The Cambridge Dictionary of English Place-Names

This alphabetical dictionary is a totally new compilation based on the archives of the English Place-Name Society and reflects recent scholarship and new research in the subject. It contains entries not only for English cities, towns and hamlets, but also geographical features such as rivers, streams and hills, as represented in the *Ordnance Survey Road Atlas of Great Britain* (1983). The dictionary provides a reflection of contemporary England, as well as its historical past.

Every place-name entry has:

- a unique National Grid reference number
- a list of historical spellings
- the age and meaning of the name and its etymology (pre-Indo-European, Indo-European, Celtic, Primitive Welsh, Anglo-Saxon, Old Norse, Old French, Middle English, Modern English)

Where appropriate, a commentary is provided on:

- comparable names
- the problems, history and significance of the name for settlement, economic and social history
- the development of the language
- its variant pronunciations and spellings

This is a major new reference work relevant to geographers, historians and historical linguists worldwide, as well as anyone interested in the history and settlement of England seeking an authoritative account of English place-names.

VICTOR WATTS was Master of Grey College, Durham University. He was Honorary Director of the English Place-Name Survey, Fellow of the Society of Antiquaries of London and Fellow of the Royal Historical Society. In addition to many articles and reviews on place-names, he compiled a *Dictionary of County Durham Place-Names* (2002), edited Book 17, *De herbis et plantis* in *On the Properties of Things: John Trevisa's Translation of Bartholomeus Anglicus' De rerum natura* (1975), and translated Boethius' *The Consolation of Philosophy* (1969, revised 1999).

The Cambridge Dictionary of

English Place-Names

Based on the collections of the
English Place-Name Society

Edited by
VICTOR WATTS

John Insley
Assistant Editor

Margaret Gelling
Advisory Editor

CAMBRIDGE
UNIVERSITY PRESS

CAMBRIDGE UNIVERSITY PRESS
Cambridge, New York, Melbourne, Madrid, Cape Town, Singapore,
São Paulo, Delhi, Dubai, Tokyo, Mexico City

Cambridge University Press
The Edinburgh Building, Cambridge CB2 8RU, UK

Published in the United States of America by Cambridge University Press, New York

www.cambridge.org
Information on this title: www.cambridge.org/9780521168557

First published 2004
Third printed 2006
First paperback edition 2010

A catalogue record for this publication is available from the British Library

Library of Congress Cataloguing in Publication Data

The Cambridge dictionary of English place-names: based on the collections of the English Place-Name Society / edited by Victor Watts, John Insley; advisory editor Margaret Gelling.
p. cm.
Includes bibliographical references and index.
ISBN 0 521 36209 1
1. England – Gazetteers. 2. Names, Geographical – England. I. Title: Dictionary of English place-names. II. Watts, V. E. (Victor Ernest) III. Insley, John. IV. Gelling, Margaret.

DA640.C36 2002 914.2′003–dc21 2001052629

ISBN 978-0-521-36209-2 Hardback
ISBN 978-0-521-16855-7 Paperback

Contents

Preface

During the course of the seventeenth annual conference of the then Council for Name Studies in Great Britain and Ireland at Christ's College, Cambridge, in March 1985, discussions took place among a group of scholars about the desirability and feasibility of compiling a new dictionary of English place-names to replace Ekwall's ageing *Concise Oxford Dictionary of English Place-Names*, the fourth and final edition of which appeared in 1960. The new work would be based on the published and unpublished collections of the English Place-Name Society and stored in electronic form. A further planning meeting was held in Cambridge in September of that year, attended by the late Professor John McNeal Dodgson of University College London, Dr Margaret Gelling of Birmingham University, Dr Peter Richards of Cambridge University Press and myself, and in December approval was given by the Syndics of Cambridge University Press for a new dictionary under the joint editorship of Dr Oliver Padel, then of the Institute of Cornish Studies, Dr Alexander Rumble of Manchester University and myself, with Professor Dodgson and Dr Gelling as advisers, to be completed within three years. Owing to pressure of work both Dr Padel and Dr Rumble were obliged to withdraw from the editorial team, to be replaced at a later date by Dr John Insley of Heidelberg University. Another grievous early blow to progress was the untimely death of Professor Dodgson in 1990.

The scope and rationale of the *Cambridge Dictionary of English Place-Names* differs from that of Eilert Ekwall's *Concise Oxford Dictionary of English Place-Names*, its standard predecessor. There is, as there must be in all selective works of reference, an element of subjectivity in the choice of names to be treated. In the case of Ekwall's *Dictionary* preference was given to names with some claim to antiquity. He aimed to include 'the chief English place-names . . . names of the country, of the counties and other important divisions, towns (*except those of late origin*), parishes, villages, some names of estates and hamlets, or even farms *whose names are old and etymologically interesting*, rivers, lakes – also names of capes, hills, bays *for which early material is available*'.[1] Ekwall included 'most of the [English] names listed in Bartholomew's *Gazetteer*', but some were omitted 'either because of the insignificance of the places or because no early forms were available'. Many hill-names in the Lake District, for example, were omitted because they were either 'comparatively late' or 'self-explanatory'. This bias, justified by the circumstance that contemporary spellings of names are unreliable guides to etymology and that names now sharing identical spellings can often be shown by early forms to have different origins, can be traced back at least to Allen Mawer's decision in his pioneering work on the place-names of Northumberland and Durham (Mawer 1920), to confine his attention 'rigidly, with some half dozen insignificant exceptions' to names found in documents dating before 1500.

By contrast, the present work aims to reflect the onomastic situation of present-day England and the selection of names for inclusion is made regardless of their antiquity or

[1] Ekwall 1960.ix (my italics). Cf. Watts 1993.7–14.

modernity. In this it reflects the practice now universal in the volumes of the English Place-Name Survey of including every name in current use as well as former names now lost or in desuetude. In this dictionary lost names are only included in so far as they are needed to make sense of contemporary names. As a basis of selection we have taken the totality of the names in England of whatever age or kind – settlements, districts, towns, villages, antiquities, rivers or other natural or man-made features – which appear in the *Ordnance Survey Road Atlas of Great Britain* first published in 1983, and we are grateful to the Ordnance Survey for making their county indices available to us. As a corollary of this bias towards the contemporary situation we have adopted the revised 1974 counties rather than the ancient historical counties, though these are given in the references at the end of each entry and every name is accompanied by its unique National Grid four-figure reference. The dictionary thus attempts a synchronic presentation of English place-names, a snapshot of the names in use today for whatever purpose of human activity, administration, industry, services, commerce, travel, planning, leisure or education.

This bias towards the present means that we include such features as new towns. It is as important and as interesting to know that that most convincing of modern place-names, Peterlee in County Durham, a new town founded in 1948, as well as conforming to the historical pattern of names ending in *-ley* or *-lee*, brilliantly preserves the name of Peter Lee, the Durham miners' leader of the 1930s, as to know that Peterborough, formerly in Northamptonshire and now in Cambridgeshire, is the **burh** or fortified enclosure built by Abbot Eadwulf in AD 963 around the reconstituted monastery of St Peter originally called *Medeshamstede*. Another such snapshot to contain the very ancient as well as the very modern could encompass Egglescliffe, a parish in County Durham, perhaps containing Primitive Welsh ***egles**, 'a church, a Christian community', and Eaglescliffe, the seventeenth-century folk-etymology of the same name adopted for the early nineteenth-century station and junction on the Stockton to Darlington railway line and thence transferred to a modern housing development.

Diachronic considerations are no less important in the Cambridge *Dictionary* than in its Oxford predecessor. But once again the presentation is different. Dr Richard Cox, in an important paper in *Nomina* (1988–9), questioned the value and validity of the term 'hybrid' in the study of the place-names of the Hebrides. He pointed out that from a linguistic point of view the term might be valid – patently many names do contain elements derived from more than one language not only in the Hebrides but also in England – but the term is not valid with regard to an *onomastic* analysis of such names. Onomastic analysis concerns itself not just with the linguistic components of a name but also with the structure of those components and how they function as names. The Hebridean name *Loch Lacsabhat*, for example, may be regarded as hybrid since the first component is the Gaelic word for lake while the second represents Old Norse *laxavatn*, 'salmon lake'. Etymologically the name may be regarded as a combination of Gaelic and Old Norse elements but this is not necessarily evidence of true hybridisation, i.e. contemporary contact between the two languages because the onomastic structure of the name is not appellative + appellative but appellative + *place-name* and the true translation of the contemporary name is not 'lake salmon lake' but 'lake Laxavatn'. There are two diachronically separate stages in the evolution of the name Loch Lacsabhat, first the Norse name *Laxavatn*, then, after the period of true language contact, the Gaelic name *Loch Lacsabhat*. Both stages participate in separate synchronic systems.

The extension of this principle of separate synchronic systems to compound place-names whose elements come from the same language has sharpened understanding of the structure of English place-names and led to new perceptions only half adumbrated in Ekwall's *Dictionary*. While Ekwall glosses the Lancashire place-names Pendlebury and Pendleton as '*burh* by *Penhill*' and '*tun* on *Penhill*' respectively, he explains Pendle itself as Welsh **pen** 'top, hill' with the addition of an explanatory Old English **hyll** 'hill'. A better presentation would have been 'Pen hill, hill called *Pen*' in which the Welsh element **pen** functions less as an appellative than as a pre-existent place-name. In the same way, we now see purely English names such as Keysoe in Bedfordshire and Cassio in Hertfordshire not as hypothetical personal name + appellative ('Cæg's hill-spur') but as *place-name* + appellative, 'the hill-spur at, of or called *Cæg*, the Key', from Old English **cæg**, 'a key', used topographically of a hill shaped in some way reminiscent of a key.

The distinction between earlier and later systems of names corresponding to changing perceptions of the landscape, its settlement, development, ownership and division is carefully observed. Names such as Great Bolas, Shropshire, are initially traced back to spellings like *Bowlas Magna* 1655, not to *Belewas* 1198. Great Bolas is a name in systemic contrast with nearby Little Bolas, *Parva Boulewas* 1342. Two names representing two separate settlements are evidenced from 1342, a different situation *onomastically* from that of 1198. A similar complexity emerges with manorial names – 'double-barrelled' or compound place-names reflecting estate division and feudal ownership – such as the Dorset Winterborne series, Winterborne Clenson, 'the W manor called Clenson, i.e. Clenston, the Clench family estate', W Houghton, 'the W manor called Houghton, Hugh's estate', W Kingston, 'the king's W estate', W Zelston, 'the de Seles W manor' together with the variants W Anderson, 'the St Andrew's W estate (from the dedication of the church)', W Stickland, 'the W estate called Stickland, earlier *Stikelane*, the steep lane', W Whitechurch, 'W with the white church': these represent a complex onomastic situation compared with that of Domesday Book (DB) where the manors are undifferentiated under the common name *Wintreborne* with the result that modern scholars cannot be certain of their precise individual identification. Later systems of such compound place-names, reflecting division and redivision of formerly unitary estates, are given separate and prior treatment to the discussion of their original simplex forms.

More importantly, compound place-names such as Beare Green in Surrey are inadequately treated if, as in the English Place-Name Society (EPNS) volume for Surrey, reference is made only to historical spellings such as *la Bere* 1263 and *Beare* 1497. These represent a simplex place-name meaning 'the woodland pasture' from OE **bǣr**. Onomastically and, indeed, historically, *la Bere* or *Beare* is a different name at a different stage from Beare Green, *Beare Green* 1816, 'the green at or called Beare'. In this dictionary, names in *End, Green, Row* and *Street*, like the manorial names, are treated as place-names in their own right and given their own run of spellings before analysis of their component elements. The attention given to these frequently late-appearing names has prompted the preparation of distribution maps which have thrown up interesting and surprising results illustrated in this dictionary and in Watts 1999.

Other systems of name-giving are also becoming increasingly recognised. It has recently been suggested, for instance, that place-names of the type Acton, 'oak-tree settlement', in Shropshire, were so named because they had a specialised function in the processing or distribution of oak timber in that county (Sa i.3). Many common place-names of this type must similarly have referred to specialised functions within a manorial estate or district economy. Shipton/Skipton, 'sheep farm', is an obvious example, Hardwick, 'herd farm', another, Ashton, 'ash-tree farm', a third. Many others deserve consideration as the implications of the perception that there is a difference between the etymological meaning of a place-name and its significance are realised. The *meaning* of the name-type Acton is 'oak-tree settlement, oak-tree estate, oak-tree farm' but the *significance* is probably not so often 'farm built of oak, farm beside an oak tree' or 'farm among oak trees' as 'place where oak is worked' or 'where oak timber is obtained'. In the same way the place-name type Stanton/ Staunton/ Stainton, 'stone farm or settlement', will frequently have been significant not primarily because of soil conditions or buildings made of stone but because it had an economic function as a place of industry, as a source of stone for building or as an estate quarry. A great deal of the town and castle of Barnard Castle in County Durham is built of locally hewn stone, some of which must have come from the local village of Stainton where to this day there is an active quarry of high-quality stone.

This dictionary attempts, therefore, to say something about the historical, administrative, commercial or economic significance of names in addition to giving their etymological meanings. Furthermore, it focuses not just on origins but includes later developments of interest. For this reason significant sixteenth-, seventeenth- and eighteenth-century spellings are cited when they illustrate demotic forms of names and local pronunciations which differ from their conventional modern spellings, e.g. the form *Femsam* 1549 for Felmersham, Bedfordshire, or *Awmsburi* 1533 for Almondsbury, Gloucestershire, both of which anticipate the modern pronunciations [femsəm] and [a:msbəri][2] by over three hundred years.

Over a period of eighty years the science of onomastics has not stood still. First, there has been a great expansion in the quantity of manuscript sources made available in published form and much detailed scholarly detective work has been applied to it, such as Sawyer's annotated list and bibliography of Anglo-Saxon charters (Sawyer). Second, there have been major advances in the understanding of place-names, in the interpretative strategies employed in their explanation, in knowledge of the languages in which they were formed, in awareness of the archaeological and historical contexts of their coinage, and especially in our understanding of the way in which the Anglo-Saxon and Scandinavian settlers perceived the land and landscape which they came to occupy and work.[3] There is thus not only a disparity in the interpretations of identical place-name types as between Buckinghamshire (published in 1925) and Cheshire (published in 1970–97), but also marked differences in the richness of evidence available and cited.

Apart from medieval writers, chronicles, early land charters and monastic records, the archives used by the sources on which this work is based are largely technical documents of local and national government in which historical place-name spellings are found. Some reference is made to a number of them in this dictionary, particularly where additional material has been used to supplement earlier work. They include Assize Rolls – legal records of Assize courts preserved in roll or scroll form; the Book of Fees – a list of feudal holdings; the Calendar of Fine Rolls – a list of enrolled legal agreements or 'fines' relating to land possession; Court Rolls – the enrolled records of manorial or hundred courts, courts convened to settle disputes within a single manor or within a subdivision of a county comprising a number of manors called a 'hundred'; Feet of Fines – one of the identical parts or 'feet' of a tripartite fine or legal settlement

[2] With ModE [a:] for lME [au] (Dobson 1968, para. 238).
[3] Cf. for example, Dodgson 1966, 1967a, 1967b and 1968; Cameron 1975, 1976; Gelling 1976, 1978 and 1998. For pre-Celtic Indo-European names, especially river-names, see Nicolaisen 1957, 1982; Krahe 1962, 1964; Tovar 1977; and Kitson 1996.

relating to land possession; *Inquisitiones post mortem* – legal investigations held upon a death concerning the ownership of land.

The materials for the dictionary are not, however, unproblematic. The English Place-Name Society was founded in 1923 and has since then published 80 annual volumes covering 26 of the 42 pre-1974 counties (Bedfordshire, Berkshire, Buckinghamshire, Cambridgeshire, Cheshire, Cumberland, Derbyshire, Devonshire, the East Riding of Yorkshire, Essex, Gloucestershire, Hertfordshire, Huntingdonshire, Middlesex, Northamptonshire, the North Riding of Yorkshire, Nottinghamshire, Oxfordshire, Rutland, Surrey, Sussex, Warwickshire, Westmorland, the West Riding of Yorkshire, Wiltshire and Worcestershire). Parts of a further six counties have also been published (Dorset, Leicestershire, Lincolnshire, Norfolk, Shropshire and Staffordshire) but there are as yet no EPNS surveys for Cornwall, Durham, Hampshire, Herefordshire, the Isle of Wight, Lancashire, Northumberland, Somerset and Suffolk. However, all of these counties except Somerset have been treated in volumes published either before the founding of EPNS or in reliable 'popular' one-volume county dictionaries published since 1940, and in addition the Society possesses unpublished material for a number of them, including the counties so far only partly published.

The earliest of these volumes, the Rev. W. W. Skeat's *Place-Names of Suffolk*, appeared in 1913. The first county to be surveyed by EPNS was Buckinghamshire, published in a single volume in 1925, while the most recent was Cheshire, published in five volumes between 1970 and 1997.

Furthermore, there are marked differences in the manner of the presentation of evidence between earlier and later EPNS county volumes.[4] Thus, for example, for Ashley Green, Buckinghamshire, *PN Bucks* gives five forms from three sources between 1227 and 1408; while for Ashley, Cheshire, *PN Ches* ii.10 gives 27 forms from 20 sources between 1086 and 1673. For Newton Blossomville and Newton Longville *PN Bucks* presents the evidence for the names and the affixes separately; for the Cheshire Newtons *PN Ches* presents the various affixes with the place-name forms to which they belong.

PN Bucks makes sparing use of medieval cartularies but when it does it takes care to distinguish the date of the deed enrolled from the date of the cartulary copy. Thus the spelling *Wlfrinton* for Wolverton in the Eynsham Cartulary is dated c.1220, the date of both deed and copy, but the form *Cubelintone* for Cublington in the same source is dated 1154(1200), which in this dictionary would appear as [1154]1200, a copy dated 1200 of a deed dated 1154. Similarly with *Asse* c.1275(1400) (in this dictionary [c.1275]1400) for Nash in the St Albans cartulary, and *Chettend'* Hy 1 (13th) (in this dictionary specifying Henry I's regnal years [1100 × 35]13th) for Cheddington in the unpublished Nostell Priory cartulary.

By contrast, the comparative absence of a rich vein of pre-Conquest evidence rightly led the editor of another early EPNS volume, A. H. Smith in *The Place-Names of the North Riding of Yorkshire* (1928), to make more extensive use of both unpublished and published registers and monastic cartularies such as the register of St Leonard's Hospital in York, a fifteenth-century manuscript (Cotton Nero D.III), the fifteenth-century register of the Honour of Richmond (Cotton Faustina B.VII), the Easby cartulary of c.1281 (MS Egerton 2827), and the fourteenth- and fifteenth-century cartularies of St Mary's Abbey, York (Harley 236 and D and Chapter, York), and of Fountains (Cotton

[4] These volumes are henceforward referred to in the text in the form PN + abbreviated county name, e.g. *PN Bucks, PN Glos*, etc. For full bibliographic details see the following in Abbreviations: iv. Sources: BdHu, Bk, Ca, Che, Gl, Nf, YN.

Tiberius C.XII and Add. MS 37770), Rievaulx and Whitby. Unfortunately, however, in citing cartulary forms *PN North Riding* gives only the date of the original deed and ignores the date of the cartulary copy. The latter can be found in the bibliography but this is not true in all cases.[5]

All this means that the quantity of evidential forms available, even from EPNS sources, varies considerably and this is inevitably reflected in the entries in this dictionary. It would have been impossible to have checked every cartulary reference or to have systematically supplemented the evidence available in these publications. And this is even more true of the counties for which only one-volume popular dictionaries are available. They necessarily present but a limited selection of forms, illustrating usually only the earliest spelling of a name and one or two representative later spellings, but generally not those of the modern period (post-1500) which are often so revealing of local pronunciations. For these counties and for counties for which no adequate volume is yet available reliance has had to be on Ekwall's pioneering collection supplemented by such manuscript collections as are held by EPNS in Nottingham or by individual scholars.

Not only the quantity but also the quality of the evidence available varies with the accuracy of transcription differing from scholar to scholar. So, for example, a random check revealed the following minor discrepancies (see table) between the DB spellings cited in the first volume of *PN Glos* and those cited in Ekwall and those in the Morris edition (*DB: Gloucestershire* 1982).

PLACE	PN GLOS	EKWALL/MORRIS
Bibury	*Begeberie –ia*	Ekwall *Begeberie*, Morris *Begeberie, Becheberie, Begabiria, Begesberi, Bercheberi*
Coln St Aldwyn	*Colne*	Ekwall *Culne*, Morris *Cvlne*
Willimastrip	*Hetrop*	Morris *Hetrope*
Cirencester	*Cire, Cyrecestr(a)'*	Ekwall, Morris *Cirecestre*
Chesterton	*Cesterton(e)*	Ekwall *Cestertone*
Pinbury	*Pennebiria –beria -buri* 1082–1359 including DB	Morris *Penneberie*
Preston	*Prestitvne*	Ekwall *Prestitune* Morris *Prestetvne, Prestitvne*
Avening	*Au- Aveninge –ynge* 1086–1587	Ekwall, Morris *Aveninge*
Bisley hundred	*Biselege hvnd'*	Morris *Biseleie Hvnd'*
Edgeworth	*Egesworde, Egeiswurde*	Ekwall *Egesworde* Morris *Egeisuurde, Egesworde*
Colesborne	*Colesborne* 1086 *-burn(e) -burn(i)a* 1086–1372	Ekwall *Colesburne* Morris *Colesborne -burne, Kolesburna*
Compton Abdale	*Contone*	Ekwall *Cuntune*, Morris *Cvntvne, Cumtona*

[5] No date is given for the Marrick Cartulary, for example, and the abbreviation *FountC* does not even appear in the bibliography and abbreviations.

In addition, because of the presentation of the material in *PN Glos* it is impossible to be sure of the exact DB spellings for Quenington (p. 44), Ashbrook (p. 52), Tetbury Upton (p. 111), Lasborough and Westonbirt (p. 114), Bisley (p. 117) and Frampton Mansell (p. 137). It would have been impossible to check every DB spelling; in those counties for which no EPNS survey volume is available, the spellings given in Ekwall have been followed, frequently checked against those in Morris. In the case of those counties for which an EPNS volume is available and a difference has been noted from the spelling given in Ekwall, the form has been checked against Morris.

In the earlier EPNS county volumes spellings are for the most part individually presented with their unique date. In later volumes, for economy of space, spellings are frequently presented in a form such as:

Donecastr(e)', -castr(i)a(m), -caster 1086 DB . . . *et passim* to 1382
Danecastr(e)', (-a, -um, -ie), -caster Hy 2 Riev . . . *et freq* to 1304

which means that in various sources between 1086 and 1382 the following spellings are recorded, *Donecastr', Donecastre, Donecastra, Donecastram, Donecastria, Donecastriam, Donecaster*; and between the reign of Henry II (1154–89) and 1304 the recorded spellings are: *Danecastr', Danecastre, Danecastr, Danecastra, Danecastrum, Danecastrie* and *Danecaster* (*PN West Riding of Yorkshire* i.29). Another convention uses the term *passim*, as in: *Lege* 1086 DB *et passim* with variant spellings *Legh', Legh, Ley(e)* to 1392. This term (meaning 'everywhere') was introduced in 1926 (*PN Beds and Hunts*), at first rather sparingly used, soon followed by *et freq* (for *et frequenter* 'and frequently', *PN North Riding of Yorkshire* 1928) although the latter was not defined until 1996 (*PN Norf* ii.xxvi) as meaning four to nine occurrences within the dates specified.[6] The presentation of historical spellings in this dictionary, although selective, necessarily follows the differing practice of the EPNS volumes, and for this reason varies in richness from name to name according to the date and nature of the source publication.

One further difficulty encountered has been with the dating of the spellings given where, again, differences of usage occur between editors and volumes. For spellings in documents which are not dated with precision from internal evidence it has been customary to give dates of the form Hy 3, for example, for a document not more closely dated than to the regnal years of a particular monarch, in this case Henry III; in this dictionary the date would be recorded as 1216 × 72 and the same convention is used for any other document for which only a *terminus post quem* and a *terminus ante quem* can be given. Thus for *Legh' sub Brokhurst* 1271–2 the date in this dictionary appears as 1271 × 2. Unfortunately in the volumes of the EPNS survey for a document of this kind the termini are presented in the form 1240–60 which is identical with the convention used for a *series* of identical spellings taken from a source which extends over a period of years. Thus in *PN Cambs* the entry *Foxton(e)* 1202–1352 FF *et passim* must refer to the fact that throughout the Feet of Fines, and elsewhere between 1202 and 1352, the regular spellings are *Foxton* and *Foxtone*, whereas, as Dr Sandred makes clear in the preliminary notes on presentation, in *PN Norf*, the entry *Martham* 1121–45, 1226–36, 1232 NCR means that there are three occurrences of this spelling in the Norwich Cathedral Register dated 1121 × 45, 1226 × 36 and 1232 respectively. Unfortunately the usage of some

[6] Occasionally other bits of Latin phraseology are encountered, such as *ter* 'three times' (*PN Beds and Hunts* p. 256) and *saepissime* 'very frequently' (*ibid.* p. 266).

EPNS volumes is not always explicit and it is sometimes unclear whether the presentation refers to the *termini inter quos* of a single spelling or to a series of spellings. In this dictionary the form 1299–1455 always refers to a series of spellings between the two dates specified and the form 1299 × 1310 to a single spelling dating from within these termini.

In the volumes of the *English Place-Name Survey* it has been customary to refer place-name etymologies to Old English forms irrespective of the likely date of coinage of the name in question. While this is defensible for names first recorded in Domesday Book or in documents of the following century, the later the attestation the more artificial and, indeed, misleading this becomes. Thus in Rugeley in Staffordshire, for example, Rugeley itself, *Rugelie* 1086, and Hagley, *Hageleia* 1130, are properly referred to the OE etyma **hrycg** and **lēah**, and ***hagga** and **lēah** respectively; but Hazel Slade in the same township, *Hazell slade* 1682, is also referred to the OE **hæsel** and **slæd** rather than to the ModE **hazel** and ModE dial **slade**.[7] The citation of etyma in OE form avoids decisions about the date of coinage. In this dictionary, however, where the evidence and balance of probability points to name coinage at a later date than the Anglo-Saxon period, etymologies are given in ME or ModE as appropriate followed by the OE source in brackets.

The vocabulary of English place-name elements – British (Brit), Primitive Welsh (PrW), Anglo-Saxon or Old English (OE), Old Norse (ON), Old French (OFr) – has been collected in A. H. Smith's *English Place-Name Elements* (PNE). In a dictionary such as this it is not possible to provide detailed discussion of such elements and constant reference is made to Smith's work for further information. However, a glossary of some of the most frequently used elements is to be found in the endmatter.

The text of the dictionary has been written by me and read in its entirety by Margaret Gelling, whose knowledge of the archaeological, diplomatic and historical contexts of English place-names and of the landscape vocabulary and of the precision with which its was used in place-names is unsurpassed. The first drafts of Cornwall, Oxfordshire and Surrey were written by Oliver Padel, Alex Rumble and John Insley respectively. Dr Insley has also read and commented on the drafts of Buckinghamshire, Cambridgeshire, Cheshire, Greater Manchester, Lancashire, Leicestershire, Lincolnshire, Merseyside, Somerset, Suffolk and Surrey, and has, out of his deep philological lore and unparalleled knowledge of Germanic personal nomenclature, improved and corrected many etymologies. Dr Michael Costen read the draft for Somerset and provided additional forms from his own collection, and Professor Richard Coates read and commented on Hampshire, Dr Paul Cullen on Kent, again providing additional forms from his own collection, as did Mr Mills on those parts of Dorset not yet covered in his EPNS volumes. Dr Wolf von Reitzenstein answered queries on Bavarian place-names and Mr David Horovitz kindly made his typescript collection of Staffordshire place-names available to me. I am also very grateful to various student friends who undertook the vital initial task of word-processing the material for me, county by county, among them Philip Atwood, Janine Brindle, Victoria Lamming, Rafael Morton, Michelle Nixon, Elisa Sarnacka and John Watts.

The compilation of a work such as this is deeply dependent upon foundations laid by previous scholars – on Mawer and Stenton, on Bradley and Wyld and Ekwall. However much those foundations have had to be modified, the later writer is but a dwarf on the shoulders of giants; and if he can see a little further than they as a result of the cumula-

[7] St i.105–7.

tive scholarship and revision of the past forty years, he still remains a dwarf in their company. To their names I should like to add the names of the two scholars who have most deeply influenced my own development, the late F. C. G. Langford, my Senior Classics master at Bristol Grammar School, and Professor G. V. Smithers, my tutor at Merton College.

This work has been unconscionably long in the gestation and has not been compiled in the soft obscurities of retirement but amid the inconvenience and distraction that is the lot of one employed today in academic bowers. It may be that the temerity of attempting such an enterprise can only expose the humble drudge to censure without hope of praise, but I cannot aver that it was written without the assistance of the learned or without patronage. Dr Gelling and Professor Kenneth Cameron, my predecessor as Hon. Director of the English Place-Name Survey, at times when the task seemed lonely or beyond hope of completion, have given me constant encouragement, Dr Insley has contributed unstintingly from his formidable philological expertise amid heavy teaching duties in Heidelberg, and the University of Durham has twice granted me periods of research-leave to advance the project. To them and to all who have lent me assistance and to my Cambridge University Press editors, Dr Peter Richards and Caroline Bundy, I am deeply grateful. If through neglect or ignorance I have shunned or misunderstood their advice, such blunders or absurdities as ignorance, inadvertency or casual eclipses of the mind have allowed to remain in this book are mine alone.

Victor Watts

Publishers' note

This Dictionary represents a major part of Victor Watts's academic output – he spent over fifteen years on its compilation. Sadly, he did not live to see it in print, dying shortly before he was due to retire from the mastership of Grey College, Durham University, and months before his magnum opus was published.

The Press is grateful to a team of specialists in the field who have helped bring the project to a conclusion. Margaret Gelling and John Insley deserve recognition, too, for advising Victor Watts throughout the long gestation of the work and contributing information in their particular areas of scholarship. We would also like to acknowledge the help of Ordnance Survey in providing the original corpus of names on which the Dictionary is based.

Research into the study of place-names is on-going. Since the inception of this Dictionary, the Institute for Name-Studies has been established within the University of Nottingham (name-studies@nottingham.ac.uk). Readers wishing to know about current or future research should contact the Institute. The English Place-Name Society, also based within the University of Nottingham, is responsible for the publication of county volumes.

Guide to dictionary entries

Dictionary entries comprise some or all of the following elements, usually but not invariably presented in the following order:

1. The head-form

Names are listed alphabetically under their main component and prefixes are ignored.

Temple Cloud will be found under Temple CLOUD

North Cliffe under the general heading CLIFFE (1) North ~ Humbs

Northcott remains NORTHCOTT

Cross references are supplied in cases of doubt

ALL STRETTON Shrops → All STRETTON Shrops

Names which occur frequently are grouped together under a single heading and ordered as shown. ACTON Shrops, ACTON ROUND Shrops, Iron ACTON Avon, etc. become

ACTON (1) ~ Shrops
(2) ~ ROUND Shrops
(3) Iron ~ Avon

Where a place-name has more than one derivation, each etymology has a separate entry:

MILTON Middle settlement or estate
(1) ~ Cambs
(2) ~ Oxon
(3) ~ ABBOTT Devon
(4) ~ REGIS Kent
(5) Great ~ Oxon
(6) Little ~ Oxon
MILTON Mill settlement
(1) ~ Cumb
(2) ~ Staffs

2. County

An abbreviated reference to the 1974 county.

3. National Grid square and unique four-figure reference number.

BRISTOL Avon ST 5973

identifies the place Bristol in the county of Avon in National Grid square ST in Easting 59 and Northing 73.

4. The translation of the name

BRISTOL 'Assembly place by the bridge'

ABINGTON PIGOTTS 'A held by the Pigott family', A being an abbreviation for Abington.

5. Dated historical spellings

These begin with the earliest attested and include all extant or lost Anglo-Saxon charter forms accompanied by their Sawyer number (S), or, in cases not included in Sawyer, a reference to Birch's *Cartularium Saxonicum*, London 1885–93 (B), or Kemble's *Codex Diplomaticus ævi Saxonici*, London 1839–48 (K).

Any forms of the name in the Anglo-Saxon Chronicle (ASC)

The Domesday Book form or forms (DB)

A selection of characteristic medieval and later spellings illustrating the development of the name to its modern form. For example:

(to, of) Brycg stowe 11th ASC(A) under year 1052, ASC(D) under year 1063, *Bristov -ou* 1086, *Bri- Brystow(e)* 12th–1675, *Bri- Brystol(l)* 1100–1675

This means that in the eleventh century the spelling *Brycg stowe* was recorded in the Anglo-Saxon Chronicle, manuscript version A, under the year 1052, and subsequently under the year 1063 in manuscript version D; the spellings *Bristov* and *Bristou* were recorded in 1086; *Bri- Brystow(e)* from the twelfth century through to 1675; and the spellings *Bri- Brystol(l)* between 1100 and 1675.

In the citation of manuscript forms, letters in brackets indicate the presence of spellings both with and without the letters in question. Thus

Coleshull(e) [1279]14th, 1340, 1345

indicates that spellings in *-hull* and in *-hulle* are found in the documents used.

The suspension mark ' is an abbreviation mark used in manuscript sources often for –*e*, sometimes for *re* or other letters:

Aketon' or Chilton(')

Dates of the form 1284–1739 mean that the spelling or spellings specified are found in documents throughout this period.

Dates of the form 1284x1300 refer to a single spelling from a source not more precisely dated than the termini given.

Dates of the form [1284]1300 indicate a copy dated 1300 of an original deed dated or purporting to date from 1284.

A spelling dated 14th indicates a single form not more precisely dated than during the 14th century (spellings dated 14th cent. occur throughout the century).

Dates preceded by an asterisk* indicate a suspicious or forged document.

Where two distinct spelling traditions exist for a name they are frequently grouped into types, as ALMONDSBURY Avon:

Type I: *Almodesberie* etc. 1086–1316
Type II: *Almundesbury* etc. 1248–1587

In the case of compound names the forms for the later compound name are given first, followed by those of the earlier simplex name, as in:

> *Acton(e) Torvill(e)* 1284–1739. Earlier simply *Achetone* 1086.

6. Etymology

This is explicit, indicating both the nominative case of noun and adjective etyma and the case form occurring in the place-name.

Italics are used when citing personal names, e.g. *Æthelræd*, and bold type when citing appellatives, e.g. **burh**, **dūn**, **tūn** etc. For classification of some common elements (e.g. **ing** and **hamm**) see the Glossary.

An asterisk* represents an unrecorded or hypothetical form.

The symbol < indicates the derivation of a name or form.

The symbol > indicates subsequent historical changes.

Th is used for the Old English thorn [þ] and eth [ð], *g* for ȝ and *w* for wynn [ƿ] except when citing manuscript sources when the original orthography is reproduced. In citing OE words ċ is used to represent the palatal assibilated sound [tʃ] and ġ the sound [j].

7. Explanatory comment

> BRISTOL The reference is probably to a crossing of the Avon. The modern form with *-ol* is an inverted spelling that arose after the development of Anglo-Norse *-ol* to *-ou-*.

8. Pronunciation

This is normally given only when recorded in the source volume used. It is presumed to be contemporary with the publication date of the volume unless otherwise stated:

> CONGRESBURY The pr is [ku:mzbri]

9. Sources

The evidence is derived from the following sources: EPNS county volumes, DEPN, non-EPNS county volumes, RN (Ekwall, *English River-Names*), RBrit (Rivet and Smith, *The Place-Names of Roman Britain*), and any important supplementary books or periodicals.

> BRISTOL Gl iii.83, TC 60

refers to *The Place-Names of Gloucestershire*, vol. 3, p. 83; and *The Names of Towns and Cities in Britain*, p. 60.

Examples of entries

STAFFIELD Cumbr NY 5442. 'Staff hill'. *Stafhole* c.1225–79, *Staffol(e)* c.1252–1777, *Staffold* 1270, *Staffeld* 1276–8, *Staffel(l)* 1307, 1508, *Stafful* 1348–68, *Staffield oth. Staffell* 1806. ON **stafr** 'a post, a pole' + **hóll** 'an isolated hill'. The reference is to a small hill marked by a post or where posts were obtained. Cu 248 gives pr [stafl], SSNNW 164, L 169.

1. headword	STAFFIELD
2. county	Cumbria
3. national grid reference	NY 5442
4. translation	'staff hill'
5. historical spellings	*Stafhole* c.1225–79, *Staffol(e)* c.1252–1777, *Staffold* 1270, *Staffeld* 1276–8, *Staffel(l)* 1307, 1508, *Stafful* 1348–68, *Staffield oth. Staffell* 1806
6. etymology	ON **stafr** 'a post, a pole' + **hóll** 'an isolated hill'
7. explanatory comment	The reference is to a small hill marked by a post or where posts were obtained
8. pronunciation	Cu 248 gives pr [stafl]
9. sources	SSNNW 164, L 169

CONGRESBURY Avon ST 4363. 'Congar's fortified place'. *Conbusburie* (sic) [688x726]17th ECW, (*on*) *Cungresbyri* [893]c.1000 Asser, *-byrig* c.1000, *Kunigresbiria* [?c.1030]lost ECW, *Cungaresbyrig* *[1065]c.1500 S 1042, *Cungresberie* 1086, *Coombesbury* 1758. Welsh saint's name *Cuncar*, OE *Congar*, genitive sing. *Congares*, + OE **byriġ**, dative sing. of **burh**. St Congar was buried here, a place mentioned by Asser as a derelict Celtic monastery. The pr is [ku:mzbri]. DEPN, CMCS 12.43.

1. headword	CONGRESBURY
2. county	Avon
3. national grid number	ST 4363
4. translation	'Congar's fortified place'
5. historical spelling	*Conbusburie* (sic) [688x726]17th ECW, (*on*) *Cungresbyri* [893]c.1000 Asser, *-byrig* c.1000, *Kunigresbiria* [?c.1030]lost ECW, *Cungaresbyrig* *[1065]c.1500 S 1042, *Cungresberie* 1086, *Coombesbury* 1758
6. etymology	Welsh saint's name *Cuncar*, OE *Congar*, genitive sing. *Congares*, + OE **byriġ**, dative sing. of **burh**
7. explanatory comment	St Congar was buried here, a place mentioned by Asser as a derelict Celtic monastery
8. pronunciation	[ku:mzbri]
9. sources	DEPN, CMCS 12.43

Abbreviations

i. Counties and geographical regions

Avon	Avon	Lancs	Lancashire
Beds	Bedfordshire	Leic	Leicestershire
Berks	Berkshire	Lincs	Lincolnshire
Bucks	Buckinghamshire	Mers	Merseyside
Cambs	Cambridgeshire	Norf	Norfolk
Ches	Cheshire	Northants	Northamptonshire
Cleve	Cleveland	Northum	Northumberland
Corn	Cornwall	Notts	Nottinghamshire
Cumbr	Cumbria	NYorks	North Yorkshire
Derby	Derbyshire	Oxon	Oxfordshire
Devon	Devon	Rutland	Rutland
Dorset	Dorset	Scilly	Isles of Scilly
Durham	Durham	Shrops	Shropshire
ESusx	East Sussex	Somer	Somerset
Essex	Essex	Staffs	Staffordshire
GLond	Greater London	Suff	Suffolk
Glos	Gloucestershire	Surrey	Surrey
GMan	Greater Manchester	SYorks	South Yorkshire
Grampian	Grampian	T&W	Tyne and Wear
Hants	Hampshire	Warw	Warwickshire
Herts	Hertfordshire	Wilts	Wiltshire
Humbs	Humberside	WMids	West Midlands
H&W	Hereford and Worcester	WSusx	West Sussex
IoW	Isle of Wight	WYorks	West Yorkshire
Kent	Kent	Y	Yorkshire

ii. Etymology

AN	Anglo-Norman	Fr	French
Angl	Anglican	G	German
AScand	Anglo-Scandinavian	Gael	Gaelic
Brit	British	Gk	Greek
CG	Continental Germanic	Gmc	Germanic
Co	Cornish	Icel	Icelandic
Dan	Danish	IE	Indo-European
Du	Dutch	Ir	Irish
E	English	LG	Low German
EFris	East Frisian	lME	Late Middle English
eME	early Middle English	lOE	Late Old English
eScand	East Scandinavian	MCo	Middle Cornish

MDu	Middle Dutch
ME	Middle English
MHG	Middle High German
MIr	Middle Irish
MLat	Medieval Latin
MLDu	Middle low Dutch
MLG	Middle Low German
ModDan	Modern Danish
ModE	Modern English
ModG	Modern German
ModSwed	Modern Swedish
MW	Middle Welsh
Nb	Northumbrian
NCy	North Country
NFr	Northern French
Norw	Norwegian
OAngl	Old Anglian dialect of Old English
OBret	Old Breton
OBrit	Old Brittanic
OCo	Old Cornish
ODan	Old Danish
ODn	Old Dutch
OE	Old English
OEScand	Old East Scandinavian
OFr	Old French
OFris	Old Frisian
OG	Old German
OHG	Old High German
OIcel	Old Icelandic
OIr	Old Irish
OKt	Old Kentish
ON	Old Norse
ONb	Old Northumbiran
ONFr	Old Northern French
OPrussian	Old Prussian
OSax	Old Saxon
OSlav	Old Slavonic
OSwed	Old Swedish
OW	Old Welsh
OWScand	Old West Scandinavian
P-Celtic	the Brittanic branch of Celtic
PIE	Primtive Indo-European
PrOE	Primitive Old English
PrCo	Primitive Cornish
PrCumb	Primitive Cumbric
PrScand	Primitive Scandinavian
PrW	Primitive Welsh
RBrit	Romano-British
Scand	Scandinavian
Scy	South Country
Skr	Sanskrit
StE	Standard English
Swed	Swedish
W	Welsh
WGmc	West Germanic
WMid	West Midlands
WSax	West Saxon dialect of Old English

iii. General abbreviations

by-n.	by-name
cent.	century
dial.	dialect
E	east, eastern
EPNS	English Place-Name Society
f.	feminine
f.n/s	footnote/s
ft.	feet (1 ft. = 0.305 metre)
in.	inch/inches (1 in. = 2.54 cm)
Kr	Kreis (German) 'district'
m.	mile (1 m. = 1.61 km)
masc.	masculine
mod.	modern
MS	manuscript
N	north, northern
n.d.	no date
n.f.	neue folge (new series)
nick-n.	nickname
n., ns.	name, names
nom	nominative
n.s.	new series
OS	Ordnance Survey map
pers.n.	personal name
pl.	plural
p.n.	place-name
pr, prs	pronunciation(s)
r.n.	river name
S	south, southern

sb.	substantive (in references to OED)
sc.	*scilicet* (that is to say)
SE	south-east
sing.	singular
s.n., s.nn.	*sub nomine* (under a specified name or names)
sp, sps	spelling, spellings
s.v.	*sub verbo* (under a word or heading)
SW	south-west
W	west, western
*	hypothetical form; also used to mark unauthentic and forged charters
[ ]	enclose phonetic script
< >	enclose graphies

iv. Sources

Unless otherwise stated, all references are to page numbers.

AA	*Archaeologia Aeliana* (cited by series, vol. and page)
A-A	Iain Taylor, *Ainmean-Àiteachan*, Aberdeen 1981
AB	*Anglia Beiblatt*
Abbo	Abbo of Fleury, *Passio Sancti Edmundi* in T. Arnold, *Memorials of St Edmund's Abbey*, Rolls Series 1890
AC	'Anonymous life of St Cuthbert' in *Two Lives of St Cuthbert*, ed. B. Colgrave, Cambridge 1940
AD	Adam of Domerham
Addenda	This refers to the published addenda to the Place-Name Survey vol. cited in the reference
ADM	A. D. Mills, personal communication
Æthelweard	*The Chronicle of Æthelweard*, ed. A. Campbell, London 1962
AEW	Jan de Vries, *Altnordisches etymologisches Wörterbuch*, Leiden 1977
AfcL	*Archiv für celtische Lexicographie*, Halle 1900–7
Age of Arthur	John Morris, *The Age of Arthur: A History of the British Isles from 350 to 650*, London 1973
AH	*The Agrarian History of England and Wales*, ed. H. P. R. Finberg (by vol., part and page)
AI	The Antonine Itinerary, cited from RBrit 150ff
Anglia	*Anglia: Zeitschrift für englische Philologie*
Antiquity	*Antiquity: a quarterly review of archaeology*
APS	*Acta Philologica Scandinavia: Tidsskrift for Nordisk Sprogforskning*
Arch J	*The Archaeological Journal* (by year and page)
Archæologia	*Archœologia: Or Miscellaneous Tracts Relating to Antiquity,* Society of Antiquaries of London
Archaeology	*The Archaeology of Somerset,* ed. Michael Aston and Ian Burrow, Somerset County Council 1982
Armstrong	A. Armstrong, *The county Palatine of Durham survey'd by Capt. Armstrong & engraved by T. Jefferys*, 1768
Arnott	W. G. Arnott, *The Place-Names of the Deben Valley Parishes*, Ipswich 1946
Årsskrift	*Sydsvenska Ortnamnssällskapets Årsskrift* (by year and page)
ASC	The Anglo-Saxon Chronicle, cited from *Two of the Saxon Chronicles Parallel*, ed. Charles Plummer, Oxford 1892–99 (different manuscript versions are shown by the letters A, B, C, etc., in brackets)
ASCharters	A. J. Robertson, *Anglo-Saxon Charters*, 2nd edn Cambridge 1956

ASE	*Anglo-Saxon England* (by vol. and page)
ASEngland	F. M. Stenton, *Anglo-Saxon England*, 2nd edn Oxford 1947
Ass	Assize Roll
Asser	*Asser's Life of King Alfred*, ed. W. H. Stevenson, Oxford 1904
ASWills	*Anglo-Saxon Wills*, ed. D. Whitelock, Cambridge 1930
Atkin 1998	M. A. Atkin, 'Places named Anstey', *Proceedings of the XIXth International Congress of Onomastic Sciences*, Aberdeen, August 4–11 1996, II.15–23
Aubrey	J. Aubrey, *The Natural History and Antiquities of the County of Surrey*, 5 vols., 1718–19
B	W. de Gray Birch, *Cartularium Saxonicum*, 3 vols. plus index, 1885–99
BAA	*Bristol and Avon Archaeology* (by year)
BAAS	British Association for the Advancement of Science
Bach	A. Bach, *Deutsche Namenkunde II: Die deutsche Ortsnamen*, 2 vols., Heidelberg 1953–4
Bahlow	Hans Bahlow, *Deutschlands geographische Namenwelt*, Frankfurt am Main 1985
Bannister	A. T. Bannister, *The Place-Names of Herefordshire*, Cambridge 1916
BAR Brit	*British Archaeological Reports*, British Series (by vol. and page)
Baring-Gould	S. Baring-Gould, *Devon*, London 1907
Baron	M. Cynthia Baron, 'A Study of the Place-Names of East Suffolk', unpublished MA thesis, Sheffield University, 1952
Barrow	G. W. S. Barrow, *The Kingdom of the Scots*, London 1973
Bath	*Two Chartularies of Bath Priory*, Somerset Record Society 7
BB	James Reed, *The Border Ballads*, London 1973
BBCS	*Bulletin of the Board of Celtic Studies*, 1921–
Bd, BdHu	Bedfordshire entries in A. Mawer and F. M. Stenton, *The Place-Names of Bedfordshire and Huntingdonshire*, Cambridge 1926
Benveniste	E. Benveniste, *Origines de la formation des noms en Indo-européen*, Paris 1973
Berger	Dieter Berger, *Geographische Namen in Deutschland*, Mannheim 1993
Berkeley	*Descriptive Catalogue of the Charters and Muniments in the possession of the Rt. Hon. Lord Fitzhardinge* [Berkeley Castle] compiled by I. H. Jeayes, Bristol 1892
BF	*The Book of Fees*. Rolls Series 1920–31 (by no. of entry)
BG	Julius Caesar, *De Bello Gallico*, ed. R. du Pontet, Oxford 1900–1
BH	*The Defence of Wessex: The Burghal Hidage and Anglo-Saxon Fortifications*, ed. David Hill and Alexander R. Rumble, Manchester and New York 1997 (numbers refer to the different MS versions: BH = A2, BH1 = B1, BH2 = B2, BH3 = B3, BH4 = B4; followed by page number)
BHA	Bede's *Historia Abbatum* in *Baedae Opera Historica*, ed. Charles Plummer, Oxford 1896
BHArch	*Brighton and Hove Archaeologist*
BHE	Bede's *Ecclesiastical History of the English People*, ed. Bertram Colgrave and R. A. B. Mynors, Oxford 1969
BHRS	*Publications of the Bedfordshire Historical Record Society*
Biddle	Martin Biddle, 'London on the Strand', *Popular Archaeology* 6, 1984, 23

Bk	A. Mawer and F. M. Stenton, *The Place-Names of Buckinghamshire*, Cambridge 1925
Blair 1988	John Blair, 'Minster Churches in the Landscape' in *Anglo-Saxon Settlements*, ed. Della Hooke, Oxford 1988, 35–58
Blair 1994	John Blair, *Anglo-Saxon Oxfordshire*, Gloucester 1994
BNJ	*British Numismatic Journal*
BONF	*Blätter für oberdeutsche Namenforschung*, München (by vol. and page)
Boniface	Works of Boniface, *Epistulae Selectae I*, ed. M. Tangl, Monumenta Germaniae Historica 1916
Bowcock	E. W. Bowcock, *Shropshire Place-Names*, Shrewsbury 1923
Bowen	E. G. Bowen, *The Settlement of the Celtic Saints in Wales*, Cardiff 1956
Boyle	J. R. Boyle, *The County of Durham*, London [1892]
BPC	Bede's prose 'Life of St Cuthbert' in *Two Lives of St Cuthbert*, ed. Bertram Colgrave, Cambridge 1940
Brentnall	M. Brentnall, *The Cinque Ports and Romney Marsh*, 2nd edn, London 1980
Britannia	*Britannia: A Journal of Romano-British and Kindred Studies*
Brk	Margaret Gelling, *The Place-Names of Berkshire*, 3 vols., EPNS 1974
Brno SE	*Brno Studies in English*
Brunner	Karl Brunner, *Altenglische Grammatik*, Tübingen 1965 (by section no.)
Bruton	*Bruton and Montacute Chartularies*, Somerset Record Society 8
BT	*The Book of Taliesin*, ed. J. Gwenogvryn Evans, Llanbedrog 1910
Buck	*Cartulary of Buckland Priory*, Somerset Record Society 25
Bülbring	K. D. Bülbring, *Altenglisches Elementarbuch. I. Teil: Lautlehre*, Heidelberg 1902 (by section)
Byn	G. Tengvik, *Old English Bynames*, Uppsala 1938
BzN	*Beiträge zur Namenforschung* (by year and page)
Ca	P. H. Reaney, *The Place-Names of Cambridgeshire and the Isle of Ely*, Cambridge 1943
Cambro-British Saints	*Lives of the Cambro-British Saints*. Welsh Manuscripts Society 1853
Camden	William Camden, *Britanniæ sive Angliæ, Scottiæ, Hiberniæ . . . descriptio*, trans. Philemon Holland, ed. E. Gibson, 2 vols., 1695
Cameron 1968	Kenneth Cameron, 'Eccles in English place-names' in *Christianity in Britain, 300–700*, ed. M. W. Barley and R. P. C. Hanson, Leicester 1968, pp. 87–92
Cameron 1998	Kenneth Cameron, *A Dictionary of Lincolnshire Place-Names*, EPNS 1998
Campbell	A. Campbell, *Old English Grammar*, Oxford 1959
Canu Llywarch Hen	*Canu Llywarch Hen* ed. Ifor Williams, 2nd edn Cardiff 1935; *The Poetry of Llywarch Hen*. Introduction, Text and Translation by Patrick K. Ford, Berkeley/Los Angeles/London 1974
Celtic Voices	Richard Coates and Andrew Breeze, *English Places, Celtic Voices: Studies of the Celtic Impact on Place-Names in England*, Stamford 2000
Chantraine	Pierre Chantraine, *Dictionnaire étymologique de la langue grecque*, 2 vols., Paris 1968
Charters I	*Anglo-Saxon Charters I: Charters of Rochester*, ed. A. Campbell, British Academy, London 1973 (by charter number)

Charters II	*Anglo-Saxon Charters II: Charters of Burton Abbey*, ed. P. H. Sawyer, British Academy, Oxford 1979
Charters III	*Anglo-Saxon Charters III: Charters of Sherborne*, ed. M. A. O'Donovan, British Academy, Oxford 1988
Charters IV	*Anglo-Saxon Charters IV: Charters of St Augustine's Abbey, Canterbury, and Minster-in-Thanet*, ed. S. E. Kelly, British Academy, Oxford 1995
Che	J. McN. Dodgson, *The Place-Names of Cheshire*, 5 vols. in 8, Cambridge 1970–81, EPNS 1998
Chronicles	R. Holinshed, *Chronicles*, London 1587
Chronique d'Egypte	*Chronique d'Egypte*, Bruxelles (by vol. and page)
CIL	*Corpus Inscriptionum Latinarum*
CMCS	*Cambridge Medieval Celtic Studies*
CoArch	*Cornish Archaeology*
Coins	*Sylloge of Coins of the British Isles* (by vol. and page)
Collinson	J. Collinson, *The History and Antiquities of the County of Somerset*, 1791
Costen	Michael Costen, *The Origins of Somerset*, Manchester 1992
CPNE	O. J. Padel, *Cornish Place-Name Elements*, EPNS 1985
Crawford	*The Crawford Collection of Early Charters*, ed. A. S. Napier and W. H. Stevenson, 1893
Cu	A. M. Armstrong, A. Mawer, F. M. Stenton and Bruce Dickins, *The Place-Names of Cumberland*, 3 vols., Cambridge 1950
Cullen	P. Cullen, 'The Place-Names of the Lathes of St Augustine and Shipway, Kent', unpublished PhD thesis, Sussex University, 1997
Cumbria	*Cumbria: A Monthly Magazine of Lakeland Life*
Cunliffe	Barry Cunliffe, *Wessex to AD 1000*, London/New York 1993
Cunnington	M. E. Cunnington, *Woodhenge*, Devizes 1929
Cur	*Curia Regis Rolls*
CW	*Transactions of the Cumberland and Westmorland Antiquarian and Archaeological Society*
D	J. E. B. Gover, A. Mawer and F. M. Stenton, *The Place-Names of Devon*, 2 vols., Cambridge 1931
DAG	J. Whatmough, *The Dialects of Ancient Gaul,* Cambridge and Edinburgh 1926, repr. 1973
DAJ	*Durham Archaeological Journal* (by vol. and page)
Dauzat-Rostaing	A. Dauzat and Ch. Rostaing, *Dictionnaire étymologique des noms de lieux en France*, Paris 1963
Db	Kenneth Cameron, *The Place-Names of Derbyshire*, 3 vols., Cambridge 1959
DB	Domesday Book. Abbreviations of the type *DB Devon* refer to the appropriate county volume of the complete edition, *Domesday Book*, text and tr. by John Morris, 34 vols., Chichester 1975–86 (by county vol. and entry no.)
DBS	P. H. Reaney, *A Dictionary of British Surnames*, 2nd edn, London, 1976
DEC	*Durham Episcopal Charters 1071–1152*, ed. H. S. Offler, Surtees Society 179, Gateshead 1968
DEPN	Eilert Ekwall, *The Concise Oxford Dictionary of English Place-Names*, 4th edn, Oxford 1960

D&CNQ	*Devon and Cornwall Notes and Queries*
DG	H. C. Darby and G. R. Versey, *Domesday Gazetteer*, Cambridge 1975
Dir	Kelly's Directory
Dixon	David Dippie Dixon, *Upper Coquetdale*, Newcastle-upon-Tyne 1903
DM	*Domesday Monachorum* in VCH Kent III
DML	R. E. Latham, *Dictionary of Medieval Latin from British Sources*, London 1975–
Do	A. D. Mills, *Dorset Place-Names: Their Origin and Meanings*, Wimborne 1986
Do i, ii, iii	A. D. Mills, *The Place-Names of Dorset* Parts I–III, EPNS 1977, 1980, 1989
Dobson	E. J. Dobson, *English Pronunciation 1500–1700*, 2 vols., 2nd edn, Oxford 1968 (vol ii. cited by para. no.)
Domesday Studies	*Domesday Studies*, ed. J. C. Holt, Woodbridge 1987
DS	Patrick Hanks and Flavia Hodges, *A Dictionary of Surnames*, Oxford 1988
DSÅ	J. Kousgård Sørensen, *Danske sø- og ånavne*, Copenhagen 1968
Duignan	W. H. Duignan, *Staffordshire Place-Names*, London 1902
Dutt	William A. Dutt, *Norfolk*, The Little Guides, London 1902
Dyer	Christopher Dyer, 'Towns and cottages in eleventh-century England' in *Studies in Medieval History presented to R. H. C. Davis*, ed. H. Mayr-Harting and R. I. Moore, London 1985
Earle	J. Earle (ed.), *Land Charters and other Saxonic Documents*, Oxford 1888
EbD	Martin Jones, *England Before Domesday*, London 1986
ECNENM	*The Early Charters of Northern England and the North Midlands*, ed. C. R. Hart, Leicester 1975
ECTV	*The Early Charters of the Thames Valley*, ed. Margaret Gelling, Leicester 1979 (by charter no.)
ECW	*The Early Charters of Wessex*, ed. H. P. R. Finberg, Leicester 1964 (by charter number)
ECWM	*The Early Charters of the West Midlands*, ed. H. P. R. Finberg, Leicester 1972
EDD	J. Wright, *The English Dialect Dictionary*, 6 vols., Oxford 1898–1905
Edda	Snorri Sturluson, *Edda*, ed. Finnur Jónsson, Copenhagen 1900
Eddi	*The Life of Bishop Wilfrid by Eddius Stephanus*, ed. Bertram Colgrave, Cambridge 1927
EDG	J. Wright, *The English Dialect Grammar*, Oxford 1905
EFlint	Hywel Wyn Owen, *The Place-Names of East Flintshire*, Cardiff 1994
EHN	O. S. Anderson (Arngart), *The English Hundred-Names*, 3 vols., Lund 1934–6
EHR	*The English Historical Review* (by vol. and page)
Ekwall 1929	Eilert Ekwall, 'Loss of a nasal before labial consonants' in *Studies in English Philology. A Miscellany in Honor of Frederick Klaeber*, ed. Kemp Malone and Martin B. Ruud, Minneapolis 1929, pp. 21–7 (repr. in SelPap)
Ekwall 1964	*Old English* wīc *in place-names*, Nomina Germanica 13, Lund 1964
Emery	F. Emery, *The Oxfordshire Landscape*, London 1974
EMEVP	*Early Middle English Verse and Prose*, ed. J. A. W. Bennett and G. V. Smithers, 2nd edn, Oxford 1968

Encyc	*The London Encyclopaedia*, ed. Ben Weinreb and Christopher Hibbert, London 1983
E&S	*Essays and Studies by Members of the English Association* (by year and page)
Ess	P. H. Reaney, *The Place-Names of Essex*, Cambridge 1935
ESt	*English Studies*
Evidence	*Place-Name Evidence for the Anglo-Saxon Invasion and Scandinavian Settlements.* Eight studies collected by Kenneth Cameron, EPNS 1975
EW	J. de Vries and F. de Tollenære, *Etymologisch Woordenboek*, Utrecht and Antwerpen 1983
Excavations	*Excavations at Mucking 2: The Anglo-Saxon Settlement*, English Heritage 1993
Exodus	*Exodus*, ed. Peter J. Lucas, London 1977
Exon	Exon Domesday, in vol. IV of the Record Commission edition of Domesday Book, London 1816
F	T. Forssner, *Continental-Germanic Personal-Names in England in Old and Middle English*, Uppsala 1916
Felix	*Felix's Life of Saint Guthlac*, ed. B. Colgrave, Cambridge 1956
FF	*Feet of Fines*
Field 1972	John Field, *English Field-Names: A Dictionary*, Newton Abbot 1972
Field 1993	J. Field, *A History of English Field-Names*, London 1993
FLH	*Folia Linguistica Historica: Acta Societatis Linguisticae Europaeae*
FmK	Ernst Förstemann, *Altdeutsche Personennamen, Ergänzungsband*, compiled by Henning Kaufmann, München/ Hildesheim 1968
FmO	Ernst Förstemann, *Altdeutsches Namenbuch*, vol. ii, *Orts- und sonstige geographische Namen*, Bonn 1913, repr. Hildesheim/München 1967 (by vol. and col.)
FmP	Ernst Förstemann, *Altdeutsches Namenbuch*, vol. i, *Personennamen*, Bonn 1900 (by col.)
FN	*Fryske Nammen*
Foerste	W. Foerste, 'Zur Geschichte des Wortes Dorf', *Studium Generale* 16, 1962, 422–33
Fordyce	William Fordyce, *The History and Antiquities of the County Palatine of Durham*, 2 vols., Newcastle-upon-Tyne 1857
Forsberg 1950	Rune Forsberg, *A Contribution to a Dictionary of Old English Place-Names*, Uppsala 1950
Forsberg 1997	Rune Forsberg, *The Place-Name Lewes: A Study of Its Early Spellings and Etymology*, Acta Universitatis Upsaliensis 100, Uppsala 1997
Fox 1989	H. S. A. Fox, 'The people of the Wolds in English settlement history' in *The Rural Settlements of Medieval England*, ed. M. Aston, D. Austin and C. Dyer, Oxford 1989, pp. 77–101
Fox 1996	H. S. A. Fox, 'Cellar settlement along the south Devon coastline' in *Seasonal Settlement*, ed. H. S. A. Fox, Vaughan Paper 39, Leicester 1996
FPD	*Feodarium Prioratus Dunelmensis*, Surtees Society 58, Durham 1872
Frid	*The Cartulary of the Monastery of St Frideswide*, Oxford Historical Society 1895–6

Frings	Theodor Frings, *Germania Romana*, 2 vols., 2nd edn, Halle Saale 1996–8
Fris	Gillian Fellows-Jensen, 'Doddington revisited', in *A Frisian and Germanic Miscellany* (in honour of Nils Århammar), Odense 1996, pp. 361–76
FT	Max Förster, *Der Flußname Themse und seine Sippe, Studien zur Anglisierung keltischer Eigennamen und zur Lautchronologie des Altbritischen*. Sitzungsberichte der Bayerischen Akademie der Wissenschaften, Philosophisch-historische Abteilung, Jahrgang 1941, München 1942
Fuller	Thomas Fuller, *The History of the Worthies of England*, London 1662
FW	Florence of Worcester, *Chronicon ex Chronicis*, ed. B. Thorpe, 2 vols., London 1848–9
Gale	Thomas Gale, *Historiœ Britannicœ Scriptores XV*, Oxford 1691
Gaz	*Cassell's Gazetteer of Great Britain and Ireland*, 6 vols., London 1896
GED	Winfred P. Lehmann, *A Gothic Etymological Dictionary*, Leiden 1986 (by letter and entry no.)
Germanic Peoples	Rolf Hachmann, *The Germanic Peoples*, London 1971
Gelling	Dr Margaret Gelling, personal communication
Gelling 1981	M. Gelling, 'The word *church* in English place-names', *Bulletin of the Council for British Archaeology Churches Committee* 15, 1981, 4–9
Gelling 1998	Margaret Gelling, 'Place-names and landscape' in *The Uses of Place-Names*, ed. Simon Taylor, Scottish Cultural Press, Edinburgh 1988, with illustrations by Ann Cole
Gildas	Gildas, *De excidio et conquestu Britanniae*, ed. Michael Winterbottom, History from the Sources, Arthurian Period Sources 7, London 1978
Gl	A. H. Smith, *The Place-Names of Gloucestershire*, 4 vols., Cambridge 1964–5
GL	J. Field, *Place-Names of Greater London*, London 1980
Glast	*Rentalia et custumaria . . . abbatum . . . Glastoniae*, Somerset Record Society 1891
Gover n.d.	J. E. B. Gover, unpublished typescript of the place-names of Cornwall
Gover 1958	J. E. B. Gover, Hampshire Place Names, unpublished typescript (1958)
GPC	*Geiriadur Prifysgol Cymru: a Dictionary of the Welsh Language*, Caerdydd, 1950– (by col.)
Gregory	*The Earliest Life of Gregory the Great*, ed. Bertram Colgrave, Cambridge 1968
Grieve	Mrs M Grieve, *A Modern Herbal*, London 1931. Cited from Harmondsworth repr. 1976
Griscom	A. Griscom, *The Historia Regum of Geoffrey of Monmouth*, London 1929
Grundy 1919	G. B. Grundy, *The Saxon Land Charters of Wiltshire*, 1st series, Arch J 76, 1919, 143–301
Grundy 1920	G. B. Grundy, *The Saxon Land Charters of Wiltshire*, 2nd series, Arch J 77, 1920, 8–126
Grundy 1935	G. B. Grundy, *Saxon Charters and Field Names of Gloucestershire*, 2 parts, Bristol and Gloucestershire Archaeological Society 1935–6
GS	H. Krahe, W. Meid, *Germanische Sprachwissenschaft* vols. i–iii, Sammlung Göschen 238, 780, 1218, Berlin 1967–9 (by vol. and para.)

Guthlac	*Das angelsächsische Prosa-Leben des hl. Guthlac*, ed. P. Gonser. Anglistische Forschungen xxvii, Heidelberg 1909
GV	B. K. Roberts, *The Green Villages of County Durham*, Durham 1978
Ha	Richard Coates, *The Place-Names of Hampshire*, London 1989
Hadfield	Charles Hadfield, *The Canals of Yorkshire and North East England*, 2 vols., Newton Abbot 1973
Harmer	F. E. Harmer, 'A Bromfield and a Coventry Writ of King Edward the Confessor' in *The Anglo-Saxons. Studies in Some Aspects of their History and Culture presented to Bruce Dickins*, ed. Peter Clemoes, London 1959, pp. 89–103
Harris	Cyril M. Harris, *What's in a Name?* London Transport 1977
Hasted	E. Hasted, *The History and Topographical Survey of Kent*, Canterbury 1778–99
HB	Nennius' *Historia Brittonum*, ed. John Morris in *Nennius: British History and the Welsh Annals*, London 1980
He	Bruce Coplestone-Crow, *Herefordshire Place-Names*, BAR Brit 214, 1989
Hellberg	Lars Hellberg, 'Kumlabygdens ortnamn och äldre bebyggelse', *Kumlabygden* III, Kumla 1967, 196–208
Heming	*Hemingi chartularium Ecclesiae Wigorniensis*, ed. T. Hearne, 2 vols., Oxford 1723
HistCleve	John Walker Ord, *The History and Antiquities of Cleveland*, London 1846
HistEl	*Historia Eliensis* or *Liber Eliensis*, Anglia Christiana Society 1848
HistKings	*Geoffrey of Monmouth: The History of the Kings of Britain*, trans. Lewis Thorpe, Harmondsworth 1966
Hodgson	J. C. Hodgson, *History of Northumberland*, 7 vols., 1896–1904
Holder	A. Holder, *Alt-celtischer Sprachschatz*, 3 vols., Leipzig 1896–1907 (by vol. and col.)
Holmberg	B. Holmberg, *Tomt och toft som appellativ och ortnamnselement*, Uppsala 1946
Holthausen	F. Holthausen, *Altenglisches etymologisches Wörterbuch*, Heidelberg 1934
Hooke	Della Hooke, *Worcestershire Anglo-Saxon Charter-Bounds*, Woodbridge 1990
Hope-Taylor	Brian Hope-Taylor, *Yeavering: An Anglo-British Centre of Early Northumbria*, Department of the Environment Archæological Reports 7, London 1977
Horovitz	Manuscript collection of Staffordshire place-names, compiled by David Horovitz
HR	*Symeonis Dunelmensis Historia Regum* in *Symeonis Monachi Opera omnia* II, ed. Th. Arnold, Rolls Series 75, 2 vols., London 1882–5
Hrt	J. E. B. Gover, Allen Mawer and F. M. Stenton, *The Place-Names of Hertfordshire*, Cambridge 1938
HSC	*Historia de Sancto Cuthberto*, Surtees Society vol. 51; Rolls Series 75
Hu	Huntingdonshire entries in A. Mawer and F. M. Stenton, *The Place-Names of Bedfordshire and Huntingdonshire*, Cambridge 1926
Humphreys	A. L. Humphreys, *Somerset Parishes: A Handbook of Historical Reference to all Places in the County*, London 1905

Hung	*Hungerford Cartulary*, Somerset Record Society
Ibis	*The Ibis: The Quarterly Journal of the British Ornithologist's Union*
ICC	*Inquisitio Comitatus Cantabrigiensis*, ed. N. E. A. S. Hamilton, London 1876
InqEl	*Inquisitio Eliensis* in vol. iv of the Record Commission edn of Domesday Book, London 1816
IF	*Indogermanische Forschungen: Zeitschrift für Indogermanische Sprache- und Altertumskunde*
Ilkow	P. Ilkow, *Die Nominalkomposita der altsächsische Bibeldichtung*, Göttingen 1968
ING	Eilert Ekwall, *English Place-Names in -ing*, 2nd edn, Lund 1962
Inn Names	Barrie Cox, *English Inn and Tavern Names*, Nottingham 1994
IntroSurvey	A. Mawer and F. M. Stenton, *The Survey of English Place-Names I.i, Introduction to the Survey*, Cambridge 1924
IPM	*Inquisitio post mortem*
IPN	Deirdre Flanagan and Lawrence Flanagan, *Irish Place-Names*, Dublin 1994
Itinerarium Kambriæ	*Itinerarium Kambriae* in Giraldus Cambrensis, *Opera*, ed. J. S. Brewer, J. F. Dimock and G. F. Warner, Rolls Series, 8 vols., London 1861–91
JCS	*Journal of Celtic Studies*
Jnl	*Journal of the English Place-Name Society* (by vol. and page)
Johansson	C. Johansson, *Old English Place-Names and Field-Names Containing* lēah, Stockholm, 1975
Jordan	Richard Jordan, *Handbuch der mittel-englischen Grammatik*, I, Heidelberg 1934
JRS	*Journal of Roman Studies*
K	*Codex Diplomaticus Aevi Saxonici*, 6 vols., ed. J. M. Kemble, London 1839–48
KPN	J. K. Wallenberg, *Kentish Place-Names*, Uppsala 1931
Kitchin	T. Kitchin's map of Durham 1763
Kitson	P. Kitson, 'Quantifying Qualifiers in Anglo-Saxon Charter Boundaries', *Linguistica Historica* xiv, 1994, 29–82
KLMN	*Kulturhistorisk Leksikon for Nordisk Middelalder*, Copenhagen 1956ff
Kluge	Friedrich Kluge, *Nominale Stammbildungslehre der altgermanischen Dialekte*, Halle (Saale) 1926 (by section)
Kluge-Mitzka	Friedrich Kluge, *Etymologisches Wörterbuch der deutschen Sprache*, 21st edn, ed. Walther Mitzka, Berlin 1975
Kluge-Seebold	Friedrich Kluge, *Etymologisches Wörterbuch der deutschen Sprache*, 22nd edn, completely revised by Elmar Seebold, Berlin/New York 1989
Krahe 1962	H. Krahe, *Die Struktur der alteuropäischen Hydronomie*, Mainzer Akademie der Wissenschaften, Abhandlung der Geistes- und Sozialwissenschaftlichen Klasse, Jahrgang 1962, nr. 5, Mainz 1962
Krahe 1964	Hans Krahe, *Unsere ältesten Flußnamen*, Wiesbaden 1964
Krahe 1966–9	Hans Krahe, *Indogermanische Sprachwissenschaft* vols. i–ii. Sammlung Göschen 59, 64, Berlin 1966–9 (by vol. and para.)
Kristensson	Gillis Kristensson, *Studies on Middle English Topographical Terms*, Lund 1970

L	Margaret Gelling, *Place-Names in the Landscape*, London 1984
La	Eilert Ekwall, *The Place-Names of Lancashire*, Manchester 1922
LancsLH	*The Lancashire Local Historian* (by issue and page)
Laur	Wolfgang Laur, *Historisches Ortsnamenlexikon von Schleswig-Holstein*, 2nd edn, Neumünster 1992
Laws	Anglo-Saxon Laws (F. L. Attenborough, *The Laws of the Earliest English Kings*, Cambridge 1922; F. Liebermann, *Die Gesetze der Angelsachsen*, 3 vols., Halle 1903–16)
LbO	Wolf-Arnim Freiherr von Reitzenstein, *Lexicon bayerischer Ortsnamen*, 2nd edn, München 1991
LDB	Little Domesday Book: DB for Essex, Norfolk, Suffolk
Lei	Barrie Cox, 'The Place-Names of Leicestershire and Rutland', unpublished Ph.D. thesis, University of Nottingham 1971
LHEB	Kenneth Jackson, *Language and History in Early Britain*, Edinburgh 1953
Li i–v	Kenneth Cameron, *The Place-Names of Lincolnshire*, EPNS 1985–97
Lincs AAS	*Lincolnshire Architectural and Archaeological Society Reports and Papers*
Lindkvist	H. Lindkvist, *Middle English Place-Names of Scandinavian Origin*, Uppsala 1912
LL	J. G. Evans and J. Rhys, *The Text of the Book of Llan Dâv*, Old Welsh Texts, Oxford 1893
LLH	*Lancashire Local Historian*
LMxAS	*Transactions of the London and Middlesex Archaeological Society*
Locus	*Locus Focus: Forum of the Sussex Place-Name Net*
Löfvenberg	M. T. Löfvenberg, *Studies in Middle English Local Surnames*, Lund 1942
Long	Brian Long, *Castles of Northumberland*, Newcastle-upon-Tyne 1967
LPN	A. D. Mills, *A Dictionary of London Place-Names*, Oxford 2001
LS	Lay Subsidy Roll
LSE	*Leeds Studies in English*
LT	*Liber Tern*
Lucerna	H. R. P. Finberg, *Lucerna: Studies in Some Problems in the Early History of England*, London 1964
Luick	Karl Luick, *Historische Grammatik der englischen Sprache*, 2 vols., 1914–40; repr. Stuttgart and Oxford 1964 (by para.)
Lund	Niels Lund, 'Thorp-names' in *Medieval Settlement: Continuity and Change*, ed. P. Sawyer, London 1976, pp. 233–5
LVD	*Liber Vitæ Ecclesiæ Dunelmenis*, Surtees Society Publications 136
LVH	*Liber Vitæ: Register and Martyrology of New Minster and Hyde Abbey*, ed. W. de G. Birch, Hampshire Record Society, Winchester 1892
M	*Montacute Chartulary*, Somerset Record Society 8
MA	*Medieval Archaeology: Journal of the Society for Medieval Archaeology*
MacBain	Alexander MacBain, *An Etymological Dictionary of the Gaelic Language*, 2nd edn, 1911; repr. Glasgow 1982
Map 1575	Saxton's county maps
Map 1610	Speed's county maps
Margary	Ivan D. Margary, *Roman Roads in Britain*, rev. edn, London 1967

Mellard	S. L. Mellard, 'The Place-Names of Muggleswick Township in the County of Durham', unpublished BA dissertation, University of Durham 1983
MI	*The Maritime Itinerary*, cited from RBrit 180ff
Middendorff	H. Middendorff, *Altenglisches Flurnamenbuch*, Halle 1902
Migne	*Patrologia Latina*, ed. J. P. Migne
Mills	David Mills, *The Place-Names of Lancashire*, London 1976
Mills 1991	A. D. Mills, *A Dictionary of English Place-Names*, Oxford 1991
Mills 1996	A. D. Mills, *The Place-Names of the Isle of Wight*, Stamford 1996
Mills 1998	A. D. Mills, *A Dictionary of English Place-Names*, 2nd edn, Oxford 1998
MLN	*Modern Language Notes*
MM	*The Mariner's Mirror, The Journal of the Society for Nautical Research*
Morgan	Richard Morgan, *Welsh Place-Names in Shropshire*, Cardiff 1997
Morris	J. E. Morris, *Northumberland*, London 1916
Morsbach	Lorenz Morsbach, *Mittelenglische Grammatik*, Halle 1896
Morton	A. Morton, *The Trees of Shropshire*, Shrewsbury 1986
Mx	J. E. B. Gover, Allen Mawer and F. M. Stenton, *The Place-Names of Middlesex apart from the City of London*, Cambridge 1942
Myres	J. N. L. Myres, *The English Settlements*, Oxford 1986
Names	*Names. Journal of the American Name Society*
NbDu	Allen Mawer, *The Place-Names of Northumberland and Durham*, Cambridge 1920
NCR	Norwich Cathedral Register
ND	The *Notitia Dignitatum*, cited from RBrit 216ff
Newm	*Chartularium Abbathiae de Novo Monasterio*, ed. J. T. Fowler, Surtees Society Publications 66, Durham 1878
Newman 1969E	John Newman, *The Buildings of England: North East and East Kent*, Harmondsworth 1969
Newman 1969W	John Newman, *The Buildings of England: West Kent and the Weald*, Harmondsworth 1969
Newman-Pevsner	John Newman and Nikolaus Pevsner, *The Buildings of England: Dorset*, Harmondsworth 1972
Newton	Robert Newton, *The Northumberland Landscape*, London 1972
Nf	Karl Inge Sandred and Bengt Lindström, *The Place-Names of Norfolk*, 2 vols., EPNS 1989, 1996
NG	*Norske Gaardnavne*, Kristiana 1897–1936
NGN	*Nomina Geographica Neerlandica*, Amsterdam/Leiden (by vol. and page)
NH	*History of Northumberland*, Northumberland County History Committee, 15 vols., 1905–40
NM	*Neuphilologische Mitteilungen* (by vol. and page)
NMS	*Nottingham Medieval Studies*
NoB	*Namn och Bygd: tidskrift för nordisk ortnamsforskning* (by vol. and page)
Nomina	*Nomina: A Journal of Names Studies Relating to Great Britain and Ireland* (by vol. and page)
Notes	E. Ekwall, *Etymological Notes on English Place-Names*, Lund 1959
NQ	*Notes and Queries*
Nt	J. E. B. Gover, Allen Mawer and F. M. Stenton, *The Place-Names of Nottinghamshire*, Cambridge 1940

Nth	J. E. B. Gover, A.Mawer and F. M. Stenton, *The Place-Names of Northamptonshire*, Cambridge 1933
NWG	*Names, Words and Graves: Early Medieval Settlement*, ed. P. H. Sawyer, Leeds 1979
O	Margaret Gelling, *The Place-Names of Oxfordshire*, 2 vols., Cambridge 1953–4
Oakden	Unpublished typescript for the survey of the place-names of Staffordshire, in possession of EPNS
OBB	*The Oxford Book of Ballads*, ed. James Kinsley, Oxford 1969
O'Donnell	*Angles and Britons*, O'Donnell Lectures, Cardiff 1963
ODEE	*The Oxford Dictionary of English Etymology*, ed. C. T. Onions, Oxford 1966
ODS	David Hugh Farmer, *The Oxford Dictionary of Saints*, 3rd edn, Oxford 1992
OEBede	*The Old English Version of Bede's Ecclesiastical History*, ed. T. Miller, Early English Text Society Original Series 95–96, London 1890–91
OED	*The Oxford English Dictionary*
OES	P. H. Reaney, *The Origin of English Surnames*, London 1967
Ogam	*Ogam: Bulletin . . . des Amis de la tradition celtique Bretagne-armorique*
OLD	*Oxford Latin Dictionary*, ed. P. G. W. Glare, Oxford 1982
O'Rahilly	T. F. O'Rahilly, *Early Irish History and Mythology*, Dublin 1984
Ord	Ordericus Vitalis, *The Ecclesiastical History*, ed. M. Chibnall, 6 vols., Oxford 1969–80
Origins	*The Origins of Anglo-Saxon Kingdoms*, ed. Steven Basset, London 1989
Ortsnamenwechsel	*Ortsnamenwechsel Bamberger Symposium 1–4 Oktober 1986*, ed. R. Schützeichel, BzN n.f. Beiheft 24, Heidelberg 1986
Orton	Harold Orton, *The Phonology of a South Durham Dialect*, London 1933 (by para.)
OS	1st edn of the One Inch Ordnance Survey Maps of England
Otium	*Otium et Negotium: Studies in Onomatology and Library Science presented to Olof von Feilitzen*, ed. F. Sandgren, Stockholm 1973
Paganism	M. Gelling, 'Place-names and Anglo-Saxon paganism', *University of Birmingham Historical Journal* VIII, 1961, 7–25
Palaestra	*Palaestra: Untersuchungen aus der deutschen und englischen Philologie und Literaturgeschichte*
Payling 1936	L. W. H. Payling, 'The Place-Names of Kesteven (South-West Lincolnshire)', unpublished MA thesis, University of Leeds 1936
Payling 1940	L. W. H. Payling, 'The Place-Names of the Parts of Holland, South-East Lincolnshire', unpublished PhD thesis, University of London 1940
Pembs	B. G. Charles, *The Place-Names of Pembrokeshire*, 2 vols., Aberystwyth 1992
Pennant 1804	Thomas Pennant, *A Tour from Alston Moor to Harrowgate and Brimham Craggs*, London 1804
Pennant 1811	Thomas Pennant, *The Journey from Chester to London*, London 1811
Perrott	Michael Perrott, 'The Place-Names of the Kesteven Division of Lincolnshire', unpublished M.Phil. thesis, University of Nottingham 1979

Persson — P. Persson, *Beiträge zur indogermanischen Wortforschung*, 2 vols., Uppsala 1912

Peut — The Peutinger Table cited from RBrit 149–50

Pevsner 1951 — Nikolaus Pevsner, *The Buildings of England: Nottinghamshire*, Harmondsworth 1951

Pevsner 1954 — Nikolaus Pevsner, *The Buildings of England: Cambridgeshire*, Harmondsworth 1954

Pevsner 1957 — Nikolaus Pevsner, *The Buildings of England: Northumberland*, Harmondsworth 1957

Pevsner 1958N — Nikolaus Pevsner, *The Buildings of England: North Somerset and Bristol*, Harmondsworth 1958

Pevsner 1958S — Nikolaus Pevsner, *The Buildings of England: South and West Somerset*, Harmondsworth 1958

Pevsner 1962NE — Nikolaus Pevsner, *The Buildings of England: North East Norfolk and Norwich*, Harmondsworth 1962

Pevsner 1962NW — Nikolaus Pevsner, *The Buildings of England: North West and South Norfolk*, Harmondsworth 1962

Pevsner 1963 — Nikolaus Pevsner, *The Buildings of England: Herefordshire*, Harmondsworth 1963

Pevsner 1964 — Nikolaus Pevsner and John Harris, *The Buildings of England: Lincolnshire*, Harmondsworth 1964

Pevsner 1965 — Nikolaus Pevsner, *The Buildings of England: Essex*, 2nd edn, rev. Enid Radcliffe, Harmondsworth 1965

Pevsner 1966 — Nikolaus Pevsner, *The Buildings of England: Yorkshire, The North Riding*, Harmondsworth 1966

Pevsner 1967 — Nikolaus Pevsner, *The Buildings of England: Yorkshire, The West Riding*, rev. Enid Radcliffe, Harmondsworth 1967

Pevsner 1968 — Nikolaus Pevsner, *The Buildings of England: Worcestershire*, Harmondsworth 1968

Pevsner 1969N — Nikolaus Pevsner, *The Buildings of England: Lancashire 2. The Rural North*, Harmondsworth 1969

Pevsner 1969S — Nikolaus Pevsner, *The Buildings of England: Lancashire 1. The Industrial and Commercial South*, Harmondsworth 1969

Pevsner 1973 — Nikolaus Pevsner, *The Buildings of England: Northamptonshire*, rev. Bridget Cherry, Harmondsworth 1973

Pevsner 1973L — Nikolaus Pevsner, *London: The Cities of London and Westminster*, 3rd edn, Harmondsworth 1973

Pevsner 1974 — Nikolaus Pevsner, *The Buildings of England: Suffolk*, 2nd edn, rev. Enid Radcliffe, Harmondsworth 1974

Pevsner 1975 — Nikolaus Pevsner, *The Buildings of England: Wiltshire*, rev. Bridget Cherry, Harmondsworth 1975

Pevsner 1992 — Nikolaus Pevsner, *The Buildings of England: Northumberland*, 2nd edn, rev. J. Grundy, London 1992

Pevsner-Hubbard 1971 — Nikolaus Pevsner and Edward Hubbard, *The Buildings of England: Cheshire*, Harmondsworth 1971

Pevsner-Lloyd 1967 — Nikolaus Pevsner and David Lloyd, *The Buildings of England: Hampshire and the Isle of Wight*, Harmondsworth 1967; repr. 1973

Piroth	Walter Piroth, *Ortsnamenstudien zur angelsächsischen Wanderung*, Frankfurter Historische Abhandlungen, vol. 18, Wiesbaden 1979
Plymley 1803	J. Plymley, *General Survey of the Agriculture of Shropshire*, London 1803
PN and History	A. Mawer, *Place-Names and History,* Liverpool 1922
PNCo	O. J. Padel, *A Popular Dictionary of Cornish Place-Names*, Penzance 1988
PNDB	Olof von Feilitzen, *The Pre-Conquest Personal Names of Domesday Book*, Uppsala 1937
PNE	A. H. Smith, *English Place-Name Elements*, 2 vols., Cambridge 1956 (by vol. and page)
PNK	J. K. Wallenberg, *The Place-Names of Kent*, Uppsala 1934
Pokorny	J. Pokorny, *Indogermansiches etymologisches Wörterbuch*, Bern 1959
Pole	Sir W. Pole, *Collections Towards a Description of the County of Devon*, 1791
Pope	M. K. Pope, *From Latin to Modern French*, 2nd edn, Manchester 1952
Problems	A. Mawer, *Problems of Place-Name Study*, Cambridge 1929
ProcHants	*Papers and Proceedings of the Hampshire Field Club and Archaeological Society*
ProcLCAS	*Proceedings of the Lancashire and Cheshire Archaeological Society*
PSA	*The Proceedings of the Society of Antiquaries of Newcastle-upon-Tyne* (by series, vol. and page)
PR	unpublished parish register
PSI	*Proceedings of the Suffolk Institute of Archaeology and Natural History* (by vol., year and part)
Ptol	Ptolemy's Geography, cited from RBrit 103ff
Pub Names	Leslie Dunklin and Gordon Wright, *A Dictionary of Pub Names*, London 1987
R	B. Cox, *The Place-Names of Rutland*, EPNS 1994
Rackham	O. Rackham, *Trees and Woodland in the British Landscape*, London 1976
Rav	The Ravenna Cosmography, cited from RBrit 185ff
Raven	Michael Raven, *A Shropshire Gazetteer*, Market Drayton 1989
RBrit	A. L. F. Rivet and Colin Smith, *The Place-Names of Roman Britain*, London 1979
RBE	*The Red Book of the Exchequer*, Rolls Series 1896
Reallexikon	*Reallexikon der germanischen Altertumskunde*, Berlin 1973–
Redin	M. Redin, *Studies on Uncompounded Personal Names in Old English*, Uppsala 1919
Reichardt 1982	Lutz Reichardt, *Ortsnamenbuch des Stadtkreises Stuttgart und des Landkreises Ludwigsburg*, Veröffentlichungen der Kommission für Geschichtliche Landeskunde in Baden-Württemberg, Reihe B, vol. 101, Stuttgart 1982
Reichardt 1983	Lutz Reichardt, *Ortsnamenbuch des Kreises Reutlingen*, Veröffentlichungen der Kommission für Geschichtliche Landeskunde in Baden-Württemberg, Reihe B, vol. 102, Stuttgart 1983
Reichardt 1986	Lutz Reichardt, *Ortsnamenbuch des Alb-Donau-Kreises und des Stadtkreises Ulm*, Veröffentlichungen der Kommission für geschichtliche Landeskunde in Baden-Württemberg, Reihe B, Forschungen, vol. 105, Stuttgart 1986

Reichardt 1993	Lutz Reichardt, *Ortsnamenbuch des Rems-Murr-Kreises*, Veröffentlichungen der Kommission für geschichtliche Landeskunde in Baden-Württemberg, Reihe B, Forschungen, vol. 128, Stuttgart 1993
RES	*The Review of English Studies*
RIB	R. G. Collingwood and R. P. Wright, *The Roman Inscriptions of Britain I: Inscriptions on Stone*, Oxford 1965
RIO	*Revue Internationale d'Onomastique*
Ritter	Otto Ritter, *Vermischte Beiträge zur englischen Sprachgeschichte*, Halle (Saale) 1922
RN	Eilert Ekwall, *English River-Names*, Oxford 1928
Robinson	T. Robinson, *An Essay towards a Natural History of Westmorland and Cumberland*, London 1709
Room 1983	Adrian Room, *A Concise Dictionary of Modern Place-Names*, Oxford 1983
Room 1988	Adrian Room, *Dictionary of Place-Names in the British Isles*, London 1988
Ross	Anne Ross, *Pagan Celtic Britain*, London 1974
Rostaing	C. Rostaing, *Essai sur la toponymie de la Provence*, Paris 1950; repr. Marseilles 1973
RRS	*Regesta Regum Scottorum*, ed. G. W. S. Barrow and others, 1–, Edinburgh 1960–
Rumble 1976	A. R. Rumble, 'Place-names and their context: with special regard to the Croydon Survey Region', Fifth C. C. Fagg Memorial Lecture, *Proceedings of the Croydon Natural History and Scientific Society* 8, 1976, 161–84
Rumble 1979	A. R. Rumble, 'The Structure and Reliability of the Codex Wintoniensis', unpublished Ph.D. thesis, University of London, 1979 (by vol. and page)
Rumble 1980	A. R. Rumble, 'HAMTVN *alias* HAMWIC (Saxon Southampton): the place-name traditions and their significance' in *Excavations at Melbourne Street, Southampton 1971–6*, Southampton Archaeological Research Committee Report 1 (CBA Research Report 33), 1980
Rumble 1987	A. R. Rumble, 'Old English *Boc-land* as an Anglo-Saxon estate name', LSE 18, 1987, 219–29
Rumble 1997	A. R. Rumble, '*Ad Lapidem* in Bede and a Mercian Martyrdom' in *Names, Places and People: An Onomastic Miscellany in Memory of John McNeal Dodgson*, ed. A. R. Rumble and A. D. Mills, Stamford 1997
Ryedale Historian	*Ryedale Historian*
Sa	Margaret Gelling, *The Place-Names of Shropshire*, 2 vols., EPNS 1990, 1995
SAC	*Sussex Archaeological Collections Relating to the History and Antiquities of the County*
Saga-book	*Saga-book of the Viking Society for Northern Research*
Saints	*Die Heiligen Englands*, ed. F. Liebermann, Hanover 1889
Salman	Arthur L. Salman, *Cornwall*, London 1903
S	P. H. Sawyer, *Anglo-Saxon Charters: An Annotated List and Bibliography*, London 1968 (by charter no.)
SBG	S. Baring-Gould, *Devon*, London 1907
Schönfeld	M. Schönfeld, *Wörterbuch der altgermanischen Personen- und Völkernamen*, 2nd edn, Heidelberg 1965

Schram O. K. Schram, 'Place-names' in *Norwich and its Region*, British Association for the Advancement of Science, 1961

Schröder E. Schröder, *Deutsche Namenkunde*, Göttingen 1938

Schwarz H. Schwarz, *Festschrift für J. Trier zum 70 Geburtstag*, ed. W. Foerste and K-H. Borck, Köln, Graz, 1964

ScotPN W. H. F. Nicolaisen, *Scottish Place-Names*, London 1976

SD *Symeonis Monachi Dunelmensis Libellus de Exordio atque Procursu Dunhelmensis Ecclesiae*, ed. Thomas Bedford, London 1732

S&DNQ *Somerset and Dorset Notes and Queries*

Secgan D. W. Rollason, 'Lists of saints' resting-places in Anglo-Saxon England', ASE 7, 61–93

SelPap Eilert Ekwall, *Selected Papers*, Lund 1963

Settlement *Medieval Settlement: Continuity and Change*, ed. P. H. Sawyer, London 1976

Settlement. Eng *English Medieval Settlement*, ed. P. H. Sawyer, London 1979

Signposts Margaret Gelling, *Signposts to the Past: Place-Names and the History of England*, London 1978

SN *Studia Neophilologica: A Journal of Germanic and Romance Philology*

SöR *Södermanlands runinskrifter*, 1924–36, by Brate-Wessén in the series *Sveriges runinskrifter*, Vitterhetsakademien (by inscription)

Speculum *Speculum: A Journal of Medieval Studies*

SPN Gillian Fellows Jensen, *Scandinavian Personal Names in Lincolnshire and Yorkshire*, Copenhagen 1968

SPNN John Insley, *Scandinavian Personal Names in Norfolk*, Uppsala 1994

Sr J. E. B. Gover, A. Mawer and F. M. Stenton in collaboration with A. Bonner, *The Place-Names of Surrey*, Cambridge 1934

SR *The Northumbrian Lay Subsidy Roll of 1296*, ed. in translation by Constance M. Fraser, The Society of Antiquaries of Newcastle-upon-Tyne, 1968 (by entry)

SRO Somerset Record Office

SS *The South Saxons*, ed. Peter Brandon, Chichester 1978

SSNEM Gillian Fellows Jensen, *Scandinavian Settlement Names in the East Midlands*, Copenhagen 1978

SSNNW Gillian Fellows-Jensen, *Scandinavian Settlement Names in the North-West*, Copenhagen 1985

SSNY Gillian Fellows Jensen, *Scandinavian Settlement Names in Yorkshire*, Copenhagen 1972

St J. P. Oakden, *The Place-Names of Staffordshire*, EPNS 1984

Staffordshire Studies *Staffordshire Studies*, Keele

Stead K. I. Sandred, *English Place-Names in -stead*, Acta Universitatis Upsaliensis, Studia Anglistica Upsaliensia 2, Uppsala 1963

Stenton 1926 F. Stenton, *The Free Peasantry of the Northern Danelaw*, Bulletin de la Société Royale des Lettres de Lund 1925–6, Lund 1926

Stenton 1947 *Anglo-Saxon England*, Oxford 1947

Stiles Patrick V. Stiles, 'Old English *halh*: slightly raised ground isolated by marsh' in *Names, Places and People: An Onomastic Miscellany in Memory*

	of John McNeal Dodgson, ed. Alexander R. Rumble and A. D. Mills, Stamford 1997, pp. 330–44
Stokes	H. G. Stokes, *English Place-Names*, London 1948
Stōw	Margaret Gelling, 'Some meanings of *stōw*' in *The Early Church in Western Britain and Ireland: Studies Presented to C. A. Ralegh Radford*, BAR Brit 102, 1982
Studies 1931	Eilert Ekwall, *Studies on English Place- and Personal-Names*, Lund 1931
Studies 1936	Eilert Ekwall, *Studies on English Place-Names*, Stockholm 1936
Surtees	Robert Surtees, *The History and Antiquities of the County Palatine of Durham*, 4 vols., London 1816–40 (by vol., part and page)
Sx	A. Mawer and F. M. Stenton with the assistance of J. E. B. Gover, *The Place-Names of Sussex*, 2 vols., Cambridge 1929–30
SWAH	*Saint Wilfrid at Hexham*, ed. D. P. Kirby, Newcastle-upon-Tyne 1974
TA	unpublished tithe award
Tacitus *Annals*	*Cornelii Taciti Annalium*, ed. C. D. Fisher, Oxford 1906
TBWAS	*Transactions of the Birmingham and Warwickshire Archaeological Society*
TC	W. F. H. Nicolaisen, Margaret Gelling and Melville Richards, *The Names of Towns and Cities in Britain*, London 1970
TCWAAS	*Transactions of the Cumberland and Westmorland Antiquarian and Archaeological Society*
Tengstrand	E. Tengstrand, *A Contribution to the Study of Genitival Composition in Old English Place-Names*, Uppsala 1940
Thomas 1976	Nicholas Thomas, *Guide to Prehistoric England*, 2nd edn, London 1976
Thomas 1980	Charles Thomas, *Christianity in Roman Britain to AD 500*, London 1980
Thorpe	B. Thorpe, *Diplomatarium Anglicum*, London 1865
TLCAS	*Transactions of the Lancashire and Cheshire Antiquarian Society*
Thurneysen	Rudolf Thurneysen, *A Grammar of Old Irish*, rev. edn, trans. D. A. Binchy and Osborn Bergin, Dublin 1946
Tomlinson	William Weaver Tomlinson, *Comprehensive Guide to the County of Northumberland*, 11th edn, Newcastle-upon-Tyne n.d.
Torrington	[J. B. Torrington] *The Torrington Diaries, Containing the Tours through England and Wales between the years 1781 and 1794*, ed. C. B. and F. Andrews, 1934–8
Townend 1998	Matthew Townend, *English Place-Names in Skaldic Verse*, EPNS 1998
TPS	*Transactions of the Philological Society* (by year and page)
Trans BGAS	*Transaction of the Bristol and Gloucestershire Archaeological Society* (by vol. and page)
Trans D&N	*Transactions of the Architectural and Archaeological Society of Durham and Northumberland* (by vol. and page)
Trautmann BSW	Reinhold Trautmann, *Baltisch-Slavisches Wörterbuch*, Göttingen 1923; 2nd edn, Göttingen 1970
TRE	*Tempore Regis Edwardi*(i), i.e. before 1086
TRHS	*Transactions of the Royal Historical Society*
Trier 1952	J. Trier, *Holz. Etymologien aus den Niederwald*, Köln, 1952
Trier 1956	J. Trier, 'Fragen und Forschungen im Bereich und Umkreis der german-

	ischen Philologie', *Festgabe für Th. Frings zum 70 Geburtstag*, Berlin, 1956, 32
TSAHS	*Transactions of the Staffordshire Archaeological and Historical Society*
TSAS	*Transactions of the Shropshire Archaeological Society* (by series, vol. and page)
TSSAHS	*Transactions of the South Staffordshire Archaeological and Historical Society* (by vol. and page)
TT	Richard Coates, *Toponymic Topics: Essays on the Early Toponymy of the British Isles*, Brighton 1988
Turner 1950	A. G. C. Turner, 'Notes on some Somerset place-names', *Proceedings of the Somerset Archaeological and Natural History Society*, 1950, 112–24
Turner 1950–2	A. G. C. Turner, 'Some Somerset place-names containing Celtic elements', BBCS 1950–2, 113–19
Turner 1951	A. G. C. Turner, 'A selection of North Somerset place-names', *Proceedings of the Somerset Archaeological and Natural History Society*, 1951, 152–9
Turner 1952	A. G. C. Turner, 'Some aspects of Celtic survival in Somerset', *Proceedings of the Somerset Archaeological and Natural History Society*, 1952, 148–51
Turner 1952–4	A. G. C. Turner, 'A further selection of Somerset place-names containing Celtic elements', BBCS 1952–4, 12–21
TYDS	*Transactions of the Yorkshire Dialect Society*
V	David Parsons and Tania Styles, *The Vocabulary of English Place-Names*, EPNS 1997– (by fascicule and page)
VB	Willibald's *Life of St Boniface* in *Vitae S. Bonifacii archiepiscopi Moguntini*, ed. W. Levison, Monumenta Germaniae Historica 1905; translated in C. H. Talbot, *The Anglo-Saxon Missionaries in Germany*, London 1954
VCH	The Victoria County History (by county, vol. and page)
Verey i	David Verey, *The Buildings of England: Gloucestershire*, vol. I, *The Cotswolds*, Harmondsworth 1970
Verey ii	David Verey, *The Buildings of England: Gloucestershire,* vol. II, *The Vale and Forest of Dean.* Harmondsworth 1970
Vising	*Mélanges de philologie offerts à M. Johan Vising par ses élèves et ses amis scandinaves*, Gothenburg/Paris 1925
Vitae SBG	*Vitae Sanctorum Britanniae et Genealogiae*, ed. A. W. Wade-Evans, Cardiff 1944
Wa	J. E. B. Gover, A. Mawer and F. M. Stenton in collaboration with F. T. S. Houghton, *The Place-Names of Warwickshire*, Cambridge 1936
Wade	G. W. Wade and J. H. Wade, *Somerset*, The Little Guides, London 1907
Wakelin	Martyn F. Wakelin, *English Dialects*, London 1972
Watts 1975	S. J. Watts, *From Border to Middle Shire: Northumberland 1585–1625*, Leicester 1975
Watts 2001	Victor Watts, 'English place-names in the sixteenth century: the search for identity', in *Sixteenth-Century Identities*, ed. A. Piesse, Manchester 2001

WATU	Melville Richards, *Welsh Administrative and Territorial Units*, Cardiff 1969
We	A. H. Smith, *The Place-Names of Westmorland*, 2 vols., Cambridge 1967 (by vol. and page)
Wealdhām	Rhona M. Huggins, 'The significance of the place-name *wealdhām*', *Medieval Archaeology* 19, 1975, 198–201
Wells	J. C. Wells, *Accents of English*, 3 vols., Cambridge 1982
Wells Cath	Wells Cathedral MSS. *Liber Alba*, Historical MSS Commission 1907, 1914
WestMids	Margaret Gelling, *The West Midlands in the Early Middle Ages*, Leicester 1992
WGÖ	Otto Kronsteiner, *Wörterbuch der Gewässernamen von Österreich*, Wien 1971
Wheeler	W. H. Wheeler, *A History of the Fens of South Lincolnshire*, 2nd edn, 1896; repr. Stamford 1990
Whitlock	Ralph Whitlock, *Historic Forests of England*, London 1979
Widén	B. Widén, *Studies on the Dorset Dialect*, Lund 1949
William of Malmesbury	William of Malmesbury, *De Antiquitate Glastonie Ecclesie*, in *The Early History of Glastonbury*, ed. J. Scott, Woodbridge 1981
Williams	*The Poems of Taliesin*, ed. Sir Ifor Williams, English version, Dublin 1975
Williamson	Tom Williamson, *The Origins of Norfolk*, Manchester 1993
Winchester	*Winchester in the Early Middle Ages*, ed. M. Biddle, Winchester Studies 1, Oxford 1976
Winton	*The Winton Domesday*, ed. and trans. Frank Barlow, in *Winchester in the Early Middle Ages*, ed. Martin Biddle, Winchester Studies 1, Oxford 1976 (by page and entry no.)
Wlt	J. E. B. Gover, Allen Mawer and F. M. Stenton, *The Place-Names of Wiltshire*, Cambridge 1939
WME	William Morley Egglestone, *Weardale Names of Field and Fell*, Stanhope 1886
Wo	A. Mawer, F. M. Stenton and F. T. S. Houghton, *The Place-Names of Worcestershire*, Cambridge 1927
Wrander	Nils Wrander, *English Place-Names in the Dative Plural*, Lund Studies in English 65, Lund 1983
Wright ii	Thomas Wright, *A Second Volume of Vocabularies* 1873 by page no. and line.
Writs	F. E. Harmer, *Anglo-Saxon Writs*, Manchester 1952; 2nd edn, Stamford 1989
WSurnames	T. J. Morgan and Prys Morgan, *Welsh Surnames*, Cardiff 1985
Wt	Helge Kökeritz, *The Place-Names of the Isle of Wight*, Uppsala 1940
WW	*Weapons and Warfare in Anglo-Saxon England*, ed. Sonia Chadwick Hawkes, Oxford University Committee for Archaeology Monograph no. 21, Oxford 1989
WYAS	*West Yorkshire: An Archaeological Survey to A.D. 1500*, ed. M. L. Faull and S. A. Moorhouse, West Yorkshire Metropolitan County Council, Wakefield 1981
YAJ	*The Yorkshire Archaeological Journal*

YCh	*Early Yorkshire Charters*, ed. W. W. Farrer, C. T. Clay, 10 vols., Edinburgh 1914–55
YE	A. H. Smith, *The Place-Names of the East Riding of Yorkshire and York*, Cambridge 1937
YN	A. H. Smith, *The Place-Names of the North Riding of Yorkshire*, Cambridge 1928
YW	A. H. Smith, *The Place-Names of the West Riding of Yorkshire*, 8 vols., Cambridge 1961–3
Zachrisson 1910	R. E. Zachrisson, *Some Instances of Latin Influence on English Place-Nomenclature*, Lund 1910
Zachrisson 1924	R. E. Zachrisson, *Some English Place-Name Etymologies* , Uppsala 1924
Zachrisson 1927	R. E. Zachrisson, *Romans, Kelts and Saxons in Early Britain*, Uppsala 1927
Zettersten	A. Zettersten, 'Middle English word studies', *Lunds Universitets Årsskrift*, n.f. Avd. I, vol. 56, no. 1, 1964, 32–4
ZONF	*Zeitschrift für Ortsnamenforschung*

Glossary of most frequently used elements

beorg OE 'barrow, mountain, hill, mound'. Cognate with OFris, OSax, OHG *Berg*, ON *bjarg*, Gothic *baírgahei* 'mountainous area' < Gmc **berga-* 'mountain', IE **bhergh-* 'height' from a verbal root meaning 'rise, grow' as in Skr *brhánt-* 'high', MIr *brī*(g) 'mountain' and the names *Brigantes*, *Birgit* 'the high one(s)'. In place-names used of small continuously curving hills, smaller than a *dūn*, with the summit typically occupied by a single farmstead or by a village church with the village beside the hill, and also of burial mounds. PNE i.29–30, L 127–8, Kluge-Seebold 75, Gelling 1998.78–81, 98.

brōc OE 'a small stream'. Cognate with LG words meaning 'marsh' (MLG *brōk*, (M)Du *broek*, OHG *bruoh*, G *Bruch*) and K Sx dialect *brook* 'marshy land'; Gmc **brōka-* 'marshland'. Related to E *brackish*, MLG, MDu *brac* 'salty', G *Brackwasser* 'salt water'; Gmc **brak-* 'marsh, stagnant water', W European/O European **m(e)r(e)g-* beside **m(e)r(e)k-* ultimately < IE **mer-* 'water, marsh' as in *mere, moor*. Not found in place-names recorded before AD 730. Characteristic term especially for streams with a visible sediment load, silt-laden, flowing over generally muddy beds with rank vegetation. PNE i.51–2, Jnl 23.26–48, L 14–16, Kluge-Seebold 108 s.v. Bruch[2].

burh OE 'a dwelling or dwellings within a fortified enclosure, a fort, castle, borough, walled town'. Cognate with OFris, OSax, OHG, G *Burg*, ON *borg*, Gothic *baúrgs* < Gmc **burg*; related to the verbs OE *beorgan*, G *bergen*, Gothic *bairgan* < **bergh-* 'defend'. In continental use the original sense was 'fortified place, place of refuge';[1] when from c.900 such fortified places became the seats of feudal lordships the sense 'castle' developed and finally 'walled town'. ON *borg* may be either an ablaut variant of *bjarg*, OE *beorg*, G *Berg* 'a hill' or of the verbal root seen in OE *beorgan*, G *bergen*, denoting a place where one can defend onself. The relationship between *burh*, *Burg* and *beorg*, *Berg* is unclear but the remoter relations seem to include Gk πύργος (*pyrgos*) 'tower' and πέργαμος (*pergamos*) 'citadel'[2] corresponding to IE zero grade[3] **bhrgh-o-* and **bherg-* possibly of non-IE origin. The basic sense in OE place-names is 'fortified place' which was applied both to prehistoric and Roman antiquities, to Anglo-Saxon fortifications, to fortified houses or manors, with later extensions to 'fortified town, town, market town' and 'borough'. The forms are complicated, the nominative sing. producing *-burgh* and *-borough*, dative sing. **byrig** *-bury* (ME *-biri*, in south-east *-beri*, in west and south-west *-buri*), genitive sing. **burge, byrh** ME *-bur* and *-bir*. PNE i.58–62, Kluge-Seebold 76, 114–15.

burna OE 'a stream'. Cognate with G *Brunnen*, Gothic *brunna*; Gmc **brunnōn* < IE **bhru-n-*, zero grade of IE **bhrēw-r/ -nt-* whence Armenian *albiwr* 'spring',

Fig. 1 Illustrations (from top to bottom) of Edlesborough, Granborough and Thornborough (Buckinghamshire), demonstrating OE **beorg**, a hill. © Ann Cole

[1] The Teutoburg forest was probably named after a tribal refuge.

[2] Cf. the name *Pergamum* for Troy.

[3] Zero grade is the term applied to the unstressed position in Indo-European vowel alternation (ablaut). At an early stage in the development of the IE vowel system it is thought that the only true vowel was /e/. Under the influence of stress, accent /e/ either remained or was lengthened to /e:/ or reduced to zero. Under the influence of pitch, accent /e/ could also become /o/ as seen in the verb *drive, drove, driven* on the IE root **dhreib-* (e-grade), **dhroibh-* (o-grade), or **dhribh-* (zero-grade).

Gk φρέαρ (*phrear*) 'well' and possibly, with loss of -r- Latin *fons* 'spring'. An early word in use before 730, the normal OE term for a stream not large enough to be called a river, used especially for streams with clear water, flowing over gravelly beds from springs, with seasonally variable flow and distinctive plant association. PNE i.63–4, Jnl 23.26–48, L 16–20, Kluge-Seebold 109.

bȳ, ODan **bȳ**, ON **býr**, 'yard, courtyard, farmhouse' < Gmc **būwi-*. Cognate with OHG, MHG *bū* 'tilled field, farm', Lithuanian *buvis* 'dwelling' and ON *bœr* 'house' < Gmc **bōwi-* on the verbal root **bheu-* seen in Skr *bhávati* 'be', Gk φύω 'bring forth', φύομαι 'become', OSlav *byti* 'grow, be', and in Gmc the lengthened grade *bōww-a- as in Gothic *bauan* 'dwell', OE, OSax, OHG *būan* 'dwell', ON *búa* 'dwell, make ready', G *bauen*. The relationship between the two senses at the base of these words, 'dwell', and 'make ready, cultivate', is not fully understood; probably they were originally two different formations on the same root. There is some evidence from Scandinavia that the basic sense of *býr* was 'dwelling, isolated secondary settlement or farmstead', but in England the full range of senses is evidenced from 'farmstead' to 'village'. PNE I.66–72, SSNEM 10–12.

ċeaster, Anglian **ċæster** or **cæster** possibly under Viking influence, Kentish and Mercian **ċester** OE '(Roman) city'. Loan word from Latin *castra* 'a camp'.

Fig. 2 Illustration of Whitcombe (Somerset), demonstrating OE **cumb**, a valley. © Ann Cole

cumb OE 'coomb, valley'. This is usually taken to be a Celtic loan, cf. W *cwm* 'valley', Brit **kumbos*, Gaulish *cumbā*, but there was also OE *cumb* 'vessel, cup, bowl' cognate with MLG *kump*, MHG, G *Kumpf* 'bowl' < Gmc **kumpa-* related to Gk κυμβη (cymbē) 'cup, vase', Skr *kumbhá*, Avestan *xumba-*, 'un mot voyageur' of non-IE origin. This like OE *byden, canne, cetel, trog* was probably used in a transferred topographical sense reinforced in western districts by *cwm*. The sense is a bowl- or trough-shaped valley shut in on three sides, short in comparison to its width and often deep for its size in contrast to the long open shape of a *denu*. PNE i.119, Nomina 6, 1982, 73–87, L 88–94, Kluge-Seebold 419, Gelling 1998.85, 90, 99.

Fig. 3 Illustrations (from top to bottom) of Standen (Wiltshire) and Vernham Dean (Hampshire), demonstrating OE **denu**, a valley. © Ann Cole

denu OE 'a valley'. Cognate with MLG *dene*. Related to OE **denn** 'a den, lair, cave, swine-pasture', MLG, MDu *denne*, MDu *dann* 'depression, bed, couch, valley', LG *danne* 'bed', MLG *dan*, Du *denne*, OHG *tan* 'forest', OHG *tenni* 'threshing place in open field', *tenar, tenra* 'flat of the hand', Gk θέναρ (*thenar*) 'palm of the hand'. The relationship of these words is very uncertain but the basic sense seems to be 'low depression, valley'. As a place-name term it is characteristically applied to long open sometimes sinuous valleys. PNE i.130, Nomina 6, 1982, 73–87, L 97–9, Gelling 1998.85.

dūn OE 'a hill'. Cognate with OFris *dūne*, OSax *dūna* (Du *duin*, ModE *dune*); WGmc **dūnō(n)* 'hill'. Probably a pre-insular loan-word from Celtic, cf. Gaulish **dūnom* in place-names like *Augustodunum* (Autun), (O)Ir *dún* 'fort', W *dinas* 'city'. If so it pre-dates the Celtic semantic development 'hill' > 'fort, ville close' and is applied in English place-names from the earliest period characteristically to low hills with fairly extensive and level summits offering good settlement sites; as Ann Cole has put it in contrast to an *ōra*, 'a *dūn* is like an upturned bowl with a limited area of flat land on top'. PNE i.138–9, L 140–58, Jnl 21.15, Gelling 1998.98.

Fig. 4 Illustrations (from top to bottom) of Billington (Bedfordshire), Stottesdon (Shropshire) and Toot Baldon (Oxfordshire), demonstrating OE **dūn**, a hill. © Ann Cole

Fig. 5 Illustrations of Athelney and Othery (Somerset), demonstrating OE **ēġ**, an island. © Ann Cole

ēa OE 'water, stream, river'. Cognate with OFris *ā*, *ē*, OHG *aha*, ON *ā*, Gothic *ahua*; Gmc **ahwō* 'river, water'. Cf. Latin *aqua* possibly < IE **akwā* replacing (?) non-IE *ap-* 'river'. The standard word in place-names for river denoting a watercourse of greater size than a **brōc** or a **burna**. PNE i.142–3, L 20–1, Kluge-Seebold, GED A60.

ēġ OE 'an island', West Saxon **īeg.** Cognate with OHG *ouwa*, G *Aue*, ON *ey*; Gmc **agwijō* 'associated with water' formed on Gmc **ahwō* as in **ēa**. The application in place-names is primarily to slightly raised dry ground offering settlement sites in areas surrounded by marsh or subject to flooding. Subsidiary senses include 'hill jutting into flat land' and 'patch of good land in moors'. PNE i.147–8, L 34–40, Kluge-Seebold 47, Gelling 1998.93, 99.

feld OE 'open or cultivated land, plain'. Cognate with OFris, OSax, OHG *feld*, OHG *fala* 'plain, heath' as in the place-name Westfalen, Latin *palam* 'open', OSlav *polje* 'field'; WGmc **felþa-* with ablaut variant *fuldō 'earth' whence OE *folde, flōr, folm.* Ultimately from the IE verbal root **pelə-/plā-* 'spread'. The sense in place-names is 'unencumbered land, land without trees or hills or fen'. The element occurs in some of the earliest place-names and its original application was probably to open land used for pasture; later, when uncultivated grazing land began to be broken in, it was applied to arable, especially to the communal arable of the open-field farming system of the later tenth century. PNE i.166–8, L 235–45, Kluge-Seebold 208.

ford OE 'a ford'. Cognate with OFris *forda*, ON *ford*, OHG *furt*, OSax *vord*; WGmc **furdu-* 'ford', IE **prtu-* 'passage, ford', Avestan *pə rə tauu-* 'ford, bridge' as in r.n. Euphrates, *hu-pərəθβ(ii)a*, 'good fording', Latin *portus* 'harbour', Welsh *rhyd* 'ford' ultimately from IE verbal root **per-* 'bring over or across'. The specifics of ford-names often relate either to the nature or position of fords (BRADFORD 'broad', DARNFORD 'hidden', LANGFORD 'long', RUFFORD 'rough', SHADFORTH 'shallow'), to the material of the track of the ford (SANDFORD 'sand', STAMFORD 'stone'), to frequenters of the ford, human or otherwise (CRANFORD 'cranes', HUNTINGFORD 'huntsmen', LATTIFORD 'beggars'), or to individual persons, the significance of which is not fully understood. PNE i.180–83, L 67–72, Kluge-Seebold 237.

hǣth OE 'heath, untilled land, waste'. Cognates OSax *hētha*, OHG *heida*, G *Heide*, ON *heithr*, Gothic *haiþi*, OE *hāth*, ModE dialect *hoath* < Gmc **haiþœi* 'waste, untilled land', Gaulish *kaito- cēto-* OW *coit*, W *coed* 'wood', possibly to be associated with Latin *caedes* 'a cutting down', *caedere* 'cut down'. The basic sense is 'uncultivated ground, uncultivated ground where heather grows, heather, heath'. The element occurs rather frequently as specific in names like Hatfield and Hedley where the sense is more likely to be the plant. PNE i.219–20, Kluge-Seebold 299.

halh OE (Anglian), WSax, Kentish **healh,** 'corner, nook, secret place'. Cognate with OE *holh* 'hole, cavity', *healoc* 'cavity, sinuosity'. The Gmc source **halhaz* is related to Gmc **hulha* 'hollow' and Bulgarian *klánik* 'space between hearth, oven and wall'. Anglian *halh* gives ModE *haugh* and dative sing. *heale* and *hale* ModE *-heale* and *-hale*. The senses detected in place-names have included 'sunken place, small valley, hollow, recess, land almost enclosed by the bend of a river, low lying land by a river, a haugh, slightly raised ground in marsh, nook or corner of land, nook or corner of a parish, piece of land projecting from or detached from its main administrative unit'. Most recently comparison has been made with North Frisian *hallig*, a term for a small, low-lying, undiked island, and it seems likely that behind many of the topographical uses in English lies the basic meaning 'slight deviation from the horizontal, slight depression or slight elevation', especially 'low-lying land higher than its surroundings but still liable to flooding'. The link with the administrative sense and with the meaning 'angle, corner' is less clear. PNE i.223–4, L 100–111, Stiles 330–44.

hām OE 'homestead, dwelling place'. Cognate with G *Heim*, Gothic *haims*; Gmc **haima-* < IE **kþoimo-*, an abstract formation in *-mo-* on the verbal root **kþei* 'to dwell' whence Skr *ksḗma* 'dwelling place, security, peace', Lithuanian *Šeima-* 'family, household', Russian, Church Slavonic *Šemija* 'family' (Lithuanian *šiemas* 'farmhouse, village' with its centum-language phonetics is probably a loan-word from Gmc). A place-name element used from the earliest times and closely associated with Roman roads, villas and settlements. The senses are 'dwelling-place, homestead, collection of dwellings, village' and later 'manor, estate'. PNE i.226–9, MA 10, 1966, 1–29, ASE 2, 1973, 1–50, Jnl 5, 1972–3, 15–73.

hamm OE 'hemmed-in land'. Cognate with LG *hamm* 'piece of enclosed land, meadow' (MLG, MDu, ModFris *ham* 'enclosed plot of pasture or meadow') and related to ModE *hem* 'hem', OFris *hemme*, MDu, MLG *hem* 'enclosed land' and OE *hemman* 'stop up, close, delete'; IE **kem-* 'press together, prevent'. It is difficult to keep this element apart from **hām** in the absence of OE forms or ME diagnostic spellings *-ham(m)e* or *-hom(m)e* and it seems likely that they were already confused in some areas in Anglo-Saxon times. The basic sense seems to be 'land hemmed in' whether by water, marsh, high ground or man-made boundaries. The relationship between this element and OE *hamm* fem. 'the inner part of the knee, the hock' possibly from Celtic **kambo-* 'bent, crooked' is disputed. From close analysis of the topography of *hamm*-names in Kent, Surrey and Sussex, Dodgson evolved the following typology of sense:

hamm 1	'land in a river bend'
hamm 2a	'a promontory of dry land into marsh or water'
hamm 2b	'a promontory into lower land even without marsh or water'; perhaps hence 'land on a hill-spur'
hamm 3	'a river-meadow'
hamm 4	'dry gound in a marsh'
hamm 5a	'cultivated plot in marginal land'
hamm 5b	'a piece of valley-bottom land hemmed in by higher ground'.

PNE i.229–31, ASE 2, 1973, 1–50, NoB 64, 1976 69–87, L 41–50.

hlāw, hlǣw OE 'mound, cairn, hill, mountain, barrow'. Cognate with OSax, OHG *hlēo*, ON runic *hlaiwa* 6th cent. 'grave', *h[l]aiwiðaR* fifth century 'buried', *hlaaiwido* fifth century 'I buried', Gothic *hlaiw* 'grave', from IE **kley-* 'lean' as in Latin *clīvus*, Lithuanian *Sleivas* 'bowlegged', OSlav *chlévina* 'dwelling' (borrowed from Gmc). The original sense seems to have been 'chamber or house partly constructed underground, burial chamber'. The usual sense in southern England is 'tumulus, burial', except in the south-west where it was replaced by *beorg*, in northern England 'hill'. PNE i.248–50, AEW 234, GED H73, Signposts 134–7, L 162–3.

hōh OE 'a hough, heel, point of land'. Cognate with ON *há* in *hámót* 'heel, ankle'. Gmc *hanha-. Related to ModE *heel* < **hāhil* < Gmc **hanh-il-* and OHG *hahsa* 'knee-joint', MLG *hesse*, Du *haas*. An anatomical term used in names for a sharply projecting spur of ground resembling in profile the shape of the foot of a person lying face-down. Usually translated in this book as 'hill-spur'. PNE i.256–7, L 167–9, Gelling 1998.83, 98.

Fig. 6 Illustrations of Ingoe (Northumberland) and Ivinghoe (Buckinghamshire), demonstrating OE **hōh**, a heel of land. © Ann Cole

-ing OE suffix. Four different uses of this suffix are identified of which 2 and 4 are most relevant to place-names. *ING passim*, BzN 2, 1967.221–45, 325–96 ; BzN 3, 1968.141–89.

1. –**ing**[1] and -**ling** OE noun suffixes denoting male persons and diminutives as *hȳrling* 'hireling' from *hȳr* 'wages', *dēorling* 'darling' from *dēor* 'dear', etc. PNE i.285.

2. –**ing**[2] OE place-name and stream-name forming suffix attached to appellatives and adjectives as DEEPING Lincs, 'deep place', OE *dēop*, DOCKING Norfolk, 'dock place, place growing with docks', OE *docce*, RUCKINGE Kent, 'rookery', OE *hrōc*, STEEPING Lincs, 'steep place', OE *stēap*, STOWTING Kent, 'hill place', OE *stūt*, and also to personal names, BOCKING Essex, 'the place called after Bocca, Bocca's (place)', CHARING Kent, 'the place called after Ceorra, Ceorra's (place)', HICKLING Notts, 'the place called after Hicel(a), Hicel(a)'s (place)'. PNE i.285–90.

3. –**ing**[3] OE patronymic as *Ecgbryht Ealhmunding* 'Ecgbriht son of Ealhmund'. PNE i.290–1.

4. –**ing**[4] OE connective particle linking a first element, appellative or personal name, to a final element. There is some uncertainty about the exact function and meaning of this element. Instances like WELLINGBOROUGH Northants which has historical forms *Wedlingeberie* 1086, *Wendlingburch* etc. 1178–1389 beside *Wendle(s)berie* 1086, *Wendlesburg* 1226 suggest that in compounds in which the first element is a personal name *-ing-* is closely similar to genitive singular *-es* and that the translation 'Wendel's fortified place, Wendel's manor' is justified. On the other hand the existence of place-names consisting of *personal name* + *ing*[2] such as Bocking, Charing and Hickling above raises the question, first proposed by Arngart (1972), of whether the correct interpretation of *-ington* and other *–ing*[4] names is not rather 'the settlement or estate called or at *X-ing'*. Such a radical departure from received practice is not adopted in this dictionary in the absence of a thorough reconsideration of *ing*-names in general. In this book we adhere to Smith's assessment in PNE (1956):

> The editors of the English place-name survey have always accepted and still adopt views . . . that in most instances *-ing-* denotes the association of a place with a particular individual and in the broadest sense has something of a genitival function without necessarily implying possession: *Teottingtun* is by this to be interpreted as 'farmstead to be associated with *Teotta'*, *Ælfredingtun* 'farmstead associated with *Ælfred'*, rather than 'farmstead belonging to *Teotta* or *Ælfred'* which would be *Teottantun* and *Ælfredestun*.

Our preferred translation is 'estate called after X', drawing attention to onomastic structure rather than possible significance. Thus, ADDINGTON Northants is 'the estate called after Adda' rather than 'Adda's estate' or 'the estate called or at Adding'; ARTHINGWORTH Northants 'the enclosure called after Earna'; HANNINGTON Northants 'the estate called after Hana', etc. Sometimes the formula 'estate associated with X' is used, which similarly leaves the exact relationship between personal name and place unspecified.

Similarly the connective function of *-ing-* in compounds in which the first element is an appellative is also in doubt. On the model of *–ing*[2] names such as THURNING Northants, 'place growing with thorns' < OE *thorn* + *ing*[2], place-names of the structure *appellative* + *ing* + *generic* have generally been taken as *place-name* or *appellative* + *generic*, an interpretation which seems to be justified by names in which the specific is a stream-name in *ing* like PIDDINGTON Northants, 'the settlement on the *Pidding'*, OE **piding* < *pide -u* + *ing*[2] or SCARRINGTON Notts, 'the settlement on the *Scearning*, the dirty stream' < OE *scearn* + *ing*[2]. So HARRINGWORTH Northants, 'the enclosure at or called *Hæring*, the stony place', OE **hæring* < *hær* + *ing*[2], HOVERINGHAM Notts, 'the estate at or called *Hofering* or at the hump of ground', OE **hofering* < *hofer* + *ing*[2], NASSINGTON Northants, 'settlement or estate called or at *Nassing*, the promontory place, OE **næssing* < *næss* + *ing*[2]. The most disputed of this category of name are the types Doddington, Donnington, Duddington and Dunnington in which the specific could be either a personal name Dunn(a) or Dudda, the usual interpretation, or an appellative, OE *dūn* 'a hill', **dod* or **dud* 'a rounded hill', 'estate called after Dunn(a) or Dudda' or 'called or at *Dūning*, the hill place' or '*Doding* or *Duding*, the place of the rounded hill'. In this dictionary preference has been given to the personal name explanation except where the topography clearly points to the appellative alternative.[4] It should be noted that some apparent *ing*-names are shown by their early spellings not to be so, for example COLLINGTREE Northants, *Colentreu* 1086, OE *Colan trēow*, 'Cola's tree'. PNE i.291–8.

-**ingas** OE plural suffix forming folk-names such as *Hæstingas* 'the people called after Hæst or Hæsta' as in the place-name HASTINGS WSusx. Most place-names which preserve such folk-names survive today in what appears to be a singular form but which is actually derived from the OE dative plural inflected form *-ingum* originally used after a preposition, for example READING Berks, 'the *Readingas*, the people called after Reada', *(to) Readingum* 871, or BARLING Essex, 'the *Bærlingas*, the people called after Bærla or Bærel', *(æt) Bærlingum* 998. The genitive plural form *-inga* also occurs in compound place-names such as BIRMINGHAM WMids, *Bermingeham* 1086, 'the homestead of the *Beormingas*, the people called after Beorma'. PNE i.298–303.

lēah OED, Anglian **lǣh**, dative sing. **lēa** and **lēage**, pl. **lēas**, dative pl. **lēum**, **lēam** and **lēagum** (Campbell § 574.2). OE 'a piece of ground, a LEA, a meadow'. There is a great variety of spellings for this element which often interchange in post-conquest sources; from nominative case **lēah** derive the ME spellings in *-legh -leigh*, from **lǣh** spellings in *-lac*, from **lēa** spellings in *-le -lee*, and from **lēage** spellings in *-ley(e)*. Cognate with OFris *lāch*, MLG *lō*, *lōch* 'copse' MDu *loo* 'woodland, scrubland', OSax, OHG *lōh* 'copse, grove, woodland, undergrowth, scrub', ON *ló* in place-names like Osló (whence ModE dialect *loo* 'open place') from Gmc **lauhō* < **louko-s* whence Latin *lūcus* 'grove', Lithuanian *laukas* 'open field'. Related to Gk λευκός (*leukos*) 'white' and so showing connection with the semantic field 'light' but not, according to Trier, in the sense 'light place in woodland'. Besides IE **leuk-* 'light' he cites a series of other extensions *leug- 'bend', **leug-* 'break', **leub- *leubh- *leup-* 'peel' of the basic IE root **leu-* 'cut off'. From the underlying sense 'woodland clearing, glade, underwood' developed the meaning 'pasture, meadow' as land-use changed. In English place-names the element seems to have been used of ancient woodland, of clearings both early and late, and of woodland pasture, becoming the usual term for woodland clearings from the mid-eighth century to a large extent in succession to **wald**. The usual translation here is 'wood, woodland

[4] The case for appellatives as the specific of these and similar names was made out by Arngart 1972, fn 12 and independently by Gillian Fellows Jensen in Årsskrift 1974, and reviewed by her in Fellows-Jensen 1996.

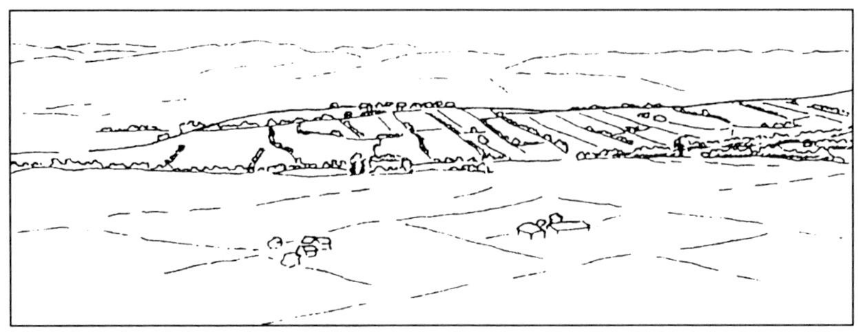

Fig. 7 Illustrations of Haselor (Warwickshire) and Wentnor (Shropshire), demonstrating OE **ofer**, a ridge. © Ann Cole

clearing' or just 'clearing'. Trier 1952, PNE ii.18–22, Schwarz, AEW 362a s.v. *ló*, Johansson, L 198–207.

mōr OE 'moor, morass, swamp'. Cognate with OFris, MLG *mōr*, OGH *muor*, G *Moor*, ON *mørr* 'fen'. Related to ModE *marsh* < WGmc **marisk*- and *mere*, OE *mere* 'sea, lake', OSax *meri* 'sea' (Du *meer*), OHG *mari, meri* (G *Meer*), ON *marr* 'sea', Gmc **mari*- < IE **mori*- whence Russian *more*, OIr *muir*, W *mor*. The basic sense in place-names is 'marsh', a kind of low-lying wetland possibly regarded as less fertile than **mersc** 'marsh'. The development of the senses 'dry heathland, barren upland' is not fully accounted for but may be due to the idea of infertility. PNE ii.42–3, L 54–6.

ofer. There are two possible sources of this element. PNE ii.53–4 gives **ōfer** 'bank, river-bank, sea-shore' and an unattested ***ofer** 'slope, hill, ridge'. **ōfer** 'border, margin, edge, brink, river-bank, sea-shore' is a well-attested[5] cognate of OFris, MLG *ōver*, MG *uover*, Du *oever*, from WGmc **ōbera*- parallel to Gk ἤπειρος, ἄπειρος (*ē- āpeiros*) < **āperjo*- 'the mainland' as opposed to the sea and the islands, cf. Skr *ápara*- 'behind, later'; the sense is 'that which lies beyond' and is clearly unrelated to the place-name term **ofer** which from field-work has been shown to be used of sites at the tip of a flat-topped hill-spur and to refer to the tip of a promontory, or to a flat-topped ridge or a level ridge with convex shoulders. As with **ōra** used in a very similar way the sense 'slope' is now abandoned. L 173–9, Jnl 22.35–41, Gelling 1998.81–5, 99.

ōra OE 'border, bank, shore'. Usually said to be cognate with OE *ōr* 'beginning, origin, front', ON *óss* 'mouth, outlet (of a river or lake)' and Latin *ōra* 'shore, coast', *ōs, ōstium* 'mouth', OSlav *usta* 'mouth'. The term is common in the south of the country and is now thought possibly to be a borrowing of Latin *ōra*. It does not occur in the North Midlands or in the North, where its function as a place-name element seems to be performed by **ofer**. Not well understood before Ann Cole's analysis in terms of a hill of characteristic profile which 'is like an upturned canoe or punt having an extensive tract of flat land terminating at one or both ends with a rounded shoulder'. The senses 'hill-slope' and 'river bank' are now abandoned and the element is usually translated 'flat-topped ridge'. PNE ii.55, L 179–82, Jnl 21.15–22, 22.26–41, Gelling 1998.85–7, 99.

stede OE 'place, site, position, station'. A difficult term to give an exact sense to, cognate with OFris *stide*, OSax *stedi, stidi*, OHG *stat*, G *Stadt* 'town, city', ON *stathr*, Gothic *staths*, Latin *statim* 'on the spot', *statio* 'standing-place', Gk στάσις (*stasis*) 'placing, something placed, something erected, a building'. The sense 'town, city' for G *Stadt* is a late development from c.1200 when the term began to replace *Burg*. Here usually translated 'place, site' or 'locality'. PNE ii.147–51, Stead *passim*.

strǣt OE, Anglian, Kentish **strēt**, 'street, high road'. A borrowing from Latin *strāta* in *via strāta* 'paved road'. The normal term in OE for a paved way or Roman road, later extended to other roads, urban streets and in SE dialects to a street of dwellings, a straggling village or hamlet. In compounds like *strǣt-tūn* and *strǣt-ford ǣ* underwent shortening to give the forms *Stratton* and *Stratford* within the WSax dialect area and *Stretton* and *Stretford* in Anglian areas although the distribution has been partially disturbed by the use of standard spelling systems. PNE ii.161–3.

thorp, throp OE 'farm, village'. Cognate with OFris *therp, thorp*, OSax *thorp*, OHG, G *Dorf* 'village, estate', ON *thorp* 'farm, estate, grave-mound', OSwed *thyrp, thørp*, Gothic *thaúrp* 'land, lived-on property' from Gmc **þurpa*- 'village, farmstead' related to Oscan *tríibúm* 'house' < **trēb*-, Lithuanian *trobà* 'house, buidling' < **trāb*-, Gk τερεμνα, τέραμνα (*teremna, teramna* < *terəb-no-) 'house, habitation', OW *treb* 'house, dwelling', W *tref- tre-* Co *tre-* in names of hamlets and villages, OIr *treb* 'habitation, family,

[5] For example in literary use in *The Battle of Maldon*.

Fig. 8 Illustration of Chinnor (Oxfordshire), demonstrating OE **ōra**, a ridge. © Ann Cole

tribe', *a-treba* 'lives', Gaulish *A-treb-ates* 'the settled ones' < IE **treb-* 'building, dwelling-place'. In continental use the later sense 'collection of dwellings, village' probably reflects the development of new settlement areas where dwellings were grouped for security. In English place-names the sense is 'outlying farm, hamlet' and as with its ON and ODan cognates it seems primarily to imply small secondary settlements. Outside the Danelaw the metathesised form *throp* (later *thrup*) is regular but in the Danelaw it is difficult to keep apart from its ON equivalent. The element does not occur in the south-east or far south-west implying that whereas the Saxons used the continental *thorp* the settlers of Kent and Sussex came from an area unfamiliar with the element and that it was obsolete by the time of the settlement of Devon and Cornwall. PNE ii.214–6, GED þ22, Kluge-Seebold 151–2.

thorp ON, ODan 'farmhouse, village'. Cognate with OE *thorp*, *throp* supra. The sense in place-names is 'secondary settlement, dependent outlying farm or hamlet' as in Denmark and Sweden where it seems to have superseded *býr*, itself a term originally denoting secondary settlements. Thorp was employed for new secondary settlements when the *býrs* had become established centres themselves for further colonisation. The element spread from Germany to Denmark where it is very common, and thence to Sweden and eastern Norway where its use is limited. Consequently it has been thought that most of the *thorp* names in the Danelaw are of Danish rather than Norwegian origin. It may, however, be that the sparseness of *thorp*-names in the NW of England is due rather to the scarcity of the kind of settlement called a *thorp*. In the Danelaw it is difficult to distinguish Scandinavian *thorps* from earlier English *thorp/throps*. English *thorps* must have existed there long before the Scandinavian settlement, but the fact that *thorp* is so much more common in the Danelaw than elsewhere is best accounted for by the supposition that the majority are Scandinavian rather than English. PNE ii.205–212, SSNEM 83–91, SSNNW 47–8, Foerste 422–33, Lund 233–5.

thveit ON, **thwēt** ODan 'a clearing, a meadow, a paddock'. Cognate with ON *thveita* 'strike, hew', OE *thwītan* 'cut, cut off'. It is unclear whether the base meaning was 'something cut off, detached piece of land' or 'something cut down, felled tree'; if the latter, the primary sense is 'woodland clearing, plot of grassland in woodland' and the sense 'detached piece of cultivated ground' secondary. The element is very common in the NW. PNE ii.218–20, SSNNW 89–91.

tūn OE 'enclosure, garden, field, yard; farm, manor; homestead dwelling, house mansion; group of houses, village, town'. Cognate with OFris, OSax, ON *tūn*, OHG *zūn*, G *Zaun* 'fence', OIr *dūn* 'fort', Gaulish *dūnon* in place-names in *-dunum*, and OE *tȳnan* 'hedge in, fence, enclose, shut' < **tūn-j-an* from Gmc **tūna-* 'fence'. From this basic sense in English developed the meaning 'place fenced in' and in place-names the senses 'enclosure, farmstead, estate and village' are all in question. Because of this uncertainty in this dictionary the translation is frequently the non-committal 'settlement' or 'farm or village'. Although *tūn* is already found as a place-name forming element before 730 its period of greatest frequency is in the centuries thereafter. Its main use seems, in fact, to coincide with the period of the break-up of large unitary estates due to land-grants made out of them by kings, noblemen and bishops to individuals of the thegnal class. It is these individuals, it is thought, whose personal names are so often compounded with *tūn* in the names of English villages, implying manorial overlordship rather than founding fathership and tillage of the soil. It is for this reason that names such as KIRTLINGTON or KNAYTON are normally translated 'the estate called after Cyrtel(a)' and 'Cengifu's estate' respectively. PNE ii.188–98, Signposts 177–85, Kluge-Seebold 806.

wald OE (Anglian), WSax and Kentish **weald** 'weald, forest, wood, grove, foliage, bushes'. The two forms survive as *wald* (unlengthened), *wold* (< ME *wōld* < *wāld* with lengthened *a* before *ld*) and *weald* (< ME *wēld* < *wǣld* < *wēald*) although the standard form *wold* has frequently intruded into SE usage. Cognate with OFris, OSax, OHG, G *Wald*, ON *vǫllr* 'untilled land, plain, meadow', Gmc **walþu* < IE **woltu-*. The original sense was 'tuft, bundle, clump' especially of foliage and twigs as cognates outside Gmc seem to show, OIr *falt, folt* 'mop of hair, foliage' < **volto-*, W *gwallt* 'hair', Lithuanian *váltis* 'panicle' < **wolət*, OPrussian *wolti* 'ear of grain, spike of flower', Russian *vóloti* 'thread, ear of grain' < **wolti-*, Gk λασιος (*lasios*) 'hairy' < *Ϝλατιος (*wlatios*) < zero grade **wlt-io-*. The sense in English place-names is 'woodland, forest', especially woodland of some extent like the Weald of Kent described as a forest, *ðæs miclan wudu eastende ðe we Andred hatað . . . se ea lið ut of ðæm walda* 'the east end of the great forest we call *Andred* . . . the river [Lympne] flows out of the forest' (*Anglo-Saxon Chronicle* under year 893). The *weald* or forest cover of the south-east was thick and impenetrable and only slowly broken in to form woodland pasture and new arable, a development probably commemorated in such names as WOMENSWOLD, 'the Wimlingas's part of the weald', SHEPERDSWELL or SIBERTSWOLD, 'Swithbeorht's part of the weald', and RINGWOULD, 'the weald called Hredling' at a period before the currency of **lēah** as the usual term for a woodland clearing. The Weald of Kent survived as wooded country until modern times, unlike the Cotswolds and the Lincolnshire and Yorkshire Wolds. Here intensive settlement and clearance brought about the emergence of open downland landscapes. Weald and wald names seem to be an early stratum of naming indicating the presence of woodland before this development took place. The sense 'open country, elevated stretch of open country or moor' is not evidenced before the thirteenth century. Trier 1956.32, PNE ii.239–42, L 222–7, Kluge-Seebold 774.

wella -e OE, WSax **wiella -e, will(a), wyll(a) -e**, Mercian **wælla -e** 'well, fountain, spring'. Cognate with MLG, MDu *welle*, ON *vella* < Gmc *wel-; and OE *wiellan* 'cause to bubble', *weallan* 'be agitated, rage seethe, bubble', OFris *walla*, OSax, OHG *wallan* 'surge, well, boil up', Norw *valla* < IE **wol-n-*; Gothic *wulan* 'be aglow, seethe', ON *olmr* 'furious' < Gmc **wulma-*, IE *wəlō*, a zero grade aorist-present form probably from the root *wel-* 'turn, roll'. The root meaning is 'that which bubbles up, a spring'. The element is poorly represented among the earliest names. PNE ii.250–3, GED W98, L 30–2.

wīċ OE 'dwelling-place, lodging, habitation, house, mansion; village, town'. Continental borrowing of Latin *vīcus* 'a group of dwellings, a village; a block of houses, a street, a group of streets forming an administrative unit', cf. Gothic *weihs* 'village', MLG *wīch* 'town', OSax *wīc*, OFris *wīk*, OHG *wīch* 'a house, a dwelling-place'. The forms of the word are difficult; nominative sing. and uninflected pl. **wīċ** give ModE *-wich*, dative pl. **wīcum** and later pl. forms **wīcu, wīcan** *-wick*. In place-names the sense is 'buildings used for a special purpose' especially dairy farming (cf. BUTTERWICK, CHESWICK, CHISWICK) but also other types of farming (BARRICK, BERWICK, CALWICK, CONRISH, COWAGE, FUGE, FUIDGE, HARDWICK, OXWICK) and trade or industry (ALDWYCH, DUNWICH, FISHWICK, HAMMERWICH, *Hamwic* (SOUTHAMPTON), *Lundenwic* (LONDON), WOOLWICH). PNE ii.257–63, Ekwall 1964, Jnl 11, 1987.87–98.

worth, **weorth**, **wyrth**, **worthiġ**, **worthiġn** OE 'court, courtyard, curtilage, farm'. Cognate with OFris, OSax *wurth* 'soil, raised house-site', MLG *wurt*, wort 'a homestead', G *Wurt* 'hillock, soil thrown up into a heap', ON *urthr* 'heap of stones' of uncertain origin. Comparison is made with OHG *werid*, MLG *werder*, MDu *waert*, *wert*, G *Werder* 'river island', Gmc **waruþa *waruþaz* 'river island' whence OE *waroth* 'shore', and with Skr *várūtha* 'protection', *vrnóti* 'defends', *varūtár* 'defender', Gmc **war-ija-*, OE *werian* 'defend' and Skr *vrti-* 'enclosure', OIr *fert* 'grave-mound'. A term in early use for a small settlement, usually translated 'enclosure'. Distinctively SW is the form **worthiġ** 'enclosure, farm, estate' (Devon and Somerset and to a lesser extent Cornwall, Dorset, Hampshire, Gloucestershire and Wiltshire and as far N as Staffordshire and Derbyshire) and distinctively W Midlands (especially Hereford and Shropshire but also S Lancashire, Cheshire, Staffordshire, Worcestershire and N Gloucestershire) **worthiġn** < *worthiġ* + *-en*, ME *-wardine*. PNE ii.273–7, Kluge-Seebold 781, 787, 801.

wudu OE 'wood, forest, grove' by back-mutation from earlier **widu** found in early place-name types like Witton. The most general OE term for a wood, cognate with OSax *widu*, OHG *witu*, ON *vithr* < Gmc **wiþu-*; and OIr *fid*, W *gwydd*, Gaulish *vidu-* as e.g. in *Viducasses*. PNE ii.279–81.

Distribution maps

The scope of this dictionary has permitted the compilation of distribution maps of other elements than those conventionally plotted, for example in A. H. Smith's *English Place-Name Elements*. The inclusion of two-component names, such as Bank Street, Beare Green, Beazley End, Beck Row and Birkdale Common, has added material which could not have been extrapolated from earlier dictionaries. The maps included here illustrate both some traditional and some novel distributions. The first six show (1) the distribution of modern spellings for OE *burh* and of *beorg* when confused with *burh*; (2) the distribution of *bury*, highlighting cases where it is applied to early earthworks especially of the Iron Age period; (3) a map of settlement in the pre-Roman Iron Age; (4) the West Midland and South West reflexes of OE *wella* 'a spring, a stream'; (5) the three variant forms of *worth*; and (6) the survival or non-survival of the OE dative plural inflectional ending *-um*.

A further five maps illustrate the significance of names in *End*, *Street* and *Green*. Recent work by historical geographers has demonstrated that in broad terms the landscape of England may be divided into three zones or provinces related to the presence or absence of woodland in the pre-conquest period. In Map 7, Professor B. K. Roberts has plotted the presence of woodland in the period c.710–1086 (see Roberts 1999. 94). The map is based on the work of H. C. Darby *et al.* (1952–7), who assembled the evidence for woodland in Domesday Book in a series of detailed maps. To this has been added the extent of rough pasture, common land and woodlands (other than plantations) in the Land Use Survey of the 1930s (Stamp 1937–44) and the evidence of Old English or Old Norse place-names indicating the former presence of woodland. Although there are many caveats concerning the interpretation of this map and those on which it is based, the synoptic view of cleared land and woodland is remarkably clear. The picture is reinforced by Map 8, which plots the density of rural settlement in the mid-nineteenth century. England may be divided into a central province of open champaign country characterised by the classic great open-field village system bounded by two provinces of a more wooded character. The distribution of *Green*, *Street* (in the sense 'row of houses, hamlet'), and to a lesser extent *End*, shown on Maps 9–11, is remarkable for its concentration in the two outer zones pointing for all their differences to the presence of similar controlling landscape features. It is argued that the *Green* names, or rather the settlements they designate (which are radically different from the 'green villages' of Northern England where the green is an element of an ancient structured planned nucleation), belong to an older landscape of former commons once surrounding ring-fenced structures or informal 'squatting' settlements. By contrast, the classic open-field system of the central zone is held to be a later planned development, overlying and obscuring earlier patterns, and to a large extent mirrored in the distribution of OE *tūn* names (Map 12).[1] Similar maps can be generated for *Common*, *Row* and other elements not usually mapped.

[1] For further detail see Roberts 1999 and Roberts *et al.* 1973; Rackham, 1986; Warner, 1987; and Hodges, 1991.

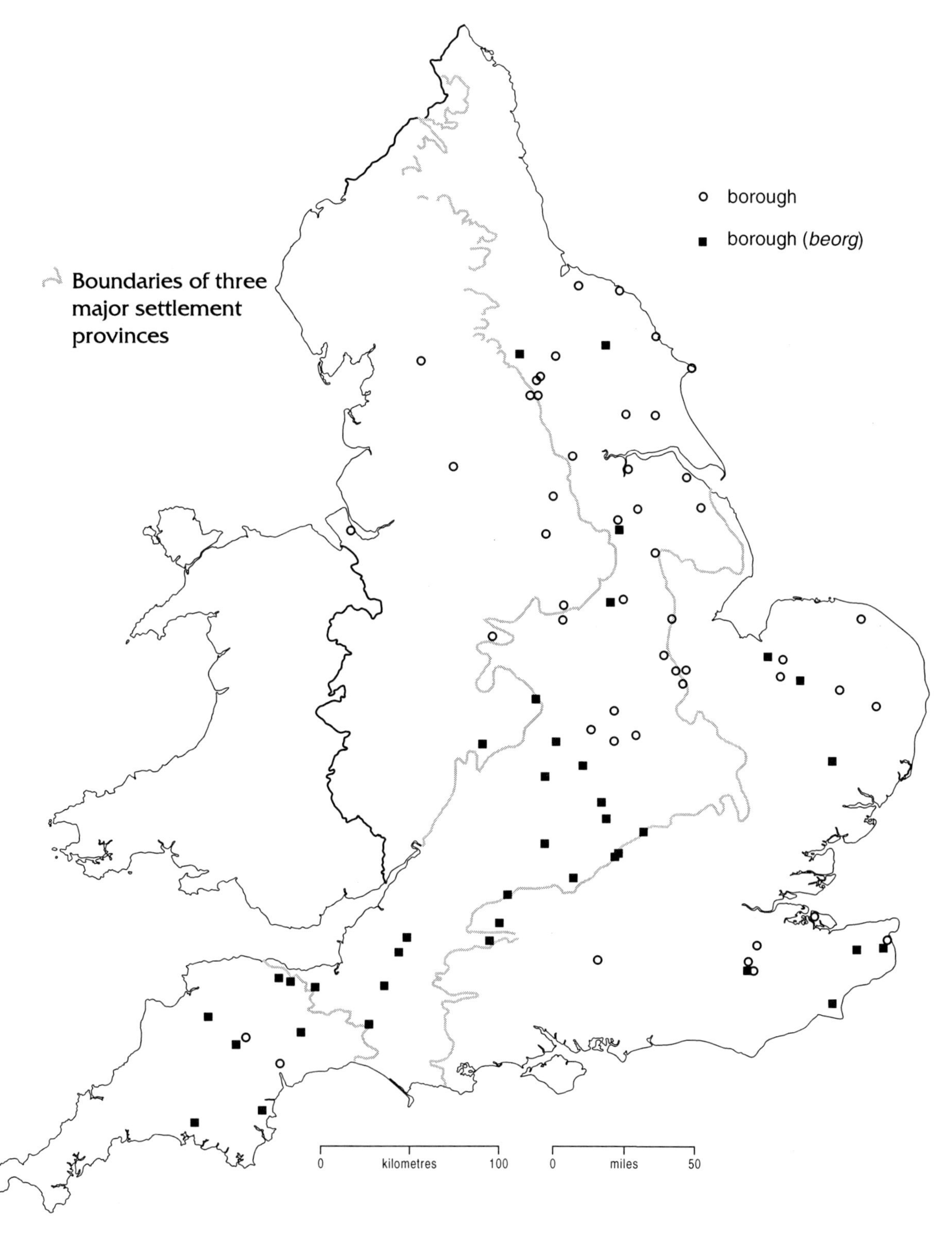

1. The distribution of the spelling 'borough' for OE *burh* and for *beorg* when confused with *burh*. © V. E. Watts; del. B. K. R.

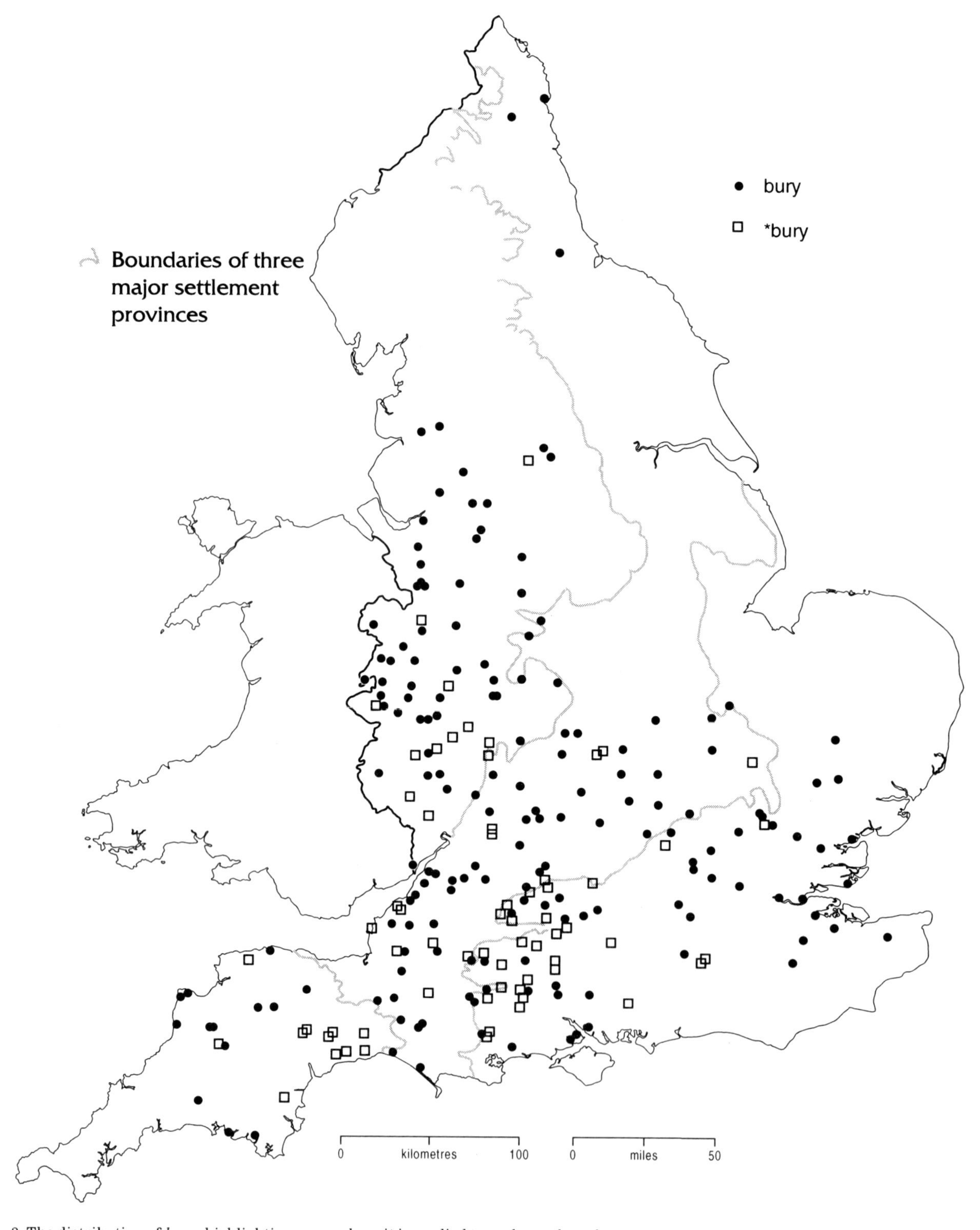

2. The distribution of *bury*, highlighting cases where it is applied to early earthworks especially of the Iron Age period (*bury). © V. E. Watts; del. B. K. R.

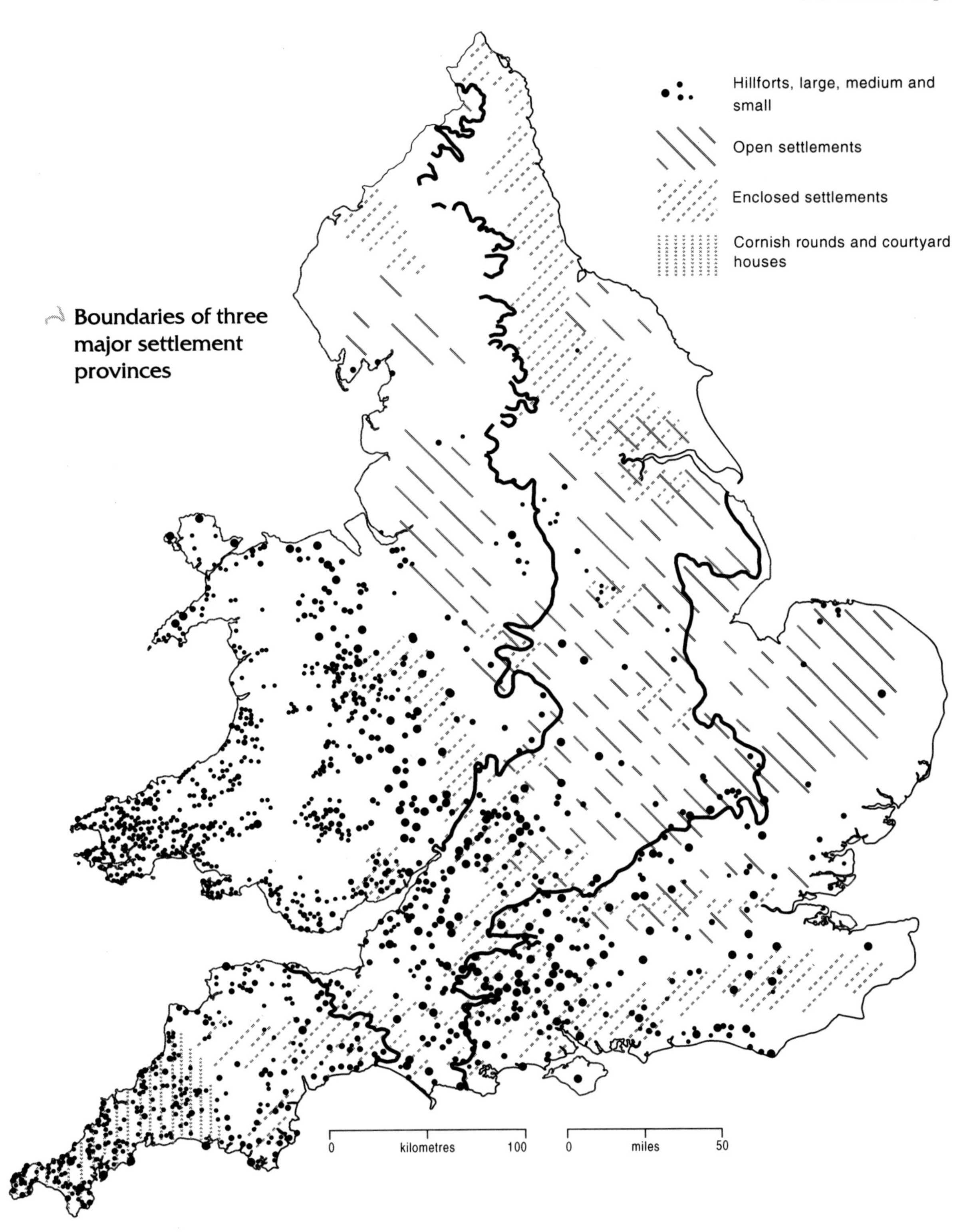

3. A map of settlement in the pre-Roman Iron Age. © B. K. R. (Roberts 1999), after Forde-Johnston 1976 and Manley 1989

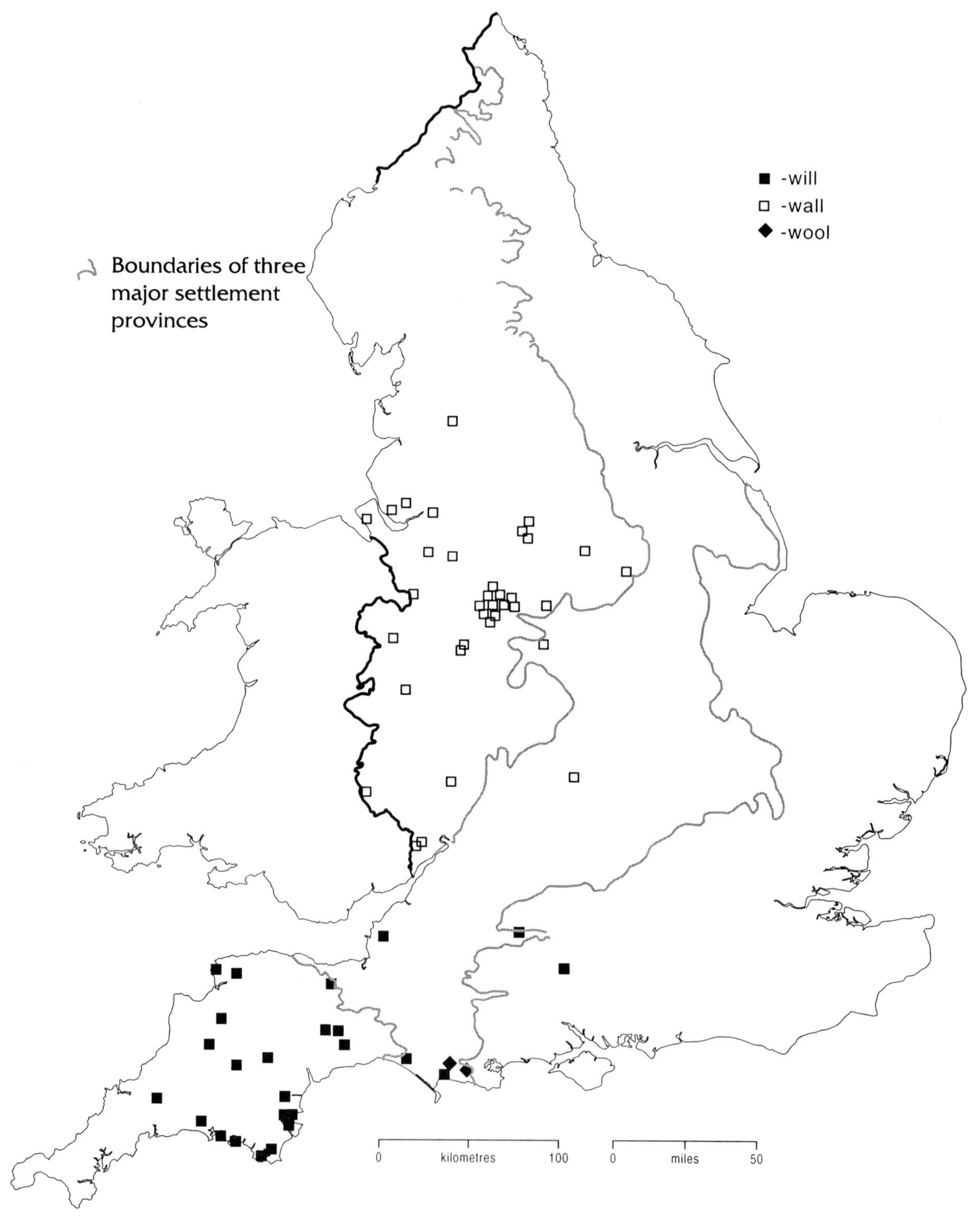

4. The West Midland and South West reflexes of OE *wella* 'a spring, a stream'. © V. E. Watts; del. B. K. R.

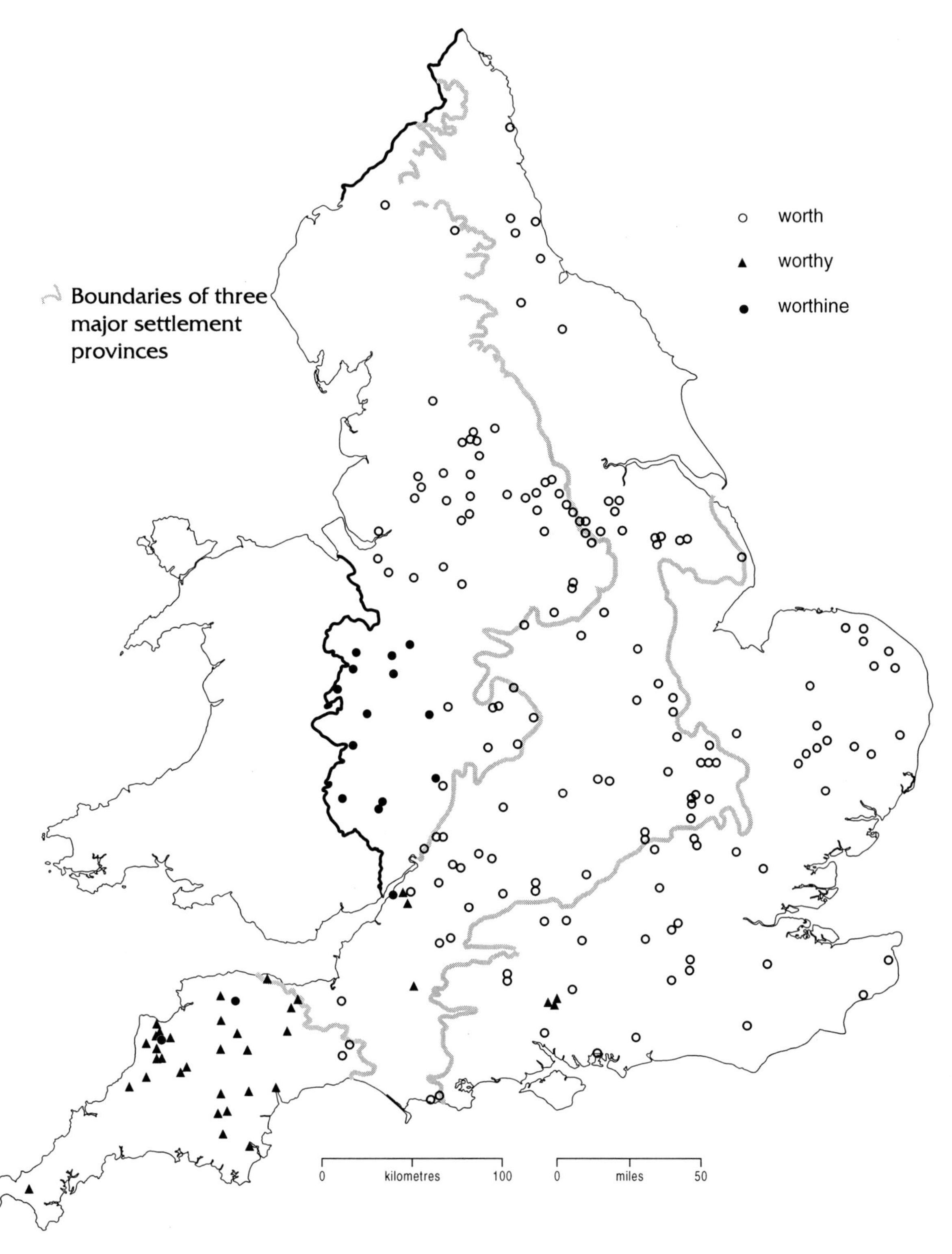

5. The three variant forms of *worth*. © V. E. Watts; del. B. K. R.

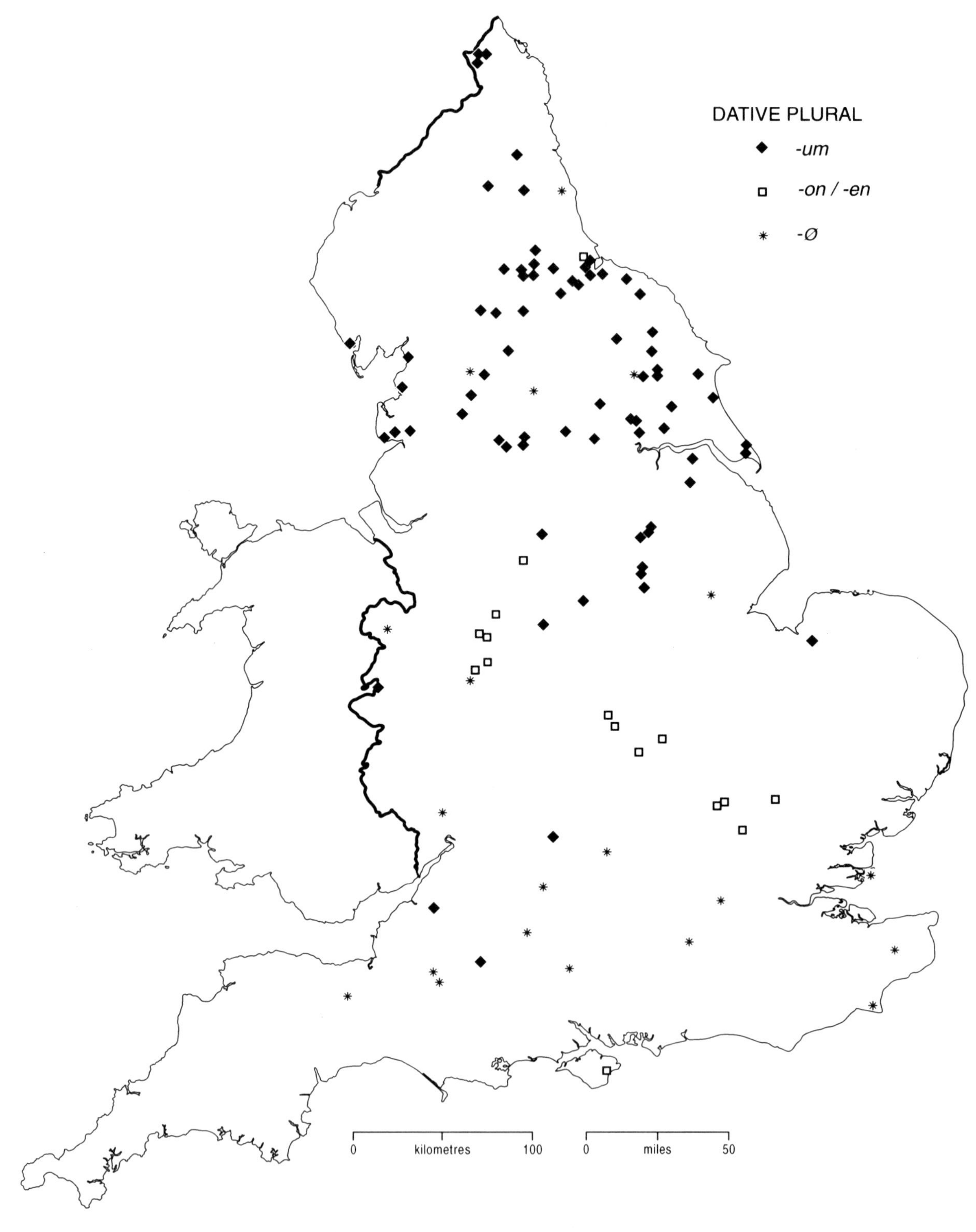

6. The survival or non-survival of the OE dative plural inflectional ending *-um*. © V. E. Watts; del. B. K. R.

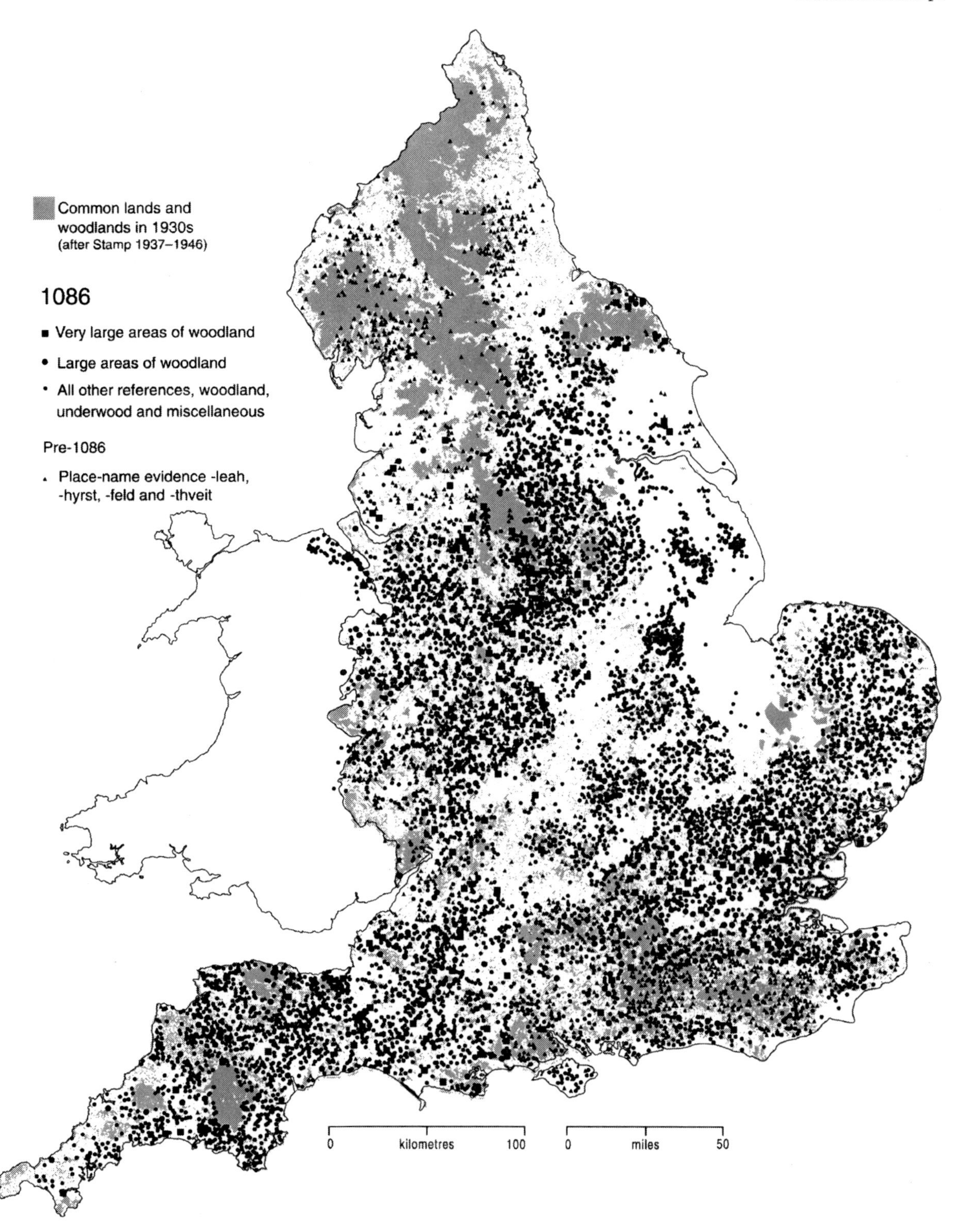

7. Presences of woodland c. 700–1086: place-names ending in *-leah, -hyrst, -feld* and *-thveit* . © B. K. R./S. W./E. H., after Darby *et al.* 1952 ff.; after Rackham 1986, fig. 5.7, Watts in Sawyer 1976, fig. 20.5, A. H. Smith 1967; OS *Gazetteer* 1992

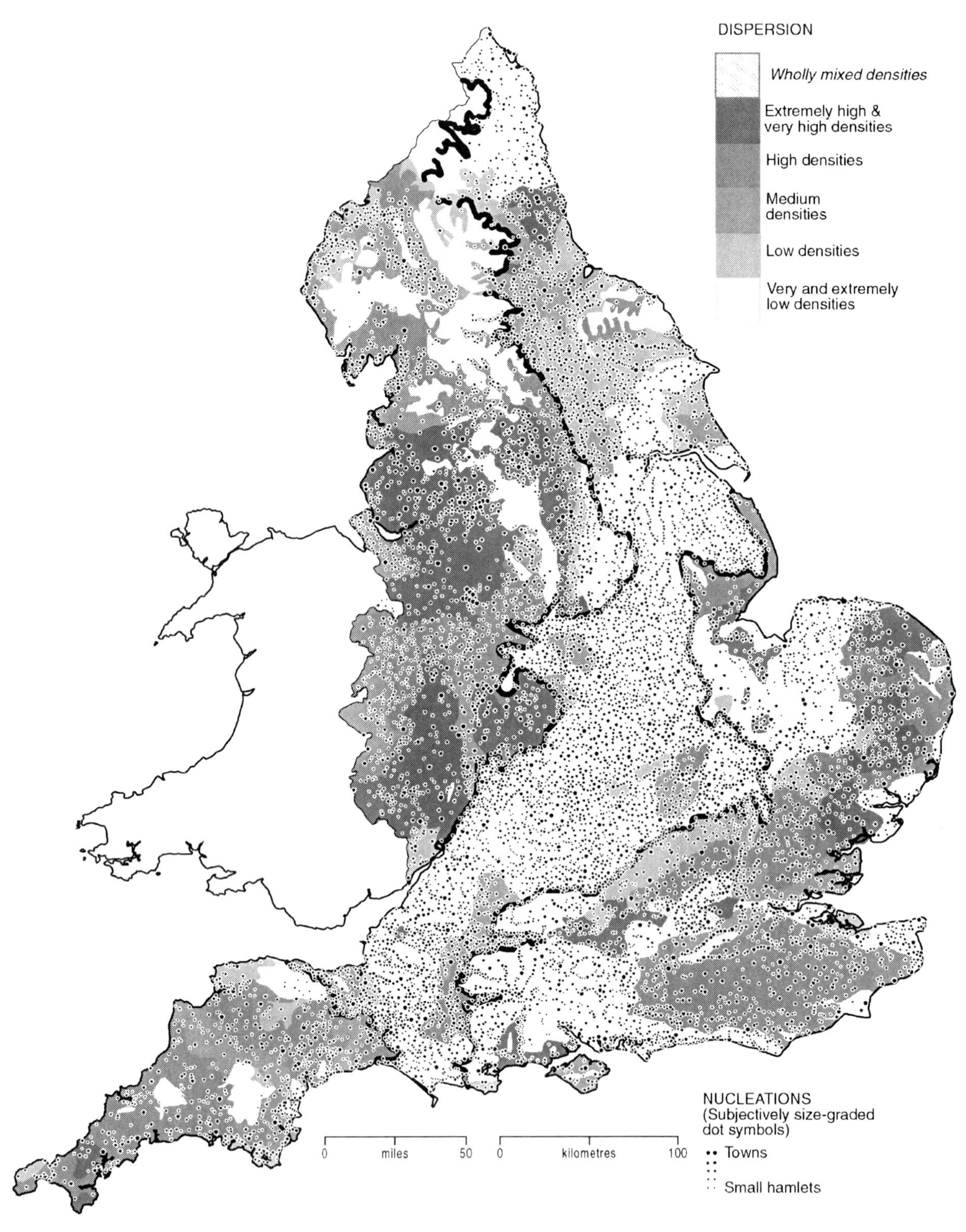

8. Rural settlement in the mid-nineteenth century. © B. K. R./S. W./E. H.

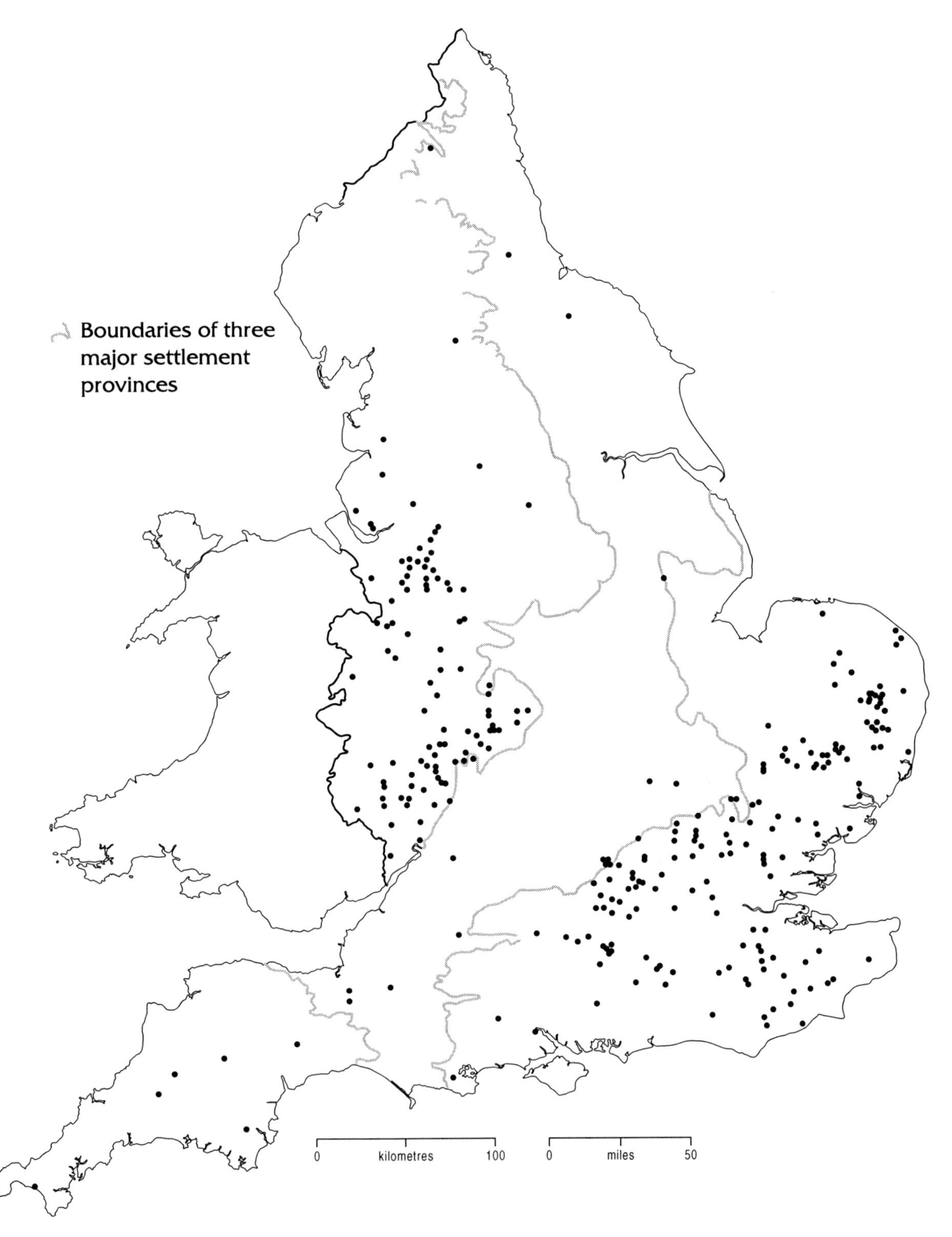

9. Selected place-name evidence for *Green* (283 locations). © V. E. Watts; del. B. K. R.

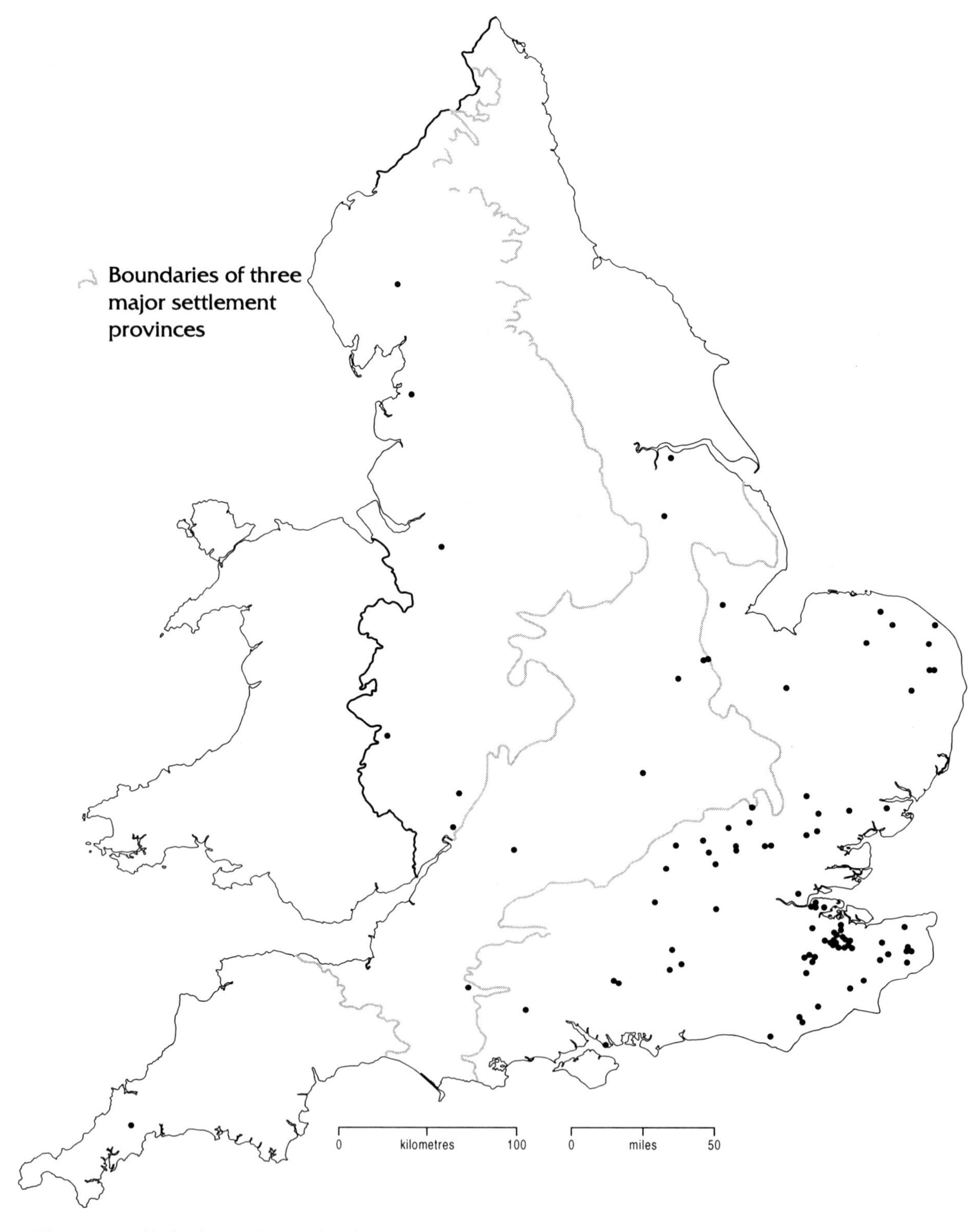

10. Place-names with the element *Street* (94 locations). © V. E. Watts; del. B. K. R.

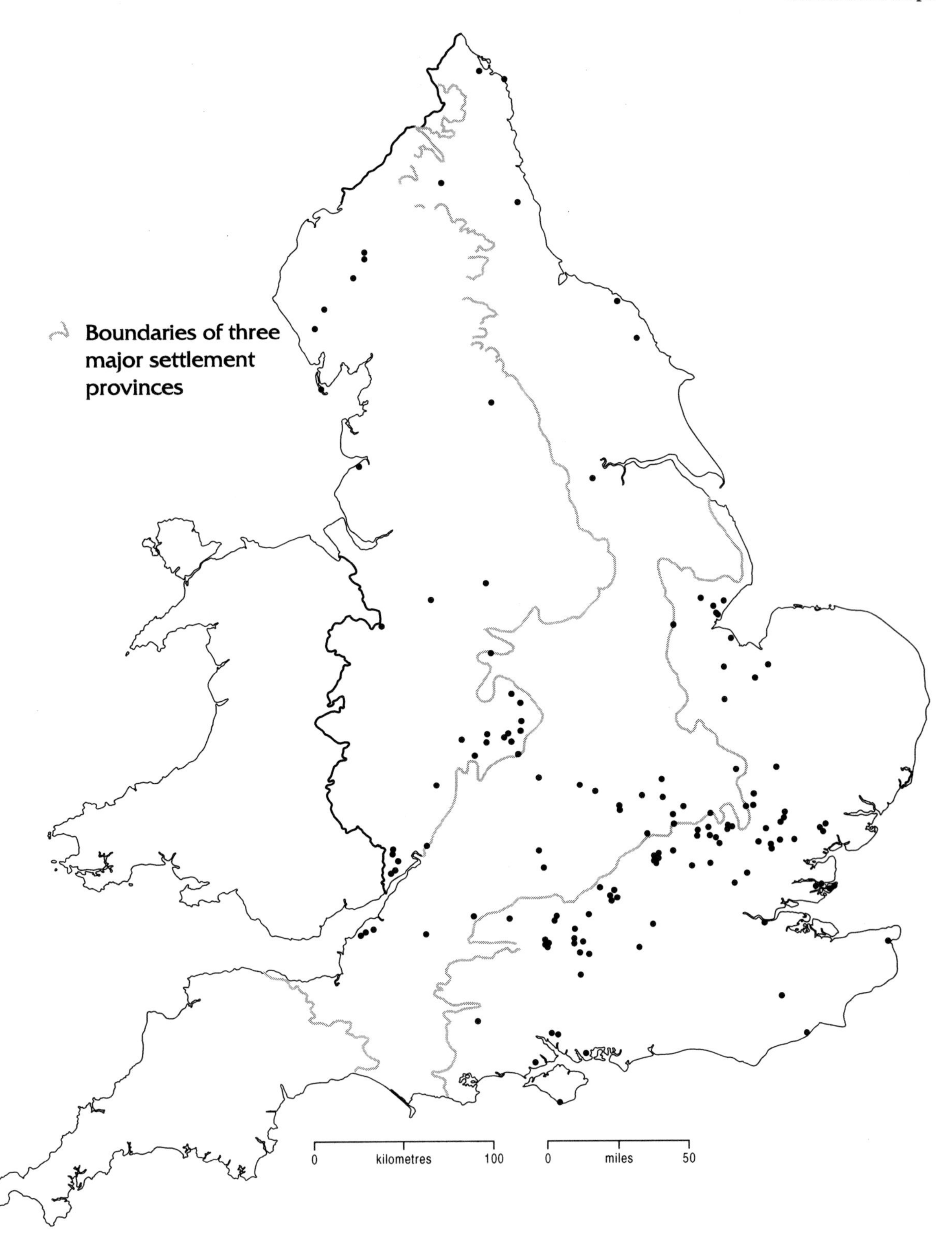

11. Selected place-name evidence for *End* (145 locations). © V. E. Watts; del. B. K. R.

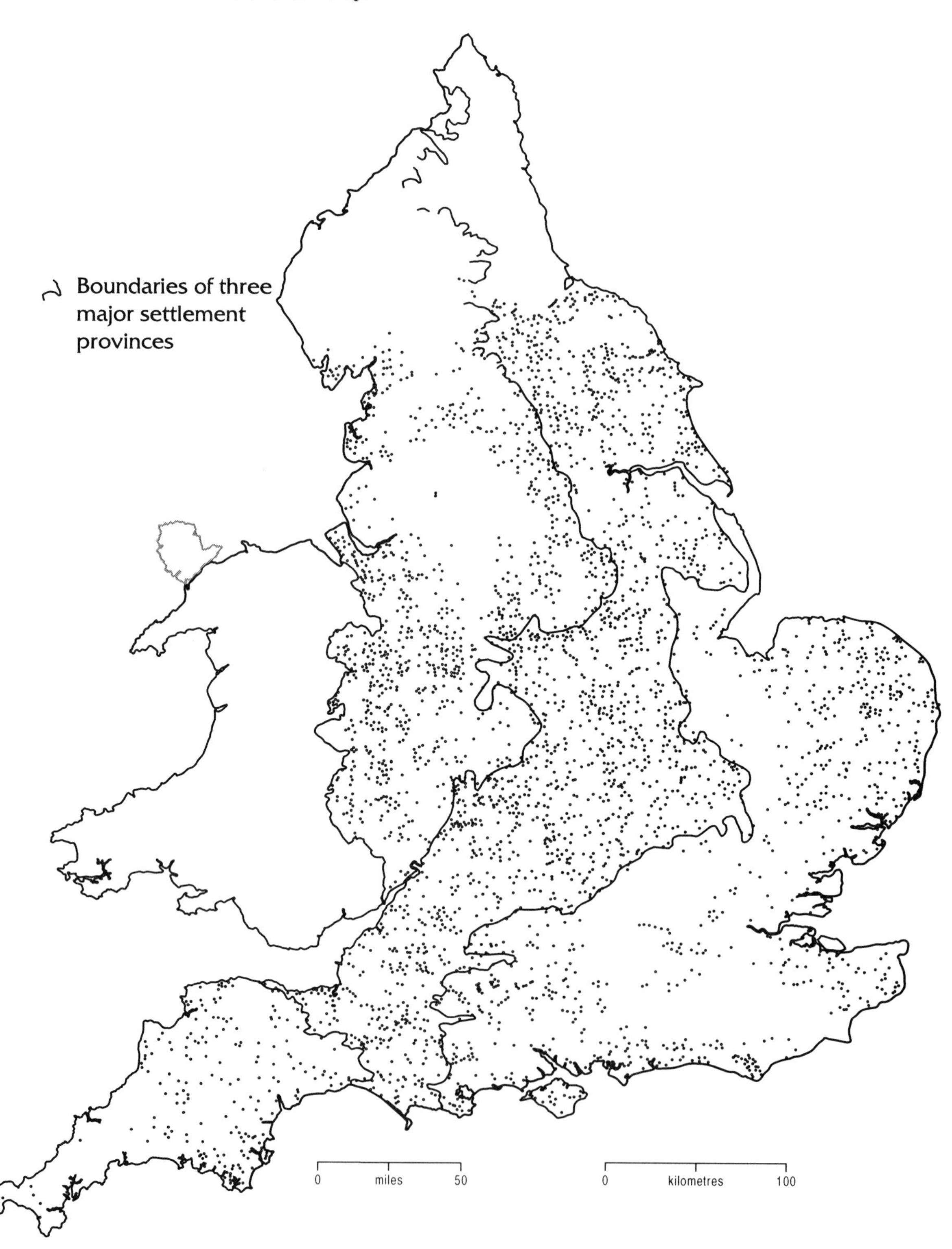

12. Place-names in *-tun* recorded in 1086. © B. K. R./S. W./E. H.

References

This bibliography contains only those works cited in the preliminary pages. The sources cited in abbreviated form in the main body of the dictionary and glossary are given in full in the List of Abbreviations (iv: Sources).

Arngart, O. (1972). 'On the *ingtun* type of English place-name', *Studia Neophilologica* 44: 263–73.
Cameron, K. (ed.) (1975). *Place-Name Evidence for the Anglo-Saxon Invasions and Scandinavian Settlements*, English Place-Name Society, Nottingham.
Cameron, K. (1976). *The Significance of English Place-Names*, Sir Israel Gollancz Memorial Lecture, British Academy.
Coates, Richard and Andrew Breeze (2000). *Celtic Voices, English Places: Studies of the Celtic Impact on Place-Names in England*, Stamford.
Cox, Richard (1988–9). 'Questioning the validity of the term "hybrid" in Hebridean place-name study', *Nomina* 12: 1–9.
Darby, H. C. *et al.* (ed.) (1952–77). *The Domesday Geography of England: Eastern England* 1952; *Midland England* 1954; *South-East England* 1962a; *Northern England* 1962b; *South-West England* 1967; *Gazetteer* 1975; *Domesday England* 1977, Cambridge University Press.
Dobson, E. J. (1968). *English Pronunciation 1500–1700*, 2 vols., Oxford.
Dodgson McNeal, J. (1966). 'The significance of the distribution of English place-names in *-ingas*, *-inga-* in south-east England', *Medieval Archaeology* 10: 1–29.
Dodgson McNeal, J. (1967a). 'The *-ing* in English place-names like Birmingham and Altrincham', *Beiträge zur Namenforschung* n.f. 2: 221–45.
Dodgson McNeal, J. (1967b). 'Various forms of Old English *-ing* in English place-names', *Beiträge zur Namenforschung* n.f. 2: 325–96.
Dodgson McNeal, J. (1968). 'Various English place-name formations containing Old English *-ing*', *Beiträge zur Namenforschung* n.f. 3: 141–89.
Ekwall, Eilert (1928). *English River-Names*, Oxford.
Ekwall, Eilert (1960). *The Concise Oxford Dictionary of English Place-Names*, 4th edn, Oxford.
Fellows-Jensen, Gillian (1996). 'Doddington revisited' in *A Frisian and Germanic Miscellany in Honour of Nils Århammar on his 65th Birthday*, Nordfriisk Institut, Odense University Press, Bredstell, pp. 361–76.
Gelling, Margaret (1976). 'Introduction' in *The Place-Names of Berkshire* III, English Place-Name Society, pp. 800–47.
Gelling, Margaret (1978). *Signposts to the Past: Place-Names and the History of England*, London.
Gelling, Margaret (1984). *Place-Names in the Landscape*, London.
Gelling, Margaret (1990). *The Place-Names of Shropshire* I, English Place-Name Society.
Gelling, Margaret (1998). 'Place-names and landscape' in Simon Taylor (ed.), *The Uses of Places-Names*, Edinburgh, pp. 75–100.
Hodges, R. (1991). *Wall-to-Wall History: The Story of Royston Grange*, London.
Kitson, P. R. (1996). 'British and European river-names', *Transactions of the Philological Society* 94: 73–118.
Krahe, Hans (1962). *Die Struktur der alteuropäischen Hydronomie*, Mainzer Akademie der Wissenschaften und der Literatur in Mainz, Abhandlung der Geistes- und Sozialwissenschaften Klasse 5, Wiesbaden.
Krahe, Hans (1964). *Unsere ältesten Flußnamen*, Wiesbaden.
Mawer, Allen (1920). *Place-Names of Northumberland and Durham*, Cambridge.
Mills, A. D. (2001). *A Dictionary of London Place Names*, Oxford.
Nicolaisen, W. F. H. (1957). 'Die alteuropäischen Gewässernamen der britischen Hauptinsel', *Beiträge zur Namenforschung* 8: 209–68.
Nicolaisen, W. F. H. (1982). 'Old European names in Britain', *Nomina* 6: 37–42.
Rackham, O. (1986). *The History of the Countryside*, London.
Rivet, A. L. F. and Colin Smith (1979). *The Place-Names of Roman Britain*, London.
Roberts, B. K. (1999). 'Of *Æcertyning*', *Durham Archaeological Journal* 14–15: 93–100.
Roberts, B. K. *et al.* (1973). 'Recent forest history and land uses in Weardale, northern England' in H. J. B. Birks and R. G. West (eds.), *Quaternary Plant Ecology*, Blackwell Scientific Publications, Oxford, pp. 207–21.
Sawyer, P. H. (1968). *Anglo-Saxon Charters: An Annotated List and Bibliography*, Royal Historical Society.
Smith, A. H. (1956). *English Place-Name Elements*, 2 vols., Cambridge.
Smith, A. H. (1967). *The Place-Names of Westmorland*, 2 vols., Cambridge.

Stamp, L. D. (ed.) (1937–44). *The Land of Britain*, Royal Geographical Society and Geographical Publications Ltd, London.

Tovar, Antonio (1977). *Krahes alteuropäische Hydronomie und die westindogermansichen Sprachen*, Sitzungsberichte der Heidelberger Akademie der Wissenschaften, Philosophisch-historische Klasse 2, Heidelberg.

Warner, P. (1987). *Greens, Commons and Clayland Colonization*, University of Leicester, Occasional Papers, 4th series, no. 2.

Watts, Victor (1976). 'Comment on "The Evidence of Place-Names" by Margaret Gelling' in P. Sawyer (ed.), *Medieval Settlement: Continuity and Change*, London, pp. 212–22.

Watts, Victor (1993–4). 'A new dictionary of English place-names', *Journal of the English Place-Name Society* 26: 7–14.

Watts, Victor (1999). 'Some place-name distributions', *Journal of the English Place-Name Society* 32: 53–72.

ABALLAVA Cumbr NY 3259. The Roman fort at Burgh-by-Sands. *Aballava* mid 2nd Rudge Cup (RBrit 232–3), *Avalana* [c.700]14th Rav, *Avalava* [c.700]13th Rav, *Aballaba* [c.408]15th ND. Possibly '(apple) orchard'. Brit ***aballā** + suffix ***-auā**. Cf. Availles, various examples, France, *Avallia*, Gaulish *aballo-* + suffix *-ia*, Avallon, Yonne, France, *Aballo* 4th, Gaulish *aballo-* + suffix *-one*, Aveluy, Somme, *aballo* + suffix *-ucium*, and Avella, Campagna, Italy, Virgil's *malifera Abella* 'abounding in apples' (*Aeneid* vii.740). RBrit 238, Dauzat-Rostaing 40–1.

ABBERLEY H&W SO 7567. 'Eadbald's wood or clearing'. *Edboldelege* 1086, *Albodeslega* c.1150, *Abbedeslegh* 1216, *Ab(b)ot(t)eley(e)* 1346–1485, *Aberley* 1480, 1484. OE pers.n. *Ēadbald* + **lēah**. The ME reflex *Abbod, Abbed* of OE *Ēadbald* led to confusion with the word *abbot*. Wo 23, DEPN.

ABBERLEY HILL H&W SO 7566. *Abberley Hill* 1832 OS. P.n. ABBERLEY SO 7567 + ModE **hill**.

ABBERTON Essex TM 0019. 'Eadburg's estate'. *Edburg(h)etunā* 1086, *(de) Eadburgetona* 1108×27, *Adbur(u)g(h)(e)ton(e)* 1208–1321, *Adburthon* 1280, *Abberton* from 1230. OE feminine pers.n. *Ēadburh*, genitive sing. *Ēadburge*, + **tūn**. Ess 314, Jnl 2.45.

ABBERTON H&W SO 9953. 'Estate called after Eadbriht'. *Eadbrihtincg tun, (in)eadbrihtincgtune* 972 S 786, *Edbritone, Edbretintune* (two separate divisions of the original manor) 1086, *Adbrighton -bry-* 1297–1377, *Abburton* 1535. OE pers.n. *Ēadbriht* + **ing**[4] + **tūn**. Wo 184, Hooke 179–90.

ABBERTON RESERVOIR Essex TL 9818. P.n. ABBERTON TM 0019 + ModE **reservoir**.

ABBERWICK Northum NU 1213. 'Alubeorht's or Alu(h)burg's dairy-farm'. *Alburwic* 1169, *Alber- Alburckwick* 1278, *Awberwyke* 1428, *Averwick* 1610, *Abberwick* from 1689. OE pers.n. *Alubeorht* masc. or *Alu(h)burg* feminine + **wīc**. For *b* > *v* cf. Great BAVINGTON. NbDu 1.

ABBEY Devon ST 1410. Cf. Dunkeswell Abbey.

ABBEY BROOK SYorks SK 1892. *Abbey Brook* 1842 OS. P.n. *Abbey* 1842 OS, ModE **abbey**, + **brook**.

ABBEY DORE H&W SO 3830 → Abbey DORE.

ABBEYSTEAD Lancs SD 5654. 'The site of the abbey'. ModE **abbey** + **stead**. The reference is to the site of a short-lived Cistercian colony from Furness Abbey founded c.1192 and removed before 1204. There are references to the *vaccary del Abbey* in 1323 and 1324. La 172, Stead 291.

ABBEYTOWN Cumbr NY 1750. *The Towne of the Abbey, Abbey Towne* 1649. So named after the Cistercian abbey of HOLME CULTRAM founded in 1150. Cu 290.

ABBEY WOOD GLond TQ 4878. *Abbey Wood* 1805 OS. A late 19th cent. residential area developed on land once owned by Lesnes Abbey, an Augustinian house founded in 1178 by Richard de Luci as an act of penance for his support of Henry II in his quarrel with Thomas Becket which led to the archbishop's murder. *Leosne* 1042×50, 1065, *Hlosnes* 11th, *Lesneis, Loisnes* 1086, *Lesnes* 1194, possibly 'the burial mounds', OE ***hlēosn**. The name was later associated with **ness** 'a headland' as in the form Lessness. DEPN, LPN 1, 135, Encyc 2, 467.

ABBOT'S FISH HOUSE Somer ST 4541. An early 14th cent. building in Weare used as a fisherman's house and a place to salt and store fish caught in the mere formerly here. Pevsner 1958 S.234.

ABBOTS LEIGH ST 5474 → Abbots LEIGH.

ABBOT'S WAY Devon SX 6266. A trackway across Dartmoor leading to Buckfastleigh.

ABBOTSBURY Dorset SY 5785. 'The abbot's manor'. *Abbedesburie* [939×46]18th S 1727, *(æt) Abbodesbyrig* [1058×66]14th S 1064, *Abedesberie* 1086, *Abbotesbir* 1227. OE **abbod**, genitive sing. **abbodes**, + **byriġ**, dative sing. of **burh**. Land here belonged to Glastonbury abbey. Do 25.

ABBOTSHAM Devon SS 4226. 'Ham held by the abbot' sc. of Tavistock. *Ab(b)edisham* 1193, 1269, *Abbodesham* 1282. Earlier simply *Hame* 1086. OE **abbod**, genitive sing. **abbodes**, + p.n. *Ham*, OE *hamm* 2b 'land on a hill-spur'. Possibly so called for distinction from LITTLEHAM SS 4323. D 83 gives pr [æpsəm].

ABBOTSIDE COMMON NYorks SD 8196. A hill-side held by the abbot of Jervaulx. YN 258.

ABBOTSKERSWELL Devon SX 8569. 'Kerswell belonging to the abbot' sc. of Horton. *Karswill Abbatis* 1285, *Abbotescharswelle* 1314, *Abbots Keswell* 1675. Earlier simply *Cærswylle* [956]12th S 601, *Carsvelle* 1086, *Kareswill* 1242, 'cress spring'. OE **cærse** + **wylle**. The manor was already held by the abbot of Horton in 1086. D 504.

ABBOTSLEY Cambs TL 2256. 'Eadbald's wood or clearing'. *Adboldesl'* [1154×89]13th, *-lee* 1216×72, *Albedesleg* 1257, *Albo(l)desle(g) -ley* 1279–1334, *Abbodesle* 1270–1313, *Abbot(t)esle(y)* 1286–1470. OE pers.n. *Ēadbald*, genitive sing. *Ēadbaldes*, + **lēah**. ME spellings with *Al-* show confusion with CG *Albold* < *Adalbald*. Hu 252.

ABBOTSWOOD Hants SU 3623. 'The abbess's wood'. *Abbys wode* 1513, *Abbes wood* 1565. ModE **abbess** + **wood**. It belonged to the abbess of Romsey. Gover 1958.185.

ABDON Shrops SO 5786. 'Ab(b)a's estate'. *Abetune* 1086, *Ab(b)eton* c.1200–1504, *Abbedon* 1301, *Abdon* from 1503. OE pers.n. *Ab(b)a* + **tūn**. Sa i.1.

ABERFORD WYorks SE 4337. 'Eadburg's ford'. *Ædburford* 1176, *Ædburgforð* 1177 P, *Ebberford* 13th cent., *Aberford* from 1208, *-forth* 1278. OE fem. pers.n. *Ēadburg* + **ford**. YW iv.97 gives pr ['abəfəθ], WYAS 294, L 69.

ABINGDON Oxon SU 4997. 'Æbba or Æbbe's hill'. *(in) Æbbandune* [687]12th S 239, [726×37]12th S 93, *Æbbanduna, Æbbendune* [811 for 815]12th S 166, *Abbandun* [931]12th S 410, *(æt) Abbandune* [968]16th S 758, *(de) Abbenduna, (in) Abbendona, (to) Abbendune* [c.931]12th S 1208, *Abbendun* [955]12th S 567, *Abbendone* 1086–14th with variants *-dun -donia -don, Abingdon(ia)* [1224]c.1250, 1535. OE masc. pers.n. *Æbba* or feminine *Æbbe*, genitive sing. *Æbban* + **dūn**. Brk 432, L 147, PNE i.10, 282.

ABINGER COMMON Surrey TQ 1145. P.n. *Abinceborne* 1086, *Ab(b)ingewurd(a)* 1191, *Abbyngesworth* 1279, *Abynchere* 1552, *Abbynsworth alias Abynger, Abyngworth alias Abinger* 1557–92, 'the enclosure at **Abbinge*, the place called after Abba', + ModE **common**. OE **worth** has been added to **Abbinġe*, the locative-dative form of a pre-existing p.n. **Abbing* from OE pers.n. *Abba* + **ing**[2]. The DB form is corrupt. The 1279 form shows an alternative *-es* genitive sing. inflexion (still remembered in the 16th cent.) in the name **Abbing* instead of the locative-dative one. For the shortened *-er* form, cf. Ember Court, Sr 91–2. Sr 259 gives pr [æbindʒə], BzN 1967.359–60.

ABINGTON PIGOTTS Cambs TL 3044. 'A held by the Pigott family'. *Pigots* first appears in 1635; the manor was held by

John Pykot and his successors from c.1427 to 1802. Abington, *Abintona -e* 1086, *-ton(e) -yn-* 1165–1352, *Abbingeton* 1198–1272, *Abbington(e) -yng-* 1273–1428, *Abington -yng-* from 1208, is the 'estate called after Abba'. OE pers.n. *Abba* + **ing**[4] + **tūn**. 'Pigotts' for distinction from Great and Little ABINGTON TL 5348, 5349. Ca 51.

Great ABINGTON Cambs TL 5348. Latin **magna** (first occurs 1218), ModE **great** (first occurs 1523) + p.n. *Abintone, (H)abintona* 1086, *Abingtuna* 1156×7, *Abintone -yn-* 1158–1428, *Abiton(e) -y-* 1199–1484, *Abington* from 1215, 'estate called after Abba', OE pers.n. *Abba* + **ing**[4] + **tūn**. Ca 99, O li.

Little ABINGTON Cambs TL 5349. Latin **parva** (first occurs 1218), ME **littel** (first occurs 1336), + p.n. Abington as in Great ABINGTON TL 5348. Ca 99.

AB KETTLEBY Leic SK 7222 → Ab KETTLEBY SK 7222.

ABLINGTON Glos SP 1007. 'Estate called after Eadbald'. *(æt) Eadbaldingtun(e)* [855]11th S 206, [899]11th S 1279, *Ablinton(a) -yn-* 1209–1501, *Ablyngton -ing-* 1286–1601. OE pers.n. *Ēadbald* + **ing**[4] + **tūn**. Gl i.26.

ABNEY Derby SK 1979. 'Abba's island'. *Habenai* 1086, *Abbeneia -ey(a) -eye* 1200–1431, *Abney* from 1416. OE pers.n. *Abba*, genitive sing. *Abban*, + **ēġ**. The meaning is probably a patch of good land in moorland. Db 25, L 36, 38.

ABRAM GMan SD 6001. 'Eadburg's homestead or village'. *Hadburham* 1190×1219, *Adburgham* before 1199–1332, *Ed-* 1212, *Abburgham* 1246, 1263, *Abraham* 1372, *Abram* 1461. OE feminine pers.n. *Ēadburg* + **hām**. La 102, Jnl 17.57 gives pr [æbrəm], 18.15.

ABRIDGE Essex TQ 4696. 'Æffa's bridge'. *Affebrigg(e)* 1203, 1328, 1485, *-brug' -bregge* 1239, 1351, *Abrigge* 1469, 1485, *Abridge or High Bridge* 1594. OE pers.n. *Æffa* + **brycg**. The same pers.n. is found in *æffan hecce* 'Æffa's hatch' *[1062]13th S 1036 in the bounds of Debden Green TQ 4398. Ess 60 gives pr [eibridʒ], 66 n.1.

ABTHORPE Northants SP 6546. 'Abba's outlying farm'. *Abetrop* 1190, 1203, *Abbetorp* 1200, 1202, *-thorp* 1230, 1241, *-throp* 1276, *Abthrop* 1255, *Ap- Abthorp* 1384, *Adthrop* 1601. OE pers.n. *Abba* + **thorp**. The DB form ascribed to Abthorpe in Nth and DEPN does not belong to this name. Nth 89 gives pr [adθrəp], SSNEM 122.

ABY Lincs TF 4178. 'The river village or farm'. *Abi* 1086, c.1115, c.1175, *Aby* from 1212. ON **á** + **bȳ**. Aby is situated on the banks of the Great Eau. DEPN, SSNEM 30, Cameron 1998.

ACASTER MALBIS NYorks SE 5945. 'A held by the Malbis family'. P.n. *Acastre -tra -ter* 1086–1541 + family name of William *Malebisse* 1167. Acaster is the 'Roman fort on the river', ON **á**, probably replacing OE **ēa**, + **cæster**. Like ACASTER SELBY SE 5741 this was a fort established to protect York against attackers sailing up the r. Ouse. The manor was held by the Malbis family 12th–15th cents. YW iv.218.

ACASTER SELBY NYorks SE 5741. 'A held by Selby abbey'. P.n. *Acastra -tre* 1086 etc. as in ACASTER MALBIS SE 5945 + p.n. *Sel(e)by* 13th cent. referring to Selby Abbey which held land here. YW iv.216.

ACCRINGTON Lancs SD 7528. 'Acorn farm'. *Akarinton* before 1194, *Akeryn(g)ton(a) -in(g)-* 1258–1446×7, *Acrin(g)ton -yn(g)-* 1287–1361, *Accryngton* 1519×21. OE **æcern** + **tūn**. The southern part of this chapelry was regarded as being in the forest where oak mast was once of great importance as a food for pigs. La 89, Jnl.17.54.

ACHURCH Northants TL 0238. Short for *Thorp Hacchurch* 1316, *Thorp Achirch* 1428. Thorp as in THORPE WATERVILLE Northants TL 0281 + lost p.n. *Asencircan, Asecyrcan* [c.980]c.1200 B 1130, *Asechirce* 1086, *-cherche -chirche -kirche -kirke* 1164–13th, *Achirche* 12th–1372, 'Asi or Asa's church'. ON pers.n. *Ási* or *Ása* (feminine) or OE **Asa* + OE **ċiriċe** with occasional substitution of Scandinavian [k] for [tʃ]. Nth 219, SSNEM 211.

ACKLAM '(The place at) the oak-tree clearings'. OE ***āc-lēah**, dative pl. ***āc-lēum**. SN 10.107–13.

(1) ~ Cleve NZ 4817. *Achelum, Aclun -um* 1086, *Aclum* [1129]15th–1247, *Acclum* c.1142–1404, *Acclam* 1399. Acklam is in systemic contrast with other dative pl. names in the district, COATHAM NZ 5925, KIRKLEATHAM NZ 5921, MARSKE NZ 6322 and UPLEATHAM NZ 6319. YN 162, SN 10.107–13, SSNY 252.

(2) ~ NYorks SE 7861. *Aclun -um* 1086, *Ac(c)lum* 1130–1365, *Acklam* 1587–1828. YE 147, SN 10.107, L 203.

ACKLETON Shrops SO 7798. 'The estate called after Eadlac'. *Aclinton* 1176, *Adelacton* 1292. OE pers.n. *Ēadlāc* + **ing**[4] + **tūn**. DEPN.

ACKLINGTON Northum NU 2201. 'The estate called after Æccela'. *Eclinton* 1176, *Ac-* 1186, *Aclington* from 1242 BF. OE pers.n. **Æccela*, genitive sing. **Æccelan*, varying with *Æccela* + **-ing**[4], + **tūn**. NbDu 1.

ACKTON WYorks SE 4121. 'Oak-tree settlement'. *Aitone, Attone* 1086, *Aike- Ayketon(a)* 12th–1497, *Acton* 1537–1638, *Ackton* 1585. ON **eik** (possibly replacing OE **āc**) + **tūn**. Cf. ACTON Sa i.1–4. YW ii.85, SSNY 114, WYAS 295.

ACKWORTH MOOR TOP WYorks SE 4316. *Ackworth Moor Top* 1771. P.n. *Ackworth Moore* 1658 + ModE **top**. See High ACKWORTH SE 4417. YW ii.94.

High ACKWORTH WYorks SE 4417. 'Acca's enclosure'. *Aceuurde* 1086, *Hac(he)wrda, Hackewrthe, Hachaworda* 12th cent., *Ackew(o)rth(e)* 1227–1551, *Ac- Akworth(e)* 1316–1638. OE pers.n. *Acca* + **worth**. YW ii.93, WYAS 296.

ACLE Norf TF 4010. 'Oak wood or clearing'. *Acle* 1086, *Achelai* 1159, *Aclee* 1197. OE **āc** + **lēah**. DEPN.

ACOCK'S GREEN WMids SP 1283. *Acock's Green* 1834 OS. Surname *Acock* + ModE **green**. Cf. John Acoc 1420.

ACOL Kent TR 3067. 'Oak-wood'. *Acholt'* 1270, *-holte* 1303, 1342, *(H)oc- Okholte* 13th. OE **āc** + **holt**. PNK 594.

ACOMB '(The settlement) at the oak-trees', OE **āc**, dative pl. **ācum**. L 218.

(1) ~ Northum NY 9366. *Aycomb* before 1769 Map.

(2) East ~ Northum NU 0564. *Akum, Acum* 1296 SR 13, 35. NbDu 1 gives pr [jekəm].

ACONBURY H&W SO 5133. 'Squirrel fort'. *Akornebir* 1213, *-bury* 1218, *Okernebur'* 1241. OE **ācweorna** + **byriġ**, dative sing. of **burh**, referring to Aconbury Iron Age hill-fort at SO 5033. DEPN, Thomas 1976.144.

ACRE Lancs SD 7824. No early forms.

Castle ACRE Norf TF 8115. 'Acre castle'. *Castelacr'* 1235. ME **castel** + p.n. Acre as in South ACRE TF 8114. 'Castle' referring to the castle here, one of the grandest motte-and-bailey castles in England, already mentioned in the late 11th cent., and for distinction from South and West ACRE TF 8114, 7815. DEPN.

South ACRE Norf TF 8114. *Sutacra* 1242. ME adj. **sūth** + p.n. *Acra -e* 1086, *Accara* 1121, 'the field', OE **æcer**. Cf. the p.ns. Aker in Norway, Åker in Sweden. The reference is probably to newly-cultivated land. 'South' for distinction from Castle and West Acre TF 8115, 7815. DEPN.

West ACRE Norf TF 7815. *Westacre* 1203. ME adj. **west** + p.n. Acre as in Castle ACRE TF 8115. DEPN.

ACRISE PLACE Kent TR 1942. P.n. Acrise + ModE **place** 'a mansion, a dwelling'. The reference is to a house of the 16th and later cents. Acrise, *Acres* 1086, *(H)acris(e)* c.1100, 1166, *Acrise* from 1226, *Ockerise* 1215, is 'oak-tree brushwood', OE **āc** + **hrīs**. PNK 430, Newman 1969E.121.

ACTON 'Farm, estate where oak is worked'. OE **āc-tūn**. The oak is well known as a timber of exceptional quality for hardness and toughness. It was especially used in shipbuilding and in all types of other construction. The roots were used to make hafts for daggers and knives and the bark in tanning leather and in dyeing. Acorns were important for feeding swine and special laws were made in Anglo-Saxon times relating to the feeding of swine known as pannage. PNE i.2, Sa i.1–4, Grieve 593ff.

(1) ~ Ches SJ 6353. *Acatone, Actune* 1086, *Acton* [c.1130]1479 and from 1253. Che iii.126.

(2) ~ Dorset SY 9878 is of different origin. See under ELSTREE GLond TQ 1895.

(3) ~ GLond TQ 2080. *Acton(e)* from 1181. Formerly a separate hamlet also known as *Chirche Acton* 'A with the church' 1347 for distinction from East Acton TQ 2180, *Estacton* 1294. Mx 81.

(4) ~ Shrops SO 3185. *Acton* from 1255. Sa i.1.

(5) ~ BEAUCHAMP H&W SO 6750. 'A held by the Beauchamp family'. Originally simply *Aactune* [718 for ?727]12th S 85, *(in) actune* 972 S 786, *Actvne* 1086. The manor was held by the Beauchamp family from the 13th cent. Wo 25, Hooke 179.

(6) ~ BRIDGE Ches SJ 5975. *pons de Actona* 13th, *Acton Brygge* 1486. p.n. *Acton* from early 13th, + ME **brigge** (OE *brycg*, Latin *pons*). Che iii.193. Addenda gives 19th cent. local pr ['akn].

(7) ~ BURNELL Shrops SJ 5302. 'That part of A held by the Burnell family'. *Akton' Burnell* 1198. P.n. *Actune* 1086, *Acton(')* from 1242 + manorial addition from the Burnell family mentioned in connection with the manor from the late 12th cent. and for distinction from ACTON PIGOTT SO 5402. Both places were probably originally part of a single estate. Sa i.4.

(8) ~ GREEN H&W SO 6950. *Acton Green* 1640. P.n. Acton as in ACTON BEAUCHAMP + ModE **green**.

(9) ~ PIGOTT Shrops SJ 5402. 'That part of A held by the Pigott family'. *Acton' Picot* 1242, ~ *Pigot* 1291 etc. P.n. *Æctune* 1086 + manorial addition from William Fitz Picot alive in 1203 and for distinction from ACTON BURNELL SJ 5302. Both places were probably originally part of a single estate. Sa i.4.

(10) ~ ROUND Shrops SO 6395. Probably 'A held by the earls of Arundel'. *Acton la Runde* 1283, *Acton(e) Rotunda* [1283]1348, 1291×2, *Acton Round* 1284×5. P.n. *Achetune* 1086, *Acton'* c.1200 etc. + manorial addition *Round*, probably a playful variant of Arundel from the earls of Arundel who possessed the manor, and for distinction from other Actons. Sa i.5.

(11) ~ SCOTT Shrops SO 4589. 'Scot's A'. *Scottes Acton* 1289, *Actone Schottes* late 13th, *Acton Scot* 1301. P.n. *Actune* 1086, *Acton* from 1255 + manorial addition from Reginald Scot who held the manor in 1255. Also known as *Acton' in longefeld'* 1261×2, OE **langa-feld** as in CHENEY LONGVILLE SO 4284 and LONGEVILLE IN THE DALE SO 5493. Sa i.6.

(12) ~ TRUSSEL Staffs SJ 9318. 'A belonging to the Trussel family'. *Acton Trussel(l)* from 1481. P.n. *Actone* 1086, *Acton'* 1206–1595 + manorial addition Trussel. St i.26.

(13) Iron ~ Avon ST 6883. 'A where iron is worked'. *Iren(e) Acton(e)* 1248–1535, *Iron Acton* from 1324. Earlier simply *Actvne* 1086, *Acton(e)* 1220–1361. ME **iren** + p.n. Acton. Gl iii.1.

ACTON Suff TL 8944. 'Aca's estate'. *Acantun* 1000×2 S 1486, *Aretona* (sic) 1086, *Acton* 1610. OE pers.n. *Āca*, genitive sing. *Ācan*, + **tūn**. DEPN.

ACTON TURVILLE Avon ST 8181. 'Acton held by the Turville or Turberville family'. *Acton(e) Torvill(e)*, ~ *To(u)rbervil(l)e* 1284–1739, earlier simply *Achetone* 1086, *Achentona Roberti* ('Robert's Acton') 1169, *Aketon'* 1236, 'Aca's estate', OE pers.n. *Aca*, genitive sing. *Acan*, + **tūn**. Gl iii.47.

ADAMTHWAITE Cumbr SD 7199. 'Adam's clearing'. *Adamthwat* 1585, *-thwait(e)* 17th cent. Pers.n. *Adam* + ON **thveit**. We ii.33.

ADBASTON Staffs SJ 7627. 'Eadbald's estate'. *Edboldestone* 1086, *Æbaldeston* 1175. OE pers.n. *Ēadbald*, genitive sing. *Ēadbaldes*, + **tūn**. DEPN.

ADBER Dorset ST 5920. 'Eata's grove'. *Eátan beares* (genitive case) [956]12th S 571, *Ate- Etesberie, Ettebere* 1086, *At(t)ebe(a)re* 1221–1498, *Adebere* 1429, *Adber* 1575. OE pers.n. *Ēata*, genitive sing. *Ēatan*, + **bearu**. Do iii.393.

ADDERBURY Oxon SP 4635. '(At) Eadburg's fortified place or manor'. *(æt) Ead(b)urggebyrig* [?c.950]10th S 1539, *(æt) Eadburgebyrig* 1015 S 1503, *Edbvrgberie* 1086, *Edburberi(a) -bir' -byri -bur(y) -ber-* 1192–1320, *Adburgebir'* 1199, *Adberbur'*, *Adbburbury* 1235×6, *Adderbury* from 1366. OE feminine pers.n. *Ēadburh*, genitive sing. *Ēadburge*, + **byriġ**, dative sing. of **burh**. O 391.

ADDERLEY Shrops SJ 6639. Partly uncertain. *Eldredelei* 1086, *Aldredelegh'* 1291×2, *-le* 1320, *Adderdeley(e) -le(gh)* 1314–60, *Adderley* from 1386. OE feminine pers.n. *Eald-*, *Ælf-* or *Æthelthrӯth* + **lēah** 'a wood, a clearing, a meadow'. Sa i.7

ADDERSTONE Northum NU 1330. 'Eadred's farm or village'. *Edredeston* 1233, *Edres-* 1234, *Hether(h)iston'* 1242 BF, *Etherstone* c.1715 AA 13.4, *Adderston* 1785. OE pers.n. *Ēadrēd*, genitive sing. *Ēadrēdes*, + **tūn**. NbDu 2.

ADDINGHAM WYorks SE 0749. Either 'the homestead of the Addingas, the people called after Adda' or 'the homestead called or at *Adding*, the place called after Adda'. *Haddincham* [972×992]11th S 1453, *Odingehem -hen, Edidha', Ediham* 1086, *Hatyng- Addingeham* c.1130, *Ad(d)ingham -yng-* 12th–1621. Either OE folk-n. **Addingas* < pers.n. *Adda* + **ingas**, genitive pl. *Addinga*, or OE p.n. **Adding* < pers.n. *Adda* + **ing**[2], locative-dative sing. *Addinge*, + **hām**. The 1086 forms seem to point to OE **Ēadgythe-hām* 'Eadgyth's estate', OE fem. pers.n. *Ēadgӯth (Edith)*, genitive sing. *Ēadgythe*, + **hām**. The alternative *-ing-* forms may well, therefore, be formations on the same pers.n., either OE folk-n. **Ēadgӯthingas* 'the people called after Eadgyth', genitive pl. *Ēadgӯthinga*, or OE **Ēadgӯthing* 'the place called after *Ēadgyth'*, dative sing. *Ēadgӯthinge* + **hām**. Originally a constituent village of the Archbishop of York's estate centred on Otley SE 2045. YW vi.57, BzN 1968.158, WYAS 297.

ADDINGTON Bucks SP 7428. 'Farm, estate called after Ead(d)a or Ead(d)i'. *Edintone* 1086, *Adinton(e)* 1175–1320, *Adington(e)* 1247. OE pers.n. *Ēad(d)a* or **Ēad(d)i* + **ing**[4] + **tūn**. Both *Ēad(d)a* and *Ēad(d)i* would give ME *Ed-* forms and with shortening of *Ēa-* > *Ǣ-* > *Æ-* the prevailing spellings in *Ad-*. Both this place and ADSTOCK SP 7330 were named after the same individual. Bk 51.

ADDINGTON Kent TQ 6559. 'Estate called after Eadda'. *Eddintvne* 1086, *Edintune, Eadintuna* c.1100, *Ed(d)in- Edingtone* 12th, *Adington'* 1256 etc. OE pers.n. *Ēad(d)a* + **ing**[4] + **tūn**. PNK 144.

Great ADDINGTON Northants SP 9675. *Adinton Major* 1284. Adj. **great**, Latin **maior**, + p.n. *Edintone* 1086, *-tune* 1348, *Adinton* 12th–1428 with variants *Add-, -en- -yn- -ing-*, 'the estate called after Æddi or Eada', OE pers.n. *Æddi* or *Eadda* + **ing**[4] + **tūn**. 'Great' for distinction from Little ADDINGTON SP 9573. Also known as *Netheraddington* 1617 'Addington lower down' sc. down-river from Little Addington. Nth 177.

Little ADDINGTON Northants SP 9573. *Adinton Minor* 1220, ~ *Parva* 1284. Adj. **little**, Latin **minor**, **parva**, + p.n. Addington as in Great ADDINGTON SP 9675. Also known as *Adynton Waterville* 1287 from its tenure by Hugh de Waterville 1199. Nth 177.

New ADDINGTON GLond TQ 3863. ModE **new** + p.n. *Eddintone* 1086, *Eddinton(e)* 1086–1291, *Adingeton* 1203, *Addinton -yn-* 1219–1330, *Adington -yng-* 1247–1544, 'the estate called after Æddi or Eadda', OE pers.n. *Æddi* or *Eadda* + **ing** + **tūn**. If the correct form of the pers.n. is *Æddi* this may be the same individual as in the p.n. Addiscombe TQ 3466, *Edescamp* 1229, *Ad(d)escomp(e)* 1279–1416, 'Æddi's enclosed land', OE pers.n. *Æddi*, genitive sing. *Æddes*, + **camp**. New Addington is a modern residential development begun c.1935. Sr 39, 49, LPN 2.

ADDISCOMBE GLond TQ 3466 → New ADDINGTON.

ADDISLADE Devon SX 7164 → ELSTREE GLond TQ 1895.

ADDLEBROUGH NYorks SD 9487. 'Authulf's fortified place'. *Otholburgh* 1153, *Authelburi -burgh* 1283, 1307. ON pers.n. *Authulfr* + OE **burh**. Roman remains have been found here. YN 262.

ADDLESTONE Surrey TQ 0464. 'Ættel's valley'. *Attelesdene* 1241, *At(e)lesdon* 1216×72–1370, *Adlesden* 1624, *Addlestone* from 1765. OE pers.n. **Ættel*, genitive sing. **Ætteles*, + **denu**. The second element has been confused with both *dūn* and *tūn*. Sr 107, L 98.

ADDLETHORPE Lincs TF 5468. 'Eardwulf's outlying farm'. *Her(d)e(r)- Harde- Arduluetorp* 1086, *Addeltorp* 1202. OE pers.n. *Eardwulf* + **thorp**. The usual DB form is *Hardetorp* with inorganic *H-*; *Herdertorp* shows AN assimilation of *r-l* to *r-r*. SSNEM 101.

ADENEY Shrops SJ 7018. 'Eadwynn's island'. *Eduney* 1212, *Edeweny* 1292, *Addeney* 1327. OE feminine pers.n. *Ēadwynn* + **ēg**. DEPN.

ADFORTON H&W SO 4071. 'Estate called after Ead- or Ealdfrith'. *Alfertintune* 1086, *Eatforton, Etferton* 1271×1307, *Alfar- Atforton* 1292, *Alferton* 1648. OE pers.n. *Ēad-* or *Ealdfrith* + **ing**[4] + **tūn**. He 23.

ADGARLEY Cumbr SD 2572 → STAINTON WITH ADGARLEY SD 2472.

ADISHAM Kent TR 2253. 'Eadi's homestead or estate'. *Edesham* 1086–1240 including *[1006 for 1001]11th S 914, *[1042×66]11th S 1047, *Adesham* 1226–54 etc. including *[616]13th S 1609, *E(a)des-Desham* c.1100. OE pers.n. **Ēadi*, genitive sing. **Ēades* + **hām**. PNK 520, KPN 8, ASE 2.29.

ADLESTROP Glos SP 2427. 'Tætel's dependent farmstead'. *Titlestrop* *[714]16th S 1250, *Tedestrop* 1086, *Tatlestrop(e)* [11th]c.1200 S 1548, 1257–1374, *-troppe* c.1130, *-throp(e)* 14th cent., *Tadelest(h)rop* 1298, 1328, *Attlesthorpe* 1330, *Addlestroppe* 1627. OE pers.n. *Tǣtel*, genitive sing. *Tǣteles*, + **throp**. The modern form is due to false analysis of the phrase 'at Tatlesthrop' as 'at Atlesthrop'. Gl i.211.

ADLINGFLEET Humbs SE 8421. 'The stream or tidal inlet of the atheling'. *Adelingesfluet* 1086, *Athelingflet* 1230. OE **ætheling**, genitive sing. **æthelinges**, + **flēot**. Adlingfleet lies on the old-channel of the r. Don. DEPN.

ADLINGTON Ches SJ 9180. 'Farm, village called after Eadwulf'. *Eduluintune* 1086, *Adelvinton* 1248, *Adelinton(e) -yn-* 1252–1287 with variants *-ing- -yng-*, *Alinton -yn-* 1286, 14th cent. with variants *-ing- -yng-* from 1337. OE pers.n. *Ēadwulf* + **ing**[4] + **tūn**. Che i.181. Addenda gives 19th cent. local pr ['adlitn].

ADLINGTON Lancs SD 6013. 'Farm, village called after Eadwulf'. *Edeluinton* [before 1190]1268, *Hedelintona* c.1190, *Adelven- Aldeventon* 1202, *Adelin- Athelington* 1246, *Adlin(c)ton*, 1202, 1295, *Adlington* from 1288. OE pers.n. *Ēadwulf* + **ing**[4] + **tūn**. La 128, Jnl 17.71, 18.21.

ADMASTON Shrops SJ 6313. 'Eadmund's estate'. *Æmundeston* 1176–8, *Edmodeston* 1177, *Æd-* 1180, *Ademon(e)ston* 1292. OE pers.n. *Ēadmund*, genitive sing. *Ēadmundes*, + **tūn**. Bowcock 20.

ADMASTON Staffs SK 0523. 'Eadmund or Eadmod's estate'. *Ædmundeston* 1176, 1178, *Edmodeston* 1177, *Ædmodeston* 1180. OE pers.n. *Ēadmund* varying with *Ēadmōd*, genitive sing. *Ēadmundes -mōdes*, + **tūn**. DEPN.

ADMINGTON Warw SP 2046. 'The estate called after Athelhelm'. *Edelmintone -a* 1086, 12th, 1221, *Ethel-* 1175, *Adel- Adilminton(a), -yn-* 1184×94–1440, *-ing-* 1251, *Admyngton* 1587. OE pers.n. *Æthelhelm* + **ing**[4] + **tūn**. Gl i.230.

ADSTOCK Bucks SP 7330. 'Ead(d)a or Ead(d)i's stock-farm'. *Edestoche* 1086, *Addestok(e)* 1221–1375, *Astocke* 1584. OE pers.n. *Ēad(d)a* or **Ēad(d)i* + **stoc**. Both this place and ADDINGTON SP 7428 were named after the same individual. Bk 52, Studies 1936.25, 39.

ADSTONE Northants SP 5951. 'Ættin's farm or village'. *A-Etenestone* 1086, *Attelestuna -ton* 1154×89–1356, *Atteneston* 12th–1330, *Atneston* 1314–1417, *Adnestone* 1522, *Adston* 1550, *Adson* 1681–1779. OE pers.n. **Ættīn*, genitive sing. **Ættīnes*, + **tūn**. Nth 38 gives pr [ædsən], Studies 1931.4.

River ADUR WSusx TQ 2118. *Adur* 1612. This seems to have been a coinage by Drayton in his *Polyolbion*, wrongly identifying the mouth of this river with *Portus Adurni* [c.400]15th ND. The original name was *bremre* 956 S 624, *Bremre, Bræmbre* c.1000, *Brembr(e)* 1249, 1438, *Bramber* 1597, itself a back-formation from the p.n. BRAMBER TQ 1910. Sx 3, RN 1, RBrit 442.

ADVENTURER'S FEN Cambs TL 5668. *Lands in Burwell common part and parcell of the 95,000 acres of Adventure land alloted and set forth by Act of Parliament for setling the draining the Great Levell of the Fenns called Bedford Levell* 1717. ModE **adventurer** 'one who undertakes or shares in commercial adventures or enterprises, a speculator'. These people *adventured* their capital on schemes to drain the fen as opposed to the *undertakers*, the contractors who undertook the actual work. In the 17th cent. the earl of Bedford went into partnership with other adventurers for this purpose. Ca 188, 292.

ADWELL Oxon SU 6999. 'Eadda's spring or stream'. *Advelle* 1086, *Ad(d)ewell(')* 1176–1523 with occasional variant *Ed(d)e-*, *Adwell* 1526. OE pers.n. *Eadda* + **wella**. O 101, L 31.

ADWICK LE STREET SYorks SE 5308. 'Adda's dairy-farm'. *(H)adeuuic, Adeuuinc* 1086, *Adewyk -wic -wyke* 1241–1428, *Atthe-* 1246–1520, *Adwick* from 13th, OE pers.n. *Ad(d)a* + **wīc**, + suffix *Le Street*, 'on the Roman road', referring to its position on the Great North Road, Margary no.28b. YW i.68.

ADWICK UPON DEARNE SYorks SE 4701. 'Adda's dairy-farm'. *(H)adeuuic* 1086, *Ad(d)ewic -wyk(e)* c.1125–1483, *Athe-* c.1160–1505, *Adwic* 1166, OE pers.n. *Ad(d)a* + **wīc**, + r.n. DEARNE SE 3408. YW i.79.

Upper AFFCOT Shrops SO 4486. ModE **upper** + p.n. *Affecot* 1274, *-cote* 1313, 1316, 'Æffa or Æffe's cottages'. OE pers.n. *Æffa* or *Æffe* feminine + **cot**. Bowcock 21.

AFFPUDDLE Dorset SY 8093. 'Æffa's Piddle estate'. *Affapidele* 1086, *Affepidel(e)* 1244–85, *-pud(e)le* 1289–1428, *Afpud(e)le* 1303–1421, *Affpuddle* from 14th. OE pers.n. *Æffa* + r.n. PIDDLE. The earliest reference to this estate, one of a series along the Piddle river, is *rus iuxta Pydelan* 'farm beside the Piddle' in a 13th cent. copy of a charter of 987 (S 1217) in which a certain Ælfrithus gave it to Cerne abbey. *Æffa* is a short form for *Ælffrith*, the name of the donor. Cf. TOLPUDDLE SY 7994. Do i.288.

AFON DYFRDWY OR RIVER DEE Mers SJ 2380 → River DEE.

AFON or RIVER WYE Glos SO 5398 → River WYE SO 5398.

AGDEN RESR. SYorks SK 2692. P.n. Agden + ModE **reservoir**. Agden, *Aykeden* 1329, *Akden* 1385, is the 'oak-tree valley', OE **āc** replaced by ON **eik** in the first form + **denu** . YW i.222, L 98.

AGGLETHORPE NYorks SE 0886. 'Aculf's outlying settlement'. *Aculestorp* 1086, *Acal- Akelthorp* 1184, 1285, 1301, *Akolfthorp* 1244, *Aglethorp(e)* 1311 etc. OE pers.n. *Ācwulf*, genitive sing. *Ācwulfes*, + **thorp**. YN 254, SSNY 54.

AIGBURTH Mers SJ 3886. 'Oakwood hill'. *Aikeberhe* [1190×1220]1268, *Aykeberh* [c.1200]1268, *-bergh -berue -berwe* 13th, *Eikebergh* 1448. ON **eiki** + **berg**. The modern form shows substitution of [θ] for the peripheral phoneme [x]. La 111, Jnl 17.62.

AIKE Humbs TA 0545. 'The oak-tree'. *Ach* 1086, *Ake* 1150×60–1650, *Aike* from 1305. OE **āc**. YE 160 gives pr [jak].

AIKETGATE Cumbr NY 4846. 'Aiket gate'. P.n. *Aik yeate* 1619, 'oak-tree gate', ME **aik** (OE *āc*, ON *eik*) + **yate** (OE *ġeat*), + pleonastic ModE **gate**. Possibly a gate to INGLEWOOD FOREST. Cu 205 gives pr [ɛ:kətjət].

AIKTON Cumbr NY 2753. 'Oak-wood settlement'. *Aiketona* c.1200, *Ayketon* 1246–1618, *Aictun* c.1200, *Ayk- Aic- Aikton* from 1232. ON **eiki** or **eik** + **tūn**, possibly replacing OE **āc** + **tūn** as in ACTON. Cu 118, SSNNW 185, L 218–9.

AILEY H&W SO 3448. 'Aileve's wood or clearing'. *Aylyuele* 1297, *Ailliveleye* 13th, *Ayley* 1335–1638. ME feminine pers.n. *Aileve* < OE *Æthelġifu*, genitive sing. *Æthelġife*, + **lēah**. Early 14th cent. spellings *Ayline- Alynleye* are probably due to confusion of *u* and *n* reinforced by OFr feminine pers.n. *Aaline*. The DB entry for Ailey, *Walelege* 1086, is a different name, 'wood or clearing of the Welsh', OE *Wealh*, genitive pl. *Wala*, + **lēah**. He 117.

AILSWORTH Cambs TL 1199. 'Ægel's enclosure'. *Ægeleswurð, (to) ægeleswurðe* [948]12th S 533, *Eglsworde* 1086, *Aileswurth* 1227. OE pers.n. *Ægel*, genitive sing. *Ægeles*, + **worth**. Nth 228.

AINDERBY MIRES NYorks SE 2592. *Aynderby in le Myre* 1498. P.n. *Endrebi* 1086, *Andrebi* 1198, 'Eindrithi's settlement', ON pers.n. *Eindrithi*, genitive sing. *Eindritha*, + **bȳ**, + 'Mires' for distinction from AINDERBY QUERNHOW SE 3481 and AINDERBY STEEPLE SE 3392. Ainderby Mires, a shrunken village, is situated in swampy ground near the river Swale. YN 239, SPN 75, SSNY 17.

AINDERBY QUERNHOW NYorks SE 3481. Short for *Aynderby juxta Querenhou* 1301, 'A by Quernhow, the mill-stone hill', ON **kvern** + **haugr**. Ainderby is *A(ie)ndrebi* 1086, *Ai- Aynderby* 1208–14th as in AINDERBY MIRES SE 2592 and AINDERBY STEEPLE SE 3392. Quernhow, *Quernhowe* 1327, *Whernehowe* 15th, is a small mound on the Roman road, Margary no. 8b, forming the boundary between the parishes of Ainderby and Middleton Quernhow SE 3378. YN 223 gives pr [ɛəndəbi kwa:nə].

AINDERBY STEEPLE NYorks SE 3392. 'A with the tower'. *Aynderby w(i)th, wythe Stepil(l)* 1316. Earlier simply *Eindre- Andrebi* 1086 as in AINDERBY MIRES SE 2595. The reference is to the conspicuous tower of the medieval church. YN 275, Pevsner 1966.56.

AINGERS GREEN Essex TM 1120. Surname Ainger, *Aunger* 1319, + ModE **green**. Cf. *Angiers* 1708, *Angel Heath* 1777, 1805. Ess 329.

AINSDALE Mers SD 3112. 'Einulf's valley'. *Einulvesdel* 1086, *-olves-* [1190–1219]1268, *Ainuluesdale -is-* 1190–1213, *Aynoluisdale* 1219–1346 with variants *-ulves- -olves- -olfes-*, *Ainoldale* 1219, *Aynollesdale* 1292, *Aynolsdale* 1451. CG pers.n. *Einulf* (from *Agin(w)ulf*), genitive sing. *Einulfes*, + **dalr**. La 125 suggests ON **Einulfr* but this is not on record. The same is true of OE **Æġenwulf*. The CG name is known from Frankish sources. La 125, SSNNW 98, L 96, Jnl 17.69.

AINSTABLE Cumbr NY 5346. 'Bracken slope'. *Ainstapillith* c.1210, *Ain- Aynstapelit(h)* 1227–1345, *Ain- Aynstaplith -lyth* 1277–1485, *Aynstaplegh, Ainstaply* 14th cent., *Aynstable* 1577, *Enstable* 1721. ON **einstapi** + **hlíth** or OE **hlīth**. Cu 168 gives pr [enstəbl], SSNNW 98, L 167.

AINSWORTH GMan SD 7610. 'Ægen's enclosure'. *Haineswrthe* c.1200, *Euenesworth* 1243, *Aynesworth* 1285–1332, *Aynsworth* 1501. OE pers.n. **Æġen*, genitive sing. **Æġ(e)nes*, + **worth**. For the pers.n. cf. OHG *Agino, Egino*. La 53, Jnl 17.41.

AINTREE Mers SJ 3798. 'Solitary tree'. *Ayntre* [before 1220]1268–1332 etc., *Aintree* from 1226. ON **einn** + **tré**. A tree that was conspicuous in the flat landscape and served to mark a boundary or meeting place. La 117, SSNNW 98, L 212, 214, Jnl 17.65.

AIRA BECK Cumbr NY 3620. 'Gravel-bank stream'. *Ayrau(c)hebeke* c.1250, *riuulum de Ayra* 1292. ON **eyrr** + **á**. Cf. the Icelandic river n. *Eyrará*. The earliest spelling retains a trace of PrON **ahwu* < Gmc **ahwō*. Cu 3, NoB 14.145–61.

AIRA FORCE Cumbr NY 3920. *Airey Force* 1789. R.n. Aira as in AIRA BECK NY 3620 + ON **fors**. Cu 254.

AIRE AND CALDER NAVIGATION Humbs SE 5920. R.ns. AIRE SE 3630 and CALDER BE 2620 + ModE **navigation**.

River AIRE WYorks SE 3630. *Yr* [959]12th S 681, *dacy* (for *ða eg*) [963]14th S 712, *Eyr(e)* 1126×35–1590, *Air'* 1130×9–1324, *Aire* from 1622. There are two theories of the origin of this name. One starts from the assumption that the spelling *dacy* is a corrupt copy of OE *tha eg* 'the island(s)' or 'the river(s)' (OE **ēġ** or **ēa**) and that the r.n. is an English one. It is topographically appropriate to the many islands formed by the different courses of the river especially between Leeds and its confluence with the Ouse. The ME form of the name would then represent a replacement of OE *ēg* by ON *eyjar*, the pl. of *ey* 'an island'. The other theory assumes that the name is of Old European origin in the form of an *r* extension of the root **is-*, **ais-* (IE **eis-*) 'move swiftly'. The posited form **Isarā* also underlies the continental r.ns. Isar, Iser and Isère, and is related to the underlying form of the r.n. URE 2085. However the British name of the Aire may well be embodied in the p.n. LEEDS SE 3034 so that the first explanation is perhaps to be preferred. YW vii.118, BzN 1957.239.

AIREDALE WYorks SE 0345. 'The valley of the river Aire'. *Air- Ayrdale* [1198]n.d.–1341, *-dall* 1534, *Aire- Ayredale* [1198]n.d.–1645, *-dall* 1340. R.n. AIRE SE 3630 + ON **dalr**, lOE **dæl**. YW iv.89.

AIRMYN Humbs SE 7225. 'Mouth of the r. Aire'. *Ermenie* 1086, *Eyreminne* 1100×8. R.n. AIRE SD 9058, SE 1938 + ON **mynni**. DEPN.

AIRTON NYorks SD 9059. 'Settlement on the river Aire'. *Air- Ayrton(e)* 1086–1676. R.n. AIRE SE 6723 + OE **tūn**. YW vi.128.

AISBY 'Asi's village or farm'. ON pers.n. *Ási*, genitive sing. *Ása*, + **bȳ**

(1) ~ Lincs SK 8792. *Asebi -by* 1086. SSNEM 30.

(2) ~ Lincs TF 0138. *Asebi -by* 1086–1560, *Ai- Aysby(e)* 16th cent. In this instance *Ási* may be a shortened form for the name of the DB tenant *Aslac*. Perrott 496, SSNEM 30.

Lower AISHOLT Somer ST 2035. *Lower Asholt* 1809 OS. ModE adj. **lower** + p.n. *Æscholt* [854]12th S 311, *Ascholt* 1197, *Aysholt* [1295]Buck, *Ashholt* 1610, *Asholt* 1809 OS, 'ash-tree copse', OE **æsc** + **holt**. DEPN.

AISKEW NYorks SE 2788. 'Oak wood'. *Echescol* 1086, *Ai- Aykescogh -schogh* 12th–1352, *Askew(e)* 1516, 1665. ON **eiki** + **skogr**. YN 236, SSNY 90, L 209.

AISLABY NYorks NZ 8608. 'Asulf's farm or village'. *Asulue(s)bi* 1086, *Aselby* c.1300–1339, *Ayslabye* 1556. ON pers.n. *Ásulfr*, secondary genitive sing. *Ásulfes*, + **bȳ**. YN 119, SPN 35, SSNY 18.

AISLABY NYorks SE 7785. 'Aslak's farm or village'. *Aslache(s)bi* 1086, *Aslakebi -by* 1167–1303, *Asle(y)- Aslaby(e)* 16th cent. ON pers.n. *Áslákr*, secondary genitive sing. *Áslákes*, + **bȳ**. YN 77, SPN 33, SSNY 18.

AISTHORPE Lincs SK 9480. 'East outlying farm'. *Æstorp, Estorp* 1086, *Esttorp* c.1115, *Astorp* 1154×89. OE **ēast** + **thorp**. E of Sturton by Stow. SSNEM 101.

AKELD Northum NT 9529. 'The oak-tree slope'. *Achelda* 1169 P, *Hakelda* 1176, *Akelde, Akil(d)* 13th cent., *Ak(h)ille* c.1320, *Akell* 1694, *Yakeld* 1733. OE **āc** + **helde**. NbDu 2, DEPN, L 162.

AKELEY Bucks SP 7037. 'Oak wood or clearing'. *Achelei* 1086, *Akeleia-lay* 1155–79, *Acle* 1220–1396, *Okelee* 1411. OE **āc**, genitive pl. **āca**, or adj. **ācen**, + **lēah**. Bk 40, L 203.

AKEMAN STREET Bucks SP 7715, Glos SP 0904, Oxon SP 3314. 'Aceman's Roman road'. *Accemannestrete* [1151×4]c.1200, *Akeman(n)estrete* c.1260, [c.1294]c.1444, 1311, *Ackeman('s) street, Ackman Street* 1724×6. OE pers.n. *Āceman*, genitive sing. *Ācemannes*, + **strǣt** referring to the Roman road from St Alban's to Cirencester, Margary nos.16a-b. The name is obscurely connected with the p.ns. *Acemannes ceastre* 973 ASC(A), *Acemannes beri* 972 ASC(F), *Akemannes ceastre* early 12th, Latin *Urbs Achumanensis* [965]12th S 735, *Civitas Aquamania* [972]12th S 785, 10th cent. innovations for Bath to dignify it as the place where Edgar was to be crowned in 973[†]. The name occurs again in the bounds of a Westminster charter for what is now Fleet Street and the Strand. The sense is probably 'road to *Acemannes ceastre*'. Gl i.15, TT 24–9, Mx 10, 223, O 1, BdHu 1–2.

[†]The latter forms may be Latinisations of 'Acemann's Roman camp' but their form suggests connection with the RBrit name of Bath *Aquæ Sulis*, the first element of which may have developed to **Acu-*, as in *Achumanensis*. A late form **Acumanenses* 'inhabitants of *Acumannā*' would give OE *Aceman(n)ēs* as **Lindenses* gave **Lindēs* in Lindsey.

AKENHAM Suff TM 1448. 'Aca's homestead'. *Achream, Ac(h)re-Acheham* 1086, *Akeham* 1214×5, *Akenham* from 1254. OE pers.n. *Āca*, genitive sing. *Ācan*, + **hām**. DEPN, Baron.

ALAVNA ROMAN FORT Cumbr NY 0337 → River ELLEN NY 1641.

ALBASTON Corn SX 4270. Partly uncertain. *Alveveston* c.1286(p), *Aliveston, Alptone* 1303, *Alpeston -is-* 1337, *Albeston* 1399. Unidentified pers.n. + OE **tūn**. If the 1286 form really belongs here a possible pers.n. is OE feminine *Ælfġifu* with secondary ME genitive sing. **-es**. PNCo 49, Gover n.d. 178.

ALBERBURY Shrops SJ 3614. 'Aluburg's manor-house'. *Alberberie* 1086, *-beria* 1155, *-bury* from 1274, *Abber-* 1291×2–1767, *Awber- Obber-* 17th–18th cent. OE feminine pers.n. *Aluburg* + **byriġ**, dative sing. of **burh**. Sa i.10.

ALBOURNE WSusx TQ 2616. Partly uncertain. *Aleburn(e)* 1176–1316, *Albourne* 1401, *Awborne* 1593, *Aburne* 1699. Unidenitifed element + OE **burna**. Possibly the specific was OE **alor** with early loss of *r* and so 'the alder-tree stream'. Sx 215 gives pr [a:bərn].

ALBRIGHTON Shrops SJ 4918. 'Eadbeorht's estate'. *Etbritone* 1086, *E(a)dburton* 1121–55, *Adbrichton(')* 1241–72, *-brighton(')* late 13th–1728 with variant *-bryght-*, *Albriton'* 1241, *Albrighton(')* 1291×2 and from 1783, *Ab(b)righton* 1585–1688, *Old Brighton* 1650. OE pers.n. *Ēadbeorht* + **tūn**. Sa i.13.

ALBRIGHTON Shrops SJ 8104. Probably 'Æthelbeorht's estate'. *Albricstone* 1086, *Albrichton* 1232, *-brighton* from 1271×2 with variant *-bryght-*, *Aylbri(g)hton* 1285, 1303. OE pers.n. *Æthelbeorht (-briht)*, genitive sing. *Æthelbrihtes*, + **tūn**. Sa i.14.

ALBURGH Norf TM 2687. 'The old barrow'. *Aldeberga* 1086. OE adj. **eald**, definite form **ealda**, + **beorg**. The reference is to the barrow near the church. DEPN, Pevsner 1962NW 73.

ALBURY Herts TL 4324. '(Place at) the old fortification'. *Eldeberie* 1086, *Audeburia -bury* 1210×12, 1291, *Aldeburia -bury* 1241–1387, *Albery* 1545, *Albury* 1552. OE **eald**, definite form oblique case **ealdan**, + **byriġ**, dative sing. of **burh**. The reference is unknown. Hrt 169 gives pr [a:bəri, ɔlbəri].

ALBURY Surrey TQ 0447. 'Place at the old fortification'. *Eldeberie* 1086, *Ældeburi* [933]13th S 420, *(et) Ealdeburi* [1062]13th S 1035, *Aldebir -biri -buri* etc. 1242–1400, *Albury* from 1487. OE **eald**, definite form dative case **ealdan**, + **burh**, dative sing. **byriġ**. The reference is probably to the RBrit camp on Farley Heath TQ 0544. Sr 219, Jnl 2.39.

ALBY HILL Norf TG 1934. *Alby Hill* 1836 OS. P.n. Alby TG 2033 + ModE **hill**. Alby, *Ala- Alebei* 1086, *Alebi* 1195ff., is 'Ali's homestead or village', ON per.n. *Áli*, genitive sing. *Ála*, + **bȳ**. DEPN, SPNN 13.

ALCASTON Shrops SO 4587. 'Alhmund's estate'. *Ælmundestune* 1086, *Al(k)hameston'* 1261×2, 1270, *Al(g)hamston'* 1271×2, *Alkamston* 1302, *Alc-* 1504, *Aulkaston* 1577, *Allcaston* 1786, *Orson* 1732. OE pres.n. *Alhmund*, genitive sing. *Alhmundes* + **tūn**. For the development of the pers.n with svarabhakti vowel, cf. DB *Alchemont* for *Alhmund*. The sequence must have been *Alhmundeston* > *Alkamundeston* > *Alkameston* > *Alkaston*. Sa i.15.

ALCESTER Warw SP 0857. 'The Roman town on the river Alne'. *Alencestre* [1138]1340, 1165–1392 with variants *Alin-* and *Alyn-*, *Alecestre* 1205–1318, *Aucestre* 1318, *Alcestre* 1349–1505, *Alincester al. Alcester* 1564, *Alsettur* 1421, *Alcettor -er* 16th cent. The RBrit name of the town was *Alauna* [c.700]13th Rav. R.n. ALNE SP 1562 + OE **ċeaster** with OFr [(t)s] for English [tʃ]. Wa 193 gives pr [ɔ:lstə].

ALCISTON ESusx TQ 5005. 'Ælfsige's estate'. *Alsi(s)ton* 1086, *Als-Alciston* 12th–1327, *Alston* 1332. OE pers.n. *Ælfsiġe*, genitive sing. *Ælsiġes* + **tūn**. Sx 414 gives prs [a:sən] and [a:stən], SS 76.

ALCOMBE Somer SS 9745. Partly uncertain. *Avcome* 1086. Unidentified element + OE **cumb**. Possibly a non-genitival variant of AWLISCOMBE Devon ST 1301 containing OE **āwel** 'hook' in some topographical sense.

ALCONBURY Cambs TL 1875. 'Alhmund's fortified place or manor'. *Acumesberie* 1086, *Alchmundesbiri -beria* 1168–1198, *Alk- Alcmundebir -bury* 1230–1428, *Alc- Alkumbury -biri* 1285–1513, *Awcum- Awcon- Awkenbury* 1535–1624. OE pers.n. *Alhmund* + **byriġ**, dative sing. of **burh**. Hu 231.

ALDBOROUGH Norf TG 1834. 'The old fort'. *Aldeburg* 1086. OE **eald**, definite form **ealde**, + **burh**. DEPN.

ALDBOROUGH NYorks SE 4066. 'The old fortification'. *Aldeburg(h)* 1145–1532, *Awdbrugh* 1598. OE adj. **ald**, definite form **alde**, + p.n. *Burc, Burg(h)* 1086–1230, referring to the site of the Roman town of ISVRIVM about 1 m. distant. Aldborough gave its name to an ancient territorial district called Burgshire, later Claro Wapentake, *Barg- Borgescire wapentac* 1086, *Burgsira* 1179399. *-shire* 1320, 1410, 'the shire of (ald)borough', p.n. Aldborough + OE **scir**. YW v.1, 80.

ALDBOURNE Wilts SU 2675. 'The stream called *Ealding*'. *(æt) Ealdincburnan* [968×71]12th S 1485, *Audingeburn'* 1226, *Aldeborne* 1086, *-burna -burne -bo(u)rne* 1165–1454, *Aldi- Audi-* 1181–1248, *Aldbourne* from 1418. R.n.**Ealding* + OE **burna**. The sense of *Ealding* is uncertain; it could be OE adj. **eald** + **ing**[3] 'the old stream' or pers.n. *Ealda* + **ing**[4] 'the stream called after Ealda'. Wlt 291 gives former pr [ɔ:bən], L 18.

ALDBROUGH Humbs TA 2438. 'Alda's fortified place'. *Aldenburg* 1086, *Aldeburh -burg(h)* 12th–1409, *Audeburg(h) -bur* 1127×40–1285, *Awbrough* 1588. OE pers.n. *Alda*, genitive sing. *Aldan*, + **burh**. YE 59 gives prs. [ɔ:bruf, ɔ:lbrə].

ALDBROUGH ST JOHN NYorks NZ 2011. 'St John's A'. P.n. *Aldeburne* 1086, *Aldeburg(he)* 12th–1556, *Awdbrough* 1606, 'the old fortification', OE adj. **ald**, definite form **alde**, + **burh**, + saint's name *John* from the dedication of the church and for distinction from ALDBOROUGH SE 4066. There are many ancient entrenchments in the parish, including Scots Dyke. YN 296.

ALDBURY Herts SP 9612. '(Place at) the old fortification'. *Aldeberie* 1086, *Aldeberi(a) -biria -bury* 1167–1317, *Audebury* 1232, *Awbury alias Awlbery* 1559, *Awbrey* 1710. OE **eald**, definite form dative case **ealdan**, + **byriġ**, dative sing. of **burh**. The reference is unknown. Hrt 26 gives pr [ɔ:b(ə)ri].

River ALDE Suff TM 4475. *Ald* 1735, *Ald(e)* 1764. A back-formation from p.n. ALDEBURGH TM 4656. RN 3.

ALDEBURGH Suff TM 4656. 'The old fortified place'. *Aldeburc* 1086, *-burg(a) -burgh(e)* [1198]1253–1674, *-borough* 1523. OE adj. **ald**, definite form **alde**, + **burh**. DEPN, Baron.

ALDEBURGH BAY Suff TM 4755. P.n. ALDEBURGH TM 4656 + ModE **bay**.

ALDEBY Norf TM 4593. Originally 'the old fort'. *Aldebury* 1086, *Aldeby* c.1180, *-bi* 13th. OE **eald**, definite form oblique case **ealdan**, + **byriġ**, dative sing. of **burh** later replaced by ON **bȳ**. DEPN.

ALDENHAM Herts TQ 1198. Either 'the old homestead' or 'Ealda's homestead'. *(æt) Ældenham* *[785]c.1100 S 124, *[1045×9]14th S 1122, *Aldenham* from 11th including *[969]13th S 774, *Ealdenham* *[1066]13th S 1043, *Eldehā* 1086, *Aude(n)ham* 1246, 1271. Either OE adj. **eald**, dative sing. definite form **ealdan**, or pers.n. *Ealda*, genitive sing. *Ealdan*, + **hām**. Hrt 59 gives former pr [ɔ:dnəm].

ALDERBURY Wilts SU 1827. 'Æthelwaru's fortified place or manor'. *(to) Æþelware byrig* [972]14th S 789, *Ædeluuaraburh* [976]12th, *Athelwarabyrig* 10th, *Alwar(es)berie* 1086, *Alwar(d)beri(e) -biri -buri -bury* 1109×20–1394, *-warde-* 1190–1356, *Alrebury* 1341, *Aldersbury* 1531, *Alderburye al. Alwardeburye* 1574. OE feminine pers.n. *Æthelwaru*, genitive sing. *Æthelware*, + **byriġ**, dative sing. of **burh**. The development of forms with *-d-* was influenced by the fact that the DB tenant is called *Alward*. Wlt 374.

ALDERFORD Norf TG 1218. 'Alder ford'. *Alraforda* 1163. OE **alor**, genitive pl. **alra**, + **ford**. DEPN.

ALDERHOLT Dorset SU 1212. 'Alder wood'. *Alreholt(e)* 1285–1332, *Altherholt* 1369, *Alderholt* from 1508 including *Cran(e)borne Alderholt* 1508, 1611, either 'Alderholt by Cranborn' or 'alder wood by C' implying that *Alderholt* was still perceived in part as a descriptive term rather than simply a p.n. OE **alor** + **holt**. Do ii.195, L 196, 220.

ALDERLEY Glos ST 7690. 'Alder wood or clearing'. *Alrelie* 1086, *-leg(e) -ley(e) -legh* 1216–1363, *Altherleye* 1312, *Alderl(e)igh -ley* from 1394. OE **alor** + **lēah**. Gl iii.23.

ALDERLEY EDGE Ches SJ 8478. *Alderley Edge* 1860. A modern development named after Alderley Edge SJ 8577, *le Hegge* early 14th, *l'Egge* 1352. P.n. Alderley as in Nether ALDERLEY SJ 8476 + OE **ecg**. Until 1779 'a dreary common containing nothing but a goodly number of Scotch firs'. Che i.225. Addenda gives 19th cent. local pr ['o:ðərli- 'ɔ:dli 'edʒ].

Nether ALDERLEY Ches SJ 8476. 'Lower Alderley'. *Aldridel' Inferior* 1285–6, *Nether Addredelegh* 1315, *Nether Alderlegh* 1275, *-ley* from 1423. ME adj. **nether**, Latin **inferior**, + p.n. *Aldredelie* 1086, *-lega -leia -le(y)* 1220–1347, *Aldirdelegh'* 13th, *Aldurlee* 1260, *Alderlegh* 1275–1653, *-ley* from 1342, probably 'Aldred's wood or clearing', OE pers.n. *Aldrēd* + **lēah**. Che i.95 prefers OE feminine pers.n. *Althrȳth* representing *Ælf- Æthel-* or *Alhthrȳth*, because this would give a genitive sing. *Althrȳthe*, + **lēah**. 'Nether' for distinction from Over Alderley SJ 8675, *Aldredelie* 1086, *Superior Aldredeleg'* [late 12th]17th, *Oure Aldredeleg'* 1281. The two places are situated below and above the escarpment known at SJ 8577 as Alderley Edge, *Alderley Edge* 1842, *le Hegge* early 14th, *l'Egge* 1352, 'the edge', OE **ecg**, and which gave its name to the 19th cent. township of ALDERLEY EDGE SJ 8478. Che i.94, 95, 99. Addenda gives pr [ald- ɔ:ldərli], formerly ['ɔ:ð- 'ɔ:dərli].

ALDERMASTON Berks SU 5965. 'The ealdorman's estate'. *Ældremanestone, Heldremanestvne, Eldremanestune* 1086, *Alderman(e)ston(e)* c.1160, 1229 etc., *Aldermaston* from 1405. OE **ealdorman**, genitive sing. **ealdormannes**, + **tūn**. Brk 198, PNE i.8.

ALDERMASTON WHARF Berks SU 6067. *Aldermaston Wharf* 1790. P.n. ALDERMASTON SU 5965 + ModE **wharf**. The wharf is situated to the N of Aldermaston on the Kennet and Avon Canal. Brk 198.

ALDERMINSTER Warw SP 2348. 'The estate of the *ealdormann*'. *Aldermanneston* 1169–1255, *Alder- Aldremeston -muston* 1327–1432, *Alderminster* from 1450 with variant *-mynster, Aldermaston* 16th–1787. OE **ealdormann**, genitive sing. **ealdormannes**, + **tūn**. The modern form is an example of popular etymology. The ealdormann in question is not known. Wo 184.

ALDERSHOT Hants SU 8650. '*Alre* corner'. *Halreshet* 1171. *Alreshete* 1248–1334, *-shute* 1316, *Aldershote* 1393. P.n. **Alre* 'the alder-tree', OE **alor**, + **scēat**. The reference of *scēat* is to the shape of the county boundary which forms an angle here protruding into Surrey as it follows the course of the river Blackwater. Ha 20, DEPN, Gover 1958.107.

ALDERTON Glos SP 0033. 'Estate called after Aldhere'. *Aldrinton(a) - yn- - tun* 1059–1412, *-ing- -yng-* 1287–1636, *Aldritone -a* 1086, 12th, 13th, *Audrington - yng-* 13th cent., *Alderton* from 1408. OE pers.n. *Aldhere* + **ing**[4] + **tūn**. Gl ii.48.

ALDERTON Northants SP 7446. 'The estate called after Ealdhere'. *Aldritone* 1086, *Aldrinton* 1184–1279, *Aldringt(h)une -ton* c.1200–1304, *Alderton* from 1316. OE pers.n. *Ealdhere* + **ing**[4] + **tūn**. Nth 96.

ALDERTON Shrops SJ 4924. 'Alder farm'. *Olreton* 1309. OE **alor**, genitive pl. **alra**, + **tūn**. DEPN.

ALDERTON Suff TM 3441. 'The alder-tree farm or village'. *al(r)etuna* 1086, *-tun* c.1150, *Alderton* from 1254 with variants *-re-, -ir-*. OE **alor**, genitive pl. **alra**, + **tūn**. DEPN, Baron.

ALDERTON Wilts ST 8483. 'The estate called after Ealdhere'. *Aldri(n)tone* 1086, *Aldrinton -yn-* 1196–1316, *-ing- -yng-* 1249–1428, *Alderton* 1603×25. OE pers.n *Ealdhere* + **ing**[4] + **tūn**. Wlt 75.

ALDERWASLEY Derby SK 3153. 'Wood or clearing belonging to *Allerwas*'. *Al(l)rewas(e)- -wasse- -leg(h) -ley(e)* 1251–1388, *Allerwas(se)le(i)gh -le(e) -ley(e)* 1273–1545, *Alderwasle(e) -legh -ley(e)* from 1407. Lost p.n. *Allerwas* + **lēah**. *Allerwas* is 'alluvial land with alders', OE **alor** + **wæsse**. Db 515, L 60.

ALDFIELD NYorks SE 2669. 'The old stretch of open country'. *Aldefelt* 1086, *-feld(e)* 1086–1512, *Awdfeild* 1587. OE **ald**, definite form **alda**, + **feld**. YW v.193, L 244.

ALDFORD Ches SJ 4259. 'The old ford'. *Aldefordia* 1153, *Ald(e)ford(e) -forthe* c.1200–1561, *Aud(e)ford* 1254, 1499, 1620, *Aldford alias Odford* 1724. OE **ald** + **ford**. A ford across the river Dee on the Roman Road from Chester to Malpas, Margary No.6a, superseded by a later ford at Aldford bridge in the 12th cent., and thereafter known as 'the old ford'. Che iv.77 gives pr ['ɔldfəd] and ['ɔ:dfəd], L 69.

ALDHAM Essex TL 9125. 'Ealda's' or 'the old homestead'. *Aldeham* 1086, *Ald(e)ham* from 1202, *Aldenham* 1167, *Oldham* 1328, 1831, *Eldeham* 1535. OE pers.n. *Ealda*, genitive sing. *Ealdan*, or adj. **eald**, definite form dative case **ealdan**, + **hām**. Ess 359.

ALDHAM Suff TM 0545. 'Alda's' or 'the old homestead'. *ealdhā, Adaldehā, Ialelham* (for *I(n) aldham*) 1086. OE pers.n. *Alda* or adj. **eald**, definite form **alda**, + **hām**. DEPN.

ALDINGBOURNE WSusx SU 9205. 'The stream called *Ealding*'. *Aldingburne* *[683]14th S 232, [688×705]14th S 45, *Ealdingburnan* [873×88]11th S 1507, *Ældyngborna* [957]13th S 1291, *Aldingeborne* 1086, *Audingburn* 1222, 1225, *Allyngbourne* 15th. OE stream-n. **Ealding* + **burna**. **Ealding* ought to mean 'Ealda's stream', OE pers.n. *Eald(a)* + **ing**[2] or 'old stream', **eald** + **ing**[2]. Sx 62 gives prs [ælnbɔ:rn] and [ældiŋbɔ:n].

ALDINGHAM Cumbr SD 2871. 'Homestead called after Alda'. *Aldingham* from 1086 with variants *-yng-* and *-hame*, *Aldingeham* 1292, *Audingham* 1587. OE pers.n. *Alda* + **ing**[4] + **hām**. La 208.

ALDINGTON H&W SP 0644. 'Ealda's estate'. *Aldintona* [709]12th S 80, *Aldintone* 1086, *Aldington* from 1227, *Aunton* 16th. OE pers.n. *Ealda* + **ing**[4] + **tūn**. Wo 260.

ALDINGTON Kent TR 0636. 'Estate called after Ealda'. *Aldinton(e)* 1086–1257, *Ealditun, (E)aldintune* c.1100, *Audinton'* 1219–41, *Audington'* 1258 etc. OE pers.n. *Ealda*, genitive sing. *Ealdan*, or *Ealda* + **ing**[4] + **tūn**. PNK 463.

ALDINGTON FRITH Kent TR 0436. 'Aldington scrub'. *Aldington-Fright corruptly so called for the Frith* 1799, *Aldington Freight* 1819 OS. P.n. ALDINGTON TR 0636 + Modern Kent dial. **frith, fright** (< OE *(ġe)fyrhth*) 'thin scrubby wood with little or no timber or with underwood only'. Earlier simply *Frith* 1222, *the fryghe* (sic) 1529. Cullen.

ALDRETH Cambs TL 1395. 'Alder-tree landing place'. *Alrehed(a)* 1169×72, *-heðe -hede -hethe -huða* 1170–c.1300, *Alderheþe -heth(e)* 1251–1345, *Aldereth* 1302, *Aldreth* 1499, *Audre* 1576, 1586. OE **alor** + **hȳth**. Aldreth lies on the bank of the river Great Ouse at the junction of dry ground and fen. Ca 232, L 62, 76–7.

ALDRIDGE WMids SK 0500. 'The alder-tree farm or dwelling'. *Alrewic* 1086, *Alrewyz* 1236. OE **alor** + **wīċ**. DEPN.

ALDRINGHAM Suff TM 4461. 'The homestead at or called *Aldring*, the alder-wood'. *Alrincham* 1086, *Alringeham* 1199, *Aldringham* from 1254. OE ***alring** < **alor** + **ing**[2] possibly used as a p.n., locative-dative sing. **alrinġe**, + **hām**. DEPN, Baron.

ALDSWORTH Glos SP 1510. 'Ald's enclosure'. *Ealdswyrðe* 1002×4, [1004]13th S 906, *-worthe* 1008, *Aldesorde -wrd(e)* 1086, *-worþe - w(u)rþe -worth(e)* 12th–c.1560, *Aul(e)sworth* 1577, 1610, 1675, *Audlesworth* 1587. OE pers.n. **Ald*, genitive sing. **Aldes*, + **wyrth**, **worth**. Gl i.23.

ALDWARK Derby SK 2257. 'The old building or fortification'. *Ald(e)werk(e) -werch(e)* c.1140–1592, *Audewerc(h) -werk* 13th cent., *Ald(e)wark(e)* from 1423. OE **ald**, definite form **alde**, + **(ġe)weorc**. The reference is unknown, unless perhaps it is to one of the tumuli in the area, eg. SK 2358. Db 339.

ALDWARK NYorks SE 2669. 'The old fortification'. *A(l)deuuerc* 1086, *Aldewerk(e)* 1175–1410, *Aldewark* 1399. OE **ald**, definite form **alde**, + **(ġe)weorc**. No traces of these fortifications remain. YN 20 gives pr [ɔ:dwa:k].

ALDWICK WSusx SZ 9198. 'The old farm or trading-place'. *Aldewyc* 1235, *-wyke* 1292, *Audewik'* 1271, *Alwicke* 1651. OE **eald**, definite form **ealde**, + **wīċ**. Alternatively this might be 'Ealda's *wīċ*'. Sx 93.

ALDWINCLE Northants TL 0081. 'Ealda's corner'. *Aldewincla* [11th]13th–1401 with variants *-wyncle -wincle* and *Aude-, Aldevincle, Eldewincle* 1086, *Alwynkell* 1539, *Aldwincle* from 1346. OE pers.n. *Ealda* + **wincel**. The reference is to the big bend in the river Nene at this place. Nth 177 gives pr [a:nikəl].

ALDWORTH Berks SU 5579. Either 'the old enclosure' or 'Ealda's enclosure'. *Elleorde* 1086, *Aldew(u)rda* 1167, 1185, *Aldew(u)rth' -worth(e)* 1220–1517. OE **eald**, definite form **ealde**, or pers.n. *Ealda*, + **worth**. Brk 495.

ALDWYCH GLond → GREENWICH GLond TQ 4077.

ALFARDISWORTHY Devon SS 2911. 'Ælfheard's enclosure'. *Alfardesworth'* 1242, *Alfardisworth(y)* 1346, *Alfrydesworth(y)* 1270, *-fredes-* 1328, *Alferysworthy* 1316, *Alfresworthe* 1379, *Alsworthy* 1525. OE pers.n. *Ælfheard*, genitive sing. *Ælfheardes*, + **worthiġ**. D 133 gives pr [ɔ:lzəri].

ALFINGTON Devon SY 1197. 'Farm, village, estate called after Ælf or Ælfa'. *Alfinton* 1244, *-yngton* 1382, *Affington* 1381. OE pers.n. *Ælf(a)* + **ing**[4] + **tūn**. Cf. ALPHINGTON SX 9190. D 604.

ALFOLD Surrey TQ 0334. 'The old fold'. *Alfold(e)* 1227–1499, *Eldefolde* 1257, *Aldefold* 1279–1313, *Ald-* 1313–1495. OE **eald**, definite form **ealda**, + **fald**. Sr 222.

ALFORD Lincs TF 4575. Partly uncertain. *Alforde* 1086, *Auford* 1175, 1202. The usual explanation is 'eel ford', ON **ál** probably replacing OE **ēl** + **ford**. Better, perhaps, 'the old ford', OE **ald** + **ford** with early loss of *d* in the sequence *-ldf-*. DEPN, PNE i.3.

ALFORD Somer ST 6032. 'Ealdgyth's ford'. *Aldedeford* 1086, *Aldicheford* (for *Aldithe-*) 1227, *Aldetheford* 1256. OE feminine pers.n. *Ealdgȳth*, genitive sing. *Ealdgȳthe*, + **ford**. *Ald-* in this name rather than *Eld-* is due to failure of lengthening before *-ld* in the consonant cluster /ldɣ/. DEPN, RN 4.

ALFRED'S TOWER Somer ST 7435. A triangular tower built c.1769–72 to decorate the grounds of Stourhead in commemoration of the place where king Alfred is supposed to have rallied the Saxons before defeating the Danes at Edington in 878. Pevsner-Cherry 1975.500.

ALFRETON Derby SK 4155. Probably 'Ælfhere's farm, village or estate'. *Elstretune* (for *Elfre-*) 1086, *Alferton* [12th]1316, 1202–17th with variants *Alver- Alfyr-* etc., *Auferton(e)* 1221×30–1631, *Alfreton(e)* from 1212, *Offerton* 1665, 1790. OE pers.n. *Ælfhere* + **tūn**. Db 187 gives pr [ɔfətən].

ALFRICK H&W SO 7553. Partly uncertain. *Alfrewike, Alfrich* 1275, *Aufrick, Awfrycke* 1577–1789. This appears to be 'Alfred's dairy farm', OE pers.n. *Ælfrǣd* + **wīc**, but an earlier spelling may belong here, *Alcredeswike* 1204×34, which must be 'Ealhred's farm', OE pers.n. *Ealhrǣd*, genitive sing. *Ealhrǣdes*, + **wīc**. If so this is an example of change of name since Alfrick cannot derive from *Alcredeswike*. Wo 28 gives prs [ɔlfrik] and [a:frik].

ALFRISTON ESusx TQ 5103. 'Ælfric's estate'. *Alvricestone* 1086, *Alf- Alvricheston* 13th–15th, *Aveston alias Alfryston* 1561, *Auston* 1574, *Auson* 1622. OE pers.n. *Ælfriċ*, genitive sing. *Ælfriċes* + **tūn**. Ælfric is the name of the tenant in 1066. Sx 415, SS 76.

ALHAMPTON Somer ST 6234. 'Settlement, estate on the r. Alham'. *Alentone* 1086, *Alemton* 1177, *Almeton'* 1243, *Alumpton* [13th]B, *Alamtone* 1365, *Alampton* 1438[†]. R.n. Alham, *Alum* [842]14th S 292, [940]14th S 462, *Alom* 1243, + OE **tūn**. R.n. *Alum* is an ancient *m*-suffix formation on the IE root **el-/*ol-* 'flow'; it is cognate with the r.ns. Alm, *Alme* 1258, a tributary of the Old Maas in North Brabant, the Lom, formerly *Almus*, a tributary of the Danube in Bulgaria, the Devon YEALME SX 6056, and other Old European r.ns. DEPN, RN 3–4, Flussnamen 36, BzN 1957.225.

ALKBOROUGH Humbs SE 8821. 'Alca's grove'. *Alchebarge* 1086, *-barua* c.1115, *-barue* 1125×8. OE pers.n. *Alca* + **bearu**. DEPN.

ALKERTON Oxon SP 3742. 'The estate called after Ealhhere'. *Alcrintone* 1086–1472 with variants *-tun' -ton(') -yn-*, *Alcretone* 1220, *Alkertone* 1526. OE pers.n. *Ealhhere* + **ing**[4] + **tūn**. O 392 gives pr [ɔ:lkətən].

ALKHAM Kent TR 2542. 'Protected, sheltered homestead'. *Ealhham* c.1100, *Aukeham* 1204, *Alkham* from 1226, *Auc- Aukham* 1263 etc. OE **ealh** + **hām**. Alkham lies in a sheltering valley. PNK 438, KPN 360, ASE 2.29, Settlements 103, GED s.v. alhs.

ALKINGTON Shrops SJ 5339. 'The estate called after Alha'. *Alchetune* 1086, *Alkinton'* 1255×6, *Alghynton* 1320, 1399, *Alkington* from 1695, *Uckington* 1703. OE pers.n. **Alha* + **ing**[4] + **tūn**. Sa i.16.

ALKMONTON Derby SK 1838. 'The Ealhmund farm or estate'. *Alchementune* 1086, *Alc- Alk(e)munton* 1243–14th, *Alk(e)monton* from 1296, *-man-* 1306–1746, *Aukmonton* 1629. OE pers.n. *Ealhmund* + **tūn**. Db 518.

ALLALEIGH Devon SX 8053. 'Ælla's wood, clearing or pasture'. *Alelege* 1238, *-legh* 1303, *Halilegh* 1238, *Allelegh* 1330. OE pers.n. *Ælla* + **lēah**. The 1238 form, however, clearly means 'holy wood or clearing' with OE **hāliġ**. D 320.

ALL CANNINGS SU 1660 → All CANNINGS.

River ALLEN Corn SX 0678. *River Allen* 1888. This name was transferred here from the nearby river Camel (formerly the *Alan*) owing to a late 19th cent. mistake, the correct historical form being *Layne* c.1470, *Laine* c.1540–1840 of unknown meaning. Farm names along the river show that it was also known as the *Dewi*, e.g. Trethevy in St Mabyn SX 0373, *Tewardeui* 1086 'house on the Dewi', OCo **ti** + **war** + ***dewy**, Pendavey in Egloshayle SX 0071, ? *Pendyfig* 949 S 552[‡], *Pendavid* 1086, *Bendewy* 1284 'foot of the Dewi', Co **ben** + ***dewy**. The meaning of **dewy* is unknown. PNCo 50, RN 6, 125.

River ALLEN Northum NY 7961. *Alwent* 1275, *Allayne, Alne, Alen, Alon* 16th cent. An *-nt* extension of the r.n. **Alaua* from IE **el-/*ol-* 'to flow, to stream'; cf. the r.ns. ALNE, AYLE BURN, ALWIN and ELLEN. NbDu 3, RN 10, Krahe 1964.35, ScotPN 187.

River East ALLEN Northum NY 8354. *Est Alon* c.1576, *East Alen* 1586. The eastern branch of the r. ALLEN Northum NY 7961. RN 10.

River West ALLEN Northum NY 7856. *West Alon* c.1576, *West Alen* 1586. The western branch of the r. ALLEN Northum NY 7961. RN 10.

ALLENDALE COMMON Northum NY 8345. P.n. Allendale as in ALLENDALE TOWN NY 8355 + ModE **common**.

ALLENDALE TOWN Northum NY 8355. P.n. *Alwentedal(e)* 1226, *Alwen(ner)dale -dall* 13th cent., *Allendaile* 1663, 'the valley of the r. Allen', r.n. ALLEN NY 7961 + ME **dale**, + ModE **town**. This form has replaced earlier *Alewenton* 1245, *Alwenton* 1479, the 'settlement on the r. Allen', r.n. ALLEN + **tūn**. NbDu 3, RN 10.

ALLENHEADS Northum NY 8545. 'The sources of the r. Allen'. *Allonheads* c.1715 AA 13.9. R.n. ALLEN NY 7961 + ModE **head**.

[†]The form *Aldamtone* [757×86]13th S 1686 may belong here; if so the spelling is corrupt.

[‡]The identification of this form is not certain.

ALLENSMORE H&W SO 4636. 'Alan's Moor'. *Aleinesmore* 1220, *Moralayn* 1265, *Mora Alani* c.1285–1346. ME (OFr, OBret) pers.n. *Aleyn*, genitive sing. *Aleynes*, + p.n. *More* 1086, *Mora* 1160×70, OE **mōr**, 'the moor'. The manor was held of the bishop of Hereford by Alan fitz Main c.1141. He 24, L 54.

ALLER Somer ST 4029. 'The alder-tree'. *(æt) Alre* 9th, 12th ASC(A,E) under year 878, *Alre* 1086, *Auler* 1610, 1750 map. OE **alor**, dative sing. **alre**. DEPN.

ALLERBY Cumbr NY 0839. 'Ailward's farm or village'. Properly short for Crosby Ailward. *Crossebyaylward* 1258, 14th cent., *Aylwardcrosseby* 1260. Later simply *Aylewardby* c.1275, *Alwardeby* 1384, *Alwarby* 1552, *Allerby* 1675. 'Ailward's Crosby' reduced to 'Ailward's *by*'. Pers.n. *Ailward* (<OE *Æthelward*) + **bȳ**. *Crosby* was ON **kross(a)-bȳ** 'farmstead, village with a cross'. *Ailward* is probably to be identified with *Ailward* son of Dolfin named in the Pipe Roll for 1163. Cu 306, SSNNW 28 no.6.

ALLER DEAN Northum NT 9947. 'Ælfhere's valley'. *Elredene* 1099×c.1122, *Aluerdane* c.1123×8 DEC, *Alre- Auredene* 1228, *Allerdene* 1539 FPD. OE pers.n. *Ælfhere* + **denu**. The name has been reformed as if containing OE **alor** 'an alder-tree'. NbDu 4.

ALLERFORD Somer SS 9046. 'Alder-tree ford'. *Alresford* 1086. OE **alor**, genitive sing. **alres**, + **ford**. DEPN.

ALLERSTON NYorks SE 8782. 'Ælfhere's stone'. *Aluresta(i)n, -ri-* 1086, *Al(l)verstain -stayn - steyn -stein* 1086–1335, *Alistan* 1316, *Ollerston* 1577. OE pers.n. *Ælfhere* + **stān**. YN 93 gives pr [ɔləstən].

ALLERTHORPE Humbs SE 7847. 'Alfward's village'. *Aluuarestorp* 1086, *Alward(e)thorp* 13th cent., *Alverthorp* 1252, *Allerthorp* 1492. ON pers.n. *Alfvarthr*, genitive sing. *Alfvarths*, possibly an adaptation of OE *Ælfweard*, + **thorp**. YE 184, SSNY 54.

ALLERTON 'Alder-tree settlement'. OE **alor** + **tūn**. L 220.

(1) ~ Mers SJ 4085. *Alretune* 1086, *-ton* [c.1200]1268–41, *Allerton* from before 1220. La 111, Jnl 17.62.

(2) ~ BYWATER WYorks SE 4227. 'A by the river'. *Alretun(e) -tona* 1086–1251, *Allerton* from 1258, *~ juxta aquam* 1258–1629, *~ by the water* 1493, *~ Bywater* from 1493. The reference is to the river Aire on which the village is situated. Originally a constituent vill of the royal estate centred on Kippax SE 4130. YW iv.89, WYAS 301.

(3) Chapel ~ Somer ST 4050. 'A with the chapel'. *Chapel Allerton* 1817 OS. ModE **chapel** + p.n. *Alwarditone* 1086, *Alwareton* 1170, *Allerton* 1610, 'estate called after Ælfweard', OE pers.n. *Ælfweard* + **ing**[4] + **tūn**. 'Chapel' for distinction from Stone ALLERTON ST 4051. There is a thirteenth cent. church here. DEPN.

(4) Chapel ~ WYorks SE 3037. 'That part of A where the chapel stands'. *Chap(p)le -el Ol(l)erton* 1576–1709. ME *Chapel* 14th + p.n. *Alretun(a) -ton(a)* 1086–14th, *Allerton* from c.13th. 'Chapel' for distinction from GLEDHOW SE 3136 and Moor ALLERTON SE 3138. Also known as *Chap(p)el(l)town(e)* 1427–1572. *Ye chapel of Allerton* is mentioned in 1365. YW iv.137, WYAS 339.

(5) Moor ~ WYorks SE 3138. 'That part of A towards the moor'. *More Ol(l)erton* 16th cent. ME *West-* 1133×44, *Mor(e) ~* from 13th + p.n. *Alretuna -ton(a)* 12th cent., *Allerton* from 1293. 'West' and 'Moor' for distinction from GLEDHOW SE 3136 and Chapel ALLERTON SE 3037. Also known as *Moretowne* 1437–1665. YW iv.137, WYAS 339.

(6) Stone ~ Somer ST 4051. *Stone Allerton* 1817 OS. ModE **stone** + p.n. Allerton as in Chapel ALLERTON ST 4050. The reference is to the old 'hundred stone' which once marked the boundary of Bempstone and Winterstoke hundreds.

ALLESLEY WMids SP 2981. 'Ælli's wood or clearing'. *Alleslega -ley(e) -lie -legh(e) -le* from 1176, *Awsley* 1782. OE pers.n. *Ælli* as in neighbouring *Alspath*, the original name of MERIDEN SP 2482, genitive sing. *Ælles*, + **lēah**. Wa 152 gives pr [ɔ:zli].

ALLESTREE Derby SK 3439. 'Æthelhard's tree'. Adelardestreu 1086, *Adelardestre* c.1200–1322, *Athelardestre* 1208–1330, *Adelastre* late 12th–1346, *Adlastre* 1216×72–1392, *Alastre* c.1200–c.1600, *Alestre* 1356–1545, *Allestre* 1607, 1609. OE pers.n. *Æthelhard*, genitive sing. *Æthelhardes*, + **trēow**. Db 423, L 212, 216.

ALLEXTON Leic SK 8100. 'Athallek's estate'. *Adelachestone* 1086, *Aðelacheston'*, *Aðelochestona* 1197, *Ad(e)lakeston(e) -acs-* c.1130–1226, *(H)athelakeston' -ax-* 1219–69, *Ad(e)- Athelokeston' -is-* 1236–1428, *(H)alakestona* late 12th, *Al(l)ex(s)ton -ax-* 1154×89–1610, *Allexton* 1653. LG pers.n. *Athallêk*, ME genitive sing. *Athallêkes*, + **tūn**. The pers.n. has been assimilated to the OE pattern of *Æthel-* + *lāc*, but the absence of ME *Ayl-* and *Ail-* spellings is against the assumption of a genuine OE **Æthelāc*. Lei 270.

ALLGREAVE Ches SJ 9767. Partially uncertain. *Awgreue* 1599, *Augreve* 1626, *Allgreave* 1831. Possibly 'hall grove', OE **hall** + **grǣfe**. Che i.168.

ALLHALLOW-ON-SEA Kent TQ 8478. A modern seaside development. P.n. Allhallow from ALLHALLOWS TQ 8377.

ALLHALLOWS Kent TQ 8377. 'All saints' sc. parish or church. *Omnium Sanctorum*, Latin genitive plural of *omnes sancti* 'all saints' 1253×4–1293 etc., *Ho All Hallows* 1285. One of six churches in Hoo lordship including HOO ST WERBURGH TQ 7872 and ST MARY HOO TQ 8076. K 122.

ALLIMORE GREEN Staffs SJ 8519. *Alley-more-green* 1661. P.n. *Aliasmore* 1401, possibly 'the marshy land with a path', ME **alley** + **more**, + **green**. St i.164, Horovitz.

ALLINGTON Lincs SK 8540. 'The princes' estate' or 'the estate called after Æthel(a)'. *Adelingetone, Adelinctune* 1086, *Athelinton, Adelington* 13th cent., *Athelington* 1276–1348×9, *Alintune -ton -y-* 1185–1301, *Alington -y-* 1226–1562, *Allington* from 1547. OE **ætheling** or pers.n. *Æthel(a)* + **ing**[4] + **tūn**. Perrott 444.

ALLINGTON Wilts SU 0663. 'The estate of the princes'. *Adelingtone* 1086, *Athelinetona* 1242, *Alingeton* 1196–1316 with variant *Alynge-*, *Alington* 1275–1316. OE **ætheling**, genitive pl. **æthelinga**, +**tūn**. Wlt 311.

ALLINGTON Wilts SU 2039. 'The estate called after Ealda'. *Aldinton(a) -yn-* 1178–1332, *-ing- -yng-* 1242–1426, 1769, *Allington* 1769. OE pers.n *Ealda* + **ing**[4] + **tūn**. Wlt 358.

East ALLINGTON Devon SX 7648. *Eastallyngton* 1581. Earlier simply *Alintone* 1086, *-yn-* 1268×75, 1312, *Alingeton* 1219, *Al(l)yngt(h)on(e)* 1242–1337, 'farm, village called after Ælla or Ælli', OE pers.n. *Ælla, Ælli* + **ing**[4] + **tūn**. 'East' for distinction from South ALLINGTON SX 7938. D 313.

South ALLINGTON Devon SX 7938. *Southalyngton juxta Chevelistone* (Chivelstone) 1320. Earlier simply *Alintone* 1086, as in East ALLINGTON SX 7648. D 320.

ALLITHWAITE Cumbr SD 3876. 'Eilif's clearing'. *Hailiuethait* 1162×90, *Alefthuayt* 1225×45, *Alithwait* 1327. ON pers.n. *Eilífr*, possibly confused with *Áleifr*, + **thweit**. La 169, SSNNW 99.

ALLONBY Cumbr NY 0843. 'Alein's farm or village'. *Alayn(e)- Alainby* 1262–1358, *Alanby* 14th cent., *Allonby* 1576. OFr(Breton) pers.n. *Alein* + **bȳ**. Cf. ELLONBY NY 4235. Cu 261, SSNNW 25.

ALLONBY BAY Cumbr NY 0742. *Allonby Bay* 1868 OS. P.n. ALLONBY NY 0843 + ModE **bay**.

ALLOSTOCK Ches SJ 7571. 'Lostock with the hall'. *Allostok* 13th, 1312 etc., *Allostock(e)* from 1534. OE **hall**, genitive sing. **halle**, + p.n. Lostock as in LOSTOCK GRALAM SJ 6975. Che ii.216 gives pr [ɔ:(l)'lɔstɔk], Addenda 19th cent. local [ɔ:'lɔstək] (sic).

ALL SAINTS SOUTH ELMHAM Suff TM 3482 > South ELMHAM.

ALLWESTON Dorset ST 6614. Uncertain.

I. *Alveston'* 1214–1870, *Allweston* 1630, *Alston, Ason* 1870.

II. *Alfetheston'* 1244.

III. *Alueueston'* 1268.

The usual spelling of the name points to 'Ælf's estate', OE pers.n. *Ælf* probably here short for Ælfhēah, Ælfwīġ or the like, genitive sing. *Ælfes*, + **tūn**. But both types II and III point to the OE feminine pers.ns. *Ælfflǣd* and *Ælfġifu* respectively, genitive sings. *Ælfflǣde, Ælfġife*, + **stān** 'Ælfflæd or Ælfgifu's stone'. Do iii.330.

ALMELEY H&W SO 3351. 'Elm-tree wood or clearing'. *Elmelie* 1086, 1137×9, *Almeleia -lie* 1165×73–1303, *Aumeley* 1234. OE **elm**, genitive pl. **elma**, + **lēah**. He 25, L 203, 220.

ALMER Dorset SY 9199. 'Eel pool'. *elmere* [943]15th S 490, 1211, *Almer(e)* from 1212. OE **ǣl** + **mere**. There is a pool beside the church. Do ii.55, L 26, Jnl 25.42.

ALMINGTON Staffs SJ 7034. 'Alhmund's estate'. *Almentone* 1086, *Alcminton* 1242. OE pers.n. *Alhmund* + **tūn**. DEPN.

ALMONDBURY WYorks SE 1615. 'Fortified place owned or maintained by the men of the village'. *Almaneberie* 1086, *Almanberia -ber(y)* 1142×54–1486 with variants *-biri -bury* etc., *Aleman(n)eburi* [c.1154]n.d., *-bir(i)* 1195×1215–early 14th, *Almanne(s)bire* 1188×1202, 1230, *Almondbury* from 1483, *Aumebery* 1471, *Ambry* 1545, *Aimbury* 1787. ON **al-menn**, genitive pl. **al-manna**, + OE **byriġ**, dative sing. of **burh**. A name contrasting with the type pers.n. + **burh** as in Dewsbury which implies ownership by an individual. The reference is almost certainly to the major Iron age hill-fort at CASTLE HILL a mile to the SW at SE 1514. YW ii.256 gives prs ['ɔ:mbri] and ['eimbri], WYAS 116, Thomas 1976.257.

ALMONDSBURY Avon ST 6084. 'Æthelmund's fortified place'.
I. *Almodesberie* 1086, *-buri -bir' -bury* c.1150–1316.
II. *Aumundesbir'* 1221, *Almundesbur(y) -bury -mondes-* 1248–1587, *Awmsburie* 1533×8.[†]
Type I is OE pers.n. *Æthelmōd*, genitive sing. *-mōdes*, probably due to confusion with OE pers.n. *Æthelmund*, genitive sing. *-mundes*, of type II but possibly an alternative name, + **byriġ**, dative sing. of **burh**. The reference is either to a fortified manor house or to the hill-fort at ST 5983. Antiquarian association of this name with Ealhmund, son of king Ecgbeorht of Wessex, is not supported by the run of forms. Gl iii.105 gives prs [ɔ:mzbəri] and [a:mənzberi].

River ALN Northum NU 1314. Ἀλαύνου (ποταμοῦ ἐκβολαί) (*Alauni fluvii ostia*) 'the mouths of the river Alaunos' [c.150]13th Ptol, *Alauna* [c.700]13th Rav, *(fluvium) Alne* [c.731]8th BHE–c.1540 including *[685]17th S 66, *be Alnes streame* [c.890]c.1000 OEBede, *A(i)le* 16th cent. A *-no-/nā-* extension of the r.n. **Alaua* from IE **el-/*ol-* 'to flow, to stream', cf. r.n. ALLEN. NbDu 4 gives prs [eil] and [jel] showing the same vowel development as in *ha'penny* and *Ralph* pronounced to rhyme with *safe*, and further northern diphthongisation to [je], cf. AYLE, RN 5, RBrit 243–7, BzN 1957.227–8.

ALNE NYorks SE 4965. *Alne, Alna* 1086 etc., *Aune* 1167, *Awne* 1402–1581. Originally the name of the r. Kyle on which Alne stands, identical with r.ns. ALN Northum NU 1214 and ALNE SP 1562. YN 21 gives prs [ɔ:n, a:n].

Great ALNE Warw SP 1259. *Magna Alne* 1584, *Great Aulne* 1627, ~ *Alne* 1658. ModE **great**, Latin **magna**, + p.n. *Alne* 1086, *Awne* 1576, identical with r.n. ALNE SP 1562 from which it is transferred. 'Great' for distinction from Little ALNE SP 1461. Also known as *Ruenale* 1169, *Rugenhelle* 1221, *Rowenale* 1247–1332, *Rownde, Round Alne* 16th cent., 'rough A', OE **rūh**, dative sing. definite form **rūwan** (ME *ru(w)en*) + Alne. Wa 194 gives pr [ɔ:n].

Little ALNE Warw SP 1461. *Parva Alna* 1221, *Luttelalne* 1315, *Little Aulne* 1540. ME adj. **lytel**, Latin **parva**, + p.n. Alne as in Great ALNE SP 1259. Wa 198.

River ALNE Warw SP 1562. *Æluuinœ fluvium* [716×737]11th S 94, *Alne stream* [n.d.]12th S 1599, *Alne* from 1221, *Aulne* 1540. On the Alne are ALCESTER SP 0857, *Alauna* [c700]13th Rav, and Great and Little ALNE SP 1259 and SP 1461. An Old European r.n. **Alaunā* on the root **al-* (IE **el-*) 'to flow'. Wa 1, RN 8 gives pr [ɔ:n], BzN 1957.225, RBrit 244.

ALNHAM Northum NT 9910. 'Village, estate centre on the r. Aln'. *Alneham* 1228–1307 AA 1912.272, *Auneham* 1242, *Aylneham* 1471×2, *Aylnam* 1507, *Aledome* 1610, *Yeldham* 1712. R.n. ALN NU 1314 + OE **hām**. Tomlinson 353 records local pronunciation *Yeldom* and NbDu 4 gives [jeldəm], RN 5.

ALNMOUTH Northum NU 2410. 'The mouth of the r. Aln'. Ἀλαύνου ποταμοῦ ἐκβολαί (Alauni fluvii ostia) 'the mouths of the river Alaunos' [c.150]13th Ptol, *Alnemuth* 1201, 1230, *Alne- Alle- Anyemue, Aunemuwe* 13th cent., *Alemuth* 1314, *Alnemouth* c.1715 AA 13.11. R.n. ALN NU 1314 + **mūth**. A new town was established here by William de Vescy in 1152 known variously as *Neubiginge* 'the new building' and *burgus de Sancto Walerico* 'the borough of St Waleric' from the dedication of the former church here. NbDu 4 gives pr [jelməθ], RN 5.

ALNMOUTH BAY Northum NU 2510. P.n. ALNMOUTH NU 2410 + ModE **bay**.

ALNWICK Northum NU 1813. 'Dairy farm, village on the r. Aln'. *Alnawic* c.1160, *Alnewic(h) -wyc* 1178–1307 AA 1912.272, *Aunewyk -wic* 1213, 1242, *Annewyk* 1269, *Anwik* 1585. R.n. ALN NU 1314 + OE **wīċ**. NbDu 5 gives pr [anik], RN 5.

ALPHAMSTONE Essex TL 8735. 'Ælfhelm's estate'. *Alfelmestuna* 1086, *-ton* 1217, 1230, *Alfameston* 1223–7, *Alph- Alfemeston* 1246–1317, *Alphameston* 1345. OE pers.n. *Ælfhelm*, genitive sing. *Ælfhelmes*, + **tūn**. Ess 405 gives pr [ælf'æmestən].

ALPHETON Suff TL 8850. 'Ælfled's estate'. *Alflede(s)ton* 1204, *Alfeton* 1254, *Alpheton* 1610. OE feminine pers.n. *Ælflēd*, genitive sing. *Ælflēde*, + **tūn**. DEPN.

ALPHINGTON Devon SX 9190. 'Farm, village, estate called after Ælf or Ælfa'. *Alfintune* 1050×73, *Alfinton(e)* 1086–1307 with variant *-yn-*, *Alphington* from 1269 with variants *Alf-* and *-yng-*, *Alfinctona* 1231, *Affynton* 1577, *-ington* 17th. OE pers.n. *Ælf(a)* + **ing**[4] + **tūn**. Cf. ALFINGTON SY 1197. D 422.

ALPORT 'The old town'. OE **ald**, definite form, **alda**, + **port**.
(1) ~ Derby SK 2264. *Aldeport(e)* 1159×90–1484, *Alport* from 1386. Db 183.
(2) ~ Derby SK 1491. *Aldeport* 1285. Db 124.
(3) ~ HILL Derby SK 3051. *Aldeport(e)* 1415–99. Possibly a fanciful name for a rock formation. Db 520.
(4) River ~ Derby SK 1292. *Alporte water* c.1556. Named after ALPORT SK 1491. Db 2.

ALPRAHAM Ches SJ 5859. 'Alhburg's village'. *Alburgham* 1086, *Alperham* 1285–1369, *Alpram* 1259–1581, *Alpraham* from 1287, *Orpram* 1719. OE feminine pers.n. *Alhburg* + **hām**. Che iii.300 gives prs ['æprəm] and ['ɔ:prəm].

New ALRESFORD Hants SU 5832. *Nova villa de Alresford* 'Alresford new town' 1236, *Nova Alresford* 1272, *Niwealreford* 1280. ME **niwe**, Latin **nova**, + p.n. *(to) Alresforda* *[701]13th S 242, *[825]12th S 273, [909]11th S 376, *Alresforde* 1086 etc. with variant *-ford, Allesford* 1408, the 'alder-tree ford', OE **alor**, genitive sing. **alres**, + **ford**. Also known as *Chepyng Alresford* 'market A' 1332 for distinction from Old ALRESFORD SU 5834. Ha 21, Gover 1958.78.

Old ALRESFORD Hants SU 5834. *Elde Alresford* 1280. ME adj. **elde** (OE *eald*) + p.n. Alresford as in New Alresford SU 5832. Gover 1958.78.

ALREWAS Staffs SK 1715. 'Alluvial land with alder-trees'. *Alrewasse* [941]14th S 479, *Alrewas* 1086, 1166, *Alderwasshe* 1485. OE **alor** + **wæsse**. Horovitz gives pr *Awl-rus* to rhyme with *walrus*. DEPN, L 59–60, 220.

ALSAGER Ches SJ 7955. 'Ælli's newly cultivated land'. *Eleacier* 1086, *Al(l)esacher -is- -as-* c.1240–1307, *Alsacher* 1285–1621 with variants *-ach(e)re, Alsager* from 1350, *Alger, Awger* 16th. OE pers.n. *Ælli*, genitive sing. *Ælles*, + **æċer**. Che iii.2 gives pr ['ɔ:lseidʒər] and older local pr ['ɔ:dʒər], L 232–3.

[†] A single spelling *Alkemundesbur'* 1287 is not significant.

ALSAGERS BANK Staffs SJ 8048. *Alsager Bank* 1833 OS. Probably surname *Alsager* from ALSAGER Ches SJ 7955 + ModE **bank** 'a hill'.

ALSOP EN LE DALE Derby SK 1655. 'A in the valley'. *Alsop in le Dale* 1535. P.n. Alsop + ModE **in** + Fr definite article **le** + ModE **dale**. Alsop, *Elleshope* 1086, *Al(e)sop(e) -opp(e)* from late 12th cent., *Ausop* 1577, is 'Ælli's valley', OE pers.n. *Ælli*, genitive sing. *Ælles*, + **hop**. Db 362, L 115, 116.

ALSTON Cumbr NY 7146. 'Halfdan's farm or village'. *Aldeneby* c.1170, *Alden(e)sto(n)n* c.1208–1479, *Aldeston* 1279–1371, *Aldston* 1628, *Auston* 1589, 1705. ON pers.n. *Halfdan* + **bȳ** replaced by OE **tūn**. The loss of *H-* is due to AN influence (Alston was a demesne manor of the great Norman family of Vipont). For a similar interchange of *tūn* and *bȳ*, cf. HOLDENBY Northants SP 6967. Cu 171 gives pr [ɔ:stən], SSNNW 25.

ALSTONE Glos SO 9832. 'Ælfsige's estate'. *Ælfsigestun* [969]11th S 1326, [1052×6]11th S 1408, *Alsiston* 1221, 1300, *Alston* 1535. OE pers.n. *Ælfsiġe*, genitive sing. *Ælfsiġes*, + **tūn**. Gl ii.41 gives pr ['ɔ:lstən].

ALSTONEFIELD Staffs SK 1355. 'Ælfstan's open land'. *Ænestanefelt* 1086, *Alfstanesfeld* 1179, *Alston Field* 1604. OE pers.n. *Ælfstān*, genitive sing. *Ælfstānes*, + **feld**. DEPN, L 240, 243, Horovitz.

ALSTON MOOR Cumbr NY 7240. *Mora de Alde(ne)ston* 1292, *Aldestonmore* 1311, *Austen, Austin, Austeyne, A(u)lston More* 16th cent. P.n. ALSTON NY 7146 + OE **mōr**. Cu 172.

ALSTON RESERVOIR Lancs SD 6136. P.n. Alston + ModE **reservoir**. Alston, *Alston* from 1226 *Aluest'-* c.1230, *(H)alles- Halues- Halfiston* 1246, *Alfston(a)* c.1256–90, *Aulston* 1567–8, is OE pers.n. **Ælf*, genitive sing. **Ælfes*, or possibly *Ælfsiġe*, + **tūn**, and so 'Ælf or Ælfsige's farm or village'. Cf. ALSTONE Glos SO 9832. La 145 gives pr [ɔlstn], Jnl 17.81.

ALSWEAR Devon SS 7222. *Alzer* 1765. The evidence is too late for an explanation to be offered, but it has the appearance of pers.n. + OE **wer** 'a weir, a fishing-weir'. D 382.

River ALT Mers SD 3403. A pre-English r.n. of unknown meaning. *Alt* [1184×90]1268, 1292, and from 1590, *Alte* [1190×1212]1268, 1238–[1328]14th. This can be compared with the continental r.ns. L'Authie, Somme, *Alteia* 723, Autise, Deux Sèvres, **Altīssa*, Elz, Baden-Württemberg, **Altia*, Autre, Ardennes, *fluviolo altro* 715, Alter, Bavaria, **Altra*, Alz, Bavaria, *Alzus* [785×98]1004, *Alzissa* 832, Oteppe, Belgium, *Altapia* 1034 with r.n. suffix *-apa*, which are regarded as *t*-extensions of Old European **al-* 'flow'[†]; or with the W Afon Aled Clwyd SH 9260, 9527, *Ughalet* 1334 'above Aled', derived from the root **pal-* 'marsh' (Latin *palus, paludem*). The latter sense would suit the topography of the Merseyside Alt which flows through flat country and cannot, therefore, be associated with Gaelic *allt* 'a stream' from OIr *alt* 'shore, cliff, glenside'. La 95, RN 9, BzN 1964.17.

ALTARNUN Corn SX 2281. 'Altar of Nonn'. *Altrenune* [c.1100]14th, *Alternon* c.1235, 1610. Co **alter** + Saint's name *Nonn* from the dedication of the church. St Nonn was believed to have been the mother of the Welsh St David and to have been buried here. PNCo 50, Gover n.d. 43, DEPN.

Great ALTCAR Lancs SD 3206. ModE **great** + p.n. Altcar. 'Great' for distinction from Little Altcar Mersey SD 3006. Altcar, *Acrer* 1086, *Altekar* 1251, *Alker* 1577, is 'carr or marshland beside the river Alt', r.n. ALT SD 3403+ ON **kjarr**. La 118, L 52.

ALTHAM Lancs SD 7732. Possibly 'swan river-meadow'. *Elvetham -u-* c.1150–1246, *Alvetham -u-* 1243–1332 etc., *Altham* from 1383. OE **ælfitu** + **hamm**. Altham stands in a bend of the river Calder. La 89, Jnl 18.15.

[†]Cf. however Reizenstein's attractive alternative explanation of the mountainous Alz as 'river flowing between high banks', Celtic **alt-* 'cliff' + suffix *-issa, -ussa*, BONF 17.24–8.

ALTHORNE Essex TQ 9098. 'The burnt thorn-bush'. *Aledhorn* 1198, *Alethorn* 1203–1323, *Althorn(e)* from 1346, *Aldthorne* 1393, *Aldern(e), Aldron* 1538–1608. OE **ǣled** + **thorn**. Ess 208.

ALTHORPE Humbs SE 8309. 'Ali's outlying farm'. *Aletorp* 1086, *-thorp* 1234. ON pers.n. *Áli*, genitive sing. *Ála*, + **thorp**. DEPN, SSNEM 101.

ALTOFTS WYorks SE 3723. 'Old building-sites'. *Altoftes* c.1090–1606, *-tofts* from 1359. OE **ald** + ON **topt**. Possibly to be identified with the lost *Westrebi* 1086, 'west or Vestarr's farm or village', ON **vestre** or pers.n. *Vestarr* + **bȳ**. YW ii.119, SSNY 40, WYAS 303.

ALTON Derby SK 3664. 'Old farm or settlement'. *Alton(e)* from 1328, *Aulton* 1577, 1610, 1767. OE **ald** + **tūn**. Db 191.

ALTON Hants SU 7239. 'Spring farm or settlement'. *Auueltona* 1080×7, *Awelton* c.1127–1291, *Aultone* 1086–1341. OE **ǣwiell** + **tūn**. There are strong springs here including the source of the river Wey. Ha 22, Gover 1958.95.

ALTON Staffs SK 0742. 'Ælfa's estate'. *Elvetone* 1086, *Alveton* 1283. OE pers.n. *Ælfa* + **tūn**. DEPN.

ALTON PANCRAS Dorset ST 6902. 'St Pancras's A'. *Aweltone Pancratii* 1226, *Alton Pancras* 1326. P.n. Alton + saint's name *Pancras* from the dedication of the church. Alton, *Awultune* 1012, *Altone* 1086, is the 'spring settlement', OE **ǣwiell** + **tūn**. The river Piddle rises here. Do 28.

ALTON PRIORS Wilts SU 1162. '(That part of) A belonging to the prior' sc. of St Swithin's, Winchester. *Aulton Prioris* 1199, *Priores Aulton* 1321. P.n. *Awelton* *[825]12th S 272, *Awelton* 1155×60–1282, *Auuil- Awl- Aultone* 1086, *Aulton* 1189, 'the estate with the copious spring', OE **ǣwiell** + **tūn**, + ModE **priors**, Latin genitive sing. **prioris**, for distinction from the other division of the manor, Alton Barnes SU 1062, *Awelton Berners* 1282, 'Alton held by the Berners family'. Wlt 317, V 35.

ALTRINCHAM GMan SJ 7688. 'Homestead, village called or at *Aldring*'. *Aldringeham* 1290, *Alt-* 1397, *Altringham* 1309–1550 with variants *-yng-* and *Ald-*, *Altrincham -yn-* from 1321, *Altrikham -c-* 1353, *-trycham* 1435. OE p.n. **Aldring* 'place named after Aldhere', locative-dative sing. **Aldrinġe*, + **hām**. Che ii.7 gives prs ['ɔltriŋəm], older local ['ɔ:tridʒəm].

ALUM BAY IoW SZ 3085. *Alum Bay* from 1720. ModE **alum** + **bay**. Alum has been obtained from here since the times of Elizabeth I. Wt 227 gives pr ['æləm bei], Mills 1996.21.

ALVANLEY Ches SJ 4974. 'Ælfwald's wood or clearing'. *Elveldelie* 1086, *Alvaldeley* 1209, *Alvand(e)ley(e)* 1292–1724 with variant *Alu-* and *-legh*, *Alvanley* from 1386, *Aulmeley* 1690. OE pers.n. *Ælfwald* + **lēah**. Che iii.219 gives pr ['ɔlvənli], older local ['ɔ:vənli].

ALVASTON Derby SK 3833. 'Alwald's farm, village or estate'. *(œt) Alewaldestune* [c.1002]c.1100 S 1536, *Aluualdestona* 1154×89, *Al(e)waldeston(e) -is-* 1154×89–1330, *Ælvvoldestrn* 1086, *-ton(e) -is-* 1154×89–1300, *Al(e)waston(e)* c.1250–1546, *Alvaston* from 1572. OE pers.n. *Alwald*, genitive sing. *Alwaldes*, + **tūn**. A late 15th cent. copy of Wulfric Spot's will of c.1002 reads *Athelwaldeston* which suggests that *Alwald* is a late form of *Æthelwald*. Db 425.

ALVECHURCH H&W SP 0272. 'Ælfgyth's church'. *Ælfgyðe cyrcan* [10th]c.1200 B 1320, *Alviethe Cyrice*, *Ælfiðe cyrce*, *Aluieuecerche* 11th, *Alvievecherche* 1086, *Aluithechirche* [c.1086]1190–1240, *Alvechirche* 1275, *Allchurch* 16th, 1675. OE feminine pers.n. *Ælfgȳth*, genitive sing. *Ælfgȳthe* varying with *Ælfġifu*, genitive sing. *Ælfġife*, + **ċiriċe**. It is impossible to say whether the pers.ns. refer to different persons or whether this is a case of confusion between the fricatives [ð] and [v]. Wo 332.

ALVECOTE Warw SK 2404. 'Afa's cottage(s)'. *Avecota -e* c.1160–1543, *Alcote* 1504, 1535, *Awcott* 1524, 1628, *Alvecote* from 1543. OE pers.n. *Afa* + OE **cot**, pl. **cotu**. The *l* in this name is a back-spelling due to the vocalisation of *Av-* > *Aw-* parallel to that of *Al-* > *Aw-* as in ALNE. Wa 24 gives pr [ɔ:kət].

ALVEDISTON Wilts ST 9723. Probably 'Ælfgeat's farm or village'. *Alfwieteston* 1165, *Alvideston -vithes- -vit(t)es-* c.1190–1395, *Alvetestun'* 1195, *-vedeston* 1272–1353, *Alvedston* 1331, *Alveston* 1428–1535, *Alstone, Awston* 16th. Probably OE pers.n. *Ælfġeat*, genitive sing. *Ælfġeates*, + **tūn**. Wlt 199 gives pr [ɔ:lstən].

ALVELEY Shrops SO 7684. 'Ælfgyth's wood or clearing'. *Alvidelege* 1086, *-leya -le(gh) -ley(e)* 1230–1390 with variants *-vith(e)- -vyth(e)-*, *Aluielea* 1160, *Alveleia* 1190, *Alvetheley alias Alveley* 1401, *Aveley* 1577–1736. OE feminine pers.n. *Ælfgȳth*, genitive sing. *Ælfgȳthe*, + **lēah**. Sa i.18 gives local pr [a:vli].

ALVERDISCOTT Devon SS 5225. 'Ælfred's cottage(s)'. *Alveredescote* 1086, *Alvredescoth* 1242, *Alvardiscote* 1320, *Alvarscot(t)e* 1449, *Alve(r)scote* 15th, *Alscote* c.1550. OE pers.n. *Ælfrǣd*, genitive sing. *Ælfrǣdes*, + **cot** (pl. **cotu**). D 112 gives pr [ɔ:lskət].

ALVERSTOKE Hants SZ 6099. 'Alwara's Stoke or outlying farm'. *Alwarestoch* 1086, *-stok(e)* c.1127–1350, *Alwarstok(e)* c.1170–1485, *Alwardstok(e)* c.1270–1412, *Alvardestok* 1351, *Alverstoke* 1530. OE feminine pers.n. *Ælfwaru* or *Æthelwaru*, genitive sing. *Ælf- Æthelware*, + p.n. *(æt) stoce* [948]12th S 532, *Stokes juxta mare* 'Stoke by the sea' 1174, OE **stoc**. According to tradition Alwara was the donor of the manor to St Swithin's at Winchester. If her name is preserved in the medieval Winchester street-name *Aylwarestrete* 1279, *Alwardestrete* 1268, *Alwarestret(e)* c.1319, c.1329, then Alwara represents OE Æthelwaru. Both manor and street show later confusion with the masc. pers.n. Æthelweard. Ha 22, Winchester i.233, DEPN, Gover 1958.26.

ALVERSTONE IoW SZ 5785. 'Ælfred's estate'. *Alvrestone* 1086, *Alvredeston(e)* 1287×90–1336, *Alvred(e)ston(e)* 1335–1401, *Aluerstone* 1271, *Averston* 1611. OE pers.n. *Ælfrǣd*, genitive sing. *Ælfrǣdes*, + **tūn**. Wt gives pr ['(h)ælvərstn], Mills 1996.21.

ALVERTON Notts SK 7942. 'The estate called after Ælfhere'. *Aluriton(e)* 1086, *Alurinton* 1154×89–c.1225, *Alv- Auvrington(a)* c.1190, 1226, *Alverton* from 1185. OE pers.n. *Ælfhere* + **ing**[4] + **tūn**. Nt 209.

ALVESCOT Oxon SP 2704. 'Ælfheah's cottage(s)'. *Elfegescote* 1086, 1185, *Elfeiscote -cot(')* 1200–1250, *Alfescot'* 1220, *Alvescot'* 1246×7. etc. OE pers.n. *Ælfhēah*, genitive sing. *Ælfhēaġes*, + **cot**, pl. **cotu**. O 298 gives prs [ɔ:lzcət], [ælvzcot] and [ælvezcot] (sic).

ALVESTON Avon ST 6388. 'Ælfwig's stone'. *Alwestan(e)* 1086–1316, *-ston'* 1287–1571, *Alveston* from 1322, *Alleston, Alla(r)ston, Alweston* 16th. OE pers.n. *Ælfwīġ*, genitive sing. *Ælfwīġes*, + **stān**. A possible alternative would be the Anglian pers.n. *Alhwīġ*. The change *Alwes-* to *Alves-* may have been due to the influence of neighbouring OLVESTON ST 6087. Gl iii.111.

ALVESTON Warw SP 2356. 'Eanwulf's estate'. *(æt) Eanulfestune* [966]11th S 1310, [985]11th S 1350, *Alvestone* 1086 etc., *All(e)ston* 1469–1515, *Aul- Awlston, Auson* 17th cent., *Alveston* from 1595. OE pers.n. *Ēanwulf*, genitive sing. *Ēan(w)ulfes*, + **tūn**. Wa 231.

ALVINGHAM Lincs TF 3691. Probably 'the homestead of the Ælfingas, the people called after Ælfa'. *Aluing(e)ha' -v-* 1086, *Alvingheheim* c.1115, *Aluingeham* 1202–18, *Alwingham* 1200. OE folk-n. **Ælfingas* < pers.n. *Ælfa* + **ingas**, genitive pl. **Ælfinga*, + **hām**. The specific could, however, be p.n. **Ælfing* 'the place called after Ælfa', locative-dative sing. **Ælfinge*. DEPN.

ALVINGTON Glos SO 6000. 'Ælfwine's estate'. *Eluinton'* 1220, *Alu- Alvinton(e) -yn-* 1221–1445, *-ing- -yng-* c.1340–1662, *Alwinton' -tune* 1221, 1282. OE pers.n. *Ælfwine* + **tūn**. Gl iii.249.

West ALVINGTON Devon SX 7243. *West Alvyngton* 1551. ModE adj. **west** + p.n. *Alvintone* 1086, *Aufinton* 1237–44, 1386, *Alfin(g)ton(e) -yn(g)-* 1232–75, *Alvyngetone* 1328, 'estate called after Ælf or Ælfa', OE pers.n. *Ælf(a)* + **ing**[4] + **tūn**. 'West' for distinction either from ALPHINGTON SX 9190 or perhaps from Easton SX 7242, 'the east settlement', *Esteton* 1333, OE **ēast**, definite form **ēasta**, + **tūn**. D 288 gives pr [ɔ:lviŋtən].

ALWALTON Cambs TL 1395. 'Estate called after Athelwold, the Athelwold estate'. *Aeþelwoldingtun* [955]12th S 566, *Alwoltune* 1086, *Aðelwoltun* 1158, *Alwalton* from 1268, *Al(l)erton* 1610–c.1750. OE pers.n. *Æthelwald* + **ing**[4] + **tūn**. Hu 180.

ALWENT BECK Durham NZ 1319 → The SOLENT Hants SZ 5098.

River ALWIN Northum NT 9208. *Alewent* [c.1200–c.1240]14th, *Alwent* [c.1240]14th. Identical with ALLEN Northum NY 7961. NbDu 5, RN 10, Krahe 1964.35.

ALWINTON Northum NT 9206. 'Settlement, village on the r. Alwin'. *Alwenton* [1233]14th, 1256–[1340]14th, *Alynntoun* 1541 AA 1912.25, *Alanton* 1539, *Allington, Allenton, Allonton* 1888 Tomlinson. R.n. ALWIN NT 9208 + **tūn**. NbDu 5 gives pr [aləntən], RN 10.

River AMBER Derby SK 4363. Uncertain. *Ambra* 1154×89, *Ambre* late 12th, 1348, c.1540, *Amber* from c.1235. Possibly Brit **Ambrā* on the IE root ***emb(h)-** 'wet, water' seen in Latin *imber* 'rain'. Db 2, RN 12, LHEB 509f, BzN 1957.228.

AMBERGATE Derby SK 3551. *Ambergate* 1836. A toll-gate on the Derby–Matlock road, named from the river AMBER SK 4363. Db 566.

AMBER HILL Lincs TF 2346. *Amber Hyll* 1575. Possibly OE **amer** 'a bunting' + **hyll**. A gravel field of 30 acres allotted under the Holland Fen Inclosure Award of 1773 for the purpose of providing materials for the repair of roads in the parishes having rights of common in Holland Fen. It was formed into a parish in 1880. Payling 1940.74, Wheeler 91 and Appendix I.1.

AMBERLEY Glos SO 8501. 'Bunting' or 'amber's wood or clearing'. *Unberleia* 1166, *Omberleia* c.1240, *Ambresleg'* 1248, *-ley* 1461, *Amberley* from 1240×50 with variant *-leie*. OE **amer**, genitive sing. *ameres*, + **lēah**. Gl i.95, Jnl 28.17, 22.

AMBERLEY WSusx TQ 0213. Probably 'bunting wood or clearing'. *Amberle* [957]14th S 1291, *Ambrelie* 1086, *Amberle(gh) -lea* 1227–1439, *Amerley* 1653. OE **amer** + **lēah**. Sx gives pr [æməli], V i.13.

South AMBERSHAM WSusx SU 9120. ModE **south** + p.n. *Æ-Embresham* [963]12th S 718, *Ambresham* 1166–1327, *Ambersham* 1575, 'Æmbri's river-bend land', OE pers.n. **Æmbri*, genitive sing. **Æmbres*, + **hamm** 1. Pers.n. **Æmbri* might be a pet-form for *Ēanbeorht* but it has also been argued that it is the OE equivalent of Latin *Ambrosius* with specific reference to Ambrosius Aurelianus. An alternative view is that the specific is the OE bird name **amer** 'bunting'. Sx 97 gives pr [æməʃəm], W xxvi, ASE 2.41, SS 81, Jnl 28.5–31, V i.13.

AMBLE-BY-THE-SEA Northum NU 2604. P.n. *Ambel(l)* 1203–96, *Ambbill* [1212]14th, *Anebelle* 1292, 'Anna's promontory', OE pers.n. *Anna* + **bile**, + ModE **by the sea**. NbDu 5, DEPN.

Chapel AMBLE Corn SW 9975. '(The part of) Amble with the chapel'. *Chapel Amble* 1664, translating earlier *Amaleglos* 1284, *(ecclesia Sancti Aldhemlmi de) Ammaleglos* 'St Aldhelm's church of A' 1302, p.n. Amble + Co **eglos** 'church' replaced with ME **chapel**. Amble, *Amal* 1086, *Ambell* 1596 Gover n.d., is 'edge, boundary', Co ***amal**. The additions arose when the estate was divided into three and distinguished this part from Lower Amble SW 9875, originally 'great Amble', *Ammalmur* 1347 (Co **meur**), and Middle Amble, *Ammalgres* (Co **cres**). PNCo 68.

AMBLECOTE WMids SO 8985. 'Æmela's cottage(s)'. *Elmelecote* 1086, *Emelecote* 1236, *Amelecot* 1242. OE pers.n. **Æmela* + **cot**, pl. **cotu**. An alternative possibility is OE **æmel** 'a caterpillar', genitive pl. **æmela**. DEPN.

AMBLESIDE Cumbr NY 3704. 'Shieling by *Amele*, the river sandbank'. *Amdeseta* (sic for *Amele-*) [1090×7]1308, *Amelsate -set(te)* 13th–1547, *Amylside -syde* 1379–1650, *Ambelsede* 1437,

Ambleside from 1564. P.n. *Amele* < ON **á** + **melr**, + **sætr**. We i.182, SSNNW 60, L 11.

AMBROSDEN Oxon SP 6019. 'Ambri's or bunting hill'. *Ameresdone* 1086, *Ambresden(') -don(e)* 1194–1581, *Ambrosdon'* [1185×9]c.1280, *Ambrosedene* 1526, *Amersden* 1581. OE pers.n. **Ambri*, genitive sing. **Ambres*, or **amer**, genitive sing. **ameres** with intrusive *b*, + **dūn** later confused with **denu**. O 161 gives pr [æmɔ:zdən], L 150, Jnl 28.21, V i.13.

AMCOTTS Humbs SE 8514. 'Amma's cottages'. *Amecotes* 1086, *Ammecotes* 1230. OE pers.n. **Amma* + **cot**, secondary (ME) pl. **cotes**. DEPN.

AMERSHAM Bucks SU 9798. 'Ealhmod or Ealhmund's homestead'. The forms fall into three types:
I. *Agmodesham* *[1066]12th S 1043, 1203–1339, *Almodesham* 1160, *Amodes-* 1165–1247, *Aumodes-* 1218–1262, *Amotesham* 1414–40.
II. *Agmondesham* *[1065]14th S 1040, *A(u)g- Aughmundesham* 13th–1398, *Agmondes-* 1348–1604.
III. *Hakmersham* 1483, *Hamersham* 1536, *Amershams* c.1600, *-ham* 1675.
The DB forms are *Elmodes- Elnodesham* 1086. Probably OE pers.n. *Ealhmund* + **hām**. OE **Ealhmōd* is not on record but seems to have been a variant in this name. Bk 209, Nt addenda.

AMESBURY Wilts SU 1641. 'Ambri's fortified place'. *(æt) Ambresbyrig* [873×88]11th S 1507, c.1000, *Amberesburg* [858]12th S 1274, *(æt) Ambresbirig -burch* [957×5]14th S 1515, [932]15th S 419, *Hambresburuh* [972]15th S 792, *Ambles- Ambresberie, Ambresberiæ* 1086, *Ambresbir(ia) -beri -buri -bury* 1130–1384, *Almesberie* 1434, *Amysbury* 1524, *Amesbury alias Ambresbury* 1703. OE pers.n. **Ambri*, genitive sing. **Ambres*, + **byriġ**, dative sing. of **burh**. Some authorities have sought to identify the pers.n. here with Ambrosius Aurelianus, the 5th cent. British leader, and suggest that this was a place fortified by him. However, this is unecessary as OE **Ambri* corresponds to OG *Ambri*. Wlt 358, Britannia 6.131f, Age of Arthur 100, English Settlements 160. For the possibility that the specific is the OE bird name **amer** see Jnl 28.5–31, V i.13–14.

AMICOMBE HILL Devon SX 5786. P.n. Amicombe, *Aunnacomb* 1346, *Ame-* 1355, *Amma-* 1364, *Amme-* 1352, + ModE **hill**. The specific of Amicombe is uncertain, the generic OE **cumb** 'a hollow, a valley, a coomb'. D 191.

AMINGTON Staffs SK 2304. 'Earma or Eamma's estate'.
I. *Ermendone* 1086, *Arminton* 1221.
II. *Aminton(a)* 1150–1468, *Amington* from 1232.
OE pers.n. **Earma* or **Eamma*, genitive sing. **Earman*, **Eamman*, + **tūn**. Both pers.ns. are possible pet forms of *Earnmund* or the like. Wa 25.

AMOTHERBY NYorks SE 7473. 'Eymundr's farm or village'. *Aimundrebi, Edmundrebia, Einde- Andebi* 1086, *Aymunderby* early 13th–1415. ON pers.n. *Eymundr*, genitive sing. *Eymundar*, + **bȳ**. YN 45 gives prs [æməbi, eməbi], SPN 77, SSNY 18.

AMPFIELD Hants SU 4023. Partly uncertain. *Anfelda* 1208, *Amfelde* 1292, *Aumfelde* 1307–33. Unidentified element + OE **feld** 'open country'. The specific could be OE **ān** 'one, single' but the sense of 'single open land' is obscure. Forms with medial *-e*, **Anefeld*, would be needed for OE *āna-feld* 'lonely open land' as in ANCROFT Northum NU 0045 'the lonely croft'. Possibly 'Anna's open land', OE masc. pers.n. *Anna*, or a reduction of **Ānsetl-feld* 'hermitage open land', OE **ānsetl**. Ha 23, Gover 1958.173.

AMPLEFORTH NYorks SE 5878. 'Ford where sorrel grows'. *Ampre- Ambreforde* 1086, *Ampilford* c.1142–1444, *Ampleford* 1167 etc. OE **ampre** + **ford**. YN 56.

AMPNEY CRUCIS Glos SP 0702. 'Ampney Holy Cross' from the dedication of the church. *Ameney(e) Sancte Crucis* 1287, *Holyrod(e)hampney* 1509, *Ampney Crucis* from 1535. P.n. Ampney + Latin **crucis**, genitive sing. of **crux** 'a cross'. Ampney, *Omenie* 1086, *Omenai - ay* 1086–1148, *Ameney(e)* 1215–1347, is 'Amma's stream', OE pers.n. **Amma*, genitive sing. **Amman*, + **ēa** confused with **ēġ**. Gl i.48.

AMPNEY ST MARY Glos SP 0802. 'St Mary's Ampney'. *Ammeneye B. Marie* 1291, *Ampney Sce' Marie* 1600, *Amney-Mary* 1700. P.n. Ampney + saint's name *Mary* from the dedication of the church. Earlier simply *Omenie* 1086, *Ameneye* 1209–1287, 'Amma's stream', as in AMPNEY CRUCIS SP 0702. Gl i.52.

AMPNEY ST PETER Glos SP 0801. 'St Peter's Ampney'. *Ameneye Petri* 1290, *Ampney Sci' Petri* 1542. P.n. Ampney + saint's name *Peter*. The estate was held by the Abbey of St Peter, Gloucester, in 1086. Ampney, *Omenie* 1086, *Omenai* 1138, *Hameneye* 1285, is 'Amma's stream' as in AMPNEY CRUCIS SP 0702. Gl i.53.

Down AMPNEY Glos SU 1097. 'The lower Ampney'. *Dunamenel* 1154×89–1361, *Dunameneie* 1221, *Dounamen(e)ye* 1300–1426, *Down(e) Ampney* 1535. OE **dūne** + p.n. *Omenel* 1086, *Amenel(l)* 1227–33, as in AMPNEY CRUCIS SP 0702. This is the most southerly of the Ampneys at the lower end of Ampney Brook. Gl i.51.

AMPORT Hants SU 3044. 'Ann held by the de Port family'. *Anne de Port* 1248, 1325, *Anne Port* 1306, *Portesanne* 1379. Earlier simply *(at) annæ, (to) anne* [901]14th S 365, *Anna* *[903]16th S 370, *Anne* 1086, either PrW district name **Onn* 'ash-trees' or a r.n. **On* cognate with G Ahne, Italian Agna, Agno etc. < IE **en-/*on-* 'water'. Amport was part of the fee of Hugh de Porth in 1086. Ha 19, RN 15, Krahe 1964.105.

AMPTHILL Beds TL 0337. 'Ant-hill'. *Ammetelle* 1086, *Amethulle* 1202, *Ampt(e)hull(e)* 14th cent., *Amptle* 17th. OE **ǣmette** + **hyll**. Perhaps a place where ant-hills abounded. Bd 67, L 171.

AMPTON Suff TL 8671. 'Amma's estate'. *hametuna* 1086, *Ametune* c.1095, *Ameton* 1196, *Ampton* 1610. OE pers.n. *Amma* + **tūn**. DEPN.

Great AMWELL Herts TL 3712. *Great Amwell* 1655. ModE adj. **great** + p.n. *Em(me)welle, Emuuella* 1086, *Eme(s)well(e)* 1199×1215–1291, *Amwell(')* from 1220 with variants *Am(m)e- Amp-* and *An-*, 'Æmma's spring', OE pers.n. **Æmma*, a pet form of names such as Ēanmǣr or Ēanmund, + **wella**. 'Great' for distinction from Little Amwell, *Parva Amwell* 1524. Hrt 211.

ANCASTER Lincs SK 9843. 'Ana's Roman fort'. *(de) Anacastro* c.1150, *Anecastre* 1185–1327, *Ancastre* 1294–1524, *-caster* from 1393. OE pers.n. *Ana* + **cæster**. Stretches of the Roman rampart and ditch survive. Identification with the RBrit settlement called *Causennis* is unlikely. Perrott 348, RBrit 305.

New River ANCHOLME Humbs SE 9718 → Old River ANCHOLME SE 9715.

Old River ANCHOLME Humbs SE 9715. Probably originally the name of the fenland through which the river flows. *Oncel* [c.1030]11th Secgan, [before 1085]12th, *Ancolna* 12th cent., *Ancolne* 12th cent.–1294, *(ueterem) Ancoln* 13th, *Ankoln* 1384, *Ancolm('), Ani- Ankholm* 13th cent., *Ancum* 1622. PrW ***an** cognate with Gaul *anam* 'a swamp', OIr *an* 'water, urine', Skr *paṅkas, paṅkam* 'mud, mire' and OE *fenn* 'swamp, fen', + r.n. COLN. The *-holm* forms are secondary and due to popular misunderstanding of *An-coln* as *Anc-holm*. Pronounced [æŋkəm]. RN 13.

ANCHOR Shrops SO 1785. Named from the *Anchor Inn*.

ANCOATS GMan SJ 8598. 'Lonely cottages'. *Einecote* 1212, *Hanekotes* 1243, *Ancoates* 1240×59, *Ancot(t)es* 1322. OE **āna** partly replaced with ON **einn**, + **cot**, secondary ME pl. **cotes**. La 35, SSNNW 193.

ANCROFT Northum NU 0045. 'The lone enclosure'. *Anacroft* c.1122 DEC, *Ane-* [c.1123×8]18th, 1208×10, 1228 FPD, *Ancroft* from [c.1180]n.d. DST. OE **āna** + **croft**. NbDu 5.

ANCTON WSusx SU 9800. 'Anneca's estate'. *Aneg(h)eton* 1279, *Ancketon* 1288, *Ancton* 1694. OE pers.n. **Anneca* + **tūn**. Sx 140.

ANDERBY Lincs TF 5275. 'Village, farm where billet-wood is worked'. *Andreby* [1123×47]13th, *Andre- Handerbi* 12th. ON **ǫndurr, andr(i)** 'board, snow-shoe' + **bȳ**. An alternative possibilty is 'Andri's village or farm', OFr pers.n. *Andri*. SSNEM 30.

ANDERBY CREEK Lincs TF 5576. P.n. ANDERBY TF 5275 + ModE **creek**.

ANDERSON Dorset SY 8897. '(St) Andrew's estate'. *Andreweston* 1331, *Andr(e)uston* 1342, *Anderston* 1428, *Anderson* from 1617. Saint's name *Andrew* from the dedication of the church in Winterborne Tomson ¼ mile away, genitive sing. *Andrewes*, + **tūn**. This is one of the manors or estates along the river WINTERBORNE, earlier referred to as *Wintreburne* 1086 (probable identification), *Winterburn* 1235×6, and *Wynterburne -born' Fif(f)(h)as(s)e* 1268, 1293, *Wynturbourne Andreston alias dict' Wynturbourne Vyueasshe* 1477, 'Winterborne with the five ash-trees'. Do ii.71.

ANDERTON Ches SJ 6475. Partly uncertain. *Andrelton'* 1182, *Aldreton'* 1183, *-er-* 1317, *Anderton'* from 1184, *(H)Enderton* 1185–6. Pers.n. + **tūn**; the pers.n. might be any of OE *Ēanrǣd*, OE feminine *Ēanthrȳth*, ON **Andrithi, E(i)ndrithi*. Che ii.95.

ANDOVER Hants SU 3645. Either 'ash-tree waters' or 'streams at or called *Onn*'. *Andeferas* [951×5]15th S 1515, *(æt, to) Andeferan* 10th Swithun, [962]12th, [962×3]c.1030, 1121 ASC(E) under year 994, *Andov(e)re* 1086, *Andev(e)re* 1156–1412, *Andover* from 1546. This appears to be an Anglicisation of PrW **Ondīƀr* 'ash-tree streams' or 'streams at or called Onn' if the first element is a district name or a r.n. as suggested under AMPORT SU 3044, PrW ***onn** (Brit **onno-*) 'ash-tree' or IE **on-* 'water' + PrW ***duƀr** (Brit **dubro-*) 'water', pl. ***dīƀr** (Brit **dubrī*). The OE forms represent nominative pl. *-as* and dative pl *-an* < *-um* after the prepositions. Such grammatical sensitivity possibly implies bilingual contact between PrW and OE. The river at Andover is now called Anton since 1801 and as in East Anton, a suburb of Andover, earlier *Andever aqua* c.1480, *Andever water* c.1540 and *water of And* 1756. The name Anton is an antiquarian identification with the mysterious p.n. *Antona* of Tacitus *Annals* xii.31, actually a corrupt spelling of the r.n. *Trisantona*, i.e. Tarrant or Trent. Its true origin is OE **ēastan tūne* 'east of the town' sc. of Andover, *Eston towne* 1582, *Estentowne* 1611. The original name of the Anton is unknown. Ha 23–4, RN 15, LHEB 273, 285, 630, Gover 1958.163, Krahe 1964.105, RBrit 510.

ANDOVER DOWN Hants SU 3946. *Andover Down F.* 1817 OS. P.n. ANDOVER SU 3645 + ModE **down**.

ANDOVERSFORD Glos SP 0219. 'Anford ford'. *Anfordes forde* 1586, *Andoversford* 1777. P.n. *Anford* influenced by ANDOVER Hants SU 3645 or by the local form *Andovere* 1266, 1275, for *Anford*, + ModE **ford**. *Anford* is *Onnan ford, æt Onnanforda* 759 S 56, *of ánna forda, in annanford* [c.800]11th S 1556, *An(n)eford(e)* 1185–1354, *Andford* 1586, *Andiford* 1779, 'Anna's ford', OE pers.n. *Anna*, genitive sing. *Annan*, + **ford**. Gl i.168 gives pr ['andəvəz'fɔ:Rd].

ANFIELD Mers SJ 3793. Possibly 'field on a slope'. *Hongfield* 1642, *Annfield* 1786, *Anfield* 1842 OS. ME **hange** + **feld**. Mills 55.

ANGERSLEIGH Somer ST 1919. 'Aunger's Leigh'. *Aungerlegh* 1354, *Angellsley* 1610. Earlier *Legh Militis* 'the knight's Leigh' 1290, OE **lēah** taken as a p.n.†. The manor was held before 1290 by John Aungier whose surname represents OFr pers.n. *Aunger* (< OHG *Ansger*). DEPN gives pr [eindʒəzli:].

ANGLE TARN Cumbr NY 4114. 'Lake with good fishing'. *Angilterne* 13th, *Angletarn(e)* 1573. ON **ǫngull** 'hook' + **tjǫrn**. Another example exists at NY 2407. We i.15.

ANGLEZARKE MOOR Lancs SD 6417. P.n. Anglezarke + ModE **moor**. Anglezarke, *Andelevesarewe* 1202, *Anlauesargh* 1224, *Anlawesaregh -arwe* 1246, *Anleshargh'* [1232–45]14th, *Anlasargh* 1339, *Anlezargh* 1514, *Anlizarke* 1686, is 'Anlaf's shieling', Scand pers.n. *Anlāf* < PrON **AnulaibaR* (ON *Óláfr*), genitive sing. *Anlāfes*, + ON **ǽrgi**. La 48, SSNNW 60, Jnl 17.38.

ANGMERING WSusx TQ 0704. '(The place) among the Angenmæringas, the people called after Angenmær'. *Angemœringum* (dative pl.) [873×88]11th S 1503, *Angemeringe -yng(e)* 1275–1397. OE folk-n. **Angenmǣringas* < pers.n. **Angenmǣr* + **ingas**, dative pl. **Angenmǣringum*. The village itself is recorded as *Angemeringatun* 'the *tūn* of the Angenmæringas' [873×88]11th S 1503. There was also a shortened form of the p.n., *Angemare* 1086–1262, *-mere* 1176–1310. Sx 163, ING 30.

ANGRAM '(The place) at the pastures'. OE **anger**, dative pl. **angrum**.

(1) ~ NYorks SD 8899. *Angram* 1195×1200. YN 272.

(2) ~ NYorks SE 5248. *Angrum* 1280–1376, *Angram* 1283, 1542. YW iv.252.

(3) ~ COMMON NYorks SD 8499. P.n. ANGRAM SD 8899 + ModE **common**.

(4) ~ RESERVOIR NYorks SE 0476. P.n. *Angrum* 1276, *Angrome* 1376, *Angram* 1822, + ModE **reservoir**. YW v.217.

ANGRY BROW Mers SD 3019. *The Angry Brow* 1842 OS. A sandbank off the coast near Southport.

River ANKER Staffs SK 2305. Unexplained. *Oncer* c.1000–1577, *Auncre* 1295 etc., *Ancre* 1332, c.1540. A possible parallel is Enkirch on the Moselle, *Ankaracha* 908, if this is *Ankar* + *aha* 'a stream' rather than **Ancariacum* from pers.n. *Ancarius*. It may belong to the IE root **ank-* 'bend' seen in Gk ἄγκος 'bend, hollow, mountain glen', Latin *uncus* 'curve, hook' and with *r*-extension ἄγκυρα 'anchor', Latin *ancrae* 'valley'. The Anker is full of bends. However, there is no independent evidence of a root **ankro-* in Celtic and the name remains an unexplained pre-English formation. St i.4, RN 14, Chantraine 10.

ANLABY Humbs TA 0328. 'Anlaf's farmstead'. *Um- Unlouebi* 1086, *Anlachbi* 12th, *Anlakeby -lacby -lauebi -lau(e)by -laweby* 13th cent., *Anlagh(e)by* c.1240–1371, *Anleby* 1251, *Anlaby* from 1392. ON pers.n. *Anlaf* (*Óláfr*) + **bȳ**. YE 216, SSNY 18.

ANMER Norf TF 7429. 'Duck pond' *Anemere* 1086, *-mera* 1177, *Anedmere* 1291. OE **æned** + **mere**. DEPN.

Abbotts ANN Hants SU 3243. 'Ann belonging to the abbot' sc. of New Minster (Hyde Abbey), Winchester. *Anne Abbatis* c.1270–1327. P.n. *(æt) Anne* [955×8]14th S 1491, *Anna* 1086, *Anne* 1201 as in AMPORT SU 3044 and ANDOVER SU 3645 + ModE **abbot**, Latin **abbas**, genitive sing. **abbatis**. The manor was given to New Minster by Edward the Elder in 901 (S 365). Ha 19, RN 15, Gover 1958.162.

ANNASIDE Cumbr SD 0986. 'Einarr's shieling'. *Ainresate* c.1145, *Ayn(n)erset* c.1150–1340, *Eynersate* 1278, *Annerside* 1503, *Annasyde* 1658, *Ann's Side* 1774, *Anderset now called Agnes Seat* 1777. ON pers.n. *Einarr* + **sǽtr**. Cu 448, SSNNW 60.

ANNA VALLEY Hants SU 3444. A modern suburb of Andover named from the DB spelling *Anna* for Abbotts ANN SU 3243. Ha 19.

ANNESLEY Notts SK 5053. 'An's wood or clearing'. *Aneslei* 1086, *-lei(a) -lega -lee -ley(e)* 1166–1287, *Annesleia -leg'* 1194–1275, *-ley(e)* from 1280×97. OE pers.n. **An(n)*, genitive sing. *Annes*, + **lēah**. Nt 112.

ANNET Scilly SV 8608. Unexplained. *Anec* [for *Anet*] 1302, *Anete* 1306, *Anet* 1336. PNCo 50.

ANNFIELD PLAIN Durham NZ 1651. P.n. Annfield + ModE **plain**. *Anfield Plain* 1857 Fordyce, *Annfield Plain* 1863 OS. *Annfield* also occurs in *Annfield House* 1863 OS. A modern colliery village.

ANNSCROFT Shrops SJ 4508. *Handcroft* 1833 OS. An early 19th cent. mining squatters' settlement. Sa ii.116.

ANSDELL Lancs SD 3428. No early forms.

†Angersleigh is not in DB.

ANSFORD Somer ST 6433. 'Ealhmund's ford'. *Almvndesford* 1086, 1219, *Anisford* 1610. OE pers.n. *Ealhmund*, genitive sing. *Ealhmundes*, + **ford**. DEPN.

ANSLEY Warw SP 3091. Possibly 'the hermitage wood or clearing'. *Hanslei* 1086, *Anesle(g)a* 1174–1241, *Onestleya* 1232, *Anest(e)leye -le(ge) -lay* 1235–1429, *Anstley* 1327, *Ansley* from 1426, *Anceley* 1550–1618. OE **ānsetl** 'lonely dwelling, hermitage' + **lēah**. Wa 75, L 206.

ANSLOW Staffs SK 2125. 'Eanswith's wood or clearing'. *ansidelege* [1008]13th S 920, *an- eansyðelege* [1012]13th S 930, *Ansedl'* c.1180, *Ansedlee* 1300, *Anzdley* 1332, *Anseley* 1546. OE feminine pers.n. *Ēanswīth*, genitive sing. *Ēanswīthe*, + **lēah** hypercorrected in unstressed position to *low* as if from *hlāw*. DEPN, Charters II, Horovitz.

ANSLOW GATE Staffs SK 1925. P.n. ANSLOW SK 2125 + ModE **gate**.

ANSTEY Herts TL 4032. 'Single path, link road'. *Anestige -stei* 1086, *Anesti -sty* 12th–1330. OE **ānstīġ**. A place where several roads join to become one. This seems the appropriate sense here but the exact sense of this p.n. type is still under discussion. Hrt 170, L 63–4, V i.18.

ANSTEY Leic SK 5408. 'Path where several ways become one'. *(H)anstige* 1086, *Anesti(a) -sty* 1183–1376, *Ansty(e) -ie* 1277–1576, *Anstey* 1537. OE **ānstīġ**. The main street of the village is a quarter of a mile long forking at either end. Lei 335, L 63.

East ANSTEY Devon SS 8626. *Estanesty* 1263, *Yestansty* 1343. ME adj. **este** 'east' + p.n. Anstey, *Anesti(n)ge* 1086, *Anesties* 1175, *(H)anestinges* 1201–5, *Anestye* 1242, *Anesteye* 1301, 'single track', OE **ānstig**. In the Midlands OE *anstiga* seems to indicate a stretch of road where several tracks become one. In this case it probably means 'single-file track up a hill' referring to the hilly track linking East and West ANSTEY SS 8527. The plural forms refer to the separate villages of E and W Anstey. D 335, L 63, Jnl 30.89.

West ANSTEY Devon SS 8527. *Westanostige* 1234, *Westanesti* 1249. ME adj. **weste** 'west' + p.n. Anstey as in East ANSTEY SS 8626. D 336, L 63.

ANSTIEBURY Surrey TQ 1544. 'The fortification at Anstie'. P.n. Anstie, *Hanstega* 1086, *Anestia* 1191, OE **ānstig** 'a stretch of road linking several routes', + OE **burh**, dative sing. **byriġ**. The reference is to an Iron Age hill-fort. Sr 270, L 64, Jnl 30.83–98.

ANSTON SYorks SK 5284. 'Single or solitary stone'. *Anestan* 1086–1379, *Anstan* 1268–1447, *Northanston* 1279 DEPN. OE **āna** + **stān**. YW i.147.

ANSTY The meaning of this term is in dispute. Probably it means 'narrow path, single-track path, an ascent', in some cases 'a stretch of road where several paths become one'. OE **ān(a)** + **stīġ**, **ānstīġ**. L 63–4, Atkin 1998, Jnl 30.83–98.

(1) ~ Dorset ST 7603. *Anesty(e)* 1219–1329, *Ansty* from 1244. The reference is to the short stretch of road forking at either end near Ansty Cross. Do iii.209 gives pr ['a:nsti], L 63–4.

(2) ~ Warw SP 3983. *Anestie* 1086, *Anesty* [c.1144]1348–1451, *Ansty* from 1249. A single road crosses a hill-spur forking at either end. Wa 96, L 64, Jnl 30.94.

(3) ~ WSusx TQ 2923. *Anstigh* 1313, *Annstie* 1603. Possibly here the reference is to the junction of tracks from Staplefield and Cuckfield leading S. Sx 261 gives pr [æn'stai], Jnl 30.92.

(4) ~ Wilts ST 9526. *Anestige* 1086, *Anesty(e) -stie -stygh* 13th cent. Wlt 183, L 64, Jnl 30.85.

River ANT Norf TG 3618. *River Ant* 1838 OS. A modern back-formation from p.n. ANTINGHAM TG 2533. The original name, *Smale*, *Smallee* 1363, *Smale Ee* 1367, 'the narrow stream', OE **smæl**, definite form feminine **smale**, + **ēa**, survives in the p.n. SMALLBURGH TG 3324. RN 372.

ANTHILL COMMON Hants SU 6512. Cf. *Ant-hill Plain* 1810 OS. ModE **ant-hill** + **common**.

ANTHORN Cumbr NY 1958. 'Solitary thorn-tree'. *Eynthorn* 1279, *Ain- Ayn(e)thorn(e)* 1279 etc., *Anthorne* 1580, *Enthorn* 1707. OE **ān**, ON **ein** + **thorn**. The tree was the site of the local court. Cu 123 gives pr [entərn], SSNNW 99.

ANTINGHAM Norf TG 2532. 'Homestead of the Antingas, the people called after Anta'. *Antingham* [1044×7]13th S 1055, 1206, *Atige- Antingham*, *Attinga* 1086, *Anting(e)ham* 1206–75, *Entingeham* 1264. OE folk-n. **Antingas* < pers.n. **Anta* + **ingas**, genitive pl. **Antinga*, + **hām**. ING 132.

ANTONY Corn SX 3954. 'Anna or Anta's farm'. *Antone* 1086, *Anton* 1289, *Antony* c.1540. OE pers.n. *Anna* or *Anta* + **tūn**. The form of the name has been influenced by the Christian name *Anthony*. PNCo 51, Gover n.d. 213.

ANTROBUS Ches SJ 6479. Possibly 'Andrithi's bush'. *Entrebus* 1086, 1281, *Anterbus* c.1250–1666, *-bush'* 1358, *Anderbusk(e)* 1295, 1306, *Antrobus* from 1457. ON pers.n. **Andrithi*, *Endrithi* + ON **buskr**. SSNNW 419 rejects this explanation on the grounds of the rarity of spellings with *-d-* and with DEPN leaves the name unexplained. Che ii.127.

ANWICK Lincs TF 1150. 'Anna's dairy-farm'. *Haniuuic*, *Amuinc* 1086, *Hanewic* 1220, *Anewic* c.1221, *Anwick* from 1241, *Amewic -wyc -k -wich* 1185–1428, *Amwic -wyc -wyk(e)* 1219–1549. OE pers.n. *Anna* + **wīċ**. The forms show assimilation of *-n-* to *-m-* before bilabial *w*. The place was a *berewic* of Ruskington. Perrott 272, Cameron 1998.

APE DALE Shrops SO 4889. 'Api's valley'. *Apedal(e)* 1277, 1283, *Ape Dale* 1833 OS. ON pers.n. *Api*, genitive sing. *Apa*, + **dæl**. DEPN.

APES HALL Cambs TL 5590. 'Ape's wood'. *Apesholt(e) -is-* 1251–1606, *Apshall* 1589, *Apeshall* 1606. ME pers.n. *Ape* < ON *Api*, genitive sing. *Apes*, + **holt**. Alternatively the specific might be the metathesised form *æpse* of OE *æspe* 'an aspen-tree'. Ca 225, PNDB 162.

APETHORPE Northants TL 0295. 'Api's outlying settlement'. *Patorp* 1086, *Apet(h)orp* from 1162, *Appelthorp* 1222. ON by-n. *Api*, genitive sing. *Apa*, + **thorp**. Nth 198 gives pr [æpθɔ:p], SSNEM 102.

APLEY Lincs TF 1075. 'The apple-tree wood or clearing'. *Apelei(a)* 1086, *Apeleia* c.1115, *Appeleye* 1219. OE **æppel** + **lēah**. DEPN.

APPERKNOWLE Derby SK 3878. 'Apple-tree hill'. *Ap(p)elknol(le) -il- -ul-* 1317–1453, *Ap(p)urknoll* 1467–87, *Ap(p)ernoll(e)* 1533×8, *-knowle* 1535. OE **æppel** + **cnoll**. Db 319.

APPERLEY Glos SO 8628. 'Apple-tree wood or clearing'. *Apperleg(a) - le(e) -lei(e) -ley(e)* c.1210–1587. OE **apuldor** or **æppel-trēow** + **lēah**. Gl ii.78, L 203, 219.

APPERSETT NYorks SD 8590. 'The apple-tree shieling'. *Appeltresate* 1280, *Apperside* 1577, *-set* 1661. ME **appeltre** + **sate** (ON *sǽtr*). YN 267.

APPLEBY 'Apple farm'. OE **æppel** possibly replacing earlier *epli*, + **bȳ**, or the whole may be a Scandinavianisation of OE **æppel-tūn**.

(1) ~ Humbs SE 9514. *Aplebi* 1086, 1130, *Appelbi* 1167. DEPN, SSNEM 30.

(2) ~ MAGNA Leic SK 3110. *Appleby Magna* 1506. P.n. *Æppelby* 1002×4 S 1536, *Apelbi -by* 1086–1292, *Appelbi* c.1160–1203, *-by* 1203–1469 + Latin **magna** 'great', for distinction from APPLEBY PARVA SK 3109. Lei 559, SSNEM 31.

(3) ~ PARVA Leic SK 3109. 'Little A'. P.n. Appleby as in APPLEBY MAGNA SK 3110 + Latin **parva**. Lei 560.

(4) ~ -IN-WESTMORLAND Cumbr NY 6820. *Appil- Appelby -bi(a)* 1132–1546, *Eppelbi -by* [1189×99]1304, 1308, *Apelby(e)* 1224–1655, *Appleby* from c.1540. Appleby was head of the Barony and County of WESTMORLAND which was incorporated into the name of the town when the county was abolished in 1974. We ii.91, SSNNW 26, L 219.

APPLEDORE 'Apple-tree'. OE **apuldor**.

(1) ~ Devon SS 4630. *le Apildore in the manor of Northam* 1335, *Apuldore* 1397, *Appelthurre* 1401, *Appledoore* 1675. The earlier

name was *Táwmuða* c.1100 ASC(D) under year 1068, *Tauemuth* 1249, *Tawe- Towemouth* 14th, 'mouth of the river Taw'. D 102.

(2) ~ Devon ST 0614. Short for Sour Appledore, 'sour apple-tree', i.e. crab-apple. *Suraple* 1086, [1173×5]1329, *-apple* [1173×5]1329, *Sureapeldor* 1242, *Sourappledore* 1303, *Southappeldore* 1797. D 547.

(3) ~ Kent TQ 9529. *Apuldre* 893 ASC(A), 968 S 1215, *Apoldre* [1032]13th S 1465, *Apuldra* 1042×66 S 1047, *Apeldres* 1086, *Appeldere* 1226 etc. KPN 298.

APPLEFORD Oxon SU 5293. 'The apple-tree ford'. *Æppelford, Appelforda* [892×9]12th S 355, *(on, of) æppelford(a)* [940]12th S 1567, *Apleford* 1086, *Appelford(')* 1175 etc. OE **æppel** possibly short for ***æppel-trēow** + **ford**. Brk 400, PNE i.3, L 67, 70.

APPLESHAW Hants SU 3048. 'Apple-tree woodland'. *Appelsag'* 1200, *Appelschage* c.1250, *Appelshawe* 1273. OE **æppel** + **sceaga**. The reference is to crab-apples since *sceaga* is not used of orchards or managed woods. Ha 24, 165.

APPLETON 'Apple orchard'. OE **æppel-tūn**. PNE i.4, L 219.

(1) ~ Oxon SP 4401. *(æt) æppeltune* [942]13th S 480, *Apletune -tone* 1086. Brk 401–2.

(2) ~ -LE-MOORS NYorks SE 7387. 'A by the moors'. P.n. *Apeltun* 1086 + Fr definite article **le** short for *en le* + ModE **moor(s)**, a late addition for distinction from APPLETON-LE-STREET SE 7373. Appleton lies at the foot of the North York Moors. YN 59.

(3) ~ -LE-STREET NYorks SE 7373. 'A on the Roman road' (Margary no. 814). P.n. *Aple- Apeltun* 1086, *Apel(l)ton(e)* 13th cent. etc. + Fr definite article **le** short for *en le* + ModE **street**, a late addition, for distinction from APPLETON-LE-MOORS SE 7387. YN 46.

(4) ~ ROEBUCK NYorks SE 5542. 'A held by the Roebuck family'. *Appleton Roebucke* 1664. Earlier simply *Æppeltune* [972×92]11th S 1453, *Apel- Apiltun(e) -ton(a)* 1086 etc. The manor was held by the Roebuck family whose surname, in the form *Rabuk*, is recorded in Appleton as early as 1379. YW iv.219.

(5) ~ THORN Ches SJ 6383. *Appleton Thorn* 1768. P.n. Appleton SJ 5186 + ModE **thorn**: a hamlet named from a thorn-tree here. Appleton, *Epletune* 1086, *Appleton* from 12th, is 'apple-tree farm, orchard', OE **æppel-tūn**. Che ii.98,96.

(6) ~ WISKE NYorks NZ 3904. P.n. *Ap(p)eton(a), Apletune* 1086 + r.n. WISKE. YN 174.

APPLETREEWICK NYorks SE 0560. 'Apple-tree Wick or dairy-farm'. *Apletreuuic -wic -wyc* 1086, *Ap(p)eltrewic -wik -wyk(e)* 12th cent., *Appetrewek* 1812. OE **æppel-trēow** + **wīc** probably used independently as a p.n. YW vi.78 gives pr ['apt(ə)rik].

APPLEY Somer ST 0721. 'Apple-tree wood or clearing'. *Ap(p)elie* 1086. OE **æppel** + **lēah**. DEPN.

APPLEY BRIDGE Lancs SD 5209. P.n. Appley + ModE **bridge**. Appley, *(boscus de) Appellae, Appelleie, Appeleye* [13th]1268, is 'apple wood', OE **æppel** + **lēah**. La 130, L 203.

APPULDURCOMBE HOUSE IoW SZ 5479. P.n. Appledurcombe + ModE **house**. An 18th cent. house, now a ruin. Appledurcombe, *Apeldurecumbe* 1216×72, *Appledurcombe* from 13th with variants *-dor-, -der-* and *-co(u)mb(e), Appeltreco(u)mb(e)* 1287×90–1327, *Apton Combe* 1611, is the 'apple-tree coomb', OE **apuldor** + **cumb**. Wt 141 gives prs ['æplikəm -ku:m] and [æpl'dərkəm], L 90, 93, 219, Pevsner-Lloyd 729, Mills 1996.23.

APSE HEATH IoW SZ 5683. *Apse Heath* 1769. P.n. Apse, *Abla* (sic for *Absa*) 1086, *Hapsa* [1100]1313, *Apse* from 1100×1, 'the aspen-tree', OE **æspe, æpse**, + ModE **heath**. Wt 168 gives pr [æps i:θ], Mills 1996.23.

APSLEY END Beds TL 1232. *Apsley End* 1834 OS. P.n. Apsley + ModE **end**. Apsley, *Aspele* 1230, is 'aspen-tree wood or clearing', OE **æpse, æspe** + **lēah**. The village lies at one end of Shillington parish. Bd 174.

APULDRAM WSusx SU 8403. 'The apple-tree enclosure'. *Apeldreham* 1100×35–1321, *Appuldram* 1440. OE **apuldor** + **hamm** 3–5 'river meadow, dry ground in marsh, cultivated plot in marginal land'. As, however, this is a parish name, OE **hām** 'a homestead' remains an alternative possibility. Sx 65, ASE 2.48, SS 81.

AQUAE SULIS Avon ST 7464. 'The springs of Sulis', the Romano-British name for BATH ST 7464. *Aquis Su- Solis* (dative pl.) [4th]8th AI. Also known as Ὕδατα Θερμά 'hot springs' [c.150]13th Ptol. The goddess name Sulis is cognate with OIr *súil* 'eye' < ***sūlis** which may or may not be related to Latin *sol* 'sun'. RBrit 255.

AQUALATE HALL Staffs SJ 7719. *Aqualate Hall* 1666. P.n. Aqualate as in AQUALATE MERE SJ 7720 + ModE **hall**. St i.147.

AQUALATE MERE Staffs SJ 7720. *Aqualate Meer* 1775. P.n. *Aguilade* 1227, *Aquilad(e) -lode* 1240–70, *Akilot(e)* 1275–1344, 'the (difficult) crossing by the oak-trees', OE **āc** + **ġelād**, + ModE **mere**. The reference is probably to a crossing of the r. Meese at Forton SJ 7521, *Forton* from 1198, 'the ford settlement', OE **ford** + **tūn**; OE *ġelād* normally means 'a river crossing', especially a difficult one, for example in a location where flooding occurs, *lād* 'a water course'. The modern sp is due to association with Latin *aqua* 'water'. St i.146, L 73, Nomina 17.10.

ARBORFIELD Berks SU 7567. Partly uncertain. *Edburgefeld* [c.1190, 1220]13th, *Erburgefeld(')* [1222]13th, 1224×5, 1247×8, *Erburghfeld(e)* 1345, 1390, *Hereburg(e)feld* 1230, 1254, *Erburfeld'* 1284, *Arburfeld* 1535. The generic is OE **feld** 'open land', the specific either the OE feminine pers.n. *Hereburh*, OE **eorthburh** 'earth fortification', or possibly an unrecorded feminine pers.n. **Eardburh*, genitive sing. *Hereburge*, **eorthburge** or **Eardburge*. Brk 123, L 239, 243.

ARBORFIELD CROSS Berks SU 7667. 'Alfild's cross'. *Alpheldecrouch* 1347, *Awfeildcross, Aufeild Crosses* 1607, *Arborfield Cross* 1790. ME feminine pers.n. *Alfild* (OE *Ælfhild*), genitive sing. *Alfilde*, + **crouche** (OE *crūċ*) later replaced by ModE **cross**. The specific shows association with ME *feld* already in the earliest recorded spelling and is replaced by the nearby p.n. ARBORFIELD SU 7567 sometime between 1670 and 1790. Brk 126.

ARBORFIELD GARRISON Berks SU 7665. P.n. ARBORFIELD SU 7567 + ModE **garrison**.

ARBOR LOW Derby SK 1663. 'Arbor tumulus'. *Harberlowe* 1533, *Harbor lowe* 1654. P.n. *Arbor*, OE **eorth-burh** 'an earthwork', + **hlāw**. OE **herebeorg** 'a military camp' would also be possible. The reference is to a large barrow built on the surrounding earth bank of a prehistoric stone circle. Db 395, Thomas 1976.76.

ARBURY HILL Northants SP 5458. *Arberry-hill* 1791. P.n. Arbury + ModE **hill**. Arbury, *Arbury, Arberry* 1712, is OE **eorth-burh**, dative sing. **eorth-byriġ**, 'an earthwork', referring to the same prehistoric camp on the hill as referred to in the name BADBY Northants SP 5659. Nth 13.

ARCHENFIELD H&W. The name of an ancient district in S Hereford bounded by the rivers Wye, Monnow and Worm meaning 'open land called or at *Erging*'. *on Ircinga felda* 10th ASC(A) under year 918, *on Iercinga felda* c.1050 ASC(D) under year 915, *Arcenefelde* 1086, *Erchenefeld* 1138, *Jercheneffeld* n.d. W p.n. *Ergin(g)* + **feld**. Some OE forms have been assimilated to the OE **ingas**, genitive pl. **inga**, p.n. pattern meaning 'people of Erging'. Erging, *Erchin, Ercincg* c.1150, is the W form of a district named from the Roman settlement of *Ariconium* at Weston under Penyard, H&W SO 6423. Nomina 10.61–74, O'Donnell 87–96, L 240, He 2.

ARCLID GREEN Ches SJ 7861. *Arclid Green* 1831. P.n. Arclid + ModE **green**. Arclid, *Erclid* 1188×1209, *Arclid* from before 1240 with variants *-lidd -lyd(d)* etc., *Arkley* 1690, may be 'Arnkell's hill-side', ON pers.n. *Arnkell* + **hlid**. SSNNW 420 rejects this explanation on account of the severe truncation of the pers.n. required and the difficult spellings *-lude -liud -luyd -luyt -leit* for the generic which look like Welsh *llwyd* 'grey'. The same name occurs at Arklid Cumbr SD 2988. Che ii.264–5.

ARDELEY Herts TL 3127. 'Earda's wood or clearing'. *Eardeleage* [924×40]13th S 453,† *Erdelei* 1086–1428 with variants *-legh -leye, Ardele(ia)* 12th, *J- Yerdele* 1203, 1287, *Yerdley* 1375, *Yardley* 1535. OE pers.n. *Earda* + **leaġe**, dative sing. of **lēah**. Hrt 151 gives former pr [ja:dli].

ARDEN WMids SP 1859, 2180. The name of a district of high ground SE of Birmingham as in HAMPTON IN ARDEN SP 2081, HENLEY-IN-ARDEN SP 1566 and TANWORTH-IN-ARDEN SP 1170. *Eardene* [1088]c.1200, *(foresta de) Ardena* 12th cent., *Ardene* 1166. *Erderna* 1174, *Arderna -e* 1189–1394, *Arden* from 1220. Probably identical with the Franco-Belgian Forest of Ardennes, the *silva Arduenna* of Cæsar's *De Bello Gallico* [c.50 BC]9th–10th 5.3.4 etc, (the wood they call) *'Αρδουένναν* Strabo 4.3.5, *Arduenna* Tacitus *Annals* 3.42, *Ardenna* Venantius Fortunatus *Carm.* 7.4.19, *Ardinna -enna* 7th, 8th. Brit ***ardu-** 'high' < PrCeltic **árdvos* cognate with Latin *arduus, ὀρθός < ορθϝμός* + suffix **-enno-**. Sps in *-erne -erna* 1174–1394 and *-ern* 1441, 1547, however, suggest an alternative explanation from OE **eard-ærn* 'a dwelling house', probably due to folk-etymology rather than primary. Wa 11, Holder I.186–7, WestMids 57, PNE i.144.

ARDEN GREAT MOOR NYorks SE 5092. P.n. Arden SE 5190, *Arden(e)* 1086, 1201 etc., *Erdene, Erden(a)* 1160–1436, usually explained as 'Earda's valley', OE pers.n. **Earda*, a short form for *Eardwulf, Eardhelm* etc., + **denu**, + ModE + **great** + **moor**; alternatively this is another instance of the Celtic n. *Ardenna* 'high district' seen in ARDEN WMids SP 1859. YN 202.

ARDINGLY WSusx TQ 3429. 'The wood or clearing of the Eardingas, the people called after Earda'. *Herdingele(ye) -legh* 1087×1100, 1240, 1279, *Erdingelega -e* 1107×18, 1205, *Ardingeleg* 1254, *Ardinglie* 1521. OE folk-n. **Eardingas* < pers.n. **Earda* + **ingas**, genitive pl. **Eardinga*, + **lēah**. Sx 251 gives pr [a:diŋˈlai].

ARDINGLY RESERVOIR WSusx TQ 3229. P.n. ARDINGLY TQ 3429 + ModE **reservoir**.

ARDINGTON Oxon SU 4388. 'The estate called after Earda'. *Ardintone* 1086, *Ardinton(a)* c.1182–1327 with variants *Ærd- Erd-, Ardington* from 1284. OE pers.n. *Earda* + **ing**[4] + **tūn**. In 1066 Ardington was divided into two estates assessed at five and nine hides respectively. The latter has been identified with an estate of nine hides granted *æt Æþeredinge tune* [961]13th S 691, 'the estate called or at Æthelræding', OE p.n. **Æthelrǣding* < pers.n. *Æthelrǣd* + **ing**[2], locative–dative sing. **Æthelrǣdinge*, + **tūn**. This name had evidently disappeared by 1066. Brk 468.

ARDLEIGH Essex TM 0529. 'Earda's wood or clearing'. *Erleiam -legam -ligam, Herlegā* 1086, *Erle(i)a -lee* 1120×35, 1201, *Erdelega* 1195, *Erdleg'* 1254, 1262, *Ardlaga -le(ga) -leia -leg(e) -lee -ley(e)* 1154×89–1313. OE pers.n. *Earda* + **lēah**. Ess 326.

ARDLEIGH RESERVOIR Essex TM 0328. P.n. ARDLEIGH TM 0529 + ModE **reservoir**.

ARDLEY Oxon SP 5427. 'Eardwulf's wood or clearing'. *Eardulfes lea* [995]13th S 883, *Ardvlveslie* 1086, *Erd- Arduluesle(g) -olv-* 1123–13th cent., *Ardulfle(e)* 13th cent., *Ardele(y)* c.1260–1526. OE pers.n. *Eard(w)ulf*, genitive sing. *Eard(w)ulfes*, + **lēah**, dative sing. **lēa**. O 196.

ARDSLEY SYorks SE 3805. 'Eored or Eared's clearing'. *Erdeslei(a) -lai(a) -lay -ley(e)* c.1160–1442, *Ardeslay -ley* 1381–1624, *Long Ardsley* 1828. OE pers.n. *Ēorēd* or *Ēa(n)rēd*, genitive sing. *Ēo- Ēarēdes*, + **lēah**. YW i.290, ii.xi.

ARDSLEY EAST WYorks SE 3025. *Ardsley east* 1641. P.n. Ardsley + ModE **east**. Ardsley, *Erdeslau(ue)* 1086, *-law(e) -lawa* before 1127–1411, *-lowe* 1219–1402, *Ardeslaw(e)* [1155×62]16th–1418, (*Est-* 1459), *Ardesley -lay* 1366, 1510, 1527, is 'Erd's mound', pers.n. *Erd* < OE *Eard, Ēanrēd* or *Ēorēd*, genitive sing. *Erdes* + **hlāw**. 'East' for distinction from Ardsley West SE 2925 which seems also to have been known variously as *West Ardsley* or *Woodkirk* 1843 OS, *Westerton* and possibly also as *Kirkham* 'the church settlement'. Cf. KIRKHAMGATE SE 2922 and WOODKIRK SE 2724. YW ii.174, WYAS 361.

ARDWICK GMan SJ 8597. 'Eadred, Eadric or Eadhere's farm or workplace'. *Atheriswyke, Atherdwic, Aderwyk* 1282, *Ardewike -wyk(e)* 1322–1422. OE pers.n. *Ēadrǣd, Ēadrīċ* or *Ēadhere* + **wīċ**. The specific has been so abbreviated by the time of the first occurrence that it is impossible to be absolutely sure of the form of the pers.n. involved. La 35 (which suggests pers.n. *Æthelrǣd*), Jnl 21.45.

ARELEY KINGS H&W SO 8070. 'The kings' Areley'. *Ardley Regis, Kyngges Arley* 1405. ME **king**, genitive sing. **kinges**, + p.n. *(H)erneleia* c.1138, *-leʒe -leie* c.1200, c.1250, *Arneley* 1275, *Arleye* 1291, 1428, *Areley* since 1453, 'eagles' wood or clearing', OE **earn**, genitive sing. **earna**, + **lēah**. 'Kings' because part of the royal manor of Martley. Also known as Nether Areley for distinction from Over or Upper ARLEY SO 7680. Wo 29 gives pr [ɛ:əli].

ARGARMEOLS → North MEOLS Mers SD 3518.

ARKENDALE NYorks SE 3860. Probably 'Eorcna's valley'. *Arche- Arghendene* 1086, *Arkenden(n) -yn-* 1200–1614, *Erkinden(e) -yn- -en-* 1246–1521, *Erkendale* 1301, *Arkendall* 1573. OE pers.n. **Eorcna*, a short form for names like *Eorconwald* etc. + **denu** later replaced by ON **dalr** as in COVERDALE SE 0682, LOTHERSDALE SD 9545, OAKDALE ST 1898. OE **earce** 'a chest', genitive sing. **earcan**, has been suggested as an alternative specific in some topographical sense. Arkendale lies on a ridge between streams and the possible sense of *earce* is not clear. The meaning 'fish weir' is too late and unsuitable. YW v.104.

ARKENGARTHDALE NYorks NZ 0002. 'The valley of Arkil's enclosure'. *Arclegarthdaile* 1557, *Archengarthdale* 1671. P.n. *Arkillesgarth* 1199, *Arkelgarth* 13th cent. etc., ME pers.n. *Archil* (ON *Arnketill*) + **garth** + **dale**. YN 295, SPN 14.

ARKENGARTHDALE MOOR NYorks NY 9405. P.n. ARKENGARTHDALE NZ 0002 + ModE **moor**.

ARKESDEN Essex TL 4834. 'Arcel's valley'. *Archesdana(m)* 1086, *Arkesden(e)* from 1241, *Harkelesheldane* 1387, *Harlesden* 1412. lOE pers.n. *Arcel* < ON *Arnkell*, genitive sing. *Arceles*, + **denu**. Ess 516.

ARKHOLME Lancs SD 5871. '(Settlement at the) shielings'. *Ergune* 1086, *Argun* 1195, *-um* 1196, *Erg(h)um* 1246–1279, *Argholme* 14th, *Erwom* 1441, *Arwyn, Erholme* 16th. ON **ǽrgi**, locative-dative pl. **ǽrgum**. La 180, SSNNW 61, Jnl 17.101.

ARKLEY GLond TQ 2295. *Arkeleyslond* 1332, *Arcleylond -lane* 1436, *Arkeley* 1547, OE **earc**, genitive pl. **earca**, + **lēah** referring to the manufacture of chests or coffers. Hrt 69.

ARKLID Cumbr SD 2988 → ARCLID GREEN Ches SJ 7861.

ARKS EDGE Northum NT 7107. *Arks Edge* 1869 OS. P.n. Arks, Borders NT 7108, + ModE **edge** (OE *ecg*) 'the sharp edge of a hill, an escarpment' here referring to a summit of 1468ft.

ARKSEY SYorks SE 5807. Possibly 'Arkel's island'. *Archesei(a)* 1086, *Arkese -ei(a) -ey(e) -ay* 1230–1597. Pers.n. *Arkel* (ON *Arnketill*), genitive sing. *Arkeles*, + OE **ēġ**. Cf. ARKESDEN. YW i.24, SSNY 152, L 35.

ARKWRIGHT TOWN Derby SK 4270. Surname *Arkwright* + ModE **town**. Sir Richard Arkwright, who invented the spinning frame, bought the manor of Sutton here in 1824. Db 310, Room 1983.4.

ARLECDON Cumbr NY 0419. Partly uncertain. *Arlauchdene* c.1130, *-lachadena* c.1185, *Arlok(e)den(e)* c.1150–1309, *Arlecden -k-* 1321, 1396, *Arladon, -leyden* 16th cent., *Arleton* 1680.

†S assigns this charter to Yardley.

Previous attempts to explain this name have suggested '*Earnlac* valley', OE river n. **Earn-lacu* 'eagle stream' < **earn** + **lacu**, + **denu**. It is difficult, however, to see how the *ă* of *lacu* could give ME spellings with *au* and *o*. Cu 335 gives prs [a:lkdən], [æərltən].

ARLESEY Beds TL 1935. 'Ælfric's island'. *Alricheseia* *[1062]13th S 1036, 1086, *Aylrices- Eluricheseye* 13th cent., *Arlicheseye* 1307, *Arlesey* 1492. OE pers.n. *Ælfrić*, genitive sing. *Æfrićes*, + **ēġ**. Bd 166.

ARLESTON Shrops SJ 6610. 'Eardwulf's estate'. *Erdelveston* 1180, *Erdulveston* 1202, 1209, *Ardelston* 1271× 1307. OE pers.n. *Eardwulf*, genitive sing. *Eardwulfes*, + **tūn**. Bowcock 28, DEPN.

ARLEY 'Eagle wood', OE **earn** + **lēah**. L 205.

(1) ~ Ches SJ 6781. *Arlegh* 1340, *-ley* from 1360. Che ii.101.

(2) ~ Warw SP 2890. *(æt) Earnlege* 1001 S 898, *Ernele* 1199, *Arnlege* 1221, *Arlei* 1086–1538 with variants *-legh -ley(e) -lege*. Wa 123.

(3) Upper ~ H&W SO 7680. *Upper Arley* 1832 OS. ModE **upper** + p.n. *Ernlege* [963]14th S 720, *Earnleie* [996 for 994]17th S 1380, *Ernlege* 1086, *Erlege -leia* 1188–1200, *Arnlege -leye* 1276–1327, *Ar(e)leye* 1332–1465. 'Upper' for distinction form ARELEY KINGS S 8070. Wo 30.

ARLINGHAM Glos SO 7111. Uncertain. *Erlingeham* 1086–c.1275, *Erlingham -yng-* c.1150–1639, *Arlingham* 1492. Possibly 'homestead' or 'water-meadow of the Eorlingas', OE folk-n. **Eorlingas*, genitive pl. **Eorlinga*, + **hām** or **hamm**. **Eorlingas* would mean 'the people called after Eorl(a)' but such an OE pers.n. is unknown. Gl ii.175.

ARLINGTON Devon SS 6140. 'Farm, village, estate called after Ælfheard'. *Alferdintona* 1086, *Alfrintone* 1258×62, *Au(v)rington* 13th, *Alrington(e)* 1284–1377, *Arlyngton* 1550. OE pers.n. *Ælfheard* + **ing**[4] + **tūn**. D 56.

ARLINGTON ESusx TQ 5407. 'Estate called after Eorla'. *Erlington, Herlintone* 1086, *Erlington* 1230–1610, *Arlyngton* 1573. OE pers.n. **Eorla* + **ing**[4] + **tūn**. Sx 408, SS 76.

ARLINGTON Glos SP 1106. 'Estate called after Alfred'. *Ælfredin(c)gtune* 1002×4, [1004]13th S 906, *Alvredintone* 1086, *Alurinton(a) -yn-* 1154×89–1363, *-ing- - yng-* 1233–1359, *Alrin(g)ton - yn(g)-* 13th–1550, *Herlingthon* 1303, *Arlington* 1584. OE pers.n. *Ælfrēd* + **ing**[4] + **tūn**. Gl i.27.

ARMATHWAITE Cumbr NY 5046. 'Hermit clearing'. *Ermiteth(w)ait* 1212, c. 1230, *Armathwayte Monialium* 'nuns' A' 1536. Probably derived from Armathwaite NY 5342, the site of a Benedictine nunnery which had property here. OFr ME **ermite** + ON **thveit**. Cu 168, SSNNW 213 no. 1(b), L 211.

ARMINGHALL Norf TG 2504. 'Nook of land of the Amer- or Ambringas'. *Hameringahala* 1086, *Ambringehale* c.1105, c.1140, *Ameringehale* 1212. Uncertain folk.n. + OE **healh**. It is unclear whether the original folk-n. was **Ameringas* < OE pers.n. *Ēanmǣr* or **Ambre* + **ingas**. DEPN.

ARMITAGE Staffs SK 0816. 'The hermitage'. *Hermitage* 13th, *Armytage* 1520. OFr, ME **ermitage**. Duignan 6, DEPN.

ARMSCOTE Warw SP 2444. 'Eadmund's cottage(s)'. *Eadmundescote* 1042 S 1394, *Edmundes-* 1166–1428, *Admundes-* 1323–32, *Admyscote* 1366, 1535, *Armscote* 1554. OE pers.n. *Ēadmund*, genitive sing. *Ēadmundes*, + **cot**, pl. **cotu**. Wo 172.

ARMTHORPE SYorks SE 6305. 'Arnulf's outlying farmstead'. *Ernulfes- Einuluestorp* 1086, *Hernoldes- Ernaldest(h)orp* 1147–1223, *Arnetorp' -thorp(e)* c.1160–1428, *Armethorp(e)* 1237–1531, *-throp* 1654. Pers.n. OE *Earnwulf* or ON *Arnulfr* alternating with CG *Ernald*, genitive sing. *Earnwulfes, Ernaldes*, + **thorp**. The change in pers.n. may reflect change of ownership or confusion in the unstressed syllable. YW i.37, SSNY 54.

ARNCLIFFE 'The eagles' cliff'. OE **earn**, genitive pl. **earna**, + **clif**.

(1) ~ NYorks SD 9371. *Arneclif(e) -cliff(e) -y-* 1086–1597. The reference is to Arnberg Scar, *Arnber Scar* 1727, 'the eagles' hill', OE **earn** + **beorg** + **sker**, a great limestone scar S of the village. YW vi.113–14.

(2) ~ NYorks SE 4599. *Erneclive, G- Lerneclif* 1086, *Erneclive* c.1170–1293, *Arneclive* c.1291. A large, steep, wooded bank. YN 178.

ARNCOTT Oxon SP 6117. 'The cottage(s) called after Earn(a)'. *Earnigcote, (æt) Earnigcotan* [983]13th S 843, *Ernicote, alias Ernicote* 1086, *Ernicote* 1123–1320 with variants *Erne- Erny-* and *-cot(a) -kote, Arnecot, Arnicote* 1278×9. OE pers.n. **Earn(a)* + **ing**[4] + **cot**, pl. **cotu**, dative pl. **cotum**. O 161.

ARNE Dorset SY 9788. 'House, building'. *Arn(e)* from 1268, *Aren* 16th, *E(a)rne* 17th. OE **ærn**. Do i.72.

ARNESBY Leic SP 6192. Possibly 'Iarund's estate'. *Erendesberie -bi* 1086, *Erendesy* 1170, 1171, *-es-* 1227–31, 1336 *Erndesby -is-* 1227–85, *Ernesbi -by -is- -ys-* c.1200–1522, *Ernebi -by* 1177–1266, 1301, *Arnesby(e)* 1610, 1695. ON pers.n. *Iǫrundr*, ODan *Iarund*, ME genitive sing. **Erendes*, + **bȳ**. Lei 425, SSNEM 31.

ARNFIELD BROOK Derby SK 0399. *Arnfield Brook* 1842. P.n. Arnfield SK 0198 + ModE **brook**. Arnfield, *Arn(e)wayesfeld -is-* 1350, 1351, *Arn(i)es(s)feld* 1360, *Arn(e)wayfeld* 14th cent., *Arnefeld* 1360, *Arnfield* 1831, is 'Arneway field', ME pers.n. and surname *Arneway* (< OE *Earnwīġ*), genitive sing. *Arneways*, + **feld**. Che i.321.

ARNFORD NYorks SD 8356. 'Ford which can be crossed by riding'. *Erneford* 1086, *Arneford(e)* 12th–1589, *Arnford* 1457. OE **ærne** + **ford**. YW vi.158.

ARNOLD Notts SK 5745. '(The settlement) at eagles' nook'. *Ernehale* 1086, *Ernes- Ærneshala, Ernehal* late 12th, *Ernhala -hal(e)* 1174–1230, *Arnhal(e)* 1212–1335, *Arnale* 1334, *Arnold* from 1474. OE **earn**, genitive pl. **earna**, + **halh**, locative-dative sing. **hale**. Nt 113 gives prs [ɑ:nəld] and [ɑ:nəl].

ARNSIDE Cumbr SD 4578. 'Arnulf's headland'. *Harnolvesheuet* 1184×90, *Arnulvesheved* 1086, *Arneshed(e)* 1472–1502, *Arn(e)sid(e) -syde* from 1517. OE pers.n. *Earnwulf*, genitive sing. *Earnwulfes*, + **hēafod** referring to a headland in the Kent estuary. We ii. 65.

ARRACOT Devon SX 4287 → ELSTREE GLond TQ 1895.

ARRAD FOOT Cumbr SD 3081. *Arrad Foot* 1864 OS. Unexplained p. or mountain name + ModE **foot**.

ARRAM Humbs TA 0344. 'At the shielings'. *Argun* 1086, *Erg(h)um* 12th–1359, *Ethome, Earham, Ar(r)am* 16th cent. ON **ǽrgi**, dative pl. **ǽrgum**. YE 79, SSNY 86.

ARRATHORNE NYorks SE 2093. 'The shieling thorn-tree'. *Erg(h)ethorn, Ergthorn* 13th cent., *Erethorn* 1285, *Arathorne* 1581, *Arrowthorne* 16th cent. ME **erg** (ON *ærgi*) + **thorn**. YN 240.

ARRETON IoW SZ 5386. 'Estate called after Eadhere'. *Eaderingtune* [873×88]11th S 1507, *Adrintone* 1086, *Arretoniæ* 1135×54, *Arreton(e)* from 1154×89, *Atherton(e) -re-* 1255–1604. OE *Ēadhere* + **ing**[4] + **tūn**. Wt 6 gives pr ['ærətn], Mills 1996.23.

ARRINGTON Cambs TL 3250. 'Settlement of the Earningas, the people called after Earna'. *(at) Earnningtone* [942×c.951]13th S 1526, *ærningetune* 1086 IE, *Erlingetona, Ernin(c)getone, Erningtune* 1086, *Erningatone* 1086×7, *Erington* 1236, *Aring(e)ton(e) -yng-* c.1205–1553. OE folk-n. **Earningas* → pers.n. **Earna* + **ingas**, genitive pl. **Earninga*, + **tūn**. Ca 69.

ARROW Warw SP 0856. Transferred from the r.n. ARROW SP 0861. *Arne* (for *Arue*) *[710]12th S 81 (a forgery of c.1097×1104), *Arue* 1086, 1230, *Ar(e)we* 1194–1390, *Arowe* 1233, *Arrowe* 1535. Wa 195.

River ARROW H&W SO 4058. 'Bright, shining one'. *(ondlong) Erge* 958 S 677, *Arewe* 1256, *Arow* c.1540. The W forms are *Arwy* [n.d.]17th, *Arro* 17th. Brit **Argwy* on the IE root **arǵ-* 'bright, shining' seen in Greek ἄργυρος 'silver', Latin *arguo* 'make clear', *argentum* 'silver'. The Celtic form was **Argouia*, cf. the Illyrian r.n. *Argua*, the Spanish Arga, a tributary of the Ebro, the East Prussian Arge (Argà in Lithuanian), the French

Argens near Cannes, *Argentios, Argenteus* in Greek and Latin sources, the Ergolz, a tributary of the Rhine near Basel, earlier *Ergenz* 1348 < OHG *Argenza* and the German Argen, *Argun* 1150, 1172, *Arguna* 773, a tributary of Lake Constance. RN 17, BzN 1957.231, Berger 40, Krahe 1964.53–4.

River ARROW Warw SP 0861. 'The swift one'. *Arwan stream* [11th]copy, *Ar(e)we* 13th cent., *Arrow(e)* from 1538. Old European r.n. **Arvā* on the IE root **er-*/**or-* 'move' seen in Latin *orior* 'rise', Gk *ὄρνυμι* 'stir oneself, move', OHG, OS *rinnan* 'run', and the French r.ns. Arve, Haute-Savoie, Avre, Eure, *Arva* c.965, Auve, Marne, *Alva* 1181, *Arva* 1132, Erve, Mayenne, *Arvam* 1060, and Italian Arvo, Calabria. Wa 1, RN 16, BzN 1957.231, Krahe 1964.45–6.

ARSCOTT Devon SS 3505, 3800 → PARKHAM ASH SS 3620.

ARTHINGTON WYorks SE 2644. 'Estate called after Earda'. *Hardinctone* 1086, *Ardinton(a)* 1086–1246, *-ing-, -yng-* 1152×71–1612, *Arthingtun(e) -ton(a) -yng-* 12th–1667. OE pers.n. **Earda* + **ing**[4] + **tūn**. YW iv.193, WYAS 307.

ARTHINGWORTH Northants SP 7581. 'The enclosure called after Earna'. *Narninworde, Arniworde, Arningvorde, Erniwade* 1086, *Earnigwurth* 1154×89, *Erning(e)w(o)rth' -wurth* 1202–39, *Arning(e)w(o)rth'* 1220–75, *Arthingworth* from 1275, *(H)ardingworth* 16th. OE pers.n. **Earn(a)* + **ing**[4] + **worth**. On formal grounds the specific could be OE **earn** 'an eagle'. Nth 109.

ARTHUR SEAT Cumbr NY 4978. *Arthur Seat* 1866 OS. Possibly identical with *Arthurt knoll'* 1552. Pers.n. *Arthur* + ModE **seat** 'lofty place, seat-shaped rock' (<ON *sǽti*). Cf. Arthur's Seat in Edinburgh. Cu 62.

River ARUN WSusx TQ 0422. *Arunus, Aron* 1577, *Arun* 1730. A back-formation from the p.n. ARUNDEL TQ 0107 probably invented by Harrison. The older name, *Tarente* [c.725]14th S 44, 1212–79, is identical with r.ns. TARRANT and TRENT from Brit *Trisanton* 'the trespasser' as the earliest reference to the mouth of the Arun shows, *Τρισάντωνος ποταμοῦ ἐκ β-ολαί* (Trisantonos potamou ekbolai) [c.150]13th Ptol. Sx 3, RN 18, 416, RBrit 476.

ARUNDEL WSusx TQ 0107. 'Hoarhound valley'. *Harundel(le)* 1086–1341, *Arundell(e)* 1087–1488, *Arondel(l)* 1093, 1273×91, 1303. OE **hārhūne** + **dell**. Sx 136, EHR 30.164, PNE i.234.

ARUNDEL PARK WSusx TQ 0108. *Arundel Park* 1813 OS. P.n. ARUNDEL TQ 0107 + ModE **park**.

ASBY 'Ash-tree farm or village'. ON **askr**, genitive pl. **aska**, + **bȳ**.

(1) ~ Cumbr NY 0620. *Asbie* 1654, *Ashby* 1713. Cu 336.

(2) Great ~ Cumbr NY 6712. *magna Askebi -by* before 1216–1415, *Myckell, Mikill Asseby(e)* 1547, *Greate ~* 1589, *Great Assby(e)* 1696. ModE **great** (Latin **magna**) + p.n. *Ascabio* 12th, *Aschaby* [1158×66]13th, *Askebi -by* c.1150–15th, *Asseby(e)* 1407–1589, *Asby* 1630. 'Great' for distinction from Little ASBY NY 6909. We ii.54, SSNNW 26, L 219.

(3) Little ~ Cumbr NY 6909. *parvo, parva Askebi -by* 1185–1415, *Lit(t)le Asby* 1428, 1567. Adj. **little** (Latin **parva**) + p.n. Asby as in Great ASBY NY 6712. We ii.54, SSNNW 26, L 219.

ASCOT Berks SU 9268. 'East cottage(s)'. *Estcota* 1177, *Ascote* 1269 etc. OE **ēast** + **cot**, pl. **cotu**. 'East' probably in relation to Easthampstead. The Royal Ascot race meeting held in June at the nearby racecourse was started by Queen Anne in 1711. Brk 88.

North ASCOT Berks SU 9069. No early forms. ModE **north** + p.n. ASCOT SU 9268.

ASCOTT-UNDER-WYCHWOOD Oxon SP 3018. *Estkote Doyliuorum sub Wicchewode* 'E held by the d'Oyly family beneath Wychwood' c.1280. Earlier simply *Esthcote* 1086, *Astcote* 1258, *Ascote* 1291, 'the east cottages', OE **ēast** + **cot**, pl. **cotu**. Ascot D'Oyly takes its suffix from Wido de Oileio who held *Escota* c.1100. See further WYCHWOOD. O 335, 386, L 229.

ASENBY NYorks SE 3975. 'Eystein's farm or village'. *Æstanesbi* 1086, *Ei- Ai- Aystenby -an-* 1198–1417, *Aysynby* 1408. ON pers.n. *Eysteinn* + **bȳ**. YN 182, SPN 78, SSNY 18.

ASFORDBY Leic SK 7019. 'Asfrøthr's village or farm'. *Osferdebie, Esseberie* 1086, *Osfordebi* 1102×6, *Asfordebi -by* 1185–1552, *As(s)h(e)ford(e)by* 1291–1537, *Esfordebi -by* 1190×1204–1263, *Asfordby* from 1232. ON pers.n. *Ásfrøthr* + **bȳ**. Some forms show that the specific was sometimes taken to be a previous p.n. **Æscabyrig* 'fort of the ash-trees' or **Æsc-ford* 'ash-tree ford'. Lei 270, SSNEM 31.

ASFORDBY HILL Leic SK 7219. P.n. ASFORDBY SK 7019 + ModE **hill**.

ASGARBY 'Asgeirr's village or farm'. ON pers.n. *Ásgeirr* with OE *-gār* for *-geirr* + **bȳ**.

(1) ~ Lincs TF 1145. *Asgarby* from late 12th, *Asger(e)bi -by* 1185–1723. Perrott 47, SSNEM 31.

(2) ~ Lincs TF 3366. *Asgerebi* 1086, 1163, *Asgerbi* c.1115, 1162, *Ansgerby* 1135–47. DEPN, SSNEM 31, Cameron 1998.

ASH '(The settlement) at the ash-tree'. OE **æsc**, dative sing. **æsce**.

(1) ~ Kent TQ 5964. *Eisse* 1086, *Aeisce* c.1100, *Esse* 1197–1232. K 32.

(2) ~ Kent TR 2858. *Æsce* c.1100, *Esse* 1226–91, *Asshe* 1270, *Aysshe* 1284. Also referred to in Latin and French forms from **fraxinus** (*(de) fraxnis* (dative pl.) 1226) and **fresne** (*ffre(s)nes* 1254). PNK 527.

(3) ~ Somer ST 4720. *Esse* 1225. DEPN.

(4) ~ Surrey SU 8950. *Essa* 1170, *Esshe* 1174×89–1272, *As(s)(c)he* 13th cent., *Ayssh(e)* 1288, 1336, *Ash* 1642. Sr 135, L 219.

(5) ASHBOCKING Suff TM 1854. 'A held by the Bocking family'. *Bokkyng Assh* 1411, *Aschebookyng* 1524, *Ashbocking* 1610. P.n. *Essa, Hassa, Ass(i)a* 1086, *Esse* 1198, 1264, *Assh(')* 1286–1338, + manorial addition from Ralph de Bocking 1338 whose family came from Bocking, Essex. DEPN, Baron.

(6) ~ BULLAYNE Devon SS 7704. 'A held by the Bullayne family'. Probably the home of William de *Asshe* 1302. D 409.

(7) ~ FELL Cumbr NY 7405. 'Ash-tree hill'. *Ashfell* [1154×89]1645, *Askef(f)ell* 1224, *Askfell, Assfell* 17th. ON **askr** + **fjall**. We ii.25.

(8) ~ MAGNA Shrops SJ 5739. 'Great A'. *Magna Asche* 1285. Latin **magna** + p.n. Ash. 'Magna' for distinction from Ash Parva SJ 5739, 'Little Ash', *Parva Asche* 1285. DEPN.

(9) ~ MILL Devon SS 7823. P.n. Ash as in Rose ASH SS 7821 + ModE **mill**.

(10) ~ PRIORS Somer ST 1529. 'Ash held by the prior' sc. of Taunton. *Esse Prior* 1263. Earlier simply *Æsce* *[1065]18th S 1042, *Aissa, Aixe* 1086. DEPN.

(11) ~ THOMAS Devon ST 0010. 'Ash held by Thomas'. *Esse Thomas* 1351. Earlier simply *Aisse* 1086, *Asshe juxta Halberton* 1339. *Thomas*. Held by *Thomas* de Esse in 1238. D 548.

(12) Bracon ~ Norf TM 1899. 'Bracon A'. *Bracon Ash* 1838 OS. P.n. *Brachene* 1175, *Brakene* 1230, 'the bracken', OE ***bræcen** or ON ***brakni**, + p.n. Ash. DEPN.

(13) Campsey ~ Suff TM 3255 > CAMPSEY ASH TM 3255.

(14) New ~ GREEN Kent TQ 6065. ModE **new** + p.n. ASH TQ 5964 + ModE **green**.

(15) Rose ~ Devon SS 7821. 'Ash held by the Ralph family'. *Esse Ra(u)f(e)* 1319, 1387, 1492, *Rowesasshe* 1400, *Roseashe* 1577. 'Rose' is a folk-etymology of *Rafe's, Raves, Rawes* from the family of Radulphus (Ralph, Rafe) who held the manor of Ash in 1198. Cf ROUSDON SY 2991. Earlier simply *Aisse* 1086, *Esse* 1198, 1242. D 391.

ASHAMPSTEAD Berks SU 5676. 'Homestead by the ash-trees'. *Essam(e)stede* 1155×8, 1348, 1355, *Ess(e)ham(p)sted'* 1212, 1220, *Asshehamstede, Ashhampstede* 1309, *Assham(p)sted(e)* 1311–1542. OE **æsc** + **hām-stede**. Brk 508, Stead 264.

ASHBOURNE Derby SK 1846. 'Ash-tree stream'. *Es(s)eburn(e) -burna -borne -borna* 1086–1330, *Ess(c)heburn(e) -bourn* 1241–1300, *Asse-* 1219–1322, *As(s)he- As(s)cheburn(e) -born(e)*

-bourn(e) from 1229. OE **æsc** + **burna**. The place takes its name from the stream on which it stands. Db 341, L 18, 219.

ASHBRITTLE Somer ST 0521. 'Bretel's Ash'. *Esse Britel* 1212, *Ashbrittell* 1610. Earlier simply *Aisse* 1086, 'the ash-tree', OE **æsc**. The tenant in 1086 was Bretel de St Clair from Saint-Clair or Saint-Clair sur l'Elle near Saint-Lô, La Manche. His name, found mainly in the SW, is probably a diminutive of OFr *Bret* 'a Breton' (OBret *Brithael*) rather than Co *Brythhael*. DEPN, PNDB 208, DBS 49.

ASHBURNHAM PLACE ESusx TQ 6814. 'Ashburnham mansion house'. P.n. Ashburnham + ModE **place**. Ashburnham, *Es(h)burnham* 1315, [12th]1432, *Ashburnhame* 1320, is the 'promontory of dry land by Ashburn, the ash-tree stream', *Esseborne* 1086. OE p.n. **Ashburn*, **æsc** + **burna**, + **hamm** 3, 6. Sx 477, ASE 2.24.

ASHBURTON Devon SX 7569. 'Settlement by Ashburn'. *Essebretone* 1086, *Aisbernatonam* [c.1150]14th, *Esperton* 1187, *Asperton(e)* 1238–1309, *As(s)hperton* 1313–56, *Aysshberton* 1483. Stream name Ashburn, *æscburne, (oð) æscburnan* early 11th S 1547, *Ays(s)heborne* 1504, 1553, 'ash-tree stream' OE **æsc** + **burna**, + **tūn**. The Ashburn, a tributary of the Dart, is now called Yeo. D 462, 17.

ASHBURY Devon SX 5097. 'Ash-tree fortified place'. *Esseberie* 1086, *-biry* 1223, *Asebiria* 1136, *Ashbiry* 1291, *Aissebyri, Aysshebyry* 13th. OE **æsc** + **byriġ**, dative sing. of **burh**. The charter forms previously assigned to Ashbury beling to UFFINGTON SU3089. There are no visible earthworks. D 126.

ASHBURY Oxon SU 2685. '(At) the fort by the ash–tree(s)'. *Eissesberie* 1086, *Æsseberia* 1187, *Esse- Asseberia -biri -bury* etc. 1188–1284, *Asschebir' -bury* 1284–1327. OE **æsc**, genitive sing. **æsces**, genitive pl. **æsca**, + **byriġ**, dative sing. of **burh**. The reference is to the Iron Age hill-fort known as Alfred's Castle. Brk 344, Thomas 1976.179.

ASHBY 'Ash-tree farm'. ON **askr** (replaced by ME *ash*), genitive pl. **aska**, + **bȳ**.

(1) ~ Humbs SE 9008. *Aschebi* 1086, c.1115. DEPN, SSNEM 31.

(2) ~ BY PARTNEY Lincs TF 4266. *Askeby iuxta Partenaye* 1228. P.n. *Aschebi* 1086, c.1115, *Askebi -by* 1154×89–1212, + p.n. PARTNEY TF 4168. Also known as *Estaskebi* 'east A' 1208 for distinction from West ASHBY TF 2672. DEPN, SSNEM 31 no.2, Cameron 1998.

(3) ~ CANAL Leic SK 3801. *Ashby de la Zouch Canal* 1795. P.n. Ashby as in ASHBY-DE-LA-ZOUCH SK 3516 + ModE **canal**. Lei 485.

(4) ~ CUM FENBY Humbs TA 2500. *Aschebi* 3×, *Achesbi* 1× 1086, *Asche- Ascbi* c.1115, *Askebi -by* 1196–1428, *Asheby* 1303, *Ashby* from 1595. Li iv.48, SSNEM 31.

(5) ~ DE LA LAUNDE Lincs TF 0555. 'A held by the de la Laund family'. *Ashby de la Launde* from 1316. Earlier simply *(H)aschebi* 1086, *Ascheby* c.1160–1303 with variants *Asshe- Ess(he)-*. The manor was held by the de la Laund family in the 12th–15th cents. One of the DB tenants was *Aschil*: it is possible that *æsc* has replaced this pers.n. Perrott 274, SSNEM 31 no.4.

(6) ~ -DE-LA-ZOUCH Leic SK 3516. P.n. *Ascebi* 1086, *Essebi(a) -by* 1154×89–1336, *As(s)hebi -by* 1277–1535 + suffix *la Z(o)u(s)ch(e)* 1203–1513, *de la Zouch(e)* 1341–1576. Alanus *la Zouche* held the manor c. 1154×89, Roger *de la Zuche* in 1202. Lei 336, SSNEM 31 no.5.

(7) ~ FOLVILLE Leic SK 7012. P.n. *Ascbi* 1086, *Essebi(a) -by* c.1130–1294, *Assheby* 1310–1507, *Ashby* from 1528 + suffix *Fol(e)vill(e)* from 1232. The manor was held by Fulco de *Foleuilla* early in the reign of Henry II. Lei 292, SSNEM 32 no.6.

(8) ~ MAGNA Leic SP 5690. P.n. *Essebi* 1086–1221, *-by* 1212–1301, *Asshebi -by* 1316–1615, *Ashby* from 1515 + affix *Magna-* 'great' late 12th–1330, *-Magna* from 1254, *Mekyll-* 1492, *Mu(t)che-* late 16th cent., *Great-* 1610 for distinction from ASHBY PARVA SP 5288. Lei 426, SSNEM 32 no.7(a).

(9) ~ PARVA Leic SP 5288. P.n. Ashby as in ASHBY MAGNA SP 5690 + Latin affix *Parva-* 'little' 1086–1502, *-Parva* from 1176, *Lytel-* 1347, *Lytell-* 16th cent. for disinction from ASHBY MAGNA SP 5690. Lei 427, SSNEM 32 no.7(b).

(10) ~ ST LEDGERS Northants SP 5768. *As(s)heby Leger* 1316, 1339. P.n. *Ascebi* 1086, *Essebi -by* 12th–1316 + affix *Sancti Leodegarri* c.1230, *St Legers* 1322, saint's n. Leger (Leodegar) from the dedication of the church. Nth 9, SSNEM 32.

(11) ~ ST MARY Norf TG 3202. P.n. *Ascebi* 1086, *Asheby* 1251, + saint's name *Mary* from the dedication of the church. DEPN.

(12) Canons ~ Northants SP 5650. *Canounes Hessheby* 1287, *Assheby Canonicorum* 1320, *Chanons Assheby* 1506. ME **canoun**, genitive pl. **canounes**, + p.n. *Ascebi* 1086, *Assheby* 1320–1506. 'Canons' from the priory of Black Canons founded here in the mid 12th cent. Nth 39, SSNEM 32.

(13) Castle ~ Northants SP 8659. *Castel Assheby* 1361. ME **castel** + p.n. *Asebi* 1086, *Esseby* 12th, *Axeby* 1235. There was formerly a castle at this place later replaced by an Elizabethan mansion. Nth 142, SSNEM 32, Pevsner 1973.138.

(14) Cold ~ Northants SP 6576. *Caldessebi* c.1150, *Chald- Kaldessebi -by* c.1160–1219, *Coldessebi -by* 1199–1235, *Cole Ashbye* 1631, *Cold Ashby* 1780. ME adj. **cald** + p.n. *Essebi* 1086, *Esseby* 12th. 'Cold' from its high and exposed situation. Nth 64, SSNEM 32.

(15) Mears ~ Northants SP 8366. 'A held by the Mares family'. *Esseby Mares* 1281, *Mares Assheby* 1297, *Maires Ashby* 1659. Family name of Robert de Mares in 1242 + p.n. *Asbi* 1086, *Essebi* 1166. Also known as *Northesseby* 1220 with reference to Castle ASHBY SP 8659. Nth 137 gives pr [mɛːəz æʃbi], SSNEM 32 no.13.

(16) West ~ Lincs TF 2672. *West Asby* 1528. ModE **west** + p.n. *Aschebi* 1086, *Asc(he)bi* c.1115. 'West' for distinction from ASHBY BY PARTNEY TF 4266. SSNEM 32 no.14, Cameron 1998.

ASHCHURCH Glos SO 9333. 'Church near the ash-tree'. *Asschirche* 1287, *Aschurch* 1487. OE **æsc** (possibly used as a p.n. Ash) + **ċiriċe**. Gl ii.52.

ASHCOMBE Devon SX 9179. 'Ash-tree coomb'. *Aissecome* 1086, *A(i)scumb(e)* 13th cent., *Ashcombe* 1291. OE **æsc** + **cumb**. D 486.

ASHCOTT Somer ST 4337. 'Cottage(s) at Ash or by the ash-tree'. *Aissecote* 1086, *Ascote* 1198, *Asshcote* 1327, *Ashton* (sic) 1610. OE **æsc** possibly used as a p.n. + **cot**, pl. **cotu**, ME *cote*. DEPN.

ASHDON Essex TL 5842. 'Hill growing with ash-trees'. *Estchentune, Æstchendune* [c.1036]12th, *Aesredune* *[1043×5]14th S 1531, *Ascendunā* 1086, *Essendun(a) -do(u)n* 12th–1387, *Ass(h)endon(e) -yn- -ing-* 1286–1390, *Ass(c)hedon(e)* 1272–1428. OE **æscen** + **dūn**. The original site of the village was at Church End TL 5841 and the present village was called *Ashdon Street* 1805 OS, 'Ashdon hamlet', ModE dial. **street**. Ess 502 gives pr [eizn], Jnl 27.22.

ASHDOWN FOREST ESusx TQ 4530. *Ashdown Forest* 1813 OS. P.n. Ashdown + ModE **forest**. Ashdown, *Hessedon* c.1200, *Essen- Ashendon* 13th cent., *Assesdune* c.1275, *Asshedo(u)n* 1293–9, is the 'hill growing with ash-trees', OE adj. **æscen** + **dūn**. Sx 2.

ASHELDHAM Essex TL 9701. 'Æschild's homestead'. *Assild(e)ham -yl-* c.1130–1291, *Asshild(e)ham* 1276–1412, *As(s)held(e)ham* 1317–1412, *As(s)heldon, Ashelham* 16th. OE feminine pers.n. **Æschild*, genitive sing. **Æschilde*, + **hām**. An earlier name was *Hain(c)tunā* 1086, *Aintune* 1198, possibly OE ***hēahing*** 'high place' + **tūn** as in HEIGHINGTON Durham NZ 2522. Ess 208.

ASHEN Essex TL 7442. '(At the) ash-trees'. *Asce* 1086, *Essa -e, Esch* c.1090–1309, *Ass(h)e, As(s)ch(e)* 1309–1412, *Asshen* 1344, 1452, *Aysshen* 1581–2. OE **æsc**, dative pl. **æscum** in the phrase *æt thǣm æscum*. Ess 406.

ASHENDON Bucks SP 7014. 'Ash-tree hill'. *Assedune -done* 1086, *Essendon* c.1218–84, *As(s)hedon* 1327–98, *Ashingdon* 1391. OE adj. **æscen** + **dūn**. Bk 101, L 150.

ASHFIELD 'Ash-tree open land'. OE **æsc**, genitive pl. **æsca**, + **feld**.

(1) ~ Suff TM 2162. *Asse- As(sa)felda* 1086, *Esfeld* 1198, *Asshe- Ashfeud -feld(e)* 1286–1470, *Ashfilde* 1568. DEPN, Baron.

(2) ~ GREEN Suff TM 2673. P.n. Ashfield + ModE **green**.

(3) Great ~ Suff TM 0068. *Aysefeld Magna* 1291. ModE **great**, Latin **magna**, + p.n. *Eascefeldā* (accusative case), *eascefelda* 1086. DEPN.

ASHFORD Usually 'ash-tree ford', OE **æsc** + **ford**.

(1) ~ Devon SS 5335. *Aisseford, Esforde* 1086, *Esforde* 1264×9, *Esse-* 1311, *Asford(e)* 1242, 1264×9, *Ayshford* 1316, 1356. A ford across the Taw estuary. D 24.

(2) ~ Kent TR 0142. Partly uncertain. *Essetesforde* 1046, *-ford* 1086, *Esseteford* 1211×2. The generic is OE **ford** 'a ford', the specific either OE ***æscet** 'a copse of ash-trees' or ***æsc-scēat** 'ash-tree corner' referring to the marked bend of the river Stour at this place. KPN 7, L 70.

(3) ~ BOWDLER Shrops SO 5170. 'A held by the Bowdler family'. *Asford Budlers* 1255, 1271×2, ~ *Boudlers* 1271×2, *Ashford Bowdler* 1675. P.n. *Esseford* 1155, 1242, *Es(ch)ford* 1203–30 + manorial addition from Henry de Bodlers 1242. Sa i.19.

(4) ~ CARBONEL Shrops SO 5270. 'A held by the Carbonel family'. *Aysford Carbonel* 1255–1308, *Assheford Carbonel* 1287, 1348, *Ashford* ~ 1303, ~ *Cardinal* 1733, *Ashford Carbonel(l)* (sic) 1356, 1361 P.n. *Esseford* 1086–1242 + manorial addition from John Carbonel 1221×2, William Carbunel 1174×85. Sa i.19.

(5) ~ HILL Hants SU 5562. *Ashforde Hill* 1600. P.n. *Ashford* 1816 OS + ModE **hill**. Ashford, the 'ash-tree ford', is a crossing of a small tributary of the r. Emborne; in the absence of earlier forms it is impossible to say how old the name is. Gover 1958.148.

(6) ~ IN THE WATER Derby SK 1969. *Ashford in the watter* 1697. Earlier simply *(æt) Æscforda* [926]13th S 397, *Aisseford* 1086, *Ays- Aisford* 1216–1348, *Esse- Asse-* 1101×8–1372, *Asshe-* 1231–1506, *Ashford* 1226×8, 1230 and from 1577. The ford crosses the river Wye. Db 27, L 70.

ASHFORD Surrey TQ 0771. 'The ford across the river Ash'. *Ecelesford* [969]c.1100 S 774, *Ecclesforde* [969]14th ibid., 1042×66, *Echelesford* 1274–1383, *(æt) Exforde* [1062]13th S 1035, *Exeforde* 1086, *Ex(e)(n)ford* 13th, *Assheford* 1488–1567. R.n. Ash + OE **ford**. The r.n. Ash, *little river Ash* 1738, is a back-formation from the later form of this p.n. after 1488. Its original name was *eclesbroc* 962 S 702, unknown element + OE **brōc**. **eċles* with palatal *ċ* is unlikely to represent PrW **eglēs* 'a church' but might represent OE **eċels* 'land added to an estate', or the genitive sing. of the OE pers.n. *Eċċel*. Alternatively, in view of the occurrence of this element in other r.ns., River Ecclesbourne Derby SK 3244, *Ecclisborne* 1298†, Ecchinswell Hants SU 4959, *Eccleswelle* 1086, and Eccleswall Court H&W SO 6523, this may be an otherwise unknown r.n. or r.n. element. Mx 1, 11, PNE i.145, RN 141, Db 7, Ha 70.

ASHILL Devon ST 0811. 'Ash-tree hill'. *As(s)hull* 1249, 1346. OE **æsc** + **hyll**. D 538.

ASHILL Norf TF 8804. 'Ash-tree wood or clearing'. *Asscelea, Essalai* 1086, *Assele* 1208. OE **æsc**, genitive pl. **æsca**, + **lēah**. DEPN.

ASHILL Somer ST 3217. 'Ash-tree hill'. *Ais(s)elle* 1086, *Esselle* 1212, *Asshulle* 1327, *Ashhill* 1610. OE **æsc** + **hyll**. DEPN.

ASHINGDON Essex TQ 8693. 'Assa's hill'. *Assandun(e)* c.1050 ASC(C) under years 1016, 1020, c.1100 ASC(D, F). c.1150 ASC(E), *Assa túnum* (dative pl.) c.1050 Knútsdrápa, *Nesenduna* 1086, *Assendon -den* 1198, 1248, *Essendon -den* 1203, 1228, *Assin(g)don -yn(g)-* 1250–1777, *Asshendon -in-* 1254–1552, *Ashyngdon -ing-* from 1325. OE pers.n. ***Assa**, genitive sing. **Assan*, + **dūn**. Alternatively the specific might be OE **assa** 'an ass'; it was early confused with *æsc* 'an ash-tree'. For full discussion of this name see Jnl 27.21–9. Ess 176.

†Ecclesbourne Glen ESusx TQ 8410 probably does not belong here, cf. the forms cited Sx 507, *Eglesbourne* 1724, probably identical with *Agnesborne'* 1706.

ASHINGTON Northum NZ 2787. 'The ash-tree valley'. *Essende* (sic) 1170, *Esinden* 1199, *Essenden* 1205–55, *-ing-* 1242 BF, *Eschedon* 1296, *Es(s)henden* 15th cent., *Ashington* 1663. OE **æscen** + **denu**. NbDu 6, DEPN.

ASHINGTON WSusx TQ 1316. 'The settlement of the Æscingas, the people called after Æsc'. *Essingeton(a)* 1073, 1268, *Esshyngton -ing-* 1315, 1319, *Asshyngton* 1397. OE folk-n. **Æscingas* < pers.n. *Æsc* + **ingas**, genitive pl. **Æscinga*, + **tūn**. Sx 183, SS 75.

ASHLEWORTH Glos SO 8125. 'Æscel's enclosure'. *Esceleuuorde* 1086, *Ess- Assel(l)esworthe -wrd -w(u)rth -wrþa -(w)urðe* 12th–1269, *Asseleworþe* 12th, *Asselworth(e) -wurthe* c.1260–1310, *Assch- As(s)helworth(e) -il-* 1316–1713. OE pers.n. **Æscel*, genitive sing. **Æsceles*, + **worth**. Gl iii.152.

ASHLEY 'Ash-tree wood or clearing', OE **æsc** + **lēah**, L 203.

(1) ~ Cambs TL 6961. *Esselei* 1086, *As(se)le(e) -leg -leie -leia -lye* 1228–1311, *As(s)h(e)le(e) -ley(e) -legh* 1260–1428. Ca 124.

(2) ~ Ches SJ 7784. *Ascelie* 1086, *As(s)heleg(h) -ley(e)* 1280–1346, *As(s)hley* from 1296. Che ii.10.

(3) ~ Devon SS 6411. *Esshe- Asselegh* 1238, *Aysshlegh* 1330. D 374.

(4) ~ Glos ST 9394. *Esselie* 1086, *-lega* 1196, 1198, *Asseleye* 1216×72, *Assheleye* 1281. Gl i.85.

(5) ~ Hants SU 3831. *Asselegh* 1275. Ha 25.

(6) ~ Northants SP 7991. *Ascele* 1086, *Estelai* 1109, *Estlaia* c.1160, *Esselega* 1195 etc. with variants *Asse- As(s)he* and *-le(e)*. Nth 155.

(7) ~ Staffs SJ 7636. *Esselie* 1086, *-legh* 1230, *Asscheleye* 1290, 1300. DEPN, Horovitz.

(8) ~ GREEN Bucks SP 9705. *Assheley grene* 1468. P.n. Ashley + ME **grene**. Cf. nearby LYE GREEN SP 9703. Ashley is *Esselega* 1193, *Esseleie* 1227, *Asscheleye* 1346. Nearby Nashleigh Farm and Hill preserve a variant of this name with prefixed *N-* from ME *atten ash-ley* 'at the ash-tree wood'. Bk 213, D 1, Jnl 2.31, L 203.

(9) ~ HEATH Dorset SU 1104. *Ashley Heath* 1811 OS. P.n. Ashley + ModE **heath**. Ashley 15 *Aisshele* 1280, *Asshele* 1317. Do 29, L 203, Ha 25.

(10) ~ HEATH Staffs SJ 7436. *Ashley Heath* 1833 OS. P.n. ASHLEY SJ 7636 + ModE **heath**.

East ASHLING WSusx SU 8207. *Estaslingge* 1287, *Estasshelyng* 1322. ME adj. **est** + p.n. *Estlinges* (for *Esc-*) 1185–93, *Eslynge* 1296, *Ashlyng* 1288, of uncertain origin, possibly representing OE folk-n. **Æscelingas* 'the people called after Æscel or Æscla' < pers.n. **Æscel*, **Æscla*, + **ingas**. The 12th cent. pl. sps may, however, stand for the two settlements of E and W Ashling. The name might therefore alternatively be OE **æsc** + **-ling** 'the ash-tree place, the ash wood'. And if the earliest sps are genuinely *Est-* and not misreadings of *Esc-* the original name would be the *Ēastlingas* 'the people who live to the E' sc. of Funtington. Sx 60, SS 65, MA 10.23.

West ASHLING WSusx SU 8107. *Westaslyng* 1288. ME **west** + p.n. Ashling as in East ASHLING SU 8207.

ASHMANSWORTH Hants SU 4157. 'Enclosure, secondary settlement of *Æscmere*, the ash-tree pool'. *æscmæreswierðe* [909]12th S 378, *æscmæres wyrðe, æscmeres weorþ* [934]12th S 427, *Esmeresworde* 1171, *Asshmansworthe* 1398. OE p.n. **Æscmere* < **æsc** + **mere**, genitive sing. *Æscmeres*, + WS **weorth, wyrth**. Forms with *-r-* survived until *Ashmersworth* 1657. *Æscmere* has been identified with a lost *escmere* [863]12th S 336 which comprised Linkenholt SU 3658 and Vernham Dean SU 3456 three miles to the W of Ashmansworth. Ha 25, Gover 1958.150.

ASHMANSWORTHY Devon SS 3317. 'Æscmund's enclosure'. *Essemvndehorde* 1086, *Asmundeswrth'* 1242, *Hasmundesworthy* 1249, *Ayshmundesworth(y)* 1316, *Ashmanesworth* 1292. OE pers.n. *Æscmund*, genitive sing. *Æscmundes*, + **worthiġ**. D 81.

ASHMORE Dorset ST 9117. 'Ash-tree pool'. *Aisemare* 1086, *Essemere* 1194–1289, *-mor(e)* 1247–50, *Assemere* 1235–1303, *Ashmere* 1314, *Asshemore* 1316, *Ashmore* 1618. OE **æsc**, genitive pl. **æsca**, + **mere** later replaced by **moor**. There is a pond at the centre of the village. Do ii.201, L 26.

River ASHOP Derby SK 1389. *acqua de Essop'* 1215, *Essope* 1250, *Eshop'* 1285, *Ashop River* 1627. Transferred from p.n. Ashop SK 1489, *Essop(e) -opp -hop(e)* 1199×1216–1285, *Eshop, Asshop(e)* 1229, 1302, 1362, *Ashop(e) -opp(e)* 1285, 1328 etc., 'ash-tree valley', OE **æsc** + **hop**. Db 3, 124, L 115–6.

ASHORNE Warw SP 3057. 'The ash-tree hill-spur'. *Hassorne* 1196, *Asshorne* 1279–1370, *Asshern* 1397–1401. OE **æsc** + **horn** in the sense 'a projecting headland, a horn-shaped hill'. Wa 257, PNE i.262.

ASHOVER Derby SK 3563. 'Ash-tree ridge'. *Essov(e)r(e) -u-* 1086–1358, *Essh-* late 12th–1552, *Assouera* 1154×9, *As(s)h(e)- Asschou(e)r(e) -ov(e)r(e)* 1299–16th. OE **æsc** + **ofer**. Db 190 gives pr [aʃə], L 175.

ASHOW Warw SP 3170. 'The ash-tree hill-spur'. *Asceshot* 1086, *Esses(c)ho* 1100×35–1291, *Assesho* 13thcent., *Ass(c)ho* 1315–1427. OE **æsc**, genitive sing. **æsces**, + **hōh**. Wa 154, L 168.

ASHPERTON H&W SO 6441. 'Ash-tree Perton'. *Sp(er)tvne* 1086, *Ashperton'* from 1224. OE **æsc** + OE **peretūn** 'the pear-orchard' used as a p.n. He 26.

ASHPRINGTON Devon SX 8157. Partly uncertain. *Aisbertone* 1086, *Asprinton(a)* 1088–1340, *A(y)springtonn(e)* 1143–1308, *A(y)sshprington(e) -yng-* 14th cent. Possibly OE pers.n. *Æscbeorht (Æscbriht)* or *Æscbeorn* + **ing**[4] + **tūn**, 'farm, village, estate called after Æscbeorht or Æscbeorn'. D 314.

ASHREIGNEY Devon SS 6213. 'Ash held by the Regny family'. *Esshereingni* 1238–1440 with variants *Rey(g)ny, Regni(e), Regnysasshe* 1381, *Ashreigney oth. Ringsaish* 1739. Earlier simply *Aissa* 1086, *Esse* 1219. OE **æsc** 'ash-tree' used as a p.n. + family name *Regny* (from Regny) first mentioned in connection with this place in 1219. D 355.

ASHRIDGE COLLEGE Herts SP 9912. P.n. Ashridge + ModE **college**. Ashridge, *Assherugge* c.1200–1540 with variants *-rigge -rudge*, is the 'ash-tree ridge', OE **æsc** + **hrycg**. Hrt 36.

ASHSTEAD Surrey TQ 1858. 'The stand of ash-trees'. *Stede* 1086, *Ested(e)* 1107×29–1291, *Axsted* 1202, *Ashsted* 1235, *Aysted* 1255, *Ashtede* 1270, *Ashted* 1605. OE **æsc** + **stede**. Sr 68, Stead 243–4, L 219.

ASHTON 'Ash-tree settlement, farm where ash-wood is obtained or worked'. OE **æsc** + **tūn**. As a timber-tree the ash was exceedingly valuable, both on account of the quickness of its growth and the toughness and elasticity of its wood which was useful for more purposes than the wood of other trees. It was employed for joists in buildings and for the manufacture of carts, wagons, axe-handles, ladders etc. PNE i.5, Grieve 65ff.

(1) ~ Ches SJ 5069. *Estone* 1086, *Ashton(e)* from 1289 with variants *Assh- Asch-*. Situated within the forest of Delamere. Che iii.268.

(2) ~ Corn SX 6028. *Ashton* 1867. A 19th cent. village and name. PNCo 51.

(3) ~ H&W SO 5164. *Estune* 1086, *Esscetun'* 1123, *Ayshton* 1308, *Ascheton* 1367, *Assheton* 1431. He 83.

(4) ~ Northants SP 7650. *Asheton* 1579. The modern form replaces earlier *Asce* 1086, *Essa, E(y)sse, Aisse* 1166–1284, *As(s)hen* 1296–1579, '(the settlement) at the ash-tree(s)', OE (**æt thǣm**) **æsce** varying with (**æt thǣm**) **æscum** (ME *atten ashen*), dative sing. and pl. respectively of **æsc** 'an ash-tree'. Nth 96.

(5) ~ Northants TL 0588. *Ascetone* 1086, *Ayston* 12th, *A(i)ston* 13th, *Ashton juxta Undele* (Oundle) 1299. Nth 210.

(6) ~ COMMON Wilts ST 8958. *Ashton Common* 1817 OS. P.n. Ashton as in Steeple ASHTON ST 9056 + ModE **common**.

(7) ~ FLAMVILLE Leic SP 4692. 'A held by the Flamville family'. P.n. *Eston(a)* 1190–1263, *Aston(')* from 1243 + suffix *Perer* 1243, *Flamvill(e)* from 1346. Hugo *de Pirar* held land in Ashton in 1243, Robert *de Flamvile* in 1247. Lei 480.

(8) ~ IN MAKERFIELD GMan SJ 5799. *Assheton in Makrefeld -er-* 1338, 1430. Earlier simply *Eston* 1212, *Aystone* 1246, *As(s)hton* 1255–1332. See further MAKERFIELD. La 100.

(9) ~ KEYNES Wilts SU 0494. 'A held by the Keynes family'. *Aysheton Keynes* 1572, *Aishen Kaynes* 1691. P.n. *Æsctun* [873×88]11th S 1507, *Essitone* 1086, *Aston* 1256, *Asshton* 1306, + manorial addition from the family of William de Keynes who held the manor in 1256. W 40 gives pr [keinz].

(10) ~ UNDER HILL H&W SO 9938. *Ashton subtus montem* 1570, ~ *Underhill* 1686. Earlier simply *Æsctun* [991]11th S 1365, *Es(s)etone* 1086, *E- Aschetona* 13th. The bounds of the people of Ashton are referred to as *Æschæma gemæru* [1042]18th S 1396. The village lies beneath Bredon Hill. Gl ii.42.

(11) ~ -UNDER-LYNE GMan SJ 9399. *Asshton under Lyme* 1305, *Assheton* ~ ~ 1355, *Asshton under Lyne* 1319. Earlier simply *Ashton* 1277. See further Lyme in LYME PARK SJ 9682. La 29, Jnl 17.30.

(12) ~ UPON MERSEY GMan SJ 7892. *Assheton super Mercy* 1421, *Ashton super Merseybanke* 1584. Earlier simply *Ayston* 1260, *Ashton* from c.1284 with variants *Assh- As(s)che- Ashe-*. Che ii.3.

(13) Cold ~ Avon ST 7572. *Cold(e) Aston* 1287–1575, ~ *As(s)h- Assheton* 1310–1767. Earlier simply *Æsctun* 931 S 416, *(æt) Æcstune* [955×7]12th S 664, *Escetone* 1086, *Ais(ch)ton* 1221, 1303. 'Cold' from its exposed position. Gl iii.62.

(14) Long ~ Avon ST 5470. ModE **long** + p.n. *Estvne* 1086, *Ayston* 1256. The village straggles along the main road for nearly two miles. DEPN.

(15) Steeple ~ Wilts ST 9056. 'A with the (church) steeple'. *Stepelaston* 1268, 1289, *Stupel A(y)shton* 1289. ME **stepel** + p.n. *(to) Æystone, (at) Aystone* [964]17th S 726, *Aistone* 1086, *Aysshton* 1333. 'Steeple' for distinction from nearby West ASHTON ST 8755. W 136

(16) West ~ Wilts ST 8755. *Westaston* 1248, 1257, *West Asshton* 1305. ME adj. **west** + p.n. Ashton as in Steeple ASHTON ST 9056. Wlt 136.

ASHTON Devon SX 8584. 'Æschere's farm, village or estate'. *Aisers- Essestone* 1086, *Ass(h)er(e)ston(e)* 1244–1356, *Aysherston* 1295, *Aysscherstone* 1317, *Asscheston* 1374. OE pers.n. *Æschere*, genitive sing. *Æscheres*, + **tūn**. D 487.

ASHURST Hants SU 3310. 'Ash-tree wooded hill'. *Assh(e)hurst* 1331. ME **ash(e)** (OE *æsc*, genitive pl. *æsca*) + **hurst** (OE *hyrst*). Ha 26, Gover 1958.194.

ASHURST Kent TQ 5038. 'Ash-tree wooded hill'. *Aeischerste* c.1100, *Eisherst'* 1220×1, *Asshehurst* 1261×2. OE **æsc** + **hyrst**, Kentish **herst**. PNK 184, L 197.

ASHURST WSusx TQ 1716. 'The ash-tree wooded hill'. *Essehurst* 1164, *Eshurst* 1296, *Aschurst* 1283, *Ayshurst* 1288. OE **æsc** + **hyrst**. The sp *Asshetehurst* 1279 shows a variant form with OE **æscett** 'an ash-tree copse'. Sx 183, SS 67, DEPN.

ASHURSTWOOD ESusx TQ 4236. *Ashurst Wood* 1819 OS, earlier *foresta de Esseherst* 1164, *boscum de Aisherst* 1189. P.n. *Escheherst* 1279, 'the wooded hill growing with ash-trees', OE **æsc** + **hyrst**, + ModE **wood**, Latin **foresta**, **boscus**. Sx 327, SS 68.

ASHWATER Devon SX 3895. 'Ash held by Walter'. *Esse Valteri* 1270, *-fitzwalteri* 1302, *Asshefitz Wauter* 1327, *Aswalter* 1285, *Aysshewater* 1385. Earlier simply *Aissa* 1086. OE **æsc** 'an ash-tree' + pers.n. *Walter*, possibly referring to *Walter* son of Ralph who was the tenant in 1165 or to *Walter* de Doneheved who held the manor in 1270. Cf. BRIDGWATER Somer ST 3037. D 127, DB 3.4.

ASHWELL Leic SK 8613 'Ash-tree spring'. *Exewelle* 1086, *Essewell(e)* 1202–74, *Asse-* 1209–1407, *Assh(e)well(e)* c.1291–1535, *Ashwell* 1610. OE **æsc**, genitive pl. **æsca**, + **wella**. R 4.

ASHWELL Herts TL 2639. 'Ash-tree spring'. *Æscewelle* c.1060, *A- Esceuuele* 1086. OE **æsc**, genitive pl. **æsca**, + **wella**. Hrt 153.

ASHWELLTHORPE Norf TM 1497. 'Outlying settlement attached to Ashwell'. *Aissewellethorp* 1254. P.n. *Aescewelle* c.1066, + p.n. *Thorp* c.1066. Ashwell is OE 'ash-tree spring', **æsc**, genitive pl. **æsca**, + **wella** and Thorp OE or ON **thorp**. DEPN.

ASHWICK Somer ST 6348. '(At) the ash-tree farm-dwellings'. *Æscwica* [1061]12th S 1034, *(æt) Æscwican* [1061×66]12th S 1427, *Escewiche* 1086, *Aswyk* c.1350, *Asshewyk(e)* 1413 Buck, *Ashwick* 1610. OE **æsc** + **wīċ**, dative pl. **wīcum**. DEPN.

ASHWICKEN Norf TF 7019. 'Ash-tree Wicken'. *Askiwiken* 1275, *Asse Wykin* 1302. ME **ash** (OE *æsc*) + p.n. *Wiche* 1086, *Wyken* 1254, 'at the dwellings', OE **wīcum**, dative pl. of **wīc**. DEPN.

ASHWORTH MOOR (RESERVOIR) GMan SD 8215. P.n. Ashworth SD 8413 + ModE **moor**. Ashworth, *Esworde* c.1200, *Assewrthe* 1236, *Hesseworthe* c.1260, *Asheworth* 1347, *Ash'orth* mid 19th, is 'ash-tree enclosure', OE **æsc** + **worth**. La 54, Jnl 17.42.

ASKERN SYorks SE 5613. Either the 'group of ash-trees' or the 'house near the ash-tree(s)'. *Askern(e)* c.1170–1428. OE ***æscern** or **æsc** + **ærn** with ON **askr** replacing OE **æsc**. YW ii.44, Kristensson 1970.59.

ASKERSWELL Dorset SY 5292. 'Osgar's spring'. *Oscherwille* 1086, *Oskereswell* 1201, *Askereswell* 1208, *Askerswill* 1346. OE pers.n. *Ōsgār*, genitive sing. *Ōsgāres*, + **wylle**. The later form of the name has been influenced by ON pers.n. *Ásgeirr* which occasionally appears as *Asker* in ME, cf. f.n. *Askereswong* c.1230–40 in Norfolk. Do 30, DEPN, SPNN 43.

ASKETT Bucks SP 8105. 'East cottage(s)'. *Astcote* c.1250, *Ascote* 1300, *Ascot(t)* 1541–c.1825. OE **ēast** + **cot**, pl. **cotu** (ME *cote*). Identical with ASCOT Berks SU 9168. Bk 171.

ASKHAM '(At) the ash-trees'. OE **æsc**, dative pl. **æscum** or ON **askr**, dative pl. **askum**.

(1) ~ Cumbr NY 5123. *Asc- Askum* 1232–1476, *Ascumbe* c.1235, *-combe* 1429, 1551, *Asc- Askom* 1291–1513, *Askham* from 1577. We ii.200, SSNNW 101.

(2) ~ IN FURNESS Cumbr SD 2177. *Askeham* 1535. The evidence is too late for certainty and earlier *Ask-holm* is also possible. 'Furness' from FURNESS SD 2087 was added for distinction from ASKHAM NY 5123. La 205.

ASKHAM 'Ash-tree homestead'. OE **æsc** replaced by ON **askr**, + **hām**.

(1) ~ Notts SK 7474. *Ascā* 1086, 1275, *Askam* 1289. 1545, *Ascham* 1167, 1276, 1317, *Askham* from 1267. Nt 44, SSNEM 212.

(2) ~ BRYAN NYorks SE 5548. 'Bryan's A'. *Askham Bryan* 1371. Earlier simply *Ascha', Ascam* 1086, *Askham* 1195–1569, *Ascam, Askam* 1231–1614. 'Bryan', a pers.n. used traditionally in the family which held Askham, e.g. *Brianus filius Alani* 1195, is for distinction from ASKHAM RICHARD SE 5348. YW iv.233.

(3) ~ RICHARD NYorks SE 5348. 'Richard's A'. *Askham Richard* 1371. Earlier simply *Ascha', Ascam* 1086, *Ask(h)am* c.1180–1684, as in ASKHAM BRYAN SE 5548. 'Richard', for distinction from Askham Bryan, refers to either *Richard Malebys de Askham* 1284 or Richard Earl of Cornwall who had great estates in Yorkshire in the 13th cent. YW iv.234.

ASKRIGG NYorks SD 9491. 'The ash-tree ridge'. *Ascric* 1086, *Askerigg* 1285 etc. ON **askr** probably replacing OE **æsc** + OE ***ric** later replaced by ON **hryggr**. YN 261, L 185, SSNY 152.

ASKRIGG COMMON NYorks SD 9493. P.n. ASKRIGG SD 9491 + ModE **common**.

ASKWITH NYorks SE 1748. 'The ash-tree wood'. *Ascvid* 1086, *Asc- Askwid* 1180×1208, 1354, *Askewith(e) -wyth* 1276–1546. ON **askr** + **vithr** possibly replacing OE **æsc** + **widu**. YW v.61, L 222, SSNY 90.

ASLACKBY Lincs TF 0830. 'Aslakr's village or farm'. *Aslachebi* 1086–1185, *Asla(c)k(e)by -bi* c.1160–1553, *Aslagby* 1327–1410, *Aslaby(e)* 16th cent. ON pers.n. *Áslákr* + **bȳ**. Perrott 95, SSNEM 32.

ASLACTON Norf TM 1591. 'Aslakr's estate'. *O- Aslactuna* 1086, *Aselaketon* 1101×7, *Aslacton* 1208. ON p.n. *Áslákr* + **tūn**. DEPN, SPPN 62–3.

River ASLAND Lancs SD 4524. 'Slow-moving ash-tree river'. *Asklone* [1195×1217]1268, 1362, *Askelon(e)* 1195×1217–1292, *Ascalon* 1223, *Ast(e)land* 1550, 1555, *Oslande* 1590. ON **askr**, possibly replacing OE **æsc**, + **lanu, lane** in the same sense as Scots dial. *lane* 'a brook where movement is scarcely perceptible, the smooth, slow-moving part of a river'. Asland is the name of the lower reaches of the river DOUGLAS Lancs SD 4812. RN 19.

ASLOCKTON Notts SK 7440. 'Aslakr's farm or village'. *(H)aslachestone* 1086, *Aslachetune, -ton* 1242, 1324, *Aslocton* 1291×1300. ON pers.n. *Áslákr* + **tūn**. Nt 219 records a former pr [aslətən], SSNEM 187.

ASPALL Suff TM 1765. 'The aspen-tree nook'. *A- Espala, Aspalle* 1086, *Asphal'* 1208, *Aspehale* 1254, *Aspal(e)* 1286–1346, *Aspall* 1524. OE **æspe** + **halh**. DEPN, Baron.

ASPATRIA Cumbr NY 1441. 'Patric's ash-tree'. *Aspatric -k(e)* c.1160–1736, *Askpatri(c)k* 13th cent., 14th cent., *As(s)patry* 15th cent., *Aspatria* 1734. ON **askr** + Goidelic pers.n. *Patric*. A compound p.n. in Celtic word order. The second element has been assimilated in modern times to Latin *patria*. Cu 262.

ASPENDEN Herts TL 3528. 'Aspen-tree valley'. *Absesdene* 1086, *Ab- Apseden(e)* 1185–1280, *Aspeden(e)* 1222–1638, *Aspyden* 1525, *Apsten* 1572. OE **æspe, æpse** + **denu**. The *en* form is a late development, *Aspynden* 1506, *Aspenden* 1698. Hrt 171 gives prs [æpsdən, a:spidən].

ASPLEY 'Aspen-tree wood or clearing'. OE **æpse, æspe** + **lēah**, dative sing. **lēa** or **lēaġe**. PNE i.4, 5, L 203.

(1) ~ GUISE Beds SP 9436. 'A held by the Guise family'. *Aspeleye Gyse* 1363. Earlier simply *(to) æpslea* [969]11th S 772, *Aspeleia* 1086. Anselm de Gyse whose family came from Guise in France held the manor in 1276. The boundary of the Aspley people is referred to as *on æpsleainga gemære* in the same charter. Bd 113.

(2) ~ HEATH Beds SP 9334. P.n. ASPLEY SP 9436 + ModE **heath**.

ASPULL GMan SD 6108. 'Aspen-tree hill'. *Aspul* 1212, *Aspull* 1262, *Asp(e)hull* 1292–1421, *Aspell* 1301. OE **æsp** + **hyll**. La 102, L 171, Jnl 17.58.

ASSELBY Humbs SE 7128. 'Askell's farmstead'. *Aschilebi* 1086, *Askelby* 1282–1379, *Astelby* c.1360, *Hasselby* 1546. ON pers.n. *Áskell* + **bȳ**. YE 248, SSNY 18.

Lower ASSENDON Oxon SU 7484. ModE **lower** + p.n. *Assundene* [late 10th]11th S 104, *Assendene* c.1240, *Asindene, Assingden* 1285, 'the ass's valley', OE **assa**, genitive sing. **assan**, + **denu**. 'Lower' for distinction from Middle ASSENDON SU 7485 and Upper Assendon Farm. O 85 suggests pers.n. *Assa* but this is not on record. L 98.

Middle ASSENDON Oxon SU 7485. ModE **middle** + p.n. Assendon as in Lower ASSENDON SU 7484.

ASSINGTON Suff TL 9338. 'Assa's estate'. *Asetona* 1086, *Asinton* 1173–4, *Assintona* 1175, *Essinton* 1219, *Assington* 1610. OE pers.n. **Assa*, genitive sing. **Assan*, + **tūn**. DEPN.

ASTBURY Ches SJ 8461. 'East fortified place'. *Astbury* from 1093 with variants *Aste-* and *buri(e) -biri -beri* etc., *Asbury* 1544. OE **ēast** + **byriġ**, dative sing. of **burh**. E of the lost *Newbold*. There is a rectangular earthwork at Glebe Farm adjoining the churchyard. Che ii.286 gives pr ['æsbəri].

ASTCOTE Northants SP 6753. 'Æfic's cottage(s)'. *Aviescote* 1086, *Auichescote* 1248, *Auescot(e)* 1300–1591, *Ascott* 1542, *Ascoat -cote* 18th. OE pers.n. *Æfic*, genitive sing. *Æfices*, + **cot**. Nth 92.

ASTERLEY Shrops SJ 3707. 'The eastern wood or clearing'. *Estrelega* 1208, *Asterlege -legh* from 1252. OE **ēasterra** + **lēah**. 'East' for distinction from WESTLEY a mile to the W at SJ

3607 and other members of a cluster of *lēah* names here. Sa ii.30.

ASTERTON Shrops SO 3991. 'The east farmstead'. *Esthampton* 1255, *Astham(p)ton* 1274,1272×1307, *Asterton* 1827. OE **ēast** + **hām-tūn**. Bowcock 29.

ASTHALL Oxon SP 2811. 'The east nook'. *(æt) Eást Heolon* early 11th, *Esthale* 1086–1354 with variants *-hal(l)a -hall(e) -halles -al(e)* - and *-all(e), Asthall(')* 1246–1428. OE **ēast** + **h(e)alh**. The earliest form (if it belongs here) is a characteristic late OE spelling of the dative plural form *healum* 'at the nooks'. O 299.

ASTLEY 'The east wood or clearing'. OE **ēast** + **lēah**. L 206.

(1) ~ H&W SO 7867. *Æstlæh* 11th, *Eslei* 1086, *Estlege* c.1150. OE **ēast** + **lēah**. 'East' in relation to ABBERLEY SO 7567. Wo 33.

(2) ~ Shrops SJ 5318. *Hesleie* 1086, *Estleg'* 1208×9–55, *Astle(yg') -leg(h)' -leye* from 1261×2. 'East' in relation to Albrighton SJ 4918. Sa i.20.

(3) ~ Warw SP 3189. *Estleia* 1086, *-lega -le(y) -legh* 1189–1316, *Astlege* from 1242 with variants *-ley(e)* and *- le(gh), Asley* 1627. E of Arley SP 2890 and Fillongley SP 2887. Wa 96.

(4) ~ ABBOTS Shrops SO 7096. 'A belonging to the abbot' sc. of Shrewsbury. *Astleye Abbatis* late 13th, ~ *Abbots* 1304. P.n. *Estleia* c.1090, *Estleg(h)'* c. 1198–1275, *Astleg' -ley(e) -legh* from 1225 + genitive singular *abbots* corresponding to Latin *abbatis*, genitive sing. of *abbas* 'an abbot'. Usually explained as 'East' in relation to Morville SO 6794, the centre of a large estate at the time of DB, but cf. NORDLEY SO 6996. Sa i.21.

(5) ~ CROSS H&W SO 8069. P.n. ASTELY SO 7867 + ModE **cross** 'a cross-roads'.

(6) ~ GREEN GMan SJ 7099. P.n. Astley + ModE **green**. Astley is *Asteleg(h)* c.1210–1332, *Asteley(e)* 1216×72–1311, *Astley* 1479. Situated E of LEIGH SD 6500, its original sense was probably 'east part of Leigh'. La 101, Jnl 17.57.

(7) ~ HALL Lancs SD 5718. No early forms. The reference is to a 16th cent. timber-framed house still extant within later façades. Pevsner 1969.96.

ASTON 'East settlement'. OE **ēast** + **tūn**. PNE i.144, ii.193.

(1) ~ Berks SU 7884. *Estun'* 1220, *-ton'* 1247×8, *Aston* 1761. E of Remenham. Brk 67.

(2) ~ Ches SJ 5578. *Estone* 1086, *-ton* 1178–1397, *Aston* from 1190. Situated E of the river Weaver in the SE corner of Runcorn parish. Contrasts with NORTON SJ 5582, SUTTON WEAVER SJ 5479 and WESTON SJ 5080. Che ii.160.

(3) ~ Ches SJ 6146. *Estune* 1086, *Aston* from 1252. E of Wrenbury SJ 5947 in which parish it lies. Che iii.102.

(4) ~ Derby SK 1884. *Estune* 1086, *-ton(')* 13th cent., *Aston* from 1285. E in relation to HOPE SK 1783. Db 610.

(5) ~ Herts TL 2722. *Eastune* 11th, *Eston(e)* 1086–1362, *Aston(e)* from 1296, *Asson* 1611. Hrt 117 gives common pr [a:sən].

(6) ~ Oxon SP 3403. *Eastun* [984]13th S 853, *Esttun* 1069, *Eston(') -tona -tun(e)* c.1130×42–15th., *Aston* from 1328. A reference to this place occurs as *(on) east hæme gemære* 'the boundary of the Aston people' [958]13th S 678. O 302.

(7) ~ Shrops SJ 5328. *Estune* 1086, *Aston(')* from 1271×2. East of Wem SJ 5129. Sa i.23.

(8) ~ Shrops SJ 6109. *Eastun* [975]12th S 802, *Eston(a)* c.1144–1249, *Aston(')* from 1231×40. Sa i.23.

(9) ~ Staffs SJ 7541. *Aston* 1298, *Aston Meyreway* 1609 referring to MAER SJ 7938.

(10) ~ WMids SP 0889. *Estone* 1086–1235, *Aston* from 1275. E of the former Staffs-Warw county boundary. Wa 28.

(11) ~ WYorks SK 4685. *Eston(e) -tona -tun* 1086–1248, *Aston'* from 12th, *Ashton* 1658. YW i.158.

(12) ~ ABBOTTS Bucks SP 8420. 'A held by the abbot' sc. of St Alban's. *Aston Abbatis* 1262. P.n. Aston + Latin **abbas**, genitive sing. **abbatis**. Earlier simply *Estone* 1086–1237×40, *Aston* from 1242. Bk 76.

(13) ~ BOTTERELL Shrops SO 6384. 'A held by the Boterell family'. *Eston Boterel(l)* 1263, 1272, *Astone Boterell* 1255. P.n. *Estone* 1086, *Eston'* 1186×90–1272, *Aston'* 1242 + manorial addition from William Boterell early 13th. E of Great Clee Hill SO 5985. Sa i.23.

(14) ~ -BY-STONE Staffs SJ 9131. *Aston near Stanes* 1276. P.n. *Eastun* [957]12th S 574†, *Estone* 1086, *Estona* 1166, + p.n. STONE SJ 9033. Also known as *Little Aston* 1266. DEPN.

(15) ~ CANTLOW Warw SP 1460. 'A held by the Cantilupe family'. *Aston Cantilup'* 1232, ~ *Cantelou* 1273. P.n. *Estone* 1086–1251, *Haston* 1254, + manorial addition from the family of John de Cantilupe who held the estate in 1205. E of Alcester SP 0857. Wa 196.

(16) ~ CLINTON Bucks SP 8712. 'A held by the Clinton family'. *Aston Clinton* 1237×40, *Arston Clinton* 1702. Earlier simply *Estone* 1086–1414, *Aston* from 1247, *Asen* 1675. A William de Clinton was connected with this place in 1196. But some forms suggest there was a lost p.n. Clinton, *Eston et Clynton* 1244. 'East' in contrast to WESTON TURVILLE SP 8511 for which a form *Weston et Clynton* is also recorded in 1262. Bk 142, Jnl 2.24.

(17) ~ CREWS H&W SO 6723. 'Aston held by the Cruze family'. *Aston Cruze* 1831 OS. P.n. Aston as in ASTON INGHAM + family name Cruze.

(18) ~ END Herts TL 2724. 'The (N) end of A'. *Aston Ende* 1592. P.n. ASTON TL 2722 + ModE **end**. Hrt 118.

(19) ~ EYRE Shrops SO 6594. 'A held by the Fitz Aer family'. *Aston Aer* 1284×5, ~ *Eyres* 1293, ~ *Ayer* 14th cent. P.n. *Estone* 1086, *Eston'* 1242 + manorial addition from Robert grandson of Aer in 1212. Surname *A(y)er* usually derives from ME, OFr *eir* 'an heir' which was occasionally used as a pers.n. DBS 17 cites *Robertus filius Aier* 1166 from Shrops. There is no connection with the name of the DB tenant *Alcher* (OE *Alhhere*). Also known as *Whetene Aston* 1255, 'Aston growing with wheat', OE **hwǣten**. Sa i.24.

(20) ~ FIELDS H&W SO 9669. *Aston Fields* 1649. Earlier simply *Eastun* [770]11th S 60, *Estone* 1086–1227, *Astone End* 1391. Wo 359.

(21) ~ INGHAM H&W SO 6823. 'A held by the Ingen family'. *Estun' Ingan* 1243. Earlier simply *Estvne* 1086, *Estone* 1256, *Astone* 1300. The manor was held by Richard Ingan in 1212, but the affix probably refers to his ancestor Hingan (OBret *Hingant, Hincant*) who was alive in 1127. Aston Ingham is the *ēast tūn* contrasting with the *west tūn* at WESTON under Penyard SO 6232. He 27.

(22) ~ JUXTA MONDRUM Ches SJ 6556. *Aston subtus Mondrem* 1290, *-iuxta-* from 1347, 'Aston by or within the forest of *Mondrem*'. Earlier simply *Estone* 1086, *Aston(a) -tun* from 1276. E of Worleston SJ 6556. Mondrem, *Mondreym* 13th, *Mondrem(e)* from 1284, is a lost name meaning 'pleasure-ground', OE **man-drēam**. It was a hunting preserve administered with Delamere Forest. Che i.10.

(23) ~ LE WALLS Northants SP 4950. *Assheton in le Walles* 1509, *Aston in the Wall* 1530, *Aston super muras* (sic) 1621. P.n. *Eston(e)* 1086–1224 + Fr definite article **le** short for *en le* + **walls**. Also known as *Aston juxta Wardon* c.1200 etc. Situated SE of Upper and Lower Boddington SP 4853, 4852. The Walls refers to local earthworks. Nth 32.

(24) ~ MAGNA Glos SP 2035. 'Great Aston'. *Aston Magna* 1626. Earlier simply *Easttune* 904 S 1281, *Eastune* [977]11th S 1333, *Estona* 1208, 1327. P.n. Aston + Latin **magna** for distinction from ASTON SUBEDGE SP 1341. Also known as *Hanging Aston* 1549, *Hangynde Aston* 1282, OE **hangende** 'steep' from its position on a steep hill. Situated E in the parish of Blockley. Gl i.234.

(25) ~ MUNSLOW Shrops SO 5186. 'A in (the parish of) Munslow. *Aston by Munslowe* 1343. P.n. *Estune* 1086, *Easton(a)* 1167–1242, *Aston* from 1255, + p.n. MUNSLOW SO 5287. East in relation to Eaton-under-Haywood SO 5090. Sa i.25.

†Identification not certain.

(26) ~ ON CLUN Shrops SO 3981. P.n. *Aston* 1833 OS + p.n. CLUN.

(27) ~ -ON-TRENT Derby SK 4129. *Aston(e) super Trente -am* 1269 etc., *As(s)on uppon Trent* 1657. P.n. Aston + r.n. TRENT. Earlier simply *Æstun, Estune* 1086, *Estuna -tona -ton(e)* late 11th–1257, *Aston(e)* from early 13th, *East Towne alias Aston* 1571. E in relation to WESTON-ON-TRENT Derby SK 4028. Db 428.

(28) ~ ROGERS Shrops SJ 3406. 'A held by Roger'. *Astone Roger* 1327, earlier simply *Aston* 1242. Held by Roger de Aston in 1242. DEPN.

(29) ~ ROWANT Oxon SU 7298. 'A held by Roald'. *Aston Roaud* 1318 etc. with variants *Rohaud, Ruaut, Ru- Rohant, Ro(uw)and*. Earlier simply *Estone* 1086. The manor was held by Rowald de Eston in 1235×6 whose name is the Romance reflex of W Frankish *Chrôdoald, (H)rôd(w)ald*, a name especially common in Brittany. E of Lewknor SU 7197. Cf. South WESTON SU 7098. O 102.

(30) ~ SANDFORD Bucks SP 7507. 'A held by the Sanford family'. *Aston' Sanford* 1242. Earlier simply *Estone* 1086, 1199. 'East' in relation to Haddenham SP 7408 which had the same lord in 1086. Bk 114.

(31) ~ SOMERVILLE H&W SP 0438. 'A held by the Somerville family'. *Aston Someruill' -vi- -vyl(l)e* from 1287, *Sumerfield Aston* 1706. Earlier simply *(ad) Eastune* [930]12th S 404, [1002]12th S 901, *Estvne* 1086, *Eston'* 1220–85. The family of Somerville held the manor from 1291. The bounds of the people of Aston are referred to as *(into) esthemmere* *[706]12th, i.e. **Ēasthǣme*, the people of *Ēast(tūn)'*, genitive pl. **Ēasthǣ ma*, + **(ġe)mǣru**. Gl ii.3.

(32) ~ SUBEDGE Glos SP 1341. 'Aston below the edge' sc. of the Cotswolds. *Aston sub Egg(e)* 1284–1535. Earlier simply *Estvne* 1086, *Eston'* 1221–74, *Aston* 1273. Contrasts with WESTON SUBEDGE SP 1240. The form *easthammore* *[709]12th S 80, a forged charter, is a corruption of OE *ēasthǣma (ġe)mǣre* 'boundary of the men of Aston'. Gl i.232 gives pr ['astən 'subidʒ].

(33) ~ TIRROLD Oxon SU 5586. 'A held by Turold de Estune' c.1150. *Aston Torald* 1376–7, *~ Tirrold* from 1709, *~ Tirrel* 1830. Earlier simply *Estone -ton' -tun* 1086–1260, *Aston* from 1317. 'Tirrold' for distinction from the other part of the village, ASTON UPTHORPE SU 5586. E of Blewbury SU 5385. Brk 510, 512.

(34) ~ UPTHORPE Oxon SU 5586. 'The upper part of A'. *Astone Upthroope* 1346, *~ Upthorpe* from 1469. Earlier simply *Upthrop* 1316 'the higher village', OE **upp** + **throp**. This part of Aston (as distinct from the other part of the village, ASTON TIRROLD SU 5586) is *(æt) Eastune* [964]13th S 725, *Estone* 1086. Brk 511.

(35) Chetwynd ~ Shrops SJ 7517. 'A held by the Chetwynd family'. *Chetwynde Aston* 1619. Earlier simply *Estona* 1155, *Eston'* 1251. Also known as *Aston Magna* 1255–1606, *Middle, Much or Great Aston* 1490 for distinction from Church or Little Aston SJ 7417, with which it probably one time constituted a single unit. East in relation to Edgmond SJ 7219. Sa i.25.

(36) Church ~ Shrops SJ 7417. *Chirche Aston* c.1300, *Churche Aston'* 1323. ME **chirche** + p.n. *Eastun* [963]12th S 723, *Aston* from 1252. Also known as *Parua Aston'* 1294×9–1422, 'little Aston' for distinction from Chetwynd or Great Aston SJ 7517 with which it was probably once a single unit. Sa i.26.

(37) Coal ~ Derby SK 3679. 'Cold A'. *Cold(e)aston* 13th, 1474–1500, *Coleaston* 16th. ME adj. **cold** + p.n. *Estune* 1086, *Aston* from late 12th. E in relation to Dronfield. 'Coal' for *Cold* refers to its exposed situation. Db 198.

(38) Cold ~ Glos SP 1319. *Cold(e)aston(e)* 1184×94–1675. Earlier simply *(æt) Eastunæ* [737×40]11th S 99, *(in) Easttune* 1033×8 S 1399, *Eston(e) -a* 1086–1209, *Aston(e)* 1184×95, 1285 etc. The situation is a cold exposed one at nearly 700ft. in the E of Bradley Hundred. Gl i.164.

(39) East ~ Hants SU 4345 → MIDDLETON Hants SU 4244.

(40) Ivinghoe ~ Bucks SP 9518. *Ivyngho Aston* 1491. P.n. IVINGHOE SP 9416 + p.n. *Estone* 1086, *Aston* 1387. Bk 97.

(41) Little ~ Staffs SK 0900. *Little Aston upon (le) Colefeld* 13th, *Little Aston* 1834 OS. Duignan 8.

(42) Middle ~ Oxon SP 4726. *Middel Aston* 1428. ME adj. **midel** + p.n. *Estone* 1086. The spellings *Midelestun'* 1220, *Middeleston'* 1242×3 and *Mid(d)el- Middalaston'* 1274×4–1366 are ambiguous as between 'the middlest *tūn*' and 'middle Aston'. Between North Aston SP 4728 and Steeple Aston 4725. O 245.

(43) North ~ Oxon SP 4728. *Nort' Eston* 1200, *North Aston'* 1246×7. ME **north** + p.n. *Estone* 1086 etc. North of Middle ASTON SP 4276. O 246.

(44) Steeple ~ Oxon SP 4725. 'A with the tower or steeple'. *Stipelestun'* 1220, *Stepelaston'* 1278×9. ME **stepel** + p.n. *Estone -tona* 1086–c.1260. 'Steeple' for distinction from North and Middle ASTON SP 4728 and 4726. O 245.

(45) Wheaton ~ Staffs SJ 8512. 'A where wheat is grown'. *Wetenaston* 1248, *Whetenaston* 1347–1571, *Wheaton Aston* 1598. ME **wheten** (OE *hwǣten*) + p.n. *Eston(a)* 1167, 1280, *Aston* 1271–1637. 'East' in relation to WESTON-UNDER-LIZARD SJ 8010, MARSTON SJ 8314 'the marsh settlement', MITTON SJ 8815 'the confluence settlement', STRETTON SJ 8811 'the Roman road settlement' and WHISTON SJ 8914 'Witi's estate'. St i.169, DEPN.

(46) White Ladies ~ H&W SO 9252. *Whitlady(e)aston* 1481, 1577. The *white ladies* were the Cistercian nuns of Whitestones who held a share of the manor from the bishop of Worcester. Earlier simply *Estvn* 1086 and *Aston Episcopi* 'the bishop's A' 1247–1318†. Wo 88.

ASTON '(Place) east in the estate'. OE **ēast in tune**. PNE i.144. ~ H&W SO 4672. *Estintun* 13th, *Aston(e)* from 1334. East in the township of Kingsland SO 4461‡. He 113.

Upper ASTROP Northants SP 5137. ModE adj. **upper** + p.n. *Estrop* 1200–69, *-torp* 1201, *Astrop* from 1269, *Astethorp* 1316, 'the eastern outlying farm', OE **ēast** + ON or OE **thorp**. Situated E of Kings Sutton SP 4936. Nth 58, SSNEM 125.

ASTWICK Beds TL 2138. 'East farm'. *Estuuiche* 1086, *Astwyk* 1316. OE **ēast** + **wīċ**. Bd 100.

ASTWOOD Bucks SP 9547. 'East wood'. *Estwode* 1151×4, 1341, *Astwode* 1242. OE **ēast** + **wudu**. Bk 30 gives pr [æstəd], L 228.

ASTWOOD BANK H&W SP 0462. *Astwood Bank* 1831 OS. P.n. Astwood + ModE **bank**. Astwood as in Astwood Court SP 1362, is *Estwude* 1244, *Astwode* 1259, the 'east wood', ME **est** (OE *ēast*) + **wode** (OE *wudu*), also *Strecches Astwode* 1319 from the 13th cent. owner Richard Strecche. It lies in the E part of Feckenham parish. Wo 318.

ASWARBY Lincs TF 0639. 'Aswarth's village or farm'. *Wardebi, Asuuardebi* 1086, *Aswardeby* 1201–1381, *Aswardbi* 1212, *-by* 1298–1576, *Aswreby* 1231, *Aserbye* 1610. ODan pers.n. *Aswarth* + **bȳ**. Perrott 50, SSNEM 33.

ASWARDBY Lincs TF 3770. 'Aswarth's village or farm'. *Asewrdeby* 1147×66, *Aswardebi -by* 1196–1212, *Asworbi* 1200. ODan pers.n. *Aswarth* + **bȳ**. SSNEM 80.

ATCHAM Shrops SJ 5409. Uncertain. *Atingeham* 1086, 1241, *-ynge-* 1315, *Ætinge-* 1194, *Hattinge-* 1291×2, *Et(t)inge-* 1149×59, 1195, *Ettingham* 1194–1443, *Etchingeham* 1195, *Attingham* 1203×4–1785, *Atyncham* 14th, *Acham* 1535–17th, *Atcham* from 1669. Possibly 'the homestead or village called *Æt(t)inġe*' in which *Æt(t)inġe* is 'the place called after Ætti or Eata', OE

†The form *æt Eastune* [977]11th S 1333 formerly identified with White Ladies Aston belongs to Aston Magna Glos.

‡DB *Hesintune* 1086 and *Herefordshire Domesday* (ed. V.H. Galbraith and I. Tait, Pipe Roll Soc. lxiii, 37) *Hesintone, id est Asciston* 1160×70 are identified with Aston, but these forms can only represent OE **Escingtūn* and *Æscestūn* 'the ash-tree estate'. *DB Herefordshire*, notes to 9.4.

pers.n. *Ætti* or *Ēata* + **ing**², locative-dative sing. **inġe**, + **hām**. However, as Sa i.27 says, everything about this name is open to dispute. The main points at issue are (i) whether the generic is **hām** or **hamm** 'land in a river-bend'; (ii) whether formations in **ing**² occur in Shropshire p.ns.; (iii) whether the pers.n. is that of St Eata, the patron saint of the parish church. On which it may be noted (i) that Atcham is situated in a bend of the r. Severn; (ii) that the form of the p.n. given by Ordericus Vitalis c.1127, who was born and christened at Atcham, *apud Etingesham*, seem to imply a genitive sing. *Etinges* of *Eting*; (iii) that elsewhere the evidence is that dedications arise from p.ns., e.g. Boston Lincs (St Botolph), Warburton Ches (St Werburg). Perhaps a quite different solution is needed, viz. '*hām* or *hamm* at the eating or grazing place', OE **et(t)ing*, **æt(t)ing*, cf. ETTINGSGHALL WMid SO 9396. Sa i.26, ING 149, BzN 1967.361.

ATHELINGTON Suff TM 2071. 'The prince or princes' estate'. *Elyngtone* [942×c.951]14th S 1526, *Alinggeton* 1219, *Athelinton* 1234, *-ing-*, *-yng-* 154, 1286, *Ayllingtone* 1327, *Alyngton(') -ing* 1336, 1524, 1568, *Allington* 1610. OE **ætheling** (genitive pl. **æthelinga**) + **tūn**. DEPN, Baron.

ATHELNEY Somer ST 3428. 'The island of the princes'. *(æt) Æþelinga eigge* 9th ASC(A) under year 878, *(æt) Æðelinga ige* 12th ASC(E) under same year, *Æthelingaeg* [c.894]11th Asser, *Adelingi* 1086, *Athelingenye* [1285]B, *Anthony* (sic) 1610. OE **ætheling**, genitive pl. **æthelinga**, + **īeġ**. DEPN, Gelling 1998.93.

ATHERFIELD POINT IoW SZ 4579. *Atherfield Point* 1769. P.n. Atherfield + ModE **point**. Atherfield, *(apud) Aderingefeldam* [959]c.1300, *Avre- Egrafel* 1086, *Aers- Ayrfeld* 1247–8, *Ath(e)refeld -er-* 1205–1533, *Ari- Ar(r)efeld* 1253–5, *Adderfield* 1744, is the 'open land of the Æthelheringas, the people called after Æthelhere', OE folk-n. **Æthelheringas* < pers.n. *Æthelhere* + **ingas**, genitive pl. **Æthelheringa*, + **feld**. The early spellings *Eg-* and *Ay-* are AN forms for OE *Æthel-*; DB *Avre-* shows AN [v] for [ð]. Wt 217 gives prs ['æðəvil] and ['æð-'aθərfild], Mills 1996.24.

Little ATHERFIELD IoW SZ 4680. *Lit. Atherfield* 1810 OS. ModE **little** + p.n. Atherfield as in ATHERFIELD POINT SZ 4579. 'Little' for distinction from Atherfield Green SZ 4679, *Atherfield Green* 1810 OS.

ATHERINGTON Devon SS 5923. 'Farm, village, estate called after Eadhere or Æthelhere'. *Hadrintone* 1272, *Adringtone* 1311, *Addrington* 1675, *Atheryngton* from 1298, *-ing-* from 1332. OE pers.n. *Ēadhere* or *Æthelhere* + **ing**⁴ + **tūn**. D 357 gives pr [æðriŋtən].

ATHERSTONE Warw SP 3198. 'Æthelred's estate'. *Aderestone* 1086–1262, *Atheredestone* 1221, *Atherestone* 1275–1402, *Atherston* 1343–1609. OE pers.n. *Æthelrēd*, genitive sing. *Æthelrēdes*, + **tūn**. Wa 77.

ATHERSTONE ON STOUR Warw SP 2051. P.n. **Eadrichestone* *[710]12th S 81 (a forgery of c.1097×1104), 12th, *-rikes-* 12th, *Eathericestum* (sic) 1070×7, *Edricestone* 1086, *Addricheston(e)* [c.1086]1190, *Aed-* 1187, *Edrichestone* 1193–1279, *Athereston* 1249–1497, *Atherston* 1318, *Adderson* 1634, 'Eadric's estate', OE pers.n. *Ēadrīċ*, genitive sing. *Ēadrīċes*, + **tūn**, + r.n. STOUR SP 2248. Wa 248.

ATHERTON GMan SD 6703. 'Eadhere's farm or village'. *Aderton* 1212, 1243, *Atherton* from 1322, with variant *-ir-*. OE pers.n. *Ēadhere* + **tūn**. As with ARDWICK SJ 8597 it is impossible to be absolutely sure of the form of the OE pers.n. Æthelrēd and Ēadrēd have also been suggested but Ēadhere seems most likely. La 102.

ATLOW Derby SK 2348. 'Eata's burial-mound'. *Etelawe* 1086, *Attelawe* late 12th–1330, *-loue -lo(u)we* 1215–1547, *At(t)low(e)* from 1382. OE pers.n. *Ēata* + **hlāw**. Atlow village lies in a valley, but no burial-mound is known. Possibly the original reference was to the hill-top at SK 2448 which rises to 814ft. Db 522.

ATTENBOROUGH Notts SK 5134. 'Adda's fortified place'. *Adinburcha* [1154×89]1316, *-burge* 1280, 1289, *Adingburg(h)* [1154×89] 1316, 1229–1428, *Adding-* 1291, 1330, *Attenborow(e)* 1617, 1679, *-borough* 1637. OE pers.n. *Adda* + **ing**⁴ + **burh**. Nt 142.

ATTINGHAM Shrops SJ 5509. A doublet of Atcham SJ 5409 without assibilation or contraction retained for the mansion of Attingham Hall built in 1783–5. The old village of Atcham was largely destroyed by the owners of the hall when the grounds were extended and landscaped in the 18th cent. Raven 18–20.

ATTLEBOROUGH Norf TM 0495. 'Ætla's fortified place'. *Atleburc* 1086, 1194. OE pers.n. *Ætla* + **burh**. DEPN.

ATTLEBOROUGH Warw SP 3790. 'Ætla's hill'. *Atreberga* c.1150, *Atteleberga* 1155×9, *At(t)lebergh(e)* 1247–1317, *Attelberewe* 1310, *Artlebury al. Attleborowe* 1604. OE pers.n. *Ætla* + **beorg**. Situated on sloping ground at the end of a hill. Wa 89, L 128.

ATTLEBRIDGE Norf TG 1316. 'Ætla's bridge'. *Atlebruge* 1086, *-brigge* 1175. OE pers.n. *Ætla* + **brycg**. DEPN.

ATWICK Humbs TA 1850. 'Farm, trading place called after Atta'. *Attingwik(e) -wyk(e) -wic(k)* 1114×24–1592, *Attingewic* 12th, *Attewik'* 1246, *Attwyk* 1491. OE pers.n. *Atta* + **ing**⁴ + **wīc**. YE 79 gives pr [atik], BzN 1968.159.

ATWORTH Wilts ST 8665. 'Ætta's enclosure'. *Attenwrðe* [1001]15th S 899, *Attewurth -worth* 1249–1369, *-warde* 1354–1475, *Ateforde* 1427, *Atford* 1480, 1615, *Afford al. Atworth* 1637. OE pers.n. *Ætta*, genitive sing. *Ættan*, + **worth**. Wlt 115 gives local pr [ætfərd].

AUBOURN Lincs SK 9262. Possibly the 'eel stream'. *Abvrne* 1086, *-burne -burn'* 1160–94ff., *Auburn* 1212–54, *Alburn* 1275. ON **ál** replacing OE **ēl** + **burna**. DEPN, PNE i.3, Cameron 1998.

AUCHOPE CAIRN Northum NT 8919. P.n. Auchope Borders NT 8521 + ModE **cairn**. A boundary marker on the border between England and Scotland. *Aucopswire* 1597 is a point on one of the thieves' roads between England and Scotland, dial. *swire* 'a neck of land, a col, a hollow on the top of a hill or ridge' BB 116. The generic of Auchope is dial. **hope** 'a small enclosed valley' (OE *hop*).

Bishop AUCKLAND Durham NZ 2029. 'Auckland held by the bishop' sc. of Durham. *Aukeland Episcopi* 1306, *Biss- Bysshopaukland* 1358–1411, *Aukland Bishop* 1420. Auckland is the 'cliff, hill on the r. Clyde'. Two Aucklands are recorded from the beginning corresponding to Bishop and West Auckland. The forms for Auckland are as follows:

I. *Alclit* [c.1040]12th, c.1104, *Alclet* 1147, c.1149×52, [1183]1382.
II. *alklint* c.1180–c.1195, *Aclent* 1213, 1239, *Auclent* 1242×3, 1252, *Auclend* 1296.
III. *Aucland* 1259–1358, *Auckland* from 1284.
IV. *Acklande* 1571, *Aickland* 1621.

PrW ***alt** + r.n. **Clūt*, identical with the Scottish r.n. Clyde, *Clota* [c.97]9th Tacitus, *Κλῶτα* *(Clota)* [c.150]13th Ptol, [c.300]8th AI, *Cled* [c.700]13th Rav, < Brit **Clouta* 'the washer, the cleanser, the strongly flowing one'. *Clota* is thought to have been the original name of the GAUNLESS NZ 1024 which is a Scand r.n. and cannot have been given before the 9th cent. The PrW name was three times remodelled, once under the influence of ODan **klint** 'a rocky cliff, a steep bank overlooking a river' (group II spellings), then, consequent upon the vocalisation of pre-consonantal *l*, under the influence of ON **auka-land** 'additional land taken into cultivation' (group III spellings), and finally, in local speech, consequent upon lME monophthongisation of *au* > *ā*, under the influence of N dial **aik** 'an oak-tree' (group IV spellings). NbDu 7, TC 53, RBrit 309.

St Helen AUCKLAND Durham NZ 1927. *Auclent sancte Elene* 1242×3, *Seint Elin Auckeland* 1303. Saint's name *Helen* from the 12th cent. church dedicated to St Helena, the mother of the emperor Constantine + p.n. Auckland as in Bishop AUCKLAND NZ 2029.

West AUCKLAND Durham NZ 1726. *Westaukland* 1357. ME adj. **west** + p.n. Auckland as in Bishop AUCKLAND NZ 2029.

AUCKLEY SYorks SE 6501. 'Alca or Alha's clearing'. *Alchelie, Alc(h)eslei* 1086, *Alkeley -lay* 1280–1590, *Aukelay* 1537, *Auckley* from 1592. OE pers.n. *Alca* or *Alha* + **lēah**. The regular appearance of medial *-e-* precludes derivation from OE **alh** 'a temple'. YW i.44.

AUDENSHAW GMan SJ 9297. 'Aldwine's copse'. *Aldenshade, Aldenesawe, Aldwynshawe -shay* [c.1200]17th, *Aldenshagh* [c.1250]17th, *Aldewainestath* 1246, *Aldewynshagh* 1422, *Awdinshawe* 1599. OE pers.n. *Aldwine* + **sceaga**. La 29, L 209, Jnl 17.30. For the *shay* form see Nomina 13.109–14.

AUDLEM Ches SJ 6644. Either 'old *Lyme*, territory formerly in *The Lyme*' or 'Alda's part of *The Lyme*'. *Aldelime* 1086–1527 with variants *-lim(a) -lym(e)*, *Aldelem* 1388–1724, *Audelym(e)* 1281–1301, *-lem* 1527–1687, *Audlem* from 1575, *Awlem* 1558, 1619. Either OE adj. **ald**, definite form **alda**, or pers.n. *Alda* + p.n. *Lyme* as in LYME PARK SJ 9682. Che iii.82 gives pr ['ɔ:dləm], locally ['ɔ:ləm].

AUDLEY Staffs SJ 7950. 'Aldgyth's wood or clearing'. *Aldidelege* 1086, *Aldithelega* 1182, *Audeyeleg* 1272. OE feminine pers.n. *Aldgȳth*, genitive sing. *Aldgȳthe*, + **lēah**. DEPN, Horovitz.

AUDLEY END Essex TL 5237. 'Audley end' sc. of Saffron Walden. *Audleyend* 1555, *Audley Inn* 1662. Family name *Audley* + ModE **end**. The reference is to the Jacobean mansion built on the lands of Walden abbey which were granted by Henry VIII to Lord Chancellor Audley in 1538. The name contrasts with SEWARDS END TL 5738. Ess 538, Pevsner 1965.61.

AUDLEY END STATION Essex TL 5136. P.n. AUDLEY END TL 5237 + ModE **station**.

AUGHTON 'Oak-tree settlement', OE **āc** + **tūn**. L 218. Cf. the p.n. type ACTON.

(1) ~ Humbs SE 7038. *Actun* 1086, *-ton'* 1180×1200–1285, *A(u)chton* 1266, *Aghton* 1288–1493, *Aughton* from 1542. YE 237.

(2) ~ Lancs SD 3905. *Achetun* 1086, *Actun* [before 1250]1268, *Acton* 1235, *Aghton* 1282–1332 etc., *Aughton* from 1499. La 121.

(3) ~ Lancs SD 5567. *Aghton* 1320×46, 1458. La 179 gives pr [aftn].

(4) ~ PARK Lancs SD 4006. P.n. AUGHTON SD 3905 + ModE **park**.

(5) ~ SYorks SK 4586. *(H)actone, Hacstone* 1086, *Acton(a)* 12th–1320, *Aghton* 1324–1604, *Aughton* from 1532. OE **āc** + **tūn**. YW i.159.

AULDEN H&W SO 4654. 'Ælfgyth's hill'. *Elvitheduna* 1158×64, *Aldon* 1547×53, *Alden* 1832 OS. OE feminine pers.n. *Ælfgȳth*, genitive sing. *Ælfgȳthe*, + **dūn**. The hamlet occupies one end of a mile-long ridge. He 123.

AUNSBY Lincs TF 0438. 'Aun's farm or village'. *Ounesbi* 1086, 1202, 1219, *-by* 1238–1428, *Ouneby* 1202–1346, *Ounsby* 1528, *Awnsby* 1535, *Outhenby* 1281. ON pers.n. *Aun* < ON *Authun*, ODan *Øthen*, genitive sing. *Auns*, + **bȳ**. SSNEM suggests alternatively that the specific could be ON ***authn** 'a deserted site', genitive sing. ***authns**, and that the reference might be to Roman remains in the area. DEPN, Perrott 52, SSNEM 33, Cameron 1998.

AUST Avon ST 5789. Unexplained. *Ætaustin* [691×9]17th S 77, *(æt, to) Austan* [794]11th S 137, *[929]11th S 401, *Augusta* c.1105, *Augst* 1375, *Aust(e)* from 1208×13. It is impossible to know whether ME *Auste* is derived from *Augusta* or whether *Augusta* is a Latinization of *Auste*. *Augusta* might be connected with the Roman Second Legion, the *Legio Augusta*, which was stationed at Gloucester until moved across the Severn to Caerleon in AD 75. The suggestion that the name derives from a Latin *Trajectus Augusti* 'passage of Augustus', a crossing of the lower Severn from Aust to Caerwent named after the same legion, is pure speculation since no such Latin name is on record and the precise location of the *Trajectus* of AI xiv is uncertain. Nor can association with Augustine's oak, *Augustinaes Ác* 731 BHE, *Augustinus aac* c.1000 OEB, be firmly established. Its site on the borders of the Hwicce and the West Saxons is not incompatible with Aust, but whether the phrase *Augustīnes āc* could have been reduced to *Auste* by the 11th cent. is very doubtful. If the OE form was *Austa* masculine or *Auste* feminine it is not of Germanic origin and for the moment must remain unexplained. Gl iii.127 gives pr [ɔ:st], RBrit 177–8.

AUSTERFIELD SYorks SK 6694. 'Open land with a sheepfold'. *Ouestraefelda, Eostrefeld* [c.715]11th ES, *Oustrefeld* 1086, *-er-* 1276, *Austerfeld(e)* 1328–1594. OE **eowestre** + **feld**. This was the venue of a synod in 702–3. YW i.46, L 238, 244.

AUSTREY Warw SK 2906. 'Ealdwulf's tree'. *(æt) Alduluestreow* [958]13th S 576, *Aldulfestreo* [1002]c.1100, *Aldulvestreu* 1086, *-tre* 1114–1288 with variants *Aldolves-* and *-tro*, *Aldestre* 1291–1545, *Austrey alias Alstrey* 1540. OE pers.n. *Ealdwulf*, genitive sing. *Ealdwulfes*, + **trēow**. Wa 13, L 212, 216.

AUSTWICK NYorks SD 7668. 'The east dairy-farm'. *Ousteuuic* 1086, *Aust(e)wic -wyc -wik(e) -wyk(e)* c.1150–1672, *Astick* 1729. ON **austr** possibly replacing OE **ēast**, + **wīc**. Austwick lies E of Clapham SD 7469. YW vi.228.

AUTHORPE 'Agi's outlying farm'. ODan pers.n. *Aghi*, genitive sing. *Agha*, + **thorp**.

(1) ~ Lincs TF 4080. *Agetorp* 1086, 1187, *Haghetorp* c.1115, *Aggetorp* late 12th, *Aghtorp* before 1219. DEPN, SSNEM 102, Cameron 1998.

(2) ~ ROW Lincs TF 5373. *Authorpe Row* 1824 OS. P.n. *Aghetorp* c.1115, *Agthorp* 1230–39, *Aggethorp'* 1272, 1282, + ModE **row**. SSNEM 102, Cameron 1998.

AVEBURY Wilts SU 0969. Possibly 'Afa's fortification'. *Aureberie* (sic) 1086, *Avesbiria -beria -byry* 1114–1256, *Aveberia -biri -biry -bery -bury* c.1180–1332, *Avenesbur'* 1255, *Avenebyr'* 1268, *Abury* 1386, *Abery* 1535, *Aubury* 1494, 1670, 1773, *Awbery alias Avebury* 1689. OE pers.n. *Afa*, genitive sing. *Afan*, + **byriġ**, dative sing. of **burh**. The reference is to the massive neolithic bank and ditch which surrounds the village and forms the most impressive prehistoric monument in Wilts. The *Aure-* and *Avene(s)-* sps are difficult and might point to the r.n. *Afon* as the specific, a supposed earlier name of the Winterbourne that flows through the village. Vocalisation of *-v-* produced ME *au-* which survives in 15th–18th cent. *Au- Aw-* sps and the former pr [ɔ:bəri]. Side by side with this, ME *au* also gave lME *ā* whence the 14th and 15th cent. forms in *A-* and the pr [eibəri]. The pr. [eivbəri] is probably a sp. pr. Wlt 293, Pevsner 1975.96, Thomas 1976.215.

AVELEY Essex TQ 5680. 'Ælfgyth's wood or clearing'. *Aluitheleam, Aluielea, Auileiam* 1086, *Alvith(e)le(e) -ley* 1202–1377, *Alviveleia -ley* 1157, 1310, *(H)awelay, Aulay* 1480, *Aweley* 1524, *Aveley* 1535. OE feminine pers.n. *Æflgȳth*, genitive sing. *Ælfgȳthe*, + **lēah**. Some curious Frenchified spellings *Auvil(l)ers -iers, Avil(l)ers* occur 1205–86, cf. the form *Auvillers* 1266 for ALVELEY Shrops SO 7584. Ess 120 gives pr [eivli].

AVENING Glos ST 8898. 'The people dwelling on the r. Avon'. *(to) Æfeningum* [896]11th S 1441 (dative pl.), *Au- Aveninge -ynge* 1086–1587, *Hav- Au- Avelinges -is* 12th cent., *Au- Aveninges* 1165×77–1297, *Avening* 1291. R.n. AVON + **ingas**. The stream running through the village is unnamed but presumed to have been Avon from PrW ***aβon** 'river'. Alternatively the *Afeningas* may have been a group of people who migrated here from one of the other river Avons. Gl i.86 gives pr ['eivniŋ].

AVERHAM Notts SK 7654. Possibly '(the settlement) at the floods'. *Aigrun* 1086, *Ægrum* c.1180, *Egrum* c.1200–1428, *Agheram* 1277, *Aghram* 1302, 1327, *Aram* 1280–1558, *Averham* from 1274. OE **ēagor**, Anglian **ēgor**, dative pl. **ēgrum**. The reference is to the river Trent Bore or *eagre*. Forms with <v> show substitution of the common fricative [v] for the peripheral phoneme [γ]. Nt 181 gives pr [ɛ:rəm].

AVETON GIFFORD Devon SX 6947. 'Aveton held by the Gifford family'. *Aveton(e) Gifford* 1276, ~ *Jefford* 1441, *Awton Gifford* 1546, 1562. Earlier simply *Avetone* 1086, 13th cent., 'settlement by the r. Avon'. R.n. AVON SX 7157 + OE **tūn**. The manor was held in 1242 by Walter *Gifford*. D 265 gives pr [ɔ:tən dʒifəd].

AVINGTON Berks SU 3768. 'Estate called after Afa'. *Avintone* 1086, *Avinton(a) -yn-* 1167–1284, *Avington* from 1294. OE pers.n. *Afa* + **ing**[4] + **tūn**. Brk 313.

AVON DAM RESERVOIR Devon SX 6765. R.n. AVON SX 7175 + ModE **dam** + **reservoir**.

River AVON. PrW ***aβon** 'river' < Brit **abona*. The base of this formation is IE **ab-* 'water, river' + suffix *-onā* as in *Devona*, the river Don in Aberdeenshire. The same root occurs in the Baltic r.ns. Abava, Abula, Abuls, the British Ἄβος (*Abus*) for the Yorks Ouse and its estuary, the Humber, the German Albe, a tributary of the Saar, and Abens, a tributary of the Danube, and is the basis of Latin *amnis* 'river' < **ab-nis*. RBrit 239, Krahe 1964.41, 63, Jnl 1.43.

(1) ~ The county name is taken from the r.n. AVON.

(2) ~ Avon ST 6966. *Abone* [4th]8th AI, *Aben* [793×6]c.800 S 139, *(in) Afene (stream)* [883]11th S 218, *(be) Āfne* 9th ASC(A) under year 652, *Au- Avene -a* 1236–1434, *Avon* from 1634. Gl i.3.

(3) ~ Devon SX 7157. *(to, upon) Afene* 847 S 289, *afne* 926, *Auene* 1238–81, *Awne, Aune, Avon* c.1540, *Owne al. Aven* 1608. D 1, RN 21.

(4) ~ Dorset SZ 1496, Hants SZ 1495, Wilts SU 1349. *(on) Afene* [892]14th S 348, [943]14th S 492, *be æfene stæþe* 'beside the Avon landing place' [892]14th S 348, *Afenan* (genitive case) [934]12th S 427, *Avene* 13th cent. Wlt 1, RN 20 where a full range of forms is given.

(5) ~ Hants SZ 1498. The settlement name, *Avere* 1086, *Auene* 1212–1397, is transferred from r. AVON SZ 1495. Gover 1958.229, RN 20, DEPN.

(6) ~ H&W SO 9238, Warw SP 2658 *afen* [699×709]11th S 64, *afene* *[780]11th S 118, *(inon, on, of) afene* [883×911]11th S 222, 972 S 786, *(neah þære éa) Afene* c.1025 Saints, *auena -am* Latin *[709]12th S 80, [714]16th S 1250, *[716]12th S 83, *(innan, on, onlong) auene* *[964]12th S 731, [996×1002]12th S 873, [n.d.]12th S 1590, [n.d.]12th S 1599, 13th cent., *Eafene* [845]11th S 198, *Avon* from c.1540. RN 21, Wa 2, Hooke *passim*.

(7) ~ CASTLE Dorset SU 1303. No early forms. The castle was built for the earl of Egmont in 1874–5. The place is named from r. AVON SZ 1496. Pevsner–Lloyd 1967.475.

AVONMOUTH Avon ST 5718. A modern port at the mouth of the river AVON begun in 1877. The mouth of the Avon is mentioned in the 10th and 11th cents., *Afenemuþan* ASC(A) under year 918, ASC(D) under years 915 and 1052. Room 1983.5.

AVONWICK Devon SX 7158. A modern name coined about 1878 from the r.n. AVON SX 7157. Earlier called *Newhouse*. D 303.

AWBRIDGE Hants SU 3325. 'Abbot ridge'. *Abedric* 1086, *Abberugge* 1207, *Abbederugge -rigge* 1218, 1238, *Abboterigge* 1256. OE **abbod** + **hrycg**. The genitive form *abbodes* might have been expected; possibly the original contained genitive pl. *abboda* 'of the abbots', sc. of St Peter's abbey, Winchester. Awbridge is on a hill-top and there is no bridge. Ha 26, Gover 1958.181.

AWKLEY Avon ST 5985. Uncertain. *Awke- Auckley* 1628–1723, *Awklers* 1631. This could be 'Alca's wood or clearing', OE pers.n. *Alca* + **lēah**, but the evidence is too late for certainty. Gl iii.120.

AWLISCOMBE Devon ST 1301. 'Valley of the fork'. *Aules- (H)ores- Holescome, Holescūbe* 1086, *Aulescoma- cumb(e)* 13th–1535, *Awelescumb'* 1221×30, *Oulis- Oulescomb(e)* 1303–56, *Alscombe* 1541, *Aulscombe alias Alliscombe* 1706. OE **āwel** 'awl, hook, fork' referring to a triple fork of the river one mile N of the village + **cumb**. D 608.

AWRE Glos SO 7008. 'Sour water-meadow'. *Avre* 1086, *Aur(e)* 1150–1595, *Awre* from 1437, *Oure, Overe* 13th, *Auerey* 1461, 1466. OE **āfor** 'bitter, sour' + **ēa, ēġ** 'water-meadow, island'. *Awre* stands in a large low-lying bend of the Severn. Gl iii.250 gives pr ['ɔ:əR].

AWSWORTH Notts SK 4844. 'Eald's enclosure'. *(æt) Ealdeswy(rðe)* [1002]c.1100, [1004]13th S 906, *Eldesvorde, Eldevrde* 1086, *Aldesworth(e)* 1280–1445, *Allesworth* 1475, 1541, *Alsworth* 1639, *Awsworth* 1703. OE pers.n. *Eald*, genitive sing. *Ealdes*, + **worth**. Nt 137.

AXBRIDGE Somer ST 4354. 'Bridge over the r. Axe'. *(to) axanbrycge* [914×9]16th BH1, *(to) oxenebrege -b'igge* [914×9]13th BH 2, 3, *Axebruge* 1084, 1168, *-brige* [12th]B, *-bridge* 1610, *Alsebrvge* (for *Aise-*) 1086, *Aucsebriges Hd* 1084, *Auxebrigge* [1172]1500. R.n. AXE ST 4647 + OE **brycg**. DEPN, RN 153, BH 112.

AXE EDGE Derby SK 0370. Partly obscure. *Axeedge* 1533. The generic is ModE **edge** 'a steep escarpment of a ridge'. Db 372.

River AXE Devon ST 3100. *(on,of) Axan* [1005]12th S 910, *Axe* from 1244. An OE r.n. **Æsce* and with metathesis **Æcse, Æxe* from Brit **Escā* ultimately < **pit-skā* (with PrCeltic loss of initial *p*) on the IE root **pid-* 'gush'. Identical with the r.ns. ESK Cumbr NY 3289, SD 1297, NYorks NZ 7407 and EXE SS 9409, 9882. D 2, RN 152.

River AXE Somer ST 4647. *Axam* (Latin accusative case) [663 for ?693]14th S 238, *Aesce* [712]14th S 1253, *Axan* (genitive sing.) [956]12th S 606, *(on) Axen* [973]14th S 793, *(on) Axa* [1068]15th, *Axe* from 1243. This is usually taken to represent an OE r.n. **Æsce* and with metathesis **Æxe*, **Eaxe* from Brit **Escā* as in r.n. ESK < **Iscā* identical with OIr *esc* 'water' < IE **pid-skā* on the root **pi-* seen in Gk πιδύειν 'to gush forth'. But certain forms for AXBRIDGE with spellings *Axan- Oxene- Aucse-* and *Auxe-* suggest an OE r.n. *Axe* perhaps to be related to the RBrit unidentified p.n. *Axium*. See further r.ns. EXE, ESK, EDEN. DEPN, RN 152, LHEB 259, 281, FT 822, Archaeologia 93, RBrit 261, HB 112.

AXFORD Hants SU 6143. 'Ash-tree ridge'. *Ashore* 1272, *Axore* 1280–1598, *Axford* 1757. OE **æsc** + **ōra**. For metathesis of *sc* > *cs* (*x*) cf. AXFORD Wilts SU 2370. Ha 26, Gover 1958.133, L 179.

AXFORD Wilts SU 2370. 'The ash-tree ford'. *Axeford* 1184–1428, *Axford* 1255, *Assheford* 1289. OE **æsc**, genitive pl. **æsca**, + **ford**. For a parallel instance of metathesis of *sc* to *cs (x)* see AXFORD Hants SU 6143. Wlt 288.

Isle of AXHOLME Humbs SE 7806. 'Haxey island'. *Haxeholm* c.1115, *Haxiholm(a)* c.1150, 1233, *Axiholm* 1179, 1233. P.n. HAXEY SK 7699 + **holmr**. Haxey is a village in Axholme and may have been the original name of the island to which pleonastic ON *holmr* was added. At the time of DB Axholme was completely surrounded by marsh and alluvium. DEPN, SSNEM 151.

AXMINSTER Devon SY 2998. 'The minster by the river Axe'. *Ascanmynster* c.900 ASC(A) under year 755, *Axanmynster* 1120 ASC(E), *Alseminstre* 1086, *Aixe- Aexe- Alseministra* 1086 Exon, *Axeministre* 1212. R.n. AXE ST 3100 + OE **mynster**. D 633.

AXMOUTH Devon SY 2591. 'The mouth of the river Axe'. *(æt) Axanmuðan* [878×888]c.1000 S 1507, *Axamuða* c.1120 ASC(E) under year 1046, *Alsemvde* 1086, *Axemuth* 1249, *-mu(w)e* 13th. R.n. AXE + OE **mūtha**. The village is a mile from the sea, but the estuary has silted up. D 636.

AYCLIFFE Durham NZ 2822. There are three types for this name:

I. *(at, æt) Aclea* early 12th ASC(E) under years 782 and 789, c.1123–1214, *Acle* [1091×2]early 12th–1441, *Ak- Acley* c.1220–1483, *Akeley* 14th cent.

II. *Heaclif* 1109, 1154×66.

III. *Aclyff(e)* 1361–1565, *Aykliffe* 1587, *Aicle alias Aicliffe* 1632, *Aycliffe* from 1717.

I is clearly 'oak wood or clearing', OE **āc-lēah**; the wood of Aycliffe is mentioned in 13th and 14th cent. records and the adjoining township is WOODHAM NZ 2826. II is probably 'the high cliff', OE **hēa**, definite declensional form of OE **hēah**, + **clif** rather than 'high Aycliffe' as has been suggested. III, 'oak bank', ME **aik** (OE *āc*) + **clif**, is probably not a continuation of II but an example of hypercorrection applied to *Acley* in view of the widespread development of original suffix *-clif* to *-cley* in northern names, e.g. SHINCLIFFE NZ 2940. NbDu 8 gives pr [jakli] corresponding to the spelling *Yakely* 1680, with N [ja] for ME *ā*, Orton 355–66.

AYLBURTON Glos SO 6202. 'Æthelbriht's estate'. *Ailbrichton'* 1186, *Ail- Ayl(e)bryghton -igh-* 1270–1546, *Ail- Aylbrit(t)on(e) -bretone* 1248–1355, *Ail(e)- Aylburton(e)* 1361, 1592. OE pers.n. *Æthelbriht -beorht* + **tūn**. Gl iii.255.

AYLE Northum NY 7149. Originally a river name identical with river ALN NU 1314. *Alne* 1258, *Ale* [1485×1509]17th AA 1912.270, *Allan water* c.1675, *Aln- Aleburn* 1777. NbDu 9, RN 5.

AYLESBEARE Devon SY 0391. 'Ægel's wood'. *Eilesberge* 1086, *Ailesberga* 1086 Exon, *Ayl- Ailesbere* 1227–74, *Ayllesbeare* 1316, 1385. OE pers.n. **Æġel*, genitive sing. **Æġeles*, + **bearu**. In the same parish is the lost n. Allen Wood, *Aylinewode* 1242, *Aylingewude* 1262, 'wood associated with Ægel', **Æġel* + **ing**[4] + **wudu**. D 580–1.

AYLESBURY Bucks SP 8113. 'Ægel's fortified place'. *Ægeles burg* 9th ASC(A) under year 571, *Ægeles byrig* 921 ibid., *Æglesbyrig* 1121 ASC(E) under year 571, *Ægelesbyrig* [c.968×71]12th S 1485, *Eilesberia* 1086, 1195, *Ailes-* 1176, 1189, *Alesbiry* 1257. OE pers.n. **Æġel*, genitive sing. **Æġeles*, + **byriġ**, dative sing. of **burh**. Bk 145.

AYLESBY Humbs TA 2007. 'Ali's farm or village'. *Alesbi* 1086–1202, *-by* 1221–1742, *Aylesby* from 1259. ON pers.n. *Áli*, genitive sing. *Ála*, later *Ales*, + **bȳ**. Cf. AILBY Lincs TF 4376. Li v.1, SSNEM 30.

AYLESCOTT Devon SS 6116 → NORTHCOTE MANOR SS 6218.

AYLESFORD Kent TQ 7359. 'Ægel's ford'. *(to) Æglesforda* c.959 S 1211, *Agelesford* [961]13th S 1212, *Elesford* 1086, *Aeilesford* c.1100[†]. OE pers.n. **Æġel*, genitive sing. **Æġ(e)les* + **ford**. PNK 145, KPN 286.

AYLESHAM Kent TR 2352. Uncertain. *Elis- Aylys- Eyles- Haylesham* 1367–1445, *Halesom* 1839 *TA*. This may be a genuinely ancient name although the evidence is late. If so, 'Ægel' or 'Hægel's *hamm* or *hām*', OE pers.n. **Æġel* or **Hæġel*, genitive sing. **(H)æġ(e)les*, + **hamm** 2b 'promontory' or 5a 'cultivated plot in marginal land' or **hām** 'homestead, estate'. PNK 534, ASE 2.34, Cullen.

AYLESTONE Leic SK 5700. 'Ægel's farm or village'. *Ai- Ayleston(e) -is-* from 1086, *Eyleston' -is-* 1234–1447, *Elston* 1549–1641, *Elson* 1725. OE pers.n. **Æġel*, genitive sing. **Æġles*, + **tūn**. Lei 428, SSNEM 373.

AYLMERTON Norf TG 1839. 'Æthelmær's estate'. *Almartune* 1086, *Adelmerton* 1199, 1208. OE pers.n. *Æthelmǣr* + **tūn**. The loss of intervocalic *th* in the first element of this pers.n. is a regular feature, cf. DB *Ailbertus, Ailuerd, Aileua*, etc. for *Æthelbeorht, Æthefrith, Æthelġifu* etc. DEPN, PNDB.

AYLSHAM Norf TG 1926. 'Ægel's homestead'. *Eilessam* 1086, *Ailesham* 1086, 1159. OE pers.n. *Æġel*, genitive sing. *Æġels*, + **hām**. DEPN.

AYLTON H&W SO 6637. 'Æthelgifu's estate'. *Ailmeton(')* 1137×9, *A(y)- Ailmeton(')* 1251×2–1334, *Eilinetona* 1160×70, *Ayleuentun'* 1250, *A(y)lyneton* 1309, 1352, *Aeltone* 1397. OE feminine pers.n. *Æthelġifu*, genitive sing. *Æthelġife*, + **tūn**. This explanation depends on the assumption that the many spellings with *-m- -in-* and *-yn-* are due to minim confusion for *-iu- -yu-*. The form of 1137×9, however, suggests that the specific may have been ME *Ailmer* from OE *Æthelmǣr*. The estate is simply called *Merchelai* 1086 as parcel of the royal manor of Much Marcle. He 29.

AYMESTREY H&W SO 4265. 'Æthelmod's tree'. *Elmodestreu* 1086, *Eilmundestro vel Bedmodestreu* 1160×70, *Aylmonedestres* c.1140, *Ailmondestre* 1291, 1353, *Ailmestre* 1419, 1428. OE pers.n. *Æthelmōd* later replaced by *Æthelmund*, genitive sing. *Æthelmōdes -mundes*, + **trēow**. The replacement is due to the rarity of the element *-mōd* compared with *-mund* and to scribal confusion due to the weakly stressed nature of the middle syllable. *Bed-* is a garbled form. He 30, DB Herefordshire notes on 1.10a, Otium 46–8.

AYNHO Northants SP 5133. 'Æga's hill-spur'. *Aienho* 1086, *Ai- Aynho* c.1185, 1243, *Ein- Eynho(o) -hou* c.1195–1375. OE pers.n. **Æġa*, genitive sing. **Æġan*, + **hōh**. Situated on a well-marked hill. Nth 48, L168.

AYOT ST LAWRENCE Herts TL 1016. 'St Lawrence's A'. *Ayete Lorencii* 1303, *Ayot Laurence* 1367. P.n. Ayot + saint's name *Lawrence* from the dedication of the church, *Eccl. Sci. Laurentii de Aete* 1291. Ayot, *Aiegete* c.1060, *Ægete* 1065, *Ægeatt* 1066×87, *Aðgiðe* [1053×66]13th S 1135, *Aiete* 1086, is 'Æga's gate or gap', OE pers.n. **Æġa* + **ġeat**. The reference is to the gap in the hills between here and AYOT ST PETER TL 2115. Hrt 118.

AYOT ST PETER Herts TL 2115. 'St Peter's A'. *Eyotte Sci Petri* 1535, *Ayott St Peters otherwise Little Ayott otherwise Ayott Montfitchet* 1770. P.n. Ayot as in AYOT ST LAWRENCE TL 1916 + saint's name *Peter* from the dedication of the church. The estate was given to William de Montfichet 1154×89, cf. *Aettemunfichet* 1255. Hrt 118.

AYSGARTH NYorks SE 0188. 'Gap, pass by the oak wood'. *Echescard* 1086, *Aykescart(h), Ai- Aykeskarth* 12th–1420, *Aykesgarth* 1374, 1388, *Ayskarth(e)* 1400–1574, *Asegarth* 1687. ON **eiki** + **skarth**. The reference is to the gap in the hills S of the river Ure forming Bishopdale. YN 262, SSNNY 90.

AYSIDE Cumbr SD 3983. Uncertain. *Aysshed* 1491, *Aysett* 1537, *Ayshead* 1573, 1592, *Aysyde* 1591. If a form *Aykeshead* 1279 belongs here, this is probably 'oak-tree shieling', ON **eik** + **sǽtr**; if not, 'river shieling', ON **á** + **sǽtr**. The generic has been confused with ME **hede** (OE *hēafod*) 'a head' and ModE **side**. La 199, SSNNW 61.

AYSTON Leic SK 8601. 'Athelstan's farm or village'. *Æðelstanestun, into Æðelstanes tune* [1046]12th S 1014, *Athestaneston* 1284, *Astaneston'* 1293, *-enes- -ones-* 1204–1300, *Atheston* 1275, *Aston(e)* 1269–1477, *Ayston* from 1535. OE pers.n. *Æthelstān*, genitive sing. *Æthelstānes*, + **tūn**. The estate was granted to Athelstan, his faithful *minister*, by Edward the Confessor in 1046. R 171.

AYTON 'River settlement'. ON **á** replacing OE **ēa**, + **tūn**.

(1) ~ NYorks SE 9985. *Atun(e)* 1086, *Aton(e) -(a)* c.1205–1385, *Ayton* 1555. Ayton lies on the river Derwent. YN 100 gives pr [jætən], SSNY 113.

(2) Great ~ NYorks NZ 5611. ModE **great** + p.n. *Atun(a)* 1086, 12th cent., *Aton(a)* 12th–1508, *Haiton* 1202. YN 165 gives prs [jætn, kæni jætn], the latter, Canny Ayton, referring to its pleasant situation.

(3) Little ~ NYorks NZ 5710. *Parva Hatona* c.1160. ModE **little**, Latin **parva**, + p.n. *Atun* 1086, as in Great AYTON NZ 5611. YN 166.

(4) West ~ NYorks SE 9884. ModE **west** + p.n. *(alia) Atune* 'the second A' 1086, *Aton(e -a)* [1200×10]15–1385, *Ayton* from 1555. YN 100, SSNY 113.

AZERLEY NYorks SE 2574. 'Atsurr's wood or clearing'. *Asserle(ia) -ley -lay* 1086–1588, *Aserla(gh) -lei -ley* 1086–1457, *Azerlagh -law(e) -lowe* 1198–1466, *Azerlay -ley(e) -le* 1276–1654. ON pers.n. *Atsurr* + OE **lēah**. YW v.199, SPN 36.

[†]The forms *Agœles þrep* ASC(A) and *Ægeles þrep* ASC(E) 'Ægel's farm', both under year 455, are often identified with Aylesford. They are, however, different names with OE generic **throp** and the identification is no more than a guess.

B

BABBACOMBE Devon SX 9365. 'Babba's valley'. *Babbecumbe* [c.1200]15th, *Babbacombe* from 1504×15. OE pers.n. *Babba* + **cumb**. D 519.

BABBACOMBE BAY Devon SX 9568. P.n. BABBACOMBE SX 9365 + ModE **bay**.

BABBINSWOOD Shrops SJ 3330. *Babyne woode* 1544×5, *Babbyes Woode alias Babbynes Wood* 1572, *Babies Wood* 1811. Earlier *foreste q'est apellee Babbyng* 'a forest called *Babbyng*' c.1320, 'the place called after Babba', OE pers.n. *Babba* + **ing**[2]. Gelling.

BABCARY Somer ST 5628 → River CARY.

BABENY Devon SX 6775. 'Babba's island'. *Babbeneye* 1260, *Babenny* 1303, *Beu- Bewbeney* 1588. OE pers.n. *Babba*, genitive sing. *Babban*, + **īeġ**. D 191.

BABWORTH Notts SK 6880. 'Babba's enclosure'. *Baburde* 1086, *-w(u)rd -wrthe -worth* c.1190–1343 etc., *Babbeuurde -wrde -w(o)rth -word* c.1190–1316. OE pers.n. *Babba* + **worth**. Nt 66.

BACH CAMP H&W SO 5460. An Iron Age hill-fort. P.n. Bach as in Upper Bach SO 5461 + ModE **camp**. Upper Bach is *Overbach* 1558, earlier simply *(La) Bache* 1269×90, 1431, OE **bæċe** 'a stream, a valley'. 'Upper' for distinction from Lower Bach SO 5360 and *Coubache* late 13th, 'cow-valley'. The same name,*Cov- Koubache*, is given to the place in Clent SO 9279 where the martyred saint-king Kenelm was secretly buried by his murderer. It is described as a deep valley between high hills named after a cow which came to lie there by the hidden body and was miraculously fed. As a result the saint's resting place was eventually revealed. The legend is as old as the 12th cent. *Vita* of St Kenelm, which preserves an alternative couplet which may be even older:

In Clent Cubeche Kenelm cunebearn
lith under haȝeþorn haudes bereafed

'In Clent Cowbach (lost, Worcestershire) Prince Kenelm lies under hawthorn beheaded'. He 111, EMEVP.96.

BACKBARROW Cumbr SD 3584. 'Hill with a backlike top'. *Bakbarowe* 1537, cf. *Bak(e)barayfell* 1538. OE **bæc** or ON **bak** + OE **beorg** or ON **berg**. Cf. Leland c.1540: 'ther was a coppe in the hille as a bakke stonding up aboue the residue of the hill'. The reference is to the hill called Old Backbarrow SD 3685. La 198.

BACKFORD Ches SJ 3971. 'Ridge-ford'. *Bacfort* 1150, *Backford* from 1186. OE **bæc** + **ford**. A ford across Backford Brook to Lea; the village stands on a 100ft. hill overlooking the ford. Che iv.172, L 69,125.

BACKWELL Avon ST 4865. 'Ridge spring'. *Bacoile* 1086, *Bacwell* 1202–1241. OE **bæc** + **wella**. DEPN, L 32, 125.

BACKWORTH T&W NZ 3072. 'Bacca's enclosure'. *Buxwurtha, Bucwortha* 1203, *Bachiswrd, Bacwrth* 1271, *Bacworth* 1296 SR, *Backworth* 1863 OS. OE pers.n. *Bacca* + **worth**. NbDu 10.

BACON END Essex TL 6018. *Beacon End* 1768, *Bacon End* 1805 OS sc. of Great Canfield. ModE **beacon** + **end**. Contrasts with Hope End TL 5720, *Offend* 1777, and Puttock's End TL 5619, *Puddocks End* 1777, *Pudhawkes End* 1805. Ess 473.

BACONSTHORPE Norf TG 1926. 'Thorp held by Bacon'. *Baconstorp* 1086, *Bacunestorp* 1203. Surname *Bacon*, originally a nickname < *bacun* 'bacon', + p.n. *Torp* 1086, 'outlying farm', OE, ON **thorp**. DEPN.

BACTON H&W SO 3732. 'Baca's estate'. *Bachetvne* 1086, *Baketon(')* 1188–1428, *Bactone* 1397, *Backingtona* c.1132, *Bakinton* 13th cent. OE pers.n. *Baca* + **ing**[4], varying with genitive sing. *Bacan*, + **tūn**. He 31.

BACTON Norf TG 3433. 'Bacca's estate'. *Baketuna* 1086, *-ton(')* 1185–1448, *Bacton* 1212, *Backton* 1623 etc. OE pers.n. *Bacca* + **tūn**. Nf ii.136.

BACTON Suff TM 0567. 'Bacca's estate'. *Bachetuna* 1086, *Baketon(')* 1198–1346, *Bakton* 1523, *Bacton* 1524. OE pers.n. *Bacca* + **tūn**. DEPN, Baron.

BACUP Lancs SD 8623. 'Ridge valley'. *Bacop(e)* 1324, 1325, a vaccary also known as *ffulebachope* 'dirty Bacup' [c.1200]14th, and *Bacopboth* 'Bacup booth' 1464, *Bacobbothe* 1507. OE **bæc** + **hop**, preceded in c.1200 by OE **fūl** and followed in 1464 by ME **bōthe** 'a temporary shelter', ODan *bōth*, N dial. *booth* 'a cowhouse, a herdsman's hut'. La 92, L 115–6, 125.

BADBURY 'Badda's fortified place'. OE pers.n. *Badda*, genitive sing. *Baddan*, + **byrig**, dative sing. of **burh**. The pers.n. *Badda* is frequently associated with hill-forts in England, suggesting that it may have been the name of some legendary hero (especially if the name is related to OE *beadu* 'war') or even an Anglicisation of some pre-English name. See also BADBY Northants SP 5659, BAUMBER Lincs TF 2274. Do ii.177.

(1) ~ Somer ST 3520.

(2) ~ Wilts SU 1980. *Baddeburi* [995]14th S 568, *-beri -biri -byr'* 1197–1255, *Badeberie* 1086, *Baddebury* 1332. The reference is to the great 5–4th cent. Iron Age hill-fort now called Liddington Castle SU 2079. Wlt 281, Thomas 1980.238.

(3) ~ CLUMP and RINGS Dorset ST 9603. *œt Baddan byrig* early 10th, 11th, *Baddebir'* 1244, *Baddebury* 1508. The site has been identified with the RBrit p.n. *Vindogladia -cladia* 4th, 'the white ditches', British ***u̯indo**- 'white, bright' + ***clādo**- 'ditch' referring to the chalk of which the hill-fort is constructed. Do ii.177, RBrit 500.

(4) ~ HILL Oxon SU 2594. *Badbery Hill* 1539, *Badbury Hill* 1710. Brk ii.362.

(5) ~ HILL Warw SP 1163.

BADBY Northants SP 5659. 'Badda's fortified place'. *baddan byrig, baddan by* 944 S 495, *(at, on) Baddanbyr(i)g* [n.d.]12th S 1565, *Badebi* [1020]12th S 957, 1086–1314 with variant *-by*, *Baddeby* 12th–1475, *Badby* from 1316. OE pers.n. *Badda*, genitive sing. *Baddan*, + **burh**, dative sing. **byriġ**, early replaced by ON **bȳ**. The reference is to the prehistoric camp on Arbury Hill SP 5458 also referred to in the phrase *þa ealdan burh œt Baddanbyrig* in S 495. See further BADBURY. Nth 10.

BADDELEY GREEN Staffs SJ 9051. *Bo- Baddeley greene* 1613×4. P.n. *Baddilige* 1227, *Badeleye* 1270, *Badilegh* 1271, *Badgley* 1572, 'Badda or Beadda's wood or clearing', OE pers.n. *B(e)adda* + **lēah**, + ModE **green**. DEPN, Horovitz.

BADDESLEY CLINTON Warw SP 2071. 'B held by the Clinton family'. *Baddesley Clynton* 1333. P.n. *Badesleia* 1166, *Baddesley(e)* from 1297, *Badgley* 1608, 1652, 'Bædi's clearing or wood', OE pers.n. **Bæd(d)i*, genitive sing. **Bæd(d)es*, + **lēah**, + manorial addition from the Clinton family who held the estate from c. 1200 and for distinction from BADDESLEY ENSOR SP 2798. Wa 53, L 203.

BADDESLEY ENSOR Warw SP 2798. 'B held by the Ensor family'. *Baddesley Endes(h)ouer(e)* 1327, 1332, ~ *Endesore* 1380, *Badgely Endsor* 1698. P.n. *Bedeslei* 1086, *Badeslega* 1169, *Baddeslei* 1216×72, 'Bædi's clearing or wood', OE pres.n. **Bæd(d)i*, genitive sing. **Bæd(d)es*, + **lēah**, + manorial addition from Thomas de Edneshoure (i.e. Edensor, Derbys) who held the estate in 1260 and for disinction from BADDESLEY CLINTON SP 2071. Wa 14 gives pr [bædʒli], L 203.

North BADDESLEY Hants SU 3919. *North Baddesley* 1558×1603. ModE **north** + p.n. *Bedeslei* 1086, *Betheslega* 1135×54, *Badeslea* 1167, *Baddesle(gh)* 1235, 1256, *Badsley* 1810 OS, 'Bæddi's wood or clearing', OE pers.n. *Bæddi*, genitive sing. *Bæddes*, + **lēah**. 'North' for distinction from South Baddesley SZ 3596 of identical origin, *Suthbadesley* 1306, earlier *Bedeslei* 1086, *Badeslea -lie* 1167, 1212, *Baddeslegh -leye -lee* 1235–1316. Ha 27, Gover 1958.34, 204, *NOWELE* 11.91–104.

South BADDESLEY Hants SZ 3596 → North BADDESLEY SU 3919.

Great BADDOW Essex TL 7204. *Magna Badewe* 1274, *Graunt Badowe* after 1420, *Mykell* ~ 1475×80. ModE **great**, Latin **magna**, Fr **grant**, ME **mikel** + p.n. *Beadewan* 975×1016 S 1487, *Baduuen* 1086, *Badewen(na)* 1163–89, *Badewe* 1199–1274 etc., *Badowe* after 1420–1545, originally a r.n. of unknown origin and meaning 'for the CHELMER', itself a late back-formation. The river itself is referred to as *(to) beadewan ea* 'to the river Beadewan' [1062]13th S 1036. 'Great' for distinction from Little BADDOW TL 7807. Ess 233 gives pr [bædə].

Little BADDOW Essex TL 7807. *Lyttil Bado* 1500. ModE adj. **little** + p.n. Baddow as in Great BADDOW TL 7204. Ess 233.

BADGER Shrops SO 7699. 'Bæcg's ridge-tip'. *Beghesovre* 1086, *Beggesoure* 1255, *Baggesour(e)* 1203×4–1502, *Badger* 1549. OE pers.n. *Bæcg*, genitive sing. *Bæcges*, + **ofer**. Sa i.29.

BADGER'S MOUNT Kent TQ 4961. A modern name. No early forms.

BADGEWORTH Glos SO 9019. 'Bæcga's enclosure'. *Beganwurþan* [862]c.1400 S 209, *Becgwirðe* [1022]15th S 1424, *Beiewrda* 1086, *-wrth'* 1221, *Begeword(ia) -w(o)rth -wurth* c.1150–1460, *Beggewurd(a) -worth(e) -wurth -uutha* 1178–1511, *Baggesworth* 1275, *Badg(e)worth(e)* 1585, 1605, *Bedgeworth* 1637. OE pers.n. ***Bæcga** + **wyrthe**, **worth**. Some spellings (*Began- Beie-*) point to the alternative pers.n. *Bēaġe* (feminine). Gl ii.115.

BADGWORTH Somer ST 3952. 'Bæcga's enclosure'. *Bagewerre* (sic) 1086, *Bægge- Beggewurda* 1158–9, *Baggeworth(y)* 1225–[1287×9] Buck, *Baddesworth* 1610. OE pers.n. **Bæcga* + **worth**. DEPN.

BADINGHAM Suff TM 3067. 'The homestead called or at *Bading*, the place called after Bada or Beada'. *Badincham* 1086, *Bad(d)ingeham* [1156×62]1396, *Bedingeham* [1189×99]1396, *Bedingham* 1203, *Badingham* from 1254. OE p.n. **Bading* < pers.n. *Bada* or *Bēada* + **ing**[2], locative-dative sing. **Badinġe*, + **hām**. DEPN, Baron, ING 129.

BADLESMERE Kent TR 0054. 'Beadel or Bæddel's pool'. *Badeles- Bedenesmere* 1086, *Ba- Bedelesmere* c.1100–1253×4, *Badenesmare -mer'* 1199, 1241. OE pers.n. **Beadel*, genitive sing. **Beadeles*, or **Bǣddel*, genitive sing. **Bǣddeles*, + **mere**. PNK 279.

BADMINTON Avon ST 8082. 'Estate called after Baduhelm'. *Badimyncgtun* 972 S 786, *Badminton' -tun -yn-* from c.1200[†]. OE pers.n. *Baduhelm* or the like + **ing**[4] + **tūn**. The exact form of the pers.n. here, which must have been a name in *Badu-*, is uncertain. Gl iii.24.

Little BADMINTON Avon ST 8084. *Parva Badminton(e) -yn-* 1274–1595, *Litell-* 1706. Latin **parva**, ModE **little** + p.n. BADMINTON ST 8082 which was also known as Great Badminton, *Magna Badminton' -yn-* 1291–1721, *Great* ~ 1706. Gl iii.29, 24.

BADSEY H&W SP 0743. 'Bæddi's island'. *Baddeseia* *[709]12th S 80, *Baddesege* [714]16th S 1250, *Baddesig* [c.860]c.1200 K 289, *Badesei* 1086, *Badsey* 1535. OE pers.n. **Bæddi*, genitive sing. **Bæddes*, + **ēġ**. The bounds of the people of Badsey are referred to as *Badesetenagemære* [840×52]12th S 203, OE **Badesǣte*, genitive pl. **Badesǣtna*, + **ġemǣre**. Wo 260, L 38, PNE ii.94.

BADSWORTH WYorks SE 4614. 'Bæddi's enclosure'. *Badesuu(o)rde* 1086, *-w(u)rth -worde -worth* 1226–1433, *Baddeswrd' -uurda* 1170×80, *-worth* 1267–1432. OE pers.n. **Bæddi*, genitive sing. **Bæddes*, + **worth**. YW ii.96, WYAS 310.

BADWELL ASH Suff TL 9969. Short for Badwell Ashfield, 'B by (Great) Ashfield', *Badewelle Asfelde* 13th, *Asshfeld Badewelle* 1320. P.n. *Badewell* 1254, 'Bada's spring', OE pers.n. *Bada* + **wella**, + p.n. Ashfield as in Great ASHFIELD TM 0068. DEPN.

West BAGBOROUGH Somer ST 1633. *West Baggebergh* 1243. ME adj. **west** + p.n. *Bacganbeorg* [899×909]c.1500 S 380, *Baggabeorc* *[1065]18th S 1042, *Bagganbeorgan* [1066×86]n.d. ECW, *Bauueb'ga*, *Bageberge* 1086, *Bagboro* 1610, 'badger or Bacga's hill', OE ***bagga**, genitive sing. or pl. ***baggan**, or pers.n. *Bacga*, genitive sing. *Bacgan*, + **beorg**. 'West' for distinction from East Bagborough ST 1732, *Eastbagborough* 1809 OS, or *Little Baggebergh* 1243. DEPN.

BAGBY NYorks SE 4680. 'Baggi's farm or village'. *Baghebi* 1086, *Bagebi -by* 1086–1400, *Baggaby* 1158×66, *Baggebi -by* 12th–1344. ON pers.n. *Baggi*, genitive sing. *Bagga*, + **bȳ**. YN 189, SPN 45, SSNY 18.

The BAGE H&W SO 2943. 'The stream valley'. *Becce* 1086, *Becha* 1160×70, *La Heche* (sic for *Beche*) 1271, *La Bache* 1537. OE **bæċe**. Another example occurs at SO 4139, The Bage Fm, *(la) Bach(e)* c.1220–1394. He 74, 140.

BAGENDON Glos SP 0106. 'Valley of the people called after Bæcga'. *Benwedene* (sic) 1086, *Bagindon' -yn-* 1211×13, 1279, *-den(e) -en-* 1291–1442, *Bagingedon(e) -den(a) -ynge* 1216–1380, *Badgendon* 1577–1813. OE folk-n. **Bæcgingas* < pers.n. **Bæcga* + **ingas**, genitive pl. **Bæcginga*, + **denu** confused with **dūn**. Gl i.55, L 99.

BAGGRAVE HALL Leic SK 6908. *Baggrave Hall* 1833 OS. Earlier simply *The Hall* 1666. P.n. Baggrave + ModE **hall**. Baggrave, *Badegrave* 1086, *-es-* 1169, *Balbe-* c.1130–13th, *Bab(b)e-* 1190–1377, *Bab-* 1299–1478, *Bag(g)rave* from 1499, is 'Babba's grove', OE pers.n. *Babba* (apparently partly confused with *Badda*) + **grāf**. Lei 305, 306.

BAGGY POINT Devon SS 4140. *Bag Poynt* 1577. D 44.

BAGINTON Warw SP 3475. 'Badeca's estate'. *Badechitone* 1086, *Badekendon* 1262, *-kynton* 1279–97, *Bathekinton(a)* [1170]1314, 1259–1401 with variants *-king- -kyng-* from 1290, *Batkinton* 1242, *Bagyn(g)ton* 1544, *Baggington alias Bathington* 1667. OE pers.n. *B(e)adeca*, genitive sing. *B(e)adecan*, + **tūn**, or *B(e)adeca* + **ing**[4] + **tūn**. Wa 155 gives pr [bægintən].

BAGLEY Shrops SJ 4027. Partly uncertain. *Bageleia* c.1090, *Baggeleg* 1225. OE ***bagga** of uncertain meaning + **lēah**. OE *bagga* means 'a bag'; this was extended to bag-like objects or creatures, as MDu *bagghe* 'a small pig'. It may have meant 'badger' in OE. DEPN, PNE i.17.

BAGNALL Staffs SJ 9251. 'Badeca's wood'. *Baggenhall* 12th, *Badegenhall* 1203, 1273, *Bagenholt* 1203–1386 with variants *-in-* and *-yn-*, *-hold -hald* c.1280×6–1400, *Bagenald* 1329–1417, *Bagnald(e)* 1470, 1538, *Bagnall* from 1547. OE pers.n. *Badeca*, genitive sing. *Badecan*, + **holt**. Oakden.

BAGSHOT Surrey SU 9163. 'Bacga's nook'. *Bache- Bagsheta* 1164, *Bacsete* 1186, *Bageset'*, *Baggeshec'* 1218, *Bagshote* 1330. OE pers.n. *Bacga* + **scēat**. Sr 153.

BAGSHOT Wilts SU 3165. 'Beocc's gate'. *Bechesgete* 1086, *-ieta* 1130, *Bekes- Bukes- Bockes- Boghkesgate* 1289–14th, *Bashette* 1525, *Bagshott* 1684. OE pers.n. *Beocc*, genitive sing. *Beocces*, + **ġeat**. The same pers.n. occurs in *Beoccesheal* [968]12th S 756, 'Beocc's nook', in the bounds of Bedwyn SU 2764, 2 m to the W. Wlt 354.

[†]The DB form *Madmintvne* 1086 is due to miscopying of Rustic or Lombardic capital B. Cf. GDB *Molebec* for *Bolebec* folio 56[v], Berks B.2, B.5, cited in Nomina 9.46; cf. *Domesday Studies* 130.

BAGSHOT HEATH Surrey SU 9161. *Bagshot Heath* 1609. P.n. BAGSHOT SU 9163 + ModE **heath**. Sr 154.

BAGTHORPE Norf TF 7932. 'Bakki or Bak's outlying settlement'. *Bachestorp* 1086, *Baket(h)orp* 1209, 1254, *Bag(g)etorp* 1198–1209, *-thorp* 1291. ODan pers.n. *Bakki*, genitive sing. *Bakka*, + ON **thorp**. The DB form with genitival *-es* suggests alternatively that the pers.n. may been the rare ON byname *Bak*. DEPN, SPNN 93–4.

BAGTHORPE Notts SK 4751. 'Baggi's outlying settlement'. *Bagthorpe* from 1490. ON pers.n. *Baggi*, genitive sing. *Bagga*, + **thorp**. Nt 131, SSNEM 123.

BAGWORTH Leic SK 4408. 'Bacga's enclosure'. *Bageworde* 1086, *Baggeworth(e)* 1270–1408, *Bagworth(e)* from 1209×35. OE pers.n. *Bacga* + **worth**. Lei 481.

BAGWY LLYDIART H&W SO 4426. 'The tip or head of L'. *Bagwy LLydiart* 1831 OS. W **bagwy** + p.n. Llydiart 'the grey ridge', PrW ***lēt** + ***garth**, as in Llwydiarth Clwyd SJ 2237, ~ Powys SN 9880, ~ Esgob Gwyn SH 4384, ~ Fawr Gwyn SH 4285, ~ Hall Powys SH 7710, ~ Powys SJ 0616, Llwydarth M Glam SS 8590, *Litgart(h)* c.1150, and as in Llidiart names in Clwyd SJ 1430, SJ 1652, SJ 0546, SJ 1143 and Gwyn SH 9012, LYDEARD ST LAWRENCE Somer ST 1232 and LYDIARD MILLICENT Wilts SU 0986. GPC 249.

BAILDON WYorks SE 1539. 'Round hill'. *(on) Bægeltune, (on) Bældune* c.1030 YCh, *Beldun(e) -don(e)* 1086, 1202×8, *Bail- Bayldun -don(a)* late 12th–1656. OE ***bæġel** + **dūn**. Probably an allusion to the circular shape of Baildon Hill. YW iv.158, L 142–3, 157, WYAS 311.

River BAIN Probably the 'helpful, useful river'. ON **beinn** 'straight, favourable, appropriate' + **á**. Cf. the ODan stream-n. **Bēn* as in p.ns. Benløse on Jutland and Binderup and Bengård in Jutland. RN 24, *DSÅ* i.131–3, *Anglia* 1996.548–9.

(1) ~ Lincs TF 2472. *(ueterem) Beinam* 'the old B' (accusative case) 1140×50, 1150×60, *Baina* 1154×89, *Baine, Bayne* [12th]14th, *Bayn* 1275, 1341, *Bane* 16th cent. RN 24, Cameron 1998.

(2) ~ NYorks SD 9390. *Bayn(e)* 1153 etc., *Bain(e), Bein* 1218. The river runs more or less straight through a narrow defile to reach Bainbridge and its junction with the Ure. The secondary sense 'hospitable, helpful, favourable' is also possible. YN 2, RN 24.

BAINBRIDGE NYorks SD 9390. 'The bridge over the river Bain'. *Bainebrig(g), Beynebrigge* 1219–1285. R.n. BAIN SD 9390 + OE **brycg**. YN 262.

BAINTON Cambs TF 0906. 'Estate called after Bada'. *Badingtun* [c.980]c.1200–1428 with variants *-yn(g)- -in-*, *Baynton* 1369. OE pers.n. *Bada* + **ing**[4] + **tūn**. Nth 229.

BAINTON Humbs SE 9652. 'Bæga's farmstead, village, estate'. *Bagenton(e)* 1086, *Baynton(e) -tona -i-* 1100×15–1600, *Bai- Bayngtun -ton(e)* 1150×60–1302. OE pers.n. *Bǣġa*, genitive sing. *Bǣġan*, + **tūn**. YE 165.

BAIT ISLAND T&W NZ 2357. *Bait Island* 1863 OS, *Bates' Island* 1888. Surname *Bates* + ModE **island**. An alternative name for ST MARY'S ISLAND. Tomlinson 58.

BAKER STREET Essex TQ 6381. *Bakerestrat -stret* 1402, 1490. Cf. *Bakers in Lugstret* 1483. ME **bakere** + **strete**. Ess 166.

BAKERS END Herts TL 3917. *Bakers ende* 1578 sc. of Thundridge parish. Earlier simply *Bakers* 1468. Surname *Baker* as in John le Bakire 1342 + ModE **end**. Hrt 206.

BAKETHIN RESERVOIR Northum NY 6391. No early forms.

BAKEWELL Derby SK 2168. 'Badeca's spring(s)'. *(to) Badecan wiellon* 924 ASC(A), *(ad) Badecanwelle* [949]13th S 548, *Beadecanweallan* before 1118, *Badequella* 1086, *Bathecwella -(e) -oc- -uc-* 1161×75–1306, *Bathekewell(e)* 1200–1345, *Bauec- Bauek- Bawekewelle* 1188×97–1397, *Bauc(h)- Bauk- Bawkwelle* 1192–1550, *Bacwell(e)* 1177–1346, *Bayequelle* 1272×1307, *Bakewell(e)* from 1330. OE pers.n. *Badeca* + **wella**, dative pl. **wellum**. Db 30.

BALCOMBE WSusx TQ 3130. Partly uncertain. *Balecumba* 1087×1100–1327 with variant *-cumb(e), Bald(e)comb(e)* 1279, 1284, *Baulcombe* 1639, *Bawcombe* 1688, *Bolkham* 1715. Possibly OE pers.n. **Bealda* + **cumb**, but the earliest sps point rather to OE **bealu** 'evil, calamity' as the specific. The village lies on a ridge between two valleys but the forms do not support derivation from *camb* 'a comb, a crest'. Sx 255 gives pr [bɔːkəm], DEPN, SS 69.

BALDERHEAD RESERVOIR Durham NY 9218. P.n. Balderhead + ModE **reservoir**. Balderhead is the 'head of the r. Balder', *Balder* 13th, *Bauder* 1577, 1626, itself possibly a back-formation from Baldersdale NY 9418, *Baldersdale* from c.1275, *Bauderdale* 1577, 'Balder's valley', pers.n. OE *Baldhere* or Dan *Baldœr*, genitive sing. *Baldheres, Baldœrs*, + lOE **dæl** or ON **dalr**. YN 2, 306, RN 25.

BALDERSBY NYorks SE 3578. 'Balder's farm or village'. *Baldrebi* 1086, *Balderbi -by* 1156–1576, *Baldersby* 1648. OE pers.n. *Baldhere* or ODan pers.n. *Baldœr* with late genitival **-s** + **bȳ**. YN 182, SSNY 19.

BALDERSDALE Durham NY 9418 → BALDERHEAD RESERVOIR NY 9218.

BALDERSTONE Lancs SD 6332. 'Baldhere's farm or village'. *Balderestone* before 1172, *Balderston* 1246–1341. OE pers.n. *Baldhere*, genitive sing. *Baldheres*, + **tūn**. La 69.

BALDERTON Notts SK 8251. 'Bealdhere's farm or village'. *Baldretune -tone* 1086, *-ton* 1287, 1291, *Baldertun* [1154×69]1327, *-ton(a)* from 1175. OE pers.n. *Bealdhere* + **tūn**. Nt 209.

BALDHU Corn SW 7743. 'Black mine'. *Baldue* 1748, 1813. Co **bal** + **du**, 'a tin mine'. The modern spelling with *dh* is an attempt to make the name look more Celtic. PNCo 51.

BALDOCK Herts TL 2434. 'Baghdad'. *Baldoce* 1135×54, *Baldok* 1248–14th, *Baudac -ak* 1214–1428. This is the OFr form of the name Baghdad, *Baldac*. The place was named after the Syrian city by the Knights Templar. Hrt 120.

Marsh BALDON Oxon SU 5699. *Mersbaldindon(e)* 1278–93, *Merschebaldon* 1428. ME **mersh** (OE *mersc*) + p.n. *Baldedone* 1086, *Baldendone* 1184, *Baldyndone -indon(e) -intone, Baudindon' -endon* 13th cent., 'Bealda's hill', OE pers.n. *Bealda*, genitive sing. *Bealdan*, + **dūn**. Also known as *Baldingtone de la Mor* 1285 'moor B'. 'Marsh' for distinction from Little Baldon Farm, *Parva Baldinton'* 1240, *Little Baldon* 1797 and Toot BALDON SP 5600. O 162.

Toot BALDON Oxon SP 5600. 'B with the look-out'. *Tot Baldington* 1428. Cf. *villa de Baldinden Sancti Laurencii cum Totbaldindon Mershbaldindon et Parva Baldindon* 'the vill of B St Lawrence with T B, Marsh B and Little B' 1316. ME **tote** (OE **tōt*) 'a look-out hill' + p.n. *Baldedone* 1086, *Baldendone -tone* 1086–1318 with variants *-dun(a) -don(a) -ton* and *-in-*, *Baudinton' -dun' -don'* 1199–1247 as in Marsh BALDON SU 5699. The boundary of the inhabitants of Baldon is *Bealddunheama gemœre* [1050]13th S 1022, *bealdanhema gemer* [1054]c.1200 S 1025. O 163, 487, L 150.

BALDWIN'S GATE Staffs SJ 7940. *Baldwin's gate* 1676, *Balding Gate* 1775. Probably surname *Baldwin* + ModE **gate**. William Baldwin was parker of Madeley Park 1293 and Baldwin's Gate lies at the S end of the former park, cf. also here *Baldewyne forlong* c.1275, *Baldwynes-pitte* 1450×1. Horovitz.

BALDWINHOLME Cumbr NY 3532. 'Baldwin's island'. *Baldewin(e)holm(e) -wyn(e)-* 1278–1366, *Baldin- Boldyngholme* 15th cent., *Balding- Bawdyn(g)- Bathenholme* 16th cent. ME (CG) pers.n. *Baldewin* + ME **holm**. Cu 145.

BALE Norf TG 0136. 'Bath wood or clearing'. *Bathele* 1086, 1177, *Bale* 1208. OE **bæth**, genitive pl. **batha**, + **lēah**. DEPN.

BALKHOLME Humbs SE 7928. 'Balki's' or 'ridge island'. *Balc- Balkholm(e)* 1199–1498, *Balterholme* 1282, *Bawkeholme* 1550. ON pers.n. *Balki* or OE **balca** + **holmr**. YE 249, L 51–2.

BALL Shrops SJ 3026. *The Ball* 1837 OS. Short for *The Original Ball*, the name of the pub whose sign depicts a globe of the earth. Raven 126.

Higher BALLAM Lancs SD 3630. Adj. **higher** + p.n. *Balholm* [1189×94]1336, *Balholme* 1324, *Balghholm* 1332, 'round-hill island', ME **balgh** (OE **balg*) + ON **holmr**. The reference is to a slight elevation in the marshlands of Lytham Moss and Brown Moss. 'Higher' for distinction from Lower Ballam SD 3630. La 151.

BALLARD POINT Dorset SZ 0481. Cf. *Ballard Hole* 1811 OS referring to caves here, and *Ballard Down* 1811 OS. Possibly surname *Ballard* + ModE **point**. But Ballard, from ME *ball* + suffix *-ard*, was originally a nick-name meaning 'bald-head', an apt description of the bare hill here. Do i.46 suggests possible derivation from OE **balg* + *hēafod* 'smooth-swelling headland', but the evidence is too late for certainty.

BALL HILL Hants SU 4263. 'Round hill'. *Ballhill* 1604. ME, ModE **ball** + **hill**. Gover 1958.158.

BALLINGER COMMON Bucks SP 9103. *Ballenger Common* 1822 OS. P.n. Ballinger + ModE **common**. Ballinger, *Baldinghore* 1195, *Beldyngore* 1297, *Baldingore* [1313]14th, *Belynger* 1504, *Bal(l)inger* 1535, 1550, is probably the 'flat-topped ridge called or at Balding, the place called after Beald(a)', OE p.n. **Bealding* < pers.n. *Beald(a)* + **ing**[4], locative-dative sing **Bealdinġe*, + **ōra**. Bk 154 gives pr [bælindʒə], BzN 1967.357, 362, Jnl 2.28, 21.19, 22.32, L 120, 180.

BALLINGHAM H&W SO 5731. 'Estate called, at or of Badelinge'. *Baldinga'* 1160×70, *Baldingesham* 1162, *Badelingeham* 1215, *Balding(e)ham* 1244×6, 1251, 1412, *Ballingham -yng-* from 1237, *Balincham* 1535. OE p.n. **Badeling* < pers.n. *Badela* + **ing**[2], locative-dative sing. **Badelinġe*, genitive sing. **Badelinges*, + **hām**. He 32, BzN 1967.362, 1968.144.

BALLOWALL BARROW Corn SW 3531. *Ballowall Cairn* 1879. P.n. Ballowall SW 3531, *Bolawall* 1302, *Bolauhel* 1396, probably 'dwelling of Louhal', Co. **bod** + pers.n. **Louhal*, + ModE **barrow**. The reference is to the composite Neolithic/Bronze Age burial mound of Carn Gluze a mile W of St Just. PNCo 51, Thomas 1976.56.

BALLS CROSS WSusx SU 9826. *Balls Cross* 1813 OS. A cross-roads.

BALMER LAWN Hants SU 3003. A modern name without early forms. Possibly a bad spelling for Barmoor in *Barmoore Wood* 1670, ? 'bare moor'. Gover 1958.207.

BALNE SYorks SE 5919. Uncertain: originally the name of a district. *Balne(a), Baln(a)* from 12th, *Baune* 1167, *Bawne* 1559–1665. Adam de la Pryme wrote in his diary in 1695: 'Mr Horatio Cay says that the Romans seeing a part of the country for a huge way round about boggy and full of quagmires, they gave it the name of Balneum', i.e. 'bath'. Late ME **balne** 'bath' is not otherwise recorded before 1471. YW ii.14 gives pr [bɔ:n], V i.45.

BALSALL COMMON WMids SP 2377. *Balsal Common* 1831 OS. P.n. *Beleshale* 1100–1327, *Balishale* 1221–1428, *Bal(e)sale* 1279–1327, *Ballsal(l)* from 1380, 'Bælli's nook', OE pers.n. **Bælli*, genitive sing. **Bælles*, + **halh**, dative sing. **hale** probably in the administrative sense 'a piece of land projecting or detached from the main unit'. Until 1863 Balsall was the SE angle of Hampton in Arden chapelry. Wa 53.4.

BALSCOTE Oxon SP 3941. 'Bælli's cottage(s)'. *Belescot(e)* c.1200–1241, *Bal(l)escot(e)* 1219–1509×47. OE pers.n. **Bælli*, genitive sing. **Bælles*, + **cot**, pl. **cotu**, ME **cote**. O 409, DEPN s.n. Balscott.

BALSHAM Cambs TL 5850. 'Bælli's homestead'. *(to) bellesham (gemære)* 'to the B boundary' [974]11th S 794, *Belesham* c.1050–1277 including [1017×35]12th S 1520 and [1042×66]12th S 1051, *Bælesham, Bælessam, Beles(s)ham* 1086, *Balesham* 1086–1284 including [c.1060]14th, *Balsham* from 1267. OE pers.n. **Bǣl(l)i*, genitive sing. **Bǣl(l)es*, + **hām**. Ca 114 gives former pr [bɔ:lsəm], O li [bɔlʃəm].

BALTERLEY Staffs SJ 7450. 'Baldryth's wood or clearing'. *Baldryðeleag* [1002×4]11th S 1536, *Baltredelege* 1086 *-legh* 1289, *Balturdeley* 1357×8. OE feminine pers.n. **Baldrȳth*, genitive sing. **Baldrȳthe*, + **lēah**. DEPN, Horovitz.

BALTONSBOROUGH Somer ST 5434. 'Bealdhun's hill or barrow'. *Balteresberghe* [744]14th S 1410, *Baltvnesberge* 1086, *Baltenes-* 1196, *Balesboro:* 1610. OE pers.n. *Bealdhūn*, genitive sing. *Bealdhūnes*, + **beorg**. DEPN.

BAMBER BRIDGE Lancs SD 5626. P.n. Bamber + ModE **bridge**. Bamber, *Bymbrig* n.d. VCH Lancs vi.290, is apparently 'Bimme's bridge', ME pers.n. *Bimme* + **brycg**. La 68.

BAMBURGH Northum NU 1834. 'Bebbe's fort'. *Bebbanburh* 10th ASC(A), c.1121 ASC(E) both under year 547, *Bebbanburg* [c.890]c.1000 OEBede, *(from) Bebban byrig* ASC(D) under year 926, *Bebba(n)- Bæbba(n)burh* c.1121 ASC(E) under years 641, 993, 1093, 1095, *Bebbanbur(c)h* 1097, *in urbem Bebban* 12th SD, *Baen- Baemburc(h) -burg* 12th cent., *Bamburg* 1199–1242 BF, *Baningburg* 1242 BF, *Bamburgh* c.1715 AA 13.4. OE feminine pers.n. *Bebbe*, genitive sing. *Bebban*, + **burh**. According to BHE iii.6 this royal city was named after a former queen, Bebba, the wife of king Æthelfrith of Bernicia (593–617). Its original name seems to have been *Dinguayrdi* or *Dinguoaroy* 9th HB 61 and 63, PrW ***din** 'a fort' + unexplained element or name. NbDu 10.

BAMFORD Derby SK 2083. 'Tree-trunk ford'. *Banford* 1086, *Baumford -forth* 1225–1501, *Bawn- Baunford -forth* 14th cent., *Bamford* from 1228. OE **bēam** + **ford**. The tree may have been used as a footbridge or may have marked the position of the ford across the river Derwent. Db 39, L 69.

BAMPTON Cumbr NY 5118. 'Tree farmstead'. *Bampton(e)* c.1160–1701, *Banton* early 13th–1699. OE **bēam** + **tūn**. The reference is either to a place beside a tree or made of beams or, perhaps more likely, where beams are manufactured and obtained. Manorial additions include the family name *Cundal(e)* 1160–1363, and personal name Patrick from *Patrick* de Culwen 12th. We ii.189.

BAMPTON Devon SS 9522. Partly uncertain. *Ba(d)entone* 1086, *Baðentune* [c.1090]12th, *Bathentona* 1156, *Baenton(a)* 1086 Exon, 1176, 1182, *Banton* [c.1156]12th, 1238, *Baunton* 1221–1368, *Bamton* 1253. The simplest explanation is OE *Bathum-tūn* 'settlement at the baths' referring to pools in the river Batherm, OE **bæth**, dative pl. **bathum**, + **tūn**. The specific could, however, be an OE **Bæth-hǣme*, genitive pl. **Bæth-hǣma*, perhaps '*tūn* of the Morebath people'. Morebath lies one mile to the N. The r.n. Batherm, *Batham* 1797, is an antiquarian back-formation from the p.n. D 530, RN 26.

BAMPTON Oxon SP 3103. Probably 'the homestead by the tree' or possibly 'made of beams'. *Bemtun* 1069, *(æt) Bemtune* 1069×72, *Bemton(i)a* 1140, c.1175, *Bamton(e)* 13th., *Bampton* from 1212. OE **bēam** + **tūn**. The precise meaning of this name is uncertain but a part of the town to the E on the edge of a large Iron Age and Roman site was known in the 14th cent. as the 'Beam'. O 304, PNE i.21, ii.194, Blair 1988.53–4, 1994.62–4.

BAMPTON COMMON Cumbr NY 4716. *Bampton Common* 1865 OS. P.n. BAMPTON NY 5118 + ModE **common** 'common land'. We ii.193.

BANBURY Oxon SP 4540. '(At) Banna's fortified place'. *Baneberie* 1086, *Bannberi(a) -ber(i) -biri(a) -biry -bur(y)* etc. 12th–1343, *Banbury* from 1285. OE pers.n. **Ban(n)a* + **byriġ**, dative sing. of **burh**. O 411.

BANHAM Norf TM 0687. 'Bean homestead'. *Benham* 1086, *Banham* from 1168. OE **bēan** + **hām**. DEPN.

BANK Hants SU 2807. No early forms. Bank is situated on a low sandstone ridge between streams. Ha 27.

BANK NEWTON NYorks SD 9153 → Bank NEWTON.

BANKS Cumbr NY 5664. 'Embankments'. *Bankys, Bankes* 1256, *les Bankes iuxta Lanercost* 1346, *Banks* 1543. ME **bank** (ODan *banke*). Apparently refers to Hadrian's wall on which Banks is situated. Cu 70.

BANKS Lancs SD 3920. *The Bank* 1713. ModE **bank**. The reference is to higher land between Martin Mere and the sea. Mills 59.

BANKS SANDS Lancs SD 3624. P.n. BANKS SD 3920 + ModE **sand**.

BANK STREET H&W SO 6362. No early forms.

BANNINGHAM Norf TG 2129. 'Homestead of the Ban(n)ingas, the people called after Ban(n)a'. *Banincham* 1086, *Banningeham* 1170, *Bani(n)gham* 1198–1242×3. OE folk-n. **Ban(n)ingas* < pers.n. *Ban(n)a* + **ingas**, genitive pl. **Ban(n)inga*, + **hām**. DEPN, ING 132.

BANNISTER GREEN Essex TL 6920. *Bunister Green* 1768, 1777, *Bannister Green* 1805 OS. P.n. *Bernestey'* 1235, *burneste* 1546 + ModE **green**. *Bernestey'* is probably 'Beorn's enclosure', OE pers.n. *Beorn*, genitive sing. *Beornes*, + **tēag**. Ess 421 gives pr [bʌnstə].

BANSTEAD Surrey TQ 2559. 'The place where beans are cultivated'. *Benstede* [727, 933, 967]13th S 420, 752, 1181, *Baenstede* [1062]13th S 1035, *Benestede* 1086, *Benestede -a* c.1150–1228, *Banested* 1196–1260×84, *Banstede* 1198 to 1796. OE **bēan**+ **stede**. Another example of the same name occurs in the nearby Merstham boundary, *(up to, of) beanstede* 947 S 528. Sr 68, Stead 244, Jnl 3.12–13.

BANTHAM Devon SX 6643. *Bantham* 1809 OS.

BANWELL Avon ST 3959. 'Murderer stream'. *Banuwille* [893]c.1000 Asser, *(æt) Bananwylle* [904]12th S 373, [905×25]lost ECW, [978 for ?968]12th S 806, *Banewelle* [?c.1030]lost ECW, *Banawelle* *[1065]c.1500 S 1042, *Banwelle* 1086, *Banewell* [1174–91]c.1240 *Wells*. OE **bana**, genitive sing. **banan**, + **wielle**. In early times criminals were sometimes ritually drowned. But Margaret Gelling suggests that the references of *bana* in this name may be to contamination of the spring. DEPN.

BAPCHILD Kent TQ 9262. 'Bacca's spring'. *Baccancelde* [696×716]11th S 22 with variants *Bac(h)an-* and *-cild -childe* from later copies, *Bakechild(')* 1204–19 etc., *Babchilde* 1572. OE pers.n. *Bacca*, genitive sing. *Baccan*, + **ċelde**. K 242, KPN 22, Jnl 8.16, L 20.

BAPTON Wilts ST 9938. 'The estate called after Babba'. *Babinton -yn-* 1221–89, *-ing- -yng-* 1249–1312, *Babeton* 1276, 1332, *Babton* 1455, *Bapton* from 1526. OE pers.n. *Babba* + **ing**[4] + **tūn**. The same pers.n., possibly referring to the same individual, occurs in nearby BAVERSTOCK SU 0232. Wlt 161.

BARBER BOOTH Derby SK 1184. *Barber Booth* 1675, 1767. Surname *Barber* + ModE **booth** 'a temporary booth or shed' used by herdsmen to shelter cattle and sheep (<ODan *bōth*). Formerly called Whitemoorley Booth, *vaccar' in Eydall vocat' Whit(e)morley* 'vaccary in Edale called W' 1590, 1607, *Whitmorlie booth* 1625–1824, 'booth at Whitemoorley, Whitemoor pasture', p.n. Whitemoor, 'white moor', ME **white** + **mor**, + **ley** (OE *lēah*). Db 87.

BAR HILL Cambs TL 3863. No early forms.

BARBON Cumbr SD 6282. 'Beaver' or 'Bera's stream'. *Berebrun(e) -brunna -e* 1086–1434, *Berburn(e)* c.1190–1461, *Barburn -borne -b(o)ron -b(r)onne* 16th cent., *Barbon* from 1609. ON **bjórr** or OE pers.n. *Bera* + OE **burna**, ON **brunnr**. Hardly 'bear stream' as the bear was not native in England. We i.23, SSNNW 102, L 18.

BARBRIDGE Ches SJ 6156. 'Bridge with, or at, a gate or barrier'. *Barbridge* c.1536, *Barr Bridge* 1621. ME **barre** + **bridge**. Che iii.151.

BARBROOK Devon SS 7147. Partly uncertain. Cf *Babbroke Mill* 1632, *Barbrick Mill* 1809 OS. The specific might be OE pers.n. *Babba* as in Babcombe SX 8677, *Babbecumb'* 1242, but the evidence is too late for certainty. The generic is OE **brōc** 'a brook'. D 65, 479.

BARBROOK RESERVOIR Derby SK 2877. R.n. Bar Brook + ModE **reservoir**. Bar Brook, *Berebroke* 1500, *Burbrok* 1577, *-brooke* 1586, *Barbrook* 1767, is of uncertain origin. Db 3.

BARBURY CASTLE Wilts SU 1576. P.n. *(æt) Beran byrg* c.890 ASC(A) under year 556, *Bereberia* 1180, *-byrie -bury* 1252, 1332, *Berbyr'* 1289, *-bury* 1391, 'Bera's fortification', OE pers.n.**Bera*, genitive ing. **Beran*, + **byriġ**, dative sing. of **burh**, + ModE **castle**. Later sps include *Barbery Down* 1653, *Barbara(h) Down*, 1673, 1759. The reference is to the great Iron Age hill-fort at SU 1476, the site of a battle between the West Saxon leaders Cynric and Ceawlin and the Britons. Wlt 278, Thomas 1980.236.

BARBY Northants SP 5470. 'The village or farm on the hill'. *Berchebi* 1086, *Beruby* 12th–1314, *Bergebi -by -we-* 1201–1314, *Barby* from 1508, *Baroughby* 1550. ON **berg** + **bȳ**. Nth 13, SSNEM 34.

BARCHESTON Warw SP 2639. 'Beaduric's estate'. *Ber(ri)cestone* 1086 (the village), *Bedric- Berricestone* 1086 (the hundred), *Bercheston* c.1190–1316, *Barchestona* [c1112]1329, *Barcheston alias Barston* 1629. OE pers.n. **Beadurīċ*, genitive sing. **Bead(u)rīċes*, + **tūn**. Wa 297, Hrt xliii.

BARCOMBE ESusx TQ 4214. 'Barley enclosure'. *Bercham* 1086–12th cent., *Ber(e)camp(e)* 1087×1100–1439, *Bercumbe* 1291, *Barcombe* 1482, *Barkham* 1588. OE **bere** + **camp**. Sx 313, Signposts 76.

BARCOMBE CROSS ESusx TQ 4216. 'Barcombe cross-roads'. *Barcombe Cross* 1813 OS. P.n. BARCOMBE TQ 4214 + ModE **cross**.

BARDEN NYorks SE 1493. 'Beorna's valley'. *Bernedan* 1086, *Berdene* 1184–1285, *Barden* 1552. OE pers.n. *Beorna* + **denu**. YN 269.

BARDEN FELL NYorks SE 0858. P.n. Barden SE 0357, *Berden(e)* c.1140–1637, *Barden in Craven* 1547, 'the barley valley', OE **bere** + **denu**, + ModE **fell**. YW vi.60.

BARDEN RESERVOIRS NYorks SE 0257. P.n. Barden SE 0357 as in BARDEN FELL SE 0858 + ModE **reservoir**.

Great BARDFIELD Essex TL 6730. *Majoris Berdefeld(e)* 12th, *Magna* ~ 1227–1459, *Myche Bard(e)feld* 1532. ModE adj. **great**, Latin **major, magna**, ME **much**, + p.n. *Bi- Byrdefeldā* 1086, *Berdefeld(e)* 12th–1459, *Bard(e)feld* 1253–1532, perhaps 'open land at *Berde*, the bank or border', OE ***byrde**, Essex dial. form ***berde**, perhaps used as a p.n., + **feld**. The reference would be to the edge or bank of the river Pant, but the existence and status of **byrde* as a p.n. element is very uncertain. Cf. STIBBARD Norf TF 9828. 'Great' for distinction from Little BARDFIELD TL 6530 and Bardfield SALING TL 6826. Ess 504, Studies 1936.163.

Little BARDFIELD Essex TL 6530. *Parva Berdefeud* 1235, *Petit Byrdefeut* 1321. ModE adj. **little**, Latin **parva**, Fr **petit**, + p.n. Bardfield as in Great BARDFIELD TL 6730.

BARDFIELD SALING Essex TL 6826 → Bardfield SALING.

BARDNEY Lincs TF 1169. 'Bearda's island'. *Beardaneu* [c.731]8th BHE, *Beardanea* [c.890]c.1000 OEBede, *(on) Bearddan igge* c.891 ASC(A) under year 716, *(on) Beardanege* c.1000 ASC(B) under same year, *Bardan ege* [c.1030]11th Secgan, *(of) Beardan igge* c.1050 ASC(C, D) under year 906, *(on) Bearðan ege* 1121 ASC(E), *Bardenai* 1086. OE pers.n. **Bearda*, genitive sing. **Beardan*, + **ēġ**. DEPN, L 38, Jnl 8.16, Cameron´1998.

BARDOLFESTON Dorset SY 7694 → BURLESTON SY 7794 footnote.

BARDON MILL Northum NY 7764. *Bardon Mill* 1867 OS. The reference is to a woollen mill established here. Tomlinson 165.

BARDSEA Cumbr SD 3074. Partly uncertain. *Berretseige* 1086, *Berdeseia* 1155–94, *Bardeseia* 1202, *Berdesey(e)* 1246–1348. The name is usually said to be 'Beornred's island', OE pers.n. *Beornrēd*, genitive sing. *Beornrēdes*, + **ēġ** in the sense 'hill-spur'. Bardsea is on the top of a hill jutting out into flat coastland. Possibly the hill-spur was called *Beard* 'the beard'. La 210, L 36.

BARDSEY WYorks SE 3643. 'Beornred's island'. *Bereleseie,*

Berdesei 1086, *Berdes(eia) -ea -ia -ee -ey(a) -ay(a)* 1158×93–1546, *Bardesey(e) -ay* 1244–1537, *Bardsey* from 1540. A reduced form of OE pers.n. *Beornrēd*, genitive sing. *Beor(nrē)des*, + **ēġ** in the sense 'high place, hill-spur'. YW iv.177, L 35–6, 38, WYAS 313.

BARDSLEY GMan SD 9201. 'Beard's wood or clearing'. *Bard(e)sley, Berdesley* 1422. OE pers.n. *Beard*, genitive sing. *Beardes*, + **lēah**. La 29.

BARDWELL Suff TL 9473. Partly uncertain. *b'deuuella, beordewella* 1086, *Berdewelle* 1190, 1197, *Berdwell* 1610. Either 'Bearda's spring', OE pers.n. *Bearda* + **wella**, or 'the spring with a brim or bank', OE **bre(o)rd** + **wella**. In the first case we have an odd DB sp for Bearda, in the second dissimilatory loss of *r* as in BEARD Derbs. DEPN.

BAREWOOD H&W SO 3856. *Bearwood* 1832 OS.

BARFORD 'Barley ford, ford where corn is carried'. OE **bere** + **ford**. PNE i.30, Bd 50–2, L 71.

(1) ~ Norf TG 1107. *Bereforda* 1086, *-ford* c.1184. DEPN.

(2) ~ Warw SP 2761. *Bereforde* 1086, *-ford(e)* 1176–1547, *Berford* 1316, *Bareford* 1535, *Barford, Barfott, Bearfoot* 16th.† Wa 248.

(3) ~ ST MARTIN Wilts SU 0531. 'St Martin's B'. *Berevord St Martin* 1304, 1380, *Berford Seint Martyn* 1400. P.n. *Bereford* 1086–1598, *Bereford al. Barford* 1598, + saint's name *Martin* from the dedication of the church. Wlt 212, L 71.

(4) ~ ST MICHAEL Oxon SP 4332. 'St Michael's B'. *Bereford Sancti Michaelis* c.1250–1346, *Bereford Seynt Michael* 1327. P.n. *Bereford(ia)* 1086–1278×9, + saint's name *Michael* from the dedication of the church and for distinction from Barford St John Oxon SP 4333, *Bereford Sancti Johannis* 1299, p.n. *Bereford* 1086–1282 + saint's name *John* and *Northbereford* 'north B' 1240×1 or *Parua Bereford* 1240×1–1349, *Little Bereford* 1364. O 393, L 71.

(5) Great ~ Beds TL 1352. ModE adj. **great** + p.n. *Bereforde* 1086 etc., *Berford* 1257, *Bareford* 1545. 'Great' for distinction from Little BARFORD TL 1857. The river forded here is the Great Ouse. Bd 50.

Little BARFORD Beds TL 1857. *Little* is a modern addition to this name for distinction from Great BARFORD TL 1352 when the *-k-* of Little Barford disappeared. Earlier simply *Bereford* 1086, *Berkeford* 1202–1581, *Berkford* 1269–1576, *Barkford* 1415–1748, *Berford* 1284, *Barford* 1539, 'birch-tree ford', OE **beorc**, genitive pl. **beorca**, + **ford**. The name has been assimilated to the common type BARFORD. Bd 100, L 220.

BARFRESTONE Kent TR 2650. Partly uncertain. *Berfrestone* 1086, *Ber(e)freston(e) -frey(e)s-* 1235–92, *Barfreistone* 1262×2. Uncertain OE pers.n., *Beorn-, Beorht-* or **Berafrith*, genitive sing. *-frithes*, + **tūn** 'estate'. PNK 578, KPN 159.

BARHAM Cambs TL 1375. 'Hill homestead'. *Bercheham* [1086]c.1180, *Bercham, Bergham* 1209–86, *Berwham* 1260, *Berewam, Ber(e)uham* 1279, 1286, *Barr(h)am* 1526–94. OE **beorg** + **hām**. Hu 233.

BARHAM Kent TR 2050. 'Beora's homestead or estate'. *Biora ham* 799 S 155, *Beorahames* (genitive case) 805 S 1259, *Beorham* 824 S 1266, *Berham* 1086, [799]13th S 155, *Bereham* 1174×5–13th including S 1266, 1414, 1616, *Be(o)reham* [809]13th S 164. OE pers.n. **Beora* + **hām**. PNK 552, KPN 87, 357, ASE 2.29.

BARHAM Suff TM 1451. 'The hill homestead'. *Bercheham* *[1042×66]12th S 1051, *Bercham* 1086, [1158×62]1331, *Bergham* 1252–1344, *Bargham* 1526, 1568, *Berham* 1286, *Berwham* 1336, *Barham* 1524. OE **beorg** + **hām**. DEPN, Baron.

BARHOLM Lincs TF 0810. 'The homestead on the hill'. *Berc(a)ham, Bercheham* 1086, *Bergham* 1242–1436, *Berham* 1138–1444, *Bargham(e)* 1460–1519, *-home* 1490, *Barham* 1882–1577, *-holme* from 1529. OE **beorg** + **hām**. The village is situated on a slight rise above low-lying fen to the E. Perrott 391, SSNEM 208.

BARKBY Leic SK 6309. 'Barki's village or farm'. *Barchebi -berie* 1086, *Barkebi(a) -by* c.1130–1553, *Barkby* from 1251. ON pers.n. *Barki*, genitive sing. *Barka*, + **bȳ**. Lei 273, SSNEM 34.

BARKESTONE-LE-VALE Leic SK 7835. *Barston in le vale* 1451, *Barkeston in le vale de Beluero* 1511. Earlier simply *Barcheston(e)* 1086–early 13th, *Barkeston(e)* from late 12th, 'Bark's farm or village', ON pers.n. *Bǫrkr, Bark*, ME genitive sing. *Barkes*, + **tūn**. The suffix referring to its location in the Vale of Belvoir is for distinction from BARKESTON Lincs SK 9241. Lei 180, SSNEM 188.

BARKHAM Berks SU 7867. 'Birch-tree meadow'. *(æt) Beorcham(me)* [952]16th and 13th S 559, *Bercheham* 1086, *Berk(e)ham* 1220–1517. OE **beorc** + **hamm**. Brk 91, L 43, 49, 220.

BARKING GLond TQ 4785. 'The Bericingas, the people called after Berica'. *Berecingas* *[677]16th S 1246, *(to) Bercongon* [685×94]8th S 1171, *(in) Berecingum* c.730 BHE (with variants *Berc-* and *-un*),*(in) Bercongum* [c.890]c.1000 OEBede (with variants *Byrc-* and *Berc-*), *(æt) Byorcingan, (into) Beorcingan* [962×91]11th S 1494, *(æt) Byorcyngan* c.1000 Saints, *Berchingas -inges -ingū* 1086, *Berking(e)* 1193 etc., *Barking* 1289. OE folk-n. **Bericingas* < pers.n. **Berica* + **ingas**. The monastery founded here by St Erkenwald c.666 was built in a place called *Beddanhaam* 'Bedda's homestead' [685×94]8th S 1171 in the district or province of the Berecingas whose territory was probably co-extensive with the later hundred of Becontree and included Dagenham. Ess 88, ING 17, TC 201.

BARKING Suff TM 0753. 'The Berecingas, the people called after Berica'. *Berechinge* [c.1030]12th, *Berchingas* *[1042×66]12th S 1051–1377×8 including 1086, with variant *-es, Berkinges* 1086, 1210×12, *Berking(') -yngg(e) -yng(e)* 1220–1442, *Barkyng* 1388. OE folk-n. **Berecingas* < pers.n. ***Berica** + **ingas**. DEPN, Baron, ING 51.

BARKINGSIDE GLond TQ 4489. *Barkingside* 1538. P.n. BARKING TQ 4785 + ModE **side**. It lies at the extreme edge of the old parish of Barking. Ess 101.

BARKISLAND WYorks SE 0519. 'Bark's newly cultivated land'. *Barkesland(e)* 1246–1609, *Barsland(e)* 1419–1777. ON pers.n. *Barkr*, secondary genitive sing. *Barkes*, + **land**. Nearby is Barsey, 'Bark's enclosure', *Barkissay, Barkeshey* 13th, named after the same individual + OE **(ġe)hæġ**. YW iii.57, L 246, 249, WYAS 314.

BARKSTON Lincs SK 9241. 'Bark's farm or village'. *Barcheston(e) -tune* 1086, *Barkestun -ton* 1212–1634×42, *Barkston* from 1212, *Barston* 1231–1766. ON pers.n. *Barkr, Bǫrkr*, secondary genitive sing. *Barkes*, + **tūn**. Perrott 486, SSNEM 188.

BARKSTON NYorks SE 4936. A 'Grimston' hybrid, 'Bark's estate'. *Barces-tune* c.1030 YCh 7, *Barchestun* 1086, *Barkestun -ton(a) -tone* 12th–1568. ON pers.n. *Barkr*, secondary genitive sing. *Barkes*, + **tūn**. YW iv.53, SPN 48, SSNY 125.

BARKWAY Herts TL 3835. 'Birch-tree way'. *Bercheuuei(g)* 1086, *Berkeweie -wey* 1210–1428, *Berc- Berkweie -wey(e)* 12th–1248, *Barkeway* 1524. OE **beorc**, genitive sing. **beorce**, genitive pl. **beorca**, + **weġ**. Hrt 172.

BARKWITH Lincs TF 1681. Uncertain. *Ba(r)cuurde, Barcourde -vorde* 1086, *Barcworda* c.1115, *Westbarkeworth* 1202, *Barkewurthe* 1252, *East, West Barkwith* 1653. Possibly OE **bearg-worth** 'pig farm' with *bearg* 'a castrated boar' early replaced by ON pers.n. *Barkr, Bǫrkr* or appellative *bǫrkr* (< **barku-*) 'bark'. The weak stressed generic was subsequently confused with ON **vithr** 'a wood'. DEPN, SSNEM 212, Cameron 1998.

BARLASTON Staffs SJ 8938. 'Beornwulf's estate'. *Beorelfestun* [1002×4]11th S 1536, *Bernulvestvne* 1086, *Berleston* 1167, 1212, *Borlaston* 1293, *Barlaston* 1300. OE pers.n. *Beornwulf*, genitive sing. *Beornwulfes*, + **tūn**. DEPN, Horovitz.

†If, however, the form *Ætberanforda* *[795 for 792]13th S 138 has any authenticity this would have to be 'Bera's ford'.

BARLAVINGTON WSusx SU 9716. 'The estate called after Beornlaf'. *Berleventone* 1086, *Berlavi(n)ton* 1242–1332, *-yng-* 1354, *Barlauinton* 1296, *Barlton* 1610, 1725. OE pers.n. *Beornlāf* + **ing**[4] + **tūn**. Sx 100 gives prs [bɑ:liŋtən] and [bɑ:ltən], SS 75.

BARLBOROUGH Derby SK 4777. '*Barley* fort or manor house'. *Barleburh* [c.1002]c.1100 S 1536, *-burc -burg(e) -burgh(e)* 1086–16th, *-boro -bor(o)w(e)* 1154×89–1428, *Barlburgh* 1440–77, *Barleyborough* 1600, 1676. Lost p.n. **Barley* + **burh**. *Barley* is either 'boar clearing', OE **bār** + **lēah**, or 'barley clearing', OE **bær** + **lēah**. There are no traces of defence works in the clearing. Db 200.

BARLBY NYorks SE 6334. 'Bardolf's farm or village'. *Bardulbi* 1086, *Bardelbi -by* 12th–1466, *Barthelbi -by* c.1163–1316, *Barlebe* 1363, *-by* 1464. OE pers.n. **Bardwulf* or CG *Bardulf*, + **bȳ**. YE 257, SSNY 19.

River BARLE Somer SS 8534. 'Hill stream'. *Bergel* 1219, *Berghel* 1298, *Bur(e)gel* 1279–80, *Burewelle* 1279, *Barle* from 1575. OE **beorg** + **wylle**. The stream rises in the hills of Exmoor. DEPN, RN 26.

BARLESTONE Leic SK 4205. 'Beorwulf's farm or village'. *Berulvestone* 1086, *Berleston(e) -is* c.1130–1417, *Barlastone* 1277, *Barleston(')* 1319–1833 OS, *Balson* 1723. OE pers.n. *Beor(w)ulf*, genitive sing. *Beor(w)ulfes*, + **tūn**. Lei 483.

BARLEY Herts TL 4038. 'Bera's wood or clearing'. *Beranlei* [c.1053]13th S 1517, *Berenleia* [11th]14th, *Beoronleam* c.1050, *Berlai* 1086, *Berleya -le(y)e* 1253–1524, *Burleye* 1303, 1435, *Barley* 1542. OE pers.n. *Bera*, back-mutated form *Beora*, genitive sing. *Be(o)ran*, + **lēah**. Hrt 173.

BARLEY Lancs SD 8240. 'Barley clearing'. *Bayrlegh* 1324, *Barelegh* 1325. Cf. *Barleboth* 'vaccary belonging to B' 1462, *Barleybothe* 1507, 1513 (ME **bōth**). OE **bere** + **lēah**. La 81, L 206.

BARLEYCROFT END Herts TL 4327 → EAST END TL 4527.

BARLEYTHORPE Leic SK 8409. 'Outlying farm where barley is grown'. *Barlithorp -y-* 1286–1412, *Barleythorpe* from 1496. ME **barli** + ON **thorp**. Some forms, *Barlicthorp(') -k-* 1334–77, show traces of OE **bærlić**. Earlier simply *Thorp juxta Ocham* c.1200–1300, 'outlying farm beside Oakham'. R 686, SSNEM 132.

BARLING Essex TQ 9289. 'The Bærlingas, the people called after Bærel or Bærla'. *(æt) Bærlingum* 998 S 1522, *Berlings* *[1042×66]13th S 1056, *Berlingā* 1086, *Berling(e)* 1179–1229, *(de) Berlingis* 1100, *-es* 1100×2, 1235, *Barling(e)* from 1240. OE folk-n. **Bǣrlingas* < pers.n. **Bǣrel* or **Bǣrla* + **ingas**. Ess 178, ING 18, Jnl 2.42.

BARLOW Derby SK 3474. Either 'boar clearing' or 'barley clearing'. *Barleie* 1086, *-lega -leg(e) -legh -leia -ley(e) -le(e) -lay* 1180–17th, *Barlow* from 1576. OE **bār** or **bær** + **lēah**. For the change of suffix cf. BARLOW T&W NZ 1560. Db 203, L 206.

BARLOW NYorks SE 6428. Either 'clearing with a barn' or 'growing with barley'. *Bernlege* c.1030 YCh 7, *Berlai(a) -ley(e) -lay(e)* 1086–1498, *Barley* 1469–1641, *Barlow(e)* 1458 etc. Either OE **bern** < *bere-ærn* or **beren** + **lēah**. YW iv.23.

BARLOW T&W NZ 1560. 'The barley clearing'. *Berleia* [1183]c.1382, *Berley(e)* 1242×3, 1369, 1380, *Barley* 1613×4, *Barlow* 1768 Armstrong. OE **bere** + **lēah**, cf. BARLEY. Weakening of the suffix *ley* to [lə] has caused confusion with Barlow < OE **bere-hlāw** 'barley hill' and consequent back-spelling. Barlow stands on a hill. NbDu 11.

BARMBY Either 'B(j)arni's village', ON pers.n. *B(j)arni*, genitive sing. *B(j)arna*, + **bȳ**, or 'the children's village or estate', ON **barn**, genitive pl. **barna**, + **bȳ**. The sense would be 'estate, village whose income is devoted to the upkeep of children' or 'secondary settlement established by the children of tenants or owners'. Jnl 17.5–13, V i.50.

(1) ~ MOOR Humbs SE 7848. *Barnebi -by in, upon the More* 1371, 1650, *Barmby super Moram* 1492. Earlier simply *Bernebi* 1086, *-by* 1201, *Barnebi* 1086–1336 with variant *-by*, *Barnby* 13th–1542, *Barmeby* 1285. YE 184, SSNY 19.

(2) ~ ON THE MARSH Humbs SE 6928. *Barmby on the Marsh* 1828. Earlier simply *Bærna-* *Barnabi* [c.1050]late 11th, *Barnebi -by* 1086–1379, *Barnby* 1342, *Barmebie* 1566. YE 249.

BARMER Norf TF 8133 Apparently 'bear pool'. *Benemara* 1086, *Beremere* 1202, 1254. OE **bera** + **mere**. DEPN, L 26–7.

East BARMING Kent TQ 7254. *Estbarmeling'* 1240, *Estbarmlinge* 1242. ME adj. **est** 'east' + p.n. *Bermelinge* 1086, *Bermeling* c.1100, 1186×7, *Bearmling(et)es* c.1100, *Barm(e)linges* 1186×7–1240, *Barm(e)ling'* 1201×2–1242×3, *Barming* 1610, of uncertain origin. 'East' for distinction from West Barming TQ 7154, *West Barmlynge*, ~ *Bramlyng* 1308, earlier simply *Bermelie* 1086. The DB form *Bermelie* for West Barming may indicate an OE p.n. **Bearm-lēah* with OE **bearm** 'bosom' but probably in the sense 'edge, berm, ditch' as its cognates ON *barmr*, LG *berm(e)*, + **lēah** 'wood or clearing', in which case the form *Bearmlinges* could represent an OE **Bearm-lē-ingas* 'the inhabitants of *Bearm-lēah*'. But this is very uncertain. The form *Bearmlingetes* survived as *Barnjet* 1819 OS for West or Little Barming, *Barmyngett* 1434, 1486, *Bermynget* 1535, apparently an example of the OFr diminutive suffix **-ette** 'little' applied to an English p.n. PNK 132–3, KPN 8, ING 8, PNE i.161.

BARMOOR CASTLE Northum NT 9939. P.n. Barmoor + ModE **castle** referring to a castellated mansion begun in 1801 on the site of an old pele-tower. Barmoor, *Beire-* *Beigermore* 1231×2, *Bey(i)rmor'* 1242 BF, *Bayremore* 1289, 1296 SR, *Bayr-* 1346, *Barmour -more* 1539, is the 'berry moor, the moor where berries are gathered', OE **beġer** + **mōr**. NbDu 11, DEPN, L 55.

BARMPTON Durham NZ 3118. 'Beorma's farm or village'. *Berme-* *Bernetun* [c.1104]12th, *Bermentun(e) -tona* 1109×14, *Bermestuna* 1154×66, *Bermeton(e)* 1141–1457, *Bermton* 1358–1422, *Bermpton* 1424, *Barm(e)ton* 1475–1614. OE pers.n. *Beorma* (possibly short for *Beornmund*) + **tūn**. NbDu 11.

BARMSTON Humbs TA 1659. 'Beorn's settlement'. *Berneston(a) -tun(e)* before 1080–1420, *Benestone -tun* 1086, *Barneston* 1441, *Barmeston* 1561. OE pers.n. *Beorn*, genitive sing. *Beornes*, + **tūn**. YE 83, SSNY 253.

BARNACK Cambs TF 0705. Uncertain. *(on) Beornican* [c.980]c.1200 B 1130, *Bernak(e)* [1052×65]c.1350, [1053]13th, 1210, *Bernac* 1086, 1200, *Bernec(a) -ek* 1162–1259, *Bernack(e)* 1205, 1209, *Barneck* 1284, *Barnicke* 1582, *Barnoak* 1779. If the B 1130 form is reliable this could be the OE folk-n. *Beornice* in the dative pl. *Beornicum* 'among the Beornice, the people of Bernicia', i.e. the people of N Northumbria. All the other forms, however, point to OE **beorn-āc** 'warrior oak-tree' with variants **ǣće** dative sing., **ǣć** plural, **ācum** dative pl. Nth 230.

BARNACLE Warw SP 3884. 'The sloping wood by the barn'. *Bernhangre* 1086, 1247, *-angre -er* 1333–1420, *Beranger* 1261–1318, *Bern(h)angel* 1299–1316, *-angul* 1314, 1545, *Barnacle* from 1547 with variant *-agell*. OE **bern** + **hangra**. Wa 101, L 195–6.

BARNARD CASTLE Durham NZ 0516. 'Bernard's castle'. *(de) castello Bernardi* 1197, 1200, *(de) Castro Bernardi* 1235–1495, *(le) Chastel Bernard* 1306–16, *Bernardcast(i)ell(')* 1365–1461, *(de) Castrobarnardi* 1412–28, *Barnardcastell* 1484–1596, ~ *castle* from 1498, *Barney Castell* 1486. CGmc pers.n. *Bernard* + ONFr **castel**. The castle was founded and built by Bernard Baliol I c.1125 as a new centre for the barony of Gainford granted to his father Guy by William the Conqueror. Its site and relationship to the ancient barony is comparable to that of Earl Alan's castle of Richmond to the ancient lordships of Catterick and Gilling. NbDu 12.

BARNARD GATE Oxon SP 4010. *Barnard Yate* 1725. Surname *Barnard* + ModE **gate**, dial. **yate** (OE *ġeat*). O 261, 447.

BARNARDISTON Suff TL 7148. 'Bernard's estate'. *Bernardeston* 1194, 1242, *Barnardston* 1610. Pers.n. CG *Bernard*, genitive sing. *Bernardes*, + **tūn**. DEPN.

BARNBURGH SYorks SE 4803. 'Barni or Bjarni's fortification' or 'fort on land held by joint inheritance'. *Berneborc -burg* 1086, *Barneburg(h) -burc(h)* 1086–1441, *Barm(e)burgh* 1379, 1495. ODan pers.n. *Barni* or *Bjarni*, genitive sing. *B(j)arna*, or ON, OE **barn** 'a child, an heir', genitive pl. **barna**, + **burh**. YW i.80, SSNY 142, Jnl 17.10.

BARNBY 'The children's village or estate '. ON or OE **barn**, genitive pl. **barna**, + **bȳ**. The reference is probably to an estate held jointly by a number of heirs. Jnl 17.5–13.

(1) ~ NYorks NZ 8112. *Barnebi* 1086. YN 135, SSNY 19 no.5.

(2) ~ Suff TM 4789. *Barnebei* 1086, *Barneby* 1086–1344, *Barnaby* 1524, *Barnbye* 1568. The reference is probably to multiple inheritance of an estate by the children of the owner. An alternative specific is the ON pers.n. *Barni*, genitive sing. *Barna*. DEPN, Baron.

(3) ~ DUN SYorks SE 6109. Short for *Barnebi on Done* early 13th, *-on Dun* 1300. P.n. *Barnebi -by* 1086–1428, *Barmby* 1379, 1641, + r.n. DON SK 4696 for distinction from BARNBY MOOR Notts SK 6684. YW i.17, SSNY 19, Jnl 17.10.

(4) ~ IN THE WILLOWS Notts SK 8652. *Barnebie in le Willowes* 1589. Earlier simply *Barnebi* 1086, c.1190, *-by* 1201–1327, *Barmeby -be* 16th cent. Nt 210, Jnl.17.5, SSNEM 34.

(5) ~ MOOR Notts SK 6684. *Barneby super le More* 1561. P.n. *Barnebi -by -beya* 1086–1364 + ModE **moor** for distinction from BARNBY DUN SE 6109 etc. Nt 67, SSNEM 20, Jnl 17.10.

BARNES GLond TQ 2276. 'The barns'. *Berne* *[939]12th S 453, 1086, *(of) Bærnun* (dative pl.) c.1000, *Bernes* 1222–1505, *Barnes* 1387. OE **bern**, pl. **bernas**, dative pl. **bernum**. Sr 11, DEPN, TC 201.

East BARNET GLond TQ 2794. *Estbarnet* 1294. ME adj. **est** + p.n. *Barneto* (Latin ablative case) [1064×77]14th, *Barnet* 1197, *(La) Barnette* 1248, *The Barnet* 1398, the 'place cleared by burning', OE **bærnet**, E of High Barnet. The reference is to a large forest area held by the abbey of St Albans on the borders of Herts and Middlesex cleared by burning. It gave names to East, Friern and High or Chipping BARNET none of which is separately mentioned in DB. They were probably not settled until the late 11th or early 12th cent. Hrt 70, TC 201, Encyc 255.

Friern BARNET GLond TQ 2892. 'The brothers' B'. *Frerennebarnethe, Frerenbarnet* 1274, *Fryeringe Barnet* 1549. ME **frere**, adjectival genitive pl. **frerene**, + p.n. *la Bernet, Barnate* 1235, 'land cleared by burning' as in East BARNET TQ 2794. The land was possessed by the knights of St John of Jerusalem. See also East BARNET TQ 2794. Mx 99, GL 50.

BARNETBY LE WOLD Humbs TA 0509. 'B on the wold'. *Barnetby le Wold* 1824. Earlier simply *Bernode- Bernete- Bernedebi* 1086, *Bernetebi -by* c.1115–1338, *Bernetby* 1180–1431, *Barnebi* 1185, *Barneby -aby* 1428–1700. This is usually explained as 'Beornnoth's farm or village', OE pers.n. *Beornnōth* + **bȳ**. The absence of the genitive ending *-es* is noteworthy: medial *-e-* may be a relic of ON genitive sing. *-a(r)*. Spellings with *-ede-* may point to OE hypocoristic pers.n. **Beornede* with possible later confusion with ME *bernet, barnet*, 'place cleared by burning'. The village lies on the W slope of the Lincolnshire wolds. Li ii.8, SSNEM 35.

BARNEY Norf TF 9932. Possibly 'Bera's island'. *Berlei* 1086, *Berneie* 1198, *Berneia* 1214. Possibly OE pers.n. *Bera*, genitive sing. *Beran*, + **ēġ**; but the specific could also be OE **beren** 'growing with barley' or **berern** 'a barn'. DEPN.

BARNHAM Suff TL 8779. 'Beorn's homestead'. *Ber(n)ham* 1086, *Bernham* 1105, 1230, 1610. OE pers.n. *Beorn* + **hām**. DEPN.

BARNHAM WSusx SU 9604. 'Beorna's river meadow'. *Berneha(m)* 1086–1316, *Bernham* 1105–1492. OE pers.n. *Beorna* + **hamm** 2a, 3, 4. A possible alternative is OE **beorn**, genitive pl. **beorna**, 'the river meadow of the warriors'. Sx 137, ASE 2.41, SS 92.

BARNHAM BROOM Norf TG 0807. *Barnham Broom* 1838 OS. P.n. Barnham + ModE **broom**. Barnham, *Bernham* 1086–1276, is possibly 'Beorn's homestead', OE pers.n. *Beorn* + **hām**. DEPN.

BARNINGHAM Durham NZ 0810. 'Homestead, estate called after Beorn(a)'. *Berningham -yng-* 1086–1406, *Berningeham* 1187, *Bernigham* 1198–[1212]14th, *Bernigeham* 1214, *Barnyngham* 1491. OE pers.n. *Beorn(a)* + **ing**[4] + **hām**. YN 302. ING 155.

BARNINGHAM Suff TL 9676. 'The homestead called or at *Beorning*, the place called after Beorn(a), or of the Beorningas, the people called after Beorn(a)'. *Bernincham, b'ninghā, b'nichā* 1086, *Beorningeham* c.1095, *Berning(e)ham, Bernicham* [1087×98]12th, *Berningeham* 1195, *Berningham* 1203, *Beringham* 1610. OE p.n. **Beorning* < pers.n. *Beorn(a)* + **ing**[2], locative-dative sing. **Beorninğe*, or folk-n. **Beorningas* < pers.n. *Beorn(a)* + **ingas**, genitive pl. **Beorninga*, + **hām**. DEPN, ING 129.

BARNINGHAM MOOR Durham NZ 0608. P.n. BARNINGHAM NZ 0810 + **mōr**.

Little BARNINGHAM Norf TG 1333. *Berningham Parva* 1254. ModE adj. **little**, Latin **parva**, + p.n. *Berning(e)ham, Bernincham* 1086, *Berningeham* 1166, *Berningham* 1203, 'the homestead of the Beorningas, the people called after Beorn', OE folk-n. **Beorningas* < pers.n. *Beorn* + **ingas**, genitive pl. **Beorninga*, + **hām**. 'Little' for distinction from North Barningham TG 1537, *Northberningham* 1291, and Town Barningham TG 1435, *Tunberningham* 1254. Also recorded here is a *Berneswrde* 1086, 'Beorn's enclosure', at Little Barningham. DEPN, ING 132.

BARNOLDBY LE BECK Humbs TA 2303. 'B on the beck or stream'. *Barnolby le Beck* 1706. Earlier simply *Bernulfbi* 1086, 1196, *Bernoluebi* 1177×8, *Bernol(e)bi -by* 1202–1526, *Barnolby* 1387–1623, *Barnoldby* from 1408, 'Beornwulf or Bjǫrnulfr's farm or village', OE pers.n. *Beornwulf* or ON *Bjǫrnulfr*, + **bȳ**. Li iv.55, SSNEM 35.

BARNOLDSWICK Lancs SD 8746. 'Beornwulf's dairy-farm'. *Bernulfesuuic* 1086, *Bernolueswic(h) -wik -wyk* before 1153–1333, *Bernoldeswick* 1155×62, *Barnoldswicke* 1576, *Bernolfwic -wik -wyk(e)* 1147×50–1415, *Bernolwic -wik -wyk* 13th–1349, *Barleweke* 1672. OE pers.n. *Beornwulf*, genitive sing. *Beornwulfes*, + **wīc**. YW vi. 34 gives pr [bɔ:lik].

BARNS GREEN WSusx TQ 1226. *Barnses Green* 1813 OS. Surname *Barns* + ModE **green**.

BARNSLEY Glos SP 0705. 'Beornmod's wood or clearing'. *(æt) Bearmodes lea* [798×822]11th S 1262, *(æt) Beorondes lea* [855]11th S 206, *Bernesleis* 1086, *-leia* 1154×89, *Berdesley(a) -leia* 1195–6, *Bardesley(e) -leg(h) -le* 1186×91–1535, *Barndesley(e) -le(gh)(e)* 1221–1494, *Barn(e)sley* 1326–1623. OE pers.n. *Beornmōd*, genitive sing. *Beornmōdes*, + **lēah**. Gl i.24

BARNSLEY SYorks SE 3406. 'Beorn's clearing'. *Berneslai(a) -lei(a) -le(y) -lay(a)* 1086–1413, *Barneslai(a) -ley(a) -lay* c.1120–1597, *Barnsley* from 1460. OE pers.n. *Beorn*, genitive sing. *Beornes*, + **lēah**. Early spellings with *ar* are AN; from late 14th they represent normal late ME *ar* < *er*. YW i.302, TC 46, L 203.

BARNSTAPLE Devon SS 5533. 'Battle-axe pillar'. *bearstaple, berdesteaple* [914×9]13th BH2,3, *B(e)arda- Bardanstapol* 979×1016 Coins, *B(e)ardastapol* 1016×66 Coins, *Beardastapol* 1018, *Berdestaple* 1155–1326, *Barnestapla -e, Bardestaplensis burg* 1086, *Bernestap(e)le* 1236–1345, *Barnestaple* 1286, *Barstaple* 1466×84, *Barstable* 1549, 1553, *Bastable, Barnstaple* 1675. This name has to be compared with Barstable Hall and Hundred, Essex, *Berdestapla* 1086, and the f.n. *Berdestapel* 1260 Herts which are compounds of OE ***bearde** 'a battle-axe' + **stapol** 'a pillar'. These are 'totem-pole' names and refer to posts or pillars set up to mark religious or administrative meeting places, cf. the p.ns. GARSTANG Lancs, GARTREE

Leics, Lincs, which signify a post with a spear. Barnstaple received its first charter as a burgh in 930 and is thus one of the earliest recorded in Britain. The town was given the right to mint in the 10th cent. and received further charters in 1154, 1189, 1201 and 1273. It has been claimed that the name of the town makes it clear that from the beginning there was an important market or staple here. However, *staple* sb. 'emporium, mart', apparently a borrowing from OFr or MDu where *stapel* seems to have meant 'raised platform' where tolls and duties were collected and justice administered, is not recorded in England before 1423. D 25, BdHu 298, Ess 140–1, Earle 466 n., PNE i.22, BH 111.

BARNSTON Essex TL 6519. 'Beorn's estate'. *Bernestunā* 1086, *-ton* 1220–1328, *Barn(e)ston'* from 1254. OE pers.n. *Beorn*, genitive sing. *Beornes*, + **tūn**. Ess 470 gives pr [bænsn].

BARNSTON Mers SJ 2883. 'Beornwulf's farm or village'. *Bernestone* 1086, *-ton* 1208–1524 with variant *-is-*, *Berleston(a)* [1096×1101]1280–1308 with variants *-is-* and *-tone*, *Berules-Bernolweston* [13th]14th, *Barneston* 1579, *Barnston* 1659. OE pers.n. *Beornwulf*, genitive sing. *Beornwulfes*, + **tūn**. Che iv.263.

BARNT GREEN H&W SP 0073. *Barn(e) Green* 1468–1789, *Brantyrene* (sic) 1535. P.n. Barnt + ME **grene**. Barnt, *Barnte* 1290, *Brante* 1315, is ME **barnet** (OE **bærnet**) 'place cleared by burning'. Wo 338.

BARNTON Ches SJ 6375. 'Beornthryth's farm or village'. *Berthinton* late 12th, *-yn(g)- -ing-* 13th cent., *Bern(e)ton* 1319–1413, *Barneton* 1517, *Barnton* from 1577. OE feminine pers.n. *Beornthrȳth* + **ing**[4] + **tūn**. Che ii.105,107.

BARNWELL Northants TL 0484. Partly uncertain. *(æt) Byrnewilla(n)* [c.980]12th B 1130, *Beornwelle* [1077]13th, *Bernewell(e)* 1086–1424. If the *y* of the earliest form is reliable this might be OE **byrġen** + **wylle** 'the spring by the burial mound'; but the later spellings point to **beorna**, genitive pl. of OE **beorn** or pers.n. *Beorna* + **wella** 'the spring of the warriors' or 'Beorna's spring'. Subsequent confusion with *bairn* (OE *bearn*) 'a child' led to popular explanations such as that of Bridges (1791): 'About the town are seven or eight wells, from which, and the custom of dipping *Bernes* or children in them, the town is supposed to be denominated'. Nth 178, DEPN.

BARNWOOD Glos SO 8618. 'Beorna's wood'. *Berneuude* 1086, *-wode* 1164×79–1378, *Bernwude -wode* 1086×7, 1221–1365, *Barn(e)wood* 1535–1617. OE pers.n *Beorna* + **wudu**. Gl ii.117, L 229.

Great BARR WMids SP 0495. *Great Barre* 1322. ME **gret** + p.n. *(æt) Bearre* [957]12th S 574, *Barre* 1086, 'the hill-top, the summit', PrW ***barr**. 'Great' for distinction from *Little Barre* 1208. The reference is to Barr Beacon SP 0697 which rises to 744ft. DEPN, WestMids 61.

BARRAS Cumbr NY 8411. Uncertain. *Barras* 1588, 1663, *Barrow(e)s* 1608, 1863, *Bar(r)house* 1661, 1761. Usually explained as 'hill house', ultimately from OE **beorg** + **hūs**. But the later forms may be a rationalisation of the derogatory term 'bare arse'. We ii.74.

BARRASFORD Northum NY 9173. 'The ford of the grove'. *Barwisford* 1242 BF, *Barewes-* 1256, *Baruys- Barue-* 1296 SR, *Barous- Barassford* 1479. OE **bearu**, genitive sing. **bearwes**, + **ford**. NbDu 12, L 70, 190.

BARRINGTON Cambs TL 3949. 'Bara's farm or estate'. *Barentona -tone* 1086–1509×47, *Barn(e)ton(a)* 1218–1459, *Barington(e) -yng-* 1236–1554. OE pers.n. **Bara*, genitive sing. **Baran*, + **tūn**. Ca 70.

BARRINGTON Somer ST 3918. Partly uncertain. *Barintone* 1086, *Barinton* 1185, 1201, *Barington* 1225, *Barrington* [1279×92]B. Possibly 'settlement at *Bæring*, the bare place, the place without vegetation', OE ***bæring** < **bær** + **ing**[2] used as a p.n. + **tūn**. Barrington lies beside a hill-spur jutting into marsh. DEPN.

Great BARRINGTON Glos SP 2113. *Majore, magna Bernin(g)ton* 1221, *Great Baryngton* 1461. Earlier simply *Berniton(e)* 1086, 1109×2, *Berninton(a) -e -tun -yn-* 1086–1485, *Bernington* 1221, *Berington -yng-* 1199–1354, *Barnton* 15th, *Barrington* 1543, 'estate called after Beorna', OE pers.n. *Beorn(a)* + **ing**[4] + **tūn**. 'Great' for distinction from Little BARRINGTON SP 2012. Gl i.193.

Little BARRINGTON Glos SP 2012. *Parua Berni(n)ton -ing- -yn(g)-* 1225–1303, *Parua Berneton'* 1401, *Little Beryngton* 1446, *Lytle Barington* 1509×47. Adj. **little**, Latin **parva**, + p.n. *Berni(n)tone* 1086 as in Great BARRINGTON SP 2113. Gl i.194.

BARRIPPER Corn SW 6238. 'Beautiful retreat'. *Beaurepere* 1397, *-repper* 1454 G. OFr **bel, beau** + **repaire**. PNCo 52.

BARROW '(Settlement at) the grove'. OE **bearwe**, dative sing. of **bearu**. L 190.

(1) ~ Lancs SD 7338. Cf. *Barowclough*, *Baroweclogbsik* 1324, 'Barrow ravine' and 'barrow ravine stream', p.n. Barrow + OE **clōh** and **sīc**. La 77.

(2) ~ Shrops SJ 6500. *Barwa* c.1200, *Bar(e)we* 13th cent., *Barowe* 1413, *Barrow* 1577. Sa i.30.

(3) ~ Suff TL 7663. *Barō* 1086, *Barue* 1201, *Bareowe* 1610. DEPN.

(4) ~ GURNEY Avon ST 5367. 'B held by the Gurnai family'. *Barwe Gurnay* 1283. Earlier simply *Berve* 1086, *Barewe* 1269. The tenant in 1086 was Nigel de Gurnai from Gournay-en-Brie, Normandy. DEPN, L 190.

(5) ~ UPON HUMBER Humbs TA 0721. *Barowe upon Humbre* 1317. Earlier simply *Adbaruae, id est Ad Nemus* 'at Barrow, that is to say at *Nemus*, the grove' [c.731]8th BHE with variant *Adbearuae*, *Æt Bearwe* [c.890]c.1000 OEBede, *Bearuwe* [737×40]11th S 99, [971]12th, [973]13th S 792, *(to) Baruwe* [971]12th S 782, *Baruue* [972]12th S 787, *Beruwe* [973]14th S 792, *Barwe* 1121 ASC(E) under year 963, [972]12th S 787, *Barewe* 1086–1368, *Barowe* 1303–1576, *Barrow* from 1535. Li ii.15.

(6) ~ UPON SOAR Leic SK 5717. *Barhou* 1086, *Baru* 1158, *Bar(e)wa -e* 1123×47–1415, *Barrow(e)* from 1413 + r.n. SOAR SK 5600, *super Sore* 1294–[1319]15th, *upon Sore* 1457–1594. Lei 275, L 190.

(7) ~ UPON TRENT Derby SK 3528. *Baru upon Trent* 1427. P.n. Barrow + ME **upon** + r.n. TRENT. Barrow is *Barewe* 1086–1291, *Barwe* 1086–1350, *Barowe* 1288, 1330, 1346, *Barrow* from 1577. Db 430, L 190.

(8) Great ~ Ches SJ 4768. *Magna Barue* 1294, *Great ~* 1391. Adj. **great** (Latin **magna**) + p.n. *Barue* [958]14th S 667, *Barwe* 1181–1445, *Barrow* from 1386. 'Great' for distinction from Little Barrow SJ 4770, *Parva Barue* 1294. Che iii.261.

(9) North ~ Somer ST 6029. *Northberwe* 1242, *Northbar(e)we* [1317] Buck, [1381×2]14th. ME adj. **north** + p.n. *Berrowene* (sic), *Berve* 1086, *Baruwe* [1187] Buck, *Barewe* 1225. 'North' for distinction from South BARROW ST 6027. DEPN.

(10) South ~ Somer ST 6027. *Sud Berwe* 1242, *Southbarewe* [1381×2]14th. ME adj. **suth** + p.n. Barrow as in North BARROW ST 6029. DEPN.

BARROW 'Burial mound, tumulus, hill'. OE **beorg**. L 127–8.

(1) ~ Leic SK 8915. *Berc* 1197–1319, *Berk* 1263–1319, *Bergh* 1316–1428, *Berugh* 1324, *Berew(e)* 1305–1405, *Barrowe* 1535–1610. There is a tumulus on the hill-crest near The Green. R 7.

(2) ~ STREET Wilts ST 8330. *Barowstreate* 1566. P.n. *la Baruwe* 1285, 'the barrow', ME **berewe** (OE *beorg*), + **strete** 'a hamlet'. A large barrow formerly stood at the end of Barrow Street Lane. Wlt 178.

Mid BARROW Essex TR 1992. 'Middle sand-bank'. ModE **mid** + Barrow as in BARROW DEEP TM 3004. Both places are now lightships.

BARROW DEEP Essex TM 3004. P.n. Barrow as in *Est, West Barowes* 1509×47 + Mod **deep** 'a deep place, deep water'. Barrow is probably OE **beorg** 'a hill' in the sense sand-bank. Ess 13.

BARROW-IN-FURNESS Cumbr SD 1969. This was originally the name of Barrow Island SD 1968, once detached from the mainland, *Barrai* 1190, 1191×8, *Barray* 1292, 1336, *Insula de Oldebarrey* 1537, *Old baro Insula* 1577, PrW ***barr** 'a top, a summit, a hill' (Brit **barro-*) + ON **ey** 'island'. The island was named from a summit on the mainland referred to as *Barrahed* 1537, *Barrohead* 1577. Identical with Barra, Scotland NF 7000, *Barru* 11th. 'In Furness' was added for distinction from other Barrows after the development of the hypercorrect form Barrow owing to the pr [barə] common to both Barrow and Barrey. La 204, SSNNW 214, L 40, 127.

BARROWAY DROVE Norf TF 5703. *Barroway Drove* 1824 OS. P.n. Barroway + ModE **drove** 'a road along which horses or cattle are driven'.

BARROWBY Lincs SK 8836. 'Hill village or farm'. *Bergebi* 1086, *Berghebi -by* 1206–1396, *Beruby* 1242–1428, *-ughby* 1296–1452, *Barowby(e)* 1406, 1594, *Bar(r)oughby(e), Barroby(e)* 16th cent. ON **berg** + **bȳ**. The specific may have been in the genitive pl. form *berga* and may have replaced OE *beorg*. The village is situated on a hill overlooking Foston Beck. Perrott 446, SSNEM 34.

BARROWDEN Leic SK 9400. 'Hill with burial mounds'. *Berchedone* 1086, *Bergedunam -dona(m) -don'* 1141–1220, *Berg(h)do(u)n* 1316–1428, *Bargh(e)don* 1263, 1459, *Beregedun* 1202, *Berewedon(e)* 1205–1364, *Berewdon* 1205, *Ber(o)ughdon* 1342–1431, *Barugdon* 1428, *Bar(r)oug(h)don* 1490–1610, *Barodon* 1471, 1537, *Bar(r)owden* from 1487, *Bar(r)adon* 1503, 1621. OE **beorg**, genitive pl. **beorga**, + **dūn**. The barrow, already ancient at the time of the Anglo-Saxon arrival, stood on a ridge overlooking the r. Welland. R 232.

BARROWFORD Lancs SD 8539. 'Grove ford'. *Barouforde* 1296, *-ford* 1324, 1325. OE **bearu** + **ford**. La 87, L 70, 190.

BARSBY Leic SK 6911. Probably 'Barn's village or farm'. *Barnesbi* 1086, 1190, *-bia -by -is- -ys-* c.1130–1502, *Barnebi -by* 1224–1308, *Baresby(e) -ys-* 1408–1610, *Barsby* from 1620. ON pers.n. *Barn*, genitive sing. *Barns*, + **bȳ**. But ON **barn** 'child' is possible as in the p.n. type BARNBY. Lei 294, SSNEM 35.

BARSHAM Suff TM 3989. 'Bar's homestead'. *Bar- Bersham* 1086, *Barsham* from 1196. OE pers.n. *Bār*, genitive sing. *Bāres*, + **hām**. DEPN.

East BARSHAM Norf TF 9133. *Estbaresham* 1254. ME adj. **est** + p.n. *Barsa- Barseham* 1086, *Barsham* from 1185, 'Bar's homestead', OE pers.n. *Bār*, genitive sing. *Bāres*, + **hām**. 'East' for distinction from North and West BARSHAM TF 9135 and 9033. DEPN, PNE i.19.

North BARSHAM Norf TF 9135. *Norbarsam* 1086, *Northbarsham* 1254. OE adj. **north** + p.n. Barsham as in East BARSHAM TF 9133. DEPN.

West BARSHAM Norf TF 9033. *Westbaresham* 1254. ME adj. **west** + p.n. Barsham as in East BARSHAM TF 9133. DEPN.

BARSTON WMids SP 2078. 'Beorhtstan's estate'. *Berces- Bertanestone* 1086, *Berstan(e)ston* 1199–1327, *Berstone* 1332–1500, *Barston* 1462, *Barson* c.1600. OE pers.n. *Beorhtstān*, genitive sing. *Beorhtstānes*, + **tūn**. A possible alternative pers.n. is *Beornstān*. Wa 55.

BARTESTREE H&W 5641. 'Beorthwald's tree'. *Bertoldestrev* 1086, *Berkwoldestre* 1206. OE pers.n. *Beorthwald*, genitive sing. *Beorthwaldes*, + **trēow**. DEPN, L 212, 217.

BARTHOMLEY Ches SJ 7652. Partly uncertain. *Bertemeleu* 1086, *Bertamelegh* 13th, *Bertumleg'* early 13th–1516 with variants *-lega -ley(e)* etc., *Bartumilegh* 1287, *Bartumley* 1507–1653, *Barthomley* 1549. Usually explained as 'wood or clearing of the dwellers at *Berton*', OE folk-n. **Bert-hǣme* < p.n. **Berton* + **hǣme** 'inhabitants', genitive pl. **Bert-hǣma*, + **lēah**. The hypothetical lost **Berton*, OE **bere-tūn** 'barley farm, demesne farm', is not identified, but the p.n. formation is a well-known one, cf. MARCHAMLEY Shrops SJ 5929, MARCHINGTON Staffs SK 1330, STRETTINGTON WSusx SU 8907, WALKHAMPTON Devon SX 5369. Doubts have been raised, however, whether the regular ME spelling with *-um-* could represent OE *hǣme* rather than an OE dative pl. ending *-um*. Possibly, therefore, this is OE p.n. **Beretūnum* 'at the granges' + **lēah**. Che iii.5 gives pr ['ba:təmli].

BARTLEY Hants SU 3112. Cf. *Bartley Ho., Bartley Lodge* 1811 OS.

BARTLOW Cambs TL 5845. 'At the birch-tree burial-mounds'. *Berk(e)lawe* 1232–1485, *-low(e)* 1260–1435, *Bertelawe* 1448, *Bart(e)lowe* 1559–99. OE **beorc** + **hlāw**. The OE form must have been in the dative-pl. *(æt) beorca-hlāwum* referring to Bartlow Hills, seven steep-sided burial-mounds of the Romano-British period. Ca 101, Pevsner 1954.233, L 220.

BARTON 'A corn farm, an outlying grange, a demesne farm, especially one retained for the lord's use and not let to tenants', OE **ber(e)tūn**, **bærtūn**. The name implies a settlement which was originally a component of a larger unit. PNE i.31.

(1) ~ Cambs TL 4055. *Barton* from *[1060]13th S 1030, *Bertona* 1086, 1168×9, *Berton(e)* 1086–1553. Ca 72.

(2) ~ Ches SJ 4454. *Berton'* 13th, 1318, 1560, *Barton'* from 13th. Part of the manor of Farndon SJ 4154. Che iv.68.

(3) ~ Devon SX 9067. *Bertone* 1333, 1414, *Barton* 1566. D 519.

(4) ~ Glos SP 1025. *Beretun* 1185, *Berton(e)* 1158–1354, *Barton(e)* 1287–1495. An outlying grange of the Knights Templar of Temple Guiting SP 0927. Gl ii.14.

(5) ~ Lancs SD 5137. *Bartun* 1086, *Barton* from 1212, *Berton* 1226. A dependent settlement of the Amounderness estate. La 148, Jnl 21.34.

(6) ~ NYorks NZ 2308. *Barton* 1086 etc. YN 285, PNE i.31.

(7) ~ Warw SP 1051. *Berton(a)* 1154×89, 1315, *Barton* 1349. An outlying farm of BIDFORD-ON-AVON SP 1052. Wa 203.

(8) ~ BENDISH Norf TF 7105. 'B within the ditch'. *Berton Binnedich* 1249. P.n. *Bertuna* 1086 + OE **binnan** + **dīċ**. The reference is to Devil's Dyke which runs N–S a mile E of the village. DEPN.

(9) ~ HARTSHORN Bucks SP 6430. Short for Barton and Hartshorn. *Barton Hartshorn* c.1450, *Barton and Herteshorne alias Beggars Barton* 1541. P.n. *Berton(e)* 1086–1241, *Barton* from 1237×40, + lost p.n. Hartshorn identical with HARTSHORNE Derbys SK 3221. Bk 58.

(10) ~ IN FABIS Notts SK 5232. *Barton in le Benes* 1388, 1451, *Berton in the beanes* 1502, *Barton in fabis* 1556. P.n. *Barton(e)* 1086–1289 + Latin **in fabis** 'in the beans'. Leland c.1540 noted the growing of beans in the district. Nt 244, xlii.

(11) ~ IN THE BEANS Leic SK 3906. *Barton in le Bean(e)s* 1591, *~ in the Beanes* 1638. Earlier simply *Barton(e)* 1086–1610, *Berton'* 1200, 1225, 1390. The affix *in the beans* alludes to the fertility of the ground. Lei 53.

(12) ~ -LE-CLAY Beds TL 0831. *Barton-in-the-Clay* 1535, earlier simply *Bertone* 1086, *Barton* 1202. The soil is a strong clay. The addition of the Fr definite article **le** (representing OFr *en le*) became fashionable in the 16th cent. By origin probably an outlying farm belonging to Shillington. Bd 145.

(13) ~ -LE-STREET NYorks SE 7274. 'B on the Roman road'. P.n. *Bartun(e) -ton(e)* 1086 + ME *in Rydale* 1280, ModE *in le Strete* 1614 as in APPLETON-LE-STREET SE 7373, Fr definite article **le** short for *en le* + ModE **street**. YN 47.

(14) ~ -LE-WILLOWS NYorks SE 7163. 'B in the willow-trees'. P.n. *Bartun* 1086, *Barton'* c.1280 etc., + ModE *in the Willos* 1574 for distinction from BARTON–LE-STREET SE 7274, Fr definite article **le** short for *en le* + ModE **willow(s)**. YN 38.

(15) ~ MILLS Suff TL 7173. P.n. *b'tona, Bertunna* 1086, *Berton* c.1235, + ModE **mill**. Also known as *Parva Bertone* 'little B' 1254 for distinction from Great BARTON. DEPN.

(16) ~ MOSS GMan SJ 7397. P.n. Barton SJ 7697, *Barton* from 1196, + ModE **moss** 'a bog'. Probably a *bere-tun* of the royal manor of Salford. La 38, Jnl 17.33.

(17) ~ -ON-THE-HEATH Warw SP 2532. *Barton on the Yethe* 1551, *~ super le Heath alias Barton Henmarsh* 1607. P.n.

Bertone 1086–1279, *Barton* from 1235, + ModE **on the heath** for distinction from BARTON SP 1051. Henmarsh, *Hennemerse* 1235, 'the marsh haunted by wild hen-birds', OE **henn**, genitive pl. **henna**, + **merse**, was an extensive area of marshland centred on MORETON-IN-MARSH Glos SP 2032. An outlying farm of Long COMPTON SP 2833. Wa 298, Gl i.230, C lvi.

(18) ~ ST DAVID Somer ST 5431. 'St David's B'. Saint's name *David* from the dedication of the church + p.n. *Bertone -tvne* 1086, *Berton* [1279] Buck, *Barton* 1610.

(19) ~ SEAGRAVE Northants SP 8977. 'B held by the Segrave family'. *Barton Segrave* 1321. Earlier simply *Berton(e)* 1086–1200, *Barton(e)* from 1179. Also known as *Northbarton'* 1242, 1303, and *Barton Hanrad* 1307. 'North' in relation to Earl's BARTON and held in 1201 by William de Henred, in 1220 by Stephen de Segrave. Nth 179.

(20) ~ STACEY Hants SU 4341. 'B held by the Stacey family', correctly the Sacy family. *Berton Sacy* 1280–1316, ~ *Stacii* 1305, *Barton Stacy* 1563. Earlier simply *(æt) bertune* [995×1006]14th S 1420, *Bertvne* 1086. The manor was held by Roger de Saci from Sacy, Marne, France in 1199. 'Stacey' is a corrupt form of this name as in NEWTON STACEY SU 4140. Ha 28, Gover 1958.173.

(21) ~ TURF Norf TG 3522. *Barton Turf* 1838 OS. P.n. *Berton* [1042 or 3]13th S 1490, *Bertuna* 1086. P.n. Barton + ModE **turf**. DEPN.

(22) ~ -UNDER-NEEDWOOD Staffs SK 1818. *Barton sub Nedwode* 1216×72. P.n. *Barton* [942]13th S 479, *Bertone* 1086, + p.n. Needwood as in NEEDWOOD FOREST SK 1624. DEPN, L 229.

(23) ~ UPON HUMBER Humbs TA 0322. *Barton super Humbram* 1202. Earlier simply *Bertune -tone* 1086, *Bartuna -tun(e) -ton(a)* from c.1115. Probably an outlying grange of Barrow upon Humber. Li ii.30.

(24) Brushford ~ Devon SS 6707. P.n. BRUSHFORD + **barton**.

(25) Earl's ~ Northants SP 8563. *Barton comitis David* 1187, 'earl David's B'. *Erlesbarton* 1261, *Barton Yerles* 1545. Earlier simply *Barton(e)* from 1086. The estate was held by David Earl of Huntingdon in the 12th cent. Nth 137.

(26) Great ~ Suff TL 8667. *Magna Bertone* 1254. ModE **great**, Latin **magna**, + p.n. *Bertuna* [945]13th S 507, *Bertun* [942×c.951]13th S 1526, *Bertunā* (accusative case) 1086. DEPN.

(27) Middle ~ Oxon SP 4326. *Mydell Barton* 1449, *Middilbarton* 1509×10. ME **middel** + p.n. Barton as in Steeple BARTON SP 4425. O 248.

(28) Steeple ~ Oxon SP 4425. 'B with the steeple'. *Stepelbertone, Stepel Bartona* 1247. ME **stepel** + p.n. *Bertone* 1086, *Barton'* 1235×6. Also known as *Magna Barton'* 'great B' 1240×1 and *Barton Seint Johan* 'St John's B' 1349 from the family *de Sancto Johanne* who held land here 12th–14th cents. The additions are for distinction from Middle BARTON SP 4326, Sesswell's Barton, 'B held by the Shareshull family', *Barton Sharshill* 1517 from the family name of William de Shareshulle 1334 and 1350, earlier *Bertone* 1086, *Bartona* c.1215 and *Bertun' Odun* 'Odo's B' 1220, *Barton Odonis* 1338 from Odo de Bertona c.1210, and Westcott BARTON SP 4225 also known as *Little Barton*. Great B and Sesswell B were constituent manors of Steeple Barton parish. O 248.

(29) Westcott ~ Oxon SP 4225. 'Barton west cottages'. *Westkote Barthona* c.1260, *West Barton* 1268, 1374, *Westcotebarton* 1509×10. P.n. *Westcot* 1246, *Wescote* 1382, 'the western cottage(s)', ME **West** + **cote** (OE *cot*, pl. *cotu*), + p.n. *Bærtune* [1050×2]13th S 1425, *Bertone* 1086 etc. Also known as *parva Bartona* 'little B' 1221 etc. O 251.

BARTON ON SEA Hants SZ 2393. A modern seaside resort named from an ancient manor called *Barton* 1810 OS, earlier *Ber- Bvrmintvne* 1086, *Berminton -yn-* 1248–1320, *(West)bermeton* c.1300–46, *Barmenton* 1627, *Barton alias Barhampton* 1670, 'estate called after Beorma', OE pers.n. *Beorma* + **ing**[4] + **tūn**. Ha 28, Gover 1958.228.

Great BARUGH NYorks SE 7479. *magna* ~ c.1200, *Great Bargh(e)* 1526. ModE **great**, Latin **magna**, + p.n. *Berg(a)* 1086–1285, *Berch* 1086, *Bergh(e)* 1219–1409, 'the hill', OE **beorg**. YN 74 gives pr [ba:f].

Little BARUGH NYorks SE 7679. *parua Berch* c.1200. ModE **little**, Latin **parva**, + p.n. Barugh as in Great BARUGH SE 7479. YN 74.

BARWAY Cambs TL 5475. 'Mound-like island'. *Bergeia -eie -eya* 1155–1230, *Ber(e)wey(e)* 1218–1494, *Barewe* 1265, 1285, *Bar(e)wey* 1359, 1448, 1564, *Barway(e)* from 1602. OE **beorg** + **ēġ**. Described in 1604 as 'a myerie island ... environed with fens'. Ca 197, L 37, 39, 127.

BARWELL Leic SP 4496. 'Boar spring'. *Barwalle* 1043–1576, *Barewell(e)* 1086–17. OE **bār** + **wælla**. Lei 503, L 31.

BARWICK Somer ST 5613. 'Barley farm, outlying grange'. *Berewyk* 1219, 1327, *Berwick* 1610, 1811 OS. OE **berewīc**. An outlying part of Yeovil. DEPN.

BARWICK IN ELMET WYorks SE 4037. *Barrewyk in Elmet* 1375, *Barwy(c)k(e) -wick(e) in Elmet(t)* 1436–1740. P.n. *Bereuuit(h)* 1086, *Berewic(a) -wik -wy(c)k -wyke* 1142×54–1473 (~ *in Elmet* 1329), *Berwik(e) -wyk(e)* 1177×93–1441, OE **bere-wīc** 'a barley farm, an outlying grange or part of an estate retained for the lord's use' + **in** + p.n. ELMET. Originally a constituent vill of a large royal estate centred on Kippax SE 4130. YW iv.106 gives pr ['barik], WYAS 315.

BASCHURCH Shrops SJ 4222. 'Bas(s)a's church'. *Bascherche* 1086, *-chirche* [1154×89]1267–1346, *-church* from 1393, *Basecherch(e)* 1086–1675 with variants *Basse-* and *- chirch(e), -church* from 1508. OE pers.n. *Bas(s)a* + **ċiriċe**. The name also appears in W translation in a mid-9th cent. source as *Eglwysseu Bassa*, in which the first element is pl. *eglwyssau* 'churches'. Sa i.31.

BASCOTE Warw SP 4063. Possibly 'Baseca's cottage(s)'. *Bachecota* 1174, *Baske- Baschecota* 1190–92, *Bacecot(e) -en* 1202–6, *Bascote* from 1206. If the 1174 form is reliable this would be OE **bæċe** 'a stream, a stream valley' + **cot**, pl. **cotu**, ME pl. *cotes*, with AN [(t)s] for English [tʃ], but the topography is not in favour. Probably, therefore, the specific is OE pers.n. *Basuca, *Baseca. Wa 134.

BASFORD GREEN Staffs SJ 9951. *Basford Green* 1810. P.n. *Bechesword* 1086, *Barclesford(e)* c.1250×60, 1253×61, 1261, *Barkesford(e) -is-* 1240–81, 1590, 1612, *Barsford(e)* 1460–1586, *Basford(e)* from 1540, 'Beorcol's ford', OE pers.n. *Beorcol*, genitive sing. *Beorcoles*, + **ford**, + ModE **green**. Oakden.

BASHALL EAVES Lancs SD 6943. 'Edge of the forest near Bashall'. *Bashull Eaves* 1608. P.n. Bashall SD 7142 + ModE **eaves**, OE *efes* 'edge of a wood', referring here to Bowland Forest. Bashall, *Bacschelf* 1086, 1251, *Bacs(c)holf(e)* 12th–1343, *Baschell* 1250, *Bashall* 1562, is 'shelf of land on the ridge called the Back', OE **bæc** + **scelf**. The reference is to the low narrow ridge S of Bashall Brook extending NE from Bashall Town SD 710420 to Backridge SD 719428, 'the ridge called (the) Back'. Bashall Hall and Town stand on the shelf formed by the broader end of the ridge. YW vi.192, L 125, 187.

BASHLEY Hants SZ 2497. 'Bægloc's wood or clearing'. *(æt) bageslucesleia* [1053]14th S 1024, *Bailocheslei* 1086, *Bailukesleia* 1152, *Bailokes- Ballokesle(y)(e)* 1272–1384, *Balloxley* 1606, *Bashley* from 1664. OE pers.n. *Bæġloc*, genitive sing. *Bæġloces*, + **lēah**. The evolution of the present-day name has been influenced by sensitivity about the 14th cent. form. Ha 28, Gover 1958.228.

BASILDON Essex TQ 7189. 'Beorhtel's hill'. *Belesdunam, Berlesduna* 1086, *Berd- Bret- Bertlesdon(') -den* 1176–1500, *Bard- Bartlesdun' -don(e) -den* 1200–1395, *Battlesden* 1240, *Bastelden* 1510, *Bastildon -yl-* 1522–52, *Basseldon* 1594, *Basildon* 1602. OE pers.n. *Beorhtel*, genitive sing. *Beorhtles*, + **dūn**. The development of this name has been influenced by association with lME *bastel* 'a fortified tower'. Ess 140.

Lower BASILDON Berks SU 6178. ModE adj. **lower** + p.n. *Bastedene* 1086–1181, *-dena -den(e)* 1155×8–1322, *Bastlesden'* 1175–1243, *Bastel(e)den(e)* 1220–1517, *Baseldene* [1335×6]1346, *Basilden* 1535, *Bassildon* (sic) 1830 OS, probably 'Bæstel's valley', OE pers.n. **Bæstel*, genitive sing. **Bæstles*, + **denu**. The same pers.n. occurs in *Bestlesforda* [688×90]12th S 252, [687]12th S 239 and *Bæstlæsford, (æt) bestles forda* [879×99]12th S 354, 'Bæstel's ford', an ancient ford of the Thames at this point. 'Lower' for distinction from Upper BASILDON Berks SU 5976. Brk 512.

Upper BASILDON Berks SU 5976. *Upper Bassildon* 1830, *Upper Basildon* 1846. ModE adj. **upper** + p.n. Basildon as in Lower BASILDON SU 6178, two miles NE on lower ground beside the Thames. Brk 514.

Old BASING Hants SU 6752. ModE **old** + p.n. *(æt) Basengum* c.890 ASC(A), *(æt) Basingum* c.1120 ASC(E), *(ad) Basingas* c.1100 ASC(F Latin) all under year 871, *Basengas* c.894 Asser, *(ad) Basyngum* (Latin form) [945]15th S 505, *Basinges* 1086–1302, *Basing -yng* from 1235, 'the Basingas, the people called after Basa', OE folk-n. *Basingas* < pers.n. **Basa* + **ingas**, dative pl. **ingum**. The Basingas must have been an important pre-Christian group of considerable antiquity possessing an outlying settlement at BASINGSTOKE SU 6352 and possibly an unlocated heathen temple at *Besinga hearh*, the 'shrine of the Basingas' [688 for 685×7]12th S 235, mentioned in a grant for the foundation of a minster by king Cædwalla of Wessex of land at Farnham SU 8446 just across the county boundary in Surrey including the unidentified *Cusanweoh* 'Cusa's shrine'. Ha 29, Gover 1958.124, ING 44, Studies 1936.37–8.

BASINGSTOKE Hants SU 6352. 'Outlying farm belonging to Basing'. *(on) embasingastocæ* (sic for *onem*, i.e. *onefn basinga stocæ*) [990]12th S 874, *Basingestoch(es)* 1086, *Basingstoke* from 1167. OE folk-n. *Basingas* as in Old BASING SU 6752, genitive pl. *Basinga*, + **stoc**. Ha 29, Gover 1958.124, Studies 1936.22, ING 44, DEPN.

BASINGSTOKE CANAL Hants SU 8453. A canal 37 miles long linking the Wey Junction Canal in Surrey with Basingstoke.

BASLOW Derby SK 2572. 'Bassa's burial mound'. *Basselau* 1086, *-lawa -e* 1156–1302, *-low(e) -lou(u)e -louwe* 1215–1460, *Baslow(e)* from 1392. OE pers.n. *Bassa* + **hlāw**. Baslow lies in the Derwent valley but no burial mound is extant. Possibly the reference is rather to the hill behind Baslow which rises to over 1000ft. at Baslow Edge SK 2673. Db 40.

BASON BRIDGE Somer ST 3445. *Bastian bridge* 1575–1680 maps, *Basingebridge* 1593 Humphreys, *Basing Br* 1750 map, *Basonbridge, ~ Bridge* from 1791 Collinson. A crossing of the r. Brue in Huntspill Level contrasting with HIGHBRIDGE ST 3147.

BASSENTHWAITE Cumbr NY 2332. 'Bastun's clearing'. *Bistunthweit* c.1160, *Bastunthweit -thwait* 13th cent., *Bastent(h)wait -thwayt -tweyt* 1241–1403, *Basyngthwaite* 1492, *Bassyngthwayte* 1540, *Basonthwaite* 1599. Surname *Bastun* (from OFr *bastun* 'a stick') + ON **thveit**. Cu 263 gives pr [basnθət], SSNNW 215, L 211.

BASSENTHWAITE LAKE Cumbr NY 2129. *Bastunwater* c.1220, *Basten-* 1260–1439, *Bastun(e)wat'* 1247, 1260, *aquam de Bastantheweyt* 1279, *Bassynt- Bassen- Basson- Basshinge-water* 16th cent., *Bassenthaitlake* 1675, *Bassenthwaite-Water* 1789. Surname *Bastun* as in BASSENTHWAITE NY 2332 + ME **water** later replaced by p.n. BASSENTHWAITE NY 2332 + **water** or **lake**. Cu 32.

BASSETT Hants SU 4216. Now an outer suburb of Southampton, Bassett was originally developed as a retreat for rich people in the 18th and 19th cents. Possibly named from the family name *Bassett*. Ha 29.

BASSINGBOURN Cambs TL 3344. 'Stream of the Bas(s)ingas, the people called after Bas(s)a'. *Basingeburna, Basingborne* 1086, *Basingb(o)urn(e) -yng-* 1237–1457, *Bassingeb(o)urn(a) -(e) -yng-* 1158–1344, *Bassingburn(e) -bo(u)rn(e) -yng-* 1218–1553. OE folk-n. **Bas(s)ingas* < pers.n. **Bas(s)a* + **ingas**, genitive pl. **Basinga*, + **burna**. Ca 52, L 16.

BASSINGFIELD Notts SK 6137. Either 'the open land of the Basingas, the people called after Basa' or 'the open land at Basing, the place called after Basa'. *Basingfelt* 1086, *Basingefeld -feud* 1201–1280, *Basingfeud -feld* 1230 etc., *Bassingfeld* 1280, 1284. Either OE folk-n. **Basingas* < pers.n. *Basa* + **ingas**, genitive pl. *Basinga*, or p.n. **Basing*, locative-dative sing. **Basinge*, + **feld**. Nt 236.

BASSINGHAM Lincs SK 9159. Probably 'the homestead of the Basingas, the people called after Basa'. *Basingeham* 1086–1231×9, *-ingham -yng-* c.1160–1634×42, *Bassingham* from 1200. The generic is probably OE **hām** 'homestead, village, estate' but the situation on the banks of the r. Witham would be appropriate to OE **hamm** 'river-bend land, a water-meadow'. The specific is either folk-n. **Basingas* < *Basa* + **ingas**, genitive pl. **Basinga*, or possibly p.n. **Basing* 'the place called after Basa', < pers.n. *Basa* + **ing**[2], locative-dative sing. **Basinge*. Perrott 245, ING 141.

BASSINGTHORPE Lincs SK 9628. 'Thorpe held by the Basewin family'. *Bas(s)uintorp* [late 12th]13th, *Basewinthorpe -wyn-* 13th cent., *Bassingthorp(e)* 1242–1819. P.n. *Torp* 1081–13th, *Thorp* 1209×35–1428, ON **thorp** 'an outlying settlement', + manorial prefix from the family of Robert Basewin who held the manor in 1202. Perrott 153, SSNEM 119.

BASTON Lincs TF 1114. 'Bak's farm or village'. *Ba(c)stune* 1086, *Baston* from 1167. ON pers.n. *Bak*, genitive sing. *Baks*, + **tūn**. Perrott 393, SSNEM 188.

BASTWICK Norf TG 4217. 'Bast or lime-tree farm'. *Bastwic* [1044×7]13th S 1055, 1202–26, *Bastuwic -uic -wic* 1086, *Bast(e)wyk* 1279–1472. OE **bæst** + **wīċ**. Bast or lime-tree bark was used for matting and making ropes. Nf ii.69.

BATCHCOTT Shrops SO 4971. 'The cottage in the stream-valley'. *Bechecot* 1212, *Bachecot* 1255. OE **beċe**, **bæċe**, + **cot**. DEPN.

BATCOMBE Dorset ST 6104. Probably 'Bata's coomb'. *Batecumbe* 1201, *-combe* 1268, *Badecombe* 1336. OE pers.n. *Bata* + **cumb**. But see BATCOMBE Somer ST 6939. Do 32, L 92.

BATCOMBE Somer ST 6939. Partly uncertain. *(at) Batecombe* [940]14th S 462, *(æt) Batancumbæ* [c.968×71]12th S 1485, *Batancumbe* [972×5]lost ECW, *Batecu͞be* 1086, *-cumbe* 1225, *Batcombe* 1610. This is usually explained as 'Bata's coomb', but the occurrence of this name-type three times in Somer and Dorset has led to the suggestion (by the late J. McN. Dodgson) that the specific is a common noun, possibly OE ***bata**, a weak side-form of *batt* 'a bat, a club, a cudgel'. This would then be a coomb where such implements were made or obtained. DEPN.

BATE HEATH Ches SJ 6879. 'Pasture heath'. *Bate Heath* 1831. In the Tithe Award (1844) there is a Bate Field, *Baytefield* 1559; ultimately from ON **beit** 'pasture'. Che ii.103.

BATH Avon ST 7464. 'The (hot) baths'. *Hat Bathu* [676]12th S 51, *in illa famosa urbe qui nominatur et calidum balneum þ' is æt þæm hatum baðum* 'in that famous city called 'at the hot baths' [864]11th S 210, *æt Hatum Baðum* [970]12th S 777, *æt Hatabaðum* c.1120 ASC(E) under year 972, *æt B(e)aþum* [781]11th S 1257, *æt Baðum* [796]11th S 148 (with variant *Baðun* in second MS also 11th), 906 ASC(A), *[931]12th S 414, *aet Bathum* [808 for 757×8]12th S 265, *æt Baðum tune* c.1050 ASC(D) under year 906, *(to, on, in to) Baðan* [914×9]16th BH, [968×71]12th S 1485, 973 ASC(A), [1061×5]12th S 1426, [1061×82]12th S 1427, *Bathancestre* late 10th Æthelweard, *(to, æt) Baðon* c.1120 ASC(E) under years 1013, 1087, 1106, *Bade -a* 1086, *(bishopric of) Baðe* 1123 ASC(E), *Bathe* 1130 ASC(E). OE **bæth**, pl. **bathu**, dative pl. **bathum**. Other forms are Latin *Bathonia* [956×61]12th S 610, 643, 661, 694, *Bathonis* [955×7] S 664, *at Acemannes beri þ' ys at Baðam* c.1200 ASC(F) under year 972 (Latin *apud Ace mannes byri .i. at Baðan*), and *in civitate Aquamania* [972]12th S 785. The reference is to the Roman baths here which are fed by hot springs. For the Romano-British name see AQUAE SULIS ST 7464 and for the alternative Anglo-Saxon name *on ðœre ealdan byrig Acemannes ceastre* 'to the ancient fortification of Aceman's Roman fort' 973 ASC(A) TT 24–30. DEPN, TC 48, L 13, BH 115.

BATHAMPTON Avon ST 7766. 'Hampton beside Bath'. *Bath Hampton* 1817 OS. Earlier simply *Hamtun* [956]12th S 627, *Hantone* 1086, 'home farm', OE **hām-tūn**. DEPN, L 13.

BATHEALTON Somer ST 0724. Partly uncertain. *Badeheltone* 1086, *Badialton -eyalton* 1196–1249, *Badelton* 1610, *Baddleston or Badialton* 1750 map, *Bathealton* 1809 OS. Unidentified specific + OE **tūn**. Ekwall suggested this might be 'Bada's *Ealdtun* or *Healhtun*, OE pers.n. *Bada* + p.n. **Eald-tūn* 'old settlement' or **Healh-tūn* 'nook settlement'. Alternatively the specific might represent OE masculine pers.n. *Beaduhelm* or feminine *Beaduhild*. DEPN gives pr [bætltən].

BATHEASTON Avon ST 7867. 'Easton beside Bath'. *Batheneston* 1258, 1263. Earlier simply *Estone* 1086, 'the eastern settlement or estate', OE **ēast-tūn**. The spelling *Bathen-* derives from the dative pl. *Bathum*. An estate of Bath Abbey contrasting with WESTON ST 7266. DEPN.

BATHFORD Avon ST 7966. 'Ford leading to or beside Bath'. *Bathford* 1817 OS. Earlier simply *(œt) Forda* [957]12th S 643, *Forde* 1086, 'the ford', OE **ford**. DEPN.

BATHLEY Notts SK 7759. 'Bath wood or clearing'. *Badeleie* 1086, *-lee -leg(he)* c.1190–1312, *Batheleg(h) -ley(e) -lay* 1242–1332, *Barlow or Bathley* 1775. OE **bæth**, genitive pl. **batha**, + **lēah**. The reference is to springs used as bathing places. Nt 182, L 13.

BATHPOOL Corn SX 2874. 'Bathing pool'. *Bathpole* 1474. ME **bath** + **pole**. A natural pool in the river Lynher. PNCo 52.

BATLEY WYorks SE 2424. Probably 'Bata's wood or clearing'. *Bathelie* 1086, *-leia -lay* 13th cent., *Batelei(a) -ley(a)* etc. 1086–1448, *Batley* from 1315. OE pers.n. *Bata* + **lēah**. Alternatively this could be 'the wood where bats are obtained', late OE **batt**, genitive pl. **batta**, + **lēah**. YW ii.179, WYAS 320, L 203.

BATSFORD Glos SP 1834. 'Bæcci's ridge'. *(œt) Bœccesore* [727×36]11th S 101, *Bœceoran* 961×72, *Beceshore* 1086, *Becheso(u)re* 1196–1273, *-ofere* 1236, *Bac(c)hesor(e)* 1235–1535, *Bat(y)sore* 1541, 1577, *Bat(t)esford(e)* 1577–1695, *Batchford* 1646, *Batsford* 1685. OE pers.n. *Bœċċi*, genitive sing. *Bœċċes*, + **ōra** partly replaced by **ofer**. Forms in *ford* are due to popular etymology although there is no ford here. Gl i.233, L 72, 179, 181, Jnl 21.21, 22.35.

BATTERSBY NYorks NZ 5907. 'Bǫthvarr's farm or village'. *Badresbi* 1086, *Batersby* 1214×22, *Batherby* 1285, *Bather(e)sby* 1301–69. ON pers.n. *Bǫthvarr* < **Bathu-harjaR*, + **bȳ**. YN 167.

BATTERSEA GLond TQ 2876. 'Beaduric's island'. *Bad- Batrices ege* [693]11th S 1248, *(Ba)doricesheah* (sic) *[677]16th S 1246, *Batriceseie* [957]12th S 645, 11th, *Badrichesey* 1198, 1204, *Batrichesegh(e)* 1350, *Batris- Batreseye* 1366–1414, *Batersey* 1408, *Battersea* 1595. OE pers.n. *Beadurīċ*, genitive sing. *Beadurīċes*, + **ēġ**. It lay along the Thames. 11th–12th cent. forms with initial *P-*, *Patricesey* c.1080, *Patricesy* 1086, *Patricheseie* 1199, show mistaken association with St Patrick. Sr 12, TC 201.

BATTERY POINT Avon ST 4677. *Portishead Point Battery* 1830 OS. ModE **battery** + **point**.

BATTISFORD Suff TM 0554. 'Bætti's ford'. *Beteforda, Betesfort* 1086, *Bat(t)esford(')* 1191–1325, *Battysforde* 1316, *Battisford* 1445, *Basford* 1524, *-forth* 1610. OE pers.n. **Bœtti*, genitive sing. **Bœttes*, + **ford**. DEPN, Baron.

BATTISFORD TYE Suff TM 0554. 'Battisford common pasture'. *Battisford Tye* 1837 OS. P.n. BATTISFORD TM 0554 + Mod dial. **tye** (OE *tēag*). Cf. CHARLES TYE TM 0252.

BATTLE ESusx TQ 7416. Short for BATTLE CHURCH or ABBEY. *ecclesia de Bello* 'church of the battle' 1082–1316. Later simply *La Batailge* 1086, *(La) Bataille* 13th–14th cents. Latin **bellum**, OFr **bataille**. The place takes its name from the abbey founded to commemorate the battle of Hastings. Sx 495.

BATTLEBURN Humbs → KIRKBURN SE 9855.

BATTLEFIELD Shrops SJ 5116. 'The battlefield'. *Bateleyfield* 1410, *Batelfeld* 1419, *Batayl- Batail- Batelfelde* 15th cent., *Battlefield* 1587. This is the site of the battle of Shrewsbury between Henry IV and the rebellious Marcher Lords in 1403. The earlier name of the place was *Hayteley* as in the original name of the battle, *Hayteleyfeld* 1406. The meaning is 'heathy clearing', OE **hæthiht** + **lēah**, to which ME *feld* was added in the sense 'battlefield'. Sa i.32.

BATTLESBRIDGE Essex TQ 7794. *Batailesbregge* 1351, *Battlesbrigge* 1519, *Battle Bridge* 1610–1805 OS. Surname *Battle* (cf. Reginald *Bataille* 1327) + ME **brigge, bregge** (OE *brycg*). Ess 193.

BATTLESBURY Wilts ST 8945. Possibly 'Pættel's fortification'. *Pattelsbury, Patelsberye* 1589, *Battlebury* 1754, *Battlesbury Castle* 1773. The evidence is late but consonant with OE pers.n. **Pœttel*, genitive sing. **Pœtteles*, + **byriġ**, dative sing. of **burh**. The reference is to the great Iron Age hill-fort at ST 8945, the inhabitants of which came to a violent end to judge by the many graves of men, women and children outside the NW entrance, whether at the hands of the Roman legions or as a result of tribal wars. The modern form is due to folk etymology no doubt as a result of this discovery. Wlt 158, Thomas 1980.236.

BATTLESDEN Beds SP 9628. 'Bæddel's hill'. *Badelesdone* 1086, *-den* 1227, *Batlesden* 1254. OE pers.n. *Bœddel*, genitive sing. *Bœddeles*, + **dūn**. Bd 115, L 151.

BATTLETON Somer SS 9127. No early forms.

BATTRAMSLEY Hants SZ 3098. Partly uncertain. *Bertamelei* 1086, *Batrameslie -le(e)* 1212–1327, *Baterhamesle* 1236–1331, *Beterhemeleg'* 1272. Unidentified specific + OE **lēah**. Possibly 'wood or clearing belonging to **Beateraham*, the homestead of the beaters', OE **bēatere** 'a beater, a boxer', genitive pl. **bēatera**, + **hām**, perhaps varying with **hǣme** 'the inhabitants of *Beateraham*', genitive pl. **hǣma**. Ha 30, Gover 1958.204.

BAUGH FELL Cumbr SD 7393. Probably 'rounded hill'. *Bawghell* (sic) 1592, *Bowe Fell* 1817. OE **balg** + **hyll** and ModE **fell** (ON *fjall*). YW vi.265.

East BAUGH FELL Cumbr SD 7491. Adj. **east** + p.n. BAUGH FELL SD 7393.

West BAUGH FELL Cumbr SD 7394. *West Baugh Fell* 1860 OS. Adj. **west** + p.n. BAUGH FELL SD 7393.

BAUGHURST Hants SU 5861. Probably 'Beagga's *Hyrst*'. *(on) beaggan hyrste* [909]12th S 378, *Baggeherst* 1175 *-hurst* 1189–1324, *Bogorste* 1558, *Baugust* 1741. OE pers.n. **Beagga*, genitive sing. **Beaggan*, + p.n. **Hyrst*, OE **hyrst** 'a wooded hill'. There is no need to see OE **bagga* 'a bag, a bag-like hill, a bag-like animal, a badger' in this name. Baughurst seems to have been a division of a wider area once called simply *Hyrst* 'the wooded hill', of which Haughurst SU 5762, *Hauekehurst* 1256, 'the hawks' *Hyrst*', and Inhurst SU 5761, *Hyneshurst* 1236, *Ineshurst* 1256–1327×77, 'Ine's *Hyrst*', were other portions. Ha 30, 100, Gover 1958.144 which gives pr [bɔ:gəst], 145.

BAULKING Oxon SU 3191. 'The playful stream of pools'. *Bedalacinge* [865×71 or 948]13th S 539, *(œt) Baþalacing* [963]12th, *Baðalacing, (œt) bada lacing* [963]13th S 713, *Bad(d)eking -yng* 13th cent., *Bad- Bathelking -yng* 1241–1376, *Baulking als Battleking* 1758. OE **bæth** 'bath, pool where people bathe', genitive pl. **batha** + stream name **Lācing* as in LOCKING Avon ST 3659. The reference is to a small tributary of the Ock E of the village containing a series of small pools originally created for fulling fleeces or woollen cloth. Brk 350, DEPN, L 13, Jnl 26.27–30.

BAUMBER Lincs TF 2274. 'Bada's fortified place'. *Badeburg* 1086, *Baburc* c.1115, *Baenburch* c.1145, *Baumbur'* 1212. OE pers.n. *Bada*, genitive sing. *Badan*, + **burh**. The pers.n. *Bada* is elsewhere associated with fortified places, especially prehistoric forts, cf. BADBURY Wilts SU 1980. DEPN

BAUNTON Glos SP 0204. 'Estate called after Balda'. *Bavdintvne -tone* 1086, *Baldinton* 1208, 1221, *Baudinton(e) -yn-* 1215–1442, *-ing- -yng-* 1248–1409, *Bawnton* 1535, *Baunton* 1682. OE pers.n. *Balda* + **ing**[4] + **tūn**. Gl i.57.

BAVERSTOCK Wilts SU 0232. 'Babba's outlying farm'. *Babbanstoc* [968]14th S 766, *Babestoche* 1086, *Bab(b)estok* 1225–91, *Baberstoke* 1535, *Baverstocke* 1558×1603. OE pers.n. *Babba*, genitive sing. *Babban*, + **stoc**. The same pers.n., possibly referring to the same individual, occurs in nearby BAPTON ST 9938. Wlt 212, Studies 1936.23.

Great BAVINGTON Northum NY 9880. *Mangna Babington* (sic) 1296 SR 19. ModE adj. **great**, Latin **magna**, + p.n. *Babington -yng-* 1242 BF-1677, *Bab(b)inton -yn-* 1242–1479, 'estate called after Babba', OE pers.n. *Babba* + **ing**[4] + **tūn**. 'Great' for distinction from Little Bavington NY 9878, *Parva Babington* 1296 SR 24, *Little Bavington* c.1715 AA 13.15. The development of *v* < *b* is modern, cf. *Averwick* for ABBERWICK NU 1213, and BAVERSTOCK SU 0232 and PAVENHAM SP 8955. NbDu 13.

BAWBURGH Norf TG 1508. 'Beawa's fortified place'. *Bauenburc* 1086, *Bauburg* c.1130. OE pers.n. *Bēawa*, genitive sing. *Bēawan*, + **burh**. DEPN.

BAWDEN ROCKS or MAN AND HIS MAN Corn SW 7053. Partly obscure.

I. *The Manrock* 1587, *the Man and his Man* 1699.

II. *little isle called Bond* 1650, *Bawden Rocks, more properly Boen Rocks* c.1720, *Boden (Rocks)* 1748, c.1870.

Type I is probably English **man** personifying the larger and smaller rocks, or less likely Co **men** 'stone'; Type II is unexplained, although the form *Boen* might be for Co **bowyn** 'a cow'. PNCo 52.

BAWDESWELL Norf TG 0420. 'Baldhere's spring'. *Baldereswella* 1086, *Baldeswell* 1208. OE pers.n. *Baldhere*, genitive sing. *Baldheres*, + **wella**. DEPN.

BAWDRIP Somer ST 3439. 'Badger trap'. *Bagetrepe* 1086, *Bakatrip* 1166, *Baggetrippe* 1243, *Bagedryp* 1294, *Baudrip* 1610. OE ***bagga** + **træppe, treppe**. DEPN, PNE i.17, ii.185.

BAWDSEY Suff TM 3440. 'Baldhere's island'. *Bald(er)eseia* 1086, *Baldreseia* 1109×31, *Baudesey(e) -eie* 1254–1408, *Balder- Baudersey(e)* 1279, 1327, 1409, *Bawdsey* 1610, 1674. OE pers.n. *Baldhere*, genitive sing. *Baldheres*, + **ēġ**. Bawdsey stands on high ground in the middle of marshland formerly flooded. DEPN, Arnott 56, Baron.

BAWTRY SYorks SK 6593. 'The ball-shaped tree'. *Baltry* [1199]1232, *Bautre* 1247–1506, *Bawtre* 1404–1555, *-try(e)* 1548–1641. OE ***ball** or adjective ***ballede** + **trēow**. YW i.47.

BAXENDEN Lancs SD 7726. 'Valley where bakestones are found'. *Bakestandene -ston-* 1305, *Bacstanden* 1324, *Baxtonden*1464. OE ***bæc-stān** + **denu**. An earlier reference is *Bastanedenecloch* [before 1194]13th, 'Baxenden ravine', p.n. Baxenden + OE **clōh**. A *bakestone* was a flat stone or slate used for baking cakes on in an oven. La 90, L 98.

BAXTERLEY Warw SP 2897. 'The baker's wood or clearing'. *Basterleia* c.1170, *Baxterley(e)* from 1221. OE **bæcestre** + **lēah**. Wa 77, SelPap 74.

BAYCLIFF Cumbr SD 2872. 'Steep slope of the bell-shaped hill'. *Bellecliue* 1212, *Beleclive -clyue* 1269, *Beelclyff* 1418, *Beacliff* 1585. OE **belle** + **clif**. Later forms, however, point to OE **bēl** + **cliff** 'signal beacon slope'. There is a Beacon Hill at SD 2771 but this can hardly be meant. La 208, L 131.

BAYDON Wilts SU 2878. 'Berry hill'. *Beidona* 1146, *-dun* 1195, c.1200, *Bei- Beydo(u)n* 1226–1412, *Baydon* 1263. OE **beġ** + **dūn**. The sps of this name rule out any identification with Gildas's *Mons Badonicus*. Wlt 285, English Settlements 159.

BAYFIELD Norf TG 0440 → BEIGHTON TG 3807.

BAYFORD Herts TL 3108. 'Bæga's ford'. *Begesford* 1086, *Begeford* c.1090, 12th, *Bei- Beyford* 1154–1428, *Bayford* 1251. OE pers.n. *Bǣġa*, *Bēaġa* or *Bēaġe* feminine, + **ford**. Hrt 214.

BAYHAM ABBEY Kent TQ 6436. *Bayham Abbey* 1819 OS. A nineteenth century Jacobean-style mansion built in 1870–72 near the site of a medieval abbey across the county boundary in Sussex. P.n. Bayham + ModE **abbey**. Bayham, *Begeha'* 1208, *Begham* c.1210, 1226, *Beg(g)(e)ham(me)* 1228–1328, *Beigham al. Bayham* 1625, is partly uncertain. The generic is OE **hamm** 3 'river meadow' or 6 'valley bottom', the specific either OE pers.n.*Bǣġa*, feminine *Bǣġe*, or **bēġ** 'a berry'. KPN 342, Sx 374, ASE 2.20, Newman 1969W.139.

BAYLES Cumbr NY 7045. 'Slopes'. *Bales* 1279, 1292, *Balles* 1301, *Bailes, Bayles* 1599, 1705. ME plural **bales** from ON *bali*. Cu 173.

BAYLHAM Suff TM 1051. 'The water-meadow at the bend' sc. in the r. Gipping. *Bel(e)ham* 1086, *Bei- Beylham* 1191–1523, *Baylham* from 1495, *Baleham* 1610. OE ***beġel** + **hamm** 1. DEPN, Baron.

BAYSDALE BECK NYorks NZ 6207. *Basedalebec* [1236]1817×30. P.n. Baysdale NZ 62–6307, *Basdale* 1189×1204, 1301, *Basedale* c.1230–1400, *Baisedale* 1483, 'cow-shed valley', ON ***báss** + **dalr**, + ON **bekkr** 'a stream'. YN 134.

BAYSTON HILL Shrops SJ 4708. *Beystaneshull* 1301, *Beystonhull'* 1364×6, *Baiston Hill* 1781. P.n. Bayston SJ 4908 + ME **hyll**. Bayston, *Begestan* 1086, *Beyeston'* 1255×6, *Beyston* 1255–1664, *-stan* 1271×2–1386, *Byston* 1585–98, *Biston* 1616, is 'Beage's stone', OE feminine pers.n. *Bēaġe* + **stān**. The reference is to a rock outcrop at The Burgs a short distance to the N. Sa i.33.

BAYTON H&W SO 6973. 'Bæga or Beage's estate'. *Beitone* 1080, *Betvne* 1086, *Bayton* 1327. OE pers.n. *Bœġa* or feminine *Beaġe* + **tūn**. Wo 38, DEPN.

BEACHAMPTON Bucks SP 7736. 'Home farm on the stream' or 'settlement of the stream dwellers'. *Bec(h)entone* 1086, *Bechamton* 1152–1284, *Bechehampton* 1230–1499, *Beachampton* 1558×1608, *Betchampton* 1654. It is impossible to say whether this is OE **beċe** + **hām-tūn** 'stream *hām-tun*' or **beċe-hǣme** 'stream dwellers', genitive pl. **beċe-hǣma**, + **tūn**. The expected form is that of 1654; the modern form is due to association with the tree-name *beech*. Bk 59.

BEACHAMWELL Norf TF 7505. 'Beacham springs'. *Bichham Welles* 1212, *Beechamwell St. Mary* 1824 OS. P.n. *Bicham* 1086, 1208, 'Bicca's homestead', OE pers.n. *Biċċa* + **hām**, + ME **welle(s)**. Nearby Devil's Dyke is *Bicchamdic* *[? 1047]14th S 1109, *Bichamdic* [1053×7]13th S 1108, 'B dyke'. DEPN.

BEACHAMWELL WARREN Norf TF 7607. *Beechamwell Warren* 1824 OS. P.n. BEACHAMWELL TF 7507 + ModE **warren**.

BEACHBOROUGH Kent TR 1638. Partly uncertain. *Belcheberche* 1262, *Bithborrow* 1690, *Beachborough* 1819 OS. Probably OE ***bylċ(e)**, Kentish dial. ***belċ(e)**, + **beorg**. Beachborough lies at the foot of Summerhouse Hill TR 166376, a considerable rounded hill. For **belċ(e)* see BELCHAMP Essex TL 8041. PNK 448, Cullen.

BEACHLEY Glos ST 5591. 'Betti's wood or clearing'. *Betesleg(a) -ley(e) -le(gh')* [1154×89]1307, 1222–1648, *Be(a)chley* 1566, 1619. OE pers.n. *Betti*, genitive sing. *Bettes*, + **lēah**. Gl iii.264.

BEACHY HEAD ESusx TQ 5895. P.n. Beachy + ModE **head** 'a headland'. Beachy, *Beuchef* 1274, *(le Forland of) Beauchief* 1404, *Bechief* 1401, *Bechy* 1546, is the 'beautiful headland', OFr **beau** + **chef**. Sx 426.

BEACON Devon ST 1705. 'Signal, beacon'. *Bekyn* 1469. ME **bēken** (OE *bēacon*). The hamlet stands high between hills rising to 819ft., one topped with a hill fort, the other with a tumulus which may be the feature referred to in the name. D 643.

BEACON END Essex TL 9524. The 'beacon end' of Stanway. ME **beken**, *le Bekyn* 1414, *le Beken* 1538, + ModE **end**. Ess 400.

BEACON FELL Lancs SD 5642. A free-standing hill rising to 873ft. overlooking the Fylde.

BEACON HILL Dorset SY 9794. *Beacon Hill* 1838, named from Lytchett Beacon, *Lechiot becon* 1575, *Litchet Beacon* 1774, a hill rising to 303ft. commanding a wide view of Poole Bay, formerly crowned with a beacon. P.n. Lytchett as in LYTCHETT MINSTER SY 9693 + ModE **beacon**. Do ii.35, 36.

BEACON HILL Hants SU 4557. *Beacon Hill* 1817 OS. ModE **beacon** + **hill**. This is the site of the boundary mark *(to) weard setle* 'the look-out place' [943]12th S 487, a hill of 859ft. commanding a wide view. Gover 1958.151.

BEACON HILL Wilts SU 2044. *Beacon Hill* 1773. Wlt 362. ModE **beacon** + **hill**.

BEACON POINT Durham NZ 4445. *Beacon Point* 1863 OS. ModE **beacon** + **point**. The reference is to a coastal navigation beacon.

BEACON POINT Northum NZ 3189. *Beacon Point* 1866 OS. The reference is to a coastal navigation beacon.

BEACON'S BOTTOM Bucks SU 7895. *Beacons Bottom* 1822 OS.

BEACONSFIELD Bucks SU 9390. 'Open land of the beacon'. *Bekenesfelde* 1184, 1204 etc., *Beckenesfeld* 1284. OE **bēacn**, genitive sing. **bēacnes**, + **feld**. Bk 214 gives pr [bekənzfi:ld], L 241, 243.

BEADLAM NYorks SE 6584. '(Settlement at) the buildings'. *Bodlun* 1086, *-um* 1086–1285, *-om* 1336, *Budelom* 1414, *Bewdlom* 1578, *-am* 1613. OE **bōthl**, dative pl. **bōthlum**. The name exhibits the dialectal change of OE *ō* to [iu] and later [iə]. The reference may have been to remains of the Roman villa here. Ryedale Historian 19.24–6, 20.12, YN 67.

BEADNELL Northum NU 2229. 'Beda's nook of land'. *Bedehal* 1161, *Bedenhala* 1176, *Bedenhal(e) -in-* 1212, 1236 BF, *Beednal* 1276, *Beednell* c.1715 AA 13.4, *Beadlin* 1753. OE pers.n. *Bēda*, genitive sing. *Bēdan*, + **halh**. NbDu 13 gives pr [bi:dlən], DEPN, L 108.

BEADNELL BAY Northum NU 2327. P.n. BEADNELL + ModE **bay**.

BEAFORD Devon SS 5514. 'Gadfly ford'. *Baverdone* 1086, *Beauford* 1238–1339, *Beu- Bew(e)ford* 1242, 1249, *Beagheford* 1281. OE **bēaw** + **ford**. The DB spelling represents OE *Bēawford-dūn* 'Beaford hill'. D 86, L 71.

BEAL Northum NU 0642. 'Bee hill', i.e. the hill where the hives are put in summer. *Behil* 1208×10 BF, *Beyl* 1228, *Behulle* 1248, *Behill, Beil* 14th cent., *Beel* c.1715 AA 13.9. OE **bēo** + **hyll**. NbDu 14.

BEAL NYorks SE 5325. 'Nook of land in the river bend'. *Begale* 1086, *Begala(m)* c.1150, *Beghal(e)* 1234–1343, *Beal(l)* 1529 etc. OE **bēag** + **halh**. The reference is to an old stretch of flat land across the river Aire bounded by Old Eye, OE **ald** + **ēa**, 'old river', i.e. the old course of the river. YW ii.55.

Great BEALINGS Suff TM 2348. *Magna Belinges -ynges* 1228–1337, *Bealings Magna* 1674. ModE **great**, Latin **magna**, + p.n. *Belinges* 1086–1316, *Bealing* 1610, possibly 'the Beolingas, the people called after Beola', OE folk-n. **Beolingas* < pers.n. *Beola* + **ingas**. 'Great' for distinction from Little BEALINGS TM 2347. DEPN, Baron, ING 51.

Little BEALINGS Suff TM 2347. *(in) paruo, parua Belinges* 1086, *litelbelings, littel belincgs* [1086]c.1180, *Parva Belinges* 13th. ME **litel**, Latin **parva**, + p.n. Bealings as in Great BEALINGS TM 2348. DEPN, Baron, ING 51.

BEAMINSTER Dorset ST 4701. 'Church called after Bebbe'. *Bebingmynster* [862]14th S 1782, *Beiminstre* 1086, *Begminister* 1091, *Beministre* 1284. OE feminine pers.n. *Bebbe* + **ing**[4] + **mynster**. The pers.n. seems to have been replaced by another feminine name, *Bēaġe*. Do 33.

BEAMISH Durham NZ 2253. 'Beautiful dwelling'. *Bellus Mansus* (Latin) 1251, *Bewmys* 1288, 1399, 1449, *Bemys* 1427, *Beamyssh* 1480, *Beamish* from 1601. OFr **beau** + **mes**. Identical with the Fr p.n. Beaumetz as in ~ lès-Aire, Pas de Calais, *Bellus Mansus* 1058, ~ lès-Loges, Pas de Calais, *Bellus Mansus* 1072, etc. OFr [ts] has been replaced by English [ʃ]. NbDu 14, DEPN, Dauzat-Rostaing 63.

BEAMSLEY NYorks SE 0752. Either 'Bedhelm's clearing' or 'clearing of the valley-bottom'. *Bedmeslei(a), Bemes- Bomeslai* 1086, *Beth(e)mesle(ia) -ley -lay* 1182×5–1695, *Bemesley -lay* before 1234–1540. OE pers.n. *Bēdhelm* or ***bethme** related to *bythme* and *bothm*, genitive sing. *Bēdhelmes*, ***bethmes**, + **lēah**. YW v.70, xi.

BEAN Kent TQ 5972. Unexplained. *Ben* 1240, 1270, 1327, *By- Bien* 1254–1332, *Been* 1278, 1347, *Bean* 1347. PNK 48.

BEANACRE Wilts SU 9066. 'Newly cultivated land where beans are grown'. *Benacre* 1276–1351, *Beneger al. Benacre* 1573, *Bineger* 1670. OE **bēan** + **æcer**. Wlt 128 gives a common pr [bınəgər].

BEANLEY Northum NU 0818. 'The clearing where beans grow'. *Benelegam* c.1150, *Beneleya -(e)* 1212, 1242 BF, *Beanley* from 1663. OE **bēan** + **lēah**. NbDu 14.

BEARE GREEN Surrey TQ 1842. *Bear Green* 1816 OS. P.n. *la Bere* 1263, *Beare* 1497, 'the woodland pasture', OE **bǣr**, + ModE **green**. Sr 267.

BEARLEY Warw SP 1860. 'The wood or clearing with the fortification'. *Burlei* 1086, *-lei(a) -ley(e)* etc. 12th–1430, *Burgelai* c.1150–1200, *Buryley* 1316, *Bereley* 1493, *Byrley* 1545, *Bearley* from 1594. OE **burh** varying with dative sing. **byriġ** + **lēah**. Wa 198 gives pr [bi:əli], L 206.

BEARPARK Durham NZ 2343. 'Beautiful retreat'. *Beaurepair(e), Beurepayr(e)* c.1270–1346, *Berpark* 1418, *Bearpark* from 1514. OF **beau** + **repaire**. An out-of-town residence of the priors of Durham emparked in 1267. The p.n. *Beaurepaire* is frequent in N and central France. Its use in the present monastic context is noteworthy when its distinctly secular connotations are remembered: the castle of the heroine of Chrétien de Troyes's courtly romance of *Perceval* was also called *Beaurepaire*. The modern form of the name is by contraction from *Beaurepaire park*. NbDu 14, Dauzat-Rostaing 64.

BEARSBRIDGE Northum NY 7857. *Bear's Bridge* 1867 OS.

BEARSTED Kent TQ 7955. 'Homestead on a hill'. *Berghamstyde* 695 Laws, *Bergamstede* [696]13th, *-rk-* [696]15th S 17, *Berge- Berhestede* 1242–1281, *Berwestede* 1226. OE **beorg** + **hāmstede**. KPN 18, Stead 89, 210, ASE 2.31, Jnl 8.29.

BEARWOOD Dorset SZ 0496. *Bear Wood* 1840. This and the nearby local names *Bear Barn* 1839, *Bear Meadow, ~ Moor* 1840, contain either OE **bearu** 'grove' or **bǣr** 'woodland pasture'. Do ii.4.

BEAUCHIEF SYorks SK 3387. 'The beautiful headland'. *Be(a)uchef in Dorhesles* c.1175, *Be(a)uchef(e) -cheff -chief* late 12th–1676, *Bechif* 1577. OFr **beau-chef**. A typical French name for a monastic site, Latinised as *Bellum Caput*. Originally called *Dorhesles* the 'hazel-trees belonging to Dore', p.n. DORE SK 3081 + OE **hæsel**. Db 208 gives pr [bi:tʃif].

BEAULIEU Hants SU 3802. 'Lovely place'. The earliest forms are Latin, *(de) Bello Loco Regis* 'the king's Bellus Locus' 1205, *(usque) Bellum Locum* 'to B L' 1208, and thereafter French, *Beulu* c.1300, *Beuleu* 1341, *Bewley* 1381–1775. Latin **bellus** + **locus**, OFr **beau** + **lieu**. 'King's' because part of the forest and therefore in the king's gift. Ha 30, Gover 1958.200.

BEAULIEU HEATH Hants SU 3400. No early forms but cf. BEAULIEU HEATH SU 4104, *Bewlie heath* 1594, p.n. BEAULIEU SU 3802 + ModE **heath**. Gover 1958.201.

BEAULIEU RIVER Hants SU 3901. *ryver of Bewlie* 1585, *Beauly River* 1759. P.n. BEAULIEU SU 3802 + ModE **river**. The earlier name was *Otere* [c.1200]17th, 1236, 'otter stream', OE **otor** + **ēa**. Ha 30, RN 313, Gover 1958.2.

BEAULIEU ROAD STATION Hants SU 3406. A railway station on the road from Lyndhurst to Beaulieu. P.n. BEAULIEU SU 3802 + ModE **road, station**.

BEAUMONT Cumbr NY 3459. *Bello Monte* 1232, 1301, *Be(u)mund*

13th cent., *Beaumont* from 1291, *Beamound* 1465, *Beamont* 1576. A common AN p.n. meaning 'beautiful hill'. Described in 1887 as a 'fair hill from whence every way lies a goodly prospect'. Cu 121 gives pr [bi:mənt].

BEAUMONT Essex TM 1725. 'Beautiful hill'. *Bealmont* 1175×80, *Beumund* 1238, *-munt* 1262, *Beaumont* from 1263, *Bemond(e)* 1427–62, Latin *Bello Monte* 1342, 1322. OFr **bel** + **mont**. What makes this name noteworthy is that it replaces an earlier pejorative name *(æt) Fulanpettæ* 1000×2 S 1486, '(at) the foul pit', OE **fūl**, dative case definite form **fūlan**, + **pytt**, referring to the marshy hollow above which the hamlet stands. Ess 327.

BEAUMONT Herts TL 1110 → BEDMOND TL 0903.

BEAUPORT PARK ESusx TQ 7813. P.n. *Beauport* 1813 OS + ModE **park**.

BEAUSALE Warw SP 2471. 'Beaw's or gadfly nook'. *Beoshelle* 1086, *Beushale* 1288, *Beausala* 12th, *-sale* from 1316 with variants *Beu-* 1242–1327, *Bele-* 1315, *Bew-* 1512, *Bau-* 1308. OE pers.n. **Bēaw* or **bēaw** 'a gadfly', genitive sing. **Bēawes*, **bēawes**, + **halh**, dative sing. **hale**, either in the sense 'valley, recess', although the actual settlement is on a small hill-spur overlooking a valley, or 'piece of land projecting from the main area of its administrative unit'; by origin Beausale was a somewhat remote chapelry of Hatton some 3 miles S at SP 2467. An alternative possibility for the specific, which was assimilated to OFr **beau**, **bel**, is OE ***bēos** 'bent grass, rough grass', cf. Bescar Notts, *Beskhale* 1207, a compound of the diminutive ***beosuc** + **halh**. Wa 200 gives prs [bju:səl] and [bouseil], Nt 89, L 105, PNE i.30.

BEAVER DYKE RESERVOIRS NYorks SE 2254. ModE p.n. Beaver Dyke, ModE **beaver** + **dyke**, + **reservoir**.

BEAWORTHY Devon SX 4699. 'Beaga's enclosure'. *Begevrde* 1086, *Beghworthy* 1242, *Beworthy -i* 1281–1345, *Beworthy vulgo Bowery* 1765. OE pers.n. *Bēaga* + **worthiġ**. For the pers.n. cf. BEIGHTON Norf 3807. D 129 gives pr [baveri].

BEAZLEY END Essex TL 1428. Cf. *Bas(e)leymede, Basalibreend* 1453–68, *Baselend* 1489, *Beazell End* c.1840, *Beazley End* 1805 OS. Possibly contains the plant name *basil*, but whatever the origin the name is in systemic contrast with BLACKMORE END TL 7430 and Rotten End TL 7229, *Rotten End* 1777, all districts of Wethersfield parish. Ess 467.

BEBINGTON Mers SJ 3384. 'Farm, village, estate called after Bebba'. *Bebinton* [1096×1101]1280, –1535 with variants *-yn-* and *-tone -tona*, *Bebington* [1154×89]1666 and from 1270 with variants *-yng-* and *-t(h)one*, *Beabington* 1646. OE pers.n. *Bebba* + **ing**[4] + **tūn**. Che iv.245 gives pr ['bebiŋtən].

BEBSIDE Northum NZ 2881. 'Bibba or Bebba's corner of land'. *Bibeshet* [1198]1271, 1204, *Bebset* 1388, 1428, *Bebside* from 1638. OE pers.n. *Bibba* or *Bebba* + **scēat** later replaced by **set** 'a fold'. NbDu 15, DEPN.

BECCLES Suff TM 4290. 'The stream pasture'. *Abecles* 1086, *Becles* 1086–1523, *Bec(c)lis* 1159–1463, *Beccles* from 1191. OE *Beclæs* < **beċe** + **lǣs**. Beccles is on the r. Waveney. DEPN, Baron.

BECCONSALL Lancs SD 4423. 'Bekan's hill'. *Bekaneshou -ho(w), -is-* 1208–1341, *Bekanshowe* 1327, *Becansaw, Becconsall* 16th. ON pers.n. *Bekan* (OIr *Beccán*), secondary genitive sing. *Bekanes*, + **haugr**. Becconsall stands on slightly elevated ground also referred to in HESKETH BANK SD 4323 overlooking the Ribble estuary marshes. La 138, Jnl 17.78.

BECKBURY Glos SP 0630. Short for Beckbury Camp, an Iron Age hill-fort. Cf. *Beckberry Lane* 1847. Gl ii.20, Thomas 1976.129.

BECKBURY Shrops SJ 7601. 'Becca's manor house'. *Becheberie* 1086, *Beckebir', Becke- Bekkeburi -bury* 1229–1376, *Beckburi* 1284×5, *Beckbury* 1535. OE pers.n. *Becca* + **byriġ**, dative sing of **burh**. Sa i.35.

BECKENHAM GLond TQ 3769. 'Beohha's homestead or enclosure'. *Bacheham* 1086, *Bekenham* 13th. OE pers.n. **Beohha*, probably a pet form of names in *Beorht-*, genitive sing. **Beohhan*, + **hām** or **hamm**. The boundary of Beckenham and of the people of Beckenham is recorded as *Biohhahema mearcæ* 862 S 331, *Beohha hammes gemæru* [955 for ? 973]10th S 671, *Beohhæma mearc* 987 S 864. DEPN, LPN 16.

BECKERING Lincs TF 1181. Uncertain. *Bechelinge* 1086, *Becheringa* c.1115, c.1160, *Bekering(e)* c.1190–1219, *-es* 1195, *Bikering(e)* 1202, [1244]14th. It is impossible to be sure whether this is the 'circle of beech-trees', OE **bēċe** with ON [k] for [tʃ] + **hring**, the most likely explanation, or whether the specific is OE **beċe** 'a stream' or the name an *ingas* formation of some kind, possibly 'the Beclingas' < pers.n. *Bec(c)el* + **ingas**. ING 65, PNE i.265, Cameron 1998.

BECKERMET Cumbr NY 0106. Usually explained as 'meeting of streams'. *Bechermet* c.1130, 13th cent., *Beckirmeth, Bekir- Bekyrmet* 12th cent., *Bekhermett* 1332, *Bechermot* 1385, *Beckermouth, Beckarmett -mond, Beck Armet* 16th cent., *Beckerment* 1722. ON **bekkr**, genitive sing. **bekkjar**, + **móti**. But the forms do not entirely support derivation from either ON **bekkjar-mót* or *bekkr* + OFr, ME *ermite* 'hermit stream'. The stressing of the pronunciation on the second element [bek'ɛrmət], however, together with the 1332 spelling, seems to show that on the analogy of other Cumbrian names like Aspatria etc. Beckermet was thought to be in Celtic word order and so to mean 'stream of the hermit'. Beckermet stands at the junction of an unnamed beck with Kirk Beck while the church stands close to the meeting of Kirk Beck with the river Ehen. Cu 337, SSNNW 103–4, L 11, 14.

BECKERMONDS NYorks SD 8780. 'The meeting of the streams'. *Becker(es)motes* 1241, *Bekirmontis* 1408, *Bekermouthe* 1540, *-monde* 1547, *-monds* 1598, *Beggermo(u)nd(es) -mons* 16th cent. ON **bekkr**, genitive sing. **bekkjar**, + **mót** later replaced by OFr **mont** 'a hill', cf. Eamont in EAMONT BRIDGE Cumbr NY 5228. This is the place where Green Field Beck and Oughterslaw Beck unite to form the river Wharfe. YW vi.115, L 11, 14.

BECKFOOT 'Foot of the beck'. ModE **beck** + **foot**.

(1) ~ Cumbr NY 0949. *Beckfoot* 1722. But there is no beck at this point. Cu 295.

(2) ~ Cumbr NY 1600. *Bekkfoote* 1578. Situated at the bottom end of WHILLAN BECK. Cu 390.

(3) ~ Cumbr SD 6196. *Beck Foot* 1864 OS. Situated at the foot of several becks. We i.31.

BECKFORD H&W SO 9735. 'Becca's ford'. *Beccanford, (to) Beccanforda* [803]11th S 1431, *Beceford* 1086, *Bekeford* 1159–1291, *Bec- Bekford* 1204–1480, *Beckford* from 1570. OE pers.n. *Becca*, genitive sing. *Beccan*, + **ford**, dative sing. **forda**. Gl ii. 43.

East BECKHAM Norf TG 1639. *Est Bekkam* 1379. ME adj. **est** + p.n. *Becche- Becham* 1086, *Becheham* 1175, 'Becca's homestead', OE pers.n. *Becca* + **hām**. 'East' for distinction from West BECKHAM TG 1339. DEPN.

West BECKHAM Norf TG 1339. *West Becham* 1300. ME adj. **west** + p.n. Beckham as in East BECKHAM TG 1639. DEPN.

BECKHAMPTON Wilts SU 0868. 'Bacca's estate' or 'the settlement of the Bæchǣme, the people who live at Bæc-hāmtūn, the *hāmtūn* at the ridge'. *Bachentune* 1086, *Bakanton* 1268, *-en- -ing-* 16th, *Bakhamtun* 13th, *Bachampton* 1242–1594, *Beckhampton alias Backhampton alias Beckington* 1730. OE pers.n. *Bacca*, genitive sing. *Baccan*, + **tūn** or folk-n. **Bæchǣme* < **bæc** + **hǣme**, genitive pl. **Bæchǣma*, + **tūn**. There is a long narrow ridge W of the village rising to Knoll Down. W 294 gives pr [bek'hæmtən], formerly [bekiŋtən].

BECK HOLE NYorks NZ 8202. No early forms. Mod dial. **beck** + **hole** 'a hollow'.

BECKINGHAM Lincs SK 8753. Uncertain. *Bekingham -yng-* c.1145–1510, *-inge* 1177–1245, *Beckingham* from 1470. Usually explained as the 'homestead of the Beocingas, the people called after Beoca', OE folk-n. *Beccingas < pers.n. Becca + **ingas**, genitive pl. *Beccinga, + **hām**, but the forms point rather to sing. p.n. *Beccing 'the place called after Becca' with **ing**[2]. The village lies on the r. Witham and an alternative suggestion for the specific is OE ***beċing** 'the stream' < **beċe** + **ing**[2] with ON [k] as in *bekkr* for English [tʃ]. The situation would also be appropriate to OE **hamm** 'river-bend land, water meadow'. Perrott 353, ING 141, Cameron 1998.

BECKINGHAM Notts SK 7790. 'The homestead of the Beccingas, the people called after Becca'. *Beching(e)hā* 1086, *Bekingeham* 1187–1252, *Bekingham -yng-* 1204–1416. OE folk-n. *Beccingas < pers.n. *Becca* + **ingas**, genitive pl. *Beccinga, + **hām**. Nt 25, ING 148.

BECKINGTON Somer ST 8051. 'Estate called after Becca'. *Bechintone* 1086, *Bekynton -in-* [1165×95–1174×95] Buck, 1186, *Beckinton* 1610. OE pers.n. *Becca* + **ing**[4] + **tūn**. DEPN.

BECKLEY ESusx TQ 8425. 'Becca's wood or clearing'. *Beccanlea* [873×88]11th S 1507, *Beckele(y)* 1261–1388, *Becklee* 1340. OE pers.n. *Becca*, genitive sing. *Beccan* + **lēah**. Sx 526, SS 66.

BECKLEY Oxon SP 5610. '(At) Becca's wood or clearing'. *Beccalege* [1005×12]14th S 943, *Bechelie* 1086, *Bekele(ye) -ley* 1123–1330. OE pers.n. *Becca* + **lēah**, dative sing. **lēaġe**. O 165.

BECK ROW Suff TL 6977. *Beck Row* 1836 OS. P.n. Beck as in *Beck Watch* 1836 OS + ModE **row**. Contrasts with nearby HOLYWELL ROW TL 7077 and WEST ROW TL 6775. A *watch* is dial. **watch** 'a piece of open water'.

BECK SIDE Cumbr SD 2382. *Beck Side* 1864 OS. ModE **beck** + **side**. Contrasts with Sandside SD 2282, *Sand Side* 1864 OS, a mile W towards Duddon Sands.

BECKTON GLond TQ 4381. A modern name from the surname of Simon Adams Beck, governor of the Gas, Light and Coke Co. which built a large works here in 1869–70. Ess 95, LPN 16, Encyc 50–1.

BECKWITHSHAW NYorks SE 2652. *Bekwithshagh* 1323, *-shaw(e)* 1511. P.n. *bec wudu* [972×92]11th S 1453, *Becvi* 1086, *Beck(e)with - wyth* 1301–1617, the 'beech-tree wood', OE **bēċe** + **wudu** later replaced by ON **bekkr** + **vithr**, + ME **shagh, shawe** 'a thin wood' (OE *sceaga*). YW v.116, L 228.

BECONTREE GLond TQ 4886. 'Beohha's tree'. *Beuentreu* (sic) 1086, *Begetreowa* [c.1150]14th, *Beg(g)entre* 1155–1242, *Bekintre -en-* 1219–1464, *Be(c)kingtre* 13th cent., *Becontre* 1594, *Beaucountry* 1777. OE pers.n. *Beohha*, genitive sing. *Beohhan*, + **trēow**. Originally the name of the hundred whose meeting place was on Becontree Heath in Dagenham. The phonological history of the name implies voicing of OE medial [x] to [γ]. Three different sound substitutions are evidenced thereafter, (a) [w] for [γ] already labialised to [γ^w] in the DB form, (b) stop [g], (c) stop [k] probably helped by association with OE *bēacn*, ME *beken* 'a beacon'. The 1777 form illustrates a piece of late fanciful etymology. Ess 87, Jordan §186.

BEDA FELL Cumbr NY 4316. Partly uncertain. *Beda Fell* 1860. Unexplained p.n. + ModE **fell**. We ii.218.

BEDALE NYorks SE 2688. Probably 'Bede's nook of land'. *Bedale* 1086 etc., *Bedhal* 1256. OE pers.n. *Bēda* + **halh**. YN 236, Redin 60, L 108.

BEDBURN Durham NZ 1031. 'Beda's stream'. *Bedeburne* 1242×3, 1283, *-bourn'* 1312, *Bedburn* from 1339. OE pers.n. *Bēda* + **burna**. NbDu 14.

BEDDINGHAM ESusx TQ 4407. 'Promontory of dry land of the Beadingas, the people called after Beada'. *Beadyngham* [801]14th S 158, *(aecclesiae) Bedingehommes* 'the church of B' [825]14th S 1435, *(þone hám æt) Beadinga hamme* 'the homestead at B' left to his nephew by king Alfred along with *þone hám æt Beadingum* 'the homestead at Beeding' [873×88]11th S 1507, *Bedinge- Bedding- Belingham* 1086, *Bedingeham* 1121–1212. OE folk-n. *Bēadingas < pers.n. *Bēada* + **ingas**, genitive pl. *Bēadinga, + **hamm** 2a. The land around Beddingham was once tidal. Beeding TQ 2109 contains the same folk-n. as Beddingham. Sx 357 gives pr [bediŋ'hæm], 203, 205, ING 123, 31, ASE 2.24, SS 82, 64.

BEDDINGTON GLond TQ 3165. 'Estate called after Beada'. *Bedintone* 12th–1318 with variant *-yn-* and including *[727]13th S 1181, *[933]13th S 420, *Beaddinctun* [900×08]12th S 1444, *-ing-* *[909]?11th S 376, *Bedington* 1229 etc., *Beddington* 1589. OE pers.n. *Bēada* + **ing**[4] + **tūn**. Sr 40.

BEDFIELD Suff TM 2166. 'Beda or Beada's open land'. *Berdefeldā* (Latin accusative case) 1086, *Bedefeld(e)* 1156–1384, *Bedfeld* 1468. OE pers.n. *Bēda* or *Bēada* + **feld**. Bedfield is about 3 m. distant from BEDINGFIELD TM 1868.

East BEDFONT GLond TQ 0873. *Estbedefont* 1235–1441 with variants *-funt(e)* and *-fount(e)*, *Bedfont east* 1593. ME adj. **est** + p.n. *Bedefunt -funde* 1086, *-font(e)* 1198–1353 with variant *-f(o)unte*, *Bedhunt* 1373, probably 'the spring with a trough', OE **byden** + ***funta**. 'East' and 'Church' as in *Chirchebedfounte* 1405 for distinction from West Bedfont TQ 0773, *Westbedefund* 1086, in Stanwell. Bedfont lies on the Roman road from London to Silchester and the reference here may be to a Roman fountain or some such building. Mx 12, Signposts 84, 86.

BEDFORD Beds TL 0449. 'Beda's ford' or possibly the 'ford leading to Biddenham'. *Bedanford* c.925 ASC(A) under year 918, *beda(n)ford* 980×90 S 1497, *Bydanford* c.1000, *Bedeford* 1086, *Bedford* 1198. OE pers.n. *Bēda*, genitive sing. *Bēdan*, + **ford**. *Bedcanford*, the site of Cuthwulf of Wessex's battle with the British in 571, is a different name and not to be identified with Bedford without corroborative evidence. Biddenham TL 0250, 'Byda or Bieda's river-bend land', lies a mile W of Bedford and it is tempting to posit the variant WS form *Bīeda* as the specific in both names although the extant spellings do not support this. Bedford might then be the 'ford leading to B(i)eda's land'. Bd 11, L 67, 69.

BEDFORD LEVEL (MIDDLE LEVEL, NORTH LEVEL and SOUTH LEVEL) Cambs TL 3393, TF 2404 and TL 5985. Bedford Level is *the Great Level (of the Fens)* 1632, 1663, *Bedford levell* 1661, so-called from Francis, earl of Bedford, who undertook the draining of the fens in 1631. In 1642 Vermuyden divided the Great Level into three areas, *North, Middle, South Levell* 1652. ModE **level**, 'a level tract of ground'. Ca 211.

New BEDFORD RIVER OR HUNDRED FOOT DRAIN Cambs TL 4987. Constructed parallel to the Old Bedford River in 1651 and so named in 1652 after Francis, Earl of Bedford, who undertook to drain the fens in 1631. Ca 210.

Old BEDFORD RIVER Norf TL 5596. A new channel for the Ouse made in 1631. *Bedford River* 1637, *Old Bedford River* 1654. The name commemorates Francis, Earl of Bedford, who undertook the draining of the fens in 1631. 'Old' for distinction from the New BEDFORD RIVER Cambs TL 4987. Ca 210.

BEDGEBURY FOREST Kent TQ 7233. P.n. Bedgebury + ModE **forest**. Bedgebury, *becgebyra* 814 S 173, *Bechebyri* c.1270, *Beggeberi -y -bury* 1278–1352, *Bedgebury* 1690, is possibly 'woodland pasture at *Becge, the bend', Kentish dial. ***becge** for OE ***bycge** probably used as a p.n. referring to the bend in the stream here, + **bēr, bǣr** 'a woodland-pasture'. KPN 126, 131.

BEDHAMPTON Hants SU 7006. Possibly 'farm, estate of the inhabitants of *Beteham*, the beet estate'. *Betametone* 1086, *Bethameton(a)* 1167, 1242, *Bedham(p)ton* from 1236. OE folk-n. **Betehǣme* the 'inhabitants of *Beteham', a hypothetical p.n. meaning 'beet estate', OE **bete** + **hām**, genitive pl. **Betehǣma*, + **tūn**. Ha 31, Gover 1958.18.

BEDINGFIELD Suff TM 1868. 'The open land of the Beadingas, the people called after Beada'. *Bedinge- Bading(h)e- Badingafelda* 1086, *Bedingefeld* c.1095–1224, *Bedingfeld(')* 1193–1610. OE folk-n. *Bēadingas < pers.n. *Bēada* + **ingas**, genitive pl. *Bēadinga, + **feld**. Bedingfeld is about 3 m. distant from BEDFIELD TM 2166. DEPN, Baron.

BEDLINGTON Northum NZ 2681. 'The estate called after Bedla

or Betla, or, called or at *Bedling*, the place called after Bedla'. *Bedlington* from [c.1040]11th, *Bet(h)li(n)gtun* c.1107, 12th, *Bellington* c.1150–1335, *Betlingetun* c.1175, *Bellingeton* 1204, *Bedelinton* 1291, *-ing-* 1315, *Beldyngton* 1507. OE pers.n. *Bēdla* or *Bētla* + **ing**[4] + **tūn**, or p.n. **Bēdling* or **Bētling* < *Bēdla*, *Bētla* + **ing**[2], locative-dative sing. **Bēd- Bētlinge*, + **tūn**. NbDu 15, BzN 1968.160.

BEDMOND Herts TL 0903. 'Fountain with a trough or tub'. *Bedesunta* (sic for *-funta*) 1331, *Bedfunte* 1433, *Bedmont -d* 1547×53. OE **byden, beden** + **funta**. The name has been reformed under the influence of nearby Beaumont TL 1110, *Beaumond* 1486, *Be(a)mond* 1487, 1535, 'beautiful hill', OFr **beau** + **mont**. Hrt 76, 78, Signposts 86.

BEDNALL Staffs SJ 9517. 'Beda's nook'. *Bedehala* 1086, *Bedenhal(e)* 1194–1428, *Bednalle* 1481. OE pers.n. *Bēda*, genitive sing. *Bēdan*, + **halh** in the sense 'small valley' or 'hollow'. St i.26, L 105.

BEDRUTHAN STEPS Corn SW 8469. *Bodrothan Steps* 1851. P.n. Bedruthan SW 8569, *Bodruthyn* 1335, 1431, *Bedrewthan* 1748, probably 'dwelling of Rudhynn', Co **bod** + pers.n. **Rudhynn*, + ModE **steps**. PNCo 52.

BEDSTONE Shrops SO 3675. 'Bedgeat's estate'. *Betietetune* 1086, *Bedeston(e)* 1176–1291×2, *-stan(e)* 1271×2, *Bedston* 1577 etc., *Bedstone* 1705. OE pers.n. **Bedġēat*, genitive sing. **Bedġēates*, + **tūn**. The pers.n. was early shortened to *Bede*. Sa i.35.

BEDWARDINE H&W → ST JOHN'S SO 8454.

BEDWORTH Warw SP 3686. 'Bēda's enclosure'. *Bedeword* 1086, *-worthe* 12th–1361 with variants *-w(u)rth(e) -wurda*, *Bedworth* from 1333. OE pers.n. *Bēda* + **worth**. Wa 97.

Great BEDWYN Wilts SU 2764. *Gretebedwyn* 1547, *Great Beden* 1655. ModE adj. **great** + p.n. *(in, æt) Bedewindan* [778]10th S 264, [873×88]11th S 1507, c.1000, *Bed(e)uuindan* [968]13th S 756 (accusative case), *Bedeuuinde* [990×1006]13th S 937, *Bedvin(d)e* 1086, *Bedewinde -wynde* 1091–1317, *Bedewin(e)* 1042×1066 Coins, *Bedewyn* 1438, of uncertain meaning. Association has been sought with OE **bedewinde* corresponding to dial. *bedwind* 'convolvulus' or with a putative Celtic stream name. The most prominent topographical feature in the neighborhood, however, is Chisbury Camp, a multivallate Iron Age hill-fort, possibly referred to here with the OE term *gewind* or a weak derivative of it, 'the winding, the circular thing'. Perhaps therefore 'Beda's ring'. Also known as *Chippingbedewynde* 1279 'B with the market' for distinction from Little or East BEDWYN SU 2965. Wlt 332, Pevsner 1975.174.

Little BEDWYN Wilts SU 2965. *Lyttelbedwyn* 1547. ModE adj. **little** + p.n. Bedwyn as in Great BEDWYN SU 2764. Also known as *Estbedewinde* 1154, *Est Bedewyn* 1437. Wlt 332.

BEEBY Leic SK 6608. 'Bee farm'. *Bebi* [966]16th S 741, 1086–1327, *Beby* 1209×19–1576, *Beebie* 1603, 1610. OE **bēo** + ON **bȳ**. Lei 278, SSNEM 35.

BEECH Hants SU 6939. 'The beech-tree'. *(la) Beche* 1239, 1462, *(atte) Beyche* 1334. Fr definite article **la** + ME **beche** (OE *bēċe*). Ha 31, Gover 1958.96.

BEECH Staffs SJ 8538. 'The beech-tree' or 'the stream-valley'. *Beche* 1199, *Le Bech* 1285, *The Breach* 1673. Fr definite article **le** + either OE **bēċe** or **beċe, bæċe**. The place lies in a steep valley running into the hill-side. DEPN, L 222, Horovitz.

BEECH HILL Berks SU 6964. 'Beech-tree hill'. *Le Bechehulle* 1384, *Bechehyll* 1572, *Beachill* 1603, *Beach Hill* 1761, *Beech Hill* 1817, 1846. OE **bēċe** + **hyll**. Brk 149, L 222.

BEECHINGSTOKE Wilts SU 0859. 'The bitches' outlying farm'. *Bichenestoch* 1086, *-stok(e)* 1289, 1303, *Bychyngstoke* 1431, *Bec(c)hingstoke* 1558×1603, *Beauchamp Stoke* 1819. OE **biċċa**, genitive pl. **biċċena**, + p.n. *Stoke* [941]15th S 478, 1260–1316, OE **stoc** 'an outlying farm'. Probably originated as a stock-farm for cattle pastured on Cannings Marsh subsequently used for keeping hounds, cf. *Stokes in Kanyngemershe* 13th. The origin of the prefix is unknown. Wlt 318, Studies 1936.19, 23.

BEEDING ESusx TQ 2109 → BEDDINGHAM TQ 4407.

Lower BEEDING WSusx TQ 2227. *Netherbetynges* 1279. ModE **lower**, ME **nether**, + p.n. *Bedynge* 1298 as in Upper BEEDING TQ 2010. Sx 203.

Upper BEEDING WSusx TQ 2010. ModE **upper** + p.n. *(þone hám æt) Beadingum* 'the homestead or estate at B' [873×88]11th S 1507, *Bedinges* 1073–1325 including 1086, *Beddinges* 1086, c.1100, 1210, *Bedingh'* 1232, *Bedyngge* 1296, 1398, *Beding* 1327, *Beedinge* 1600, 'the Beadingas, the people called after Beada', OE folk-n. **Bēadingas* < pers.n. *Bēada* + **ingas**, dative pl. *Bēadingum*. Upper and Lower Beeding belong together, Lower B probably marking the swine-pastures in the Weald of the original settlement at (Upper) Beeding. Paradoxically Lower B is on higher ground and further up country than Upper B. The adjectives must refer to status. Sx 205, ING 31, SS 64.

BEEDON Berks SU 4877. '(At the) tub-shaped valley'. *Bydene* [965]12th S 732, *Bedene* 1086–1517 with variants *-en(a) -one*, *Budene* c.1200–1315 with variants *-en(') -one'*, *Budon als Bedon* 1535, *Beeden* 1761. OE **byden**, dative sing. **bydene**. The term 'tub' was apparently used of the steep narrow valley below the village. The place is also referred to in *on beden weg* 'to Beedon way' [948]13th S 542 and *andlang byden hæma ge mæres* 'along the boundary of the Beedon people' [951]16th S 558. Brk 232, 651–3, L 89.

BEEFORD Humbs TA 1254. '(The settlement) beside the ford'. *Biuuorde* 1086, *Beford(e)* 1147×54–15th, *Biford(a)*, *Byford(e)* c.1160–1260, *Beforth(e)* 1180×97–1527, *Beafurth* 1567. OE **bi** + **ford**. Alternatively this could be the 'bee ford', OE **bēo**. YE 76 gives pr [bi:fəθ], V i.82.

BEEFSTAND HILL Northum NT 8214. P.n. Beef Stand NT 8213, *Beef Stand* 1869 OS, + ModE **hill**. The hill rises to 1842ft. on the boundary between England and Scotland. Beef Stand, a hill of 1672ft., marks a traditional place where cattle were grazed. Cf. Scott, *Redgauntlet* (1824): 'Ye ken the place they call the Beef-stand, because the Annandale loons used to put their stolen cattle in there? ... it looks as though four hills were laying their heads together to shut out the daylight from the dark, hollow space between them.'

River BEELA Cumbr SD 5181. 'Embankment river'. *Betha* c.1195–1823, *Bethey* 1647, *Beetha* 1770, *Bela* 1702, *Beela* 1745. ON *****beth** + **á**. The reference is to banks made to control flooding. RN 31, We i.2.

BEELEY Derby SK 2667. 'Bega or Beage's clearing'. *Begelie* 1086, *-leg' -lei -ley* 1206–1307×27, *Bei- Beyleie -leia -ley(e)* late 12th–1303, *Beleg(e) -ley(e) -legh -lee -lay(e) -ly* 1210–1610, *Beelege* 1216×72, *Beeley* from 1565, *Bylegh(e) -ley(e)* 1215–1592. OE pers.n. *Bēga* or *Bēaġe* feminine + **lēah**. Alternatively the specific might be OE **beġ** 'a berry'. Db 44.

BEELSBY Humbs TA 2001. 'Beli's farm or village'. *Belesbi* 1086–1259, *-by* 1202–1594, *Bilesbi* 1086, *Bielsby* 1576, *Beelesby* 1601. ON pers.n. *Beli*, genitive sing. *Bela*, secondary (ME) genitive sing. *Beles*, + **bȳ**. Li iv.57, SSNEM 35.

BEENHAM Berks SU 5968. 'Homestead where beans are grown'. *Benham* [1142×84]12th–1316, *Beneham* 1242–1552, *Bi(e)nham* 13th cent., *Beenham* 1761. OE **bēan** + **hām**. Brk 150, 875.

BEER Either 'pasture, woodland pasture for feeding swine', OE **bǣr** or **bearu** 'a wood, a grove'.

(1) ~ Devon SY 2289. *Bera* 1086, *Bere* 1242–91, *Beare* 1303, 1337. D 620, PNE i.23.

(2) ~ CROCOMBE Somer ST 3220. 'B held by the Craucombe family'. *Cracombesbere* 1325, *Beare Crockam* 1610. Earlier simply *Bere* 1086. The manor was granted to Godfrey de Craucombe in 1227. See CROWCOMBE Somer ST 1336. DEPN.

(3) ~ HACKETT Dorset ST 5911. 'Beer held by Haket'. *Berehaket* 1362–1483, *-ha(c)ket(t)* 16th, *Beer Hackett* 1667. Earlier simply *Bera* 1175–1203, *Bere* 1244–1431, either OE **bǣr** 'woodland pasture' or **bearu** 'wood, grove'. The manor was held by Haket de Bera 1176–86. Do iii.290.

(4) ~ HEAD Devon SY 2287. P.n. BEER SY 2289 + ModE **head** 'a headland'.

BEESANDS Devon SX 8140. *Base Sande* 1514. Situated close to BEESON SX 8140 by which the name has probably been influenced. D 333.

BEESBY Lincs TF 4680. 'Besi's village or farm'. *Bize- Besebi* 1086, *Beseby* c.1155, [1182]1409. ON pers.n. *Besi*, genitive sing. *Besa*, + **bȳ**. SSNEM 35 prefers the explanation 'village or farm where bent-grass grows', OE ***bēos** + **bȳ**, but compounds with this element do not show regular spellings with medial *-e-*. DEPN, Cameron 1998.

BEESON SX 8140. 'Bædi's farm or village'. *Bedeston* 1377×99, *Beaston* 1509×47. OE pers.n. *Bǣdi*, genitive sing. *Bǣdes*, + **tūn**. D 333.

BEESTON 'Bent-grass farm'. OE ***bēos** + **tūn**. Studies 1931.55.

(1) ~ Norf TF 9015. *Bestone* 1254. DEPN, Studies 1931.55.

(2) ~ Notts SK 5236. *Bestune* 1086, *-ton(a)* 1169–1434, *Beeston'* from 1182, *Beeston oth. Beeson* 1753. Nt 139 gives a local pr [bi:sən].

(3) ~ WYorks SE 2830. *Bestone* 1086, *-tun(a) -ton(a)* 12th–1418, *Beeston(e)* [1175×89]copy–1733, *Beas- Beys- Beiston* 16th cent. YW iii.217, WYAS 322.

(4) ~ REGIS Norf TG 1642. 'The king's B'. *Beeston Regis* 1838 OS. P.n. *Besentuna, Besetune* 1086, *Besenton'* 1185, *Bestone* 1254, *Beestone* 1379, OE ***bēos** varying with ***bēosen** 'growing with bent-grass', + Latin **regis**, genitive sing. of **rex**. DEPN.

BEESTON Beds TL 1648. 'River bend' or 'trade settlement'. *Bistone* 1086, *Buistona* 12th, *Beston* 13th–14th cents., *Beeston* 1219. OE **byġe** 'a bend' or **byġe** 'commerce, traffic', genitive sing. **bȳġes, byġes**, + **tūn**. Beeston is situated at a well-marked bend of the river Ivel. Bd 107 gives pr [bi:sən].

BEESTON Ches SJ 5458. 'Rock where a market is held'. *Buistane* 1086, *Bestan* 1170–1353, *Beston* 1240–1561, *Beeston* from 1237, *B(e)ustan'* 1247–82, *-ston* 1284–1398. OE **byġe** + **stān**. The reference is to the prominent hill on which the remains of Beeston Castle stand SJ 5359. Che iii.302 gives prs ['bi:stən] and [''bistən].

BEETHAM Cumbr SD 4979. '(Amongst the) embankments'. *Biedvn* 1086, *Bethum* [1090×7]1308, c.1125–1534, *Bethom(e)* c.1190–1537, *Beetham* from 1612. ON ***beth**, dative pl. ***bjǫthum**. See River BEELA. We i.xiv and 66 which gives pr [bi:ðəm], SSNNW 104.

BEETLEY Norf TF 9717. 'Mallet wood or clearing'. *Betellea* 1086, *Betel'* 1204. OE **bētel** + **lēah**. The reference is to the manufacture and supply of wooden tools. DEPN.

BEGBROKE Oxon SP 4613. 'Becca's stream'. *Bechebroc* 1086, *Bekebroc(h) -brok(e)* 1188–1476, *Beckbroc* 1235×6, 1297, *Begbroke* from 1551×2. OE pers.n. *Becca* + **brōc**. O 251, L 15, 16.

BEIGHTON Norf TG 3807. 'Beaga's estate'. *Begetuna* 1086, *Begeton* 1186, *Beghetun* 1202. OE pers.n. *Bēaga* + **tūn**. OE *Beaga*, a short form for a name like *Bēagmund*, not certainly evidenced in independent use, is required to explain the form with medial *-gh-* as also in BEAWORTHY Devon SB 4699. OE pers.n. *Bæġa* or *Bēaġe* (feminine) would normally give something like Bayton, as in Bayfield Norf TG 0440, *Baiafelda* 1086, *Baifeld* 1180, 1200, *Beinfeld* 1200, pers.n. *Bæġa*, genitive sing. *Bæġan*, + **feld**; cf. BAYHAM ABBEY Susx TQ 6436, BAYFORD Herts TL 3108, BAYSTON HILL Shrops SJ 4808, BAYTON H&W SO 6973. DEPN.

BEIGHTON SYorks SK 4483. 'Brook settlement'. *(æt) Bectune* [1002×4]c.1100 S 1536, 1086, *-tone(e)* 1233–1328, *Beghton* c.1270–17th, *Beighton* from 1347. OE **beċe** + **tūn**. The reference is to Ochre Dyke, a tributary of the r. Rother. Db 209 gives pr [beitən].

BEKESBOURNE Kent TR 2055. 'Bec's (manor called) *Bourne*'. *Bekesburn'* 1270, *Bekesbourn'* 1292. Earlier simply *Bvrnes, Borne* 1086, *Burnes* c.1100–1201, *Burn(e)* 1250, 1254, OE **burna** 'a stream'. The manor was held from 1198 by William de Beche or Bec. Also known as *Livinge(s)burn* 13th, 'Leofwine's *Bourne*' from the 1066 tenant Lēofwine, *Levine* DB. PNK 540.

BELAUGH Norf TG 2818. Partly uncertain. *Belhae, Belahe* *[1044×7]13th S 1055, *Belaga* 1086, *Belhag* [1147×9]13th, *Belhagwe* 1249. OE *Bel-haga*, unidentified element + **haga** 'enclosure'. The identical name occurs at Bylaugh Norf TG 0319, *Belega* 1086, *Belag* 1203, 1208, *Belhawe* 1254, *Belhaye* 1275, and in a lost f.n. *Belahaye* 1161 in Glos†. It seems likely that we have to do with an unknown OE appellative **bel-haga**, possibly a compund of *bǣl* 'a fire, a funeral pyre' and **haga** meaning 'enclosure where dead are, or were, cremated'. See further BELTON TG 4802. DEPN, Studies 1936.160 which gives pr [bilæ], Glos iv.102, 134, V i.78.

BELBROUGHTON H&W SO 9277. *Bellebrocton* 1292, *Belne-* 1298–1323, *Bellebroughton* 1368, 1431. A compound of the p.ns., Bell as in Bell Hall SO 9377 and Broughton. Bell, *Beolne* [817]11th S 181, 11th, 1300, *Bellem* 1086, *Belna -e* c.1150–1292, *Bolne* 1346, is by origin the name of the brook flowing through this place, *aqua que vocatur Beolne* 'stream called B' 1300, possibly related to OE **beolone** 'henbane' and so 'stream where henbane grows'. Henbane was widely used in ancient medicine. Broughton, *Broctun* [817]11th S 181, 11th, *Brotvne* 1086, 1292, is 'brook settlement' OE **brōc** + **tūn**. Wo 274, RN 32.

BELCHAMP OTTEN Essex TL 8041. 'Otto's B'. P.n. Belchamp + pers.n. *Oton* 1255 representing Latin genitive sing. *Ottonis*, of Otto, 1154×89, a descendant of Otto the goldsmith who held Gestingthorpe in 1086 and whose family were hereditary masters of the mint. Belchamp, *Bylcham* *[c.940]13th S 453, [c.1000]c.1125, *(at) Belhcham, Bolcham* [1035×44]13th S 1521, *Belcham* 1086–1431×2, *Belcamp* 1086, *Belchamp(e)* from 1100, is the 'homestead at (the hill called) Bylc', OE ***bylċ** probably used as a p.n. + **hām**. This was early assimilated to the Fr pattern *bel* + *champ* 'beautiful country'. In this interpretation **bylċ* (or **bylċe*) is an i-mutation derivative of OE **bulc* cognate with ON *bulki*, Norw dial. *bulk* 'a lump', and refers to the ridge of high ground on which the Belchamps lie. However, DEPN and PNE i.27 prefer OE *belċ* 'a beamed or vaulted roof' which occurs once in the OE poem *Exodus* 73 (*bælce* dative sing.) referring to the pillar of cloud covering the Israelites: this *belċ* is an i-mutation derivative of OE *balca* 'a beam'‡. The meaning would be 'house with a roof made of beams' or possibly 'house at (the hill called) *Belc*, the vaulted roof'. In this case the spelling *Bylcham* would have to be an inverse spelling for *Belcham* due to the unrounding of OE *y* in the Essex dial. Ess 409 f.n. cites the K p.ns. BEACHBOROUGH TR 1638, *Belcheberche* 1262, Bilchester and Belce Wood, Sturry, *Bilechewode* 1249, *Bylychewode* 1292, as other possible candidates for OE **bylc* or **bylce*, PNK 447, 574.

BELCHAMP ST PAUL Essex TL 7942. 'St Paul's B'. *Belchampe of St Paul of London* 1451, *Belching Paule* 1535. P.n. Belchamp as in BELCHAMP OTTEN TL 8041 + saint's name *Paul* (*Sci Pauli* 1235) referring to possession by St Paul's cathedral. Ess 408 gives pr [pɔ:lz belʃəm], 409.

BELCHAMP WALTER Essex TL 8240. 'B held by Walter' sc. de Tey, a descendant of William de Beauchamp 1135×54 whose name is also recorded here. *Water Bechampe alias Walter Belchampe* 1548. P.n. Belchamp as in BELCHAMP OTTEN TL 8041 + pers.ns. *Will(elm)i* 1235, 1255 and *Water* 1297, 1305. Ess 409.

†Neither Bilhaugh Notts SK 6368 nor Bealeys in Lockington Humbs SE 9947 belong here, the former exhibiting an overwhelming preponderance of *Bil(e)- Bille* spellings which Nt 76 take to be the OE pers.n. *Billa*, and the latter early spellings in *Begh-* unknown to Ekwall in his investigation in Studies 1936.159 (see YE 161).

‡ Exodus, note to line 73, takes *bælc* to be a descendant of Gmc **balkuz* 'partition' (related to *balca* < Gmc **bal(u)kōn*) without breaking or retraction. This is difficult.

BELCHFORD Lincs TF 2975. 'Belt's ford'. *Beldeforda* [1075]14th, *Bades- Beltesford* 1086, *Beltesford(a)* c.1155, 1183×4, *Beldforda* c.1155, *Belchesford* 1200, [1199]1300. OE pers.n. **Belt*, genitive sing. **Beltes*, + **ford**. SSNEM 373, Cameron 1998.

BELFORD Northum NU 1033. 'The ford by the bell-shaped hill'. *Beleford* 1242 BF-1313, *Belle-* 1300, 1301, *Belford* from 1296 SR, *-forth* 1323–1550. OE **belle** + **ford**. The reference is to a round hill just N of the village. NbDu 16, Studies 1936.160, PNE i.27.

BELL BUSK NYorks SD 9056. Probably the 'bell-shaped bush'. *Bel(l)busk(e)* 1585–1660. ME **belle** + **busk** (ON *buskr*). YW vi.128.

BELL HALL H&W SO 9377 → BELBROUGHTON SO 9277.

BELL HILL Northum NT 8410. *Bell Hill* 1869 OS. 'The round or bell-shaped hill'. An isolated peak in the Cheviots rising to 1612ft.

BELLEAU Lincs TF 4078. 'Helgi's meadow'. *Elgelo* 1086, *Hel(e)gelo* c.1160, 1191, *Helekheluue* c.1180, *Helg(h)elowe -lawe* 1242×3, 1251, *Hellowe* 1311, *Bellowe* 1536. ON pers.n. *Helgi*, genitive sing. *Helga*, + **ló**. Belleau stands on the Great Eau and the name has been remodelled as if Fr **bel** + **eau** 'beautiful eau, beautiful water'. SSNEM 151, Cameron 1998.

BELLERBY NYorks SE 1192. 'Belgr's farm or village'. *Belgebi* 1086, *Belgerby -re-* 12th–1244, *Bellerby* 1285 etc. ON by-n. *Belgr* 'skin-bag', genitive sing. *Belgjar*, + **bȳ**. YN 252, SPN 51, SSNY 20.

BELLEVER Devon SX 6577. Possibly 'stream ford'. *Welford* 1355, *Wele-* 1477, 1489, *Wella- Willa-* 1579, *Bellab(o)ur* 1608–9, *Bellaford* 1663, 1702, *Bellefor* 1736. OE **wella** + **ford**. If this is correct the name shows an unusual series of sound substitutions W- (> V-) > B-. The reference is to a crossing of the East Dart River where there is now a clapper bridge. D 192.

BELLINGDON Bucks SP 9405. Probably 'Billa's hill' or 'hill called Billing'. *Bil(l)endon* 1196, *Belindene* 1198–1227. OE pers.n. *Billa*, genitive sing. *Billan*, or p.n. **Billing* < **bill** 'sword' or **bile** 'bill, beak' + **ing**[2], + **dūn** early replaced with **denu**. Bellingdon is situated on a long narrow sword-like ridge. This may not, however, have been the earliest site and *denu* referring to one of the sinuous dry valleys either side of the ridge may be original. Bk 224, Wa xli, Jnl 2.32.

BELLINGHAM GLond TQ 3772 → COULSDON TQ 3059.

BELLINGHAM Northum NY 8383. 'The settlement called or at *Belling*, the bell-shaped hill'. *Bainlingham* c.1170, *Belinge-Bellingham, Belingjam* 13th cent., *Belyncham* 1332, *Bellingeham, Bellyngeam* 1524. OE ***belling**, locative-dative sing. ***bellinġe**, + **hām**. The village name is still pronounced [belindʒəm]. NbDu 16, BzN 1967.331.

BELLS YEW GREEN ESusx TQ 6136. *Bells Yew Green* 1819 OS.

BELLSHILL Northum NU 1230. *Bellshill* 1866 OS. Probably 'the hill of the bell', ModE **bell** (OE *belle*) 'a bell, a bell-shaped hill'. The reference is to a rounded hill rising to 578ft. due W of the village at NU 1130.

BELMESTHORPE Leic TF 0410. Probably 'Beornhelm's outlying farm'. *Beolmesđorp* [1042×55]13th S 1481, *Belmestorp(e)* 1086–1233, *-thorp(e)* from c.1200, *Belstropp* 1573. OE pers.n. *Beornhelm*, genitive sing. *Beornhelmes*, + ON **thorp**. The pers.n. is not recorded in so contracted a form elsewhere, but no alternative explanation has been proposed. R 161, SSNEM 103.

BELMONT Lancs SD 6716. 'Beautiful hill', OFr **bel** + **mont**. A fashionable name type in the 19th cent.

BELMONT RESERVOIR Lancs SD 6717. P.n. BELMONT SD 6716 + ModE **reservoir**.

BELOWDA Corn SW 9661. Probably 'dwelling of Loude'. *Boloude* c.1240, *Bolloude* [late 13th]14th, *Belleuse* 1302, *Belowdy commonly and not unproperly called Beelowzy* 1602, *Belovedy alias Belowsey* 1609, *Belouda or Belovely Beacon* 1903 Salmon. Co **bod** + pers.n. **Loude*. Medially and finally Co *d* regularly became *z* causing English re-interpretation and a contrasted alternative re-interpretation in 1609. PNCo 53.

BELPER Derby SK 3547. 'Beautiful retreat'. *Be(a)urepeir -repeyr -repair(e) -repayr(e)* 1231–1482, *-reper(e)* 1291–1624, *Beaupeyr* 1269, *-per* 1493, *Belper* from 1577. OFr **beau-repaire** reformed as **bel** + **repaire**. Cf. BEARPARK Durham NZ 2343. Db 525, PNE i.28.

BELSAY Northum NZ 1078. 'Spur of the bill-shaped hill'. *Bilesho* 1162, *Billes-* 1203, *Bel(l)es(h)o(u)* 1166–1296, *Belsou* 1242 BF, *-ow* 13th cent., *Belsey* 1663, c.1715 AA 13.5. OE **bile**, genitive sing. **biles**, later replaced by **belle**, genitive sing. **belles**, + **hōh**. Alternatively the specific might be OE pers.n. *Bill*, a shortened form of *Bilfrith* or the like, genitive sing. *Billes*, and so 'Bill's hill-spur'. NbDu 17, L 167–8.

BELSFORD Devon SX 7659. Possibly 'Bægel's ford'. *Bailleford* 1347, *Baillesford* 1383. OE pers.n. **Bæġel*, genitive sing. **Bæġ(e)les*, + **ford**. The evidence is late, however, and the other possibilities are OFr **baillie** 'a bailiff' or OE ***bēġel** 'a bend'. The exact site of the crossing of the river Harbourne is unclear, but the river goes through a marked bend at SX 7658. D 325.

BELSTEAD Suff TM 1341. Possibly the 'place of the (funeral) pyre'. *Bel(e)steda* 1086, *Belsted(e)* 1198–1610. There is no certainty about the specific but it could be OE **bēl** 'a fire, a funeral pyre'; the specifc is OE **stede**. An alternative suggestion is OE **bel* cognate with ON *bil*, Swed dial *bil*, Dan dial *bil, bœl* 'an interval, a space' in the sense 'open land in forest' or 'patch of dry land in marsh', the appropriate sense here. DEPN, Stead 186, Studies 1936.160, PNE i.28.

BELSTONE Devon SX 6193. 'Bell-stone'. *Belesthā* 1086 DB, *Bellestam* 1086 Exon, *Belstan(a)* 1166–1254, *Belestone* 1259, *Belston* 1292. OE **belle** + **stān**. 'The bellstone was a remarkably fine logan-rock that rolled like a ship in a gale and boys were wont to make it crack nuts for them. It has been thrown down and broken up by quarrymen' (S. Baring-Gould in the *Little Guide* to Devon, 1907). Perhaps it was shaped like a bell or made a noise like a bell. D 131.

BELTHORN Lancs SD 7124. Cf. *Belthorn-moor* 1771. Possibly 'thorn-tree on the beacon', OE **bēl** + **thorn**. The site of Belthorn is on the slope of a prominent hill rising to over 900ft. La 19.

BELTINGE Kent TR 2068. Uncertain. *Beltinge* from 1189×99, *Byltyng* 1498×9. Formally this could be an *ing*[2] derivative of OE *belt* 'a belt, a girdle'. An OE **Belting*, locative-dative sing. **Beltinġe* 'at Belting', would be 'the belt-like place', but what that might have meant is not clear. It might have been the name of the small stream in the nearby valley, 'belt-like stream', or it may be the Kentish dial. form for an OE **bylting* from **bylte* 'a heap, a small hill' and so 'hill place' referring to the ridge on which the village stands, cf. Belting Hill in Herne, *Beltinghill* 1461 and BILTING TR 0549. K 509, ING 203, Cullen.

BELTOFT Humbs SE 8006. Uncertain, possibly 'curtilage by the funeral pyre' or 'at the glade'. *Beltot* 1086, *Beltoft* 1202. OE **bēl** or ***bel** + **toft**. Cf. BELTON SE 7806. DEPN, PNE i.26, SSNEM 213, V i.77.

BELTON Partly uncertain. The generic is OE **tūn** 'settlement, farm, village', the specific possibly OE **bēl** 'a fire, a funeral pyre, a cremation'. The significance of such a compound is unknown. Alternatively the element *Bel-* in this and similar names may be an unknown element **bel* found also in continental names like Beelen in Münster, *Bele* 1146, and Beelen in Gelderland, *Bele* 1188, for which a connection with Norw dial. *bali* 'a hill' has been suggested. The sense 'patch of dry ground in marsh' has been suggested for English examples. FmO 386, Studies 1936. 159–63, PNE i.26, V i.77, DEPN.

(1) ~ Humbs SE 7806. *Beltone* 1086, *Bealton* 1179, 1224, *Beltona* 1224. Cf. BELTOFT SE 8006. DEPN.

(2) ~ Leic SK 4420. *Beltona* c.1130, *Belton(e)* from 1199, *Beleton(e)* early 13th–1282. Lei 343.

(3) ~ Leic SK 8101. *Belton(e)* 1066×87–1610, *Bealton* 1167,

Beautone c.1200, 1237. SK 4420 The first element has been partly confused with OFr **bel**, **beau** 'beautiful'. The village lies on raised ground between two streams in what would have originally have been densely wooded ground. R 68.

(4) ~ Lincs SK 9339. *Beltone* 1086, *-ton(a)* from 1163. DEPN, Perrott 487, SSNEM 213 (under Beltoft).

(5) ~ Norf TG 4802. *Beletuna* 1086, *Beleton'* 1198–1219, *Belton* 1229, 1278. DEPN.

(6) ~ HOUSE Lincs SK 9239. P.n. BELTON SK 9339 + ModE **house**. A notable late 17th cent. country house.

BELVEDERE GLond TQ 4978. *Belvedere* 1805. The name of a mansion built c.1740 and since demolished. OFr **obel** + **vedeir** 'beautiful view'. LPN 19, Encyc 57.

BELVIDE RESERVOIR Staffs SJ 8610. *Belvide Reservoir* 1895. P.n. Belvide + ModE **reservoir**. Built in *Belvide Field* in 1834 as a feeder for the Shropshire Union Canal. Belvide may be OFr **bel** + **vedeir** 'beautiful view'. St i.39, Horovitz.

BELVOIR Leic SK 8233. 'Beautiful view'. *Belveder* 1130, *Beluer(o) -v-* c.1130–1534, *Bellouidere -v-* 1145–[1198]1301, *Belueeir, Balveir* 1154×89, *Bealue(i)r -v-* 1174×82–1250, *-voir* 1464–75, *Beauer -v-* 1252, 15th cent., *Belvoyr(e)* 16th cent. OFr **bel**, **beau** + **vedeir**. Belvoir Castle is situated on a high hill-top with commanding view to the N and E. Lei 138 gives pr ['bi:və].

BEMBRIDGE IoW SZ 6488. '(Land, place) within the bridge'. *Bynnebrygg* 1316, *-brigge* 1345, 1468, *Bynbryg(g)e, Binbrigge -bridge* 1441–1688, *Bembridge* 1775, *Bimbridge* 1781. OE **binnan** + **brycg**. A Latin form of the name, *infra pontem*, occurs 1324, 1440. Until the 19th cent. Bembridge was cut off from the rest of the island by extensive marshes along the river Yar and could only be reached by boat or by the bridge at Yarbridge. Wt 34 gives prs ['bim-] and ['bembridʒ], Mills 1996.27.

BEMBRIDGE POINT IoW SZ 6488. *Bambridge Point* 1769, 1775, *Bimbridge Point* 1781. P.n. BEMBRIDGE SZ 6488 + ModE **point**. Wt 36, Mills 1996.28.

BEMPTON Humbs TA 1972. 'Tree farm or village'. *Benton(e)* 1086–1525, *Bempton(a)* 1114×24–1595, *Bemton(a)* 1128×32–1285. OE **bēam** + **tūn**. The sense may be 'farm where beams are made and obtained'. YE 106.

BENACRE Suff TM 5184. 'The bean field'. *Benagra* 1086, *Beanacer* c.1095, *Benaker* 1229, *Benacre* from 1254. OE **bēan** + **æcer**. DEPN, Baron.

Lower BENEFIELD Northants SP 9988. *Benyfeld Netherthorp* 1481. P.n. Benefield + **netherthorp**, the 'lower village', in contrast to Upper BENEFIELD SP 9889. Benefield, *Beringafeld* [10th]12th S 1566, *Berifeld* 1201, *Benefeld* 1086–1346, *Beni(ng)feld -y(ng)-* 12th–1482, is the 'open land of the Beringas, the people called after Bera', OE folk-n. *Beringas* < pers.n. **Bera* + **ingas**, genitive pl. *Beringa*, + **feld**. The change *r* > *n* is an AN substitution. Nth 211 gives pr [benifi:ld].

Upper BENEFIELD Northants SP 9889. *Benefield Overthorp* 1641. P.n. Benefield as in Lower BENEFIELD SP 9988 + **overthorp**, the 'upper village'. It is situated uphill from the *netherthorp*. Earlier simply *Upthorp* c.1220–1376. Nth 212.

BENENDEN Kent TQ 8032. 'Bynni or Bynna's woodland-pasture'. *Benindene* 1086, *-den(n)(e)* 1215×28–1251, *Binningdaenne, Bennedene* c.1100, *Bynyndenn' -in-* 1242×3, 1253×4. OE pers.n. *Byn(n)i* + **ing**[4] or *Bynna*, genitive sing. *Bynnan*, + **denn**. In the dial. of Kent the pers.n. *Bynni, Bynna* would have been **Benni, *Benna*. PNK 347, KPN 348.

North BENFLEET Essex TQ 7590. *North Bemflet(e)* 1337. ME adj. **north** + p.n. Benfleet as in South BENFLEET TQ 7785. Also known as *Parva Benflet* 'little B' 1255 etc. Ess 142.

South BENFLEET Essex TQ 7785. *Suth Benfleete* 1281. ME adj. **suth** + p.n. *(to, æt) Beamfleote* c.900 ASC(A) under years 894, 1067, *(æt) beamfletan* [995×1005]11th, *Beamflete* 1066×87, *-fleotam* c.1250, *Benfleota* [1068]1309, *-flet* 1086–1255 etc., *Bem- Bamflet(e)* 1157–1319, 'the tree-marked creek', OE **bēam** + **flēot**. Also known as *Magna Benflet* 'great B' 1248 for distinction from North or Little BENFLEET TQ 7590. South Benfleet overlooks the great creek or inlet from the Thames which cuts off Canvey Island and where the Danish warlord Hæsten built a raiding fortress in 894. North Benfleet lies several miles inland. Ess 142, Jnl 2.42.

BENGATE Norf TG 3027. 'The bean road'. *Benegate* 1327×77, *Bengate* 1838 OS. ME **bene** (OE *bēan*) + **gate** (ON *gata*). Nf ii.205.

BENINGBROUGH NYorks SE 5357. 'The fortified place called after Benna'. *Benniburg* 1086, *Beningburg(h) -yng-* 1180–1317, *Beningeburc* 1167, *-burg* 1223. OE pers.n. *Benna* + **ing**[4] + **burh**. YN 19, BzN 1968.161.

BENINGTON Herts TL 3023. 'Settlement on the river Beane'. *Belintone* 1086, *Beninton(e)* 1086–1230, *Benington(e)* from 1248 with variants *-yng, -ig-*. R.n. Beane, *Bene* c.925 ASC(A), c.1000 ASC(D) under year 913, *Beane* 1577, + **ing**[2] or [4] + **tūn**. Beane is a pre-English r.n. of uncertain origin appearing in early sources as *Bene ficcan* c.925 ASC(A) under year 913, *Bene ficean* c.1000 ASC(D) under same year, *Beneficche* 13th, in which *ficcan*, (?) the oblique case of **ficce*, has been explained as representing PrW **bicc* 'small', and *bene* the Celtic word **benā*, OIr *ben*, W *bun*, Co *benen* 'wife, woman, goddess': but this is very uncertain. Hrt 1, RN 27, FT 163, LHEB 567.

BENINGTON Lincs TF 3946. 'The estate called after Beonna'. *Benigtun* 1154×89, *Benington* from 1166. OE pers.n. *Beonna* + **ing**[4] + **tūn**. DEPN, Cameron 1998.

BENNACOTT Corn SX 2992. Probably 'Bynna's cottage'. *Bennacote* 1201, *Bunacote* 1302, *Benecote* 1443 G. OE pers.n. *Bynna* + **cot**. G cites a form *Buningcote* c.1150 which points to *Bynna* + **ing**[4] + **cot**. PNCo 53.

Long BENNINGTON Lincs SK 8344. *Longebeniton* 1274. ME **long** + p.n. *Beningtone, Beninctun* 1086, *Benington -yng-* 1150–1602, *Bennington* from c.1163, 'the estate called after Beonna', OE pers.n. *Beonna* + **ing**[4] + **tūn**. Perrott 356, Cameron 1998.

BENNIWORTH Lincs TF 2081. Probably 'the enclosure of the Beonningas, the people called after Beonna'. *Beningvrde* 1086, *Beningeorde* c.1135, *Beningewurða* 1171, *-wurda* 1195. OE folk-n. **Beonningas* < pers.n. *Beonna* + **ingas**, genitive pl. **Beonninga*. Alternatively this might be the 'enclosure called or at *Beonning*, the place called after Beonna', p.n. **Beonning* < pers.n. *Beonna* + **ing**[2], locative-dative sing. **Beonninge*, + **worth**. DEPN, Cameron 1998.

BENOVER Kent TQ 7048. Partly uncertain *Dan(e) Street* 1782, *Denover Street* 1819 OS, *Benhover Street* 1801. For *street* see BROAD STREET TQ 8356. The generic of Benover is ultimately from OE **ofer** 'a flat-topped hill', the specific is unclear. PNK 169.

BENSON Oxon SU 6191. 'The estate called after Benesa'. *Bænesing tun* 9th ASC(A) under year 571, *Benesing tún* 9th ASC(A) under year 777, *Benesingtun* 977×1000 ASC(B), 1121 ASC(E) under years 571 and 777, *(from) Bynsincgtune, (in) Beonsincgtune* [887]11th S 217, *Bensinton(a) -tun(a) -yn-* 1140–1315, *Benston* 1526. OE pers.n. *Benesa* + **ing**[4] + **tūn**. O 116.

BENTHALL Shrops SJ 6702. 'Bent-grass nook'. *Benetala* [1120]copy, *Benethala* 1167, *-hal(e)* 1221×2–1421, *Benthale* 1271×2–1421, *-hall* from 1698, *Bental* 1577, *Bentall -oll* 17th. OE **beonet** + **halh**. Sa i.37.

BENTHAM Glos SO 9116. 'Bent-grass homestead' or 'water-meadow'. *Benetham* 1220–1322, *Bentham* from 1378. OE **beonet** + **hām** or **hamm**. Gl ii.115.

High BENTHAM NYorks SD 6669. *Over Bentham* 1404–1681, *Upper ~* 1598. ModE **high, upper**, ME **over**, + p.n. *Benetain* 1086, *-haim* 1202×8, *-ham* 1214–90, *Bentham* from 1204, 'the homestead in the bent grass', OE **beonet** + **hām**. 'High' for distinction from Low BENTHAM SD 6469. YW vi.237 gives pr ['bentəm].

Low BENTHAM NYorks SD 6469. *Nether Bentham* 1598, *Lower ~*

1674. ModE **low, lower**, ME **nether**, + p.n. Bentham as in High BENTHAM SD 6669. YW vi.237.

BENTLEY 'Clearing where bent-grass grows'. OE **beonet** + **lēah**. L 204.

(1) ~ Hants SU 7844. *(æt) beonetleh* [963×75]12th S 823, *(to) beonet legæ (gemære)* 'to the Bentley boundary' [973×4]12th S 820, *Benedlei* 1086, *Benetlea -leg(a) -leia -lie* 1167–1407. Ha 32, Gover 1958.112.

(2) ~ H&W SO 9866. *Beneslei* 1086, *Benetlega -e -ley* 1185–1281, 1499, *Bentley* 1578. Also known as Bentley Pauncefote, *Benetelege Pancevot* 1212, from Richard Panzeuot who held land here in 1185. Wo 366.

(3) ~ Humbs TA 0236. *Benedlage* 1086, *Benetlee -ley(e) -leg(a)* 1163×89–1269, *Bentley* from 1281. YE 204.

(4) ~ SYorks SE 5605. *Benedleia, Beneslei -laie, Benelei* 1086, *Benetleia -ley(e)* etc. c.1200–1453, *Bentley* from 1285. YW i.24, L 204.

(5) ~ Warw SP 2895. *Benechelie* (sic) 1086, *Benetleia* 1221–1301 with variants *-legh* and *-leye, Ben(e)tleye* 1316, 1335. Wa 78.

(6) ~ HEATH WMids SP 1676. *Bentley Heth* 1401. P.n. Bentley + ME **heth**. Wa 73.

(7) Fenny ~ Derby SK 1750. *Fennibent(e)ley(e) -y-* 1271–1310 etc. Earlier simply *Benedlege* 1086, *Benetlea -legh -ley* 1175–1260, *Bent(e)leg(h) -le(ye)* 1251–1291 etc. 'Fenny', from OE **fennig** 'marshy', is for distinction from Hungry Bentley SK 1783, one of the lost villages of Derbyshire, *Hungrybentele* 1269, *--bentley* 1464, *Beneleie* 1086 with OE **hungrig** used of poor unproductive land. Db 344, 530.

(8) Great ~ Essex TM 1121. *Magna Ben(e)leg(h)* 1227, *Moche Bentley* 1552. ModE adj. **great**, ME **much**, Latin **magna**, + p.n. *Benetleye* [1035×44]13th S 1521, *Benetleiam -leã, Menetleam* 1086, *Benetleg(a) -leghe -ly -leye* 1212–1303, *Bentley(e)* from 1323. 'Great' for distinction from Little BENTLEY TM 1125. Ess 328.

(9) Little ~ Essex TM 1125. *Parva Ben(e)leg(h)* 1254, 1295. ModE adj. **little**, Latin **parva**, + p.n. Bentley as in Great BENTLEY TM 1121. Ess 328.

BENTON Devon SS 6536. 'Farm, village called after Bunta'. *Botintone* 1086 DB, *Bontintona* 1086 Exon, *Buntingeton* 1270, *Bot- Bodyngdon* 1301–46, *Bunton* 1582. OE pers.n. *Bunta* + **ing**[4] + **tūn**. D 30.

BENTWORTH Hants SU 6640. 'Binta's enclosure'. *Bintewrde* 1100×35, *Binte(w)orda -wurða -wurd* 1130–1223, *Bynteworth* 1277–1431. OE pers.n. **Binta* + **worth**. Ha 32, Gover 1958.130.

BENWICK Cambs TL 3490. 'Bean farm'. *Beymwich* (sic) 1221, *Benewich -wik(e) -wyk(e)* 1244–1331, *Benwyk(e)* 1256, 1399, 1549, *Beynwyk(e) -wik(e)* 1251–98, *Ben(n)ick* 1645, 1763. OE **bēan** + **wīc**. Ca 246 gives pr [benik].

BEOLEY H&W SP 0669. 'Bee wood or clearing'. *Beoleah* 972 S 786, *Beolege* 1086, *Buleg -ley* 1244–1428, *Beeley* 1346, 1481, 1741, *Beoley* from 1431. OE **bēo** + **lēah**. Wo 186 gives pr [bi:li:], L 205.

BEPTON WSusx SU 8518. 'The estate called after Bæbba, Bæbbi or Bæbbe'. *Babintone* 1086–1303, *Beb(b)in(g)ton -yn(g)-* 1240–1584, *Bebiton -y-* 1296–1357, *Bebton* 1527–1726, *Bepton* 1583. OE pers.n. *Bæbba, Bæbbi* or *Bæbbe* feminine + **ing**[4] + **tūn**. Sx 15, SS 74.

BERDEN Essex TL 4629. 'Cow-house valley'. *Berden(e)* from 1085, *-dane* 1086, *Bierden* 1428. OE **bȳre** with Essex dial. *e* for *y* + **denu**. Ess 548, DEPN.

BERE Either 'pasture, woodland pasture for feeding swine', OE **bǣr**, or **bearu** 'a wood, a grove'. PNE i.16, 22.

(1) ~ ALSTON Devon SX 4466. *Berealmiston* 1433×72, *Berealbeston* 1542, *Berealston* 1549, *Alson* 1675. The earliest reference to this place, *Alphameston que est hamella de Byrfferers* 'Alston which is a hamlet of BERE FERRERS' 1339, reveals its origin, viz. 'Ælfhelm's farm or villlage', OE pers.n. *Ælfhelm*, genitive sing. *Ælfhelmes*, + **tūn**. D 223.

(2) ~ FERRERS Devon SX 4563. 'Bere belonging to the Ferrers family'. *Byr Fer(r)ers* 1306–40, *Bere Ferers* 1389. Earlier simply *Ber* 1242–1311, *Bere* 1301, 1331, *Byr* 1281–1377. The form *Birland* 1086, 1318, 1319, and the frequent ME spellings *Byr(e), Bire*, call in question the derivation from *bǣr* or *bearu*. However, no satisfactory alternative has been proposed. OE *bȳre* 'a byre, a cow-shed' is a North country element. The manor was held by William de Ferers in 1242. D 223.

(3) ~ REGIS Dorset SY 8494. 'The king's Bere'. *Bi- Byre Regis* 1495, *Beare Regis* 1549. P.n. Bere + Latin **regis**, genitive sing. of **rex** 'a king'. Bere, *Bere* from 1086, *Beer(e)* 1242–1617, *By- Biere* 1259–1410, *Bi- Byre* 1327–1447, is either OE **bǣr** 'woodland pasture' or **bearu** 'wood, grove'. There was formerly a royal forest here. Do i.273.

(4) Forest of ~ Hants SU 6711. *Foresta de Ber'* 1237. ME **forest** + p.n. *la Bera -e* 1168–1232, 'the swine-pasture', Fr definite article **la** + ME **bere** (OE *bǣr*). A forest in which rights of pannage were exercised. Mislingford in the forest at SU 5814, *Mestlingeford* 1290, *Maslyngefeld* 1280, *Maslyngefford* 1298, the 'ford of the people of *Mastley', contains a lost p.n. OE **Mastlēah* 'beech-mast wood or clearing', beech-mast being a favourite food of foraging pigs. Ha 32, Gover 1958.7, 51.

BEREPPER Corn SW 6522. 'Beautiful retreat'. *Beauripper* [1443]19th, *Breepar* 1699. OFr **bel, beau** + **repaire**. PNCo 53.

East BERGHOLT Suff TM 0735. *Estbergholt* 1394–1622. ME adj. **est** + p.n. *Bercolt* 1086, 1212, *Bercholt* 1130–1209, *Berholt* 1199–1327, 1511, *Bergholt -hout* 1228–1507, 'the hill copse', OE **beorg** + **holt**. 'East' for distinction from West BERGHOLT Essex TL 9627. DEPN reports an old pr with medial [rf]. DEPN, Baron.

West BERGHOLT Essex TL 9527. *Westbergholt Sakevyle* 1491. ME adj. **west** + p.n. *Bercolt(a)* 1086, *Bercolt* 1183, *Berch(h)olt(e)* 1190–1272, *Bergh(h)olt(e)* from 1248, *Barfold(e)* 16th cent., 'hill wood', OE **beorg** + **holt**. 'West' for distinction from East BERGHOLT Suffolk TM 0734. Ess gives former prs [ba:fould, ba:fl].

BERINSFIELD Oxon SU 5796. No early forms. The name was invented by a local historian for the airfield here to commemorate Birinus, the apostle of the West Saxons. Signposts 140.

BERKELEY Glos ST 6899. 'Birch wood or clearing'. *(æt) Be(o)rclea* [824]11th S 1433, *(æt) Berclea* 833 S 218, *(on) Beorclea* 1121 ASC(E), *Berchelai(a) -le -lay -ley(a)* 1086–1190, *Berkelai -lay(a) -lei(a) -ley(a) -le(ye)* 1080–1512, *Bark(e)ley* 1492–1675. OE **beorc** + **lēah**. The form *Berclinga* [804]11th S 1187 is the genitive pl. of OE *Beorclingas* 'the men of Berkley' referring to the monks of Berkley Monastery. Gl ii.211 gives prs ['baRkli] and ['ba:kli].

BERKHAMSTED Herts SP 9907. 'Hill homestead'. *Beorhðanstædæ*[966×75]12th S 1484, *Beorh ham stede* c.1100 ASC(D) under year 1066, *Berch(eh)ãstede* 1086, *Berkhamsted(e) -ch-* 1133–1428, *-kam-* 12th–1377. OE **beorg** + **hāmstede**. Also known as *Great Barkhamsteed* 1580 for distinction from Little BERKHAMSTEAD TL 2907. Hrt 27 gives former pr ['ba:kəmstid], Stead 237.

Little BERKHAMSTED Herts TL 2907. *Parva Berchamstede* 1254. ModE **little**, Latin **parva** + p.n. *Berchehãstede* 1086, *Berchamsted(e)* 1208–1428, the 'birch-tree homestead', OE **beorc**, genitive sing. **beorce**, genitive pl. **beorca**, + **hāmstede**. 'Little' for distinction from (Great) BERKHAMSTED SP 9907. Hrt 217, Stead 238.

BERKLEY Somer ST 8149. 'The birch-tree wood or clearing'. *Berchelei* 1086. OE **beorc**, genitive pl. **beorca**, + **lēah**. DEPN.

BERKSHIRE 'The county of *Bearroc*'. *Bearrocscire, Berrocscire* [893]11th Asser, *Bearrucscire* c.900 ASC(A) under year 860, *Berchesire* 1086. OE **scīr** 'county, shire' + p.n. *Bearroc*, reflecting Brit **barrāco-* 'hilly'. This Celtic p.n. may have referred originally to the region dominated by the Berkshire Downs, but was early transferred to a wood in the south-west of the county. Asser states that the shire was named from the wood, which is possible. The wood is now lost. Brk i.1–2, V i. 52.

BERKSWELL WMids SP 2479. 'Beorcol's spring'. *Berchewelle* 1086, *Bercles- Berk(e)leswell(e)* 1221–14th, *Berkeswelle* 1235–1535, *Barkeswell* 1504–1657, *Barswell* 1518–1657. OE pers.n. *Bercul, Beorcol*, genitive sing. *Beorc(o)les*, + **wella**. There is a spring on the S side of the church known as Berk's Well. Wa 56 gives pr [ba:(k)swəl], L 31.

BERMONDSEY GLond TQ 3579. 'Beornmund's island'. *Vermundesei* (sic) [708×15]c.1200 B 133, *Bermundesye* 1086 etc., *Bermonsey* 1450, *Barmonsey* 1540, *Barmsey* 1617. OE pers.n. *Beornmund*, genitive sing. *Beornmundes*, + **ēġ**. The reference is to dry ground in marshland beside the Thames before the river was embanked. A form of the original spelling has survived and ousted the natural development illustrated by the 1617 form. Sr 16 gives obsolete pr [ba:mzi], TC 202, Encyc 59.

BERNARD WHARF Lancs SD 3451. Pers.n. *Bernard* + ModE **wharf** 'a sand-bank'.

BERNEY ARMS STATION Norf TG 4605. Inn-name *Berney Arms* 1837 OS + ModE **station**. The Berney Arms is named after the Berney family.

BERNICIA, the ancient name of the kingdom of Northumbria N of the Tees, is a Latinisation of OW *Beornica, id est Berneich, Birneich* 9th HB, OE *Beornice* 9th ASC(A), *Bœrnice* 12th ASC(E), MW *Bryn- Byrneich, Brenn(e)ych, Breennych*. The people of Bernicia are the *Bernicii* [c.731]8th BHE. The name is usually associated with the RBrit tribal name Brigantes and derived from PrW **Breȝent(e)ich, *Brɩȝent(e)ich*, < **Brigantaccī* 'the high ones' from Brit **brigant-* related to **brigā* 'a hill' (W *bre*). There are serious phonological difficulties with this derivation, however, and LHEB 701–5 suggests instead PrW **Bernecc, *Bɩrnecc* 'the land of the mountain passes' < Brit **Bern- Birnacciā* from Celtic **bernā, *birnā*, as in OIr *bern*, Gaelic *bearn* 'a gap, a mountain pass'.

BERRICK SALOME Oxon SU 6293. 'B held by the Suleham family'. *Berwick Sullame* 1571, ~ *Sallome* 1737. Earlier simply *Berewich* 1086, 'the outlying farm', OE **berewīc**. 'Salome' from Almaric de Suleham 1235×6 from Sulham Berks SU 6424 and for distinction from other Berwicks. O 120, 106.

BERRIER Cumbr NY 4029. 'Hill shieling'. *Berghgerge* 1166, *Berherge* 1167, *Bergher* 1242, *Berier(e), Beryer(e)* 1285–1361, *Beeryer* 1610. ON **berg** + **ǣrgi**. Cu 181, SSNNW 62, L 127.

BERRINGTON Northum NU 0043. 'Berry hill'. *Berigdon'* 1208×10 BF, *Bering-* 1296 SR, *Beryngdon* 1342, *-yngton* 1370, *Barrington* 1610, c.1715 AA 13.9. OE **beriġe** + **dūn**. The reference is probably to Berrington Law NU 9843. This seems more likely than the alternative suggestion that the specific might be OE **byriġ**, locative-dative sing. of **burh** 'a fort', referring to a doubtful camp half a mile S of the village. NbDu 18, L 144.

BERRINGTON Shrops SJ 5307. 'The settlement associated with a fort'. *Beritune* 1086, *-ton(a)* 1121, 1382, 1577, *Bi -Byriton -yton* c.1090–1449, *Beringtun'* 1226, *Berrington* from 1596. OE **byriġ**, dative sing of **burh**, + **tūn**. Possibly part of an organized system of defence. Sa i.38–41, WW 150.

BERROW Somer ST 2952. 'The hills or mounds'. *Berges* 1196, *Bergh* 1249, *Berrowgh* 1610. OE **beorg**, pl. **beorgas**. The form *(at) Burgh'* [973]14th S 793 is a late unreliable spelling. The reference is to sand-dunes. DEPN.

BERROW FLATS Somer ST 2854. 'Berrow mud-flats'. P.n. BERROW ST 2952 + ModE **flat**.

BERROW GREEN H&W SO 7458. *Berrow Green* 1832 OS. P.n. Berrow + ModE **green**. Berrow, *(de) Berga* 1275, *(atte) Berewe* 1327, is OE **beorg** 'a hill' referring to Berrow Hill SO 7458 which dominates the village to the W. Wo 63.

BERRY '(At the) fortification'. OE **byriġ**, dative sing. of **burh**. PNE i.58.

(1) ~ HEAD Devon SX 9456. Identical with *Byri pointe* c.1550. There are old earthworks on the headland. D 508.

(2) ~ POMEROY Devon SX 8261. 'B held by the Pomeroy family'. *Bury Pomerey* 1281, 1294, *Burgh Pomeray* 1303, *Piry Pomeray* 1413. Earlier simply *Berie* 1086 DB, *Beri* 1086 Exon, 1242, *Berri* 1267, *Byry* 1275. The reference may be to an earlier fortification on the site of the medieval castle. The manor was held by Ralph de Pomerei in 1086. Pomerei is OFr *pommeraie* 'apple orchard' as in La Pomeraye, Calvados, Normandy. D 505.

BERRY HILL Glos SO 5712. 'Hill with a fortification'. *Burchull'* 1221, *Burg- Borouhulle* 1270–93, *Burlhill, the Burl* 1612, *Bury Hill* 1830. OE **burh** + **hyll**. The form *Burl* is a contraction of *burh-hyll*. Gl iii.226.

BERRYL'S POINT Corn SW 8467. Partly uncertain. *Berryls Point* c.1870. Possibly from the surname *Burrell* or *Birrell* + ModE **point**. PNCo 53.

BERRYNARBOR Devon SS 5646. 'B held by the Nerebert family'. *Bery Narberd* 1244, *Ber(r)i Neyberd* 1281, 1315, *Bery Nerberd* 1288, *Byry in Arberd* 1394. Earlier simply *Biri* 1167, 1261, *Bery* 1209, 1234, *Bury* 1281, p.n. BERRY + family name *Nerebert* 1209. The 1086 spellings *Hvrtesberie* DB, *-ia* Exon, and *Hertesberie* 1121 point to an earlier form of the name, 'Heort's or stag fortified place or manor', OE pers.n. *Heort* or **heort**, genitive sing. *Heortes*, **heortes**, + p.n. **byriġ** as in BERRY. D 27, DEPN.

BERSTED WSusx SU 9200. Partly uncertain. *Beorgan stede* *[680 ?for 685]10th S 230, *Berkestede* 1248, *Bergested* c.1250, *Northber(e)wested* 1267, *(Suth)berg(h)ested(e)* 1275–1347, *Burgested* 1271, *Bursted* 1375, 1640, *Bersted(e)* from 1398, *Barsted* 1535, *Birsted* 1649. This might be 'Beorge's estate', OE feminine pers.n. **Beorge*, genitive sing. **Beorgan*, + **stede**, or it might represent OE **Beorg-hāmstede* 'the tumulus homestead'. The boundary of the Bersted people is *(to) beorgansteding a mearce* [988]14th S 872. Sx 90, Stead 251, ASE 2.34, SS 87.

The BERTH Shrops SJ 4323. 'The fort'. *Berth Hill* 1833 OS. OE **burh** 'a fort' referring to an Iron Age earthwork set in marshland. The modern form shows substitution of [θ] for the ME peripheral phoneme [x] as also in Berth Hill Staffs SJ 7839, *Berth Hill* 1833 OS. Sa i.31.

BERWICK 'A barley farm', later 'an outlying part of an estate, the outlying lands of a manor retained for the lord's use'. OE **bere-wīc**. PNE i.31.

(1) ~ ESusx TQ 5105. *Beruice, Berewice* 1086, *Berwicke al. South Berwick al. Barrike* 1571. Sx 411, SS 78.

(2) ~ BASSETT Wilts SU 1073. 'B held by the Bassett family'. *Berewykbasset* 1321, 1409, *Berwyk Basset* 1325, *Barwyk Basset* 1449. P.n. *Berwicha* 1168, *Berewic* 1185, + manorial addition from the family of Alan Basset who held the estate in 1211. Wlt 254 gives pr [barik].

(3) ~ HILL Northum NZ 1775. Short for Berwick on the hill, *Berewic super montem* 1428, *Barricke of the hill* 1595. Earlier simply *Berewic* 1205 Cur, 1242 BF, *-wyc* 1242 BF. NbDu 18, DEPN.

(4) ~ ST JAMES Wilts SU 0739. 'St James's B'. *Berewyk Sancti Jacobi* c.1190–1291, *Barwike St. James* 1595. P.n. BERWICK + saint's name *James* from the dedication of the church. W 232.

(5) ~ ST JOHN Wilts ST 9422. 'St John's B'. *Berewyke S. Johannis* 1265, *Barwyke Seynt John* 1536, *Barrike St John* 1606. P.n. *Berwicha* 1167, *Berewic(h') -wicke* 1196–1255, + saint's name *John* from the dedication of the church. Wlt 201.

(6) ~ ST LEONARD Wilts ST 9233. 'St Leonard's B'. *Berewyk Sci Leonardi* 1291, *Berwyk Seynt Leonarde* 1534, *Barwyke Sci Leonard'* 1544. P.n. *Berewica* 1100×35, *Berewyk* 1249, + saint's name *Leonard* from the dedication of the church. Also known as *Cold Barwyke* 1545. Wlt 184.

(7) ~ -UPON-TWEED Northum NT 9953. *Berewicum super Twedam* 1229. P.n. *Berwich* 1136, *-wyc -wic(h) -uvich -wyk'* 12th cent. RRS + r.n. TWEED NT 9551. Also known as *Suthberwych* 1287. By origin an outlying vill of an early multiple estate formerly centred upon perhaps Swinton Borders NT 8147 or Kimmerghame Borders NT 8151, subsequently raised to burgh status by David I. It became the chief seaport of

Scotland and after changing hands many times was finally incorporated into England in 1482. *Suth* for distinction from North Berwick Borders NT 5585. DEPN, ScotPN 78, Barrow 1973.30,

BESCAR LANE STATION Lancs SD 3914. P.n. Bescar Lane + ModE **station**, a halt on the line from Wigan to Southport. Bescar Lane is the 'lane leading to Bescar, the birch marsh', *Birchecar* 1331, 1359, *Birchcarre* 1546, ME **birche** (OE *birċe*) + **ker** (ON *kjarr*). La 124.

BESFORD H&W SO 9145. 'Betti's ford'. *Bettesford* 972 S 786, *Beford* 1086, *Bez(e)- Bezce- Best- Bes(c)e- Besseford(e) -fort* [c.1086]1190–1322, *Besforde* 1327. OE pers.n. *Betti*, genitive sing. *Bettes*, + **ford**. Wo 187.

BESSACARR SYorks SE 6101. 'Newly cultivated ground by the rough grass'. *Besacla -e* c.1160–1202, *Besacra -e* c.1190–1605, *Besakell'* 1379, *Bezacle* 1573, *Bessacre* 1731, *Bessacle* 1771. OE **bēos** + **æcer**. Forms with *l* are due to AN confusion of unstressed *-re* and *-le*. Bessacarr is in the marshes SE of Doncaster. YW i.40, L 232.

BESSELS LEIGH Oxon SP 4501 → Bessels LEIGH SP 4501.

BESSINGHAM Norf TG 1636. 'Homestead of the Basingas, the people called after Basa'. *Basingeham* 1086, 1176, *Basingham* 1205–54. OE folk-n. **Basingas* < pers.n. **Basa* + **ingas**, genitive sing. **Basinga*, + **hām**. ING 132.

BESTHORPE Norf TM 0695. 'Bøsi's outlying farm'. *Besethorp* 1086, *Bestorp* 1198. ODan pers.n. *Bøsi*, genitive sing. *Bøsa*, + **thorp**. Not identical with BESTHORPE Notts SK 8264 or BEESTHORPE Notts SK 7260, both of which contain OE **bēos* 'bent-grass'. DEPN, SPN 70.

BESTHORPE Notts SK 8264. 'The outlying settlement in the bent-grass'. *Bestorp* 1147–1203, *-thorpe* from 1236, *Beisthrope -thorp(e)*, *Beastropp* 16th. OE ***bēos** + **thorp**. Bent-grass thrives in boggy areas, and the village lies in a marshy area by the Trent. Nt 201, SSNEM 103.

BESWICK GMan SJ 8697. Possibly 'Beac's farm or work-place'. *Bexwic* 1200×30, *-wycke -wyk* 1322, *-wik* 1359. OE pers.n. **Bēac*, genitive sing. **Bēaces*, + **wīċ**. La 35.

BESWICK Humbs TA 0148. 'Bessi or Bōsi's dairy-farm'. *Baseuuic -wic* 1086, *Besewic -wik(e) -wyk(e)* 1177×99–1406, *Bessewyk(e)* 1254–1332, *Beswyk(e)* 1371–1562. ON pers.n. *Bessi* or *Bōsi*, genitive sing. *Bessa, Bōsa*, + **wīċ**. YE 159, SSNY 143.

BETCHWORTH Surrey TQ 2150. 'Becci's settlement'. *Becesworde -uuorde* 1086, *Becheswurd -wrde -worth* etc. c.1170–1420, *Bacheswrthe* c.1215, *Becheworth'* 1215–1302, *Bettisworth* 1448. OE pers.n. **Beċċi*, genitive sing. **Beċċes* + **worth**. Sr 282.

BETHERSDEN Kent TQ 9240. 'Beaduric's woodland-pasture'. *Beatrichesdenne* 1086×7, c.1194, *Bed(e)richesdenne* 1142×8, 1146, *Baedericesdaenne* c.1100, *Baderigges- Bet'ichesden(n)* 1226, *Bederesden(e)* 1270. OE pers.n. **Beadurīċ*, genitive sing. **Beadurīċes*, + **denn**. PNK 405, KPN 248.

BETHNAL GREEN GLond TQ 3583. *Blethenalgrene* 1443, *Bethnall alias Bednall grene* 1576. P.n. *Blithehale* 13th cent., *Blithenhale in Stebenhethe* 1341, 'Blitha's' or 'happy nook', OE pers.n. *Blītha*, genitive sing. *Blīthan*, or adj. **blīthe**, definite form dative case **blīthan**, + **halh**, dative case **hale**, + ME **grene**. Originally a corner, nook, or angle of Stepney parish. Mx 83.

BETLEY Staffs SJ 7548. 'Beta or Bette's wood or clearing'. *Betelege* 1086, *Bettelega* 1175, *Betle* 1605. OE pers.n. *Beta* or feminine *Bette* + **lēah**. DEPN, Horovitz.

BETSHAM Kent TQ 6071. 'Bæddi's estate'. *Bedesham* 1278–1450 with variants *-is- -ys-*† *Bedsham* 1450, 1486, *Bettisham* 1487, *Bettsham* 1493, *Betsam* 1505, 1522. OE pers.n. *Bæddi*, Kentish form *Beddi*, genitive sing. *Beddes*, + **hām**, though *hamm* 'enclosure' is also possible. PNK 47, ASE 2.47.

†DB f.11b has the form *Bedesham* for a place in Eastry hundred usually identified with Betteshanger but actually for Beauchamp Wood, Lane and Bottom in Nonington; Betsham is in Axton hundred. Cullen.

BETTESHANGER Kent TR 3152. Probably 'Byttel's wooded slope'. *Betleshangre* 1176–1261×2, *Betlesengre* 1198. OE pers.n. **Byttel*, Kentish **Bettel*, genitive sing. **Bettles*, + **hangra**. PNK 578, ASE 2.35, Cullen.

BETTISCOMBE Dorset SY 3999. 'Betti's coomb'. *Bethescomme* 1129, *Bet(t)escumbe* 1244, 1288, *Bettyscombe* 1436. OE pers.n. *Betti*, genitive sing. *Bettes*, + **cumb**. Do 35, L 92.

BETTON Shrops SJ 3102. *Betton* 1836 OS.

BETTON Shrops SJ 6937. Partly uncertain. *Baitune* 1086, *Beitona* 1087–c.1190, *Becton(a)* 1121–1292×5, *Betton* from 1255×6. Also known as *Bectone sub Lima* 'B under Lyme' and *Betton in Hales* 1318. Either 'the stream settlement', OE **beċe** + **tūn**, or 'the beech-tree settlement', OE **bēċe** + **tūn**. The additions, for distinction from BETTON SJ 3102 and Betton Abbots SJ 5107, are Lyme as in LYME PARK SJ 9682, and the district name Hales, as in NORTON IN HALES SJ 7038. Sa i.46.

River BEULT Kent TQ 7747. Unexplained. *Beule* 1612, *Beult* 1819 OS. Another example of this r.n. is Bewl TQ 6834 as in BEWL BRIDGE RESERVOIR TQ 6832. RN 33 gives pr [bʌlt], now [belt], Cullen.

BEVENDEAN ESusx TQ 3406. 'Beofa's valley'. *Bevedene* 1086, *Bevenden* 1230–88, *Bevinden -yng-* 15th cent. OE pers.n. *Beofa*, genitive sing. *Beofan*, + **denu**. Sx 308, SS 71.

BEVERCOTES Notts SK 6972. 'The beaver huts'. *Beu'cate* (sic) 1192, *Beuerc(h)ote* 1189×99, *-cot* 1201–1300, *-cotes* 1203–81 etc., *Beaver Coates* 1665, *Berecotes* 1327. OE **beofor** + **cot**. The reference is possibly to dams built by beavers. Nt 45.

BEVERLEY Humbs TA 0339. 'Beaver stream'.

I. English forms:

(a) *Beferlic* [c.1030]11th *Secgan*, *Beofor-* c.1100, *Beuer- Bever-* c.1066–12th, *Beureli -v-* 1086, *Beuerlye* 13th, *Beverlec* 1115×23, 1202;

(b) *Beuirleg(e)* 12th, *Beu- Beverlei(a) -lai(a) -y* 1166–1828, *Beu-Beverle* 1196–1476;

(c) *Beverlac(')* 1200–1530;

II. Latin forms: *Beverlacum* 12th cent., adj. *Beverlicensis* c.1066, *-lacensis* 12th cent.

English forms (a) point to OE **beofor** + ***licc** 'a stream'. English forms (b) show confusion with OE **lēah** which also produced AN spellings in *-lac*. English forms (c) and the Latin forms may either represent this AN spelling or possible monastic etymologising, *Beverlac quasi locus vel lacus castrorum*. The reference is to the r. Hull: beaver bones have been found at nearby Wawne. The Anglo-Saxon monastery here was earlier referred to by Bede as *Inderauuda* 'in the wood of the Deirans'. YE 192, PNE ii.24, Forsberg 1950.102.

BEVERSTON Glos ST 8693. 'Beaver's stone'. *(æt) Byferes stane* 12th ASC(E) under year 1048, *Beurestane* 1086, 1194, *Beverstan* 1236, *-ston* 1287, *Beverston(e)* c.1175–1764. OE **beofor**, genitive sing. **beofores**, + **stān**. The reference may have been to a stone resembling a beaver, or the word may be used as a pers.n. Gl ii.213.

BEVINGTON Glos ST 6697. 'Estate called after Beofa'. *Beu-Bevinton -yn(g)-* 1234–1577. OE pers.n. *Bēofa* + **ing**[4] + **tūn**. Gl ii.224.

BEWALDETH Cumbr NY 2134. 'Homestead of Aldgyth'. *Bualdith* 1255, 1260, *Bowaldef -if* 13th cent., *Boaldith* 1278, *-eth* 1704, *Bowaldeth* 1292–1777. OE, ON **bū** + OE feminine pers.n. *Aldgȳth* in Celtic word order. Cu 264, SSNNW 20 gives pr [biwɔ:dəθ].

BEWCASTLE Cumbr NY 5674. 'Roman fort with a booth'. *Buchastre*, *Buch(e)castre* c.1177, *Buthecastra* [c.1178]1348, *Buth(e)- Both(e)castr(e)* late 12th–1372, *Bothcastel(l)* 14th cent., *Beau- Bewcastell* late 15th cent., *Bewcastle* from 1580. ON **búth** + OAngl **cæster** referring to the Roman fort within which the *búth* was built presumably for use as a shieling. The development to Bewcastle instead of **Bewcaster* took place because of the medieval castle built within the earlier defences (cf. the form *Bothe Castell castr'* 1379). Cu 60, SSNNW 201.

BEWCASTLE FELLS Cumbr NY 5681. *Bewcastle Fells* 1868 OS. P.n. BEWCASTLE + ModE **fell(s)** (ON *fjall*).

BEWDLEY H&W SO 7875. 'Beautiful place'. I. (French forms) *Be(a)uleu, Beaulieu* 1275–1424. II. (Anglicised forms) *Buleye* 1316, *Beudle* 1335, *Bewdeley* 1547. There is also a Latin form *Bellum Locum* 1308. OFr **bel, beau** + **lieu**. Both Leland and Camden comment on the aptness of this name. Wo 40.

BEWERLEY NYorks SE 1564. 'The beavers' wood or clearing'. *Beurelie* 1086, *Beuerl(a)i -leia -lay -ley* c.1142–1553, *Bewerley* 1575, *Bureley* 1577. OE **beofor**, genitive pl. **beofra**, + **lēah**. YW v.142.

BEWHOLME Humbs TA 1650. Uncertain. 'At the bends', or 'at the rings'. *Begun* 1086, *Begum* 12th, 13th cent., *Beghum* [1144×54]16th, 1159×82–1349, *Beyom* 1316, *Bewham -ho(l)me* 16th cent. ON ***bjúgr**, dative pl. ***bjúgum**. The village itself stands on a small hill; the reference is thought to be to the nearby streams with twisted courses. The name might, however, represent OE **bēag**, dative pl. **bēagum** 'at the rings', referring to some lost feature such as a ring of stones or trees. YE 77 gives pr [biuəm], SSNY 90.

New BEWICK Northum NU 0620. *New Bewick* 1868 OS. ModE adj. **new** + p.n. Bewick as in Old BEWICK NU 0621.

Old BEWICK Northum NU 0621. *Old Beawick* c.1715 AA 13.7. ModE adj. **old** + p.n. *Beuuiche* [c.1136]n.d. DST, *Be- Bowich* 12th cent., *Bowic* 1203, *Bewyk* 1296 SR 103, 'the bee-farm' sc. of Eglingham parish, the place where the people of Eglingham took their bees in summer to the heather-covered hills. 'Old' for distinction from New BEWICK NU 0620. NbDu 19 gives pr [bju:ik], Newton 56.

BEWL BRIDGE RESERVOIR Kent TQ 6832. P.n. Bewl Bridge, *Beauldbridge* 1576, *Beaulbridge* 1596, + ModE **reservoir**. Bewl Bridge is unexplained r.n. *Bewl, Beaul* 1576, 1596, *The Bewle* 1782, + ModE **bridge**. Cf. River BEULT TQ 7747. RN 33.

BEXHILL ESusx TQ 7407. 'Box-tree wood or clearing'. *bex- bixlea* [772]13th S 108, *Bexelei* 1086, *Byx- Bix- Buxle* 1186–15th cent., *Bixel* 1278, *Bex(h)ill* 1535.† OE **byxe** + **lēah**. The land of the Bexhill inhabitants is *bæxwarena land* [772]13th S 108, abbreviated p.n. **Bex* + OE **ware** 'inhabitants', genitive pl. **warena**. Sx 489 gives pr [bæxhil] (sic).

West BEXINGTON Dorset SY 5386. *West Bexington* 1810 OS. ModE **west** + p.n. *Bessintone* 1086, *Buxinton* 1212, *Bexinton* 1243, *Bexingtun* 1285, 'settlement where box-trees grow', OE **byxen** + **tūn**. According to Do 35 East and West Bexington are separately recorded as early as the 13th cent.

BEXLEY GLond TQ 4775. 'Box-tree wood or clearing'. *Bixle* [765×85]12th S 37, *Byxlea* [814]?10th S 175, *Bixle* 12th, *Bexle* 1314. OE **byxe** 'a box wood' + **lēah**. DEPN, LPN 21.

BEXWELL Norf TF 6305. 'Beac's stream'. *Bekeswella* 1086, *-well* 1177, 1196. OE pers.n. *Bēac*, genitive sing. *Bēaces*, + **wella**. DEPN.

BEYTON Suff TL 9362. 'Beaga or Beage's estate'. *begatona* 1086, *Beketon* 1208, *Beighton* 1610. OE pers.n. *Bēaga* or feminine *Bēaġe* + **tūn**. DEPN.

BIBURY Glos SP 1106. 'Beage's manor house'. *began byrg* [899]11th S 1279, *Beagan byrig* 11th S 1254, *Begeberie -a* 1086, 1186×9, *Becheberie* 1086, *Behebiria* 1221, *Bi- Bybur'* 1154×89–1303, *-bury* from 1291. OE feminine pers.n. *Bēaġe* + **burh**, dative sing. of **byriġ**. The reference is to Beage the daughter of Earl Leppa to whom this estate was granted by the Bishop of Worcester in the 8th cent. Gl i.26, Brk 825.

†A series of unusual spellings, *Bealsa* 1174, *Beause* 1226–7, *Bease* 1248, *Bewes* 1357, *Bause* 1289, are best explained as Anglo-Norman influence, Vising 188–9.

BICESTER Oxon SP 5922. 'Beorna's Roman fort' or 'Roman fort of the warriors'. *Bernecestre* 1086–1320 with variants *-cestr(ia), Bere- Burcestria* [1151×2]c.1444–1574 with variants *-ce(s)tre* and *-ce(s)ter, Bysseter* 1517, *Bister* 1685. OE pers.n. *Beorna* or **beorna**, genitive pl. of **beorn**, + **ċeaster**. O 198 gives pr [bistə].

BICKENHALL Somer ST 2818. Probably 'Bica's nook'. *Bichehalle* 1086, *Bikehal* 1201, *Bikenhal* 1243, *Bicknell* 1610. OE pers.n. *Bica*, genitive sing. *Bican*, + **healh** although the DB spelling possibly points to **heall** 'a hall'. Other spellings, *Bikehilla* 1186, *-hull* 1243, show an alternative form with OE **hyll** 'a hill'. DEPN.

BICKENHILL WMids SP 1882. 'Bica's' or 'beak hill'. *Bichehelle* 1086, *-hul* 1100×35, *By- Bikenhull(e)* 1202–1492, *Bigenhull* 1354, *Bygnell* 1461, 1535, *Byknell* 1525, *Bickenhill oth. Bicknell oth. Bignell* 1770. OE pers.n. *Bica* or ***bica**, genitive sing. *Bican*, ***bican**, + **hyll**. Wa 59 gives former pr [bignəl], V i.96.

BICKER Lincs TF 2237. Probably 'the village marsh'. *Bichere* 1086, 1212, *Bikere* 1176, 1194, *Bicre* 1206. ON **bȳ** + **kjarr**. Other possibilities are OE *bī* + *kjarr* '(the place) beside the marsh' or ME *biker* 'quarrel, dispute' used in a concrete sense for land subject to dispute. DEPN, SSNEM 152, L 52–3, Cameron 1998.

BICKER HAVEN Lincs TF 2533. P.n. BICKER TF 2237 + ModE **haven**.

BICKERSTAFFE Lancs SD 4404. 'The bee-keeper's landing-place'. *Bikerstad* [before 1190]1268, *Bickerstat* 1212, *Bi- Bykerstat(h) -ir-* 1226–1385, *Bikerstaff* 1267. OE **bīcere** + **stæth**. The etymology of this name is clear, but the reference is problematical as there are just two small brooks here. La 121, Jnl 17.67, L 80, SSNNW 420.

BICKERTON Ches SJ 5052. 'Bee-keepers' estate'. *Bicretone* 1086, *Bikerton* 1180–1724 with variants *Byker- -ir- -yr- -ur-* and *-tone, Bickerton* from 1350. OE **bīcere**, genitive pl. **bīcera**, + **tūn**. Che iv.4.

BICKERTON NYorks SE 4550. 'The bee-keepers' estate'. *Bic(h)retone* 1086, *Bi- Bykerton* 12th–1519, *Bickerton* 1521. OE **bīcere**, genitive pl. **bīcera**, + **tūn**. YW iv.247.

BICKINGTON Devon SX 7972. 'Estate called after Beocca'. *Bechintona* [1107]1300, *Buketon* 1219, 1228, *Bukyngton* 1303, 1428, *Byketon* 1330. OE pers.n. *Beocca* + **ing**[4] + **tūn**. D 465.

BICKINGTON Devon SS 5332. Possibly 'estate called after Bucca'. *Buckyngton* 1570, *Bukington* 1606. OE pers.n. *Bucca* + **ing**[4] + **tūn**. But cf. BICKINGTON SX 7972 and Abbots BICKINGTON SS 3813. D 114.

Abbotts BICKINGTON Devon SS 3813. *Abbots Bekenton* 1580, *Abbotsbekington* 1636. Earlier simply *Bichetone* 1086 DB, *Bicatona* 1086 Exon, *Bechatona* [1189]1465, *Bukyngton(e)* 1291, 1340, 'Bica's farm or village', OE pers.n. *Bica*, genitive sing. *Bican*, + **tūn**. The manor was granted to Hartland Abbey some time before 1189. D 124.

High BICKINGTON Devon SS 6020. *Heghebuginton* 1423. Earlier simply *Bichentone* 1086, *Bykanthon'* 1242, *Bykynton* 1373, *Bukinton* 1212–1350 with variants *-yng- -e(n)-*, 'Bica or Beocca's farm or village', OE pers.n. *Bica* or *Beocca*, genitive sing. *Bican, Beoccan*, + **tūn**. D 358.

BICKLEIGH Partly uncertain. Either 'Bica's wood or clearing', OE pers.n. *Bica* + **lēah**, or 'wood or clearing of the point', OE ***bīca** + **lēah**. PNE i.33–4, L 205.

(1) ~ Devon SS 9407. *Bichelie* 1086, *Bi- Bykelegh(e)* 1238–1302. It is uncertain whether the *villa venatoria quae Saxonica dicitur Bicanleag* of S 372 ([904]12th) belongs here. D 554.

(2) ~ Devon SX 5262. *Bichelie* 1086, *Bi- Bykeleg(he)* 1225–91. D 224.

BICKLETON Devon SS 5031. 'Bicela's farm or village'. *Picaltone* 1086, *By- Bikelton* 1281–1330. OE pers.n. **Bicela* + **tūn**. D 117.

BICKLEY GLond TQ 4268. Either 'Bica's wood or clearing' or 'wood or clearing at the pointed ridge'. *Byckeleye* 1279, *Bykeleye* 1292. OE pers.n. *Bicca* or ***bica** + **lēah**. DEPN.

BICKLEY MOSS Ches SJ 5449. *Byckeley Mosse* 1554, also called *mora de Bykeley* 1394 and *Bykleye Heath* 1537. P.n. Bickley SJ 5348 + ModE **moss** (OE *mos*) 'peat bog'. Bickley, *Bichelei* 1086, *Bikeleg'* early 13th–1579 with variants *Byk(k)e- Bicke-* and *-ley(e) -lee*, *Bicklegh* from 1272 with variants *Bik- Byck-* and *-ley*, *Bekelegh -ley* 1329–1487, is probably 'glade of the bee-hives', OE ***bīc**, genitive pl. ***bīca**, + **lēah**. This part of Cheshire seems to have specialised in bee-keeping and honey-gathering, cf. BICKERTON SJ 5052. On formal grounds, however, there are alternatives which cannot be ruled out, viz. OE pers.n. *Bic(c)a*, or OE ***bica**, ***bice** 'a woodpecker'. Che iv.6, 8, L 205.

BICKNACRE Essex TL 7802. 'Bica's newly cultivated land'. *Bikenacher* 1186–7, *-acre -a -akere* 1190–1309. OE pers.n. *Bica*, genitive sing. *Bican*, + **æcer**. Ess 275, L 232–3.

BICKNOLLER Somer ST 1139. 'Bica's alder-tree'. *By- Bikenalre* 1291, 1334, *Bicknaler* 1610, *Bicknoller* 1809 OS. OE pers.n. *Bica*, genitive sing. *Bican*, + OE **alor**, dial. **oller**. DEPN.

BICKNOR Kent TQ 8658. 'Flat-topped hill of the beak or point' or 'Bica's flat-topped hill'. *Bikenora* 1185×6, *Bi- Bykenor(e)* 1232–78 etc. OE ***bica** or pers.n. *Bica*, genitive sing. ***bican**, *Bican*, + **ōra**. PNK 231.

English BICKNOR Glos SO 5815. 'Bicknor on the English side' of the river Wye. *Englise Bi- Bykenor(e)* 1248. Earlier simply *Bicanofre* 1086, *Bi- Bykenour(a)* 1190–1221, 'Bica's ridge' or 'ridge with a point'. OE pers.n. *Bica* or ***bica**, genitive sing. *Bican*, ***bican**, + **ofer**. The village is located at the end of a marked ridge. 'English' for distinction from Welsh BICKNOR H&W SO 5817. Gl iii.211, PNE i.33, L 174.

Welsh BICKNOR H&W SO 5917. *Welsh Bicknor* 1831. ModE adj. *Welsh* + p.n. Bicknor as in English BICKNOR Glos SO 5815 on the English side of the Wye. He 33.

BICKTON Hants SU 1412. Partly uncertain. *Bichetone* 1086, *Bi- Byketon* 1227–1412. It is uncertain whether this is the 'settlement at the point or beak' or 'Bica's estate', OE ***bica** or pers.n. *Bīca* + **tūn**. ME *bike* 'a bees' nest', a late northern back-formation from OE *bıcere*, is not possible here. Ha 33, Gover 1958.214, Anglia 103.1–25, V i.95.

BICTON Shrops SJ 4415. 'Beak hill'. *Bicheton* 1086, *Bi- Byketon(')* 1255–1404, *Bykton* 1535, *Bicton* 1670, *Bikedoun* c.1200×12, *Bykedon(e)* 1203×4–1362. OE ***bic** 'a point' + **dūn**. Bicton is situated at the waist of a long narrow hill with sharp projections N and S. Probably two forms of the name originally existed, **bic-dūn* in which the *d* was early devoiced after voiceless stop *c* > *bic-tūn*, and **bica-dūn* in which *bica* is the genitive pl. 'hill of the beaks' and *d* was not devoiced. Sa i.47.

BICTON Shrops SO 2982. Uncertain. *Biggeton* 1271×2, *Biketon* 1284, *-don* 1302, *Bykidon* 1397. It is unclear from the forms whether this is a compound of ME *bigge* 'big' + *tūn* (or *dūn*) or identical with BICTON SJ 4415. Sa i.47.

BIDBOROUGH Kent TQ 5643. 'Bitta's hill'. *Bitteberga* c.1100, *Bi- Bytteberg(h)* 1248–1344, *Bytberghe* 1346, *Bitteborugh* 1367. OE pers.n. *Bitta* + **beorg**. PNK 184.

BIDDENDEN Kent TQ 8438. 'Bida's woodland-pasture'. *Bidingden*, *Bydyngdene* [993 for 996]14th S 877, *Bidindaenne* c.1100, *Bi- Bydenden(n)e* 1204, 1262. OE pers.n. *Bida* + **ing**[4] + **denn**. PNK 330, KPN 341.

BIDDENHAM Beds TL 0250. 'Byda or Bieda's river-bend land'. *Bide(n)ham* 1086, *Beydenham* 1240, *Budeham* 1247. OE pers.n. *Byda*, *Bīeda*, genitive sing. *Bydan*, *Bīedan*, + **hamm** 1. Biddenham lies a mile W of BEDFORD TL 0449 which is possibly 'Bieda's ford', i.e. the ford leading to his land. Bd 26.

BIDDESTONE Wilts ST 8673. 'Biedin's estate'. *Bedestone* 1086–1337, *Bedeneston'* 1187, *Bud(d)eston(e)* 1215–1464, *Bydesden'* 1274, *Bi- Byd(d)eston* 1339, *Bi- Bydston* 1487, 1552, *Bitt- Byd- Bedson* 16th. OE pers.n. **Bīedīn*, genitive sing. **Bīedīnes*, + **tūn**. Wlt 82 gives pr [bidstən].

BIDDISHAM Somer ST 3853. 'Biddi's water-meadow'. *Biddesham quod Tarnuc proprie appellatur* 'B properly called T' *[1065]18th S 1042, 1203, *Bides-* 1209, Biddesham, *Bilsham* 1610. OE pers.n. **Biddi*, genitive sing. **Biddes*, + **hamm**. The earlier name was *Ternuc* [663 for ?693]14th S 238, [946×75]13th S 1740, *Tarnuc* *[1065]18th S 1042, *Ternoc* 1086, *Tornuc* [1129×39]14th, *Tornok* 1370, 'dry place', ultimately from Brit **tarnāco-*. Cf. Tarnac, Corrèze, France, though Dauzat-Rostaing 669 prefer derivation from pers.n. *Tarinus*. Biddisham lies on a tributary of the Axe. DEPN, Turner 1950.122.

BIDDLESDEN Bucks SP 6340. 'Valley of the building'. There are three types:

I. *Beches- Betesdene* 1086, *Bet(e)lesden(a)* c.1145–1394.

II. *But(t)lesden(a)* 1154×89–1522.

III. *Bit(t)lesden* 1224–1394, *Byldesden* 1553, *Bidleston* 1766. OE **(ġe)bytle**, genitive sing. **(ġe)bytles**, + **denu**. Bk 41 gives pr [bitəlzdən], Jnl 2.22, L 99.

BIDDLESTONE Northum NT 9608. 'The valley of the building'. *Bitlesden* [1181]14th, *Bi- Bydlisden(e)* 13th cent., *Bidelston* 1313, *Bittleston* 17th cent. OE **(ġe)bytle**, genitive sing. **bytles**, + **denu**. Alternatively the specific might be unrecorded OE pers.n. **Byttel*, genitive sing. **Bytt(e)les*, and so 'Byttel's valley'. NbDu 21, L 99.

BIDDULPH Staffs SJ 8857. '(The settlement) by the quarry'. *Bidolf* 1086–1227, *Bydulph* 1291. OE **bī** + ***dulf** or ***dylf**. Stone quarries still exist here. DEPN, PNE i.140, L 53.

BIDDULPH MOOR Staffs SJ 9508. *Biddulph Moor* 1842 OS. P.n. BIDDULPH SJ 8857 + ModE **moor**.

BIDEFORD Devon SS 4526. Partly uncertain. *Bedeford* 1086 DB, *-ford(e)* 13th cent., *Bediforda* 1086 Exon, *Bediford -y-* 1201–2, 1356, *Budiford -e-* 1232–91, *By- Bidiford -e-* 1238–1314. The generic is OE **ford** 'a ford', the specific uncertain. OE pers.n. *Bȳda* + **ing**[4], or pre-English r.n. **Bȳd* indentical with the Gloucestershire BOYD have been proposed. Better perhaps is Dr Gelling's suggestion, OE **byden** 'bucket, tub' used literally or in a transferred topographical sense of one of the side-valleys here or even of the deep trough of the r. Torridge itself. There is a settlement called Bidna (from *byden*) 2 miles N of Bideford by the Torridge estuary. D 87 gives pr [bidifəd], 102.

BIDEFORD BAR Devon SS 4333. P.n. BIDEFORD SS 5031 + ModE **bar** 'a bank of sand across the mouth of a river'.

BIDFORD-ON-AVON Warw SP 1052. P.n. *Budiford(e)* 1155×6–1327 including *[710]12th S 81, *Bedeford* 1086, 1155–62, *Bide-* 1141–1246, *Bude-* 1291–1454, *Bidford* 1371, perhaps 'the ford in the hollow', OE **byden** 'a trough' in the topographical sense 'depression, hollow, valley' + **ford**, + r.n. AVON. Bidford lies on the line of Ryknild Street (Margary no.18a) which both N and S of the Avon descends from high ground to make the crossing. Wa 201.

BIELBY Humbs SE 7843. 'Beli's farm or village'. *Belebi -by* 1086–1421, *Belby* 1359, *Beylby*, *Byelby* 16th cent. ON pers.n. *Beli*, genitive sing. *Bela*, + **bȳ**. YE 232, SSNY 21.

BIERLEY IoW SZ 5178. Partly uncertain. *Berlay* 1341, *Beer ley* 1565, *Beerloy* 1781, *Bere Lay* 1810 OS. Possibly the 'woodland pasture clearing', OE **bǣr** + **lēah**, or the 'barley clearing', OE **bere** + **lēah**. The forms are too late for certainty. Wt 252 gives pr [b(i)ər'lai], Mills 1996.28.

BIERTON Bucks SP 8315. 'Settlement, farm belonging to the *burh*' sc. of Aylesbury. *Bortone* 1086, *Burton(a)* 1133×40–1342, *Berton(')* 1184–1218, 1672, *Byrton* 1347–82, 1616, *Beerton* 14th–1627, *Bierton* from 1382, *Bear- Byerton* 1672. OE **burh**, genitive sing. **byrh**, + **tūn**. Bk 146 gives pr [bi:ətən], Jnl 2.25, WW 149.

BIGBURY Devon SX 6646. 'Bica's fortified place'. *Bicheberie* 1086, *Bickeberi* 1201–1332 with variants *Bik(k)e- Byke-* and *-biry -bury*, *Byggebury* 1485. OE pers.n. *Bica* + **burh**, dative sing. **byriġ**. D 266.

BIGBURY BAY Devon SX 6342. *Bigberie Baie* 1588. P.n. BIGBURY SX 6646 + ModE **bay**. D 19.

BIGBURY-ON-SEA Devon SX 6544. A modern holiday resort named after BIGBURY SX 6646.

BIGBY Lincs TA 0507. 'Bekki's village or farm'. *Bechebi* 1086, c.1115, 1233, *Bekebi -by* 1191–1576, *Bigby* from 1526. ODan pers.n. *Bekki*, genitive sing. *Bekka*, + **bȳ**. Li ii.48, SSNEM 36.

BIGGAR Cumbr SD 1966. 'Triangular piece of land where barley is grown'. *Bigger* 1292, *Bygger* 1537, 1539. ON **bygg** + **geiri**. La 205, SSNNW 104.

BIGGES'S PILLAR Northum NU 1207. *Bigges Pillar* 1868 OS. Surname *Biggs* + ModE **pillar**. The reference is to a prominent hill-top cairn on Alnwick Moor at 885ft., possibly originally a medieval beacon. PSA 5.i.123.

BIGGIN '(The) building'. ME **bigging** < ON *byggja* 'to build'. PNE i.35.

(1) ~ Derby SK 1559. *le Byggyng* 1415, 1417, *Byging* 1547, short for *New(e)- Neubbigging(e)* 1244, 1265, *New Byggyng* 1340. Db 368.

(2) ~ Derby SK 2648. *Bugyng* 1329, *Bi- Bygging -yng(e)* 1330–1635, *Byggyn* 1556, *Biggin* 1668. Also known as *New(e)bigging* 1223×39–1262, *-biggin* 1298, *Nubiggyng* 1346. Db 531.

(3) ~ NYorks SE 5434. *Neuebiggynge, le Biggyng* 13th. YW iv.63.

BIGGIN HILL GLond TQ 4158. 'Hill with or near a building'. *Byggunhull* 1499, *Biggin Hill* c.1762. ME **bigging + hill**. LPN 22.

BIGGINS 'Buildings'. Pl. of ME **bigging** from ON *byggja* 'to build'.

(1) ~ Cumbr SD 6078. *Bi- Bygginges -yng(es)* 13th–1537, *-yn(e)s* 16th, *(the) Biggins* 1558–1725. We i.43.

(2) Friar ~ Cumbr NY 6309 → SUNBIGGIN NY 6508.

BIGGLESWADE Beds TL 1944. 'Biccel's ford'. *Pichelesuuade* 1086, *Bikeleswade* 1183, *Bygelswade* 1486. OE pers.n. **Biccel*, genitive sing. **Bicceles*, + **wæd**. Bd 101, L 83.

BIGHTON Hants SU 6134. 'Estate called after Bica'. *Bicincgtun* [959]15th S 660, *Bighetone* 1086, *Bic- Bikenton* 1165–1280, *Byketon* 1229–1541, *Bighton* 1535. OE pers.n. *Bīca* + **ing**[4] + **tūn**. Ha 33, Gover 1958.86.

BIGNOR WSusx SU 9814. 'Bicga's ridge'. *Bigeneure* 1086, *Byg(g)e- Biggeneu(e)re -oure* 1249–1398, *Bignore* 1397. OE pers.n. **Bicga*, genitive sing. **Bicgan*, + **yfer** varying with **ofer**. Sx 124.

BILBERRY Corn SX 0160. 'Billa's mound'. *Billebery* c.1280, *Billibyry* [c.1280]14th. OE pers.n. *Billa* + **beorg**, but OE **byriġ**, dative sing. of **burh** 'fortified place' is also formally possible. No such fortification survives. Probably refers to a lost barrow or to the hill on which the hamlet stands. PNCo 53.

BILBOROUGH Notts SK 5141. 'Billa's fortified place'. *Bileburch* 1086, *-burg* 1086–1203, *Billeburg(h) -burc* c.1180–1243, *Billburgh* 1242–1353, *-borou* 1293. OE pers.n. **Bil(l)a* + **burh**. Nt 140.

BILBROOK Staffs SJ 8803. 'Bilders brook'. *Bilrebroch* 1086, *Bilrebroc* 1166, 1271, *Billebroc* 1160–13th. OE ***billere**, a water-plant of some kind, 'bilders', + **brōc**. Duignan 14, Horovitz.

BILBROUGH NYorks SE 5346. 'Billa's fortification'. *Mileburg* (sic) 1086, *Billingeburgh* 1246, *Bi- Byleburc -g* before 1166–1310, *Bi- Bylleburc -burg(h)* 1167–1286. OE pers.n. **Bil(l)a* + **burh**. The same pers.n. occurs in BILTON SE 4750, 4 miles away. There is no trace of any fortified place here. YW iv.235.

BILDESTON Suff TL 9949. 'Bildr's estate'. *Bilestunā* 1086, *Billestona* 1130, *Bildeston(e)* 1166–1242. ON pers.n. *Bildr*, ODan *Bild*, seondary genitive sing. *Bildes*, + **tūn**. DEPN.

BILHAM SYorks SE 4906. Either 'Billa's homestead' or 'homestead at the bill-shaped hill'. *Bil(e)ham, Bilam* 1086, *Bileham -y-* c.1150–1312, *Bilham* from 1243. Either OE pers.n. **Bil(l)a* or **bile** or **bill** + **hām**. YW i.86.

BILLERICAY Essex TQ 6794. 'Dye house, tannery'. *Byllyrica* 1291, *Billirica -er-* 1274–1551, *Bellerica* 1382. Latin ***bellerīca**. This rare word ultimately of Persian origin may have entered use in the early 13th cent. as a result of the importation of myrobalan, a source of tannin and of black dye, from the east in the wake of the third and fourth crusades. Identical forms are recorded for Bellerica Fm in Witham Friary, Somerset, *Billerica* 1535, a lost Kentish *Billirica* 1278, *Bellirica* 1293 in Lympne, and a field-n. *Billerica* 1549, *Bellericary* (sic) 1732, in Longbridge Deverill, Wilts. Ess 146 gives pr [bil'riki], Wlt xxxvii, S&DNQ 30.353–4, Jnl 15.20–3.

BILLESDON Leic SK 7102. 'Bil's hill'. *Billesdone* 1086–1610, *Byllesdon(') -doun* 1289–1516, *-den* 1326–1580. OE pers.n. *Bil*, genitive sing. *Bil(l)es*, + **dūn**. Lei 205, L 156.

BILLESLEY Warw SP 1456. 'Bil's wood or clearing'. *Billes lœh* [699×709]11th S64, *Billeslei* 1086 etc. with variants *-ley(e)* and *-legh(e)*. OE pers.n. *Bil*, genitive sing. *Billes*, + **lēah**. Wa 203.

BILLING Northants SP 8062. Probably the 'bell-shaped hill, the prominent hill'. *Bel(l)inge, Bellica* (sic) 1086, *Bellinges* 1179, 1252, *Billynges -inges* 1223–1318, *Billing* from 1220. OE ***belling** < **belle** + **ing**[2], later replaced by ***billing** < **bile**, **bill** + **ing**[2]. The plural forms may represent OE **Billingas* 'the people of the *bill*' or refer to the two hamlets of Great and Little Billing, *Magna, Parva Biling(g)e* 12th. The former suggestion is supported by the form *Bilmiga land* (for *Billinga*) 'the territory of the *Billingas*', 7th Tribal Hidage. Two other forms, *Bethlinges* [1185×8]1329, 13th, are difficult to equate with this explanation, and may not belong here. Great Billing is situated on the edge of a prominent hill rising to 385ft. overlooking the Nene valley. The situation recalls that of BILLINGHAM Cleve NZ 4623. Nth 132, ING 68, BzN 1967.326.

BILLINGBOROUGH Lincs TF 1134. 'The fortified place called or at *Billing*, the ridge'. *Bil(l)inge- Bellinge- Bolinburg* 1086, *Billingeburc* 1167–1219, *Bil(l)ingburg(h)* 1218–1519, *Billing- Bullingborough* 1551. OE ***billing**, locative-dative sing. ***billinge**, perhaps varying with folk-n. **Billingas* 'the people of the *bill* or *bile*', OE **bill** 'a sword, an edge' or **bile** 'a bill, a beak' + **ingas**, genitive pl. **Billinga*, + **burh**. Billingborough is situated at the end of a ridge of higher ground by the edge of the Holland fens. Cf. HORBLING TF 1135. Perrott 100, V i.101.

BILLINGE Mers SD 5300. '(Settlement at) the hill'. *Billinḡ* 1202, *Billing* 1206, 1246, *Billinge* from 1284, *Bulling* [c.1200]1268–78 etc., *Bullynth* (for *-ynch*) 1292, *Bullinge -yngge* 1332, *Billindge* 1580, 1585. OE ***billing**, locative-dative sing. ***billinġe**. La 104 gives pr [bilindʒ] (1911), BzN 1967.326–32, Jnl 17.59.

BILLINGFORD Norf TG 0120. Perhaps 'ford at Billing'. *Billingeforda* 1086, *-ford* 1212. OE p.n. **Billing* < ***billing** 'a hill', locative-dative sing. **Billinġe*, + **ford**. Alternatively this could be the 'ford of the Billingas, the ford leading to the land of the Billingas, the people who live at Billing or beside the *billing*', OE folk-n. **Billingas*, genitive pl. **Billinga*, + **ford**. The ford crosses the river Wensum beside a prominent hill-spur. DEPN, BzN 1967.329.

BILLINGFORD Suff TM 1678. *Billingford* 1610. Formally identical with BILLINGFORD Norf TG 0120, *Billingeford* 1086, 'the ford of the Billingas'. DEPN.

BILLINGHAM Cleve NZ 4623. 'Homestead, estate called or at (the) *Billing*'. *Billingham* from [c.1040]12th., *Bellingahā* 1087×93, *-yngham* 1378, 1381, *Billingeham* [c.1123]12th, [c.1190×5]n.d. OE ***billing** 'a hill, a prominence, a ridge', locative-dative ***billinge**, + **hām**. Billingham stands on a prominent ridge overlooking low ground. BzN 1967.326, TC 52.

BILLINGHAY Lincs TF 1555. 'The island called or at *Billing*'. *Belingei* 1086, *-eie -eia* early 13th, *Billingeie -eye* before 1189–1513, *-heia -hey(e)* 1185–1526, *Billinghay* from 1560. OE ***billing** + **ēġ**. Billinghay is situated in fenland ½ m. from the edge of a ridge of high ground. Perrott 308, V i.100.

BILLINGLEY SYorks SE 4304. Either the 'wood or clearing of the Billingas, the people either called after Billa or from Bilham', or 'wood or clearing at *Billing*, the bill-shaped hill'. *Bilingelei(a) -lie, Bingelie* 1086, *Billing(e)lea -leg(a)* 1167–96,

Billingley from 1316. OE folk-n. **Billingas* < pers.n. **Billa* or p.n. BILHAM SE 4906 + **ingas**, genitive pl. **Billinga*, or OE ***billing**, locative-dative sing. ***billinge**, + **lēah**. YW i.94, V i.100.

BILLINGSHURST WSusx TQ 0825. 'The wooded hill at or called *Billing*, the bill-shaped hill'. *Bellingesherst* 1202, *Bi- Byllyngeherst -hurst -inge-* 1225–1397, *Billingeshurst -yng-* 1249–1379. OE hill-n. ***billing** probably used as a p.n., genitive sing. ***billinges**, + **hyrst**. The village stands at the end of a prominent hill-ridge. Sx 147, BzN 1967.329–30.

BILLINGSLEY Shrops SO 7085. 'The wood or clearing of the hill called *Billing*'. *Billingesle* 1087×94, *Bi- Byllingesleya -leg -ley(e)* 1138–1577, *By- Billingele(gh) -leye* 13th cent., *Billinsley* 1751. OE p.n. **Billing* referring to a ridge running NW from the village, OE ***billing**, genitive sing. **Billinges*, + **lēah**. Sa i.48.

BILLINGTON Beds SP 9422. 'Billa's hill'. *Billendon* 1196, *Billesdon* 1202, *Billingdon* 1276, *Billington* 1798. OE pers.n. *Billa*, genitive sing. *Billan*, + **dūn**. Bd 116, L 151.

BILLINGTON Lancs SD 7235. 'Hill called *Billing*'. *Billingduna* 1196, *Bil(l)in(g)don* 1203–1242, *Billinton* 1208–59, *Billington* from 1243, *Belington -yng-* 1296–16th. OE ***billing** 'edge, hill' + **dūn**. The SE boundary of this place is formed by Billington Moor, a long ridge earlier called *Billingahoth* c.1130, either the '*hōh* or hill-spur of the **Billingas*, the people of the *bill(ing)*', genitive pl. **Billinga*, or the 'hill-spur at or called *Billing*' taking the medial *a* as a sign of a locative-dative pronunciation [indž] as in BILLINGE Mersey SD 5300. La 71, BzN 1967.326, Jnl 17.45, 18.19.

BILLOCKBY Norf TG 4213. Probably 'Bithill-Aki's farm'. *Bit(h)lakebei* 1086, *Billokebi -by* 1198–1332×3, *Bi- Byllokby* 1316–1428. ON pers.n. **Bithil-* or *Bithill-Áki* 'wooer Áki', genitive sing. **Bithil(l)-Áka*, + **bȳ**. Nf ii.46, Jnl. 19.10, SPNN 94.

BILLSMOOR PARK Northum NY 9496. *Billsmoor Park* 1869 OS. P.n. Billsmoor + ModE **park**.

BILLY MILL T&W NZ 3369 → BILLY ROW Durham NZ 1637.

BILLY ROW Durham NZ 1637. 'Row of dwellings at *Billy*'. *Billyraw* 1425, *Billerawe* 1484, 1549, *-rowe* 1498, *Billie Rawe* 1621. P.n. *Billy* + OE **rāw**. *Billy* is *Billey* 1334, 1349×50, *Billy* 1371, OE ***billing** 'a ridge, a hill edge' referring to the prominent ridge of high ground rising to 1017 ft. at NZ 1538 and still called Billy Hill. For the development ***billing** > *billy* compare ModE *penny* < OE **pening**. The same development occurs in the p.n. Billy Mill, Tyne & Wear NZ 3369 *Molendinum de Billing* 1320. NbDu 22.

East BILNEY Norf TF 9419. *East Bilney* 1838 OS. ME **est** + p.n. *Billneye* 1254, *Bylneye* 1316, 'Billa's island', OE pers.n. *Billa*, genitive sing. *Billan*, + **ēġ** in the sense 'dry land between streams, hill-spur'. The alternative suggestion that the specific is OE *bile* 'a bill, a beak, a promontory' suits the topography but *bile* is only recorded as a masculine i-stem with genitive sing. *biles*. 'East' for distinction from West BILNEY TF 7115. East Bilney is three miles W of Billingford TG 0120, the ford leading to the Billingas, either 'the people of the *billing*' or 'the people called after Billa'. DEPN, L 38–9.

West BILNEY Norf TF 7115. *Westbilneye* 1272×1307. ME **west** + p.n. *Be- Binelai, Bilenei* 1086, *Bilneia* 1205, 'Billa's island', OE pers.n. *Billa*, genitive sing. *Billan*, + **ēġ** referring to a raised patch of higher ground in marshland. 'West' for distinction from East BILNEY TF 9419. DEPN.

BILSBORROW Lancs SD 5140. 'Fortified place of the promontory'. *Billesbure* 1187, *Billisburg* [c.1200]1268, *-borgh* 1268–80, *-borowe* 1509, *Billesburgh* 1212–1332 etc., *Bilsborough* 1508. OE **bill**, genitive sing. **billes**, + **burh**. The village is situated on a low promontory overlooking a tributary stream of the river Brock. La 162, Jnl 17.94.

BILSBY Lincs TF 4776. Possibly 'Billi's village or farm'. *Billesbi* 1086, 1193, *Bilesbi* c.1150×60. ON pers.n. *Billi*, ME genitive sing. *Billes*, + **bȳ**. Pers.n. *Billi* is rare. A possible alternative specific is OE **bill** 'a sword, an edge' or **bile** 'a bill, a beak', genitive sing. **bil(l)es**. SPN 53, SSNEM 36, Cameron 1998.

BILSINGTON Kent TR 0434. 'Bilswith's estate'. *Bilsvitone* 1086, *Bilswithetun, Bilsindetune* c.1100, *Bilsinton'* 1186×7–1226. OE feminine pers.n. *Bilswīth*, genitive sing. *Bilswīthe*, + **tūn**. PNK 469.

BILSTHORPE Notts SK 6460. Either 'Bildr's outlying settlement' or 'the outlying settlement of the promontory'. *Bildestorp* 1086, 1215, 1242, *-thorp(e) -is-* 1272–1346, *Billesthorp* 1265, *Billistrop* 1545, *Bilstropp alias Bilsthorp* 1600. ON pers.n. *Bildr*, secondary genitive sing. *Bildes*, or **bildr** '(blood-letting) knife' used in a topographic sense, + **thorp**. The pers.n. (which is by origin a by-name from the appellative **bildr**) is very rare. The old village stands on a promontory which may have been called *Bildr*. Nt 45, SSNEM 103.

BILSTON WMids SO 9597. 'The settlement of the Bilsætan, the inhabitants of the *Bill*'. *Bilsetnatun* [996 for 994]17th S 1380, *Billestune* 1086. OE folk-n. **Bilsǣtan* < OE **bill** 'a sword, an edge, a ridge' or **bile** 'a bill, a beak' + **sǣtan**, genitive pl. **Bilsǣtna*, + **tūn**. The reference is to a ridge of high ground. The boundary of the Bilsætan is mentioned in S 860, *Bilsatena gemæro* [985]12th. DEPN, V i.99–100.

BILSTONE Leic SK 3605. Either 'Bildr's village or estate', or 'promontory village'. *By- Bildeston(e) -is-* 1086–1434, *By- Billeston -is-* 1277–1422, *By- Bilston(')* 1430–1610. ON pers.n. *Bildr*, genitive sing. *Bild(e)s*, + **tūn**; but the pers.n. is rare and the specific may rather be the Scand appellative **bildr** used in Scandinavian p.ns. of angle-shaped topographical features such as the promontory on which Bilstone stands. Lei 535, SSNEM 188.

BILTING Kent TR 0549. Probably the 'hill place'. *Belting'* 1270, 1272×1307, *Beltinge -yng(e)* 1272×1307–1346, *Byltyng(')* 1346, 1438. This name presents the same problems as BELTINGE TR 2068 with which it is identical. Either the 'place like a belt or girdle', OE ***belting**, referring to the Great Stour river which cuts through the North Downs here, or more likely ***bylting**, Kentish dial. ***belting** 'hill place' < ***bylte, belte** + **ing**^{2} referring to Bilting Hill TR 049499, a remarkable round hill. The element *belt* or *belte* (for **bylte*) is established in this name by the form *Beltesburne* 1272×1307, 'the stream of Belt'. PNK 377, ING 9.

BILTON Humbs TA 1633. 'Billa's settlement'. *Bil(l)eton(e)* 1086, *Bi- Bylton(a)* 1199×1216–1563. OE pers.n. *Bil(l)a* + **tūn**. YE 46.

BILTON Northum NU 2210. 'The settlement at the hill-edge'. *Bilton('), Bylton* from 1242 BF. OE **bill** + **tūn**. The hamlet lies on the 200ft. contour of a ridge of high ground overlooking the river Aln. NbDu 22.

BILTON NYorks SE 4750. 'Billa's farmstead'. *Biletun -ton(e)* 1086, *Bilton* 1219–1597. OE pers.n. *Bil(l)a* + **tūn**. The same pers.n. occurs in BILBROUGH SE 5346, 4 miles away. YW iv.248.

BILTON Warw SP 4873. Partly uncertain. *Ben- Beltone* 1086, *Belton(e)* [c.1155]1235–1329, *B(e)ulton* 1255–1329, *Beel- Boel- Beolton* 1316–86, *Bylton* 1492, *Bilton* 1608. One suggestion is that the specific is a lost stream name *Beolne* which also occurs in Worcestershire [871]11th S 181 and forms the first el. of the name BELBROUGHTON SO 9277, or OE **beolone** 'henbane' from which the r.n. derives. The generic is OE **tūn** 'an estate'. Wa 125, V i.83.

BILTON PARK NYorks SE 3157. *Parcum de Bilton* 1409, *Bilton park* 1540. P.n. BILTON SE 4750 + OFr, ME **park** 'an enclosed tract of land for beasts of the chase'. One of the parks in the Forest of Knaresborough. YW v.105.

BINBROOK Lincs TF 2193. 'Bynna's brook'. *Binnibroc* 1086, *Binnabroc* c.1115, *By- Binnebroc -brok(e)* 1099–1332, *Binbroke* 1244–1610, *-brook(e)* from 1587. OE pers.n. *Bynna*, genitive sing. *Bynnan*, + **brōc** possibly varying with *Bynna* + **ing**^{4} + **brōc**. Other possibilities canvassed include OE **binnan brōce** '(the place) within the brook', or 'the valley brook', OE

binn(e) 'a basket, a manger, a bin' used in the transferred sense 'valley'. The village is situated on sloping ground overlooking a small tributary of Tetney drain. Part of the village lies within a fork of two streams and may have been the original site. Li iii.2, V i.102, 103.

BINCOMBE Dorset SY 6884. Possibly 'bean coomb'. *Beuncumbe* [987]13th S 1217, *Beincome* 1086, *Biencomme -comb(e)* 1157–1321, *Bincumbe* 1244, *Byncomb(e)* 1327–1459. OE **bēan** + **cumb**. The earliest spellings, however, including S 1217 which is a careful copy, are difficult. An alternative might possibly be **Being-cumb*, OE **bēo** + **ing**[2] 'bee place coomb', or pers.n. *Bēaġe* + **ing**[2] 'coomb called after Beage'. Do i.197, L 93.

BINEGAR Somer ST 6149. 'Beage's wooded hill-side'. *Beaganhangran* [852×74]18th S 1701, *Begenhangra* *[1065]18th S 1042, *Behenhanger* 1243, *Benhangre* 1176, *Benager* 1610. OE feminine pers.n. *Bēaġe*, genitive sing. *Bēaġan*, + **hangra**. DEPN, L 195–6.

BINFIELD Berks SU 8471. 'Open land where bent-grass grows'. *Benetfeld'* c.1160–1220, *-feld(e)* 1176–1400, *Bent(e)feld(')* 13th cent., *Benefeld(')* 1185–1389 with variants *-feud*, *Benfelde* 1388, *Bynfeld* 1517. OE **beonet** + **feld**. Brk 76, L 243.

BINFIELD HEATH Oxon SU 7478. *Bynfeildeheath* 1589. Hundred name Binfield + ModE **heath**. Binfield, *Benifeld* 1175×6–1205, *Benetfeld(')* 1187×8–1201, is 'the open land where bent grass grows', OE **beonet** + **feld**[†]. The heath was probably the meeting place of Binfield hundred. O 65, 81, DEPN, EHN ii.219.

BINGFIELD Northum NY 9772. Possibly 'open land by a hollow'. *Bingefeld* 1180, *Bynge-* 1298, *Bingfield* c.1715 AA 13.13. OE ***bing** + **feld**. OE **bing* is not on record but there are parallel formations in other Gmc languages, ON *bingr* 'part of a room, a bed', Faroese *bungur* 'chest, coffer', OSwed *binge* 'grain-box', and especially MLG *binge* 'a kettle-shaped hollow in hills'. Bingfield stands on a hill-side overlooking the valley of the Erring Burn. Alternatively the specific might be a contraction of either an OE **Bynning* 'the place called after Bynna' < OE pers.n. *Bynna* + **ing**[2], or of OE **Bynninga*, genitive pl. of **Bynningas* 'the people called after Bynna' < *Bynna* + **ingas**. NbDu 22, L 244.

BINGHAM Notts SK 7039. Probably 'the homestead of the Bynningas, the people called after Bynna'. *Binghehā* 1086, *Bingeham* 1164–1227, *Bynge-* 1237, 1272, *Bingham Byng-* from 1212, ~ *in the Vale* 1375. OE folk-n. **Bynningas* < pers.n. *Bynna* + **ingas**, genitive pl. **Bynninga*, + **hām**. Alternatively the specific might be OE ***bing** 'a kettle-shaped hollow'. Cf. BINGFIELD Northum NY 9772. Bingham lies on the edge of the gentle valley of SAXONDALE SK 6839. Nt 220.

BINGHAM'S MELCOMBE Dorset ST 7702 → Bingham's MELCOMBE.

BINGLEY WYorks SE 1139. Probably the 'wood or clearing of the Bynningas, the people called after Bynna'. *Bingheleia*, *Bingelei* 1086, *Bi- Byngelay -lai(a) -l(e)'* etc. 1156×85–1346, *Bingley* from 1303. Probably OE folk-n. **Bynningas* < OE pers.n. *Bynna* + **ingas**, genitive pl. *Bynninga*, + **lēah**. But it is also possible that there was an OE ***bing** meaning either 'a rounded lump' related to ON *bingr* 'bed, bolster, heap', or 'a hollow' related to MHG *binge* 'kettle shaped hollow in the hills'. Either sense would fit the topography of Bingley. YW iv.161, Nt 220–1, WYAS 323.

BINHAM Norf TF 9839. 'Bynna's homestead'. *Binneham*, *Benincham* 1086, *Binham* from 1156. OE pers.n. *Bynna* + **hām**. The second DB form points to an alternative formation *Bynna* + **ing**[4] + **hām**. DEPN.

BINLEY Hants SU 4253. Partly uncertain. *Benelega* 1184, *Benleg' -le(ye)* 1248–1359, *Bynle(ghe) -lygh* 1269–1333, *Beenley* 1483, *Bindley* 1610. This is probably the 'bean clearing', OE **bēan** + **lēah**, although 'bees' wood or clearing', OE **bēona**, genitive pl. of **bēo**, + **lēah**, or 'Bynna's wood or clearing' are also possible. Ha 33, Gover 1958.155.

BINLEY WMids SP 3777. 'Billa's promontory'. *Bilnei*, *Bilueie* (sic) 1086, *Bilneye* [c.1155]1251–1424, *Binlea* 1189×99, *Bynle* 1279, *Bingley* 1603×15. OE pers.n. *Billa*, genitive sing. *Billan*, + **ēġ** 'an island, a promontory'. This is preferable to 'ridge promontory', OE *bile* + *ēġ* since *bile* is believed to be masculine gender with genitive sing. *biles*. Wa 156, L 36.

BINLEY WOODS Warw SP 3977. P.n. BINLEY WMids SP 3777 + ModE **wood(s)**.

BINNIMOOR FEN Cambs TL 4597. *Binnimoor Fen* 1824 OS. P.n. Binnimoor + ModE **fen**. Binnimoor is *Benimore* 1621, *Binnymore* 1636. The forms are too late for certainty but this might represent OE *binnan mōre* 'within the fen'. Ca 253.

BINSEY Cumbr NY 2235. Unexplained. *Binsay Hawse* 1742, *Binsa*, *Binsay Fell*, *Binsell fell* 1777. The feature is a steep conical hill with a tumulus marked on the summit. The generic is probably, therefore, ON **haugr** 'a mound' which can develop to *-ay*, cf. Calva Hill near Workington, *Calvey* 1717, Greenah NY 1545 *Grenehou* 1295, *Greney* 1608, Lowsay NY 1148, *Lowsehowe* 1604, + N dial. **hause, hawse** (ON *hals* 'a neck, a col') referring to the gap through which the road from Bewaldeth to Uldale passes, and **fell** (ON *fjall*). The specific is probably a pers.n. such as OE *Byni*. Cu 300, 456, 255, 297.

BINSTEAD IoW SZ 5791. 'Bean place'. *Benestede* 1086–1507, *Bensted(e)* 1225–1595, *Byn- Binsted(e)* 1276–1559, *Binstead* 1781. OE **bēan** + **stede**. Wt 42, Stead 279, Mills 1996.29.

BINSTED Hants SU 7740. 'Bean place'. *Benestede* 1086–1324, *Bensted(e)* 1212–1359, *Bi- Bynstede* 1269–1375. OE **bēan** + **stede**. According to Stead 171 the reference is probably to the horse-bean, 'an essential element in the poor man's diet' but also used as horse and cattle fodder. Ha 33, Gover 1958.97, PNE i.21, Stead 171, 270.

BINTON Warw SP 1454. 'The estate called after Byni or Bynna'. *(æt) Bynningtune* [c.1010×23]11th S 1460, *Buninton(e) -y(n)- -yng-* 1215–1427, *Beninton* 1086, 1194, *-ing-* 1227, 1416, *Benitone* 1086–1288, *Bynton* c.1384. OE pers.n. *Byni* or *Bynna* + **ing**[4] + **tūn**. Wa 204.

BINTREE Norf TG 0123. 'Bynna's tree'. *Binnetre* 1086, *Binetre* 1180, OE pers.n. *Bynna* + **trēo**. DEPN.

BINWESTON Shrops SJ 3004. 'Binna's Weston'. *Binneweston* 1255, 1292, *Bynne Weston* 1327. OE pers.n. *Binna* + p.n. WESTON, *Westun* 1255. DEPN, Sa i.308.

BIRCH Essex TL 9417. 'The birch-tree'. *(Parva) Bric(ce)iam* 1086, *Birch(e) (Magna, Parua)* 1194–1243 etc., *Berche Magna, Parva* 1594, *Great, Little Burch* 1608. Latin **parva, magna**, ModE **little, great**, + p.n. Birch, OE **birċe**. Ess 362.

BIRCH GMan SD 8057. 'Birch-trees'. *Birches* 1246, *(del) Byrches* 1277–1322. OE **birċe**, pl. **birċas**. La 31.

BIRCH GREEN Essex TL 9418. P.n. BIRCH TL 9419 + ModE **green**. Known in 1805 OS as *Layer Breton Heath*.

BIRCH VALE Derby SK 0286. A modern name. ModE **birch** as in Birch Hall, *Birch Hall* 1538×44 + **vale**. Db 115.

Little BIRCH H&W SO 5131. *Little Byrche* 1523. ModE adj. **little** + p.n. *Birche(s)* 1160×70–1535, *La Birich* 1162, 'the birch-trees', OE **birċe**, pl. **birċas**. 'Little' for distinction from Much Birch SO 5030, *Muchelbirches* 1340. These two manors were also known as *Seyntmariebirches* 1302×3 'St Mary's B' and *Byrches Sancti Thomas* 1334 'St Thomas's B' respectively from the dedication of the churches. In DB the estate is called *Mainavre* 1086, OW **mainaur** 'manor, multiple estate', possibly referring to a *maenor wrthir* or upland division of Archenfield centred on the hill-fort at Aconbury SO 5033 with a corresponding *maenor fro* or lowland division at Hentland SO 5426. He 33–4, Jnl 10.65, Ag. Hist. 1.ii.307.

Much BIRCH H&W SO 5030 → Little BIRCH SO 5153.

Great BIRCHAM Norf TF 7632. *Great Bircham* 1824 OS. ModE

[†]The form *(on) beonan feld* [964]13th S 722, 'Beona's open land', does not belong here but to a place S of the Thames.

adj. **great** + p.n. *Brecham* 1086–1190ff, *Breccham* 1195, 'homestead at the newly cultivated land', OE **brēċ** + **hām**. 'Great' for distinction from *Parva Brecham* 'little Brecham' 1254. DEPN, PNE i.48.

BIRCHANGER Essex TL 5122. 'Birch-covered wooded slope'. *Bilichangrā, Blichangram, Becangrā* 1086, *Byl(y)changr(e), Bil(e)cha(u)ngr(e)* 12th–1503, *Birecengre* 1123–33, *Birich(e)angr(e)* c.1160–1322, *Birch(h)angre* 1222–1334, *Belchangre -er* 1288–1589, *Birchanger alias Belchanker* 1536. OE **birċe** + **hangra**. Ess 518 gives pr ['bə:tʃæŋgə].

BIRCHER H&W SO 4765. 'Birch-tree ridge'. *Birchour'* 1173×86, *Byrchore* 1388. OE **birċe** + **ofer**. He 217.

BIRCHINGTON Kent TR 2969. Partly uncertain. *Birchil- Birchenton'* 1240, *Bru- Bir- Berchinton -yn-, Bircheton* 1354–78, *Bercel- By- Birchel- Berchelton* 1264–92. The generic is OE **tūn** 'village, estate', the specific uncertain: perhaps 'birch hill', OE **birċe** + **hyll**, or 'birch copse', OE ***birċel**, a formation parallel to *brēmel*, or adjective **birċen** 'growing with birches'. PNK 594.

BIRCHOVER Derby SK 2462. 'Birch-tree ridge'. *Barcovere* 1086, *Bi- Byrchou(e)r(e) -over(e)* from 1226. OE **birċe** + **ofer**. The DB spelling suggests there was an alternative form with OE **berc**. Db 45, L 175–6, 220.

BIRCHWOOD Lincs SK 9369. A modern name with no early forms. ModE **birch** + **wood**.

BIRCOTES Notts SK 6391. A modern settlement name.

BIRDBROOK Essex TL 7041. 'Brook of the young birds'. *Bridebroc* 1086, *Bri- Bryd(e)brok(e)* 1248–1428, *Bri- Bryddebrok(e)* 1292–1331, *Birdbrok(e)* 1432–1559, *Burbrooke or Bridbroke* 1594. OE **bridd**, genitive pl. **bridda**, + **brōc**. Ess 411 gives pr [bʌbruk].

BIRDHAM WSusx SU 8200. 'The promontory or river meadow of the young birds'. *Bridham* 12th–1492 including *[683]14th S 232, *Brideham* 1086–1248 with variant *Bridde-, Byrdham* 1492, *Burdham* 1595. OE **bridd**, genitive pl. **bridda**, + **hamm** 2a, 3, 4, 5b. Perhaps a natural hatching ground later subject to human exploitation. Sx 80, ASE 2.41, SS 81.

BIRDINGBURY Warw SP 4368. 'Fortified place of the Byrdingas, the people called after Byrda'. *Byrtingabirig* [1043]17th S 1000, *Burtingbury* *[c.1043]17th S 1226, *Berdingeberie, Derbingerie* 1086, *Burthingberi* c.1195–1333 with variant *-bury*, *Burdingbury* 1316, 1363, *Birdingeburie* 1580, *Burthebire* 1199, *-bury* 1316, *Burdebery* 1373, *Burbury* 1627. OE folk-n. *Byrdingas* < pers.n. *Byrda* + **ingas**, genitive pl. *Byrdinga*, + **byriġ**, dative sing. of **burh**. Wa 126 gives pr [bə:bəri].

BIRDLIP Glos SO 9214. Partly uncertain. *Bri- Brydelep(e)* 1221, 1287, 1376, 1480, *Brudelep* 1240, 1295, *Bri- Brydlep(e)* 1295, 1423, 1494, *-lipp(e) -lyppe* 1494, 1529, *Bi- Byrd(e)lyp(pe) -lip* 1537–1637. The generic is OE **hlēp** 'leap, a steep place', the specific either OE **bridd** 'bird', genitive pl. **bridda**, or **brȳd** 'bride', genitive sing. **brȳde** or pl. **brȳda**. Birdlip is situated at the edge of the Cotswolds escarpment; the sense is 'leap such as only a bird could make' or a reference to some forgotten incident or folk-tale about a 'bride's leap'. Gl i.156.

BIRDSALL NYorks SE 8165. 'Bridd or the bird's nook of land'. *Briteshale -a* 1086, *Brideshal(a) -hale* 1086–1300, *Bi- Byrdsal(l)* c.1155–1531. OE pers.n. *Bridd* or **bridd**, genitive sing. *Briddes*, **briddes**, + **halh**. The reference is to a small indentation in the high ground of Birdsall Brow. YE 141, L 108.

BIRDSGREEN Shrops SO 7685. *burdes greean* 1593, *Birds Green* 1833 OS. Surname *Bird* + ModE **green**.

BIRDWELL SYorks SE 3401. 'Spring frequented by birds'. *The Birdwell* 1642. ModE **bird** + **well**. YW i.294.

BIRDWOOD Glos SO 7418. 'Bird wood'. *Bridwode* 1250×60–1342, *Byrdwood* 1532. OE **bridd** + **wudu**. Gl iii.196.

BIRK BECK Cumbr NY 5907. 'Birch stream'. *Bi- Byrk(e)be(c)k* 1279–1777. ON **birki** + **bekkr**. We i.3.

BIRKDALE Mers SD 3215. 'Birch-copse valley'. *Birkedale* [c.1200]1268, 1305 etc., *Berkdale* 1311, *Birkedene* [c.1200]1268. ON **birki** + **dalr** possibly replacing OE **denu**. La 125, SSNNW 105, L 94–5.

BIRKDALE COMMON NYorks NY 8302. P.n. *Birkedale* 1301, 'dale where birch-trees grow', ON **birki** + **dalr**, + ModE **common**. YN 272.

BIRKENHEAD Mers SJ 3288. 'Headland growing with birch-trees'. *Bircheveth* 1190–1216, *Birkheued* 1260–1579 with variants *Byrk(e)-* and *-heuid -hevet -hed* 1347–1535, *Birkenhed* 1278–1666 with variants *Byrk-*, *-yn-* and *-hedd(e)*, *Berkenhead* 1673, 1724. OE **birċe** varying with **birċen** + **hēafod** influenced by ON **birki** and **hǫfuth**. Che iv.313 gives prs [bɛ:kən'ed] and ['bərkən'ed], SSNNW 216, L 161, 220.

BIRKENSHAW WYorks SE 2028. 'Birch wood'. *Birkenestawe* (for *-scawe*) 13th, *Birkens(c)hawe* 1274, 1604. OE **birċen** with Scand /k/ for /tʃ/ + **sceaga**. YW iii.13, L 209.

BIRKER Cumbr NY 1800. 'Birch-tree shieling'. *Birkergh* 1279, *Byrker* 1432, *Birker* 1560. ON **birki** + **ærgi**. Cu 342, SSNNW 62.

BIRKER FORCE Cumbr SD 1899. P.n. BIRKER NY 1800 + N dial. **force** (ON *fors*) 'a waterfall'.

BIRKIN NYorks SE 5327. 'The birch-tree copse'. *Byrcene* c.1030 YCh 7, *Berchi(n)ge -ine* 1086, *Birkin(e) -y-* c.1160–1683, *Byrkinga* c.1165, *Birkinge -y-* 1193. OE **birċen** and ***birċing** with Scandinavian /k/ for /tʃ/. YW iv.16.

High BIRKWITH NYorks SD 8077. ModE **high** + p.n. *Bircwid* c.1190, *Birk(e)with -wyth* 1190–1692, *Bir(k)quith -yth* 1535–1778, 'the birch-tree wood', ON **birki** + **vithr**. YW vi.218.

BIRLEY H&W SO 4553. 'Fortified place clearing'. *Burlei* 1086, *Burleg(e)* 1230, 1242, *Birley* 1832 OS. OE **burh** + **lēah**. DEPN, L 206.

BIRLING Kent TQ 6860. 'The Bærlingas, the people called after Bærla'. *Boerlingas* [788]12th S 129, *Bœrlingas, (of) Bœrlingan* [973×87]12th S 1511, *Berlinge* 1086, *Berlinges* 1200, 1240, *Berling(')* 1229–64, *Berlyng', Burlinge, Byrling'* 1278 etc. OE folk-n. **Bǣrlingas* < OE pers.n. **Bǣrla* + **ingas**. PNK 146, KPN 71, ING 9.

BIRLING Northum NU 2406. Possibly the 'place called after Byrla'. *Berlinga* 1186, *Berling'* 1188, *Byrlyngs* [c.1210]14th, *Birling(')* from 1242 BF. OE pers.n. **Byrla* + **ing**[2]. NbDu 23, ING 78.

BIRLINGHAM H&W SO 9343. 'River-bend land of the Byrlingas'. *(in) Byrlingahamme* (dative case) [972]11th S 786, *Berlingeham* 1086, 1221, 1330–4, *Burlingeham* [c.1086]1190–1245, *Birlingham* from 1212. OE folk-n. **Byrlingas* 'the people of Byrla' < OE pers.n. **Byrla* 'the cup-bearer' or **Byrla* < **Byrgla* + **ingas**, genitive pl. **Byrlinga*, + **hamm** 1. Birlingham lies in a great bend of the Avon. Wo 188, ING 128, L 43, 47, 49.

BIRMINGHAM WMids SP 0787. 'The homestead at or called *Berming*, the place called after Beorma'.

I. *Bermingehā* 1086, *-ham* 1214–1476, *Bi- Byrmingeham -ynge-* 1206–1510, *Burmingeham -ynge-* 1221–1438, *Birmingham* from 1227, *Burmingham* 1235–1387;

II. *Bermincham -yn-* 1245–1417, *Burmincham -yn-* 1260–1652, *By- Birmincham -yn-* 1326–1534;

III. *Bermingham -yg-* 1247, 1298;

IV. *Brimingham* 1200, *Brym(m)yngham -ing-* 1424–1617;

V. *Brymecham -ych-* 1420–1542, *Bromech(h)am -ychehame -wicham -icham* 16th cent., *Bromegem* 1650.

OE p.n. **Beorming* < pers.n. *Beorma* + **ing**[2], locative-dative sing. **Beorminġe*, + **hām**.

Type II sps show that the assibilated pr of *ing* in this name is ancient. Together with *r*-metathesis it produced the familiar *Brummagem* sb. and adj. recorded in OED from 17th cent. and abbreviated to *Brum* [brʌm]. Type V has been influenced by the form and pr of nearby Bromwich [brʌmidʒ] as in Castle BROMWICH SP 1590. Wa 34 gives popular prs [brʌmədʒəm] and [bɛ:nigəm], TC 52, BzN 1967.221–45, 363–4.

BIRMINGHAM INTERNATIONAL AIRPORT WMids SP 1784. P.n. BIRMINGHAM SP 0787 + ModE **international airport**.

BIRSTALL Leic SK 5909. 'Site of a stronghold, disused stronghold'. *Burstell(e)* 1086, *-stal(l)(e)* c.1130–1610, *Birstal* 1270. OE **burh-stall.** Lei 345 gives pr ['bɜ:stʌl].

BIRSTALL WYorks SE 2326. 'The site of a fort'. *Birstale* 12th, *Bir- Byrstall* 1195×1211–1822. OE **byrh-stall.** YW iii.14, WW 147.

BIRSTALL SMITHIES WYorks SE 2225. *Birstall Smythies* 1642. P.n. BIRSTALL SE 2326 + ModE **smithy**. YW iii.15.

BIRSTWITH NYorks SE 2359. 'The site of a farm'. *Beristade* 1086, *Bri- Brystath* 1386, 1508, *-stithe* 1563, *Birstwith* 1663. ON **bȳjarstathr** possibly replacing OE ***byriġ-stede**. YW v.131, Stead 87, SSNY 143.

BIRTLEY 'The bright clearing', OE **be(o)rht, birht**, definite form **beorhta, birhta**, + **lēah**.

(1) ~ Northum NY 8778. *Birtleye* 1229, *Birteley, Brutteleg* 13th cent., *Britelay* 1346. NbDu 23.

(2) ~ T&W NZ 2756. *Britleia* [1183]c.1382, *-ley* 1344–1451, *By- Birt(e)lei(a) -ley(a) -ley(e)* c.1190–1411, *Birtley* from 1369, *-lee* 1417–1628, *Bruteley* [1286]1332. NbDu 23.

BIRTLEY H&W SO 3669. Partly uncertain. *Brit- Birdleia* c.1183, *Berkley* 1833 OS. It is just possible that this might be 'birch-tree wood or clearing', OE **berc** + **lēah**, but the forms are too few and too inconsistent for certainty. Another form, *Brerehelde* 1292, cited under this name, is of quite different origin, OE **brēr, brere** 'briar' + **helde** 'slope'. He 128.

BIRTSMORTON H&W SO 8035 → BIRTS STREET SO 7836, CASTLEMORTON SO 7937.

BIRTS STREET H&W SO 7836. *Birts Street* 1831 OS. Birts as in Birtsmorton SO 8035 + ModE **street**. Birtsmorton, *Morton le Bret* 1241, *Brittes Morton* 1250, is 'Morton, the marsh settlement, OE **mōr** + **tūn**, held by the le Bret family' from Brittany to whom the manor was granted in 1166. Wo 213.

BISBROOKE Leic SP 8899. 'Beetle' or 'Byttel's stream'. *By- Bitlesbroch* 1086, 1087×1100, *-broc -broke* early 12th–1286, *Bittel(l)isbroc -es- -brok* 14th cent., *By- Bissebrok(e)* c.1291–1534, *By- Bisbrok(e)* 1394–1557. OE ***bitel**, a strong variant of *bitela* 'an insect, a beetle, a water-beetle', genitive sing. ***bit(e)les**, or pers.n. *Byttel*, genitive sing. *Bytt(e)les*, + **brōc**. R 238, V i.106.

BISHAMPTON H&W SO 9951. 'Bisa's farm'. *Bisantvne* 1086, *Bisshantune* [c.1086]1190, *Bissamtona, Bi- Byshampton* from 1275. OE pers.n. ***Bisa**, genitive sing. ***Bisan**, + **hām-tūn**. Wo 97 gives pr [biʃəmtən].

BISHOPDALE BECK NYorks SD 9885. P.n. *Biscop(p)edale* 1202–1279×81, *Bis(s)hopdale* 1289–1519, 'the bishop's valley', OE **biscop** + **dæl**, + ModE **beck**. YN 265, L 96.

BISHOP ROCK Scilly SV 8006. *Maenenescop'* 1302, *The byshop and hys clerkys* 1564, *Bishop* 1689. A fanciful name for a rock in the sea, cf. The Bishop, Breage (named from its shape), Mullion, St Columb Minor, and Bassenthwaite, Cumbria. Translated from Co **men an escop** 'rock of the bishop'. In 1302 a certain Muriel de Trenywith and her two daughters were abandoned here to drown for theft. PNCo 53.

BISHOPSBOURNE Kent TR 1852. 'Bourne belonging to the (arch)bishop' sc. of Canterbury. *Biscopesburne* c.1090, *Bissopesburn(e)* before 1223, 1248, *Byshops Borne* 1596. OE **biscop**, genitive sing. **biscopes**, + p.n. *Burnan* 799 S 155, *Burnes* 1086, 'the stream(s)', OE **burna**, pl. **burnan**, referring to the Little Stour or Nail Bourne. 'Bishop's' for distinction from other manors on this stream, BEKESBOURNE TR 2055 'Bek's *Bourne*', KINGSTON TR 1951 'the king's *Bourne* manor', LITTLEBOURNE TR 2057 and PATRIXBOURNE TR 2055 'William Patricius's *Bourne*'. PNK 555, KPN 85, Cullen.

BISHOP'S CASTLE Shrops SO 3288.

I. Latin forms, *Castrum Episcopi* 1255–1316;

II French forms, *Chastel Eveske* 1318×19, ~ *Episcopi* 1394;

III English forms, *Bisshopescastel* from 1282, *Bishop's Castle* 1577;

IV Welsh form, *Trefesgob* from late 15th.

The castle was built c.1127 in the Bishop of Hereford's estate of Lydbury North and the town planted in the late 12th cent. Sa i.49.

BISHOPSTEIGNTON Devon SX 9073. 'Teignton held by the bishop' sc. of Exeter. *Teygtone Episcopi* 1262, *Teynton Bishops* 1341. Earlier simply *Tantone* 1086 DB, *Taintona* 1086 Exon, *Teinton* [c.1150]14th, 1185, 1224, *Teyngton* 1245, 'settlement on the river Teign', r.n. TEIGN + OE **tūn**. The manor was already held by the bishop of Exeter in 1086 in contrast to KINGSTEIGNTON SX 8773 held by the king. D 487.

BISHOPSTOKE Hants SU 4719. 'Outlying farm belonging to the bishop' sc. of Winchester, already in DB. *Stoke Episcopi* c.1270–1301, *Bishopestoke* 1379. Latin **episcopus**, genitive sing. **episcopi**, ME **bishop** + p.n. *Stoches* 1086. Identical with *(æt) yting stoce* 'at the farm called *Yting*, the Jutes' place' [960]12th S 683, OE folk-n. *Ȳte* + **ing**[2]. Ha 34, Gover 1958.67.

BISHOPSTONE 'Bishop's estate'. OE **biscop**, genitive sing. **biscopes**, + **tūn**.

(1) ~ Bucks SP 8010. *Bissopeston* 1227. This estate was parcel of the manor of Stone held by Odo, bishop of Bayeux, in 1086. Bk 165, Jnl 2.23 gives local pr [biʃtən].

(2) ~ ESusx TQ 4701. *Biscopestone* 1086, *Bisschopeston* 1279, *Bushopston*. The manor belonged to Chichester already in 1086. Sx 365, SS 76.

(3) ~ H&W SO 4143. *Bissopeston(a)* 1135×54–1328, *Biscopestone* 1166, 1210×12. The DB name is *Malveshille* 1086, this being one of the divisions of the tripartite estate of Mansell as in MANSELL GAMAGE and MANSELL LACY SO 3944 and 4245. He 35.

(4) ~ Wilts SU 0725. *Bissopeston* 1166, 1190, *Bysschopestone* 13th, *Busshopeston(e)* 1399, 1406. An estate belonging to the bishop of Winchester and carved out of Ebbesbourne ST 9924, and so known alternatively as *Ebleburn Bissopiston* 1244 or *Ebbelesbourne Episcopi* 1310, 'the bishop's (part of) Ebbesbourne'. Wlt 392.

(5) ~ Wilts SU 2483. *Bissopeston* 1186, 1223, *Bisshopeston* 1247, *Busshopeston* 1332, *Bussheton* 1534. Originally part of Ramsbury Parish SU 2771 which was a manor held by the bishop of Salisbury. Wlt 286 gives former pr [buʃtən].

BISHOP'S WOOD Staffs SJ 8309. *Bishoppes wood* 1598, *Bishop's Wood* from 1661. Short for *Bishopes Brewode* 1302, ME **bishop**, genitive sing. **bishopes**, + **wode** as in BREWOOD SJ 8808. St i.39.

BISHOPSWOOD Somer ST 2512. *Bishops Wood* 1809 OS.

BISHOPSWORTH Avon ST 5768. 'The bishops' enclosure'. *Bichevrde, Biscopewrde* 1086, *Biscopewurth* 1243. OE **biscop**, genitive pl. **biscopa**, + **worth**. In 1086 this was an estate belonging to the Bishop of Coutances. DEPN.

BISHOPTHORPE NYorks SE 5947. 'The bishop's outlying estate'. *Biscupthorp* 1275. Earlier *Badetorp(es)* 1086, *T(h)orp* c.1180–1373, and ~ *Archiepiscopi* 1225. The DB form represents either 'the two outlying farmsteads', ON **báthir** + **thorp**, or 'Bada's outlying farmstead', OE pers.n. *Bad(d)a* + **thorp**. The estate was purchased by archbishop Gray c.1225 and is still the residence of the archbishops of York. YW iv.225, SSNY 55.

BISHOPTON Durham NZ 3621. 'Estate held by the bishop' sc. of Durham. *Biscoptun* c.1104, 1144×52, 1189×1212, *-tune* 1339, *Bissopton'* 1174×93, 1195×1221, 1315, *Bissopeston'* 1239, *Bishopton* from c.1220, *Bus(s)hopton'* 1342, 1593–1637. OE **biscop** + **tūn**. NbDu 23.

BISLEY Glos SO 9006. 'Bisa's wood or clearing'. *(to) Bislege* [896]11th S 1441, *Bi- Byseleg(e) -lei(a) -legh(e) -le(ye)* 1086–1424, *Bi- Bysle(y) -legh* 1211×13–1600. OE pers.n. *Bisa* + **lēah**. The same pers.n. occurs in the neighbouring Bismore, 'Bisa's marsh', *Bismore* 1609, Bussage SO 8803, 'Bisa's ridge', *Bisrugge* 1287, Buscombe, 'Bisa's coomb', *Biscombe* 1380, and the lost *Bisendune* 1250×60, 'Bisa's hill'. Gl i.117, 127, 139, 147.

BISLEY Surrey SU 9559. 'The woodland clearing with bushes'.

Bisselegam c.1103, *-le(gh)* 13th cent., *Busselegh(e)* [933, 967]13th S 420, 752, 1263–1402, with variant *-l(e)y(e)*, *Bisscheleye* 1279, *Busshele(y)* 1283–1471, *-legh* 1493, *Bisley* from 1587. OE ***bysc** + **lēah**. Alternatively the specific might be the OE pers.n. *Byssa*. Sr 103 gives pr [bizli], DEPN.

BISPHAM Lancs SD 3040. 'The bishop's manor'. *Biscopham* 1086–1196, *-heym* [early 13th]1268, *Bischopeham* 1094, *Bischopham* 1155, c.1190, *Bisbhaym* c.1270, *Bi- Bysp(e)ham* from 1327, *Byspham in ye Fyle* 1577, *By- Bissham* 16th. OE **biscop** + **hām**. For *Fyle* see The FYLDE. La 156, Jnl 17.91, 18.15.

River BISS Wilts ST 8656 → BLISSFORD Hants SU 1713.

BISSOE Corn SW 7741. 'Birch-trees'. *Bedou* c.1250, *Bedow* c.1260, *Besow* 1480. Co **bedewen** 'a birch-tree', pl. ***bedew** > ***bezow**. PNCo 54, CPNE 18.

BISTERNE CLOSE Hants SU 2302. 'Bistern enclosure'. P.n. Bisterne + ModE **close**. Bisterne, *Betestre* 1086, *Bettest(h)orn'* 1187, 1349, *Budes- Bedenestorn* 1190, *Butestorna -thorne* 1219, 1362, *Budesthorne* 1300, is 'Bytti's thorn-tree', OE pers.n. **Bytti*, genitive sing. **Byttes*, + **thorn**. Ha 34, Gover 1958.219.

BITCHFIELD Lincs SK 9828. 'Bil's open land'. *Bille(s)felt* 1086, *Billesfeld* before 1160–1445, *Bylshe-* 1490, *Bylche-* 1501–45, *Biche- Byche-* 1535–71, *Bytchfeild* 1583, *Bitchfield* 1634. OE pers.n. *Bil(l)*, genitive sing. *Billes*, + **feld**. N and EMidl dial. pronunciation of [s] as [ʃ] led to false etymological association with ModE *bitch*. Perrott 152, Dobson §373.

BITTADON Devon SS 5441. 'Bitta's valley'. *Bedendone* 1086, *Bet(t)enden(n)* 1205, 1219, *Bi- Bytteden -e* 1242–1884, *Byttedon* 1346. OE pers.n. **Bitta*, genitive sing. **Bittan*, + **denu**. *Be-* spellings may point to an alternative form of the pers.n., **Beotta*. D 29.

BITTAFORD Devon SX 6657. Uncertain. *Bittiford* 1677, *Bittaford* 1699. The evidence is too late but this might be 'Bitta's ford' with the same OE pers. n. as in BITTADON SS 5441. The ford crosses Lud Brook. D 287.

BITTERING Norf TF 9317. Probably the 'place called after Brihthere'. *Britringa* 1086, *Bit(t)ringe* 1203, *Bitering* 1252, *-e* 1275, *Bitteryng* 1476. The evidence is actually insufficient to say whether as seems likely this is OE *Brihthering* 'the place called after Brihthere', OE pers.n. *Brihthere* (*Beorhthere*) + **ing**², or a pl. folk-n. **Brihtheringas* 'the people called after Brihthere', pers.n. *Brihthere* + **ingas**. ING 55.

BITTERLEY Shrops SO 5677. 'Butter pasture'. *Buterlie* 1086–1349 with variants *-le(e) -leg(h) -ley(e)*, *Butter-* 1172×7–1655, *Byterleye* 1306×7, *Bitterley* from 1346. OE **butere** + **lēah**. Sa i.50.

BITTERNE Hants SU 4413. Partly uncertain. *Bi- Byterne* c.1090–1354, *Bitterne* from 1208. Probably a compound of OE **bita** 'a bit' and **ærn** 'a house'. A 'house for bits' probably referred to tools or horse-tackle. Ha 34, Gover 1958.35.

BITTESWELL Leic SP 5385. 'Stream or spring of the broad valley'. *Betmeswell(e) -wel* 1086, *By- Bitmeswell(e) -is-* 1199–1369 with variants *Butmes- -us-*, *By- Bit(t)eswell(') -is- -ys-* from 1416. OE **bytme**, genitive sing. **bytmes**, + **wella**. Lei 430, L 32, 87, Jnl 20.43, 45.

BITTESWELL AERODROME Leic SP 5184. P.n. BITTESWELL SP 5385 + late ModE **aerodrome**.

BITTON Avon ST 6869. 'Settlement, estate on the river Boyd'. *Betvne -ton(e) -tun* 1086–1205, *Be- Button(e)* c.1150–1535, *Bi- Bytton(')* from 1248. R.n. Boyd + OE **tūn**. Boyd, *(andlang, on tha ealdan) byd* [950]19th S 553, *(on) byd* 972 S 786, *Byde* 1287, *Boyd* 1712, is the name of an affluent of the Bristol Avon of uncertain origin, connected either with W *budr* 'dirty' or *budd* 'profit, benefit'. S 553 refers to *tha ealdan byd'* 'the old (course of the) Boyd' and also to *Bydincel* 'the little Boyd', a small eastern affluent of the Boyd. Gl iii.75, i.4, RN 46, GPC 344, 345.

BIX Oxon SU 7285. 'The box-tree' or 'box-wood'. *Bixa* 1086–[1240]c.1280, *Bi- Byxe* 1201 etc. OE **byxe**. Also known as *Byx Gybewyne* [1240]c.1280 and *Bixe Brond* 1285 from Ralph Gibbewin 1165 (OFr pers.n. *Gibouïn*) and Robert Brand 1254×5 (a shortened form of MLG pers.n. *Hildebrand*). Cf. MARSH GIBBON Bucks SP 6423. O 66, L 222.

BLABY Leic SP 5697. Probably 'Blar's village or farm'. *Bladi* (sic) 1086, *Blabi* 1175–1316, *-by* from 1209×19, *Blayby* 1518. ON by-n. *Blár* + **bȳ**. The ON adjective *blár* 'black, dark', with the topographical sense 'cheerless, bleak, exposed' from which the pers.n. is derived does not suit the site here. Lei 431, SSNEM 37.

BLACK BECK NYorks SE 1093. No early forms. ModE **black** + **beck**.

BLACK BUOY SAND Lincs TF 4139. A modern name for a mooring buoy + ModE **sand** 'a sand-bank'.

BLACK BURN Cumbr NY 6940. *Black Burn* 1762. ModE **black** + **burn**. Also known locally as *Dirtpot Burn*. Cu 5.

BLACK COMBE Cumbr SD 1486. *Blackcoum* 1671, *-comb* 1777. ModE **black** + **comb** (OE *camb*) 'a crest'. So named from the blackness of the heath growing on it. Cu 449.

BLACK DOG Devon SS 8009. A hamlet named after a public house which appears as *Black Boy* 1809 OS. D 397.

BLACK DOWN 'Black or dark hill'. ME **blak(e)** + **doun** (OE *blæc*, definite form *blaca*, + *dūn*). L 142–3, 147.

(1) ~ Devon ST 0906. *Blackdowne* 1631, 1700. High ground rising to 929ft. and part of the same ridge referred to in BLACKBOROUGH ST 0909. D 558.

(2) ~ Devon SX 5081. *Blakedon* 1345, *Blackdowne* 1563. D 201.

(3) ~ HILLS Devon ST 1616. *Blakedoune* 13th, *Blacke Downe* 1566. These hills form the boundary between Devon and Somerset. D 18.

(4) ~ Dorset SY 6187. *Black Down* 1822 OS.

BLACK DUB Cumbr NY 0845. N dial. **dub**, first recorded 1500×20, means 'a muddy or stagnant pool, a deep, dark pool in a river'. The river was originally called *Polneuton* [1189]n.d., 1201, 'Newton brook' in Celtic word order, OIr *poll* 'a pool, a stream' + p.n. Newton, referring to WESTNEWTON NY 1344. Cu 5.

BLACK FELL Northum NY 7173. 'Black mountain'. *Black Fell* 1868 OS. ModE **black** + **fell** (ON *fjall*).

BLACK HEAD 'Black headland'.

(1) ~ Corn SX 0348. *Blake Hedd, Blak-Hed* c.1540. PNCo 54.

(2) ~ Corn SW 7716. *Peden due alias Blackhead* 1699. A translation of Co **pen-du** with late Co *dn* for *n*. PNCo 54.

BLACK HEATH Wilts SU 0751. *Blacke Heath* 1610, 1695. W 240.

BLACK HILL. 'Black or dark hill'. ModE **black** + **hill**, (OE *blæc*, definite form *blaca*, + *hyll*).

(1) ~ Ches SJ 9982. *Blacke hill end(e)* 1611. Che i.278.

(2) ~ Derby SE 0704. *Black Hill* 1842. A summit rising to 1908ft. on the boundary with West Yorkshire. Che i.322.

(3) ~ Devon SY 0285. *Black Hill* 1809 OS.

(4) ~ NYorks SD 7760. *Black Hill* 1845.

(5) ~ Shrops SO 3279. *Black Hill* 1833 OS.

(6) ~ WYorks SE 0704. *Black Hill* 1843.

BLACK KNOWE 'Black summit'. ModE adj **black** + N dial. **knowe** (OE *cnoll*).

(1) ~ Northum NY 6481. A summit of 1615ft.

(2) ~ Northum NY 5891. *Black Knowe* 1868 OS.

BLACK MARSH Shrops SO 3299. *Black Marsh* 1836 OS.

BLACK MOUNTAIN Shrops SO 1983. *Black Mountain* 1836 OS.

The BLACK MOUNTAINS H&W SO 2427. *Black Mountains* 1832 OS. Formerly the mountains of Talgarth, Itinerarium Kambriæ 36.

BLACK ROCKS NYorks TA 0486. No early forms. ModE **black** + **rock**.

BLACK SAIL PASS Cumbr NY 1811. Unexplained p.n. *Le Blacksayl, Blacksol* 1322, + ModE **pass**. Cu 26, 391.

BLACKA BURN Northum NY 7878. *Blacka Burn* 1868 OS.

BLACKAWTON Devon SX 8051. 'Black Awton'. *Blakeaueton(e)* 1281, 1308, *Blakaueton* 1284, 1338, *Blackawton alias Blackaveton* 1671. ME adj. **blak** + p.n. *Auetone* 1086–1266, either 'Afa's farm or village', OE pers.n. *Afa* + **tūn**, or 'settlement on the river Afon', PrW **aβon** taken as a r.n. + OE **tūn**. *Avon* would be the earlier name of The Gara, an unexplained r.n. of recent origin. The precise significance of the affix *Black* is uncertain: it may refer to soil or vegetation. D 315, 6.

BLACKBOROUGH Devon ST 0909. 'Black hill'. *Blacheberie -berge* 1086, *Blakeburgha* [1173×5]1329, *-berg(he)* 1242–74. OE **blæc**, definite form **blaca**, + **beorg**. Also known as *Butyesblakeburgh* and *Blakeberghesboty* 1302, 1357, from the family name of Ralph Buty who held the manor in 1242. The hill is part of the high ground referred to in BLACK DOWN ST 0906 reaching to 929ft. D 564, L 128.

BLACKBOROUGH END Norf TF 6613. Hill name *Blakeberge* c.1150, *-berg* 1205, 'the black hill', OE **blæc**, definite form **blaca**, + **beorg**, + ModE **end**. DEPN.

BLACKBOYS ESusx TQ 5220. Short for 'Blackboy's farm'. *Blakeboy(e)s* 1437, 1547. From the surname *Blackboy*: a Richard Blakeboy is recorded in 1398. Sx 394.

BLACKBROOK Staffs SJ 7639. Transferred from the stream name here, *Black Brook* 1833 OS. ModE **black** + **brook**.

BLACKBROOK RESERVOIR Leic SK 4517. *Blackbrook Reservoir* 1806. P.n. Blackbrook + ModE **reservoir**. Blackbrook, *le Blakebrok* c.1280, *Blackbrooke* 1610, is 'the dark brook', OE **blæc**, definite form **blaca**, + **brōc**. Lei 353, 398.

BLACKBURN Lancs SD 6828. Originally a stream name, the 'dark stream'. *Blacheburne* 1086, *Blakeburn* 1187–1332 etc., *Blakburn* 1349×50, *Blackbo(u)rne* 1547–1608, *Blagburne* 1590, *Blegburn* 1864. Blackburn lies on a river called Blackwater, formerly Blackburn, *Blak* [12th]14th. The local pronunciation is ['blegbən]. La 74, 66, Jnl 17.46.

BLACKBURN COMMON Northum NY 8092. *Blackburn Common* 1869 OS. R.n. Blackburn + ModE **common**.

BLACKBURY CASTLE Devon SY 1892. An Iron Age hill-fort on *Blacke downe* 1525. P.n. Blackbury, ultimately representing OE **blæc** 'black' + **burh** 'fortified place', dative sing **byriġ**, + pleonastic ModE **castle**. D 632.

BLACKDEN HEATH Ches SJ 7871. P.n. Blackden SJ 7870 + ModE **heath**. Blackden, *Blakedene* 1308, *Blak(e)den(e)* 1414–55, *Blackden* from 1548, *Blagden* 1621, is 'dark valley', OE **blæc**, definite form **blace**, + **denu**. Che ii.222, L 98.

BLACKFIELD Hants SU 4401. A modern estate at Fawley named from *Blackfield Common* 1810 OS. ModE **black** + **field**.

BLACKFORD 'Black ford'. OE **blæc**, definite form **blaca**, + **ford**. L 68.

(1) ~ Cumbr NY 3962. *Blakforthe* 1384, *Blackford* 1561. Cu 118.

(2) ~ Somer ST 6526. *Blacaford* [955×9]18th S 1757, *Blakeford* [959×75]18th S 1768, 1276, *Blacheford* 1086, *Blakford* 1610. DEPN.

(3) ~ Somer ST 4147. *Blacford* 1227, *Blakeford* 1257. DEPN.

BLACKFORDBY Leic SK 3318. 'Farm, village at Blackford'. *Blakefordeb'* c.1130, *-by* 1276, 1392, *Blacfordebi -k-* 1199–1398, *Blaugherby(e) -ou-* 1525–1692. P.n. **Blackford* ('the black ford', OE **blæc** + **ford**) + ON **bȳ**. The village is in a coal area. Lei 339, SSNEM 37.

BLACKGANG IoW SZ 4876. No early forms. Cf. BLACKGANG CHINE SZ 4876.

BLACKGANG CHINE IoW SZ 4876. *Blackgang Chine* 1781. Blackgang appears to be 'black path' referring to the path along the bottom of the cliffs. The chine, a great 400ft. deep ravine leading down to the sea, is Hants/IoW dial. **chine** (OE *ċinu*) 'a ravine'. Wt 114 gives pr ['blæ(k)gæŋ tʃain], Mills 1996.29.

BLACKHALL COLLIERY Durham NZ 4539. P.n. *Black Hall* 1863 OS + ModE **colliery**. See further BLACKHALL ROCKS NZ 4638.

BLACKHALL ROCKS Durham NZ 4638. P.n. Blackhall as in BLACKHALL COLLIERY NZ 4539 + ModE **rock**. The reference is to a cave in the sea cliffs here.

BLACKHAM ESusx TQ 4939. '(At the) black promontory or shelf of land'. *Blacheham* c.1095, *Blakeham* 1176, *-hamme* 1288, *Blakenhamme* 1279. OE **blæc**, definite form dative case **blacan**, + **hamm** 2 or 7. Alternatively the specific could be OE pers.n. *Blaca*, genitive sing. *Blacan*. If the original site was Blackham village the sense is 'promontory', if Blackham Court, 'enclosure on a shelf of ground'. Sx 370, ASE 2.24–5, SS 82.

BLACKHEATH Essex TM 0021. *le blakeheth* 1435. Cf. *Blackheathe Corner* 1563. ME **blak** + **heth**. Ess 378.

BLACKHEATH Surrey TQ 0446. 'The black or dark heath'. *le Blakheth* 1500, *Blackheath alias Churtheath* 1581. ME **blak** + **heth**. Once the site of the meeting place of the Blackheath Hundred, *Blacheatfeld, Blache(d)- Blachetfelle* 1086, *Blakehethe* 1332. The alternative first element in 1581 is OE **ċert** 'rough ground'. Sr 254, 219, EHN 3.63–4, L 245.

BLACKHEATH WMids SO 9786. Probably an ancient name but no forms have been found earlier than *Black Heath* 1834 OS. ModE **black** + **heath**.

BLACKLAND Wilts SU 0168. 'The black newly cultivated land'. *Blakeland* 1195–1332 with variant *-lond, Blaklond* 1232. OE **blæc**, definite form **blace**, + **land**. Wlt 260.

BLACKLEY GMan SD 8602. 'Black wood or clearing'. *Blakeley* 1282–1577, *Blackeley* 1577. OE **blæc** + **lēah**. La 37 gives pr [bleikli].

BLACKMAN'S LAW Northum NY 7498. Surname *Blackman* + dial. **law** 'a hill, a conical hill' (OE *hlāw*). *Blackman* may originally have been the nickname of the hill itself.

BLACKMOOR Hants SU 7733. Probably '(at the) black pool'. *Blachemere* 1168, *Blakemere* 1200–1300[†]. OE **blæc**, dative sing. **blacan**, + **mere**. Ha 35, Gover 1958.93.

BLACKMOOR 'Black moor'. OE **blæc** + **mōr**. L 55.

(1) ~ GATE Devon SS 6443.

(2) ~ VALE Dorset ST 7315. *The Vaile of Whithart alias Blackemore* 1575, *The Vale or Forest of Blakemore or White-Hart* 1774, earlier *boscum de Blakemore* 'Blackmoor wood' 1212, 1328, and *foresta de Blakemor(e)* 1217–1547, named from *Blakemor(e)* 1258–1445, 'the dark moor or marshy ground', OE **blæc**, definite form **blaca**, + **mōr**. Do iii.274, 268.

BLACKMORE Essex TL 6001. 'The black marshland'. *(la) Blakemore -a* 1213, 1232 etc., *Blakamore* 1374, *Blakmore* 1234, 1471. OE **blæc**, definite form **blaca**, + **mōr**. As parish name this replaced earlier *Phinghería* 1086, *Fingery(e)* 1234–68, *Fi - -Fyngrith(e) -reth(e)* 1204–1310, possibly the 'stream of the Finningas, the people called after Fin', OE folk-n. **Finningas* < pers.n. *Fin* + **ingas**, genitive pl. **Finninga*, + **rīth**. Ess 236.

BLACKMORE END Essex TL 7430. *Blakemore Ende* 1534 sc. of Wethersfield parish. Contrasts with BEAZELEY END TL 7428. Ess 467.

BLACKNEST Hants SU 7941. *Black Nest* 1816. Possibly identical with *Blakenhurste* 1272, 'at the dark wood', OE **blæc**, dative sing. definite form **blacan**, + **hyrst**. Gover 1958.98.

BLACKO Lancs SD 8641. *Blackow(e)* 1514, 1575. Named from Blacko Hill SD 8642, *Blacho* 12th, *Blakhow -hou* 1329, 1335, *the Blackoo* 1540, 'the black hill', OE **blæc** + ON **haugr**. La 87, 67 gives pr [blakə].

BLACKPOOL Lancs SD 3035. *Blackpoole* 1602, *Le poole commonly called Black poole* 1637. The town took its name from a peaty-coloured pool of water called *Pul* 1268, *Le Pull* 1416, which drained Marton Mere. Described in 1786 as 'Mr Forshaw's Bathing Place'. OE **blæc** + **pull**. La 156, Mills 64, L 28.

BLACKPOOL AIRPORT Lancs SD 3281. P.n. BLACKPOOL + ModE **airport**.

BLACKPOOL GATE Cumbr NY 5377. *Blackpool Gate* 1796. P.n. *Blackpool* (adjective **black** + **pool** in the sense 'stream,

[†]The form *Blackemere* [931×9]14th S 1206 cited in DEPN does not belong here.

rivulet') + ModE **gate**. A gate into Kershope Forest on the Black Lyne. Cu 62, PNE ii.68.

BLACKROD GMan SD 6110. 'Black clearing'. *Blacherode* c.1188×9, *Blakerode* 1201–1332 etc., *Blacrode* 1219, 1226, *Blakrode* 1414. OE **blæc** + ***rod(u)**. La 45, PNE ii.86, Jnl 17.37.

BLACKSMITH'S CORNER Essex TM 0131. No early forms. ModE **blacksmith** + **corner**.

BLACKSTONE WSusx TQ 2316. 'The estate called after Blæcsige'. *Blexinton* 1262, *Blaxinton* 1288, *Blaxton* 1373, 1605, *Blackson* 1610, *Blackstone* 1813 OS. OE pers.n. **Blæcsiġe* + **ing**[4] + **tūn**. Sx 221, SS 77.

BLACKTHORN Oxon SP 6219. 'The blackthorn or sloe-tree'. *Blaketorn* 1190–1309 with variants *-thurn(e)* and *-thorn(e)*, *Blackthorne* 1316. OE **blæc**, definite form **blaca**, + **thorn** and **thyrne**. O 167, PNE i.37.

BLACKTHORPE Suff TL 9063. *Blackthorpe* 1837 OS.

BLACKTOFT Humbs SE 8424. 'Black homestead' or 'Blakki's curtilage'. *Blaketoft(e)* [1153×90]copy, 1197×1206–1398, *Blaktoft -c-* [13th]15th, 1325–1537, *Blachestoft* [1195]14th. OE **blæc** + **toft** or ON pers.n. *Blakki*, genitive sing. *Blakka*, secondary (ME) genitive sing. *Blakkes*, + **toft**. YE 244.

BLACKTON RESERVOIR Durham NY 9418. P.n. Blackton + ModE **reservoir**. Blackton, *Blakedene* 1301, is 'the black valley', OE **blæc**, definite form **blace**, + **denu**. YN 306.

BLACKWATER Corn SW 7346. 'Black stream'. *Black water coomb* c.1696. ModE **black** + **water**. Contrasts with CHACEWATER SW 7544. PNCo 54, Gover n.d. 362.

BLACKWATER Hants SU 8460. *Blakwater* 1588. The village is named from the BLACKWATER RIVER SU 8360. Ha 35, Gover 1958.112.

BLACKWATER IoW SZ 5086. Originally a r.n., 'black stream'. *Blakwater* 1548. ModE **black** + **water**. Wt 8, Mills 1996.30.

River BLACKWATER Berks SU 7364. 'Black water'. *la Blakewat'e* 1279, *la Blakewater(e)* 1272×1307, *atte Blakewatere* 1327, 1327×77. ME **blak** + **water**. The spellings with medial *e* may point to the OE definite form *blace* of *blæc*. The Blackwater rises near Farnham Surrey and runs along the border between Surrey and Hants and Berks and Hants to the Loddon below Swallowfield. According to RN 36 the Blackwater has clear water but a dark bed, and is so named for contrast with the Whitewater in Hants which has greyish-white somewhat milky coloured water. Brk 7, Sr 2, RN 35.

River BLACKWATER Essex TL 8322. *le Blak Water* 1477, *Blackwater* 1576. ME **blak** + **water**. This name superseded the ancient name PANT TL 6631 as in the *Battle of Maldon* [c.1000]18th which is still used but now confined to the river above Bocking. Ess 4, 9, RN 35.

BLACKWATER RIVER Berks SU 7564, Hants SU 8360. *(la) Blakewater(e)* 1256, *Blakwatere* 1298. Blackwater, ME **blak** + **water**, is a replacement of original *Swalewe* 1272 which survives in SWALLOWFIELD Berks SU 7264. Cf. SWALLOW Humbs TA 1703. Ha 35, Gover 1958.2.

BLACKWELL 'Black, dark spring'. OE **blæc** + **wella**. L 31.

(1) ~ Derby SK 1272. *Blachewelle* 1086, *Blakewell(e)* 1101×8–1535, *Bla(c)k- Blacwell(e)* from 1234, *Bla(c)k- Blacwall(e)* 1328–1577. Db 46.

(2) ~ H&W SO 9972. *Blakewell* 1216×72. Wo 362, L 31.

BLACKWOOD HILL Staffs SJ 9255. *Blackwood(e) Hill* 1598–1842. P.n. *Blacwode* 1299–1361, *Blackwood* 1538×44–1617, 'the dark wood', ME **blake** (OE *blæc*, definite form *blaca*) + **wode** (OE *wudu*), + ModE **hill**. Oakden, Horovitz.

BLACON Ches SJ 3868. 'At the black hill'. *Blachehol* 1086, *Blachenol* 1093, *Blachenot(h)* [1096×1101]1150, [1096×1101]1280, *Blakene* c.1200–1387, *Blaken* 1262–1709, *Blacon* from 1536. OE **blæc**, dative sing. definite form **blacan**, + **cnoll** alternating with **cnotta**. The reference is to the headland of Blacon Point (cf. Point Farm SJ 3766) overlooking the river Dee. Che iv.168 gives pr ['bleikən].

BLADON Oxon SP 4414. Transferred from *Bladene*, the old name of the river EVENLODE SP 3220. *Blade* 1086, *Bladene* 1141–1797 with variants *-en(a), Bladon(')* from 1208. The river is *(innon, andlang) Bladene* [1005]12th S 911, –1363 with variants *-en(a)*, an unexplained pre-English r.n. O 7, 252.

BLAGDON 'Black hill'. OE **blæc**, definite form **blaca**, + **dūn**. L 143, 147.

(1) ~ Avon ST 5058. *Blachedone* 1086, *Blakedune* 1189×99, *Blakedon* 1212 Fees. DEPN, L 143, 147.

(2) ~ Devon SX 8560. *Blakedune -don(e)* 1242–1436. D 518.

(3) ~ HILL Somer ST 2118. P.n. Blagdon + ModE **hill**. Blagdon, *Blakedona* [1155×8]1334, *-done* 1225, *Blagdon* 1809 OS. Blagdon lies N of the BLACKDOWN HILLS. DEPN, L 143, 147.

(4) ~ LAKE Avon ST 5159. A large artificial lake formed in the early 20th century as a reservoir for Bristol. P.n. BLAGDON ST 5058 + ModE **lake**.

BLAISDON Glos SO 7017. 'Blecci's hill'. *Blechedun(a) -don(e)* 12th–1509×47, *Blechesdun(a) -don(e)* c.1200–1358, *Bleysdon* 1535, 1556, *Bleasdon* 1563. OE pers.n. **Bleċċi*, genitive sing. **Bleċċes*, + **dūn**. Gl iii.195, L 143, 149.

BLAKEBROOK H&W SO 8176. No early forms.

BLAKEDOWN H&W SO 8878. *Blake Down* 1831 OS.

BLAKE HALL Essex TL 5305. 'The black hill'. *la Blakehull(e)* 1212, c.1218, *-hall* 1222, 1329, *Blackehall* 1511. OE **blæc**, definite form **blaca**, + **hyll**. The present hall is a Queen Anne house and much post-dates the first appearance of *hall* in this name. Ess 53, Pevsner 1965.90.

BLAKEHOPE FELL Northum NY 8494. *Blakehope Fell* 1869 OS. P.n. Blakehope NY 8594, *Blakehope* 1869 OS, + ModE **fell** (ON *fjall*). Blakehope is 'the black valley', ME adj. **blake** + dial. **hope** (OE *hop*).

BLAKEMAN'S LAW Northum NY 8795. *Blakeman's Law* 1869 OS. Surname *Blakeman* + N dial. **law** 'a hill' (OE *hlāw*). Cf. BLACKMAN'S LAW NY 7498.

BLAKEMERE H&W SO 3641. 'The black mere'. *Blakemere* 1249. OE **blæc**, definite form **blaca**, + **mere**. DEPN.

BLAKENEY Glos SO 6707. '(At the) black island' or 'Blaca's island'. *Blakeneia -ey(e)* 12th–1662. Either OE **(æt thǣm) blacan ēġe**, OE adj. **blæc**, oblique case definite declension **blacan**, or pers.n. *Blaca*, genitive sing. *Blacan*, + **ēġ**. Gl iii.251, L 39.

BLAKENEY Norf TG 0243. Either the 'black isle' or 'Blaca's isle'. *Blakenye* 1242, *Blakene* 1248. OE **blæc**, definite form dative case **blacan**, or pers.n. *Blaca*, genitive sing. *Blacan*, + **ēġ**. The earlier name of this place was *Esnuterle, Snuterlea* 1086, *Sniterle* 1242, which presents the same problem as SNITTERBY Lincs SK 9894 and the other names cited there. The generic is **OE lēah** 'a wood, a clearing'. DEPN.

BLAKENEY POINT Norf TG 0046. P.n. BLAKENEY TG 0243 + ModE **point**.

BLAKENHALL Ches SJ 7247. 'At the black nook'. *Blachenhale* 1086–1581 with variants *Blaken-* (from c.1200) and *-ale*, *Blaken(h)all* from c.1188, *Blaknall* 1396–1584. OE **blæc**, dative sing. definite form **blacan**, + **halh**, dative sing. **hale**. The place lies in a valley opening into that of Checkley Brook. Che iii.51 gives prs ['blækənɔ:l] and ['blæknɔ:l], L 106, 111.

BLAKENHALL WMids SO 9197. No early forms have been found but the name shows traces of OE inflectional *-an* and might be either 'Blaca's nook', OE **Blacan-halh*, or '(the settlement) at the black nook', OE *(æt thǣm) blacan hale*.

Great BLAKENHAM Suff TM 1250. *Blakenham Magna* 1291. ModE **great**, Latin **magna**, + p.n. *Blach(eh)am* 1086, *Blakeham* 1180–1253, *Blakenham* from 1254, 'Blaca's homestead', OE pers.n. *Blaca*, genitive sing. *Blacan*, + **hām**. 'Great' for distinction from Little BLAKENHAM TM 1048. DEPN.

Little BLAKENHAM Suff TM 1048. *Blakenham parva* 1254. ModE **little**, Latin **parva**, + p.n. Blakenham as in Great BLAKENHAM TM 1250. Also known as *Blakenham sup' mont* 'B on the hill' 1610. DEPN.

BLAKESHALL H&W SO 8381. 'The black mire'. *Blakesole* [c.1190]c.1250, 1327, *Blakesal(l)* 1275, 1575, *Blackshall -soll* 17th. OE **blæc**, definite form **blace**, + **sol**. There is a deep hollow filled with black mud on the opposite side of the road from the old farm here. Wo 257.

BLAKESLEY Northants SP 6250. 'Blæcwulf's wood or clearing'. *B(l)aculveslea, Blaculveslei, Blachesleuue* 1086, *Blaculfesle(a)* 1190–1241, *-ulueslea -ley -lee* 1185–1352, *Blacoslegh* 1330, *Blakeslee* 1386, *-ley* from 1468, *Blaxley* 18th. OE pers.n. **Blæcwulf*, genitive sing. **Blæcwulfes*, + **lēah**. Nth 39 gives pr [breiksli] (sic).

BLAKEY RIDGE NYorks SE 6897. No early forms. Perhaps identical with Blakey Moor SE 8794, Blakey Topping SE 8793, *Blakehou* 1223, *Blakay* 1577, 'the black mound', ME **blake** (OE *blæc*) + **howe** (ON *haugr*). YN 94.

BLANCH FELL Lancs SD 5760. No early forms.

BLANCHLAND Northum NY 9650. 'The white launde'. Latin *Alba Landa* 1203, 1270, French *Blanchlande* [1165]copy. OFr **blanche** 'white' + **lande** 'a glade, untilled ground' whence ModE *laund, lawn*, Latin **alba** + **landa**. Cf. LAUNDE ABBEY Leic SK 7904 and Whitland Dyfed SN 1916. *White* probably refers to the bareness of newly cleared land. Blanchland was a house of Premonstratensian canons founded in 1165; it is not impossible that the name was transferred from another Premonstratensian house at Blanchelande near Coutances in Normandy. NbDu 25.

BLANCHLAND MOOR Northum NY 9553. P.n. BLANCHLAND NY 9650 + ModE **moor**.

BLAND HILL NYorks SE 2053. *Bland Hill* 1606 etc. Local family name *Bland* 1593 etc. + ModE **hill**. YW v.124.

BLANDFORD 'Gudgeon ford'. OE **blǣġe** 'a blay, a gudgeon' (a small freshwater fish), genitive pl. **blǣġna**, + **ford**. Studies 1931.62, L 71.

(1) ~ CAMP Dorset ST 9108. A modern military camp rebuilt 1964×71. Newman-Pevsner 101.

(2) ~ FORUM Dorset ST 8806. 'B with the market'. *Blaneford(e) Forum* 1297–1470, *Blandford* ~ from 1506. Also known as *Cheping Blaneford* 1288 etc. P.n. Blandford + Latin **forum**, ME **cheping**. Blandford is *Blaneford* 1086–1458, *Blandeford* 1377. 'Forum' for distinction from BLANDFORD ST MARY ST 8805, BRYANSTON ST 8706 (*Blaneford Brian* 1268, 1270), and Langton Long Blandford ST 8905, *Bleneford* 1086, *Blæneford* 1086 Exon, *Longa Blaneford(')* 1212–1428, *Lang(e)blan(e)ford(e)* 1244–1548 with addition *alias, otherwise Lang(e)ton Boteler, Botyler, Botiller* 1420–66, 'the long village held by the le Botiller family'. Do ii.87 gives pr ['bla:nvəRd], 106.

(3) ~ ST MARY Dorset ST 8805. 'St Mary's B'. *Blaneford(e) St Mary, ~ Sancte Marie* 1254–1426, *Seyntmaryblandford* 1450. P.n. Blandford + saint's name *Mary* either from the dedication of the church or because the manor belonged to the nuns of St Mary in Clerkenwell. Also known as *Parva Blaneford* 'little B' 13th cent. Earlier simply *Bleneford(e)* 1086, *Blaneford(e)* 1086–1332. Do ii.73.

BLANKNEY Lincs TF 0706. Probably 'Blanca's island'. *Blachene* 1086, *Blankenei(e) -ey* 1185–1535, *Blanken(ey)* 1206–1395, *Blaunkeney(e)* 1249–1395, *Blaunkney(e)* 1272–1556, *Blankney* from 1476. OE pers.n. **Blanca*, genitive sing. **Blancan*, + **ēġ**. An alternative suggestion is 'horse island', OE **blanca**, genitive pl. **blancena**, + **ēġ**. OE *blanca* (ME *blonke*), however, is used exclusively in poetry in the surviving literature; an OE **Blanca* would have parallels in ON *Blakki*, OHG *Blanka*. Perrott 310, L 38.

BLASHFORD Hants SU 1506. 'Blæcca's ford'. *Blachford* c.1170, *Blac(c)he- Blec(c)heford* 1248–1406, *Blaschforde* 1429. OE pers.n. **Blæċċa* + **ford**. Ha 36, Gover 1958.213.

BLASTON Leic SP 8095. Apparently 'Blath's estate'. *Bladeston(e), Blavestone* 1086, *Blaston(a)* from 1086, *Blaeston'* 1165–1473, *Bla(t)heston'* 1254–1291. The specific is best interpreted as an unrecorded pers.n. from ON *blath* 'a leaf, a blade' used as a by-name, + OE **tūn**. Lei 206, SSNEM 188.

East BLATCHINGTON ESusx TQ 4800. ModE adj. **east** + p.n. *Blechinton(e) -yng-* 1169, 13th cent., *Blachington* 1225, 14th cent., either 'estate called after Blæcca', OE pers.n. *Blæċċa* + **ing**[4] + **tūn**, or 'estate or settlement at or called **Blæċing*, the black place' < OE adj. **blæc** + **ing**[2]. 'East' for distinction from West BLATCHINGTON TQ 2706. Sx 362, SS 76.

West BLATCHINGTON ESusx TQ 2706. ModE adj. **west** + p.n. *Blacinctona* 1121, *Blech- Blachin(g)ton* 1242–1476, *Blatchington or Bletchington* 1795, either 'estate called after Blæcca', OE pers.n. *Blæċċa* + **ing**[4] + **tūn** or 'estate or settlement at or called **Blæċing*, the black place' < OE adj. **blæc** + **ing**[2]. 'West' for distinction from East BLATCHINGTON TQ 4800. Sx 290, SS 75.

BLATHERWYCKE Northants SP 9795. 'Bladder farm'. *Blarewic(he)* 1086–1175, *Blatherwyk* 12th–1436 with variant *-wik(e)*. OE **blǣdre** + **wīc**. The exact sense of *blædre* is uncertain; in a 15th cent. glossary *bledder* glosses Latin *berula*, a water plant also called bilders, one of the names of watercress. Blatherwycke lies beside Willow Brook. Nth 156.

BLAWITH Cumbr SD 2888. 'Dark wood'. *(foresta de) Blawit* 1276, *Blawith* from 1341, *Blathe* 1600. ON **blár** + **vithr**. La 214 gives prs [bla:ð] and [bla:θ], SSNNW 106, L 222.

BLAXHALL Suff TM 3657. 'Blæc's nook'. *Blac(c)hessala, Blaccheshala* 1086, *Blakeshal(e)* 1254–1327, *Blaxhal(e)* 1286–1346, *Blaxhall* 1524, *Blaxall* 1610. OE pers.n. *Blæc*, genitive sing. *Blaces*, + **halh**. DEPN, Baron.

BLAXTON SYorks SE 6700. 'The black stone'. *Blacston* 1213, *-stan* 1294, *Blak(e)stan(e)* 14th cent., *Blackston(e)* 1566–1620. OE **blæc** + **stān**. Possibly a stone marking the county boundary. YW i.45.

BLAYDON T&W NZ 1762. Uncertain. *Bladon(e)* 1303, 1340, 1613×4, 1723×4, *Bla(i)den* 1723, *Blaydon* from 1576 Saxton. This is usually taken to be 'the dark' or perhaps 'the bleak hill', ON **blá** + OE **dūn**. Such a compound, however, could hardly arise before the ME period when the ON word was naturalised in English as N dial. *bla* 'lead coloured, bleak'. This would fit the chronology of recorded sps, but possibly this is really a lost r.n. as suggested by TC identical with BLADON Oxon SP 4434 of unknown origin but pre-English. The reference would be either to Barlow Burn or possibly to the Derwent itself. NbDu 25, TC 55.

BLAZE MOSS Lancs SD 6153. No early forms but this is probably p.n. Blaze from ON **blesi** 'bare spot' + **mos** 'a bog'.

BLEA MOOR NYorks SD 7782. 'Blue moor'. *Blemor* 1293. OE **blēo** + **mōr**. YW vi.245.

BLEADON Avon ST 3456. 'Coloured hill.' *(æt) Bleodune -done, (to) Bleodunæ* [956]12th S 606, *Bleodonam* [1053]lost ECW, *Bledone* *[975]14th S 804, 1086. OE **blēo** + **dūn**. The reference is to Bleadon Hill which rises immediately N of the village and may have presented a variegated appearance due to outcrops of white limestone in the surface or to shadows cast by the evening sun. DEPN, L 142–3, 147.

BLEADON HILL Avon ST 3557. *Bleadon Hill* 1809 OS. The hill was originally simply Bleadon and when this name became transferred to the village BLEADON ST 3456 pleonastic ModE **hill** was added.

BLEAKLOW HILL Derby SK 1096. *Bleaklow Hill* 1843. P.n. Bleaklow SK 0996 + ModE **hill**. Bleaklow, *Blakelowe* 1216×72, *Blacklow* 1627, is 'the black hill', OE **blæc**, definite form **blaca**, + **hlāw**. Db 70.

BLEAN Kent TR 1261. Short for *Blean Church* 1819 OS or St Cosmas and St Damian in the Blean, *Cosmerysblen -is-* 1432, *Cosme and Dannyane Le Bleen* 1525. Blean, a district name for the old forest of Blean, is *(in, on) Blean* [724]15th S 1180, [786]12th S 125, 814 S 177, [956]12th S 608, *(to) Blean ðem wiada* 'to the wood B' 858 S 328, *(silva que dicitur) Blean* 'the wood called B' [850]13th S 300, *Blea hean hrycg* 785 S 123,

Bleanheanric [791]13th S 1614 'the high ridge of B', *Blehem* 1086, *Blien* 1190, *Bleen* 1230, *Blen* c.1090–1243, *the Blee* 14th Chaucer *Manciple's Prologue* 3, *þe Ble* 15th Lydgate *Thebes* 1047 rhyming with *se* 'see'. The OE form was *Blēa* nominative, *Blēan* oblique case, from OE ***blēa** 'coarse, rough ground' corresponding to OHG *blacha* 'coarse cloth' used topographically in the G p.ns. Blee near Düsseldorf, *Blee* 12th. It is possible that the forms in S 123 and S 1614 (which is dependent on S 123) usually taken as containing the OE adj. *hēah*, oblique case strong form *hēan*, and the DB spelling show traces of the spirant [x] from Gmc **blahwōn* and that the correct interpretation of S 123 is simply 'Blean ridge'. PNK 357, Studies 1931.60, PNE i.38, Jnl 8.17, Cullen.

BLEASBY Notts SK 7149. Either 'Blesi's village' or 'the village at or by the bare ground'. *Bleseby* 13th–1386, *Bley- Blay- Bleesby* 16th, *Bleasby* 1652. Either ON pers.n. *Blesi* or ON **blesi**, 'a blaze, the white spot (on a horse's forehead)' used topographically of a bare spot on the ground, genitive sing. *Blesa*, **blesa**, + **bȳ**. The pers.n. is not independently recorded in L or Y and is rare in Scandinavia. The issue is complicated by the existence of two OE forms for this place, *Blisetune* [958]14th S 659, and *Blisemere* [958]14th ibid. The first can be regarded as a Grimston hybrid, the second as a case of dittography due to the preceding entry *Gypesmere*. Nt 155.

BLEASDALE Lancs SD 5745. 'Valley with a bare spot'. *Blesedale* 1228, 1297, *Blesdale* 1348, *-tale* c.1540. ON **blesi** + **dalr**. Alternatively the specific could be the ON pers.n. *Blesi*. La 165, NoB 8.85, L 94, Jnl 17.97.

BLEASDALE MOORS Lancs SD 5648. P.n. BLEASDALE SD 5745 + ModE **moor(s)**.

BLEATARN 'Dark pool'. ON **blá** + **tjǫrn**. Such names are also found in Norway, *Blaatjernet*, and Sweden, *Blåtjärn*.

(1) ~ Cumbr NY 2914. *Bleaterne* 1587. Cu 32.

(2) ~ Cumbr NY 7313. *Blatern(a) -terne* 12th–1577, *-tarne* 1540–1648, *Bletarne* 1547, *Blay- Bleytarne* 1604–52, *Bleatarn* from 1621. The pool, possibly a fishpond, no longer exists. We ii.82, S 106 no.2.

BLEDINGTON Glos SP 2422. 'Settlement on the river Bladen'. *Bladintvn -yn- -ton(a)* 1086–1428, *-ing- -yng-* 1251–1771, *Bledington* from 1424. R.n. Bladen + OE **tūn**. Bladen, *Bladon* *[675]13th S 1245, *Bladaen, on bleadene* [718 for ?727]11th S 84, *Bladen* [772 for 775]11th S 109, *(innon, of, andlang) Bladene* *[777]12th S 112, [779]12th S 115, [969]11th S 1325 *(on, of) blædene* [949]c.1600 S 550, *Bladon'* 12th, *Bladen'* 1154×89–1363, is the old name of the river Evenlode of unknown origin and meaning. Gl i.3, 213, RN 36.

BLEDLOW Bucks SP 7702. 'Bledda's tumulus'. *Bleddanhlæwe* [966×75]12th S 1484, *Bleddehlæwe* [1020×38]11th S 1464, *Bledelai* 1086, *Bledelaw(e)* 1175–1361, *Bledlow* 1485. OE pers.n. **Bledda*, genitive sing. **Bleddan*, + **hlǣw**. Bk 167.

BLEDLOW RIDGE Bucks SU 7997. *Bledelowerigge* 1247, *-rugge* 1216×72. P.n. BLEDLOW SP 7702 + ME **rigge** (OE *hrycg*). Bk 168.

BLENCARN Cumbr NY 6331. 'Hill-top with a cairn'. *Blencarn(e)* from 1159. PrW ***blain** (W *blaen*) + ***carn**. Cu 214, L 128, GPC 279.

BLENCATHRA Cumbr NY 3127. *the Rackes of Blenkarthure* 1589, *Blencarter*, *Blenkarthur* 1794. PrW ***blain** (W *blaen*) 'point, top' + unexplained second element at some time associated with the pers.n. *Arthur*. *Rackes* is dial. **rack** (ON *reik*) 'a way, a path'. The name of a mountain rising to 2847ft. Also called SADDLEBACK. Cu 253, L 128, GPC 279.

BLENCOGO Cumbr NY 1948. 'Hill-top of cuckoos'. *Blencoggou* c.1190, *-cogo(u) -kogow -cogow -cogoh* 1278–1500, *Blen(e)coghou -how* c.1220–1306. PrW ***blain** (W *blaen*) + W **cog**, pl. **cogow**. Some spellings show assimilation to ON *haugr* 'a hill'. The hamlet is situated on a low hill in the Waver mosslands. Cu 122, SSNNW 420, O'Donnell 80, L 128, GPC 279, 540.

BLENCOW Cumbr NY 4532. Possibly 'hill called *Blain*'. *Blenco* 1231, *Blencou -kou -cow(e) -kowe* 1278–1608, *Blenkhaw* 1255, *-howe* 1285, 1382 *-hou* 1310, *Blenchow* 1278, *-hou(gh)* 1279, 1324. Thought to be a compound of PrW ***blain** (W *blaen*) 'a point, a summit' and ON **haugr** 'a hill', but the substitution of [k] for the [h] of *haugr* is difficult. Cu 186, SSNNW 217, L 128, GPC 279.

BLENDWORTH Hants SU 7013. 'Blædna's enclosure'. *Blednewrthie* c.1170, *Bledenewrth* 1256, *Blendeword' -worth* 1207–1341. OE pers.n. **Blǣdna* + **worthiġ** later replaced by **worth**. Ha 36, Gover 1958.54, DEPN.

River BLENG Cumbr NY 1008. 'Dark one'. *Bleng* 1576. ON **blæingr**, a derivative of ON **blár** 'dark'. This is confirmed by the form *Bleyng*[.]*it* 1391 representing *Bleyngfit*, river n. Bleng + ON **fit** 'a meadow'. Cu 5, RN 37.

BLENHEIM PALACE Oxon SP 4416. P.n. Blenheim, *rectius* Blindheim near Dillingen in Bavarian Swabia, the site of the Duke of Malborough's victory over the French and Bavarians in 1704, + ModE **palace**. O 293.

BLENNERHASSET Cumbr NY 1741. 'Hay shieling called or at *Blenner*'. *Blennerheiseta* 1188, *Blenreheyset(e) -er- -yr- -ar- -hayset -sat* 1230–1400, *Blynroset*, *Blanrasset* 17th, *Blennerhasset(t)* from 1353. P.n. **Blenner* (possibly W *blaen-dre* 'hill-farm', PrW **blain* + **treβ*) + ON **hey sætr**. Cu 265 gives pr [blin'reisit], L 128, GPC 279.

BLETCHINGDON Oxon SP 5017. 'Blæcci's hill'. *Blecesdone* 1086, *Blec(c)hesdune -don(a) -den -doun* 1123×33–1428, *Blechyndene* 1526. OE pers.n. **Blæċċi*, genitive sing. **Blæċċes*, + **dūn**. O 201, L 149–50.

BLETCHINGLEY Surrey TQ 3250. 'The wood or woodland clearing called or at *Blæccing*, the place called after Blæcci, or of the Blæccingas, the people called after Blæcca or Blæcci'. *Blachingelei* 1086–1307 with variants *-le(ha) -leg* and *-leye*, *Blecchingeleg'* 1195–1327 with variants *Blech-* and *-ynge-*, *Blescingele(ie)* 13th cent., *Blecchyng Lee* c.1482. OE p.n. **Blæċċing*, locative-dative sing. **Blæċċinge* < pers.n. *Blæċċa* or *Blæċċi* + **ing**[2], or folk-n. **Blæċċingas* < pers.n. *Blæċċa* or *Blæċċi* + **ingas**, genitive pl. **Blæċċinga*, + **lēah**. Sr 308.

BLETCHLEY Bucks SP 8634. 'Blæcca's wood or clearing'. *Blechelai* 1106×9, 1152×8, *Blacchelai* 1155, *Bletchele* 1227, 1380. OE pers.n. *Blæċċa* + **lēah**. Bk 17, DEPN, L 203.

BLETCHLEY Shrops SJ 6233. 'Blecca or Blecci's wood or clearing'. *Blecheslee* 1222, *Blec(c)heleg* 1254–5, *Blecheley* 1309, 1534×5. OE pers.n., *Bleċċa* + **lēah**. The earliest spelling suggests that a by-form of the pers.n. **Bleċċi*, genitive sing. *Bleċċes*, may have existed. Bowcock 46, DEPN.

BLETSOE Beds TL 0258. 'Blecci's hill-spur'. *Ble- Blacheshou* 1086, *Blettesho* c.1230, *Blech- Blek- Blethenesho* 13th. OE pers.n. **Bleċċi*, genitive sing. **Bleċċes*, varying with **Bleċċīn*, genitive sing. **Bleċċnes*, + **hōh**. The development of *t* in this name is due to simplification of the cluster [tʃs] to [ts]. Bd 26, Studies 1936.5, L 168.

BLEWBURY Oxon SU 5385. '(The settlement) at the hill-fort on bluish-white soil'. *Bleoburg, (to) bleobyrig* [944 ? for 942]13th S 496, *Bleobirie -buria -beria -bir' -bury* [1144]13th–1380, *Blidberia*, *Blitberie* 1086, *Bleubiri* [1091]13th, *Bleberi(a) -ber(y) -bir(i) -biry -byry -byre -ber(y) -bur(y)* 1144–1377. OE **blēo** + **byriġ**, dative sing. of **burh**. The reference is to the hill-fort on Blewburton Hill. Brk 151.

BLICKLING Norf TG 1728. 'The Bliclingas, the people called after Blicla'. *Blikelinga*, *Bliclinga*, *Blikelinges* 1086, *Bliccling* 1166, *Blicling'* 1218. It is not absolutely clear whether this is actually OE *Blicling* 'the place called after Blicla', OE pers.n. **Blicla* + **ing**[2], or as seems likely, OE *Bliclingas* 'the people called after Blicla', pers.n. **Blicla* + **ingas**. ING 56.

BLIDWORTH Notts SK 5956. 'Blitha's enclosure'. *Blideworde* 1086, *-wurda* 1186, *Bleðwurða*, *Blede- Bleðewurda* 1162–5, *Blithewurth' -wrth -worth'* 1240–1332, *Blythworth* 1335, *Blideth*

1670. OE pers.n. *Blītha* + **worth**. Nt 115 gives local prs [blidəθ, blidəf].

BLINDBURN Northum NT 8210. 'The hidden stream'. *Blindburn* 1869 OS. ModE **blind** + **burn**.

BLINDCRAKE Cumbr NY 1434. 'Hill-top of the rock'. *Blenecreyc* 1154×89, *-crayc* 1268, *Blencraic -creic -crayc -crake -kerck* 12th–1610, *Blyncrake* 1555, *Blindcraick* 1567. PrW ***blain** (W *blaen*) + ***creig** (W *craig*, OW *creic*). Possibly an earlier name of MOOTA HILL. Cu 266 gives pr [blin'kreiək], L 128, GPC 279, 578.

BLINDLEY HEATH Surrey TQ 3646. *Blyndley Heathe* 1559, *Blyndlye* ~ 1585. P.n. Blindley, possibly a compound of OE **blind** 'blind, concealed, dark, and **lēah**, + ModE **heath**. Sr 317.

BLISLAND Corn SX 1073. Partly obscure. *Glustone* (sic for *Blus-*) 1086, *Bloiston* 1177, 1198, *Bl(i)eston* 1195, *Bliston* 1291–1438 G, *Bleselond* 1284, *Blislonde* 1300. Unidentified element + OE **tūn**, 'farm, village', later replaced with **land**. PNCo 54, DEPN.

BLISS GATE H&W SO 7472. *Bliss Gate* 1832 OS. A gate onto Rock Common.

BLISSFORD Hants SU 1713. A corruption of earlier *(la) Bisseford* 1199×1216, *Byseford* 1298, *Blesford alias Blisford* 1670, 'ford across the r. Biss'. This r.n. is not on record here but would be identical with r. Biss Wilts ST 8656, *(on) bis* [964]14th S 727, *Byssi, Bissy* [1001]15th S 899, *le Bisse* 1468, *The Biss* 1575, of unknown origin, possibly related to W *bys*, Brit **bissi-* 'finger'. Gover 1958.214, Wlt 2, RN 34, GPC 367.

BLISWORTH Northants SP 7253. 'Blithi's enclosure'. *Blidesworde* 1086, 1205, *-worth* 1216, *Blythes- Blithesworth* 13th cent., *Blise- Blyseworth* 1166–1499, *Blisworth* from 1428. OE pers.n. **Blīth(i)*, genitive sing. **Blīthes*, + **worth**. Nth 143.

River BLITHE Staffs SK 0822. 'The gentle one'. *bliðe, (up æfter) bliþe* 996 Charters II.27, *Blithe* from 12th, *Blythe* 1284, 1577. OE **blīthe**. St i.5, RN 38.

BLITHFIELD RESERVOIR Staffs SK 0524. P.n. *Blidevelt* 1086, *Blithefeld* 1257, 1388, 'the open land by the r. Blithe', r.n. BLITHE SK 0822 + OE **feld**, + ModE **reservoir**. RN 38, L 240, 243.

BLITTERLEES Cumbr NY 1052 → LEES SCAR LIGHTHOUSE NY 0953.

BLOCKLEY Glos SP 1635. 'Blocca's wood or clearing'. *Bloccanleah* [855]11th S 207, *-lea* [978]11th S 1337, *Bleccelea* (sic) [964]12th S 731, *Blochelei* 1086, *Blockele(ia) -ley(e)* 12th–1348. OE **Blocca*, genitive sing. **Bloccan*, + **lēah**. Gl i.234.

BLOFIELD Norf TG 3309. Possibly 'blue, i.e. bleak open land'. *Bla(uue)felda, Blawefelle* 1086, *Blafeld* 1156, *Blofeld* 1294. OE adj. *blāw* is not independently attested. It occurs as a noun meaning 'pigment' in Erfurt Gloss no. 1152 *blata, pigmentum* 'purple dye, dye, grey dye': *haui-blauum* 'iron-coloured, bluish, grey dye'. The adjectival form is *blǣwen* WW 162 which is probably a contracted form of *blǣhǣwen* Lev 8 'light blue'. The cognate ON adj. *blá(r)*, which seems to have influenced this name, means 'dark blue, livid'. PNE i.38 suggests the sense 'cold, cheerless' for the word in p.ns. but this is only a guess. DEPN suggests the noun *blāw* may have had the sense 'woad' and compares the G p.n. Blaufelden near Crailsheim, Baden-Württemberg, *Blauelden* 12th. Blaufelden, however, lies on the same stream as a village called Blaubach of which the specific is clearly the stream name **Blāwa* 'the blue one', as also in Blaubeuren, Baden-Württemberg, *Blabivron* [1175×8]c.1300. Possibly, therefore, Blofield is the 'open land by the river **Blāwe*'. Cf. the stream names *Blabec* discussed in RN 36. DEPN, PNE i.38, Reichardt 1986.56ff, V i.109.

BLOODYBUSH EDGE Northum NT 9014. *Bloodybush Edge* 1869 OS. One of the highest hills of Kidland rising to 2001ft. The *Alnwick Gazette* for July 1893 comments: 'The name is suggestive, and almost certainly points to a sanguinary encounter having taken place there in those bygone days which are spoken of in the locality as "times of trouble". Who were the combatants, which side was victorious, are problems which probably defy all historic research.' Possibly related to the events of 28 July 1585 when Lord Francis Russell was killed at a meeting of the Border Wardens at Windy Gyle and the Scots chased the Englishmen for several miles into their own country. Dixon 78, 83.

BLORE Staffs SK 1349. Unexplained. *Blora* 1086–1203, *Blore* from 1199. OE ***blōr** of unknown meaning. This element has been linked to ME *blure* 'a blister' and thought to be used as a term for a hill, but this is very uncertain. Oakden, PNE i.39, V i.117.

BLOXHAM Oxon SP 4235. 'Blocc's homestead'. *Bloc(c)esham* 1086–1183, *Bloxham* from 1227. OE pers.n. **Blocc*, genitive sing. **Blocces*, + **hām**. O 394.

BLOXWICH WMids SK 0002. 'Blocc's dairy farm or trading place'. *Blocheswic* 1086, *Blokswich* 1271. OE pers.n. **Blocc*, genitive sing. **Blocces*, + **wīċ**. DEPN, Nomina 11.93.

BLOXWORTH Dorset SY 8894. 'Blocc's enclosure'. *(in) Blacewyrðe* *[987]13th S 1217, *Blocheshorde* 1086, *Blo(c)kesw(u)rth(e) -worth(e)* 1200–1465, *Bloxworth(e)* from 1440. OE pers.n. **Blocc*, genitive sing. **Blocces*, + **worth**. Do ii.75.

BLUBBERHOUSES NYorks SE 1755. Either 'houses associated with the Bluber family' or 'houses by the bubbling spring'. *Bluberh(o)usum* 1172, 1203×15, *Bluberhus(e) -hous* 1172–1244, *-huses* 1279×81. By-n. *Bluber* as Walter Bluber 1229 or ME ***bluber** + **hus**, dative pl. **husum**, ME nominative pl. **huses**. YW v.120, xii.

BLUE ANCHOR Somer ST 0243. *Blew Anchor* 1750 map, *Blue Anchor* 1809 OS. The name of an inn here.

BLUE ANCHOR BAY Somer ST 0145. P.n. BLUE ANCHOR ST 0243 + ModE **bay**.

BLUE BELL HILL Kent TQ 7462. This seems to be what is marked as *Up^r Bell* 1819. ModE **bell** 'a rounded hill'.

BLUNDELL SANDS Mers SJ 3099. Named from INCE BLUNDELL SD 3203.

BLUNDESTON Suff TM 5195. 'Blund's estate'. *Blundeston* from 1203, *-is-* 1275, *Blunteston -tun* 1205, 1229. ME nick-n. *Blund* < AN *blunt, bl(o)und* 'fair-haired, blond', genitive sing. *Blundes*, + **tūn**. DEPN, Baron.

BLUNHAM Beds TL 1551. 'Bluwa's homestead'. *Blunham* 1086. OE pers.n. **Bluwa*, genitive sing. **Bluwan*, + **hām**. Bd 88.

BLUNSDON ST ANDREW Wilts SU 1389. 'St Andrew's B'. *Bluntesdon Seynt Andreu* 1281, *Blounesdon Andrew* 1429, *Androblunsdon* 1544. P.n. *Blu- Blontesdone* 1086, *Bluntesdun(a) -don* 1204–42, 'Blunt's hill', pers.n. **Blunt*, genitive sing. **Bluntes*, + **dūn**, + saint's name *Andrew* from the dedication of the church. The origin of this pers.n. is uncertain. There appears to have been an OE pers.n. **Blunt* as in *Bluntesig* [988]11th S 1358, possibly related to ON *blundr* 'dozy' used as a nick-n., which is the source of ME *blunt*, and there is OFr *blund, blond* 'fair-haired', the source of the ModE surname Blunt, Blount. Any one of these three, OE **Blunt*, ME *blunt* used as a nickname or OFr *blund*, similarly used, could be at issue here. 'St Andrew' for distinction from Broad BLUNSDON SU 1590. Wlt 30 gives former pr [blʌnsən].

Broad BLUNSDON Wilts SU 1590. *Bradebluntesdon(e)* 1234, *Brode-* 1263, 1279. ME adj. **brode** 'great' or 'chief' + p.n. Blunsdon as in BLUNSDON ST ANDREW SU 1389. Wlt 31.

BLUNTISHAM Cambs TL 3647. 'Blunt's homestead'. *Bluntesham* from 1086 including [1042×66]12th S 1051, *Bluntsome* 1545, *Blunsham* 1558×1603. OE pers.n. **Blunt*, genitive sing. **Bluntes*, + **hām**. The same person is probably referred to in *Bluntesdiche* 'Blunt's dike' in nearby Neddingworth. Hu 204 gives prs [blʌntsəm, blʌntʃəm].

BLYBOROUGH Lincs SK 9394. 'Blitha's fortified place'. *Blibvrg* 1086, c.1200, *-burc* c.1115, *Blieburc* 1181, 1203, *Blitheburc'*

-burgh' 12th, 1286. OE pers.n. **Blītha* + **burh**. DEPN, Cameron 1998.

BLYFORD Suff TM 4276. 'Ford across the r. Blyth'. *Blitleford* (for *Blithe-*) [mid 11th]13th S 1516, *Blideforda* 1086, *Bliford(e)* 1254–1336, *Blyford* 1568. R.n. BLYTH TM 2967 + OE **ford**. RN 38, DEPN, Baron.

BLYMHILL Staffs SJ 8112. 'Wild plum-tree hill'. *Brumhelle* 1086, *Blumenhull(e)* 1218–1420, *Bly- Blimenhul(l)* 1236–1375, *Blymehull -hill* 1428, *Plymylle* 1423, *Blymhill* from 1485. OE **plȳme** varying with ***plȳmen** 'growing with plum-trees' + **hyll**. Voicing of initial *p* > *b* is not uncommon in the dial. of Staffs. St i.128, L 171, Horovitz.

BLYTH Northum NZ 3181. Transferred from the r.n. BLYTHE NZ 1877. The village was properly *Snoc de Bliemue* 1208, *le Blithsnoke* 1386, 'Blythemouth snook', p.n. Blythemouth + OE ***snōc** 'a point, a projection, a tongue of land at the mouth of a river'. Blythemouth, *Blithemuthe* [1236]14th is the 'mouth of the r. Blythe', r.n. BLYTHE + OE **mūtha**. NbDu 26.

BLYTH Notts SK 6286. A transfer from the old name of the river Ryton, *Blide* 1086–1335 with variants *Bli(d)a, Blith* 1226, 1233, *Blythe* 1241–57, the 'pleasant river', OE **blīthe**. Nt 68.

River BLYTH Suff TM 2967. 'The gentle or merry one'. *Blith* 1586. OE **blīthe**. Although the evidence is late the r.n. is an early one as is shown by the p.ns. BLYFORD TM 4276 and BLYTHBURGH TM 4575 and the name of the district around the river, Blything Hundred, *Blidinga* 1086, *Bliðingehundr'* 1176, *Blythyng* 1275, '(the hundred of) the dwellers by the r. Blyth', r.n. Blyth + **ingas**, or 'the r. Blyth district', r.n. Blyth + **ing**[2]. RN 38, DEPN, Baron.

BLYTHBURGH Suff TM 4575. 'The fortified place on the r. Blyth'. *blideburh -burc, bledeburc* 1086, *Blieburc* 1159–1209, *Bliburgh(')* 1235–1453, *Blythburg* 1523×4. R.n. BLYTH TM 2967 + OE **burh**. RN 38, DEPN, Baron.

River BLYTHE Northum NZ 1877. 'The gentle one'. *Blida* 1130 P, *Blitha* [1133×40, 1153×9]14th, 1204 Ch, *Blye* 1257, *Blythe* from 1267. OE **blīthe**. Forms without *th* are due to Anglo-Norman influence. NbDu 26, DEPN, RN 39.

River BLYTHE Warw SP 2185. 'Merry, pleasant, gentle one'. *Blitha* 1154×89, *Blithe* 1276, 1316, c.1540, *Blythe* 1299, 1365. OE **blīthe**. Wa 2, RN 38.

BLYTHE BRIDGE Staffs SJ 9541. 'The bridge over the r. Blithe'. *Blythbryge* 1475. R.n. BLITHE SK 0822 + ME **brigge**.

BLYTHE SANDS Kent TQ 7580. No early forms.

BLYTON Lincs SK 8594. Partly uncertain. *Blit(t)one* 1086, *Blituna* c.1115, *Bliton* 1223. Blyton is 5 m. E of BLYBOROUGH SK 9394 with which it probably shares the same specific. Alternatively this could be 'Bligr's farm or village', ON pers.n. *Blígr* + **tūn**. DEPN, SSNEM 188.

BOARHUNT Hants SU 6008. 'The spring of the fortification or town'. *(æt) Byrhfunt'* [10th]12th S 1821, *Bor(e)hunte* 1086–1316, *Burhunt(e)* 1170–1316, *Burrant* 1544. OE **burh**, genitive sing. **byrh**, + ***funta**. The spring is that at Offwell farm SU 6207. Ha 36, Gover 1958.18, DEPN, Signposts 84, 86.

North BOARHUNT Hants SU 6010. ModE adj. **north** + p.n. BOARHUNT SU 6008.

BOARSHEAD ESusx TQ 5332. *Boreshead* 1556. ModE **boar's** + **head**. The reference is apparently to a rock resembling an animal head in the grounds of the house here. Sx 377.

BOARSTALL Bucks SP 6214. 'Fort or manor house site'. *Burchestala* 1158–9, *Borestall* 1175, *Bostall* 1545. OE **burh-steall**. Bk 115.

BOASLEY CROSS Devon SX 5093. P.n. Boasley + ModE **cross** 'a cross-roads'. Boasley, *(æt) borslea* c.970 B 1247, *Borsleia -legh* 1316, *Boslie* 1086 DB, *-leia* 1086 Exon, *Borseley* 1733, is 'clearing where prickly shrubs grow', OE ***bors** + **lēah**. D 174 gives pr [bouzli].

BOBBING Kent TQ 8864. 'The Bobbingas, the people called after Bobba'. *Bobinge* 11th, [1234]1329, *Bobbing'* 1205–26, *Bobbinges* 1179, 1234, 1240, *Bobing'* 1243. OE folk-n. **Bobbingas* < OE pers.n. *Bobba* + **ingas**. PNK 243, KPN 82, ING 9.

BOBBINGTON Staffs SO 8090. 'The estate called after Bubba'. *Bubintone* 1086, *Bubington* 1236, *Bovington otherwise Bubbington otherwise Bublington* 1600. OE pers.n. *Bubba* + **ing**[4] + **tūn**. DEPN, Horovitz.

BOBBINGWORTH Essex TL 5305. 'Enclosure called or at Bubbing, the place called after Bubba'. *Bubingeordā* 1086, *Bu- Bob(b)ing(e)w(u)rth(e) -worth(e)* 1243–1474, *Bovingworth* 1561, *Bovenger -inger* 1607–1706. OE p.n. **Bubbing* < pers.n. *Bubba* + **ing**[2], locative-dative sing. **Bubbinġe*, + **worth**. Ess 52 gives pr [bʌvindʒə], BzN 1967.364, 1968.144.

BOCADDON Corn SX 1758. 'Dwelling of Cadwen'. *Bodkadwen* 1315, *Bocadwen* 1507, *Bocaddon* 1636 G. Co ***bod** + pers.n. **Cadwen*. PNCo 55.

BOCKHAMPTON Dorset SY 7291. Probably 'settlement of the *Bochœme*, the people living at **Bochamtun* or **Bocland*'. *Bochehamtone* 1086, *Boc- Bokham(p)ton(e)* 1228–1407, *Bu(c)khampton* 1244, *Bock- Book- Beak- Boakhampton, Bockington* 17th.† OE folk-n. **Bōchǣme* (OE **bōc** + **hǣme**), genitive pl. **Bōchǣma*, + **tūn**. *Boc-* must be short for an unrecorded p.n. *Bochamtun* 'beech-tree farm' or *Bocland* 'land granted by charter'. Do i.367.

BOCKING Essex TL 7623. 'Boccing, the place called after Bocca' and 'the Boccingas, the people called after Bocca or living at Boccing'. *(on, æt) Boccinge* [961×95]11th S 1501, 995×9 S 939, *Boccinges* [995×9]12th S 1218, *Boccing* 1042×66 S 1047, *Bochinges* 1086, 1155×8, *Bocking* from 1255. This seems to be OE p.n. **Boccing* < pers.n. **Bocca* + **ing**[2] varying with folk-n. **Boccingas* < **Bocca* + **ingas**. Modern Bocking at TL 7623 is not the original site of the settlement which was rather at BOCKING CHURCHSTREET TL 7525. Modern Bocking is actually short for *Bocking Street* 1805 OS, 'Bocking hamlet'. Ess 413 gives pr [bɔkən], lxi, ING 189, BzN 1967.332.

BOCKING CHURCHSTREET Essex TL 7525. P.n. BOCKING TL 7623 + p.n. *le Cherchestrete* 1389, *Church Str.*[t] 1805. This is the original settlement site called Churchstreet, the 'church hamlet', ME **chirche** + **strete**, for distinction from *Bocking Street* 1805, now simply BOCKING. Ess 415.

BOCONNOC Corn SX 1460. 'Dwelling of Conec or Kenec'. *Bochenod, Botchonod* 1086, *Botkennoc* [1241]14th, *Boskennec* 1282, *Bokonnecke, Boccunnek* 1310ff, *Bodkonok* 1331. Co ***bod** + pers.n. *Conek* or *Kenec*. PNCo 55, DEPN.

BODDINGTON Glos SO 8925. 'Estate called after Bota'. *Botingtvne* 1086, *Botintun' -ton(e) -yn-* 1086–1385, *-yng-* 1271, 1287, *Bodin(g)ton -yn(g)-* 1248–c.1560. OE pers.n. *Bōta* + **ing**[4] + **tūn**. Gl ii.76.

Lower BODDINGTON Northants SP 4852. ModE adj. **lower** + p.n. *Botendon(e)* 1086–1235, *Botindon -y-* 1199–1368, *Budinton* 1244, *Bodynton* 1309, *Botyngdon* 1358, *-ton* 1428, 'Bota's hill', OE pers.n. *Bōta*, genitive sing. *Bōtan*, + **dūn**. 'Lower' for distinction from Upper BODDINGTON SP 4853.

Upper BODDINGTON Northants SP 4853. *Uvrebotindon* 1261, *Over Botingden* 1289. ME adj. **over**, **upper** + p.n. Boddington as in Lower BODDINGTON SP 4852. Nth 33.

BODENHAM H&W SO 5351. Probably 'Boda's river-bend land'. *Bodehā -ham* 1086, 1180, *Bodenham* 1249. OE pers.n. **Boda*, genitive sing. **Bodan*, + **hamm** 1, or possibly **hām** 'a homestead'. Bodenham lies in a large bend of the river Lugg. Also known as *Bodenham Devereux* 1588 from the 1086 tenant William, ancestor of the Devereux family from Évreux, Eure, Normandy. DEPN, He 36, L 47.

BODENHAM Wilts SU 1626. 'Bota or Botta's homestead or promontory'. *Boterham* 1209, *Bote(n)ham* 1249–1370, *Botte(n)ham* 1255–1345, *Botnam* 1480, *Bodenham* 1695. OE

†Identification of the form *(æt) Buchœmatune* 1001×12 S 1383 with this place is uncertain. Writs 484.

pers.n. *Bōta* or *Botta*, genitive sing. *Bōtan, Bottan,* + **hām** or **hamm**. Bodenham's location on a spur in the angle between the rivers Ebble and Avon suggests *hamm* 2a 'a promontory of land into marsh or water' or 2b 'land on a hill-spur'. Wlt 393 gives pr [bɔd(ə)nəm].

BODHAM Norf TG 1240. 'Boda's homestead'. *Bod(en)ham* 1086, *Bodeham* 1175. OE. pers.n. *Boda*, genitive sing. *Bodan*, + **hām**. DEPN.

BODIAM ESusx TQ 7826. 'Boda's promontory'. *Bodeham* 1086, *Bodiham -y-* 1166–16th cent. with variant *-hamme* 13th, 14th cents., *Bodgiham* 1610. OE pers.n. *Boda*, genitive sing. *Bodan*, + **hamm** 2. Beside the genitive sing. compound **Bodanhamm*, ME *Bode(n)- Bodi(n)ham*, an *-ing-*[4] formation may also have existed, **Bodinghamm* '*hamm* called after Boda'. The reference is to a promontory of land between Kent Ditch and the River Rother. Sx 518, ASE 2.25, SS 83.

BODICOTE Oxon SP 4638. 'The cottage(s) called after Boda'. *Bodicote* from 1086 with variants *-cot(a) -kot'*. OE pers.n. *Boda* + **ing**[4] + **cot**, pl. **cotu**. O 395.

BODIEVE Corn SW 9973. 'Lord's dwelling'. *Boduff* 1302, *Bodyuf* 1323, *Bodeff* 1425, *Bedeve* 1584 G, 1610. Co ***bod** + ***yuf**. PNCo 55.

BODLE STREET GREEN ESusx TQ 6514. P.n. Bodle Street + ModE **green**. Bodle Street, *Bodylstret* 1539, *Bodell Strete* 1588, is 'Bodle hamlet', surname *Bodle* + ModE dial. **street** 'a hamlet'. A Potemen Bothel is recorded in the 13th cent. and the surname recurs 1318–46. Sx 480.

BODMIN Corn SX 0767. 'Dwelling by church-land'. *Bodmine* c.975, 1086–15th with variant *-myne* G, *Botmenei* [c.1100]c.1200, *Bodmen* 1253, *Bodman* 1337, *Bodmyn* 1552. Co ***bod** + ***meneghy**. The development of this name is due to shift of stress from the penultimate syllable (**bod-menéghy*) to the first syllable Bódmin. Also known as *Petrocys stowe, Paetrocysstow* 'St Petroc's holy place' in 10th cent. manumissions (G): his relics were translated here c.1000. PNCo 55.

BODMIN MOOR Corn SX 1876. *Bodmin Moor* 1813. P.n. BODMIN SX 0767 + **mōr**. An invented name replacing *Foy Moor* 1844, *Fawimore* 1185, *Fowymor(e)* 1347, 1355, r.n. FOWEY SX 0962 + **mōr**. Also simply *The Moares* 1610. PNCo 55.

BODMIN PARKWAY STATION Corn SX 1164. A modern euphemism for a railway station distant from the town it serves. Formerly *Bodmin Road Station*, 'railway halt at the road to Bodmin'. PNCo 55.

BODNEY Norf TL 8398. 'Beoda's island'. *Bu- Bodeneia* 1086, *Bodeneie* 1199, *Be- Bodeneye* 1254. OE pers.n. **Beoda*, genitive sing. **Beodan*, + **ēġ** probably in the sense 'island of dry ground in marsh'. DEPN, L 38.

The BOG Shrops SO 3598. *The Bog* 1836 OS.

BOGNOR REGIS WSusx SZ 9399. 'The king's B'. P.n. *Bucgan ora* *[680? for 685]10th S 230, *Bug(g)enor(e)* 1275, *Bogenor(e)* 1270–1405, 'Bucge's flat-topped hill', OE feminine pers.n. *Bucge*, genitive sing. *Bucgan*, + **ōra**, + Latin **regis**, genitive sing. of **rex** alluding to the convalescence of George V here in 1929. Sx 92, SS 73, Jnl 21.18, Room 1983.11.

BOHORTHA Corn SW 8632. 'Cow-yards'. *Byhurthw* 1284, *Behorthou* 1302, *Bohurthowe* c.1500, *Bohurra -hurrowe* 1604. Co ***buorth**, pl. ***buorthow**. PNCo 56, Gover n.d. 433.

BOJEWYAN Corn SW 3934. Possibly 'dwelling of Uyon'. *Bosuyon* 1302, 1327, *Bojewyan alias Bosuyan* 1699. Co ***bod** + pers.n. **Uyon*. The name exhibits the normal development of medial *d* to /dj/ and /z/. PNCo 56.

BOLAM Durham NZ 1922. '(At the) rounded hills'. *Bolum* 1235, 1242×3, *Bolom* 1303–1519, *Bolam* from 1366. OE ***bol**, dative pl. ***bolum**. Bolam is situated on undulating ground overlooking the Tees lowlands. On formal grounds, however, the specific could also be the dative pl. of OE ***bola** 'a tree-trunk' and the name refer to woodland clearance. DEPN.

BOLAM Northum NZ 0982. Either '(the settlement) at the rounded hills' or 'at the tree-trunks'. *Bolum* c.1155–1339, *Boolun -on* 12th cent., *Bolom* 1324, 1507. OE **bolum**, dative plural of either OE ***bol** 'a smooth rounded hill' or **bola** 'a bole, a tree-trunk'. Either explanation is possible for this name. DEPN, NbDu 27.

BOLAM LAKE Northum NZ 0881. P.n. BOLAM NZ 0982 + ModE **lake**.

Great BOLAS Shrops SJ 6621. *Bowlas Magna* 1655, *Greate Bolas* 1672, *Big Bolas* 1755. ModE **great**, Latin **magna** + p.n. *Belewas* 1198, *Boulewas* [1199]1265–1443, *Boulwas* 1316–1437, *Bolas* 1520, *Bolus* 1750, of uncertain origin, possibly OE ***bogel** 'a little bend' + **wæsse** 'riverside land which floods and drains rapidly'. The reference is to a former bend in the river Tern now straightened. 'Great' for distinction from Little Bolas SJ 6421, *Parva Boulewas* 1342, *Little Bolas* 1705. Sa i.51.

BOLBERRY Devon SX 6939 → BOLT HEAD SX 7236.

BOLD HEATH Mers SJ 5389. P.n. Bold + ModE **heath**. Bold, *Bolde* 1204–12 etc., *Bold* from 1257, is OE **bold** 'dwelling, house, palace'. La 107, Jnl 17.60.

BOLDERFORD Hants SU 2904 → BOLDRE SZ 3198.

BOLDERWOOD Hants SU 2407 → BOLDRE SZ 3198.

BOLDON T&W NZ 3561. 'The round hill'. *Boldune -(a)* 1133×44–[1312]15th, *Boldon(')* from 1211. OE ***bol** + **dūn**. Boldon village lies half a mile N of a distinctive round hill at NZ 3560. NbDu 27.

BOLDON COLLIERY T&W NZ 3462. A modern mining settlement developed after 1863. P.n. BOLDON NZ 3561 + ModE **colliery**.

BOLDRE Hants SZ 3198. Uncertain. *Bolra* 1135×54, *Bolre* 1152–1341, *Balre* 1236, *Boldre* 1324, 1610. The origin of this name is very uncertain. Connection with Mod dial. *bo(u)lder* sb.2 'bulrush' is hazardous but possibly suggests that *Bolre* was originally a name of the Lymington river also preserved in Bolderford SU 2904, *Bovreford* 1086, *Bolreford* 1291, the 'ford leading to Boldre', and in Bolderwood SU 2407, *Bolrewode* 1331. On this assumption the DB forms *Bovre* (2×) and *Bovreford* must show early vocalisation of *-l-* to *-u-* rather than representing OE *bī ofre* 'by the flat-topped ridge'. Ha 36, Gover 1958.203–4.

East BOLDRE Hants SU 3700. ModE **east** + p.n. BOLDRE SZ 3198. East Boldre is at the extreme E of the parish of Boldre. Ha 36, Gover 1958.203.

BOLDRON Durham NZ 0314. 'Forest clearing where steers are kept'. *Bolrum* [1175×88]15th, *Bolerum* 1204, [1258]13th, *Bolron* 1285, [1302]15th, 1577, *Boldron* from 1285, *Boldram* 1564. ON **boli** + **rúm**. YN 303.

BOLE Notts SK 7987. Probably '(the settlement) at the tree-tunks'. *Bolun* 1086, 1212, 1299, *-um* 1202–1373, *Bole* from 1290, *Boole* 1379, *Boyl(l)e, Booll* 16th cent., *Boalle* 1638. OE ***bola** or ON **bolr**, locative-dative pl. **bolum**. On formal grounds the specific could be OE ***bol** 'a smooth rounded hill', dative pl. ***bolum**. Nt 25, SSNEM 152.

BOLEHILL Derby SK 2955. *Boale Hill* 1599, *Bolehill* 1620. ME ***bole** 'place where ore was smelted' (? < OE *bolla* 'a bowl') + **hill**. Before the invention of furnaces ore was often smelted in a round cavity on the top of a hill. Db 416, 673.

BOLHAM Devon SS 9514. 'Bola's homestead'. *Bolehā* 1086, *Bolleham* 1175–1285, *Bolham* 1284, *Boldham* 1675. OE pers.n. *Bola* + **hām**. D 541.

BOLHAM WATER Devon ST 1612. P.n. Bolham + ModE **water** 'stream, river'. Bolham, *Boleham* 1086, 1227, *Bolleham* 1215–56, *Bolham* from 1228, *Bollam* 1290, is 'Bola's homestead', OE pers.n. *Bol(l)a* + **hām** or possibly **hamm** 'a water meadow'. D 610.

New BOLINGBROKE Lincs TF 3057. *New Bolingbroke* 1824 OS. ModE **new** + p.n. Bolingbroke as in Old BOLINGBROKE TF 3565. A new settlement founded c.1817 by John Parkinson 'upon that part of the West Fen which was allotted at the enclosure to Bolingbroke parish'. Room 1983.76.

Old BOLINGBROKE Lincs TF 3565. '*Bolin* stream'. *Bolinbroc -im-* 1086, *Bulincbroc* c.1156, *Bulingbroc* 1202. R.n. *Bolin* as in River BOLLIN Chesh SJ 7785 + OE **brōc**. DEPN.

BOLINGEY Corn SW 7653. 'Mill-house'. *Mellingy* 1516 G, *Velingey* 1566, *Melinge* 1650, *Bollingey* 1732 G. Co ***melyn-jy** (<*melin* + *chy*) with hyper-correct de-mutation of *V-*, *v* being the mutated form of both *b* and *m*. PNCo 56.

BOLLIHOPE COMMON Durham NY 9834. P.n. Bollihope + ModE **common**. Bollihope NY 0034 is *Bothelinghoppe* [c1294]18th, *Bolyhop* 1416, *Baly- Bailihop* late 15th. In Bollihope were *Bolyhopsheles* 1377, *Bolyopshele* 1382, 'Bollihope shieling(s)'. Possibly ME ***botheling**, an unrecorded diminutive of ME *bothe* 'a hut', + **hop**, 'side-valley with a temporary shelter or herdsman's hut'. NbDu 27.

River BOLLIN Ches SJ 7785. Uncertain. *Bolyne* 13th–1415, *Bolin -yn* 1210–1621, *Bollin* from 13th. Possibly 'eel river', OE **bōl** + **hlynn** 'a torrent, a noisy stream'. However, *bōl* occurs only once glossing MedLat *mūrenula* 'little eel or lamprey' and its meaning and status is very uncertain. Che i.15, RN 40, ES 40.236–7.

BOLLINGTON 'Settlement on the river Bollin'. R.n. BOLLIN + OE **tūn**.

(1) ~ Ches SJ 7286. *Bolintun* 13th, *-ton(a)* c.1233, *-yn-* 1312–1431, *Bolington -yng-* 1287, 1353, *Bollington* from 1584. Che ii.43.

(2) ~ Ches SJ 9377. *Bolynton* 1288–1454, *Bolington(e) -yng-* 1285 etc., *Bollington* 1559. The river on which Bollington stands is now the Dean, a tributary of the Bollin whose name it must once have shared. Che i.187. Addenda gives 19th cent. local pr ['bɔlitn], now ['bɔlintən, 'bɔlin̥tən].

BOLNEY WSusx TQ 2623. 'Bola's island'. *Bolneye* 1263, 1280, *Bolene(y) -ye* 1271–1332, *Boney* 1621. OE pers.n. **Bola*, genitive sing. **Bolan*, + **ēġ** in the sense 'patch of dry gound in marsh'. Sx 257 gives pr [bouni], SS 70.

BOLNHURST Beds TL 0859. 'Bulls' wooded hill'. *Bo- Bulehestre* 1086, *Bollenhirst* [1066×87]14th, *Bolnhurst* 13th. OE **bula**, genitive pl. **bulena**, + **hyrst**. Bd 13, L 197.

BOLSOVER Derby SK 4770. 'Bull or Boll's promontory'. *Belesovre* 1086, *Bellesores* 1199, *Bol(l)is- Bol(l)es(h)ou(e)r(e) -v- -(h)oura -(h)or(e) -(h)o(u)res* 1149×53 etc., *Bules(h)o(u)res* 1197–1209, *Boulsover* 1577, 1610, 1675. OE pers.n. **Bul(l)* or **Bol(l)*, genitive sing. **Bulles* or **Bolles*, + **ofer**. Db 214 gives pr [bauzə], L 175.

BOLSTERSTONE SYorks SK 2796. 'The pillow stone'. *Bolstyrtone* 1375, *-er-* 1402–1739, *Bolstirston* 1419, 1426. OE **bolster** + **stān**. A stone on the village green on which felons are said to have laid their head for execution. Forms without *s* are due to the difficulty of the sequence *st–st*. YW i.257, iii.261.

BOLSTONE H&W SO 5532. 'Bola's stone'. *Boleston* 1193, *Bolestan* 1194, 1200. OE pers.n. *Bola* + **stān**. DEPN, He 39.

BOLTBY NYorks SE 4986. 'Bolti's farm or village'. *Boltebi -by* 1086–1399, *Boutebi -by* 1176–1271, *Boltby* 1316. ON pers.n. *Boltr* or **Bolti*, genitive sing. *Bolta*, + **bȳ**. YN 198, SPN 60, SSNY 22.

BOLTER END Bucks SU 7992 → CADMORE END SU 7892.

BOLT HEAD Devon SX 7236. 'The top end of Bolt'. *Bolt Head* 1809 OS, *Bult poynt* 1577. P.n. Bolt + ModE **head** 'a headland'. Bolt represents OE **bolt** 'a bolt, a missile, an arrow' used topographically of the stretch of high coast land rising to 432ft. as viewed from the sea. Bolberry SX 6939, *Boltes- Boteberie* 1086, *Boltebyry -bir(y) -bure -bury -byr(e)* 1224–1333 is 'fortified place at Bolt', p.n. Bolt as in BOLT HEAD + OE **burh**, dative sing. **byriġ**. Cf. also BOLT TAIL SX 6639 at the N end of the ridge. D 307.

BOLTON 'Enclosure with buildings, a collection of buildings'. OE ***bōthl-tūn**. In some cases *bōthl* may signify the superior hall or court of an estate and *bōthl-tūn* imply a settlement dependent on a royal vill.

(1) ~ Cumbr NY 2344. *Boulton'* 1200 etc., ~ *in Allirdale* 1335, *Bolton* from 1212 (~ *Alverdale* 1278), *Bothelton -il-* 13th cent. Cu 268 gives pr [boutən].

(2) ~ Cumbr NY 6323. *Botelton* 12th, *Bothelton* c.1200, *Boelton* c.1120, *Boulton* 1176–1617, *Bolton* from 1317. We ii.139.

(3) ~ GMan SD 7108. *Boelton* 1185, *Bothelton* 1212, *Botun'* 1236, *Bowelton* 1248, *Boulton* 1285–1332. Also known as *Magna Boulton* 1285 and *Bolton on the Mores* 1331, *Bolton (in) le Moors* 1547, 1843 OS. La 45 gives pr [boutn], TC 56, Jnl 17.37.

(4) ~ Humbs SE 7752. *Bodelton* 1086, *Bouelton* [12th]copy, *Boulton* 13th–1655, *Bolton* from 1200. YE 174.

(5) ~ Northum NU 1013. *Bolton* from 1200, *Bo(d)el- Bowil- Bowyl- Boulton, Boyltun* 13th cent. NbDu 27.

(6) ~ ABBEY NYorks SE 0754. P.n. *Bodelton(e)* 1086, *Botheltona* 12th, *Boulton(a)* 1120×47–1689, *Bolton* 1246×66–1607, + *Abbey* 1586. The reference is to the ruins of the 12th cent. Augustinian priory. YW vi.63.

(7) ~ -BY-BOWLAND Lancs SD 7849. *Bolton in Bouhelant* 12th, ~ *by Bowland* from 1415. P.n. *Bodeltone* 1086, *Bothelton -il-* 1189×99–c.1250, *Boulton* 12th–1642, *Bolton* from c.1200, + p.n. Bowland as in Forest of BOWLAND SD 6652. YW vi.184.

(8) ~ FELL Cumbr NY 4868. *Bolton Fell* 1384, 1592. Cu 91.

(9) ~ FELLEND Cumbr NY 4768. P.n. BOLTON FELL NY 4868 + **end**.

(10) ~ GATE Cumbr NY 2340. 'Gate leading to Bolton'. *Bolton yate, -ga(i)te* 1578. P.n. BOLTON NY 2344 + OE **ġeat** influenced by ON **gata**. Cu 269.

(11) ~ HALL NYorks SE 0789. P.n. Bolton as in Castle BOLTON SE 0391 + ModE **hall**.

(12) ~ -LE-SANDS Lancs SD 4868. *Bodeltone* 1086, *Boelton(e)* 1094, 1219, 1249, *Bothelton* c.1190–1246 etc., *Boulton(e)* 1206–1310, *Bolton* from 1265. *Le-Sands* refers to the situation of the village near the sands of Morecambe Bay. La 186, Jnl 17.103.

(13) ~ -ON-SWALE NYorks SE 2599. *Bolton oppon Swale* 1403. Earlier simply *Boletone* 1086, *Bo(h)eltona* 1184, 1280, *Bolton* c.1300. YN 277.

(14) ~ PERCY NYorks SE 5341. 'B held by the Percy family'. *Bolton Peyrssy* 1509. Earlier simply *Bodel- Bade- Bodetone -tune* 1086, *Bo(w)elton* 1205, 1276, *Bolton* 1226–1621. The manor was part of the extensive estates received by William de Percy from the Conqueror. YW iv.221.

(15) ~ PRIORY NYorks SE 0754 > BOLTON ABBEY SE 0754.

(16) ~ UPON DEARNE SYorks SE 4502. P.n. *Bode(l)tone* 1086, *Bowoltona* early 12th, *Boelton* 1164×81, 1236, *Bo(u)lton* 1248–1605, + suffix *upon Dyrne* 1300, r.n. DEARNE SE 3408. YW i.83.

(17) Castle ~ NYorks SE 0391. 'B with the castle'. P.n. *Bodelton(a) -tun* 1086–1252, *Boelton* 1160–1208, *Bolton* 1396 etc. + ModE **castle**. The castle was built by Richard le Scrope in 1379. YN 256.

(18) East ~ MOOR NYorks SE 0094. P.n. Bolton as in Castle BOLTON SE 0391 + ModE **east** + **moor**.

(19) West ~ NYorks SE 0290. *(Alia) Bodelton* 'the second B' 1086, *(Little) Boulton* 1296–1316. The additions are for distinction from Castle BOLTON SE 0391. YN 266, DG 538.

BOLT'S LAW Durham NY 9545. Uncertain. *Boltislawe* 13th. Nearby are Bolts Burn, Bolts Fine (ModE **fine** 'a lease') and Bolts Walls, the latter probably identical with *Boldshele* 1382, *Baltshele* 1418. Probably surname *Bolt* or *Bold* + OE **hlāw** 'a hill' (rising to 1773ft.). However, the occurrence of the identical name in Northumberland at NY 6981 rising to 1259ft. in the wastes of Wark Forest casts some doubt on the surname explanation. NbDu 28.

BOLT TAIL Devon SX 6639. 'The bottom end of Bolt'. *Bolt Tail* 1765. P.n. Bolt as in BOLT HEAD SX 7236 + ModE **tail**. D 307.

BOLVENTOR Corn SX 1876. *Boldventure* 1844. The 'bold venture' was a mid-19th cent. attempt to found a farming settlement in the middle of the moorland. PNCo 56.

BOMERE HEATH Shrops SJ 4719. *Bolemeresheth* 1325×46, *Bomer Heath* 1833 OS. Identical with Bomere SJ 4908,

Bole- Bulemar(') 'the bull pond', OE **bula** + **mere**, + ME **heth**. Gelling, Sa ii.105.

BONBY Humbs TA 0015. 'Bondi's farm or village' or 'village of the peasant proprietors'. *Bundebi* 1086, 1219, *Bondebi -by* c.1115–1535, *Bondby* 1294–1723, *Bonby* from 1383. ON pers.n. *Bóndi*, genitive sing. *Bónda*, or **bóndi**, genitive pl. **bónda**, + **bȳ**. Li ii.56.

BONDLEIGH Devon SS 6504. Uncertain. *Boleneia* 1086, *Bonlege* 1205, *Bonelegh* 1242–1390, *Bondleigh* 1620, *Bundleigh* 1809 OS. Possibly 'Bola's island', OE pers.n. *Bola*, genitive sing. *Bolan*, + **īeġ**, with metathesis of *Boleney* > *Boneley* as if a name in *lēah* 'wood, clearing'. Alternatively *Bolan-lēah* 'Bola's clearing' may have been the original form. D 359.

BONEHILL Staffs SK 1902. 'Bula's or bull hill'. *Bolenhull* 1230, 1271, *Bulenhull* 1230, *Bonhill* 1601, *Bonehill* 1608. OE pers.n. *Bula*, genitive sing. *Bulan*, or **bula**, genitive sing. **bulan**, genitive pl. **bulena**, + **hyll**. DEPN, L 171, Horovitz.

BONHUNT Essex TL 5033 → WICKEN BONHUNT.

BONINGALE Shrops SJ 8102. Partly uncertain. *Bolingehal'* c.1200, *Bolinghale -hall -yng-* 1255×6–1446, *Bolinghal* 1276, *Bonyngale* 1577. The generic is OE **halh**, dative sing. **hale**, the specific either OE pers.n. *Bola* + **ing**[4] 'the nook of land called after Bola' or OE ***boling**, an **ing**[2] derivative of OE ***bol** 'a smooth, rounded hill' or **bola** 'a tree-trunk, a bole', cf ModE *bolling* ' a pollard tree'. Sa i.52.

BONNINGTON Kent TR 0535. Possibly the 'estate called after Buna'. *Bonintone* 1086, *Buningtun, Bonintone* c.1100, *Bu- Boninton -yn-* 1207×8–1242×3, *Bu- Bonington* 1251 etc. OE pers.n. *Buna* + **ing**[4] + **tūn**. Another possibility might be the 'reed-bed farm', OE ***buning** < **bune** + **ing**[2] + **tūn**. PNK 464.

BONSALL Derby SK 2858. 'Bunt's nook of land'. *Bunteshale* 1086, *-hal'* 1195, 1295–8, *-ale* 1306, *Bontishal(e) -al(e) -is-* 1251–1474, *-hall* 1519–1758, *Bondeshal(e) -al(e) -is-* 1330–1428, *Bondsale* 1340–1412, *Bon(e)sale* 1427–1577, *Bonsall* from 1505. OE pers.n. **Bunt*, genitive sing. **Buntes*, + **halh**. Db 345 gives pr [bɔnsəl], L 106.

BOOKER Bucks SU 8491. Unexplained. *Bokar* 13th, *(G'.) Booker F.* 1822 OS. Bk 206.

Great BOOKHAM Surrey TQ 1354. *Magna Bockam* c.1270, *Magna Bukham* 1255, *Greate Bookeham* 1588. ModE **great**, Latin **magna**, + p.n. *Bochehā* 1086, *Beche-* 1165–99, *Bocham* 1225–1508 including [933, 967, 1062]13th S 420, 752, 1035 with variant *Bo(c)k-*, 'the settlement growing with beech-trees', OE **bōc** + **hām**. 'Great' in contrast to Little Bookham. Sr 99, ASE 2.32, L 222.

BOOLEY Shrops SJ 5725. 'The bull pasture'. *Boleley* c.1100, *Boley(e)* 1285, 1300. OE **bula** + **lēah**. Bowcock 48, DEPN.

BOOSBECK Cleve NZ 7619. 'Stream near the cowshed'. *Bosbek* 1375. OE **bōs** + ON **bekkr**. YN 145 gives pr [biuzbek].

BOOT Cumbr NY 1701. 'The bend'. *Bout, the Bought* 1587, *Boot* 1791. ME **bought** (OE **buht*). Refers to the acute bend in the valley of the Esh at this point. Cu 389, PNE i.56.

BOOTHBY 'Booth farm or village'. ON **bóth**, genitive pl. **bótha**, + **bȳ**.

(1) ~ GRAFFOE Lincs SK 9859. *Boothby Graffoe* 1693. Originally two separate names, *Bodebi* 1086, 1135×54, *Bothebi -by* 1202–1642, *Boothby* from 1538, *Bobi -by* c.1160–1428, and *Grafho* 1166–1200, *-ho(u)* 1167–1281, *Graff(h)o(u)* 1202–1576, *Graffoe* 1603, the 'hill-spur with or by a grove', OE **grāf** + **hōh**, referring to a lost site on the high ground of the Cliffe Range, probably the meeting place of Boothby Graffoe wapentake, the place where the temporary booths for the meeting were erected. The two names are first combined as *Boby et Grafhow* 1298. Perrott 218, SSNEM 38.

(2) ~ PAGNELL Lincs SK 9730. 'B held by the Paynell family'. *Botheby Paynell* 1388, *Boothby Pannell* 1600. Earlier simply *Bodebi* 1086, *-by* [c.1150×60]1409, *Boebi -by* 1138–1327, *Bothebi -by* 1202–1551. The manor was held by the Norman family Paynel in the 14th cent. Perrot 491, SSNEM 38.

BOOTHSTOWN GMan SD 7200. A 19th cent. development named after Booths Hall (lost) SD 7300, *Bothes man.* (for *manor*) 1500, *The Bouthe* c.1540, *Boothes hall* 1577, 'the booths', ODan **bōth**, ME pl. **bōthes**. La 40, SSNNW 62.

BOOTLE 'Building'. OE **bōthl**. PNE i.43.

(1) ~ Cumbr SD 1088. *Bodele* 1086, *Botle* c.1135–1321, *Botel(l) -yl -il* c.1170–1513, *Butile* 1369, *Butle* 1571. Cu 345.

(2) ~ FELL Cumbr SD 1488. *Bootle Fell* 1864 OS. P.n. BOOTLE SD 1088, *Boothill* 1777, + ModE **hill** and **fell** (ON *fjall*). Cu 345.

(3) ~ STATION Cumbr SD 0989. P.n. BOOTLE SD 1088 + ModE **station**.

(4) ~ Mers SJ 3394. *Boltelai* 1086, *Botle* 1212–57, *Botel* 1284, *Bot(h)ull* 1322, 1332, *Bolde* 1226. La 116.

BORASTON Shrops SO 6170. 'Fort Aston'. *Bureston* 1188–1271×2, *Buraston* 1255×6, 1274, 1633, *Boraston(')* from 1271×2. OE **burh** + p.n. Aston 'the east settlement', OE **ēast** + **tūn**. E of Burford SO 5868. Sa i.52.

BORDEN Kent TQ 8862. 'Woodland pasture where boards are made' or 'by the hill'. *Bordena* 1176×7, *Borden(e)* from 1190. OE **bord** or ***bor** + **denn**. In the case of the first explanation the name is to be compared with the lost p.n. *bord-dæne* 824 S 1434 (OE **denu**). PNK 243, PNE i.42, V i.126.

BORDLEY NYorks SD 9364. Either the 'wood where boards are obtained' or 'Borda's wood or clearing'. *Borelaie* 1086, *Bordelei(a) -lay -ley* 1134×52–1596. OE **bord**, genitive pl. **borda**, or pers.n. *Borda* from *Brorda* by dissimilation, + **lēah**. YW vi.81, PNDB 208, L 205.

BORDON CAMP Hants SU 7936. A modern military camp at Bordon, *Burdunesdene* c.1230, 'Burdun's valley', surname *Burdon* + ME **dene** (OE *denu*). The manor belonged to the de Burdon or Burdun family. Ha 37, Gover 1958.103.

BOREHAM Essex TL 7509. 'Hill homestead'. *Borham* 1086–1265 etc. including [1035×44]13th S 1521, *Boreham* 1594. OE ***bor** + **hām**. Ess 238, Studies 1936.131, PNE i.42.

BOREHAM Wilts ST 8844. Possibly '(the settlement) at the fortifications'. *Burton Delamare al. Booreham* 1573, *Boreham, Boram, Bowram* 16th. OE **burgum, burhum**, dative pl. of **burh** 'a fort', referring to the Iron Age hill-forts of Battlesbury and Scratchbury Camps. An alternative possibility is OE ***borum** 'at the hills', dative pl. of ***bor**, referring to the hills on which these camps are located. The earlier name Burton, *Buriton* 1241–2, *Bouri- Buryton* 14th, means 'the camp settlement', OE **byriġ**, dative sing. of **burh**, + **tūn**. Wlt 158, Thomas 1980.236, 239.

BOREHAM STREET ESusx 6611. 'Boreham village'. P.n. Boreham + ModE dial. **street** 'a village'. Boreham, *Boreham* from 12th cent., is 'hill promontory', OE ***bor** + **hamm** 2 or 3. Sx 483, ASE 2.49, SS 85.

BOREHAMWOOD Herts TQ 1996. *bosci de Borham* 1188, *Burhamwode* 13th, *Borehamwode* 1329, *Barramwoode* 1554. P.n. *Boreham* [1189×99]1301, 'hill homestead or enclosure', OE ***bor** + **hām** or **hamm**, + ME **wode** (OE *wudu*). Hrt 74, Studies 1936.131, V i.125.

BORLE BROOK Shrops SO 7088. *Borebroke* 1569×70, *Borle Brook* 1833 OS. P.n. Borle as in *aq'm de Borle* 1271×2 and *Borle Mylne* 1585, *The Burle Myll* 1592, late ME **burl** 'to dress cloth', + ModE **brook**. Alternatively this could be late ME **burl** 'to bubble, to flow gently'. Gelling.

BORLEY Essex TL 8442. 'Boar wood or clearing'. *Barlea* 1086, *Barle(e) -leg' -ley* 1203–1314, *Borle(gh) -lei -leye* 1224–1340. OE **bār** + **lēah**. Ess 415.

BOROUGH GREEN Kent TQ 6057. *Borrow Grene* 1575. The origin of this name is unclear. PNK 156.

BOROUGHBRIDGE NYorks SE 3966. 'The bridge of *Burgh*'. *Pontem de Burgo* 1155, *Pons Burgi* 1175×1203–1428, *Pont de*

Burc 1171, *Burghbrig(g) -brygg* early 13th–1549, *Burrowbrig(ge)* 1508, 1519. P.n. *Burgh* for ALDBOROUGH SE 4066 + ME **brigg**. The bridge carried the Great North Road across the river Ure. YW v.82.

BORROWASH Derby SK 4134. '*Burgh* ash-tree'. *Burghas'* 1269, *-asshe* 1330, *Boroughashe* 1535, *Borowe Ashe* c.1600. P.n. *Burg(e)* c.1200–1362, *Burgh* 1269–1417, OE **burh** 'the fort, manor house', + **æsc**. Db 488.

BORROW BECK Cumbr NY 5205. *Borrow Beck* 1836. Earlier *Borra watter* c.1180, *Burg(h)ra* 1195–1278, *Borghra* 12th, 1235, *watter of Borgherey, Borough watter* 16th, *Borrow R.* 1777. R.n. *Borghra* 'river of the fortification', ON **borg**, genitive sing. **borgar** + **á**, + OE **wæter**, ModE **beck**. The reference is to the Roman fort at Low Borrow Bridge NY 6102. We i.4, SSNNW 107.

BORROWBY NYorks SE 4289. 'Hill farm'. *Ber(g)(h)ebi -by* 1086–1333. ON **berg** + **bȳ**. Borrowby lies on a hill. YN 205 gives pr [barəbi], SSNY 20.

BORROWDALE 'Fortress valley'. ON **borg**, genitive sing. **borgar**, + **dalr**.

(1) ~ Cumbr NY 2516. *Borgordale* [c.1170]n.d., *Borudale* 1209–1337, *Borhe- Borcher(e)dale* 13th cent., *Borowdale* 1398, *Bor(r)adell* 16th cent. Instead of ON genitive sing. *borgar*, the specific might in this case be the r.n. *Borghra* 'the river of the fort', ON **borgar-á** as in BORROW BECK NY 5205, here an alternative name for the upper Derwent. There is a fort at Castle Crag NY 2515. Cu 349, SSNNW 107 no.1, L 94.

(2) ~ Cumbr NY 5603. *Borgher(e)dal* 1154–89, *Burwerdale* 1199, *Borwedale* 1282, *Borghdale* 1362, *Borowdale* 1523. Instead of genitive sing. *borgar*, the specific might be the r.n. *Borghra* 'the river of the fort', ON **borgar-á** as in BORROW BECK NY 5205. The reference is to the Roman fort at Low Borrow Bridge. We i.138,SSNNW 107 no.2, Jnl 2.70, L 94.

(3) ~ FELLS Cumbr NY 2613. *Borowdale Fells* 1509×47. P.n. BORROWDALE NY 2516 + ModE **fell(s)** (ON *fjall*). Cu 349.

BORWICK Lancs SD 5273. 'Demesne farm'. *Bereuuic* 1086, *-wyk -wi(c)k* 1285–1347, *Berwik* 1228, 1332, 1518, *Borwyc* 1255, *Barwick* 14th. OE **bere-wīc**. La 188, Jnl 17.104.

BOSAVERN Corn SW 3630. Probably 'dwelling of Avarn'. *Bosavarn* 1302. Co **bos** (from earlier **bod*) + pers.n. **Avarn*. PNCo 56.

BOSBURY H&W SO 6943. 'Bosa's fortified place or manor house'. *Bosanbirig* before 1118, *Boseberge* 1086, *-bir'* 1230. OE pers.n. *Bōsa*, genitive *Bōsan*, + **byriġ**, dative sing. of **burh**. DEPN, He 39.

BOSCASTLE Corn SX 0990. 'Boterel's castle'. *Castrum Boterel* 1284, *Castel Botereaus* 1287, *Boterelescastel* 1302, *Botrescastell* 1343, *Boscastel* c.1540, *Boscastle* 1610 Map. Latin **castrum**, OFr **castel** + family name *Boterel, Botreaux*, an important Norman family in Cornwall which probably came from Les Bottereaux in Normandy. The modern form has been assimilated to Cornish names in *Bos-* (from **bod* 'a dwelling'). PNCo 56.

BOSCOBEL HOUSE Shrops SJ 8308. P.n. Boscobel + ModE **house**. The place originated as a hunting lodge in Brewood Forest built about 1580. Boscobel, *Baskabell* 1707, *Boscobel* 1784, is 'the beautiful wood', an adaptation of Italian *bosco bello*. Sa i.53, Raven 31.

BOSCOMBE Dorset SZ 1191. Possibly 'coomb where spiky plants grow'. *Boscumbe* 1273, *Bascombe* 1593. OE **bors** + **cumb**. Do 39, L 93.

BOSCOMBE Wilts SU 2038. 'The coomb growing with prickly plants'. *Boscumbe* from 1086 with variant *-comb(e)*, *Borrescumb'* 1256, *Borscumbe -combe* 1275–1332, *Borscome* 1537. OE ***bors** + **cumb**. Wlt 361, L 93.

BOSHAM WSusx SU 8004. 'Bosa's promontory of dry land'. *Bosanhamm* [c.731]8th BHE, *-ham* c.1121 ASC(E) under year 1049, *Bosenham* c.1121 ibid. under years 1046 and 1048, *Boseham* 1086–1407, *Bosham* from 1306, *Bozam* 1628. OE pers.n. *Bōsa*, genitive sing. *Bōsan*, + **hamm** 2, 4, 5. The same pers.n. occurs in the neighbouring hundred of Bosmere in Hants, *Boseberg* 1086, which included Hayling Island, Warblington and parts of Havant. Bosa's territory straddled and probably preceded the Sussex-Wessex boundary. Sx 57–8, SS 81, ASE 2.25, NoB 48.147.

BOSKEDNAN Corn SW 4434. Probably 'dwelling of Kennon'. *Boskennon* 1327, *-kennan* 1597, *Boskednan* 1623. Co **bos** (from earlier **bod*) + pers.n. **Kennon* (with late Co *dn* for *nn*). PNCo 57.

BOSLEY Ches SJ 9165. 'Bot's wood or clearing'. *Boselega* 1086, *-leg(h) -le(e) -ley(e)* 1275–1351, *Bozeley* 1275, *Bothis- le Botesleg'* 1286, 1287, *Botesle* 1322, *Bosley(e)* from 1306. OE pers.n. **Bōt*, genitive sing. **Bōtes*, + **lēah**. Addenda gives 19th cent. local pr ['bɔ:zli]. Che i.54.

BOSSALL NYorks SE 7160. 'Bot's nook of land'. *Boscele, Bosciale* 1086, *Boz(h)al(e)* 1225–89, *Bossal(e)* 1295–1416. OE pers.n. **Bōt*, genitive sing. **Bōtes*, + **halh**. The village stands on a low spur in a loop of the r. Derwent. YN 36, L 108, 110.

BOSSINEY Corn SX 0688. 'Dwelling of Kyni'. *Botcinnii* 1086, *Boccyny, Bocciny* 1236, *Boscini* 1291. Co ***bod** (later *bos*) + pers.n. *Kyni*. PNCo 57, DEPN.

BOSSINGHAM Kent TR 1549. 'Open land called after Bosa'. *Bossingcamp* 1226, *Bosingkomp* 1264. OE pers.n. *Bōsa* + **ing**[4] + **camp**. The element *camp* has not been thoroughly studied, but many examples of *camp*-names seem to be associated with Roman settlements. PNK 542, Signposts 76–7.

BOSTOCK GREEN Ches SJ 6769. *Bostock Green* 1666. P.n. Bostock SJ 6768 + ModE **green**. Bostock, *Botestoch* 1086, *Bostoc* c.1200–1536 with variants *-sto(c)k(e)*, is 'Bota's outlying farm'. OE pers.n. *Bōta* + **stoc**. Che ii.202.

BOSTON Lincs TF 3244. 'Botwulf's stone'. *Botelvestan* (for *Botolve-*) [after 1114]17th, *Botuluestan* 1130, *Botulfstan* 1281, *Botelstane* 1323, *Botelston -ul-* 1347–86, *Botolph- Botolfston* c.1360, 1363, *Boston* from 1235. The Latin name is *villa Sancti Botulfi* 'St Botulf's vill' 1093–1351 with variant *Botulphi*. OE pers.n. *Bōtwulf*, genitive sing. *Bōtwulfes*, + **stān**. The place was early associated with the English saint Botulf d. 680 who studied in Germany and founded a monastery at a place called Icanho, *(æt) Icanho* 891 ASC(A) under year 654, *(æt) Icanhoe* c.1122 ASC(E) under year 653, *Icheanog* [675×90]13th S 1798, 'Ica's hill-spur', OE pers.n. *Ica*, genitive sing. *Ican*, + **hōh**, usually assumed to have been the earlier name of Boston. Boston, however, was in the territory of the Middle Angles while we know from the anon. *History of the Abbots* §4 that St Botulf's foundation lay *ad Anglos Orientales* 'among the East Angles', perhaps at IKEN Suffolk TM 4155. There is no reason to believe that the *stān* of Boston was a preaching stone associated with St Botulf rather than a boundary marker or that the dedication of the great church at Boston was not inspired by the popular etymology of the place which preceded it rather than vice-versa. Payling 1940.115, 116, TC 57, Cameron 1998.

BOSTON SPA WYorks SE 4245. A well-known mineral spring was discovered here in 1744. Originally known as *Thorp Spaw* 1771 from nearby THORP ARCH. The village was constructed in an open field to exploit the spring in 1753. The origin of *Boston* in this name is unknown, cf. *Bostongate* 1799, *Boston* 1822. YW iv.82, B.M.Scott, *Boston Spa* 1976, Langdale 1822.238, Room 1983.11.

Great BOSULLOW Corn SW 4133. *Great Bosullow* c.1870, earlier *Bossowolo-meour* 1517. ModE adj. **great**, Co **meur**, + p.n. Bosullow. 'Great' for distinction from *Botuolo bichan* 1244, *Boschiwolou-bigha* 1302, 'little Bosullow', p.n. Bosullow + Co **byghan**. Bosullow appears to be 'dwelling of a cottage of light', Co ***bod** + **chy** + **golow** with mutation of *g* to *w*. The significance is unknown. PNCo 57.

BOSWINGER Corn SW 9941. 'Dwelling of Wengor'. *Boswengar* 1301. Co **bos** (earlier **bod*) + OCo pers.n. *Wengor*. PNCo 57.

Husbands BOSWORTH Leic SP 6484. *Husband Boreswurth* 1548, *Husbands Boseworth* 1555. ModE **husband** for *husbandman* 'a farmer' referring to the situation of the village in a farming district, + p.n. *Bareswsord(e)* 1086–1214×21, *Baresw(o)rth(e) -is-* 1166–1278, *Boresw(o)rth(e) -is-* 1189–1720, 'Bar's or boar enclosure', OE pers.n. *Bār* or **bār** 'a boar', genitive sing. *Bāres*, **bāres**, + **worth**. 'Husband' for distinction from Market BOSWORTH SK 4003. Lei 208.

Market BOSWORTH Leic SK 4003. *Market(t) Bosworth* from 1518. ME **market** + p.n. *Boseword(e)* 1086–late 12th, 1364, *-worth(e)* 1232–1496, *Bosworth* from 1265, 'Bosa's enclosure', OE pers.n. *Bōsa* + **worth**. 'Market' for distinction from Husbands BOSWORTH SP 6484. Lei 484.

BOTALLACK Corn SW 3632. Probably 'dwelling on a steep brow'. *Botalec* 1262, *-ik* 1327–1447 G, *Botalleck* 1610. Co ***bod** + ***talek** 'steep-browed' (<*tal* + *-ek*). The situation of the place, near the cliff, makes this solution more likely than the alternative 'dwelling of Talek', ***bod** + pers.n. **Talek*. PNCo 57.

BOTANY BAY GLond TQ 2999. *Botany Bay* 1819. This hamlet only developed after the enclosure of Enfield Chase in 1777. Its name is transferred from Botany Bay, Australia, a term found fequently in field-names for a remote or inaccessible spot in a township. The original Botany Bay, an inlet in the coast of New South Wales, was so named by the naturalists of Captain Cook's expedition of 1770. Botany Bay was used as a penal settlement in 1787–8. Mx 74, Field 1972.25, GL 32, Encyc 81.

BOTCHESTON Leic SK 4804. 'Bochard or Botcher's estate'. *Borchardeston* 1265–1363, *Borcherd(e)ston* 1416–1512, *Bocherston' -ar-* 1340–1492, *Bocheston' -as-* 1360–1541, *Botcheston* 1610. OFr pers.n. *Bochard* (OG *Burchard*) or surname *Botcher* < ME *botcher* 'mender, patcher, cobbler', genitive sing. *Bochardes, Botcheres*, + **tūn**. Lei 492.

BOTESDALE Suff TM 0575. 'Botwulf's valley'. *Botholuesdal* 1275, *Botulfesdale* 1313, *Botesdade* (sic) 1437, *Botisdale* 1568, *Buddesdale* 1610. OE pers.n. *Bōt(w)ulf*, genitive sing. *Bot(w)ulfes*, + **dæl**. DEPN, Baron.

BOTHAL Northum NZ 2386. 'Bota's nook of land'. *Bothale* 1212 BF, *Bothal(a) -halle, Bot(t)ehal(e)* 13th cent. including [1154×89]1271, *Bottal* 1346, *Bottell* 1428. OE pers.n. *Bōta* + **halh**. NbDu 28 gives pr [bɔtl], L 108.

BOTHAMSALL Notts SK 6773. 'Valley bottom spring'. *Bodmescel(d)* 1086, *-kil* c.1175, *-hil(l) -hel -'hull* c.1190–1335, *Both(o)meskill* 1269, 1281, *Bothumsil -el(l)* 1356–1400, *Bothamsall* 1689, *Bottomsall* c.1825. Probably OE **bothm**, genitive sing. **bothmes**, + OE **celde**, ON **kelda**, later replaced by **hyll**. There is a strong spring in the side of the hill on which the village stands. Nt 69 gives prs [bɔtəmsəl] and [bɔðəmsəl].

BOTHEL Cumbr NY 1839. 'Building'. *Bothle* c.1125, c.1135, *Bothel* from 1230, *Bothil(l) -yll* 1279–1777, *Bold* 1580, *Bole* 1675, *Boal(d)* 18th. OE **bōthl**. Cu 271 gives pr [bou(ə)l].

BOTHENHAMPTON Dorset SY 4691. 'Home farm in the valley bottom'. *Bothehamton* 1268, *Bothenamtone* 1285, *Bothenhampton* 1322, *Baunton* 1497. OE **bothm** + **hām-tūn**. Do 39, Jnl 20.42, 45.

BOTLEY Either 'Bota's wood or clearing', OE, pers.n. *Bota* + **lēah**, or 'remedy, help wood', OE **bōt**, genitive sing. **bōte**, + **lēah**, a wood in which tenants had a right to take timber. PNE 1.43, V i.129–30.

(1) ~ Bucks SP 9802. *Bottlea Helie* 'Elias's Botley' 1167, *Bot(t)ele* 1227. Bk 224.

(2) ~ Hants SU 5113. *Botelie* 1086, *Boteleia -leg*, 1195, 1236, *Botlai* 1184 *-legh* 1349, *Bottel(e)(y)e* 13th–1346. Ha 38, Gover 1958.35, DEPN.

(3) ~ Oxon SP 4805. *(apud) Boteleam* [before 1170]12th, *Botele(gh) -ley* 1227–1401. Brk 451.

BOTOLPHS WSusx TQ 1909. Short for 'St Botolph's church or parish'. *Sancto Botulpho* (Latin prepositional case) 1288, *St Botulph* 1442, *Botoulf* 1395, *Botulphes* 1453, *Butt(el)les* 1565, 1610, *Buttols* 1726. Saint's name *Botolph* from the dedication of the church. Sx 222 gives prs [bʌtəlls] (sic) and [bʌtəlfs].

BOTTESFORD 'Ford belonging to the *bōtl* or house'. OE **bōtl**, genitive sing. **bōtles**, + **ford**.

(1) ~ Humbs SE 8907. *Budlesforde* 1086, *Botlesforda* c.1115, *-ford* 1202, *Botnesford* 1272. DEPN.

(2) ~ Leic SK 8039. *Bot(t)lesford(e) -is-* 1086–1518, *Bot(t)esford(e) -is- -ys-* 1086 and from 1368. Lei 142, L 72.

BOTTISHAM Cambs TL 5460. 'Boduc's homestead'. *Bodekesham* *[1062]13th S 1030, –1494 including [1077]17th, *-is-, -ys-* 1284–1492, *Bodi(s)chesham, Bodichessham* 1086, *Boding(e)s(h)am* 1155–1227, *Bodesham* c.1210–1527, *Bot(h)ekesham -is- -ys-* 1218–1468, *Bot(t)esham* 1298–1605, *Bottisham -ys-* from 1434. OE pers.n. **Boduc*, genitive sing. **Boduces*, + **hām**. Ca 129, O li gives pr [bɔtsəm].

BOTTOM FLASH Ches SJ 6667. A lake formed by subsidence of a reach of the river Weaver. *Flash Meadow* is mentioned in the Tithe Award (1838). Mod dial. **flash** (ME *flasshe*) 'flooded grassland; a sheet of shallow water'. Che ii.214.

BOTTON HEAD Lancs SD 6661. 'The head of Botton'. P.n. Botton + ModE **head**. Botton, *Bottun* [c.1230]1268, *Botine* 1246, *Botten* 1341, is ON **botn** 'flat wet land beside a river in a valley bottom'. La 182, L 86–7, Jnl.20.40.

BOTUSFLEMING Corn SX 4061. Partly obscure. *Bothflumet* 1259, *Boflumiet* 1261, *Botflumyet, -fleming* 1291, *Bodflumiet* 1318, *Bodflemy* 1336, *Botflemyng* 1348, *Blowflemyng* 1553, *Blosflemin* 1610. Co ***bod** 'dwelling' (later *bos*) + unknown element later replaced by the family name *Fleming*. PNCo 57, DEPN.

BOUGH BEECH RESERVOIR Kent TQ 4948. P.n. Bough Beech + ModE **reservoir**. Bough Beech, *Boubeche* 1396, *Bowbeche* 1440–4, is probably 'bent beech-tree', OE **boga** or **bogen** 'bowed' + **bēċe**. PNK 78.

BOUGHSPRING Glos ST 5597. *Bowspring* 1830. Gl iii.266.

BOUGHTON 'A manor held by book', i.e. by written title. OE **bōc-tūn**. Formally in some of these names *bōc* could be 'a beech-tree'.

(1) ~ ALUPH Kent TR 0348. 'Aluf's B'. *Boton Alou* 1237, *Bocton' Alulphi, Alulfi* 1270 etc. P.n. Boughton + pers.n. *Aluf*, an early tenant of the manor recorded in 1199 as *Alolf' de Bouton'*, representing OE *Ælf-* or *Æthelwulf*. Earlier simply *Boctun* [1013×20]11th S 1220, *Boltvne* (sic) 1086, *Boctune* c.1100, *Bouton'* 1199–1240, *Boc(h)tone* 1211×2, 1253×4. Also known as *Earlesboctune* c.1100–1240; it was held by Earl Godwin c.1020 and subsequently by Count Eustace. PNK 380, DEPN.

(2) ~ LEES Kent TR 0247. 'B pasture'. *Boctone(s)lese* 1313. Earlier simply *la Lese* 1240, *la Lewse* (sic for *Leswe*) 1254, *(atte) Lese de Bocton' Alulphi* 1313. P.n. BOUGHTON ALUPH TR 0348 + ME **lese, leswe**, OE *lǣs*, oblique case *lǣswe*. PNK 381.

(3) ~ MALHERBE Kent TQ 8849. 'B held by the Malherbe family'. *Bectone Malherbe* 1253×4, *Boctun' Malerbe* 1275. P.n. Boughton + family name *Malherbe*. Robert de Malherbe held the manor 1199×1216. Earlier simply *Boltone* (sic) 1086, *Boctun(e)* c.1100–1199. PNK 203.

(4) ~ MONCHELSEA Kent TQ 7651. 'B held by the Montchensie family'. *Bocton' Monchansy* 1278. P.n. Boughton + family name *Montchensi, de Monte Canisio*. The manor was held by William de Montecan' or Muntchanesy 1242×3–1257. Earlier simply *Bo(c)ton'* 1242×3–1273, *Bouton'* 1226. Also known as *Westbocton* 1275. PNK 207.

(5) ~ STREET Kent TR 0657. *Bocton Street* 1690 for 'B on the street', i.e. the Roman road from London to Canterbury, Margary no.1. Earlier simply *Boltvn(e)* (sic) 1086, *Boulton' -tun'* 1166×7, *Bouton'* 1240, *Bocton(')* 1226–78. This is Chaucer's *Boghtoun under Blee, Canon's Yeoman's Prologue* 556, *Boctone*

iuxta Bleen 1292 etc. from its position beneath the forest of BLEAN. Also known as *Biscopestune* 'the bishop's estate' c.1100 and *Bisshoppesbocton'* 1292 from its possession by the (arch)bishop of Canterbury. PNK 300.

(6) ~ HOUSE Northants SP 9081. P.n. *Boctone* 1086, *Boketon, Bo(c)htun* 12th, *Buke- Bu(c)h- Bugh- Bu(t)ton* 1201–80, *Bouhton* 1246, *Boughton* from 1316, p.n. BOUGHTON + ModE **house**. A substantial house of c.1500 remodelled c.1690–1700. Nth 173 gives pr [bautən], Pevsner 1973.110.

BOUGHTON 'A goat farm'. OE **bucc**, genitive pl. **bucca**, + **tūn**. Some of these names may, however, contain OE pers.n. *Bucca*.

(1) ~ Norf TG 1709. *Buchetuna* 1086, *Buche- Bugeton* 1180, *Buketon* 1197. DEPN.

(2) ~ Northants SP 7565. *Boche- Buchetone -done, Buchenho* 1086, *Bucton* 1175–99, 1305, 1597, *Bok(e)- Buketon(e)* 12th–1349, *Bouhton* 1216×72, *Boughton* from 1302. Nth 133 gives pr [bautən].

(3) ~ Notts SK 6768. *Buchetun -tone* 1086, *Buketon* c.1190–1472, *Bucton* 1208–1280, *Bughton* 1276–1377, *Boghton* 1332. Nt 70.

BOULBY Cleve NZ 7619. 'Boli or Bolli's village or farm'. *Bol(l)ebi* 1086, *-by* 1204–1363, *Bolby* 1407, 1412, *Bowlby* 1575, 1665. ON pers.n. *Boli* or *Bolli*, genitive sing. *Bol(l)a*, + **bȳ**. YN 140 gives pr [boulbi], SSNY 22.

BOULDON Shrops SO 5485. Partly uncertain. *Bolledone* 1086, *-don* 1255–1346 with variant *Bole-, Bul(l)ardone* 1166, *Bul(l)ar(d)dun -don* 1203×4, 1205, *Bulledon(')* 1255×6–1294×9, *Boldon* 1431, *Boulden* from 1495. The generic is OE **dūn** 'a hill'. The village is situated in a valley between hills and must be named from the high ground to the NW or SW. If the *-ard-* forms are primary the specific may be a lost W name for these hills ending in PrW ***arth** 'a height', perhaps PrW ***bulch** + ***arth** 'height of the gap' referring to the gap in the hills cut by Clee Brook. Cf TOLLARD ROYAL W ST 9417. Sa i.54 gives pr ['bouldən], CPNE 9, 26.

BOULMER Northum NU 2614. 'The bull pond'. *Bulemer* 1161, *Bulmer* 1296, *Boomer* 1663. OE **bula** + **mere**. The reference is to one of the shallow lagoons found on the sea-shore, possibly to that now called Boulmer Haven. NbDu 28 gives pr [bu:mə], L 27.

BOULMER HAVEN Northum NU 2613. *Boulmer Haven* 1868 OS. P.n. BOULMER + ModE **haven**. The reference is to an inlet between the rocks of North Reins and Marmouth Scars, possibly the original port of Boulmer.

BOULSWORTH HILL Lancs SD 9336. P.n. Boulsworth + ModE **hill**. Boulsworth, *Bulswyre* 14th, *Bulswar(r)e* 1618, 1620, is 'bull's neck', ME **bule**, genitive sing. **bules**, + **swire** (OE *swīra* or ON *sviri*). The reference is probably to the shape of the long massive ridge which stretches from SD 9235 to SD 9736. La 67.

BOULTHAM Lincs SK 9669. 'The village or homestead' or 'the water-meadow where ragged robin grows'. *Bvletham* 1086, 12th, *Bulteham* 1202, 1218, 1396, *Bultham* 1205–1597, *Boultham* 1675. OE **bulut** + **hām** or **hamm**. Boultham is situated on low-lying ground by the Witham river. Perrott 248, Studies 1936.106.

BOURN Cambs TL 3256. 'The stream'. *Bron(n)a, Brone, Brunna, Brunam -e* 1086, *Brunna -e, Bronne* 1185–1445, *Burne* 1441–1506, *Bourne* 1540–53. ON **brunnr** or a metathesied form of OE **burna**. The reference is to Bourn brook which joins the Cam at Grantchester.

BOURNE Lincs TF 0920. '(The settlement) at the springs'. *Brune* 1086–1239, *Brunum* [c.1150×60]1409, 1314, *Brun(na)* 1138–1486, *Bourn'* 1382–1604. ON **brunnr**, dative pl. **brunnum**. Bourn Eau river, *Brunne* 1327, *Burn' Ee* 1376, OE **ēa**, rises SE of the church from a copious spring called St Peter's Pool. RN 42, Cameron 1998.

River BOURNE Wilts SU 2038. Not really a r.n. at all, but the OE appellative **burna** 'a stream'. It is referred to variously as:

I. Collingbourne 'the stream of Colla's people', *(of) Collengaburnan* [921]14thS 379, *Colebourne* 1348, cf. COLLINGBOURNE DUCIS SU 2453 and COLLINGBOURNE KINGSTON SU 2355;

II. Winterbourne, *(fram) winter burnan* [972]14th S 789, *Winterburne* 1216×72, a stream that flows only in winter as in the Wilts WINTERBOURNE village names;

III. Simply 'the stream', *(andlang) burnan* [949]14th S 543, *la Bourne* 1518. Wlt 2, RN 41.

St Mary BOURNE Hants SU 4250. 'St Mary's Bourne'. *Maryborne* 1476, *Bourne Sce Marie* 1483. Saint's name *Mary* + p.n. *Borne* 1185, *Burna* 1188 'the stream', OE **burna**. The present church, the *capella* mentioned in 1185, is dedicated to St Peter but the p.n. implies an earlier dedication to the Virgin Mary. 'St Mary' for distinction from the HURSTBOURNES SU 3853, 4346. Ha 143, Gover 1958.155.

BOURNE EAU Lincs → BOURNE TF 0920 and River GLEN TF 2427.

BOURNE END Beds SP 9644. 'Stream end' sc. of Cranfield. *le Burnehende* 1306. OE **burna** (*la Burne* 1227) + **ende**. Bd 69.

BOURNE END Bucks SU 8987. 'The stream end' sc. of Wooburn. *la Burnhende* 1222, *Burnend'* 1236, *Brone End* 1766, *Bone End* 1822 OS, 1826. ME **burne** + **ende**. The reference is to the spot where the Wooburn or Wye flows into the Thames, a purely local name until the arrival of the railway, contrasting with Cores End SU 9087, *Coars End* 1822 OS. Bk 198, YE addenda.

BOURNE END Herts TL 0206. 'Stream end' sc. of Bovingdon. *le Bournend* 1357. ME **bourne** (OE *burna*) + **end** (*ende*). Hrt 29.

BOURNEMOUTH Dorset SZ 0991. 'The mouth of the stream'. *la Bournemowþe* 1407, *B(o)urnemouthe* 16th, *Bourne Mouthe* 1811 OS. ME **bourn** (OE *burna*) + **mouth** (OE *mūtha*). *Bourne*, the name of the stream which enters the sea near the pier, is all that appears on maps of this area as late as 1817, and this was the original name of the town which stems from a villa built here in 1812 by Captain Lewis Tregonwell for himself and his family. In 1836 Sir George Jervis laid out a marine village round Tregonwell's house. The stream is still called the Bourne. The 1407 reference is the record of 'a great fish, called in English a whale, washed up on the shore near the Bournemowthe'. Do 40, Room 1983.12, Ha 38.

BOURNEMOUTH (HURN) AIRPORT Dorset SZ 1198. P.n. BOURNEMOUTH SZ 0991 + ModE **airport**. The alternative name Hurn, *Herne* 1086, *Hern* 1811 OS, *(la) Hurne* 1242, 1367, is 'the corner of land', OE **hyrne** referring to the piece of land between the r. Stour and the Moors River. Do 90.

BOURNES GREEN Glos SO 9104. *Bournes Green* 1830. P.n. Bourne + **green**. Bourne, *Bourne (in parochia de Biseleye)* 1394, 1647, is OE **burna** 'a stream'. Gl i.118.

BOURNHEATH H&W SO 9574. *Bourne Heath* 1831 OS.

BOURNMOOR Durham NZ 3151. *Bourn Moor* 1833. A modern colliery village named from the nearby Moor or Herrington Burn.

BOURNVILLE WMids SP 0481. A modern invention for George Cadbury's chocolate factory developed here in the 1870s. The original name was to have been Bournbrook from Bournbrook Hall, *Burnebrock* c.1250, *Byrnebroc* 1275, *Barnebrok, Barnbrook* 1511–1831 OS, 'Beorna's brook', OE pers.n. *Beorna* + **brōc**†. It was changed to Bournville 'because it had a French sound, and French chocolate was looked upon as the best'. Room 1983.12 citing I. A. Williams, *The Firm of Cadbury: 1831–1931*, Wo 351.

BOURTON 'Fortified enclosure or village'. OE **burh-tūn**. WW 145–53.

(1) ~ Avon ST 3864. *Burton* 1274. DEPN, WW 152.

(2) ~ Dorset ST 7730. *Bureton(')* 1212–1310, *Burton(')* 1244–1332,

†Room 1983.12 erroneously derives Bournbrook from the r. Bourne more than 1 m to the N.

Borton(') 1268–92, *Bourton* 1774. Do iii.2, WW 147, 151 Dorset no.1.

(3) ~ Oxon 2387. *Burgtun* [821]12th S 183, *Burgh(e)ton(')* 13th–1336, *Borton* 1412, *Burton* 1517, *Bourton* from 1535. Brk 352, WW 150.

(4) ~ Shrops SO 5996. *Burtune* 1086–1833 with variant *Burton('), Burg(h)ton* 1255×6, 1316, *Bourton* from 1291. Sa i.55, WW 150.

(5) ~ ON DUNSMORE Warw SP 4370. *Burton super Dunnesmore* 1235. P.n. *Bortone* 1086, *Burgton* 1232, *Bourton* from 1298, + p.n. *Dunesmore* 1189×99, *Dunnes-* 1235–1315, *Dounesmore* 1330–1424, probably 'Dunn's moor', OE pers.n. *Dun(n)*, genitive sing. *Dunnes*, + **mōr**, an open stretch of land to the S and SW of Rugby. Cf. DUNSMORE Bucks SP 8605. Wa 127 gives pr [bo:tən], 12, L 54–5.

(6) ~ ON-THE-HILL Glos SP 1732. *Burton super montem* 1535, ~ *upon the hill* 1552. Earlier simply *Bortvne* 1086, *Burton'* 1173–1536, *Bourton* c.1195–1495. Also known as Bourton in Henmarsh, *Burton' in Hennemeyrs'* 1274. Gl i.236, WW 149.

(7) ~ ON-THE-WATER Glos SP 1620. *Burton super aquam* 1535, ~ *upon the water* 1690. Earlier simply *Burchtun* *[714]16th S 1250, *(to) Burhtune, (into) Burghtune* [949]c.1600 S 550, *Burghton* 1221, 1375, *Bortune* 1086, *Burton* 1195–1235. The reference is probably to the large four-sided Iron Age hill-fort at Salmonsbury SP 1620, *Sulmonnesburg* 779 S 114, 'ploughman's camp', ie. where he keeps his oxen, OE ***sulh-man**, genitive sing. ***sulh-mannes**, + **burh**. Gl i.195, WW 149, Verey i.131.

(8) Black ~ Oxon SP 2804. ModE adj. **black** supposed to refer to the black habit of the Austin canons who held the estate + p.n. *Bor- Bvrtone* 1086, *Burt(h)ona -tun(') -ton(')* 1122–1385, *Bourton* from 1360. O 306 gives pr [bɔ:tən] as commoner than [bɛ:tən], WW 150.

(9) Flax ~ Avon ST 5069. 'Bourton where flax grows'. ModE **flax** + p.n. *Buryton* 1260 Ass, *Bricton* 1276, *Bratton* 1316, *Boryton* 1327, uncertain element + OE **tūn**. The forms are too varied for explanation although those of 1260 and 1327 point to OE *byriġ*, dative sing. of *burh*: the prefix may allude to Flaxley Abbey which possessed the principal estate in the parish. DEPN, Wade 136.

(10) Great ~ Oxon SP 4545. *Magna Burton(')* c.1265–1346, *Mucheleburton* 1323. ModE **great**, Latin **magna**, ME **muchel** + p.n. *Burton'* [1208×13, 1209×12]1300. 'Great' for distinction from adjoining Little Bourton, *Burton Parvum* 1278×9, also simply *Bourton* 1317. O 414.

BOUTS H&W SO 0358 → INKBERROW SO 0157.

BOVENEY Oxon SU 9377. '(The place) above the island'. *Boveni(a)e* 1086, *Bo- Buuenia -eie -eia -eya -eye* 1155–1226. OE preposition **bufan** + **ēġe**, dative sing. of **ēġ**. Boveney is just above a small island in the Thames. Bk 215, PNE i.56, L 39, DEPN.

North BOVEY Devon SX 7483. *Northebovy* 1199. Earlier simply *Bovi* 1086. ME **north** + r.n. BOVEY. 'North' for distinction from BOVEY TRACEY SX 8178, *Sutbovi* 1219. D 470.

River BOVEY Devon SX 7583. Uncertain. *aqua de Boui* 1238, *Bouy, Bouie* 1577, 1586, *the Bovey* c.1620. A pre-English r.n. comparable with Bobbio, Italy, *Bobium*. D 2 gives pr [bʌvi], RN 44.

BOVEY TRACEY Devon SX 8178. 'Bovey held by the Tracy family'. *Bovitracy* 1309, *Boveytracy* 1401. Earlier simply *Buui* before 1093, *Bovi* 1086, *(in) Boveio* 1088, *Bovy* 13th cent. R.n. BOVEY + family name Tracy who held the manor from 1219. Also known as *Sutbovi* 1219 'south Bovey' for distinction from North BOVEY SX 7483 and the lost *Adonebovi* 1086. This represents OE *Ofdune Bovi* '(that part of) Bovey lower down the stream', probably to be identified with Little Bovey SX 8376, *parua Bouy* 1303. Ancient estates often took their name from a river. Affixes were added as and when they were sub-divided. Cf. AFFPUDDLE SY 8093, BRIANTSPUDDLE SY 8193, PIDDLETRENTHIDE SY 7099, PIDDLEHINTON SY 7197, PUDDLETOWN SY 7594, TOLPUDDLE SY 7994 and TURNERS PUDDLE SY 8293. D 466 gives pr [bʌvi].

BOVINGDON Herts TL 0103. 'Bofa's hill'. *Bovyndon -in- -en-* c.1200–1321, *Buvindon* c.1200, *Buvendon(e)* 1216×72, 1222. OE pers.n. *Bōfa*, genitive sing. *Bōfan*, or *Bōfa* + **ing**[4], + **dūn**. Hrt 29.

BOVINGTON CAMP Dorset SY 8388. A modern army camp established in the 1920s. P.n. Bovington + ModE **camp**. Bovington, *Bovintone* 1086, *Bovinton -yn-* 1236–1335, *Bovyngton* 1288–1403, is the 'estate called after Bofa', OE pers.n. *Bōfa* + **ing**[4] + **tūn**. Do i.190, Newman-Pevsner 497.

BOW Devon SS 7201. 'The arch'. *la Bogh* 1281, *Boghe* 1343, *Bowe* 1400. French definite article **la** + ME **bogh** (OE *boga*) believed to refer to an arched bridge over the river Yeo, the earlier name of which was probably *Nymet*, whence the forms *Nymetbogh(e)* 1270, 1311, *Nimete- Nymetbowe* 1281. D 360.

BOW GLond TQ 3738. *Bow(e)* 1594. Short for *Stratford atte Bowe* from 1279, p.n. Stratford + ME *at the bowe* referring to the arched bridge built here in the 12th cent. replacing the original ford to the N. Earlier *Stratford* 1177–1273, *Stretford* 1265–1593, the 'ford (over the river Lea) on the Roman road' to Colchester, Margary no. 3a, OE **strǣt** + **ford**. Mx 134, Encyc 82.

BOWBANK Durham NY 9423. Uncertain. *Bowbanck(e)* 1561, 1575. Probably 'bent, steep hill', ON **bogi** + ODan **banke**; but the specific might be ON **bógr** 'shoulder'. YN 308, PNE i.40.

BOWBURN Durham NZ 3038. 'Crooked stream'. *Bowburn* 1863. ModE **bow** + **burn**. The modern colliery village is named after Bowburn Beck which flows through two tight loops at NZ 3037 and 3137.

BOWCOMBE IoW SZ 4686. '(Place) above the valley'. *Bove- Bouecome* 1086, *Bouecumb(a) -e -comb(e)* 1185–c.1484, *Boecumba -e* 1154×89, c.1200, *Bowcombe* from 1299, *Buccomb* 1635, 1823. OE **bufan** + **cumb**. This explanation is preferred to 'Bofa's coomb' because the tumuli on Barcomb Down above the valley probably mark the meeting place of Bowcombe Hundred. Wt 96 gives prs ['bæukəm] and ['bakəm], L 90, Mills 1996.32.

BOWD Devon SY 1090. 'Curved wood'. *Boghewode* 1281, 1330, *Bowewode* 1345, *Bowood* 1566. OE **boga** + **wudu**. D 590.

BOWDEN Devon SX 8449. 'Settlement in or on a curve of land'. *Bogheton(e)* 1333, 1390, *Boweton* 1392. OE **boga** + **tūn**. The precise significance is unclear. D 331.

BOWDEN HILL Wilts ST 9367. *Bowden Hill* 1817 OS. P.n. *Bovedon(e)* c.1240–70, 1561, *Bouedon* 1374, 1411, *Bowdon* 1561, *Bowden* 1539, '(the place) above the down', OE **(on) bufan dūne**, + pleonastic ModE **hill**. Wlt 103.

Great BOWDEN Leic SP 7588. Latin prefix **magna**, *Mangna* c.1180×1200, *Magna* before 1250–1374, ME **mikel**, *Me- Mikell* late 15th cent., **much**, *Meche, Mich', Moch, Mych* early 16th cent., ModE **great**, *Great* from 1515 + p.n. *Bugedon(e)* 1086†, 1208, *Buggedon' -den(e)* 1173–1227, *Buggenden* 1181–96, *Bughedon'* 1229–47, *Bu(w)edon(e)* 1230–1242, *Boudon(e)* before 1250–1514, *Bowdon'* 1430–1576, 'Bucga's hill', OE pers.n. *Bucga*, genitive sing. *Bucgan*, + **dūn**. The modern form with *-ow-* [au] must develop from ME *ū* < *-uw-* < *-ug-* and shows that the name has been reformed as if from OE *boga-dūn* 'bow or bend hill'. 'Great' for distinction from Little BOWDEN SP 7486. Lei 226.

Little BOWDEN Leic SP 7486. Latin prefix **parva**, *Parva* 1220–1465, ME **litel**, *Litill, Litull, Lytyll* 1509–18 + p.n. Bowden as in Great BOWDEN SP 7588. Lei 228.

BOWDERDALE Cumbr NY 6704. 'Valley of *Buthar-á*, the river of the booth or shelter'. *Butheresdal* 1224, *Boudirdal* c.1270, *Bowthirdale* 1377, *Bowdardall* 1594, *Bootherdale* 1764. ON river n. **Búthar-á*, ON **búth**, genitive sing. **búthar**, + **á**, ME

†DB also has one spelling *Bigedone*.

genitive *Butheres, + **dale** (ON *dalr*). Cf. the Icelandic and Danish r.ns. *Búða(r)á* and *Bodeå*. There is another Bowderdale, *Beutherdal(bek), Boutherdal(beck)* 1322, of identical etymology at NY 1607. We ii.31, Cu 440, SSNNW 108, L 94.

BOWDON GMan SJ 7586. 'Curved hill'. *Bogedone* 1086, *Boudun -donia* 1189×99, *Bou- Bowdon* from 1278, *-den(e)* 15th, *Baw(e)- Baudon* 15th. OE **boga** + **dūn**. The reference is to Bowdon Downs which curve sharply to enclose a depression opening to the NW. Che ii.15 gives pr ['boudən], L 143, 155.

BOWER Northum NY 7583. 'The dwelling'. *Boure* 1296 SR 173 (p), *The Bowre: an ancient pile* c.1715 AA 13.14, *the Bower* pre 1769 Map. OE **būr** 'a dwelling, a chamber' or ON **búr** 'a storehouse'. It was a residence of the Charltons in the Middle Ages and later. Tomlinson 220.

BOWERCHALKE Wilts SU 0223. 'Bower Chalke'. *Burchelke* 1225–68, *-chalke* 1268–1316, *Bourchalk(e)* 1312–80, *Bowerchalke* 1547, *Burg(h)chalke* 1279–1456. Either OE **(ġe)būr** 'a peasant', genitive pl. **būra**, or **būr** 'a dwelling, a chamber' or OE **burh** + OE p.n. *Ċealce as in *to cealcan gemere* *[826]12th S 275 'to the bounds of Chalke', *(æt) Ceolcum* [955]14th S 582, *Cheolca* [974]14th S 799, *Chelche* 1086, *Chelk(e)* 1175–1305, *Chalke* 1242, 1316, *Chawke* 1547, 'the chalky place', OE ***ċealċe** possibly replaced by *ċealc*. 'Bower' for distinction from Broad CHALKE SU 0325. The meaning and origin of Bower are uncertain. In the 16th cent. tenants in Chalke paid grain rents called 'bower corn' which varied according to 'bower custom'. The name might mean 'peasants' Chalke' or 'borough Chalke'. Wlt 203, PNE i.87, Jnl 19.47.

BOWERS GIFFORD Essex TQ 5788. 'B held by the Gifford family'. *Burisgiffard* 1315. P.n. Bowers + family name of William Giffard 1243. Bowers, *Bure* 1065, *Bura* 1086, *Bures* 1157–1292 etc., *Bowers* from 1291, is OE **būr** the sense of which in p.ns. is uncertain. In literary texts it meant 'inner room, chamber' implying a superior dwelling as well as a possible sense 'cottage'. 'Gifford' for distinction from *Bures Tany* 1260 from the family of Richard de Tany 1289. Ess 144 gives pr [dʒifəd], Jnl 2.42.

BOWES Durham NY 9913. 'The bends' sc. in the river Greta. *Bogas* 1148, *Boghes* 12th–1333, *Bouuys* 1241, *Bowes* from 1283. OE **boga**, ON **bogi**, secondary plural **bogas**. YN 304, DEPN.

BOWES MOOR Durham NY 9211. *Bowes Moor* 1863 OS. P.n. BOWES NY 9913 + **mōr**.

Forest of BOWLAND Cumbr SD 6652. *chacea, foresta de Bouland* 1343, 1423, *the forest of Bolland* 1583, 1650. OFr, ME **forest** + p.n. *Bochland(e)* 12th, 1194–1211, *Bohland* c.1150, *Bogelanda* 1192×9, *Boghland* 1355, 1361, *Boeland(am)* 1102×14, c.1150, *Boweland(e)* 1232×4, 1313, *Bouland* c.1140, 1254–1367, *Bowland(e)* 1311–1652, *Bolland* 1548–1613, the 'district characterised by bends', OE **boga** + **land**. The reference is to the sharp bends of the river Hodder which runs through the district. YW vi.209, 112 gives pr [bɔlənd], La 142, L 246, Jnl 17.80.

BOWLAND BRIDGE Cumbr SD 4189. *Bowland Bridge* 1632. Possibly a surname from BOWLAND SD 6548 + ModE **bridge**. We ii.83.

BOWLEY H&W SO 5452. 'Tree-stump or Bola's wood or clearing'. *Bolelei* 1086, *Bol(l)ey(e)* 1305–34. OE ***bola** or pers.n. *Bola* + **lēah**. He 36.

BOWLHEAD GREEN Surrey SU 9138. *Boveled Grene* c.1550, *Bowlled* ~ 1647, *Bowlers Green* 1823. P.n. *Buuelipe* 1216×72, *Bovelith* 1294–1428 with variants *-lythe* and *-lithe*, OE locative phrase **bufan hlithe* 'above the slope', **bufan** + **hlith**, + ModE **green**. Sr 212.

BOWLING GREEN H&W SO 8251. *Bowling Green* 1831 OS. ModE **bowling green**.

BOWMANSTEAD Cumbr SD 3096. No early forms. Probably surname *Bowman* + ModE **stead** 'farmstead' (OE *stede*).

BOWNESS-ON-SOLWAY Cumbr NY 2262. 'Rounded headland'. *Bounes* c.1225, *Bou- Bownes(s)* 1279–1582, *Bogh(e)nes -neys -nesse* 1292–1319, *Bo(a)nes(e), Bo(u)nes(se)* 16th cent., *Bulnes(se)* 17th cent. Either OE **boga** 'bow' + **næss** or ON **bogi** + **nes**. For the modern affix (for distinction from BOWNESS-ON-WINDERMERE) see SOLWAY. Cu 123, SSNNW 108, L 172.

BOWNESS-ON-WINDERMERE Cumbr SD 4096. 'Bull headland'. *Bulebas* (sic for *nas*) c.1200, *Bulnes* 1282, 15th cent., *Bowlnes* 1566, *Bownas* 1675, *-nes(s)* 1706, 1718. OE **bula** + **næss**. For the modern affix (for distinction from BOWNESS-ON-SOLWAY) see WINDERMERE. We i.185 gives pr ['bounes], L 173.

BOWOOD HOUSE Wilts ST 9770. P.n. Bowood + ModE **house**. An 18th cent. country mansion. Bowood, *Bunewode* (sic for *Buue-*) 1216×72, *Bouewode* 1304, is '(the place) above the wood', OE **(on) bufan wuda**. Wlt 257, Pevsner 1975.121.

BOWSDEN Northum NT 9941. 'Boll's valley'. *Bolesden* 1196 P, *Bollesden(e), Bollisdun -don* 13th cent., *Bowsdenn* 1579, *Bowsdon* c.1715 AA 13.12. OE pers.n. **Boll*, genitive sing **Bolles*, + **denu** partly replaced by **dūn**. NbDu gives pr [bauzən], DEPN.

BOWTHORPE Norf TG 1709. 'Outlying farm at the bow or bend'. *Bo(w)ethorp* 1086, *Boytorp* 1183, *Bugetorp* 1230. OE **boga** or ON **bogi** + **thorp**. The reference is to a sharp bend in the river Yare. Only *Boytorp* 1183 points to ON pers.n. **Bōi*. DEPN, SPNN 97.

BOX Glos SO 8600. 'The box-tree'. *Boȝ* 1234, *la Boxe* 1260–1377×99, *Box* from 1374, *the Box* 1737. OE **box**. Gl i.96, L 222.

BOX Wilts ST 8268. 'The box-tree'. *Boczam* (Latin accusative) 1144, *la Boxe -a* 1181–1259, *(la) Box* from 1340. OE **box**. Wlt 82.

BOXBUSH Glos SO 7413. No early forms. Gl iii.205.

BOXFORD Berks SU 4271. 'Box-tree ridge'. *Boxoran* 960 S 687, [968]12th S 761, *(æt) Boxoran, (to) Boxorran* [958]13th S 577, *Boxhora -e* 1167–1402, *Boxore* 1199 etc. with variants *-or(a)*, *Boxforth* 1517. OE **box** + **ōra**. Brk 133, L 222, Jnl 21.17, 19.

BOXFORD Suff TL 9640. 'The box-tree ford'. *Boxford* from 12th. OE **box** + **ford**. DEPN.

BOXGROVE WSusx SU 9007. 'The box-tree grove'. *Bosgrave* 1086, 1280, *Boxgrave* 1225 etc., *Boxgrove* from 1337, *Bosegroue, Bosgrave* 16th. OE **box** + **grāf(a)**. Sx 66.

BOXLEY Kent TQ 7758. 'Box-tree wood or clearing'. *Boselev* (sic) 1086, *Boxlea, Boxelei* c.1100, *Boxley(e) -lega* 1100×7 etc. OE **box**, genitive pl. **boxa**, + **lēah**. PNK 133.

BOXTED Essex TM 0033. 'Beech-tree place'. *Bo- Bucchestedā, Bocstede* 1086, *Boc- Boxsted(e)* 1163–1428, *Boxted* from 1257. OE **bōc** + **stede**. Ess 363, Stead 195.

BOXTED Suff TL 8251. 'The beech-tree or box-tree place'. *Boesteda* 1086, *Bosted* 1196, *Bocsted(e)* 1154–1280, *Boxsted(e)* c.1190–1568, *Boxted(e)* from 1298. OE **bōc** or **box** + **stede**. Beech is more likely than box which is native only in S counties from Glos to Kent. DEPN, Stead 187.

BOXWORTH Cambs TL 3463. 'Bucc's or the buck's enclosure'. *Bochesuuorde -worth* 1086, *Bokesw(o)rth(e)* 1199–1479, *Buckesw(u)rth(e)* 1185, 1228, *Box(e)w(o)rth* 1272, 1332, 1479. OE pers.n. **Bucc* or **bucc**, genitive sing. **Bucces*, **bucces**, + **worth**. Ca 164.

BOYLESTONE Derby SK 1835. Possibly '*Boghyll* farm'. *Boilestun* 1086, c.1166, *Boi- Boyleston(e)* from c.1190. P.n. **Boghyll*, genitive sing. **Boghylles*, + **tūn**. **Boghyll* is not recorded, but would mean 'rounded hill', OE **boga** + **hyll**. Db 532.

BOYNTON Humbs TA 1368. 'Estate called after Bofa'. *Bouinton(e) -tona, Bov-* 1086, *Bouinton(a) -v-* [1114×24]c.1300, 1128×32–1249, *Bouington(a) -tun, Bovyng-* [12th, 13th]c.1300–1353, *Boventon(a)* 1180×90, 1200, *Boi- Boyngton(a)* 1259, 1297, *Boynton* from c.1265. OE pers.n. *Bofa*, genitive sing. *Bofan*, varying with *Bofa* + **ing**[4] + **tūn**. YE 99, BzN 1981.280f.

BOYTON Corn SX 3191. 'Boia's farm'. *Boietone* 1086, *Boyton* from 1291 G. OE pers.n. *Boia* + **tūn**. PNCo 58, DEPN.

BOYTON Suff TM 3747. 'The estate assigned to the servant'.

Boi- Bohtuna 1086, *Boiton(')* 1187–1232, *Boyton(')* from 1204. OE **boia**, ME **boie** 'servant, churl' or pers.n. *Boia* + **tūn**. DEPN, Baron, BzN 1981.365.

BOYTON Wilts ST 9539. 'Boia's farm or village'. *Boientone* 1086, *Boynton* 1366, *Boiton'* 1166, *Boi- Boyton* from 1242. OE pers.n. *Boia*, genitive sing. *Boian*, + **tūn**. Wlt 162, BzN 1981.364, 375–403.

BOZEAT Northants SP 9059. 'Bosa's gate'. *Bosiet(e)* 1086, 1216, 1284, *Bosiat(e) -yat(e)* 1162–1389, *Bozeatt* 1632. OE pers.n. *Bosa* + **ġeat**. The reference is probably to the low pass over the high ground on the boundary between Northants and Bucks. Nth 189 gives pr [bouʒət].

BRABLING GREEN Suff TM 2964. *Babylon Green* 1837 OS. Cf. Babylon Clwyd SJ 3260, Kent TQ 8406, Dorset ST 5816, *Babylondwey* 1531, *Bablynwaye* 1563, *Babylon Hill* 1838. Biblical n. *Babylon* recorded from 1634 in rhetorical reference to a great and luxurious city, used ironically in p.ns. Do iii.295, Field 9.

BRABOURNE Kent TR 1041. 'The broad stream'. *(æt) Bradanburnan* [844×64]10th S 1198, [993 for 996]14th S 877, *Bradeburnan* 863 S 332, *Brade- Brebvrne* 1086, *Braburne* from 1166. OE **brād**, dative case definite form **brādan**, + **burna**. KPN 205.

BRABOURNE LEES Kent TR 0840. 'Brabourne pasture'. *Braborne Lees* 1674, 1799. P.n. BRABOURNE TR 1041 + ModE **lease** (OE *lǣs*). Cullen.

BRACEBOROUGH Lincs TF 0813. Partly uncertain. *Breseburc -burg, Braseborg* 1086, *Bresseburc -burg* 1180–1202, *Bressenburc* 1191, 1194, *Bressing- Brassingburg(h) -yng-* 1212–1505, *Brasborough* 1495, 1538×9, *Brase-* 1551, *Brace-* from 1562×7. There are several possibilities:

(1) 'the fortified place of the bold one(s)', OE **bræsne, bresne**, genitive pl. **bræsna**, + **burh**;

(2) 'Bræsna's fort', OE pers.n. **Brœsna < brœsne*;

(3) 'the brazen fort', OE **bræsen**, definite form feminine **bræsne**, + **burh**;

(4) 'the gadfly fort', OE **brēosa**, genitive pl. **brēosna**, + **burh**. DEPN, Studies 21–2.

BRACEBRIDGE HEATH Lincs SK 9867. P.n. Bracebridge + ME **hethe**, cf. the surname *del heth'* 1392, *the Heath Farm* 1762. Bracebridge, *Brachebrige, Bragebruge* 1086, *Bracebrig(ge)* 1154–1597, *-bridge* from 1549, is unidentified element + OE **brycg** varying with ON **bryggja**. This is probably not a 'brushwood causeway', OE ***bræsc**, carrying the Foss Way but a reference to a bridged crossing of the r. Witham. The pr with [s] for [ʃ] is an AN sound substitution. Lincs i. 190–1, L 66.

BRACEBY Lincs TF 0135. 'Breithr's village or farm'. *Bre(i)zbi* 1086, *Brei- Breyceby -bi* 1206–1327, *Braceby* from 1226×8. ON pers.n. *Breithr*, genitive sing. *Breiths*, + **bȳ**. Perrott 492, SSNEM 38.

BRACEWELL Lancs SD 8648. 'Breith's spring'. *Braisuelle* 1086, *Brai- Braycewell(e)* 1147×50–1428, *Bracewell* from 1303. ON pers.n. *Breithr* possibly replacing OE **Brœġd*, genitive sing. *Breiths, Breiz*, + OE **wella**. YW vi.38, SSNY 153, L 31.

BRACKENFIELD Derby SK 3759. *Bra(c)ken- Brakynfeld* 16th. A late reformation of earlier *Brachentheyt* (sic) 1269, *Brakenth(w)eyt -thweit -thwayt(e) -thwait -twytt -in- -yn-* 1275–16th cent., 'bracken clearing', ON ***brækni** + **thveit**. Db 217, SSNEM 166.

BRACKLESHAM BAY WSusx SZ 8195. P.n. *Braclœshamstede* [714]14th S 42, *Brakelesham* [945]14th S 506, [1150]1233, 1428, *Brac(c)- Braklesham* 1278–1428, 'Braccol's estate', OE pers.n. *Braccol*, genitive sing. *Braccoles*, + **hām** varying with **hāmstede**, + ModE **bay**. Sx 87, Stead 253, ASE 2.34, SS 86.

BRACKLEY Northants SP 5837. 'Bracca's wood or clearing'. *Brachelai* 1086, *Braccalea* c.1170, *Brakele* [1156]1318–1415 with variants *Bracke- Brakke* and *-ley(e) -lee, Brackleye* 1316. OE pers.n. **Bracca* + **lēah**. The traditional etymology first noted by Camden 1610, 'a place full of *Brake* or *Ferne*', does not suit the forms very well but is not entirely impossible. Nth 49.

BRACKNELL Berks SU 8668. '(At) Bracca's nook'. *(on) braccan heal, (of) braccan heale* [942]12th S 482, *Brackenhal(e)* 1185–1241, *Brecknoll* 1607, *Bracknell* 1800. OE pers.n. **Bracca*, genitive sing. **Braccan*, + **healh**, dative sing. **heale**. The reference is probably to Bracknell's situation in the SW corner of Warfield parish. Brk 116, L 101, TC 57.

BRACON ASH Norf TM 1899 → Bracon ASH TM 1899.

BRADBOURNE Derby SK 2052. 'The broad stream'. *Bradeburne* 1086–1340 with variant sps. *-burn(a) -born(e) -bourn, Bradburn(e) -born(e) -bourne* from 1199×1216. OE **brād**, definite form **brāda**, + **burna**. Db 349.

BRADBURY Durham NZ 3128. 'Plank fortification'. *Brydbyrig* [c.1040]12th, c.1104, *Bridbirig* [1072×83]15th, *Bredbir' -beri -bury* 1153×95–1374, *Bradberi -y* c.1240–1534, *Bradbury* from 1381×2. OE **bred** (later replaced by **brād**) + **byriġ**, dative sing. of **burh**. The spellings with *i/y* may point to OE **briden** 'made of boards' instead of **bred**. Cf. BREDBURY GMan SJ 9291. NbDu 28.

BRADDEN Northants SP 6448. 'The broad valley'. *Braden(e)* 1086, 1195, 1203, *Brade(n)den(e)* 1185–1272, *Bradden(e)* from 12th. OE **brād**, definte form **brāde**, + **denu**. The forms with medial *-e(n)-* point to OE *æt thǣm brādan dene* 'at the broad valley'. Nth 40.

BRADDOCK Corn SX 1661. Possibly 'broad oak' or 'broad hook'. *Brodehoc* 1086, *Brothac* 1201, *Bretohk* (sic) 1224, *Brothec* 1284, 1317, *-ok* 1291, *-ek* 1316, *Brodhoke* 1342, *Brodok(e)* 1422 G, 1552, *Brodock* 1553, 1610, *Broadoake* 1658 G, *Bradocke* 1563. OE **brād** + **āc** 'an oak-tree', or OE **hōc** 'a hook of land'. The church is situated on a broad spur or 'hook' of land. Forms with *th* show the results of the E Cornish sound-change of medial *d* > *th* of which the 1201 spelling is the earliest example. The spelling *Brod-* for OE *brād* in the DB form must be added to the handful of other examples of incipient rounding of OE *ā* found in DB p.n. forms. PNCo 58, 31, PND 45 fn.1, DEPN.

BRADENHAM Bucks SU 8297. 'The broad homestead or enclosure'. *Bradeham* 1086–1253, *Bradenham* from 1227, *-hamme* 1535. OE adj. **brād**, definite form dative case **brādan**, + **hām** or **hamm** 6. The 1535 spelling is too late to be decisive in favour of *hamm* 6 'piece of valley bottom hemmed in by high ground'. Bk 175 gives pr [brædnəm].

BRADENHAM Norf TF 9208. '(At) the wide homestead or enclosure'. *Brade(n)ham* 1086, *Westbradeham* 1197, *Estbradeham* 1242. OE adj. **brād**, definite form dative case **brādan**, + **hām** or **hamm**. The Bradenhams occupy a broad valley. DEPN.

BRADENSTOKE Wilts SU 0079. 'The outlying farm belonging to Braydon'. *Bradenestoche* 1086, *-stok(e)* 1198–1255, *Bradenstok* 1224, 1349, *Bradstock(e)* 1636. P.n. Braydon + OE **stoc**. Braydon is the pre-English and unexplained name of an ancient forest near Cricklade, *(silva) Bradon* *[688]12th S 234, 1243, 1611, *(silva) Braden* [796]13th S 149, 1240, 1328, *(on) Bradene* c.925 ASC(A) under year 905, [956]13th S 1577–1321, *Brœdene* c.1100 ASC(D) under year 905, *Braedone* 1272, *Breden -on* 1316, *Braddon* 1424, *Braydon* 1589, 1634. Known also simply as *Stoche* 1086. Wlt 270 gives prs [breidənstouk] and popular [brædstok], 11, Studies 1936.23.

BRADFIELD 'Broad open land'. OE **brād**, definite form **brāda**, + **feld**.

(1) ~ Berks SU 6072. *(æt) Bradanfelda* 990×2 S 1435, *Bradefelt* 1086, *-feld(') -feud* 1167 etc., *Bradfeld(e)* 1316, 1517. Brk 200, L 242.

(2) ~ Essex TM 1430. *Bradefelda* 1086, *-feld* 1198–1339. Ess 329.

(3) ~ Norf TG2733. *Bradefeld* 1177–1347, *Bradfeld(e)* 1316, 1548. Formerly in Suffolk. Nf ii.145.

(4) ~ SYorks SK 2692. *Bradesfeld* 1188, *Bradefeld* 1268–1432, *Brad-* 1290–1604. YW i.221, L 238–9, 242.

(5) ~ COMBUST Suff TL 8957. 'Burnt B'. *Burnebradfeld* 1610. P.n. *bradefella -felda(m)* 1086, *Bradefelde* c.1095, *-feld* 1197, + ME **burned**, ModE **combust** (Latin *combusta*) 'burnt', for distinction from BRADFIELD ST CLARE TL 9057 and BRADFIELD ST GEORGE TL 9160. DEPN.

(6) ~ GREEN Ches SJ 6859. *Bradfield Green* 1831. P.n. Bradfield + ModE **green**. Bradfield, *Bradfield within Minshull Vernon* 1652, earlier *Ligh Bradfield* 1595. *Ligh* represents the OFr definite article *le* which is occasionally confused with OE **lēah**. Che ii.249.

(7) ~ MOORS SYorks SK 2392. *Bradfield Moor* 1771. P.n. BRADFIELD SK 2692 + ModE **moor**. YW i.221.

(8) ~ ST CLARE Suff TL 9057. B held by the St Clare family'. *Bradefeud Sencler* 1254, *St Cleres Bradfeld* 1610. P.n. Bradfield as in BRADFIELD COMBUST TL 8957 + manorial addition from John le Seyncler 1253. DEPN.

(9) ~ ST GEORGE Suff TL 9160. 'St George's B'. P.n. Bradfield as in BRADFIELD COMBUST + saint's name *George*.

(10) High ~ SYorks SK 2692. ModE **high** + p.n. BRADFIELD SK 2692.

BRADFORD 'The broad ford'. OE **brād**, definite form **brāda**, + **ford**. L 67–8.

(1) ~ Devon SS 4207. *Bradeford* 1086, 1242, *Bradafford* 1346. The ford crosses the Torridge. D 131.

(2) ~ GMan SJ 8698. *Bradeford* 1196–1359, *Bradford* 1282. The ford crosses the river Irwell. La 35.

(3) ~ Northum NU 1532. *Bradeford* 1212 BF 203–1296 SR 141, *Bradfurth* 1460. The ford crosses the Warrenburn. NbDu 29.

(4) ~ WYorks SE 1633. *Bradeford(e)* 1086–1412, *-forth(e)* 1190×1220–1547, *Bradford* from 1252, *-forth* 1428–1630. The ford carried a wide road across Bradford Beck in the centre of the town. YW iii.241 gives pr ['bratfəθ], WYAS 331.

(5) ~ ABBAS Dorset ST 5814. 'The abbot's B'. *Braddeford Abbatis* 1386, *Bradforde Abbat* 1486. P.n. Bradford + Latin **abbas**, genitive sing. **abbatis**. Bradford is *Bradford'* [839×55]14th, *apud Bradeford, æt bradan forda* [933]12th S 422, *bradanford* [998]12th S 895, *Bradeford(')* 1086–1460, *Bradford* 1163. The manor belonged to Sherborne abbey. Do iii.294.

(6) ~ LEIGH Wilts ST 8362 > Bradford LEIGH ST 8362.

(7) ~ -ON-AVON Wilts ST 8260. *(æt) Bradanforda be Afne* c.900 ASC(A) under year 652, *Bradeforda* [1001]15th S 899, *Bradeford* 1086–1244, *Bradford* from 1415, *Brodeford* 1340–1485. P.n. Bradford + r.n. AVON ST 91171. Wlt 116.

(8) ~ -ON-TONE Somer ST 1722. P.n. Bradford + r.n. TONE. Earlier simply *(fram) Bradan forda* [882]12th S 345, *Bradefor(d)* 1086, *Bradford* 1610. DEPN.

(9) ~ PEVERELL Dorset SY 6592. 'B held by the Peverell family'. *Bradeford Peverel(l)* 1244–1431, *Bradford Peverell* from 1352, earlier simply *Bradeford(e)* 1086–1543. Various members of the Peverel(l) family are recorded here 1200–14th. Do i.334.

(10) West ~ Lancs SD 7444. *West Bradford* from 1614. ModE **west** for distinction from Bradford WYorks SE 1633, + p.n. *Bradeford(e)* 1086–1348, *Bradford* 1285–1438, *Bradforth(e)* 1497–1641. YW vi.194.

BRADING IoW SZ 6386. 'The Brerdingas, the people living by the *Breord* or ridge'. *Brerdinges* [683]c.1300, 1253–96, *Brerardinz* 1086, *Brerdinge -ynge* 1235–1487, *Brarding(e) -yng(e)* 1218–1583, *Bradinge* 1235–1613, *Bradyng -ing* from 1507. OE folk-n. **Breordingas* < **breord** 'a brim, a margin, a border, a bank' + **ingas**. The reference is to the high ground of Brading Down which may have been called the *Breord*. Wt 49 gives pr [breidn], Mills 1996.32.

BRADLEY 'The broad clearing'. OE **brād**, definite form **brāda**, + **lēah**. L 205.

(1) ~ Derby SK 2246. *Braidelei* 1086, *Brad(e)leia -le(e) -leg(e) -legh(e) -ley(e) -ly* from c.1192. The 1086 form may show influence of ON **breithr**. Db 533, SSNEM 200.

(2) ~ Hants SU 6341. *Bradelie* 1086†, *Bradelega -legh -ligh -ley(e)* 1167–1341. Ha 38, Gover 1958.131.

(3) ~ H&W SO 9860. *Bradanlæh* [723×37]11th S 95, *-læg(e)* [789]11th S 1430, [962]11th S 1301, *-leage* (dative case) [803]11th S 1260, *Bradinleah* 11th, *Bradelege* 1086, *-ley(e)* 1275, 1376, *Bradley* 1418. Wo 166.

(4) ~ Humbs TA 2406. *Bredelou* 1086, *Bredelai* c.1115, *-la(y) -le(a) -ley(e)* 1170–1558, *Bradley* from 1385. Li v.6.

(5) ~ Staffs SJ 8818. *Bradelia -lie* 1086, *Bradeleia -lega -legh -leya -ley(e)* 1154×89–1608, *Bradley(e)* from 1308. St i.133.

(6) ~ GREEN H&W SO 9961. *Bradley Green* 1831 OS. P.n. Bradley + ModE **green**.

(7) ~ IN THE MOORS Staffs SK 0641 → BRADLEY IN THE MOORS SK 0641.

(8) Great ~ Suff TL 6653. *Bradeleya Magna* 1254, *Bradley mag* 1610. ModE **great**, Latin **magna** + p.n. *Bradeleia* 1086. DEPN.

(9) Little ~ Suff TL 6852. *Parva Bradel'* 1199, *Bradeleya Parva* 1254, *Bradley p'ua*. ModE **little**, Latin **parva**, + p.n. *Bradeleia* 1086. DEPN.

(10) Low ~ Lancs SE 0048. *Netherbradley* 1592. Adj. **low**, earlier **nether**, + p.n. *Bradelei -lay -ley* 1086–1372, 1606, *Bradlei -lay -ley* c.1190–1545. 'Low' for distinction from High Bradley SE 0049 *Overbradley* 1579, the two together being known as Bradleys Both. YW vi.11.

(11) Maiden ~ Wilts ST 8038. 'B held by the nuns' sc. of Amesbury. *Maydene Bradelega* 1199×1216, *Maydenbradley* 1547. OE **mæġden** 'a maiden', genitive pl. **mæġdena**, + p.n. *Bradelie* 1086, *-leia* 12th. Also known as *Deverall Puellarum* 1178, 'Maidens' Deverill', Latin **puella**, genitive pl. **puellarum**, + r.n. Deverill. Wlt 172.

(12) North ~ Wilts ST 8555. *North Bradlegh* 1352, *North Broadley* 1632. Earlier simply *Bradlega* 1174, *Bradelege* c.1210. 'North' for distinction from Maiden BRADLEY ST 8038. Wlt 138.

(13) West ~ Somer ST 5536. *West Bradley* 1610, 1817 OS. ModE adj. **west** + p.n. *Bradelega* 1196. DEPN.

BRADLEY IN THE MOORS Staffs SK 0641. P.n. *Bretlei* 1086, *Bredlege* 1274, 1327, *Bradley* 1834 OS, 'wood or clearing where boards are obtained', OE **bred** + **lēah**, + ModE **in the moors**. The name has been assimilated to the form of the numerous other BRADLEY names. DEPN.

BRADMORE Notts SK 5831. 'The broad lake'. *Brademere* 1086–1346, *-mar(e)* 1227–1316, *Bradmore* 1612. OE **brād** + **mere**. The reference is to a large expanse of marshy land SW of the village (Bradmore Moor, Bunny Moor SK 5730, Gotham Moor SK 5430–5530) now drained. Cf. BUNNY SK 5829. Nt 245, L 27.

BRADNINCH Devon SS 9903. 'At the broad ash-tree'. *Bradenese* 1086–1348 with variants *-esse -eis -eys(h) -es(s)h, Brahanies* 1200, *Braeneis, Braeles, Braneys* early 13th, *Bradenasse* 1244, *-assh(e)* 1342, 1349, *-ych -iche* 1306, 1330, *-ynge* 1380, *Bradenynch* 1404, *Brodninche -nidge -niche* 16th. OE *(æt thǣm)* **brādan** (dative sing. definite inflection of **brād**) **æsce** (dative sing. of **æsc**). Cf. WHITNAGE ST 0215. The second *n* is intrusive and is never heard in pronunciation. Occasional forms *Bradanech(e)* 1238–1389 point to confusion with OE **ǣċ**, dative sing. of **āc** 'an oak-tree'. D 555 gives pr [bræniʃ].

BRADNOP Staffs SK 0155. '(The settlement) at the broad marshland enclosure'. *Bradenhop(e)* 1219–1395, *Bradnap(p)e* 1227–1606, *Branop(e)* from c.1240. OE **brād**, definite form dative case **brādan**, + **hop** in the earlier sense 'a plot of enclosed land, especially in marshes'. Much of the land here is marshy. Oakden, PNE i.259.

BRADPOLE Dorset SY 4794. 'The broad pool'. *Bratepolle* 1086, *Bradepol(e)* 1212, 1219, *Bradpole* 1457. OE **brād**, definite form **brāda**, + **pōl**. Do 42, L 28.

†The OE form *(æt) Bradanleage* [909]12th S 377 cited for this name in DEPN does not belong here.

BRADSHAW GMan SD 7312. 'The broad copse'. OE **brād**, definite form **brāda**, + **sceaga**. *Bradeshawe* 1246, *-shagh* 1246–1332, *Bradsha* 1577. La 46, L 209, Nomina 13.109–14.

BRADSTONE Devon SX 3880. '(At) the broad stone'. *(of) brada, (æt) bradan stane* c.970 B 1246, *Bradestone* 1086 DB, *Bradastana* 1086 Exon, *-stona* 1174×83, *Bradstone* 1699, *Braston* 1539. OE **brādan** (dative sing. definite form of **brād**) + **stāne** (dative sing of **stān**). The reference is probably to the capstone of a cromlech now preserved as a stile in a hedge (Baring-Gould 115). D 173 gives pr [brasən].

BRADWALL GREEN Ches SJ 7563. *Bradwall Green* 1831. P.n. Bradwall + ModE **green**. Bradwall, *Bradwall* from before 1226 with variants *Brade-* and *wal(e) -walle, Brad(e)well(e)* 1281–1385, 1724, is 'the broad spring', OE **brād**, definite form **brāda**, + **wælla**. Che ii.265, 267.

BRADWELL 'The broad spring'. OE **brād**, definite form **brāda**, + **wella**. L 31.

(1) ~ Bucks SP 8339. *Bradewelle* 1086–1378, *Brodewelle* 1086. Bk 17, L 31.

(2) ~ Derby SK 1781. *Bradewell(e) -wella* 1086–1335, *Bradwell(e)* from 1285, *Brad(e)wall(e)* 1339–1660. Db 48.

(3) ~ Essex TL 8023. *Bradewell(e)* 1238 etc., *Brodwell* 1594. Ess 282 gives pr [brædl].

(4) ~ GROVE Oxon SP 2408. P.n. Bradwell as in BROADWELL SP 2504 + ModE **grove**. Cf. *Grove Piece, Grove Enclosures* 1776, O 309.

(5) ~ -ON-SEA Essex TM 0006. *Bradfelt in mari* (sic) 1285. Earlier simply *Brad(e)well(e)* 1212–1552 with suffixes *juxta Mare* 1509×47, *next the See* 1522. Ess 209 gives pr [brædəl].

(6) ~ WATERSIDE Essex TL 9907. P.n. BRADWELL TM 0006 + ModE **waterside**. Earlier *Bradwell Wharf* 1805 OS. Referred to in 1285 as *quay in Bradewelle called Hokflet. Hokflet* or *Hacflet* 1086 means 'hook-shaped creek', OE **haca, hōc** + **flēot** referring to Bradwell Creek. Ess 209.

BRADWORTHY Devon SS 3214. 'Broad enclosure'. *Brawordine* 1086 DB, *Bravordina* 1086 Exon, *Braworthy* 1233, *Bradeworthi* 13th, *Bradworthy* 1461. OE **brād** + **worthiġn**. This was a large manor containing many dependent settlements and the sense may be 'broad estate'. D 132.

BRAFFERTON Durham NZ 2921. 'Farm, village at or called *Bradford*'. *Bradfortuna* [1091×2]12th, *Bradfordt'* c.1190, *Bradfertona* [1183]1382, *Brafferton(')* from early 13th, *Braf(f)irton* c.1290–1418. P.n. **Bradford* 'the broad ford', OE **brād** + **ford**, + **tūn**. The p.n. *Bradford* is not independently recorded here. NbDu 29.

BRAFFERTON NYorks SE 4370. 'Settlement by the broad ford'. *Brad- Bratfortune, Bratfortone* 1086, *Braf(f)erton* 1226 etc. OE **brād** + **ford** possibly used as a p.n. *Braford* + **tūn**. Brafferton is near the river Swale. YN 23.

BRAFIELD-ON-THE-GREEN Northants SP 8258. *Brayfield de le Grene* 1503, *Brafeld of the Grene* 1539, *Brayfeild super le Greene* 1663. Earlier simply *Brach(e)sfeld* 1086, *Bragefeld(a)* 1086–1166, *Braunfeld* 12th–1333, *Branfeld(e)* c.1150–1397, *Brafeld* 1241, *Braifeld* 1316, 'open land on *Bragen*', p.n. *Bragen* as in Cold BRAYFIELD Bucks SP 9252 + **feld**. Nth 144 gives pr [breifi:ld], L 241, 243.

BRAGBURY END Herts TL 2621. *Blakborweend* 1451, *Bragberg End* 1598, *Brackberrie end* 1638. P.n. Bragbury + ME **end** sc. of Datchworth parish. Bragbury, *Brakeburue* 1294, *Blakeberwe* 1296, *Brageberwe* 1307, *Brack(e)burgh(e)* 1314, 1429, is 'bracken hill', OE **bracu** + **beorg** confused with **burh**, dative sing. **byriġ**. Hrt 123.

BRAGENHAMM Bucks SP 9028 → DAGENHAM GLond TQ 5084.

BRAIDES Lancs SD 4451. No early forms.

BRAIDLEY NYorks SE 0380. 'The broad clearing'. *Bradeleie* 1270. OE **brād**, definite form **brāda**, + **lēah**. YN 254.

BRAILES Warw SP 3139. 'The hill court'. *Brailes* from 1086 with variant *-ay-*, *Breyles* 1224–1319. PrW ***breʒ** + ***lis** 'a hall, a court, the chief home in a district'. The reference is to the prominent hill rising to 761ft. E of the village. Wa 276, Jnl.1.44, 49.

BRAILSFORD Derby SK 2541. '*Brailes* ford'. *Brailesfordham* 1086, *Brai- Braylesford(e) -forth(e) -is-* 1086 etc. Lost p.n. *Brailes* identical with BRAILES Warw SP 3139 + **ford**. Alternative suggestions for the specific are OE pers.n. **Bræġel*, genitive sing. **Bræġles*, or **bærġels**, **bræġels**, a secondary form of OE **byrġels** 'a burial-place, a tumulus'. Db 536.

River BRAIN Essex TL 7918. *Brain River* 1848. A modern back-formation from the p.n. BRAINTREE TL 7622. Earlier names were *Pod's Brook* 1777 from the family name *Pod*, *Barus* 1586, *ripa de Nottele* 1385 from Black and White NOTLEY TL 7620, 7818, and *aqua de Wyham* 1254 from WITHAM TL 8114. See also RAYNE TL 7222 for a further possibility. Ess 4.

BRAINTREE Essex TL 7622. 'Branca's tree'. *Branchetreu* 1086, *Bran(e)k(e)tre(e)* 1199–1550, *Brangtree* 1412–52, *Braintre(e)* from 1486, *Bayntre* 1513, *Baintree* 1546. OE pers.n. **Branca* + **trēow**. The earlier name was *Great Reyne* 1311–21, *Rein'* 1202, *Resnes* 1240, *Magna Reines* 1248, ModE **great**, Latin **magna**, + p.n. Rayne as in RAYNE TL 7222, properly *Little Rayne* 1557. Ess 415 gives prs [breintri] and [bra:ntri].

BRAISEWORTH Suff TM 1371. 'Gadfly enclosure'. *Briseworde -wrda -uolda* 1086, *wurd(a) -wurða* 1191–1201, *-worth(e)* 1286–1346 including [1189×99]1336, *Bruswrthe* 1254, *Bresworth* 1327, 1367, *Braisworth* 1610. OE **brīosa** + **worth**. DEPN, Baron.

BRAISHFIELD Hants SU 3723. Partly uncertain. *Braisfelde* c.1235, 1346, *Bray(e)sfeld* 1282–1447. This might be 'brushwood open land', OE ***bræsc**, the ancestor of Mod dial. *brash* 'twigs, undergrowth' or 'small stones' (OED brash sb.), + **feld**. This word, however, probably does not occur in Bracebridge in BRACEBRIDGE HEATH Lincs SK 9867 which contains an unknown specific perhaps also appearing here. Ha 38, Gover 1958.181, DEPN.

BRAITHWAITE 'Broad clearing'. ON **breithr** + **thveit**.

(1) ~ Cumbr NY 2323. *Braithait* c.1160, *Brai- Braythwayte -thwait(e) -thweit -thwa(y)t* 1230–1569. Cu 369, S 109 no.2, L 211.

(2) Low ~ Cumbr NY 4242. Adj. **low** + p.n. *Braythweyt* from 1285 with variants *-thueyt -twaythe* etc., *Braithwait in Inglewod forest* 1350. 'Low' for distinction from High Braithwaite, Braithwaite Hall NY 4141 and Braithwaite Shields NY 4241. Cu 225 gives pr [brɛ:θət], SSNNW 109 no.1, L 211.

BRAITHWELL SYorks SK 5394. 'The broad well or stream'. *Bradeuuelle* 1086, *-well(e)* c.1170–1324, *Breith- Braith(e)- Brayth(e)well(e)* 1199–1446. OE **brād**, definite form **brāda**, replaced by ON **breithr**, + **wella**. YW i.132, SSNY 139.

BRAMBER WSusx TQ 1910. 'The bramble thicket'. *Bremre* 956 S 624, *Brenbria -bre* before 1075, [1150]1253, c.1270, *Brembre -a* before 1080–1438 including 1086, *Bremble* 11th–1234, *Brambre* 1295–1505, *Brambrough* 1535. OE **brēmer, brember**. Sx 222, DEPN.

BRAMCOTE Notts SK 5037. 'The cottages in the broom'. *Broncote, Brun(e)cote* 1086, *Bramcote* from c.1175. OE **brōm** + **cot**. In the compound name OE *ō* was shortened to *o/a*. Nt 140.

BRAMDEAN Hants SU 6128. 'Broom valley'. *(of, andlang) bromdene* [924 for ? 824]12th S 283, *(to, on) brómdœne* [932]12th S 417, [1045]12th S 1007, *Bron- Biondene* 1086, *Bramdene* 1436. OE **brōm** with ME shortening to *Bram-* + **denu**. Ha 30, Gover 1958.87, DEPN, L 98.

BRAMERTON Norf TG 2904. 'Settlement by the bramble-thicket'. *Brambretuna* 1086, *Bramerton* 1254. OE **brēmer, brember** + **tūn**. DEPN.

BRAMFIELD Herts TL 2915. Partly uncertain. *Brandefelle* 1086, *Brantefeld(e)* 1151–1318, *Brantfeld(e)* 1310–1546, *Branfeld* 1502, *Brampfeld* 1535, *Bramfield oth. Brantfield* 1766. This is usually taken to be the 'steep open land', OE **brant**, definite form **branta**, + **feld**, but the land slopes only gently from 400ft. W of the village to 288ft. An alternative would be 'burnt open land', OE **brende**, **brænde** + **feld**. Hrt 108 gives common pr [bræmfəl].

BRAMFIELD Suff TM 4074. 'Broom open land'. *Bu'felda', brunfelda* 1086, *Bramfeld* 1163–1346, 1523×4, *-feild* 1610. OE **brōm** + **feld**. DEPN, Baron.

BRAMFORD Suff TL 1246. 'Broom ford'. *Bromford* [1035×44]14th S 1521, *Brum- Bromford* 1212–1338, *Brun- Branfort* 1086, *Bramford* from 1198 with variants *Braun- Braum-* 1219–1507. OE **brōm** + **ford**. DEPN, Baron.

BRAMHALL GMan SJ 8985. 'Broom nook'. *Bramale* 1086, *-hall* from 1612, *Bramall* 1578, *Brom(h)ale* 1181–c.1280, *Brom(m)all -(e)hall* 15th. OE **brōm** + **halh**, dative sing. **hale**. Che i.258.

BRAMHAM WYorks SE 4243. 'Homestead in the broom'. *Bram(e)ha', Braham* 1086, *Brameham* 1126×75–1280, *Bramham* from 1147×53. OE **brōm** + **hām**. YW iv.82, WYAS 332.

BRAMHOPE WYorks SE 2543. 'Broom valley'. *Bra(m)hop* 1086, *Bramhop(a) -hop(p)e* 1156–1561, *Bramup(pe)* 1594, 1665, *Bramop* 1604. OE **brōm** + **hop**. The reference is to a small side-valley of Wharfedale. YW iv.195 gives pr ['braməp]. L 115–6.

BRAMLEY Hants SU 6559. 'Broom clearing'. *Brumelai* 1086, *Bromelega* 1167, *Bromleye -legh -ligh -lygh* 1241–1327. OE **brōm** with ME shortening to *Bram-* + **lēah**, dative sing. **lēaģe**. Ha 39, Gover 1958.125.

BRAMLEY Surrey TQ 0044. 'The woodland clearing growing with broom'. *Brunlei -lege, Bronlei, Brolege* 1086, *Bromlega -leg(e) -le(e)* etc. 1172–1539, *Brumlea -lega -legh(e)* etc. 1173–1313, *Brunlygh* 1335, *Bramleg'* 1235, *-leye* 1382, *Bramley* from 1473. OE **brōm** + **lēah**. Sr 225, L 205.

BRAMLEY SYorks SK 4992. 'Clearing growing with broom'. *Bramelei(a) -le -lia -ley(e) -lay* 1086–1318, *Bramlei -ley -lay* 1199–1596. OE **brōm** + **lēah**. YW i.134, L 205.

BRAMPFORD SPEKE Devon SX 9298, 'Brampford held by the Speke family'. *Bramford Spec* 1275, 1285. Earlier simply *Brenteforlond* 'land at *Brenteford*' [944]14th S 498, *Branfortvne, Bran- Brenford* 1086, *Branford* c.1170–1249, *Bramford* 1194. Although this looks like OE *brōm-ford* with normal shortening to *Bram-*, 'ford where broom grows', the identification of S 498 shows that originally the specific must have been *Brente* of unknown origin. It can hardly be PrW r.n. **Breȝunt -ent* as in BRENTFORD GtrLond TQ 1778 since this is a crossing of the Exe. Perhaps identical with *Brent* in BRENT KNOLL ST 3350, East BRENT ST 3452 etc. referring to high ground at Upton Pyne SX 9197. The manor was held by Richard de Espec c.1170. D 422, Nt xxxvii.

BRAMPTON 'Broom settlement'. OE **brōm** + **tūn** with early shortening of *ō* to *ă*. In many cases the reference is not simply to a landscape growing with broom but to its economic and medicinal value. Not only was it used to make brooms, baskets and a fibre like flax, but it had other uses in consolidating banks and dunes, providing shelter for game and young plantations, as a source of dye and as a material for thatching. It was also used in medicine. Broom buds were a delicacy in salads and a winter fodder for sheep. PNE i.52, Grieve 367.

(1) ~ Cambs TL 2171. *Brantune* 1086, *Branton* c.1155–1291, *Bramtona* c.1150–1242, *Brampton* from 1227. Hu 233.

(2) ~ Cumbr NY 5361. *Brantun* c.1169, *-ton* 1197–1552, *Brampton* from 1169. Cu 65.

(3) ~ Cumbr NY 6723. *Branton(a)* 1178–1627, *Braunton* 1279, *Brampton* from 1283. We ii.114.

(4) ~ Lincs SK 8479. *Branthon* [1054×7]12th, *-tune* [1075×92]12th, 1086, *-tuna* c.1115, *Bramton'* 1206. DEPN, Cameron 1998.

(5) ~ Norf TG 2224. *Brantuna* 1086, *Bramptone* 1254. DEPN.

(6) ~ SYorks SE 4101. *Brantone* 1086, *Bramtun -ton(a)* c.1143, *Brampton* from 1198. Also known as Brampton Bierlow, *-birlagh* 1307, *Biarlawe* 1583, from ON **bȳjar-lǫg** 'a township'. YW i.106.

(7) ~ Suff TM 4381. *Bram- Brantuna* 1086, *Brampton(')* from 1242. DEPN, Baron.

(8) ~ ABBOTS H&W SO 6026. 'B held by the abbot' sc. of Gloucester abbey. *Brvntvne* 1086, *Bromtun* 1242. DEPN.

(9) ~ ASH Northants SP 7987. Short for *Brampton in the Ash* 1712, so called 'on Account of its Preheminence for this Sort of Tree' (Morton 1712). Earlier simply *Branton(e)* 1086–1219, *Brampton* from 12th. Nth 158.

(10) ~ BRYAN H&W SO 3772. 'B held by Brian', probably Brian Unspac fl. 1157×8. *Bramptone Brian* 1275. Earlier simply *Brantune* 1086. DEPN, He 41.

(11) ~ STATION Suff TM 4183. P.n. BRAMPTON + ModE **station**.

(12) Chapel ~ Northants SP 7266. *Chappell Brampton(e)* 1474. ME **chapel** + p.n. *Branton(e)* 1086–1255, *Bramton(e)* 1194–1203, *Brampton(e)* from 12th. There is no trace of the chapel implied by the name which is also for distinction from Church BRAMPTON SP 7165. Nth 79.

(13) Church ~ Northants SP 7165. 'B with the church'. *Chyrche Brampton(e)* 1287. ME **chirche** + p.n. Brampton as in and for distinction from Chapel BRAMPTON SP 7266. Nth 79.

(14) Little ~ Shrops SO 3781. *Little Brampton* 1833 OS. ModE **little** + p.n. *Brompton'* 1255×6, *Bromton* 1346. 'Little' for distinction from BRAMPTON BRYAN H&W SO 3772. Gelling.

(15) Old ~ Derby SK 3371. *Brandune* 1086, *-tune -ton(a)* 1086–1325, *Bramton(e)* 1179–1560, *Brampton(a) -(e)* from 1237. 'Old' by contrast with Brampton SK 3670, *New Brampton* 1803. Db 220, 235.

BRAMSHALL Staffs SK 0633. 'Broom shelf'. *Branselle* 1086, *Brumeshel* 1195, *Bromschulf* 1327. OE **brōm** + **scylf**. The village lies at the edge of an enormous flat boggy field on top of a hill. DEPN, L 187.

BRAMSHAW Hants SU 2615. 'Bramble woodland'. *Bramessage* 1086, *Bremscaue* 1158, *-shawe* 1280, *Brumesaghe* 1186, *-schawe* 1280, *Bremblessath* 1212, *Brambelshagh* 1272, *Brembelshawe* 1342, *Bramshawe* 1642, *Bramshire -sheere* 1670. OE **brēmel** varying with **brōm** + **sceaga**. Ha 39, Gover 1958.205, DEPN.

BRAMSHILL Hants SU 7461. Partly uncertain. *Bromeselle* 1086, *Bromeshelle -hill(e) -hull(e)* 1167–1346, *Bramshill* 1449, *Bramcell* 1530, *Bramzell* 1738. The specific is OE **brōm**, genitive sing. **brōmes** with ME shortening to *Brams-*, + uncertain element, possibly **hyll**, or **scelf** later replaced by **hyll**. Ha 39, Gover 1958.119 which gives pr [bræmzəl], DEPN.

BRAMSHILL PLANTATION Hants SU 7562. P.n. BRAMSHILL SU 7461 + ModE **plantation**.

BRAMSHOTT Hants SU 8432. 'Bramble corner'. *Brenbresete* 1086, *Brembels(h)ete* 1207–91, *-chete* 1230, *-shute* 1316–64. OE **brēmel** + **scēat** varying with **scīete**. The reference is to the same corner of the county as is referred to in GRAYSHOTT SU 8735. Ha 40, Gover 1958.98, DEPN.

Kirk BRAMWITH SYorks SE 6211. 'B with the church'. ME prefix *Kyrk- Kirke-* 1341, + p.n. *Branuuat -uuit(h)e -uuode -uuet* 1086, *Branwyth* 1200, 14th, *Bramwik* 1154, *-wich* 1291, *Bramwith* from 12th, 'the clearing growing with broom', OE **brōm** + ON **thveit** or **vithr** varying with OE **wudu** and **wīc**. 'Kirk' for distinction from South or Sand Bramwith SE 6211, *alia Branuuat* 1086, *Bramwith* 1324–1842, *Sand ~* 1771. YW ii.29, i.13, SSNY 153.

BRAN END Essex TL 6525. 'Burnt end' sc. of Stebbing. *Brandend* 1565, 1617, *Braneend* 1612. ME **brende** + **ende**. Possibly short for *Brandonande* 1422, cf. *Brandon (ende) cross, mylle* 1517, 1536, p.n. *Brandon* 'burnt hill', OE **brende** + **dūn**, + **ende**. Ess 458.

BRANCASTER Norf TF 7743. 'Roman camp where broom grows'. *Bramcestria* [955×1001]14th S 1810, *Broncestra* 1086, *Bramcestre* c.1110, *Brancestr'* 1170. OE **brōm** + **cæster**. This is the Roman fort of Branodunum, *Brano- Branaduno* [c.400]15th ND, from British ***bran(n)o-** 'raven', possibly used as a pers.n. + **duno-** 'hill fort'. It is tempting to see the first element of the RBrit name as preserved in the English name; if this is so, RBrit *Bran-* was early taken as a form of OE *brōm* by the process of folk-etymology. DEPN, RBrit 274.

BRANCEPETH Durham NZ 2238. 'Brand's path'. *Brandespethe* 1155, *Brantes- Brentespethe* 12th cent., *Braundespath'* 1311, *Bra(u)ncepath'* c.1230–1612, *Branspath -peth(e)* 1242×3, *Braunspith -pyth* 1345, 1384, *Braunsepeth* 1418–1564, *Brancepeth* from 1585. ON pers.n. *Brandr*, genitive sing. *Brands*, + OE **peth** varying with **pæth**. Northern dial. *peth* means 'a sunken path, a path steeply descending', while OE *pæth* is frequently used of a Roman road. The Roman road at Brancepeth (Margary no.83) makes a sharp oblique descent to cross Stockley Beck. The legend of the great 'brawn' or boar slain by Hodge of Ferry (Surtees iii.284) is a romantic invention which has led to the naming of a nearby farm *Brawn's Den*. NbDu 29, Arch.J cxviii.141, Nomina 12.33.

New BRANCEPETH Durham NZ 2241. ModE **new** + p.n. BRANCEPETH NZ 2238. A modern colliery village.

BRANDESBURTON Humbs TA 1147. 'Brand's Burton'. *Brandisburtune -tone* 1086, *Branzbortune* 1086, *Brandesburton(a)* [13th]c.1300 etc. ON pers.n. *Brandr*, genitive sing. *Brands*, + p.n. *Bur- Borton* 1086, *Burton* 1228, OE **burh-tūn** 'a fortified enclosure', partly influenced by ON *borg*. YE 74, WW 151.

BRANDESTON Suff TM 1460. 'Brant's estate'. *brandes- Brantestuna* 1086, *Branteston* 1195–1326, *Brandeston* from 1248. OE pers.n. **Brant*, genitive sing. **Brantes*, + **tūn**. DEPN, Baron.

BRANDISTON Norf TG 1321. 'Brant's estate'. *Brantestuna* 1086, *-ton* 1203. OE pers.n. **Brant*, genitive sing. **Brantes*, + **tūn**. DEPN.

BRANDON 'Broom hill'. OE **brōm** + **dūn** with shortening of *ō* to *ă*. L 142, 153–4, 157.

(1) ~ Durham NZ 2339. *Bramdon(e)* 12th, 1260, *Prandon(a)* c.1220, c.1240, *Brandon(')* from c.1250.

(2) ~ Northum NU 0417. *Bremdona* [c.1150]copy, *Breme- Bramdon, Bromdun* 13th cent., *Brandon* from 1350, *Braundon* 1480. Spellings with *-e-* may be by mistake for *o* or may point to an alternative specific OE ***brēmen** 'growing with broom' later replaced by *brōm*. Cf. BRANTON NU 0416. NbDu 30.

(3) ~ Suff TL 7886. *Bromdun* 10th Thorney fragm, *Brandune* [1042×66]12th S 1051, *Brandona, brantona* 1086, *Brandon* 1610. DEPN.

(4) ~ Warw SP 4076. *Brandune* 1086, *-don(e)* from 12th cent., *Braundon* c.1185–1656. Wa 157.

(5) ~ BANK Norf TL 6288. P.n. Brandon as in BRANDON CREEK TL 6091 + ModE **bank**.

(6) ~ CREEK Norf TL 6091. cf. *Brandon Creek Bridge* 1824 OS.

(7) ~ PARK Suff TL 7784. *Brandon Park* 1836 OS. P.n. BRANDON TL 7886 + ModE **park**. The park of Brandon House, a Georgian building. Pevsner 1974.113.

(8) ~ PARVA Norf TG 0707. 'Little Brandon'. *Brandon Parva* 1838 OS. P.n. Brandon + Latin adj. **parva**. Earlier simply *Brandun* 1086, *-don* 1199. There is no 'great' Brandon surviving. DEPN.

BRANDON Lincs SK 9048. 'The high ground by the river Brant'. *Brandune* 1086, *-done* 1166–1642. R.n. BRANT SK 9457 + OE **dūn**. Perrott 377.

BRANDSBY NYorks SE 5872. 'Brandr's farm or village'. *Branzbi* 1086, *Brandesby* 1221×5–1458. ON pers.n. *Brandr*, genitive sing. *Brands*, + **bȳ**. YN 28, SPN 62, SSNY 22.

BRANDS HATCH Kent TQ 5764. Partly uncertain. *Bronkeshach' -esch* 1292, *Bronkeshech'* 1327, *Brontyshecche* 1334, *Barnshatch* 1819 OS. The generic is OE **hæċċ**, Kentish **heċċ** 'a hatch, a gate'. The specific is unclear: an OE **branc* 'a steep slope' has been proposed, cognate with G *Brocken* 'lump, chunk', OHG *brocko*, if this is to be derived from **bronk-* by *n*-assimilation, and ON *brekka* 'steep hillock' < Gmc **brenkōn* of unknown origin. Brands Hatch is on a steep slope. PNK 44.

BRAND SIDE Derby SK 0468. P.n. Brand + **side**. Brand, *Le Brent* 1555, *The Brande* 1614, is 'burnt place, place destroyed or cleared by burning', OE **brende**. Db 372.

BRANE Corn SW 4028. 'Dwelling of Bran'. *Bosvran* 1323–4, *Borrane* 1386, *Brane* 1588. Co ***bod** (later *bos*) + pers.n. *Bran* with mutation of *B-* to *V-*. Bran was a common pers.n. as well as the name of a Welsh mythological figure associated with hill-forts, as in Dinas Bran, Wales, and *Branodunum*, the Roman fort at Brancaster Norf TF 7844. This may be the explanation in this name with reference to Caer Bran Round, the hill-fort at SW 4029, *Caer Brân* 1754, 'fort near Brane'. PNCo 58, Thomas 1976.62.

BRANKSOME Dorset SZ 0692. The name comes from a house called *Branksome Tower* built in 1852 to the designs of the Scottish romantic architect William Burn. The house was named after Branksome Tower, the setting of Sir Walter Scott's *Lay of the Last Minstrel* published in 1805, itself in turn modelled on Branxholm Castle near Hawick, Border NT 4611. Do ii.1, Room 1983.13, Newman-Pevsner 329.

BRANKSOME PARK Dorset SZ 0590. P.n. BRANKSOME SY 4794 + ModE **park**.

BRANNEL Corn SW 9551 → ST STEPHEN SW 9453.

BRANSCOMBE Devon SY 1988. 'Branoc's coomb'. *(æt) Branecescumbe* [878×888]11th S 1507, *(æt) brances cumbe* 1050×72, *Branchescome* 1086, *Brankescumb(e)* 1219, 1310, *Brannescumbe* 1342, *Brauns- Braynescombe* 15th. OW pers.n. *Branoc*, secondary genitive sing. *Branoces*, + OE **cumb**. D 620, L 92.

BRANSDALE NYorks SE 6296. 'Brandr's valley'. *Brannesdale* c.1150, *Brauncedale* 1276, 1301, *Brandesdal'* 1279×81. ON pers.n. *Brandr*, genitive sing. *Brands*, + **dalr**. YN 65, SPN 62.

BRANSFORD H&W SO 7952. 'Ford leading to *Bragen*'. *Bregnesford* [963]11th S 1304, *Bradnesford* 1086, *Branesforde* [c.1086]1190–1316, *Brantesford* 1255, *Bransford* from 1275 with variants *Brauns- Braunce-*. *Branesforde* [716]12th S 83 is spurious; but the *Bragenmonna broc* [963]11th 'brook of the men of *Bragen*' in the bounds of S 1303 referring to a stream flowing into the Teme opposite to Bransford seems to confirm that the specific of Bransford is an old name possibly derived from OE **bræġen**, ***bragen** 'brain, crown of the head' used in a topographic sense of high ground but more probably of pre-English origin. There is a hill-spur at Bransford rising some 60ft. above the Teme. See also BRAFIELD-ON-THE-GREEN Northants SP 8258 and Cold BRAYFIELD Bucks SP 9252. Wo 189.

BRANSGORE Hants SZ 1897. 'Brand's triangle of land'. Cf. *Bransgoer Common* 1757, *Bransgore* 1811 OS, *Bransgrove* (sic) 1817. ME pers.n. or surname *Brand*, genitive sing. *Brandes*, + **gore** (OE *gāra*). The reference is to an area of raised ground between valleys. Gover 1958.223.

BRANSTON Leic SK 8129. 'Brant or Brandr's farm or village'. *Branteston(e)* 1086–1276, *Brandeston' -is-* 1184–1378, *Braundeston' -is-* 1284–1413, *Branston(')* from 1263, *Braunston'* 1216×72–1618. OE pers.n. *Brant*, genitive sing. *Brantes*, or ON pers.n. *Brandr*, genitive sing. *Brands*, + **tūn**. Lei 160.

BRANSTON Lincs TF 0267. 'Brandr's farm or village'. *Branztun(e) -tone* 1086–[1254]13th, *Brancetuna -ton* 1185–1279, *Branston* from 1185. ON pers.n. *Brandr*, genitive sing. *Brands*, + **tūn**. As the pr with [s] shows, this is a Scandinavian coinage, probably for an Anglo-Saxon settlement taken over and partly renamed by the Danes. Perrott 315, SSNEM 188.

BRANSTON Staffs SK 2221. 'Brant's estate'. *Brontiston* [941]14th S 479, *Brantestone* 1086, *-ton* 1230. OE pers.n. *Brant*, genitive sing. *Brantes*, + **tūn**. DEPN.

BRANSTONE IoW SZ 5683. 'Brand's farm or estate'. *Brandestone* 1086–c.1300, *Brondes- Raundeston(e)* c.1240–c.1450, *Brendestone* 1299, *Brens(t)on* 1769–1844 including 1810 OS. Pers.n. *Brand*, genitive sing. *Brandes*, + **tūn**. The same place as *Heantune* [982]14th S 842, 'at the high settlement', OE **hēah**, definite form dative case **hēan**, + **tūn**. Wt 169 gives pr [bræns(t)ən], 145, Mills 1996.33.

BRANT FELL Cumbr SD 6795. 'Steep hill'. *Brant Fell* 1863 OS. N dial. **brant** (OE *brant*) + **fell** (ON *fjall*). YW vi.266.

River BRANT Lincs SK 9457. 'The deep one'. *Brante* [1216×72]1301, 1295, *Brant* from 1316, *Brent* 1503. OE **brant** + **ēa**. Perrott 39, RN 48, Cameron 1998.

BRANTHAM Suff TL 1034. 'The steep village or enclosure'. *Brantham* from 1086, *Braham* 1198–1313. OE **brant** + **hām** or **hamm**. The ground drops steeply away from the church to the r. Stour. DEPN.

BRANTHWAITE 'Clearing where broom grows'. ME **brame** (OE *brōm*) + **thwaite** (ON *thveit*).

(1) ~ Cumbr NY 0525. *Bromthweit* 1210, *Bramtweit -thuait -thwayt* 1212–1369, *Branthwayt(e)* 1230–1570, *Bramptweyt* 1279, *-thwayte* 1544, *Branthate* 1602. Cu 366 gives pr [branθət], SSNNW 218 no.2, L 211.

(2) ~ Cumbr NY 2937. *Braunthwait* 1332, *-thwayt* 1345, *Branthwaite* from 1560. Cu 276 gives pr [branθət], S 218 no.1, L 211.

BRANTINGHAM Humbs SE 9429. 'Village at the steep place'. *Brentingeha'* 1086, *-ham'* 1167, 1202, *Bre(n)dingham, Bre(n)tingha'* 1086, *Brentingeham* 1224, *-ingham -yng-* [1080×6]17th, 1153–1296, *Brantingham* from [1160×80]n.d., *Brantingeham* 1202, 1246. OE ***branting**, an *ing*[2] derivative of *brant* 'steep', locative-dative sing. ***brantinge**, varying with **Brantingas* the 'dwellers at *Branting*, the steep place', genitive pl. **Brantinga*, + **hām**. Situated at the S end of the Yorkshire Wolds in very steep country. YE 221, ING 152 which prefers derivation from *Brant-* or *Brentingas*, 'the people called after Brant or *Brenti'.

BRANTON Northum NU 0416. 'Broom settlement'. *Bremetonam* [c.1135]copy, *Bremtun -ton(e)* 13th cent., *Bromb- Brampton* 14th cent., *Braunton* 1490, *Branton* from 1498. OE ***brēmen**, later replaced by **brōm**, + **tūn**. Cf. nearby BRANDON Northum NU 0417. NbDu 30.

BRANTON SYorks SE 6401. 'Settlement in the broom'. *Branton(e) -tun(a)* 1086–1207, *Bramtun(a) -ton* c.1160–1265, *Brampton* 1246–1771. OE **brōm** + **tūn**. YW i.40.

BRANXTON Northum NT 8937. 'Branoc's farm or village'. *Brankeston* [1195]1335, 1242 BF, 1249, *Branckiston* 1296 SR, *Branxston* 1346, *Branckston* c.1715 AA 13.5. OW pers.n. *Branoc*, secondary genitive sing. *Bran(o)ces*, + **tūn**. NbDu 30.

BRASSINGTON Derby SK 2354. 'Settlement, estate called after Brandsige'. *Branzinctun* 1086, *Brassinton* c.1100, 1371, *Bracintun -tona -yn-* c.1141–1325, *Bras(s)ington(e) -yng-* from 1296, *Brasson, Braston* 1601. OE pers.n. **Brandsiġe* + **ing**[4] + **tūn**. Db 351 gives pr [bræsən].

BRASTED Kent TQ 4755. 'The broad place'. *Briestede* 1086, *Bradesteda* c.1100, 1184, *-sted(e)* late 11th–1399, *Bradsted(e)* 1291–1455, *Brasted(e)* from 1361. OE **brād**, definite form **brāda**, + **stede**. PNK 70, Stead 210.

BRASTED CHART Kent TQ 4653. 'Brasted rough ground'. *Brastead Chart* 1819 OS. Cf. *bosco de Chert* 'Chart wood' 1226. P.n. BRASTED TQ 4755 + ME **chert** (OE *ċert, ċeart*, Surrey dial. *chart* 'rough common overgrown with gorse and bracken'). PNK 71, PNE i.90.

BRATOFT Lincs TF 4764. 'The broad curtilage'. *Bre(ie)toft* 1086, *Breitoft* c.1115, *Braitoft* 1156, 1166. ON **breithr** + **toft**. SSNEM 148.

BRATTLEBY Lincs SK 9480. 'Brot-Ulfr's village or farm'. *Brotulbi* 1086, *Brotol- Brotulebi* c.1115. ON pers.n. **Brot-Úlfr* + **bȳ**. SSNEM 38.

BRATTON Partly uncertain. Occasional spellings in *Brett-* and the ease of confusion between *c* and *t* suggest OE *Bræc-tūn* as the most likely source, 'settlement by the brushwood or uncultivated land', OE **brǣc** + **tūn**. OE *Brāda-tūn* 'the broad farm or village' is unlikely in the absence of ME spellings in *Brade-*. D 30.

(1) ~ Wilts ST 9152. *Bracton(e)* 1196–1333, *Bratton* from 1177. Wlt 145.

(2) ~ CAMP Wilts ST 9051. *Bratton Castle* 1817 OS. P.n. BRATTON ST 9152 + ModE **camp, castle**. The reference is to an Iron Age hill-fort with massive defences. Pevsner 1975.140, Thomas 1980.237.

(3) ~ CLOVELLY Devon SX 4691. 'B held by the Claville family'. *Bratton Clavyle* 1279, ~ *Clovelleigh* 1625, *Bracton Clavile* 1331. Earlier simply *Bratone* 1086, *Bratton(e)* 1242–1427, *Bretton* 1367. Roger de Clavill' held land here in 1254 by right of marriage. The manorial affix has been influenced by the p.n. CLOVELLY. D 173.

(4) ~ FLEMING Devon SS 6437. 'B held by the Fleming family'. *Bratone -œ, Brotone* 1086 DB, *Bratona* 1086 Exon, *Bratton(e)* from 1242, *Bretton* 1285. Baldwin le Fleming held the manor in 1242. D 29.

BRATTON SEYMOUR Somer ST 6729. 'B held by the St Maur family'. P.n. Bratton + family name St Maur. The manor was held by Roger de St Maur c.1400. Earlier simply *Broctvne* 1086, *Brocton* 1195–14th, *Bratton* 1252, 'brook settlement', OE **brōc-tūn**. DEPN.

BRAUGHING Herts TL 3925. 'The Breahhingas, the people called after Breahha'. *Breahingas* [after 825]17th S 1791, *(æt) Bra(c)hingum* c.975×8, *(ad) Brahcingum* [c.975×8]13th S 1497, *Brachinges* 1086, *Brack- Bracchinges* 12th, *Braching(e) -(c)k-* 1135–1251, *Braghing(e) -yng* c.1200–1428, *Brauhing(e)* 1268–1303, *Brawinge -ynge* 1307–39, *Braughing* 1545, *Braffyng* 1571, *Braffin* 1683. OE folk-n. **Breah(h)ingas* < p.n. **Breahha* + **ingas**. Hrt 189 gives prs ['bra:fiŋ, bræfiŋ], ING 25.

BRAUNSTON Leic SK 8306. 'Brant or Brandr's farm or village'. *Branteston(e)* 1167–1304, *Braundeston'* 1231–1410, *Braunston(')* from 1300, *Braunson, Brawnson* 1566. OE pers.n. *Brant*, genitive sing. *Brantes*, or ON *Brandr*, genitive sing. *Brands*, + **tūn**. R 74.

BRAUNSTON Northants SP 5266. 'Brant's farm or village'. *Brantestun, to brantes tune* 956 S 623, *-ton(e)* 1175–1291 with variant *Braunt-, Brandeston(e)* 1086–1294 with variant *Braund-, Braunston* from 1304. OE pers.n. **Brant*, genitive sing. **Brantes*, + **tūn**. Nth 14.

BRAUNSTONE Leic SK 5502. 'Brant or Brandr's farm or village'. *Branteston(e) -is-* 1086–c.1350, *Brandeston(e) -is-* 1242–1381, *Braunteston(e) -d-* 1239–1395, *Braunston(e)* from 1302. OE pers.n. *Brant*, genitive sing. *Brantes*, or ON pers.n. *Brandr*, genitive sing. *Brandes*, + **tūn**. Lei 486.

BRAUNTON Devon SS 4836. Partly uncertain. *Brantone, Bracton* 1086, *Bramton(a)* 1167–1229×35, *Brampton(e)* 1157–1302, *Branton* 13th, *Brauntone* 1269. To these must be added the name of the hundred of Braunton, *Bran(c)tone, Bractona* 1084. If these forms really belong here and are trustworthy, this may be 'settlement on the Branoc', PrW r.n. *Branoc* < *bran* 'cow' + suffix *-oc* + OE **tūn**. If not this would be 'broom settlement, farm or village', OE **brōm** shortened to **brăm** + **tūn**. The first interpretation is supported by the lost name *Brannocminster* 855×60 S 1695, *Brancminstre* [973]14th S 791 for an estate of 10 hides granted out of the royal manor of Braunton by King Æthelbald of Wessex to Glastonbury Abbey. A 14th cent. tradition makes Braunton the burial place of St Brannoc. D 32 gives pr [bra:ntən], 24, DB Devon ii.1, 5.

BRAUNTON BURROWS Devon SS 4535. 'Braunton (sand-)hills'. *Burgh in Brampton* 1570, *Braunton Boroughes* 1609. P.n. BRAUNTON SS 4836 + SW dial **burrow** (< OE *beorg*). D 34.

BRAWBY NYorks SE 7378. 'Bragi's farm or village'. *Bragebi -by* 1086, *Brahebi* 1165, *Brauby* 1301. ON pers.n. *Bragi*, genitive sing. *Braga*, + **bȳ**. YN 57, SPN 61, SSNY 23.

Great BRAXTED Essex TL 8614. *Magna Bracsted* 1206. ModE **great**, Latin **magna** + p.n. *Brac(c)hestedam, Bracstedā* 1086, *Brac- Brak(e)sted(e) -stude* 1228–1313, *Braxsted(e)* 1214–1513, possibly 'fern-brake place', OE **bracu** + **stede**, although **bræc** 'brake, brushwood, thicket' and ***brǣc** 'newly cultivated land' are also possible. 'Great' for distinction from Little Braxted TL 8314, *Parva* 1254. Ess 283, Stead 196, PNE i.45–6.

BRAY Berks SU 9079. Probably 'the marsh'. *Brai(o), Bras* 1086, *Brai, Bray(e)* from 1156. The usual explanation has been OE **brēġ** 'brow of a hill'. However this does not fit the topography and further the absence of *e* spellings is a difficulty unless we posit a West Saxon form **brœġ*. An alternative suggestion is OFr **brai** 'mud' as in the common Fr p.n. *Bray(e)*: the sense would be 'marsh', but OFr elements are rare in DB p.ns. DEPN, Brk 43, L 34, Anglia 102.415.

River BRAY Devon SS 6932. Best explained as a back-formation from BRAYFORD SS 6834, High Bray SS 6934 and South Bray SS 6624, *Brai* 1086. *aqua de Bray* 1249, *Bray* 1577. D 2, RN 49.

BRAY SHOP Corn SX 3374. 'Bray's workshop'. *Bray's Shop* 1728. Surname *Bray* found in the parish registers from 1619 (G) + ModE **shop** 'workshop, smithy'. PNCo 58.

BRAYBROOKE Northants SP 7684. 'The broad brook'. *B(r)adebroc, Baiebroc* 1086, *Brai- Braybroc -broke* 1147–1428. OE **brād**, definite form **brāda**, + **brōc**. The brook here is normally insignificant but 'upon sudden rains swells to a great depth' (Bridges 1791). For this reason, no doubt, the 15th cent. bridge is three-arched and built with massive cutwaters. Nth 110, Pevsner 1973.122.

BRAYDON HOOK Wilts SU 2167 → THEYDON BOIS Essex TQ 4598.

Cold BRAYFIELD Bucks SP 9252. ModE adj. **cold** since the 16th cent. alluding to the bleak exposed situation + p.n. *(æt) Bragenfelda* [967]13th S 750, *Brau- Brakefeld* c.1175, *Brainfeud* c.1218, *Braunfeld* 1247–1376, *Brafeud* 1242, 'open ground by a raised area', OE **bræġen**, ***bragen** 'brain, crown of the head, a hill' + **feld**. The village occupies a raised circular site surrounded on three sides by the river Ouse. The identification of the first form is not quite certain. Bk 3, PNE i.46, ECTV no. 151, L 241, 243.

BRAYFORD Devon SS 6834. 'Ford at or leading to Bray'. *Brayford* 1651. P.n. Bray as in High Bray SS 6934 + **ford**. Bray, *(æt) Brœg* c.970 B 1253, *Braia* 1086, *brai* 1121, is OW ***breȝ** 'a hill'. D 58, 57, RN 49, L 129.

BRAYSTONES Cumbr NY 0005. '(The) broad stones'. *Bradestanes* 1247, 1279, *Brayde-* 1278–94, *Brai- Brey- Braystanes* 1279–1391, *Braystones* from 1279, *Braythestanes* 1294, *Breith-* 1300. OE **brād** replaced by ON **breithr** + OE **stān**. Possibly refers to stepping stones or a ford across the river Ehen. Cu 413, SSNNW 218.

River BRAYTHAY Cumbr NY 3203. 'Broad river'. *Brayza* (for *-tha*) 1154×89, *Braitha* 1157–1235, *Braythay(e) -ey* from 1443. ON **breithr** + **á**. The same name *Breiðá* occurs in Iceland and in the Faroes. We i.5, RN 49.

BRAYTON NYorks SE 6030. Either the 'broad farmstead' or 'Breithi's farmstead'. *Breiðe-tun* c.1030 YCh 7, *Braiþatun* c.1050 YCh 9, *Brai- Brayton(a) -tun* c.1070–1588. ON **breithr** possibly replacing OE **brād**, definite form **breitha**, or ON pers.n. *Breithi*, genitive sing. *Breitha*, + **tūn**. YW iv.24, SPN 64, SSNY 116.

BRAZACOTT Corn SX 2690. 'Brosya's cottage'. *Brosiacote* 1330, *Brassacott* 1618. Surname *Brosia* + ME **cot** (OE *cot*). PNCo 58.

BREACHWOOD GREEN Herts TL 1522. *Brachewoddgrene* 1327. P.n. Breachwood + ME **grene**. Breachwood, *Brachewode* 1327, *le Brachewodde* 1547×53, is the 'wood beside the land newly broken in for cultivation', ME **brach**, **brech** (OE *brǣċ*) + **wode** (OE *wudu*). Hrt 23.

BREADSTONE Glos SO 7100. '(At the) broad stone'. *Bradelestan'* 1236, *Bradeneston* 1273, *Bradeston(e)* c.1250–1522, *Bradston(e)* 1479–1640, *Bredston* 1561. OE **(æt thǣm) brādan stāne** from OE **brād**, oblique case definite form **brādan**, + **stān**. The form *Bradele* shows AN *-l-* for *-n-*. Gl ii.214 gives pr ['bredstən].

BREAGE Corn SW 6128. 'Church of St Breage'. *Egglosbrec* [c.1170]13th, *Vicaria Sancte Breace* 1264, *Eglosbrek* 1380 G, *Seyntbreke* 1495 G, *Breke* 1522, *Brege* 1610. Co **eglos** + saint's name *Breage* from the dedication of the church. St Breage was believed to have been born in Ireland and to have come to Cornwall together with other saints commemorated in the names CROWAN SW 6434, GERMOE SW 5829 and SITHNEY SW 6328. PNCo 59 gives pr [bri:g], older [breig], DEPN.

BREAM Glos SO 6005. 'The rough ground'. *le, la Breme* 1282–1576, *Bream(e)* 17th cent. ME **breme**, possibly a derivation of OE **brōm** 'broom'. The settlement lies on the edge of the forest of Dean in a parish whose name, Newland, indicates 'newly cultivated land'. Gl iii.236.

River BREAMISH Northum NT 9215. 'The roaring river'. *Bromic* [c.1040]12th, *Bromis(h)* 16th cent., *Bremyz -iz* 1293 Ass, *-ish* 1532, *Bramish* 1755. A British r.n. formed with the suffix *-īkā* on the root **brōm-*, the lengthened grade of IE **bhrem-* seen in βρέμω, Latin *fremo*, OHG *breman*, OE *bremman, brēme* (< **brōmi-*), W *brefu*, all with underlying sense 'rumble, roar'. Cf. BREMENIUM Northum NY 8398. Dixon writes that in its upper reaches the Breamish is a 'brawling torrent' well known for the suddenness with which its spates descend and for its destructive power. NbDu 30, RN 49, Dixon 79, BzN 1957.219.

BREAMORE Hants SU 1519. 'Broom-covered moor'. *Brumore* 1086, 1208, *Brum- Brommore* 1196–1332, *Bremmora* 1245, *Brymmore* 1299–c.1490, *Bremmer* 1544, *Breamore* 1558. OE **brōm** + **mōr**. Ha 40, Gover 1958.217 which gives pr [bremə], DEPN.

BREAMORE HOUSE Hants SU 1519. A large late Elizabethan house. P.n. BREAMORE SU 1519 + ModE **house**. Pevsner-Lloyd 143.

BREAN Somer ST 2955. Uncertain. *Brien* 1086, *Bren* 1212, *Breene* 1243, *Broen* 1254, *Breon* 1334, *Brayne* 1610. The run of spellings points to an OE **Brēon* which might be the nominative or dative pl. of an otherwise unknown **brēo*. It has been suggested that the original reference was to BREAN DOWN ST 2858 and that the name is related to PrW ***breȝ** (Brit **brigā*) 'a hill'. DEPN.

BREAN DOWN Somer ST 2858. *Brean Down* 1809 OS. P.n. BREAN ST 2955 + ModE **down**.

BREARTON NYorks SE 3261. 'Briar settlement'. *B(r)aretone* 1086, *Braherton* 1173×85, *Brerton* 1198–1415, *Brearton* 1597. OE **brēr** + **tūn**. YW v.106.

BREASTON Derby SK 4633. 'Brægd's farm or village'. *Braidestune -tone* 1086–1431 with variant *Bray-*, *Brey- Breideston(e) -is-* 1208×15–1428, *Breyston* 1350, *Bra(y)ston(e)*, *Brayson* 16th cent. OE pers.n. **Brœġd*, genitive sing. **Brœġdes*, + **tūn**. Db 431.

BREAST SAND Norf TF 5427. *Breast Sand* 1824 OS. ModE **breast** + **sand**. A sand-bank in The Wash.

BRECKLES Norf TL 9494. 'Meadow by newly-cultivated land'. *Brecc(h)les* 1086, *Brecles* 1254. OE **brēċ** + **lǣs**. The settlement lies at the edge of extensive heathland. DEPN, L 233.

BREDBURY GMan SJ 9291. 'Fortified place built of planks'. *Bretberie* 1086, *Bredburi -y* from c.1190, *Bradbury* c.1270–1638 with variants *-bur(i)e*. OE **bred** + **byriġ**, dative sing. of **burh**. Cf. BRADBURY Durham NZ 3128. Che i.262.

BREDE ESusx TQ 8218. 'Broad valley'. *Bretda* [11th]18th, *Brada* 1166, *Brede* from c.1220. OE **brǣdu**. The reference is to the broad valley called Brede Level. Sx 514.

River BREDE ESusx TQ 8217. A back-formation from the village name BREDE TQ 8218. *Brede* 1801. A possible earlier reference to this river may be *aqua de Bradeham* [c.1200]13th., 'the water of the *hamm* or *hām* of Brede'. Sx 4, RN 50.

BREDENBURY H&W SO 6156. 'The fortification or manor house of boards'. *Brideneberie* 1086, *Briden(e)biria -burch -bury* 1160×70–1341, *Brydenburye* 1585. OE **briden**, **breden**, definite form oblique case, **bridenan**, + **byriġ**, dative sing. of **burh**. He 41.

BREDFIELD Suff TM 2653. 'The open land at or called *Brede*, the plain'. *Brede- Brade- Berdesfella, Brede- Berde- Bradefelda* 1086, *Bredefeld(')* 1286–1402, *Bradefeld -feud* 1275, 1291, *Bredffeld* 1524. OE **brǣdu** probably used as a p.n. + **feld**. DEPN, Baron.

BREDGAR Kent TQ 8860. 'The broad triangle'. *Bradegare* c.1100–1240, *Bradgare* 1205, *Bredgar* 1610. OE **brād**, definite form **brāda**, + **gāra**. The reference is to the triangular point of high ground on which the village stands. PNK 245.

BREDHURST Kent TQ 7962. 'Plank wood'. *Bredehurst* 1240, *Bredhurst* from 1278, *Brid- Bradhurst* 1336, 1416. OE **bred**, genitive pl. **breda**, + **hyrst**. Bredhurst is the 'wooded hill where boards or planks are made or obtained'. Forms in *Brid-* and *Brad-* show assimilation to the more familiar OE *bridda* 'a bird' and *brād* 'broad'. PNK 208, L 198.

BREDON H&W SO 9236. Originally a hill name 'Bre hill'. *Breodun* [772 for 775]11th S 109, [780]11th S 117 with variant *Brea-* [780]11th S 116, [841]11th S 193, *-dvn* 1086, *Bredon* from 1208. PrW ***breȝ** probably used as a proper name by English speakers + **dūn**. The reference is to Bredon Hill SO 9639 which rises to 961ft. The bounds of the people of Bredon are recorded as *Breoduninga gemære* 984. In two charters, S 109 and 116, Bredon is specifically said to be *in provincia Hwicciorum* 'in the province of the Hwicce'. Wo 101, L 129, 142, 145, 154.

BREDON HILL H&W SO 9639. *Bredon Hill* 1828 OS. P.n. BREDON SO 9236 + pleonastic ModE **hill**. The original Welsh name of the hill was simply *Bre* to which the Anglo-Saxons added *dūn* and in later times *hill*: all three words mean 'hill'.

BREDON'S NORTON H&W SO 9339 → Bredon's NORTON.

BREDWARDINE H&W SO 3344. 'Board enclosure'. *Brocheurdie* 1086, *Brodewordin* 1160×70, *Bredewr(t)hin -werdin* 1185×9–c.1200, *-wordin* 1291, *Brad(e)w(o)rthin* 1200, 1260, *Bradew(a)rdin* 1219, *Bradwardyn* 1397. OE **bred**, genitive pl. **breda**, + **worthiġn**. Spellings with *Brad-* are due to false association with ME *brode, brad-* 'broad'; the two earlier spellings are aberrant (unless there was an alternative form for this name). He 42.

Long BREDY Dorset SY 5690. *Langebride* 1086, *-bridie* 1244. OE adj. **lang**, definite form **langa**, + p.n. *Bridian* [987]13th S 1217, by origin a Celtic r.n., now the r. Bride, *Brydie* 1288, meaning 'boiling, gushing stream' from an OW **Brydi* related to Co *bredion* 'to boil'. The r.n. occurs again in BURTON BRADSTOCK SY 4889 and LITTLEBREDY SY 5889. The BH form *(to) brydian* [914×9]16th may belong here or under BRIDPORT SY 4692, LITTLEBREDY or Bredy Farm SY 5089. Do 42, RN 52, BH 108.

BREEDON ON THE HILL Leic SK 4022. P.n. Breedon, 'hill (called) *Bre*', *Briudun* [c.731]8th BHE, *Breodun* [c.890]c.1000 OEBede, *Bre(o)dune* c.1121 ASC(E) under years 675 and 731, –1425, *Bredon(e)* 12th–1610, *Breedon* from 1572. Hill name **Bre*, PrW ***breȝ** (Brit **brigā* 'a hill') + OE **dūn**. The village lies on the S slope of a hill which has been a settled site from antiquity and is crowned with an Iron Age hill-fort within which an Anglo-Saxon monastery was later established. Pleonastic *on the hill* has been added since 1610 to the simplex p.n. Lei 346, L 129, 142, 145, 156.

BREIGHTON Humbs SE 7034. 'Bright farmstead or enclosure'. *Bristune -tone, Bricstune* 1086, *Brigh- Bryghton* [1195×1214]copy, 1298–1567, *Breighton* from 1636. OE **briht, beorht** + **tūn**. YE 239 gives pr [bretən].

Lower BREINTON H&W SO 4739. ModE adj. **lower** + p.n. *Br(e)untune* 1200×19, *Bre- Brahintone* c.1215, 1252, *Brei- Breynton(e)* from 1218, possibly the 'settlement of Bræge', OE p.n. **Bræġe* < PrW ***breȝ** 'a hill' referring to the hill which rises to 277ft. N of the village, genitive sing. **Bræġan*, + **tūn**. 'Lower' for distinction form Upper Breinton SO 4640. He 43.

Upper BREINTON H&W SO 4640. *Upper Brienton* (sic) 1831 OS. ModE adj. **upper** + p.n. Breinton as in Lower BREINTON SO 4739.

BREMENIUM Northum NY 8398. 'Fort by the roaring stream'. Βρεμένιον (Bremenion) [c.150]13th Ptol, *Bremenio, Bremænio* [c.300]8th AI, *Bremenium* [c.700]14th Rav, *Bre(g)uoin* [944]c.1100 HB, *Brewyn* 14th BT. The name of the Roman fort at High Rochester. Brit ***brem-** 'to roar' from the root seen also in the r.n. BREAMISH NT 9215. The reference is to the nearby Sills Burn which was probably originally called **Bremiā* 'the roaring one'. The W forms of the name show the expected development **Breμēn* > [brev̆uin] > [brewuin]. RBrit 276, Antiquity 23.48.

BREMHILL Wilts ST 9873. 'The bramble-bush'. *Brœmel* [937]14th S 434, *Breomel* [937]12th S 436, *Bremel* [937]14th S 434, 1190–1428, *Bremele* 1233–1342, *Bremhill* from 1430, *Bremble* 1561, 1637, *Breme* (sic) 1086. OE **brēmel**. Wlt 86 gives obsolete pr [brimbl].

BRENCHLEY Kent TQ 6741. 'Brænci's wood or clearing'. *Braencesle* c.1100, *Brencheslega -le(ie) -leg(h')* 1184×5–1254, *Bre- Branchelee* 1254 etc. OE pers.n. **Brænċi*, genitive sing. **Brænċes*, + **lēah**. PNK 189, KPN 33, DEPN.

BRENDON 'Broom hill'. OE **brōm** + **dūn** influenced by **brēmel** 'bramble'.

(1) ~ Devon SS 7648. *Brandone* 1086, *-don* 1205, *Bramdon* 12th *Bremdon(e)* 12th–1328, *Brendon* 1275, *Bryndon* 1544. D 58.

(2) ~ COMMON Devon SS 7645. Identical with *la More de Bremdon* 1425. P.n. BRENDON SS 7648 + OE **mōr** and ModE **common**. D 60.

BRENDON HILLS Somer ST 0135. *Brendon Hill* 1809 OS. P.n. Brendon + ModE **hill**. Brendon, *Brunedun* [1204]1313, *Brundon* 1227, is 'hill called Bruna, the brown one', OE **brūn** used as a p.n. *Brūna* or *Brūne* as in Brown Farms ST 0231, *(in) Brunan* 854 BCS 476, *Brune* 1086, + **dūn**. DEPN, L 145, 147.

East BRENT Somer ST 3451. *Estbrente* 1196, *Est Brunte* 1305, *East Brent* 1610. ME adj. **est** + p.n. *Brente* [663 for ?693]14th S 238, [725]12th S 250, [13th]14th, 'the steep place', OE ***brente**, a derivative of the adj. *brant* 'steep' referring to Brent Knoll. Cf. the hundred name *Suðbrenta* 'south Brent' 1086 Exon. Occasional spellings with *-u-* (*Est Brunte* 1305 above) and with *-ie-* (*Brienta* 1228 for BRENTOR Devon SX 4881) led Ekwall to suggest a British origin, **Brigantia*, an *-nt-* suffix formation on **brigā* 'a hill'. There is, however, no certainty about the meaning of such a formation and the only sense attested for W *braint* (OW *bryeint* < Celtic **brigant-i̯on*, cf. *Brigantes*) is 'privilege, prerogative, title, status' etc. The DB form *Brentemerse* 'Brent marsh' 1086 possibly suggests a lost Old European r.n. here as in BRENTFORD GLond TQ 1778, *Brœgenet* 951 etc. S 1450, *Brainte* 1203, the Anglesey Braint, *Bremt* (for *Breint*) [1198]14th and the Swiss *Prin(t)ze* at Brienz, all from the goddess name *Brigantiā*. DEPN, RN 51, Krahe 1964.60, GPC 307.

River BRENT GLond TQ 1682. 'Holy river'. *(of) brægentan* 972×8 S 1451, *(andlang, into) Brægente, Brœingte* c.955 S 1450, *Brainte, Breynte* 1203–1599, *Brent(e)* from 1274, *Brentbrook* 1593. Brit **Brigantiā*. Mx 1, RN 51, GL 32.

South BRENT Devon SX 6960. *Southbrynt* 1497, *Southbrent* 1714. ME adj. **south** + p.n. *Brenta* 1086, 1238, *Brente* 1086–1337, 'a steep place', OE ***brente** referring to Brent Hill SX 7061 rising to 1020ft. 'South' for distinction from Easter and Wester

Overbrent further up the river Avon at SX 6961, *Ouera Brenta* 1304. For an alternative explanation see North BRENTOR SX 4881. D 290.

BRENT ELEIGH Suff TL 9447 → Brent ELEIGH.

BRENTFORD GLond TQ 1778. 'Ford across the river Brent'. *Breguntford* 705 B 115, *(æt) Bregentforda* [781]11th, *Brægent- Bre(ge)ntforda* c.1050 ASC(C, D) under year 1016, *Brain- Breinford* 1222–1341, *Braynt- Brentford* 16th. R.n. BRENT TQ 1682 + OE **ford**. Mx 31, TC 202.

BRENT KNOLL Somer ST 3350. 'Brent hill'. *Brenteknol* 1289. P.n. Brent as in East BRENT ST 3451 + OE **cnoll**. The reference is to an isolated prominent hill rising to 449ft. DEPN.

North BRENTOR Devon SX 4881. *Northbrienta* 1322, *North Brentor* 1809 OS. Contrasts with *South Brentor* 1809 OS, both places lying N and S of a conspicuous rocky hill crowned with a medieval church called Brent Tor SX 4780, *(apud) Brentam* 1154×89, *Brenta* 1238, 1244, *Brienta, Brente* 1228, *Brentetov(re)* 1232–1309, *Brentor* 1423, *Brenter* 1648, OE ***brente** 'a steep place' later taken as a p.n. **Brente*, + **torr** 'a rocky peak'. For possible derivation from Brit **Brigantiā* see East BRENT Somer ST 3451 and BRENT KNOLL ST 3350. D 213, DEPN, LHEB 70, Anglia 44.110.

BRENT PELHAM Herts TL 4330 → Brent PELHAM.

BRENTWOOD Essex TQ 5993. 'The burnt wood'. *(Le) Brend(e)wode* 1254–1376, *Brentwood* 1532. ME **brende** + **wode**. The earliest references are Latin, *Bosco arso* 1176–1264, *boscum arsum* 1222, Latin **boscus** + **arsus**. The wood was formerly parcel of the ancient parish of South Weald TQ 5793. Ess 123.

BRENZETT Kent TR 0027. 'Burnt building, burnt folds'. *Brensete* 1086–1291, *Branzet* 1278. OE **brende** + **(ġe)set**, pl. **setu**. PNK 476, PNE ii. 120.

BRERETON Staffs SK 0516. 'Briar hill'. *Brerdun* 1292, *Brer(e)dun -don* 13th–1483, *Brereton* from 1412, *Brewarton* 1547, *Bruerton* 1577–1801. OE **brēr** + **dūn**. St i.106.

BRERETON GREEN Ches SJ 7764. *Brereton alias Bruerton Grene* 1559, *Breerton Green* 1656, p.n. Brereton + ModE **green**. Brereton, *Bretone* 1086, *Brerton* from c.1100 with variants *Brer- Bruer-* and *-tune -toun*, is 'briar farm', OE **brēr-tūn**, probably referring to the construction of the enclosing fence. Che ii.274 gives pr ['briətən], locally ['brɛətən].

BRERETON HEATH Ches SJ 8164. *Brereton Heath* 1831. P.n. Brereton as in BRERETON GREEN SJ 7764, + ModE **heath**. Che ii.276.

BRESSINGHAM Norf TM 0881. 'Homestead or promontory of the Briosingas, the people called after Briosa or the gadfly'. *Bresing(a)- Brasincham* 1086, *Bry- Brisingeham* [1087×92]12th, 1198, *Brisingham* 1202–54, *Bresingham* 13th. OE folk-n. **Brīosingas* < OE pers.n.**Brīosa* or **brīosa**, genitive pl. **Brīosinga*, + **hām** or **hamm** 2a. Bressingham occupies a hill-spur abutting on fenland. DEPN, ING 133.

BRETBY Derby SK 2923. 'Village or farm of the Britons'. *Bretebi* 1086, *-by* from c.1232, *Bret(t)ebi -by* c.1150–1403. ON **Bretar**, genitive pl. **Breta**, + **bȳ**. Db 623, SSNEM 39.

BRETFORD Warw SP 4277. 'The plank ford'. *Bretford* from [1100×35]copy, *Bredford* 1199–1315. OE **bred** + **ford**. The reference may be to a plank foot-bridge or a post marking the ford. Wa 157, L 69.

BRETFORTON H&W SP 0944. 'Settlement at Bredford'. *Bretfertona* *[709]12th S 80, *Brotfortun* [714]16th S 1250, *Bradferdtuna* [c.860]12th K 289, *Bratfortvne* 1086, *Brat- Bradforton* 1235–1346, *Bretforton(e)* from 13th. OE p.n. **Bredford* 'plank ford' < **bred** + **ford**, + **tūn**. The *Brat- Brad-* forms are due to confusion with the more common name Bradford. Wo 261.

BRETHERDALE Cumbr NY 5705. 'The brothers' valley'. *Bri- Brytherdal(e)* [1154×89]1247, 1292, 16th cent., *Bretherdale* from 12th, *Breredale* 14th cent. ON **bróthir**, genitive pl. **brœthra**, + **dalr**. ON *œ* normally gives ME *ē* which can be raised to *ī*. We ii.42, SSNNW 110, L 96.

BRETHERTON Lancs SD 4720. 'Brothers' or brother's village or estate'. *Bretherton* from [before 1190]1268, *-ir-* 1268–79, *Brotherton* 1292, *Breder- Brodderton* 1501. ON **bróthir**, genitive pl. **brœthra**, + OE **tūn**. La 137 compares the ON p.n. *Brœðragarðr* and (with ON genitive sing. *brœthr*) modern Norwegian Brödre-Aas (Buskerud) and Brörby (Kristians Amt). The significance is either that of a joint inheritance by brothers, or, as in the case of Brödre-Aas, the transfer of an estate by its owner to a younger brother. Jnl 17.78.

River BRETT Suff TM 0145. *Bret* 1735. A back-formation from BRETTENHAM TL 9653. An earlier form of the back-formation is *Breton* 1577, 1586, 1610. RN 52.

BRETTENHAM Norf TL 9398. 'Bretta's homestead or river-bend land'. *Bretham* 1086, 1201, *Breteham* 1170, *Bretenham* 1257. OE pers.n. **Bretta*, 'the Briton', genitive sing. **Brettan*, + **hām** or **hamm** 1. Brettenham lies in a bend of the river Thet. DEPN.

BRETTENHAM Suff TL 9653. 'Bretta's homestead'. *Bret(t)ham, Bretenhama* 1086, *Brethenham* c.1095, 1233, 1275, *Bretenham* 1275, 1610. OE pers.n. **Bretta*, genitive sing. **Brettan*, + **hām**. *Bretta* is a name meaning 'Briton'. DEPN, RN 52.

BRETTON 'Settlement of the Britons'. OE **Brettas**, genitive **Bretta**, + **tūn**.

(1) Monk ~ SYorks SE 3607. *Munk(e)bretton(a)* 1225, *Monk(e)-* from 1300, *Monk(e)burton* 16th cent. ME **munke** + p.n. *Bret(t)one* 1086, *Bretton(a) -tun* 12th–1651. 'Monk' refers to the monks of Bretton Priory founded c.1154 for distinction from West BRETTON. YW i.273.

(2) West ~ WYorks SE 2813. *Westbretton(a)* from 1193×1211. OE **west** + p.n. *Bretone* 1086, *Bretton(a) -tun* 1155–1666. 'West' for distinction from Monk BRETTON. YW ii.99.

BREWHAM Somer ST 7236. 'Homestead, estate on the r. Brue'. *Briweham* 1086, [1162×70]Bruton, *Bru(w)ham* [1139×61–c.1280]Bruton, *Briu- Bruuham* 1218, *Brywham, Bruham* 1610. R.n. BRUE + OE **hām**. DEPN, RN 55.

BREWINNEY Corn SW 4527 → PAUL SW 4627.

BREWOOD Staffs SJ 8808. 'Bree wood'. *Brevde* 1086, *Breoda -e* 1139–66, *Broude* c.1150, *Brewde* 1201–1559, *Brewod(e) -wud(e)* 1200–1496, *Bre(e)wood(e), Braywoode* 16th cent. P.n. *Bree, Bray* < PrW ***breȝ** 'a hill' + OE **wudu**. The forest of Brewood, *foresta de Brewuda* 1187, was a large area of S Staffs extending into Shrops subject to forest law from at least the 12th cent. and disafforested in 1204. Horovitz gives pr [bru:d], St i.35, L 129, 228.

BREYDON WATER Norf TG 4907. *aqua de Breything* c.1450, *Breydon Water* 1838 OS. P.n. *Breþingh'* 1269, *Breything* 1304, *Breyding* 1462, ODan ***breithing** 'a place where a narrow piece of water widens out' + ModE **water**. The reference is to the mouth of the river Yare. RN 478, Nf ii.29.

BRIANTSPUDDLE Dorset SY 8193. 'Brian's Piddle estate'. *Brianis Pedille* 1465, *Bryans, Brians Pud(d)ell, Puddle* 1480–1607, *Briantspuddle* 1687. Pers.n. *Brian* from Brian de Turbervill 1316, 1327, whose family held the manor from 1205, + p.n. Piddle as in r. PIDDLE. Also known as *Turberuile Pudele* 1268, *Turbervilespudel(l)(e) -vyles- -uylys-* 15th, and *Prestepidel(a)* [1093×1100]1313, 1220–1372, the 'priests' Piddle', OE **prēost**, genitive pl. **prēosta**, from its tenure by Godric *presbiter* 'the priest' in 1086. Also known simply as *Pidele* 1086, 1205. Do i.289.

Great BRICETT Suff TM 0350. *Magna Brisete* 1235, *Briset mag:* 1610. ModE adj. **great**, Latin **magna**, + p.n. *Brieseta* 1086, *Brisete* 1188–1344, *Breset* 1203–1402, *Brissete* 1212, 'gadfly fold(s) or stable(s)', OE **brīosa** + **(ġe)set**, pl. **-setu**. 'Great' for distinction from Little B, *Parva Briset* 1212. DEPN gives pr [braiset], Baron.

BRICKENDON Herts TL 3207. 'Brica's hill'. *Brikandun* *[959]12th S 1293, *(æt) Brycandune* [975×1016]11th S 1487,

Birkendune *[1062]13th S 1036, *Birchendone* 1086, *Bri- Brykendon(e)* 1176–1402. OE pers.n. **Brica*, genitive sing. **Brican*, + **dūn**. Hrt 218.

BRICKET WOOD Herts TL 1301. *Briteyghtwod* 1505, *Bricket Wood* 1655. P.n. Bricket + ME **wode** (OE *wudu*). Bricket, *Bruteyt* 1228, *-eghte* 1272×1307, *Brygteyght* 1436, is the 'bright island', OE **beorht, briht** + **ēġeth** in the sense 'piece of dry ground in marshland'. The wood is reported to have occupied a patch of heavy wet boulder-clay amidst surrounding dryer loams. Hrt 97.

Bow BRICKHILL Bucks SP 9034. 'Bolle's Brickhill'. *Bolebrichill, Bolle Brichulle* 1198, *Bowebrykhyll* 1472–80. ME pers.n. *Bolle* (< OE *Bolla* or *Bolli*) + p.n. *Brichell(a)e* 1086, *Brichille* 1152×8, 1160×65, 'hill called Brig', PrW ***brig** (Brit **brīco-*) taken as a p.n. 'hill top' + tautologous OE **hyll**. For a parallel instance of a PrW element taken as a p.n., see BRILL SP 6513. 'Bow' for distinction from Great and Little BRICKHILL SP 9030, 9132. Bk 30 gives pr [bou brikəl], Wa addenda xli, L 129, 171.

Great BRICKHILL Bucks SP 9030. *Magna Brikhelle* 1197, *Moche Brikell* 1505, ModE **much, great**, Latin **magna**, + p.n. *Brichelle* 1086, identical with Brickhill in Bow BRICKHILL SP 9034. Bk 31, Ca, D addenda.

Little BRICKHILL Bucks SP 9132. *Parva Brichulle* 1198. Latin adj. **parva**, ModE **little**, + p.n. *Brichella* 1086, as in Bow BRICKHILL SP 9034. Bk 32.

BRICKLEHAMPTON H&W SO 9842. 'Brihthelm's estate'. *Bricstelmestvne* 1086, *Bricht(th)elmentona* [c.1086]1190, *Brihtellementon* 1204, *Britelhampton* 1327, *Bri(c)klanton* 1340, 1577, *Bricklehampton* from 16th. OE pers.n. *Briht- Beorhthelm* + **tūn**. Wo 190.

BRIDEKIRK Cumbr NY 1133. 'Church of St Bride'. *Bridekirk(e)* from c.1210. Irish saint's name *Bride* + ON **kirkja**. Cu 272, SSNNW 53 no.1.

BRIDESTOWE Devon SX 5189. 'Holy place of St Brigid or Bride'. *Bridestov* 1086, *-stowe* from 1221×30 with variants *Byrighte- Brightes- Brigide-*. 13th cent. Irish saint's name *Brigid* + OE **stōw**. D 177 gives pr [bridistou].

BRIDESTOWE AND SOURTON COMMON Devon SX 5588. P.ns. BRIDESTOWE SX 5189 and SOURTON SX 5390 + ModE **common**.

BRIDFORD Devon SX 8186. Partly uncertain. *(of) Bridafordes (gildscipe)* (genitive case) 1072×1103, *Brede- Brideford* 1086, *Brideford(a)* 1086–1438, *Brudeford* 1253, *Bredeforde* 1399, *Bridford* 1285. The generic is OE **ford** 'a ford', the specific uncertain. Among the possibilities canvassed are **brȳda**, genitive pl. of OE **brȳd** 'brides' ford', **bridda**, genitive pl. of OE **bridd** 'bird ford', ***brȳde** 'gushing, surging stream', ***bryd** 'board, plank'. D 423, NMS 39.12–8.

BRIDGE Kent TR 1854. 'The bridge'. *Brige* 1086, *Brygge* c.1100, *Brigge* 1235, *Brugge(s), Bregg'* 1240. OE **brycg** (Kentish **brecg**) referring to a crossing of Nail Bourne or the Little Stour by the (Roman) road from Canterbury to Dover, Margary no.1a. Cullen.

BRIDGE END Lincs TF 1436. 'The end of the causeway'. *Bridgend(e)* 1255–1556, *Bridge end* from 1607. ME **brigge** + **end**. The reference is to Holland Bridge, *Hoilondebrige* 1199, *pontem de Holond* 1329, *-olandbrigge* 14th cent., a causeway across the fens to Donington maintained by Bridgend Priory. Perrott 123.

BRIDGEFOOT Cumbr NY 0529. No early forms but the bridge across may be identical with that called *Briggethorfin* 'Thorfinn's bridge' c.1260. ON **bryggja** + pers.n. *Thorfinnr* in Celtic word order. Cu 360, SSNNW 111.

Great BRIDGEFORD Staffs SJ 8827. *Great Bridgeford* 1836 OS. ModE adj. **great** + p.n. *Brigeford* 1086, *Brugeford* 1246, either 'the ford by a bridge' or 'with a (foot-)bridge', OE **brycg** + **ford**. 'Great' for distinction from Little Bridgeford SJ 8727, *Little Bridgeford* 1836 OS. DEPN, L 65, 72.

BRIDGE GREEN Essex TL 4636. *Bruggegrene(strete)* 1497. ME **brugge** (OE *brycg*) + **grene**. Ess 527.

BRIDGEMARY Hants SU 5803. 'St Mary's bridge'. *Bridge Mary* 1759. ME **brigge** + saint's name *Mary* from the dedication of Alverstoke church. Ha 41, Gover 1958.27.

BRIDGEND Cumbr NY 3914. *Bridgend* 1859. ModE **bridge** + **end**. Refers to a bridge over Deepdale Beck. We ii.224.

BRIDGERULE Devon SS 2702. 'Bridge held by Ruald'. *Briggeroald* 1238, *Bruge Ruardi* 1242, *Bryggeruwel* 1393. Earlier simply *Brige* 1086, *Brugia* 1121, 'the bridge', OE **brycg**, referring to a crossing of the Tamar. The manor named after the bridge was held by Ruald Adobed in 1086. D 135.

BRIDGES Shrops SO 3996. *Bridges* 1833 OS.

BRIDGE SOLLERS H&W SO 4142. 'B held by the Sollers family'. *Bruge . . . de Solers* 1160×70, *Bruges Solers* 1291. Earlier simply *Bricge, Brigge* 1086. P.n. Bridge, OE **brycg**, + family name *Solers* from Soliers near Caen. Henry de Solers is mentioned in connection with this holding in 1160×70. DEPN, DB Herefordshire 2.48.

BRIDGE STREET Suff TL 8749. *Bridge Street* 1836 OS. ModE **bridge** (over an affluent of the Stow) + dial. **street** 'a hamlet'.

BRIDGETOWN Somer SS 9233. *Bridgetown* 1809 OS. ModE **bridge** + **town** referring to the settlement in Exton township which grew up beside the bridge over the r. Exe.

BRIDGEWATER CANAL Ches SJ 7287. GMan SJ 7287. Originally formed in 1758 at the expense of the Duke of Bridgewater to transport coal from his Worsley mine to Manchester and extended to Runcorn in 1776. Pevsner 1969.25.

BRIDGE YATE Avon ST 6873. 'The gate beside Breach'. *Brech(e)yate* 1554, *Breach(e)yate* 1612–38, *Bridge Yate* 1779. P.n. Breach + Mod dial. **yate** (OE *ġeat*). The p.n. Breach survives in the farm name Breaches near Siston, *la Brech* 1301, 'land newly broken up for cultivation', OE **brēċ**. Gl iii.71, 67.

East BRIDGFORD Notts SK 6943. *Estbrug(g)eford* 1240–1327, *Estbrig(g)eford* 1291–1335, *Eastbridgford* 1786. ME adj. **est** + p.n. *Brugeford(e)* 1086–[1100×35]1241, *Brig(g)eford* 1192–1295, 'the ford by the bridge', OE **brycg** + **ford**. 'East' for distinction from West BRIDGFORD SK 5837. Also known as *Briggeford juxta Neuton* 1291 referring to Newton SK 6841 and *Brygeforthe on Hyll* 1558. Nt 222.

West BRIDGFORD Notts SK 5837. *Westburgeforde* 1572, *West Bridford al. Bridford at the Bridge ende* 1595. ModE adj. **west** + p.n. *Brigeforde* 1086, *Brigeford juxta pontem de Notingham* 1238, *Brygeforde ad pontem* 1298, *Briggeford (atte Briggend)* 1302, 1361, 'the ford by (Nottingham) bridge', OE **brycg** + **ford**. The name was transferred to the settlement S of the river Trent and the additions are for distinction from East BRIDGFORD SK 6943. Nt 231.

BRIDGHAM Norf TL 9686. 'The homestead or water-meadow by the bridge'. *Brugeham* [1042×66]12th S 1051, *Briggeham* 1230. OE **brycg** + **hām** or **hamm** 1. DEPN.

BRIDGNORTH Shrops SO 7193. 'The bridge in the North'. *Brugg' Norht* 1282–1550 with variants *Brug(g)e* and *North(e)*, *Bri- Bryggenorth(e)* 1322–1506, *Bridgenorth* c.1540–1834, *Bridgnorth* 1661. P.n. *Bridge* + pseudo-manorial addition **north** for distinction from other bridge names in England which acquired manorial suffixes, e.g. BRIDGERULE Devon SS 2702, BRIDGE SOLLERS H&W SO 4142, BRIDGWATER Somer ST 3037. Bridge, *(apud) Brugiam* [1100×35]1267, *Bruges* 1121–1459, *Brugg(e)* 1261–1349, is 'the bridge', OE **brycg** frequently + AN nominative sing. inflectional ending *-s*. The name is short for Quatbridge, *(æt) Cwatbrucge* c.924 ASC(A) under year 895, *(æt) Bry(g)cge* 11th ASC(D) under year 896, *Quatt- Cwatbrycge, Brycge* c.1130, 'the bridge in or leading to Quatt', p.n. QUATT SO 7588 + OE **brycg**. Sa i.56.

BRIDGTOWN Staffs SJ 9808. 'Bridge town', a district of Cannock developed beside Churchbridge SJ 9808, cf. *a pasture called*

Chirchebrigge 1538, *Church Bridge* 1834 OS. Land adjoining the bridge belonged to a Lichfield guild which probably maintained the bridge. St i.58, Duignan 39–40.

BRIDGWATER Somer ST 3037. 'Bridge belonging to Walter' sc. de Douai. *Brigewaltier* 1194, *Bridgewater* 1610. Earlier simply *Brvgie* 1086, *in Brugiis* [13th] Buck, 'the bridge', OE **brycg**. Part of the fee of Walter of Douai in 1086 who is said to have built a bridge here across the Parrett. DEPN, PNE i.54, VCH vi.192.

BRIDGWATER BAY Somer ST 1852. *Bridgewater Bay* 1809 OS. P.n. BRIDGWATER ST 3037 + ModE **bay**.

BRIDLINGTON NYorks SE 6296. 'Estate called after Brehtel'. *Bretlinton* 1086, *Bredlin(g)ton(a) -(e)* 1120×9–1618, *Bridlin(g)ton(a) -yng-* 1119×30–1828. OE pers.n. **Brehtel* metathesized from *Berhtel*, + **ing**[4] + **tūn**. For loss of *h* and assimilation of *t* to *d*, cf. BASILDON Essex TQ 7088. YE 100 gives pr [bɔlitən], Redin 139.

BRIDLINGTON BAY Humbs TA 2065. P.n. BRIDLINGTON + ModE **bay**.

BRIDPORT Dorset SY 4692. 'Harbour or market town belonging to or called Bredy'. *Brideport* 1086, *Bridiport* 1157, *Brudiport* 1207, *Bredeport* 1266, *Britport* 1426, c.1540. P.n. Bredy as in Long BREDY SY 5690 + OE **port**. Do 43, BH 108.

BRIDSTOW H&W SO 5824. 'St Brides holy place'. *Bridestowe* 1277. Saint's name *Bride* + OE **stōw**. Earlier W forms are *Lann San Freit*, ~ ~ *Bregit* c.1130 with PrW ***lann** 'enclosure, churchyard'. DEPN, Stōw 192.

BRIERFIELD Lancs SD 8436. A 19th cent. textile community, no early forms. It is either derived from the name of one of the old town-fields of Little MARSDEN SD 8536, an ancient township absorbed into Brierfield and Nelson, or a new creation partly modelled on nearby Briercliffe, *Brerecleve* before 1193, *-clive -clyf -cliffe* 1258–1311, *Brerclif* 1332, 'steep hill where briars grow', OE **brēr** + **clif**. La 85, 86, Room 1983.13, Mills 66.

BRIERLEY Glos SO 6215. 'Briar wood or clearing'. *Brierley* 1830, cf. *Brierlies brooke* 1669. ME **brēr** + **ley**. This place lies in the Forest of Dean. Gl iii.221.

BRIERLEY H&W SO 4956. Partly uncertain. *Bredege* (for *Bretlege*) 1086, *Bradelega* 1086 DBH, *Bradelega* 1160×70, *Brerel(e)y* 1553, 1599. The DBH and 1160×70 forms are clearly 'the broad wood or clearing', OE **brāda** + **lēah**, the 16th cent. forms 'briar clearing', ME **brere**. The DB form points to an original OE **bred** + **lēah**, 'wood or clearing where boards are obtained', in which the rarer element *bred* was early replaced by *brāda*. The identification of the DB spellings with Brierley, and their original forms, remains uncertain. He 123, DB Herefordshire notes to 1.10a.

BRIERLEY SYorks SE 4110. 'Wood or clearing where briars grow'. *Breselai -lie* 1086, *Brerleia -ley -lay* 13th cent., *Brerelay -lai -leia -ley* 1194–1589, *Brirleye -lay* 1271, 1304, *Bryerley* 1566, *Brearley* 1588. OE **brēr** + **lēah**. YW i.268.

BRIERLEY HILL WMids SO 9287. *Brierley Hill* 1834 OS. P.n. *Brereley* 14th, 'the briar wood or clearing', OE **brēre** + **lēah**, + ModE **hill**. DEPN.

BRIGG Humbs TA 0007. 'The bridge', short for 'Glanford bridge'. *punt, pontem de Glanford* 1218–37, *Glanford Brigg* 1235, *pontem de Glamford* 1203, *Glaumfordbrig(ges)* 1331–1475, *Brigg alias Glanford Brigg* 1674, *Brigg(e)* 1681, 1734. P.n. Glanford + ON **briggja**. Glanford is the 'ford where sports are held', *Glanford* 1183, OE **glēam** (influenced by ON *glaumr*) + **ford**. The bridge, and earlier the ford, carries the road across the r. Ancholme. Li ii.117.

BRIGHAM Cumbr NY 0830. 'Homestead near or village at the bridge'. *Brig(g)ham* from c.1175, *Brugeham* 13th cent., *Bridgeham* 1610, *Brigholme* 1649. OE **brycg** influenced by ON **bryggja**, + **hām**. Cu 355, SSNNW 193, L 65.

BRIGHAM Humbs TA 0753. 'The village or farmstead by the bridge'. *Bringeha'* 1086, *Bri- Brygham* 1187×1207–1828. OE **brycg** (influenced by ON *bryggja*) + **hām**. Probably named in relation to North Frodingham with which, together with Church End, it may have formed a continuous settlement. YE 90.

BRIGHOUSE WYorks SE 1423. 'Houses by the bridge'. *Brig(g)huses* 1240, 1298, *-houses* 13th, 14th cents., *Brighouse* from 1324. OE **brycg** with Scand /g/ for English *dʒ* + **hūs**. The reference is to an ancient crossing of the r. Calder. YW iii.76 gives pr ['brigəs].

BRIGHTGATE Derby SK 2659. '*Breach* gate'. *Briche yate* 1543, *Breech yate* 1620, *Breach Gate* 1833. P.n. *Breach* + OE **ġeat**. Breach, *le Breche* 1415, is Fr definite article **le** + ME **brech** (OE *brēċ*) 'land broken up for cultivation'. Db 412.

BRIGHTHAMPTON Oxon SP 3803. 'The estate called after Brihthelm'. *Byrhthelmingtun* [984]14th S 853, *Brist(h)elmeston(e)* 1086–1304, *Briht- Bric(h)t- Bricthtelmeston'* 1161–1202, *Brichelminton'* 1246×7, *Brithelminton* 1268×81, *Brychampton* 1284, *Brighthampton* from 1316. OE pers.n. *Brihthelm* + **ing**[4] + **tūn**. The replacement of *ing*[4] by the genitive sing. *-es-* between 984 and 1086 implies possession by rather than just association with Brihthelm. Cf. BRIGHTWALTON Berks SU 4279. O 329.

BRIGHTLING ESusx TQ 6821. 'The Birhtlingas, the people called after Birhtela'. *(æt) Byrhtlingan* [1016×20]18th S 1461, *Brislinga* 1086, *Bresling* 12th, *Brictling* 1236, *Brichelinges* 1248, *Brightling(e)* 1273–1535, *Brykelyng* 1550. OE folk-n. **Birhtlingas* < pers.n. *Birhtela* + **ingas**, dative pl. **Birhtlingum*. The development of the name shows metathesis of *r*, *Birht-* > *Briht-*, and in some forms substitution of stop [k] for fricative [ç]. The DB form shows the common AN spelling *s* for [ç]. Sx 471, MA 10.23, SS 65.

BRIGHTLINGSEA Essex TM 0816. 'Brihtric or Brihtling's island'.

I. *Brictriceseiā* 1086, *Bric(h)tricheseye -eie* 1212.

II. *Brichtlinges(s)eya -e(ie)* 1096–1253, *Bri- Bryghtling(e)seia -ey(e) -yng-* 1100–1513.

III. *Bryk(ke)l(e)s(h)ey* 1458–1535, *Brykyl- Brikilsey* 1512–13, *Brightlesea* 1613.

Type I is clearly 'Brihtric's island', OE pers.n. *Beorht- Brihtrīċ*, genitive sing. *Brihtrīċes*, + **ēġ**. Type II is 'Brihtling's island', OE pers.n. **Brihtling*, genitive sing. **Brihtlinges*, + **ēġ**. This may refer to a son of Brihtric, it may be a pet form of that name, or it may be a linguistic distortion of it to avoid the sequence *r - r*. Ess 330 gives prs [britlzi, briklzi].

BRIGHTON ESusx TQ 3105. 'Brihthelm's estate'. *Bristelme(s)tune* 1086, *-ton* 12th–13th cents., *Brictelmestune* c.1100, *Bright(h)elmeston* 14th–15th cents., *Bryghteston* 1437, *Brighton* from 1813. OE pers.n. *Brihthelm* (< *Beorhthelm*), genitive sing. *Brihthelmes*, + **tūn**. A strange pseudo-etymological form occurs, *Brighter Limeston* 1636. Sx 291.

BRIGHTON Corn SW 9054. *Brighton* 1888. Probably transferred from BRIGHTON ESusx TQ 3105. PNCo 60.

New BRIGHTON Mers SJ 3093. *New Brighton* 1841. A seaside resort developed in the 19th cent. and named after BRIGHTON ESusx TQ 3104. Che iv.327, Room 1983.76.

BRIGHTSTONE IoW SZ 4282. 'Brihtwig's estate'. *Brihtwiston(e)* 1212–25, *Bristes- Bric(h)tes- Brit(t)es- Brighteston(e)* 1232–1478, *Brixton* 1399–1810 OS, *Brisson* 1720, *Brightstone* 1295, 1431. OE pers.n. *Brihtwīġ*, genitive sing. *Brihtwīġes*, + **tūn**. Possibly to be identified with DB *Weristetone* 1086 owing to *W/B* confusion. Wt 63 gives prs ['braistən, braisn] and ['bri(k)stən], lxii, Mills 1996.33.

BRIGHTSTONE BAY IoW SZ 4380. *Brixton Bay* 1769, 1781, 810 OS. P.n. BRIGHTSTONE SZ 4282 + ModE **bay**. Wt 64, Mills 1996.34.

BRIGHTSTONE FOREST IoW SZ 4384. P.n. BRIGHTSTONE SZ 4282 + ModE **forest**.

BRIGHTWALTON Berks SU 4279. 'Estate called after Beorhtwald'. *(æt) Beorhtwaldingtune* [939]13th S 448,

Brichtwaldi(n)t' [1066×87]1312, *Bristwoldintona* 1087, *Bristoldestone* 1086, *Brichtuuoldestun* c.1100, *Bri(c)th- Bricht- Brig(h)twalton' -ton(e)* 1220–1338, *Brickleton* 1574, *Brightwalton alias Brickleton* 1667, 1755, *Brightwaltham* 1761. OE pers.n. *Beorhtwald* + **ing**[4] + **tūn**. Brk 237, PNE i.294, Signposts 180.

BRIGHTWELL 'Bright spring'. OE **be(o)rht**, definite form **be(o)rhta**, + **wella**.

(1) ~ BALDWIN Oxon SU 6595. 'B held by Baldwin' sc. de Bereford who received the manor in 1373. Earlier simply *(æt) Berhtanwellan, (æt) Byrhtan wellan* [887]11th S 217, *Brete- Britewelle* 1086, *Bric(h)t- Brict(t)e- Brycte- Bryhtwell(e)* 1166–1371, *Bri- Bryght(e)well(e)* 1241–1371. 'Baldwin' and also 'east' (*Estbritwell* 1275) for distinction from BRIGHTWELL-CUM-SOTWELL Oxon SU 5890. O 120, L 31.

(2) ~ -CUM-SOTWELL Oxon SU 5890. Originally two distinct settlements, Brightwell and Sotwell.

I. *Beorhtawille, (æt) Brihtanwylle* [854]12th S 307, *(æt, to) Beorhtanwille* [945]12th S 517, *Bricsteuuelle, Bristowelle* 1086, *Brightewell'* 1220, *Brightwell* from 1517. Also known as *Westbrightwell* 1315 for distinction from BRIGHTWELL BALDWIN or *Estbritwell* 1275 SU 6595.

II. Sotwell, probably 'Sutta's spring or stream', *(æt) Suttanwille* [945]12th S 517, *(æt) Suttan uulle* [948]12th S 536, *(æt) Stottanwille* [957]15th S 641, *-welle* [957×c.958]14th S 1496, *Sotwelle* 1086, OE pers.n. **Sutta*, genitive sing. **Suttan*, + **wella, wylle**. Brk 515, 529, L 31.

(3) ~ Suff TM 2543. *Briðwelle* [1042×66]12th S 1051, *Brihtewella* 1086, *Brightwell(')* from 1336 with variant *Bryght-*. DEPN, Baron.

BRIGNALL Durham NZ 0612. Partly unexplained. *Bring(en)hale* 1086, *Bringenhale* 1280, *Brig(g)enhal(e)* 1227–1335, *Bryngnell* 1544. The generic is OE **halh** 'a nook of land'. The specific has the form of a pers.n. **Bringa* or **Brynga*, genitive sing. **Bringan*, but no such name is known. An OE folk-name **Bryningas*, genitive pl. **Bryninga*, has been suggested, but the forms do not support this and the supposed parallel with BRINNINGTON GMan SJ 9192 no longer holds. Whatever its origin, it has been reshaped under the influence of Northern dial. **brigg** 'a bridge'. YN 302, SSNY 253, L 108, 111.

BRIGSLEY Humbs TA 2501. 'The forest clearing of or by the bridge'. *Brige(s)lai* 1086, *Brig(h)e(s)la* c.1115, *Briggele, Brickelai* 1202, *Brigesle(a) -lay -ley(e)* 1196–1620, *Briggesle(e) -ley -lay -ly(e)* 1202–1679, *Brig(g)elega -ley(a) -le(y)e* 1210–1454. OE **brycg** varying with genitive sing. **brycges** + **lēah**. The bridge carried the road from Watham to Ravendale across Tetney Drain. Li iv.60, SSNEM 202.

BRIGSTEER Cumbr SD 4889. 'Causeway used by steers'. *Brig(g)(e)stere -y-* 1227–1612. OE **stēor** + **brycg** influenced by ON **bryggja** in Celtic word order. Refers either to a causeway across marshland or a bridge across the river Pool to pastures on Helsington Moss. We i.109, SSNNW 219.

BRIGSTOCK Northants SP 9485. 'The bridge settlement'. *Bricstoc* 1086, *Brichestoch* 1095–1312 with variants *Bric(ke)- Brik(k)e-* and *-stok(e), Birckestoc', Birggestok, Berkestok'* 13th, *Briggestok(e)* 1237–1343 with variants *Bryg(g)e- Brige-* and *Brugge-, Brigstoke* 1466. OE **brycg** + **stoc**. This explanation assumes an early assimilation of OE voiced assibilated [dʒ] to unvoiced stop [k] before the following *s*. An alternative view would be that the *Brigge-* form is an instance of popular etymology working on an earlier *bierce-stocc* 'birch-tree stump' with r-metathesis. There is a bridge across Harper's Brook. Nth 158, L 65.

BRILL Bucks SP 8940. 'Hill called Bree'. *Burhella* [1072]1225, *Bruhella* [1072]1200, *Brunhelle* 1086, *Brehill(a) -hulle* 1154×89–1489, *Bryll super montem* 'Brill on hill' 1535. PrW ***bre[ʒ]** 'brow of a hill' taken as a p.n. + OE **hyll**. Bk 118, L 129, 170–1.

BRILLEY H&W SO 2649. 'Burnt clearing'. *Brunleg(e)* 1219–52, 1397, *Bro- Brumlegh(e) -leye -leie* 1233–1337, *Brynlegh* 1259×60, *Brilley* 1535. OE **bryne** partly replaced by **brōm** 'broom' + **lēah**. He 45.

BRIMFIELD H&W SO 5268. 'Open land where broom grows'. *Bru- Bromefeld* 1086, *Bru- Bromfeld(e)* 1174×86–1397, *Brimfield* from 1174×86 with variants *Bry-* and *-feld -feud, Bremell alias Bremfeld* 1576. OE **brōm** + **feld**. He 46.

BRIMINGTON Derby SK 4073. 'Farm, village, estate called after Bremi'. *Brimintune* 1086, *-inton' -enton(e)* 1199–1288, *Brimington(e) -yng-* from 1219, *Brimiton'* 1200, 1243, *Bremington(e) -yng-* 1290–16th cent. OE pers.n. *Brēmi* + **ing**[4] + **tūn**. Db 227.

BRIMPSFIELD Glos SO 9412. 'Bremi or Brymi's open land'. *Bri- Brymesfeld(e)* 1086–1496, *Brumes- Bremesfeld* c.1180–c.1300, *Bri- Brymmesfeld(e)* 1279–1492, *Brymsfeld* 1494–1577, *Brimpsfeld* 1760. OE pers.n. *Brēmi* or **Brymi*, genitive sing. *Brēmes, *Brymes*, + **feld**. Gl i.144 gives pr ['brimzfi:ld].

BRIMPTON Berks SU 5564. 'Estate called after Bryni'. *(æt) Bryning tune* [944]13th S 500, *Brinton(e)* 1086–1305, *Brimtona -ton(e)* 13th cent., *Brimpton* from 1220. OE pers.n. *Brȳni* + **ing**[4] + **tūn**. Brk 239.

BRIMSCOMBE Glos SO 8703 → EASTCOMBE SO 8904.

BRIND Humbs SE 7431. 'The burnt place'. *Brend(e)* 1188×91–1362, *La, Le Brend(e)* 1254–1360, *Brynd(e)* 1432–1612. ME **brende**. The reference is to land cleared by burning, or to a place so destroyed. YE 243.

BRINDLE Lancs SD 5924. 'Brook hill'. *Brumhull* 1203–4, *Burnhull(e)* 1204–92, *Burnul* 1212–51, *Burnhil* 1246, *Burnehill* 1332, *Birne-* 1448, *Brun(e)hill -hull* 1227–54, *Brynhill* 1480, *Bryndill* 1509, *Brundell* 15th, 16th cent. OE **burna** + **hyll**. A small stream rises south of the village. La 134, Jnl 17.76.

BRINDLEY FORD Staffs SJ 8854. *Brindley Ford* 1872. A new settlement to house workers in the extensive coal and iron works here. For the name cf. BRINDLEY HEATH Staffs SJ 9914.

BRINDLEY HEATH Staffs SJ 9914. *Brindley Heath* 1708. P.n. Brindley as in *Brindley Slade* 1735 (dial. **slade** 'a valley'), *Brinsy Coppice* 1698, probably the 'woodland cleared by burning' ultimately < OE **brende** 'burnt' + **lēah**, + ModE **heath**. St i.106, Horovitz.

BRINETON Staffs SJ 8013. Possibly the 'estate called after Bryni'. *Brunitone* 1086, *Brintona* 1116, 1181×4, *Bri- Brynton* 1305–1563, *Brininton* 1211, *Brunton* 1223–1360, *Bruynton* 1310–1408, *Brineton* 1775. OE pers.n. *Brȳni* + **ing**[4] + **tūn**. St i.129.

BRINGEWOOD H&W SO 4673. Probably 'Burrington wood'. *Buringwode* 1301, *Boryngwode* 1333. The place lies in Burrington parish and is probably a compound of the shortened form **Buring* as in BURRINGTON SO 4472 + ME **wode** (OE *wudu*). He 51.

BRINGEWOOD CHASE Shrops SO 4673. *Bringwood Chase* 1832 OS. P.n. *Buringwode* 1301, *Boryngwode* 1333, + ModE **chase**. Bringwood is short for Burrington Wood, p.n. BURRINGTON H&W SO 4472 + ME **wode** (OE *wudu*). He 51.

BRINGHURST Leic SP 8492. 'Wooded hill of the Bryningas, the people called after Bryni'. *Bren(n)ingehurst* 1199, *Brin(n)ingehurst(e)* 1220, 1229, *Brin(n)- Bren(n)inghurst -herst -hirst* [1041×57]12th–1310, *Bry- Bringherst -hyrst -hurst* from 1312. OE folk-n. *Brȳningas* < OE pers.n. *Brȳni* + **ingas**, genitive pl. *Brȳninga*, + **hyrst**. The village is sited on a small hill overlooking the River Welland in what was formerly heavily wooded country. Lei 209, L 198.

BRINGTON Cambs TL 0875. 'Estate called after Bryni'. *Breninctune* 1086, *Bri- Brynin(g)ton* 1253–1286, *Brington* from 13th. OE pers.n. **Bryni* + **ing**[4] + **tūn**. Hu 235.

Great BRINGTON Northants SP 6665. *Moche Bryngton* 1559, *Greate Brinton* 1657. ModE adj. **much, great** + p.n. *Brinintone*

1086, *Bri- Brynton(e)* 1086–1334, *Bry- Brington* from 1325, *Brynkton* 1330, 'the estate called after Bryni', OE pers.n. *Brȳni* + **ing**[4] + **tūn**. 'Great' for distinction from Little BRINGTON SP 6663. Also known as *Chyrchebrynton* 1312. Nth 79.

Little BRINGTON Northants SP 6663. *Little Brinton* 1284. ME adj. **litel** + p.n. Brington as in Great BRINGTON SP 6665. Nth 79.

BRININGHAM Norf TG 0334. 'Homestead of the Bryningas, the people called after Bryni'. *Bruning(a)- Burningham* 1086, *Brini(n)gham* 1254, 1275. OE folk-n. **Brȳningas* < pers.n. *Brȳni* + **ingas**, genitive pl. **Brȳninga*, + **hām**. ING 133.

BRINKBURN PRIORY Northum NZ 1198. *Brenkburn abbey* c.1715 AA 13.8. P.n. Brinkburn + ModE **abbey, priory** (founded c.1135). Brinkburn, by origin a river name, *Brinkeburn(e)* [c.1120]copy, 1158–14th, *Brenkeburn* 1313, 1507, *Brenkborne* 1542, 1663, is 'Brynca's stream', OE pers.n. *Brynca* + **burna**. One early spelling, *Brincaburch* c.1175, 'Brynca's fortified place', may indicate substitution of original OE **burh** with **burna** as in the names NEWBURN NZ 1965 and SOCKBURN NZ 3407. NbDu 31, DEPN.

BRINKHILL Lincs TF 3773. 'Brynca's wood or clearing'. *Brincle* 1086, 1200, *Brincla* c.1115, *Brinckell'* 1212, *Brenkel* 1314. OE pers.n. *Brynca* + **lēah**. DEPN.

BRINKLEY Cambs TL 6254. 'Brynca's wood or clearing'. *Brinchel'* [1154×89]1508, *Bri- Brynkelai -leia -leie -lay -le(e) -leg(h)* -leye 1174×99–1551, *Brinklow* 1576, 1610. OE pers.n. *Brynca* + **lēah**. Ca 115.

BRINKLOW Warw SP 4379. 'Brink' or 'Brynca's tumulus'. *Brinchelau* [1100×35]copy, *-lawa* 1173–1279, *Bry- Brinkelaw(e)* 12th–1298, *-lowe* 1275–1546, *Brynklow* 1502. OE ***brince** or pers.n. *Brynca* + **hlāw**. The village is situated overlooking a steep slope. Wa 98.

BRINKWORTH Wilts SU 0184. 'Brynca's enclosure' or 'the enclosure at the brink'. *Brinkewrða* [1065]14th S 1039, *-worþe -w(o)rth -wurth* 1191–1479, *Brenchewrde, Brecheorde* 1086, *Brenkewrtha -w(u)rth -worth* 1156–1370. OE pers.n. *Brynca* or ***brince** + **worth**. The village lies at the E end of a ridge of higher ground overlooking Brinkworth Brook. Wlt 65.

BRINNINGTON GMan 9192. 'Farm, village, estate called after Bryni'. *Brinintona* [1154×89]17th, *Bry- Brinyn(g)ton -in(g)-* 1308–48, *Brinnington* from 1285. OE pers.n. *Brȳni* + **ing**[4] + **tūn**. Che i.268.

BRINSCALL Lancs SD 6221. 'Burnt huts'. *Brendescoles* [c.1200, 13th]14th, *-schales* 1246. ME **brende** + ON **skáli**. La 132.

BRINSLEY Notts SK 4649. Probably 'Brun's wood or clearing'. *Brunesleia -lega -leya -leg(h) -ley(e)* 1086–1236, *Brunnesley(e)* 1198–1303, *Brynesley* [c.1185]15th, *Brines-* 1280, *Brinnes-* 1316, *Brynnes-* c.1350, 1490. OE pers.n. *Brūn*, genitive sing. *Brūnes*, + **lēah**. A man named *Brun* who held four bovates here in the time of Edward the Confessor may or may not be referred to in this name. Nt 117 records a one-time pr [brʌnzli].

New BRINSLEY Notts SK 4650. *New Brinsley* 1836 OS. ModE adj. **new** + p.n. BRINSLEY SK 4649.

BRINSOP H&W SO 4444. 'Bryni's enclosed valley'. *Brun(e)shop(e)* 1103×7–1305, *Brun(e)- Brinhope* 1143×55–1308, *Bryn(e)shope* 1446–1535. Earlier simply *Hope* 1086. OE pers.n. *Brȳni*, genitive sing. *Brȳnes*, + **hop**. B lies in a small side-valley. He 47, L 113, 116.

BRINSWORTH SYorks SK 4290. 'Bryni's ford'. *Bri- Brynesford'* 1086–1366, *Brynesforth* 1348, *Bri- Brynsford* 1303, *-furth* 16th cent., *Brindsworth* 1822. OE pers.n. *Brȳni*, genitive sing. *Brȳnes*, + **ford**. YW i.177.

BRINTON Norf TG 0335. Partly uncertain. *Bruntuna* 1086, *Brinton* 1197, 1252, *Bryneton* 1291. The most likely etymology is 'Bryni's estate', OE pers.n. *Brȳni* + **tūn**. But the proximity to BRININGHAM TG 0334 suggests that this may have been the 'farm of the Bryningas', OE folk-n. **Brȳningas*, genitive pl. *Brȳninga*, + **tūn**, or the 'estate called or at *Bryning*, the place called after Bryni', OE pers.n. *Brȳni* + **ing**[2], + **tūn**. For a parallel case cf. Great BRINGTON Northants SP 6665. DEPN.

BRISLEY Norf TF 9521. 'Gadfly infested wood or clearing'. *Bruselea* c.1105, *Brisele* 1199, 1254, *Brissele* 1270. OE **brīosa** + **lēah**. DEPN.

BRISLINGTON Avon ST 6270. 'Brihthelm's estate'. *Bristelton -le-* 1194–1243, *Brihtelmeston* 1199. OE pers.n. *Beorht- Brihthelm*, genitive sing. *Brihthelmes*, + **tūn**. The name shows early substitution of *s* for the peripheral phoneme *h*, a substitution common in Anglo-Norman. DEPN.

The BRISONS Corn SW 3431. 'Shoal of rocks, reef'. *Bresan island* 1576, *Breezam* 1588 G, *Bresoms* 1732 G, *Breezon Island* 1750. French **brisant** 'breaker, reef, shoal'. PNCo 60.

BRISTOL Avon ST 5973. 'Assembly place by the bridge'. *(to, of) Brycg stowe* 11th ASC(D) under year 1052, ASC(A) under year 1063, *Bristov -ou* 1086, *Bri- Brystow(e)* 12th–1675, *Bri- Brystol(l)'* 1100–1675. OE **brycg** + **stōw**. The reference is probably to a crossing of the Avon. The modern form with *-ol* is an inverted spelling that arose after the development of AN *-ol-* to *-ou-*. Gl iii.83, TC 60.

BRISTOL AIRPORT Avon ST 5065. A modern development opened after the second world war. P.n. BRISTOL ST 5973 + ModE **airport**.

BRISTON Norf TG 0632. 'Settlement at the gap'. *Burstuna* 1086, *Birston* 1191, *Birs- Burstone* 1254. OE **byrst** + **tūn**. The reference is to a gap in the surrounding hills through which the river Bure flows. Cf. BURSTON TM 1383. DEPN.

River BRIT Dorset SY 4795. A back-formation from the form *Britport* for BRIDPORT SY 4692. *Bruteport water* 1577, *Bride* 1577, 1586, *Brute* 1586, *Bert* 1612. Its original name, *Woch* (for *Woth*) 12th, *Woth* 1288, survives in the p.n. Wooth SY 4795, *Woth* 1207, ~ *Fraunceys* 1276, *Wooth Fraunces* 1405, either 'vocal, sounding stream', OE **wōth** 'sound, voice, melody, song', or 'sweet, pleasant stream', OE ***wōth**, an unmutated variant of *wēthe* 'sweet, mild, pleasant'. Do 43, 162, RN 52, 469.

BRITANNIA Lancs SD 8821. A late 19th cent. name presumably taken from an inn. It is the highest point on the former railway line from Rochdale to Bacup and Ravenstall which opened in 1881. Room 1983.14.

BRITFORD Wilts SU 1628. Apparently 'the bride ford'. *Brutford* *[826]12th S 276, 1243–1428, *Bredford* 1086, *Bretford* 1086–1325, *Brytford* c.1100 ASC(C) under year 1065, 1399, *Britford* from 1115, *Birtford* 1488, 1754. The earliest occurences are *Brytfordingea landscære* [c.670]12th S 229, *Brytfordingealand sceare* [948]11th S 540, 'the boundary of the people of Britford'. OE **brȳd** + **ford**. The significance of such a compound is unknown and the specific may be a different unidentified element. Wlt 220 gives pr [bɔ:fəd], PNE i.55.

BRITWELL SALOME Oxon SU 6793. 'B held by the Sulham family'. *Brutewell(e) Sol(e)ham* 1320–57, *Britwell Salome* from 1613. P.n. *Brutwelle* 1086, *Brutewell'* 1235×6, *Bretewelle* 1220, *Brittewilla* [c.1090]13th, formally 'the spring of the Britons', OE folk-n. *Bryttas*, genitive pl. *Brytta*, + **wella, wylle**, + family name of Almaric de Suleham (Sulham Berks SU 6424) who held the manor in 1235×6. However, the usual term for Britons in p.ns. is *Wealas*, genitive pl. *Weala*; furthermore occurrences elswhere of this name-type, Britwell SU 9582, *Brute- Brittewelle* 13th, Brightwells Farm in Watford, Herts TQ 1097, *Brute- Brittewell(e)* 13th cent., suggest the existence of a hitherto unrecognised compound p.n. of uncertain meaning. The charter form *Brutuwylle* [1042×66]11th S 1047 belongs to Britwell Prior, ECTV 301. O 105 gives local pr [sɔləm], Bk 217, Hrt 104.

BRIXHAM Devon SX 9255. Probably 'Brioc's enclosure'. *Briseham* 1086, 1143, *Brisehamme* 1088, *Brixa- Brixeham* 1143, *Brixham* from 1242, *Brik(k)esham* 1205–1347. OW pers.n. *Brioc*, secondary genitive sing. *Brioces*, + **hamm**.

Alternatively the specific might be OE pers.n. *Beorht- Brihtsiġe*. Cf. BRIXTON SX 5552. D 507.

BRIXTON Devon SX 5552. Probably 'Brioc's farm or village'. *Brisestone* 1086, *Brikeston'* 1200–1376, *Brixton* from 1284, *Bryokeston* 1346. OW pers.n. *Brioc*, secondary genitive sing. *Brioces*, + **tūn**. Alternatively the specific might be OE pers.n. *Beorht- Brihtsiġe*. Cf. BRIXHAM SX 9255. D 249 gives pr [briksən].

BRIXTON GLond TQ 3175. 'Brihtsige's stone'. *(æt) brixges stane, brixes stan* [1062]13th K 813, *Brixi(e)stan(e)* 1086–1279, *Bricsiston* 1169, *Briston* 1316. OE pers.n. *Brihtsiġe* < *Beorhtsiġe* + **stān**. The neighbouring manor of Hatcham was held in 1066 by one *Bricsi*, i.e. *Brihtsige*. The stone may, therefore, have been his boundary stone. Alternatively it may have marked the meeting place of Brixton hundred. Sr 11, 23.

BRIXTON DEVERILL Wilts ST 8638. 'Brihtric's r. Deverill estate'. *Britricheston* 1229, *Bri(g)htricheston* 1242, 1291, *Bri- Bryghteston(e)* 1316, 1332, ~ *Deverell* 1402, *Brixston Deverell* 1555. OE pers.n. *Beorh- Brihtrīċ*, genitive sing. *Brihtrīċes*, + **tūn** + r.n. DEVERILL. This subdivision of what may have once been a single estate along the r. Deverill, originally called simply *Devrel* 1086, was held in 1066 by a man called *Brictric* and subsequently became known as 'Britric's estate or division of Deverill', cf. *Deverill quod fuit Bristricii* [1087]15th, '(that part of) Deverill that was Brictric's'. For other divisions of this land unit see KINGSTON DEVERILL ST 8437, LONGBRIDGE DEVERILL ST 8640 and MONKTON DEVERILL ST 8537. Wlt 165.

BRIXWORTH Northants SP 7470. 'Bricel or Brihtel's enclosure'. *Bricleswore* 1086, *Brihteswrðe* 1198, *Brythtesworth* 1253, *Brikelesworth* 12th–1426 wth variants *Bryk-* and *-wurth*, *Bryxworth* 1571. OE pers.n. **Bricel* or *Brihtel*, genitive sing. **Bric(e)les, Briht(e)les*, + **worth**. Nth 122.

BROAD BENCH Dorset SY 8978. *Broad Bench* 1811. A coastal shelf of rock. Do i.104.

BROADBOTTOM GMan SJ 9894. 'The broad valley-bottom'. *(le) Brodebothem* 1286, *le Brodbothum* 1360, *Broadbottom* 1831. OE **brād** + **bothm**. Che i.314, L 86.

BROADBRIDGE HEATH WSusx TQ 1431. *Bradbruggesheth* 1441. P.n. or surname *Broadbridge* + ME **heth**. Sx 240.

BROADBRIDGE WSusx SU 8105. 'The long bridge or causeway'. *Bradebrig(g)'* 1192, 1202. OE **brād**, definite form feminine **brāde**, + **brycg**. The road from Fishbourne to Havant crosses low-lying land and three streams here. Sx 58.

BROADBURY Devon SX 4697. 'Great fort'. *Brodebury* 1440, *Broadbury Down* 1791. OE **brād** + **burh**, dative sing **byriġ**. The name is now applied to a wide sweep of downland but originally applied to Broadbury Castle SX 4895, a prehistoric encampment at the highest point (917ft.) of the downs. D 130.

BROADCLYST Devon SX 9897. 'Great Clyst'. *Brodeclyste* 1372. Earlier simply *(æt) Glistune* 11th ASC(A) under year 1001, *Clistone* 1086, 13th cent., 'Clyst estate or manor', r.n. Clyst + OE **tūn**. ME **brode** (OE *brād*) + r.n. CLYST SY 0098. 'Broad' or 'great' for distinction from other divisions of the the river Clyst estate, CLYST HONITON SX 9893, CLYST HYDON ST 0301, CLYST ST GEORGE SX 9888, CLYST ST LAWRENCE ST 0200 and CLYST ST MARY SX 9790. D 573.

BROAD DOWN Devon SY 1793. *Broad Down* 1809 OS.

Gedney BROADGATE Lincs TF 4022. *Gedney Broadgate* 1824 OS. P.n. GEDNEY TF 4024 + p.n. *Brod(e)gate* 1272×1307–1526, *Brad(e)gate* 1380, 1382, *Broadgate* from 1637, the 'wide road', OE **brād**, definite form **brāda**, + ON **gata**. Payling 1940.23.

BROAD GREEN H&W SO 7756. *Broad Green* 1832 OS. ModE **broad** + **green**. Possibly the home of the family whose name is *de la Grene, atte Green* 1275–1327. Wo 104.

BROADHEATH GMan SJ 7689. *Broad Heath* 1831. Che ii.21.

BROAD HEATH H&W SO 6665. *Broad Heath* 1578. ModE **broad** + **heath**. Earlier *Hanleyesheth* 1377 'heath belonging to Hanley' as in HANLEY CHILD SO 6465. Wo 51.

BROADHEATH H&W 8056. *Broad Heath* 1646. Earlier simply *Hethe* 1240 and *le Brode* 1327, 1418. ME **brode** + **hethe**. Wo 90.

BROADHEMBURY Devon ST 1004. 'Great Hembury'. *Brodehembyri -biri -bery -buri* 1273–1334. ME adj. **brode** 'great' + p.n. *Hā- Hanberie* 1086, *Hamber(ia) -biry -byr* 1166–1272, *Hembir(i) -byr(ia)* 1272–49, '(at the) high fort', OE **hēan**, dative sing. definite form of **hēah**, + **byriġ**, dative sing. of **burh**. The reference is to Hembury Fort, an Iron Age hill-fort, the largest earthwork in Devon. 'Broad' or 'great' for distinction from PAYHEMBURY ST 0801. D 557, Thomas 1976.95.

BROADHEMPSTON Devon SX 8066. 'Great Hempston'. *Hemmeston Magna* 1232, *Brodehempstone* 1362. ME adj. **brode** 'great', Latin **magna**, + p.n. *Hamistone* 1086, *Hameston'* 1179, *(H)emmeston(e)* 1175–1281, 'settlement at the river Hems', r.n. Hems, *Hemese* 1287, *Hemse* 1467, 'Hæmi's stream', OE pers.n. *Hǣmi*, genitive sing. *Hǣmes*, + **ēa**, + **tūn**. Also known as *Hemmiston Cauntelu* 1285 from William de Cantelupo who held the manor in 1242. 'Broad' for distinction from LITTLE-HEMPSTON SX 8162. D 509,7.

BROAD HILL Cambs TL 5976. *Broad Hill ground* 1814. ModE **broad** + **hill**. Ca 201.

BROADLANDS HOUSE Hants SU 3520. A 17th and 18th cent. mansion. P.n. *Brodeland* 1541, *Brodelandes* 1547, 'broad open land', ME **brode** (OE *brād*) + **land**, + ModE **house**. Ha 41, Gover 1958.183.

BROAD LAYING Hants SU 4362. *Broad Layings* 1817 OS, *Broad Lane* 1859. This is marked in 1817 as an area of rough pasture and is to be associated with ModE *lay down* in the sense 'convert arable to pasture, put under grass'.

BROADLEY COMMON Essex TL 4207. *Bradley Com*[n]. 1805 OS. P.n. *Bradeleye* 1216×72 'the broad wood or clearing', OE **brād**, definite form **brāda**, + **lēah**, + ModE **common**. Ess 50.

BROADLEY Lancs SD 8816. No early forms. Apparently 'broad clearing or pasture', OE **brād** + **lēah**.

BROADMAYNE Dorset SY 7286. 'Great Mayne'. *Brademaene* 1202, *Brodemayn(n)(e)* 1288–1575. OE adj. **brād** + p.n. *Maine* 1086, *Mayne* 1236–1348, PrW ***main** 'a rock, a stone', referring to the large sarsen stones NE of the village. Also known as *Mayn(e) Martel(l)* 1244–1450 from the Martel(l) family who held the manor from 1202. 'Broad' for distinction from Fryer Mayne SY 7386, *Mayne (Hospital(l)'), Ospitalis* 1244–1332, *Frarenemayne* 1337, *Freremayn(e)* 1449–1533, 'Mayne held by the Knights Hospitaller', ME **frere**, genitive pl. **frerene**, and Little Mayne SY 7287, *Maine* 1086, *Parva Maene* 1202, *Lyttlemayne* 1306. Do i.337, 208.

BROADMERE Hants SU 6247. Cf. *Broad Mear Pond* 1728. ModE **broad** + **mere**. Gover 1958.132.

BROAD OAK Dorset ST 7812. No early forms. Do iii.193.

BROADOAK Dorset SY 4396. 'The large oak-tree'. *Brode Woke* 1493. ME **brode** + **oke**. On the development of *w* before ME *o* see Dobson § 431. Do 44.

BROAD OAK ESusx TQ 8320. *Broadoke formerly Motts* 1567, *Combers farm al. Broad Oak* 1653. ModE **broad** + **oak**. The 16th cent. form is from the surname *Mott*. Sx 516.

BROADOAK ESusx TQ 6022. 'The broad oak-tree'. ModE **broad** + **oak**.

BROAD OAK H&W 50/4821. *Broad Oak* 1831 OS. ModE **broad** + **oak**.

BROAD OAK Kent TR 1661. *Broderhok, (del) Brodhok* 13th cent., *Brodeoke* 1473, 1479. The modern form is clearly 'broad oak-tree' but the frequency of the 13th cent. form with medial *-er-* suggests a different origin, perhaps 'the broader oak', ME **broder** + **ok** (OE *brādra* + *āc*), as in the lost p.n. *Broderehoke* c.1453 in Pluckley. On the other hand OE, ME **hōc** 'a hook, an angle, a corner' is added to other names in this parish, e.g.

Bokwelleshok 14th (Buckwell), *Caldecoteshok* 14th (Calcott), suggesting alternatively that the specific of Broadoak may have been an earlier p.n. **Brāda-ōra* 'the broad hill-slope'. PNK 514, Cullen.

BROAD SOUND Scilly SV 8309. 'The broad channel'. *The brode sownde* c.1540. ModE adj. **broad** + **sound**. PNCo 60.

BROADSTAIRS Kent TR 3967. *Brodestyr* 1479, *Brodestayer* 1565, *Broadstairs* 1819 OS. Cf. *Brodsteyr Lynch* 1434×5. ME **brode** + **stair**. According to Room 1983.14 the reference is to a ledge (*Lynch*, modern dial. **linch**) cut in the cliff-face to lead down to the sea. A gateway to the sea was built here in 1440. The name Broadstairs has superseded the original parish name ST PETERS TR 3868, *borgha scī Petr'* 'St Peter's borough' 1254, *villa scī Petri* 1270, so named from the dedication of the church. PNK 602, TC 60.

BROADSTONE Dorset SZ 0095. No early forms. An ecclesiastical parish formed in 1906. Do ii.4.

BROADSTONE Shrops SO 5489. *Broadstone* 1833 OS.

BROAD STREET Kent TQ 8356. 'The broad street'. *Bradestrete* n.d., *Brodestret* 1327. OE **brād**, definite form **brāde**, + **strēt**. One of a system of similar names including Chestnut Street TQ 8763, cf. *in bosco de Castayner'* 1214, wood called *la Chastenere* 1278, OFr **chastaigniere** 'chestnut plantation', EYHORNE STREET TQ 8354 'hawthorn-tree street', Hazel Street TQ 6939, TQ 8559, Key Street TQ 8864, *Kaystrete* 1254, ME **key** 'a wharfe', Moor Street TQ 8265, *Morestrete* 1278, on the Roman road from London to Canterbury, Nevington Street TQ 8564 on the same road, OAD STREET TQ 8662 'old street', Swanton Street TQ 8759, Ware Street TQ 7956, OE **wer** 'a weir', WARREN STREET TQ 9253, Water Street, Way Street and WEST STREET TQ 9054, in which *street* has acquired the local meaning 'village, hamlet'. Similar systems appear elsewhere in Kent. Cf. CHURCH STREET TQ 7174. PNK 218, 227, 236, 244–5, KPN 115.

BROAD TOWN Wilts SU 09978. 'The great settlement'. *Bradetun* 12th, *Labradetun, La Bradetune* 1205, 1220, *Brodetone* 1206, 1272, *Brodtowne* 1503. OE **brād**, definite form **brāda**, + **tūn**. This remained a descriptive term long enough to attract use of the French definite article *la* in the 13th cent. and follow the normal phonetic developement of OE *tūn*, ME *toun* to ModE *town*. It contrasts with Little Town 0978, 'the small settlement', *Lytelton* 1247, *la Lyteletoune* 1349, *Liteltowne* 1497. Wlt 265.

BROADWAS H&W SO 7655. 'The broad alluvial land'. *Bradeuuesse -wassan* [786 or 589 for 789×90]11th S 126, *Bradewas* [c.1086]1190, 1275–1327, *Brodwas* 1535–1611, *Braddis* 1595. OE **brād**, definite form **brāda** (or **brāde**), + ***wæsse**. Broadwas is beside a meandering stretch of the river Teme. The boundary of the people of Broadwas is *Bradsetena gemœre* [961×72]11th S 1370. The DB form is *Bradeweshā* 1086 with either OE **hām** or **hamm** affixed to the p.n. Broadwas, i.e. 'the estate or river-meadow of Broadwas'. Wo 103, L 59–60.

BROADWATER WSusx TQ 1404. 'The broad water'. *bradan wœtere* (dative case) [946×7]12th S 1504†, *Bradewatre -er* 1086–1451, *Brodewater* 1584, *Brawater* 1242–68, 1763. OE **brād**, definite form neuter **brāde**, + **wæter**. The reference is probably to an area liable to flash flooding between Worthing and Lancing. Sx 192.

BROADWAY H&W SP 1037. 'The broad road'. *to bradan þege, in bradanuuege* 972 S 786, Hooke 226, *Bradeweia -wega -weye* 1086–1275, *Brodwey* 1554. OE **brād**, definite form oblique case, **brādan**, + **weġ**. The boundary of the people of Broadway is *bradsetena gemœre* [c.860]12th S 1591a, Hooke 377. The reference is to one of the routeways crossing the estate, possibly the saltway leading from the vale of Evesham into Gloucestershire on the line of the A 44. Wo 191, Hooke 229, L 83.

†Identification not certain.

BROADWAY Somer ST 3215. 'The wide road'. *Bradewei* 1086, *-weie* 1225, *Broadway* 1610. OE **brād**, definite form **brāda**, + **weġ**. DEPN.

BROADWAY HILL H&W SP 1136. *Broadway Hill* 1828 OS. P.n. BROADWAY SP 1037 + ModE **hill**.

BROADWELL '(The) broad spring'. OE **brād**, definite form **brāda**, + **wella**. L 31.

(1) ~ Glos SP 2027. *Bradewell(e)* [11th]c.1200, 1086–1428, *Bradwell* 1433–1675, *Broadwell* 1684. Gl i.214 gives pr ['bradəl].

(2) ~ Glos SO 5911. *Brodwelles* 1635. Gl iii.214.

(3) ~ Oxon SP 2504. *Bradewelle* 1086–1369 with variants *-well(')* and *Brodewell* 1517. O 308.

(4) ~ Warw SP 4565. *Bradewell(a)* 1130–1379, *Bradwell* 1419, 1656, *Brodwell* 1533, *Broadwell* 1625. Wa 139 gives pr [brædəl].

(5) ~ HOUSE Northum NY 9153. P.n. Broadwell 'the broad stream or spring', ModE **broad** + **well**, + **house**.

BROADWEY Dorset SY 6683. 'Great Wey'. *Brode Way -waye -waie* 1243–1483. ME adj. **brode** + p.n. *Wai(a)* 1086†, *Waie* 1166–1227, transferred from the r. WEY SY 6681. 'Broad' for distinction from the other manors named from the r. Wey, Creketway (lost), *Kriketesweie* 1371, held by the Cricket family from CRICKET ST THOMAS Somer ST 3708, Rowaldsway (lost), *Wayeruaud* 1249, *Rowaldesweу(e)* 1299, 1436, held by Rualet de Waie in 1166, Southway (lost), *Sutwaye* 1285, *South(e)wey* 15th, Wayhoughton (lost), *Wayehogheton, Wayhouton* 1288 'Hugh's manor of Wey', one of the three manors called *Waia* held by Hugh fitz Grip in 1086, Radwey (lost), *Radewey(e)* 1280–1386 'red Wey'††, Causeway SY 6681, *Caucesweie* 1371, *Caus(e)wey(e)* 1381–1473, UPWEY SY 6684, Stottingwey SY 6684, *Stottingewaie* 1212, *Stottingway(e) -yng- -wey(e)* 1288–1682, possibly 'Stotting's Wey', and WEYMOUTH SY 6779. Do i.199–201, 233, 247.

BROADWINDSOR Dorset ST 4302. 'Great Windsor'. *Magna Wyndesor* 1249, *Brodewyndesore* 1324. Latin adj. **magna**, ME **brode**, + p.n. *Windesore* 1086, *Windlesor* 1202, 'river-bank with a windlass', OE **windels** + **ōra**. 'Broad' for distinction from Littlewindsor ST 4404, *Parua Windesoria* 1189, *Parva Windlesor* 1209, *Little Windesore* 1279, earlier simply *Windresorie* 1086 (with AN confusion of *l* and *r*). Do 44, 99.

BROADWOODKELLY Devon SS 6105. 'B held by the Kelly family'. *Brawode Kelly* 1261, *Brodwode Kelly* 1291. Earlier simply *Bradehode* (sic) 1086, *Bradewode* 1175, '(the) broad wood', OE **brād**, definite declensional form **brāda**, + **wudu**. The manor was held by William de Kelly in 1242, cf. KELLY SX 3981. The affix is for distinction from BROADWOODWIDGER SX 4189. D 136.

BROADWOODWIDGER Devon SX 4189. 'B held by the Wyger family'. *Brod(e)wode Wyger* 1306, 1322. Earlier simply *Bradewode* 1086, *Bradwode* 1282, *Brawod'* 1242, '(the) broad wood', OE **brād**, definite declensional form **brāda**, + **wudu**. The manor had passed to the Wyger family from the Vypunds by 1273. D 179 gives pr [bradud].

BROBURY H&W SO 3444. Partly uncertain. *Brocheberie* 1086, *Brogebir' -beria* 1158×65, *Brokeburi -bury* 1183×5–1317, *Brocbury* 1291, 1317. Usually taken to be OE **brōc** 'stream' + **byriġ**, dative sing. of **burh** 'a fortified place, a manor house', but the 'stream' here is the river Wye for which the term *brōc* is unlikely. Either, therefore, OE **brocc** 'badger', genitive pl. **brocca**, or pers.n. **Broca*. He 47, PNE i.52.

BROCHALL Northants SP 6362. 'Badger hole'. *Brocole* 1086, *Broc- Brokhole* 13th, *Brockhall* 1567. OE **brocc** + **hol**. Nth 80.

River BROCK Lancs SD 5140. 'The brook'. *Broc, Brok* 1228, *Broc* [1190×1212, 1230×68]1268, *Broke* [1230×68]1268, *Brock*

†There are eight DB manors so named which are difficult to identify individually. Almost certainly one of them represents the settlement which later became Broadwey.

††Or possibly OE **rād-weġ** 'road suitable for riding'.

[1190×1212]1268, 1292, *the Brooke rill* 1577, *Brook* 1590. OE **brōc**. La 140, RN 54.

BROCKBRIDGE Hants SU 6118. *Brock-bridge* 1810 OS. Gover 1958.46.

BROCKDAM Northum NU 1624. *Brockdam* 1868 OS.

BROCKDISH Norf TM 2079. 'Enclosure by the stream'. *Brodise* 1086, *Brochedisc* [1087×98]12th, *Brokedis* 1166. OE **brōc** + **edisc**. Brockdish is situated on the river Waveney. DEPN.

BROCKENHURST Hants SU 2902. 'Broken up, uneven wooded hill'. *Broceste* 1086, *Brokenhurst* 1100–1490, *-herst* 1182, 1184, *Brokeherst* [1135×54]1331, *Brocheherst* 1158, *Broknest* 1546. OE **brocen** + **hyrst**. The alternative suggestion 'Broca's wooded hill', OE pers.n. **Broca*, genitive sing. **Brocan*, is unnecessary. Ha 41, DEPN.

BROCKFORD STREET Suff TM 1166. *Brockford Street* 1837 OS. P.n. *Brocaforde* 1044×65 ASCharters, *Brokeford(e)* 1065×97, 1446, *-ford(')* 1209–1327, *Brocfort* 1086, *Brokford(e)* 1286, 1346, 1524, *Brockford* from 1316, 'the badger ford', OE **brocc**, genitive pl. **brocca**, rather than 'the brook ford', OE **brōc**, + ModE **street** 'a village, a hamlet'. DEPN, Baron.

BROCKHAM Surrey TQ 1949. 'The river-meadow either by the brook or frequented by badgers'. *Broc(k)ham* 1241, *Brok-* 1243–1347, *Brooke-* 1589, 1615. OE **brōc** or **brocc** + **hamm** 3. Brockham lies at the confluence of Tanners Brook and the river Mole. Sr 282, ASE 2.39.

BROCKHAMPTON Glos SP 0322. 'Estate of the brook-dwellers' or 'homestead by the brook'. *Broc- Brokhamtone* 1166, 1361, *Broc- Brokhampton* 1248–1535, *Brock(e)hampton* from 1572. Either OE **brōc-hǣme**, genitive pl. **brōc-hǣma**, + **tūn**, or **brōc** + **hām-tūn**. Gl i.178.

BROCKHAMPTON H&W SO 5932. 'Brook settlement' or possibly 'settlement of the brook-dwellers'. *Brochamtona* 1160×70, *Brochampton* 1274. Either OE **brōc** + **hām-tūn** or folk-n. **Brōchǣme* 'the brook dwellers', genitive pl. *Brōchǣma*, + **tūn**. The DB name of the place was *Caplefore* (for *-ford*) 1086 'ford leading to Caple' as in How CAPLE SO 6030 and King's CAPLE SO 5628. A later spelling is *Capulfford* 1420. He 48, DB Herefordshire notes to 2.15.

BROCKHOLES WYorks SE 1511. 'Badger holes'. *Brockholis -es* 1275, 1297. OE **brocc-hol**, ME plural **brocc-holes**. YW ii.272.

BROCKLESBY Lincs TA 1311. 'Broklauss's farm or village'. *Brach- Brochelesbi* 1086, *Broclesbi -by* c.1115–1442, *Broclous(e)bi -by* 1143–1338, *Broclausbi -by* c.1150–1291, *Brocklesby* from 1339. ON pers.n. *Bróklauss*, a nickname meaning 'breechless', + **bȳ**. Li ii.6, SSNEM 39.

BROCKLEY Avon ST 4666. 'Badgers' wood or clearing'. *Brochelie* 1086, *Broclegha* 1199×1216 *Berkeley*, *Brockeleg* 1225, *Brokkeley* 1278 Ass. OE **brocc**, genitive pl. **brocca**, + **lēah**. DEPN.

BROCKLEY GREEN Suff TL 8254. P.n. *brocle, broc-lega, Brode* (for *Brocle*) 1086, *Brochelee* 1196, *Brocleye* 1254, *Brokley* 1610, 'badger wood' or 'the brook wood or clearing', OE **brocc** or **brōc** + **lēah**, + ModE **green**. DEPN.

BROCKTON 'Brook settlement', OE **brōc-tūn**. Sa i.59.

(1) ~ Shrops SJ 3104. *Brockton* 1272. DEPN.

(2) ~ Shrops SJ 7203. *Broctone* 1086, *Brocton(')* from 1212 with variants *Broch- Brok-* and *-tun*, *Brockton* 1544. Situated by Mad Brook. Sa i.60.

(3) ~ Shrops SO 3285. *Brochton* 1252, *Bruthon, Bruchton* 1259. Bowcock 53.

(4) ~ Shrops SO 5793. *Broctune* 1086, *Broc(h)- Brokton* 13th cent., *Brockton* 1548. Situated by a small tributary of the r. Corve. Sa i.60.

BROCKWEIR Glos SO 5401. 'Brook weir'. *Broc- Brokwere* c.1145–1395, *Brockwer(e)* 1559, 1616. OE **brōc** + **wer**. A weir in the river Wye where it is joined by a stream called *Brocweresbroc* 1355. Gl iii.233.

BROCKWOOD PARK Hants SU 6226. P.n. Brockwood + ModE **park**. Brockwood is *Brookwood* 1810 OS which may or may not be reliable: either 'badger' or 'brook wood', ModE **brock** or **brook** + **wood**.

BROCKWORTH Glos SO 8916. 'Brook enclosure'. *Brocowardinge* 1086, *Broc- Brokw(u)rthin -wurdin -wordin* 1183–1248, *Broc- Bro(c)kwrd -w(o)rth(e) -wurth(e)* c.1210–1692. OE **brōc** + **worthiġn**, later **worth**. Gl ii.118.

BROCOLITIA Northum NY 8571. 'The rocky' or 'heathery place'. *Brocoliti* [c.700]14th Rav, *Procolitia* [c.400]15th ND. Either Brit ***broc-** 'a pointed rock, a sharp peak', or ***u̯roico-** 'heather', + adj. suffix ***-litu-** 'abounding in'. The name of the Roman fort at Carrawburgh. RBrit 284.

BROCTON Staffs SJ 9619. 'The brook settlement or estate'. *Broctone* 1086, *Broctuna -e* 1167, 1176, *Brocton* from 1199. OE **brōc** + **tūn**. St i.33.

BRODSWORTH SYorks SE 5007. 'Brodd's enclosure'. *Broches- Brodesuuorde* 1086, *Broddesworde* 1156–1300, *-worth* 13th–1545, *Brodesworth -wurth* 1281–15th. OE pers.n. **Brodd* or **Brord* or ON *Broddr*, genitive sing. *Broddes*, + **worth**. YW i.71, SSNY 143.

BROGBOROUGH Beds SP 9638. 'Broken hill'. *Brockebergh* 1222, *Brokeneberwe* 1263, *Brockeborewe* 1308. OE **brocen** + **beorg**. Bd 83, Ca lii.

BROKEN CROSS.

(1) ~ Ches SJ 6873. *Broken Cross Farm* 1831. From a wayside cross formerly standing at the intersection of King Street, the Roman Road to Middlewich, Margary no. 70a, and Penny's Lane. It was broken in the 17th cent. *The Cross(e) feild* is recorded 1650. Che ii.199.

(2) ~ Ches SJ 8973. *Brokencrosse* 1539. At a meeting of medieval roads. Che i.118.

BROKENBOROUGH Wilts ST 9189. 'The broken hill or barrow'. *(in) brokene beregge, Brokeneberge* [956]13th S 629, *Brokeneberg* [1065]14th S 1038, *Brocheneberge* 1086, *Brokeneberga -e -berwe* 1156–1268, *Brokenburgh* 1232, *Brockenborough* 1558×1603. OE **brocen** + **beorg**. The reference is to a hill detached from the main body of high ground or to a barrow. According to Migne lxxxix 309A there was an earlier name of this place, *Kairdurberg* which appears to be PrW *cair* 'a fortified place' + late Brit **duro-* 'fort, a walled town' + OE *beorg*. Wlt 53–4, L 128.

BROMBOROUGH Mers SJ 3582. 'Bruna's fortified place'. *Brunburg* [1100×35]1285–1421 with variants *-bur(g)h*, *Brumburg'* [1153]1280–1535 with variants *-burgh(e) -bur(h) -brugh*, *Bromborough* from 1277. OE pers.n. *Brūna* + **burh**. The place may be identical with the *Brunanburh* ('Bruna's *burh*' with genitive sing. *Brūnan*) near which Athelstan defeated a great invasion of Norsemen and Scots in 937 at the battle celebrated in heroic verse in ASC. Che iv.237 gives prs ['brɔmbərə], locally ['brumbərə].

BROMCOTE Warw SP 4088. 'The broom cottage(s)'. *Brancote* 1086–1301×18, *Bramcote* 1297–1390, *Bromcote* from 1208. OE **brōm** shortened to *bram* + **cot**, pl. **cotu**. Wa 101.

BROME Suff TM 1376. 'Broom'. *Brum, Brom* 1086, *Brom* 1156–1384, *Brome* from 1195. OE **brōm**. DEPN, Baron.

BROME STREET Suff TM 1576. 'Brome hamlet'. P.n. BROME TM 1376 + ModE **street**.

BROMESWELL Suff TM 3050. Either 'the stream belonging to Brome' or 'broom hill'. *Bromes- Brumeswella, Brumesuuelle* 1086, *Bromeswell* from 1254, *Brumswall* 1610. OE **brōm** possibly representing the p.n. BROME TM 1376, genitive sing. **brōmes**, + **wella**, or **brōm** + **swelle** 'a swelling, a hill'. DEPN, Baron.

BROMFIELD Cumbr NY 1747. 'Brown open land'. *Brounefeld* c.1125, *Brunfeld'* c.1200–1438, *Brumfeld* [c.1155]c.1200–1576, *Bromfeld(e)* c.1210–1417, *Bromefeild, Broomefielde* 1578. OE **brūn** with early assimilation of *n* to *m* before *f* and confusion with *brōm*, + **feld**. Cu 272.

BROMFIELD Shrops SO 4877. 'Open land where broom grows'.

I. *Bromfeld* 1060×1 ECWM, *-feld* 1203×4–1535 with variants *-feud -fel -fild* and *-fylde*, *Bromfield* 1684, *Bromfeld'* 1209×9.
II. *Brunfelde* 1086, *-feld* 1194–5, *Brumfeld -feud* 1203×4–1319.
OE **brōm** + **feld** probably used in this name in contrast with adjacent higher ground. Sa i.60.

BROMHAM Beds TL 0051. 'Bruna's homestead'. *Bruneham* 1086, *Bruham* 1164, *Bromham* 1227. OE pers.n. *Brūna* later confused with **brōm** 'bramble', + **hām**. Bd 28 gives pr [brumәm].

BROMHAM Wilts ST 9665. 'The broom homestead', *Bromham* from 1086. OE **brōm** + **hām**. Wlt 254.

BROMLEY 'Wood or clearing where broom grows'. OE **brōm** + **lēah**, dative sing. **lēaġe**.
(1) ~ GLond TQ 4069. *Bromleag* 862 S 331, *Brom leage* [955 for ? 973]10th S 671, *Bronlei* 1086, *Bromlega* 1178. DEPN, LPN 31.
(2) ~ COMMON GLond TQ 4266. *Bromley Common* c.1762. P.n. BROMLEY + ModE **common**. The common was enclosed by Acts of Parliament in 1764 and 1821. LPN 31, Encyc 98.
(3) ~ GREEN Kent TR 0036. *Bromley Green* 1819 OS. P.n. Bromley + ModE **green**. Bromley, *Brūlegh* 1240, *Brumlegh'* 1254, *Bromleye* 1292, is the 'clearing where broom grows'. Contrasts with CHEESEMAN'S GREEN TR 0238. PNK 364.
(4) Abbots ~ Staffs SK 0824. 'B held by the abbot' sc. of Burton abbey. *Bromleia Abbatis* 1203, *Pagets Bromley formerly called Bromley Abbots* 1749. ModE **abbot** + p.n. *(æt) bromleage* [996]14th S 878, [1002×4]11th S 1536, *Bromlege* 14th. DEPN.
(5) Great ~ Essex TM 0826. *Great Brummeley* 1397, *Moche Brymley* 1506. ME adjs. **grete, much** + p.n. *Brūbeleiam, Brumleā -leiam* 1086, *Brum(e)legh(a) -leg(he) -ley(e) -le(ya) -lee* 1232–1345, *Bromleg(h)* 1258–1336. 'Great' for distinction from Little BROMLEY TM 0928. Ess 332.
(6) King's ~ Staffs SK 1216. *Bramlea Regis* 1167. ModE **king**, Latin **rex**, genitive sing. **regis**, + p.n. *Bromleage* [942]13th S 479. An estate granted by king Edmund to Wulfsige Maur' in 941. DEPN.
(7) Little ~ Essex TM 0928. ModE adj. **little**, Latin **parva** (from 1238) + p.n. Bromley as in Great BROMLEY TM 0826. Ess 332.

BROMPTON 'Settlement where broom grows'. OE **brōm** + **tūn**.
(1) ~ Kent TQ 7768. *Brompton* 1819 OS. It is uncertain whether this is a genuine Brompton or if it is to be associated with the name of Benedict de Brunstun, a tenant in Gillingham in 1206; this surname would be 'Bruni's estate', OE pers.n. *Brūn(i)*, genitive sing. *Brūnes*, + **tūn**. PNK 129.
(2) ~ NYorks SE 9482. *Bruntun(e) -ton* 1086–1665, *Brum(p)ton(e)* 1219–1399, *Brompton(e)* 1285 etc. YN 96.
(3) ~ NYorks SE 3796. *Bruntun -tone* 1086, *Bromtune* 1088, c.1130. YN 209.
(4) ~ -ON-SWALE NYorks SE 2199. P.n. *Brunton* 1086, 1231, *Brum(p)ton(a)* 1160 etc. + r.n. SWALE. YN 286.
(5) Patrick ~ NYorks SE 2290. 'Patrick's B'. Manorial prefix *Pateryke* 1154×89, + p.n. *Bruntun -ton(e)* 1086. The OIr pers.n. *Patric* alluding to an (unknown) feudal owner, was probably introduced by Irish-Norwegian settlers in this area. YN 240.

BROMPTON RALPH Somer ST 0832. 'Ralph's B'. *Brompton Radulphi* 1274, *Brūptonrafe* 1610. P.n. Brompton + pers.n. *Ralph* from Ralph FitzWilliam who held the manor after William de Mohun. Earlier simply *Bvrnetone* 1086, *Brumton* c.1235, *Bruneton* c.1250, 'settlement by or on *Bruna* or *Brune*', the original name of the BRENDON HILLS ST 0135†. DEPN, DBSomerset 25.7.

BROMPTON REGIS Somer SS 9531. 'The king's B'. *Brompton Regis* 1291, *Brūpton regis* 1610. P.n. Brompton + Latin **rex**, genitive sing. **regis**. Brompton, *Brvne- Bvrnetone* 1086, is the 'settlement on or by *Bruna* or *Brune*', the original name of the BRENDON HILLS ST 0135. The manor was held by the king in 1086. DEPN.

†*Brunantun* [729×44]13th S 1677 and *Brunham* [824]n.d. ECW may also belong here.

BROMSASH H&W SO 6424. 'Bremi's ash-tree'. *Bromesais, Bremesese, Bremesse, Bromesesce* 1086, *Bromesheff* (sic for *-hess*) 1228, *Bromesasse* 1282. OE pers.n. **Brēmi*, genitive sing. **Brēmes*, + **æsc**. The ash-tree marked the site of the hundred court. He 129, L 219.

BROMSGROVE H&W SO 9670. 'Bremi's grove'. *Bremesgraf* [803]11th S 1260, *-grefan* (dative case) ibid., [804]11th S 1187, *Bremesgrave* 1086–1387, *Bromesgrava* 1167, 1232, *Brom(m)esgrove* 1441. OE pers.n. **Brēmi*, genitive sing. **Brēmes*, later replaced by **brōm**, + **graf(a)**. Wo 336, L 193–4.

Castle BROMWICH WMids SP 1590. 'B with the castle' as opposed to Little Bromwich SP 1086 and West BROMWICH SP 0092. *Castelbromwic, Chastelbromwix* 13th, *Chastelbrunwyz* 1272, *Castel Bromwich* 1322–1549, *Castle Bromage* 1667. ME **castel** + p.n. *Bramewice* 1168, *Bromwic* 1192, 'the broom farm', OE **brōm** + **wīċ**. The reference is to a 12th cent. motte N of the church. Wa 40 gives pr [brʌmidʒ].

West BROMWICH WMids SP 0092. *Westbromwich* 1322. ME adj. **west** + p.n. *Bromwic* 1086 as in Castle BROMWICH SP 1590. 'West' for distinction from Little Bromwich SP 1086 and Castle BROMWICH SP 1590. DEPN.

BROMYARD H&W SO 6554. 'Broom enclosure'. *Bromgeard* [840×52]10th S 1270, *Bromgerde* 1086, *Bromiarde* 1160×70. OE **brōm** + **ġeard**. DEPN, DB Herefordshire notes on 2.49.

BROMYARD DOWNS H&W SO 6655. *Bromyard Downs* 1832 OS. P.n. BROMYARD SO 6554 + ModE **downs**.

BRONYGARTH Shrops SJ 2637. 'The breast of the hill'. *Brongarth* 1272, *Bronagard* 1302. OW **bronn** + definite article **y** + **garth**. O'Donnell 101.

BROOK 'Brook, marshy stream'. OE **brōc**. L 15, Jnl 23.26–48.
(1) ~ Hants SU 3428. *le Broke* 1362, 1447, *Brooke* 1586. Ha 42, Go 186.
(2) ~ Hants SU 2714. *Brook* 1811 OS.
(3) ~ IoW SZ 3883. *Broc* 1086, before 1189, *(la) Brok(e)* c.1200–1600, *Brook(e)* from 1348. Wt 73 gives pr [bruk], Mills 1996.34.
(4) ~ Kent TR 0644. *Broc* 1066×87, *Broke* 1226. Brook lies in the lowlands beside the Great Stour river. The sense is probably 'water meadow' rather than 'marshy land' referring to exploitation as pasture rather than mere description. Alternatively the reference could simply be to the r. Stour. PNK 382, PNE i.51.
(5) ~ Surrey SU 9337. Short for *Broke(strete)* 1548, 'the hamlet at B'. ME **broke**, sometimes combined with **strete** 'a street, a hamlet'. Cf. *Broklond* 1549. Sr 218.

BROOK STREET Essex TQ 5792. Either 'brook village' or 'Roman road by the brook'. *Brok(e)stre(e)t(e)* 1479–1535. ME **broke** (OE *brōc*) + **strete** (OE *strēt*) in either the local sense 'hamlet' or 'Roman road' referring to the road to Colchester, Margary no.3a. The name of the brook is *Sed(e)bur(ghe)brok(e)* 1270–1376, *Sidyngbournebroke* 1341, 'brook by *Sideburh*, the wide fortification', OE p.n. **Sīdeburh* < adj. **sīd**, definite form feminine **sīde**, + **burh**, + **brōc, burna**. The reference is to South Weald Camp, an early Iron Age circular enclosure now in poor condition. Ess 136, Pevsner 1965.361.

BROOKE Leic SK 8505. 'The stream'. *Broc* 1176–1298, *Brok(e)* 1242–1551, *Brooke* 1607. OE **brōc** referring to the upper reaches of the River Gwash. R 78.

BROOKE Norf TM 2899. 'The stream'. *Broc* 1086, 1254. OE **brōc**. The reference is to a small stream that flows into the river Chet. DEPN.

BROOKHOUSE GREEN Ches SJ 8161. *Brookhowse Green in Smalewoode* 1591. P.n. *Brookhowse* 'house by a brook', ME **broke** + **hous**, + ModE **green**. Che ii.316.

BROOKHOUSE Lancs SD 5464. No early forms. Apparently 'house by the brook', ModE **brook** + **house**, referring to Turn Brook, a tributary of the Lune.

BROOKLAND Kent TQ 9825. *Broklande* 1262. Earlier simply *Broke* 1253×44, 'watermeadow', OE **brōc**, Kent–Sussex dial. **brook** 'marshy land, water meadow' referring to the pasture-land of Walland Marsh. *Land* is a late addition in the sense 'newly cultivated land (at *Broke*)' probably referring to coastal reclamation. Brookland lies on a long tongue of alluvium. PNK 477, L 248–9, Cullen.

BROOKMANS PARK Herts TL2404. *Brockman Park* 1822 OS. Earlier simply *Brokemanes* 1468, i.e. Brokeman's manor. Surname Brookman, cf. John *Brok(e)man* 1405, 1437, + ModE **park**. Hrt 66.

BROOKTHORPE Glos SO 8312. 'Outlying farmstead near the brook'. *Brostorp* 1086, *Brocht(h)rop* 1096×1112–1148×63, *Broc- Brokþrop(e) -throp(e)* 12th–1354, *Brock(e)thorpe* 1294, *Brook(e)throp(e)* 1535, 1676. OE **brōc** + **throp**. Gl ii.161.

BROOKWOOD Surrey SU 9557. 'The wood by the brook'. *Brocw(u)de* 1225, 1234, *Brocwod* 1263, *Brookewood* 1598. OE **brōc** + **wudu**. Sr 156, L 15, 228.

BROOM 'Bramble brake, broom'. OE **brōm**. PNE i.52.

(1) ~ Beds TL 1743. *Brume* 1086, *Brome* 12th. Bd 97.

(2) ~ Warw SP 0953. *Brome* 1086–1656. Also known as *Kinges Brome* 1285, *Lutlebrome* 1315 'Little Broom', and *Bromeburnell* 1436 from Alina of Burnel who held the manor in 1280. BCS 127 (S 81) cited in DEPN is not a genuine Anglo-Saxon charter but a forgery made at Evesham c.1097×1104. Wa 202, ECNENM 74.

BROOME H&W SO 9078. 'The broomy place'. *Brome* 1169, *Broome* 1343. OE **brōm**. Wo 278.

BROOME Norf TM 3491. 'The place growing with broom'. *Brom* 1086, [1087×98]12th, 1190. OE **brōm**. DEPN.

BROOME Shrops SO 4081. 'The place growing with broom'. *Brome* 1255×6, 1272 etc., *Broome* 1832 OS. OE **brōm**. Gelling.

BROOMEDGE GMan SJ 7086. 'Hill growing with broom'. *Brown Egge* 1519, ~ *Edge* 1666, *Broomedge* 1831. Associated with *Broomhull* c.1220 and *le Brom(e)* 1289–1528, *Broome* 1559, 'the broomy place', OE **brōm**. Che ii.37.

BROOME PARK Northum NU 1112. *Broome Park*. ModE **broom** + **park**.

BROOMER'S CORNER WSusx TQ 1221. *Brommer Corner* 1813 OS.

BROOMFIELD 'Open land growing with broom'. OE **brōm** + **feld**.

(1) ~ Essex TL 7009. *Brumfeldam* 1086, *-feld -feud* 1200, 1239, *Brimfeld(e) -feud* 1224, 1255, 1602, *Bromfeld(e)* 1232 etc. OE **brōm** + **feld**. Ess 241 gives pr [brʌmfəl].

(2) ~ Kent TQ 8352. *Brvnfelle* 1086, *Brumfeld* c.1100, 1243×4, *Bromfeud -feld* 1240, 1246. PNK 209.

(3) ~ Kent TR 1966. *Bromefeld* 1470, *Bromfield* 1482×3, *Broomefield* 1647. PNK 509.

(4) ~ Somer ST 2231. *Brvnfelle* 1086, *Bromfeld* 1243, *Brumfeld* 1610. DEPN.

BROOMFLEET Humbs SE 8827. 'Brungar's stretch of river'. *Brungareflet* 1150×4, *Brungar(a)flet -gari-* 12th cent., *Brūgaresfleta* 1200, *Brung(e)flete* 1294, 1379, *Brun- Broun(e)- Brownfleet* 14th cent., *Brumflet(e)* 1322, 1600, *Bromeflete* 16th cent. OE pers.n. *Brūngār* + **flēot**. The reference was probably to possession of fishing rights in Broomfleet Hope, a side channel of the Humber, cf. YOKEFLEET. YE 222.

BROOMHEAD RESERVOIR SYorks SK 2696. P.n. Broomhead + ModE **reservoir**. Broomhead, *Bromyheued* c.1280–1335, *(le) Bromyhed(e), Brom(e)heued(e)* 14th cent., *-he(a)d(e)* 1379–1625, *Broomhead* 1822, is the 'headland overgrown with broom', OE **brōmig** later replaced by **brōm** + **hēafod**. The reference is to the prominent hill rising up from Ewden beck. YW i.223.

BROOM HILL Dorset SU 0302. 'Hill growing with broom' or 'where broom is obtained'. *Bromehill* 1591. ME **brome** + **hill**. Do ii.153.

BROOMHILL Northum NU 2401. Short for *Broomhill Colliery* 1868 OS. P.n. Broom Hill, ModE **broom** + **hill**, + **colliery**.

BROOMLEE LOUGH Northum NY 7969. *Broomlee Lough* 1868 OS. P.n. Broomlee, no early forms, identical with BROMLEY, OE ***brōm-lēah** 'the clearing overgrown with broom', + dial. **lough** (ONb *luh*, OW *loch*, W *llwch*) 'a lake'. GPC 2234.

BROOMY LODGE Hants SU 2111. *Bromehill Lodge* 1596, *Broomy Lodge* 1664. P.n. *Bromehill* 'broom hill', ModE **broom** + **hill**, + **lodge**. An extraparochial area in the New Forest. Ha 42, Gover 1958.212.

BROSELEY Shrops SJ 6702. 'The wood or clearing of the fort-guardian'. *Burewardeslega, Burwarleg'* 1177, *Bur(e)- Bor(e)wardesle -leg(h) -ley(e)* 1203×4–1701, *Brosley* 1535–1706, *Broseley* from 1577. OE **burhweard**, genitive sing. **burhweardes**, + **lēah**. Sa i.63, WW 146.

BROTHERS WATER Cumbr NY 4012. A local story says the lake was named after two brothers drowned in a skating accident. *Brother-water* 1671, *Broad Water* 18th cent., *Brothers Water* 1802. We i.15.

BROTHERTOFT Lincs TF 2746. Possibly 'Brothir's curtilage'. *Brothertoft* from 1531. ODan pers.n. *Brōthir* + ON, ME **toft**. Cameron 1998.

BROTHERTON NYorks SE 4826. 'Brothir or the brother's estate'. *Broðer-tun* c.1030 YCh 7, *Broðortun* c.1050 YCh 9, *Brotherton* 1198–1822. ON pers.n. *Brothir*, genitive sing. *Brothur*, or **bróthir** 'a brother', genitive sing. **bróthur**, + **tūn**. The reference might be to property handed over by or to a brother. YW iv.45, SPN 65, SSNY 126.

BROTTON Cleve NZ 6819. 'Stream settlement'. *Bro(c)tune* 1086, *Brocton* 1272, *Brotton* from 1181. OE **brōc** + **tūn**. YN 142.

BROUGH 'Fortified place'. OE **burh**. Metathesis of *ur* > *ru* is not uncommon with this element, as is also substitution of [f] for the ME peripheral phoneme [x], cf. *tough* [tʌf] < OE *tōh* [to:x].

(1) ~ Cumbr NY 7914. *Burc* 1174–1204, *Burg(us -um)* 1198–1362, *Burgh* 1228–1671, *Brough(e)* from 1594. The reference is to the Roman fort of VERTERAE within which the medieval castle was built. We ii.63 gives pr [bruf].

(2) ~ Derby SK 1882. *Burc* 1195–1206, *Burg', Burgum* (Latin accusative form) 1196–1362, *Burgh* 1330–18th cent., *Brugh* 1348, 1532, *Brough* from 1658. The reference is to the Roman fort of *Navio*, for which see River NOE SK 1385. Db 50, RBrit 423–4 (where the grid reference is incorrectly given).

(3) ~ Humbs SE 9426. *Burg(h)o* [late 12th]15th, 1202, *Burgh(e)* 1239–1610, *Brough* 1650. Site of a Roman military and naval station on the Humber. YE 220 gives pr [bruf], RBrit 437–8.

(4) ~ Notts SK 8358. *Burgh* 1525, *Bruff* 1677. The reference is to the Roman settlement of *Crococalana*. Nt 204 gives pr [bruf], RBrit 327.

BROUGHALL Shrops SJ 5641. Uncertain. *Burn(t)hale* 1255×6, *Bungale, Bughhale* 1271×2, *Worchal'* 1327, *Burhalle* 1483×5, *Broughall* 1833 OS. Possibly OE *burh-halh* 'nook of land with or belonging to a *burh*, a fortified place or manor house'; but the variety of forms is difficult. Gelling.

BROUGHTON 'Brook settlement'. OE **brōc** + **tūn**. Cf. BROTTON. L 12, 15.

(1) ~ Bucks SP 8940. *Brotone* 1086, *Broctun'* [1156×66]14th, *Broctone* 1237, 1216×72, *Brouton* 1241, 1284, *Broughton(')* from 1300×20. Bk 32, Jnl 2.25.

(2) ~ Cambs TL 2878. *Broctune* 1086, [n.d.]14th, *Brocton* 1202×7, 1245, *Broucton* 1254×67, 13th, *Broughton* from 1303. Hu 206.

(3) ~ Cumbr NY 0731. *Broctuna* 12th, *Broc- Brogton'* 13th cent., *Broghton* 1286, *Broughton* from 1332. The reference is to the river Derwent. Cu 274 gives pr [brɔ:tn, brautn].

(4) ~ Lancs SD 5235. *Broctun* 1086, *Brocton(a)* before 1160–1226, *Brochton* 1261, *Broucton* 1262, *Brouton* 1269, *Brog(h)ton* 1297–1332, *Broughton* 1490. Blundel or Woodplumpton Brook flows through the village. La 147, Jnl 17.84.

(5) ~ Northants SP 8375. *Burtone* 1086, *Brocton(a), Bructon(e)* 1191–1281, *Brohtune* 1125×8, *Brouton* 1235. Nth 123 gives pr [brautən].

(6) ~ NYorks SD 9451. *Broctuna -e -ton(e)* 1086–1254, *Broghton* 1278–1444, *Broughton* 1316. YW vi.42.
(7) ~ NYorks SE 7673. *Broctun(e), Brostone* 1086, *Broctuna -ton* 1145×53–1285, *Broghton* 1328, 1369. YN 46.
(8) ~ Oxon SP 4238. *Brohtune* 1086, 1285, *Broctun -ton* c.1166–1268, *Brog(h)ton(')* 1246×7–1382, *Broughton* from 1336. O 396.
(9) ~ ASTLEY Leic SP 5292. 'B held by the Astley family'. *Broch- Broghton Astleye(ye)* 1322, *Broughton Astley* from 1535. Earlier simply *Brocton(e)* 1086–1303, *Bro(h)- Brostone* 1086, *Brotton* c.1291, 1301, *Broghton'* 1286–1385, *Broughton'* from 1327. The manor was held by Thomas de Estle in 1220. Lei 432.
(10) ~ BECK Cumbr SD 2882. *Broctunebec* c.1246, *Brockton- Broghtunbec* c.1272. P.n. *Broghtona* 1332–1412, + ON **bekkr**. La 213.
(11) ~ GIFFORD Wilts ST 8763. 'B held by the Gifford family'. *Brocton Giffard* 1288, *Broughton Gyffard* 1409. P.n. *Broctun* [1001]15th S 899, *-ton(e)* 1086–1288, + manorial addition from the family of John Gifford who held the manor in 1281. Also known as *Magna Browton* 'great B' 1401. Wlt 119 gives pr [brɔ:ton dʒifərd] (sic).
(12) ~ HACKETT H&W SO 9254. 'B held by the Hackett family'. *Broctone Haket* 1275, *Haggetts Broughton* 1544. Earlier simply *Broctun* 972 S 786, *-tvne* 1086, *-ton(e)* c.1150–1316, *Broghton* 1514. The Hackett family held land here before the end of the 12th cent. Wo 192.
(13) ~ IN FURNESS Cumbr SD 2187. *Broughton in Furness* 1864 OS. Earlier simply *Brocton* 1196, 1235, *Broghton(a)* c.1300, 1378. La 222.
(14) ~ MILLS Cumbr SD 2290. *Broughton Mills* 1864 OS. P.n. Broughton as in BROUGHTON IN FURNESS + ModE **mills**.
(15) ~ MOOR Cumbr NY 0533. *Broghton more* c.1187. P.n. BROUGHTON NY 0731 + **mōr**. Cu 275.
(16) ~ POGGS Oxon SP 2304. 'B held by the Pugeys family'. *Broughton Pouges* 1526. P.n. *Brotone* 1086, *Brocton(a) -tun'* 1192–c.1300, *Broghton* 1316–47, *Broughton* from 1330 + manorial suffix from Imbert le Pugeys, also called *de Salvadya* 'of Savoy' and *de Bampton* (Oxon SP 3103). He came to England, with Eleanor of Provence who married Henry II in 1236, from Puy-en-Velay, Haute-Loire, and his name means 'a man from Puy' or 'a dweller by the hillock' (OFr *puy*). His byname is also found in STOKE POGES Bucks SU 9884. O 309, DBS.
(17) Church ~ Derby SK 2033. 'B with the church'. *Chirchebroghton(e)* 1327, 1330, *Kyrk(e)- Kirk(e)bro(u)ghton* 1383–c.1600. ME **chirche** + p.n. *Broctun(e)* 1086, c.1166, *-ton(e)* 1211×13–1369, *Broghtone* 1333, *Broughton* 1577, 1610. 'Church' for distinction from West Broughton SK 1433, *West Brocton* 1328, 1330, *West Bro(u)ghton(e)* 1330, 1415, 1692. Db 541, 550.
(18) Drakes ~ H&W SO 9348. 'B held by the Drake family'. Earlier simply *Broctun* 972 S 786, *Broctune* 1086. The reference is to Bow Brook. 'Drakes' from the Drake family here in 1275 for distinction from BROUGHTON HACKETT SO 9254, both manors held by Pershore abbey. Wo 218.
(19) Field ~ Cumbr SD 3881. *Field Broughton* 1864 OS. ModE **field** + p.n. *Brocton* 1276, *Broghton* 1314–1429. 'Field' for distinction from Wood Broughton SD 3781. Field and Wood Broughton are situated on parallel arms of Ayside Pool (River Eea). La 198.
(20) Great ~ NYorks NZ 5406. *Magna Broctun* 1086, *Mekil Broghton* 1481, *Great Broughton* 1665. ModE **great**, ME **mikel**, Latin **magna** + p.n. BROUGHTON. YN 168.
(21) Little ~ NYorks NZ 5406. *Parva Brocton(a)* 1302. ModE **little**, Latin **parva**, + p.n. *Broctune* 1086 as in Great BROUGHTON NZ 5406. YN 168.
(22) Upper ~ Notts SK 6826. *Overbroughton* 1601. ModE **over, upper** + p.n. *Brotone* 1086, *Brocton(a) -tun* 1108–1291, *Brohton* 1205, *Broghton* 1292–1346, *Broutton* 1323. Also known as *Brocton' Sulleny* 1242, *Broghton Sulny* 1292, *Broughton Sulney* 1604 from Alfred de Sulenie who held the manor in 1205. 'Upper' for distinction from Nether BROUGHTON Leic SK 6925. Nt 230.
(23) West ~ Derby SK 1433 → Church BROUGHTON SK 2033.

BROUGHTON 'The fortified enclosure'. OE **burh-tūn**. WW 150.
(1) ~ GMan SD 8201. *Burton* 1177, 1201, *Burghto(u)n* 1323–52. OE **burh-tūn**. Modern Broughton lies across the Irwell from a narrow peninsula on which a *Castle Irwell* is marked 1843 OS. La 32, WW 151.
(2) ~ Humbs SE 9608. *Burtune* 1086, *Bructun* 1185, *-tone* 1209×19, *Brendebrocton* 1250. DEPN, WW 150.
(3) Brant ~ Lincs SK 9154. 'Burnt B'. *Brent Broughton* 1245, *Brendebrocton* 1250–1607, *Brant Broughton* from 1447. ME **brende** influenced by the r.n. BRANT SK 9457 on which the village stands, + p.n. *Burtune* 1086, *Bructun -ton* 1185–1329, *Bruh- Brocton(e)* 13th cent., *Broughton* 1332–1607. Perrott 360, WW 150, Cameron 1998.

BROUGHTON Hants SU 3033. Partly uncertain. *Brestone, Brocton (Hvndret)* 1086, *Burchton* 1173, 1191, *Burghton* 1272–1412, *Berc(h)ton* 1176, 1191, *Bergton* 1236–1331, *Bereweton* 1219, *Borgheton* 1235, *Broughton* 1485, *Brawghton* 1593. The majority of forms point to OE **beorg** 'hill, barrow' + **tūn** with some confusion of the specific with **brōc** 'a brook' and **burh** 'fortified place'. There are barrows on Broughton Down but the village also extends along the valley of Wallop Brook. Ha 42, Gover 1958.188 which gives pr [brɔ:tən], DEPN.

BROWN BANK HEAD NYorks SE 1058. No early forms. P.n. Brown Bank, ModE **brown** + **bank**, + **head**.

BROWN EDGE Staffs SJ 9053. Uncertain. *Browenage* 1298, *Browene Edge* 1572, *Brown Edge* 1630. This might be 'the brown edge', ME **broun** (OE *brūn*) + **egge** (OE *ecg*); alternatively the specific might be OE **brū** 'an eyebrow, a brow, an edge', genitive pl. **brūwena**, and so 'the brows' edge'.

BROWNHILL Lancs SD 6830. No early forms. ModE **brown** + **hill**. A suburb of Blackburn presumably named from the hill that rises to 774ft. at SD 7031.

BROWNHILLS WMids SK 0505. *Brown Hill* 1749 Map. ModE **brown** + **hill**.

BROWNIESIDE Northum NU 1623. *Brownyside* 1868 OS. Perhaps ultimately from OE **brūning** 'the brown place' + **sīde** 'hill-side'.

BROWNLOW HEATH Ches SJ 8360. *Brownlow Heath* 1842. P.n. Brownlow + ModE **heath**. Brownlow, *(the) Browne Low(e)* 1586, is the 'brown mound', OE **brūn** + **hlāw**. Che ii.287.

BROWNSEA ISLAND Dorset SZ 0288. *Branksea Island* 1773, *Brownsea Island* 1811. P.n. Brownsea + ModE **island**. Brownsea itself means 'Brunoc's island', *Brunkes'* 1235, *Brunkesey(e)* 1241–c.1540, *Brownsey alias Brunckesey* 1558×1603, *Branksey* 1575, 1695, OE pers.n. **Brūnoc*, genitive sing. **Brūnoces*, + **ēġ**. Do i.43.

BROWNSHOLME HALL Lancs SD 6845. P.n. Brownsholme + ModE **hall**. A Jacobean house rebuilt in the early 19th cent. There are no early forms for Brownsholme but it appears to be a late analogical formation in **holm** < ON *holmr*. In Sweden post-medieval manor houses were given names in *holm* even when not on islands or raised ground in marsh. Pevsner 1967.150.

BROWNSOVER Warw SP 5177. 'Bruno's Waver'. *Bruneswauvre* 1216×72, *-wavere* 1235, c.1245, *Brounesware* 1324–5, *Brouneswovere* 1313, *Brownes Over* 1548. Pers.n. *Brūne* (ON *Brúni*, ODan *Bruni* or OE *Brūna*), genitive sing. *Brūnes*, + p.n. *Gavra* 1086, *Wavere* 1318, an earlier name of the river Swift, identical with river WAVER Cumbr NY 1950 from OE **wæfre** 'wandering'. Held in 1086 by Bruno whose name is retained for distinction from CHURCHOVER SP 5180. Wa 100, PNDB 209.

BROWNSTON Devon SX 6952. 'Brunweard's farm or village'. *Brunardeston* 1219, 1242, *Brunewardestuna* 1231, *Brouniston*

1290. OE pers.n. *Brūnweard*, genitive sing. *Brūnweardes*, + **tūn**. The modern form probably develops from an alternative form with pers.n. *Brūn* short for *Brūnweard*, genitive sing. *Brūnes*. D 279.

BROWN WILLY Corn SX 1579. 'Hill of swallows'. *Brunwenely* [1239]15th, *Brounwellye hill* 1576. Co **bronn** + **gwennol**, pl. **gwennili**, with mutation of *gw* to *w*. PNCo 60.

BROW TOP Lancs SD 5258. No early forms. ModE **brow** + **top**. Refers to the 625ft. ridge stretching NE from Lower Brow Top SD 5257 to Fell End Farm SD 5359.

BROXBOURNE Herts TL 3707. 'Badger's or Brocc's stream'. *Brochesborne* 1086, *Brokesbo(u)rne -burne* 1203–1494, *Broxborn* 1511. OE **brocc** or pers.n. *Brocc*, genitive sing. **brocces**, *Brocces*, + **burna**. Hrt 219 gives common pr [brɔgzbən].

BROXTED Essex TL 5727. 'Badger's head'. *Brocheseued* [1042×66]12th S 1051, *Broc(c)hesheuot* 1086, *Brox(e)- Bro(c)kesheved* 1218–1428, *Brokeshede* 1542, *Broxsted* 1546. OE **brocc**, genitive sing. **brocces**, + **hēafod**. Ess 471.

BROXWOOD H&W SO 3654. 'Badger's wood'. *Brokeswo(o)de* c.1250×72, 1339, *Brockeswode* 1304. OE **brocc**, genitive sing. **brocces**, + **wudu**. He 156.

River BRUE Somer ST 4243. 'Brisk, vigorous one'. *Briuu* [681]?10th S 236, *Bru* [744]14th S 1410, [842]14th S 292, *Briu* [c.1200]13th, *Bruw* [1204×13]B, *Bryw(e)* 1346–1540, *le Brewe* 1550. Cf. W **bryw** < Celtic **briwo-* 'lively, vigorous, powerful'. The river has a brisk current, especially in its upper reaches. RN 54, LHEB 373, GPC 342.

Temple BRUER Lincs TF 0053. 'B held by the Knights Templar'. *(in) templo Bruerie* 1185, *Temple Bruer(e)* from 1236 with variants *brewer, Bruar, brower*. Earlier simply *(la) Bruere* 12th–1428 with variant *Bruiere* and Latin *Brueria(m)* 1163–1302, 'the heath', OFr **bruiere**, Latin **brueria**. The parish is situated on the Lincoln Cliff or Heath and is surrounded by heath names, *Caythorp Heath, Frieston Heath, Rauceby Heath, Welbourne Heath*, all 1824 OS. A house of the Knights Templar was established here in the reign of Henry II and passed to the Knights Hospitaller c.1312. Perrott 305.

BRUERA Ches SJ 4360. 'Heath'. *Bruera* from 1141×57, *Brewera* 1620. OFr **bruiere**, Med Latin **bruer(i)a**. The location of a chapel, *ecclesia sancte Marie de Bruera* 12th, anglicised as *Churchenheath* 1294, *Churton Heath* 1547. Also known simply as *Heeth* 1157×94, *le Heth* 1317, OE **hǣth**. Che iv.115.

BRUERN ABBEY Oxon SP 2620. P.n. Bruern + ModE **abbey**. A Cistercian monastery founded in 1147. Bruern, *Bruaria* 1159, *Bruir', Bruera -(e)* 1162–1447, *Bruern(e)* from c.1200, is 'the heath', Latin **brueria**, OFr **bruiere**. The form *Bruern* may be due to popular etymology as if from OE **brēowærn** 'a brewery'. O 337, DEPN.

BRUISYARD Suff TM 3266. Probably the 'bower or farmer's yard'. *Buresiart* 1086, *-iard* 1191, 1203, *-yard* 1229, *Bursierd -yerd* 1254–1346, *Bruysyerd* 1316, 1336, *Brusyerd -ierd* 1327, 1346, *Brusyard(e)* 1568, 1610, *Brushard* 1612, *Brusard* 1674. OE **būr** or **(ġe)būr**, genitive sing. **(ġe)būres**, + **ġeard**. A third possibility might be ME *brushe* 'brushwood' < OFr *broce* although this would be unusually early for a Fr loan-word. DEPN, Baron.

BRUMBY Humbs SE 8909. Either 'Bruni's farm or village' or 'spring farm or village'. *Brunebi* 1086, 1395, *Brunneby* 1271–1360, *Brouneby* 1340, *Brunby* 1316–1436, *Brumby* from 1388. Either ON pers.n. *Brúni*, genitive sing. *Brúna*, or **brunnr**, + **bȳ**. Cf. BURNBY SE 8346 and Brøndby, Denmark. SSNEM 39.

BRUND Staffs SK 1061. *Brunde* 1598 Horovitz, *Brund* 1840 OS. W Midl dial. **burned, brund** (ME *berned, brend*) 'burnt' referring to a place cleared by burning.

BRUNDALL Norf TG 3208. Partly uncertain. *Brundala* 1086, *Brundhal* 1257, *Brundale* c.1180. Uncertain element + OE **halh**, dative sing. **hale**, 'a nook'. The specific may represent OE ***brōmede** 'growing with broom'. DEPN, L 110.

BRUNDISH Suff TM 2669. 'The stream *edish* or enclosed pasture'. *Burnedich* 1177, *-edis* 1204, 1208, *Brundishe* 1610. OE **burna** + **edisc**. DEPN.

BRUNDISH STREET Suff TM 2671. 'Brundish hamlet'. *Brundish Street* 1837 OS. P.n. BRUNDISH TM 2669 + Mod dial. **street**.

BRUNE'S PURLIEU Hants SU 1815 → DIBDEN PURLIEU SU 4106.

BRUNTON 'Brook settlement'. OE **burna** + **tūn**.

(1) ~ Northum NU 2024. *Burndon, Burneton (Batayll')* 1242 BF, *Burnton* 1296 SR, *Burneston* 1377. Held in 1242 by *Walter Bataill'*. NbDu 33.

(2) Low ~ Northum NY 9270. *Low Brunton* 1868 OS. ModE **low** + p.n. BRUNTON as in *Brunton House* 1868 OS.

BRUSHFORD 'Causeway ford'. OE **brycg** in the sense 'causeway' + **ford**. It is unclear whether the change [dʒ] > [ʃ] is due to unvoicing before [f] or whether it is due to folk etymology because of the frequent use of brushwood for causeways. L 65,72.

(1) ~ Devon SS 6707. *Brigeford*(3×), *Brisforde*(1×) 1086, *Brig(g)eford(e)* 1219–49, *Brigford(e)* 1269–91, *Bryxford* 1322–92, *Brushford* from 1330, *Brishford* 1333, *Burshford* 1687. D 361.

(2) ~ Somer SS 9225 (under Battleton Brushford on map) *Brige- Brucheford* 1086, *Brige- Brugeford* 13th cent., *Brissheford* 1427, *Brussheford* 1428. D 361.

(3) ~ BARTON Devon SS 6707. Identical to BRUSHFORD (1).

BRUTON Somer ST 6834. 'Settlement on the r. Brue'. *Brvme- Brave- Briwe- Briuuetone* 1086, *Bruwton* [1135×66–1177×94]Bruton, *Briuton* [1142×66]Bruton, *Briweton* c.1150, *Bruton* from [1139×61]Bruton. R.n. BRUE ST 4243 + OE **tūn**. DEPN.

BRYANSTON Dorset ST 8706. 'Brian's estate'. *Brianeston(e)* 1268–1451 with variants *Bryanes- Bri- Bryanis-*. ME pers.n. *Brian*, genitive sing. *Brianes*, + **toun** (OE *tūn*). The manor was held c.1220 by Brian de Insula. Earlier known as *Blaneford Brian* 1268, 1270, the 'Blandford estate held by Brian', probably to be identified with one of the DB manors called *Blaneford* 1086, for which see BLANDFORD FORUM ST 8806. Do ii.90.

BRYHER Scilly SV 8714. Probably 'the hills'. *Braer* 1319, *Brayer* 1336. Co ***bre** + plural ending **-yer**. The hilliest part of the original single island of Scilly before it sank and became subdivided. PNCo 60.

BRYN GMan SD 5600. Possibly an area of 'scorched land'. *Burnal* 1212, *Brunne* 1276, *(del) Brynne* 1432, *the Bryn* 1491. The spellings of 1212 and 1276 suggest original [y] which rules out the explanation from PrW **brïnn* favoured by most commentators. More likely is OE **bryne** 'burning, fire' referring to land scorched through natural causes or through forest clearance. La 100, Jnl 1.44, 17.57.

BRYN Shrops SO 2985. 'The hill'. *Bren* 1272. PrW ***brïnn**. DEPN, Jnl 1.44.

BRYN GATES GMan SD 5901. No early forms. Presumably refers to gates on the turnpike between Ashton in Makerfield and Hindley. L 129.

BUBBENHALL Warw SP 3672. 'Bubba's hill'. *Bvbenhalle* 1086, *-hull* 1211, 1242, *Bobenhulle* c.1225–1656, *Bubbenhull* 1371, *Bubnell* 1524–1651. OE pers.n. *Bubba*, genitive sing. *Bubban*, + **hyll**. Wa 158, L 170.

BUBWITH Humbs SE 7136. 'Bubba's wood'. *Bobewyth* [1066×9]copy, *Bobwyth* 1268, *Bubuid -vid* 1086, *-with -wyth(e)* 12th–1573, *Bubbewith -wyth(e)* 1206–1414, *-wych(e)* 1279×81, 1317. OE pers.n. *Bubba* + ON **vithr** probably replacing OE *wīc* 'dairy-farm'. YE 239, SSNY 144, L 222.

BUCKABANK Cumbr NY 3749. Partly uncertain. *Bucothebank'* c.1345, *Bukobank* 1408, *Buccabank* 1488. This appears to be unknown element + *of the bank*, ME **banke** 'bank, slope'. Cu 131, SSNNW 220.

Long BUCKBY Northants SP 6267. *Longe Bugby* 1565. ModE adj.

long + p.n. *Buchebi* 1086, 1190, *Bucke- Bukkebi -by* 1175 etc., *Buckby* 1284, of uncertain origin. Either ON pers.n. *Bukki*, genitive sing. *Bukka*, + **bȳ** possibly replacing OE **burh**, dative sing. **byriġ**; or ON **bukkr** 'a buck', genitive pl. **bukka**, possibly replacing OE **bucca** 'a he-goat' or pers.n. *Bucca*. 'Long' refers to the length of the village. Nth 65, SSNEM 39.

BUCKDEN Cambs TL 1967. 'Bucga or Bucge's valley'. *Bugedene* 1086–1286 with variant *Bugge(n)-* 1167–1238, *Bo- Bukeden* 1245–1497, *Buckden* 1279. OE pers.n. *Bucga* or feminine *Bucge*, genitive sing. *Bucgan*, + **denu**. Hu 252, L 98.

BUCKDEN NYorks SD 9477. 'The bucks' valley'. *Buckeden(e) -dena* 12th–1316, *Bukden(e)* c.1300–1545, *Buckden* 1576. OE **bucc**, genitive pl. **bucca**, + **denu**. YW vi.115.

BUCKDEN PIKE NYorks SD 9678. P.n. BUCKDEN SD 9477 + ModE **pike** 'a pointed or peaked summit'. PNE ii.63.

BUCKENHAM Norf TG 3505. 'Bucca's homestead or promontory'. *Buc(h)anaham* 1086, *Bokenham Ferye* 1451. OE pers.n. *Bucca*, genitive sing. *Buccan*, + **hām** or possibly **hamm** 2b. Buckenham overlooks the river Yare. DEPN.

New BUCKENHAM Norf TM 0890. *Nova Bukham* 1286. ModE adj. **new**, Latin **nova**, + p.n. *Buch(eh)am* 1086, *Bucheham* 1151, 'Bucca's homestead', OE pers.n. *Bucca*, genitive sing. *Buccan*, + **hām**. 'New' for distinction from Old BUCKENHAM TM 0691, *Vetus Bokenham* 1343, ModE adj. **old**, Latin **vetus**, + p.n. Buckenham as in New BUCKENHAM TM 0890. DEPN.

BUCKERELL Devon ST 1200. Unexplained. *Bucherel* 1165, *Bukerell(e) -el* 1221×30 etc., *Bokerel(le)* c.1220–1326. Cf. CHICKERELL Dorset SY 6480. D 610.

BUCKFAST Devon SX 7467. 'Buck fastness'. *Bucfæsten* [1046]12th S 1474, *-festen -fasten* 13th, *Bvcfestre* 1086, *Buc(k)festr(e)* 1228, c.1270, *Buckefast'* 1286, *Bucfast(e) -k-* 1286–1357, *Buffestr(e)* 1257, 1263. OE **buc(c)** 'a buck, a male deer' + **fæste(r)n**. The specific is clearly singular and the reference may have been to a particular thicket where a buck once took shelter rather than to a place of shelter for bucks for which genitive pl. *bucca* might be expected. D 293.

BUCKFASTLEIGH Devon SX 7366. 'Buckfast wood or clearing'. *Leghe Buc(c)festre* 13th, 1310, *Bucfasten(es)legh* 1306, 1353, *Bok- Bucfastlegh(e)* 1334, 1349, *Bookefasteligh* 1538. P.n. BUCKFAST SX 7467 + OE **lēah**. D 293.

BUCKHORN WESTON → Buckhorn WESTON.

BUCKHURST HILL Essex TQ 4193. *Bu(c)kkershyl(l)* 1496, 1529, *Bokert hill* 1498, *Bokehurste Hill* c.1528, *Bookers Hill* 1563, *Buckets Hill* 1739, *Buckott's Hill* 1770–7. P.n. *(la) Bocherst(e)* 1135, 1203, *(la) Bukhurst* 13th, 'the beech-tree hill', OE **bōc** + **hyrst**, + ModE **hill**. Ess 53.

BUCKINGHAM Bucks SP 6933. 'River-bend land of the Buccingas, the people called after Bucc(a)'. *(to) Buccingahamme*, 918 ASC(A), *Buccinga hámme* 12th ASC(D) under year 915, *Buccyngaham* c.1000, *Bochingeham* 1086, *(to) bucch- buckingeham* [914×9]13th BH, *Bukingeham* 1152×8, *(to) Bukyngham* [914×9]14th BH, *Buckyngham* 1508. OE folk-n. **Buccingas* 'the people called after the buck or the he-goat or Bucc(a)' < OE **bucc**, **bucca** or pers.n. *Bucc(a)* + **ingas**, genitive pl. **Buccinga*, + **hamm** 1. Bk 60, xxxii, YE addenda, TC 61, L 43, 47, 49, BH 118.

BUCKLAND 'Estate granted by royal charter'. OE **bōc-land**. A technical term for land granted by charter, to which certain rights or privileges were attached making it free from services; freehold land, the equivalent of Latin *libera terra*. The word survives as a p.n. in southern counties where the vowel of the first syllable, originally long in ME *bokeland*, underwent shortening to [u] (spelt *o*) and lowering to [ʌ] in the 17th cent. Contrast COPELAND and FA(U)LKLAND. PNE i.39, L 246, LSE 18.219–29, Reallexicon iii.103–4.

(1) ~ Bucks SP 8812. *Bocheland* 1086, *Bocha- Bokelanda* [1133×47]14th, *Bochland* 1157, *Bucland* 1265, *Bokland* [1278×86]14th. The manor was held in 1066 by Godric brother of Wulfwig, bishop of Dorchester. Bk 148 speculates that the estate was given to an early bishop of Dorchester by royal charter (*bōc*), Jnl 2.25, LSE 18.223.

(2) ~ Devon SX 6743. *Bocheland* 1086, *Bocland'* 1242. D 312.

(3) ~ Glos SP 0836. *Bochelande* 1086, *Boc- Bokland* 1154×89–1363, *Buc- Bukland(e)* 1267, 1428. The reference is to land granted to St Peter's Gloucester by Coenred king of the Mercians (704–9). Gl ii.3.

(4) ~ Herts TL 3533. *Bochelande* 1086, *Bocland(e) -lond(e)* 1198 etc. Hrt 175 gives pr [bɔkəldən].

(5) ~ Kent TR 3042. *Bocheland(e)* 1086, *Bokeland* [1087]13th, 1292, *Bocland(e)* 1226, 1268. PNK 567.

(6) ~ Oxon SU 3498. *(æt) Boclande* [957]12th S 639, *Bocheland(e)* 1086, *Buklond* 1412. Brk 385.

(7) ~ Surrey TQ 2250. *Bochelant* 1086, *Bocland(e)* from 1225 with variant *-lond'* to 1323, *Bukelonde* 1293, *Bukkelond* 1448. The shape of the parish, formed by two detached parts of the manorial hundred of Reigate, reflects the carving out of a new settlement from an existing one. Sr 285, Rumble 1987.227, L 246.

(8) ~ ABBEY Devon SX 4866. A mansion house once owned by Sir Francis Drake converted from the buildings of the abbey founded here in 1278. See further BUCKLAND MONACHORUM.

(9) ~ BREWER Devon SS 4120. 'B held by the Brewer family'. *Boclande Bruere* 1290, *Bruweresbocland* 1360. Earlier simply *Bocheland* 1086, *Bockland* 1219. William Briwerre held the manor in 1219. Also known as *Northboclaunde* 1312 for distinction from BUCKLAND FILLEIGH. D 88.

(10) ~ COMMON Bucks SP 9206. *Buckland Comm^{n}.* 1822 OS. BUCKLAND SP 8812 + ModE **common**.

(11) ~ DINHAM Somer ST 7551. 'B held by the Dinant family'. *Bokelonddynham* 1329. Earlier simply *Boclande* [951]14th S 555, *Bochelande* 1086, *Buckland* 1610. Contrasts with *folcland* at nearby FAULKLAND ST 7354. The manor was held by Oliver de Dinant in 1205. The family name comes from Dinan, Brittany. DEPN, LSE 18.227.

(12) ~ FILLEIGH Devon SS 4609. 'B held by the Filleigh family'. *Bokelondefilleghe* 1333. Earlier simply *Bochelan* 1086, *Bokland* 1242. Nicholas de Fyleleye held the manor in 1285. Also known as *Suht Bokland* 1249, *Southboclande* 1268 for distinction from BUCKLAND BREWER. D 90.

(13) ~ IN THE MOOR Devon SX 7273. *Bokelaund in the More* 1318. Earlier simply *Bochelande* 1086, *Bocland'* 1242. The village lies on the edge of Dartmoor; the affix is for distinction from the many other Devon Bucklands. D 512.

(14) ~ MONACHORUM Devon SX 4868. 'The monks' B'. *Boclande Monachorum* 1291, *Monkyn Bockeland* 1525. Earlier simply *(of) bóc lande* c.970 B 1247, *Bocheland* 1086, *Boclaund* 1234. *Monachorum* is the genitive pl. of Latin *monachi* 'monks' and refers to the monks of the original Buckland Abbey founded here in 1278. *Monkyn* is a fossilized form of southern ME *munken(e)* 'monks'. D 225 gives pr [mən'ækərəm], PNE ii.45.

(15) ~ NEWTON Dorset ST 6905. 'B associated with Sturminster Newton'. *Newton Buckelond* 1533, *Buckland Abbas otherwise Newton* 1703. P.n. *Boklond toun* [854]14th S 303, *Boclond(e) -lande* [941]14th S 474, [946×55]18th S 1737, *Bokeland(e)* [966]14th S 742(2), 1196–1387, *Bochelande* 1086, *Buklonde* 1330, + Newton as in STURMINSTER NEWTON ST 7814. The manor was held by Glastonbury abbey. Do iii.239.

(16) ~ ST MARY Somer ST 2713. 'St Mary's B'. *Bokeland S. Marie* 1346. P.n. *Bocheland(e)* 1086, *Buckland* 1610, + saint's name *Mary* from the dedication of the church. DEPN, LSE 18.227.

(17) East ~ Devon SS 6731. *Estbokland'* 1242, *Yestbokeland* 1346. ME adj. **este** 'east' + p.n. *Bocheland* 1086. 'East' for distinction from North and West BUCKLAND SS 4740, 6531. D 34.

(18) North ~ Devon SS 4740. *North Buckland* 1809 OS. ModE adj. **north** + p.n. *Bocheland* 1086, *Boclaunde Dyn(e)ham* 1301, 1561. The affix *Dyn(e)ham*, presumably a manorial addition, has not been identified. 'North' for distinction from East and West BUCKLAND SS 6731, 6531.

(19) West ~ Devon SS 6531. *West Boclaunde* 1242. ME adj. **west** + p.n. *Bochelant* 1086. 'West' for distinction from East BUCKLAND and North BUCKLAND SS 6731, 4740. D 35.

(20) West ~ Somer ST 1720. *West Buckland* 1610. Earlier simply *Bocland* [899×909]c.1500 S 380, *[1065]18th S 1042. 'West' for distinction from Buckland ST 1819, *Buckland* 1809 OS. DEPN, LSE 18.227.

BUCKLEBURY Berks SU 5570. '(At) Burghild's fortified place'. *Borch- Borge(l)deberie* 1086, *Burchildeburia* [1142×84]12th, *-beri* 1176, *Burghildesbir'* 1284, *B(o)urghuldesbury* 1340, *Burgul(le)bury* 1307, 1571, *Buckelbery* 1556, *Bugglebury* 1651. OE feminine pers.n. *Burghild*, genitive sing. *Burghilde*, + **byriġ**, dative sing. of **burh**. Brk 154.

Upper BUCKLEBURY Berks SU 5468. No early forms. ModE adj. **upper** + p.n. BUCKLEBURY SU 5570.

BUCKLERS HARD Hants SU 4000. *Bucklers Hard* 1789, *Buckleshard* 1810 OS. Surname *Buckler* + Mod dial. **hard** 'firm foreshore, landing place'. Earlier known as *Montague Town vulgar Bucklesbury* 1759 and as *Montaguetown* 1727. This name commemorates John Montagu, second duke of Montagu c.1688–1749, who drew up a prospectus for developing the site in 1724. The later name refers to the Buckler family recorded as living here since 1664. Bucklers Hard has a quay on the river Beaulieu and was a centre of shipbuilding in the 18th cent., many of Nelson's warships coming from here. Ha 43, Gover 1958.201, Room 1983.15, Pevsner-Lloyd 149.

BUCKLESHAM Suff TM 2441. 'Buccel's homestead'. *Bukelesham* 1086, 1286, 1314, *Buclesham* 1086, 1286, *Bokelesham* 1279, *Bucclesham* 1344, *Buckelsham* 1356, 1568, *Bucklesham* 1610, *Bucklesome* 1674. OE pers.n. **Buccel*, genitive sing. **Bucceles*, + **hām**. DEPN, Arnott 27, Baron.

BUCKLOW HILL Ches SJ 7383. *Buckley-Hill* 1750, *Bucklow Hill* 1831. P.n. Bucklow + pleonastic ModE **hill**. Bucklow, *Boclou'* 1240, *Buckelawe* 1260, *Bucklowe* 1511, is 'Bucca's mound', OE pers.n. *Bucca* + **hlāw**, the meeting place of the hundred of Bucklow at the junction of Mere, Millington and Rostherne townships on Watling Street. Che ii.51. Addenda gives 19th cent. local pr ['bukli], now ['buklou].

BUCKMINSTER Leic SK 8722. 'Bucca's minster'. *Bucheminstre* 1086, *Bukeminstre -y- -er* 1242–1399, *Buc- Bukminstre -y-* 1254–1451, *-ministre* 1266–1314, *Bukminster -y-* 1311–1535. OE pers.n. *Bucca* + **mynster**. Bucca was either the founder or the owner of the church. Lei 148.

BUCKNALL Lincs TF 1769. 'Bucca or the he-goat's nook'. *Bokenhale* [806]17th S 162, *Buchehale* 1086. OE pers.n. *Bucca*, or **bucca**, genitive sing. *Buccan*, **buccan**, + **halh** probably in the sense 'nook of dry ground in marsh'. DEPN, L 103.

BUCKNALL Oxon SP 5625. 'Bucca's' or 'he-goat's hill'. *Buchehelle* 1086, *Bu(c)kehull(a) -hell' -hill'* 1123–1284×5, *Bu(c)kenhull(e)* 1194–1378, *Buknell* 1526. OE pers.n *Bucca*, or OE **bucca**, genitive sing *Buccan*, **buccan**, + **hyll**. O 203, L 171.

BUCKNALL Staffs SJ 9047. 'Bucca's nook'. *Buchenhole* 1086, *-hale* late 12th, *Buccenhal* 1227. OE pers.n. *Bucca*, genitive sing. *Buccan*, + **halh** in the sense 'small valley, hollow'. DEPN, L 105.

BUCKNELL Shrops SO 3574. 'Bucca's or the he-goat's hill'. *Buchehalle* 1086, *Bu(c)k- Bockenhale* 1333–4, *Bu(c)k- Bo(c)kenhull(e)* 1208–72, *Bucknell* from 1535. OE pers.n. *Bucca* or **bucca**, genitive sing. *Buccan*, **buccan**, + **hyll**. Sa i.64.

BUCKNOWLE Dorset SY 9481 → Church KNOWLE SY 9381.

BUCK'S CROSS Devon SS 3422. Identical with *East Buckish* 1809 OS. P.n. Buckish + ModE **cross** 'a cross-roads'. Buckish, *Bochewis* 1086, *Bochiwis* 1167, *Buchiwises* 1238, *Bokish* 1311, is 'estate granted by charter', OE **bōc** + **hīwisc** 'a household, a family, land to support a family'. In the time of Edward the Confessor the estate was held jointly by three thanes. D 81.

BUCKS GREEN WSusx TQ 0732. *Bucks Gr.* 1813 OS. Probably surname *Buck* recorded here in 1725 + ModE **green**. Sx 158.

BUCKS HILL Herts TL 0500. *Bucks Hill* 1822 OS. Nearby is *Bucks Bottom* ibid. Cf. *Buckeslandes* 1556. Surname *Buck* + ModE **hill**. Hrt 107.

BUCKS HORN OAK Hants SU 8041. *Buckshorn Oak* 1859. The reference is to a particular oak-tree in Alder Holt Wood so named from its shape. Gover 1958.98.

BUCK'S MILLS Devon SS 3523. *Buckish Mill* 1809 OS. P.n. Buckish as in BUCK'S CROSS SS 3422 + ModE **mill**. D 81.

BUCKTON H&W SO 3873. Partly uncertain. *Buctone* 1086, *Buctun* 1252, *Buketon* 1292, *Bocton* 1199–1272×1307. The specific is uncertain, either OE **bucc** 'a buck, a male deer', **bucca** 'he-goat', or a pers.n. derived from either, + **tūn** 'settlement, estate'. He 49, DEPN, PNE i.56.

BUCKTON Northum NU 0838. Either 'deer' or 'goat farm' or 'Bucca's farm or village'. *Buketon* 1208×10 BF 28, *Bucketon* 1296 SR 146, *Bukton* 1344, *Buckton* c.1715 AA 13.9. OE **bucc** 'a buck, a male deer', genitive pl. **bucca**, or **bucca** 'a he-goat', or pers.n. **Bucca*, + **tūn**. The first explanation is assumed by a traditional rhyme, *The Black Sow of Rimside and the Monk of Holy island*:

From Goswick we've geese, from Cheswick we've cheese,
From Buckton we've venison in store,
From Swinhoe we've bacon but the Scots have it taken
And the Prior is longing for more.

NbDu 33, Tomlinson 541–2.

BUCKWORTH Cambs TL 1576. 'Bucc's or the buck's enclosure'. *Buchesworde* 1086, *Bu(c)keswrda -worth -w(u)rth* 1180–1294, *Buckew(o)rth* 1225, 1272×1307, *Buk- Bucworth* 1327–1428. OE pers.n. **Bucc* or **bucc**, genitive sing. **Bucces*, **bucces**, + **worth**. Hu 235.

BUDBROOKE Warw SP 2665. 'Budda's brook'. *Bvdebroc* 1086–1205, *Buddebroc -k* 1100×35–1295, *Boudebrok* c.1185, *Budbrook* 1487. OE pers.n. *Budda* + **brōc**. Wa 204, L 15–16.

BUDBY Notts SK 6270. 'Buti's village or farm'. *Butebi* 1086, *-by* 1275–1335, *Buttebi -by* 1168–1330, *Boteby -bi* 1252–1335, *Budby* from 1433. ON pers.n. *But(t)i*, genitive sing. *But(t)a*, + **bȳ**. Nt 91, SSNEM 40.

BUDE Corn SS 2106. Unexplained. *Bude* from 1400, *Bedebay* 1468, *Bede baye* 1576. The stream is *the Bedewater* 1577, 1586: *Bude* might, therefore, be a r.n. related to W *budr* 'dirty', as possibly in the r.n. Boyd in BITTON Glos ST 6869, *Byd* 10th. In both cases, however, the absence of *r* is a difficulty. The 15th and 16th cent. spellings show Co *u* > *i*. PNCo 60, RN 56, 46, Gl i.4.

BUDE HAVEN Corn SS 1906. *Beedes haven* c.1605, *Beeds Haven* 1610. P.n. Bude + ModE **haven**. PNCo 60.

BUDLAKE Devon SS 9800. Unexplained. *Buddlelake* 1779, *Budlake* 1809 OS. D 576.

BUDLE Northum NU 1535. 'The building(s)'. *Bolda* 1165, 1177, *Bodle* 1196, 1212 BF, *Bodlum* 1205, *Bodhill'* 1242 BF, *Buddill* 1538, *Bewdle* c.1715. OE **bōthl, bōdl, bōld**. The 1205 form is the dative pl. *bodlum* 'at the buildings' and suggests that the other forms are also plural: in this word the sing. and pl. forms are indistinguishable in the nominative and accusative cases. NbDu 33 gives pr [bʌdl].

BUDLE BAY Northum NU 1535. *Budle Bay* 1866 OS. P.n. BUDLE + ModE **bay**.

BUDLEIGH SALTERTON Devon SY 0682 → Budleigh SALTERTON SY 0682.

East BUDLEIGH Devon SY 0684. *Estbodelegh* 1327×77, *East Budleigh* 1671. ME adj. **est** + p.n. *Bodelie* 1086 DB, *-leia* 1086 Exon, *Bud(d)elega -le(ghe)* 1125×9–1269, probably 'Budda's wood or clearing', OE pers.n. *Budda* + **lēah**. OE pers.n. *Budda* is a nick-name from **budda** 'a dung-beetle' which might be the specific in this name. Alternatively the specific might be OE **bōthl** 'dwelling, house'. 'East' for distinction from West Budleigh, a detached part of East Budleigh Hundred, referred to in 1333 as *Westbuddelegh*. D 582, 414, Löfvenberg 19.

BUDOCK WATER Corn SW 7832. A 19th cent. hamlet, *Budockwater* 1884. P.n. Budock + ModE **water**. Budock, *ecclesia Sancti Budoci* 1208–1342 G, *Plu vuthek* c.1400, *Budok* 1535 G, *Bythick* 1727, *Beaudock* 1749, is saint's name *Budoc* preceded in the c.1400 form by Co **plu** 'a parish' and with mutation of *B*- to *V*-. St Budoc was believed to have been born while his mother, a Breton princess, was floating to Ireland in a barrel after being cast adrift for alleged unfaithfulness to her husband. Budoc returned to Brittany in a stone coffin and became bishop of Dol. Cf. ST BUDEAUX Devon SX 4558 and St Botolphs Dyfed SM 8907. The 1727 form shows Co *u* > *i* and medial *d* > [ð]. PNCo 60, DEPN.

BUDSHEAL Devon SX 4560. See under ST BUDEAUX SX 4558.

Great BUDWORTH Ches SJ 6677. *Magna, Great Budworth* 1554. Adj. **great** (Latin **magna**) + p.n. *Budewrde* 1086, *-worth* 1394, *Buddewrtha* 1190×1211, *-worth(e)* 13th–1375, *Budworth* from 1209 with variants *-wurth(e) -worthe*, 'Bud(d)a's enclosure', OE pers.n. *Bud(d)a* + **worth**. 'Great' for distinction from Little BUDWORTH SJ 5965. Che ii.107 gives local pr ['budə:θ].

Little BUDWORTH Ches SJ 5965. *Little Boddeworth* 1353, ~ *Budworth(e)* from 1582, *Budworth Parva* 1668. Adj. **little** (Latin **parva**) + p.n. *Bodeurde* 1086, *Bod(d)eworth* c.1300–1403, *Bud(d)eworth(e)* [1153×81]1353×7–1456 with variants *-w(u)rth(e) -word*, *Budworth* from 1295, as in Great BUDWORTH SJ 6677. Also known as *Buddew(o)r(th) in foresta (de Mara)* 1290–1311 and *Budworth en le Frith* 1295–1860 (*in the Frith*), OE **fyr(h)th** 'wooded country'. Che iii.184 gives local pr ['budə:θ].

BUERTON Ches SJ 6843. 'Enclosure belonging to the fortified place'. *Burtune* 1086, *-tun -ton* c.1200, 1260, *Buerton* from 1272. OE **byrh-tūn**. The reference is not known. Che iii.87 gives prs ['bju:ərtən] and ['bu:ərtən], WW 149.

BUGBROOKE Northants SP 6757. 'Bucca's brook' or 'brook of the bucks or he-goats'. *Buchebroc* 1086–1359 with variants *Bucke- Bukke-* and *-brok*, *But(t)ebroc -dd-* c.1175–1206, *Bukbrok* 1332, *Bugbroke al. Budbroke* 1598. OE pers.n. *Bucca*, or **bucca** 'a he-goat' or genitive pl. of **bucc** 'a buck', + **brōc**. Nth 80, L 15, 16.

BUGLAWTON Ches SJ 8863. 'Haunted Lawton'. *Buggelauton* 1287–1359, *Buglauton* 1357, *-lawton* 1468. ME **bugge** 'a hob-goblin' + p.n. *Lautune* 1086, *Lauton* 1278–1561, 'settlement by a mound', OE **hlāw** + **tūn**. 'Bug-' for distinction from Church LAWTON SJ 8255. Che ii.290.

BUGLE Corn SX 0159. A 19th cent. village, *Bugle* 1888. In 1840 the inn, then newly built, was said to have been so named in honour of a local bugle-player. The inn-sign was a coaching horn. PNCo 61.

BUGTHORPE Humbs SE 7758. 'Buggi's outlying farm'. *Buchetorp* 1086, 1194×8, *Bughetorp* 1086, *Bugetorp* 1086–1298, *Bugget(h)orp(e)* 1154×89–1579. ON pers.n. *Buggi*, genitive sing. *Bugga* + **thorp**. YE 149, SSNY 56.

BULBARROW HILL Dorset ST 7705. P.n. Bul Barrow + ModE **hill**. Bul Barrow, *Buleberwe* 1270, *Bulbarowe* 1545, is 'Bula's barrow', OE pers.n. *Bula* (or possibly **bula** 'a bull') + **beorg**. The reference is to a Bronze-Age bowl-barrow. Do iii.233 gives pr ['bu:bəR], Newman-Pevsner 498.

BULBY Lincs TF 0526. 'Boli's village or farm' or 'bull village or farm'. *Bolebi* 1086, *Boleby -bi* 1160–1534, *Bolby* 1275–1506, *Boolby* 1481×2, *Bulby* from 1506. ON pers.n. *Boli*, *Bóli* or *Bolli*, genitive sing. *Bol(l)a*, *Bóla*, or OE ***bula** or ON **boli**, genitive pl. **bola**, + **bȳ**. Perrott, SSNEM 40.

BULFORD Wilts SU 1743. Partly uncertain. *Bultisford* 1178, 1286, *-es-* [1154×89]1270, *Bu- Bolteford* 1249–1428, *Bulti(ng)ford* 1249, 1268, *Bultford* 1341–1541, *Bulford al. Bultford* 1603×25. The generic is OE **ford** 'a ford', the specific perhaps an OE pers.n. **Bult* cognate with OE *bolt* 'a bolt', ON *boltr* (source of ON pers.n. *Boltr*), ModIcel *bolti* 'a ball, an iron nail', ModNorw and Dan dial. *bolt*, ModSwed *bult* 'an iron nail', Shetland *bolt, bult* 'a big, clumsy figure'. The alternative suggestions, OE ***bulutiġ** 'growing with ragged robin' from *bulut* 'ragged robin, cuckoo flower' or, since the river Avon here divides into more than one stream, OE *bulut-īeġ* 'ragged robin island', leave the *-s-* of the earliest forms unexplained. It seems likely that later popular etymology associated the name with the verb *bolt, boult* 'sift flour'. Wlt 362, SPN 60.

BULFORD CAMP Wilts SU 1843. P.n. BULFORD SU 1743 + ModE **camp**. The reference is to a military barracks begun in the First World War and mostly built in 1935–9. Pevsner 1975.151

BULKELEY Ches SJ 5354. 'Bullock's clearing'. *Bulceleia* 1170, *Bulkeleh*, *Bulkileia -ley(e)* etc. from 1180, *Buckley -legh* 1308–1783. OE **bulluc**, genitive pl. **bulca**, + **lēah**. Che iv.17 gives prs ['bulkli] and ['bukli].

BULKINGTON Warw SP 3986. 'The estate called after Bulca'. *Bochintone* 1086, *Bulkyngton(a) -kin- -kyn-* 1143–1721, *Buckington al. Bulkington* 1682. OE pers.n. **Bulca* + **ing**[4] + **tūn**. Wa 100.

BULKINGTON Wilts ST 9458. 'The estate called after Bulca'. *Boltintone* (sic) 1086, *Bo- Bulkinton(e) -yn- -en-* 1211–1395, *Bu- Bolkyngton -ing-* from 1332, *Buckington -yng-* 16th. OE pers.n. *Bulca* + **ing**[4] + **tūn**. Wlt 126.

BULKWORTHY Devon SS 3914. Possibly 'Bulca's enclosure'. *Bvcheswrde* 1086, *Bulkewrth(y) -worth(y) -wurthi* 1219–1426. OE pers.n. **Bulca* + **worthiġ**. The specific might, however, be OE ***bulca** related to ON *bulki* 'a heap, a lump' used topographically of the hill on which Bulkworthy stands or even of one of the tumuli here. OE **bulluc** 'a bullock', genitive pl. **bull(u)ca**, is also possible, and so 'bullock farm'. D 91.

BULLDOG SAND Norf TF 6027. No early forms. A sand-bank in The Wash. According to Wheeler 124 and Appendix I.7 Bull Dog Bank and Sluice in Holbeach, Lincs approx TF 4335, were so named because navvies building the enclosure bank in 1838 seized a bulldog which a bailiff had brought with him to assist in the arrest of one of the men and, having killed it, buried it in the bank. It seems strange that another bulldog name should exist in such proximity.

BULLEY Glos SO 7619. 'Bull pasture'. *Bvlelege* 1086, *Bulleg(a) -ley(a) -legh* c.1105–1592. OE **bula** + **lēah**. Gl iii.190, L 206.

Lower BULLINGHAM H&W SO 5238. *Neathere Bullynghope* 1265, *Netherbolynghop* 1315, *Lower Bullingham* 1831 OS. ModE **lower**, ME **nether** + p.n. *Boniniope*, *Boninhope* 1086, *Bolingehop* 1236, *Bullingehope* 1242, *Bullyngeshope* 1341, *Bolynghope* 1396, 'enclosure in marshland' or 'promontory jutting into marshland at or called *Bulling*, the place called after Bulla', OE p.n. **Bulling* < pers.n. *Bulla* + **ing**[2], locative-dative sing. **Bullinġe*, + **hop**, later replaced by *ham* < OE *hamm* 'a meadow, a riverside meadow, an enclosure'. 'Lower' for distinction from Bullinghope SO 5137, *Bollynghope Superior* 'upper B' 1316, *Bullingham* 1831 OS. He 49, Bannister 35, ING 149, 171 noting pronunciation with [ndʒ], BzN 1967.366, 1968.145, L 117–8.

BULLINGHOPE H&W SO 5137 → Lower BULLINGHAM SO 5238.

BULL POINT Devon SS 4646. *Bull Point* 1809 OS.

BULLPOT FARM Cumbr SD 6681. 'Bull hole'. *Bullpot* 1823. Also called *Bull Pot of the Witches*. ModE **bull** + **pot**. The reference is to a pit in the limestone. We i.28.

BULMER 'Bulls' pool'. OE **bula**, genitive pl. **bulena**, + **mere**.

(1) ~ Essex TL 8440. *Bulenemera* 1086, *Bul(e)mera -e* 1178–1262. Ess 417.

(2) ~ NYorks SE 6967. *Bolemere* 1086, *Bulemer'* 1159–1285, *Bulmer(e)* 1190 etc. YN 39.

(3) ~ TYE Essex TL 8438. 'B enclosure'. *Bulmere Tye* 1310. P.n. BULMER TL 8440 + OE **tēag**. Ess 419.

BULPHAN Essex TQ 6385. 'Fort marshland'. *Bulgeuen* 1086, *Bur(e)g(h)efen* 1243, 1247, *Bulchfen* 1248, *Buluefen* 1255, *Bulewephen* 1291, *Bulephen -fan* 1291, 1300, *Bulfen* 1310, *Burghfanne* 1342, *Bulfanne* 1375, 1399, *-vand -uan* 1547, 1594. OE **burh** apparently with svarabahkti vowel **buruh**, genitive sing. **bur(u)ge**, + **fænn, fenn**. A similar substitution of *l* for *r* also occurs in BULVERHYTHE Kent TQ 7809. Ess 144 gives pr [bulvən].

BULVERHYTHE ESusx TQ 7809. 'Landing-place of the borough people', viz. the people of Hastings. *Bulwareheda* 1135×54, *Bu- Bolewar(e)heth(e) -hith(e) -huth* 1250–1335, *Burewarehethe* 1229, *Bu- Bolverhethe* 1430–1519, *Bulverhyde* 1672. OE **burhware**, genitive pl. **burhwara**, + **hȳth**, SE dial. form **hēth**. Sx 535.

BULWELL Notts SK 5345. 'The bull spring'. *Buleuuelle* 1086–1255 with variants *-welle* and *-wella*, *Buluuelle* 1086, *Bolewella -(e)* 1174–1255, *Bulwill* 1461. OE ***bula** + **wella**. The specific could also be the pers.n. *Bula*. Nt 141 gives a local pr [bulil].

BULWICK Northants SP 9694. 'The bull farm'. *Bulewic -wik* 1162–1301, *Bulwic* 1248, *Bullicke* 1622. OE **bula** + **wīc**. Nth 161 gives pr [bulik].

BUMBLE'S GREEN Essex TL 4005. *Brummers Green* 1777, *Brimmers Green* 1805. Uncertain surname + ModE **green**. Ess 27.

Helions BUMPSTEAD Essex TL 6541. 'Hellean's B', i.e. B held by Tihel the Breton, otherwise Tihel de Herion from Helléan, Morhiban, Brittany. *Bunsted(e) Helyon* 1234, locally 'Helen's B'. Bumpstead, *Bumesteda, Bumme- Bunstedā* 1086, *Bumsted(a)* [1042]c.1170–1452, *Bunsteda -(e)* 1178–1307, *Bumpsted(e)* 1262–1561, probably represents *Bun-hāmstede* < OE **bune** 'reeds' possibly used as a r.n. + **hāmstede** 'homestead'. But OE *bune-stede* is perhaps also possible with *n* > *m* dissimilation as in ModE *brimstone* < OE *brynstān*. Reeds grow in the stream which flows through Steeple Bumpstead. 'Helions' for distinction from Steeple BUMPSTEAD TL 6741. Ess 508, Stead 196, PNE i.57.

Steeple BUMPSTEAD Essex TL 6741. 'B with the church tower'. *Bunstede atte Tour* 'B at the tower' 1285, ~ *alatour* 1284, ~ *ad Turrim* 1428, *Stepilbumsted(e) -el-* 1260–92. ME **stepel, tour**, Fr **tour**, Latin accusative case **turrim**, + p.n. Bumpstead as in Helions BUMPSTEAD TL 6541. Ess 508.

BUNBURY Ches SJ 5658. 'Buna's fortified place'. *Boleberie* 1086, *Boneburi* c.1170–1674 with variants *Bon(ne)-* and *-bury -biri* etc., *Buneburi -beri* etc. 1180–1387, *Bunbury* from 1283. OE pers.n. *Buna* + **byriġ**, dative sing. of **burh**. Che iii.305.

BUNGAY Suff TM 3389. 'Bung island'. *Bongeia, Bunghea* 1086 *Bungheia* 1174, *Bungeia* 1175, 1191, *Bungey(e)* 1173–1615, *Bungay* from 1235. OE ***bung** + **ēġ**. Bungay occupies a ridge site at the neck of a great loop in the r. Waveney which it stops like a bung. The modern word is not evidenced before c.1440 and is said to be of MDu origin. However, such a word very likely existed in OE cognate with ON *bunga* 'a lump, a swelling, an arch', OHG *bungo* 'a swelling, a tuber', Mod Norw *bunge* 'a bump, a boil' whence Shetland *bungi* 'a bump, a swelling' and the ON by-n. *Bungi*. DEPN.

BUNNY Notts SK 5829. 'Reed island' or 'the island on the r. *Bune*'. *Bonei* 1086, *-ei(a) -ai -ey(e)* 1175–1302, *Buneya* 1227, *Bunney(a)* 1231 and from 1576. Either OE **būne** or r.n. **Bune* + **ēġ**. If the specific is a r.n. it is identical with the early name of Claydon Brook Bucks SP 7427, *(into) Bunon, (andlang) Bunan* [995]c.1250 S 883, *Buna* 1228, *Bune* 1228, 1242, which may be the same as OE **būne**. It may have been the earlier name of Fairham Brook which flows by Bunny. There is flat marshland to the W of the village. Nt 245, RN 56, Bk 2, O 197, L 39.

BUNTINGFORD Herts TL 3629. 'Buntings' ford'. *Buntingeford* 1185, 13th, *-ynge-* 1308, *Buntiford* 1226, *-ing-, -yng-* from 1255. OE ***bunting** (ME *bounting*), genitive pl. ***buntinga**, + **ford**. Hrt 182 gives common pr [bʌn(t)ifəd], PNE i.57.

BUNWELL Norf TM 1193. 'Reed spring'. *Bunewell* 1198, 1254. OE **bune** + **wella**. DEPN.

BURBAGE 'Fort ridge'. OE **burh** + **bæċ**. L 12, 125–6.

(1) ~ Derby SK 0472. *Burbache* 1417, *Burbage* from 1482. The reference of *burh* in this name is unknown. The generic could alternatively be OE **beċe, bæċe** 'a stream'. See L 125–6.

(2) ~ Leic SP 4492. *Burhbeca* 1043, *Burbece* 1086, *-bech(e)* 1202–1339, *Burbach(e)* 1203–1433, *-bage* from 1445, *Burbach or Burbage* 1805 OS. The village lies on a low ridge. Lei 503, L 126.

BURBAGE Wilts SU 2361. Partly uncertain. *Burhbece, (andlang) burgbeces* [961]12th S 688, *Burhbec* [990×1006]12th S 937, *(wið) byr bœces* [968]12th S 756, *(wið) burhbeces* [968]13th ibid., *Burberge -betc(e) -bed* 1086, *Burbech(e)* 1158–1227, *-bach(e)* 1178–1286, *Burgbeche* 1249, *Burghbach* 1352, *Burbage* from 1353. This is usually said to be 'the fort stream', OE **burh** + **bæċe, beċe**. On the analogy of Burbage Derbs SK 0472 and Leics SP 4492, however, it has been suggested that the generic is rather OE **bæċ** 'a ridge'. The topography is indecisive; the village lies at the edge of a broad flat shelf of land, but the bounds between Burbage and Bedwyn follow the line of a nearby stream. The reference to a **burh** is also unclear; the bounds of S 688 refer to an *eorðburh* or earthwork immediately before *wiðbyr bœces* but if that is correctly identified with Godsbury SU 2157 it cannot be the reference in this name. Wlt 336–7, L 126, Arch J 77.78.

BURCOMBE Somer SS 7538. No early forms.

BURCOMBE Wilts SU 0730. Apparently 'Bryda's coomb'. *Brydancumb* [937]14th S 438, *Bredecumbe* 1086, *Brudecumba -e -combe* 1225–1428, *Bridecumbe -combe* 1268–1302, *Birdecombe* 1526, *Burcombe* 1535. OE pers.n. **Bryda*, genitive sing. **Brydan*, + **cumb**. However, this pers.n. is unattested and the specific may be an otherwise unidentified element. Wlt 213.

BURCOT Oxon SU 5696. 'The cottage(s) called after Bryda'. *Bridicot(e)* 1198–1256, *Bri- Bry- Brudecot(e)* 1200–1428, *Byrdcote* 1551×2, *Burcott* 1761. OE pers.n **Brȳda* + **ing**[4] + **cot**, pl. **cotu**. O 149.

BURDALE NYorks SE 8762. Either the 'nook of land where planks are obtained' or the 'house made of planks'. *Bredhalle* 1086, *-hale* 1272, *Bred(d)al(e)* 1086–1297, *Bridale* 12th, 1281, *Byrdale* 1500, *Burdall* 1650. OE **bred** + either **halh** or **hall**. The situation is a deep side-valley NW of Fimber at the extremity of Wharram Percy parish. YE 132 gives pr [bɔdl].

Great BURDON Durham NZ 3116. *magna Burdon* 1326–1506, ~ *Magna* 1367–1624, *Great Burdon* from 1583. ModE adj. **great**, Latin **magna**, + p.n. *Burdune* 1109–1282, *-dun(a)* 1149×53–1282, *Burdon* 1326–1872, 'fortification hill', OE **burh** + **dūn**. The reference is possibly to nearby Sadberge hill. 'Great' for distinction from Little Burdon NZ 3216 *Burdon Parva* 1439–1559×63, *Lytle Burdon* 1557. NbDu 34, L 144, 158.

River BURE Norf TG 2521, 4709. *Bure* 1577, 1586. One of Harrison's inventions, a back-formation from BRISTON TG 0632, *Burstuna* 1086, *Burstone* 1254, *Buresdune* 1586, Burgh Hall TG 2127, *Bure towne* 1586, or BURGH NEXT AYLSHAM TG 2125, *Bure* 1586. RN 57.

BURES Suff TL 9034. 'The cottages'. *Adburā, bure, bura* 1086, *Buren* c.1095, *Buras* c.1180, *Bures* from 13th, *Buers* 1610. OE **būr**, pl. **būras**. DEPN.

Mount BURES Essex TL 9032. 'Bures at the mound'. *Bures atte Munte* 1328. ME **munt** + p.n. *Bura* 1086, *Bures* from 1198, 'the dwellings', OE **būr**, pl. **būras**. The affix refers to the mound of the Norman castle near the church and serves to distinguish this place from nearby BURES or Bures St Mary, Suffolk TL 9034. Ess 363 gives pr ['bjuəs].

BURFORD Oxon SP 2512. 'The ford at the fortified place'. *Bureford* 1086–1380 with variants *-forde* and *-forda*, *Burghforde* 1285, *Burford* from 1316. OE **burh** + **ford**. Burford is situated on the river Windrush. O 310, L 72.

BURFORD Shrops SO 5868. 'The fort ford'. *Bureford* 1086–1386,

Bireford 1255×6, *Burford* from 1284. OE **burh**, dative sing. **byriġ**, + **ford**. One of a series of 'fort' names along the river Teme. By the time of DB Burford was the *caput* of the Scrupe, later the Mortimer, barony in Shropshire. The mound of the Norman *motte* may be the successor of the Anglo-Saxon *burh*. Sa i.67, Raven 42.

BURGESS HILL WSusx TQ 3119. *Burges hill* 16th. A modern settlement developed since the opening of the Brighton railway in 1846; probably named from the family of John Burgeys which occurs in Clayton 1296–1332. Sx 260.

BURGH 'A fort, a fortified place, a manor house'. OE **burh**.

(1) ~ Suff TM 2351. *Burc(g), Burh(e)* 1086, *Burg* 1254, *Burgh* from 1286. The reference is to a banked enclosure called Castle Field just N of the church containing a Roman villa overlying an important Belgic Iron Age site. DEPN gives pr [bʌrə], DEPN, Baron, Pevsner 1974.128.

(2) ~ BY SANDS Cumbr NY 3259. 'The fort by the sands'. *Burgh super le Sablunes* 1271, ~ *by the Sandes* 1356. Earlier simply *Burg(um)* c.1160–1307, *Burgh* from c.1220, OE **burh** Latinised as *burgus, burgum* referring to ABALLAVA NY 3259, the fort on Hadrian's Wall at this place. Cu 126 gives pr [brʌf].

(3) ~ CASTLE Norf TG 4804. *Borough-Castell* 1281. P.n. *Burch* 1086, *Burc* 1168, + pleonastic ME **castel**. The reference is to the Roman fort Gariannum on the river Yare formerly in Suffolk. DEPN gives pr [bʌrə].

(4) ~ LE MARSH Lincs TF 5046. 'B in the marsh'. *(de) Burgo in Marisco* 1275. P.n. *Burch, Burg* 1086, *Burc* c.1115 + Fr definite article **le** for *en le* + ModE **marsh**. This was a major fenland settlement with prolific evidence of RBrit occupation and a pagan Anglo-Saxon inhumation grave on Cock Pit Hill. DEPN, Cameron 1998.

(5) ~ NEXT AYLSHAM Norf TG 2125. P.n. *Burc* 1086 + p.n. AYLSHAM TG 1926. DEPN.

(6) ~ ON BAIN Lincs TF 2286. *Burg super baine* [1154×89]13th. P.n. *Burg* 1086, *Burc* c.1115 + r.n. BAIN TF 4080. There is a long barrow on Burgh Top near Burgh Farm. DEPN, Cameron 1998.

(7) ~ ST MARGARET or FLEGGBURGH Norf TG 4414. 'St Margaret's B' or 'B in Flegg'. *Burg Sancte Margarete* 1254, *Flegburg* 1232, *Burgh in Fleg(ge)* 1271–1470. Earlier simply *Burc, Burh* 1086, *Burg'* 1196, 1275. 'St Margaret's' from the dedication of the church. The reference of *burh* is unknown. See further FLEGGBURGH TG 4414. Nf ii.47.

(8) ~ ST PETER Norf TM 4693. 'St Peter's B'. *Burgh St Peter* 1837 OS. Saint's name *Peter* from the dedication of a church now disappeared. Earlier known as *Whettaker Borough* 1515 from *Qwetacre Sancti Petri* 'St Peter's Wheatacre' 1254. See further WHEATACRE TM 4594. DEPN.

BURGHCLERE Hants SU 4761. 'Clere with the fortification'. *Burclere* 1171, *Burh- Burgclere* 1218, 1307, *Burghclere* 1327, *Borough Clere* 1565. ME **burgh** (OE *burh*) referring either to the hill-fort on Beacon Hill or to the manor or palace of the bishop of Winchester, + p.n. *cleran* [749]12th S 258, *(æt) clearan* [873×88]11th S 1507, [959]12th S 680, *(æt) clere* [931]12th S 412, *(æt) west clearan* [953]12th, *clearas* [951×5]15th S 1515, *Clere* 1086, *Clara* 1167, of unknown meaning and origin. The name, which was later associated with ME *clere* 'clear', Latin *clarus*, feminine *clara*, occurs again in HIGHCLERE SU 4360 and KINGSCLERE SU 5258 (to which some of the charter forms cited above belong). *Clere* must have been an ancient district name perhaps for the range of hills on which the villages stand, perhaps derived from a former name of the river Enborne to the N, cf. the stream-name *cleara(n) flod* [900]11th S 360, [909]12th S 377 from PrW **clījár* 'bright' < Brit **clisáro-*, W *claer*. Ha 55, Gover 1958.151, DEPN, GPC 487.

BURGHFIELD Berks SU 6668. 'Open land by a hill'. *Borgefel(le)* 1086, *Burgefeld', Burg(h)efeld(e)* c.1160–1365, *Bergefeld(a) -feud, Bereg(h)efeld'* 1167–13th cent., *Berwefeld -ue-* 1185–1328, *Bur(u)ghfeld* 1305–1517. OE **beorg** + **feld**. The boundary of the Burghfield people is mentioned in *to Beorhfeldinga gemære* [946×51]13th S 578. Brk 204, L 127, 239, 243.

BURGHFIELD COMMON Berks SU 6566. *Burfield Common* 1761, *Burghfield Common* 1817. P.n. BURGHFIELD SU 6668 + ModE **common**. Brk 207.

BURGH HEATH Surrey TQ 2457. *Borow heth* 1545, *Borough Heath* 1680. P.n. *Berge* 1086, *Berges* 1196–1206, *Burgh(e)* 1237, 1276, '(the settlement) at the barrow' or simply 'the barrows', OE **beorg**, dative sing. **beorge**, nominative pl. **beorgas**, + ModE **heath**. Some ten tumuli were recorded here in former times. Sr 69 gives pr [bʌrə] for Burgh, which is reflected in the 1680 form for Burgh Heath. Sr 71.

BURGHILL H&W SO 4844. 'Fort hill'. *Bvrgelle* 1086, *Burch(e)- Burghull(e)* 1103×7–1495, *Burghyll* 1535. OE **burh** + **hyll**. The reference is possibly to Credenhill Iron Age hill-fort. He 50, Thomas 1976.146.

BURGH ISLAND Devon SX 6443. *The Boro' Island* 1667, *Burough or Bur Isle* 1765, *Burr Island* 1907 SBG. SW dial. **burrow** 'heap, mound' (OE *beorg*). On the hill once stood a chapel of St Michael de *la Burgh* 1411. D 267.

BURGHLEY HOUSE Cambs TF 0406. P.n. Burghley + ModE **house** referring to the Elizabethan mansion house of the Cecil family. Burghley, *Burchle* [10th]13th S 68, *Burglea* 1086, *Burgelai -le* 1177–1227, *Bourle* 1285, *Burle(e)* 1301, is the 'wood or clearing of or near the borough' sc. of Stamford, OE **burh** + **lēah**. Nth 243.

BURGHWALLIS SYorks SE 5311. 'Burgh held by the Walleys family'. P.n. *Burg* 1086–1272, *Burgh* 1246–1606, OE **burh** 'the fortified place', + manorial suffix *Wal(l)eys* 1272, *Wallis* from 1571, 'Welsh', the name of an important family in SYorks in the 12th cent. YW ii.35.

BURHAM Kent TQ 7262. The present village is a modern development and the name short for *Burham Street* 1819 OS, 'Burham hamlet', p.n. Burham + Kent dial. **street**. Burham, *burhham* [c.975]12th B 1322, *Burhham* [995]12th S 885, c.1045 S 1471, *Burham* from 1205 including [1016×20]18th S 1461, *Borham* 1086, *Burgham, Bur(u)ha'* 13th, is the 'fort homestead, estate or meadow', OE **burh** + **hām** or **hamm**. The original site of the settlement was at Burham Court TQ 7161 beside a meander of the r. Medway, a classic site for a *hamm* 3 'river meadow'. This is not, however, supported by the spellings which point to *hām*. The reference of *burh* is not known. PNK 302, 305, ASE 2.29.

BURITON Hants SU 7320. 'Settlement at *Bury*, the fort'. *Buriton* from 1227 with variant *Bury-*, *Bergton* 1229. OE **byriġ**, dative sing. of **burh**, probably already used as a p.n. for the Iron Age earthworks on Butser Hill SU 7120, + **tūn**. The name Buriton began as a descriptive term and eventually replaced the earlier name of the parish which survives now only as the farm name Mapledurham SU 7321, *Malpedreshā* 1086, *Mapelderesham* 1190, 1200, *Mapeldorham* 1287, 'maple-tree estate', OE **mapuldor**, genitive sing. **mapuldres**, + **hām**. Ha 44–5, Gover 1958.55, DEPN, Thomas 1976.140.

BURLAND Ches SJ 6153. 'Peasants' newly cultivated land'. *Burlond* 1260, *Burland* from 1288 with variants *Bour- Boor-* and *-londs -londe*. OE **(ġe)būr** + **land**. Che iii.134, L 9, 247, 249.

BURLAWN Corn SW 9970. Either 'happy dwelling' or 'dwelling of Lowen'. *Bodolowen* 1243, *Bodlouen* 1277, *Bodeloweyn* 1317, *Borlawerne* 1469 G, *Burlawren* 1748 G. Co ***bod** + **lowen** 'glad, happy' or pers.n. *Lowen* derived from it. The earlier explanation from ***elowen** 'an elm-tree' is less likely. PNCo 61.

BURLESCOMBE Devon ST 0716. 'Burgweald's coomb'. *Berlescome* 1086, *Bur(e)woldescumbe* 12th, *Borwoldiscumbe*

1374, *Burlescombe* 1219, *Bordlescumb'* 1242, *Burdelescombe -is-* 1291–2, 1377, *Brodelyscom, Bri- Brudelescombe* 14th, *Burlescom al. Buscombe al. Buddlescom* 1586. OE pers.n. *Burgweald*, genitive sing. *Burgwealdes*, + **cumb**. D 546 gives pr [bə:leskəm], L 92.

BURLESTON Dorset SY 7794. 'Burdel's estate'. *Bordelestone* [843 for 934]17th S 391, *Burdeleston* 1212–1546 with variants *-als -els-* and *-les-*, *Burleston* from 1535†. OFr pers.n. *Burdel*, secondary genitive sing. *Burdeles*, + ME **toun** (OE *tūn*). This was originally one of the r. Piddle estates referred to in DB simply as *Pidele* 1086. Do i.301.

BURLEY 'Wood or clearing by or containing a fortified place or belonging to a *burh*'. OE **burh-lēah**. L 206.

(1) ~ Hants SU 2103. *Burgelea* 1178, *Burglee -lye -ley* 1256–1306, *Burle* 1251. The *burh* is Ringwood which is recorded in 1086 as having four hides in the New Forest where 14 villagers and 6 smallholders dwelt with 7 ploughs although the place is not named. Ha 45, Gover 1958.218, DEPN, DB Hampshire 1.30 and note.

(2) ~ Leic SK 8810. *Burgelai* 1086, 1179, *Burghle* 1294–1383, *Burle* 1218–1406, *Burley(e)* from 1316. The reference is uncertain but could be to Oakham two miles away. R 10.

(3) ~ GATE H&W SO 5947. *Burley Gate* 1832 OS.

(4) ~ IN WHARFEDALE WYorks SE 1646. P.n. *burhleg* [972×92]11th S 1453, *on Burhleage* c.1030 YCh, *Burg(h)elei -lai -lay* 1086–1311, *Burley -lay* 13th–1616 + p.n. WHARFEDALE SE 2646, *Querfildale* 1318. There is no trace of any fortification today. YW iv.196, L 206.

(5) ~ LODGE Hants SU 2405. *le Loge de Burley* c.1490, *Burley Lodge* 1811 OS. P.n. BURLEY SU 2103 + ME **loge** 'a temporary dwelling, a hut'. Gover 1958.218.

(6) ~ STREET Hants SU 2404. 'Burley hamlet'. P.n. BURLEY SU 2103 + ModE **street**.

BURLEYDAM Ches SJ 6042. *Burldame* [1621]1656, *Burleydam* from 1643. P.n. Burley + ModE **dam** referring to an old milldam. Burley, *Burley* from c.1130–17th, *Burle* 1253, is the 'peasants' clearing', OE **(ġe)būr** + **lēah**. Che iii.95.

North BURLINGHAM Norf TG 3610. *Northbirlingham* 1254. OE adj. **north** + p.n. *Berling(e)ham -unge-*, *Berningeham* 1086, *Berlingeham* 1177, 1207, *Birlingeham* 1193ff, *-ingham* 1198, 1202, *Burlingham* 1204, 'homestead of the Byrlingas, the people called after Byrla', OE **Byrlingas* < pers.n. **Byrla* + **ingas**, genitive pl. **Byrlinga*, + **hām**. The exact form of the pers.n. here is uncertain; it may alternatively have been **Bǣrla* as in BARLING Essex TQ 9289, BIRLING Kent TQ 6860. Cf. also BIRLING Northum NU 2406 and BIRLINGHAM H&W SO 9343. 'North' for distinction from South BURLINGHAM TG 3707. DEPN, ING 133.

South BURLINGHAM Norf TG 3707. *Sutberlingeham* 1086, *Suthbirlingham* 1254. OE adj. **sūth** + p.n. Burlingham as in North BURLINGHAM TG 3610. DEPN, ING 133.

BURLTON Shrops SJ 4526. 'The settlement at *Burh-hyll*'. *Burghelton* 1241, *-hulton* 1285. OE P.n. **Burh-hyll* 'the fort-hill' + **tūn**. The reference is to a large multivallate enclosure at SJ 453263. According to Raven 42 the village was moved from its original site after the Black Death. DEPN, Gelling.

BURMARSH Kent TR 1031. 'Marsh belonging to the people of the *burh*' sc. of Canterbury. *Burwaramers* [c.848]13th S 1193, *-mersce* [1016×20]18th S 1461, *Borewarmers* [c.848]14th S 1651, *Borchemeres, Burwarmaresc* 1086, *Burewaremarais -eis* [1087]13th, 1226, *Borwarmershe* 1240. OE **burhware** 'the town dwellers', genitive pl. **burhwara**, + **mersc**. Burmarsh belonged to St Augustine's Abbey, Canterbury, from an early date. KPN 270, Charters IV.91–2.

†The spellings *Burdalueston* [843 for 934]17th S 391, *Burdufues- Burdolveston* 1292 are due to confusion with nearby Bardolfeston SY 7694.

BURMINGTON Warw SP 2638. 'The estate called after Beorma'. *Burdintone* 1086, *Burminton(e) -ing-* late 12th–1428. OE pers.n. **Beorma* + **ing**[4] + **tūn**. *Beorma* is a pet form of *Beornmǣr* or *Beornmund* which might account for the DB form with *-d-*. Wa 298.

BURN NYorks SE 5928. 'The place cleared by burning'. *Byrne* c.1030 YCh 7, *Birne -y-* 1194×1214–1495, *Burne* 1285, 1511–1673, *Bryne* 1374. OE **bryne**, metathesised **byrne**. YW iv.26.

BURNAGE GMan SJ 8692. Unexplained. *Brona(d)ge, Bronnegge, Brounegg* [1322]n.d. This looks like OE **brūn-ecg** 'brown edge' but this does not suit the topography. The specific could alternatively be OE **hecg**, but the evidence is too little for certainty. La 30.

BURNASTON Derby SK 2832. 'Brunwulf's farm or village'. *Burnulfestune* 1086, *Brunolviston -ues-* 13th, *Brun(n)aldeston(e) -is-* 1276–1397, *Brunalston* 1277, 1300, 1393, *Brunastone* 1330, *Burnaston* from 1495. OE pers.n. **Brūnwulf*, genitive sing. **Brūnwulfes*, + **tūn**. Since **Brūnwulf* is not recorded, the specific might alternatively be ON pers.n. *Brynjólfr*. Db 543, SSNEM 189.

BURNBY Humbs SE 8346. 'Stream farm or village'. *Brunebi* 1086, *-by* [1150×60]1200–1310, *Brunnebi -by* [12th]13th–1377, *Brunbi -by* 15th cent., *Brumby* 1402, *Burn(e)by(e)* 1466–1828. ON **brunnr** + **bȳ**. YE 180 gives pr [bɔmbi], SSNY 23.

BURNESIDE Cumbr SD 5095. 'Brunulf's headland'. *Brunolvesheved* c.1180–1292, *Brunol(de)sheved* 13th cent., *Burnol(l)esheued* 1333, *-hed(e)* 15th cent., *Burneshe(a)d* 1552–1777, *Burn(e)syd(e) -side* from 1576. Pers.n. *Brunulf* (OE **Brūnwulf*, OHG *Brunulf* or ON *Brynjolfr*), genitive sing. *Brunulfes*, + **hēafod** referring to the hill SE of Burneside Hall at 5159. We i.153 gives pr ['bɛ:nisaid], L 161.

BURNESTON NYorks SE 3085. Either 'Bryning's farm' or 'farm at the place cleared by burning'. *Brennigston* 1086. Either OE pers.n. *Brȳning*, ON *Brýningr* or OE ***bryning**, genitive sing. *Brȳninges, Brýnings* or **bryninges**, + **tūn**. YN 226, Redin 165, SPN 67, SSNY 126.

BURNETT Avon ST 6665. 'Burnt place'. *Bernet* 1086 Exon, [1107]1300, *Burnet* 1227, 1327. OE **bærnet**. The reference is to land cleared by burning. DEPN.

BURNHAM 'Homestead by a stream'. OE **burna** + **hām**. L 16–7.

(1) ~ Bucks SU 9382. *Burneham* 1086, 1164, *Burnham* from 1175. According to Leland and Harrison in the 16th cent. a stream called the Bourne ran past Burnham. Bk 215.

(2) ~ BEECHES Bucks SU 9585. P.n. BURNHAM SU 9382 + ModE **beeches** 'beech-trees'.

(3) ~ DEEPDALE Norf TF 8043. P.n. *Brun(e)ham* 1086, *Brunham* 1158, *Burnham* 1191, + p.n. *Depedala* 1086, *-dale* 1381, 'the deep valley', OE **dēop**, ON **djúpr**, definite form **dēopa**, + **dæl**, ON **dalr**. The Norfolk Burnhams all take their name from the river Burn which flows North from the Creaks into the sea at TF 8546. There are no less than 9 settlements, the core of them clustering around what is now called Burnham Market. DEPN.

(4) ~ GREEN Herts TL 2616. Partly uncertain. *Burnehamgrene* 1409, *Burn- Brenamgrene* 1451, cf. *Burnhamstret* 1380. The difficulty here is that although the name seems to be p.n. *Burnham* 'stream homestead', OE **burna** + **hām**, + ME **grene**, the hamlet is far from any water. Possibly *Burnham* is a surname. Contrasts with WOOLMER GREEN TL 2518 and HARMER GREEN TL 2516. Hrt 124.

(5) ~ MARKET Norf TF 8342. A modern name which has superseded BURNHAM WESTGATE. P.n. Burnham as in BURNHAM DEEPDALE + ModE **market**.

(6) ~ NORTON Norf TF 8243 → Burnham NORTON.

(7) ~ -ON-CROUCH Essex TQ 9496. P.n. *Burn(e)ham* 1086, the 'stream homestead' referring to a tributary of the Crouch, + r.n. CROUCH TQ 8095. Ess 211.

(8) ~ OVERY now divided into BURNHAM OVERY STAITHE TF 8444 and BURNHAM OVERY TOWN TF 8443. Originally simply *Overhe* 1457, the place 'across the stream', ME **over** (OE *ofer*) + **ee** (OE *ēa*) referring to the river Burn. Burnham Overy was the land E of the river from Market Burnham and was divided into a quay settlement, *Overy Staith* 1824 OS, and a 'town' settlement around the Norman church of St Clement, *Burnham Overy* 1824 OS. DEPN.

(9) ~ THORPE Norf TF 8541→ Burnham THORPE TF 8511.

(10) ~ ULPH Norf TF 8341. 'Ulf's B'. *Burnham Ulph* 1824 OS p.n. Burnham + ON pers.n. *Úlfr* probably referring to *Uluus Bunting* recorded 1161×77 in connection with Brancaster, Burnham and Burnham Deepdale. The church is 12th cent. SPNN 438.

(11) ~ WESTGATE Norf TF 8342. 'B western hamlet'. *Brunham Westgate* 1276. The original name for Burnham Market. For the sense of *Westgate* cf. EASTGATE TG 1423.

BURNHAM '(The settlement) at the springs'. ON **brunnr**, dative pl. **brunnum**, influenced by OE *burna*, dative pl. *burnum*.

(1) High ~ Humbs SE 7801. ModE **high** + p.n. *Brune* 1086, *Brunham* 1154×89 etc., *Burnham* from 1185 possibly representing OE **burna** + **hām** or **hamm** rather than *burnum*. SSNEM 152.

(2) Low ~ Humbs SE 7802. ModE **low** + p.n. *(in altero) Brune* 1086 as in High BURNHAM SE 7801. SSNEM 152.

BURNHAM Humbs TA 0517. Uncertain. *Brune* 1086, *Brunum* c.1115, *Brunnum* 1156–1343, *Brunham* 1242–1431, *Burnham* from 1428. DEPN and SSNEM 152 take this to be the dative pl. *burnum* of OE *burna* 'a stream' or ON **brunnr* 'a spring'. But there are no streams marked at this place which lies on a hill-brow. It is better, therefore, to think of ON **brūnum** 'at the brows, at the moors', dative pl. of **brūn**, or **brunum** 'at the places cleared by burning', dative plural of **bruni**. Li ii.280.

BURNHAM-ON-SEA Somer ST 3049. The addition 'on sea' is a recent innovation. Earlier simply *Burnhamm* [873×88]11th S 1507, *Bvrnehā* 1086, *Burnham* from 1170, 'water-meadow by the burn', i.e. the r. Brue, OE **burna** + **hamm**. DEPN.

BURNHILL GREEN Staffs SJ 7900. *Burnell Grene* 1566, *Burnhill Green* 1833 OS. P.n. *Byrnhill* 1279, *Birnhull(e)* 1309, 1413, possibly 'the tumulus hill', OE **byrġen** + **hyll**, + ModE **green**. Horovitz.

BURNHOPE Durham NZ 1848. Partly uncertain. *Bromhope* 1284, *Brunhop* 1307, 1385, *Burnhop* 1382. The later forms point to OE **burna** + **hop** 'valley with a stream' referring to Whiteside Burn NZ 2048. But the original specific may have been OE **brōm** 'broom' and so 'valley where broom grows'. NbDu 34.

BURNHOPE RESERVOIR Durham NY 8438. P.n. Burnhope + ModE **reservoir**. Burnhope, *Burnehope* 1580, *Burnhope in Weardale* 1647, *Burnope, Burnup* 1685, is OE **burna** + **hop** 'side-valley with a stream'.

BURNHOPE SEAT Durham NY 7837. P.n. Burnhope as in BURNHOPE RESERVOIR NY 8438 + N dial. **seat** < ON *sǽti* 'a seat, a lofty place'. Burnhope Seat rises to 2448ft. and marks the boundary between Durham and Cumbria.

BURN HOWE RIGG NYorks SE 9198. No early forms. The ridge is named from a tumulus or 'howe' at SE 9199. P.n. Burn Howe, ModE **burn** + **howe** (ON *haugr*), + Mod dial. **rigg** (OE *hrycg*).

BURNISTON NYorks TA 0193. Either 'Bryning's farm' or 'farm at the place cleared by burning'. *Brinctun, Brinnistun -ton* 1086, *Brini(n)gstun -tona* 1091×5–1185×95, *Bri- Bryningeston(a)* 1224×38–1408, *Briniston(a)* 1108×14–1322, *Burnysshton* 1550. Cf. BURNESTON SE 3085. YN 107 gives pr [bɔnistən].

BURNLEY Lancs SD 8432. Probably 'clearing, pasture beside the river Brun'. *Brunlaia* 1124, *-lay* 1305, 1309, *-ley(a)* [1155×8] 1230–1533, *Brumley(e)* 1292–1341, *Bro(u)nlay* 14th, *Bromley* 1416, *Burnde- Brundelegh* 14th cent., *Burneley* 1533, 1577. R.n. Brun + OE **lēah**. Brun, *Browne* 1469, a name also preserved in Brownside SD 8632, *Brownes Wode, Brownesyd* 1542 and Brunshaw SD 8532, *Brunschaghe* 1296, ME **shagh** 'copse' (OE *sceaga*), is a stream name derived from OE **brūn** 'brown'. The modern form Brun instead of Brown is due to the influence of the names Brunshaw and Burnley, earlier *Brunley*, in which [u:] was shortened to [u] before the date of the Great Vowel Shift diphthongization of [u:] > [au]. La 83, RN 56, Jnl 17.51.

BURN MOOR NYorks SD 6964. 'The brown moor'. *Brune- Bronemor* 1165×77, *Broun(e)mor(e)* 1177, 1307. OE **brūn**, definite form **brūna**, + **mōr**. YW vi.233.

BURN MOOR FELL NYorks SD 7064. No early forms. P.n. BURN MOOR SD 6964 + ModE **fell** (ON *fjall*).

BURNMOOR TARN Cumbr NY 1804. Partly unexplained. *Burman, Burneman Tarne* 1570, *Burnmoor Tarn* 1784. Probably identical with *Burmeswater* 1322, *Br(o)um(e)berwater* 16th cent. P.n. Burnmoor, *Burnam, Burmer Moore* 16th cent., unexplained element + ModE **moor**, + ModE **tarn** (ON *tjǫrn*). Cu 32.

BURNOPFIELD Durham NZ 1656. 'Field at Burnop'. *Burnopfield* 1863 OS. P.n. Burnop + ModE **field**. Burnop is OE **burna** + **hop** 'side-valley with a stream' referring to the steep dene leading down to the river Derwent at NZ 1658.

BURNSALL NYorks SE 0361. 'Bryni's nook of land'. *Brineshal(e), Bri- Brynshale* 1086–1304, *Bry- Brinsal(e)* 1140–1428, *Brunisale* 1179×1202, *Brunshale* 1348, *Burn(e)sall* 1409 etc. OE pers.n. *Brȳni*, genitive sing. *Brȳnes*, + **halh**, referring to the site of the village in a nook or bend of the river Wharfe. YW vi.83, L 107.

Mabe BURNTHOUSE Corn SW 7634. 'Burnthouse in the parish of Mabe'. *Burnthouse* 1748 G. P.n. MABE SW 7533 + **burnt house**. 'Mabe' for distinction from Burnthouse SW 7636, *Burnthouse* 1699, in St Gluvias. PNCo 113.

BURNTWOOD Staffs SK 0509. 'The wood cleared by burning'. *Brendewode* 1298, 16th, *Brundwood* 1571, *Burndwood* 1680. ME **brende** + **wode** (OE *wudu*). Duignan cites a Forest Jury of 1262 which found 'that a certain heath was burnt by the vill of Hammerwich (SK 0607) to the injury of the king's game'. Land was frequently cleared for cultivation by burning. DEPN, PNE i.49, L 228.

BURNT YATES NYorks SE 2561. 'Burnt gates'. *the Burnt Yate* 1576, *Burntyeates* 1581. ModE **burnt** + **yate** (OE *ġeat*). Probably a gate in the wall separating the Forest of Knaresborough from the estates of Fountains Abbey. Nothing is known of the burning of the gates. YW v.98.

BURPHAM Surrey TQ 0152. 'The river-bend land containing a fortification'. *Borham* 1086, *Bur(g)ham* 1255–1346, *Burfham* 1592. OE **burh** + **hamm** 1 or 2a. No earthwork survives. Sr 162, ASE 2.39.

BURPHAM WSusx TQ 0408. Partly uncertain. *Burhham* [c.920]17th Gale, *Bercheham* 1086, *Burgham* 1229–98, 1631, *Burfeham* 1558. Probably OE **burh** + **hām** 'the homestead or estate by the fortified place' referring to ancient earthworks here perhaps guarding the passage of the Arun. DB already shows confusion with *beorg* 'a hill' referring to the marked eminence 1 m NE. The generic could also be *hamm* senses 1–3, 'river-bend land, promontory, land on a hill-spur'. OE *burhhamm* might be 'the fortification enclosure'. The modern pr shows substitution of [f] for peripheral phoneme [x]. Sx 166, ASE 2.33, SS 86.

BURRADON Northum NT 9806. 'Fortification hill'. *Burhedon* c.1200, *Burwedon -ton* 1242 BF 1117–8, *Borudoune* 1296 SR 174, *Burghdon* 1323, *Burrowdon* 1628. OE **burh**, genitive sing. **burge**, + **dūn**. NbDu 34, L 144, 158.

BURRADON T&W NZ 2772. 'The fort hill'. *Burgdon* c.1150, *-dunie* before 1162, *Bu- Borudon', Buruedon'* 1242 BF, *Burudon* 1296 SR, *Boroudon* 1346, *Burroden* 1638, *Burradon* 1662. OE **burh** + **dūn**. NbDu 34, L 144, 158.

BURRAS Corn SW 6734. Probably 'short ford'. *Berres* 1337, *Burras* 1625. Co **ber** + **rid** (later **res*). PNCo 61.

BURRATOR RESERVOIR Devon SX 5568. P.n. Burrator SX 5567 + ModE **reservoir**. Burrator, *Burrator* 1907 SBG, is the name of a small hill at the S end of the early 20th cent. reservoir. The whole parish of SHEEPSTOR SS 5567 abounds in tors or granite outcrops.

BURRIDGE Hants SU 5110. *Burridge F., Burridge Wood* 1810 OS. Probably identical with *Burhegge* 1315, 'fort or manor house hedge', ME **burgh** (OE *burh*) + **hegge** (OE *hecg*). Gover 1958.32.

BURRILL NYorks SE 2387. 'Fortification hill'. *Borel(l)* 1086, 1184, 1285, *Burel(l)* 12th–1572. OE **burh** + **hyll**. YN 237 gives pr [bɔril].

BURRINGHAM Humbs SE 8309. 'Village, estate of the Burningas, the stream-dwellers', or 'village at or called *Burning*, the stream place'. *Burringham* 1199, *Buring(e)-Burningham* 1218, *Burnyngham* 1281. OE folk-n. **Burningas* < OE **burna** + **ingas**, genitive pl. **Burninga*, or p.n. **Burning* < **burna** + **ing**[2], + **hām**. In either case OE *burna* seems to be tributary stream of the r. Trent. ING 143.

BURRINGTON Avon ST 4859. 'Fortified enclosure'. *Buringtone* [1100×35]13th AD, *Buringtune* 1189×99. OE **byriġ-tūn** 'enclosure, settlement at the fort'. **byriġ** is the dative sing. of **burh**. DEPN, WW 152.

BURRINGTON Devon SS 6316. 'Farm, village, estate called after Beorn(a)'. *Bernintone* 1086, *Berington* 1249, *Burnintone -y(n)g- -yn-* 1265–1305, *Buringtune* 1275, *Burryngtone* 1304. OE pers.n. *Beorn(a)* + **ing**[4] + **tūn**. D362.

BURRINGTON H&W SO 4472. 'Fort settlement'. *Boritvne* 1086, *-ton* 1333, *Buryntone* 1397, *Beryngton* 1535. OE **byriġ-tun**. He 51, WW 149.

BURROUGH GREEN Cambs TL 6355. 'Burgh green'. *Borough(e)grene* 1571. P.n. Burgh + ModE **green**. Burgh, *(at) Burg* [1043×45]13th S 1531, 1234–1336, *Burch* 1086, *Burgh* 1285–1449, *Borw* c.1278, *Borow* 1501, *Borough* 1529, is OE **burh** 'a fortified place'. The reference is to earthworks in Park Wood. Ca 115.

BURROUGH ON THE HILL Leic SK 7510. *Burrough-on-the-hill* 1641. Earlier simply *Burg(-o -um)* 1086–1316, *Burgh* 1242–1536, *Burc -k* 1232×46–[1216×72]1404, *Bor(r)ow(e), Borough(e), Burrow(e)* 16th cent., 'the fort', OE **burh**. Also known as *Erd(e)- Erth(e)burg(h)* 1201–1514, *Ard(e)brogh* 1480, *-borough* 1510, 'fort with earthen ramparts', OE **eorthburh**. The reference is to the great Iron Age hill-fort on Burrow Hill SK 7611, possibly the tribal capital of the Coritani. The present village is situated on a different hill ¾ m. to the SW, whence the affix *on the hill*. Lei 187, Thomas 1876.163.

BURROW BRIDGE Somer ST 3530. *Burough Bridge* 1809 OS. P.n. Burrow as in *Michaels Burro* 1610 + ModE **bridge**. Earlier simply *(æt þam) Beorge* *[1065]18th S 1042, *Bergh* 1325, 'the hill', OE **beorg**. The reference is to Burrow Mump or Mount, a natural eminence visible for miles around, reputedly the place of Alfred's camp in 879, crowned with the tower of a church dedicated to St Michael. DEPN, Pevsner 1958S.110.

Nether BURROW Lancs SD 6175. *Nethirburgh* 1370. Adj. **nether** + p.n. *Borch* 1086, *Burg(')* 1212, 1252, *Burgh* 1259, 1332, OE **burh** 'a fortified place' referring to the Roman fort of *Galacum* at Over Burrow SD 6175, *Overburgh* 1370. The fort lies on a promontory between Leck Beck and the river Lune. *Galacum* is an error for **Calacum* standing for British **Calācon* with the sense 'noisy stream', literally 'loud-calling one'. It is probably the original name of LECK BECK SD 6376 which is of Scandinavian origin. The two manors together are referred to as *Burras, Burros* 1577 and the township as *Burrow with Burrow*. La 184, RBrit 288.

BURROWHILL Surrey SU 9763. 'Burrow hill'. *Borohill* 1542, *Burroughhill* 1609. ModE **burrow** (OE *beorg*) probably used as a p.n. + **hill** (OE *hyll*). Sr 118.

BURSCOUGH BRIDGE Lancs SD 4411. P.n. BURSCOUGH + ModE **bridge**. The bridge carries the road from Liverpool to Preston over the Leeds and Liverpool canal.

BURSCOUGH Lancs SD 4310. 'Wood belonging to *Burh*, wood beside the (old) *burh*'. *burgechou* c.1190, *Burgaschoe* before 1200, *Burscogh'* [1189×91]14th, *Burscogh* c.1190, 1327, 1519, *Burreschoc* c.1224–56, *Burscho* 1232–56, *Burscou* [1199–1315]14th, *Birscogh -scow -skou* 13th, 14th cent. OE **burh**, genitive sing. **burge**, + ON **skógr**. The site of the lost **burh** or fortified place or manor is unknown but is referred to in the forms *Burgastude -stud'* 1189–99, *Burchistude* c.1224–42, *Bourcghestude* c.1230–42, *Burgestude* 1232–42, OE **burh-styde** 'the site of a **burh**'. La 123, Stead 292, L 209, Jnl 17.68.

BURSEA Humbs SE 8033. 'Shed pool'. *Bi- Byrsay -sey(e)* 1259–1416, *Bursai -sey* 1303–1567. OE **byre** + **sǣ**. Situated in marshy land by the r. Foulness. YE 234 gives pr [bɔsi].

BURSHILL Humbs TA 0948. 'Broken hill'. *Bristehil* [12th]c.1536, [c.1200,1293]15th, *Bri- Brystil(l) -yl(l)* 1200–1571, *Bristhil(l)* [1172]15th, 1246–1510, *Bursall, Bursill, Brustill* 16th, *Boshill* 1786. OE **byrst** + **hyll**. The reference is to a small hill forming part of the narrow ridge called Barff Hill, *le Bergh* [12th]c.1536, OE **beorg** 'a hill'. YE 74 gives pr [bɔsil].

BURSLEDON Hants SU 4809. 'Beorsa's hill'.
I. *Brixendona -tona* c.1170, *Burxedun'* 1218, *Brexheldene* 1228.
II. *Bursedona* 1208, *-dene* 1292, *Bercildon -sil- -sel-* 1245–1333, 1610, *Bursindene -don* 1248, 1292, *Bursuldon -sel-* 1263–1330, *Bursyngdon* 1288, *Bruseldon* 1343, *Brisselden* 1619.
Type II represents OE pers.n. **Beorsa*, genitive sing. **Beorsan*, + **dūn** with later substitution of *l* for *n*. Type I, however, may represent the full form of the name of which **Beorsa* would be a pet form, OE *Beorht- Brihtsiġe* + **ing**[4] + **dūn**. Ha 45, Gover 1958.36, DEPN.

BURSLEM Staffs SJ 8749. 'Burhweard or the fort-guardian's (part of) Lyme'. *Barcardeslim* 1086, *Burewardesleg' Lime, Borewardeslyme* 1242, *Burewardeslime* 1252, *Burseleym* 1485, *Burslem* 1599. OE pers.n. *Burhweard* or appellative **burhweard**, genitive sing. *Burhweardes*, **burhweardes**, + district name Lyme as in LYME PARK Chesh SJ 9682. DEPN. gives pers.n. *Burgheard* with OE *g* [ɣ] > [ɣʷ] > [w], WW 146, Horovitz.

BURSTALL Suff TM 0944. 'The fort site'. *Burg(h)estala* 1086, *Burcstal* 1194, *Burstal* 1193–1208, *-stall* from 1275. OE **burh-stall**. DEPN.

Great BURSTEAD Essex TQ 6892. *Great* 1208–88, *Magna* 1189, 1274, *Much* 16th + p.n. *Burgestede* 975×1016–1274, *Burghestedā* 1086, *Burg(he)sted(e)* 1208–1333, *Bur(e)stude -sted(e)* 1212–16th, 'site of a stronghold', OE **burh**, genitive sing. **burge**, + **stede**. 'Great' for distinction from Little BURSTEAD TQ 6691. Ess 145, Stead 197.

Little BURSTEAD Essex TQ 6691. *Parva* 1228 + p.n. Burstead as in Great BURSTEAD TQ 6892. Also known as *West* Burstead 1241. Ess 145.

BURSTOCK Dorset ST 4202. 'Burgwynn or Burgwine's outlying farm'. *Burewinestoch* 1086, *Burgestoch(e)* 1179, 1180, *Burg(h)estok* 1200, 1244, *Burstok* 1291. OE pers.n. *Burgwynn* feminine or *Burgwine* masculine + **stoc**. The modern form may either be a worn-down form of the longer name or derive from a short form *Burge* or *Burga* for Burgwynn or Burgwine respectively. Do 47, Studies 1936.20.

BURSTON Norf TM 1383. 'The settlement by a landslip'. *Borstuna* 1086, *Birston* 1196–9, *Burston(e)* 1212, 1254. OE **byrst** + **tūn**. Cf. BRISTON TG 0632. DEPN.

BURSTON Staffs SJ 9430. Possibly 'Burgwulf's estate'. *Burouestone* 1086, *Bureweston* 1242, 1252, *Burcheston* 1278. OE pers.n. *Burgwulf*, genitive sing. *Burgwulfes*, + **tūn**. DEPN.

BURSTOW Surrey TQ 3041. 'The fortified place'. *Burestou* 1121, 1135×54, *-stowe* 1235, 1241, *Burgstowe* 1252, *Burgh-* 1308, *Burstow alias Bristow* 1727. OE **burh-stōw**. Sr 286.

BURSTWICK Humbs TA 2227. 'Brusti's dairy farm'. *Brocste- Brostewic* 1086, *Brustewic -wyc -wyk* 1203×21–1392, *Brustwic -wyk(e) -wik(e)* [1170×5]c.1300, 1199×1216–1461, *Burstwi(c)k -wyk* 1280–1650. ON pers.n. *Brusti*, genitive sing. *Brusta*, + **wīc**. YE 33 gives pr [bɔstwig].

BURTERSETT NYorks SD 8989. 'The elder-tree shieling'. *Beutresate* 1280, 1283, *Birtresatte* 1285, *Butterside* 1577, *Burtersett* 1608. ME **burtre** + **sate** (ON *sǽtr*). YN 267.

BURTLE Somer ST 4043. Uncertain. Cf. *Burtle house* 1610, *West Burtle, Burtle Hill, Burtle Heath* 1817 OS. Burtle Hill is a hill in the Somerset levels. Burtle is therefore probably uncertain element + **hill**.

BURTON 'A fortified enclosure, a farm or village by a fort'. OE **burh-tūn**. A secondary form **byrh-tūn** with *byrh*, genitive sing. of *burh*, lies behind some examples below. WW 145–53.

(1) ~ Ches SJ 3174. *Burton* from 1152. The reference is to the earthworks at Burton Point SJ 3073. Che iv.211.

(2) ~ Ches SJ 5164. *Burtone* 1086, *Burton* from 13th. Che iii.270.

(3) ~ Dorset SZ 1694. *Buretone* c.1100, *Buriton* 1236, *Burton* 1248. The sense here is possibly 'farm of the borough' referring to Christchurch a mile to the S. Ha 45.

(4) ~ Lincs SK 9674. *Burton(e)* 1086, *-tuna* c.1115. DEPN, WW 150.

(5) ~ Northum NU 1633. 'The settlement by the fort'. *Burton* from 1242 BF. There are earthworks to the NW at NU 1533. NbDu 34.

(6) ~ Somer ST 1944. *Burton* 1809 OS.

(7) ~ Wilts SU 0730. *Burton* 1204. Wlt 81, WW 146–7, 152.

(8) ~ AGNES Humbs TA 1063. 'Agnes's B'. *Bortona* 1086, *Burtun(e) -ton* 1086–1265, *Burton Agnetis* 1255, ~ *Agnes* from 1316. From Agnes de Percy or Agnes de Albemarle c.1175. YE 88 gives pr [bɔtn, bɔʔn]. WW 151.

(9) ~ CONSTABLE Humbs TA 1836. 'B held by the Constable family'. *Burton Con(e)stable* c.1265–1407. P.n. *Burton* + family name *Constable*. Also known as *Santriburtone* 1086, unexplained, and *Erneburgh Burtona* 1190 from *Erneburgh*, widow of Gilbert de Alost, early 12th. YE 61, WW 151.

(10) ~ FLEMING Humbs TA 0872. 'B held by the Fleming family'. *Burton(a) Flandrensis, (le) Flamang -ing* 12th cent., ~ *Fleming* from 1240. P.n. *Burton(e)* 1086, *Burtun -ton(a)* 12th–1828, + family name *Fleming* who held Burton in the 12th cent. YE 112 gives pr [bɔtn], WW 151.

(11) ~ GREEN Warw SP 2675. *Burton greene* 1585, 1650. P.n. BURTON + ModE **green**. Wa 58.

(12) ~ HASTINGS Warw SP 4189. 'B held by the Hastings family'. *Burugton de Hastings* 1313, *Burton Hastinges* 1540. P.n. *Burhtun* [1002×4]c.1100 S 1536, *Bortone* 1086, *Burton(a)* from 1220, *Bourton* 1332, *Broughton* 1436, + manorial addition from the family of Henry de Hasteng who held the estate in 1242. Wa 102, WW 150.

(13) ~ -IN-KENDAL Cumbr SD 5376. *Bortun* 1086, *Burton* since 1120×30. 'In Kendal', *in Kendal(e)* 12th, for distinction from BURTON IN LONSDALE SD 6572. We i.57, WW 151.

(14) ~ IN LONSDALE NYorks SD 6572. P.n. *Borctune* 1086, *Burton(a)* 1130–1428, + p.n. *Lanesdale* 1086, *Lonesdale* 1198–1443, the 'valley of the river Lune', r.n. LUNE SD 3653, 6075 + OE **dæl**. YW vi.248, 218, WW 151.

(15) ~ JOYCE Notts SK 6443. 'B held by the Jorz family'. *Burton Joyce* 1726. P.n. *Bertune* 1086, *Birton(a)* c.1170–1291, *Burton* 1222, + manorial affix *Jorce* 1327, *Jorse* 1356, *Joys, Joce, Jorse* 16th cent. Geoffrey de Jorz held the manor in 1235. Possibly the manor originally belonged to nearby Woodborough SK 6247, or the reference might be to the earthwork at SK 6344. Nt 157, WW 150.

(16) ~ LATIMER Northants SP 9074. *Burton(e) Latymer* 1482, 1512. Earlier simply *Burton(e)* from 1086. Held by William le Latimer in 1280. Nth 180, WW 150.

(17) ~ LAZARS Leic SK 7616. P.n. *Burton(e)* from 1086, *Burghtun'* 1237, + affix *de Sancto Lazaro* 1200×1240, *Sancti Lazari* 1254–1520, *Lazars* from 1449 referring to the hospital of St Lazarus for lepers founded here c.1135. The village stands on a hill-top, a suitable site for an early defensive work. Lei 150, WW 149.

(18) ~ -LE-COGGLES Lincs SK 9725. *Burton in le Coggles* 1583, ~ *le Coggles* 1641. P.n. *Bertune -tone* 1086, *By- Birtun(e) -ton(e)* 1212–1571, *Burton* from 1185, + 16th cent. affix *in le Coggles*, ModE **coggle** 'a rounded water-worn stone', an antiquarian addition for distinction from other Burtons and referring to local soil conditions. Perrott 155, WW 150.

(19) ~ LEONARD NYorks SE 3263. 'St Leonard's B'. P.n. *Burton(e) -tun* 1086 etc., + saint's name *Leonard* 1276 etc., probably from the former dedication of the church, now St Helen's. YW v.93, WW 151.

(20) ~ ON THE WOLDS Leic SK 5820. P.n. *Burton(e)* 1086–1610 + affix *juxta Prestwo(u)ld* 1295, *super Waldas* 1301, *othe Wold* 1413, *super Olds, on the Owles* 1604 referring to the village's site on high ground NE of Loughborough. Cf. Prestwold SK 5721, *Prestewald(e)* 1086–1282, 'the priests' part of the wolds', WALTON ON THE WOLDS SK 5919, WYMESWOLD SK 6023. Leic 282, 313, L 227, WW 149.

(21) ~ OVERY Leic SP 6798. 'B held by the Noveray family'. *Burton(e) Ouerey -ay* 1317. Earlier simply *Burtone* 1086. The manor was held by Robert de Noveray in 1261. There are earthworks, probably not defensive, in the village which lies in a valley between two spurs of land. There is no trace of fortification on the spurs. Lei 210, WW 149.

(22) ~ PEDWARDINE Lincs TF 1142. 'B held by the Pedwardine family'. *Burton Ped(e)wardyn(e)* 1369–1642. Earlier simply *Burton -tun* 1086–1675. The manor was held through marriage by Roger Pedwardine of Pedwardine in Brampton Bryan H&W. Perrott 56, WW 150.

(23) ~ PIDSEA Humbs TA 2531. A combination of two p.ns., *Bortun(e)* 1086, *Burton(a)* 13th cent., *Burton(a) Pi- Pydse* 13th–1389, + *Pidsea* as in *Piddese (mere)* 1260, *the water of Pidsey* 1550, the 'marsh pool', OE ***pid(e)**, ***pidu** 'fen, marsh' + **sǣ**, the names of a lost mere and fishery. The reference is probably to Owstwick Drain. YE 55 gives pr [bɔtn], WW 151.

(24) ~ UPON STATHER Humbs SE 8717. *Burtonstather* 1275. P.n. *Burtone* 1086 + ON **stathar**, pl. of **stǫth** 'a landing place'. Burton upon Stather lies on the banks of the r. Trent. DEPN, WW 150.

(25) ~ UPON TRENT Staffs SK 2423. *Burton super Trente* 1234. P.n. *Byrtun* [1002×4]11th S 1536, [1012]13th S 929, *Byrtune* 1066, *Bertona* 1086, + r.n. TRENT SJ 9231. DEPN, WW 150.

(26) ~ WOOD Ches SJ 5692. *Burtoneswod* 1228, *B(o)urtonewod* 1251–1341 'wood belonging to (a lost) Burton'. Earlier simply *Burton* 1200. La 96.

(27) Bishop ~ Humbs SE 9839. 'B held by the (arch)bishop' sc. of York. *Bi- Bys(s)hop Burton(e)* from 1349. Earlier simply *Burton(e)* from 1086. Part of the fee of the Archbishops of York in 1086 and after. YE 192.

(28) Cherry ~ Humbs SE 9942. *Cheri -y Burton(e)* 1444–1518, *Cherry* ~ from 1562. Earlier simply *Burton(e)* 1086, *-ton -tun* 1166–1828, and *North* ~ 1289–1562. *North* and 'Cherry' for distinction from Bishop BURTON SE 9839. YE 191 gives pr [bɔtn], WW 151.

(29) Constable ~ NYorks SE 1690. 'B held by the constable' sc. of Richmond. P.n. *Burton* from 1279 + *Conestabel* 1279×81. YN 247, WW 151.

(30) Gate ~ Lincs SK 8383. 'Goat B'. *Gaiteburton* [1199]1330,

Geiteburtone 1219. ON **geit**, genitive pl. **geita**, + p.n. *Bortone -a* 1086, *Burtuna* c.1115, *Burtun* 1163 etc. This must have been a place where goats were kept. SSNEM 209, WW 150.

(31) East ~ Dorset SY 8386. *Estburton(e)* 1280–1332 etc. 'East' for distinction from West Burton SY 8285, *Westburton(e)* from 1280, a deserted medieval village. This large settlement is simply referred to as *Burton(e)* 1210–80. There is no evidence for a fortification here. Do i.176, WW 152.

(32) Little ~ Dorset ST 6412 → LONGBURTON ST 6412.

(33) West ~ Dorset SY 8285 → East BURTON Dorset SY 8386.

(34) West ~ NYorks SE 0186. ME prefix *West* 1254 + p.n. *Burton* 1086. YN 265, WW 151.

(35) West ~ WSusx SU 9913. 'The settlement to the W of and belonging to Bury'. *Westburgton* 1230, *Westburton* 1279, *Bury Westburton* 1296. ME adj. **west** + p.n. **burhtun** < p.n. BURY TQ 0113 + **tūn**. West Burton is situated in Bury parish one m NW of Bury. But cf. WW 152. Sx 125, SS 77.

BURTON BRADSTOCK Dorset SY 4889. 'B held by Bradenstoke'. P.n. Burton + p.n. BRADENSTOKE Wilts SU 0079, referring to possession by Bradenstoke abbey. Burton, *Bridetone* 1086, *Briditona* 1244, *Brytton* 1327, *Berton* 1381, is the 'settlement on the r. Bride', r.n. *Bride* as in Long BREDY SY 5690, + OE **tūn**. Do 48.

BURTON SALMON NYorks SE 4927. 'Salamon's B'. *Burton Salamon* 1516. Earlier simply *Breiðe-tun* c.1030 YCh 7, *Bretton(a)* 1154×63–1428, *Burton* 1425 etc., 'the Britons' settlement', OE *Brettas*, genitive pl. *Bretta*, + **tūn**. The c.1030 YCh 7 form is probably erroneous by confusion with BRAYTON SE 6030 mentioned in the same charter. 'Salmon' is from *Salamone de Brettona* who witnessed a charter c.1230. YW iv.40.

BURWARDSLEY Ches SJ 5156. 'The *burhweard*'s wood or clearing'. *Burwardeslei* 1086, *-ley(a) -legh* etc. 1186–1819, *Burwardsley* from 1653, *Borosley* 1499, *Bursley* 1524–1671. OE **burhweard** 'the guardian of a town or stronghold', genitive sing. **burhweardes**, + **lēah**. The derivative OE pers.n. *Burgweard* is also possible, but this may have been the estate of an officer charged with the defence of Chester or a fortification on the Welsh border. Che iv.93 gives prs ['bɔ:wə(d)zli] and locally ['bouəzli], ['bu:əzli], L 206.

BURWARTON Shrops SO 6285. 'The fort-guardian's estate'. *Burertone* 1086, *Burwardton'* 1194–5, *Burwarton* from 1225, *Bur(e)warston* 13th. OE **burhweard** + **tūn**. Alternatively the specific might be the OE pers.n. *Burgweard*. Sa i.67, WW 146.

BURWASH ESusx TQ 6724. 'Ploughed land by a fortified place or belonging to a borough'. *Burgersa* 12th., *Burhercse* before 1170, *Burgherse -ers(s)h(e) -ersch(e)* 13th–14th cents., *-es(s)e* 1224, *Burgasshe* 1378, *-wassh* 1428, *Burwash* 1440, *Bur(r)ish* 1586, 1620. OE **burh** + **ersc**. The reference is unknown. A Sussex rhyme preserves the local pronunciation,

To love and cherish
From Battle to Berrish.

Sx 461 gives prs [bʌriʃ] and [beriʃ], SAC 35.68.

BURWASH COMMON ESusx TQ 6423. Marked as *Burwash Wheel* 1813 OS. P.n. BURWASH TQ 6724 + ModE **common**.

BURWELL Cambs TL 5866. 'Fort spring'. *Burcwell* [1062] 14th S 1030, *Burghwell(e)* 1298–1509×47, *Bur(e)uuella -e* 1086, *Burwell(e)* from 1227. OE **burh** + **wella**. The Norman castle here may have occupied the site of an earlier earthwork. Ca 187, L 32.

BURWELL Lincs TF 3579. 'The spring by the fortified place'. *Buruelle* 1086, *Burewelle* c.1110, *Burwell* from c.1115, *Burgwelle* 1292. OE **burh** + **wella**. DEPN, L 32.

BURY '(At the) fortified place'. OE **byriġ**, dative sing. of **burh**.

(1) ~ Cambs TL 2883. *Byrig* [974]14th S 798, *(æt) Byryg* c.1000, *Biria, Bi- Byri(g), Biry* [1100×35]14th, 1253–1336, *Bury* from 1359. The reference is to an earthwork SW of the village described as the 'Roman Camp'. Hu 206.

(2) ~ GMan SD 8010. *Bury* from c.1190 with variants *Biri, Buri, Biry, Byry, Berye*. Neither the date nor the nature of the fortification are known. La 61, TC 63, Jnl 17.43.

(3) ~ Somer SS 9427. *Bury* 1809 OS. Although the evidence is late this can hardly be other than OE **byriġ**, dative sing. of **burh** 'a fortified place'. The hamlet lies by a hill-spur between valleys.

(4) ~ WSusx TQ 0113. *Berie* 1086, *Bery* 1203–91, *Bur(r)y* 1271–1332. The reference is unknown but the place was important enough to give its name to Bury Hundred. Sx 124.

(5) ~ DITCHES Shrops SO 3283. *Bury Ditches* 1836 OS. The reference is to the great Iron Age hill-fort on the summit of Sunnyhill. Ultimately from OE **byriġ**, dative sing. of **burh** 'a fort', + **dīċ**. Thomas 1976.183.

(6) ~ GREEN Herts TL 4521. 'The manor green'. *le Beregrene* 1369, *Berygrene* 1413. Fuller, *Worthies of England* ii.17, 1662, writes 'Surely no County can shew so fair a *Bunch of Berries*, for so they term the fair *Habitations* of *Gentlemen* of *remark*, which are called *Places, Courts, Halls* and *Mannors* in other Shires'. Hrt 179, 243.

(7) ~ HILL Hants SU 3443. *Bury Hill* 1817 OS. The *Bury* (ultimately from OE *byriġ*, dative sing. of *burh*) is a hill-fort with three lines of bank and ditch dating from the 3rd cent. BC to 1st cent. AD. Pevsner-Lloyd 632, Thomas 1976.140.

(8) ~ ST EDMUNDS Suff TL 8564. 'St Edmund's Bury'. *Sancte Eadmundes Byrig* 1038. Saint's n. *Ēadmund*, genitive sing. *Ēadmundes*, + p.n. *(on) Byrig* c.1035 Wills, *Burye* 1610. OE **byriġ**, dative sing. of **burh**. St Edmund, king of the East Angles, was martyred and buried here in 870. An early name of the place is *Sanctæ Eadmundes stow* 'the holy place of St Edmund' [962×91]11th S 1494. These forms replaced the original name *(æt) Bæderices wirde* [945]13th S 507, *Beodrichesworth* [962]13th S 1213, *Beadriceswyrð* c.1030 *Secgan*, 'Beaduric's enclosure', OE pers.n. *Beadurīċ*, genitive sing. *Beadurīċes*, + **worth**, **wyrth**. DEPN.

(9) ~ WALLS Shrops SJ 5727. *Bury Walls* 1833 OS. The reference is to the 20-acre Iron Age hill-fort with triple earthen ramparts. Ultimately from OE **byriġ**, dative sing. of **burh** 'a fort', + **wall**. Thomas 1976.183.

BURYTHORPE NYorks SE 7964. 'Bjǫrg's outlying settlement'. *Berg(u)etorp* 1086, *Bergert(h)orp* 1180×90, 1268, *Berwerthorp* 1252, *Berg(h)thorp(e)* 1289–1360, *Berethorp* 1409, *Burithorp(e) -y(e)-* 1519–1650, *Berrythorpe* 1554. ON feminine pers.n. *Bjǫrg*, genitive sing. *Bjargar*, + **thorp**. YE 142, SPN 54, SSNY 56.

BUSBRIDGE Surrey SU 9842. Partly uncertain. *Bursebrig(g)e -brugg(e)* etc. c.1252–1394, *Bussebrigge* 1398, *-brugge* 1448, *Busbridge* from 1619. Unidentified element or pers.n. + OE **brycg** 'a bridge'. Cf. Bushblades Durham NZ 1653, *Bursebred -brades* 12th cent., *-blades* 1243, *Bushblades* 1717, possibly 'broad strip(s) of land where prickly shrubs grow' with OE ***burs** cognate with MDu *bors, burs*. An alternative possibility is 'Burgsige's bridge', OE pers.n. *Burgsiġe* + **brycg**. Sr 196.

Great BUSBY NYorks NZ 5205. *Magna Buskebi(a), Buskeby* 1180–1327, *Magna Busbye* 1587. ModE **great**, Latin **magna**, + p.n. *Buschebi* 1086, 1202, *Buskby* 1369, 'the farmstead or village in the scrub', ON ***buskr**, ***buski**, + **bȳ**. YN 169, SSNY 23.

Little BUSBY NYorks NZ 5104. ModE **little** + p.n. *Buschebi(a)* 1086 as in Great BUSBY NZ 5205. A lost village whose site is probably marked by Busby Hall. YN 169, SSNY 23.

BUSCOT Oxon SU 2397. 'Burgweard or the fortress warden's cottage(s)'. *Boroardescote* 1086, *Bur(gh)wardescota -e* 1130–1474, *Burscote* 1412, *Busketts* 1666, *Buscot* 1736. OE pers.n. *Burgweard* or **burh-weard**, genitive sing. *Burgweardes*, **burh-weardes**, + **cot**, pl. **cotu**. Brk 353, WW 146.

BUSHBURY WMids SJ 9203. 'The bishop's manor'. *Byscopesbyri* [996]17th, *Biscopesberie* 1086. OE **biscop**, genitive sing. **biscopes**, + **byriġ**, dative sing. of **burh**. DEPN.

BUSHEY Herts TQ 1395. 'Bush, thicket enclosure'. *Bissei*

1086–1330 with variants *Byss-* and *-eg' -eie -eye, Bi- Bys(s)heye* 1230–1359, *Bussheye* 1330, 1383. OE **bysc** + **(ġe)hæġ**. Cf. nearby OXHEY TQ 1193. Hrt 64.

BUSH GREEN Norf TM 2187. *Bush Green* 1838 OS. ModE **bush** + **green**. There is a network of similar names in this part of Norf including GREAT GREEN TM 2789, HEMPNALL GREEN TM 2493, NORTH GREEN TM 2277, SHELTON GREEN TM 2390, Bedingham Green TM 2892, Denton Green TM 2788, Earsham Green TM 3284, Silver Green TM 2593, Hollands Green TM 2991, Wacton Green TM 1890, and *Coblers Green, Culham Green, Denton Great Green, Darram Green, Little Green, Stratton Wood Green, Rhees Green, Thorpe Green, Twainton's Green* etc. all 1838 OS.

BUSHLEY H&W SO 8734. 'Clearing where bushes grow'. *Biselege* 1086, *Bisselega -le(y)* 1159–1221, *Busse- Bi- Bys(s)che- Bishele(ye)* 1212–1417, *Bussheley* 1415, 1575. OE ***bysc**, genitive pl. ***bysca**, + **lēah**. Wo 104.

BUSHTON Wilts SU 0677. 'The bishop's estate'. *Bi- Byssopeston* 1242, 1268, *Bushepeston* 14th, *Bushton* 1558×1603, *Bushen* 1753. ME **bishop**, genitive sing. **bishopes**, + **tūn**. The reference is to that part of the manor of *Clive* as in CLYFFE PYPARD SU 0776 held by the bishop of Winchester in 1086. It is a late formation in *tūn*. Wlt 266.

BUSHY PARK GLond TQ 1569. *Bushie Park* 1650. ModE adj. **bushy** + **park**. The evidence is too late for certainty: so the specific might be 'bush enclosure', ME **bush** + **hay** (OE *(ġe)hæġ*). Mx 15, Encyc 113.

BUSSAGE Glos SO 8803 → BISLEY SO 9006.

High BUSTON Northum NU 2308. *Uuerbuttesdun* 1186, *Bud- Butlisdon Superiore* 1242 BF, *Botilisdon superior* 1296 SR. ModE adj. **high**, ME **ofer**, Latin **superior** + p.n. *Buttesdon -ton* 1166, *Butlesdon, Bot(e)leston -don* 13th cent., *Botilston* 1307, *Butels- Bot(el)eston* 1346, *Buston* 1428, possibly 'Buttel's hill', OE pers.n. **Buttel*, genitive sing. **Butt(e)les*, + **dūn** varying with **tūn**. OE *dūn* is original as the Bustons lie on the edge of high ground beside a hill rising to 356ft. It is possible, therefore, that the specific is rather an OE appellative ***butt** 'a hill' varying with ***buttel** 'a little hill'. Alternatively the name may contain OE **bōtl**, genitive sing. **bōtles**, 'the hill belonging to the *bōtl*' referring to nearby SHILBOTTLE NU 1958. Cf. BUTSFIELD Durham NZ 1045. 'High' for distinction from Low Buston NU 2207. NbDu 35.

BUTCHER'S PASTURE Essex TL 6024. Surname *Butcher* + ModE **pasture**. No early forms.

BUTCOMBE Avon ST 5161. 'Buda's coomb'. *Budancumb* [984×1016]12th S 1538, *Bvdicome* 1086, *Budecumb* 1225, *Budicumbe* 1259×60 Ass. OE pers.n. *Buda*, genitive sing. *Budan*, varying with *Buda* + **ing**[4], + **cumb**. DEPN, L 92.

BUTLEIGH Somer ST 5233. 'Budeca's wood or clearing'. *Budecalech* [725]12th S 250, *Bodecanleighe* [801]14th S 270a, *Budeclega* *[971]13th S 783, *Bucticanlea* (for *Budican-*) [984]15th S 850, *Bodvchelei, Bodechelie* 1086[†], *Butley* 1610. OE pers.n. **Budeca*, genitive sing. **Budecan*, + **lēah**. DEPN, Turner 1952.150–1.

BUTLEIGH WOOTTON Somer ST 5034 → Butleigh WOOTTON.

BUTLEY Suff TM 3650. Possibly 'Butta's wood or clearing'. *Butelea -lai* 1086, *Buttele* 1186–1336, *Butele* 1191, 1291, *Butley* from 1235. OE **Butta* + **lēah**. Another possibility is 'the wood or clearing of the butts', OE **butt**, genitive pl. **butta**, + **lēah**. DEPN.

BUTSER HILL Hants SU 7120. *Butser Hill* 1810 OS. P.n. Butser + ModE **hill**. Butser, *(on) bryttes oran* [956]13th, *byrhtes oran* [959×63]12th S 811, *Britteshore* 1448, *Butser* 1593, is 'Briht's ridge', OE pers.n. *Briht*, genitive sing. *Brihtes*, + **ōra**. Ha 46, Gover 1958.55.

[†]Another spelling *Bodeslege* 1086 may also belong here, DB Somerset 8.12.

BUTSFIELD Durham NZ 1045. 'Open land of the dwelling house'. *Botelisfield* 1284, *Buttelesfeld* c.1300, *Butlesfeld* 1334, *Botesfeld* 1312, *But(t)esfeld* 1382, 1385. OE **bōtl**, genitive sing. **bōtles**, + **feld**. Alternatively the specific may have been an OE pers.n. *Bōtel* or **Buttel*. NbDu 35.

BUTTERBURN Cumbr NY 6774. 'Stream surrounded by good grazing land'. *Butterburne* 1603. ME **butter** + **burn**. A tributary of this burn is called Cheese Burn. Cu 6, 97.

BUTTERCRAMBE NYorks SE 7358. 'Piece of rich land in the bend' sc. of the river Derwent. *Butecram(e)* 1086, 1234, *Botercram(e)* c.1155–1416, *But(t)ercram* 13th cent. OE **butere** + ***cramb**. YN 36.

BUTTERKNOWLE Durham NZ 1025. 'Butter hill'. *Butterknowle* 1647, 1652. ModE **butter** + **knoll** (OE *cnoll*). The hill in question is described as a rock or stones in 1652, probably colliery waste. *Butter* in minor names usually refers to good land producing rich butter: the name must antedate the industrial developments here and have applied to a natural feature – unless its use is ironic.

BUTTERLEIGH Devon SS 9708. 'Butter pasture'. *Bvterlei* 1086, *-lega -lea* 1161×84, 1187, *Boterleg(he)* 1251–91. OE **butere** + **lēah**. The reference is to rich pasture. D 577, L 206.

BUTTERLEY WYorks SE 0410. 'Clearing with butter producing pasture'. *Inn & Out Butterly* 1850. Ultimately OE **butere** + **lēah** but the antiquity of the name is unclear. YW ii.278.

BUTTERMERE (Lake) Cumbr NY 1815. 'Lake surrounded by good grazing land'. *Water of Buttermere* 1343. OE **buttere** + **mere** Cu 33.

BUTTERMERE Cumbr NY 1716. Transferred from the lake name BUTTERMERE NY 1815. *Butermere* 1230, *Buttermer'* 1260. Cu 355, L 26.

BUTTERMERE Wilts SU 3461. 'Butter pond'. *Butermere* [863]12th S 336, *(æt) Buter mere* 931 S 416, *Bwtermære* 1066×87, *Butremare -mere* 1086, *Buttermere* from 12th. OE **butere** + **mere**. The sense is probably 'pond by good pasture'. Wlt 339, L 26.

BUTTERMERE FELL Cumbr NY 1915. *Buttermere Fell* 1868 OS. P.n. BUTTERMERE (Lake) NY 1815 + ModE **fell** (ON *fjall*).

BUTTERSHAW WYorks SE 1329. 'Copse near the butter producing pasture'. *Buttershawe* 1567. ME **butere** + **shawe**. YW iii.10.

BUTTERTON Staffs SK 0756. 'Butter hill'. *Buterdon* 1200, *Butterdon* 1223, *Boturton* 1422. OE **butere** + **dūn**. A hill with rich pasture producing good butter. DEPN, L 144, Horovitz.

BUTTERWICK 'Dairy-farm, farm where butter is made'. OE **butere** + **wīċ**.

(1) ~ Humbs SE 8305. *Butreuuic* 1086, *Buterwic* 1219. DEPN.

(2) ~ Lincs TF 3845. *Butruic* 1086, *Buterwic* 1198, 1202. DEPN, Cameron 1998.

(3) ~ NYorks SE 9971. *Butruid* (sic) 1086, *Butreuic -w-* 12th, early 13th, *Butterwic -wyc -k(e)* 1120×35–1370. YE 114.

(4) ~ NYorks SE 7377. *Butruic* 1086, *Buttrewyc, But(t)erwic -wyk(e)* 1145×8 etc. YN 47.

(5) ~ LOW Lincs TF 4243. P.n. BUTTERWICK TF 3845 + dial. **low** 'a river, an inlet'.

BUTT GREEN Ches SJ 6651. *Butt Green* 1831, ModE **butt** 'a strip of land abutting on a boundary' + **green**. Butt Green is at the edge of Stapeley township. Che iii.72.

BUTTONOAK Shrops SO 7578. *Buttonoak* 1802. Gelling.

BUXEY SAND Essex TM 1103. P.n. Buxey + ModE **sand** 'a sandbank'. Buxey, *Bokesey(e)* 1308×23, *Bokesehey* 1323, cf. *Bucksey Marsh* 1598, must once have been an island before the encroachment of the sea. The name means 'Bucc's island', OE pers.n. *Bucc*, genitive sing. *Bucces*, + **ēġ**. Ess 18.

BUXHALL Suff TM 0057. 'Bucc or the buck's nook'. *(æt) Bucyshealæ* [1000×2]11th S 1486, *Boccheshale* 1050 Thorpe, *Buckeshala, Bukes(s)alla* 1086, *Buchessala* 1165, *Bukkeshal* 1254, *Bux(h)ale* 1254–1327, *Buxhall* 1568. OE pers.n. *Bucc* or

bucc from which it is derived, genitive sing. *Bucces*, **bucces**, + **halh**, dative sing. **hale**. DEPN, Baron.

BUXTED ESusx TQ 4923. 'Beech-tree or box-tree place'. *Boxted'* 1199, *Boxstude -stede* 1210×12–1405, *Boc- Bok- Bux- Buckstед(e)* 1229–1412. OE **bōc** or **box** + **stede**. Sx 389, Stead 253.

BUXTON Derby SK 0673. 'Rocking stone(s) or buck stone(s)'. *Buchestanes* before 1108, *Buc(k)stanes -is* 1101×8–13th, *Bu(c)k- Bucstones* 1257–1357, *Buc- Bukston* 1277–1357, *Buxton* from 1577. OE **būg-stān**, pl. **būg-stānas**, or **bucc** 'a buck, a male deer', + **stān**, pl. **stānas**. Db 52.

BUXTON Norf TG 2322. 'Bucc's estate'. *Bukes- Buchestuna* 1086, *Buxstone* 1254. OE pers.n. *Bucc*, genitive sing. *Bucces*, + **tūn**. DEPN.

BYERHOPE RESERVOIR Northum NY 8546. *Byerhope Reservoir* 1866 OS. P.n. Byerhope NY 8646 + ModE **reservoir**. Byerhope, no early forms, appears to be the 'valley with a cow-shed', ModE **byre** + dial. **hope** 'a side-valley' (OE *hop*), here a side-valley opening off the river East Allen.

BYERS GREEN Durham NZ 2234. 'Village green at *Byres*'. *Byers greine* 1562, ~ *Green* from 1647. P.n. *Bires* 1197×1209–1418, *(les) Byres* 1228×37–1514 (~ *Geffrey* 1382, 1498), *Byars* 1306, *the Byarsse* 1506, *Byers* 1547, OE **bȳre**, plural **bȳras** 'byres, cow-sheds', + **grēne**. The settlement began as a clearing on SPENNYMOOR NZ 2534 (*assart de Byres* [1183]c.1320) where herdsmen's sheds were established. The green was built over in modern times. NbDu 36, GV 28 and fig.10.

BYFIELD Northants SP 5253. '(The settlement) beside the open land'. *Bivelde -f-* 1086, *Bifeldia* 12th. OE **bī**+ **feld**. Nth 33, L 241, 244.

BYFLEET Surrey TQ 0661. '(The settlement) by the river'. *æt Bifleote* [1062]13th S 1035, *Biflet* 1086–1200, *Biflete* [933, 967]13th S 420, 752, *Byflete* 1284, 1308, *Byfleet* from 1270. OE locative phrase ***bi flēote** with the element *flēot*, usually a tidal inlet, here referring to the river Wey. Sr 104, Jnl 22.44–6.

BYFORD H&W SO 3943. Perhaps 'ford at a river-bend'. *Buiford* 1086–c.1170, *Bu- Biford(')* 1217×19–1317, *Byford* from 1291. OE **byġe** + **ford**. There are more pronounced bends in the river Wye than at Byford, but the parish boundary follows an old, very sharply curving course of the river. The alternative, therefore, OE **byġe** + **ford**, 'commerce ford', is unnecessary. He 51, PNE i.72.

BYGRAVE Herts TL 2636. '(Place) by the entrenchments'. *Bigravan* [973]15th S 792, *(æt) Biggrafan* 11th, *Bigrave* 1086, *Bygrave* from 1204. OE preposition **biġ** + **græf**, dative pl. **grafum**, or ***grafa**, dative sing. ***grafan**. The reference is to ancient earthworks here. Hrt 155.

BYKER T&W NZ 2764. '(The settlement) beside the marsh'. *Biker'* 1219 BF–1298, *Byker* from 1242 BF. ME **bi** + **ker** (ON *kjarr*). NbDu 36.

BYLAND ABBEY NYorks SE 5478. P.n. *Beghland(a)* 1142×3–1231, *Beiland(a), Beyland* c.1150–1242, *Biland(a), Byland* 1285 etc., 'Bega's land', OE pers.n. *Bēġa* + **land**, + ModE **abbey** referring to the Cistercian monastery removed here in 1177. YN 194.

Old BYLAND NYorks SE 5585. *Veteri Bella Landa* (Latin) 1293, *Old Bylande* 1541. ME **old**, Latin **vetus**, dative case **veteri**, + p.n. *Begeland* 1086, 'Bega's newly cultivated land', OE pers.n. *Bēġa* + **land**. Reinterpreted in 1293 as 'the beautiful glade' as if from OFr *bel* + *launde*. The original site of BYLAND ABBEY SE 5478 in 1143. Also known as Byland-on-the-Moor. YN 197, L 247, 249.

BYLAUGH Norf TG 0319 → BELAUGH TG 2818.

BYLEY Ches SJ 7269. 'Beofa's wood or clearing'. *Bevelei* 1086, *-ley* 1291, *By- Bivele(ye) -leg(h)* etc., late 12th–1499, *Bygele, Byle* 1304, *Byley* from 1410. OE pers.n. *Bēofa* + **lēah**. Che ii.233 gives pr [baili].

BYRNESS Northum NT 7602. *Byrness* 1869 OS. A new village for Kielder forestry workers.

Castle BYTHAM Lincs SK 9819. 'B with the castle'. *Castelbytham -i-* 1309–1554, *Castle Bytham* from 1331. Cf. *castellum de Biham* 1219. ME **castel** + p.n. Bytham as in Little BYTHAM TF 0117. The earthworks of the castle which was built in the 11th cent. and demolished in 1221 still dominate the village. Formerly also known as *Westbitham* 1086–1321 and *Magna Byham* (sic) 'great B' 1230–1439, *Graunt Byham* 1268, *Great Byt(h)am* 1275–1577. Perrott 158.

Little BYTHAM Lincs TF 0117. *Litilbitham* 1464, *Parva Bi- Bytham* after 1391–1537. ME adj. **litel**, Latin **parva**, + p.n. *Bytham* [1066×8]c.1200, *Bi(n)tham* 1086, *By- By(h)hamel(l)* c.1150–1291, 'the homestead or settlement by the valley bottom', OE **byt(h)me** + **hām**. Also known as *Est Bi- Bytham* 1288–1821 for distinction from Castle or Great BYHTAM SK 9819. The village is situated at the junction of a side-valley where it descends into the valley of the West Glen river. The forms *Bi- Byhamel* show AN loss of medial [ð] and addition of Fr diminutive suffix *-el*. Perrott 161.

BYTHORN Cambs TL 0575. 'By the thorn-bush'. *Bitherna* [10th]14th S 1808, *Bierne* 1086, *By- Bierne* 1252–c.1350, *Bi- Bytherne* [1127]14th, 1248–1451, *Bithorne, Bythorn* 1285, 1545. OE **bī** + **thyrne**. Hu 236, DEPN.

BYTON H&W SO 3764. 'Bend settlement'. *Boitvne* 1086, *Bu(y)ton(e)* 1287–1410, *Byton* from 1479×80. OE **byġe** + **tūn**. The reference is said to be to a bend in the river Lugg, but the more prominent feature here is the bend in the hills in which the village lies. He 51.

BYWORTH WSusx SU 9821. 'Bæga's enclosure'. *Byworthe, Be(h)gworth* 1279. OE pers.n. *Bǣġa* + **worth**. One of a group of five neighbouring *worths* from E to W, LODSWORTH SU 9223, PETWORTH SU 9721, BYWORTH, Hesworth and FITTLEWORTH both TQ 0119. Sx 116, SS 79.

C

CABOURNE Lincs TA 1401. 'Jackdaw stream'. *Caburne* 1086–1610, *-burn(')* 1199–1707, *-borne* 1319–1741. OE **cā** + **burna**. Li iv.65.

CADBURY Devon SS 9104. 'Cada's fort'. *Cadebirie* 1086, *-beria -biri -byre -bury* 1093–1356 with variant *K-*. OE pers.n. *Cada* + **byriġ**, dative sing. of **burh**. The reference is to Cadbury Castle hill fort SS 9105 (*Cadbury Castell* 1397). Cf. CADELEIGH SS 9107. For the possibility that *Cada* was a mythological name see South CADBURY Somer ST 6325. D 559.

CADBURY BARTON Devon SS 6917. 'Cadbury farm'. P.n. Cadbury + SW dial. **barton** 'a farmyard, a demesne farm'. Cadbury, *Cadebire* 1238, 1242, *Caddebir* 1238, *Cadebury* 1303, is 'Cada's fortified place', OE pers.n. *Cada* + **byriġ**, dative sing. of **burh**. D 377.

CADBURY CAMP Avon ST 4572. *Cadbury Camp* 1830 OS. An Iron age hill-fort. No early forms are known but the camp must have been originally called Cadbury whether independently or as a transfer from CADBURY Devon SS 9105 or South CADBURY Somer ST 6325.

CADBURY CASTLE Somer ST 6225. *Cadbury Castle* 1811 OS. P.n. Cadbury as in South CADBURY ST 6325 + ModE **castle**. The reference is to South Cadbury Iron Age hill-fort on a site occupied from Neolithic times until the reign of Ethelred the Unready who set up an emergency mint here. The original name of the fort was simply Cadbury 'Cada's fort'. Thomas 1976.195.

North CADBURY Somer ST 6327. *Northkadebir'* 1212, *-cadeb'* [1239]B, *Northcadebury* 14th. ME adj. **north** + p.n. Cadbury as in South CADBURY ST 6325. Earlier simply *Cadeberie* 1086. DEPN.

South CADBURY Somer ST 6325. *Svdcadeberie* 1086, *South Cadbury* 1610. OE adj. **sūth** + p.n. *Cadanbyrig* c.1000, 'Cada's fort', OE pers.n. *Cada*, genitive sing. *Cadan*, + **byriġ**, dative sing. of **burh**. The reference here, as at CADBURY Devon SS 9104 and CADBURY CAMP Avon ST 4572, is to an Iron Age hill-fort, CADBURY CASTLE ST 6225. It seems unlikely that the same pers.n. should occur in association with two such distinctive places; we should probably consider that Cada was a traditional figure of myth or folklore. Cf. also Cadson Bury Corn SX 3467, p.n. Cadson, *Cadokestun* [1175]1378, *Cadeston* 1327 + **bury**, earlier *Cadebyre* 1262, *Cadebury* 1365, *Cadbery* 1408, Cadbury Farm Hants SU 3127, Cadbury Shrops SO 6173, and the p.n. type BADBURY SU 1980. DEPN, Gover n.d. 187.

CADDINGTON Beds TL 0619. 'Cada's hill'. *Cadandun -en-* [c.1053]13th S 1517, *Cadendone* 1086, *Cadingdun* 1195, *Cadington* 1247. OE pers.n. *Cada*, genitive sing. *Cadan*, + **dūn**. A 16th and 17th cent. local pr is shown by the spelling *Carrington* 1563, 1690. Bd 145, DEPN.

CADEBY Either 'Kati's village or farm', ON pers.n. *Káti*, genitive sing. *Káta*, + **bȳ**, or 'the boys' farm', ON **kati**, genitive pl. **kata**.

(1) ~ Leic SK 4202. *K- Catebi* 1086–early 13th, *-by* c.1176–1476, *Cadeby(e)* from 1516. Lei 488, SSNEM 40.

(2) ~ Lincs TF 2695. *Cadebi* 1086, *Cadeby* from c.1200. Cameron 1998.

(3) ~ SYorks SE 5100. *Catebi -by* 1086–1521, *Cadeby* from 1479. YW i.63, SSNY 23.

CADELEIGH Devon SS 9107. 'Cada's wood or clearing'. *Cadelie* 1086–1334 with variants *-leye -legh(e)*. OE pers.n. *Cada* + **lēah**. Probably the same Cada as gave his name to CADBURY SS 9104. D 559.

CADGEWITH Corn SW 7214. *Cadgewith* 1748, short for *Porthcaswith* 1358, *Por Cadgewith* 1699, 'harbour of the thicket', Co **porth** + ***caswyth** 'thicket, bramble-brake'. PNCo 62.

CADISHEAD GMan SJ 7092. '*Cadwall* shieling'. *Cadwalesate* 1212, *Cadwalsete* 1212, 1271, *Cadewallessiete* 1226, *-set* c.1300, *-wallisete* 1329, *-walle(s)heved* 1322, 1346, *Cadyswalhede* 1538. OE p.n. **Cadawœlla* 'Cada's spring', OE pers.n. *Cada* + **wælla**, + ON **sǽtr**. Cadishead is an isolated settlement on the Mersey at the S end of Chat Moss. *Cadwall* may have been an old name of Glaze Brook SJ 6894. The 1322 form with ME **heved** (OE *hēafod*) 'a headland' may refer to the point of raised ground in the junction of Glaze Brook and the Mersey SJ 7091. The alternative suggestion that the specific is the pers.n. *C(e)adwalla* is unlikely since this would normally give **Chadwalle* in this area. La 39, Jnl 17.34.

CADLEY Wilts SU 2066. Possibly 'Cada's wood or clearing'. *Cadley* 1817 OS. OE pers.n. *Cada* + **lēah**. But the evidence is far too late for certainty and the name might be manorial. Cadley is at the edge of Savernake Forest. Wlt 354, 244.

CADMORE END Bucks SU 7892. *Cadmore End* 1822. P.n. Cadmore + ModE **end**. Contrasts with Bolter End SU 7992, *Bolter End* 1822, Wheeler End SU 8093, *Wheeler End* 1822, and LANE END SU 8092, *Lane End* 1822. Cadmore, *Cademere* 1236, is 'Cada's pond', OE pers.n. **Cada* + **mere**. Bk 177.

CADNAM Hants SU 2913. 'Cada's estate or hemmed-in land'. *Cadenham* 1272–1331. OE pers.n. *Cada*, genitive sing. *Cadan*, + **hām** or **hamm**. *Cada* may be short for *Cademan* as in the name of a lost ford in the parish, *Kademannesforde* 1279. If so, this might be an instance of the PrW pers.n. **Cadμān* (OE *Cœdmon*) and could date back to the earliest contacts between Welsh and English in this region. But it could also mean 'Cada's man'. *Cada* itself is ultimately derived from Brit **catu-*, PrW **cad* 'battle'. Ha 47, Gover 1958.195.

CADNEY Humbs TA 0103. 'Cada's island'. *Catenai -ase* 1086, *Cadenai(e) -(a) -ay -ei(a) -eya -eye* c.1115–1526, *Cadney* from 1318. OE pers.n. *Cada*, genitive sing. *Cadan* + **ēġ**. Li ii.75.

CAER CARADOC 'Fort of Caradoc'. PrW **cair** + pers.n. *Caradọ̄g* (Brit *Caratācos*). A name type applied to hill-forts on the W border commemorating the name of Caratācus (Caractacus), the British leader against the Romans.

(1) ~ Shrops SO 4795. *Caer Caradoc* 1833. This seems to be a folk-etymology of earlier *Cordokes* 1271×2, *Cordakes* 1291×2, *(The) Cordock* 1636–64, *Cordock Hill* 1695, probably a pre-English hill-n. for the highest peak in the Church Stretton Hills. A hill-fort on a summit rising to 1506ft. Thomas 1976.183, Gelling.

(2) ~ Shrops SO 3175. *Caer Caradoc* 1833 OS. An Iron Age hill-fort. Thomas 1976.183.

CAERHAYS CASTLE Corn SW 9441 → ST MICHAEL CAERHAYS SW 9642.

CAESAR'S CAMP Hants SU 8350. *Caesars Camp* 1816 OS. For other examples of this name type cf. Caesar's Camp Beds TL 1748, GLond TQ 2271 and TQ 4263, SYorks SK 3995, all Iron Age hill-forts with modern names, Thomas 1976.49, 133, 254.

CAINHOE Beds TL 1036. Partly uncertain. *Chainehou, Cainou* 1086, *Kaino, Keinho* 13th cent. Possibly 'spur, heel of the hill called *Cæge*, the key-shaped height'. OE hill-n. **Cǣġe* from either OE *cǣġ* 'a key' or **cæġ* 'a stone', genitive sing. *Cǣġan*, + **hōh**, dative sing **hō**. A pers.n. **Cǣġa* masc. or **Cǣġe* feminine is also possible. Cf. KEYSOE TL 0763. Bd 147, ESt 43.41, L 168.

CAISTER-ON-SEA Norf TG 5112. A modern form for what was formerly called *East Caister* 1837 OS. 'East' for disinction from West CAISTER TG 5111. Earlier simply *Castra* 1086–1183 and *[1044×7]16th S 1055, *Castre* 1186–1547, *Caistor* 1248×53, *Cayster* 1469, *Castor* c.1750, the 'Roman town', OE **cæster**. A 30 acre Roman harbour town has been excavated here. Romano-Saxon pottery in use c. AD 300 points to trade or the presence of *foederati*. Some 150 burials of the later Saxon period (c.650–850) were found outside the walls. Nf ii.3, Pevsner 1962NE.109.

West CAISTER Norf TG 5111. *West Caister* 1837 OS. Adj. **west** + p.n. Caister as in CAISTER-ON-SEA TG 5112.

CAISTOR Lincs TA 1101. 'The Roman town'. *CASTR* 975×8 Coins, *Castre* 1070×87–1556 including 1086, *Castra* 1090–1254, *Castor* c.1275, 1392, 1555, *Caster* 1331–1697, *Caister* 1485–1576, *Caistor* from 1634. OE **cæster**. For possible identification with the Roman town of *Bannovalium* see HORNCASTLE TF 2669. Li ii.87.

CAISTOR ST EDMUND Norf TG 2303. 'St Edmund's C'. *Castre Sancti Edmundi* 1254. Saint's name *Edmund* from the dedication of the church added to p.n. *Castre* [c.1025]13th S 1528, *Castrum* 1086, 'the Roman town', OE **cæster**. The Roman town of *Venta Icenorum*, the 'market town of the Iceni'. Excavation has produced evidence of occupation into the 3rd century AD followed by signs of burning and sudden death and a Saxon settlement beyond the town belonging to the troubled period c.360–400. The settlement was occupied until c.500 and overlies some 2–4th cent. Roman huts. DEPN, RBrit 492, Pevsner 1962NW.108.

CAISTRON Northum NT 9901. 'The *Cers* or marsh thorn-tree'. *Kersten* 1184, *Kerst(h)irn, Ke(r)s- Cres- Castern(e)* 13th cent., *Kestryn* 1428, *Kaistrin* 1663. ME **kers** + ME **thirn** (OE *thyrne*). The simplex *Cers* occurs c.1160, possibly the plural of ME **ker** (ON *kjarr*) 'a marsh'. It survives as a singular noun *carse* in Scottish dial. and p.ns. Caistron lies in the valley of the river Coquet. NbDu 37.

CALBOURNE IoW SZ 4286. 'Kale stream'. *(to, on, andlang) Cawelburnan* [826]12th S 274, *(apud) Cawelburnam* 828]c.1300 S 1581, *Cawelburn(a) -e -bourne* 1220–1338, *Cavborne* 1086, *Calburna -e -bourn(e)* from 1232. Probably OE **cawel** + **burna** referring to sea-cabbage, *brassica oleracea*, or sea kale, *crambe maritima*, or even the ordinary cultivated cabbage. Cf. however, the r.n. CALE Somer ST 7126, earlier *Cawel*, for an alternative explanation. Wt 75 gives prs ['kə:bərn] and ['kæł bərn], Mills 1996.36.

CALCOT Berks SU 6671. 'Cold cottages'. *Caldecote* 13th cent., *Colecote* 1325, *Cal(di)cote* 1552, *Calcot* 1544, 1846. OE **cald** + **cot**, pl. **cotu**. Cf. CALDECOTE. Brk 222.

CALDBECK Cumbr NY 3239. 'Cold stream'. *C- Kalde- C- Kaude- Cawdebe(c)k -bec(k)* from [11th]13th, *Cawlbek* 1506. OE (Anglian) **cald** or ON **kaldr** + **bekkr**. Cu 275 gives pr [kɔ:bek], SSNNW 113, L 14.

CALDEBERGH NYorks SE 0985. 'The cold hill'. *Caldeber* 1086, *-bergh* 1184–1311, *Cauldeberg(h)* 1269, 1293. OE **cald**, definite form **calda**, + **beorg**. YN 253 gives pr [kɔ:dbə], L 128.

CALDECOTE 'Cold cottages'. OE **cald**, pl. **calde**, + **cot**, pl. **cotu**. Cf. CALCOT.

(1) ~ Cambs TL 1488. *Caldecote* from 1086, *Caudecote* 1248, *Cawcott* 1576. Hu 181.

(2) ~ Cambs TL 3456. *Caldecote* from 1086, *Caudecot(e)* 1227–75, *Cawket* c.1570–1607, *Cowcote* 1644, 1690. Ca 156. O li gives pr [kɔ:dikʌt].

(3) ~ Herts TL 2338. *Caldecota* from 1086 with variant *-cote*, *Calcott* 1526. Hrt 155.

CALDECOTT 'Cold cottages'. OE **cald** + **cot**, pl. **cotu**.

(1) ~ Leic SP 8693. *Caldecot(e)* 1086–1410, *-cott* from 1535, *Calcot(e)* 1426–1553. R 244.

(2) ~ Northants SP 9668. *Caldecote* from 1086, *Caudecote(s)* 1234, 1269, *Calcote* 1461, *Caucot* 1675, *Caldecott* 1689. Nth 190.

River CALDER 'Hard, rapid, violent stream'. PrW ***caled-duβr** (Brit **caleto-dubro*).

(1) ~ Cumbr NY 0609. *Kalder* [c.1200]1318, *aquam de K- Calder* 1292, *River of Cawder* c.1675. Cu 7, RN 60.

(2) ~ Lancs SD 5346. *Keldir* c.1200, *Caldre* 1228–c.1350, *Calder* 1228, c.1540, *Kaldir* 1324. La 140, RN 59.

(3) ~ Lancs SD 7833. *Caldre* before 1193–14th, *Calder* from c.1200, *Kalder, Caldyr* [1294]14th, *Kelder* 1295–6. La 66, RN 60.

(4) ~ WYorks SE 2620. *Kelder* 12th–1467, *Calder* from 13th. Spellings with *e* may represent a weakening of the unstressed vowel at a time when the main stress was still on the generic **duβr**, cf. ModW *Clettwr* where the unstressed vowel has actually disappeared. YW vii.121.

(5) ~ BRIDGE Cumbr NY 0305. *Pons aque de Calder* 1509×47, *Calder Bridge* 1644. R.n. CALDER NY 0609 + **brycg**.

(6) ~ VALE Lancs SD 5345. R.n. CALDER SD 5346 + ModE **vale**.

CALDERBROOK GMan SD 9418. *Calderbrook* 1843 OS. Cf. Caldermoor SD 9316, *Caldermoor* 1843 OS. These two names suggest that the original name of the river Roach was CALDER. Roach itself is probably a back-formation from the name Rochdale.

CALDON CANAL Staffs SJ 9453. *Caldon Canal* 1836 OS. P.n. Caldon as in Caldon Grange and Low SK 0848, a variant of CAULDON SK 0749, + ModE **canal**. A canal extending from Apedale Hall in NW Staffs to Handford and thence to the Grand Trunk Canal at Stoke-on-Trent.

CALDWELL 'The cold spring'. OE **cald**, definite form **calda**, + **wella**, Mercian **wælla**.

(1) ~ Derby SK 2517. *(æt) Caldewællen* [942]14th S 1606, *Caldewelle* 1086, *Cald(e)well(e)* from 1114, *Cald(e)wall(e)* 1086–1552, *K- Caud(e)well(e)* 1236–1338, 1651. The earliest spelling points to a dative plural form *æt Caldewællum*, 'at Coldwells'. Db 625.

(2) ~ NYorks NZ 1613. 'So caullid from a lattle font or spring, by the ruines of the old place, and so rennith into a becke halfe a quarter of a mile off'. Leland. YN 299 gives pr [kɔ:dwel].

CALDY Mers SJ 2285. 'Cold arse'. *Calders* 1086–1287, *Kalders* c.1350, *Caldei(e)* 1182–1283, *Caldey* 1237–1724 with variant *Kald-*, *Caldy* 1553, 1819. OE **cald** + **ears**, referring to the prominent hill on which the two manors of GRANGE SJ 2286 (formerly Great Caldy) and Little Caldy lie. The later forms mean 'cold island', OE **cald** + **ēġ**, but precisely how they relate to the earlier forms is uncertain. Che iv.282 gives pr ['kɔ:ldi], L 39.

River CALE Somer ST 7126. A pre-English r.n. of unknown origin and meaning. *Cawel, Wricawel* (for *Win-* 'white Cawel', PrW ***wïnn**) [956 for 953×5]14th S 570, *Caul* [1216×72]14th, *Cale Water* 1577, 1586. Possibly identical with the first element of Caul Bourne I o W SZ 4188, *Cawelburnan* [826]12th S 274, and *Caweldene* [940]13th S 461 in White Waltham, Berks, although these are usually taken to be compounds of OE *cawel* 'colewort'. A better suggestion is Co **cawal**, W **cawell** 'a basket, a creel, a pannier' and so 'creel river'. Creels were used for trapping fish and the Cale was renowned for its fishing. 'White' and 'Black' were regularly used to distinguish different arms of a river, cf. Whiteadder and Blackadder, Borders NT 8555, 8452. RN 63, Wt 75, Brk i.74, *Charters* V.76, Turner 1952–4.13. See also WINCANTON Somer ST 7128.

CALF ModE **calf** usually used in p.ns. of some smaller object near a large one. Cf. RICCAL NYorks SE 6237.

(1) The ~ Cumbr SD 6697. *The Calf* 1859. A mountain name. We ii.45, i.55.

(2) ~ TOP Cumbr SD 6685. 'Top of the calf'. *Calf Top* 1865 OS. Hill name *Calf* 1865 OS + ModE **top**. A summit of 1999ft. perhaps seen as the 'calf' to the loftier summit of Crag Hill SD 6983 (2250ft.).

CALIFORNIA Norf TG 5114. No early forms. Transferred from California, USA, as a name suitable for a remote place. The original California was a poetic name for an island in the ocean, peopled by Amazon-like women and rich in gold. Cortez applied it to the Californian peninsula whence it was exported N along the coast by Spanish settlers in 1542–3. S.R. Stewart, *A Concise Dictionary of American Place Names*, Oxford 1986 s.n., Nf ii.14.

CALKE Derby SK 3722. 'Chalk, limestone'. *C- Kalc, C- Kalk(e)* from 1132, *Chalke* 1195–1230, *Cauke* 1577, 1610. OE **calc**. The reference is to outliers of the carboniferous limestone here some sixteen miles south of the main outcrop. Db 626 gives pr [kɔ:k], Jnl.19.49.

CALLALY Northum NU 0509. 'The calves' pasture'. *Calualea* 1161, *Calvele(ya)* 1212, 1226×8, *Caluley* 1236, 1296 SR 163, *Calele* 1425. OE **calf**, genitive pl. **calfa**, **lēah**. NbDu 37, BF 204, 370, 598, L 206.

CALLERTON 'Calves' hill'. OE **calf**, genitive pl. **calfra**, + **dūn**.

(1) Black ~ T&W NZ 1769. *(Blac)kalverdon'* 1242 BF, *Black Callirdon* 1311, *Black Callerton* 1865 OS. ME adj. **blak** + p.n. *Calverdona* 1212 BF, *Calverdon* 1296 SR. 'Black' for distinction from High Callerton Northum NZ 1670, ModE adj. **black** + p.n. *Calverduna* 1100×35, *Calverdon'* 1242 BF, 1346, and also known as *Calverdon de Valenc'* 1296 SR, *Callerton Valkens* 1428, from its sale to Sir William de Valence 1264×8. NbDu 37.

(2) High ~ Northum NZ 1670. *High Callerton* 1864 OS. ModE **high** + p.n. *Calverdon'* 1242 BF 1115 as in Black CALLERTON Tyne and Wear NZ 1769.

CALLESTICK Corn SW 7750. Unexplained. *Calestoch* 1056, *Kalestoc* c.1250, *Kellestek* 1302, *Kelestok* 1306. It is unknown whether this is an English name in **stoc** or a Cornish name in suffix **-ek** (OCo *-oc*), and no satisfactory solution for the first element has yet been proposed. Cf. CALSTOCK SX 4368. PNCo 62, Notes 15.

CALLINGTON Corn SX 3669. 'Settlement at or of the bare hill'. *Calwetone* 1086, *Calwinton* 1187, *Calyton* 1306, *K- Cal(l)yngton* 1318–45 G, *Killitton* 1610. Hill-name **Calwe* or **Calwa* 'the bare one' < OE **calu** 'bald, bare', dative or genitive sing. **Calwan*, + **tūn**. There is no evidence that the forms *Cælling* [905]12th BCS 614, *Cællwic* 980×8 and *Kelli wic ygkernyw* 13th Mabinogion 'forest grove in Cornwall', which have usually been identified with Callington, belong here; they are incompatible with the early forms for Callington and are therefore disregarded. The later form with *-ing-* is probably hyper-correct owing to the frequent reduction of *-ing-* to *-i-* in p.ns. The reference is to Kit Hill 1096ft., a mile NW at SX 3771 which dominates the town. PNCo 62, PNE i.79, Notes 15.

CALLOW H&W SO 4934. 'The bare one'. *Calue* 1175, *(La) Caluwe -ewe* 1285–1346. OE **calu**, dative sing. **calwe**, definite form **calwan**. Originally the name of one of the hills here. For further examples see CALLOW HILL SO 7473. He 52.

CALLOW END H&W SO 8349. Called *Callow Green* in 1832 OS, *Callaway* in 1820. Possibly a hill-name identical with CALLOW SO 4934 or surname derived from such a hill-name + ModE **end**. Wo 224.

CALLOW HILL H&W SO 7473. 'Hill called Callow'. *Callow Hill* 1832 OS. Hill-name identical with CALLOW SO 4934 + ModE **hill**. Further examples include Callow Hill SP 0264, *Kalewan hulle* 1221, *Callow Hill* 1613, Callow Hill Wood SO 3928, *Calowe* 1300, *Caluhull* 1334, *Calewehull* 1360, Callow Farm SO 5721, *(La) Calewe* 1302, 1313, *La Calowe* 1473. Wo 318, He 83, 198.

CALLOW HILL Wilts SU 0384. 'The bare hill'. *(le) Calewehulle* 1300–1343×77, *Callowe Hill* 1558×1603, *Calley Hill* 1815. OE **calu**, dative sing. definite form **calewan**, + **hyll**. Wlt 65.

CALLOWS GRAVE H&W SO 5967. *Callows Grave* 1832.

CALMORE Hants SU 3414. 'Sea-kale marshland'. *Cauwelmor* 13th, *C- Kaulemor(e)* 1276, *Colmore* 1557. OE **cawel** + **mōr**. The place is adjacent to the former tidal marshes of the r. Test. The sea-kale of Southampton Water was famous in the 18th cent. as a wild food as far afield as London. Ha 47, Gover 1958.200, Jnl 21.6.

CALMSDEN Glos SP 0408. 'Calumund's valley'. *Kalemundes-, to Calmundesdene* [852]13th S 202, *Calemundesdene* 12th–1327, *Calmondesden(e)* 1220–1403, *Calmesden(e)* 1379–1584. OE pers.n. **Calumund*, genitive sing. **Calumundes*, + **denu**. Gl i.148.

CALNE Wilts SU 0070. Probably originally a r.n. *Calne* from c.1120 ASC(E) under year 978 *(Cálne)* including [951×55]15th S 1515, *(et) Calnæ* [997]12th S 891, *Calna* 1091–1144, *Cauna -e* 1086–1228, *Cawne* 1556,*Cane* 1672. The r.n. occurs as the specific of CALSTONE WELLINGTON SU 0268 and is identical with Kale Water, Roxburghshire, *Kalne* 1165×1214, and COLNE Lancs SD 8940, *Calna* 1124, all pre-English r.ns. of uncertain origin. Formally they could derive from **Caluna* parallel to *Alauna*, the origin of the r.ns. Alne and Ale Water. The root **cal-* 'to call' seems to occur in *Calagum* for **Calacum*, the name of the RBrit fort at Burrows-in-Lonsdale, Lancs SD 6175, possibly also in *Crococalana*, the RBrit settlement at Brough, Notts SK 8358, and in several Gaulish names. The sense would have been 'noisy stream, loud calling one'. The river in question is now called Marden, *Merkedene* 1245, which is itself of OE origin and not originally a r.n. but a p.n., OE *mearc-denu* 'the boundary valley'. It formed one of the boundaries of Chippenham Forest. Wlt 256 gives prs [kɑ:n] and popular [kan], 8, RN 90, RBrit 288, 327.

CALOW Derby SK 4171. 'Bare-hill nook'. *C- Kalehal(e)* 1086, *C- Kalahala -(e)* before 1108–1269, *C- Kalal(e)* 1278–1459, *Calall(')* 1430–1484, *Calo* 1561, *-low(e)* from 1558. OE **calu** + **halh**. It is suggested that *calu* in this name refers to, or was the name of the small sub-circular hill on which the church stands overlooking a very slight valley. Db 229 gives pr [keilou], L 107, 110.

CALSHOT CASTLE Hants SU 4802. *Calshot castel* 1579. P.n. CALSHOT SU 4701 + ModE **castle**. One of Henry VIII's castles built to defend the S coast against invasion from France. Ha 47, Pevsner-Lloyd 158.

CALSHOT Hants SU 4701. Uncertain.

I. *(æt) celcesoran* [980]12th S 836, *Calchesores* c.1300.

II. *celces(h)ord* [11th]14th, *Calchesord(e)* 1347, 1428, *Calshorde* c.1450.

III. *kelcheford* 1344, *Calcheford* 1353.

IV. *Calshotespoynt* 1539, *Calshot castel* 1579, *Cawshot* 1610.

The specific cannot be OE *ċealc* 'chalk' because of the initial [k]. Possibly, therefore, 'Cælic or Cælci's shore', OE pers.n., OE pers.n. **Cæliċ* or *Cælċi*, genitive sing. **Cælċes*, + **ōra** 'shore, flat-topped hill' varying with **ord** 'point, spit of land' referring to the spit of land extending into Southampton Water, itself later replaced with **shot** as if from OE *scēat* 'a nook, a corner of land'. An alternative suggestion is OE *caliċ, cæl(i)ċ, celċ* 'a cup, a chalice' perhaps for a cup-shaped harbour or landing place and so 'cup shore, cup point'. Ha 47, Gover 1958.198, Mx 86, Jnl 21.21, 22.29.

CALSTOCK Corn SX 4368. Partly obscure. *Calestoch* 1086, *-stoc* 1208, *K- Calistoke* [1208]17th, 1329, *Calestoke* 1265, *Kalstock* 1270, *Calstock* 1610. The generic is OE **stoc** 'an outlying farm', the specific possibly a reduced form of the p.n. Callington SX 3669 and so 'outlying farm of Callington'. But this is very uncertain. PNCo 62, Notes 15, DEPN.

CALSTONE WELLINGTON Wilts SU 0268. 'C held by the Willington family'. *Calston Wely* 1500, *Caulston Wellington*

1568, *Cawston Willington* 1607. P.n. *Calestone* 1086–1316 with variant *-ton, Calston* 1242–1326, probably 'the settlement belonging to Calne', r.n. *Calne*, genitive sing. *Calnes*, + **tūn**, + manorial addition from the family of Ralph de Wilinton 1228. Wlt xl cites an undated form *Calveston* which is probably a transcription error for *Calneston*. Wlt 257 gives popular pr [kɑ:sən].

CALTHORPE Norf TG 1831. 'Kali's outlying farm'. *Caleðorp* *[1044×7]16th S 1055, *Cala- Catetorp* 1086, *Caletorp'* 1197, 1209. ON pers.n. *Kali*, genitive sing. *Kala*, + **thorp**. DEPN, SPNN 244.

CALTHWAITE Cumbr NY 4640. 'Calves' clearing'. *Caluethweyt -thwayt* 1272–1619, *Calthwayt(e)* 1348, 1460, *Calthatt*, *Cawthwayte* 16th cent. OE (Anglian) **calf**, genitive pl. **calfa**, or ON **kalfr**, genitive pl. **kalfa**, + **thveit**. Cu 202 gives pr [kɔ:θət], SSNNW 113, L 211.

CALTON NYorks SD 9159. 'The calf farm'. *Caltun -ton* 1086–1610, *Calveton* 1304, *Cawl- Caulton, Cawton* 16th cent. OE **calf** + **tūn**. YW vi.130 gives pr [kɔ:tn].

CALTON Staffs SK 1050. 'The calf farm'. *Caldon* 1228, *Calton(e)* from 1229, *C- Kouton'* 1236, *Calveton* 1340, *Cowton* 1599, 1607, *Cawton* 1558–1689. OE **calf**, genitive pl. **calfa**, + **tūn**. Oakden gives prs ['kɔ:ltə] and ['koutə], Horovitz.

Great CALVA Cumbr NY 2931. ModE adj. **great** + p.n. *Calva* 1749, probably 'calf hill', OE (Anglian) **calf** or ON **kalfr** + **haugr**. 'Great' for distinction from Little Calva NY 2831. Cu 326.

CALVELEY Ches SJ 5958. 'Calves' pasture'. *Kaluileia -leh*, *Caluilee -leg'* c.1180–1474 with variants, *Calveley* from 1282. OE **calf**, genitive pl. **calfa**, + **lēah**. Che iii.307 gives pr [ka:vəli], L 206.

CALVER Derby SK 2474. 'Calf ridge'. *Calvoure* 1086, *Kalv- Caluou(e)r(e) -over(e)* 1200–1409, *Calu- Calvore* 1199–1490, *Cawver als. Calver* 1578. OE **calf** + **ofer**. Db 54 gives pr [ka:və], L 175–6.

CALVER HILL H&W SO 3748. *Calver Hill* 1833 OS.

CALVERHALL Shrops SJ 6037. 'The calves' nook'. *Cavrahalle* 1086, *Chalurehal'* 1219, *Calfrehall', K- Caluerhal(e)* 13th cent., *Calverhalle* 1284×92, *-hall* from 1490, *Callerall* 1427, *Corrar* 1843. OE **calf**, genitive pl. **calfra**, + **halh** in the sense 'dry ground in marsh', dative sing. **hale**. Sa i.69 says the local form is still *Corra*, Raven 42.

CALVERLEIGH Devon SS 9214. '*Calwood* clearing'. *Calodelie* 1086, *Calewudelega* 1194–1535 with variants *Cal- Kal(e)-*, *-wod(e)-* and *-legh(e), Calwoodley al. Calverley* 1557. P.n. *Calwood* 'bare wood' < OE **calu** + **wudu**, + **lēah**. Pole comments c.1630, *Kawoodlegh nowe called corruptlye Calverley*. The change of medial *w* > *v* is seen in other SW names, e.g. LIFTON SX 3885, MEAVY SX 5467. D 539, i.xxxv.

CALVERLEY WYorks SE 2036. 'Calves' pasture'. *Cau- Caverlei(a)* 1086, *Kalu- Kalverlaia -leia -lay(e)* 12th cent.–1261, *Calverley* from 1202×8, *Cauverlay* c.1250, *Cawverley* 1404. OE **calf**, gen.pl. **calfra**, + **lēah**. YW iii.224 gives prs ['ka:]- and ['kɔ:vələ], a modern sp-pr. YW iii.224, L 206.

CALVERT Bucks SP 6824. No early forms.

CALVERTON Bucks SP 7939. 'Calf farm'. *Calvretone* 1086, *Cauverton* 1206–7, *Caluerton* 1227. **calf**, genitive pl. **calfra**, + **tūn**. Bk 18 gives pr [kælvətən], D xlix.

CALVERTON Notts SK 6149. 'The calves' farm'. *Calvretone* 1086, *Calverton(a)* from c.1190, *Caverton* 1583, *Coverton* 1663. OE **calf**, genitive pl. **calfra**, + **tūn**. Nt 158 gives prs [kɑ:vətən] and [kɔ:vətən].

CAM Glos ST 7599. From the r.n. Cam. *Camma* 1086–1236, *C- Kamme* 1195–1575, *Cam* from 1472. The r.n. is not so early evidenced, *Cammes Brooke* 1566, *Cam amniculum* 'the little stream Cam' 1590, but probably represents PrW ***camm** (British **cambo-*) 'crooked'. The stream has a winding course. Gl ii.215, i.4, RN 65.

Lower CAM Glos SO 7400. *Lower Cam* 1830 OS. ModE adj. **lower** + p.n. CAM ST 7599.

CAM BECK NYorks SD 7978. P.n. Cam as in CAM FELL SD 8180 + ModE **beck**.

River CAM Cambs TL 5268. A back-formation from Cambridge. *Cam* from 1610. Ca 2, RN 64.

CAMBEAK Corn SX 1296. *Cambeak* 1732 G, 1789. Probably 'comb beak', ModE **comb** 'a comb, a crest' + **beak**. The reference is to a sharply-crested promontory. Co **cam** + ***pyk** (with mutation of *p* to *b*) 'crooked point' is formally possible, but ***pyk** is extremely rare and an English name is more likely in this part of Cornwall. PNCo 63.

CAMBER ESusx TQ 9619. 'The chamber'. *Camere* 1375, *Portus Camera* 'Camber harbour' 1397, *(le) Ca(u)mbre* 15th cent., *the Cambre* c.1450, 1512. Cf. *Camber Point* 1747. AN **cambre** (OFr *chaumbre*) applied to the confined space of the harbour. An anchorage in Finisterre is similarly called *La Chambre*. Sx 538.

CAMBER CASTLE ESusx TQ 9619. P.n. CAMBER TQ 9218 + ModE **castle**. One of Henry VIII's 'artillery castles' built in 1539.

CAMBERLEY Surrey SU 8860. Developed in the early 19th cent. as a place for retired army officers from the Royal Military Academy at Sandhurst nearby. Originally called *Cambridge Town* after the newly-appointed Commander-in-chief of the army, the Duke of Cambridge, who built the army staff college here in 1862. On the establishment of a railway station here in 1877, to avoid confusion with Cambridge, its name was changed to Camberley, a name invented by a Dr E. Atkinson and probably suggested by other local names in *-ley* e.g. Frimley, Eversley, Yately. Sr 127, Ess lix, Room 1983.18.

CAMBERWELL GLond TQ 3376. Partly uncertain. *Cambrewell(e)* 1086–1373, *Camerwell(e)* 1174–1611, *Camberwell(e)* from 1241. Unidentified element + OE **wella** 'a spring, a stream'. Professor Coates raises the possibility that the specific might be a borrowed form of Latin *camera* 'vault, room' referring to a building at the spring. Sr 17 gives pr [kæməwəl], TC 202.

CAMBLESFORTH NYorks SE 6426. Probably 'Camel's ford'. *Cameles- Canbes- Ca'bes- Gamesford(e)* 1086, *Camelesford(e)* 1154×81–1468, *Camelsford* 1311–1536, *-forth* 1458, *Camblesforth* 1641. OE pers.n. **Camel*, genitive sing. **Cameles*, + **ford**. Although the pers.n. *Camel* is not on record, the alternative suggestion, W r.n. *Camlais* 'the crooked stream', PrW **camm* + **gle(i)s*, is equally conjectural and seems an unlikely survival for what can only have been a minor stream in the marshland here. There is no identifiable stream in or near the village. YW iv.5.

CAMBO Northum NZ 0285. 'The hill-spur with a crest or ridge'. *Camho* 1230, *Kamho, Cam(b)hou -hogh* 13th cent., *Camhow* 1296 SR 82, *Cambow* 1346, *Cammo* 1583, *Camma* 1715. OE **camb** + **hōh**. NbDu 38 gives pr [kamə], L 129, 168.

CAMBOGLANNA Cumbr NY 6166. 'Curved bank, bank at a bend'. *Camboglans* mid 2nd Ridge Cup, *Cambog[lani]s* 2nd Amiens patera, *Gabaglanda, Cambroianna* [c.700]13th Rav. Brit ***cambo-** + ***glanno-** (W *glann* 'bank, shore'). The Roman fort at Castlesteads NY 5163. RBrit 293.

CAMBOIS Northum NZ 3083. 'The bay'. *Cammes* [c.1040]12th HSC, *Kambus* c.1150, *Camb(h)us, Camh(o)us, Cammus* 13th cent., *Cambois* 1363, *Cambhous* 1377×8 32.292, *Cammosse* 1551. Identical with W *camas, camais*, Irish *cambas*, early Irish *cammas* 'a bay', from British ***cambo-** 'crooked'. This name early underwent folk-etymological association with OE *hūs* 'a house' and OFr. *bois* 'a wood'. Cf. KEMPSTON Beds TL 0347. NbDu 38 gives pr [kaməs], GPC 397.

CAMBORNE Corn SW 6440. 'Crooked hill'. *Camberon* 1182, *Cambron* 1291–1558 G including [c.1230]14th, 1755, *Camborne* from 1431. Co **cam** + **bron** influenced by OE **burna**. PNCo 63.

CAMBRIDGE Cambs TL 4658. 'Bridge across the river Granta'. I. *grontabricc* [c.745]9th Felix, *Grontebrugae* 12th, *(to) Grantanbrycge* c.925 ASC(A) under year 921, c.1000 ASC(B) under year 875, c.1100 ASC(D), c.1150 ASC(E) both under same year, *(to) Grantebrycge -bricge* c.890 ASC(A) under year 875, [c.970]12th, c.1050 ASC(C) under year 876, 1170, *Grantabricge* c.1050 ASC(C) under year 1010, c.1100 ASC(D), *Grentebrige* 1086.
II. *Cantebrigie -a* 1086, *-brigge* 1185–1540 with variants *-brig(g)(e) -brug(g)e* etc., *Cauntebrig'* 1230–1351 with variants *-brig(g)e -breg(g)e* etc.
III. *Caumbrig(g)e* 1348–1458 with variants *-brygge -brege, Cambrugge* 1378, *-bregge -brig(g)e -bryg(g)e* etc. 1412–1552. R.n. GRANTA TL 5051, 5231 + OE **brycg**. The earliest references are to *Grantacaestir* c.731 BHE, *Grantacester* c.1000 OE Bede, *Granteceaster* c.1050 Guthlac, 'Roman fort on the Granta', r.n. GRANTA + OE **ċeaster**. This was the Roman town of *Duroliponte*. The forms *Kair-Grant* 12th, *Caergrant* 15th and *Caergraunte* 15th, found in Henry of Huntingdon, Ralph Higden and John Trevisa are antiquarian formations. The development of the name shows that initial *Gr-* became *Cr-* and was then simplified by dissimilation in the sequence *Cr-n-r* in **Crantebrigge* to *C-n-r*. For the development *Gr- > Cr-* which is not actually evidenced here, see the forms for GRANTCHESTER TL 4355. Ca 36, RN 64, TC 66, RBrit 351, L 65.

CAMBRIDGE Glos SO 7503. 'Bridge over the r. Cam'. *Cambrigga -brig(g)e* 1200×10–1516, *Cambridge* 1535. R.n. Cam as in CAM ST 7599 + OE **brycg**. Gl ii.248.

CAMDEN TOWN GLond TQ 2784. So named in 1795 after Charles Pratt, first Earl Camden of Camden Place, Kent and Bayham Abbey, Sussex, who came into possession of the manor of Kentish Town by marriage and granted leases here for building houses in 1791. Hence also the names Bayham Street and Place and Pratt Street. Mx 141, GL 35, Encyc 47, 119.

Queen CAMEL Somer ST 5924. 'The queen's Camel'. *Camel Reginæ* (Latin) 1280, *Quene Cammell* 1431. ME **quene** + p.n. *Cantmæl* 995 S 884, *Cantmel* [939×46]13th S 1718, [955×9]13th S 1755, [959×75]13th S 1764, *Camel(le)* 1086, possibly Brit **canto-* as in QUANTOCK HILLS + PrW ***mę̄l, moil** 'bare hill'. The Camels lie in a valley due S of Camel Hill ST 5925 which rises from low ground to 258ft. 'Queen' commemorating the donation of the estate by Edward I to queen Eleanor before 1280 and for distinction from West CAMEL ST 5724. DEPN, Turner 1952–4.18.

River CAMEL Corn SX 0168. The name originally referred only to the upper section of the river and probably means 'crooked one'. *Camel* 1602, *Camel or Alan River* 1748. Earlier forms as in CAMELFORD SX 1083. Co **cam** + adj.suff. ***-el** or diminutive suffix **-ell**. The name of the main river was *Alan* 1199–1803, *Aleyn* [1285–1381]14th, a common Celtic r.n. discussed under River ALLEN. It was incorrectly transferred to the River ALLEN SX 0678 by the Ordnance Survey. PNCo 63, RN 6, 66, Gover n.d. 680, BzN 1957.226.

West CAMEL Somer ST 5724. *Wescammel* 1291, *West Camell* 1610. ME adj. **west** + p.n. Camel as in Queen CAMEL ST 5924. DEPN.

CAMELFORD Corn SX 1083. 'Ford of the River Camel'. *Camelford* from [c.1256]c.1260, *Camleford* 1256. R.n. CAMEL SX 0168 + OE **ford**. Camelford first occurs in Layamon's *Brut* 14240, an Arthurian poem based on Geoffrey of Monmouth's *History of the Kings of Britain*, where it is substituted for Geoffrey's *Camblam*, the place of Arthur's last battle. *Camblam* is in fact the unlocated *Camlann* of the older Welsh Annals, and Layamon's identification of it with Camelford is simply a piece of creative guesswork. PNCo 63, Gover n.d. 67.

CAMELSDALE Surrey TQ 8932. No early forms. Probably surname *Camel* + ModE **dale**.

CAMERTON Avon ST 6857. 'Settlement or estate on the river Camelar'. *Cam(e)lartone* [946×55]13th S 1738–9, *Camelertone* 1086, *Camelarton* 1227. R.n. *Camelar* + **tūn**. *Camelar*, to-day Cam Brook, is *(innon) Cameler, (andlang) Camelar* [961]12th S 694. The name Cam is probably a shortened form of *Camelar* although it could represent PrW ***camm** 'crooked'. *Camelar* is obscure but might represent the Celtic divinity name *Camulos, Camalos* + r.n. suffix *-ra, -ro*. Cf. CAMPTON Beds TL 1238. DEPN, RN 65–6.

CAMERTON Cumbr NY 0331. Partly obscure. *Camerton(a)* from c.1150, *Camberton(e)* c.1150–1686. The generic is OE **tūn** 'farm, village, estate', the specific unexplained. Cu 281.

CAM FELL NYorks SD 8180. P.n. Cam + ModE **fell**. *Camp(e)* 1190, 1220, 1401, *Cambe* 1307–56, *Came* 1547, *Cham* 17th cent., is 'the comb, the crest', OE **camb**, referring to the great ridge between Cam Beck and Gayle Moor crossed by a Roman road, Margary no. 73, referred to as *Cambesgate* 1338, 'Cam's way', p.n. Cam + ON **gata** 'a road'. YW vi.218, L 129.

CAMMERINGHAM Lincs SK 9482. Possibly 'the homestead of the Cameringas, the people called after Cafmær or Cantmær'. *Came(s)lingeham* 1086, *Camring- Cameryngham* c.1115, *Cambrigeham* 1126, *-inge-* c.1175, 1192, *Cameringham* 1202, 1275, *Kamerincham* 1212. OE folk-n. **Cameringas* < pers.n. **Cǣf-* or **Cantmǣr* + **ingas**, genitive pl. **Cameringa*, + **hām**. Alternatively the specific could be a p.n., 'the homestead of the people who live at *Cā-mere* 'the jackdaw pond', or pers.n. **Cǣf- *Cantmǣr* + **ing**[2], 'the homestead called or at **Cǣf-Cantmǣring*, the place called after Caf- or Cantmær'. Cf.CAMBERWELL Surrey TQ 3376, CAMERTON Cumbr NY 0331 and a lost *Cameringcroft* 1257 in Lincs. DEPN, ING 143.

The CAMP Glos SO 9109. *The Camp* 1779. The place lies N of two long barrows. Gl i.131.

Broad CAMPDEN Glos SP 1637. *Bradecampeden(e)* 1224–73, *Brode Caump(e)dene* 1313, 1327. Also known as *Parva Campdene* 1216 'little Campden' (sic) and *Large Campeden* 1291. ME adjs. **brāde**, **brode**, **large** and Latin **parva** + p.n. Campden as in Chipping CAMPDEN SP 1539. Gl i.238.

Chipping CAMPDEN Glos SP 1539. 'Campden with the market'. *Chepyngcampeden(e)* 1287–1446, *Chipping Campden* 1689. ME **cheping** + p.n. *Campedene -a* 1086–1492, 'valley of the enclosures', OE **camp**, genitive pl. **campa**, + **denu**. The sense of *camp* in this name is unknown but some connection with Latin *campus* seems possible, perhaps a reference to cultivation encroachment on the former *campus*. The situation is a well marked stream valley. 'Chipping' for distinction from Broad CAMPDEN SP 1637. A market here was granted to Hugh de Gondeville c.1180 and confirmed in 1247. The *Campsæte* of *on Campsætena gemære* 'to the boundary of the *Campsæte*' [1005]12th S 911 are 'the inhabitants of Campden'. Gl i.237, Signposts 78.

Castle CAMPS Cambs TL 6343. *Castlecamps* 1384. ME **castel** referring to the castle built by Aubrey de Vere, earl of Oxford, shortly after 1086 + OE **camp** used as a p.n. probably referring to a Roman site at Horseheath. Earlier *Magna Ca(u)mpes* 'great Camps' 1236–1314, and *Caumpes Comitis* 'the earl's Camps' 1218. 'Castle' for distinction from Shudy CAMPS TL 6244. Ca 102, Signposts 76.

Shudy CAMPS Cambs TL 6244. Originally Little or Wood Camps, *Parva Campes* 1316, *Woode Campes* 1570. ModE **little**, Latin **parva**, ModE **wood** + p.n. Camps as in Castle CAMPS TL 6343. The element Shudy appears as *Sude-* 1218, 1236, *S(c)hude- S(c)hode-* 1256–85, *S(c)hudi-* 1286–1345, *Shudy-* 1331, possibly representing OE **scydd** 'a shed'. Sometimes it appears as *Suth(e)-* 'south' in the 13th cent. Ca 103, Signposts 76.

CAMPSALL SYorks SE 5413. Partly uncertain. *Cansale* 1086, 1196, 1207, *Camshale* 1142×86, *Camsala -(e)* c.1165–1535, *K-Cames(h)al(e)* 1227–1335, *Camps(h)ale* c.1165–1428, *Campsall* from 1402. The generic is OE **halh** 'a nook of land, land in a river bend', the specific uncertain, possibly pers.n. OE *Cam* or *Cammi* or a topographical term derived from PrW ***camm** 'crooked'; Campsall is situated at a sharp bend of a stream. YW ii.45, DEPN, L 108, 111.

CAMPSEY ASH Suff TM 3255. *Campesche in Ashe* 1286, *Campese et Ahs* 1346, *Campsy Ash* 1674. A combination of two p.ns., *Campsea, Campeseia* 1086, *Campes(s)e* 1185–1524, *Campeseye* 1254, *Campseye* [1433]16th, 'the island of the open land or of the camp', OE **camp**, genitive sing. **campes**, + **ēġ**, + p.n. *Esce* 1086, *Eysse* 1249, *Assh* 1308, *Ays(s)h(e)* 1336–1568, *Ashe* 1610, 'the ash-tree', OE **æsc**. The element *camp* in this name may be related to Roman settlement. DEPN, Baron, Signposts 76–7.

CAMPTON Beds TL 1238. 'Settlement on the river *Camel* or *Camelar*'. *Chambeltone* 1086, *Camelton* c.1150, *Kamerton* 1152×8, *Campton* 1526. Lost r.n. + **tūn**. The stream on which Campton stands has no name to-day. The r.n. suggested as the first element may be compared with Cam Brook, *Cameler -ar* [961]12th S 694, on which stands CAMERTON Avon ST 6857. Bd 167.

CAMULODUNUM Essex TM 0025. 'Fortress of Camulos', the Roman name for Colchester. *CA, CAM, CAMV, CAMVL, CAMVLODVNO* coins of Cunobelinus c.5 BC×AD 40, *(a) Camaloduno* [1st AD]11th Pliny NH ii.187, *Colonia Camulodunum* [1st AD]11th Tacitus *Annals*, *Camuloduno* [4th]13th Peutinger, *Καμουδόλανον (Camudolanum)* [c.150]c.1200 Ptol MS U (with variants *Καμρυδόλανον (Camrudolanum)*, *Καμβρυδόλανον (Cambrudolanum)*, *Καμουδόλαν (Camudolan)*, *Καμουλόδουνον (Camulodunum)* [43]15th Dio Cassius, *CAMULODUNI -O, CAMALODUNI -O* inscriptions, *Manulodulo Colonia* (sic for *Chanu-*) Rav [8th]13th. Celtic war-god name *Camulos* + **dunum**. The divine name Camulos occurs in numerous continental pers.ns. such as *Camulogenus, Andecamulos* in Gaul and in other p.ns. such as *Camulodunum*, the Roman fort at Slack, Yorks SE 0817 transferred from Almondbury hill fort SE 1514 or Old Lindley Moor SE 0918, *Camuloessa*, an unidentified fort in southern Scotland. The cult of this war god was probably brought to Britain where it became widely extended by the Belgae. Camulodunum was also simply known as *Colonia* 'the colony' referring to the legal staus of the city. Against Ess 368–9 and DEPN s.n. Colchester RBrit 312–3 argues that the specific of the modern city name is in fact *Colonia* rather than the r.n. Colne on the model of Lincoln < *Lindum Colonia*, Köln (Cologne) in Germany < *Colonia Agrippina* and the many names in which *chester* was added to a surviving first part of the RBrit name, e.g. Manchester, Rochester, Winchester etc. It is unlikely that this difference of view can be resolved. Berger 153, Ess 367.

River CAN Essex TL 6807. No early forms. *Can* is given as the name of the Wid in 1622, 1720. It is a back-formation from Great CANFIELD TL 5917 which is actually on the Roding. See also WRITTLE TL 6606. Ess 5.

CANADA Hants SU 2918. A modern name for a new intake of marginal land in a remote part of the parish of West Wellow projecting into the New Forest. Ha 48.

CANAL FOOT Cumbr SD 3177. *Canal Foot* 1864 OS. ModE **canal** + **foot**. The foot of the canal built by the engineer John Rennie in 1793–6 to connect Ulverston to the sea. Pevsner 1969.252.

CANDLESBY Lincs TF 4567. 'Calnoth's farm or village'. *Calnodesbi* 1086, *Kanleby* 1196, *Candlouebi, Kandelesbi* 1202. OE pers.n. **Cal(u)nōth*, genitive sing. **Cal(u)nōthes*, + **bȳ**. The same pers.n. occurs in the Lincs hundred name Candleshoe, *Calnodeshou* 1086, c.1115. DEPN, SSNEM 40.

Brown CANDOVER Hants SU 6041 → PRESTON CANDOVER SU 6041.

Chilton CANDOVER Hants SU 5940 → PRESTON CANDOVER SU 6041.

CANE END Oxon SU 6779. 'The canons' end'. *Canonende, Cannone ende* 1534, *Cane end* 1551×2. ME, OFr **canoun** + **ende**. The place belonged to the canons of Notley abbey, Bucks. Contrasts with Kidmore End SU 6979. O 76.

CANEWDON Essex TQ 8994. Possibly 'hill of the Caningas, the people called after Cana'. *Carendunā* 1086, *Cagnenda* (for *-euda*) 1087, *Canuedon'* 1181, *Canu(u)don* 1194, 1494, *C- Kanewedun -don* 1224 etc., *Caneudon* 1303, 1558, *Canydon* 1498, *Can(n)yngdon* 1502, 1539. OE folk-n. **Caningas* < pers.n. *Cana* + **ingas**, genitive pl. **Caninga*, + **dūn**. Although the evidence is lacking in this name the development *-inga-* > *-ega-* > *-ewe-* is posited on the model of p.ns. DANBURY TL 7805, DENGIE TL 9801, East HANNINGFIELD TL 7601, HONINGTON Suff TL 9174, MANUDEN TL 4926, MONEWDEN Suff TM 2358 and Coney WESTON Suff TL 9758. Ess 179 gives prs ['kænədən, 'kænjudən] and [kænəndən], Zachrisson 9, Jnl 2.43.

Great CANFIELD Essex TL 5917. *magna Caneueld* 1285, *Magna Kanewell* 1254, *Much Canwell* 1663. ModE **great**, Latin **magna**, ME **much**, + p.n. *Chene-Canedfeldam, Canefelda* 1086, *C- Kan(e)feld(a) -e* 12th–1347, *C(h)an(e)uel(l), Kan(n)evell* 1100–1235, 'Cana's open country', OE pers.n. *Cana* + **feld**. 'Great' for distinction from Little Canfield TL 5821, *Parva Kanewell* 1254. Also known as *Childe(s)* and *Childene Canefeld* 1285–1347, OE **ċild**, genitive sing. **ċildes**, ME genitive pl. **childene**. Ess 472.

CANFORD CLIFFS Dorset SZ 0689. P.n. Canford + ModE **cliff(s)**. Canford, *Cheneford* 1086, *Kene-* 1181, 1196, *K- Caneford* 1195–1440, *Canford(e)* from 1307, is 'Cana's ford', OE pers.n. *Cana* + **ford**. The reference is to a crossing of the r. Stour at SZ 0398. Do ii.2, 39.

CANFORD HEATH Dorset SZ 0295. *Canford Heath* 1811. Earlier simply *the hethe* 1542, *the Lawndes and Heath of Canford* 1558×1603, and also *the greate waste of Canford* 1612 and *launds of Caneford* 1275. P.n. Canford as in CANFORD CLIFFS SZ 0689 + ModE **heath, laund**. Do ii.4, 6.

CANN Dorset ST 8721. '(The) cup'. *Canna* [1100×35]15th, *Canne* 1202–1564, *Can* 1575–1655, *Cann* 1744. OE **canne** 'a can, a cup' referring to the deep valley in which the village lies. Do iii.90.

CANN COMMON Dorset ST 8820. *Cann Common* 1811. P.n. CANN ST 8721 + ModE **common**. Do iii.95.

All CANNINGS Wilts SU 1660. 'The old C'. *Aldekanning'* 1205, *Alle Kanynges* 1258, *Alcanninges* 1316. OE **eald**, definite form **ealda**, + p.n. *Caninge* 1086, *Caninges* [1091]13th, 1194–1249, *Caningas* 1148, 1175, 'the Caningas, the people called after Cana', OE folk-n. **Caningas* < pers.n. *Cana* + **ingas**. The earliest reference to the *Caningas* is in ASC(E) under year 1010 where the progress of a marauding Danish host is traced from Northampton to Wessex where they came to Canning Marsh, *wið Caningan mærsces* (*Canegan* ASC(C,D)). The territory of the *Caningas* must have been extensive; they gave their name to Cannings Hundred, *hund' de caniga, Canege hundred* 1086 Exon. 'Old' for distinction from Bishops CANNINGS SU 1660. Wlt 249, ING 47.

Bishops CANNINGS Wilts SU 0364. *Canyng Episcopi* 1294, *Bisshopescanyngges* 1314. This was an estate of the bishop of Salisbury already in 1086 when it is distinguished from All CANNINGS SU 1660 as *Caninghā* 'the Canning estate' sc. of the Bishop of Salisbury. Wlt 249, ING 47.

CANNINGTON Somer ST 2539. 'Settlement, estate by the Quantock Hills'. *Cantuctune* 873×88 S 1507, *Cande- Cantetone* 1086, *Cantoctunæ* (Latin genitive case) 1086 Exon, *Cantinton* 1187, *Caninton* 1178, *Camyngton* [1295] Buck, *Cannington* 1610. P.n. Quantock as in QUANTOCK HILLS ST 1537 + OE **tūn**. DEPN.

CANNOCK Staffs SJ 9810. 'Hill'. *Chenet* 1086, *Cancia, Chenot, Chnoc, C(h)not, Knot* 12th cent., *K- Can(n)oc -ok(e)* c.1135×40–1428, *le Cank(e)* 15th cent., *Cannock* from 1493. OE **cnocc** modified in Anglo-Norman pronunciation to *canoc* just as *Cnut* became *Canute*. Earlier derivations from Brit ***cunāco** supposedly meaning 'a hill' (more correctly PrW ***cönyg** of unknown meaning) depended on the form *Canuc*

[956]12th S 608, but the identification of this place is uncertain. The reference is to Shoal Hill 650ft. NW of the town centre. St i.56, TC 66, SelPap.46, Jnl 16.15.

CANNOCK CHASE Staffs SK 0016. *Le Cank chase* 1322×58, *Cannock Chase* 1834 OS. P.n. CANNOCK SJ 9810 + ME **chace** 'a hunting ground'. Duignan 28.

CANNOCK WOOD Staffs SK 0412. *Canckwood* 1623, *Cannock Wood* 1666. P.n. CANNOCK SJ 9810 + ModE **wood**. St i.58.

CANNON STREET STATION GLond TQ 3280. P.n. Cannon Street + ModE **station** built in 1863–6. Cannon Street, *Cannon Street* from 1667, was earlier *Candelwrichstrete* c.1180, 'candle-makers street', OE **candelwyrhta** + **strǣt**. GL 35–6, Encyc 122.

CANONSTOWN Corn SW 5335. *Canons Town* 1839. A 19th cent. village named after John Rogers (1778–1856), canon of Exeter Cathedral. PNCo 64, Room 1983.19.

CANT BECK Lancs SD 6374. Uncertain. *Kant* 1202. British ***canto**- of obscure origin and meaning, cf. OCo ***cant**. Probably to be compared with the Gaulish r.n. **Cantia* which lies behind Cance, Ardèche, France, a tributary of the Rhône, Chanza on the Spanish-Portuguese border, a tributary of the Guardiana, and the Kinz of Kinzweiler near Aachen, Germany, a tributary of the Ruhr, possibly meaning 'river at the edge, corner stream'. La 169, RN 69, PNE i.80, CPNE 37 and references there cited.

CANTERBURY Kent TR 1557. 'The *burh* or town of the people of Kent'. *Cant wara burg* 9th ASC(A) under year 754, [c.890]c.1000 OEBede, *Cont wara burg* 9th ASC(A) under year 851, [c.890]c.1000 OEBede, *(to) Cantuare beri* 9th ASC(A) under year 870, *(into) Cantware byri* c.1050 ASC(D) under year 1023, *Cantware -a burh* c.1120 ASC(E) under years 851, 1011, *Cantwara byrig* c.1120 ASC(E) (7×), *(to) Cantwareberig* 1070 ASC(A), *Cantwarbyrig* c.1120 ASC(E) (7×), *-beri* 12th ASC(E) under year 1140, *Canterburie* 1086, *Caunterbury* 14th Chaucer. OE *Cantware* 'the people of Kent', genitive pl. *Cantwara*, + **byriġ**, dative sing. of **burh**. Originally a descriptive term, this form subsequently replaced the RBrit name of the city, *Durovernum Cantiacorum* '*Durovernum* of the *Cantiaci*', *Δαρόυερνον (Darvernum)* [150]13th Ptol, *Duroruerno, Duraruen(n)o* [4th]8th AI, *Duro Averno Cantiacorum* [c.700]13th Rav, Brit ***duro**- + ***u̯erno**- 'alder fort, walled town by the alder-swamp' a name which survived in early English official usage as *(ciuitas) Dorobernie -uernie -uernis* [7th–8th]13th Charters IV passim, *civitas Doruuernis, Doruuernensis* [c.731]8th BHE and *Dorwitceaster* c.1122 ASC(E) under year 604. The medieval Latin form *civitas Cantuaria* 1086 etc. is the source of the abbreviation *Cantuar* used by the Archbishops of Canterbury. DEPN, RBrit 353, TC 66, Jnl 8.30.

CANTLEY Norf TG 3803. 'Canta's wood or clearing'. *Cantelai* 1086, *Cantelea* 1196, *Kantele* 1212. OE pers.n. **Canta* + **lēah**. DEPN.

CANTLEY SYorks SE 6202. 'Canta's wood or clearing'. *Cathalai* 1086, *Canteleia -lie -lay -lai(a) -ley(e)* 1086–1544, *Cantley* from 1441. OE pers.n. **Canta* + **lēah**. YW i.39.

CANTLOP Shrops SJ 5205. Probably 'sing wolf'. *Cantelop* 1086–1694 with variants *-hop(e) -ope*, *Cantlopp* 1579 etc., *Cantlop* from 1613. No satisfactory English etymology has been proposed, although the generic might be OE **hop** in the sense 'enclosure in waste'. Probably this is an instance of the common French place-name type C(h)anteloup 'sing wolf' composed of the imperative of the verb *c(h)anter* 'to sing' and *loup* 'a wolf' and used of places where wolves could be heard howling. Sa i.69, Dauzat-Rostaing 142 s.n. Canteleu.

CANTSFIELD Lancs SD 6273. 'Open land of the river Cant'. *Cantesfelt* 1086, *C- Kancefeld* 1208–43, *C- Kaunsfeld* 1327–76, *Cauntefeld* 1341. R.n. Cant as in CANT BECK SD 6374, genitive sing. OE *Cantes*, + **feld**. La 183 gives pr [kansfi:ld], L 243, Jnl 17.102.

CANVEY ISLAND Essex TQ 7783. *Canwaie Iles* 1586. P.n. Canvey + ModE **isle**. Canvey, *Caneveye* 1254, 1343, *Kaneveye -weye* 1259, 1264, is obscure. Comparison is made with CANEWDON TQ 8994 for which the sense 'hill of the Caningas' is posited, OE **Caninga-dūn*. Possibly, therefore, the 'island of the Caningas, the people called after Cana', OE folk-n. **Caningas* < pers.n. *Cana* + **ingas**, genitive pl. **Caninga*, + **ēġ**. *Caninga* is thought to have developed *Caninga* > *Caniga, Canega* > ME *Canewe* with subsequent *w/v* confusion. Ess 148.

CANWELL HALL Staffs SK 1400. *Canwell Hall* 1834 OS. P.n. *Canewelle* 12th, *-well* 1209×35, either 'Cana's spring', OE pers.n. *Cana* + **wella**, or 'the spring with a cup', OE **canne** + **wella**, + ModE **hall**. The reference is to an aluminous spring called St Modwen's Well famed in the Middle Ages for cures. DEPN, Duignan 32, PNE i.80, L 32.

CANWICK Lincs SK 9869. 'Cana's dairy-farm'. *Canuic, Caneuuic* 1086, *Canewic -wyke(e) -wik(e)* 1146–1539, *Canwyk(e)* 1303–1560. OE pers.n. *Cana* + **wīċ**. Li i.209.

CANWORTHY WATER Corn Sx 2291. *Kenworthy Water* 1748. P.n. *Canworthy* + ModE **water** referring to the River Ottery by which the hamlet is situated. Canworthy, *Carneworthy* 1327, 1366 Gover n.d., *Carnary* 1613 Gover n.d., *Kennery (bridg)* 1699, is 'farm called or at *Carn*', unknown p.n. *Carn*, Co **carn** 'rock-pile, tor', + OE **worthiġ** with usual Co reduction to *-ery*. PNCo 64.

CAPE CORNWALL Corn SW 3431. *Cape Cornwall* 1589 Gover n.d., *Cap Cornwall* 1699. ModE **cape** 'a headland' (ME *cap*) + county name CORNWALL. Its Cornish name is unexplained *Kulgyth* [1580]18th, *The Kilguth(e)* c.1605, 1610 Map. PNCo 64.

CAPEL Surrey TQ 1440. 'The chapel'. *Capelle* 1235, *la Capele* 1356, 1403, *Chapell* 1531. Capel was a chapelry of Dorking. ONFr **capel**. Sr 264 gives pr [keipəl].

CAPEL-LE-FERNE Kent TR 2538. Short for *Capell' in le Ferne* 1535, 'chapel in the fern-covered land', ME, ONFr **capel** + Fr **en** + Fr definite article **le** + OK ***ferne** 'area growing with ferns', a Frenchification of earlier English *Capel ate Verne* 1377. Also known as *(villa) scē Marie in the Verne* 1369, *seynte Marie in the fferne* 1377, 'St Mary's vill in the Ferne' from the dedication of the church. PNK 441–2.

CAPEL ST MARY Suff TM 0938. 'St Mary's chapel'. P.n. *Capeles* 1254, 1291, *Capel(l)e* 1275–1336, *Capell* 1610, 'the chapel', ME, ONFr **capel** + saint's name *Mary* for distinction from Capel St Andrew TM 3748, *Capeles* 1086, *Capel(l)e* 1316–44, + saint's name *Andrew*. DEPN, Baron gives pr [keipəl], DEPN.

CAPENHURST Ches SJ 3673. Possibly 'Capa's wooded hill'. *Capeles* 1086, *Capenhurst* from 13th with variants *-in- -yn-* and *-hyrst -hirst*. OE pers.n. **Capa*, genitive sing. **Capan*, + **hyrst**. The alternative suggestion, OE **cape* 'a look-out place' is unlikely on topographical grounds, but **capa* 'a look-out', which would be the source of the suggested pers.n., is also possible. Che iv.200 gives pr [keipənərst].

CAPERNWRAY Lancs SD 5371. 'Merchant's corner'. *Coupmanwra* c.1200–1227×36, *-mannes-* 1229, *Koupemoneswra* 1212, *Copenwra* 1235, *Caponwra, Capernwray* 14th. ON **kaupmathr** 'merchant' possibly used here as a pers.n., + **vrá**. La 187 gives pr [ke:pnre:], Jnl 17.103.

CAPHEATON Northum NZ 0380. 'Great or chief Heaton'. *Heton' Magnam, Magnam Heton'* 1242 BF 1114, 1115, *Magna Heaton* 1276, *Manga Hetton* 1296 (sic) SR 22, *Cap(p)itheton* 1454, 1465, *Capheton* 1536. Latin **magna** 'great', later **caput** 'head' + p.n. HEATON, OE **hēa** + **tūn** 'the high farm, village or estate'. 'Chief' or 'Great' to distinguish the Lord's settlement here from the church settlement at KIRKHEATON, a separate place in the same estate or parish. NbDu 38.

How CAPLE H&W SO 6130. 'Hugh's Caple estate'. *Huwe Caples* 1316, 1328, *Hugh Caple* 1334, *Howcaple* 1535. ME pers.n. *Hugh* probably from Hugh de Caple fl. c.1115×25 + p.n. *Capel* 1086, *Caple(s)* 1166–1291 of uncertain meaning, possibly OE ***capel**, a *-ul/-ol* derivative of ***cape** 'a look out place' referring to CAPLER CAMP, the Iron Age hill-fort at SO 5932. Caple was the name of a large area in late Anglo-Saxon times. Cf. also King's CAPLE SO 5629. He 53, Pevsner 1963.128, Thomas 1976.145, Nomina 10.65.

King's CAPLE H&W SO 5629. 'The king's Caple estate'. *Chingestaples* (sic for *-caples*) 1155, *Kyngescaple* c.1250. lOE **cing**, genitive sing. **cinges**, + p.n. *Cape* 1086, *Caple* 1285, 1325, OE ***cape** or ***capel** 'a look out place' as in How CAPLE SO 6130. The estate was once held by King Edward the Confessor. He 53, Nomina 10.65.

CAPLER CAMP H&W SO 5933. P.n. Capler as in *Capler Wood, Capler Farm* 1831 OS + ModE **camp** referring to the Iron Age hill-fort here. Capler, *Capullore* 1298, *Capellar* 1732 is 'Caple ridge', p.n. Caple as in How CAPLE SO 6030 and King's CAPLE SO 5628 + OE **ofer** or **ōra**. An alternative name of the fort is preserved in the f.ns. Upper and Lower Walboro, *Wobury* 1732, *Woldbury* n.d. with OE **burh**, which may be the same as the nearby farm name Oldbury SO 6332, *Oldbury* 1831 OS. He 214, Thomas 1976.145.

CAPLESTONE FELL Northum NY 5988. *Caplestone Fell* 1868 OS. Possibly ModE **capel, caple** 'a composite stone of quartz, schorl and hornblende' + **stone**, + **fell**.

CAPSTONE Kent TQ 7665. 'Cybbel's estate'. *Kebbeliston'* 1254, *Kebleston'* 1338, *Capson* 1819 OS. OE pers.n. **Cybbel*, Kentish dial. form **Cebbel*, genitive sing. **Cebbeles*, + **tūn**. PNK 129.

CAPTON Devon SX 8353. 'Capia's farm'. *Capieton* 1278, *Capinton* 1285, *Capyatone* 1330, *Capeton* 1351, 1475. Surname *Capia* as in Roger Capia 1330, Will Capia 1308, + ME *toun*. D 322.

CAR DYKE Cambs TL 4769. ModE dial. **carr** + **dike**. A broad deep artificial channel of Roman origin forming part of a canal system from Cambridge to Lincoln. Its old name is *Tillinge* [c.1160]1348, 1235, *Tyllinge* 1279, *Tillyng* 1430, *the Old Tillage or Twilade* 1863. OE ***tilling** 'the stretcher', referring to the length of the channel from Waterbeach to Benwick TL 3490, King's Delph TL 2595, and Peterborough. Ca 33.

CAR DYKE Lincs TF 1437. 'Karr or Kari's ditch'. *Caredic, Cardik* [1154×89]13th, *Karesdic* 1199×1215, *le Caresdik'* 1381, *Carr Dyke* 1771. ON pers.n. *Kárr* or *Kári*, genitive sing. *Kára*, ME *Kares*, + **dīċ**. The reference is to a Roman drainage ditch extending from Lincoln to the r. Nene. Perrott 42.

CARADON HILL Corn SX 2770. *Caradon Hill* c.1605. Named from Caradon SX 2971: *Carnetone* 1086, *Carnedune* [c.1160]15th, *Carnedon* 1234. This is 'hill at *Carn*'. OE **dūn** + p.n. *Carn*, representing Co **carn** 'tor, rock'. PNCo 64.

CARBIS BAY Corn SW 5238. *Carbis Bay* 1884. A 19th cent. village named from the bay. P.n. Carbis SW 5238 + ModE **bay**. Carbis is *Carbons* 1391, *Carbisse* 1584 Gover n.d., *Carbouse alias Carbease* 1617 Gover n.d., from Co ***car-bons** 'causeway' (< **car* 'a cart' + *pons* 'a bridge') if the 1391 form is correct and not to be read as *Carbous*; cf. the other names Carbis SX 0059, SX 0255 and Carbows SW 5533. The Cornish name of the bay is Barrepta (Cove), *Parrupter* c.1499, *Porthreptor* [1580]18th, *Porriper* 1621 G, Co **porth** 'cove, harbour' + unidentified element. PNCo 64.

CARBROOKE Norf TF 9402. 'Brook call Kere'. *Chere-Weskerebroc* 1086, *Kerebroc* 1195, 1202. R.n. *Kere* probably identical with River KEER Lancs SD 5372, *Kere* [1262]13th, Brit **kēro-* 'dusky, dark', + OE **brōc**. DEPN, RN 223.

CARBURTON Notts SK 6173. Partly uncertain. *Carbertone* 1086, *-ton* 1173–1332, *-bir- -byr-* 1330, 1335, *Carbarton -bur-* 1422×71. Apparently **car** + p.n. BARTON, OE **bere-tūn**, later confused with BURTON, OE **byrh-tūn**. *Car* might be OE **carr** 'a rock' but no such topographical feature exists here today. The settlement lies in a valley and ME *ker* 'marsh' would be possible if early *e* spellings were recorded. Nt 71.

CARCLEW Corn SW 7838. Apparently 'barrow of colour, coloured tumulus'. *Crucleu* 1314, *-lewe* 1388 Gover n.d., *Crukleu* 1327, *Crekelewe* 1540 Gover n.d. Co **cruc** 'barrow, hillock' + **lyw**. Alternatively the second element might be a pers.n. **Lew* or a stream-name identical with r. LEW Devon SS 5301. PNCo 65, CPNE 151.

CARCROFT SYorks SE 5410. 'Marsh enclosure'. *Kercroft -kroft* 12th–1366, *Kar- Carcroft(e)* 1365–1586. ME **ker** (ON *kjarr*) + OE ME **croft**. YW ii.32.

CARDEN Chesh SJ 4553 → CLUTTON SJ 4654.

CARDESTON Shrops SJ 3912. 'Card's estate'. *Cartistune* 1086, *Cardeston(')* from 1255×6, *Karston* 1535, *Cars(t)on* 1577–1750, *Caerdeston* 1800. OE pers.n. **Card*, genitive sing. **Cardes*, + **tūn**. The 1800 form shows association with W **caer** 'a fort'. Sa i.70 reports a modern pr *Carson*.

CARDINGTON Beds TL 0847. 'Cærda's farm,village estate'. *Chernetone* 1086, *Kerdinton* c.1190, *Kardinton* 1220. OE pers.n. *Cærda*, genitive sing. *Cærdan*, + **tūn**. Bd 88 gives pr [kæriŋtən].

CARDINGTON Shrops SO 5095. 'The estate called after Card(a)'. *Cardintune* 1086, *Cardinton'* 1170, *Kardinton(a)* 1185–1271×2, *Kardington'* 1271×2, *Cardyngton* 1397, 1535, *Cardington* from 1559×60. OE pers.n. **Card(a)* + **ing**[4], or genitive sing. **Cardan* + **tūn**. Sa i.71.

CARDINHAM Corn SW 1268. 'The fort *Dinan*'. *Cardinan lebiri* c.1180, *Cardinan* 1194, 1229, *Cardinam* 1251, *Cardinnum* 1643 G. Co ***ker** + *Dinan*. This is a pleonastic name since *Dinan* itself means 'fort', Co ***dynan**, a diminutive of ***dyn**. The reference is to Bury Castle SX 1369, an Iron Age hillfort 1 m to the NW. In the form of c.1180 *lebiri* is 'the fort', Fr definite article **le** + OE **byriġ**, dative sing. of **burh**. It may refer again to Bury Castle, or to Cardinham Castle SX 1268, a motte-and-bailey castle which was the Norman estate centre. The later form of the name in *-ham* has been influenced by the possession of the manor by the Dynham family in the late 13th cent., so that it looks as if it were 'fort of the Dynham family'. PNCo 65.

CARDURNOCK Cumbr NY 1758. 'Pebbly fort'. *Cardrunnock -ok(e)* 13th–1469, *Cardonock -dunnok* 14th cent., *Cardron(n)ok(e) -ock* 1485–1589, *Cardornoc(k) -durnock* 18th cent. PrW ***cair** + ***durnọg**. If PrW **durn* (W *dwrn*) 'a fist', from which **durnōg* is derived, can also mean 'fist-sized lump of stone' the reference may be to the cobble-stones of the structure of the Roman coastal fortlet here. Cu 123, Jnl.1.47, GPC 1106.

CAREBY Lincs TF 0216. 'Kari's farm or village'. *Careby -bi* from c.1226, *Karby -bi* 1202–56, *Carby* 1256–1626, *Cairby* 1506. ON pers.n. *Kári*, genitive sing. *Kára*, + **bȳ**. Perrott 162, SSNEM 81.

CAREY H&W SO 5631. 'Stony stream'. *Cari* 1162, *Kary* 1237, *Cary* 1272×1307, 1523. An Old-European r.n. identical with the rivers CAREY Devon SX 3687 and CARY Somer ST 4530. Old European **Karīsā* < **kar-* 'hard, stony' + suffix **-īs-*. He 32. Krahe 1964.59.

River CAREY Devon SX 3687. *Kari* 1238, *Kary* 13th. A Brit r.n. possibly meaning 'hard, stony stream'. D 3, RN 71.

CARGO Cumbr NY 3659. 'Crag hill'. *Cargaou* [c.1178]1348, *Kargho(we)* 1195, 1339, *Cargo(we)* 1303–1374. PrW ***carreg** perhaps taken as a p.n. *Carg* (cf. p.n. CARK SD 3676) + OE **hōh**, ON **haugr**. The reference is to a long hill rising above the marshes. Cu 94, GPC 431.

CARGREEN Corn SX 4362. 'Seal rock', literally 'rock of a seal'. *carrecron* late 11th S 951, *Kaer- Corgroyn -growne* 1478, *Cargrene* 1508 G, *Cargreen* 1610. Co **karrek** + **ruen** 'a seal'. The same name occurs again in the Cornish name of Black Rock SW 8331, *Caregroyne, 'n garrek ruen* c.1400. Later reinterpreted as if containing ***ker** 'a fort, a round'. PNCo 65, 67, CPNE 203.

CARHAM Northum NT 7938. '(The settlement) at the rocks'. *Carrum* [c.1040]12th, c.1107, *Karham* 1242 BF, 1255, 1296 SR, *Karram -um* 1251, 1252. OE **carr**, dative pl. **carrum**. NbDu 39.

CARHAMPTON Somer ST 0042. 'Settlement, estate called or at

Carrum'. *Carumtune* 873×88 S 1507, *Carintune* [904×25]lost ECW, *Care(n)tone* 1086, *Carhampton* 1610. P.n. *(æt) Carrum* 9th ASC(A) under years 833, 840, 'at the rocks', OE **carrum**, dative pl. of **carr**, + **tūn**. The village lies between two marked hills. DEPN.

CARHARRACK Corn SW 7341. Possibly 'fort of the high place'. *Carathek* 1408, *Carharthek* 1423, *Carharracke* 1590. Co ***ker** + *adj. **arthek** (<Co **arth* 'height' + adj. suffix *-ek*). PNCo 65.

CARINES Corn SW 7959. 'The part of *Crou* near the island'. *Crouwortheynys* 1348, *Crowarthenes* 1473, *Correynes* 1553 Gover n.d., *Carrenys* 1620 Gover n.d. P.n. *Crou* 1302, 1327, Co **krow** 'a hut', + **worth** 'at, against' + **enys**, 'island'. The reference is to its position near marshy ground. The addition is for distinction from the other parts of the original settlement, Colgrease, *Cowgres* 'middle Crou', Co **cres** 'middle', and Carevick SW 7958, *Crowarthevycke* 1567, Co **worth** + uncertain element. PNCo 65.

CARISBROOKE IoW SZ 4888. Partly uncertain. *Caresbroc* 1071–1352, *C- Karesbrok(e)* 1100–1583, *Carisbrooke* from 1327, *Cas(e)broke* 1393–1600. Possibly Brit r.n. *Cari(c)* as in Castle CARY Somer ST 6332 + OE **brōc**. But OE *carr* 'a rock' referring to the hill on which Carisbrooke Castle stands or an element **cear* 'hollow, gorge' as in G Karbach have also been proposed. Wt 93 gives prs ['keiz- 'kiəz- 'kæridʒ-] and ['kæzbruk], RN 71, LBO 200, Mills 1996.36.

CARK Cumbr SD 3676. 'The stone, the rock'. *Karke* 1491, 1587, *Carke* 1537. PrW ***carreg**. Cark is situated on the S slope of a rocky ridge. La 197, GPC 431.

CARLAND CROSS Corn SW 8453. *Carland Cross* 1972, a road junction named from the adjacent farm, *Cowland* 1813, *Carland* 1840, of uncertain origin. PNCo 65.

CARLBY Lincs TF 0414. 'The peasants' farm or village'. *Carlebi* 1086–1219, *-by* 1146–1602, *Carlby* from 1282. ON **karl**, genitive pl. **karla**, + **bȳ**. Perrott 401, SSNEM 40.

CARLECOTES SYorks SE 1703. 'The churls' cottages'. *Carlecotes* 13th–1614, *-coytes* 1379, 1647. ON **karl** probably replacing OE **ceorl**, genitive pl. **karla** + OE **cot**. YW i.339.

CARLETON 'Settlement of the free peasants'. ON ***karlatún** possibly replacing OE **ċeorlatūn**. Lucerna 144–60. See also CARLTON, CHARLTON, CHORLTON.

(1) ~ Cumbr NY 4252. *Karleton'* 1212, 13th cent., *Carleton* from 1231. Cu 148, SSNNW 186 no.5.

(2) ~ Lancs SD 3440. *Carlentun* 1086, *Carlton* [before 1190–1260]1268, *Karltun, Karl(e)ton(a)* [1190]1268–1256, *Carleton* from 1327. A dependent settlement of Earl Tostig's manor of Preston-in-Amounderness. La 157, Lucerna 151, Jnl 17.91, 21.34.

(3) ~ NYorks SD 9749. *Carlentone* 1086, *C- Karleton(e)* 1119×47–1441. YW vi.30, SSNY 114.

(4) ~ FOREHOE Norf TG 0905. 'C with the four hills'. *Karleton Fourhowe* 1268. Earlier simply *Carletuna* 1086. Forehoe is ME **four** + **howe** (ON *haugr*). DEPN.

(5) ~ RODE Norf TM 1192. 'C held by the Rode family'. *Carleton Rode* 1201. Earlier simply *Carletuna* 1086. The manor was held by Walter de Rede in 1302, by Robert de Rode in 1346. DEPN.

(6) East ~ Norf TG 1702. *Est Karleton* 1311. Earlier simply *Karltun* [1042×53]13th S 1535, *Carletuna* 1086. DEPN.

CARLINGCOTT Avon ST 6958. 'Cottage(s) called after Creodela'. *Credelincote* 1086, *Credlingcot* 1199, *Crudelincot* 1225, *Credlincote* [1239]14th *Mont*, *Carnicot* 1817 OS. OE pers.n. **Creodela* + **ing**[4] + **cot**, pl. **cotu**. DEPN.

CARLISLE Cumbr NY 4055. 'The Roman fort of Luguvalium'. There are three types:

I. *Luel* 9th–14th, *Cair Ligualid* [9th]10th, *Cairleil* [1129]12th, *Kaer-Leil* c.1200, *Car- Karleolum* 1131 etc., *Karlisle* 1318.

II. *Cardeol* [1092]1122, *Car- Karduil(l) -doyl -doil(le)* c.1125–1417.

III. *Luercestre* [c.1050]12th.

PrW ***cair** + p.n. LUGUVALIUM NY 4055. Cu 40 gives a pr [kɛrəl], O'Donnell 80, 81.

Port CARLISLE Cumbr NY 2462. *Port Carlisle* 1869 OS. A modern name for the harbour built by the earl of Lonsdale in 1819. The canal to Carlisle was built in 1823. ModE **port** + p.n. CARLISLE NY 3955. Pevsner 1967.74.

CARLOGGAS Corn SW 8765, 9554 > MUSBURY Devon SY 2794.

CARLTON 'Settlement of the free peasants'. ON **karl**, genitive pl. **karla** often replacing OE **ċeorl(en)a**, + **tūn**. The name type implies that the settlement was originally a constituent of a larger, often royal, estate. See CHARLTON. Lucerna 144–60.

(1) ~ Beds SP 9555. *Carlentone* 1086, *Carleton* 1198, *Carlton* 1240. Bd 30.

(2) ~ Cambs TL 6453. *Carletunes* (genitive case) [975×1016]11th S 1487, *-tun(e)* 1154×89, 1218, *Carletone* 1086–1551, *Carlentone* 1086, *-tona -tuna -tune* [1086×87]1417, [c.1098]15th 1221, 1121–60, *Carlton* from 1135×54. Ca 116, Lucerna 150.

(3) ~ Cleve NZ 3921. *Carlentune* 1109, 1154×66, *Carleton* [1183]14th, 1242–1818, *Carltun* 12th, *Carlton* from 1349. Nomina 12.34.

(4) ~ Leic SK 3904. *Karlintone* 1202, *Karleton'* 13th cent., *Carleton(e)* 1277–1576, *Carlton(')* from 1387. Lei 489.

(5) ~ NYorks SE 6186. *K- Carlton* early 13th, 1414, *Carletona* 1301. YN 72.

(6) ~ NYorks SE 6424. *Carleton(a) -tun* 1086–1641, *Karlinton'* 1169. YW iv.3.

(7) ~ NYorks SE 0684. *Carleton (in Coverdale)* 1086 etc. YN 253.

(8) ~ Notts SK 6141. *Carentune* 1086, *Karleton* 1182, 1276×88, ~ *juxta Notingham* 'C by Nottingham' 1294, *Carleton* 1197–1507. Nt 159, SSNEM 182.

(9) ~ SYorks SE 3609. *Carlentone* 1086, *Karleton(a) -tun* 1086–1684. YW i.276, SSNY 113.

(10) ~ Suff TM 3864. *Carletuna* 1086, *Carleton* 1254–1346, *Carlton* 1524. DEPN, Baron.

(11) ~ WYorks SE 3327. *Carlentone* 1086, *Carletone* 1243–1606. YW ii.137.

(12) ~ COLVILLE Suff TM 5190. 'C held by the Colville family'. *Carleton Colvile* 1346. P.n. *Karletun, Carletuna* 1086, *Carleton* 1229–1610, + manorial addition from Robert de Colevill 1230 whose family came from Colleville, Seine-Maritime, 'Kolli's vill', ON pers.n. *Kolli*, genitive sing. *Kolla*, + *ville* (Latin *villa*). DEPN, Baron.

(13) ~ CURLIEU Leic SP 6997. C held by the Curly family'. *Carleton(e) Curly* 1272–1457. Earlier simply *Cherletona* [c.1055]13th, *Cherlentonœ* c.1131 under year 1081, *Carlintone* 1086, *Carleton(e)* 1086–1576. William de Curly held land here in 1253. The place was a dependency of the royal manor of Great Bowden. Lei 211, Lucerna 151.

(14) ~ HUSTHWAITE NYorks SE 4976. 'C by HUSTHWAITE SE 5174'. *Carlton Husthwat* 1516. Earlier simply *Carletun, C- Karleton* 1086 etc. YN 190.

(15) ~ IN CLEVELAND NYorks NZ 5004. *Carletun* 1086. YN 176.

(16) ~ IN LINDRICK Notts SK 5884. *Carlton in Lindric* 1227. P.n. *Careltune, Car(l)etone* 1086, *Carletuna -e* c.1150, *Carle- Karleton* 1194–1572, + p.n. *(bosco, fossatum de) Lindric* c.1150–1227, *-ri(c)k* 1234–80, *Limbric* 1230, probably the 'limetree ridge', OE **lind** + ***ric**. Nt 72, 12, SSNEM 182.

(17) ~ -LE-MOORLAND Lincs SK 9057. 'C in the marsh'. *Carleton in Moreland* 1293–1565, ~ *in le Moreland(e)* 1546–7, *Carlton le Moorland* from 1734. P.n. *(æt) Carlatune* [1066×8]c.1200, *Carletune* 1086, *-ton* 1180–1607, + Fr definite article **le** for *en le* + ME **moreland**. The village is situated in uncultivated marshland of the r. Brant floodplain. Perrott 249, SSNEM 182.

(18) ~ MINIOTT NYorks SE 4081. 'C held by the Miniott fam-

ily'. *Carleton Mynyott* 1579. Earlier simply *Carletun -ton* 1086. The family of *Miniott* held land here in the 14th cent. YN 187.

(19) ~ MOOR NYorks SE 0384. P.n. CARLTON SE 0684 + ModE **moor**.

(20) ~ -ON-TRENT Notts SK 7963. P.n. *Carentune, Carlentun(e), Carletun(e) -ton(e)* 1086, *Karletun* c.1190, *-ton'* 1198 etc. with variants *Carle-* and *-ton*, + suffix *on Trent* 1297. Also known as *Nordkarleton'* 1242, *North-* 1245, *Northcarleton* 1289, and ~ *super Trentam* 1349 for distinction from Little or South CARLTON SK 7757. Nt 182, SSNEM 182.

(21) ~ SCROOP Lincs SK 9444. 'C held by the Scope family'. *Carleton Scrop(e)* 1434–1607, *Carlton Scro(u)p(e)* 16th cent., ~ Scroop from 1702. Earlier simply *Carletun(e)* 1086, *Carlentona* c.1115, *Carleton* 1242–1556. Land in the parish was held by the Scope family in the 14th cent. Perrott 364, SSNEM 182.

(22) East ~ Northants SP 8389. *(æt) Carlatune* [11th]13th, *Carlintone* 1086, *Carlenton* c.1115, *K- Carleton* c.1115–1349. 'East' is a modern addition for distinction from CARLTON CURLIEU Leic SP 6997. Nth 162, Lucerna 152.

(23) Little ~ Notts SK 7757. *Lytel Carleton* 1425. ME **litel** + p.n. *Karlet'* c.1180, *-ton* 1276, *Carleton* 1281×8. Also known as *Sutkarleton* 1278, *Suthcarleton* 1300 for distinction from CARLTON-ON-TRENT or North Carlton SK 7963. Nt 193.

(24) Great ~ Lincs TF 4185. *magna Carleton* [1154×89]1272×1307 *Maior* ~ 1294. ModE **great**, Latin **major** + p.n. *Carletune -a* c.1115. 'Great' for distinction from Little Carlton TF 4085, *parua carl'* [1183×4]1272×1307, *Parva Karletona* 1209–19, earlier simply *Carletone* 1086, and Castle Carlton TF 4083, *castre Karleton* 1253, *(de) castro de Carleton, Castel Carleton* 1266, referring to the site of the castle of Hugh Bardolf, Justiciar under Richard I. DEPN, SSNEM 182, Cameron 1998.

(25) North ~ Lincs SK 9477. *Nortcarletone* 1086. OE **north** + p.n. *Carletuna* c.1115. 'North' for distinction from South CARLTON SK 9576. SSNEM 182.

(26) South ~ Lincs SK 9576. ModE **south** + p.n. *Carlentone -tun, Carletvne* 1086, *Carletuna* c.1115. 'South' for distinction from North CARLTON SK 9477. SSNEM 182.

CARL WARK SYorks SK 2681. 'man's work'. *Carles Work* 1789. An ancient fort. Db 112.

CARN BREA VILLAGE Corn SW 6841. 'Village beside Carn Brea, the tor above Brea'. *Carnbree* 1348, *Kernbray* 1610. Co **carn** + p.n. Brea, Co. ***bre** 'hill'. The reference is to a granite pile of 740ft. at SW 6840. PNCo 66.

CARN TOWAN Corn SW 3626. 'Tor of the sand-dunes'. *Carn Towan* 1888 (of a rock), 1906 (of a farm). The hamlet is even more recent. Co **carn** + **towan**. PNCo 66.

CARNABY Humbs TA 1465. 'Keyrandi or Kærandi's farm or village'. *Cherendebi* 1086, *Kerendeby* 1155×7, *-an-* 1312, *Kern(t)tebi -by* [1154×89]c.1300, 1190×6–1312, *Kernetbi -by* 12th–1577, *Kerneby* 1285, *Car-* 1448, *Carnaby* from 1481. ON pers.n. **Keyrandi*, genitive sing. **Keyranda*, or **Kærandi*, genitive sing. **Kæranda*, + **bȳ**. Metathesis of *Kerend-* to *Kerned-* and subsequent confusion with OE names in *-nōth* and ME names in *-et, -ot* such as *Bennet, Bagot* etc. explain the development. YE 86, SSNY 23.

CARNE Corn SW 9138. 'Tor, pile of rocks'. *Kern* 1396 Gover n.d., 1513, *Carne* 1562. Co **carn**. The reference is probably to the barrow called Carne Beacon SW 912386 where the 8th cent. king Gerent of Cornwall was believed to lie buried in a boat, although *carn* is normally used of natural features rather than man-made ones. PNCo 66.

CARNFORTH Lancs SD 4970. 'Cranes' ford'. *Chreneforde* 1086, *Corneford* 1212, c.1388, *Carneford(e)* 1212–1383, *Kerneford* 1246–1332 etc. OE **cran**, **corn**, genitive pl. **corna**, + **ford**. La 187, L 71, Jnl 17.103.

CARNHELL GREEN Corn SW 6137. *Carnhell Green* 1813. An 18th or 19th cent. hamlet. P.n. Carnhell SW 6137 + ModE **green**. Carnhell, *Karnhell* 1249, *Carnhel* 1302, 1351 Gover n.d., appears to be either 'tor near a hall', Co **carn** + **hel**, or, if the *h* is not original, possibly 'rocky place', Co **carn** + adjective suffix ***-el**. PNCo 66.

CARNON DOWNS Corn SW 7940. (Moorland of) *Carnon* 1569, *Carnon Downes* 1683, *Goon Carnon* 1782. P.n. Carnon + ModE **downs** alternating with Co **goon** 'downland, enclosed pasture'. Carnon, *Karnen* 1284, *Carnan* 1301, 1378 Gover n.d., *Carnon* 1485 Gover n.d., is probably 'little rock' (Co **carn** + **-ynn**) or else 'rocky place' (Co **carn** + ***-an**). PNCo 66.

CARNYORTH Corn SW 3733. 'Roebuck crag', literally 'crag of a roebuck'. *Carnyorgh* 1334 Gover n.d., *Carnzorgh* 1418, *Carnyorke alias Carnyorth* [1550]1572. Co **carn** + **yorch**. In the 1418 form *z* is a graphy for *ȝ* = [j]. PNCo 66, Gover n.d. 631.

CARPERBY NYorks SE 0089. Partly unexplained. *Chirprebi* 1086, *Kerperby(a)* 1137×46–1420, *C- Karperbi(a)* 1168, [1172×81]late 13th. The generic is ON **bȳ**. The first element *Kerper* does not represent OIr pers.n. *Cairpre* and has not been satisfactorily explained. YN 266, SSNY 23.

CARPLEY GREEN NYorks SD 9487. No early forms. Unexplained p.n. Carpley + ModE **green**.

CARR END Northum NU 2232. 'The end of the rock'. *Carr End* 1866 OS. N dial. **carr** (OE *carr*) used of rocks and reefs in coastal waters + ModE **end**. The *carr* at Seahouses helps to form the defences of the small harbour as do the Muckle Car and the Little Car at Craster.

CARR SHIELD Northum NY 8047. *Carr Shield* 1866 OS. Probably 'the rock shieling', N dial. **carr** (OE *carr*) + **shield** (ME *shele*) 'a hut, a shieling'.

CARR VALE Derby SK 4669. A modern name, possibly related to *the Carrclose* 1657, 'the marsh close', ModE **car** (ME *ker*, ON *kjarr*) + **close**. Db 216.

The CARRACKS Corn SW 4640. 'Rocks'. *Carrocks* 1732 Gover n.d., *Carracks* 1748. Co **karrek** with English plural. PNCo 66, Gover n.d. 667.

CARRICK ROADS Corn SW 8335. 'Shelter for ships at *Carrick*', a name given to the estuary of the Fal. *Caryk Rood* c.1540. P.n. Carrick (Co **karrek** 'rock' referring to Black Rock SW 8331) + ModE **road** 'a sheltered place of water for ships; a roadstead, an anchorage'. Described by Leland c.1540 as 'a sure herboro for the greatest shyppes that travayle the occean'. PNCo 66.

CARRINGTON GMan SJ 7492. 'Farm, estate called or at *Carring*, the rock place'. *Carrintona* [1154×89]17th, early 13th, *Carintun -ton(a)* 12th–1294, *Carrington(a)* from early 13th. OE p.n. **Carring* < **carr** + **ing**[4], + **tūn**. Che ii.17.

CARRINGTON Lincs TF 3155. A township founded in 1812 after the draining of Wildmore Fen and E and W Fens, taking its name from the principal land-owner, the banker Robert Smith, Lord Carrington. DEPN, Room 1983.19.

CARRINGTON MOSS GMan SJ 7491. *Carrington Moss* 1831. P.n. CARRINGTON SJ 7492 + ModE **moss** 'a bog'. An extensive bog on the S side of the Mersey. Che ii.18.

The CARRS NYorks SE 9779. The name of the marshes along the river Hertford between Yedingham and Flixton. ON **kjarr**, ME **ker** 'marsh'. Flixton Carr is *marisco de Flixton* 1162×75, Willerby Carr *marisco de Willardby* 1170×85, East Heslerton Carr *Hallow Carr* 1584. YE 117, 118, 121.

CARRVILLE Durham NZ 3043. *Carrville* is called by Fordyce (1857) a village of 'recent existence'. Mod dial. **car** (ON *kjarr*, ME *ker*) + **ville**, an element frequently used for new settlements in the 19th cent. Compare BOURNVILLE SP 0481, CHARTERVILLE SP 3110, COALVILLE SK 4213, EASTVILLE TF 4057 etc.

CARRYCOATS HALL Northum NY 9279. P.n. *Carricot* before 1245, *Carre Cottes* 1542, *Cary(e) Coats* 1663, 1769 Armstrong, 'the cottages beside Carry Burn', r.n. Carry + **cot**, + ModE **hall**. The origin of the stream name *Carry Burn* 1866 OS is unknown. Cf. however r.ns. CAREY, CARY and CARRISBROOKE. NbDu 40.

CARSHALTON GLond TQ 2764. 'Cress Aulton'. *Chershautone* 1199, *Kers- Cresaulton* 13th cent., *Carshaulton* 1279, *Carshalton* 1323. ME **kerse, cresse** 'water-cress' + p.n. *Æuueltone* *[727]13th S 1181, *Aultone* 1086, 'settlement by a spring', OE **ǣwiell-tūn**. ME *kerse* was added for distinction from other Aultons. Water-cress was produced here in the 13th cent., and a grant of property dated 1216×72 mentions a *kersenaria*, a 'watercress-bed'. Watercress beds still exist by the river Wandle which rises here and is the spring referred to. Sr 41, TC 202.†

CARSINGTON Derby SK 2553. 'Cress farm'. *Ghersintune* 1086, *Kercin- Kersin- Kersynton(e)* 1251–1469, *Kersyngton(e) -ing-* 1269–1516, *Carsyngton(e) -ing-* from 1416, *Kerstone* 1437, *Carston* 1577–1675. OE ***cærsen** + **tūn**. Db 355 gives pr [ka:sən].

CARSINGTON RESERVOIR Derby SK 2552. P.n. CARSINGTON SK 2553 + ModE **reservoir**.

CARSWELL MARSH Oxon SU 3299. *Carsewell Merssh* 1467, *Corswell Marsh* 1761. P.n. Carswell, *Chersvelle* 1086, *Karswelle* 1233×3, 'the spring or stream where watercress grows', OE **cærse** + **wella**, + ME **mersh**. Brk 386.

CARTER BAR Northum NT 6906. 'The cart-drivers' gate'. *Cartergate* 1842 BB 119, earlier *The Carter* 1695 Map. This is the boundary on the turn-pike road from England into Scotland opened in 1776. The earlier crossing was known as *the Red Sewer* [1375]15th Barbour's *Bruce*, *the Read Squire* 1695 Map, *the Reidswire* 1803, r.n. REDE NY 8298 + dial. **swire** (OE *swīra*) 'a neck of land, a col, a hollow on the top of a hill or ridge'. It was the site of a famous Border skirmish in 1575.

CARTER'S CLAY Hants SU 3024. No early forms. The surname *Carter* occurs 1840. Gover 1958.189.

CARTERTON Oxon SP 2807. A modern settlement named after its founder, William Carter, in 1901. Mills 68.

CARTERWAY HEADS Northum NZ 0451. 'The head of the cart-drivers' road'. *Carterway Head* 1863 OS. The place marks the summit of a long climb on the turn-pike road from Allensford NZ 5007 on the river Derwent to Corbridge.

CARTHEW 'Black fort'. Co ***ker** + **du** with mutation of *d-* to *th-*.

(1) ~ Corn SX 0055. *Cartheu* 1201 Gover n.d., *Carduf* 1327, *Carthu* 1367, 1431. The spelling of the 1327 form is remarkably archaic, cf. PrW *duβ*. PNCo 67, Gover n.d. 382.

(2) ~ Corn SW 6836. *Kaerdu* 1315, *Cardu* 1338, *Carthu* 1356. Gover n.d. 535.

(3) ~ Corn SW 9571. *Cartheu* 1201, *Kardu* 1327. Gover n.d. 339.

CARTHORPE NYorks SE 3083. 'Kari's outlying farm'. *Caretorp* 1086, *Karethorp* 1246–1322, *Carthorp* 1161×70 etc., *Carethropp* 1558. ON pers.n. *Kári*, genitive sing. *Kára*, + **thorp**. YN 226, SPN 161, SSNY 56.

CARTINGTON Northum NU 0304. 'Cretta's valley or hill'. *Cretenden* 1220 Cur, *Kertindun -don'* 1236, 1242 BF 598, 1120, *-den -ton, Cartindune* 1296 SR 53, 110, 115, *-yngdon* 1314, *-yngton* 1346. OE pers.n. *Cretta*, **Certa*, genitive sing. *Crettan*, + **denu** early replaced by **dūn**. Metathesis of *r* to give a form **Certa* is not unusual. NbDu 40, DEPN.

CARTMEL Cumbr SD 3878. 'Rough, rocky, sterile sand-bank'. *Ceartmel* 12th, *Cart-* 12th–1215 etc., *Car- Karmel* 1188, 1190, *Kartmel* 1206–70, *Kertmel(l)* 1157–1279 etc. ON ***kartr** + **melr**. *Ker-* forms are inverted spelling owing to the ME sound change *er* > *ar*. In DB 1086 this place is *Cherchebi* 'village with a church', ON **kirkju-bȳ**. La 195, SSNNW 114.

CARTMEL FELL Cumbr SD 4188. *Cartmel Fell* 1865 OS. P.n. CARTMEL SD 3878 + ModE **fell** (ON *fjall*).

CARTMEL SANDS Cumbr SD 3376. *Cartmell Sands* 1864 OS. P.n. CARTMEL SD 3878 + ModE **sands**.

CARTMEL WHARFE Cumbr SD 3668. *Cartmel Wharfe* 1852 OS. P.n. CARTMEL SD 3878 + dial. **wharf** 'an embankment, a sand-bank'. The reference is to a sand-bank in Morcambe Bay.

River CARY Somer ST 4530. 'Hard, stony stream'. *Kari* [725]14th S 251, *Cari* [729]14th S 253, [966]14th S 743, *Cary* from 1279. An Old European r.n. **Karīsa* on the root **kar-* 'hard, stone, stony' as also River CAREY Devon SX 3687 and the W r. Afon Ceri Dyfed SN 3247. Several ancient estates take their name from this river. RN 70, Krahe 1964 59.

(1) BABCARY Somer ST 5628. 'Babba's Cary'. *Babachan* (sic for *-chari*), *Babecari* 1086, *Babbe Cari*, ~ *Kari* [before 1196]14th, *Babbekari* 1200, *Babekary* 1212, *Bab(b)ecary* [14th]B. OE pers.n. *Babba* + r.n. CARY. 'Bab-' for distinction from the other estates named after the r. Cary, Castle CARY ST 6332, CARY Fitzpaine ST 5427 and Lytes CARY ST 5326. DEPN, DB Somerset 22.18, 45.1–2, RN 71.

(2) ~ FITZPAINE Somer ST 5427. Held by Margery Fitz Payn in 1243, earlier simply *Cari* 1086, and also known as *Stipelkari* 'steeple C' 1225 and *Lytilkary* 1439, *Little Care* 1610. In the 16th cent. the manor house was called *Phippens Cary*, 'Fitzpaine's C'. DEPN, VCH iii.99.

(3) Castle ~ Somer ST 6332. *Castellum de Cari* 1138, *Castelkary* 1237, *Castelcari, Caricastel* [1318×9]14th, *castle Caree* 1610. OFr, ME **castel** + p.n. *Caric, Cary* [c.680]13th, [672 for 682]13th ECW (S 237), *Cari* 1086, *C- Kari* [1152×8]14th, *C- Kary* [13th]B, transferred from the r.n. CARY.

(4) Lytes ~ Somer ST 5326. An estate held by William de or le Lyte in 1284×5. Earlier simply *Cvri* 1086, *Kari* 1284×5. Also known as Tucks or Tuckers Cary, *Towkerekary* 1255×6, from the surname *Tucker*, Little Cary, *Lytilkary alias Tokeryskary* 1439, and Cooks Cary from Thomas Cooke to whom the manor was sold in 1720 by Thomas Lyte IV. DEPN, VCH iii.100–2.

CASHMOOR Dorset ST 9713. Possibly 'pool where water-cress grows'. *Cashmore* 1774. Cf. *Cashmere Field* 1841. The evidence is late but this is probably ultimately from OE **cærse** + **mere**. Do iii.117.

CASSINGTON Oxon SP 4510. 'The homestead where cress grows' or 'called or at *Cærsing*, the place growing with cress'. *Cerse- C(h)ersitone* 1086, *Ch- Kersin(g)t(h)on(a) -ton(e) -tune -yn-* before 1123–1252, *K- Carsinton(e) -yn-* 1197–1320, *Cassington(')* from 1246×7. OE adj. ***cærsen** or p.n. **Cærsing* < **cærse** + **ing**², + **tūn**. O 252, PNE i.76.

CASSIOBURY PARK Herts TQ 0897. *Cashiobury Park* 1822 OS. P.n. Cassiobury + ModE **park**. Cassiobury, *Cayeshoobry* 1509, *Cassh- Cayshobury* 1545, is 'Cassio manor', p.n. Cassio + Mod dial. **bury**. Cassio, *(æt) Cægesho* [793]13th S 136, *Cagesho* [11th]14th, *C(h)aissou* 1086, *Cais- Cays- Kaiesho* 1154×89, *Key- Kayshou* 13th, is the 'hill-spur of or called *Cæg*, the key', OE **cǣġ** 'a key' used as a p.n. for a key-shaped hill, genitive sing. **cǣġes**, + **hōh**. This explanation is preferable to positing an unrecorded pers.n. **Cǣġ*. Cf. KEYSOE Beds TL 0762. Hrt 104 gives pr [kæsjou], L 168.

CASSWELL'S BRIDGE Lincs TF 1627. No early forms. A bridge over the South Forty Foot Drain. Surname *Casswell* as in Thomas Casswell 1777 + ModE **bridge**. Payling 1940.43.

CASTALLACK Corn SW 4525. Probably 'rocky' or 'fortified place'. *Castellack* 1284, *Castalak* 1356 Gover n.d. Co **castell** 'castle, village', also 'tor, rock' + adjective suffix **-ek**. PNCo 67, Gover n.d. 654.

CASTERTON Cumbr SD 6279. 'Settlement near the fortification'. *Castretun(e) -ton'* 1086–1447, *Castertun -ton'* from 1202, *Casterington* 1279. OE **cæster** + **tūn**, the N equivalent of CHESTERTON. The reference is unknown but Casterton lies beside a Roman road, Margary No. 7c. We i.27.

Great CASTERTON Leic TF 0009. Latin prefix **magna**, *Magna* 1218–1428, suffix **maior** 'greater', *Maior* 1254–1428, ME prefix **brigge**, *Bry- Brigg(e)* 1265–1553, *Bridge* 1610, 1642, + p.n. *Castreton(e)* 1086–1371, *Casterton(')* from 1202, the 'settlement

†Another charter form sometimes cited here, *(æt) Aweltune* [873×88]11th S 1507, belongs to Alton Hants or Alton Priors Wilts.

at or by a roman fort', OE **cæster** + **tūn**. There are extensive traces of a RBrit settlement here. 'Great' for distinction from Little CASTERTON TF 0109. 'Bridge' referring to a crossing of the river Gwash on the line of Ermine Street. R 130.

Little CASTERTON Leic TF 0109. Latin prefix **parva**, *Parva* 1262–1537, suffix **minor** 'lesser', *Minor* 1254–1428, ModE **little**, *Little* 1610, + p.n. Casterton as in Great CASTERTON TF 0009. R 134.

CASTLE AN DINAS Corn SW 9462. 'Fort at Dennis'. *castel an dynas* c.1504, *Castell Dennyse* c.1582, *Castle Andenas* 1610. Co **castell** + p.n. Dennis SW 9463, *Dynes* 1428, Co ***dynas** 'hill-fort', also *An dynas* with the definite article 'the hill-fort'. The reference is to the great, late Iron Age triple-ringed earthwork on Castle Downs, said to have been the site of the death of Cador, duke of Cornwall and husband of king Arthur's mother. PNCo 67, Gover n.d. 320, Thomas 1976.62.

CASTLE CARROCK Cumbr NY 5455. 'Castle called *Cairog*'. *Castelcairoc -kayrok -c -keyrok -c -cayrok -kairoc* c.1165–1530, *Castelcarroc* 1212, 1221, *-karrok'* 1255. PrW ***castell** + ***cairōg**, a derivative corresponding to W *caerog* 'fortified' from PrW **cair*. There is a fort half a mile E of the village. Cu 74, GPC 384.

CASTLE COMBE Wilts ST 8477 → Castle COMBE ST 8477.

CASTLE DITCHES Hants SU 1219. *Castle Ditches* 1811 OS. The reference is to the three lines of rampart and ditches of WHITSBURY Iron Age camp. Cf. *Castell Coppice* 1545. Wlt 224, Thomas 1976.143.

CASTLEFORD WYorks SE 4225. 'Ford by the Roman fort'. *æt Ceaster forda* [948]11th ASC(D), *Casterford -re-* c.1130–1241, *Castelford(e)* 12th–1528, *-forth* 1276–1506, *Castleford* from 1290. OE **ċeaster** later replaced by OFr **castel** + **ford**. At this place Ermine Street from Doncaster to the north, Margary no.28b, crosses the Aire. The fort in question is the Roman station *Lagentium*. YW ii.69, RBrit 383.

CASTLE HILL Suff TM 1547. ModE **castle** + **hill**.

CASTLE HILL WYorks SE 1514. *Castell hill* 1582. The name of the great Iron age hill-fort from which ALMONDBURY SE 1615 takes its name. The ancient earthworks were later temporarily re-used as a manorial administrative centre and it is probably to the mediaeval castle that the name refers; the castle is referred to as *castrum de Almondbury* 1399, and in the family name *del Castle* 1333, *del Castell* 1338. YW ii.258, WYAS 302, 735, 737, Thomas 1980.257.

CASTLE HOWARD NYorks SE 7170. An early 18th cent. mansion built by the Howard family, replacing, on a changed site, the earlier Hinderskelfe Castle, *Hildre- Ilderschelf* 1086, *Hi- Hylderskelf* 1207–1418, *Hi- Hynderskelfe* 1316, 1483, 'the shelf of land growing with elder', OE ***hylder**, ***hyldre** + **scelf** later replaced by ON **skjalf**. YN 40, SSNY 158.

CASTLEMORTON H&W SO 7937. 'Morton with the castle'. *Castel Morton* 1346, 1428. ME **castel** + p.n. *Mortun* 1235, the 'moor settlement or estate', OE **mōr** + **tūn**. 'Castle', referring to a castle erected here during the anarchy of king Stephen's reign, is for distinction from Birtsmorton SO 8035, *Brittes Morton* 1250, *Birch- Burchmorton* 16th and 17th cents., earlier *Morton le Bret* 1241, 'Morton held by the le Bret family' to whom it was granted in 1166: they came from Brittany. Cf. BIRTS STREET SO 7836. The two Mortons, presumably originally a single estate, lie on the edge of an extent of wet-land called Longdon Meadows which include *Wet Field*, *Broad Marsh*, *Pig Marsh*, *Langdon Marsh*, *Marsh End*, and *Pendock Marsh*, all 1831 OS, a marsh at one time of 10,000 acres extent, the last remains of the tidal estuary of the Severn above Gloucester. Wo 214, 213, 209.

CASTLE POINT Northum NU 1441. ModE **castle** + **point** 'a promontory'. The reference is to Lindisfarne Castle.

CASTLE RING Staffs SK 0412. *Castle-hill* 1686, 1834 OS, *Castle Ring* 1907 OS. ModE **castle** + **ring** referring to one of the most striking Iron Age hill-forts in England. St i.58, Thomas 1976.198, Horovitz.

CASTLESHAW MOOR GMan SD 9911. P.n. Castle Shaw SE 0009 + ModE **moor**. Castle Shaw, *Castylshaw* 1544, *Castleshaw(e)* 1581, 1657 is 'copse near the fort', ME **castel** + **shawe** (OE *sceaga*), referring to the Roman fort *Rigodunum* at SD 9909. YW ii.311, RBrit 448.

CASTLESIDE Durham NZ 0748. 'Castle hill-side'. *Castleside* 1864 OS. ModE **castle** + **side**. The reference is unknown.

CASTLETHORPE Bucks SP 7944. 'Outlying farm by the castle' sc. of Hanslope. *Castelthorpe* 1252–1447, *Castelthrope* 1486. Also simply *Throp* 1255, *Thrupp* 1616. ME **castel** + **thorp**. Bk 14.

CASTLETON Derby SK 1583. 'Castle farmstead'. *Castelton(e)* 1216×72–1625×49, *-il-* 1275–1577, *Castleton* from 1577. OE **castel** + **tūn**. The reference is to nearby PEVERIL CASTLE SK 1482. Db 55.

CASTLETON NYorks NZ 6807. 'Castle farm'. *Castleton* 1577. Named from Danby Castle, *castro de Daneby* 1242. YN 131.

CASTLETOWN T&W NZ 3558. A modern housing development named from Hylton Castle, a late 14th cent. castle now in ruins.

CASTON Norf TL 9597. 'Catt's estate'. *Ca(s)testuna* 1086, *Catestuna* 1121, *Cattestun -ton* 1191, 1194ff. OE pers.n. **Catt*, genitive sing. **Cattes*, + **tūn**. DEPN gives pr [ka:stə].

CASTOR Cambs TL 1298. 'The Roman town' sc. of *Durobrivae*. *Castre* 1086–1428, *Castor* from 1227 including [1189]1332. OE **ċeaster**, **cæster**. Earlier known as *(to) Kyneburga cæstre, (be) Cyneburge cæstre* [948]12th S 533, 'Cyneburg's fort', OE feminine pers.n. *Cyneburg*, genitive sing. *Cyneburge*, + **cæster**. The church is dedicated to St Kyneburga, daughter of King Penda of Mercia, who married Alcfrith, son of Oswiu of Northumberland, and subsequently became abbess of the convent here. She died c.680. Her relics, and those of her sister Cyneswith and another kinsman Tibba were translated to Peterborough and subsequently to Thorney. Her name is preserved in the local name Lady Conyburrow's Way. Nth 232 gives pr [ka:stə], Saints 119.

CAT AND FIDDLE Ches SK 0071. *Cat and Fiddle Inn* 1831. An important inn on the highest point on the Macclesfield-Buxton turnpike. Che i.173.

CATBRAIN Avon ST 5780. 'Cat-brain'. *Catbrain* 1830. ModE dial. **cat(s)brain** is a term for soil of 'rough clay mixed with stones'. Gl iii.133.

CATCLEUGH RESERVOIR Northum NT 7303. P.n. Catcleugh NT 7403, *Cattechlow* 1279, 'the ravine of the wild cats', OE **catt**, genitive pl. **catta**, + **clōh**, + ModE **reservoir**. NbDu 41.

CATCLIFFE SYorks SK 4288. 'Wild-cats' bank'. *Catteclíue* 13th, *-clif -clyf* 1255, 1316, 1390, *Catclif(f) -clyf(f)(e)* 1322–1593. OE **catt**, genitive pl. **catta**, + **clif**. The reference is to the steep bank of the r. Rother. YW i.179.

CATCOTT Somer ST 3939. 'Cada's cottage(s)'. *Caldecote* 1086, *Cadicot* 1225, *Katicote* 1243, *Caldecot* 1251. OE pers.n. *Cada* varying with adj. **cald** 'cold', + **cot**, pl. **cotu** (ME *cote*). DEPN.

CATER'S BEAM Devon SX 6369. *Cater's Beam* 1809. Surname *Cater* + **beam** 'a beam-engine, a pump used for mining purposes'. D 192.

CATERAN HILL Northum NU 1023. *Cateran Hill* 1868 OS.

CATERHAM Surrey TQ 3455. Partly uncertain. Perhaps 'Catta's enclosed land' or 'the enclosed land associated with the wild cat'. *Cathe- Catenham* 1179–1235, *Katrehamme* 1263, *Katerham -re-* 13th cent., *Catere-* 1279, *Caterham* from 1372, *Catriham* 1398, *Katri-* 1422×71, *Catter-* 1552–1710. Perhaps OE pers.n. *Catta* or *Catte* feminine or ***catte** 'a wild cat', genitive sing. *Cattan*, ***cattan**, + **hamm** 2 'a promontory of dry land' or 6 'a piece of valley bottom hemmed in by higher ground'. The earlier derivation from PrW **cadeir* 'a chair' is unlikely; the medial *r* is probably intrusive as in DUNSFOLD TQ 0036. Sr 311 gives former pr [kætərəm], ASE 2.22–3.

CATESBY Northants SP 5259. 'Katr or Kati's village or farm'. *Catesbi* 1086–1471 with variants *Kates-* and *-by*, *Cate- Kateby* 1246–1330. ON pers.n. *Kátr* or *Káti*, secondary genitive sing. *Kates*, + **bȳ**. Nth 16, SSNEM 41.

CATFIELD Norf TG 3821. 'Kati's open land'. *Catefelda* 1086, *-feld(e)* 1101–1432, *Cattfeld* 1197, *Catfeld(e)* 1269–1535. ON pers.n. *Káti*, genitive sing. *Káta*, + **feld**. Nf ii.86.

CATFORD GLond TQ 3872. 'Wild-cat ford'. *Catford* 1254, *Cateford* 1311. OE **catt**, genitive pl. **catta**, + **ford**. GL 36.

CATFORTH Lancs SD 4735. 'Cat ford'. *Catford* 1332, *Catforthe* 1514. OE **catt** + **ford**. The forms are too late to be sure whether the specific is OE singular *catt* or OE genitive pl. *catta*. La 162, L 71.

CATHERINGTON Hants SU 7014. 'Settlement of the Cateringas, the people called after Cator'. *Cateringatun* 1014 S 1503, *C- Kateringeton* 1176, 1199, *C- Katerin(g)ton -yn(g)-* 1218–1382, *Katherington* 1596. OE folk-n. **Cateringas* < pers.n. **Cator* + **ingas**, genitive pl. **Cateringa*, + **tūn**. OE pers.n. **Cat(t)or* is explained as a pet form of **Cador* < PrW **Cadur* < Brit **Catuu̯iros*. But this is by no means certain and the specific may be an unidentified p.n. *Cater* and so 'settlement of the dwellers at (the) *Cater*', perhaps a hill-name from PrW ***cadeir** < Latin *cathedra* 'a seat'. The modern form seems to have influenced dedication of the church to St Katherine. Ha 49, Gover 1958.56, LHEB 555, FT 800.

CATHERTON Shrops SO 6578. 'The settlement at *Cadeir*'. *Caderton'* 1291×2, *Carderton* 1316, *Cader(e)ton* 1439–c.1540, *Catherton* from 1619. PrW ***cadeir** probably used as a hill-n. + OE **tūn**. DEPN, Gelling.

CATLOW FELL NYorks SD 7160. P.n. *Catlowe in Bolland* 'C in Bowland' 1625×49, *Catlow* 1665, 'wild-cat hill', ME **cat** (OE *catt*) + **low** (OE *hlāw*), + ModE **fell**. Cf. Forest of BOWLAND SD 6455. YW vi.201.

CATLOWDY Cumbr NY 4576. 'Dirty stream'. *Kackledy* 1275, *Catlody* 1509×47, *Cat(h)lowthy*, *Cat(t)lowd(e)y* 17th, 18th. This seems to be 'cack-lady' used as a nickname of the small stream that rises here and transferred to the settlement. Mod dial. **cack** (OE *cac(c)* 'dung') + **lady**. Also known as *Lairdstown* 1697–1821. Cu 105.

CATMORE Berks SU 4580. 'Wildcat pond'. *(be eastan) catmere* [935×38]13th S 411, [1042]13th S 993, *(oþ) catmeres (gemære)* [951]16th S 558, *Catmere* 1086–1428 with variants *-mer(a)*, *Catimor'* 1196. OE **catt** + **mere**. Brk 496.

CATON Lancs SD 5364. Probably 'Kati's farm or village'. *Catun* 1086, *Caton* from 1185, *Catton* 1186–1273, *Katon* 13th, *Kaiton* 1664. ON pers.n. *Káti* + **tūn**. This seems to fit the forms better than the alternative, OE pers.n. *C(e)atta*. La 177 gives pr [ke:tn], Jnl 17.100.

CATOR COURT Devon SX 6877. P.n. Cator + ModE **court**. Identical with *H^r Cator* 1809 OS. Cator is *Cadatrea* 1167, *Cadetrewe* 1270, *-tru* 1412, 'Cada's tree', OE pers.n. *Cada* + **trēo**. D 526.

CATON MOOR Lancs SD 5763. P.n. CATON SD 5364 + ModE **moor**.

CATSFIELD ESusx TQ 7213. 'Catt's' or 'the cat's open land'. *Cedesfeld -felle* 1086, *Cat(t)esfeld(e)* 12th–15th cents. OE pers.n. *Catt* or **catt** 'a wildcat', genitive sing. *Cattes*, **cattes**, + **feld**. Sx 485, L 244.

CATSHILL H&W SO 9674. 'Cat's hill'. *Catteshulle* 1221–75, *Catshill* 1427. OE **catt** 'wild cat' or nick-name or surname *Catt*, genitive sing. *Cattes*, + **hyll**. Wo 339, L 171.

CATTAL NYorks SE 4454. Probably the 'nook of land haunted by wild-cats'. *Catale -ala* 1086, *Catal(l)* 1165, 1307, 1488, *K- Cahal(a) -hale* 1150×1200–1236, *C- Kathal(e)* 1175×1205–1328, *Cattall* 1421–1502. It has been argued that the forms with medial *-h-* are best explained as an AN reflex of medial [ð], but there are also instances of the loss of *t* before an element beginning with *h*, e.g. STRETHALL Essex TL 4839, *Strahal(e)* 1212–70, STRETHAM Cambs TL 5714, *Straham* 1170–1302. Probably, therefore, 'the nook of land haunted by wild-cats', OE **catt** + **halh**, rather than 'Katha's nook of land', ON pers.n. *Katha* + **halh**. The reference is to land in a bend of the river Nidd. YW v.17, Ess 534, C 238, SPN 159, XCIII, SSNY 253, L 107, 111.

CATTAWADE Suff TM 1033. 'The cat ford'. *Cattiwad* 1247, 1286, *-wade* 1256, *Catiwade* 1610. OE **catt** + **ġewæd**. DEPN, Baron.

CATTERALL Lancs SD 4942. Uncertain. *Catrehala* 1086, *C- Katerhale* 1212–1332, *Kateral(l)* [1200]1268–1346, *Caterall* 1346 etc. The generic either is or has been assimilated to OE **halh** here in the sense 'promontory of dry ground reaching into marsh'. This may be a reformation of ON **kattar-hali** 'cat's tail' referring to an elongated ridge along which the village stretches. La 162, L 107, 111, Jnl 17.95.

CATTERICK NYorks SE 2497 → CATTERICK BRIDGE SE 2299.

CATTERICK BRIDGE NYorks SE 2299. P.n. Catterick + ModE **bridge**. Catterick, *Κατουρ(ρ)ακτόνιον*, *Τα(κ)τουρακτόνιον* (*Catur(r)actonium, Ta(c)turactonium*) [c.150]13th Ptol, *Cataractoni -e* [c.300]8th AI, *Cactabactonion* [c.700]13th Rav, *Catraeth* c.600 Gododdin, *Cataracta, Cataractone* [c.731]8th BHE, *Cetreht(tun), Cetrihtun* [c.950]c.1000 OEBede, *Catrice* 1086, *Catrik* 1362, 1400, 1441, *C- Katerik* 1283–1390, is the 'place of the battle ramparts', Brit ***catu-** 'battle' + ***ratis**, ***racte** 'rampart, fortification'. This name subsequently underwent popular reinterpretation as if from Latin *cataracta* 'a waterfall, rapids' referring to the rapids in the Swale at Richmond. The place is the site of a Roman town. YN 242, LHEB 564, RBrit 302–4.

CATTERICK GARRISON NYorks SE 1897. P.n. Catterick as in CATTERICK BRIDGE SE 2299 + ModE **garrison**. A military camp established in 1923. Pevsner 1966 120.

CATTERLEN Cumbr NY 4833. Unexplained. *C- Kaderleng* [1158, 1165, 1190]c.1200, *Katir- C- Katerlen -lyn* 13th cent., *Catterlen -ley* 16th cent., *-land* 1722. Possibly a compound of PrW ***cadeir**, an element sometimes used of a hill or lofty place, with an unknown second element. The reference here would be to the hill at High Dyke NY 4732. Cu 182.

CATTERTON NYorks SE 5146. 'Settlement at or called *Cader*'. *Cadreton(e) -tune* 1086, 1230, *Cadartona -tuna* c.1140×8, 1156, *K- Cathertuna -ton* 1173–1355, *K- Caterton -ir-* 1211–1531. PrW p.n. *Cader* 'the chair, the hill' < Brit **cateira* + **tūn**. YW iv.236 gives prs ['katə-] and ['kaθətən].

CATTHORPE Leic SP 5578. 'Thorpe given by Isabel le Cat' sc. to Leicester Abbey. *Kattorpt* 12th, *Catthorp(e)* from 1218, *Torpkat, Thorp le Cat* 13th. ME surname *Cat* as in *Ysabelle le Cat, Ysabelle Chat*, the 12th cent. donor of a virgate of land to Leicester Abbey, + p.n. *Torp* 1086–1243, *Thorp(e)* 1243–1361, ON **thorp** 'an outlying farm'. Lei 435, SSNEM 119 no. 24.

CATTISTOCK Dorset SY 5999. 'Catt's outlying farm'. *Stoke, Cattesstok* *[843 for 934]17th S 391, *K- Cattestok* 1288, also simply *Stoche* 1086. OE pers.n. *Catt*, genitive sing. *Cattes*, + **stoc**. Do 50, Studies 1936.21.

CATTON Either 'Catta or Káti's farm or village' or 'farm or village frequented by wildcats'. Pers.n. OE *Catta*, ON *Káti*, or OE **catt** + **tūn**.

(1) ~ Norf TG 2312. *Cattuna, Catetuna* 1086, *Catton* 1212. DEPN.

(2) ~ NYorks SE 3778. *Catune* 1086, *C- Katton* from 1199. YN 183.

(3) ~ HALL Derby SK 2015. *Catton Hall* 1833. P.n. Catton + ModE **hall**. Catton, *Canton* (sic) [942]13th, *Chetun* 1086, *Catton(e)* from c.1141, is probably 'Catta's farm', OE pers.n. *Catta* + **tūn**, although **catta**, genitive pl. of **catt** 'a cat' is also possible. Db 627.

(4) High ~ Humbs SE 7153. *High-Catton* 1828. Earlier simply *Caton* 1086, *Cattuna -ton(a)* 1170×85–1398 and *Over Catton* 1355. High for distinction from Low CATTON SE 7053. YE 186.

(5) Low ~ Humbs SE 7053. *Low Catton* 1828. Earlier simply *Cattune* 1086 and *Nether Catton* 1583. 'Low' for distinction from High CATTON SE 7153. YE 186.

CATTON Northum NY 8257. 'The wild-cats' valley'. *Cattedene*

1225, 1298, *Catton* 1343, *Caddon* 17th cent. OE **catt**, genitive pl. **catta**, + **denu**. NbDu 41.

CATWICK Humbs TA 1345. 'Catta's dairy-farm' varying with 'dairy-farm of the Cattingas, the people called after Catta', or 'dairy-farm called or at *Catting*, the place called after Catta'. *Catin(ge)- Cotingeuuic* 1086, *Cattingewic* 1120×40, *C- Kattewic -wy(c)k* 1120×7–1377, *C- Katwic -wyk(e)* [12th]c.1536, 1259–1531, *C- Kattingwic -wyk -yng-* [late 12th]c.1536, 13th cent. OE pers.n. *Catta*, genitive sing. *Cattan*, or folk-n. **Cattingas* < pers.n. *Catta* + **ingas**, genitive pl. **Cattinga*, or p.n. **Catting* < *Catta* + **ing**[2], locative dative **Cattinğe*, + **wīc**. The specific could also represent OE *catte* 'a she-cat'. YE 73 gives pr [katik], BzN 1968.163.

CATWORTH Cambs TL 0873. 'Catt's or the (wild)cat's enclosure'. *Catteswyrð* [983×95]c.1200 B 1130, *Parva Cateuuorde, alia Cateuuorde* 'little C, the other C' 1086, *Catteswurda* 1163, *Cattewurda* 1167, *Cat(t)eworth* 1199–1428, *Catw(o)rth* 1288, 1325. OE pers.n. **Catt*, genitive sing. **Cattes*, apparently varying with **catt**, genitive pl. **catta**, + **worth**. It is suggested that genitival *-s* could be lost by confusion with the weak pers.n. *Catta*. Hu 236 gives pr [kætəθ].

CAULCOTT Oxon SP 5024. 'The cold, inhospitable cottage(s)'. *Caldecot' -cot(e)* 1199–1285, *Coldecote* 1285, *Caucot, Cauld Cott, Caulcut* 1669. OE **cald** + **cot**, pl. **calde cotu**. O 219 gives pr [kɔ:kət] or [kɔlkət], PNE i.77.

CAULDON Staffs SK 0749. 'Calf hill'. *Celfdun* [1002×4]11th S 1536, *Celfdune* [1002]n.d. K 1298, *Caldone* 1086–1311, *Caluedon'* 1196, *Calfdon(a)* 1185, c.1200, *Caldon(')* 1224–1838, *Coldon* 1332, *Calden* 1562×6, 1651, *Cawdon* 1598, *Cauldon* 1599. OE **celf**, Mercian **cælf**, genitive pl. **celfa, cælfa**, + **dūn**. Oakden, L 144, 155.

CAULDON LOW Staffs SK 0848 → WEAVER HILLS SK 0946.

CAULDRON SNOUT Durham NY 8128. *Cauldron Snout* 1863 OS. ModE **cauldron** (ME *caudron*) 'a cauldron, a place where water boils', + **snout** (OE *snūte*) 'a snout, a nozzle'. The reference is to spectacular rapids on the upper Tees.

CAUNDLE An unexplained stream-name, name early associated with OE **candel** 'a candle'. It occurs in the names of several estates along the course of the brook.

(1) Bishop's ~ Dorset ST 6913. *Kaundele Episcopi* 1260, 1280, *Candel(l) Episcopi* 1280, 17th, *Caundel Bishops* 1294, 1297, *Busshopescaundell'* 1497. ModE **bishop**, Latin **episcopus**, genitive sing. **episcopi**, + p.n. *Candel* 1224–8, *Caundele* 1294, *Caunnell* 1454. 'Bishop's' because the estate belonged to the bishop of Salisbury and for distinction from Purse and Stourton CANDLE ST 6917, 7115. Do iii.311

(2) Purse ~ Dorset ST 6917. 'C held by the Purse family'. *Pursca(u)ndel* 1241–1341, *Ca(u)ndel Purs* 1270–1340, *Candle Purse* 1664. Family n. *Purse* + p.n. *Candel* 1086–1203, *Caundel* 1241. 'Purse' for distinction from Bishop's and Stourton CANDLE ST 6913, 7115. Do iii.319.

(3) Stourton ~ Dorset ST 7115. 'C held by the Lords Stourton'. *Sturton Candell* 1569×74. Family name *Stourton* from the Lords Stourton who held the manor from the 15th cent. to 1727 + p.n. *Candel(le), Candele* 1086, *Candel* 1212. Also known as *Candel Malherbe* 1202 and *Candel Hadden* 1270: Robert Malherbe granted 2½ hides of land here to Henry de Haddon in 1202. Do iii.276 gives prs ['stɔ:tən 'kɔ:ndl] and ['stɔ:tən 'kændl].

CAUNSALL H&W SO 8581. 'Conn's nook'. *Conneshale* 1240, 1327, *Counsall* 1614, *Cornsall* c.1830. OE pers.n. **Conn* from a W name such as *Conhouarn* or *Conmael*, genitive sing. **Connes*, + **hale**, dative sing. of **healh** in the sense 'land in a river bend' referring to a bend of the river Stour. Wo 257.

CAUNTON Notts SK 7460. 'Calnoth's farm or village'. *Calnestone -tune* 1086, *Calnodeston -noð-* 1166–7, *Kalnadatun* c.1160, *Kalnadton* c.1220, *Calnat(t)on(')* 1175–c.1260, *C- Kalneton* 1226–1428, *Caunton* from 1392. OE pers.n. **Cal(u)nōth*, genitive sing. **Cal(u)nothes*, + **tūn**. Nt 183.

CAUS CASTLE Shrops SJ 3307. P.n. *Caus* from 1134, 1255 etc. with variants *Kaus, Caws, Cause* (1577), *Cawse* (1672), *Caos* 1217, *C- Kauz* 1246–1328, *Caur(e)s* 1255–1636, *Caux* 1330–1453, + ME **castel**. Although this is said to be a Fr name transferred from Pays de Caux in Normandy there is no secure evidence that Robert fitz Corbet who built the castle originated from there. This is likely rather to be an independent use of OFr **caus** 'chalk' for a building characterised to an unusual extent by the use of lime or chalk. The *Caur(e)s* spellings may indicate folk-etymology based on W *cawres* 'a giantess' used as a nick-name for the castle. The earlier name of the manor was not *Alretone* 1086 as frequently stated since this is now known to be Trewern Powys SJ 2811. Sa ii.56.

CAUSEWAY Dorset SY 6681 → BROADWEY SY 6683.

CAUSEY PARK Northum NZ 1795. *Cawsee Park* 1517, *Cassapark* 1705. P.n. *(Capella de) Calceto* 'Causey chapel' 1221, *La C(h)auce* 1242 BF–1346, *le Cawse* 1455, OFr **caucie**, Latin **calcetum** 'a raised way', + ModE **park**. The reference is to a paved way on the E boundary of the Park on the line of the Great North Road. NbDu 42, HN 2.2.131, Pevsner 113.

CAUSEY PARK BRIDGE Northum NZ 1994. *Causey Park Bridge* 1868 OS. P.n. CAUSEY PARK NZ 1795 + ModE **bridge**.

CAUSEY PIKE Cumbr NY 2120. Named from an unknown causeway. *Cawsey-pike* 1784. OFr **caucie** + ModE **pike** (OE *pīc*) 'point'. Cu 372.

CAUTLEY Cumbr SD 6995. Possibly 'clearing with a trap'. *Cawtley(e)* 1574, 1603, 1655, *Cawteley* 1649, *Co(w)tley* 1661. ME **cautell** 'trick, crafty device' (as, for instance, a device for catching fish) + OE **lēah**. YW vi.266.

CAVE Humbs. OE **cāf** 'swift, quick'. Originally a r.n., the name of Mires Beck which flows with rapid course from the Wolds through North Cave and South Cave.

(1) North ~ Humbs SE 8932. *North Cave* from 1148×58. Earlier simply *Cave, Cava* 1086 etc. YE 224.

(2) South ~ Humbs SE 9231. *S(o)uthcave* from 1228. Earlier simply *Cave, Cava* 1096–1529. YE 223.

CAVENDISH Suff TL 8046. 'Cafa or Cafna's enclosed pasture'. *Kauana- Kanauadis(c), Rana uadisc* 1086, *Kafenedis, Kaftnedich* 1219, *Cavenedis* 1242, *Cauenedess* 1229, *Candishe* 1610. OE pers.n. *Cāfa* or **Cāfna*, genitive sing. *Cāfan*, **Cāfnan*, + **edisc**. The exact form of the pers.n. here is uncertain. Cf. CAVENHAM TL 7669. DEPN.

CAVENHAM Suff TL 7669. Possibly 'Cafnoth's homestead'. *Canauathā, Ranauahā* (for *Cauana- Kauana-*) 1086, *Cauenham* 1198, 1291, *Caveham* 1210, *Caneham* 1610. OE pers.n. **Cāfnōth* or *Cāfa* or **Cāfna*, genitive sing. *Cāfan*, **Cāfnan*, + **hām**. Cf. CAVENDISH TL 8046. DEPN.

CAVERSFIELD Oxon SP 5825. 'Cafhere's open land'. *Cavrefelle* 1086, *C- Kaveresfeld(e)* 1222–1391. OE pers.n. **Cāfhere*, genitive sing. **Cāfheres*, + **feld**. O 204, L 243.

CAVERSHAM Berks SU 7175. 'Cafhere's homestead or river meadow'. *Cavesham* 1086, *Caueresham* 1086–1285, with variants *Kau- Kav-*, *Cauersham* 13th cent. OE pers.n. **Cāfhere*, genitive sing. **Cāfheres*, + **hām** or **hamm**. Brk 175.

CAVERSWALL Staffs SJ 9543. 'Cafhere's spring'. *Cavreswelle* 1086, *Caureswell(e) -v-* 1167–1369, *Cavereswall(e)* 1242–1359, *Caverswall* from 1272, *Careswelle -walle* 1333–1553. OE pers.n. **Cāfhere*, genitive sing. **Cāfheres*, + **wella**, Mercian **wælla**. Oakden.

CAW Cumbr SD 2394. Probably 'the calf', referring to a hill-top rising to 1735ft. Cf. *Calfheud* 1170×84 'head or top of Caw'. Caw is a minor peak S of the main group consisting of Coniston Old Man (2633ft.), Wetherlam (2502ft.) and others; Caw is as it were the 'calf' of the more northern group. Cf. CALF. La 194, CW 1918.94.

CAW FELL Cumbr NY 1310. 'Calf field'. *Cawffelde* 1578. ME **calf** + **feld**. Cu 387.

CAWOOD NYorks SE 5737. 'Jackdaw wood'. *(on) Kawuda*

[963]14th S 712, *Cawuda* [972×92]11th S 1453, *Cauda, (on) Cawuda* c.1030 YCh 7, *Cawod(e)* 1135–1543. OE **cā** + **wudu**. YW iv.38.

CAWSAND Corn SX 4350. Possibly '*Cow* sand'. *Couyssond* 1405, *Cawson* 1619 Gover n.d., *Cawsand* 1634 Gover n.d.; cf. Cawsand Bay SX 4450, *Causet Bay* 1583, *Causam Bay* 1602, *Cousham Baye* 1610. Rock name *Cow* + **sand**. *Cow* is a frequent name for coastal rocks, e.g. Cow and Calf SW 9680 and 8370, The Cow and The Calf Do i.111, Cow Corner Do i.125, The Bull and The Blind Cow Do i.132, Cow and Calf Rocks YW iv.214. There is, however, a phonological problem since [kau] rather than [kɔ:] would be expected from ME *cow*. PNCo 67, Gover n.d. 233.

CAWSTON Norf TG 1323. 'Kalfr's estate'. *Caluestune, Cau(e)stuna* (13×), *Caustona* (2×), *Gaus- Caups- Caustituna, Castune -tona* 1086, *Causton'* 1162×74, 1176×87, 1275. ON pers.n. *Kálfr*, secondary genitive sing. *Kalfes*, + **tūn**. DEPN, SPNN 243.

CAWTHORNE SYorks SE 2802. 'The cold, exposed thorn-tree'. *Caltorn(e)* 1086–1194, *Calthorn(a) -(e)* 1090–1486, *Cawlthorne* 1426, 1561, *Cawthorne* from 1546. OE **cald** + **thorn**. YW i.323.

CAWTON NYorks SE 6476. 'The calf farm'. *Caluetun -tone* 1086, *C- Kalueton(a)* 1160×75–1416, *Caulton* 1418, *Caw(e)ton* 1538, 1579, 1665. OE **calf**, genitive pl. **calfa**, + **tūn**. YN 52 gives pr [kɔ:tən].

CAXTON Cambs TL 3058. 'Kak's estate'. *Caustone* 1086, 1559, *Kachestona* 1135×54, *Ka(c)kestune -ton(a)* c.1150–1330, *Chachestun, Chakeston, Cakeston(a)* c.1150–1324, *Caxton(a) -(e)* from c.1150. ON pers.n. *Kakkr*, secondary genitive sing. *Kakkes*, + **tūn**. Ca 157.

CAYNHAM Shrops SO 5573. Uncertain. *Caiham* 1086, *K- Cayham* 1255×6–1314, *Kainham* 1255×6, *Cainham* 1255, 1686ff., *Caynham* from c.1291 with variant *Kayn-*. The specific is uncertain but possibly OE **cǣġ(e)** 'a key' used as a topographical term for a hill, genitive sing. **cǣġan**, + **hām** 'a village' or **hamm** 'land between two streams'. The reference would be to the hill crowned by Caynham Camp and the village name mean either 'key village or promontory', or, 'village or promontory of Key'. In the 14th cent. French romance of *Fouke Fitz Warin Keyenham* is said to be named from *Chastel Key* built by King Arthur's seneschal Kay. Sa i.72.

CAYNHAM CAMP Shrops SO 5473. P.n. CAYNHAM SO 5573 + ModE **camp** 'an earthwork'. The reference is to a Bronze- and Iron Age settlement with formidable rock-cut defences. Thomas 1976.184, Raven 44.

CAYTHORPE 'Kati's outlying farm.' ON pers.n. *Káti*, genitive sing. *Káta*, + **thorp**.

(1) ~ Lincs SK 9348. *Catorp* 1086–1202, *-thorpe* 1218–1748, *Catt(h)orp(e)* 1180–1434, *Caythorpe* from 1576. Two aberrant spellings, *Carltorp* 1086, *Carletorp* 1202, are probably mistakes due to the proximity of Carlton Scroop. Perrott 366, SSNEM 106.

(2) ~ Notts SK 6845. *Cathorp(e)* c.1170–1394, *Caitorp'* 1178, *Caythorppe* 1528, *Cathrop(e)* 1562, *Catrope* 1674. Earlier called *Alwoldestorp* 1086, 'Ælfwald or Æthelwald's outlying farm'. Nt 159 records a one-time pr [kat(ə)rəp], SSNEM 124, 102, DB Notts 18, 6.

CAYTON NYorks TA 0583. 'Cæga's farm'. *Caitun(e), Caimton(a)* 1086, *C- Kaiton(a), C- Kayton* from 12th cent. OE pers.n. **Cǣga* + **tūn**. YN 103.

CAYTON BAY NYorks TA 0684. P.n. CAYTON TA 0583 + ModE **bay**.

CEFN EINION Shrops SO 2886. 'Ridge of the anvil'. *Cefn Einion* 1836 OS. W **cefn** + **einion**.

CELLARHEAD Staffs SJ 9547. *Cellarhead* 1734, *Sellarhead* 1776, *Cellar Head* 1810. ModE **cellar** 'a beer-cellar' + **head**. A modern name of a former site for fairs, cock-fighting, and bull-baiting at an ancient cross-roads. There were formerly no less than four inns here, one at each of the four road-crossings. Oakden.

CERNE The name of several Dorset estates along the course of the r. Cerne, the 'rock or stony stream', OE **ċēarn* < **ċearn* < PrW **carn*. The modern pr with [s] instead of [tʃ] is usually assumed to be due to AN influence, but the stream-name also occurs with expected [tʃ] in CHARMINSTER SY 6892. For another example of the r.n. see CHARMOUTH SY 3693. RN 72, LHEB 272.

(1) ~ ABBAS Dorset ST 6601. 'C held by the abbot' sc. of Cerne abbey. *Cerne Abbatis* 1288, *Serne Abbas* 1447. P.n. *Cernel* 1086, *Cerne* 1175 + Latin **abbas**, genitive sing. **abbatis**. Do 51.

(2) Nether ~ Dorset SY 6698. '(The) lower Cerne estate', i.e. downstream from Cerne Abbas. *Nudernecerna* 1206, *Nudercerne* 1244, *Nethercerne* 1288. OE **neothera**, oblique case **neotheran**, + p.n. CERNE. Do 51.

(3) Up ~ Dorset ST 6502. '(The) upper Cerne estate', i.e. upstream from Cerne Abbas. *Obcerne* 1086, *Upcerne* 1202. OE **upp** + p.n. CERNE. Do 51.

North CERNEY Glos SP 0208. *Nort Cerney* 1269, *North(e)-* 1277–1564. ME adj. **north** + p.n. *(into) Cyrnea* [852]13th S 202, *Cernei -ey(a) -eye -ay(a)* 1086–1564, *Sarney(e)* 1392, as in South CERNEY SU 0597. Gl i.148.

South CERNEY Glos SU 0597. *Sud- Suth Cerney* 1274, 1285, *South(e)-* 1287–1584. ME adj. **suth** + p.n. *(æt) Cyrne, (terra) Cyrne* [999]13th S 896, [990×1006]13th S 937, *Cernei -ey(a) -ai -ay(a)* 1086–1512, *Sarney(e)* 1286, 1492, 'the Churn stream', r.n. CHURN SP 0108 + **ēa**. Gl i.58 gives prs ['sɔ:ni] and ['saRni].

CERNEY WICK Glos SU 0796. 'Cerney outlying farmstead'. *Cernewike* 1220, *Cerneywyke -wike* 1398, 1402, *la Wyke (de Cerney)* 1240–1327. P.n. Cerney as in South CERNEY SU 0597 + OE **wīċ**. Gl i.58.

CHACELEY Glos SO 8530. Partly uncertain. *Ceatewesleah* *[972]11th S 786, *Chad(d)eslega -leia -ley(e)* c.1086–1354, *Ched(d)eslega* 1154×89, *Chaseleia* 1185, *-ley* 1632. The generic is OE **lēah** 'wood, clearing', the specific obscure, possibly a pers.n. **Ceatwe* or a PrW p.n. **Cēdïw* 'yew wood' or **Cēdou* 'wood' (PrW **cę̄d* + suffix **-ou* < British **-ouiā*). Gl ii.56.

CHACEWATER Corn SW 7544. 'Chace or hunting-ground on the stream'. *Chasewater* 1613, 1732. OFr, ME **chace** 'a tract of ground for breeding and hunting wild animals' + ME **water** 'a stream'. The chace belonged to the manor of Goodern or Blanchland. It is mentioned in the 12th cent. legend of Tristan as a hunting ground of King Mark of Cornwall (Beroul, *Tristan* 4085) and in the later Life of St Kea as the hunting ground of another king of Cornwall, Teudar or Theodoricus of the saint's Lives. Contrasts with BLACKWATER SW 7346. PNCo 67, Gover n.d. 462.

CHACKMORE Bucks SP 6835. 'Ceacca's marsh'. *Chalkemore* (sic) 1229, 1272×1307, *Chacke- Cha(k)kemore* 1241–1379, *Chackmore* 1590. OE pers.n. *Ċēac(c)a* + **mōr**. There is no chalk here and the first two forms must be regarded as errors. The pers.n. is an original by-n. from OE *ċēace* 'cheek, jaw'. Bk 47, L 55, Anglia 113.373.

CHACKRELL Devon ST 0716 → CHICKERELL Dorset SY 6480.

CHACOMBE Northants SP 4943. 'Ceawa's valley'. *Cewecumbe* 1086, *Chaucumba -cumb(e)* 1178–1371, *Chacombe* from 12th. OE pers.n. **Ċeawa* + **cumb**. The reference is to a broad side-valley opening on to the river Cherwell. Nth 50 gives pr [tʃeikəm], L 91–2.

CHAD VALLEY WMids SP 0485 takes its name from the Chad Valley Company established here in 1860. The stream known locally as *Chad Brook* is the *Horeburn* of HARBORNE SP 0284.

CHADDERTON GMan SD 9005. 'Settlement at or called *Chader*'. *Chaderton* c.1200–1332, *K-* c.1250, *Chaterton* 1224, *Chadreden -ton, Chad- Chathirton -er-* 14th cent., *Chadderton* 1468.

Possibly PrW ***cadeir** 'a chair' + OE **tūn**. PrW **cadeir* seems often to have been used with reference to a hill or lofty place, perhaps in this instance to the hills called Chadderton Heights at SD 8907. La 50, PNE i.75, Jnl 17.40.

CHADDESDEN Derby SK 3737. 'Ceadd's valley'. *Cedesdene* 1086, *Chad(d)esden(e) -a -is-* 1154×9–1315, *Chad(d)esdon* 1309, 1577, 1610, *Chadsden* 1577. OE pers.n. *Ċeadd*, genitive sing. *Ċeaddes* + **denu**. Db 544.

CHADDESLEY CORBETT H&W SO 8973. 'C held by the Corbet family'. *Chad(d)esley(e)* 1275–1431 when *Corbet* first appears. The manor came into the possession of the Corbets at the end of the 12th cent. Chaddesley, *ceadresleahge* [816]11th S 179, Hooke 107, *(æt) Ceadresleage* [816]11th S 180, Hooke 113, with variants *Cedres- Cead(d)es-* all 11th, *Cedeslai* 1086, is 'Ceadder's wood or clearing', OE pers.n. **Ċeadder* varying with *Ċeadd*, genitive sing. *Ċeadd(er)es*, + **lēah**. Wo 234 gives pr [tʃædʒli].

CHADDLEWORTH Berks SU 4177. 'Ceadela's enclosure'. *Ceadelanwyrð* 960 S 687, *Cedeneord, Cedeledorde* (sic) 1086, *Chedeleswrtha* [1100×35]12th, *Chadel(e)worth(e) -wurth' -wrdh'* 1167 etc., *Chaddelworth* 1517, *Chaddleworth* 1550. OE pers.n. **Ċeadela*, genitive sing. **Ċeadelan*, + **worth**. Brk 289.

CHADLINGTON Oxon SP 3322. 'The estate called after Ceadela'. *Cedelintone* 1086, *Chedelint(ona) -ton(e)* c.1130–1208, *Chadelin(g)ton' -tun -t(h)one -ton(a) -yn(g)-* 1196–1346, *Chadlington* from 1304. OE pers.n. **Ċeadel(a)* + **ing**[4] + **tūn**. O 338, Signposts 172.

CHADSHUNT Warw SP 3453. 'Ceadel's spring'. *Ceadeles funtan* [949]16th S 544, *Chaddeleshunt* [c.1043]15th S 1226, *Chadeleshunt* [1043]15th S 1000, c.1057–1384, *Cedeleshvnte* 1086, *Chedeleshunte* 1175–99, *Chadelesfunz -funt* 1195–1242, *Chaddeshunte* 1291, 1535, *Chadson* 16th cent., *Chadshunt* 1689. OE pers.n. **Ċeadel*, genitive sing. **Ċeadeles*, + ***funta**. Wa 249, Signposts 84, 86, L 22.

CHADWELL GLond TQ 4888 → CHADWELL ST MARY Essex TQ 6478.

CHADWELL ST MARY Essex TQ 6478. 'St Mary's C.' P.n. Chadwell + saint's name *Mary* from the dedication of the church. Chadwell, *Celdeuuellā* 1086, *C(h)aude- C(h)ald(e)well(e)* 1205–1361, *Chad(e)well* 1272, 1285, 1678, *Shadwell* 1665–6, is the 'cold spring', OE **ċeald**, definite form **ċealda**, + **wella**. 'St Mary' for distinction from Chadwell (Heath) in Dagenham GLond TQ 4888, *Chaudewell'* 1254, and Chardwell in Arkesden TL 4734, *Chaldewell(e)* 1361–1452. Ess 150 gives pr [tʃædl].

CHADWICK END WMids SP 2073. 'The Chadwick end' sc. of Balsall. *Shadditch end* 1667. P.n. *Chedelesuuich* 1100×35, *Cheddeleswyke* 1221, *Chadeleswiȝ -wyc* 1199–1247, *Chadelwic* 1224, *Chaddeswich* 1285, 1460, 'Ceadel's dairy-farm', OE pers.n. *Ċeadel*, genitive sing. *Ċeadeles*, + **wīċ**, + ModE **end** as in nearby *Bedlams End* 1831 OS, FEN END SP 2275, MEER END SP 2474, Needlers End, *Nelders* 1540, possibly '(the settlement) at the elder-trees', ME *at than ellers* misdivided, and Oldish End, *Oldedich* 1288, *Oldysh* 1789, 'the old ditch', OE **ald**, definite form **alda**, + **dīċ**. Wa 54, 55.

CHAFFCOMBE Somer ST 3510. Usually explained as 'Ceaffa's coomb'. *Caffecome* 1086, *Chaffacombe* [1204]1313, *Chaffecumbe* 1236, *Chafcombe* 1610. OE pers.n. **Ċeaffa* + **cumb**. Formally, however, the specific could be OE **ċeaf** 'chaff' perhaps referring to fragments and shavings from the working of wood in this wooded area of piecemeal felling and cutting. DEPN, VCH iv.121.

CHAGFORD Devon SX 7087. 'Brushwood ford'. *C(h)ageford* 1086, *Chageford(e)* 1196–1429, *Chaghhford* 1275, 1285. OE ***ceacga** + **ford**. D 424 gives prs [tʃægfəd] and [tʃægivɔrd].

CHAILEY ESusx TQ 3919. 'Brushwood clearing'. *Cheagele, Chaglegh* c.1095, *Chaggele(ye) -legh -ly(e)* 1268–1442, *Chayly* 1588. OE ***ceacga** + **lēah**. The expected modern form would be *Chagley: the development to Chailey is unexplained. Sx 296, SS 66, Nomina 13.111.

CHAINHURST Kent TQ 7347. Uncertain. *Chainhurst* 1819. This name has been associated with the surname *de Che(y)ney, Cheuenye* 1278 from Great Cheveney TQ 7342 and assumed to be an old manorial genitive sing. **Cheyneyes* > **Chainers* > *Chainhurst*. But the place stands on a low hill beside the river Beult and may be an ancient *hyrst* name. PNK 315.

CHALBURY COMMON Dorset SU 0206. P.n. Chalbury + ModE **common**. Chalbury is *(on) cheoles burge (eastgeat)* 'to the east gate of Chalbury' [946]14th S 519, *(in, an) cheoles byris (east gete)* [956]14th S 609, *Chelesbyr' -bur(y)* 1244–1448, *Chalesbury* 1361, *Chalbury* from 1448, *Chawbery, Chabury* 16th, 'Ceol's fort', OE pers.n. *Ċēol*, genitive sing. *Ċēoles*, + **byrig**, dative sing. of **burh**. Chalbury is a prominent univallate Iron Age hill-fort. Do ii.134, Thomas 1976.111.

CHALDON Dorset 'Calves' hill, hill where calves are pastured'. *Celvedune, Calvedona* 1086, *Chauvedon* 1199–1285, *Chalvedon(e)* 1224–1489, *Chaldon* from 1501. OE(WS) **ċealf**, genitive pl. **ċealfa**, + **dūn**. Do i.108 gives pr ['tʃ ɔ:ldən].

(1) ~ BOYS or West CHALDON Dorset SY 7782. 'C held by the Boys family'. *Chau(e)don Boys* 1270–99, *Chalvedon(e) Boys* 1280–1428 from its tenure by the *Boys* or *de Bosco* family. Also known as *West Chavedon* 1269–1459, *Westchaldon* from 1454 for distinction from East CHALDON or CHALDON HERRING SY 7983. Do i.108.

(2) ~ DOWN Dorset SY 7882. *Chaldon Down* 1811. P.n. CHALDON + pleonastic ModE **down** 'hill'. Do i.108.

(3) ~ HERRING or East CHALDON Dorset SY 7983. 'C held by the Harang family'. *Chavedon Hareng* 1235×6, *Chalvedon Hareng, Harang, Heryng* 1244–1489, *Chaldon Hearing* 1587. P.n. CHALDON + family name *Harang*, holders of the manor from 1154×89. Also known as *Est Chalvedon* 1269–1489, *Est Chaldon* 1535. 'East' in contrast to West CHALDON or CHALDON BOYS SY 7782. Do i.108.

(4) East ~ Dorset SY 7983 → CHALDON HERRING SY 7983.

(5) West ~ Dorset SY 7782 → CHALDON BOYS SY 7782.

CHALDON Surrey TQ 3255. 'The upland pasture for calves'. *Cealuadune* 967 S753, *Cealfadun'* [1062]13th S 1035, *Calvedone* 1086, *Chaluedune* [727, 933, 967]13th S 420, 752, 1181, *-don(e)* 1235–1445, *Chaldon* from 1539. OE(WS) **ċealf**, genitive pl. **ċealfa**, + **dūn**. For the significance of the name cf. MERSTHAM TQ 2953. Sr 42 gives pr [tʃɔ:ldən], L 144, 146.

CHALE IoW SZ 4877. 'The throat'. *Cela* 1086, *Chele* 1181–1255, *Chale* from 1117, *Shale* 1534, 1591. OE **ċeole** 'throat' in the sense 'gorge, ravine'. The reference is to the 400ft. deep cleft of Blackgang Chine, one of the principal sights of the island. The name illustrates the local dial. development of OE *eo* > *ea*. Wt 112 gives prs [tʃiəɫ, tʃeəɫ], xciv, Mills 1996.38.

CHALE BAY IoW SZ 4677. *Chall bay* 1544, *Chale bay* 1583, *Shale baye* 1591. P.n. CHALE SZ 4877 + ModE **bay**. Wt 113, Mills 1996.38.

CHALE GREEN IoW SZ 4879. P.n. CHALE SZ 4877 + ModE **green**. The modern name corresponds to *North End* 1775, *Chale Street* 1810 OS, dial. **street** 'a hamlet', *North Green* 1812, and earlier *Stroad Green* 1769 from *Stoude* 1345, 'scrubby marshland', OE **strōd**. Mills 1996.38.

CHALFONT AND LATIMER STATION Bucks SU 9997. See CHALFONT ST GILES SU 9893 and LATIMER TQ 0099.

CHALFONT COMMON Bucks TQ 0091. P.n. Chalfont as in CHALFONT ST GILES SU 9893 + ModE **common**.

CHALFONT ST GILES Bucks SU 9893. 'St Giles's C'. *Chalfund Sancti Egidii* 1237, *saincte Gyles Chaffyn* 1557, *St Giles Cha(l)funt* 1675. Saint's name *Giles* (Latin *Egidius*) from the dedication of the church, + p.n. *Celfunte -funde* 1086,*Chalf(h)unt(e)* 1220× 34–1509, *Chafhunte* 1196, 1379,

Chawfount 1538, *Chafforn* 1675, 'calf spring', OE **ċealf** + **funta**[†]. Jnl 3.33. Bk 218, YE addenda, Signposts 84, 86, L 22.

CHALFONT ST PETER Bucks TQ 0090. 'St Peter's C'. *Chalfhunte Sancti Petri* 1237×40. P.n. Chalfont as in CHALFONT ST GILES SU 9893 + saint's name *Peter* (Latin *Petrus*) from the dedication of the church. Bk 219.

Little CHALFONT Bucks SU 9997. ModE adj. **little** + p.n. Chalfont as in CHALFONT ST GILES SU 9893.

CHALFORD Glos SO 8902. 'Chalk ford'. *Chalford(e)* c.1250–1598, *Chelke- Chalkford* 13th, *Chawford* 1634, 1643. OE **ċealc** + **ford**. Gl i.127.

CHALGROVE Oxon SU 6396. '(The settlement) at the chalk pit'. *Celgrave* 1086, *Chalgraue -a -grave* 11th–1577, *Chalcgraua -e* 1235–85, *Chalgrove* from 1285. OE **ċealc** + **græf**, dative sing. **grafe** later replaced by ME *grove*. The reference is to an isolated patch of chalk amidst the surrounding clay dug out for marling. A series of hollows so created are still visible here. O 122, L 193–4, Jnl 19.50.

CHALK Kent TQ 6772. 'The chalk'. *(of) cealce* [c.975]12th B 1322, *Celca* 1086, *Chalcha* 1164×5, *Chalke* 1215 etc., *Chauk'* 1219. OE **ċealc**. Chalk lies at the junction of the chalk and marshland where it outcrops bordering on the Thames. This could have been a place where chalk was loaded to be taken upstream for use on the clayey fields around London, perhaps being landed at Chelsea, the *cealc-hyth*, 'the chalk landing place'. KPN 302, 306, Jnl 19.52.

Broad CHALKE Wilts SU 0325. 'Great Chalke'. *Magna Chelk* 1289, *Grete Chalke* 1312, *Brode Chalk* 1380. ME **brode**, Latin **magna**, ME **grete**, + p.n. Chalke as in BOWERCHALKE SU 0223. Wlt 203–4, Jnl 19.47.

CHALLACOMBE Devon SS 6940. '(The) cold coomb'. *Celdecome* 1086, *Cheldecombe* 1438, *Caldecumba* 1167, *Chaude- Chaldecumb(e)* 1242, 1244, *Choldecomb(e)* 1326, 1346, *Cholcombe* 1428, *Cholycombe* 1620. OE **ċeald**, definite declensional form **ċealda**, + **cumb**. D 60 gives pr [tʃɔləkum], L 93.

CHALLOCK Kent TR 0150. 'The calf enclosure'. *(ad) cealfa locum* (Latin form) [824]11th S 1434, *(et) cealflocan* 833×9 S 1482, *Cealueloca* c.1090, *cheafloce* 12th, *Chalfeloc* 1240, *Chalfloke* 1254 etc., *Chalewelok'* 1247, *Challok* 1610. OE **ċealf**, genitive pl. **ċealfa**, + **loca**. KPN 146, 154, Cullen.

East CHALLOW Oxon SU 3888. *Estchaulo* 1316, *Est Challowe, Eschallawe* 1284. ME **est** + p.n. *(to) Ceawan Hlewe* [947]12th S 529, [958]13th S 657, *(to) ceawan hlæwe* [947]13th S 529, *Ceveslane* (sic) 1086, *Chawelawe* 1220 etc., *Chalowe* 1382, 'Ceawa's tumulus', OE pers.n. **Ċeawa*, genitive sing. *Ċeawan*, + **hlǣw, hlāw**. 'East' for distinction from West CHALLOW SU 3688. Brk 292.

West CHALLOW Oxon SU 3688. *Westchau(e)lawe, West Chaulowe, Weschallawe* 1284. ME **west** + p.n. Challow as in East CHALLOW SU 3888. Brk 292.

CHALTON Beds TL 0326. 'Calf farm'. *Chaltun* [1131]13th, *Chalton* 1195, *Chalfton* 1227. OE **ċealf** + **tūn**. Bd 136.

CHALTON Hants SU 7316. 'Chalk settlement'. *cealc- cealhtun* 1015 S 1503, *Chalc- Chalgh(e)- Chalk(e)ton* 1218–1346, *Chaulton* 1167, *Chauton* 1229–75, *Chalton* from c.1195. OE **ċealc** + **tūn**. The place lies on the chalk downs at a spot where the top-soil is denuded causing the white to show through. The DB form is *Ceptune* (5×). Ha 50, Gover 1958.57, DEPN.

CHALVINGTON ESusx TQ 5109. 'Estate called after Cealfa' or 'calf farm'. *Cavel- Calvintone* 1086, *Chalin(g)ton(a)* 12th–14th cents., *Chavinton'* 1242, *Chalington* 1416, *Chalvington al. Chaunton* 1576, *Chalton* 1580. Either OE pers.n. *Ċealfa* + **ing**[4] + **tūn**, or OE **ċealf** + **ing**[4] + **tūn**. Sx 398, SS 76.

CHANDLER'S CROSS Herts TQ 0698. *Chandres Cross* 1822 OS. Surname *Chandeler, Chaundler* 1659 etc., + ModE **cross**. Hrt 107.

CHANDLER'S FORD Hants SU 4320. *Chandlers Ford* 1759. This is actually a modern corruption of *searnægles ford* [909]11th, *Sarnayllesford* 1280, 'Searnægel's ford', OE pers.n. **Searnæġel*, genitive sing. **Searnæġles*, + **ford**. The modern form has been influenced by the surname *le Chaundler* known in nearby South Stoneham since the 14th cent. The ford carries the Roman road from Winchester to Bitterne, Margary no.422, over Monks Brook. Ha 50, Gover 1958.37.

SOUTH CHANNEL Humbs SE 9231. ModE **south** + **channel**. The name of that part of the r. Humber that flows south of Read's Island.

CHANSTONE H&W SO 3635 → WALTERSTONE SO 3425.

CHANTRY Somer ST 7146. No early forms.

CHANTRY Suff TM 1443. *Chauntry* 1805 OS. The modern estate is named after The Chantry, an early 18th cent. mansion house here. Pevsner 1974.308.

CHAPEL-EN-LE-FRITH Derby SK 0580. 'Chapel in the scrubland'. *capellam de Frich'* (for *-Frith'*) 1241, *Capella de le Frith* 1272, *-del Frithe* 1330–1535, *Capellam in le Fryth* 1484, 1570. ME **chapel** + Fr preposition **en** + Fr definite article **le** + p.n. Frith < OE **(ġe)fyrhth** 'wooded country' referring to the Peak Forest. The chapel is first mentioned as *Ecclesia de Alto Peccho* 1219, *Ecclesia de Peck'* 1226–8, '(High) Peak chapel'. Db 60, L 191.

CHAPEL HILL Lincs TF 2054. *Chapel Hill* 1612. ModE **chapel** + **hill**. The reference is to the *capella Sancti Nicholai de Docdyke* 1343, *Dokdyk capella* 1526, a chapel founded to serve the settlement which grew up round the terminus of Dogdyke Ferry. Perrott 321, Cameron 1998.

CHAPEL LAWN Shrops SO 3176. *Chapel Lawn* 1833 OS.

CHAPEL-LE-DALE NYorks SD 7377. 'The chapel in the valley'. *the Chappell ith Dale* 1677, *Chapel in Dale* 1807. Also known as the *Fells Chapp.* 1692. ModE **chapel** + Fr definite article **le** from names such as Chester-le-Street and short for *en le* + **dale**. A chapel of ease of Ingleton. YW vi.243.

CHAPEL POINT Corn SX 0243. *Chapelle Land or Point* c.1540. ME **chapel** + **point**. The chapel is mentioned in 1327 but no remains of it are known. PNCo 68.

CHAPEL ROW Berks SU 5769. *Chapel Rewe* 1617, *Chappel Row* 1689, *Chapel Row* 1761. ModE **chapel** + **row**. The reference is to Magdalen Chapel the ruins of which were removed in 1770. Brk 156.

CHAPEL ST LEONARDS Lincs TF 5572. *the chapell of seint Leonard in Mumby* 1503. ModE **chapel** + saint's name *Leonard*. The chapel is first mentioned in 1257. Cameron 1998.

CHAPEL STILE Cumbr NY 3205. *Chapel Stile* 1865 OS. Named from *Langdale Chappell* 1660. ModE **chapel** + **stile**. We i.206.

CHAPELEND WAY Essex TL 7039. No early forms but cf. *Chapell howse* 1598. The sense is the chapel end of Stambourne. Ess 457.

CHAPELFELL TOP Durham NY 8734. *Chapel Fell Top* 1866 OS. Named from the village of ST JOHN'S CHAPEL in Weardale NY 8838.

CHAPELGATE Lincs TF 4124. 'The road to the chapel'. *Chapelgate* from 1272×1307. ME **chapel** + **gate** (ON *gata*).Payling 1940.25.

CHAPELTON Devon SS 5726. 'Chapel settlement'. *Chapelton* 1809 OS. ModE **chapel** + **ton** (OE *tūn*).

CHAPELTOWN Lancs SD 7315. No early forms. The church belongs to 1840–1. Pevsner 1969.249.

CHAPELTOWN SYorks SK 3696. *Chap(p)eltown* 1707. Named from a chapel of ease of Ecclesfield recorded as *Capella* 1260, *(le) Chap(p)el(l)* before 1279–1649. ME **chapel** + ModE **town**. YW i.246.

CHAPMAN SANDS Essex TQ 8383. 'Merchant's sand-bank'.

[†]The form *Ceadeles funtan* [949]16th S 544 belongs to Chadshunt Warw, not here.

Chapmansand 1402, *Chapnassand* 1412. ME **chapman** + **sand**. Ess 17.

CHAPMANS WELL Devon SX 3593. *Chapmans Well* 1809 OS. Surname *Chapman* (cf. Walter Chepman 1330) + **well**. D 164.

CHAPMANSLADE Wilts ST 8247. 'The valley of the chapmen or merchants'. *Chepmanesled'*, *Chipmannesled* 1245, *Chy- Chipman(ne)slade* 1268–1456, *Chapman(n)(e)slade* from 1245. OE **ċēap- ċȳpman**, genitive pl. **ċēap- ċȳpmanna**, + **slæd**. Wlt 147.

CHAPPEL Essex TL 8928. *Chap(p)ell (Paris(s)he)* 1528–94, *Alba Capella* Latin 'white chapel' 1536, *Chappell alias Pontesbright* 1603×25. ME **chapel** from a chapel existing here since at least 1285 *in capellam de Ponte Brichrich*. The original name was *Britesbrig* 1272, *Pontbritrich* 1437, *Pontesbright -is* 1486–1596, 'Beohrtric's bridge', OE pers.n. *Beorhtriċ, Brihtriċ* + **brycg**, OFr **pont**. Ess 364.

River CHAR Dorset SY 3894. A back-formation from CHARMOUTH SY 3693. *Charebroke* c.1540, *Chare* 1577, *Char* 1750. The original name was *(aqua de) Cerne* 1288, a r.n. identical with Cerne in CERNE ABBAS ST 6601. RN 73.

CHARD Somer ST 3208. 'House on the chart or rough ground'. *Cerdren* *[1065]18th S 1042, *Cerdre* 1086, *Cerda* 1166, *Chard* 1610. OE **ċeart** + **renn** (a side form of *ærn*). DEPN.

South CHARD Somer ST 3205. *Sutcherde* 1261. ME adj, **suth** + p.n. CHARD ST 3208, possibly the *frusa* (v.l. *frisa*) *Cerdren* of S 1042. Another part of Chard is Crimchard ST 3109, 'Cynemær's Chard estate', *Cynemerstun* *[1065]18th S 1042, *Kinemerscherd* 1196, *Crim Chard* 1809 OS. DEPN.

CHARDSTOCK Devon ST 3104. 'Outlying farm belonging to Chard'. *Cerdestoche* 1086, 1154×89, *C(h)erdestok(e)* 1166–1297, *Chardestock* 1278, *Church Stoke, Cherstocke* 16th. P.n. CHARD Somer ST 3208 + **stoc**. D 654, Studies 1936.16.

CHARDWELL Essex TL 4734 → CHADWELL ST MARY TQ 6478.

CHARFIELD Avon ST 7292. Probably 'chert open land'. *Cirvelde* 1086, *Certfeld* c.1200, *Cherfeld* 1248–1535 with variants *-veld' -vylde*, *Char(e)feld(e)* 1274–1615 with variant *-vild*. Although *ċert* is primarily a SE word well attested in Kent and Surrey, if the 1200 spelling is significant this must be 'chert open land, rough-surfaced open land', OE **ċert** + **feld**. Otherwise the specific might be OE **ċerr** 'a bend' possibly referring to a bend in the stream running up into the valley beside Charfield Hall ST 7291. Gl iii.26, xi, L 240, 242.

North CHARFORD Hants SU 1919. *Northchardeford* 1248, *N. Charford* 1811 OS. ME adj. **north** + p.n. *Cerdeford, Cerdifort* 1086, *Sherdiford* c.1200, *Shardeford* 1236, *Cherdeford* 13th, 1325, *Chardeford* 1242–1327, unknown element + OE **ford**. The occasional spellings with medial *-i-* suggest specific **Ċerding* of unknown meaning but possibly related to the specific of CHARD Somer ST 3208. The name is usually associated with the ASC form *Cerdices ford* 9th (A), *Certices ford* c.1121 (E) under years 508 and 519, and with *Cerdices leaga* (A), *Certices ford* (E) under year 527 where the West Saxon prince Cerdic and his son Cynric fought a series of battles against the Britons. However, the ASC account of the early conquest of Wessex is fictitious and unreliable. On the one hand Cerdic is a genuine OE pers.n. (of W origin, PrW **Car´dig* < Brit **Caratīcos*) and the chronicler Ethelwerd locates *Cerdicesford* 'Cerdic's ford' on the Avon like Charford; on the other the genitival *-es-* of a form *Cerdicesford*, if genuine, would be expected to survive in ME. In reality we do not know the location or the genuineness of the p.n. *Cerdicesford* or its associated names *Cerdices ora, Cerdices leaga* and *Natanleag*. Ha 51, Gover 1958.212, DEPN, RN lxvii fn.1, Origins 84ff, LHEB 203, 237, 554, 613.

South CHARFORD Hants SU 1618. *Southchardeforde* 1248. ME adj. **south** + p.n. *Cerdeford* 1086 as in North CHARFORD SU 1919. Gover 1958.212.

CHARING Kent TQ 9549. Either '*Ceorring*, the place called after Ceorra' and '*Ceorringas*, the people called after Ceorra, or who live at *Ceorring*', or 'the turning' and 'the people who live at *Cerring*, the turning'. *(æt) Ciorrincge* 799 S 155, *Cerringges* [799]13th S 155, *Cirringe* 940 S 464, *Cherringes* 1086, 1176, *Cyrringe, Cearringe, Cerringes -is* c.1100, *Cheringe(s)* 1154×89, *Charr- Cherringes* 1182, *Charing* 1610. Either OE p.n. **Ċeorring* < pers.n. *Ċeorra* + **ing**[2] varying with folk-n. **Ċeorringas* < pers.n. *Ċeorra* + **ingas**, or OE ***ċerring** < OE **ċerr** + **ing**[2] and folk-n. **Ċerringas*. PNK 388, ING 184, BzN 1967.333.

CHARING CROSS GLond TQ 3080. *The stone cross of Cherryngge* 1334, *La Charryngcros* 1360. P.n. Charing, *(to) cyrringe* [c.1000]16th, *(La) Charryng* 1263, 'the turning', OE **ċierring**, + ME **cross**. The reference is either to the bend in the river Thames here or to a bend in Akeman Street as it approached the river near what is now Trafalgar Square; and to the cross, of which the present one is a modern replica, erected by Edward I to mark one of the resting places of the coffin of his queen Eleanor on its way from Harby in Notts to Westminster Abbey in 1290. Mx 10, 167, 223, GL 37, Encyc 141.

CHARING HEATH Kent TQ 9249. *Charing Heath* 1798. P.n. CHARING TQ 9549 + ModE **heath**. Cullen.

CHARINGWORTH Glos SP 2039. Possibly 'enclosure of the Ceaforingas'. *Chevringavrde* 1086, *Chaveringewurd -wrth'*, *Chev(e)ringew(o)rth* 1190–1220, *Che- Chaveringworth(e) -yng-* 1277–1328, *Charyngworth(e) -ing-* 1353–1677. OE folk-n. **Ċeaforingas* 'the people called after Ceafor' < OE pers.n. **Ċeafor* + **ingas**, genitive sing. **Ċeaforinga*, + **worth**. An alternative possibility is an OE p.n. **Ceaforing* 'beetle place' < OE **ċeafor** + **ing**[2], locative dative sing. **Ceaforinge*. Gl i.242.

CHARLBURY Oxon SP 3519. '(At) the fortified place called after Ceorl'. *Ceorling(c)burh* c.1000, *Cherleberi(am) -bery -bir(y) -buri(a) -bury -byri* 12th–1325, *Charlebury* 1320, *Chorlbry* 1429. OE pers.n. *Ċeorl* + **ing**[4] + **burh** varying with dative sing. **byriġ**. O 415.

CHARLCOMBE Avon ST 7467. 'The free peasants' coomb'. *Cerlecvme* 1086, *Cherlecumba-e* 1156, 1225. **ċeorl**, genitive pl. OE **ċeorla**, + **cumb**. DEPN, L 93.

CHARLECOTE Warw SP 2656. 'The cottage(s) of the peasants'. *Cerlecote* 1086, *Cherlecote* c.1185–1302, *Sherlecote* 1217–1361, *Charlecote* 1484. OE **ċeorl**, genitive pl. **ċeorla**, + **cot**, pl. **cotu**. Wa 250 gives pr [ʃɑ:lkət].

CHARLECOTE PARK Warw SP 2656. The 16th cent. mansion house of the Lucy family. P.n. CHARLECOTE SP 2656 + ModE **park**.

CHARLES Devon SS 6832. Uncertain. *Carmes* 1086, *Charnes -is -ys* 1242–1317, *Charles* from 1244 with variants *-is -ys, Charells* 1616. Possibly Co **carn** 'rock-pile, tor' + ***lys** 'court'; but this is very uncertain. D 61, CPNE 40.

CHARLES' TYE Suff TM 0252. 'Charles common pasture'. *Charles Tye* 1837 OS. P.n. Charles + Mod dial. **tye** (OE *tēag*). Cf. nearby BATTISFORD TYE TM 0254, Kersey Tye TL 9843, *Kersey Tye* 1837 OS, Lindsey Tye TL 9845, *Lindsey Tye* ibid., NEDGING TYE TM 0149, Noakes Tye TM 0043, *Nokes Tye* 1837 OS.

CHARLESTOWN Corn SX 0351. *Charles-town* 1800. A port and village founded in 1790 or 1791 to serve the growing china-clay industry and named after their sponsor, Charles Rashleigh of Menabilly (1747–1825). The older Cornish name of the place survives in Polmear SX 0853, *Portmoer* 1354, *Porthmuer* 1403 G, *Polmere* 1610, 'great harbour', Co **porth-meur** later anglicised as if from OE **pōl** 'pool' + **mere** 'lake'. 'Great' for distinction from PORTHPEAN SX 0350 'little harbour' (Co **byghan**). PNCo 68, Gover n.d. 386, Room 1983.22.

CHARLESTOWN Dorset SY 6579. *Charlestown* 1893. On the 1811 OS this place appears as *Furzeland*.

CHARLESWORTH Derby SK 0092. 'Ceafl's enclosure'. *Cheueneswrde* 1086, *Chauelesworth(e) -with -is-* 1285–15th cent.,

Chal(l)esworth 1552–1758, *Charlesworth* 1767. OE pers.n. **Ċeafl*, genitive sing. **Ċeafles*, + **worth**. *Ceafl* represents OE **ċeafl** 'jaw' and is here used either in the topographical sense 'ravine' as the name of the valley of the river Etherow below Charlesworth, or as an OE nick-name. Db 68.

CHARLTON 'A settlement of free peasants'. OE **ċeorla** varying with **ċeorlena**, genitive pl. of **ċeorl**, + **tūn**. The name-type implies that the settlement was originally a constituent of a larger, often royal, estate. See CARLTON. Lucerna 144–60.

(1) ~ GLond TQ 4278. *Cerletone* 1086. GL 37.

(2) ~ H&W SP 0145. *Ceorletun* *[780]11th S 118, 1086, *Ceorlatuna* 11th, *Cherleton* 1208–1431, *Charlton* 16th. One form, *Cherlinton* 1346, 1428, goes back to genitive pl. *ceorlena*. A member of the royal manor of Cropthorne. Wo 105, Lucerna 154.

(3) ~ Northants SP 5235. *Cerlintone* 1086, *Cherlington* 12th, *Cherlenton* 1297, *Cherleton(a)* c.1190 etc., *Cherleton* 1316. A constituent of the royal manor of Kings Sutton 2 miles to the W. Nth 56, Lucerna 152.

(4) ~ WSusx SU 8812. *Cherleton* 1271–1311, *Churletone* 1302. Sx 54, SS 76.

(5) ~ Wilts SU 1156. *Cherlentona* 1154×89, *Cherleton* 1198–1332, *Charleton* 1368, *Chorleton* 1559. Wlt 319.

(6) ~ Wilts ST 9689. *Cherletune* [680 for 681]13th S 71, *-ton* 1225–1332, *Cerletone* 1086, *(æt) Ceorlatunæ* [10th]c.1150, *Cheorletun* [1065]14th S 1038, *Chorletone* 1216×72, *Cherelton* 1268. Wlt 54 gives pr [tʃ ɔ:ltən], Lucerna 154.

(7) ~ Wilts ST 9022. *Cherlton* 1249, *Cherleton* 1256–1382, *Charlton* 1412. Wlt 188.

(8) ~ Wilts SU 1723. *Cherleton* 1209–1316, *Churleton* 1279, 1281, *Charleton* 1358, *Charelton* 1524. Wlt 397.

(9) ~ ABBOTS Glos SP 0342. 'Charlton held by the abbot' sc. of Winchcomb. *Charleton Abbatis* 1535, ~ *Abbottes* 1592. Earlier simply *Cerletone* 1086, *Cherleton(a)* 12th–1327, *Chorletona* 1175. Gl ii.5, Lucerna 151.

(10) ~ ADAM Somer ST 5328. 'C held by the fitz Adam family'. *Cherleton Adam* 13th cent., *Cherleton Fitz-Adam*, ~ *Fizadam* [before 1269]B. Earlier simply *Cerletone* 1086, *Cherleton* [1142×66–before 1257]Bruton. Also known as *Estcherleton* [c.1252]Bruton, *East Carlton* 1610. Held in 1206 by William fitz Adam; the family name was commemorated when the manor was given to Bruton priory and served to distinguish it from nearby CHARLTON MACKRELL ST 5228. DEPN, DB Somerset 19.43.

(11) ~ DOWN Dorset ST 8700. *Charlton Down* 1863. Formerly simply *le Downes* 1508. P.n. Charlton as in CHARLTON MARSHALL ST 9003 + ModE **down(s)**. Do ii.12.

(12) ~ HORETHORNE Somer ST 6623. 'C in the hundred of Horethorne'. *Charleton Horethorne* 1811 OS. Earlier simply *(æt) Ceorlatune* ? c.950 S 1539. Horethorne is *Hareturna* 1184, 'the grey thorn-tree', OE **hār**, definite form **hāra**, feminine **hāre**, + **thyrne**. *Horethorne Down* 1811 OS lies between this place and Corton Denham ST 6322. Also known as *Cherleton Kanvill* 1225 from the 12th cent. tenant Gerard de Camvile. DEPN.

(13) ~ KINGS Glos SO 9620. 'The king's Charlton'. *Kynges Cherleton(e)* 1245, 1270, *Charleton Regis* 1520–1675, ~ *Kinges* 1535. Earlier simply *Cherleton(e)* 1160–1327, *Charl(e)ton* c.1300–1557. Gl ii.96, Lucerna 156.

(14) ~ MACKRELL Somer ST 5228. 'C held by the Makerel family'. *Cherletun Makerel* 1243, *Cherleton Makerell* [1381×2]14th. Earlier simply *Cerletvne* 1086. Also known as *West Carlton* 1610. 'Mackrell' for distinction from nearby CHARLTON ADAM ST 5328. DEPN.

(15) ~ MARSHALL Dorset ST 9003. 'C held by the Marshall family'. *Cherleton(e) Marescal* 1288, *Charlton Marshall* 1571. Earlier simply *Cerletone* 1086, *Cerlentone -tonia* 1087×1100, c.1165, *Cherlenton(a) -tunia* 1167×87, 1188, *Cherleton(e)* 1201–1428, *Charletone* 1337–1571. 'Marshall' from the family which gave its name to STURMINSTER MARSHALL and for distinction from the lost Little Charlton, *Charleton Parva* 1462–71, also known as *Charleton Rectoris de Stourmynstre Marchall* 1447, *Charleton Sturmynstr'* 1448, 'C held by the rector of Sturminster Marshall'. Do ii.11.

(16) ~ MUSGROVE Somer ST 7231. 'C held by the Mucegros family'. *Cherleton Mucegros* 1225, *Cherlton Mucegros* [1329]Bruton, *Charlton Musgroue* 1610, *Charlton Musgrave* 1811 OS. Earlier simply *Cerletone* 1086. The identification of the DB form is not entirely certain. DEPN.

(17) ~ -ON-OTMOOR Oxon SP 5912. *Cherleton' super Ottemor'* 1278×9, *Cherleton on Ottemore* 1315. P.n. *Cerlentone* 1086, *Cherlenton(a)* [1108]12th, [c.1190]c.1200, *Charlentone* [1081]12th *-yn-* 1336, *Cherleton(e) -tun'* [c.1190]c.1280–1316, *Cherlton* 1245, *Charleton* 1351, + p.n. OTMOOR SP 5614. An alternative view is that this is 'Ceorla's homestead', OE pers.n. **Ċeorla*, genitive sing. **Ċeorlan*, + **tūn**, rather than ***ceorlena** 'of the ceorls'. O 205.

(18) North ~ Northum NU 1622. *Charleton de(l) North* 1242 BF 1117–8, *Northcharlton'* 1296 SR 143. ME **north** + p.n. *Cherletona* 1166. 'North' for distinction from South CHARLTON NU 1620. NbDu 42.

(19) Queen ~ Avon ST 6367. 'C held by the queen'. Originally simply *Cherleton* 1291. An estate of Keynsham abbey given to queen Catherine Parr by Henry VIII. DEPN.

(20) South ~ Northum NU 1620. *Charleton' del Suth'* 1242 BF, *Sutcharleton'* 1296 SR. ME **suth** + p.n. Charlton as in North CHARLTON NU 1622.

(21) West ~ Devon SX 7542. ModE **west** + p.n. *Cheletona* (sic) 1086, *Cherleton(e)* 1242–1311, *Churleton(e)* 1267–1328, *Charleton* from 1396. 'West' for distinction from East Charleton SX 7642 which was called *Eastown* in 1765 while West Charleton is *West Town* 1813 OS. D 319.

CHARLWOOD Surrey TQ 2441. 'The free peasants' wood'. *Cherlewde*, *Chierlewode* 12th cent., *Cherlewod'* 1199, *Cherlewode* 1218–1459, *Charle-* c.1310–1488, *Chorl-* 1372, *Charley-* 1553. OE **ċeorl**, genitive pl. **ċeorla**, + **wudu**. Sr 287 gives pr [tʃɑ:liwud] as in the 1553 spelling, L 229.

CHARLYNCH Somer ST 2337. 'Chard's Lynch, Lynch belonging to Chard'. *Cerdesling* 1086, *Cherdelinch(e)* 1291, [1295] Buck, 1316, *Charsynch* (sic) 1610. P.n. CHARD ST 3208 + **hlinċ** 'a terrace' possibly used as a p.n. The reference is to a terrace on the hill on which the village lies. DEPN.

CHARMINSTER Dorset SY 6892. 'Church on the r. Cerne'. *Cerminstre* 1086, 1200×10, *Chermyn(y)str(e) -minster* 1289–1435, *Charminstr' -mynstr(e)* 1376–1449, *-myster -mister* 1549–1642. R.n. CERNE + OE **mynster**. Do i.338 gives pr ['tʃa:RmistəR].

CHARMOUTH Dorset SY 3693. 'Mouth of the r. Cerne'. *Cernemude* 1086, *-muth* 13th, *Charnemothe* 1394, *Chermouth* 1432. R.n. *Cerne* as in CERNE ABBAS ST 6601 + OE **mūtha**. By normal processes of sound change OE *Cerne* became ME *Chern*, late ME *Charn* and, with loss of *-n-* in the combination *-rnm-*, *Char* whence the modern r.n. CHAR. Do 54.

CHARNDON Bucks SP 6724. Possibly 'hill called Carn'. *Credendone* 1086, *Charendone* 1227–47, *Chardon(e)* 1255, 1316, 1491, *Charndone* 1316. Possibly PrW ***carn** 'a heap of stones' probably used as a p.n., + OE **dūn**. If trustworthy the DB form points to 'Creoda's hill', OE pers.n. *Creoda*, genitive sing. *Creodan*, + **dūn**, but this should have given present day *Crendon as Long CRENDON SP 6908 with which perhaps there is some confusion, not Charndon. Bk 52 gives pr [tʃa:dən], L 145, 150.

CHARNEY BASSETT Oxon SU 3895. 'C held by the Bassett family'. *Charney otherwise Cerney otherwise Cerney Basses and Weeks* 1833. P.n. Charney + unknown family name *Basses* later assimilated to the well known name Bassett and p.n. *Weeks* referring to Charney Wick, *wika de Cerneia* c.1240. Charney, *Ceornei* *[821]12th S 183, *Cernei -eya -ey(e)* 1086–1323,

Cherneye 1284, *Charney(e)* from 1284, is 'Cern island', r.n. *Ce(a)rn* + OE **ēġ**. The *Ce(a)rn* is the stream which rises on the W boundary of Pusey SU 3596 and flows to Race Farm and S and SE to the Ock. The boundary of the Cearningas, 'the people of Charney or of the *Ce(a)rn*', is *(on) cearninga gemære* [958]13th S 654. The base of the stream-n. is Brit ***carn-** 'a rock, a heap of stones' as in the Gloucestershire CERNEYS. Brk 389, RN 73, LHEB 272.

CHARNOCK RICHARD Lancs SD 5515. 'C held by Richard'. *Charnoc Ricardum* 1204, *Schernoc* 1242×3, *Chernock Ricard* 1288, ~ *Richard* 1324, *Ricardeschernok* 1292. P.n. *Chernoc* + pers.n. *Richard* referring to Richard de Chernok who held the manor in 1242. 'Richard' for distinction from Heath Charnock SD 5914, *Hethechernoce* 1270, *Heth Chernok* 1332, ME **heth** + p.n. *Chernoc*, also known as *Chernock Gogard* 1284 from a family of the name. Charnock, *Chernoc* before 1190, is a W name, possibly r.n. **Cern* + suffix ***-ǭg**. La 129, 130, Jnl 17.73.

CHARNWOOD FOREST Leic SK 4814. P.n. Charnwood + ModE **forest**, cf. *the foreste of Charley* c.1545, *Charley Forest* 17th–18th cents. Charnwood, *Cernewoda* 1129, *Charnewod(e)* 1242–1427, *Charnwod(e)* 1266–1470, *-wood* from 1576, is 'the wood at or called *Charn*', PrW ***carn** (Brit **carno-, carnā*) 'a heap of stones' referring to the rugged outcropping of granite often appearing as heaps of broken rock, + OE **wudu**. The 16th–18th cent. forms *Charley* derive from an alternative form in OE **lēah** 'a wood or clearing', *Cernelega* 1086, *Cherlega* c.1130, *Charley(e)* 1240–1610. Lei 103, 351, L 228.

CHARSFIELD Suff TM 2556. 'The open land of the river *Cear*'. *Cer(r)es- Cer- Caresfella, C- Keres- Car(e)sfelda* 1086, *Caresfeld* c.1150, *Charesfeld -feud* 1252–1524, *Charsfelde* 1610. OE r.n. **Ċear*, genitive sing. **Ċeares*, + **feld**. R.n. **ċear* is from the same root as River CARY Somer ST 4530 and the Car in Wales, SO 0112. DEPN, Baron.

Great CHART Kent TQ 9842. *Great Chart* 1271, Latin *Magna Chert* 13th, also known as *Michelchert -chart* and *East-cert* 'east Chart' 1042×1066 S 1047, *Estchert* 1270. ME **grete, michel**, Latin **magna** + p.n. *Cert, Cherd* [762]13th S 25 Charters IV, *Cert* 867×70 S 1200, [839]13th S 1625, 858 S 328, *Ceart* 1006 for 1002 S 914, c.1100, *Certh* 1086, OE **ċert** 'rough ground', K Sr dial. *chart* 'rough common overgrown with gorse and bracken'. The Chartland is one of the sharply differentiated *pays* or zones of settlement in Kent, stretching across the county from W to E between the Downland and the Weald. It was originally heavily wooded and marked the boundary of the original primary settlement of the county. 'Great' for distinction from Little CHART TQ 9446. One of the Chart estates was known as Seleberht's Chart, *Seleberhtes cert* 799 S 155, *Selebertes ceart* [799]13th ibid. PNK 409, KPN 46, Nomina 3.98.

Little CHART Kent TQ 9446. *Litelcert* 1086, c.1100, *Litlechert* 1206×29, Latin *parua Chert* 1233, referred to as *oðer Cert* 'the other C' 1042×66 S 1047. OE **lytel**, Latin **parua**, + p.n. Chart as in Great CHART TQ 9842. PNK 392, KPN 46.

CHART SUTTON Kent TQ 7950. Short for *Chert near Sutton* 1280. Earlier *silva quae dicitur Cært* 'wood called C.' 814 S 173, *Certh* 1086, *Cert* c.1100, *Chert* 1278, 1291, OE **ċert** 'rough ground'.

CHARTER ALLEY Hants SU 5957. *Charter Alley* 1817 OS, *Chatter Alley* 1859. Gover 1958.141.

CHARTERHOUSE Somer ST 4955. *Chartuse* 1243, *Charterhouse* 1610. OFr **chartrouse** 'a house of Carthusian monks' altered to *-house* by popular etymology. DEPN, PNE i.92.

CHARTERVILLE ALLOTMENTS Oxon SP 3110. Named from the Chartist movement of the 1840s. In 1847 the National Land Company founded by the Chartist leader, Feargus O'Connor, bought a farm of 300 acres here and divided it into allotments with cottages as part of an unsuccessful scheme to settle 81 North Country mechanics here. ModE **charter** + **ville**. O 365.

CHARTHAM Kent TQ 1055. 'Estate in the chart, the rough ground'. *Certham* [870×89]15th S 1202, *Certeham* 874 for 844 S 319, *Cert(a)ham* [1042×66]11th S 1047, *Certeham* 1086, *Certaham -e-* c.1100, *Chartham* 1610. OE **ċert** + **hām**. PNK 369, ASE 2.29.

CHARTHAM HATCH Kent TQ 1056. Called *Hatch Green* 1727. The hatch or grating or gate (OE **hæċċ**) is probably referred to in the lost p.n. Bow Hatch, *Bouehecce* 1313×4–1348 etc., *Bow Hatch* 1790, 'above the hatch', OE *bufan hæċċe*. PNK 369, Cullen.

CHARTRIDGE Bucks SP 9303. 'Cærda's ridge'. *Charderuge* 1191×4, 1227, *Chardrugge -rigge* 1200–41, *Chartrugge* c.1200. OE pers.n. **Ċærda* + **hrycg**. Bk 224.

CHARWELTON Northants SP 5356. 'Settlement on the Cherwell'. *Cerweltone* 1086, *Cheruoltona* 1094×1100, *-walton(e)* 1175, 1229, *Charwelton(e)* from 1175, *Charlton* 1605. R.n. CHERWELL SP 4947 + **tūn**. Nth 17 gives pr [tʃa:ltən].

CHASE TERRACE Staffs SK 0409. A modern name for a mining settlement developed here by 1870. ModE **chase** as in CANNOCK CHACE SJ 9816 + **terrace**. Cf. CHASETOWN SK 0508.

CHASETOWN Staffs SK 0508. A mid-19th cent. development to house miners. ModE **chase** as in CANNOCK CHACE SJ 9816 + **town**.

CHASEWATER WMids SK 0307. A modern reservoir. ModE **chase** as in CANNOCK CHASE SJ 9917 + **water**. Cf. CHASE TERRACE SK 0410 and CHASETOWN SK 0508.

CHASTLETON Oxon SP 2429. 'The homestead by the heap (of stones)'. *Ceastelton* [777]12th S 112, *Cestitone* 1086, *Chestelton(')* [c.1100]12th, 1509×10, *Chastleton* from 1286. OE **ċeastel** + **tūn**. The reference was probably to a heap of stones forming the remains of Chastleton Burrow Camp, an Iron Age plateau fort. O 341, PNE i.85, Signposts 153, Studies 1936.136, Thomas 1976.179.

CHAT MOSS GMan SJ 7096. 'Ceatta's bog'. *Catemosse* 1277, *Chatmos* 1322, *Chatmosse* 1577. OE pers.n. *Ċeatta* + OE **mos**. A very extensive bog between Glaze Brook and the Mersey. La 39, L 57.

CHATBURN Lancs SD 7644. 'Ceatta's stream'. *Chatteburn* 1242–1332 etc., *Chatburn* 1341, *Chadburn* 1416. OE pers.n. *Ceatta* + **burna**. La 79, Jnl 17.48.

CHATCULL Staffs SJ 1934. 'Ceatta's kiln'. *Ceteruille* (sic) 1086, *Chatculne* 1199, 1327, *Chatkull* 12th. OE pers.n. *Ċeatta* + **cyln**. DEPN, Horovitz.

River CHATER Leic SK 8503. Unexplained. *Chatere* 1263, 1286, *Chater* c.1545, 1610. Lei 93. RN 74 gives pr ['tseitə].

CHATHAM Kent TQ 7567. 'Estate at or called *Chet*'. *(æt) cetham* [880]12th S 321, *Cætham* c. 975 B 1321, *Ceteham* 1086, *Caetham*, *Cettaham* c.1100, *Chatham* from 1195. PrW ***cę̄d** 'wood' probably used or interpreted as a p.n., + **hām**. The boundary of the people of Chatham is referred to as *Ceðæma mearce* (sic for *Cethæma*) [995]12th S 885, OE **Cēthǣme*, genitive pl. **Cēthǣma*, + **mearc**. KPN 225–6, PNK 127, ASE 2.29, TC 70.

CHATHILL Northum NU 1827. *Chathill* 1866 OS. A small isolated hill rising to 166ft.

CHATSWORTH HOUSE Derby SK 2870. P.n. Chatsworth + ModE **house**. A great classical mansion, 1687–1707, the seat of the Dukes of Devonshire. Chatsworth, *Chetesuorde* 1086, *Catesworth* c.1192, *-wrðe* 1199, *Chatteswrthe -worth(e) -is-* late 12th–1625×49, is 'Ceatt's enclosure', OE pers.n. **Ċeatt*, genitive sing. **Ċeattes*, + **worth**. Db 73.

CHATTENDEN Kent TQ 7572. Partly uncertain. *Chetindunam*, *Chatendune* n.d., *Chatindone* 1281, *Chetyndone* 1287, *Chatyndon* 1535. Uncertain element + OE **dūn** 'hill'. Proximity to Chatham TQ 7567 has led to the suggestion that the specific is *Ċethǣma*, genitive pl. of *Ċethǣme* 'the people of Chatham' and so 'hill of the Chatham people', but 'Ceatta's hill' is perhaps better, OE pers.n. *Ċeatta* + **ing**[4] + **dūn**. PNK 115, DEPN.

CHATTERIS Cambs TL 3986. 'Ceatta's ridge'. *Cæateric* *[974]14th

S 798, *Ceateric* *[1060]13th S 1030, 1235, *Ceatrice* [1178]1334, *Chatric'* 1199–1251, *Chateriz* 1086–1473 including *[974]14th S 798, *Chatriz* 1086–1302, *C(i)etriz* 1086, *Cetricht* 1109, *Chatteriz -yz -ysse* 1258–82. OE pers.n. *Ċeatta* + ***riċ**. The reference is to a straight narrow ridge of land beside the road running NW from the settlement. Ca 247, PNE ii.83, L 184.

CHATTERIS FEN Cambs TL 3979. *Chatteris Fen* 1838 OS. P.n. CHATTERIS TL 3986 + ModE **fen**.

CHATTISHAM Suff TM 0942. 'Ceatt's homestead'. *Cetessam* 1086, *Chettesham* 1190, *Chat(t)esham* 1190–1610, *Chattisham* 1275, 1346. OE pers.n. **Ċeatt*, genitive sing. **Ċeattes*, + **hām**. DEPN, Baron.

CHATTON Northum NU 0528. 'Ceatta's farm or village'. *Chetton* 1177, 1342, *Chatton* from 1242 BF. OE pers.n. *Ċeatta* + **tūn**. NbDu 43 gives pr [ʃatən] with Northumbrian dial. [ʃ] for [tʃ].

Great CHATWELL Staffs SJ 7914. *Much Chatwall* 1509×40, *Gt ~* 1604, 1755. ModE adjs. **much**, **great** + p.n. *Chat(t)ewell* 1203, 1275, *Chatwell* from 1430, *Chat(t)ewalle* 1275–1397, *Chat(t)wall* 1447–1679, 'Ceatta's spring', OE pers.n. *Ċeatta* + **wella**, Mercian **wælla**. There is a spring here called St Chad's Well, a piece of folk-etymology from the p.n. 'Great' for distinction from Little Chatwell SJ 7814, *Lytel Chatwalle* 1472, *Little Chatwell* 1834 OS, now erroneously simply *Chadwell*. Horovitz records a 19th cent. pr *Chattle*. St i.155, L 31.

CHAWLEIGH Devon SS 7112. 'Calf pasture'. *Calvelie* 1086, *Chaflege* 1201, *Cheluelega* 1227, *Chauveleg* 1228, *Chalvelegh(e)* 1285–92. OE **ċealf**, genitive pl. **ċealfa**, + **lēah**. D 363.

CHAWSTON Beds TL 1556. 'Cealf's thorn-tree'. *Chaueles-Calnestorne* (sic for *Calues-*) 1086, *Chaluesthorn* 1180, *-ton* 1227, *Chalston* 1418, *Chawson* 1826. OE pers.n. *Ċealf*, genitive sing. *Ċealfes*, + **thorn**. Bd 65 gives pr [tʃɔ:sən].

CHAWTON Hants SU 7037. Probably 'chalk settlement'. *Celtone* 1086, *Cautona* 1167, *Chaltun* c.1195, *Chaulton'* 1214, *Chauton* 1242–1412, *Chawton* 1548[†]. OE **ċealc** + **tūn**. Formally the specific could alternatively be OE *ċealf* 'a calf' and so 'calf farm', but Chawton lies on the chalk downs. Ha 52.

CHEADLE GMan SJ 8688. Partly uncertain. *Cedde* 1086, *Chedle* 1153–1326 with variants *-legh -lee*, *Chelle* c.1185–c.1292, *Cheddle* early 13th–early 14th, *Chedel(l)* 1281–15th with variants *-il(l) -yll -ul(l) -hill -hull*, *Cheadle* from 1294. Usually taken to be '*Ched* wood', PrW ***cę̄d** (Brit *cēto-* W *coed*) 'a wood' taken as a p.n. + OE **lēah**, but *Ched* usually becomes *Chet* as in CHEETHAM HILL SD 8400. On formal grounds OE **ċēod(e)** 'a bag, a bag-like hollow' would be possible: Cheadle lies by a depression at the confluence of the Mersey and Micker Brook. Che i.246 gives prs ['tʃi:dəl], older local ['tʃedəl].

CHEADLE Staffs SK 0143. Probably '*Ched* wood'. *Celle, Cedla* 1086, *Chedele* 1197, *Chedle* 1227, 1253. P.n. *Ched* < PrW ***cę̄d** 'a wood' + OE **lēah**. Cf. however, CHEADLE SJ 8688 DEPN, Jnl 1.45, L 191, Horovitz.

CHEAM GLond TQ 2463. 'River bend land by stumps or brushwood'. *Cegeham* *[727]13th S 1181, *Cheham* *[933]13th S 420, *Cegham* [946×7]12th S 1504, *[967]13th S 752, *Ceigham* *[1041×66]11th S 1047, *Ceiham* 1086, *Chei- Cheyham* 1199–1428, *Chayham* 1226, *Chey-*, *Chayme* 16th cent., *Cheam* 1680. OE ***ċeġ** + **hamm.** Sr 43, TC 202.

CHEARSLEY Bucks SP 7110. 'Ceored's wood or clearing'. *Cerdes- Cerleslai* 1086, *Cherdesle(e) -ley* [1154×89]1313, 1235–1397, *Chardesleye -le(e)* 1296–1542, *Charesley* 1526, *Chers(e)ley(e)* 1552. OE pers.n. *Ċeored* (< *Ċēolrǣd*), genitive sing. *Ċeor(e)des*, + **lēah.** Bk 103, Jnl 2.23.

CHEBSEY Staffs SJ 8628. 'Cebbi's island'. *Cebbesio* 1086, *Chebbesee* 1222, *-ey* c.1272. OE pers.n. *Ċebbi*, genitive sing. *Ċebbes*, + **ēġ**. Chebsey is situated on the r. Sow. DEPN, L 38, Horovitz.

CHECKENDON Oxon SU 6683. Uncertain. *Cecadene* 1086, *Chakenden(e) -don(') -in- -yn-* [c.1175×80]c.1444–1768, *Chekenden* 1258, *Checkingdon* early 18th. This might be 'Ceacca's valley', OE pers.n. **Ċeacca*, genitive sing. **Ċeaccan*, + **denu**. Alternatively the specific might be OE ***ċeacce** 'a lump' referring to some landscape feature such as a small hill, genitive sing. ***ċeaccan**, perhaps 'the valley of the hillock or with a hillock', or an onomatapoeic bird–n. as Scots *chack* for the Wheat-ear in Orkney. O 44, PNE i.83.

CHECKLEY Ches SJ 7346. 'Ceaddica's wood or clearing'. *Chackileg* 1252, *-ylegh -lee -ley* 1329–1440, *Chadkeleg'* 1275, *Chatkileg* c.1300, *Chat(t)kylegh -ley* 1323–1510, *Checke- Chekkeley(e)* 1321–1584, *Checkley* from 1519. OE pers.n. **Ceaddica* + **lēah**. Che iii.56.

CHECKLEY Staffs SK 0237. Possibly 'Ceacca's wood or clearing'. *Cedla* (sic) 1086, *Checkeleg* 1196, *Chekelee* 1189×99. OE pers.n. **Ċeacca* + **lēah**. The same pers.n. may occur in CHECKENDON Oxon SU 6682 although an unrecorded element **ċeacce* 'a lump, a hill' has also been proposed for these names. DEPN, PNE i.83.

CHEDBURGH Suff TL 7957. 'Cedda's hill or barrow'. *Cedeberia* 1086, *Cheddeberg* 1254, *Chedeberwe* 1275, *Chedber* 1610. OE pers.n. *Ċedda* + **beorg**. DEPN.

CHEDDAR Somer ST 4553. Uncertain. *(æt, in, on) Ceodre* 873×88 S 1507, [904×25]lost ECW, [960 for 941]12th S 511, [956]n.d. S 611, [978 for ?968]12th S 806, *(apud) Ceod(d)rum, Ceoddri* (genitive sing.) c.1000, *Cedre, Ceder* 1086, *Chedre* 1157–[1299]c.1500, *Chedder* 1389, 1610, *Cheddar* 1817. The Anglo-Saxon monastery here is referred to as *Ceoddor mynster* [1067]c.1500. The gorge is *(andlang) ceodder cumbes* [1068]c.1500, the caverns *Chederhole* c.1125–30. This has been explained as OE ***ċēodor**, a derivative of *ċēod(e)* 'a bag' in a topographical sense such as 'bay, hollow, ravine', the reference being either to the deep limestone gorge or to the caves at the foot of the cliffs. The gorge, however, is not bag shaped. Another p.n. apparently with this element is Chitterley, Devon SS 9404, *Chederlia, Chiderlie* 1086, *Chedderlegh* before 1226, *Chudderlegh* [n.d.]15th. Turner suggests a Celtic origin, W **cau, ceu** 'hollow', Co ***kew**, 'hollow, enclosure', + **dor** 'door'. DEPN, D 555, Studies 1931.68, PNE i.89, Turner 1952–4.14, CPNE 57, GPC.

CHEDDAR GORGE Somer ST 4754. P.n. CHEDDAR + ModE **gorge**. Cf. *Cedder rocke* 1610.

CHEDDAR RESERVOIR Somer ST 4453. P.n. CHEDDAR + ModE **reservoir**.

CHEDDINGTON Bucks SP 9116. Probably 'Ceta's hill'. *Cete(n)done, Cetedene* 1086, *Chet- Chedendon* c.1200–62, *Chetyngdon* 1291–1407, *Cehdyngton* 1383–1429. OE pers.n. *Ċeta*, genitive sing. *Ċetan*, + **dūn**. Bk 90, L 150.

CHEDDLETON Staffs SJ 9752. 'The settlement in the deep valley'. *Celtetone* (sic) 1086, *Chetelton' -il- -yl- -ul-* 1203–1509×47, *Chedelton'* 1352–1755 *-le-* 1775. OE **ċetel** + **tūn**. DEPN, PNE i.91.

CHEDDON FITZPAINE Somer ST 2427. 'C held by the Fitzpaine family'. *Cheddon Fitzpaine* 1809 OS. Earlier simply *(of) twam Cedenon* 'from the two Cheddons' 1066 ECW, *Vbcedene, Succedene* (for *Sut-*), *Opecedre, Cedre* 'up' and 'south Cheddon' 1086, *Cheddon* 1610, uncertain element + OE **denu** 'a valley'. The specific might be PrW ***cę̄d** used as a p.n., '*Ced* valley'. The two Ceddons are, as the DB spellings show, Upper Cheddon ST 2328 and South Cheddon, later Cheddon Fitzpaine. DEPN, DB Somerset 22.22.

CHEDGRAVE Norf TM 2699. 'Ceatta's pit or grove'. *Scatagraua* 1086, *Chategrave* c.1165×70, *Chattegraua* 1158, *Chatte- Schategraue* 1268×9, *Chedgrave* 1396. OE pers.n. *Ċeatta* + **græf** or **grāf(a)**. DEPN, RN 77.

CHEDINGTON Dorset ST 4805. 'Estate called after Cedd(a)'. *Chedinton* 1194, *Chedington* 1230. OE pers.n. *Ċedd* or *Ċedda* + **ing**[4] + **tūn**. Do 54.

CHEDISTON Suff TM 3577. 'Cedd's stone'. *Cidestan(es),*

[†]The form *Chalvedone* c.1230 probably does not belong to this name.

Cedestan, Sedestana 1086, *Chedestan* 12th–1346, *Chedeston* 1203, 1275, *Chestone* 1327, *-stan* 1446, *-sten* 1524–1610, *-ston* 1674. OE pers.n. *Ċedd*, genitive sing. *Ċeddes*, + **stān**. DEPN, Baron.

CHEDWORTH Glos SP 0512. 'Cedda's enclosure'. *(æt) Ceddanwryde* [862]14th S 209, *Cedeorde* 1086, *Cheddewrda -wurðe -w(u)rth(a) -worth(e)* 1194–1509×47, *Ched(d)worth(e)* 1276–1712. OE pers.n. *Ċedda*, genitive sing. *Ċeddan*, + **wyrthe, worth**. Gl i.150.

CHEDZOY Somer ST 3337. 'Cedd's island'. *Chedesie* [729]14th S 253, *Cheddes(e)ia* 1175, [1186–1275×92] Buck, *Chedesia* 1194, *Cheddesey* [1363×86] Buck, *Chedsey* 1610, *Chedzoy* 1809 OS. OE pers.n. *Ċedd*, genitive sing. *Ċeddes*, + **ēġ**, WS **īeġ** in the sense 'dry ground in marsh'. Chedzoy is an island of higher ground in the Parrett marshes. DEPN, L 38.

CHEESEMAN'S GREEN Kent TR 0238. *Cheeseman's Green* 1761, 1819 OS, *Chismans Green* 1798. Surname *Cheeseman* + ModE **green**. Cullen.

CHEETHAM HILL GMan SD 8400. P.n. Cheetham + ModE **hill**. Cheetham, *Cheteham* late 12th, *Chetam* 1212, *Chetham* 1226, 1332 etc., *Cheetham* 1254, is '*hām* called or at *Ched*', PrW ***cęd** (Brit *cēto-* W *coed*) 'a wood' taken as a p.n., + OE **hām**. Cf. nearby Cheetwood SD 8300, *Chetewode* 1489, 1522, *Chetewood* 1597, '*Ched* wood', identical with CHETWODE SP 6429. La 33, Jnl 17.32, 18.15.

CHEETWOOD GMan SD 8300 → CHEETHAM HILL SD 8400.

West CHELBOROUGH Dorset ST 5405. *Westchelbergh* 1346. ME adj. **west** + p.n. *Celberge* 1086, *Cheleberg* 1204, 'Ceola's hill', OE pers.n. *Ċēola* + **beorg**. Alternatively the specific might be OE **ċeole** 'a throat, a channel, a gorge' referring to the valley between West and East Chelborough ST 5505, *Estchelberewe* 1343. Do 54.

CHELDON Devon SS 7313. 'Ceadela's hill'. *Chele- Cadeledone* 1086, *Chedeladon -ledune -don* c.1185–1285, *Ched(d)eledon(e)* 1285, 1321, *Chydeladon* 1281. OE pers.n. *Ċeadela* + **dūn**. D 376.

CHELFORD Ches SJ 8174. 'Ceola's ford' or 'throat ford'. *Celeford* 1086, *Chelleford* c.1200, *Chelford* from 1210, *Cholleford* c.1250, *Cholford* 1341. OE pers.n. *Ċēola* or *Ċeolla*, or **ċeole**, + **ford**. The landscape has been changed but *ceole* in a topographical sense 'throat, narrow valley' may have applied to the original site of the ford. Che i.75, L 69.

CHELKER RESERVOIR NYorks SE 0551. Uncertain. *Chelchar* 1651, *Chelker* 1730, + ModE **reservoir**. Possibly OE ***ċelce** 'a chalky place' referring to the Mountain limestone of this area + ME **ker** (ON *kjarr*) 'a marsh'. YW vi.66.

CHELLASTON Derby SK 3730. 'Ceolheard's farm or village'. *Celerdestune -ar-* 1086, *Chelardeston(e) -is-* 1197–1535, *Chellarston* 1230×50, *Chel(l)astun -ton(e)* from 1262. OE pers.n. *Ċēolheard*, genitive sing. *Ċēolheardes*, + **tūn**. Chellaston Hill is referred to in S 922 as *Ceoleardesbeorge* 1009 (OE **beorg**). Db 628.

CHELLINGTON Beds SP 9656. 'Ceolwynn's estate'. *Chelewentone* 1219, *Cheluinton* 1247, *Chelewynton* 1273, *Chelinton* 1242, *Chillington* 1393. OE feminine pers.n. *Ċēolwynn*, genitive sing. *Ċēolwynne*, + **tūn**. Bd 30 gives pr [tʃiliŋtən], DEPN.

CHELMARSH Shrops SO 7288. 'Pole Marsh'. *Cheilmers* [674×704]13th S 1799, *Celmeres* 1086, *Chelmersh* 1331–1651 with variants *Chel(l)e-*, *-merssh* and *-marsh(e)* (from 1577), *-mish, -mash, -mesh* 18th cent., *Cheilmere, Cheylmers(e) -mers(s)h* 1228–1494. OE ***ċeġel** + **mersc**. The marsh no longer exists but must have been marked out with posts. The site is probably that now occupied by Chelmarsh Reservoir SO 7387. Sa i.74, Studies 1936.164.

CHELMARSH RESERVOIR Shrops SO 7387. P.n. CHELMARSH SO 7288 + ModE **reservoir**.

River CHELMER Essex TL 6520, 7709. A back-formation from the p.n. CHELMSFORD TL 7006. *Water of Shelmereford* 1239, *Chelmer* 1576, *Chelme* 1577, 1586. Ess 5.

CHELMONDISTON Suff TM 2037. 'Ceolmund's estate'. *Chelmundeston(')* 1174–1275, *-mond-* 1219, *Chelmyng(s)ton* 1316, 1327, *Chelmeton(')* 1286, 1524, *Chempton* 1568, 1610. OE pers.n. *Ċēolmund*, genitive sing. *Ċēolmundes*, + **tūn**. DEPN, Baron.

CHELMORTON Derby SK 1170. 'Ceolmær's hill'. *Chelmerdon(e) -dona* 1101×8–1587, *Chelmar(e)don* 1196–1346, *Chelmerton* 1415, 1564–18th cent. OE pers.n. *Ċēolmǣr* + **dūn**. Db 74, L 143, 155.

CHELMSFORD Essex TL 7006. 'Ceolmær's ford'. *Celmeresfort* 1086, *Chelmaresford* 1200–63, *Chelmer(e)sford* 1225–1594, *Chelm(e)sford* from 1272, *Chen(ne)sford* 1470, 1638, *Chemsford* 1514, 1793. OE pers.n. *Ċēolmǣr*, genitive sing. *Ċēolmǣres*, + **ford**. Ess 245 gives prs [tʃemsfəd, tʃensfəd, tʃenʃfəd] and [tʃendʒfəd].

CHELSEA GLond TQ 2778. 'Chalk landing-place'. *Caelichyth* 764 for 767 S 106, *Celchyth* [789]12th S 131, *celichyð* [816]11th B 358, *Cealchyþe* c.900 ASC(A) under year 785, *-hithe* 1071×5, *Chelc(he)hethe* 1231–1428, *-huthe(e)* 1300–82, *Chels(e)hithe* 1499, *Chelsyth* 1556, *Chelsey(e)* 1523. OE ***ċ(i)elċe** + **hȳth**. The name is actually puzzling: two early forms point to OE *cælic* 'a cup, a chalice' which offers no satisfactory sense. This was either a place where chalk or limestone was landed or a chalky landing place. Mx 85, PNE i.278, TC 202, GL 38.

CHELSFIELD GLond TQ 4864. 'Ceol's open country'. *Cillesfeld* 1086 (DB), *Chilesfeld* 1086, *Chelesfeld* 1190–1242. OE pers.n. *Cēol*. genitive sing. *Cēoles*, + **feld**. DEPN, LPN 46.

CHELSWORTH Suff TL 9848. 'Ceorl's or the peasant's enclosure'. *Ceorleswyrðe* 962 S 703, *-weorð* 1000×2 S 1386, *Cæorlesweorþ* 962×91 S 1494, *Cerleswordã* 1086, *Chelsworth* 1610. OE pers.n. *Ċeorl* or **ċeorl**, genitive sing. *Ċeorles*, **ċeorles**, + **wyrth, worth**. DEPN.

CHELTENHAM Glos SO 9522. 'Well-watered valley of (the hill called) *Cilta* or *Celta*'. *Celtan hom, (æt) Celtanhomme* [803]11th S 1431, *(of) Ciltan ham* 11th, *Chinteneham* 1086, *Chilt(e)ham* 1155×9–1316, *-en-* c.1240–1675, *Cheltenham* from c.1250, *Cheltnam* 1562–1631. Pre-English ***ciltā** 'steep hill' probably taken as a p.n., secondary OE genitive sing. **ċiltan, ċeltan**, + **hamm**. The reference is to Cleeve Hill which dominates the site. Gl ii.101 gives pr ['tʃeltnəm], PNE i.88, Jnl 6.10–14.

CHELVESTON Northants SP 9969. 'Ceolf or Ceolwulf's village'. *Celuestone* 1086, *Chelveston(e)* from 1206, *Cheston(e)* 1461, 1675, *Chelston* 1603×25–1823. OE pers.n. *Ċēolf* or *Ċēolwulf*, genitive sing. *Ċeol(wul)fes*, + **tūn**. Nth gives pr [tʃelsən].

CHELVEY Avon ST 4668. 'Calf farm'. *Calviche* 1086, *Caluica* 1086 Exon, *Chalvy* 1285, *Chelvey* 1326 FF. OE **ċealf** + **wīċ**. DEPN.

CHELWOOD Avon ST 6361. 'Ceola or Ceolla's enclosure'. *Ceolanwirthe* [n.d.]13th William of Malmesbury, *Celeworde, Cellewert* 1086, *Chel(l)eworth* 1225, 1243. OE pers.n. *Ċēola* or *Ċeolla* + **wyrth**, later replaced by **worth**. DEPN.

CHELWOOD GATE ESusx TQ 4130. *Chellworth Gate* 1564, *Chellwood Gate* 1579, *Charlwood Gate* 1813. P.n. Chelworth, *Chelworthe* 1546, + ModE **gate**. Chelworth is probably 'Ceola's enclosure', OE pers.n. *Ċeola* + **worth**. Sx 335.

CHENIES Bucks TQ 0198. 'Cheyney's place or estate'. *Cheynes* 1536, *Cheyney* 1675. Family name *Cheyne*, genitive sing. *Cheynes*. Short for *Ysenamsted Cheyne* 13th, *Ismansted Chenies* 1552, '*Isenhamstede* held by the Cheyne family'. Earlier simply *Hisenstude* 1161, *Isenhamstede* 1196–1352, 'Isa's homestead', OE pers.n. **Īsa*, genitive sing. **Īsan*, + **hāmstede**. See also LATIMER TQ 0099, another manorial name for a division of *Isenhamstede*. The Cheyne family were associated with this division from at least 1232. Bk 221 gives prs [tʃeini] and [tʃi:ni], Stead 263, Jnl 2.32.

CHERHILL Wilts SU 0370. Unexplained. *Ciriel* 1155–1275, *C(h)eriel* 1156–1324, *Chiriel(l) -yel(l)* 1221–1446, *Cheryell, Chirrell* 16th cent., *Cher(r)ell* 1637, 1668. The explanation in Wlt 261 is impossible since the supposed element *ial* 'fertile upland' is a ghost word, cf. CPNE 138.

CHERINGTON Glos ST 9098. 'Church village'. *Cerintone* 1086,

Chederintone 1166, *Cherinton(a) -yn-* [12th]1267–1308, *Chiri(n)- Chyry(n)ton* 1195–1455, *-ing- -yng-* 1234–1377×99, *Cheryngton -ing-* c.1300–1583. OE **ċyriċe** + **tūn**. The 1166 form may, however, point to OE ***ċeador** + **ing**[2] + **tūn** 'settlement at **ċeadoring*, the bag-like depression' referring to the deep valley here and perhaps already used as a p.n. Gl i.89, Studies 1931.89.

CHERINGTON Warw SP 2936. 'The church village'. *Chiriton(e) -y-* 1199–1349, *Cherinton* 1200, *Scheryngton* 1512, *Cheryngton* 1535. OE **ċyriċe** + **tūn**. Wa 279, Studies 1931.38.

CHERITON 'Church settlement'. OE **ċyriċe** + **tūn**. Studies 1931.33ff.

(1) ~ Devon SS 7346. *Ciretone* 1086, *Chirinton* 1198, *Ceriton* 1205. This is the site of a church dedicated to St Brendan subsequently moved to the village of Brendon. D 59.

(2) ~ BISHOP Devon SX 7793. 'C held by the bishop' sc. of Exeter. *Cheritone Episcopi* 1316, *Bisshopps Cheriton al. South Cheriton* 1675. Earlier simply *Ceritone* 1086, *Cheriton(e)* 1270–1316, *Churiton(e)* 1242–91. Also known as *Suthchiriton* 1314 in relation to CHERITON FITZPAINE SS 8606. One acre here was granted to the bishop of Exeter by Eleanor de Melewis before 1280. D 427.

(3) ~ FITZPAINE Devon SS 8606. 'C held by the Fitzpaine family'. *Chiriton Fitz Payn* 1334, *Cheryton Phezpayn* 1510, ~ *Fyshepyne* 1555. Earlier simply *Cerintone* 1086, *Churiton(e)* 1242–1328, *Cheriton* 1428. Also known as *Churitone Santone* 1274. Thomas de Santon' held three parts of a fee here in 1242, and Roger son of Pagan held the manor in 1256. D 414.

(4) North ~ Somer ST 6925. *Northchiriton* 1243. ME adj. **north** + p.n. *Eiretone* (for *Cire-*), *Cherintone, Ciretvne* 1086, *Ciretona -tuna, Cherintona* 1086 Exon, *Cheri- Chirintone* 1198, *Che- Chiriton(e)* 1225, *Chyriton'* 1243, *Cheriton* 1610. 'North' for distinction from South CHERITON ST 6925. DEPN, Studies 1931.38.

(5) South ~ Somer ST 6924. *Suthchuryton* 1329. ME adj. **suth** + p.n. CHERITON. DEPN, Studies 1931.38.

CHERITON Devon ST 1001. 'Settlement of the free peasants'. *Cherletone* 1086, *-ton* 1242, 1330, *Brodechurleton(e)* 14th. OE **ċeorl**, genitive pl. **ċeorla**, + **tūn**. D 566.

CHERITON Hants SU 5828. Probably 'church settlement'. *Chiriton* 1162–1408, *Cheriton(a)* from 1208. OE **ċiriċe** + **tūn**. An alternative specific might be PrW ***crüg** 'a hill, a barrow, a mound' referring to the conspicuous long barrow ¾ m E of the village. Ha 52, Gover 1958.68.

New CHERITON Hants SU 5827. A modern develpoment. ModE **new** + p.n. CHERITON SU 5828.

CHERRINGTON Shrops SJ 6620. 'River-bend settlement'. *Cerli(n)tone* 1086, *Cherinton(a) -yn-* 1155–1428, *Cheryngton -ing-* 1376–1765, *Cherrington* 1786, *Charington* 1535. OE ***ċerring** referring to a bend in the river Meese + **tūn**. Alternatively this could be 'the estate called after Ceorra', OE pers.n. *Ċeorra* + **ing**[4], or genitive sing. *Ċeorran*, + **tūn**. Sa i.76.

CHERRY COBB SANDS Humbs TA 2221. So named 1786. A cherry-cob is a cherry-stone. YE 38.

CHERTSEY Surrey TQ 0466. 'Cerot's island'. *Cerotaesei id est Ceroti insula* [c.731]8th BHE, *Cirotesige -ege -egt* [672×4]13th S 1165, *Ceroteseg id est Cerotis insula* [871×99]13th S 353, *Ciroteseg* [1062]13th S 1035, *Certeseg* [827]13th S 285, *Ceortes ege* [871×89, 967]13th S 752, 1508, *Ceortesige* 10th ASC(A) under year 964, 11th[c.1200], *Ceoreteseye* [1006×12]13th S 940, *Ceortesege -æge* 1121 ASC(E) under years 1084 and 1110, *Ceortes(e)ig, Cyrtesig* c.1000, *Certeseige* [1053×66]13th S 1094, *Certesi* 1086, *Certeseye* c.1130–1466 including [933]13th S 420, *Chierteseye* 1297, *Cherte- Charteseye* 1311–87, *Chertsey* from 1559. OE pers.n. *Ċeor(o)t* < Brit *Cerotus*, genitive sing. *Ċeor(o)tes*, + OE **ēġ**. Sr 105, Jnl 8.19, L 37.

River CHERWELL Northants SP 4947. Unknown element + OE **wella, wylle** 'a spring, a stream'. *Ceruelle* [681]12th S 1167, *-welle* [904]11th S 361, *-willan* [1004]1313, *Cearwellan* [864]11th S 210, [929]11th S 402, *-wyllan* [904]11th S 361, [c.968]13th S 1569, *Cear wyllun* 944 S 495, *-wylle* [1005]c.1200, *Charewell(e)* 1185–1401, *Charwell(')* 1220–1577, *Cherwell* from c.1540. Attempts to establish an OE **ċear*, **ċær*, 'a hollow, a valley' cognate with OHG *kar* 'a container, a bowl' as in the G p.n. Karbach, *Carabach* [9th]12th, and the numerous Karbachs in Austria, are quite uncertain. A better suggestion is OE *ċ(i)err* or **ċearr(e)* 'a turn, a bend', but the specific of all these names may well in fact be a Celtic water or river name. Nth 2, RN 75, LbO 200, WGÖ 93.

CHESELBOURNE Dorset SY 7699. 'Gravelly stream'. *(be) Chisleburne, (juxta) Cheselburneam* [859 probably for 870]15th S 334, S 342, *(ad, to) Cheselburne* [942]15th S 485(1), [1019]15th S 955(1), 1212–1428. OE **ċisel, ċeosol** + **burna**. Do iii.202.

CHESFORD BRIDGE Warw SP 3069 → SEIGHFORD Staffs SJ 8825.

CHESHAM BOIS Bucks SU 9699. 'C held by the Bois family'. *Chesham Boys* 1339. P.n. CHESHAM SP 9501 + family name *De Bosco, De Bois* from the family which held this part of Chesham from 1213. The earlier form *Boys in Cesteresham* 1276 appears to be intended as 'wood in C' (OFr **bois**). Bk 226 gives pr [tʃesəm boiz].

CHESHAM Bucks SP 9501. 'Hemmed-in valley bottom of the stone heap'. *(æt) Cæstæleshammæ* [966×75]12th S 1484, *Cestreham* 1086, 1201, *Cestresham* 1154×89–1209×19. OE **ċeastel**, genitive sing. **ċeasteles**, + **hamm** 6. The reference is to the group of puddingstones now forming the foundation of the parish church tower. *Ċeastel* was early misunderstood as if ME *castel* and so partly replaced by *ċeaster*. Bk 223 gives pr [tʃesəm], L 44, 47, 49, Jnl 2.32.

CHESHIRE 'The province of the city of Cheshire'. *Legeceaster scir* 11th ASC(C) under year 980, *Ceasterscire* 12th ASC(E) under year 1085, *Cestre Scire* 1086, *Cheschire* 1430. P.n. CHESTER SJ 4066 + OE **scīr** 'shire, county'. Ches i.1.

CHESHUNT Herts TL 3502. Probably 'Roman town spring'. *Cestrehunt(e)* 1086–1304, *-honte* 1086, *Chest(e)hunte* c.1150–1428, *Cheston* 1478–1675, *Chesson* 1700, *Cheshunt* 1775. OE **ċeaster** + **funta**. The place lies on Ermine Street but no Roman town is known here. Hrt 222 gives pr [tʃesənt], Signposts 84, 96.

CHESIL BEACH Dorset SY 6180. 'Shingle beach'. *the Chisil, bank of Chisil, Chisille bank* 1535×43, *The Beach of Pebbles* 1710. ME **chisel** (OE *ċisel*) + ModE **beach**. A ridge of pebbles extending for 16 miles between Burton Bradstock and Portland. Do i.218.

CHESLYN HAY Staffs SJ 9807. 'Cheslyn enclosure'. *Haya de Chi- Chyst(e)ling* 1252–90, ~ *Chest(e)lyn -lin* 1256–1570, *Cheslinhay* 1695–1834. P.n. *Chistlin* 1236, 'coffin ledge or terrace', OE **ċest, ċist**, + **hlinċ**, + ME **hay** 'part of a forest fenced in for hunting'. The reference may have been to a place where a coffin was found. St i.67, Horovitz.

River CHESS Herts TQ 0695. A modern back-formation from CHESHAM SP 9601.

CHESSINGTON GLond TQ 1863. 'Cissa's hill'. *Cisendone, Cisedune* 1086, *Chissendon(am)* 1129×35–1291, *Chessendone* 1255, *-ing-* 1279, *Chesyngton* 1563. OE pers.n. *Ċissa*, genitive sing. *Ċissan*, + **dūn**. Sr 72.

CHESTER Ches SJ 4066. 'The Roman city'. *Cestria* [c.1043]15th, early 12th–1661, *Cestre* 1086–1454, *Ceastre, Cæstre* 12th ASC(E), *Chester* from 1152. OE **ċeaster**. The original name was *Dēvā* 'place on the river Dee' (*Δηοῦα* [c.150]13th Ptol, *Deva* [4th]8th A1, [7th]13th Rav). It was a Roman fortress, headquarters of the 20th Valeria Victrix Legion whence the W name *Caerlleon* 'fortress-city of the legions', PrW ***cair** + ***legion**, *Carlegion* c.731 BHE, c.1118, *Cair Legion* c.800 HB, *Caerleon* 14th, Latinised as *civitas Legionum* c.731 BHE, c.1118, *urbs Legionis* c.880 HB, and anglicised as *Legacæstir, Legaecaester* c.731 BHE, *Legeceaster* 894 ASC(A), *Legeceaster* [c.890]c.1000 OE Bede, 1042×66, c.1118, *Legceaster -re* c.1000 ASC(B), 12th ASC(E) and *þe citee of Legiouns* 1387 Trevisa. The reduction of the OE name *Legacæstir* to the simplex *ċeaster* was due to the political and commercial importance of the place in the Middle Ages. Che v.1.i.2–7.

CHESTER-LE-STREET Durham NZ 2751. 'Roman fort on the Roman road'. *Cæstre, Ceastre* c.1104, *Cestre, Cestria* 12th–18th cents, *Cestria in strata* 1406, *Chestre (in) le Strete* 1411, 1419, *Chester le Street* from 1607. OE **ċeaster** + Fr definitie article **le** (short for *en le*) + **strēt**. The RBrit name, *Concangis* (*Co-Ceganges* [c.700]13th Rav, *Conca(n)gios* [c.420]15th ND) of unknown meaning but possibly containing the same element as CONSETT NZ 1051, seems to be remembered in the earliest English forms *Kuncacester* [c.700]c.900 AC, *Cunceceastre* [c.1040]12th HSC, *Cuncacester* 1104×7 SD. NbDu 43, RBrit 314, Jnl 16.16.

CHESTERBLADE Somer ST 6641. Partly obscure. *Chesterbled* [802×39]17th ECW, 1395, *Cesterbled* [1065]18th S 1042, *Chestreblad(e)* 1259, 1327, *Chesterblade* 1610. OE **ċeaster** referring to the entrenchments on Small Down $\frac{1}{2}$ mile S of the village + either **blæd** 'blade, leaf' in some unknown application or **bledu** 'bowl, cup' referring to the hollow valley below the earthworks. DEPN, PNE i.18, Charters III.81.

CHESTERFIELD Derby SK 3871. 'Open land by a Roman town'. *(ad) Cesterfelda* [955]14th S 569, *Cestrefeld* 1086, *Chestrefeld(e) -feud -feild -field -ur- -er-* from [1166]c.1250. OE **ċeaster** + **feld**. There was a minor Roman settlement at Chesterfield on the line of Ryknild Street. Db 231, L 240, 243.

CHESTERFIELD Staffs SK 1005. 'Open land by a Roman town'. *Cestrefeld Alani* 'Alan's C' 1167, *Cestrefeud* 1218, *Chesterfeld* 1332. Roman remains have been found here but the reference may rather be to the Roman town of *Letocetum* at Wall $\frac{1}{2}$ m to the N. DEPN, Duignan 39, L 240, 243, Signposts 152, Horovitz.

CHESTERFIELD CANAL Notts SK 7284. P.n. CHESTERFIELD Derby SK 3871 + ModE **canal**.

Great CHESTERFORD Essex TL 5042. *Magna Cesterford* 1238, *Moche Chesterford* 1467×72, *Great Chestover* 1570×2. ModE adj. **great**, Latin **magna**, ME **much**, + p.n. *Cestrefordā -fort* 1086, *-ford* 1226–1329, *Chesterford* from 1303, 'the ford by the Roman town', OE **ċeaster** + **ford**. The Icknield Way crosses the Cam or Granta here. A Roman town lies in the fields NW of the village. The earliest reference is to the bounds of Æthelweard's Chesterford, *Æþelwardes cesterforda gemæra, Casterforda gemerae* [1004]12th S 907. 'Great' for distinction from Little CHESTERFORD TL 5141. Ess 519.

Little CHESTERFORD Essex TL 5141. *Parva Cesterford* 1231, *Little Chestover* 1570×2. ModE adj. **little**, Latin **parva**, + p.n. Chesterford as in Great CHESTERFORD TL 5042. Ess 519.

CHESTERTON 'Settlement beside a Roman town'. OE **ċeaster-tūn**.

(1) ~ Cambs TL 1295. *Cestretuna* 1086, 1184, *Cestreton -er-* 1192–1326, *Chesterton* from 1345. The reference is to the Roman town of *Durobrivae*, Water Newton. The boundary of the people of Chesterton is referred to as *Ceastertuninga gemærie* [955]12th S 566. Hu 181, RBrit 348.

(2) ~ Cambs TL 4560. *Cestretone* 1086, *-ton(a) -(e)* 1086–1394, *Chesterton(e)* from 1206. The reference is to the Roman town of *Duroliponte*, Cambridge. The earliest forms for CAMBRIDGE are *Grantacæstir* etc. 'the Roman town on the river Granta'. Ca 147.

(3) ~ Oxon SP 5521. *Cestertune* [1005]12th S 911, *Cestretone* 1086–1309 with variants *-ton(')* and *-tun'*, *Chesterton* 1316. Chesterton is $\frac{3}{4}$ of a mile NW of the Roman town of Alchester in Wendlebury SP 5619. O xvii, 206, Signposts 152.

(4) ~ Staffs SJ 8349. *Cestretone* 1201, *Cestreton* 1214, *Chesterton* 1276. The reference is to the Roman fort site on the projected line of the road from Derby to Stoke on Trent, Margary no.181, p.310. DEPN, Horovitz.

(5) ~ Warw SP 3558. *Cestretune* [1043]15th S 1000, *-ton(e)* 1086–1326 with variant *Cester-*, *Chesterton* from 1350. There is a Roman settlement on the Fosse Way 1$\frac{1}{4}$ miles NW of the village at SP 3459. Wa 251.

CHESTFIELD Kent TR 1366. A manorial name from the family of William de Cestevile or Cestuill' 1242×3. *Chestevile* 1491, *Chestefeld* 1486. The family name may represent OFr *chastel, chestel vieil* 'old castle' although subsequently associated with *ville* 'a town'. PNK 493.

CHESWARDINE Shrops SJ 7229. 'Cheese-producing settlement'. *Ciseworde* 1086, *Chesewordin* 1159×60–1428 with variants *-yn-* and *-wurth-*, *-worth-*, *-werd-*, *-ward-*, *Cheswardine* from 1662, *Cheshardine* 1586, *Chezadine* 1671. OE **ċēse** + **worth** varying with **worthiġn**. Sa i.78 gives local pr *Chezadine*.

CHESWICK Northum NU 0346. 'The cheese farm'. *Chesewic* 1208×10 BF, 1228 *-wyk'* 1296 SR, *Cheswike* 1539, *Chesswick* 1639. OE **ċēse-wīc**. For the local rhyme about villages formerly part of the estate of the monks of Holy Island, see BUCKTON NU 0838. The village of Cheswick was destroyed by the Scots in 1400. NbDu gives pr [tʃizik].

CHESWICK BLACK ROCKS Northum NU 0347. *Cheswick Black Rocks* 1865 OS. P.n. CHESWICK NU 0346 + ModE **black** + **rocks**.

CHESWICK GREEN WMids SP 1376. *Cheswick Green* 1831 OS. P.n. *Chesewych* 13th, *-wik -wyk* 1267, 1301, 'the cheese farm', OE **ċēse** + **wīċ**, + ModE **green** as in nearby Shelly Green SP 1476, *Shelley greve* (sic) 1580, *Shelley* 1311, 'the wood or clearing on the shelf of land', OE **scelf** + **lēah**, and Waring's Green SP 1274, *Warings Green* 1831 OS. Wa 293.

River CHET Norf TM 3799. A back-formation from CHEDGRAVE TM 3699 before the change *t* > *d*. For the original name see LODDON TM 3698. RN 77.

CHETNEY MARSHES Kent TQ 8871. *Chattenemersch* 1371, 1380, *Chitney Marsh* 1819 OS. P.n. Chetney TQ 8969 + ME **mersh**. Chetney, *Chattenea* 1192, *Cheteneye* 1235, is 'Ceatta's island', OE pers.n. *Ċeatta*, genitive sing. *Ċeattan*, + **ēġ** in the sense 'dry ground in marsh'. K 251.

CHETNOLE Dorset ST 6007. 'Ceatta's hill'. *Chetenoll* 1242, *Chattecnolle* 1268, *Chetecnolle* 1288, *Chetnoll* 1535. OE pers.n. *Ċeatta* + **cnoll**. The reference is to a hill SE of the village. Do 55, L 137.

CHETTISCOMBE Devon SS 9614. 'Coomb of the kettle-shaped hollow'. *Chetelescome* 1086, *Chetescumbe* 1281, *Chettiscome* 1285. OE **ċietel**, genitive sing. **ċieteles**, + **cumb**. The coomb running N from here opens into a kettle-like hollow after $\frac{1}{3}$ mile. D 541, L 93.

CHETTISHAM Cambs TL 5483. Probably 'homestead of (the wood called) Chet'. *Chetesham* c.1170–1475, *Chedesham* 1221–77, *Checham* 1606, *Churcham* 1638–1808. OE p.n. ***Ċet** <PrW ***cę̄d**, genitive sing. ***Ċetes*, + **hām**. There are several references to woodland at this place (13th, 1277, 1606). Ca 217.

CHETTLE Dorset ST 9513. 'The kettle'. *Ceotel* 1086, [1107]1300, *Chetel* [1100×35, 1154×89]1496, [c.1183]14th, *Chettel* 1233, *Chettle* 1575. OE ***ċeotel**, a variant of *ċetel, cytel* ' kettle, a kettle-shaped valley'. The village lies in a deep valley surrounded by hills. Do ii.290, L 89.

CHETTON Shrops SO 6690. 'The estate called after Ceatta'. *Catinton* 1086, *Chatinton* 1167, *Chetinton(e) -yn-* 1210×12–1334, *Chetyton' -i-* 1255–95, *Chetyngton -ing-* 1270–1421, *Cheton(')* 1271×2, 1316, *Chetton* from 1397. OE pers.n. *Ċeatta* + **ing**[4], or genitive sing. *Ċeattan*, + **tūn**. Sa i.78.

CHETWODE Bucks SP 6429. 'Wood called Chet'. *Cetwuda* (dative case) [949]16th S 544, *Ceteode* 1086, *Chetwude -wode* 1223–31, *Chitewode* 1465, *Chickwood* 1615. PrW ***cę̄d** 'a wood' taken as a p.n. + pleonastic OE **wudu**. Bk 62 gives pr [tʃitwud], Jnl 2.33 local pr [tʃitu:d], L 191, 228.

CHEVELEY Cambs TL 6760. 'Chaff wood'. *(æt) Cæafle* 1000×2 S 1486, *Ceaflea* [1022]12th S 958, *Cheaflea* [1022]18th ibid., *Ceauelai, Chauelai* 1086, *Cheueleie -lea -le(e) -ley(e)* 1086–1539. OE **ċeaf** in the sense 'rubbish, fallen twigs' + **lēah**. This was a royal wood, presumably where wood-working left off-cuts and shavings behind. Ca 125 gives pr [tʃi:vli].

CHEVENING Kent TQ 4857. 'The ridge place' or 'the people at *Cefn*, the ridge'. *Ciuilinga* c.1100, *Chiueling'* 1226, *Chivening(es)* 1199–1250, *Chi- Chyvening(e)* 1240–71, *Chevening(e) -yng'* from 1254. PrW ***cevn** (Brit **cemno-*) 'a ridge' + OE **ing**[2] and **ingas**. Chevening lies beneath a ridge rising to over 700ft. which may have been known as *Cefn* to the Anglo-Saxon settlers as if a p.n. and not an appellative. PNK 52, ING 10, Jnl 1.45.

Great CHEVERELL Wilts ST 9854. *Magna Cheverel* 1227, ~ *Chiverel* 1178, *Grete Cheverell* 1522, ModE adj. **great**, Latin **magna**, + p.n. *Chevrel* 1086, *Cheverel* 1179–1242, *Chiverel* 1178–1287, identical with Keveral Corn SX 2954, *Keverel* 1299, probably 'the joint-village place, the place where land is ploughed jointly', cf. Co ***kevar** + ***-yel**. The practice of co-tillage is mentioned in the Welsh Laws. The form *Capreolum* 1103 shows that in the middle ages the name was associated with Fr *chevreuil* 'a goat, a roe-buck'. 'Great' for distinction from adjoining Little CHEVERELL ST 9853. See also CHICKERELL Dorset SY 6480. Wlt 238.

Little CHEVERELL Wilts ST 9853. *Parva Cheverel* 1236, ~ *Chiverel* 1242, *Lytill Cheverell* 1522. ModE adj. **little**, Latin **parva**, + p.n. Cheverell as in Great CHEVERELL ST 9854. Wlt 238.

CHEVINGTON Suff TL 7859. 'Ceofa's estate'. *Ceuentunā* 1086, *Cheventon* 1201, 1254, *Cheuington* 1610. OE pers.n. *Ċeofa*, genitive sing. *Ċeofan*, + **tūn**. DEPN.

CHEVINGTON DRIFT Northum NZ 2699. P.n. Chevington as in East Chevington NZ 2699, *Chivi(n)gton del Est* 1242 BF, + ModE **drift** 'a track, a grass lane'. Cf. West CHEVINGTON NZ 2297.

West CHEVINGTON Northum NZ 2297. *(West) Chi- Chyvington -yng-* 1236 BF–1430, *Chivigton' del West* 1242 BF. ME adj. **west** + p.n. *Chiuingtona* 1212 BF, *Chevyngton* 1335, 1428, *Chiventon* 1724, 'the estate called after Cifa', OE pers.n. **Ċifa* + **ing**[4] + **tūn**. Forms with *Chev-* may indicate the variant form of the pers.n. *Ċeofa* with back mutation of OE *i* > *eo*. NbDu 44 gives pr [tʃivəntən].

The CHEVIOT Northum NT 9020. Unexplained. *Chiuiet* 1181, *Chiuiet(t) -iot(t) -yot* [12th, 13th]14th, *Chyviothe -iat, Chiveot -eat -iat -yat* 16th cent. BB 116, *Cheviot* 1842 BB 119, *Chevy (Chace)* [15th]16th. Meaning and etymology unknown, but possibly a compound – or understood to be a compound – in OE *ġeat* 'a gate, a gap in the hills'. Pronounced [tʃiviət]. NbDu 44.

The CHEVIOT HILLS Northum NT 8212. *Cheviot Hills* 1595 Watts 1975.39, 1695 Map, *The Cheviots* 1842 BB 119. P.n. CHEVIOT NT 9020 + ModE **hills**.

CHEVITHORNE Devon SS 9715. 'Cifa or Ceofa's thorn-tree'. *Chiveorne, Chenetorne* (sic) 1086, *Chiveorna, Chevetorna* 1086 Exon, *Chevethorn* 1198–1285, *Chi- Chyvethorn(e)* 1238–1316, *Cheathorne* 1630. OE pers.n. *Ċifa* varying with its back-mutated form *Ċeofa* + **thorn**. D 542.

CHEW 'The gap, fissure, cleft'. OE ***ċēo** with accent shift to ***ċeó**, recorded only as a plural *cian* in OE glossing Latin *branciae* 'gills of a fish'. The word is related to OE *ċinu* 'a fissure, a ravine' used in p.ns. and clearly possessed the topographical sense 'gap, fissure, cleft'. La 33, 45, 75, Jnl 17.45.

(1) ~ Derby SJ 9992. *Chew Wood* 1842, referring to a deep narrow valley. Db 80.

(2) ~ Lancs SD 7136. *(le) Cho* 13th–1355, *Choo* 1325, referring to a small valley widening out to join the r. Calder. La 71.

(3) ~ MOOR GMan SD 6607. *Chow More* 16th. Situated at the head of several small valleys. La 45.

(4) ~ RESERVOIR GMan SE 0301. P.n. Chew as in Chew Brook SE 0202, *Chew Brook* 1843 OS, + ModE **reservoir**. Chew Brook flows through a deep narrow trough which widens out into Greenfield Brook.

CHEW MAGNA Avon ST 5763. 'Great Chew', also known as 'Bishops Chew'. *Chew Magna* 1817. P.n. Chew + Latin **magna**. Chew, *Ciw* *[1065]c.1500 S 1042, *Chyw* [1061×6] 13th S 1113, *Chiu* 1084, 1086 Exon, *Chivve* 1086, *Chiw* 1225, is by origin the r.n. Chew ST 5762, 6565, found also as the specific of Chewton Mendip ST 5953, *Ciwtune* [873×88]11th S 1507, *Ciwetvne* 1086, *Chiwton* 1281, 'settlement, estate on the river *Chew*'. It represents W *cyw* 'young of an animal'. Several examples of W r.ns. identical with animal names are known, e.g. *pant ciu, foss ciu* in the *Liber Landavensis*, Aber Gwybedyn, Dyfed (*gwybedyn* 'a gnat'), Nant Ceiliog, Powys SN 8114 (*ceiliog* 'a cock'). 'Magna' and 'Bishops' for distinction from Chew STOKE ST 5561. RN 77, GPC 451, 828, 1744.

CHEW STOKE Avon ST 5561 → Chew STOKE.

CHEW VALLEY LAKE Avon ST 5659. A modern reservoir named from the river Chew as in CHEW MAGNA ST 5763.

CHEWTON MENDIP Somer ST 5953. 'Chewton by the Mendip hills'. *Cheuton by Menedep* 1313, *Chewton Mendip* 1610. P.n. Chewton + Mendip as in MENDIP HILLS ST 5255. Chewton, *Ciwtun* [873×88]11th S 1507, *Ciwetvne* 1086, *Chuyton* 1243, *Chiwton* 1281, is the 'settlement or estate on the r. Chew', r.n. Chew as in CHEW MAGNA Avon ST 5763 + OE **tūn**. DEPN, RN 77.

CHICHELEY Bucks SP 9045. 'Cica's wood or clearing'. *Cicelai* 1086, *Chi- Chechele(i)* 1151×4–1485, *Chickes- Chi(t)chley, Chick- Chicherly, Chi(t)chly* 1680–93. OE pers.n. **Ċiċa* + **lēah**. Bk 33 gives prs [tʃitʃili, tʃetʃli], Jnl 2.22.

CHICHESTER WSusx SU 8604. 'Cissa's Roman town'. *Cisseceastre* 9th ASC(A) under year 895, c.1100, *Cycester* [988]14th S 872, *Ciceastre* c.1121 ASC(E) under year 1086, *Cicaestre* 1130 ibid., *Cicestre* 1086–1428, *Chichestr'* 1417, *Sesettyr* 1475×85. OE pers.n. *Ċissa* + **ċeaster**. According to ASC under year 477 Cissa was the third son of Aelle who in that year led the invasion of what was to become Sussex. It is impossible to know whether *Cisseceastre* was really named after Cissa or whether the pers.n. is derived from the p.n. which may have had some quite different origin, e.g. OE ***ċisse** 'a gravelly place' < **ċis* + *-e* (Gmc *-jōn*). Sx 10 gives pr [tʃidəstə], PNE i. 95, 142.

CHICHESTER HARBOUR WSusx SU 7600. P.n. CHICHESTER SU 7600 + ModE **harbour**.

CHICKERELL Dorset SY 6480. Unexplained. *Cicherelle* 1086, *Chi- Chykerel(l)* 1227–1481. A small group of mainly SW names of this type have not been satisfactorily explained, viz. BUCKERELL Devon ST 1200, Chackrell in Birlescombe, Devon ST 0716, *boscum de Chakerell* 13th, Cheverell Wilts ST 9854, *Chevrel* 1086, *Cheverel(l)* from 1166, *Chiverell* 1175–1291, Keveral Corn SX 2955, *Kewerel* 1226, *Keverel* 1299, Petteril Corn, *Peterel* 1268, River Peterill Cumbr NY 4352, *Peterel* 1268–1330, Tregatherall Corn SX 1189. Probably like the r.n. DEVERILL Wilts and its Cornish parallels Deveral SW 5935 and Derval SW 4130, these all contain a Brit adjective suffix represented by Co **-(y)el*, OBreton, W *-(i)ol*. Do i.204, CPNE 138, Cu 23, D 548, RN 323, W 238.

CHICKLADE Wilts ST 9134. Uncertain. *Cytlid* 899×924 S 1445, *(boscus de) Chitlad* 'C wood' 1300, *Ciclet* 1211, *(in bosco de) Siclat* 1232, *Chicled, Chi- Chykled(e)* 1232–1369, *Chik(k)elade* 1279–1442, *Chicklad(e)* from 1289. From the 13th cent. the generic was interpreted as OE *(ġe)lād* 'a river crossing' but this is incompatible both with the earliest form and the topography. It has been suggested that the specific is PrW ***cẹd** 'a wood' with interchange of *c* and *t* before *l* as in the name WATLING STREET. Wlt 184.

CHICKSANDS Beds TL 1239. 'Cica's sandy land'. *Chichesane* 1086, *Chikesand* 1190. OE pers.n. **Ċica* + **sand**. Bd 168 gives prs [tʃiksəndz, tʃiksən].

CHICKSGROVE Wilts ST 9729. 'Chick's grove'. *Chichesgrave -a* 1154×89, 1225, *Chikesgrave* 1225–1312, *Chekesgrave* 1452, *Cheekesgrove al. Cheesgrove* 1636, *Chisgrove* 1648. OE **ċicen** possibly used as a nickname or surname, genitive sing. **ċic(n)es**, + **grāf**. W 194

CHIDDEN Hants SU 6517. Partly unexplained. *(æt) cittandene* [956]12th S 598, *Chitedene* 1218–57, *Cheddene* 1236, 1280, *Chi- Chydden(e)* from 1242. Other early forms include references to the boundary of the people of Chidden, *(on) citwara mearce* [956]12th in S 598 and 1042 in S 994, OE folk-n. **Ċitware* 'the inhabitants of Chidden', genitive pl. *Ċitwara*, and to the ridge or ridges (OE *bæc*) of the *Citware, (to) cittanwara becun, (of) citwara beca, cytwara bæcce* [956]12th S 598. Were it not for these forms *Cittandene* would be 'Citta's valley', OE pers.n. **Ċitta*, genitive sing. *Ċittan*, + **denu**, but *Citware* seems to be a reduced form of *Cittanware* which can only be 'the dwellers at *Cittan*', an otherwise unknown p.n. Ha 53, Gover 1958.52.

CHIDDINGFOLD Surrey SU 9635. Probably 'the open land of the Ceodelingas, the people called after Ceodel'. *Chedelingefelt* c.1130, *Chideringfald* c.1180, *Chidingefeld* c.1185–1596 with variants *Chyd- -ynge-*, *-fald* and *-fold*, *Chedingefeld* c.1185, 1229, *Chudingefaud* c.1250, *Chiddingfold* 1597. The generic shows occasional confusion of **feld** 'open land' with **fald** 'a fold for animals'. Sr suggests the specific is the genitive pl. **Ċeoderinga* of a folk-n. **Ċeoderingas* 'the dwellers in the hollow', from OE *ċeodor* 'a bag, a hollow' + **ingas**. A better suggestion is folk-n. **Ċēodelingas* < pers.n. **Ċēodel*, a hypocoristic form of a by-n. from OE *ċēod(e)* 'a bag, a pouch'. Sr 186.

CHIDDINGLY ESusx TQ 5414. Probably 'wood or clearing of the Cittingas, the people called after Citta'. *Cetelingei* (sic) 1086, *Chit(t)ingeleg(h)e* c.1230–1345, *Citinghleye* 1273, *Chiddelingeli* 1279, *Chidyng(e)le(gh)* 14th cent., *Chittingly* 17th cent. OE folk-n. **Ċittingas* < pers.n. *Ċitta* + **ingas**, genitive pl. *Ċittinga*, + **lēah**. For the folk-n. cf. Chiddingly Wood TQ 3532, *Citangaleaghe* [765]13th S 50, *Cytinglegh* 1288. Sx 398, 271.

CHIDDINGSTONE Kent TQ 4945. Partly uncertain. *Cidingstane* c.1100, *Cid(d)ingstan(e)* 1218–80, *Chy- Chidingestone(e)* 1240–88. The generic is OE **stān** 'a stone', the specific uncertain, possibly an OE p.n. **Ċid(d)ing* 'the place called after Cidda', OE pers.n. *Ċidda* + **ing**[2], or the verbal noun **ċīding** 'brawling, quarreling' from OE **ċīdan** 'to quarrel'. PNK 77.

CHIDDINGSTONE CAUSEWAY Kent TQ 5246. *Chiddingstone Causeway* 1819 OS.

CHIDEOCK Dorset SY 4292. 'Wooded place'. *Cidihoc* 1086, *Cidioc* 1240, *Chidioc* 1268. OE p.n. **Ċīdioc* < PrW ***cęd** + ***-ịọg**. Do 56, LHEB 229, 295, 327, 554, 557, L 190.

North CHIDEOCK Dorset SY 4294. *N. Chideock* 1811 OS. ModE **north** + p.n. CHIDEOCK SY 4292.

CHIDHAM WSusx SU 7903. 'The bay promontory'. *Chedeham* 1193–1271 etc., *Chudeham* 1247–1312 etc., *Chideham* 1223, *Chyd(e)ham* 1248, 1428. OE **ċēode** + **hamm** 2a. Chidham lies on a peninsula in Chichester Harbour which forms a bay or inlet here. Sx 59, NoB 48.147, Studies 1931.70, PNE i.89, ASE 2.42, SS 81.

CHIEFLOWMAN Devon ST 0015 > UPLOWMAN ST 0115.

CHIEVELEY Berks SU 4773. 'Cifa's wood or clearing'. *(æt) Cifanlea* [951]16th S 558, 960 S 687, *Ciuenlea* [965]13th S 732, *Civelei(a)* 1086–13th, *Chiuelai* 1167 etc. with variants *Chyue- Chive- Chyve-* and *-l(eg)' -le(y) -lye*, *Chevelee* 1401×2. OE pers.n. **Ċifa*, genitive sing. *Ċifan*, + **lēah**, dative sing. **lēa**. Brk 241, 936.

CHIGNALL SMEALY Essex TL 6611. 'The Smealy (part of) Chignall'. *Chikenale Smythle* 1481, *Smeley Chyggenhale* 1548. P.n. Chignall as in CHIGNALL ST JAMES TL 6709 + p.n. *Smetheleye* 1254–1387, *Smelie* 1544, 'the smooth wood or clearing', OE **smēthe** + **lēah**. Ess 247.

CHIGNALL ST JAMES Essex TL 6709. 'St James's C'. *Chigehal(e) Sancti Jacobi* 1255, *Chyng- Chingenhall alias Chiggenhall alias Chignall St Mary and St James* 1545, 1563. P.n. Chignall + saint's name *James*, Latin *Jacobus*, from the dedication of the church. Chignall, *Cingehala, Cinguehellam* 1086, *Chikehala* 1179, 1254, *Chikenhale* 1202–60, *Chi- Chyge(n)hal(e)* 1203–1339, *Chignall* 1640, is 'Cicca's or chicken nook', OE pers.n. *Ċicca*, genitive sing. *Ċiccan*, or **ċiċen**, genitive pl. **ċicna**, + **halh**. 'St James' for distinction from CHIGNALL SMEALY TL 6611. Ess 246.

CHIGWELL Essex TQ 4493. 'Cicca's spring'. *Cingheuuella -ā* 1086, *Chi- Chyngewell(')* 1235, 1376, *Chi- Chyg(ge)well(e)* 1158–1333, *Chi- Chyk(e)well(e)* 1225–1331. OE pers.n. *Ċicca* + **wella**. The *-ing-* spellings may point to an earlier form **Ċiccingwella* 'spring at or called *Ċiccing*, Cicca's place' or 'spring called after Cicca' or simply to a genitival compound *Ċiccan wella* 'Cicca's spring' with metathesis. Ess 54.

CHIGWELL ROW Essex TQ 4693. *Chigwell Rowe* 1518. P.n. CHIGWELL TQ 4493 + ModE **row** 'a row of cottages'. Ess 56.

CHILBOLTON Hants SU 3939. 'Estate called after Ceolbald'. *Ceolboldingtun* [909]11th S 376, *(æt, in) Ceolboldinc(g)tun(e), Ceolbaldinctuna* [934]12th S 427, *Cilbode(n)tvne* 1086, *Chilboltinton* c.1190, *Chilbolton(e)* from 1284. OE pers.n. *Ċēolbald* + **ing**[4] + **tūn**. Ha 53, Gover 1958.174, DEPN.

CHILCOMB Hants SU 5029. 'Valley called or at *Cilta*, the steep slope'. *Ciltacumb* *[855×8]12th S 325, *Ciltancumb* [909]11th S 376, *(of, æt) Ciltancumbe* [963×75]12th S 817, [984×1002]12th S 946, *Ciltecūbe* 1086, *Chiltecumbe* 1171–1341, *Chilcum* 1579. Brit ***ciltā** probably taken as a p.n. + OE **cumb**. The reference is to the abrupt scarp of Deacon Hill or Magdalen Hill Down or both. Ha 53, Gover 1958.69, Jnl 16.7–15.

CHILCOMBE Dorset SY 5291. 'Valley at or called *Cilta*, the steep hill'. *Ciltecombe* 1086, *-cumb* 1268, *Childecumbe* 1198, *Chylcombe* 1558. Pre-Brit ***ciltā** 'a steep hill' possibly taken as a p.n. + OE **cumb**. Chilcombe lies at the foot of a steep slope crowned with a hill-fort. Do 56, Jnl 16.10, 14.

CHILCOMPTON Somer ST 6451. 'The children's Compton'. *Childecumpton* 1227, *Chilcompton* 1610. OE **ċild**, genitive pl. **ċilda**, + p.n. *Contone -tvne* 1086[†], 'coomb settlement', OE **cumb** + **tūn**. OE *ċild* can mean 'young nobleman' or 'young monk'. The sense here is probably 'estate used to provide for the upkeep of the offspring of a landowning family'. DEPN.

CHILCOTE Leic SK 2811. 'Cottage or cottages of the younger sons or of the young men or retainers'. *Cildecote* 1086, *Childecot(e)* 1195–1440, *Chilcote -y-* from 1428. OE **cild**, genitive pl. **ċilda**, + **cot**, pl. **cotu**. This name may indicate an estate given to the younger sons of a family as a joint posession. Lei 561.

CHILDREY Oxon SU 3687. 'The stream (of the spring) called Cille'. *(to) Cillariðe* [?c.950]10th S 1539, *Celrea -rey(a)* 1086–1242×3, *Chil(d)re* 1206, *Chilrey* 1517, *Childrey* 1766. The name of the village is derived from that of the stream, Childrey Brook, which occurs as *(andlang) cilla riðe* [940]12th, *(andlang) cilla riþe* [940]13th S 471, *(on) cilla(n) riðe* [947]12th S 529, *(on) cyllan rið* [956]12th S 597 etc. OE ***ċille** 'a spring' + **rīth**. An alternative suggestion is that the specific is OE feminine pers.n. *Ċille*, 'Cille's spring'. Brk 470, NoB 61.49, Signposts 176.

CHILDSWICKHAM H&W SP 0738. 'The child's *Wigwen*'. *Childeswicwon* *[706]12th S 1174, *Eildesuucque* (sic for *Cildes-*) [714]16th S 1250 with variant *Cildesuicoque* 16th, *Chi- Chyldes Wi- Wy(c)kewan(ne) -wane* 1220–1585, *Childes wican* 1577, *Childeswi(c)kham* 1637–84. ME **child** (OE *ċild*) 'a noble born youth', genitive sing. **childes**, + p.n. *Wicwone* *[706]12th S 1174, *Wicwona* *[709]12th S 80, *(in) uuiguuennan* 972 S 786, *Wicvene* 1086, *Wi- Wykewan(n)(e)* 1220–1474 of unknown origin and meaning. The name also occurs in nearby WICKHAMFORD H&W SP 0641 and also in a Wiltshire stream-name in Bradford on Avon, *Wigewen broke* [1001]15th S 899. DEPN suggests that *Wigwenne* in these names might represent W **gwig-gwaun*, OW *gwic* + *guoun* with the possible sense 'village' or 'forest pasture'. Gl ii.6, DEPN, Hooke 23–4, 40, 178, Grundy 1920.103, CPNE s.vv. *goon*, **gwyk*.

[†] *Cumtun* [970]12th S 777 may belong here or under Compton Dando.

CHILDWALL Mers SJ 4189. 'Children's spring'. *Cildeuuelle* 1086, *Childewelle* 1094–1302, *-walle* 1212–1376, *Childwall* 1423, *Cheldewell* 12th. OE **ċild**, genitive pl. **ċilda**, + **wella**, **wælla**. La 112, Jnl 17.63.

CHILFROME Dorset SY 5898. 'The Frome estate held by the noble-born sons'. *Childefrome* 1206, *Childesfrome* 1268, *Chillfroome* 1653. Earlier simply *Frome* 1086. OE **ċilda**, genitive pl. of **ċild**, + p.n. Frome transferred from the r.n. FROME SY 8089. Do 56.

CHILGROVE WSusx SU 8314. 'The throat grove'. *Chelegrave* 1200–1303 etc., *Chilgrave* 1341, *Chilgrove* 1428. OE **ċēole** + **grāf(a)**. The reference is to the deep-cut valley in which Chilgrove lies. It is possible, however, that the generic was originally OE **græf** 'a trench, a ditch' and that the name means '*Ceole* trench, the trench called *Ceole*, the throat'. Sx 49.

CHILHAM Kent TR 0653. 'Cilla or Cille's estate'. *Cylleham* *1016×35 S 981, *Cilleham* 1086–1214, *Chilham* from c. 1140. OE pers.n. *Ċilla* or feminine *Ċille* + **hām**. PNK 372, KPN 332, ASE 2.29.

CHILLATON Devon SX 4381. 'Farm, estate of the young (noble)men'. *Childeton'* 1242, *Chillatun* 1284. OE **ċild**, genitive pl. **ċilda**, + **tūn**. D 217.

CHILLENDEN Kent TR 2753. 'Ci(o)lla or Ci(o)lle's valley'. *Ciollan dene* 833×9 S 1482, *Cilledene* 1086, *Cillin(g)den'* 1226–54, *Chilinden'* 1270. OE pers.n. *Ċi(o)lla* or feminine *Ci(o)lle*, genitive sing. *Ċi(o)llan*, + **denu**. PNK 579, KPN 175.

CHILLERTON IoW SZ 4884. 'The valley settlement'. *Celertvne* 1086, *Chelierton(e) -yer-* 1279–1409, *Chel(l)erton* 1189–1583, *Chillerton* 1708. OE **ċeole** 'throat, valley' + **ġeard-tūn** as in YORTON Shrops SJ 5023, *Iartune* 1086. There is a marked valley at Chillerton Farm. Wt 133 gives prs ['tʃil- 'tʃelərtn], Mills 1996.39.

CHILLESFORD Suff TM 3852. 'Gravel ford'. *Cesefortda* 1086, *Chiselford* 1184–1344, *Chi- Cheselesford* 1254, 1275, *Chyseford'* 1286, *Chyllysford* 1523, *Chillesforde* 1610. OE **ċeosol** + **ford**. The soil here is mixed overlying crag (shelly sand). DEPN, Baron.

CHILLINGHAM Northum NU 0626. 'The homestead or village called or at *Chevelinge*, the place called after Ceofel'. *Chauringeham* before 1167, *Cheulingeham* 1186, *Chevelingham -yng-* 1231–1346, *Chillyngham* 1348, 1507. OE **Ċeofeling* < pers.n. *Ċeofel(a)* + **ing**[2], locative-dative sing. **Ċeofelinge*, + **hām**. Pronounced [ʃiliŋəm] with Northumbrian [ʃ] for [tʃ] as CHATTON NU 0528. Like the other Northumberland *-ingham* names probably originally pronounced with assibilated [indʒ]. This assibilated pronunciation was probably lost through dissimilation after initial [ʃ]. NbDu 45, BzN 1968.164, ING 157.

CHILLINGTON Devon SX 7942. 'Estate called after Ceadela'. *Cedelintone* 1086, *Chedelington -yng-* 1238–1310, *Chedlyngton* 1478, *Chidlington al. Chillington* 1597. OE pers.n. *Ċeadela* + **ing**[4] + **tūn**. D 332.

CHILLINGTON Somer ST 3811. 'Estate called after Ceola'. *Cherinton* 1231, *Cheleton(e)* 1261, 1285, *Chellington* 1610. OE pers.n. *Ċēola* + **ing**[4] + **tūn**. DEPN.

CHILLINGTON HALL Staffs SJ 8607. *Chillington Hall* 1834 OS. P.n. *Cillentone* 1086, *Cildentona* 1129, *Chi- Chylinton -yn-* 1175×82–1562, *Chy- Chillington -yng-* from 1236, 'the childrens' estate', OE **ċild**, genitive pl. **ċildena**, + **tūn**. For the gen. pl. form cf. Brunner §289 A.3. St i.36, Horovitz.

CHILMARK Wilts ST 9732. 'The signpost'. *(æt) Chieldmearc* [924×939]14th S 458, *Childmerk(e)* 1195, 1242, *cigel marc, to cigelmerc broce* 'to Chilmark brook' [983]15th S 850, *Chilmerc* 1086–1206, *-merk* c.1190, 1289, *Chilmark* 1306. OE***ċiġel** 'a pole' + **mearc**. The earliest sps show folk-etymological association with OE *ċild* 'child'. Wlt 185, Studies 1936.165.

CHILSON Oxon SP 3119. 'The estate of the young nobleman'. *Cildestuna* c.1200, *Childestune -a -tona -ton(e)* early 13th–1390, *Childston* 1401×2. OE **ċild**, genitive sing. **ċildes**, + **tūn**. O 378, PNE i.93.

CHILSWORTHY Corn SX 4172. 'Ceol's enclosure'. *Chillesworthy* 1337, *Che- Chyllesworthy* 1471 G. OE pers.n. *Ċeol*, genitive sing. *Ċeoles*, + **worthiġ**. PNCo 68, Gover n.d. 179.

CHILSWORTHY Devon SS 3206. 'Ceol's enclosure'. *Chelesworde* 1086, *Chilles- Chellis- Cheleswrth'*, *Chellesworth* 1238–78, *Chullesworthi* 1310. OE pers.n. *Ċēol*, genitive sing. *Ċēoles*, + **worth**, later **worthiġ**. D 147.

CHILTERN HILLS Oxon SU 7799. District n. Chiltern + ModE **hill(s)**. Chiltern, possibly 'the high ones', is *Cilternes efes* 'the edge of Chiltern' [1006 for 1002]11th S 914, *Ciltern* 1121 ASC(E) under year 1009, *Ciltre* 1241–1335, pre-Brit ***ciltā-** 'high' + suffix **-erno-**. The Chiltern dwellers are the *Ciltern sætna* (genitive pl.) in the 7th cent. Tribal Hidage. Bk 174, Hrt 7, DEPN. Jnl 16.10, 13, 14.

CHILTERN HUNDREDS Bucks SU 9588. P.n. Chiltern + ModE **hundred(s)**. Chiltern, *Ciltern* [7th]11th Tribal Hidage, c.1121 ASC(E) under year 1009, *Cilternes efes* 'the eaves of Chiltern' [1005]17th S 911, *Ciltria* 12th, *Chiltre* 1100×35, is an ancient hill-name **Cilt-ern-* from pre-British ***ciltā** 'steep hill'. The inhabitants of the Chilterns are mentioned as *Ciltern sætna feoþer þusend hyda* 'the 4000 hides of the Cilternsæte' [7th]11th Tribal Hidage†. Bk 2, 174, Jnl 2.31, 16.10, 13.

CHILTHORNE DOMER Somer ST 5218. 'C held by the Dumere family'. *Chilterne Dunmere* 1280, *Chilterne Dummer* 1610. Earlier simply *Cilterne -a* 1086–1204, [n.d.] Buck, a name identical with Chiltern in CHILTERN HILLS SU 7596 from pre-British ***ciltā** 'steep hill'. The manor was held by Henry de Dummere c.1200 and is named after him for distinction from its other division Chilthorne Vagg, *Cilterne Fageth* 1100×35, ~ *Fageth, Faghet* [1107×22, 1152×8]14th, *Chylterne Fag* 1276. Earlier simply *Ciltern* [1091×1106, 1100×18]14th. Robert Faget occurs [c.1100]14th whose surname represents OFr *fagot* 'a bundle of firewood'. DEPN, Jnl 6.13, VCH ii.381.

East CHILTINGTON ESusx TQ 3715. ModE adj. **east** + p.n. *Childel- Childe(n)tune* 1086, *Chilting'* 1212, *Chiltigton juxta Lewes* 1283, 'estate, settlement called or at Chilting, the hill place', OE p.n. **Ċilting* < pre-English **ċiltā** + **ing**[2], + **tūn**. 'East' for distinction from West CHILTINGTON WSusx TQ 0198. Sx 299, SS 75, Jnl 16.8–9.

West CHILTINGTON WSusx TQ 0918. *West Chilton* 1675. ModE **west** + p.n. *Cillingtun* [969]13th S 774, *[1066]13th S 1043, *Cilletune -tone* 1086, *Chi- Chyltin(g)ton* 1247–1342, 'the settlement at or called *Ciltine*', Brit p.n. **Ciltinā* < ***ciltā** 'a slope' + - **ing**[2], + **tūn**. *Ciltine* is almost certainly the *Ciltine* of Eddius's *Life of Wilfrid* [c.710×20]11th ch. 42, and the *Ciltinne* 764 for 767 of S 106. 'West' for distinction from East CHILTINGTON ESusx TQ 3715. Sx 174, Jnl 16.8–10, 13.

CHILTON 'The children's estate'. OE **ċild**, genitive pl. **ċilda**, + **tūn**. The reference is to ownership or use by young people of some kind, e.g. as a source of revenue for the support of the young monks of a monastery or the young heirs of a landowner. PNE i.93, Signposts 125, 184.

(1) ~ Bucks SP 6811 *Ciltone* 1086, *Chilton(')* from 1152×8. Bk 120.

(2) ~ Durham NZ 2829. Short for *Chilton Buildings* 1863 OS. A modern colliery village named after Great Chilton NZ 2930, *ciltona* [1091×2]12th, *Chilton* from 1187 (*Great* ~ from 1586), ~ *Magna* 1350–1634. 'Great' for distinction from Little Chilton NZ 3031 (now Ferryhill Station), *Chilton Parva* 1354~1580, *Little Chilton* 1675. NbDu 45.

(3) ~ Oxon SU 4885. *(emb) cylda tun* [879×99]12th S 354, *Cildatun* [1015, 1052]13th S 934, 1023, *Cil(le)tone* 1086, *Chilton(e)* from 1218×21. Brk 497.

†Eddi's *Ciltine* [c.680]11th belongs not here but under East and West CHILTINGTON Sussex TQ 3715, 0918.

(4) ~ CANDOVER Hants SU 5940 → PRESTON CANDOVER SU 6041.

(5) ~ CANTELO Somer ST 5721. 'C held by the Cantelu family'. *Chiltone Cauntilo* 1361, earlier simply *Childeton* 1201‡. The manor was held by Walter de Cantelu in 1201. DEPN.

(6) ~ LANE Durham NZ 3031. 'Lane leading to C'. P.n. CHILTON NZ 2829 + ModE **lane**.

(7) ~ STREET Suff TL 7546. 'Chilton hamlet'. *Chilton Street* 1805 OS. P.n. *Chilton* from 1254, + ModE **street**. DEPN.

(8) ~ TRINITY Somer ST 2939. 'Holy Trinity's C'. *Chilton Sancte Trinitatis* 1431. Earlier simply *Cille- Cildetone* 1086, *Chileton* 1208, *Chilton* 1610. The addition is from the dedication of the church. DEPN.

CHILTON CHINE IoW SZ 4082. *Chilton Chine* 1769. P.n. Chilton + Hants/IoW dial. **chine** (OE *ċinu*) 'cleft, gorge'. Chilton, *Celatvne* 1086, *Cheletuna -ton(e)* c.1192–1305, *Chelton(a) -e* 1173–1608, *Chilton* 1306, 1781, is the 'gorge farm or estate', OE **ċeole** 'a throat, gorge, valley, referring to the deep chasm or valley of the chine + **tūn**. Wt 65 gives pr [tʃɪltn], Mills 1996.39.

CHILTON POLDEN Somer ST 3739. Short for Chilton upon Polden, p.n. Chilton + p.n. Polden as in Polden Hills. Earlier simply *Ceptone* 1086, *Cahalton* 1285, *Chauton* 1303, *Cheltone* 1327, *Chelton* 1610, of uncertain origin as the variety of forms shows. DEPN suggests 'settlement on the limestone hills', OE **ċealc** + **tūn**, which at least fits the topography. Polden is *Poeldune* [725]14th, *Poldone* [after 729]13th, *Poweldone, Poldone* [n.d.]13th, *Pouldon* 1241, *Poweldun* 1235×52, lost p.n. *Bouel(t)* [705]13th, *Pouelt* [705]n.d. S 248, *Poeld, Poelt .i. Grentone* [725]13th, *Poelt, Poholt* *[725]12th S 250, [725]14th S 251, *Pouholt* [729]14th S 253, *Poponholt* [after 729]13th, *Poholt* [754]13th S 1680, possibly OW **pou** 'country' + OE **holt** 'wood', + OE **dūn**. But *holt* looks like a piece of folk-etymology. Comparison has been made with the W name Builth Powys SO 0350, originally a district name, *Buelt* c.800 HB, c.1282, 1301×2, *Buhelt, Buheld* 13th, Brit **bou̯o-gelt-* 'cow pasture'. DEPN, Turner 1950–2.117.

CHILTON FOLIAT Wilts SU 3270. 'C held by the Foliot family'. *Cilton Roberti Foliot* 'Robert Foliot's C' 1167, *Chilton Foliot* 1227. P.n. *Cilletone* 1086, *Chilton(e)* 1155 etc., 'Cilla's farm or village', OE pers.n. *Ċilla* + **tūn**, + manorial addition from the surname of Robert Foliot 1167. Wlt 340.

CHILWORTH Hants SU 4118. Partly uncertain. *Celeorde* 1086, *Cheleuuorth -worth* 1230–1341, *Chuleworthe* 1334, *Chilworth* from 1412. The name is usually explained as 'Ceola's enclosure', OE pers.n. *Ċēola* + **worth**, but there is a marked 'throat' or valley here between Castle Hill and Chilworth Ring so that this is very likely the 'throat or valley enclosure', OE **ċeole** + **worth**. Ha 53, Gover 1958.37.

CHILWORTH Surrey TQ 0247. Either 'Ceola's settlement' or 'the settlement at the throat-like valley'. *Celeorde* 1086, *Chelewurth -wrth(e) -worth* c.1172–1283, *Chelworth* 1383, *Chylworthe* 1556. Either OE pers.n. *Ċēola* or **ċeole** 'a throat' + **worth**. Sr 245.

CHIMNEY Oxon SP 3500. 'Ceomma's island'. *Ceommanyg* 1069, *(æt) Ceommenige* c.1080, *Chimeneye* 1285, *Chymneye* c.1360, *Chimley* 1797. OE pers.n. **Ċeomma*, genitive sing. *Ċeomman*, + **īeġ**. O 302, L 37–8.

CHINEHAM Hants SU 6554. 'Ravine homestead or estate'. *Chineham* from 1086, *Chi- Chynham* 1154×89–1341. OE **ċinu** + **hām**. The road to Basingstoke passes between low hills here. Ha 53, Gover 1958.146, DEPN.

CHINGFORD GLond TQ 3893. 'Gravel or shingle ford'. *Cingeford* [1042×66]12th/13th S 1056, *Cing(h)efort* 1086, *Ching(e)ford(e)* 1181–1543, *Chingelford* 1242–1349, *S(c)hingelford* 1327 etc., *Shyngleford* 1535. OE ***ċingel**, ***singel**, + **ford**. The ford crosses the river Lea. The soil is gravelly. There are many variant spellings of this name, one of which, *Chagingeford* 1219, gave rise to a possible alternative explanation, 'ford of the dwellers by the stumps', OE **Cæġingaford* from the element **ċeġ* seen in CHEAM. Pile-dwellings have been excavated nearby. However, although early reduction of such a name is not unparalleled this explanation is hazardous. Ess 18, TC 203, L 69, Encyc 156.

CHINLEY Derby SK 0482. 'Wood or clearing at the deep valley'. *Chyn- Chinley(e) -le(ge)* from 1285. OE **ċinu** + **lēah**. Db 76.

CHINLEY HEAD Derby SK 0584. *Chinleyhead* 1624. P.n. CHINLEY SK 0482 + ModE **head**. Db 78.

East CHINNOCK Somer ST 4913. *Estcinnok* 1243, *Estchinnok -y-* [1318×9, 1381×2]14th. ME adj. **est** + p.n. *Cinnuc* ?c.950 S 1539, 1091–1106, *Cinioch* 1086, *Cynnuc* [1100×18]14th, *Ci- Cynnoc -k* [1107×22–1302]14th, *Chi- Chynnuc* [1152×8]14th, *Chinnoc(h)* c.1155, 1230×1, *Cynnock* 1284×5, *Chinnock* [c1227]14th, probably OE ***ċinnuc** 'little chin' (OE *ċinn* + *-uc*) in some topographical sense of the hills around the place. The alternative, OE ***ċinuc** 'little valley' (< *ċinu* 'fissure, ravine' + *-uc*), equally well fits the topography but not the early spellings. 'East' for distinction from *Middle Chenock* 1610, *Mid Cinnoc* [c.1250]14th, and West CHINNOCK ST 4613, *West Cinnoc* [c.1227]14th. Also known as *Kinnoc Monachorum* 'monks' C' 1284 referring to possession by Montacute priory. DEPN, DB Somerset 19.44, Turner 1950–2.114 who prefers a Welsh origin comparing Welsh *cunnog, cynnog* 'milk-pail' found as a stream-name, Nant-y-Cynog, Gwyn SH 5804, *Nant Kynnog* 1592, and Gael *cuinneag* in the same sense found as a mountain name, Quinag, Highld NC 2028.

CHINNOR Oxon SP 7500. 'Ceonna's flat-topped ridge'. *Chennore* 1086–1285 with variants *-or(a), Chinnor(a) -ore* from 1194 with variant *Chy-*. OE pers.n. **Ċeonna*, genitive sing. **Ċeonnan*, + **ōra**. O 106, L 179–80, Jnl 21.17, 21, 22.35, Gelling 1998.85.

CHIPCHASE CASTLE Northum NY 8875. *Chipchase Castle* 1866 OS. P.n. Chipchase + ME **castel**, a large-scale mid-14th cent. tower with later additions. The meaning and origin of Chipchase, *Chipches* 1229, 1242 BF, 1296 SR, *Chip(e)ches(se)* 1255–1346, *Chipchase* 1298, are unknown. One suggestion is OE **ċipp** 'a beam, a log' + ***ċeas** 'a heap', referring to a structure of logs, possibly an animal trap. The generic has been assimilated to OFr, ME **chace** 'a tract of ground for breeding and hunting wild animals'. NbDu 45, DEPN, Pevsner 1992.230.

CHIPNALL Shrops SJ 7231. Probably 'Cippa's knoll'. *Ceppecanole* 1086, *Chi- Chyppe(k)nol(l)* c.1240×60–1337, *Chipnall* from 1577. OE pers.n. *Ċippa* + **cnoll**. Sa i.79.

CHIPPENHAM Cambs TL 6669. 'Cip(p)a's homestead'. *Cy- C(h)ipeham* 1086, *Chip(p)eham* 1086–1302, *Chip(p)enham* from 1086. OE pers.n. *Ċip(p)a*, genitive sing. *Ċip(p)an*, + **hām**. Ca 189, cf. TC 72–3.

CHIPPENHAM Wilts ST 9173. 'Cippa's promontory'. *(to, æt) Cippanhamme* 891 ASC(A) under year 878, [873×88]11th S 1507, c.1000 Asser under year 853, *(æt) Cippan homme* 899×924 S 1445, *Cyppan hamm* [930]?10th S 405, *Chi- Chepehã* 1086, *Chy- Chippenham* 1177–1219, *Chippenham* from 1227. OE pers.n. *Ċippa*, genitive sing. *Ċippan*, + **hamm**. Chippenham is situated on a promontory of higher ground in a sharp bend of the river Avon. Wlt 89.

CHIPPERFIELD Herts TL 0401. 'The traders' field'. *Chiperfeld* 1375, *Chi- Chy- Cheperfeld* 1382–1437. Cf. also *Chepervillewode* 'Chipperfield wood' 1315, *Chepervilerowe* 1437. ME **cheper** (OE *ċēapere*, genitive pl. *ċēapera*) + **feld**. Hrt 45.

CHIPPING 'A market'. OE (Angl, Kt) **ċēping**, (WS) **ċī(e)ping**.

(1) ~ Lancs SD 6243. *Chippin* 1203, *Chypyn* 1225×6, *Chypping* 1241, *Chipping* 1242, *Che- Chip(p)in, Chypyn* 13th cent., *Schi- Schypen* 1311. Forms in *-in -yn* are due to the influence of p.n. Chippingdale in the pronunciation of which *n* developed by assimilation to following *d*. Chippingdale, *Chipinden* 1086, *Chippendal, Chipindale, Chippingdale* 13th cent., 'Chipping valley', p.n. Chipping + OE **denu** later replaced with ON **dalr**,

‡The DB form *Citerne* 1086 has been associated with this name but formally it must belong to the Chilthornes, DB Somerset 26.7.

was formerly the name of the district around Chipping. La 143, Jnl 17.80.

(2) ~ Herts TL 3532. Short for New Chipping, 'the new market', *le Neuchepyng* 1360, *New Chepyng in Buckland* 1518–29, *New Chipping* 1533–8, *Cheppen* 1700. A market was established here in 1252. Hrt 176.

(3) ~ HILL Essex TL 8215. 'Market hill'. *Chippinghill* 1608. Ultimately from OE **ċīeping** + **hyll**. Ess 301.

(4) ~ SODBURY Avon ST 7382 → Chipping SODBURY.

CHIPSTABLE Somer ST 0427. Partly uncertain. *Cipestaple* 1086, *Chippestapel* 1251, *Chipstable* 1610. The generic is OE **stapol** 'post, pillar, platform' but it is impossible to say whether the specific is OE **ċipp** 'a log', genitive pl. **ċippa**, referring to some sort of log foundation or platform, or pers.n. *Ċippa*. DEPN, PNE i.94, ii.146.

CHIPSTEAD Surrey TQ 2757. 'The place where a market is held'. *Tepestede* 1086, *Chepstede* c.1115, [727, 933, 967, 1062]13th S 420, 752, 1035, 1181, *Shepstede* 1226, 1428, *Chipsted(e)* 1231–1449, *Cheapstid* 1711. OE **ċēap-stede**. The parish occupies a central position in a complex of Anglo-Saxon estates in the Banstead-Coulsdon region. Sr 290, Stead 244–5, Rumble 1976.175.

CHIRBURY Shrops SO 2698. 'Church fort or manor'. *(æt) Cyricbyrig* c.1000, c.1050 ASC(B, C) under year 915, *Cireberie* 1086, *Chirebiri -bury -bir(y)* 1227–1334, *Chyrbury* 1272, *Chirbury* from 1287. OE **ċiriċe** + **byriġ**, dative sing. of **burh**. It is uncertain whether the name was newly coined in 915 when Æthelflæd, the lady of Mercia, built a fort in this area in her war against the Norsemen or whether it already existed. The site of the fort is unknown. Sa i.80.

CHIRDON BURN Northum NY 7683. *Chirdon Burn* 1866 OS. P.n. *Chirden* 1255–1610, *Chirdon* 1663, *Cherdon* c.1715 AA 13.14, unknown element or name + OE **denu** 'a valley', + OE **burna**. OE **ċiriċe** has been suggested as the specific giving 'the church valley, the valley with or belonging to a church', but this cannot be established from the surviving forms. The alternative, OE **ċerr** 'a bend', is topographically apt but normally gives ME *cher, char*. NbDu 45, Studies 1931.53.

CHIRTON Wilts SU 00757. 'Church estate or village' *Ceritone* 1086, *-ton(a)* 1242, 1259, *Chi- Chyriton* 1230–86, *Chur(u)ghton* 1316–41, *Chirghton* 1349. OE **ċiriċe** + **tūn**. Wlt 312.

CHISBURY Wilts SU 2766. 'Cissa's fortification'. *Cheseberie* 1086, *Chessebure -byr' -bury* 1247–92, *Chisseberia -byr' -bur(y)* 1166–1282, *Chussebur(ia) -bury* 1257–1408, *Cheesbury* 1819. OE pers.n. *Ċissa* + **burh**, dative sing. **byriġ**. The reference is to Chisbury camp, a multivallate Iron Age hill-fort. According to a late unreliable tradition there was in the days of Kinuinus, king of the West Saxons, a *regulus* of parts of Wilts and Berks whose dominium included the bishop's seat at Malmesbury and whose capital was at Bedwyn in the S part of which he built a *castellum* now called *Cysseburi* after him. In itself this does not prove that *Ċissa* is the original specific rather than ***ċise** 'a gravelly place' referring to the gravelly subsoil here, but Mills 1998 cites an early 10th cent. charter form *Cissanbyrig*, which seems to substantiate the pers.n. Cf. CISSBURY RING WSusx TQ 1308. Wlt 334, Pevsner 1975.174.

CHISELBOROUGH Somer ST 4614. 'Gravel hill'. *Ceolseberge* 1086, *Ciselberg* [1091×1106–1135×7]14th, *Chiselberge* [1135×66, 1152×8]14th, 1253, *Chysselborough* 1610. OE **ċisel, ċeosol** + **beorg**. The village lies beside several large oval hills. The soil is loam overlying gravel. DEPN, Gaz s.n., L 128.

CHISELBURY Wilts SU 0128. *Chiselbury* 1721. Possibly 'the gravel fortification', OE **ċiṡel** + **byriġ**, dative sing. of **burh**, but the evidence is very late. W 215 cites two charter forms, *cester slæd byrg* 'camp valley fort' [901]14th S 364 and *ceaster blæd byrig* [994]14th S 881 'camp ledge fort', both of equal authority; it is not impossible that the present name is a direct descendant of the former. The reference is to Chiselbury Camp, a univallate Iron Age hill-fort. Pevsner 1975.251.

CHISELHAMPTON Oxon SU 5998. 'Hampton on gravelly soil'. *Chiselentona* [1147]18th, *Chiselhamt'* c.1170–1230 with variants *-ton(a), Chiselhampton(')* from [1209×13]c.1300. ME **chisel** (OE *ċisel, ċeosol*) + p.n. *Hentone* 1086, '(At) the high settlement', OE **(æt thǣm) hēan tūne** (dative sing. of **hēah** + **tūn**). 'Chisel-' for distinction from the adjacent Brookhampton SU 6098. O 155, PNE i.95.

East CHISENBURY Wilts SU 1452. *Estchesyngebur'* 1289, *Est Chussyngbury* 1376. ME adj. **est** + p.n. *Chesigeberie* 1086, *Chesingbiria, Chisingburi* 1202, *Chesingebiria -byre -bir'* 1211–52, *Chysyngbury* 1304, 'fort of the Cisingas, the people who live on the gravel, or at Cising, the gravel place'. OE folk name **Ċisingas* < ***ċis** + **ingas**, genitive pl. **Ċisinga*, or p.n. **Ċising* < ***ċis** + **ing**[2], locative dative sing. **Ċisinge*, + **burh**, dative sing. **byriġ**. The reference is to Chisenbury Camp. 'East' for distinction from West Chisenbury or Chisenbury Dallyfolly SU 1352, *Westchesyngebur'* 1289, *Westchisingbury* 1313, *Chesingbur' voc. Folye, Chesingebir qui dicitur la Folye* 1275, *Chisenbury de la Folly* 1544×53, held by Roger de la Folie in 1202. Wlt 328.

Great CHISHILL Cambs TL 4238. *Magna Chiselhelle* 1248, *Graunt Chisull* after 1420. ModE adj. **great**, Latin **magna**, Fr **grand** + p.n. *Cishella(m) -ā -e* 1086, *Chishull(e) -hell* 1212–1327×33, *Chishill* from 1222, 'gravel hill', OE ***ċis** + **hyll**. The soil, however, is reported to be chalk and clay overlying chalk. Ca 373.

Little CHISHILL Cambs TL 4237. *Little Chyshull* 1269. ME adj. **litel** + p.n. Chishill as in Great CHISHILL TL 4238. Ca 373.

CHISLEDON Wilts SU 1879. 'The gravel valley'. *(æt) Cyseldene* [873×88]11th S 354, 925×33 S 1417, *Ciseldenu* [879×99]12th S 354, *(æt) Ceolseldene* (sic) [900]11th S 359, *Ceoseldene* [901]14th S 366, *Chiseldene* 1086, *Cheselden(e)* 1242–1306, *Chusselden(e)* 1281–1393, *-done* 1316. OE **ċisel, ċeosel**, + **denu**. Wlt 281.

CHISLEHURST GLond TQ 4470. 'Gravel wooded hill'. *Cyselhyrst* [955 for ? 973]10th S 671, [998]12th S 893, *Chiselherst* 1159. OE **cisel** + **hyrst**. DEPN.

CHISLET Kent TR 2264. Uncertain. *Cistelet* *[605]13th S 4 with variant *Cistelei*, 1086–1204, *Chistelet* [1087]13th, 1242, *Chisteled* 1199, *Cislested* 1202, *Chislet* 1610. There is no independent evidence for OE ***ċistelett** 'chestnut copse'. A better suggestion is *ċist-ġelǣt* 'water-conduit from a cistern'. A hypocausted Roman building has been found here. PNK 503, KPN 6, PNE i.96, Nomina 17.27–8, Cullen.

CHISWELL GREEN Herts TL 1303. *Chiswell Green* 1782. P.n. Chiswell + ModE **green**. If Chiswell is an ancient name it may be the 'gravelly stream', OE ***ċis** or **ċisel** + **wella**. Hrt 93, PNE i.95.

CHISWICK GLond TQ 2077. 'Cheese farm'. *(of) Ceswican* [c.1000]12th, *Chesewic -k* 1181–1470, *Chiswyk* 1537, *Cheswyke* 1566. OE **ċ(ī)ese** + **wīċ**, dative pl. **wīcum**. Mx 88, TC 203.

CHISWORTH Derby SJ 9891. 'Cissa's enclosure'. *Chisewrde* 1086, *-worth* 1330, *Chissewrde -wrthe -worth(e)* 1199–1360, *Chisworth* from 1285. OE pers.n. *Ċissa* + **worth**. Db 80.

CHITHURST WSusx SU 8423. 'Citta's wooded hill'. *Titesherste* 1086, *Chite(s)herst* 1248, 1288, *Chyteherst -hurst* 1279–14th cent., *Chydhurst* 1296. OE pers.n. **Ċitta* + **hyrst**. Sx 33.

CHITTENING Avon ST 5382 > PILNING ST 5585.

CHITTERING Cambs TL 4970. 'Quivering place, quaking bog'. *Chit(t)eringe* 1426, 1434, 1621, *Chet(t)ering(e) -yng* 1423–1539. Dial. **chittering**. Ca 185, ING 175.

CHITTERLEY Devon SS 9404. See under CHEDDAR ST 4553.

CHITTERNE Wilts ST 9944. Uncertain. *Che(l)tre* 1086, *Cet(t)ra -e* 1166–1279, *Chy- Chittern(e)* 1208–1381. Possibly a derivative of PrW ***cę̄d** 'a wood' with suffix **-erno** or OE **ærn** 'a house'; but this is very uncertain. Wlt 163, LHEB 327.

CHITTLEHAMHOLT Devon SS 6420. 'Wood of the people of Chittlehampton'. *Chitelhamholt(e)* 1288–1481, *Chetilmeholt* 1377, *Chittenholt* 1577, *Chittleholt al. Chittlehamholt* 1708. OE **ċietel-hǣme** 'dwellers in the hollow', genitive pl. **ċietel-hǣma**, + **holt**. D 337, L 196.

CHITTLEHAMPTON Devon SS 6325. 'Settlement of the dwellers in the hollow'. *Cvremtone* 1086, *Curemetona* 1086 Exon[†] *Chitelimtona* [1107]1300, *Chitelhamton(e)* from 1176 with variants *-hampton(e), Chitelmetun* 1219. OE **ċietel-hǣme**, genitive pl. **ċietel-hǣma**, + **tūn**. D 338.

CHITTOE Wilts ST 9566. Uncertain. *Chetewe* 1167–1305, *Chi-Chytewe* 1202–1356, *Chitue* 1409. *Chittowe* 1558×1603. Possibly a derivative of PrW ***cęd** 'a wood' with name forming suffix **-ouįā**. Wlt 252 gives pr [tʃitu:], LHEB 327, 376ff.

CHIVENOR Devon SS 5034. 'Cifa's flat-topped hill'. *Chyvenor(e)* 1284–1311, *Chivenore* 1285. OE pers.n. *Ċifa*, genitive sing. Ċifan, + **ōra**. The reference is to the flat-topped ridge stetching from Heanton Punchardon to Tutshill SS 5035–5055. D 45, Jnl 21.19.

CHOBHAM Surrey SU 9762. 'Ceabba's settlement'. *Cebbeham* [1053×66]13th S 1094, *Cebehā* 1086, *Cabbeham* [c.1156]1285, *Chebe-* [672×4, 933]13th S 1165, 420, *Chabbe-* [672×4, 1062]13th S 1165, 1035, *Chabe-* [967]13th S 752–1432, *Chobham* from 1241. OE pers.n. *Ċeabba* + **hām**. Sr 114–15, ASE 2.31.

CHOBHAM COMMON Surrey SU 9665. P.n. CHOBHAM SU 9762 + ModE **common**.

CHOBHAM RIDGES Surrey SU 9159. *Chobham Ridges* from 1765. P.n. CHOBHAM SU 9762 + p.n. *the Rudges* 1607, ultimately OE **hrycg** as shown by the form *Ruggestrate dun* 'ridge-street down' in the later bounds attached to [672×4]13th S 1165, referring to a track along the sand-hills here. Sr 116, ECTV nos. 309, 354.

CHOLDERTON Wilts SU 2342. Possibly 'the estate called after Ceolthryth'. *Celdretone* 1086, 1170×5, *Celdrintone* 1086, *Cheldrington* [1154×89]1270–1324, *Childreton* 1170×5, *Chaldrington* 1341–1472, *Chauldrington* 1557, *Choldrington* 1656, *Chalderton* 1676, *Cholderton oth. Choldrington* 1764. OE feminine pers.n. *Ċēolthrȳth* (or *Ċēolhere* or *Ċēolrǣd*) + **ing**[4] + **tūn**. Wlt 362.

CHOLESBURY Bucks SP 9306. 'Ceolweald's fort'. *Chelewoldesbye* (for *-berye*) 1227, *Chelewoldesbyr'* 1262, *Chelwoldesbury* 1363–77, *Chollisbury* 1526. OE pers.n. *Ċēolweald*, genitive sing. *Ċēolwealdes*, + **byriġ**, dative sing. of **burh**. The reference is to Cholesbury Iron Age hill-fort SP 9307. Bk 91, Thomas 1976.53.

CHOLLERTON Northum NY 9372. 'The settlement at *Ceolford*, the ford in a gorge, or *Ceolanford*, Ceola's ford'. *Choluerton* 1154×95–1316, *Chelverton, Chelreton* 13th cent., *Chollerton* from 1241×6. OE p.n. **Ċeolford*, **ċeole** 'a throat, a channel, a gorge' + **ford**, or *Ċeolanford*, OE pers.n.**Ċeola*, genitive sing. **Ċeolan* + **ford**, + **tūn**. Reduction of *ford* in medial position is regular, cf. BRAFFERTON Durham NZ 2921, DULVERTON Somer SS 9127 and STAVERTON Glos SO 8923. The reference is probably to the gorge of the N Tyne valley between Chollerton and Wall. NbDu 46.

CHOLMONDELEY CASTLE Ches SJ 5351. Built in 1801–30 W of the Old Hall, a 16th cent. half-timbered building remodelled by Vanbrugh c.1713–15 and later demolished: *the newly erected Mansion house called Cholmondeley Castle* 1787. P.n. Cholmondeley + ModE **castle**. Cholmondeley, *Calmundelei* 1086, *Chelmundeleia* 1180–1369 with variants *C(h)ele-* and *-leg(he) -ley, Cholmondeley* from 1400, *Cholmley -ly -leighe* 1397–1724, *Chomley* c.1420–1579, is 'Ceolmund's wood or clearing', OE pers.n. *Ċēolmund* + **lēah**. Che iv.21 gives pr ['tʃoumli] and ['tʃʌmli], Pevsner-Hubbard 1971.176.

[†]Not *Citremetona* as printed by Ellis, DB Devon ii.52, 10.

CHOLSEY Oxon SU 5886. 'Ceol's island' sc. of dry ground in marsh. *Ceolesig, (to) ceolsige* [879×99]12th S 354, *(to) ceolesige* [945]12th S 517, *(æt) Ceolsige* [?after 975]11th S 1494, *(into) Ceolesige* [1003×4]13th S 1488, *(æt) Ceolesege* c.1050 ASC(D) under year 1006, *Celsei -ea* 1086, *Chel(e)seia -eye* 1176–1401×2, *Chausi(e) -(e)ia -y(e) -ey(e)* 1167–1332, *Cholsey* from 1316. OE pers.n. *Ċēol*, genitive sing. *Ċēoles*, + **īeġ**. Brk 162, L 38.

CHOLSTREY H&W SO 4659. 'Ceorl's tree'. *Cerlestreu* 1086, *Scholestre* 1397, *Cholstrey* 1599. OE pers.n. *Ċeorl*, genitive sing. *Ċeorles*, + **trēow**. He 124.

CHOPPINGTON Northum NZ 2583. 'Estate called after Ceabba'. *Cebbington* [c.1040]12th HSC, *Chabiton* 1181, *Chabyn(g)ton -in-* [1183]c.1320–1381, *Chayn(g)ton, Chepynton* 14th cent., *Cheaping- Cheppington* 17th cent. OE pers.n. *Ċeabba* + **ing**[4] + **tūn**. The name was subsequently associated with OE **ċēap** 'trade, merchandise, a market' and underwent the same development as CHOPWELL Durham NZ 1158 and CHOBHAM Surrey SU 9761. NbDu 46.

CHOPWELL T&W NZ 1258. 'The spring where commerce takes place'. *Cheppwell(e)* [1153×9–1317]14th, 1242, 1316, *Chapwell(')* late 12th–mid 14th, *Chopwell* from 1564. OE **cēap** + **wella**. Nomina 12.65ff.

CHORLEY 'Peasants' wood or clearing'. OE *ċeorl*, genitive pl. **ċeorla**, + **lēah**. L 206.

(1) ~ Ches SJ 5751. *Cerlere* 1086, *Chorleg'* 1280, *Chorley* from 1398. Che iii.115.

(2) ~ Lancs SD 5817. *Chorle* early 13th, *Cherleg* 1246, *Cherle(gh) -lag* 1252–78, *Chorley* from 1257. La 131, Jnl 17.74.

(3) ~ Shrops SO 6983. *Cherleye* 1291×2, 1307, *Chorley* 1833 OS. Gelling.

(4) ~ Staffs SK 0711. *Cherlec* 1231, *Churleye* 1306, *Scherleg'* c.1310, *Chorley* from 14th, *Chorley alias Charley* 16th. DEPN, Duignan 39, Horovitz.

CHORLEYWOOD Herts TQ 0396. *(in) bosco de Cherle* 1278, *Charle* 1433, *Charleywood* 1603×25, 1822 OS, *Chorley Wood* 1730. P.n. Chorley 'the wood or clearing of the peasants', OE **ċeorl**, genitive pl. **ċeorla**, + **lēah**, + pleonastic ModE **wood**. Hrt 73, L 206.

CHORLTON 'The peasants' settlement'. OE **ċeorl**, genitive pl. **ċeol(en)a**, + **tūn**. Cf. CARLTON, CHARLTON.

(1) ~ Ches SJ 7250. *Cerletune* 1086, *Cherl(e)ton* 1295–[1342]1438, *Chorl(e)ton* from 1288. Che iii.59 gives prs ['tʃ ɔ:rltən], older ['tʃɔ:rtən].

(2) ~ LANE Ches SJ 4547. 'Lane leading to Chorlton', p.n. Chorlton SJ 4648 + ModE **lane**. Chorlton is *Cherl(e)ton* 1283–1378, *Chorlton* 1284–1588, *Chorlton* from 1330, *Charlton* 1635, *Chalton* 1727, *Chatton* 1769. Che iv. 27.

(3) ~ -UPON-MEDLOCK GMan SJ 8496. *Chorlton upon Medlock* 1843 OS. P.n. *Cherleton* 1177–1201, *Cherlton* 1226, 1278, *Chorelton* 1212, *Chorlton* 1278, + r.n. MEDLOCK. Originally a dependency of the king's manor of Salford. La 32, Lucerna 157, Jnl 17.31.

(4) Chapel ~ Staffs SJ 8138. 'C with the chapel'. *Chapel Chorlton* 1838 OS. ModE **chapel** + p.n. *Cerletone* 1086, *Cherleton* 1267, *Chawlton* 1559. 'Chapel' for distinction from Hill Chorlton SJ 7939. DEPN.

CHORLTON-CUM-HARDY GMan SJ 8193. *Chorlton cum Hardy* 1842. Originally two separate settlements, *Cholreton* 1243–1314, *Chollerton* 1322–1561, *Chorleton* 1551, *Chelverton* 1259–60, 'Ceolferth's farm or village', OE pers.n. *Ċēolferth -frith* + **tūn**, and *Hardey* 1555, 1558, *Hardy Farm* 1842 OS, ME **hard** 'a raised piece of firm land' + **ey** (OE *ēġ*) 'an island', referring to a spur of dry land between Chorlton Brook and the r. Mersey. The spellings for Chorlton are difficult to keep apart from those of CHORLTON-UPON-MEDLOCK SJ 8496 with which it was early confused. La 31. An alternative explanation of p.n. Hardy is 'Hearda's island', OE pers.n. **Hearda* + **eg**. NQ n.s. 44.168.

CHOSEN HILL HOUSE Glos SO 8818 → CHURCHDOWN SO 8820.

CHOWLEY Ches SJ 4756. 'Ceola's wood or clearing'. *Celelea* 1086, *Chelleia -leye* 1208–90, *Scholley* 1284–1463, *Cholley* 1272–1073 with variants *-le(gh)*, *Chowley* from 1551, *T(h)ouley*, *Towley* 1690. OE pers.n. *Ċēola* + **lēah**. Che iv.84 gives prs ['tʃouli] and ['tʃauli].

CHRISHALL Essex TL 4439. 'Christ's nook'. *Cristeshala* [1068]1309, *-halā* 1086, *-hale* 1199–1248, *Criss(h)alle* 1313, 1346. Divine name *Crist*, genitive sing. *Cristes*, + OE **halh**. The reference is to a nook of land at the county boundary. Cf. STRETHALL TL 4939. Ess 521 gives pr [krisəl].

CHRISTCHURCH Cambs TL 4996. So named in 1862 when the church was built. Ca 291.

CHRISTCHURCH Dorset SZ 1592. 'The church of Christ' of Twinham (which it has replaced as the name of the place). *Christi ecclesia de Twinham* 1087×1100, *Cristeschirche of Twynham* 1318, also simply *Christecerce* c.1125, *Christescherche* 1176, *Crischurch* 1354. OE pers.n. *Crīst*, genitive sing. *Crīstes*, Latin *Christi*, + **ċiriċe**. The original name of the place was *(æt) Tweoxn eam* 10th ASC(A) under year 901, *(æt) Tweoxnám* c.1050 ASC(D) under same year, *(æt) twynham* [843 for 934]17th S 391, *Thuinam* 1086, '(place) between streams', OE **(be)tweoxn, (be)twēonan** + **ēam**, dative pl. of **ēa** 'a stream'. The place lies between the rivers Stour and Avon. Do 56, Ha 54.

CHRISTCHURCH Glos SO 5713. *Christ Church* 1830 OS. The church was built in 1816. Verey 1970.104.

CHRISTCHURCH BAY Dorset SZ 2292. P.n. CHRISTCHURCH SZ 1592 + ModE **bay**.

CHRISTIAN MALFORD Wilts ST 9678. *Cristine Malford* 1374, *Christian Malford* 1611, *Curst Mavord* 1585. These forms are all due to popular etymology of the historically correct forms *Cristemal(l)eford* [955]14th S 1575–1330, *(at) Cristemalford* [940]14th S 466, 1330, *Cristemeleford* [937]14th S 434, 1086–1275 with variants *Cristes-* and *-meles-*, 'the ford with a crucifix', OE **cristel-mǣl** < *Crist* + **-el** + **mǣl** varying with **cristesmǣl** < *Crist*, genitive sing. *Cristes*, + **mǣl**, + **ford**. Wlt 67.

CHRISTLETON Ches SJ 4465. '(At) the Christian or the Christians' enclosure or settlement'. *Cristetone* 1086, *Cristentune -tona* 1096–c.1311, *Cristelton* 1157–1653 with variants *Christ- C(h)ryst- -il- -yl-* and *-tone -tun*, *Christleton* from 1200, *Chrisilton* c.1480, *Chrisleton* 1570–1724, *Kysterton*, *Kyrsylton* 16th, *Chris(t)lington* 1620–1792. OE **(Æt) cristenan tune* or **(Æt) cristena tūne*, OE adjective **cristen**, dative sing. **cristenan**, or genitive pl. **cristena**, + **tūn**, dative sing. **tūne**. The changes *r - n* > *r - l* or *r - r* are common in place-names. Che iv.xiii and 107 which gives prs ['krisltən], older local ['krislitən].

CHRISTMAS COMMON Oxon SU 7193. ModE **christmas** + **common**. No early forms but cf. *Christmas Coppice* 1603×25, *Christmas Green* early 18th. Possibly a coppice where holly grew with its associations with Christmas. O 95.

CHRISTON Avon ST 3757. 'Settlement, village, estate of *Cryc*'. *Crucheston* 1197, *Chricheston* 1204. OE p.n. *Cryċ* (from PrW ***crüg** 'a hill'), genitive sing. *Cryċes*, + **tūn**. The OE p.n. is found in *Cyrces gemæro* 1068 'the bounds of *Cyrc*' (with *r*-metathesis for *Cryc*). The reference is to Bleadon Hill which rises to 575ft. half a mile W of the village. DEPN.

CHRISTON BANK Northum NU 2122. *Christon Bank* 1868 OS. Family name *Christon* (1698) + ModE **bank**.

CHRISTOW Devon SX 8384. 'The Christian holy place'. *Cristinestowe* 1244–91 with variants *-e(i)n-*, *Cristowe* 1361, *Christow al. Christenstow* 1618. OE **crīsten**, definite form **crīstena**, + **stōw**. D 430.

CHRIST'S HOSPITAL WSusx TQ 1428. A modern institution.

CHUDLEIGH Devon SX 8679. Partly uncertain. *Ceddelegam* [c.1150]14th, *Ched(d)eleghe -le* [c.1200]15th, 1236–63, *Chedley* 1456, *Chuddeleghe* [c.1200]15th, 1281–1484, *Chi- Chyd(d)eleghe* 1249–1421. Probably 'Ciedda's wood or clearing', OE pers.n. **Ċiedda* + **lēah**, philologically the easiest explanation of the *-e-*, *-u-*, *-i-* spellings. But the specific might be OE **ċēod(e)** 'a bag' referring to the hollow beside which the village lies now traversed by the M5, or to the bag-like outcrop of Chudleigh Rocks. D 489, Studies 1931.70.

CHUDLEIGH KNIGHTON Devon SX 8477 → Chudleigh KNIGHTON.

CHULMLEIGH Devon SS 6814. 'Ceolmund's wood or clearing'. *Calmonlevge* 1086, *Chamundesleg* 1219, *Chaumun(d)leg(h)* 1228, 1254, *Chulmeleg(h)* 1274–97, *Chy- Chilmelegh(e)* 1281–1334, *Chymley* 1600, *Chimleigh or Chulmleigh* 1675. OE pers.n. *Ċēolmund* genitive sing. *Ċēolmundes*, + **lēah**. Forms in *Cal- Cha(u)-* are difficult and point to OE **Ċealmund*, cf. **Calumund* in CALMSDEN Glos SP 0408. D 377 gives pr [tʃʌmli].

CHUNAL Derby SK 0391. 'Nook of land by the throat'. *Ceolhal* 1086, *Chelhala -e* 1185, 1186, 1216×72, *Cholhal* 1216×72, *-ale* 1309 etc., *Chunall* 1767. OE **ċeole** + **halh**. The reference is to nearby Long Clough (OE **clōh** 'a deep valley, a ravine'). The modern form shows an unusually late example of the dissimilation *l - l* > *n - l*. Db 69, L 106.

CHURCH Lancs SD 7428. 'The church'. *Chirche* 1202–1332 etc., *Chyrche* 1202–85, *Churchkyrk* 1536. OE **ċiriċe**. Still called locally Church Kirk with pleonastic ModE **kirk** (ON **kirkja**). La 90, Mills 73, Jnl 17.54.

CHURCHAM Glos SO 7618. 'Ham with the church'. *Chi- Chyrch(e)hamme -homme* 12th–1329, *Churchehamme* 1200, *-ham* 1425. ME **chirche** + p.n. *Hamme* 1086, 'river meadow', OE **hamm**. 'Church' for distinction from HIGHNAM SO 7919 'Ham belonging to the monks (OE *hiġna*)'. Gl iii.196, L 43.

CHURCH COVE Corn SW 7112. *Church Cove* 1888. ModE **church** + **cove**. The reference is to nearby Landewednack church. PNCo 68.

CHURCHDOWN Glos SO 8820. Probably 'church hill'. *Circesdvne* 1086, *Chi- Chyrchesdon(e) - d(o)un* 12th–1400, *Chi- Chyrchedon(e)* 1221–1458, *Churchdon'* 1221, *Church(e)down(e)* 1543–90, *Chorsdown* 1719, *Churson* 1577. OE **ċiriċe**, secondary genitive sing. **ċiriċes**, + **dūn**. OE *ċiriċe* is normally a weak feminine noun with genitive sing. *ċiriċan* but the *-es* genitive is found in the 10th cent. boundary survey of S 345. The parish church is on top of a hill which does not have the distinctive profile of the kind of hill called *cryc* from PrW **crüc*, the usual explanation of this name. The 1577 local form *Churson* survives in the name Chosen Hill Ho SO 8818. The reference is to Churchdown Hill rising to 500ft. Gl ii.119.

CHURCH END ModE **church** + **end**.

(1) ~ Beds SP 9921. 'Church end' sc. of Totternhoe. No early forms.

(2) ~ Beds TL 1937. *Church End* 1834 OS, sc. of Arlesey. Contrasts with *Dean End* and *South End* 1834 ibid.

(3) ~ Cambs TF 3909. Sc. of Parson Drove. No early forms. The site is marked as *Parson Drove* in 1824 OS. The church is 13th cent. Contrasts with Gate End, *Gatysende* 1480, and RING'S END TF 3902. Ca 279, Pevsner 1954.365.

(4) ~ Cambs TL 4857. Sc. of Cherry Hinton. No early forms.

(5) ~ Hants SU 6756. No early forms. Short for Church End of Sherfield on Loddon SU 6758.

(6) ~ Warw SP 2992. *Church End* 1840 sc. of ANSLEY SP 3091.

(7) ~ Wilts SU 0278. *Church end* 1828 OS. Short for Church End of Lyneham as opposed to Townsend SU 0079.

CHURCHEND 'The church end' of a village. ModE **church** + **end**.

(1) ~ Essex TL 6323. *Cherche ende* 1529, *le Churche Yende* 1549. The church end of Great Dunmow. Ess 477.

(2) ~ Essex TR 0092. The church end of Foulness in contrast to COURTSEND TR 0293.

CHURCH FENTON NYorks SE 5136 → Church FENTON.

CHURCH HOUSES NYorks SE 6697. ModE **church** + **houses**.

CHURCHILL 'Hill called *Cruch, Crich,* the prominent hill or tumulus'. PrW ***crüg** + **hyll** frequently taken as 'church hill', OE **ċiriċe** + **hyll**. Signposts 139, L 139.
(1) ~ Devon ST 2901. *Churchill* 1809.
(2) ~ Oxon SP 2824. *Cercelle* 1086–15th with variants *-ell(a) -hil* and *-hull(a), Churchehull(') -hill(e)* 15th cent. The old church is overlooked by a large hill. O 343.
(3) ~ H&W SO 8879. *Circhul* 11th, *Cercehalle* 1086, *Chyrchull* 1275. Wo 278, L 139.

CHURCHILL Avon ST 4459. 'Church hill'. *Cherchille* 1201, *Chyrchehull* 1243. OE **ċyriċe** + **hyll**. If, however, the form *Crichell* [n.d.]12th in Bath Abbey Chartulary belongs here the specific is OE p.n. *Cryċ* from PrW ***crüg** 'a hill'. DEPN.

CHURCHINFORD Somer ST 2112. 'Churcham ford'. *Suthchurchamford* 1386, *Churchamford* 1499, *Churchingford* 1809 OS. P.n. *Churcham* 'church homestead or meadow', OE **ċiriċe** + **hām** or **hamm**, + **ford**. There are remains of an ancient church here, now part of a farm building. D 619.

CHURCH KNOWLE Dorset SY 9381 → Church KNOWLE.

CHURCHOVER Warw SP 5180. 'Waver with the church'. *Chirchewavre* [1100×35]1397, 1325, *Chirche Wavere* 1247–1397, *Chirchewover* 1427, 1468, *Church Over al. Waver* 1535. ME **chirche** + r.n. *Wavre, Wa(v)ra* 1086, *Waver(e)* 13th cent., *Wover* 1445, an earlier name of the Swift, OE **wæfre** 'wandering' as in river WAVER Cumbr NY 1950. 'Church' for distinction from BROWNSOVER SP 5177 'Brun's Waver'. Wa 103.

CHURCHSTANTON Somer ST 1914. 'Cherry Stanton'. *Cheristontone* 1258×79, *Chery Sta(u)nton* 1282–5, *Churestaunton al. Churchestaunton* 1512. ME **cheri** + p.n. *Stantone* 1086–1291 with variant *Staun-*, 'stone settlement', OE **stān** + **tūn**. The substitution of Church for Cherry is a piece of late folk-etymology. D 618.

CHURCHSTOW Devon SX 7145. 'Place with a church'. *Churechestowe* 1242, *Chyri- Chyrestowe* 1244, 1281, *Chur(e)stowe* 1276, 1308, *Churche Stoke* 1459, *Chestow* 1577. OE **ċiriċe** + **stōw**. D 295 gives pr [tʃə:stou].

CHURCH STREET Kent TQ 7174. ModE **church** + Kent dial. **street** 'village, hamlet'. The original name of this settlement was Higham as in HIGHAM TQ 7171. It was supplanted at the time when Higham Upshire was commonly shortened to its first element only. The reference is to the settlement beside the parish church, St.Mary's, on the edge of the marshes. In the neighbourhood there are numerous examples of other street names, Chequers Street in the same parish from the name of the inn replacing *Upper Higham* 1805 OS, Cooling Street TQ 7474 in Cooling, *Cowling Street* 1805, West Street TQ 7376 in Cliffe, Fen Street and Sharnal Street TQ 7974 (from the *Sharnwelle* 1332, the 'filthy spring', but recorded as *Hog Street* 1805) in High Halstow, Bill Street, *Bilstrete* 1372, Home Street, *Westhollestrete* 1292, *Holme* 1332, OE **holeġn** 'a holly tree', Oak Street, *Haven Street* 1819 OS, otherwise Haydan Street, *Haddun* 889 S 1276, *Hathdune, œdune* c. 975 B 1321–2, *Hadone* 1086, *Hey- Hayton(e)* 14th, *Heyton Fee* 1572, *Havonfee* 1613, 'heath hill', OE **hǣth** + **dūn**, Silver Street TQ 8760 and Sole Street, *la sole* 'muddy place', OE **sol**, all in Frindsbury. Similar systems appear elsewhere in Kent, cf. under BROAD STREET TQ 8356. PNK 115, 120, KPN 228.

CHURCHTOWN Lancs SD 4843. *Church Town* 1786. The church settlement of Kirkland township in Garstang parish, site of St Helen's church, the old church of Garstang. Kirkland is *Kirkelund* [c.1230]1268, *-land* 1362, 'the church grove', ON **kirkja** + **lundr**. Room 1983.24, La 163, Jnl 17.95.

CHURCHTOWN Mers SD 3518. A suburb of Southport formerly called North MEOLS. *Church Town* 1895. St Cuthbert's church was built here between 1730 and 1739. Pevsner 1969.99.

CHURN CLOUGH RESERVOIR Lancs SD 7838. P.n. Churn Clough + ModE **reservoir**. There are no early forms for Churn Clough, but it may be dial. **churn** 'a daffodil' + **clough** 'a ravine' (< OE *clōh*).

River CHURN Glos SP 0108. Unexplained. *(innon, on) Cyrnéa, Cirnea* [c.800]11th S 1556, [852]13th S 202, *(innan, on, andlang) Cyrne* [999]13th S 896, *Cheern* 15th, *Churn(e)* c.1540–1695. OE **Ċiren*, lOE *Ċyren* + OE **ēa**. The meaning of *Ċiren* which occurs again in CIRENCESTER SP 0201 and ultimately derives from the Romano-British name CORINIUM is unknown. Gl i.5, RN 78.

River CHURNET Staffs SK 0345. Unexplained. *Chirned* 1239, *Chirnet* 1240–1318, *Churnet(t)* c.1275 and from 1540. A pre-English r.n. of unknown origin and meaning. St i.7, RN 79, Horovitz.

CHURNSIKE LODGE Northum NY 6677. *Churnsike Lodge* 1866 OS. P.n. *Churn Sike* + ModE *lodge*. Churn Sike may be ModE **churn** + **sike** 'a small stream' referring to the turbid nature of the water.

CHURT Surrey SU 8538. 'The rough ground'. *Cert* [685×7]c.1135 S 235, *Chert* 1210–1561, *(la) Churt* from 1206, *Ch(e)arte* 1583–1765. OE **ċert**. Sr xix, 178.

CHURTON Ches SJ 4156. Possibly 'hill-settlement'. *Churton* from c.1170, *Chi- Chyrton* c.1190–1535. OE ***cyrc**, ***crȳċ** < PrW **crüg* + **tūn**. The village occupies a small circular elevation. Che iv.70.

CHURWELL WYorks SE 2729. 'The peasants' spring'. *Cherlewell -wall* 1226, *Chorl(l)ewel(l)e* 13th cent., *Churlwell* 1418, 1489, *Chorwell* 1503, *Charlewell* 1565, *Charwell* 1608, *Churwell* from 1560. OE **ċeorl**, genitive pl. **ċeorla** + **wella**. The 1226 form with *-wall* points to the Mercian dial. form **wælla**. YW iii.222, L 32.

CHUTE CAUSEWAY Wilts SU 2955. The Roman road from Winchester to Mildenhall, Margary no. 43. P.n. Chute as in Upper CHUTE SU 2953 + ModE **causeway**. The road here is a magnificent *agger* 27ft. wide and up to 4ft. high.

Lower CHUTE Wilts SU 3153. ModE **lower** + p.n. Chute as in Upper CHUTE SU 2953.

Upper CHUTE Wilts SU 2953. ModE adj. **upper** + p.n. *Cett'* 1235, *Chuth'* 1268, *Cheut* 1289–1343×77. The village takes its name from Chute Forest, *(silva que vocatur) Cetum* 'the wood called C' 1086, *Cet(te)* 1210–53, *Chet(t)(e)* 1231–72, *Chute* from 1283, PrW ***cę̄d** 'a wood', borrowed into OE as ***cę̄t**, WS *ċīet*, later *ċīt* and *ċȳt*. Wlt 12, 340, LHEB 327.

CHYANDOUR Corn SW 4731. 'Cottage of the stream'. *Chiendour* 1452, *Chyendower* 1504. Co **chy** + **an** + **dour**. PNCo 69, Gover n.d. 623.

CHYSAUSTER Corn SW 4535. 'Cottage of Silvester'. *Chisalwester* 1302, *Chisalvestre* 1313. Co **chy** + pers.n. *Silvester*. PNCo 69 givs pr [tʃe'zɔistə].

CILURNUM Northum NY 9170. 'The cauldron'. *Cilurno* [c.400]15th ND, *Celunno, Celumno* (sic) [c.700]14th Rav. OW **cilurnn** 'a bucket, a pail' (Brit **cilurno-*, ModW *celwrn*) referring to a deep natural pool in the N Tyne river known as The Ingle Pool. The RBrit name for Chesters Roman fort. RBrit 307, GPC 458.

CINDERFORD Glos SO 6514. 'Cinder ford'. *Si- Synderford* 1258–1320, 1612. OE **sinder** + **ford**. The ford carries the Roman road from *Ariconium* to Lydney, Margary no.614, across Soudley Brook at SO 6572. The reference is to slag from iron-smelting which took place in the Forest of Dean from Roman times to the late Middle Ages. Gl iii.217.

CIRENCESTER Glos SP 0201. 'The Roman town *Corinium*'. *Cirenceaster, œt, to Cirenceastre* late 9th ASC(A) under years 577, 628, 879, 880, c.1120 ASC(E) under years 879, 880, *Cirrenceastre, Cairceri* 894 Asser, *Cair Ceri* 11th Nennius, *Cirneceastre* c.1120 ASC(E) under year 628, [999]13th S 896, *(œt) Cyrenceastre* early 12th ASC(D) under year 1020, *Cyrneceaster* [990×1006]13th S 937, *(on) Cyrnceastre* c.1120 ASC(E) under year

1020, *Cire- Cyrecestr(a)'* 1086–1413, *Ci- Cyrencestr(e)'* 12th–1759, *Syrencestre* 1439–c.1560, *Cisetur* 1453, *Cicet(t)er* c.1490, 1718, *Sissetur* c.1500, *Sisator* 1685. Romano-British p.n. CORINIUM + OE **ċeaster**. RBrit *Corinium* would become PrW **Cerin*, OE **Ċiren*, lOE *Ċyren* whence the r.n. CHURN SP 0108 with normal [tʃ] and Ciren- and CERNEY with abnormal [s]. Gl i.60 gives prs ['sairən'sestə, 'saiRən, 'zaiRnsestə, 'sisitə], the latter a common 15th–18th cent. pr but not used locally today.

CISSBURY RING WSusx TQ 1308. 'Cissa's fort'. *Sieberie hill* c.1588, *Sissabury* 1610, *Ciss(i)- Sizebury* 18th. Earlier simply 'the old *byry*' 1477. A 16th cent. antiquarian invention in order to associate this prominent Iron Age hill-fort with Cissa, third son of Aelle who, according to ASC, led the Anglo-Saxon invasion of what was to become Sussex in the year 477. Cf. also CHICHESTER SU 8604 and CHISBURY Wilts SU 2766. Sx 197, Thomas 1980.210.

The CITY Bucks SU 7896. No early forms. *City* occurs occasionally in modern field and minor names mostly sarcastically of a small settlement. Cf. Brk 284.

Great CLACTON Essex TM 1716. *Great Claketon* 1286, *Magna* ~ 1335, *Mochel Clacton* 1431×2, *Much Clafton* 17th cent. ME adjs. **great, much(el)** + p.n. *(of) Claccingtune* [c.1000]c.1125, *Clachintuna* 1086, *Clachentona* 1127, *Clakinton -tun* 1206×7–1241×3, *Claketon* 1204–1335, *Clacton* from 1318×46 with variant *Clak-*, *Clatton* 1426–1607, *Claston* 1539–1685, *Clackton Magna corruptlie Claston* 1594, *Clafton* 1550–1727, *Claffhen* 1688, either 'estate called after Clacc', OE pers.n. **Clacc* + **ing**[4] + **tūn**, or 'settlement at or called Claccing, OE ***claccing** 'hill, hill-like place' < ***clacc** + **ing**[2], + **tūn**. The spelling *Clafton*, for which *Claston* is a corrupt form owing to *f/ſ* confusion, represents the development [k] > [x] > [f] as also in CLAUGHTON Lancs SD 5666. 'Great' for distinction from Little CLACTON TM 1618. Ess 334, Jnl 2.45.

Little CLACTON Essex TM 1618. *Claketon Parva*, *Little Claston* 1545, ~ *Claffhen* 1688. ModE adj. **little**, Latin **parva**, + p.n. Clacton as in Great CLACTON TM 1716. Ess 334.

CLACTON-ON-SEA Essex TM 1715. An early Victorian resort development named after Great CLACTON TM 1716. Pevsner 1965.126.

CLAIFE HEIGHTS Cumbr SD 3797. P.n. Claife + ModE **height(s)**. Claife, *Clayf* 1272×80–1400, is ON **kleif** 'steep hillside up which there is a path' referring to the steep slope rising from Lake Windermere. Cf. the Norwegian p.n. Kleiv. La 219, SSNNW 114.

CLAINES H&W SO 8558. 'Clay headland'. *Cleinesse* 11th, *Clei- Cleynes* 1234–91, *Claynes* 1283, 1428. OE **clæġ** + **ness**. The reference is to a slight headland on which the church stands. Wo 110, L 173.

East CLANDON Surrey TQ 0551. *East Cleyndon* 1243. ME adj. **est** + p.n. *Clanedun* 1086, *Clandone* 1240, 1327, *Clendon* 1201–1325, *-done* [727, 933]13th S 1181, 420, *-dune* [967]13th S 752, *(æt) Clenedune* [1062]13th S 1035, *-don'* 1206, 'the clean or clear upland', OE **clǣne** + **dūn**. Also known as *Clendon Abbatis* 'abbot's C' 1281–1428. The reference is to the smoothness of the chalk downs above the village, which belonged to Chertsey Abbey before the 16th cent. 'East' for distinction from West CLANDON TQ 0452. Sr 137–8, L 144, 146.

West CLANDON Surrey TQ 0452. *Westclanedon* 1225, *West Clandon alias Kynges Clandon* 1565. ME adj. **west** + p.n. *Clanedun* 1086, *altera Clenedone* [727]13th S 1181, ~ *Clendone* [933]13th S420, ~ *Clenedune* [1062]13th S 1035, *Glen- Cleendon* 1241, *Clyn-* 1263, 'the clean or clear upland', as in East CLANDON TQ 0551. Also known as *Clendon Regis* 1290–1428, *Kynges* ~ 1313. Sr 138–9.

CLANDOWN Avon ST 6855. *Clan Down* 1817. Probably identical with CLANDON Surrey, 'clean hill', i.e. one free from vegetation or harmful growth such as thorn-bushes and the like. Cf. L 144, 146.

CLANFIELD Hants SU 6916. 'Clean open land'. *Clanefeld* 1207–1341, *Clenefeld* 13th cent. OE **clǣne** + **feld**. This name-type is in contrast with names such as Driffield, Horfield 'dirty open land'. 'Clean' may refer to absence of weeds or undergrowth. Clanfield adjoins the Weald and marks the boundary between open and wooded countryside. Cf. CLANVILLE SU 3149. Ha 55, Gover 1958.58, L 226.

CLANFIELD Oxon SP 2802. 'Clean open land', i.e. uncluttered by undergrowth or weeds. *Chenefelde* (sic) 1086, *Clenefeld(e)* 1196–1320, *Clanefeld(e) -feud(') -felt* 1204–1395, *Clanfeld(e) -fyld* from 1275×6. OE **clǣne** + **feld**. O 312, L 242.

CLANNABOROUGH BARTON Devon SS 7402. 'Clannaborough farm'. P.n. Clannaborough + SW dial. **barton** 'farm, farmyard' (OE *bere-tūn*). Clannaborough, *Cloenesberg* 1086, *Cloueneberge* 1239, *-burg(he)* 1242–1334, is 'cloven hill', OE **clofen** + **beorg**. The precise topographical feature in question is unclear. The church lies below a ridge which dips at this point; several valleys run up to it. D 364, L 88, 128.

CLANVILLE Hants SU 3149. 'Clean open land'. *Clavesfelle* 1086, *Clanefeud* 1256, *Clanefeld* 1263–1316, *Clanfeld* 1504, *Clanvell* 1654, *Clanvill* 1695. OE **clǣne** + **feld** as in CLANFIELD SU 6916. Clanville adjoins Chute Forest, PrW **cę̄d* 'forest', cf. Upper and Lower CHUTE Wilts SU 2953, 3153; the name marks the boundary between open and wooded countryside. Ha 55, Gover 1958.170, DEPN, L 242.

CLAPGATE Dorset SU 0102. '*Clapgate* 1811. ModE **clap-gate**, 'a small gate which shuts when slammed or swings to of itself'. This place lies on the boundary of Colehill parish and the name probably refers to a gate of the forest of Holt. Do ii.138.

CLAPHAM Beds TL 0252. 'Homestead on or by a hillock'. *Clopham* [986]n.d. ECTV, *Cloppham* *[1060]14th S 1030, *Clopeham* 1086, *Clapham* 1247. OE ***clopp(a)** + **hām**. Bd 22, Studies 1936.136.

CLAPHAM GLond TQ 2875. 'Hillock enclosure'. *Cloppaham* 871×89 S 1508, *Clop(p)eham* 1086–1325, *Clopham* 1184–1500, *Clapham* 1503. OE ***clopp(a)** + **hamm**. The form *Clophammesgrave* 1255 'wood belonging to Clapham' seems to prove that this is a *hamm* name rather than a *hām*. Unrounding of *o* to *a* is regular in non-standard varieties of English in the ME period. Sr 21, Jordan §272, Dobson §87.

CLAPHAM NYorks SD 7469. 'Homestead on *Clæpe*, the noisy stream'. *Clapeham* 1086–1559, *Clapham* from 12th. OE stream-n. **Clæpe* related to OE *clappettan* 'throb, palpitate', ModE *clap*, OHG *chlaffon* 'to sound' and G stream-n. Klaffenbach, + **hām**. YW vi.232.

CLAPHAM WSusx TQ 0906. 'The hill homestead'. *Clopeham* 1073, 1086, *Clopham* 1230–1428, *Clopham al. Clapham* 1625×49. OE ***clopp(a)** + **hām**. Sx 195, Studies 1936.137, ASE 2.33, SS 86.

CLAPPERSGATE Cumbr NY 3603. 'Road of the rough bridge'. *Clap(p)ergate* 1588, 18th cent., *Clappersgate* from 1588. Dial. **clapper** 'a rough bridge, esp. one made of planks laid on piles of stones' + ON **gata**. Possibly identical with Brathay Bridge. We i.209.

CLAPTON Somer ST 4106. 'Hill settlement'. *Clopton* [1174×9]14th, 1243, 1274, *Cloptune* [1246]14th, *Clapton* 1811 OS. OE ***clopp(a)** + **tūn**. The reference may be to Shave Lane Hill ½ mile E of the village although this is rather larger than the sense 'hillock' usually assumed for **clopp(a)*. DEPN, PNE i.99–100.

CLAPTON-IN-GORDANO Avon ST 4774. 'Clapton in *Gorden*, the triangular valley'. Earlier simply *Clotvne* 1086, *Cloptun* 1225, 'settlement, village by the hill', OE ***clopp(a)** + **tūn**. Gordan, *Gordeyne* 1270, *Gordon* 1293, *(in) Gordano, (in) Gorden(e)* 1333, refers to the long gore-shaped valley stretching from Walton-in-Gordano ST 4273 NE to the river Avon, OE **gār** + **denu**. One form, *(in) Gordenlond* 1271, may show that the suffix was interpreted as a form of ME **gardin** 'a garden'. DEPN.

CLAPTON-ON-THE-HILL Glos SP 1618. *Clopton super montem* 1590, *Clapton on the hill* 1641. Earlier simply *Clopton* 1171×83–1611, *Clapton* 1577, 1679, 'hill settlement', OE ***clopp(a)** + **tūn** with SW dial. *a* for *o*. The village is situated on a round hill rising to 725ft. Gl i.198, Studies 1936.136–40, PNE i.99.

CLAPWORTHY Devon SS 6724. Partly uncertain. *Clobworthy* 1731, *Clapworthy Mill* 1809 OS. The specific is ModE **worthy** 'an enclosure' (OE *worthiġ*), but the evidence is too late for further explanation. D 347 gives pr [klaperi].

CLARBOROUGH Notts SK 7383. 'The fortified place where clover grows'. *Claueburch, Claureburg, Clavrebvrg* 1086, *Claverburg(h) -burc(h)* 1154×89–1316×24, *Clareburg(h)* 1242–1343, *Clarburgh* 1286. OE **clǣfre** + **burh**. Nt 27.

CLARE Suff TL 7745. Uncertain. *Claram* (Latin accusative case) 1086, *Clara* c.1145, *Clare* from 1198. Possibly identical with the r.n. in p.ns. BURGH- HIGH- and KINGSCLERE Hants SU 4660, 4358 and 4258. The forms for the Cleres, however, are much more varied. DEPN.

Goodworth CLATFORD Hants SU 3642. 'The Goodworth part of Clatford'. *Goodworth Clatford* 1579. P.n. *Godorde* 1086, *Gudeworthe* 1235, 'Goda's enclosure', OE pers.n. *Gōda* + **worth**, + p.n. Clatford as in Upper CLATFORD SU 3543. The manor of Clatford developed three different foci, at Upper CLATFORD SU 3543, Goodworth Clatford and Lower Clatford to which this name subsequently transferred. Ha 83, Gover 1958.171.

Upper CLATFORD Hants SU 3543. *Upclatford* 1306. ME **uppe**, ModE **upper**, + p.n. *Cladford* 1086, *Clatford* from 1156, 'burdock ford', OE **clāte** + **ford**. 'Upper' for distinction from Goodworth CLATFORD SU 3642 further downstream from here. Ha 55, Gover 1958.165.

CLATWORTHY Somer ST 0530. 'Burdock enclosure'. *Clatevrde* 1086, *-wurth* 1227, *Clatewurthy* 1243, *Clatworthy* 1610. OE **clāte** + **worth**, later **worthiġ**. DEPN.

CLAUGHTON 'Settlement on or by the hill', OE ***clacc** + **tūn**.

(1) ~ Lancs SD 5342. *Clactune* 1086, *Clacton* 1185, 1208 etc., *Clahton* 1252, *Claghton* 1285–1332 etc., *Clawton* 1348, *Clayghton* 1554. Claughton lies on sloping ground near several marked hills, e.g. Sullom Hill SD 5244, rising to 525ft. The 1554 form shows the results of lME *ā* < *au* surviving in the present-day pronunciation ['klaitən]. La 162, Jnl 17.95.

(2) ~ Lancs SD 5666. *Clactun* 1086, *Clahton* 1208–55, *Klacton* 1250×75, *Clauton* 1241, *Clafton* 1246, *Claghton* 1297–1332 etc., *Claughton* 1297. Claughton lies at the foot of Claughton Moor which rises steeply to 1184ft. at SD 5863. The spellings of this name and the modern pronunciation [klaftn] exhibit two parallel developments, (1) ME [ax] > [aux] > [au], and (2) substitution of neighbouring fricative phoneme [f] for peripheral [x]. La 178, Jnl 17.101.

CLAVERDON Warw SP 1964. 'Clover hill'. *Clavendone* 1086, *Claverdona -e* 1123–1547, *Claredon* 1316–1625, *Claverdon* from 1521. OE **clǣfer** + **dūn**. Wa 206.

CLAVERHAM Avon ST 4466. Partly uncertain. *Clivehā* 1086, *Claverham* 1248. Probably OE **clāfre** 'clover' + **hamm** 2a 'promontory of dry ground into marsh': but OE **hām** is not impossible. The DB form is anomalous. DEPN.

CLAVERING Essex TL 4832. 'Place growing with clover'. *Clǣfring* c.1000, *Clauelinga* 1086, *Clauering(a), Clavering(e) -yng(g)* 1154–1335, *Claveringes* 1177–1306. OE **clǣfre** + **ing**². Ess 548, ING 189, Jnl 2.48.

CLAVERLEY Shrops SO 7993. 'Clover clearing'. *Claverlege* 1086–1636 with variants *-lai -lay -le(ga) -ley(e)* etc., *Clarel(e)y* 1361–1667, *Clearly* 1698. OE **clǣfre** + **lēah**. Sa i.82 gives local pr [cla:li].

CLAVERTON Avon ST 7864. 'Settlement at Clatford'. *(æt) Clatfordtune* [984×1016]12th S 1538, *Claftertone* 1086, *Clav- Claferton* from 1212. P.n. **Clatford* 'ford where burdock grows', OE **clāte** + **ford**, + **tūn**. A ford across the river Avon. DEPN.

River CLAW Devon SX 3498. A back-formation from CLAWTON SX 3599. *Claw* c.1620. D 3, RN 80.

Long CLAWSON Leic SK 7227. *Long Claxton* 1632, *Long Clauston* 1725, *Long Clawson* 1710. ModE adj. **long** + p.n. *Clachestone* 1086, *Clacstun(e)* 12th, *Claxtun* 1236, *Claxton(e)* c.1130–1718, *Claucstuna* [12th]early 15th, *Clauxton'* 1216×72, early 14th, 1564, *Clau- Clawson* from 1539, either 'Klakk's farm or village' or 'settlement of the hill or peak', ON pers.n. *Klakkr*, genitive sing. *Klakks*, or ODan **klakk**, OE ***clacc**, genitive sing. ***clacces**, + **tūn**. 'Long' refers to the linear shape of the village. Lei 154, SSNEM 189.

CLAWTON Devon SX 3599. 'Settlement on the claw'. *Clavetone* 1086, *Clavatona* 1088, *Clauton'* 1242 etc. OE **clawu** + **tūn**. The reference is to a 'claw' of higher ground between two streams on which the church stands. D 138.

CLAXBY 'Klak's village or farm'. ON pers.n. *Klakkr*, ODan *Klak*, genitive sing. *Klaks*, + **bȳ**.

(1) ~ Lincs TF 4571. *Clachesbi* 1086, *Clakesbi* 1176–1202. DEPN, SSNEM 41, Cameron 1998.

(2) ~ Lincs TF 1194. *(æt) Cleaxbyg* [1066×8]c.1200 Wills, *Clachesbi* 1086, *Clakesbi -by* 1150–[1190]1302, *Claxbi -by* from 1209. Li iii.17, SSNEM 41.

CLAXTON 'Clacc's. farm or estate'. Late OE pers.n. *Clacc* (probably representing ODan **Klak*), genitive sing. *Clacces*, + OE **tūn**.

(1) ~ Norf TG 3303. *Clake- Clarestona* (sic) 1086, *Clakeston'* 1200, *Claxtone* 1254, 1275. DEPN, SPNN 264.

(2) ~ NYorks SE 6960. *Claxtorp* 1086, *Claxton(a)* from 1282. The earlier spelling, if reliable, is an alternative form with ON **thorp** 'outlying farm'. YN 37.

CLAY CROSS Derby SK 3963. *Clay-Cross* 1734. The name of the Civil Parish in Clay Lane, *Cley Lane* 1573, from the surname *Clay*, a family well established in the area, + ModE **cross** 'a cross-roads'. Db 237.

CLAYBROOKE MAGNA Leic SP 4988. 'Great Claybrooke'. Latin prefix **magna**, *Magna* 1261–1316 (suffix 1428), varying with ME **nether** 'lower', *Nether* 1399–1725, + p.n. *Claibroc* 1086–1210, *Cley- Claybrok(e)* 1265–1576†, the 'stream with a clay bed', OE **clæġ** + **brōc**. The soil is light clay, and loam and sand overlying clay. 'Magna' for distinction from Claybrooke Parva SP 4987, *Parva* 1261–1316, 'little C', also known as Over Claybrooke 'upper Claybrooke', *Over-* 1596, 1721. Lei 436–7, L 15.

East CLAYDON Bucks SP 7325. *Est Cleydon* 1247. ME adj. **est** + p.n. *Clai(n)done* 1086, *Cleindona* c.1200, *Cleydon* 1220, '(at) the clay hill', OE **clǣġiġ**, definite form dative case **clǣġan**, + **dūn**. 'East' for distinction from Botolph Claydon SP 7324, *Botle Cleidun* 1224, *Claydon St Botolph* c.1825, 'Claydon with the buiding' (OE **botl** later misinterpreted as saint's name *Botolph*) and Middle and Steeple CLAYDON SP 7225, 6926. Bk 131 gives pr [bɔtl], 133, DEPN, L 144, 150, 153.

CLAYDON Oxon SP 4550. '(At) the clayey hill'. *Clein- Claindona* 12th cent., *Clayndone* 1241, 1247, *Claydon* from 1215. OE adj. **clǣġiġ**, dative sing. definite form **clǣġiġan** + **dūn**, dative sing. **dūne**. O 418.

CLAYDON Suff TM 1350. 'The clay hill'. *Clainduna -e* 1086, *Cleidun* 1198, *Claidon'* 1166, 1168, *Claydon* from 1315, *Cleydon* 1610. OE ***clæġen** + **dūn**. DEPN.

Middle CLAYDON Bucks SP 7125. *Middelcleydon* 1242. Earlier simply *Claindone* 1086 as in East CLAYDON SP 7325. Bk 133.

Steeple CLAYDON Bucks SP 6926. 'C with the church tower'. *Stepel Cleydon* c.1220–1357. Earlier simply *Claindone* 1086, *Cleindon* 1200, c.1215, *Claydon* 1235, as in East CLAYDON SP 7325. The village stands on Oxford clay. Bk 53.

†The form *clæg broc, (on) clæg broce* [962]13th S 833 does not belong here, TBWAS 90.85.

CLAYGATE CROSS Kent TQ 6155. 'Claygate crossroads'. P.n. Claygate + ModE **cross**. Claygate, *Cleygate* 1270, is the 'clay gate', OE **clǣġ** + **ġeat**. PNK 154.

CLAYGATE GLond TQ 1563. 'Gate leading to clayey ground'. *Clæigate* [1066]12th S 1043, *Cleigate* [1066]c.1400 ibid., *Claigate* 1086. OE **clǣġ** + **ġeat**. The place lies on the London clay. Sr 91, DEPN.

CLAYHANGER 'Clay wooded slope'. OE **clǣġ** + **hangra**. L 195.
(1) ~ Devon ST 0222. *Clehangre* 1086, *Cley(h)angre* 1271, 1283, *Clayhangre* 1322. D 533.
(2) ~ WMids SK 0404. *Cleyhungre* 1216×72. Cf. CLAYHANGER Devon ST 0223, Clayhanger Lane, Wadeford, Somer. DEPN, L 195.

CLAYHIDON Devon ST 1615. 'Clay Hidon'. *Cleyhidon* 1485, ME **cley** + p.n. *Hidone* 1086–1428 with variant *-don*, 'hay hill', OE **hīeġ** + **dūn**. The parish which is on the Black Down Hills is remarkable for its clay. D 610, L 144, 148.

CLAYPOLE Lincs SK 8549. 'The pool on or by the clay ground'. *Claipol* 1086, *Clai- Clay- Clei- Cleypol(e)* 1185–1642. OE **clǣġ** + **pōl**. The reference could be to a 'pool' or deep place in the r. Witham. Perrott 368.

CLAYTON 'Settlement on clayey soil', OE **clǣġ** + **tūn**.
(1) ~ Staffs SJ 8543. *Claitone* 1086, *Cleyton* 1254. DEPN.
(2) ~ SYorks SE 4507. *Clai- Clayton(e) -tun* 1086–1620. Also known as *Clayton in the Clay* 1771 for distinction from CLAYTON WEST SE 2510. YW i.89.
(3) ~ WSusx TQ 3014. *Glai- Glaytone* 1073, 1274, *Claitune* 1086, *Clayton(e)* from 1284 with variant *Clei-* 12th–14th. Part of the parish lies on the gault, a clay formation. Sx 259, SS 75.
(4) ~ WYorks SE 1231. *Claiton(e) -tun* 1086, 13th, *Clayton(a)* 1177×93–1549. YW iii.255.
(5) ~ -LE-MOORS Lancs SD 7431. 'C on the moors'. *Clayton super Moras* 1284, *~ on the Moors* 1390. Earlier simply *Cleyton* 1243, *Clayton* 1263, 1277. La 89.
(6) ~ -LE-WOODS Lancs SD 5622. 'C in the woods'. Earlier simply *Cleitonam* 1160, *Claiton* 1180×95–1332, *Cleyton* [c.1200]1268 etc., *Clayton* 1246. For the suffix cf. nearby WHITTLE-LE-WOODS SD 5821. La 134, Jnl 17.76.
(7) ~ WEST WYorks SE 2510. *Clayton West* 1822. P.n. *Clac- Claitone* 1086, *Cleitun -ton* 12th, 1204×9, *Cleyton* 1284, 1526, *Clai- Claytun -ton(e)* 1194–1549, *Claton* 1297, *Cleeton* 1444, + ModE **west**. YW i.320.

CLAYWORTH Notts SK 7388. 'The enclosure on the claw'. *Clauorde -v-* 1086, *Claworth -wurða -w(u)rda* etc. c.1130–1672, *Clauew(u)rda* 1163–4, *Clauworth* 1299–1347, *Clayworthe* 1317. OE **clawu** + **worth**. The village stands on a low curving hill which projects into the flat land along the r. Idle and was probably described as 'the claw'. The modern form probably owes its development to the early ME monophthongisation of *au*, cf. Nomina 13.109ff. Nt 28.

CLEADON T&W NZ 3862. 'Cliff hill'. *Cliuedon(')* [1183]c.1320, 1242×3–1307, *-den* 1292, 1387, *Clyuedon(')* [1183]c.1382, 1280–1350, *Cleuedon(')* [1183]c.1382–1471, *Cledon(e)* 1242×3–1675, *Cleadon* from 1399. OE **clif**, genitive pl. **clifa**, + **dūn**. The reference may be to coastal cliffs. NbDu 47, L 134–5, 143, 158.

CLEARBURY RING Wilts SU 1625. *Clearbury Ring* 1773. P.n. *Clereburu* 1632, unknown element, possibly OE *clǣfre* 'clover', + **burh** + ModE **ring**. The reference is to a small univallate Iron Age hill-fort. Wlt 394, Pevsner 1975.365.

CLEARWELL Glos SO 5708. 'Clover spring'. *Clouerwalle* c.1282, *Clourewall* 1435, *Clorewall'* 1385, *Clowerwall* 1515×29–1692. ME **clovre** (OE *clāfre*) + **walle** (OE Mercian *wælla*). The modern form is due to popular etymology. Gl iii.236.

CLEASBY NYorks NZ 2513. Partly uncertain. *Clesbi -by* 1086, *Clesebi -by* 1184–c.1300. The first element is possibly a pers.n. derived from ODan *klēss* (ONorw *kleis*) 'inarticulate', + **bȳ**. YN 284, SSNY 24.

CLEASBY HILL NYorks NY 9707. P.n. CLEASBY NZ 2513 + ModE **hill**.

CLEATLAM Durham NZ 1118. '(Settlement at the) clearings where burdock grows'. *Cletlum* c.1200–1348, *Clete-* 1235, c.1300, *Clet(e)lam* 1421–1564, *Cleatlam* from 1616. OE **clǣte** + **lēah**, dative pl. **lēum**. The earliest form, *Cletlinga* [c.1040]12th, [1072×83]15th, appears to be a Latinisation of OE **clǣte** + suffix **-ling**, 'place where burdock grows'. NbDu 47, ING 78.

CLEATOR Cumbr NY 0113. 'Burdock shieling'. *Cletertha* c.1185, *C- Kletern(e) -erwe* c.1220–1322, *Cleterhe -ergh* c.1225–1338, *Cleter(e)* c.1250–1418. ME **clete** (OE **clǣte*) + ME **erg** (ON *ǽrgi*). Forms in *-erne* are transcription errors for *-erue*, the regular ME development of *-erg*. Cu 357, SSNNW 63.

CLEATOR MOOR Cumbr NY 0214. *Cleator Moor* 1732. P.n. CLEATOR NY 0113 + **mōr**. Cu 357.

CLECKHEATON WYorks SE 1825. 'Heaton beside the round peaked hill'. *Claketon* 1285, *Clakheton* 1299–1581, *-heaton* 1466. OE **clacc** + p.n. HEATON, *Hetun -tone* 1086–1413, 'the high settlement', OE **hēah**, definite form **hēa**, + **tūn**. YW iii.16.

Brown CLEE HILL Shrops SO 5985. One of the Clee Hills, *Lesclives alias Cley alias Cleys alias Browne Clee* c.1612. Adj. **brown** + hill name Clee as in CLEEHILL SO 5975. Sa i.86.

CLEE ST MARGARET Shrops SO 5684. 'St Margaret's C'. *Clya Sancte Margarete* 1255×6, *Cle St Margaret's* 1280, *Clea St Margaret* 1421. P.n. *Cleie* 1086, *Cleia* 1221×2, *Clie* 1199, *Clia* c.1210, *Clye* 1259, hill-name Clee as in CLEEHILL SO 5975 + saint's name *Margaret* from the dedication of the church. Sa i.88.

CLEEDOWNTON Shrops SO 5880. *Clee Downton* 1832 OS. Hill-name Clee as in CLEEHILL SO 5975 + p.n. *Downton* 1291, 'the hill settlement', OE **dūn** + **tūn**. 'Clee' for distinction from Downton SJ 5412 and Downton Hall SO 5279, *Dounton* 1291, 1320. DEPN.

CLEEHILL Shrops SO 5975. The name does not appear in 1832 OS but is clearly transferred from Clee Hill SO 6076, *(montem qui dicitur) Clie* [674×704]13th S 1799, *Clivas* 1320, *Cleya (est quidam mons)* 1292, *the Cle Hills* c.1540, *Lesclives alias Cley alias Cleys alias Browne Clee* c.1612, *The Clee* 1650–93, *Clea* 1743, *Clayhill* 1766, 1770, OE ***clēo** 'the ball-shaped massif' in contrast to the long narrow massifs of The Long Mynd, Wenlock Edge etc. For other forms see CLEE ST MARGARET SO 5684, CLEETON ST MARY SO 6178, CLEOBURY MORTIMER SO 6776 and CLEOBURY NORTH SO 6287. In the dial. of Shropshire **clēo* would > ME *clø* (spelt *cleo, cleu, clue, clu*), *clē* (spelt *clee*) and *clī* (spelt *cli(a), clie, cly(a)*). This rare word was frequently replaced by OE **clif** 'a cliff', Latin **clivus** 'a slope' (*clive, clyve* spellings), and OE **clǣġ** (*clei, clai, clay* spellings). The reference is to the sub-circular shape of the massifs of Browne Clee and Clee Hill. Sa i.82–7.

CLEETHORPES Humbs TA 3008. 'The outlying settlements belonging to Clee'. *Clethorpes* 1588–1695, *Clee thorpes, Cleethorpes* from 1598. P.n. *Cleia* 1086, *Cle* c.1115–1594, *Clee* 1232–1351, *Clee als Cley* 1315–1759, 'the clay place', OE **clǣġ**, + ModE **thorps** referring to Hole and Itterby, now parts of Cleethorpes. Originally the reference was to Itterby alone, *Thorpe* 1406, *Clethorpe* 1552, *Cleethorp(e)* 1593–1840, p.n. Clee + ON **thorp**. Li v.15–6.

CLEETON ST MARY Shrops SO 6178. 'St Mary's C'. P.n. *Cleoton* 1241–1398, *Cleton(e)* 1255–1410, *Cleeton(')* from 1291×2, 'the settlement by Clee Hill', hill-name Clee as in CLEEHILL SO 5975 + OE **tūn**, + saint's name *Mary* from the dedication of the church first recorded in 1885. Sa i.88.

CLEEVE '(Place at) the cliff, the steep slope'. OE **clif**, dative sing. **clife**.
(1) ~ Avon ST 4565. '(At the) steep slopes'. *Clifan, Cylfan* [950×75] 13th S 1766, *Clive* 1243, *la Clive* 1242×3 Ass, *Clyve* 1327. OE **clif**, dative pl. **clifum** (lOE *clifan*). The ground slopes steeply upwards E of the village. DEPN, L 131, 133–4.

(2) ~ HILL Glos SO 9826. *Clevehill* 1564. ME **clive** 'cliff, steep hill' (OE *clif*) + **hill**. The hill rises to over 1000ft. Gl ii.90.

(3) ~ PRIOR H&W SP 0849. 'C held by the prior' sc. of Worcester. *Clive Prioris* 1240, *Priours Cleve* 1535. P.n. *Clive* 1086 + ME **priour**. The place stands on a conspicuous ridge. 'Prior' for distinction from Bishop's CLEEVE Glos. SO 9527. Wo 314.

(4) Bishop's ~ Glos SO 9527. 'Cleeve held by the bishop' sc. of Worcester. *Bissopes Clive* 1284. Earlier simply *(æt, in, into) Clife* [768×79]11th S 141, [889]11th S 1415, [899×904]10th S 1283, *Clive, Cliue -a -y-* 1086–1561, *Cleve* 1534, OE **clif** 'steep hill' referring to the steep escarpment of Cleeve Hill beneath which the village lies. The lands of the monastery of Cleeve were granted to the bishop of Worcester in 889 (S 1415). Gl ii.86.

(5) Old ~ Somer ST 0341. *Old Cleue* 1610. ModE adj. **old** + p.n. *Clive* 1086, 1227, [1258]Bruton. It is uncertain whether *(at) Clifan* [959×75]13th S 1766 (apparently dative pl. **clifum**) belongs here or under CLEEVE ST 4566. DEPN, L 131–4.

CLEHONGER H&W SO 4638. 'Clay sloping wood'. *Clevnge* (sic) 1086, *Clahungra* 1160×70, *Cleyhongre* 1163×7, *Clehungr(e)* 1320–1535. OE **clæġ** + **hangra**. He 54, L 195, DB Herefordshire note on 21.7.

CLEMSFOLD WSusx TQ 1131 → CLIMPING SU 9902.

CLENCH COMMON Wilts SU 1765. *Clench Common* 1817 OS. P.n. *Clenchia* n.d., *Clenche* 1289, 1314, *Cleynche* 1330–72, *Clynche* 1572–96, OE ***clenċ**, the source of Mod dial. *clench, clunch* 'a lump, a mass', referring to Martinsell Hill SU 1763, a big rounded hill, + ModE **common**. *Clenċ* or its ablaut variant **clinċ* occurs again in Clinghill, Bromham ST 9665, *Klynghulle, le Clynche* 1409, *Clynghill* 1569. Wlt 349, 254, PNE i.98.

CLENCHWARTON Norf TF 5920. 'Settlement of the Clencware, the people who live at Clenc'. *Ecleuuartuna* 1086, *Clenchewarton* 1196, *Clencwarton* 1205. OE folk-n. **Clenċware* < p.n. **Clenċ* + **ware**, genitive pl. *Clenċwara*, + **tūn**. The folk-n. occurs again in the name *Clencware hundred* 11th. P.n. *Clenċ* is identical with Clench in CLENCH COMMON Wilts SU 1765, from OE ***clenċ** 'a lump, a mass, a hill'. DEPN, Wlt 349.

CLENT H&W SO 9379. 'The rock' or 'hill'. *Clent* from 1086. OE ***clent**. The reference is to the Clent Hills with the summit of Walton Hill at 1036ft. Wo 279, PNE i.98.

CLENT HILLS H&W SO 9479. *Clent Hills* 1831 OS. P.n. CLENT SO 9379 + ModE **hill(s)**.

CLEOBURY MORTIMER Shrops SO 6776. 'C held by the Mortimer family'. *Clebury Mortimer* c.1270–1385, *Cleybur' Mortem'* 1255, *Cleobury Mortymer* 1315–82, ~ *Mortimer* from 1666. P.n. *Clai- Cleberie* 1086, *Claiberi* 1201, *Cleybur'* 1242, *Cleobr'* 1235×6, 'the manor or fortified place by Clee', hill-name Clee as in CLEEHILL SO 5975 + OE **byriġ**, locative-dative sing. of **burh**, + manorial addition from Ralph de Mortimer from Mortemer, Seine-Maritime, the holder in 1086. Sa i.89.

CLEOBURY NORTH Shrops SO 6287. *North Claibury* 1221–1714 with variants *Cley- Cle-* and *-biry -buri*, *Clebury North* 1283–1334, *Cleobury* ~ from 1753. P.n. *Cleoberie* 1086 as in CLEOBURY MORTIMER SO 6776 + pseudo-manorial ME **north**. The sp *Clibbury* 1586 represents the local pr still reported by Bowcock 74 in 1923. Sa i.90.

CLEVANCY Wilts SU 0575. 'Cleve held by the Wancy family'. *Clif Wauncy* 1231, *Clyve Auncy* 1428, *Cleveansey* 1580, *Cleeve Ansty al. Cleeve Ancey* 1726. P.n. *Clive* 1086, c.1220, '(the settlement) at the steep slope', OE **clife**, dative sing. of **clif**, referring to the steep scarp of the downs lying between Cherhill and Liddington, + manorial addition from the family of Robert de Wancy who held the estate c.1220 and for distinction from CLYFFE PYPARD SU 0776. Wlt 268 gives pr [kli'vænsi], L 131, 134.

CLEVEDON Avon ST 4074. 'Hill of the cliffs'. *Clivedone* 1086, *Cliu- Clivedon* 1172–1242. OE **clif**, genitive pl. **clifa**, + **dūn**. The church occupies an isolated hill with sheer cliffs to seaward. DEPN, L 131, 134–5, 143, 145, TC 75.

CLEVEDON COURT Avon ST 4271. *Clevedon Court* 1830 OS. An early 14th cent. manor house. P.n. CLEVEDON ST 4074 + **court**. Pevsner 1958N.186.

CLEVELAND Cleve NZ 6213. 'Cliff district'. *Clive- Clyveland(a)* 1104×14–1452, *Cleveland* from [c.1270]15th. OE **clif**, genitive pl. **clifa**, + **land**. YN 128.

CLEVELAND HILLS NYorks SE 5899. P.n. Cleveland + ModE **hill(s)**. Cleveland, *Cli- Clyveland(a)* 1104×14–1452, *Cleueland* from c.1270, is the 'steep precipitous district', OE **clif**, genitive pl. **clifa** with subsequent opem syllable lengthening to ME *clēve*, + **land**. YN 129.

CLEVELEYS Lancs SD 3143. 'Cliff wood, clearing or pasture'. *Cliueleie* [c.1180]1268, *Cliveley* c.1270, *Klillegh* c.1380. OE **clif**, genitive pl. **clifa**, + **lēah**. La 166, L 134–5.

CLEVERTON Wilts ST 9785. 'Clover hill'. *Claverdon(e)* 1216×72, 1257, *Cleverdon(e)* 1284–1332, *Cleverton* 1627. OE **clǣfre** + **dūn**. Wlt 61.

CLEWER Somer ST 4351. Originally a folk-n., 'the hill dwellers'. *Cliwere* [676×85]13th S 1668, [705×9]13th S 1674, *Cliveware* 1086, *Clifwere* 13th, *Clywar* 1276, *Clewer* 1817 OS. OE **Clifware* < **clif** + **ware**. The village lies at the N edge of the Wedmore Hills in the Somerset Levels. DEPN.

CLEY HILL Wilts ST 8344. 'The hill called Cley'. *Cley hill* from 1540. P.n. *Cly* 1316, *Cley* 1560, *Clye* c.1560, probably identical with Clee in CLEEHILL Shrops SO 5975 from OE ***clēo** 'a ball shaped massif', + ModE **hill**. In WMidl and SW dialects OE *ēo* sometimes developed through the stage *ø* to *y* and the resultant ME *clī* was frequently assimilated to the word *clay*. Association with the Shropshire hill name is suggested by the 1812 forms *Clee or Clay Hill*. Cley Hill is a prominent isolated hill-mass. Wlt 152, PNDB 65 and f.n.2, Sa i.82–7.

CLEY NEXT THE SEA Norf TG 0443. *Cley Next the Sea* 1838 OS. Earlier simply *Claia* 1086, *Claya* 1242, 'the clay, the clayey place', OE **clæġ**. According to Gaz, however, the soil is light overlying chalk. DEPN gives pr [klai].

Cockley CLEY Norf TF 7904. *Coclikleye* 1324. Unexplained element + p.n. *Cleia* 1086, *Claia* 1086, 1199, *Cleye* 1254, 'the clay, the clayey place', OE **clæġ**. Cockley is obscure: it might represent OE *cocc*, genitive pl *cocca*, + *lēah*, 'cocks' wood or clearing'. The soil is sandy, overlying loam, according to Gaz. DEPN.

CLIBURN Cumbr NY 5824. 'Stream by the bank'. *Clibbrun* 1133–47, *Cli- Kli- Clybburn(e)* 1204–1703, *Cli- Clyburn(e)* 1250–1366. OE **clif** + **burna** varying with ON **brunnr**. The *clif* is either the slope on which the village stands or one of the scars alongside the river Leith. We ii.136 gives prs ['klibən, 'klaibə:n], the latter a spelling pronunciation. SSNNW 193, L 18, 134–5.

CLIDDESDEN Hants SU 6349. 'Clyde valley'. *Cleresden* 1086, *Cledesdene* 1194–1341, *Cludes-* 1219–1428, *Clydes-* 1240, *Clides-* 1274, *Clisden* 1665. Hill-name **Clȳde*, genitive sing. **Clȳdes*, + **denu**. **Clȳde* is a derivative of OE *clūd* 'a rock, a rocky hill', either an unrecorded *i*-declension form **clȳde* < **clūdiz* or an old locative-dative sing. **clūdi*, referring to Farleigh Hill SU 6247. Ha 57, Gover 1958.126 which gives pr [klizdən], DEPN.

CLIFF END ESusx TQ 8813. 'End of the cliff'. *Cliveshende* c.1197. OE **clif**, genitive sing. **clifes**, + **ende**. The line of cliffs extending from Hastings ends here in the marshes of Pett Level. Sx 513.

CLIFFE 'A bank, a steep slope'. OE **clif**.

(1) ~ Kent TQ 7376. *Clive* [830 for 833]13th S 1414, 1086 etc., *cliua, cliue* c.975 B 1321–2, *Clyffe* 1610. The boundary of the

Clifware, the dwellers at Cliffe, is referred to in 778 as *clifwara gemære* S 36. KPN 57.
(2) ~ NYorks SE 6632. *Clive* 1086, *Clif(f), Clyf(f)* 1200–1542. The reference is to the bank of the r. Ouse. YE 258, L 132.
(3) ~ HILL ESusx TQ 4310. 'Hill called Cliff'. P.n. Cliffe + ModE **hill**. Cliffe, *la Clyve* 1248, *Clyf* 1296, *Clyve by Lewes* 1345, is OE **clif** 'steep slope' used as a simplex p.n. to which pleonastic *hill* was subsequently added. Sx 354.
(4) King's ~ Northants TL 0097. *Kyngesclive* 1305, *Clyve Regis* 1347. ME **king**, genitive sing. **kinges**, Latin **regis** + p.n. *Clive* 1086–1328 with variants, *Clyve and Cleve, Clif* 1349, *Clyf* 1393, 'the hill-slope', referring to the slope above Willow Brook. The manor was already held by the king at the time of DB. Nth 199.
(5) North ~ Humbs SE 8737. *North- Nordclif* early 13th. ME adj. **north** + p.n. *Cliue* 1086, *Clif* 12th. The reference is to the steep W scarp of the Wolds. YE 227.
(6) South ~ Humbs SE 8736. *Sudclif* early 13th, *South-* from 1303. ME adj. **suth** + p.n. *Cliue* 1086, *Cli(f) -y-* 1259–1466 as in North CLIFFE SE 8737. YE 225.
(7) West ~ Kent TR 3444. *Wesclive* 1086, *Westcliua -clive -clyve* 1172×3–1242×3. Earlier *(in loco qui dicitur) æt clife* 'place called At Cliffe', *(to) clife* 1042×4 S 1044[†]. 'West' for distinction from ST MARGARET'S AT CLIFFE TR 3644. KPN 322.
(8) ~ WOODS Kent TQ 7373. P.n. CLIFFE TQ 7376 + ModE **wood(s)**.

CLIFFORD 'Ford at the a steep bank'. OE **clif** + **ford**. L 69, 131–5.
(1) ~ WYorks SE 4244. *Cliford* 1086, *Clif- Clyfford(e)* 1160×80–1620 (~ *in Elmette* 1298), *-forth(e)* 1343–1657. The ford crossed Carr Beck. YW iv.87.
(2) ~ H&W SO 2445. *Cliford* 1086, *Clifford* 1230. DEPN.
(3) ~ CHAMBERS Warw SP 1952. 'C belonging to the chamberlain'. *Chaumberes Clifford* 1388, *Clifford Chambers* 1723. P.n. *(æt) Clif(f)orda* [922]17th S 1289, [966]11th S 1311, *Clifort* 1086, *Clifford(e) -ia(m), Clyfford(e)* 1086–1625, + ME **chamberere**. The manor was given in 1099 to St Peter's Abbey, Gloucester for the maintenance of the *camerarius* or chamberlain. Gl i.239.

CLIFFORD'S MESNE Glos SO 7023. 'Meend held by the Clifford family'. *Cliffords Meend -mynde* 1749, ~ *Mine* 1830. Family name *Clifford* + p.n. Meend, ME ***munede** 'forest-waste' (PrW **mïnïth*, W *mynydd* 'mountain'). The modern spelling is due to false association with AFr *mesne* 'demesne land'. Gl iii.175, PNE ii.34.

CLIFFS END Kent TR 3464. 'The end of the cliff'. *Cliuesende* c.1200–50, *Clives(h)end'* 1236–1346 with variants *Clyues-* and *-(h)ende*. OE **clif**, genitive sing. **clifes**, + **ende**. PNK 600, Cullen.

CLIFTON 'Cliff settlement'. OE **clif** + **tūn**. PNE i.99, L 131–4.
(1) ~ Avon ST 5773. *Clistone* (sic for *Clif-*) 1086, *Cli- Clyfton(e)* from 1167[‡]. The reference is to the cliff in the Avon Gorge at the edge of Clifton Down. Gl iii.97.
(2) ~ Beds TL 1739. *cliftune* 980×90 ECTV 171 (S 1497), *Cliftone* 1086. Bd 169.
(3) ~ Cumbr NY 5326. *Cli- Clyfton* 1196–1703. The village stands on two cliffs overlooking the river Lowther. We ii.187, L 131–4.
(4) ~ Derby SK 1644. *Clif- Cliptune* 1086, *Clif- Clyfton(e)* from early 13th. Db 432.
(5) ~ H&W SO 8446. *Clifton* from 1256. The 'cliff' is a gentle rise from 40ft. on the Severn to 55ft. Wo 227.
(6) ~ Lancs SD 4630. *Clistun* [sic] 1086, *Clifton* from 1226. Clifton lies on a fairly steep slope above the marshland along the Ribble. La 150.
(7) ~ Northum NZ 2082. *Clifton* from 1242 BF. NbDu 48.
(8) ~ Notts SK 5534. *Cliftun(e)* 1086, *Clifton(a)* from 1165. The reference is to the steep slope overlooking the r. Trent. Nt 246, Jnl.2.53.
(9) ~ Oxon SP 4831. *Cliftona* from c.1170 with variants *Cliff- Clyf-* and *-ton(e)*. There is a steep bank overlooking the Cherwell. O 256.
(10) ~ CAMPVILLE Staffs SK 2510. 'C held by the Campville family'. *Clifton Caunvil* 1284. P.n. *Clyfton* [941]14th S 479, *Clistone* 1086, + manorial addition from Richard de Camvill who held the manor in 1231. His family came from Canappeville, Canapville or Canville in Normandy. DEPN, Duignan 41.
(11) ~ HAMPDEN Oxon SU 5495. 'C held by the Hampden family'. P.n. *Cliftona* [1146]18th etc. with variants *Cliff- Clyf-* and *-ton(e)*, + suffix *Hampden* first noted on a map of 1836. No family of this name is known in association with this place. The church lies on a small cliff overlooking the Thames. O 149.
(12) ~ REYNES Bucks SP 9051. 'C held by the Reynes family'. *Clyfton Reynes* 1383. Earlier simply *Clys- Clif- Clistone* 1086. The manor was held by Ralph de Reynes in 1302. Bk 34.
(13) ~ UPON DUNSMORE Wa SP 5376. *Clifton super Donesmore* 1306. P.n. *Cliptone* 1086, *Clifton(a)* from 1169, + p.n. DUNSMORE SP 5476. 'Standeth upon the top of an indifferent hill... *Cliffe* with the Saxons, signifying... any shelving ground' (Dugdale). Wa 127.
(14) ~ UPON TEME H&W SO 7161. *Cliftun ultra Tamedam* 'C beyond the T' *[930]11th S 406, *Clistvne* 1086. The situation is high ground overlooking the Teme. Wo 43 gives pr [klifn].
(15) Great ~ Cumbr NY 0429. *Great Clifton* 1300, ModE adj. **great** + p.n. *Cliftona* [c.1160]n.d., *Cliftun', Cli- Clyfton* 1210 etc. Also known as *Alta Clyton* 1284 'high Clifton', and *Kirkeclifton'* c.1260 'church Clifton', all for distinction from Little Clifton NY 0528, *Parua Clifton* 1278. The reference is to the steep bank overlooking the river Derwent. Also known as *Clifton Game*l 1212, from the holder *Gamel* in 1212. 359, L 131–4.
(16) North ~ Notts SK 8272. *Nort Clifton* c.1160, *North* ~ from 1280, *Norclifton* 1455. OE **north** + p.n. *Cli(f)- Clistone* 1086. Also known as *Clifton juxta Dunham* 1275. 'North' for distinction from South CLIFTON SK 8270. Nt 202.
(17) South ~ Notts SK 8270. *Suth Clifton* 1280 etc. ME **south** + p.n. Clifton as in North CLIFTON SK 8272. Nt 202.

CLIMPING WSusx SU 9902. 'The Climpingas, the people called after Climpa'. *Clepinges* 1086, *Clenpi(n)ges* c.1086, *Clinpingh(es)* c.1194, *Cly- Climpinges* 1228–1345, *Climping* c.1260. OE folk-n. **Climpingas* < pers.n. **Climp(a)* + **ingas**. The same pers.n., originally a nick-n. meaning 'lumpy', occurs in nearby Clemsfold TQ 1131, *Climpesfaude* 1285, *Cli- Clympesfold* 1327–1442. Sx 138, 159, ING 32, MA 10.23, SS 64.

CLIMSON Corn SX 3674 → STOKE CLIMSLAND SX 3674.

CLINGHILL Wilts ST 9665 > CLENCH COMMON SU 1765.

CLINT NYorks SE 2559. 'The steep bank'. *Cli- Clynt(e)* from 13th. ODan **klint** 'a rocky bank, a steep bank' here with reference to the r. Nidd. YW v. 98.

CLINTBURN Northum NY 7179. *Clintburn* 1866 OS. Dial. **clint** 'a hard rock projecting on the side of a hill or river-bank' (ODan *klint* 'a rocky cliff') + **burn**.

CLINT GREEN Norf TG 0210. No early forms. The topography hardly supports derivation from dial. **clint** 'a hard rock projecting on the side of a hill or riverbank' (ODan *klint*).

CLIPPESBY Norf TG 4214. 'Clip's village or farm'. *Clipesby, Clepesbe(i), Clepebei* 1086, *Cly- Clip(p)esby* 1101–1535, *Clep(p)esby* c.1160, 1396. lOE pers.n. *Clip* < ON *Klyppr* or **Klippr*, genitive sing. *Clipes*, + **bȳ**. Nf ii.51, Jnl. 19.11, SPNN 268.

CLIPSHAM Leic SK 9716. Possibly 'Cylp's homestead'. *Ky- Kilpesham* 1203–1375, *Clip(p)esham -is-* 1286–1472, *Cly- Clipsham* from 1329. OE pers.n. **Cylp*, possibly a by-n. related to ON *kylp* 'a small sturdy person', genitive sing. **Cylpes*, + **hām**. R 80.

[†]Possibly *(æt) Clife* [c.985×1006]11th S 1455 also belongs here. Charters IV.119.

[‡]The form *(æt) Cliftune* [970]12th S 777 does not belong here.

CLIPSTON Northants SP 7181. 'Klyppr's farm or village'. *Clipestune* 1086, *-ton(e)* 1086–1313 with variant *-y-*, *Clipston(e)* from 12th with variant *-y-*, *Clypson* 1572, *Clipson* 1702. ON pers.n. *Klyppr*, genitive sing. *Klypps*, + **tūn**. Nth 111, SSNEM 189.

CLIPSTON Notts SK 6334. 'Klyppr's farm or village'. *Clipestvne* 1086, *Clipeston* 1278, 1287, *Clipston* from 1315. Also known as ~ *juxta Pluntre* 1278, ~ *super le Hill* for distinction from Old CLIPSTONE SK 6064. ON pers.n. *Klyppr*, genitive sing. *Klypps*, + **tūn**. Nt 232, SSNEM 189.

New CLIPSTONE Notts SK 5863. A modern coal-mining settlement. ModE adj. **new** + p.n. Clipstone as in Old CLIPSTONE SK 6064.

Old CLIPSTONE Notts SK 6064. Adj. **old** + p.n. *Clipestvne* 1086, *-ton(a) -e* 1130 etc., *Cli- Cly- Klyppeston(a)* 1088–1287, *Clipston(a)* 1173–1305. ON pers.n. *Klyppr*, genitive sing. *Klypps*, + **tūn**. Also known as *Kyngesclipston* 1290 for distinction from CLIPSTON SK 6334. 'Old' for distinction from New Clipstone SK 5863, a modern coal-mining settlement. Nt 73, SSNEM 189.

CLITHEROE Lancs SD 7441. 'Hill with loose stones'. *Cliderhou -(h)owe* 1102–1416, *Clitherow* 1124, *Clithero* 1356. OE ***clȳder**, ***clider** + ON **haugr**. The crag on which the castle stands consists of loose crumbling limestone. La 78, PNE i.101, Jnl 17.48.

CLIVE Shrops SJ 5124. 'The cliff'. *Clive* from 1255 with variants *(La) Clyve, Klive, The Clive, (La) Cleve* 1291×2–1577. OE **clife**, dative sing. of **clif** 'a cliff' referring to the steep slope on the S side of Grinshill Hill. Sa i.91 gives pr [kliv].

CLOCK FACE Mers SJ 5291. A district of St Helens around Clock Face Colliery.

CLODOCK H&W SO 3227. Short for 'St Clydog's church'. *Cladoc* 1266, *S. Cleddoc'* 1540, *Cludock* 1778. The earliest reference is *Merthirclitauc* [early 8th]c.1130 with W **merthyr** 'saint's grave'. OW saint's name *Clitauc* < Brit **Clutācos*. He 56, LHEB 677.

CLOPHILL Beds TL 0838. 'Clop hill, hill called *Clop*'. *Clopelle* 1086, *Clophull* 1227, *Clophill* 1247. OE ***clopp(a)** 'a lump, a hillock' probably used as a p.n. + **hyll**. Bd 146, DEPN, Studies 1936.136, PNE i.99.

CLOPTON GREEN Suff TL 7654. *Clapton Green* 1836 OS. P.n. Clopton + ModE **green**. There are no early forms for this Clopton but the settlement occupies a marked hill E of Wickhambrook and this must be another example of OE ***clopp(a)** + **tūn**.

CLOPTON Northants TL 0680. 'The farm or village on the hill'. *Cloptun, Cloptona* [c.960]14th S 1808, *Clotone* 1086, 1201, *Clopton* from 1175, *Clapton* 1227, 1683, 1936. OE ***clopp(a)** 'a lump, a hillock, a hill' + **tūn**. The reference is to the ridge that rises from the river Nene. Nth 217, Studies 1936.137, PNE i.99.

CLOPTON Suff TM 2252. 'The hill settlement'. *Clop(e)tuna* 1086, *Clopton(')* from 1186. OE ***clopp(a)** + **tūn**. DEPN.

CLOTHALL Herts TL 2732. 'Burdock nook'. *Clatheala* c.1060, *Cladhele* 1086, *Clothalla -hal(e)* 1185–1314. OE **clāte** + **healh**. Hrt 155.

CLOTTON Ches SJ 5263. 'Dell settlement'. *Clotone* 1086, [1096×1101]1280, *Clottona* 1157–94, *Clotton* from 1246. OE **clōh** + **tūn**. The hamlet is situated at the mouth of a small low valley. Che iii.271 gives pr [klɔtən].

Temple CLOUD Avon ST 6257. '(That part of) Cloud held by the Temple family'. Originally called simply *la Clude* 1199, *Clude -a* 1204, *la Cloude* 1311×12 FF, OE **clūd** 'the hill'. The same element occurs in nearby CLUTTON ST 6258 'settlement at Cloud' and refers to the high ground E of the village which rises to 511ft. at ST 6058. There is no record of property held by the Knights Templar here but a Richard de Templo occurs in this area in the Buckland Priory chartulary. DEPN.

CLOUGH FOOT WYorks SD 9023. 'Foot of Clough, the ravine' (ultimately OE **clōh**). The reference is to the steep narrow valley through which the road from Todmorden to Bacup passes. YW iii.180.

CLOUGHA Lancs SD 5459. 'Hill-spur with a ravine'. *Clochoch* 1199, *Clochehoc, Cloghou* 1228, *Clough ho hill* 1577. OE **clōh** + **hōh**. Clougha Pike, 1355ft., is a projection from the mass of Grit Fell with the ravine of Rowton Brook to the S. La 169 gives pr [klɔfə], L 88.

CLOUGHTON NYorks TA 0094. 'Valley settlement'. *Cloctune -ton(a)* 1086, *Clochton* 1231, *Cloghton* 1368, 1404, *Cloughton* 1577. OE **clōh** + **tūn**. YN 108 gives pr [kloutən].

CLOUGHTON NEWLANDS NYorks TA 0195. P.n. CLOUGHTON TA 0094 + ModE **newland** (OE *nīwe-land*) 'land newly taken in to cultivation'.

CLOUGHTON WYKE NYorks TA 0295. P.n. CLOUGHTON TA 0094 + ModE dial. **wick** 'a creek, an inlet, a small bay' (ON *vík*).

CLOVE LODGE Durham NY 9137. *Clove Lodge* 1863 OS. Probably N dial. **clof** 'a cleft between hills' + **lodge**.

CLOVELLY Devon SS 3124. Probably 'crevice at Velly, the wheel-rim'. *Clovelie* 1086, *-leia* 1086 Exon, *Clovely -i* 1242–1361. OE ***clof** 'crevice, cleft' referring to the cleft in the coastline where the present village lies, + p.n. Velly SS 2924, *(la) Felye* 1287, 1333, *Velye* 1301, *Velly* 1566, OE **felġ** 'a felly or wheel rim' used topographically of the hilly ground here which rises shaped like a wheel around the marked *clof* or indentation of the site. A recent alternative suggestion is Co **cleath** (PrCo [klɔ:ð]) 'a dike' referring to the Iron Age hill-fort of Clovelly Dikes + pers.n. *Felec* or **Fili* as in PHILLACK Corn SW 5539, PHILLEIGH SW 8639 and W Caerphilly ST 1587, and so 'Felec's earthworks'. D 70 gives pr [klou'veli], 76, L 87, Jnl 28.36–43.

CLOVELLY DYKES Devon SS 3123. P.n. CLOVELLY SS 3124 + ModE **dyke(s)**. The nearby hamlets of East and West DYKE SS 3123 [dik], *E^t Dicks* 1809 OS, *Westdich* 1333, *W^t Dicks* 1809 OS, are named from this feature which is an Iron Age hill-fort called *Ditchen Hills* in 1809 OS. The form *Ditchen* may represent OE dative pl. *dīcum* 'at the ditches'. D 71, Thomas 1976.94.

CLOWBRIDGE RESERVOIR Lancs SD 8228. P.n. Clow Bridge SD 8228 + ModE **reservoir**. Clow Bridge, no early forms, is probably dial. **clow, clough** 'a ravine' (OE *clōh*) + **bridge**, referring to the ravine of Limy Water crossed here by the road from Burnley to Rawenstall.

CLOWNE Derby SK 4975. Originally a r.n., PrW **Colūn* meaning 'water, river' <Brit **Colaunā* of unknown meaning. *(æt) Clune* [c.1002]c.1100 S 1536, 1086–1316, *Cloune* 1269–16th cent. The original r.n. was later replaced by r.n. POULTER. **Colūn* is preserved also in the name CLUMBER PARK. Cf. r.n. COLNE. Db 238, LHEB 308–9, Jnl.1.45.

CLOWS TOP H&W SO 7171. *Clowes Toppe* 1663. P.n. Clows + ModE **top**. The reference is to the summit of a 760ft. massif on the road from Tenbury Wells to Bewdley. Clows, *Cluse* 1275, *(la) Clouse* 1294, 1328, is ME **clouse** (OE *clūs(e)*) 'an enclosure, a close, a narrow passage'. Probably a late medieval enclosure on the hill originally called MAMBLE SO 6971 and then PENSAX. Wo 39.

CLUBMOOR Mers SJ 3893. *Club Moor* 1842 OS.

CLUBWORTHY Corn SW 2792. Partly uncertain. *Clobiry* 1322, 1330, *Clobery* 1748, *Clubbery* 1842, *Clubworthy* 1813. An English name, but the forms are too late for certainty. The specific might be Cornish dial. **clob** 'clod of earth, mud', OE **clōh** 'a deep valley, a dell' or ***clop(p)** 'a lump, a hillock, a hill'; the generic OE **byriġ**, dative sing. of **burh** 'fort, fortified place', or **beorg** 'a hill'. However, there is no known fort nearby and the name probably refers to the hill just W of the hamlet. The modern form with *-worthy* is a hyper-correct one due to the fact that *-ery* elsewhere often stands for older *-worthy*, cf. Canworthy in CANWORTHY WATER SX 2291. PNCo 70.

CLUMBER PARK Notts SK 6274. P.n. Clumber + ModE **park**. A modern Country Park, first enclosed as a park for the use of Queen Anne by John Holles, Duke of Newcastle, in 1707. Clumber, *Clun- Clūbre* 1086, *Clumber* from 1227, is '*Clun* hill', r.n. *Clun* as in CLOWNE Derby SK 4975 + PrW ***breȝ** (Brit **brigā*). Nt 106, Jnl.1.44.

CLUN FOREST Shrops SO 2286. *Clun Forest* 1836 OS. R.n. Clun as in CLUN SO 3081 + ME **forest**.

CLUN Shrops SO 3081. By origin a r.n. of unknown meaning. *Clune* 1086–1540, *Cloune* 1242–1343, *Clunn(e)* 1500–1774, *Clun* from c.1540. PrW **Colūn*, **C'lūn* (Brit **Colaunā*) as in CLOWNE Derby SK 4975, CLUMBER PARK Notts SK 6274 and the r.n. COLNE. Spellings with *-nn-* show shortening of *ū* > *ŭ* possibly transferred from the compound names Clunbury, Clungunford and Clunton. The W form is Colunwy, *Colunwy* 13th, < Brit **Colaunouiā*. This and the following names were made famous by the old rhyme quoted by A.E.Housman in *A Shropshire Lad* (1896):

> Clunton and Clunbury,
> Clungunford and Clun,
> Are the quietest places
> Under the sun.

Sa i.91, LHEB 688.

CLUNBURY Shrops SO 3780. 'The fortified place on the river Clun'. *Cluneberie* 1086, *Clumbire -bury* 1221×2 1716 including [c.1158]1348, with variant *Clom-, Clunbury* from 1316. R.n. Clun as in CLUN SO 3081 + **byriġ**, dative sing. of OE **burh**. Sa i.92.

CLUNGUNFORD Shrops SO 3978. 'Gunward's Clun'. *Clongunford* 1272–1668, *Cloune Goneford* 1242, *Clungunford* from 1291×2. P.n. *Clone* 1086, r.n. Clun as in CLUN SO 3081 used as a p.n. + pers.n. *Gunward*, the name of the tenant at the time of King Edward the Confessor. Sa i.93 cites a local form [gʌnəs] corresponding to the sps *Clungunnas* 1691 and *Clungunnus* 1778.

CLUNTON Shrops SO 3381. 'The settlement on the river Clun'. *Clutone* 1086, *Cluntune* [c.1155]1348, *Clunton* from 1302, *Clin- Clynton* 1225×6–1291×2. R.n. Clun as in CLUN SO 3081 + OE **tūn**. Sa i.93.

CLUTTON Avon ST 6259. 'Settlement at Cloud'. *Cluttone* [839×58]13th S 1694, [939×46]13th S 1732, *Clvtone* 1086, *Clotton* 1205. P.n. Cloud as in Temple CLOUD ST 6257 + **tūn**. DEPN.

CLUTTON Ches SJ 4654. 'Settlement by the rocky hill'. *Clutone* 1086, *Clutton* from 1275. OE **clūd** 'a rock, a rocky hill' + **tūn**. The reference is to the scar of Cliff Hill (370ft.) SJ 4653, ¾ of a mile SE of the village in Carden, *Caworthin* 1300, *Caurdyn* 1363, 'enclosure at the rock', OE **carr** + **worthiġn**. Che iv.72, 53.

CLYFFE PYPARD Wilts SU 0776. 'C held by the Pypard family'. *Cliva Pipard* 1242, *Pippardesclyve* 1282, *Cleve Pepper* 1570, *Cleeve alias Clieff Pypard* 1603. P.n. *(æt) Clife* [983]12th S 848, *Clive* 1086–1242, '(the settlement) at the steep slope', OE **clife**, dative sing. of **clif**, referring to the steep scarp of the downs lying between Cherhill and Lydington, + manorial addition from the family name of Richard Pipart who held the estate in 1231 and for distinction from CLEVANCY 'Cleve held by the Wancy family' SU 0575. Wlt 266 gives pr [kli:v paipəd], formerly [pipər] or [pepər], and cites a local rhyme

> White Cleeve, Pepper Cleeve, Cleeve and Cleveancy,
> Lyneham and lousy Clack, Chris Malford and Dauntsey.

L 131

River CLYST Devon SY 0098. Perhaps 'clear stream'. *(on, of) clyst* [937]11th S 433, [951 for ? 961]11th S 669, *Clyst* 1281, *Clist* c.1200, 1321. A Celtic r.n., OE, OW **Clist*, earlier **Clust* < Brit **klūst-*, **kloust-* < IE **kleu-* 'to wash, to swill' seen in the r.n. **Clyt*, **Clit* in AUCKLAND Durham and cognate with Latin *cluo* 'to cleanse'. Along the river Clyst lies a series of estates named from the river including MONKERTON SY 9693 'monks' C' and SOWTON SY 9792 'south C'. D 3 gives pr [klist], RN 81.

(1) ~ HONITON Devon SX 9893. Properly Clyst Hinton, 'the Clyst estate belonging to a religious community'. *Clysthynetone* 1281–1330, *Hynetonisclyst* 1340, *Honiton Clyst* 1472, *Clist Honyton* 1652. Earlier simply *(æt) Hina tune* c.1100, *Hinetun -ton* 1219, 1276, OE **hiġna**, genitive pl. of **hiwan** 'a household, members of a family, a religious community', + **tūn**. The manor belonged to the religious community of Exeter cathedral. *Hinton* was later assimilated to the form of the Devon p.n. Honiton. D 584.

(2) ~ HYDON Devon ST 0301. 'The Clyst estate held by the Hydon family'. *Clyst Ydone* 1258×77, *Clisthydon* 1285. Earlier simply *Clist* 1086. The manor was held by Richard de Hidune in 1242. D 577.

(3) ~ ST GEORGE Devon SX 9888. 'St George's C'. Short for Clystwick St George. *Clystwik Sc̄i Georgii* 1327, *Clyst Sancti Georgii* 1334. Earlier simply *Clyst wicon* [951 for ?961]11th S 669, *Clistwike* 1072×1103, *Chisewig -wic* 1086, *Clisewic* 1086 Exon, *(of) Clistwich' -wike* 1200, 1259, 'Clyst dairy-farm', OE **wīċ**. The S 669 spelling *Clyst wicon* looks like an OE dative pl. in *-um* and may include Clyst St Mary as well as Clyst St George. Saint's name *George* from the dedication of the church. D 585.

(4) ~ ST LAWRENCE Devon ST 0200. 'St Lawrence's C'. *Clist Sancti Laurencii* 1203, *Clistlauraunz* 1319, *Clist Seint Laurenz* 1361. Earlier simply *Clist* 1086. Saint's name *Lawrence* from the dedication of the church. D 578.

(5) ~ ST MARY Devon SX 9790. 'St Mary's C'. *Clist(e) Sancte Marie* 1242–1344, *Seynt Mary Clyst* 1544. Earlier simply *Cliste* 1086. Saint's name *Mary* from the dedication of the church. D 586.

(6) Broad ~ Devon SX 9897 → BROADCLIST SX 9897.

(7) West ~ SY 9795 → SOWTON SY 9792.

COAD'S GREEN Corn SX 2976. *Coades Green* 1813. An 18th or 19th cent. hamlet, local surname *Code* + **green**. PNCo 70, Gover n.d. 167.

COALBROOKDALE Shrops SJ 6604. 'Coalbrook valley'. *Coalbrook Dale* 1833 OS. P.n. *Caldebrok* 1250, 'the cold stream', OE **cald**, definite form **calda**, + **broċ**, + ModE **dale**. The modern spelling of the name has been influenced by the coal-field developed here in the 18th century. DEPN.

COALCLEUGH Northum NY 8045. Either 'the cool ravine' or 'the ravine where charcoal is burnt'. *Coalcleugh* 1866 OS. OE **cōl** or **col** + N dial. **clough, cleugh** 'a deep valley or ravine' (OE **clōh*).

COALEY Glos SO 7701. 'Cove wood or clearing'. *Coue- Coveleg(e) -le(e) -legh -ley(e)* 1086–1574, *Cow(e)ley* 1472–1695, *Coley* 1672, *Coaley* 1731, 1777. OE **cofa** + **lēah**. The precise meaning of *cofa* 'hollow, recess' in p.ns. is uncertain. Gl ii.219.

COALPIT HEATH Avon ST 6780. *Colepitt Heath* 1702. ModE **coal-pit** (ME *cole-pit*) + **heath**. Gl iii.69.

COALPORT Shrops SJ 6902. 'Coal port'. *Coalport* 1833 OS. ModE **coal** + **port**. Planned and created by the iron-master William Reynolds in 1796 as a canal–river interchange port linking the river Severn with the Shropshire canal by means of Hay Incline. Coal from the coal-pits at nearby Blist's Hill was transported via the canal and the incline for loading on Severn barges. Porcelain was made here in the 18th cent., but Coalport is now only a trade-name for ware manufactured at Stoke-on-Trent. Raven 62.

COALVILLE Leic SK 4213. Short for *Whitwick-Coalville* 1838, a 19th cent. coal mining town which grew up in WHITWICK Leic SK 4316. ModE **coal** + **ville**. Lei 354.

COALWAY Glos SO 5910. Named from a road leading to Coleford on which coal was carried. Gl iii.229.

COATE Wilts SU 0461. 'The cottages'. *Cotes* 1255–1324, *la Cote* 1289, cf. *Coate Grove* 1686. ME **cote** (OE *cot*), pl. **cotes**. Wlt 251.

COATES 'The cottages'. OE **cot**, pl. **cotu** replaced by ME *cotes*. Cf., COATE, COATHAM.
(1) ~ Cambs TL 3097. *Cotes* c.1280, *Cootes, Coats* 1603. Ca 264.
(2) ~ Glos SO 9801. *Cota* 1175, *Cotes* 1220–1584. Gl i.68.
(3) Great ~ Humbs TA 2310. *Magna Cotes* 1242. ModE adj. **great**, Latin **magna**, + p.n. *Cotes* 1086, *Cotis, Cotun* c.1115. 'Great' for distinction from South or Little Coates TA 2408, *Sudcotes* 1086, *Sut Cotun* c.1115, *Parva Cotes* 1242. The *Cotun* spellings of c.1115 represent an alternative form of the name derived from OE dative pl. *cotum* 'at the cottages'. DEPN.
(4) South or Little ~ Humbs TA 2408 → Great COATES TA 2310.

COATHAM '(At) the cottages'. OE **cot**, dative pl. **cotum**. The significance of OE *cot* names is explored by C. Dyer, 'Towns and cottages in eleventh-century England' in H. Mayr-Harting and R. I. Moore eds., *Studies in Medieval History presented to R. H. C. Davis*, London 1985, 91–106.
(1) ~ Cleve NZ 5925. *Cotum* [1123×8]15th–1404, *Cotom* 1231, 1443. The reference is to fishermen's huts. The name is in systemic contrast with other dative pl. names in the district, ACKLAM NZ 4817, KIRKLEATHAM NZ 5921, MARSKE NZ 6322 and UPLEATHAM NZ 6319. YN 156.
(2) ~ MUNDEVILLE Durham NZ 2820. 'C held by the Mundaville family'. *Cotum Mundeville* 1235, *Cotom Mo(u)nd(e)vill -vyll* 1374–1549, *Coatham Mundeville* from 1647. Earlier simply *Cotum* 1214×18–1333. The family name Mundaville derives either from Émondeville, Manche, *Amundavilla* 1050×60, or Mondeville, Calvados, *Amundi villa* 990. Members of this family held various estates in Durham in the 12th cent. The manorial addition of the name *Mundeville* is for distinction from COATHAM STOB NZ 4016. NbDu 48, AA4 24.60–70, DEC 77, Dauzat-Rostaing 262, 464.
(3) ~ STOB Cleve NZ 4016. 'C with the tree-stump(s)'. *Cottam Stubbs* 1768. Earlier simply *Cotum* 1336 and also *Cotone super Tesiam* 'C on Tees' 1235×6, *Cotom next Langnewton* 1390, *Coatham Conyers* 1768. OE ***stobb** 'a tree-stump'. The various additions are for distinction from COATHAM MUNDEVILLE NZ 2820. Coatham Stob was held by the Conyers family (from Cogners, Sarthe or Cuignières, Oise, *Cuneriis* 1165) from the 14th cent. to 1569. NbDu 48, Dauzat-Rostaing 199.

COBBATON Devon SS 6126. 'Cobba's farm'. *Cobetone* 1308, *Cobbeton* 1330. OE pers.n. *Cobba* + **tūn**. D 351.

COBBIN'S BROOK Essex TL 4001. '*Cobbing* brook'. P.n. or stream-n. *Lobbing'* (sic for *Co-*) 1228, *(water of) Cobbinge* 1547, 'place or stream called after Cobba', OE pers.n. *Cobba* + **ing**², + pleonastic ModE **brook**. Ess 5, RN 210.

COBERLEY Glos SO 9616. 'Cuthbert's wood or clearing'. *Culberlege, Coberlie* 1086, *Cuthbrithleya* 1148×79–1215, *Cuthberleya -leg'* 1182, 1220, *Cuberley(a) -leg(e) -lei(a) -le(gh)* 12th–1706, *Coberley(a) -le(ye) -legh* 13th–1610, *Cowberley, Cow Berkeley* 16th. OE pers.n. *Cūthbeorht* + **lēah**. Gl i.152 gives pr [kubəli].

COBHAM Kent TQ 6768. Probably 'Cobba's hill-spur'. *(to) cobba hammes (mearce)* 'to the boundary of Cobham' 939 S 447, *Cob(b)eham* c. 1100–1225, *Cobham* from 1226. OE pers.n. **Cobba* + **hamm** in sense 2b. The reference is either to the site of Cobham hill-fort or to that of the village on an isolated island contour of 350ft. The latter site has suggested that the specific might rather be OE ***cobb(e)** 'a round lump, a cob' used topographically. The spelling *hammes* of S 447 is not absolutely decisive against *hām* 'homestead', cf. *wichammes* 974 for West Wickham, Cambs. PNK 110, KPN 244, ASE 2.21, Settlements 9.

COBHAM Surrey TQ 1060. 'Cofa's homestead or river-bend land'. *Covenhā* 1086, *Covenham Couen-* 1218–1604 including [1062]13th S 1035, *Cove-* [727, 933, 967]13th S 420, 752, 1181–1517, *Cobbe-* 1434, *Cobe-* 1465–1583, *Cobham* from 1570. OE pers.n. *Cofa* + **hām** or **hamm** 1. Parts of the parish were distinguished by the affixes *Chirch-* and *Stret-* from 1298. For the sound development cf. colloquial *sebm* for *seven*. Sr 87, ASE 2.48, MLN 50.523.

COBNASH H&W SO 4560. *Cobnash* 1832. Possibly 'Cobba's ash-tree', OE pers.n. *Cobba*, genitive sing. *Cobban*, + **æsc**, as suggested in Bannister 46.

COCK BECK NYorks SE 4739. Possibly '(wood)cock stream'. *Co(c)k'* 1293, 1590, 1622, *Koc* 1348, *Coket* early 14th, *Cokbek* 1543, *Cock River* 1771. ME **cock** (OE *cocc*) + **beck** (ON *bekkr*). YW vii.123.

COCK CLARKS Essex TL 8102. *Cocklarkes* 1545, *Cokeclarkes* 1509×47. Possibly ME **cok-lake** 'place where cocks play' (OE *cocc* + *lāc*). Ess 223.

COCKAN Cumbr NY 0718 → COGRA MOSS NY 0919.

COCKAYNE NYorks SE 6198. ME *Cokaygne*, the name of an imaginary land of luxury and idleness, used jocularly in p.ns. PNE i.105.

COCKAYNE HATLEY Beds TL 2549 → Cockayne HATLEY.

COCKAYNE RIDGE NYorks NZ 6100. *Cockayne Ridge* 1861 OS. P.n. COCKAYNE SE 6198 + ModE **ridge**.

COCKEN Durham NZ 2847 → HART Cleve NZ 4634.

River COCKER Cumbr NY 1523. 'Crooked river'. *Koker* c.1170, *Coker* 12th–1305. Brit ***kukrā-**. Cu 9.

River COCKER 'Winding river'. British r.n. **Kukrā* < adjective ***kukro-**'crooked'.
(1) ~ Cumbr NY 1523. *Koker* c. 1170, *Coker* 12th–1305. Cu 9.
(2) ~ Lancs SD 4651. *Cocur* [etc. to '1590']. La 168, RN 83.

COCKERHAM Lancs SD 4652. 'The homestead by the r. Cocker'. *Cocreham* 1086, *Kokerham* [1190]1268, 1246, *Cokerham* [1204×5]1268, 1332, *-heim* c.1155, *Kokerheim -haim* 1206, 1246. R.n. Cocker as in River COCKER SD 4651 + OE **hām**, partly replaced with ON **heimr**. La 170, Jnl 17.98, 18.17.

COCKERINGTON Lincs TF 3790. 'Settlement on the *Cocering*'. *Crochinton, Cocrinton(e)* 1086, *Cocringtuna* c.1115, *Cockeringtune* 1197, *K- Cokerington'* [c.1163]13th, 1202. R.n. **Cocering*, PrW **Cocr* < Brit ***Kukrā** 'the crooked stream', + **ing**², + **tūn**. North Cockerington is situated on the r. Lud, South Cockerington on Greyfleet. It is suggested that one or other of these streams had the alternative name *Cocering* 'the crooked stream'. DEPN, Cameron 1998.

COCKERMOUTH Cumbr NY 1230. 'Mouth of the river Cocker'. *Cokyrmoth* c.1150, *Cokirmowth, Cocre- C- Kokermuth'* 13th cent., *K- Cocermu(e)* 1194–1310, *Cockermouth* 1547. R.n. COCKER NY 1523 + OE **mūtha**. The Cocker joins the Derwent here. Cu 361.

COCKERNHOE Herts TL 1223. 'Cock-house hill-spur'. *Cokernho* 1221, 1314. OE **cocc** + **ærn** + **hōh**. Hrt 19.

COCKFIELD Durham NZ 1224. Either 'Cocca's open land', or 'open land frequented by cocks'. *Kokefeld* 1223, *Co(c)kefeld* 1242×3–1458, *Cokfeld* 1314–1516, *Cockfield* from 1607. OE pers.n. *Cocca* or **cocc**, genitive pl. **cocca**, + **feld**. The birds would be woodcock or the like. NbDu 49.

COCKFIELD Suff TL 9054. 'Cohha's open land'. *Cokefeld* [946×c.951]13th S 1483, *Cohhanfeldœa* 962×91 S 1494, *Cochanfelde* [962×91]13th ibid., *Cothefeldā* (for *Coche-*) 1086, *Cochefelde* c.1095, *Cockefeld* 1196, *Cockfeld* 1610. OE pers.n. *Cohha*, genitive sing. *Cohhan*, + **feld**. DEPN.

COCKFOSTERS GLond TQ 2896. *Cokfosters* 1524, *(house called) Cockefosters* 1613. Possibly named from the house of the *cock* or *chief foster* or *forester*. Cockfosters lies at the edge of Enfield Chase. Mx 73, GL 39, Encyc 192.

COCKING WSusx SU 8717. 'The Coccingas, either the people called after Cocca or the people living at or by the *Cocc*, the hillock'. *Cochinges* 1086, *C- Kokinges* 1189, 1261, *Kocking'* 1199.

OE folk-n. **Coccingas* < pers.n. *Cocc(a)* or OE ***cocc** possibly used as a p.n. + **ingas**. Sx 16, ING 32, MA 10.23, SS 64.

COCKINGTON Devon SX 8963. 'Estate called after Cocca'. *Cochintone* 1086, *Kokington* from c.1180 with variants *Cok-*, *-in(g)-* and *-yng-*. OE pers.n. **Cocc(a)* + **ing**[4] + **tūn**. D 511.

COCKLAKE Somer ST 4449. *Cocklake* 1817 OS. This may be 'cock lake' referring to part of the marshes along the river Axe, but the evidence is too late for certainty.

COCKLEY BECK Cumbr NY 2401. *Cockley Beck* 1865 OS. P.n. *Cockley*, probably 'wood-cock clearing', + ModE **beck**.

COCKPOLE GREEN Berks SU 7981. *Cockpole Green* 1846. P.n. Cockpole + ModE **green**. Cockpole, *Cockpole* 1607, is the 'pool frequented by woodcocks', ultimately from OE **cocc** + **pōl**. Brk 63.

COCKS HILL Devon SX 5678. *Cockshull* 1573. Formerly also known as *Mewyburgh(e)* 1240†. The hill rises to 1642ft. D 192.

COCKSHUTT Shrops SJ 4329. 'Cock-shoot'. *La Cockesete* 1270, *Kackeshute* 1427, *Cockshutt(e)* from 1557. OE ***cocc-scīete** 'a woodland glade where nets are stretched to catch a woodcock'. More usual as a minor name, this instance is unique in becoming a parish name (1872). The earliest form probably belongs here, but may refer to one of the other Cockshutts in Shrops. Sa i.94.

COCKTHORPE Norf TF 9842. Possibly 'cock Thorp, Thorp where cocks are reared'. *Coketorp* 1254. Earlier simply *Torp* 1086, 'outlying settlement', ON **thorp**. It is uncertain whether the prefix is OE *cocc* 'a cock' or pers.n. *Cocc(a)*. Another possibility is *cock* in the sense 'chief'. Cf. COCKFOSTERS Essex TQ 2796. DEPN.

COCKWOOD Devon SX 9780. 'Cock wood'. *Est- Westcokwode* 1514. ME **cock** + **wode**. D 494.

COCKYARD H&W SO 4134. Partly unexplained. *Cochard* 1327. The specific is probably OE ***cocc** 'a hill' referring to the marked hill called *Cockyard Tump* 1831 OS possibly + W **ardd** 'a height'. The modern form has been assimilated to ModE **cock** + **yard**. Bannister 47.

COD BECK NYorks SE 4277. 'Cotta or Kotti's beck'. *Cotesbec* [13th]copy, *Cottebec* 1279, *Coddebek* c.1540, *Codbeck* 1562. OE pers.n. *Cotta* or ODan pers.n. *Kotti*, genitive sing. *Kotta*, + **bekkr**. YN 2, RN 85.

CODDA Corn SX 1878. Unexplained. *Codda* 1459, 1671, short for *Stingede-lace* [1239]15th, *Stumcodda* 1280, *Stymkodda* 1385, 1401. Co ***stum** 'bend' + unidentified element. *lace* in 1239 is OE **lacu** 'stream, watercourse', presumably referring to a tributary of the nearby river Fowey. PNCo 70, Gover n.d. 44, L 23.

CODDENHAM Suff TM 1354. 'Codda's homestead'. *Code(n)- Cadenham* 1086, *Codeneham* 1154×89, *Codham* 1195, 1206, *Codeham* 1204–1335, *Codenham* 1254–1523, *Codden-* from 1568. OE pers.n. *Codda*, genitive sing. *Coddan*, + **hām**. DEPN, Baron.

CODDINGTON Ches SJ 4555. 'Farm, village called after Cot(t)a'. *Cotintone* 1086, *Cotituna* 1150, *Cot- Cothinton -yn-* 1188–1390, *Codinton* 1157–1535 with variants *-yn- -ton(a)*, *-tun*, *Codington* 1258–1724, *Codd-* from 1289. OE pers.n. *Cot(t)a* + **ing**[4] + **tūn**. In the township was *le Codyngeheye* 1284×7 'Cot(t)a's fenced-in enclosure'. Che iv.85.

CODDINGTON H&W SO 7242. 'Farm, village called after Cot(t)a'. *Cotingtvne* 1086, *K- Cotinton(e) -yn(g)-* 1276–1428, *Codynton* 1302. OE pers.n. *Cot(t)a* + **ing**[4] + **tūn**. He 58, Bannister 47.

CODDINGTON Notts SK 8354. Partly uncertain.

I. *Cotintone* 1086, 1250, *Cotington -yng-* early 13th, 1335.

II. *Chodingtona* 1163, *Codinton* 1226, *-ing- -yng-* 1252–1343.

This could be 'the estate called after Codda or Cotta', OE pers.n. *Cod(d)a* or *Cot(t)a* + **ing**[4] + **tūn**. Or the specific might rather be an **ing**[2] derivative of OE **codd** 'a bag, a sack, a hollow', or of OE **cot** 'a cottage'. Coddington is situated on a small hill-spur overlooking a hollow. Nt 211, Årsskrift 1974.54.

CODFORD ST MARY Wilts ST 9739. 'St Mary's C'. *Codeford Sce Marie* 1291, *Codford Marys* 1552. P.n. *Codan ford* [901]12th S 362, *Coteford* 1086–1327, *Codeford* 1212–1428, 'Coda's ford', OE pers.n. *Coda*, genitive sing. *Codan*, + **ford**, + saint's name *Mary* from the dedication of the church and for distinction from CODFORD ST PETER ST 9640. Wlt 164.

CODFORD ST PETER Wilts ST 9640. 'St Peter's C'. *Codeford Sci Petri* 1291, *West Codford al. Codford St Peter* 1619. P.n. Codford as in CODFORD ST MARY 9739 + saint's name *Peter* from the dedication of the church. Wlt 164.

CODICOTE Herts TL 2118. 'Cottages associated with Cuthere or Cutha'. *(æt) Cuðeringcoton, Cuðingcoton* [1002]13th S 900, *Codicote* from 1086 with variants *Cody- Cud(d)i-*. OE pers.n. *Cūthhere* and pet form *Cūtha* + **ing**[4] + **cot**, dative pl. **cotum**. Hrt 109 gives pr [kʌdikət].

CODNOR Derby SK 4149. 'Cod(d)a's ridge'. *Cotenoure* 1086, *Cod(d)enoura -(h)ou(e)r(e) -or(e)* 1182–1519, *Codno(u)r(e)* from 1316. OE pers.n. **Cod(d)a*, genitive sing. **Cod(d)an*, + **ofer**. Db 434, L 175–6.

CODRINGTON Avon ST 7278. 'Estate called after Cuthhere'. *Godriton* c.1100, *Cu- Cod(e)rin(g)ton -yn(g)-* 1221–1763, *Co- Cutherington* 1303, 1307. OE pers.n. *Cūthhere* + **ing**[4] + **tūn**. Gl iii.57.

CODSALL Staffs SJ 8703. 'Cod's nook'. *Codeshale* 1086, 1271, *Coddeshal* 1167, 1248. Pers.n. *Cōd*, genitive sing. *Cōdes*, + **halh** probably in the sense 'a corner of land'. Codsall lies on high ground partly surrounded by streams. DEPN, L 104–5.

CODSALL WOOD Staffs SJ 8405. *Codsall Wood* 1834 OS. P.n. CODSALL SJ 8703 + ModE **wood**.

COFFINSWELL Devon SX 8968. 'Coffin's well'. *Coffineswell* 1249, *Wylle Coffyn* 1281, *Coffynswille* 1301. Surname *Coffin* + p.n. *Welle* 1086, *Willa* 1086 Exon, *Wille* 1272×13, 'the spring, the stream', OE **wylle**. Hugo Coffin held the manor in 1185. D 512.

COFTON HACKETT H&W SP 0075. 'C held by the Hackett family'. *Corfton Hakett* 1431, *Korfen Hackett* 1650, *Coston Hackett* (sic) 1831 OS. Earlier simply *(æt) Coftune* [780]11th S 117, 849 S 1272, [934]11th S 428, *Costone* 1086, *Kofthon, Coftone* 1280, 1295, 'cove settlement', OE **cofa** + **tūn**. OE *cofa*, which variously means 'closet, chamber, cave, den' is probably here used of a cove or recess in the Lickey Hills beside which the settlement lies. 'Hackett' for distinction from another division of the manor, Cofton Richards SP 0175, *Corfton Richart* 1431, also known as *Cofton Walteri* 'Walter's Cofton' in 1256. William Haket held one Cofton in 1166, a certain Richard the other also in 1166, and a Walter in 1256. Wo 346, PNE i.104.

COGENHOE Northants SP 8260. 'Cugga's hill-spur'. *Cugenho* 1086–1428 with variants *Cuge- Cugge(n)-* and *-hou*, *Cogenho* 1313–47, *Cogenhoe* 1639, *Kukenho* 1247, *Cook(e)nhoe* 17th. OE pers.n. **Cugga*, genitive sing. **Cuggan*, + **hōh**. Nth 144 gives pr [kuknou].

High COGGES Oxon SP 3709. ModE adj. **high** + p.n. *Coges* 1086, *Kogis* 1195, *Cogges* from 1201×3 with variants *K-*, *Coggis -ys*, 'the cogs', OE ***cogg** 'a cog of a wheel', pl. ***coggas**. The reference is unclear. O 333, L 124.

COGGESHALL Essex TL 8522. 'Cocc's nook'. *Coggashaele* [1042×66]12th S 1646, *-eshale* c.1060, *Kockeshale* [1052×66]13th S 1519, *Coghes(s)hala* 1086, *C(h)oggeshal(a) -e -hall(e) -is-* 1181–1578, *Coksale* 1412, *Cox(h)all* 1558×1603. OE pers.n. **Cogg* or *Cocc*, genitive sing. **Cogges, Cocces*, + **healh**. Also possible would be OE **cocc** 'a heap, a hillock' or **cocc** 'a cock' and so 'nook of the hillock' or 'cock's nook'. Ess 365 gives prs [kɔksl, kɔksɔ:l] and now usually [kɔgiʃ ɔ:l].

COGRA MOSS Cumbr NY 0919. Also known as Congra Moss. Cf.

†Not the r. Meavy as RN 282 unless this name formerly also applied to the Walkham.

Cockan NY 0718, *Cockan* 1636, *Cockin* 1650, Cockleygill, *Cocklay-gill* 1636, and *Cockall* and *Moss Beck* 1867 OS. Cu 406.

East COKER Somer ST 5412. *Estkoker* 1243, *-cocre* 1275. ME adj. **est** + p.n. *Cocre* 1086, *Cokre* [1152×8, 1216×72]14th, *Coker* from [1280]14th with variants *Cocra, Koker*, originally a stream n., cf. Leland's *Coker water* c.1540, identical with Cocker in COCKERHAM Lancs SD 4651 and COCKERMOUTH Cumbr NY 1230, < Brit **kukrā* 'crooked one'. DEPN, RN 84.

West COKER Somer ST 5113. *Westcocre* 1227, *West Coker, Koker* [late 13th]14th. ME adj. **west** + p.n. Coker as in East COKER ST 5412. DEPN, RN 84.

COLAN Corn SW 8661. 'St Collen's (church)'. *(decenn' de) Sancto Culano* 1201, ~ *Colano* 1205, 1296, 1303, *(Ecclesia) Sancti Coelani* 1270, *(Sien) Colan* 1302, 1429, *Cullan* 1610. Saint's name *Collen* or *Colan* from the dedication of the church. Nothing is known of this saint whose name probably recurs in Langolen 'church site of Collen' in Finistère, Brittany, and in LLangollen, Clwyd SJ 2141, *Lancollien* 1234, *LLangollen* 1391. PNCo 70, Gover n.d. 318, DEPN.

COLATON RALEIGH Devon SY 0787. 'C held by the Raleigh family'. *Coleton Ralegh* 1316, *Colyton Rawley* 1680. Earlier simply *(of) Colatune* 1072×1103, *Coletone* 1086–1395, 'Cola's farm or village', OE pers.n. *Cola* + **tūn**. Wimund de Ralegh held the manor in 1242. D 587.

COLBURN NYorks SE 1999. 'Cool' or 'coal-black stream'. *Corburne* 1086, *Colebrun(n) -burn* 12th–1260, *Colburn(e)* from 13th, *Cowburne* 1574. OE **cōl**, definite form **cōla**, or **col**, ON **kol**, + OE **burna** or ON **brunnr**. YN 243 gives pr [koubən], SSNY 253.

COLBY Cumbr NY 6620. 'Farmstead or village by a hill'. *C-Kollebi -by* 1100–1351, *Colebi -by* 1132–1277, *Colby(e)* 1369–1777, *Coweby* 1337, *Cou- Cowby* 17th cent. ON **kollr** + **bȳ**. ON pers.n. *Kolli* is also possible, but the village lies beside a marked hill. We ii.96 gives pr ['koubi], SSNNW 27.

COLBY, COLEBY 'Koli's farm or village'. ON pers.n. *Koli*, genitive sing. *Kola* + **bȳ**.
(1) ~ Humbs SE 8919. *Colebi* 1086, 1202. DEPN, SSNEM 41.
(2) ~ Norf TG 2231. *Colebei* 1086, *Colebi -by* 1191–1202. DEPN, SPNN 275.

COLCHESTER Essex TM 0025. 'Roman town on the river Colne'. *(to) Colneceastre* 921 ASC(A), *Colenceaster* [931]12th S 412, *Coleceastra -cestr(i)e -cestr(i)a* 1067–1291, *(in) cole castro* 1086, *Colchestre* 1223–1428, *-er* 1491. R.n. COLNE TL 9327, TM 0616, + OE **ċeaster**. The earliest record is W *Cair Colun* [9th]10th Nennius, an exact equivalent of OE *Colneceaster*; later it was *Caer Colyne* 1594. There are also two folk etymologies: in the 12th cent. Henry of Huntingdon associated the name with Old King Cole – *Kair-Collon, id est Coleceastria* and *regis Britannici de Colecestre, cui nomen erat Coel* 'a British king of Colchester called Coel' – whence in the following century Robert of Gloucester could write *Cole was a noble man... & Colchestre after is name icluped is.* And in the 15th cent. the form *Golchester* was associated both in English and Welsh (*Caercolden, Caergolden*) with gold. For the alternative view that the specific is rather Latin *colonia*, see under CAMULODUNUM TM 0025. Ess 367, TC76.

COLD ASH Berks SU 5170. *Cold Ash* 1761. A parish carved out of the civil parish of Thatcham for ecclesiastical purposes in 1865 and for civil purposes in 1894. Brk 245.

COLD FELL Cumbr NY 6055. 'Cold mountain'. *Caldefell pyke* 1589, *Calfeildpike, Calefell* 1603. ME **cald** (ON *kald*) + **fell** (ON *fjall*). Cu 104.

COLD LAW Northum NT 9523. 'Cold hill'. ModE adj. **cold** + N dial. **law** 'a hill' (OE *hlāw*). Nearby Coldlaw Burn NT 9417, *Caldelauburne* 1255, is named from another exposed shoulder of The Cheviot. NbDu 50.

COLDBLOW Kent TQ 5173. *Cold Blow* 1799. A fanciful name, cf. Coldblow Farm TR 3550 called *Wadling Court* 1819 OS. There are many examples of this name type in E Kent e.g. in Aldington, Bilsington, Hastingleigh, Lyminge, Temple Ewell, Wye, and its pl. variant Cold Blow(e)s in Brabourne, Challock, Upper Hardres, Monks Horton, Ickham, Kingsnorth and Ruckinge. A Cool Blows occurs in Elham. An early recorded instance is *Kalblowesande* 1317×8. Cullen.

COLDEAN ESusx TQ 3408. No early forms. Possibly 'charcoal valley', i.e. a valley where charcoal is made, ultimately OE **col** + **denu**.

COLDEAST Devon SX 8174. Unexplained. *Choleyest* 1430, *Coldyest* 1452, *-east* 1609. Formally this could be OE **ċeald** + **ēast** but such a compound would be unparalleled. The farm is in the E of the parish. The specific might alternatively be OE **ċeole** 'a throat' if this was formerly the name of the valley running NW to Ilsington at the SE end of which Coldeast stands. D 475.

COLDEN COMMON Hants SU 4822. *Golding Com*^{n}. 1810 OS. Cf. *Coledown Heath* 1540. P.n. Colden + ModE **heath, common**. Colden, *Colvedene* 1208–1333, *Coldend* 1759, *Golding* 1810 OS, is unexplained element + OE **denu** 'a valley'. Ha 58, Gover 1958.76.

COLDFAIR GREEN Suff TM 4361. *Cold Fair Green* 1837 OS. Possibly an ironic name, ModE **cold fare** 'cold food' + **green**.

COLDHARBOUR Surrey TQ 1543. 'Inhospitable dwelling'. *Coldharbor-Hill* 1675, *Cole Harbour* 1749. ModE **coldharbour**. Sr 268, 406, Nomina 8.73–8.

COLDMEECE Staffs SJ 8532 → MILLMEECE SJ 3333.

COLDRED Kent TR 2746. Probably 'charcoal clearing'. *Colret* 1086, *Colrede* c.1180–1226, *Colred* 1235. The boundary of the people of Coldred is *(andlang) colredinga gemercan* [944]13th S 501. OE **col** 'coal, charcoal' + ***ryde**, Kentish ***rede** 'a clearing'. Although this became a coal-mining area the original sense is probably 'clearing where charcoal is made'. KPN 265, PNE ii.89, Charters IV.104.

COLDRIDGE Devon SS 6907. 'Charcoal ridge'. *Colrige* 1086–1316 with variants *-rigge -rugge*. OE **col** + **hrycg**. D 365.

COLDSMOUTH HILL Northum NT 8528. *Coldsmouth Hill* 1865 OS.

COLDWALTHAM WSusx TQ 0216 → Cold WALTHAM TQ 0216.

COLE Somer ST 6733. Originally a r.n. *Colna* 1212, *Colne* 1219, *Kolle* 1285. Pre-English r.n. **Colün* of unknown meaning identical with r. COLNE Essex TL 9237, originally applied to the r. Brue which may be a later name. DEPN.

River COLE Wilts SU 2294. *River Cole* 1828. The lateness of the evidence naturally suggests that this is a late back-formation from COLESHILL Oxon SU 2393. However, the existence of a messuage in Wanborough called *Colne* 1336, may be evidence that the r.n. is identical with r.ns. COLN Glos SP 1305, COLNE Essex TL 9327, Cambs TL 3775 and as in COLNEY HEATH, STREET and London COLNEY TL 2005, 1502 and 1704, and in COLNBROOK TQ 0277. An earlier name of the river is preserved in Lynt Bridge SU 2198 and in the bounds of Wanborough, *(innan, of) Lentan* *[854]12th S 312, with later forms *Lente* 1221, 1227, *Liente, Lyenthe* 1216×72 and *Leynt* 1482. This is linked by Ekwall with the old name of a brook in King's Norton WMids SP 0579, *(of, on) Leontan, Liontan* [699×709]11th S 64, implying an OE form **Lēonte* connected with W *lliant* 'a torrent, a flood, a stream'. Yet a third name is preserved in the bounds of Wanborough and Little Hinton, *of smitan* *[854]12th S 312 and *innan smitan stream*, [1043×53]12th S 1588, from OE **Smīta* or *Smīte* 'the gliding one' on the root of the verb *smītan* in the sense 'glide, slip'. Wlt 5, Brk 9, RN 249, 373.

River COLE WMids SP 1787. 'The hazel-tree stream'. *(in, on, of) Colle* [849]10th S 1272, 972 S 786, 1282, 1457, 1540, *Cole broke* c.1460–1603. PrW ***coll** < **koslo-* cognate with Co **coll* 'hazel-trees', W *coll* 'hazel, sapling, twig', Latin *corylus* and *hazel* itself. Wa 2, RN 85, LHEB 273, CPNE 62, GPC 546.

COLEBATCH Shrops SO 3287. Partly uncertain. *C- Kolebech(e)* 1176–1272, *Colebach'* 1271×2, 1334, *Colbach(e)* 1306×7, 1416, 1641, *Cowbach* 1577, 1675. The generic is OE **bæċe** 'a stream-valley', the specific either OE pers.n. *Cola*, OE **col** 'charcoal', or **cōl**, definite form **cōla**, 'cool'. Sa i.94.

COLEBROOK Devon ST 0006. Either 'the cool brook' or 'Cola's brook'. *Colebroche* 1086, *-broc* 1175, *-brok* 1314, *Colbroke* 1390, 1488. OE **cōl**, definite form **cōla**, or pers.n. *Cola*, + **brōc**. D 560.

COLEBROOKE Devon SS 7700. Either 'the cool brook' or 'Cola's brook'. *Colebroc(h)* 1154×89–1291 with variants *-brok*, *Colbrok(e)* 1280. OE **cōl**, definite form **cōla**, or pers.n. *Cola*, + **brōc**. D 403.

COLEBY Lincs SK 9760. 'Koli's village or farm'. *Colebi -by* 1086–1576, *Colleby* 1245–1340, *Colby* 1276–1642. ON pers.n. *Koli*, genitive sing. *Kola*, + **bȳ**. Perrott 221, SSNEM 41.

COLEFORD Devon SS 7701. Short for *Colbrukeford* 1330, *Colbrokforde* 1333, 'ford over Colebrooke', p.n. COLEBROOKE SS 7700 + **ford**. D 404.

COLEFORD Glos SO 5710. 'Ford where coals are carried'. *Coleford(e)* from 1282. OE **col**, genitive pl. **cola**, + **ford**. Coal was mined in the Forest of Dean from the 13th cent. and also in Roman times. Gl iii.214, VCH Glos ii.215, L 71–2.

COLEFORD Somer ST 6849. 'Coal ford'. *Culeford* 1234, *Colford* 1291. OE, ME **col** + **ford**. Coleford is in an old mining area so that the name may be quite late and refer to a ford across which coal was carried rather than charcoal. Coalpits are marked here in 1610 (*Colepitts*). DEPN, PNE i.105, L 71–2.

COLEHILL Dorset SU 0200. 'Charcoal hill' or 'hill called Cole'. *Colhulle* 1341, *-hill* 1518, *Collehill* 1547, *Colehill'* 1578. OE **col** or **coll** 'a hill' taken as a p.n. + **hyll**. The hill referred to is that NW of Wimborn Minster called *Cole Hill* 1847. Do ii.137.

COLEMAN'S HATCH ESusx TQ 4533. 'Coleman's hurdle' or 'fish-trap'. *Colmanhacche* 1495, *Colmans-* 1535, *Comans Hatch*. Surname *Coleman* recorded here in 1279 and 1327 + ME **hache** (OE *hæċċ*). Sx 317.

COLEMERE Shrops SJ 4332. 'Cula's lake'. *Colesmere* 1086, *Colemere* from 1157×72, *Culmere* 1279–1440, *Coomer* 1655, *Coumer* 1704. OE pers.n. *Cūla* + **mere**. Sa i.96 gives local pr [ku:mə].

COLEORTON Leic SK 4017. ME prefix **col** 'coal', *Col(l)e* 1443–1610, *Coal* 1719, + p.n. *Ovretone* 1086, *Ouer- Overton(e)* 1086–1666, *Owerton -or-* 1346, 1347, 1527, *Orton* from 1456, the 'settlement on the flat-topped ridge', OE **ofer** + **tūn**. The original name was simply Overton to which the affix Cole was added for distinction from other Overtons. The earliest reference to a coal-mine here is *Colpitt close* 1539, but the evidence above points to mineral exploitation at least a century earlier. Lei 362.

COLERNE Wilts ST 8171. Probably 'the charcoal house', a place where charcoal was made, stored or used. *Colerne* from 1086, *Cul(l)erne* 1156–1728. OE **col** + **ærn**. The main difficulty is to explain the *Cul-* sps. There seems to have been a variant form OE **cul* 'charcoal' which occurs in the Wilts minor name *Culputtehalve* 1375, 'charcoal pit half'. Wlt 93 gives pr [kʌlərn], 434.

COLESBOURNE Glos SO 9913. 'Col's stream'. *(to) Colesburnan (forda)* 'to the ford of C' [800]11th S 1556, *(æt) Col(l)esburnan* [798×822]11th S 1262, [n.d.]11th B 1320, *Colesborne -burn(e) -burn(i)a* 1086–1372, *Collesb(o)urne* 1171×83–1535. OE pers.n. **Col(l)*, genitive sing. **Col(l)es*, + **burna**. Gl i.154.

COLESDEN Beds TL 1255. 'Col's valley'. *Colesden(e)* 1195–1220. OE pers.n. **Col* < ON *Kolr*, genitive sing. **Coles*, + **denu**. Bd 65.

COLESHILL Bucks SU 9495. Probably 'Coll's hill'. *Coleshull(e)* [1279]14th, 1340, 1345, *Colshull* 1304. OE pers.n. *Col(l)* from ON *Kol(l)r*, genitive sing. *Col(l)es*, + **hyll**. Bk 227 gives pr [kousəl].

COLESHILL Oxon SP 2393. Probably 'Coll's hill'. *(æt) Colleshylle* [?c.950]10th S 1539, *Coleselle -halle* 1086, *Colleshill* 1278, *Coleshel' -hill(') -hull(e)* from 1185. OE pers.n. **Coll*, genitive sing. **Colles*, + **hyll**. The uncertainty about this pers.n. has led to other suggestions such as OE ***coll** 'hill' or an unrecorded stream name from PrW ***coll** 'hazel-trees'. Brk 356, L 170, Löfvenberg 43.

COLESHILL Warw SP 2089. Partly uncertain. *Colleshyl* [799]11th S 154, *Coleshelle* 1086, *-hull(e)* 1161–1625, *-hill* 1718, *Colsull* 1365, *Colsell* 1540, 1626. The generic is OE **hyll**, the specific probably the r.n. COLE SP 1787 and so 'the hill on the r. Cole'. Wa 42, Brk 356–7.

COLETON Devon SX 9051 → COLLATON SX 8760.

COLGATE WSusx TQ 2332. Probably 'the charcoal gate'. *la Collegate* 1279. ME **col** or pers.n. *Colle* (Nicholas) + **gate**. The reference is to one of the gates into St Leonard's Forest and probably to the manufacture and supply of charcoal there. Cf. FAYGATE TQ 2134. Sx 203.

COLKIRK Norf TF 9126. 'Koli or Cola's church'. *Colechirca -kirka* 1086, *Colechircha* 1161, *-kerca* 1168, *Kolekirch'* 1209. ON pers.n. *Koli*, genitive sing. *Kola*, or OE pers.n. *Cola*, + **ċiriċe**, ON **kirkja**. DEPN, SPNN 275.

COLLATON ST MARY Devon SX 8760. 'St Mary's C'. P.n. Collaton + saint's name *Mary* from the dedication of the church. Also known as *Collaton Kirkham* 1809 OS 'C church village', p.n. Collaton + ModE **kirkham**. Earlier simply *Coletone* 1261, 'Cola's farm or village', OE pers.n. *Cola* + **tūn**. A common name in S Devon, cf. Shiphay Collaton SX 8965, *Coletone* 1086, and Coltone SX 9051. D 517.

COLLEGE BURN Northum NT 8825. *College Burn* 1865 OS. P.n. *Colledge* 1542, probably the 'cold stream', OE, ME **cōl** + N dial. **letch** 'a stream flowing through boggy land' (OE **læcc*, **lecc*, **lece*), + ModE **burn**. Alternative specifics might be OE, ME **col** '(char)coal' and so 'the stream where charcoal is burnt', or OE pers.n. *Cola*, 'Cola's stream'. RN 86–7.

COLLIER LAW Durham NZ 0141. *Cloyer Lawe* (sic for *Colyer*) 1647. Probably surname *Collier* + OE **hlāw** 'a hill'. Collier Law rises to 1694ft. and marks the boundary between three parishes.

COLLIER ROW GLond TQ 4991. 'Charcoal-burners' row'. *Colyer(s) Rowe* 1453. ME **colier** + **row** (OE *rāw*). Ess 117.

COLLIER STREET Kent TQ 7145. 'Collier hamlet'. *Collier Street* 1819. Surname *Collier* + Kent dial. **street**. Other *street* names in the vicinity include *Denover Street* 1819, now simply BENOVER TQ 7048, Haviker Street TQ 7246 from the surname *He(a)uakere* 13th. Cf. BROAD STREET TQ 8356. PNK 170.

COLLIERS END Herts TL 3720. *Colyersend* 1526. Surname *Collier* + ModE **end** sc. of Standon parish. Hrt 199.

COLLIFORD LAKE RESERVOIR Corn SX 1772. P.n. Colliford SX 1870 + ModE **lake** + **reservoir**. A lake and reservoir constructed in the early 1980s. Colliford, *Colaford* mid 13th, *Coleford(e)* 1291, 1370, *Culeford* 1296, is probably 'the cool ford', OE **cōl** + **ford**, but it could alternatively be 'Cola's ford', OE pers.n. *Cola*. PNCo 71, Gover n.d. 285.

COLLINGBOURNE DUCIS Wilts SU 2453. 'C held by the duke'. *Colingburn' Comitis* 1256, *Colyngbourne ducis* 1507. Originally a stream name, Colingbourne, + Latin **ducis**, genitive sing. of **dux** 'a duke', earlier **comitis**, genitive sing. of **comes** 'a count, an earl'. The manor was held by the earls, later dukes, of Lancaster and the additions are for distinction from Collingbourne Kingston SU 2355. Colingbourne, *Colengaburnam* [903]16th S 370, *(at) Colingburne, (of) Collengaburnan* [921]14th S 379, *(æt) Collinga burnan* [931×9]10th S 1533, *Coleburna* 1086, *Colingeburne* 1086–1257, has usually been explained as the 'stream of the Collingas, the people called after Col(l)a', OE **Collingas* < pers.n. *Colla* + **ingas**, genitive pl. **Collinga*, + **burna**. Another possibility is that the specific is the original name of the stream on which the two Collingburns stand. The name would then be either 'stream of the folk who live by the river *Coll*', OE **Collingas* < r.n. **Coll* as in COLESHILL Warw SP 2089 + **ingas**, genitive pl. **Collinga*, + **burna**, or r.n. **Colling* as possibly in COWAN BRIDGE Lancs SD 6376 + **burna**. Wlt 342.

COLLINGBOURNE KINGSTON Wilts SU 2355. 'The king's Collingbourne estate'. *Colingeburne Kingeston* 1306, *Colyngborne Kyngston* 1372. The manor was held by the king in 1086 and the addition 'king's estate', lOE **cynges tūn**, is for distinction from neighbouring Collingbourne Ducis SU 2453. For Collingbourne see COLLINGBOURNE DUCIS. Wlt 342.

COLLINGHAM Notts SK 8361. 'The homestead of the Colingas, the people called after Cola'. *Colingeham* 1086–1229, *Colingham* c.1155–1284, *Collingeham* 1182, 1187, *Collingham* 1269. OE folk.n. **Colingas* < pers.n. *Cola* + **ingas**, genitive pl. *Colinga*, + **hām**. Alternatively the specific might be the stream name *Colling* discussed under COWAN BRIDGE Lancs SD 6376, here referring to the Fleet, *le Flete* 1348, OE **flēot** 'a creek, a stretch of water', an old channel of the Trent on which Collingham stands. Nt 203, ING 148.

COLLINGHAM WYorks SE 3845. 'The homestead of the Collingas, the people called after Colla'. *Collingham* from 1185, *Col(l)ingeham* 1167–1275, *Colingham -yng-* 1173–1428. OE folk-n. **Collingas* < pers.n. *Colla* + **ingas**, genitive pl. **Collinga*, + **hām**. Alternatively the specific might be the stream-name *Colling* discussed under COWAN BRIDGE Lancs SD 6376 or an *ing* derivative of OE **coll* 'a hill'. In the first case the reference would be to the unnamed stream which flows through the village to join the Wharfe at SE 3846, in the latter to the high ground just E of the village at SE 3945. YW iv.174, ING 155.

COLLINGTON H&W SO 6560. 'Estate called after Cola'. *Col(l)intvne* 1086, *Colinton* 1305. OE pers.n. *Cola* + **ing**[4] + **tūn**. A Great and Little C were formerly distinguished, *Collinton Major'* 1291, *Parva Colintone* c.1250×72, the former now deserted. He 58.

COLLINGTREE Northants SP 7555. 'Cola's tree'. *Colentreu* 1086, *Colintre -en- -un- -yn-* 1199–1587, *Colyng(e)tre -ing- -trowe* 14th cent. OE pers.n. *Cola*, genitive sing. *Colan*, + **trēow**. Nth 145, L 212–3.

COLLOWAY Lancs SD 4459. *Collingeswelle* [c.1200]1268. Doubt has been cast on whether *Collingeswelle* is the ancestor of the modern name. It has been explained as 'Colling's well or brook', OE pers.n. *Colling*, genitive sing. *Collinges*, + **wella**. The specific might, however, be an old stream name, as suggested for COWAN BRIDGE Lancs SD 6376. A small stream rises close to Colloway Farm. La 175.

COLLYHURST GMan SD 8500. 'Wooded hill grimy with coal dust'. *Colyhurst* 1322, 1556, 1586, *Collyhurst* 1843 OS. OE **coliġ** + **hyrst**. La 35, Jnl 17.32.

COLLYWESTON Northants SK 9902. 'Colin's Weston'. *Colynweston* 1309–47, *Colines Weston* 1322, *Colyweston* 1374, *Collywesson* 1611. *Colin* is said to be a pet-form of the name of the Nicholas who held the manor in the 13th cent. Earlier simply *Weston(e)* 1086–1330, the 'west settlement' in contrast to EASTON ON THE HILL TF 0104. Nth 200 gives pr [kɔli'wesən].

COLMWORTH Beds TL 1058. 'Culma's enclosure'. *Co-Culmeworde* 1086, *Culmw(u)rth* 1202, 1240, *Colmworth* 1247. OE pers.n. **Culma* + **worth**. Bd 53.

River COLN Glos SP 1305. Unexplained. *(fluvium...) Cunuglae*, (of) *Cunuglan (sulhforda)* 'from the ford of the sunk road on the Coln' [718×45]11th S 1254, *(bi) Cunelgan* [855]11th S 206, *(juxta) Cunelgnan* [899]11th S 1279, *(be) Culne* [n.d.]11th B 1320, *Culna* 1257, *Colne*, *Cowne* 1590, 1637, *Cole* 1631. Together with these forms must be taken those for the estates along the river, COLN ROGERS SP 0809, COLN ST ALDWYNS SP 1405 and COLN ST DENNIS SP 0810. The OE form *Cunugle*, prepositional case *Cunuglan*, has not been explained and may be pre-Celtic. The development seems to be *Cunugle* > *Cungle* > **Cunle* > *Culne* by metathesis. Gl i.5, RN 87, Grundy 1935.42.

COLN ROGERS Glos SP 0809. 'Coln once belonging to Roger de Gloucester' who gave the manor to St Peter's Abbey, Gloucester, c.1100. *Culne Roger(i)* 1200–1354, *Colne Roger(s)* from 1287. Earlier simply *Culna(m) -e* 1100–1284. R.n. COLN SP 1305 + pers.n. *Roger*. Gl i.165.

COLN ST ALDWYNS Glos SP 1405. 'St Athelwine's Coln'. *Culna*, *Culn(e) Sancti Ail- Aylwyn(i) -win(i)* 1154–1542, *Coln(e) Sancti Aylwyn(i)* 1276–87, ~ *Allwyns* 1587, *Cowne Allens* 1620. P.n. *Colne* 1086, *Chulna* c.1127 as in River COLN SP 1305 + saint's name *Athelwine* from the dedication of the church. St Athelwine was bishop of Lindsey from 679. Two charter forms referring to this Coln and to land at Ablington SP 1008 are respectively, *(æt) Enneglan* (sic for *Cuneglan*) [862]14th S 209 and *Cungle* [962]11th S 1298. Gl i.29.

COLN ST DENNIS Glos SP 0810. 'Coln held by (the church of) St Denis' of Paris to which the estate was given in 1069. *Colne Sci' Dionis' -isi*, ~ *Seint Denys* 1287–1610. Earlier simply *Colne* 1086. R.n. COLN SP 1305 + Saint's name *Denis* (Latin *Dionisius*). Gl i.166.

COLNBROOK Bucks TQ 0277. 'Cola's' or 'the cool brook'. *Colebroc* [1146]c.1225, *-brok(e)* 1337–64, *Colbrok(e)* 1392–1526, *Col(e)brook* 1675, *Colnbrook* 1822 OS. OE pers.n. *Cola* or adj. **cōl**, definite form **cōla**, + **brōc**. The modern spelling shows the influence of the r.n. Colne on which Colnbrook stands, but the absence of any early spellings with *-n-* seems to show that this is not the origin of the p.n. Bk 242, L 16.

COLNE Cambs TL 3776. An old stream-name identical with River COLNE TL 9327. *Colne* from 1086 including [1042×66]12th S 1051, *Cone* c.1160. Hu 208, RN 88.

COLNE Lancs SD 8940. Obscure. *Calna* 1124–[1155×8]1230, *Kaun* 1242, *Caln(e)* 1246–55, *Caune* 1251, 1305, *Colne* from 1296. By origin a pre-English river name of unknown origin. La 87, RN 90.

COLNE ENGAINE Essex TL 8530. 'The river Colne estate held by the Engaine family'. *Colum Engayne* 1255, *Colne (de) Engayne* 1288, 1303, *Gaynescolne* 1475–1554. P.n. Colne + family name *Engaine* from Viel Engayne 1218. Colne, *(at) Colne* [946×c.951]13th S 1483, 1000×2 S 1486 and from 1223, *Colun* 1086–1321, *Coles* 1086, 1154×89, is transferred from the r.n. COLNE TL 9327. Also known as *Parva Colun* 'little C' 1086, *Lytel Colum* 1272, *Engaynes also called Little* 1594, and *Colum Viel* 1254, *Golun Vital* 1235, *Culn Vital* 1238. Vital is the Latinate version of pers.n. Viel. Ess 379 gives pr [kɔːən], RN 88.

Earls COLNE Essex TL 8528. 'C held by the earl' sc. of Oxford. *Colne Comitis* 'count's C' 1303, 1372, *Cunte Colum* 1274, *Erlescolne* 1358. ME **erle**, **cunte**, Latin **comes**, genitive sing. **erles**, **comitis**, + p.n. *Colne* [1035×44]13th S 1521 as in COLNE ENGAINE TL 8530. Also known as *Aubrey's Colum* 1141×52 and *Golun, Culn, Colum de Ver* 1235–54 from Alberic or Aubrey de Vere to whom the manor was granted by William I, and *Chepingge Colne* 1309, *Magna Colne* 1331, and *Monks Coln* 1449 or *Colne Priore* 1513. Ess 381 gives pr [aːlz kɔːən].

River COLNE Essex TL 9327, TM 0616. A British r.n. of unknown origin and meaning. *Colnewater* 1256, 1484, 1544, *Colne* 1362 and from 1576. Brit *Colūn* < earlier **Colaun-* of unknown origin. On the river are COLNE ENGAINE TL 8530, Earls COLNE TL 8528 and Wakes COLNE TL 8928. Ess 5, RN 88.

Wakes COLNE Essex TL 8928. 'C held by the Wake family'. *Colne Wake* 1375, *Colwake* 1381, *Weakscone* 1688. Family name *Wake* from Baldwin de Wake d.1282 + p.n. Colne as in COLNE ENGAINE TL 8530. Also known as *Colum Malot* 1227, *Colum Saer* 1254 and *Colum Quency* 1255, *Colne Quincy* 1428 from Robert Malet, the DB tenant, and Saer de Quincy whose niece married Baldwin de Wake. Ess 382.

COLNE POINT Essex TM 1012. R.n. COLNE + ModE **point**. Earlier names are *St Osyth Point* 1805 OS from ST OSYTH TM 1215 and *the Nesse commonly called Westnesse* 1703 contrasting with *(le) Estness* 1326, 1366. Ess 336.

COLNE VALLEY Essex TL 8529. R.n. COLNE TL 9327 + ModE **valley**. In the Colne Valley are COLNE ENGAINE TL 8530, Earls COLNE TL 8528 and Wakes COLNE TL 8928.

COLNEY Norf TG 1707. 'Cola's island'. *Coleneia* 1086, *Colneia* 1175, 1197. OE pers.n. *Cola*, genitive sing. *Colan*, + **ēġ**. Colney occupies a ridge of high ground in a bend of the river Yare. DEPN.

COLNEY HEATH Herts TL 2005. *Colneyhethe al. Tytenhangrehethe* 1427. Earlier *bruera de Colne* c.1275. R.n. Colney as in London COLNEY TL 1603 + ME **hethe**, Latin **bruer(i)a**. Cf. also TYTTENHANGER TL 1805. Hrt 96.

COLNEY STREET Herts TL 1502. *regiam viam que dicitur Colnee* 'the king's highway called C' 1275, *Colneystrete* 1477, *Conystreete* 1659. R.n. Colney as in London COLNEY TL 1603 + ME **strete** 'Roman road' referring to WATLING STREET TL 1110. Hrt 100.

London COLNEY Herts TL 1603. 'C on the road to London'. *London Colney* 1555. 'London' for distinction from COLNEY STREET TL 1502 and, between the two, Colney Chapel, *capella de Colnea* 1209–35. Colney is r.n. Colne as in COLNBROOK TQ 0277 + OE **ēa**. Hrt 67.

COLQUITE Corn SX 0570 → CULCHETH Chesh SJ 6595.

COLSTERDALE NYorks SE 1281. 'Coalman valley'. *Colserdale* 1301, *Coster-* 1330, *Kwustardhall* 1416, *Cowsterdale* 1616, *Colster-* 1705. ME **colster** 'one who has to do with coals' + **dale**. There is reference to a coal-mine here in 1330, but *colster* could also refer to charcoal burning. YN 230.

COLSTERWORTH Lincs SK 9324. 'The enclosure of the charcoal-burners'. *Colstewrde -vorde* 1086, *-worth* 1276–1371, *Colsterw(o)rth(e)* 1290–1725. OE **colestre**, genitive pl. **colestra**, + **worth**. Colsterworth is the site of a Roman blast furnace for iron smelting. Perrott 167.

COLSTON BASSETT Notts SK 6933. 'C held by the Basset family'. *Coleston Basset(t)* 1228–77, *Colston Basset* 1265. P.n. *Coletone* (sic) 1086, *Colestune* [1120]c.1195, *-ton(a)* [c.1155]c.1195–1244, 'Kolr's farm or village', ON pers.n. *Kolr*, genitive sing. *Kols*, + **tūn**, + family name Basset. The estate was held by Ralf Basset, justiciar under Henry I, in 1120. Nt 232, SSNEM 189.

Car COLSTON Notts SK 7242. 'Church Colston'. *Kyrcoluiston* 1242, *-colston* 1277–1346, *Kirkolston* 1277–86, *Kyrkecolston* 1364, *Kercolston* 1271–1346, *Carecolston* 1409, *Carr Colson* 1650. ME **kirk** later replaced by ME **ker**, ModE **car** 'marshland' + p.n. *Colestone* 1086, *-ton* 1235, 1269, 'Kolr's farm or village', ON pers.n. *Kolr*, genitive sing. *Kols*, + **tūn**. *Kirk* for distinction from COLSTON BASSET SK 6933. Nt 223, SSNEM 189, Mx xxxii gives local pr [kousən].

COLT CRAG RESERVOIR Northum NY 9378. P.n. Colt Crag + ModE **reservoir**. Colt Crag, *Colt Crag* 1868 OS, 1 m S of Old Colt Crag, a hill rising to 785ft, probably contains *colt* with reference to a natural feature or rock used in the same way as calf (as in the Cow and Calf rocks near Ilkley WYorks SE 1246) to designate a lesser rock or stone.

COLTISHALL Norf TG 2720. Partly uncertain, possibly 'Cohhede or Coccede's nook'. *Cokeres- Coketeshala* 1086, *Couteshal* 1200, 1207, *-hale* 1219, 1254. The recorded forms are too few and contradictory for satisfactory explanation. The DB sps suggest OE pers.n. **Cohh- *Coccede*, genitive sing. **Cohh- *Coccedes*, + **halh**. DEPN.

COLTON Cumbr SD 3186. Partly uncertain. *Coleton* 1202, *Colton* from 1332. This could be 'settlement where charcoal is made', OE **col** + **tūn**; 'Cola's farm or village', OE pers.n. *Cola* + **tūn**; or 'settlement on Colton Beck', stream name **Cole* 'the cold stream' + **tūn**. La 216, SSNNW 350.

COLTON 'Koli or Cola's farm or estate'. ON pers.n. *Koli*, genitive sing. *Kola*, or OE pers.n. *Cola*, + **tūn**.

(1) ~ Norf TG 1009. *Coletuna* 1086, *-tun(e)* 1182–1202, *Coleton'* 1198–1209. DEPN, SPNN 275.

(2) ~ NYorks SE 5444. *Coletun(e)* 1086, 1232, *C- Kolton* from 1249. YW iv.223.

COLTON Staffs SK 0520. 'The settlement where charcoal is made'. *Coltone -tvne* 1086, *-tun(a)* 1201, 1227, *-ton* 1176. OE **col** + **tūn**. DEPN, PNE i.105.

COLWALL GREEN H&W SO 7541. *Colwall Green* 1831 OS. P.n. Colwall as in COLWALL STONE SO 7542 + ModE **green**.

COLWALL STONE H&W SO 7542. P.n. Colwall as in Old Colwall SO 7431 + ModE **stone**. In 1831 OS the site is named *Tan house*. The modern development is due to the arrival of the railway here. (Old) Colwall is *Colewelle* 1086–1241, *Col(l)ewall(e)* 1140×8–1506, 'the cool spring', OE **cōl**, definite form **cōla**, + **wella**, Mercian form **wælla**. He 59.

COLWELL Northum NY 9575. 'The cool' or 'coal-black spring'. *Colewel(l)* 1236–55, *Collewell* 1296 SR, *Col(l)well* from 1318, *Collell* 1663. OE **cōl**, definite form **cōla**, or **col** + **wella**. NbDu 51.

COLWELL BAY IoW SZ 3288. *Colwell Bay* 1769. P.n. Colwell + ModE **bay**. Colwell, *Colewhelle* 1417, *Collwell(s)* 1608, 1630, *Colwell* 1720, is the 'cool spring', ultimately representing OE **cōl**, definite form **cōla**, + **wella**. Wt 126, Mills 1996.40.

COLWICH Staffs SK 0121. 'Cola's farm or estate'. *Calewich* 1166, *Colewich* 1240, *-wyz, Colwich* 1247. OE pers.n. *Cola* + **wīċ**. An alternative possibility is the 'place where charcoal is made', OE **col** + **wīċ**. DEPN gives pr [kɔli:].

COLWORTH WSusx SU 9102. 'Cola's enclosure'. *(æt) Coleworð* [988]14th S 872, *Coleworþe* 12th *-worth* 1288. OE pers.n. *Cola* + **worth**. Sx 75, SS 78.

River COLY Devon SY 2595. Unexplained. *cullig* [1005]12th S 910, *Coley river* c.1550, *Coly* 1577. Identical with the lost Yorks r.n. *Coli* 1307 (near Whernside and Appletreewick). OE **Culi* of uncertain origin. D 3 gives pr [kɔli], RN 91.

COLYFORD Devon SY 2492. 'Ford over the river Coly'. *Culy- Culiford* 1244–74, *Culli-* 1356, 1675, *Colyford* 1282, *Colyvert* 1538–44. R.n. COLY + OE **ford**. D 622, RN 91.

COLYTON Devon SY 2494. 'Settlement on the r. Coly'. *Culinton(am)* [946]12th Laws, *-ton(a)* 1086–1229, *Colintun -ton* 1219–30, *Cullingthon'* c.1225, *Cvlitone* 1086, *Culi(g)tun* 1193×1205, 1238, *Coulitone* 1237×74, *Culliton* 1356, *Colyton* 1292, *Cooliton* 1590. R.n. COLY + OE **tūn**. Some spellings show evidence of OE **-ing** used as a stream name suffix. Cf. CORYTON SX 4583, DARTINGTON SX 7862, TAVISTOCK SX 4774. D 621, RN 91.

COMB FELL Northum NT 9118. *Comb Fell* 1869 OS. ModE **comb** 'the crest of a hill, a ridge' + **fell**. A mountain rising to 2132ft.

COMBE 'Coomb'. OE **cumb** 'vessel, cup, tub' influenced by PrW **cumm* (Brit **cumbo-*, ModW *cwm*) 'valley', Particularly common in SW counties (except Cornwall). In E areas, where it is primarily of English origin, it is characteristically used of short, broad valleys, usually bowl- or trough-shaped with three steeply rising sides. In SW areas, where it is primarily of W origin, it is also used of narrower, deeper, longer valleys often winding between steep sides. L 88–94.

(1) ~ Berks SU 3761. *Cvmbe* 1086, *Cumb(e)* 1219. Brk 294.

(2) ~ H&W SO 3465. *La Cumbam* 1244, *(La) C(o)umbe* 1263–1307, *Combe* 1399. The reference is to the great hollow on the N side of Wapley Hill. He 60, L 91–2.

(3) ~ Oxon SP 4116. *Cvmbe* 1086, *Cumb(e)* 13th cent., *Combe* from 1399. O 254.

(4) ~ FLOREY Somer ST 1431. 'C held by the Flury family'. *Cumbeflori* 1291, *Combefloryе* 1610. Earlier simply *Cumba* [1155×8]1334. Hugh de Flury held the manor c.1155; his name comes from one of the Fleurys in France. DEPN, DBS s.n. Fleury 130.

(5) ~ HAY Avon ST 7359. 'Combe held by the Hayway family'.

Cumbehawya 1249. Earlier simply *Cvme* 1086. The manor was held by Thomas de Ha(i)weie in 1225. The 1086 identification is not absolutely certain. DEPN.

(6) ~ HILL Glos SO 8827. *Coombe Hill* 1778. As there is no coomb at this place, this may be ME *camb, comb* 'a comb, a crest'. Gl ii.83.

(7) ~ MARTIN Devon SS 5846. 'C held by Martin'. *Cumbe Martini* 1265, *Coumb' Martin* 1276. Earlier simply *Cvmbe* 1086, *Cumbe* 1198. The manor was held in 1133 by Robert son of *Martin*. D 36, DEPN.

(8) ~~ BAY Devon SS 5748. *Combe Martin Cove* 1809 OS. P.n. COMBE MARTIN SS 5846 + ModE **bay**.

(9) ~ MOOR H&W SO 3663. *Combe Marsh* 1833 OS. P.n. COMBE SO 3463 + ModE **marsh** and **moor**.

(10) ~ RALEIGH Devon ST 1502. 'C held by the Raleigh family'. *Comberalegh* 1383. Earlier simply *Cumb(a)* 1237, 1242, *la Cumbe* 1300. Henry de Ralegh held land here in 1292. In 1086 this estate was called *Otri* from the r. Otter, and later known as *Otercomb* held by John Ralegh. Also known as *Cumbe Sancti Nicholai* 'St Nicholas's Combe' 1260 from the dedication of the church, and as *Combe Banton* 1285, *Comb Matthei* 1303, *Combebampton* 1331 from its tenure by Mattheus de Banton (Bampton) in 1242. D 638, DB Devon ii.23, 21.

(11) ~ ST NICHOLAS Somer ST 3011. 'St Nicholas's C'. *Combe St Nicholas* 1610. P.n. *Cuma* [1065]18th S 1042, *Cvmbe* 1086, + saint's name *Nicholas* from the dedication of the church. DEPN.

(12) Abbas ~ Somer ST 7022. 'C held by the abbess' sc. of Shaftesbury. *Coumbe Abbatisse* 1327, *Abbas Combe* 1610. P.n. *Cvmbe* 1086 + Latin **abbatisse**, genitive sing. of **abbatissa**. The manor was held by Shaftesbury abbey already in 1086. DEPN.

(13) Castle ~ Wilts ST 8477. *Castelc(o)umbe* [1270]1383, 1322, 1330. ME **castel** + p.n. *Come* 1086, *Cumbe* 1186–1270. A motte and bailey castle was built here c.1140. Wlt 78, Pevsner 1975.159.

(14) East ~ Somer ST 1631. Cf. *West Combe* 1809 OS. ModE adjs. **east, west** + p.n. Combe as in COMBE FLOREY ST 1321.

COMBEINTEIGNHEAD Devon SX 9071. 'Coomb in Teignhead, the ten hides district.' *Cumbe in Tenhide* 1227, ~ ~ *Tynhid* 1259, 1277, *Comyngteynhede* 1417, *Comyng Tynid* 16th. Earlier simply *Cūbe* (for *Cvmbe*) 1086, *Comba* 1086 Exon, *Combe* 1227. OE **cumb** used as a p.n. + **in** + p.n. Teignhead 'ten hides' OE **tīen-hīde** as in TINHEAD Wilts. Teignhead was a detached part of Wonford hundred lying south of the Teign estuary containing some thirteen manors amounting to about ten hides. The name has been altered to Teignhead through association with the r.n. Teign. D 459.

COMBERBACH Ches SJ 6477. 'Stream of Cumbra, the Welshman, or of the Welshmen'. *Cambrebech* 1172×81, *Combrebeche* 1190, *Comberbach(e)* from c.1230. OE *Cumbra* or folk-n. *Cumbre*, genitive pl. *Cumbra*, + **bæċe**. Che ii.111 gives pr [ˈkʌmbərbætʃ].

COMBERTON Cambs TL 3856. Probably 'Cumbra's estate'. *Cumbertone* 1086, *-tuna* 1155, 1218, *Cumberton(e)* 1176 etc., *Commertona* 1086, *Comerton(e)* 1279–1404, *Comberton(e)* from 1286. OE pers.n. *Cumbra* + **tūn**. The specific could, however, be OE **Cumbra*, genitive pl. of **Cumbre* 'the Cymry, the Welsh'. Ca 73, PNE i.119.

Great COMBERTON H&W SO 9542. *Magna Comberton* 1428. ModE **great**, Latin **magna**, + p.n. *Cumbrincgtun* 972 S 786, *Cūbrintvne, Cūbritone* 1086, *Cu- Combrington* 1270, *Cumbreton* 1198, 'estate called after Cumbra', OE pers.n. *Cumbra* 'the Welshman' + **ing**[4] + **tūn**. 'Great' for distinction from Little COMBERTON SO 9643. Wo 193.

Little COMBERTON H&W SO 9643. *Parva Comberton* 1428. ModE **little**, Latin **parva**, + p.n. Comberton as in Great C SO 9542. Wo 193.

COMBPYNE Devon SY 2992. 'Coomb held by the Pyne family'. *Combpyn* 1377, *Compine* 1675. Earlier simply *Come* 1086. OE **cumb** 'a coomb' used as a p.n. + family name *Pyne*. Also known as *Cumba Ricardi Coffin* 1175, 'Richard Coffin's Coomb'. Sir Thomas Pyn was patron of the church in 1278. D 637.

COMBROOK Warw SP 3051. 'The valley brook'. *Cumbroc(e)* 1217–33, *-brok* 1288–1656, *-brocke* 1591. OE **cumb** + **brōc**. The reference is to the stream which flows down the *combe* of Compton Verney SP 3152, 'the valley settlement', *Contone* 1086, *Cum(p)ton* 13th cent., OE **cumb** + **tūn**, and passes Combrook described by Dugdale as 'lying neer unto a narrow and deep valley'. Wa 252.

COMBS Derby SK 0478. 'The valleys'. *Cumbes* 1251, *Coumbe(s)* 1285–1346, *Combes* from 1375. OE **cumb**, pl. **cumbas**. The hamlet is situated in the valley of Meveril Brook and at the entrance to the valley of Pyegreave Brook. Db 61, L 9, 91.

COMBS Suff TM 0456. 'The crests'. *Cambas* 1086, *Cambes* 1130–1286, *Combes* 1212–1610. OE **camb**, pl. **cambas**. The reference is to the two hill-spurs between which the settlement lies. DEPN, Baron.

COMBS FORDS Suff TM 0577. *Combs Ford* 1837 OS. P.n. COMBS TM 0456 + ModE **ford**.

COMBS RESERVOIR Derby SK 0379. P.n. COMBS SK 0478 + ModE **reservoir**.

COMBWICH Somer ST 2542. 'Coomb trading place'. *Cōmiz, Comich* 1086, *Cumwiz* 1178, *Comwidg* 1610. OE **cumb** + **wīċ**. 'Trading place' is the most likely sense here; Combwich Pill provided a natural harbour at the mouth of the Parrett. It was an important Roman settlement connected by road to Ilchester and beyond. DEPN, Archaeology 72.

The COMMON Wilts SU 2432.

COMMONDALE NYorks NZ 6610. 'Colman's valley'. *Colemandale* 1273. OIr pers.n. *Colmán* + ON **dalr**. YN 148.

COMMON MOOR Corn SX 2469. *Common Moor* 1867. A 19th cent. hamlet situated on the common moor of St Cleer parish. ModE adj. **common** + **moor** 'rough grazing'. PNCo 71.

COMMON SIDE Derby SK 3375. 'Beside the common'. Cf. *Commonside House* 1818. The reference is to Barlow Common. Db 205.

COMPSTALL GMan SJ 9691. *Compstall (Bridge)* from 1608. A 19th cent. civil parish formed out of Werneth SJ 9592. Perhaps 'valley fishing place', OE **cumb** + **stall**. It lies beside the r. Etherow. Che i.303.

COMPTON 'Coomb settlement'. OE **cumb** + **tūn**. PNE i.119, L 1, 89–92.

(1) ~ Berks SU 5379. *Contone* 1086, *Cumpton(a)* 1195–1243. Brk 498.

(2) ~ Devon SX 8664. *Cumpton* 1244, *Compton* 1330. D 516.

(3) ~ Hants SU 4625. *Cvntvne* 1086, *Cumton* c.1195, *Compton* from 1158 with variant *Cump-* 1202–1316. Ha 58, Gover 1958.174, DEPN.

(4) ~ Surrey SU 9547. *Contone* 1086, *-ton* 1196, *Cumpton* 1190–1339, *Compton* from 1287. The village lies in a depression below the Hog's Back. Called *Estcompton* in 1467 to distinguish it from Compton SU 8546, *Cumpton* 1241. Sr 194, 176, L 1, 89–92.

(5) ~ WSusx SU 7714. *Cumtun* 1015 S 1503, *Contone* 1086, *Cumpton* 1224. Sx 47 gives pr [kʌmtən].

(6) ~ Wilts SU 1352. *Contone* 1086, *Cumpton* 1256. Wlt 328.

(7) ~ ABBAS Dorset ST 8718. 'C held by the abbess' sc. of Shaftesbury. *Cum(p)ton Abbatis(se)* 1293, 1428, *Compton(') Abbatisse* 1340. Earlier simply *(to, atte) Cumtune* [956]14th S 630, *Cuntone* 1086, *Cum(p)ton'* 1268–1342. The manor belonged to Shaftesbury abbey from 956. 'Abbas' is a reduced form of Latin **abbatisse**, genitive sing of **abbatissa** 'an abbess'. Do iii.99 gives pr [ˈkɔmtən].

(8) ~ ABDALE Glos SP 0616. 'Compton called Abdale' for distinction from Cassey Compton SP 0415 'Compton held by the

Cassey family' at the bottom of the same coomb. *Compton Apdale* 1504, ~ *Abdale* 1535. Earlier simply *Contone* 1086, *Cumton'* 1221, 1285, *Cumpton(a)* 1221–83, *Compton* from 1291. The source of *Apdale* is unknown, probably a surname from Apedale Staffs SJ 8149, NYorks SE 0194, or Ape Dale Shrops SO 4889. Gl i.167.

(9) ~ BASSETT Wilts SU 0372. 'C held by the Bassett family'. *Cumptone Basset* 1228. P.n. *Contone* 1086, *Comtona* 1182, + manorial addition from the family name of Fulke Basset who held the estate in 1242. Wlt 262.

(10) ~ BAY IoW SZ 3684. *Compton Bay* 1550. P.n. *Cantvne* 1086, *Cumptune -ton(e)* 1155×61–1302, *Compton(')* from 1154×89, + ModE **bay**. Wt 126 gives pr [kʌmtn], Mills 1996.41.

(11) ~ BEAUCHAMP Oxon SU 2887. 'C held by the Beauchamp family'. *Cumton' Beucamp* 1235×6, *Compton Bechaump* 1379. P.n. *(æt) Cumtune* [955]13th S 564, *Contone* 1086, *Compton(')* from 1327, + family name of Walter de Beauchamp who held the manor in 1220. Brk 360.

(12) ~ BISHOP Somer ST 3955. 'C held by the bishop' sc. of Wells. *Compton Episcopi* 1332. Earlier simply *(into) Cumbtune* 1067, 1068†. DEPN.

(13) ~ CHAMBERLAYNE Wilts SU 0229. 'C held by the Chamberlain family'. *Compton Chamberleyne* 1316. P.n. *Contone* 1086, *Cumpton* 1208–49, + manorial addition from the family name of Robert Camerarius and Geoffrey le Chaumberleng who held the estate in 1208 and 1242 respectively. Wlt 399.

(14) ~ DANDO Avon ST 6464. 'C held by the Dando family'. *Cumtun Daunon* 1256. Earlier simply *Cumtun* [970]12th S 777††, *Contone* 1086, *Cumton* 1225. Compton was held by Alexander de Alno in the later 12th cent. His family name, variously *de Alno, Auno, Alneto, Dauno* is from Aunou in Normandy, *Alnetum* 12th. DEPN.

(15) ~ DOWN Wilts SU 1051. *Compton Down* 1817 OS. P.n. COMPTON SU 1352 + ModE **down**.

(16) ~ DUNDON Somer ST 4932. Short for 'C near Dundon'. *Cumpton by Dunden* 1289, *Cumpton Dundo* 1610. P.n. *Contone* 1086, *Compton* 1610‡, + p.n. DUNDON ST 4732. DEPN.

(17) ~ MARTIN Avon ST 5457. 'C held by Martin (de Tours)'. *Cumpton Martin* 1226×8. Earlier simply *Contone* 1086. DEPN.

(18) ~ PAUNCEFOOT Somer ST 6425. 'C held by the Pauncefoot family'. *Cumpton Paunceuot* 1291, *Cumpton Pauncefford* 1610. Earlier simply *Cvntone* 1086. *Pauncefote* is a Norman nick-name meaning 'round belly'. DEPN, DBS 266.

(19) ~ VALENCE Dorset SY 5993. 'C held by the Valence family'. *Compton Valance* 1280, earlier simply *Contone* 1086, *Cumton* 1212, *Cumpton* 1252. 'Valence' from William de Valencia, earl of Pembroke, who was granted the manor in 1252. Do 60.

(20) ~ VERNEY Warw SP 3152 > COMBROOK SP 3051.

(21) ~ WYNYATES Warw SP 3342. 'C at Windgate'. *Cumpton Wincate* 1242, *atte Wingate* 1247, *Compton Wind(e)gate* from 1268 with variants ~ *Windȝate* 1279, ~ *(atte) Wynyate(s)* 1315 etc. P.n. *Contone* 1086 + p.n. Wynyate(s) 'the wind gate, a gap through which the wind blows', ME **wind-yate** (OE *wind-ġeat*), referring to the point where the combe narrows and forms a windy or draughty passage. Wa 279.

(22) East ~ Somer ST 6141. *East Compton* 1817 OS. ModE adj. **east** + p.n. *Coumpton* 1327. DEPN.

(23) Easter ~ Avon ST 5782. 'The more easterly C' sc. than Compton Greenfield ST 5782. *Estore Compton* 1305, *E(a)ster ~* 1584, 1622, *Eastward ~* 1777. Earlier simply *Compton* 1291. Gl iii.106.

†The forms *Cumbtune* [904]12th S 373 and [904×25]lost ECW may also belong here.

††This form may alternatively belong under Chilcompton.

‡The form *Cumtun* [762]13th S 1685 may also belong here.

(24) Fenny ~ Warw SP 4152. *Fennicumpton(e)* 1221–1535 with variants *Fenny-* and *-compton*, *Venny Cumpton* 1317, *Veny Compton* 1594. ME adj. **fenny** 'muddy' + p.n. *Contone* 1086, *Cumton* 1173, *Cumpton* 1235. Wa 269.

(25) Little ~ Warw SP 2630. *(in) Litlan-Cumtune* [1005]16th S 911, *Contone parva* 1086, *Parva Compton* 1301. 'Little', Latin **parva**, for distinction from Long COMPTON SP 2833. Wa 299.

(26) Long ~ Staffs SJ 8522. *Long Compton* late 18th.

(27) Long ~ Warw SP 2833. *Long Compton* 1299, *Compton Longa* 1535. ME **long** + p.n. *Cuntone* 1086, *Cumtona* 1123, *Cumpton(a)* 1218. Also known as *Magna Cumpton* 'great C' 1275, *Cumpton in Hennemersh* 1278, ~ *Garynges* 1305, *Over Compton* 1520 and *Much Compton* 1600, all for distinction from Little COMPTON SP 2630. Wa 299, DEPN.

(28) Nether ~ Dorset ST 5917. 'Lower C'. *Nethecumton'* 1268, *Nethercumpton(')* 1288, *Nether(e) Compton* from 1297. Earlier simply *Cuniton'* (for *Cum-*) [860×6]14th, *(on, to) Cumtun (bricgge)* [903 for 946×51]12th S 516, *Cumbtun* [998]12th S 895, *Com- Cumtona* 12th, *Cumton(e)* 1201–85, *Compton'* 1280, 1428. 'Nether', ME **nether** (OE *neothera*) for distinction from Over Compton ST 5916, *superior Cumtona* [1145]12th, *Ouerecumpton'* 1268, 'upper C', OE **uferra**. Nether Compton lies in the valley of the r. Yeo and is the original site; Over Compton lies on high ground to the W. Do iii.324.

(29) Over ~ Dorset ST 5916 → Nether COMPTON ST 5917.

(30) West ~ Dorset SY 5694. *West Compton* 1811. Also known as Compton Abbas West, *Cumpton Abbatis* 1291, 'C belonging to the abbot' sc. of Milton abbey. Earlier simply *Comptone* [834 for 934]17th S 391, *Contone* 1086. 'West' for distinction from COMPTON ABBAS. Do 60.

(31) West ~ Somer ST 5942. *West Compton* 1817 OS. ModE adj. **west** + p.n. Compton as in East COMPTON ST 6141.

River CONDER Lancs SD 5158. 'Winding river'. *Kondover* [1190×1220]1268, *Kondoure* [1225×50]1268, *Gondour' -douere* 1228, *Candovere* 1246, *Condofre* 1292, *Condar* c.1540. PrW ***camm** (Brit **cambo-*) + ***duβr** (Brit **dubro-*). La 168, RN 91, Jnl 1.46, 47.

CONDERTON H&W SO 9637. 'Estate of the Kent dwellers'. *Cantuaretun* [875]11th S 216, *C- Kanterton* 1201–58, *Conterton* 1269–1327. OE folk-n. *Cantware*, genitive pl. *Cantwara*, + **tūn**. The presence in Worcestershire of an estate of migrants from Kent is surprising but not unparalleled. The stream name Whitsun Brook SO 9951, *Wixenabroc* 972 S 786, is 'the stream of the Wixan', a tribal grouping otherwise only known in Lincolnshire. Cf also EXTON SU 6121, an estate of the East Saxons in Hants. Wo 115, xix, 16, PN and History 10.

CONDICOTE Glos SP 1528. '(At) Cunda's cottages'. *Cundicotan* [1051×3]11th S 1475, *Connicote* 1086, *Cunnecote* 1154×89, *Condicote* from 1086, *Cundicol(te) -y-* 1154×89–1675. OE pers.n. *Cunda* + **ing**⁴ + **cot**, dative pl. **cotum**. Gl i.216 gives pr ['kundikət].

CONDOVER Shrops SJ 4906. 'The flat-topped ridge overlooking the river Cound'. *Cone(n)dovre* 1086, *Conedovera -e* c.1090–1366, *Cunedoura -our(e) -ov(e)r(e)* 1130–1276, *Condover* from 1494, *Condor(e), Conder* 1416–1586, *Cunder* 1597, 1675. R.n. Cound as in COUND SJ 5504 + **ofer**. Sa i.97.

CONDURROW Corn → PRESTON CANDOVER Hants SU 6041.

CONEYHURST WSusx TQ 1023. 'Rabbit wood'. *Coneyhurst* 1574. ModE **coney** (ME *coni*) + **hurst** (OE *hyrst*). Sx 175.

CONEYSTHORPE NYorks SE 7171. 'The king's outlying settlement'. *Coninges- Covngestorp* 1086, *Cun(n)i(n)gestorp* 1167–1204, *Coningesthorp -yng-* 1251–1436, *Conisthorp -ys-* [1285]16th, 1577, *Conistropp* 1615. OEScand **kunung**, secondary genitive sing. **kununges**, + **thorp**. YN 48, SSNY 56.

CONGERSTONE Leic SK 3605. 'The king's estate'. *Cunningeston(e)* 1086–early 13th, *Coning(g)eston(e) -yng-* 1232–1397, *-ig(g)-* 1154×89–1270, *Cun- Congeston' -is-* 1243–1547, *Konestone* 1339, *Cun- Conston* 1435–1610, *Kyngerston* 1277,

Congerstone 1833 OS. OE **cyning**, genitive sing. **cyninges**, influenced by ODan **kunung**, + **tūn**. Lei 536.

CONGHAM Norf TF 7123. Partly uncertain. *Congre- Conghe- Concham* 1086, *Cungheam* 1121, *Congham* 1197, *Cangham* 1199. Uncertain element + OE **hām** 'a homestead'. Comparison has been made with the Yarmouth street called The Conge, *le Conge* 1286, *in vno Cong* 1290, *Kingeskonch* 1305 etc. The 1290 form suggests that this obscure element was still a living appellative. Cf. MedLat *conchus* 'a shell', *conch(a), conca* 'the basin of a fountain'; but no clear sense is apparent here. DEPN, Nf ii.31.

CONGLETON Ches SJ 8663. 'Settlement at *Cunghill*'. *Cogeltone* 1086, *Congulton* c.1262–1498 with variant *-el-*, *Congleton* from 1322, *Connkelton'* 1307, *Congerton* 1547–1690. P.n. **Cunghill* + **tūn**. OE ***cung** 'a protuberance' is suggested as referring to one of the big hills in the district such as The Cloud SJ 9063 (1125ft.), Congleton Edge SJ 8860 or The Mount SJ 8562. Che ii.294 gives prs ['kɔŋltən] and ['kɔŋgltən], addenda Ch. ii.ix 19th cent. local ['kɔŋərtn], Årsskrift 1978.24–31.

CONGRESBURY Avon ST 4363. 'Congar's fortified place'. *Conbusburie* (sic) [688×726]17th ECW, *(on) Cungresbyri* [893]c.1000 Asser, *-byrig* c.1000, *Kunigresbiria* [?c.1030]lost ECW, *Cungaresbyrig* *[1065]c.1500 S 1042, *Cungresberie* 1086, *Coombesbury* 1758. W saint's name *Cuncar*, OE *Congar*, genitive sing. *Congares*, + OE **byriġ**, dative sing. of **burh**. St Congar was buried here, a place mentioned by Asser as a derelict Celtic monastery. The pr is [ku:mzbri]. DEPN, CMCS 12.43.

CONINGSBY Lincs TF 2257. 'The king's village or farm'. *Cuningesbi* 1086, c.1200, *Coni(n)ghes- Coningesbi* c.1115, [1198]1328. ODan **kunung**, genitive sing. **kunungs**, + **bȳ**. Held by the king in 1086. SSNEM 42, Cameron 1998.

CONINGTON 'Royal estate'. ON **konungr** replacing OE **cyning** + **tūn**.

(1) ~ Cambs TL 1785. *(æt) Cunictune* 957 S 649, *Cunistone* [1017×35]14th S 1523, *Coninctune* 1086, *Cun(n)ington -yng-* 1214–96, 1662, *Conyngton -ing-* from 1290. Hu 182 gives pr [kʌniŋtən].

(2) ~ Cambs TL 3266. *(æt) Cunningtune* 975×1016 S 1487, *Cunington -yng-* 1214–1348, *Cun(i)tone* 1086, *Conyngton(e) -ing-* 1265–1570. The boundary of Conington is mentioned *be Cunigtunes gemæra* [1012]12th. Ca 165, PSA iii.49.

CONISBROUGH SYorks SK 5098. 'The king's stronghold'. *(æt) Cunugesburh* [1002×4]c.1100 S 1536, *Cu- Coningesburg, Coningesborc* 1086, *C- Kuningeburg(h), Cunnyngburgh'* 12th cent., *Cuningesburc(h), Coningesb(o)urg(h) -burc' -ynges-* 1121–1428, *Cun- Conesburg(h)* 1201–1466, *Connysburgh -is-* 1441–1521, *-borow(e), Cunsburgh* 16th cent., *Conisbrough* from 1771. ON **konungr** probably replacing OE **cyning**, genitive sing. **konungs**, + OE **burh**. One of a series of forts along the Don valley. Mentioned in the will of Wulfric Spot, but certainly a royal estate long before its first recorded royal association among the lands of king Harald in 1066. Geoffrey of Monmouth says that its earlier name was *Kaerconan* 'Conan's fort', but this looks like an imaginative extrapolation from *Cuningesburh*. The name *Conan* has been revived in Conanby, the name of a housing estate in Conisbrough. YW i.125, TC 76, SSNY 144.

High CONISCLIFFE Durham NZ 2215. *Conesclive Superiore* 1313, *Overcunscliff* 1359, 1501, *Over Consclife* 1564, *High Conscliff(e)* from 1717, *High Coneyscliffe* 1801, 1804. ModE **high** + p.n. *æt Cininges clife* c.1121 ASC(E) under the year 778, *Cingcesclife* [c.1040]12th, *Cingesclyffe* [1072×83]15th, *Cun(e)scliue* [c.1210×20]14th, 1242×3, 1408, *Conesclyue -cliue -cliff* 1235–1376, *Conysclyf'* 1314, *Coney(s)cliff* 1809, 1811, 'king's cliff or hill', OE **cyning**, genitive sing. **cyninges**, + **clif**. OE *cyning* was subsequently replaced by ODan **kunung**. The early 19th cent. spellings show association with ModE *coney* 'a rabbit'. The Laud MS of the Anglo-Saxon Chronicle records the murder here in the year 778 of Eadwulf son of Bosa, a king's high reeve. 'High' for distinction from Low CONISCLIFFE NZ 2413. Pronounced [kunizklif]. NbDu 51, Nomina 12.34, L 134, 136.

Low CONISCLIFFE Durham NZ 2413. *Nether Conescliue, Conyscliff* [early 14th]15th, *Nethircunscliff, -con(y)scliff -clyf(f)* 1359–1525, *Low Con(y)scliff(e)* 1717. ModE **low**, earlier **nether**, + p.n. Coniscliffe as in High CONISCLIFFE NZ 2215.

CONISHOLME Lincs TF 4095. 'The king's island'. *Cunyngesholme* [c.1155]1409, *Cuninggesolm* 1195, *Cuningesholm* before 1185, 1196. ODan **kunung**, genitive sing. **kunungs**, + **holmr**. The vill owes its exstence to the sea-bank on which it is built; the bank may have been in existence by the early 11th cent., but the village may not have been founded until after the compilation of DB. SSNEM 166, Jnl 7.56, Cameron 1998.

CONISTON 'The king's manor'. Scandinavianised form of KINGSTON with ON **konungr**, genitive sing. **konungs**, for OE *cyning*.

(1) ~ Cumbr SD 3097. *Coningeston'* 1157×63, 1257, *Koningeston* 1196, *Kunyngston* 1336. The name possibly preserves the memory of a small Scandinavian mountain kingdom. La 215, SSNNW 186.

(2) ~ Humbs TA 1535. *Cuningeston* 1190, *Coningeston -yng* 1260–1421, *Coniston(a) -y-* 1297–1585. The earliest form is *Co(i)ningesbi* 1086 with ON **bȳ**. Either an original *bȳ* name has been replaced by later *tūn* or an original OE *cyningestūn* has been Scandinavianised in the DB record. YE 47, SSNY 24.

(3) ~ COLD NYorks SD 9055. 'Cold C'. *Calde Cuningeston'* 1202, *Cold(e)coniston* 1509×47, 1586, ~ *Cold(e)* from 1577. ME **cald**, ModE **cold** + p.n. *Co- Cuninges- Conegheston(e)* 1086, *Coning(e)stun -yng- -ton(e)* 1155–1475, *Coniston(a) -ys-* from 1303. A royal estate in DB. YW vi.45.

(4) ~ WATER Cumbr SD 3097. *Coniston water* 1774. P.n. CONISTON SD 3097 + ModE **water**. Earlier *turstiniwatra* 1157×63 (Latinised), *Thurstainewater* 1196, *Thurstane Water* 1539, *Thurston water* 1774, ON pers.n. *Thórsteinn* + OE **wæter**. La 192.

CONISTONE NYorks SD 9867. 'Royal estate'. *Cunestune* 1086, *Cunig(g)estun -ton(a)* c.1130–late 12th, *Coning(e)ston -yngs-* 1198–1379, *Coniston(e) -ys-* from 1303. ODan **kunung** possibly replacing OE **cyning**, genitive sing. **kunungs**, + **tūn**. Held by the king's thanes in DB. YW vi.85

CONISTONE MOOR NYorks SE 0170. *Conyston morez* 1540. P.n. CONISTONE SD 9867 + ModE **moor**. YW vi.87.

CONNOR DOWNS Corn SW 5939. *Conner Down* c.1870. A 19th cent. village on former downland. The downs are mentioned as *Conerton Down* [1580]18th, *Conner Down* 1813, lost p.n. Connerton SW 5841 + OE **dūn**. Connerton, *Conarditone* 1086, *Conarton* [c.1155]16th, *Cunarton* [c.1200]16th, *-tone* 1185, *Connerton* 1284–1342, is partly obscure, consisting of an unidentified element or pers.n. + OE **tūn**. The DB form may point to OE *Cyneheardingtun* 'estate called after Cyneheard'. PNCo 72, Gover n.d. 599.

CONOCK Wilts SU 0657. Unexplained. *Cunet* 1211, *Kunek* 1242, *Cunnuc* c.1250, *Conok* 1279, *Connok* 1372, *Conke* 1522, 1527. Possibly PrW ***cönyg** of unknown origin but probably meaning 'hill'. Although Conock itself lies on flat ground the reference could be to a nearby hill. Cf. CONSETT Durham NZ 1051. Wlt 312, Jnl 16.17.

CONONLEY NYorks SD 9847. Partly uncertain.

I. *Cutnelai* 1086.

II. *Conanlia* 1155×87, *Conynley -unlay -on-* 1362–1638.

III. *Conendeley* 1171×81, 1175×84, *Conondlay -ley* 1405–1787.

IV. *Cuneteleia* 1191, *Cunetlay* 1246, *Conotlay* 1254, *Cunedlay* 13th, *Conedley* 1287, 1295.

V. *Cunigley(e)* 1254, 1273, *Coningley -lay -yng-* 1303, 1433.

It is impossible to say whether type II or type IV is primary.

Type II is 'Conan's wood or clearing', OIr pers.n. *Conán* cognate with W *Cynan*, Breton *Conan* + OE **lēah**. Type III could be a development of this with intrusive *d* between *n* and *l*. Type IV appears to contain unexplained p.n. **Cuned*, **Cunet* representing PrW **cönēd* < Brit **cunētjū* of unknown origin and meaning. Type V is due to folk etymological association with ODan *kunung* 'a king'. YW iv.27.

CONSALL Staffs SJ 9848. Partly uncertain. *Cuneshala* 1086, *Cu- Coneshale* 1227–1421, *Conleshale* 1227–8, *Cunsall* 1369, 1386, 1529 and 1606–1841, *Co(u)nsall* 1561–1837. Uncertain element or name + OE **halh** 'a nook or corner of land'. The forms do not support derivation from OE *cyning* replaced by ON ***kon-ingr** as in High and Low CONISCLIFFE Durham NZ 2215, 2514. Oakden gives prs [kɔns]- and [kunsəl].

CONSETT Durham NZ 1051. 'Head of the hill called *Cunec*'. *Conekesheued* [1183]c.1382, *Conkeshevet -d* c.1200–1382, *Konkeshed* 1385, *Consheved* 1422, *Conset* 1415×16–1592, *Consett* from 1443, *Conseid* late 15th, *Conside* 1580, 1725. OE hill-name **Cunec* < pre-English ***cönyg** of unknown origin, probably meaning 'hill', genitive sing. **Cuneces*, + OE **hēafod**. Cf. CONOCK Wilts SU 0657. NbDu 51, L 161, Jnl 16.15–7.

CONSTANTINE BAY Corn SW 8574. *Colstenton Bay* 1699, *Constantine Bay* 1813. Saint's name *Constantine* as in St Constantine's Church SW 8674 + ME **baye**. The church is recorded as *chapel of Sanctus Constantinus* 1390, and *Eglosconstantyn* c.1525, Co **eglos** + saint's name as in CONSTANTINE SW 7329. PNCo 72.

CONSTANTINE Corn SW 7329. 'Church-site of Constantine'. *Sanctus Constantinus* 1086, *(ecclesia) Sancti Constantini* 1291, *Langustenstyn* 1367, *Costentyn* 1441, *Constantin* 1610. Saint's name *Constantine* from the dedication of the church. His identity is unknown – possibly the 6th cent. Constantine who was king of Devon and Cornwall, possibly the royal *Custennyn Gorneu* 'Constantine the Cornishman' mentioned in Welsh pedigrees. PNCo 72 records a former pr [kɔ'stentən], Gover n.d. 501.

CONYER Kent TQ 9664. Short for *Conyers Quay* 1819 OS. Surname Conyer, *Coneyire* 1324, 'money stamper, coiner', OFr *coignier*, + ME **key** 'a landing place'. PNK 278.

COOKBURY Devon SS 4006. 'Cuca's fortified place or manor'. *Cukebyr'* 1242, *Cokebiry* 1291, *-bury* 1420. OE pers.n. **Cuca* + **byriġ**, dative sing. of **burh**. D 139.

COOKHAM Berks SU 8985. Either 'cook' or 'hill settlement'. *Coccham* [798]13th S 1258, [968×71]12th S 1485, *(to) Cócham* c.997 S 939, *Cocham* *[1052×66]13th S 1477, *Cocheham* 1086–1164, *Cokeham* 1265 etc., *Coukham* 1327, *Cookham* 1399. OE **cōc** 'cook' or ***cōc(e)** 'lump of earth, hillock' + **hām**. The two earliest forms might be thought to point to OE ***cocc** 'a hill' but this name clearly had *ō* throughout and they must be scribal mistakes. Brk 79, Do i.229, L 137.

COOKHAM DEAN Berks SU 8785. *Cookham Dean* 1761, 1790. P.n. COOKHAM SU 8985 + ModE **dean**. Cf. *Le Dene* 1344, *Le Deane, Deanefeild* 1608, *The Deane* 1699, from OE **denu** 'valley'. Brk 81.

COOKHAM RISE Berks SU 8885. No early forms. P.n. COOKHAM SU 8985 + ModE **rise** 'a piece of rising ground'. The reference is to the upper end of Cookham SU 8985. Brk 81.

COOKHILL H&W SP 0558. 'Cocc hill'. *Cochehi* (sic) 1186, *Cokehelle* [c.1086]12th, *K- Coc- Cokhull* 1227–1494, *Cokehill* 1542. OE ***cocc** 'a hill' probably taken as a p.n., + pleonastic **hyll**, 'hill called *Cocc*'. This is the highest point at the S end of a ridge of high ground stretching N from here to Headless Cross. Wo 326, PNE i.103.

COOKLEY H&W SO 8480. Apparently 'Culna's cliff'. *Culnan clif* [964]17th S 726, *Culleclive, Culla clife* 11th, 1240, *Colecliff* 1275, *Cookley* from 1608, *Cookecliffe* 1649. OE pers.n. **Cūlna*, genitive sing. **Cūlnan*, + **clif**. Wo 258 regards this unattested pers.n. as an *n*-extension of *Cūla* like *Tilne* from *Tila* and *Wilne* from *Willa*, Redin 160–1. But these forms are very doubtful; perhaps a name of W origin lies behind it. The reference is to an escarpment overlooking the r. Stour, also referred to in the name Austcliff to the N, *Astenes Clive* 1240, *Alstanesclive* 1275, *Ausclif* 1589, 'Ælfstan's cliff'. For the interchange *-clif* and *-cley* cf. AYCLIFFE Durham NZ 2822 and SHINCLIFFE Durham NZ 2940. Wo 258, 257, L 131, 135.

COOKLEY Suff TM 3575. 'Cuca's wood or clearing'. *Cokel(e)i* 1086, *Kukeleia* [1154×89]1268, *C- Kokeleye* 1275–1346, *Cokeley* 1610. OE pers.n. *Cuca* + **lēah**. DEPN, Baron.

COOKLEY GREEN Oxon SU 6990. P.n. Cookley + ModE **green**. Cookley, *Cokelea* [c.1183]13th, *Coclee* 1246, *Cookley* from 1608×9, is possibly 'the hillock wood or clearing', OE ***cōc** or ***cōce** 'a lump of earth, a hillock' + **lēah**. O 137, Do i.229.

COOKSBRIDGE ESusx TQ 4013. *Cooke's Bridge* 1590. Surname *Cook* probably to be associated with the family of Thomas Coke of Hamsey 1543 + ModE **bridge**. Another place in Hamsey was called *Cookspike al Parkfield* 1635. Sx 315.

COOKSMILL GREEN Essex TL 6305. *Cokysmylgreene* 1500. P.n. Cooks Mill, surname *Cook* + ME **mill**, + ModE **green**. A windmill was built here by one Richard Cook in 1274. Ess 280.

COOLHAM WSusx TQ 1222. Short for *Coolham Green* 1813 OS.

COOLING Kent TQ 7576. 'The Culingas, the people called after Cul(a)'. *Culingas* 808 S 163, *[1006 for 1001]c.1200 S 914, 1042×66 S 1047, *(æt) Culingon* c. 959 S 1211, *Culinges* [961]13th S 1212, 1203–1242×3, *Colinge(s)* 1086, *Colinga -es* 11th, *C(o)uling(e)* 1205–84×5, *Cowling* 1610. The boundary of the people of Cooling occurs as *culinga gemære* 778 S 35. OE pers.n. *Cūl(a)* +**ingas**. PNK 112, KPN 57, ING 10.

COOMBE 'Coomb, valley'. OE **cumb**. L 88–94.

(1) ~ Corn SS 2011. *Come* [1439]15th, *Combe* 1520. PNCo 72, Gover n.d. 20.

(2) ~ Corn SW 9551. *Combe* 1813. PNCo 72.

(3) ~ ABBEY Warw SP 4079. *Comb Abby* 1721. P.n. *Cumba* 1156–1316, *Coumbe* 14th cent., + ME **abbeie**, 'an abbey'. Wa 105.

(4) ~ BISSETT Wilts SU 1126. 'C held by the Bisset family'. *Cumba Maness' Biset* 1167, *Coumbe Byset* 1288, *Combissett* 1639. P.n. *Come* 1086, *Cumbe* 1086, 1158, + manorial addition from Manasser Biset, cup-bearer to Henry II, who held the manor in the later 12th cent. Wlt 221.

(5) ~ KEYNES Dorset SY 8484. 'C held by the Keynes family'. *Cumbe Willelmi de Cahaignes* 1199 'W de Keynes's C', ~ *Chaynes, Kaynes* 13th, *Coumb(e) Caynes, Kaynes, Keynes* 1302–1443. Earlier simply *Cume* 1086, *Comb(e) -a* 1280–1434. The manor was held by William de Cahaignes in 1199 and by his family until at least 1386. Do i.114.

COOMBES WSusx TQ 1908. 'The valley(s)'. *Cumba* 1073, *Cumbe* 1086–1320 (*La* ~ 1280, 1320), *Cumbes* 1202–1327, *Co(u)mbes* 1265–1405. OE **cumb**, pl. **cumbas**. The boundary of the people of Coombes is *cumbhæma gemære* 956 S 624. Sx 224, SS 69.

COPDOCK Suff TM 1141. 'The peaked or pollarded oak-tree'. *Coppedoc* 1195–1275, *-ac* 1254, *-ok* 1286–1327, *Copdoke* 1417, *Copdock* 1656. OE **coppede** + **āc**. DEPN, Baron.

COPELAND 'Purchased land'. ON **kaupaland**.

(1) ~ Cumbr. *Caupalandia* c.1125, c.1135, *Coupland(a)* c.1125–1500, *Copeland(i)a -e* c.1140–1279. An ancient barony or district in W Cumbria embracing the land between the rivers Derwent and Duddon. The reference is probably to an unrecorded 10th cent. Norse land purchase. Coleridge's diary for 26th August 1802 contains a nice suggestion – 'a great national convention of mountains which our ancestors called Copland, that is the Land of Heads'. Cu 2, Jnl 2.56, Cumbria 1983.168–71, L 246.

(2) ~ FOREST Cumbr NY 1507. *Foreste de Coupland* [c.1282]1594, *Copland Forest* 1495. P.n. COPELAND + OFr **forest**. Cu 37.

COPFORD GREEN Essex TL 9222. *Copford Green* 1805 OS. P.n.

Copford TL 9223 + ModE **green**. Copford, *(into, of) Coppanforda* [961×95]11th S 1501, [c.1000]c.1125, *Copefordā* 1086, *C- Kop(e)- Coppeford* 1225–1428, *Cofford* 1248, is 'Coppa's ford', OE pers.n. *Coppa*, genitive sing. *Coppan*, + **ford**. Ess 385.

COPLEY Durham NZ 0825. Possibly 'hill-top clearing' or 'Coppa's clearing'. *Koppeleyker* 'Copley carr' 1315. OE **copp** or pers.n. *Coppa* + **lēah**. The form cited is for a *carr* or marsh at Copley. NbDu 52.

COPLOW DALE Derby SK 1679. *Coplow Dale* 1658. P.n. Cop Low + ModE **dale**. Cop Low, no early forms, is 'hill with a peak', ModE **cop** (OE *copp*) + **low** (OE *hlāw*). Db 133.

COPMANTHORPE NYorks SE 5547. 'The merchant's outlying farm'. *Copeman Torp* 1086, *Coup(e)- Caup(e)mant(h)orp* 1276–1657. ON **kaup-mann**, possibly genitive pl. **kaup-manna**, + **thorp**. YW iv.227, SSNY 56.

COPPATHORNE Corn SS 2000. 'Copped, pollarded thorn-tree'. *Copelle horno* 1699, *Coppet Thorn* 1748. ModE adj. **copped** + **thorn**. The first form is corrupt. However, if the forms cited by Gover n.d., *Cobbethorne* 1360, *Cobthorne* 1612, belong here, this is 'Cobba's thorn-tree', OE pers.n. **Cobba*. PNCo 72, Byn 305.

COPPENHALL Staffs SJ 9119. 'Coppa's nook'. *Copehale* 1086, *Cop(p)enhale* 1166–1483, *Cop(p)nall* 12782×1307, 1586–1657, *Coppenhall* from 1240. OE pers.n. *Coppa*, genitive sing. *Coppan*, + **halh** in the sense 'promontory into marsh'. St i.82, L 105.

COPPERHOUSE Corn SW 5738. No early forms.

COPPINGFORD Cambs TL 1680. 'Traders' ford'. *Copemaneforde* 1086, *Copmanford* 1272×1307–1584, *Coppyngford* 1535, 1564. lOE **coupmanna**, genitive pl. of **coupman** < ON *kaupmann*, + **ford**. Hu 237.

COPPLESTONE Devon SS 7702. 'Peaked or peak-like stone'. *(on) copelan stan* 974 S 795, *Copelaston -le-* 1281, 1294, *Coplestone* 1384. OE **copel**, definite form **copela**, + **stān**. The reference is to a 10ft. high granite cross-shaft marking the boundary of three parishes. The sense of *copel* is uncertain but the usual gloss 'unsteady, rocking' hardly suits this instance. It seems to be a derivative of *copp* 'a summit, a peak' and perhaps meant 'towering'. Cf. ModE *copple* 'a (bird's) crest'. D 403, PNE i.106.

COPPULL Lancs SD 5614. 'Cop hill'. *Cophill* 1218, *Cophull* 1243–1322 etc., *Coppel* 1276, *Coppull* 1386. OE **cop(p)** 'the top of a hill, a summit, a peak' which may have been the original name of the hill, + **hyll**. La 129, L 137, 170.

COPSALE WSusx TQ 1725. 'Cobb's nook'. *Cobsale* 1650, *Copsale* 1795, *Cobshill* 1823. Surname *Cobb* + ME **hale** (OE *healh*). Sx 231.

COPSTER GREEN Lancs SD 6733. P.n. Copster + ModE **green**. Copster, no early forms, might be identical with Copster Greater Manchester SD 9203, *the Coppedhyrst* 1422, *Copthirst* 1507, 'the peaked hill', OE **coppede** + **hyrst**, or Copster SYorks SE 2802, *Copstorth* 14th, 1422 the 'hill-top plantation', OE **copp** + **storth**. La 51, YW i.314.

COPT HEATH WMids SP 1777. 'Copp heath'. *Cophethe* 1498, *Coppeheathe* 1600, *Copt Heath* 1831 OS. P.n. *le Coppe* 1353, OE **copp** 'the summit, the crest', + ME **heth**. The heath lies on high ground between Knowle and Solihull. Wa 69.

COPT HEWICK NYorks SE 3471 → Copt HEWICK SE 3471.

COPT OAK Leic SK 4812. 'The pollarded oak-tree'. *le Coppudhok* c.1230, *(le) Coppedoke* 1343×71, *Coppyd Oke* 1550, *Coptoke* 1578, *the Copt Oak* 1642. ME **copped** 'without a top, pollarded' (OE *coppod*) + **oke** (OE *āc*). Lei 518.

COPTHORNE WSusx TQ 3139. 'The pollarded thorn-tree'. *Coppethorne* 1437. ME **copped** + **thorn**. Sx 284.

COPYTHORNE Hants SU 3014. 'The pollarded thorn-tree'. *Coppethorne* 1327×77, *Copped Thorne* 1754. ME **coppede** + **thorn**. Ha 59, Gover 1958.194.

COQUET ISLAND Northum NU 2904. *Cocþædesæ* [699×705]c.900 AC, *Cocuedes(ce)* [699×705]13th ibid., *Concuedesee* [699×705]14th ibid., *cocuedadedes eu* (sic) c.1150, *Cocuedeseu* 1350×1400 marginalia in AC, *Insula Coket* [1135×54]12th, *Coketeland* [1347]14th, *Cokette Elande* 1471×2. R.n. COQUET NX 1699 + OE **ēġ** later replaced by **ēġland** and ModE **island**. RN, Jnl 8.19.

River COQUET Northum NZ 1699. 'The red river'. *Coccuveda -neda* [c700]14th Rav, *Coquedi fluminis* [699×705]c.900 AC, [c.721]10th BPC, *cocuedi fluminis* [721]12th ibid., *Cocwuda* [c.1040]12th, *Coqued* c.1107, *Koket* 1139×52, 1242 BF, *Coket* 1200–1250 BF. Brit ***cocco-** + W **wedd** as in W *cochwedd* 'red appearance'. The upper reaches of the river are filled with red porphyritic detritus from the Cheviot. The form *Cocwuda* appears to be a folk etymology as if the name were 'cock wood'. Pronounced [koukit]. NbDu 52, RN 93, RBrit 311, Jnl 8.19, GPC 8.

COQUETDALE Northum NU 0800. *Cokedale* [1150×62]14th, 1279, *Kokedal(e)* c.1160, 1275, 1430, *Cokdale* 1471×2, *Cookdale* 1479, *-daill* 1552 Dixon 30, *Cockda(y)le* 1541 ibid. 46, *Choketdale* c.1160. R.n. COQUET NZ 1699 + ME **dale**. RN 93.

CORBRIDGE Northum NY 9964. 'The bridge (over the river Tyne) by CORSTOPITUM'. *Corebricg* [c.1040]12th, *-brugia -brig(g)(e)* 1139×43 RRS–1507, *Cornbrigg, Cornebrig'* 1250. *Cor-* from CORSTOPITUM + OE **brycg**. 12th and 13th cent. spellings *Colebruge, Colbrige* also occur showing AN dissimilation of *r - r* to *l - r*. NbDu 52.

CORBY A name-type of uncertain origin, usually explained as 'Kori's village or farm', ON pers.n. *Kori*, genitive sing. *Kora*, + **bȳ**. This is, however, a rare pers.n., the only certain instance being an Irish thrall in Iceland. The specific might be OE ***corf** 'a cutting, a gap, a pass'.

(1) ~ Northants SP 8888. *Corbei -bi* 1086 etc. with variant *-by*, *Corebi -by* 1166–1252. OE ***corf** + **bȳ** with early loss of *f* in the sequence *-rfb-*. Nth 162, SSNEM 42.

(2) ~ GLEN Lincs SK 9924. *Corbi -by* 1086–1642, *Corebi -by* 1156–1418. *Glen* is a modern addition (1956). Corby lies in the valley of the r. Glen where it narrows to form a gap. Perrott 170, SSNEM 42.

Great CORBY Cumbr NY 4754. *Corkeby Magna, Mekill Corkby* 1348, *Mikle Corbye* 1580, ModE adj. **great** (Latin **magna**, ME **micel**) + p.n. Corby. Also known as *Cunscorkeby, Kunskorkeby* 1289, *Comscorkeby, Conis Corkby* 1348, 'king's Corby'. 'Great' for distinction from Little Corby NY 4756, *parva Corkeby* c.1170–1332, *Litilkorby* 1450. Corby, *Chorkeby* [c.1115]17th, c.1160, c.1175, *C- Korkeby -bie* 1130–1621, *Corkby* 1235, *Corbye* 1580, is either 'Corc's village or farm', OIr pers.n. *Corc(c)* + **bȳ**, or 'oat farm', ON **korki** + **bȳ**. Cu 161, SSNNW 27.

CORBY PIKE Northum NT 8401. *Corby Pike* 1869 OS. Probably N dial. **corbie** 'a raven, a carrion crow' + **pike** 'a pointed hill'.

CORELEY Shrops SO 6174. 'The cranes' wood or clearing'. *Cornelie* 1086, *-le(ye)* 13th cent., *Cornleg(h) -leye -lay* 1242–1399, *Corle(y)* 1255–1710, *Coreley* from 1577. OE **corn**, genitive pl. **corna**, + **lēah**. Sa i.97.

CORFE 'A cutting, a gap, a pass'. OE ***corf**. See also CORSCOMBE Dorset ST 5105. L 88.

(1) ~ Somer ST 2319. *Corf* 1243, *Corfe* 1610. The reference is to a natural cutting into the Black Down Hills through which the road from Taunton to Honiton passes. DEPN.

(2) ~ CASTLE Dorset SY 9681. The name of the borough is *Corf(f)(e) Castel(l) -castel(l)(e)* 1302–1575, of the castle itself *castellum de Warham* 'Wareham castle' 1086, *castellum de Corf(f)(e)* 1196–1435. Also known simply as *Corf* 1162–1456 including [955]14th S 573, [956]14th S 632, *Corfe* from 1242, *Corft* 1399, 1427, and as *(æt) Corfesgeate* 11th ASC(D), 12th ASC(E), *(at) Corfgeate, Porta Corf* (Latin) early 12th ASC(F) all under year 979, 'the pass or gate at or called Corfe', p.n. CORFE + OE

ġeat, Latin **porta**. The reference is to the gap in the central ridge of the Purbeck Hills at this place. Do i.5.

(3) ~ MULLEN Dorset SY 9896. 'C with the mill'. *Corf le Mulin* 1176×7, *Corfmulin -molyn(e), Molyn* 1216×72–1428, *Corfe Mullen* from 1545. P.n. CORFE + OFr **(le) molin**. Earlier simply *Corf* 1086–1327. The reference is to its situation between two hills. 'Mullen', for distinction from CORFE CASTLE, refers to a valuable mill here already mentioned in DB. The spelling *Croft molendina* 1199×1216 shows development of final *-t* as in *Corft* for Corfe Castle, and metathesis. Do ii.15, Dobson §437.

(4) ~ RIVER Dorset SY 9685. Flows past Corfe Castle.

CORFTON Shrops SO 4985. 'The settlement by the river Corve'. *Cortune* 1086, *Corfton* from 1221×2. R.n. CORVE SO 5083 + OE **tūn**. Sa i.99.

CORHAMPTON Hants SU 6120. 'Corn estate'. *Cornhamton* 1201, *-hampton* c.1225–1428, *Corhamtunne* 1232, *Corehampton* 1598. OE **corn** + **hāmtūn**. Ha 59, Gover 1958.49.

CORINIUM Glos SP 0201. The Romano-British name of Cirencester. *Κορίνιον* (Corinium) [c.150]13th Ptol (with variants *Κορίννιον* (Corinnium), *Κορόνιον* (Coronium), *Cironium Dobunorum* [c.700]13th Rav. The authentic form appears to be *Corinium*, identical with Karin, a city on the coast of Illyria (Yugoslavia), *Κορίνιον* [c.150]13th Ptol, *Corinium* [77]9th/10th Pliny NH 3.105. This appears to be an unknown element *Corin-* or *Corinn-* + suffix **-io-*. Corinium was a *polis* of the Dobuni. Gl i.60, RBrit 321.

CORLEY Warw SP 3085. 'The cranes' wood or clearing'. *Cornelie* 1086, *-legha -le(g)* 1220–79, *Cornlea -leg(e) -ley(e)* 1182–1313, *Corlee* 1213, *-ley(e)* form 1291. OE **corn**, genitive pl. **corna**, + **lēah**. Wa 81.

CORLEY ASH Warw SP 2986. *Corley Ash* c.1840. P.n. CORLEY SP 3085 + ModE **ash** 'an ash-tree'. Wa 81.

CORLEY MOOR Warw SP 2885. *Corleymore* 1411. P.n. CORLEY SP 3085 + ME **more**. Wa 82.

Great CORNARD Suff TL 8940. *Cornherth Magna* 1254, *Cornerd mag:* 1610. ModE adj. **great**, Latin **magna**, + p.n. *Corm-* (for *Corni-*) *Corn(i)erda* 1086, *Cornerde* c.1095, *Cornerth* 1196, *Cornherd* 1197, 'the cultivated land where corn is grown', OE **corn** + **erth**. 'Great' for distinction from Little Cornard TL 9039, *Cornherth Parva* 1254, *Cornerd p'ua* 1610. DEPN, L 233.

CORNEY Cumbr SD 1191. 'Crane island'. *Cornai(e) -ay(e) -eie -ey(e) -ye* c.1160 etc. OE **corn** + **ēġ**. Cu 364, L 36, 39.

CORNFORTH Durham Nz 3134. 'Ford frequented by cranes'. *Corneford* c.1116–c.1450, *-forth* 1382–1647, *Cornford* c.1292–1436, *-forth* from 1382, *-furth* 16th cent. OE **cran**, **cron**, genitive pl. **corna**, + **ford**. Metathesised forms of *cran, cron* are frequent in p.ns. NbDu 54.

CORNHILL-ON-TWEED Northum NT 8539. P.n. *Cornehale* c.1180, 1208×10 BF, 1335, *Cornhale* 1228, *Cornale* 1229 FPD, *Cornell* 16th cent., 'the cranes' haugh, the river-bend land where cranes are seen', OE **corn**, genitive pl. **corna**, + **halh**, locative-dative sing. **hale**, + r.n. TWEED NT 9452. NbDu 54 gives pr [kɔ:nəl].

CORNHOLME WYorks SD 9026. No early forms. Apparently 'corn meadow', but *corn* in p.ns. often stands for *crane*. YW iii.180.

CORNISH HALL END Essex TL 6836. *Cornish End* 1805 OS. 'Cornish hall end' sc. of Finchingfield TL 6832. Cornish Hall is *Corners Hall* 1395, *Cornertheshalle* 1475, *Cornerdeshall* 1490, *Corner Hall* 1536, *Cornish Hall* 1768, family name *Cornerde* 1303 from Cornard, Suffolk, + ME **hall**. Cornish Hall, sometimes simply called *Cornerthys* 1381, *Cornerdes* 1428, *Cornett* 1768, lay at the extreme N end of the parish at a place called *Norton* 1235, the 'north settlement', OE **north** + **tūn**, and *Nortonstrete* 1425, 1561, with ME **strete** 'a hamlet', whence the original name of the hall, *Nortonhalle* 1378 and *North Hall alias Cornerdes* 1395, *Northhalle* 1425. Ess 426.

CORNRIGGS Durham NY 8441. 'Ridges where corn grows'. *Corn Riggs* 1886 WME. Writing in 1804 Pennant commented on the hills of Weardale, 'the hills that bound this vale are low, cultivated almost to their tops'.

CORNSAY Durham NZ 1443. 'Crane's hill-spur'. *Cornesho* [1153×95]15th, 1190×5, *-hou(h)* 1189×1212, 1242×3, *-how(e)* c.1225–1498, *Croneshowe* 1303, *Corns(h)ow(e)* 1301–1521, *Corneseye* 1549, *Cornsey alias Cornseyrawe* 1629. OE **cran**, **cron**, genitive sing. **cornes**, + **hōh**. Metathesised forms of OE *cran, cron* are frequent in p.ns. NbDu 54, L 167–8.

CORNWALL The county name derives from OE *Cornwalas* 'the *Corn*-Welsh' in which *Corn-* comes from the original tribal name **Cornowii* 'the people of the horn' referring to their situation at the end of a long peninsula, the 'horn' of Britain, and *-walas* the plural of OE **w(e)alh** 'a foreigner, a Welshman'. *On Corn walum* 891 ASC(A), *-wealum* c.1121 ASC(E) under year 997, 'among the *Corn*-Welsh' (dative plural), *Cornualia* 1086, *Cornwal* c.1198. The form *Cornubia* [c.705]mid 9th is an artificial latinisation of the name; *Cornugallia* 1086 shows reinterpretation of the second part as *Gallia* 'Wales'; *Kernow* c.1400, the Cornish form of the name, derives directly from the original **Cornowii*. PNCo 72.

CORNWELL Oxon SP 2727. 'The spring where cranes are seen'. *Cornewelle* 1086–1428 with variant *-well(a), Cornwell(e)* from 1148×61. OE **corn**, genitive pl. **corna**, + **wella**. Cranes were once widespread in England. O 346, L 31, Ibis 140 (1998) 882–500.

CORNWOOD Devon SX 6059. 'Crane wood'. *Cornehvde* 1086, *Curnwod'* 1242, *Cornewud -wode* 1249–1337, *Cornwode* 1263–97, *Cournewode* 1404, *Cornwood vulg. Curlewood* 1675. OE **cran**, **corn**, genitive pl. **corna**, + **wudu**. L 228. D 268.

CORNWORTHY Devon SX 8255. 'Crane enclosure'. *Corneorde* 1086, *-worthi -wrth(e) -wordi -wrthy* 1205–91, *Cornworth(i)* 1250, 1291. OE **cran**, **corn**, genitive pl. **corna** + **worthiġ**. The specific could be OE **corn** 'corn, grain' and so 'corn farm', but the persistent medial *-e-* suggests rather genitive pl. **corna**. D 320.

CORPUSTY Norf 1130. 'Korpr's path' or 'pigsty'. *Corpestih -stig, Corpsty* 1086, *Corpesti* 1196, 1202. ON pers.n. *Korpr*, genitive sing. *Korps*, + OE **stīġ**. DEPN, SPNN 280.

CORRINGALE Essex TL 5216 → CORRINGHAM TQ 7183.

CORRINGHAM Essex TQ 7183. 'Homestead of the Curringas, the people called after Curra'. *Currincham* 1086, *Culingham* 1201, *Curingeham -yge-* 1204–48, *Cur(r)ingham* 1206–c.1250, *Coringham -yng-* 1242–1428. OE folk-n. **Curringas* < pers.n. **Curra* + **ingas**, genitive pl. **Curringa*, + **hām**. *Curra* occurs also in *Curringtun* [786]12th S 125, a *vicus* in Canterbury, and in Corringale TL 5216, *Curing(e)hal' -halle* 1217–84, *Cor(r)ing(e)hal(e)* 12678, 1291, the 'nook of the Curringas'. Ess 151 gives pr [kɔrənəm], Wlt xxxvii, ING 121.

CORRINGHAM Lincs SK 8791. Partly uncertain. *Coringeham* 1086–1162, *Coringheham* c.1115, *Corincham* 1212, *Coringham* 1086, 1242×3. Either the 'homestead of the Coringas, the people called after Cora', OE folk-n. **Coringas* < pers.n. **Cora* + **ingas**, genitive pl. **Coringa*, + **hām**, or the 'homestead called or at *Coring*, the place called after Cora', OE p.n. **Coring* < pers.n. **Cora* + **ing**[2], locative-dative sing. **Coringe*, + **hām**. OE **Cora* is paralleled by OScand *Kori*. Cf. CARRINGTON TF 3155. DEPN, ING 143, SPN 180.

CORSCOMBE Dorset ST 5105. Partly uncertain. *Corigescumb* [1014]12th S 933, [1035]12th S 975, *Cor(ie)scumbe* 1086, *Corescumbe* 1244. The generic is OE **cumb** 'a coomb, a valley', the specific perhaps the genitive sing. of an old stream-name identical with Curry in CURRY MALLET Somer ST 3221. It seems unlikely to be a reduction of OE **corf-weġ* 'road in a pass' at this early date. Do 61, L 88, 93.

CORSHAM Wilts ST 8770. 'Cosa or Cossa's homestead'. *Coseham* [1001]15th S 899, 1336, *Cossehā* 1086, *-ham* 1130, 1134, 1424, *Corsham* from 1160, *Cosham* 1185–1572, *Cossam*

1196–1558×1603, *Cossum* 1741. OE pers.n. *Cosa* or *Cossa* + **hām**. Wlt 95 records former pr [kɔsəm].

CORSLEY Wilts ST 8246. 'The wood or clearing at or called Cors'. *Corselie* 1086, *-le(g)a -legh(e) -le* 1179–1316, *Corsley(e)* from 1211. PrW ***cors** 'a marsh, a bog' probably taken as a p.n. + OE **lēah**. *Cors* may have been the name of the stream subsequently called Rodden Brook. Cf. CORSTON ST 9284. Wlt 152, CPNE 66.

CORSTON Avon ST 6965. 'Settlement, estate on the river *Corse*'. *(æt) Corsantune* [941]12th S 476, *(æt) Corsatune* [956]12th S 593, *Corsantun* [972]12th S 785, *Corstvne* 1086. OE r.n. *Corse*, genitive sing. *Corsan*, + **tūn**. The r.n. is independently recorded (in the oblique case) as *Corsan* [941–72]12th S 476, 593, 711, 735, 785. It derives from PrW ***cors** 'marsh, bog' as also in CORSHAM Wilts ST 8770 and CORSLEY Wilts ST 8246. DEPN, RN 95, GPC 566.

CORSTON Wilts ST 9284. 'The settlement on the stream called Cors'. *Corstuna* [1065]14th S 1038, *Corstone* 1086, *Corston* since 1177, *Corsen* 1673. PrW stream name **Cors* < PrW ***cors** 'a marsh, a bog' + OE **tūn**. The reference is to Gauze Brook, properly *Cors Brook*, as the following forms show, *Corsaburna* [701]12th S 243, *Corsborne* *[854]14th S 305, *Corsbrok* [956]13th S 1577, *Gosbrooke* 1599, *Gauze Brook* 1773. An additional meaning of **cors* was probably 'reeds' as in Co **cors* and Breton *korz*. A later alternative name of the stream is preserved in Rodbourne Rail Farm ST 9484 and RODBOURNE ST 9383, *Reodburna* [701–982]14th S 243, 260, 841, *Rodborne* 1535, 'the reed stream', OE **hrēod** + **burna**. Wlt 50 gives popular pr [kɔːrsən], 51.

CORSTOPITUM Northum NY 9864. The Roman town at Corbridge. *Corstopitum -pilum, Cor stopitu* [c.300]8th AI, *Corie lopocarium* [c.700]13th Rav. These are all corrupt spellings for a name which was probably **Coriosopitum* or **Coria Sopitum* 'the tribal centre of the *Sopites*', Brit ***corio-** 'host, army', ***coria** 'hosting place, tribal centre', + folk-n. **Sopites*, genitive pl. **Sopitum*, a tribal name of unknown meaning. The same name, *Coriosopites*, is recorded of a people in SW Britany c.875 and, also in France, the name of a people called the *Coriosolites* survives in the p.n. Corseul. One problem in connecting the *Cor-* of Corbridge with *Coriosopitum* is that it might have been expected to have undergone either PrW internal *i*-affection > **Ceirio-* or OE *i*-mutation > **Cere-*. The absence of either of these changes could be due to adoption of the name by the incoming Anglo-Saxons after completion of *i*-mutation in the sixth cent. (Luick §201) and before the period of PrW *i*-affection (seventh or eighth cent. LHEB 611, 616). RBrit 322.

CORTON DENHAM Somer ST 6322. 'C held by the Dinan family'. *Corftan Dynham* 1308. Earlier simply *Corfetone* 1086, *Corftona* 1168, *Corton* 1610, 'settlement by the cutting', OE ***corf** + **tūn**, referring to a pass on the road from Corton to Charlton Horethorn over Horethorn Down which rises here to 644ft. DEPN.

CORTON Suff TM 5497. 'Kari's estate'. *Karetuna* 1086, *C- Korton* from 1226. ON pers.n. *Kári*, genitive sing. *Kára*, + **tūn**. DEPN.

CORTON Wilts ST 9340. Partly uncertain. *Cortitone* 1086, *-ton* 1227, *Cortinton -yn(g)-* 1259–1348, *Corton(e)* 1242, 1316, *Corton al. Cortington* 1570. Uncertain element or pers.n. + OE **tūn**. The specific might be pers.n. *Cort* or *Corta* + **ing**[4] and so 'the estate called after Cort(a)', or OE ***corte** + **ing**[4] and so 'the settlement at or called **Corting*, the hurdle or fence place' referring to some sort of enclosure or possibly a fishing weir in the river Wylye. Wlt 162, Nomina 7.38.

CORVE DALE Shrops SO 5488. 'The valley of the river Corve'. *Corvedale* from 1255×6, *Corfedale* 1402, *Cordale* 1675. R.n. CORVE SO 5083 + ME **dale**. Sa i.99, RN 96.

River CORVE Shrops SO 5083. *Corf* [674×704]13th S 1799, *Corve* from 1250. OE ***corf** 'a valley, a pass' transferred to the river that runs through the long valley between Wenlock Edge and Clee Hill. The loss of the pre-English name for this major river is surprising. Sa i.99, RN 96.

CORYTON Devon SX 4583. 'Settlement on the r. Cory'. *(æt) cur(r)itune* c.970 B 1246, *Curiton(e)* 1238–1339, *Coriton* 1086–1284, *Corryton* 1293, *Coryngton* 1547. R.n. **Cory* + OE **tūn**. **Cory* is thought to be the old name of the river Lyd as in LYDFORD SX 5185, itself an English name (OE **Hlyde* 'the loud one'). On the Lyd in Coryhill SX 4683 was the lost *Curiford* 1242, *Coryfforde* 1346. The meaning and origin of **Cory* are unknown. Cf. CURRY MALLET Somer ST 3221, CURRY RIVEL Somer ST 3824, and West CURRY Corn SX 2893 for possible other examples. D 181 gives pr [kɔritən], RN 97.

CORYTON Essex TQ 7482. Surname *Cory* + ModE **ton**. An industrial village at the oil refining and storage plant opened here in 1922 on Fobbing Marsh by Messrs Cory Brothers and Co. The chairman was Sir Clifford Cory d.1941, second son of John Cory of Glyn Cory. Earlier called *Kynochtown* from Messrs Kynoch's munition works on land bought by Kynochs in 1896 and subsequently acquired by the Cory brothers. Ess 152, Room 1983.27.

COSBY Leic SP 5494. Probably 'Cossa's village or farm'. *Cosbi* 1086, *Cossebi* 1086, 1325, *-by* c.1130–1406, *Cosby(e)* from 1308. Probably OE pers.n. **Cossa* + **bȳ**. Derivation from the Scand pers.ns. *Kofsi* or *Kopsi* is doubtful in view of the absence of supportive spellings. Lei 437, SSNEM 42.

COSDON HILL Devon SX 6391. *Cawson-hill* 1797, *Cawsand or Cawsorn Hill* 1809 OS, *Cawsand Hill* 1982 OS Atlas. P.n. Cosdon + ModE **hill**. Cosdon, *the Hoga of Cossdoune* [1240]15th, *Costendoune* 1240, *Cosdon al. Cosson* 1608, is 'Costa's hill', OE pers.n. **Costa*, genitive sing. **Costan*, + **dūn**. *Hoga* is a Latinised form of OE *hōh* 'hill-spur'. D 448 (under Cawsand Beacon) gives prs [kɔsdən] and [kɔːzən].

COSELEY WMids SO 9494. Partly uncertain. *Colseley* 1357, *Coseley* 1834 OS. The generic is OE **lēah** 'a wood or clearing', the specific possibly OE pers.n. *Col*, genitive sing. *Col(l)es*. DEPN.

COSFORD STATION Shrops SJ 7905. P.n. *Cospelford* 1189, 1286, *Cospesford* 1271×2, *Cosford* 1833 OS, 'the fetter ford', OE ***cospel** or **cosp**, genitive sing. **cospes**, + **ford**, + ModE **station**. A halt on the railway line between Telford and Wolverhampton. Cosford is a crossing of the river Worf. The exact sense of the name is unknown. Gelling.

COSGROVE Northants SP 7942. 'Cof's grove'. *Covesgrave* 1086–1328, *Colesgrave* 1199, 1338, *Coges-* 1253, *Cosegrave* 1275, 1445, *G- Cosgrave* 1341, 1401. OE pers.n. **Cōf*, genitive sing. **Cōfes*, + **grāf**. Nth 97, L 194.

COSHAM Hants SU 6505. Possibly 'Cossa's estate'. *Cosham* from early 12th ASC(E) under year 1015, *Cos(s)eham* 1086–1265, *Cor(e)sham* 1250, 1280. OE pers.n. *Cossa* + **hām**. Cf. CORSHAM Gover 1958.19. DEPN.

COSSALL Notts SK 4842. 'Cott's nook of land'. *Coteshale* 1086, *Cotesale* [c.1156]15th, *Cotsale* c.1250, *Cossall'* from 1242. OE pers.n. **Cott*, genitive sing. **Cottes*, + **halh**. The reference is to land in a bend of the r. Erewash. Nt 143.

COSSINGTON Leic SK 6013. Probably the 'estate called after Cusa'. *Cosinton(e) -yn-* 1086–1389, *Cosenton* late 13th, *Cusinton' -yn-* 1176–1299, *Cosington -yng-* 1209×35–1566, *Cusington -yng-* 1221–1504, *Cossyngton* 1591, *Cussington* 1604–25. OE pers.n. *Cusa* + **ing**[4] + **tūn**. Lei 284.

COSSINGTON Somer ST 3540. 'Estate called after Cosa or Cusa'. *Cosingtone* [729]14th S 253, *Cosintone* 1086, *Cusinton* 1196, 1225, *Cussington* 1610. OE pers.n. **Cosa* or *Cusa* + **ing**[4] + **tūn**. DEPN.

COSTA BECK NYorks SE 7682. Possibly the 'choice stream'. *(aqua de) Costa* from [1157×8]13th, *Costey* 1536. ON **kostr**, genitive pl. **kosta**, + **á**. YN 2, RN 99.

COSTESSEY Norf TG 1711. 'Cost's island'. *Costeseia* 1086, c.1184,

1196, *Costesseia* c.1130. OE pers.n. *Cost*, genitive sing. *Costes*, + **ēġ**. Like COLNEY TG 1707 this place lies on a ridge of high ground in a bend of the river Wensum. DEPN.

New COSTESSEY Norf TG 1809. A new suburb of Norwich. ModE adj. **new** + p.n. COSTESSEY TG 1711.

COSTOCK Notts SK 5726. Possibly 'the outlying farm of the Cyrtlingas, the people called after Cyrtel or Cyrtla'.
I. *Cotingestoche, Cortingestoche(s)* 1086, *Cortingstoc(a) -stok(a) -yng-* 1158–1335, *Cordingstoch' -stok(a)* 1165–1202.
II. *Chirtlingastoca* 1158, *Kertlingestok'* 1219, *Kirtlingstoc* 1253.
III. *Curtlingestoke* 1211, *K- Cort(e)lingstok(e) -yng-* 1235–1428.
IV. *Costoke* 1539, *Costock(e)* from 1559.
The three forms correspond to OE **Cortingas* 'the people called after Corta' < OE pers.n. **Corta* + **ingas**, genitive pl. **Cortinga*, OE **Cyrtlingas* 'the people called after Cyrtla or Cyrtel' < OE pers.n. **Cyrtla, Cyrtel*, itself a diminutive of **Corta*, + **ingas**, genitive pl. **Cyrtlinga*, and OE **Cortlingas* 'the people called after Cortla or Cortel' < OE pers.n. **Cortla, *Cortel*, an unmutated diminutive of **Corta*, + **ingas**, genitive pl. **Cortlinga*, each compounded with OE **stōc**. Nt 251, Studies 1936.28.

COSTON Leic SK 8422. Possibly 'Katr's village or farm'. *Caston(e)* 1086–1294, *C- Kauston'* 1219, *Coston(')* from 1227, *Coaston, Coasen* 17th cent. ON pers.n. *Kátr*, genitive sing. *Káts*, + **tūn**. Lei 169 gives pr ['koustən].

COTEBROOK Ches SJ 5765. *Coatbrook* 1828, *Cote Brook* 1831, apparently identical with *le Brokh* 1360, *Colbroke* 1476, *Colebrooke* 1603×25, 'cool or cold stream', OE **cōl** or **cald** + **brōc**. Che iii.292.

COTEGILL Cumbr NY 6504. 'Cottage ravine'. *Cotegill* 1615, 18th cent. ModE **cot** 'cottage' + **gill**. We ii.45.

COTEHELE HOUSE Corn SX 4268. P.n. Cotehele + ModE **house**, an important late 15th cent. mansion house. Cotehele, *Cotehulle* c.1286, *Cotehele* 1305, *Cuttayle* 1602, appears to be 'cottage hill', OE **cot** + **hyll**, but the stressing of the pr [kə'ti:l] suggests that this is a reinterpretation of OCo **cuit** + ***heyl** 'wood on an estuary'. The banks of the Tamar estuary are still well-wooded. PNCo 73.

COTEHILL Cumbr NY 4650. 'Cottage hill'. *Cotehill* 1550, *Cotthil* 1547×53. ME **cot** + **hill**. Cu 164.

COTEN END A street name in Warwick SP 2865. *Cawton End* 1525, *Cottenende* 1545, *Coten end* 1656. Earlier simply *Cotes* 1086–1257, *Coton* 1486, 'the cottages', OE **cot**, secondary ME plurals **cotes, coten**. The cottages were the homes of suburban small-holders engaged in trading and industrial activity in Warwick. Wa 264, Dyer 96–100.

COTES 'The cottages'. ME **cotes**, pl. of **cot** (OE *cot, cotu*). See also COATES. Dyer 1985.
(1) ~ Staffs SJ 8434. *Cota* 1086, *Cotes* 1251. DEPN.
(2) ~ Leic SK 5520. *Chotes* 1135×54, *Cotes* from late 12th, *Coat(e)s* 1558–1694. Lei 285.
(3) ~ North Lincs TA 3500. 'The north huts'. *Nordcotis* c.1115, *Northcotes* 1202. OE **north** + **cot**, pl. **cotu**, ME **cotes**. 'North' for distinction from Somercotes TF 4193 and 4296. A daughter settlement from North Thoresby TF 2998 originating as herdsmen's huts on land reclaimed from the sea. DEPN, Jnl 7.56.

COTESBACH Leic SP 5832. Probably 'Cott's stream'. *Cotesbece* 1086, *Cottebec' -bech(e) -is- -ys-* 1254–1608, *Cotes-* 1274–1507, *Cotesbach(e)* from 1274. OE pers.n. *Cott*, genitive sing. *Cottes*, + **beċe, bæċe**. Lei 438, L 12.

COTGRAVE Notts SK 6435. Probably 'Cotta's grove'. *Godegrave* (sic) 1086, *Cotegrava -e* 1094–1355, *Cotgrave* from 1205. OE pers.n. *Cotta* + **grǣfe**. Nt 233, L 194.

COTHAM Notts SK 7947. '(The settlement) at the cottages'. *Cotun(e), Cotes* 1086, *Cotum* 1232–1351, *Cotun(')* 1235–1316, *Cotom* 1334, *Cot(h)am* 16th, *Cotton al. Cotham* 1693. OE **cotum**, dative-pl. of **cot**. Also known as *Southcoton -om* 1387, 1528 for distinction from COTTAM SK 8180. For the possible significance of this name in relation to Newark, see Dyer 1985.91–106. Nt 212.

COTHELSTONE Somer ST 1831. 'Cuthwulf's estate or farm'. *Cothelestone* 1327, *Cuthelstone* 1333, *Cothelston* 1610. OE pers.n. *Cūth(w)ulf*, genitive sing. *Cūth(w)ulfes*, + **tūn**. DEPN.

COTHERSTONE Durham NZ 0119. 'Cuthhere's farm or village'. *Codrestune -ton* 1086, *Cothereston* 1279×81–1354, *Cotherston* from c.1250, *Cudderston* 1576. OE pers.n. *Cūthhere*, genitive sing. *Cūthheres*, + **tūn**. Despite frequent statements to the contrary in older literature the pers.n. Cuthbert does not occur in this name. YN 306.

COTHERSTONE MOOR Durham NY 9136. *Cotherston Moor* 1863 OS. P.n. COTHERSTONE NZ 0119 + **mōr**.

COTHILL Oxon SU 4699. 'The cottage well'. *Cotwell* 1738, *Cothill* 1783, 1812. ModE **cot** + **well**. Brk 415.

COTLEIGH Devon ST 2002. 'Cotta's wood or clearing'. *Cotelie* 1086, *-leia* 1086 Exon, *Cotteleg(he) -le(ye)* 1195–1321. OE pers.n. *Cotta* + **lēah**. D 625.

COTON '(The place at) the cottages'. OE **cot**, pl. **cotu**, dative pl. **cotum**; also Me pl. **cotes, coten**. PNE i.108, Dyer 1985. See also COTHAM, COTTAM.
(1) ~ Northants SP 6771. *Cota, Cote* 1086, *Cotes* 12th–1428, *Cotene* 1285, *Estcoton* 1369, *Cotton* 1581. Nth 67.
(2) ~ Staffs SJ 9832. *Cote* 1086. DEPN. A form *Cottin* 1598 cited by Horovitz may belong here.
(3) ~ CLANFORD Staffs SJ 8723. 'C beside Clanford'. P.n. *Cote* 1086, *Coton* 1291 + p.n. *Clanford* 1679, 'the clean ford', OE **clǣne** + **ford**, for distinction from Coton near Gnosall SJ 8120, *Coton* from c.1260. DEPN, St i.155.
(4) ~ IN THE ELMS Derby SK 2415. *Cotun(e)* 1086, *-ton* from 1295, *K- Cotes -is* 1086–1284×6. The form *Cotuhalfne* [942]14th S 1606 is 'half of Coton'. Db 630.
(5) Clay ~ Northants SP 5977. 'C on clay soil'. *Claycoten* 1330, *-ton* from 1427. ME **clay** + **coten**, weak plural of **cot**. 'Clay' for distinction from COTON SP 6771. Earlier simply *Cotes* 1175–1285, and *Cleycotes* 1284, 1360. Nth 66.

COTSWOLD HILLS Glos SO 9707. Otherwise The Cotswold(s), 'Cod's forest'. *(montana de) Codesuualt* 12th, *Coddeswold* 1269, 1294, *Coteswaud* 1250, *-wold* 1305, *Cottyswolde* 1440, 1480, *Cotssold* 1541, *Cotsall* 1602. OE pers.n. **Cōd*, genitive sing. **Cōdes*, + **wald**. Gl i.2, L 224.

COTT Devon SX 7861. *Cott* 1809 OS. ModE **cot** 'a cottage'.

COTTAM '(The place) at the cottages'. OE **cotum**, dative pl. of **cot**.
(1) ~ Lancs SD 5032. *Cotun* 1227, *Cotum* [c.1230]1268–1284, *Kotum, Coton* 1246, *Cotam* 1292, *Cotham* 1577. La 147, Jnl 17.84.
(2) ~ Notts SK 8180. *Cotum* 1280–1346, *Cotom* 1334, 1374, *Cotton* 1522, *Cottam* 1593. Also known as *Northcotom* 1374 for distinction from COTHAM SK 7947. Nt 29.

COTTENHAM Cambs TL 4567. 'Cotta's homestead or promontory'. *Cotenham* [948]14th S 538, c.1050, [c.1071]12th, [1042×66]13th S 1051, 1086–1500, *Coteham* 1086–1315, *Cottenham* from 1246. OE pers.n. *Cotta*, genitive sing. *Cottan*, + **hām** or **hamm** 2a. Ca 149.

COTTERDALE NYorks SD 8394. 'Cotter valley'. *Cottesdale* 1266–7, *Cotterdale* from 1280. P.n. Cotter as in Cotter End SD 8293, Cotter Force SD 8492 and Cotter Riggs SD 8392 + ME **dale** (ON *dalr*). Cotter is thought to be ON **kotar** 'the huts', pl. of **kot** as in the Norwegian p.n. Kaater (ONorw *Kotar*). YN 258.

COTTERED Herts TL 3129. Partly uncertain. *Chodrei* 1086, *Codreye* 1236, *Codreth(e)* 1220–1402, *Coddreth -red* 1248–1350, *Cotered al. Codreth* 1553, *Cotterhead* 1663. Unidentified element + OE **rith** 'a stream'. The absence of spellings with medial *-e-* shows that the specific cannot be OE pers.n. *Codda*. OE ***cōd** 'spawn of a fish' cognate with ON *kóth* has been suggested and there are occasional spellings reflecting *ō*,

Coudreya 1254, *Coudrede* 1287, *Coodred* 1509; better, perhaps, might be OE **codd** 'bag, sack' used once of a hollow in a ME f.n. and so 'stream in a hollow, stream called *Codd*'. OED s.v. cites occasional spellings with *oo* for *cod*. Hrt 157 gives pr [kɔtred].

COTTERSTOCK Northants TL 0490. 'The dairy-farm'. *Codestoch* 1086, *-stoc -stok(e)* [1100×20]c.1200–1232, *Coþestoche* c.1175, *Cother(e)stok(e)* 12th–1428, *Cot(t)erstok(e)* 1316, 1346, *Cotterstocke* 1615. OE **corther** 'a place of churning, a churn, a dairy' + **stoc**. Nth 200, DEPN, ESt 77.375–8.

COTTESBROOKE Northants SP 7173. 'Cott's brook'. *Cotesbroc* 1086–1342 with variant *-brok, Codesbroc* 1086, *Codebrok* 1251, *Cottesbroc* 1219, *-broke* 1575. OE pers.n. **Cott*, genitive sing. **Cottes*, + **brōc**. Alternatively the specific might be OE pers.n. **Codd* with unvoicing of *d* to *t* before following *s*. Nth 67, L 15–16.

COTTESMORE Leic SK 9013. 'Cott's moor'. *Cotesmore* 1086–1535, *Cottesmor(e)* from 1202[†]. OE pers.n. *Cott*, genitive sing. *Cottes*, + **mōr**. R 16, L 55.

COTTINGHAM Humbs TA 0532. 'Homestead of the Cottingas, the people called after Cott(a)'. *Cotingeham* 1086–1244, *Cotinghā* 1086, *-ham -yng-* [1150×60]1201–1523, *Cottingham* from 1196. OE folk-n. **Cottingas* < pers.n. **Cott(a)* + **ingas**, genitive pl. *Cottinga*, + **hām**. It seems unlikely that the specific could be an *-ing* derivative of OE *cot* 'a cottage' in composition with *hām*. YE 205, ING 153.

COTTINGHAM Northants SP 8490. 'The homestead of the Cottingas, the people called after Cott(a)'. *Cotingeham* 1086–1227, *Cotingham* [1137]12th–1343 with variant *Cotyng-*. OE folk-n. **Cottingas* < OE pers.n. **Cott(a)* + **ingas**, genitive pl. **Cottinga*, + **hām**. Nth 163, ING 146.

COTTINGWITH 'Cotta's dairy farm'. OE pers.n. *Cotta* + **ing**[4], or genitive sing. *Cottan*, + **wīč**, replaced by ON **vithr**.

(1) East ~ Humbs SE 7042. ME adj. **est** (1276) + p.n. *Coteuuid, Cotinuui* 1086, *C- Kottingwith -y-* 1225×30, *Coting- Cattingwic* 1231. It lies E of the r. Derwent. YE 237.

(2) West ~ NYorks SE 6942. ME adj. **west** (1309) + p.n. *Coteuuid* 1086, *Cotingwith* 1100×15–1309, *Cottingwic* 1214. It lies W of the r. Derwent. YE 264.

COTTISFORD Oxon SP 5931. 'Cott's ford'. *Coteford* [1081]12th, *Cotesforde* 1086–1504 with variant *Cot(t)esford, Cotisford* 1312, 1675, *Cottisford* 1797. OE pers.n. **Cott*, genitive sing. **Cottes*, + **ford**. O 206.

COTTON Staffs SK 0646. 'Coda's estate'. *Codetune* 1086, *Codinton* 1194. OE pers.n. *Coda*, genitive sing. *Codan*, or *Coda* + **ing**[4], + **tūn**. DEPN.

COTTON Suff TM 0767. 'Coda's estate'. *Code- Cote- Coti(n)tuna, Kodetun* 1086, *Cotton(')* from 1195. OE pers.n. *Coda* + **ing**[4] + **tūn**. DEPN, Baron.

Far COTTON Northants SP 7558. *Far Cotton* 1779. ModE adj. **far** + p.n. *Cotes* 1199–1316, *Cotun* 13th, *Coten* 1294–1325, *Coton juxta Norhampton* 1394, 'the cottages', OE **cot**, secondary ME plurals **cotes, coten**. Also known as *West Cotton* 1779 for distinction from Cotton End SP 7559, *Cotten End* 1686. Cf. COTEN END Warw SP 2865. The cottages were the dwellings of suburban small-holders engaged in trading and industrial activity in Northampton. Nth 147, Dyer 98.

COTWALTON Staffs SJ 9234. Partly uncertain. *Cotewaltun* [1002×4]11th S 1536, *Cotewoldestune* 1086, *Codewalton* 1176. Unidentified specific + **tūn**. The specific seems to be a p.n. **Cotewal*, possibly 'Cotta's' or 'the cottages' spring', OE pers.n. *Cotta* or **cota**, genitive pl. of **cot**, + Mercian **wælla**. DEPN.

COUGHTON H&W SO 5921. 'Settlement at *Cocc*, estate called *Cocc*'. *Cocton(e)* 1216×72, 1286, *Co(c)kton* 1280–1381, *Cottona* 1328, *Coughton* 1542. OE ***cocc** + **tūn**. The reference is to the pyramidal hill occupied by Chase Wood SO 6022. He 198, Bannister 51.

COUGHTON Warw SP 0860. 'The hill settlement'. *Coctvne* 1086, *-ton(a)* 1186–1293, *Co(c)kton* 1294–1428, *Cohton* 1200, *Coghton* 1410, *Cowton* 1676, *Coughton* from 1472×8. OE ***cocc** + **tūn**. Situated near a prominent hill. Wa 207 gives pr [koutən].

King's COUGHTON Warw SP 0859. 'The part of C held by the king'. *Kyngescoketon* 1262, 1306, *Kingscoughton* 1653. ME **king**, genitive sing. **kinges**, + p.n. COUGHTON SP 0860. King's Coughton was within the royal manor of Alcester. Wa 194.

COULSDON GLond TQ 3059. Partly uncertain. **Curedesdone* [672×4]13th S 1165, *Cudredesdone* [933]13th S 420, *-dune* [967]13th S 752, *Cuðredesdune* [1062]13th S 1035, *Colesdone* 1086–1374, *Cul(l)esdon(a)* 1100×29–1498, *Coulesdon* 1346, *Coulsdon* 1597, *Couldisdon* 1610. This is usually said to be 'Cuthræd's hill', OE pers.n. *Cūthrǣd*, genitive sing. *Cūthrǣdes*, + **dūn** with Anglo-Norman substitution of [l] for [r] as in e.g. Bellingham GLond TQ 3772, *beringa hamm* 973 B 1295, *Bellingeham* [late 12th]13th, 'enclosure of the Beringas, the people called after Beora or the bear people', OE folk-n. **Beringas* < OE pers.n. *Be(o)ra* or **bera** + **ingas**, genitive pl. **Beringa*, + **hamm**. In Coulsdon it would have to have taken place in a reduced OE form **Cur(r)esdun* for which there is no evidence: indeed a hypocoristic form **Curra* for *Cūthrǣd* would not have an *-es* genitive sing. form. A possible alternative suggested by Professor Coates might be 'hill at or called *Cull*' in which *Cull* would be PrW ***cull** < Late Latin **cullus* < Latin *culleus*, 'leather bag, scrotum', ModW *cwll*, used in a topographical sense. In this case either the identification of *Cuthredesdun* with Coulsdon is wrong, or there were two different names for the same place. Sr 44, PNK 6, DEPN, ASE 2.20, TC 203, GPC 640.

COULSTON Wilts ST 9554. 'Cufel's farm or village'. *Covelestone, Cuuleston(e)* 1086, *Cov-, Cou-, Cuveleston* 1206–1332, *Couleston* 1316, *Cowston* 1613, *Cowlson* 1637. OE pers.n. *Cūfel*, genitive sing. *Cūfeles*, + **tūn**. Wlt 140 gives prs [koulstən] and popular [koulsən].

COULTON NYorks SE 6374. 'Charcoal-burning settlement'. *Col(e)tun(e)* 1086–c.1285, *Colton* from 1086. OE **col** + **tūn**. YN 50 gives pr [koutən].

COUND Shrops SJ 5504. Originally a r.n. of unknown meaning as in COUND BROOK SJ 5305. *Cuneet* 1086, *Conet* 1242–94, *Conede* 1255–1397, *Cunde* 1351–1750, *Counde* 1397–1809, *Cound* from 1571. Brit **Cunētịu* of unknown meaning found also in the r.ns. KENNET(T) and KENT. Sa i.102 gives pr [kund].

COUND BROOK Shrops SJ 5305. *Cound Brook* 1833 OS. R.n. Cound as in COUND SJ 5504 + ModE **brook**.

COUNDON Durham NZ 2429. 'Cows' hill'. *Cundun'* 1197, *-don'* 1197, 1310×11, *Cundum* 1242×3, *Condon -om* 14th cent., *Coundon* from [1183]c.1320, *Cowndon* 1522. OE **cū**, genitive pl. **cūna**, + **dūn**. NbDu 55, L 144, 157–8.

COUNTERSETT NYorks SD 9187. 'Constance's shieling'. *Constansate* 1280, 1283. OFr pers.n. *Constance* + ME **sate** (ON *sǣtr*). YN 263.

COUNTESS WEAR Devon SX 9489. *Countess Wear* 1809 OS, *Countess Weir* 1907. A modern parish named from a weir built by Isabella de Fortibus, Countess of Devon, in 1284 to obstruct navigation to the city of Exeter in return for offences against her by the citizens of Exeter. S. Baring-Gould, *Devon*, The Little Guides, 157.

COUNTESTHORPE Leic SP 5895. 'Outlying farm held by the countess' sc. of Leicester. *Cuntass(e)thorp* 1242–1337, *Countesthorp(e) -is- -ys-* from 1395. Earlier simply *Torp* [1156]1318, 1209×35, *Thorp(e)* 1276–1535. ME **cuntesse** + **thorp**. Simon de Montfort, earl of Leicester, died seised of lands in Countesthorpe in 1256. Lei 439, SSNEM 132 no. 6.

[†]The form *(æt) Cottesmore* [971×83]14th S 1498 does not belong here but to a lost place in Bradwell, Oxon. O 308, ECTV 133.

COUNTISBURY Devon SS 7449. 'Fortified place of Cunet'. *Contesberie* 1086, *Cuntesberia -bir(y) -byr'* 1177–1275, *Cunisbere, Consberye* 16th, *Cunsbery* 1629×57. P.n. **Cunet* (from PrW ***cönuid** < British **cunētiū* of unknown meaning) + **byriġ**, dative sing. of **burh**. Identical with the *arx Cynuit* 'citadel called Cynuit', a place near the Devon coast where according to Asser Ubba, the brother of Ivarr the Boneless, suffered defeat in 878 (*Life of King Alfred* c.54). D 62.

COUPLAND Northum NT 9331. 'The purchased land'. *Coupland* 1242 BF, *Coupeland* 1296 SR, *Copeland* 1663. ON **kaupa-land**. NbDu 52, L 246.

COURTEENHALL Northants SP 7653. 'Curta's nook' or 'the nook of the fence'. *Cortenhale* 1086–1316, *Curtehala* 1100×35, *Cortehalle* [1109×22]1356, *Cortenhalle* 1299, *Cortnall, Cawtnoll, Curtenhall* 16th., *Courtenhall -in-* 1657. OE pers.n. **Curta*, genitive sing. **Curtan* + **halh**, or OE ***cort(e)**, ***curt(e)**, genitive sing. ***cortan**, ***curtan**, + **halh**. Courteenhall lies in a small secluded valley. Nth 145 gives pr [kɔ:t(ə)nɔ:l], PNE i.108, Nomina 7.38.

COURTSEND Essex TR 0293. The 'cottage end' sc. of Foulness. *Cotes End* 1777, 1805 OS. ModE **cote**, pl. **cotes**, + **end**. The settlement included *Pekiscote* and *Sacriscote* in 1614 and *London coate* in 1643. Contrasts with CHURCHEND TR 0092. Ess 184.

COURTWAY Somer ST 5243. *Courtgrove* 1809.

COUSLEY WOOD ESusx TQ 6533. *Corslewode* 1437, *Cousley wood* 1627. P.n. *Coreslie* 1296, *Co(u)rsley, Coursly* 1547, + ME **wode**. The p.n. Cousley is partly obscure; another Cousley in Withyham ESusx (TQ 4935) is *Coresle* 1285, *Coseley* 1758, unidentified pers.n. or element **Cor* or ***cor**, genitive sing. **Cores*, ***cores**, + OE **lēah**. Cf. CORRINGHAM Lincs SK 8791. Sx 386.

COVE Devon SS 9519. 'Hollow'. *La Kove* 1242, *(aqua de) Cove* 1249. There is a marked hollow in the hills with a water-course just N of the hamlet. D 542.

COVE Hants SU 8455. Uncertain. *Cove* from 1086, *la Cove* 1305. OE **cofa**. The etymology is clear but the sense uncertain. The basic meaning of *cofa* and its Gmc cognates is 'enclosed chamber' (in various specialised senses, 'bedchamber, storeroom, stable, pigsty, hut shed'). In English it also developed the sense 'hollow in a rock, cave, den' and eventually 'sheltered recess on the coast' (16th cent.). The application here is uncertain. Cove stands on a hill-spur overlooking a small narrow valley. Ha 60, Gover 1958.108.

North COVE Suff TM 5281. *North Cove* 1285. ME adj. **north** + p.n. *Cove* from 1204, OE **cofa** 'an inner chamber, a cave, a den, a hollow, a cove'. The exact sense is unclear; DEPN suggests that N Cove was an outlying part of S Cove. Baron.

South COVE Suff TM 4980. *Suth Coue* 1327. ME adj. **south** + p.n. *Coua* 1086, *Cove* from 1203, OE **cofa** as in North COVE TM 4689. The exact sense is unclear; DEPN suggests that the reference is to a former cove or creek on the coast, possibly the indent in the hills N of Potters Bridge. Baron.

COVEHITHE Suff TM 5281. 'Cove harbour'. *Coofythe, Coveheith alias Nerthale* 1523, *Coveheyth alias Northales* 1524, *Couehith* 1610. P.n. Cove as in North and South COVE TM 4689, 4980, + ModE **hithe**. The original name was *Northhala, Nordhalla, Nor(t)hals* 1086, *Northales* 1253–1633, 'the north nook or nooks', OE **north** + **halh**, pl. **halas**, or 'the north spit of land', **north** + **hals**. DEPN, Baron.

COVEN Staffs SJ 9107. Uncertain. *Cove* 1086, *Covena -e* 1116–1433, *Coven* from 1199. Possibly OE **cofum**, dative pl. of **cofa** 'an inner chamber, a cell, a shelter, a hut' and so perhaps '(the settlement) at the huts'. Other senses of this word, 'cave, cove, recess in a hill-side' are either too late or inappropriate to the topography here. Horovitz cites the gloss *cofa: pistrinum* 'bakehouse' (Wright ii.117, 30) which suggests that the reference may be to rounded kilns for burning charcoal perhaps to supply nearby ironworks. St i.37.

COVENEY Cambs TL 4882. 'Cofa's island' or 'bay island'. *Coueneia* [1017×35]12th S 1520, 1170, *Coueney(e)* from 1251. OE pers.n. *Cofa* or **cofa** 'bay, creek', genitive sing. *Cofan*, **cofan**, + **ēġ**. Possibly the island originally lay in a deep bay represented by West Fen. Ca 230.

COVENHAM RESERVOIR Lincs TF 3496. P.n. Covenham as in COVENHAM ST BARTHOLOMEW TF 3394 + ModE **reservoir**.

COVENHAM ST BARTHOLOMEW Lincs TF 3394. 'St Bartholomew's C'. *Covenham Sancti Bartholomei* 1254. P.n. *Covenham* from 1086 including [before 1067]13th, *Conam* 1383–1625, 'Cofa's homestead', OE pers.n. **Cofa*, genitive sing. **Cofan*, + **hām**, + saint's name *Bartholomew* from the dedication of the church. The topography does not support derivation from OE **cofa** 'a recess in a hill-side'. Li iv.4.

COVENHAM ST MARY Lincs TF 3394. 'St Mary's C'. *Couenham B. Marie* 1265. P.n. Covenham as in COVENHAM ST BARTHOLOMEW TF 3394 + saint's name *Mary* from the dedication of the church. Li iv.7.

COVENTRY WMids SP 3379. 'Cofa's tree'. *Couentr'* [1043×53]13th Harmer 1959, *Couentria* (Latin) [1043×53]16th S 1099, *Cofantreo* c.1060 ASC(C) under year 1053, *Cofentreo* c.1060 ASC(D) under same year, *Cofentreium* (Latin) c.1070, *Coventrev* 1086, *Couintre(a) -en-* 1151–1442, *Coventry* from 1249. OE pers.n. **Cofa*, genitive sing. **Cofan*, + **trēo**. The reference may have been to a prominent tree or to a cross erected by the unknown Cofa. The charter forms of S 1000, S 1098, S 1226 occur in late forged documents and are not cited here. Wa 160, TC 78, L 215, 218.

COVENTRY AIRPORT Warw SP 3574. P.n. COVENTRY WMids SP 3379 + ModE **airport**.

COVENTRY CANAL Warw SP 3196, 3786. P.n. COVENTRY WMids SP 3379 + ModE **canal**.

River COVER NYorks SE 1186. Either 'the stream in the hollow' or simply 'the stream'. *Cobre* c.1150, *Couer* 1279, *Cover* from 1336, *Cour* 1565. A British r.n., either PrW **cou* 'hollow' (W *cau*) + **ber* or PrW **gober* (W *gofer*) 'a stream' < Celtic **u̯obero* on the root **bheru-* 'boil' as in the p.n. *Voberna*, modern Vobarno near Brescia. The river runs through a deep valley. YN 2, RN 100, LHEB 372, 434, Holder iii.422, GPC 441, 1429.

COVER HEAD BENTS NYorks SE 0078. P.n. Cover Head, *Coverhede* 1405, 'the head of the river Cover', r.n. COVER SE 1186 + ME **hede**, + ModE **bent** (OE *beonet*) 'bent grass'. YN 254, PNE i.28.

COVERACK Corn SW 7818. Unexplained. *Covrack* 1588, short for earlier *Porthkoverec* 1284, *Pordcofrek* 1302, Co **porth** 'cove, harbour' + p.n. *Covrec* of unknown origin. Coverack Bridges SW 6630 lies on the river Cobar, *Coffar* 1284, 1286, *Cofar* 1323, *Chohor* 1336–54, also of unknown origin. A similar stream name may lie behind this name + adj. ending **-ek**. Cf. River COVER NYorks SE 1186 and *cofer fros* 960 S 684, the name of a stream in the bounds of Ladrock and Perranzabuloe with Co *frot* 'stream' already exhibiting the change *d* > *z*. PNCo 73, RN 82, 100, Gover n.d. 553, 680, CPNE 100–1, DEPN.

COVERACK BRIDGES Corn SW 6630 → COVERACK SW 7818.

COVERDALE NYorks SE 0683. 'Valley of the river Cover'. *Coverdale* from 1202. R.n. COVER SE 1186 + ON **dalr**. YN 3.

COVERHAM NYorks SE 1086. 'Homestead on the r. Cover'. *Covreham* 1086, *Coverham* from 12th. R.n. COVER SE 1186 + OE **hām**. YN 254.

COVINGTON Cambs TL 0570. 'Estate called after Cofa'. *Covintune* 1086, *Couyngton* 1260, 1331, 1493. OE pers.n. *Cofa* + **ing**^{4} + **tūn**. Hu 238.

COW GREEN RESERVOIR Durham NY 8030. P.n. Cow Green + ModE **reservoir**.

COW RIDGE NYorks SE 5496. *Cow Ridge* 1858 OS. Mod **cow** + **ridge**.

COWAN BRIDGE Lancs SD 6376. *Collingbrigke, Colligbrige* [c.1200]1268. This has usually been explained as 'Colling's bridge', OE pers.n. *Colling* + **brycg** influenced by ON **bryggja**, *Colling* being a patronymic formed on **Coll*. OE *ing*[2], however, is regularly used to form stream names. Cowan Bridge carries the road from Skipton to Kirby Lonsdale over LECK BECK, a Scandinavian stream name (ON **lœkr**). *Colling* could well, therefore, be an earlier name of the stream, possibly an *ing*[2] derivative of PrW ***coll** 'hazels'. Cf. COLLOWAY Lancs SD 4459, COLESHILL Warw SP 2089, COLLINGBOURNE DUCIS and COLLINGBOURNE KINGSTON Wilts SU 2453, SU 2355. La 184.

Little COWARNE H&W SO 6051. *Parva Coura* 1145, 1148×63, ~ *Cour(e)* 1160×70, 1291. ModE adj. **little**, Latin **parva**, + p.n. *Cogre* [1017×41]17th, *Colgre* 1086, *Coura -e* 1088–1160×70 with variant *Cura* c.1150, *Cowra* c.1250, possibly OE **col** 'charcoal', **cōl** 'cool', **cole** 'a hollow' or **cū** 'a cow' + **grēd** 'pasture' or **grene** 'a green piece of grassland'. 'Little' for distinction from Much COWARNE SO 6245 to the form of which this name has been assimilated. He 60, DB Herefordshire 7,8 note.

Much COWARNE H&W SO 6245. *Covene Majori* 1088, *Magne Co(w)ern(e)* 1243–1373 with variant *Coure* 1291, *Cowarn Magna* 1397, *Muchel Cowarne* 1429. ME adj. **muchel**, Latin **major** 'greater', **magna**, + p.n. *Cuure* (for *Cuuren*) 1086, 'cow-house', OE **cū** + **ærn**. 'Much' for distinction from Little COWARNE SO 6051, originally apparently a quite different name. He 60, DB Herefordshire 19,10.

COWBEECH ESusx TQ 6114. 'The pollarded beech-tree'. *Coppetebeche* 1261, *Kopped(e)beche* 1296, 1316, *Cop(pe)beche* 16th., *Cobbeach* 1622, *Cobeech* 1724. OE **coppede** + **bēċe**. Sx 481.

COWBIT Lincs TF 2617. 'The river-bend frequented by cows'. *Coubith* c.1250, *Coubiht* 1267, *Coubyt* 1316, *-bi(g)ht -by(g)ht -bith* 1331–92, *Coubit, Cowbyt(t)* 1488, *Cubbit* 1572, *Cubbett* 1618, *Coobitt* 1764. OE **cū** + **byht**. The reference is to a cow-pasture in a bend of the r. Welland. Payling 1940.10 gives prs [kubət -it].

COWDEN Kent TQ 4640. 'Cow pasture'. *Cudena* c. 1100, *K-Cudenn(e)* 1240–1258, *Couden(')* 1237, 1259, *Cowden* 1610. OE **cū** + **denn**. PNK 87.

COWDEN STATION Kent TQ 4741. P.n. COWDEN TQ 4640 + ModE **station**.

COWES IoW SZ 4995. 'The cows'. *(le) Estcowe, Westcowe* 'the east' and 'west cow' 1413, *the Cow(e)* 1512–25, *the Kowes* 1545, *Cowes* 1622. ME **cow** 'a cow' originally referring to sand-banks off the coast of the Medina and later transferred to the coast itself. Wt 120 and Mills 1996.42 cite other examples of animal names applied to coastal sand-banks.

COWES ROADS IoW SZ 5097. P.n. COWES SZ 4995 + ModE **road** 'a sheltered piece of water where ships may ride at anchor'. Cf. *the tone of the roads called the Esturley or the Westerly Cowe* 1539. Wt 119, Mills 1996.43.

East COWES IoW SZ 5095. *East Cowes* 1769. ModE adj. **east** + p.n. COWES SZ 4995. Mills 1996.42.

COWESBY NYorks SE 4689. 'Kausi's farm or village'. *Cahosbi* 1086, *Cousebi -by* 1199–1407. ON pers.n. *Kausi* + **bȳ**. YN 201 gives pr [kouzbi], SSNY 24.

COWFOLD WSusx TQ 2122. 'The cow fold'. *Coufaud* 1232, *Coufold* 1336, *Cowfold(e)* from 1589. ME **cou** (OE *cū*) + **fald** (OE *falod*). Sx 209, DEPN, SS 79.

COWGILL Cumbr SD 7587. Probably 'dam ravine'. *Callgill* 1592, *Gaw- Cawegill* 17th, *Coegill* 1732, *Cowgill* 1847. Dial. **caul** 'a dam' + **gill**. The place lies beside the river Dee. YW vi.255.

COWICK Humbs SE 6521. 'Dairy-farm'. *Cuwic -wik* 12th–1246, *Cowyk(e) -wik -wick(e)* 1229–1531. OE **cū** + **wīċ**. YW ii.26.

COWLEY Devon SX 9095. 'Cofa or Cufa's wood or clearing'. *conanleygh* (for *couan-*) [944]14th S 498, *Cuueleghe* c.1200, *Couelegh(e) -v-* 1237–1389, *Cowlegh* 1537. OE pers.n. **Cofa* or *Cufa* + **lēah**. D 455, Nt xxxvii, L 203. But see COWLEY Oxon SP 5504.

COWLEY GLond TQ 0582. 'Cofa's clearing or wood'. *Cofenlea* [959]12th S 1293, *-an-* *[998]14th S 894, *Covelie* 1086, *Coueleg' -ley(e)* 1204–1440, *Cowlee* 1294, *-ley* 1535. OE pers.n. *Cofa*, genitive sing. *Cofan*, + **lēah**. Mx 32. But see COWLEY Oxon SP 5504.

COWLEY Glos SO 9614. 'Cow pasture'. *Kulege* 1086, *Culeg(a) -ley(e)* 13th cent., *Couley(e) -legh* 1253–1357, *Cowley* 1535. OE **cū** + **lēah**. Gl i.156, L 206.

COWLEY Oxon SP 5504. This deceptively simple name, one of at least five examples, is in fact of uncertain origin. *Couelea* [1004]1312×3 S 909–1428 with variants *-le(ya) -leia -lai -lay -ley* and *-lie, Cueueleia ley(e) -le -v-* [1199]c.1320, 1230–94, *Cofleya -lye -le(a)* 13th cent., *Cowley* from 1509×10. The generic is OE **lēah** 'a wood or clearing', the specific either an OE pers.n. **Cufa*, an OE word cognate with Norw *kuv* 'a round top', an OE **cufl* 'a block of wood, a stump', or **cofa** possibly in the sense 'shelter, hut'. O 27, DEPN, L 203.

COWLING NYorks SD 9643. 'The hill-top place'. *Collinghe* 1086, *-ing(e) -yng* 1202–1605, *Cullyng* 1315, *Cowling(e)* from 1562. OE ***colling** < **coll** + **ing**[2]. The reference is to Cowling Hill of 1000ft., *Collinge* 1202. YW vi.12 gives pr ['kauliŋ], ING 77.

COWLING NYorks SE 2387. *Collyng(e)* 1400, 1538, *Cowling(e)* from 1572. Short for *Thornton' Colling'* 1202, *Thorneton Collinge -ynge* 1270–1328, 'Thornton held by the Colling family'. Earlier simply *Torneton* 1086, 'farm with or near a thorn-tree or trees', OE **thorn** + **tūn**. YN 237 gives pr [koulin], ING 77, DBS 86, 80.

COWLINGE Suff TL 7254. 'Culing, the place called after Cul or Cula'. *Cvlinge* 1086, *Culing(es) -a, Coelingia* 12th, *Culinges* 1195–1258, *Culing(')* 1206–28, *Culingge* 1257, *Cowlidge* 1610. OE pers.n. *Cūl(a)* + **ing**[2], locative-dative sing. **inġe**. The *Culinges* sps may represent OE folk-n. **Cūlingas* 'the people called after Cula', pers.n. *Cūla* + **ingas**. DEPN gives pr [ku:lindʒ], ING 210, BzN 1967.334.

COWM RESERVOIR Lancs SD 8818. P.n. Cowm + ModE **reservoir**. Cowm, *magnam Cumbam, paruum Cumbe* [13th]14th, *le Mikelcoumbebrok, Litelcumbe* c.1300, is 'great and little valley', Latin **magna, parvum**, ME **mikel** + OE **cumb** used as a p.n. La 59.

COWPEN BEWLEY Cleve NZ 4824. '(At) the fish-traps in the manor of Bewley'. *Cowpan-, Coopan Bewley* 17th cent., *Coopen Bewley* 1749–1872. Earlier simply *Cupum* 1154–1244, *Cupun* 1296, 1300, *Coupon* 1339–1692, *Cowpen* 1611, OE **cūpe**, dative pl. **cūpum**, + p.n. Bewley, *grangia de Bello Loco* 1233×44, ~ ~ *Beuleu* 1296, *Buley grange* 1576, the moated manor house of the Prior of Durham called *Beaulieu* 1446, 'beautiful place' like BEAULIEU Hants SU 3801, now represented by Low Grange NZ 4625.

COWPLAIN Hants SU 6911. *Cow-plain* 1859. ModE **cow** + **plain** in the technical sense 'open space in a forest'. Before development Cowplain was one of a number of *plains* in the Forest of Bere. Ha 60, Gover 1958.57, Rackham 157.

COWSHILL Durham NY 8540. Partly uncertain. *Coushille* 1336, *Coueshill* 1685, *Cowshill* 1886 WME. This may simply be 'cow's hill', OE **cū**, genitive sing. **cūs**, + **hyll**, but the specific might be ME **cove** (OE *cofa*) in the sense 'recess in the steep side of a hill'. Cowshill lies at the foot of a steep narrow valley.

East COWTON NYorks NZ 3003. *Est Coutona* 1314. ME adj. **est** + p.n. *Corketune, Cottvne, Cotun(e)* 1086, *C- Kuton(a)* 1154×89–1243, *Coutona* 1184–1316, originally 'Corc's estate', OIr pers.n. *Corc(c)* as in CORBY Cumbria NY 4754 + **tūn**. The name was early reformed as the 'cow farm' as in North COWTON NZ 2803, OE **cū** + **tūn**. YN 281 gives pr [ku:tən].

North COWTON NYorks NZ 2803. Prefixes ME *North* 13th, Latin *Magna* 'great' 1273, + p.n. *(alia) Cudtun -tone* 'the other C' (besides South Cowton) 1086, *Coutona* [1184]15th, the 'cow farm', OE **cū** + **tūn**. 'North' for distinction from East COWTON NZ 3003 and South Cowton NZ 2902, *Cudtun -tone* 1086. YN 281.

COX COMMON Suff TM 4082. No early forms.

COXBANK Ches SJ 6541. 'Wood-cock's hill'. *Coxbank* 1831, *Cock Bank* 1842. ModE **cock** + **bank**. Che iii.85.

COXBENCH Derby SK 3743. Probably 'cock's slope'. *Cokkesbenche* 1395, 1543, *Cocksbench(e)* 1533, 1633, *Coxbench* 1634. ME **cok** (OE *cocc*), genitive sing. **cokkes**, + **bench** (OE *benč*). The specific might, however, be the nickname or surname *Cock*. Db 570.

COXHEATH Kent TQ 7451. 'Cock's heath'. *Cokkyshoth* 1489, *Coxhoth* 1585. Surname Cock or Cook, *le Cok'* 1339, + ME **hoth** (OE **hāth*) later replaced by **heath**. Nearby is Cock Street TQ 7750, *Cock Street* 1819 OS 'Cock hamlet'. Cf. BROAD STREET TQ 8356. PNK 138.

COXHOE Durham NZ 3136. Possibly 'hill-spur of (the hill called) *Cocc*'. *Cokeshow* 1233×44, *-hov* 1235×6, *-howe* 1293, early 14th, *Coxhowe* 1298–1717, *Coxhoe* from 1586×7, *Coxsay, Cokseye, Coxeaye* 16th cent. Hill name **Cocc*, genitive sing. **Cocces*, + OE **hōh**. The village stands at the end of a very prominent hill ridge which may have been known as the *Cocc* < OE ***cocc** 'a heap, a lump, a hillock'. However, the specific could alternatively be either OE **cocc** 'a cock' or the pers.n. *Cocc*. NbDu 56.

COXLEY Somer ST 5243. 'The cook's wood or clearing'. *Cokesleg* 1207, *Cokesleye, Kockesley* 1269, *Cokkesleghe* 1327. OE **cōc** used as a by-name, genitive sing. **cōces**, + **lēah**. One of the tenants of land in the manor of Wells in 1086 was the wife of Manasses the Cook, *uxor Manasses coqui*. Manasses was a cook in the royal household and probably held this land. DEPN, Turner 1951.155.

Great COXWELL Oxon SU 2793. *magna Cokewell* 1225, *Magna Cogeswell'* 1284. ModE adj. **great**, Latin **magna**, + p.n. *Cocheswelle* 1086, *Kokeswell'* 1201, *Cokeswell(e)* [1217]n.d.–1327, 'Cocc's spring', OE pers.n. **Cocc*, genitive sing. **Cocces*, + **wella**. Doubt, however, has been cast upon this explanation because of the absence of *-cc-*, *-ck-*, *-kk-* sps. 'Great' for distinction from Little COXWELL SU 2893. Brk 362.

Little COXWELL Oxon SU 2893. *magna et parva Cokewell* 1225. Latin **parva** + p.n. Coxwell as in Great COXWELL SU 2793. Brk 362.

COXWOLD NYorks SE 5377. Partly uncertain. *Cuha walda* [757×8]17th B 184, *Cucualt* 1086, *Cuk(e)wald* 1154×89–1406, *Cookwold, Cuckwould* 16th, *Coxwo(u)ld* from 1627. Unknown element or name + **wald**. The specific could be OE ***cucu** 'a cuckoo' and so 'cuckoo forest', but *wald* is not otherwise compounded with words for wild creatures. The earliest spelling is not trustworthy, coming from a 17th cent. transcript of a papal letter and there is no evidence for an OE pers.n. **Cuha*. The current spelling replacing *Cockwold* 1612, *Cuckwold* 1616, is a reformation possibly due to the sensibilities of the local landowner, William Bellasyse ennobled in 1627, to a name homophonous with *cuckold*; indeed the form *Cuckoldie* actually occurs in 1579.
YN 191 gives pr [kukud], L 226–7, Jnl 27.43–7.

CRAB ROCKS Humbs TA 2074. *The Crab Rocks* 1856 OS. ModE **crab** + **rock**.

CRABADON Devon SX 7555. 'Crab-apple farm'. *Crabbeton* 1306, 1330, *Crabbaton* 1809 OS. ME **crabbe** + **toun** (OE *tūn*). D 300.

CRABBS CROSS H&W SP 0464. No early forms. Probably surname Crabbe + ModE **cross** 'a cross-roads'.

CRABTREE WSusx TQ 2225. 'The crab-apple tree'. No early forms. ModE **crab** + **tree**.

CRACKENTHORPE Cumbr NY 6622. 'Cracand's outlying farm'. *Crac- Crakant(h)orp(p)* 1185–15th, *Crakent(h)orp* 1206–1678, *Crackenthorp(e)* from 1510. ON pers.n. **Krakandi* + **thorp**. Two other examples of this name type are known, a lost vill in Beetham SD 4979, *Crakintorp* 1254, and a lost *Cracanethorpe* 1267×8 in Caton SD 5364. This suggests that the specific of these names is perhaps rather OE **crācena*, genitive pl. of **crāca* 'a crow, a raven' and that they mean accordingly 'outlying farm of the crows'. We ii.101, SSNNW 202.

CRACKINGTON Corn SX 1595. Partly uncertain. *Cracumtona* c.1170, *Crakenton* 1181, *Craketon* 1182, *Crac- Crak(h)am(p)ton* 1302–69, *Crackington* 1610. With these forms must be taken *Crachemua -nwe* [for *-mue*] 1086, *Crakemude* 1196, and the forms for Crackington Haven SX 1496, *Crakamphavene* 1358, *Crackington Horn* (for *Hawn*, the dial. form of ME **havene**) 1813. P.n. **Crak*, Co ***crak** 'sandstone', or **Craken*, Co ***cragen** 'rock', + OE **tūn** 'farm' and **mūtha** 'mouth'. At an early date *tūn* was replaced by *-hamtūn* on the analogy of other English *-hampton* names, and ME *havene* was added to the reduced form of this as if it were **Crackham*. The reference is uncertain; however Crackington is situated on a stream which may have been called *Crak*, cf. the river Crake Cumbr SD 2987 *Crec, Crayke* c.1160, *Craic* 1196 'rocky stream', PrW ***creig** cognate with Co **crak*. PNCo 73 gives local pr [krækan ɔ:n], Gover n.d. 59–60, DEPN, Jnl 1.45, CPNE 68, RN 101–2.

CRACKLEY BANK Shrops SJ 7611. *Crackley Bank* 1833 OS.

CRACKPOT NYorks SD 9796. 'Limestone rift where crows abound'. *Crakepot(e)* 1298, 1301, 1377×99. ON **kráka** + ME **potte**. YN 271.

CRACOE NYorks SD 9760. 'Crow hill'. *Crakehou -how(e)* 12th–1380, *Crac- Crakhou -how(e)* 1179–1428, *Crakoe, Cracoe* from 1588. ON **kráka** + **haugr**. YW vi.88 gives pr ['kre:kə].

CRADLEY H&W SO 7347. 'Criddi's wood or clearing'. *Cyrdesleah* [1016×35]11th S 1462, *Credelaie* 1086, *Cradele(i)a -leye* 1166, 1283, 1346, *Credele(i) -leye* 1189×93, 1241, 1291. OE pers.n. **Criddi*, genitive sing. **Criddes*, + **lēah**. **Criddi*, related to pers.n. *Creoda*, occurs again in KERSOE SO 9940. Here it has been reformed as if identical with Cradley WMids SO 9484, *Cradeleie* 1086, *Cradlega -ley(e)* 1180–1485, either 'Crada's wood or clearing' or the 'cradle clearing', OE pers.n. **Crada* or **cradol** + **lēah**, perhaps a place where cradles or hurdles were made. He 61, Wo 294 which gives pr [kreidli] for the WMids example, DEPN.

CRADLEY WMids SO 9484 → CRADLEY H&W SO 7347.

CRAFTHOLE Corn SX 3654. Uncertain. *Crofthol* 1348, *Crofthole* 1610, short for *Croftilberwe* 1314, *Croftholburgh* 1420, 'Crafthole borough'. Hardly 'croft hollow' since the place is situated on a hill; probably, therefore, as the 1314 form suggests, 'croft hill', OE **croft** + **hyll**. PNCo 74.

CRAG HILL Cumbr SD 6983. No early forms. ModE **crag** 'a crag, a rock' + **hill**. We i.25.

CRAG LOUGH Northum NY 7668. 'The lake by the crag(s)'. *Crag Lough* 1868 OS. ModE **crag** + N dial. **lough** (OE *luh*, itself a loan-word from PrW, cf. W *llwch*). The reference is to the rocky sill of Hotbank Crags which the Roman wall follows at this point. GPC 2234.

CRAGDALE MOOR NYorks SD 9182. P.n. Cragdale + ModE **moor**. Cragdale, *Cragdal* 1218, *Crakedale* 1307, is either the 'rock valley' or 'Kraki's valley', ME **cragge** or ON pers.n. *Kraki*, genitive sing. *Kraka*, + **dale** (ON *dalr*). The valley is rocky and scarred. YN 263 gives pr [krægdil].

CRAGG VALE WYorks SE 0123. *Cragg Vale* 1839. P.n. Cragg SD 9923, *(the) Cragg(e)* 1449–1642, ME **cragge** 'a crag, a rock' (W *craig*), + ModE **vale**. YW iii.164, 160, GPC 578.

CRAGHEAD Durham NZ 2150. *Craghead* 1857 Fordyce, *Crag Head* 1864 OS. ModE **crag** + **head**. A modern colliery village.

CRAKEHALL NYorks SE 2490. Partly uncertain. *Crachele* 1086, *K- Crakehale* 1157–1298, *-hall* from 1231, *Crakall* 1364, *Crakell* 1663. Either 'Craca's nook of land', OE pers.n. *Craca* + **halh**, or 'nook of land where crakes are seen', ON **kráka** 'a crow' or **krákr** 'a raven' + **halh**. The reference is to some member of the rail or *Ralidae* family, e.g. the corncrake. YN 237, L 108, 111.

CRAMBE NYorks SE 7364. '(Settlement) at the bends' sc. in the river Derwent. *Crambom -un, Cranbon(e)* 1086, *Crambum*

1086–1391, *Cramb(e)* from 1577. OE ***cramb**, dative pl. ***crambum**. YN 38.

CRAMLINGTON Northum NZ 2676. Partly uncertain. *Cramlingtun(a) -intun(e)* c.1130, *Cremelington, Cramlingtuna, Cramilintona* c.1160, *Cram(e)lington* 13th cent., *Cramelton* 1292, *Cramlyngton* 1430. The specific could be either OE pers.n. **Cramel* or p.n. **Cranwella* 'the spring frequented by cranes'. The name seems to show alternation between **Crameltūn* 'Cramel's farm', **Cranwell-tūn*, and an *ing*[2] formation **Cranwellingtūn* 'the *tūn* (called or at) **Cranwelling*, the *Cranwella* stream'. NbDu 56, BzN 1968.165, Årsskrift 1974.29.

East CRAMLINGTON Northum NZ 2876. Adj. **east** + p.n. CRAMLINGTON NZ 2676. The settlement grew up around *Cramlington Colliery* 1864 OS. 'East' for distinction from West Cramlington NZ 2575, *West Cramlington* 1867 OS.

CRANAGE Ches SJ 7568. 'Crow's boggy stream'. *Croeneche* 1086, *Cranlach(e)* 12th–c.1259, *Croulach(e)* 1188–1260, *Crau(e)-Craw(e)nach(e)* late 12th–1312, *Cran(n)ach(e)* 1246–1346, *Crannage* 1318, *Cranage* from 1399, *Carnage* 1775. OE **crāwe**, genitive pl. **crāwena**, + **læċ(ċ)**. Che ii.223 gives pr ['krænidʒ].

CRANBERRY Staffs SJ 8236. A modern name, no early forms.

CRANBORNE Dorset SU 0513. 'Crane stream'. *Creneburne* 1086, *-burna(m) -(e)* 1163–1543, *Cranbourn(e)* from 1252. OE **cran**, genitive pl. **crana**, + **burna**. Do ii.205, L 18.

CRANBORNE CHASE Dorset ST 9417. *Chaceam de Craneburn'* 1236, *Cranburne Chace* 1618. P.n. CRANBORNE SU 0513 + ME **chace** 'a tract of ground for breeding and hunting wild animals'. Do ii.193.

CRANBOURNE Berks SU 9272. 'Brook frequented by cranes or herons'. *Crampeburn'* 1279, *Cranbourn* 1337, *Cranburn* 1761, *Cranbourne* 1800. OE **cran**, genitive pl. **crana**, + **burna**. Brk 37.

CRANBROOK Kent TQ 7736. 'Crane stream or bog'. *Cranebroca* c. 1100, *Cranebroc -k* c.1200–70. OE **cran**, genitive pl. **crana**, + **brōc**, K Sx dial. *brook*. PNK 318.

CRANBROOK COMMON Kent TQ 7938. P.n. CRANBROOK TQ 7736 + ModE **common**.

CRANE ISLANDS Corn SW 6344. *Crane Island* c.1870. Uncertain, but related to the antiquity called Crane Castle SW 6343. A similar coastal name occurs at SW 6912, Crane Ledges. PNCo 74.

CRANFIELD Beds SP 9542. 'Open country frequented by cranes'. *Crangfeld* *[1060]14th S 1030, *Cranfelle* 1086, *Cranefeld* [c.1125]14th. OE **cranoc** later replaced by **cran**, genitive pl. **crana**, + **feld**. *Crangfeldinga dic* 'the dyke of the people of Cranfield' and *crancfeldinga gemære* 'the boundary of the Cranfield people' are mentioned in a charter of [969]11th S 772. Bd 68, Wo xl, Sx xlii, L 244.

CRANFORD GLond TQ 1077. 'Crane or heron ford'. *Cranford(e)* from 1086 with variants *Cram-* 1199, 1236, *Craum-* 1247, 1542 and *Cramp-* 1575. OE **cran**, genitive pl. **crana**, + **ford**. Mx 32.

CRANFORD ST ANDREW Northants SP 9277. P.n. *Craneford* 1086–1332, *Cranford(a)* from [c.1150]1267 with occasional variant *Craun-*, the 'cranes' ford', OE **cran**, genitive pl. **crana**, + **ford**, + saint's name *Andrew* from the dedication of the church. Nth 180, L 71.

CRANFORD ST JOHN Northants SP 9276. P.n. Cranford as in CRANFORD ST ANDREW SP 9277 + saint's name *John* from the dedication of the church. Nth 180, L 71.

CRANHAM GLond TQ 5787. 'Crows' hill-spur'. *Craohv* (sic) 1086, *Crawenho* 1201–91, *Cran(e)ham* 1486–93, *Crainham* 1535. OE **crāwe**, genitive pl. **crāwena**, + **hōh**. Sps in *-ham* may represent a variant name of the place with OE **hām** 'a homestead' or possibly OE **hōum**, the dative pl. of **hōh**. There was an alternative name of this place, *Wochendunā* 1086, *Wokindon* 1254×5, *Woke(n)den -don* 1254, 1338 as in North and South OCKENDON TQ 5984, 5982. Ess 124, LPN 57.

CRANHAM Glos SO 8913. 'Homestead or meadow where cranes are seen' or 'Crana's homestead'. *Craneham* 1148×79–1544, *Cranham* from 1327, *Cron(e)ham* 1189–1598, *Crunham* 1287. OE **cran** + **hām** or **hamm**, or pers.n. **Crana* + **hām**. Gl i.157.

CRANK Mers SJ 5099. A district named from *Crank Hall* 1843 OS. This could be OE **cranuc** + **halh** 'crane nook' but early forms are needed.

CRANLEIGH Surrey TQ 0638. 'The cranes' wood or clearing'. *Cranelega* 1166–1553 with variants *-legh* and *-le(y)(e), Cranlea* 1167–1871 with variants *-leghe* and *-le(y)(e), Crannelegh* 1488, *Crandley* 1657. OE **cran**, genitive pl. **crana**, + **lēah**. The present spelling is a modern adoption by the GPO to avoid confusion with CRAWLEY WSusx TQ 2736. Sr 229.

CRANMORE IoW SZ 3990. 'Crane marsh'. *Cranmores* 1559, *Cranmore* 1781. The name is probably referred to in the surnames of Nicholas de *Cranemore* 1235, Thomas de *Cranemore* 1285, and in the boundary point *to þæs móres heafde* 'to the head of the moor or marshland' [949]14th S 543. OE **cran**, genitive pl. **crana**, + **mōr**. Wt 209, Mills 1996.43.

CRANMORE Somer ST 6643. 'Cranes' pool'. *Cranemere* [955×9]13th S 1746, [959×75]lost ECW, 1196, 1241, *Crenemere* 1084, *Crenemelle* 1086, *East Cranmere, West Cranmer* 1610. OE **cran**, genitive pl. **crana**, + **mere**. DEPN.

CRANOE Leic SP 7695 'Crows hill-spur'. *Craweho -v-* 1086, *Crawenho(u)* 1306–1519, *Cranow(e)* 1385–1619, *Craynowe, Creyno* 15th. OE **crāwe**, genitive pl. **crāwena**, + **hōh**. Cranoe lies on a small spur of land rising steeply from the broad valley of the River Welland. Lei 212, L 168.

CRANSFORD Suff TM 3164. 'The cranes' ford'. *Crane- Crenefort, Crane(s)forda* 1086, *Craneford* 1203–1336, *Cranesford(')* 1254–1336, *Cransforde* 1610. OE **cran**, genitive sing. **cranes**, pl. **crana**, + **ford**. DEPN, Baron.

Great CRANSLEY Northants SP 8376. ModE adj. **great** + p.n. *Cranslea* [956]12th S 592, 1086, *Cranesleg* 1086–1458 with variants *Cranys- -le(ye)* and *-legh*, 'the crane's wood or clearing', OE **cran**, perhaps used as a nickname, genitive sing. **cranes**, + **lēah**. 'Great' for distinction from Little Cransley SP 8276, *Little Crannesley* 1555. Nth 124.

CRANSWICK Humbs TA 0252. Partly uncertain. *Cransuuic, Cranzuic -vic* 1086, *Crancewic -wik(e)*, 12th, 13th cents., *Cranke- Cranchewic* 1166, *Crauncewik(e) -wyk(e)* 1235×49–1492, *Craneswi(c)ke* 16th cent. DB *z* and regular ME *ce* point to [ts] and a Scandinavian pers.n. **Krant*, genitive sing. **Krants*, but no such name is known. Perhaps, therefore, a Scandinavianisation of OE *Cranoces-wīċ*, 'the crane's dairy-farm', with OE **cranoc**, possibly used as a proper name. The village is now called Hutton Cranswick from nearby HUTTON Humbs TA 0253. YE 156.

CRANTOCK Corn SW 7960. '(Church of) St Carantoc'. *(Canonici) S' Carentoch* 1086, *(ecclesia de) Sancto Carentoco* [1100×35]1270, *Seint Karentoc* 1234, *Crantocke* 1546, *Carantack* 1610. Saint's name *Carantek* (early W *Carantoc(us)*, W *Carannog* 'lovable one') from the dedication of the church. He was believed to be of royal birth in Cardiganshire; on his way to Cornwall he assisted King Arthur by taming a dragon. He was the leader of the group of monks who evangelised central Cornwall in the 6th–7th cents. He is also patron saint of Llangranog Dyfed SN 3154 and Carhampton Somer ST 0042. The name of the church-town survives as Langurra, a house in the village, *Langorroc, Langoroch* 1086, *Langor(r)ou* 1201–1336, *Lancorru'* 1302, 'church-site of Correk', Co **lann** + pers.n. *Correk*, probably a pet-form of Carantek. Cf. the Breton p.n. Carantec. PNCo 74, Gover n.d. 367, DEPN.

CRANWELL Lincs TF 0349. 'The spring frequented by cranes'. *Cranewellam* [1051]13th, *Craneuuelle -welle* 1086, *-well(a)* 1149–1551, *Cranwell* from 1358. OE **cran**, genitive pl. **crana**, + **wella**. There are springs S of the village. Perrott 281, Cameron 1998.

CRANWICH Norf TL 7894. 'Cranes' meadow'. *Cranewisse* 1086, *Crenewiz* 1200, *Kernewiz* 1254. OE **cran**, genitive pl. **crana**, + **wisse**. DEPN records a local pr [krænis] for 1883.

CRANWORTH Norf TF 9804. 'Cranes' enclosure'. *Crana-Craneworda* 1086, *Craneworth* 1211. OE **cran**, genitive pl. **crana**, + **worth**. DEPN.

CRAPNELL Somer ST 5945 → CROPTHORNE H&W SO 9944.

CRAPSTONE Devon SX 5067. *Crap Stone* 1678. D 226.

CRASTER Northum NU 2519. 'Crow fort, the fort frequented by crows'. *Craucestre* 1242 BF, 1428, *Craucestre* 1244–1346, *Crasestir* 1415 Morris, *Craister* 1460, 1663, *Crawstor* 1538. OE **crāwe** + **ċeaster**. The specific could also be the feminine pers.n. *Crāwe* derived from **crāwe**. The reference is to Craster Heugh Camp, a double ramparted earthwork S of the village. NbDu 57 gives pr [kreistə], a pronunciation which results from the early monophthongisation of ME *au* > *ā* beside retention of the diphthong into the 16th cent. Cf Nomina 13.109ff for this phenomenon. Pevsner 1957.135.

CRASWALL H&W SO 2836. 'Cress spring'. *Cressewell* 1231, *Crassewalle* 1255. OE **cærse**, **cresse** + **wella**, Mercian **wælla**. DEPN, Bannister 53.

CRATFIELD Suff TM 3175. 'Cræta's open land'. *Cratafelda* 1086, *Cratefeld(')* *-feud* 1165–1524, *Cratfeild* 1610. OE pers.n. **Crǣta* + **feld**. DEPN, Baron, L 243.

CRATHORNE NYorks NZ 4407. 'Thorn-tree in the corner of land'. *Gra- Cratorne* 1086, 1279×81, *Crathorn* from c.1160×75. ON **krá** + OE, ON **thorn**. The reference is to a bend in the river Leven. YN 174.

CRAVEN An ancient regional name for the district between Ribblesdale and Great Whernside. *Cravescire* 'Craven shire' 1086, *Craven(a)* from 1134, *Crawyn*, *Cravyn* 15th. A Celtic name probably related to the Italian p.n. Cremona in the sense '(wild) garlic district', cf. W *craf*, Ir *creamh* earlier *crem*, Gk *κρόμμυον* 'onion', OE *hramse* 'ramsons' < IE **qremus-*. YW vi.1, Chantraine I.586, Krahe, *Sprache der Illyrier* i.104, Pokorny 580, GPC 575.

CRAVEN ARMS Shrops SO 4382. *The Craven Arms* 1833 OS. Originally an inn named after the Earl of Craven who created and planned the present town as a result of the railway junction developed here in the late 1840's. Raven 64.

CRAWCROOK T&W NZ 1363. 'The nook of land frequented by crows'. *Crau- Crawcrok(e)* [1183]c.1320–1530×1, *Kraukruke* 1242×3, *Craucruc* 1303, *Crauwecrok'* 1311, *Crawecroke* 1369–1498, *Crawcrooke* 1616–47. OE **crāwe** + ***crōc**. NbDu 57.

CRAWLEY Hants SU 4234. 'Crow wood or clearing'. *Crawanlea* [909]11th S 376, [963×75]12th S 827, *Crawelie* 1086, *Craweleia -le(ye) -legh(e)* 1208–1316, *Craule* 1316. OE **crāwe**, genitive pl. **crāwan**, + **lēah**. The expected genitive pl. would be *crāwena* but traces of an *-an* genitive pl. are found in late WS. This might be genitive sing *Crāwan* of the OE feminine pers.n. *Crāwe* and so 'Crawe's wood' but the bird-name seems more likely. The boundary of the inhabitants of Crawley is *Craweleainga mearce* 909 S 376, 'the mark of the *Craweleaingas*, the people who live in Crawley'. Ha 60, Gover 1958.175, Brunner §276 A.5.

CRAWLEY Oxon SP 3412. 'Crow wood or clearing'. *Croule* 13th cent., *Craule* 1227–c.1384, *Craw(e)le* 1285–1387. OE **crāwe** + **lēah**. O 314, L 205.

CRAWLEY WSusx TQ 2736. 'Crow wood or clearing'. *Crauleia* 1203, *Crawele(y)* 1248, 1279, *Crawley* from 1272. OE **crāwe** + **lēah**. Sx 261, SS 66.

North CRAWLEY Bucks SP 9244. *North Crawele* 1388. ME adj. **north** + p.n. *Crauelai* 1086, *Craule* 1151×4–1404, *Crawley* from 1432, 'crow wood or clearing', OE **crāwe** + **lēah**. Also known as Great Crawley, *Magna Crawele* 1197, for distinction from Little Crawley SP 9245, *Parua Crawle* 1202. The explanation of 'North' is unclear since North or Great Crawley lies S of Little Crawley. Bk 34, L 205.

CRAWLEY DOWN WSusx TQ 3437. 'The downland belonging to C'. *Crauledun'* 1272, *Crawleydone* 1437. P.n. CRAWLEY TQ 2736 + ME **doun** (OE *dūn*). Sx 284.

CRAWLEYSIDE Durham NY 9940. 'Crawley hill-side'. P.n. Crawley + OE **sīde**. Crawley, *Crawlawe* 1418, *le Crawlawe* 1528, is 'crow hill', OE **crāwe** + **hlāw**.

CRAWSHAWBOOTH Lancs SD 8125. 'Crawshaw vaccary'. *Crawshaboth* 1507. P.n. Crawshaw + ME **bōthe** 'a temporary shelter, a cow-house, a herdsman's hut'. Crawshaw, *Croweshagh* 1324, 1325, 'crow wood', OE **crāwe** + **sceaga**, was already referred to as a vaccary in 1324. La 92.

CRAY NYorks SD 9479. *Crei* 1202, 1241, *Cray(e)* from 1499. Originally a stream name from PrW ***crei** 'fresh', W *crai*. YW vi.116, Jnl 1.45, GPC 578.

Foots CRAY GLond TQ 4770. 'Fot's Cray'. *Fotescraei* c.1100, *Votescray(e)* 1242×3. ME nick-name *Fōt*, referring to *Goduin' fot* who held the manor in 1066, genitive sing. *Fōtes*, + r.n. *Cray*. S 864 mentions the enclosure of the people who live on the Cray, *Cræg sætena haga* 987. KPN 83, PNK 18, RN 103, SPN 85, GPC 578.

North CRAY Kent TQ 4972. *Northcraei* c. 1100, *Northcray* from 1254. OE adj. **north** for distinction from St Mary CRAY TQ 4767, *Sudcrai* 1086, + p.n. *Craie* 1086, as in River CRAY. PNK 19.

River CRAY Kent 'Fresh water'. *Cræges æuuelma* 'source of the Cray' [798]12th S 1258, *Cræges æuulma* [791]12th S 1613, *(of, on, andlang) Crægean* [814]10th S 175, *Crale* [c.1200] c.1260, *Crey*, *Cray*, *Crea* 16th. PrW ***crei**, W *crai* 'fresh, new' in the sense 'pure, clean'. This word is thought to be identical with Middle Breton *crai* 'sour, trop fermenté' and may mean rather 'fermenting, rising river' referring to its aptness to flood. Identical with Afon Crai as in Abercrai Powys SN 8928, *Cray* 1578, and CRAY NYorks SD 9479. KPN 83, RN 103, Jnl 1.45, GPC 578.

St Mary CRAY GLond TQ 4767. 'St Mary's Cray'. *Creye scē Marie* 1257, *Seynte Mary Crey* 1270. Saint's name Mary from the dedication of the church + p.n. Cray as in River CRAY Kent. Also known as *Svdcrai* 'south Cray' 1086 for distinction from North CRAY TQ 4972. PNK 20.

St Paul's CRAY GLond TQ 4768. *Craye scī pauli* 1254, *Craye Paulin(i)*, *Paulinescreye*, *Craypaulin* 13th., *Paulscray* 1610. Saint's name *Paul*, short for Paulinus, from the dedication of the church, + p.n. *(aliam) Craie* 'the other C' 1086, as in River CRAY Kent. Also known as *Rodulfes craei* c. 1100. PNK 23.

CRAYFORD Kent TQ 5175. 'Ford across the river Cray'. *Creiford'* 1199, *Craiford* 1202, 1350, *Crayford* from 1354, *Creford* 16th. R.n. CRAY + OE **ford**. Crayford has been identified since the 16th century[†] with *Crecganford* 9th ASC(A) under year 457, c. 1120 ASC(E) under year 456, the site of the battle with the British in which Hengest and Æsc won the overlordship of Kent. On linguistic grounds, however, the identification cannot be upheld unless the Chronicle forms are corrupt. OE *Crecganford* would be expected to give ME **Creggeford*, ModE **Creg-* or **Credgeford*. PNK 29, KPN 83, Palæstra 147–8.47, TC 79, Origins 60–61.

CRAYKE NYorks SE 5670. 'The rock'. *Crec* [685]17th S 66, *Creic* [10th]11th S 1660, c.1000, 1086–1229, *Creik*, *Creyk* 1227–1364, *Crake* 1440, 1470, 1577. PrW ***creig**. The village site is a commanding one on the summit of a conspicuous outlying hill of the Howardians. It was given to St Cuthbert by king Egfrid in 685 to provide a resting place on the road to York. Crayke remained an outlying part of County Durham until 1844. YN 27, Jnl 1.45.

CRAYS HILL Essex TQ 7192. *Crays Hill* 1805 OS. Surname *Cray* + ModE **hill**. The same surname occurs in Ramsden Cray (lost ½ mile N of Crays Hill), *Rammesden(e) Cray*, *Greye*, *Creye*

[†]The identification is first made in MS notes by Robert Talbot (d.1558) in MS C of the Anglo-Saxon Chronicle, 'nunc Creford non longe a Dartford'.

1254–1333, p.n. Ramsden as in RAMSDEN BELLHOUSE and HEATH TQ 7194, 7195, + family name of Simon de Craye 1248. Ess 168.

CRAY'S POND Hants SU 6380. *Gray's Pond* 1812. Surname *Gray* + ModE **pond.** O 54.

CRAZE LOWMAN Devon SS 9814 → UPLOWMAN ST 0115.

CREACOMBE Devon SS 8119. 'Crow valley'. *Crawecome* 1086, *Craucomb* 1254, *Creu- Crew(e)comb(e)* 1284–1422, *Croucombe* 1291. OE **crāwe** + **cumb.** D 379 gives pr [kreikəm] which derives from an early monophthangisation of ME *au* > *ā*, cf. Nomina 13.109–14.

North CREAKE Norf TF 8538. *Northcrec* 1211. OE adj. **north** + p.n. *Creic(h)* 1086, *Crech* 1190, *Cre(i)c* 1196, 'the cliff or rock', PrW ***creig**. The reference is unclear; the two Creakes lie on the river Burn which here forms a narow defile between high ground. Perhaps the reference is to some exposure of the underlying chalk in the river valley, or to the much ploughed out Iron Age camp on Bloodgate Hill TF 8536. Much levelling of the ancient landscape has occurred. 'North' for distinction from South CREAKE TF 8536. DEPN.

South CREAKE Norf TF 8546. *Suthcreich* 1086. OE adj. **sūth** + p.n. Creake as in North CREAKE TF 8538. DEPN.

CREATON Northants SP 7072. Partly uncertain. *Crap- Creptone* 1086, *Creton(e)* 1086–1484, *Creiton* 1202, *Creyton* 1526, 1563, *Great Creaton* 1657. Possibly PrW ***creig** used as a p.n. + **tūn** 'the settlement at Creig, the hill'. The village lies on a ridge. The difficulty is to account for the DB spellings with *p*; this is sometimes written erroneously for wynn (w), þ (th), f, r, or w, none of which are likely graphies for the voiceless palatal spirant [ç] which would have developed in this position and subsequently been lost. Nth 68, Jnl 1.45.

CREDENHILL H&W SO 4543. 'Creoda's hill'. *Credehulle* 1067×71–1274, *Cradenhille, Credenelle* 1086, *Credenhulla -(e)* 1160×70–1385. OE pers.n. *Creoda*, genitive sing. *Creodan*, + **hyll**. The reference is to the large hill that stands alone N of the village rising to 720ft. crowned with an Iron Age hill-fort occupied c.400 BC–c.AD 75. He 63, Thomas 1976.147.

CREDITON Devon SS 8300. 'Settlement on the river Creedy'. *Cridie* [739]11th S 255, *(æt) Cridiantune* 930 S 405, *Crideton* 933, *Crydiatun* 974 S 795, *Cridiantun(e)* c.1000 ASC(C) under year 977, 980×8 K 1334, 1046, *(in to) crydian tune* 1008×12 S 1492, *Chridia- Crediatone* 1084, *Critetone* 1086, *Cridinton* 1175, *Crieton* 1181, *Cridi(n)- Cride- Crydy- Criddington* 13th cent., *Credington* 1274, *Crediton* 1284, *Kyrtone* 1380, *Kirton* c.1550, 1637, *Curton, Kerton* 17th. Crediton was originally the bishop of Devon's seat and the church is referred to as *Cridiensis ecclesia* 933 S 421. R.n. Creedy as in CREEDY PARK SS 8301 + OE **tūn**. D 404 gives pr [kəːrtən], 402.

CREECH 'A mound, a hill, a barrow'. PrW ***crüg**.

(1) ~ GRANGE Dorset SY 9182 > GRANGE HEATH SY 9083.

(2) ~ HEATHFIELD Somer ST 2726 → MONKTON HEATHFIELD ST 2526.

(3) HILL Dorset SU 0413. *Critchill* 1838. P.n. CREECH + pleonastic ModE **hill**. A 300ft. hill on the boundary of three parishes.

(4) ~ ST MICHAEL Somer ST 2725. 'St Michael's Creech'. *Creech S^t. Michael* 1809 OS. P.n. *Crice* 1086, *Criche* [1091×1106–1269]14th, *Cruche* 1157, *Criz* [1303]14th, *Crich, Crych(e)* [14th]14th, [1369] Buck, *Creche* 1610, PrW ***crüg** 'a prominent hill', + saint's name *Michael* from the dedication of the church. As Dr Gelling points out, the name is a puzzle since the hill on which the village stands is hardly striking enough to be called a *crüg*. The charter form *collem qui dicitur britannica Cructan apud nos Crycbeorh* 'the hill called *Cructan* ('Creech on the river Tone') in Welsh, *Crycbeorh* ('hill called *Cryc*') by us' [672 for 682]16th S 237 does not refer to Creech St Michael but to a prominent hill at ST 2525 on the S side of the Tone and Black Brook, currently occupied by the Creech Castle Hotel, a 19th cent. construction on the site of a previous house called Creechbarrow. 'St Michael' for distinction from EVERCREECH ST 6438. According to VCH vi.17 the 17th cent. form of the name, *Michael Creech*, is a variant of *Muchel* or *Michel Creech* 'great Creech' for distinction from *Little Creech*. The village occupies a low headland overlooking the Tone. DEPN, Gelling 1998.82.

(5) East ~ Dorset SY 9382. *Est(e)crich(e) -crych(e)* 1337–1546, *Estecreche* 14th, *East Cretch* 1586. ME adj. **est** + p.n. *Crist, Criz, Cric* 1086, *Cri- Crych(e)* 1224–1545, *Crech(e)* 1204, 1244. The reference is to the remarkable cone-shaped hill 637ft. high called Creech Barrow, *Crechbarrow* 1610 with pleonastic ModE **barrow** (ultimately from OE *beorg*), also referred to as *magnum montem* 'the great hill' 14th. 'East' for distinction from West Creech SY 8982, *Crich(e)* 1319–1509, *West(e) Crych(e)* [1324]17th, 1459, *West Creche* 1577. Do i.89, 91, 96.

CREED Corn SW 9347. '(Church of) St Crite'. *Sancta Crida* c.1250, 1310, *(Sancte) Cride* 1275, 1291, 1378, *Crede* 1509, *Creede* c.1570. Saint's name *Crite* (Latin *Crida, Creda*) from the dedication of the church. Cf. SANCREED SW 4229. PNCo 74, Gover n.d. 439, DEPN, CMCS 12.60.

CREEDY PARK Devon SS 8301. *Creedy House* 1809. R.n. Creedy + ModE **park, house**. Creedy, *(on, oþ) Crydian* [739]11th S 255, *Crydian (Bricge)* 997, *(on, oþ) Cridian* [739]11th S 255, *Cridia* 1018–1244, *Cri- Credie* 1086, *Cride* 16th cent., is OE *Cridie*, a weak feminine noun on OW **Crīdī* or **Criðī* from Brit **Critīo-* or **Cridīo-* 'winding' or 'swinging river'. The root is probably cognate with Latin *cardo* 'hinge', W *cerdded* 'go', *cerdd* 'a going', OCo *kerd*, ultimately from a root **(s)qer-* 'jump, leap, stride', cf. Greek σκαίρω 'dance'. D 4, RN 103, LHEB 286, 673, GPC 465.

Lower CREEDY Devon SS 8402 → UPTON HELLIONS SS 8403.

CREEKMOUTH GLond TQ 4582. 'Mouth of the creek'. ModE **creek** + **mouth**. A modern industrial development at the mouth of Barking Creek. In 1323 the creek is referred to as *Fletesmouthe de Barkinnge*, 'the mouth of Barking fleet', ME **flete** (OE *flēot*), genitive sing. **fletes**, + **mouthe** (OE *mūtha*). Ess 90.

CREETING ST MARY Suff TM 0758. 'St Mary's C'. *Creting Sancte Marie* 1254. P.n. *Cratingh -inga(s), Gratinga, (in) Cra- Gratingis* 1086, *Gratingis* [1157]1378, *Cretinges* 1199–1219, 1568, *Cr- Gretinge* 1210×12, *Cretinges, aliam Cretinges* 1212, *Cretingges* 13th, *Creting* 1610, 'the Crætingas, the people called after Cræta', OE folk-n. **Crǣtingas* < pers.n. **Crǣta* + **ingas**, + saint's name *Mary* from the dedication of the church and for distinction from Creeting St Peter TM 0758, *Creting Sancti Petri* 1254, the lost Creeting All Saints, *Creting Omnium Sanctorum* 1254, earlier *Stepelcreting* c.1200, 'C with the tower', and Creeting St Olave, *Creting Olaui* 1254. DEPN, ING 53, Baron.

CREETON Lincs TF 0120. 'Cræta's farm or village'. *Cretone -tun(e)* 1086–1241, *Crectone* 1166, *Cretton'* 1202, *Criton* 1219. OE pers.n. **Crǣta*, **Crēta*, + **tūn**. DEPN, Perrott 175.

Long CRENDON Bucks SP 6908. *Long Crindon* 1626. ModE adj. **long** + p.n. *Credendona* 1086, *Crehendon(a)* [c.1145]c.1300, 1186, *Creendon* 1175–1230, *Creindon* 1204, *Crendon(a)* from 1155 with variants *Crun- Cro(y)n- Crayn-*, 'Creoda's hill', OE pers.n. *Creoda*, genitive sing. *Creodan*, + **dūn**. The modern affix is merely descriptive of the shape of the village but may have been adopted for distinction from GRENDON UNDERWOOD SP 6820. Bk 122, Ca addenda, L 150.

CRESSAGE Shrops SJ 5904. '(The settlement) at Christ's oak-tree'. *Cristesache* 1086, *Cristesech(e)* c.1200–1322 with variants *-ess- -is-* and *-hech -eck, Cry- Crisseg(g)e - ech -egh* 1271–1627, *Cressedge* 1590, 1723, *Cressage* from 1535. OE *Crist*, genitive sing. *Cristes*, + **ǣċ**, dative sing of **āc**. Popular tradition has grown up around this name maintaining that it was an oak-tree, no longer extant, under which St Augustine once preached, part of a medieval forest already cleared in the 17th cent. It is said that a cutting from the original tree survives at Lady Oak 5603, a young oak supporting the hulk of an ancient one, but there were other trees in the parish called *the Cursed Oaks* 1433 which may be a survival of *Cristes æc*. Sa i.102, ii.139, Raven 64, Morton 50–3.

CRESSING Essex TL 7920. 'Place or brook where cress grows'. *Cressyng(e)* 1136, 1323, *Cressing* from 1227, *Ki- Kersing(e) -inges -yng(es)* 1185–1414, *Cursing(e)* 1602–78, *Crissing* 1313, *Cressen* 1706. OE **cærsing**, ***cresing**. Ess 285 gives pr [krisn], ING 189.

Great CRESSINGHAM Norf TF 8501. *Great Kersingham* 1264. ME adj. **gret** + p.n. *Cressin(c)ga- Cresinga- Gresingham* 1086, *Kersingeham* 1168, 1200, *Cressingham* 1200, 'the homestead of the Cærsingas or of the cress-beds' or 'homestead called or at Cærsing, the cress-bed'. OE folk-n. **Cærsingas* 'the people of the cress-beds' < ***cærsing**, genitive pl. **Cærsinga*, or ***cærsing** used as a p.n., locative sing. ***cærsinġe**, + **hām**. 'Great' for distinction from Little CRESSINGHAM TF 8700. DEPN, ING 133, 171 which gives pr [-ndʒ-], BzN 1967.367.

Little CRESSINGHAM Norf TF 8700. *Parva Cresingham, (in) parvo Cressingaham* 1086. ModE adj. **little**, Latin **parva, parvus**, + p.n. Cressingham as in Great CRESSINGHAM TF 8501. ING 133.

CRESSWELL 'Cress spring'. OE **cresse** + **wella**. L 31.

(1) ~ Derby SK 5274. *Kressewella -(e)* 1176–16th cent., *Kersewella* 1154×89, *Cres(s)well(e)* from 1290. Db 256.

(2) ~ Northum NZ 2993. *Kereswell* 1234, *Kercewell* 1255, *Carswell* 1450, *Cres(s)e- Crassewell* 13th cent. NbDu 57, L 31.

(3) ~ Staffs SJ 9739. *Cressvale* 1086, *Cressewella* 1190, *Cresswall* 1611. DEPN, L 31.

CRETINGHAM Suff TM 2260. 'The homestead of the Greotingas, the gravel people'. *Gretingaham, gratinge ham* 1086, *Gretin(c)geham* [1086]c.1150, *Gretingham* 1142–1336, *Cretyngham -ing-* 1327–1639. OE folk-n. *Grēotingas* < **grēot** + **ingas**, genitive pl. *Grēotinga*, + **hām**. The soil and subsoil here is clay: the name *Grēotingas* means 'the gravel people, the people who live at *Greoting*, the gravelly place', OE ***grēoting** < *grēot* + *ing*[2]. Whether or not this was some location on the r. Deben, it is interesting to note that *Grēotingas* is an exact parallel to the 4th cent. folk-n. *Greutingi*, a by-n. for the Ostrogoths from the sandy steppes of Southern Russia. DEPN, Schönfeld 113.

CREWE 'Weir, fish-trap'. PrW ***criu**. Che iii.9–10.

(1) ~ Ches SJ 4253. *Creuhalle* 1086, *Cryu* [1090×1101]1280, 1150, *Crue* 13th–1609, *Crewe* from 1326. From a fish-trap or weir in the river Dee. Che. iv.73 gives pr [kru:].

(2) ~ Ches SJ 7055. *Creu* 1086, *Crewe* from 1297, *Cru(w)e* 1220–1586. The modern borough takes its name from a fish-trap or weir in Valley Brook or its tributary near CREWE HALL SJ 7354. Che iii.9 gives pr [kru:].

(3) ~ HALL Ches SJ 7354. *Crewe Hall* 1579. Also known as *Great Crewe* 1573 for distinction from Crewe Green SJ 7255, *Little Crewe* 1573. P.n. CREWE SJ 7055 + ModE **hall**. Che iii.10.

CREWKERNE Somer ST 4409. 'House at or called Cruc'. *Crucern* [873×88]11th S 1507, *Crukerne* 1266, *Crokethorne, Crewkern Hund.* 1610, *Crewkern or Crook horn* 1750 map. P.n. *Crvche* 1086, *Cruke* 1225, PrW ***crüg** 'the hill', + OE **ærn**. The reference is either to the hill site of the town itself or to nearby Bincombe Hill. DEPN, VCH iv.4.

CREWS HILL GLond TQ 3199. *Crews Hill* 1819. Surname *Crew* as in Sarah Crew 1742 + **hill**. Mx 74.

CRIBBA HEAD Corn SW 4022. Probably 'crests, ridges'. *Cribba Head* 1888. Co **krib** 'comb, crest', pl. **kribow**. Frequent in the names of coastal rocks. PNCo 75, CPNE 70.

CRICH Derby SK 3554. 'The hill'. *Cryc* 1009, *Crice* 1086, *Crich(e) -y-* from 1228, *Cruch(e)* 1149×59–1544. PrW ***crüg** < Brit **crouco-*, late Brit **crūgo-*. This name was borrowed after Brit *ū* had become PrW *ü*, that is early 6th cent. at the earliest. Db 436 gives pr [kraitʃ], Jnl.1.46, L 138.

CRICHEL 'Hill called *Crich*'. PrW ***crüg** 'a mound, a hill, a barrow' taken as a p.n. + pleonastic OE **hyll**. Identical with CHURCHILL Avon ST 4459 and CREECH HILL SU 0413. Studies 1931.45.

(1) Long ~ Dorset ST 9710. *Langecrechel* 1208, *Longakerchel* 1258, *Langecruchel -crich-* 1268, *Lang, Longe Kurchel* 1280, *Langekurchill* 1348, *Lang(e) Crychell* 1494, *Long Creechill*. ME adj. **lang, long** + p.n. *Circel* 1086, *Crechel* 1204, *Kerchel* 1208–51, *Curchel* 1244, *Crouchull'* 1332. The reference is to Crichel Down which rises to 347ft. SW of the village. Do ii.274, 268.

(2) Moor ~ Dorset ST 9908. 'The part of C at the marshy ground'. *Mor Kerchel(l)* 1212–after 1290, *Mour Curchil* 1328–1564 with variants *Mor(e)-* and *-kurchyl* etc., *Moore Crechill* 1541. ME **more** (OE *mōr*) + p.n. CRICHEL as in Long CRICHEL. Moor Crichel lies downstream from Long Crichel and close to the low lands beside the r. Allen. Do ii.140.

CRICK Northants SP 5872. 'The hill'. *Crec* 1086–1331 with variant *Crek, Kreic* 1202, *Creyk* 1300–1330, *Creek(e), Creake* 1340–1583, *Cricke* 1598. PrW ***creig**. The reference is either to Crack's Hill SP 5973, *Crick Hill* 1839, ½m NW of Crick, or to the ridge on which the village itself stands. Nth 68, Jnl 1.45.

CRICKET ST THOMAS Somer ST 3708. 'St Thomas's C'. *Cruk Thomas* 1291, *Creket Thomas* 1610, *Cricket St Thomas* 1811 OS. P.n. Cricket, *Cruche* 1086, *Cruket* [n.d.] Buck, PrW ***crüg** 'hill' + OFr diminutive suffix **-ette** + saint's name *Thomas* from the dedication of the church and for distinction from Cricket Malherbe ST 3611, 'C held by the Malherbe family', *Cryket Malherbe* 1320, earlier simply *Crvchet* 1086, *Cruket* 1201, 1242, PrW ***crüg** + OFr diminutive suffix **-ette**. The original names of the two places were *Cruc* and *Crucet* 'little *Cruc*'. The topography suggests that *crüg* here refers to the extensive area of high ground reaching to 730ft. at Windwhistle ST 3809. DEPN.

CRICKHEATH Shrops SJ 2923. Possibly 'the heath at Crick'. *Cruchet* 1272, *Crukin* 1302, *Cru(c)keyth* 1303, 1392, *Cruckheyth* 1400, *Krukieth* c.1510–23, c.1565–81, *Crickieth* 1586, 1609, *Crickett* 1672, *Crickheath* late 16th. The earliest sps represent PrW ***crüg** 'a hill' + diminutive suffix OE **et**, W **yn** cf. MW *crucyn* 'a little hill'. But the other forms bear comparison with Criccieth, Gwyn SH 5038, 'the hill of the bondmen', W *crug* + *caeth*, pl. *ceith*. The modern form has been influenced by ModE **heath**. DEPN, O'Donnell 101, Morgan 1997.24.

CRICKLADE Wilts SU 1093. Partly uncertain. *(to) crecgelade* [914×9]1562 BH1, *(To) croccegelate* [914×9]1204×15 BH2, *(to) croccegelade* [914×9]c.1225 BH3, *Crecca gelad* c.925 ASC(A), *Creaccgelad* c.1000 ASC(B), *Creocc gelad* c.1050 ASC(D) all under year 905, *Crecalad* c.1200 ASC(F) under year 1015, *Cricgelad(e)* c.1050 ASC(D), *(æt) Cræcilade* c.1120 ASC(E), *Crecalade* c.1050 ASC(C) all under year 1016, *(into) Cracgelade* [c.975]14th, *Crocgelad* [1008]13th S 918, *Croc(i), Croccel, Crocglad* 1016×35 Coins, *Crec(elā), Croc, Creccelad, Crecca, Crecel, Crecla* 1042×66 Coins, *Cri(c), Criic* 1066×87 Coins, *Crichelade* 1086, 1177, *Crikelade* 1190–1290, *Crickelade* 1205–1397 with variants *Kricke-* and *Crikke-, Creyklade* 1311, *Creklade* 1446, *Cricklett* 1637. Unidentified element + OE **ġelad** 'a river crossing', particularly one liable to be difficult owing to flooding as happens here on the Thames. The specific is usually said to be PrW **creig* 'a rock' possibly referring to Common Hill SU 0893, an isolated hill rising to 373ft., but the sps are too varied to support this. Drayton gives an interesting folk-etymology *Greeklade* in *Polyolbion* relying on the tradition dismissed by Selden that a colony of Cambridge scholars at one time taught Greek here before migrating to Oxford, a folk-etymology which may, however, go back to the 11th cent. form *Crecalade*, OE *Crēcas* 'the Greeks', genitive pl. *Crēca*. Wlt 42, Nomina 17.10–11 and map p.17, Watts 2001.

CRIDLING STUBBS NYorks SE 5221. *Credlingstubbes* 1480–1607,

~ *Stubbs* 1675. A combination of two earlier p.ns., *Cred(e)ling, Cridelinc -yng* 1155×7–1341, *Cridlinc* [1173]14th, *Cri- Cryd(e)ling(e) -yng* from 1202, 'Cridela's place', OE pers.n. **Cridela* + **ing**², + p.n. *Stobbes* 1296, *Stubbes -ez* 1368, 1486, 'the tree-stumps, the clearing where tree-stumps are left', OE, ME **stubb** 'a tree-stump', ME pl. **stubbes**. YW ii.61, ING 77.

CRIGDON HILL Northum NT 8605. *Crigdon Hill* 1869 OS. In the absence of earlier forms no certainty is possible, but this might be PrW ***crüg** 'a hill' (Brit **crouco-*, late Brit **crūgo-*) + OE **dūn**. Studies 1931.47.

CRIGGLESTONE WYorks SE 3116. 'The settlement of *Cryc-hyll*'. *Crigest' -tone* 1086, *Crikeleston'* 12th–1462, *Cri- Cryg(e)leston* 1188×1202–1377. OE p.n. **Crȳc-hyll*, genitive sing. **Crȳc-hylles*, + **tūn**. For *Crȳc-hyll* see Long CRICHEL Dorset ST 9710. Studies 1931.47, YW ii.101.

CRIM ROCKS Scilly SV 8009. Crim + ModE **rock(s)**. Crim, *Crim* 1689, is possibly Co **krybynn** 'little ridge', diminutive of *krib* 'a comb, a crest', as in GRIBBIN HEAD Corn SX 0949, misunderstood as if late Co *Cribm* and hyper-corrected to *Crim*. PNCo 75.

CRIMCHARD Somer ST 3109 → South CHARD ST 3205.

CRIMPLESHAM Norf TF 6503. 'Crympel's homestead'. *Crepelesham* 1086, *Crimplesham* 1200, *Crimplelesham* 1203, *Crunplisham* (for *Crim-*) 1291. OE pers.n. **Crympel*, genitive sing. **Crympeles*, + **hām**. DEPN.

CRINGLEFORD Norf TG 1904. 'Ford at Cringel, the round hill'. *Cringelforð* 1043 or 4, *Kringelforda* 1086, *Cringelford* 1191. OE ***cringol** or ON **kringla** possibly used as a p.n. + OE **ford**. ON **kringla** means 'a circle' and is used of the circular sweep of a river, a round hill, or some similar circular feature. The reference here is probably to the hill on which Cringleford stands. There may, however, have been an OE cognate adj. **cringol* 'twisting' which may have been applied to the ford itself, the road to which takes a marked twist. DEPN, PNE i.112, ii.7.

CRIPPLESEASE Corn SW 5036. *Cripples-ease* 1884. A 19th cent. inn and hamlet named from its position near the top of a hill. PNCo 75.

CRIPP'S CORNER ESusx TQ 7821. Surname *Cripps* recorded here c.1540 + ModE **corner**. Sx 330.

CROASDALE FELL Lancs SD 6857. P.n. Croasdale + ModE **fell** (ON *fjall*). Croasdale, *Crossedale* 1194×1211, 1423, *Crosdale* 14th, is the 'valley with a cross', OIr, late OE **cros** + **dalr**. The pedestal of an ancient cross remains near Woodhouse in Slaidburn on the road up the valley. YW vi.211.

CROCKENHILL Kent TQ 5067. 'Pottery slope'. *Crokornheld* 1388, *Crokerneheld* 1390, *Crokkenhill* 1471. OE **crocc-ærn** as in CROCKERNWELL Devon SX 7592 + **helde**. PNK 39.

CROCKERNWELL Devon SX 7592. 'Crockern spring'. *Crochewelle* 1086, *Crokkernewell* 1284, *Crok(k)ernewill* 1330, 1412. P.n. **Crockern* + **wylle**. **Crockern*, which occurs again in Crockern Tor and Crockern Farm in Lydford, *Crock- Crokkerntor* 1481, is OE **crocc-ærn** 'a pottery'. D 428, 193.

CROCKERTON Wilts ST 8642. 'Crocker's farm'. *Crokerton* 1249–1351, *Crock-* from 1369. The evidence suggests this is surname *Crocker* + ME **toun** (OE *tūn*). A John *le Crocker* occurs in the parish in 1268 and a Stephen *le Crokker* in 1289. The surname Crocker is ME *crokkere* 'a potter' (OE **croccere*). If the settlement goes back to OE times it may represent OE **crocc-era-tūn* 'the settlement of the potters'. Wlt 166.

CROCKHAM HILL Kent TQ 4450. *Cockham Hill* (sic) 1819 OS. Without more forms it is impossible to offer an etymology or to know which is the correct spelling. The 1819 form is suspiciously like Coakham Farm TQ 4449 the early forms of which are given as *Cob(b)ecumbe* 1232, 1313, 'Cobba's' or 'hill coomb', OE pers.n. *Cobba* or ***cobb(e)** 'a round lump, a cob'. PNK 75.

CROCKLEFORD HEATH Essex TM 0426. *Crokylfeld heth* 1538, *Crackleford Heath* 1768. P.n. Crockleford + ModE **heath**. Crockleford, *Crockeresford* 1206, *Cro(c)kel(e)ford(e)* 1308, 1323, 1544, is the 'potter's ford', OE **croccere** possibly used as a surname, genitive sing. **crocceres**, + **ford**. Ess 326.

CROCKLEY HILL NYorks SE 6246. No early forms. P.n. Crockley + ModE **hill**.

CROFT 'A small enclosed field; a small enclosure of arable or pasture land near a house, a curtilage'. OE **croft**.

(1) ~ Ches SJ 6393. *Croft* from 1212. La 98.

(2) ~ Lincs TF 5061. *Croft* 1086, c.1115. An ancient enclosure in the marshes. DEPN.

(3) ~ MARSH Lincs TF 5360. P.n. CROFT TF 5061 + ModE **marsh**.

(4) ~ -ON-TEES NYorks NZ 2809. *Croft super Teyse* 1252. P.n. *Croft* from 1086, + r.n. TEES. YN 282.

CROFT Leic SP 5195. 'The machine, the engine'. *Craeft* 836 S 190, *Crec, Crebre* (sic) 1086, *Creft'* 1139×47–1250, *Craft(e)* 1136×53–1590, *Croft* 1610. OE **cræft** 'strength, power, art, skill' used in the concrete sense 'a work of art' applied either to some lost engine or device, perhaps some kind of water mill, or to some feature of Croft Hill, the prominent ridge of ground N of the village. Lei 490.

CROFT AMBREY or AMBERY H&W SO 4466. *Croft Ambrey* 1832 OS. Apparently p.n. Croft SO 4565 + ModE **ambry** 'a store house, a treasury' used of the Iron Age hill-fort which contains a late 2nd cent. AD mound, perhaps the site of a sanctuary. Croft, *Crofta* 1086, *Croft* 1163, 'the enclosure', OE **croft**, may refer to the same hill-fort or to the village near the fort. *Ambry* seems an unlikely term for a hill-fort and may conceal an earlier name in *-bury* (OE *byriġ*, dative sing of *burh*) such as 'Anna's fortification'. Local tradition wrongly associates the camp with the British king Ambrosius. DEPN, VCH Herefordshire I.208, Thomas 1976.148, Bannister 54.

Old CROFT RIVER Norf TL 5098. *the river of Croft* 1606, *the Old Craft River* 1830. Named from the crofts or small-holdings abutting the river bank, cf. Croft Farm TF 4900 and *Littleport Crofts* 1606, *Lyttleport Craftes* 1672. ModE **croft**. The earlier name of the river was Well as in OUTWELL TF 5103, UPWELL TF 5002 and WELNEY TL 5293. Ca 9 gives pr [ould kra:ft], 226.

CROFTON WYorks SE 3817. 'Farmstead with a croft'. *Scroftune, Scrotone* 1086, *Croftun(e) -ton(a)* 12th–1634. OE **croft** + **tūn**. YW ii.113.

CROGLIN Cumbr NY 5747. 'Crooked torrent'. *Crokelyn* c.1140, 1361, *Krokelin, Crokeling* 1279, *Croglin(e) -lyn* from [c.1155]1200, *Crogelin(e) -lyn* 1195–1363. Originally the name of CROGLIN WATER NY 5646. OE ***crōc** + **hlynn**. Cu 183, SSNNW 421.

CROGLIN WATER Cumbr NY 5646. *aquam de Croglyn* c.1210, ~ *Crocling, Crokelin* 1279, *Croglyn beck* 1568, *watter of Croklinge* 1603. P.n. CROGLIN NY 5747 + ModE **water**. Cu 9.

CROMER Herts TL 2928. 'Crow pond'. *Croumere* 1191, *Crawemere* 1255–6, *Crowmere* 1259. OE **crāwe** + **mere**. Hrt 151.

CROMER Norf TG 2142. 'Crow pool'. *Crowemere* 13th, *Crowmere* 1297. OE **crāwe** + **mere**. No such pool now exists. DEPN.

CROMER POINT NYorks TA 0392. *Cromer Point* c.1855 OS. P.n. Cromer + ModE **point**.

CROMFORD Derby SK 2956. 'Ford by a bend'. *Crunforde* 1086, *Crumford -forth* 1204–1330, 16th cent., *Cromford(e) -forth(e)* from 1276. OE ***crumbe** + **ford**. The reference is to a right-angled bend in the river Derwent. Db 358.

CROMHALL Avon ST 6990. 'Bent nook'. *Cromhal(a) -hale* 1086–1592, *Cromale* 1086, 12th, 1227, *Crumhal(a) -hale -hall* 1187–1704. OE **crumb** + **halh**. The reference is to a small valley within which the stream makes a number of bends. *Crum(b)e* or *Crom(b)e* may in fact have been the name of the stream, 'the twisting one'. Gl iii.3, xi, RN 106, Jnl 1.46, L 103, 107, 110.

CROMHALL COMMON Avon ST 6989. *Cromhall Common* 1839. P.n. CROMHALL ST 6990 + ModE **common**. Gl iii.4.

CROMWELL Notts SK 7961. 'The crooked spring'. *Crunwelle* 1086, *-wella* 1178, *Crumwelle -well(a)* 1166–1675, *Crumbwell(e)* 1216–1346, *Cromwell* 1280. OE **crumb** + **wella**. Nt 185 records a local pr [krʌməl].

CRONDALL Hants SU 7948. '(At) the (chalk-)pits'. *(æt) Crundellan* [873×88]11th S 1507, *(æt) Crundelan* [955×8]14th S 1491, *(æt) Crundelom* [968×71]12th S 1485, *Crundelas* [973×4]12th S 820, *Crundele* 1086, 1205, *-dal(l)e* 1246–1332, *Crondall* from 1428. OE **crundel**, dative pl. **crundelum**. Ha 61, Gover 1958.108, DEPN which gives pr [krundəl].

CRONKLEY FELL Durham NY 8427. *Cronkley Fell* 1863 OS. P.n. Cronkley NY 8628 + ME **fell**. Cronkley, *Cronkley* 1863 OS, is named after Cronkley Scar, a steep crescent of rocky cliff overlooking a bend in the river Tees at NY 8329–30. It is identical with Cronkley Northum NZ 0252, *Crombeclyve* 1268, *Crumcliffe* 1298, *Cronkley* 1663, and Crunkly Gill NYorks. NZ 7057, *Crūbeclif -clive* 1086, 'crooked cliff', OE **crumb** + **clif**. NbDu 57, YN 133.

CRONKLEY Northum NZ 0252 → CRONKLEY FELL Durham NY 8427.

CRONTON Mers SJ 4988. Partly uncertain. *Crenton* 1198, *Growynton* 1242, *Crohinton* 1243, *Croenton* [13th]14th, *Cronington* 1251, *-in-* c.1310, *Croynton* 1322–32, *Crouington* 1246, *Crowynton* c.1301, *Crouwenton* 1333, *Crawynton* 1292, *Craunton* 1341. This might be 'farm, village where saffron grows', OE ***croging** 'saffron place' or ***crogen** 'growing with saffron' < *croh* + *-ing²* or *-en*, + **tūn**. La 107, Jnl 17.60.

CROOK 'Bend, nook, secluded corner of land'. OE ***crōc**, ON **krókr**.

(1) ~ Cumbr SD 4695. *Crok(e)* 1170×84–1647, *Crook(e)* from 1220. The reference is to the secluded location of the church and hall at SD 4594. We i.177, SSNNW 117.

(2) ~ Durham NZ 1635. 'Secluded corner of land'. *(le) Croke next Bra(u)ncepath* 1378–1457×8, *Croke* 1498, *Croyke* 1549, *Crooke* 1564, 1621. Crook is situated in a fold of hills remote from the mother church at Brancepeth. The early forms cited in NbDu 58 do not belong to this name.

CROOKFOOT RESERVOIR Cleve NZ 4331. P.n. Crookfoot + ModE **reservoir**. The reservoir is a modern construction on land referred to as *Crookfoot Hill, Crookfoot Dean* 1838 TA. Crookfoot means 'the foot of the crook' in which crook refers to an indentation or hollow in surrounding high ground.

CROOKHAM Berks SU 5464. 'Homestead, village at the bend or bends in the river'. *Croc(he)ham* 1086–1517 with variants *Crok(e)ham* 1170, *Crokam* 1517, *Crookham* 1713, 1761. OE ***crōc** + **hām**. Brk 188.

CROOKHAM Northum NT 9138. '(The settlement) at the bends' sc. in the r. Till. *Crutun* (sic for *Crucum*) 1242 BF, *Crucum* 1244, 1296 SR, *Crukum, Crocum* 14th cent., *Croukham* 1542. OE ***crōc**, locative-dative pl. ***crōcum**. NbDu 58.

CROOKHAM VILLAGE Hants SU 7952. A modern parish carved out of Crondall and Church CROOKHAM SU 8152, ModE **church** + p.n. *Crocham* 1200–75, *Crokham* 1257, *Croukham* 1327, *Croham* 1362, *Crookham* 1598, of uncertain meaning, either OE **crōc** 'a crook, a bend', **crōh** 'a nook' or even **croh** 'saffron' + **hām** 'village, estate' or **hamm** 'enclosure'. The precise origin and meaning of this name is obscure; if *crōh* is correct the reference might be to the valley S of Crookham Village, but the forms suggest *crōc*. Ha 61, Gover 1958.109, Brk i.188–9. DEPN.

Church CROOKHAM Hants SU 8152. 'C with the church'. A modern development. ModE **church** + p.n. Crookham as in CROOKHAM VILLAGE SU 7952.

CROOKLANDS Cumbr SD 5383. Possibly 'secluded lands'. *Crokelandes* c.1320, *Crockelands* 1692, *Crooklands* 1731. Alternatively this could be 'lands belonging to a family surnamed *Croke*'. The combination of *land* in the pl. with a surname is not infrequent in medieval records as a descriptive term for a land-holding but does not often survive into the post-medieval name stock. We i.96.

CROOME COURT H&W SO 8844. P.n. Croome + ModE **court**, an 18th cent. mansion by Lancelot Brown. Croome is the name of a large area of land W of the Severn including Croome D'Abitot SO 8844, 'C held by the d'Abitot family', *Crombe Dabetoth* 1275, Earls CROOME SO 8742, and Hill Croome SO 8840, *Hylcromba* [1038]18th S 1392, *Hilcrūbe* 1086, which stands on a steep hill. Earlier simply *cromman, (æt) crom(b)an* [969]11th S 1322, *Crūbe* 1086, *Crumba* 1208. Spellings in S 1322 may point to OE *(æt þæm) crumbum* 'at the bends', dative pl. of OE ***crumbe**, used as a district name referring to bends in the course of the Severn, or to OE **Cromban*, dative sing. of *Crombe*, from an assumed British r.n. **Crumbā*, **Crombā* 'crooked stream' referring to the brook which rises in Croome d'Abitot and runs through the other Croomes to join the Severn NW of Tewkesbury. Since, however, Brit **crumb* would normally give PrW **crumm* and since the medieval landscape here has been much affected by drainage canalisation the choice must remain uncertain. Croome D'Abitot was also known as *Molde Crombe* [1182]18th, 1340, 'Maud's C' from Maude de Crombe who held land here in 1182, *Cromb Osbern* 1349, 'Osbert's C' from one Osbert who held the manor in 1182, *Abbots Croome* 1535, a rationalisation of *Abitots*, the family who held the manor c.1150–1450, and *Clares Crome* 1584 from the Clare family who possessed it in the 16th cent. The identification and descent of the various Croome manors is a complex picture and may be followed in Hooke 179, 188, 274–77 and Forsberg 212. Wo 119, RN 106, PNE i.116, Pevsner 1968.126.

CROOME D'ABITOT H&W SO 8844 → CROOME COURT SO 8844.

Earls CROOME H&W SO 8742. 'C held by the earl' sc.of Warwick. *Erles Crome* 1495. Earlier simply *Crūbe* 1086, *Cromba* [1086]1190. ME **erle**, genitive sing. **erles**, + p.n. Croome as in CROOME COURT SO 8844. The manor passed to the Earls of Warwick before 1369. Also known as *Crombe-, Crumb Adam* 1255–1340 and *Crombe Simon* or *Simondis Crombe* 1310–97 from earlier 12th cent. owners. Wo 118 gives pr [krûm], Hooke 188.

Hill CROOME H&W SO 8840 > CROOME COURT SO 8840.

CROPREDY Oxon SP 4646. Partly uncertain. *Cropelie* 1086, *Cropperi(a) -r(y) -ri -rey(e) -rri* 1109–1367, *Croprithi* [c.1275]c.1450, *Cropredy* from 1390. The generic is OE **rīthiġ** 'a small stream', the specific either OE pers.n. **Croppa*, or **cropp** 'a cluster, a bunch, a sprout, a flower' perhaps referring to water-plants or the same word in the sense 'hump, hill, hill-top'. O 419 gives prs [krɔprədi, krɔpərdi], DEPN, PNE i.113.

CROPSTON Leic SK 5511. 'Kroppr's estate'. *Cropeston'* c.1130, 1299, 1301, *Croppeston* 14th cent., *Cropston(e)* from 1194, *Cropson* 1605. Scand by-n. *Kroppr*, ME genitive sing. **Croppes*, + OE **tūn**. Lei 408, SSNEM 374.

CROPSTON RESERVOIR Leic SK 5410. P.n. CROPSTON SK 5511 + ModE **reservoir**.

CROPTHORNE H&W SO 9944. Partly uncertain. *Cropponþorn, Croppeþorne* *[780]11th S 118, *Croppanþorn* [841]11th S 196, *Cropetorn* 1086, *Croppethorne* 1305–1428. Probably 'Croppa's thorn-tree', OE pers.n. **Croppa*, genitive sing. **Croppan*, + **thorn**. It has been suggested, however, that the specific is OE *cropp(a)* 'the sprout of a plant, a bunch of blooms, a cluster of berries', also 'the crop of a bird' and possibly 'hump, hill, hill-top' like ModE *crop* 'round swelling', ON *kroppr* 'lump on the body' used as a local name in *Landnámabók*. Possibly, therefore, 'thorn-tree of (the hill called) *Croppa*', perhaps referred to in the nearby lost name *Croppendune*, and comparable with Crapnell Farm Somer ST 5945, *Croppanhull* [705]14th S 274. Wo 119, PNE i.113.

CROPTON NYorks SE 7589. 'Hill-top settlement'. *Croptun(e)* 1086, 1167, *Cropton(a)* from c.1200. OE **cropp** + **tūn**. YN 78.

CROPWELL BISHOP Notts SK 6835. '(The portion of) C belonging to the (Arch)bishop' sc. of York. *Bishopcroppehill, Bissop Croppehull* 1280, *Croppehull Episcopi* 1316, *Crophilbisshop* 1330, *Cropwell Byshop* 1562. P.n. *Crophille -helle* 1086, *Crophulla -hille -hylle* 1178 etc., *Cropp(h)ill* 1226, 1230, 'the humped hill', OE **cropp**, ON **kroppr**, + **hyll**, + ME **bishop**. The reference is to a small but prominent round hill called Hoe Hill at SK 6736 (OE **hōh**). Nt 234–5, L 170.

CROPWELL BUTLER Notts SK 6837. '(The portion of) C belonging to the Butler family' of Warrington. *Croppill Boteiller* 1265, *Cropphil Boteler* 1276, *Cropwell Botler* 1538, ~ *Butler* 1684. P.n. Cropwell as in CROPWELL BISHOP SK 6835 + surname *Butler*. The manor was held by Ricardus *Pincerna* 'the butler' already in 1187. Nt 234–5.

CROSBY 'Village or farmstead marked by crosses'. ON **kross**, genitive pl. **krossa**, + **bȳ**. SSNNW 28.

(1) ~ Cumbr NY 0738. *Crosseby* 1123–1540. Cu 282.

(2) ~ Mers SJ 3198. Modern Crosby is a 19th cent. resort which developed at what was called in 1842 *Crosby-sea-bank* near the modern Waterloo Marine. The ancient settlements were at Great and Little CROSBY SJ 3299 and SD 3101.

(3) ~ COURT NYorks SE 3991. P.n. Crosby + ModE **court**. Short for *Crosseby et Cotunam* 1252, *Crosby Cote* 1928. Earlier simply *Cotun* 1086, *Cotem* 1088, '(the settlement) at the cottages', OE **cot**, dative pl. **cotum**. YN 208.

(4) ~ GARRET Cumbr NY 7209. 'Gerard's C'. *Crosseby Gerard* 1206, *Crosby Garret(t)* from 1585. Earlier simply *Crossebi -by* 1200–1388. The feudal affix is OFr(OG) pers.n. *Gerard* but the identity of this man is unknown. The church contains Anglo-Saxon fabric but there are no traces of cross sculpture. We ii.39.

(5) ~ -ON-EDEN Cumbr NY 4459. *Crosby Eden* 1642 from the river n. EDEN NY 5830. Earlier simply *Crossebi* c.1200, *Crosseby* 1278–1310. Cu 76.

(6) ~ RAVENSWORTH Cumbr NY 6214. 'Hrafnsvart's C'. *Crosbi(e) -by Rau- Raven(e)swart -su(u)art -svart* 12th–1366, *-swarth(e)* 1203–1394, *-swath* 1369–1652, *Crosby Ravensworth* from 1625. Earlier simply *Crossebi -by* 12th–1361, *Crosbi(e) -by* from 1294. The affix is the ON pers.n. *Hrafnsvartr*. We ii.154.

(7) ~ RAVENSWORTH FELL Cumbr NY 6010. *Crosby Ravensworth Fell* 1859. Earlier simply *the, ye Fells* 17th cent., *Crosby Fell* 1823. P.n. CROSBY RAVENSWORTH NY 6214 + ModE **fell** (ON *fjall*). We ii.159.

(8) Great ~ Mers SJ 3299. *magnam Crossby* c.1190, *Great Crosseby* 1246 etc. ME adj. **great**, Latin **magna**, + p.n. *Crosebi* 1086, *Crossebeyam* (Latin accusative) 1094, *-bi -by* 1177–1332, *Crossby* c.1190, 'Great' for distinction from the other half of the divided manor, Little Crosby SD 3101, *Little Crosseby* 1243, 1332, *Crosseby parua* 1322. La 118, SSNNW 28, Jnl 17.65.

(9) Low ~ Cumbr NY 4459. *Nethyr Croseby* 1422, *Low Crosbye* 1667. ModE adj. **low** (ME **nether**) + p.n. CROSBY-ON-EDEN NY 4459. 'Low' for distinction from High Crosby, *High Crosseby* 1292. Cu 76.

CROSBY Humbs SE 8711. 'Krok's village or farm'. *Cropesbi* 1086, *Crochesbi* c.1115, 1130, *Crosseby* 1206. ON pers.n. *Krókr*, genitive sing. *Króks* + **bȳ**. DEPN, SSNEM 43.

CROSCOMBE Somer ST 5944. Partly uncertain. *Correges cumb* [705]n.d. S 248, *Coristone* 1086, *Coriscoma* 1086 Exon, *Croscombe* 1610. Neither the forms of this name nor those of CORSCOMBE Devon ST 5105 support the usual explanation, OE **corfweġ** 'road through a pass or cutting' (< *corf* + *weġ*), genitive sing **corfweġes**, + **cumb**. The specific is probably rather the genitive sing. of an old r.n. identical with Curry in CURRY MALLET ST 3221. DEPN, L 88, 93.

CROSDALE FELL Lancs > CROASDALE FELL Lancs SD 6857.

CROSS Somer ST 4154. *Cross* 1817 OS.

CROSS DRAIN Lincs TF 1613. A drain in Deeping Fen running at right angles to the North and South Drove drains.

STONE CROSS ESusx TQ 6104. *the cross called Stonecrosse* 1563. ME **stone** + **cross**. SX 449.

CROSS FELL Cumbr NY 6834. *Cross Fell* 1608. Earlier *Fendesfeld* 1340, *Fendesfell* 1479, 'fiend's mountain', ME **fend** (OE *fēond*), genitive sing. **fendes**, + **fell** (ON *fjall*). 'Formerly called *Fiends-Fell* from evil spirits which are said in former times to have haunted the Top of this mountain' Robinson (1709). The fiend may well have been the Helm Wind which comes down from Cross Fell. The cross from which the modern name derives was erected to give a Christian association to a mountain under the influence of evil powers. The same name also occurs in Tweeddale. Cu 243, lxxix.

CROSS GREEN Devon SX 3888. *Cross* 1809 OS. Possibly the home of *Johanna atte Crosse* 1333. ME **cross** possibly in the sense cross-roads + ModE **green**. D 180.

CROSS GREEN Suff TL 9852. *Cross Green* 1837 OS. Contrasts with nearby *Cooks Green* TL 9753 and *High Street Green* TM 0055, both 1837 OS.

CROSS HOUSES Shrops SJ 5407. *Cross Houses* 1675. A squatter settlement on the parish and township boundary at a T-junction, *the Crosse* 1635, 1653, 1758, of the Shrewsbury-Bridgnorth road and the road to Atcham. Sa ii.97.

CROSS IN HAND ESusx TQ 5621. *via cruce manus* 'street, road called *cruce manus*, hand's cross' 1547, *Crosse atte Hand, Crosse in Hand* 1597. Sx 406.

CROSS LANES NYorks SE 5264. *Cross Lanes* 1858 OS.

CROSSCANONBY Cumbr NY 0739. 'Crosby held by the canons'. *Crosbycanon'* 1393, *Crosbycannonby* 1535, *Croscanonby* 1552. Land here was granted along with the church to the canons of Carlisle. For earlier forms see CROSBY NY 0738. Part of the shaft of a 10th cent. stone cross carved with biting dragons still survives in the church. Cu 282, SSNNW 28.

CROSSDALE STREET Norf TF 2239. 'C hamlet'. *Crossdale Street* 1838 OS. P.n. Crossdale + dial. **street**. Contrasts with nearby *Church Street* 1838 OS at Northrepps TG 2439.

CROSSENS Mers SD 3719. 'Cross headland'. *Crossenes* c.1250, 1323, *Crosnes* 1327, 1341, *Crossons* 1550. ON **kross**, genitive pl. **krossa**, + **nes**. La 126, SSNNW 118.

CROSSGILL Lancs SD 5562. No early forms but this is a compound of OIr, late OE **cros** 'a cross' + **gil** 'a ravine'. The remains of a cross survive on the road to Brookhouse.

CROSSINGS Cumbr NY 5076. No early forms. The reference is to a cross-roads.

CROSSLANES Shrops SJ 3218. *Crosslanes* 1836 OS. A cross-roads.

CROSSMOOR Lancs SD 4438. No early forms.

CROSSWAY GREEN H&W SO 8468. *Crossway Green* 1831 OS.

CROSSWAYS Dorset SY 7688. A modern development at a cross-roads on Knighton Heath.

CROSTHWAITE Cumbr SD 4391. 'Clearing with a cross'. *Crosthwait(e)* from 1187×1200 with variants *-tweit -tw(h)eyt -thwayt(e) -thweyt*. ON **kross** + **thveit**. We i.80, SSNNW 118, L 211.

CROSTON Lancs SD 4819. 'Farm, village with a cross'. *Croston* from 1094, *Croxton* [1189×1202]1268. OIr, late OE **cros** + **tūn**. La 136, Jnl 17.77.

CROSTWICK Norf TG 2515. 'Cross clearing'. *Crostueit* 1086, *Crosthweyt* 1302. ON **kross** + **thveit**. DEPN gives pr [krɔsik].

CROSTWIGHT Norf TG 3430. 'Cross clearing'. *Crostwit* 1086, *Crostweit -tweyt* 1208–1460, *Crosthueit -thweyt* 1198–1433, *Crostwight -wick* 1650. ON **kross** + **thveit**. The preservation of the diphthongal spelling suggests that the name was coined before c.900. Nf ii.148 gives local prs [kɔsit, kɔːsit].

River CROUCH Essex TQ 8095. *Crouch* from 1576. A back-formation from a lost p.n. in *crouch* 'a cross', cf. *Crouchefeld* 1375 in Rawreth TQ 7793 and *Whoulue(r)crouch* 1375 near Hullbridge

TQ 8194. The river was also known as *Borneham Water* 'Burnham water' 1535 and *the Burne* 1586. The original name seems to have been *Huolue* (*Huolne* c.1200) as in *Whoulue(r)crouch* supra and HULLBRIDGE, *Whoulnebregg'* for *Whoulue-* 1375, probably representing ME *wholve*, dial. *hull, hulve* 'a passage for water' < OE *hwealf* 'arch, vault, concave, hollow'. Ess 6, RN 107.

CROUGHTON Northants SP 5433. 'The settlement in the fork'. *Crevel- Criwel- Cliwetone* 1086, *Creulton* 1174×83–1292, *Crou(e)lton* 12th–1428, *Crow(l)- Crofton* 16th cent., *Croughton, Crolton* 1618. OE ***creowel** + **tūn**. The village is situated on a hill in the fork of two streams. Nth 51 gives pr [kroutən].

CROW HILL H&W SO 6427. *Crow Hill* 1831 OS.

CROW SOUND Scilly SV 9312. *Crawe Sound* 1570. P.n. Crow Rock at the entrance to the sound, *The Crow* c.1585, + **sound** 'a channel'. PNCo 75.

CROWAN Corn SW 6434. 'Church of St Crowan'. *Eggloscrauuen* [c.1170]13th, *Ecclesia Sancte Crawenne* 1238, *ecclesia de Sancto Crewano* 1201, *Sancta Crouwenna* 1269, *(Ecclesia de) Crewenne* 1291, *Crowan* 1610 Map. Co **eglos** 'church' + saint's name *Crowan* (Latin *Cruenna*) from the dedication of the church. She was believed to have come from Ireland along with St Breage and others. PNCo 75, Gover n.d. 587, DEPN.

CROWBOROUGH ESusx TQ 5130. 'Crow hill'. *Cranbergh* (for *Crau-*) 1292, *Crowbergh* 1390, *Croweborowghe, Crowbar(r)owe* 16th. OE **crāwe** + **beorg**. Sx 372.

CROWCOMBE Somer ST 1336. 'Crow coomb'. *Crauuancumb* [904]12th S 373, *Crawancumb* [978 for ? 968]12th S 806, *Crawecūbe* 1086, *Crauecombe* [late 12th]14th, *Crokam* 1610. OE **crāwe**, genitive pl. **crāwena, crāwan**, + **cumb**. DEPN, L 93.

CROWDECOTE Derby SK 1065. 'Cruda's cottage(s)'. *Crudecote* early 13th–1306, *Croudecot(e)* 1251–1417, *Crowth(e)cote* 1445, 1555, *Crowdecote* 1482. OE pers.n. **Crūda* + **cot**, pl. **cotu**. Db 365.

CROWDUNDLE BECK Cumbr NY 6530. P.n. Crowdundle + ModE **beck**. Crowdundle, *Crawdunda(i)le* 1601, *Crawdundale* 1616, *Craudundel* 1750, *Crowdundale* 1777, is probably 'Crawdon valley', p.n. **Crawdon*, 'crow hill', OE **crāwe** + **dūn**, + ON **dalr**. The earlier name of the beck was *Bocblencarn* (for *Bec-*) 1228, 'Blencarn beck', p.n. BLENCARN NY 6331 + **bekkr** in Celtic word-order. Cu 5, 215.

CROWDY RESERVOIR Corn SX 1483. P.n. Crowdy + ModE **reservoir**. The reservoir was built in 1973 on the site of Crowdy Marsh, *Crowdy Marsh* 1813, earlier *Crowdy upon the moore* c.1613, 'hovel in the marsh', Co ***krow-jy** 'hut, cottage' (*krow* + *chy*). PNCo 75, Gover n.d. 53, CPNE 73.

CROWFIELD Northants SP 6141. 'Open land frequented by crows'. *Crowefeld* 1287. OE **crāwe** + **feld**. Nth 60.

CROWFIELD Suff TM 1557. Probably 'the open land by the nook'. *Crofelda* 1086 OS, *Cropfeld* 1212, *Crofeld'* 1214×5, *Crosfeld* 1219, *Croffeld* c.1230, 1294, *Crofeud* 1273, *Crowefeld* 1441. OE **crōh** + **feld**. The place lies at the head of a shallow but marked valley. An alternative possibility would be OE **croh** 'saffron'. DEPN, Baron, Studies 1936.167.

CROWHURST ESusx TQ 7512. 'Wooded hill at or with a corner'. *(on) Croghyrste* [772]13th S 108, *Croherst -hurst* 1086–13th, *Crawe(n)hurst -hirst -herst* 1274–1316, *Crouherst -hurst* 1278–1342, *Crow(e)hurst -herst* 14th–16th cents. OE **crōh** + **hyrst**. The name was early re-interpreted as if 'crow wooded hill' sometimes with ME *crawene* < OE *crāwena*, genitive pl. of *crāwe*. Crowhurst lies in a well-marked bend of a river valley. Sx 502, Studies 1936.167, PNE i.113, SS 68.

CROWHURST Surrey TQ 3947. 'The wooded hill frequented by crows'. *Crouhurst* 1154×89–1321, *Cro(we)- Crau- Craw(e)-* 13th cent. OE **crāwe** + **hyrst**. Sr 315, L 197.

CROWLAND Lincs TF 2410. 'The land in a bend'. There are two types;

I. *Cruglond* [c.745]c.800, *Crugland, Cruulond, Cruwland* [c.745]9th–10th Felix, *Cruwland* [10th]c.1050 Guthlac, *Cruland* c.1030 Secgan–1331, *Crouland(e)* 1264–1303, *Crowland(e)* from 1402.

II. *Croiland* 1086–1526, *(insula) Croyland(ie)* 1179–1418, *Croylond* 1254–1535.

OE ***crūw, *crūg** 'a bend' + **land** in the sense 'land taken in from marsh'. The reference may have been to a bend in the r. Welland. Although OE **crūw, *crūg* is not attested it seems to have been an ancient p.n. element formed on the root **krū- /*krau-* seen in cognate Gmc languages, G *Kräuel*, Du *krauwel* 'a claw', Norw *kryl* 'a hump' < **krūli-* < **krūw-ila-*. The spelling *Crug-* has been explained as a Latinised form, the spellings *oi, oy* as representing ME [ui] derived from it. Both types I and II seem to have existed side by side for a long time, development of the name being influenced by the existence of the Med Lat word *croia, croa, crovus, crowus* 'a cruive, a fish-trap'. Payling 1940.12, Studies 1936.168, PNE i.117, Kluge-Seebold 410, EW 176, L 248–9, BzN 1981.285–92, 334–7.

CROWLAS Corn SW 5133. Uncertain. *Crourys -es* 1327, *Creulys* 1345, 1360, *Croulys* 1361, *Crowlis* 1564. Either 'weir-ford', Co ***crew** + **rys** (OCo *rid*), or 'hovel ford', Co **krow** + **rys**; or if the *l* form is original, 'hovel court, hovel ruins', Co **krow** + ***lys**. The situation at a stream crossing makes the 'ford' more likely. PNCo 75, Gover n.d. 644.

CROWLE H&W SO 9256. Partly uncertain. *Croglea* (dative case) 836 S 190, *Crohlea* [840×48]11th S 205, *Crohlea, Croelai* 1086, *Croule(ga) -leia* c.1150–1241, *Croule* 1232–1540×4. The boundary of the people of Crowle is *crohhæma gemære* [943 for 963]11th S 1297 and nearby Bow Brook *crohwællan* [840×8]11th S 205. Crowle is situated on a hill in a large bend of Bow Brook; the specific is OE ***crōh** probably meaning 'nook, corner' or 'valley' referring either to the stream or a recess in the hill itself. The generic is OE **lēah** 'wood, clearing'. Wo 315, Studies 1936.167, PNE i.113.

CROWLE Humbs SE 7712. 'The winding river'. *Crull* [c.1080]14th, 1232, *Crul(e)* 1086. OE ***crull**. The river from which this name is derived, *(super) Crullam* c.1100, *Crull* 1352, *Crowle* 1540, has disappeared owing to drainage. RN 108.

CROWMARSH GIFFORD Oxon SU 6189. 'C held by the Giffard family'. *Crowmershe Giffard* 1334 etc., *Crowmarsh Gifford* from 1636×7. P.n. *Cravmares* 1086, *Craumers* 1174–1338 with variants *-mersa -mers(s)he -mersch(e)*, 'the crow marsh', OE **crāwe** + **mersc**, + family name of Walter Gifard who held this place in DB. O 47, L 53.

CROWNHILL Devon SX 4857. *Crownewyll Wood* 1547. P.n. *Crownewyll*, uncertain element + OE **wylle** 'a spring, a stream', + ModE **wood**. D 255.

CROWNTHORPE Norf TG 0803. Partly uncertain. *Congrethorp, Cronkethor* (sic) 1086, *Crunge(l)thorp* 1252–83, *Crunkelthorp* 1316. Uncertain element or pers.n. + ON **thorp** 'outlying farm'. The specific is related in some way to ON *krungr* 'a hump', or Norwegian dial. *krungla* 'a crooked tree', *krunglutt* 'crooked', but whether as appellative or pers.n. is uncertain. DEPN.

CROWTHORN SCHOOL Lancs SD 7418. No early forms.

CROWTHORNE Berks SU 8364. 'Thorn-tree frequented by crows'. *Crowthorne* from 1607. ModE **crow** + **thorn**. In 1607 this was a single tree at the junction of the Bracknell and Wokingham road. The hamlet grew up after the arrival of Wellington College in 1856 and of Broadmoor Asylum in 1863. It became a separate ecclesiatical parish in 1874 and a civil parish in 1894 having been formerly in Sandhurst. Brk 125, Arch J 22.85, Room 1983.29.

CROWTON Ches SJ 5774. 'Crow farm or village'. *Crouton* 1260–1605, *Crowton* from 1330, *Crawton, Croton* 15th, 16th. OE **crāwe** + **tūn**. Che iii.195 gives pr ['krɔ:tən].

CROXALL Staffs SK 1913. 'Croc's hall or nook'. *Crokeshalle*

[942]14th S 1606, *Crocheshalle* 1086, *Croxhale* c.1200–59, *Croxall* 1577. OE pers.n. *Crōc*, genitive sing *Crōces*, + **hall** or **halh**. DEPN, Duignan 47, Horovitz.

CROXDALE Durham NZ 2636. 'Tail of land of the river bend, or, belonging to Croc'. *Crocestail* c.1190, *-teil* [1189×99]1336, *Crocestail, Croxtelle* 1242×3, *Croxstall(e)* 1259×60, 1329×30, *Croxdale* from 1354 with variant spellings *-dall(e) -dell -dayl(e) -dail(e)*. OE ***crōc**, genitive sing. ***crōces**, or pers.n. *Crōc*, genitive sing. *Crōces*, + **tæġel**. The reference could be to a bend in Tursdale Beck which in former times made a long loop to the north before joining the Wear, or to a bend in the Wear itself. NbDu 58.

CROXDEN Staffs SJ 0629. 'Croc's valley'. *Crochesdene* 1086, *Crokesdene* 1212. OE pers.n. *Crōc*, genitive sing. *Crōces*, + **denu**. DEPN, L 98.

CROXLEY GREEN Herts TQ 0795. *Crokesleigrene* 1349–96. P.n. Croxley + ME **grene**. Croxley, *Crokesleya* 1168–1280 with variants *Crockes-* and *-lege -leye* is 'Croc's wood or clearing', OE pers.n. *Crōc*, genitive sing. *Crōces*, + **lēah**. Hrt 81.

CROXTETH Mers SJ 4096. 'Landing place at a bend'. *Crocstad* 1257, *Croxstath* 1297, *Crokstath* 1372, *Crostoffe* c.1540. ON **krókr** referring to a bend in the river Alt + **stǫth** 'a landing place' (genitive sing. **stathar**). An alternative possibility for the generic is ON **stathir** 'place, site' not certainly evidenced in English p.ns., but a possible introduction from the Isle of Man. La 114, SSNNW 56.

CROXTON 'Croc's estate'. Late OE pers.n. *Crōc* (probably representing ON *Krókr*), genitive sing. *Crōces*, + **tūn**. Alternatively, since *Crōc/Krókr* is of rare occurrence in England, it has been suggested that the specific may be OE ***crōc** 'a nook of land', genitive sing. ***crōces**.

(1) ~ Cambs TL 2459. *Crochestone -tune, Crocestona* 1086, *Croxton(e)* from 1199. Ca 158.

(2) ~ Humbs TA 0912. *Crocestone, Crochestune -tone* 1086, *Crochestuna -tune -ton* c.1115–67, *Croxtun(e) -a -ton(')* 1155–1243. The place lies in a triangular shaped valley. L ii.98, SSNEM 375.

(3) ~ Norf TL 8786. *Crokestuna* 1086, *Croxtun'* 1202, *Crokestone* 1254. DEPN, SPNN 280.

(4) ~ Staffs SJ 7832. *Crochestone* 1086, *Croxton* 1327, *Croxson* 1583. DEPN.

(5) ~ KERRIAL Leic SK 8329. 'C held by the Cryoll family'. *Croxton(e) Kiriel* 1247–1516, *~ Kernal* from 1500. Earlier simply *Crohtone* 1086, *Crocheston(e)* 1195–early 15th, *Crocston(e)* 1223–1328, *Croxton(e)* from 1198. This may well be 'farm, village of the nook of the land', with OE ***crōc**, genitive sing. ***crōces**. The reference would be to the distinct nook in the hills in which Croxton Park lies. The manor was granted to Bertramus de Cryoll in 1239. Lei 158 gives pr ['krousən], SSNEM 375 no 3.

(6) ~ PARK Leic SK 8227. P.n. CROXTON SK 8329 + ME **parc**. *Parco de Croxton* [1189]1290, *Croxton Park(e)* 1610, 1612, 1806. Lei 159.

(7) ~ South Leic SK 6810. OE adj. **sūth** 'south', *Sut-* 1199, *Sud-* 1201–1224, *Suth-* 1202–1324, *South(e)* from 13th + p.n. *Crocheston(e)* 1086, c.1130, *Croptone* (sic) 1086, *Crokeston(e)* 1201–1423, *Croxton(e)* from 1199, *Croson* c.1570–1701, *Crowson* 1607, 1629, possibly 'farm, village of the nook of land', OE ***crōc**, genitive sing. ***crōces**, + **tūn**. The village lies in a hollow between ridges. 'South' for distinction from CROXTON KERRIAL SK 8329. Lei 286 gives pr ['krousən]. SSNEM 375 no. 4.

CROYDE Devon SS 4439. Uncertain. *Croyde* 1765. Earlier spellings are for Croyde Hoe SS 4240, *Crideholde* 1086, *-ho* 1242–1329, *Cride al. Cridehoe* 1670, 1713. Originally thought to have been the name of the stream that flows through the village, identical with the r.n. Creedy in CREEDY PARK SS 8301. Alternatively there may have been an OE ***crȳde** meaning 'that which thrusts out, a headland', formed on the root of OE *crūdan* 'to press, to drive', with explanatory OE **hōh** 'a hill-spur' and so 'hill-spur called *Cryde*, the Thruster'. The modern form is due to confusion of ME *ī* and *oi* in the 17th and 18th cents. D 43, L 168.

CROYDE BAY Devon SS 4239. *Cride Bay* 1577. P.n. CROYDE SS 4439 + ModE **bay**. D 43.

CROYDON Cambs TL 3149. 'Crow valley'. *Crauedena, Crauuedene* 1086, *Crauden(e)* 1199–1457, *Craw(e)den(e)* 1203–1541, *Crou(e)den(e) -w-* 1195–1362, *Croyden* 1577, 1645, *Croydon* 1668. OE **crāwe** + **denu**. For the diphthong *-oy-* cf. Croyland for CROWLAND Lincs TF 2310. Ca 53.

CROYDON GLond TQ 3365. 'Saffron valley'. *(æt) Crogedena* [809]13th S 164, *Crogdene* [c.871]15th S 1202, *-dæne* [973×87]c.1200 S 1511, *Croindene* 1086–, *Croyndenne* 1347, *Croindon(e)* 1229–1351, *Croydon(e)* from 1233×52. OE **croh** + **denu**. Forms with medial *n* probably derive from a side-form of the name with OE adj. **crogen* 'growing with saffron' or adj. **crogiġ* 'saffrony', definite form dative case **crogigan* in the phrase *æt thǣm crogigan dene* 'at the saffrony valley'. OE *croh* is a loan-word from Latin *crocus* probably referring to the autumnal crocus, *Crocus sativus*, grown by the Romans for the manufacture of saffron dye. It is not a native plant but may have continued to flourish in the wild after the end of the Roman period long enough to attract the attention of the English settlers. When the plant was re-introduced to western Europe by the crusaders the OE word had disappeared and a new loan-word *saffron* (OFr *safran* < Arabic *zaçfarān*) became current. The reference is to the valley of the river Wandle. Sr 47, DEPN, TC 23, 203, BzN 16.293–4.

CROYDON HILL Somer SS 9640. *Croydon Hill* 1809. P.n. Croydon as in Croydon House SS 9640 + ModE **hill**. Croydon, *Craudon* 1243, *Croudon* 1331, is 'crow hill', OE **crāwe** + **dūn**. DEPN.

CRUCKMEOLE Shrops SJ 4309. '*Cruc*, the cruck-framed building, on the r. Meole'. *Crok(e)mele* 1291×2, *Crokkemele* 1340, *Cruckmele* 1577, 1600 etc. P.n. *Cruc* as in CRUCKTON SJ 4310 + r.n. Meole, *Mele* 1255, 'the cloudy one', OE **melu**, **meolu** 'meal' used of a stream with cloudy water. Sa ii.32.

CRUCKTON Shrops SJ 4310. 'The settlement at *Cruc*, the cruck-framed building or river bend'. *Crotton'* 1271×2, *Crukton'* 1271×2, 1291×2, *Cruckton* from 1597, *Crok- Crocton* 1271–1339, *Croketon* 1490, 1631. P.n. *Cruc* < OE ***crōc** + **tūn**. Sa ii.34.

CRUDGINGTON Shrops SJ 6318. 'The settlement at Cruc hill'. *Crugetone* 1086, *Crugelton(a) -tun'* 1138–late 14th, *Crudgelton'* c.1260, *Crucheltuna* c.1140, *Crouging- Crogen- Crugging- Crogiton* 16th, *Crudginton* 1606, *Crudgington* 1667. P.n. **Cruc hill* < PrW ***crūg** 'a pointed hill' + OE **hyll**, + **tūn**. Sa i.103.

CRUDWELL Wilts ST 9592. 'Creoda's spring'. *Croddewell(e)* *[854]14th S 305, 1362, 1383, *Cruddewell(e)* [901]14th S 1579, 1270, *Creddewilla* [1065]14th S 1038, *Credvelle* 1086, *Criddanuille* c.1125 S 305, *Crudwell* 1624. OE pers.n. *Crēoda*, genitive sing. *Crēodan*, + **wylle**. The same pers.n. occurs in the lost names *Cruddemores lake* [974]14th K 584, *Credemore, Credehemefeld* 'the open land of the inhabitants of Crudwell' 1360, *Crudehamwlleslake* [n.d.]13th S 1577 and in the bounds of the inhabitants of Crudwell, *Cruddesetene imere* [956]14th S 629. In form it seems to vary between *Crēoda* and **Crudda*, cf. OE *crūdan* 'to press', preterite pl. *crudon*, p.pl. *croden* and ME *crudde* 'curds' from the same root. Wlt 56, L 31.

CRUGMEER Corn SW 9076. 'Great barrow or hill'. *Crucmur* 1336, *Crukmur(e) -moer* 1350–1467, *Crigmer* 1610, *-meare* 1702. Co **cruc** 'barrow, hillock' + **meur** 'great'. Situated on an isolated hill-top. PNCo 76 gives pr [krigmi:ə], Gover n.d. 355.

CRUMMOCK WATER Cumbr NY 1519. *lake of Crumbokwatre* 1230, *Crombocwater* 1307, *Cromack water* 1570. R.n. *Crummock* as in Crummock Beck, the name of the upper Cocker which flows through the lake, + ME **water**. Crummock Beck, *Crombok' -boc* after 1150, 1189, *Crumboc(h) -bok* 1189–1285, *Cromocke* 1578, is PrW ***crummōg**, Brit **crumbāco* 'crooked' as in the Irish r.ns. Cromoge and Crummoge. Both Crummock and Cocker have the same meaning. See River COCKER NY 1523. Cu 33, 10, RN 108.

CRUMPSALL GMan SD 8402. Partly uncertain *Cormeshal* 1235, *Curmisale* 1282, *Curmeshale* 1322, 1444, *-al(l)e* 1322, *Cormesall* 1500, *Cromsall* 1548, *Crumpsall* 1552. The early spellings do not well support the usual explanation 'Crumb's nook', although a pers.n. **Crum(b)* is possible < OE *crumb* 'crooked, bent'. The generic is OE **halh** referring to a small valley opening off the r. Irk. La 37, L 107, Jnl 17.33.

CRUNDALE Kent TR 0749. 'The gully'. *Crundala* c. 1100, *Crumdale* 1226–1272×1307, *Crundale* from 1242×3. OE **crundel** 'a gully, a chalk-pit, a quarry'. Crundale lies in a marked valley, but it is impossible to say whether this was the feature that gave rise to the name or whether there was a quarry here. Minor names in the area suggest the latter. PNK 382, Cullen.

CRUNKLY GILL NYorks NZ 7057 → CRONKLEY FELL Durham NY 8427.

CRUWYS MORCHARD Devon SS 8712. 'Morchard held by the Crues family'. *Cru(w)ys Morchard* 1257×80, *Crewsmorchard* 1603×25, *Morcestr(e) Crues* 1279–91. Earlier simply *Morceth, Morchet* 1086, 'great wood', PrCo ***mör** + ***cę̄d**. The manor was held by Alexander de Crues in 1242. The affix is for distinction from MORCHARD BISHOP SS 7607. D 380, L 190.

CRYSTAL PALACE GLond TQ 3470. ModE **crystal** + **palace**. A huge glass conservatory designed to house the Great Exhibition of 1851. Originally erected in Hyde Park it was lated removed to Sydenham where it was destroyed by fire in 1936. Encyc 221.

CUBBINGTON Warw SP 3468. 'The estate called after Cubba'. *Cobintone* 1086–1504 with variants *-yn-* and *-ton, Cvbi(n)tone* 1086, *-intone* 1086–1265, *Cubbintona* [1170]1314, *Cubbyngton* 1535. Probably OE pers.n. **Cubba* + **ing**[4] + **tūn**, or OE **Cubban*, genitive sing. of **Cubba*, + **tūn**, 'Cubba's estate'. Wa 169.

CUBERT Corn SW 9076. '(Church of) St Cuthbert'. *(Vicaria) Sancti Cuberti* 1269, *(Ecclesia) Sancti Cutberti* 1291, 1374, ~ ~ *Cuthberti* 1342, c.1400, *Eglos Cutbert* 1402, *St Kibberd* c.1605, *Kibbert -d* 1604, 1610, *Egloscubert* 1622. Co **eglos** + OE saint's name *Cūthbeorht* from the dedication of the church to St Cuthbert of Lindisfarne and Durham. Cf. Gwbert Dyfed SN 1650. PNCo 76, Gover n.d. 369, DEPN.

Great CUBLEY Derby SK 1638. Adj. **great** + p.n. *Cobelei* 1086, *Cob(b)elegh -le(ye)* 1269–c.1350 *-lag, Cub(b)eleia -legh(e) -ley(e)* c.1166–1354, *Cubley(e)* from 1302–45. 'Cubba's wood or clearing', OE pers.n. **Cubba* + **lēah**. Db 547.

CUBLINGTON Bucks SP 8322, 'Estate called after Cubbel'. *Cubelintone* [1154]1200–1247, *Cublintone* 1237×40, *Coblington* 1238–1347. OE pers.n. **Cubbel* + **ing**[4] + **tūn**. The DB spelling *Coblincote* 1086 points to an alternative form, 'cottages associated with Cubbel', OE pers.n. **Cubbel* + **ing**[4] + **cotu**, pl. of **cot**. Bk 77. Jnl 2.33 gives local pr [kubəltun].

CUCKFIELD WSusx TQ 3024. 'The open land where cuckoos are heard'. *Kuku- Kukefeld* 1087×1100, *Cucufeld(a)* 1114×22, 1121, *Cu- Cokefeld* 1220–1497, *Cukfeld* 1245, *Cookfield* 1422. OE ***cucu** + **feld**. Sx 261 gives prs [kukfi:ld] and [kukful].

CUCKLINGTON Somer ST 7527. 'Estate called after Cuca or Cucol'. *Cocintone* 1086, *Cukelingeton* 1212, 1274, *Cucklington* 1610. OE pers.n. **Cuca* or its diminutive **Cucol(a)* + **ing**[4] + **tūn**. DEPN.

CUCKMERE RIVER ESusx TQ 5408. *river called Cokemere Haven* 1423. Earlier simply *Cokemer(e)* 14th, *Cookemere* 1335, *The Cuckmer* 1586, either 'Cuca's lake', OE pers.n. *Cuca* + **mere** or 'living lake', OE **cucu** < *cwicu*, referring to the running water of the stream. The pre-English name of the river is possibly preserved in the name Exceat [ek'set] near Westdean TQ TV 5199, *Essete -a* 1086, *Exete -a, Exsetas -es*, which has been interpreted as 'the settlers by the river Exe', r.n. *Exe* as in River EXE SS 9409, SX 9882 + OE **sǣte**. Sx 4, RN 109, ZONF 3.203–4.

CUCKNEY Notts SK 5671. 'Cuca's island'. *Cuchenai* 1086, *-eia -eya -eie -eye -aye* c.1150–1295, *Cokeneye* 1211–1327, *Coknay* 1510, *Cuckney* 1684. OE pers.n. **Cuca*, genitive sing. **Cucan*, + **ēġ**. Nt 75, L 38.

CUDDEN POINT Corn SW 5427. *Cuddan poynt* 1576. Unidentified p.n. + ModE **point**. Co **cudin** 'tress of hair' is possible, but the meaning would be obscure. PNCo 76.

CUDDESDON Oxon SP 5903. 'Cuthen's hill'. *(æt) Cuþenes dune* [956]10th S 587, *Codesdone* 1086–1428 with variants *-don(a)* and *-duna, Cudesdon* [1122–1346]c.1425 with variants *-duna -e -dona -dena* and *dene, Cuddesdone* 1285, 1526. OE pers.n. **Cūthen*, genitive sing. **Cūthenes*, + **dūn**. **Cūthen* (**Cūthin*) would be a hypocoristic form of names in *Cūth-*, but might alternatively represent a reduced form of *Cūthwine*. O 167 gives pr [kʌdzdən], L 150, DEPN, Gelling 1998.79.

CUDDINGTON Bucks SP 7310. 'Estate called after Cud(d)a'. *Cudintuna* 1115×25, 1176, *Cudington* c.1218, *Codyngton* 1339–1539. OE pers.n. *Cud(d)a* or *Cuddi* + **ing**[4] + **tūn**. Bk 158.

CUDDINGTON Ches SJ 5971. 'Estate called after Cud(d)a'. *Codynton* c.1235, *Codington* 1270–1579 with variants *-in- -yn(g)-*, *Cudinton* 1260, *Cudington* 1289–1508 with same variants, *Cuddington* 1585. OE pers.n. *Cud(d)a* + **ing**[4] + **tūn**. Che iii.198 records older local pr ['kuditn].

CUDDINGTON HEATH Ches SJ 4746. *Cudyngton Hethe* 1532, *Kiddington Heath* 1692. P.n. Cuddington SJ 4546 + **hǣth**. Cuddington, *Cuntitone* 1086, *Kydinton* 1284, *Kydd- Kiddinton -yng- -ing-* 1475–1860, *Cudin(g)ton* 1288–1567 with variants *Cod-* and *-yn(g)-*, *Cuddington* from 1350, is the 'estate called after Cyd(d)a or Cud(d)a', OE pers.n. *Cud(d)a* with *u>i* before dentals as in DIDCOT Berks SU 5189, DIDDINGTON Cambs TL 1965 and DINNINGTON SYorks SK 5386, + **ing**[4] + **tūn**. Che iv.28, Addenda.

CUDDY HILL Lancs SD 4937. No early forms. Northern dial. **cuddy**, a diminutive of *Cuthbert*, is a pet name for a donkey.

CUDHAM GLond TQ 4459. 'Cuda's homestead'. *Codeham* 1086, *Cudeham* 1278, *Codam* 1458. OE pers.n. *Cuda* + **hām**. DEPN, LPN 62.

CUDLIPTOWN Devon SX 5279. 'Cudlip hamlet'. *Cudliptowne* 1641. P.n. Cudlip + ModE **town**. Cudlip, *Cudelipe* 1114×19, *Codelip(p) -lepe* 1238–1428, *Cuddelypp* 1474, is 'Cudda's leap or steep place', OE pers.n. *Cuda* + **hlīep**. The village lies at the foot of steep ground rising to over 1500ft. D 232.

CUDWORTH Somer ST 3710. 'Cuda's enclosure'. *Cvdeworde* 1086, *Cudewurth* 1243, *Cudworth* 1610. OE pers.n. *Cuda* + **worth**. DEPN.

CUDWORTH SYorks SE 3808. 'Cutha's enclosure'. *Cutheworth(e) -wrth* 12th–1342, *Cothew(o)rth(e)* 13th–1434, *Cudeuurdia* c.1185, *-wrth(e)* 13th cent., *Cudworth* from 1371. OE pers.n *Cūtha* + **worth**. YW i.280.

CUFFLEY Herts TL 3002. 'Cuffa's wood or clearing'. *Kuffele* 1255, *Co- Cuffele(ye)* c.1275–c.1430. OE pers.n. **Cuffa* + **lēah**. Hrt 114.

CULBONE HILL Somer SS 8247. P.n. Culbone SS 8448 + ModE **hill**. Culbone, *Culbone* 1610, is said to derive from the dedication of the church to St Culbone or Columbanus. The original name of the hill is *Chetenore* 1086, *Kitenore* 1236, *Oare Hill* 1809 OS, 'kite ridge', OE **cȳta**, genitive sing. **cȳtan**, + **ōra**. Cf. Oare SS 8047, *Are* 1086, *Oure* 1610, referring to the same feature. DEPN.

CULCHETH Ches SJ 6595. 'Nook of a wood'. *Culchet* 1201, *Kulcheth* 1246, *C- Kulchit(h)* 1246–1311 etc., *Culcheth* from 1322, *Kilchith -cheth* 1246, 1577, *-chif* 1303, *Culsheth, Kilshay* 16th. PrW ***cül** (Brit **coilo-*) + ***cę̄d**. Cf. Colquite Corn SX 0570, *Kilcoit* 1308, Blaencilgoed Dyfed SN 1410, *Blanculcoyt* 1325, Culcoed Dyfed SN 3121, ~ Gwyn SH 4250, CULGAITH Cumbr NY 6129. La 97, Jnl. 1.45, 46, CPNE 58.

CULFORD Suff TL 8370. 'Cula's ford'. *Cule- Coleford* c.1040 and [c.1040]14th S 1225, *Culeford* 11th, 1197, *-fordā* 1086, *Culfurth* 1610. OE pers.n. *Cūla* + **ford**. DEPN.

CULGAITH Cumbr NY 6129. 'Nook of a wood'. *Culgait* c.1140–1290 with variants *-gayt -geyt, Culgayth -gaith* from 1232 with variant *Kul-*. PrW ***cül** + ***cę̄d**. Identical with CULCHETH Ches SJ 6595. Cu 184, Jnl 1.45, 46, L 190, GPC 629.

CULHAM Oxon SU 5095. 'Cula's river-bend land'. *Culeham* [821]12th S 183, 1225, *Culanhom* [821]13th S 183, *-ham* [940]12th S 460, *Coleham* [1152]13th., [1190×1200]c.1245, *Cullum* 1675. OE pers.n. **Cūla*, genitive sing **Cūlan*, + **hamm** 1. Culham is situated on a bend in the river Thames. O 150, L 43, 49.

CULKERTON Glos ST 9396. Possibly 'estate called after Culcere'. *Cvlcortone -torne* 1086, *C- Kulkerton(a) -tune* 12th–1587, *Culc- Culkrinton* 1195, 1292. OE pers.n. **Culcere* (+ **ing**[4]) + **tūn**. The alternative suggestion, OE **culcor* 'waterhole' (related to OE *ōden-colc* 'hollow used as a threshing floor') is unlikely as this place is on a hill. Gl i.105, xii.

CULLERCOATS T&W NZ 3670. 'The dove-cotes'. *Culvercoats* c.1600, *Cullercoats* 1693. ModE **culver** (OE *culfre*) + **cote**. NbDu 59.

CULLINGWORTH WYorks SE 0636. 'Enclosure of the Culingas, the people called after Cula'. *Colingauuorde* 1086, *Culingeworth', Cullingewrthe* 1208, *Cul(l)ingw(u)rth(e) -worth* 1180×4–1687. OE folk-n. **Cūlingas* < pers.n. **Cūla* + **ingas**, genitive pl. **Cūlinga*, + **worth**. For the folk-n. **Cūlingas* cf. COOLING Kent TQ 7576. YW iv.162.

CULLOMPTON Devon ST 0107. 'Settlement on the river Culm'. *(æt) Columtune* [873×88]11th S 1507, *(æt) Culumtune* before 1097, *Colitone* 1086, *Cu- Columton* 1212, 1230, *Culmeton* 1257, *Columpton(e)* 1263, 1291, *Culmpton* 1321, *Columpton al. Colehampton, Columbton, Collupton vulg. Culliton* 1675. R.n. CULM + **tūn**. D 561.

River CULM Devon ST 0105. 'Looped river'. *(on, up, of) culum* [925×39]11th S 386, *Culum* 1238, *Cul(u)mp* [1291]1408, *Columb* 1577. OW **culm** 'knot, tie'. The course of the river is a succession of loops and frequently divides into two. Along it were various estates named from the river, Culm Davy ST 1215, originally 'coomb held by David de Wydeworth' 1242, *Cumbe juxta Culum* 'Coomb by Culm' [c.760]13th S 1691, *Cūbe* 1086 subsequently influenced by the r.n., Culm Pyne ST 1314, 'C held by Herbert de Pyne' 1242, *Colvn* 1086, and UFFCULME ST 0612. D 4, RN 109.

CULM VALLEY Devon ST 1013. 'Valley of the river Culm'. R.n. CULM ST 0105 + ModE **valley**.

CULMINGTON Shrops SO 4982. Probably 'the estate called after Cuthhelm'. *Comintone* 1086, *Cu- Colminton(e) -yn-* 1159×60–1772, *Colmyngton* 1350, *Culmington* from 1688. OE pers.n. *Cūthhelm* + **ing**[4] + **tūn**. Sa i.104.

CULMSTOCK Devon ST 1013. 'Outlying farm on the river Culm'. *Culumstocc, (to) Culumstoce* *[670 for 925×39]11th S 386, *-stok* 1244, *(æt) culmstoke* [1050×73]n.d. K 940, *Cvlmestoche* 1086, *Columbstoke* 1675. R.n. CULM ST 0105 + OE **stoc**. D 612, RN 109.

CULVER CLIFF IoW SZ 6385. 'Pigeon cliff(s)'. *Culver Cleues* 1550, *Couluer clyffes* 1591, 1611, *Culver Cliff* 1700. ModE **culver** (OE *culfre*) + **cliff**. Wt 262.

CULVERSTONE GREEN Kent TQ 6363. 'Culverstone common'. *Culverson Green* 1819 OS. P.n. Culverstone + ModE **green**. Culverstone appears to be a compound of *culver* 'dove' and *stone* or *toun* but the form *Culversole* 1381 shows it must originally have been OE **sol** 'miry place' and so 'miry place of the doves'. The specific is probably OE **culfre**, genitive pl. **culfra**. PNK 104.

CULVERTHORPE Lincs TF 0240. Partly uncertain. *Cal(e)warthorp(e)* 1271–1562, *Kilwardthorp* 1338, *Calverthorp* 1506. Uncertain element prefixed to earlier simplex *Torp* 1086, *Thorp* 1212–1346, OE, ON **thorp**, 'the outlying farm'. The prefix may represent OE **calf-weard* 'a herdsman' used either as an occupation term or as a surname. Perrott 59, SSNEM 119.

River CULVERY Devon > TEDBURN ST MARY SX 8194.

CULWORTH Northants SP 5446. 'Cula's enclosure'. *Culeorde* 1086, *-wurda -worth(e)* 1184–1371, *Coleworth* 14th cent., *Colworth* 1331. OE pers.n. **Cūla* + **worth**. Nth 52 gives pr [kɔləθ].

CUMBERWORTH Lincs TF 5073. 'Cumbra's enclosure'. *Combreuorde* 1086, *Cumberworda* c.1115, *-worth* 1209×19. OE pers.n. *Cumbra* + **worth**. DEPN.

CUMBRIAN MOUNTAINS Cumbr NY 2716. From the Latin form of the name of Cumberland, *(comitatu) Cumbrie* 1230, *Cumbria* 1231. Cumberland (abolished in 1974 when it was united with Westmorland to form the new county of Cumbria), *Cumbra land* c.960 ASC(A) under year 945, *(to) Cumer lande* 1121 ASC(E) under year 1000, *(to) Cumber lande* late 11th ASC(D) under year 1000, *Cumberland* from c.1150, *Cumbreland'* 1191, 1459, is 'the land of the *Cömmri*, the Welsh', OE folk-n. **Cumbre* 'the *Cömmri*, the Welsh, the Cumbrian Britons', genitive pl. **Cumbra*, + **land**. Cf. WESTMORLAND. Cu 1, PNE i.119.

CUMMERSDALE Cumbr NY 3953. 'Valley of the *Cömmri*, the Cumbrian British'. *Cumbredal* 1227, *-er-* 1279, *Cumbersdale -re-* 1285–1345, *Commersdaill* 1540, *Cummersdale* 1622. OE folk-n. **Cumbre*, genitive pl. **Cumbra*, + ON **dalr**, lOE **dæl**. Cu 130, SSNNW 225, L 96.

CUMNOR Oxon SP 4604. 'Cuma's ridge'. *Cumenora(n)* [821]12th, [821]13th S 183, *(ad) Cumanoran* [931]12th S 410, *Cumenoran* [968]12th S 757, *Cumenor(e)* 1201–52, *Comenore* 1086–1412 with variants *-or(')*, *Cumner* 1830. OE pers.n. **Cuma*, genitive sing. **Cuman*, + **ōra**. The pers.n. here is *Cuma* < OE *cuma* 'a guest, a stranger' and not *Cumma* as in Cumma, an 8th cent. abbot of Abingdon whose name is a regular hypocoristic form of OE *Cūthmund*. Brk 445.

CUMREW Cumbr NY 5450. 'Valley of the slope'. *Cumreu* c.1200–1291, *Cumrew* from 1290. PrW ***cumm** (W *cwm*) + ***riu** (W *rhiw*). The slope in question is Cumrew Fell. Cu 77, Jnl 1.46, 50, O'Donnell 80, GPC 640.

CUMWHINTON Cumbr NY 4552. 'Valley of Quintin'. *Cumquintin(a) -yntyn(e)* [c.1155]1200–1540, *Cumwhyntyn* 1485, *cum Whynton* 1551. PrW ***cumm** (W *cwm*) + OFr pers.n. *Quintin*, a late coinage in Celtic word order. Cu 161, Jnl 1.46, O'Donnell 81, L 92, GPC 640.

CUMWHITTON Cumbr NY 5052. '*Whittington* valley'. *Cumwyditon* 1278, *-whytyton -iton* 1279, *-w(h)ytington -quiting-* 1292, *Com- Cumquityngton* 1332, 1363, *Comwhytton* 1359, *Cumwhitton* 1485. PrW ***cumm** + lost p.n. *Whittington* 'farm, village, estate called after Hwita', OE pers.n. *Hwīta* + **ing**[4] + **tūn**. This form of p.n., a late coinage in Celtic word order, is strongly suggestive of territory recovered by the Cumbrian British from its Anglian occupiers. Cu 78 gives pr [kum'h-witən], xxi, O'Donnell 81–2, L 92, GPC 640.

CUNDALL NYorks SE 4272. Partly uncertain. *Cun-, Goindel* 1086, *C- Kundale* 12th–1457, *Cundall* from 1418. Unidentified element or name + ON **dalr**, lOE **dæl**, 'a valley'. The specific is uncertain, possibly OE **cūna**, ONb **cȳna**, genitive pl. of **cū** 'a cow' and so 'cow valley'. YN 181, SSNY 154.

CURBAR Derby SK 2574. 'Corda's fortified place'. *Cordeburg -burgh(e)* 1203–1389, *Cordborgh -burg -burgh(e) -bourg* 1356–1531, *Corburg -borowe -borough* 1365–1643, *Corber -bar* 1577. OE pers.n. **Corda* + **burh**. Db 80.

CURBRIDGE Hants SU 5211 'Quern bridge'. *Cur- Kernebrugge, Kerebrigge* 1236, *Cornebrigge* 1272, *Kernebregge* 1286, *Corebrugge* 1281, *Currebrigge* 1314. OE **cweorn** + **brycg**. Possibly a bridge made of or supported on quernstones. Ha 62, Gover 1958.32.

CURBRIDGE Oxon SP 3308. 'Creoda's bridge'. *(æt) Crydan brigce* [956×7]12th S 1292, *Crudebrigg' -brug(g)'* 1200, 1240×1, *Credebrigge -brugg'* 1209, 1268, *Curbrigge* 1517. OE pers.n. *Creoda*, genitive sing. *Creodan*, + **brycg**. O 315, L 65.

CURDRIDGE Hants SU 5213. 'Cuthræd's ridge'. *cuðredes hricgæ* [900]11th S 360, *Curderigge* 1208, 1350, *Courdridge* 1593. OE pers.n. *Cūthrǣd*, genitive sing. *Cūthrǣdes*, + **hrycg**. Ha 62, Gover 1958. 45.

CURDWORTH Warw SP 1893. 'Creoda's enclosure'. *Credeworde* 1086, *-wrth(e)* 1154×89–1359, *Cruddewrth* c.1200, *Croddewurthe* c.1235, *Crudworth* 1316–1670, *Cord-* 1547, *Curd-* from 1656. OE pers.n. *Crēoda* + **worth**. Wa 44.

CURLAND Somer ST 2717. 'Newly cultivated land belonging to Curry'. *Curiland* 1252. P.n. Curry as in CURRY MALLET ST 3221 + OE, ME **land**. Either 'outlying land belonging to Curry' or to an owner with a surname derived from the p.n. DEPN.

CURRY MALLET Somer ST 3221. 'C held by the Mallet family'. *Curi Malet* 1225, 1284, *Cory malet* [13th]Buck, *Curry mallet* 1610. Earlier simply *Cvri* 1086. Originally a stream-n. of unknown meaning and origin occurring in the other Curry estate names. A possible suggestion is 'border river', related to W *cwr* 'corner, border', Co *cor* 'hedge, boundary'. The manor was held by William Malet 1189×99 whose name is either a patronymic, a diminutive of *Malo* or a nickname meaning 'hammer' or 'cursed one'. DEPN, DB Somerset 21.1–2, Turner 1950.115, CPNE 65, DBS 229.

CURRY RIVEL Somer ST 3925. 'C held by the Revel family'. *Curry Revel* 1225, 1610. Earlier simply *Curi* [934×9]13th S 455, *Chori*, *Couri* 1084, *C(h)vri* 1086. P.n. Curry as in CURRY MALLET ST 3221 + family name of Richard Revel to whom the manor was granted by the king in 1194. His surname is a nickname from OFr *revel* 'pride, rebellion, sport'. Pronounced locally [raivəl]. DEPN, Turner 1950.115, DBS 293.

North CURRY Somer ST 3125. *Nortcvri* 1086, *North Cury* 1262×3, *Northcory* [1329] Buck. OE adj. **north** + p.n. *(on) curig ie mære* 'to the boundary of Curry island' *[854]12th S 309, *(usque ad) Curig* [904]12th S 373, *(æt) Curi* [1066×8]12th K 897, *Cori* 1084, as in CURRY MALLET ST 3221. Another Curry mentioned in DB is Currypool ST 2238, *Cvriepol* 1086. DEPN, DB Somerset 21.13, Turner 1950.115.

West CURRY Corn SX 2893. *Westacory* c.1220, *Wescori* 1300, *West Cory* 1543, *West Curry* 1842. ME adj. **west** + p.n. Cory, *Chori* 1086, *Kery* 1284, *Curry* 1654, of unknown meaning, but identifical with Cory SS 2116, Cory and CORYTON Devon SX 4583 and CURRY MALLET ST 3221 and CURRY RIVEL ST 3824 in Somerset, possibly all originally names of rivers or streams. 'West' for distinction from *East Correy Parke* 1654. Also known as *Great Cory* 1748, 1813. PNCo 76, RN 98, Gover n.d. 2.

CURRYPOOL Somer ST 2238. → North CURRY ST 3125.

West CURTHWAITE Cumbr NY 3248. *West Kirthwaite* 1578. ModE adj. **west** + p.n. *Kyrkehuaite* 1272–1619 with variants *Kirke-* and *-thwayt -thweyt -thwait(e)*, *Kirthate* 1616, 'church clearing', ON **kirkja** + **thveit**. The reason for the name is unknown. 'West' for distinction from East Curthwaite, *Estkirk . . . wayt* 1367, *Est Kirthwaite* 1578, *East Curthat* 1699. Cu 329 gives pr [kərθət], SSNNW 118. L 210.

CURTISDEN GREEN Kent TQ 7440. *Courtson Gr.* 1819 OS. The generic is ModE **green** but the available evidence is too late to offer a full explanation. PNK 309.

CURTISKNOWLE Devon SX 7353. 'Curt's hill'. *Cortescanole* 1086, *Curtescnolle -knolle* 1242, 1297, *Goosnoll al. Courtesknoll* 1767. OE pers.n. **Curt*, genitive sing. **Curtes*, + **cnoll**. D 300.

CURY Corn SW 6721. 'Church of St Cury'. *Egloscuri* 1219, *(Ecclesia de) Sancto Corentino* 1284, *Seyntcorentyn* 1369, *Cury* from 1473. Co **eglos** 'church' + saint's name *Curi*, a pet form of *Corentin*, from the dedication of the church. A Breton saint, patron of the cathedral at Quimper, and believed to have been the first bishop of Cornouaille. PNCo 77, Gover n.d. 544, DEPN.

CUSHAT LAW Northum NT 9213. 'Wood-pigeon hill'. *Cousthotelaw* (sic for *Cousch-*) [c.1200]14th, *Cowshotlaw* 1536. OE **cūsc(e)ote** + **hlāw**. OE *cūsc(e)ote* survives as N dial. *cushat*. NbDu 59.

CUSHUISH Somer ST 1930. *Cows Huish* 1809 OS. The evidence is too late for explanation. The generic of the name, however, is ultimately OE **hīwisc** 'measure of land sufficient to support a family'.

CUSOP H&W SO 2341. Partly uncertain. *Cheweshope* 1086[†], *Kiweshope* 1160×70–1198, *Kywishope* 1292, *C- Kusop* 1302, 1317. Uncertain element + OE **hop** 'a secluded valley'. Cusop lies in a small side-valley of a tributary of the r. Wye. The specific may, therefore, be the old name of the stream, identical with the r.n. Chew in CHEW MAGNA Avon ST 5763. He 63, Bannister 55.

CUT HILL Devon SX 5982. *Cut Hill* 1809 OS. Named after nearby Cut Lane; a cut lane is 'a track or way formed by removing the soft surface soil and leaving exposed the harder and often stony subsoil; a way artificially cut through'. The reference is to the North West Passage Peat Pass on Dartmoor. D 193 (citing Trans. Devon. Assoc. 58.362).

CUTNAL GREEN H&W SO 8868. *Cuttenhall Green* 1642. P.n. *Cutnall* 1644, + ModE **green**. The forms are too late for certainty, but this could be 'Cudda's nook or hall', OE pers.n. *Cudda*, genitive sing. *Cuddan*, + **halh** or **hall**. Wo 241 gives pr [kʌtlənd].

CUTSDEAN Glos SP 0830. 'Cod's valley'. *Cottesdena -den(e)* 12th–1626, *Cuttesdowne als Cutson* 1583, *Cuts Dean* 1718. OE pers.n. **Cōd*, genitive sing. **Cōdes*, as in COTSWOLD HILLS, + *denu*. This was a hamlet of Temple Guiting in 1354; but another part of the same estate granted to Worcester Priory was known as *Codestun(e)* [977]11th S 1353, 1086, 1196, *Cot(t)eston(e)* 1221–1419, 'Cōd's estate'. Gl ii.7.

CUTTHORPE Derby SK 3473. 'Cutt's outlying farm'. *Cutthorp(e)* from 1417. Surname *Cutt* + ME **thorpe**. The Cutt family are evidenced in the area from the 14th cent. Db 221, SSNEM 125.

CUXHAM Oxon SU 6695. 'Cuc's river-meadow'. *Cuces hamm, (to) Cuces Hamme* [995]c.1000 S 1379, *Cukesham* 1203–81, *Cuchesham* 1086, *Cuxham* from 1278×9. OE pers.n. **Cuc*, genitive sing. **Cuces*, + **hamm** 1. The bounds of the people of Cuxham are *(to) Cuceshæma gemære* [887]11th S 217. O 125, L 43, 49.

CUXTON Kent TQ 7066. 'Cucola's stone'. *Cucolanstan* [880]12th S 321, *Cuclestana*, *Cucclestane* c. 975 B 1321–2, *Coclestane* 1086, *Cukston* 1610. OE pers.n. **Cucola*, genitive sing. **Cucolan*, + **stān**. PNK 113.

CUXWOLD Lincs TA 1710. 'Cuca's forest'. *Cvcvalt*, *Cucualt* 1086, *Cucuwald* c.1115, 1163, *Cukewald(')* 1146–1281, *Cokewald(')* 1242–1431, *-wold* 1291–1387, *Cokeswold* 1519–1610, *Cuxwold* from 1706. OE pers.n. *Cuca* + **wald**. A recent alternative suggestion for this n. and COXWOLD NYorks SE 5377 is 'cuckoo forest' with OE ***cucu**, but *wald* is not otherwise compounded with words for wild animals. Li iv.73, L 275, Jnl 27.47.

[†]The identification is not certain, DB Herefordshire 1.3.

DACRE Cumbr NY 4526. *Dacor* c.1125, *Dacre* from c.1200, *Daker(e) -ir -yr* 1212–1318 transferred from the stream-name Dacre Beck, *amnem Dacore* c.731 BEH, *(big) Docore ðære éa* [c.890]c.1000 OEBede, *riuulus de Daker* 1292. PrCumb ***dagr** (Brit **dacrū*) 'tear' in the sense 'trickling stream'. Cu 186 gives pr [dɛəkər], 10, RN 111, Jnl 1.46, GPC 921 s.v. *deigryn, deigr*.

DACRE NYorks SE 1960. *Dacre* from 1086, *Daker* 1156–1608. Originally a r.n. from PrW **deigr* 'a tear, a drop', for one of the tributary streams of the river Nidd at this point. It appears to be a name used for an insignificant stream. YW v.139, Jnl 1.46.

DACRE BANKS NYorks SE 1961. *Dacreban(c)kes* 1557–1615, *Banks (in Netherdale)* 1795, 1817. P.n. DACRE SE 1960 + ME **bank(e)** 'a bank, the slope of a hill or ridge'. YW v.140.

DADDRY SHIELD Durham NY 8937. Possibly 'shieling held by the Daudry family'. *Danterre Sheele* (for *Dauterre*) 1615, *datherie she(a)le* 1685, *Daddry Shield* 1866 OS. Surname *Daudry* + ME **shele**. Members of this family variously spelled *de Audre(i), Daudri, Daldre*, are frequent witnesses of charters in the 12th and early 13th cents. in Durham.

DADFORD Bucks SP 6638. 'Dod(d)a's ford'. *Dodeford* 1086–1379, *Doddeforde* 1227–47, *Dadef'* 1237, *Dudford* 1685, *Dadford* 1693. OE pers.n. *Dod(d)a* + **ford**. Bk 48, Jnl 2.23.

DADLINGTON Leic SP 4098. 'Estate called after Dædela'. *Dadelinton(e) -yn-* 1216–1369, *-ing- -yng-* 1266–1419, *Dadlington' -yng-* from 1274. OE pers.n. **Dǣd(e)la* + **ing**[4] + **tūn**. Lei 547.

DAGENHAM GLond TQ 5084. 'Dæcca's homestead'. *deccanhaam* [685×94]8th S 1171, *Dæccanham* *[677]16th S 1246, *Dake(n)ham* 1194–1335, *Daginham* 1262, *Dagenham* from 1274, *Dagnam* 1499. OE pers.n. **Dæcca*, genitive sing. **Dæccan*, + **hām**. An exact parallel is Daknam, East Flanders, *Dackenham* 1156. Voicing of [k] to [g] is paralleled at Bragenham Bucks SP 9028, *Brag(n)ham* 1178–1512, *Bracham* 13th cent., *Brakenham* 1241, 'Bracca's homestead', OE pers.n. *Bracca*, genitive sing. *Braccan*, + **hām**. Ess 91, Bk 83, TC 203.

DAGLINGWORTH Glos SO 9905. 'Enclosure called after Dæggel'. *Daglingworth -yng- -wurth* c.1150–1535, *Dag(g)(e)lingew(o)rth -ynge-* 1200–1335. OE pers.n. **Dæggel* + **ing**[4] + **worth**. Gl i.69.

DAGNALL Bucks SP 9916. 'Dagga's nook'. *Dagehale* [1192]c.1300, 1235–1300, *Dagenhale* 1196–1372, *Dagnall* 1539. OE pers.n. **Dagga*, genitive sing. **Daggan*, + **halh** referring to a nook in the slope of the Chiltern Hills. **Dagga* is a hypocoristic form of names in *Dæġ-*, cf. OSax *Dago*, OHG *Taggo*. Bk 94, L 102.

DALBURY Derby SK 2634. 'Dalla's fortification'. *Delbebi* (sic for *-beri*) 1086, *Dalebiry* c.1141, *Dal(e)bury* 1281–1333 etc., *Dalby(e)* 16th. OE pers.n. *D(e)alla* + **byriġ**, locative-dative sing. of **burh**, partly replaced by ON **bȳ**. DB also has the form *Dellingeberie* 1086, which appears to be 'fortification called or at *Dallinge*, the place called after Dalla', locative-dative sing. *Dallinġe*, + **byriġ**. Db 548.

Great DALBY Leic SK 7414. Latin prefix **magna**, *Magna* c.1130–1537, **maior** 'greater', *Major* 1450, 1535, + p.n. *Dalby* from 1086, with variants *-bi* 1086–1404 and *Dale-* c.1130–1404, *Daubi -by* late 12th–1404, 'valley village or farm', ON **dalr** + **bȳ**. The name of the valley in which the Dalbys lie was *Crumdale* [late 13th]1449, 'the crooked valley', OE **crumb** + **dæl**, from which nearby Crown Hill is named. Also known as *Chacombe- Chakun- Checom Dalby* 1272–1535 and *Dalby Chal- Cha(u)combe* 1233–1724 from Huge de Chaucumbe who held the manor in the 12th cent. The affixes are for distinction from Little DALBY SK 7713. Lei 152, SSNEM 43 no. 2(a).

Little DALBY Leic SK 7713. Latin prefix **parva**, *Parva* 1212–1494, OFr **petit**, *Petit* 1266 and ME **litel**, *Litle, Litul, Litylle* 1444–1552, *Little* from 1610 + p.n. Dalby as in Great DALBY Leic SK 7414. Also known as *Dalby Paynal* 1242–1367, ~ *Perer* and ~ *Tateshale -is-* 1242 from manorial tenants. Lei 153.

Old DALBY Leic SK 6723. 'Wold D, D on the wolds'. *Dalbia super Waldas* c.1130, *Dalby* ~ 1316, *Dauby de Wauz* 1209, ~ *super (le) Wold(e)* 1316–1529, *Dalby (vp)on (the) Oldas* 1610, 1688, *Old Dalby* 1718. Earlier simply *Dalbi* 1086, 1208, *Dalby(e)* 1209×35–1610, *Daubi -by* 1206–1340, *Dawbye* 1543, 'the valley village', ON **dalr** + **bȳ**. The village lies in a small valley on the edge of the Wolds. For loss of *w* before ME *ō* see Dobson §421. Lei 281.

River DALCH Devon SS 8011. 'Dark stream'. *(on, oþ) doflisc* [739]11th S 255. PrW ***duβ** + ***gles**. Identical with DAWLISH SX 9676, DOUGLAS SD 4619 etc. D 4, RN 130.

DALDERBY Lincs TF 2466. 'The farm or village of the small valley'. *Dalbi* c.1115, *Dauderby* late 12th, *Dalderby* from 1221 including [1147×51]14th. This is either ON **dalr** or **dæld**, genitive sing. **dældar**, + **bȳ** referring to a small valley running down to the r. Bain. DEPN, SSNEM 43, Cameron 1998.

DALE Derby SK 4338. 'The valley'. *(la, le) Dal(e), Dala, Dalle* late 12th etc. Short for Deepdale, *Depedala -dal(e)* 1158–1272×1307, OE **dēop** + **dæl** (ON *dalr*). Db 442, L 95.

DALE DIKE RESR. SYorks SK 2491. Dale Dike, 'the valley ditch', is the name of the upper part of the r. Loxley in Bradfield Dale, *Bradfield Dale* 1770×9, p.n. BRADFIELD SK 2692 + ME **dale**, + ModE **dike** (OE *dīc*). YW i.233, L 94.

DALE HEAD Cumbr NY 4316. 'Head of the dale', sc. of Martindale. *Dailehead* 1573, *Dalehead(e)* 1588–1823. Earlier *Martendaleheved* 1363, *Martindalehead* 1589. P.n. MARTINDALE NY 4319 + ME **heved** (OE *hēafod*). We ii.218.

DALES HEAD DIKE Lincs TF 1563. *Dales Head Dyke* 1824 OS. P.n. *Dale head* 1842 + ModE **dyke**. The reference is to The Dales, an area subject to flooding between Martin TF 1259 and the r. Witham, ultimately from OE **dāl**, ON **deill** 'a share, an allotment' of common meadow. Cf. Martin DALES TF 1761. Perrott 330, Wheeler 183.

DALHAM Suff TL 7261. 'The valley homestead'. *dalham* 1086, *Dalham* 1610. OE **dæl** + **hām**. DEPN.

Field DALLING Norf TG 0039. 'Dalling in the open country'. *Fildedalling* 1272. OE **(ġe)filde** or adj. ***filden** + p.n. *Dalli(n)ga* 1086, *Dallenges* 1138, *Dallinge, Dalliges* 1198, *Daullinges* 1223×4, 'the Deallingas, the people called after Dealla', OE folk-n. **Deallingas* < pers.n. *Dealla* + **ingas**. 'Field' for distinction from Wood DALLING TG 0827. DEPN, ING 56.

Wood DALLING Norf TG 0827. 'Dalling at the woodland'. *Wdedallinges* 1198, *Wode Dallinges* 1199. OE **wudu** + p.n. *Dallinga* 1086, *Dallinges* 1199, *Dauligg'* 1236, *Dallinge* 1242×3, 'the Deallingas, the people called after Dealla', OE folk-n. **Deallingas* < pers.n. *Dealla* + **ingas**. 'Wood' for distinction from Field DALLING TG 0039. Possibly this place was an offshoot of Field Dalling some ten miles to the N. DEPN, ING 56.

DALLINGHOO Suff TM 2655. 'The hill-spur of the Deallingas, the people called after Dealla'. *Dal(l)inga- Dalinge- Delingahou* 1086, *Dalingeho* c.1150, *Dal(l)ingho -yng-* 1275–1589, *Dallinghoo* 1674. OE folk-n. **Deallingas* < pers.n.

Dealla + **ingas**, genitive pl. **Deallinga*, + **hōh**. The name contrasts with simplex HOO at TM 2558. DEPN, ING 56, Baron.

DALLINGTON ESusx TQ 6519. 'Estate called after Dealla'. *Dalinton(e)* 1086–14th, *Dalington* 13th–16th cents., *Dolinton'* 1232. OE pers.n. *Dealla* + **ing**^{4} + **tūn**. Sx 473, SS 76.

DALLOW NYorks SE 1971. 'The valley enclosure'. *Dalhagh(a)* 1175–1221, 1461, *Dalagh(e)* 1198, 1535, *Dal(l)a* 1379, 1618. OE **dæl** + **haga**. YW v.212.

DALLOWGILL MOOR NYorks SE 1671. P.n. DALLOW SE 1971 + ModE **gill** (ON *gil* 'a ravine') + ModE **moor** (OE *mōr*).

DALSTON Cumbr NY 3650. Probably 'Dall's farm or village'. *Daleston'* 1187–1298, *Dalston'* from 1187, *Dalles-* 1190–1292, *Dawes-* 1464, *Dauston* 1576. OE pers.n. **D(e)all* (<adj. *deall* 'foolish'), genitive sing. **D(e)alles*, + **tūn**. The weak form *Dealla* is on record. Cu 130 gives pr [dɔ:stən].

DALTON 'Valley settlement' OE **dæl** + **tūn**. PNE i.126.

(1) ~ Lancs SD 4908. *Daltone* 1086, *-ton* from 1212, *Daltun* [before 1225]1268. The reference is to the valley of the Douglas. La 105.

(2) ~ Northum NZ 1172. *Dalton* from 1201 FF, *Dawton* 1436. NbDu 59, DEPN.

(3) ~ Northum NY 9158. *Dalton* from 1256 Ass. NbDu 59, DEPN.

(4) ~ NYorks NZ 1108. *Daltun, altera Daltun* 1086, *Daltona et alia Daltona* 1184. Also known as *Dalton Travers* 1258, *Dalton Michel* 1259, *Dalton Norreys* 1285, *Dalton in le Gayles* 1559, and the *Dawtons* 1577. Two and later three manors are distinguished by various manorial additions alluding to possession by the Travers family, a tenant called Michel and a tenant called John Norris. For *in le Gayles* see GAYLES NZ 1207. Dalton lies in a narrow glen. YN 290 gives pr [dɔ:tən].

(5) ~ NYorks SE 4376. *Deltune* 1086, *Dautona* c.1200, *Dalton* from 13th, *Daweton* 1573. YN 183 gives pr [dɔ:tən].

(6) ~ NYorks NZ 2907. *Dalton super Tese* 1221×6, 'D on Tees'. YN 283.

(7) ~ SYorks SK 4694. *Daltun(e) -ton(e)* 1086–1822, *Dauton* 1197, 1260. The reference is to the valley of Dalton Brook. YW i.179.

(8) ~ -IN-FURNESS Cumbr SD 2373. *Dalton in fournais* 1332. Earlier simply *Daltune* 1086, *Dalton(am)* from 1189×94. 'In Furness' from FURNESS SD 2087 for distinction from other Daltons. La 201, L 95–6.

(9) ~ -LE-DALE Durham NZ 4048. *Dalton in valle* 1333, 1343, *Dalton in le Dale* 1420, *Dawton in le Hole* 1609. Earlier simply *Daltun* [c.716]10th, [c.1040]12th–1344×5, *Dalton(a)* 1154×66–1609. The addition *le Dale* < French definite article **le** (short for *in le*) + ME **dale** is for distinction from DALTON PIERCY Cleve NZ 4631. NbDu 59.

(10) ~ PIERCY Cleve NZ 4631. *Dalton Percy* [1381×2]–1564, ~ *Piercy* from 1729. Earlier simply *Daltun* [1150]15th, *Dalton(e)* 1235–1407. The manor was held by the Percy family in the 13th cent.

(11) North ~ Humbs SE 9352. *Northdalton(a)* from 1150×60. ME adj. **north** + p.n. *Dalton(a)* 1086–1400, *Dauton* 1246. N in relation to South DALTON SE 9645. YE 168 gives pr [dɔltn].

(12) South ~ Humbs SE 9645. *S(o)uthdalton* from 1259. ME adj. **suth** + p.n. *Delton* 1086, *Dalton -tun* 1166 etc. S in relation to North DALTON SE 9352. YE 190.

DALWOOD Devon ST 2500. 'Valley wood'. *Dalewude -wde -uuode* 1195–1231, *Dalw(u)de -wod(e)* 1201–1344, *Dal(l)ad* 1545. OE **dæl** + **wudu**. D 638 gives pr [daləd].

DAMERHAM Hants SU 1015. 'River meadow of the judges'. *(æt) Domra hamme* [873×88]11th S 1507, *(æt) Domerhame* [944×6]14th S 513, c.1100 ASC(D) under year 946, *Domarham* [975×91]11th S 1494, *Dobreham* (sic) 1086, *Domerham* 1186–1428, *Damerham* 1231, 1564, *Damerum* 1675. OE **dōmere**, genitive pl. **dōmera**, + **hamm** 3. Damerham was a royal estate. Its revenue was presumably appropriated to the support of judges. Ha 64, Wlt 400.

DAMFLASK RESR. SYorks SK 2791. P.n. Damflask + ModE **reservoir**. Damflask, *Damflask* 1849, is the 'damm pool or marsh', ModE **dam** (ON *dammr*) + dial. **flask** (ODan *flask*). YW i.233.

DAMGATE Norf TG 4009. *Damgate* 1838 OS. Probably 'gate by the dam'. *Dam* also had the sense 'causeway through fens'. OED s.v. sb. 1.c.

DANBURY Essex TL 7805. The 'stronghold of the Dænningas, the people called after Dæne'. *Danengeberiam* 1086, *Dan(n)ing(e)bir(y) -yng- -byry -bury* 1233–1306, *Daingeb' Dainghebury* 1235, 1290, *Dan(e)bury* from 1249, *Daneweberi -bury* 1300–1428. OE folk-n. **Dænningas* < pers.n. *Dæne* + **ingas**, genitive pl. **Dænninga*, + **byriġ**, dative sing. of **burh**. The remains of an ancient camp are said to have been visible here in the 18th cent. Ess 248 which takes *Dæningas* to be the 'woodland pasture people', OE **dænn**, **denn** 'a swine-pasture' + **ingas**, but this element seems not to occur outside Kent, Sussex and Surrey, DEPN.

DANBY 'Farm, village of the Danes'. ON *Danir*, genitive pl. *Dana*, + **bȳ**.

(1) ~ NYorks NZ 7008. *Danebi -by* 1086–1328, *Danby* from 1285. YN 131.

(2) ~ LOW MOOR NYorks NZ 7110. P.n. DANBY NZ 7008 + ModE **low** + **moor**.

(3) ~ WISKE NYorks SE 3398. P.n. *Danebi -by* 1086 + r.n. *Wiske* late 13th. The village lies on the river WISKE NZ 4300, SE 3497. YN 276.

River DANE Ches SJ 6868, 9664. Possibly 'slow river'. *Dau-Daven(e)* 12th–1724, *Daane* 1295, 1392, *Dane* from 1443. OW **dafn** 'a drop, a trickle'. If this is correct, the topography of the river, especially in its upper reaches, suggests that the appellation was given ironically. Che i.20, RN 112.

DANE END Herts TL 3321. *Dane End* 1404, 1556. Possibly surname *Dean* + ME **end** sc. of Little Munden parish. Hrt 136.

DANE'S BROOK Somer SS 8331. *Dunsbrook flud* 1610, *Daine's Brook* 1809 OS.

DANEBRIDGE Ches SJ 9665. 'Bridge over the river Dane'. *Dauenbrugge* 1357, *Dane Bridge* 1611. R.n. DANE SJ 6868 + OE **brycg**. Che i.166.

DANEBURY Hants SU 3237. An outstanding Iron Age hill-fort, the first phase dating from the 4th cent. BC, and the final refurbishment to the Roman advance of AD 43–4. The modern form of the name is misleading: it has nothing to do with the Danes as the early forms show, *Duwnebury Hill* 1491, *Dunbery hill* 1593, *Dunbury* 1637, *Deanbury Hill* 1817 OS, probably 'hill fort', OE **dūn** + **byriġ**, dative sing. of **burh**. For a similar example of popular etymology cf. Norsebury Ring Hants SU 4940, another hill-fort, *(to) næsan byrig* [900]11th S 360, 'the ness fort', later *Nosebury* 1272×1307, 'the nose fort'. Ha 64, Gover 1958.192, 82, Thomas 1976.141.

DANEHILL ESusx TQ 4027. 'Denn hill'. *Denhill* 1437, *Danell* 1558×1603, *Danehill* 1760×1820. P.n. *Denne* 1279, OE **denn** 'swine-pasture', + ME **hill**. The historically correct pronunciation is illustrated by the spelling *Danell*; in Sussex OE *an* when subject to i-mutation usually becomes [æn], cf. East DEAN TV 5998. Sx 335.

DANES' DYKE Humbs TA 2171. Originally called Flamborough Dyke, *Flayn(e)burghdyk -dike* 1392, 1446, 1452. It is in fact a pre-Roman earthwork separating Flamborough Head from the mainland. YE 106, Thomas 1976.155.

DARDEN LOUGH Northum NY 9795. *Darden Lough* 1869 OS. P.n. Darden + Northumberland dial. **lough** 'a lake' (OE *luh*, itself a loan-word from PrW, cf. W *llwch*). There are no early forms for Darden, but it could be a compound of OE **dēor** + **denu** 'deer valley'. GPC 2234.

River DARENT Kent TQ 5466. 'Oak-tree river'. There are three types;

I. *diorente* 822 S 186, *De- Dærentam* [983]12th S 849, *Derente* [c.1200] c.1260, *Dernthe* 1399, *Darnt* 1577.

II. *Derguentid* [800]11th HB with variants *Derguint, Der(e)uent, Derguent* [c.1071] c.1400.
III. *Derwent* [1147]12th, *Derwent(e)* 1292, [c.1350] c.1400.
PrW ***derwïnt**, Brit **deruentiū* < **deruā* 'oak-tree'). KPN 144, RN 113, Jnl 1.46.

DARENTH Kent TQ 5671. Transferred from the river name DARENT TQ 5466. *Darente, Daerintan* [940]12th S 1210, *(æt) Dærœntan, Dœrente* [973×87]12th S 1511, *Tarent* 1086, *Derente* 1185–1240, *Darente* 1193×1205 etc., *Darent vulgo Darne* 1769. KPN 144, RN 113.

South DARENTH Kent TQ 5669. *South Darent* 1819 OS. ModE **south** + p.n. DARENTH TQ 5671.

DARESBURY Ches SJ 5782. 'Dēor's fortified place'. *Deresbiria* [1154×89]17th, *Deresbury* 1194–1599 with variants *-beri -biri -buri* etc., *Darisburi* c.1200, *Daresbury* from 1434, *Dasbury* 1660, *Doresburi* c.1200. OE pers.n. *Dēor*, genitive sing. *Dēores*, + **byriġ**, dative sing. of **burh**. Che ii.148. Addenda gives 19th cent. local pr ['da:zbri], now ['dɛ:rzbəri], older local ['da:rzbri].

DARFIELD SYorks SE 4104. 'Open land frequented by deer'. *Dereuueld -uuelle* 1086, *Derfeld(e)* 1155–1531, *Darffeld* 1380, *-feld(e)* 1433–1567. OE **dēor** + **feld**. YW i.95, L 244.

DARGATE Kent TR 0761. 'Deer gate'. *Deregate* 1275, *Dergate* 1348, 1458, *Dargate* 1535. ME **dere** (OE *dēor*) + **gate** (OE *ġeat*). PNK 304.

DARITE Corn SX 2569. Obscure. *Daryet* 1506, *Daryth* [1510]c.1595, *Daryte* [1530]c.1595. Possibly from the surname Daryte, *Daryth* 1391, *Daryte* 1569. The surname is not explained. PNCo 77.

DARLASTON WMids SO 9897. 'Deorlaf's estate'. *Derlaveston* 1262, *Derlaston* 1316. OE pers.n. *Dēorlāf*, genitive sing. *Dēorlāfes*, + **tūn**. DEPN.

DARLEY 'Deer wood or clearing'. OE **dēor**, genitive pl. **dēora**, + **lēah**. The reference is probably to a glade where deer were captured or corralled.
(1) ~ DALE Derby SK 2763. P.n. Darley + ModE **dale**. Darley is *Derelei(e)* 1086, *Derl(e)ia -leie -le(i)gh -le(e) -ley(e)* [1120×6]1329, 1121×6–1575, *Darley* from 1448. Db 81.
(2) Little ~ Derby SK 3438. *parua Derl'* c.1140. ModE **little**, Latin **parva**, + p.n. *Derleg(a) -leia -leie -ley(a) -leye* 1154–1570. Db 443.

DARLEY GREEN WMids SP 1874 > DORRIDGE SP 1775.

DARLINGSCOTT Warw SP 2342. 'Darling's cottage(s)'. *Derlingescot* 1210, *Derlingiscote* 1272, *-es-* 1275–1331, *Darlingscot* 16th. ME surname *Derling* (OE *dēorling* 'darling'), genitive sing. *Derlinges*, + **cote** 'cottage(s)' (OE *cot*, pl. *cotu*). Wo 173, DEPN.

DARLINGTON Durham NZ 2914. 'Estate called after Deornoth'. The forms fall into three types:
I. *Dearthingtun* [c.1040]12th, *Derington* 1300, *Darington* 1558.
II. *Dearningtun -ton* c.1104, *Dernington* [1070×83]15th, c.1200–c.1250, *Derninton* 1217×26, *Dernton* [1443]15th, *Darn(e)ton* 1574–1675.
III. *Derlington(a -e) -yng- -ig-* c.1175×1200–1589×90, *Darlington* from [c.1297]1329.
OE pers.n. *Dēornōth* + **ing**[4] + **tūn**. *Dēornōthingtūn* early underwent two independent shortenings, one to *Derthingtun*, the other to *Derningtun*. The forms are almost exactly parallel to those for DARRINGTON WYorks SE 4820, I. *Derth- Darthington* c.1170–1495, *Darington* 1243–1615; II. *Darnintone* 1086, *Dernington* 1307; III. *Derlington(e)* 1279×81, *Darlyngton* 14th. II is the spoken form which still survives in the local pr [da:ntən], III is the chancery form exhibiting AN substitution of *r-l* for *r-n*. NbDu 60, YW ii.63.

DARLISTON Shrops SJ 5833. 'Deorlaf's estate'. *Derloueston* 1199, *Dorlaveston* 1224, *Derlawstun* 1249, *Derlaston* 1327. OE pres.n. *Dēorlāf*, genitive sing. *Dēorlāfes*, + **tūn**. Bowcock 84, DEPN.

DARLTON Notts SK 7773. 'Deorlufu's farm or village'. *Derluveton* 1086, *Derlintun' -ton* 1156–1227, *-ing- -yng-* 1216×72–1345, *Darlyngton* 1471, *Derleton* 1172–1316, *Darleton* 1458. OE feminine pers.n. **Dēorlufu*, genitive sing. **Dēorlufe*, + **tūn**. The development of *ing-* forms in the 13th cent. is noteworthy. Nt 46.

DARNBROOK FELL NYorks SD 8872. R.n. Darnbrook SD 8871 + ModE **fell** (ON *fjall*). Darnbrook, *Dern(e)broc -broke* 12th–1461, *Darn(e)bro(o)ke* 1541–1695, is the 'hidden stream', OE **derne** + **brōc**. YW vi.138.

DARRAS HALL Northum NZ 1571. *Dareshall* 1695 Map, *Darras Hall* 1864 OS. Short for Callerton Darras, 'the part of the manor of Callerton held by the Darrayns family', *Calverdon' Araynis* 1242, ~ *Darayns* 1296, ~ *Darreyne, de Arreyns* 1346, *Calverton Darrays* 1360, *Callerton Darres* 1428. The manor was held by Wydo de Araynis in 1242, whose family originated from Airaines, Somme, France. For Callerton see Black CALLERTON T&W NZ 1769 and High CALLERTON Northum NZ 1670.

DARRINGTON WYorks SE 4820. 'Estate called after Deornoth'. *Darni(n)tone* 1086, *Dernington* 1307, *Dardin(g)ton(a)* c.1090–1235, *Darthin(g)ton(a) -yng-* c.1170–1445, *Derthington -yng-* 1364–1495, *Derlington(e)* 1279×81, *Darlyngton* 14th, *Darington -yng-* 1243–1615, *Darrington* from 1558. OE pers.n. *Dēornōth* + **ing**[4] + **tūn**. In the compound p.n. OE pers.n. *Dēornōth*, ONb *Dēarnōth*, was reduced to *Darth-* or *Darn-* and the latter shows occasional AN substitution of *l* for *n* as in DARLINGTON Durham NZ 2914. YW ii.63.

DARSHAM Suff TM 4169. 'Deor's homestead'. *Ders(h)am, Diresham* 1086, *Dersham* 1224–1524, *Darsham* from 1469×70. OE pers.n. *Dēor*, genitive sing. *Dēores*, + **hām**. DEPN, Baron.

River DART Devon SX 7565. 'Oak-tree river'. *to Dertan, on dertan stream* 10th B 1323, *Derte -a* 1162–1326, *Dert* 1360, 15th, *Dart* from 1575, *Darnt* 1577, *Darent* 1586. PrW r.n. **Derwïnt* < PrW ***derw** 'an oak-tree' + r.n. suffix ***ïnt** (*-ent-*). The early loss of *w* is surprising; the derivation, however, is substantiated not by the 16th cent. spellings here but by the earliest spelling for DARTMOUTH SX 8751. D 4, RN 114.

East DART RIVER Devon SX 6676. ModE adj. **east** + r.n. DART.

Little DART RIVER Devon SS 8316. *Little Dart* 1765. Earlier simply *Darte* 1544. A tributary of the Taw, identical with River DART SX 7565. Earlier spellings are found in the forms for Dart Raffe SS 7915, 'Dart held by Ralph', *Derta* 1086, *Derth'* 1242, *Derte Rauf* 1329. The manor was held by Radulfus in 1086 and by Ralph de Derthe in 1242. D 5, 398.

West DART RIVER Devon SX 6373. ModE adj. **west** + r.n. DART.

DARTFORD Kent TQ 5474. 'Ford across the river Darent'. *Tarenteford* 1086, 1154×4, 1158×9, *Darenteford* 1089–1154×89, *Derteford* c.1100–1219 etc., *Dartfoorde* 1610. R.n. DARENT TQ 5466 + OE **ford**. PNK 31.

DARTINGTON Devon SX 7862. 'Settlement on the river Dart'. *Dertrintone* (sic) 1086, *Dertinton(ia)* 1162–1282 with variant *-ing-*, *Dertyngton* 1326. R.n. DART Devon SX 7565 + **ing**[2,4] + **tūn**. The form *Derentunehomm* 'river-bend land at or called Derwent *tūn*' [833]14th S 277 has been associated with Dartington but the ascription is uncertain and hardly fits the run of genuine forms. D 297, RN 114.

DARTINGTON HALL Devon SX 7962. An Elizabethan mansion with ruins of a 13th cent. hall. P.n. DARTINGTON SX 7862 + ModE **hall**.

DARTMEET Devon SX 6773. *Dartameet* 1616. R.n. DART SX 7565 + ModE **meet**. The meeting point of the East and West Dart rivers. D 198.

DARTMOOR Devon SX 6276. *Dertemora -e* 1181–1359, R.n. DART SX 7565 + OE **mōr**. D 18.

DARTMOOR FOREST Devon SX 6180. *Forest of Dertemore* 1232×9. P.n. DARTMOOR SX 6276 + ME **forest**. D 18.

DARTMOUTH Devon SX 8751. 'Mouth of the river Dart'. *(to) Dærentamuðan* 11th ASC(C) under year 1049, *(to) Dertamuðan* 11th ASC(D) under year 1050, *Dertemuthe -mue* 1231–1377,

Dartemuth 1265. R.n. DART SX 7565 + OE **mūtha**. D 321, RN 114.

DARTON SYorks SE 3110. 'Deerpark or deer farm'. *Dertun(e) -tone -ton(a)* 1086–1433, *Darton* from 1333. OE **dēor-tūn**. YW i.317.

DARWELL RESERVOIR ESusx TQ 7121. P.n. *(la) Derefold(e)* 1294–1320, *Darvoll* 1603, 'the deerfold', OE **dēor** + **fald**, + ModE **reservoir**. Sx 472, 475.

DARWEN Lancs SD 6922. This is the river name DARWEN transferred to the settlement on its banks. *Derewent* 1208, 1246, *Darrun* 1868. Also known as Over Darwen, *(in) superiori Derwent* 13th, *Overderewente* 1276, *Overderwyn -darwyn* 14th–16th cent., *Overdarwen* 1518×19. 'Over' for distinction from Lower DARWEN SD 6825. La 75, Jnl 17.47 gives local pr 'Darren' [dærən].

Lower DARWEN Lancs SD 6825. *Netherderwent* 1311, 1335, *Nether Derwyn, Darwine* 16th, 17th cent. Adj. **nether** 'lower' + r.n. DARWEN. 'Lower' for distinction from DARWEN SD 6922. La 75, Jnl 17.47.

River DARWEN Lancs SD 6129. 'Oak river'. *Derewente* 1227, *Derwent* 1240–1362, *-wint* before 1290, *Darwent* c.1540, *Darwen* 1656. PrW ***derwïnt** (Brit **deru̯entiū*), a r.n. derived from Brit ***deru̯ā** 'an oak-tree'. La 66, RN 121, Jnl 1.46, 17.44.

Avon DASSET Warw SP 4050. 'The part of D on the r. Avon'. *Afnedereceth* 1185, *Auene Dercete* 1202, *Aven(e)dercet -s(c)et(e)* 1227–1428, *Awen Darset* 1496, *Aven Dassett* 1535, *Avon Dossett* 1616. R.n. AVON + p.n. *Derceto* 1086, *-chet* 1173–1214, *-cet* 1233, 1241, probably 'the oak-tree wood', PrW ***derw** + ***cę̄d**, referring to the Notts, Leics, Northants and Warwicks wolds on the edge of which the settlement is situated like DOSTHILL Staffs SP 2199 with the same origin. See the map in Fox 1989. An alternative suggestion is 'the deer-shelter', OE **dēor** + **ċēte**. Also known as Little Dasset, *Dersete parva* 1291. 'Avon', for distinction from Burton DASSET SP 3951, must have been a former name of the stream that flows here and joins the Cherwell at SP 4542. Wa 267, DEPN.

Burton DASSET Warw SP 3951. 'The part of D by Burton'. *Burton Dassett* 1604, *Burton Dorsett* 1800. P.n. *Buriton* 1327, 1332, *Birton* 1497, *Burton* 1538, 'the fort settlement', OE **byriġ**, dative sing. of **burh**, + **tūn**, + p.n. *Dercetone* 1086, *Derceth* 12th, *Derset* 1291–1509, *Dorset* 1326, 1496, either 'the oak-tree wood' or 'the deer-shelter' as in Avon DASSET SP 4050. Also known as *Magna Dercet* 1242, *Great Dorcestre* (sic) 1267, *Greate Dasset* 1625, and *Chepynge Dorset* 1397, *Chipping Dorset* 1650, 'D with the market', ME **chēping** (OE *ċ(i)ēping*). Wa 268, WW 150.

DATCHET Berks SU 9877. Unexplained. *Deccet* 990×2 S 1454, *Daceta* 1086, *Dachet(te)* 1181–1338, *Dechet(te)* 1237–1449, *Dotchatt* 1626. Possibly a compound name of unknown element + PrW ***cę̄d** (Brit **cēto-*, W *coed*) 'wood'. Bk 234, LHEB 327.

DATCHWORTH Herts TL 2619. 'Dæcca's enclosure'. *Decewrthe* *[969]12th S 774, *(in) Deccewyrððe* *[1066]13th S 1043, *Deceswrþe* [1049]14th, *Daccewrth'* [1049]13th S 1123, *(æt) Tæccingawyrðe* [11th]12th K 1354, *Dæccewrðe* c.1060, *Dæcceuuyrthe* 1065, *Daceuuorde* 1086, *-w(o)rth(e)* 1193–1428, *Thacheworth(e)* 1248–1531, *Thatchworth al. Datchworth* 1558×1603. OE pers.n. **Dæċċa* + **worth**. The form *Tæccingawyrðe* shows assimilation of initial *D-* to *T-* after preposition *æt* and a unique form in *-inga-*, genitive pl. of **ingas**, 'the enclosure of the Dæccingas, the people called after Dæcca'. Later forms show assimilation to the word *thatch*. One of two neighbouring *worths*, Datchworth and Knebworth. Hrt 122 gives pr [dætʃər(θ)].

DAUNTSEY Wilts ST 9982. 'Domgeat's island'. *Dometesig* [850]14th S 301, [1065]14th S 1038, *(at) Domeccesige* *[854]14th S 305, *Dauntes- Dameteseye* [850]14th S 11580, *Dantesie* 1086, 1160, *Dauntesa* [1154×89]1270, *-eye* 1268–1316, *Daundesey* 1407, *Daunsey* 1516, *Dancy* 1655. OE pers.n. **Dōmgeat*, genitive sing. **Dōmgeates*, + **īeġ**. The Avon still forms two channels here. Wlt 68 gives pr [da:nsi] as in the rhyme cited under CLYFFE PAPARD SU 0776.

DAVENHAM Ches SJ 6671. 'Village on the river Dane'. *Deveneham* 1086, *Davenham* from 1178, *Daveham* 1325, 1536, *Daneham* 1430–1599, *Denham* 1402, 1739. R.n. DANE SJ 6868 + OE **hām**. Che ii.203 gives pr ['deivnəm], Addenda 19th cent. local ['deinəm].

DAVENTRY Northants SP 5762. 'Dafa's tree'. *Daventrei* 1086, *Dauentre -in-* 1199–1537, *Daventry* from 1320, *Dantr(e)y, Daintree* 1610–57, *Dauntrye, Dawntrie* 1639–40. OE pers.n. **Dafa*, genitive sing. **Dafan*, + **trēow**. Medial [v] was vocalised to give ME [au] which survives in the 17th cent. *au, aw* spellings and in the pronounciation [do:ntri], while lME [a:] < [au] gave the parallel 17th forms with *a, ai* and the pronounciation [deintri]. Now usually [dævəntri]. Nth 18, L 215, cf. Nomina 13.109ff.

DAVENTRY RESERVOIR Northants SP 5863. P.n. DAVENTRY + ModE **reservoir**.

DAVIDSTOW Corn SX 1587. 'Holy place of St David'. *(ecclesia) Sancti David alias Dewstow* 1269, *Dewystowe* 1311, *Dewestowe* 1313, *(par.) Sancti David* 1377, *Dewstowe alias Davystowe* 1423, *Dustow* 1678, *Davidstowe* 1610. Saint's name *Dewy* (later replaced with its English equivalent *David*) + OE **stōw**. St David, patron saint of Wales, was also venerated in Cornwall and Brittany. The church at nearby ALTARNUN is dedicated to his mother. Cf. Dewstow Gwent ST 4688 and St David's Dyfed SM 7525. Co 77, Gover n.d. 53, DEPN.

DAWLEY Shrops SJ 6807. Probably 'the woodland clearing called after Dealla'. *Dalelie* 1086, *Dalilea -leg(h) -ley(e) -y-* c.1200–1428, *Daghele* 1228, *Dahlegh'* 1301, *Dawley* from 1559×60. OE pers.n. *D(e)alla* + **ing**[4], varying with genitive sing. *D(e)allan*, + **lēah**. Sa i.106 gives local pr [douli].

DAWLISH Devon SX 9676. Originally a Welsh r.n., 'black stream'. *Doflisc* 1044 S 1003, 1069, 1050×73, *Dovles* 1086, *Duvelis* 1148, *Douelis* c.1200, 1253, *Dovelish* 1302, *Doulissh* 1323, *Dawlisshe* 1468. PrW ***duβ** + ***gles**. Identical with DALCH SS 8011, DOUGLAS SD 4619 etc. D 491, RN 130.

DAWLISH WARREN Devon SX 9778. *Warenna in Manerio de Douelis* c.1280, 'warren in the manor of Dawlish'. P.n. DAWLISH SX 9676 + ME **wareine** 'a game preserve', later 'a piece of ground for the breeding of rabbits, a warren'. A sandy spur of land projecting into the Exe estuary. D 494.

DAWPOOL BANK Mers SJ 2381. A sand-bank covering a silted-up channel in the river Dee formerly called *Dawpool Deep* 1842. P.n. Dawpool (lost at SJ 2383), *Dalpole* 1454, *Dawpoole* 1707, 'pool or creek at a valley', lOE **dæl** or ON **dalr** + **pōl**, + ModE **bank**. The reference is to a break in the coastal hills at SJ 233833 where the creek is marked on the 1842 1" OS map. Dawpool SJ 2484 is the name of the 19th cent. mansion (now destroyed) built for the ship-owner Thomas Henry Ismay. Che iv.280, Pevsner 1971.363.

DAWS HEATH Essex TQ 8188. *Daws heath* 1563, *Doesheathe* 1579, *Dores Heath* 1558×1603. Surname *Daw* < ME *dawe* 'a jackdaw' or *Dore* + ModE **heath**.

DAWSMERE Lincs TF 4430. Perhaps 'jackdaw pond'. *Daws Mere Creek* 1824 OS. ModE **daw** + **mere**.

DAYLESFORD Glos SP 2426. 'Dægel's ford'. *Dæglesford (vadum)* 'D ford' [718 for ? 727]11th S 84, [841]11th S 194, [875]11th S 215, [979]11th S 1340, *Dæiglæsford* *[964]12th S 731, *Degilesford* [979]11th S 1340, *Deilesford* *[777]12th S 112, *[1061×5]12th S 1238, *Dagelesford* *[1061×5]12th S 1238, *Eilesford* (sic for *Deiles-* misunderstood as *D'Eiles-*) 1086. OE pers.n. **Dæġel*, genitive sing. **Dæġeles*, + **ford**. Gl i.217, Wo 121.

DE LANK RIVER Corn SX 1376. P.n. *Dymlonke -lanke* c.1650 + ModE **river**. The evidence is too late for certainty; one possibilty is Co ***dyn** + ***lonk** 'gorge, gully' and so 'fort of the ravine'. The river flows through a gorge but no hill-fort is known near it. PNCo 78.

DEADWATER Northum NY 6096. *Deadwater* 1868 OS. According to Hodgson the North Tyne here 'runs in a most sluggish manner along a level plain, from which circumstance it is called Deadwater'. Tomlinson 229.

DEAL Kent TR 3752. Uncertain. *(In) Addelā -am* 1086, *(ad) dale* [1087]13th, *Dele -a* 1154×5, 1158×9, *Dale* 1240 etc., *Deale* 1610. This could either be the oblique case **dele** of OE Kentish **del** 'a pit, a hollow, a valley' or **del** 'a share of land, a district.' Phonology favours the former in which OE *ĕ* > *ę* > ModE *[ei]* later replaced by [i:]. The earliest form has the prefixed preposition *At*. PNK 577, KPN 299, PNE i.126, TC 81, L 94.

DEAL HALL Essex TR 0097. No early forms.

DEAN 'The valley'. OE **denu**, ME **dene**. In some names ME *eo, oe* and *u* spellings point to an OE back-mutated variant **deonu**. L 97.

(1) ~ Cumbr NY 0725. *Dene* c.1170–1385, *Deane* 1559. Cu 366, L 97.

(2) ~ Devon SX 7364. *Dene* 1086, *La Dene* 1244, later *Niderdenam* 1154×89, *Netherden* 1326, 'lower Dean', and *Dean Prior* 1809 OS. See also DEAN PRIOR. D 298.

(3) ~ Hants SU 5619. *Dean* 1810 OS. The surname *atte Dene* 'at the valley' 1292 may belong here. Gover 1958.49.

(4) ~ Somer ST 6744. *Dene* *[1065]18th S 1042, *Dean* 1817 OS.

(5) ~ HILL Wilts SU 2526. *Dean Hill* 1811 OS, cf. *Deane hill close* 1637. P.n. Dean as in West DEAN SU 2527 + ModE **hill**. Wlt 390 under Deanhill Fm.

(6) ~ PRIOR Devon SX 7363. 'The prior's Dean'. *Dene Prioris* 1316, *Dene Pryour* 1415. Earlier simply *Dene* 1086. Also known as *Over(e)dene* 1242, 1244, 'upper Dean', and *Church Dean* 1809 OS, 'Dean with the church'. According to BF 1242 it is Lower Dean (*Nitheredene*) that was held by the Prior of Plympton and the name Dean Prior is thus correctly applied in the 1809 OS. Since then the names DEAN and DEAN PRIOR have been erroneously interchanged. D 298, DB Devon ii.20, 13.

(7) ~ ROW Ches SJ 8781. 'Row of houses at Dean'. *Denerawe -rowe* 1477, *-row* 1512, *Dean Row* 1724. P.n. Dean + OE **rāw**. Dean, *le Dene* 1286–1357, is 'the valley', OE **denu**. The hamlet lies between the valleys of the Bollin and the Dean: it may well be named from the river DEAN. Che i.221.

(8) East ~ Hants SU 2726. *Estdena* 1167, *-dene* 1195–1334, *-dune* 1256–1327, *-deone -doene* 1332–90. ME adj. **est** + p.n. *Dene* 1086, *Dune* 1212. 'East' for distinction from West DEAN Wilts SU 2527. Ha 64, Gover 1958.188, DEPN.

(9) East ~ ESusx TV 5598. *Esdene* 1086, *Estden(n)'* 1279, *Estdeyn* 1545. OE adj. **ēast** + p.n. *Dene* 1086. For the pr [dein] implied by the 1545 form cf. DANEHILL TQ 4027. 'East' for distinction from WESTDEAN ESusx TV 5299. Sx 419, xlv.

(10) East ~ WSusx SU 9012. *Estdena* [1150]1227. ME adj. **est** + p.n. Dean as in West DEAN SU 8612. Dean is probably to be identified with the royal estate called *Edelingedene* [1002]13th S 904, 'the princes' Dean', OE **ætheling**, genitive pl. **æthelinga** + p.n. Dean, granted by Ethelred the Unready to Wherwell Abbey and with *(æt) Dene* [775 for ? 725]10th S 43 granted by Nunna to the bishop of Selsey. Sx 47 gives pr [i:zdi:n], xlv, SS 71.

(11) Forest of ~ Glos SO 6311. *Dana sylva, Dane(i)um nemus* 12th, *(forest' de) Dena -e* 1100–1488, *(forest of) Dean, Deen(e)* 1456–1669. ME **forest** + p.n. Dean, referring to the valley of Cannop Brook which runs through the forest into the Severn near Lydney. Gl iii.209.

(12) River ~ GMan SJ 8881. Short for Dean Water, 'valley stream'. *Deyne Water* 1552, *Deanwater* 1632. ME **dene** (OE *denu* or a back-formation from DEAN ROW SJ 8781) + **water** (OE *wæter*). In 1291 it is called *aqua de Honford*, 'Handforth water', see HANDFORTH SJ 8583. Che i.20.

(13) Upper ~ Beds TL 0467. *Overdene* 1430, *Overdeane* 1539. Adj. **over, upper**, + p.n. *Dene* from 1086. 'Upper' for distinction from Lower Dean TL 0569 lower down the valley, *Dene juxta Tillebroke* 1307, *Netherdeane* 1539. Bd 14.

(14) West ~ Hants SU 2527. *West Duene* 1319, 1324, *Westdune* 1270, *Westdene* 1279, *Westdoene* 1297, 1327, *-deone* 1345. 1351. ME adj. **west** + p.n. *(æt) Deone* [873×88]11th S 1507–1332, *Duene* 1086, *Dune* 1242–1316, *Dene* 1275, *Doene* 1289, 1428, *Deene* 1467, *Dean* 1547. The reference is to the valley of the r. Dun. 'West' for distinction from East DEAN Hants SU 5619. Wlt 377, Studies 1931.65.

(15) West ~ WSusx SU 8612. *Westdena* [1150]1227. ME **west** + p.n. Dean as in East DEAN SU 9012. Sx 49 gives pr [wezdi:n], SS 71.

(16) West ~ Wilts SU 2527. *Westdune* 1270, *Westdoene* 1297, *Westdeone* 1345, *Westden* 1275, *-dene* 1279, *-deene* 1310. ME adj. **west** + p.n. *(æt) Deone* [873×88]11th S 1507, 1275–1332, *Duene* 1086, 1315, *Dune* 1242–1316, *Doene* 1289, 1428, possibly representing OE ***deonu** < *denu* 'a valley' by *u*-mutation. The reference is to the valley of the river Dun. The twin villages may represent the estate of the Roman village round which West Dean has grown. 'West' for distinction from East DEAN Hants SU 2726. Wlt 377, Studies 1931.65, Ha 64.

DEANE Hants SU 5450. 'The valley'. *Dene* 1086–1544, *Deane* 1610. OE **denu** referring to the dry upper reach of the valley of the river Test. Ha 65, Gover 1958.136.

DEANSCALES Cumbr NY 0926. 'Dean shielings'. *Deneschall -scal'* 1278–9, *Deenskalis* 1391. P.n. DEAN NY 0725 + ON **skáli**. Cu 366.

DEANSHANGER Northants SP 7639. 'Dynni's sloping wood'. *Dinneshangra* [937]14th S 437, *Dune(s)- Daneshanger* 1227–1330, *Deneshangre* 1252–1336, *Deanshanger* 1640. OE pers.n. *Dynni*, genitive sing. *Dynnes*, + **hangra**. Nth 101 gives prs [dens- dʌns- dinsæŋə], L 195.

DEARHAM Cumbr NY 0736. 'Deer village'. *Derham* c.1160–1456, *-heim* 1212, *Dereham(e)* c.1212–1429, *Deareham* 1580. OE **dēor** + **hām**. An ancient settlement where deer were kept or hunted. Cu 283.

DEARNE SYorks SE 4604 → r. DEARNE SE 3408.

River DEARNE SYorks SE 3408. Unexplained. *Di- Dyrne, Dirna* c.1154–1309, *Derna* 1154×9, 1186, *Derne* 1340–1588, *Darne* 1577, *Dearne* from 1588. The early forms are against derivation from OE **dierne** 'hidden, secluded' since this word would have been **derne** in the OE dial. of this area. Nevertheless it seems to have influenced the later development of the name. YW vii.125 gives pr [diən, də:n].

DEBACH Suff TM 2454. 'The ridge overlooking the r. Deben'. *Depebek -be(c)s, De(ben)beis* 1086, *Debech(')* 1201, 1286, *Debbeche* 1270, *Debach* from 1250, *-bage* 1524–1612, *-bache* 1610, *Debbadge* 1668. This cannot be 'the deep stream' as the place stands on a ridge of high ground; the generic must be OE **bæċ** 'a back, a ridge'. The place overlooks Potford Brook, a tributary of the Deben, which may have shared the main river's name in former times. DEPN, Baron gives pr [debidʒ].

DEBDEN Essex TL 5533. 'The deep valley'. *Deppedanā* 1086, *Depeden(e)* 1227–1422, *Depden* 1348, 1421, *Debeden* 1428. OE **dēop**, definite form **dēope**, + **denu**. Ess 523.

DEBDEN GREEN Essex TL 5832. *Debden Gr.* 1805 OS. P.n. DEBDEN TL 5533 + ModE **green**.

River DEBEN Suff TM 2061. *Deue* 1577, 1586, *Deane* 1618, *Deben* 1735. A back-formation from DEBENHAM TM 1763. RN 117, Arnott 2.

DEBENHAM Suff TM 1763. Partly unexplained. *Debham* [10542×66]13th S 1051, 1377×8, *Dephenham, Depbe(n)ham, Depham -heam* 1086, *Deb(b)eham* 1168–1275, *Debenham* from [1086]c.1180. Unidentified element + **hām**. Not 'the homestead by the r. Deben' since the r.n. is a back-formation from the p.n. Nor 'the homestead on the *Dēope*, the deep river' as the Deben is not such a river. RN 117, Baron.

DEDDINGTON Oxon SP 4631. 'The estate called after Dæda'. *Dædintun* [1050×2]13th S 1425, *Dadintone* 1086–1526 with variants *-ton(a) -tun -thone -ing-* and *-yn(g)-, Dedinton'* 12th cent., *Dedyngton* 1580. OE pers.n. *Dǣda*, a short form of names like *Dǣdhēah*, + **ing**[4] + **tūn**. O 256.

DEDHAM Essex TM 0533. 'Dedda's homestead'. *Delham* 1086, *Di- Dyham* 1142–1303, *Dedham* from 1166. OE pers.n. **Dydda*, Essex dial. form **Dedda*, + **hām**. Ess 386 gives pr [dedəm].

River DEE Ches SJ 4056. 'The goddess, the holy one'. *Δηοῦα* [c.150]13th Ptol, *Deva* [4th]8th AI, [c.700]13th Rav, *Dee* from [1043]17th S 1000, [c.1043]17th S 1226, 1188, *De* 1086. The Welsh forms are *Dubr duiu* [10th]c.1200, *Deverdoe(u)* [1191]c.1200, [c.1214]13th, *Dyfrdwy* from 14th. Brit **Dēua** borrowed into OE as **Dēw*, later *Dē*, before the 7th cent. development of the W diphthong *ui* in *Dwy*. The prefix *Dubr- Dever- Dyfr-* in the W forms is W *dyfr*, a weakened form of W *dwfr* 'a river'. An alternative name occuring in early W poetry is *Aerfen* 'battle goddess'; this may have been the original name of the river dropped because too sacred and potent for everyday use and replaced by the allusive and less specific *Deua*. Giraldus Cambrensis records a tradition that the Dee was still supposed in his time to predict the outcome of wars between the Welsh and the English by eating away its bank on either the Welsh or the English side. The same name occurs in the Scots Dee NO 8198, *Δηοῦα* [c.150]13th Ptol, the Irish Dee and the Spanish Deba and Deva. Che i.21, RN 117.

DEENE Northants SP 9492. 'The valley'. *Den* 1065–1285, *Deen* 12th–1316. OE **denu**. The reference is to the valley of Willow Brook. Nth 163.

DEENETHORPE Northants SP 9591. 'Thorpe, the secondary settlement, belonging to Deene'. *Denetorp* 1169, *Deenthorp* 1246, 1382, *Dyn(g)thorp* 16th. P.n. DEENE SP 9492 + p.n. *Torp* 13th, *Trop* 1235, ON **thorp**. Nth 163, SSNEM 133.

DEEPCUT Surrey SU 9057. The name of a cutting on the Basingstoke Canal, at the head of a series of locks, where the ground rises to 288ft. ModE **deep** + **cut**.

DEEP DALE Durham NY 9715. *Deep Dale* 1862 OS. ModE **deep** + **dale**. A deep narrow valley cutting up into moorland.

DEEPDALE Cumbr SD 7284. 'Deep valley'. *Depedale* 1433, *Depdal(l), Dib(b)dale* 16th cent., *Depe- Deeb- Deepdale* 17th cent. OE **dēop** + **dæl**. YW vi.253, L 96.

DEEPING GATE Cambs TF 1509. 'The road to Deeping'. *Depynggate* 1390, 1422. P.n. Deeping as in DEEPING ST JAMES and ST NICHOLAS and Market and West DEEPING TF 1609, 2115, 1310, and 1009 + ME **gate** (ON *gata*). Nth 234.

DEEPING Lincs TF 1310. 'The deep place, the deep fen'. *Depinge* 1086, *-ing -yng* c.1128–1526. OE ***dēoping**. Already by the time of DB this area was divided into *Estdeping(e)* 1086–1509, later DEEPING ST JAMES, *Deping (Sancti) Jacobi* c.1221, 1526, *Deeping St James* 1649, with saint's name *James* from the dedication of the Benedictine priory church founded here in 1139, and WEST DEEPING, *West Depinge* 1086, *Westdeping(a) -yng-* 1138–1392, *West Deeping* 1603. Perrott 402, 407, PNE i.130.

DEEPING ST NICHOLAS Lincs TF 2115. P.n. DEEPING + saint's name *Nicholas* from the dedication of the church built in 1845–6.

Market DEEPING Lincs TF 1310. *Markyddepyng* 1412, *Market(t) Deping(e) -yng* 1457–1554, ~ *Deeping* from 1576. ME **market** referring to the market granted here in 1220 + p.n. DEEPING TF 1008 etc. Also known as *Deping Sancti Guthlac* 'St Guthlac's D' 1291, *Deeping(e) Guthlac's* 1570–1607 with saint's name *Guthlac* from the dedication of the late 12th cent. church. Perrott 405.

River DEER Devon SS 3200. A back-formation from the p.n. DERRITON SS 3303. *Deer river* 1765. It is earlier referred to as *aqua de Dyraton* 'the water of Derriton'. D 5.

DEERHURST Glos SO 8729. 'Wooded hill frequented by deer'. *Deorhyrst* [804]11th S 1187, *-hyrst(a) -hurst* 1042×66, *(on) Deor hyrste* *[1066]13th S 1043, 11th ASC(D) under year 1016, 1053, *Deorhirstan* 1062×6, *-hyrste* 1066, *Derherst* 1086, *-hurst(e)* 12th–1443, *Dear(e)hurst* 1587, *Durhurst(e)* 1221–1407, *Durest* 1451, *Durist* 1719. OE **dēor** + **hyrst**. The ground rises rapidly to 100ft. behind the low-lying village. Gl ii.78, L 197–8.

DEERPLAY MOOR Lancs SD 8627. P.n. Deerplay + ModE **moor**. Deerplay, *Derplaghe* 1296, 1305, *Derpelawe* 1324, *Dirpley* 1736–1828, is the 'place where deer play', OE **dēor** + **plaga**. La 93, 263, PNE ii.67, Jnl 17.55.

DEFFORD H&W SO 9243. 'The deep ford'. *Deopanforda* (dative case) 972 S 786, *Depeforde* 1086, *Defforde* 1327. OE **dēop**, dative case definite form **dēopan**, + **ford**, dative case **forda**. A ford across Bow Brook (*Le Bowne* in 1393). Wo 194, L 68.

DEIGHTON 'Ditch settlement, settlement by or surrounded by a ditch'. OE **dīc** + **tūn**. *Dis-* spellings are AN for [di:tʃ] or [di:ç] and depend on the OFr sound change [s] > [ç] before consonants.

(1) ~ NYorks NZ 3801. *Dictune -ton* 1086–1285, *Di- Dyghton* 1316–1536. A large moat is marked on the map beside the village. YN 209 gives pr [di:tən].

(2) ~ NYorks SE 6244. *Diston(e)* 1086, *Dicton -k-* 1176–1291, *Di- Dyghton* 1285–1400, *Deighton* 1828. YE 267.

(3) Kirk ~ NYorks SE 3950. 'D with the church'. P.n. *Distone* 1086, *Dystone* 1280, *Dihtona* 1194, *Dicton(a)* 1154–1293, *Di- Dyghton* 1246–1661, + prefix *Kirk(e)* from 1361, ON **kirkja**. 'Kirk' and *South* 1303 for distinction from North DEIGHTON SE 3951. YW v.23 gives pr [kək 'di:tən].

(4) North ~ NYorks SE 3951. *Nordictun, Northdictun -ton* 1197, *North Ditton* 13th cent., ~ *Di- Dyghton* 1303–1542. OE **north** + p.n. *Distone* 1086, as in Kirk DEIGHTON SE 3950. YW v.25.

DEIRA (lost) NYorks. *Deiri, Deri, Deirorum prouincia* 'The Deiri, the province of the Deiri' [c.731]8th BHE, *Dere* c.1000 OEBede, 11th. In OE times Deira was the name applied to the kingdom of South Northumbria centred on York. It included Beverley which in the 8th cent. was known as *In Derauuda* 'in the wood of the Deiri, the men of Deira'. The meaning of Deiri, from PrW **deir*, is unknown. YE 12, LHEB 419–20.

DELABOLE Corn SX 0683. '(The portion of) Deli with the pit'. *Delyou Bol* 1284, *Delyoubol -i-* 1302–46, *Delebole* 1596. P.n. Deli SX 0883 + Co **pol** 'a pit', probably referring to the great slate quarry which must therefore be over 700 years old. Deli, *Deliov* 1086, is 'leaves', Co **delyow**, double plural of **deyl**. The addition 'with the pit' is for distinction from the other part of Deli, Delamere SX 0683, *Delyoumur* 1284, 'great Deli', p.n. Deli (*Deliav* 1086) + Co **meur**. PNCo 78, Gover n.d. 135.

DELAMERE Ches SJ 5669. A parish created by Act of Parliament in 1812 out of the last remaining part of the old royal Forest of Delamere 'to be known as Delamere Forest'. See DELAMERE FOREST SJ 5571. Che iii.211 gives prs ['deləmi:r] and older local ['dæləmər].

DELAMERE FOREST Ches SJ 5571. 'Forest of The Mere'. *Foresta de la Mare* 1233×7–1527 with variants *la Mar(a), forest of Delamere* from 1308. Earlier simply *foresta* 1086, *foresta de Mara* 1153×60–1439 with variants *Ma(i)re, Maris*. OFr **forest** + **de** + **la** + MedLatin **mar(a)** < OE *mere* 'a lake'. The reference is to one of the lakes at Eddisbury, either Blakemere SJ 5571, *Blakemere* 1359, 'black lake', or Oakmere SJ 5767, *Ocmare* 1277, *Okemere* 1347, 'oak-tree lake'. Che i.8, iii.217.

DELAMERE STATION Ches SJ 5570. A halt on the railway line from Chester to Manchester. P.n. DELAMERE + ModE **station**.

DELPH GMan SD 9808. 'The quarry'. *Delf(e)* 1544–1733, *Delph* 1817. OE **(ġe)delf**. YW ii.311.

DELPH BANK Lincs TF 3821. P.n. Delph + ModE **bank**. Delph, *le Delffe* 1272×1307, *le Delf* 1416, 1456, *Gednay delph* 1572, *Gedney Delf* 1723, is 'the drainage ditch', ME **delf** (OE *(ġe)delf*, dial. *delf, delft*). Cf the Dutch p.n. Delft. Payling 1940.24, Wheeler 102.

DELPH RESERVOIR Lancs SD 7015. P.n. Delph + ModE **reser-**

voir. Delph, no early forms, is ModE **delf** 'a digging, a trench, a pit, a quarry'. PNE i.128.

River DELPH Norf TL 5596. *The Delph or Thirty Feet* 1821. An artificial drain between the Old and New Bedford rivers, perhaps already referred to as *the Delph* 1617. ModE **delf**, **delph** 'a drainage channel'. Ca 6.

The DELPH Suff TL 7080. No early forms. ModE **delf** 'a ditch, a trench, a quarry'.

DEMBLEBY Lincs TF 0437. 'The village or farm by the pool'. *Den- Del- Dembelbi* 1086, *Dembelby* 1212–1553, *Dembleby* from 1242×3. ON ***dembil** + **bȳ**. A stream forms a pool in the middle of the village. Perrott 53, PNE i.129, SSNEM 44, Cameron 1998.

DENABY 'Village of the Danes'. OE folk-n. *Dene*, genitive pl. *Dena* and *Deniġea*, + **bȳ**. A name given by English speakers in an English dominated area.

(1) ~ MAIN SYorks SK 4999. A modern coal-mining locality named after OLD DENABY SK 4899 + ModE **main** 'a main seam of coal'.

(2) Old ~ SYorks SK 4899. ModE **old** + p.n. *Denege- Degenebi* 1086, *Deningebi -by* 1166–1240, *Deningby -yng-* 1240–1408, *Deneby* 13th cent., *Denyby* 1388–1585, *Dennabye* 1598. Medial *-ing(e)-* is an unetymological alteration of the genitive pl. suffix *iġa*. 'Old' for distinction from the modern mining settlement of DENABY MAIN SK 4999. YW i.122, SSNY 25.

DENBURY Devon SX 8269. 'Fortified place of the *Defnas*, the men of Devon'. *Deveneberie* 1086, *-bire -byr(') -byri -bury* 1228–1336. OE folk-n. *Defnas*, genitive pl. *Defna*, + **byriġ**, dative sing. of **burh**. The reference is to Denbury Camp SX 8169, a large earthwork, perhaps used as a strongpoint by the Devon Britons during the Saxon take-over. D 523, i.xiv note 1.

DENBY 'Village of the Danes'. OE **Dene**, genitive pl. **Dena**, + **bȳ**.

(1) ~ Derby SK 3946. *Denebi* 1086, *-by(e)* 1234–1395, *Denby(e)* from 1308. Db 444, SSNEM 44.

(2) ~ DALE NYorks SE 2208. *Denby Dale* 1851. P.n. Denby + OE **dæl**. Denby is *Denebi -by(e)* 1086–1611, *Denby(e)* 1304–1581. YW i.326, SSNY 25.

(3) Upper ~ WYorks SE 2207. *Overdeneby* 1261, ~ *Denby* 1573. YW i.326.

DENCHWORTH Oxon SU 3891. 'Denic's estate'. *Denichesuurde* *[821]12th S 183, *(æt) deniceswurþe, (to þam norðran) Deneceswyrðe* [947]12th, *(æt) Deneceswurþe, (to þam norðran) denceswurþe* [947]13th S 529, *Deniceswyrð* 960 S 687, *(ad) Denceswyrðe* *[811]12th S 166, *(æt) Dencesuurþe* [958]13th S 657, *Denceswurthe* [965]12th S 733, *Denchesworde* 1086, *Denchewurth'* 1221. OE pers.n. **Deniċ*, genitive sing. **Deniċes*, + **worth**. Brk 472.

DENE MOUTH Durham NZ 4540. *Dean Mouth* 1862 OS. ModE **dene** + **mouth**. Marks the point where Castle Eden Dene opens to the sea.

DENFORD Northants SP 9976. 'The valley ford'. *Deneford(e)* 1086–1428, *Denford* from 1241, *Derneford* 1241, 1249. OE **denu** + **ford**. Denford lies on the Nene at the mouth of a small side-valley. Nth 180, L 98.

DENGE BEACH Kent TR 0718. *Denge Beach* 1816 OS. P.n. Denge, as in DENGE MARSH TR 0419, + ModE **beach**.

DENGE MARSH Kent TR 0419. *denge mersc* [774]10th S 111, *Di- Dengemareis -eys -ais* 1225–1275, *Dengemersse* 1253, *Denge Marshe* 1610. P.n. Denge, OE **dyncge**, Kentish **dencge** 'manured land', + **mersc**. Cf. DUNGENESS TR 0916. PNK 482, KPN 55.

DENGIE Essex TL 9801. Partly uncertain. *Deningei* [c.706×9]17th S 1787, *(on) Denesige* [942×c.951]13th S 1526, *Daneseiã* 1086, *-eiam -heia -ey(e)* 1123–1354, *Den(e)seie -eye* 1274, 1369, *danningam* 1222, *Danengeye -ynge-* 1276–1305, *Dange(ye)* 1251–1428, *Daungey* 1395, *Dengeye* 1344. The generic is OE **ēġ** 'an island' probably referring to the whole peninsula between the Blackwater and the Crouch E of Maldon. The specific seems to vary between two forms, an *ing*[2] formation **Den(n)ing*, **Dæn(n)ing*, and a genitive sing. form *Denes, Dænes*, which point to OE pers.n. *Dæni*, 'Dæni's island' and the 'island called or at *Dæning*, the place called after Dæni'. The modern pr [dendʒi] is best explained as coming from an OE assibilated locative–dative form **Denninġe* 'at Denning' in subsequent forms of which [z] was widely substituted for [dʒ] through false association with the Danes. The name is to be associated with DANBURY TL 7805 between Chelmsford and Maldon which has the same pers.n. specific. In ancient times this whole area was regarded as a *regio* which subsequently became the hundred of Dengie, *Danesie -ia -heia -ee* 1185–1264, *Deneseye* 1274, *Danseie* 1586, *Dancy* 1713, *Denegeia* 1189×99, *Dangeye* 1251, *Dansing -yng* (in rhyme) 14th, 1324, 1594, *Denge* 1594. The hundred reached inland as far as Maldon and the Woodhams, *wood* names to be associated with *lēah* names at Hazeleigh TL 8203, *Halesleiam* 1086, *Hai- Heylesle(y)* 13th cent., 'Hægel's wood or clearing', and PURLEIGH TL 8301, where it abutted on the *feld* or open land of the Haningfields. Ess 213.

DENGIE FLAT Essex TM 0404. 'D sand-bank'. P.n. DENGIE TL 9801 + ModE **flat**.

DENHAM 'Valley homestead'. OE **denu** + **hām**.

(1) ~ Bucks TQ 0486. *Denham* *[1065]12th S 1040, *Deneham* *[1066]12th S 1043, 1195–1227, *Daneham* 1086, *Denham* from 1233. Bk 235, L 98.

(2) ~ Suff TM 1974. *deha', denham, Delham* 1086, *Denham* 1610. DEPN.

(3) ~ Suff TL 7561. *denhã* 1086, *Denham* 1610. DEPN.

(4) ~ CASTLE Suff TL 7452. *Denham Castle* 1836 OS. P.n. DENHAM TL 7561 + ModE **castle**. A motte and bailey castle with wet ditches. Pevsner-Radcliffe 1974.227.

(5) ~ GREEN Bucks TQ 0388. P.n. DENHAM TQ 0486 + ModE **green**.

(6) ~ STREET Suff TM 1872. 'D hamlet'. *Denham Street* 1837 OS. P.n. DENHAM TM 1974 + Mod dial. **street**.

DENHOLME WYorks SE 0734. 'The water-meadow in the valley'. *Denholm(e)* from 1252. OE **denu** + ON **holmr**. YW iii.257.

DENMEADE Hants SU 6611. 'Valley meadowland'. *Denemede* 1205–1327, *Denmede* 1412. OE **denu** + **mǣd**. The reference is to the valley of the headwater of the Wallington river. Ha 65, Gover 1958.53.

DENNE PARK WSusx TQ 1629. P.n. *Den* 1813 OS + ModE **park**. Sx 229.

DENNINGTON Suff TM 2867. 'Denegifu's estate'. *Binneuetuna'* (for *Dinueue-*), *dinguiet'* (for *dingiue-*), *dingifetuna -giue-*, *dingiuetona* 1086, *Dingieueton* 1169, *Dinnieueton* 1190, *Dinnyngton* 1610. OE feminine pers.n. *Deneġifu*, genitive sing. *Deneġife*, + **tūn**. DEPN, Baron.

DENNY ISLAND Avon 4581. *The Denny Island* 1830 OS. The evidence is too late for explanation.

DENNY LODGE Hants SU 3305. *Dinney Lodge* 1664, *Denny Lodge* 1810 OS. P.n. Denny + ModE **lodge** 'a house in a forest' (OED sense 2). Denny is *Dunie* c.1300, *la Dunye* 1331, *Dinne* 1347, *le Dony* c.1490, 'the hill island', OE **dūn** + **īeġ**, referring to a large area of raised land forming a low *dūn* amidst marshland. There is still marshland between Pond Head SU 3007 and Decoy Pond SU 3507. Ha 65, Gover 1958.202, Rackham 144, 157, 163.

DENSHAW GMan SD 9710. 'The valley copse'. *Denshaw* 1635, *Deanshaw* 1771, 1822. Ultimately OE **denu** + **sceaga**. YW ii.311.

DENSOLE Kent TR 2141. *Lower, Vpper Dens Hall* 1698, *Densell* 1819 OS, *(Little) Densole* 1841 *TA*. Cf. *Densall Minnis, Densall Bushes* 1539×40. This has been explained as 'miry pool at the pasture', OE **denn** + **sol**. The evidence is too late for certainty but there is still a pool at Densole Farm. PNK 455, Cullen.

DENSTON Suff TL 7652. 'Deneheard's estate'. *Danardes-*

danerdestuna, Damardestunā 1086, *Denardeston* 1220, *Denston* 1610. OE pers.n. *Deneheard*, genitive sing. *Deneheardes*, + **tūn**. DEPN.

DENSTONE Staffs SK 0940. 'Deni's estate'. *Denestone* 1086, *-ton* 1208. OE pers.n. *Deni*, genitive sing. *Denes*, + **tūn**. DEPN.

DENT Cumbr SD 7086. Unexplained. *Denet* 1202–52, *Deneth* 1279–81, *Dent(e)* from 1278. Possibly PrW **Dinnéd* from a supposed Brit **Dindeto-* or **Dindētiō* related to OIr *dinn, dind* 'a hill', ON *tindr* 'a point, a crag'. This explanation is influenced by Dent in Cleator NY 0313, *mons Dinet* c.1200, *Denthill* 1576, *Dint* 1690. But the name may be an old r.n. of unknown origin and meaning. YW vi.252.

DENTDALE Cumbr SD 7186. 'Dent valley'. *Dentdale* 1577. P.n. DENT SD 7086 + ME **dale** (ON *dalr*).

DENTON 'Valley settlement'. OE **dentūn** < *denu* + *tūn*. PNE i.130, L 98, 150.

(1) ~ Cambs TL 1487. *Dentun* [963×92]12th B 1130, *Dentone* 1086 etc. Hu 183.

(2) ~ Durham NZ 2118. *Denton* from 1200, *-tone* 1235, 1242×3. The village is situated in the shallow valley of Cocker Beck. NbDu 61.

(3) ~ ESusx TQ 4502. *Denton* from [801]14th S 158, *Deanton* [825]14th S 1435. Sx 365, SS 76.

(4) ~ GMan SJ 9295. *Dentun* c.1220, *Denton* from 1255. A small brook rises close to the church and runs in a slight valley SW. La 30, Jnl 17.30.

(5) ~ Kent TR 2146. *dene tun* [799]10th S 156, *Danetone* 1086, *Deni(n)tone -īg- -e(n)- -y(n)-* 1203–1304, *Denton* from 1441. Some spellings suggest an alternative specific **dening** 'valley place', an **ing**[2] derivative of **denu**, possibly an analogical development on the pattern of other *-ington* names. PNK 556, KPN 88.

(6) ~ Lincs SK 8632. *Dentone -tune* 1086, *Denton* from 1174. Perrott 451.

(7) ~ Norf TM 2788. *Dentuna* 1086, *Denton* 1199. DEPN.

(8) ~ NYorks SE 1448. *dentun* [972×92]11th S 1453, *(on) Dentune* c.1030 YCh 7, *Denton(a)* from 1086. YW v.63.

(9) ~ Oxon SP 5902. *Denton(e)* from [1122]c.1425, *Dinton, Dyntone* 1285. O 169.

(10) ~ FELL Cumbr NY 6262. P.n. DENTON NY 6165 + ModE **fell** (ON *fjall*).

(11) Nether and Upper ~ Cumbr NY 5862, 6165. *Denton* from 1169, *duas Dentonas* c.1184, *Overdenton* 1363, *Nethirdenton* 1457. The reference is to the valley of the river Irthing. Cu 81, L 98, 150.

DENTON Northants SP 8358. Probably 'the estate called after Dodda'. *Dodintone* 1086–1376, *Dodington -yng-* 12th–1563, *Dudintun -ton* 1195–1229, *Denynton* 1371, *Deynton* 1563, *Denton, Doddington Parva* 1749. OE pers.n. *Dodda* + **ing**[4] + **tūn**. An OE ***dod(d)** 'a rounded hill' has been suggested as the specific of this name, but the topography does not securely bear this out: Denton is situated in a small valley on sloping ground, although there is an isolated piece of high ground to the NW at SP 8459. Nth 146, Årsskrift 1974.37.

DENVER Norf TF 6101. 'The Danes' crossing'. *Danefella -faela* 1086, *Denever(e)* 1200, 1254. OE folk-n. *Dene*, genitive pl. *Dena*, + **fær**. The reference is probably to the Fen Causeway which carries the Roman road from Peterborough to Denver, Margay no. 25, across the Nene and Great Ouse fens. DEPN.

DENWICK Northum NU 2014. 'Valley farm'. *Den(e)wyc* 1242 BF, *Denewick* 1278, *-wike* 1296 SR, *Dennek* 1538. OE **denu** + **wīc**. NbDu 62.

DEOPHAM Norf TG 0500. 'The homestead at or called Deop'. *D(i)epham* 1086, *Depham* 1227. OE P.n. **Dēop* 'the deep place' + **hām**. The reference is to a deep place in the nearby lake called Sea Mere. DEPN.

DEOPHAM GREEN Norf TM 0499. *Deopham Green* 1838 OS. P.n. DEOPHAM TG 0500 + ModE **green**.

DEPDEN GREEN Suff TL 7757. P.n. *Depdanā* 1086, *Depedene* 1198, *Debden* 1610, 'the deep valley', OE **dēop**, definite form feminine **dēope**, + **denu**, + ModE **green**. DEPN.

DEPTFORD GLond TQ 3676. 'The deep ford' across the river Ravensbourne. *Depeforde* 1293, 1344, *Depford(e)* 1313, 1344. OE **dēope** + **ford**. Forms with medial *t* appear from the 15th cent. Cf. DEPTFORD Wilts SU 0537. KPN 2, TC 204.

DEPTFORD Wilts SU 0138. 'The deep ford'. *Depeford* 1086–1316, *Dup(p)eford* 1242–81, *Doepe- Deopeford* 1341–1422, *Depford* 1432×43, *Detford* 1558×1603, *Debtford* 1630. OE **dēope** + **ford**. A ford across the river Wylye. Wlt 231.

DERBY Derby SK 3536. 'Deer farm'. *Deoraby* c.1000 Æthelweard under year 871, c.1050 ASC(C) under 917, 924×39 coins, c.955 ASC(A) under year 942, 978×1016 coins, *Deorby -bi* 957–1066 coins, *Derabi* 924×39 coins, *Derby* 1086, *-bi(a) -by* 1129×38–1340, *Darbiam -by(e) -bie* 1360–1675. ON **djúr**, genitive pl. **djúra**, + **bȳ**. This ON name replaced earlier OE *Northuuorthige* c.1000 Æthelweard, *Norðweorðig* c.1020, 'northern enclosure', OE **north** + **worthiġ**. Possibly the sense was N in relation to Tamworth. The deer of Derby may have come from Little DARLEY SK 3438 2 m. N of Derby. Db 446, SSNEM 43.

West DERBY Mers SJ 3993. *Westderbi* 1177, *-derebi* 1201, *West Derby* 1330. Adj. **west** + p.n. *Derbei, Derberie* 1086, *Derbeia* 1153, *Derby* 1094–1332, *(de) Derebi (Wapentachio)* 1188, 1202, *-by* 1226×8, 'deer-park, deer-farm', ON **djúr**, genitive pl. **djúra**, + **bȳ**. In the DB form *Derberie* there has been confusion with OE **byriġ**, dative sing. of **burh**. 'West' for distinction from DERBY SK 3536. La 93, 114, SSNNW 28, Jnl 17.55, 64.

DERE STREET Durham NZ 2119. Northum NY 9278, NZ 0757. NYorks SE 4363. 'Roman road of the wild animals'. *Deorestrete* [c.1040]12th, c.1104, *Derestrete* 15th. The Roman road from York to Newstead, Scotland, Margary No. 8a-f, OE **dēor**, genitive pl. **dēora**, + **strēt**. It is possible that **dēor** has replaced OE *Dere* 'the people of Deira', genitive pl. *Dera*, in this name and that it was originally 'the Roman road of the Deirans'. The OE folk-name *Dere* is derived from PrW **Deir* < British **Dobrịā* or **Dubrịā* 'well watered land' referring to the district around York. YE 12, LHEB 419–20.

East DEREHAM Norf TF 9813. *Estderham* 1428. ME adj. **est** + p.n. *Derham* 1086, the 'deer-enclosure', OE **dēor** + **hamm**. 'East' for distinction from West DEREHAM TF 6500. DEPN.

West DEREHAM Norf TF 6500. *Westderham* 1203. ME adj. **west** + p.n. *Deorham* c.1200 ASC(F) under year 798, *Der(e)ham* 1086, *Derham* [1087×98]12th, 1193, *Dierham* 1197, the 'deer-enclosure', OE **dēor** + **hamm**. 'West' for distinction from East DEREHAM TF 9813.

DERRINGSTONE Kent TR 2049. 'Deoring's estate'. *Dieringestune, de Dieringestun', de Diering'* before 1223, *Deringeston'* 1262×3. OE pers.n *Dēoring*, genitive sing. *Dēoringes*, + **tūn**. KPN 189, Cullen.

DERRINGTON Staffs SJ 8922. Partly uncertain. *Dodintone* 1086, *-ton* 1242, *-ington* 1288, *Duddinton* 1203, *Dudington* 1236, *Derington* 1601. Probably 'the estate called after Dudda', OE pers.n. *Dudda* + **ing**[4] + **tūn**. An alternative possibility is OE ***dodding**, ***dudding** 'a small hill' referring to the site of the village on a low hill beside Doxey Brook. DEPN, Årsskrift 1974.37.

DERRITON Devon SS 3303. 'Dyra's estate'. *Direton* 1238, *Dyreton* 1278. OE or ME pers.n. *Dȳra* + **tūn**. OE *Dȳra, Dīera* survived in Devon as a surname *Dyra, Dira* until the 14th cent. D 153.

DERRY HILL Wilts ST 9670. 'Dairy hill'. *Derry Hill* 1695, ModE **dery**, a 17th cent. variant of *dairy*. Wlt 257.

DERRYTHORPE Humbs SE 8208. 'Outlying settlement at or called *Dudding*, the place called after Dudda'. *Dudingthorp'* [c.1184]15th, *Dodithorp'* 1263, *Dudythorp* 1278×9, *Dodingthorp* 1316. OE p.n. **Dudding* < pers.n. *Du(d)da* + **ing**[2], + **thorp**. For the sound change *d* > *r* cf. DORRINGTON Shrops. DEPN, SSNEM 125.

DERSINGHAM Norf TF 6830. 'Homestead of the Deorsigingas, the people called after Deorsige'. *Dersincham* 1086, *Dersingeham* 1166, 1244, *Dersingham* 1203, 1236. OE folk-n. **Deorsiġingas* < pers.n. *Dēorsiġe* + **ingas**, genitive pl. *Deorsiġinga*, + **hām**. DEPN, ING 134.

DERVAL Corn SW 5935 → CHICKERELL Dorset SY 6480.

DERVENTIO ROMAN FORT Cumbr NY 1131. *Derwentione* [c.700]13th Rav. The fort is named from the Brit river n. DERWENT NY 2514. RBrit 334.

River DERWENT 'Oak river, river in the oakwood'. PrW ***derwïnt**, a derivative of Brit **deru̯ā* 'an oak-tree'. Possibly a reshaping by popular etymology of an earlier Old European participial formation on the IE root **dreu̯* 'to run' seen in continental r.ns. such as Drewenz, East Prussia, *Drawanta* 1243, Durance, a tributary of the Rhone, *Druantia* 1st Pliny, and other examples in France, Drän, Carinthia, Austria, *Trewina* 890, Drava, Croatia, *Dravos* Strabo, Traun, Bavaria, *Druna* 788 etc. LHEB 282, 502–3, TPS 1996.78.

(1) River ~ Cumbr NY 2514. *Dorvantium* [c.700]13th Rav, *Deruuentionis, Diorwentionis fluuii* c.731 BHE, *Deorwentan streames* [c.890]c.1000 OEBede, *Dyrwenta* 1104×8, *Derewent(e)* [c.1140]n.d.–1316, *Derwent* from [c.1210]n.d., *Darwent* 1465 etc., *Darwen* 1576, 1664. Cu 11, RN 122, RBrit 334 s.n. Derventio.

(2) ~ Derby SK 2664. *(in, neah) Deorwentan* 1009 S 922, c.1020, before [1085]12th, *(super) Derewentam* 12th cent., *Derewent(e)* 1175–1390, *Derwent* from 1154×89, *Darwent* 1443, 16th cent., *Darwen* 1505–1630. Db 5 gives pr [darən].

(3) ~ Northum NZ 0450. *Dyrwente* [c.1040]12th HSC, *Derewent(a) -e* 1259–1371 including [1138×59]14th, *Derwent* from 1312 including [1138×59]14th, *Darwayne* 1565, *Darwyn -in* 17th cent., *Darwen* 1764. NbDu 62, RN 122.

(4) ~ NYorks SE 7035, 8578. *Der- Doruuentionem* [c.731]8th BHE (accusative case), *Deorwentan* [959]12th (oblique case) S 681, *Derewent(a) -(am)* [1109×14]15th–1339, *Derwent* from 1177. The Roman fort at Malton on the upper Derwent SE 7971 was *Derventione* [300]8th AI, [c.420]15th ND. YN 3, YE 2, RN 121, RBrit 333.

(5) ~ RESERVOIR Derby SK 1791. R.n. Derwent as in River DERWENT + ModE **reservoir**.

(6) ~ RESERVOIR Durham NZ 0152. R.n. Derwent + ModE **reservoir**. Derwent, *Dyrwente* [c.1040]12th, *Dirwencionis fluminis* 12th, *Derewentam -e* [1138×59]15th, [1183]c.1320, 1259–1303, *Derwent* from 1312, *Darwent, Derwen, Darwayne* 15th cent., *Darwin, Darwyn* 16th, *Darwen* 1764. NbDu 62, RN 122.

(7) ~ RESERVOIR Northum NZ 0152. R.n. DERWENT NZ 0450 + ModE **reservoir**.

(8) ~ WATER Cumbr NY 2420. *Derewentewatre* [1199×1215]–1279, *Derwentwater* from [1209×10]n.d., *Darwent- Daren- Darrantwater* 16th. R.n. DERWENT + OE **wæter**. The lake is named from the river DERWENT NY 2514 which flows through it. Cu 33.

DESBOROUGH Northants SP 8083. 'Deor's fortified place'. *Dies- Dereburg* 1086, *Dereburc'* 1200, *Deresburc -burg(h)* 1166–1255, *Deseburg* 12th–1393, *Desbrow al. Desborough* 1705. OE pers.n. *Dēor*, genitive sing. *Dēores* + **burh**. Nth 111, DEPN.

DESFORD Leic SK 4703. 'Deer' or 'Deor's ford'. *Deresford* 1086–1362, *Diresford* 1086, *Dersford(e)* 1209–1387, *Desford(e)* from 1322. OE **dēor** or pers.n. *Dēor*, genitive sing. **dēores**, *Dēores*, + **ford**. Lei 491.

DETCHANT Northum NU 0836. 'The ditch end'. *Dichende* 1166, *Di- Dychend'* 1242 BF–1296 SR, *Dychent -ant* 14th cent., *Ditchin* 1570, *Detchon* 1715. OE **dīc** + **ende**. NbDu 62 gives pr [detʃən].

DETLING Kent TQ 7958. 'The Dettlingas, the people called after Dettel'. *Detlinges* 1066×87–1264, *Detling(e) -ynge* 1198–1278, *Detling* from 1275, *Dytlinge* c. 1100. OE folk-n. **Dyttlingas*, Kentish form **Dettlingas*, < pers.n. **Dettel*, **Dyttel* + **ingas** PNK 136, ING 11.

DEVA Ches SJ 4066. 'Goddess'. The Roman name of the River DEE and the city of CHESTER, q.v.

DEVERAL Corn SW 4130 → CHICKERELL Dorset SY 6480.

River DEVERILL Wilts, an old name for the upper part of the river Wylye as in BRIXTON DEVERILL ST 8638, KINGSTON ~ ST 8436, LONGBRIDGE ~ ST 8640 and MONKTON ~ ST 8537. *Deferael -eal* [968]13th, *Deverel* c.1300, *stream called Deverell water* 1736. As a p.n. referring to one or other of the above villages it occurs as *Devrel* 1086, *Deverel(l)* c.1160–1213, *Deurel, Deyreals* 1164, 1166. PrW ***duβr** 'water' + adjectival suffix ***-(i)ol**, Co *-(y)el*. Identical with Deveral Corn SW 5935, *Deverel* 1324, *Defriel* 1356, and Derval Corn SW 4130, *Deverel* 1326, *Dyfryl* 1340. Wlt 6, CPNE 82, 138, RN 123.

DEVIL'S CAUSEWAY Northum NU 1202. *The Devil's Causeway* 1867 OS. The name of the Roman road from Bewclay to Berwick upon Tweed, Margary no.87.

DEVIL'S DITCH Cambs TL 6062. *Dæmonis fossam* (Latin) 1574, *Deuilsdike* 1594, *(The) Devil's ditch* 1604–1789. Large ancient earthworks are often ascribed to the devil or to supernatural forces ever since the Anglo-Saxons called Roman buildings *eald enta geweorc* 'ancient work of the giants'. But this is a modern appellation and earlier this great earthwork, probably a boundary wall of Anglo-Saxon origin between the East and Middle Angles or the Angles and the Mercians, was referred to simply as 'the ditch', *dicum* (dative pl. 'at the ditches') c.925 ASC(A) under year 905 referring to Devil's Ditch and Fleam Ditch, *(le) Dych(e)* 1336, 1450; as 'Reach ditch', *fossam de Reche* 12th, *Reach(e) dytch, ditch* 1591×1601; 'St Edmund's ditch' *fossatum de Sancto Edmundo* 1220, *fossam Sancti Edmundi* 1354–1574, *Seynt Edmond his diche, the dych of Seynte Edmunde* 14th; and as the 'great ditch' *magnum fossatum* 13th–1491, *le Micheldyche* 1315. OE **dīc**, Latin **fossa(tum)** with prefixed adj. Latin **magnus**, ME **michel**, p.n. REACH TL 5666, saint's name *Edmund*, and ModE **devil**, Latin **daemo**. Ca 34, O li.

DEVIL'S DYKE Norf TF 7408. *The Devils Dyke* 1824 OS. Earlier known as *Bicchamdic* *[? 1047]14th S 1109, *Bichamdic* [1053×7]13th S 1108, p.n. Beacham as in BEACHAMWELL TF 7505 + OE **dīc**. An early Saxon linear earthwork with a ditch on the E controlling the Fen road, mentioned in the Ramsey Abbey Charter of 1053. It may have been a temporary frontier between hostile goups in the early phase of the Anglo-Saxon settlement. R. Rainbird Clarke, *Norwich and its Region*, BAAS 1961.103, Pevsner 1962NW 265.

DEVIL'S DYKE WSusx TQ 2611. *Devils Dyke* 1813 OS. An example of the common association of the devil with ancient earthworks. Sx 287.

DEVIL'S WATER Northum NY 9356. 'The black stream'. *Divelis* 1233, 1269, *Deueles, Dyvils* 13th, 14th cents., *le Ewe Devyls* 1464, *Devell, Douols, Deuilles brooke, Dill* 16th cent., *Devil's Water* 1869 OS. PrW ***duβ**, ***dü** + ***gles**, ***gleis**. Identical with r.ns. DALCH SS 7510, DAWLISH WATER SX 9676 etc. The 1464 form is preceded by ME **ewe** 'a river' (OFr *eve*) possibly replacing OE *ēa*. NbDu 62, RN 130.

DEVIZES Wilts SU 0061. 'The boundaries'. *(apud) Divisas* (Latin) 1141–1233, *Divises* 1152–1280, *Devises -y-* 1195–1316, *le Devisez* 1485, *the Devizes* 1675. Latin **divisæ** varying with OFr **devises** referring to the boundary between the hundreds of Potterne and Cannings. Devizes Castle was known as *Castrum Divisarum* (Latin genitive pl.) 1223–42, and also as *Castrum de Vises* 1330 whence the forms *la Vyses* 1335, *Vise(s), The Vyse* 15th cent. and *the Vies, the toune of Vyes* 16th cent. The town of Devizes grew up around the Norman Castle built by bishop Roger of Salisbury in the 12th cent. Wlt 242 gives pr [divaiziz].

DEVOKE WATER Cumbr SD 1596. 'Devoke lake'. *Duvokeswater*

[c.1205]n.d., *Duuokwat'* 1279, *Devoke* 1626. Possibly MW pers.n. *Dyfog*, genitive sing. *Dyfoges*, + OE **wæter**. *Dyfog* represents OW **Dibauc*, OBrit **Dubāco-* 'black'. The specific might also represent an original stream name, Brit **Dubācā* 'the black one'. Cu 33.

River DEVON Notts SK 7847. Probably the 'black river'. *Dyvene* 1252–1433, *Deven* 1330–1458, *Devon* from 1582, *Dene* 1576–1622. Brit **Dubona* < ***dubo-** 'black', PrW ***duβ, dū**, + suffix **-no/-na**. The reference is to the steep ravine in which the upper Devon runs. Identical with the Scottish r.n. Devon, Perthshire. Nt 3 gives prs [di:vən], formerly [di:n], RN 124, ScotPN 177.

DEVON The county name ultimately preserves the name of the British kingdom of Dumnonia. The earliest reference to the shire is *(mid) Defena scire* 9th ASC(A) under year 851 with later spellings *Defna scir* under year 894, *Defna scire* (genitive sing.) 11th ASC(C) under year 977, and *(on) Dæfenan scire* early 12th ASC(E) under year 1017, OE folk-n. *Defnas* or *Defene* 'the people of Devon', genitive pl. *Defna*, + **scīr**. The people of Devon are *Defna* (genitive pl.) in ASC(A) under year 823, *Defenum* (dative pl.) [951×5]14th S 1515, and Devon is *(on) Defnum* ASC(A) under year 894 and *(on) Defenum* ASC(A) under year 897. OE *Defnas* (or *Defene*) is derived from the Romano-British tribal name *Δαμ- Δουμνόνιοι* (*Dam- Dumnonii*) [c.150]13th Ptol, *Dumnonii* 3rd AD Solinus, and territory name *Damnonia* c.540 HB, *Dibnenia* c.1150 *Vita S. Gildae*, *Domnonia* [893]c.1000 Asser, *Dyfneint* Mabinogion. This last is a reformation with OW *naint*, the pl. of *nant* 'a valley' as if the name meant 'deep valleys'. The original name is unexplained but recurs again in the SW Scottish tribal name *Damnonii* and in the root *Dumno-* found in Celtic pers.ns. of the form *Dumnorix* etc. Possibly 'men, worshippers of *Dumnōnus*', cf. the Irish *Fir Domnann* 'worshippers of *Dumnū* or *Dumnōnū*'. The Irish *Domnainn* seem to have been a branch of the *Dumnonii* of Devon and Cornwall. *Dumnonii* stands for earlier **Dubnonii* on the root **dubno-* 'deep, world'. Exeter was known as *Isca Dumnoniorum* 'Isca of the Dumnonii' 4th Peut, *Isca Dumnuniorum* [3rd]8th AI, and *Isca Dumnamorum* [c.650]13th Rav. D 1, O'Rahilly 93–4, RBrit 342.

DEVONPORT Devon SX 4554. A district of Plymouth originally known as *Plymouth Dock*. The name was changed to *Devonport* in 1824. County name DEVON + ModE **port**. D 240.

DEVORAN Corn SW 7939. 'Waters'. *Dephryon* 1275, *Deffrion* 1278, *Dofryoun* 1327, *Deveryon* 1435, *Devrian or Devoran* 1683. Co ***devryon**, pl. of ***devr-** 'water'. The village is situated where three streams merge into a wide creek. PNCo 78, Gover n.d. 447.

Much DEWCHURCH H&W SO 4831. 'Great D'. *Dewchurch Magna* 1535. ModE adj. **much** + p.n. *Dewischirch(e) -es-* 1148×55–1341, 'Dewi's church', W saint's name *Dewi* (St David) + ME **chirche**. W and Latin forms are *Lann Deui* [7th, mid 8th]c.1130 and *Podii Deui* [early 8th]c.1130, PrW ***lann** 'churchyard, church' and L **podium** 'district'†. He 65, O'Donnell 90, Jnl 10.68.

Little DEWCHURCH H&W SO 5331. *Lytel Deuchurche* 1397, *Dewchurch Parva* 1535. ME adj. **litel**, Latin **parva**, + p.n. Dewchurch as in Much DEWCHURCH SO 4831. He 63.

DEWLISH Dorset SY 7798. Transferred from the stream now called Devil's Brook: a Celtic r.n. meaning 'black stream'. *Devenis* 1086, *Deueliz -is -ys* 1194–1300, *-ich -is(s)(c)h(e) -ys(s)(c)h(e)* 1288–1575, *Dewelisshe* 1481, *Du(e)- Diulish(e)* 17th. PrW ***duβ, dü** + ***gles, gleis**. Identical with DAWLISH Devon SX 9676, DEVIL'S WATER Northum NY 9455 etc. Do i.303 gives prs ['dju:liʃ] and ['du:liʃ].

†This does not correspond to W *pau* 'country, land' as sometimes stated since *pau* derives from Latin *pagus*, GPC 2703.

DEWSALL COURT H&W SO 4833. 'Dewi's spring'. *Dewiswell(e) -ys-* 1160×70, *Deuswelle* c.1174, *Dewshall* 1625. W saint's name *Dewi* (St David) + ME **welle**. Latin forms are *Fonte David, ~ Dauid* 1148×9–1269. Part of the composite manor of Westwood, *Westevde* 1086. Dewsall is 1 mile N of Much DEWCHURCH SO 4831. He 68, L 31, 104, Jnl 10.68.

DEWSBURY WYorks SE 2422. 'Dewi's fortified place'. *Deusberia -ie* 1086, *Dewesbiri -y -byry* 1091×7–1323, *-bury* 14th cent., *Deuhesbir'*, *Deuwythbiris* 12th, *Dewsbury* from 1364. W pers.n. *Dewi* (Latin *David*), secondary genitive sing. *Dewes*, + OE **byriġ** dative sing. of **burh**. YW ii.184, TC 82.

DIAL POST WSusx TQ 1519. Short for *Dial Post Farme* 1702. Sx 188.

River DIBB NYorks SE 0563. Uncertain. *Dibe* 1154×77, *Dib* 13th, *Dibb* 1717. YW vii.126 derives the name from OE **dybb* 'a pool', but this seems an unlikely name for such a narrow stream. Possibly, therefore, a back-formation from Dibble's Bridge SE 0563, *Dib(b)les Bridge* 1654 YW vi.78. Dibble might be derived from OE **dēop** + **wella** 'deep spring, deep stream', or ON **djúpr** + **hylr** 'the deep hollow', or surname *Dibble*. See also DIBBLE BRIDGE NZ 6707. RN 125, DBS 101.

DIBBLE BRIDGE NYorks NZ 6707. *Depilbrigge* 1301, *Dybell Brigge* 1539. P.n. Dibble + OE **brycg**. Dibble, *Depehil* [1119]15th, *Dephil* [1129]15th, 1170×90, [1239]15th, is the 'deep hollow', OE **dēop** or ON **jupr** + **hylr**. YN 148.

DIBDEN Hants SU 4007. Probably 'valley at Deep'. *Depedene* 1086–1413 with variant *-den*, *Diepedena* 1165, *Dupedene* 1201, 1291, *Dibbden* 1491, *Dybden* 1516. OE **dēope** + **denu**. The reference of the name is uncertain; Dibden itself occupies a ridge between Southampton Water and the low but hardly 'deep' ground of a tributary of the Beaulieu River at Dibden Bottom. This may be the *denu* but the specific probably refers to a zone of deep water in Southampton Water called locally Deep Lake in which Deep will have been an early p.n. OE **Dēope* 'the deep'. Ha 65, Gover 1958.203.

DIBDEN PURLIEU Hants SU 4106. Short for *Dibden in purlieu* 1486. P.n. DIBDEN SU 4007 + late ME **purlewe** 'tract of land on the fringe of a forest', AN *puralé* 'a perambulation' sc. of a boundary especially to establish the bounds of a forest where disafforestation has taken place. The reference is to an area round the nucleus of the New Forest removed from the forest in the 14th cent. when the forest bounds were established by perambulations c.1300 and in 1327, and in which the king retained or claimed ancient rights. *Purlewe* occurs again at Brune's Purlieu SU 1815 (surname *Brewen* 1670), Holbury Purlieu SU 4202, Ogden's Purlieu SU 1811 (surname *Okeden* c.1490) and Purlieu SZ 1899. Ha 65, Gover 1958.211.

Upper DICKER ESusx TQ 5510. ModE adj. **upper** + p.n. *Dikere* 13th–1359, *(la) Diker* 1229–91, *Deker* 1535, which seems to be the ME word **dyker** meaning the number ten. In the Gloucestershire entries in DB certain rentals are expressed in *dickers* or tens of rods of iron. Possibly the rent or toll of the iron-works here was similarly expressed and the land subsequently named after it. Sx 439.

DICKLEBURGH Norf TM 1782. Partly uncertain. *Dicclesburc* 1086, *Dikelburg -le-* 1254. The generic is OE **burh** 'a fortified place, a manor house'. The specific may be pers.n. **Diċel* or **Dicla* and so 'Dicel or Dicla's fortified place'. Dickleburgh, however, lies on the Pye Road, the Roman road from Baylham to Caistor St Edmund, Margary no. 3d, which has a marked *agger*. The specific may, therefore, be a p.n. **Dīċ-lēah* 'the dyke wood or clearing' referring to this feature. DEPN.

DIDBROOK Glos SP 0531. 'Dydda's brook'. *Duddebrok'* 1248, *Di- Dydebroc -broke* 1257–1327, *Di- Dydbrok(e)* 1275–1588. OE pers.n. **Dydda* + **brōc**. Gl ii.9, L 16.

DIDCOT Oxon SU 5189. 'Dudda's cottage(s)'. *Dudecota* 1206–1335 with variants *-cot(e) -cothe* and *-kote, Duddecot'* 1208, *Dudcote* 1390–1517, *Didcot or Dudcot* 1657. OE pers.n. *Dud(d)a* + **cot**, pl. **cotu**. Brk 517.

DIDDINGTON Cambs TL 1965. Partly uncertain. *Dodinctun* 1086, *Dodintone, -ynton* 1086–1318, *Dudington -yng-* 1227–1497, *Dodington -yng-* 1252–1551. Usually explained as 'estate called after Dudda', OE pers.n. *Dudda* + **ing**[4] + **tūn**. But the place lies on raised ground above the Ouse and the specific might be OE ***dodd**, ***dudd** 'a rounded hill'. Hu 254, Årsskrift 1974.37.

DIDDLEBURY Shrops SO 5085. 'Dudela's fortified place or manor'. *Dodeleberia* c.1090, 1121, *Dodelbury* 1318, *Dudeleberi(a) -bire -bur(y)* 1138–1298, *Diddelbir'* 1231, *Diddlebury* 1688, *Delbury* 1796–1833 OS. OE pers.n. **Dudela* + **byriġ**, dative sing. of **burh**. Sa i.108.

DIDDLINGTON Dorset SU 0007 → GRITTLETON Wilts ST 8680.

DIDLEY H&W SO 4532. 'Dudda's wood or clearing'. *Dodelegie -lige* 1086, *Dud(d)ele(ia) -ley(a)* 1166–1431, *Dod(d)eleye* 1210×12, 1316. OE pers.n. *Dudda* + **lēah**. For the change *u* > *i* cf. DIDDINGTON Hunts TL 1965, DIDDLEBURY Shrops SO 5085, and Didlington Norf TL 7797, *Dudelingatuna, Dodelintona* 1086, *Dudelington* 1254. He 175, DEPN.

DIDMARTON Glos ST 8287. Partly uncertain. *Dydimeretune* [972]10th S 786, *Dedmertone* 1086, *Dudmerton'* 1220–1498, *Di-Dydmarton* 1380. This is usually taken to be 'Dydda's Marton or boundary settlement', OE pers.n. **Dyd(d)a* + **(ġe)mǣre-tūn**. The parish is on the Wiltshire boundary. But ponds have been found here so that *meretūn* 'pond settlement' is also possible. The 972 form may in fact point to original OE *Dyddingmeretūn* 'settlement at *Dyddingmere*, Dydda's pond', OE p.n. **Dyddingmere* < pers.n. *Dydda* + **ing**[4] + **mere**, + **tūn**. Gl iii.28, Jnl 25.45–6.

DIDSBURY GMan SJ 8591. 'Dyddi's fortified place'. *Dedesbiry* 1246, *Diddisbiry -es-, Didesbyri* 1276, *Diddesburye, Dutesbure* 1322, *Doddesbury* 1577, *Duddesburye* 1593, *Didsbury* c.1280. OE pers.n. **Dyddi*, genitive sing. **Dyddes*, + **byriġ**, dative sing. of **burh**. La 31, Jnl 17.31.

DIDWORTHY Devon SX 6862. 'Dudda's enclosure'. *Duddewrth(e)* 13th, 1238, *Dodeworthi* 1333. OE pers.n. *Dudda* + **worthiġ**. For unrounding and fronting of *u* > *i* cf. DINWORTHY SS 3115. D 291 gives pr [didəvər].

DIGBY Lincs TF 0854. 'Farm, village by the ditch'. *Dicbi* 1086, *Di-Dycby* 1405–38, *Dig(g)- Dyg(ge)by* 1160–15th. OE **dīċ** + **bȳ**. The reference is to an old drainage channel. Perrott 283, SSNEM 44.

DIGGLE GMan SE 0007. Probably 'ditch hill'. *Diggel* late 12th, *Dighil(l)* early 13th, 1468, *-hull* c.1272, *Dikele* 1249, *Deghall* 1638, *Diggle* 1822. OE **dīċ** + **hyll**. YW ii.311.

DILHAM Norf TG 3325. 'Homestead or promontory where dill grows'. *Dilham, Dillam* 1086, *Dilham* from 1101, *Dilam* 1610. OE **dile** + **hām** or **hamm** 2b. Nf ii.150.

DILHORNE Staffs SJ 9743. 'Quarry house'. *Dulverne* 1086–1386 with variant *-u-, Duluerne* c.1187, 1200, *Di- Dylverne* 1236–1604, *Delverne* 1281–1327, *Dyl(l)ron* 1414–1581, *Dillerne* 1534–1607, *Dyllon* 1583, *Dilhorne* 1665, *Dilhorn als Dillerne* 1786. OE ***dulf**, ***dylf**, + **ærn**. Oakden, PNE i.140, Horovitz.

DILSTON Northum NY 9763. 'The settlement on Devil's Water'. *Diulestuna* 1139×42 RRS, *Dovelestone* 1166, *Deuelestune, Develstone* 1171, *Di- Dyvel(e)ston -is-* 1176–1296 SR, *Dileston* 1298, *Devyleston* 1346. R.n. **Divles* as in DEVIL'S WATER NY 9356 + OE **tūn**. NbDu 63, RN 130.

DILTON MARSH Wilts ST 8549. P.n. *Dulinton* 1190, *Dultun'* 1221, 1265, *-ton* 1236–1587, *Dilton* from 1516, probably 'the estate called after Dulla', OE pers.n. **Dulla* + **ing**[4] + **tūn**, + ME **mersh**, *Mersshe* 1332. Wlt 147.

DILWYN H&W SO 4154. 'Shady or secret place(s)'. There are four types:

I. *Dilge* 1086 (2×), *Diliga* 1123, *Dile* c.1130.

II. *Dilven* 1086, *Dilu(i)n* 1137–1275, *Dilwin* 1654.

III. *Di- Dylum* 1193, 1272, *Di- Dylon(e) -iam* 1205–1251×2.

IV. *Dil(u)ve* 13th, 1372, *Dylewe* 1334.

OE **dīġle** (type I), dative pl. **dīġlum** (type II). Type II has been remodelled as if a W name in *gwyn* 'white'. He 68.

Sollers DILWYN H&W SO 4255. 'D held by the Sollers family'. *Solersdylewe* 1344, *Dylwesolers* 1345. Earlier simply *Dilge* 1086, *Dilege* 1160×70, *Dilewe* 1303. Family name Sollers as in BRIDGE SOLLERS SO 4142 + p.n. DILWYN SO 4145. He 68.

DINCHOPE Shrops SO 4584. Partly uncertain. *Dudingehope* c.1180, *Dodinghope* 1534×5, *Dynchop* 1503. This might be 'the valley called after Dudda', OE pers.n. *Dudda* + **ing**[4] + **hop** with unexplained assibilation of *ing* > *indʒ*; or 'the valley at or called Dudding, the place called after Dudda', p.n. **Dudding*, locative-dative sing. *Duddinġe*, + **hop**. In the latter case it is possible that the specific might be OE **dud(d)*, a variant of **dod(d)*, 'the rounded summit of a hill'. The meaning would then be 'the valley by the **dudding*', referring to one of the surrounding hills. The valley is one of a series of funnel-shaped side-valleys opening from Dale Hope into Corve Dale or in this case W into Ape Dale. Cf. EASTHOPE SO 5695, WESTHOPE SO 4786. Bowcock 87, DEPN, Nomina 6.35, 9.105, L 113, 117–121.

DINDER Somer ST 5744. 'Valley house'. *Denrenn* *[1065]18th S 1042, *Dinre -a* 1174–6, *Dinder* 1610. OE **denu** + **renn**, an early metathesised form of *ærn*. D lies in the upper valley of one of the tributaries of the r. Brue. DEPN.

DINEDOR H&W SO 5336. 'Fort hill'. *Dur(r)a* 1067×71, c.1130, *Dunre* 1086, *Dunre* 1176–1357, *Dinra* 1170, *Duyndre* 1350, *Dyndure* 1453. PrW ***dīn** + ***breʒ**. The reference is to Dinedor Iron Age hill-fort of the 2nd cent. BC. Tudor antiquaries later developed the name as if W *din dwr* 'hill by the river'. He 71, LHEB 320, Thomas 1976.148, Bannister 64.

DINES GREEN H&W SO 8255. *Dynes Green* 1831 OS. Probably surname *Dyne(s)* + ModE **green**. Wo 92.

DINGLEY Northants SP 7787. Partly uncertain. *Dinglei, Tinglea* 1086, *Dingele(ye) -lea -y-* 1166 etc. Possibly 'Dynni's clearing', OE pers.n. *Dynni* + **ing**[4] + **leah** with the same loss of syllable as in *king* < OE *cyning*; or OE **dyngel*, ME **dingle** 'a deep hollow', + **lēah**; there are marked valleys to the E and W of Dingley. Nth 164.

DINMORE H&W SO 4850 → HOPE UNDER DINMORE SO 5052.

DINNINGTON Somer ST 4012. 'Settlement at Duning, the hill-place'. *Dinni- Dvnintone* 1086, *Doniton* 1201, *Dunington* 1254, *Dunnington* 1610. OE ***dūning** 'hill' < *dūn* + *ing*[2] possibly used as a p.n. + **tūn**. The first DB spelling is probably an error for *Dunin-* due to minim confusion. The village lies beside a prominent hill. DEPN, Årsskrift 1974.38.

DINNINGTON SYorks SK 5386. Probably 'farmstead or settlement associated with Dunna'. *Dunin- Dunni- Domnitone* 1086, *Dunin(g)ton(e) -tun* 1091×7–1305, *Donington -yng-* 1147–c.1230, *Dinington -y-* 1271–1526, *Dynnyngton -i-* 1379–1591. OE pers.n. *Dunna* + **ing**[4] + **tūn**. However, Dinnington lies on high ground so that the specific might be OE ***dūning** 'hill place', an *ing*[2] derivative of *dūn* 'a hill'. For the sound change *u* > *i* cf. DINNINGTON T & W NZ 2073 and DINTON Bucks SP 7610 and Wilts SU 0131. YW i.146, Årsskrift 1964.38.

DINNINGTON T&W NZ 2073. Possibly 'the settlement at the high ground'. *Donigton'* 1242 BF, 1296 SR, *-yngton* 1346, *Dunington* 1255, *Dunnyngton* 1650, *Dinnington* 1663. OE ***dūning** < *dūn* + *ing*[2] + **tūn**. Dinnington lies at the edge of an area of hilly ground. Alternatively 'the estate called after Dunn or Dunna', OE pers.n. *Dunn(a)* + **ing**[4] + **tūn**. The sound change [u] > [i] is regular in NE dialects, cf. SHILLMORE NT 8807 and *Wissington* for WOOLSINGTON etc. NbDu 63, Årsskrift 1974.38.

Low DINSDALE Durham NZ 3410. *Low Dinsdale* 1861 OS, earlier *Nethir Dinsdale* 1624. ModE adj. **low**, earlier **nether**, + p.n. *Di- Dytneshal(e) -all'* 1174×95–1366, *Di- Dytineshale, -en-* 13th cent., *Di- Dytensale* 1283–1378, *Dedinsale -all, -y-* 1406–1624, *Dinsdell* 1573, *-dale* from 1590, 'Dyttin's nook of land', OE pers.n. **Dyttin*, genitive sing. **Dyttines*, + **halh**. Alternatively the 'nook of land belonging to Deighton', OE p.n. **Dīctūn*, genitive sing. **Dīctūnes*, + **halh**. 'Low' for distinction from Over Dinsdale NZ 3411 on the opposite bank of the Tees in Y, *Di(g)nes- Dirneshale* 1086, *detnisale* 1088, *Dinneshall, Dydensale, Ditneshal(l), Ditensala* 12th cent., *Dit(t)ensale* 15th cent., *Dynsda(i)ll* 16th cent. The Y Dinsdale formed a detached part of Allerton wapentake the nearest township of which was DEIGHTON NYorks NZ 3801, the *Dīctūn* of the compound p.n. Dinsdale. NbDu 63, YN 279, L 108, 110.

DINSDALE STATION Durham NZ 3413. P.n. Dinsdale as in Low DINSDALE NZ 3410 + ModE **station**.

DINTON Bucks SP 7610. 'Estate called after Dunn(a)'. *Danitone* 1086, *Donentona* c.1180, *Duninton* 1208–1316, *-ing-* c.1218, 1241. Probably OE pers.n. *Dunn* or *Dunna* + **ing**[4] or genitive sing. *Dunnan* + **tūn** rather than 'hill settlement', OE ***dūning** < **dūn** + **ing**[2], + **tūn**. Bk 159, YN addenda, Årsskrift 1974.38.

DINTON Wilts SU 0131. 'The estate called or at Duning'. *Domnitone* 1086, *Dunyngtun -ing-* 1154×89–1268, *Dunniton* 1184–1242, *Duniton -y* 1198–1249, *Dynton* 16th cent. OE ***dūning** 'the hill-place' (OE *dūn* + *ing*[2]) + **tūn**, 'the settlement at or called *Duning* or beside the hill formation', referring to the prominent nearby hill of Wick Ball (OE **ball*, ME *balle* 'a ball, a rounded hill') with its Iron Age hill-fort. Alternatively this could be OE pers.n. *Dunna* + **ing**[4] + **tūn**, 'the estate called after Dunna'. Wlt 160–1, Årsskrift 1974.38, Pevsner 1975.220.

DINWORTHY Devon SS 3115. 'Dunna's enclosure'. *Dunneworth'* 1242, *Donnewrth'* 1279, *Donnaworthi* 1333, *Dunworthy* 1499. OE pers.n. *Dunna* + **worthiġ**. For unrounding and fronting of *u* > *i* cf. DIDWORTHY SX 6862. D 134.

DIPPENHALL Surrey SU 8146. '(The settlement) at the deep nook of land'. *Depehal, Dupehale, Duppehalle, Depen- Dupenhale* 1224–99, *Dyp(pe)nall* 1548, *Dipnell* 1823. OE **dēop**, locative-dative sing. definite form **dēopan**, + **halh**, dative sing. **hale**. Sr 170, L101, 110.

DIPTFORD Devon SX 7256. '(The) deep ford'. *Depeforde* 1086–1428, *Dippeforda* 1084, *Dup(p)eford* 1267, 1303, 1385, *Ditford* 1671. OE **dēope**, definite form **dēopa**, + **ford**. A ford across the river Avon. D 288, 299.

DIPTON Durham NZ 1553. 'Deep valley'. *Depedene* 1189×95, *Depeden* 1339, 1349. OE **dēope** + **denu**. NbDu 64, L 98.

DISEWORTH Leic SK 4524. 'Digoth's enclosure'. *Digþeswyrþe* [967]14th S 749, *Diwort* 1086, *Digaðeswrð, Digðeswrthia* c.1180, *Dig(e)theswurth -worth(e)* 1184–1227, *Dy- Digeswrthe -worth'* 1193–1324, *Dy- Dichesw(o)rth(e) -is-* 12th–1296, *Dy- Dithesworth (e) -is-* late 12th–1546, *Dy- Dis(s)eworth(e)* 1242–1610. OE pers.n. **Digoth*, genitive sing. **Digethes*, + **worth**. Lei 415.

DISHFORTH NYorks SE 3873. 'The ford by the ditch'. *Di- Dysford(e)* 1086–1403, *-forthe* 1541–1612, *Disseford* 1157, 1198, *Diceford* 1202, 1208, *Di- Dysceford* 1208–1350, *Dishford* 1665. OE **dīċ** + **ford**. For other examples of AN substitution of *s* for OE *ċ* cf. DISS, DISSINGTON. The reference is probably to a drainage channel dug in the carrs beside the r. Swale. YN 184.

DISLEY Ches SJ 9784. Partially obscure. *Destesleg'* c.1251, *Destlegh* 1394, *Di- Dystislegh -ley(e) -es-* 1274–1495, *Di- Dysteleg(h) -ley(e)* 1286–1533, *Distley* 15th, *Disley* from 15th. Unknown element + **lēah** 'a wood, a clearing'. The solutions so far proposed are rather desperate, OE pers.n. **Dȳstig* 'Dusty', OE *dæġ-wist* 'food, a meal' and so 'wood where provisions must be taken, where meals are eaten, where a living is had from day to day', **dǣge-wist* 'dairy-maid's living' and so 'place where a dairy-maid makes a living'. The name reamins unexplained. Che i.269 gives pr ['dizli], addenda 19th cent. local ['disli].

DISS Norf TM 1179. 'The ditch'. *Dice* 1086, *Dic* 1130, *Dize* 1158, *Disze* 1190, *Disce* 1191. OE **dīċ** with AN spellings *z, sz* representing [ts] for OE [tʃ], subsequently simplified to [s]. The reference is unknown but may have been to the valley of the river Waveney as also, perhaps, in DITCHINGHAM TM 3491. DEPN.

DISTINGTON Cumbr NY 0023. Partly uncertain. *Distingtona* [before 1230]n.d., *Distington* from 1256. The prevalence of *-ing-* spellings makes explanation from OE ***dȳsten** 'dusty' + **tūn** unlikely. Possibly the specific was an OE ***dȳsting** 'dusty place' from OE *dūst* + *ing*[2]. Cu 375.

South DISTRICT Norf TL 5298. A modern name for former *Londoner Fen* 1824 OS.

DITCHEAT Somer ST 6236. 'Dike gate'. *Dichesgate* [842]14th S 292, *-gete* [855×60]13th S 1699, *Dicesget* 1086, *Dichesgete* 1196, *Dichiat* 1610. OE **dīċ**, genitive sing. **dīċes**, + **ġeat**. The reference may be to the Foss Way, Margary no. 5b, which runs to the W of the village. Dr Gelling points out that both this and LAMYATT ST 6535 lie either side of the gap in the horseshoe of hills surrounding Evercreech which is probably the *ġeat* in question here. In S 292 there is mention of *Dich, Dichforde* and a *strete yate* all referring to the same road and dike. DEPN.

DITCHINGHAM Norf TM 3491. 'Homestead of the Dicingas, the people who live by the dike or ditch'. *Dicingaham* 1086, *Dichingeham* 1178, 1194, *Dikingeham* 1196, *Dichingham* 1212, *Dikingham* 1230. OE folk-n. **Dīċingas* < **dīċ** + **ingas**, genitive pl. **Dīċinga*, + **hām**. Ditchingham lies on the river Waveney which may be the *dīċ* of this name, cf. DISS TM 1179. DEPN, ING 134.

DITCHLING ESusx TQ 3215. 'The Diccelingas, the people called after Diccel'. *Dicelinga* (genitive pl.) [c.765]13th S 50, *(þone ham æt) Diccelingum* [873×88]11th S 1507, *Dicen- Dicel- Digelinges* 1086, *(in) Dicelingis* 1121, *Dicheling* 13th–16th cents., *Dichlinge* 1589. OE folk-n. **Diċċelingas* < pers.n. **Diċċel* + **ingas**, dative plural **Diċċelingum*. Sx 300, ING 33.

DITCHLING BEACON ESusx TQ 3313. *Ditchling Beacon* 1813 OS. P.n. DITCHLING TQ 3215 + ModE **beacon**. It stands on a spur of the South Downs that rises to 813ft.

DITTISHAM Devon SX 8655. 'Dydi's promontory'. *Didasham* 1086, *Didisham* 1230–1316 with variants *Dy-* and *-es-*, *Dyteshamme* 1340, *Dytsham* 1570, *Dedishome* 1462. OE pers.n. *Dydi*, genitive sing. *Dydes*, + **hamm** 2b. The reference is to the promontory N of the village protruding into the river Dart – 'a thumb of high land stretches into the Dart and forces it to make a great sweep about it' (Baring-Gould). D 322 gives prs [ditsəm] and [ditʃəm], L 43, 49.

DITTON 'Ditch settlement'. OE **dīċ** + **tūn**.

(1) ~ Ches SJ 4985. *Ditton* from 1194 with variant *Dy-*, *Dutton* 1202–1341 by confusion with Dutton Ches SJ 5579. The reference is probably to a drainage ditch in the low-lying land near the Mersey. Contrasts with Upton SJ 5087, *Upton* from 1251, 'upper settlement' on higher ground. La 106.

(2) ~ Kent TQ 7157. *(to) dictune* c. 975 B 1322, [995×1005]12th S 1456, 1086, *Ditton(')* from 1163×4. The reference may have been to the stream called Bradbourne which flows past the church, or to a ditch around the original settlement. KPN 302, 306.

(3) ~ GREEN Cambs TL 6658. *Ditton Green* 1836 OS. P.n. Ditton as in WOODDITTON TL 6559 + ModE **green**. Contrasts with nearby *Houghton Green* 1836 OS, *Heighton Green* 1821, *Hoghton Green* c.1825, *Cross Green* 1836 OS, BURROUGH GREEN TL 6355 and KIRTLING GREEN TL 6855. Ca 128, 118.

(4) Fen ~ Cambs TL 4860. *Fen* 1281, *Fenny-*1285 + p.n. *(æt) Dictunæ* 962×991 S 1494, *(æt) Dictune* [1000×2]11th S 1486, [1042×66]13th S 1051, *(at) Dittone* [946×c.951]13th S 1483, *Dittune* [1042×66]13th S 1051, *Dittona* 1086, *Ditton(e)* from 1200, 'settlement by (Fleam) dyke', OE **dīċ** + **tūn**. 'Fen' for distinction from from WOODDITTON TL 6559. Ca 142.

(5) Long ~ Surrey TQ 1665. *Longa Dittone* 1242. ME adj. **long** +

p.n. *Ditune* 1086, *Ditton* 1233. 'Long' alludes to the relative length of the parish compared with its width and is for distinction from nearby Thames DITTON TQ 1567. Sr 57.

(6) Thames ~ Surrey TQ 1567. *Temes Ditton* 1235. R.n. THAMES + p.n. *Dictun* [1005]12th S 911, *-ton(a)* 13th, *Ditone* 1086. The village lies nearer to the r. THAMES than does Long DITTON TQ 1665. Sr 90–1.

DITTON PRIORS Shrops SO 6089. 'D held by the prior' sc. of Wenlock Priory. *Dodyton Prioris* 1346, *Prioursduditon* 1406, *Ditton Prioris, Priorz Dytton* 1535. P.n. Ditton + Latin **prior**, genitive sing. **prioris**. Ditton, *Dodintone -en-* 1086, *Dodintun' -ton(') -yn-* 1161–1334 with variant *Dud-, Do- Dudington' -yng-* 1190–1331, *Dodyton'* 1334, *Dytton* 1510×11, is either 'Dudda or Dodda's estate', OE pers.n. *Dud(d)a, Dod(d)a* + **ing**[4], or genitive sing. *Dud(d)an, Dod(d)an*, + **tūn**; or possibly OE ***dodding**, a derivative of OE **dod(d)* 'the rounded summit of a hill', referring to the site of the church, + **tūn**. 'Priors' for distinction from Earls DITTON SO 6275. Sa i.109.

Earls DITTON Shrops SO 6275. 'D held by the earl'. *Erlesdodinton* 1398, *Erles Dytton* 1566, *Earls Ditton* 1770. ME **erle**, genitive sing. **erles**, referring to the Mortimer family who were earls of March and held the manor from 1086, + p.n. *Dodentone* 1086, *Doddington al's Ditton al's Erles Ditton* 1603×25, possibly 'Dudda's estate', OE pers.n. *Dud(d)a*, genitive sing. *Dud(d)an*, + **tūn**. However the forms for nearby DODDINGTON SO 6176, apparently a doublet of this place, point rather to OE *Dodding-tūn* 'the estate at Dodding', OE ***dodding**, a derivative of ***dod(d)** 'the rounded summit of a hill', + **tūn**, which suits the topography here. It is noteworthy that another place in the district, Dudnil SO 6074, *Duddenhill or Dudnil* 1833 OS, apparently 'Dudda's hill' also occupies a marked hill-site. 'Earls' for distinction from DITTON PRIORS SO 6089. Sa i.113.

DIXTON Glos SO 9830. Possibly 'dike-hill down'. *Di- Dyc(c)lesdon(e), Dickles-, -duna* 1059–1327, *Dricledone* 1086, *Di- Dycleston(e)* 1212, 1418, 1593, *Dichestone* 1166, *Di- Dyxton* 1487–1591, *Dixon* 1611, 1640. OE p.n. **Dīc-hyll* 'dike-hill', genitive sing. **Dīc-hylles*, + **dūn**. The reference is to The Knolls, an Iron Age camp NW of the village at SO 9731. This is more probable than the alternative suggestion of unknown pers.n. **Diccel*. Gl ii.49.

DIZZARD POINT Corn SX 1699. *Dazard Point* 1813. P.n. Dizzard SX 1698 + ModE **point**. Dizzard, *Lisart, Disart* 1086, *Dysert* 1238, *Disard* 1324, *Dyzade* 1610, is Co ***dy-serth** 'very steep'. There is no evidence in Cornish for **dyserth* 'wilderness, hermitage' equivalent to Welsh *diserth*, Latin *desertum*. Diserth Clwyd SJ 0579 is also at the foot of a steep hill. PNCo 78, Gover n.d. 60, CPNE 85.

DOBWALLS Corn SX 2165. 'Dobb's walls or ruins'. *Dobwalls or Hogswall* 1607, *Dobbewalles* 1619. Surname Dobb (*Dobbe* 14th, 16th) + ModE **wall**, plural **wall(e)s**. PNCo 78.

DOCCOMBE Devon SX 7786. 'Valley where docks grow'. *Dockumb'* 1221×30, *Doccombe* from 1330. OE **docce** + **cumb**. D 485.

DOCKING Norf TF 7637. 'The place growing with dock'. *(et) Doccyncge* 1035×40 S 1489, *Dochinga -e* 1086, *Dochinge -a -k-* 1157–1212, *Docking* from 1208. OE ***doccing** < **docce** + **ing**[2]. DEPN, ING 199.

DOCKLOW H&W SO 5657. 'Dock hill or tumulus'. *Dochelowe* 1277×92, *Dockelawe* 1291, *Dok(k)lowe* 1341, 1397. OE **docce** + **hlāw**. He 72.

DOCKRAY Cumbr NY 3921. 'Nook where dock or sorrel grows'. *Doc- Dokwra* 1278–1478, *Dockewra* 1279, *Dockray* from 1577. ME **dokke** (OE *docce*) + ME **wra** (ON *vrá*). Cu 333, SSNNW 225.

DOCTOR'S GATE Derby SK 0794. *Docto Talbotes gate* (sic) 1627, *The Doctor's Gate* 1789. ModE **doctor** + **gate** in the sense 'road' (ON *gata*). The name of the Roman road from Brough to Melandra Castle, Margary No.711. It is not known who Dr Talbot was. Db 21.

DODD FELL NYorks SD 8484. *Dodd Fell* 1860 OS. Mod dial. **dod** (ME *dodde* 'the rounded summit of a hill') + ModE **fell** (ON *fjall*). PNE i.133.

DODDINGHURST Essex TQ 5998. 'Wooded hill of the Duddingas, the people called after Dudda'. *Doddenhenc* 1086, *Dutingehest* 1181, *Duddingeherst -hurst* 1218–76, *-ing- -yng-* 1253–1300, *Dod(d)inghirst -herst -hurst* from 1260. OE folk-n. **Duddingas* < pers.n. *Dudda* + **ingas**, genitive pl. **Duddinga*, + **hyrst**. Ess 152.

DODDINGTON 'Estate called after Dudda'. OE pers.n. *Dudda* + **ing**[4] + **tūn**. OE *Dudda* is a common pers.n. Occasionally, especially in the N country, however, where the topography favours it, the specific may be OE ***dodd, dudd** 'a rounded hill' or ***dodding, dudding**. PNE i.133, Årsskrift 1974.37ff, Fris 361–76.

(1) ~ Cambs TL 4090. *Dundingtune* (sic) [c.975]13th LibEl, *Dudintone, Doddintona, Dodinton(e), Dodincgtune, Dodingetone* 1086, *Dudingtune -tone -yng-* 1170–1559, *Dodington(e) -yng-* 1256–1600. Although this place is on slightly raised ground above the fens it seems safer to take it as the 'estate called after Dudda'. Ca 251, Årsskrift 1974.37.

(2) ~ Kent TQ 9357. *Duddingtun, Dodintuna* c. 1100, *Du- Dodinton(e) -yn* c. 1180–1278, *Du- Dodington -yng-* 1254–78. OE pers.n. *Dudda* + **ing**[4] + **tūn**. PNK 274, Årsskrift 1974.37.

(3) ~ Lincs SK 9070. *Dodinctone, Dodintune* 1086, *-inton* 13th cent., *-ington -yng-* 1177–1607, *Doddington* from 1561. The village is situated on a low circular hill. Perrott 251, Årsskrift 1974.37.

(4) ~ Northum NT 9932. *Dodinton* 1207, 1281, *Do- Dud(d)ington -yng-* 1242 BF-1428, *Dorrington* 18th cent. OE ***dodding** < ***dod**, ***dud** 'a rounded hill or summit' + **ing**[2], + **tūn**. The reference is to nearby Dod Law. NbDu 65, Årsskrift 1974.37.

(5) ~ Shrops SO 6176. This place is not marked on the first edition of the one inch Ordnance Survey and seems to be a doublet of nearby Earls DITTON SO 6275 to which the forms probably apply. *Dodington* 1285, 1566, *Dodinton* 1316, *Dodyton* 1431. Sa i.113 suggests that this and Earls Ditton were both members of the same land unit called *Dodingtūn*. DEPN.

(6) Dry ~ Lincs SK 8546. *Dry Dodyngton* 1325. ME adj. **dry** + p.n. *Dudintun'* ?1085×9, [1186×8]14th, *Dodintune -tone* 1086–1283, *-ington -yng-* 1157–1531, *Doddington* 1570×1, 'Duda's estate', OE pers.n. *Duda*, genitive sing. *Dudan*, varying with *Duda* + **ing**[4] + **tūn**. Although the village is situated on a circular hill the earliest forms (not known in 1974) are against derivation from ***dodding**. Perrott 387, Årsskrift 1974.37, Cameron 1998.

(7) Great ~ Northants SP 8864. *Great Dodington* 1290, *Dodington Magna* 1346. ME adj. **great**, Latin **magna**, + p.n. *Dodintone* 1086–1397 with variants *-ing- -yn(g)-, Dorrington* 1558×1603, 1675. Doddington is situated on a hill-slope above the river Nene. Nth 138, Årsskrift 1974.38.

DODDISCOMBSLEIGH Devon SX 8586. '*Leigh* held by the Doddescombe family'. *Doddescumbeleghe* 1309, *Dascomley* 1628. Earlier simply *Levge* 1086, *Legh* 1303, 'the wood or clearing', OE **lēah**. Also known as *Leghe Gobol(de)* c.1260, *Gobaldeslegh* 1289, and *Leghe Peverel* 1313. The manorial additions refer to tenure by Godebold 1086, Hugh Peveril 1242, and Ralph de Daddescumb or Doddescambe c.1260. D 494 gives pr [daskəmzli], DEPN.

DODFORD H&W SO 9373. 'Dodda's ford'. *Doddeford* 1232–1327. OE pers.n. *Dodda* + **ford**. Wo 340, L 69.

DODFORD Northants SP 6160. 'Dodda's ford'. *doddanford* 944 S 495, *Dodeford(e)* 1086–1326, *Dodde-* 1218–1377, *Dadford* 1702, 1730. OE pers.n. *Dodda*, genitive sing. *Doddan*, + **ford**. The territory and the bounds of the people of Dodford with Newnham SP 5859, *doddafordinga land, doddafordung* *gemære*, are mentioned 1021×23 in S 977. A 17th cent. tradition held erroneously that the place was named from a water-weed called *dod* which grows plentifully here. Nth 20, L 69.

DODINGTON Avon ST 7580. OE ***doding + tūn.** 'Settlement by the rounded hill'. *Dodinton(e) -yn-* 1086–1382, *Do- Duddintone* 1166, *Dudin(g)ton'* 12th–1248, *Dodington* from 1221. Dodington lies at the foot of the indented Cotswold escarpment. Gl iii.48, Årsskrift 1974.38.

DODLESTON Ches SJ 3661. 'Dod(d)el's farm or village'. *Dodestune* 1086, *Dodleston* from 1153 with variants *Dodel(i)s- -es-, Dod(d)les -is-* etc. OE pers.n. **Dod(d)el*, genitive sing. **Dod(d)eles*, + **tūn**. Che iv.156 gives pr ['dɔdlstən].

DODMAN POINT Corn SX 0039. *Dudman Poynt* 1512 Gover n.d., *Dudman foreland* c.1540, *~ poynt* 1610, *Dodman* 1564, *Deadman Point* 1699. Surname Dodman, *Dudemann* 1201, *Dudeman* 1206, *Dudmann* 1469, + ModE **point**. The original Cornish name survives in PENARE SW 9940. PNCo 79 gives pr [ded-man], an English reinterpretation as seen in the 1699 form.

DODWORTH SYorks SE 3105. 'Dod's enclosure'. *Dodesuu(o)rde* 1086, *-worth* 1300, *Doddewrd(a)* c.1090–c.1232, *-worth* 1301, 1307, 1414, *Dodwurth* 1349, *-worth* 1379, *Dudewurða* 1170. OE pers.n. *Dod(d)i*, genitive sing. *Dod(d)es*, or *Dod(d)a* + **worth**. YW i.306.

DOE LEA Derby SK 4566. By origin a r.n., 'valley river'. *(aquam de) Dal* 1149×53–1229, *Dalhe(e)* 1154×89, *Dalle hee* 1349, *Dalle* 1496, *Dawley* 1540, 1767. OE **dæl** (ON *dalr*) + **ēa**. The modern form Dawley, *Dawl-ey*, has been wrongly understood as if *Daw-ley*, a name in *ley* (OE *lēah*) and then reinterpreted as *Doe Lea*. Db 5.

DOG VILLAGE Devon SX 9896. No early forms.

DOGDYKE Lincs TF 2055. 'Ditch where docks grow'. *Dockedic* 1154×89, *-dyk* 1289, 1531, *Do(c)kdic -dyk(e) -dike* 1275–18th, *Dogdyke -dike* from 1484. OE **docce** + **dīċ**. Both Water Dock, *rumex aquaticus*, and Great Water Dock, *rumex hydrolapathum*, were valued for herbal and medicinal properties. Although attractive, it is unlikely that the specific is identical with ModE *dock* sb 'the bed in the sand or ooze in which a ship lies dry at low tide, an artificial inlet to admit a boat' first recorded in England in 1513 and believed to be a borrowing from MDu *docke*, itself not recorded before 1436 and of unknown origin, possibly from late Latin **ductia* 'an inlet' (from *dūcere* 'to lead'). Perrott 320, Grieve 259.

DOGMERSFIELD Hants SU 7852. Possibly 'open land of *Doccemere*, the dock pond'. *Dochemeresfelda* 1106, *Docche- Dokemeresfeld* 1167, 1188×9, *Dog(ge)meresfeld* 1198–1352. P.n. **Doccemere* 'dock pond' < OE **docce**, the plant 'dock', + **mere**, genitive sing. **Doccemeres*, + **feld**. The corrupt DB form *Ormeresfelt* 1086 has recently been revived in the name of Ormersfield Farm SU 7852 (*Dormersfield F.* 1816 OS). The pool mentioned may have been the ancestor of Tundry Pond SU 7752 at Tundry Green, *Tundry greene* 1629, OFr **tonderie** 'a cloth-shearing workshop'. Ha 66, Gover 1958.113–4, DEPN, L 26, 243.

DOLEBURY Somer ST 4558. *Dolberry* 1817 OS. Ultimately from OE **byriġ**, dative sing. of **burh** 'fort', referring to the great Iron Age hill-fort here. The specific is unknown. Pevsner 1958N.165.

DOLES WOOD Hants SU 3852. *Boscus de Doles* 1216×72, *Doles Wood* 1817 OS, cf. *Dowles Heath* 1650. P.n. *Doles*, ME **dole** (OE *dāl*)'a share', pl. **doles**, probably referring to apportionments in a common woodland, + **wode** (OE *wudu*). Gover 1958.160.

DOLPHINHOLME Lancs SD 5153. 'Dolfin's island'. *Dolphineholme* 1591, *Dolphinhoulme* 1621. Pers.n. *Dolfin* (ON *Dólgfinnr* rather than OFr *Dalfin, Daufin*) + ON **holmr**. La 164.

DOLTON Devon SS 5712. Partly uncertain. *Ovel- Duveltone* 1086, *Du(v)eltona* 1086 Exon, *Duwel- Dyvil- Du(gh)el- Dewelton(e)* 13th cent., *Deul- Doghelton* 14th, *Dowelton* 1477, *Doulton* 1675. The generic is OE **tūn** 'farm, village, estate' the specific unexplained. Connections have been sought with PrW **duβ* 'black' or an OE p.n. **Dūfe-feld* 'open country frequented by doves', but neither suggestion adequately accounts for the spellings recorded which point rather to a specific something like **dugol, *dugel*. Cf. DOWLAND SS 5610 which has the same specific. D 366.

Little DON or THE PORTER RIVER SYorks SK 2399. *Litteldon* 13th, *Little Dun River* DATE. See r. DON SK 4696, THE PORTER OR LITTLE DON RIVER SK 2399.

River DON SYorks SK 4696. *Don* from before 1135, *Done* c.1175–1439, *Doon* 1275, 15th cent., *Dun(ne)* 1300–1633. PrW **Dōn* < Brit **Dānu* from the IE root **da-* 'flowing' seen also in the r.ns. Danube, **Dānou̯i̯os*, and Rhône, **Ro-dānos*. See further DONCASTER SE 5702. YW vii.126, FT 145 note 2, BzN 1957.245, Krahe 1964.93, 103, RBrit 329.

DONCASTER SYorks SE 5702. 'Fortification on the river Don'. *Dano* [c.300]8th AI, [c.420]15th ND, *Cair Daun* [c.800]828×9 HB, *(æt) Doneceastre* [1002×4]11th S 1536, *Donecastr(e) -castria -caster* 1086–1382, *Donacastre* c.1160, *Doncastr(e) -caster* 1119×47–1822, *Dane-* c.1160–1304, *Dene-* 1165–1268. R.n. DON SK 4696 + OE **cæster**. The Roman station here was called *Danum*, originally the name of the r. Don. *Dane* spellings represent the N dial. treatment of OE *ŏ* before nasals after shortening from *ō* in the compound p.n. *Dene* spellings are due to mistaken association with the Danes, ON *Danir*, OE *Dene*. YW i.29, TC 83, RBrit 329.

DONHEAD ST ANDREW Wilts ST 9124. 'St Andrew's D'. *eccl. Sancti Andree de Dunheved* 'the church of St Andrew of Donhead' 1298, *Do(u)nhefde Sci Andree* 1346, *Donhead Andreu* 1373, *Donnett St Andrew* 1690. P.n. *Dunheved, Dunehefda* [871×99]c.1400 S 356, *Dunheued -v-* [956]14th S 630, 1242, 1249, *(to) Dun heafdan* [955]14th S 582, *Dunheve* (sic) 1086, '(the place at) the end of the down', OE **dūn** + **hēafod**, referring to the edge of the ridge of high ground stretching E of Shaftesbury beneath which the village is located, + saint's name *Andrew* from the dedication of the church and for distinction from nearby DONHEAD ST MARY; both villages are probably referred to in the dative pl. form *Dun heafdan* (OE *dūn-hēafdum*) in S 582. Wlt 187 gives popular pr [donət].

DONHEAD ST MARY Wilts ST 9024. 'St Mary's D'. *Donheved Sancte Marie* 1298, *Dunhed(d) Mary* 1588×1603. P.n. Donhead as in DONHEAD ST ANDREW ST 9124 + saint's name *Mary* from the dedication of the church. Wlt 187.

DONINGTON 'Settlement called or at *Duning*, the high ground'. OE ***dūning** < **dūn** + **ing**[2] + **tūn**.

(1) ~ Lincs TF 2035. *Dvninc- Donninctune, Dunninc hund'* 1086, *Doninton'* 1167, *Duningetona* [1189×99]1290, *Dunington* 1202, *Duninge-, Dunninton* 1203. Donington is situated on raised ground in the fens. DEPN, Årsskrift 1974.39, Cameron 1998.

(2) ~ ON BAIN Lincs TF 2382. *Donyngton super Beyne* 1216×72. P.n. *Duninctune* 1086, *Duningtun'* c.1150, *Dunnington* 1202 + r.n. BAIN TF 2472 for distinction from DONINGTON TF 2035. Donington is situated on the edge of the Wolds overlooking the Bain. DEPN, Årsskrift 1974.39.

DONISTHORPE Leic SK 3114. 'Durand's outlying farm'. *Durandestorp' -is-* 1086–1273×1307, *-thorp* 13th–1497, *Donasthorp(e)* 1278–1549, *-is-* from 1535, *Dunnistrop* 1715. CG pers.n. *Durand*, genitive sing. *Durandes*, + **thorp**. Lei 565, SSNEM 108.

DONKEY TOWN Surrey SU 9360. Unexplained.

DONNA NOOK Lincs TF 4399. *Donna Nook* 1824 OS.

DONNINGTON Berks SU 4669. 'Estate called after Dunn(a)'. *Dunintona* 1167–1334 with variants *-i(n)ton' -ynton -y(e)ton'*, *Dony(n)gton(e)* 1284–1394, *Dodynton'*. OE pers.n. *Dunn* or *Dunna* + **ing**[4] + **tūn**. Two forms, however, *Deritone* 1086, *Derinton'* 1253, unless due to *n*/*r* confusion, apparently contain OE pers.n. *Dēor(a)* not *Dunn(a)*. The coexistence of the two names probably reflects pre-Conquest tenurial development. The isolated form *Dodynton'* is the result of confusion with the common p.n. Doddington. The alternative view of Årsskrift 1974.39 that we are concerned with OE **dūn** + **ing**[2] + **tūn** is unnecessary in view of the frequency of the OE pers.n. Dunna. Brk 264, Signposts 179.

DONNINGTON Glos SP 1928. 'Estate called after Dunn(a)'. *Doninton -yn-* c.1195–1359, *-ing- -yng-* 1262–1590, *Dunninton(e) -yn-* early 13th, *-yng- -ing-* 1585, *Donnington* 1597. OE pers.n. *Dunn(a)* + **ing**[4] + **tūn**. Nearby Duncombe House SP 1727, *(innon, of) dunnen cumbe* [779]12th S 115, 'Dunna's coomb' and a series of local names in *Dunnen, Dunnes* in the Anglo-Saxon Bounds of Donnington establish the pers.n. *Dunn(a)* rather than the appellative ***dūning**. Gl i.217, Årsskrift 1974.39.

DONNINGTON H&W SO 7134. Partly uncertain. *Dvnninctvne* 1086, *Duni(n)tona -tuna* 12th, *Dunnington* 1231, *Donyngton* 1341. Donnington Hall occupies a hill site which suggests the specific might be OE ***dūning** 'a small hill' and so 'estate called or at *Dūning*'; otherwise, OE pers.n. *Dunna* + **ing**[4] + **tūn**, 'estate called after Dunna'. He 72, Årsskrift 1974.39.

DONNINGTON Shrops SJ 5807. Probably 'the estate called after Dunna'. *Dunniton* 1180, *Duninton* 1201, *Donynton* 1303, *Donyngton by Wroxcestre* 1348. OE pers.n. *Dunna* + **ing**[4] + **tūn**. Donnington is situated on a marked hill so that the alternative OE ***dūning** 'that which is like a **dūn**, a hill' might be possible except for the early *-nn-* spelling. DEPN, Årsskrift 1974.39, Sa i.111.

DONNINGTON Shrops SJ 7013. Probably 'the estate called after Dunna'. *Donnyton* 1271×2, *Dunnyton* 1577, *Dunnington* 1740. OE pers.n. *Dunna* + **ing**[4] + **tūn**. The spellings with *-nn-* are against derivation from an OE ***dūning** 'the hill-place'. Sa i.111.

DONNINGTON WSusx SU 8502. 'The estate called after Dunnic or Dunneca'. *Dunne- Dunketone* [966]12th S 746, *Cloninctune* (for *Don-*) 1086, *Duneketon* 1219, 1248, *Donegh(e)ton* 1291, 1388, *Donghton al. Donnyngton* 1558. OE pers.n. *Dunnic, Dunneca* + **ing**[4] + **tūn**. Identical with DUNCTON SU 9617. Sx 69, SS 74.

Castle DONNINGTON Leic SK 4427. ME prefix **castel**, *Castel(l)-* 1302–1565, *Castle-* from 1571 + p.n. *Duniton(e)* 1086, 1242, *Duninton(e)* 1086–1242, *Dunington(e) -yng-* 1193–1624, *Dorrington(e) -yng-* from 1227, either the 'farm, village at or called *Duning*', OE **Dūning* 'that which has the characteristic of a *dūn*, hill-place' + **tūn**, or 'estate called after Dunn', OE pers.n. *Dunn* + **ing**[4] + **tūn**. Donnington lies on the N slope of a ridge of high ground in an arc of *ington* settlements. Lei 363, Årsskrift 1974. 39.

DONYATT Somer ST 3314. 'Dunna's gate'. *(on) Duuneȝete* (for *Dunne-*) [725]13th S 249, *Doniet* 1086, 1610, *Duneiet* 1176, *Dunniete* 1212, *Dunnechete* before 1270 Humphreys, *Dounyate* [1362×2]14th, *Dunyatt* 1603×25 Humphreys. OE pers.n. *Dunna* + **ġeat**. The reference is probably to a gap in the hills here through which the river Isle flows and the road from Chard to Taunton passes. In the same early source occurs the name *Dunnepool* 'Dunna's pool' [725]13th S 249. DEPN.

DONYLAND Essex TM 0222. 'Newly cultivated land called after Dunna'. *(on) Dunninclande, Dunninglande -æ* [962×91]11th S 1494, *Duning Lond* [962×91]13th ibid., *Dunulanda, Dunilanda(m) -ā* 1086, *Duni- Doniland(a) -lond(es)* from 1147. OE pers.n. *Dunna* + **ing**[4] + **land**. Ess 387, L 247, 249.

DORCHESTER Dorset SY 6890. 'The Roman town of *Dorn* (*Durnovaria*)'.

I. Romano-British forms: *Durno(no)varia* [c.300]8th AI, *Duriarno* [8th]13th Rav.

II. Welsh form: *Durngueir* [893]11th Asser.

III. Full English forms: *Dornwerecestre* [833]14th S 277, *Dornuuarana ceaster* 847 S 298, *Dornwaracestre -ceaster -ceastræ* [863, 864, 868]12th S 336, S 333, S 340.

IV. Reduced English forms: *Doracestria* [843 for 934]17th S 391, *Dorn(e)ace(a)ster* [937]12th, 13th S 434–5, *Dorecestre* 1086–1244, *Dorcestr(e)* 1194–1447, *Dorchester* from 1273.

The Romano-British name is a compound of Brit ***durno-** 'fist' found also in Scottish *Dornock, Dornoch* < **durnāco-* 'place covered with fist-sized pebbles' + ***u̯ariā**, an element of uncertain meaning found also in *Argentovaria* and *Varar* (the r. Farrar, Scotland NH 2438), possibly pre-Celtic **var-* < IE **u̯er-/ u̯or-* 'wet'. If so *Durnovaria* was by origin a water-name of some kind rather than the name of Maiden Castle. The RBrit name was anglicised in full form as *Dornwara ċeaster* 'the city of the inhabitants of Dorn', shortened p.n. *Dorn-* (as in the county name *Dornsǣte* 'the Dorn people', modern DORSET) + OE **wara**, genitive pl. of **ware** 'inhabitants', + **ċeaster**, and in short form as *Dorn-* + **ċeaster** 'the Roman town of *Dorn*'. Do i.347 gives prs ['dɔːtʃistə, 'dɔːdʒestə, 'daːRtʃistəR] and ['daːdistəR], BzN 1957.235, SPN 180–3, RBrit 345.

DORCHESTER Oxon SU 5794. 'The Roman town Dorcic'. *Dorcic, Dorciccaestræ* [c.731]8th BHE, *Dorce(s)ceastre* c.900 ASC(A) under year 635, *Dorche- Dorkecestre* 1086, *Dorchestre* 1331 etc. A Latin adjectival form *Dorcocensis* occurs [995]c.1000 S 1379. P.n. *Dorcic* + OE **ċeaster**. The origin of *Dorcic* is not certain: either British **Duro-c-*, a diminutive of **duro-* 'a fort' or the root **derk-/dork-* as in Breton *derch*, W *drych* 'mirror, aspect, appearance', OIr *dercaim* 'I see', OE *torht* 'bright'. O 152, RBrit 513, GPC 1091.

DORDON Warw SK 2600. 'Deer hill'. *Derdon* 12th–1398, *Dere-* 1230, *Dordon* from 1292. OE **dēor** + **dūn**. Wa 21, L 144, 154.

DORE SYorks SK 3081. 'The door or narrow pass'. *Dore* from c.900 ASC(A) under year 827, 1086, *Dor* c.955 ASC(A) under year 942, *Doer, Dawre, Dawer* 16th cent. OE **dor**. A pass on the ancient boundary between Mercia and Northumbria. Db 240.

Abbey DORE H&W SO 3830. *Abbey Dore* 1831. A new name which came into use in the 18th cent. for the ruins of the Cistercian abbey founded here in 1147. ModE **abbey** + r.n. *Dour* [early 8th]c.1130, *Do(w)r* c.1130, *Dore* from 1213, PrW ***duβr** 'water'. The early spellings *Dour, Dowr* and *Dor* probably stand for *Dovr* and *Dôr* respectively < late Brit **Dobra* < Brit **Dubrā*. The abbey site was known as *Blanchebershale* c.1120×5, *Blancharbesal* 1232, *Blak Berats Haulle* c. 1540, possibly 'White *Ebershale*', OFr adj. **blanche** + p.n. **Ebershale* 'Eadberht's nook', OE pers.n. *Ēadberht*, genitive sing. *Ēaberhtes*, + **halh**, dative sing. **hale**. He 20, LHEB 418.

DORKING Surrey TQ 1648. Either 'the Deorcingas, the people called after Deorc', or 'the Dorcingas, the dwellers on the r. *Dorce*'. *Dorchinges* 1086, *Dork-* 1180–c.1270, *Dorkingg* from 1219 with variants *-ing(g)(e)* and *-yng(g)(e), Derkyng* 1431, *Darking* 16th, *Darken* 1701. OE **ingas** folk-n. with first element either the pers.n. *Deorc* or r.n. **Dorce*, a postulated old name for the upper MOLE identical with Wilts stream name *(innan) Dorcan* [1043×53]12th S 1588 from IE root **derk-* 'glance' seen in OE *torht* 'bright', OIr *derc* 'eye', and so meaning 'clear, bright stream'. Sr 269, ING 29, RN 129.

DORMANSLAND Surrey TQ 4042. 'Deorman's or Dereman's land'. *Deremanneslond* 1263, 1288, *Dermannysland* 1489, *Dormans* 1593. Either the OE pers.n. *Dēorman* or its ME derivative *Dereman*, genitive sing. *Deremanes*, + **land**. Richard *Dereman* lived nearby in 1418. Sr 328.

DORMANSTOWN Cleve NZ 5823. Planned as a company town for the firm of Dorman Long, bridge builders and steel manufacturers, in 1918, but not fully developed. Room 1983.32.

DORMINGTON H&W SO 5840. 'Estate called after Deorma'. *Dorminton* 1206, 1242, *-ing-* 1290. OE pers.n. *Deorma*, short for *Dēormōd* or *Dēormūnd*, + **ing**[4] + **tūn**. DEPN.

River DORN Oxon SP 4420. A late back-formation from DORNFORD SP 4139.

DORNEY Oxon SU 9379. 'Humblebee island'. *Dornei* 1086, 1185, *Dorney* from 1209×19. OE **dora**, genitive pl. **dorena**, + **ēġ**. Brk 228, L 39.

DORNFORD Oxon SP 4319. 'The hidden ford'. *Deorneford* *[777]12th S 112, *Dœrneford* 1109, *Derneford(e)* [c.1100]late 12th–1401, *Darneford'* c.1160, *Darnford* 1676, *Danford* 1606×7, *Durneford'* 1268. OE **d(i)erne** + **ford**. O 293.

DORRIDGE WMids SP 1775. 'The deer ridge'. *Derrech* 1400, *Dorrech* 1401, *Doryge* 1411, *Dorridg* 1608. OE **dēor** + **hrycg**. The settlement occupies a small promontory of high ground. One m E is Darley Green SP 1874, *Derley* 1278, *Dorley* 1608, 'the deer wood or clearing', OE **dēor** + **lēah**. Wa 63.

DORRINGTON Lincs TF 0852. Probably the 'estate called after Deor or Deora'. *Derintone* 1086–1180, *Dir- Dyrington -yng-* c.1160–1539, *Derington -yng-* 1276–1428, *Durrington -yng-* 1515–1791, *Dorrington* from 1563. OE pers.n. *Dēor(a)* + **ing**[4] + **tūn**. A possible alternative specific might be OE ***dēoring** 'deer place' < **dēor** + **ing**[2] + **tūn** and so the 'deer farm'. Perrott 284, Årsskrift 1974.50.

DORRINGTON Shrops SJ 4703. 'The estate called after Doda'. *Dodinton'* 1198–1274, *Dodyton(')* 1291–1591, *Dodyngton -ing-* 1483–1629, *Doddington* 1545–1694, *Dorington* 1584–1619, *Dorrington* from 1589. OE pers.n. *Doda* + **ing**[4] + **tūn**. For the late change [d] > [r]. Sa ii.112.

DORSET The county name was originally a folk-n., 'the Dorn people'. *Dorset* [891]14th, *(on) Dor sætum* (dative pl.) 11th ASC(C) under years 978 and 982, *(mid, on) Dorsætum, Dorn sætum, (into) Dorsætan* 12th ASC(E) under years 837, 845, 998 and 1015, *(on) Dorsǣtan* 12th ASC(D) under year 1078, *Dorsete* 1086. P.n. *Dorn-* from RBrit *Durnovaria* (DORCHESTER SY 6890) + OE *sǣte* 'inhabitants'. Do 67.

DORSINGTON Warw SP 1349. 'The estate called after Deorsige'. *Dorsitune* [1058×62]12th S 1479, *Dorsinton(e) -tvne* 1086–1535, *-ing- -yng-* 1234–1753, *Dersinton(')* 1203–1359 with variants *-ing-* and *-yng-*, *Dorston* 1525. OE pers.n. *Dēorsiġe* + **ing**[4] + **tūn**. Gl i.241.

DORSTONE H&W SO 3141. 'Estate called after Deorsige'. *Dorsin(g)ton -yn-* 1137×9–1309×24, *Dorston(e)* from 1250. OE pers.n. *Dēorsiġe* + **ing**[4] + **tūn**. An alternative name appearing in DB, *Dodintvne* 1086, is either the 'estate at or called *Doding* 'the rounded hill', OE ***dodding** + **tūn**, or the 'estate called after Dodda', OE pers.n. *Dodda* + **ing**[4] + **tūn**. The survival of the present name may have been influenced by the place's location on the river Dore. He 73, DB Herefordshire 23.2.

DORTON Bucks SP 6814. 'Door settlement'. *Dortone* 1086–1237×40, *D(o)urton* 1291–1806, *Dorton* from 1343. OE **dor** + **tūn**. The reference is to the 'door' or narrow pass through the Brill Hills now taken by the railway line from Bicester to High Wycombe. Bk 123.

DOSTHILL Staffs SK 2100. '*Dercete* hill'. *Dercelai* (sic) 1086, *Dercetehull -hille* 1195–1247, *Derstill* 1273, *Dorsthull* 1316, *Dostell* 1526, *Dastyll* 1550, *Dosthill* 1618. P.n. *Dercete* probably representing PrW ***derw** + ***cę̄d** 'oak-tree forest' referring to the Notts, Leics, Northants and Warwicks wolds on the edge of which the village lies, + OE **hyll**. Cf. Avon and Burton DASSET Warw SP 4050, 3957 and the map in Fox 1989. An alternative suggestion is 'the deer shelter', OE **dēor-ċēte**, + **hyll**. DEPN, Wa 17.

DOUBLEBOIS Corn SX 1964. 'Double wood'. *Dobelboys* 1293, *(in bosco de) Doubleboys* 1337, *Dublebois* 1359. Fr **double** + **bois**. The English form of this name is found in Twelvewood SX 2065, *Twyfeldewode* 1375, *Twyvalewode* 1498, ME **twifeld** (OE *twifeald*) + **wode** (OE *wudu*). PNCo 79 gives pr [dʌbl boiz], Gover n.d. 273, 278.

DOUGH CRAG Northum NY 9795. *Dough Crag* 1869 OS. Probably identical with DOW CRAG NY 8418.

DOUGHTON Glos ST 8791. 'Duck farm'. *(æt) Ductune* *[775×7]11th S 145, *-tun(e) -ton*, *Duchtune* before 1241, *Dughton(e)* 1287–1305, *Doughton* from 1327, *Duf(f)ton* 17th. OE **dūc-tūn** (*dūce* + *tūn*). Gl i.112 gives pr ['dautən] but ['dʌftən] is still current locally.

River DOUGLAS Lancs SD 4812. 'Dark stream'. *Duglas* [1147]12th, *-is* [1199×1220]1268, *Dug(g)les* 1200×20, [1200×33]1268, [1212×35]1268, *Dug(e)les* 1212×35, [1221×32]1268, 1577, *Dowles* 1577. PrW ***duβ**, **dū** + ***gle(i)s**. La 126, RN 129, Jnl 1.47, 17.70.

DOULTING Somer ST 6443. Originally a r.n. of uncertain origin and meaning. *Dulting* 1125–1267 including [725]12th S 250, [688×726]13th S 1672, [946×55]13th S 1740, *Doulting* [946×55]13th S 1742, *Doltin* 1086, *Duulting* 1267, *Daulting* 1610. Forms for the r.n. (now the Sheppey) are *Doulting(strem)* [705]14th S 247, *Duluting* [705]? S 248, apparently a compound of unknown element ***dulut** + r.n. formative suffix **ing**[2]. *Dulut* might represent PrW **dū* 'black' + an element related to the root of W *llud* 'slime' if this really is a derivative of Brit **louto-* or **loutā* rather than a late variant of *glud*. DEPN, GPC 2218.

DOUR HILL Northum NT 7902. 'Bleak, unfriendly hill'. *Dour Hill* 1869 OS. ModE **dour** + **hill**.

DOUSLAND Devon SX 5368. 'Dove's land'. *Dovesland* late 13th, *Douseland* 1653. Surname *Dove*, genitive sing. *Doves*, + ME **land**. D 244 gives pr [dauzlənd].

River DOVE Derby SK 1431. 'Dark river'. *(an, and lang) Dufan* [951]14th S 557, [1008]14th S 920, *Duve* late 12th, 1305, *Dove* from 1225, *Douve* 1255–1438. PrW ***duβ**, < Brit ***dubo-**. Db 6, RN 134, Jnl.1.47.

River DOVE NYorks SE 6793. 'The dark river'. *Duue* 1100×3–[13th]14th, *Dow* 1577, 1626, *Dovebeck* 1614. PrW ***duβ**. YN 3, RN 134.

River DOVE Suff TM 1370. No early forms. In 1577 and 1586 Harrison calls it *Eie* from EYE TM 1473.

DOVE DALE Derby SK 1452. 'Valley of the r. Dove'. *Duvesdale* 1269, *Duvedale* 1296, 1297, *Douuedale* 1332. R.n. DOVE + **dæl** (ON *dalr*). Db 398.

DOVE HOLES Derby SK 0778. *Dove Holes* 1836. R.n. DOVE + **hole**. Db 399.

DOVE STONE RESERVOIR GMan SE 0103. Cf. *Dove stone wood* 1771. From *Dove Stones* 1843 OS near Dean Rocks SE 0203 + ModE **reservoir**. Nearby is *Fox Stone* 1843. YW ii.314.

DOVENBY Cumbr NY 0933. 'Dufan's farm or estate'. *Dunaneby* (for *Duuane-*) 1230, *Du(v)ane- Douvanne- Dovanues- Douan(e)by* 13th, *Dovenby* from 1541, *Dolphinby* 1671. ON pers.n. *Dufan* (< OIr pers.n. *Dubhān*, a diminutive of *dub* 'black') + ON **bȳ**. Cu 284 gives pr [dɔfnbi], SSNNW 29.

Low DOVENGILL Cumbr SD 7299. Adj. **low** + p.n. *Devingill -en-* (for *Dov-*) 1578–97, *Dov- Douengill* from 1628, 'Duvan's ravine', ON pers.n. *Dufan* (< OIr pers.n. *Dubhān*, a diminutive of *dub* 'black') + **gil**. We ii.31.

DOVER BECK Notts SK 4963. *Douerbec -k -v-* 1154×89–1480, *Dourbek* [115×65]14th, *Dorebekke* 1330, *Dorbeck(e)* 1577, 1663. PrW ***duβr** (Brit **dubro-*) 'water' used as a r.n. + ON **bekkr**. Nt 3, RN 135, Jnl.1.47.

DOVER Kent TR 3141. 'The waters'.

I. Latin forms; *(ad portum) Dubris* [4th]8th AI, [c.700]13th Rav, [c.425]10th ND.

II. English forms; *Dofras* [696×716]11th S 22, *(at) Dobrum* [844]18th S 1439, *(on) Doferum* c. 1000, *(of, æt, to) Doferan* [1016×20]18th S 1461, c.1050 ASC(D) under year 1052, c.1200 ASC(E) under year 1048, *(to) dofran* 1042×4 S 1044, *Dov(e)re, dou(e)re, apud Douerā* 1086.

Dubris is the locative-dative case of Latin **Dubrœ* from Brit ***dubro-** 'water'. The English form *Dofras* probably represents late Brit **doβras* from a Latin feminine pl. **Dubrās* (also from Brit **dubro-*), of which *Doferum* is the normal OE dative pl. form. The waters at Dover are referred to as *(be) doferware broce* 'beside the brook of the inhabitants of Dover' 1042×4 S 1044. This is probably the river Dour which flows into the sea at Dover, *Dubris* [c.650]13th Rav, *the river of Dovar* c.1540, *Dour* 1577. The use of the descriptive phrase *doferware broc* suggests that the original r.n. was forgotten in the Middle

Ages to be re-invented by antiquarians in the 16th century. RBrit 341, RN 135, DEPN, TC 84, Jnl 1.47, 8.20.

DOVERDALE H&W SO 8666. 'Valley of the river Dover'. *(on) Douerdale, (andlang) Douerdœles* [706]12th S 54, *(andlang) Doferdœles* [817]11th S 1596, *Lvnvredele* (sic for *Dvuvre-*) 1086, *Douerdale -v-* from 1262, *Dar- Dordall* 1558. R.n. **Dover* < PrW **duβr** 'water' + OE **dæl**. Wo 239, RN 136, L 94–5.

DOVERIDGE Derby SK 1134. 'Bridge over the river Dove'. *Dub(b)rig(e) -(b)rigge* 1086–1454, *-(b)rug(g)(e)* 1275–1562, *Dunebrug* (for *Duue-*) 1229, *Duuebrug(e)* 1252, 1275, *Du(e)brig(g)* 1252, 1306, *Douuebrig(ge) -brug(ge)* 1271–1392, *Doubrig(ge) -brug(ge)* 1275–1382, *Doveridge* from 1306. R.n. DOVE + OE **brycg**. Db 549.

DOW GRAG Durham NY 8418. *Dow Crag* 1863 OS. Probably N dial. **dow** 'a dove' (< OE *dūfe*) + ME **cragge**.

DOWDESWELL Glos SP 0019. Possibly 'Dogod's spring(s)'. *Dogodeswellan, Dogedes wyllan* [781×800]11th S 1413, *Dodesuuelle* 1086, *Doudeswell(a) -e* 12th–1458, *Dowdeswell(e)* 1185–1691. OE pers.n. **Dogod*, genitive sing. **Dogodes*, + **wella, wylle**, plural **wyllan**, dative pl. **wyllum**. Gl i.167.

DOWLAND Devon SS 5610. Partly unexplained. *Dvvelande* 1086, *Du(g)heland(a)* [1173×5]1329, 1242, *Duelonde* 1269, 1316, *Doweland* 1428. The generic is OE *land* 'open country', the specific unexplained but identical with that of DOLTON SS 5712. D 367 gives pr [daulənd], L 248–9.

DOWLISH WAKE Somer ST 3712. 'D held by the Wake family'. *Duueliz Wak* 1243. Earlier simply *Dovles* 1086, *Duueliz* 1196, *Doueliz* [n.d.] Buck, originally the name of the stream here, *Douelish* [705]n.d., PrW ***duβ-gles** 'black stream' possibly referring to black algae on the bed. Ralph Wac held the manor in 1189 and in 1284 Andrew Wak held *Westdouewyz* (West Dowlish) and Ralph Wake *Estdunelitz* (Dowlish Wake), *East Daulish* 1610. DEPN, DB Somerset 5.1.

DOWN 'Hill'. OE **dūn**.

(1) ~ ST MARY Devon SS 7404. 'St Mary's Down'. *Dune St Mary* 1297. P.n. Down + saint's name *Mary* from the dedication of the church. Down is *Done* 1086, *Doune* 1284. Possibly identical with *Vlwardesdone* 1086 'Wulfweard's estate at Down'. D 368, DEPN, DB Devon ii. 1, 72, L 143, 146, 148.

(2) East ~ Devon SS 6041. *Estdoune* 1260. ME adj. **est** + p.n. *Dvne* 1086. 'East' for distinction from West DOWN SS 5142. D 37.

(3) West ~ Devon SS 5142. *Westdo(u)ne* 1273, 1288. ME adj. **west** + p.n. *Dvne* 1086, *Dune* 1242. 'West' for distinction from East DOWN SS 6041 at the opposite end of the high ground either side of BITTADON SS 5441, 'Bitta's valley'. D 39.

DOWN PARK WSusx SU 7822 → UPPARK SU 7717.

Lower DOWN Shrops SO 3384. *Lower Downe* 1645, *Lower Down* 1836 OS. ModE **lower** + p.n. *Downesay* 1577, *Down* 1641, 'the hill (belonging to the Say family)', ModE **down** (OE *dūn*) + family name *Say* as in HOPESAY SO 3983. Gelling.

WEST DOWN Wilts SU 0548. Cf. *le Downes* 1509–47. ModE **west** + **down**.

DOWNDERRY Corn SX 3154. Unexplained. *Downderry* 1699, 1706. Another example at SW 9550 is first recorded in 1685 so that the name type cannot be connected with the unsuccessful siege of Londonderry in 1689. Connection with Co ***dery**, pl. of *dar* 'an oak-tree', is possible. PNCo 79, Gover n.d. 220.

DOWNE GLond TQ 4361. 'The hill'. *Dona* 1283, *(la) Dune* 1304, 1308. OE **dūn**. The place lies on the edge of the North Downs. PNK 25.

DOWNEND Berks SU 4775. 'Down end' sc. of Chieveley, the district of Chieveley near the downs. *Downend* 1749. ModE **down** (OE *dūn*) + **end**. Brk 243.

DOWNEND IoW SZ 5287. 'The end of the down' sc. of Arreton Down. ModE **down** sb. + **end**.

DOWNGATE Corn SX 3672. 'Gate leading on to the downs'. *Down Gate* 1748. ModE **down** + **gate**. A 19th cent. hamlet on the edge of Hingston Downs. PNCo 79, Gover n.d. 207.

Little DOWNHAM Cambs TL 5284. *Lytle Dounham* 1357. ME adj. **litel** + p.n. *Dun(e)ham, Donham* 1086, *Dunham* 1086–1294, *Dounham* 1286–1506, 'hill homestead', OE **dūn** + **hām**. The site is a low hill (50ft.) in the fens. 'Little' for distinction from DOWNHAM MARKET TF 6003. Ca 224.

DOWNHAM Essex TQ 7395. 'Hill homestead'. *Dunham* 1168–1276, *Dounham* 1316–1424, *Di- Dynham* 1221–1307. OE **dūn** + **hām**. Ess 154.

DOWNHAM Lancs SD 7844. '(The settlement) at the hills'. *Dunun* 1188–9, *Dunum* 1194, *Dounum* 1251, 1276, *-am* 1296, 1361, *-om* 1311, 1332 etc., *Dunham* 1246. OE **dūnum**, locative-dative pl. of **dūn**. Downham is situated on the lower slopes of Pendle Hill SD 7941 near Worsaw Hill SD 7743 and Gerna SD 7744, *Grenehou* [c.1300]14th, 'green hill', **grene** + ON **haugr**. La 79, L 143, 157–8, Jnl 17.49.

DOWNHAM Northum NT 8633. '(The settlement) at the hills'. *Dunum* 1186 P–1296 SR, *Dunhum* 1255, *Downeham* 1542. OE **dūn**, locative-dative pl. **dūnum**. The village is situated in the foothills of the Cheviot overlooking the Tweed valley. NbDu 66, DEPN, L 143, 158.

DOWNHAM MARKET Norf TF 6103. This is the ModE version of earlier references to a market here, *market æt Dunham, mercatum de Dunham* *[1042×66]14th S 1109, *Forum de Dunham* c.1110, *Mercatus de Dunham* 1130. P.n. *Dunham* 1086, 'the hill homestead', OE **dūn** + **hām**, + lOE **market** (ONFr *market*). Downham lies on an escarpment rising to 127ft. overlooking the river Great Ouse and the fens. DEPN.

DOWNHEAD Somer ST 6945. 'End of the hill(s)'. *Duneafd* [854]14th S 303, *Dvnehefde* 1086, *Duneheued* 1196, *-hefd* 1244, *Dunhaved* [n.d.] Buck, *Dunnyet* 1610 (by confusion with Donyatt ST 3314). OE **dūn**, genitive pl. **dūna**, + **hēafod**. D stands on flat ground at the E end of the Mendip Hills. DEPN.

DOWNHOLLAND CROSS Lancs SD 3607. P.n. Downholland + ModE **cross** 'a crossroads'. Downholland, *Dunholand* 1298, *Doun(e)holand(e)* 1325–1422, is 'lower Holland', ME adj. **doun** + p.n. *Holand* 1086, 1298, *Hoiland(a)* 1194, 1226, 'newly-cultivated land by the hill-spur', OE **hō-land**, *hōh* + *land*. Spellings with *oi* are AN graphies for *ō*. 'Down' for distinction from Up HOLLAND SD 5105. Downholland stands on the slope of a low ridge overlooking Downholland Moss and the river Altcar marshes. The name refers to land taken in at the limits of cultivable ground on the edge of marshland. La 120, L 248, Jnl 17.67.

DOWNHOLME NYorks SE 1197. '(The settlement) at the hills'. *Dune* 1086, *Dunum* 12th, 13th, *Downhum* 1535. OE **dūn**, dative pl. **dūnum**. YN 270 gives pr [du:nəm].

DOWNLEY Bucks SU 8495. No early forms.

North DOWNS Surrey SU 8147. 'The northern uplands' sc. of SE England'. *Northdownes* c.1570. OE **north** + **dūn**. 'North' for distinction from South DOWNS ESusx TQ 4902, WSusx SU 8913, TQ 3109. Sr 8.

South DOWNS WSusx SU 9214. ModE **south** + **downs** in the sense 'open expanse of elevated land, upland pasture', cf. Evelyn 1646, *downs of fine grass, like some places in the south of England* (*Memoirs* I,229 (1857)).

DOWNSIDE ABBEY Somer ST 6550. A modern Benedictine monastery founded at Douai in N France in 1607 and moved here in 1814 to Downside House (built c. 1700) at Norton Down, *Norton Down* 1817 OS. Pevsner 1958N.182.

DOWNTON Hants SZ 2792. Partly uncertain. *Dunchinton* 1164, *Do- Duneketon* c.1200–1397, *Donkton alias Dunckton* 1544, *Downton* 1691. This could be OE **Duneccingtūn* 'estate called after Dunecca', OE pers.n. *Dunecca* + **ing**[4] + **tūn**, or **Dunnocatun* 'hedge-sparrow settlement', OE **dunnoc**, genitive pl. **dunnoca**, + **tūn**. The modern form has been influenced by the p.n. DOWNTON Wilts SU 1820 on the opposite side of the New Forest. Ha 66, Gover 1958.224.

DOWNTON Wilts SU 1820. 'The hill settlement'. *Duntun*

*[672]12th S 229, *[826]12th S 275, *(in, æt) Duntune* [948]?11th S 540, [951×5]14th S 1515, [997]12th, *Duntone* 1086, *Dounton* 14th cent., *Est Downton* 1547, *Dunkton or Dounton* 1675. OE **dūn** + **tūn**. Situated in the Avon valley between downland on either side. Wlt 394.

DOWNTON ON THE ROCK H&W SO 4273. *Downton on the Rock* 1832 OS. Earlier simply *Dvntvne* 1086, *Do- Dunton(a)* c.1100–1527, *Dounton(e)* 1328–1535, the 'hill settlement', OE **dūn** + **tūn**. He 75, L 143.

DOWSBY Lincs TF 1129. 'Dusi's village or farm'. *Dusebi* 1086–1212, *-by* 1226–1327, *Douseby* 1280–1539, *Dousby* 1334, *Dowsby* from 1535. ON pers.n. *Dúsi*, genitive sing *Dúsa*, + **bȳ**. Perrott 113, SSNEM 44.

DOWSDALE Lincs TF 2810. 'Dusi's allotment'. *Dousedale* 1331, 1723, *Dowsdale* from 1535, *-daill* 1572. ON pers.n. *Dúsi*, genitive sing. *Dúsa*, + **deill**. Payling 1940.14.

DOWTHWAITEHEAD Cumbr NY 3720. 'Upper end of Dowthwaite'. *Dowthate Head* 1563, *Dowthwaite Head* 1568. P.n. Dowthwaite + OE **hēafod**. Dowthwaite is *Dowethweyt* 1285, *Dowthwate in Maderdale* 1487, 'Dufa's clearing', ON pers.n. *Dúfa* + **thveit**. The specific could also be the rare OE pers.n. *Dūua* or the ON or OE bird-name, **dūfe, dúfa** 'a dove', from which the pers.ns. derive. Cu 222, SSNNW 120.

DOXEY Staffs SJ 9023. 'Docc's island'. *Dochesig* 1086, *Dokeseia* 1168, *Doteshay* 1203, *Dokeseye* 1355. OE pers.n. **Docc*, genitive sing. **Docces*, + **ēġ**. DEPN, L 38, Horovitz.

DOYNTON Avon ST 7274. 'Village, estate called after Dyd(d)a'. *Didintone* 1086, *Deinton* 1194, 1613, *Doy- Doington'* 1221–1314, *Duy- Doi- Doynton(e)* 1248–1753, *Dynton* 1535, 1638. OE pers.n. *Dyd(d)a* + **ing**[4] + **tūn**. Gl iii.77.

DOZMARY POOL Corn SX 1974. *Dosmerypole* [13th]15th, 1516, *Dosmery Poole* 1610. P.n. Dozmary + ME **pole** (OE *pōl*). Dozmary, *Thosmery* [c.1241×4]15th, *Tosmeri* c.1300, is unexplained. PNCo 80, Gover n.d. 285.

DRAKE'S ISLAND Devon SX 4652. *Drake or St Nicholas Island* 1907. A modern renaming through association with Sir Francis Drake of St Nicholas Island, *isle of St Nicholas* 1396, *Sent Nicholas Ilond* 1573, so called from a chapel of St Nicholas there. D 236.

DRAKELAND CORNER Devon SX 5758. *Drakeland Corner* 1622. P.n. Drakeland, surname Drake + ME **land**, + ModE **corner** 'a corner, a nook'. Probably to be associated with the family of Ralph *Drake* 1374. Situated in a small valley opening off Tory Brook. D 252.

DRAKELOW Ches SJ 7070. 'Dragon mound'. *Drakelow(e)* from 1310×30. OE **draca** + **hlāw**. An allusion to the ancient folk-belief in the dragon which guards the burial treasure, as in *Draca sceal on hlæwe, frod, frætwum wlanc* 'It is for the dragon to be on the burial mound, old and wise, resplendent with treasure' (the Old English *Maxims II*, 26–7). Che ii.198.

DRAUGHTON Northants SP 7676. 'The settlement on a hill needing a good pull'. *Dractone, Bracstone* 1086, *Dra(c)h- Drai- Drayton* 12th cent., *Dracton(')* 1184–1228, *Drau(c)hton* 13th cent., *Draughton* from 1317, *Drawton* 1715. ON **drag** probably replacing OE **dræġ** apparently preserved in the 12th cent. *Drai- Dray-* spellings, + **tūn**. Draughton is on a hill that rises approximately 130ft. in half a mile. Cf. the p.n. type DRAYTON. Nth 112, lii, NoB 20.54, 64, SSNEM 183.

DRAUGHTON NYorks SE 0352. 'Settlement on a track along which loads are dragged'. *Dracton(e) -tona* 1086–1487, *Drauthona* before 1207, *Draghton* 1276–1587, *Draughton* from 1423, *Draython* 1287. OE **dræġ** replaced by ON **drag**, + **tūn**. YW vi.65 gives pr ['draftən].

DRAX NYorks SE 6726. Either 'the portage' or 'the fishery'. *æt Ealdedrege* 'at old *Dreg*' [959]12th S 681, *Drac* 1086, *Drax* from 1100×35, *Dracas* [1147×55]1329. Either OE **dræġ** 'a portage, a place where boats are dragged overland or pulled up from the water', or OE **dræġe** 'a drag-net'. Drax is very close to an old channel of the river Aire; a fishery is known to have existed here in 1090×1100 in which drag-nets were used. In either case ON **drag** 'a portage' was early substituted and a new ME plural form *drages* introduced. This was subsequently reduced to *drags* and assimilated to *Drax* under the influence of nearby Long DRAX SE 6828. YW iv.8.

Long DRAX NYorks SE 6828. 'The long reach'. *Langrak(e)* 13th–1616, *Lang Drax, Longdrax* 18th cent. OE **lang** + **racu**. This was originally the name of a stretch of the river Ouse. The 18th cent. forms are influenced by nearby DRAX SE 6726. The original sense survives in Langrick Reach, cf. LANGRICK TF 2648. *Racu* 'a hollow, a stream, the bed of a stream, the straight stretch of a river' survives in Modern dial. *rake* 'a path, a track, a course'. YW iv.11.

DRAYCOTE Warw SP 4470. 'The cottage(s) at the place where loads are dragged' or 'where drays or sledges are kept'. *Draicote* 1203, *Draycote* from 1298. OE **dræġ** + **cot**, pl. **cotu**. Wa 127.

DRAYCOTE WATER Warw SP 4670. P.n. DRAYCOTE SP 4470 + ModE **water**.

DRAYCOTT 'Cottages where loads are dragged or drays are kept'. OE **dræġ** + **cot**, pl. **-cotu**.

(1) ~ Derby SK 4433. *Draicot* 1086, *Drai -Draycot(e)* 1188–1243, etc. The reference is probably to a narrow neck of land in a meander of the r. Derwent. Alternatively, 'shed where drays are kept'. Db 456.

(2) ~ Glos SP 1835. *Draicota* 1208, *Draycote* 1275. The sense is probably 'shed where a dray is kept'. Gl i.235.

(3) ~ Somer ST 4750. *Draicote* 1086, *Draycot* 1227. D lies at the foot of a steep track over the Mendip Hills. DEPN.

(4) ~ IN THE CLAY Staffs SK 1528. *Draycott in the Clay* 1836 OS. P.n. *Draicote* 1086, 1251 + ModE **in the clay** for distinction from DRAYCOTT IN THE MOORS SJ 9840 and referring to the same conditions as Coton in the Clay SK 1629. The road runs steeply through the original settlement at SK 1529. DEPN.

(5) ~ IN THE MOORS Staffs SK 9840. *Drycote* (sic) *in le More* 1420. P.n. *Draicot* 1251, *Draycote* 1291 + ModE **in the moors**. DEPN, Horovitz.

DRAYTON 'Settlement at a portage or at a place where loads have to be dragged'. OE **dræġ** + **tūn**. PNE i.134. *Germanska Namnstudier tillägnede Evald Liden* NoB.20.46ff.

(1) ~ Hants SU 6705. *Drayton* from 1242. Gover 1958.21.

(2) ~ H&W SO 9076. *Dreiton* 1200, *Drayton* 1255. A road rises steeply from the village due S to Barrow Hill. Wo 237.

(3) ~ Leic SP 8392. *Dreitun* [1041×57]12th ECNENM 20, *-ton -ey-* 1186–1216×72, *Draiton(e)* 1163–1384, *Drayton(e)* from c.1150. The village lies at the foot of a hill rising evenly 250ft. in half a mile; the modern road from Drayton to Neville Holt takes this gradient directly in two straight stages probably following the original trackway. Lei 213.

(4) ~ Norf TG 1813. *Draituna* 1086. DEPN.

(5) ~ Oxon SU 4797. *(æt) Draitune* [958]13th S 650, *Drætun* [983]13th S 851, *Draigtun, Drægtun* [960]12th, 13th S 682, [1000]13th S 897, *Draitune -tone* 1086, *Drayton* 1517. Brk 406.

(6) ~ Oxon SP 4341. *Draitone* 1086 etc., *Drei- Dreyton(')* 1268–1316, *Drayton* from 1228. O 397.

(7) ~ Somer ST 4024. *Draitvne* 1096. The precise reference is unclear. The marshland between D and Muchelney may have been a route where goods were dragged on sledges. DEPN.

(8) ~ BASSET Staffs SK 1900. 'D held by the Basset family'. *Drayton Basset* 1301. P.n. *Draitone* 1086 + manorial addition from the 12th cent. tenants, the Basset family. The reference is to a steep ascent on the Roman road from High Cross to Wall, Margary no. 1g. DEPN.

(9) ~ CAMP Hants SU 4343. P.n. *Draitone* *[903]14th S 370, 1086, *(into) Drægtune* [995×1006]14th S 1420, 1019 S 956, *Drayton* from 1284, + ModE **camp**. Gover 1958.173.

(10) ~ PARK Northants SP 9680. P.n. *Drayton* from 12th with variants *Drai- Drey-* + ModE **park**. The road to Drayton House rises steadily from Lowick. Nth 186, NoB 20.46–70.

(11) ~ PARSLOW Bucks SP 8428. 'D held by the Parslow family'. *Dreyton Passelewe* 1268, 1283, *Drayton Paslow* 1526, 1552, ~ *Passeleiu* 1552. Earlier simply *Drai(n)tone* 1086, *Northdreitune* [c.1230]14th, and *Drayton Monachorum* 'monks' Drayton' 1237×40. The road through the village rises from 320ft. to over 420ft. Held in 1086 by Ralf Passaquam (OFr *Passe l'ewe*). 'North' for distinction from Drayton Beauchamp Herts SP 9111. The monks of Woburn abbey also claimed rights here. Bk 66 gives pr [pa:zlou], Jnl 2.23.

(12) ~ ST LEONARD Oxon SU 5996. 'St Leonard's D'. P.n. *Drœtona* [1146]18th, *Draitun' -ton'* [1199]c.1425, 1204–5, *Drei- Dreyton'* 13th cent., *Drayton* from 1235, + saint's name *Leonard* from the dedication of the church. O 153.

(13) Dry ~ Cambs TL 3862. *Dreiedraiton* 1227, *Dry(e)drayton* from 1272. Earlier simply *Draitone* 1086–1362. Two roads rise steeply to the village which stands on a dry ridge above the fens. 'Drye-Drayton, so called *not* from the Dryenesse of the *Soile*, but for that it *standeth* in the *Upland* and Champion Countrie, thereby to *distinguish* it from the other *Drayton*, which taketh Appelation from the *Fenne*' (BM Cole MSS xlvii f.156). Ca 152.

(14) East ~ Notts SK 7775. *Estdrayton* 1280, 1316, ~ *Dreyton* 1304, *East Drayton* 1606, *Estdraton* ~ *Dretton -dreaton* 16th. ME adj. **est** + p.n. *Draitone* 1086, *Dreitona* 1168, *Drayghton* 1287. 'East' for distinction from West DRAYTON SK 7074. Both places may have lain at the ends of a portage between the Trent and the Idle, although they could also have arisen independently referring to steep slopes. Also known as *Drayton in the Clay* 1340 and *Mikill, Grett* or *Magna Drayton* 1486×1515–1606. Nt 47.

(15) Fen ~ Cambs TL 3468. *Fendreiton' -drey-* 1188–1279, *Fendrayton(e)* from 1218, *Fenny Drayton* 1836 OS. Earlier simply *Draitone* 1086. The earliest reference is to the bounds of Drayton, *(be) Drœgtunes gemœra* [1012]12th S 926/1562. The reference is possibly to a portage from the river Ouse to the ascent to Fenny Stanton on the Roman road from Cambridge to Godmanchester, Margary no.24. Ca 166.

(16) Fenny ~ Leic SP 3597. *Fen(n)y Drayton(')* from 1327, *Drayton in the Clay* 1655. Earlier simply *Draiton(e)* 1086–1352, *Drayton(')* from 1227. As the affix indicates, the village lies on a heavy wet ground, once marshland, astride the Roman road from Manchester to Leicester, Margary no. 57 b. In Anglo-Saxon times the road must have been in bad condition necessitating the hauling of carts and wagons. The place of the *drœg* is probably marked by the sharp deviation of the modern road through the village from the line of its antecedent. 'Fenny' for distinction from DRAYTON. Lei 556–7.

(17) Market ~ Shrops SJ 6734. *Draiton Market* c.1540. ModE **market** + p.n. *Draitune* 1086, *-ton(')* 1203×4–1452, *Drayton(')* from 1271×2. The reference here is probably to the use of sledges on marshy ground. Also known as *Magna Drayton(')* 1271×2, 1529, *Muche Drayton* 1529 and *Great Drayton* 1549. 'Market' and 'Great' for distinction from Little Drayton SJ 6633. Sa i.198.

(18) West ~ GLond TQ 0679. *Westdrayton* 1465. ME adj. **west** + p.n. *Drœgtun* *[939]13th S 453, *Draitone* 1086. West Drayton lies in a large bend of the river Brent; however, it hardly seems such as to have necessitated dragging boats overland. Mx 33, GL 97.

(19) West ~ Notts SK 7074. *West Draytone* 1269, ~ *Draitone* 1305, ~ *Dreyton* 1689. ME adj. **west** + p.n. *Draitun* 1086, 1191×3. 'West' for distinction from East DRAYTON SK 7775. Also known as *Little Drayton* 1297, 1689, ~ *Dreaton* 1675 and *Drayton juxta Gamelston* referring to GAMSTON SK 7176. Nt 48.

DREWSTEIGNTON Devon SX 7390. 'Drew or Drogo's Teignton'. *Teyngton Drue* 1275, *Druesteynton* 1272×1307, *-teyngton* 1327, *Drewstenton* 1623. Pers.n. *Drew*, genitive sing. *Drewes*, + p.n. *Taintone* 1086, *Teintun* 1235, *-ton* 1238, 'settlement on the river Teign', r.n. TEIGN + OE **tūn**. The manor was held by a man called in Latin Drogo, in French Drew, in the late 12th cent. D 431 gives pr [-teintən], SBG 168.

DRIBY Lincs TF 3874. 'Dry village or farm'. *Dribi* 1086–1200, *Drebi* c.1115, *Driebi* 1130, *Driby* from late 12th. OE **drȳġe** + ON **bȳ**. The village is situated on chalky soil. SSNEM 44.

DRIFFIELD Glos SU 0799. Partly uncertain. *Drifelle* 1086, *Driffeld* 1190–1587. Possibly 'stubble field', OE **drīf** + **feld** or like DRIFFIELD Humbs TA 0257 'open land characterised by dirt', OE **drit** + **feld**. Gl i.70.

Great DRIFFIELD Humbs TA 0257. ME adj. **grete** (from 1466) + p.n. *(on) Driffelda* c.1100, c.1121, *Drifelt -d* 1086, *Driffeld(a) -y-* 1149×54–1546, 'open land characterised by dirt', OE **drit** + **feld**. 'Great' for distinction from Little Driffield TA 0057, *Parva Driffield* 1290, *Lit(t)le* 1367 etc. YE 153 gives prs [drifil, dðrifil], 154, L 238–9, 242.

DRIFT Corn SW 4328. 'The village'. *Dref* 1302, *Driffe* 1610, *Drift* 1748. Also known as *Drek bichan* 1244, *Drefbygan* 1262, 'little Drift' (Co **byghan**). Co ***tref** with lenition of *t* to *d* after lost definite article *an*, and late (English) excrescent *t* as in dial. *clift* for *cliff*. PNCo 80, Gover n.d. 658.

DRIFT RESERVOIR Corn SW 4329. P.n. DRIFT SW 4328 + ModE **reservoir**. Constructed in 1961. PNCo 80.

DRIGG Cumbr SD 0698. 'Portage'. *Dreg* 1175×99–1363, *Dregg(e)* 1279–1574, *Drigg* from 1572. OScand ***dræg** as in the Swedish p.n. Dräg and Danish p.n. Drejø. This element is related to English *drag* and refers to a place where boats were dragged overland, here between the sea and the river Irt. Cu 376, SSNNW 120.

DRIGHLINGTON WYorks SE 2228. 'Estate called after Dryhtel'. *Dreslin(g)tone* 1086, *Drichtlington* 1202, *Dri- Dryght(e)lington -yng-* 1401–1526, *Drytlington* 1251, *Drigh- Dryghlington -yng-* 1303–1641. OE pers.n. **Dryhtel* or **Dryhtla* + **ing**[4] + **tūn**. YW iii.19 gives pr ['driglintən].

DRIMPTON Dorset ST 4105. 'Dreama's estate'. *Dremeton* 1244, *Dremintun* 1250, *Dremyton* 1268, *Drempton* 1288. OE pers.n. **Drēama* + **ing**[4] + **tūn**. Do 67.

DRINKSTONE Suff TL 9561. 'Dreng's estate'. *Drincestune* [1042×66]12th S 1051, *renges-* (sic) *Drencestuna, Drincestona* 1086, *Drencestun* c.1095, *Drencheston* 1192, *Drenchistone* 1254, *Drenkeston* 1610. ON pers.n. *Drengr*, ME genitive sing. *Drenges*, + **tūn**, rather than the unrecorded and phonologically difficult OE pers.n. **Drēmiċ* suggested by DEPN.

DRINKSTONE GREEN Suff TL 9760. P.n. DRINKSTONE TL 9561 + ModE **green**.

DROINTON Staffs SK 0226. 'The drengs' settlement'. *Dregetone* 1086, *Drengetone* 1199, 1284, *Dreynton* 1619, *Drointon* 1836. Late OE **dreng** (ON *drengr*), genitive pl. **drenga**, + **tūn**. DEPN, PNE i.136, Horovitz.

DROITWICH H&W SO 8963. 'Dirty' or 'Noble Wich'. *Driht- Dryghtwych* 1347–1466, *Dryt- Dyrt- Dertwych(e)* 1353–1491, *Droitwich* from 1466. ME **drit** or, if the spellings *Driht- Dryght-* are not back-spellings due to the loss of [ç] before [t], OE *dryht* which in compounds has the sense 'noble, princely', + p.n. Wich, *Wiccium emptorium* 'market Wich' *[716]12th S 83, *(in) Wico emptorio salis quem nos Saltwich vocamus* 'in the Wich where salt is sold which we call Saltwich' [716×7]12th S 97, *Wich* (39×) *Wic* (1×) 1086, OE **wīċ** 'a trading place' sc. where salt is manufactured and sold. The salt-houses (*salinae*) here are mentioned throughout DB and the place was also known in Anglo-Saxon times as Saltwich or the salt-wich, *Saltwic* [888]13th S 220, *(æt þære) sealtwic* 'at the salt-wich' [1017]17th S 1384, OE **sealt-wīċ**. The forms of this name are complicated and can be elucidated as follows; *Driht- Dryghtwych* are probably back-spellings owing to the loss of [ç] before [t] rather than evidence of OE *dryht* 'noble, splendid' at so late a date. The *oi* form is also a back-spelling owing to the convergence of ME *ī* and *oi* under early ModE /əi/ and *Drīt-* for expected *Drit-* < OE *drit* has the vowel of the associated verb *drīten* < OE *ġedrītan*. 'Dirt' because the place is low-lying – Leland c. 1540, Torrington 1781 and *The Beauties of England and Wales* early 19th all call Droitwich dirty – and for distinction from the lost names Upwich, *Upwic* [962]11th S 1301, 'the upper *wīc*', Middlewich, *Middelwic* *[972]17th S 788, 'the middle *wīc*' and Netherwich, *neodemestan wic* (accusative case) *972 S 786, 'the nethermost *wīc*', i.e. lowest down the river Salwarpe. Cf. also Higher and Lower Wych SJ 4943, 4843, *Fulewich* [1096×1101]1150 'the foul *wic*', later itself also called *Dritwyche* 1482, *Dirtwich* 1485 with 'dirt' replacing 'foul', again with *drīt* for *drit* and back-spelled *Droytewhicche* 1530. Wo 285–7, Che iv. 51, BzN 16.328–9, Pevsner 1968.134.

DRONFIELD Derby SK 3578. 'Open land infested by drones'. *Dranefeld(e) -feud* 1086–1330, *Dranfeld(e) -fyld(e) -feild* 1330–17th, *Drone-* early 12th–17th, *Dron-* 1282–1312 etc. OE **drān**, genitive pl. **drāna**, + **feld**. Db 243.

DROVE ModE **drove** used in the Fens in two senses, (1) 'a road along which cattle are driven to and from the fenland pastures', (2) 'a channel for drainage'.

(1) North ~ DRAIN Lincs TF 1817. *North Drove Drain* 1824 OS. P.n. North Drove + ModE **drain**.

(2) South ~ DRAIN Lincs TF 2114. *South Drove Drain* 1824 OS. P.n. *South Drove* 1817 + ModE **drain**. Payling 1940.17.

DROXFORD Hants SU 6018. 'The ford of *Drocen*, the dry place'. *Drocenesforda* [826]12th S 275, [939]12th S 446, *Drokenesford(e)* c.1127–1342, *Drocelesford* [10th]12th S 1821, *Drocheneford* 1086, *D- Trokenesford* 13th cent., *Droxenford alias Droxford* 1722. OE ***drocen** 'a dry place' possibly used as a p.n., genitive sing. ***drocenes**, + **ford**. Ha 67, Gover 1958.46.

DROYLSDEN GMan SJ 9098. Probably 'Drygel's valley'. *Drilisden* [c.1250]17th, *Drilsden* c.1290, 1502, *Dri- Drylesden* 16th. The generic is OE **denu** 'a valley'. The specific, ME *Drīles*, has been explained as the genitive sing. of a pers.n. **Drȳġel* from OE *drȳġe* 'dry' denoting a small person of withered appearance. La 35.

DRUMBURGH Cumbr NY 2659. 'Buck ridge'. *Drumbogh* [1171×5]1333–1610, *-boc(k)* 13th, *Drombogh* [c.1225]n.d.–1496, *-beugh* 1610, *-burgh* 1300, *Drumburgh* from 1485. OW **drum** 'ridge' + W **bwch** 'a buck' later replaced by OE *burh* through association with the castle built here by Robert le Brun in 1307. Cu 124, GPC 351.

DRURIDGE BAY Northum NZ 2896. *Druridge Bay* 1866 OS. P.n. Druridge NZ 2795, *Dririg* 1242 BF, *Dri- Dryrig(e) -rygge* c.1250–1428, *Drurigg* 1354, *Druridge* 1663, 'the dry ridge', OE **drȳġe** + **hrycg** partly influenced by ON **hryggr**, + ModE **bay**. NbDu 66, L 169.

DRYBECK Cumbr NY 6615. 'Stream which sometimes dries up'. *Dri- Drybek(e) -beck(e)* 1256–1777. ME **dry** (OE *drȳġe*) + **beck** (ON *bekkr*). We ii.98, SSNNW 226.

DRYBROOK Glos SO 6417. 'Dry stream'. *Dry(e)- Dri(e)bro(o)k* 1338, 1339. Cf. *Druy(e)brokeswalle -forde* 1282. 'Drybrook spring' and 'ford'. OE **drȳġe** + **brōc**. Gl iii.217.

DUBMILL POINT Cumbr NY 0745. 'Dubmill promontory'. P.n. Dubmill + ModE **point**. Dubmill is *Dubmylne* 1332, *Dubmill* 1538, 'mill on the Dub', river n. BLACK DUB NY 0845 + OE **mylne**. Cu 296.

DUBRIS Kent TR 3540. The RBrit form of the name DOVER TR 3141.

DUBWATH Cumbr NY 1931. Uncertain; probably identical with *Dobwra* 1292, 'nook of land by the stream', ME, N dial. **dub** + ME **wra** (ON *vrá*). Cu 435.

DUCK'S CROSS Beds TL 1156. No early forms. Possibly surname *Duck* + ModE **cross** 'a cross roads'. This is in fact a T-junction for which *cross* is also used regularly in Shropshire.

DUCKINGTON Ches SJ 4952. 'Estate called after Ducc(a)'. *Dochintone* 1086, *Dokin(g)ton* 1288–1656 with variants *Doc- -yn(g)-*, *Duc- Dukinton(a) -yn-* 1216–1426, *Duckington* from 1430. OE pers.n. **Ducc(a)* + **ing**^{4} + **tūn**. Che iv.30.

DUCKLINGTON Oxon SP 3507. 'The estate called after Ducel'. *Duclingtun, (to) Duclingtune, (to) Duclingdune* [958]13th S 678, *Duceling dune* [1044]12th S 1001, *Dochelintone* 1086, *Duklyn(g)ton* 14th., 1428. OE pers.n. **Ducel* + **ing**^{4} + **tūn**. O 317.

Long DUCKMANTON Derby SK 4471. *Long Duckmanton* 1840 OS. Adj. **long** + p.n. *(æt) Ducemannestune* [c.1002]c.1100 S 1536, *Dochemanestun* 1086, *Duk(e)man- Ducman(e)- Duck(e)mantun(e) -ton(e)* 1154×89–1200 etc., *Dogmanton* 1299–1392, 'Ducemann's farm or village', OE pers.n. **Dūcemann*, sometimes in the genitive sing. form **Dūcemannes*, + **tūn**. 'Long' for distinction from *Middle Duckmanton* 1840 OS (*hodie* Duckmanton SK 4472) and *Far Duckmanton* 1840 OS (lost). Db 309.

DUDDENHOE END Essex TL 4636. *Dudney End* 1768. P.n. Duddenhoe + ModE **end** sc. of Elmdon parish. Duddenhoe, *Dud(d)enho* 1189×99, 14th, *Dodenho(o)* 1212–1416, is 'Dudda's hill-spur', OE pers.n. *Dudda*, genitive sing. *Duddan*, + **hōh**. Ess 526.

DUDDINGTON Northants SK 9800. Probably the 'estate called after Dudda'. *Dodinton(e)* [1029]14th, 1086–1391 with variants *-ing- -yn(g)-* and *-tun*, *Dudinton(a)* 1174 etc. with variants *-ing- -yng-*, *Duddington* from 1246, *Dorington* 1713. OE pers.n. *Dudda* + **ing**^{4} + **tūn**. OE ***dod(d)**, ***dud(d)** 'a rounded hill', has been suggested as the specific in this name, but although Duddington is on a slope the topography does not support this sense. Nth 201, Årsskrift 1974.38.

DUDDO Northum NT 9342. Usually explained as 'Dudda's hill-spur' but possibly rather 'hill-spur called *Dod*, the rounded hill'. *Dudehou* 1208×10 BF, *-ho* 1228, *Duddowe* 1366×7 32.279. Either OE pers.n. *Dudda* or OE ***dod(d)**, ***dud(d)** 'a rounded hill' + **hōh**. The reference could be to one of the surrounding hills, Duddo Hill (275ft.), Grindonrigg (262ft.) or Mattilees Hill (3275ft.). NbDu 66.

DUDDON Ches SJ 5164. 'Dud(d)a's hill'. *Dudedun* 1185, *Duddon* from 1288. OE pers.n. *Dud(d)a* + **dūn**. Che iii.273 gives prs ['dud(ə)n] and ['dʌd(ə)n].

River DUDDON Cumbr SD 2398. Uncertain. Possibly by origin the name of the river valley, 'Dudd's valley' or 'valley of the black river'. *Dudun* before 1140, *Dudenam* 1157×63, *Duden(e)* [1170]n.d.–1500, *Doden* 1279–1459, *Dudden* 1509×47. OE pers.n. *Dudd* or river n. from Brit **dubo-* 'black' + **denu**. Cu 11, RN 137, SSNNW 226.

DUDDON BRIDGE Cumbr SD 1988. *pontem de Duden, Doden* 1292, *Dodenbrig* 1332, *Duddonbridge* 1610. R.n. DUDDON SD 2398 + **brycg**. Cu 415.

DUDDON SANDS Cumbr SD 1775. *Dudden Sandes* c.1540, 1600. R.n. DUDDON SD 2398 + **sand**. Cu 39.

DUDLESTON HEATH Shrops SJ 3636. P.n. *Dodeleston* 1267, 'Duddel's estate', OE pers.n. *Duddel*, genitive sing. *Duddeles*, + **tūn**, + ModE **heath**. DEPN.

DUDLEY T&W NZ 2673. A modern mining settlement. *Dudley Colliery* 1863 OS.

DUDLEY WMids SO 9490. 'Dudda's wood or clearing'. *Dudelei* 1086, 1199, *Duddel(a)ege -leye* c.1140–1275. OE pers.n. *Dud(d)a* + **lēah**. Wo 289, L 203.

DUDNIL Shrops SO 6074 → Earls DITTON SO 6275.

River DUDWELL ESusx TQ 6824. 'Dudda's spring'. *Dudewell(e)* 1200×5, 1446. OE pers.n. *Dudda* + **wella**. Sx 461.

DUFFIELD Derby SK 3443. 'Open country frequented by doves'. *Duvelle* 1086, *Duffeld(e)* late 11th–12th etc., *Dovefeld* 1535×43. OE ***dūfe** + **feld**. Db 553, L 237, 244.

North DUFFIELD NYorks SE 6837. *Nortdufelt* 1086, *Northduffeld* 1185×1205 etc. OE **north** + p.n. *Duffeld* [1070×83]14th–1363, the 'open land where doves are seen', OE ***dūfe** + **feld**. YE 260, 262, L 237, 244.

South DUFFIELD NYorks SE 6833. *Suddufel(d) -felt* 1086, *Southduffeld* 1226–1529. OE **sūth** + p.n. Duffield as in North DUFFIELD SE 6837. YE 260, 262, L 237, 244.

DUFTON Cumbr NY 6825. 'Dove farm or village'. *Duston'* (for *Duf-*) 1176, *Dufton* from 1256. OE ***dūfe** + **tūn**. We ii.108.

DUFTON FELL Cumbr NY 7628. *Dufton Fell* 1588. P.n. DUFTON NY 6825 + ModE **fell** (ON *fjall*). We ii.110.

DUGGLEBY NYorks SE 8767. 'Dubgilla's farm or village'. *Difgeli-Dighelibi* 1086, *Di- Dyuegilbi -v- -by* late 12th–1336, *Dukelby* late 12th, *Dugilby* 1246, *Duggelby* 1280–1356, *Duggleby* 1303, 1828. OIr pers.n. *Dubgilla* ['duvjilə], in which the first vowel became [i] before the following front vowel, later influenced by pers.n. *Dughel* (OIr *Dubgall*), + **bȳ**. YE 124, SSNY 25–6.

DUGGLEBY HOWE NYorks SE 8866. P.n. DUGGLEBY SE 8767 + ModE **howe** (ON *haugr*) 'a mound'. The most spectacular Neolithic round barrow in Britain. PNE i.235, Thomas 1976.241.

DUKINFIELD GMan SJ 9497. 'Ducks' open land'. *Dokinfeld* 1285–1620 with variants *-en- -ing- -yn(g)-* and *-feld(e) -feeld -feild -felt*, *Dukenfeld* 1285–1819. OE **dūce**, genitive pl. **dūcena**, + **feld**. Che i.276.

DULCOTE Somer ST 5644. '(At) the cottages at or on the Doulting'. *Dulticotan* *[1065]18th S 1042, *Dulcot* 1817 OS. R.n. *Dulting* [725]12th S 250 as in DOULTING ST 6443 + OE **cot**, dative pl. **cotum**.

DULFORD Devon ST 0606. Possibly 'place hollowed out'. *Dylfytt* 1525, *Delvett* 1540. OE ***dylfet**. The reference may be to an old marl-pit. D 558.

DULLINGHAM Cambs TL 6357. 'Homestead of the Dullingas, the people called after Dulla'. *(at) Dullingham* from 1086 including [1043×5]13th S 1531, *Dullingeham* 1086–1272×98, *Dol(l)ing(e)ham -yng(e)-* 1086–1478. OE folk-n. **Dullingas* < pers.n. **Dulla* + **ingas**, genitive pl. **Dullinga*, + **hām**. Ca 118, ING 128.

DULOE Beds TL 1560. Possibly 'peg-shaped hill-spur'. *Diuelho* 1167, *Duvelho* 1211, *Deuelho* 1227. OE ***dyfel**, a word perhaps cognate with ModE *dowel*, or, in a different sense, with ***dyfe** 'hollow, valley, deep place'. Formally the specific could, however, be OE **dēofol** 'devil'. Bd 57, Sr xl, PNE i.140, L 168.

DULOE Corn SX 2358. 'Two Looes'. *Dulo* 1283–1610, *Doulo* 1291, *Duloo* 1342, 1467, *Dewlo(o)* 1321, 1504. Co **dew** + river name Looe as in LOOE SX 2553. Duloe is situated between the two rivers West and East Looe. PNCo 80, Gover n.d. 258, DEPN.

DULVERTON Somer SS 9127. Possibly 'settlement by the hidden ford'. *Dolertrne -tone* 1086, *Dulverton* from 1212, *Dilvertone* 1225, *Delverton* 1291. OE **dīegol** + **ford**. DEPN.

DULWICH GLond TQ 3373. 'Marshy meadow where dill grows'. *Dilwihs* [967]14th S 747, *Dilwich(e)* 1127, *Dilewis(se)* 13th, *-wyssh(e)* 1277–1316, *Dyl(l)ewyche* 1369, 1431, *Dulwich* 1530, 1603, *Dullidge* 1675. OE **dile** + **wisce**. The plant was widely cultivated for its carminative and other properties. Sr 19 gives pr [dʌlidʒ], TC 204, Grieve 255.

DUMBLETON Glos SP 0136. Possibly the 'toe of Dumel'. *(ad) Dumolan, Dumol(a)tan* [930]11th S 404 with variants *Domelten* and *Dumoltun -tan*, *(æt) Dumeltan* [1002×4]c.1200, *Dumbeltun*, *(æt) Dumaltun* [995]16th S 886, *Dumbeltune* 995, *Dumoltun*, *(to) Dumeltun* [1002]13th S 901, *(æt) Dumeltun* [1003×4]13th, *Dv(n)bentvne* 1086, *Dumbelton' -t(h)one -tun* 1206–1326, *Dumbleton(a)* 1202–1708. The specific is probably an older Celtic name for Dumbleton Hill (551ft.) SP 0035 beneath which Dumbleton is situated, PrW ***duβ-mēl** 'black, bare hill' + OE **tā** 'toe', dative sing. **tān**, later confused with OE **tūn**, referring to the northen spur of the main hill. Gl ii.9.

DUMMER Hants SU 5846. 'Hill pond'. *Dvmere* 1086, *Dunmere* 1196–1236, *Dummer(e)* from 1204. OE **dūn** + **mere**. There is a small pond in this hilltop village and other ponds on the hillside to the S. Ha 67, Gover 1958.131.

DUMPLINGTON GMan SJ 7697 → GRITTLETON Wilts ST 8680.

DUNBALL Somer ST 3040. There is some confusion over this name. Dunball ST 3040, *Dunball* 1875 Dir, is a hamlet in Puriton parsh which developed at *Dunball Clyce* 1809 OS as the result of the building of a wharfe on the Parrett here in 1844 with a railway goods station for the transfer of goods and minerals from vessels to trucks. The *clyce* (Somerset Levels dial. **clyse, clice**, ME *clüse*) was the sluice built at the seaward end of King's Sedgemoor Drain as a result of the Draining Act of 1791. But there is another Dunball downstream at The Island ST 2945 in Huntspill parish, *Dunbal Isle* 1791 Collinson, *Dunball* 1809 OS, opposite *Dunbal Point* 1750 map (North Clice ST 2644). This island was, according to Collinson, made early in the 18th cent. by making a 40 yards cut across the isthmus of the Parrett no doubt to aid navigation. The relationship between the two Dunballs and their etymology is obscure in the absence of further evidence. VCH vi.274.

DUNCHIDEOCK Devon SX 8787. 'Wooded fort'. *Dvnsedoc* 1086, *Dunsidioc(h') -ok(e)* 1187–1261, *Dunchidyoc -k* 1291, 1396. PrW ***din** (influenced by OE *dūn*) + ***cę̄dǭg** (Brit **caito-* + adjectival suffix *-āco*). The reference is to a large entrenchment on Penhill. D 495 gives pr [dʌntʃidik], Baring-Gould 169.

DUNCHURCH Warw SP 4871. 'Dunn's church' or 'Church of *Dun*, the hill'. *Donecerce* 1086, *-chir(e)che* 1143–1343, *Duneschirche* c.1150–1207, *Dunnes-* 1201, 1208, *Duncherch -chirche* 1250–1316. OE pers.n. *Dunn* or p.n. **Dūn* < **dūn**, genitive sing. **Dunnes*, **Dūnes*, + **ċiriċe**. Cf. DUNSMORE SP 5476. Wa 128.

DUNCOMBE HOUSE Glos SP 1727 → DONNINGTON SP 1928.

DUNCOTE Northants SP 6750. 'Dunna's cottages'. *Dunecote* 1227, 1305, *Don(e)cote* 1253–1316, *Duncote* 1302. OE pers.n. *Dunna* + **cot**, pl. **cotu**. Nth 43.

DUNCTON WSusx SU 9617. 'The estate called after Dunnic or Dunneca'. *Donechitone* 1086, *Dunecketuna* 1135×54, *Doneketon* 1261–1352, *Donkton* c.1260, *Don(e)gheton* 1281–1428, *Dunken* 1670. OE pers.n. *Dunnic, Dunneca* + **ing**[4] + **tūn**. Identical with DONNINGTON SU 8502. Sx 101, SS 75.

DUNDON Somer ST 4732. 'Hill valley'. *Dundene* [959×75]lost ECW, *Dondeme* (sic) 1086, *Dunden* 1236, 1243. OE **dūn** + **denu**. The village lies between two hills including Dundon Beacon 337ft to the E. The 1817 OS erroneously marks D as *Dundon Compton* and Dundon Compton simply as *Compton*. DEPN.

DUNDRAW Cumbr NY 2149. 'Steep slope called Drum'. *Drumdrahrigg* [1194]n.d., *Drumdrayf -dragh -drake -dra(ye)* 13th cent., *Drumdraw(e)* 1308, 1316, *Dundrah(e)* [c.1230]13th cent., *-dragh(e)* 1255–1610, *-draw(e)* from 1278. W **drum** 'ridge' + ON **drag** (possibly replacing OE *dræg*) with ME **rigg** (OE *hrycg*) in 1194. Cu 139, SSNNW 226.

DUNDRY Avon ST 5566. 'Steep ascent of the down'. *Dundreg* *[1065]c.1500 S 1042, *Dundrey -dray* 1227, 1230. OE **dūn** + **dræġ**. The N and S approaches to Dundry, which is situated just below the summit of Dundry Hill (764ft.), are by steep ascents. DEPN.

DUNDRY HILL Avon ST 5766. A long steep ridge. P.n. DUNDRY ST 5566 + ModE **hill**.

DUNFORD BRIDGE SYorks SE 1502. P.n. Dunford + ModE **bridge**. Dunford, *Dunneford* 1282, is the 'ford across the r.Don', r.n. DON SK 4696 + OE **ford**. YW i.339.

DUNGENESS Kent TR 0916. *Dengeness(e)* 1335, *Denge Nasse* 1610, *Denge Ness* 1819 OS. P.n. Denge as in DENGE MARSH TR 0520 + ME **ness** 'a headland'. KPN 56.

DUNGEON BANK Ches SJ 4580. *Dungeon Banks* 1843 OS. A

sand-bank in the river Mersey named from *Old Dungeon* at Hale Heath SJ 4582.

DUNGEON GHYLL FORCE Cumbr NY 2906. *Dungeon Ghyll* 1800. According to Dorothy Wordsworth's Journal *dungeon* was a local topographical term for a fissure or cavern, *ghyll* (ON *gill*) 'a short and, for the most part, narrow valley, with a stream running through it' and *force* 'the word universally employed for waterfall'. P.n. Dungeon Ghyll (ME **donjon** + **gill**) + N dial. **force** (ON *fors*). We i.206, xix fn.2.

DUNHAM 'Hill village'. OE **dūn** + **hām**. L 143, 153, 155.

(1) ~ -ON-THE-HILL Ches SJ 4773. *Dunham de Hill'* 1344, ~ *super Montem*, ~ *on the Hill* 1534. Earlier simply *Doneham* 1086, *Don(e)ham* 1310–1559, *Dunham* from 1283. The pleonastic addition *on the hill* is for distinction from DUNHAM TOWN SJ 7488. Also known as *Stony Dunham* 1327–1616 and *Stanry Dunham* 1348, *Stanredunham* 15th cent., OE adj. **stāniġ**, ME **stanry**, from the outcropping rock of the hill on which the village stands. Che iii.253.

(2) ~ ON TRENT Notts SK 8174. *Duneham* 1086–1280, *Dunham* from 1155, ~ *super Trent* 1410. There is a low hill to the W of the village. Nt 48.

(3) ~ TOWN GMan SJ 7488. *Dunham Town* 1841. P.n. Dunham + ModE **town**. Dunham, *Doneham* 1086–1274, *Dunham* from 1274, is 'village, homestead at a hill', OE **dūn** + **hām**. Also known as Dunham Massey, *Dunham Mass(e)y, Mascy, Masci* c.1280–1564, from the surname of the DB tenant *Hamo de Masci* and for distinction from DUNHAM-ON-THE-HILL SJ 4773. Che ii.19 gives pr ['dunəm].

(4) Great ~ Norf TF 8714. *Magna Dunham* 1242. ModE **great**, Latin **magna** + p.n. *Dunham* 1086. Great Dunham lies in an area of higher ground called Dunham Hill. 'Great' for distinction from Little DUNHAM TF 8612. DEPN.

(5) Little ~ Norf TF 8612. *Little Dunham* 1824 OS. ModE adj. **little** + p.n. Dunham as in Great DUNHAM TF 8714.

DUNHAMPTON H&W SO 8466. 'Hill farm'. *Dunhampton* from 1222, *Dudhampton* 1582. OE **dūn** + **hām-tūn**. Wo 270 gives pr [dʌnən].

DUNHOLME Lincs TF 0279. Either 'Dunna's' or 'the hill homestead'. *Duneham* 1086, c.1115, 1202, *Dunham* c.1115. c.1155. OE pers.n. *Dunna* or **dūn** + **hām**. The village is situated at the edge of rising ground. DEPN.

DUNK'S GREEN Kent TQ 6152. *Dunks Green* 1819 OS. Kentish surname *Dunk* + ModE **green** 'common'. Cf. Dunk's Farm. PNK 191.

DUNKERY HILL Somer SS 9042. Cf. *Dunkery Beacon* 1750 map, 1809 OS. A hill rising to 1704ft. Earlier simply *Duncrey* 13th, *Dunnecray* 1298, perhaps PrW ***dīn** + **creig** 'rock hill'. DEPN.

DUNKESWELL Devon ST 1407. 'Hedge-sparrow or Dunnoc's spring'. *Dodvcheswelle* 1086 DB, *Doduceswilla* 1086 Exon, *Dunekeswell(a)* 1219–78, *Dunekewill' -kys-* 1221×30, *Dunkeswell* 1234, *Dunxwell* 1605. OE **dunnoc** 'a hedge-sparrow' or pers.n. *Dunnoc* derived from it, genitive sing. **dunnoces**, *Dunnoces*, + **wylle**. The DB forms suggest an alternative pers.n. *Duddoc*. D 614.

DUNKESWICK WYorks SE 3046. 'Lower Keswick'. *Dunkeswy(c)k(e) -wic(k)* 1135×50–1615. OE **dūne** 'down, lower' + p.n. *Chesuic* 1086, *Kesewych -wik -wyk(e)* 1235–1550, 'dairy-farm producing cheese', OE **ċēse** + **wīc** with /k/ instead of /tʃ/ due to ON influence. 'Down' or 'lower' from its position in the valley bottom and for distinction from East KESWICK SE 3644. YW v.50 gives pr [duŋ'kezik], SSNY 147.

DUNKIRK Kent TR 0759. *Dunkirk* 1790. Transferred from Dunkerque, France, *Dunkerka* 1067, 'the church of the dunes', MDutch *dûne* 'dune' + *kerke* 'church'. There were lively commercial connections with Dunkerque which was actually an English possession for a short period in the 17th century. The term is usually used in English p.ns. with overtones of remoteness and difficult management especially after the Duke of York's unsuccessful siege in 1793. Dunkirk in Kent was a wild place, a tract of thinly populated woodland, an extra-parochial vill from which no tithes were payable, renowned as a haunt of law-breakers. Sir William Courtenay, the 'Messiah' Mad Tom, hid here in 1837–8. There was no church or parson until 1840. Field 1993.153, Cullen.

DUNLEY H&W SO 7969. 'Dunna's wood or clearing'. *Dunelege* 1221, *Donesley* c.1225, *Dunley* from 1275 with variant *Donne-* 1327. OE pers.n. *Dunna* + **lēah**. Wo 29.

DUNMAIL RAISE Cumbr NY 3211. 'Dunmail's cairn'. *Dunbalrase stone* 1576, *Dunmail-raise* 1610. OW pers.n. *Dunmail* + ON **hreysi**. The last but one king of Strathclyde bore the name *Dunmail*, but no connection with the place is known. A lost *Dumball Raise* near Milburn NY 6529 is said to have been the meeting place of the Lordships of Long Marton, Melburn and Knock. Cu 312, We ii.120.

DUNMOOR HILL Northum NT 9618. *Dunmore Hill* 1869 OS. P.n. Dunmoor + ModE **hill**. Dunmoor, no early forms, is possibly 'the hill moor', OE **dūn** + **mōr**, but 'brown moor', ModE **dun** + **moor** is also possible.

Great DUNMOW Essex TL 6221. *Magna Dunmaue* 1225, *Muchdunmo* 1546. ModE **great**, Latin **magna**, ME **much**, + p.n. *(at) Dunemowe* [942×c.51]13th S 1526, *(of) Dunmæwan* [c.1000]c.1125, *Dunmawe* 1119–1352, 'the hill meadow(s)', OE **dūn** + ***māwe**, ***mǣwe**, dative pl. ***māwum**. 'Great' for distinction from Little DUNMOW TL 6521. Ess 474 gives pr [dʌnmə], Studies 1931.72, Cu lxxiii, PNE ii.37.

Little DUNMOW Essex TL 6521. *Parva Dunmawe* 1237. ModE adj. **little**, Latin **parva**, + p.n. *(at) Dunmawe* [1043×5]13th S 1531 as in Great DUNMOW TL 6221. Ess 474.

DUNNERDALE Cumbr SD 2093. Probably 'valley of the river Duddon'. *Dunerdale* 1293, 1300, *Doner-* 1300, *Dunerdall* c.1550. Nearby is Dunnerholme SD 2179, *Dunreholm* c.1220, *Duner- Donnerholme* 1252. Formally the specific could be OE pers.n. *Dunhere*, but composition with ON *dalr* 'a valley' and *holmr* 'an island' suggests a Scand specific, possibly **Duthn*, a Scandinavianized form of the river n. DUDDON SD 2398, actually recorded once in the spelling *Duthen* 1196, genitive sing. **Duthnar*; or **Duthn-á* with ON *á* 'a river', genitive sing. **Duthn-ár*, with subsequent loss of intervocalic *-th-*. Dunnerdale and Hall DUNNERDALE SD 2195 both lie in the valley of the river Duddon. SSNNW 226.

Hall DUNNERDALE Cumbr SD 2195. ModE **hall** + p.n. DUNNERDALE SD 2093.

DUNNINGTON Humbs TA 1552. 'Estate called after Dudda'. *Dodinton(e)* 1086, 1190, *Dudingtun -ton -yng-* 12th–1349, *Dodington(a) -yng-* 1160×80–1512, *Donnyngton* 1566. OE pers.n. *Dud(d)a* + **ing**[4] + **tūn**. The sound change in this name is paralleled in DUNTON Bucks but may have been influenced by DUNNINGTON NYorks. For a different interpretation (OE **dodd* 'a hill') cf. Årsskrift 1974.38. YE 77 gives pr [dunitən].

DUNNINGTON NYorks SE 6752. Either the 'settlement called or at **Duning*, the hill-place', or the 'estate called after Dunn(a)'. *Domni- Donni- Do'niton* 1086, *Duninton* 1200, 1204, *Dunington* 1225–1307, *Donington* 1225–1465. OE ***dūning** < **dūn** + **ing**[2] or pers.n. *Dunn* or *Dunna* + **ing**[4], + **tūn**. Dunnington lies on raised ground. YE 273, Årsskrift 1974.39.

DUNNINGTON Warw SP 0653. Either 'the estate called after Dunn(a)' or 'the settlement at *Duning*, the hill-place'. *Donyngton* 1315, 1343, 1545, *Donnington Hethe* 1547. OE pers.n. *Dunn(a)* + **ing**[4] or OE ***dūning** < OE *dūn* + *ing*[2], + **tūn**. Wa 221, Årsskrift 1974.39.

DUNNOCKSHAW Lancs SD 8127. 'Hedge-sparrow copse'. *Dunnockschae* 1296, *Donocshay* 1536, *Dunnockschagh(e)* 1305, 1328, *Dunnokschaw* 1323. OE **dunnoc** + **sceaga**. La 80, Jnl 17.50.

DUNNOSE IoW SZ 5878. 'Nose or promontory of the down'. *Donnose* 1544, *Dounnose* 1550, *Donnesse* 1595, *Douñose* 1600,

Dunnose 1769. ME **doun** (OE *dūn*) + **nose**. Wt gives pr ['dʌnouz], Mills 1996.46.

DUNSBY Lincs TF 1026. Probably 'Duni's village or farm'. *Dunesbi* 1086, 1190×1200, *Dun(n)esby* 1223–1554, *Dounes-* 1283–1549, *Dunsby* from 1291. OE pers.n. *Duni*, genitive sing. *Dunes* + ON **bȳ**. The village stands on the edge on high ground overlooking fenland and the specific might alternatively be OE **dūn** 'a hill', genitive sing. **dūnes**. Perrott 114, SSNEM 45.

DUNSCROFT SYorks SE 6508. Either 'Dun's enclosure' or 'croft by the r. Don'. *Donescroft* 1404, *Dunscroft(e)* 1592–1614. OE pers.n. or ME surname *Dun(n)*, genitive sing. *Dunnes*, or r.n. DON SK 4696 + OE **croft**. YW i.9.

DUNSDEN GREEN Oxon SU 7377. *Donsden grene* 1589, *Dunsden Greene* c.1605. P.n. Dunsden + ModE **green**. Dunsden, *Dunesdene* 1086–1341 with variants *Dunnes-* and *-den(a) -denne -don'*, *Denesden* 1231–1368, is 'Dyni's valley', OE pers.n. *Dyni*, genitive sing. *Dynes*, + **denu**. O 69–70.

DUNSFOLD Surrey TQ 0036. 'Dunt's fold'. *Dunte(s)fold, Duntesfaud(e)* 1241–1428 with variants *-fald* and *-feld*, *Dunterfeld* 1272, *Duntresfolde* 1307, *Dunnesfold* 1430, *Dunsfold* from 1480. OE pers.n. **Dunt* as in DUNTISBOURNE ABBOTS etc. Glos SO 9708, genitive sing. **Duntes*, + **fald**. The 1272 and 1307 forms have unexplained intrusive *r*, as in ME forms of other Surrey p.ns., CATERHAM TQ 3455, FRENSHAM SU 8411, SANDERSTEAD GLond TQ 3361, *(an) sonden stede, (on) sondenstyde* 871×89 S 1508, *Sandestede* 1086–1220, *Sandersted(e) -re* 1221–1325, 'the sandy place', OE **sanden** + **stede**. Nearby was *D(o)unteshurst* 1294–1332, 'Dunt's wooded slope' from OE **hyrst**. Sr 53, 234–5.

DUNSFORD Devon SX 8189. 'Dunn's ford'. *Dvnesford* 1086–1290 with variant *Dun-*. OE pers.n. *Dunn*, genitive sing. *Dunnes*, + **ford**. D 434.

Lower DUNSFORTH NYorks SE 4464. *Nether Dunford* 1331, *Nether Dunsford* 1251. ModE **lower**, ME **nether**, + p.n. *Doneford(e)* 1086, *Duneford* 1199–1210, *Dunesford(e)* 1086–1251, *Dunsford* 1180–1577, *-forth* from 1577, *Duncefordes* 1208, 'Dunn's ford', OE pers.n. *Dunn*, genitive sing. *Dunnes*, alternating with *Dunna*, genitive sing. *Dunna(n)*, + **ford**. YW v.83, 84.

Upper DUNSFORTH NYorks SE 4464. *Uverdunesford* 1251, *Over Dunsford* 1577, *upper* ~ 1732. ME **over**, ModE **upper** + p.n. *Dunesford* 1089×1118, *Dunsford* 1154×89, as in Lower DUNSFORTH SE 4464. YW v.83, 84.

DUNSLEY NYorks NZ 8511. 'Dunn's forest clearing'. *Dunesla* 1086, *-le* 1086–1227, *-lac* 1100×c.1115, 1133. OE pers.n. *Dunn*, genitive sing. *Dunnes*, + **lēah**. YN 124.

DUNSMORE Bucks SP 8605. *Dunsmore* 1822 OS. This could be 'Dunn's moor', OE pers.n. *Dunn*, genitive sing. *Dunnes*, + **mōr**, as Dunsmore in RYTON and STRETTON-ON-DUNSMORE Warw SP 3874, 4072.

DUNSMORE Warw SP 5476. 'Dunn's moor' or 'Moor of *Dun*'. *Dunesmore* 1189×99, *Dunnes-* 1235–1315, *Do(u)nes-* 1235–1424, *Dunsmore heathe* 1588. OE pers.n. *Dunn*, genitive sing. *Dunnes*, or p.n. **Dūn* 'the hill', OE **dūn**, genitive sing. **Dūnes*, + **mōr**. Wa 12.

DUNSOP BRIDGE Lancs SD 6550. *Dunsop Bridge* 1704. P.n. Dunsop + ModE **bridge**. Dunsop, *Duleshop(p)* [930]13th, *Dunsuppe* 1652, is a problem. If the OE form is reliable, *Dules-* may represent PrW ***dū-gles** 'black stream' as in the r.n. DOUGLAS; if *Dules-* is a mistake for *Dunes-* the name means 'Dun's valley', OE pers.n. *Dunn*, genitive sing. *Dunnes*, + **hop**. Sound substitution of *n* for *l* occurs sporadically, cf. HINDERWELL NYorks NZ 7916, STITTENHAM WOOD NYorks SE 67 67. YW vi.212.

DUNSTABLE Beds TL 0221. 'Dunn or Dunna's post' or 'post at or on the hill'. *(æt) Dunestaple* 1123 ASC(E), *Dunstaple* [1187]13th, *Dunstable* 1287. OE pers.n. *Dunn(a)*, genitive sing. *Dunnes*, or **dūn** + **stapol**. The post may have been a boundary mark of Dunn(a)'s estate; or it may have marked the intersection of Icknield Way and Watling Street in the gap in the hills around which the town subsequently developed. Bd 120.

DUNSTALL Staffs SK 1820. 'The farmstead'. *Tunstall* 13th, *Donestal* 1272. OE ***tūn-stall** with voicing of initial *t*, cf. *p* > *b* in BLYMHILL SJ 8112. DEPN, PNE ii.198.

DUNSTALL GREEN Suff TL 7460. P.n. Dunstall + ModE **green**. There are no early forms for Dunstall but it is identical with DUNSTALL Staffs SK 1820 < OE **tūn-stall** 'a homestead'.

DUNSTAN Northum NU 2419. 'The hill stone'. *Dunstan* from 1242 BF. OE **dūn** + **stān**. The reference is to the rocky outcrop on which Dunstanburgh Castle was subsequently built. NbDu 67, L 158.

DUNSTANBURGH CASTLE Northum NU 2521. P.n. Dunstanburgh + ME **castel**. Dunstanburgh, *Dunstanburgh* from 1321, is 'Dunstan fortified place', p.n. DUNSTAN NU 2419 + OE **burh**. The present castle, probably overlying an Iron Age fort, was begun in 1314. NbDu 67, Pevsner 1992.257.

DUNSTER Somer SS 8843. 'Dunn's Torr'. *Dunestore* 1138, *-tor(ra)* c.1150, [1186] Buck, *Donstorre* [late12th]14th, [1237×42]Bruton, *Dunestere* 1238, *Dunsterr* 1242, *Dones- Dunstore* [1346×7]Bruton, *Dunster* 1610. OE pers.n. *Dunn*, genitive sing. *Dunnes*, + p.n. *Torre* 1086, OE **torr** 'a rock, a rocky outcrop'. The reference is to the prominent hill on which the castle was later built. DEPN.

DUNSTON Lincs TF 0662. 'Dunn or Duni's farm or village'. *Dunestune* 1086–[1163]13th, *-tun(a)* 13th cent., *-ton* 1188–1249, *Dunnestun -ton(a)* 1264, 1291, *Dunston(e)* from 1185. OE pers.n. *Dunn* or *Duni*, genitive sing. *Dunnes, Dunes*, + **tūn**. Perrott 322, Cameron 1998.

DUNSTON Norf TG 2202. 'Duni or Dunn's estate'. *Dunestun* 1086, *-ton* 1186. OE pers.n. *Duni* or *Dunn*, genitive sing. *Dun(n)es*, + **tūn**. DEPN.

DUNSTON Staffs SJ 9217. 'Duni or Dunn's estate'. *Dunestone* 1086, *Du- Doneston* 1201–1587, *Dunston* from 1220, *Dunson* 1552. OE pers.n. *Duni* or *Dun(n)*, genitive sing. *Dun(n)es*, + **tūn**. St i.84.

DUNSTON T&W NZ 2262. Uncertain. *Dunston* 1647, 1652, cf. *Dunstanfield* 1647, a field in Whickham, and *Dunstill staithe* 1590, *Dunstan Staith, Banks* and *Hall* 1768 Armstrong. Possibly identical with DUNSTAN Northum NU 2419, referring to some feature of Dunston Hill NZ 2261.

DUNSVILLE SYorks SE 6407. A modern creation from *Dun* as in nearby BARNBY DUN SE 6109 and DUNSCROFT SE 6508 referring to the r. Don + ModE **ville**.

DUNSWELL Humbs TA 0735. Uncertain. *Doncevale* 1349, 1353, *Dounceuall* c.1362, *Downeswall* 1546, *Dunswell* 1828. By origin possibly a Fr name in **val** 'a valley' perhaps with surname Dunn or Dunce from John Duns Scotus, the schoolman, not originally derogatory. Also known as *B(e)arhouse* 17th cent., *Beer-houses* 1828. YE 206.

DUNTERTON Devon SX 3779. Partly uncertain. *Dondritone* 1086, *Dunterdune* 1242, *-tone* 1242–84. Possibly an earlier Celtic name, PrW **din-treβ*, 'fort village' influenced by OE **dūn**, + **tūn**, 'village, settlement at **Dintref*'. There is a small fortification on a promontory here. D 182.

DUNTISBOURNE 'Dunt's stream'. An original stream name that gave its name to a number of estates N of Cirencester. OE pers.n. **Dunt*, genitive sing. **Duntes*, + **burna**. L 18.

(1) ~ ABBOTS Glos SO 9708. 'D held by the abbot' sc. of St Peter's Abbey, Gloucester. *Duntesburn(e) Abbatis* 1287, *Dountesborne Abbottes* 1580, *Dunsburn -borne Abbot(te)s* c.1560–1657. Earlier simply *Duntesburn(e)* 1055–1322, *Dantes- -is-, Tantes- Dvntesborn(e)* 1086, *Duntesbourne* 1291–1471. Gl i.71 gives pr ['dunzbɔRn].

(2) ~ LEER Glos SO 9707. 'D held by (the abbey of) Lire' in Normandy, to whom it was given by Roger de Laci before 1086. *Duntesb(o)urn(e) Lyre* 1307, 1540, *Dunsborne Lerr* 1795. Earlier simply *Tantesborne* 1086. R.n. DUNTISBOURNE + Fr p.n. Lire. Gl i.72.

(3) ~ ROUSE Glos SO 9806. 'D held by the Rous family'. *Duntesb(o)urn(e) R(o)us* 1287–1474, *Dunsburn Rowse* 1533. Earlier simply *Dvntesborne* 1086. R.n. DUNTISBOURNE + surname *Rous* as in Sir Roger le Rous 13th. Gl i.73.

DUNTISH Dorset ST 6906. 'Hill pasture'. *on dounen tit* [941]14th S 474, *Dunetes -isse -ys(h') -edish(')* 13th cent., *Duntisse -disse* 1264, *Duntish(')* from 1299. OE **dūn** + ***etisc** in some forms confused with **edisc**. Do iii.242, L 143, Studies 1931.33.

DUNTON 'Hill settlement'. OE **dūn-tūn**.

(1) ~ Norf TF 8830. *Dontuna* 1086, *Dunton* 1198–1236. Dunton overlooks the valley of the river Wensum. DEPN.

(2) ~ BASSET Leic SP 5490. 'D held by the Bassett family'. *Dunton(e) Basset(t)* from 1409. Earlier simply *Donitone* 1086, *Dunetunam* 12th, *Dunton(e)* c.1130–1591. Radulfus Basset held the manor in 1166. It is perhaps possible that the DB form is a reduction of OE **dūning-tūn*. Lei 440, Årsskrift 1974.39.

(3) ~ WAYLETTS Essex TQ 6590. 'Dunton cross-roads'. *Dounton Wey- Waylate* 1395, 1451. P.n. Dunton + ME **weylate** (OE *weġ-ġelǣte*). Dunton, *Dantuna* 1086, *Duntona* 1109×24, *Dunton(a)* from 1200, is the 'hill settlement, farm or estate'. Ess 155, Jnl 2.42.

DUNTON Beds TL 2344. 'Estate called after Dunna'. *Donitone* 1086, *Duniton* 1185, *Dunton* 1286. OE ***dūning** or pers n. *Dunna* + **ing**[4], + **tūn**. Although Dunton lies on a steep escarpment overlooking the river Rhee, the alternative explanation, OE *Dūning-tūn* 'settlement called or at **Dūning*, the hill-place', < OE *dūn* + *ing*[2], is unnecessary in view of the frequency of the pers.n. *Dunna* in OE. Bd 103.

DUNTON Bucks SP 8224. 'Settlement called after Dudda'. *Dodintona* 1086, 1242, *Dudinton* 1198, 1237×40, *Do- Dudington(e) -yng-* 1221–1465, *Donyngton* 1480×96, *Donington al. Dunton* 1522×47. OE pers.n. *Dudda* + **ing**[4] + **tūn**. Although Dunton is situated on a hill this explanation is preferable to OE ***doding**, ***duding** < ***dod**, ***dud** + **ing**[2], + **tūn** since *Dudda* is a common pers.n. and *dod* characteristically a northern word. Bk 67, Årsskrift 1974.38.

DUNTON GREEN Kent TQ 5157. *Dunton Green* 1819 OS. P.n. Dunton + ModE **green** 'common'. Dunton, *Dunington'* 1244, 1292, *Denyngton* 1346, is probably the 'estate called after Dunna', OE pers.n. *Dunna* + **ing**[4] + **tūn**. Alternatively the specific might be an OE ***dūning** 'a small hill' < *dūn* + *ing*[2]. In this case the reference is to an isolated round hill NE of the hamlet of the kind usually called a *beorg*. PNK 55, Årsskrift 1974.38–9.

DUNWICH Suff TM 4770. Possibly 'the trading-place at or called Dumnoc'. *Duneuuic* 1086, *-wic* 1156–1258, *-wich(')* 1173–1218, *Dunwich* from 1230. P.n. *(in civitate) Dommoc, Domnoc* [c.731]8th BHE, *(in) dommoc ceastre* [c.721]11th, 12th ibid., *Domnoc* 12th ASC(F) under years 636ff, 'the deep place, the port with deep water', Brit ***dubno-** + suffix ***-āco-**, + OE **wīċ**. Other early references are *Dommocceaster* [c.890]c.1000 OEBede, *Dummucæ civitas* 803 B 312, *Dammace civitas* [803]copy ibid. The identification of *Dumnoc* with Dunwich is not, however, universally accepted. DEPN, Baron, J. M. Wallace-Hadrill, *Bede's Ecclesiastical History of the English People: A Historical Commentary*, Oxford 1988.78.

DURDAR Cumbr NY 4051. Possibly 'oak-tree copse'. *Derdarre* 1336, *Durdar* 1794. Gaelic **doire-darach**, cf. Dardarroch D & G NX 8586 and Irish names like Darragh Co. Clare, Aghadarragh Co Tyrone, Clondarragh Co. Wexford, all with *darach* 'abounding in oaks'. Cu 149, IPN s.nn.

DURDLE DOOR Dorset SY 8080. *Dirdale Door* 1811, *Duddledoor* c.1825, *Durdle or Dudde Door* 1826. P.n. Durdle + ModE **door** referring to an arched rock projecting into the sea. Durdle is probably 'hill with a hole', ultimately from OE **(ġe)thyrlod** 'pierced' (< *thyrlian* 'to pierce') or ***thyrlede** 'having a hole' + **hyll**. Cf Durlston in DURLSTON BAY SZ 0477. Do i.132.

DURHAM Durham NZ 2642. 'Hill island'. The forms fall into two types:

I. *Dunholm* c.1000–c.1230, *-holme* 1068–1128, *Dunolm(um)* [1183]c.1382–1412×3, *Dunelm(um)* 1146–1386, 1625.

II. *Durelme, Dureaume* [c.1191×1200]13th, *Dure(s)m(e)* 1303–1432, 1620, *Duram* 1297 Rob Gl, *Durham* from 1334, *Durram* c.1500.

OE **dūn** + lOE **holm** (ON *holmr*). Type II shows Anglo-Norman substitution of *r-l* for the sequence *n-l* and false analysis as if it were a name in OE *hām*. The city is situated on a lofty peninsula surrounded on three sides by the river Wear. NbDu 67, DEPN, L 57, 157.

DURLEIGH Somer ST 2736. 'Deer wood or clearing'. *Derlege* 1086, *-leya -lega -lege -legh* 13th, *Durlay* 1610. OE **dēor** + **lēah**. DEPN.

DURLEIGH RESERVOIR Somer ST 2636. P.n. DURLEIGH ST 2736 + ModE **reservoir**.

DURLEY Hants SU 5116. 'Deer wood or cleaning'. *(to) deorleage* [900]11th S 360, *Durlea* *[903]16th S 370, *Derleie* 1086, *-leia -legh -lygh(e)* 12th–1421, *Durlee* 1256. OE **dēor-lēah**. Probably not just a place where deer were seen but where they were coralled or slaughtered as in the deer-hunt in *Sir Gawain and the Green Knight*. Cf. DARLEY. Ha 68, Gover 1958.47.

DURLEY Wilts SU 2364. 'The hidden wood or clearing'. *Durnley* 1229, *Durnelyghe* 1264, *Durle(ye)* 1235–1306, *Derley, Dirlay* 1278–9, *Dor- Dyerlegh* 1307×27. OE **dierne** + **lēah**. Wlt 338.

DURLSTON BAY Dorset SZ 0477. *Durlston Bay* 1774, *Durleston Bay* 1826. P.n. Durlston + ModE **bay**. Durlston is probably 'rock with a hole in it', ultimately from OE **thyrel** 'pierced' + **stān**. The original reference was probably to some coastal feature rather than later quarrying. Do i.57 gives pr ['da:Rlstən].

DURLSTON HEAD Dorset SZ 0377. *Durlston Head* 1774. P.n. Durlston as in DURLSTON BAY SZ 0477 + ModE **head** 'a headland'. Do i.57.

Great DURNFORD Wilts SU 1338. *Magna Derneford* 1270, *Moche Durneford* 1412. ModE adj. **great**, ME **much**, Latin **magna**, + p.n. *D(i)arneford* 1086, *Durneford* 1158–1474, *Derneford(e)* 1158–1316, 'the hidden ford', OE **dierne** + **ford**. 'Great' for distinction from Little Durnford SU 1234 in the extreme S of the parish. Wlt 363.

DURNOVARIA Dorset SY 6890. The Romano-British name of DORCHESTER q.v.

DURRINGTON Wilts SU 1544. 'The estate called after Deora'. *Derintone* 1086, *-ton* c.1160, 1179, *-ing- -yng-* 1281–1428, *Durenton(a)* 1178, [1154×89]1270, *-in- -yn-* 1190–1319, *Duryngton* 1291, 1398, *Durryngton* 1552. OE pers.n. *Dēora* + **ing**[4] + **tūn**. Wlt 364.

DURRINGTON WSusx TQ 1105. 'Deora's estate'. *Derentune* 1086, *Derinton(am)* [1150]1230, 1290, *Deryngton* 1542, *Durington(e)* 1219–1327, *Durrington* from 1248. OE pers.n. *Dēora*, genitive sing. *Dēoran*, varying with *Dēora* + **ing**[4], + **tūn**. Sx 195, SS 75.

DURSLEY Glos ST 7698. 'Deorsige's wood or clearing'. *Dersilege* 1086, *Derseleg(a) -leie -leye* c.1153–1331, *Durseleg(h) -ley(e)* 1211×13–1558 with variant *Dures-, Dursleg(a) -legh -ley(e)* 1220–1645. OE pers.n. *Dēorsiġe* + **lēah**. Gl ii.222.

DURSTON Somer ST 2928. 'Deor's estate or farm'. *Destone* (sic for *Ders-*) 1086, *Derstona* 1086 Exon, *Dirston* [1165×95–99] Buck, *Durston* [1165×95–1369] Buck, 1610, *Dourston* [1329] Buck. OE pers.n. *Dēor*, genitive sing. *Dēores*, + **tūn**. DEPN.

DURWESTON Dorset ST 8508. 'Deorwine's estate'. *Derwinestone -vin-* 1086, *Durwineston(a) -wyn-* 1166, 1242–1399, *Durweston* from 1412. OE pers.n. *Dēorwine*, genitive sing. *Dēorwines*, + **tūn**. Do ii.92 gives pr ['dʌrestən].

DUSTON Northants SP 7260. Partly uncertain. *Duston(e)* 1086 etc. Possibly 'the dust farm or village', OE **dūst** + **tūn**. But the specific might rather be OE ***dus** 'a heap' referring to the hill on which the village is situated. Nth 82.

New DUSTON Northants SP 7162. Adj. **new** + p.n. DUSTON SP 7260.

DUTCH RIVER Humbs SE 7121. So named in 1817 from the Dutch engineers who drained the Yorkshire marshlands. This drainage channel was planned in 1632. YW vii.127.

DUTON HILL Essex TL 6026. *Duton Hill* 1805 OS. P.n. *Dewton* 1570 of unknown origin + ModE **hill**. Ess 486.

DUTTON Ches SJ 5779. Partially uncertain. The forms fall into two types:
I. *Duntune* 1086, *-ton* 1247, 1254, *Duninton* 1183, *-yng-* 1223×7, *Doncton* 1269–1300.
II. *Dutton -ton* from 1150, *Dudton* c.1233, *Ditton -ton(e) -y-* 13th–1390, 1653.
Type I is 'hill settlement', OE **dūn** + **tūn** varying with **dūning** + **tūn**. The reference is to the isolated hill which rises to 214ft. near Aston at SJ 5678. Type II has been explained as 'Dudda's farm or village', OE pers.n. *Dudda*, genitive sing. *Duddan*, + **tūn** with early loss of inflexional *-an* and assimilation of *dt* to *tt*. Although the existence side by side of two different name-forms for the same place is not unusual, it is tempting to reconcile the two types, by regarding *Duntune* also as representing a contracted **Dudntūne* < **Duddan-tūne*. Much better, however, is the assumption of a side-form ***dud(d)** beside ***dod(d)** 'a rounded hill'. Che ii.112, PNDB 224–5.

DUXFORD Cambs TL 4846. 'Ducc's enclosure'. *Dukeswrthe* [942×51]13th S 1526, *-w(o)rth(e)* 1211–1320, *Du- Dochesuurda, Dochesurda- -uuorde -uurðe* 1086, *Dokesw(o)rth(e)* 1251–1435, *Duxwurth* 1442, *Duxforth(e)* 1535–64, *Duxford* 1548. OE pers.n. **Ducc*, genitive sing. **Ducces*, + **worth**. Ca 92.

DYE HOUSE Northum NY 9358. *Dye Ho.* 1867 OS. No early forms. Possibly ME **dey-house** 'a dairy'.

DYKE Devon SS 3123. *Westdich* 1333. OE **dīċ**. The reference is to CLOVELLY DYKE SS 3132, a large Iron Age hill fort. D 71 gives pr [dik], Thomas 1976.94.

DYKE Lincs TF 1022. 'The ditch'. *Dic* 1086, 1219, 1327, *Dik* 1234–1329, *Dyke* from 1346. OE **dīċ**, ON **dík**. The reference is to CAR DYKE TF 1437 on which the village stands. DEPN.

DYMCHURCH Kent TR 1029. 'Dema or the judge's church'. *Deman c.* c.1100, *Demechirche -cherche* 1240–91, *Dymechirch'* 1261×2, *Deemchurche* 1585. OE **dēma** 'judge' possibly used as a by-name *Dēma*, genitive sing. *Dēman*, + **ċiriċe**. The surname Deem occurs 1279, *Richard Deme*. In favour of derivation from *dēma* is the fact that in the 16th cent. at least Dymchurch was the seat of government of Romney Marsh. It was the meeting-place of the Lords of the Level – 'the Lords, Bailiffs and Jurats of Romney Marsh' – where they met to maintain the efficient drainage of the marsh and upkeep of the walls with power to enforce maintenance. The jurisdiction is likely to have been of considerable antiquity. PNK 462, DBS 98, Brentnall 1980.111, Cullen.

DYMOCK Glos SO 7031. Partly uncertain. *Dimoch* 1086–1159, *Di- Dymmoc(k)' -ok* 1156–1777, *Di- Dymoc -ok(e) -ocke* 1221–1730. Probably 'swine fort', PrW ***din** + **moch**, but the first element might be PrW ***tiȝ** 'house' with initial *d-* owing to lenition after a preceding consonant and so 'pig-sty'. Gl iii.168 gives pr ['diməkl].

DYRHAM Avon ST 7475. 'Deer enclosure'. *Deorham* 9th–11th ASC under year 577, *(on) Deorham, (of) Deorhamme* [950]19th S 553, *(into) Deor hamme* 972 S 786, *Dirham* 1086, *Derham* 1220–1695, *Deerham* 1659, *Dyrham* from 1540. OE **dēor** + **hamm**. Gl iii.49 gives pr ['diərəm], L 44, 49.

DYRHAM PARK Avon ST 7475. *Dyrham Park* 1830. A country mansion of the 16th–18th cents. Gl iii.49, Verey i.230.

E

EAGLAND HILL Lancs SD 4345. No early forms. The name refers to a small rise in the level of Pilling Moss.

EAGLE Lincs SK 8767. 'The oak-tree wood or clearing'. *Aclei -ey, Aycle, Akeley, Achelei* 1086, *Eicla* 1135×9, 1141, *Aicle* 1220×35–1449, *Eycle* 1212–1428, *Ec(c)les -is* 1155–1449, *Egle* 1442–1602, *Eagle* from 1553. OE **āc** + **lēah** partly Scandinavianised by substitution of ON **eik**. Perrott 223, SSNEM 216, Cameron 1998.

EAGLESCLIFFE Cleve NZ 4215. *Eaglescliffe* is first recorded as a spelling in 1639 PR. It is an alteration of nearby EGGLESCLIFFE NZ 4113 due to popular etymology. It was established locally as a spelling long before its use for the station of Eaglescliffe Junction on the old Stockton and Darlington railway in 1825.

EAGLESFIELD Cumbr NY 0928. 'Open country at *Egles*'. *Eglesfeld* [c.1170]–1357, *Eggles-* 1333. P.n. **Egles*, PrW ***eglēs** 'church, Celtic Christian centre' + **feld**. Cu 378, Signposts 82, 96–7, L 243.

EAKRING Notts SK 6762. 'The circle of oak-trees'. *Ec(he)ringhe* 1086, *Ekering(a)* 1174–1322, *Ecrynge* 1563, *Eckrin* 1716, *Aik(e)ring(a)* c.1190–1508 with variants *Ay-* and *-yng*, *Eikering(e)* 1156–1553 with variants *Ey-* and *-yng*. ON **eik**, genitive pl. **eika**, + **hringr** (or OE **hring**). Nt 49, SSNEM 153.

EALAND Humbs SE 7811. The 'island'. *Aland* 1316, 1372. OE **ēaland**. DEPN.

EALING GLond TQ 1781. 'The Gillingas, the people called after Gilla'. *Gillingas* [693×704]18th S 1783, *Yllinges* 12th cent., *Illing' -yng* 1130–1563, *Yilling(g) -yng(ge)* 1294–1399, *Elyng* 16th cent., *Ealing al. Yealing* 1622. OE folk-n. **Ġillingas* < pers.n. **Ġilla* + **ingas**. *Chircheyllinge* 'church Ealing' occurs 1274 for distinction from *Little* or *West* Ealing. Mx 90.

EAMONT BRIDGE Cumbr NY 5228. *Eamontbridge* 1865 OS. R.n. EAMONT NY 4725+ **brycg**.

River EAMONT Cumbr NY 4725. 'Junction of streams'. *(æt) Ea motum* 11th ASC(D) under year 926, *Amont* 12th, *Emot(t)* [1220×47]15th, *Amot(e)* 1285–1408, *Eamont* from 1558. OE **ēa-(ġe)mōt**. The reference is to the junction of the Lowther and the outflow from Ullswater. In some forms OE *ēa* has been replaced by ON *á*; the intrusive *-n-* is due to French influence, as in BECKERMONDS. Cu 12 gives prs [i:mənt] and [jæmen], RN 139, L 11, 21.

EARBY Lancs SD 9046. Probably the 'upper farmstead'. *Eurebi* 1086, *Eu- Everby* 1260–1373, *Ereby* 1526, *Ear(e)by* from 1557. ON **øfri**, **efri** + **bȳ**. Identical with Norwegian *Øverby*. 'Upper' relative to Thornton in Craven lower down the valley of Earby Beck. However, the ON pers.n. *Iofurr* is also formally possible so that this might alternatively be 'Iofurr's village or farm'. YW vi.33, SSNY 26.

EARCROFT Lancs SD 6824. No early forms.

EARDINGTON Shrops SO 7290. 'The estate called after Earda'. *Eardigton* 1012×56 ECWM, *Ardintone* 1086, *Erdinton -yn-* 13th cent., *Erdington(') -yng-* 1228–1705, *Eardyngton* 1535, *-ing-* 1746. OE pers.n. *Earda* + **ing**^{4} + **tūn**. Sa i.112.

EARDISLAND H&W SO 4258. The modern spelling is a corruption of 'earl's Leen', lOE **eorl** 'an earl', genitive sing. **eorles** + the district name LEEN. The forms are confusing: *Werlesluna* 1067×71, *Orleslen(a)* 12th, *Erleslen(e)* 1230, 1291, *Erleslonde* 1321, 1326, *Ereslond* 1529×30, *Erislonde* 1535, *Aresland* 1577, 1660, *Eardsland* 1786. The estate was held by earl Morcar, son of earl Algar of East Anglia, who had the estate before 1066. Leen is *Lene* 1086 as in The LEEN SO 3859. He 76, Bannister 64, DEPN.

EARDISLEY H&W SO 3149. 'Ægheard's wood or clearing'. *Herdeslege* 1086, *Eirdesleg'* 1137×9, *Eierdesl'* 1249, *Eiardeleye* 1252, *Erdesley(e)* 1219–75, *Erdisley* 1292–1535, *Ardesle, Ardelai* 12th, *Jerdisley* 1413. OE pers.n. *Æġheard*, genitive sing. *Æġheardes*, + **lēah**. He 77, DEPN.

EARDISTON H&W SO 6968. 'Eardwulf's estate'. *Eard(g)ulfestun* [?781×96]11th S 1185, *Ardolvestone* 1086, *Erdelstone* 1295, *Erdeston* 1275, *Adris- Aderestone* 1349, *Yeardiston* 1787. OE pers.n. *Eardwulf*, genitive sing. *Eardwulfes*, + **tūn**. Wo 58 gives pr [ja:distən].

EARDISTON Shrops SJ 3725. Partly uncertain. *Erdeston* 1203× etc., *Eardiston* 1837 OS. Uncertain pers.n., possibly *Ēorēd*, genitive sing. *Ēorēdes*, + **tūn**. Gelling.

EARITH Cambs TL 3875. 'Gravel or muddy landing place'. *Herhethe* [1244]14th, *E(a)rheth* 1260–86, *Eryth(e)* 1557, 1616. OE **ēar** + **hȳth**. Hu 204, L 62, 76–7.

EARLE Northum NT 9826. 'The hill where rods are obtained'. *Yherdhill* 1242 BF, *Yerd(h)il(l)* 1255–1428, *Yerdle* 1542, *Earlle* 1579, *Yardhill, Yerle, Erle* 18th cent. OE **ġerd** + **hyll**. NbDu gives pr [jerl], L 171.

EARLESTOWN Mers SJ 5795. A railway town which developed slowly after 1826 and was named after Sir Hardman Earle, director of the Liverpool and Manchester Railway Company at the time. Room 1983.34.

EARLHAM Norf TG 1908. 'Earl homestead or estate'. *Erlham* 1086–1198, *Herlham* 1196, 1242. OE **eorl** + **hām**. DEPN.

EARLSDON WMids SP 3278. A modern refashioning due to popular etymology of p.n. *Eldsone* 1675, *Elsdon* 1834 OS, 1841, of unknown origin and meaning, possibly the same as *Alsdenfeld* 1423. The evidence is too late and thin. Wa 166.

EARL SEAT NYorks SE 0758. *Earl Seat* 1858. ModE **earl** + **seat** 'a high place' (ON *sǽti*. YW vi.61.

EARL'S GREEN Suff TM 0366. *Earl's Green* 1837 OS. Contrasts with nearby *Bacton Green* TM 0365, *Haughley Green* TM 0264, *Hestley Green* TM 1567 and *Smith's Green*, all 1837 OS.

EARLSHEATON WYorks SE 2521. 'Heaton held by the earl' sc. of Warren. *Herleshetone* 1284, *Erles-* 1316 etc. ME **erle**, genitive sing. **erles** + p.n. HEATON 'the high settlement', *Et(t)one* 1086, *Hetun -tone* early 13th–1554, YW ii.194.

EARL'S SEAT Northum NY 7192. *Earl's Seat* 1869 OS. A peak in Kielder Forest. N dial. **seat** 'a lofty place' (ON *sǽti*). The reference is to the Earl of Northumberland who owned the land.

EARLSWOOD Warw SP 1174. 'The earl's wood'. *le Erlyswode* 1461×83, *Erlys wood* 1475, *Urleswoode* 1544. ME **erle**, genitive sing. **erles**, + **wode** (OE *wudu*). Belonged to the earls of Warwick as overlords of the manor of Tanworth SP 1170 in which the wood lies. Wa 294.

EARNLEY WSusx SZ 8197. 'Sea-eagle wood or clearing'. *Earnaleach* 780 S 1184, *Earneleagh* [780]14th ibid., *Earneleia* [930]14th S 403, *Erneleia* c.1197, *-le(y)e* 1291–1438. OE **earn**, genitive pl. **earna**, + **lēah**. Sx 82, SS 65, LSE 18.176, 177.

EARSDON T&W NZ 3272. 'Ea(n)red or Eored's hill'. *Hertesdona* 1203, *Erdisdunam* 1271, *-ton* 1296 SR, *Erdesdon* 1363, *Eresdon* 1428. OE pers.n. *Ēa(n)rēd* or *Ēorēd*, genitive sing. *Ēa(n)rēdes, Ēorēdes*, + **dūn**. NbDu 69.

EARSHAM Norf TM 3289. Partly uncertain. *Ersam* 1086, *Earesham* [1087×98]12th, *Eresham* 1158, 1212, *Erlsham* 1248.

The generic is either OE **hām** 'homestead, estate' or **hamm** 2a 'promontory into marsh'. The specific might be OE **eorl** 'nobleman, earl' as the isolated 1248 spelling suggests. Otherwise a much reduced form of a pers.n. such as *Ēanhere* is possible. Another solution might be OE **ears** 'buttock' referring to the projecting hill on which the village stands. A similar problem occurs with the name Eastbury Manor in Hollow, H&W SO 8258, *Earesbyrig* 11th, *Eresbyrie* 1086, *Esebyre -bire* 1240, 1275, *Eselbyre* 1240, *Eylesbyri* 1255 and a lost *Earesbroca* in Arley for which Wo 130 suggests an OE pers.n. *Ēar* or *Ēare*. DEPN.

EARSWICK NYorks SE 6257. 'Ætheric's dairy-farm'. *Edresuuic, Edrezuic* 1086, *Ethericewyk* early 13th, *(H)everswyk'* 1292, 1295, 1301, *Herswyk* 1295, *Etherswik* 1322, *Ereswick* 1577. OE pers.n. *Ætheric*, genitive sing. *Ætherices*, + **wīċ**. YN 12 gives pr [i:əzwik].

New EARSWICK NYorks SE 6155. ModE **new** + p.n. EARSWICK SE 6257. A modern suburb of York.

EARTHAM WSusx SU 9409. 'The ploughed enclosed land'. *Ercheham* (for *Erthe-*) 1100×35, *Urtham* 1279–80, *Ertham* 1279–1402. OE **erth** + **hamm** 6b 'a piece of land in a bay of higher ground'. Sx 70, ASE 2.49, SS 81.

EASBY NYorks NZ 5708. 'Esi's farm or village'. *Esebi -by* 1086–1369. ON pers.n. *Ēsi*, genitive sing. *Ēsa*, + **bȳ**. YN 167, SSNY 26.

EASEBOURNE WSusx SU 8922. 'Esa's stream'. *Eseburne* 1086–1346. OE pers.n. *Ēsa* + **burna**. Some sps, *Estburn* 1219, *Eastborn* 1447, show confusion with EASTBOURNE TV 6199. Sx 16 gives pr [ezbɔ:n], SS 72.

EASEDALE Cumbr NY 3208. 'Asi's valley'. *Asedale* 1332, 1375, *Aisedale* 1706, *Easedale* 1801, *Aysdale* 1812. ON pers.n. *Ási*, genitive sing. *Ása*, + **dalr**. We i.199, SSNNW 120.

EASEDALE TARN Cumbr NY 3008. *Easedale Tarn* 1802 OS. P.n. EASEDALE NY 3208 + N dial. **tarn** (ON *tjǫrn*). We i.200.

EASENHALL Warw SP 4679. 'Esa's hill'. *Esenhull* 1221–1789 with variants *Esyn-* and *-hill*, *Eysenell* 1445, *Easnell* 1546. OE pers.n. *Ēsa*, genitive sing. *Ēsan*, + **hyll**. Wa 107, DEPN.

EASINGTON 'Estate called after Esa or Esi'. OE pers.n. *Ēsa* or *Ēsi* + **ing**[4] + **tūn**.

(1) ~ Bucks SP 6810. *Hesintone* 1086, 1152×8, *Esinton* 1185–1302, *Esington -yng-* 1241–1366. Bk 121.

(2) ~ Cleve NZ 7417. *Esingetun -ton* 1086, *Esinton(a)* 1154×61–1369, *Esington -yng-* 12th–1371. The single *-inge-* spelling is probably to be disregarded rather than taken as pointing to OE *Ēsinga-tūn* 'settlement of the Esingas, the people called after Esa'. YN 140.

(3) ~ Humbs TA 3919. *Hesinton* 1086, 1260, *-ing-* 1175×95, *Esinton(e)* 1086–1339×49, *Esinctun* 1115, *-ington(a) -yng-* 1227–1546, *Easington* 1695. YE 17, BzN 1968.166.

(4) ~ Lancs SD 7050. *Esintune* 1086, *Hesington* 1216×72, *Esington -yng-* 1285–1546. YW vi.201.

(5) ~ COLLIERY Durham NZ 4343. A modern mining village, p.n. Easington + ModE **colliery**. Easington NZ 4143 is *Esingtun* [c.1040]12th, *-ton(')* [1183]c.1320, 1242×3–1580, *Esin- Esynton(a)* 1147–1228, *Esyngton* [1183]c.1400, 1357–1570, *Easington* from 1597. NbDu 70.

(6) ~ FELL Lancs SD 7249. *Easington Fell* 1846. P.n. EASINGTON Lancs SD 7050 + ModE **fell** (ON *fjall*). YW vi.202.

(7) ~ LANE T&W NZ 3645. 'The lane leading to Easington'. P.n. Easington as in EASINGTON COLLIERY Durham NZ 4343 + ModE **lane**.

EASINGTON Oxon SU 6697. Esa's hill. *Esidone* 1086, *Esendon(e)* 1199–1227, *Esindon' -don(e) -den -yn-* [c1200]c1280–1428, *Es(s)in(g)ton' -don(e) -yng-* 1240–85, *Esyngton* 1408. OE pers.n. *Ēsa*, genitive sing. *Ēsan*, + **dūn**. O 125, L 150.

EASINGTON Northum NU 1234. 'The settlement on or called *Yesing*, the gushing stream'. *Yesington* 1242 BF 1119, *Y(h)esyngton -ing-* 1278–1346, *Easengtoun* 1579. OE ***ġēosing** < **ġēose*, cognate with ON *gjósa* 'to gush', + **ing**[2], + **tūn**. Cf. JESMOND Tyne & Wear NZ 2566. NbDu 69.

EASINGWOLD NYorks SE 5369. 'The forest of the Esingas, the people called after Esa'. *Eisicewalt, Eisincewald* 1086, *Esingwald -yng-* 1167–1451, *Esingewald(e)* 1169–1247, *Easingwould* 1666. OE folk.n. **Ēsingas* < pers.n. *Ēsa* + **ingas**, genitive pl. **Ēsinga*, + **wald**. The reference is probably to part of the same forest as COXWOLD SE 5377. YN 24 gives pr [i:əzinud], L 226–7.

EASOLE STREET Kent TR 2652. *Hazle St.* (sic) 1819 OS. This form is a false realisation of Easole Street, 'Easole village', p.n. Easole + Mod dial. **street**. It is in systemic contrast with nearby Holt Street TR 2550, *Holstrete* 1547, *Old Street formerly called Holt-Street* 1790 (the place is close to Ackholt TR 2451, *Akholte* 1226, 'oak wood', OE **āc** + **holt** though the earliest spelling of Holt Street point to OE **hol(?)** 'a hole, a hollow') and Frogham TR 2550, *Frogham Street* 1819 OS. Easole, *Eswalt, Essewelle* 1086, *Easole* from 1242×3, *Eswalle -well' -wole, Estwel(l), Heasuele, (H)essole* 13th cent., *E(a)swole* 1254–1318, is a p.n. which gave great difficulty to the medieval scribes as the variety of spellings shows. Probably identical with *(æt) Oesewalum* 824 S 1434 and *Oesuualun* [830 for 833]13th S 1414, which appears to be a compound of OE **ōs** 'a god', genitive pl. **ēsa** (earlier *ǣsa* < *ansi-*), + **walu** 'ridge of earth or stone', dative pl. **walum**, and so 'at the ridges of the gods'. The reference is unexplained, but on the face of it this appears to be a p.n. which alludes to pagan Anglo-Saxon religious belief. PNK 534, KPN 147, PNE i.159, ii.245.

EASTBOURNE ESusx TQ 6199. 'East Bourne'. *Estburn* 1279. ME adj. **est** + p.n. *Borne, Burne* 1086–13th, 'the stream', OE **burna**, referring to the stream which rises near the parish church and flows into the sea. One of the streets of Eastbourne, Bourne Street, marks its course. 'East' for distinction from WESTBOURNE WSusx SU 7507. Sx 426.

EASTBRIDGE Suff TM 4566. *East Bridge* 1837 OS. E of Backford Bridge over the Minsmere River. Earlier simply *Bringas, briges* 1086, *Brigge, Bregge* 1275–1344, 'the bridge', OE **brycg**. Baron.

EASTBURN Humbs SE 9955 → KIRKBURN SE 9855.

EASTBURY Berks SU 3477. '(At the) manor to the east' sc. of Bockhampton. *(æt) Eastbury* [c.1000]c.1300, *Estberi* 1165 etc. with variants *-buria -bir(i) -byri -bere -bery -bur(y), Eastbury* 18th. OE **ēast** + **byriġ**, dative sing. of **burh**. Brk 335, Signposts 124.

EASTBURY GLond TQ 0991. 'East manor'. *Estbury* 1290. 1456, *Estberies* 1556. OE **ēast** + **byriġ**, dative sing. of **burh**. Hrt 107.

EASTBURY MANOR H&W SO 8258 → EARSHAM Norf TM 3289.

EASTCHURCH Kent TQ 9871. 'East church'. *Eastcyrce* c.1100, *Estcher(i)che -churche* etc. 1226 etc. OE **ēast** + **ċiriċe**. The church is E of Minster on the Isle of Sheppy. PNK 246.

EASTCOMBE Glos SO 8904. 'East coomb'. *E(a)stcomb(e)s* 1633, 1763. ModE **east** + **coomb**. 'East' possibly for distinction from Brimscombe SO 8702 two miles SW at the mouth of the same valley, *Bremescumbe* 1306, 'Brēmi's valley'. Gl i.123, 141.

EASTCOTE GLond TQ 1188. 'East cottages'. *Estcot(t)e* 1248–1323, *Ascote* 1248–1323, *Ascote* 1356–1710, *Eastcote* 1819. OE adj **ēast** + **cot**, pl. **cotu**. 'East' for distinction from a lost *Westcott* in Ruislip. Mx 47, GL 45.

EASTCOTE WMids SP 1979. Partly uncertain. *Erescote* 1307–69, *Erscot* 1331, *Ernes- Ermescote* 1332, *Escote* 1434, 1454, 1725, *Estcourt* 1628. The 17th cent. and the modern forms are rationalisations of an earlier obscure name, possibly 'Earn's cottage(s)', OE pers.n. *Earn*, genitive sing. *Earnes*, rather than **earn** 'an eagle', + **cot**, pl. **cotu**. Wa 55.

EASTCOTT Corn SS 2515. 'East cottage(s)'. *Yestecote* 1302 Gover n.d., *Es(t)cote* 1327, *Estecote* 1432. ME **est** (OE *ēast*) + **cot**, plural **cote** (OE *cotu*). Situated NE in the parish of Morwenstow SS 2015. PNCo 80, Gover n.d. 20.

EASTCOTT Wilts SU 0255. 'The east cottage(s)'. *Estcota* 1167, 1187, *Escote* 1298, *Escott* 1640. ME **est** + **cote** (OE *cot*, pl. *cotu*). Situated E of Easterton, istelf an eastern settlement of Market Lavington. Wlt 315.

EASTCOURT Wilts ST 9792. 'The east cottage(s)'. *Escote* [901]14th S 1579, 1332, 1638, *Eastcotun* [974]13th S 796, *Estcote* 1222–79, *Estcott otherwise Eastcourt* 1736. OE **ēast** + **cot**, dative pl. **cotum**. The modern folk-etymology is due to the building of Eastcourt House in the late 17th cent. Wlt 57 gives pr [eskət].

EAST END

(1) ~ Avon ST 4870. 'East end' sc. of NAILSEA ST 4470. *East End* 1830 OS. Contrasts with WEST END ST 4469.

(2) ~ Dorset SY 9998. *East End* 1811. A hamlet in the E part of Corfe Mullen parish. Do ii.17.

(3) ~ Hants SU 4161. Short for the East End of East Woodhay SU 4061. *Estende* 1447, *East End* 1817 OS. In contrast with Heath End SU 4162, *Heath End* 1817 OS and *North End* 1817 ibid., *Northende* 1441. Gover 1958.158.

(4) ~ Hants SZ 3697. Short for the East End of Baddesley SZ 3496 in Boldre parish. *East End* 1810 OS. Gover 1958.205.

(5) ~ Herts TL 4527. *East End* 1676 sc. of Furneux Pelham. Contrasts with Barleycroft End TL 4327, *Barleycroft End* 1676, and Patient End TL 4227, *Payston End* 1557, *Patient End* 1676 from *Paynottys al. Paynestone* 1434, 'Payn's estate', all in the same parish. Hrt 186.

(6) ~ Kent TQ 8335. The east end of Benenden parish. No early forms.

(7) ~ Oxon SP 3914. *le Estend* 1529×30 sc. of North Leigh. ME **est** + **end**. O 275.

Good EASTER Essex TL 6212. 'Godgyth or Godgiefu's Easter'. *Godith(e) -yth(e) Estre* 1208–91, *Godyve Estre* 1296, *Godes(h)estre* 1272–1315, *Goodestre* 1428–96. OE feminine pers.n. *Gōdgȳth* or *Gōdġiefu*, genitive sing. *Gōdgȳthe* or *Gōdġiefe*, + p.n. *Æstre* [1017×35]12th, *Estre* [1042×66]12th S 1051–1434, *Estram* 1086, *Ester* [1068]1309, 1227, 'the sheepfold', OE **eowestre**. The forms *Estern'* 1275, *estren* 1285, represent either nominative or dative pl., OE *in thā eowestran, to eowestrum*. 'Good' for distinction from High EASTER TL 6214. The forms point overwhelmingly to pers.n. *Godgyth* but *Liber Eliensis* reports that the estate was given to Ely by the will of a widow called Godiva, i.e. Godgiefu, in the reign of Cnut. Ess 478 gives pr [estə], Studies 1931.74.

High EASTER Essex TL 6214. *Alt Estre* 1236, *Haut-* 1273, *Hegh-* 1274, *High Eastre* 1300, 1302. OFr adj. **halt, haut**, ME **hegh**, + p.n. Easter as in Good EASTER TL 6212. Ess 478.

EASTERGATE WSusx SU 9505. 'The eastern gate'. *Estergat(e) -re-* 1263–1331. ME **ester** (OE *ēasterra*) + p.n. *Gate* 1086–1332, *Gates* 1248–1327, 'the gate', OE **ġeat**, pl. **gatu**. The reference here and at WESTERGATE SU 9305 is unknown. Cf. also WOODGATE SU 9304. Sx 140.

EASTERN GREEN WMids SP 2879. *Eastern Green* 1821, 1834 OS, a modification of earlier *Estend grene* 1528. Probably so called in relation to Flint's Green SP 2680 since Eastern Green lies at the W rather than the E end of Allesley parish. Wa 153.

EASTERN ISLES Scilly SV 9514. *Eastern Isles* 1892. The easternmost group of the Scillies. PNCo 80.

EASTERTON Wilts SU 0154. 'The eastern settlement'. *Esterton (juxta Stepellavynton)* 1348–1481, *Easterton Garnham* 1591. ME **ester** (OE **ēastor, ēasterra*) + **toun** (OE *tūn*). East of Market Lavington. Wlt 239.

EASTERTOWN Somer ST 3454. *Easter Town* 1809 OS. ModE **easter** (OE *ēasterra*) + **town**. The reference is to the eastern part of Lympsham ST 3354.

EAST FEN Lincs TF 4055. *The East Fenne* 1661, *East Fen* 1824 OS. ModE **east** + **fen**. The eastern part of the Fens between the Witham and the sea-coast drained and enclosed 1801–18. Wheeler 197ff.

EASTFIELD HALL Northum NU 2206. *Eastfield Hall* 1868 OS. P.n. Eastfield + ModE **hall**. Probably a reference to the ancient eastern open field of the village of Sturton NU 2107. Contrasts with the Northfield NU 2407 of Birling NU 2506.

EASTFIELD NYorks TA 0384. Cf. *High* and *Low East Field* c.1855 OS. ModE **east** + **field**. The reference is to the old open east field of Cayton.

EAST GARSTON Berks SU 3676. 'Esgar's estate'. *Es(e)gar(e)ston(a)* 1180–1284, *Edgar(e)ston* 1232, 1412, *Estgarston* 1275×6 etc., *Estgreston'* 1284, *Est Garston, Eastgarston* 1535, *Ergaston* 1401×2, *Azgaston* 1559. Pers.n. *Esgar* (< ODan *Æsgēr*), genitive sing. *Esgares*, + **tūn**. It is generally accepted that the reference is to Edward the Confessor's staller Esgar who held 30 hides at Lambourn in 1066. These 30 hides are identical with East Garston. Esgar was of Scandinavian descent being the grandson of the Cnutian magnate Tovi the Proud, the founder of Waltham Abbey. The forms of 1232 and 1412 show association with ME pers.n. *Edgar* (OE *Ēadgār*). The forms from 1275 show reinterpretation of the name as if OE *ēast gærstūn* 'eastern grass enclosure'. The local pr at the beginning of the century is said to have been [a:gæstən]. Brk 330, Signposts 124, 229, Writs 560.

EASTGATE Durham NY 9538. '(The) east gate' of the forest of Weardale. *Estyat* 1508, *-yate in Wardell* 1547, *Est Gate* 1540, *the East Gate* [1569]1792. An *Estyatshele* 'Eastgate shieling' is mentioned in 1457. ME **est** + **yate** (OE *ġeat*). The east gate of the hunting park of the Bishop of Durham in Weardale. NbDu 70.

EASTGATE Norf TG 1423. 'The eastern hamlet' sc. of Cawston. *Eastgate* 1838 OS. ModE **east** + **gate**. ModE *gate*, ON *gata*. In Norf, ModE gate, On gata, is used (like dial. *street*) for hamlet settlements. Cf. nearby Southgate TG 1324, *South gate* 1838 OS (actually to the N of Eastgate). For other examples cf. Lingate TG 2731, *Lyngate* 1588, and Tungate TG 2629, *Tungate* 1563–91, in North Walsham, and Bengate TG 3027, *Benegate* 1345, Briggate TG 3127, *Briggate* 1237–51, Lyngate TG 3126 and Withergate TG 2927, *Wytegate* 13th, in Worstead. Nf ii.178, 180–1, 205–6.

EASTHAM Mers SJ 3580. 'East village or estate'. *Estham* 1086–1657, *Eastham* from 1539, *Estam* 1343, 1432, *Estom* 1599, *Estem* 1671. OE **ēast** + **hām**. The village is situated at the E end of the great Domesday Manor that stretched the length of the Wirral from Moreton SJ 2690. It was originally a chapelry of Bromborough parish. Che iv.187 gives pr ['i:stəm], 239.

EASTHAM SANDS Mers SJ 3981. *Eastham Sands* 1842, cf. *Eastham Bank* 1831. P.n. EASTHAM SJ 3580 + ModE **sand(s)**. A sand-bank in the Mersey estuary. Che iv.188.

EASTHAMPSTEAD Berks SU 8667. 'Homestead at the gate'. *Lachenestede* (sic) 1086, *Yezhamesteda* 1167, *Yethamsted(e)* 1176 etc., *Yeshamsted(e)* 1224–1327, *Yes(s)hampstede(e)* 1316–43, *Essamested'* 1216, *Es(t)hamstede* 1284, *Estehampsted* 1535. OE **ġeat**, genitive sing. **ġeates**, + **hāmstede**. The DB form seems to be corrupt unless it contains OE **lacu** 'stream' or ***læċċ** 'boggy stream' subsequently replaced with *ġeat*. This in turn gave way to ME **est** 'east' as a result of popular etymology. Brk 23, Stead 265.

EASTHOPE Shrops SO 5695. 'The east valley'. *East hope* 901 S 221, *Stope* 1086, *Est(h)op(e)* 1166–1685, *Yestope* 1780. OE **ēast** + **hop**. The reference is to one of a series of funnel-shaped side-valleys opening from Hope Dale, the valley between the parallel ridges of Wenlock Edge and the Aymestrey Limestone escarpment, into Corve Dale. Cf. WESTHOPE SO4786, DINCHOPE SO 4584. Sa i.113.

EASTINGTON 'Place east in the estate or village'. OE **ēast in tune**. PNE i.144.

(1) ~ Devon SS 7409. *Estyngton* 1330, 1384, *Eastown* 1809 OS. A settlement at the east end of Lapford parish. D 370, PNE i.144.

(2) ~ Glos SP 1313. *Estinton(e) -yn-* 1119–1328, *Estington -yng-* 1529–1605. Gl i, 171.

EASTINGTON Glos SO 7705. Possibly 'Eadstan's estate'. *Esteueneston'* 1220, *Estanes- Estones- Esteneston* c.1230–87, *Estinton - yn-* 1227–1393, *-yng-* 1287, 1509–71, *Eastington* 1422, 1577, 1695. The 13th cent. forms may be explained from an OE *Ēadstānes-tūn* later assimilated to the form of EASTINGTON SP 1313, but the first form, if trustworthy, seems to point to 'Stephen's farm' or to a possible OE pers.n. **Ēadstefn*. Gl ii.194.

EASTLEACH 'East Leach'. *Estleche -lech* 1138–1372. ME **est** + p.n. *Lec(c)e* 1086, *Lech(e)* 1221, 1227, *Lecche* [862]14th S 209, from the River LEACH SP 1707. The name of an estate on the river Leach surviving in EASTLEACH MARTIN SP 2005 and EASTLEACH TURVILLE SP 1905. 'East' for distinction from NORTHLEACH SP 1114 higher up the river. Gl i.31,33.

(1) ~ MARTIN Glos SP 2005. 'St Martin's Eastleach'. *(Est)leche Sancti Martini* 1291, *Eastleach Marten* 1587. P.n. EASTLEACH + saint's name *Martin* from the dedication of the church. Gl i.31.

(2) ~ TURVILLE Glos SP 1905. 'E held by the Turville family'. *Estleche Rob' de Tureuill'* 1221, *Eastleach Turvile* 1587. P.n. EASTLEACH + family name *Turville*. The manor was held by Robert de Turevill in 1200. Also known as *Lech(e) Sci Andree* 1291, 1535, 'St Andrew's Leach' from the dedication of the church. Gl i.33.

EASTLEIGH Hants SU 4519. 'East wood or clearing'. *east lea* [932]13th, *Estleie* 1086, *Estleg(a) -le(ya)* 1218–71, *Estleighe* 1582. OE **ēast** + **lēah**. Ha 69, Gover 1958.37.

EASTLING Kent TQ 9656. 'The Eslingas, the people called after Esla'. *Eslinges* 1086–1242×3, *Eastlinges, Æslinge* c.1100, *Asling'* 1231, *Esling'* 1242×3, *Estling'* 1275. OE folk-n. **Ēslingas* < pers.n *Ēsla* + **ingas**. Forms with medial *-t-* show association with *ēast* 'east'. Forms with *Æs-* are back-spelling due to the equivalence of *e* and *œ* in the Kentish dial. and forms in *As-* may be due to shortening of *ēa* in *ēast*. The same folk-n. occurs in *Islingham Farm* 1819 OS (lost in Frindsbury TQ 7469), the 'homestead of the Eslingas', *Æslingaham* [761×5]12th S 33, [764]12th S 105, *Æslingeham* ibid., c.1100, *Eselingham* 1100–37, *Eslingeham* 1242×3. PNK 284, ING 11, 119.

EAST MIDLANDS AIRPORT Leic SK 4526. P.n. East Midlands + ModE **airport**.

EAST MOOR Derby SK 2868. *le Eastmore* 1575. Fr definite article **le** + p.n. Eastmore, ModE **east** + **moor**. E of Beeley SK 2667. Db 45.

EASTNEY Hants SZ 6799. '(Place) on the east side of the island' sc. of Portsea. *Esteney* 1247, *-ey(e) -y(e) -e* 1244–1382. OE **ēast** + **īeġ** in the phrase *be ēastan īeġe*. Ha 69, Gover 1958.25.

EASTNOR H&W SO 7337. '(Place) east of the ridge'. *Astenofre* 1086, *Est(e)nover(e)* 1140×8, 1166, c.1317, *Esteno(u)r(e)* 1241–1368, *Estnor* 1538. OE **ēastan** + **ofer**. He 78, Bannister 65, L 176, Jnl 22.36.

EASTOFT Humbs SE 8016. 'Homestead by the ash-wood'. *Eschetoft(h)* 1164×77, 1293, *Esketoft* 1199×1209, 1304, *Estoft(e)* 1252–1594, *Eastoft(e)* from 1572. ON **eski** + **topt**. YW ii.4.

EASTON 'East settlement'. OE **ēast-tūn**. PNE i.144.

(1) ~ Cambs TL 1371. *Estone* 1086 etc., *Eston al. Esson* 1578. 'East' as opposed to OLD WESTON TL 0977 either side of the hundred centre at LEIGHTON BROMWOLD TL 1175. Hu 238 gives pr [i:sən], cf. the spelling *Wessen* 1594 for WESTON TL 0972.

(2) ~ Cumbr NY 4372. *Estuna* [c.1155]c.1200, *Eston* 1267–1509×47, *Easton* from 1584. Situated in the E of Arthuret parish. Cu 53.

(3) ~ Devon SX 7188. *Shutt al. Eston* 1578. East of Chagford SX 7087. The alternative name *Shutt* represents OE **scīete**, an i-mutation variant of **scēat** 'a corner' referring to situation in a corner or projection of a parish. D 426.

(4) ~ Devon SX 7242 → West ALVINGTON Devon SX 7243.

(5) ~ Dorset SY 6971. *Easton* from 1608, *East Town* 1774. On the E side of the Isle of Portland contrasting with WESTON SY 6871. Do i.219.

(6) ~ Hants SU 5132. *Eastun* [871×7]12th S 1275, *(to) eastune* [963×75]12th S 827, *Estune* 1086, *Eston(e)* 1167–1341. The purported earliest reference is *to Eastuninga mearce* 'to the boundary of the *Eastuningas*, the people of Easton' *[825]12th S 273. Easton lies E of the Worthy estates of King's and Headbourne WORTHY SU 4933. Ha 69, Gover 1958.69.

(7) ~ Lincs SK 9226. *Estone* 1086, *-ton* 1174–1548, *Easton* from 1601. The village lies E of Stoke Rochford. Perrott 176.

(8) ~ Norf TG 1310. *Estone* *[1044×7]13th S 1055, *Estuna* 1086. E of Honingham. DEPN.

(9) ~ Somer ST 5147. *Eastun* *[1065]18th S 1042, *Easton* 1817 OS. E of Westbury ST 5048.

(10) ~ Suff TM 1858. *Estuna* 1086, *Eston(')* 1219–1336, *Easton* 1610. DEPN, Baron.

(11) ~ GREY Wilts ST 8887. 'E held by the Grey family'. *Eston Grey* 1289, *Aston(e) Gray, Grey* 1311–1460. P.n. *Estone* 1086 + manorial addition from the family name of John de Grey who held the estate in 1243. The most easterly village in Dunlow Hundred. Wlt 76.

(12) ~ HILL Wilts SU 2159. *EastonHill* 1817 OS. P.n. Easton as in EASTON ROYAL SU 2060 + ModE **hill**.

(13) ~ -IN-GORDANO Avon ST 5175. *Eston in Gordon* 1293, *~ in Gorden* 1330. Earlier simply *Eastun* *[1065]c.1500 S 1042, *Estone* 1086. Contrasts with WESTON-IN-GORDANO ST 4474. For the affix see CLAPTON-IN-GORDANO. DEPN.

(14) ~ MAUDIT Northants SP 8858. 'E belonging to the Mauduit family'. *Estonemaudent* 1298, *Estonmauduyt* 1377. Earlier simply *Estun* 12th ASC(E) under year 656, 1239, *Eston(e)* 1086–1389. E with reference to Whiston or Denton or Yardley Chase in general. Held by John Maled', Latin *Maledoctus*, in 1166 whose family-name was also spelt *Maudu(i)t* in the 13th cent. Nth 190.

(15) ~ ON THE HILL Northants TF 0104. *Easton on the hill* 1791. Earlier simply *Eston(e)* 1086 etc. E for distinction from COLLYWESTON SK 9902. The village is situated on the brow of a hill. Nth 201.

(16) ~ ROYAL Wilts SU 2060. *Estone* 1086, *-ton(e)* 1175–1324, *Aston* 1242, 1282, *Eastone* 1522. 'Royal' is a common modern addition probably because the east half of the parish was in the King's forest of Savernake. Wlt 345.

(17) Crux ~ Hants SU 4256. 'Croc's Easton'. *Eston Croc* 1242, *Crockes Estone* 1307, *Crux Easton* 1607. OE pers.n. *Crōc*, genitive sing. *Crōces*, + p.n. *Eastun* [801]13th S 268, [821]12th S 183[†], *(æt) Eastune* [961]12th S 689, *Estune* 1086. The estate was held by Croc the hunter in 1086. It lies in a detached part of Hurstborne hundred E of the main block. *Crux* is a late development influenced by Latin **crux** 'a cross'. Ha 62, Gover 1958.152.

(18) Great ~ Leic SP 8493. *Easton Magna* 1619, *Great Easton* 1717. ModE **great**, Latin **magna** + p.n. *Estun* [1041×57]12th ECNENM 20, *Estone* 1086, *Astuna* mid 12th, *Eston(e)* 1146–1548, *Easton* 1576. The village lies 3 miles E of Medbourne. Lei 214.

(19) Ston ~ Somer ST 6235. 'Stony Easton'. *Stonieston* 1230, *Stonyeston* [1348]B, *Stonaston* 1610. ME adj. **stoni** + p.n. *Estone* 1086. DEPN.

Great EASTON Essex TL 6125. *Great Eyston* 1556. ModE adj. **great** + p.n. *E(i)stanes* 1086, *Eystanes* 1267–1373, *Eastuna* 1119, *Estunam -tune -ton(a)* 1121–1485, *Ei- Eystan(e)* 1141–1428, *Eyston* 1219–1556, *Easton, Aston* 1669, 'Æga's stone(s)', OE pers.n. *Æga* + **stān**, pl. **stānas**, early assimilated to the common p.n. type EASTON, ASTON. 'Great' for distinction from Little EASTON TL 6023. Ess 484.

Little EASTON Essex TL 6023. *Little Eyston* 1556, earlier *Parva* 1303, *Petyt* 1272. ModE adj. **little**, Latin **parva**, Fr **petit**, + p.n. Easton as in Great EASTON TL 6125. Ess 484.

[†]Identification not absolutely certain.

EASTREA Cambs TL 2997. 'Eastern island' in relation to Whittlesey. *Estereie* [c.1020]12th LibEl, *Estrey(e)* 1285–1459. OE **ēasterre** + **ēġ**. Cf. WESTRY TL 3998. Ca 259, L 37, 39.

EASTRINGTON Humbs SE 7930. 'Farm or village of the easterners'. *Eastringatún* [959]c.1200, *Estrin(c)ton* 1086, *Estrington -yng-* 1169–1579, *Eastrington* from 1583. OE folk-n. **Ēastringas* < **ēasterra** + **ingas**, 'the easterners, the people living E of Howden', genitive pl. **Eastringa*, + **tūn**. YE 246.

EASTRY Kent TR 3155. 'The eastern district'. *Éastorege* 805×7 S 1264 (with varant spellings *Easterege, Eoster(e)ge), Eastrœge* 825×32 S 1268, *Eastrege* [824]13th S 1266, *Eastrige* *[1006 for 1001]11th S 914, *Eastryge* 1042×66 S 1047, *Estrei(a)* 1086, *Eastreie* 12th. OE ***ēastor**, comparative form **ēasterra**, + ***ġē**. The word *ġē* is cognate with Go *gawi* and G *Gau*, and refers to an ancient administrative division of the early kingdom of Kent, perhaps consisting originally of the 4 lathes of Lyminge, Wye, Canterbury and Eastry itself. In S 128 the district is called a *regio, in regione eastrgena* 788, in which *eastrgena* appears to be the genitive pl. of a word meaning 'the people of Eastry'. Traces of a western *ġē* may exist in the p.n. Wester TQ 7450 although the evidence is very late, *Westerey* 1254, *West(e)re, Westry* 1278, *West(e)re* 1292–1491. KPN 73, PNK 139, Origins 69–74.

EAST-THE-WATER Devon SS 4626. The name of that part of Bideford on the east bank of the river Torridge, on the opposite side from the centre of town. Settled at least since the 15th cent. D 88.

EASTHORPE Essex TL 9121. The 'east dependent settlement'. *Estorp* 1086–1228, *Est(t)horpe* 1218–1328. OE **ēast** + **thorp**. E of Blackwater. Ess 388 gives pr [i:stəp].

EAST VILLAGE Devon SS 8405. Identical with *East Sandford* 1809 OS. So called for distinction from West Sandford SS 8102, *Westsaunford* 1347, and the village of SANDFORD SS 8202. D 413.

EASTVILLE Lincs TF 4057. *East Ville* 1812, *Eastville* 1824 OS. ModE **east** + **ville**. A modern coinage for one of several new villages created by Act of Parliament in 1812 on newly drained fenland, the former East Fen, *Estfen* before 1150–1296. E of MIDVILLE TF 3857. Cf. also FRITHVILLE TF 3150 and Westville TF 3051. Wheeler 229, Cameron 1998.

EAST WEAR BAY Kent TR 2537. *East Ware Bay* 1798, *East Wear Bay* 1819 OS. P.n. East Wear + ModE **bay**. Cullen.

EASTWELL Leic SK 7728. 'The eastern spring'. *Estewell(e)* 1086–1449, *Estwell(e)* 1086–1610, *Eastwell* 1603. The source of the river Devon. 'East' possibly for distinction from Caldwell SK 7825 or Holdwell SK 7323, or from one of the springs on the W side of the ridge overlooking the village. Lei 162, L 32.

EASTWELL PARK Kent TR 0147. *Eastwell Park* 1819 OS. P.n. Eastwell + ModE **park**. Eastwell, *Estwelle* 1086, 1100, also simply *Welle(s)* 13th cent., is the 'east spring', OE **ēast** + **wella**, contrasting with nearby WESTWELL TQ 9947. PNK 383.

EASTWICK Herts TL 4311. 'East farm'. *Esteuuiche* 1086, *Estwic -wik -wyk(e)* 1204–1318. OE **ēast** + **wīċ**. E of Stanstead. In 1342, 1391, it is *Estwyk atte Flore* with ME **flor** 'a floor' which sometimes refers to a Roman pavement. Hrt 191.

EASTWOOD Essex TQ 8588. 'East wood'. *Nestuda, Estuudā* 1086, *Estwode* 1232–1216×72. OE **ēast** + **wudu**. E of Prittlewell. The form *Nestuda* represents ME *atte nestwude* < *atten estwude* (OE *æt thǣm ēastwuda* 'at the east-wood'). Ess 181.

EASTWOOD Notts SK 4646. 'The east clearing'. *Estewic* 1086, *Estweit* 1165, c.1200, *Estwette* 1546 with variants *-wet -w(h)aite -weith -w(e)yt, Estthwet* c.1185, *-thweyt -thwayt(e) -thwait* 1223–1332, *Eastwood* from 1575. OE **ēast** + ON **thveit**. Nt 144, SSNEM 216.

EASTWOOD WYorks SD 9625. 'The east wood'. *(del) Estwode* 1286, *(del) Estewode* 1364, *East(e)wood* from 1572. OE **ēast** + **wudu**. E of Todmorden SD 9234. YW iii.175.

EATHORPE Warw SP 3969. 'The river hamlet'. *Ethorpe* 1232–1656, *Ethrop(e)* 1520–1656, *Ethrupp* 1741. OE **ēa** + **thorp**. The place lies in a bend of the r. Leam. Wa 129.

EATON 'River settlement or estate'. OE **ēa** + **tūn**. Sa i.114–5, L 20.

(1) ~ Ches SJ 8765. *Yei- Yeyton* c.1262, 1318, *Yay- Yaiton, Yeaton, Yat(t)on* 14th., *Ayton* 1365, *Eton* 1549, *Eaton alias Yayton* 1666. The place is on the river Dane. *Y-* spellings are due to stress-shift in the initial diphthong whereby *ēa* > *eá* > *ya* instead of *ēa* > *ē*. Che i.61.

(2) ~ Norf TG 2106. *Ettune -a* 1086, *Etona* 1147×9, *Eton(e)* 1232, 1254. On the river Yare. DEPN.

(3) ~ Notts SK 7178. *Æt- Etune, Etunae, Ettone* 1086, *Eton(a)* 1168–1524, *Eaton* 1793. The reference is to the r. Idle. Nt 50.

(4) ~ Oxon SP 4403. *Eatune* [815]12th S 166, *Eatun* [821]12th S 183, *Eatone* 1212. Brk 402, L 20.

(5) ~ Shrops SO 3789. *Eton(')* 1252–1291×2. The village lies on the r. Onny. DEPN, Sa i.115.

(6) ~ Shrops SO 5090. *Eton(')* 1227–1747, *Eaton* from 1631, ~ *under-Haywood* 1737. The village lies on a large stream now called variously Byne, Eaton or Lakehouse Brook. Sa i.117.

(7) ~ BISHOP H&W SO 4439. 'E held by the bishop' sc. of Hereford. *Eton Episcopi* 1341. P.n. *Etvne* 1086, *Eton(e)* 1241–1506, + Latin **episcopus**, genitive sing. **episcopi**, ModE **bishop**. He 80.

(8) ~ CONSTANTINE Shrops SJ 5906. 'E held by the Constantine family'. *Eatton Constentine* [1217]1285, *Eaton Constantine* 1730. P.n. *Etune* 1086, *Eton(')* 1235–1399×1413 + manorial addition from the *(de) Constentin* family who appear in connection with this manor in the first half of the 13th cent. The parish is bounded by the river Severn and the road S from Eaton leads to the only crossing between Buildwas SJ 6304 and Atcham SJ 5409. Sa i.115.

(9) ~ HALL Ches SJ 4161. *the haule of Eton* 1566, *Eaton Hall* 1656. P.n. *Eaton* [1043]17th S 1226, *Etone* 1086–1150, *Eton* 1178–1549, *Eaton* from 1313, + **hall**. The reference is to the 17th century Grosvenor house remodelled by Alfred Waterhouse for the first Duke of Westminster in 1870–3. Che iv.148, Pevsner-Hubbard 1971.207.

(10) ~ HASTINGS Oxon SU 2597. 'E held by the Hastings family'. *Eton' Willelmi de Hasting'* 1220, *Eton Hastinges* 1298. P.n. *Etone* 1086 + manorial affix from the Hastings family who held land here in the 12th and 13th cents. probably as descendants of the DB holder, Walter FitzOther, castellan of Windsor. Brk 364.

(11) ~ SOCON Cambs TL 1658. Properly *Eaton cum Soca* 'Eaton and its soke' 1645, cf. *Soka de Eton* 'Eaton soke' 1247. P.n. *Etone* 1086–c.1360, *Eaton* 1208, 1247, + OE **sōcn** 'a jurisdiction, a liberty'. Eaton lies on the Great Ouse. Bd 54.

(12) ~ UPON TERN Shrops SJ 6523. *Eton(')* c.1223–1255×6. The village lies on the river Tern. DEPN, Sa i.115.

(13) Castle ~ Wilts SU 1495. 'E with the castle'. *Castel(l) Eton* 1469, *Castle Eaton al. Eton Meysye* 1601. ME **castel** + p.n. *Ettone* 1086, *Etton(a)* 1195–1249, *Etune* 1086, *Eton(e)* 1228–1332. The village is situated on the river Thames. 'Castle', referring to the 12th cent. castle built on a Saxon earthwork at Kempsford SU 1595 to defend the crossing of the Thames, and for distinction from other Eatons. The manor was held in 1242 by Robert de Meysey. Wlt 23 gives former pr [jetən], cf. the form *Easte Yetton* 1442.

(14) Water ~ Oxon SP 5112. *Watereaton'* [c.1220×30, 1227×8]c.1425, *Water Eton'* [1227]c.1425. ME **water** + p.n. *Eatun* [864]11th S 210, [900 for 904]11th S 361, *Etone* 1086. 'Water' for distinction from WOODEATON SP 5312. O 266.

EATON 'Island settlement'. OE **ēġ-tūn** with OE **ēġ** 'an island, a dry place in marsh or moorland, hill jutting into flat land'. Frequently interchanges with **ēa-tūn** but ME *ei- ey-* and *ay* spellings are characteristic for OE *ēġ-tūn*. L 37.

(1) ~ Ches SJ 5763. *Eyton* 1240–1671, *E(i)ton* 1240, *Ayton* 1304–1724, *Eaton* from 1240. Situated at the end of a spur running down between two streams. Che iii.289 gives prs ['i:tən] and ['eitən].

(2) ~ Leic SK 7929. *(H)aitona* c.1130, *Ayton'* 1229, *Eiton(e)* early 13th–1539, *Eton(e)* 1216×72–1520, *Eaton* 1576. The village lies on a wedge of land formed by two branches of the river Devon. Lei 161.

(3) ~ BRAY Beds 9720. 'E held by the Bray family'. Earlier simply *Eitona* 1086, *Eton* 1241. The manor was granted to Sir Reginald Bray in 1490. Bd 121.

(4) Little ~ Derby SK 3641. *Little Eton* 1392, ~ *Eaton* 16th. Earlier simply *Detton* (sic) 1086, *Eyton(e)* 1232–1330, *Eaton* 1577. The village occupies a hill-spur site between rivers. 'Little' for distinction from Long EATON SK 4933. Db 457, L 36.

(5) Long ~ Derby SK 4933. *Long(e) Eyton(e)* 1288–1330, *Longeaton* 1577, 1610. Earlier simply *Aitune* 1086, *Eitun* 1176, -*ton(e)* 1220–1495. The settlement occupies a site between the river Erewash and an unnamed tributary. Db 458.

Great EAU Lincs TF 4484. ModE **great** + dial. **eau** 'a feeder of a drainage channel, a water-course', *A* c.1200, c.1210. Also called *Adic* '*A* ditch' c.1200, *Withern Eau* 1824 OS. *Eau* derives from ON *á* 'a river'; this became ME *ō* which was identified with Fr *eau*. 'Great' for distinction from Little Eau TF 4085, probably identical with *Le Seventowne Aa* 1430 in Saltfleetby. RN 140, Stenton 1926.120, 136.

River EAU Lincs SK 9098. *River Eau* 1824 OS. Dial. **ea** 'a river', **eau** 'a drainage channel', both ultimately from OE *ēa* partly replaced by ON *á*. RN 140.

EBBERSTON NYorks SE 8982. 'Eadbriht's settlement'. *Edbriztun(e)* 1086, *E- Ædbri(c)hteston* 12th cent., *Edbriston* 1185×95–1301, *Ebreston(a)* 1114×9–1359, *Eberston* 1316, 1408. OE pers.n. *Ēadbriht*, genitive sing. *Ēadbrihtes*, + **tūn**. YN 95.

EBBESBOURNE WAKE Wilts ST 9924. 'E held by the Wake family'. *Ebleburn' Simonis Wach* 1167, *Eblesburne Wake* 1285. P.n. *Eblesburna* *[826]12th S 275, *(æt) Eblesburnan* [902]12th S 1285, [947]12th S 522, [986]12th S 861, *Eblesborne* 1086, *Ebsborne* 1571, 'Ebbel's stream', OE pers.n. **Ebbel*, genitive sing. **Ebbeles*, + **burna**, + manorial addition from the family name of Simon Wach who held the estate in 1167. One of several estates along the course of the river EBBLE. Wlt 207.

River EBBLE Wilts SU 1126. Short for Ebblesbourne, 'Ebbel's stream', *(on) ebblesburnon, Ybbles burnan* *[672]12th S 229, *(on) Eblesburnan* *[825]12th S 275, *[905 for 931×4]12th S 393, [947]12th S 522, [957]12th S 640, *(innan) Ebbeles burnan* [955]14th S 582, *(on) Yblesburnan* *[948]12th S 540, *Ebelesburne* 1288, *Ebbelesborne* 1289. OE pers.n. **Ebbel*, genitive sing. **Ebbeles*, + **burna**. Also known as *Chalkbourn* c.1540 from Broad CHALKE SU 0325. Wlt 7, RN 140.

EBCHESTER Durham NZ 1055. 'Ebba or Ebbe's Roman fort'. *Ebbecestr'* 1230, *Ebecestre* 1242×3, *Ebcestre* c.1332×45, *-chestr(e)* 1384–1533, *Ebchester* from [1153×94]1349. OE per.n. *Ebba* or *Ebbe* feminine + **ċeaster**. The Roman fort is that of *Vindomora* 'bright waters', a name which did not survive. The medieval tradition that St Ebba, the daughter of king Ethelfrith of Northumbria and sister of kings Oswald and Oswi, the foundress of the monastery at Coldingham and remembered in the name *St Abb's Head*, also founded a monastery at Ebchester, is almost certainly a late rationalisation of the p.n. NbDu 71, RBrit 502.

EBFORD Devon SX 9887. 'Ebba's enclosure'. *Ebworthy* 1465, *Ebworthie* 1525, *Ebbeford* c.1630. OE pers.n. *Ebba* + **worthiġ**. The modern form with substitution of *ford* for *worthy* is probably due to the fact that the Clyst is fordable here at ebb tide. D 601.

EBRINGTON Glos SP 1840. 'Eadbriht's estate'. *Bristentune* 1086, *Edbricton'* 1200–1248 with variants *-brich-*, *-brigh-*, *Ebrihton* *-yh- -igh- -ygh-* 1274–1692, *Ebrington* from 1382. OE pers.n. *Ēadbeorht -briht* (+ **ing**[4]) + **tūn**. The dedication of the church is to St Eadburg but this saint's name is incompatible with forms of the p.n. Gl i.242 gives pr ['jæbətən].

EBURACUM NYorks SE 6052. The Roman legionary fortress and *colonia* at YORK. Ἐβόρακον (Eburacum) [c.150]13th Ptol, *Eburacum* [4th]8th AI, [c.670]13th Rav, [c.731]8th BHE, *ab Eboraci* 237, *Col(oniae) Ebur- Ebor(acensis)* Inscrs, *Eboraci* (locative) 210 *Codex Justinianus*, 4th Aurelius Victor, Eutropius, *Historia Augusta*, St Jerome *Interpretatio Chronicae Eusebii*, *Eboracum* c.417 Orosius, before 519 Cassiodorus, *Eburaci -or-* (locative) c.450 Prosper Tiro. Brit p.n. **Ebŭrācon*, either the 'place abounding in yews', Brit ***ebŭro-*** 'yew', cf. OIr *ibar* 'yew-tree', or some other plant such as ModW *efwr* 'cow-parsley, hogweed', Breton *evor* 'bourdaine, hellebore', + suffix ***-āco-***; or 'the estate of Eburos', cf. Gaulish pers.n. *Ebŭros* identical with OIr *Ibar*, OW *Ebur*, MW *Efwr*, + suffix ***āco-***. RBrit 355–7, YE 275–80, CPNE 96.

ECCHINSWELL Hants SU 5060. Partly uncertain.

I. *Eccleswelle* 1086, *Echeleswelle* 1256.

II. *Echeneswolle* (for *-welle*) 1172, *-well(e)* 1186–1327, *Egeneswel* 1176, *Icheneswelle* 1379, *Itchingswelll* 1817 OS, *Echinswell* 1573. To the first group of forms must be added the name of a tributary here of the Enborne, *(æt) ec(e)lesburna* [931]12th S 412. These are either 'Eccel's spring and stream', OE pers.n. **Eċċel*, genitive sing. **Eċċeles*, + **wella, burna**, or 'spring and stream of the *Ecel*', an unrecorded r.n. possibly related to Latin *agilis* 'swift, agile'. The later *-n-* forms may have been influenced by the r.n. ITCHEN. Ha 70, Gover 1958.145.

ECCLES 'Church, Celtic Christian centre'. PrW ***eglēs***.

(1) ~ GMan SJ 7798. *Eccles* from c.1200 with variants *-is*, *Hecles*, *Hekkeles*, *Eck(el)les* 13th, *ecils* 1590. La 37, Jnl 1.47, Cameron 1968.87, TC 88.

(2) ~ Kent TQ 7360. *(of) æcclesse* c. 975 B 1322, *Aiglessa* 1086, *Eccles* from 1208. KPN 305, Cameron 1968.87.

(3) ~ ROAD Norf TM 0190. The name of a station on the Thetford to Norwich railway. P.n. *Eccles* from 1086, *Ecclis* 1254, later with ModE **road**. DEPN, Cameron 1968.87

River ECCLESBOURNE Derby SK 3244 > ASHFORD Surrey TQ 0771.

ECCLESFIELD SYorks SK 3594. 'Open country at *Eccles*'. *Eclesfelt -feld* 1086–1291, *Ecclesfeld* c.1155–1532, *Eg(g)lesfeld(e)* 1161–1528. PrW ***eglẹ̄s*** 'a Celtic Christian centre or church community' probably used as a p.n. + OE **feld**. YW i.244, Evidence 3, Jnl 1.47, L 243.

ECCLESHALL Staffs SJ 8329. 'Nook at a Celtic Christian centre'. *Ecleshelle* 1086, *Eccleshale* 1227, *-hall* 13th. PrW ***eglẹ̄s*** + **halh** probably in the sense 'land not included in the general administrative arrangement of a region'. DEPN, Cameron 1968.88, Jnl 1.47, Signposts 97, Thomas 1980.262–5, L 109, Horovitz.

ECCLESTON 'Settlement at or with a British Church'. PrW ***eglēs***, perhaps used as a p.n., + **tūn**.

(1) ~ Ches SJ 4162. *Eclestone* 1086, *Eccleston(a)* from c.1188 with variants *Ec(c)lis- (H)Ecles- Ekles-* etc., *Egleston -is-* 1506. Che iv.151, Cameron 1968.88, 89.

(2) ~ Lancs SD 5116. *Aycleton* 1094, *Ecclestun* c.1180, [before 1212]1268, *Etcheleston* c.1190, *Ekeles- Eckeles- Ec(c)lis- Ec(k)leston* 13th, *Eccleston -is-* 14th cent. If this is the church settlement of a larger estate, Leyland would be the estate in this case. La 131, Cameron 1968.88, Jnl 17.74, 21.34.

(3) ~ Mers SJ 4895. *Ecclestun* 1190–c.1220, *Eccliston* before 1220, 1243, *Ec(c)leston* 1246–1332 etc. La 108, Cameron 1968.88, Jnl 17.61.

(4) Great ~ Lancs SD 4340. *Great Ecleston* 1285, ~ *Eccleston* 1296, *Magna Eccleston* 1346. Adj. **great**, Latin **magna**, + p.n. *Eglestun* 1086, *Eccliston* 1212, 1243. 'Great' for distinction from Little Eccleston SD 4139, *Parua Eccliston* 1261, *Little Eccleston*

1331. There is no record of an old church but the name may signify the church settlement of a larger estate: in this case it would be Amounderness. La 161, 154, Cameron 1968.88, Jnl 17.93, 21.34, Thomas 1980.262ff.

ECCLESWALL COURT H&W SO 6523 → ASHFORD Surrey TQ 0771.

ECCUP WYorks SE 2842. 'Ecca's valley'. *Echope* 1086, *-hopa* 12th cent., *Ecop(p), Ecopa, Eccope* 1150×60–1625, *Ecupe* 1590, *Eccupp* 1620. OE pers.n. *Ecca* + **hop**. YW iv.190, L 115, 116.

ECKINGTON Derby SK 4279. 'Farm, village, estate called after Ecca or Ecci'. *(æt) Eccingtune* [c.1002]c.1100 S 1536, *Echintune* 1086, *-ton(e) -yn-* 1188–17th, *Eckenton(e)* 13th cent., *Ekyngton(e) -ing-* 1306–16th, *Eckington* from 1577. OE pers.n. *Ecca* or *Ecci* + **ing**[4] + **tūn**, possibly varying with *Eccan* + **tūn** in which *Eccan* is the genitive sing. of *Ecca*, 'Ecca's farm or village'. Db 247.

ECKINGTON H&W SO 9241. 'Estate called after Ecca or Ecci'. *Eccyncgtun* 972 S 786, *Aichintvne* 1086, *Ekinton -yn-* 1233–97, *Ekkyngton* 1542. OE pers.n. *Ecca* or *Ecci* + **ing**[4] + **tūn**. Wo 195.

ECTON Northants SP 8263. 'Ecca's farm or village'. *Echentone* 1086, *Eche- E(c)keton -i-* 1164–1428, *(H)ekinton* 1175–90, *Ekyngton* 1355, *Ecton* from 1585. OE pers.n. *Ecca*, genitive sing. *Eccan*, + **tūn**. Nth 138.

EDALE Derby SK 1285. 'Island valley'. *Aidele* 1086, *Ei- Eydale* 1275–1529, *Edall* 1550–late 18th, *Edale* 1732. OE **ēġ** + **dæl** (ON *dalr*). Edale stands on Grinds Brook and several small streams enter the brook around the village giving it an island character. Db 87, L 36–7, 95–6.

EDALE CROSS Derby SK 0786. *Edowe Crosse* 1640. P.n. EDALE SK 1285 + ModE **cross**. The reference is to a boundary cross. Db 88.

EDBURTON WSusx TQ 2311. 'Eadburg's estate'. *Eadburgeton* 12th, *-tun* c.1246, *Adburg(h)ton* 1261, *Ebberton, Abburton* 14th. OE feminine pers.n. *Ēadburh*, genitive sing. *Ēadburge*, + **tūn**. Sx 206 gives pr [æbətən], SS 75.

EDDISBURY HILL Ches SJ 5569. P.n. Eddisbury + ModE **hill**. Eddisbury, *æt Eades byrig* 914 ASC(C), *Edesberie* 1086, *-bury -buri* 1331–1666, *Eddisbury* from 1404, is 'Ēad's fortified place', OE pers.n. *Ēad*, genitive sing. *Ēades*, + **byriġ**, dative sing. of **burh**. The reference is to the Iron Age hill-fort on Eddisbury Hill, garrisoned by Æthelflæd, lady of the Mercians, in 914, subsequently capital of Eddisbury Hundred and chief lodge of the Forest of Delamere. Che iii.213 gives prs ['edizbəri] and older local ['edzbəri] and ['edʒbəri].

EDDYSTONE ROCKS Corn SX 3834. P.n. Eddystone + ModE **rocks**. Eddystone, *Ediston* 1405, *Edestone* 1478, *Ideston* 1587, is ModE **eddy** 'a small whirlpool' + **stone**. The reference is to water impeded and breaking round the rock. The first reference is to a shipwreck here and provides evidence for the word *eddy* earlier than the first independently recorded instance c.1450. PNCo 80.

River EDEN Kent TQ 4645. *Eden* 1577, 1586. A back-formation from EDENBRIDGE TQ 4446. The original name of the river was probably Avon, *(on norþan) auene* [814]10th S 175, PrW ***aβon** 'river'. KPN 137, RN 143, 22.

River EDEN. A British river n. PrW **Idon* from Brit **Itunā*, with Brit loss of initial *p*, from the root seen in Skr *pitú-* 'sap', Gk πίτυς 'pine', πιδύειν 'gush', πίδαξ 'spring', OE *fǣtte*, ON *feitr* 'fat', ON *fit* 'wet meadow'. PrW **Idon* would have undergone back-mutation in OE to produce OE **Iodon*, **Eodon* > ME *Eden*. RN 142.

(1) River ~ Cumbr NY 5830. Ἰτούνα *(Ituna)* [c.150]13th Ptol, *Eden* from [1130]n.d. with variants *Edene -a, Eaden* 1589. Cu 12, RN 142.

(2) CASTLE ~ Durham NZ 4237. 'Eden with the castle'. *Castelleden* [1239]c.1410–1561, *Casteleden(')* 1248–1472, *Castle Eden* from 1561. ONFr **castel** + p.n. Eden. Robert de Brus had a castle here c.1150. There were originally two Edens, *duas Geodene* [c.1040]12th. Castle Eden is identical with *Ioden australis* [c.1040]12th 'south Eden' for distinction from the lost village of *Yoden* NZ 4341, *Iodene* 1104×7, *Edena -e* 12th cent-1242×3, *Eden* from c.1170. Both ultimately derive from the name of the stream that runs through Castle Eden Dene. NbDu 71, LHEB 554, 578, 673, RBrit 380.

EDENBRIDGE Kent TQ 4446. 'Eadhelm's bridge'. *Eadelmesbrege* c.1100, *Edelmesbrugg' -brig' -breg'* 1226–1279, *Edelmebrigg* 1214, *Edelnebrigg'* 1221, *Edenebrigge* 1314. OE pers.n *Ēadhelm*, genitive sing. *Ēadhelmes*, + **brycg**, Kentish **brecg**. The r.n. EDEN TQ 4645 is a 16th cent. back-formation from Edenbridge. PNK 73, KPN 137, RN 143.

EDENFIELD Lancs SD 8019. 'Open land by *Ayton*'. *Aytounfeld* 1324, 1443, *Ayten-* 1509, *Aton-* 1579, *etenfelde* 1591, *edenfeld* 1615. Lost p.n. *Ayton* + **feld**. *Ayton* is probably OE **ēġ** + **tūn** 'island settlement' referring to a patch of dry ground surrounded by marsh near the river Irwell. La 64, L 243.

EDENHALL Cumbr NY 5632. 'Nook of land beside the river Eden'. *Edenhal'* 1159–1212, *-hall* from 1313, *Ednale, Ednell* 15th cent., *Eednell* 1606. R.n. EDEN NY 5830 + OE **halh**. Cu 190 gives old pr [i:dnəl], L 108–9, 111.

EDENHAM Lincs TF 0621. 'Eada's homestead'. *Ed(en)eham* 1086, *Edenham* from 1125×30, *Edingham* 1316, *Ednam(e)* 1466–1607. OE pers.n. *Ēada*, genitive sing. *Ēadan*, + **hām**. Perrott 177.

EDENHOPE HILL Shrops SO 2588. P.n. *Etenehope* 1086, *Edenehope* 1284, *Edenhope* 1272, 1641, *Ednop(p)* 1577–1672, 'Eada's valley', OE pers.n. *Ēada*, genitive sing. *Ēadan*, + **hop**, + ModE **hill**. Sa i.117 reports the local pr as in the sp *Ednop*.

EDEN PARK GLond TQ 3868. P.n. *Eden* 1819 OS + ModE **park**. An estate acquired by William Eden, first Lord Auckland, in 1807. LPN 73.

EDENSOR Derby SK 2570. 'Eadin's ridge'. *Edensoure* 1086, *Edenesou(e)r(e) -hou(e)re -or(e)* 1199–1376, *Edins- Ednis- Ednes-* 1154×59–late 17th, *Edynso(u)re* 1453, 1547, *Edinsor* 1577, *Ensoure* 1567. OE pers.n. **Ē(a)dīn*, genitive sing. **Ē(a)dīnes*, + **ofer**. The settlement was moved here from its original site in the 19th cent. to improve the view from Chatsworth. Db 90 gives pr [ensə] as in the surname *Ensor*. L 175–6.

EDENTHORPE SYorks SE 6106. A modern name containing the surname *Eden* replacing earlier STREETTHORPE. YW i.22.

EDGBASTON WMids SP 0584. 'Ecgbald's estate'. *Celboldestone* (sic) 1086, *Eg(e)baldestone* 1184–1262, *Eg(e)baston* 1279–1553, *Edgebaston* 1553. OE pers.n. *Ecgbald*, genitive sing. *Ecgbaldes*, + **tūn**. The DB form is either corrupt or evidence of an earlier form of the name, 'Ceolbald's estate'. Wa 45 gives prs ['edʒbəstən] and [edʒ'ba:stən].

EDGCOTT Bucks SP 6722. Partly uncertain. *Achecote* 1086–1461, *Echecote* 1162–1535, *Edgecote* 1598. The generic is OE **cot** 'a cottage', pl. **cotu**, the specific either OE adj. **ǣċen** 'made of oak', or pers.n. **Æċċa*. Bk 54.

EDGE Shrops SJ 3908. 'The escarpment'. *Egge* 1255–1308, *Edge* from 1555. OE **ecg**. The village lies at the E tip of a long narrow summit. Sa ii.36.

EDGE END Glos SO 5913. Cf. MILE END SO 5911.

EDGE HILL Mers SJ 3689. *Edge Hill* 1842 OS. ModE **edge** 'an escarpment' + **hill**.

EDGE HILL Warw SP 3847. 'Hill called Edge'. *Edg(e)hill* 1656. P.n. *(le) Hegge* c.1250, 1275, *(le) Egge* c.1270, 1272, OE **ecg** 'a steep slope' used as a p.n. + ModE **hill**. The reference is to the steep edge of the long narrow ridge stretching from SP 3646 to SP 3948. Wa 12.

EDGEBOLTON Shrops SJ 5721. 'Ecgbald's homestead or village'. *Egebaldesham* c.1190×98, *Egge-* c.1230×35, *Egebalde(n)ham* c.1220×40–1292×5, *Eg-* 1291×2, 1377×99, *Egebaldon'* 1291×2, *Edgebalton* 1577, *Edgbolton* 1645, 1687. OE pers.n. *Ecgbald*, genitive sing. *Ecgbaldes*, varying with *Ecgbalda*, genitive sing. *Ecgbaldan*, + **hām**. With the weakening of the generic under weak stress to [ən] reinterpretation of the name as a compound in *-ton* took place. Sa i.118.

EDGEFIELD Norf TG 0934. 'Open land with enclosed pastures'. *Edisfelda* 1086, *Edichfeld* 1191, *Edesfeld* 12th, *Egesfeld* 1189×99, 1197. OE **edisc** + **feld**. DEPN, L 159, 242, 244.

EDGEFIELD STREET Norf TG 0933. 'Edgfield hamlet'. P.n. EDGEFIELD TG 0934 + ModE **street**. Otherwise known as *Ramsgate Street* 1838 OS.

EDGEWORTH Glos SO 9406. 'Ecgi's enclosure'. *Egesworde, Egeiswurde* 1086, *Egesw(u)rth(e)* 1222–85, *Egewurð -w(u)rth* 1138–1236, *Egge(s)worth(e)* 1220–1427, *Edgeworth* 1539, 1602. OE pers.n. *Ecgi*, genitive sing. *Ecges*, + **worth**. The specific has been assimilated to ME **egge** (OE *ecg*) 'an edge' owing to the situation on one of the steep slopes of the upper Frome valley. Gl i.128.

EDGMOND MARSH Shrops SJ 7120. *Edgmond Marsh* 1833 OS. P.n. EDGMOND SJ 7129 + ModE **marsh**.

EDGMOND Shrops SJ 7219. 'Ecgmund's hill'. *Edmendune* 1086, *-dona* c.1090, 1138, *Eadmundona* 1121, *Eigmendona, Egmundun'* 1155, *Egmendon' -dune -ton'* 1167–1306×7, *Egmundon -don'* 1227–1316, *Eggemendon'* 1177, 1230, *-munden* 1317, *-mundoun* 1401, *Egmonde* 1535, *Edgemonde* 1549, *Edgemond alias Edgmonden* 1601. OE pers.n. *Ecgmund* + **dūn**. Some early sps suggest that the name was sometimes believed to contain the commoner pers.n. *Ēadmund*. The loss of the final element *-don*, which is characteristic of Shrops, is probably due to the influence of the stress pattern of W names, in which the stress falls on the penultimate syllable. Cf. Child's ERCALL SJ 6625 and FITZ SJ 4417. Sa i.119, Nomina 11.107.

EDGTON Shrops SO 3885. 'Ecga's hill'. *Egedune* 1086, *-dun' -don(')* 1232–1431, *Egge-* 1232–1346, *Egeton'* 1272, *Egge-* 1316, 1327, *Edgeton* 1577, *Edgton* from 1667, *Edgon, Edgedon, Edgs(t)on, Edson* 17th cent. OE pers.n. *Ecga* + **dūn**. Sa i.121.

EDGWARE GLond TQ 2092. 'Ecgi's weir or fishing pool'. *Ægces wer* 972×8 S 1451, *Eggeswere* 1176–1411, *Egg(e)were* 1270–1389, *-ware* 1310, 1475, *Edgware* 1495. OE pers.n. *Ecgi*, genitive sing. *Ecges*, + **wer**. The location of the weir was at the point where Edgeware brook (formerly the *Stanburna* 'the stony stream') crosses the High Street. Mx 50, GL 46.

EDGWORTH Lancs SD 7416. 'Enclosure beneath the escarpment'. *Eggewrthe* 1212, *-worth* 1276–1332. OE **ecg** + **worth**. Edgworth is situated at the foot of Edgworth Moor which rises steeply to a height of over 1500ft. Alternatively the specific could be the OE pers.n. *Ecga*. La 47, Jnl 17.38.

EDINGALE Staffs SK 2112. Partly uncertain. *Ednunghal(l)e* 1086, *Edeling(e)hale* 1208, *Ederingehal* 1181, *Edenynghale* c.1170, 1272 etc. The variety of forms is due to AN confusion of the liquids *l-r-n*. Probably 'the nook of the Edeningas, the people called after Edin or Eadwine', OE folk-n. **Ēdeningas* < pers.n. **Ēdīn* or *Ēadwine*, + **ingas**, genitive pl. **Ēdeninga*, + **halh** in the sense 'land in a river bend' or 'promontory into marsh'. DEPN, L 105, 110.

EDINGLEY Notts SK 6655. 'Eddi's clearing'. *Ed(d)yngleia* [c.1180]14th, *Eddingley(e)* 1270–1392, *Eding-* from 1302. OE pers.n. *Eddi* + **ing**[4] + **lēah**. Nt 160.

EDINGTHORPE Norf TG 3232. 'Eadgyth's outlying farm'.

I. *Ædidestorp* 1177, *Edith(e)thorp(e)* 1234–1429, *Edythorp(e)* 1250–1535.

II. *Edinest(h)orp -ynes-* 1198–1347, *Edinetorp(e), Edynethorp* 1227–1324.

III. *Edyngthorp(e)* 1419–1548.

OE feminine pers.n. *Ēadġȳth*, genitive sing. *Ēadġȳthe*, + **thorp**. ME *Edithe* < *Ēadġȳth* is regularly confused with *Ediue* < *Ēadġifu*. This seems to have happened here with subsequent misreading of *u* as *n*. Nf ii.153.

EDINGTON Somer ST 3839. 'Eadwine or Eadwynn's estate or farm'. *Eduuinetone* 1086, *Edinton, Edingtone* 1243, *Edington* 1610. OE pers.n. *Ēadwine* or *Ēadwynn* feminine, genitive sing. *Ēadwynne*, + **tūn**. DEPN.

EDINGTON Wilts ST 9253. 'Etha's hill'. *(æt) Eðandune* [880×5]11th S 1507, *Eþandun* 891 ASC(A) under year 878, *Eðandun* 957 S 646, *Ethandun* [894]c.1000 Asser, *Edendone* 1086, *Edindun -don(e)* 1236–97, *Edyngdon(e)* 1291–1512, *Edyngton* 1428, *Eddington* 1546. OE pers.n. *Ētha*, genitive sing. *Ēthan*, + **dūn**. An alternative possibility is OE *(æt thǣm) ēthan dune* '(at the) bare downland' from the rare OE adj. *ēthe* 'waste, bare, uncultivated', referring to the chalk downs below which the village lies. Wlt 140, L 149.

EDITHMEAD Somer ST 3249. *Eddy Mead* 1791 Collinson, *Edymead* 1809 OS. The generic is ModE **mead** (OE *mǣd*) 'a meadow', but no suggestion can be made for the specific without earlier forms.

EDLESBOROUGH Bucks SP 9719. 'Eadwulf's hill or barrow'. *Edulf(u)esberga* 1175, 1185, *Aedulf(u)esberga* 1181–9, *Edelesbergh* 1227, 1247, *Edgeborough* 1716[†]. OE pers.n. *Ēadwulf*, genitive sing. *Ēadwulfes*, + **beorg**. Bk 91 gives pr [edʒbərə].

EDLINGHAM Northum NU 1109. 'The homestead at or called *Eadwulfing*, the place called after Eadwulf'. *Eadwulfincham* [c.1040]12th HSC, *Eadulfingaham* c.1130 SD referring to year 737, *E(a)dulf- Eduluing(e)ham* 12th cent., *Edelingham* 1236, 1242 BF, 1122, *Edlincham* 1296 SR, *Edlyngeham* 1346. OE p.n. **Ēadwulfing* < pers.n. *Ēadwulf* + **ing**[2], locative-dative sing. **Ēadwulfinġe*, + **hām**. Pronounced [edlindʒəm] from the assibilated locative-dative form. NbDu 71, ING 157, BzN 1967.370.

EDLINGTON Lincs TF 2371. 'The estate called after Edla or Eadulf'. *Ellingetone* 1086, *(H)edlingtuna* c.1115, *Edlington(')* from 1202 including [1125]13th. OE pers.n. **Ēdla* or *Ēadulf* < *Ēadwulf* + **ing**[4] + **tūn**. DEPN, Cameron 1998.

New EDLINGTON SYorks SK 5498. A modern coal-mining locality named from (Old) Edlington SK 5397, *Eilin- Ellintone* 1086, *Edelington'* c.1195–1316, *Edlingtin -ton -yng-* c.1200–1641, 'estate named after Edla', OE pers.n. **Ēdla* + **ing**[4] + **tūn**. YW i.129.

EDMONDSHAM Dorset SU 0613. 'Eadmod's homestead or enclosure'. *(A)medesham* 1086, *Æd- Edmodesham* 1176–1303, *Edmundesham -mond-* 1195–1438, *Edmonsam* 1536, *-mond-* 1618, *Ensom* 1664. OE pers.n. **Ēadmōd* later replaced by the commoner *Ēadmund*, genitive sing. **Ēadmōdes*, + **hām** or **hamm**. Occasional spellings *Age- Au(g)modesham* 1177–1244 point to a third form with pers.n. *Ealhmōd*. Do ii.217 gives prs ['edmənʃəm] and ['enʃəm].

EDMONDSLEY Durham NZ 2349. 'Edemann's clearing'. *Edemannesleye* c.1190, *Edman(ne)slege -ley* [1183]c.1382, 1242–1421, *Edmundesle(y)* [1183]c.1320, 1338, *Edmondesley* [1183]c.1400, 1418. OE pers.n. *Edemann*, genitive sing. *Edemannes*, + **lēah**. *Edemann* may either represent OE ***ēdemann** 'a shepherd' (< OE *ēde* 'a flock of sheep' + *mann*) or possibly mean 'Eadu's man', OE pers.n. *Ēadu* fem., genitive sing. *Ēade*, + *mann*. By the early 14th cent. *Edemann* became confused with the much more common pers.n. *Edmund*. NbDu 72, DEPN.

EDMONDTHORPE Leic SK 8517. 'Edmund's outlying farm'. *Torp Edmundi* 1298, *Thorp(e) Edmun(d)* 1298–1610, *Edmon(d)thorpe* from 1487. Earlier 'Eadmer's outlying farm', *Edmerestorp* 1086, 1165, *Edmer(e)thorp* 1344–72, *Thorp(e) Ed(d)mer(e)* 1290–1617. ME pers.n. *Edmund* replacing OE pers.n. *Ēadmǣr* + **thorp**. The manor belonged to the fee of Edmund Crouchback, brother of Edward I in 1298. Lei 200, SSNEM 108.

EDMONTON GLond TQ 3493. 'Ealdhelm's estate'. *Adelmetone* 1086, *Edelmeton(e)* 1202–1550, *Edelminton -yn(g)-* 1211–1406, *Edmenton* 1369, *Edmonton* 1464. OE pers.n. *Ēadhelm* + **ing**[4] + **tūn**. Mx 67.

EDMUNDBYERS Durham NZ 0150. 'St Edmund's *Byres*'. *Edmindbires* (for *Edmund-*) [1183]c.1382, *Edmundbire*

[†]The DB form *Eddinberge* 1086 is aberrant.

[1183]c.1320, *Edmundbyr(e)s -bir(e)s* 1261×74–late15th, *Edmond-* 1369–1448, *Edmondbyers* 1580, *Edmund-* from 1644×5. Saint's name *Edmund* from the dedication of the church + OE **bȳre**, pl. **bȳras**, 'the cow-sheds', used as a p.n. as in BYERS GREEN NZ 2234. NbDu 72.

EDNO-VEAN Corn SW 5429 → PERRANUTHNOE SW 5329.

EDSTASTON Shrops SJ 5232. 'Eadstan's estate'. *Stanestune* 1086, *Edis- Edestaneston'* 1255×6, *Edenestanton'* 1291×2, *Edstanton* 1376, *Estaston* 1577, *Edstaston* from 1672. OE pers.n. *Ēadstān*, genitive sing. *Ēadstānes*, + **tūn**. Sa i.121.

EDSTOCK Somer ST 2340 → MARTOCK ST 4619.

EDSTONE Warw SP 1861. 'Eadric's estate'. *Edricestone* 1086, *Eadricheston* 1154×89, *Edricheston(e)* 12th–1288, *Edreston* 1379–1701, *Edeston* 1266, 1403, *Edston* 1530, *Edstone* 1538. OE pers.n. *Ēadrīċ*, genitive sing. *Ēadrīċes*, + **tūn**. Wa 243.

Great EDSTONE NYorks SE 7084. *Micheledestun* 1086. ModE **great**, ME **michel**, + p.n. *Edestun -ton* 1140–1285, *Edneston'* 1231, 'Eadīn's farm or village', OE pers.n. **Ēadīn*, genitive sing. **Ēad(ī)nes*, + **tūn**. YN 58.

Little EDSTONE NYorks SE 7184. *Parva Edestun* 1086. ModE **little**, Latin **parva**, + p.n. *Edenston'* 1167, as in Great EDSTONE SE 7084. YN 76.

EDVIN LOACH H&W SO 6658. 'Edvin held by the Loges family'. *Ydefen Loges* 1286, *Edven Loche* 1576. P.n. Edvin + family name Loges for distinction from EDWYN RALPH SO 6457. The church of St Giles at Edwin was given c.1158×63 to St Guthlac's priory, Hereford, by Gerwy de Loges whose family name is a common p.n. in Normandy. Edvin, *Edevent* 1086, *Gedesfenna* 1123, *G- Yedeuenna* 1148–67, *G- Yedefen(n)* 1160–1265, is 'Gedda's fen', OE pers.n. *Ġedda* + **fenn** with W voicing of initial [f] > [v]. He 80, L 41.

EDWALTON Notts SK 5935. 'Ealdwald's farm or village'. *Edvvoltone, Edwoltun* 1086, *-ton* 1298, *Edwalton* from 1227, *Eddolton, Edowlton* 17th. OE pers.n. *Ēadwald* + **tūn**. Nt 246 gives pr [edəltən].

EDWARDSTONE Suff TL 9442. 'Eadweard's estate'. *Eduardestuna* 1086, *Edwardston* 1610. OE pers.n. *Ēadweard*, genitive sing. *Ēadweardes*, + **tūn**. DEPN.

EDWINSTOWE Notts SK 6266. 'Edin's holy place'. *Edenestou* 1086, *-stoua -stow(e)* 1146–1330, *Ednestowe* 1230, *Edenstow(e)* 1275–1577 with occasional variants *Eddens-*, *capella Sci Edwini in Haya de Bircwrde* 'St Edwin's chapel within the fence of Birchwood' 1205, *Edwynstow* 1300. OE pers.n. **Edīn*, genitive sing. **Edīnes*, + **stōw**. The proximity of the chapel of St Edwin in Birklands (*Bircwrde*), a part of Sherwood Forest, probably caused the reformation of the name to its modern form. Nt 75, Introd i.171 fn.5, Studies 1931.7–8, BAR Brit 102.192.

EDWORTH Beds TL 2241. 'Edda's enclosure'. *Edeuuorde* 1086, *Eadwrthe* 1185, *Eddeworth* 1202, *Edworth* 1355. OE pers.n. *Edda* + **worth**. *Edda* is a pet form of names in *Ēad-*: Simeon of Durham mentions one *Edwine, qui et Eda dictus est* 'Edwine also called Eda' HR 63. Bd 104, Wa xlii, Redin 65.

EDWYN RALPH H&W SO 6457. 'Edvin held by Ralph'. *Yeddefen(n)e Radh* 1291, ~ *Rauf* 1349, *Edvyn Rauf* 1562. Earlier simply *Gedeuen* 1086. P.n. Edvin as in EDVIN LOACH with vocalisation of [v] > [w] + pers.n. Ralph as in Ralph de Yedefen, lord of the manor in 1176, or one of his synonymous successors. He 80.

EFFINGHAM Surrey TQ 1253. 'The homestead of the Æffingas, the people called after Æffa or Æffe'. *Epingehā* 1086, *Effingeham* 1180–1314 including [933, 1062]13th S 420, 1035, *Effing-* from 1283. OE folk-n. **Æffingas* < pers.n. *Æffa* or *Æffe* feminine, hypocoristic forms of names in *Ælf-* + **ingas**, genitive pl. **Æffunga*, + **hām**. The 1086 spelling (DB) is a misreading of insular minuscule <f> as <p>. Sr 102, ING 122, ASE 3.32.

EFFINGHAM JUNCTION STATION Surrey TQ 1055. 'The railway station at Effingham Junction'. P.n. EFFINGHAM TQ 1253 + ModE **junction** and **station**.

EFFORD Devon SS 8801. *Effords* 1809 OS.

EGERTON GMan SD 7114. *Egerton* 1843 OS. From surname *Egerton*, the family name of the Earls of Bridgewater who held property here. La 47, Room 1983.36.

EGERTON Kent TQ 9047. 'Estate called after Ecgheard'. *Eardingtun* c.1100, *Egarditton* 1202×3, *Egyardyng- E(gge)yardington* 1270, *Ad- Edgarinton* 1206, 1240, *Egerton* 1610. OE pers.n. *Ecgheard* + **ing**[4] + **tūn**. The spelling *Estiardinton* 1240, however, seems to be *est iardinton*, i.e. east + *Eardingtun* of c.1100 which appears to be a compound of **earding* < OE **eard** 'native place, dwelling place' + **ing**[2], + **tūn**, cf. ME *erdingstow*, lOE *eardungstōw* 'dwelling place'. K 392.

EGERTON FORSTAL Kent TQ 8946. *Egerton Fostal* 1819 OS. P.n. EGERTON TQ 9047 + Kent and Sussex dial. **fostal** 'a paddock near a farmhouse or a way leading to a farmhouse' (OE **foresteall*). PNE i.184.

EGGARDON HILL Dorset SY 5494. P.n. Eggardon + pleonastic ModE **hill**. Eggardon, *Giochresdone* 1086, *Jekeresdon* 1204, *Ekerdon* 1244, *Egerdone* 1316, is said to be 'Eohhere's hill', OE pers.n. *Eohhere*, genitive sing. *Eohheres*, + **dūn**. Eggardon Hill is one of the most spectacular and undamaged Iron Age hill-forts in Dorset. The specific is possibly, therefore, the name of an unknown mythical figure, perhaps connected with OE *ġeocor* 'full of hardship' related to Gothic **jiuka* 'anger' < IE **yewĝ-* 'be stirred up', **yewdh-* 'move vigorously, battle', Sanskrit *yudhmá-* 'warrior'. Do 70.

Low EGGBOROUGH NYorks SE 5623. ModE **low** + p.n. *Eg(e)- Eche- Acheburg* 1086, *Egeburgh* c.1185, 1234, *Egburg(h)* 1156–1593, *-browghe* 1552, 'Ecga's fortified place', OE pers.n. *Ecga* + **burh**. YW ii.57.

EGGESFORD STATION Devon SS 6811. A halt on the line from Exeter to Barnstaple. P.n. Eggesford + ModE **station**. Eggesford, *Egkeneford* 1238, *Eg(g)ene(s)ford* 1242–1428, *Eggesford* from 1377, is probably 'Ecgen's ford', OE pers.n. Ecgīn, *Ecgen*, genitive sing. *Ecgenes*, + **ford**. D 369 gives pr [egzvəd], Studies 1931.8.

EGGINGTON Beds SP 9525. 'Ecca's hill'. *Ekendon* 1195, *Egynton* 1304. OE pers.n. *Ecca*, genitive sing. *Eccan*, + **dūn**. Bd 121, PNE i.2.

EGGINTON Derby SK 2628. 'Estate called after Ecga'. *(æt) Ecgintune* [1012]14th S 928, *Eghintune* 1086, *Eg(g)inton(e) -yn- -en-* 1207–1524, *Ekentone* 1219, *-yng-* 1234, 1428, *Eg(g)ington(e) -yng-* 1303–1586. OE pers.n. *Ecga* + **ing**[4] + **tūn**. Db 459.

EGGLESCLIFFE Cleve NZ 4113. Probably 'church-community cliff'. *Eggescliva* 1155×89, *Eggasclif* before 1172, *Eggescliue* c.1175–1400 with variants *-clif -clyf -clyve*, *Eggisclife* 1554, 1625, *Eggleclif* c.1190, *Ecclescliue* 1197, *Eg(g)lescliue -clyf* 13th cent., *Eg(g)lisclif(f) -clyff* 1374–1527, *Egglescliffe* from 1605×6, *Eaglescliff(e)* 1639–1874. A difficult name. It has been claimed as a compound of PrW ***eglẹ̄s** 'a church, a church community' and OE **clif** 'a hill', but the earliest spellings do not obviously support this. On the other hand derivation from OE pers.ns. *Ecgi* or *Ecgel*, genitive sing. *Ecges*, *Ecg(e)les*, 'Ecgi or Ecgel's cliff', might have been expected to give a modern pr with *Edge-* rather than *Eg-* as in EDGEWORTH Glos SO 9406, *Ege(s)w(u)rth(e)* 1222–87, *Eggesw(o)rthe)* c.1230–1427, and EDGEWARE GtrLond TQ 2091, *Eggeswere* 1176–1411, *Eggewere* 1294–1389. The village is situated on a steep brow overlooking the r. Tees. Cameron 1968.88, Evidence 3.

EGGLESTON Durham NY 9923. 'Ecgwulf or Ecgel's farm or village' or possibly 'church-site settlement'. *Egleston(')* 1197–1620, *Egliston(')* 1313–1604, *Eggleston(')* from 1211, *Egeston* 1378. OE pers.n. *Ecgwulf*, *Ecgel*, genitive sing. *Ecg(e)les*, + **tūn**. The pr [eg-] rather than [edʒ-] is probably due to Scandinavian influence. But this might be an example of PrW ***eglẹ̄s** + **tūn**. NbDu 72.

EGGLESTON COMMON Durham NY 0027. *Eggleston Common* 1862 OS. P.n. EGGLESTON NY 9923 + ModE **common**.

EGGLESTONE ABBEY Durham NZ 0615. P.n. Egglestone + ME **abbey**. Egglestone, *Eghistun -ton* 1086, *Egleston* [1157]15th, 1208–1400, *Egliston* [c.1200]1399×1413, 1398–1421, *Eggleston* 1226, 1234, is 'Ecgi or Ecgel's farm or village', OE pers.n. *Ecgi* or *Ecgel*, genitive sing. *Ecg(el)es*, + **tūn**. The pr with [eg-] rather than [edʒ-] is probably due to Scandinavian influence but this might be another instance of Pr ***eglęs** + **tūn** 'church-site settlement'. YN 301.

EGHAM Surrey TQ 0171. 'Ecga's homestead'. *Egehā-* 1086, *Egeham* 1225–1453 including [672×4, 933, 967, 1062]13th S 1165, 420, 752, 1035, *Egge-* 1203–1342 including [1053×66]13th S 1094, *Egham* from 1201, *Eggen-* 1241, 1272. OE pers.n. *Ecga*, genitive sing. *Ecgan*, + **hām**. Sr 119–20, ASE 2.31.

EGLETON Leic SK 8707. 'Ecgwulf's estate'. *Egoluestun'* 1218, *Egiltun' -ton(') -el-* 1209–1461, *Egleston* 1319, 1334, *Egleton* from 1565. OE pers.n. *Ecgwulf*, genitive sing. *Ecg(w)ulfes*, + **tūn**. R 84.

EGLINGHAM Northum NU 1019. 'The homestead at or called *Ecgwulfing*, the place called after Ecgwulf'. *Ecgwulfincham* [c.1040]12th HSC, *Ecgwulf- Eguluing- Egwiluinge- Egulinge- Eggleuingeham* 12th cent., *Egling(e)ham* 1200–1346, *Eglinjham* 1596. OE p.n. **Ecgwulfing* < OE pers.n. *Ecgwulf* + **ing**[2], locative-dative sing. **Ecgwulfinğe*, + **hām**. Pronounced [eglindʒəm] from the assibilated locative-dative form. NbDu 72, ING 157, BzN 1967.370.

EGLOSHAYLE Corn SX 0072. '*Heyl* church'. *Egloshail* 1166, *Eg(g)losheil* 1201, c.1210, *-heyl* 1258, *Egloshayl -haille -hale* 1291–1428 Gover n.d., *Egleshall* 1610. Co **eglos** + p.n. *Heyl* 1284 Gover n.d., the former name of the Camel estuary, as in *Hægelmutha'* 11th, p.n. *Heyl* + OE **mūtha** 'mouth, estuary', and Hayle Bay SW 9379. *Heyl* is Co ***heyl** 'an estuary'. PNCo 81, DEPN.

EGLOSKERRY Corn SX 2786. 'Church of Keri'. *Egloskery* [c.1145]15th, 1291–1610, *Eglescheria* [c.1170]15th, *Egloskury* [c.1280]15th. Co **eglos** + saint's name *Keri*, one of the 24 children of Broccan and female. Nothing else is known of her. Cf. St Quiry in Plounévézel near Carhaix-Plouguer in Brittany. PNCo 81, Gover n.d. 143, DEPN.

EGMANTON Notts SK 7368. 'Agmundr or Ecgmund's farm or village'. *Agemuntone* 1086, *Egmanton* from 1244. ON pers.n. *Agmundr*, or OE pers.n. *Ecgmund* with Scandinavian [g] for English [dʒ], + **tūn**. The DB form may have AN *a* for OE *e*. Nt 50, SSNEM 190.

EGREMONT Cumbr NY 0110. 'Pointed hill'. *Egremont* from [c.1125]n.d., *Egremunt* [c.1135]12th–1281, *-mund* [c.1160]n.d.–1317, *Eggermuth* 1260. OFr **aigre** + **mont**. A feudal invention for the site of Egremont Castle, *castellum de Egremundia* [c.1125]15th, ~ *de Acrimonte* c.1205 (Latinised form). Identical with Fr p.n. Aigremont. Cu 379 gives occasional pr [egə'məθ/egə 'mət].

EGTON NYorks NZ 8006. 'Ecga's settlement'. *Egetune -ton* 1086, 13th cent., *Eggeton'* c.1170×95–1410, *Egton* 1285. OE pers.n. *Ecga* + **tūn**. YN 129.

EGTON BRIDGE NYorks NZ 8006. P.n. EGTON NZ 8006 + ModE **bridge**.

EGTON HIGH MOOR NYorks NZ 7701. P.n. EGTON NZ 8006 + ModE **high** + **moor**.

River EHEN Cumbr NY 0112. Unexplained. *Egre* [c.1125, c.1180]15th, *Eger* c.1203, *Eghene -a* [c.1160]15th, 1334, *Eghen* 1322, [1327]15th, *Ehen* from c.1205, *End* 17th cent., *Enn* 1794, *Eigne* [c.1170]15th, *Eygne* 1322, *Eyne* 1292–1539. The common early form *Egre, Eger* is irrelevant being a back-formation from or modification of the original name under the influence of the p.n. EGREMONT NY 0110. The original form must have been an OE *Ēgen(e)* of unknown origin. Cu 13, RN 143 gives pr [e:n].

EIGHT ASH GREEN Essex TL 9425. *Eight Ash Green* 1777. ModE **eight** + **ash** 'an ash-tree' + **green**.

ELBERTON Avon ST 6088. 'Ealdbriht's estate'. *Eldbertone* 1086, *Elbri(g)hton(a), Albric(h)ton* 1167–1248, *Ail- Aylberton(e)* 1248–1675, *Elberton als. Elverton* 1601. OE pers.n. *Ealdbeorht -briht* + **tūn**. The earliest reference is to the ditch of the inhabitants of Elberton, *œwelburhe leme dich* (sic for *œþelburhe heme* with OE thorn misread as wynn) [986]c.1400 S 862. Here the pers.n. is the feminine pers.n. *Æthelburg* but all the other forms point clearly to masculine *Ealdbeorht*. Gl iii.113.

ELBURTON Devon SX 5353. Probably 'Æthelbeorht's farm or village'. *Aliberton* 1254, *Ailberton* 1423, *Elberton* 1485. OE pers.n. *Æthelbeorht* + **tūn**. Cf. AYLBURTON Glos SO 6101. D 256.

ELCOMBE Wilts SU 1380. 'Ella's' or 'the elder-tree coomb'. *Elecome* 1086, *Ellecumba -e* 1168–1281, *Elecumbe* 1242–75. OE pers.n. *Ella* or **elle(n)** + **cumb**.Wlt 279, L 92.

ELDERNELL Cambs TL 3298. 'Alder-tree nook'. *Alrenhale* 1315, *Aldernhale* c.1350, *Eldernall* 1563. OE **ælren** + **halh**. Ca 259.

ELDERSFIELD H&W SO 8031. Partly uncertain. *Yldresfeld* 972 S 786, *Edresfelle* 1086, *Heldresfelde* 1167, [c.1086]1190, *Eldresfeld* 1262–1493, *El(le)sfeld* 1431, 1493. The generic is OE **feld**, 'open land', the specific uncertain. OE ***hyldre**, genitive sing. ***hyldres**, 'an elder-tree' has been suggested but the rarity of forms with initial *h* and unlikeliness of *e*-forms for OE *y* in this area render this improbable. The specific is more likely to be a pers.n., perhaps *Ealdhere* varying with *Ēadrēd*. Wo 197, L 240, 243–4, PNE i.274.

ELDROTH NYorks SD 7665. 'Alder hill or headland'. *Ellerhowyth* 1383, *El(l)droth(e)* 16th cent. ON **elri** + **hǫfuth**. YW vi.226.

ELDWICK WYorks SE 1240. 'Helgi's dairy-farm'. *Helguic, Heluuic* 1086, *Hele(s)wyk* 13th cent., *Hel(l)wick(e) -wycke* 16th cent. ON pers.n. *Helgi*, genitive sing. *Helga*, + OE **wīċ**. YW iv.162, SSNY 145.

Brent ELEIGH Suff TL 9447. 'Burnt E'. *Illeya Combusta* 1254, ~ *Arsa* 1260, *Brendeylleye* 1312, *Burnteylie* 1610. ME adj. **brende**, Latin **combusta, arsa**, + p.n. *(œt) Illanlege* [1000×2]11th S 1486, *Illeleia, Ilelega, lelegā* 1086, 'Illa's wood or clearing', OE pers.n. **Illa*, genitive sing. **Illan*, + **lēah**. 'Brent' for distinction from Monks ELEIGH TL 9647. DEPN.

Monks ELEIGH Suff TL 9647. *Illeya Monachorum* 1254, *Monkesillegh* 1304, *Munkesilye* 1610. ME **monke**, Latin **monachus**, genitive pl. **monkes, monachorum** + p.n. Eleigh as in Brent ELEIGH TL 9447. DEPN.

ELERKEY Corn SW 9139 → VERYAN SW 9139.

ELFORD Northum NU 1830. Possibly 'the eel ford'. *Eleford* 1255, *Elsford* 1268, 1280, *Elford* from 1296 SR 134. OE **ēl** + **ford**. The specific could also be a pers.n. such as *Elli* or *Ella* as in nearby ELLINGHAM NU 1725, or OE ***ēl** 'a strip of land'. The reference is to a ford across Crackerpool Burn. NbDu 73.

ELFORD Staffs SK 1910. Probably 'Ella's ford'. *Elleford* [1002×4]11th S 1536, 1179, *Eleford* 1086, *Elford* 1413. OE pers.n. *Ella* + **ford**. DEPN, Horovitz.

ELHAM Kent TR 1743. Possibly the 'eel-pond'. *Alham* 1086, [1100×35]1275, *Aelham* c.1100, *Aleham* 1202, *Elham* from 1100×35 with varants *Ele- Hel(e)-, Elhamme* c. 1195. OE **ǣl, ēl** + **hamm** in the sense 'enclosure'. Hasted writing in 1778×99 mentions deep ponds here in which quantities of eels were found. Alternatively *hamm* could be used here in either of senses 3 or 6, 'river meadow, piece of valley-bottom hemmed in by higher ground'. PNK 431, KPN 194, ASE 2.21.

ELING Hants SU 3612. 'The Edlingas, the people called after Edla'. *Edlinges* 1086, *Eilling(as -es)* [1100×35]14th, 1130, *Elinges* 1158–1327, *Ailinges* 1186, *Eling' -ynge* 1158–1341, *Eeling* 1655. OE folk-n. **Ēdlingas* < pers.n. **Ēdla* + **ingas**. The *Ei-* and *Ai-* spellings may point to alternative pers.n. *Ēthla*. Ha 71, Gover 1958.196, ING 42.

ELISHAW Northum NY 8695. 'Illa's wood'. *Illishawe* 1254, *Illescagh -schawe* 1278, 1291, *-shawe* 1411, *Ellyshawe* 1534,

Lishaw 1894 Heslop. OE pers.n. *Illa + **sceaga**. The change in the initial letter may be due to the influence of nearby Elsdon. NbDu 73 gives pr [(e)liʃə], DEPN.

ELKESLEY Notts SK 6875. Possibly 'Ealac's wood or clearing'. *Elchesleig -l(e)ie* 1086, *Elkesle(y) -lei(e) -lay* 1227–1380 etc., *Elsley* 1541–1675. OE pers.n. *Ēalāc*, genitive sing. *Ēalāces*, + **lēah**. Nt 78 gives pr [elzli].

ELKINGTON Northants SP 6276 → ELTHAM GLond TQ 4274.

North ELKINGTON Lincs TF 2890. *Northalkinton* 12th, 1205. ME adj. **north** + p.n. *Archintone, Alchinton* 1086, *Helchingtuna* c.1115, 'the estate called after Ealac or Eadlac', OE pers.n. *Ēalāc* or *Ēadlāc* + **ing**[4] + **tūn**. DEPN.

South ELKINGTON Lincs TF 2988, *Suthelkintone* [1160×9]17th. ME adj. **suth** + p.n. Elkington as in North ELKINGTON TF 2890. DEPN, Cameron 1998.

ELKSTONE Glos SO 9612. 'Ealac's stone'. *Elchestane* 1086, *Elkestan(e)* 1177–1291, *Elkeston* 1291–1580, *Elston* 1540–1675. OE pers.n. *Ēalāc*, genitive sing. *Ēalāces*, + **stān**. Gl i.159.

Upper ELKSTONE Staffs SK 0559. *Over Elkesdon* 1272–1337, ~ *Elkeston* 1293, *Upper & Lower Elkston* 1689, *Upper Elkstone* 1840. ME **over**, ModE **upper** + p.n. *Helkesdon'* c.1175, *Elkesdon'* 1227–1341, *-dun* 1251, 'Ealac's hill', OE pers.n. *Ēalāc*, genitive sing. *Ēalāces*, + **dūn**. 'Upper' for distinction from Lower Elkstone SK 0658, *nether Elkysdon* 1286, ~ *Elkeston* 1337–1506. Oakden.

Kirk ELLA Humbs TA 0229. *Ki- Kyrkellay -ey* from 15th, *Kirkella* 1594. 'Ella with the church' for distinction from West ELLA TA 0029. ME **kirk** + p.n. *Aluengi* 1086, *Heluiglei* 1156×7, *Eluele(y -e) -lay -v-* 1189–1496, *Ellay -ey* 16th, 'Ælfa's clearing', OE pers.n. *Ælfa*, genitive sing. *Ælfan*, varying with *Ælfa* + **ing**[4], + **lēah**. YE 217.

West ELLA Humbs TA 0029. *Westeluelle* 1305, *-eluele(gh) -ley* 14th cent., *Westella* 1594. ME **west** + p.n. Ella as in Kirk ELLA TA 0229. YE 218.

ELLAND WYorks SE 1121. 'Newly-cultivated land by a river'. *Elant, Elont* 1086, *Eland(a) -(e)* 12th–1588, *Yeland* 1203×30, 1545, *Elland* from 1541. OE **ēa** + **land**. The place lies on the r. Calder. YW iii.43 gives prs ['elənd] and ['jelənd], L 21, 246, 249.

ELLASTONE Staffs SK 1143. 'Eadlac's estate'. *E(de)lachestone* 1086, *Adelakeston* 1197, *Adlacston* 1236, *Athelaxton* 1242, *Ethelaston* 1327. OE pers.n. *Ēadlāc*, genitive sing. *Ēadlāces*, + **tūn**. DEPN, Duignan 56, Horovitz.

River ELLEN Cumbr NY 1641. A Brit r.n. identical with ALNE. *Alne* [1171×5]14th–1610, *Al(l)en* 13th cent., *Aylne* 1410, *Elne* 1777. Cu 13.

ELLEN'S GREEN Surrey TQ 0935. *Ellen's Green* from 1789. Earlier simply *Aylweynes, Al(l)eynes* 16th cent. Surname *Allen*, recorded here in 1599, < ME pers.n. *Aylwin* (OE *Æthelwine*), + ModE **green**. Sr 239.

ELLENHALL Staffs SJ 8426. Partly uncertain. *Linehalle* 1086, *Ælinhale* c.1200, *El(l)inhale* 12th–1258. The generic is OE **halh** 'a nook' in the sense 'tongue of land between streams', the specific uncertain. DEPN, L 105, 110.

ELLERBECK NYorks SE 4396. 'Alder stream'. *Elrebec* 1086, *Alrebec* 1086, 1088. ON **elri** + **bekkr**. YN 212, SSNY 93.

ELLERBY NYorks NZ 7914. 'Ælfweard's farm or village'. *Elworde- Alwardebi* 1086, *Elverdeby* 13th–1316, *Elred(d)eby* 1301, 1316, *Ellerby* 1369. OE pers.n. *Ælfweard* + ON **bȳ**. YN 136, SSNY 26.

New ELLERBY Humbs TA 1639. ModE adj. **new** + p.n. Ellerby as in Old ELLERBY TA 1636.

Old ELLERBY Humbs TA 1636. ModE adj. **old** + p.n. *Aluuarde- Alu(u)erdebi* 1086, *Elward(e)by -uard-* c.1265–1563, *Ellerbye als. Ellwerbye* 1583, *Elerdeby* 1650, 'Ælfweard's farm or village', OE pers.n. *Ælfweard* + **bȳ**. YE 47, SSNY 26.

ELLERDINE HEATH Shrops SJ 6122. Cf. *Stanton Heath, Heath House* 1833 OS. P.n. *Elleurdine* 1086, *-wurd' -wurth(in)' -wardyn -worthyn* 1195–late 14th, *Elw(o)rthin -wardine -wardyne* 1234–1680, *Ellerdon* 1577, *Ellerdine* from 1589, 'Ella's enclosure', OE pers.n. *Ella* + **worthiġn**, + ModE **heath**. Sa i.122.

ELLINGHAM Hants SU 1408. Partly uncertain. *Adelingeham* 1086, *Haslingueham* c.1165, *El- Alingeham -ynge-* 1167–1327, *Aylingham* c.1210, *Elingham -ynge-* 1204–1341, *Ellingham* 1579. Probably the 'estate of the noblemen', OE **ætheling**, genitive pl. **æthelinga**, + **hām**; but the location beside the r. Avon suggests that the generic may be OE **hamm** 'a water-meadow'. Gover 1958.213, ING 42.

ELLERKER Humbs SE 9229. 'The alder-tree marsh'. *Alrecher* 1086, *-ker* 1204, *Elreker* 13th cent., *Ellerker* from 1196. OE **alor** replaced by ON **elri** + **kjarr**. YE 222.

ELLERTON Humbs SE 7039. 'The alder-tree farm or village'. *Elreton(e)* 1086, *-tun(a) -ton* 1198–1298, *Ellerton* from 1225. ON **elri** + **tūn**. YE 238.

ELLERTON Shrops SJ 7125. 'Æthelheard's estate'. *Athelarton* 13th, *Ethelarton* 1285, *Alarton* 1203. 1312. OE pers.n. *Æthelheard* + **tūn**. Bowcock 96, DEPN.

ELLESBOROUGH Bucks SP 8306. 'Ass hill'. *Esenberge* 1086, *-bergh* 1241, *Hesel'bga* [133]14th, *Eselbergh* 1195–1382, *-burgh* 1237–1470, *Eseburgh* [1227×37]14th, *Elisborough* 1491. OE **eosol** + **beorg**. Probably a hill where asses were pastured. Bk 149, Jnl 2.25.

ELLESMERE Shrops SJ 4035. 'Elli's lake'. *Ellesmeles* 1086, *Ailes- Ellismera* 1177, *Ellesmere* from 1204 with variant *-is-*, *Eylesmere* 1271×2, *Elsmere* 1516–1750 with variants *-mire -meere -m(e)are*. OE pers.n. *Elli*, genitive sing. *Elles*, + **mere**. Sa i.122.

ELLESMERE PORT Ches SJ 3976. A town which grew up at the junction of the Ellesmere (or Shropshire Union) Canal with the River Mersey. The Dee and Mersey branch was opened in 1795 and the port takes its name from the canal, itself named from ELLESMERE Shrops SJ 4035. Che iv.198, Pevsner 1971.216, Room 1983.36. Addenda gives prs ['elzmi:r], older local ['elzmər].

ELLINGHAM Norf TM 3592. 'Homestead at or called *Eling*'. *Elinc(g)ham* 1086, *Elingham* 1201–1254. OE p.n. *Eling* 'place called after Eli', OE pers.n. *Eli* + **ing**[2], or 'place called *Ēling*, the eel place', OE **ēl** + **ing**[2], + **hām**. Ellingham lies on a tributary of the Waveney and may have had eel-traps. DEPN, ING 134.

ELLINGHAM Northum NU 1725. 'The homestead called or at *Elling*, the place called after Ella'. *Ellingeham* c.1130, *El(l)ing(e)ham* c.1160–1296 SR, *Elling(c)ham, Ellincham* 1242, 1278, *Elyngham* 1346, *-yngeham* 1507. OE p.n. *Elling < pers.n. *Ella* + **ing**[2], locative-dative sing. *Ellinġe, + **hām**. Pronounced [elindʒəm] from the assibilated locative-dative form. NbDu 73, ING 157, BzN 1967.371.

Great ELLINGHAM Norf TM 0197. *Magna Elingham* 1242×3. ModE adj. **great**, Latin **magna**, + p.n. *Ellincham, Ailincham, (H)elincham, Helingham* 1086, *Elingeham* 1207, 'homestead of the Elingas, the people called after Eli or Ella'. OE folk-n. *El(l)ingas < pers.n. *Eli* or *Ella* + **ingas**, genitive pl. *El(l)inga, + **hām**. 'Great' for distinction from Little ELLINGHAM TM 0099. ING 134, BzN 1967.371.

Little ELLINGHAM Norf TM 0099. *Parva Elingham* 1242×3. ModE adj. **little**, Latin **parva**, + p.n. Ellingham as in Great ELLINGHAM TM 0197.

ELLINGSTRING NYorks SE 1783. 'Water-course called *Eling*, the eel-stream'. *Elingestrengge* 1198, *El(l)yngstrynge, Elling- Ellyngstring(e)* 1285–1571. OE r.n. *Ēling < OE **ǣl, ēl** 'an eel', + **ing**[2], + ON **strengr** 'a water-course'. YN 231.

ELLINGTON Cambs TL 1671. 'Estate called after Ella'. *Elintune* 1086, 1106×13, *Elint(h)on -ynton* 1207–85, *Elyngton -ing-* 1267–1346 etc. OE pers.n. *Ella* + **ing**[4] + **tūn**. Hu 239.

ELLINGTON Northum NZ 2791. 'The settlement on the eel-stream'. *Elingtuna* 1166 RBE, *Helingtone* 1167, *El(l)ington* 13th cent. OE ***ēling** < **ēl** + **ing**[2], + **tūn**. The reference would be to the river Lyne. Alternative explanations are OE pers.n. *Eli* or *Ella* + **ing**[4] + **tūn** 'the estate called after *Eli* or *Ella*', or an *ing*[2] derivative of OE ***ēl** 'a strip of land'. NbDu 73, Årsskrift 1974.34.

High ELLINGTON NYorks SE 1983. ModE **high** + p.n. *Ellintone* 1086, *Ellington -yng-* from 12th, *Elington -yng-* 1219–85, the 'estate called after Ælla', OE pers.n. *Ælla*, Northumbrian *Ella*, + **ing**[4] + **tūn**. YN 231.

Low ELLINGTON NYorks SE 2083. ModE **low** + p.n. Ellington as in High ELLINGTON SE 1983.

ELLIS CRAG Northum NT 7401. *Ellis Crag* 1969 OS. Surname *Ellis* + ModE **crag**.

ELLISFIELD Hants SU 6345. 'Ielfsa's open land'. *Esewelle* 1086, *Elsefeld* 1167–1392, *Ulsefeld* 1219–1428, *Ellesfeud* 1236, *Ilsfeld* 1579. OE pers.n. **Ielfsa* + **feld**. Ha 71, Gover 1958.131 which gives pr [elsfi:ld], DEPN, L 243.

ELLISTOWN Leic SK 4311. Surname *Ellis* + ModE **town**. A coal-mining village named after the colliery owner Joseph Joel Ellis. The first turf on the Ellistown new colliery estate was cut by Mrs J. J. Ellis in 1873. Room 1983.36.

ELLONBY Cumbr NY 4235. 'Alein's farm or village'. *Alaynby* c.1220, *Alain-* 1306, 1381, *Alem- Alein-* 13th cent., *Allanby* 1580, 1783, *Elenby* 1619, *Ellonby* 1706. OFr pers.n. *Alein* + ON **bȳ**. Cu 240, SSNNW 25.

ELLOUGH Suff TM 4486. Possibly '(at) the pagan temple'. *Elga* 1086, *Elgh* 1286–1336, *Elecg* [1199]1319, 14th, *Heleg* 1254, *Elegh* 1286, *Ellough* from 1316, *Ellow(e)* 17th. OE **ealh**, dative sing. ***ealge**. DEPN, Baron.

ELLOUGHTON Humbs SE 9427. Partly uncertain.
Type I. *Elgendon* 1086, *Elgedon* 1185, 1196, *Elgton* 1290, *Elgh-* 1339, *Ellegton* 1191, *Elegh- El(l)ughton* 1381, *Eloughton* 1544;
Type II. *Helgedon* 1196, *-ton* 1200, 1201;
Type III. *Elv(h)eton(a), Helvehet'* 1233.
Possibly 'temple upland', ON ***elgr**, genitive sing. ***elgi(ar)**, + **dūn**. Alternative possibilities are 'Helgi's upland', ON pers.n. *Helgi*, genitive sing. *Helga* + **dūn** or, if the *n* of the DB form is to be trusted, OE **on thǣm hǣlgan dūne* 'at the holy hill' with *dūn* later assimilated to OE *tūn*. YE 220, SSNY 254.

ELLWOOD Glos SO 5908. Probably 'elder wood'. *Ellwood* 1618. ME **eller** (OE *ellern*) + **wode** (OE *wudu*). Gl iii.229.

ELM Cambs TF 4607. '(At) the elm-tree(s)'. *Ælm* c.1150 ASC(E) under year 656, *Elm(e)* from c.1213 including [c.973]c.1253, *Eolum* [973]15th S 792, *Elym* 1221. This is almost certainly **elmum**, dative pl. of OE **elm**, or possibly dative sing. **elme**. However, the forms *Eolum* and *Elym* if reliable are difficult to explain. Accordingly it has been suggested that this might be an old folk-n. **Eole*, dative pl. **Eolum*, perhaps related to the folk-n. *Elvecones* < Gmc **elwekaz* < **elwaz* + *ika-*, cf. OHG *elo* 'brown, yellow'. Ca 266, L 220, Schönfeld 74.

Great ELM Somer ST 7449. ModE adj. **great** + p.n. *Telvve* 1086, *Telma* 1086 Exon, *Theaumes* 1247, *Elme* 1327, 1610, *Elm* 1817 OS. OE **elm** 'an elm-tree' with prefixed *t-* by misdivision of *æt Elme*. Great and Little Elm are sometimes distinguished. DEPN.

ELM PARK GLond TQ 5385. An underground station opened in 1935. No early forms. ModE **elm** + **park**. For *Park* in railway names see Jul 23, 18–25.

ELMBRIDGE H&W SO 9067. 'Elm-tree ridge'. *Elmerige* 1086, *Elmrugge, Aumbrug'* 1212, *Elm(e)brug(g)e* 1270–1492. OE pers.n. **elm**, genitive pl. **elma**, + **hrycg** later replaced by ME *brugge*. Wo 290, L 169, 220.

ELMDON Essex TL 4639. 'Hill growing with elm-trees'. *Elm̄dunā* 1086, *Elmedon(a) -e* 1141 etc., *E(a)umeden' -don(e)* 1198–1254, *Elmendon* 1301–73. OE ***elmen** + **dūn**. Ess 525 gives pr [eləmdən].

ELMDON WMids SP 1783. 'The elm-tree hill'. *Elmedone* 1086, *-don(a)* 1217–1656. OE **elm**, genitive pl. **elma**, + **dūn**. Occasional sps *Ilmedon* 1297 and *Elmendon* 1316, *-yn-* 1458, point to alternative forms with OE ***ylme** 'an elm-tree, an elm wood' and ***elmen** 'growing with elms'. Wa 60, L 154, 220.

ELMDON HEATH WMids SP 1681. *Elmedon heath* 1609. P.n. ELMDON SP 1783 + ModE **heath**. Wa 60.

ELMESTHORPE Leic SP 4696. 'Ailmer's outlying farm'. *Ai- Aylmerestorp' -ar-* 1199–1371, *-merst(h)orp'* 1196×1208–1458, *Ai- Aylmest(h)orp(e)* 1209×35–1467, *Elmesthorp(e)* from 1458. lOE pers.n. *Ailmer* (OE *Ælhelmǣr*), genitive sing. *Ailmeres*, + **thorp**. Lei 493, SSNEM 494.

ELMET An ancient district name in WYorks occurring in the names BARWICK IN ELMET SE 4037 and SHERBURN IN ELMET SE 4933. *Elmed -t (sætna)* 'the people who live in E' [7th]c.1000, *(in silva) Elmete* 'E forest' [c.731]8th BHE, *Elmet* c.800 HB, *Elmeth* 1431. Identical with Welsh cantref name *Elfed* of obscure origin. YW iv.1, WATU 66.

All Saints South ELMHAM Suff TM 3482. Short for All Saints parish of S Elmham. *parochia vocata All Hallows* 1523, *All Seyntes* 1524, *Alsainct* 1610, + p.n. *Suth Elmham* 1229–1346, ME **south** + p.n. *Alme- Alma- Halme- Elmeham* 1086, *Elmham* from c.1105, 'the elm-tree homestead', OE **elm** + **hām**. There were six parishes here. 'South' for distinction from North ELMHAM Norf TF 9820. DEPN, Baron.

North ELMHAM Norf TF 9820. *Northelmeham* 1252. ME adj. **north** + p.n. *Ælmham* [1047×70]13th S 1499, *Elmenham* 1086, *Elmham* 1167, 'elm-tree homestead', OE **elm** varying with ***elmen** 'growing with elm trees' + **hām**. Elm-wood was much used in carpentry for keels, coffins, wagons, wheel-barrows, funiture and buildings on account of its extreme toughness. This may have been an estate where elm was worked. 'North' for distinction from South ELMHAM Suffolk TM 3083. DEPN.

St Cross South ELMHAM Suff TM 2984. One of the six S Elmham parishes. *Sancroft* 1254, *Sand(e)croft* 1286–1391, *St Crost* 1610, 'the sandy croft', OE **sand** + **croft**, + p.n. South Elmham as in All Saints South ELMHAM TM 3482. DEPN, Baron.

St James South ELMHAM Suff TM 3281. One of the six S Elmham parishes. *parochia Sancti Jacobi* 1523, *Seynt Jamys* 1524, *St Iames* 1610. Saint's name *James* from the dedication of the church + p.n. S Elmham as in All Saints South ELMHAM TM 3482. Baron.

St Margaret South ELMHAM Suff TM 3183. *ecclesie Sancte Margarete de South Elmham* 1229–88, *Seynt Margarettes* 1524, *St Marget* 1610. One of the six S Elmham parishes, cf. All Saints South ELMHAM TM 3482. Baron.

St Michael South ELMHAM Suff TM 3483. *parochia Sancti Michaelis* 1523, *Seynt Mighelles* 1524, *St Michael* 1610. One of the six S Elmham parishes, cf. All Saints South ELMHAM TM 3482.

ELMHURST Staffs SK 1112. 'The elm-tree wooded hill'. *Elm- Elme(s)hurst* from 1259, *Elinghurst* c.1275. OE **elm** + **hyrst**. Duignan 57, Horovitz.

ELMLEY CASTLE H&W SO 9841. 'Elmley below the castle' referring to the vanished Despenser castle built in the late 11th cent. ½ m. S of the village and for distinction from ELMLEY LOVETT SO 8769. *Almeley sub castellum* 1313, *Castel Elmeleye* 1327, p.n. Elmley + ME **castel**. Elmley, *Elmlege* (dative case) *[780]11th S 118, *Elmlæh, into Elmlea* [1046]18th S 1396, *Elmelege* 11th–1275, is the 'elm-tree wood or clearing', OE **elm** (genitive pl. **elma**) + **lēah**. Wo 122, L 199, 203, 220, Pevsner 1968.142, Hooke 362.

ELMLEY ISLAND Kent TQ 9468. *Elmley Island* 1819 OS. P.n. Elmley + ModE **island**. Elmley, *Elmele* 1226–77, *-leg'* 1238, *-ley* 1270, is the 'elm-tree wood or clearing', OE **elm** + **lēah**. K 248.

ELMLEY LOVETT H&W SO 8769. 'Elmley held by the Lovett family'. *Almeleye Lovet* 1275, *Awmeley ~* 1428, *Elmeleye ~* 1327. P.n. Elmley + family name Lovett from the early 13th cent. lords of manor and for distinction from ELMLEY CASTLE SO 9841. Elmley, *Ælmeleia* 1086, is the 'elm-tree wood or clearing', OE **elm**, genitive pl. **elma**, + **lēah**. The boundary of the people of Elmley is referred to as *Elmesetene gemœre* [817]18th S 1597, *elmsetena gemœre* [980]11th S 1342. Wo 240, L 199, 203, 220, Hooke 305.

ELMORE Glos SO 7815. 'Elm-tree ridge'. *Elmoura -our(e) -ovre -ouere* 1137–1252, *Elemor(e)* 1263–1358, *Elmor(e)* 1249–1557. OE **elm** + **ofer**. Gl ii.162, L 176–7, 220.

ELMORE BACK Glos SO 7716. 'The back part of Elmore'. *Elmore Back* 1830. P.n. ELMORE SO 7815 + ModE **back**. Gl ii.163.

South ELMSALL WYorks SE 4711. *Suthelmeshal(e)* 1230, 1268, *Southelmesale* 1253–1366. ME adj. **suth** + p.n. *Ermeshale* 1086, *Emsala* 1137×9, *Elmeshale* 1170×80, 1243, 'nook of land of the elm-tree', OE **elm**, genitive sing. **elmes**, + **halh**. 'South' for distinction from North Elmsall SE 4812, *Ermeshala -hale* 1086, *Elmesale* 1264–1365, *North ~* from 1320, *Elmesall* 1316–1459. YW ii.36, 39, L 108, 220.

ELMSCOTT Devon SS 2321. 'Ielfmund's cottage(s)'. *Ilmundescote* 1281, *Ylmondes- Ylmannescote* 1301, *Yelmes- Eme(n)scott* 1566. OE pers.n. **Ielfmund*, genitive sing. **Ielfmundes*, + **cot**, pl. **cotu**. D 73 gives pr [emskət].

ELMSETT Suff TM 0546. 'The dwellers or the houses at the elm-tree copse'. *(œt) Ylmesœton -un* [962×91]11th S 1494, [1000×2]11th S 1486, *Elmesetā* 1086, *Elmeset* 1610. OE ***elme** + **sǣte** 'dwellers' or **sǣte** 'house', dative pl. **sǣtun**. *Ylme-* is a hyper-correct sp for *Elme-*. DEPN.

ELMSTEAD MARKET Essex TM 0624. *Elmested Market* 1475. P.n. Elmstead + ME **market**. Elmstead, *Elme- Almestedā* 1086, *Elmested(e)* 1233–1593, is 'elm-tree place', OE **elm**, genitive pl. **elma**, or ***elme** 'elm wood', + **stede**. Ess 337, Stead 198.

ELMSTED Kent TR 1145. 'Elm-tree place'. *elmanstede* 811 S 168, [811]13th S 1617, *elmes stede* [811]13th ibid., *Elmested(e)* 11th–1354 including [1087]13th, *Elmsted(a)* 1166, 1610. OE ***elmen** varying with **elm**, genitive pl. **elma**, + **stede**. Alternatively the generic could be a reduced form of OE *hāmstede* 'a homestead' which is frequently compounded with tree names. KPN 118, Stead 88–9, 214–5, PNE i.232.

ELMSTONE Kent TR 2660. 'Æthelmær's estate'. *Elm'est'* 1154×89, *A(i)lmarestone(e) -meres-* 1202–3, *E(y)lmereston'* 1240–78, *Elmestone* 1610. OE pers.n. *Æthelmǣr*, genitive sing. *Æthelmǣres*, + **tūn**. PNK 517.

ELMSTONE HARDWICKE Glos SO 9226. Originally two separate places, Elmstone and Hardwicke. Elmstone, *Almvndestan* 1086, *Eilmundestan* 1221, *Eil- Ayl- Eylmundeston -mond-* 1221–1319, *Elmeston(e) -ys-* 1506–97, is 'Alhmund's stone', OE pers.n. *Alhmund*, genitive sing. *Alhmundes*, + **stān**. The spellings *Eil- Ayl-* point to alternative pers.n. *Æthelmund*, but *Alhmund* is confirmed by *Al(c)hmunting tu(u)n* [889]11th S 1415, *(on) Alhmundingtune* and *se Alhmunding snœd* [899×904]18th S 1283, 'estate' and 'detached piece of land called after Alhmund', both probably with reference to this place. Hardwicke, *Herdeuuic* 1086, *-wik-wyk(e)* 1212–1525, *Hard(e)wi(c)k -wyk(e) - wike* c.1260, 1451–1597, is 'the herd farm', OE **heorde-wīc**. Gl ii.81, 82, PNDB 185.

ELMSWELL Suff TL 9863. 'The elm-tree spring'. *Elmeswella(m)* 11th, 1086, *-well* 1200–54, 1610. OE **elm** + **wella**. DEPN.

ELMTON Derby SK 5073. 'Elm-tree farm or village'. *Helmetune* 1086, *Elmeton(e)* 1176–17th, *Elmton* from 1286. OE **elm**, genitive pl. **elma**, + **tūn**. Probably a place where elm wood was prepared and worked. Because of its extreme toughness elm wood was widely used for wheels, furniture, carts, sheds, buildings and water pipes. Db 256.

ELSDON Northum NY 9393. 'Elli's valley'. *Eledene* 1226, *Hellesden* 1236, *Elisden'* 1242 BF, *Ellesden(e) -is-* 1244–1336, *Elsden* 1663. OE pers.n. *Elli*, genitive sing. *Elles*, + **denu**. The first form can be explained from the variant form of the pers.n. *Ella*. Two late spellings, *Helvesden* 1324, *Eluesden* 1432, offer evidence of a tradition of an alternative pers.n. OE *Ælf*. NbDu 74, DEPN.

ELSECAR SYorks SE 3800. 'Ælfsige's marsh'. *Elsecar -ker* 1746, 1822. OE pers.n. *Ælfsīġe* + ON **kjarr**. OE *Ælfsige* became ME *Elsi* which also occurs in nearby lost *Aylsi rode* 1259×66, *Elsy Royd* 1753×71, 'Ælfsige's clearing', OE ***rodu**. YW i.113.

ELSENHAM Essex TL 5425. 'Elesa's homestead'. *Elsenham* from 1086 with variants *-in(g)-*, *-yn-*, *Alsenham* 1086. OE pers.n. *Elesa*, genitive sing. *Elesan*, + **hām**. Elesa is a very ancient pers.n. Ess 527 gives pr [elznəm].

ELSFIELD Oxon SP 5410. 'Elesa's open land'. *Esefelde* 1086, *Elsefeld(e) -feud* 1123–1428. OE pers.n. *Elesa* + **feld**. O 170, L 243.

ELSHAM Humbs TA 0312. 'Elli's village'. *Ele(s)ham* 1086, *Elesham* c.1115–1382, *Ellesham* c.1160–1529, *Elsham* from 1247×8. OE pers.n. *Elli*, genitive sing. *Elles*, + **hām**. Li ii.105.

ELSING Norf TG 0516. 'The Elesingas, the people called after Elesa or Ælesa'. *Helsinga* 1086, *Alsinges* 1197, *Ausinges* 1268, *Ausinge* 1242×3, *Elsyng* 1347. OE folk-n. *Elesingas* < OE pers.n. *Elesa* varying with side-form **Ælesa*, + **ingas**. DEPN gives pr [elziŋ], ING 56.

ELSLACK NYorks SD 9349. Either 'Elli's stream' or 'Ella's valley'. *Eleslac* 1086, 1267, *Elselak* 1240, *Elslac(k)* from 1219, *Elleslak(e) -lac* 1231–1400. OE pers.n. *El(l)i*, *Æl(l)i*, genitive sing. *Elles*, *Ælles*, + OE **lacu** referring is to a small side stream of Earby Beck, or pers.n. *Ella* + **slakki** 'a shallow valley'. Since *lacu* is often used of a side channel or a river, a sense not applicable here, the second interpretation is probably to be preferred. YW vi.44, L 23, 123.

ELSTEAD Surrey SU 9143. 'The elder-tree place, stand of elder-trees'. *Helestede* [1128]1318, *El(l)e-* 1222–1336, *Elsted* 1258, *-de* 1294–1390, *Elstead* from 1593. OE **ellen-stede**. Sr 167, PNE i.150, Stead 245, L 222.

ELSTED WSusx SU 8119. 'The elder-tree place'. *Halestede* 1086, *Ellested(a)* 1180–1428, *Elnestede* 1212–1456, *Elsted al. Elnested* 1618. OE **ellen** or genitive pl. **el(le)na** + **stede**. Sx 34, Stead 254.

ELSTON Notts SK 7548. Probably 'Eilafr or Eilifr's farm or village'. *Elvestune* 1086, *Eluestun* 1165, *-tona* c.1190, *Heylueston'* c.1200, *Eyleston(e)* 1236–1583, *Ayles-* 1236–1362, *Elleston* 1270, *Elson* 1574. ON pers.n. *Eiláfr* or *Eilifr*, secondary genitive sing. *Eiláfes*, *Eilifes*, + **tūn**. Nt 212 records a former pr [elsən], SSNEM 190.

ELSTONE Devon SS 6716. *Elson* 1809 OS.

ELSTOW Beds TL 0547. 'Ælna's meeting-place'. *Elnestou* 1086, *Elvestoue* c.1150, *Alnestow* 1182, *Aunestow* 1202, *Helenstoe* c.1270, *Elstow* 16th cent., *Elvestow* 1766. The spellings imply a form *Ælnestowe* c.1050 probably representing OE **Ælnan stow*, OE pers.n. **Ælna*, genitive sing. **Ælnan*, + **stōw**. **Ælna* would be a pet form of names such as *Æthelnōth* or *Ælfnōth*: another possibility would be **Ællen*, earlier **Ællīn*, genitive sing. **Ællenes*, and so *Ællenes stōw*. The *Elv-* forms, though long-lived, derive from an error in transcription: the *Helen* form is influenced by the dedication of the parish church, earlier the Chapel of St Helena. Bd 70.

ELSTREE AERODROME Herts TQ 1596. P.n. ELSTREE TQ 1895 + ModE **aerodrome**.

ELSTREE GLond TQ 1895. 'Tidwulf's tree'. *Tiðulfes treow* [785]12th S 124, *Tydolvestre* 13th cent., *Idolvestre* 1254, *Idelestre* 1320, *Illestre(e)* 1487–1536, *Elstre*. OE pers.n. *Tīdwulf*, genitive sing. *Tīdwulfes*, + **trēow**. Initial *T-* is sometimes lost from p.ns. through contact with preceding *at* and false analysis whereby *at Tidolvestre* > *at Idolvestre*. Cf. Acton Dorset SY 9878, originally *Tacatone* 1086 'sheep farm', OE ***tacca** 'a young sheep', Arracot Devon SX 4287, originally *Tad(i)ecote* 1327–30 'toad cottage', OE **tādiġe** 'toad', Addislade Devon SX 7164, originally *Tadyeslade* 1306, 'toad valley', ADLESTROP Glos SP 2427. Hrt 74, DEPN, D 208, 299.

ELSTRONWICK Humbs TA 2232. 'Ælfstan's dairy-farm'. *Asteneuuic* 1086, *Elstainnewic* [late 12th]c.1536, *Elstanwik -wyk(e)* c.1265–1546, *Elsternwyk -wicke* 1535, 1609, *Elstramwick* 1650. OE pers.n. *Ælfstān* + **wīc**. YE 53 gives pr [elstθrənwig].

ELSWICK Lancs SD 4238. 'Ethelsige or Ethel's farm'. *Edelesuuic* 1086, *Hedthelsiwic* c.1160, *Ethel(i)swic -wyke* 1200–1346, *Etheleswic -wyk* 1212–98 etc., *Elleswyk -wike* 1302×3, 1549, *Elswyk* 1539. The 1160 form points to a compound of OE (Northumbrian) pers.n. *Ethelsiġe* + **wīc**. Otherwise the specific is OE (Northumbrian) pers.n. **Ethel*, genitive sing. **Etheles*, a shortened form of compound names in *Ethel-*. La 161, Jnl 17.94, BARBrit 180.249 for 9th cent. Northumbrian moneyers' names in *Edel- Edil-*.

ELSWORTH Cambs TL 3163. 'Elli's enclosure'. *Eleswurth(e)* [974]14th S 798, 1249, *-worth* [974]13th S 798, 1337, *-worð* *[1060]14th S 1030, *-uuorde* 1086, *Elleswurth(e) -worth(e)* 1218–1473. The bounds of Elsworth are mentioned as *(be) Elleswyrðe gemœra* [1012]12th S 1562. OE pers.n. *El(l)i*, genitive sing. *El(l)es*, + **worth**. Ca 167.

ELTERWATER Cumbr NY 3204. 'Swan lake'. *Helt'- Heltewatra, Elterwat'* 1157×63, *Helterwatra* [1154×89]n.d., *Elterwater* from 1608. ON **elptr**, genitive sing. **elptar**, + OE **wæter** (perhaps replacing ON *vatn* 'water'). Whooper swans overwinter on the lake. We i.16.

ELTHAM GLond TQ 4274. 'Elta's homestead'. *A- Elteham* 1086, 13th cent., *Healte- Aeltheham* c.1100, *El(l)tham* from 1210×12. OE pers.n. **Elta* + **hām**. The same pers.n. occurs in Elkington Northants SP 6276, *Eltetone* 1086, *Elte(n)- Eltesdon* 13th cent., 'Elta's hill'. KPN 2, DEPN, ASE 2.29, TC 204.

New ELTHAM GLond TQ 4573. ModE adj. **new** + p.n. ELTHAM TQ 4274. A 20th- cent. development in a place formerly called *Pope Street* 1805, surname *Pope* + ModE **street** 'a hamlet'. LPN 76.

ELTISLEY Cambs TL 2759. 'Elti's wood or clearing'. *Hecteslei* (sic) 1086, *Helteslay -l(e)* 1202–c.1330, *Eltisle(e) -ys-* 1218–1434, *-ley* 1500. OE pers.n. **Elti*, genitive sing. **Eltes*, + **lēah**. Ca 158.

ELTON 'Eel farm or village'. OE **ēl** + **tūn**.

(1) ~ Cleve NZ 4017. *Eltun* 12th., *Eltone* 13th cent., *Elton* from 1406.

(2) ~ Ches SJ 4575. *Eltone* 1086, *Elton* from early 13th., *Ealton* c.1617. Che iii.255.

(3) ~ Derby SK 2261. *Eltune* 1086, *-ton(e)* from 1269, *Eleton* 1254. The village is situated above a small stream. Othewise 'Ella's farm or village', OE pers.n. *Ella* + **tūn**, although the dearth of forms with medial *-e-* makes this unlikely. Db 364.

ELTON Cambs TL 0893. 'The prince's estate'. *Æþeling- Ælintun* [972×92]12th B 1130, *Adelintune* 1086, *Aethelyngtone* [1123×36]14th, *Ail(l)inton* 1207–9, *Ayl(l)ington -yng-* 1215–1355, *Aileton, Aylton, Ailton* 1517–93. OE **ætheling** + **tūn**. Hu 184 prefers OE pers.n. **Æthel* + **ing**[4] as specific, 'estate called after Æthel'. PNE i.7.

ELTON Glos SO 7014. 'Ælfwine or Ælfwynn's estate'. *Elwinton* 1201, *Elue- Elveton(a) -tune* 1221–1420, *Elton* from 1542. OE pers.n. *Ælfwine*, feminine *Ælfwynn*, genitive sing. *Ælfwynne*, + **tūn**. Gl iii.203.

ELTON H&W SO 4571. 'Eel-fishery estate'. *Elintvne* 1086, *Elintona* 1160×70, *El(l)eton* [1175×9]14th, 1199, 1333, *Elton(e)* 1397, 1535. OE ***ǣling** + **tūn**. Elton lies beside a stream called Wigmore Lake. He 81.

ELTON Notts SK 7438. Uncertain. *Ailetone* 1086, *Eylton* 1280, *Elleton(am)* [1088]13th–c.1280, *Elton* from 1195, *Eleton* 1242, 1259. The two first spellings cast doubt on the otherwise satisfactory explanation 'Ella's farm or village', OE pers.n. *Ella* + **tūn**. OE pers.ns. in *Æthel-* frequently appear in DB as *Ail-* (beside many other variant spellings); possibly 'Æthel's farm or village', OE pers.n. *Æthel* varying with *Ella* + **tūn**. Nt 224.

ELVASTON Derby SK 4132. 'Ælwald's farm or village'. *Ælwoldestune* 1086, *Aylwoldeston(e)* 13th cent., *Ail- Aylwardeston(e) -is-* 1219–1331, *Ail- Aylwaston(e)* 1208×26–1431, *Elwaston* 1450, 1475, *-vas-* from 1493. OE pers.n. *Æthelwald*, lOE *Ælwald*, genitive sing. *Ælwaldes*, + **tūn**. Db 461.

ELVEDEN Suff TL 8279. Probably 'elf valley'. *Eluedenā, Heluedana -dona, Haluedona* 1086, *Eluedene* c.1095, *-den* 1179, *Elveden* 1242, *Elden* 1610. OE **elf**, genitive pl. **elfa**, + **denu**. Alternatively this might be 'swan valley', OE *elfet-denu* < **elfitu** + **denu**. DEPN, cf. E&S 19.156.

ELVINGTON Kent TR 2750. 'Ælfgyth's estate'. *Elventune* c.1100, *Elfgethetun* [1087]13th, *Eluīgton'* 1240, *Eluinton(') -yn(g)-* 1262–1731. OE feminine pers.n. *Ælfġȳth* + **ing**[4] + **tūn**. PNK 582.

ELVINGTON NYorks SE 7047. 'Ælfwine or Ælfwynn's settlement'. *Aluuinton(e)* 1086, *Eluinton -v-* 1176–1279×81, *Eluington(a) -v- -y-* 1180×97–1546. OE pers.n. *Ælfwine* or *Ælfwynn*, + **tūn**. YE 272.

ELWICK Cleve NZ 4532. 'Ella or Ægla's farm or village'. *Ailewic* c.1150, *Ellewic* 1174–1441 with variants *-wyc -wyk, Elwyk(e) -wik(e)* c.1250–1558, *Elwick* from 1622. OE pers.n. *Æġla* or *Ella* + **wīċ**.

ELWICK Northum NU 1136. 'Ella's farm'. *Ellewich* 1154×66, *-wic* 1203, *Elwyc -wyk* 1242 BF, 1296 SR, *Ellick* 1637. OE pers.n. *Ella* + **wīċ**. NbDu gives pr [elik].

ELWORTH Ches SJ 7461. 'Ella's enclosure'. *Ellewrdth* 1208, *-worth(e)* 1305–18, *Elworth* late 13th and from 1527. OE pers.n. *Ella* + **worth**. Che ii.266.

ELWORTHY Somer SR 0834. 'Ella or Elli's enclosure'. *Elwrde* 1086, *Elleswurða* 1166, *Elleworthe* 1166, 1225, *Elworthye* 1610. OE pers.n. *Ella* varying with *Elli*, genitive sing. *Elles*, + **worth** later replace with **worthy** (OE *worthiġ*). DEPN.

ELY Cambs TL 5380. 'Eel district'. *(In regione quæ uocatur) Elge* 'in the district called Ely' c.731 BHE, c.1150, 1170, *elgae* 8th BHE, *(æt) Elige* c.900 ASC(A) under year 673, *(in þæm þeodlonde þe is geceged) Elige, hélige* c.1000 OE Bede, *(in regione, on, into) Elig* [970]13th S 778, *Elyg* 1086, [1004]18th S 907, *(æt, into) Helig(e)* [957]18th S 646, 1015 S 1503, 11th ASC(D) under year 1072, 12th ASC(E) under year 673, *(into) Ælig* [c.1000]11th, *ǽl cǽg* c.1040 OE Bede, *Ylig* 11th, *(in) Hely* 1108–c.1400 including [1045×66]12th S 1100, *(into, in) Heli* c.1130, c.1260, *(æt) Eli* c.1100 ASC(F) under year 673–1203, *(in) Ely* from 1086 including [956]12th S 572 and [970]18th S 776. OE **ǣl**, **ēl** + **ġē** early confused with **īeġ** 'island'. Already Bede states, *Elge...regio...in similitudinem* insulæ *uel paludibus, ut diximus, circumdata uel aquis, unde et a copia* anguillarum *quæ in eisdem paludibus capiuntur nomen accepit*, 'Ely resembles an *island* in that it is surrounded by marshes or by water. It derives its name from the large number of *eels* which are caught in the marshes'. Later writers suggest that the name is a compound of two Hebrew words, *quoniam dicitur* 'el' *Deus*, 'ge' *terra, quod simul dei terra sonat*, '*el* is God and *ge* land, which together makes "God's land"'. Bede gives the size of the Ely *regio* as 600 hides. Eel rents were a fruitful source of income for the abbots and bishops of Ely in the middle ages. Ca 213.

EMBERTON Bucks SP 8849. 'Æmbriht's estate'. *Ambri- Ambretone* 1086, *Embertone* c.1215×47, *Embertom al. Emmerton* c.1450. OE pers.n. *Æmbriht* (an attested form for *Ēanbeorht*) + **tūn**. Bk 35 gives pr [emətən].

EMBLEHOPE MOOR Northum NY 7495. *Emblehope Moor* 1869 OS. P.n. Emblehope NY 7494, *Emelhope* 1325, *Hemelhop -il-* 14th cent., *Emlopp* 1686, apparently 'the caterpillar valley', OE **emel** + **hop**, + ModE **moor**. NbDu 75 gives pr [emləp], L 116–7.

EMBLETON Cumbr NY 1730. Partly uncertain. *Emelton* 1195–1407, *Embelton'* 1233–1322, *Embleton* from 1243. Apparently OE **emel** 'caterpillar' + **tūn**, or perhaps OE pers.n., e.g. *Ēanbald*, + **tūn**. Cu 384.

EMBLETON Northum NU 2322. 'Æmele's or caterpillar hill'. *Emlesdune* c.1200, *-done* 1212, *Emelesdona* 1212 BF, *Emil- Emeldon* 1242 BF–1346, *Emylton* 1538, *Embleton* from 1507. OE pers.n. *Æmele* or **emel**, genitive sing. *Æm(e)les*, **em(e)les**, + **dūn**. NbDu 75, DEPN.

EMBLETON BAY Northum NU 2423. *Embleton Bay* 1868 OS. P.n. EMBLETON NU 2322 + ModE **bay**.

EMBOROUGH Somer ST 6151. 'The flat hill'. *Amelberge* 1086, *Emeneberge* 1200, *Eueneberia* 1194, *Emnebergh* 1238, *Enboro* 1610, *Emborrow* 1817 OS. OE **efn, emn**, definite form **efna, emna**, + **beorg**. The reference is to Red Hill ST 6050 SW of the church which has a very low-profile curve. DEPN.

EMBSAY NYorks SE 0053. 'Embi's island'. *Embesie* 1086, *-ay(a) -ai -ey(a)* 1131×40–1305, *Emmesey(e) -ay(e)* 1120–1363, *Embsay(e) -ey* 1305, 1591, *Emsey* 16th cent. OE pers.n. *Embi*, genitive sing. *Embes*, + **ēġ** in the sense 'hill jutting into flat land; patch of good land in moors'. YW vi.67 gives pr ['emsə], L 35.

EMBSAY MOOR NYorks SE 0056. P.n. EMBSAY SE 0053 + ModE **moor**.

EMBSAY RESERVOIR NYorks SE 0054. P.n. EMBSAY SE 0053 + ModE **reservoir**.

EMERY DOWN Hants SU 2808. 'Hill belonging to the Emmory family'. *Emerichdon* 1376, *Emeryesdowne* c.1490. Family-n. *Emmory* recorded here in 1389 + ME **doun** (OE *dūn*). The family-n. is from Fr *Emaurri*, ultimately OG *Amalric*. Ha 72, Gover 1958.208, DBS s.n. Amery.

EMLEY WYorks SE 2413. 'Eama's forest clearing'. *Amelai -leie* 1086, *Emelei(a) -le(y) -lay(e)* 1150×70–1551, *Emmelei(a) -le(y) -lay* 1190×1220–1361, *Emley* from 1328. OE pers.n. *Ēama* + **lēah**. Alternatively the pers.n. may have been an OE **Emma*. YW ii.218.

EMMANUEL HEAD Northum NU 1343. *Manwell Head* 1610. P.n. Manwell of unknown origin + ModE **head**. A coastal headland on Holy Island.

EMMER GREEN Berks SU 7276. *Ember Green* 1705, *Emmir Green* 1840, 1846. Obscure element, possibly a surname, + ModE **green**. Brk 178.

EMMINGTON Oxon SP 7402. 'The estate called after Eama'. *Amintone* 1086–1428 with variants *-ing-, -yng-, -tun* and *-ton('), Eminton(')* 1199–1402 with variants *-ing-, -yn(g)- -tone* and *-tune*. OE pers.n. *Eama* + **ing**[4] + **tūn**. O 107.

EMNETH Norf TF 4906. Uncertain. *Anemeða* 1170, *Enemeða* 1171, *-meth* 1203, *-methe* 1251, 1361. Attempts to explain this name have focused on the possibility that it contains an old name of the Nar[†] cognate with r.n. Emel, itself an old name of the r. Mole, Surrey TQ 1263, 2347, *(on, andlang) Emenan* [983]c.1325 S 847, *(on) Æmenan, Emenan* [1005]c.1200, representing OE **Ǣmene* 'the misty one' < OE ***ǣmen**. This sense would admirably fit a river in the fenlands. Emneth may, therefore, represent *Ǣmene*, genitive sing. *Ǣmenan*, + **(ġe)mȳthe** 'the mouth of a river where it runs into another' or **hȳth** 'a landing place'. Emneth is situated at the classic location for a *hȳth* at the junction of fen and firmer ground. The recorded sps, however, are difficult to reconcile with OE **Ǣmenan-mȳthe* or **Ǣmenan-hȳth* which should give ME **Emnemeth* or **Emneth*. Other possibilities are probably, therefore, preferable, 'Eana's river mouth', OE pers.n.*Ēana*, genitive sing. *Ēanan*, + **(ġe)mȳthe** or 'Eana's mowing or meadow', *Ēanan* + **mǣth**. DEPN, RN 146.

EMNETH HUNGATE Norf TF 5107 → Emneth HUNGATE TF 5107.

EMPINGHAM Leic SK 9508. 'Homestead of the Empingas, the people called after Empa'. *Epingeham* 1086, before 1118, *Empingaham* 1106–1259, *Empingham* from 1140. OE folk-n. **Empingas* < pers.n. **Empa* + **ingas**, genitive pl. **Empinga*, + **hām**. R 138, ING 147.

EMPSHOTT Hants SU 7531. 'Corner of the bee-swarm(s)'. *Hibesete* 1086, *Himbeset* c.1170, *Y- Imbes(c)hete* 1218–1316, *Imbeshute* 1297–1341, *Impshoote* 1515, *Emshot* 1535, *Empshot* 1579. OE **imbe** + **scēat** or **scīete**. Probably originally a corner of the Greatham estate. Ha 72, Gover 1958.90.

EMSWORTH Hants SU 7405. 'Æmele's enclosure'. *Emeleswurth* 1224, 1239, *Emlesworth* 1304, *Emnesw(o)rth* 1224, 1570, *Empnesworth* 1268–1364, *Emmesworth(e)* 1303, 1327. OE pers.n. *Æmele*, genitive sing. *Æmeles*, + **worth**. Ha 72, Gover 1958.17.

EMSWORTH CHANNEL Hants SU 7401. *Emsworth Channel* 1810 OS. P.n. EMSWORTH SU 7405 + ModE **channel** referring to a navigable channel between Hayling Island and Thorney Island.

ENBORNE Berks SU 4365. 'Duck stream'. *Aneborne -burn(a)* 1086–1223, *Ened(e)burn(e) -borne* 1220–1353, *Eneburna* 1170 etc., with variants *-burn(e) -bern -borne, Westenborne alias Enborne Cheney* 1615. OE **ened** + **burna**. The stream in question is not the present Enborne river which had a different name till at least 1761. East and West Enborne were different manors, the latter being granted to John Cheney in 1542. Brk 9, 294, L 18, RN 148.

ENCHMARSH Shrops SO 5096. *Enchmarsh* 1833 OS.

ENDERBY 'Eindrithi's village or farm'. ON pers.n. *Eindrithi*, genitive sing. *Eindritha*, + **bȳ**.

(1) ~ Leic SP 5399. *Andretesbie* 1086, *Andrede(s)bi* 1188–95, *Endredeby* 1204–1330, *Endrebi* 1086, *-by* 1207–1381, *Enderby* from 1254. Forms with genitival *-es* are probably due to Anglicisation of the pers.n. Lei 494, SSNEM 45 no 1.

(2) Mavis ~ Lincs TF 3666. 'E held by the Malebisse family'. *Malebisse Enderby* 1229. Family n. *Malebisse* + p.n. *Endrebi* 1086, c.1115, *Enderbi* 1142×53, *Andrebi* 1154×89, usually taken to be 'Eindrithi's village or farm', ON pers.n. *Eindrithi*, genitive sing. *Eindritha*, + **bȳ**. The pers.n. is not found independently in Lincs or Yorks although occurring in at least 8 p.ns. The specific may have been an older district name but no satisfactory alternative explanation is known. 'Mavis' for distinction from Wood ENDERBY TF 2763 represents the surname of William Malebisse who held the manor in 1202, cf. ACASTER MALBIS NYorks SE 5945. *Malebisse* is a Norman surname meaning 'evil beast' < Latin *mala bestia*. DEPN, SSNEM 45.

(3) Wood ~ Lincs TF 2763. 'E near the wood'. *Wodenderby* [1198]1328. ME **wode** (OE *wudu*) + p.n. *Endrebi* 1086, 1195, as in Mavis ENDERBY TF 3666. SSNEM 45.

ENDMOOR Cumbr SD 5485. 'The end moor'. *Thendmore* 1586, *Endmoore* 17th cent. Definite article **the** + ME **ende** + **mōr**. We i.96.

ENDON Staffs SJ 9253. Partly uncertain. *Enedun* 1086, 1227, *-don* 1203–1322, *Endon otherwise Yondon* c.1621, *Yenn* 1586, *Yen* 1665. Probably 'lamb hill, the hill where lambs are reared', ***ēan** 'a lamb', genitive pl. ***ēana**, + **dūn**, but the specific could be pers.n. *Ēana*. Oakden, Studies 1936.70, Horovitz.

ENFIELD GLond TQ 3296. 'Eana's open land'. *Enefeld(e)* 1086–1393, *Enfeld* 1293, *Endfelde -field* 1323–1638, *Envill* 1460. OE pers.n. *Ēana* + **feld**. The reference is to cleared space in the woodland of Enfield Chase. Mx 71.

ENFIELD CHACE GLond TQ 2998. *Enefeld Chacee* 1325. P.n. ENFIELD TQ 3296 + ME **chace** 'a tract of open country in which game is bred and hunted'. Mx 73.

ENFORD Wilts SU 1351. 'Duck ford'. *Enedford* [960 for 941]12th S 511, 1290, 1311, *-forde* 1086, *Endford* 1376, *Eneford* 1200–52, *Enford* 1345. OE **ened** + **ford**. Wlt 328, L 71.

ENFORD DOWN Wilts SU 1149. P.n. ENFORD SU 1351 + ModE **down**.

ENGINE COMMON Avon ST 6984. No early forms. Named from an engine used at Yate Colliery formerly worked here.

ENGLEFIELD Berks SU 6272. 'Open land of the Angles'. *(on) Englafelda* c.900 ASC(A) under year 871, *Inglefelle, Englefel* 1086, *Englefeldia* 1154×89 etc. with variants *feld(e) -feud'*. OE folk-n. *Engle*, genitive pl. *Engla*, + **feld**. The name denotes an isolated settlement of Angles in Wessex. Brk 211, 839, PNE i.153.

ENGLEFIELD GREEN Surrey SU 9970. 'The green at Englefield'. *Enfield Green* 1695, 1728, *Englefield or Enville Green* c.1800. The form *Englefield* occurs from 1644 and was earlier *Hingefelda* [967]13th S 752, *Ingefeld* 1291–1434, *Ingel-* 1282, 1625,

[†]Cf. *aqua de Emenhus* 1250 referring to a lost place *Emenhouse* on the Nar.

Ingle- 1586, 1609, *Engelfeud* 1272×1307, 'Inga's open space', OE pers.n. *Inga* + **feld**. The first *l* is intrusive, perhaps by association with ENGLEFIELD SU 6272. Sr 120.

ENGLISHCOMBE Avon ST 7162. 'Engel's coomb'. *Engliscome* 1086, *Ingeliscuma* 1086 Exon, *Inglescumbe* 1227. OE Pers.n. **Engel*, genitive sing. **Engeles*, + **cumb**. DEPN.

ENGLISH STONES Avon ST 5285. *English Stones* 1830. Rocks in the Severn. Gl iii.138.

ENHAM-ALAMEIN Hants SU 3649. P.n. Enham + p.n. Alamein. A rehabilitation centre for ex-servicmen established here after World War I, re-endowed after the Second World War by the Egyptian government in gratitude for the defeat of the German army at El Alamein in 1942. It was then called *Alamein Village*, but the original name was Knight's Enham, *Knytheneshenehem* 1216×72, *Enham militis* 1316, *Knyghtesenham* 1389, for distinction from the royal part of Enham, King's Enham, *Enham regis* 1316. Knight's Enham was assessed as 1½ knight's fees. Enham, *Eanham* 11th, *Enham* from 1167, is the 'lamb estate', OE ***ēan** + **hām**†. The forms *Knytheneshenehem* and *Cnyghteneseham* 1298 show illogical double genitive, ME **kniht** + adjectival **-ene** < OE weak declension genitive pl. *-ena*, + **-es**. Ha 73, Gover 1958.166, DEPN, Studies 1936.70.

ENMORE Somer ST 2335. 'Duck pond'. *Animere* 1086, *Enemere* 1200, [1295] Buck, *Enedemere* 1315, *Enmore* 1610. OE **ened**, genitive pl. **eneda**, + **mere**. Possibly the forerunner of the lake in the grounds of Enmore Castle. DEPN, VCH vi.36.

ENNERDALE Cumbr NY 0715. Originally 'Anund's valley', later understood as 'valley of the river Ehen'. *Anenderdale* [c.1135]12th, *Ananderdale* [1189×99]1308, *Enderdale* 1303, *Eghner- Eynor- Eghenner- Eynerdale* 14th cent., *Enerdale* 1395, *Ennerdale* 1777. ON pers.n. *Anundr*, genitive sing. *Anundar*, later partly replaced by river n. EHEN NY 0112, + **dalr**. Cu 385, SSNNW 121, L 94.

ENNERDALE BRIDGE Cumbr NY 0715. P.n. ENNERDALE NY 0715+ ModE **bridge**. *Bridge in Enerdall* 1656. Cu 385.

ENNERDALE FELL Cumbr NY 1313. *Ennerdale Fell* 1868 OS. P.n. ENNERDALE NY 0715 + ModE **fell** (ON *fjall*).

ENNERDALE WATER Cumbr NY 1015. *Ennerdale Water* 1868 OS. P.n. ENNERDALE NY 0715 + ME **water**. Earlier names were *Eyneswater* 1322 (river n. EHEN NY 0112 + **water**), and *Brodwater* 1610, *ye Broadwater* 1650. Cu 34.

ENSBURY Dorset SZ 0996. 'Ægen's fortified place'. *Eynesburgh* 1463, *Ensbury* early 17th. OE pers.n. *Æġen*, genitive sing. *Æġnes*, + **burh**, dative sing **byriġ**. Do ii.23.

ENSDON Shrops SJ 4117. *Ensdon* 1833 OS.

ENSIS Devon SS 5626. 'The farm, the enclosures'. *Hayne* 1809 OS. ME **heghen**, pl. **heghenes**.

ENSTONE Oxon SP 3724. 'Enna's stone'. *Hen(n)estan -stona* 1086–1309, *En(n)estan(e) -a -ston(e) -stan(n)(')* [1185]13th–1349, *Enston* 1381, 1428. OE pers.n. *Enna* + **stān**. The reference is to remains of a Neolithic stone burial chamber now known as Hoar Stone. O 347, Thomas 1976.177.

ENTWISTLE STATION Lancs SD 7217. P.n. Entwistle + ModE **station**, a stop on the railway line from Bolton to Blackburn. Entwistle, *Hennetwisel* 1212, *(H)en(n)etwysel, Ennu- Emmetwesille* 1276, *Entwysel* 1292, 1341, is 'Enna's river fork', OE pers.n. *Enna* + **twisla**, referring to the junction of Broadhead Brook and a tributary to form what is now Wayoh Reservoir. La 47, Jnl 17.38.

ENVILLE Staffs SO 8286. 'The level open ground'. *Efnefeld* 1086, *Euenfeld* 1183, *Evenefeud* 1240, *-feld* 1332ff. OE **efn**, definite form **efna**, + **feld**. DEPN, Horovitz.

EPPERSTONE Notts SK 6548. Possibly 'Eorphere's farm or village'. *Ep(re)stone* 1086–1330, *Aperstone* c.1090, *Eperiston -es-* 1242–1327, *Eperson* 18th cent. OE pers.n. *Eorphere*, genitive sing. *Eorpheres*, + **tūn**. Nt 162.

EPPING Essex TL 4602. 'The Eppingas, the upland people'. *Ep(p)ingam, Eppinḡ, Epingā* 1086, *Eppinges* 1086–1225, *Epping(e) -a* from c.1144, *Upping* 1227, *Yppynge* 1455, *Ippyng -ing* 1508, 1509×47. OE folk-n. **Yppingas*, Essex dial. form **Eppingas*, < **yppe** + **ingas**. The reference is to the high ground at the N end of Epping forest. The original site was near Epping Church at Epping Upland TL 4404. The present town of Epping is at a place originally called *Eppinge streete* 'Epping hamlet' 1594. Ess 22, ING 19.

EPPING FOREST Essex TQ 4197. *Epping Forest* 1662. P.n. EPPING TL 4602 + ModE **forest**. Also known as *foresta de Wautham* 'Waltham forest' 1261 and *foreste de Essex* 1170, *foresta regis Essex* 'the royal forest of Essex' 1248. At its fullest extent it stretched from Waltham Abbey, Harlow, Hainault and Havering atte Bower to Navestock and Chelmsford. Ess 1.

EPPING GREEN Essex TL 4305. *Epping Long Green* 1805 OS. P.n. EPPING TL 4602 + ModE **green**.

EPPING GREEN Herts TL 4305. *Epping Green* 1822 OS. P.n. EPPING TL 4602 + ModE **green**.

EPPING UPLAND Essex TL 4404. No early forms. The site is at Epping Church, *Epping churche* 1594. P.n. EPPING TL 4602 + ModE **upland**. Ess 22.

EPPLEBY NYorks NZ 1713. 'Apple-tree farm or village'. *Aplebi* 1086, *Appelbi -il- -by* 1157–1440, *Eppilby* 1421. OE **æppel** possibly replacing ON **epli**, + **bȳ**. YN 298, SSNY 26.

EPSOM DOWNS STATION Surrey TQ 2259. 'The railway station at Epsom Downs'. Cf. *Downelande Grove* 1574. P.n. EPSOM TQ 2160 + ModE **down** + **station**. Sr 74.

EPSOM Surrey TQ 2160. 'Ebbi's homestead'. *Ebbesham* [late 10th]early 12th S 1457–1719, *Eveshā* 1086, *Ebesham* [727, 933, 967, 1062]13th S 1181, 420, 752, 1035–1375, *Ebs-* 1297–1581, *Epsam* 1404, *-some* 1680, 1734, *Epsom* from 1718. OE pers.n. *Ebbi*, genitive sing. *Ebbes*, + **hām**. Sr 74, ASE 2.32.

EPWELL Oxon SP 3540. 'Eoppa's spring'. *(on, of) Eoppan wyllan (broc)* [956]12th S 617, [956]16th S 584, *(on) Eoppan welles (stream)* [956]12th S 611 with variant *wylles* [956]13th ibid., *Ep(p)ewella -well(e)* 1187–1506, *Epwell* 1537. OE pers.n. *Eoppa*, genitive sing. *Eoppan*, + **wylle**. O 421, L 31.

EPWORTH Humbs SE 7803. 'Eoppa's enclosure'. *Epeurde* 1086, *Appe(l)wurda* 1179, *Eppeworth* 1233. OE pers.n. *Eoppa* + **worth**. DEPN.

Child's ERCALL Shrops SJ 6625. 'E held by the young nobleman'. *Childes Erkelewe, Erkalew* 1249×4, *Childes Ercall* 1410, *Child(e)s Arcoll* 1577, 1699. ME **child**, genitive sing. **childes**, + p.n. *Arcalun* 1086, *Arkeluu* 1198, *Arkelawe* 1200, *Ercalewe, Erkelawe, Erkalow* 14th cent., *Archall* 1708, possibly a district name from OE **ēar** 'gravel, mud, earth' + an OE **Calwe*, **Calwa* 'bare hill, bare one' < **calu** 'bald, bare, lacking vegetation' possibly originally applied to the site of High Ercall SJ 5917 or to the hill called The Ercall SJ 6409. Subsequently partly assimilated to ME **lawe** (OE *hlāw*) 'a tumulus, a hill'. 'Child's', for distinction from High ERCALL SJ 5917, may refer to a junior member of the Strange family which held the manor. Also known as *Parva Erkalawe* 1242, *Little Ercall* 1449. For loss of the final syllable cf. EDGMOND SJ 7219. Sa i.124 gives pr *Arcall* but 18th cent. sps point to a pr with [tʃ], Nomina 11.107.

High ERCALL Shrops SJ 5917. *Hegh Arkall'* 1390, *High Arcall -coll* 16th, *High(e) Ercall* from 1569, ME adj. **hegh** (OE *hēah*) + p.n. *Archelou* 1086, *Archalou* 1121, *Archalue* 1155, *Ercalu(we) -lowe -lewe, Erkalow -loe -lewe* 13th–14th cent., *Erkall* 1394, *Ercall, Arcoll, Archaule* 16th cent., *Archall* 1703, as in Child's ERCALL SJ 6625. Also known as *Magna Ercaluwe* 1315, *Miche Ercall* 1535, *Greate Ercoll* 1548×9. Sa i.124.

ERDINGTON WMids SP 1192. 'The estate called after Earda'. *Hardintone* 1086, *Erdinton(e)* c.1180–1357, *Erdingtona* 1230, *-yngton* 1346–1581, *Yardington* 1517–71, *Yarnton* 1576–1649. OE

†DB gives the form *Etham* for both manors; this may be a mistake but may alternatively represent OE **ete** + **hām** 'pasture estate'.

pers.n. *Earda + **ing**[4] + **tūn**. Wa 30 records a former pr [ja:rntən].

River EREWASH Notts SK 4743. Originally 'the wandering stream'. *Irewys -wis* [c.1145]14th, c.1175, 1314, *Yrewis* c.1175, *-wys* 1317, 1330, *Irrewysa* c.1300, *Irwys(se)* 1330, c.1500, *Erwys -wis* 17th, *Yrewas* 1272×1307, *Erwash* 1576, 1577, 1586. OE **irre** + **wisce** later replaced by **(ġe)wæsc** 'a washing, a flood'. The Erewash was a very winding stream and had many different channels, cf. the form *Holdeerwys* 1330 'old Erewash'. Alternatively the specific could be OE pers.n. *Íra*. Nt 4 gives pr [eriwɔʃ], RN 148–9.

ERIDGE GREEN ESusx TQ 5535. P.n. Eridge + ModE **green**. Eridge, *Ernerigg* 1203, *-regge* 13th cent., *-rugge* 14th–15th cents., *Er(r)egge* 1296, 1382, is 'eagles' ridge', OE **earn**, genitive pl. **earna**, + **hrycg**. Sx 374.

ERISWELL Suff TL 7278. 'The boar's spring'. *Hereswella* 1086, *Ereswell* 1183, 1242, *Evereswell* 1249, *E*^{r}*sewell* 1610. OE **eofor**, genitive sing. **eofores**, + **wella**. In the absence of further evidence this explanation depends on the 1249 sp. Otherwise possibly 'Her or Heri's spring', late OE pers.n. *Her* or **Heri*, genitive sing. *Heres*, + **wella**. DEPN.

ERITH GLond TQ 5177. 'Gravel landing-place'. **Earhyð* [677]16th S 1246, *-hið* c.960 S 1458, *Erhede* 1086, *Herhetthe, Erhether -huthe* 13th cent. OE **ēar** + **hȳth**. The sense is probably 'gravelly landing place' rather than 'place where gravel is landed'. KPN 17, DEPN, TC 204.

ERLESTOKE Wilts ST 9654. 'The earl's Stoke'. *Erlestok(e)* from 12th. Also known as *Littlestoke* 1310, 1344, and simply *Stokes* 1190, *Stoke* 1242, OE **stoc** 'an outlying farm'. The reference is probably to Earl Harold who held Melksham to which Stoke belonged at the time of King Edward the Confessor. Wlt 126, Studies 1936.23.

River ERME Devon SX 6355. Unexplained. *Irym* 1240, *Hyrm* 1281, *Irm* 1346, 1355, *Erm(e)* from 1240, *Arme* c.1550. The form *Ermyn* 1425 has been influenced by the p.n. ERMINGTON SX 6353, but it seems unlikely that the name as a whole can be a back-formation. RN 149.

ERMIN WAY Glos SO 9611. The Roman road from Silchester to Cirencester and Gloucester, Margary no. 41(b)-(c). Identical with *Ermingestrete* 13th, *the Irminstreet* 1779, a transferred name from ERMINE STREET named after the *Earningas*, an Anglo-Saxon people whose name survives in Armingford, Cambs, *Ernin(c)gaford* 1086 'the ford of the Earningas', and ARRINGTON Cambs TL 3250. Gl i.16, Ca 50.

ERMINE STREET Cambs TL 1593, 2667, Humbs SE 9418, Herts TL 3720, Lincs SK 9687. 'Roman road of the Earningas, the people called after Earna'. *Earninga strǣt(e)* [955]12th S 566, 957 S 649, [1012]12th S 1562, *Erningstret(e)* [before 1152]13th, c.1230, c.1300, *Arningstrat* 1251, *Ermingestrete* [c.1090]c.1230, c.1400, 1586. OE folk-n. *Earningas* < pers.n. *Earna* + **ingas**, genitive pl. *Earninga* + **strǣt**. The name of this people is preserved in ARRINGTON TL 3250 and in *Ærningaford* [970]13th S 776, *Earmingaford, Earnigaford* [970]c.1100 S 779, the ford which once carried Ermine Street across the r. Cam at TL 3448. BdHu 2, Herts 6, Ca 22.

ERMINGTON Devon SX 6353. 'Settlement on the river Erme'. *Ermen- Ermtone* 1086, *Erminton(a)* from 1131 with variants *-ing-* and *-yng-*, *Armyngton* 1614. R.n. ERME SX 6355 + OE **tūn**. D 272, 264, RN 149.

ERPINGHAM Norf TG 1931. 'The homestead of the Eorpingas, the people called after Eorp'. *(H)erpincham, Erpingeham Nord, Erpingaham Nord, ~ Svd, Erpingham Svd* 1086, *Erpingham* from 1210×12 including *[1044×7]13th S 1055, *Erpingeham* 1201. OE folk-n. **Eorpingas* < pers.n. **Eorp* + **ingas**, genitive pl. **Eorpinga*, + **hām**. DEPN, ING 134.

ERRWOOD RESERVOIR Derby SK 0175. P.n. Errwood + ModE **reservoir**. Errwood is *Errwood* 1693. Che i.173.

ERWARTON Suff TM 2134. 'Eoforweard's estate'. *Alwartunā, Eure Wardestuna* 1086, *Euerewardeston* 1196, *Euerwar(d)ton(e)* 1254–1346, *Erwarton* from 1346, *Arwarton -wer-* 1568, 1610. OE pers.n. *Eoforweard*, genitive sing. *Eoforweardes*, + **tūn**. DEPN, Baron.

ERYHOLME NYorks NZ 3208. '(The settlement) at the shielings'. *Argun* 1086, *-um* 1179, *Erg(h)um* 12th–1346, *Eryom* 1285–1404, *Eriholme* 1665. OWScand ***ǽrgi**, dative pl. ***ǽrgum**. YN 280 gives pr [erium], SSNY 86.

ESCOMB Durham NZ 1830. '(At the) enclosures'. *Ediscum* [c.994]11th S 1659, [c.1040]12th, [1072×83]15th, c.1104, *Escum* 1242×3, 1303, *Escombe* [1183]c.1320–1647, *East Combe* 1647, *Escomb* from 1647. OE **edisc**, dative pl. **ediscum**. NbDu 76.

ESCRICK NYorks SE 6342. 'Ash-tree *Ric*, the strip of land'. *Ascri* 1086, *Ascric* 1156×7, *Esk(e)rik(e) -c- -y-* 1169–1524, *Estcrick, Eskirke, Eskrigg* 1607. OE **æsc** 'an ash-tree' influenced by ON **eski** 'growing with ash-trees', + OE ***riċ** with Scand [k] for English [tʃ] perhaps already used as a p.n. The reference is to a narrow ridge of land running from Stillingfleet to WHELDRAKE SE 6845 *Queldric* 14th. Escrick is halfway along it. YE 267–9, SSNY 152, L 184.

ESH Durham NZ 1944. Probably 'the ash-tree' or 'ash-tree thicket'. *Es* 1153×95, 1185×95, c.1270, *héés* 1180×96, *Ess(e)* 1188×96–1322, *Eys* c.1204, *(H)eyss(c)he, Eissche* early 14th, *Es(s)che* 1298–1356, *Essh(e)* 1312–1567, *Esh* from 1382, *Assh* 1382. There is some uncertainty about the etymology of this name: the majority of forms point to OE **æsc, esc** 'an ash-tree' or possibly an OE collective ***esce** 'a clump of ash-trees, an ash thicket' with Anglo-Norman *s* for *sh*. OE **esc** and forms with *ei/ey* spellings could be the result of *ś*-umlaut (Morsbach §87 Anm.3), but *ei/ey* can also be graphies for a long vowel and together with the spelling *héés* 1180×96 possibly point to OE ***hǣs** 'brushwood'. NbDu 77.

ESH WINNING Durham NZ 1941. 'Esh coal-mine'. P.n. ESH NZ 1944 + ModE **winning**. Coal was first won at Esh Winning only in the second half of the 19th cent.

ESHER Surrey TQ 1464. 'The share of land with ash-trees'. *(to) Æscæron* [1005]12th S 911, *Aissela -e* 1086, *Assere* 1242, *A(ys)sher* 1509, *Esshere* 1100×29–1623 including [1062]13th S 1035, ~ *Episcopi* 1404, *Esere-Wateville* 1284, *Esher* from 1680. OE **æsc-scearu**. The 1086 (DB) form is corrupt. Part belonged to the bishop of Winchester from 1245, the other was held by William de Vatteville (from Vatteville, Eure, France) in 1086. Sr 92.

ESHER COMMON Surrey TQ 1362. *Esher Com*^{n} 1816 OS. P.n. ESHER TQ 1464 + ModE **common**.

ESHOTT Northum NZ 2097. 'The clump of ash-trees'. *Esseta* 1186, *Esset'* 1242 BF, *Es(sc)het(te)* c.1200–1428, *Eshott* 1638. OE ***æscett**. NbDu 77, PNE i.5.

ESHTON NYorks SD 9356. 'Ash-tree settlement'. *Estune* 1086, *-ton(a)* 1200×16–1316, *Escheton* 1287, 1379, *Es(s)heton* 1303–1597, *Es(s)hton* 1428–1654. OE **æsc** + **tūn**. YW vi.47, L 219.

River ESK An Old European river n. **Eis-kā* formed by a *k*-suffix extension of the IE root **is-/*eis-* 'move swiftly, strongly'. BzN 1957.241.

(1) ~ Cumbr NY 3666. *Esch* after 1165, *Ask* 1194×1214, *Eske* [c.1205]n.d., 1279, *Esk* from early 14th. Cu 13, RN 152, BzN 1957.241.

(2) ~ Cumbr SD 1297. *Esc* [c.1140, 1180]n.d., *Esk* from c.1180, *Hesk, Esch, Eske* 13th cent. Cu 14, RN 151, BzN 1957.241.

(3) ~ NYorks NZ 7207, 8708. *Esch* 1109×14–[1199]15th, *Esk(e)* from 1204. YN 3.

ESK DALE NYorks NZ 7407. 'Valley of the r. Esk'. *Eschedale -a* 1086, *Eskedal(a)* 12th cent., *Eskdale* 1336. R.n. ESK NZ 7207, 8708 + ON **dalr**. YN 119.

ESKDALE Cumbr NY 1700. 'Valley of the river Esk'. *Eskedal(e)* [1285]n.d.–1368, *Asshdale* 1461, *As(s)he- Esshedaille -dale* 16th cent., *Eshdell* 1606. R.n. ESK SD 1297, probably identified by Viking settlers with ON *eski* 'ash-trees', + ON **dalr**. Cu 388 gives occasional pr [eʃdəl], SSNNW 227.

ESKDALE GREEN Cumbr NY 1300. P.n. ESKDALE NY 1700 + **green**. A halt on the Eskdale railway.

ESPLEY HALL Northum NZ 1790. This is *High Espley* 1866 OS. P.n. *Es(s)peley* 1242 BF, 1257, *Aspele* 1252, 'the aspen-tree wood or clearing', OE **æspe** + **lēah**, + ModE **hall**. NbDu 78, L 204.

ESPRICK Lancs SD 4036. 'Ash-tree slope'. *Eskebrec* [c.1210]1268, *Ester- Escebrec* 1249, *Askebrek* 1332, *Esbrek(e)* 15th, 16th. ON **eski** + **brekka**. The terrain slopes gently up from Esprick on the banks of Thistleton Booth to High Moor at SD 3936. La 154, Jnl 17.90.

ESSENDINE Leic TF 0412. 'Esa's valley'. *Esindone* 1086, *Esenden'* 1222–1316, *Es(s)ingden(e) -y-* 1286–1509, *Ezenden, Easondyne* 17th, *Is(s)enden(e)* 1185–1263. OE pers.n. *Ēsa*, genitive sing. *Ēsan*, + **denu**. R 147, L 98.

ESSENDON Herts TL 2708. 'The valley of the Eslingas, the people called after Esla'. *(into) Eslingadene* [11th]12th K 1354, *Esendene* [1077]14th, 1342, *Essendon(e)* 1172–1362. OE folk-n. **Ēslingas* < pers.n. **Ēsla* + **ingas** as in EASTLING Kent TQ 9656, genitive pl. **Ēslinga*, + **denu**. The later forms point rather to 'Esa's valley', pers.n. *Ēsa*, genitive sing. *Ēsan*, + **denu** as in ESSENDINE Leics TF 0412. **Ēsla* is a diminutive form of *Ēsa*. Hrt 223.

ESSEX The county name. *(regis) East Saxonorum* 'king of the east Saxons' [685×94]8th S 1171, [704]8th S 65, *East Se(a)xe* 9th–12th ASC, *(to) ast sexū* 10th B 1335, *Essexe syre* [1086]12th. OE **ēast** + folk.n. *Seaxe*, Latin *Saxi*. Jnl 2.39.

ESSINGTON Staffs SJ 9603. 'The settlement of the Esningas, the people called after Esni'. *Esingetun* [996 for 994]17th S 1380, *-ton(e)* 1166, 1271, *Eseningetone* 1086, *Esenington(a) -yng-* c.1201×23–1324, *Esington -yng-* 1205–1436, *Essington* from 1271. OE folk-n. **Esningas* < pers.n. *Esni* + **ingas**, genitive pl. **Esninga*, + **tūn**. St i.49.

ESTHWAITE Cumbr SD 3595. 'East clearing'. *Estwyth* 1539, *Easthwaite* 1670. ME **est** + **thweit** (ON *thveit*). La 218 gives pr [estwət].

ESTHWAITE WATER Cumbr SD 3696. *Esthwaite Water* 1845 OS. P.n. ESTHWEAITE SD 3595 + ModE **water** replacing earlier *Estwater* 1537, 'east lake', ME **est** + **water**. Also known as the *Mere of Hawkshed* 1539, p.n. HAWKSHEAD SD 5398 + **mere**. La 218, Mills 82.

ESTON Cleve NZ 5519. 'East settlement'. *Astun(e)* 1086, *Eston(a)* from 1160×72. OE **ēast** + **tūn**. Eston is situated in the east of the parish of Ormesby. YN 157.

ESTOVER Cambs TL 4298 → WESTRY TL 3998.

ETAL Northum NT 9239. Either 'Eata's nook of land in the river bend' or 'river-bend land used for grazing'. *Ethale* 1232–1371, *Etale* 1296 SR, *Etal* 1346, *Eatle* 1655. OE pers.n. *Ēata* or **ete** + **halh**. NbDu 78 gives pr [i:təl].

ETCHILHAMPTON Wilts SU 0460. Partly uncertain.
I. *Ec(h)esatingetone* 1086, *Hechesetingeton* 1206.
II. *Echehamt', Ehelhampton'* 1196, *Echelham(p)ton* 1236–1332, *Echilhampton* 1622, 1647, *Ashlington* 1622, 1773.
The generic is OE **tūn** 'settlement, village, estate'. The best explanation of the specific is Ekwall's. Form I represents OE **Ǣċ-sǣtingas*, short for **Ǣċ-hyll-sǣtingas*, 'the dwellers at **Ǣċ-hyll*, the oak-tree hill' referring to Etchilhampton Hill due W of the village, OE **ǣċ**, dative sing. of **āc**, + **hyll** + **sǣtingas**, genitive pl. **Ǣċ-sǣtinga*; Form II **Ǣċ-hǣme*, again short for **Ǣċ-hyll-hǣme* with the same meaning, OE **ǣċ** + **hyll** + **hǣme**, genitive pl. **Ǣċ-hǣma*, **Ǣċ-hyll-hǣma*. Wlt 313 gives pr [æʃəltən], formerly [æʃliŋtən], DEPN.

ETCHINGHAM ESusx TQ 7126. 'Promontory or river meadow of the Eccingas, the people called after Ecci'. *Hechingehā* 1158, *E(c)chingeham* 12th–14th cents., *-hamme* 1176, *Echingham* 13th–16th cents. OE folk-n. **Eċċingas* < pers.n. *Eċċi* + **ingas**, genitive pl. *Eċċinga*, + **hamm** 2, 3, 6. Sx 455, ING 124, ASE 2.25, SS 83.

ETCHINGHILL Kent TR 1639. Partly uncertain. *Tettingehelde, Tettingheld'* 1240, *Tetingheld' -yng-* 1278, 1327. The generic is OE **helde** 'a slope, a hill'. The specific might represent OE genitive pl. **Tettinga* of **Tettingas* 'the people called after Tetta or Tette', OE pers.n. **Tetta* or feminine *Tette* + **ingas** or locative-dative **Tettinge* of **Tetting* 'the place called after Tetta or Tette'. Initial *T-* became lost by false division of the phrase *at Tettingheld* as if *at Ettingheld*. PNK 435.

ETCHINGHILL Staffs SK 0218. Partly unexplained. *Eychilhill* 1504, *Echynge hill* 1554, *le Echin- Ichinhill* 1584, *Itching Hill* 1698, *Eaching Hill* 1798, *Hitching Hill* 1834. Unidentified specific + ModE **hill**. St i.106, Horovitz.

ETHERLEY Durham NZ 1628. Partly uncertain. *Ederlee* 1392, *-ley* 1407, 1437×45, *Edirley* 1408, 1437, *Etherley* 1647. The generic is OE **lēah** 'a clearing', the specific possibly OE **eodor** 'a fence, an enclosure' or a pers.n. such as *Æthelrǣd, Ēadhere* or *Ēadrǣd*. The forms are too late for certainty. NbDu 78.

River ETHEROW Derby SJ 9791, GMan SJ 9791. Originally a hill name 'water-course hill-spur' transferred to the river. *Ederau* c.1216×20, *Ederhou* 1226, *-ou* 1216×72, c.1251, 1285, *Edderowe* 1290, *Ederow* 1386, *Etherow* 1767. OE **ēdre** 'water-course' + **hōh** 'a hill-spur'. An older name of this river may be preserved in the p.n. TINTWISTLE Ches SK 0297. Db 7, Che i.23.

ETON Berks SU 9677. 'River settlement'. *Ettone* 1086–1255, *Eitun -ton* 1155–1211, *Eton* from 1207. OE **ēa** + **tūn**. Forms in *Ei-* show confusion with OE *ēġ-tūn* 'island settlement'. The river in question is the Thames. The famous public school, Eton College, was founded by Henry VI in 1440. Bk 236, L 20, DEPN.

ETTERSGILL Durham NY 8829. Probably 'Eitri's ravine'. *Ethresgilebec* [c.1160×83]16th, *Ettresgile, Etresghilebec* 1333. Pers.n. ON *Eitri* or possibly *Edred, Eder* < OE *Ēadrǣd, Ēadhere* or *Æthelrǣd*, secondary genitive sing. *Etres, Edres*, + **gil**. ON *gil* is a diagnostic for Irish-Norwegian settlement in upper Teesdale. Nomina 12.30.

ETTINGSHALL WMid SO 9396. Probably 'the nook of land of the grazing place'. *Ettingeshale* 996, 1261, *Etinghale* 1086, *Eltingehal(e)* 1175, 1196. OE ***et(t)ing**, genitive sing. ***et(t)inges**, + **halh**. DEPN.

ETTINGTON Warw SP 2749. 'Eata's' or 'pasture hill'. *Ete(n)done* 1086, *Etendon(e) -in- -yn-* 1247–87, *Etyngton* 1485, *Eatington* 1587, *Ettington* 1651. OE pers.n. *Ēata*, genitive sing. *Ēatan*, or **ete(n)** + **dūn**. Wa 253 gives pr [etiŋtən], L 154.

ETTON Cambs TF 1306. Partly uncertain. *Etton(a)* from 1125×8. The generic is **tūn** 'farm, village, estate', the specific uncertain, perhaps a pers.n. *Ēata*, perhaps OE **ēa** 'river'. Etton now lies on a small stream for which *ēa* would be inappropriate but which may have once been an old course of the Welland. Nth 234.

ETTON Humbs SE 9843. Possibly 'Eata's farm or village'. *Eton* 1086, 13th cent., *Etton(e)* 1086, *Ettun -ton(a)* 1179×89–1212, *Ecton* 13th cent. OE pers.n. *Ēata* + **tūn**. YE 190.

ETWALL Derby SK 2732. 'Eata's or pasture spring'. *Etewell(e)* 1086–1436, *Ettewella -(e)* 1199–1328, *Et(t)ewall(e)* 1275–1399, *Et(t)wall* from 1351. OE pers.n. *Ēata* or **ete** + Mercian **wælla**. Db 559, L 31.

EUSTON Suff TL 8978. Probably 'Efi's estate'. *Eu(e)stuna* 1086, *Euuestun* c.1095, *Euston* 1242, 1610. OE pers.n. *Efi*, genitive sing. *Efes*, + **tūn**. The pers.n. is possibly that of an 8th cent. East Anglian moneyer. DEPN, BNJ 55.30, Signposts 259.

EUSTON STATION GLond TQ 2982. P.n. Euston transferred from EUSTON Suff + ModE **station**. The station, opened in 1837 adjacent to Euston Grove and Euston Square, was built on land owned by the Duke of Grafton whose seat was at Euston Hall, Suffolk. Harris 27.

EUXIMOOR FEN Cambs TL 4799. *Tuxmore Fen* (sic for *At Euxmore Fen*) 1654. P.n. Euximore + pleonastic ModE **fen**. Euximore, *Yekeswellemoor* 1431, *Ȝekeswellmore* (for *Ȝekes-*) 1434, *Ixwellmoor* 1605, *Euxmoor* 1562, *Eusimore* 1618, *Eximore* 1654, is 'Yekeswell moor'. P.n. Yekeswell 'cuckoo's spring',

ME **yeke** (OE *ġeac*), genitive sing. **yekes**, + **welle** (OE *wella*), + **moor**. Ca 289.

EUXTON Lancs SD 5519. 'Eofoc's farm or village'. *Eueceston* 1187, *Euekeston* 1188, *Eu(c)keston* 1212–1332 etc. OE pers.n. **Eofoc*, genitive sing. **Eofoces*, + **tūn**. The usual explanation of the specific as OE pers.n. *Æfic, Efic* is impossible for phonological reasons. La 133, Jnl 17.76, Anglia 109.541.

EVEDON Lincs TF 0947. 'Eafa or Eafe's hill'. *Euedune* 1086, 1185, *-done* 1166, *-don(')* from 1296. OE pers.n. *Eafa* or feminine *Eafe* + **dūn**. Perrott 62, Cameron 1998.

EVENLEY Northants SP 5834. 'The level wood or clearing'. *Evelaia, Avelai* 1086, *Euenlai -lee -ley -leg* from 1147, *Yevynle* 1427, *Imley, Emley* 17th. OE **efen** + **lēah**. Nth 52 gives prs [imli, emli, evənli].

EVENLODE Glos SP 2229. 'Eowla's river-crossing'. *(æt) Eu(u)langelade* [772 for 775]16th, 11th S 109, *Eowengelad* *[784]17th S 122, *(æt, to) Eowlangelade* [969]11th S 1325, *Eunelade* *[777]12th S 112, *[1044×59]12th S 1057, 1221, *-lode* [c.1050]12th S 1548, *Eownilade* [779]12th S 115, *-ig-* [c.1000]11th S 1534, *Eowenland* *[964]12th S 731, *Eunilade* 1086, 1221, *Evenlade* 1185, 1275, *-lod* 1369, 1712, *Emlade* 1378, *-lode* 1428–1723. OE pers.n. **Eowla*, genitive sing. **Eowlan*, + **ġelād**. The later *Even-* is an inverted spelling due to the development of OE *efen* 'even' to *em-* as in ME *emcristen* for *even-cristen* 'fellow Christian'. Gl i.219 gives prs ['emloud], ['i:vənloud], L 73–4.

River EVENLODE Oxon SP 3220. *Euenlod(e)* 1577, 1586, *(Y)enload* 1612. A late back formation from p.n. EVENLODE Glos SP 2229. The original name of the river was *(to) Bladene* [1005]late 12th etc., *Bladon* *[675]13th S 1245, *Bladene* [718 for ?727]11th S 84 –1298 including [777]12th S 112, [779]12th S 115, [969]11th S 1325, [979]11th S 1340, S 1548 & S 1550, *Bladen* [772 for 775]16th S 109, 13th cent., a Brit r.n. of unknown origin and meaning. O 7, RN 36, 157.

EVENWOOD Durham NZ 1524. 'Level wood'. *Efenwuda* [c.1040]12th, 1104, *Ewenwod(e)* 1242–1522, *Evenwod(e)* 1295–1522, *Evenwood* from 1422. OE **efen** + **wudu**. NbDu 79.

EVERCREECH Somer ST 6438. Partly uncertain. *Evorcric* *[1065]18th S 1042, *Evrecriz* 1086, *Euercriz* 1086 Exon–1310, *Euercrich'* 1327. This might be 'cow-parsnip or bogweed hill', W **efwr**, Co ***evor**, + PrW ***crüg**. Evercreech stands in a horseshoe of hills of which Creech Hill is the highest but no one specific *crūg*-type hill relates directly to Evercreech. Dr Gelling suggests that Creech here may have become a district name and that Evercreech was a specialised settlement in the district, '(the settlement in) Creech where boars are bred', OE **eofor** + p.n. Creech < ***crüg**. DEPN, Turner 1952–4.16, CPNE 96, GPC 1173.

EVERDON Northants SP 5957. 'Wild boar hill'. *eferdun* 944 S 495, *Everdon(e)* from 1086. OE **eofor** + **dūn**. The boundary of the people of Everdon is mentioned in 1021×3 S 977 as *eofordunenga gemære*. Nth 21.

EVERINGHAM Humbs SE 8042. 'The homestead of the Eoforingas, the people called after Eofor'. *Yferingaham* [972×92]11th S 1453, *Euringha'* 1086, *Everingeham* 1185×95–1288, *Everingham* from 1192×8. OE folk-n. **Eoforingas* < pers.n. *Eofor* + **ingas**, genitive pl. **Eoforinga*, + **hām**. YE 233, ING 153.

EVERLEIGH Wilts SU 2053. 'The boar wood or clearing'. *Eburlea(g)h* [704]13th S 245, *Everlegh -le(ye)* 1249 etc. OE **eofor** + **lēah**. Probably a place where swine were pastured. Wlt 329, L 205.

EVERLEY NYorks SE 9789. 'Wild boar clearing'. *Eurelai -lag* 1086, *Everle* 1177×89–1328, *Yereley* 1577. OE **eofor** + **lēah**. YN 115 gives prs [evələ], and [jiələ] for which cf. YEARSLEY SE 5874, YORK SE 6052.

Great EVERSDEN Cambs TL 3653. *Micheleueresdona, Euerisdone magna* 1272. ME adj. **michel**, Latin **magna**, + p.n. *Euresdone* 1086, 1199, *Au(e)resdone* 1086, *Evresdon* 1200×3–1382, *Eversdon(e)* 1239–1503, *-den* from 1303, 'boar's hill', OE **eofor**, genitive sing. **eofores**, + **dūn**. 'Great' for distinction from Little EVERSDEN TL 3752. Ca 159.

Little EVERSDEN Cambs TL 3752. *Everesdone Parva* 1240. ModE adj. **little**, Latin **parva**, + p.n. Eversden as in Great EVERSDEN TL 3653. Ca 159.

EVERSHOLT Beds SP 9933. 'Boar wood'. *Eures(h)ot* 1086, *Euresholt* 1185, *Eversoll* 1518. OE **eofor**, genitive sing. **eofores**, + **holt**. Bd 123, L 196.

EVERSHOT Dorset ST 5704. 'Corner of land frequented by wild boar'. *Teversict* (sic) 1202, *Theuershet* 1268, *Evershet* 1286, *Evershute* 1432. OE **eofor** + **scēat(a)**, WS ***scīete**. The spellings with *T-* and *Th-* derive from misdivision of the prepositional form *at Evershot*. The reference of *scīete* in this case is to the valley of the upper reaches of the r. Frome. Do 71.

EVERSLEY CROSS Hants SU 7961. 'Eversley cross-roads'. Cf. *Cross Green* 1759, 1816 OS. P.n. EVERSLEY SU 7762 + ModE **cross**. Ha 73, Gover 1958.120.

EVERSLEY Hants SU 7762. 'Boar' or 'Eofor's wood or clearing'. *(æt) Everslea* [1053×66]13th S 1129, *Evreslei* 1086, *Evereslea -le(ge) -le(gh) -ligh* 1174–1291, *Everle* 1341, 1412. OE **eofor** or pers.n. *Eofor*, genitive sing. **efores**, *Eofores*, + **lēah**. If this is the common noun *eofor* the reference may be to pig husbandry. Ha 73, Gover 1958.119.

EVERTON 'Boar farm'. OE **eofor** + **tūn**.

(1) ~ Beds TL 2051. *Euretune -tone* 1086, *Euerton* 1227. Bd 104.

(2) ~ Mers SJ 3592. *Evretonam* 1094, *Everton* from 1201, *Ouerton* 1226, 1346, *Earton* 1577. La 115, Jnl 17.64.

(3) ~ Notts SK 6991. *Evretone* 1086, *Everton* from 1184. Nt 29.

EVERTON Hants SZ 2994. 'Settlement on the r. *Gifl*'. *Iveletona* 13th, *I- Yvelton* 1272–1428, *Yevelton -il-* c.1300–1661, *Evelton* 1646, 1810. OE r.n. **Ġifl* as in r. IVEL TL 1938 and YEO in YEOVIL and YEOVILTON Somer ST 5515, 5422, + **tūn**. *Ġifl* must have been the old name of Danes Stream. Ha 73, Gover 1958.227.

EVESBATCH H&W SO 6848. 'Esa's stream'. *Sbech* (sic) 1086, *Esbec(he)* 1160×70, 1195×9, *Es(se)bach(e)* 1268×75–1495, *Eastbatch* 1652, *Evesbatch* 1757. OE pers.n. *Ēsa* + **bæċe**. The modern spelling is probably a back-spelling due to the pronunciation of names such as Evesham as [i:zəm]. He 82, Bannister 71, L 12.

EVESHAM H&W SP 0343. 'Eof's Ham or river bend'. *Eveshomme* *[709]12th S 79, *Euesham* *[714]16th S 1250, [1016]12th S 935, *Eoues- Eofeshamme* 1017×23, *Eoueshom* 1033×8, *Heofeshamm* 12th ASC(C) under year 1037, *Eofeshamm* 12th ASC(D) under year 1045, ASC(C) under year 1054, *Evesham* from 1086, *Esam* 1675. OE pers.n. *Ēof*, genitive sing. *Ēofes* + p.n. *Etham* i.e. 'at Ham' *[706]12th S 1174, *[710]12th S 81, *[716]12th S 83, *homme* *[709]12th S 80, OE **hamm** 1 'land in a river bend' referring to the large loop here in the river Avon in which the town is situated. Also known as *Cronuchomme* *[706]12th S 1174, *[708]12th S 78, [716×7]12th S 97, *cronochomme* *[709]12th S 80, *Cronuchamme* *[n.d.]12th S 226, 'crane Ham', OE **cranoc** + **hamm**, for distinction from other *hamms* along the river, and once as *(æt) Ecguines hamme* [840×52]12th S 203, 'Ecgwine's Ham' from Bishop Ecgwine of Worcester, 693–717, founder of the monastery at Evesham. Wo 262 gives prs [i:vʃəm, i:viʃəm, eisəm, i:ʃəm, i:səm] and obsolete [i:zəm], L 43, 50, Hooke 46–7.

EVINGTON Leic SK 6203. 'Estate called after Eafa'. *Avinton(e)* 1086–1207, *Ev- Euinton(e) -yn-* c.1130–1435, *-en-* c.1200–1338, *Ev- Euington(') -yng-* from 1250. OE pers.n. *Eafa* + **ing**[4] + **tūn**.

EWDEN VILLAGE SYorks SK 2796. P.n. *Uden(e)* 1290–1442, 1822, *Over Ewden* 1608, 'yew-tree valley', OE **īw** + **denu**, + ModE **village**. YW i.223.

EWELL Surrey TQ 2262. 'The spring, the source'. *Eauuelle* 1066, *Etwelle* 1086, *Æwella* 1158–90, *Ewelle -a* 1155 etc., *Euuelle* [727,

933]13th S 420, 1181, *Aiwella -e* 1158–1202, *Awell(e)* 1196–1325, *Yewell* 1603, 1710. OE **ǣwell**. One of the sources of the Hogs Mill river. The 1086 (DB) form is corrupt. Sr 75 gives pr [ju:əl], Nomina 9.6, L 12.

EWELL MINNIS Kent TR 2643. *Ewell Minnis* 1819 OS. P.n. Ewell as in Temple EWELL TR 2844 + Kent and Sussex dial. **minnis** 'common land' (ME *menesse*, ultimately OE *(ġe)mǣnnes*). Cf. nearby RHODES MINNIS TR 1542, STELLING MINNIS TR 1446, SWINGFIELD MINNIS TR 2142, and cf. The Minnis TR 2743. *Meunesa* (sic for *Menn-*) 1204, *La Mannesse* 1226. PNK 562.

Temple EWELL Kent TR 2844. 'Ewell held by the Knights Templar'. *Templ' de Ewell* 1213, *Ewell' Templers* 1270. ME **temple** + p.n. *E(t)welle* 1086, *Aewellan, Eawelle* c.1100, *Ewell* 1219, 'the stream, the source', OE **ǣwylle**, Kentish **ēwell**. The boundary of Ewell is referred to as *(be) æwillemearce* [765×92]15th. PNK 560.

EWELME Oxon SU 6491. 'The powerful spring'. *Auuilma -e* 1086, *Ewelma -e* from c.1183. OE **ǣwelm**. O 126, L 12.

EWEN Glos SU 0097. 'Spring, source'. *(at) Awilme, Awelm* *[931]14th S 415, *Euulme* [n.d.]14th S 1552, *Ewulm* [937]12th, *Ewelme* 1289, 1428, *Ewen* from 1621, *Yewelme als. Yewen* 1736. OE **ǣw(i)elm**. The reference is to the source of the river Thames at Thames Head SO 9899. Gl i.76, L 12.

EWERBY Lincs TF 1247. Probably 'Ivar's village or farm'. *Geres- Ieres- Grene(s)bi* 1086 DB, *I- Ywar(e)by -bi* 1196–1723, *I- Ywardeby* 13th–1652, *Ewarby* 1530, *Ewerby* from 1534. ODan pers.n. *Ivar* + **bȳ**. The DB forms are difficult. Pers.n. *Ivar*, ON *Ívarr*, derives from **Inhu-harjaz* with stress on the first element. The side-form **Ingu-harjaz*, with stress on the second element, produced *Yngvarr*; possibly the p.n. was wrongly analysed as preposition *in* + *Gvares-by* instead of *Yngvares-by*, cf. the lost vill Ingarsby Lincs SK 6805 which has DB forms *Inuuaresbie* and *in Gerberie*. Perrott 61, SSNEM 46, 54, SPN 153.

EWHURST Surrey TQ 0940. 'The wooded hill growing with yew trees'. *Iuherst* 1179, *Ewerste* 1207, *Ewehurst* 1462, *Iwirste, Iwerste* 1206, *Iwhurst* 1253–c.1430, *Highhurst* 1789, 1816. OE **īw** + **hyrst**. Sr 237 gives pr [ju:əst], L 197, 222.

EWHURST GREEN ESusx TQ 7924. P.n. Ewhurst + ModE **green**. Ewhurst, *Werste* 1086, *Hyerst, Yherst* 1195, *Ywehurst -herst -hirst* 13th–16th cents., *Ewehirst* 1509, is 'yew-tree wooded hill', OE **īw** + **hyrst**. Sx 518.

EWORTHY Devon SX 4495. 'Yew enclosure'. *Yworthy* 1468, *Yewry* 1765. OE **īw** + **worthiġ**. D 183.

EWSHOT Hants SU 8149. 'Yew-tree corner'. *Hyweshate* 1236, *Yweset* 1256, *Iweshete* 1305, *Ushotte alias Eweshotte* 1579. OE **īw** + **scēat**. Possibly a division of the same *scēat* or indentation in the boundary between Hants and Surrey as ALDERSHOT SU 8650. Ha 74, Gover 1958.109, L 222.

EWYAS HAROLD H&W SO 3928. 'Ewyas held by Harold'. *Ewias Haroldi* c.1160×70, *Euuiasharold'* 1176, *Euwyas Harold* 1300, *Haraldesewyas* 1371. District name Ewyas + pers.n. Harold as in *Castelli Harralldi de Ewas* 'the castle of Harold of Ewyas' 1198×1214, referring to Harold, son of the first earl of Hereford, Ralph of Mantes d.1057, son of Drogo, count of Mantes and Edward the Confessor's sister Godgifu. Harold founded the priory of Ewyas Harold c.1100 as a cell of St Peter's, Gloucester. Ewyas, *Ewyas* [mid 10th]c.1130, 1148×52, *Ewias* 1086–1166, *Eugias* c.1130, is OW **eguic, euic**, W *ewig* 'hind doe', < Celtic **ou̯īkā* and so cognate with Ir *ói*, Latin *ovis* 'sheep', + suffix **-as** and so 'hind' or 'sheep district'. This district is included as a commote in the earliest list of W cantrefs and commotes and stretched from Cusop Hill SO 2341 in the N to the river Monnow SO 4716 and from the valley of the Grwyne Fawr SO 2331 in the W to Golden Valley SO 3537. 'Harold' for distinction from Ewyas Lacy, *Ewyas Lascy* 1219, 'Ewyas held by the Lacy family', a separate lordship centered on LONGTOWN SO 3229. He 82, 6, 58, DB Herefordshire 19.1, GPC 1262.

EXBOURNE Devon SS 6002. 'Cuckoo stream'. *Eche- Hechesbvrne* 1086, *Yekesb(o)urn(e)* 1242–92, *Yeksbourne* 1297, *Ekesburn(e) -borne* 1285–1303. OE **ġēac**, genitive sing. **ġēaces**, + **burna**. D 140.

EXBURY Hants SU 4200. Probably 'Eohhere's fortified place'. *Teocreberie* 1086, *Ykeresbir(ie)* 1196–1201, *Ocresbiry* c.1205, *Ek(e)res- Hukeres- Eukeresbiri -bury* 1291–1412, *Eukeresbir' -ber' -bury* 1212–1339, *Ekesbury* 1316–67, *Exbere* 1484, *-burye* 1579. OE pers.n. *Eohhere*, genitive sing. *Eohheres*, + **byriġ**, dative sing. of **burh**, probably referring to the earthwork on a low promontory at the mouth of the Beaulieu river. This explanation assumes early substitution of [k] for the peripheral phoneme [x]. The DB spelling shows initial *T-* from preceding preposition *æt*. Ha 74, Gover 1958.196, DEPN.

Nether EXE Devon SS 9300. 'Settlement lower down the Exe'. *Niresse* 1086, *Nitherexe* 1196–1238, *Netherexe* 1238. OE **neothera** + r.n. EXE SS 9409. 'Nether' for distinction from Up EXE SS 9402. D 506.

Up EXE Devon SS 9402. 'Settlement higher up the Exe'. *Vlpesse* 1086, *Uphexe* 1238, *Uppe Esse* 1242, *Upexe* 1346. OE **upp(e)** + r.n. EXE SS 9409. 'Up' for distinction from Nether EXE SS 9300. D 445 gives pr [ʌbəks].

River EXE Devon, Somer SS 7341, 9409. *Ἴσκα* [c.150]13th Ptol, *Uuisc* c.1000 Asser, *(to, andlang) Eaxan* [739]11th S 255, *(on, andlang) exa, (on) exan stream* 11th S 433, *(ut, on, andlang) exan* 11th S 389, 1044, *Exe* from 1238. A British r.n. *Iscā* on the IE root **eis-/*is-* 'move swiftly'. On the Exe are Nether EXE SS 9300 and Up EXE SS 9402. RN 153, BzN 1957.241.

EXE VALLEY Devon SS 9415. R.n. EXE SS 9409 + ModE **valley**.

EXEBRIDGE Somer SS 9224. 'Bridge over the river Exe'. *Exebrigge* 1255, *Exbridge* 1610. R.n. EXE SS 7341, 9409, + ME **brigge** (OE *brycg*). DEPN.

EXELBY NYorks SE 2986. 'Eskil's farm or village'. *Aschilebi* 1086, *-e-* 1161×70, *Eskelby* 12th–1316, *Exkilby* 1372, *Exilby -y-* 15th cent. ON pers.n. *Ásketill, Eskil* + **bȳ**. YN 226 gives pr [eʃəlbi], SSNY 18.

EXETER Devon SX 9292. 'The Roman city of *Isca*'. *Ἴσκα (Iska)* [c.150]13th Ptol, *Isca Dumnoniorum* 'Isca of the Dumnonii' 4th Peut, ~ *Dumnuniorum* [3rd]8th AI, ~ *Dumnamorum* [c.650]13th Rav, *Adescancastre* c.750 VB, *Escanceaster* c.890 ASC(A) under year 876, *Exanceaster* c.900 ASC(A) under year 894, 11th S 386, *-ceastre* c.1000, [900×25]c.1140, *-cestre* 932 B 639, c.1000, *-cester* c.1100 ASC(D) under year 1067, *(to) Eaxanceastre* [914×9]16th BH1, *-ceaster* 925×40, *-cestre* coins 1014×35, *Eaxeancestre* c.1000, *Eaxeceaster* 1121 ASC(E) under year 1003, coins 978×1013, *Execeaster* [928]12th S 400, *Exaceaster* 11th S 389, *Exceaster* coins 1014×35, *Ecxeceaster* coins 1035×40, *Execester* coins 1042×66, *excestre* 1050×73, *Exacestre* 1070, *Execestre* 1086, 12th ASC(E) under year 1135, 1196, *Excetur* 1456, *Excyter* 1523, *Exeter* 1547, *Exiter* 1549, Latin *(in Civitate) Exonia* 1086. R.n. EXE, RBrit *Isca*, + OE **ċeaster**. Cf. also DEVONSHIRE. D 20, BH 109.

EXETER AIRPORT Devon SX 9993. P.n. EXETER SX 9292 + ModE **airport**.

EXFORD Somer SS 8538. 'Ford across the river Exe'. *Aisseford(e)* 1086, *Exeford* 1243, *Exford* 1610. R.n. EXE SS 7341, 9409 + OE **ford**. DEPN.

EXHALL Warw SP 1055. 'The nook of land at a Celtic Christian centre'. *Eccleshale* *[710]12th S 81, *Ecleshelle* 1086, *Ec(c)leshale* 1194–1316, *Exhall* from 1535. PrW ***eglẹ̄s** + OE **hale**, dative sing. of **halh**. Wa 208, Cameron 1968.88, L 106, 109.

EXMINSTER Devon SX 9487. 'The minster by the river Exe'. *(æt) Exanmynster* [873×88]11th S 1507, *Axe- Esse- Aisseminstre* 1086, *Exemenistre* 1208, *Exemystere* 1447, *Axmister* 1675, *Exminster al. Exmister* 1742. R.n. EXE SS 9409 + **mynster**. D 496.

EXMOOR FOREST Somer SS 7642. *Exmoor Forest* 1809 OS. P.n. Exmoor + ModE **forest**. Exmoor, *Exemora* 1204, *Ex more* 1610, is 'the wet land where the r. Exe rises', r.n. EXE SS 7641, 9409, + OE **mōr**. DEPN.

EXMOUTH Devon SY 0080. '(Settlement at the) mouth of the Exe'. *Exanmuða* c.1025 ASC under year 1001, *(on) Exanmuðan* [1042]12th S 998, *(of) Examuða* 1072×1103. R.n. EXE SS 9409 + **mūtha**. D 591, TC 91.

EXNING Suff TL 6165. 'The Gyxningas, the people called after Gyxa or Gyxin'. *Essel(l)inge* 1086, *Esselinga* 1086 ICC, *Heselinges* [1086]c.1180, *Ixninges, (in) Exningis* 1158, *Exning'* 1159, *Exening' -ingis, Exining* 1161, *Exninga -e* 1170–1269 including [c.975]12th, *Exninges* 1185, *Ixlinges* 1212, *Yxninge* 1210×12, *Ixning(')* 1220, 1226×8. OE folk-n. **Ġyxningas* < pers.n. **Ġyxa(n)* or *Ġyxīn* + **ingas**. The medial *n* may be explained either as a fossilised survival of the *n* of the n-stem pers.n. **Ġyxa* or as the *n* of a diminutive side-form **Ġyxīn*. Both forms would be nick-ns. related to OE *ġe(o)csa, ġihsa* 'hiccup'. Cf. HEVENINGHAM TM 3372. DEPN, ING 50.

EXTON Devon SX 9886. 'Farm, village by the river Exe'. *Exton'* 1242, 1305, *Exeton* 1244. R.n. EXE SS 9409 + OE **tūn**. D 602.

EXTON Hants SU 6121. 'Settlement, estate of the East Saxons'. *(æt) east seaxnatune, east seaxena tunes (bōc)* [940]12th S 463, *Essessentune* 1086, *Estsexentun* c.1127, *Eseteneton* 1199–1230, *Exton* from 1182. OE folk-n. *Ēast Seaxe*, genitive pl. *Ēast Seaxna*, + **tūn**. The name implies an isolated settlement or estate of East Saxons in the territory of the *Meonware*, the dwellers by Meon. Ha 74, Gover 1958.69, DEPN.

EXTON Leic SK 9211. 'The ox farm'. *Exentune* 1086, *Exton(e)* from 1107. OE **oxa**, genitive pl. ***exna**, + **tūn**. R 19, DEPN.

EXTON Somer SS 9233. 'Settlement on the river Exe'. *Essetvn* 1086, *Exton* from 1216. R.n. EXE SS 7341, 9409, + OE **tūn**. DEPN.

EYAM Derby SK 2176. '(Settlement at the) islands'. *Aiune* 1086, *Ayum, Ai(h)um* 13th cent., *Ei- Eyum* 1225–1520, *Eyam* from 1446, *Eam(e)* 16th cent. OE **ēġum**, locative-dative pl. of OE **ēġ**. The hamlet lies between Hollow Brook and Jumber Brook. Db 92 gives pr [i:m], L 36–7.

EYDON Northants SP 5450. 'Æga's hill'. *Egedone* 1086, *Eindune -don(a), Eyndon(a)* c.1200–1253, *Eidon(e)* from c.1220 with variant *Ey-, Edon, Eadon* 16th. OE pers.n. **Æġa*, genitive sing. **Æġan*, + **dūn**. Nth 35 gives pr [i:dən], L 153.

EYE Cambs TF 2202. 'The island'. *Ege* 12th ASC(E) under year 963, c.1115, *Eia, Eya* 1125×8, [1189]1332, 1227, *Eye* from 1284. OE **ēġ**. The place is situated on rising ground and before draining of the fen was surrounded by water in winter time. Nth 234, L 36–7.

EYE H&W SO 4964. 'The island'. *Heya* 1123, *Eia, Eya* c.1150–1242×9, *Eye* from 1431, OE **ēġ** here in the sense 'raised ground in marsh'. Eye is situated on raised ground between Main Ditch and another water channel. He 84, L 36–7.

EYE Suff TM 1473. 'The island'. *Heia* 1086–1231 with variants *Haia, Hea* and *Heye, Eia* 1086–1194, *Eye* from 1103. OE **ēġ**. DEPN, Baron.

River EYE Leic SK 8018. *Eye* from c.1545. OE **ēa** 'river'. On the river is Eye Kettleby SK 7316 discussed under Ab KETTLEBY SK 7223. Lei 94, RN 157.

EYE BROOK Leic SK 7702. The original name is *Litelhe* 1218–99, *Lytele* 1269, *Littleye* 1276, *Little Ey* 1610, 'Little Eye' in which *Eye* is OE **ēa** 'river'. 'Little' for distinction from the River Eye. Lei 94, RN 157.

EYEBROOK RESERVOIR Leic SP 8595. R.n. Eye as in EYE BROOK SK 7702 + ModE **reservoir**.

EYEWORTH Beds TL 2545. 'Island enclosure'. *Ai(ss)euuorde* 1086, *Eyworth* 1232. OE **ēġ** in the sense 'island' or possibly 'hill-spur' + **worth**. The settlement lies at the end of a low ridge between streams. Bd 105, L 279.

EYHORNE STREET Kent TQ 8354. *Eyhorn Street* 1819 OS. P.n. Eyhorne + Mod dial. **street** 'a village'. Contrasts with nearby Broad Street TQ 8356, Ware Street TQ 7956, Weavering Street TQ 7856 etc. Eyhorne, the meeting place of Eyhorn Hundred, *Haihorne -borne, Aihorde* 1086, *Hai(h)ornh'dr, Aihornehundredo* 12th, *Eyhorn(')* 1219–29 with variants *(H)ay(t)horn(e), Ay(t)horn(e), Eythorn* and *Heyhorn(e)*, is the 'hill-spur with an enclosure' or 'with a thorn-tree', OE **hæġ** 'a fence, an enclosure' + **horn** 'a projecting headland' varying with **hæġthorn** 'a hawthorn-tree'. The village stands by a low spur of land extending from the North Downs. PNK 218, KPN 101.

EYKE Suff TM 3151. 'The oak-tree'. *Eik* 1185, *Eyk* 1270–1416, *Eyck* 1291, *Ike* 1605, *Eyke* 1610. OE **eik**. DEPN, Arnott 59, Baron.

EYNESBURY Cambs TL 1859. 'Eanwulf's fortified place'. *Eanulfesbirig* c.1000, *Einuluesberie* 1086, *-biri* c.1125, *Ainesbiri* 1163, *Eynesbyr -bir -bury* 1234–1504. OE pers.n. *Ēanwulf* later replaced by CG *Einulf*, genitive sing. *Ēanwulfes, Einulfes*, + **byriġ**, dative sing. of **burh**. Hu 255.

EYNSFORD Kent TQ 5365. 'Ægen's ford'. *Ænes ford* 960×88 S 1458, *Æinesford(am)* [960×88]12th ibid., *Elesford* 1086, *Aeinesford* c.1100, *Eines- Eyne(s)- Aine(s)ford(e)* 1164×5–1223 etc. OE pers.n. *Æġen*, genitive sing. *Æġenes*, + **ford**. K 39, KPN 289.

EYNSHAM Oxon SP 4309. 'Ægen's river-meadow'. *(ad) Egenes homme* [864]11th S 210, *Egonesham* c.900 ASC(A) under year 571, *Egnesham -ham(i)e* [1105]12th–1367, *Ei- Eynesham* 1185–1471, *Eynsham* from 1428. OE pers.n. **Æġen*, genitive sing. **Æġenes*, + **hamm** 1. O 258.

EYPE Dorset SY 4491. 'Steep place'. *Yepe* 1365–1677, *Yep* 1406, *Ype* 1676, *Eype* 1839. The earliest references are to *Estyep* 1300–30, *Uuerȝep* 1320 'upper Eype', *Estrhep* 1329, 'east Eype', OE **ēast** and **ēasterra** + **ġēap**. The development of the name is difficult; ME *Yepe* [jɛ:p] should give ModE *Yeap* or *Yepe* pronounced [ji:p] whence present-day [i:p] with assimilation of initial [j] to following [i:]. If the spelling *Eype* is really significant it would point to an unusual raising of ME [ɛ:] to [i:] prior to the Great Vowel Shift to PDE [ai]. This seems to have occurred sporadically in some SW dialects. Do 71, Notes 46–7, Widén 34.1.(b) note.

EYTHORNE Kent TR 2849. There are two types;

I. *(æt) Heagyðe ðorne, (æt) Hægyðe ðorne* 805×7 S 41 with endorsements *He(a)gyþe ðornes boc* 11th cent., *heageþorne* 12th cent., *Aegyðe ðorn, Eagyðe ðornes* (genitive) [824]10th S 1266, *Egethorn(e)* 13th ibid., 1210×12–1270, *Egedorn* c.1100, *Eythorn* 1256, *Aythorn'* 1270 etc., *Eyhorne* 1610.

II. *hageþorne* 12th cent. endorsement of S 1266, *Heythorn'* 1249, *Hagthorn'* 1254.

Type I represents OE *Hēahgȳthe thorn* 'Heahgyth's thorn-tree', type II OE **hagu- hæġthorn** 'a hawthorn-tree', a rationalisation of I. PNK 582, KPN 100.

EYTON 'Island settlement, settlement on an area of dry ground in marsh', sometimes 'promontory settlement'. OE **ēġ-tūn**. Sa i.128.

(1) ~ Shrops SO 3787. *Eton'* 1251, *Eyton(')* from 1255, *Eyton alias Yeton* 1547. Sa i.128.

(2) ~ H&W SO 4761. *Eiton(')* 1186×99, 1308, *Eytune* 1238×58, *Eytone*†. Eaton Hall W of the village occupies a hill-top overlooking the Lug River. He 84, L 36–7.

(3) ~ UPON THE WEALD MOORS Shrops SJ 6514. *Eyton in Wydemore* 1255×6, *Eton' sub Wyldemor'* 1291×2, *Eyton super Wildemore* 1346, *Iton Wilmore* 1589, *Eyton-in-the-Wildmore* 1646. P.n. *Etone* 1086, *Eiton* 1200, *Eyton(')* from 1231 + ME **wilde more** 'the wild, uncultivated marsh' (OE *wilde* + *mōr*), with unusual failure of ModE diphthongisation of ME *ī* > [ai] and assimilation to *weald* (OE *weald* 'woodland'). Sa i.129, 244.

†*Ettone* 1086 probably refers to Eaton near Leominster SO 5058. DB Herefordshire 1,10 c.

F

FACCOMBE Hants SU 3958. 'Facca's coomb'. *(andlang) faccan cumbes* [863]12th S 336, *Faccancumb* ? 950 S 1539, *Facūbe* 1086, *Faccombe* from 1316, *Fackham* 1579, 1697. OE pers.n. **Facca*, genitive sing. **Faccan*, + **cumb**. The village overlooks a small trough-like valley opening off the larger valley referred to in the p.n. COOMBE SU 3761. Ha 76, DEPN.

FACEBY NYorks NZ 4903. 'Feit's farm or village'. *Fe(i)zbi, Foitesbi* 1086, *Fayceby* c.1160–1367, *Faceby* from 1285. ON pers.n. *Feitr*, genitive sing. *Feits*, + **bȳ**. YN 176 gives pr [fɛəsbi], SSNY 26.

FADDILEY Ches SJ 5953. 'Fadda's wood or clearing'. *Fadilee* c.1220, *Fadileg(h)* from 1295 with variants *Faddi- Faddy-* and *-ley(e) -ly, Faddelee* 1260, *Fadelegh* 1288–1724 with variants *Fadde-* and *-ley -leigh* etc. OE pers.n. *Fad(d)a* + **lēah**. Forms with medial *-i-, -y-*, probably represent a reduced form of **ing**[4]. Che iii.142 gives prs [fadili], [fadəli] and [fadli].

FADMOOR NYorks SE 6789. 'Fada's moor'. *Fademor(a)* 1086–1231, *Faddemor* c.1150–1219, *Fadmore* 1285–1462. OE pers.n. **Fad(d)a* + **mōr**. YN 62, SSNY 254.

FAILAND Avon ST 5271. Partly uncertain. *Failaund* 1280 Ass (p), *Feylond* 1446×7 FF. Unidentified element possibly from OE *(ġe)fæġe* 'popular, acceptable' + ME **land**.

FAILSWORTH GMan SD 8901. Partly uncertain. *Fayleswrthe* 1212, *Failesw(o)rth(e)* c.1200–46, *Felesworde* 1226, *Faylesworde* 1451, 1461, *Faylsworth* 1537, *Failsworth* 1843 OS. The generic is OE **worth** 'an enclosure', the specific an unidentified element, possibly an unknown pers.n. **Feġel* or **Fæġel* or an unrecorded OE **fēġels* from OE *fēġan* 'to join', perhaps in the sense 'hurdle, fence'. Cf. HURWORTH NZ 3010. La 36, PNE i.166, Jnl 17.32.

FAIRBURN NYorks SE 4727. 'Fern stream'. *Faren-burne* c.1030 YCh 7, *Fareburn(a) -e* 1086–1287, 1617, *Farburn(e)* 1216×56–1602, *Fairburne* 1616. OE **fearn** + **burna**. YW iv.48, L 18.

FAIRFIELD H&W SO 9575. 'Hog open land'. *forfeld* [817]11th S 181, *For(e)feld(e)* 11th–1316, *-field* 1616, 1820, *Fairfield* 1831 OS. OE **fōr** + **feld**. The original name meant 'open land where pigs are kept'; the present form is a 19th cent. 'improvement'. Wo 275, Studies 1936.76–7, PNE i.180, L 244.

FAIRFIELD Mers SJ 3791. *Fairfield* 1842 OS.

FAIRFORD Glos SP 1501. 'Fair, beautiful ford'. *(æt) Fagranforda* [862]14th S 209, *Fareford(e)* 1086, 1190, 1498, *Feir(e)- Feyr(e)ford(e)* 1100–1349, *Fair- Fayrford* from 1221. OE **fæġer** + **ford**. The earliest form is OE *(æt thǣm) fagran forda* 'at the fair ford', OE **fæġer**, dative sing. definite form oblique case **fæġran**, + **ford**, dative sing. **forda**. The ford carried the road from Lechlade to Cirencester across the r. Coln. Gl i.34, L 68.

FAIRHAM BROOK Notts SK 5532. *Fairham Brook* c.1825. P.n. Fairham as in Fairham Bridge SK 5633 + ModE **brook**. Fairham Bridge is *Feareholme Bridge* 1690 of unknown origin. Earlier known as *Ke(u)worthbroke* 1346, p.n. KEYWORTH SK 6130 + ME **broke**. Nt 4, 249.

FAIRLIGHT ESusx TQ 8612. 'Fern clearing'. *Farleg(h)(e) -leye* 1176–17th cent., *Farnleg(h)* 13th cent., *Fairlight* 1673. OE **fearn** + **lēah**. For the development to *-light* see East HOATHLY TQ 5216. Sx 507 gives pr [fe:rlai].

FAIRMILE Devon SY 0897. *Le faire mile* c.1425. Probably referred to an especially good stretch of the main road between Exeter and Honiton. D 605.

FAIR OAK Hants SU 4918. 'Beautiful oak-tree'. *Fereoke* 1596, *Fair Oak* 1695. ME **fair** + **ok**. No fair is known to have been held here so the sense must be 'beautiful'. Ha 76, Gover 1958.70, Room 1983.38.

FAIROAK Staffs SJ 7632. 'Beautiful oak-trees'. *Fair Oak* 1838 OS.

FAIRSEAT Kent TQ 6261. *Fairseat, Farsee Street* 1782, *Facy Street* 1819 OS. So called for the pleasantness and the extensiveness of the view from the place. The form *Facy Street* represents *Fairseat Street* 'Fairseat hamlet', Modern dial. **street**. PNK 155.

FAIR SNAPE FELL Lancs SD 5947. A steep mountain rising to 1673ft. P.n. Fair Snape + ModE **fell** (ON *fjall*). Fair Snape, (Higher and Lower Fair Snape SD 5845), a vaccary on the lower slopes of the fell, *Fayrsnape* 1323–4, *Fairsnap* 1341, is 'beautiful pasture', ME **fair** (OE *fæġer*) + **snape** (OE **snæp*, ON *snap*) 'a patch of grass for sheep to nibble at in snow covered fields, poor pasturage'. La 166, SSNNW 122, Nomina 10.174.

FAIRSTEAD Essex TL 7616. 'Pleasant place'. *Fairstedam* 1086, *Faire- Fayrested'* 1185–1400, *Fayrsted(e)* 1246–1470, *Firsted, Fristed* 1594. OE **fæġer** + **stede**. Ess 286, Stead 199.

FAIRWARP ESusx TQ 4626. No early forms.

FAIRY CROSS Devon SS 4024. *Fairy Cross* 1809 OS. A cross-roads.

FAKENHAM Norf TF 9230. 'Facca's homestead'. *Fachenham, Fagan(a)ham* 1086, *Fakeham* 1212, *Fakenham* from 1254. OE pers.n. *Faca*, genitive sing. *Facan*, + **hām**. DEPN gives pr [feik-].

Little FAKENHAM Suff TL 9076. *Litla Fachenhā* 1086, *Fake(n)ham Parva* 1254. OE adj. **lȳtel**, Latin **parva**, + p.n. *Fakenham* c.1060, *Fachenham* 1086, *Fakeham* 1242, 'Facca's homestead', OE pers.n. *Facca*, genitive sing. *Faccan*, + **hām**. 'Little' for distinction from Great Fakenham TL 9876, *Fake(n)ham Magna* 1254, ModE **great**, Latin **magna**, + p.n. *Fachenhā -ham* 1086. DEPN.

River FAL Corn SW 9246. Unexplained, possibly pre-Celtic. *(to) fæle* [969]11th S 770, 1049 S 1019, *Fale* [c.1210]late 13th, *Fal* 1378, *Vale* 1576. PNCo 83, Gover n.d. 681, RN 157.

FALDINGWORTH Lincs TF 0684. Partly uncertain. *Fal(d)ingeurde* 1086, *Faldinguorda* c.1115, *-wrd -word(a)* 12th, *-worth* from 1202. OE **falding* 'the act of folding animals' has been posited to account for ME forms Fald- Fold- Follingworth of frequent occurrence in minor and late names in Yorks. The presence of medial *-e-*, however, points rather to OE ***falding** 'a folding place' < **fald** 'a fold for animals' + **ing**[2], locative-dative sing. ***faldinge**, + **worth**, 'the enclosure called or at the *falding*, the folding place', or folk-n. **Faldingas* < pers.n. *Falda* + **ingas**, genitive pl. **Faldinga*, + **worth**, 'the enclosure of the Faldingas, the people called after Falda'. DEPN, PNE i.164.

FALFIELD Avon ST 6893. 'Fallow-coloured open land'. *Falefeld(e)* 1227–1497, *Falfeld* 1232, 1445, *-field* 1570, *Fawfyld* 1510. OE **fealu** + **feld**. Gl iii.6, L 240, 242.

FALKENHAM Suff TM 2938. 'Falta's homestead'. *Faltenham* 1086–1416, *Falceham* 1200, *Falcenham* 1254–1408, *Falkenham* from 1327, *Ffaultenham* 1613. OE pers.n. **Falta*, genitive sing. **Faltan*, + **hām**. DEPN, Arnott 43, Baron.

FALLOWFIELD GMan SJ 8593. Either 'fallow field' or 'fallow-coloured open land'. *Fallufeld* 1317, *Falofeld* 1417, *Falowfelde* 1530, *Fallowfield* 1843 OS. The generic is OE **feld**, the specific uncertain as both OE **falh** 'fallow land' and **fealu** 'fallow-coloured, pale brown' give ME *fallu*. La 30, L 242.

FALLOWLEES BURN Northum NZ 0092. *Fawleyburne* before 1265, *Fauleyburn* 14th. P.n. Fawley + OE **burna**. Fawley is the 'variegated clearing', OE **fāh** + **lēah**. The name has been influenced by nearby Fallowlees NZ 0194, *Falalee* 1388, *Falowleys* 1436, *Fallowlees* 1663, the 'clearing where land has been broken up for ploughing', OE **falh** + **lēah**. NbDu 80.

FALMER ESusx TQ 3508. 'Dark lake'. *Falemere* 1086–1327, *Falmere* from 1332, *Fawmer* 1607, *Famer* 1727. OE **fealu** + **mere**. Sx 308 gives pr [fa:mə].

FALMOUTH BAY Corn SW 8130. P.n. FALMOUTH SW 8032 + ModE **bay**.

FALMOUTH Corn SW 8032. 'Mouth of the Fal'. *Falemue* 1225, 1297, *Fale(s)muth* 1234, *Falemouth* 1403, *Fallmouth* 1478, *Falmouth* 1610. R.n. FAL SW 9246 + OE **mūtha** 'mouth, estuary'. Described by Leland as a 'havyn very notable and famose, and in a manner the most principale of al Britayne'. The name was transferred to the growing town in the 15th century, the original settlements being Smitheck, *Smythwyck* 1370, 'the smith's workplace', OE **smith** + **wīċ** and Pennycomequick, *Penicomequicke* 1646, an English nick-name meaning 'get-rich-quick'. PNCo 83, Gover n.d. 497, RN 157.

FALSTONE Northum NY 7287. 'The yellow stone'. *Faleston* 1255, *Faustan* 1371, *-ston* 1610, 1695 Map. OE **fealu** + **stān**. NbDu 80.

South FAMBRIDGE Essex TL 8694. *S(o)uthfambreg(ge)* 1291, 1336. ME **suth** + p.n. Fambridge as in FAMBRIDGE STATION TQ 8597 at North Fambridge. Ess 215.

FAMBRIDGE STATION Essex TQ 8597. P.n. Fambridge + ModE **station**. Fambridge, *Fanbruge* [1042×66]12th S 1051, 1086, *Phenbruge* 1086, *Fambrugge -brig(ge) -breg(ge)* 1086–1331, is 'fen bridge', OE **fænn, fenn** + **brycg**. Also known as North Fambridge, *Northambregg* 1274, *-brige* 1535, for distinction from South FAMBRIDGE TQ 8694. Ess 214 gives pr [fa:mbridʒ].

FANGDALE BECK NYorks SE 5694. *Flandgedalebec* (sic) 1201. P.n. Fangdale + ON **bekkr**. Fangdale, *Fangedala* c.1160, is the 'hunting valley, the valley where game is taken', ON **fang** 'hunting, fishing' + **dalr**, rather than the 'valley of the Fangá, the river where fish are taken' since the stream here is too minor. YN 68, DEPN, SSNY 155.

FANGFOSS Humbs SE 7653. 'Ditch where fish are taken'. *Frangefos(s)'* 1086, 1199, *Fangefos(s)(e)* 1120×9–118, *Fangelfosse* 1200, *Fangfoss* from 12th. ON **fang** 'hunting, fishing' + **foss**. The reference is possibly to fish-traps. The form *Fangel-* may show an *-el* derivative form parallel to ODan **fangæl* found in the p.n. Fangel referring to a river rich in fish, and in West FINGEL Devon SX 7491. YE 185, SSNY 155.

FAR FOREST H&W SO 7274. The reference is to WYRE FOREST SO 7576.

FARCET Cambs TL 2094. '(At) the bull's head'. *Fearresheafde* (dative case) [956]14th S 595, c.1000, *-heafod* [973]15th S 792, *Farresheafde* [963]12th S 1448, *-heued* [c.1150]14th, *Faresheued* [963]12th S 1448–1353, *Far(i)sheued* 1260–1327, *Fasset* 1526–95. OE **fearr**, genitive sing. **fearres**, + **hēafod** in the sense 'headland'; the village occupies a low ridge in the fens. Hu 185 gives pr [fæsət].

FARCET FEN Cambs TL 2392. *Faresheved Fen* 1279. P.n. FARCET TL 2094 + ME **fen**. Hu 185.

FARDEN Shrops SO 5776. *Farden* 1832 OS.

FAREHAM Hants SU 5706. Probably 'fern homestead or estate' or 'fern river-bend land'. *Fearnham* [963×75]12th S 822, *Fernhā* 1086, *-ham* 1164–1208, *Fer(e)- Ferreham* 1136–1294, *Farham* 1195–1327, *Fareham* from 1233, *Far(e)nham* 1207–1328, *Fayram* 1541, *Pharam* 1609. OE **fearn** + **hām** or **hamm**. The topography supports either *hām* or *hamm* 1 'land in a river bend' or 2 'promontory of dry land into marsh or water'. S 636 refers to the 'brook of the *Fearningas*, the fern people or the people of Fareham', *Fearninga broc* 956. On formal grounds, however, *Fearningas* could also be explained as 'the people called after *Fearn*', a pers.n. which might be related to a name such as PrW *Farinmail* or *Farinmagil*, one of three Welsh kings slain by Cuthwine and Ceawlin at the battle of Dyrham in 577. Ha 76, Gover 1958.28, DEPN.

FAREWELL Staffs SK 0811. 'The beautiful spring'. *Fager-Faierwelle* 1200, *Fayr- Fager(e)- Farewell* 13th, *Faurewell* 1251. OE **fæġer** + **wella**. DEPN, Horovitz.

FARINGDON Oxon SU 2895. 'The fern covered hill'. *(æt) Færndunæ* [c.968×71]12th S 1485, *Ferendone* 1086–1272 with variants *-don(a)* and *-duna, Farendon(e) -dune -duna* [1158]13th etc., *Farndon'* 1242×3, *Faringdon* 1509×47. OE **fearn** + **dūn**. Confusion with *ing*[4] is common in this type of name. Brk 365.

Little FARINGDON Oxon SP 2201. *Parva Ferend(una)* 1156, *Parva Farendon -in-* 1220, 1316, *Little Farendon'* 1359. ME adj. **litel**, Latin **parva**, + p.n. Faringdon as in FARINGDON SU 2895 of which it was an outlying possession. O 319.

FARINGTON Lancs SD 5324. 'Fern farm or village'. *Farinton* before 1149–1242, *-tunā* 1153×60, *Farington -yng-* from 1246. OE **fearn** + **tūn**. La 135, Jnl 17.76.

FARLAM Cumbr NY 5558. '(Settlement) at the fern clearings'. *Farlam* from 1166, *Farlaham* 1279, 1316. OE **fearn** + **lēah**, dative pl. **fearn** + **lēam**. *Farlam* is the overwhelmingly frequent form, but the two spellings in *-ham* could point to OE **fearn-lēah** + **hām** 'homestead at *Farnley, the fern-clearing'. Cu 83, cf. L 204.

FARLEIGH 'Clearing where ferns grow'. OE **fearn** + **lēah**.

(1) ~ Avon ST 4969. *Farley* 1817 OS. Probably 'fern covered clearing', but the evidence is too late for certainty. Cf. L 204.

(2) ~ Surrey TQ 3660. *Ferlega* 1086, *Farle(g)* 1215–1314, *-legh* 1314, *Farnleg(h), Farnelegh* 1255–79, *Varl(y)e* 1431–9. Sr 316, L 204.

(3) ~ HUNGERFORD Somer ST 8057. 'F held by the Hungerford family'. *Farlegh Hungerford* 1404. Earlier simply *Fearnlæh* [987]12th S 867, *Farnleghe* [1001]15th S 899, *Ferlege* 1086, *Farlegh* 1362, cf. *Fa'ley Cast* (i.e. *Castle*) 1610. Sir Thomas Hungerford acquired the manor in 1369. DEPN.

(4) ~ WALLOP Hants SU 6246. 'F held by the Wallop family'. *Farley Wollop* 1617. P.n. Farleigh + family name Wallop as in John Wallop who obtained the manor in 1487. Previously known as *Farle Mortymer(e)* 1390, 1412, from its feudal overlords c.1297–1487, and simply *Ferlege* 1086, *Farnlegh* 1256, *Farlegh* 1291. 1316. Ha 76, Gover 1958.132.

(5) East ~ Kent TQ 7353. *East Farlegh* 1291. ME adj. **est** + p.n. *(on) Fearnlege* 871×89 S 1508, *Fearnleag(e)* 898 S 350, *Fernlege* [961]13th S 1212, 1042×66 S 1047, *Fearnlega -e* c.975, *Fernleah* [1006 for 1001]11th S 914, *Ferlaga* 1086, *Farleg(h) -le(g)e* 1202–26 etc. 'East' for distinction from nearby West FARLEIGH TQ 7152. KPN 227, DEPN.

(6) Monkton ~ Wilts ST 8065. 'The monks' F estate'. *Farleye Monketon* 1370, *Mounton Farley* 1773. ME **monkton** (OE *munuc*, genitive pl. *muneca*, + *tūn*) + p.n. *Farnleghe* [1001]15th S 899, *-lege* 1196, 1249, *Farlege* 1086. Also known as 'monks' Farleigh', *Farley Monachorum* (Latin genitive pl.) 1316, *Monkene- Munkesfarlegh* 14th. A Cluniac priory was founded here in 1125. Wlt 120, DEPN.

(7) West ~ Kent TQ 7152. *West Farlegh* 1291. ME adj. **west** + p.n. *Ferlaga* 1086, as in nearby East FARLEIGH TQ 7353. KPN 227, DEPN.

FARLESTHORPE Lincs TF 4774. Possibly 'Faraldr or Farulf's dependent settlement'. *Farlestorp'* 1160–1202, *Fareslestorp* 1202. ON pers.n. *Faraldr* or OSwed *Farulf*, genitive sing. *Faralds, Farulfs*, + **thorp**. DEPN, SSNEM 126.

FARLETON Cumbr SD 5381. Partly uncertain. *Farelton* 1086, *Farleton* from 1190, *Farlton* 1184×90–1610. The generic is OE

tūn 'farm, village', the specific any of OE pers.n. **Færela*, ON *Faraldr* or *Farle*, or even an older p.n. **Fearn-hyll* or **Fearn-lēah* 'fern hill or clearing'. The first alternative is best. We i.68, SSNNW 190.

FARLEY 'Clearing where ferns grow'. OE **fearn** + **lēah**.

(1) ~ Shrops SJ 3807. *Fernelege* 1086, *Fere- Farleye* 1291×2, *Farley* from 1581, *Fareley(e)* 1563–1694, *Fairley* 1923 Bowcock. Sa i.130.

(2) ~ Staffs SK 0744. *Fernelege* 1086, *Farleye* 1273, *Farley* from 1274. DEPN, Horovitz.

(3) ~ Wilts SU 2229. *Farlege* 1086, *Fernelega -le(y)e* 1109×20–1327×77, *Farnlee -leg'* 1257, 1263, *Farle(y)gh* 1285, 1351. Wlt 378, L 204.

(4) ~ GREEN Surrey TQ 0645. *Farley Greene* 1690. P.n. *Farleye* 1305, *Farnlegh* 1332, + ModE **green**. Sr 220, L 204.

(5) ~ HILL Berks SU 7564. *Far(e)ley Hill* 1550–1790, *Fawley Hill* 1761. P.n. Farley + ModE **hill**. Farley, *Fer(n)lega* 1167–1225, *Farle(g)a -legh* 1169–1345, is the 'fern clearing'. Brk 110, 109.

(6) ~ MOUNT Hants SU 4029. A modern country park centred on a house called Farley Mount, p.n. *Ferlege* 1086, *Ferley* 1212, + ModE **mount**. Also known as Farley Chamberlayne, *Ferlega Camerarii* 1167, *Farlegh Chaumberlayn* 1297, from its tenure by the family of Robert Camerarius 'the chamberlain' 1212. Ha 77, Gover 1958.179.

FARLEYS END Glos SO 7715. *Farleigh End* sc. of Elmore 1830. P.n. Farleigh with pseudo-genitival **-s** + ModE **end**. Farleigh, *Far(e)nleye* 12th, *Farelehe -ley(e) -leigh* 1248–1638, is 'fern clearing', OE **fearn** + **lēah**. Gl ii.162, L 204.

FARLINGTON NYorks SE 6167. 'Estate called after Færela'. *Fer- Farlintun* 1086, *Ferlinton(a)* 1167–1310, *Ferlington* 1249–1316, *Farlington* from 1316. OE pers.n. **Færela* + **ing**^{4} + **tūn**. YN 31.

FARLOW Shrops SO 6480. 'Fern tumulus'. *Ferlau* 1086, *Ferlaue* 1206, *Ferlowe* 1399, *Farlawe* 1255, *Farlowe* 1273–1410, *-low* from 1403, *Ferulawe* (sic for *Fern-*) c.1188, *Fern- Farnlawe -lowe* 1221×2–1293. OE **fearn** + **hlāw**. A seated skeleton was found in a Bronze-Age mound when the present church was built 1857×8 on its dominating hill site. Sa i.131.

FARMBOROUGH Avon ST 6660. 'Fern-clad hill(s)'. *Fearnberngas* (sic) [901]c.1400 S 363, *Ferenberge* 1086. OE **fearn** + **beorg**, pl. **beorgas**. DEPN, L 128.

FARMCOTE Glos SP 0629. 'Fern cottage(s)'. *Fern(e)cote -a* 1086–1487, *Farn(e)cote -kote* 1248–1560, *Farm(e)cote -cott* 1569–1742. OE **fearn** + **cot** (pl. **cotu**). Gl ii.20.

FARMINGTON Glos SP 1315. 'Settlement at Thornmere'. *Tormentone* 1086, *Tor- Thormerton(a)* 1183–1621, *-mar-* 1287, 1351, *Farmynton* 1577, *Farmington* 1632, 1719. P.n. **Thornmere*, 'thorn-tree pool', OE **thorn** + **mere** 'pool', + **tūn**. The reference is probably to a lake $\frac{1}{2}$ mile E of the village at SP 1415. Substitution of initial *Th-* by *F-* is widespread in local varieties of English. Gl i.172, Wakelin 98, Jnl 25.45.

FARMOOR Oxon SP 4407. 'The far moor'. *Ear Moor* (sic) 1761, *Far Moore Common* 1808. ModE **far** + **moor**. Brk 447.

FARMOOR RESERVOIR Oxon SP 4406. P.n. FARMOOR SP 4407 + ModE **reservoir**.

FARNBOROUGH 'Fern hill'. OE **fearn** + **beorg**.

(1) ~ GLond TQ 4464. *Farberghe* 1147×82, *Ferenberga* 1180, *Farnberg(a)* 1185–1226. The boundary of the people of Farnborough is *fearnbiorginga mearc* 862 S 331, *Fearn beorghginga mearc* [987]10th S 864. KPN 208, 211, DEPN.

(2) ~ Hants SU 8754. *Ferneberga* 1086, *(ecclesia de) Ferenbergo* (Latin form) 1230, *Farnburge* 1243, *Farneburewe* 1284. In this name *beorg* was later replaced by *burh*. Ha 77, Gover 1958.110, DEPN, L 128.

(3) ~ Warw SP 4349. *Feornebeorh* [990×1006]13th S 937, *Ferneberg(e)* 1086–1245, *Farnburgh* 1221–1535 with variants *Farne-* and *-borough, Farnborough* 1632. OE **fearn** + **beorg** referring to the little hill on which the town stands. Wa 270, L 128.

FARNCOMBE Surrey SU 9745. 'The valley growing with ferns'. *Fernecome* 1086, *Fern- Farnecumbe* 1225–80, *Farncombe* from 1332, *Francom(b)(e)* 1325–1642. OE **fearn** + **cumb**. Sr 197, L 90, 93.

FARNDALE NYorks SE 6697. 'Fern valley'. *Farnedale* c.1154×63, 1276, 1416, *Farendale -dal(a)* late 12th–1301, *Farndale* from 1279×81. OE **fearn** + **dæl**. YN 63, L 96.

FARNDALE MOOR NYorks NZ 6600. P.n. FARNDALE SE 6697 + ModE **moor**.

FARNDISH Beds SP 9263. 'Enclosed pasture growing with ferns'. *Fernadis -edis* 1086, *Farnedisch* 13th cent., *Frendish al. Frendich* 1509. OE **fearn** + **edisc**. Bd 38.

FARNDON 'Fern hill'. OE **fearn-dūn**. L 142–3, 153, 155.

(1) ~ Ches SJ 4154. *(æt) Fearndune* [924]11th ASC(C), *-dun* c.1118, c.1130, *(æt) Farndun* [924]11th ASC(D), 1333, *Farndon* from 1245, *Ferentone* 1086, *Farendun -don* 1194–1529 with variants *-in- -un-*, *Faryngton* 1391–1508. Che iv.73 gives pr [fa:rndən].

(2) ~ Notts SK 7752. *Farendune* 1086, *-don* 1163–1543, *Farandon* with occasional variants *-in-* and *-yn-*, *Farnedon* 1270–1329, *Farndon* from 1316. Originally the name of a wedge of raised ground between the rivers Trent and Devon further S at SK 7649 where the late p.ns. Thorpe and East Stoke are found. Nt 213, L 155.

(3) East ~ Northants SP 7185. *E(a)st Farnedoun* 1616, 1617, *East Farrington al. Farndon* 1702. ModE adj. **east** + p.n. *Ferendon(e)* 1086, 1203, *Farendon(e)* 12th–1337, *Farndon* from 1288. 'East' for distinction from West FARNDON SP 5251. Nth 113.

(4) West ~ Northants SP 5251. *Westfarindon* 1275, *Westfarndon* 1300. ME adj. **west** + p.n. *Ferendon(e)* 1086–1269, *Faryndon* 1294, 1316, *Farndon* 1284. 'West' for distinction from East FARNDON SP 7185. Also known as *Ferendon juxta Wodeford* 1394, *Faryndon juxta Hynton* 1294. Nth 37.

FARNE ISLANDS Northum NU 2337. Uncertain. *insula Farne* [699×705]c.900 AC 96, [c.721]10th BPC 214, c.1160, 1204, *Farnea* [699×705]13th AC 96, 14th, *Farnealond* 1257. The islands are covered with fern and the traditional explanation has been 'fern islands', ONb **farn** (OE **fearn**) + **ēġ** later replaced by ME **eland**. The proximity and similarity of Lindisfarne, however, suggests that in Farne we may have to do with a hitherto unidentified element of pre-English origin. NbDu 80, Jnl 8.20, Celtic Voices 255.

FARNHAM Dorset ST 9515. 'Homestead or enclosure where ferns grow'. *Fernham* 1086–1486, *Farnham* from 1199. OE **fearn** + **hām** or **hamm**. Do ii.221.

FARNHAM Essex TL 4724. 'Fern homestead'. *Phern(e)ham* 1086, *Fernham* 1194–1281, *Farnham* from 1206. OE **fearn** + **hām**. Ess 550.

FARNHAM NYorks SE 3560. 'Fern homestead'. *Farneham* 1086–1609, *Farnham* from 1204. OE **fearn** + **hām**. YW v.91.

FARNHAM Suff TM 3660. 'Homestead where ferns grow'. *Farn- Ferneham* 1086, *Farnham* from 1198. **fearn** + **hām**. DEPN, Baron.

FARNHAM Surrey SU 8446. 'River meadow where ferns grow'. *Fernham* [685×7, 801×14, 909]c.1135 S 235, 1263, 382, 1206–17, *Fear(na)ham* [858]c.1150 S 1274, *(æt) Fearnhamme* c.900 ASC(A) under year 894, *Fearnham* [963×75]c.1150 S 823, *Ferneham* 1086, *Farnham* from 1233. OE **fearn** + **hamm** 3. Sr 169, ASE 2.23, L 42–3, 47, 49.

FARNHAM COMMON Bucks SU 9685. P.n. Farnham as in FARNHAM ROYAL SU 9683 + ModE **common**.

FARNHAM GREEN Essex TL 4625. *Farnham Green* 1805 OS. P.n. FARNHAM TL 4724 + ModE **green**.

FARNHAM ROYAL Bucks SU 9683. *Fernham Riall* 1477, *Farnham Roiall* 1478. P.n. Farnham + ME **riall** < OFr *reial* 'royal'. Farnham, *Ferneham* 1086, *Farnham* from 1200, is 'fern homestead', OE **fearn** + **hām**. Bk 229.

FARNINGHAM Kent TQ 5466. Partly uncertain. *Frinningaham* 1042×66 S 1047, *Fo- Ferninge- Ferlingeham* 1086, *Ferni(n)ge- Fremga- Frenigaham, Faerningeham* c.1100, *Freni(n)g(e)ham* 1176×7–1242×3, *Ferningeham* 1198, 1201×2, *Framingham* 1203, *Fremingeham* 1211×2. Unidentified folk-n. < unidentified element or name + **ingas**, genitive pl. **inga**, + **hām** 'homestead, estate'. Whatever the original form of the specific of this name it was later assimilated to **fearning* 'ferny place', and **fearningas* 'dwellers in the ferny place', genitive pl. **fearninga*. A full study is needed of this and associated names, Farningham Farm TQ 8035, *ffrenyng(e)ham* 1327, 1338, Frinningham TQ 8158, *Frimingeham* after 1264, *Frenyngham* 1316, FRING Norfolk TF 7334, *Ffainges, Frenga, Frenge* 1086, and Franjum in Frisia, *Franingaheim* n.d. PNK 40, KPN 326–7, ING 118–9, 57, ASE 2.30.

FARNLEY 'Forest clearing overgrown with fern'. OE **fearn** + **lēah**. L 204.

(1) ~ NYorks SE 2148. *(on) Fearnleage* c.1030 YCh 7, *Fernelai -ley -lay* 1086–1577, *Farnelay -lai -ley* 12th–1658, *Farnlaia* 1170×80, *-le(y)* 1240–1542. YW v.58, L 204.

(2) ~ TYAS WYorks SE 1612. 'F held by the Tyas family'. P.n. *Fereleia, Ferlei* 1086, *Ferneleia -lay -ley* 12th–1599, *Farlag'* 1202, *-lay* 1244, *Farneley(e) -lay* 1267–1605, *Farnley* from 13th, + manorial addition *-tyes* 1361, *Tyas* 1499. Land was held here in the 13th cent. by an owner called *le Tyeis* 'the German'. YW ii.267.

(3) New ~ WYorks SE 2431. A new settlement named after Farnley SE 2532, *Fernelei* 1086, *-ley -lay* 1379–1516, *Farnelei(a) -lay -ley* 12th–1597. YW iii.211.

FARNSFIELD Notts SK 6456. 'Open land of *Fearn*, the area growing with fern'. *Fearnesfeld, (on) Fearnes felda* [958]14th S 659, *Farnesfeld* 1086–1333, *Franesfeld* 1086, *Farnefeld* 1187–1344. OE **fearn**, genitive sing. **fearnes**, + **feld**. Nt 163, L 243.

FARNWORTH Ches SJ 5187. 'Fern enclosure'. *ffarneword* 1324, *Farneworth* 1518. OE **fearn** + **worth**. La 106.

FARNWORTH GMan SD 7305. 'Fern enclosure'. *Farnewurd* 1185, *Ferneworthe* c.1200, *Farneworth* 1278–1448, *Farnworth* 1278×9, *Farnorth* 1586. OE **fearn** + **worth**. La 43, Jnl 17.36.

FARRINGDON 'Fern-clad hill'. OE **fearn** + **dūn**. PNE i.166.

(1) ~ Devon SY 0191. *Ferentone, Ferhendone* 1086, *Ferendon(e)* 1234–91, *Ferndon* 1242–1327, *Farndon(e)* 1310–72, *Faryngdone* 1310×15. D 588, L 148.

(2) Lower ~ Hants SU 7035. A modern development, ModE adj. **lower** + p.n. Farringdon as in Upper FARRINGDON SU 7135.

(3) Upper ~ Hants SU 7135. ModE **upper** + p.n. *Ferendone* 1086, *-don(a)* 1186, 1208, *Farendon(e)* 1200–74, *Far(i)ndon(e)* c.1270–1341, *Faryngdon* 1428. Ha 76, Gover 1958.91, DEPN.

FARRINGTON GURNEY Avon ST 6255. 'F held by the Gurney family'. *Farrington Gournay* 1817 OS. P.n. Farrington + family name *Gurney* from Gournay-en-Brai in Normandy. Earlier simply *Ferentone* 1086, *Ferenton* 1225, 'fern village or estate', OE **fearn** + **tūn**. Ferns were prized and cultivated for their medicinal properties, especially as a vermifuge. The ashes were used in both soap- and glass-making. The leaves were sometimes eaten and used as fodder for sheep and goats. DEPN, Grieve 300ff.

FARSLEY WYorks SE 2135. 'Heifer clearing'. *Fersellei(a)* 1086, *Ferselee -lay -lai -lei(a) -ley* 12th–1468, *Fersleg -ley(a) -lay* 1208–1505. OE ***fers**, genitive pl. ***fersa**, + **lēah**. YW iii.228, xiii.

FARTHINGHOE Northants SP 5339. 'Hill-spur called or at *Fearning*'. *Ferningeho* 1086–1232, *Ferningho -ling-* 1195–1229, *Farningho -yng-* 1220–1406, *Ferthingo* 1198, *Farthinghoe* from 1580, *Farnigo* 1595. OE p.n. **Fearning* 'ferny place' < **fearn** + **ing**[2], locative-dative sing. **Fearninge*, + **hōh**. Nth 53 gives pr [fa:nigou], L 169.

FARTHINGSTONE Northants SP 6154. 'Farthegn's farm or village'. *Fordinestone* 1086, *Fardeneston* 1166, *Farding(e)stun -ton -yng-* 1166–1478, *Farthingeston(e) -yng-* 1261–1428. ON pers.n. *Farthegn*, secondary genitive sing. *Farthegnes*, + **tūn**. Nth 22 gives pr [farəkstən], SSNEM 190.

FARWAY Devon SY 1895. Uncertain. *Farewei* 1086–1346 with variants *-weie* and *-weye, Farweye* 1297. Usually explained as OE **fær-weġ**, but the precise significance of a compound of OE *fær* ' a going, a road, a passage, a ford' + *weġ* 'path, road' is unclear. Dr Gelling suggests rather *fǣr-weġ* 'danger road' pointing to the steep declivities on the roads into the village. D 625, L 66, 83.

FATFIELD Durham NZ 3053. 'Rich, luscious field'. *Fatfield* 1863 OS. ModE **fat** + **field**.

FAUGH Cumbr NY 5054. 'The fallow'. *Faughe* 1589, *The Faugh* 1603, *le Faugh* 1626. ME dial. **faugh** (OE *falh*). Cu 89.

FAULKBOURNE Essex TL 7917. 'Falcon stream'. *Falcheburnā* 1086, 1185, *Falkeburn(a)* 1198, 1256, 1346, *Faukeburn(e)* 1236–1303, *Faw(e)bo(u)rn(e)* 1352–61, 1468. OE ***falca** + **burna**. Ess 287 gives pr [fɔbən], Studies 1931.23–4.

FAULKLAND Somer ST 7354. 'Land held by folk-right'. *Fouklande, Falclond* 1243. OE **folcland**. This an Anglo-Saxon legal term for land from which the king drew food-rents and customary service as distinct from *bōcland* which was exempt by charter (*bōc*) from such services. So here Faulkland is distinguished from *bōcland* at nearby BUCKLAND DINHAM ST 7551. DEPN, PNE i.179, AS England 306–8, Reallexikon s.v. Folkland.

FAULS Shrops SJ 5932. Short for *Faul's Green* 1805, *Faux Green* 1806. P.n. *Le Faall* 1301, *(le) Falles* 1363, 1383, *Fawles* 1672, OE **(ġe)fall** 'a felling of trees, a forest clearing' + ModE **green**. Gelling, PNE i.165.

FAVERSHAM Kent TR 0160. 'The smith's estate'. *Fefres hám* 811 S 168, *Fefresham* 812, 858 S 169, 328, *Febresham* 811, 815, 858 S 168, 178, 328, *Feferesham* [850]13th S 300, *Fæfresham* c.935, *Favreshant* 1086, *Faveres- Faures- Fevresham* 1154×5–1162×3 etc., *Feversham* 1610, 1819 OS. OE ***fæfer**, Kentish ***fefer** < Latin *faber*, perhaps used as a by-n., genitive sing. ***fæf(e)res**, ***fef(e)res**, + **hām**. Faversham is close to a small RBrit settlement at Syndale TQ 9960, and the grave goods from Anglo-Saxon burials near Faversham suggest that there was a centre of fine metal-working here during the pagan period. It is possible that the metal-working was already established here at the time of the English settlement, and that the industry continued in the ensuing period. KPN 117, PNE i.163, TC 94, ASE 2.30, Signposts 80, Charters IV.82.

FAWFIELDHEAD Staffs SK 0765. 'The head of Faw Field'. *Fawfeild Hill Heydde, Fawfild Heade* 1558×1603, *Fairfield Head* 1695, *Faw Field Head* 1840 OS. P.n. *Faufeld* 1308, 'the many coloured open land', OE **fāh**, definite form **fāga** (ME *fawe*), + **feld**, + ModE **head**. Oakden, Horovitz.

FAWKHAM GREEN Kent TQ 5865. *Fawkham Green* 1819 OS. P.n. Fawkham TQ 5966 + ModE **green** 'a common'. Fawkham, *Falcheham* [973×87]12th S 1511, *Fealcnaham* [10th]12th S 1457, *Falchenham* c.975, *Fachesham* 1086, *Falkeham* 1210×12, 1240, 1610, *Falkham* 1819 OS, *Faukeham* 1215, *Faukham* 1245, is probably 'Fealcen's estate', OE pers.n.**Fealcen*, genitive sing. **Fealcnes*, + **hām**. The same pers.n. occurs again in *(to) fealcnes forda* 898 S 350. KPN 295, 358, 232.

FAWLER Oxon SP 3717. '(The estate containing a building with) the flagstone floor'. *Flauflor* [1066×87]13th, *Flagaflora* [1100×35]12th., *Fauflur -flor(e)* 1204–1242×3, *Fau(e)lor(e)* 1284–1380, *Fawlour* 1428, *Fuller* 1535, *Fallow* 1761, 1830. OE ***flage** + **flōr**. Brk 372.

FAWLEY Berks SU 3981. Partly uncertain. *Faleslei -lie -le(g)a* 1086–1190, *Faleweslega* 1177, *Falelea* 1197–1241 with variants *-l(egh)', Falleye* 1301, *Great Fawley* 1761. Uncertain element + OE **lēah** 'wood, clearing'. Fawley is usually regarded as having the same specific as FAWSLEY Northants SP 5656, either OE **fealg** 'fallow(-land)' confused here with the adj. **fealu** 'lightish red, fallow', the name then denoting a 'clearing of woodland belonging to some larger area of fallow land'; or a noun **fealu*, the name of a forest, 'fallow-coloured wood', or, more likely, an animal name, 'fallow deer'. Neither of these suggestions is convincing and it may be preferable to see the first element as an OE pers.n. derived from the adjective *fealu* 'fallow, yellow, tawny, dun-coloured'. North or Great Fawley and South or Little Fawley were distinct manors. South FAWLEY SU 3980 is three-quarters of a mile S of Fawley. Brk 298, Nth 23, PNE i.165.

FAWLEY Bucks SU 7586. Partly uncertain. *Falelie* 1086, *Fauley* c.1225. The generic is OE **lēah** 'wood or clearing', the specific either of OE **falh** 'ploughed land' and so 'clearing that has been ploughed' or **fealu** 'fallow coloured, pale brown'. Bk 175, Studies 1936.171.

FAWLEY Hants SU 4603. Partly uncertain. *Falegia, Falelie* 1086, *Falesleia* 1130, *Faleleia -legh -leye -le(e)* 1194–1327, *Fawele* 1291, *Fawleigh* 1675, *Valile, Vallie* 1311, 1316. Either 'fallow coloured clearing', OE **fealu** + **lēah**, or 'clearing with land broken in for arable', OE **fealg** + **lēah**, would be possible although some ME spellings with *-w-* (**Falweleye*) might have been expected. If *(to) Faleðlea* [10th]12th in S 1821, a list of lands belonging to Winchester cathedral, belongs here, then the name would mean 'hay clearing', OE **fælethe** + **lēah**, a possibility indicated by the 1130 spelling, although *fælethe* is probably the Mercian dial. form and WS *filethe* would be expected. The name remains problematic. Ha 77, Gover 1958.197.

South FAWLEY Berks SU 3980. *Suthfalalea* [1154×89]1270, *Suffalleleg'* 1224×5, *Suth Falele, ~ Fauele, ~ Fal(l)e(s)leg(h)'* 1284, *Southfalle* 1401×2. ME adj. **suth** + p.n. FAWLEY SU 3981. Brk 298.

FAWLEY CHAPEL H&W SO 5929. *Fawley Chapel* 1831 OS. P.n. Fawley + ModE **chapel** referring to the Norman church here. Fawley, *Fæliglæh* [1016×35]11th S 1462, *Falileam* 1142, *Filileam* 1158, *Felileie -leyam* 1166, 1198, *Faliley* 1292, *Falley(e)* 1284, 1334, *Folley* 1641, is possibly 'the wood or clearing where felloes are made', OE **f(i)elġe**, Mercian **fælġe**, + **lēah**. Also known as *Much Fawley* 1831 OS for distinction from *Little Ffawley* 1586, referring to Fawley Court SO 1730. He 48–9.

FAWTON Corn SX 1668. 'Settlement on the river Fowey'. *Fauuitone* 1084, *Fauuitona* 1086 (Exon), *Fawintone* 1086 (DB), *Fawyton* 1229, *Foweton* 1311. R.n. FOWEY SX 0962 + OE **tūn**. PNCo 84, RN 164, DEPN.

FAXFLEET Humbs SE 8624. Uncertain. *Flaxflet(e)* 1185, 1202, 16th cent., *Faxflet(e)* 1190–1566. If the forms with *Flax-* are original, 'stretch of water, tidal creek where flax grows', OE **fleax** + **flēot(e)** with loss of the first *l* by dissimilation. Alternatively OE **feax** + **flēot(e)** with *feax* 'hair' in the sense 'coarse grass'; or 'Faxi's creek or stretch of water', with ON, OE pers.n. *Faxi*, a name occurring once in a literary text as the name of a hairy monster. YE 224, lx, L 21, Jnl 29.81.

FAYGATE WSusx TQ 2134. *Fay Gate* 1614. P.n. *Feye* 1380, possibly a contraction of OE **fēo-heġe* 'cattle-enclosure', + ModE **gate**. Sx 233.

FAZAKERLEY Mers SJ 3896. 'Wood or clearing at *Fasacre*, the newly-cultivated border strip'. *Phasakyrlee* c.1250, *ffasacrelegh* 1325, *Fazakerley* 1509. P.n. *ffasacre* 1325 'border strip', OE **fæs** 'border, fringe' + **æcer** 'newly-cultivated land', + **lēah**. La 116, L 232–3.

FAZELEY Staffs Sk 2002. 'The bull's pasture'. *Faresleia -is-* c.1142, *Faresleye* 1335. OE **fearr**, genitive sing. **fearres**, + **lēah**. DEPN.

FEARBY NYorks SE 1981. Partly uncertain. *Federbi* 1086, *Fetherby* 1184, *Fe(g)herbi -by* 1193–1316, *Faireby* 1231, *Feyerby* 1369, *Fearby* from 1537. Possibly 'four villages', PrScand ***fethuru** 'four' + ON **bȳ**, cf. SEVENHAMPTON SP 0312, SU 2090; or 'feather village' perhaps with reference to production of down, OE **fether** + ON **bȳ**. YN 232, SSNY 27.

FEARNHEAD Ches SJ 6390. 'Headland covered with fern'. *Ferneheued* 1292, *(del) Fermhed* (sic for *Ferni-*) 1332, *Fernyhed* 1467. OE **fearn** or adj. **fearnig**, + **hēafod**. La 106.

FEATHERBED TOP Derby SK 0991. *Featherbed Top* 1850. P.n. Featherbed + ModE **top** referring to a summit of 1785ft. Featherbed occurs again in *Featherbed Moss* 1843 SK 0892 and *Featherbed Moss* 1837 near Buxton SK 0669 and refers to soft soil. Db 126, 70, 373, Field 1972.75.

FEATHERSTONE Staffs SJ 9405. 'The tetralith'. *Feoþer(e)stan* [996]17th S 1380, *Ferdestan* 1086, *Fetherstan* 1280×90, *-ston* 1244–1411, *-stone* 1294–1775. OE **feother** + **stān**. No trace of the cromlech or tetralith remains. St i.122.

FEATHERSTONE WYorks SE 4221. 'The four stones'. *Fredestan, Ferestan(e)* 1086, *Federstan(a)* 1108×14–1601, *Fetherstan(a) -stane* 12th–1428, *Fedderstone, Featherson* 16th cent. OE **feother** + **stān**. The reference may have been to a cromlech or tetralith no longer extant though no such feature has been found at any of the places so named. YW ii.86.

FEATHERSTONE CASTLE Northum NY 6760. *Featherstone Castle* 1869 OS. P.n. Featherstone + ME **castel**, a 14th cent. tower much rebuilt. Featherstone, *Fetherstan* 1255, OE **feother-stān** 'a tetralith', presumably refers to a lost cromlech consisting of three uprights and a headstone. Also recorded are *Fetherstanehalg* 1204, *Fetherstanishalu* 1236 BF, *Fetherstan(e)halcht* c.1215, *-halwe* 1242 BF, *-halgh* 1346, *Ferstonehale* 1222, *Feyrstanhalth* (sic for *-halch*) 1296 SR, 'land in a river bend at Featherstone', p.n. Featherstone + OE **halh**. NbDu 82, DEPN, Pevsner 1992.278.

FECKENHAM H&W SP 0161. 'Fecca's river-meadow'. *feccan homm* [804]11th S 1187 with variant *feccanhorn, feccan ham* 1046×52 S 1227, *Fec(c)hehā* 1086, [1154×89]1266, [1189×99]1326, *Fecke- Fek(k)eham* [c.1086]1190–1275, *Fe(c)kenham* 1233, 1312, *Fecnom* 1699. OE (W Mercian) pers.n. **Fæcca, *Fecca*, side form of **Facca*, genitive sing. **Fæccan, *Feccan*, + **hamm** 3. Wo gives pr [feknəm], L 43,39, Hooke 92–3, 348.

FEERING Essex TL 8720. 'The Feringas, the people called after Fera'. *Feringes* 1066–1274, *Feringe* 1075–1274, *Ph- Feringas, ferigens* 1086, *Ferring(es)* 1203, 1206. OE folk-n. **Fēringas* < pers.n. **Fēra* + **ingas**. Ess 389, ING 19.

FEETHAM NYorks SD 9898. '(The settlement) at the meadows'. *Fytun -on* 1242–1298, *Fethom* 1645. ON **fit**, dative pl. **fitjum**. YN 271 gives pr [fi:təm].

FEIZOR NYorks SD 7967. 'Fiach's shieling'. *Fegesargh, Fehhesherge, Fegheserche* 12th, *Feysergh(e)* 1294, 1599, *Feiser* 1538, 1643, *Feazor* 1597, *Feisor* 1658. Pers.n. *Feghe* < OIr *Fíach*, genitive sing. *Feghes*, + ON **ǽrgi**. YW vi.226 gives pr ['fe:zə] and ['fi:zə].

FELBRIDGE Surrey TQ 3839. 'The bridge at the open country'. *Feltbruge, Feldbrigge* 12th cent., *Felbridge* from 1749. OE **feld** + **brycg**. Sr 318, L 66, 242.

FELBRIGG Norf TG 2039. 'The plank bridge'. *Felebruge* 1086, *-brigge* 1207. ON **fjǫl**, genitive pl. **fjala**, + **bryggja**. Cf. ON *fjala-brú* 'a bridge of planks'. DEPN, PNE 1.55, 174.

FELCOURT Surrey TQ 3841. 'The cottage(s) in the open country'. *Feldecote* 1403, *Felcot -corte* 1535, 1596. OE **feld** + **cot**, pl. **cotu**. Sr 329.

FELDEN Herts TL 0404. *Felden* 1588, *Felddane* 1667. Although the evidence is too late to offer a secure explanation this appears to be a name in ME **dene** (OE *denu*) 'a valley'. There are valleys both N and S of the hamlet. Hrt 42.

FELIXKIRK NYorks SE 4684. 'St Felix's church'. *Ecclesia S. Felicis* 1210, *Felicekyrke -kirk* 1293, 1316, *Fillyxchurche* 1578. Saint's name *Felix* from the dedication of the church + ON **kirkja**. YN 199 suggests that DB *Fridebi* (327v) 'Frithi's farm or village' is an earlier name of this place. Cf. FIRBY SE 2686. YN 199, SSNY 27.

FELIXSTOWE Suff TM 2934. 'Filica's meeting place or holy place'. *Filchestou* 1254, *-stowe* 1286–1508, *Fylthestowe* 1366,

Fylstowe 1405, *Fylcestowe* 1493, *Felyxstowe* 1503–55, *Felixstowe* from 1503, *Fylchestow al. Felixstowe* 1521, *Felixstowe al. Fylstow* 1538, *Fylstowe al. Fylchestowe* 1547. OE pers.n. *Filica* + **stōw**. The sense of *stōw* is uncertain here. There does not seem to be any connection with St Felix. Modern Felixstowe is a late 19th and early 20th cent. development, the original site being at Old FELIXSTOWE TM 3135. The earlier name of the place was *burg, burch* 1086, 'the fort', OE **burh**, referring to a Roman fort washed away by the encroaching sea at Walton Castle. DEPN, *Stōw* 192, Pevsner 1974.474.

Old FELIXSTOWE Suff TM 3135. This is the original site of Felixstowe where the Benedictine priory used to be. Modern Felixstowe is a late 19th and early 20th cent. development. Pevsner 1975.210.

FELKINGTON Northum NT 9444. 'Feoluca's hill'. *Felkindon* 1208×10 BF–c.1250, *Felkenden* 1238, *Felkyngton* 1441. OE pers.n. **Feoluca*, genitive sing. **Feolucan*, + **dūn**. NbDu 83, L 158.

FELL END Cumbr SD 7298. *Fell End* 1669, 1690, 1739, *Felend* 1686, named from *ye Fells* 1651. ModE **fell** (ON *fjall*) + **end**. We ii.35.

FELL SIDE Cumbr NY 3037. 'Fell hill-side'. *Felside* 1560. ModE **fell** (ON *fjall*) + **side**. Cu 278.

FELLING T&W NZ 2762. 'The clearing'. *Felling* 1325 and from 1580, *(del) Fellyng* 1371–1512. OE ***felling**. NbDu 83 records that it was still known locally as '*the* Felling'.

FELMERSHAM Beds SP 9957. 'Feolumær's homestead or land in a river bend'. *Falmeres- Flammeresham* 1086, *Faumerisham* 1189×99, *Felmer(e)sham* 13th cent., *Femsham al. Femsam* 1549. OE pers.n. *Feolumǣr*, genitive sing. *Feolumǣres*, + **hām** or **hamm**. The name *Feolumǣr* occurs in two lost charters (*[709]12th S 79, [963]11th S 1307), once in the spelling *Fealamær* which would account for the *Falm- Faum-* forms: the name is paralleled by Gothic *Filumar*. Felmersham lies at a bend in the river Great Ouse. Bd 31 gives pr [femsəm], L 47.

FELMINGHAM Norf TG 2529. 'The homestead of the Felmingas, the people called after Felma or Feolma'. *Felmi(n)cham* 1086, *Felmingham* from 1101, *Felmingeham* 1175–1257. OE folk-n. **Felmingas* < pers.n. **Felma* or **Feolma* + **ingas**, genitive pl. **Felminga*, + **hām**. Nf ii.156, ING 134.

FELPHAM WSusx SZ 9699. Probably 'the fallow enclosure'. *Felhhamme* [873×88]11th S 1507, *Felhham* [953]14th S 562, *Falcheham* 1086, *Falk- Falgham* 1230–1428, *Felk(h)am, Felcham, Felgham* 1177–1428, *Phelpham, Felffam, Feltham, Felgham* 1575–93. OE **fealh** + **hamm** probably in senses 2, 4 or 5, 'promontory of dry land in marsh, cultivated plot in marginal land' (the original topography is lost). The *Falk-* sps are normal with substitution of [k] for peripheral phoneme [ç], those in *Felk-* difficult. 16th cent. sps show substitution of [θ] and [f] for the peripheral phoneme [ç]. Sx 140, Ritter 139, ASE 2.25, SS 82.

FELSHAM Suff TL 9457. 'Fæli's homestead'. *fealshā* 1086, *Fealsam, Felesham* c.1095, *Fales- Felsham* 1203, 1610. OE pers.n. **Fǣli*, genitive sing. *Fǣles*, + **hām**. DEPN.

FELSTED Essex TL 6720. 'Site in open country'. *Felestedā, Felstede, Phenstedā* 1086, *Feldsted(a) -e* 1085–1208, *Felsted(e)* 1203–1552, *Fenested* 1246. OE **feld** + **stede**. Ess 421, Stead 199.

FELTHAM GLond TQ 1072. 'Homestead at the open land'. *Feltham* from [969]c.1100 S 774, *Feltehā* 1086, *Feltam* 1655, *Feltem* 1668. This explanation depends on the unvoicing of *d* to *t* as also in UPWALTHAM WSusx SU 9413. Alternatively the specific might be OE *felte* 'mullein', a plant-name derived from OE *felt* as in FELTWELL Norf TL 7190. Mx 14, GL 49.

FELTHORPE Norf TG 1617. Possibly 'Fæla's outlying farm'. *Felethorp, Faltorp* 1086, *Feletorp* 12th cent., *Felestorp* 1254. OE pers.n. **Fǣla* + **thorp**. DEPN.

FELTON 'Settlement in open ground'. OE **feld** + **tūn**.

(1) ~ Avon St 5265. *Felton* from 1243. DEPN, L 240, 242.

(2) ~ H&W SO 5848. 'Settlement in open land'. *Feltone* 1086, *Felton(a)* from 1148×63. He 85, L 240, 242.

(3) ~ Northum NU 1800. 'The settlement in the open land'. *Feltona* 1166, *-tunia* 1215, *-ton* from 1242 BF. *Parva Felton* 'little F' 1242 BF refers to Old Felton NU 1802. NbDu 84, L 240, 242.

(4) ~ BUTLER Shrops SJ 3917. 'F held by the Butler family'. *Felton Buttler* c.1270. P.n. *Feltone* 1086, *Felton(')* from 1241 + manorial addition from Hamo le Butiler, the tenant in 1242. Sa i.131.

(5) West ~ Shrops SJ 3425. *West Felton* from 1397, *Welch, Welsh Felton* 17th cent. ME adj. **west** + p.n. *Feltone* 1086, *Felton* 1303, 1344 etc. 'West' for distinction from FELTON BUTLER SJ 3917. Sa i.307.

FELTWELL Norf TL 7190. 'The spring where wild marjoram or mullein grows'. *Feltwelle* [1042×66]12th S 1051, *Feltuuella, Fatwella* 1086, *Feltewell* 1169, 1196, *Fautewelle* 1162. OE **felte** + **well**. The precise botanical species referred to is uncertain. The soil is sand and chalk which would favour marjoram. DEPN, PNE i.169.

FELTWELL ANCHOR Norf TL 6589. An inn name. No early forms.

FEN END WMids SP 2275. 'The fen end' sc. of Balsall. *Fennend* 1540, *Fenny End* 1667. ModE **fen** + **end**. Wa 54.

FEN ROAD Cambs TL 4698. The Roman road across the fens from Upton to Denver via Peterborough, Margary no. 25. ModE **fen** + **road**.

FENBY Humbs TA 2500. 'Fen village or farm'. *Fen(de)bi* 1086, *Fembi* c.1115, *Fenbi -by* from 1190. OE **fenn** + **bȳ**. Li iv.49.

FENCE Lancs SD 8337. 'The enclosure'. *Fens in Penhill* 1402, *del Fence* 1425, *the Fence* 1515. ME **fence** (OFr *(de)fence*). La 82, Jnl 17.51.

FENCE HOUSES T&W NZ 3250. 'The houses by the fence'. *Fence Hos.* 1863 OS. A modern colliery settlement. The reference is unknown and the location seems to be too distant from the *Lumley parke paile* 'Lummley park fence' mentioned in a will of 1597.

FENCOTE NYorks SE 2893. 'Cottages in the fen'. *Fencotes* from 1270. OE **fenn** + **cot**. The reference is to settlements in the swampy ground beside the river Swale at Great and Little Fencote. Cf. KIRKBY FLEETHAM SE 2894, AINDERBY MIRES SE 2592. YN 239.

Thorpe FENDYKES Lincs TF 4560. 'Fendyke in T'. Cf. *Thorp Fendike Sholter* 1661. P.n. Thorp as in THORPE ST PETER + p.n. *Fendyke* 1824 OS, 'the fen dyke', ModE **fen** + **dike**. The Fendike is a pre-17th cent. drainage channel taking the waters of Steeping River from Firsby Clough to Wainfleet and the sea. Wheeler 199.

FENISCOWLES Lancs SD 6425. 'Dirty huts'. *Feinycholes* 1276. ME **fenny** (OE *fennig̣*) + **scoles** (ON *skáli*). Probably referred to muddy ground beside the river Darwen. La 74.

FENITON Devon SY 1099. 'Settlement on Vine Water'. *Finetone* 1086, *Finetuna* 1185–1476 with variants *Fyne-* and *-ton(e)*, *Vinetone* 1309, *Feneton* 1169, 1311, *Venyton* 1585, 1637. R.n. Vine + OE **tūn**. Vine Water, *(on) Finan* [1061]1227, *Fynee* 1553, *Vine* 1797, may be identified with Co **fyn** 'end, boundary'. The stream here forms a boundary between Feniton and Buckerell. The form *Fynee* looks like a compound of the r.n. + OE **ēa** 'a stream, a river'. D 563,15, CPNE 98.

FENNY BRIDGES Devon SY 1198. *Saint Annes Bridge al. Fynee Bridge* 1553. The main Exeter-Honiton road crosses bridges over the river Otter and Vine Water for which see FENITON SY 1099. D 608.

FENROTHER Northum NZ 1792. 'The clearing or assart where wood is stacked'. *Finrode* 1189, *Fin- Fynrothre -rother* 1232–1428, *Fenrother* from 1256 Ass. OE **fīn** + ***rod(u)** later replaced by ***rother**. Alternatively the specific could be OE **fīna** 'a woodpecker' or ***finn** 'coarse grass'. It was later misunderstood as **fenn** 'marshland'. NbDu 84, DEPN.

FENSTANTON Cambs TL 3168. 'Fen Stanton'. *Fenstanton* from 1260, *Feni- Fennystanton* 1344–1526, 1836 OS. Earlier simply

Stantun [1012]12th S 926, *Stanton(e)* 1086 etc., *Staunton* 1227–86, 'stone settlement', OE **stān** + **tūn**. 'Fen' or 'Fenny' for distinction from LONGSTANTON TL 3966. Hu 267.

FENTON 'A marsh settlement'. OE **fenn** + **tūn**. L 40–1.

(1) ~ Cambs TL 3279. *Fentun* 1236, *Fenton* 1279. Hu 211.

(2) ~ Lincs SK 8476. *Fentuna* c.1115, *Fenton* from c.1225. DEPN, Cameron 1998.

(3) ~ Lincs SK 8750. *Fentona* [c.1145]after 1269, *-ton* before 1219. Perrott 370, Cameron 1998.

(4) ~ Northum NT 9733. *Fenton(')* from 1242 BF. NbDu 84.

(5) ~ Staffs SJ 8944. *Fentone* 1086, *-ton* from 1273. DEPN, L 40–1.

(6) Church ~ NYorks SE 5136. 'F with the church'. ME prefix *Kirk(e)* from 1338 (ON *kirkja*), ModE **church**, + p.n. *(on) fentune* [963]14th S 712, *Fen(n)tún* c.1030 YCh 7, *Fentun, Fenton(a)* from 1086. YW iv.63.

(7) South or Little ~ NYorks SE 5235. *S(o)uthfenton* c.1270, *Lit(t)lefenton* from 1512. ME **suth**, ModE **little**, + p.n. Fenton as in Church FENTON SE 5136. YW iv.64.

FENWICK 'A dairy-farm in or by marshland'. OE **fenn** + **wīċ**. L 40–1.

(1) ~ Northum NZ 0572. *Fenwic* 1242 BF, *-wik* 1296 SR, *Fennewyk* 1346. The place is situated near MATFEN NZ 0371. NbDu 84.

(2) ~ Northum NU 0540. *Fenwic* 1208×10 BF, *Fennewick* 1312, *Fenneck* 1579. NbDu 84.

(3) ~ SYorks SE 5916. *Fenwic* 1166–1226, *-wyk(e)* 13th–1496, *Fennicke* 1622. YW ii.47.

FEOCK Corn SW 8238. 'Church of St Fioc'. *Lan- Lamfioc* [c.1165]17th, *ecclesia(m) Sancte Feoce* 1264–1350, *de Sancto Feoko* 1291, (church of) *Seyntfeok* 1392. Co ***lann** 'church-site' + saint's name *Fioc* from the dedication of the church. Nothing is known of this saint, including the gender; cf. St Fiac, Brittany. The form *Lanfioc* survives in the farm name La Feock pronounced [la'veig]. PNCo 83, Gover n.d. 447.

FERNDOWN Dorset SU 0700. P.n. Fern + ModE **down**. Fern, *Fyrne* 1321, *Ferne* 1358, is probably 'fern brake', OE ***(ġe)fierne** rather than *fierġen* 'a wooded hill' which scarcely fits the topography with its flat surface and only a low hill. Do ii.225, Studies 1936.140.

FERNHAM Oxon SU 2992. 'River meadow where ferns grow'. *Fernham* from [821]12th S 183, *Færhom* [821]13th S 183, *Farnham* 1241–1517. OE **fearn** + **hamm** 1. Fernham lies at the end of the block of higher ground on which FARINGDON SU 2895 is situated to the S of which is a level area traversed by head-streams of the Ock, the *hamm* of the p.n. Brk 371.

FERNHILL HEATH H&W SO 8659. *Fernall Heath* c.1830. P.n. Fernhill + ModE **heath**. Fernhill, *Fernhull* 1275, 1327, is 'fern hill', OE **fearn** + **hyll**. Wo 111.

FERNHURST WSusx SU 8928. 'Wooded hill where bracken grows'. *Fernurst* c.1195, *-herst* c.1200, *Farnhurst* from 1269, *Farnyst* 1313. OE **fearn** + **hyrst**. Sx 19 gives pr [fɑ:nəst], SS 67.

FERNILEE Derby SK 0178. 'Fern(y) clearing'. *Ferneley(e) -le(i)gh* 1101×8–1714, *Fernileg(h)(e) -ley(e) -le(e) -y-* 1236–72 etc. OE **fearn, fearniġ** + **lēah**. Db 98.

FERNILEE RESERVOIR Derby SK 0176. P.n. FERNILEE SK 0178 + ModE **reservoir**.

FERNWORTHY RESERVOIR Devon SX 6684. P.n. Fernworthy + ModE **reservoir**. Fernworthy, *Vernaworthy* 1355, *Ferne-* 1377, is 'fern enclosure', OE **fearn** + **worthiġ**. D 198.

FERRENSBY NYorks SE 3760. 'Farm or village of a Faroe Islander'. *Feresbi* 1086, *Feringebi* 1239, *Feringesby(e) -ynges-* 1269–1576, *Ferinsby -bie* 1606. ON **færeyingr**, possibly used as a by-name, + **bȳ**. YW v.92, SSNY 27.

FERRIBY 'Village near the ferry'. ON **ferja** + **bȳ**. Cf. the Swedish p.n. Färjeby, *j Färioby* 1383. SSNY 27.

(1) North ~ Humbs SE 9825. *North Feriby* 1284 etc. ME adj. **north** + p.n. *Ferebi* 1086, *-bi -by* 1088×93–1297, *Feribi -y-* 1150×3–1546. The village lies at the N end of an important passage of the Humber. Cf South FERRRIBY SE 9820. YE 218, SSNY 27.

(2) South ~ Humbs SE 9820. *Suthferiby* [1100×35]14th–1309. ME adj. **suth** + p.n. *Ferebi* 1086, 1202, *Feribi -by* 1167–1431. The village lies at the S end of an important passage of the Humber. Cf. North FERRIBY SE 9825. DEPN, SSNEM 46.

FERRING WSusx TQ 0902. 'The Feringas'. *Ferring* from 1269 including [762 for 765]14th S 48 and [711 for 791]14th S 1178 with variant *-yng(e), Feringes* 1086–1305, *Farryng* 1660. OE folk-n. *Fēringas* as in FEERING Essex TL 8720 with vowel shortening in the trisyllabic form as in READING Berks SU 7173. Sx 167, ING 33, MA 10.23, SS 64.

FERRYBRIDGE WYorks SE 4824. 'The bridge at the ferry'. Latin *Pontem ferie* 12th, *pontis de Feria* 1227, English *Feribrig(ge) -bryg(g)* 1198–1597, *-bridge* 1545. ON **ferja** + ME **brigge** (OE **brycg**). The ferry, *Ferie, Fereia* 1086, *Feri(a)* 12th cent., *Fery* 1290–1541, which carried the traffic of the Great North Road across the Aire, was replaced by a bridge in the 12th cent. The present structure is 18th cent. YW ii.66, Pevsner 1967.200.

FERRYHILL Durham NZ 2932. 'Ferry on the hill'. *Ferry on the Hill* 1421–1587, *Ferry Hill* from 1422. P.n. Ferry, *æt Feregenne* [c.994]early 11th S 1659, *Ferie -a* 1154×61–c.1350, *Fery* [1183]c.1320–1550, *Ferry* 1575, 1580, the 'wooded hill', OE **ferġen**, + pleonastic ME *on the hill*. NbDu 85.

FERSFIELD Norf TM 0683. Either 'the open land growing with furze' or 'the heifer open land'. *Fersafeld* 1035×40 S 1489, *Ferseuella* 1086, *Fersfelde* 1212. Either OE **fyrs**, SE dial. form **fers**, 'furze', or ***fers** 'a heifer', genitive pl. ***fersa**, + **feld**. DEPN, PNE i.190.

FETCHAM Surrey TQ 1455. 'Fecca's settlement'. *Fecehā* 1086, *Fecheham* c.1170–1285, *Fecham* [973×87]early 12th S 1511–1436, *Fecchen-* 1252, *Fek-* 1259. OE pers.n. **Feċċa*, genitive sing. **Feċċan*, + **hām**. Sr 76, ASE 2.32, PNDB 250.

FEWSTON NYorks SE 1954. 'Fot's estate'. *Fostun(e) -ton(e)* 1086–1582, *Foteston* 13th, *Fosceton(e)* 1280, 1282, 1388, *Fooston* 1414, *Fuston(e)* 1441, *Fui- Fuyston* 1454–late 17th, *Feweston(e)* 1628, 1658. Pers.n., either OE *Fōt* or ON *Fótr*, genitive sing. *Fōtes, Fóts*, + **tūn**. YW v.122 gives prs ['fiustən] and ['foustən], SSNY 127.

FEWSTON RESERVOIR NYorks SE 1854. P.n. FEWSTON SE 1954 + ModE **reservoir**.

FIDDINGTON Glos SO 9231. 'Estate called after Fita'. *(æt) Fittingtune* [1004]13th, *Fitentvne -tone* 1086, *Fi- Fytinton(a) -yng-* 1220–91, *Fi- Fydin(g)ton(e) -yn(g)- -tune* 1248–1713, *Fiddington* from 1579. OE pers.n. **Fita* + **ing**[4] + **tūn**. Gl ii.53.

FIDDINGTON Somer ST 2140. 'Estate called after Fita'. *Fitintone* 1086, *-ton* 1236–43, *Fytyngton* [1295] Buck, *Fidington* 1304. OE pers.n. **Fita* + **ing**[4] + **tūn**. DEPN.

FIDDLEFORD Dorset ST 8013. 'Fitela's ford'. *Fitelford(e)* 1244–1456, *Fi- Fyttleford* 1575–1870, *Fiddleford* 1795. OE pers.n. *Fitela* + **ford**. Do iii.181, L 69.

FIDDLERS HAMLET Essex TL 4701. No early forms.

FIELD Staffs SK 0233. 'The open land'. *Felda* 1114×8, 1130, *Field* 1686. OE **feld**. DEPN, L 240, 242.

FIELD HEAD Leic SK 4909. The field in question is probably that in *Markfield Field* 1833 OS, alluding to one of the open fields of the village, or possibly to the *field* of Markfield itself. See MARKFIELD Leic SK 4810.

FIFEHEAD '(Estate of) five hides'. OE **fīf** + **hīd**. In the 11th cent. five hides was regarded as the usual size of a thegnly holding and it was also a unit for the assessment of military obligations.

(1) ~ MAGDALEN Dorset ST 7821. 'St Mary Magdalen's F'. *Fifyde Maudaleyne* 1388, *Vifide* ~ 1393, *Fifhead Magdalen* 1644. Earlier simply *Fifhide* 1086, *Fyf(h)ide -hyde* 1285–8, *Vifhide* 1316. The manor was assessed at 5 hides in DB. Do iii.6.

(2) ~ NEVILLE Dorset ST 7610. 'F held by the Neville family'. *Fi- Fyf(h)id(e) -(h)yd(e)* 13th etc., *Fifehead Nevile* from 1650.

Earlier simply *Fifhide* 1086, 1235×6. The manor was assessed at 5 hides in DB. It was held by William de Nevill' in 1235×6. Do ii.95.

FIFIELD Berks SU 9076. 'Five hides'. *Fifhide -hyde* 14th cent., *Fifhydefeld* 1396, *Fyfeelde Greene* 1573, *Fifeild Feild* 1636, *Fyfield Green and Lane* 1800. OE **fīf** + **hīd**, genitive pl. **hīda**. Fifield is a representative of a fairly common p.n. compound (cf. FIFEHEAD, FITZHEAD, FIVEHEAD, FYFIELD). Brk 45, 21, PNE i.172, 246, Stenton 1926.487, 583.

FIFIELD Oxon SP 2418. 'The five-hide estate'. *Fifhide* 1086–1400 with variants *Fyf-* and *-(h)yde*, *Fifide -yde* 1241–1399, *Fyfield* 1797. OE **fīf** + **hīd**. Cf. p.n. types FIVEHEAD, FYFIELD. O 351.

FIGHELDEAN Wilts SU 1547. 'Fygla's valley'. *Fisgledene* 1086, *Fichel- Fy- Fikeldene* 1115–1263, *Fig(h)- Fy(g)helden(e)* 1227–1321, *Figheldean* 1616, *Fyeldean* 1572, *Feildeane -den* 1640, 1645, *Filedean* 1718. OE pers.n. **Fyġla* + **denu**. Wlt 365 gives pr [fai(ə)ldi:n], formerly [fikəldi:n], L 98.

FIGSBURY RING Wilts SU 1833. *Clorus's Camp or Figbury Ring* 1773. P.n. Figsbury + ModE **ring** referring to a 5–4th cent. BC Iron Age hill-fort with a concentric ditch within the main rampart. The forms for Figsbury, *Frippesbury* 1695, *Fripsbury* 1721, are too late for elucidation. Wlt 384, Pevsner 1975.590, Thomas 1976.238.

FILBY Norf TG 4613. Possibly 'the farm or village where boards are made' or 'with a plank bridge'. *Phileb(e)y, Filebey* 1086, *Fil(l)ebi -by* 1165–1340, *Fy- Filby* from 1346. ODan ***fili** + **bȳ**. Alternatively the specific might be OE pers.n. *Fila*. Nf ii.7.

FILEY NYorks TA 1180. 'Monster island'. *Fiuelac* 1086, *Fiue- Fyue- Fivelei(a) -le(y) -lay(e)* before 1080–1402, *Fi- Fyley* 1447–1650. OE **fīfel** + **ēġ**. This poetic formation refers to Filey Brigg, a long ridge of rock projecting into the sea which has the appearance of a monster swimming into land. DB *-lac* normally represents late Northumbrian *-lǣh* < OE *-lēah*; for *-ac* as a possible spelling for OE *ēġ*, however, see Gate HELMSLEY SE 6955. YE 110 gives prs [fa:lə, faⁱlə], PNE i.172, SSNY 254.

FILEY BAY NYorks TA 1379. *Fyley Bay* 1651. P.n. FILEY TA 1180 + ModE **bay**. YE 111.

FILEY BRIGG NYorks TA 1282. 'Jetty belonging to Filey'. *Filey-Bridge* 1828. P.n. FILEY TA 1180 + ON **bryggja**. YE 111, L 66.

FILGRAVE Bucks SP 8748. Partly uncertain. *Firigraue* c.1218, *Fi- Fylegraue* 1240–1328, *Filgrave* 1316. Either 'Fygla's pit' or 'Fygla's grove', OE pers.n. **Fyġla* + **græf** or **grāf(a)**. Bk 15, L 193, 195. Jnl 2.33 gives local pr [filgru:v].

FILKINS Oxon SP 2404. 'The Filicingas, the people called after Filica' or '*Filicing*, Filica's place'. *Filching* 12th cent., *Filkinges* 1180, *Fi- Fylking(') -yng(e)* 1185–1390, *Filkynges* 1383, *Filkinch* 1185, *Fylkynche* 1316, 1336, *Filechinge* 1269, c.1270, *Fylekinge* 1316, *Filkins* 1397. OE folk-n. **Filicingas* < pers.n. **Filica* + **ingas**, or p.n. **Filicing* < pers.n. **Filica* + **ing**², locative-dative sing. **Filicinġe*. O 320.

FILLEIGH Devon SS 6627. 'Hay clearing, clearing where hay is made'. *Filelie* 1086, *Filelei* 1086–1297 with variants *Fyle-* and *-legh(e)*, *Fillilegh* 1278, *Philleigh* 1736. OE ***filith-lēah** from *filethe* + **lēah**. D 42.

FILLEIGH Devon SS 7410. *Filleigh Farms* 1809 OS. Possibly the same as FILLEIGH SS 6627.

FILLINGHAM Lincs SK 9384. 'The homestead or village of the Fyglingas, the people called after Fygela'. *Fil- Fel- Figelingeham* 1086, *Figlingaham -ingheim* c.1115, *Fil(l)ingeham* 1170–1219, *Fugelingam, Figelingham* 1202, *Fillinge- Figlincham* 1212, *Fulingeham* 1218. OE folk-n. **Fyġlingas* < pers.n. *Fyġ(e)la* + **ingas**, genitive pl. **Fyġlinga*, possibly varying with p.n. **Fyġling* 'the place called after Fygela' < pers.n. *Fyġ(e)la*, locative-dative sing. **Fyġlinge*, + **hām** partly replaced by ON **heim**. Cf. FYLINGDALES NYorks SE 9199. DEPN, ING 143, Cameron 1998.

FILLONGLEY Warw SP 2887. 'The wood or clearing of the Fyglingas, the people called after Fygla'. *Filinge- Filunge- Felingelei, Filunger* (sic) 1086, *Fillingeleia -unge- -legh- -ley(e)* 1206–1546, *Filinglegh* 1251, *Filongele* 1265, *Fyl(l)ongley* 1345–1564. OE folk-n. **Fyġlingas* < pers.n. *Fyġla* + **ingas**, genitive pl. **Fyġlinga*, + **lēah**. Wa 82.

FILTON Avon ST 6079. 'Hay farm'. *Fi- Fylton* from 1187. OE **filethe** + **tūn**. Gl iii.103.

FIMBER Humbs SE 8960. Either the 'pond amidst rough coarse grass' or the 'wood-pile pond'. *Fi- Fymmar(a)* 1121×7–1207, *-mer(e)* 1205×15–1489, *Fi- Fymara -mare* 12th cent., *Fi- Fymber* from 1541. OE ***finn** or **fīn** + **mere**. There are still two pools in the village, the lower one of considerable size. YE 128 gives pr [fimə], Problems 92.

Great FINBOROUGH Suff TM 0157. *Fineberg Magna* 1254, *Finbarow magna* 1568, *Finboro magna* 1610. ModE adj. **great**, Latin **magna**, + p.n. *Fineberga* 1086, 'woodpecker hill', OE **fīna** + **beorg**. If however *fīna* wer used as a nick-n. this might be 'Fina's barrow'. 'Great' for distinction from Little Finborough TM 0154, *Parva Fineberg'* 1226×8. DEPN, Baron.

FINCHALE PRIORY Durham NZ 2947. P.n. Finchale + ME **priorie**. Finchale, *Finchale* from 1153×95, *ffynckhal'* c.1220×30, *Fynk(h)all(')* c.1220–1456, *-halgh* 14th cent., is 'finch haugh, nook of land in a river bend frequented by finches', OE **finc** + **halh**. Under the year 788 in the Anglo-Saxon Chronicle a synod is mentioned at *Pincan heale* (*Wincanheale* MS D) in the land of the Northumbrians. Although OE ***pinca** also means 'finch, chaffinch', there is no proof that Finchale was the site of this synod or any evidence of settlement at this place before St Godric established a hermitage here in the early 12th cent., later replaced by the priory in 1197. NbDu 85.

FINCHAM Norf TF 6806. 'Finch homestead'. *P(h)incham* 1086, *Fincham* from [1087×98]12th, *Fincheham* c.1150. OE **finċ**, genitive pl. **finċa**, + **hām**. DEPN.

FINCHAMPSTEAD Berks SU 7963. 'Homestead frequented by finches'. *Fincham(e)sted(e)* 1086–1307, *-stæde* c.1122 ASC(E) under year 1098, *Heamstede* c.1122 ASC(E) under year 1103, *Fynchamsted(e)* 1316–1458, *Fync(h)hampstede* 1412. OE **finċ** + **hāmstede**. Brk 95, Stead 265.

FINCHDEAN Hants SU 7312. 'Finc's' or 'finch valley'. *Finchesdene* 1167–1265, *-den* 1182, 1190, *Fincheden'* 1230. OE pers.n. *Finċ* or **finċ**, genitive sing. *Finċes*, **finċes**, + **denu**. Ha 78, Gover 1958.54, 58.

FINCHINGFIELD Essex TL 6832. 'Open country of the Fincingas, the people called after Finc'. *Fincingafeldā, Fincinghefelda, Phincing(h)efelda(m)* 1086, *Finch(el)esfeld'* 1190–4, *Fy- Finching(e)feld -ing(g)(e)- -yng- -feud* 1121–1337. OE folk-n. **Finċingas* < pers.n. **Finċ* + **ingas**, genitive pl. **Finċinga*, + **feld**. Ess 425 gives pr [fintʃinfl], Jnl 2.46.

FINCHLEY GLond TQ 2791. 'Finch clearing or wood'. *Finchlee -ley(e)* from c.1208, *Vynchelay* 1407. OE **finċ** + **lēah**. The variant forms *Finchesleg' -ley(e)* 1235–1483 and *Finching(e)ley(e)* 1260, 1581 are compatible with this explanation and do not necessarily point to OE pers.n. *Finċ*. Mx 92.

FINDERN Derby SK 3030. Unexplained. *Findre* 1086, *Fi- Fyndern(e)* c.1100–1204 etc. If this is a case of intrusive *-d-* we might think of OE *fīn-ærn* 'house with a heap'. Cf. DINDER Somer ST 5744. Db 464.

FINDON WSusx TQ 1208. 'The hill with a heap or wood-pile' or 'called or at *Fin*, the heap, the heap-shaped hill'. *Fintona* 1073, *Findune -tune* 1086, *Findon* from 1100×35. OE **fīn** + **dūn**. The church is situated at the foot of a well-marked spur with ground rising steeply behind it. The name-type *Findon* is found elsewhere, e.g. Findon Hill, Durham NZ 2446. Sx 197.

FINEDON Northants SP 9172. 'Valley where the *thing* assembled'. *Tingden(e)* 1086–1274 with variant *Tyng-, Tindena -e* 1167–1241, *Thing- Thyngden(e)* 12th–1363, *Fyndon* 1603×25, *Finedon* 1685. OE **thing** + **denu**. Nth 181 gives pr [findən].

FINGAL STREET Suff TM 2369. 'Fennel street' possibly in the sense 'stinking street'. *Fincle Street* 1837 OS. Nearby are *Coal*

Street and *Shop Street* both 1837 OS with ModE **street** 'a hamlet'. The possible meaning of the common and puzzling name Finkle Street is exhaustively examined in Nomina 18.7–31.

West FINGEL Devon SX 7491. *West Fangle* 1610. ModE **west** + p.n. *Fenghyl* 1317, *Feynghel* 1330, *Fingle* 1765, 'Stream where fish are caught'. OE ***fengel**, a derivative of the stem *fang* 'to hold, catch'. The reference may be to traps. A stream name *fengel* occurs in the land boundaries of Culmstock, [925×39]11th S 386. D 432, 612.

FINGEST Bucks SU 7791. 'Assembly wooded-hill'. *Tingehurst* 1163, *-herst* 1233, 1246, *Ti- Tynghurst(e)* 1209–1342, *Thinghurst* c.1240, 1535, *Thynchehurst* 1402, *Thingest* 1552, *Fingest* 1572, 1660. OE **thing** + **hyrst**. Bk 176 gives prs [fi- vindʒəst], Wlt addenda, L 198, Jnl 2.31. Jnl 2.33 gives pr [viŋist].

FINGHALL NYorks SE 1889. 'Nook of land at **Fining* or at the place where wood is heaped'. *Finegal(a)* 1086, *Fingala -e, Fyngale* 1086–1406, *Fi- Fynyng(h)ale* 12th cent., *Fynghall* 1361, *Fyngell* 16th cent. OE ***fining** probably used as a p.n. + **halh**. Alternatively this might be the 'nook called after Finn', OE pers.n. *Finn* + **ing**[4] + **halh**. For pers.n. formations with *ing* and *halh* cf. also BONINGALE Shrops SJ 8102, KILLINGHALL NYorks SE 2858. The reference is to a small valley. YN 247 gives pr [fiŋgəl], L 108.

FINGRINGHOE Essex TM 0220. 'Hill-spur of the Fingringas, the people who dwell on the finger of land'. *(æt) Fingringaho* 1000×2 S 1486, [962×91]11th S 1494, *-inge-* 1202, *Fi- Fyngringho(o) -yng-* 1202–1428, *ffinrigo* 1581. OE folk-n. **Fingringas* < **finger** + **ingas**, genitive pl. **Fingringa*, + **hōh**. The reference is to the broad finger of land which thrusts out between Roman river and Geeton creek. Cf. nearby LANGENHOE TM 0018. Ess 315 gives pr [fiŋriŋhou].

FINMERE Oxon SP 6332. 'Woodpecker pool'. *Finemere* 1086–1331 with variants *Fyne-* and *-mer(a)*, *Fi- Fynmere* from 1251. OE **fīna** + **mere**. O 208.

FINNINGHAM Suff TM 0669. Possibly 'the homestead of the Finningas, the people called after Finn'. *Fin(n)inga-Felincham* 1086, *Finegeham* [1087×98]12th, *Fyningham* 1170×5, *Finingeham* [1169×87]1268, 1191, *Fynnyngham* 1610. OE folk-n. **Finningas* < pers.n. *Finn* + **ingas**, genitive pl. **Finninga*, + **hām**. Alternatively the specific may have been folk-n. **Fīningas* 'the people called after Fina', a nick-name < OE *fīna* 'a woodpecker' occurring again in Great FINBOROUGH TM 0157 6 m to the SW. DEPN suggested that *Fīningas* might be elliptical for 'the people of Finborough' instead of **Finebergingas*. ING 130, Baron.

FINNINGLEY SYorks SK 6799. 'Wood or clearing of Fenningas, the fen-dwellers' or 'at Fenning, the fenny place'. *Feniglei* 1086, *Feningelay -lea* 12th cent., *Fenyngley* 15th cent., *Finhingley* c.1190, *Finingelay -yngeleye, Finyngley -le -ing-* 13th cent. OE folk-n. **Fenningas* < **fenn** + **ingas**, genitive pl. **Fenninga*, or p.n. **Fenning* < **fenn** + **ing**[2], locative-dative **Fenninge*, + **lēah**. Nt 79.

FINSBURY GLond TQ 3282. 'Finn's manor'. *Finesbir' -bury* 1235–1475, *Vinisbir' -es-* 1231, 1235, *-bury* 1397. Anglo-Scand pers.n. *Fin(n)*, genitive sing. *Finnes*, + ME **bury** (OE *byriġ*, dative sing. of *burh*). Mx 93.

FINSTHWAITE Cumbr SD 3687. 'Finn's clearing'. *Fynnesthwayt* 1336. ON pers.n. *Finnr*, genitive sing. *Finns*, + **thveit**. La 217, L 211.

FINSTOCK Oxon SP 3616. 'The place frequented by woodpeckers'. *Finestochia* with variants *Fyne-* and *-stoc(hes) -stok(e)* 1135×50–1300, *Fi- Fynstok(e)* from 1208. OE **fīna** + **stoc**. O 422.

FIRBANK Cumbr SD 6294. 'Scrub bank'. *Frebanc* 1215–54, *Frethebank(e) -banc* 1225–1406, *Fri- Frytheban(c)k* 1230–1585, *Firbank* from 1528. ME **frith** (OE *fyrhthe*) + **banke**. We i.32, L 191.

FIRBECK SYorks SK 5688. 'Frithi's' or 'scrubland stream'. *Friebec* 1171×9, *Fridebec(h) -t-* 1190, *Fri- Fryth(e)bek(e)* 1276–1535, *Fir- Fyrbe(c)k(e)* 1489–1598. ODan pers.n. *Frithi*, genitive sing. *Fritha*, or OE **fyrhthe** + **bekkr**. YW i.140, L 191.

FIRBY NYorks SE 2628. 'Frithi's farm or village'. *Fredebi* 1086, *Fritheby* 1184, 1252, *Fryth- Freth- Frithby* 1285–1400, *Fi- Fyrthby* 1485, 1566, *Fyrby* 1566. ON pers.n. *Frithi*, genitive sing. *Fritha*, + **bȳ**. YN 237, SSNY 27.

FIRE BEACON POINT Corn SX 1092. *Fire Beacon Point* 1813. ModE **fire-beacon** + **point**. PNCo 84.

FIRGROVE GMan SD 9213. A modern name. No early forms.

FIRLE BEACON ESusx TQ 4806. *Firle Beacon* 1813 OS. P.n. Firle as in West FIRLE TQ 4707 + ModE **beacon**.

West FIRLE ESusx TQ 4707. *Westferles* 1255, *Westfarles* 1309, *Westfrylles, West frille* 16th cent. ME adj. **west** + p.n. *Ferla -e(s)* 1086, *Ferles* [1189]14th–1327, *Firle(s)* 1271–1407, *Virle* 1412. This name must be taken with Frog Firle TQ 5101, *Ferle(s)* 1086, *Froggeferle* 1288, a lost Pig Firle, *Pyggeferl* 1300, and a lost *Firoland* [772×87]14th S 1183, apparently meaning 'the land of the Firolas', OE folk-n. **Firolas* possibly representing **Fierelas* 'the oak people' on the root **ferh-* seen in OHG *fereh-eih*, Lombardic *fereha*, 'winter oder Speiseeich' and OE *fiergen* 'wooded hill' + suffix *-el* (*-il*). 'West' for distinction from *Estfirle* 1235. Sx 359 gives pr [fʌrəl]. For an alternative view suggesting derivation from Latin *feralia* 'uncultivated land' see now Jnl 30.5–15.

FIRSBY Lincs TF 4563. 'The village or farm of the Frisians'. *Frisabi* c.1115, *Friseby -bi* 1125–1254, *Frisby* [1115]14th. ON folk-n. *Frísir*, genitive pl. *Frísa*, + **bȳ**. DEPN, SSNEM 81, FN iii.62.

FIR TREE Durham NZ 1434. *Fir Tree* 1857 Fordyce. A modern mining village named from *The Fir Tree*, a local public house.

FISHBOURNE IoW SZ 5592. 'Fish stream'. *Fisseburne* 1267, *Fishbourn Creek* 1769. OE **fisc**, genitive pl. **fisca**, + **burna**. The reference is to Wootton Creek on the E side of the entrance to which lay a place called *Fisshehous* 14th. Wt 43 gives pr [ˈviʃbərn], Mills 1996.49.

FISHBOURNE WSusx SU 8304. 'The fish stream'. *Fiseborne* 1086, *Fissaburna* c.1090, *Fyssheborne* 1289, 1315. OE **fisc**, genitive pl. **fisca**, + **burna**. Sx 58, 70, SS 72.

FISHBURN Durham NZ 3631. 'Fish stream'. *fisseb'* c.1170×80, *Fisseburn(e)* 1180×86–1256, *Fysse-* 1296, 1321, *Fiss(c)he- Fyssheburne(e)* 1233×44–1530, *Fishburn* from 1403×4. OE **fisc**, genitive pl. **fisca**, + **burna**. NbDu 87.

FISHER'S POND Hants SU 4820. *Fishers Pond* 1813. Surname *Fisher* + ModE **pond**.

FISHERSTREET WSusx SU 9531. Cf. *Fisher Street Westdean* 1813 OS. Surname *Fisher* recorded in Petworth in 1296 + dial. **street** 'a row, a hamlet'.

FISHER TARN RESERVOIR Cumbr SD 5592. P.n. Fisher Tarn + ModE **reservoir**. Fisher Tarn is probably named after a member of the family of James *Fisher* of Scalthwaiterigg mentioned in 1669, or could simply be ModE **fisher** + N dial. **tarn** 'a small lake' (<ON *tjǫrn*). We i.132.

FISHLAKE SYorks SE 6513. 'Fish stream' or 'side channel where fish are taken'. *Fiscelac, Fixcale* 1086, *Fislac* 1147–1300, *-lak(e)* c.1150–1428, *Fi- Fyschelake -sshe-* 1194×9–1525. OE **fisc**, genitive pl. **fisca**, + **lacu**. YW i.14, L 23.

FISHPOOL GMan SD 8009. Cf. *Fish Pool Brook* 1843 OS.

FISHTOFT Lincs TF 3642. 'Fish curtilage'. *Fishtoft* from 1416 with variants *Fy- Fisshe-* and *-tofte*. ME **fish** + p.n. *Toft* 1086–1526, *Toftes, Toftum* [late 12th]14th, ODan, lOE **toft**, pl. **toftas**, dative pl. **toftum**. The prefix indicates a connection with fishing since Fishtoft is situated near the coast; but it could be the surname *Fish*. Payling 1940.119.

FISHTOFT DROVE Lincs TF 3149. P.n. FISHTOFT TF 3642 + dial. **drove** 'a fen road'.

FISKERTON Lincs TF 0572. 'The settlement of the fisherman or -men'. *Fiskertuna* [1060]12th S 1029, *Fiscartune -tone* 1086, *Fischertune* c.1115. OE **fiscere** or genitive pl. **fiscera** replaced by ON **fiskari**, genitive pl. **fiskara**, + **tūn**. SelPap 73, SSNEM 183.

FISKERTON Notts SK 7351. 'The settlement of the fishermen'. *Fiscertune* [958]14th S 659, *Fiscartune* 1086, *Fiskerton* from 1236. OE **fiscere**, genitive pl. **fiscera**, replaced by ON **fiskari** or with Scandinavian [sk] for English [ʃ], + **tūn**. The settlement is situated on the bank of the river Trent. Nt 164, SSNEM 183.

FISTRAL BAY Corn SW 7862. Unexplained. *Fistal Bay* 1813, *Fistral Bay* c.1870. PNCo 84, Gover n.d. 325.

FITTLETON Wilts SU 1449. 'Fitela's farm or village'. *Viteletone* 1086, *Fiteletune* 1250, *Fi- Fytelton* 13th–1394, *Fitlington* 1233, 1243, *Fiddleton* 1645, 1725. OE pers.n. *Fitela* + **tūn**. The same pers.n. occurs in the bounds of nearby Enford, *on fitelan slœdes crundœl* 'to the quarry of Fitela's valley' [934]12th S 427. Wlt 330, Grundy 1919.232.

FITTLEWORTH WSusx TQ 0119. 'Fitela's enclosure'. *Fitelwurða* 1168, *Fitelew(o)rth* 1200, 1256, *Vy- Vitelworth* 1296, 1450. OE pers.n. *Fitela* + **worth**. Sx 126, SS 79.

FITTON END Cambs TF 4312. *Fitten End* 1824 OS. Probably for 'Fitton end of Newton'. P.n. Fitton + ModE **end**. Fitton occurs in *Fittonehall* 1366, either as an independent p.n. + **hall** or from the surname of Alan de *Fittun* c.1213, de *Fittune* c.1254, which probably represents ON **fit** 'meadowland by a river' either in the dative pl. *fitum* or + **tūn** 'farm, enclosure'. The place lies in Newton Fen. Ca 273, 276.

FITZ Shrops SJ 4417. Short for Fitz Hoe, 'Fitt's spur of land'. *Witesot* (sic) 1086, *Phitesoth* 1128, *Phitesso* 1138, *Fit(t)esho* c.1175–1275, *Fi- Fyttes* 1255–1811, *Fitts* 1535–1722, *Fittz* 1577, *Fitz* 1658. OE pers.n. *Fitt*, genitive sing. *Fittes*, + **hōh** referring to the long narrow spur of land at the end of which the church and manor house are situated. For the loss of the final element, cf. EDGMOND SJ 7219. Sa i.132, Nomina 11.107.

FITZHEAD Somer ST 1228. '(Estate of) five hides'. *Fifhida* *[1065]18th S 1042, *Fifida* 1178, *Fyfhide* 1330, *Fitshead* 1610, *Fitzhead* 1809 OS. OE **fīf** + **hīd**, pl. **hīde**. The modern form may have been influenced by the nearby p.ns. Norton Fitzwarren ST 1925 and Cheddon Fitzpaine ST 2427. Cf. FIVEHEAD ST 3522, FYFETT ST 2314. DEPN.

FITZWILLIAM WYorks SE 4115. A mining village built c.1900 to house the miners of Hemsworth colliery worked by the Fitzwilliam Hemsworth Colliery Company. The name commemorates the family of the 3rd Earl Fitzwilliam (1786–1857) which owned the land. Room 1983.39.

FIVE ASHES ESusx TQ 5525. 'The five ash-trees'. *Five Ashes* c.1512. ModE **five** + **ash**, pl. **ashes**. Sx 396.

FIVE OAK GREEN Kent TQ 6445. *Five Oak Green* 1819 OS.

FIVE OAKS WSusx TQ 0928. 'The five oak-trees'. *Five Oaks* 1740. ModE **five** + **oak(s)**. Sx 151.

FIVEHEAD Somer ST 3522. '(Estate of) five hides'. *Fifhide* 1086, 1225, *Fyfhead* 1610. OE **fīf** + **hīd**, pl. **hīde**. Cf. FITZHEAD ST 1228, FYFETT ST 2314. DEPN.

FLACKWELL HEATH Bucks SU 8989. *Flackwell Heath* 1822 OS. P.n. Flackwell + ModE **heath**. Flackwell, *Flac- Flakewelle* 1227, *Flakwell* 1537, is probably 'hurdle stream', OE ***flaca** + **wella**. Bk 202 gives pr [flækəl].

FLADBURY H&W SO 9946. 'Flæde's fortified manor'. *Fledanbyriġ* [691]17th Hooke 20, *Fledanburg* [691×9]18th S 76 with variant *flœdanburh* [691×9]11th, *fladeburg* *[709]12th S 80, *(on) Flaedanbyrg* [777×81]18th S 62, *Fledebyrig*, *flœdan- fledanburh* [798×821]11th S 185, *Fledebirie* 1086, *Fladebury -bure* [c.1086]1190, 1275, 1291. OE feminine pers.n. **Flǣde*, genitive sing. **Flǣdan*, + **byriġ**, dative sing. of **burh**. Wo 126, Hooke 21–4, 47, 95, Anglia 110.156.

FLAG FEN Cambs TL 2894. *Flagg(e) Fen(n)* 1666, 1677. ME **flagge** 'a reed, a rush' + **fen**. Nth 226.

FLAGG Derby SK 1368. '(Place, settlement at) the turves'. *Flagun* 1086, *Flagg(e)* from 1230. ON **flag**, locative–dative pl. **flagum**. The meaning is 'place where turves are cut'. Db 100, SSNEM 153.

FLAMBOROUGH Humbs TA 2270. 'Fortification on the promontory'. *Flaneburc -burg* 1086, *Fleynesburg(h) -ai- -ei-* 12th–1251, *Fleynburg(h) -ai- -ay- -ei-* [1114×24]c.1300, 1244–1518, *Flaymburgh* 1461, *Flamburgh(e)* 1511, 1552, *-borough* from 1573. ON **fleinn** 'hook, barb' used topographically in the sense 'spit of land, tongue of land' as in the Danish p.ns. Flenø and Flensborg, + **burh**. The reference is to the massive pre-Roman entrenchment known erroneously as DANES' DYKE. There is a tradition that Flamborough is named after Flayn, the brother of the Skarthi who founded SCARBOROUGH on a similar site further up the coast. It is more likely, however, that Flayn is an eponymous hero created in medieval romance much like the Grim of GRIMSBY. YE 105 gives pr [flɛ:mbrə], APS i.320, SSNY 145, Thomas 1976.155.

FLAMBOROUGH HEAD Humbs TA 2570. 'Flamborough headland'. *Vlem- Vlamberger hovede* [14th]15th, *Flambrough Head* 1651. The earliest spellings are from a continental source, *Das Seebuch* ed. K.Koppmann, Bremen, 1876. YE 106.

FLAMSTEAD Herts TL 0814. 'Place of refuge, sanctuary'. *Fleamstede* [1005]13th S 912, *Flāmestede* 1086, *Flamsteda -(e)* 1166–1488. OE **flēam-stede**. The place is close to the Herts-Beds border; the manor was held by the tenure of providing protection for travellers. Cf. FLIMWELL WSusx TQ 7131. Hrt 32, Stead 95, 239.

FLANSHAM WSusx SU 9601. Partly unexplained. *Flennesham* 1220, *Flo- Flemesham* 1279, *Flem-* or *Flenisham* 1398, *Flansham* 1688. The original form of the name is unclear; possibly OE pet-n. **Flœmmi*, genitive sing. **Flœmmes*, + **hamm** 4 'dry ground in marsh'. Sx 140, ASE 2.44, SS 84.

FLASBY NYorks SD 9456. 'Flat's farmstead'. *Flatebi* 1086, *Flatteby* 12th, *Flatesby* c.1160, *Flasceby -bi* 1155–1400, *Flasby* from 1379. ON pers.n. *Flatr* or *Flati*, genitive sing. *Flats, Flata*, + **bȳ**. YW vi.48, SSNY 28.

FLASH Staffs SK 0267. 'The swamp'. *The Flasshe* 1586, 1605, *The Flash(e)* 1598–1601, *Flash* 1682. ME **flasshe**. Oakden, Duignan 61, PNE i.175, Horovitz.

The FLASHES Cleve NZ 6125. ModE **flash** 'a breaker'. The reference is to an extensive reef of rocks off Redcar submerged at high tide.

FLAT HOLM Avon ST 2265. 'Fleet island'. *Flotholm* 1375, *Floteholmes* 1387, *Flat Holmes* 1809 OS. ON **floti**, OE **flota** + ON **holmr**, lOE **holm**. The reference is to the use of the island as a base by Viking fleets one of which was starved out here in 918. The modern form has been influenced by nearby STEEP HOLM Avon ST 2260. The original name was *(œt) Bradan Relice* 918 ASC(A), *(into) Bradan Reolice* 1067 ASC(D), 'the broad Relic', OE **brād**, definite form dative case **brādan**, + p.n. *Relic*, OIr **re(i)lic(c)** 'graveyard, cemetery' (ulimately < Latin *reliquiœ* 'relics'). The name survived as *Reoric* in Florence of Worcester (c.1116). It may be identical with *insula Echni qui modo Holma uocatur* 'the island of Echnus now called Holm' in Cambro-British Saints 63, possibly referring to St Éogan of Ardstraw. DEPN.

The FLATT Cumbr NY 5678. 'Level piece of ground'. *The Flatt* 1618. Definite article **the** + ME **flat** (< ON *flot*). Cu 63.

FLAUNDEN Herts TL 0100. Partly uncertain. *Flawenden(e)* 13th, *Flauden(e)* from 1279. Perhaps 'Flaha's valley', OE pers.n. **Flaha*, genitive sing. **Flahan*, + **denu**. Alternatively 'flagstone valley', OE ***flage**, genitive sing. ***flagan**, + **denu**. Hrt 34, PNE i.174–5.

FLAWBOROUGH Notts SK 7842. 'Stone hill'. *Flodberge* 1086, *Flouberge -bergh* c.1190–1343, *-berewe* 1252, *-burgh* 1280, *Flaubergh* 1316–51, *Flawborrowe* 1590. OE **flōh** + **beorg**. The *beorg* is a small round knob on one side of the hill. Nt 214, L 128.

FLAWITH NYorks SE 4865. Either the 'troll's ford' or the 'ford by the flat meadowland' or 'where water-lilies grow'. *Flathwayth* c.1190, *Flathewath(e)* 1207–1301, *Flawith* from [1316]16th, 1582. ON **flagth** 'a giantess' or ***flatha** or OE **fleathe** + ON **vath**. YN 21, DEPN, PNE i.175, 176, L 82.

FLAXBY NYorks SE 3958. 'Flat's farm or village'. *Flatesbi* 1086, *Flasceby* 1158–1379, *Flaxby* from 1407. ON by-n. *Flatr*, genitive sing. *Flats*, + **bȳ**. The modern form shows dial. [ks] from [ts]. YW v.15 gives prs ['flazbi, 'flaksbi], SSNY 28.

FLAXFLEET Humbs → FAXFLEET SE 8624.

FLAXLEY Glos SO 6915. 'Flax clearing'. *Flax(e)ley(a) -le(a) -leg(a)* 1160–1662. OE **fleax** + **lēah**. Gl iii.232, L 206.

FLAXPOOL Somer ST 1435. *Flaxpool* 1809 OS. Apparently ModE **flax** + **pool**.

FLAXTON NYorks SE 6862. 'Settlement where flax is grown'. *Flaxtune -ton(a), Flastun -tona* 1086, *Flacstune* c.1160. OE **fleax** + **tūn**. YN 37, SSNY 255.

FLEAM DYKE Cambs TL 5553. 'The fugitives' dyke'. The forms for Fleam Dyke must be taken with those for Flendish Hundred which takes its name from the dyke. For the dyke they are *Flemesdich* c.1260, *Flemdich* 1279–1366, *Fleam Dyke* c.1825, for the Hundred *Flamingdice, Flammindic, Flammidinc -ding* 1086, *Flamminc- Flammi(c)gedic, Flammingedich, Flammedigedig, Flamencdic* 1086 InqEl and ICC, *Flam(m)edich(e)* 1155–14th, *Flem(e)dich(e)* 1188–1523, *Flendiche* 1428, 1570, *Flendishe -yshe* 1422×61, 1560. This seems to be OE **flēmingas** 'fugitives', genitive pl. **flēminga**, + **dīċ**, probably with variant **flēmena**, genitive pl. of **flēma** also meaning 'fugitive'. The name may further have been influenced by ME **flēme** 'a stream, a river, an artificial watercourse or mill-race'. The existence of full and reduced forms of the same name is similar to Castle and Sible HEDINGHAM Essex TL 7835, 7734, *He- Hidingham* 1086, *Hethingaham* [1100×35]13th, beside the hundred name Hinckford, *Hidingaforda -inge-* 1086, *He- Haingeford* 1166–7. The earthwork is post-Roman and probably constructed to mark the boundary between the East and Middle Angles: as such it would have been an important place for fugitives from justice, although there is also evidence of fierce fighting on the dyke which may take its name 'from some remarkable fight at this place' (Camden, Britannia, 1772, i.390). Other references to the dyke include *dicum* 'at the dykes' c.925 ASC(A) under year 905 referring to The Devil's Dyke as well as Fleam Dyke, *(on þa) dic* [974]11th S 794 (in the bounds of West Wratting), *magnum fosse* 13th, *(in) magno fossato de magna Wylburgham* 'in the great ditch of Great Wilbraham' 1289, and *fossatum de Balsham* 1285, *Balsham ditch* 1632, 1801, 1812. Cf WRAKENDIKE NZ 3162. Ca 35, 140, ING 121. O.li gives pr [flem ditʃ].

FLECKNEY Leic SP 6493. Possibly 'Flecca's island'. *Flechenie* 1086, *-eia* c.1160, *Fleckney(e)* c.1130–1528, *Flekney* 1209×35–1576, *Fleckney* 1467. OE pers.n. **Flecca*, genitive sing. **Fleccan*, + **ēġ** 'dry ground in marsh'. An alternative suggestion is an OE ***fleca** 'hurdle' and so 'island where hurdles are obtained'. Cf. FLECKNOE Warw SP 5136. Lei 216, L 39.

FLECKNOE Warw SP 5163. 'Flecca's' or 'hurdle hill-spur'. *Flechenho* 1086, *Flek(k)enho(u)* 1227–1361, *Flecknall* 1518, *Flecknowe* 1535. OE pers.n. **Flecca*, genitive sing. *Fleccan*, or ***fleca**, genitive sing. ***flecan**, + **hōh**. Wa 150.

FLEET 'an estuary, an inlet, a creek'; 'a river, a stretch of river (sometimes implying fishing rights), a reach'. OE **flēot(e)**. Jnl 29.79–87.

(1) ~ Hants SU 8154. *Flete* 1313, *le Flete* 1505. This is an unusual instance of ME **flete** (OE *flēot*) 'a creek, a stretch of water flowing through flat land' applied to a large natural pond here, Fleet Pond, *Fletepondes* 1505. Ha 78, Gover 1958.110.

(2) ~ Lincs TF 3823. *Fleot, Flec* 1086, *Flet* 1175×91–1275, *Flete* 1203–1495. The village now lies in the heart of the fens near the Roman sea-bank but formerly at the head of an arm of the sea. A fishery is recorded here in DB and *flēot* may have had the technical sense 'stretch of river with fishing rights'. Payling 1940.18.

(3) ~ HARGATE Lincs TF 3924 → Fleet HARGATE TF 3914.

(4) ~ HAVEN Lincs TF 4129 → GEDNEY DYKE TF 4126.

FLEETHAM NYorks SE 2894. 'Homestead by the stream'. *Fleteham* 1086–14th, *Fletham* 1270–1400. OE **flēot** + **hām**. The reference is probably to Hill Beck rather than to the river Swale itself. YN 239, L 21.

FLEETWOOD Lancs SD 3247. The town is an early 19th cent. development named in 1836 after Sir Peter Fleetwood, a developer who lived at Rossall Hall (later Rossall School) and saw the possibilities for constructing a harbour and docks at the mouth of the river Wyre to serve as an outlet for the manufacturing industries of the hinterland. From 1840–47 it was the northern terminus from Euston whence passengers embarked for Scotland, but the completion of the railway to Glasgow in 1847 finished the boom at Fleetwood. La 158 footnote, Room 1983.39, Pevsner 1969.120.

FLEGGBURGH or BURGH ST MARGARET Norf TG 4414. 'Fort or manor in (the hundred of) Flegg'. *Flegburg* 1232, *Burgh in Fleg(ge)* 1271–1470. Earlier simply *Burc, Burh* 1086. Flegg, *Flec* 1086, c.1100, *Fleg* 1107×7–1333, *Flegge* 1175, 1310, *Flegg* from 1302, is 'the sedge district', an ODan name for an area of dense Scandinavian settlement NW of Yarmouth between the Bure and the Thurne or Hundred Stream, referring to its marshy nature and cover with such vegetation. The exact etymon is uncertain but must be the source of ME **flegge** 'sedge', cf. ModDan *flæg*, ModE *flag* 'an iris'. Nf ii.1, 47, Jnl 19.6–7.

FLEMPTON Suff TL 8170. 'The Fleming estate'. *flemingtuna* 1086, *Flameton* 1195, *Fleminton* 1197, *Flempton* 1610. OE *Fleming* 'a Fleming, a native of Flanders' + **tūn**. Another possibility is OE **flēming** 'a fugitive'. ON *Flæmingr* is found as a pers.n. in FLIMBY Cumbria NY 0233 and Fleming is recorded as a surname from 1228. DEPN, PNE i.176.

FLETCHING ESusx TQ 4323. 'The Fleccingas, the people called after Flecci'. *Flescinge(s)* 1086–13th, *Flec(c)hinges* 13th cent., *Flec(c)hing(e)* 1268–16th cent. OE folk-n. **Flećċingas* < pers.n. **Flećċi* + **ingas**. Sx 345, ING 34, SS 65.

Old FLETTON Cambs TL 1997. ModE **old** + p.n. *Fletun* 1086, *Fletton* from 1227, 'the settlement on the fleet or inlet or stream', OE **flēot** + **tūn**. The precise reference is unknown but references to the *Fleete* occur in 1578 and to a piece of land called *the fleets* in the Enclosure Award. 'Old' for distinction from New Fletton TL 1997, a new suburb of Peterborough immediately S of the river. Hu 186, x1.

FLEXBURY Corn SS 2107. 'Flax mound or fort'. *Flexberi* 1201, *Flexbury* from 1306. OE **fleax** + **beorg** or **byriġ**, dative sing. of **burh**. No fort is known here and there is no distinctive natural hill; **beorg** probably refers, therefore, to a lost tumulus. PNCo 84, Gover n.d. 25.

FLEXFORD Surrey SU 9350. 'The weir where flax grows'. *Flexwere* 1317, 1386, *Flexewer, Flaxverd* c.1440, *Flexworth(ye)* 16th, *-wood* 17th, *Flaxford* 1749, 1816 OS. OE **fleax** + **wer**. The second element was variously reinterpreted as *worth, wudu* and *ford*. Sr 135.

FLIMBY Cumbr NY 0233. 'Village or farm of the Flemings'. *Flemyngeby* [1171×5]1333, *-inge-* 1201, *-ing-* c.1174. ON *Flǽmingr*, genitive pl. *Flǽminga*, + **bȳ**. Cu 286, SSNNW 30.

FLIMWELL ESusx TQ 7131. 'The fugitives' spring'. *Flimenwelle* 1210, *Flemenewelle* 1288, *Flemyngwell* 1309, *Flimwell* 1409. OE **flīma**, genitive pl. **flīmena**, + **wella**. The place is on the county boundary and would have received fugitives from the neighbouring jurisdiction. Sx 452.

FLINTHAM Notts SK 7446. 'Flinta's homestead or village'. *Flint(e)hā* 1086, *-ham* 1205, *Flintham* from 1184. OE pers.n. **Flinta* + **hām**. Nt 224.

FLINTON Humbs TA 2236. 'Flint farm or village'. *Flentun*,

Flintone 1086, *Fli- Flynton(e)* 1163×5–1828. OE **flint** + **tūn**. YE 54.

FLITCHAM Norf TF 7226. 'The homestead where flitches of bacon are produced'. *Flicham* 1086–1227, *Flitcham* 1207, 1824 OS. OE **flicce** + **hām**. DEPN.

FLITTON Beds TL 0536. Uncertain. *flittan* 980×90 S 1497, *Flichtham* 1086, *Flitte* 1166–14th, *Flete* 1183, *Flitten* 13th cent., *Flitton* 1318. Bd 148 interprets this name as OE **flēotum** 'at the streams', dative pl. of **flēot** with early shortening to *flit*. But such shortening is hardly likely by the time of the earliest form and ignores the DB spellings both for Flitton and FLITWICK TL 0335, *Flichtham* (for ?**flihtum*) and *Flicteuuiche*. S 1497 is a contemporary record which preserves the dative pl. ending *-um* unchanged (*westwicum, brahingum, welingum, twingum*). The form *Flittan* might otherwise be regarded as the dative pl. of OE *(ġe)flit* 'dispute, strife' in the sense 'disputed land(s)'. The spelling with double *-tt-* might be analogical with nouns ending in a double consonant after a short vowel or might represent an OE **flitta* with consonant gemination and transfer to the *n*-declension. The DB spellings may show association with OE *flyhte* 'a patch' as in *flyhte-clāth* 'a patch of cloth' used in the Rushworth Gospels at Mk 2.21 and Mt 9.16 to translate *commisura* 'joint'. Such a noun would be a *-jôn* formation on the root of *fleohta* 'hurdle', *fleohtan* 'weave, plait' meaning something like 'fence, hurdle'. Formally *Flittan* could be nominative-accusative pl. of an OE **flitte* of unknown meaning. Whatever the right answer the feature in question gave its name to the hundred of Flitt, *Flictham* 1086, *Flete* 1185, *Flitte* 13th–14th cents, the rural deanery of Fleete, *Flitte, Flute* 1291, and the river Flitt. Bd 148, Kluge 81–2, L 22.

FLITWICK Beds TL 0335. Partly uncertain. *Flicteuuiche* 1086, *Flit(te)- Flet(te)wik* 13th–14th cents., *Flotewyk* 1276. Unexplained element **flitte* as in FLITTON TL 0536 + OE **wīc**. Bd 72 gives pr [flitik], L 22.

FLIXBOROUGH Humbs SE 8715. 'Flik's fortification'. *Flichesburg* 1086, *-burc* c.1115, *Flickesburc* 1202. ON pers.n. *Flík* or *Flikkr*, genitive sing. *Flíks, Flikks,* + **burh**. DEPN, SPN 83.

FLIXTON 'Flik's settlement or estate'. ON pers.n. *Flik* or *Flikkr*, genitive sing. *Fliks, Flikks,* + OE **tūn**.

(1) ~ GMan SJ 7494. *'Flixton* from 1177 with variant *Flyx-, Fluxton(a)* 1228–1506. La 37, SSNNW 190, Jnl 17.33.

(2) ~ NYorks TA 0479. *Fleuston(e)* 1086, *Flixton(a)* 12th–1828. YE 116, SSNY 126.

(3) ~ Suff TM 3286. *Flixtuna* 1086, *Flixton* from 1254. DEPN, Baron.

FLOCKTON WYorks SE 2415. 'Floki's settlement'. *Flocheton(e)* 1086, *Floc- Floktun(a) -ton(a)* 12th–1536, *Floketon(a) -tun* 1145×60–1339, *Flocktun -ton* 1150×70–1607. ON pers.n. *Flóki*, genitive sing. *Flóka,* + **tūn**. YW ii.203, SSNY 126.

FLODDEN Northum NT 9235. 'The hill by the water-channel'. *Floddoun* 1517, *Flowdoun* 1521, *Flodden* 1695 Map. OE **flōd(e)** + **dūn**. Flodden lies beside a tributary stream of the river Till. DEPN.

FLOOKBURGH Cumbr SD 3675. Probably 'Floki's fortified place'. *Flokeburg* 1246, *Flokesburgh* 1394. ON pers.n. *Flóki*, genitive sing. *Flóka,* + OE **burh**. Other explanations are possible, such as 'fluke fortified place or borough', OE **flōc**, ON **flóki** 'a flat fish, a fluke', or 'fortified place in flat terrain', again ON **flóki**. Both are plausible since Flookburgh is a fishing village where flukes are (or were) caught and is also situated on flat low-lying ground. Pers.ns, however, are of great frequency as specifics of names in *-burh*. La 197, Mills 84, SSNNW 203, VCH Lancs 8.270.

FLORDON Norf TM 1897. 'Floor hill'. *Florenduna* 1086, *Florendone* 13th cent., *Flordone* 1291. OE **flōre**, genitive sing. **flōran**, + **dūn**. The reference is unknown, but it may have been to a paved floor perhaps of Roman origin. DEPN, PNE i.178.

FLORE Northants SP 6460. 'The floor'. *Flora* 1086, *Flore* from 1086, *Floure* 1330, *Flower* 1535–1779. OE **flōr**. This could be a reference to a Roman pavement. No such feature is known here although a fine floor mosaic was discovered in nearby Nether Heyford SP 6658 in 1669. Alternatively **flōr** might mean 'valley bottom'; Flore is situated in the valley of the r. Nene at the foot of higher ground. Nth 82 gives pr [flu:ə].

FLOTTERTON Northum NT 9902. Possibly the 'settlement by the *flot-way*, the road that floods'. *Flotweyton(')* c.1160–1346 with variants *-wai- -way-, Flot(t)e-* 13th cent., *Flotwarton* 14th., *Flotterton* 1538. OE ***flot-weg** + **tūn**. The exact sense of OE **flot-weg* is unknown. It may have been a floating road, i.e. on rafts. OE *flot* is recorded in the sense 'deep water, sea', but its basic sense is 'state of floating, that which floats'. Presumably the reference is to flooding of the road bewteen Flotterton and Warton which crosses a tributary of the river Coquet. The name was subsequently influenced by that of WARTON NU 0002 and remodelled as if **Flot Warton*. NbDu 87, DEPN.

FLOWTON Suff TM 0846. 'Floki's estate'. *Flochetuna* 1086, *Floketon(e)* 1201–1357, *Floweton* 1503, *Flowton* 1524. ON pers.n. *Flóki*, genitive sing. *Flóka,* + **tūn**. DEPN, Baron.

FLUSHING Corn SS 8033. *Flushing* 1698, 1699. Transferred from Flushing in Holland (Vlissingen, *Vlisseghem* 1220, pers.n. *Flisse* + suffix *-inghem*). Described in the early 18th cent. as 'lately built by the Dutchmen', the village was founded in 1661. PNCo 84, Gover n.d. 522, Pevsner 1951.56.

FLYFORD FLAVELL H&W SO 9855. *Fleford Flavell,* ~ *Fluvell* 16th, 17th, is a combination of two p.ns., Flyford + Flavell, *Flavel* 1190–1428, an AN form of the name Flyford added to it for distinction from GRAFTON FLYFORD SO 9656. Flyford, *Fleferth* [930]16th S 404, *fleferð* [956]11th S 633, *flæferth, fleferð* 972 S 786, *(æt) fleferht* [1002]13th S 901, *Flefrith* 1316–7, *Fleford* 1420 is the name of an old wooded district of central Worcestershire which included the estates of Flyford Flavell, Grafton Flyford and Dormston SO 9857, all of them alongside the Piddle Brook which may explain the modern form with *ford*. Flyford Flavell was also called *ælflæde tūn* 972 S 786, 'Ælflæd's estate', OE feminine pers.n. *Ælfflǣd*, genitive sing. *Ælfflǣde,* + **tūn**. A possible short form of names in *-flǣd* is **Flǣde* as in FLADBURY. Flyford might, therefore, be regarded as 'Flæde's wooded countryside', OE feminine **Flǣde* + **(ġe)fyrhth**, but this is very uncertain, the recorded forms do not really support it, and the name is best left unexplained. Wo 199, Hooke 161, 168, 179, 193, 195.

FOBBING Essex TQ 7183. Either the 'place called after Fobba' or 'the Fobbingas, the people called after Fobba'. *Fobbing(e)* from 1200 including [1068]1309, *Phobinge* 1086, *Fob(b)inges* 1125–1243, *Vobbing -yng* 1392×3. Either OE p.n. **Fobbing* < pers.n. *Fobba* + **ing**[2] or folk-n. **Fobbingas* < pers.n. *Fobba* + **ingas**. Ess 156, ING 19.

FOCKERBY Humbs SE 8419. 'Folcward's village or farm'. *Fulcwardby(e)* 1164×77, 1293, *Folquard(e)by* 1194×1203–1422, *Folk- Folcard(e)by* 1241–1304, *Folkeby* 1362, *Fockerby* from 1573. ON pers.n. *Folkvarthr* + **bȳ**. This pers.n. is actually borrowed from OHG *Folcward*; the p.n. may therefore be a post-conquest formation directly with the continental pers.n. YW ii.5, SSNY 28.

FODDER FEN Cambs TL 5287. *Fodder Fen* c.1840. This name occurs in various places in Cambridgeshire eg. Fodder Fen TL 5980, *Foderfen* 1325, Fodder Fen, Wicken, *fodder fen* 1345, *the fother fenn* 1541, Fodder Fen Common TL 4893, *Fodder Fen* c.1840. ME **fodor** (OE *fōdor, foddor*) + **fen**. Ca 202, 203, 236.

FOGGATHORPE Humbs SE 7537. 'Fulcard's outlying settlement'. *Fulcartorp* 1086, *Folkwarethorp* [1154×89]14th, *Folcware-* 1156–7, [1189–99]1308, *Folkerthorp(e)* 12th–1493,

Foker- 1419, *Foggerthorp* from 1610. CG pers.n. *Fulcard* + **thorp**. YE 240, SSNY 58.

FOLE Staffs SK 0437. 'Cattle spring'. *Fowall* c.1260, 1332, *Fo(o)wale* c.1272, *Fowell* 1290, *Fole* 1538. OE **feoh** + Mercian **wælle**. Horovitz.

FOLESHILL WMids SP 3582. 'The people's or Folc's hill'. *Focheshelle* 1086, *Folkeshulla -hull(e) -hill* c.1144–1451, *Folxhull, Foxhul, Foxull, Foxehall* 16th, *Fulsall, Folshull* 1535, *Fossell, Focell* 17th. Either OE **folc** or pers.n. **Folc*, genitive sing. **folces**, **Folces*, + **hyll**. The reference may be to some early meeting-place. Wa 109.

FOLKE Dorset ST 6513. 'The people', i.e. presumably 'land held by the people, land held in common'. *Fulk* 1166, *Folk(')* from 1244, *Foke* 1575, *Fooke* 1569×74. OE **folc**. Do iii.330.

FOLKESTONE Kent TR 2235. 'Folca's stone'. *Folcanstan* [696×716]11th S 22, *(æt) Folcanstanœ* 824 S 1434, *(to) Folcanstane* 845 for 830 S 282, 833×9 S 1482, 1042×4 S 1044, *Folcanstan* [927]12th S 398, *(to) folces stane* 946 S 510, c.1121 ASC(E) under year 1052, *Folcstane* 12th ASC(F) under same year, *Fvlchestan* 1086, *Folkston* 1610. OE pers.n. **Folca*, genitive sing. **Folcan*, + **stān**. The change to *folces stan* is probably due to false association with OE *folc* 'people'; the DB form suggests association with OG *Fulco*. The territory of the men of Folkestone at Burmarsh is *terra folcanstaninga* [c.848]13th S 1193. In ASC(A) Folkestone is simply *(to) Stane* under year 993. PNK 445, KPN 23, Jnl 8.21, TC 95, Charters IV.92.

FOLKINGHAM Lincs TF 0733. 'The village or homestead of the Folcingas, the people called after Folca'. *Folching(e)- Fulchingehā* 1086, *Folkingham* from before 1176, *Fu(c)k- Fokingeham* late 12th, *Foukingham* 1274. OE folk-n. **Folcingas* < pers.n. **Folca* + **ingas**, genitive pl. **Folcinga*, + **hām**. Perrott 116, ING 141.

FOLKINGTON ESusx TQ 5603. 'Estate called after Folca'. *Fochintone* 1086, *Fokin(g)ton* 1121–14th cent., *Fockington, Foginton* 1579, *Foyngton* 1610. OE pers.n. **Folca* with early assimilation to **Focca* + **ing**[4] + **tūn**. Sx 411 gives pr [fouiŋtən], SS 76.

FOLKSWORTH Cambs TL 1490. 'Folc's enclosure'. *Folchesworde* 1086, *Fulkeswurþe -w(o)rthe* 1152–1316, *Folkesw(o)rth(e)* c.1200–1322 etc., *Fokesworth* 1239, *Foxworth* 1526. OE pers.n. **Folc* (partly influenced by CG *Fulc*), genitive sing. **Folces*, + **worth**. Hu 186 gives former pr [foxwə:θ] (sic for [foks-]).

FOLKTON NYorks TA 0579. 'Folki or Folca's estate'. *Fulcheton* 1086, *Fulketun -ton'* c.1165, 1220, *Folketun -ton(a)* 12th–1418, *Folkton* from c.1170, *Folton* 1525, *Foul- Fowlton* 17th cent. ON pers.n. *Folki* possibly influenced by CG *Fulco*, genitive sing. *Folka*, or OE **Folca*, + OE **tūn**. YE 115 gives pr [fautn], SSNY 126.

FOLLIFOOT NYorks SE 3452. 'Place where horse-fights are held'. *Pholifet* 12th cent., *Folifait(h) -feit(h) -y-* c.1200–1505, *Folifate* 1444–1585, *Follifoot* from 1580. OE **fola** + **(ge)feoht**. Horse-fighting was a sport practised by the Vikings. YW v.27, P. G. Foote, D. M. Wilson, *The Viking Achievement*, 1970, 402.

FOLLY GATE Devon SX 5797. *Folly Gate* 1809 OS. A gate on the turnpike road from Oakhampton to Hatherleigh. D 150.

FONTBURN RESERVOIR Northum NZ 0493. R.n. Font Burn + ModE **reservoir**. The river Font, *Funt* c.1200–14th cent., *Font* from 1261, is PrW **funtōn* 'a spring, a stream' < Latin *fontāna*. NbDu 88, RN 160.

FONTHILL BISHOP Wilts ST 9332. 'F held by the bishop'. *Fontel Epi* 1291, *Bishop Funthill* 1695. P.n. *Funtial* 899×924 S 1445, *Funtgeall* [900]12th S 1284, *Funteal* [963×75]12th S 818, *Fontel* 1086–1361, *Funtel(l* 1166–1257, by origin a stream-n. *(to) Funtgeal, (on) Funtal* [983]15th S 850, 'the spring place, the place abounding in streams', PrW ***font**, ***funt** + ***iol** (Co **-(y)el*) + Latin **episcopus**, genitive sing. **episcopi**, ModE **bishop**, referring to tenure of the manor by the bishop of Winchester in 1086, and for distinction from FONTHILL GIFFORD ST 9231. For the formation cf. River DEVERILL. Wlt 190 gives pr [fʌnt(h)il], RN 161, LHEB 345, Signposts 84.

FONTHILL GIFFORD Wilts ST 9231. 'F held by the Gifford family'. *Fontel Giffard* 1291, *Funtel Giffard* 1297, *Fountell Gifford* 1316. P.n. Fonthill as in FONTHILL BISHOP ST 9332 + manorial addition from the family name of Berengar Gifard who held the estate in 1086. Wlt 190.

FONTMELL MAGNA Dorset ST 8616. 'Great F'. *Magnam Funtemell'* 1391, *Fontmell Magna* 1795. Latin adj. **magna** + p.n. *F- ffuntemel* [871×7]15th S 357(1, 2), *ff- Funtemel, (in, to) funtmel* [932]15th S 419, *Fontemale* 1086, *Fu- Fontemel(l)(')* 1201–1406, originally a r.n., 'stream, spring by the bare hill', PrW ***funtōn** + ***mēl**. 'Great' for distinction from Fontmell Parva ST 8214, 'little F', *Parva Funtemel(l)* 1250, 1360, *Litel Fontemell* 1308. Do iii.103 gives pr ['fɔntməl].

FONTMELL PARVA Dorset ST 8214 → FONTMELL MAGNA ST 8616.

FONTWELL WSusx SU 9507. No early forms. Possibly a made-up modern name.

FOOLOW Derby SK 1976. 'Multi-coloured hill'. *Fou- Fowlowe* 1269–1451, *Fulowe* 1329, *Folow(e)* 1354–17th cent., *Foolowe* 1461×83, 1610. OE **fāg** + **hlāw**. The phonology of this name illustrates a dial. development of ME *ǭ* > [u:]. Db 101 gives pr [fu:lə].

FORCETT NYorks NZ 1712. 'Fold by the ford'. *Forsed* 1086, *Forset(a)* 1086–1367, *Forsett(e)* 1285–1519. OE **ford** + **(ġe)set**. YN 299, SSNY 255, L 68.

FORD 'Ford'. OE **ford**. PNE i.181, L 67–72.

(1) ~ Bucks SP 7709. *Forda* [c.1200]14th, *Donyngtonsford* 'Dinton's ford' 1262. The village lies one mile SE of Dinton. Bk 160.

(2) ~ Devon SX 7840. *Forde* 1086, *Ford* 1422. D 319.

(3) ~ Glos SP 0829. *Forda(m)* 1154×89–1237, *la Forde* 1216×72, 1287, *Ford* from 1227. Gl ii.15.

(4) ~ Mers SJ 3398. *la Forde* 1323, *the Forde* 1408, *Forde* 1547. La 117.

(5) ~ Northum NT 9437. *Forda* 1225, *Ford* from 1242 BF, *Furde* 1507. A crossing of the river Till. NbDu 88, L 68.

(6) ~ Shrops SJ 4113. *Forde* 1086 etc. with variants *Ford(a), La Forde* 1291×2, 1453, 1456, *Furde* 1255, *Fourd, Foord(e)* 1562–1672, *Foard* 1703. A crossing place of Cardeston Brook called locally Welshman's Ford. Sa i.133.

(7) ~ Staffs SK 0654. *Forde* 1240–1592, *Fourd* 1558, 1599, *le f(f)o(o)rd(e)* 1363–1618, *Ford* 1558×1603, 1836 OS. A crossing of the r. Hamps. Oakden.

(8) ~ WSusx TQ 0003. *Fordes* c.1194–1310, *Vorde* 1279, *Forde* 1316, 1329, *Foord, Fourde* 16th. Either a ford of the Arun where Ford Ferry is marked on the first ed. OS or a N-S crossing of the tributary of the Arun here. Sx 141.

(9) ~ Wilts ST 8474. *(la) Forde* 1249, 1297, *Foorde* 1422. Wlt 113.

(10) ~ END Essex TL 6716 sc. of Great Waltham. *ffordende* 1376, *le fforthynde* 1377, *Forthende* 1399. ME **ford** + **ende**. Ess 272.

(11) ~ STREET Somer ST 1518. *Ford Street* 1809 OS. ModE **ford** + dial. **street** 'a hamlet'. The reference is to a former ford at Blacken Bridge ST 1519. For other nearby *street* names cf. Silver Street ST 1721 and 5342.

FORDCOMBE Kent TQ 5240. Possibly the 'fir-tree coomb'. *ffyrecoumbe* 1313, *Fercombe* n.d., *Fordcomb Green* 1819 OS. The evidence is too scanty for certainty but this may be OE ***fyre** (ME *firre*) + **cumb**. Another *(to) fyrcumbe* occurs in the bounds of Farnborough, Berks SU 4381, in a charter of 1042 (S 993), identical with *(to) furcumbe* [931]13th S 411, possibly containing OE **furh** 'a furrow, a trench' used as a p.n., genitive/dative sing. **fyrh**. In the Kentish dial. this would appear as **ferh*. The sense would be 'coomb at *Ferh*, the furrow or trench' referring to the narrowness of the Medway valley here. PNK 91, Brk iii.671–2.

FORDE ABBEY Dorset ST 3505. A Cistercian abbey founded in

1141. P.n. Forde + ME **abbeye**. Forde, *Ford* 1189, 1227, *Forde* 1291, is the 'ford' across the r. Axe, OE **ford**. Do 74.

FORDER GREEN Devon SX 7867. Possibly the home of John *atte Forde* 1333. D 510.

FORDHAM Cambs TL 6370. 'Ford homestead' or 'promontory'. *(æt) Fordham* from c.972, *Fordeham* 1086–1428. OE **ford** + **hām** or **hamm** 2. Ca 191.

FORDHAM Essex TL 9228. 'Homestead by the fords'. *Fordehā, Forhā -ham, Infordehā* 1086, *Fordham* from 1087. OE **ford**, genitive pl. **forda**, + **hām**. Ess 391.

FORDHAM Norf TL 6199. 'The ford homestead'. *For(d)ham* 1086, *Fordham* 1175×86. OE **ford** + **hām**. The reference is to a crossing of the Wissey. DEPN.

FORDHAM ABBEY Cambs TL 6369. *Fordham Abbey* 1829. P.n. FORDHAM TL 6370 + ModE **abbey**. A 13th cent. priory replaced by an early 18th cent. house. Ca 191, Pevsner 1954.310.

FORDINGBRIDGE Hants SU 1414. 'Bridge of the *Fordingas*, the Ford people'. *Fordingebrige* 1086, *Fordingebrig' -brug'* 1227, 1255. OE folk-n. **Fordingas* 'the people who live at Ford', the original name of the place before the bridge was built, p.n. *Forde* 1086–1345, OE **ford**, + **ingas**, genitive pl. **Fordinga*, + **brycg**. Ha 78, Gover 1958.213.

FORDON Humbs TA 0475. '(Village) in front of the hill'. *Fordun(e) -a* 1086–1331, *Fordon(e)* from 12th. OE **fore** + **dūn**. Fordon stands at the foot of a steep hill. YE 108 gives pr [fɔdn].

FORDSTREET Essex TL 9227. 'Ford village'. *Fordstreet* 1728. ModE **ford** referring to the same ford of the river Colne as FORDHAM TL 9228 + dial. **street** 'a hamlet'. Earlier called *Oldford* 1542. Ess 359.

FORDWELLS Oxon SP 3013. 'The springs beside the ford'. No early forms but cf. *Fordwell Poole als. Duckpoole als. Sewkeford* 1641. Duckpool is ModE **duck** + **pool**, *Sewkeford', Sewkeden'* 1300, from OE pers.n. *Seofeca* + **ford** and **denu**. O 300.

FORDWICH Kent TR 1859. 'Trading place by the ford'. *(juxta) Fordeuuicum* (Latin) [675]15th S 7, *Forduuic* [747]13th S 1612, *Fordwic -wik* [c.763 or 4]13th S 29 (with variants *-wich(t)* 14th, *Fordewik* 15th), *Fordƿic* [1053×66]13th S 1092, *Forewic* 1086, *Fordwik* [1087]13th, c.1100 etc. OE **ford** + **wīc**. Fordwich was the port of Canterbury. In 747 Eadberht, king of Kent, granted Reculver the toll due on a ship at Fordwich and in c.761 the same king granted the abbess of St Peter's Minster, Thanet, remission of toll due on two ships at Fordwich and Sarre. The ford crossed the river Stour. KPN 9, Jnl 8.33, Charters IV.26, 135, 180–1.

FORELAND IoW SZ 6687. 'The promontory'. *East foreland* 1591, *Foreland* 1769. ModE **foreland** 'a cape, a headland, a promontory'. Wt 37, Mills 1996.50.

North FORELAND Kent TR 4069. *North Foreland* 1596, 1690. Earlier simply *Forland* 13th, *the Forland* 1432, ME **forlonde** 'cape, headland, promontory'. 'North' for distinction from SOUTH FORELAND TR 3643. PNK 603, Cullen.

South FORELAND Kent TR 3643. *South Foreland* 1710. ModE **south** + **foreland** 'a cape, a headland, promontory'. 'South' for distinction from NORTH FORELAND TR 4069. Cullen.

The FORELAND or HANDFAST POINT Dorset SZ 0582. *Foreland* 1811. ModE **foreland**, 'a cape, a headland'. See also HANDFAST POINT SZ 0582. Do i.44.

FORELAND POINT Devon SS 7551. *The Foreland* 1907 Baring-Gould.

FOREMARK Derby SK 3326. 'Ancient building work or fortification'. *Fornvverche* 1086, *Forn(e)werc -werk(e)* 1228–1489, *-warke* c.1275–1529, *Formewarke* 1486, 16th cent., *Fornemerch* 1271, *-merk(e)* 1337, *-mark* 1552, *Farmark(e)* 1577, 1610. ON **forn** + **verk** later replaced by ME **merk, mark** (OE *mearc*). The reference is either to old foundations or to the earth formations a little to the E of Wall Hill. Db 634, SSNEM 149.

FOREMARK RESERVOIR Derby SK 3324. P.n. FOREMARK SK 3326 + ModE **reservoir**.

FORENESS POINT Kent TR 3871. *Foreness P.*[t] 1819 OS. P.n. *Fayre nasse* 1596, 1610, *Fair Nasse* 1719, 'the beautiful headland', ME **fair** (OE *fæġer*) + **nass** (OE *næsse*) + ModE **point**. The form of the name has been influenced by the proximity of North Foreland. Cullen.

FOREST GATE GLond TQ 4085. A modern name commemorating the gate which once stood in Woodgrange Road to prevent cattle straying from Epping forest into the highway. Ess 96, GL 49.

FOREST GREEN Surrey TQ 1241. Partly obscure. *Follis Green* 1680 etc., *Forrest* ~ 1738, *Folles* ~ 1807. Unexplained element, possibly a variant of the surname *Follows*, + ModE **green**. Sr 263, DS 188.

FOREST HALL Cumbr NY 5401. *Forest Hall* 1718. The forest is Fawcett Forest NY 5203, *(ye, the) Forrest* 1581–1618, *Forrast* 1592, *Fawsedforast* 1596, *Fauside Forest* 1777, ME **faugh** 'fallow' + **side** 'hill-side'. We i.137–9.

FOREST HEAD Cumbr NY 5857. 'Head of the forest'. *Forest Head* 1768. ModE **forest** + **head**. The reference is to the King's Forest of Geltsdale. Cu 86.

FOREST HILL Oxon SP 5907. 'The ridge-like hill'. *Fostel* 1086, *Forsthull(e) -hell(a) -e -hil(le) -hill'* [1122]c.1425–1428, *Foresthull(e) -hill* 1219 etc., *Foresthill alias Fersthill alias Fosthill* 1762. OE ***forst** + **hyll**. Later forms with *Forest* are due to popular etymology. O 171.

FOREST MOOR NYorks SE 2256. OFr, ME **forest** as in *foresta de C- Knare(s)burc -burg(h)* 1168, 1299, *the For(r)est of* ~ 15th–1540, + ModE **moor**. The Forest of Knaresborough was one of the royal forests N of the river Trent, extending some 20 miles W of Knaresborough as far as Barden SE 1493 and Appletreewick SE 0560, and including three great parks, Bilton Park SE 3157, Hay-a-Park SE 3758, and Haverah Park SE 2254. YW v.133, 77.

FOREST ROW ESusx TQ 4235. *Forstrowe* 1467, *Forrest Rowe beside Grynsted* c.1540. ME **forest** + **row** probably referring to a row of cottages in Ashdown Forest. Sx 327.

FOREST TOWN Notts SK 5662. A modern suburb of Mansfield named from SHERWOOD FOREST SK 6060.

FOREST-IN-TEESDALE Durham NY 8629. *The forest of Teisdaile* 1569. OFr, ME **forest**, 'a forest, unenclosed woodland devoted to the hunting of game', + p.n. *Tes- Thesedale* [12th]1640, 1303, *Tesedale* 1235–1485, *Tesdale* 1283–1408, r.n. TEES NY 7733 + ME **dale** (ON *dalr*).

FORESTBURN GATE Northum NZ 0696. *Forestburn Gate* 1868 OS. R.n. *Forest Burn* 1868 OS + ModE **gate**. The Forest Burn flows out of Rothbury Forest.

FORESTSIDE WSusx SU 7512. *Forest Side* 1810 OS. ModE **forest** + **side**. The reference is to Stansted Forest.

FORMBY Mers SD 2907. 'Forni's village' or 'old village'. *Fornebei* 1086, *Fornebi(a) -by* 1177–1298, *Formby* 1509. Either ON pers.n. *Forni*, genitive sing. *Forna*, or ON adjective **forn** 'old', + **bȳ**. If the second explanation is correct, the allusion may be to an earlier settlement abandoned because of erosion or burial by sand. La 125, SSNNW 30.

FORMBY HILLS Mers SD 2708. Short for Formby Sand Hills, p.n. FORMBY SD 2907 + *Sand Hills* 1842 OS.

FORNCETT ST MARY Norf TM 1493. 'St Mary's Forncett'. *Forncett St Mary* 1838 OS. P.n. *Fo(r)nesete, Fornes(s)eta* 1086, *Fornesset* 1199, *Fornesete* 1254, either 'Forni's fold', ON pers.n. *Forni*, genitive sing. *Forna*, + OE **(ġe)set**; a purely Scandinavian formation, 'Forni's or the old dwelling', ON pers.n. *Forni* or adj. **forn** + **sǣti**; or OE folk-n. **Forn-sǣtan* 'settler from FORNHAM' Suffolk TL 8367, 8566, + saint's name *Mary* from the dedication of the church. DEPN, SPNN 124–5.

FORNCETT ST PETER Norf TM 1692. 'St Peter's Forncett'.

Forncett St Peter 1838 OS. P.n. Forncett as in FORNCETT ST MARY TM 1693 + saint's name Peter from the dedication of the Anglo-Saxon church.

FORNHAM ALL SAINTS Suff TL 8367. 'All saints' F'. *Fornham Omnium Sanctorum* 1254. ModE **all saints** from the dedication of the church, Latin **omnes sancti**, genitive **omnium sanctorum**, + p.n. *Fornham* from 11th including 1086, 'the trout homestead', OE **forna** + **hām**. 'All Saints' for distinction from Fornham St Genevieve, 'St Genevieve's F', *Genonefœ fornhā* (for *Genoue-*) 1086, *Fornham Sancte Genovefe* 1254. The meaning is probably a place where trout were caught; however, the Fornhams lie on both sides of the Lark, a back-formation from Lackford: perhaps the river's original name was *Fornēa* 'the trout river'. In this case Fornham would be 'the homestead on the *Forne*, the trout river'. DEPN.

FORNHAM ST MARTIN Suff TL 8567. 'St Martin's F'. *Fornham Sancti Martini* 1254. P.n. *Fornhā* 1086 as in FORNHAM ALL SAINTS TL 8367 + saint's name *Martin* from the dedication of the church. DEPN.

FORSBROOK Staffs SJ 9641. 'Fot's brook'. *Fotesbroc* 1086–1276, *-brok(e)* 1185×90–1327, *Focebroc* 1200, *Fossebrok(e)* 1282×130–1567, *Foss(e)brook* 1682–1795, *Forsebrocke* 1586, *Forsbrook* 1837. OE pers.n. *Fōt*, genitive sing. *Fōtes*, + **brōc**. The /r/ is a late intrusion and local pr [fɔsbru:k] is still heard. Oakden, L 16.

FORSTON Dorset SY 6695. 'Forsard's manor'. *Fos(s)-Forsardeston* 1236–1371, *Forston* from 1431. William Forsard held this manor in 1285 but his family must have been here before then. By origin this was one of the nine manors called *Cernel* 1086 after the r. CERNE like Herrison SY 6794, 'Hareng's manor', *Harengestun* 1224, earlier *Cernel* 1086, and Pulston SY 6695, 'Pullein's manor', *Pulleinston* 1236, earlier *Cernel* 1086 and *Cerna Pulli* 1166. Do i.340.

FORTHAMPTON Glos SO 8532. 'Estate called after Forthelm'. *Fortemel- Forhelmentone* 1086, *Forthelme(n)ton(a) -tune* 1100–1229, *Forthhampton(e)* 1248–1710. OE pers.n. *Forthelm* + **ing**[4] + **tūn** assimilated to the common p.n. type in *-hampton*. Gl ii.57.

FORTON 'Ford settlement'. OE **ford** + **tūn**.

(1) ~ Hants SU 4143 → MIDDLETON SU 4244.

(2) ~ Lancs SD 4851. *Fortune* 1086, *fforton(a)* [1170–1206]1268, *Forton(a)* from [1207]1268. Named from two early fords of the river Cocker, *Langwathforde* 1268, ON **langa** + **vath** 'the long ford', and *Scamwath* as in *Scamwathlithe* [1220×40]1268, ON **skamma** + **vath** 'the short ford'. La 166, Jnl 17.97.

(3) ~ Shrops SJ 4316. *Fordune* 1086, *Forton(') -e* from 1240. The reference is to a crossing of the r. Severn later replaced by Montford Bridge. Sa i.133.

(4) ~ Somer ST 3307. *Forton* 1809 OS. Early forms are needed but this probably the 'ford settlement' referring to the stream crossing at the E end of the village.

(5) ~ Staffs SJ 7521. *Forton* from 1198. The reference is to a crossing of the r. Meece. St i.146, L 68.

FORTUNESWELL Dorset SY 6873. 'Lucky well, well or spring where fortunes can be told'. *Fortunes Well* 1608. ModE **fortune** + **well**. Do i.219.

FORTY FOOT or VERMUDEN'S DRAIN Cambs TL 3888. This is *Vermu(y)dens Eau* 1654, and *Adventurers Fortyfoot Dreyne* 1664. It was made for Sir Cornelius Vermuyden in 1651. Surname *Vermuyden* + ModE **eau** (OE **ēa** influenced by Fr. *eau*); ModE **adventurer** as in ADVENTURERS' FEN TL 5668. Ca 209.

South FORTY FOOT DRAIN Lincs TF 1633. *Forty feet Drain* n.d., *Fourty foot* 1633. This is the main drain in the Black Sluice District, extending from Boston Haven TF 3242 to Gutheram Cote TF 1820, first cut by the Adventurers who drained Lindsey Level in the middle of the 17th cent. It was later opened out and improved under the Black Sluice Act of 1765, and again in 1846. 'South' for distinction from the North Forty Foot cut in the mid 18th cent. which runs from Chapel Hill TF 2054 to Boston, the last two miles on a parallel course. Payling 1940.107, Wheeler 244, 252ff, 265ff.

FORTY GREEN Bucks SP 7603 → PITCH GREEN SP 7703.

FORTY HILL GLond TQ 3398. 'Hill at or called Forty'. *Fortyehill* 1610, *Fortee hill* 1686. Forty Hill overlooks the marshes of the river Lea; it is alluded to in the names John *atte Fortey* 1327×77 and Hugh *Fortey* 1420 whose surname represents OE **forth** + **ēġ** 'island of higher ground standing out from surrounding marsh'. Mx 73, PNE i.185.

FORWARD GREEN Suff TM 0959. *Forward Green* 1837 OS. ModE adj. **forward** + **green**. The reference is to the green's location vis-à-vis Earl Stonham. Contrasts with nearby *Broad Green* ibid. TM 0959.

FOSBURY Wilts SU 3158. 'The fortified place of the ridge'. *Fostesberge, Fistesberie* 1086, *Forstebyri -bury* 1268, 1307×27, *Forstesbery* 1270, *Fossebury* 1509×47, *Fostbury* 1721. OE ***forst**, genitive sing. ***forstes**, + **byriġ**, dative sing. of **burh**, referring to Fosbury Camp, a bivallate Iron Age hill-fort on a ridge S of the village. Wlt 356, Pevsner 1975.250.

FOSDYKE Lincs TF 3133. 'Fótr's ditch'. *Fotesdic* 1183–1212, *-dich* 1196–1200, *-dik* 1202. ON pers.n. *Fótr*, genitive sing. *Fót(e)s*, + OE **dīċ**, ON **dík**. SSNEM 226, Cameron 1998.

The FOSS WAY Counties between Devon and Lincs SK 8462, SP 4786, ST 9495 etc. 'The ditch road'. *le Fossewey* 1422, *(the) Foss(e)way* 1457, 17th cent. Earlier simply *(and lang) fosse* [705]14th S 247, *Foss* [779–c.1055]12th S 115, 550, 935, 1550, 1553, *(stratum que vocatur) fos* [852]13th S 202, *stratam publicam que ab antiquis stret nunc fos nuncupatur* 'the public way called *stret* by the ancients, now *Fos*' [956]13th S 1577. [978]11th S 1337, *Fosse streat* [970]12th S 777, *(chiminam, viam de) Foss(e)* c.1132–1779. OE **foss** + **weġ**. The Roman road from Axmouth to Lincoln, Margary nos. 5 a-f, which frequently ran upon a conspicuous *agger* with ditches on either side. Gl i.17, Nt 11, Wa 7, Wlt 15, Wo 3.

FOSSDYKE NAVIGATION Lincs SK 9274. P.n. *Fossedic* [1154×89]17th, c.1155–66, *-dik'* 1272, OE **foss** + **dīċ**, + ModE **navigation** 'a canal'. The Fossdyke, a Roman canal linking the r. Witham to the Trent, frequently silted up. It was re-opened for navigation by Richard Ellison in 1741. Cameron 1998, Wheeler 10, 138, 159, 430.

FOSSEBRIDGE Glos SP 0811. *Foss Bridge* 1779. ModE **fosse** as in FOSS WAY SP 0811 + **bridge**. A bridge carrying the Fosse Way over the river Coln. Gl i.166.

FOSTER STREET Essex TL 4909. *Forsterstrete* 1440. Surname *Fo(r)ster* + ME **strete** 'a hamlet'. There is mention here in 1659 of *land late ffosters*. Ess 38.

FOSTON Derby SK 1831. 'Fot's farm or village'. *Fostun -ton(e)* before 1138–1283 etc., *Foteston'* early 14th. OE pers.n. *Fōt*, genitive sing. *Fōtes*, + **tūn**. The forms *Farulveston* 1086 and *Farleston'* 1331 also belong to this place and represent a separate name, 'Farulf's farm or village', OGer pers.n. *Farulf*, genitive sing. *Farulfes*, + **tūn**. Db 560, SSNEM 190.

FOSTON Lincs SK 8542. 'Fotr's farm or village'. *Foztun(e)* 1086, *Foston(')* from 1199, *Foteston'* 1205, *Fotstun* 1212, 1218. ON pers.n. *Fótr*, genitive sing. *Fóts*, + **tūn**. Probably an Anglo-Saxon settlement taken over and partly renamed by the Danes. SSNEM 191 no.1.

FOSTON NYorks SE 6965. 'Fotr's estate'. *Fostun(a)* 1086, 1167, *Fotestun* 1231, *Fotston* 1233, *Foston* from 12th. ON pers.n. *Fótr*, genitive sing. *Fóts*, + OE **tūn**. YN 39, SSNY 127.

FOSTON ON THE WOLDS Humbs TA 1055. *Foston on le Wolde* 1609. Earlier simply *Fodstone* 1086, *Fotston* 1248, *Fostun -ton(a)* 13th–1549, 'Fotr's farm or village', ON pers.n. *Fótr*, genitive sing. *Fóts*, + OE **tūn**. YE 91, SSNY 127.

FOSTON BECK Lincs SK 8740. P.n. FOSTON SK 8542 + ModE **beck**

(ON *bekkr*). Referred to in 1577 as 'a brooke that riseth about Denton and goeth by Sydbrooke'. Perrott 372, SSNEM 273.

FOTHERBY Lincs TF 3191. 'Fotr's farm or village'. *Fo(d)rebi* 1086, *Fotrebi* c.1115–1288, *Foterby* 1154×89–1495, *Foderby* 1242–1551, *Fothrebi* 1207, *Fotherby* from 1272. ON pers.n. *Fótr*, West Scand genitive sing. *Fótar*, + **bȳ**. There is no need to see the specific as ON **fótar**, genitive sing. of **fótr** 'a foot', used in some topographical sense such as 'village of the foot' sc. of the Wolds. Li iv.16, SSNEM 47.

FOTHERINGHAY Northants TL 0693. 'The grazing island'. *Fodringeya* [c.1060]13th, *-eia* 1086–1457 with variants *-eie -eye -ay* and *-yng-*, *Fotheringeia* 1212, c.1250. OE ***fōdoring** < *fōdor* + *ing*[2], + **ēġ**. The village lies on low ground between the Nene and Willow Brook. Nth 202 gives pr [fɔðəriŋgei], DEPN.

FOUL MILE ESusx TQ 6215. *Fowle Myle* 1653, *Foul Mile* 1740. ModE adj. **foul** + **mile** referring to a bad stretch of road. Sx 482.

FOULDEN Norf TL 7699. 'Bird hill'. *Fugalduna* 1086, *Fugeldona* 1166. OE **fugol** + **dūn**. DEPN.

FOULHOLME SANDS Humbs TA 1921. *Foulholme Sand* 1786. P.n. Foulholme, ME **fūl** 'dirty, foul' + **holm** 'island', + ModE **sand**. An island at low tide in the Humber estuary. YE 38.

FOULNESS Norf TG 2341. *Foulness (Rock)* 1838 OS. Possibly identical with FOULNESS Essex TR 0495.

River FOULNESS Humbs SE 7839. 'The foul or filthy river'. *(on, andlang) Fulanea* [959]12th S 681, *Fulna* [1165×85]13th, 1268, 1325, *Fulnach* 1285, *Fulne(y)* 14th cent., *Fowlney* 1577, *Foulness* 1695. OE **fūl**, definite form oblique case **fūlan**, + **ēa** partially replaced by ON **á**, earlier **ahh*, and ModE **ness**. YE 4, RN 163.

FOULNESS ISLAND Essex TR 0192. *Foulness Island* 1805 OS. P.n. Foulness + ModE **island**, *insulam* 1412. Foulness, *Fulenesse* c.1190, *Fulnes(se)* 1235–74, *Fughelnes(se)* before 1219, 1324, *Foulnes(se)* 1273–1412, is the 'birds' headland', OE **fugol**, genitive pl. **fugla**, + **næss**. Ess 183.

FOULNESS POINT Essex TR 0495. P.n. Foulness as in FOULNESS ISLAND TR 0192 + ModE **point**.

FOULNESS SANDS Essex TR 0997. *Foulness Sand* 1805 OS. P.n. Foulness as in FOULNESS ISLAND TR 0192 + ModE **sand** 'a sand-bank'.

FOULNEY ISLAND Cumbr SD 2464. P.n. Foulney + ModE **island**. Foulney, *Fowley* 1537, 1667, *Foulney, the Fola* 1577, is usually taken to be 'bird's island', OE **fugol**, genitive pl. **fugla**, + **ēġ**, or ON **fugl** + **ey**. But this does not explain the modern form with *-n-*. Either this is due to the influence of nearby WALNEY ISLAND SD 1768 or the correct derivation is OE *(æt thǣm) fūlan ēġe* 'at the foul island'. A description of the island in 1537 mentions that it 'bred innumerable foul of divers kind'. La 204 gives pr [fɔ:lni].

FOULRICE NYorks SE 6270 → SAUNDERTON Bucks SP 7901.

FOULRIDGE Lancs SD 8942. 'The ridge where foals graze'. *Folric* 1219–21, *-rig(ge) -ryg(g)(e)* 1246–1395, *Fulrigge* 1542, *Folrige* 1551. OE **fola** + **hrycg** reformed with ON **hryggr**. La 88 gives pr [fɔ:lridʒ], L 169, Jnl 17.53.

FOULSHAM Norf TG 0324. Probably 'Fugol's estate or homestead'. *Folsham* 1086, *Folesham* 1156, 1168, *-is-* 1254. OE pers.n. *Fugol*, genitive sing. *Fugoles*, + **hām**. DEPN gives pr [foulsəm].

FOUNTAINS ABBEY NYorks SE 2768. A Cistercian abbey established in 1132 simply called *Fontibus* (Latin dative pl.) 1131×3–1428, *(apud) Fontes* 12th, 13th cents., *Fonteyns -ayns* 1275, *Fountain(e)s -eyns -ayn(e)s* 1294–1680, '(at) the springs', OFr, ME **fontein** 'a fountain, a spring'. The reference is to the springs at the site of the abbey. YW v.191.

FOUNTAINS FELL NYorks SD 8670. *Fontance Fell* 1540, *Fountaynes -ain(e)s Fell(ys)* 16th cent. P.n. Fountains as in FOUNTAINS ABBEY SE 2768 + ModE **fell** (ON *fjall*) 'a fell, a mountain'. The abbey owned extensive sheep pastures here. YW vi.138.

FOUR ASHES Suff TM 0070. 'The four ash-trees'. ModE **four** + **ash**. Nearby are BADWELL ASH TL 9969 and Great ASHFIELD TM 0068.

FOUR CROSSES Staffs SJ 9509. *the 4 Crosses* 1674, *Fower Crosses* 1682. Described in 1786 as an ancient inn where two roads intersect on Watling Street. Duignan 66, Horovitz.

FOUR ELMS Kent TQ 4648. *Four Elms* 1819 OS.

FOUR FORKS Somer ST 2336. *Four Forks* 1809 OS. ModE **four** + **forks** referring to a crossroads where four roads meet.

FOUR GOTES Cambs TF 4516. 'Four channels or drains'. *the four gotes* 1438, 1570, *Le Quatuor Goates* 1597, *Four Gouts* c.1825. ME **four** + **gote**. The meeting place of *the four gotes* of Wisbech Drain, Levenington High Lode, Newton Lode and Tydd Drain. Ca 285.

FOUR LANES Corn SW 6838. *Four Lanes* 1872. A late 19th cent. mining village at a crossroads. PNCo 84.

FOUR MARKS Hants SU 6734. 'Four boundaries'. *Fowrem'kes* 1548. ME **four** + **merk** (OE *mearc*). The site is the junction of four parishes, Chawton, Faringdon, Medstead and Ropley, and owes its development to the advent of the railway. Ha 79, Gover 1958.79.

FOUR OAKS 'The four oak-trees' ModE **four** + **oaks**.

(1) ~ ESusx TQ 8624. *Four Oaks* 1795. Earlier known as *Brownsmiths Oaks* 1724 from the manor of *Bromenesmyththe* 1320, *Brownsmyth* 1469, unknown element + OE **smiththe** 'a smithy'. Cf SEVENOAKS Kent TQ 5355, WHITESMITH TQ 5213. Sx 529.

(2) ~ WMids SP 2480. *Four Oaks* 1830. Wa 58.

(3) ~ WMids SP 1099. *Four Oaks* 1834. Cf. *Four Oaks Hill* 1725. Wa 51.

FOUR THROWS Kent TQ 7729. 'Four trees'. *Fourtrowes* 1790, *Four Throws* 1819 OS. Ultimately from OE **fēower** + **trēow**. PNK 339.

FOURLANES END Ches SJ 8059. *(Smallwood) Four Lane Ends* 1831. A crossroads near Smallwood SJ 8060. Che iii.19.

FOURSTONES Northum NY 8867. *Fourstanys* 1236, 1242 BF, *Four(e)stanes* 1256–1346, *Fourstones* c.1536. OE **fēower** + **stān**, pl. **stānas**. The place is said to be so named from the four stones, Roman altars, which marked the boundaries of the township, but it may rather be a reference to a lost tetralith. NbDu 88, Tomlinson 150, PNE i.171.

FOVANT Wilts SU 0029. 'Fobba's spring'. *Fobbefunte*, *Fobbanfuntan (boc)* 'the Fovant charter' [901]14th S 364, *(æt) Fobbafuntan* [994]14th S 881, *Febefonte* (sic) 1086, *Fobbefunta -e* 1154×89, 1224, *Fofunte* 1574, 1267, *Fovehunte* 1279–89, *Fovunte* 1305, *Fovent* 1574, 1592. OE pers.n. **Fobba*, genitive sing. **Fobban*, + ***funta**. Wlt 214 gives pr [fɔvənt], Signposts 84, 86.

FOWBERRY TOWER Northum NU 0329. P.n. Fowberry + ModE **tower**. An original tower-house of the 15th cent. rebuilt in 1666 and again in 1776. Fowberry, *Folebir' -byr'* 1242 BF, *Follebiri -birr' -bery* 1288, 1296 SR, 1349, *Folbury* 1346, 1428, *Foulbery* 1538, *Fowberye* 1542, is the 'foal fortified place', OE **fola** + **burh**, dative sing. **byriġ**, presumaby a stud-farm. NbDu 89, Pevsner 1992.286.

FOWEY Corn SX 1251. *Fawi* c.1233, *Fawy* 1255–1330, *Fawe* 1257–1325, *Fow(e)y* from 1301, *Foy* 1330, 1602–94. Transferred from the r.n. FOWEY SX 0962. PNCo 82 gives pr [foi], Gover n.d. 396, RN 164, DEPN.

River FOWEY Corn SX 0962. 'Beech-tree river'. *Fawe* [c.1210]late 13th, 1276–1303, *Fawy* [1241] early 14th, 1284–1339, *Fow(e)y* from 1344, *Foy* 1576–1602. Co ***faw** pl. 'beech-trees' + name suffix ***-i**. Cf. r.n. INNY, and for an earlier form of the r.n. FAWTON SX 1668. PNCo 84 gives pr [foi], Gover n.d. 681, RN 164, DEPN.

FOWLMERE Cambs TL 4245. 'Wild birds' mere'. *Fuglemære*, *Fugelesmara* 1086, *Fulemere* 1086–1258, *Fulmere* 1086–1428,

Foulmer(e) 1313 etc., *Fowlemer(e)* 1494–1583. OE **fugol**, genitive pl. **fugla**, + **mere**. According to Conybeare (*Highways and Byways in Cambridgeshire and Ely* 1827, 230) a mere noted for its wealth of wild fowl formerly existed here, but by his time it was a worthless patch of land 'full of springs and rivulets'. Ca 83.

FOWNHOPE H&W SO 5834. '(At) the multi-coloured Hope'. *Faghehop, Fauue Hope* 1243, *Fowe(n)hop(e)* 1275–1341, *Fowelppe* c.1540. OE **fāg**, dative case definite form **fāgan**, + p.n. *Hopa* 1067×71, 1142×6, *Hope* 1086, 1142, 'the secluded valley', OE **hop**. The reference is to a small side-valley of the river Wye. The prefix is for distinction from other nearby *hop* names at Sollers HOPE SO 6033 and WOOLHOPE SO 6135. He 85, Bannister 76, L 113, 116.

FOXCOTE RESERVOIR Bucks SP 7136. P.n. Foxcote + ModE **reservoir**. Foxcote, *Foxescote* 1086, *Foxcote* 1197, *Foscote* 1486, 1526, is 'fox cottages', OE **fox**, genitive sing. **foxes**, + **cotu**, pl. of **cot**. Bk 43.

FOXEARTH Essex TL 8344. 'Fox-hole'. *Focsearde* 1086, *Fox(h)erth(e) -(h)erd(e)* 1198–1274 etc., *Foxearth* 1594. OE **fox** + **eorthe**. This name shows that the meaning 'fox-hole' for *earth* was current some 500 years before the first example given in OED (1575). Ess 429 gives pr [foksəθ].

FOXFIELD Cumbr SD 2085. *Foxfield* 1864 OS. ModE **fox** + **field**.

FOXHAM Wilts ST 9777. 'The fox homestead or enclosure'. *Foxham* from 1219 including [1065]14th S 1038, *Voxham* 1553. OE **fox** + **hām** or **hamm**. Wlt 87.

FOXHOLE Corn SW 9654. *Foxhole* 1686, 1748. The name of a tin-work and associated hamlet. Co 85.

FOXHOLES NYorks TA 0173. 'Fox-earths'. *Foxele, Foxohole* 1086, *Foxhol(e) -hola -holo* 1086–1353, *Foxholes* from c.1130. OE **fox-hol**. YE 115.

FOX LANE Hants SU 8557. No early forms. Possibly related to the name Foxlease Farm, cf. *Foxleye corner* 1516. *Foxley* is 'fox wood or clearing' and was used as a surname. Ha 79, Gover 1958.111.

FOXLEY Norf TG 0321. 'Fox wood or clearing'. *Foxle* 1086, 1254. OE **fox** + **lēah**. DEPN.

FOXLEY Wilts ST 8986. 'Fox wood or clearing'. *Foxelege* 1086, *Foxlegh -le(e) -ley(e)* from 1227, *Voxlee* 1509×47. OE **fox**, genitive pl. **foxa**, + **lēah**. Wlt 70, L 205.

FOXT Staffs SK 0348. 'The fox burrow'. *Foxwiss* 1176, *Foxwist -wyst* 1293–1533×8, *Foxt* from 1578, *Fox* 1682–1755. OE **fox** + **wist**. Oakden.

FOXTON Cambs TL 4148. 'Fox farm'. *Foxetune* 1086, *Foxtona -ton(e)* from 1086. OE **fox**, genitive pl. **foxa**, + **tūn**. Ca 83.

FOXTON Leic SP 7089. 'Fox's farm or village'. *Foxestone* 1086, *Foxton(e)* from 1086. OE **fox**, genitive sing. **foxes**, + **tūn**. OE *fox* in p.ns. is said to refer to a fox's earth or to a place infested by foxes. *Fox* was, however, used as a nick-name from at least the 12th cent., cf. Toue fox of Saleby, Lincs, 1154×89 (ON *Tófi*, ODan *Tovi*). Lei 217, DBS 352.

FOXUP NYorks SD 8676. 'Fox valley'. *Foxhop(e)* 1457–1587, *Foxup(p)* 1632, 1675. OE **fox** + **hop**. A small side-valley leading into Littondale. YW vi.122.

FOXWIST GREEN Ches SJ 6268. *Fox's Green* 1831. P.n. *Foxwyste* 1475, *Foxwist* 1488, 'fox's lair', OE **fox** + **wist**, + ModE **green**. Che iii.182.

FOY H&W SO 5928. There are two types for this name;

I. *Lann Timoi* [860×86]c.1130, *Lanntiuoi* [1056×1104, 1063×6]c.1130, 'church of St Timoi or Tifoi', PrW ***lann** + W saint's name *Tifoi* or *Tyfwy*, a hypocoristic name formed by prefixing the particle **ti** 'thy' to the name *Moc* or *Mwy*. An exact pararel is Lamphey Dyfed SN 0100, *Lantefei* 12th.

II. *Sancte Fidis* (genitive case) 1139×48, *Sancta Foa* c.1187×98, 1205×1216, *Foy* from 1148×61, 'St Faith', Latin *sancta Fides*, due to confusion of the W name *Moe, Foe* with French *foi*, Latin *fides*.

The present dedication of the church is to St Mary. He 86, O'Donnell 89, Pembs 694.

FRADDON Corn SW 9158. 'Place of streams'. *Frodan* 1321, c.1510, *Fraddon* 1702. OCo **frot** 'stream' + name suffix ***-an**. PNCo 85.

FRADLEY Staffs SK 1613. 'Fodder clearing'. *Fod(e)resleye* 1262, *Frodele* 1269, *Fradley* 1834 OS. OE **fōdor** + **lēah**. Duignan 63.

FRADSWELL Staffs SJ 9931. 'Frod's spring'. *Frodeswelle* 1086, *-uella* 1155, *-well* 1177, *-wall* 13th. OE pers.n. *Frōd*, genitive sing. *Frōdes*, + **wella**, Mercian **wælla**. DEPN, L 31.

FRAISTHORPE Humbs TA 1561. 'Freisting or Freysteinn's outlying settlement'. *Frestint(h)orp* 1086, 1212, *Fraistingt(h)orp -ay- -ei- -ey-* 12th, 13th cent., *Fraist(h)orp -ei- -ey- -ay-* 13th–1828, *Fraistropp* 1561, *Frast(h)rup* 1650. ON pers.n. **Freistingr* or *Freysteinn* + **thorp**. YE 87 gives pr [frɛ:zθrəp], SSNY 58.

FRAMFIELD ESusx TQ 4920. Partly uncertain. *Framelle* 1086, *Fremisfeld -es-* 13th cent., *Fremedfeld* 1265, *Fremifeld(e)* 1296–1340, *Fren(e)feld* 1305, 1395, *Frantfeild* 1673. Some forms point to OE pers.n. **Fremi* as the specific, genitive sing. **Fremes*, but the 1265 spelling with *-d-* and the modern pronunciation suggest OE *frem(e)de* 'foreign, alien, strange, unfriendly'. Perhaps, therefore, 'unfriendly open land', OE **feld**. Sx 392 gives pr [fræntfi:ld], SS 70.

Upper FRAMILODE Glos SO 7510. 'Framilode further up the river Severn' sc. than Framilode or Framilode Passage SO 7410. *Upper Framilode* 1831 OS. Framilode, *Framilade* 1086, 1138, *Framilod(e) -y-* 1139×48–1575, *Fremelad(e) -lode* 12th–1367, is the 'Frome crossing', a ferry across the river Severn ¼ mile below the outfall of the river Frome. R.n. FROME SO 7706 + OE **ġelād**. ME *Frem-* spellings seem to show *i-* mutation of *From* caused by medial *-i-* < *ġe-*. Gl ii.179.

FRAMINGHAM EARL Norf TG 2702. 'F held by the earl' sc. of Norfolk. *Framelingham Comitis* 1254. P.n. Framingham + Latin **comes** 'count, earl', genitive sing. **comitis**. Earlier simply *Framinga- Framincham* 1086, *Framingeham* 1130–94, 'the homestead of the Framingas, the people called after Fram', OE folk-n. **Framingas* < pers.n. *Fram* + **ingas**, genitive pl. **Framinga*, + **hām**. The form *Framelingham* 1254 points either to a variant form with diminutive pers.n. *Framela* or to influence from the Suffolk p.n. FRAMLINGHAM TM 2863. DEPN, ING 135.

FRAMINGHAM PIGOT Norf TG 2703. 'F held by the Picot family'. *Framelingham Picot* 1254. The manor was held by Ralph Picot in 1254. *Framelingham*, 'the homestead of the Framelingas, the people called after Framela', seems to be a variant with diminutive pers.n. Framela for the Frama of Framingham in FRAMINGHAM EARL TG 2702. DEPN, ING 135.

FRAMLINGHAM Suff TM 2863. 'The homestead of the Framelingas, the people called after Framela'. *Fram(a)linga- Frameling(a)ham* 1086, *Framillingeham* 1175, *Frameling(e)ham* 1194–1327, *Frem(e)lingeham* 1180×1, 1218, *Framlingham* from 1336. OE folk-n. **Framelingas* < pers.n. **Framela* + **ingas**, genitive pl. **Framelinga*, + **hām**. DEPN, ING 130, Baron.

FRAMPTON 'Settlement on the river Frome'. R.n. FROME + OE **tūn**.

(1) ~ Dorset SY 6295. *Frantone* 1086, *Fromton* 1188, *Frompton* 1253, *Frampton* 1264. R.n. FROME SY 8089 + OE **tūn**. Do 74.

(2) ~ COTTERELL Avon ST 6682. 'F held by the Cotel family'. *Franton' Ade Cotelli* 'Adam Cotel's F' 1167, *Frampton Cote(l) -ele* 1257–1497, ~ *Cott(e)rell* from 1535. Earlier simply *Frantone* 1086, *Frompton* 1220–1316, *Franton'* 1236, 1378, 'settlement on the river Frome'. Gl iii.116.

(3) ~ MANSELL Glos SO 9202. 'F held by the Mansell family'. *Frompton Maunsel* 1368, *Frampton Mauncell* 1466. Earlier simply *Frantune -tone* [c.1075]1367, 1086, 1240×55, *Frompton(a)* 1211×13–1377×99, *Framton'* 1221, *Frampton* from 1305. The manor was held by John Mansell in 1285, by William Mauncell in 1303. Gl i.138.

(4) ~ ON SEVERN Glos SO 7407. P.n. *Frantone* 1086, *Framtone* c.1180–1312, *Frompton(e)* 1216×18–1414, *Frampton(a)* from 1235, + affix *super Sabrinam* 1279–1733, *(up)on Seu- Severne* from 1248. Gl ii.196.

FRAMPTON Lincs TF 3239. 'Frani, Franki or Frameca's farm or village'. *Franetone, Frantvne* 1086, *Franton(a)* 1154×89–1346, *Fraunton* 1259–1371, *Fra(u)ncton* 1183–1355, *Fra(u)nk(e)ton* 1294–1357, *Framton* 1202–55, *Frampton* from 1241. Pers.n. ON **Fráni*, ODan *Franki* or OE **Frameca* + **tūn**. Payling 1940.82.

FRAMPTON WEST END Lincs TF 3041. *West end(e) of Frampton* 1553–1707. P.n. FRAMTON TF 3239 + ModE **west** and **end**. Payling 1940.85.

FRAMSDEN Suff TM 1959. 'Fram's valley'. *Framesdena* 1086, *-den(')* 1213–1568, *Framsden(')* from 1275. OE pers.n. *Fram*, genitive sing. *Frames*, + **denu**. DEPN.

FRAMWELLGATE MOOR Durham NZ 2644. P.n. Framwellgate + **mōr**. Framwellgate, *Framuuelgate* c.1275, *-welgate* 1309–1777, *Framwellgate* from 1635, *Framaygate* 1583×4, is the 'road leading to *Framwell*, the vigorous spring', OE **fram** + **wella**. The well-head, no longer in its exact original spot, but still preserved at NZ 271429, formerly supplied a cistern or pant in the Market Place at Durham. Pronounced locally [fræməgeit].

FRANCHE H&W SO 8278. 'Frea's ash-tree'. *Frenesse* 1086, *Frenysse* 1249, *Freynes* 1275, *Fraynsh* 1307, *Fraunche* 1587. OE pers.n. **Frēa*, genitive sing. **Frēan*, + **æsc**. Wo 249, L 219.

FRANKBY Mers SJ 2486. Probably the 'Frenchman's farm'. *Frankeby* [1230]17th, 1278–1526, *Fraunk(e)bi -by* 1346–1434, *Frankby* 1546. ME **Franke** (OE *Franca* 'a Frenchman, a Frank') + **bȳ** (ON *býr*). This explanation was preferred in Che iv.287 because the alternative specific, pers.n. *Franki*, was believed to be an ODan pers.n. whereas this part of Cheshire is a Norse-Irish district. Furthermore in the DB account of nearby Caldy it is recorded that *unus Francigena cum i serviente habet ii carucas* 'a Frenchman with one serf has two ploughs'. This Frenchman may well be the eponym of Frankby. The name *Franki* is, however, found in Man c.1000 and therefore the question remains open. Che iv.287, SSNNW 30.

FRANKLEY H&W SO 9980. 'Franca's wood or clearing'. *Franchelie* 1086, *Frankele(ge)* 1166–1278. OE pers.n. *Franca* + **lēah**. Wo 346.

FRANKTON Warw SP 4270. 'Franca's estate'. *Francton* [1043]1267, *Franchetone* 1086, *-ton(e)* 12th–1332, *Frankton* from 1279, *Franckton* 1604. OE pers.n. *Franca* + **tūn**. Wa 129, DEPN.

English FRANKTON Shrops SJ 4529. *Englyshe Francton* 1577. ModE adj. **English** + p.n. *Franchetone* 1086, 1166, *Franketon(')* 1221×2–1428, *Frankton(')* from 1255×6, 'Franca's farm or village', OE pers.n. *Franca* + **tūn**. 'English' for distinction from Welsh Frankton SJ 3633. Sa i.134.

Lower FRANKTON Shrops SJ 3732. *Lower Frankton* 1724. ModE adj. **lower** + p.n. Frankton as in English FRANKTON SJ 4529. Sa i.135.

Welsh FRANKTON Shrops SJ 3633. *Welsch Frankton* 1577. ModE adj. **Welsh** + p.n. *Franckton'* 1544×5 as in English FRANKTON SJ 4529. Sa i.135 suggests that English Frankton had a detached area of pasture on the Whittington–Ellesmere–Hordley boundary and that Welsh Frankton was one of several small settlements that grew up on this common.

Great FRANSHAM Norf TF 8913. *Fransham Magna* 1254. ModE adj. **great**, Latin **magna**, + p.n. *Frande(s)ham* 1086, *Fran(e)sham* 1197–8, unkown element or pers.n. (possibly from Dan *Frœndi*) + **hām** 'a homestead, a village'. 'Great' for distinction from Little FRANSHAM TF 9012. DEPN, SPNLY 88.

Little FRANSHAM Norf TF 9012. *Fransham Parva* 1254. ModE adj. **little**, Latin **parva**, + p.n. Fransham as in Great FRANSHAM TF 8913. DEPN.

FRANT ESusx TQ 5835. 'Place overgrown with bracken'. *Fernet(te)* c.1105, 13th cent., *Fern(e)th(e)* 1230–1439, *Farn(e)th(e)* 1286–1349, *Frenthe* 1332, *Fraunt* 1526, *Franthe* 1537, *Fant(e)* 17th cent. OE **fyrnthe**, SE dial. form **fernthe**. Possibly to be identified with the *denn* or 'swine-pasture' of Annington called *fyrnþan* (dative case) in 956 S 624. Sx 373 gives pr [fænt].

FRATTON Hants SU 6500. 'Estate called after Froda'. *(œt, to) Frodin(c)gtune* [982]14th S 842, *Frodintone* 1086–1530 with variants *-yn-*, *-en-* and *-ton*, *Froditonia* c.1160, *Frodyton(e) -i-* 1235–1327, *Fradington alias Fratton* 1699. OE pers.n. *Frōda* + **ing**[4] + **tūn**. Ha 79, Gover 1958.25, DEPN.

FREATHY Corn SX 3952. Unexplained. *Vridie* 1286, *Freathy* 1699. The surname *Fridia* is recorded in 1327, *Fredea* in 1428, but whether they represent this p.n. or the surname type Friday is unknown. PNCo 85, Gover n.d. 225.

FRECKLETON Lancs SD 4329. 'Settlement at *Frecwœl*'. *Frecheltun* 1086, *-tuna* 1153×60, *Frekelton -il-* [c.1190]1268–1428, *Freke(n)ton -quen-* 1201–70, *Frequelton* 1212, *Freculton* 15th, 16th cent. OE p.n. **Frecwœl*, < **frec** 'greedy, dangerous' + **wǣl** 'a deep place in a river', + **tūn**. In former times there was a large bend in the Ribble at Naze Mount SD 4327 where the river was dangerously deep. La 150, Jnl 17.87.

FREDDEN HILL Northum NT 9526. No early forms.

FREEBY Leic SK 8020. 'Fræthi's village or farm'. *Fredebi* 1086, *Fretheby* 1227–1560, *Freyth(e)by* 15th cent., *Fritheby* 1273–1467, *Frayby* 1539, *Freby(e)* 1578–1666, *Freebie* 1604. ODan pers.n. *Frœthi*, genitive sing. *Frœtha*, + **bȳ**. Lei 164, SSNEM 47.

FREELAND Oxon SP 4113. *Freelandes* 1605, *Freeland* 1833 OS. ModE **free** + **land**. A squatter settlement on the boundary between Eynsham and Handborough parishes which did not pay rent. Emery 163.

FREETHORPE Norf TG 4005. Probably 'Fræthi's outlying farm'. *Frietorp* 1086, *Frethetorp* 1219, 1253, *Frethorp* 1234–86. ON pers.n. *Frœthi*, genitive sing. *Frœtha*, + **thorp**. OE *frith, freothu* 'refuge, protection' is an alternative possibility for the specific of this name. DEPN, SPNN 127.

FREISTON Lincs TF 3743. 'The Frisian's farm or village'. *Fristune* 1086, *-ton(a)* 1168–1386, *Frestuna* 1158, *-ton* 1200–1535, *Frieston(a)* 1208, 1235, 1350, *Freeston* 1353, *Freiston* from 1526. OE *Frīsa* 'a Frisian' + **tūn**. Payling 1940.121 gives prs [fri:stən] and [fri:sən], FN iii.58.

FREMINGTON Devon SS 5132. 'Estate called after Fremi'. *Framintone* 1086, *Framigtona* [1107×28]1307×27, *Fremigtun* [1107×28]1307×27 etc. with variants *-in(g)- -yng-* and *-ton*. OE pers.n. **Fremi* + **ing**[4] + **tūn**. The same name probably occurs in the nearby lost *Fremesmore* 'Fremi's moor' 1227. D 113.

FREMINGTON NYorks SE 0499. 'Estate called after Frema'. *Fremin(g)ton -yng-* from 1086. OE pers.n. **Frema* + **ing**[4] + **tūn**. YN 273, BzN 3.168.

FRENCHBEER Devon SX 6785. 'Friendship's Beer'. *Fryncepisbere* 1346, *Freinschebeare* 1504×15. Surname *Friendship* (cf. John *Frencheyppe* 1525) + p.n. Beer, OE **bearu** 'a grove'. D 425.

FRENSHAM Surrey SU 8441. Possibly 'Freomund's settlement'. *Fermesham* [963×75]c.1150 S 818, *Fremes-* [967]13th S 752–1583, *Farmes-* 1241–1545, *Fermers-* 1246, *Fernes-* 1258–1395, *Frinse-Frenes-* 16th, *Frensham* from 1553. OE pers.n. *Frēomund*, genitive sing. *Frēomundes*, + **hām**. Sr suggests pers.n. OE **Feorhmǣr* or *Feorhmund* but names in *Feorh-* are not on record. Sr 177–8, ASE 2.32.

FRESHFIELD Mers SD 2908. A district of Formby. *Freshfield* 1895. According to Room 'the district was originally a village on a site called Church Mere. This was buried under sand for approximately a hundred years from the mid-18th century, after which the land was said to have been cultivated by a Mr Fresh who laid top-soil over the sand. When this was used for further building, the new village was named after him'. Room 1983.41.

FRESHFORD Avon ST 7860. 'Ford across the *Fersc*'. *Ferscesford* [984×1016]12th S 1538, *Fersefor*ð [1001]15th S 899, *ferscforda* [n.d.]12th Bath, *Fersshford* 1327. OE **fersc** used as a noun 'fresh or running water' in contrast to the tidal Avon downstream at SALTFORD ST 6867. A ford across the river FROME ST 7643. The Frome may here have had the alternative name *Fersc* 'freshwater'. Cf. the unidentified *Fresca* [c.720]10th BHA. DEPN, RN 165.

River FRESHNEY Humbs TA 2308. 'Fresh-water river'. *Fresken* 1258, *Freskeney* 1275, 1280, *Freshney* from 1280. OE ***fresc, fersc**, definite declensional form oblique case **frescan**, + **ēa**. The reference is to fresh as opposed to salt water. Li v.43, RN 165.

FRESHWATER IoW SZ 3487. Originally a r.n., 'river with fresh water'. *Frescewatre* 1086, *Freskewatre -water(e)* c.1145–1312, *Freschewater(e)* 1194–1379, *Freshwater* from 1250. OE ***fresc** + **wæter**. Forms with *-sk-* probably represent the definite declensional form oblique case **frescan* in the phrase *æt thǣm frescan wætre* referring to the highest place reached by the tidal waters of the Yar and where fresh water began. Wt 122 gives pr ['freʃwɔ:dər], RN 165, Mills 1996.51.

FRESHWATER BAY IoW SZ 3485. *Freshewater bay* 1550. P.n. FRESHWATER SZ 3487 + ModE **bay**. Wt 124.

FRESSINGFIELD Suff TM 2677. Partly uncertain. *Fessefelda* (sic) 1086, *Fresingfeud -feld* [1100×35]1396, 1197–1286, *Frisingefeld* 1185, *Fressingfeld* 1229. The generic is OE **feld** 'open land', the specific possibly **fyrs** 'furze' as in FERSFIELD Norf TM 0682 with *r*-metathesis varying with ***fyrsen, *frysen** 'growing with furse'. If the 1185 sp is primary, however, this would be 'the open land of the Frisingas, the people called after Frisa, the Frisian', OE folk-n. **Frīsingas* < pers.n. *Frīsa* + **ingas**, genitive pl. **Frīsinga*, + **feld**. The DB form might represent a variant with genitive pl. *Frīsa* 'of the Frisians'. DEPN, FN 3.59, 66, Baron.

FRESTON Suff TM 4160. 'The Frisian's estate' or 'the settlement of the Frisians'. *Fresantun* 1000×2 S 1486, *Fresetuna* 1086, *Freston(')* from 1242. OE *Frēsa* possibly used as a pers.n., genitive sing. *Frēsan*, possibly genitive pl. **Frēsan*, + **tūn**. DEPN, FN 3.59, Baron.

FRETHERNE Glos SO 7309. 'Sanctuary thorn-tree'. *Fridorne* 1086, *Frethorn(e)* 1236–1535, *Frethern(e)* from 1368. OE **frith, freothu** or **freothen** + **thorn**. Gl ii.178 gives pr ['freðən].

FRETTENHAM Norf TG 2417. 'Fræta's homestead'. *Fretham* 1086, *Freteham* 1174, *Fretenham* from 1202. OE pers.n. *Frǣta*, genitive sing. *Frǣtan*, + **hām**. DEPN.

FRIAR'S GATE ESusx TQ 4933. Sx 372.

FRIDAY BRIDGE Cambs TF 4605. *Fridayesbrugg'* 1298, *Frydaybrigge* 1340. Friday was the name of a fishery belonging to the monks of Ely, *fridai* 1086, *Frideiwere* 1251 'Friday weir'. There was a *Fryday lake* 1570 in Elm and another near Whittlesey, *Fridaylake* 1244. There was also a *Frydaye weyr* 1549 in Haddenham. All these names allude to the provision of fish for Friday consumption. ME **fryday** (OE *Frīġedæġ*) + **brigge** (*brycg*). Ca 268.

FRIDAYTHORPE Humbs SE 8759. 'Fridag's outlying settlement'. *Fridag(s)- Frida(r)storp* 1086, *Fridait(h)orp -y-* 1196–1546, *Fredathorpe* 1562. CG pers.n. *Frigdag* or OE **Frīgedæġ* or ON *Frjádagr* + **thorp**. YE 129, SSNY 58.

FRIESTHORPE Lincs TF 0783. 'Outlying farm of the Frisians'. *Frisetorp* 1086, *Frisa- Frisætorp* c.1115, *Fristorp* 1146, [1154×89]1329, *Frestorp* c.1200, 1202, *Frisethorp* 1266, 1281. ON folk-n. *Frísir*, genitive pl. *Frísa*, + **thorp**. Possibly a colony from the lost vills of East and West Firsby TF 0085, SK 9985, *Frisebi* 1086, *Frisaby* c.1115. SSNEM 109, 46, FN iii.63, 62.

FRIETH Bucks SU 7990. 'Scrubland'. *ffrith* c.1307, *The Frith* 1548, *Freeth* 1766, 1826, *Frieth* 1822 OS. OE **(ġe)fyrhth**. According to Bk 178 this place was still known locally as *the Frieth*, L 191, Jnl 2.31.

FRILFORD Oxon SU 4497. 'Frithela's ford'. *Frieliford* 1086, *Fryleford(')* [1066×87]13th–1396×7, *Fridlef(ord)'* 1220, *Fritleford* 1327, *Frylford* 1517. OE pers.n. **Frithela* + **ford**. This pers.n., a hypocristic form of names in *Frithu-* and *-frith*, occurs in *friðelabyrig* [955×7]12th S 663 in the bounds of Hinksey, Seacourt and Wytham. Cf. also FRILSHAM SU 5473Berks. Brk 407.

FRILSHAM Berks SU 5473. 'Frithel's homestead'. *Fril(l)esham* 1086–1203, *Frid(e)lesham* 1174–1327, *Frydelsham* 1409. OE pers.n. **Frithel*, genitive sing. **Fritheles*, + **hām**. Brk 246.

FRIMLEY Surrey SU 8858. 'Fremma's wood or woodland clearing'. *Fremelegh -le(ye) -lye* 13th–1349, *Fremeley* [933] and *Fremesleya* [967]both 13th S 420, 752, *Fremle* 1203, 1279, *-ley* 1445, *Frym(e)ley* 1535, 1543, *Frimley* from 1575. OE pers.n. **Fremma* + **lēah**. **Fremma* may be a hypocoristic form of the pers.n. in FRENSHAM SU 8441. Sr 126–7.

FRINDSBURY Kent TQ 7469. 'Freond's fortified place or manor'. *Freondesberiam* [764]12th S 105, *Freondesberia*, *Frinondesbyrig* (sic) c.975 B 1321–2, *Frandesberie* 1086, *Frendesberiam* 1100×35. OE pers.n. **Frēond*, genitive sing. **Frēondes*, + **byriġ**, dative sing. of **burh**. KPN 49.

FRING Norf TF 7334. 'The Freingas, the people called after Frea'. *Frainghes, Frenga -e* 1086, *Frainges* 1136×45, *Fréėinge* [c.1160]13th, *Freng(es)* 1198, *Freinge(s)* 1205–18. OE folk-n. **Frēingas* < pers.n. **Frēa* + **ingas**. DEPN, ING 57.

FRINGFORD Oxon SP 6029. 'Ford called or at *Fering*, the place called after Fera'. *Feringeford* 1086–1246×7, *Feringford(e) -yng-* 1243–1359, *Fringeford(')* 1205–41, *Fryngeford* 1522, *Frinchford* 1622. OE p.n. **Fēring* < pers.n. **Fēra* + **ing**[2], locative–dative sing. **Fēringe*, + **ford**. This explanation (rather than 'the ford of the Feringas, the people called after Fera', OE folk–n. **Fēringas* < pers.n. **Fēra* + **ingas**, genitive pl. **Fēringa*, + **ford**) is adopted to account for the assibilated pronunciation implied by the form *Frinchford*. O 209.

FRINSTED Kent TQ 8957. 'Protected, fenced-in place'. *Fredenestede* 1086, *Fredenesstede*, *Fridenastede* 11th, *Frethenested(e)* 1223–1323, *Frethne- Frednestede* 1243, *Frenestede* 1223, 1242×3, *Frensted* 1610. OE ***frithen, freothen** + **stede**. PNK 210, KPN 21, Stead 215.

FRINTON-ON-SEA Essex TM 2319. P.n. Frinton + ModE **on** and **sea**, a twentieth cent. addition marking the development of Frinton as a resort by Sir Richard Cooker 1890–1900. Earlier simply *Frietunā, Frientunam* 1086, *Frienton'* 1158–1218, *Fri- Frynton(e)* 1230–1428, possibly 'Fritha's estate', OE pers.n. **Fritha*, genitive sing. **Frithan*, + **tūn**. An alternative specific is OE ***frithen** 'protected, safe, secure' perhaps meaning 'fenced in'. Ess 339, PNE i.188, Pevsner 1965.185.

FRISBY ON THE WREAKE Leic SK 6917. 'The Frisian's village'. P.n. *Frisebi(e)* 1086–1404, *-by* 1100×35–1444, *Freseby* 1244–1367, *Fri- Frysby* 1304–1610, ON *Frísir*, genitive pl. *Frísa*, + r.n. WREAKE SK 6616, *super Wrethek, Wreke* 1329–1541, for distinction from (Old) Frisby SK 7001. Lei 287, FN 3.62.

FRISKNEY Lincs TF 4655. Partly uncertain. *Frischenei* 1086, *Freschena* c.1115, *Freschenei* c.1150, *Freskenei -ay* late 12th. This could be OE *(æt thǣre) frescan īe* '(at the) freshwater stream', OE ***fresc, fersc**, dative sing. definite form ***frescan**, + **ēa**, dative sing. **īe**; but the majority of forms point to **ēġ** 'an island'. DEPN, Cameron 1998.

FRISKNEY FLATS Lincs TF 5051. P.n. FRISKNEY TF 4655 + ModE **flat** 'a level tract of land over which the tide flows'.

FRISTON ESusx TQ 5498. Possibly 'Freo's estate'. *Friston* from 1200, *Frison* 1610. OE pers.n. **Frēo*, genitive sing. **Frīġes*, + **tūn**. It has been claimed that this is a name in OE *Frīsa* 'a Frisian' but the complete absence of spellings with medial *-e-* is a difficulty. Sx 420 gives pr [frisən], FN 3.59.

FRISTON Suff TM 4160. 'The settlement of the Frisians'. *Frisetuna* 1086, *Freston* 1225–1346, *Fryston* 1524, *Friston* 1568. OE folk-n. genitive pl. *Frīsa* + **tūn**. DEPN, PNE i.187, FN 3.59, Baron.

FRITCHLEY Derby SK 3553. Partly obscure. *Furcesley(e)* 1216×72, 1330, *Furchesleye* 1350, 1362, *Fyrchislegh'* 1309, *Fyrtesley* [for *Fyrcesley*] 1340, *Frytcheley* 1493, 1544×7, *Fritchley* from 16th, *Frithley* 1610–1767. The specific has not been explained, the generic is OE **lēah** 'a clearing'. A possible solution might be an OE masc. or neuter *i*- stem ***fyrċ**, genitive sing. ***fyrċes**, related to *forca* 'a fork'. Db 437.

FRITH BANK Lincs TF 3147. *Frith bank* from 1648. P.n. Frith + ModE **bank**. Frith, *le Frith* 1322–4, *le Fryth* 1415, *an enclosed several marsh by Boston called Fryth* 1416, is ME **frith** (OE *(ġe)fyrhth*) in the sense 'fenland overgrown with brushwood'. Payling 1940.117, Wheeler appendix I.15, PNE i.190.

FRITH COMMON H&W SO 6969. *Frith Common* 1832 OS. ModE **frith** (OE *fyrhth*) + **common**.

FRITHAM Hants SU 2314. 'Cultivated plot in scrubland'. *Friham* 1212, *Fry- Fritham* from c.1280. OE **fyrhth(e)**, ***fryth(e)** + **hamm** 5a. Also known as *Frythambayly* 'Fritham jurisdiction' 1452. Ha 80, Gover 1958.206, L 191.

FRITHELSTOCK Devon SS 4619. 'Frithulac's outlying farm'. *Fredelestoch* 1086, *Fredeletestoc* 1086 Exon, *Frithelaghestok -lakes-* 1223–38, *Fry- Frithelestoke* 1269, 1287, *Frithelstocke al. Fristocke* 1605×23. OE pers.n. *Frithulāc*, genitive sing. *Frithulāces*, + **stoc**. D 92 gives pr [fristɔk].

FRITHVILLE Lincs TF 3150. *Frithville* 1824 OS. A modern coinage for one of several new townships created by Act of Parliament in 1812 on newly drained fenland. P.n. *le Frith* 1322–5 as in FRITH BANK TF 3147 + ModE **ville**, an element briefly in vogue in new 19th cent. p.n. formations. Cf. EASTVILLE TF 4057, MIDVILLE TF 3857. Wheeler 229.

FRITTENDEN Kent TQ 8141. 'Wooded pasture'. *Friððingden* [?804]13th S 159, [850]13th S 300, *Fri- Frythin(g)den(n)* 13th cent., *Fritindenne* 13th., *Frytenden* 1610. OE ***fyrhthen**, Kentish ***ferhthen** or ***fyrhthing, ferhthing** 'wooded place' possibly used as a p.n., + **denn**. PNK 324, KPN 95, PNE i.190, Charters IV.64.

FRITTON 'Enclosed, fenced-in settlement'. OE **frith(u)** + **tūn**.

(1) ~ Norf TM 2293. *Fride- Frede- Frithetuna* 1086, *Fretone* 11th, *Freton* 1199. *Friþetun* [1042×53]13th S 1535 may belong here or under FRITTON TG 4600. DEPN, Nf ii.115.

(2) ~ Norf TG 4600. *Fridetuna* 1086, *Freton* 1224, *Fretone* 1254. DEPN.

FRITWELL Oxon SP 5229. Possibly 'the spring used for divination'. *Fert(e)welle* 1086, *Fretewill -well(e)* [1166]c.1310–1428, *Frit(t)ewell(e)* 13th cent., *Frytwell* 1526. OE ***freht** + **wella, wylle**. The absence of sps reflecting OE *ht* remains, however, a difficulty. O 211, Jnl 29.65–70.

FRIZINGTON Cumbr NY 0317. 'Settlement of Frisa's people'. *Frisingaton* c.1160, *Frisington* from c.1206 with variants *Frys-, -yng-*. OE folk.n. **Frīsingas* < OE pers.n. *Frīsa* + **ingas**, genitive pl. **Frīsinga*, + **tūn**. Cu 336, FN 3.57.

FROCESTER Glos SO 7803. 'Roman camp by the river Frome'. *Frowecestre* 1086, *Froucestr(ia) -cestre -er* 1154×89–1378, *Frou- Frowcet(t)or -er* 1535–1662, *Froster* 1601, 1673. R.n. FROME SO 7706 + OE **ċeaster**. The reference is to remains of a Roman villa here. Gl ii.197 gives pr ['froustə], but locally ['frɔstə] is usual. For the development of *Frōm* to *Frow* cf. LHEB 491. Frocester, a mile distant from the Frome, is less closely associated with the r.n. than Frampton SO 7407 and Framilode SO 7510.

FRODESLEY Shrops SJ 5101. 'Frod's clearing'. *Frodeslege* 1086 etc. with variants *-lega -leg(h)* and *-ley(e), Frodsl(e)y* 1569–1781, *Frodgley* 1584–1710. OE pers.n. *Frōd*, genitive sing. *Frōdes*, + **lēah**, dative sing. **lēaġe**. There are interesting variant *e, a,* and *i* spellings as for FRODSHAM Chesh SJ 5278, *Fredesleg(h) -ley(e)* 1242–1297, *Fridesleye* 1271×2, 1292×5, *Fragl(e)y* 1642, 1674, 1684. Sa i.136 gives local pr [frʌdʒli] and [frɔdzli].

North FRODINGHAM Humbs TA 0953. *North Frothingham -yng-* 1285 etc., *North Froddingham* 1521, 1650. ME adj. **north** + p.n. *F(r)otingha'* 1086, *Fro(h)ing(e)ham* 12th cent., the 'homestead of the Frodingas, the people called after Froda', or 'called or at *Froding*, the place called after Froda', OE folk-n. **Frōdingas* < pers.n. *Frōda* + **ingas**, genitive pl. **Frōdinga*, or p.n. **Frōding* < pers.n. *Frōda* + **ing**[2], locative-dative **Frodinġe*, + **hām**. The *-th-* forms are due to the influence of ON pers.n. *Fróthi*. 'North' for distinction from South Frodingham Humbs TA 3126, *Forthingham* c.1265, *Frothingham -yng-* 1290–1572, *Sowth, South* ~ from 1285, *Frodingham* from 1293, of similar origin. YE 75, 27, ING 153, BzN 1968.169, SPN 87.

FRODSHAM Ches SJ 5278. 'Frod's village'. *Frotesham* 1086, *Frodesham -is- -ys-* 1150–1619, *Frothes- Fradesham* 13th, *Fradsham* 1437–17th with variants *Fradson -som(e) -same*. OE pers.n. *Frōd*, genitive sing. *Frōdes*, + **hām**. Che iii.221 gives prs ['frɔdsəm -ʃəm] and older local ['frad- 'fratsəm] obsolescent by c.1886.

FROG FIRLE ESusx TQ 5101 → West FIRLE TQ 4707.

FROGGATT Derby SK 2476. 'Frog cottage(s)'. *Froggecot(e)* 1225, 1330, 1424, *Frogcot(e)* 1272×1307, 1348, *Froggot(e)* 1319–1416, *Froggat(t)* 1339. OE **frogga** + **cot**, pl. **cotu**. A reference to a damp low-lying situation, possibly ironic rather than literal, cf. Field 84 s.n. Frog Hall. Db 102.

FROGHALL Staffs SK 0247. 'The hollow infested with frogs'. *Frogholle* 1434, 1481, *-hole* 1445–72, 1769, *Froggall* 1639, *Froghall* from 1558×1603. OE **frocga** + **hol**. Oakden.

FROGMORE Hants SU 8460. 'Marshy place', literally 'frog marsh'. *(marsh called) Frogmore* 1567. ModE **frog** + **moor**. Ha 80, Gover 1958.112, Jnl 28.61–70.

FROGMORE HOUSE Berks SU 9776. P.n. Frogmore + ModE **house**. Frogmore, *Frogmore* from 1573, is probably 'frog pond' ultimately from OE **frogga** + **mere**, the latter subsequently confused with **mōr**. Frogmore House lies in the southern section of the Home Park S of Windsor Castle. This type of name is discussed in Jnl 28.61–70. Brk 29.

FROLESWORTH Leic SP 5090. 'Freothulf's enclosure'. *Frel(l)esworde* 1086, 1209×19, *-worth(a)* 13th cent., *Fredleswurð(a)* 1175, 1176, *Frollesworth(e) -is-* 1235–1549, *Frolesworth* from c.1291. OE pers.n. *Freothulf* (*Freothuwulf*), genitive sing. *Freothulfes*, + **worth**. Lei 441.

FROME A common r.n. and p.n. transferred from the r.n.

(1) ~ Somer ST 7747. Transferred from the r.n. FROME ST 7743. *Froom* [705]12th B 114, *(ón) Frome* 955 ASC(A) and from 1086, *Frowme* 1610. The enclosure of the people of Frome is referred to as *(on) Fromesetinga hagen* [964]14th S 727, folk-n. *Fromesǣtingas*, genitive pl. *Fromesǣtinga*, + **hæġen**. DEPN, RN 166.

(2) ~ ST QUINTIN Dorset ST 5902. 'F held by the St Quintin family'. *Fromequintin* 1288, *Fromeseyntquynteyn* 1452, earlier *Litelfrome* 1086, *Littlefrome* 1202, the 'little Frome estate'. R.n FROME SY 8089 + family name St Quintin, earlier OE **lȳtel** + r.n. FROME. Do 75.

(3) Bishop's ~ H&W SO 6648. 'The Frome estate held by the bishop' sc. of Hereford. *Frome Episcopi* c. 1285, *Byschopusfrome* 1428. Earlier simply *Frome* 1086. ME **bischop**, genitive sing. **bischopes**, Latin **episcopus**, genitive sing. **episcopi**, + p.n. Frome from the r.n. FROME SO 6544. 'Bishop's' for distinction from Canon and Castle FROME SO 6543 and 6645. He 88.

(4) Canon ~ H&W SO 6543. 'The Frome estate held by the canons' sc. of Llanthony priory. *Froma Canonicorum* 1160×70–1243, *Canonffrome* 1397. Earlier simply *Frome* 1086. ME **canon**, Latin **canonicus**, genitive pl. **canonicorum**, + p.n. Frome transferred from r.n. FROME SO 6544. Hugh de Lacy gave the manor to his foundation of Llanthony prima before c.1115. Also known as *Parva Frome* 1131 or *Frome Minor* c.1132, 'little' or 'lesser F' for distinction from Bishop's and Castle FROME SO 6648 and 6645. He 89.

(5) Castle ~ H&W SO 6645. 'Frome with the castle'. *Froma Castri* 1243, *Castyl Frome* 1428. P.n. Frome transferred from

r.n. FROME SO 6544 + Latin **castrum**, ME **castel**. Earlier called *Bricmarifrome* 1074×85, *Brismerfrvm* 1086, 'Brictmer's Frome', OE pers.n. *Brihtmǣr* referring to the TRE tenant, and also *Majoris Frome* 1101×2 'greater F' and *Frome Herberti* 'Herbert's F' referring to the 12th cent. tenant Herbert de Castello from Castle Holgate, Shrops. A motte and traces of the bailey survive. He 89, Pevsner 1963.231.

(6) Halmond's ~ H&W SO 6747 → River FROME SO 6544.

River FROME Uncertain, perhaps simply 'running water'. PrW ***frōm** < Brit **frām-* whence ModW *ffrau* 'flood, stream' and *ffraw* 'fine, brisk'. GPC 1311.

(1) ~ Avon ST 6377. *(andlang) Frome, Fromes* [950]19th S 553, *Frome* from early 13th, *Froom* 1712. RN 166.

(2) ~ Dorset SY 8089. Welsh forms, *Frauu* [894] Asser, *Fraw* [before 1118]12th, c.1540; English forms, *Frome* from 12th including [859 for ? 870]15th S 334, [966]15th S 744, *Frume* [? 870]15th S 342. RN 166.

(3) ~ Glos SO 7706, 9303. *Frome* from 1248, *Fraw als. Frome* c.1540, *Froome or Stroudwater* 1779. The r.n. forms the specific of FRAMPTON and FROCESTER. Gl i.7, RN 167.

(4) ~ H&W SO 6544. *From* [840×52]10th S 1270, 1577. The river gave its name to several estates along its length, Bishop's FROME SO 6648, Canon FROME SO 6543, Castle FROME SO 6645, Halmonds Frome SO 6747, *Froma Hamunde* 1205×16, *Haymondsfrome* 1399, from the 12th cent. tenant Hamo, earlier *Nerefrvm* 1086, *Nederefroma* 1160×70, 'lower F', Priors Frome SO 5739, *From' Prioris* 1243 held by St Guthlac's Priory, Hereford, earlier simply *Frome* 1086, and Larport SO 5738, *Lorteport* 1148×63, 'dirty market', OE **lort** + **port**, earlier *Frome* 1086, *Froma Henrici* 1243 'Henry's F' from Henry de Monmouth who held the manor in 1243, and *Little Froma* 1281. RN 167, He 88–9, 149.

(5) ~ Somer ST 7743. *From* [701]12th B 105, *Frón* [701]10th B 106, *Fromœ* [987]12th S 867, *Frome* from 1218, *Frowme flu:* (i.e. *flumen* 'river') 1610. DEPN, RN 166.

FROMES HILL H&W SO 6846. *Fromes Hill* 1832 OS. P.n. Frome as in Castle FROME SO 6645 + ModE **hill**.

FROSTENDEN Suff TM 4881. 'Frog valley'. *Froxedena* 1086, *Frostenden(e)* from 1225, *Frosteden* 1242. The DB sp points clearly to OE **frosc, frox**, genitive pl. **frosca, froxa**, + **denu**. The later sps seem to suggest that **Frox(e)den > Froxten > Frosten* to which a second pleonastic *denu* was added. DEPN, Baron.

FROSTERLEY Durham NZ 0337. 'The forester's Lea'. *ffrosterley* [1183]14th, *Frosterley* from 1296, *-le(e)* 1405–1530, *Forsterley* 1382. ME **forester** + **ley** (< OE *lēah* probably used as p.n.). A holding of the Bishop of Durham's forester in Weardale. The apparent combination of OE *lēah* with a specific of ME origin is not to be taken as evidence that *lēah* was still a living element in the 12th cent. Frosterley adjoins Rogerley NZ 0137, *Rogerleia* [1183]14th, *Rogerley* from c.1320, 'Roger's clearing or pasture', OG pers.n. *Rodger* + ME **ley** (OE *lēah*). This is probably an instance of 12th cent. subdivision of an earlier Anglo-Saxon woodland estate called *Lēah* 'the clearing, the pasture'. NbDu 90.

FROXFIELD Wilts SU 2967. 'The frogs' open land'. *Forscanfeld* [801×5]12th S 1263, *Froxefeld* 1212–1341, *Froxfeld* 1242, 1289, 1604, *Vroxfeld* 1297, *Throxfeld* 1377. This is probably to be taken with the stream name *Forsca burna* 'the frog stream' [778]?10th S 264, if this is the brook on which Froxfeld lies; in which case *Forscanfeld* is probably an error for **Forscafeld*, OE **frosc, frox, forsc** 'a frog', genitive pl. **forsca, froxa, frosca**, + **feld**. Alternatively *Forscan* may be a genuine form representing the genitive sing. of OE **Forsce* 'the frog stream'. Wlt 346, Grundy 1919.154, L 244.

FROXFIELD GREEN Hants SU 7025. P.n. Froxfield + ModE **green** 'a common', an element frequently used of secondary settlements. Froxfield, *(œt) Froxafelda* [c.968×71]12th S 1485, *Froxefeld(e)* 1207, 1245, *Froxfeld* 1242, is 'frogs' open land', OE **frosc, frox**, genitive pl. **froxa**, + **feld**. Ha 80, Gover 1958.61, DEPN, L 244.

Lower FROYLE Hants SU 7644. *Lower Froyle* 1817 OS. ModE **lower** + p.n. Froyle as in Upper FROYLE SU 7543. Also known as *Northfroille* 1348. Gover 1958.100.

Upper FROYLE Hants SU 7543. ModE **upper** + unexplained p.n. *Froli* 1086, *Froila -e* 1167, 1185, *Froyle* from 1236, *Frohill' -hull* 1199, 1205. This might be any of 'free hill', i.e. free from service or charge, or 'Friga or Frig's hill', OE **frēo**, pers.n. **Friġa* or pagan goddess name *Frīġ, Frēo* + **hyll**. The *hill* spellings, however, look suspiciously like a folk-etymology for an earlier name that was not understood. Ha 80, Gover 1958.100, DEPN, BzN 1981.304.

FRYER MAYNE Dorset SY 7386 → BROADMAYNE SY 7286.

FRYERNING Essex TL 6400. 'Ing held by the brothers' sc. of the hospital of St John of Jerusalem. *Friering* 1469, *Fryer Inge* 1539, *Gynge Freren* 1542, *Fryerning* 1558×1603. ME **frere**, genitive pl. **frerene**, + p.n. *Ingā* 1086, *Ging(es), Gynge(s)* 1291, 1306, as in INGRAVE TQ 6292. Ess 254.

FRYSTON 'The settlement of the Frisians'. OE folk-n. *Frīsa* + **tūn**. FN 3.60.

(1) Ferry ~ WYorks SE 4824. 'F by the ferry'. *Ferry Freiston* 1535, *Ferry Freeston* 1645. ModE **ferry** + p.n. *Friston(e) -tona, Fryston* 1154–1522. The ferry, *Fereia* 1086, is mentioned in DB. YW ii.65, DB Yorks ii.9W 57, FN 3.60.

(2) Monk ~ SYorks SE 5029. 'F held by the monks' sc. of Selby Abbey. Prefix *Muneches-* 1166, *Monk(e)-* from 1398, + p.n. *(on) fryyetune* [963]14th S 712, *Fristun* c.1030 YCh, *Friston* from c.1070 with variants *Frys-* and *-tun(a) -tona, Freis- Frieston* 16th cent. King Edgar's charter of 963 survives only in late copies and the form *fryyetun* must be an error for *frysetun*. YW iv.41, FN 3.60.

(3) New ~ WYorks SE 4527. *Newton Fryston* 1522. ModE **new**, short for Newton as in NEWTON SE 4427 on the opposite side of the Aire. YW ii.67.

(4) Water ~ WYorks SE 4626. 'F by the water' sc. of Aire. *Water Fryston* 1532. ModE water + p.n. *Fristone* 1086, *Friston(a), Fryston* 1155×8–1532. Earlier forms of the affix are *on Ayr(e)* 1289 etc. and *by the Water* 14th. YW ii.66, DB Yorks ii.9W 56, FN 3.60.

FRYTON NYorks SE 6875. Probably 'Frithi's settlement'. *Frideton, Fritun* 1086, *Fri- Fryton(a)* from 12th. ODan pers.n. *Frithi*, genitive sing. *Fritha*, + OE **tūn**. For an alternative explanation see FRITTON Norfolk TG 4600, TM 2293, from OE **frith(u)** + **tūn** 'an enclosed place, a fenced-in settlement'. YN 50 gives pr [fritən], SSNY 127.

FULBECK Lincs SK 9450. 'The dirty stream'. *Fulebek* 1086–1247, *Fulbec* 1212–1535, *-beck* from 1275. OE **fūl**, definite form **fūla**, + ON **bekkr**. An isolated form *Fulebroc* 1194 suggests that *bekkr* may have replaced original OE **brōc**. Perrott 372, SSNEM 217.

FULBROOK Oxon SP 2613. 'The foul, dirty brook'. *Fvlebroc* 1086, *Fulebroc(h) -brok(e)* 1156×66–1316, *Fulbroc* 1167, *-brok* 1326 etc. OE **fūl**, definite form **fūla**, + **brōc**. O 352.

FULFORD NYorks SE 6149. 'The dirty ford'. *Fuleford* 1086–1230, *Fulford* from 1150×61. OE **fūl**, definite form **fūla**, + **ford**. The York–Doncaster road crosses a small stream now bridged at this point. YE 275.

FULFORD Somer ST 2029. 'The dirty ford'. *North- Southfuleford* 1327. OE **fūl**, definite form **fūla**, + **ford**. A Latin translation, *sordidum vadum* *[854]12th S 311, is the earliest form. DEPN.

FULFORD Staffs SJ 9538. 'The dirty ford'. *Fuleford* 1086, 1167, *Fulford* 1583. OE **fūl**, definite form **fūla**, + **ford**. DEPN, Horovitz.

FULHAM GLond TQ 2576. 'Fulla's river-bend land'. *Fulanham* [c.705]17th S 1785, *(œt, on, of) Fullanhamme* 9th ASC(A) under years 879 and 880, *-homme* 12th ASC(E) under year 880,

Fulham from 1274. OE pers.n. *Fulla*, genitive sing. *Fullan*, + **hamm** 1. Fulham lies in a bend of the river Thames. Mx 101 gives pr [fuləm].

FULKING WSusx TQ 2410. 'The Folcingas, the people called after Folca'. *Fochinges* 1086, *Folkinges* 1091–1267, *Fulkyng* 1327, 1427. OE folk-n. **Folcingas* < pers.n. **Folca* + **ingas**. Sx 284, ING 34, MA 10.23, SS 65.

FULL SUTTON Humbs SE 7455 → Full SUTTON.

FULLER STREET Essex TL 7415. 'Fullwood hamlet'. *ffullere-street* 1415, *Fullwood Street* 16th, *Fuller Street* 1805 OS. P.n. Fullwood + ME **strete**. Fullwood, *Folewode* 1323, is the 'foul or dirty wood', OE **fūl**, definite form **fūla**, + **wudu**. Ess 287.

FULLER'S MOOR Ches SJ 4954. 'Fuller's marsh'. *Fuller's More* 1668×71, *Fuller's Moor* 1842. ModE **fuller** (OE *fullere*) or surname *Fuller*, + **moor**. Che iv.15.

FULLERTON Hants SU 3739. 'Settlement, estate of the wild-fowler(s)'. *Fugelerestune* 1086, *Fu- Foghelerton(e)* 1234–1327, *Foulerton* 1410, *Fullherton al. Fughlerton* 1653. OE **fuglere**, genitive sing. **fugleres** varying with pl. **fuglera**, + **tūn**. Fullerton at the confluence of the Anton, Dever and Test must have offered excellent opportunities for wild-fowling. Ha 81, Gover 1958.172.

FULLETBY Lincs TF 2973. Probably the 'village or farm by the muddy road-junction'. *Fullobi* (2×, 1× corrected from *Fulnodebi*) 1086, *Fuledebi -t-* c.1115, *Fulleteby -bi* [1153×62]1409, 1175×81, *Foletheby* [c.1160]13th. OE p.n. **Ful-læte* < **fūl** + **(ġe)lǣt(e)**, + **bȳ**. SSNEM 47, PNE ii.11.

FULMER Bucks SU 9985. 'Wild fowl pond'. *Fugelmere* 1198, *Fulmere* 1247–1509, *Foulmere* 1296–1375. OE **fugol** + **mere**. Bk 237, L 26.

FULMODESTON Norf TF 9931. 'Fulmod's estate'. *Fulmotestuna* 1086, *Fulmodeston* 1242, *-e* 1254. lOE pers.n. **Fulmōd* (OHG *Folcmōd* or possibly OSwed *Folkmōdh*), genitive sing. **Fulmōdes*, + **tūn**. DEPN, SPNN 124.

FULNETBY Lincs TF 0979. Possibly 'Folcnoth's village or farm'. *Fulnedebi* 1086, *-by* 1234, *-netebi* c.1115, [1154×89]1406, *Fullotebi* 1154×89, *Fulnotebi* before 1187, *-nethebi* 1196, *Fullethe-Fulnathebi* 12th, *Fulnotesbi* 1200. Possibly OE pers.n. **Folcnōth* or ON **full-nautr* 'one who has a full share' used as a by-n., + **bȳ**. DEPN, SSNEM 48, Cameron 1998.

FULSTOW Lincs TF 3297. 'The meeting-place of the birds'. *Fvgelestov* 1086, 1183, *-stow(e)* 1180–1219, *Fulestou(e) -stowe* 1147–1311, *Fulstow(e)* from 1239. OE **fugol**, genitive pl. **fugla**, + **stōw**. Possibly a gathering place before migration. SSNEM 376, Cameron 1998.

FULWELL T&W NZ 3959. 'The dirty spring'. *fulewellā* 1154×66, *Fulewell(e)* 1189×1212–1296, *Fulwell* from 1345. OE **fūl**, definite form **fūla**, + **wella**. NbDu 90.

FULWOOD 'The dirty wood', OE **fūl**, definite form **fūla**, + **wudu**. L 228.

(1) ~ Lancs SD 5331. *Fulewde* 1199, *ful(e)wude* 1228, *Fuluuode* 1252, *-wode* 1297, *-wood* 1551, *Foghel(l)wo(o)d* 1346–8. Part of the forest of Lancaster. The 14th cent. forms suggest that the specific may have been OE **fugol** 'a bird' and so 'wood where fowling takes place'. La 148, Jnl 17.84.

(2) ~ SYorks SK 3085. *Folwod* 1284×1327, *Fulwode* 1332, 1441, *-wood* 1586. YW i.194.

FUNTINGTON WSusx SO 8008. 'The settlement called or at *Funting*, the fountain place'. *Fundintune* 12th, *Funtington* from 1252. OE ***funting** < **funta** + **ing**², + **tūn**. Sx 60 gives prs [fʌndən] and [fʌniŋtən], SS 74.

FUNTLEY Hants SU 5608. 'Spring wood or clearing'. *Funtelei* 1086, *-lye -legh -leg' -lighe* 1251–1305, *Fontele(g')* 1242–1428, *Funtle* 1305. OE **funta** + **lēah**. There is a copious spring here which may have been used in Roman times. Ha 78, Gover 1958.33, Signposts 84, 86, Nomina 9.6.

FURNESS Cumbr SD 2087, 1969, 2374. 'Headland of (the island called) *Futh*'. *Futhpernessa* [c.1150]13th, *ff- Fudernesium* 1127, [1127×33]1398, *Furnesio* c.1155, *Furneis -eys -ays* 1194×9–1295, *Furnes* from 1170. P.n. **Futh* < ON **futh** 'cunt, anus', genitive sing. **futhar**, + **nes**. ON *futh* is used in the names of skerries and small islands in Norway. Here it was used in the original name of PIEL ISLAND SD 2363, *Fotherey* c.1327, ON **futhar** + **ey** 'island', with reference to the long deep depression which runs from N to S through the island. The original application of the name Furness was to the point opposite this island at what is now called Rampside Point SD 2466. From there it was transferred to refer to the whole region and divided into two parts, Low or Plain Furness, *Lowfurnes* 1546, *Playne Furneys* 1582 and High Furness, *Heigh Furnes* 1584. La 200, Mills 1976.85, SSNNW 124, L 173.

FURNESS FELLS Cumbr NY 3000. *Fournes-fell'* 1338. P.n. FURNESS + ModE **fell(s)**. Also referred to as *Montanis de Furnesio* 1196, *Heigh Furnes* 1584. La 201.

FURZEBROOK Dorset SY 9384. *Furzebrook* 1811. ModE **furze** (OE *fyrs*) + **brook** (OE *brōc*). Do i.91.

FURZEHILL Devon SS 7244. 'Furze hill'. *Fershull* 1198, *Furshille* 1242, *Forshull* 1301. OE **fyrs** + **hyll**. D 65.

FYFETT Somer ST 2314. *Fyfet* 1809 OS. Early forms are wanting but this might be OE **fīf** + **hīde** 'five hides'. Cf. FITZHEAD ST 1228, FIVEHEAD ST 3522.

FYFIELD '(Estate of) five hides', the standard holding of a thegn. A hide was originally the unit of land needed for the support of a single family. The five-hide unit was the basis of the OE obligation to provide one soldier for the *fyrd*. OE **fīf-hīda**, dative case **fīf-hīdum**. AS England 638, Speculum 36.61–74, EHR 63.453ff.

(1) ~ Essex TL 5707. *Fifhidam -ā* 1086, *Fi- Fyfhide -hyde* 1092–1332, *Fyf(f)eld(e)* 1412, 1523–94. Ess 57, Jnl 2.40.

(2) ~ Glos SP 2004. *Fif(h)ida -e* [1100×35]1313, [1189×99]1372, 1207–1327, *Fyfeld* 1566, *Fifield* 1777. Gl i.31.

(3) ~ Hants SU 2946. *(to) fif hidan, (æt) fif hidon, fif hida (land)* [975]12th S 800, *Fifhide* 1086–1346, *Fyfeld* 1535, *Fifield al. Fiffhead* 1641. The manor is assessed at five hides in DB. Ha 81, Gover 1958.166.

(4) ~ Oxon SU 4298. *(æt) Fif Hidum* [956]13th S 603, *(æt) Fifhidan* [968]12th S 758, *Fivehide* 1086, *Fi- Fyfhide -hyde* c.1180–1284, *Fyfeld* 1517. Brk 408.

(5) ~ Wilts SU 1468. *Fifhide* 1086–1242, *Fyfilde al. Fyfilt* 1501, *Phiphild* 1553, *Fyfelde* 1558. The manor paid geld for five hides in DB. Wlt 295.

The FYLDE Lancs SD 3439. 'The plain'. *Filde* 1246, *Fylde* 1293. OE **(ge)filde**. The reference is to the flat W part of Amounderness Hundred, the district between the Ribble and the Cocker. The term is a collective formation on OE *feld* and meant 'an area of open land', cf. G poetic *Gefilde* (OHG *gefildi*). La 139, L 240, Kluge-Seebold 251.

FYLINGDALES NYorks SE 9199. 'Fyling valleys'. *Ffilingdales* 1395. P.n. *(Nort)figelinge, Fig(c)linge* 1086, c.1175, *Figelingam et aliam Figelingam* 1100×35, *Philinch* 1132×7, *Tribus Figelinges* 'the three Fylings' 1181, *Fieling(an)* 1133–1308, *(North)filinge -fyling(e)* c.1280 etc., OE folk-n. **Fyġ(e)lingas* 'the people called after Fygela', OE pers.n. **Fyġ(e)la* + **ingas**, + ME **dale**. *North* for distinction from *South Fyling* or FYLINGTHORPE NZ 9404. YN 116, ING 74.

FYLINGDALES MOOR NYorks SE 9199. P.n. FYLINGDALES SE 9199 + OE **mōr**.

FYLINGTHORPE NYorks NZ 9404. '*Fyling* outlying settlement'. A name compounded from Fyling as in *aliam Fielingam* 1133, *Sutfieling* 1140×65, 'South Fyling' for distinction from 'North Fyling' or FYLINGDALES SE 9199, and *thorpe* as in *Prestethorpe* 1280, 'the outlying settlement of the priests', OE **preost**, genitive pl. **preosta**, + ON, ME **thorp**. The land was held by the monks of Whitby. YN 117, ING 74.

G

GADDESBY Leic SK 6813. Either 'Gadd's village or farm' or 'village of the spur of land'. *Gadesbi(e)* 1086, *-by* c.1130–1502, *-berie* 1200, 1201, *Gaddesbi* c.1130–1404, *-by* from 1209×35 with variants *-is-*, *-ys-*. ON pers.n. *Gaddr*, ME genitive sing. *Gaddes*, or ON **gaddr** 'goad, spur' used topographically, + **bȳ**. Gaddesby is situated on a spur of land between streams. Lei 291, SSNEM 48.

Great GADDESDEN Herts TL 0211. *Magna Gatesdene* 1254, *Gaddysden the More* 1432–43. ModE adj. **great**, **more** 'greater', Latin **magna**, + p.n. *Gǣtesdene* [944×6]13th S 1497, *Gatesdene* 1086–1428 with variant *-den*, *Gadesden(e) -ys-* 1228–1405, *Gaddesden* 1402, *Gadson* 1679, 'Gæti's valley', OE pers.n. **Gǣti*, genitive sing. **Gǣtes*, + **denu**. 'Great' for distinction from Little GADDESDEN SP 9913. The r. Gade which flows past here, *Gadus* 1577, is a back-formation from the p.n., invented by Harrison. Hrt 34 gives pr [gædzdən], RN 168.

Little GADDESDEN Herts SP 9913. *Parva Gatesdene* 1204. ModE adj. **little**, Latin **parva**, + p.n. Gaddesden as in Great GADDESDEN TL 0211. Hrt 34.

GAGINGWELL Oxon SP 4025. Apparently 'the kinsmen's spring'. *Gadelingwell(e)* [c.1173]13th–c.1280, *Gadelingewelle* [1193]13th., *Gageinwell* early 18th, *Gogingwell* 1791. OE **gædeling** 'a relative, a kinsman, a comrade', genitive pl. **gædelinga**, + **wella**. O 348, L 32.

GAILEY Staffs SJ 9110. 'The clearing growing with bog-myrtle'.
I. *Gageleage* [1002×4]11th S 1536, [1004]11th S 906, *Gragelie* (sic) 1086, *Gaghley* 1251.
II. *Gaeleg'* 1200, *Gaele* 1257, *Galey* 1775, 1834, *Gailey* 1880.
III. *Gau(e)ley(e) -legh -v-* c.1158×65–1470, *Gaw(e)ley* 1459, 1505.
The full range of the forms of this name occurs in the 1569 record *Gauvell als Gaveley als Gawley als Galeigh als Galey*. III shows the normal development of OE **gagol-lēah** with labialisation of *g* [ɣ] > [ɣʷ] > [w]. Forms with *-v-* if not purely orthographic may show substitution of the peripheral phoneme [ɣ] by the neighbouring fricative [v]. The current form is the result of early ModE monophthongisation of ME *aw* [au] to [a:] as in ModE *gale* 'the wild myrtle'. Gale was used for animal feed, as a substitute for hops in brewing, and in dying wool and tanning. The presence of bog-myrtle here is recorded in *The Gentleman's Magazine* 1786, 'At a place called Foulmire, about a mile from the Four Crosses, an aromatic shrub of the myrtle kind grows spontaneously. It is called gale or sweet gale, and gives its name to a hamlet near it. Where it flourishes is a black morassy ground between two copses, greatly sheltered from the bleak winds... It thrives not anywhere else, and seems confined to this spot of a few acres'. St i.89, Duignan 65, Horovitz, Nomina 13.109.

GAINFORD Durham NZ 1716. 'Direct ford'. *Geg(e)nford, Geagenforda* [c.1040]12th, *Gegen- Geinforde* c.1104, *Gaineford* 1197, 1242×3, *Gain- Gaynford(e)* 1235–1540, *Gaynesdford'* 1306, 1313, *Gaynforth* 1306–1587, *-furth* 1449. OE **ġeġn** + **ford**. A direct as opposed to an oblique crossing of the Tees. OE **ġeġn** would normally produce ME *yein*; the hard [g] in this name is due to Scandinavian influence. NbDu 90, Nomina 12.34.

GAINSBOROUGH Lincs SK 8189. 'Gegn's fortified place'. *Gegnes burh* c.1122 ASC(E) under year 1013, *Genesburuh* 11th ASC(C), *Gæignesburh* c.1050 ASC(D), *Gainesburg* 1086, *Gleinesburc* (sic) c.1115. OE pers.n. *Ġeġn* with ON [g] for English [j], genitive sing. *Ġeġnes*, + **burh**. DEPN, TC 97.

GAINSFORD END Essex TL 7235. P.n. Gainsford + ModE **end**. Gainsford, *Gaynesfordes* 1429, 1534, *Gensforde* 1573, *-fords* 1768, *Gainsford* 1777, is short for Gainsford's homestead from the surname *Gaynesforde* 1465–70. Ess 464.

GAISGILL Cumbr NY 6305. 'Wildgoose ravine'. *Gagesgill -gylle, Gas(s)egill(e)* 1310, *Gaisgill* 1572. ON **gás**, genitive pl. **gása**, + **gil**. The second *g* in the first two forms is probably erratic. We ii.50, SSNNW 124, L 99.

GAITSGILL Cumbr NY 3846. 'Shieling(s) where goats are kept'. *Geytescall* 1278, *Gayt(e)sc(h)ales* 1279–1567 with variants *Gait-skales, Gaitskaile al. Gateskill* 1670, *Gateskale sometimes written Gatesgill* 1816. ON **geit**, genitive pl. **geita**, + **skáli**. Cu 133 gives pr [gɛ:tskil], SSNNW 63 s.n. Gatesgill.

GALBY Leic SK 6901. 'Village, farm on poor soil'. *Galbi* 1086–1306, *Gaubi -by* 1154×89–1313, *Galby* from 1232. ON ***gall** 'barren spot' + **bȳ**. The village lies on raised ground where the soil is stiff clay and loam; it is surrounded by Anglo-Saxon *tūn*-names. Lei 218 gives pr ['gɔ:lbi], SSNEM 48, Gaz.

GALGATE Lancs SD 4855. 'The Galloway road'. *Gawgett* 1605. This explanation derives less from the form of the name, which is indecisive, than from the tradition of cattle drovers from Galloway using the road. La 170 compares *Galwaithegate* [1190×1220]1268, the old name of a N–S road on the borders of Lambrigg and Firbank (SD 5993) known locally as Scotch Lane on account of its being a road along which cattle were driven from Scotland to England; p.n. *Galweithia* Galloway, literally, 'the foreign Gael' + ON **gata**. La 170, We i.21 under Road VI.

GALHAMPTON Somer ST 6329. Probably 'settlement of the rent-paying peasants'. *Galmeton* 1199, *Galampton* 1303. OE ***gafolmann**, genitive pl. ***gafolmanna**, + **tūn**. DEPN, PNE i.192.

GALLEY COMMON Warw SP 3192. *Galley Common* 1591, 1655. This might be 'gallows common', ME **galowe** + **common**, but the evidence is too late for certainty. Wa 91.

GALLEYWOOD Essex TL 7102. 'Wood which pays a tax'. *Gauelwode* 1250–1345, *Gawel(l)- Gawlewod(e)* 1307, 1419–21, *Gall(e)wo(o)dheth -end* 1450, 1633×5, *Gawel Wood, Gallwood* 1612, *Gallywood* 1777. OE **gafol** + **wudu**. The sense is 'woodland from which a rent is derived'. Ess 234.

GALMPTON 'Settlement of the rent-paying peasants'. OE ***gafolman**, genitive pl. ***gafolmanna**, + **tūn**. D 305.
(1) ~ Devon SX 8956. *Galmentone* 1086, *Galmentūn* 1198, *-ton* 1285–1309, *Gaumenton* 1249, *Gampton* 1756. D 511 gives pr [gamtən].
(2) ~ Devon SX 6840. *Walementone* 1086, *Galmeton'* 1211, *Yalmeton* 1238, *Gaumeton* 1242, *Galmaton -me-* 1244–1325, *Gealmeton* 1428, *Gaylmington* 1281. D 304 gives pr [gæmptən].

GALPHAY NYorks SE 2572. 'Gallows enclosure'. *Galghagh(e) -haga* 12th–1469, *-ha(y)* 1457, *Gallwhae* 1467, *Gal- Gawhay* 16th cent., *Galfay* 1822. OE **galga** + **haga** with [f] replacing the peripheral phoneme [x] in *galga*. YW v.199 gives pr ['gɔ:fə].

GALTRES NYorks (lost SE 5663). 'Boar thicket(s), brushwood where boars lurk'. *Galtrys -is* 1171–1451, *Galtres* 1177–1577, *Gautris -es* 1227–1577. ON **gǫltr** 'a boar' (PrON **galtuR*) + OE **hrīs**. The name of an old royal forest N of York alluded to in the names STOCKTON and SUTTON ON THE FOREST, SE 6555, 5864. YN 8 gives pr [gɔ:triz].

GAMBLESBY Cumbr NY 6139. 'Gamel's village or farm'. *Gamelesbi* 1177–1564 with variant *-by*, *Gamblesby* 1580. ON pers.n. *Gamall*, secondary genitive sing. *Gam(a)lles*, + **bȳ**. The origin of this name and that of GLASSONBY is revealed by a writ of Henry I stating that the king has given to Hildred of Carlisle and his son Odard *terram que fuit Gamel filii Bern et terram illam que fuit Glassam filii Brictrici drengorum meorum* 'the territory of my drengs Gamel son of Bern and Glassam son of Brictric'. Hildred and Odard are known to have made payment for this grant in 1130 which thus dates the coinage of these two names to the first quarter of the 12th cent. Cu 192 gives pr [gaməlzbi], SSNNW 31.

GAMLINGAY Cambs TL 2452. 'The island of the Gamelingas, the people called after Gamela'. *Gamelinge(i)* 1086, *-ingeheia -ingee -ei(e) -eia -ey(e)* 1086–1432. OE folk-n. **Gamelingas* < pers.n. **Gamela* + **ingas**, genitive pl. **Gamelinga*, + **ēġ**. Ca 160.

GAMSTON 'Gamel's farm or village'. ON pers.n. *Gamall*, secondary genitive sing. *Gamalles*, + **tūn**.

(1) ~ Notts SK 7176. *Gamelestun(e)* 1086, *-ton* 1192–1337, *Gameston* 1541, *Gamston* 1569, *Gambston* 1586, *Gampstone* 1686. One of the three DB land-holders here was called *Gamel*. Nt 51, SSNEM 191.

(2) ~ Notts SK 6037. *Gamelestune* 1086, *-ton* 1246, 1297, *Gamelston* 1302, *Gamston* 1595, *Gamson* 1710. Nt 231, SSNEM 191.

GANAREW H&W SO 5316. 'Mouth of the hill or of the pass'. *Genoreu* c.1136, *Guenerui* 1186, *Genoire -oyre* [c.1220]13th Layamon, *Genereu -w* 1325, 1345, *Generrywe* 1397, *Gannerew* 1831 OS. W **genau** + OW **riu** 'steep slope; way'. In Geoffrey of Monmouth's *History of the Kings of Britain* the castle of Genoreu in 'Erging country . . . beside the River Wye, on a hill called Cloartius' was the site of Vortigern's fatal last stand against Aurelius Ambrosius. The reference is to Little Doward Camp, an Iron Age hill-fort ½ mile E of Ganarew†. The name Ganarew probably refers to the pass between the hill-fort and the high ground to the W partly shared by the river Wye‡. He 90, DEPN, HistKings 187–8, Thomas 1976.149, GPC 1391.

GANSTEAD Humbs TA 1434. 'Gagni's farmstead'. *Gagenestad* 1086, *Gag(h)en(e)sted(e)* 1196–1334, *Gaunested(e)* 1260, *Gaunstede* [1150×60]1347, 1421, *Ganste(a)d* from 1572. ON pers.n. **Gagni*, genitive sing. **Gagna*, + **stede**. YE 48, Stead 295, SSNY 145.

GANTHORPE NYorks SE 6870. 'Galm's outlying settlement'. *Gameltorp*, *Galmetona* 1086, *Galmet(h)orp* 1202–1344, *Gametorp'* 1200, *Ganthorpe* from 1577. ON pers.n. *Galmr* + **thorp**. The DB substitution of the commoner pers.n. *Gamel* may have been helped by the fact that a TRE tenant of this place was called *Gamel*. YN 34 gives pr [gɔnθrəp], SSNY 58.

GANTON NYorks SE 9877. 'Galma's settlement'. *Galmeton* 1086–1352, *Gaunton* 16th cent., *Ganton or Galmpton* 1822. OE pers.n. *Galma* or ON *Galmr* + **tūn**. YE 118 gives pr [gantn], SSNY 127.

GARBOLDISHAM Norf TM 0081. 'Gærbald's homestead'. *Gerboldesham* 1086, 1233, *Garboldesham* 1254. OE pers.n. **Gǣrbald*, genitive sing. **Gǣrbaldes*, + **hām**. DEPN gives pr [ga:bəlsəm].

GARE HILL Wilts ST 7840. *Gear Hill* 1817 OS.

†Little Doward, *Lyttledowarth* 1413, earlier *Douwarth* is OW **deu-arth** 'two heights' referring to Great and Little Doward. The W name was misread by Geoffrey who mistook initial *d* as if *cl* to produce his *Cloartius* c.1136 whence Layamon's *Cloard* [c.1220]c.1260, *Gloarþ* [c.1220]c.1275. Ganarew may have been the site of *Lanndougarth* [7th]c.1130, 'the church by the two hills'. He 90, O'Donnell 95.

‡The Welsh forms cited by B. G. Charles, O'Donnell 92, from the Welsh versions of Geoffrey, *castell genorwy, kastell Genorwyw, castell Goronw* and *castell Goronvy* are explained by him as 'Gwynwarwy's castle'; this must be a re-interpretation. See further, however, Jnl 31.113–4 and Celtic Voices 182.

GARFORD Oxon SU 4296. 'Gara's ford' or 'the ford at the triangular piece of land or strip of raised ground in the marsh'. *(æt) Garanforda* [940]13th S 471, *Garanford* 960 S 687, *Gareford(') -fordia -forde* [1066×87]13th, 1175–1396×7, *Garford(')* 13th and from 1517. OE pers.n. **Gāra* or **gāra**, genitive sing. *Gāran*, **gāran**, + **ford**. L 70 prefers derivation from the appellative *gāra* but **Gāra* would be a regular hypocoristic formation from such names as Gārwulf or Wulfgār possibly occurring also in GORING SU 6081. Brk 410.

GARFORTH WYorks SE 4033. 'Gæra's ford'. *Gereford(e)* 1086–1305, *Gerford* 1241–1428, *-forth* 1379, *Garford* 1336–1459, *-forth* 1632. OE pers.n. *Gǣra* + **ford**. Alternatively the specific may have been ON **geiri**, OEScand **gēre**, 'a triangular piece of ground' probably replacing OE **gāra**, referring to land between the roads that intersect at SE 393324. YW iv.95, xi.

GARGRAVE NYorks SD 9354. 'The triangular copse'. *G(h)eregraue* 1086–1276, *Gair- Gayrgrave -graue -a* c.1160–1466, *Gargrave* from 1182×5. ON **geiri** possibly replacing OE **gāra** + OE **grāf**. Alternatively the original specific may have been OE **gār** 'a spear' and so 'wood where spear-shafts are obtained'. YW vi.53 gives pr [ga:griv].

GARMSLEY CAMP H&W SO 6261. P.n. Garmsley + ModE **camp**, an undated probably Iron Age hill-fort. Garmsley is *Garmesley* or *Wrathes* 17th cent. The evidence is too late for certain explanation but Garmsley seems to be a pers.n., perhaps *Garmund*, + OE **lēah** 'wood or clearing'. Wo 80, Thomas 1976.148.

GARNETT BRIDGE Cumbr SD 5299. *Garnet bridge* 1651. Named from the Garnett family whose name also occurs in Garnett House in Strickland Ketel SD 4996, *Garnettes house* 1517. The family name can be traced back to *Gernet*, *Garnett* 1189×1200, 1446. We i.158, 155.

GARRAS Corn SW 7023. Probably 'rough moorland'. *Garros* 1571 Gover n.d., *Garrows Common* c.1696, *Garras* 1748. Co **garow** 'rough' reduced to **gar-** + ***ros** 'promontory, hill-spur, moor'. Identical with Breton Garros. Apparently an 18th cent. settlement built on common grazing-land. PNCo 85, Gover n.d. 571, CPNE 102, Dauzat-Rostaing 311.

High GARRETT Essex TL 7726. *High Garret* 1768. Earlier simply *Barrets* from the family of William Baret 1413×8. The original house, later demolished, was called *Low Barrets* when c.1428 John Baret built a new tall manor house called *High Barrets*, later *High Garrets*. Ess 414.

GARRIGILL Cumbr NY 7441. 'Gerard's ravine'. *Gerardgile* 1232, 1285, *Garrardgylle* 1443, *Garey- Garrygill* 1631. CGmc pers.n. *Gerard* + **gil**. The reference may be to the devil several times recorded with this name in ME literature. Cu 174, SSNNW 229.

The GARRISON Scilly SV 8910. *The garison in the Hugh or New Towne* 1642. ModE **garrison** (OFr *garison*). The reference is to the fort built in 1593–4 and the main fort of 1715–46. PNCo 85, Pevsner 1951.191.

GARROW TOR Corn SX 1478. P.n. Garrow SX 1477 + dial. **tor** 'a rocky outcrop'. Garrow, *Garroy* 1353, *Garrow* 1681, *Garrah* 1748, is unexplained. The form *Garros* in *Garros Moors* c.1640 can be compared with GARRAS SW 7023, but the earlier forms and the surname *Garra* 1327 are incompatible. PNCo 85, Gover n.d. 104.

GARSDALE Cumbr SD 7489. 'Garth's valley'. *Garcedale* c.1240–1331, *Garsedal(e)* 1241–51, *Garstall* 1399, *Garsda(y)le* 1568. ON pers.n. *Garthr*, genitive sing. *Garths*, + **dalr**. The earliest spelling represents ON *-ths-* by *-ce-* showing early simplification of *Garths-* to *Gars-*. YW vi.261.

GARSDALE HEAD Cumbr SD 7892. *Garsdale Head* 1858. P.n. GARSDALE SD 7489 + ModE **head**. YW vi.262.

GARSDON Wilts ST 9687. 'Grass hill' *Gersdune* [701]14th S 243, *-don* 1279–81, *Garsduna* [1066×87]13th, *-don* from 1248. OE **gærs** + **dūn**. Wlt 58, L 144.

GARSHALL GREEN Staffs SJ 9634. *Garshall Green* 1836 OS. P.n.

Garshall as in Garshall House SJ 9633, *Garnonshale* 1310, *Geringeshalgh -halew -ow* 14th, *Garingshall otherwise Gashall* 1601. 'Garnon's nook', surname Garnon, originally a nick-n. from OFr *gernon* 'a moustache', + ME **hale** (OE *halh*) + ModE **green**. DEPN, DBS s.n. Garnon, Horovitz.

GARSINGTON Oxon SP 5802. 'The grass hill'. *Gersedvn* 1086, *Gersendona -dun(a) -don(e) -den' -tuna -tun(e) -ton(')* [c.1110]13th–1267, *Gersindon(e) -dun(e) -tona -ton(e) -yn-* [1122]c.1325–1357, *Gersyngton -ing- -don(e)* [1122]c.1425–1428, *Garsin(g)ton -yn(g)- -den' -don* 1224–1476. OE ***gærsen** + **dūn**. O 174, L 144, Gelling 1998.79.

GARSTANG Lancs SD 4945. 'The spear-shaft', marking the site of some public function such as a legal assembly. *Cherestanc* 1086, *Gair- Geir- Gayrstang* [1170×84]1268–1292, *Gerstang* [1189×1200]1268–1278, *Garstange* 1494, *Garstan* 1577. ON **geirr** 'spear'+ **stǫng** (< ***stangu-**) 'a pole'. La 163 compares Girstenwood in Scotland which was *Gairstang* 1305. Cf. Langobardic *gairethinx* used to denote a legal assembly in the Edictus Rothari (643) and the laws of Liutprand (712–44). La 163, Lindkvist 47, Jnl 17.95, SSNNW 125, Nomina 10.174.

GARSTON Mers SJ 4084. Partly uncertain. *Gerstan* 1094–1367 etc., *Gerhstan* 1122, *Gerestan(am)* 1142, 1212, *Grestan* c.1155–1325, *Gerstun* 1297, *Gerston* 1202, 1324, *Garston* from 1262. Possibly 'great stone', OE **grēat**, N dial. **gert** 14th cent., + **stān**. Cf. the lost hundred name Greston in Gloucestershire, *Gretestan(e)* 1086, *Grestan* 1170–1265, *Greston* 1248–1568. No such stone is known today. La 111, Gl ii.1, Jnl 17.62.

East GARSTON Berks SU 3676 → EAST GARSTON.

GARSWOOD Mers SJ 5599. Partly uncertain. *Grateswode* 1367, *Gartiswode* 1479, *Garteswodde* 1508. Unidentified element + ME **wode** (OE *wudu*) 'a wood'. La 100.

GARTHORPE Humbs SE 8419. 'Gerulf's outlying settlement'. *Gerulftorp* 1086, *Geroldtorp* 1180. Pers.n. ON *Geirulfr* or CG *Gairulf* + **thorp**. DEPN, SSNY 99.

GARTHORPE Leic SK 8320. Possibly the 'outlying farm at the gore'. *Garethorp(e)* c.1130, 1199–1580, *-throp(p)* 1530–1714, *Geretorp* 1180–1209. OE **gāra** 'gore, triangular piece of land' + **thorp**. The reference is probably to the angle between two streams in which the village lies. The alternative explanation, ON pers.n. *Geiri*, genitive sing. *Geira*, anglicised to **Gāra*, would require spellings with *-ei- -ey-*. Lei 169, SSNEM 109, SPN LXXIX, 98.

GARTON 'Farm or village at the triangular piece of land'. OE **gāra** + **tūn**.

(1) ~ Humbs TA 2635. *Gartun, Garton* 1086 etc. The reference is possibly to the angle formed at the road junction here. YE 58.

(2) ~ -ON-THE-WOLDS Humbs SE 9859. *Garton in Waldo* 1208, *~ super Waldas* 1301, *~ on the Wold* 1538, *~ (in) le Woold* 1610. P.n. *Gartun(e), Garton* 1086 etc. + ModE **on the wold** referring to the high tract of chalk hills extending in a crescent from the Humber near Wauldby and Cave to the North Sea at Flamborough Head. YE 96, 13.

GARVESTON Norf TG 0107. 'Gerulf's estate'. *Ge- Girolfestuna* 1086, *Gerolvestone* 1254. lOE pers.n. *Gerulf* < ON *Geirúlfr*, ODan *Gērulf*, OG *Gairulf* or OE **Gǣrulf*, genitive sing. *Gerulfes*, + **tūn**. DEPN, SPNN 137.

GARWAY H&W SO 4522. 'Guoruoe's church'. *Garou* 1137, *Gar(e)wi -wy* 1160×70–1338, *Gorewy* 1320, *Langarewi* 1199. PrW ***lann** + pers.n. *Guoruoe*. *Lann Gu(o)rboe, Lann Gu(o)ruoe* c.1130 LL, is the same name probably located not here but at Eaton Bishop SO 4439. The later *Lann Guorboe* is named after its first incumbent as LL makes clear: *Gvoruodu rex . . . dedit . . . agrum . . . deo et sancto Dubricio . . . fundavit locum in honore sancte trinitatis, et ibi guoruoe saccerdotem suum posuit* 'King Gvoruodu gave the land to God and St Dyfrig . . . he founded the place in honour of the Holy Trinity and placed there his priest Guoruoe'. He 90, 14, CPNE 144, Nomina 10.69.

GASTARD Wilts ST 8868. 'The goats' tail of land'. *Gatesterta -e* 1154–1279, *Gastard* from 1428, *Gadsteed* 1601. OE **gāt**, genitive pl. **gāta**, + **steort**. Wlt 96.

GASTHORPE Norf TL 9780. 'Gadd's outlying farm'. *Gades- Gatesthorp* 1086, *Gaddesthorpe* [1087×98]12th, *Gatestorp* 1244, *Gadisthorp* 1275. lOE pers.n. *Gadd* < ON pers.n. *Gaddr*, genitive sing. *Gaddes*, + **thorp**. DEPN, SPNN 128.

GATCOMBE IoW SZ 4984. 'Goats' valley'. *Gatecome* 1086, *Gat(h)ecumb(e) -co(u)mb(e)* c.1220–1611, *Gatcumbe* c.1300, *Gatcombe* 1535. OE **gāt**, genitive pl. **gāta**, + **cumb**. Wt 132 gives pr ['gækəm], Mills 1996.53.

GATEACRE Mers SJ 4287. *Gateacre* 1842 OS. Possibly 'goat acre'. Mod dial. **gait** (ON *geit*).

GATEBECK Cumbr SD 5586. 'Stream by the road'. *Gatebeck* 1685. ModE **gate** + **beck**. We i.63.

GATEFORTH NYorks SE 5628. 'Goat ford'. *Gæite-ford* c.1030 YCh 7, *Geiteford'* 1166, *Gayte- Gaiteford* 1251–1428, *Gateforth* from 1470. ON **geit**, genitive pl. **geita**, possibly replacing OE **gāt, gāta**, + OE **ford**. A ford regularly used for moving goats. YW iv.27, SSNY 156.

GATEHOUSE Northum NY 7889. *Gatehouse* 1869 OS. The hamlet contains remains of several 17th cent. fortifications. Pevsner 1992.287.

GATELEY Norf TF 9624. 'Goats' pasture'. *Gatelea* 1086, *-leia* 1156, *Gotele* 1202. OE **gāt**, genitive pl. **gāta**, + **lēah**. DEPN, L 206.

GATENBY NYorks SE 3287. Possibly 'Gaithan's farm or village'. *Ghetens- Chenetesbi* 1086, *Gaitaneby* 1184, *Gaitenebi* 1228, *Gey- Gaytenby* 13th cent., *Gaittyngby, Gatonby* 16th. OIr pers.n. *Gáethín(e)* + ON **bȳ**. This explanation depends on the assumption that this rare pers.n. was early assimilated to ON *geit* 'a she-goat', genitive pl. **geitna* as in *geitna-njóli* (aegopodium) and *geitna-skóf* (lichen proboscideus). YN 227 gives pr [gɛətənbi], SSNY 28.

GATESCARTH PASS Cumbr NY 4709. Possibly 'pass frequented by goats'. *Gaitscarthe, Gaytskarthe* 1578, *Gatescar* 1823, ~ *Scarth* 1859. ON **geit** + **skarth**. But this could equally well be ME **gate-scarth** 'road pass' or an adaption of GOATSCAR NY 4706 from the other side of the pass in Longsleddale. The pass takes Brant Street over the mountains from Haweswater to Longsleddale. We ii.174.

GATESHEAD T&W NZ 2560. 'The goat's headland'. *Gateshevet* 1144×53, *-heued* [1183]c.1320, 1197–1508 with variant *-is-* 1296–1514, *Gateshed(e)* 1385–1587, *-head* from 1598, *Gate- Gait- Gaytsid(e) -syd* 1433–1733×4. The Latin forms are *Ad Caprae Caput* [c.731]8th BHE with literal translation *æt Rǣge heafde* [c.890]c.1000 OEBede, and *Ad Caput Caprae* c.1107. OE **gāt**, secondary genitive sing. **gātes**, + **hēafod**. OE *gāt* was originally a feminine noun with genitive sing. *gāte*. The Latin forms of Bede (731) and Symeon of Durham (c.1107) are learned translations of OE *æt Gāte-hēafde* and the form in the OE translation of Bede's *History*, *æt Rǣge heafde*, is in turn a translation of *Ad Caprae Caput* with OE feminine noun *rǣge* 'a hind'. The reference is to the headland overlooking the river Tyne at NZ 255636 where St Mary's Church now stands, on which a goat or goats must have been seen. The theory that the 'goat's head' refers to a pagan custom of exposing animal heads on poles is discussed Sr 403–6 and dismissed by Ekwall *Namn och Bygd* 14.129ff. The name *Gateshead* is not connected with the RBrit p.n. *Gabrosentum* 'goat's path' which is now identified with the Roman fort at Moresby Cumbria NX 9821. NbDu 92, TC 97, RBrit 364, Jnl 8.21, L 160–1.

GATESHEATH Ches SJ 4760. 'Heath near or allotted into cattle-walks'. *Gates Heath* 1831. ModE **gate** (ON *gata*) in the sense 'allotment of pasture' + **heath**. Che iv.96.

GATHURST GMan SD 5407. 'Goats' wooded hill'. *Gatehurst* before 1547. OE **gāt**, genitive pl. **gāta**, + **hyrst**. La 128, L 197.

GATLEY GMan SJ 8488. 'Goats' bank'. *Gatescliue* [12th]17th, *Gateclyue* 1290, *-clyf(fe)* 1381, 1383, *Gatley* from 1602. OE **gāt**, genitive pl. **gāta**, + **clif**. Che i.244, L 136.

GATTON Surrey TQ 2752 > MERSTHAM TQ 2953.

GATWICK WSusx TQ 2640 → LONDON (GATWICK) AIRPORT TQ 2640

River GAUNLESS Durham NZ 1024–2130. 'Profitless river'. *Gauhenles* c.1185, *Gawenless(e)* 1242×3, 1420, *Gaunles(se)* 1292–1647, *-less* from 1647. ON **gagn-lauss**. The possibity that this Scandinavian r.n. replaced an earlier British r.n. is discussed under the entry for Bishop AUCKLAND NZ 2029. NbDu 93, RN 169.

GAUNT'S COMMON Dorset SU 0205. *Gaunt Common* 1811. Cf. *landes namyd the great Gawntz* 1535, *Gantts farme* 1646. Probably named from John of Gaunt, duke of Lancaster 1372–99, who was granted a fair at nearby Holt in 1368 and who according to tradition had a house here. The neighbouring manor of Kingston Lacy belonged to the Duchy of Lancaster. Do ii.147.

GAUTBY Lincs TF 1772. 'Gauti's village or farm'. *Ganteby* (for *Gaute-*) [before 1129]1336, *Goutebi* 1195×6, *Gautebi* 1212. ON pers.n. *Gauti*, genitive sing. *Gauta*, + **bȳ**. SSNEM 81, Cameron 1998.

GAUZE BROOK Wilts ST 9083 > CORSTON ST 9284.

GAWBER SYorks SE 3207. 'Gallows hill'. *Galgbergh* 1304, *Galbergh* 1379, 1415, *-ber* 1607, *Gawber(d)hall* 1550, 1590. OE **galga** + **beorg**. YW i.316.

GAWCOTT Bucks SP 6831. 'Rent cottages'. *Chauescote* 1086, *Gavecote* 1255–1331, *Gauekote -cote* 1284, 1316, *Galcote* 1480, *Gocot* 1675. OE **gafol** + **cotu**, pl. of **cot**. Bk 60. Jnl 2.33 gives local pr [ga:kut].

GAWSWORTH Ches SJ 8969. 'Smith's enclosure'. *Govesurde* 1086, *Gous(e)worth* c.1265 etc. with variants *-w(u)rth(e)*, *Gowesworth* 1274, *Gaus(e)worth(e)* 1279, *Gawsworth* from 1389. PrW ***goϸ** 'a smith' + secondary genitive ending **-es** + **worth**. The specific may be used here as a pers.n., cf. the surnames *Goff(e)*, *Gough*. Che i.66 gives pr [gɔ:zwərθ], formerly ['gɔ:zuθ].

GAWTHORPE HALL Lancs SD 8034. P.n. Gawthorpe + ModE **hall**. An early 17th cent. house built for the Shuttleworth family on the outskirts of Padiham. Gawthorpe, *Gouthorp'* [1256]1439, *Goukethorp* 1324, *Gawthrop* 1472, is either 'cuckoo farm', ON **gaukr** + **thorp**, or 'Gaukr's outlying farm', ON pers.n. *Gaukr* + **thorp**. ON *gaukr* was also used to denote a fool or simpleton. La 83.

GAWTHROP Cumbr SD 6988. 'Cuckoo farm'. *Gawthorpe* 1592, *Gathrope* 1612, *Gauthrope* 1679. ON **gaukr** 'cuckoo' + **thorp** 'outlying farmstead'. YW vi.256.

GAWTHWAITE Cumbr SD 2784. Partly uncertain. *Golderswatt* 1552. The specific is ON **thveit** 'a clearing' possibly used in the sense 'the shelving part of a mountain'. The village lies on the eastern slope at the head of a pass on the Lowick-Broughton road. La 214.

GAYDON Warw SP 3653. 'Gæga's hill'. *Gaidon(e)* 1194–5, *Gaydon* from 1248, *Geadon* 1615. OE pers.n. **Gǣġa* + **dūn**. Wa 254.

GAYHURST Bucks SP 8446. 'Wooded hill where goats are kept'. *Gateherst* 1086, *Gathurst* 1237–55, *Gothurst(e)* 1290–1526, 1806, *Gaherst -hurst* 1167–1389, *Geyhurst* 1526, *Gayhurst* 1806. OE **gāt**, genitive pl. **gāta**, + **hyrst**. Bk 4 gives pr [geiə:st], L 197.

GAYLE NYorks SD 8889. 'The ravine'. *Ga(y)le* 1606. Short for earlier *Seldalegile* 1280, *Sleddalgayle* 1285, p.n. SLEDDALE SD 8587 + ON **geil**. The reference is to the narrow gap leading into the dale. YN 267.

GAYLE MOOR NYorks SD 7982. P.n. GAYLE SD 8689 + OE **mōr**.

GAYLES NYorks NZ 1207. 'The ravines'. *Gales* 16th cent., *Gailes* 1576. ON **geil**. A 1258 form *Austgail*, ON **austr**, refers to the easternmost of the three inhabited ravines in the township. YN 290.

GAY STREET WSusx TQ 0820. *Gay Street* 1813 OS. Situated on Stane Street, the Roman road from Chichester to London, Margary no. 15.

GAYTON LE MARSH Lincs TF 4284. 'G in the marsh'. *Gayton' in le Mersshe* 1378. P.n. *Gaiton'* 1202, *Geiton'* 1206, *Gayton* 1236, 'the goat farm', OE **gāt** replaced by ON **geit** + **tūn**, + Fr definite article **le** short for *en le* + ModE **marsh** for distinction from Gayton le Wold TF 2385, 'G on the Wold', *Gedtune*, *Gettune* 1086, *Gertuna* c.1115, *Gattun* 1154, *Gaitun(a)* c.1155–c.1180, *Gayton* 1156×8. Alternatively the specific could be OE pers.n. **Gǣġa*. DEPN, SSNEM 197, 184, Cameron 1998.

GAYTON Mers SJ 2780. Probably 'goat farm'. *Gaitone* 1086, *Gayton* from 1237 with variants *Gai- Gey-* and *-tone*, *Geaton* 1615–1727. ON **geit** 'a she-goat' possibly replacing OE **gāt** + **tūn**. Alternatively the specific might be the OE pers.n. **Gǣġa*. Che iv.275, SSNNW 187.

GAYTON Norf TF 7219. Partly uncertain. *Gaituna* 1086, *Geitun* c.1150, *-ton* 1198. This is normally taken to be 'the goat farm', OE **gāt**, genitive pl. **gāta**, + **tūn**. The proximity of GAYWOOD TF 6320, however, suggests that this might rather be 'Gæga's estate'. ME *ai* spellings might reflect this etymolgy or be due to the influence of ON *geit*. DEPN.

GAYTON Northants SP 7054. 'Gæga's farm or village'. *Gaiton(e)* from 1086 with variant *Gay-*, *Gainton* 1166, *Gauton* (for *Gan-*) 12th, *Garton* 1269×71. OE pers.n. **Gǣġa*, genitive sing. **Gǣġan*, + **tūn**. Nth 90.

GAYTON Staffs SJ 9828. Probably 'Gæga's estate'. *Gaitone* 1086, *ton* 1227, *Gayton* 1272. OE pers.n. **Gǣġa* + **tūn**. DEPN.

GAYTON SANDS Mers SJ 2578. *Gayton Bank or Big Ben* 1842. P.n. GAYTON SJ 2780 + ModE **sands(s)** and **bank**. A sandbank in the Dee estuary. Che iv.275.

GAYTON THORPE Norf TF 7418 → Gayton THORPE TF 7418.

GAYWOOD Norf TF 6320. 'Gæga's wood'. *Gaiuude* 1086, *Geywode* c.1105, *Gaiwde* c.1140. OE pers.n. **Gāega* + **wudu**. DEPN.

GAZELEY Suff TL 7264. 'Gægi's wood or clearing'. *Gaysle* 1219, *Gasel(e)* 1248, *Gaisle* 1254, *Gaiesley* 1610. OE pers.n. **Gǣġi*, genitive sing. **Gǣġes*, + **lēah**. DEPN.

GEDDING Suff TL 9458. 'The Gyddingas, the people called after Gydda'. *ge(l)dinga* 1086, *Geddingis* 1182×5, *Gedding'* 1185–1253, *Geddinges* 1190ff., *Gedding* 1610. OE folk-n. **Gyddingas* < pers.n. **Gydda* + **ingas**. DEPN, ING 53.

GEDDINGTON Northants SP 8983. 'The goat farm'. *Geitenton(e)* 1086–1356 with variants *Gait- -in-* and *-yn-*, *Geyt- Geitington -yng-* 12th–1337, *Gadinton(e)* C1086, 1195, *Gaidintun'*, *Geidinton* 1196–1227 with variants *Geyd- -yng-* 1376, 1397. OE ***gāting** 'goat place' < **gāt** + **ing**[2] influenced by ON **geit** 'a goat' or by the pers.n. *Geitr*, + **tūn**. Nth 165, Årsskrift 1974.50 comparing the Danish p.n. Geding (ODan *gēt*) and Swedish Getinge, SSNEM 184.

GEDNEY Lincs TF 4024. 'Gæda's island'. *Gadenai -ay* 1086, 1130–1344, *Gedenei* 1201, *-ey(e)* 1231–1432, *Gedney(e)* from 1227. OE pers.n. **Gǣda*, genitive sing. **Gǣdan*, + **ēġ**. Payling 1940.22, L 38.

GEDNEY BROADGATE Lincs TF 4022 → Gedney BROADGATE TF 4022.

GEDNEY DROVE END Lincs TF 4629. *Drove End* 1824 OS. P.n. Gedney Drove, *magna draua de Gedeneye* c.1360, p.n. GEDNEY TF 4024 + dial. **drove** 'a fen road' + **end**. The village is situated on the Wash, five m. NE of Gedney across Gedney Marsh. Payling 1940.26.

GEDNEY DYKE Lincs TF 4126. 'G ditch'. *Gedney dyk* 1272×1307. P.n. GEDNEY TF 4024 + ME **dike**. The reference is to Fleet Haven TF 4129, *portum de Flet* 1275, 1313, *Flete(s)havene* 1338, 1342, 'Fleet harbour', OE **flēot** 'a creek' + **hæfen**, formerly a tidal creek running across the foreshore up to the sea bank at Holbeach. Payling 1940.26, 29, Wheeler 127, Jnl 29.81.

GEDNEY HILL Lincs TF 3311. *Gedney Hill* 1604, 1786. P.n. GEDNEY TF 4024 + ModE **hill**. Also known as Fen End Chapelry, cf. *Gedney Hill Chapel* 1824 OS. Payling 1940.27, Wheeler Appendix I.17, Pevsner 1964.537.

GEDNEY MARSH Lincs TF 4429. *Mariscum de Gedeney* 1313, *le Mers(c)h(e)* 1294, 1350, *Ged(e)ney(e) fenne* 1379–1509, *Gedney Marsh* 1824 OS. P.n. GEDNEY TF 4024 + ModE **marsh**.

GEE CROSS GMan SJ 9593. *Gee Crosse* 1629, *Gee Cross* 1831. The site of a stone cross at a crossroads, associated with the family of *Dicon Gee* 1494 and *Robert Gee of Gee Crosse* 1629. Che i.303.

GELDESTON Norf TM 3991. 'Gyldi's estate'. *Geldestun* 1242, *-ton(e)* 1242, 1273. OE pers.n. *Gyldi*, Norfolk dial. form **Geldi*, genitive sing. *Gyldes*, **Geldes*, + **tūn**. DEPN.

River GELT Cumbr NY 5654. 'Madman, wild man'. *Gelt* from c.1210, *Kelt* c.1235. OIr **geilt**. Cu 14, RN 170.

King's Forest of GELTSDALE Cumbr NY 6053. *forresta mea de Geltesdale* c.1210, 1589, 1609, *Foreste de Guiltesdale*, *Guyltsdall* 1609, *Forest of Gweltesdale* 1610. ME **forest** 'a tract of land set aside for the preservation and hunting of wild animals' + p.n. Geltsdale as in GELTSDALE MIDDLE Cumbr NY 6051. Cu 38.

GELTSDALE MIDDLE Cumbr NY 6051. P.n. Geltsdale + ModE **middle** 'middle ground'. Geltsdale, *Geltesdale* 1285–1603, *Gelsdale* 1603, is the 'valley of the river Gelt', river n. GELT NY 5654 + ON **dalr**. Cu 87, SSNNW 229.

GENTLESHAW Staffs SK 0511. 'Gentle's grove'. *Gentylshawe* 1505, 1529, *Gentleshore* 1788. Surname Gentle as in John Gentyl 1341 + ModE **shaw** (OE *sceaga*). Originally a grove of ancient trees on Cannock Chase. Duignan 66, Horovitz.

GEORGEHAM Devon SS 4639. 'St George's Hamm'. *Hamme Sci Georgii* 1356, *Georgeham* 1535. Earlier simply *Hame* 1086, *Hamme* 1242. Saint's name *George* from the dedication of the church + **hamm** 'an enclosure', used as a p.n. D 43 gives pr [dʒɔːdʒ'hæm].

GEORGIA Corn SW 4836. Probably 'broken-down hedge'. *Gorga moor*, *The Gorga Craft* c.1696, *Georgia (Croft)* 1841. Co ***gor-ge** 'low or broken-down hedge' (< **gor-* + *kee* 'hedge, bank'), dial. **gorgoe**, **gurgey** 'low hedge, rough fence for waste land'. Subsequently influenced by the name of the state of Georgia named after George II in 1732. *Craft* for *croft* is frequent in dial. Gover n.d. 665 mentions a mine here called *Great Georgia* in 1765. PNCo 86, CPNE 44, 110.

GERMANSWEEK Devon SX 4394. 'St German's Week'. *Wyke Germyn* 1458, *Wykejarmen* 1474, *Germaneswyk* 1468, *Germans Weeke* 1699. Earlier simply *Wiche* 1086. Saint's name *Germanus* + OE **wīċ** 'a dairy farm' used as a p.n. 'German's' for distinction from WEEK SS 7316 and Southweek SX 4393, *Wiche* 1086, *Sudwik'* 1242, S of Germansweek, and Westweek SX 4293, *Westwyke* 1411 W of Germansweek. D 183, 181.

GERMOE Corn SW 5829. 'St Germoe's (church)'. *(capella) Sancti Germochi* [c.1176]1300, *Germogh* 1283–1404, *Germow* 1535, *Sent Germowe* 1549, *Garmow* 1610. Saint's name *Germoch* from the dedication of the church. This is the only occurrence of the saint's name who may have been connected with St Breage. PNCo 87, Gover n.d. 509.

GERNA Lancs SD 7744 → DOWNHAM SD 7844.

GERRANS Corn SW 8735. 'St Gerent's (church)'. *Seint Geren* 1201, *(ecclesia) Sancti Gerenti* 1202, ~ ~ *Gerendi* 1261, *Seynt Gerens* 1386, *St Gerance* 1578, *Gerans* 1610. Saint's name *Gerent* from the dedication of the church. He is probably the historical Cornish king of the early 8th cent. PNCo 87, Gover n.d. 450, DEPN, CMCS 12.45.

GERRANS BAY Corn SW 9037. *Gerrans Bay* 1813. P.n. GERRANS SW 8735 + ModE **bay**. PNCo 87.

GERRARDS CROSS Bucks TQ 0088. *Gerards Cross* 1692, *Gerrard's Cross* 1822 OS. Family name *Gerrard* (*Jarrard* 1552, *Jarret* 1566) + ModE **cross** 'a crossroads'. Bk 238, D l, O xlv, Room 1983.44.

GESTINGTHORPE Essex TL 8138. 'Outlying settlement of the Gyrstlingas, the people called after Gyrstla'. *(œt) Gyrstlingaþorpe* [975×1016]11th S 1487, *Gristlyngthorp* [1035×44]13th S 1521, *Ghestingetorp*, *Glestingethorp* 1086, *Gesting(e)thorp(e)* 1218–1325 etc. with variants *Gestling(e)-* 1231–1678, *Gestning(ge) -yng-* 1238–1339, *Gesthorp(e)* 1458, 1559. OE folk-n. **Gyrstlingas* < pers.n. **Gyrstla* + **ingas**, genitive pl. **Gyrstlinga*, + **thorp**. An alternative possibility is that the Gyrstlingas were the 'inhabitants of the gorse place', OE ***gyrstling** < **gorst** + **ling**. An OE *gyrstling* may have been used as a p.n. like CLAVERING TL 4832, the 'clover place'. Ess 430 gives pr [gestəp], ING 35.

The GIANT Dorset ST 6601. The reference is to the 180ft. long turf-cut figure of a giant holding a knobbed club resembling a Roman Hercules figure (possibly carved out as a memorial to the Emperor Commodus who declared himself Hercules incarnate after defeating the Scots c. AD 187). Newman-Pevsner 1972 1972.135.

GIBBET HILL Surrey SU 9035. 'Gallows hill'. ModE **gibbet** + **hill**. The hill rises to 895ft. near the Devil's Punch Bowl. No early forms.

GIBBON HILL NYorks SE 0196. *Gibbon Hill* 1860 OS.

GIBRALTAR Lincs TF 5558. *Gibraltar*, *Gibraltar Point* 1824 OS. Transferred from the British Colony on the S tip of Spain to a similar location at the extreme N boundary of the Wash. Gibraltar, Arabic *Jabal Tariq* 'the rock of Tarik', is named after Tariq ibn Ziyad, the leader of the Moors who seized and fortified it in AD 711. Annemarie Schimmel, *Islamic names*, Edinburgh 1989, 33.

Great GIDDING Cambs TL 1183. *Magna Geddinge* 1252. ModE adj. **great**, Latin **magna**, + p.n. *Redinges* (sic), *Gedelinge*, *Geddinge* 1086, *Geddinge* 1147–1399, *Gidding -yng* from 1285, 'the Gydelingas or Gyddingas, the people called after Gydela or Gydda'. OE folk-n. **Gydelingas*, SE dial. form **Gedelingas*, < pers.n. **Gydela*, **Gedela* + **ingas**, or **Gyddingas*, **Geddingas*, < pers.n. **Gydda*, **Gedda* + **ingas**. 'Great' for distinction from Little GIDDING TL 1382 Hu 240.

Little GIDDING Cambs TL 1382. *Parva Gydding* 1272×1307. ModE adj. **little**, Latin **parva** + p.n. Gidding as in Great GIDDING TL 1183. Hu 240.

Steeple GIDDING Cambs TL 1381. 'G with the tower'. *Stepelgedding* 1260, *Stepel Guiddyng* 1291. ME **stepel** + p.n. Gidding as in Great and Little GIDDING TL 1183, 1382. Hu 241.

GIDEA PARK GLond TQ 5390. *Guydie hall parke* 1668, *Gidea Park* 1881. P.n. Gidea + ModE **park**. Gidea, *(la) Gid(i)ehall'*, *Ged(i)ehall(e)* 1258, *Gydihall by Ramford* (sic) 1466, *Gedyhall* 1478, *Geryhall* 1510, is 'the giddy hall', ME **gidi** (OE *gydiġ*) + **hall**. Two other examples are recorded, in Little Clacton, Essex, *Gyddyhall alias Eynegaynehall* 1589 (from the surname *Engaine*), *Geddyehall* 1643, and Giddeahall in Yatton Keynell, Wilts, *Giddy Hall* 1773. The term is perhaps used for a foolish building like the common minor name Folly. Spelling variation between *e* and *i* is regular in names derived from OE *y*. Ess 117, 336, Wlt 114, TC 205.

GIDLEIGH Devon SX 6788. 'Gydda's wood or clearing'. *Geddelegœ* 1158, *Gi- Gyddeleia -legh(e) -ley* 1167–1324, *Gud(d)eleg(h) -leghe* [c.1200]15th, 1284, 1330. DB Devon 15.7 identifies *Chiderleia* 1086 with Gidleigh, but the identification has not been independently established. OE pers.n. **Gydda* + **lēah**. D 438.

GIGGLESWICK NYorks SD 8063. 'Gikel's dairy-farm'. *Ghigeleswic* 1086, *Gic- Gik(e)leswic(h) -wik(e) -y-* 12th–1428, *Gig(g)(e)leswi(c)k(e) -y-* 1221–1659. Pers.n. *Gikel*, genitive sing. *Gikeles*, + OE **wīċ**. YW vi.144, SSNY 145.

GILBERDYKE Humbs SE 8329. 'The dike by the land of Gilbert'. *Gilbertdike -dyke* 1376–1619, *Gilberdyke* from 1349. Pers.n. *Gilbert* + p.n. *Dyc*, *Dyk(e) -i-* 1234–1559, 'the dike', OE **dīċ**. YE 246.

GILBERT STREET Hants SU 6532 → NORTH STREET SU 6433.

GILBERTSTONE H&W SO 3727 → WALTERSTONE SO 3425

GILCRUX Cumbr NY 1138. 'Retreat by a hill'. *Killecruce* c.1175, *Gillecruz -cruce* 1230–c.1280, *Gilcrux* from 1247, *-cr(o)uce -crous -crowse -crewse* c.1280–1558, *-cross* 1378. OW **cil** replaced at an early date by ON **gil**, + **crüg** later associated with Latin *crux* 'a cross'. Cu 287 gives pr [gilkru:s], L 138, GPC 478 s.v *cil*, 631 s.v. *crug*.

GILDERDALE FOREST Cumbr NY 6844. P.n. Gilderdale + ModE **forest**. Gilderdale, *Gilderdale* from 1279, *Gilder(e)sdale* 1285, 1292, 1695, is 'valley of the trap or snare', ON **gildri**, genitive sing. **gildra**, + **dalr**. Cu 175.

GILDERSOME WYorks SE 2429. '(The place at the) guild-houses'. *Gildehusum* 1181, *Gild(h)us* 13th cent., *Gildesom* 1315, *Gyldosum* 1323, *Gildersome* from 1504. ON **gildi-hús**, dative pl. **(í) gildi-húsum**. YW iii.223.

GILDINGWELLS SYorks SK 5585. 'Gushing springs'. *(la) Gildanwell* 13th, 14th cents., *-welles* 1345, *Gyldyn- Gyldenwelles* 1403–1588, *Gildingwell(e)s* 1546–1641. OE ***gildande** + **wella**. YW i.149.

GILGARRAN Cumbr NY 0323. 'Garran's ravine'. *Gillegarran* before 1230, *Gilgarran* from 1321. ON **gil** + surname *Garran* (<Gaelic nickname *gearrán* 'gelding', cf. Cumbrian dial. *garron* used of an ungainly horse or person) in Celtic word order. Cu 375, SSNNW 126.

GILKICKER POINT Hants SZ 6097. P.n. Gilkicker + ModE **point**. Gilkicker, *Gilkicker* 1679, 1759, is named after Gilkicker Tower, a landmark for ships built in 1669 and destroyed in 1965. The name is unexplained. Ha 82, Gover 1958.27, Pevsner-Lloyd 257.

GILLAMOOR NYorks SE 6890. 'Gedling moor'. *Gedlingesmore* 1086, *Gillingamor* late 12th, *-inge-* c.1170, 1231, *Gillingmore -yng-* 1195–1399, *Gillemore* 1282, *Gillimore* 1577. OE p.n. **Gedling*, genitive sing. **Gedlinges*, varying with locative–dative sing. **Gedlinge*, + **mōr**. **Gedling* is 'the place called after Getla', OE pers.n. **Gētla* + **ing**[2]. YN 64 gives pr [giləmuə], BzN 1967.336.

GILLING EAST NYorks SE 6177. P.n. *G(h)ellinge* 1086, *Gy- Gilling'* from c.1140, *Gilinge* [1090]11th, *Gillinges* 1239, the 'place called after Getla', OE p.n. **Gēdling* < pers.n. **Gētla* + **ing**[2], locative–dative **Gēdlinge*, varying with folk-n. **Gētlingas* 'the people called after Getla', pers.n. **Gētla* + **ingas**, + ModE **east**. This folk-n. also occurs in GILLING WEST NZ 1805, YETLINGTON Northumb NU 0209, and in Bede's *Ingetlingum* [c.731]8th BHE and *Ingætlingum* HA. It appears to be a parallel to OSwed *Getlinge* (Hellqvist, *Svenska ON på Inge* 34). YN 53, SSNY 146, BzN 1967.336.

GILLING WEST NYorks NZ 1805. P.n. *G(h)elling(h)es* 1086, *Gellyng(h)es* 12th, *Gwyllingues* 1137×46, *Gillinges* 1200, 1241, *Gi- Gylling(e) -ynge* from 1166, as in GILLING EAST SE 6177, + ModE **west**. Bede records (BHE 3.14) that king Oswin of Deira was murdered in the home of a nobleman at *Ingetlingum* after disbanding an army that he had raised at *Wilfaresdun* ten miles NW of the village of Catterick. This has generally led to the identification of *Ingetlingum* '(the place) among the Getlingas' (dative pl. *Getlingum*) with Gilling West. YN 288, BzN 1967.337.

GILLINGHAM Dorset ST 8026. 'Homestead of the Gyllingas, the people called after Gylla'. *Gillinga hám* 11th ASC(D) under year 1016, *Gillingaham* 12th†, *Geling(e)ham* 1086, *Gy- G(h)illingham* 1152–1310, *Gillingham* from 1198. OE folk-n. **Gyllingas* < pers.n. **Gylla* + **ingas**, genitive pl. **Gyllinga*, + **hām**. Do iii.9 gives pr ['giliŋəm].

GILLINGHAM Kent TQ 7767. 'Homestead of the Gyllingas, the people called after Gylla'. *gillingeham, (to) gyllinge ham* [c.975]12th B 1321–2, *Gelingehã* 1086, *Gel(l)ingeham* 11th, *Gillingeham* c.1100–1228 etc., *Gillingham* from 1202. OE folk-n. **Gyllingas* < pers. **Gylla* + **ingas**, genitive pl. **Gyllinga*, + **hām**. PNK 128, KPN 303, ING119, ASE 2.30, TC 98.

†*Gillingaham* [993]10th S 876 may also belong here.

GILLINGHAM Norf TM 4191. 'Homestead of the Gyldingas or Gyllingas, the people called after Gylda or Gylla'. *Kildincham, Gillingaham* 1086, *Gelingeham* 1107×18, *Gil(l)ingham* from 1198, *Gelyngham* 1375. OE folk-n. **Gyldingas* or **Gyllingas* < pers.n.**Gylda* or **Gylla* + **ingas**, genitive pl. **Gyldinga*, + **hām**. Both *Gylda* and *Gylla* may be forms of **Gȳthla*. DEPN, ING 135.

GILLOW H&W SO 5328 → MICHAELCHURCH SO 5225.

GILLOW HEATH Staffs SJ 8858. *Gylloowe Heht* 1427, *Gilloe heath* 1675. P.n. ? *Gilleloh* 1227, *Gillowe* 1279 + ME **hethe** (OE *hǣth*). Cf. Gillow H&W SO 5328, *Gilhou* 1228, 1350 for which a W origin has been proposed, OW **cil** + **luch**, 'the lake nook'. Horovitz.

GILMORTON Leic SP 5787. 'Golden Morton'. *Aurea Morton* 1248, *Guldenemorton* 1293, *Gy- Gilden(e)morton* 1303–1509, *Gy- Gild(e)morton* 1266–1641, *Gy- Gil(e)morton* from 1515×18. ME **gilden** (OE *gylden*) 'golden, i.e. wealthy, prosperous' + p.n. *Morton(e)* 1086–1576, 'the marsh settlement'. Lei 442.

River GILPIN Cumbr SD 4687. Uncertain. *the watt^r of Gylpyne* 16th. Probably a back-formation from Gilpin Beck, *Gilpin beck(e)* 1614, or Gilpin Bridge, *Gilpin Bridge* 1718. *Gilpin* is a well evidenced but unexplained surname in S Westmorland from the 13th cent. There may be a link with Swed dial. *gölpa* 'deep place in a river, small lake'. We i.7, RN 172.

GILSLAND Cumbr NY 6366. 'Gille's land'. *Gillesland(e) -lond -laund* c.1165–1589, *Gilsland* 1618. Pers.n. *Gille*, genitive sing. *Gilles*, + ME **land**. Probably named from Gille son of Bueth mentioned in the foundation charter of Lanercost Priory c.1166. Cu 2, SSNNW 126.

GILSLAND SPA Cumbr NY 6367. *Gilsland Spa* 1794. P.n. GILSLAND + ModE **spa**. Chalybeate and sulphuric springs were discovered here in 1812 and the village became a spa for a while. Cu 3, Pevsner 1967.127, Gaz s.n.

GIMINGHAM Norf TG 2836. 'Homestead of the Gymmingas, the people called after Gym(m)i or Gymma'. *Giming(h)eham* 1086, *Gemingheam* 1121, *Gimmingeham* 1188, *Gummingeham* 1192ff, *Gimingham* from 1206. OE folk-n. **Gymmingas* < pers.n. **Gymi*, **Gymmi*, or **Gymma* + **ingas**, genitive pl. **Gymminga*, + **hām**. DEPN, ING 135.

East GINGE Oxon SU 4486. *Estgeyng'* 13th cent., *Estgeynch* 1325, *Estgynge* 1517, *Est–Gins* 18th. ME adj. **est** + p.n. *(ad) Gainge* [815]12th S 166, *Geinge* [821]12th S 183, *Gaincg* [955]12th S 567, *Gainge, (to) Gæinge* [956]12th S 583, *Gaing* [959]12th S 673, 1212, *Gæging* [959]13th S 673, *Gainz* 1086, 1157, *Geing'* 1241×3, *Ginge* from 1217, *Gindge* 1617, by origin a stream name, OE **Gǣġing* from OE *gǣġian* 'to turn aside'. It meant 'the bending stream' and survives as Ginge Brooke for which there are early forms *(iuxta riuulam) Geenge* 'beside Ginge brook' [726×37]12th S 93 and *(on) Gæing broc* [956]12th S 583. The present form and earlier spelling *Gainz* represent the locative–dative form **Gǣġinġe* 'at Ginge'. Brk 10, 469, RN 172.

West GINGE Oxon SU 4486. *Westgenge* 1247×8. ME adj. **west** + p.n. Ginge as in East GINGE SU 4486. Brk 469.

GIPPING Suff TM 0763. 'The Gyppingas, the people called after Gyppa or Gyppi'. *Gippinges* [1154×89]13th, *Gypping, Gippingneweton* 1272×1307, *Gippinge* 1316, *Gyppinges Newtone* 1318, *Gipping* 1610. OE folk-n. **Gyppingas* < pers.n. **Gyppa* or *Gyppi* + **ingas**. DEPN, ING, 54 RN 172.

River GIPPING Suff TM 1152. A back-formation from the p.n. GIPPING TM 0763. *Gipping* from 1586, *Gippen* 1764. RN 172.

GIPSEY BRIDGE Lincs TF 2850. *Gipsey Br.* 1824 OS. Cf. GYPSEY RACE Humbs TA 0970, OE **ġips** + **ēa** 'an intermittent stream'.

GIRDLE FELL Northum NT 7001. *Girdle Fell* 1869 OS.

GIRSBY NYorks NZ 3508. 'The pig-farm'. *Grisebi -by* 1086 etc. ON **gríss**, genitive pl. **grissa**, + **bȳ**. Alternatively the specific might be ON pers.n. *Gríss*. YN 280, SSNY 28.

GIRTON Cambs TL 4262. 'Gravel settlement'. *Gri- Gre- Gryttune* *[1060]13th S 1030, *Gri- Gryttona -(e)* 1135–1541, *Grettona -(e)*

1086–1382, *Gretone* 1086, *Gerton* 1399, *Gi- Gyrton* from c.1460. OE **grēot** + **tūn**. Ca 176.

GIRTON Notts SK 8266. 'The settlement on the gravel'. *Gretone* 1086, *Grettuna, Gretton(a)* 1145–1576, *Girtone* 1525, *Gyrton* 1538, *Girton oth. Gretton* 1604. OE **grēot** + **tūn**. Nt 204.

GISBOROUGH MOOR Cleve NZ 6213. P.n. GUISBOROUGH NZ 6015 + **mōr**. Cf. entry for GLAISDALE MOOR below.

GISBURN Lancs SD 8348. Probably 'rushing stream'. *Ghiseburne* 1086, *Gi- Gyselburn(e)* 12th–1338, *Gi- Gyseburn(e)* [12th]14th–1305, *Gi- Gysburn(e) -borne*[1176]14th–1587. Probably OE ***gysel** + **burna**. Alternatively the specific is a pers.n., OE **Gysla* or **Gīsla* or ON *Gísli*. YW vi.164, L 19.

GISBURN FOREST Lancs SD 7457. *foresta de Gi- Gyseburn* [c.1150–1269]14th, *forrest of Gysborne* 1560, *Gisburne Forrest* 1585. P.n. GISBURN + ME **forest**. YW vi.167.

GISLEHAM Suff TM 5188. 'Gysela's homestead'. *Gisleham* from 1086 with variants *Gy-* and *-el-*, *Gyslam* 1568, *Gislam* 1610. OE pers.n. **Gys(e)la*, a hypocoristic form of *Gūthsiġe* or the like, + **hām**. DEPN gives pr [gizləm].

GISLINGHAM Suff TM 0771. 'The homestead of the Gyselingas, the people called after Gysla'. *Gyselingham* [1043×7]13th S 1470, *Gisling(a)- Gisling(h)e- Gisling- Gis(i)linc- Gissilincham* 1086, *Gislingeham* [1087×98]12th, *Gis(e)lingham* from 1189×99. OE folk-n. **Gyselingas* < pers.n. *Gys(e)la* as in GISLEHAM TM 5188 + **ingas**, genitive pl. **Gys(e)linga*, + **hām**. DEPN, ING 130, Baron.

GISSING Norf TM 1485. 'The Gyssingas, the people called after Gyssa or Gyssi'. *Gers- Gessinga* 1086, *(ad) Gessinge* [1087×98]12th, 1198, 1242×3, *Gessinges* 1180, 1190, *Gissinges* 1211, *Gissing(e)* from 1196. OE folk-n. **Gyssingas* < pers.n. **Gyssa* or **Gyssi*, + **ingas**. DEPN, ING 57.

GITTISHAM Devon SY 1398. 'Gyddi's enclosure or homestead'. *Gideshā* 1086–1375 with variants *Gy-* and *-ham, Giddesham* 1238, 1242, *Gitesham* 1249, *Gitsam(e)* 1545, 1675. OE pers.n. **Gyddi*, genitive sing. **Gyddes*, + **hamm** or **hām**. D 589 gives pr [gitsəm].

Great GIVENDALE Humbs SE 8153. *Great Gevydale* 1564, *Great Givendaile* 1650. ModE adj. **great** + p.n. *Ghiuedale* 1086, *Geueldal(e) -v-* 1120×9–1363, *Gaveldal* 1198, *Ganedale* (for *Gaue-*) 1203×4, 1212, *Giv- Gi- Gyueldal(e)* 13th cent., *Gevendale* 1231–1421, the 'valley of the r. **Gævul*', ODan r.n. **Gævul* 'generous, rich in fish' + ON **dalr**. 'Great' for distinction from Little Givendale *Geuedale* 1086, *Estgeveldale* 1296, *parua Geueldale* 1342, *Little Gevydale* 1564, *Little Gevendaile* 1565. YE 177 gives prs [gindəl, gi:ndəl] and [geldən], 179, SSNY 94, L 96.

GLAISDALE NYorks NZ 7603, 7705. 'Valley of the r. *Glas*'. *Glasedale* 12th–1665, *Glasdale* 1223, 1227, 1369. R.n. **Glas* representing PrW ***glas** 'blue, green' or OE ***glæs** 'clear, bright, shining', + ON **dalr**. YN 132, L 96.

GLAISDALE MOOR NYorks NZ 7201. P.n. GLAISDALE NZ 7603, 7705 + ModE **moor** (OE *mōr*).

GLAISDALE RIGG NYorks NZ 7404. P.n. GLAISDALE NZ 7603, 7705 + Mod dial. **rigg** (ON *hryggr*).

GLANDFORD Norf TG 0441. 'The ford where sports are held'. *Glam- Glanforda* 1086, *Snitesle Glaumford* 1254, *Glamford* 1257, 1275. OE **glēam** + **forda**. For similar names cf. BRIGG Lincs TA 0007, formerly *Glan- Glaumford Bridge* 13th, PLAITFORD Hants SU 2719, PLAYFORD Suff TM 2148. DEPN.

GLANTON Northum NU 0714. Probably the 'look-out hill'. *Glentendon* 1186, 1278, *Glentedone -dun* 1210, 1212, 1236 BF, *Glante(n)don* 1200 Ch, 1219, 1278, *Glatendon'* 1242 BF, *Glantoun* 1320. OE ***glente**, genitive sing. ***glentan**, + **dūn**. The village lies beside a prominent hill offering extensive views N and S along the Roman road from Corbridge to Berwick. An alternative possibility is OE ***glente** 'a hawk' and so 'hawk hill'. NbDu 94, DEPN, PNE i.203.

GLANTON PIKE Northum NU 0514. *Glanton Pyke* 1868 OS. P.n. GLANTON NU 0714 + N dial. **pike** 'a pointed hill'.

GLAPTHORN Northants TL 0290. Partly uncertain. *Glapthorn* from 1185, *Glape- Clapethorn(e)* 13th cent. The forms point to an OE ***glæp-thorn**, an otherwise unrecorded plant name, rather than to *Glappan thorn* 'Glappa's thorn-tree'. Cf. GLAPWELL Derby SK 4765. Nth 203.

GLAPWELL Derby SK 4765. 'Buck-bean spring'. *Glapewelle* 1086, *Glapwell(e)* 1186–1243 etc., *-walle* 1227, 1586, *Clapwell(e)* 1236–1399. OE **glæppe** 'buck-bean' (*Menyanthes trifoliata*), also called Marsh Trefoil or Marsh Clover, a water-plant common in bogs, having medicinal properties, + Mercian **wælla**. Alternatively the specific might be OE pers.n. *Glæppa*. Db 258, Grieve 117 s.n. Bogbean.

GLARAMARA Cumbr NY 2410. *Gleuermerghe* 1211, *Glaramara* 1784. Also *Hovedgleuermerhe* 1209×10, 'Glaramara head', p.n. Glaramara + OWScand **hǫfuth**. The mountain name Glaramara is probably '*Glaram* shieling', in which *Glaram* is ON **gliúfrum** '(at) the ravines', locative–dative pl. of **gliúfr**, and the generic ON **ǽrgi** 'a shieling, a mountain pasture'. Cu 350.

GLASCOTE Staffs SK 2303. Partly uncertain. *Glascote* from 1206 including [1154×89]1398. Uncertain element + OE **cot** 'a cottage', pl. **cotu**. The specific could be OE **glæs** and the sense 'huts where glass is made'. Wa 26.

GLASSHOUSE HILL Glos SO 7020. P.n. Glasshouse SO 7021, *the Glasshouse* 1755. ModE **glasshouse** + **hill**. A glasshouse was a building where glass was made. Gl iii.188.

GLASSHOUSES NYorks SE 1764. 'Houses where glass is made'. *Glassehouse(s)* 1387, 1533, *Glashows* 1526. OE **glæs** + **hūs**. YW v.149.

GLASSON Cumbr NY 2560. Uncertain. *Glassan* 1259–1363, *Glasson* 1580. Possibly originally a river name, 'green river', from Irish *glass*, W *glas* 'grey, blue'. Cu 125, GPC 1401.

GLASSON Lancs SD 4456. Uncertain. *Glassene* [c.1265]1268, *Glasson* 1552. Old Glasson SD 4455 is situated on a slight rise overlooking both the Lune and the Conder. It may be from an OE ***glǣsne** 'bright, shining place', a derivative of **glǣs* 'clear, shining, bright' referring to the water of either of these rivers. In G names, however, it has been suggested that *glas(en)-* derives from *calasna*, a word meaning 'boundary'. Cf. GLAZENWOOD Essex TL 8022. La 171 gives pr [glazən], PNE i.203 s.v. ***glǣs**, BONF 17.51.

GLASSONBY Cumbr NY 5738. 'Glassan's village or farm'. *Glassanebi -by* 1177–1273, *Cl-* 1197–1279, *Glassenbi -by -an-* 1202–1432, *Glassonby* from 1548. OIr, Old Celtic pers.n. *Glas(s)ān* + **bȳ**. Possibly the same person as is mentioned in a writ of Henry I quoted in a plea of 1201, stating that the king has given to Hildred of Carlisle *terram que fuit Gamel filii Bern et terram illam que fuit Glassam filii Brictrici drengorum meorum*, 'the lands formerly held by my drengs Gamall son of Beorn and Glassan son of Beorhtric'. Cu 194, 192, SSNNW 31–2.

GLASTON Leic SK 8900. Probably 'Glathr's estate'. *Gladestone* 1086, *Glathestun -ton* c.1100, 1273, *Glaston(e)* from 1225, *Glaceton* 1254–1302, *Glaiston -ay-* 1506–1724, *Glason* 1515, 1620. ON pers.n. *Glathr*, genitive sing. *Glaths*, + **tūn**. R 248 gives pr ['gleistən], SSNEM 191.

GLASTONBURY Somer ST 4938. 'The fortified place of the Glæstingas'. *(in) Glastingaburghe* [705]14th S 247, *Glastingburi* [725]14th S 251, *Glestingaburg* [732×55]n.d., *Glastingaburh* [744]14th S 1410, *(æt, on) Glæstinga byrig* 9th ASC(A) under year 688, c.1000 Saints, 12th ASC(E) under year 1016, *(into) Glæstyngabyrig* 971×83, *Glastingberie* 1086, *Glassenburye* 1610. The church of Glastonbury is *Ecclesia Glastingberiensis* 1086. OE folk-n. **Glæstingas*, genitive pl. **Glæstinga*, + **byriġ**, dative sing. of **burh**. Other relevant forms are *Glastingai* [678]13th S 1666, *Glastingaea* *[704]13th S 246, *Glasteie* *[725]12th S 250, *Glastingei* [745]14th S 257, 'the island of the Glæstingas', **Glæstinga* + **īeġ**. The Glæstingas are 'the people of Glastonia', the British-Latin name of the place, *(in) Glaston'* [798]14th B 284, *villa Glaston'* [944]14th S 499, *Glastonia* [1125]12th, 1199×1200 etc., *Glastonia, i.e. Urbs Vitrea* 'G, that is, the city of glass', Caradoc, *Life of Gildas*. This is probably a misunderstanding since Glastonia is almost certainly a derivative of OCeltic **glasto-* as in Gaulish *glastum* 'woad'. Cf. Treglasta Corn SX 1886, *Treglastan* 1086 Exon, 1226–72, *Tregalsta* [1300×1]14th. Alternatively the name has been associated with Co **glastan** 'oak-trees'. However, Latin *vitrum* can also mean 'woad'. The W forms of the name, *Ineswytrin* [601]14th B 835, *Ineswitrin* [601]12th B 836, *Ynisgustrin* Caradoc, *Ineswitrim* [1125]12th, explained as 'glassy island', PrW ***ïnïs** (Brit **enistī* 'island') + W **gwydrin** (earlier *gutrin*) 'of glass', are probably folk-etymological translations into W of the English name **Glæstinga-īeġ*. DEPN, Turner 1950 116, 1952–4.16, CPNE 104.

GLATTON Cambs TL 1586. 'Pleasant settlement, farm or estate'. *Glatune* 1086, *Glatton* from 1167. The etymology is proved by the form *glædtuninga weg* 'the road of the Glædtuningas, the people of Glatton' 957 S 649 in the bounds Conington. OE **glæd** + **tūn**. Hu 187.

River GLAVEN Norf TG 0540. Possibly a back-formation from GLANDFORD TG 0441. Glaven appears to be a hyper-correct form for *Glam, Glaum*, owing to the regular loss of intervocalic *-v-* in forms like *sen* for *seven* in *sennight, Denshire* for *Devonshire, e'en* for *even* and p.ns. like Lawley Shropshire, *Lavelei* 1086, Rainow Ches, *Ravenhoh* 1288, Ranskill Notts, *Ravenschel* 1086 etc. Cf. also River WAVENEY TM 4691. Harrison's forms *Glow, Glowy* 1577, *Glow, Glowie* 1586 are probably artificial unless they represent a local pronunciation **Gloford* like STOFORD Wilts SU 0835, Somer ST 5613, for *Stanford*. RN 174 gives prs [glævn] and [gleivn].

GLAZEBURY Ches SJ 6797. A modern name. R.n. Glaze as in Glazebrook SJ 6592 + p.n. Bury as in *Bury Lane* 1843 OS. Glazebrook, *Glasebroc* [[1190–1245]1268, is 'brook called *Glas*, the blue-green stream', W **glas** (Brit **glasto-*) + OE **brōc**. RN 175.

GLAZELEY Shrops SO 7088. Possibly 'the wood or clearing by the stream called *Glæs*'. *Gleslei* 1086, *Gleseleg(e)* 1221–1274, *Glasele* 1255–1711 with variants *-ley(e)* and *-leg(h), Glazeleye* 1308, *Glasleg(h) -ley* 1261–1733. R.n. **Glæs* < OE ***glæs** 'clear, bright, shining', + **lēah**. The reference is thought to be to one of the streams flowing into Borle Brook. Sa i.136.

GLAZENWOOD Essex TL 8022. *Glas(e)newode* 1291, 1323, *Glasonwode* 1424. P.n. **Glasen* + OE **wudu**. **Glasen, (on) Glæsne* 961×995 S 1501, *Glas(e)n(e)* 1198–1363, is probably OE ***glǣsne** 'bright, shining', a derivative of **glǣs*, referring to a stream which flows E from the wood to join the Blackwater at Bradwell. In G names, however, it has been suggested that *glas(en)-* derives from *calasna*, a word meaning 'boundary'. Ess 283, PNE i.203 s.v. ***glǣs**, BONF 17.51.

GLEADLESS VALLEY SYorks SK 3783. P.n. *(le) Gladeleys, Gledeleys* 13th., *Gleadleys* 1584, *-les* 1692, *-less* 1822, + ModE **valley**. The generic is OE **lēah** 'a forest clearing', the specific any of OE **glǣd** 'bright', ***glād**, ***glǣd**, or ***glēd** 'a glade' or **gleoda** 'a kite'. YW i.165.

GLEADSMOSS Ches SJ 8268. 'Hawk moss'. *Glead Moss* 1831. Dial. **glead** (OE *gleoda*) + **moss** 'a bog'. Che i.90.

GLEASTON Cumbr SD 2570. Partly uncertain. *Glassertun* 1086, *Gleston(a)* 13th–c.1540, *Cle(y)ston -don* 1246, *Glay- Glaiston* 1577. The specific is probably a form of OE *glǣs-* on the root seen in OE *glisnian* 'to gleam'. The reference may have been to the brook at Gleaston or to the light sunny aspect of the village, or to the beacon fire that periodically blazed on Beacon Hill SD 2170. The generic is OE **tūn** 'farm, village'. La 209 gives pr [gli:stn].

GLEDHOW WYorks SE 3136. 'Kite hill'. *Gledhou* 1334×7, *-(h)owe -oo, Gleado(we)* 16th cent. ME **glede** (OE *gleoda*) + ME **howe** (ON *haugr*). Earlier part of Allerton township and known as East Allerton or Allerton Gledhow, *Alreton -tuna* 12th, 13th (*Est-* 1172), *Estallerton* 13th cent., *Gledhawe Allerton* 1285, *Allerton -ir- Gledhow(e) -hou* 1285–1527, for distinction from Chapel and Moor ALLERTON SE 3037, 3138. YW iv.136, WYAS 339.

Great GLEMHAM Suff TM 3461. *magna Glemham* 1336, ~ *ma:* 1610, *Great Glemham* 1837 OS. ModE adj. **great**, Latin **magna** + p.n. *Gl(i)em- Glaimham* 1086, *Glemmeham, Glamessam* 1086 Inquisitio Eliensis, *Glemham* 1180, 'the homestead where revelry or sports take place', OE **glēam** + **hām**. Also known as *Northglemham* 1254 for distinction from Little GLEMHAM TM 3458. DEPN.

Little GLEMHAM Suff TM 3458. *Parva Glemham* 1254. ModE **little**, Latin **parva**, + p.n. *Gl(i)em- Glaimham* 1086, *Glemmeham, Glamessam* 1086 Inquisitio Eliensis, *Glemham* from 1180 including [1156×92]1396, *Glameham* [1189]1253, as in Great or North GLEMHAM TM 3461. DEPN, Baron.

GLEMSFORD Suff TL 8348. 'The play ford'. *Glemesford* [1042×66]12th S 1051, c.1125, *Clamesford(a)* 1086, *Glammesforda* 1086 Inquisitio Eliensis, *Glamesford* c.1160, *Glemeford* 1232, *Glemysforde* 1610. OE **glēam** 'revelry, joy', genitive sing. **glēames**, + **ford**. The reference is to the celebration of games or sports. Cf. BRIGG Lincs TA 0007, *Glaumford Bridge* 1294, GLANDFORD Norf TG 0441, PLAYFORD TM 2147. DEPN.

River GLEN 'Clean, holy or beautiful river'. PrW ***Glen'** < Brit **Glanịo-* or **Glanịā*, a mid-sixth cent. loan into English. The root **glano-* meant 'clean, holy, beautiful' (ModW *glan*), and occurs in the Fr p.n. Glaignes, Oise, *Glana* 1253, originally the name of a stream. RN 177, LHEB 589, 602, Dauzat-Rostaing 321, Jnl 8.44.

(1) ~ Lincs TF 2427. *Glenye* 1275, *le, la Glene* 1390, 1435, *Glean* 1500, 1653, *Glen* from 1366. Also known as Bourne Eau, *aqua que vocatur Brun(n)e* 1240, 1331, *water of Brunne* 1327, *Brun(ne)he(e)* 1331, *Burn(e) Ee in Surflete* 1375, *Burne Eu* 1553, *Bourne Eay* 1663, p.n. BOURNE TF 0920 + dial. **eau** (OE *ēa* replaced by ON *á*). Perrott 39, Payling 1940.2, RN 42, 140.

(2) ~ Northum NT 9430. *fluvio Gleni* [c.731]8th BHE, *Glene* [c.890]c.1000 OEBede–1293, *Glyne* c.1540, *Glin 1586*. Cf. Glendale, *Glendale* from [c.1122]13th. Possibly identical with Nennius's river *Glein* at which one of the twelve battles of Arthur was fought. NbDu 94, RN 177, HB 56.

(3) West ~ River Lincs TF 0022. ModE adj. **west** + r.n. GLEN TF 2427.

Great GLEN Leic SP 6597. *Magna Glen* 1238–1519, ~ *Magna* 1294–1629, *Mikel* ~ 1406, 1410, *Much* ~ 16th cent., *Great(e)* ~ from 1598. Latin **magna**, ME **micel, much**, ModE **great**, + p.n. *Glen* 1086–1610, *Glenne* 1199, *Gleen* 1350, *Gleane* 1582. The earliest form of all is *(æt) Glenne* [849]11th S 1272 which may belong here or under GLEN PARVA. The simplest explanation would be to assume that this was originally a r.n. identical with GLEN Lincs TF 0711, 2427 and Northum NT 9030, Brit **Glanịā* from **glano-* 'clean, holy, beautiful', and that this was the earlier name of the River SENCE SP 6096, whose current name is of English origin. Spellings with a lengthened vowel in the open syllable (*Gleen, Gleane*) would support this. However, the forms in S 1272 and 1199 with *-nn-* point rather to Brit **glennos* 'a valley', W *glyn*, Gaelic *gleann*. The evidence is contradictory. Lei 219, RN 177, L 99, GPC 1914.

GLEN PARVA Leic SP 5798. 'Little Glen'. *Glen Parva* from early 13th., *Little Glen* 1352. Latin **parva**, ME **litel**, + p.n. *Glen* early 13th–1610, *Gleen* 1323–89, as in Great GLEN SP 6597. 'Parva' for distinction from Great GLEN Leic SP 6597. Lei 444, RN 177.

GLENDHU HILL Cumbr NY 5686. P.n. Glen Dhu + ModE **hill**. A summit of 1685ft. Glen Dhu, *Glendeu* 1339, is 'black glen', PrW ***glïnn** + **dū**, ModW *glyndu*, spelt in a pseudo-Gaelic way. A late borrowing into English from the 10th cent. immigrants from Strathclyde. Cu 61, Jnl 1.48, 47, O'Donnell 82, GPC 1414, 1097.

GLENDHU HILL Northum NY 5686. *Glendhu Hill* 1868 OS. Cf. GLENDUE FELL NY 6455.

GLENDON HALL Northants SP 8481. P.n. *Clen(e)done* 1086, *Clendon* 1175–1388, *Glendon* from 1205, the 'clean hill', i.e. bare of vegetation, OE **clǣne** + **dūn**, + ModE **hall** referring to an 18th cent. mansion. Nth 113, L 144, 153.

GLENDUE FELL Northum NY 6455. *Glendue Fell* 1869 OS. P.n. *Glendew* 1239, the 'black glen', PrW ***glïnn** (Brit **glennos*) + ***dū** (Brit **dubo-*), + ModE **fell** (ON *fjall*). Glendue is one of the narrowest and darkest valleys in S Tyndale. The treatment of PrW *ü* shows that the name cannot have been borrowed into English before c.1000. NbDu 94, FT 27ff, LHEB 311.

GLENFIELD Leic SK 5306. 'Clean open ground'. *Clanefelde* 1086, *Clenefeld(e)* c.1131 Ord under year 1081–1361, *Glene-* 1254–1386, *Glenfeld(e)* 1302–1576, *-f(e)ild(e)* 1517–1610. OE **clǣne** + **feld**. The reference is to lack of vegetation or to clean soil. Lei 496, L 241–2.

GLENRIDDING Cumbr NY 3717. 'Bracken valley'. *Glenredyn* 1292, *Glen Roden* 1577, *Glenrhodden* 1777, *Glenridding(e) -ynge* from 1426, *Glenridden* 1777. PrW ***glïnn** + **redin** (ModW *rhedyn*) later replaced by **rydding** 'a clearing'. We ii.222 gives pr [glen'ridən], L 100.

GLENTHAM Lincs TF 0090. 'Village, homestead at *Glente*, the look-out hill'. *Gland- Glant- Glentham* 1086, *Glentheim* c.1115, *Glentham* 1197. OE ***glente** as in GLENTWORTH SK 8488 and GLANTON Northum NU 0714. Glentham and Glentworth lie at the E and W edges respectively of the limestone ridge which stretches N from Lincoln and offers extensive views in either direction. The OE name of the ridge may have been **Glente* 'the look-out place'. DEPN, PNE i.203.

GLENTWORTH Lincs SK 8488. 'Enclosure at *Glente*, the look-out place'. *Glenteuurde -wrde -urde* 1086, *-worda* c.1115, *-wurðða* 1166. OE ***glente** as in GLENTHAM TF 0090 + **worth**. DEPN, PNE i.203.

GLEVUM Glos SO 8318. The RBrit name of GLOUCESTER meaning 'bright', either literally referring to the reflection of the setting sun here in the waters of the Severn or transferred in the sense 'noble, famous place'. *Clevo* [c.300]8th AI, *Glebon Colonia* [c.700]13th Rav. The adjective *Glevensis* occurs in RIB 161 *Dec(urio) Coloniæ Glev(ensis)*. Brit. **Glēu̯on* < **Glaiu̯on* on the root **gleivo-* as in OW *gloeu*, ModW *gloyw* 'bright, shining, sparkling, clear, limpid', Ir *glé* 'clear, bright, pellucid'. RBrit 368, LHEB 325, Archaeologia 93.35, Britannia 1.70, GPC 1411.

GLEWSTONE H&W SO 5622. 'Gleaw's estate'. *Gleanston* (sic for *Gleau-*) 1212, *Glewston* 1568, *-stone* 1722. OE pers.n. *Glēaw*, genitive sing. *Glēawes*, + *tūn*. He 145, Bannister 83.

GLINTON Cambs TF 1506. Partly uncertain. *Clinton* *[1060]14th S 1030, 1285, *Glinton(e)* from 1086. The generic is OE **tūn** 'farm, village, estate', the specific uncertain, perhaps OE ***glind** 'a fence, enclosure', perhaps a lost r.n. identical with the River GLYME Oxon SP 4418. Nth 235.

GLOOSTON Leic SP 7595. 'Glor's farm or village'. *Glorstone* 1086, *Gloreston(e)* c.1130–1629, *Gloureston* 1515, *Glouston(')* 1269–1465, *Glooston(')* from 1405. OE pers.n. *Glōr*, genitive sing. *Glōres*, + **tūn**. Lei 221.

GLOSSOP Derby SK 0394. 'Glott's valley'. *Glosop* 1086, *-hop(e)* 1285–1402, *Glotsop'* 1219, *Glossop(e)* from 1223. OE pers.n. **Glott*, genitive sing. **Glottes*, + **hop**. Db 103, L 115–6.

GLOSTER HILL Northum NU 2504. *Gloucester Hill, Glowster-hill* 17th cent., *Gloster Hill* c.1715 AA 13.16. P.n. Gloster + ModE **hill**. Gloster, *Gloucestre* [before 1178]14th, could be the 'Roman fort used for sport', OE **glēow** + **ċeaster**, or could be identical with GLOUCESTER SO 8318. No Roman fort is known in the area but a Roman altar set up by the first cohort was dug up here in 1856. NbDu 94, Tomlinson 416.

GLOUCESTER Glos SO 8318. 'The Roman fort *Glevum*'. *Gloecester* c.800 HB, *Gleawan ceaster* late 9th ASC(A) under year 577, *(ad, of, to, on) Gleaweceasdre -c(e)astre* [671 for 674×9]14th S 70–12th ASC(E) under year 1122, *Gleweceaster -re* [984]11th S 1346, c.1000, 12th ASC(E) under years 1103, 1123, *to Glew cestre* 11th ASC(D) under years 1124–7, *Glowe- Glouuecestre -iam* 1086–1185, *Gloucestr(e)' -ia -er* from 1086, *Glow- Gloucetur -er* c.1165–1416, *Gloster* 1666. The coin forms are *Gleovv, Gleoce, Gleo(v)* 1040×42 and *Gleiwe, Glie* 1042×66, the W forms *Cair Gloiu* c.800 HB, *Cairclau* c.1125, *Kaerglov* c.1150. RBrit p.n. GLEVUM SO 8318 + OE **ċeaster**. The RBrit p.n. seems to have been borrowed in the form *Glēwe*; forms in *Gleaw(e)-*, genitive sing. *Gleawan-*, show folk-etymological assimilation to OE *glēaw* 'wise, prudent', genitive sing. definite declension **glēawan**, whence the later *Glewe* spellings, as if the sense were 'Roman fort of the wise-man'. The form *Gleow*, whence the later *Glow-* spellings due to accent shift *Gléow-* > *Gleów-* and possible association with the root of the OE verb *glōwan* 'to glow', may have been influenced by the W form *Gloiu*. Gl ii.123 gives prs ['glɔstə] and ['gla:stəR]. LHEB 327, Coins 19.113.

GLOUCESTER & CHELTENHAM (STAVERTON) AIRPORT Glos SO 8821. P.ns. GLOUCESTER SO 8314, CHELTENHAM SO 9522 and STAVERTON SO 8923 + ModE **airport**.

GLOUCESTER & SHARPNESS CANAL Glos SO 7406. P.ns. GLOUCESTER SO 8314 and SHARPNESS SO 6702 + ModE **canal**. Opened in 1827. Verey 1970V.273.

GLUSBURN NYorks SE 0044. Originally a stream name referring to Glusburn Beck, 'the glittering, shining stream'. *Glus(e)brun* 1086, *Glusebrunna* 1170, *-burn(a) -burne* 1182×5–1343, *Gluceburne* [1154×81]1412, *Glusburn(e)* from 1316. OE or ON ***glus(s)** + ON **brunnr** varying with OE **burna**. YW vi.16, L 18.

River GLYME Oxon SP 4418. 'The bright one'. *(to, andlang) Glim* [958]12th S 675, *Glime* 1229, *Glyme* from [1298]c.1400. Brit ***glīmo-**. O 7, RN 180.

GLYMPTON Oxon SP 4221. 'The settlement on the river Glyme'. *Glimtuna* [1050×2]13th S 1425, *-tone* 1086–1664 with variants *-tun(')* and *-ton(')*, *Gli- Glympton* from 1252. R.n. GLYME SP 4418 + OE **tūn**. O 7, 265, RN 180.

GLYNDE ESusx TQ 4509. 'Fence, enclosure'. *Gli- Glynde* from 1210, *Gline* 1587. OE ***glind** cognate with Du *glind, glint* as in p.ns. *Glinthuis, Glinthorst*. Sx 352 gives pr [glain], SS 66.

GLYNDEBOURNE ESusx TQ 4510. 'Glynde stream'. *Glindborne* 1662. Earlier simply *Borne* 1403 and *Burne juxta Glynde* 1288. P.n. GLYNDE TQ 4509 + ME **burne** (OE *burna*). Sx 353.

GNOSALL Staffs SJ 8321. Partly uncertain. *Geneshale* 1086, *Gnowesala* 1140, *Gnod(w)eshall* 1199, *Gnow- Gnou(e)shal(e)* c.1149–1452, *Gnousal* c.1255, *Gnosall* from 1610. Uncertain element or pers.n. + **halh** 'a nook' in the sense 'small valley, hollow'. DEPN's suggestion, pers.n. **Gnēath* < OE *gnēath* 'niggardly' with accent shift **Gnéath* > **Gneáth* > *Gnāth* > **Gnōth*, is implausible although this element did participate in the Gmc personal nomenclature as shown by the runic pers.n. *Knauþimanr* SöR 46. A better proposal is that the specific is a p.n. identical with ME *genow, gannow* < OW *genou* 'mouth, opening of a valley' referring to the narrow stream-valley beyween Gnosall and Gnosall Heath. Cf. GANAREW H&W SO 5316. St i.153, PNE i.194, L 105, 111, NQ 238(NS 40).13–4, CPNE 101, Horovitz.

GNOSALL HEATH Staffs SJ 8220. *Gnosall Heath* 1775. P.n. GNOSALL SJ 8321 + ModE **heath**. St i.160.

GOADBY Leic SP 7598. 'Gauti's village or farm'. *Goutebi* 1086, 1182, *-by* 1259–1517, *Gauteby* 1232, 1445, *Goudeby* 15th cent., *Goadeby* 1641. ON pers.n. *Gauti*, genitive sing. *Gauta*, + **bȳ**. Lei 221, SSNEM 49 no. 1.

GOADBY MARWOOD Leic SK 7826. 'G held by the Malreward family'. *Godeby Morwode* 1576, *Goadby Maurewood* 1725. P.n. *Golte- Goutebi* 1086, *Goutebi -by* c.1130–1428, *Gawteby* 1428, *Goudeby* 1346–1428, *Godeby(e)* 1462–1610, as in GOADBY SP 7598 + family name *Malreward*. The heiress Ada de

Quatermars married Geoffrey Maureward 1216×72. Lei 163, SSNEM 49 no.2.

GOATACRE Wilts SU 0177. 'The goat cultivated land'. *Godacre* 1242, *Got-* 1268–1345. OE **gāt** + **æcer**. Wlt 269.

GOATHILL Dorset ST 6717. 'Goat hill, hill where goats are pastured'. *Gatelme* (sic) 1086, *Gathulla* 1176, *Gothull(e)* 1254–1320, *-hill* 1284. OE **gāt** + **hyll**. The DB form may be a mistake or point to an alternative form in OE **helm** 'a helmet, a summit of a hill'. Do iii.382.

GOATHLAND NYorks NZ 8301. 'Goda's newly-cultivated land'. *Godeland(ia)* c.1110–1240, *Gotheland(e)* c.1180–1408, *Gote-Goutland* 17th. OE pers.n. *Gōda* with substitution of Scand *th* for *d*, + **land**. YN 81 gives pr [gɔːədlənd].

GOATHLAND MOOR NYorks SE 8598. P.n. GOATHLAND NZ 8301 + ModE **moor** (OE *mōr*).

GOATHURST Somer ST 2534. 'Goat wooded hill'. *Gahers* (sic for *Gatherst*) 1086, *Gothurste* 1292, *Gotehirst* [1295] Buck, *Gotehurst* 1610. OE **gāt** + **hyrst**. DEPN.

GOATSCAR Cumbr NY 4706. *Goatscar* 1857, an extensive rocky hill-side and scar. Nearby was (the lost) *Galtecoue* 1238×46, *Goytcowe* 1577, *Goatecoue* 1578, and *Carlegowtecowghe, Carlgoutcave, the carle of Gawtcawe* 1578, 'the carl of Goatscar' probably referring to a prominent rock pile. ON **gǫltr** 'a wild boar' + **sker** 'a scar' and OE **cofa** 'a cove, a recess in the hills'. We i.161, 165.

GOBOWEN Shrops SJ 3033. *Goebowens* 1699, *Goebowen* 1765, *Gobowen* 1770. 'Owen's embankment' has been suggested, W **cob** + pers.n. *Owen*, but *cob* seems to be a 19th cent. borrowing of English *cob* 'a mole, a pier' and this etymology is very uncertain. Gelling, WS 172, GPC 523.

GODALMING Surrey SU 9743. The Godhelmingas, the people called after Godhelm'. *æt Godelmingum* [873×88]early 11th S 1507, *Godelminge* 1086, *-ing(e)* 12th–1476 with variants *-yng(g)(e)* and *-ingg(e), Godelminges -helming(es)* 1154–1227, *God(h)alminge(s)* 13th cent., *Godalmyn(e)* 1485–1622. OE folk-n. **Godhelmingas* < pers.n. **Godhelm* + **ingas**, dative pl. **ingum**, possibly alternating with **ing**[2]. Sr 195 gives pr [gɔdəlmiŋ], ING 29.

GODINGTON Oxon SP 6327. 'Goda's hill'. *Godendone* 1086–1285 with variants *-dune -(a) -don(a)* and *-den('), Godinton' -yn--don(e) -dun' -tone* 1200–1428, *Goding(e)don(')* 1215, 1316, *Godington* from 1241. OE pers.n. *Gōda*, genitive sing. *Gōdan*, + **dūn**. O 212, L 149.

GODMANCHESTER Cambs TL 2470. 'Godmund or Guthmund's Roman fort'. *Godmundcestre* 1086, 1173, *-cestria* 1168, 1176, 1286, *Gutmuncetre* 1146×54, *Gum(m)uncestre* 1175, 1177, *Gumecestre* 1197–1467, *Gom(m)ecestre* 1267, 1334, *Godmanchester (al. Gunecestre)* 1535, 1597. OE pers.n. *Gōdmund* or *Gūthmund* + **ċeaster**. The reference is to the Roman settlement probably called *Durovigutum*. Hu 255, RBrit 354.

GODMANSTONE Dorset SY 6697. 'Godman's estate'. *Godemaneston(e)* 1166, 1201, *Godmanneston* 1251, *Godmanston* 1268. OE pers.n. *Godmann*, genitive sing. *Gōdmannes*, + **tūn**. Do 78.

GODMERSHAM Kent TR 0650. 'Godmær's homestead or estate'. *Godmeresham* 822 S 1620, *Godmæres hám* 824 S 1434, 1042×66 S 1047, *Gomersham* 1086. OE pers.n. *Godmǣr* genitive sing. *Gōdmǣres*, + **hām**. KPN 145, ASE 2.29.

GODNEY Somer ST 4842. 'Goda's island'. *Godenie* [?670×672]13th ECW, *Godeneia* [971]15th S 783, *Godnye* [1154×89]1227, *Gedney more* 1610[†]. OE pers.n. *Gōda*, genitive sing. *Gōdan*, + **īeġ**. DEPN.

GODOLPHIN CROSS Corn SW 6031. *Godolphin Cross* 1888. P.n. Godolphin as in GODOLPHIN HOUSE SW 6031 + **cross** 'a crossroads'. A 19th cent. village. PNCo 88.

GODOLPHIN HOUSE Corn SW 6031. *Godolphin Hall* 1610. P.n. Godolphin + ModE **house** with reference to the 16th cent. hall of the Godolphin family. Godolphin, *Wotholca* 1166, *Gludholghan* 1186, *Wulgholgan* 1194, *Woldholgan* 1201, *Godholkan* c.1210, *God- Gotholg(h)an* 1296–1451, *Godolphin* 1613, *-fyn* 1620, is unexplained; possibly Co ***go-** 'little' + unidentified word *tolghan* with lenition of *t* to *d*. PNCo 87, Gover n.d. 490.

GODREVY ISLAND Corn SW 5743. *Godrivie island* 1528 Gover n.d., *The rokket Godryve* c.1540, *Gudreny Isle* (for *-reuy*) 1610. P.n. Godrevy SW 5842 + ModE **island**. Godrevy, *Goddrevi* c.1250 Gover n.d., *Godrevy* 1297, is 'small-holdings, little farms', Co ***godre** (**go-* + **tref*), plural ***godrefi**. Early ModE **rocket** is a SW dial. diminutive of *rock*. PNCo 88, Gover n.d. 600.

GODSHILL Hants SU 1715. 'God's hill'. *Godeshull(e)* 1235–1363, *-hill(e)* 1251, *Goddeshyll* 1411. ME **god**, genitive sing. **godes**, + **hull** (OE *hyll*). The reason for this name is unknown. Gover 1958.211.

GODSHILL IoW SZ 5281. 'God's hill'. *Godeshul* 1142×7, *-hella* 1135×54, *-hull(e)* 1183–1410, *-hille* 1255–1441, *Godshull* 1307, *-hyll* 1535. OE **god**, genitive sing. **godes**, + **hyll**. The hill on which the church stands was probably a site of pagan worship. Wt 139 gives pr ['gadziɫ], Mills 1996.54.

GODSTONE Surrey TQ 3451. Possibly 'Codd's settlement or stone'. *Godeston* 1248(p), 1434, *-stone* 1308, 1331 both (p), 1446, *Codeston(e)* 1279–1415 mostly (p), *Coddeston(e)* 1288–1347(p), *Goddeston* 1294(p), *Godstone* from 1548. OE pers.n. *Codd*, genitive sing. *Coddes*, + either **tūn** or **stān**. According to Aubrey 1718–19 the place was so called 'from its excellent Quarries of Free-stone'. However, almost all the early sps come from personal names and this may not be a local name at all but a manorial substitution of the family name *Coddestone* possibly representing CUTSDEAN Glos SP 0830.Sr 317, Ess lx.

GODSTONE STATION Surrey TQ 3648. P.n. GODSTONE TQ 3451 + ModE **station**.

South GODSTONE Surrey TQ 3648. ModE **south** + p.n. GODSTONE TQ 3451.

GODWAY H&W SO 3540 → WALTERSTONE SO 3425.

GOG MAGOG HILLS Cambs TL 4954. *Gogmagoghil(l)s* from 1576, *Hogmagogge Hill* 1667. Gogmagog was one of the giants according to Geoffrey of Monmouth's 12th cent. mythical *History of the Kings of Britain* who inhabited Britain prior to the arrival of Brutus. He was a particularly repulsive creature, 12ft. tall, who led an attack on the Britons but was finally captured, challenged to a wrestling match and cast onto the rocks at Gogmagog's Leap, Plymouth Hoe. According to another tradition[†] Gog and Magog were two giants captured by Brutus and brought in chains to London to serve in the royal palace. Their original effigies in the Guildhall were destroyed in the Great Fire of London and again in 1940, the modern ones 14ft. high dating from after World War II. Wickerwork models of them were carried in the Lord Mayor's shows. The legend concerning the Cambridge hill told by Drayton, *Polyolbion* (1612) XXI, is that Gogmagog fell in love with the nymph Granta. She, however, would have no truck with him and he was metamorphosed into the hill. Camden (1586) attributed the name to scholars of the University while the local historian John Layer c.1635 attributes to them the cutting of 'a high and mighty portraiture of a giant' in the trench of Vandlebury Camp which was renewed from time to time 'but is now of late discontinued'. The antiquary William Cole also saw it c.1724, but all trace has since been lost. Ca 35.

[†]A bad form *Coneneie* *[725]12th S 250 may belong here.

[†]Ultimately of biblical origin. In Ezechiel 38–9 Gog comes from Magog and is a sinister demonic leader against the people of God and is ultimately annihilated with his vast hosts in a terrifying conflagration; in Revelations 20 Gog and Magog are two nations who attack the city of God and also earn destruction.

GOLANT Corn SX 1254. Probably 'fair in a valley'. *Gulnant* 1299, *Golananta* 1342, 1370, *Gol(l)enanta* 1385–c.1475, *Golenance* c.1462, *Gollant* 1478, *Golanit, Glant* 17th. Co **gol** + **nans**, older *nant*. The parish was otherwise known as St Sampson, *(capelle) Sancti Sampsonis* c.1280, 1338, from the patron saint, St Samson of Dol. PNCo 88, Gover n.d. 419.

GOLBERDON Corn SX 3271. Partly uncertain. *Golberton or Goberton* 1620, *Goberdon* 1627, *Gol-* 1659, 1679. Unidentified element + **don** < OE *dūn*, 'hill'. All the above forms refer to rough pasture land; the 18th cent. village which developed on it is first recorded as *Golborne* 1679, *Goldburne* 1691, apparently a stream-n. in ModE **burn** (OE *burna*). PNCo 88, Gover n.d. 204.

GOLBORNE GMan SJ 6198. 'Marsh-marigold stream'. *Goldeburn* 1187–1332 etc., *Goldburn(e)* 1203–1390, *Golburn* 1259, *Golborne* 1468, *Gowborne -burn* 16th. OE **golde** + **burna**. Transferred from the stream now called Millingford Brook on which the village stands but which must once have been called *Goldburn*. La 99, Jnl 17.56, 18.24.

GOLCAR WYorks SE 1016. 'Guthlac's shielings'. *Gudlagesarc -argo* 1086, *Gouthelaghcharthes* 1272, 1308, *Guthlacharwes* 1306, *G(o)ulakarres* 13th, *Goulekar* 1451, *Golcar* from 1534, *Gowkar* 1715. Pers.n. *Guthlac* (ON *Guthleikr* or *Guthlaugr*), genitive sing. *Guthlaces*, + **erg**. YW ii.292, SSNY 86.

GOLDEN CROSS ESusx TQ 5312.

GOLDEN GREEN Kent TQ 6348. *Golden Green* 1819 OS.

GOLDEN POT Hants SU 7143. *Golden Pot* 1816 OS. Named from an inn at the crossroads here. Gover 1958.97.

GOLDEN VALLEY Glos SO 9022. *Golden Valley* 1830. Gl ii.84.

GOLDEN VALLEY H&W SO 3636. A translation of the W name *Ynis Stratdour* (for *yn Istratdour*), *Istratour, Estrateur* c.1130, 'valley of the river Dore', OW **(i)strat** 'broad valley' + r.n. *Dour* as in Abbey DORE SO 3830. The occurrence of the Latinised surname *(Richard) de aurea valle* c.1130 shows that the W name was already interpreted as 'golden valley' owing to ambiguity of the spelling *ou* as between [ov] in the r.n. *Dour* [dovr] < PrW **Dobra* 'the waters' and [oy] in OW *our* 'gold'. The alternative name Straddle, *Stradelei, Stratelie* 1086, *Strad(d)el(l)e* c.1100–1517, is either 'Roman-road wood', OE **strǣt** + **lēah**, or 'Strat valley', OW **(i)strat** + OE **dæl**. He 17, DB Herefordshire 2,54, Nomina 10.70.

GOLDENHILL Staffs SJ 8553. *Golden Hill* 1670 Horovitz. Contrasts with nearby (lost) *Hunger Hill* 1836 OS at SJ 8455.

GOLDERS GREEN GLond TQ 2488. *Golders Greene* 1612. Surname *Golder* + ModE **green**. Mx 58, GL 51.

GOLDHANGER Essex TL 9009. 'Gold-wooded slope'. *Goldhangram* 1086 *-e* 1253–1348, *Goldhanger* from 1248. OE **gold** 'gold, treasure' or **golde** 'marigold', + **hangra**. There is a typical *hangra* slope here. The sense of *gold* is uncertain. Ess 302, L 195.

GOLDING Shrops SJ 5403. 'Gold valley'. *Goldene* 1086–1733 with variant *Golden, Goldyng* 1569, *Golding* 1730. OE **gold** + **denu**. The reference is either to yellow-coloured flowers or to the valley as a source of wealth in some way. Sa i.137.

GOLDSBOROUGH NYorks NZ 8314. 'Golda's fortification'. *Golborg* 1086, *Goldeburg(h)(e)* 1086–1301, *Goldesburgh* 1303, 1402. OE pers.n. *Golda* with later ME genitive sing. **-es** + **burh**. YN 137.

GOLDSBOROUGH NYorks SE 3856. 'Godel's fortification'. *Gode(ne)sburg* 1086, *Godel(l)esbur(g) -burc* 12th–1250, *Godelsburg'* 12th cent., *Goldesburg(h) -burc* 1154–1543, *Gould(e)sbrough(e)* 1564–1641. OE pers.n. **Godel*, genitive sing. **Godeles*, + **burh**. YW v.15.

GOLDSITHNEY Corn SW 5430. 'Fair of St Sithny'. *Pleyn-goylsithny* 1399, *Goylsithny* 1403, *Golsithny* 1409, *Golsury* [for *-siny*] 1610. Co **gol** + saint's name *Sydhni*. *Pleyn* in the first form is Co **plen** 'arena, field'; it means 'field of St Sithny's fair'. The fair moved here some time before 1284 from Merthersithney in SITHNEY SW 6328 ('grave of Sithny', Co ***merther**). PNCo 88, Gover n.d. 610.

GOLDTHORPE SYorks SE 6404. 'Golda's outlying farm'. *Gulde-Go(l)detorp* 1086, *Goldethorp(p)* 1307, 1386, 1528, *Goldthorp(e)* from 1285, *Goltorp* 1227, *-thorp(e)* 1276–1400, *Gowlthorp* 1572. OE pers.n. *Golda* + ON **thorp**. YW i.83, SSNY 59.

GOMELDON Wilts SU 1835. 'Gumela's hill'. *Gomeledona* 1189, *-don(e)* 1275–1312, *Gomeldon* from 1279, *Gombledon -ton* 16th cent. OE pers.n. **Gumela* + **dūn**. Wlt 80.

GOMERSAL WYorks SE 2026. 'Gumer's nook of land'. *Gome(r)shale* 1086, *Gumereshale, Gumersale* 13th cent., *Gomersall(e)* 1285–1604. Pers.n. *Gumer* (OE **Gūthmǣr*), genitive sing. *Gumeres*, + **halh** in the sense 'little valley'. YW iii.21 gives pr ['gumǝsǝl], L 107, 110.

GOMSHALL Surrey TQ 0847. 'Guma's shelf of land'. *Gomeselle* 1086, *Gumesele* 1156, *Gomeselve* 1154, *-shelve -shulve -shelf* 13th–14th, *Gomshulf* 1388, *Gumeselva -silva -sselua -sselue -ssolve* 1167–1210×12, *Gumishill* 1172, *-eshull'* 1233, 1255, *Gumshelf -shulve* 1287–1316, *Gumshull* 1609, *Gunshal* 1675. OE pers.n. *Guma* + **scelf** with occasional reinterpretation of second element as *hyll*. Sr 248 records pr [gʌmʃǝl], L 187.

GONALSTON Notts SK 6747. 'Gunnulf's farm or village'. *Gunnvlvestvne, Gunnuluestone -fes-* 1086, *Gunnoluiston' -ves-* 1192–1299, *Gunnileston* 1231, *Gonol(e)ston* 1280, 1293, *Gunnolston* 1302, *Gonalston* 1458, *Gonaston* 1549, *Gunneston* 1634, *-is-* 1738. Pers.n. ON *Gunnolfr*, ODan *Gunnulf*, secondary genitive sing. *Gunnulfes -olfes*, + **tūn**. Nt 166 gives pr [gʌnǝlst(ǝ)n, gʌns(ǝ)n], SSNEM 192.

Great GONERBY Lincs SK 8938. *Gunwarbie Magna* 1634, *Gt. Gunnerby* 1808. ModE **great**, Latin **magna**, + p.n. *Gunforde-Gouerdebi* 1086, *Gun(n)(e)fordebi -by* 1130–1263, 'Gunnfrøthr's farm or village', ON pers.n. *Gunnfrøthr* + **bȳ**. Subsequent forms, *Gun(e)wardebi -by* 1173–1610, *-wardby* 1292–1634×42, *-reby* 1299, *-erby(e) -bie* 1553–1685, *Gunnerbie -by* 1566–1821, *Gonerbie -bye* 1531–1607, have been influenced by those for Little Gonerby (lost) SK 9036, *Little Gonerby* 1605, ~ *Gunnerby* 1778, earlier simply *Gunnewordebi* 1086, 'Gunnvarthr's farm or village', ON pers.n. *Gunnvarthr* + **bȳ**. Perrott 452, SSNEM 49, Cameron 1998.

GOOD EASTER Essex TL 6212 → Good EASTER TL 6212.

GOODBER COMMON Lancs SD 6263. Cf. Goodber Fell SD 6262. P.n. Goodber + ModE **common**, **fell**. There are no early forms for the p.n. Goodber.

GOODERSTON Norf TF 7602. 'Guthhere's estate'. *Godestuna* 1086, *Gurreston* 1177ff, *Gutherestone* 1254, *Gutherstun* 1267. OE pers.n. *Gūthhere*, genitive sing. *Gūthheres*, + **tūn**. DEPN.

GOODLEIGH Devon SS 5934. 'Goda's wood or clearing'. *Godelege* 1086–1394 with variants *-leg(he)*. OE pers.n. *Gōda* + **lēah**. D 44.

GOODMANHAM Humbs SE 8843. 'Village of the Godmundingas, the people called after Godmund'. *Godmund(d)ingaham* [731]8th BHE, [c.890]c.1000 OEBede, *Gudmundham* 1086–14th, *Guthmundham* 1191×1203–1404, *Goodmad(h)am* 1487, 1592, *Goodmanham* from 1504. OE *Gōdmundingas* < pers.n. *Gōdmund* + **ingas**, genitive pl. *Gōdmundinga*, + **hām**. In the 13th and 14th cents. the OE pers.n. *Godmund* was partially assimilated to ON *Guthmundr*. Goodmanham is the site of the shrine destroyed by the priest Coifi in Bede's dramatic account of the conversion of Northumbria in AD 627. Along with nearby SANCTON it seems to have formed the cult centre of the kingdom of Deira in its pagan phase. It is striking, therefore, that the pers.n. on which the folk-n. *Godmundingas* is formed might alternatively be regarded as a compound of OE *god* 'a god, an image of a god', and *mund* 'protection'. Perhaps the *Gudmundingas* were 'the people under divine protection'. YE gives pr [gudmǝdǝm], HE ii.13, Settlements 189, 196.

GOODNESTONE Kent TR 2554. 'Godwine's estate'. *Godwineston'* 1179×80, 1240, *Godwineston(e) -wynes-* 1196–1270,

Goodwinston 1610. OE pers.n. *Gōdwine*, genitive sing. *Gōdwines*, + **tūn**. PNK 532.

GOODNESTONE Kent TR 0461. 'Godwine's estate'. *Godwinestoñ* 1207×8, 1215, *Godwyneston(e)* 1253×4, 1260. OE pers.n. *Gōdwine*, genitive sing. *Gōdwines*, + **tūn**. PNK 286.

GOODRICH H&W SO 5719. Short for 'castle Goodrich' from the name of Godric Mapson, the holder of the estate in 1086. *Castelli Odrici* 1101×2, *Castello Godrici* 1146, *Goderychescastell* 1372, *Goderih* 1322×6, *Goderich* 1538, *Gotheridge* 1671. The original name of the estate was *Hulla* 'the hill' 1086, OE **hyll**. He 91, DB Herefordshire 1,60, Bannister 84, Studies 1936.143, PNE i.267.

GOODRINGTON Devon SX 8858. 'Estate called after Godhere'. *Godrintone* 1086, *God(e)rington -yng-* 1199–1414, *Godel- Gothel- Guther- Gotherington -yng-* 13th cent., *Gorrenton (sands)* 1667. OE pers.n. *Gōdhere* + **ing**[4] + **tūn**. D 517 gives prs [gɔrintən] and [gʌrintən].

GOODWOOD HOUSE WSusx SU 8808. P.n. *Godiua- Goddiuewuda* c.1200, *Godivewod -eyue- -yeue-* 13th cent., 'Godgiefu's wood', OE feminine pers.n. *Gōdġiefu*, genitive sing. *Gōdġiefe*, + **wudu**, + ModE **house**. Nearby was *Godiuemere* 1209, 'Godgiefu's pool'. Sx 66.

GOODYERS END Warw SP 3486. Family name Goodyer + **end**. Short for 'Goodyer's end of Exhall'. Amicta Godyer of Monks Kirby SP 4683 is mentioned 1327. Wa 109.

GOOLE Humbs SE 7423. 'The ditch'. *Gulle in Houk'* 1362, *Gowle* 1535–1165, *Goyll* 1535, *Goole* from 1540. ME **goule** (OE **gūl*, **gŭl-*) surviving as dial. *gool* 'a channel made by a stream'. For *Houk'* see HOOK SE 7625. Originally a stream name. The pronunciation [gu:l] is dialectal since ME *ū* normally becomes ModE diphthong [au]. YW ii.16.

GOOLE FIELDS Humbs SE 7519. *Goole Field Houses* 1771. A modern township carved out of GOOLE SE 7423. It is named from *Feildhouses* 1555, houses established in the old fields of Goole. YW ii.16.

GOONBELL Corn SW 7249. 'Far downs'. *Goonbell* 1813. Co **goon** 'downland, enclosed pasture' + **pell** 'far, distant' with lenition of *p* to *b*. 'Distant' in relation to St Agnes SW 7150. PNCo 89, Gover n.d. 363.

GOONHAVERN Corn SW 7853. 'Downs of summer-ploughed land'. *Goenhavar* 1300, *Gunhaver* 1338 Gover n.d., *Goonhavern* 1748. Co **goon** + ***havar**. Refers to an area of rough grazing which had a piece of summer-ploughed land in or near it. The village is 19th cent. PNCo 89, Gover n.d. 376.

GOONHILLY DOWNS Corn SW 7120. P.n. Goonhilly + ModE **down(s)**. Goonhilly, *Goenhili* [c.1240]14th, *Gonhely* 1284, *Gonelgy* 1315 is either 'downs of brackish water', Co **goon** + **hyly** 'brine, salt water' or more likely 'downs of hunting', Co **goon** + verbal noun ***helghy** 'hunting'. PNCo 89.

GOOSEHAM Corn SS 2216. 'Goose meadow'. *Gosham* 1201–1340, *Gousham* 1417, 1430, *Ham* 1610. OE **gōs** + **hamm**. PNCo 89, Gover n.d. 21.

GOOSETREY Ches SJ 7769. 'Gorse-tree'. *Gostrel* 1086, *Gostre* 1190–1553, *Gosetre* 13th cent., *Gorestre(e)* 1192–1208, *Goostre* 1258, *Goosetrey* from 1579. OE ***gorst-trēow** from OE *gorst* 'gorse, juniper, broom' + *trēow*. Che ii.226 gives pr ['gu:stri], Addenda 19th cent. local ['gu:(ə)stri].

GOOSEY Oxon SU 3691. 'Goose island'. *Goseie* [811]12th S 166, *Gosige* [955]12th S 567, *Gosei* 1086, *Goseya -ey(e)* 1220–1316. OE **gōs** + **ēġ, īeġ**. Brk 411, L 39.

GOOSNARGH Lancs SD 5536. 'Gosan or Gusan's shieling'. *Gusansarghe* 1086, *Gosenhar(egh) -ar(g)e -arch, Gosnarhe -arwe* 13th cent., *Gosenargh* 1284–1332 etc. OIr pers.n. *Gosan* or *Gusan* + **ǣrgi**. La 149, Jnl 17.86.

GORE SAND Somer ST 2851. 'Triangular sand-bank'. *Gore Sand* 1809 OS. ModE **gore** (OE *gāra*) + **sand**. A spit of sand at the mouth of the river Parrett.

GOREFIELD Cambs TF 4112. 'Muddy open land'. *Gorefeld* c.1190, *Gordefelde* 1520, *Gorefeild greene* 1620, *Gorfield Green* 1824 OS. OE **gor** + **feld** with late addition of ModE **green**. Ca 272.

GORING Oxon SU 6081. 'The Garingas, the people of the gore' or 'called after Gara'. *Garinges* 1086–1366 with variants *-ingis -inghes -ingges* and *-ynges, Garing(e) -ingg' -yngg'* c.1181–1316, *Goringe* 1278×9, *Goryng* 1405. OE folk-n. **Gāringas* < OE **gāra** or pers.n. **Gāra* + **ingas**. There is a large triangle of level ground here which favours the appellative explanation. O 51, ING 46.

GORING-BY-SEA WSusx TQ 1103. P.n. *Garinges* 1086–1301, *Garing(e)* 1202–1327, *Goryng* 1280, 'the Garingas, the people called after Gara, or the dwellers at the gore or triangle of land', OE folk-n. **Gāringas* < pers.n. *Gāra* or **gāra** + **ingas**, + ModE **on sea**. Sx 168, ING 34, MA 10.23, SS 64.

GORLESTON-ON-SEA Norf TG 5203. P.n. *Gorlestuna* 1086, *Gurleston(a)* 1130, 1235, unknown element or pers.n. + OE **tūn**, + ModE **on sea**. It is suggested that the specific is a pers.n. related to ME *gurle, girle, gerle* 'a girl' from OE **gyrela, *gyrele*. DEPN.

North GORLEY Hants SU 1611. *Northgarle -gorley* early 13th. ME adj. **north** + p.n. *Gerlei* 1086, *Garleia* c.1210, *Gorley* 1227, 'wedge-shaped wood or clearing', OE **gār(a)** + **lēah**. The reference may be to the wedge of high ground E of the village possibly known as 'the gore' and so the better translation may be 'wood or clearing at the gore or at Gore'. 'North' for distinction from South GORLEY SU 1610. Ha 83, Gover 1958.215.

South GORLEY Hants SU 1610. *Suthgorley* 1279, *South Goreley* 1811 OS. ME adj. **south** + p.n. Gorley as in North GORLEY SU 1611. Gover 1958.215.

GORPLE RESERVOIRS WYorks SD 9231. P.n. Gorple + ModE **reservoir**. Gorple, *Gorpawle* 1575, *Gorpelin* 1817, may contain dial. **gor(p)**, **gorpin**, **gorblin** 'an unfledged bird' (hardly compounded with **play**) referring to an activity of adult birds. YW iii.193.

GORPLEY RESERVOIR WYorks SD 9123. P.n. Gorpley + ModE **reservoir**. Gorpley, *Gorpley* 1783, is unexplained but comparable to GORPLE SD 9231. YW iii.181.

GORRAN CHURCHTOWN Corn SX 0042. P.n. Gorran + ModE **churchtown**. Gorran, *Sanctus Goranus* 1086, *ecclesia de Sancto Gorano* 1270, *Seynt Goron -an* 1302, 1306, 1428, *Langoron* 1374, *Goran* 1591, *Laworran* 1717, *Gurren* 1757, is properly Co ***lann** 'church-site' + saint's name *Guron* or *Vuron*, a saint unique to Cornwall. St Guron was believed to have lived a religious life at Bodmin until St Petrock crowded him out and he moved here. PNCo 89, Gover n.d. 398, DEPN, CMCS 12.60.

GORRAN HAVEN Corn SX 0141. *Gurran hone* 1699, *Gorranhaven* 1748. P.n. Gorran as in GORRAN CHURCHTOWN SX 0042 + dial. **hawn** (ModE *haven*). This name replaces earlier *Porthjust* 1374, *Porteuste* 1576, *Porth East* 1792, 'harbour of St Yust', Co **porth** + saint's name *Iust*. PNCo 89, Gover n.d. 399.

GORSLEY Glos SO 6826. 'Gorse clearing'. *Gorst(e)leg(h) -le(y)* c.1220–1424, *Gors- Gosseleye* 1287. OE **gorst** + **lēah**. Gl iii.176.

GORTON GMan SJ 8896. 'Mud settlement'. *Gorton* from 1282. OE **gor** + **tūn**. The stream running through the township is Gore Brook 'the muddy stream'. La 35.

GOSBECK Suff TM 1655. 'The goose stream'. *Gosebech* 1179, *-bec -bek* 1203–1346. OE **gōs**, genitive pl. **gōsa**, + **beċe** with [k] from ON *bekkr*. DEPN, Baron.

GOSBERTON Lincs TF 2331. *Gosburton* 1487, *Gosberton* from 1535. Whatever these forms were thought to represent, presumably 'goose Burton', they are a reformation with ME **toun** (OE *tūn*) of earlier *Gozeberdecherca, Goseb'techirche, Gosb't cherche* 1086, *Gosebertcherche* 1180–5, *Gosberchirch(e) -kirk(e) -kyrk(e)* 1202–1526, 'Gosbert's church', CG pers.n. *Gosbert* + ON **kirkju** possibly replacing OE *ċiriċe*. Payling 1940.86, SSNEM 213.

East GOSCOTE Leic SK 6413. A modern development resurrecting an old name, *Eastgoscott(e)* 1604–80, *East Goscoate* 1610, ModE **east** + p.n. *Gose(n)cote* 1086, *Gosecot(e)* 1169–1363, *Goscot(e)* 1276–1580, either 'Gosa's cottage' or 'geese shelter', OE pers.n. *Gōsa*, genitive sing. *Gōsan*, or OE **gōs**, genitive pl. **gōsa**, + **cot**. Goscote was the name of one of the Leicester hundreds divided into East and West divisions in the 14th cent. Lei 268.

GOSFIELD Essex TL 7829. 'Wild-goose open country'. *Gosfeld* 1198–1346, *-feud* 1232–85, *Gosfel* 1315, *Gossefeld* 1349, 1412. OE **gōs** + **feld**. Ess 431 gives pr [gɔ:sfl].

GOSFORTH Cumbr NY 0703. 'Goose ford'. *Goseford* c.1150–1396, *Gosford* c.1170–1440, *Gosseford* c.1225–1514, *Gosforth* from 1388. OE **gōs**, genitive pl. **gōsa**, + **ford**. The final *-th* is due to Scandinavian influence. Cu 393, L 71.

GOSFORTH T&W NZ 2368. 'The goose ford'. *Goseford(')* 1166–1296, *Gosford* 1296, 1663, *Goseforth* 1278, 1378, *Gosworth* 1699. OE **gōs**, gentitive pl. **gōsa**, + **ford**. The ford took the Great North Road across the Ouse Burn at NZ 2469. NbDu 95, SR 28, 39, L 71.

GOSMORE Herts TL 1927 'Goose pool'. *Gosmere* 1283, *Gosemere* 1354, *Gossmoore* 1609. OE **gōs**, genitive pl. **gōsa**, + **mere**. Hrt 15.

GOSPORT Hants SZ 6199. 'Goose town, goose market'. *Goseport* 1251–1379, *Gosporte* 1466. OE **gōs**, genitive pl. **gōsa**, + **port**. Ha 83.

GOSWICK Northum NU 0545. 'The goose farm'. *Gossewic* 1202 FF, *Gosewic* 1208×10 BF, *Gosewic(he)* 13th cent., *Gossewyk* 1327. OE **gōs**, genitive pl. **gōsa**, + **wīc**. Pronounced [gɔzik]. See CHESWICK Northum NU 0346. NbDu 95.

GOTHAM Notts SK 5330. 'Goat farm or village'. *Gathā* 1086, *-ham* 1158, [1189×99]1335, 1310, *Gataham* 1148×53, *Gaham* [1154×89]1318–1302, *Gotham* from 1269, *Goteham* 1477–1657, *Goatum* 1750. OE **gāt**, genitive pl. **gāta**, + **hām**. Nt 247 gives pr [goutəm].

GOTHERINGTON Glos SO 9629. 'Estate called after Guthhere'. *Godrinton* 1086–1561 with variant *-yn(g)-* from late 12th, *Goderin(g)ton -tune -yn(g)-* early 13th–1535, *Gutherin(g)ton(a)* 1209, 1254, 1604, *Gotheryn(g)ton -ing-* 1306 and from 1561, *Gotherton* 1684, 1709. OE *Gūthhere* + **ing**[4] + **tūn**. Gl ii.87.

GOUDHURST Kent TQ 7237. Partly uncertain. *Guithyrste* c.1100, *Guthurst, Guhtherste* c.1200, *Gudherst(e) -hurst* 13th cent., *Gouth(h)erst* 1304, 1316, *Goodherst* 1610. The generic is OE **hyrst** 'wooded hill', the specific OE **gūth** 'combat, battle' or pers.n. *Gūtha*. PNK 306, L 198.

GOULCEBY Lincs TF 2579. Probably 'Kolkr's farm or village'. *Colchesbi* 1086, *C- Kolkesbi* 1193–4, *Gol(c)kesbi* 1185–1212, *Golcebi* 1202. ON pers.n. **Kolkr*, genitive sing. **Kolks*, + **bȳ**. The pr with [s] is a survival of the Scand genitive sing. in *-s* showing that the name was coined by Scandinavian speakers. DEPN, SSNEM 49.

GOUTHWAITE RESERVOIR NYorks SE 1369. P.n. *Gowthwaite* 1598, possibly 'cuckoo clearing' or 'Gaukr's clearing', ON **gaukr** or pers.n. *Gaukr* + **thveit**, + ModE **reservoir**. YW v.216.

GOVETON Devon SX 7546. 'Gova's hill or farm'. *Gouedon* 1244, *-ton(e)* 1330, 1372, 1500, *Goton al. Goveton* 1686. Surname *Gova* as in William Gova living at Sigdon SX 7346 in 1372 probably from Co *gof* 'a smith', + **dūn** replaced by **tūn**. D 318.

GOWDALL Humbs SE 6222. 'Nook of land where marigolds grow'. *Goldale* 12th–1543, *Gowdale* 1546, *Goudle* 1602, *Gowdall* from 1641. OE **golde** + **halh**. The *halh* was a piece of ground within one of the loops of the r. Aire. YW ii.17.

River GOWY Ches SJ 4765. *Gowy* from 1577. Possibly 'winding river with a bend in it', a late borrowing, or rather one of Harrison's inventions, from W ***gwy** 'curving' with pr [gu:i] > [gaui]. The existence of **gwy*, however, is extremely doubtful. The earlier name was *Tervin* 1209, *Teruen* 1209–1279, 'boundary river', PrW ***tervïn** (W *terfyn*) as in TARVIN SJ 4967. ***tervïn** is cognate with Latin *terminus*; it must have been borrowed into English after PrW lenition of *-m-* to *-v-*, viz. seventh century or later. Che i.26 and Addenda.

GOXHILL '(Place at) the gusher, the gushing spring'. ON ***geysill**, dative sing. ***gausli**. The modern spelling is a back-spelling due to the use of [x] for [z] as in ROXBY NYorks NZ 7616.

(1) ~ Humbs TA 1021. *Golsa -e* 1086–1192, *Goxa* 1147×68, *Go(u)sla -e* 1135–1549, *Gousel -il* 1127–1475, *Goushill* 1263–1566, *Gouxill* 1349, 1462, 1528, *Goxhill* from 1462. Li ii.119, SSNEM 217, Jnl 27.7.

(2) ~ Humbs TA 1844. *Golse* 1086, *Gosla* 1135×9–1155×7, *Gousle* 1197–c.1400, *Gousell' -ow-* 1209–1580, *Gous(h)ill -(h)yll* c.1265–1504, 1881, *Gouxhill* 1375–1610, *Goxhill* from 1567. YE 66 gives pr [gouzəl], SSNY 156, Jnl 27.6.

River GOYT Derby SK 0172, SK 0178, GMan SJ 9589. 'A rush of water, a water-course'. *Guit* [1208×29]1608, [1202×29]1611, 1244, *Guyt* 1244–[1385]17th, *Gwid, Gwyth* 1285, *Goyt* from c.1257, *le Goyte* 1503. OE ***gȳte**, ***gōte** > dial. **goit** possibly replacing PrW ***gwuith** 'channel, conduit', W *gwyth*. It is unlikely that this name could derive directly from ***gwuith** (<Celtic **u̯eido-*) since Brit *w-* is normally taken over as *w-* in English borrowings. Che i.27, Db 8, GPC 1790.

GOYT'S MOSS Derby SK 0172. 'Goyt's bog'. *Goyt's Moss* 1831. R.n. GOYT SK 0172 + N dial. **moss** (<OE *mos*). Che i.173.

GRADDON MOOR Devon SS 4602. P.n. Graddon + **moor**. Graddon, *Gratedon* 1244, *Graddon* 1492, is 'the great hill', OE **grēat**, definite form **grēate**, + **dūn**. D 171.

GRAFFHAM WSusx SU 9217. 'The grove homestead'. *Grafham* 1086–1314, *Grefham* 1281–1409, *Grofham* 1292–1428, *Graffham* from 1248. OE **grāf** + **hām**. OE *grāf* gave expected ME *grof* and also *graf* with early shortening of *ā* > *a* and consequent confusion with ME *graf, gref* 'a pit' < OE *græf*. Sx 21, ASE 2.33, SS 86.

GRAFHAM Cambs TL 1669. 'Grove homestead'. *Grafham* from 1086, *Grofham* 1342–1548. OE **grāf** or **grāfa** + **hām**. Grafham lay in an ancient forest commemorated in the name Weybridge Farm TL 1873, *Wauberge nemus* 'Waldbeorg wood' c.1110, OE **wald** 'wooded hill, forest' + **beorg** 'hill'. The Ramsey Cartulary records six *gravetae* ('groves') in Grafham. Hu 232, 241.

GRAFHAM WATER Cambs TL 1468. A modern man-made lake. P.n. GRAFHAM TL 1669 + ModE **water**.

GRAFTON 'Settlement in or at the wood'. OE **grāf** + **tūn**. Possibly a name indicating special concern with the management of woodland within a larger unit.

(1) ~ H&W SO 4937. *Crafton* 1303, *Grafton* 1316. DEPN.

(2) ~ H&W SO 5861. *La Grafton* 1251, *Graphne* 1674. Wo 42.

(3) ~ NYorks SE 4163. *Graftune -ton(a)* from 1086. YW v.88.

(4) ~ Oxon SP 2700. *Graptone* 1086, *Grafton(a) -tun -ton'* from 1130. O 320.

(5) ~ FLYFORD H&W SO 9656. Short for 'Grafton under Flyford'. *Grafton juxta Flavell* 1285, ~ *sub Fleuarth* 1317, ~ *Fleford* 1509×38. Earlier simply *Graftun* [884]18th S 219, 972 S 786, *Garstvne* 1086. 'Flyford' for distinction from other Graftons; for the forms and meaning see FLYFORD FLAVELL SO 9946. Wo 200.

(6) ~ REGIS Northants SP 7546. 'G held by the king' for distinction from GRAFTON UNDERWOOD SP 9280. P.n. *Grastone* (for *Graf-*) 1086, *Grafton(e)* from 1166 + **regis**, genitive sing. of Latin **rex** 'a king', and also with manorial addition *Wydevyle* 1465. The manor was held by Robert de Wivill in 1204. Nth 99.

(7) ~ UNDERWOOD Northants SP 9280. *Grafton(e) Underwode* 1367. Earlier simply *Grastone* (for *Graf-*) 1086, *Grafton(e)* from 1166. 'Underwood' for distinction from GRAFTON REGIS SP 7546. The place lies near Rockingham Forest. Nth 181.

(8) Ardens ~ Warw SP 1154. 'The part of G held by the Arden

family'. *Ardens Grafton* 1565. Family name Arden, who held land here as early as c.1200, + p.n. *alia Graftun* 'the other Grafton' 1070×7, *Graston* 1086, *Grafton* 1208. The 12th cent. form *Greftone* in the spurious charter S 1214 for Temple Grafton shows contamination with the mutated side-form *grǣfe*. 'Ardens' for distinction from Temple ~ SP 1255. Also known as *Grafton Minor* c.1200 and *Nelthere* or *Inferior Grafton* 1289, 1316. Wa 209, L 193–4.

(9) East ~ Wilts SU 2560. *East Grafton* 1817 OS. ModE adj. **east** + p.n. *Graf- Grastone* 1086, *Grafton(a)* from 1130. 'East' for distinction from West GRAFTON SU 2460. Wlt 347.

(10) Temple ~ Warw SP 1255. 'The part of G belonging to the knights Templar'. *Temple Grafton* from 1363. ME **tempel** + p.n. *Grastone* 1086, *Grafton(e)* 1189–1275[†], *Greftone* *[962]12th S 1214, as in Ardens ~. 'Temple' for distinction from Ardens ~ SP 1154. The Knights Templar are not known to have held this place, but the Hospitallers did, cf. *Grafton hospitelarium* 'Grafton of the Hospitallers' 1306. Also known as *Grafton Major* c.1200 and *Over Grafton* 1344. Wa 209.

(11) West ~ Wilts SU 2460. *West Grafton* 1817 OS. ModE adj. **west** + p.n. Grafton as in East GRFTON SU 2560.

GRAFTY GREEN Kent TQ 8748. *Grafty Green* 1819 OS. P.n. Grafty + ModE **green** 'common'. Grafty, *Gras(i)tegh, Grestheg(h)e* 13th, *Grastye* 1270, *la Grauetegh'* 1327, is the 'grass enclosure', OE **græs** + **tēag**, either alternating with or replaced by **græf** 'pit, grave' or **grāfa** 'grove' + **tēag**, probably due to the orthographic similarity between long *s* and *f*. PNK 205.

GRAIGNANT Shrops SJ 2535. *Craignant* 1837 OS. Possibly a frenchified W p.n., **crai** 'fresh' + **nant** 'brook'.

GRAIN Kent TQ 8876. 'Sandy ground'. *Grean* c.1100–1278, *Gren(e)* 1189–1280, *Greyn, Grane* 1535. OE ***grēon** cognate with MLG *grēn* 'sand (on the sea shore)'. Situated on alluvium at the far top of the Isle of Grain. PNK 131.

Isle of GRAIN Kent TQ 8776. *Ile of Greane* 1610. ModE **isle** + p.n. GRAIN TQ 8876. Formerly a true island.

GRAINSBY Lincs TF 2799. 'Grein's farm or village' or 'the farm or village of the fork'. *Grenesbi* 1086–1208, *Greinesbi* c.1115–1218, *Greynesby* 1242–1503, *Grainesbi(a)* 1147–1212, *Granesby* 1390–1554, *Grainsby(e)* from 1562. ON pers.n. *Grein* or **grein**, genitive sing. *Greins*, **greins**, + **bȳ**. The village is situated on a small promontory between two valleys. Li iv.98, SSNEM 50.

GRAINTHORPE Lincs TF 3897. 'Germund's farm or village'. *Germund(s)torp* 1086, *Ghermudtorp(sic)* c.1115, *Germundthorp* 1156×8. CG pers.n. *Germund* + **thorp**. A daughter settlement from Covenham on land reclaimed from the sea. Germund was enfeofed with the estate by count Alan of Britanny. SSNEM 109, Jnl 7.56, Cameron 1998.

GRAIZELOUND Humbs SK 7798. *Graiselound* 1824 OS. Earlier simply *altero Lvnd* 'the other Lound' 1086 beside East LOUND SK 7899. ON **lundr** 'a grove'. SSNEM 157.

GRAMPOUND Corn SW 9348. 'Great bridge'. *Grauntpount* 1302–99, *Graundpont* 1373, *Graunt-* 1422, *Grampond* 1426 Gover n.d., *-pont* 1466 Gover n.d., *Grompont* 1610. Fr **grand** + **pont**. The reference is to the crossing of the river Fal at the foot of the village. Recorded as *Ponsmur* 1297, 1301, Co **pons** 'a bridge' + **meur** 'great'. It is uncertain whether the French or Cornish name is primary. PNCo 89.

GRAMPOUND ROAD Corn SW 9150. *Grampound Road* c.1870. P.n. GRAMPOUND SW 9348 + ModE **road**. A 19th cent. village. PNCo 90.

GRANBY Notts SK 7536. 'Grani's farm or village'.

I. *Granebi* 1086, *-by* [c.1155] 14th, 1242–1346, *Granby* from 1284.

II. *Grenebi* 1086, *-by* [1189×99] 1308, 1170×84, *Grenesbi* c.1170.

Form I is ON pers.n. *Grani*, genitive sing. *Grana*, + **bȳ**, form II probably shows DB <e> for [a], but might possibly point to a mutated by-form, **Grœni*, genitive sing. **Grœna*, + **bȳ**, or to confusion with ON **grein** 'a fork' or OE **grēne** 'green'. Granby is situated on a spur close to the fork of the r. Whipling and a stream called The Grimmer. A lost minor name in the next parish of Barnstone SK 7335 is *Grane- Grenehou* [c.1200] 14th, the *haugr* or burial mound of Grani. Nt 225, SSNEM 50.

GRAND UNION CANAL Bucks SP 9220, Warw SP 2466. Cut in 1810–14 from the Soar at Aylestone, Leics, 45 miles to the Grand Junction Canal at Welton, Northants. Gaz.

GRANDBOROUGH Bucks SP 7625. 'Green hill'. *Grenebeorge* [1042×9]13th S 1228, *Grenesberga* 1086, *Greneburne -berne* (for *-burue -berue*) 1284–5, *Grenborough* 1535, *Grandborow* 1653, *Granborough* 1766. OE **grēne** + **beorg** sometimes confused with **burh**. Bk 134, L 128.

GRANDBOROUGH Warw SP 4966. Probably 'the green hill'. *(G)rane- Greneberge* 1086, *Grenebergh* 1199–1547 with variants *-berge -berwe -burgh -borough, Grenberge* 1300, *-berow* 1326, *-bo(u)rgh* 1358–1439, *Grend(e)burgh* 1390, *Granborough* 1538, *Graundborowe* 1585. Two charter forms are also preserved in unreliable late copies, *Greneburgan* [1043]17th S 1000, *Grœnesburgh* [c.1043]17th S 1226. OE **grēne** + **beorg** showing usual ME confusion with **burh**. Wa 130, L 128.

GRANGE 'A grange, an outlying farm of an abbey; a mansion house'. ME **grange**.

(1) ~ Cumbr NY 2517. *grangia nostra de Boroudale* 1396, *The Grange* 1576, *Grange-in-Borrodell* 1669. A grange of Furness Abbey. Cu 350.

(2) ~ Mers SJ 2286. *Graunge* 1519, *(The) Grange* from 1656, *graunge vocat' le Hall in Westkirkeby* 'grange called *le Hall*', *Grangia de Hall* 1547, *grangia de Wyrrall* 1547. Earlier *Caldaygrange* 1341, *Caldygrange* 1553 'Caldy grange'. The manor was a grange of Basingwerk Abbey in Clwyd and took its later name from the old hall demolished in 1819. Earlier known as Great Caldy, *Calders* 1086–[1096×1101]1280, *Magna Caldey(e)* 1281–1641, *Magna Cawedy* 1606, *Great Caldey* [1552]17th–1656, ModE adj. **great**, Latin **magna**, + p.n. CALDY SJ 2285. 'Great' for distinction from Little Caldy, *Parua Caldey* 1280 etc., *Little Calday* 1552. Che iv.288, 282.

(3) ~ NYorks SE 5796. *Grange* 1861 OS.

(4) ~ HEATH Dorset SY 9083. *Grange Heath* 1811. Named from Creech Grange SY 9182, *grangia apud Crich* 1319, *Criche grange* 1535, an outlying farm or grange at one time belonging to Bindon abbey SY 8586 in Wool. See CREECH. Do i.98, 96.

(5) ~ HILL GLond TQ 4492. *Graungehill* c.1534, earlier simply *la Graunge* 1274, a grange or outlying farm of the Cistercian abbey of Tilty near Thaxted, founded in 1153. ME **grange** + ModE **hill**. Ess 55.

(6) ~ MOOR WYorks SE 2216. *Grangemo(o)re* 1659, 1660, 1753. ME **grange** as in Denby Grange, *grangia de Denby* 1331, + **more** (OE *mōr*). YW ii.234.

(7) ~ -OVER-SANDS Cumbr SD 4077. *Grange* 1491. A grange of Cartmel Priory. Grange Farm gave its name to the village as it developed into a coastal resort. It was the terminus of one of the routes across the sands of Morecambe Bay and was usually approached in the 19th cent. by coach across the sands. La 198, Mills 88, Room 1983.47.

(8) ~ VILLA Durham NZ 2352. A modern mining village named from *Pelton Grange* (lost), p.n. PELTON NZ 2553 + ME **grange**. For *Villa*, a variant of *ville*, cf. CARRVILLE NZ 3043.

(9) High ~ Durham NZ 1731. No early forms. ModE **high** + **grange**.

(10) The ~ Hants SU 5636. *Grange* 1610, *The Grange* 1816 OS. A 17th cent. house redesigned by William Wilkins for the Drummond family in 1804–9. ModE **grange**. Gover 1958.83, Pevsner-Lloyd 258.

GRANGETOWN Cleve NZ 5520. A settlement which grew up

[†]The charter S 81 is spurious and therefore not cited here.

with the development of iron and steel works at the mouth of the Tees in the 1830's. It is named after *Eston Grange* 1861×3 OS, probably by origin a grange or outlying farm of Guisborough Priory.

Great GRANSDEN Cambs TL 2756. *Magna Granteden* 1272×1307, *Mekel Grantesdene* 1339. ModE adj. **great**, ME **michel**, Latin **magna** + p.n. *Grantesden(e)* 1086–1485, *Gra(u)ncenden* 1200, 1245, *Gransden* 1598, 'Granti's valley', OE pers.n. *Granti*, genitive sing. *Grantes*, + **denu**. 'Great' for distinction from Little GRANSDEN TL 2755. Hu 258.

Little GRANSDEN Cambs TL 2755. *Little Grantesdene* 1294. ME adj. **litel** + p.n. *Grantandene* [973]13th, *Grantendene -(a)* 1086–c.1185, *Grentedene* [1042×66]13th S 1051, *Grantedene* 1086–c.1278, *Grantsedene -es-* 1199–1415, *Gran(n)(e)sden* from 1485, 'Granta's valley', OE pers.n. *Granta*, genitive sing. *Grantan*, later replaced by *Granti*, genitive sing. *Grantes*, + **denu**. 'Little' for distinction from Great GRANSDEN TL 2756. Ca 161.

GRANSMOOR Humbs TA 1259. 'Grante or Grente's moor or marsh'. *Grentesmor(a)*, *Grenzmore* 1086, *Grancemor(e)* 12th–1265, *Graunce-* 1254–1506, *Grans(e)more* 1583, 1650. Pers.n. ON **Grentir*, genitive sing. *Grents*, or OE *Granti* or *Grenti*, genitive sing. *Grantes*, *Grentes* with *-es* replaced by ON *-s* (to account for the *-z-* and *-ce-* spellings) + OE, ON **mōr**. YE 88, SSNY 95, L 55.

River GRANTA Cambs TL 5231. Possibly 'fen' or 'muddy river'. *Gronte fluminis* (genitive case) [c.745]9th Felix, *Grantan stream* 'course of the Grante' [c.890]c.1000 OEBede, *Grante ēa* [10th]c.1050 Guthlac, *Grant(e)* 1285–1669, *Granta* from 1576. *Granta* is a learned Latin form for OE *Grante* feminine, genitive sing. *Grantan*, a pre-English r.n. possibly from IE **ghrn̥-tu-*, Gmc **grunþa* 'sea bottom' as in ON *grunn* 'a shallow spot', OE *grund*, OHG *grunt* 'ground, bottom', which may be related to Danish *grums*, Swedish *grummel*, OFrisian *grum* 'sediment' < IE **ghrem-*, **ghrendh-*, **ghren-* 'grind, scrape, gnash'. Ca 6, RN 183.

GRANTCHESTER Cambs TL 4355. Properly 'the dwellers by the river Granta'. *Grantesete* 1086–1426 with variants *-set(a)*, *Gransete* 1199–1393, *Grancestr(e)* 1208–1559, *Grauncestre* 1322–1480, *Gransete -ceter* 1349–1576, *Granchester* 1643. R.n. GRANTA TL 5231 + OE **sǣte** early confused with ME **chester** < OE *ċeaster*. Ca 75.

GRANTHAM Lincs SK 0135. 'The homestead on the sand'. *Grant- Gran(d)- Graham* 1086, *Grantham* from 1227. OE ***grand** + **hām**. Cf. ON *grandi* 'a sand-bank'. The soil and subsoil is mostly clay and sand. Perrott 455, PNE i.208.

GRANTHAM CANAL Leic SK 7431. *Canal from Nottingham to Grantham* 1824 OS. P.n. GRANTHAM + MadE **canal**. The Grantham and Nottingham canal was cut in 1793 linking Grantham to the river Trent near Nottingham. Gaz.

GRANTLEY NYorks SE 2369. 'Granta's clearing'. *Grante-lege* c.1030 YCh 7, *Grentelai(a)* 1086–1167, *Grantelei(a) -le(y) -lay* 12th–1535, *Graunt(e)ley* 13th, 16th cents., *Grantle -lay* 1207–1468. OE pers.n. **Granta* + **lēah**. YW v.197.

GRAPPENHALL Ches SJ 6386. 'Nook of the ditch or drain'. *Gropenhale* 1086–1519 with variants *-(h)al -ale*, *-hall* 1448–1677, *Grapenale* 1216×72, *Grappenhall* from 14th, *Grapnall* 1554–1698. OE ***grōpe**, genitive sing. ***grōpan**, + **hale**, dative sing. of **halh**. Che ii.140 gives pr ['grapənɔ:l] and ['grapnəl], Addenda 19th cent. local ['grɔpnə(l)].

GRASBY Lincs TA 0804. Partly uncertain. *Gros(e)bi* 1086, *Grossebi -by* c.1115–1208, *Gressebi -by* 1165–1384, *Gresby* 1238–1762, *Grisby* 1387–1625, *Grassebi* 1202, *Grasby* from 1697. The *Gresse- Gras-* forms probably represent OE **gærs, græs, gres** or ON **gras, gres** 'grass' + **bȳ**. The original specific may have been a pers.n. derived from OFr *gros* 'big, stout'. Li ii.135, SSNEM 50.

GRASMERE (the lake) Cumbr NY 3306. 'Lake with grassy shores'. *Grissemere* 1375, 1383, *Grismyre* 1580, *Gresmyer* 1570, *Gresmier Tarne* 1615×30, *Grassmeer-water* 1671, *-mere-* 1787. The modern name is ME **gres** + OE, ME **mere**, but the early forms point to ON **gríss** 'a young pig'. See further GRASMERE (the village) NY 3307. We i.16, L 26.

GRASMERE (the village) Cumbr NY 3307. *Ceresmere* 1203, *Gressemer(e)* 1245–1530, *Gresmer(e)* 1274–1777 with variants *-myer*, *-myre* 1558, 1650, *Grasmere* from 1247, *Gri- Grysmer(e)* 1254–1569. The village takes its name from GRASMERE (the lake) NY 3306. We i.198.

GRASMOOR Cumbr NY 1720. *Grasmire* 1784. Apparently ModE **grass** + **mire** 'a bog'. Cu 354.

GRASSCROFT GMan SD 9704. 'Grass field'. *Grasscroft* 1728. ModE **grass** + **croft**. YW ii.314.

GRASSENDALE Mers SJ 3985. 'Grazing valley'. *parvam Gresyndale* [13th]14th, *Gresselond Dale* n.d., *Garstan Dale* 1842 OS. ME **gresing** 'grazing, feeding' + **dale**, ON *dalr*. In the 19th cent. this name was assimilated to that of nearby GARSTON SJ 4084, but the limited evidence available does not point to a common origin of the two names despite their proximity. La 111, SSNNW 230.

GRASSHOLME Durham NY 9221. 'Watermeadow used for pasture'. *Grassholm* 1863 OS. ModE **grass** (OE *gærs*, ON *gres*) + dial. **holm** (ON *holmr*).

GRASSHOLME RESERVOIR Durham NY 9422. P.n. GRASSHOLME NY 9221 + ModE **reservoir**.

GRASSINGTON NYorks SE 0064. 'Pasture settlement'. *Gh- Chersintone* 1086, *Gersington -yng-* 12th–1336, *Karsynton* 1320, *Garsyngton -ing-* 1379–1594, *Gressington -yng-* 1373–1626, *Gris(s)ington -y-* 1453–1697, *Gryston* 1452, *Griston* 1630–1700, *Grasinton* 1746. OE **gærsing** + **tūn**. The reference is to the great moorland pastures N of the village. YW vi.97 gives prs ['grasintən, 'gə:stən].

GRASSINGTON MOOR NYorks SE 0368. *Grassington Moor* 1761. Earlier *the Owte Moore* 1611. P.n. GRASSINGTON SE 0064 + ModE **moor** (OE *mōr*).

GRASSMOOR Derby SK 4066. 'Grass moor'. *Gresmore* 1549, *Grassmore* 1568, *-moore* 1630. OE **gærs** + **mōr**. The reference is to good grazing. Db 260.

GRASSTHORPE Notts SK 7967. Probably 'the outlying farm in the grass, the grazing farm'. *Grestorp* 1086–1242, *-thorp(e)* 1266–1428, *Grysthorp(e)* 1388, 1535, *Christhorpe* 1678, 1720. OE **gærs, græs** or **gres**, or ON **gres**, + **thorp**. The reference is to meadowland in the Trent valley, cf. the Danish and Swedish p.sn. *Grædstrup* and *Grästorp*. Nt 186, SSNEM 110.

GRATELEY Hants SU 2641. 'The great wood or clearing'. *Greatteleiam* (Latin accusative form) 929 B 1341, *(æt) Greatanlea* c.935, *Greteleia* 1130, *Grateleya -le(ye) -lye* from 1242. OE **grēat**, definite declensional form **grēata**, + **lēah**. In OE *grēat* normally means 'tall, thick, stout, massive' as opposed to *smæl* 'thin, slender, narrow'. Ha 83, Gover 1958.166, DEPN.

GRATWICH Staffs SK 0231. 'Dairy farm by the gravelly stream'. *Crotewich* 1086, *Grotewic* 1176, *-wis* 1242, *Gretewyz* 1236, 1242. OE ***grēote** + **wīċ**. DEPN, PNE i.209.

GRAVELEY Cambs TL 2564. Partly uncertain. *Greflea* *[974]14th S 798, [964, 1077]17th, *Grauel'* *[974]13th S 798, *Græflea* *[1060]14th S 1030, *Gravele* 14th–1448 including [974, 1317]17th and [1078]14th, *Gravelei* 1086. The generic is OE **lēah** 'a wood, a clearing', the specific either **græf** 'a grave, a pit' or **grǣfe** 'a grove, a copse, a thicket'. L 193 gives explanation 'grove wood' but perhaps 'clearing containing a pit' is preferable. Ca 167.

GRAVELEY Herts TL 2328. 'Copse wood or clearing'. *Gravelai* 1086, *-le -ley(e)* from 1185, *Gravene* 1212, *Gravenl'* 1220. OE **grāfa** + **lēah**. Hrt 125.

GRAVELLY HILL WMids SP 1090. *Gravelly Hill* 1834 OS. ModE **gravelly** + **hill**.

GRAVELS Shrops SJ 3300. *Gravel Mine* 1836 OS.

GRAVENEY Kent TR 0562. 'Ditch stream'. *(aet) Grafon aea* 811 S 168, *Grafonea* [811]10th and 13th ibid., *Grafon eah, (Æt) grafon œa* 812 S 169, 814 S 177, *gravenea* 11th S 169, [830 for 833]13th S 1414, [946]13th S 515, *Gravanea* 815 S 178, 1006 for 1001 S 914, *Gravenel* 1086. OE ***grafa**, genitive sing. ***grafan**, + **ēa**. A very ancient name containing both an element unrecorded in English but corresponding to OHG *grabo* and a very archaic spelling for *ēa* < **ahwō*. Exact parallels are the two German names Grebenau on the Fulda near Melsungen, *Grabanowa* 786, 1057, and Grabenau near Miesbach, *Grabanouua* 1114. Cf also Grabe near Mühlhausen in Thuringia, *Grabaha* 997, *Grabaho* 1472. KPN 117, PNE i.143, 208, DEPN, FmO i.1086.

GRAVESEND Kent TQ 6473. 'The end of the grove'. *Graveshā* 1086, *Grauesand(')* -*v*- c.1100, 1215, *Grauesend(')* -*v*- from 1226. OE **grāf**, genitive sing. **grāfes**, + **ende**. PNK 100, TC 100.

GRAYINGHAM Lincs SK 9396. Partly uncertain. *Gra(i)ngeham* 1086, *Greingheham* c.1115, *Graingeham* c.1146, *Grahingaham* 1157, *Greingeham* 1196, *Graincham* 1212, *Greyngham* 1242×3. This may be the 'homestead or village of the Grægingas, the people called after Græga', OE folk-n. **Grǣġingas* < pers.n. **Grǣġ(a)*, genitive pl. **Grǣġinga*, + **hām**, or the 'homestead at or called *Grœging*', p.n. **Grǣġing* < pers.n. **Grǣġ(a)* + **ing**[2], locative–dative sing. **Grǣġinġe*, + **hām**. DEPN, ING 144, Cameron 1998.

GRAYRIGG Cumbr SD 5797. 'Grey ridge'. *Grarig(e) -rigg(e)* 1160×70–1597, *Gray- Grey(e)rig - rigg(e) -rygge* 1398–1758. ON **grár** (influenced by OE **grǣg**) + **hryggr**. Grayrigg is a small area of grey carboniferous limestone. We i.140, L 169.

GRAYS Essex TQ 6177. *Grayes* 1399, *Grace* 1547. Short for Grays Thurrock, *Grace Thurrock* 1552. Earlier *Turrokgreys* 1248, *Thurroc de Grey* 1254, and simply *Turruc* 1086 as in Little THURROCK TQ 6477. The manor was granted to Henry de Grai from Graye in Normandy in 1195. Ess 125.

GRAYSHOTT Hants SU 8735. 'Grove corner'. *(boscus de) Grauesseta* 'Grayshott wood' 1184×5, *Graveschete -shette* c.1200–1333, *Graveshote* 1548, *Grayshott* 1544. OE **grāf**, genitive sing. **grāfes**, or **grāfa** + **scēat**. The reference is to a piece of Hampshire surrounded on three sides by Surrey. The grove in question is referred to explicitly in the sources and may have been a virtual p.n. *Grave*. Ha 83, Gover 1958.102, DEPN.

GRAYSWOOD Surrey SU 9134. Partly unexplained. *(le) Grasewode* 1479, 1518, *Greyes- Grace Wood* 16th. Unknown element + ME **wode** (OE *wudu*). Possibly the same as a lost *Gerardeswode* 1332–1448 in which the first element is the OFr pers.n. *Gerard*. Sr 204.

GRAYTHORPE Cleve NZ 5127. A modern industrial development. Surname *Gray* + **thorp**.

GRAZELEY Berks SU 6966. 'Wolf or badger wallowing place'. *(on) grœgsole (burnan andlang burnan on) grœgsole (hagan)* 'to Grazeley Brook, along the brook to Grazeley enclosure' [946×51]13th S 578, *Greyshull(') -(h)ulle* 1241–1659, *Greyshall* 1539×40, *Greseley* 1284, *Grazeley* 1790. OE **grǣġ** + **sol**. When the name became opaque in ME the generic was variously reinterpreted as if from **hyll**, **hall** or **lēah** as a result of popular etymology. Brk 166, 647, PNE i.297, ii.134, L 58.

GREASBROUGH SYorks SK 4295. Either 'gravelly' or 'grassy stream'. *Gersebroc, Gres(s)eburg* 1086, *Gresbroc -brok(e)* 1156–1593, *Grise- Grysebroc(k) -brok* 13th cent., *Greisbrocke* 1511, *Greesborough* 1653. OE **grēosn** or **gresen** + **brōc**. YW i.181, xi.

GREASBY Mers SJ 2587. Probably 'wood fortification'. *Gravesberie* 1086, *Grauesbyri* [1096×1101]1150, *-biri -beri*, *Gravisby* [1096×1101]1280, *Gravesbi(a) -y* 1153×81–c.1310, *Greuesby* 1249–1432, *Greseby* 1271–1374, *Gresby(e)* 1271–1639, *Greas(e)by* from 1579. OE **grǣfe**, genitive sing. **grǣfes**, + **byriġ**, dative sing. of **burh**, later replaced by ON **bȳ**. The specific could alternatively be OE **græf** 'a digging, a pit, a trench'. Che iv.291, SSNNW 194.

GREAT BANK Lancs SD 3323. The name of a sand-bank at the mouth of the Ribble.

GREAT BORNE Cumbr NY 1216. *Great Borne* 1868 OS. A summit of 2020ft. Perhaps ModE **bourn** 'a boundary'. Earlier apparently known as *Hardecnut* 1230, 'hard crag', ON **harthr** + **knútr**, identical with HARDKNOTT PASS NY 2310. Cu 385.

GREAT DODD Cumbr NY 3420. *Dod Fell* 1783. ModE adj. **great** + ME **dod** 'a rounded hill'. Cu 222.

GREAT END Cumbr NY 2208. Short for 'Wasdale great end'. *the Wastall great ende* 1578, *hill called Great End* 1805. ModE **great** + **end**. Refers to the 2995ft. high mountain which blocks the end of Wasdale. Cu 353.

GREAT FEN Cambs TL 5978. *Great Freffen* 1640. Earlier simply *Frythfen* 1343 'fenland overgrown with brushwood'. OE **(ġe)fyrhth** + **fenn**. One of the farms here is Frith Farm, *le Fryth* 1397. Ca 202.

GREATFORD Lincs TF 0811. 'The gravelly ford'. *Griteford(e)* 1086, 1207, 1375, *Gret(e)ford* 1178–1613, *Grat- Grafford* 1192×8, *Grettford* 1242×3, 1327, 1597. OE **grēot** + **ford**. The earliest reference is simply *(œt) Forde* [852]12th S 1440. The ford crosses the r. Glen. The soil is loam overlying gravel. Perrott 408.

GREAT GABLE Cumbr NY 2110. *Great Gavel* 1783. A modern translation of *Mykelgavel* 1338, ON **mikill** 'great' + **gafl** 'gable'. Refers aptly to the triangular shape of the mountain. Cu 389.

GREAT GREEN Suff TL 9155. Short for *Cockfield Great Green* 1837 OS, p.n. COCKFIELD TL 9054 + ModE **great** and **green**. Contrasts with nearby *Colchester Green* TL 9255, *Elm Green* TL 9156, *Parsonage Green*, all 1837 OS, and THORPE GREEN TL 9354.

GREATHAM 'Gravel estate or enclosure'. OE **grēot** + **hām** or **hamm**.

(1) ~ Cleve NZ 4927. *Gretham* 1196–1622, *Grytham* 1382, *Greatham* from 1581, *Greetham* 1630. Greatham occupies a patch of gravel in the marshes of the Tees estuary. Pronounced [griːtəm].

(2) ~ Hants SU 7731. *Greteham* 1086, *Grietham* 1167, *Gretham* 1179–1316, *Grutam* 1235×6, *-ham* 1280–1397. The village is situated on the occasionally pebbly Folkestone Beds. Ha 84, Gover 1958.91 which gives pr [gretəm].

GREATHAM WSusx TQ 0415. 'The gravel meadow'. *Gretha(m)* 1086, *Gruteham* 1135×54, *Gret(e)ham* 1121–1619, *Gritham* 1724. OE **grēot** + **hamm** 1 or 3 'river-bend land'. Greatham lies in a bend of the r. Arun, but as a parish and DB manor may alternatively be an example of *hām* 'a homestead'. In Sussex the two forms *hamm* and *hām* seem to have fallen together at an early time. Sx 151 gives pr [gritəm], ASE 2.49, SS 81–2.

GREAT LAKE Notts SK 5773. A lake created in the grounds of Welbeck Abbey in the late 18th cent. and later enlarged. Pevsner 1957.196.

GREAT RIDGE Wilts ST 9336. *Great Ridge* 1773. Earlier *Chicladrygh'* 1348, *Chikeladerugge* 1362, *Chicklade Ridge* 1635, p.n. CHICKLADE ST 9134 + ME **rugge** (OE *hrycg*). A ridge of high ground near Chicklade on the route of the Roman Road from Old Sarum to the Mendip Hills, Margary no. 45b. Wlt 185.

GREATSTONE-ON-SEA Kent TR 0822. A modern coastal development dating between the First and Second World Wars named from the *Great Stone* 1745, 1819 OS, that once lay on the south side of Romney Sand at *Stone End* [1617]1737. Contrasts with LITTLESTONE-ON-SEA TR 0824. Room 1983.48, Newman 1969W.368, Cullen.

GREATWORTH Northants SP 5542. 'The enclosure on the gravel'. *Grentevorde* (for *Greute-*) 1086, *Gret(t)ew(o)rth -wurth* 12th–1312, *Grutte-* c.1200–1383, *Gretworth* 1316, 1657, 1826, *Greetworth* 1284, 1702, *Gritworth* 1651, 1712. OE **grēot** + **worth**. Nth 35 gives pr [gretwɔːθ].

GREEB POINT Corn SW 8733. P.n. Greeb + ModE **point**. Greeb,

The Greeb 1748, is Co **krib** 'crest, ridge' with lenition of *k* to *g* after a lost Co definite article *an*. PNCo 90.

The GREEN Cumbr SD 1784. *Green* 1781, *The Green* 1864. Cu 418.

GREEN ORE Somer ST 5750. 'Green flat-topped hill'. *Grenor* 1610, *Green Ore* 1817 OS. ModE adj. **green** + **ore** ultimately < OE *ōra*. Green Ore lies on the Roman road across the Mendip Hills, Margary no.45b.

GREEN ROAD STATION Cumbr SD 1883. A railway station on the coastal line from Whitehaven to Barrow in Furness at the road to The GREEN SD 1784.

GREEN STREET Herts TQ 1998. *Grenestrate* c.1250, *Grenestreate* 1544. OE **grēne** + **strǣt, strāt**. Hrt 69.

GREEN STREET GREEN GLond TQ 4563. *Green Street Green* c.1762. P.n. *La Grenestrete* c.1290, 'green or grassy hamlet', ME **grene** + **strete**, + ModE **green** 'a village green'. LPN 95, Encyc 343.

The GREEN Wilts ST 8731. *Green* 1811 OS. Cf. the local f.n. Green Hayes, *Grynehayes* 1609, 'the green enclosures' or 'the enclossures at the green'. ModE **green**. Wlt 483.

GREENBOOTH RESERVOIR GMan SD 8515. P.n. Greenbooth + ModE **reservoir**. Cf. *Greenbooth Mill* 1843 OS.

GREENDYKES Northum NU 0628. No early forms. ModE adj. **green** + **dyke(s)**.

GREENFIELD Beds TL 0534. 'Green open country'. *Grenefelde* 1286, *Grenfield* 1766. ME **grene** + **feld**. Bd 150.

GREENFIELD GMan SD 9904. 'Green open-land'. *Grenefeld* 1323, *Greenfield* 1642. OE **grēne** + **feld**. YW ii.314.

GREENFIELD Oxon SU 7191. 'The open land by the green'. *Grenefeld* 1479–1522×3, *Greenfield* from 1603×25. ME **grene** + **feld**. The settlment adjoins Cookley Green. O 97.

GREENFORD GLond TQ 1382. 'The green ford'. *(et) grenan forda* 845 S 1194, *Greneford(e)* 1086–1448, *Grynford* 1561. OE adj. **grēne**, definite form dative case **grēnan**, + **ford**. Mx 33.

GREENGATE Norf TG 0116 → WOODGATE TG 0216.

GREENHAM Berks SU 4865. 'Green river-meadow'. *Greneham* 1086–1517. OE **grēne** + **hamm**. Brk 247.

GREENHAUGH Northum NY 7987. 'The green nook'. *le Grenehalgh* 1325, *Greenhaug^h* n.d. Map. OE **grēne** + **halh**. The reference is to the location of the hamlet in a small side-valley of the N Tyne. NbDu 96, L 108, 110.

GREENHEAD Northum NY 6665. 'Green hill-end'. *le Greneheued* 1289. OE **grēne** + **hēafod**. The reference is to the end of the high ground along which the Roman wall runs overlooking Tipalt Burn. NbDu 96.

GREEN HILL Northum NY 8647. No early forms. ModE adj. **green** + **hill**.

GREEN HILL Wilts SU 0686. *Greenehills* 1630. ModE **green** + **hill**. Wlt 40.

GREENHILL GLond TQ 1688. *Grenehulle* 1334, *-hill* 1563, *Green Hill* 1675. There is no hill here. The name is probably, therefore, transferred from that of one of the early lords of the manor, such as Henry de Grenehulle 1282 or Richard de Grenhull 1307. Mx 52

GREENHILL SYorks SK 3481. ModE **green** + **hill**.

GREENHITHE Kent TQ 5974. 'Green landing place'. *Grenethe* 1264–93, *Greneheth(e)* 1278–1405, *Grenehith(e)* 1342 etc. ME **grene** (OE *grēne*) + **hethe** (OE *hȳth*, Kentish **hēth*). One of a series of *hithe* names on the lower Thames including CHELSEA GLond TQ 2778, ERITH GLond TQ 5177, LAMBETH GLond TQ 3078, PUTNEY GLond TQ 2274, Rotherhithe GLond TQ 3577, *Retherhith* 1127, the 'landing-place for cattle', OE **hrīther**, and STEPNEY GLond TQ 3581. PNK 51, DEPN.

GREENHOLME Cumbr NY 5905. 'Green water-meadow'. *Greenholme* 1770. ModE **green** + **holm** (<ON *holmr*). We ii.46.

GREENHOW HILL NYorks SE 1164. *Greenhow-hill* 1701. P.n. Greenhow, 'the green hill', ME **grene** (OE *grēne*) + **howe** (ON *haugr*), + ModE **hill**. Cf. *Grenehoo-Morez* 'G moors' 1540. The reference is to a natural hill. YW v.143.

GREENLEE LOUGH Northum NY 7769. *Greenley Lough* n.d. Map, *Greenlee Lough* 1868 OS. P.n. *Greenleye* 1285, 'the green clearing or pasture', OE **grēne** + **lēah**, + N dial. **lough** (OE *luh* < OW *loch*, Brit **luk-su-*) 'a lake'. NbDu 96, GPC 2234.

GREENMOUNT GMan SD 7714. A modern name partly modelled on nearby *Greenhalgh Fold* 1843 OS.

GREENODD Cumbr SD 3182. 'Green promontory'. *Green Odd* 1774. ModE **green** + dial. **odd** 'a small point of land' (< ON *oddi*). La 213.

GREENSIDE HILL Northum NT 9716. *Greenside Hill* 1869 OS. P.n. Greenside 'the green hill-side' + ModE **hill**.

GREENSIDE T&W NZ 1462. Probably 'the green hill-side'. *Green(e)side* 1647. OE **grēne** + **sīde**.

GREENSTEAD GREEN Essex TL 8227. *Grinstead Green* 1768. P.n. Greenstead + ModE **green**. Greenstead, *Grensted* 1497, is 'green place', ME **grene** + **stede**. Ess 437, *Stead* 200.

GREENSTED Essex TL 5302. 'Green place'. *Gernestedam* 1086, *Grenested(e)* 1185–1412, *Grynstede* 1320–1408. OE **grēne** + **stede**. Ess 57 gives pr [grinsted], Stead 200.

GREENWICH GLond TQ 4077. 'Green trading-place'. *Gronewic* 918 B 661, *Grenewic* [964]13th S 728, c.1120 ASC(E) under year 1014, *Grena wic* c.1120 ASC(D, E) under years 1013, 1016, *Grenwic* c.1200 ASC(F) under year 1016, *Greenwic* *[1044]copy S 1002, *Grenviz* 1086, *Grenewych* 1291. OE **grēne** + **wīċ** in the sense 'port, trading place' as also in Aldwych GLond, Latin *Vetus vicus* 1199, *Aldewic(h)* 1211 'the old trading place', OE **eald**, definite form **ealde**, + **wīċ**, DUNWICH TM 4470, *Hamwich* under SOUTHAMPTON SU 4211, IPSWICH TM 1744, *Lundenwic* under LONDON TQ 3079, NORWICH TG 3208, SANDWICH TR 3358, WOOLWICH TQ 4478 etc. It is suggested that the Danish army which encamped here in 1013 chose this place because it was already an established trading place with facilities for landing goods from the river. PNK 237–8, DEPN, TC 205.

GREET Glos SP 0230. 'The gravelly place'. *Grete, Greta(m)* 1185–1582, *Greete* 1477–1722. OE ***grēote**. Gl ii.33.

GREETE Shrops SO 5770. 'The gravelly place'. *Grete* 1196–1577, *Greete* 1552. OE ***grēote**, a derivative of **grēot** 'gravel'. The soil is 'mostly stiff gravel and strong loam overlying clay' (Gaz). Sa i.138.

GREETHAM Leic SK 9214. 'The homestead on the gravel'. *Gretham* 1086–1610, *Greteham* 1398–1535, *Greatham* 1608, *Greetham* 1610. OE **grēot** + **hām**. R 24.

GREETHAM Lincs TF 3070. 'The homestead or village on the gravel'. *Grand- Gretham* 1086, *Gre- Graham* 12th, *Greteham* 1233, *Gretham* 1259. OE **grēot** + **hām**. Identical with GREATHAM Cleve NZ 4927. The first DB form may represent OFr *grand* by false analysis of *Gret-* as if ME **grēt** 'great' or OE ***grand** 'gravel' as in GRANTHAM SK 9135. The soil is varied, overlying white clay and gravel. DEPN.

GREETLAND WYorks SE 0821. 'Newly cultivated land by the rocks'. *Greland* 1086, *Greteland(e)* 13th–1518, *Gretland* 1277–1675, *Greetland* from 1589. ON **grjót** + OE **land**. The reference is to the steep rocky hill-side overlooking the Calder. ON *grjót* may have replaced OE *grēot* 'gravel'. YW iii.47, SSNY 95, L 246, 249.

North GREETWELL Lincs TF 0173. ModE **north** + p.n. *Grentewelle* 1086, *Gretwella* c.1115, *Gretewelle* 1120–1586 with variants *Grethe- Grette-* and *-well'*, *Greetwell* 1612, 'the gravel stream', OE **grēot** + **wella**. North Greetwell is a modern development N of the DB vill. DEPN, Li i.70.

GREGSON LANE Lancs SD 5926. No early forms. Probably from the surname *Gregson*.

GREINTON Somer ST 4136. 'Grǣga's farm or estate'. *Graintone* 1086, 1166, *Greinton* 1201, 1202, *Grendon* 1610. OE pers.n. **Grǣġa*, genitive sing. **Grǣġan*, + **tūn**. The earlier name of this place was *Poeld, Poelt .i. Grentone* [725]15th S 250. See further Polden under CHILTON POLDEN ST 3739. DEPN, Turner 1950–2.117.

GRENDON 'Green hill'. OE **grēne** + **dūn**, L 143, 150, 153–4.
(1) ~ Northants SP 8760. *Grendon(e)* from 1086, *Gryndon* 1313, *Grindon* 1622. Nth 146.
(2) ~ Warw SP 2799. *Grendone* 1086, *Grendon* from 1235. Wa 15, L 143, 150, 153–4.
(3) ~ BISHOP H&W SO 5956 → GRENDON GREEN SO 5957.
(4) ~ COMMON Warw SP 2798. P.n. GRENDON SP 2799 + ModE **common**.
(5) ~ UNDERWOOD Bucks SP 6820. Short for Grendon under Bernwood, *(be tweox) Byrne wuda* c.950 ASC(A) under year 921, *bernewude* 1189×99, *Barnwood* 1610, unidentified element + OE **wudu**. Grendon, *Grennedone* 1086, *Grendon* 1231, *Gryndon* 1626, is the 'green hill'. See also WOTTON UNDERWOOD SP 6816. Bk 103 gives pr [grindən], 132, L 143, 150, 153–4, 229.

GRENDON GREEN H&W SO 5957. P.n. Grendon, as in Grendon Bishop SO 5956, + ModE **green**. Called *Little Common* 1832 OS. Grendon Bishop is *Grendone Episcopi* 1316, earlier simply *Grenedene* 1086, *Grendon* 1269, 'green valley', OE **grēne** + **denu**. The estate was held by the bishop of Hereford from 1241. He 94, L 98.

GRENOSIDE SYorks SK 3393. *Granhowside* 15th, *Grennall Syde* 1617, *Grenowside* 1646. P.n. Greno + ME **sīde**. *Greno* occurs also in Greno Knoll, *Gravenhou -howe* before 1279, 1329, 1410, *le Grenow* 1277, 'excavated hill-spur', OE **græfen** + **hōh** or ON **haugr**. The reference is to ancient quarries. YW i.246.

GRESHAM Norf TG 1638. 'Grazing farm'. *Gersam, Gressam* 1086, *Grasham* 1194, 1242, *Gresseham* 1254. OE **gærs** + **hām**. DEPN.

Castle GRESLEY Derby SK 2818. 'G with the castle'. *Castelgresele* 1252, *-ley(e)* 1330–1488. ME **castel** + p.n. Gresley, as in and for distinction from Church GRESLEY SK 2918. Db 635.

Church GRESLEY Derby SK 2918. 'G with the church'. *Churchegreseleye* 1363. ME **churche** + p.n. *Gresele(a) -lee -leia -leg(h)(e) -ley(e)* c.1125–1177 etc., *Gresleia -le(i) -ley(e)* 1166–1281 etc., *Griseleia -leya* 1194–c.1280, probably 'gravel clearing', OE **grēosn** + **lēah**. The forms do not entirely substantiate this explanation, but the soil overlies sand and gravel. 'Church' for distinction from Castle GRESLEY SK 2818. Db 636.

GRESSENHALL Norf TF 9616. 'Grassy or gravelly nook'. *Gressenhala* 1086, *Gresenhal* 1203, *Gressinhale* 1254, *Gressingehal* 1195, *Gressinghal* 1196, *Grossenhale* 1289. OE ***gærsen** or **grēosn** + **healh**. The soil is reported to be loam and clay overlying clay and gravel. DEPN, Gaz.

GRESSINGHAM Lancs SD 5669. 'Homestead at *Gærsing*, the grazing place'. *Gersingeham* 1183, 1194, *Gersingham* 1204–85 etc., *Gressingham* from 1206. OE ***gærsing** + **hām**. The earliest occurrence is compounded with OE **tūn**, *Ghersinctune* 1086, but the simplex *Gærsing* also occurs, *Gersinch* 1177×1200, *Gersinges, Kersing'* 1226×8, the *-inch* spelling possibly reflecting a locative–dative form **Gærsinġe*. That the generic is *hām* rather than *hamm* 'water meadow', which would fit the topography well, seems to be confirmed by the isolated spelling *Gersinghaim* 1204×12 with ON **heimr** for **hām**. La 178, Jnl 17.100, 18.18.

River GRETA 'Boulder river'. ON **grjót** + **á**. (Identical with Icelandic r.n.)
(1) ~ Cumbr NY 3225. *Greta* from 1278, *gratha* 1578, *gretowe* 1589. *Grjótá* and Norwegian *Grjota, Gryta, Grøta*. Cu 16, RN 185, L 11.
(2) ~ NYorks SD 6271. *Greta* from [1200]1268. YW vii.128, RN 185.
(3) ~ BRIDGE Durham NZ 0813. *Greta Bridge* 1862 OS. R.n. Greta + ModE **bridge**. The present bridge was built in 1773. Greta, *Gretha* 1279, 1280, *Gretay* 1341, *Gretey* c.1540, *Greta* 1577, is the 'stony river'. Its bed is strewn with boulders and the name extremely apt. YN 4, RN 185.

GRETTON Glos SP 0130. Possibly the 'gravel settlement'. *Gretona* 1175, *Grettun -ton(a)* 1185–1703. OE **grēot** + **tūn**. Gl ii.33 reports that the soil here is clay not gravel and therefore prefers the explanation 'settlement near Greet', p.n. GREET SP 0230 + **tūn**.

GRETTON Northants SP 8994. Partly uncertain. *Gretone* 1086, 1215, *Grettone -tun* 1163–1346. The forms suggest 'grit or gravel settlement', OE **grēot** + **tūn**, but OE **grēat** 'great' might be an alternative since the soil is reported as being 'clayey, overlying clay' (1896). Nth 166, Gaz.

GRETTON Shrops SO 5195. 'The gravelly settlement'. *Grotintune* 1086, *-ton(')* 1214–1285, *Gretton(')* from 1255×6, *Gretinton' -ing-* 1269, 1271×2. OE ***grēoten**, an adjectival derivative of **grēot** 'gravel', + **tūn**. The place is situated on a small patch of morainic deposits consisting of clayey gravel. Sa i.139.

GREWELTHORPE NYorks SE 2376. 'Gruel's outlying farmstead'. *Gruelthorp(e)* c.1280–1695, *Growel-* 1290–1460, *Grewel-* from 1457. Surname *Gruel*, a nickname for a miller or baker from OFr *gruel* 'fine flour', + p.n. *Torp* 1086, 12th, *Thorp(e)* 12th–1424, ON **thorp** 'an outlying farm'. YW v.xii, 206 gives pr ['griuwilθrəp].

GREYGARTH NYorks SE 1872. 'Enclosure belonging to the Grey family'. *Low Gray Garth* 1619. Surname *Grey* + ModE **garth**. YW v.213.

GREYSOUTHEN Cumbr NY 0729. 'Suthan's rock or cliff'. *Graykesuthen* c.1187, *creik- Craicsuthen* 1230, 1258, *G- Craykesothen* 1292, *Cray- Creysuthen -sothen* 1292–1307, *Graysothen(e)* 1299–1406, *Graysone* 1505, *Greysoon* 1765. OIr pers.n. *Suthán* + OW **creic**, MIr **craicc** in Celtic word order. Cu 397 gives pr [greisu:n].

GREYSTOKE Cumbr NY 4431. 'Outlying farm called or at *Creic*'. *Creistoch'* 1167, *Creystok', Kreystoc(k)* 13th cent., *Crekestoc* [1171×5]1333, *Craysto(c)k* 1245–1359, *Graystok(e)* from 1259. The family name is *Graistoc* c.1180–c.1237, *Greystoke* 1253, 1307 showing early influence of ME adj. **grei** 'grey'. But the p.n. seems to be OW **creic** 'a hill' taken as a p.n. + OE **stōc**. The reference is to the 1124ft. peak NW of the village at NY 4231. This explanation depends on the reliability of the one form *Crekestoc*. DEPN suggests that the specific may be the r.n. CRAY. Cu 195, Studies 1936.30, DEPN.

GREYWELL Hants SU 7151. Possibly 'wolf spring'. *Graiwella* 1167, *Grei- Greywell* from 1235, *Grewell* 1235, *Gruell* 1579. OE ***græġ** + **wella**. But *grǣġ* 'grey' is also possible. Ha 84, Gover 1958.114 which gives pr [gru:əl], NM 96.361–5.

GRIBBIN HEAD Corn SX 0949. *Gribbin Head* c.1870. P.n. Gribbin + ModE **head** 'a headland'. Gribbin, *Grebin* 1699, is probably Co **kriban** 'little ridge' (*krib* + diminutive suffix) with lenition of *k* to *g* after a lost Co definite article *an*. Cf. Gribin, Clwyd SJ 2146. The older name of the promontory is *Pennarthe* 1525, *Penarth-Point* c.1540, Co ***pen-arth** 'promontory, headland'. PNCo 90, CPNE 70.

GRIFF Warw SP 3588. 'The valley'. *Griva* 12th, *Gryva, (la) Gryve* 1233–1473, *la Greve* 13th–1606, *Gryff* 1488–1654, *Greefe* 1603×25. ON **gryfja**. The reference is to Griff Hollow. Wa 80.

GRIFFE GRANGE VALLEY Derby SK 2556 → IBLE SK 2557.

GRIKE Cumbr Ny 0814. *Grike* 1867 OS. A peak rising to 1594ft.

GRIME'S GRAVES Norf TL 8189. *Grimes Graves* 1824 OS. A collection of over 360 flint mine shafts and shallow workings of the Neolithic period subsequently named after **Grīm* or *Grímr*, a byname of Óthinn. Near Grime's Graves was *Grimeshou* 1086, 'Grímr's mound' (ON **haugr** or OE **hōh**), the meeting place of the Grimshoe hundred. SPNN 144, 146–8, EHN I.75, *SelPap* 82.

GRIMEFORD VILLAGE Lancs SD 6112. P.n. Grimeford + ModE **village**. Grimeford is *Grindford* 1840 (in *Grindford Farm*). The reference is to a ford of the river Douglas, but the evidence is too late for explanation. Indeed in 1577 it was called

Andertonford 'ford leading to Anderton (SD 6113)' and in 1786 *Headless Cross* referring to a cross still surviving. Mills 89.

GRIMESTHORPE 'Grim's outlying farmstead'. ON pers.n *Grímr*, genitive sing. *Gríms*, + **thorp**.

(1) ~ SYorks SE 4109. *Grimestorp* late 12th, *-thorp* 1369, *Grynthorp* 1535, *Grimethorp* 1735. YW i.268.

(2) ~ SYorks SE 3587. *Grimestorp* 1297, *Gri- Grymesthorp(e)* 1346–1822, *Grymsthorpe* 1623, *Grimstrop* 1657. A *Grimeshou* 1086, 'Grim's burial mound', ON **haugr**, is mentioned in DB. For a similar pairing cf. GRANBY Notts SK 7536, *Haggenby* in HACKENBY DYKE NYorks SE 4941, CADEBY SYorks SE 5100, LEGSBY Lincs TF 1385. YW i.210, SSNY 59.

GRIMLEY H&W SO 8360. 'Grima's wood or clearing'. *griman-lea(ge)* *[851]11th S 201, *Grimanlæge* [964]12th S 731, *(Æt) grimanlege, (into, æt) grimanleage* [961×72]11th S 1370, *Grimanleh* 1086, *Grimley* from 1542. OE pers.n. **Grīma*, genitive sing. **Grīman*, + **lēah**. The boundary of the people of Grimley is *grim setene gemære* [969]11th S 1323. Wo 126, Hooke 115, 278, 286–7.

GRIMOLDBY Lincs TF 3988. 'Grimolfr's farm or village'. *Grimaldbi -old-* 1086, *Grimolbi* c.1115, *Grimmoldbi* 1189×99. ON pers.n. *Grímólfr* with common ME substitution of *-olf* by *-old -ald* possibly reflecting the Frankish pers.n. *Grimoald* + **bȳ**. DEPN, SSNEM 51, Insley.

GRIMSARGH Lancs SD 5834. 'Grim's shieling'. *Grimesarge* 1086, *-arg(h) -erg -harch -harwe* 1246, 1262, 1341 etc., *Grymsar* 15th, *Grimsar* 1638×9. ON pers.n. *Grímr*, genitive sing. *Gríms*, + **ǽrgi**. La 145, SSNNW 64 gives pr [grimzə], Jnl 17.82.

GRIMSBY Humbs TA 2709. 'Grim's village or farm'. *Grimesbi* 1086–1392, *-by* 1155–1499, *Grimsby* from 1328. ON pers.n. *Grímr*, genitive sing. *Gríms*, + **bȳ**. That Grim was popularly held to be the founder of this place is seen in the ME poem *Havelok the Dane* 744ff.

And for that Grim that place aute
The stede of Grim the name laute,
So that Grimesbi it calle
That theroffe speken alle.

'Because Grim possessed that place, the place took the name of Grim so that all who speak of it call it Grimsby'. Li v.46, SSNEM 51.

New GRIMSBY Scilly SV 8815. *New Grymsey* c.1540, c.1585, *Newe Grynssey* 1570, *New Grimsbye* 1652. ModE adj. **new** + p.n. **Grimsey* 'Grim's island', ON pers.n. *Grímr*, genitive sing. *Gríms*, + **ey** later assimilated to the p.n. GRIMSBY Humbs TA 2810. *Grimsey* is the former name of Tresco or for the whole large island of Scilly before its subdivision. 'New' for distinction from Old Grimsby SV 8915, *Old Grymsey* c.1540, *Olde Grynssey* 1570, *Old Grimsbye* 1652. PNCo 90.

GRIMSCOTE Northants SP 6553. 'Grim's cottage(s)'. *Grimescote* 12th–1349 with variant *Grymes-*, *Grimestorp* 1235. ON pers.n. *Grímr*, secondary genitive sing. *Grímes*, + **cot**, occasionally replaced by *thorp*. Nth 91.

GRIMSCOTT Corn SS 2606. 'Grim's cottage(s)'. *Gri- Grymescote* 1284, 1394. ON pers.n. *Grímr*, secondary genitive sing. *Grímes*, + ME **cot**, pl. **cote**, OE *cotu*. PNCo 90, Gover n.d. 13.

GRIMSPOUND Devon SX 7080. 'Grim's pound'. First referred to in 1797, this is a striking circular walled Bronze Age settlement on the edge of Dartmoor. For the use of Grim as a name for the Devil frequently associated with prehistoric remains, cf. Grims Dyke Wilts SU 0831, *grimes dic* [956]14th S 612, Grims Ditch Herts SP 9309, TL 0009, *Grymesdich* 1291, Grimes Dike GLond TQ 1190, 1392, *Grim(m)esdich* 13th. D 482, W 15–6, Hrt 7–8, Mx 11, Thomas 1976.87–8.

East GRIMSTEAD Wilts SU 2227. *Estgremsted* 1270, *-grymsted(e)* 1310–72, *Istgrinstede* 1318. ME adj. **est** + p.n. *Gremestede* 1086–1281, *Gremsted(e)* 1263–1327, *Gramestede* 1086, *Grimesteda -e* 1186–1236, *Grymsted(e)* 1275–1428, *Grenested(e) -a* 1160–96, *Grensted(e)* 1196–8, 'the green homestead', OE **grēne** + **hām-stede**. Alternatively this is OE **grēne-stede** and the specific has been influenced by nearby Grim's Ditch along the county boundary with Hants. But the *grem-* sps remain a difficulty. 'East' for distinction from adjoining West GRIMSTEAD SU 2126. Wlt 379, Stead 283–4.

West GRIMSTEAD Wilts SU 2126. *Westgrimstede* 1249. ME adj. **west** + p.n. Grimstead as in East GRIMSTEAD SU 2227. Wlt 379, Stead 283–4.

GRIMSTHORPE Lincs TF 0422. 'Grim's outlying farm'. *Grinestorp* 1166, *Gri- Grymestorp(e)* 1166–1413, *Grimsthorpe* from 1371. ODan pers.n. *Grīm*, genitive sing. *Grīms*, + **thorp**. Perrott 179, SSNEM 127.

GRIMSTON 'Grimr's farm or village'. ON pers.n. *Grímr*, genitive sing. *Gríms*, ME genitive sing. *Grimes*, + **tūn**. In many of these names rather than the Scandinavian pers.n. it seems likely that *Grímr* (or its OE equivalent *Grīm*), originally a by-name for Óthinn (or Wōden), is used for a settlement in a poor situation, one fit to be occupied by a devil or a hobgoblin or haunted or plagued by such. SSNY 203, Signposts 233.

(1) ~ Leic SK 6821. *Gri- Grym(m)eston(e) -is-* 1086–1426, *Gri- Grymston(e)* from c.1240. Lei 294, SSNEM 192 no. 2.

(2) ~ Norf TF 7221. 'Grim's estate'. *(æt) Grimastune* 1035×40 S 1489, *Grimes- Erimestuna* (sic) 1086, *Grimeston'* 1198. ON pers.n. *Grímr*, secondary genitive sing. *Grímes*, + **tūn**. As this village is in a favourable situation it is probably an English estate taken over by a Danish lord and renamed after him rather than the mythological name from Grim, the by-n. of Óthinn, used elsewhere in a derogatory sense for settlements in poor situations. DEPN, SSNY 203, SPNN 144, 147.

(3) North ~ NYorks SE 8467. ModE prefix *North* from 1574 + p.n. *Grimetona, Grimeston(e)* 1086, *Gri- Grymeston(e) -tun* 12th–1404, *Gri- Grymston(e)* from 1084. 'North' for distinction from Grimston Garth near Garton TA 2635, *Grimestun -tone* 1086. YE 141, SSNY 128, 202–3.

GRIMSTONE Dorset SY 6394. 'Grim's estate'. *Gri- Grymeston* 1212–1344, *Gry- Grimston(e)* from c.1226, *Grym(e)stan(e)* 1324–1409, *Grymstede* 1425. ON pers.n. *Grímr*, secondary genitive sing. *Grímes*, + **tūn** occasionally replaced by **stān** and once by ME **stede**. The pers.n. *Grímr* was common even outside the Danelaw by the time of DB. There are three examples of its use as a surname (which may be its use here) in the Dorset Lay Subsidy roll of 1332. Do i.373.

GRIMWITH RESERVOIR NYorks SE 0664. P.n. *Grimwith* 1858, possibly 'the spectre wood', ultimately OE **grīma** + ON **vithr**, + ModE **reservoir**. YW vi.79.

GRINDALE Humbs TA 1371. 'Green valley'. *G(e)rendele* 1086, *Grendale -dal(a)* 1086–1342, *Gryndale* 1333. OE **grēne** + **dæl**. YE 104 gives pr [grindl], L 96.

GRINDALYTHE NYorks SE 8967–9971. *Grendalith* 1367, *Grendalyth* 1479. P.n. *K- Crandala -dale* 1123–1345, *C- Krendale* 1192–1504, *Grendal(e)* 1180–1399, 'crane valley', OE **cran** + **dæl**, + ME **hlith** 'a slope'. The original name of the valley in which KIRBY GRINDALYTHE SE 9067 lies. YE 12.

GRINDLE Shrops SJ 7503. 'The green hill'. *Grenhul* c.1190–1250. OE **grēne** + **hyll**. Bowcock 108.

GRINDLEFORD Derby SK 2477. Possibly 'bar ford, ford with a bar'. *Grundelford* 1248, 1330, 1380, *Gryndelford -il- -ul-* 1216×72–1386, *Grindelford(e) -ul-* 1330, 1356, *Gryndlefordbridge* 1584, 1585. OE **grindel**, ***gryndel** 'a bar, a bolt', + **ford**. The sense is uncertain and there are other possibilities; e.g. 'ford where grindstones are found', OE ***gryndel**, ***grindel**, an instrumental formation on the root of *grindan* 'to grind', short for **gryndel-stān* as in dial. *grindlestone*; or dial. **grindle** 'a ditch'. A ford of the river Derwent. Db 94.

GRINDLETON Lancs SD 7545. 'Farm, village beside the *Grendling* or gravelly stream'. *Gretlintone* 1086, *Grellinton*

1177×93, *Grillington(a) -tun -y-* [late 12th]14th, 1212×19, 1379, *Gren- Grinlington -y-* 1258–1343, *Grenedelington* 1303, *Gryndlyngton* 15th, *Grenel- Grynleton* 1402–1594, *Grindel- Gryndilton* 1497–1594, *Grindleton* from 1560. OE r.n. **Grendling* (<OE ***grendel** + **ing**[2]) + **tūn**. YW vi.195, BzN 1968.181, Årsskrift 1974.30.

GRINDLEY BROOK Shrops SJ 5243. *Grenley Broke* 1534, *Grindley Brooke* 1665. P.n. *Grenleg'* 1230, 1260, *Grenlegh -ley(e) -le(e)* 1308–1338, *Grynleye* 1299, *Gryndley* 1543, 'the green wood or clearing', OE **grēne** + **lēah**, + ModE **brook**. The stream is referred to in 1334 as *le Brock*. Che iv.47.

GRINDLOW Derby SK 1877. 'Green hill or burial mound'. *Greneslaw* 1199, *Grenlawe* [1199]1285, *Gren(e)low* 1259–1767, *Greenlow, Grinlow* 1577. OE **grēne** + **hlāw**. No burial mound is known here but the name may be compared with that of Green Low chambered tomb at SK 2358. Db 105.

GRINDON 'Green hill'. OE **grēne** + **dūn**. L 143, 155, 158.

(1) ~ Northum NT 9144. *Grandon* (sic) 1208×10 BF, *Grendona* [1183]14th, *Gryndone* 1539. NbDu 96.

(2) ~ Staffs SK 0854. *Grendone* 1086, *Grenedun* 1236, *Grendon(a)* 1116×35–1584, *Gryndon* 1577–1610, *Gryn* 1577, *Grynn* 1618, *Grinne* 1586, *Gryn(e) als Gryndon* 1564, 1583. Oakden.

GRINGLEY ON THE HILL Notts SK 7390. *Grenele(y) super Montem* 1517, 1541, *Grenly of the Hill* 1535. P.n. Gringley + ModE **on the hill**, Latin **super montem** for distinction from Little GRINGLEY SK 7380. The village stands on a lofty eminence (Beacon Hill SK 7490 which rises to over 250ft.) overlooking marshes. Gringley, *Gringeleia* 1086, *-lai -lay -ley(e) -leg(e)* 1154×89–1380 with variant *Grynge-, Gringaleia* 1154×89, *Grynley* 1272×1307, 1375, *Grynley* 1539, *Grenley(e)* 1303, 1316, is problematical. Possibly the 'clearing of the dwellers on the green hill', OE folk-n. **Grēningas* < OE **grēne** + **ingas**, genitive pl. **Grēninga*, + **lēah**. Nt 30–1.

Little GRINGLEY Notts SK 7380. *Parva Gryngley* 1585, *Little Gringley* 1587. ModE adj. **little**, Latin **parva**, + p.n. *Grenelei -leig(e)* 1086, *-ley(e)* 1290–1597, *Grenlay* 1225, *-ley* 1242–1384, *Grynley(e)* 1305, 1557, the 'green clearing', OE **grēne** + **lēah**. 'Little' for distinction from GRINGLEY ON THE HILL SK 7390 by the form of which this name has been reshaped. Nt 59.

GRINSDALE Cumbr NY 3658. Probably 'green-headland valley'. *Grennesdal(e)* c.1180, 1399, *Grenesdale* 12th–1360, *Grin(n)esdale -is-* c.1200–1402, *Grynsdale* 1485. ON p.n. **Groennes* 'green headland' **groen** + **nes** + **dalr**. The reference is to the promontory formed by a loop of the river Eden on which the church stands; it is luscious green pasture-land. Cu 140, SSNNW 127, L 94.

GRINSHILL Shrops SJ 5223. Partly uncertain. *Grivelesul* (for *Grin-*) 1086, *Grineleshul(l)* 1242–1428, *Gril(l)eshill(') -hull(')* c.1224×32–1431, *Cryneshull* 1271×2, *Greneshull'* 1334, *Grynshill* 1577, *Grinsell* 1587–1779, *Grinshill* from 1746. Possibly OE ***grynel**, a derivative of **grin, gryn** 'a noose, a snare', genitive sing. ***gryneles**, + **hyll**. It is suggested that there may have been animal snares on the hill, or that the outline of the hill itself was thought to resemble some sort of snare (in which case **Grynel* would be the name of the hill). Sa i.139.

East GRINSTEAD WSusx TQ 3938. *Estgrenested(e)* 1271–1505, *Est(e)grinstede* 1275, 1316. ME adj. **est** + p.n. *Grenestede* (hundred) 1086, *Grenesteda* 1091–c.1170, *-sted(e)* 1091–1476, *Grinsted(e)* c.1140–1302, 'the green place', OE **grēne** + **stede**. 'East' for distinction from West GRINSTEAD TQ 1720. Sx 331, Stead 254, SS 87.

West GRINSTEAD WSusx TQ 1720. *West Grensted(e)* c.1255–1457, *Westgrenested(e)* 1272–1500, *Westgrimstede* 1330, *-in-* 1378. ME adj. **west** + p.n. *Grenestede* (hundred) 1086, *Granestede* c.1230, *Grensted* 1261, 'the green place, the grazing pasture', OE **grēne** + **stede**. Sx 184, Stead 255, SS 87.

GRINTON NYorks SE 0498. 'Green enclosure'. *Grinton* from 1086, *Grenton(e)* c.1180–1397. OE **grēne** + **tūn**. YN 270.

GRISEDALE 'Pig valley', ON **gríss**, genitive pl. **grísa**, + **dalr**. Cf. GRIZEDALE. L 94.

(1) ~ Cumbr NY 3715. *Crisdale* 1291, *Gri- Grys(e)dal(e)* 1292–1823. We ii.223.

(2) ~ Cumbr SD 7793. *Grisedale* [13th]1307 and from 1292, *Gris- Grysdale* 1400, 1621. YW vi.261.

(3) ~ PIKE Cumbr NY 1922. P.n. Grisedale as in Grisedale Gill NY 2023 + ModE **pike**, referring to a 2591ft. peak. The stream in Grisedale Gill is *Grisebek* 1331, 'stream of the young pigs', ON **grísa** + **bekkr**. Cu 16.

(4) ~ TARN Cumbr NY 3412. *Grisdale Tarn* 1787. P.n. GRISEDALE NY 3715 + dial. **tarn** 'a small lake' (<ON *tjǫrn*). We ii.225.

GRISTHORPE NYorks TA 0882. Probably 'pig farm'. *Grisetorp* 1086, *Gris- Grysthorp(p)* from 1175, *Gri- Grysethorp* 1181 etc. ON **gríss**, genitive pl. **grísa**, + **thorp**. Alternatively the specific could be ON pers.n. *Gríss*. YN 104 gives pr [grisθrəp], SSNY 59.

GRISTON Norf TL 9499. Probably 'Gyrth's estate'. *Gris- Grestuna* 1086, *Gerdestuna* [c.1150]1391, *Greston* 1166, *Gristone* 1254. ODan pers.n. *Gyrth*, genitive sing. *Gyrths*, + **tūn**. DEPN, SPNN 168.

GRITTENHAM Wilts SU 0382. 'The gravelly promontory'. *Gruteham* [850]14th S 1580, 1288–9, *Greteham* 1156, *Grutenham* [1065]14th S 1038, 12th–1349, *Gretenhem* 1291, 1362, 1558×1601, *Gritnam* 1595. OE **grēote** varying with ***grīeten** + **hamm**. The soil is gravelly. The place lies on a spur of land by Brinkworth Brook. Wlt 66.

GRITTLETON Wilts ST 8680. Partly uncertain. *Grutelington(e)* [940]14th S 472, 1242, 1354, *Grutlyngton* 1327, 1360, 1601, *Gretelintone* 1086, *-ing-* 1186–1269, *Gritelington -yng-* 1279, 1330, *Gritilton* 1337, *Grittleton al. Grittlinton* 1687. Possibly OE ***grēoteling**, ***grīeteling** 'a gravelly place' (OE *grēot* + *el* + *ing*[2]) + **tūn**, or OE pers.n. **Grytel* + **ing**[4] + **tūn** 'the estate called after Grytel'. For the formation cf. Diddlington Dorset SU 0007, *Di- dydelingtune* [946]14th S 519, either 'the estate called after Dydel' or 'the settlement called or at *Dydeling*, the hill place', OE **dydeling* + *tūn*, Dumplington GMan SJ 7697, *Dumplington* 1229, 'the settlement called or at *Dympling*, the hollow place', OE **dympel* + *ing*[2] + *tūn*, PEOPLETON H&W SO 9350, *Piplincgtun* [972]10th S 786, 'pebble place', OE **pybbel* + *ing*[2] + *tūn*, etc. Wlt 79, Do ii 134, Årsskrift 1974.39–40, 45.

GRIZEBECK Cumbr SD 2385. 'Pig stream'. *Grisebek* 13th. ON **gríss**, genitive pl. **grísa**, + **bekkr**. La 221, SSNNW 128.

GRIZEDALE 'Pig valley'. Identical with GRISEDALE. L 94.

(1) ~ Cumbr SD 3394. *Grysdale* 1336, 1537. La 220, SSNNW 128.

(2) ~ FOREST Cumbr SD 3394. P.n. GRIZEDALE SD 3394 + ModE **forest**.

GROBY Leic KS 5207. 'Village, farm by the pit or hollow'. *Grobi* 1086–c.1200, *-by* from 1205, *Groubi -by* c.1140, 1403, *Grooby(e)* 1576, 1675. ODan ***grōf** 'something hollowed out' + **bȳ**. The reference is to nearby Groby Pool which lies in a deep hollow with streams draining into it. Lei 497 gives pr ['gru:bi:], SSNEM 51.

GROMFORD Suff TM 3858. No early forms.

GROOMBRIDGE ESusx TQ 5337. 'The servants' bridge'. *Gromb'gg'* 13th, *Gromenebregge* 1318, *Grombrugg* 1355, *Grumbrygge* c.1480, *Groomebridge* 1601. ME **grome**, genitive pl. **gromene**, + **brigge, bregge** (OE *brycg, brecg*). Sx 371.

GROSMONT NYorks NZ 8205. Transferred from Grandmont near Limoges, France. *Grosmunt'* 1226×8, *Grosmont* 1540, *Grandimont(e)* 13th cent., *Grauntmont, Gromunde* 1301, 'the great hill', OFr **gros, grand** + **mont**. A priory founded in 1200 and given to the order of Grandmont near Limoges. YN 120.

GROTON Suff TL 9541. 'The sandy or gravelly stream'. *Grotena* 1086, *Grotene* 11th–1254, *Groton* 1610. OE ***groten** or possibly **grota**, genitive pl. **grotena**, + **ēa**. DEPN, PNE i.210.

GROVE 'The grove'. OE **grāfa, grǣfe**. L 193.

(1) ~ Dorset SY 6972. *(la, le) Groue* 1321×4, *Grove* 1608. Do i.219.
(2) ~ Kent TR 2362. *(la) Graue* 1254, 1292, *Groue* 1332. PNK 527.
(3) ~ Oxon SU 4090. *la Graue* 1188, *Grove* from 1316. Brk 474, L 193.
(4) ~ Notts SK 7379. *Grave* 1086–1428, *Grove* 1303–1332 and from 1512. In this name ME **grove** (OE *grāf*) replaces **grǣfe**. Nt 51.
(5) ~ PARK GLond TQ 4172. *Grove Park* 1905. LPN 96.

GROVELY WOOD Wilts SU 0534. *Grovely Wood* 1817, earlier *foresta de Gravelinges* 'G forest' 1086–1281, *forest de Gravele* 1154–1279. P.n. *(on) grafan lea* [940]14th S 469, *Grofle* 1317, *Grovely* 1448, 'the grove wood', OE **grāfa**, genitive sing. **grāfan**, + **lēah**, + ModE **wood**. The forms *Gravelinges* 1086–1281, *Graueling(a) -inge* 1190–1352, represent OE folk-n. **Grāfanlēahingas*, **Grafanlēingas*, 'the people of Grovely'. Wlt 13, ING 48, L 193.

GRUMBLA Corn SW 4029. 'The cromlech'. *Gromlogh* (for *-legh*) 1503, *Grumbler* 1872. Co ***cromlegh** 'cromlech, dolmen, quoit' (< *crom* 'curved' + *legh* 'flat stone, slab' referring to the capstone of a dolmen) with lenition of *c* to *g* after a lost Co definite article *an*. There are remains of a barrow including a retaining wall of very large stones. Gover n.d. suggests identification with *kestelcromlegh* [943 for 925×39]14th S 450 and *Kestelcomleghe* 1238, 'cromlech castle or earthwork', Co **castell**. PNCo 91, Gover n.d. 42–3, CPNE 72.

GRUNDISBURGH Suff TM 2250. 'Grund's fortified place or manor'. *Grundesburc(h)* 1086, *-burg(h)* 1086–1346, *Groundisburgh* 1327, *Grundesboro* 1610. OE pers.n. **Grund*, genitive sing. **Grundes*, + **burh**. DEPN, Baron.

GUARLFORD H&W SO 8145. 'Gærla's ford'. *Garlford* 1275, *Garleford* 1291–1541, *Galvert* 1820. OE pers.n. **Gǣrla* + **ford**. Wo 211 gives pr [gɔ:lfəd].

GUESTLING GREEN ESusx TQ 8513. P.n. Guestling + ModE **green**. Guestling, *Gestelinges* 1086, *Gestlinges* 1194–1206, *Grestling* 1198, *Gestling(e)* 1220–17th is 'the Gerstelingas, the people called after Gerstel', OE folk.n. **Gerstelingas* < pers.n. *Gyrstel*, SE dial. form *Gerstel* + **ingas**. Sx 508, ING 34–5, SS 65.

GUESTWICK Norf TG 0627. 'Cleared ground or pasture belonging to Guist'. *Geg(h)estueit* 1086, *Geistweit, Geystweyt* 1203–44, *Geystethweyt* 1254. P.n. GUIST TF 9925 + ON **thveit**. DEPN gives pr -[tik].

GUGH Scilly SV 8908. Uncertain. *Agnes Gue* 1652 refers to common land belonging to St Agnes. Possibly Co ***keow** 'hedges', the plural of **kee**, meaning 'home-field'; but there is also ***kew** 'a hollow, enclosure' found in the WCo field name *gew* or *gews*. PNCo 91, CPNE 44, 57.

GUIDE POST Northum NZ 2585. *Guide Post* 1867 OS.

GUILDFORD Surrey SU 9949. 'Golden ford'. *(æt) Gylde forda* [873×88]early 11th S 1507, *Gyld(a)* 978–1066, *Gilfrd* 1066–87 Coins, *Gilde- Geldeford* 1086, *Gelde- Gylde- Gulde-* 12th–1636, *Geu(l)de-* 13th, *Gylford* 1477–1546. OE ***gylde** + **ford**. The reference is to the colour of the sand at the original site of the ford rather than to yellow flowers. Sr 9–10, PNE i.211, Jnl 2.38, L 67, 70.

GUILE POINT Northum NU 1340. *Guile Point* 1866 OS. A coastal name, presumably meaning the 'treacherous headland'. ModE **guile** + **point**.

GUILSBOROUGH Northants SP 6773. 'Gyldi's fortification'. *Gisleburg* 1086, *Gilleburc* 1160, *Gildeburch -burg* 1194–1226, *Gildesburc -burg(h)* late 12th etc., *Gylesburgh* 1367, *Gilsborowe* 1610. OE pers.n. **Gyldi*, genitive sing. **Gyldes*, + **burh**. According to an 18th cent. authority the place is named from an encampment whose remains were then still visible. Nth 70.

GUISBOROUGH Cleve NZ 6015. 'Gigr's fortification'. *Ch- Ghigesburg, Gighesborc, Ghigesborg, Giseborne* 1086, *Gisebur(g)h* c.1130–15th, *Gi- Gysburgh* 1285–1577, *Gysborow -borough* 1530, *Gi- Gyseburne(e)* [1119]15th–1430, *Gi- Gysburn* 1228–1483. ON pers.n. *Gígr*, secondary genitive sing. *Gíges*, + OE **burh** varying with ON **borg** and OE **burna**, cf. NEWBURN Northum. YN 149 gives prs [gi:zbrə, gizbrə], SPN 100.

GUISELEY WYorks SE 1942. 'Gislic's forest clearing'. *Gislicleh* [972×92]c.1050 S 1453, *Gi- Gysele(i) -ley(e) -lai -lay* 1086–1526, *Gui- Guysley* 16th cent. OE pers.n. **Ġīsliċ* + **lēah**. From the 15th cent. the specific was (wrongly) regarded as the genitive sing. form of pers.n. *Guy*. YW iv.146 gives prs ['gɑ:zlə] and ['gaizli].

GUIST Norf TF 9925. Uncertain. *(et) Gæssæte* 1035×40 S 1489, *Geysete* [1035×40]13th ibid., *Gegeseta* 1086, *Geiste* 1200, 1254. Possibly 'the settlers on the river Gæge', OE r.n. **Gæġe* 'the turning or wandering one' < vb. *gæġan* as in *forgǣġan* 'pass by' and *ofergǣġan* 'transgress' + **sǣte**. This would have been a nickname for the Wensum which makes a big turn at this point. The formation would be parralel to *Tomsetan* [849]11th S 1272, 'the settlers on the river Thame (Staffs)'. Another example of a r.n. formed on the base of *gæġan* is Ginge Brook Oxon SU 4487, *rivum Geenge* [726×37]13th S 93, *Gæing broc* [958 for 959]13th S 673, and the settlement name Ginge SU 4486, *(ad) Gainge* *[811 for 815]12th S 166, *Geinge* [821]12th S 183, *Gaging* [821]13th ibid., *Gain(c)g* *[955]12th, 13th S 567, *Gainge, Gæinge* *[956]12th, 13th S 583, *Gaing, Gæging* [958 for 959]12th, 13th S 673, r.n. **Gǣġe* + **ing**². Alternative suggestions are OE pers.n. *Gǣġa* or *Gǣġi* + **(ġe)set** 'a fold' or ON **sǣti**, lOE **sæte** 'a seat, a residence'. DEPN gives pr [gaist], RN 172, ING 207, Hooke 135, 141.

GUITING POWER Glos SP 0924. 'G held by the Poer family'. *Gettinges Poer* 1220, *Gutinge Po(u)er* 1287, *Gui- Guyting - yng(e) Po(w)er* 1327–1491. Earlier simply *(bi) Gythinge* [814]11th S 171, *(on) Gytinc -ing* [977]11th S 1335, [987]11th S 1353, *Getinge* 1086, *Guting(e) -yng(e)* 1265–1354, *Gi- Gyting(e) -yng(e)* 1220–1696, *Gui- Guyting - yng(e)* 1248–1742, 'the spring', OE **gyting** < *gyte* 'a pouring forth, a flood' + *ing*². The spring is referred to as *fontanum quod nominatur Gytingbroc* 'spring called Guiting Brook' [780]11th S 116, and *Gytin(c)ges æwylm* [974]11th S 1335, [987]11th S 1353, 'the spring of Guiting'. The manor was held by Roger le Poer 1220, William le Poer 1287 etc. Also known as *Nether Gutinge* 'lower Guiting' from its location lower down the Guiting stream than Temple GUITING SP 0928. Gl i.12 gives pr ['gaitiŋ 'pauəR].

Temple GUITING Glos SP 0928. 'G held by the Knights Templar'. *Guttinge(s) Templar'* 1221, *Temple Guting(e)* 1290–1414. Earlier simply *Geting(e) -inges* 1086–1274, *Gutinge(s)* 1139×48–1354 as in GUITING POWER SP 0924. Gl ii.13.

East GULDEFORD ESusx TQ 9321. *Est Guldeford* 1517. ModE adj. **east** + p.n. *Gilford* 1598 from the family name *Guldeford*. Also known as *Newguldford* 1508, the earliest reference. It is recorded in 1507 that 'the church of New Guldeford, within the brooke (*marisco*) commonly called *Guldeford Inning*, now reclaimed from the sea and made dry land by Richard Guldeford, Kt., having been newly built at his expense' was consecrated by bishop Fitzjames and made into a parish church. 'East' for distinction from GUILDFORD Surrey TQ 0049. Sx 530.

GULLAND ROCK Corn SW 8778. *Gullond Rok* 1545 Gover n.d., *The Gull rock* 1576. P.n. Gulland + ModE **rock**. Gulland, *Gull land* 1694 Gover n.d., *Gulland* 1748, is 'gull-land', ModE **gull** + **land**. PNCo 91, Gover n.d. 355.

GULVAL Corn SW 4831. 'St Gwelvel's (church)'. *(ecclesia) Sancte Welvede* 1302, ~ *Sancte Welvele de Lanystly* 1327, *Sanctus Weluelinus* 1377, *(ecclesia) Sancte Gwelvele alias Wolvele* 1413, *Gulvall* 1522. Saint's name *Gwelvel* from the dedication of the church. Nothing is known about him or her. The churchtown was called *Landicle* 1086, *Lanestly -i* 1244–91, *Lanisley Common* 1888, Co ***lann** 'church-site' + unknown element **ystly**. PNCo 91, Gover n.d. 622, DEPN.

GUMLEY Leic SP 6890. 'Godmund's wood or clearing'. *Godmundeslaech* [749]12th S 92, *-leah* 779 S 114, *Godmvndelai*,

Gvtmvndeslea 1086, *Gu- Gothmundele(y)* 1109–1364, *Gom(m)undele(y)* 1087×1100, 13th cent., *Gummeley* 1518, *Gumley* from 1535. OE pers.n. *Gōdmund*, genitive sing. *Gōdmundes*, later replaced by ON pers.n. *Guthmundr*, + **lēah**. Lei 222, SSNEM 217.

GUNBY Humbs SE 7135. 'Gunna, Gunni or Gunnhildr's village or farm'. *Gunelby* 1066×9, *Gundeby* 1070×83, *Bonnebi* (sic) 1086, *Gunneby* 1154–1354, *Gunby* from 1154. ON feminine pers.n. *Gunnhildr* varying with shortened form *Gunna*, or *Gunni* + **bȳ**. YE 239 gives pr [gumbi], SSNY 29.

GUNBY Lincs SK 9121. 'Gunni's farm or village'. *Gvnnebi* 1086, 1200, 1202, *-by* 1212–1428, *Gunby* from 1327. ON pers.n. *Gunni*, genitive sing. *Gunna*, + **bȳ**. Perrott 187, SSNEM 51, Cameron 1998.

GUNDLETON SU Hants 6133. No early forms and not recorded in 1817 OS.

GUNFLEET SAND Essex TM 2611. *Gunfletsond* 1320, *the Gonfle(e)te* (sand-bank) 1509×47. P.n. Gunfleet + ME **sand** 'a sand-bank'. Gunfleet, *les Gunflethavyn* 1504, *Ganfleete* (a port) 1586, is the original name of Holland Brook TM 1226, unknown element + OE **flēot**. The haven was probably at Holland Creek. Ess 17.

GUNN Devon SS 6333. No early forms. Probably Co **goon** 'down-land, unenclosed pasture'. The hamlet is situated on the edge of Hutcherton Down SS 6333.

GUNNERSIDE NYorks SD 9598. 'Gunnar's shieling'. *Gunnersete* 1301, *Goner-* 14th, *Gonnerside* 1655. ON pers.n. *Gunnarr* + **sǽtr**. YN 271 gives pr [gunəsit].

GUNNERTON Northum NY 9075. 'Gunware's farm or village'. *Gunwarton* 1169–1428. OE feminine pers.n. *Gunnwaru* (ON *Gunnvǫr*), genitive sing. *Gunnware*, + **tūn**. NbDu 97, SPN 118.

GUNNESS Humbs SE 8411. 'Gunni's headland'. *Gunnesse* 1199–1250, *Gunnes* 1219. ON pers.n. *Gunni* + **nes**. Gunness lies in a bend of the r. Trent. DEPN, SSNEM 169.

GUNNISLAKE Corn SX 4271. 'Gunna's stream'. *Gonellake* 1485, *Gunnalake* 1608, 1614, *Gunnislake* from 1796. OE pers.n. *Gunna* + **lacu**. The reference is probably to the side channel by Weir Head. Later reformed under the influence of Co dial. **gunnis** 'crevice in a mine-lode'. A 19th cent. village named after an earlier tin-mine. PNCo 91, Gover n.d. 180.

GUNTHORPE Norf TG 0134. 'Gunni's outlying farm'. *Gune(s)-Gunatorp* 1086, *Gunethorp'* 1219. ON pers.n. *Gunni*, genitive sing. *Gunna*, + **thorp**. DEPN, SPNN 165.

GUNTHORPE Notts SK 6844. Probably 'Gunni's outlying farm'. *Gvlne- Gūnetorp* 1086, *Gunthorp(e)* from 1088, *Gunthrop(p)* 1606, *Guntrop* 1709. ON pers.n. *Gunni*, genitive sing. *Gunna*, + **thorp**. Nt 167, SSNEM 111.

GUNTON HALL Norf TG 2234. P.n. *Gunetune* 1086, *-ton'* 1206–20, *Guntone* 1275, *Gunton* from 1316, 'Gunni's estate', ON pers.n. *Gunni*, genitive sing. *Gunna*, + **tūn**, + ModE **hall** referring to the mansion house built in 1742. DEPN, SPNN 165.

GUNTON STATION Norf TG 2535. P.n. Gunton as in GUNTON HALL TG 2234 + ModE **station**.

GUNVER HEAD Corn SW 8977. *Gunver Head* 1813. Unexplained p.n. Gunver + ModE **head**. PNCo 92.

GUNWALLOE FISHING COVE Corn SW 6522. *Gunwalloe Fishing Cove* 1888. Gunwalloe, *(eccl.) Sancti Wynwoluy* 1291, *(capella) Sancti Wynwolai* 1332, *St Wynwolay* 1449, *Wynwala* 1499, *Gonwallo* 1584, is saint's name *Gunwalloe* or *Winwaloe* from the dedication of the church. Believed to have been born in Brittany where he is patron saint of Landévennec where he was abbot, *Landevennoch* 818, as also of Landewednack in Cornwall, in both of which names he is commemorated in the pet form **To-winn-oc*. Earlier known as *ecclesia de Winiton* 1219, *parochia de Wynanton* 1333, ~ ~ *Wynianton* 1337, 1380, from Winnianton SW 6620, the chief manor of the parish, *Winetone* 1086, *Winne(n)- Uinne- Uuingetona* 1086 Exon, *Wi- Wyni(e)nton* 1195–1346, *Gwynyon* 1439, *Winnington* 1678, apparently 'estate called after Wine', OE pers.n. *Wine* + **ing**[4] + **tūn**. However, this kind of formation is unlikely this far W in Cornwall and the specific may be an unidentified Cornish p.n., perhaps the district name seen in Pedngwinian Point SW 6521, *Pengwenyon point* 1610, 'the end of Gwenyon', Co **pen**. PNCo 92, Gover n.d. 548, 550, Dauzat-Rostaing 382, CPNE 218–9.

GURNARD IoW SZ 4796. Uncertain. *Connore, Gom(m)ore* 1279, *Gornore* 1280–1327, *Gurnore* 1333, *Gurnard(e) -er(d)* 16th, *Gurnet* 1781, 1821. Possibly the '(place) at the muddy shore', OE **gyru** or ***gyre**, definite form dative case **gyr(w)an**, + **ōra**. The later forms of the name have been influenced by the fish-name *gurnard, gurnet*. Wt 187 gives pr ['gəːrnər], Mills 1996.56.

GURNARD BAY IoW SZ 4795. *Gurnard Bay* 1769, *Gurnet Bay* 1781–1829. P.n. GURNARD SZ 4796 + ModE **bay**. Wt 187, Mills 1996.56.

GURNARD'S HEAD Corn SW 4338. *Gurnard's Head* 1732, 1748. ModE **gurnard**, a fish with a heavily armoured head, + **head** 'a headland'. From the shape, resembling the head of a gurnard fish. Locally *Izner, the Isnarl* 1935 from nearby Chapel Jane SW 4338, *Innyall Chappell* [1580]18th, possibly Co ***enyal** 'desert, wild'. PNCo 92, Gover n.d. 667.

GURNEY SLADE Somer ST 6249. → Gurney SLADE.

GUSSAGE ALL SAINTS Dorset SU 0010. 'All saints' G'. *Gersich(e) Omnium Sanctorum* 1155, *Gussich(e)* ~ ~ 1242×3, *Gussehich All* Saints 1425, *All Hallows Gussage* 1466. The simplex name is recorded in the following forms, *Gyssic* [966×72]early 12th, *Gessic* 1086, *Gessich(e)* 1183–14th, *Gersic* [1091–1158]after 1305, *-sich(e)* 1155–1233, *Gussich(e)* 1211–1445, *Gusshigge, Gussege* 1432–1500, *Gussage* from 1509×47. This is said to be a compound of OE ***gyse** 'a gush of water' and **sīċ** 'a water course, a small stream', but *sīċ* is otherwise used only of very minor water-courses and would be inappropriate in composition with *gyse* or for the long swift-flowing stream here. The alternative suggestion, ***gysiċ** 'little gusher' < *gyse* + diminutive suffix *-iċ*, might be an ironic or affectionate name. 'All Saints' from the dedication of the church. Do ii.276 gives prs ['gʌsidʒ] and ['gisidʒ].

GUSSAGE ST MICHAEL Dorset ST 9811. 'St Michael's G'. *Gyssich(e) St Michael* 1273, *Gussych(e) Michaelis* 1340, *Mighelys Gussech* 1393, *Michaels Gussage* 1618, *Gussage St Michael* 1795. P.n. Gussage as in GUSSAGE ALL SAINTS SU 0010 + saint's name *Michael* from the dedication of the church. Do ii.143.

GUSTON Kent TR 3242. 'Guthsige's estate'. *Gocistone* 1086, *Gusistune, guthistun* [1087]13th, *Guts(i)eston* 1208, 1218, *Guthsieston* 1229, *Gussiston'* 1254, *Guston* 1610. OE pers.n. *Gūthsiġe*, genitive sing. *Gūthsiġes*, + **tūn**. PNK 562.

GUY'S HEAD Lincs TF 4825. A modern name dating from after the channelling of the r. Nene whereby Guy's Hospital in London gained land in Sutton Marsh in 1733 and subsequently made more enclosures in 1747, 1805 and 1865. Payling 1940.52, Wheeler 129.

GUY'S MARSH Dorset ST 8420. 'Guy's marsh'. *Gyesmersch* 1401, *Gay's Marsh* 1841, *Guy's Marsh* 1869. Pers.n. or surname *Guy* + ME **mersh** (OE *mersc*). The person referred to is unknown. Do iii.93.

GUYHIRN Cambs TF 3903. Possibly 'the guide corner'. *le Gyerne* 1275, *(le) Gy(e)herne -hyrne -hirne* 1275–1513, *Guy(e)hyrne -herne -hirne -hurne -horne* 1438–1579. OFr **guie** + ME **hirne** (OE *hyrne*). The reference is probably to channelling to guide the waters flowing up the Wisbech river here along Waldersea Bank. It was from here that the channel called Moreton's Leam was cut in 1478. Ca 293.

GUYZANCE Northum NU 2103. 'The *Guines* manor'. *Gysnes*

[1240]14th, 1242 BF–1346, *Gynes* 1252, *Gysinis -ys* 1266, 1296 SR, *Gysyns* 1428, *Guisons* 1586. A manorial name from a Norman family which originated from Guines near Calais, *Gisna* 807. NbDu 97, Dauzat-Rostaing 337 s.n. Guignecourt.

River GWASH Leic SK 8306. Uncertain. *le Whasse* c.1230, *Wesse* 1263, *Wasse* 1276–1307, *Washe* c.1545, *Gwash* 1586. Possibly OE adj. **hwæss** 'sharp', related to *hwœt* 'sharp, lively' or ***wæsse** 'land by a meandering river which floods and drains quickly' transferred to the river itself. Neither explanation is entirely satisfactory; cf. however River MEASE SK 2711. In the 16th cent., the name was associated with the more famous WASH TF 5041. The form Gwash is pseudo-Welsh possibly introduced by the antiquarian Camden. Lei 95, RN 436.

GWEEK Corn SW 7026. Uncertain. *Wika* 1201, *Wyk* 1300, *Wi-Wyke* 1302–37, *Wick* 1610, *Gwyk* 1358. This could be OE **wīċ** 'village, trading post' assimilated into Cornish by substitution of *Gw-* for *W-*, or the original Co ***gwyk** 'village' or 'forest'. Gweek is at the SW extremity of Constantine parish. Another possibility might be ON **vík** 'a bay, an inlet' referring to Helford river estuary at the head of which the village stands. PNCo 92, Gover n.d. 503, DEPN.

GWENNAP Corn SW 7340. 'St Wenep's (church)'. (Church of) *Lamwenep* 1199, *Lawenep* 1444, *ecclesia Sancte Weneppe* 1269–1342, ~ ~ *Wenep* 1291, *Seynt Gwenape* 1520, *Gwennape, Gwynep* c.1600 G. Co ***lann** 'church-site' + saint's name *Wenep* from the dedication of the church. Nothing is known of her; possibly identical with *Wynup*, one of the 24 sons and daughters of King Broccan of Breconshire. PNCo 92, DEPN.

GWENNAP HEAD Corn SW 3621. *Gwennap Head* 1888. Surname *Gwenappe* + ModE **head** 'a headland'. PNCo 93.

GWENTER Corn SW 7417. Uncertain. *Wynter* 1263, 1315, *Wenter* 1324, *Gwynter* 1519. Possibly 'windy-land', Co **guyns**, earlier **gwint*, + **tyr**, or 'white water', Co **guyn** + **dour**, but that should have medial *-th-* as Gwenddwr, Powys SO 0643. Another possibility might be OE **winter** 'a vine-yard'. PNCo 93, Gover n.d. 554.

GWINEAR Corn SW 5937. 'St Winier's (church)'. *(ecclesia) Sancti Wyneri* 1258, ~ *de Sancto Wyniero* 1286, ~ *Sancti Wynieri* 1291, *Seynt Wynyer* 1350, *Gwynnyar* c.1535, *Gwynyer* 1610. Saint's name *Winier* (Gwinear) from the dedication of the church. He was reputed to have been the son of a pagan Irish king who came to Cornwall and was martyred with his 777 companions by the pagan Cornish king Theodoric. He has a dedication at Pluvigner (*Ple Guinner* 1259, 'parish of Guigner') in Morbihan, Brittany. PNCo 93, Gover n.d. 596, DEPN, Dauzat-Rostaing 535.

GWITHIAN Corn SW 5841. 'St Gwythian's (church)'. *(parochia) Sancti Goziani* 1327, *parochia Sancti Goi- Goythiani* 1335–1405, *Gothian* 1524, *Gwithian* from 1563. Saint's name *Gwythian* from the dedication of the church. Nothing certain is known of him, but he was honoured also in Brittany. Initial *Gw-* is probably due to the influence of nearby GWINEAR SW 5937. PNCo 93, Gover n.d. 599, DEPN.

GYPSEY RACE Humbs TA 0970. 'The intermittent stream or stream that flows partly underground'. *Gipse, Gypse* c.1198, 1327. OE ***ġips** + **ēa**. A local dial. term for streams that spring intermittently from the Yorkshire Wolds in wet seasons. William of Newburgh describes this one in 1198: 'In the province of Deira, not far from the place where I was born, a wonderful thing occurs... There is a village some miles distant from the North Sea, beside which those famous waters commonly called *Gipse* spring from the ground from numerous springs, not continuously, but at intervals of years. They form a torrent of some magnitude and flow through the lower ground into the sea. When they dry up it is a good sign, for their appearance is said to be a reliable portent of impending famine.' *Race* is from ON **rás** 'a rush of water, a channel, a water-course'. YE 4 gives pr [gipsə] with Scandinavian [g] for English [j].

HABBERLEY H&W SO 8077. 'Heahburg's wood or clearing'. *Harburgelei* 1086, *Haberlega -ley(e), Haburley* 1183–1550. OE feminine pers.n. *Hēahburg*, genitive sing. *Hēahburge*, + **lēah**. Wo 249.

HABBERLEY Shrops SJ 3903. Probably 'Heathuburh's clearing'. *Habberleg'* 1242, *-le(ye)* 1299–1309, *-ley* from 1577, *Hatburleg'* 1242, *Haberle(gh) -ley(e)* 1271–1774. OE feminine pers.n. *Heathuburh* + **lēah**. An alternative possibility for the pers.n. might be OE *Hēahburh*. Sa i.140.

HABBLESTHORPE Notts SK 8181 A lost vill in North LEVERTON WITH HABBLESTHORPE Notts SK 7882. Possibly 'Hæppel's outlying farm'. *Happelesthorp* [1153×4]14th, 1267–1364, *Appeles-* 1268, *Hap(e)les-* 1267–1330, *Habylthorp* 1527, *Habylstrope* 1539, *Hablesthorpe* 1594, *Happelstrappe* (sic) 1602. OE pers.n. **Hæppel*, genitive sing. **Hæpp(e)les*, + ON **thorp**. Nt 34, SSNEM 127.

HABITANCUM Northum NY 8986. 'Place belonging to or associated with Habitus'. *Habitanci* 4th Inscr., *Evidensca* (sic for *Avitanco)* [c.700]13th Rav. Latin pers.n. *Habitus* + p.n. forming suffix **-anc-**. The RBrit name of the Roman fort at Risingham. RBrit 371.

HABROUGH Humbs TA 1413. 'The high fortification'. *(H)aburne* 1086, *Haburc* c.1115–1252, *-burg(')* before 1180–1300, *-burgh(')* 1259–1817. ON **há(r)** probably replacing OE *hēa(h)* + **burg**. The place stands on an island of relatively high ground. Li ii.139.

Great HABTON NYorks SE 7576. *Great Habton* 1365. ME **great** + p.n. *Habetun, Ab(b)etune* 1086, 'Habba's settlement', OE pers.n. **Hab(b)a* + **tūn**. 'Great' for distinction from Little HABTON NYorks SE 7477. YN 74, SSNY 256.

Little HABTON NYorks SE 7477. *(parva) Hab(b)eton* c.1163–1333, *parva Habton* 1365. ModE **little**, Latin **parva**, + p.n. Habton as in Great HABTON NYorks SE 7576. YN 74.

HACCOMBE Devon SX 8970. Possibly 'Hacca's coomb'. *Hacome* 1086, *Haccumbe, Haccombe* from 1238. OE pers.n. *Hacca* + **cumb**. The specific might alternatively be OE *haca* 'a hook', but the topographical application is not clear. D 459.

HACEBY Lincs TF 0336. 'Haddr's village or farm'. *Hazebi* 1086, 1165×6, 1218, *Hatsebi* 1115, *Hacebi* 1172, 1198, *-by* from 1201. ON pers.n. *Haddr*, genitive sing. *Hadds*, + **bȳ**. Perrott 129, SSNEM 52, Cameron 1998.

HACHESTON Suff TM 3059. 'Hæcci's estate'. *He(t)ce- Haces- Hece(s)tuna* 1086, *Hecetune* c.1095, *Hascheton* 1197, *Hacheston* 1292, *Hachston* 1610. OE pers.n. *Hæċċi*, genitive sing. *Hæċċes*, + **tūn**. DEPN.

HACKENBY DYKE NYorks SE 4941. Lost village name *Hagede- Haghedene- Hagendebi* 1086, *Hag(g)andeby* 12th–1388, *Hagenby* 1310, 'Haggande's village or farm', ON pers.n. *Hǫggvandi* 'the hewer, the executioner' < PrScand ***haggwan**, genitive sing. *Hǫggvanda* + **bȳ**. In *Haggenby* was *Hagandehou* 12th, 'Haggandes's burial mound', ON **haugr**. For a similar pairing see GRIMESTHORPE SYorks SE 3587. YW iv.239, SSNY 29, SPN 148.

HACKFORD Norf TG 0502. Partly uncertain. *Hakeforda* 1086, *-ford* 1203, *Hacford* 1254. Possibly OE **haca** 'a hook' used topographically of a bend of some kind or as a pers.n. *Haca*, + **ford**. Both valley and road make a bend at this place. DEPN, PNE i.213.

HACKFORTH NYorks SE 2493. 'Ford at the bend'. *Acheford(e)* 1086, *Hac- Hakford* 1184–1350, *Ha(c)keford* 13th cent., *Hackfourth* 1565. OE **haca** + **ford**. YN 240.

HACKINSALL Lancs SD 3447. 'Hakon's spur of land'. *Hacunesho* c.1190–1262, *Hacunshou -owe* [1199×1200]1268–1335, *Hacinshou* 1262×8, *Hakensowe -awe* 1324. ON pers.n. *Hákon*, genitive sing. *Hákons*, + OE **hōh** later replaced with ON **haugr**. Hackinsall is situated on slightly raised ground in a low and level district. La 160, SSNNW 129, Jnl 17.92.

HACKLETON Northants SP 8055. 'The estate called after Hæcel'. *Hachelintone* 1086–1405 with variants *Hakelin(g) -yng-*, *Hakilton* c.1265, *Hakelton* 14th cent. OE pers.n. **Hæċel* with [k] for [tʃ] before [l] + **ing**[4] + **tūn**. Nth 146.

HACKNESS NYorks SE 9690. 'The hook-shaped headland'. *Hacanos* [c.731]8th BHE, *Heaconos* [c.890]c.1000 OEBede, *Hagenesse* 1086, *Hakenes(se)* c.1081–1354. OE **haca** + **nōs(e)** later replaced by **næss**. Hackness lies at the foot of a very prominent ridge projecting between the river Derwent and Lowdales Beck. YN 112, YE xlix.

HACKNEY GLond TQ 3585. 'Haca's' or 'bend island'. *Hakeneia -ey(e)* 1198–1443, *Hackney* 1535, *Haqueneye* 1593. OE pers.n. *Haca* or **haca**, genitive sing. *Hacan*, **hacan**, + **ēġ** in the sense 'dry land in marsh'. Mx 105, L 39, GL 54.

Upper HACKNEY Derby SK 2961. *Overhackney alias Hackney lane* 1586. The evidence is too late for explanation unless the name is to be associated with the form *Hackinhale* 1216×72, 'Haca's or hook corner of land', OE **haca** or pers.n. *Haca*, genitive sing. **hacan**, *Hacan*, + **halh**, dative sing. **hale**. Db 83.

HACKPEN HILL Devon ST 1112. 'Hook hill'. *(on) hacapenn* *[670 for 925×39]11th S 386, *Hakepen(ne)* 1249–91. OE **haca** + ***penn** < PrW **penn*. A prominent hook-shaped hill. D 538.

HACKPEN HILL Wilts SU 1375. *Hackpen Hill* 1828 OS, earlier *Hacpendowne* 1570, *Hackpinn Downes* 1695. P.n. *(on) hacan penne* 939 S 449, *Hakepen* 1245, probably a compound of OE **haca** 'a hook', genitive sing. **hacan**, and PrW **penn** 'a head, an end, a top, a height, a hill' adopted as a loan-word in OE, + ModE **hill**. The reference is to the 'hook' or projecting end of the hill at Monkton Down SU 1172. Wlt 310.

HACKTHORN Lincs SK 9982. 'The hawthorn-tree'. *Haggethorn* [968]13th, 1193, *Age- Hageto(r)ne* 1086, *Hage- Hakethorn* 1202, *Hacatorn, Hachethorna* c.1115. OE **hagu-thorn** varying with ***haca-thorn**. DEPN, PNE i.213, Cameron 1998.

HACKTHORPE Cumbr NY 5422. Possibly 'outlying farm on the hook or promontory'. *Haka- Hacatorp* c.1150, 1170×80, *Haketorp -thorp(e)* c.1150–1401, *Hac- Hakthorp* 1295–1429. ON **haki** or OE **haca** + **thorp**. The reference would be to the pointed ridge of land protruding into a junction of the river Leith with a tributary stream at NY 5521. Alternatively the specific could be ON pers.n. *Haki*, genitive sing. *Haka*. We ii.182, SSNNW 57.

HACONBY Lincs TF 1025. 'Hakon's village or farm'. *Hacunes- Hacone(s)bi* 1086, *Haconebi -by* 1219–1372, *Hacunebi* 1135–1201, *Hacunbi* 1185, *-by* 1254–1539, *Hacumby* 1267–1547, *Hacomby* 1316–1556, *Haconby* from 1316. ODan pers.n. *Hākon* sometimes with secondary genitive sing. **-es**, + **bȳ**. Perrott 118, SSNEM 52.

HADDENHAM Bucks SP 7408. 'Hæda's homestead'. *Nedreham* (sic) 1086, *Hedreham* 1087×1100, *Hedenham* 1142–1247,

Hadenham 1196–1373. OE pers.n. **Hǣda*, genitive sing. **Hǣdan*, + **hām**. Bk 161.

HADDENHAM Cambs TL 4695. 'Hæda's homestead'. *(on) hædan ham* [970]13th S 780, *Hadenham* 1086–1484, *Haddenham on the hill* 1537, *Had(d)re- Hæderham* 1086. OE pers.n. *Hǣda*, genitive sing. *Hǣdan*, + **hām**. Ca 231.

HADDISCOE Norf TM 4496. 'Haddr's wood'. *Had- Hatescou -scov* 1086, *Had(d)esco* 1209, 1236. ON pers.n. *Haddr*, secondary genitive sing. *Haddes*, + **skógr**. DEPN, SPNN 172.

Chapel HADDLESEY NYorks SE 5826. *Chappel(l) Had(d)lesey* 1605 etc. ModE **chapel** + p.n. *Hæðel-sǣ* c.1030 YCh 7, *Mediana Haþelsay* 'middle H' 1190×1210, *Hathelsay(e) -sey(e)* 1279×81–1428, *Haddelsay* 1426, *Hausay(e) -sey(e) -sai* 1196–1268, as in East or Goose HADDESLEY SE 5925. YW iv.18.

East or Goose HADDLESEY NYorks SE 5925. *Esthathilsay(e) -el- -sey(e)* 1288–1379, *Est Haddlesey* 1542, *Goose Hadlesey* 1650, *Esthausay* 1264. ME **est**, ModE **goose**, + p.n. *Haðel-sǣ* c.1030 YCh 7, OE *Hathel* of unknown meaning + **sǣ** 'a lake'. *Hathel* has been explained either as a derivative of OE *hāth* or *hǣth* 'heath' with the sense 'heathland' like *brēmel* beside *brōm*, *thȳfel* beside *thūf* etc., or perhaps as a word related to *hatho-* in OE *hatholitha* 'elbow' and *heathor* 'enclosure, prison, chain', the sense of which is believed to be 'hollow, cavity' as in Greek κοτύλη 'a bowl, a basin, a cup, a cavity'. The reference would be to a marshy hollow in or near the course of the river Aire. Both attempts to explain this name are very uncertain. YW.iv 18–19 and xi, Notes 62–3, Chantraine 573.

West HADDLESEY NYorks SE 5626. *West(e)hathel(e)say(e) -ey(e)* 1280–1387, *Westhadylsay* 1460, ~ *Haddysley* 1538. ME **west** + p.n. *þriddan Haðel-sǣ* 'the third H' c.1030 YCh 7, *Hausey -ay* 1234, as in East or Goose HADDESLEY SE 5925. YW iv.20.

HADDON 'Heath-hill'. OE **hǣth** + **dūn**. L 153, 155.

(1) ~ HALL Derby SK 2366. *Haddon Hal* 1610. Pn. Haddon as in earlier *castrum Haddon* 'Haddon castle' c.1460 + ME **halle**. The reference is to a medieval manor house dating from about 1370. Haddon is *Hadun(e) -a* 1086, 1189×99, *Haddon(e) -a* 1101×8–1202 etc., *Heddon'* 1176–1206. Db 106.

(2) ~ HILL Somer SS 9828. *Haddon beacon* 1610, *Haddon Down Hill* 1809 OS.

(3) East ~ Northants SP 6668. *Esthaddon* 1265, 1279, 1305. OE **ēast** + p.n. *Ed(d)one, Hadone* 1086, *Haddun* 1185, *Haddon* 1195, *Hed(d)on(e)* 1230. 'East' for distinction from West HADDON SP 6271. Nth 83, L 153, 155.

(4) Over ~ Derby SK 2066. *Hadd(on)' Superior* c.1199, 1210, *Uuer Haddon(e)* 1206, 1230 etc. with variants *Ufre, (H)oure* etc. OE **uferra** 'upper', Latin **superior**, for distinction from Nether Haddon SK 2366, *Underhaddon* 1206, *Thunre-Haddone* 'the lower H' 1212, *Nethir -er Haddon(e)* 1248, 1257 etc., earlier forms as HADDON HALL SK 2366. Over Haddon is situated on high ground at 750ft. overlooking the river Lathkill, Nether Haddon on low ground in the Wye valley. Db 107, 106.

(5) West ~ Northants SP 6271. *Westhaddon* from 12th. OE **west** + p.n. *Ecdone, Ed(d)one* 1086. 'West' for distinction from East HADDON SP 6668. An 18th cent. writer on the natural history of Northamptonshire comments on the healthy sheep 'upon the heaths' at West Haddon. Nth 71.

HADDON Cambs TL 1392. 'Hædda's hill'. *(æt) Haddedune* [951]14th S 556, *Haddune* [c.1150]14th, *Adone* 1086, *Haddon* from 1268. OE pers.n. *Hǣdda* + **dūn**. Hu 188.

HADEMORE Staffs SK 1708. *Hademore* 1834 OS.

HADFIELD Derby SK 0296. 'Heathland'. *Hetfelt* 1086, *-feld* 1216×72, *Hedfeld'* 1232, *Haddefeld'* 1185–6, 1353, 1439, *Hadfeld* 1372, 1378. OE **hǣth** + **feld**. Identical with HATFIELD. Db 103, L 242.

HADHAM CROSS Herts TL 4218. 'Hadham cross-roads'. *Hadham Cross* 1653. P.n. Hadham as in Much HADHAM TL 4319 + ModE **cross**. Hrt 178.

HADHAM FORD Herts TL 4321. *Hadham Ford* 1805 OS. Earlier simply *le Forthe* 1363, *Ford* 1582. P.n. Hadham as in Much HADHAM TL 4319 + ME **ford**. Hrt 178.

Little HADHAM Herts TL 4422. *Parva Hadā* 1086, *Little Hadham* 1805 OS. ModE **little**, Latin **parva**, + p.n. Hadham as in Much HADHAM TL 4319. Hrt 176.

Much HADHAM Herts TL 4319. *Magna Hedham* 1278, *Muchel Hadham* 1373, *Myche* ~ 1552. ME adj. **much**, Latin **magna**, + p.n. *Hædham* [946×75]17th S 1795, *Hedham* [975×91]11th S 1494, [1000×2]11th S 1486, *Hadham* from [1042×66]13th S 1051, *Hadam* 1086, 'Hæda's homestead', OE pers.n. *Hǣda* or *Hædda*, + **hām**. The absence of the genitival ending *-an* is noteworthy; for this reason the name has usually been taken to be 'heath homestead', OE *hǣth* + **hām**. The difficulty here is the lack of any trace of *-th-*. Hrt 176 gives pr [hædəm].

HADLEIGH Essex TQ 8087. 'Heath wood or clearing'. *Hæþlege* [c.1000]c.1125, *Hadleg(a) -lea -legh(e) -le(y)e* 1121–1330, *Hathleg(h)* 1237–40. OE **hǣth** + **lēah**. This was simply *Leam* in 1086, probably part of the *leigh* of LEIGH-ON-SEA TQ 8385. Ess 185, Jnl 2.43.

HADLEIGH Suff TM 0242. 'The heath covered clearing'. *(into) Hedlæge* [962×91]11th S 1494, *Hæðleh* 1042×66 S 1047, *Hetlegā* 1086, *Hadlega* 1183, *Hadley* 1610. **hæth** + **lēah**. The boundary of Hadleigh is referred to as *Hædleage gemære* 1000×2 S 1486. DEPN.

HADLEY Shrops SJ 6712. 'Heath clearing'. *Hatleye* 1086, *Hathlegh'* 1271×2, *Hedleia -leg(a) -ley* c.1138–14th, *Hadlega -ley(e)* from 1191. OE **hǣth** + **lēah**. Sa i.141.

HADLEY END Staffs SK 1320. *Hadley End* 1836 OS. Cf. *Hadley End Gates, Hadley Cottage* ibid. Contrasts with *Snail's End* ibid. at SK 1419.

HADLOW Kent TQ 6350. Partly uncertain. *Haslow* 1086, *Haslo* c.1100, *Hadlo* 1214×26, 1248, *Haudlo(u)* 1270, 1280, 1292. This might be the 'heather-covered mound' or the 'chief mound', OE **hāth** + **hlāw** or **hēafod** + **hlāw**. The reference is to a low hill overlooking the PNK Bourne. K 176.

HADLOW DOWN ESusx TQ 5324. *Hadledowne* 1333. P.n. Hadlow + ME **doun** (OE *dūn*). Hadlow, *Had(de)le(gh)* 1254–1457, is 'Headda's wood or clearing', OE pers.n. *Headda* + **lēah**. Sx 395.

HADNALL Shrops SJ 5220. 'Headda's nook'. *Hadehelle* 1086, *Hedenhale* 13th, *Hadenhal(e)* 13th–1431, *Hadnall* from 1450. OE pers.n. *Headda*, genitive sing. *Headdan*, + **halh**, dative sing. **hale**, referring to the slight valley running from E to W. Sa i.142.

HADRIAN'S WALL Cumbr NY 4661. See The ROMAN WALL Cumbr NY 4661.

HADRIAN'S WALL Northum NY 9669. *The Wall of the Picts, Vallum Severi* 1695 Map, *Roman Wall* 1869 OS. Named after the Roman emperor Hadrian who in AD 122 visited Britain and *murum per octoginta milia passuum primus duxit qui barbaros Romanosque divideret* 'first built a wall 80 miles long to divide the barbarians and the Romans' *Vita Hadriani, Scriptores Historiae Augustae*. Cf. WALL NY 9169.

HADSTOCK Essex TL 5645. 'Hada's outlying farm'. *Hadestoc* *[1042×66]13th S 1051, [1042×66]12th, *-stok(e)* 1209–1346, *Haddestok(e)* 1270, 1280, *Hadstok* 1303. OE pers.n. *Hadda* + **stoc**. An earlier name of this place seems to have been *Caddanno, Cadenho* [1008]12th, *Cadenhov* 1086, 'Cada's hill-spur', OE pers.n. *Cada*, genitive sing. *Cadan*, + **hōh**. Ess 510.

HADSTON CARRS Northum NU 2800. 'Hadston rocks'. P.n. *Hadeston* 1189, 1242 BF, 1296 SR, *Hadistonam* 1236 BF, *Haddeston* 1255, *Hadsen* 1676, 'Hæddi's farm or village', pers.n. OE *Hæddi*, genitive sing. *Haddes*, + **tūn**, + N dial. **carr** (OW **carr*, cf. Gaelic *cárr* 'rocky shelf', Ir *carr, carraig*, W *careg*, OW *carrecc*). NbDu gives pr [hadsən].

HADZOR H&W SO 9162. 'Headd or Headdi's ridge-tip'. *Headdesofre* 11th, *Hadesore* 1086, *Had(d)esouere* c.1215–1428,

Haddesor 1327–1486, *Hadser, Hadsore* 1533, 1535. OE pers.n. *Headd(i)*, genitive sing. *Headdes*, + **ofer**. Hadzor House is situated on the tip of a low ridge. Wo 291, L 177.

HAFFENDEN QUARTER Kent TQ 8840. *Haffenden Quarter* 1798, ~ *Quar.*ʳ 1819 OS. Contrasts with Further Quarter TQ 8939, *Further Quarter* 1819 OS, and Middle Quarter TQ 8938, *Middle Quarter* 1819 OS, divisions of High Halden parish. P.n. Haffenden + ModE **quarter**. If the form *efreðingdenn* 863 S 332 is correctly identified with this place, Haffenden, otherwise *Heuerthinge, Hereuerthing'* 1244, *Herferthige -thyng'* 13th, is the 'woodland pasture called after Herefrith', OE pers.n. *Herefrith* + **ing**[4] + **denn**, a woodland pasture belonging to Mersham. KPN 216, Cullen.

East HAGBOURNE Oxon SU 5288. *Esthakeburn'* 1275×6, *East Hagborn* 1752. ME adj. **est** + p.n. *Haccaburne, (æt) hacceburnan* [879×99]12th S 354, *(æt) Hacceburnan* 990×2 S 1454, *Hacheborne* 1086, *Hakeburn' -b(o)urne* 1199–1454×5, *Hagbourne* from 1454×5, 'Hacca's stream', OE pers.n. *Hacca* + **burna**. 'East' for distinction from West HAGBOURNE SU 5187. Brk 519.

West HAGBOURNE Oxon SU 5187. *Westakeburn'* 1241, *Westhakeburn'* 1284, *Westhagbourn* 1517. ME **west** + p.n. Hagbourne as in East HAGBOURNE SU 5288. Brk 519.

HAGGBECK Cumbr NY 4773. No early forms. Apparently 'stream by the clearing', ON **hǫgg** (< **haggu-*) + **bekkr**.

HAGGERSTON Northum NU 0443. 'Hagard's farm or village'. *Agardeston* 1196 P, *Hagardestun -ton* 1208×10 BF–1268, *Haggarston* 1278. Family name *Hagard* 'wild hawk, falcon' (OFr *hagard* 'wild, untamed'), genitive sing. *Hagardes*, + **tūn**. NbDu 98, DS 233 s.n. Haggard.

HAGLEY 'Haw wood'. OE ***hagga** + **lēah**. In some cases an OE pers.n. *Hæcga* is also possible. PNE i.222, L 203.

(1) ~ H&W SO 9180. *Hageleia* 1086, *Haggele(ge) -leg -ley* 1212–1346. Wo 292.

(2) ~ H&W SO 5641. *Hagley* 1831 OS.

HAGWORTHINGHAM Lincs TF 3469. Partly uncertain. *Hacberding(h)am -ingeham -incham, Haberdingham* 1086, *Ag- Hagwordingheheim, Hawordingeham* c.1115, *Hagwurdinge- Aburdingeham* 1167, *Hacwrdhingham* 1197, *Hagwrðingham* 1198, *Hagworthingham* from 1202. Possibly a compound of OE **hām** and a p.n. **Hagworthing*, 'the village at or called *Hagworthing*, the hawthorn-tree enclosure' < OE ***haga-worth** + **ing**[2]. The DB forms may point to an alternative form with ON pers.n. *Hagbarthr*. ING 144, SPN 122, Cameron 1998.

HAIGH GMan SD 6009. 'The enclosure, assart, messuage'. *Hage* 1194, *Hagh(e)* 1298–1332, *Hawe* 1330, c.1540, *Hay* 1539, *Haighe* 1548, *haigh* 1581, *Thaigh al. Le Haigh* 1628. OE **haga**. La 102, TT 76 for the sense, Nomina 13.109–14 for the form, Jnl 17.58.

HAIGHTON GREEN Lancs SD 5634. P.n. Haighton + ModE **green**. Haighton, *Halctun* 1086, *Halicton* 1212, *Halech-* 1226, *Halton* [before 1268]1268, *Halghton* 1278, 14th, *Haughton, Ha(i)ghton* 1560–1600, is 'village, settlement in a nook'. OE **halh** + **tūn**. La 147, Jnl 17.84.

HAILE Cumbr NY 0308. 'Nook, corner'. *Hale* c.1180–1584, *Hayll* 1509×46, *Hayle* 1569. OE **halh**, dative sing. **hale**. The reference is to the valley of Kirk Beck as it cuts into the hills at NY 0309. Cu 398, L 109.

HAILES Glos SP 0530. Uncertain. *Heile* 1086, *Helis* [12th]1253, *Hei- Heyles -is* 1114–1387, *Hai- Hayles* 12th–1714. An ancient settlement with nearby hillforts and a salt-way. A British folk-n. **Salenses* either 'the people living by the stream called **Saliā*, the dirty one' (< **sal-* 'dirty') or 'the salt folk' would account for ME *Hailes*. Gl ii.15.

HAILEY Herts TL 3710. 'Hay clearing'. *Hailet* 1086, *Heilet* 1100×35, *Heile(e)* 1235–1303, *Heyle(gh) -le(g)e* 1241–1374, *Hayley* 1675. OE **hēġ** + **lēah**. The *-t* of the earliest forms is either a mistake or the Fr diminutive suffix **-ette**. Hrt 212.

HAILEY Oxon SP 3512. 'The hay clearing'. *Haylegh'* 1240×1, *Hayley* 1517, *Heyley* 1797. OE **hēġ** + **lēah**. O 321.

HAILSHAM ESusx TQ 5909. 'Hægel's homestead'. *Hamelesham* (sic) 1086, *Ei- Eyles- (H)e(i)lesham* 1189–17th including [1078×1100]15th, *Hay- Hailesham* 1230–1494. OE pers.n. *Hæġel*, genitive sing. *Hæġ(e)les*, + **hām**. Sx 435 gives pr [helsəm], ASE 2.33, SS 86.

HAINAULT GLond TQ 4691. '(The) household's wood'. *Henehout* 1221, *Hyneholt(e)* 1239–1334, *(le) Henneold(e)* 1475, 1552, *Heinault* 1590, *Hainault* 1654. OE **hīġna**, genitive pl. of **hīwan** 'a household, the members of a community' here referring to the nuns of Barking abbey, + **holt**. The modern spelling is due to erroneous association with queen Philippa of Hainault. Ess 2.

HAINFORD Norf TG 2218. 'The enclosure ford, the ford by the enclosure'. *Han- Hamforda* 1086, *Heinford* 12th, 1199, *Henford* 1206. OE ***hæġen** + **ford**. The earliest reference is to a 'homestead at Hainford', *Hemfordham* (for *Hein-*) [1052×66]13th S 1519. DEPN.

HAINTON Lincs TF 1884. 'The hedge farm or village'. *Haintone -tun, Gaintone* 1086, *Heintuna* c.1115, *-ton* 1193, *Hainton* 1197. OE ***hæġen** varying with ON ***hegn** + **tūn**. DEPN, PNE i.215, 241.

HAISTHORPE Humbs TA 1264. 'Haskel's outlying settlement'. *Ascheltorp -il-, Haschetorp* 1086, *Haschelthorp* 1190, *Hastorp* 1246, 1283, *-thorp(e)* 1265–1364, *Hastropp* 1601, *Haysthorp* 1294, 1828. ON pers.n. *Haskel* (from *Hǫskuldr*) + **thorp**. Some of the DB forms show influence of ON pers.n. *Ásketill*. YE 89 gives pr [eəsθrəp], SSNY 59.

HALAM Notts SK 6754. '(The settlement) at the nooks'. *Healum* [958]14th S 659, *Halum* 1197–1340, *Halom(e)* 1392–1535, *Halam* from 1604. OE **halh**, locative-dative pl. **halum**. The reference is to short valleys running into hills W and SW of the village. Nt 167 gives pr [heiləm].

HALBERTON Devon ST 0012. Possibly 'Haligbeorht's farm or village'. *Halsbretone* 1086, *Halberton(e)* c.1200–1340, *Hauberton(a)* c.1200, 1247, *Haulberton* 1675. OE pers.n. *Hāliġbeorht* + **tūn**. D 548.

High HALDEN Kent TQ 8937. *High Halden* 1610. ModE adj. **high** + p.n. *Hadinwoldungdenne* c.1100, *Hathwoldindanna* 1157, *Hathewaldingedenne* 1261, *Hathewoldenn(e)* 13th cent., 'woodland pasture called after Heathuweald', OE pers.n. *Heathuweald* + **ing**[4] + **denn**. PNK 361.

Little HALDON Devon SX 9176. *Lytlehaldon al. Letheweldowen* 1564, *Little Haldon al. Leughwelldoune* 1614, ~ ~ *al. La Well Doune* 1692. A hill-range so named for distinction from Great Haldon SX 8983, *Hagheledon* 1281, *Hagheldon* 1301–31, *Hauldon* 1296, *Heweldon* 1475, *Hayeldoune* 1329, 'hail hill', OE **hagol** + **dūn**. The reference may have been to some famous storm. D 18–19.

HALE '(Place at) the nook'. OE **hale**, dative sing. of **halh**. L 100–107.

(1) ~ Ches SJ 4682. *Halas -es* 1094, *Hale* from 1201. OE **halh**, plural **halas**. Refers to raised patches of dry ground in the marshland along the Mersey shore. La 110, L 107.

(2) ~ GMan SJ 7786. *Hale* from 1086. Refers to its situation in a hollow E of Bowdon SJ 7586. Che ii.24.

(3) ~ Hants SU 1919 *Hala* 1158, *(la) Hale* from 1219. The reference is to the E end of Fordingbridge Hundred abutting on the W limit of the New Forest. Ha 85, Gover 1958.216, L 101.

(4) ~ Lincs TF 1442. *Hale* from 1086, *Hayll, Haile* 16th cent., *Hole* 1339–1585. Two villages are distinguished, *Magna* ~ 1204, *Great* ~ from 1234, and *Lit(t)le* ~ from 1180, both lying on islands of gravel amid the boulder clay. Perrott 66, L 103, Cameron.

(5) ~ BANK Ches SJ 4883. 'Hill leading to Hale'. *Halebonke* c.1240, 1426, *-bank* 1509. P.n. HALE SJ 4682 + ME **banke**. La 110, Mills 1976.90.

(6) ~ BARNS GMan SJ 7985. *Halebarns* 1616, *Hale Barn* 1831. P.n. Hale SJ 7786 + ModE **barn** referring to a tithe-barn here demolished in 1848. Che ii.24.

(7) ~ STREET Kent TQ 6749. *Hale Street* 1819 OS. P.n. *la Hale* 1278, *the hale* 1327–48, 'the nook or dry ground in marsh', OE **healh**, dative sing. **hēale**, + dial. **street** 'a village'. PNK 165.

(8) Holme ~ Norf TF 8807. *Holmhel* 1267, *Holm Hale* 1838 OS. Apparently short for Holm and Hale as in *Hale Hall* 1838 OS, *Hale* 1254. For Holme → HOLME HALE TF 8807. DEPN.

(9) Upper ~ Surrey SU 8449. ModE **upper** + p.n. *Hale* from 1222, '(the settlement) at the nook', OE **halh**, dative sing. **hale**. Sr 176, L 100–01, 103, 106–7.

HALES Norf TM 3897. 'The nooks'. *Hals* 1086, [1087×98]12th, *Hales* 1236. OE **halh**, pl **halas**, in the sense 'valleys'. DEPN.

HALES Staffs, Shrops. Originally the name of a district incuding lands on both sides of the Staffs-Shrops border probably meaning 'the raised patches of ground isolated by marsh'. *Halas* 1086, *Hales* from 1291. OE **halh**, pl. **halas**. Cf. NORTON IN HALES Shrops SJ 7038, Market DRAYTON Shrops SJ 6734 formerly Drayton-in-Hales. DEPN, Sa i.198, L 104.

HALESOWEN WMids SO 9683. 'Hales held by Owen'. *Hales Owayn* 1272, *Halesowing* 1690. P.n. *Hala* 1086, 'the nooks', OE **halh**, pl. **halas**, + manorial addition from the Welsh prince Owain ap Dafydd, nephew of Henry II by his sister Emma, who became lord of Hales in 1204. The manor was earlier forfeited to the crown and so known in the 12th cent. as *Hales Regis* (Latin), 'the king's H'. Wo 293, VCH Wo 3.141–2.

HALES PLACE Kent TR 1459. *Hales Place* 1839 *TA*. Surname *Hales* + ModE **place** 5.b. 'a dwelling house, a mansion'. The place is named after Sir Edward Hales whose home was here in the 18th cent. PNK 498, Cullen.

HALESWORTH Suff TM 3877. 'The nook enclosure'. *Healesuurda, Halesuuorda* 1086, *Haleswurde -word -wurth -worth(e)* 1195–1446. OE **healh** perhaps used as a p.n., genitive sing. **heales**, + ´**worth**. The reference may be to a nook in the encircling hills. DEPN, Baron.

HALEWOOD Mers SJ 4485. 'Wood belonging to Hale'. *Halewode* [c.1200]1268, *Halewood* 1509. P.n. HALE SJ 4682 + OE **wudu**. La 110, Jnl 17.62.

HALFORD Shrops SO 4383. 'The nook ford'. *Hauerford* 1155, *Haleford* 1221×2, *Hancforde* (for *Hauc-*) c.1260×70, *Halegford* 1291×2, *Hawkeford* 1535, *Hawford* 1535–1744. In spite of the variety of forms there can be no doubt that this is a compound name of OE **halh** + **ford**. The spellings *Hanc-* (for *Hauc-*) and *Hawkeford* probably represent a side form in which [k] was substituted for [x] in the ME form *haugh* thus making it identical with *hawke* 'a hawk'. OE **halh** in this name probably means either 'a detached or projecting part of an estate' (Halford formerly belonged to Bromfeld Priory 5 miles SE) or possibly 'land in a river bend' on the assumption that the river Onny formerly made a meander around the village here. Sa i.142.

HALFORD Warw SP 2645. 'The nook ford'. *Halc(h)ford* 1154×89, *Halec- Haleghford* 1190, 1221, *Hale-* 1176–1316, *Halford* from 1329, *Haw(l)ford(e)* 16th, *Hauford* 1675. OE **halh** + **ford**. Wa 254, L 69–70, 101, 105.

HALFPENNY GREEN Staffs SO 8292. *Halfpenny Green* from 1448. ME **halfpenny** + **green**. Probably a reference to payment charged to drovers for overnighting their animals. Field 1993.193, Horovitz.

HALFWAY HOUSE Berks SU 4068. *Half Way House* 1675, *Halfway House* 1761. Situated approximately half-way between Newbury and Hungerford. Brk 275.

HALFWAY HOUSE Shrops SJ 3411. *Halfway House* 1836 OS. Originally the name of an inn half-way between Shrewsbury and Welshpool.

HALFWAY HOUSES Kent TQ 9372. *Halfway House* 1819 OS. Halfway between Queenborough TQ 9172 and Minster TQ 9573.

HALIDON HILL Northum NT 9654. *Halidone hill* c.1535. P.n. Halidon + pleonastic ModE **hill**. Halidon itself, *Halydon* 1338, 1357, *Haly(ng)doun* c.1390, is the 'holy hill', OE **hāliġ** + **dūn**.

HALIFAX WYorks SE 0925. Possibly 'area of coarse grass in the nook of land'. *Hali- Halyfax* from 1091×7. OE **halh** + **ġefeaxe**. In the 16th cent. the name was wrongly interpreted as OE *hāliġ-feax*, 'holy hair', and a story invented of a maiden killed by a lustful priest whose advances she refused. The tree on which her head was suspended became an object of pilgrimage and when its bark was finally stripped for relics the fibres beneath were held to be the maiden's holy hair. Another story based on the same false etymology makes Halifax the final resting-place of the head of St. John the Baptist. YW iii.104.

West HALLAM Derby SK 4341. *Westhal'* 1210, 1221×30, *Westhalum* 1230–1431, *-halam* 1272×1307, 1442, *West Hallam* 1601. Adj. **west** + p.n. Hallam as in Kirk HALLAM SK 4540. Db 468.

Kirk HALLAM Derby SK 4540. 'Church H'. *Kyrkehallam* 1154×89, *Kyrk- Kirkhalam* 1442, 1577, 1610. ME **kirk** + p.n. *Halun* 1086, 1195, 1258, *Halum* 1230–1417, *Hallom* 1291, 'at the nooks', OE **halh**, locative-dative pl. **halum**. Also known as *Burhhalum* [1011]13th, 'Hallam with the fortification', OE **burh**. *Kirk* was added to the name for distinction from the daughter settlement at West HALLAM SK 4341. Db 466, L 103, 108.

HALLAND ESusx TQ 5016. 'Hall land'. *Hollond* 1375, *Halland* 1376. Surname *Hall* + ME **land**. Land here was acquired by Richard ate Halle of Lullington in 1343. Sx 400.

HALLATON Leic SP 7896. 'Nook settlement'. *Alctone* 1086, *Halcton(e)* 1229, 1285, *Haleghton* 1285–1387, *Halec(h)ton(e)* 1167–1285, *Halug(h)ton(e)* 1289–1451, *Haluton'* c.1246–1317, *Halouton(') -ow-* 1296–1524, *Hallaton* 1576. OE **halh** + **tūn**. Hallaton is situated in a narrow side-valley. Lei 223, L 103.

HALLATROW Avon ST 6357. 'The holy tree'. *Helgetrev* 1086, *Halghetre* 1259. OE **hāliġ**, definite form **hālga**, + **trēow**. DEPN, L 212, 214.

HALLBANKGATE Cumbr NY 5859. *Hallbank Gate* 1754. P.n. Hall Bank + ModE **gate**. Hall Bank, *Hallebanke* c.1200, *haubanke, hawle banke* 1589, *Hall Bank* 1822, is 'hall slope', OE **hall** + ME **banke**. Cu 85.

HALLEN Avon ST 5580. 'End of the *halh*'. *Hale yende* 1537, *Hallend* 1690, *Hallen* 1830. ME **hale** (from the dative sing. *hale* of OE *h(e)alh*) + **ende**. The sense of *halh* in this name is probably 'dry ground in marsh'. Gl iii.133.

HALL GREEN WMids SP 1181. *Hall Green alias Hawe green* 1619 Deed, *Hall Green* 1834 OS. Surname *Hawes* + ModE **green**. Hall Green Hall was the home of the Hawes family, cf. William Hawes d. 1610 commemorated in Solihull church.

HALLING Kent TQ 7064. OE folk-n. of uncertain meaning and origin. *Hallingas* [765×85]12th S 37, *Hallingas, (of) heallingan, (de) hallingis* (Latin) c.975 B 1321–2, *Hallinges* 1086, 1182×4, *Hallyng(e) -ing(e)* from 1100×7. The boundary of the Hallingas is *Heallingwara mearc* [880]12th S 321. In the same manor was *hallesmeri* [765×85]12th S 37, *halles meres* (genitive sing.) [880]12th S 321 'the lake of Halling'. The *Hallingas* were probably the 'people called after Heall', *Healling* the 'place called after Heall' and *halles mere* 'Heall's lake', although the specific might be OE **heall** 'a hall'. KPN 75, PNK 116, ING 12, BzN 1967.337.

Great HALLINGBURY Essex TL 5119. *Hallyngbery Magna* 1335. ModE adj. **great**, Latin **magna**, + p.n. *Halingeb(er)iam, Halingheberia, Hallingeberiam* 1086, *Hal(l)ing(h)eb(er)i(a) -bir(ia) -bery* 1096–1464, *Hallingbury* from 1303, 'fortified place of the Hallingas, the people called after Heall or Healla', OE folk-n. **Hallingas* < pers.n. **H(e)all(a)* + **ingas**, genitive pl. **Hallinga*, + **byriġ**, dative sing. of **burh**. The reference is to Wallbury Camp Iron Age hill-fort at TL 4917. 'Great' for distinction from Little HALLINGBURY TL 5017. Ess 34 gives pr [hɔliŋbri], ING 12, Thomas 1976.123.

Little HALLINGBURY Essex TL 5017. *Hallyngbury Parva* 1303. ModE **little**, Latin **parva**, + p.n. Hallingbury as in Great HALLINGBURY TL 5119. Ess 34.

HALLINGTON Northum NY 9875. 'The holy valley'. *Halidene* 1247–1479, *Hallidene* 1547, *Hallendon* 1608, *Hallington* 1663. OE **hāliġ** + **denu**. Leland records a tradition that the name refers to the battle of Heavenfield in 634 in which St Oswald defeated the British king Cadwallon. NbDu 99.

HALLINGTON RESERVOIRS Northum NY 9776. P.n. HALLINGTON NY 0875 + ModE **reservoir(s)**.

HALL OF THE FOREST Shrops SO 2183. *Hall of the Forest* 1836 OS. The reference is to Clun Forest.

HALLOUGHTON Notts SK 6851. 'The settlement in the nook of land'. *Healhtune* [958]14th S 659, *Halghton* 1280–1330, *Haluton* 1235, *-ow- -we-* 1330, *Halughton* 1336, *Hallaughton al. Hawton* 1549, *Halloughton oth. Haughton* 1772. OE **halh** + **tūn**. Halloughton lies in a small shallow side-valley off Halloughton Dumble (dial. **dumble** 'a wooded valley'). Nt 168 gives prs [hɔ:tən] and [hɔ:'n], L 103.

HALLOW H&W SO 8258. 'Enclosures in the nook'. *(De) Halhegan, (territoriis . . . de) heallingan, (lond ge mœra of) halhegan* 'the boundaries of H', *(œt, into) halheogan* *[816]11th S 179, *Hallege* *[964]12th S 731, *Halhegan* 1086, *Hallaga -e* [c.1150]c.1240, *Hallauwe* c.1250–98, *Hallowe* 1291, 1327, *Hallewe* 1428. Hallow is also mentioned in the bounds of S 180 in the non-compound form *(in) hagan* [876]11th. This, apparently the most authentic form, represents OE **hagum**, 'at the enclosures', dative pl. of **haga**. The reference is to enclosures in the ancient woodland W of the Severn called *Weogorena leage* ibid., 'the wood or clearing (possibly 'woodland-pasture') of the Wigoran', the tribal name preserved in the p.n. WORCESTER SO 8555. To specify which enclosures were meant the prefix **halh** 'a nook' was added to form the compound p.n. 'enclosures in the tongue of land' referring to the tongue or nook of land between the Laughern Brook and the Severn. The form *heallingan* probably represents OE **healhingum*, dative case of folk-n. **Healhingas* 'the *halh* people, the people who dwell in the *halh* or nook of land'. Wo 129 gives pr [hɔlou], ECWM 184–96, Hooke 107–15.

HALL ROAD STATION Mers SD 3000. A halt on the railway line from Liverpool to Southport at Hall Road, the road leading to Little Crosby Hall at SD 0132.

HALLSANDS Devon SX 8138. 'Sandy shore by the hole or cave'. *Hole Sande* 1514, *Helys(l)ands* 1540, *Hall Sands* 1809 OS. The earliest reference is to *Le Hole* c.1347, OE, ME **hol** 'a hole, a hollow'. Hallsands is a late developed fishing village at the mouth of Hollowcombe, *Holecombe* 1377×99. D 333, Fox 1996.

HALL'S GREEN Herts TL 2728. *Hallsgreene* 1710. Surname *Hall* + ModE **green**. Hrt 147.

HALLTHWAITES Cumbr SD 1785. *the hall of Thaytis* 1449, *Halthwett(es)* 1608, *Hawthwaites* 1657. ME **hall** + p.n. *Thueites* c.1170, *Thuaites* 1183×1216, *Tweites in Copland'* 1466, *Thwaits up in the head of Millome* 1675, ON **thveit**, secondary ME pl. **thweites**, 'clearings'. Cu 417.

HALLWORTHY Corn SX 1887. 'Marsh of Gorgi'. *Halworgy* 1439, *Haleworthy alias Halldrunkard* 1748. Co **hal** 'moor, marsh' + pers.n. *Gorgi* (OCo *Wurci*) with lenition of *G-* to *W-*, assimilated to English names in *-worthy*. Halldrunkard, *Haldronket* 1415, 1439, is 'marsh of the promontory wood', Co **hall** + ***trongos** 'nose-wood, wood on a spur of land' (< *tron* 'nose' + *cos*, OCo *cuit*) with lenition of *t-* to *d-*. PNCo 94, Gover n.d. 54.

HALMER END Staffs SJ 8049. *Halmoreende* 1541, *Halmer End* 1579. P.n. *Halmore* 1493 + ModE **end**. Horovitz.

HALMORE Glos SO 6902. 'Pool in the nook of land'. *Halmare* 1287, 1370, *-more* from 1456, *-mer* 1521–1635, *Hamer -or* 17th. OE **halh** + **mere**. Situated at the edge of Hinton township; there is a Pool Farm there. Gl ii.229.

HALNAKER WSusx SU 9108. Probably '(the settlement) at the half acre'.

I. *Helneche -ache* 1086, *Hannac* 1105, *Hannak(e)* 1272–1302, *Halnac -k'* 1166–1301, *Halfnac -k'* 12th–1372, *Halnakere -acre* 1316, 1386, *Halvenaker* 1452, *Hanycarr* 1605, *Hanekar* 1650.

II. *Halfnaked(e)* 1274–1425, *Halnaked(e)* 1274–1336, *Haunaked(e)* 1288, 1329.

OE **healf**, definite form dative case **healfan**, + **æcer**, dative case **æcre**. The forms are not conclusive but this may have been a small area of arable in primarily heathland. Sx 67.

HALSALL Lancs SD 3710. 'Hæle's nook'. *Heleshale, Herleshala* 1086, *Haleshale* 12th, 1280, *Halsale* 13th cent., *Halsall* from 1346. OE pers.n. **Hœle*, genitive sing. **Hœles*, + **halh**, dative sing. **hale**. *Halh* here probably has the sense 'promontory into a marsh'. La 120, L 107, Jnl 17.67.

HALSE Northants SP 5640. There are two forms for this name:

I. *Hals* 1284–1349, *Halse* from 1485, *Hawes* 1657–1823, 'the neck of land' referring to the high neck of land between valleys on which the village stands.

II. *Hasou* 1086, *Hasho* 1229, *Halsou -ho* c.1160–1346, *Haus(h)o* 12th 1236, the 'hill-spur called or at Halse', OE **heals** used as a p.n. + **hōh**. Nth 49 gives pr [hɔ:z], L 168.

HALSE Somer ST 1427. 'The pass'. *Halsa -e* 1086, *Alse, Hausa* [12th] Buck, *Halse* from 1243 including [c.1152–1423] Buck, *Hause* [1152, 1329×46] Buck, *Hawse* [1281×90] Buck, *Haulse* 1610. OE **heals** 'neck, pass' referring to the situation of the village between two hills. DEPN.

HALSETOWN Corn SW 5038. *Halse Town* 1839 commemorates the name of James Halse of Truro (1768–1838), MP for St Ives, who built the town for the accommodation of miners, providing, it is said, in order to secure his parliamentary seat, each house with just enough land to qualify its tenant for the vote, and selecting the tenants from his supporters. PNCo 94, Gover n.d. 627, Room 1983.49.

HALSHAM Humbs TA 2727. 'Homestead on the neck of land'. *Halsaham* [1033]14th, *Halsem* 1086, 1288, 1294, *Halsam* 1086–1482, *-ham* from [1190]1301, *Hausaim* early 13th. OE **hals** + **hām**. The reference is probably to a rise of ground between drainage streams on which the church now stands. YE 30.

HALSINGER Devon SS 5138. 'Hazel' or 'neck wood'. *Halsangra* 1167, *Halshangre, Halesangre* 1244. OE **heals** or **hæsel**, Devon/Somerset dial. **halse**, + **hangra**. With *hangra* 'hazel' is perhaps more likely; otherwise the reference is to a neck of high ground forming a pass between valleys. D 33, lviii, BH 109.

HALSTEAD Essex TL 8130. 'Shelter place'. *Hal(te)steda* 1086, *Halsted(e)* c.1180–1592, *Hausted(e)* c.1180–1287. OE **hald** in the sense 'protection, shelter for animals' + **stede**. Ess 433 gives pr [hælstid], Stead 201, Studies 1936.52–3, PNE i.222.

HALSTEAD Kent TQ 4861. Probably 'place of shelter, hiding place'. *Haltesteda* c.1100, *Aldestede* 1210–12, *Halt(e)sted(e)* 1226–1291, *Halsted(e)* 1201–1601. OE **hald** + **stede**. This solution fits the topography. PNK 56, PNE i.222, Ess 433, Stead 216.

HALSTEAD Leic SK 7405. 'Place of refuge'. *Elstede* 1086, *Hal(l)ested(e)* c.1130–1466, *Halsted(e)* 1200–1610, *Hau-Hawsted(e)* 1167–1351, *Haustead* 1620, *Haldsted'* 1230, 1236. OE **hald** + **stede**. It is uncertain whether the sense is 'fortified place' or 'animal shelter'. Lei 321, Studies 1936.52, Stead 289.

HALSTOCK Dorset ST 5308. 'The holy outlying-farm'. *(in) Halganstoke* [998]12th S 895, *Halgestok -stoch* c.1192, 1212, *Halghestok* 1285, *Hallestok* 1399. OE **hāliġ**, definite form oblique case **hālgan**, + **stoc**. Identical with Halstock in Oakhampton, Devon, *Halghestok* 1240. The Dorset Halstock is so named because the estate belonged to Sherborne abbey. Do 81, Studies 1936.21.

HALSTON Shrops SJ 4107 → HINTON no.6.

High HALSTOW Kent TQ 7875. *High halsto* 1610, *High Halstow* 1805 OS. ModE adj. **high** + p.n. *Halgesto* c.1100, *Halghestowe* 1270, *Halewestoue -stowe* 1240, 1278 'the holy place'. OE **hāliġ**, definite form **hālge**, + **stōw**. 'High' for distinction from Lower HALSTOW on the opposite side of the Medway TQ 8567. PNK 119, Stōw 192.

Lower HALSTOW Kent TQ 8567. *Lower Halstow* 1819 OS. ModE adj. **lower** + p.n. *Halgastaw* c.1100, *Halegestowe* 1199, 1200, *Halewestowe* 1199, 'the holy place'. OE **hāliġ**, definite form **hālge**, + **stōw**. 'Lower' for distinction from High HALSTOW on the opposite side of the Medway TQ 7875. PNK 248, Stōw 192.

HALSTOW MARSHES Kent TQ 7778. P.n. Halstow as in High HALSTOW + ModE **marshes**.

HALTHAM Lincs TF 2463. 'Wood village or homestead'. *Holtham* 1086–1254, *Holteim* c.1115. OE **holt** + **hām** partly replaced by ON **heim**. DEPN.

HALTOFT END Lincs TF 3645. P.n. Haltoft + ModE **end** with reference to Freiston village; in 1824 OS it is *Freiston Ultra End*, i.e. that part of Freiston beyond the end of the village. Haltoft, *(H)alketoft* 12th cent. is the 'nook or corner curtilage' again referring to a nook or end of Freiston, OE **halc**, ME **halke** 'a corner, a nook, a hiding place, + ON **toft**. Cf. SCRANE END TF 3841. SSNEM 223, PNE i.222.

HALTON 'Farm or village in or by a nook'. OE **halh** + **tūn**.

(1) ~ Bucks SP 8710. *Healtune* 1020×38 S 1464, *Haltone* 1086, 1195, *Halkhton* 1237×40, *Haulton* 1766. The village is situated near an indent in the Chiltern hills. Bk 162, L 103–4, 107–8, 153.

(2) ~ Lancs SD 5064. *Haltune* 1086, *-tun -ton(a)* [c.1225]1268, *Halg(h)- Halehton* 1246–51, *Halton* from 1243. *Halh* here probably refers rather to the valley of the Lune than to water-meadow as there is no meadow-forming bend in the river at this place. La 179, L 107, Jnl 17.101.

(3) ~ EAST NYorks SE 0454. *Est Halton* 1314. ME adj. **est** + p.n. *(H)altone* 1086, *Halton(a) -tun* 1120–1597. The reference is to the side-valley opening from the river Wharfe, which Halton East overlooks. 'East' for distinction from HALTON WEST SD 8454. YW vi.70.

(4) ~ GILL NYorks SD 8776. *Haltongyll -gill* 1457–1685, *Haughton Gill', Hautongill* 17th cent. P.n. Halton + ME (ON) **gil** 'a ravine'. The village lies at the head of the deep valley of Littondale. YW vi.121.

(5) ~ HOLEGATE Lincs TF 4165. 'Halton hollow way'. *Hauton Holgate* 1576, *Halton Holegate* 1824 OS. P.n. *Haltun* 1086, c.1190, *Hauton* c.1135, *Haltona* c.1150, *Halton* 13th, the 'nook settlement', + **holegate** (OE *hol*, ON *holr* + *gata*). The reference is to a small valley in the rising ground on the N edge of the fens and to the road which pierces the sandstone cliff there. DEPN, Gaz s.n., Cameron 1998.

(6) ~ GATE Northum NY 6558. *Haltonlee Gate* 1869 OS. P.n. *Halton Lee* 1869 OS + ModE **gate**.

(7) ~ WEST NYorks SD 8454. *West Halghton* 1303. P.n. *Halctun* 12th, *Halton* from c.1200, + ME **west** for distinction from HALTON EAST SE 0454. YW vi.157, L 107.

(8) East ~ Humbs TA 1419. *Esthouton* 1331, *Esthalton* 1415–1601. ME adj. **est** + p.n. *Haltune* 1086, *-tun* c.1115–1338, *-ton* from 1143, *Hauton(e)* 1190–1331. The reference is to a piece of dry ground reaching into the Humber marshes. 'East' for distinction from West HALTON SE 9021. Li ii.148.

(9) East ~ SKITTER Humbs TA 1422. *Halton Skitter* 1824. 'Halton sewer'. P.n. HALTON + *Scitra* c.1155, *Schitere* 1155–60, *Skyter* 14th cent. 'the sewer', OE ***scitere**, with ON initial *sk*. Li ii.149, PNE ii.112.

(10) West ~ Humbs SE 9021. ModE adj. **west** + p.n. *Haltone* 1086, *Halghton* 1219, *Halton* from c.1115. The village lies in a tiny indentation of the 100ft. contour. DEPN, L 103.

HALTON Ches SJ 5482. 'Settlement at the heathery place'. *Heletune* 1086, *Heltun -ton* 1199–1305, *Halton(a)* from 1185, *Hau(l)ton* c.1200–14th, *Hethelton* 1174, *Hadeltona* c.1200, *Hau(l)ton* c.1200–14th, *Hathelton* [c.1250]17th. OE ***hāthel**, ***hǣthel** + **tūn**. Che ii.166 gives pr ['hɔltən], Addenda 19th cent. local ['hɔ:ttn].

HALTWHISTLE Northum NY 7064. 'High *Twisel*, the fork'. *Hautwisel -twysel(l) -twisill* 1240–1372, *Hoatewhisle* 1655. OFr **haut** + OE **twisla** probably already used as a p.n. referring to the fork between Haltwhistle Burn and the river S Tyne. The village lies on high ground in the fork. NbDu 99 gives pr [hɔ:təsəl].

HALVERGATE Norf TG 4106. Uncertain. *Halfriate* 1086, *Haluergiata* 1158, *-iet* 1177, *-gata* 1182. The generic appears to be OE **ġeat** 'a gate' influenced by ON **gata** 'a road'. But the whole name might represent OE **healf** 'half' + **hereġeatu** 'the feudal service or payment called heriot' and mean 'land for which a half heriot was paid'. DEPN.

HALVERGATE MARSH Norf TG 4707. P.n. HALVERGATE TG 4106 + ModE **marsh**. Simply *The Marsh* 1838 OS.

HALWELL Devon SX 7753. 'The holy well'. *(to) halganþille* [914×9]16th, 12th BH1, 2, *-þylla* [914×9]13th BH3, *Halgewill(e)* c.1240–1322, *Halgh(e)wille* 1330, 1368, *Hallewell* c.1400, *Holwell* 1675. OE **hāliġ**, definite declensional form oblique case **hālgan**, + **wylle**. D 323 gives pr [hɔ:lwel].

HALWILL Devon SX 4299. 'The holy well'. *Halgewelle* 1086, *-willa* 1086 Exon, *Hal(e)ghewill(e)* 1242–1318. OE **hāliġ**, definite form **hālge**, + **wylle**. D 141.

HALWILL JUNCTION Devon SS 4400. P.n. HALWILL SX 4299 + ModE **junction**.

HAM 'Enclosed land' of some kind. OE **hamm** for which the following senses have been identified: 1 'land in a river-bend', 2a 'a promontory of dry land into marsh or water', 2b 'a promontory into lower land, even without marsh or water; land on a hill-spur', 3 'a river meadow', 4 'dry ground in a marsh', 5a 'a cultivated plot in marginal land', 5b 'an enclosed plot, a close', 6 'a piece of valley-bottom land hemmed in by higher ground'. ASE 2.26–7.

(1) ~ GLond TQ 1672. *Hama* c.1150, *Hamme* 1154 etc., *Hammes* 1235, 1241, 1383, *Ham* 1532. Sr 57.

(2) ~ Glos ST 6898. *Hamma, Hamm(e)* [1108]1300, 1195–1547, *Homme* 1315–56, *Ham* 1472. Gl ii.223.

(3) ~ Kent TR 3254. *Hama* 1086, *Hamme(s)* c.1100–1254. The site is a hill-spur overlooking North Stream, a *hamm* 2b. PNK 584.

(4) ~ Wilts SU 3362. *(æt) Hamme* 931 S 416, 1229–1363, *Hame* 1086. OE **hamm** 5b 'an enclosed plot, a close' or 2b 'land on a hill-spur'. Wlt 348.

(5) ~ GREEN Avon ST 5357. 'Green at Ham'. *Hampne Green* 1830 OS, cf. *Hampne House* 1830 ibid.

(6) ~ GREEN H&W SP 0163. *Ham Green* 1831 OS. P.n. *Hamm(e)* 1240–80, *Homme* 1275–1346, 'the water-meadow', OE **hamm** 3, + ModE **green**. Wo 319.

(7) ~ HILL Somer ST 4816. Short for *Hamden hill* 1610 as in NORTON SUB HAMDON ST 4615 and STOKE SUB HAMDON ST 4717. Hamdon, *Hamedone* [1091×1106–1189×99]14th, *Homedon* 1244, *Hamedon* 1248, is 'the river meadow hill', OE **hamm**, genitive pl. **hamma**, + **dūn**. It is a famous archaeological site consisting of an Iron Age hill-fort and Roman settlement. In the Middle Ages it was the site of a fair. DEPN, Thomas 1976.193.

(8) ~ STREET Somer ST 5534. 'Ham village'. *Ham Street* 1817 OS. P.n. Ham + dial. **street**. The *ham* may, however, refer specifically to either nearby Tilham ST 5535 or *Lattisham* 1817 OS.

(9) East ~ GLond TQ 4283. *Eastham* 1206, *Esthammes* 13th cent. ME adj. **est** (OE *ēast*) + p.n. *Hamme* [958]12th S 676, 11th–12th cents., *Hame* 1086, *Hammes* 13th. The sense here is 'promontory of dry land into marsh', OE **hamm** 2a, referring to dry

land projecting into the Thames marshes between Bow Creek TQ 3980 and Barking Creek TQ 4582. Ess 94, Jnl 2.41.

(10) High ~ Somer ST 4231. *Heyghe Hamme* 1380, *Highham* 1610. ME adj. **hegh** + p.n. *Hom* [702]13th S 244, *Hamme* [965]n.d. ECW, [973]14th S 791, *[1065]18th S 1042, *Hame* 1086. The sense here is 'promontory of high ground reaching into wet land' **hamm** 2. The 'lake of Ham-land' is *Hamelondesmere* [725]14th S 251. Contrasts with Low HAM ST 4329. DEPN.

(11) Low ~ Somer ST 4329. *Nitherhamm* 1264. ME adj. **nether** + p.n. Ham as in High HAM ST 4231. DEPN.

River HAMBLE Hants SU 5241. 'Crooked one'. *Homelea* [c.731]8th BHE, *Hamalea* [8th]9th, *(innan, andlang) Hamele* [900]11th S 360, *Hamel* 1369, *Hamble* 1586. OE r.n. **Hamel-ēa* < ***hamol**, ***hamel** 'maimed, mutilated, crooked' + **ēa**. The Hamble has a very winding course. Cf. the G r.n. Hamel, *Hamele* 1309, a tributary of the Weser at Hameln. RN 189, Gover 1958.3, Berger 124.

HAMBLE Hants SU 4607. Transferred from the r.n. HAMBLE SU 5241. *Amle* 1147, *Hamele* 1165–1391 with addition *de Rys* 1386, *in the Rice* 1391, 'in the brushwood', OE **hrīs**. Ha 85, Gover 1958.38.

HAMBLEDEN Bucks SU 7886. 'The misshapen valley'. *(æt) hamelan dene* 1015 S 1503, *Hanbledene* 1086, *Hameleden* 1182–1240, *Hammulden* 1552, *Hambleden* 1654. OE ***hamol**, definite form dative case ***hamelan**, + **denu**. The reference is probably to the sharp bend in the valley, an unusual feature for a *denu*. Bk 177 gives pr [hæməldən], L 98, Jnl 2.31.

HAMBLEDON '(The) mutilated or scarred hill'. OE ***hamol**, ***hamel** (definite form oblique case ***hamelan**) + **dūn**.

(1) ~ Hants SU 6414. *(to) hamelandunæ* [956]12th S 598, *Hamledune* 1086, *Hameldon(a)* 1154–1341. Ha 85, Gover 1958.52.

(2) ~ Surrey SU 9638. *Hameledune* 1086–1291 with variant *-done, Hameldon(e)* c.1270, 1304, *Hemeldon* 1422×71, *Hambeldon* 1552, *Hambledon* from 1606. Sr 202, L 143–4, 146, 156–7, 226.

HAMBLETON 'The maimed or irregularly shaped hill'. OE ***hamol** + **dūn**.

(1) ~ Lancs SD 3742. *Hameltune* 1086, *-ton* 1177–1332 etc., *Hambleton* 16th. The reference is to one of the small hills standing out of the Wyre estuary flats. La 155, YW iv.29, Jnl 17.90.

(2) ~ HILLS NYorks SE 5089. P.n. *Hamelton(a)* c.1160, 1350, 1452, *Hameldon'* 1301, + pleonastic ModE **hill**. YN 198.

(3) Upper ~ Leic SK 9007. ModE **upper** + p.n. *Hameleduna* 1067, *Hamedun(e)* 1086, 1232, *-don(e)* 1202–1556, *Hambleton* 1346, 1684. Also known as *Magna* 'great' 1288, 1344 and *Great* 1684, for distinction from Nether Hambledon (lost), *le Nethertowne of Hambledon* 1549. Earlier *Parva-* c.1200–1442, *Little-* 1655, *Nether-* 1549. The reference is to a flat-topped hill that looks as if it has been sliced off. R 179, L 143–6, 156–7, 226.

HAMBLETON 'Settlement at *Hamel*, the scarred or misshapen hill'. OE p.n. **Hamel* (OE ***hamol**) + **tūn**.

(1) ~ HILL NYorks SE 1473. P.n. *Hameldun* 1175, + pleonastic ModE **hill**. YW v.213.

(2) ~ NYorks SE 5530. *Hameltun -ton(a)* 1086–1546. YW iv.xi, 28.

HAMBRIDGE Somer ST 3921. 'Bridge at Ham, the promontory of dry land'. *Hambridge* 1811 OS. ModE **ham** < OE *hamm* 2a + **bridge**.

HAMBROOK Avon ST 6479. 'Stone brook'. *Hanbroc* 1086, *Hambrok(e)* 1227–1552, *-brook(e)* from 1598. OE **hān** + **brōc**. A stone is marked on the 6in. OS map near the grounds of Hambrook House. Gl iii.124.

HAMBROOK WSusx SU 7806. 'The water-meadow stream'. *Hambrook* 1327. OE **hamm** 1 + **brōc**. Sx 60.

HAMELDON HILL Lancs SD 8128. P.n. Hameldon + ModE **hill**. A hill rising to 1309ft. separated from Great Hameldon SD 7928 by a deep gully. Great Hameldon (1343ft.) is *Hameldon* before 1194, 'maimed, notched or cut-off hill', OE ***hamol** + **dūn**. La 67, PNE i.231, L 143–6, 156–7.

HAMELDOWN TOR Devon SX 7080. P.n. Hamel Down + ModE **tor**. Hamel Down is the *great waste called Hameldon* 1566, *Hammeldon* 1652, 'mutilated hill', ultimately from OE ***hamol** + **dūn**. The shape of the hill resembles a hog's back. D 527.

HAMERINGHAM Lincs TF 3068. 'The homestead or village at or called *Hamering*'. *Hamerīgā* 1086, *Hamringhe- Hamrigeheim* c.1115, *Hamerig- Hamringham* c.1175, *Hameringeham* 1188ff., 1218, *Hamringham* 1212. OE p.n. **Hamering* 'that which is like a hammer, a hammer-shaped hill' < OE **hamor** + **ing**[2], locative–dative sing. **Hameringe*, + **hām**. ING 144.

HAMERTON Cambs TL 1379. 'Hammer farm or village'. *Hambertun* 1086, *Hamertun* 1152, [c.1155]13th, *Hamerton* from 1199. The *-b-* of the DB form is regarded as inorganic, leaving OE **hamor** + **tūn**. The sense of *hamor* is unclear; the simplest solution is 'place where hammers are made or used, place with a smithy', but *hamor* might be short for a plant name such as *hamor-secg* 'hammer-sedge' or *hamor-wyrt* 'hammer-wort, wall pellitory', or it might possess a topographical sense 'hammer-shaped rock or crag' as ON *hammarr*. Hu 242, PNE i.231.

HAMMERSMITH GLond TQ 2279. 'Smithy where a hammer is used'. *Hamersmyth'* 1294–1393, *Hamyrsmyth in the parish of Fulham* 1535, *Hammersmith* 1675. OE **hamor** + **smyththe**. Hammersmith, a hamlet of Fulham, did not gain independent status until 1834. Mx 108.

Green HAMMERTON NYorks SE 4657. 'H with the green'. ME **grene**, *Gren(e)* 1176, + p.n. *(H)ambretone* 1086, *Hamertun -on(a)* 1121–1586, 'settlement where hammer-sedge (*Carex hirta*) or hammer-wort (wall-pellitory) grows', OE **hamor** for *hamor-secg -wyrt* or *dūth-hamor*, + **tūn**. The precise botanical species is uncertain. 'Green' for distinction from Kirk HAMMERTON SE 4655. YW v.xi, 8.

Kirk HAMMERTON NYorks SE 4655. 'H with the church'. *Kyrkehamerton* 1226. ON **kirkja** + p.n. *Ambre- Hambretone* 1086, *Hamertun -tona* 1118–1539, as in Green HAMMERTON SE 4657. 'Kirk' for distinction from Green HAMMERTON SE 4657. DEPN, YW v.11.

HAMMERWICH Staffs SK 0607. 'The hammer work or trading place'. *duæ Humeruuich* 'the 2 Hs' 1086, *Hamerwich* 1191, *-wic* 1220. OE **hamor** + **wīċ**. Probably a reference to industrial activity. DEPN.

HAMMOND BECK Lincs TF 2038. *Hamundebek* [1315]c.1331, *Hamondbek* 1315. ME pers.n. *Hamond* + ME **bek** (ON *bekkr*). RN 172, Cameron 1998.

HAMMOND STREET Herts TL 3304. 'Hammond village'. *Hammonds Street* 1805 OS. Surname *Hammond*, cf. *Hamund* 1296, + Mod dial. **street** 'a hamlet'. Hrt 222.

HAMMOON Dorset ST 8114. 'Ham held by the Moun family'. *Ham Galfridi de Moiun* 'Geoffrey de Moun's Ham' 1194, *Hamme Moun* 1280, *Ham(m)oone* 1611 etc. Earlier simply *Hame* 1086, *Hamme* 1202–1331, 'the enclosure or river meadow', OE **hamm**. Hammoon is surrounded on three sides by the r. Stour. The Moun or Mohun family derives its name from Moyon in Normandy, *Moion* 1027. Do ii.99 gives pr [hə'mu:n], Dauzat-Rostaing 482 s.n. Mouais.

Great HAMPDEN Bucks SP 8401. *Magna Hamdene* 1284. ModE **great**, Latin **magna**, + p.n. *Hamdenam* 1086, *Hampdene* [c.1200]14th, 'valley of the enclosures', OE **hamm** 5a 'a cultivated plot in marginal land', genitive pl. **hamma**, + **denu**. 'Great' for distinction from Little HAMPDEN SP 8503. Bk gives pr [hæmdən], L 99, Jnl 2.26.

Little HAMPDEN Bucks SP 8503. *Parva Hamdene* 1247. ModE **little**, Latin **parva**, + p.n. Hampden as in Great HAMPDEN SP 8401. Bk 151.

HAMPDEN PARK ESusx TQ 6002. An early 20th cent. development at Eastbourne named after Lord Hampden, the owner of the manor of Ratton, *Radin- Radetone* 1086, 'Ræda's estate', on which it was built. Sx 424.

HAMPNETT Glos SP 1015. '(At the) high settlement'. *Heamtun* 1062×66 S 1480, *Hanton(e)* 1086, [12th]1267, *Hampton(a)* early 12th–1330, *Hamtonet(t)* 1211–82, *Hamptonet(t) -en-* 1212–1500, *Hampenet* 1458, *Hampnet(t)* from 1447. OE **hēah**, dative case definite form **hēan**, + **tūn** with later addition of French diminutive suffix **-et**. Gl i.173.

East HAMPNETT WSusx SO 9106 → WESTHAMPNETT SU 8806.

HAMPOLE SYorks SE 5010. Either 'cock pool' or 'Hana's pool'. *Honepol* 1086, *Hanepol(e)* 1086–1345, *Hampol(e)* 1202×8–1822. OE **hana** or pers.n. *Hana* + **pōl**. YW i.70, L 28.

HAMPRESTON Dorset SZ 0598. 'The priests' estate of or called Ham'. *Hamme Preston(e)* 1244–1431, *Hamme et Prestone* 1327, *Hampreston(e)* from 1372. P.n. *Hame* 1086, *Hamme* 1204–1583, 'enclosure or river-meadow', OE **hamm**, + p.n. PRESTON. The reference is probably to land belonging to the college of Wimborne Minster. Do ii.224 gives pr ['ha:m 'prestn].

HAMPSHIRE 'The shire or county of Hampton', i.e. Southampton. *(buton, mid, on, of) Hamtunscire* c.900 ASC(A) under year 755–c.1121 ASC(E) under year 1011, *Hantescire* 1086. OE p.n. *Hamtun* 'Southampton' + **scīr**. The 1086 spelling represents a reduced form of Norman Fr **Hantunescire* with assimilation of *-mt-* to *-nt-*. From it derives the modern county abbreviation 'Hants'. Rumble 1980.10, 16–8.

HAMPSTEAD 'Homestead'. OE **hāmstede**. Stead 65–7, 88–9, PNE i.232.

(1) ~ GLond TQ 2485. *(and lang) Hemstedes (mearce)* 'along the boundary of H' *[959]10th S 1450, *(in loco qui celebri æt) Hamstede (nuncupatur vocabulo)... hamstede* [978 for ?974]18th S 805, *hamstede...(to medeman) hemstede* *[986]10th S 1450, *Heamstede* *[1066]13th S 1043, *Hamestede* 1086, *Hamsted(e)* 1223–1428, *Hampsted(e)* 1253–1566. The 959 and 1066 forms may represent the side-form **hǣmstede**. Mx 111, DEPN, PNE i.217, 232, Stead 242f.

(2) ~ NORREYS Berks SU 5276. 'H held by the Norreys family'. *Hampstede Norreys* 1517. P.n. Hampstead + family name *Norreys*. Earlier simply *Hanstede* 1086, *Hamsteda -sted(e)* 1167 etc., *Ham(p)stede* 1284–1410. Also known as *Hamesteda Willelmi de Sifrewast* 'William Sifrewast's H' 1167 and *Hampsted Ferrers* 1375. The estate was held by William de Sifrewast in 1166–7 from whom it descended to the Ferrers family and the trustees of John Norreys in 1448. Brk 249, Stead 266, VCH iv.74.

(3) ~ MARSHALL Berks SU 4165. 'H held by the (Lord) Marshal'. *Hamstud' Mareschal, Ham(p)sted(e) Marchal',* ~ *Mareschal(l)* 1284–1400. Earlier simply *Hamesteda -e* 1086, *Hamsted(e)* [1154×89]1270 etc. Hampstead Marshall first appears in the hands of the Lords Marshal early in the 13th cent. Brk 299, Stead 266, VCH iv.179.

HAMPSTHWAITE NYorks SE 2658. 'Hamal's clearing'. *Hamethwayt* 12th cent., *Hameleswaith'* 1208, *Ham(e)st(h)weit -t(h)wayt(e) -thwait(e)* 1222–1406, *Hampsthwait* from 1339, *-twit* 1663. ON pers.n. *Hamall*, secondary genitive sing. *Hamales*, + **thveit**. YW v.133 gives pr ['amstwit].

HAMPTON '(At) the high settlement'. OE **hēan-tūne**, dative sing. of **hēah**, definite form **hēa**, + **tūn**.

(1) ~ H&W SP 0234. *hamtona* *[709]12th s 80, *hantun* *[714]16th S 1250, *(æt) haeantune* *[780]11th S 118, *Hantvn(e)* 1086, *Hamton* [c.1086]1190, *Hampton* 1327. The village occupies raised ground between Merry Brook and the r. Isebourne. Wo 133, Hooke 46, 23, 30.

(2) ~ Shrops SO 7486. *Hempton* 1391. The village lies on high ground overlooking the river Severn. DEPN.

(3) ~ WMids SO 9198 → WOLVERHAMPTON SO 9198.

(4) ~ HEATH Ches SJ 4949. *Hampton Heth* 1476, ~ *Heath* 1582. P.n. *Hantone* 1086, *Hanton* 1259–1420, *Hampton* from 1260, + ME **hēth** (OE *hǣth*). Che iv.34.

(5) ~ IN ARDEN WMids SP 2081. *Hamton in Ardena* 1100×35, *Hantuna in Ardena* 1154×89, *Hamptone in Arderne* 1200–1375, *Hampton in Ardren* 1534, ~ *in Arding* 1667. P.n. *Hantone* 1086 + forest-n. ARDEN. The village occupies a plateau of high ground. Wa 61.

(6) ~ ON THE HILL Warw SP 2564. *Hampton super montem* 1443, 1649, ~ *uppon the Hill* 1483×1509. P.n. *Hamtone* 1154×89 + ME **upon the hill**, Latin **super montem**. Also known as *Hampton Curly* 1275–1501 from the family of William de Curly who held the manor in 1235. Wa 205.

HAMPTON 'Settlement, estate in a river bend or between rivers'. OE **hamm** 1 + **tūn**. L 43, 50.

(1) ~ GLond TQ 1369. 'River-bend settlement'. *Hamntone* 1086, *Hantune -tona* 12th cent., *Hamton* 1202–94, *Hampton* from 1237. The reference is to a great bend of the river Thames. Mx 14.

(2) ~ BISHOP H&W SO 5538. 'H belonging to the bishop' sc. of Hereford. *Hampton Episcopi* 1592. Earlier simply *Hantvne* 1086, *Hamtona* 1186×1200, *Hampton(e)* 1246–1535, *Hompton(e)* 1334, 1376. The estate lies on low ground between the rivers Lug and Wye. He 95, L 43, 50.

(3) ~ COURT GLond TQ 1568. 'Hampton manor'. *Hampton Courte* 1476. P.n. HAMTON TQ 1369 + ME **court**. The original reference was to the building that preceded Wolsey's palace. Mx 15.

(4) ~ LUCY Warw SP 2557. 'H held by the Lucy family'. *Hampton Lucy* 1606. P.n. *Homtune* [781]11th S 120, *Hantone* 1086, *Hamtone* 1251, + manorial addition from the Lucy family. Also known as *Hampton Episcopi* 1270–1535 and *Bisshopeshampton* 1315, *Busshopes* ~ 1492. The Bishops of Worcester held the manor till 1556 when it came into the hands of the Lucys of Charlecote. It lies in a large bend of the river Avon. Wa 233.

HAMPTON 'Home farm'. OE **hām-tūn**.

(1) ~ LOVETT H&W SO 8865. 'H belonging to the Lovet family'. *Hamton Louet* 1315. Earlier simply *Hamtone iuxta Wiccium* 'H by Droiwich' *[716]12th S 83, *hatun* (for *hamtun*) [817]11th S 181, *Hamtun* 11th Heming, *Ham- Hantvne* 1086, *Hampton'* 1242. Wo 303.

(2) ~ POYLE Oxon SP 5015. 'H held by the Poyle family'. *Hampton Poile,* ~ *Poyle(y),* ~ *Poyell* from 1428. P.n. *Hantone* 1086–1208, *Hamptone* 13th, + family n. of Walter de la Puile who obtained the manor by marriage in 1252. O 213.

(3) Meysey ~ Glos SP 1200. 'H held by the Meysey family'. *Hamtone Rogeri de M(o)eisi* 1221, *Hampton Meys(e)y, Meisi* 1243–1587, *Meseishampton* 1287, *Meysey Hampton* from 1437. Family name *Maisi, Moisi* + p.n. *Hantone* 1086, *Hamtun* 1185, Gaufrid de Maisi occurs 1134, Robert de Meisi 1185. The affix is for distinction from MINCHINHAMPTON SO 8700. Gl i.74.

HAMSEY ESusx TQ 4112. Properly Ham Say, *Hammessay* 1342, 'Ham held by the de Say family' which is first mentioned in connection with the manor in 1222. Ham, *(æt) Hamme wiþ Læwe* 'Ham by Lewes' c.959 S 1211, *Hamme juxta Læwes* [961]12th S 1212, *Hame* 1086, *Hammes* c.1155–1439, is 'land in a river bend' referring to the river Ouse. OE **hamm** 1. Sx 315, SS 82.

HAMSTALL RIDWARE Staffs SK 1019 → Hamstall RIDWARE.

HAMSTEAD IoW SZ 3991. 'The homestead'. *Hamestede* 1086, 1284, *Hampstede* 1154×89–1540, *Hamsted(e)* 1202–1468, *Hampstead* 1769. This was a grange of Quarr abbey during the Middle Ages. Wt 210 gives pr ['(h)æmstid], *Stead* 280, Mills 1996.56.

HAMSTEAD WMids SP 0493. 'The homestead'. *Hamsted(e)* 1213–1356, OE **hāmstede**. *Hampsted(e)* 1289–1555. Stead 290.

HAMSTERLEY 'Clearing infested with corn-weevil'. OE ***hamstra** + **lēah**. PNE i.232.

(1) ~ Durham NZ 1131. *Hasteleia* (for *Hā-*, i.e. *Ham-*) 1130, *Hamsterleia* c.1200, *-ley* from c.1200, *Hampsterle(y)* 1242×3, 1474–89. NbDu 100.

(2) ~ Durham NZ 1156. *Hansterlege* 1242×3, *Hamsterley* from 1359×60, *Hampsterley* 1367–1621. NbDu 100.

(3) ~ FOREST Durham NZ 0428. ModE **forest**. A modern Forestry Commission plantation.

HAMSTREET Kent TR 0033. *Homstrate* 1256×65, *Ham Street* 1816 OS. P.n. *Hame* 1086, *(la) Hamme* c.1100–1254, OE **hamm** 2b 'a promontory into lower ground', + ME **strate, strete** 'a hamlet'. Also known as *Easthamme* 1253×4. The site is a hill-spur overlooking Romney Marsh. PNK 473, Cullen.

HAMWORTHY Dorset SY 9991. 'Ham enclosure'. *Hamworthy* 1463. P.n. *Hamme* 1236–1476, 'the enclosure, the promontory', OE **hamm**, + ME **worthy** (OE *worthiġ*). The reference is to the situation of Hamworthy on a marked peninsula between Lytchett Bay and Holes Bay, a site 'enclosed' by water. Do ii.20.

Long HANBOROUGH Oxon SP 4114. ModE adj. **long** + p.n. Hanborough as in Church HANDBOROUGH SP 4212.

HANBURY H&W SO 9663. '(At) the high fortification'. *Heanburg* [657×74]18th S 1822, *(in) heanbyrg, heanbirige* (genitive case), *(aet) heanbyrig* 836 S 190, *Hambyrie* 1086, *Hambury* 1275, 1327, *Hanbury juxta Wych* 'by Droitwich' 1379. OE **hēah**, oblique case definite form **hēan**, + **byriġ**, dative sing. of **burh**, referring to the ramparts of the Iron Age hill-fort within which the church stands in an elevated position. Also known as North Hanbury, *Norð Heanbyrig* 11th B 1320, *Northeanburh* 12th, for distinction from HENBURY Avon ST 5678, also an estate of the bishop of Worcester. Wo 321, Hooke 18–19, 97, ECWM 37.

HANBURY Staffs SK 1727. '(The settlement) at the high fortified place'. *Hamb'* c.1185, *Hambur(y)* 1251, 1430, *Hanbury* 14th. OE **hēah**, definite form dative case **hēan**, + **byriġ**, dative case of **burh**. The village stands on a steep hill overlooking the valley of the r. Dove. DEPN, Duignan 72.

HANCHURCH Staffs SJ 8441. '(The settlement) at the high church'. *Hancese* (sic) 1086, *-churche* 1212. OE **hēah**, dative case definite form **hēan**, + **ċiriċe**. The village occupies high ground overlooking the valley of the r. Trent. There is no church here but a square plot of ground is supposed to be the site. DEPN, Horovitz.

Church HANDBOROUGH Oxon SP 4212. 'H with the church' for distinction from Long HANBOROUGH SP 4114. ModE **church** + p.n. *Haneberge* 1086–1370 with variants *-berg(a) -burgh -berwe* etc., *Hageneb'ga* 1156, *Haunneberg'* 1268, *Hanboroughe -borowe* 1428, 'Hagena's hill', OE pers.n. *Hagena* + **beorg**. O 268, L 128.

HANDBRIDGE Ches SJ 4165. 'Rock bridge'. *Honebrugge* [c.1150]1400, 1202–1450 with variants *-brug -brig(g)(e) -brygg*, *Hondbrigg* 1285–1656 with similar variants, *Hornebrugge* 1350, *Houndbryge* 1482, *Handbridge* from 1544. ME **hōne** (OE *hān*, genitive sing. *hāne*, 'a rock, a boundary stone') + OE **brycg**. The reference is to the prominent outcrop of sandstone rock on which the hamlet stands and which fronts the River Dee with bold cliffs. The variety of forms shows a great deal of folk-etymology connecting the name with *hand, hond* 'a hand', *hund* 'a dog' and *horn* 'a horn, a projecting piece of land'. Originally simply called *Bruge* 1086 'bridge', *Brugge, Bryg(g)e* 16th. Che v.1.i.53.

HANDCROSS WSusx TQ 2629. *Handcross(e)* 1617, 1703, *Hancrosse* 1629. Possibly a crossroads with a one-handed sign-post or a cross with a sign. In the latter case the name may be a descendant of *cruce' Alex'* 'Alexander's cross' 1272. Sx 278.

HANDFAST POINT or the FORELAND Dorset SZ 0582. *Hanfast Point* 1575, c.1586, *Handefaste -feste Pointe* 1583, *Handfast Point* 1861. P.n. Handfast + ModE **point**. Handfast, *Hanfast* late 16th is probably 'rock stronghold', OE **hān** + **fæsten**, but '(at the) high stronghold' is also possible, OE **hēan**, definite form dative case of **hēah**, + **fæsten**. The reference is to Studland Castle, *castellum de Studlande* [1381]16th which formerly stood here. See also The FORELAND SZ 0582. Do i.44.

HANDFORTH Ches SJ 8583. Partially uncertain. *Haneford* [c.1153–81]1285, *Hanford* 1394, 16th cent., *Hon(e)ford* late 12th–14th, *Handford -forth* 1576. OE *hana* + **ford** in which *hana* could be either OE pers.n. *Hana*, 'Hana's ford', OE **hana** 'a cock', 'cock's ford', or genitive pl. of OE **hān** 'a stone', 'ford of the stones', possibly referring to markers placed at the ford. Che i.254, L 71.

HANDLEY '(At) the high wood or clearing'. OE **(æt thǣm) hēan lēage**, dative sing. of **hēah**, definite form **hēa**, and **lēah**. L 207.

(1) ~ Ches SJ 4657. *Hanlei* 1086, *-ley(e) -leg(h) -leigh* etc. 1161–1703, *Handlegh* 1360, *-leye* 1557. Che iv.90 gives pr ['handli] and ['hanli].

(2) ~ SIXPENNY Dorset ST 9917. 'Handley in Sixpenny Handley hundred'. *Sixpennyhanley* 1575, *Soxpenny Handley* 1741. Earlier simply *(at) Hanlee, (in) henlee* [871×7]15th S 357(1, 2), *(at) heanlegen, Hanlee* [956]14th S 630, *Hanlege* 1086, *Han- Henle(gh) -ley(e)* 1230–1618, *Handeleygh* 1496. 'Sixpenny', for distinction from other Handleys, refers to Sixpenny Handley hundred which takes its name from this Handley and Sixpenny ST 8416, *Sexepenne* 1340, *Seppens* c.1540. The first reference occurs in the bounds of Fontmell Magna in S 419, *on ðæs lutlen seaxpennes suð eke* 'to the S edge of the little *Seaxpenn*' [932]15th. *Seaxpenn* is Pen Hill which rises to 356ft. at ST 8517: the name means 'hill of the Saxons', OE folk-n. *Seaxe* + PrW ***penn**, like PENSAX H&W SO 7269, and probably marks an ancient boundary. It was the meeting place of the hundred. Do iii.113, 104.

(3) West ~ Derby SK 3977. *West Hand(e)ley* 1468–9, 1596, for distinction from Middle and Nether Handley SK 4078, *Mydlehanley, Netherhanley* 1586. Handley is *Henlei* 1086, *Hanleg' -le -ley(e)* 1230–17th, *Hand(e)ley* 1470, 1481, 1562. Db 302.

HANDSACRE Staffs SK 0916. 'Hand's newly cultivated land'. *Hadesacre* 1086, *Hendes- Hundes- Handesacra* 1167–96, *Hondesacr'* 1242. OE pers.n. **Hand*, genitive sing. **Handes*, + **æcer** 'cultivated land on the margin of a settlement'. DEPN, L 232–3.

HANDSWORTH SYorks SK 4186. Partly uncertain. *Handesuuord -uurde* 1086, *-wrth -wrd(a) -worth(e) -wurth(e)* 12th–1428, *Handelesworth* 1193×9, *Hannesworth(e)* 1367–1573, *Hansworth* 1473–1641, *Handsworth* 1692. First element probably pers.n. OE **Hand*, **Handel* or a reduced form of **Handwulf*, genitive sing. **Handes*, + **worth**. YW i.165.

HANDSWORTH WMids SP 0390. 'Hun's enclosure'. *Honeswor de* 1086, *-worthe* 1242, *Huneswordne* 1209, *-wurth* 1222. OE pers.n. *Hūn*, genitive sing. *Hūnes*, + **worth** varying with **worthiġn**. DEPN.

HANFORD Staffs SJ 8742. Probably 'the cocks' ford'. *Heneford* 1086, 1307, *Honeford* 1212, 1234×40, *Hanneford* 1250, *Hon- Handford* 1327. OE **hana**, genitive pl. **hanena**, + **ford**. Alternatively the specific might be pers.n. *Hana*. DEPN, L 71, Horovitz.

HANGINGSTONE HILL Devon SX 6186. No early forms. ModE **hanging stone** + **hill**.

HANHAM Avon ST 6472. '(At) the rocks'. *Hanvn* 1086, *Hanum* c.1150–1189, *Hanam* 1154–1400, *Hanham* from 1535. OE **hān**, dative pl. **hānum**. Hanham is bounded on the SW by the deep rocky gorge of the Avon along which are many rocks and old quarries. Gl iii.78.

HANKELOW Ches SJ 6745. 'Haneca's mound'. *Honcolawe* late 12th, *Honkelawe, Hun(c)ke- -i- -y- -low(e)* 1260–1542, *Hankylowe* 1520, *Hankelow* [1621]1656. OE pers.n. **Haneca* + **hlāw**. Che iii.89 gives pr ['hæŋkəlou] and ['hæŋkəlɔ:].

HANKERTON Wilts ST 9790. 'Hanceca's estate'. *Hanekyntone* [680]14th S 1578, [901]14th S 363, 1279, *Haneketon* 1249–79, *Hanketon* 1428, *-er-* 1535. OE pers.n. **Haneca*, genitive sing. **Hanecan*, + **tūn**. Wlt 59.

HANKHAM ESusx TQ 6105. 'Haneca's dry ground'. *(æt) Hanecan hamme* [947]14th/15th S 527, *Henecham* 1086, *Hanekham* 1293.

OE pers.n. *Haneca*, genitive sing. *Hanecan*, + **hamm** 4. Cf. HANKERTON Wilts ST 9790, HANKELOW Ches SJ 6745. Sx 447, ASE 2.26, SS 83.

HANLEY '(At) the high wood or clearing'. OE **hēah**, oblique case definite form **hēan**, + **lēaġe**, dative sing. of **lēah**. L 207.

(1) ~ Staffs SJ 8847. *Henle* 1212, *Hanlih* 1227, *Hanley* 1332. The site is a headland of 500ft. DEPN, Duignan 73, L 207.

(2) ~ CASTLE H&W SO 8442. 'H with the castle'. *Hanley Castrum* 1535 referring to the castle built by king John 1206×13. Earlier simply *Hanlie -lege* 1086, *Hanley* from 1314. 'Castle' for distinction from HANLEY SWAN SO 8142. Wo 201, Pevsner 1968.187.

(3) ~ CHILD H&W SO 6465. 'Children's H'. *Cheldreshanle* 1255, *Chuldrenehanle* 1265, *Hanley Chylde* 1581. OE **ċild** 'young person, young monk, noble born youth', genitive pl. **ċildra** (ME *childres, childrene*), + p.n. *Hanlege* 1086. The estate may have been one whose income was used to support young monks, but the exact reference is unknown. Wo 50.

(4) ~ SWAN H&W SO 8142. *Hanley Swan* 1831 OS. 'Swan' from the Swan Inn here for distinction from Hanley CASTLE SO 8442.

(5) ~ WILLIAM H&W SO 6766. 'H held by William' sc. de la Mare in 1242. *Williames Henle* 1275. Earlier simply *Hanlege* 1086. Also known as *Hanley Thome* 1242 from Thomas de la Mare who held the manor in 1212 and *Ouer Hanley* 1577. 'William' for distinction from Hanley CHILD SO 6465. Both Hanleys occupy the same tract of high ground. Wo 51.

HANLITH NYorks SD 9061. 'Hagni's slope'. *Hangelif* 1086, *Haghnelit -enlid* 12th, *Haghenlith -lyth(e)* 13th–1349, *Haun- Hawn(e)lith -lyth* 1260, 1373, *Hannlith* 1407, *Handleth(e)* 16th cent. ON pers.n. *Hagni*, genitive sing. *Hagna*, + **hlíth**. The reference is to the steep hillside E of the river Aire. The phonological development was *-agh-* > *-au-* > *-a-*. YW vi.130, SSNY 96.

East HANNEY Oxon SU 4192. *Esthenneya* 1220, *Esthanney* 1242×3. ME adj. **est** + p.n. *Hannige* [956]13th S 597, *Hannie* [968]12th, *(æt) Hanige* [968]13th S 759, *Hannei, Hanlei* 1086, *Hann(e)ia -ega -eya -ey(e)* from 1155, 'the wildfowl island', OE **hana**, genitive pl. **hanena**, + **īeġ, ēġ**. The island was probably the strip of land between Letcombe and Childrey Books. A further reference to this land is probably *on aniges ham, of haniges hamme* [958]13th S 651. 'East' for distinction from West HANNEY SU 4092. Brk 477, L 37, 39.

West HANNEY SU 4092. *Westhannea* c.1200, *West Henneye* 1324. ME **west** + p.n. Hanney as in East HANNEY SU 4192. Brk 479.

East HANNINGFIELD Essex TL 7601. *Esthannyngefeld* 1255. ME adj. **est** + p.n. *Hamniggefelde -inge-* [c.1036]12th, *Haningefeld(ā)* 1086, the 'open country of the Haningas, the people called after Han or Hana', OE folk-n. **Haningas* < pers.n. *Hān* or *Hana* + **ingas**, genitive pl. *Haninga*, + **feld**. Later forms include *Han(e)feld(e)* 1248–1387, *Han(e)wefeld* 1296, 1323, < **Hanegefeld* < *Hani(n)gafeld*, and *Hanvil(l)e* 1685, 1768. The Hanningfields mark the open country W of the woodland of the Dengie peninsula. 'East' for distinction from South and West HANNINGFIELD TQ 7497, 7399. Ess 250 gives former pr [hænvəl].

South HANNINGFIELD Essex TQ 7497. *Suhaningefeud* 1232, *Suthannyngfeld* 1329. ME adj. **suth** + p.n. Hanningfield as in East HANNINGFIELD TL 7601. Ess 250.

West HANNINGFIELD Essex TQ 7399. *Westhanege(s)feld* 1203–12, *Westhaningefeld* 1247. ME adj. **west** + p.n. Hanningfield as in East HANNINGFIELD TL 7601. Ess 250.

HANNINGFIELD RESERVOIR Essex TQ 7398. P.n. Hanningfield as in East HANNINGFIELD TL 7601 + ModE **reservoir**.

HANNINGTON Hants SU 5455. 'Estate called after Hana'. *Hanningtun* [1023]12th S 960, *Hanitune* 1086, *Hanyton(e) -i-* 1171–1302, *-ing -yng-* 1237–1382. OE pers.n. *Hana* + **ing**[4] + **tūn**. Ha 86, Gover 1958.139.

HANNINGTON Northants SP 8171. 'The estate called after Hana'. *Hani(n)tone* 1086, *Hanin(g)tone -yn(g)-* 1195–1349. OE pers.n. *Hana* + **ing**[4] + **tūn**. Nth 125.

HANNINGTON Wilts SU 1793. Either 'Hana's hill' or 'cock hill'. *Hanindone* 1086–1398 with variants *-yn-* and *-don, Hanedon(e)* 1211–89, *Hanyngdone* 1300–1564, *Hannyngton* 1576. OE pers.n. *Hana*, genitive sing. *Hanan*, or **hana** 'a cock', genitive pl. **hanena**, + **dūn**. Wlt 24.

HANNINGTON WICK Wilts SU 1795 → Hannington WICK SU 1795.

HANSLOPE Bucks SP 8046. 'Slope of the enclosures'. *Hammescle, (H)anslepe, Hamslape* 1086, *Ham(e)slap(e)* 1104–1398, *Hamslope* from 1488, *Hanslep* 1480, 1509. OE **hamma**, genitive pl. of **hamm**, + ***slǣp**. *Hamma* seems preferable to the proposed OE heroic name *Hǣma* in view of the absence of *e*-spellings in the forms of the name. Hanslope is on a hill with roads leading up to the church. Bk 6 gives pr [hænsləp], Studies 1936.186, L 57.

HANTHORPE Lincs TF 0823. 'Hermothr's outlying settlement'. *Hermodestorp* 1086, *Hermet(h)orp* 1166–1652, *Harmetorp -thorpe* 1219–1649, *Hornethorpp* 1653, *Hanthorpe* 1819. ON *Hermóthr* + **thorp**. Perrott 127, SSNEM 111.

HANWELL Oxon SP 4343. 'Hana's road'. *Hanewege* 1086, *-weie -wey(e)* 1220–1242×3, *Hanewell(e)* 1235×6–1476, *Hanwell* from 1428. OE pers.n. *Hana* + **weġ** subsequently replaced by *welle*. O 398.

HANWOOD Shrops SJ 4409. Partly uncertain. *Hanewde* 1086–1398 with variants *-wud'* and *-wod(e), Hanwode* 1535, *(Greate) Hanwood* from 1559. The generic is OE **wudu** 'a wood', the specific uncertain; OE **hana** 'a cock' would fit but this is mostly used of the domestic cock; OE **hān** 'a rock' is also possible but the reference would be unexplained; perhaps, therefore, OE pers.n. *Hana*, 'Hana's wood'. Sa i.143.

HANWORTH GLond TQ 1271. 'Hana's' or 'cock enclosure'. *Haneworde -worth(e)* 1086–1359, *Hanworth* 1428. OE pers.n. *Hana* or **hana** + **worth**. Mx 152.

HANWORTH Norf TG 1935. 'Hagena's enclosure'. *Haganaworda* 1086, *Haneworth* 1270. OE pers.n. *Hagena* + **worth**. DEPN.

Cold HANWORTH Lincs TF 0383. *Calthaneworth* 1322. OE, ME **cald** + p.n. *Haneu(uo)rde* 1086, *Haneworda* c.1115, *-worth'* c.1215. 'Hana's enclosure', OE pers.n. *Hana* + **worth**. 'Cold' for distinction from POTTERHANWORTH TF 0566. DEPN, Cameron 1998.

HAPPISBURGH Norf TG 3831. 'Hæp's fortified place'. *Hapesburc* 1086, *Hapesburg(h)* 1100×35–1430, *Happesburg* 1229, 1291, *Hap(p)isburgh* from 1353. OE pers.n. **Hæp(p)*, genitive sing. **Happes*, + **burh**. Happisburgh was the chief place in the hundred of Happing, *Hapincha -inga* 1086, *Happinggehundred* [1135×54]13th, *Happing'* 1168–1332, *Happinges* 1185, 1230, 1266, representing OE folk-n **Happingas* 'the people called after Hæp'. *Hæp* must have been a chieftain of some sort, the *Happingas* his retainers or followers, Happisburgh the stronghold of the district of Happing, isolated from the surrounding country by rivers and marsh. There is no clear evidence that it was the site of a Roman fort. DEPN gives pr [heizbrə], Nf ii.84, 92, ING 57, 114–5, Schram 142.

HAPPISBURGH COMMON Norf TG 3729. *Happisburgh Great Common* 1838 OS. P.n. HAPPISBURGH TG 3831 + ModE **common**.

HAPSFORD Ches SJ 4774. 'Hæp's ford'. *Happesford* 1216×72–1641 with variants *Hapes- Hap(p)is- -ys- -us- Apes- Appis-* and *-forde, Harp(e)sford(e)* 1317–1671, *Hapsford* from 1348. OE pers.n. **Hæp(p)*, genitive sing. **Hæppes*, + **ford**. Che iii.257.

HAPTON Lancs SD 7931. 'Settlement on or by the heap'. *Upton* 1241, *Apton* 1243, *Hapton* from 1246. OE **hēap** + **tūn**. The reference is either to a hill of 575ft. at SD 7831 on which Hapton Castle once stood or to Great Hameldon SD 7928 (1343ft.). The earliest spelling, if reliable, suggests OE **up** + **tūn**, the 'higher settlement'. La 80, Jnl 17.50.

HAPTON Norf TF 1796. 'Heabba's estate'. *Habetuna -i-* 1086, *Habeton* 1198, *Hapetun* 1242. OE pers.n. *Heabba* + **tūn** possibly varying with *Heabba* + **ing**[4] + **tūn**. DEPN.

HARBERTON Devon SX 7758. 'Settlement by the Harbourne river'. *Herburnat' -ber-* 1108, 1109, *Herburton* 1158, *-berton(e)* 1220, 1360, *Hurbertun'* 1212–1364, *Harberton* 1603×25. R.n. HARBOURNE + **tūn**. D 325.

HARBERTONFORD Devon SX 7856. Partly uncertain. *Hurbertonforde* 1571, *Harbertonford* 1628. Probably 'ford leading to Harberton', p.n. HARBERTON SX 7758 + **ford**; but the original may have been 'ford over the Harbourne river', r.n. HARBOURNE + **ford** influenced by the nearby p.n. HARBERTON. D 327.

HARBLEDOWN Kent TR 1357. 'Herebeald's hill'. *Herebaldune* 1086×7, *Herbald(o)une* c.1180–c.1206, *Herboldon(a)*, *Herebolddune* 1174×5, *Herboldon'* 1175×6, *Herbaudon' -baldon* 13th cent. OE pers.n. *Herebeald* + **dūn**. Only a single form shows genitive inflexion, *Herbaldesdon'* 1270. PNK 499, L 146, Cullen.

HARBORNE WMids SP 0284. 'The dirty stream'. *Horeborne* 1086, *-burn* 1221. OE **horu** + **burna**. DEPN, L 17–8.

Market HARBOROUGH Leic SP 7387. *Mercati de Hauerberghe* 1219, *Market Har(e)berowe* 1553. ME **market** + p.n. *Haverberga(m) -berg(h)(e) -burg(h)e* 1153–1501, *Har(e)berg(h)* before 1250–1350, *-burgh(e)* 1347–1610, *Harborowe* 1459–1553, 'hill where oats are grown', OE ***hæfera** cognate with ON *hafri* + **beorg**. This place, originally an outlying part of the fields of Great Bowden, lay on the main Leicester-Northampton road at a crossing of the r. Welland, where a market developed. Lei 224, Studies 1936.106.

HARBOROUGH MAGNA Warw SP 4779. 'Great H'. *Herdeborough Magna* 1316, *Hardeburgh Magna* 1498, *Harborough Magna al. Church Harborowe* 1609. P.n. *Herdeberge* 1086–1371 with variants *-ber(e)gh -ber(o)we* and *-borogh*, *Hardebarwe* 1262, 1359, 'the hill of the flocks', OE **heord**, genitive pl. **heorda**, + **beorg**, + adj. Latin **magna** (first recorded 1232) for distinction from Harborough Parva SP 4778, 'little H', *Herdeberge* 1259, *Parva* ~ 1291, *Lyttle Hereberwe* 1296. Wa 111, 116.

HARBOTTLE Northum NT 9304. Partly uncertain. *Hir- Hyrbotle -botil* 1220–1479, *Herbotill* 1430, *Harbotell* 1539. The generic is OE **bōthl** 'a house, a manor', the specific apparently either **hȳra** 'a hireling' or **hȳr** 'hire, wages'. In late Saxon times Harbottle was the focal point of the 'ten towns of Coquetdale'. The sense seems to be 'place to which others are subject' or 'where payment is made'. Alternatively the specific might be as in HARLOW HIll NZ 0768 and so 'magpie house'. NbDu 101, PNE i.276,

HARBOURNE RIVER Devon SX 8056. Partly uncertain. *Hurburn(e)* 1244, 1315, *Hareborne, Hartburn* 1577. The generic is OE **burna** 'a stream', the specific possibly OE **hēore** 'gentle, mild, pleasant' or **heort** 'a hart, a stag'. However, this looks like one of Harrison's 16th cent. antiquarian etymologies. D 7, RN 191.

HARBURY Warw SP 3759. 'Hereburg's fortified place'. *(æt) Hereburgebyrig* [1002×4]c.1100 S 1536, [1004]c.1100 S 906, *Erbur(ge)- Edburberie* 1086, *Herberburia* 12th–1397 with variants *-bury -biri* and *-bire*, *Herberg* 1299, *-bury* 1334, *Harberg al. Harburbury* 1553, *Harbury al. Hungry Harbury* 1740. OE feminine pers.n. *Hereburh*, genitive sing. *Hereburge*, + **byriġ**, dative sing. of **burh**. *Hungry* 1740 alludes to the poor and unproductive quality of the land here. Wa 170.

HARBY 'The farm or village of the herdsmen'. OE **hierde**, genitive pl. **hierda**, + **bȳ**. Very often a designation of a small dependent settlement attached to a larger tenurial unit. SN 70.11–23.

(1) ~ Leic SK 7431. *Hertebi* 1086, *-by* [1277]15th, *Hertheby* 1282, *Herdebi* 1086–1268, *-by* c.1130–1472, *Herdbi -by* c.1130–1553, *Hardeby* 1363–1475, *Hardby* 1413–1548, *Harby* from 1494. Alternatively the specific might be a pers.n., ON *Herrøthr* or *Hjǫrtr*, genitive sing. *Hjartar*. Lei 156, SSNEM 52 no. 2.

(2) ~ Notts SK 8870. *Herdrebi* 1086, *Herdebi -by* 1266–1365, *Hertheby* [1154×89]1316, 1284×90–1332, *Hardeby* 1383, *Harby* 1615. Nt 205, SSNEM 52.

HARCOMBE Devon SY 1590. 'Hare coomb'. *Haracumb'* c.1200, *Harecumbe juxta Sidemuth* 1340. OE **hara** + **cumb**. D 597.

HARDEN WYorks SE 0838. 'Hare valley'. *Heredene* 1180×4, *Hareden(e)* late 12th–1333, *Harden(a) -dene* 1207×27–1676. OE **hara** + **denu**. YW iv.164.

HARDENDALE Cumbr NY 5814. Possibly 'Hardwine's valley'. *Harendale* 1235, *-dall* 1579, *Harndall -dale* 1574–1759, *Hardingdall'* 1242–56, *Hardenesdale* 1247–1544, *Hardendale* from 1544. Possibly OE pers.n. *Hardwine*, genitive sing. *Hardwines*, + **dæl** (ON **dalr**). If this is correct *Hardwine* was early reduced to *Harden*, and spellings with *Harding-* are analogical rather than the local surname *Harding*. We ii.168.

HARDHAM WSusx TQ 0417. 'Heregyth's river-bend land or estate'. *Heriedeham* 1086, *Herietham* 1300, 1307. OE feminine pers.n. *Heregȳth*, genitive sing. *Heregȳthe*, + **hamm** 1 or **hām**. Hardham is situated in a great bend of the r. Arun and is also a parish. Probably an instance of the early coalescence of *hamm* and *hām* in Sussex. Another set of sps for this name survives, *Eringeham* 1135×54, 1283×4, *Helming(e)ham* c.1130, 1189, *Heringham -yng-* 1189–1300, *Hiringham* 1285, *Herringham* 1724, which probably represent a reduced form of *Heregȳthinge-hamm* or *-hām* in which *Heregȳthing*, locative–dative sing. *Heregȳthinge*, is a p.n. meaning 'the place called after Heregyth', pers.n. *Heregȳth* + **ing**[2]. Sx 128, ING 125, MA 10.24, BzN 3.170, ASE 2.49, SS 81.

HARDHORN Lancs SD 3538. 'The store-house'. *Hordern* 1298–1327, *Hordorn* 1332. OE **hordærn**. It must have been a store-house belonging to the lords of Staining or to Whalley Abbey. La 157, Jnl 17.91.

HARDINGHAM Norf TG 0403. 'Homestead of the Heardingas, the people called after Hearda'. *Hardingeham* 1161, *Hardingham* from 1242 ×3. OE folk-n *Heardingas* < pers.n. **Hearda* + **ingas**, genitive pl. *Heardinga*, + **hām**. DEPN, ING 135.

HARDINGSTONE Northants SP 7657. 'Hearding's thorn-tree'. *Hardingestorp* 1086, *-trop* c.1120, *Hardingestone* 1086, *Herd- Harding(e)ston'* 1235–6, *Hardighestorne* c.1145, *Hardingesthorn(a) -t(h)orn(e) -yng-* 1155×8–1428. OE pers.n. *Hearding*, genitive sing. *Heardinges*, + **thorn**. The forms in *-torp* are erroneous. Nth 147, SSNEM 91.

HARDINGS WOOD Ches SJ 8254. *Hardingeswood* 1597. Surname Harding + ModE **wood**. Che ii.321.

HARDINGTON 'Estate called after Hearda'. OE pers.n. **Hearda* + **ing**[4] + **tūn**.

(1) ~ Somer ST 7452. *Hardintone* 1086, *Hardington* from 1225. DEPN.

(2) ~ MANDEVILLE Somer ST 5111. 'H held by the Mandeville family'. *Hardintone* 1086, [late 12th]14th, *Herdinton(e)* 1166, 1243, *Hardin(g)ton* 1243, *Hardington* 1610. DEPN.

(3) ~ MARSH Somer St 5009. *Hardington Marsh* 1811 OS. P.n. Hardington as in HARDINGTON MANDEVILLE ST 5111 + ModE **marsh**.

HARDKNOTT PASS Cumbr NY 2301. Hill name Hardknott NY 2302 + ModE **pass**. Hardknott, 1803ft., is *Hardecnuut* c.1210, *Ardechnut* 1242, *Hard-knot* 1610, 'hard knot', i.e. 'craggy hill', ON **harthr** + **knútr**. The original name of the pass was *Wainscarth* 1242, 'wagon gap, cleft through which wagons could go', ME **wain** (<OE *wæġen*) + ON **skarth**. Cu 343.

HARDLEY Hants SU 4204. 'The hard wood or clearing'. *Hardelie* 1086–1327 with variants *-lega -le(ye)*, *Hardleg'* 1201. OE **heard**, definite form **hearda**, + **lēah**. The reference is probably to hard land contrasted with nearby marshland bordering Southampton Water. Ha 86, Gover 1958.198.

HARDLEY STREET Norf TG 3801. P.n. *Hardale* 1086, 1286, *Hardele* c.1115, 13th, *Hardeleygh* 1268, of uncertain origin, possibly 'the hard clearing, the clearing on hard ground', OE **heard** used as a noun + **lēah**, + dial **street** 'a village'. The reference would be to the firm ground of Broom Hill intruding into Hardley marshes. DEPN.

HARDMEAD Bucks SP 9347. 'Heoruwulf's meadow'. *Horel- Herulf- Herouldmede* 1086, *Harewemede* 1194–1235, *Haremede* 1233–1366, *Hardmede* 1284–1385. OE pers.n. *Heoruwulf* + **mǣd**. The first two DB spellings show *e/o* variation for OE [ø] and [e] < *eo*; the third DB spelling shows confusion with the pers.n. *Herold* < CG *Hairold*. The specific was early misinterpreted as the adjective *hard*. Bk 36 gives pr [ha:mi:d], L 250.

HARDRAW NYorks SD 8691. 'The shepherd's row' sc. of dwellings. *Hardrawe* 1606. Ultimately OAngl **herde** + **rāw**. YN 259.

Upper HARDRES COURT Kent TR 1550. P.n. Upper Hardres + ModE **court**. Upper Hardres is *Hec(h)hardis* 'high H' 1226, *upper Hardres* 1596, ME **hegh**, ModE **upper** + p.n. *(in) haredum* (dative pl.) [785]?9th S 123, *(in) haraðum* [786]?12th S 125, *haredā* [809]13th S 164, *(to, fram) Harþan* [933]c.1400, *Hardes* 1086–16th, *Hardres* from 1191. OE ***harath**, ***harad** 'a wood', pl. ***harathas**, dative pl. ***harathum**, cognate with OHG *hard* 'mountain forest' in the mountain names Harz, *Hart* 781, *Harz* c.870, and Haardt, *Hart* 1262. The area is characterised by stony soil which was left wooded and uncultivated. Also known as *Great Harden* 1570. KPN 67, Palæstra 147–8.35, StN 1.99, PNE i.234, Middendorff 66, Berger 120–1, 126, Cullen.

Lower HARDRES Kent TR 1553. *Nider Hardre* c.1205(p), *Nethere Hardres* 1241, *Nether hardres* 1610, *Lower Hardres* 1819 OS. ME adj. **nither** (OE *nithera*) and ModE adj. **lower** + p.n. Hardres as in Upper HARDRES COURT TR 1550. Also known as *Little Harden* 1570. KPN 67, Cullen.

HARDSTOFT Derby SK 4463. 'Hjort's curtilage'. *Hertestaf* 1086, *Hertistaf -toft(e)* 1257–1453, *Hertstofte* 1476, *Har(e)stoft(e)* 1439–1575. ON pers.n. *Hjǫrtr*, secondary genitive sing. *Hertes*, + **toft**. Db 269, SSNEM 149.

HARDWAY Hants SU 6001. No early forms. ModE **hard** 'a firm beach or foreshore', locally 'a street adjoining a landing', and so 'the road to the hard'.

HARDWAY Somer ST 7234. *Hardway* 1817 OS. Possibly this simply means 'hard, cheerless road'. ModE **hard** + **way**. It is a track that crosses high exposed downland.

HARDWICK 'Herd farm'. OE **heorde-wīc**. The reference is to that part of a manor devoted to livestock as distinct from the **bere-tūn**, the barton or arable farm (see BARTON).

(1) ~ Beds TL 2857 → Kites ~ Warw SP 4768.

(2) ~ Bucks SP 8019. *Hardwich -vic -uich* 1086, *Herdewyc(h) -wic -wik* 1208–84, *Hardwyke* 1435. Bk 79 gives pr [ha:dik].

(3) ~ Cambs TL 3578. *Hardwic* [1042×66]13th S 1051, *-uic(h) -wic(h) -wyk(e)* etc. from 1086. Ca 162.

(4) ~ Norf TM 2289. *Herdeuuic, Hierduic* 1086, *Herdwick* 1254. DEPN.

(5) ~ Northants SP 8569. *Heordewican* [c.1067]12th, *Herdewiche -wik(e) -wyk(e)* 1086–1313, *Hardewiche* 1086, *-wyke* 1220, *Hardwyk* 1428. The first form, if it belongs here, is the dative pl. and suggests a group of 'hardwicks'. Nth 125.

(6) ~ Oxon SP 5729. *Hardewich* 1086, *Herdewic -wyc -wik(e) -wyk'* 1123–1309, *Hardwic'* 1205, *Hardwick Audley* 1580. The manor was held by William de Audele in 1345. O 216.

(7) ~ Oxon SP 3706. *Herdewic'* 1199–1285 with variants *-wike* and *-wyk(e), Herdwic(h)' -wik(') -wick(e)* 1200–95. O 324.

(8) ~ Warw SP 3447 → Kites ~ SP 4768.

(9) ~ HALL Derby SK 4663. *Hardwyck Hall* 1431. P.n. Hardwick + ME **halle**. A late 15th cent. ruined manor house. Hardwick is *Herd(e)wik(e) -wyk(e)* 1257–1453. Db 269.

(10) Blind ~ WYorks SE 4523. *Blynd- Blind(e) Hardwick(e)* 1584, 1631, 1722. Earlier simply *Herd(e)wic* c.1196, 1251, and also known as *Spitle Hardwick(e)* 1294, 1556. 'Blind', OE **blind**, because there is no thoroughfare through the hamlet and *Spitle*, ME **spitel** 'a hosptial', because it belonged to the Hospital of St. Nicholas in Pontefract and for distinction from East ~ SE 4618 and West ~ SE 4118. YW ii.79.

(11) East ~ WYorks SE 4618. ME prefix *Est* 15th cent. + p.n. *Herdewica* 1120×8, *Herthewic -wyk(e)* 1121×7, 1258, 1328, *Herdwik -wyk(e)* 1243–1402, *Hardwic(ke) -wyk(e) -wik(e)* 1215–1568. The spellings with *-th-* may be due to ON influence. 'East' for distinction from West ~ SE 4118 and Blind ~ SE 4523. YW ii.72.

(12) Kempston ~ Beds TL 0244. *Kempston Hardewik* 1485. Earlier simply *Hardwyke* 1276. A livestock farm belonging to KEMPSTON TL 0347. Bd 77.

(13) Kites ~ Warw SP 4768. 'H where kites are seen'. *Kytherdewyk* 1387, *Kytehardwik* 1529, *Kytes Hardwicke al. Hardwicke Grombold* 1661. ME **kyte** (OE *cȳta*) + p.n. *Herdewice* 1208, *Herdwic* 1236, *Hardwyk* 1464. Other examples of Hardwick prefixed with a bird name are Hardwick Farm SP 3447 *Herdewic in porroch. de Thiesho* c.1136, *Kuyte herdewyk* 1316 (ME *kyte*) and Hardwicke Farm Beds TL 2857, *Herdwich* 1209, *Puttokesherdwyke* 1286 (ME *puttok* 'kite'). Wa 139, 285, Hu 258.

(14) Priors ~ Warw SP 4756. 'H belonging to the prior' sc. of Coventry as in DB. *Herdewyk Priour* 1310, *Prioures Herdywyk* 1410, *Hardewyk Prior* 1488. ME **priour** + p.n. *Herdewiche* 1086–1195, *-wic* 1236, 1316. Wa 270.

(15) West ~ WYorks SE 4118. *West Hardwick* from 1559. ME prefix *West* 1390 + p.n. *Harduic, Arduwic* 1086, *Hardewic(h)* 12th cent., *Herdewic(a) -wyk(e)* 1119×35–1400, *Herthewyke -wic(k)* 13th, 1343. 'West' for distinction from Blind ~ SE 4523 and East ~ SE 4618. YW ii.88.

HARDWICKE 'herd-farm'. OE **heorde-wīc**.

(1) ~ Glos SO 9127. *Herdeuuic* 1086, *-wik -wyk(e)* 1212–1525, *Hard(e)wi(c)k -wyk(e) -wike* c.1260–1597. The herd-farm of the manor of Elmstone as in ELMSTONE HARDWICKE SO 9226. Gl ii.82.

(2) ~ Glos SO 8012. *Herdewik(e) -wice - wyk(e)* late 12th–1469, *(la) Hard(e)wi(c)ke -wyk(e)* 1222–1704. Gl ii.180.

(3) ~ H&W SO 2743. *Herdwyc* 1272, *Herdewyk* 1309×24, 1328. The herd-farm of Clifford parish SO 2445. He 55.

HAREBY Lincs TF 3365. 'Hari's farm or village'. *Harebi* 1086–1202. ON by-n. *Hari* from *hari* 'a hare', genitive sing. *Hara*. + **bȳ**. DEPN, SSNEM 53, Cameron 1998.

HAREDEN Lancs SD 6450. 'Hare valley'. *Harden* 1343–1817. OE **hara** + **denu**. YW vi.212.

HAREFIELD GLond TQ 0590. 'Army open land'. *Herefelle* 1086, *-feld* 1206–1342, *Harefeld* 1223–94, *Harfeld* 1393–1428, *-vil(l)* 17th cent. OE **here** + **feld**. The reason for the name is unknown. Mx 35.

HARE HATCH Berks SU 8077. Perhaps 'trap for catching hares'. *Hare Hatch* from 1607. ME **hare** + **hacche**. Brk 121, PNE i.213.

HAREHOPE Northum NU 0920. 'Hare valley'. *Harop* c.1150, 1236 BF, *Harhop* 1242 BF, 1308, *Hayropp* 1289, *Hareupp* 1628. OE **hara** + **hop**. NbDu 101, L 116.

HARESCOMBE Glos SO 8410. Probably 'Heresa's coomb'. *Hersecome* 1086, *-cumbe* 1148×79, 1253, *Harsecumbe -co(u)mbe* 1222×8–1597, *Haresco(u)mbe* 1294–1535, *Hascombe* 1577, 1655. OE pers.n. **Her(e)sa* + **cumb**. Cf. HARESFIELD Gl ii.164.

HARESFIELD Glos SO 8110. 'Heresa's open land'. *Hersefel(d) -felde* 1086–1255, *Harsefeld(e)* 12th–1438, *Haresfeld(e)* 1100×35–1509. OE pers.n. **Her(e)sa* + **feld**. Gl ii.182, L 240, 243.

HARE STREET Herts TL 3929. Partly uncertain. *Harestrete* 1498, *-streete* 1568. This is usually said to be OE **here-strǣt** 'army street', but the forms are too late for certainty. The reference seems to be to an ancient side road from or parallel to Ermine

Street, the Roman road from Braughing to Chesterton, Margary no. 2b, rather than to Ermine Street itself, or to Stane Street since there is another reference in Great Hormead, *Herestrete* 1472, *Hare Street* 1593. Hrt 151, 179.

HAREWOOD '(The) grey wood'. OE **hār**, definite form **hāra**, + **wudu**.

(1) ~ WYorks SE 3245. *æt harawudu* 10th, *Hareuuode* 1086, *-wud(e) -w(o)d(e)* 1190~1377, *Harwod(e)* 1135×50–1545, *-wood(e)* 1545–1678. It has been suggested that the specific might be OE **hær* 'a rock, a heap of stones' referring to the large rock in the parish called the Greystone. But the forms point clearly to OE *hāra* which could only otherwise be the inappropriate genitive plural form if derived from *hær*. YW iv.180 gives prs ['hɛəwud] and ['hɑ:wud].

(2) ~ FOREST Hants SU 4044. *Hare Wood Forest* 1817 OS. P.n. *Harewode* 1198–1379, *Horwud(e)* 1222, 1238. Ha 87, Gover 1958.8.

HARFORD Devon SX 6359. Probably 'street ford'. *Hereford* 1086–1301, *Herforde* 1291, 1299. OE **here** short for **herepæth** + **ford**. D 275.

Fleet HARGATE Lincs TF 3924. 'Fleet road'. *(pontem de) Flete Hergate* 1316, *Flete Hargate* 1364. Also simply *Herregate* 1276, *Hargate* 1488, 1572. P.n. FLEET TF 3823 + ME **hergate** 'army road, road' (ON **hergata*). The reference is to the road from Sutterton TF 2835 to King's Lynn TF 6119. Payling 1940.18, Cameron 1998.

HARGRAVE 'Hoar wood' or 'hare wood'. OE **hār**, definite form **hāra**, or **hara** 'a hare' + **grǣfe** or **grāfa**. L 193. The habitat of the hare is field and heath rather than woodland. While it is impossible to distinguish between **hāra** and **hara** in most names, **hāra** is probably to be preferred in compounds with woodland terms.

(1) ~ Ches SJ 4862. *le Haregreve* early 13th, *Hargreve* late 13th–1545 with variants *Hare-* and *-grave* (from 1287) *-greave(s) -greeve*. Che iv.105.

(2) ~ Northants TL 0370. *Haregrave* 1086–1330, *Heregrave* 1086, 1282, *Hargrave* from 1242. Nth 191, PNE i.234, L 193.

(3) ~ Suff TL 7759. *Haragraua* 1086, *Haregraue* c.1150, 1610, *Hargrave* 1254. DEPN.

HARKER Cumbr NY 3960. Partially obscure. *Herker* 1589, *Harkar* 1617. The specific is uncertain and the forms are too late, but the generic is probably ME **ker** 'marsh' (<ON *kjarr*). Cu 147.

HARKSTEAD Suff TM 1834. Probably 'Hereca's place'. *Herchesteda* 1086, *Herkested(e)* 1198–1534, *Horkestede* 1331, *Harkisted* 1506, *Harksted* 1567, 1610. OE pers.n. *Hereca* + **stede**. Stead 188.

HARLAND HILL NYorks SE 0284. No early forms. Possibly the 'tract of land marked by a heap of stones', OE ***hær** + **land**, + ModE **hill**.

HARLASTON Staffs SK 2110. 'Heorulf's estate'. *Heorelfestun* [1002×4]11th S 1536, *Horvlvestone* 1086, *Herlaueston* 1165, 1242, *Herlaweston* 1288, *Herlaston* 1393. OE pers.n. *Heorulf* (*Heoruwulf*), genitive sing. *Heorulfes*, + **tūn**. Some later sps point to pers.n. *Heorulāf*. DEPN, Horovitz.

HARLAXTON Lincs SK 8832. 'Hiǫrleifr's farm or village'. *Herlavestvne* 1086, *-ton(e)* 1174–1240, *Herlowes- -lawes- -loucs- -laghes- -laceston* 13th cent., *-laston* 1242–1432, *-laxton* 1265–1587, *Harlacston* 1291, *-laxton* from 1310. ON pers.n. *Hiǫrleifr*, secondary genitive sing. *Hiǫrleifes*, + **tūn**. The pers.n. has been reformed as if an OE name in *-lāf*. Perrott 468, Insley.

HARLE SYKE Lancs SD 8635. No early forms. The generic is OE **sīċ**, ON **sík** 'a ditch, a trench, a small stream'.

HARLESDEN GLond TQ 2283. 'Heoru- or Herewulf's estate'. *Herulvestune* 1086, *Herleston* 1195–1330, *-don* 1291, *Harleston* 1365–1795, *-don* 1564, *-den* 1606. OE pers.n. *Heoruwulf* or *Herewulf*, genitive sing. *Heoru- Herewulfes*, + **tūn**. Mx 162.

HARLESTON Devon SX 7945. Partly uncertain. *Harliston* 1252, *Harleston* from 1333. Unknown pers.n. + **tūn**. The evidence is too late for certainty; possibilities are OE *Herel, Herewulf* or *Heoruwulf*. D 314.

HARLESTON Norf TM 2483. 'Here- or Heoruwulf's estate'. *Heroluestuna* 1086, *-tun* [1087×98]12th, *Harolveston* 1228. OE pers.n. *Here-* or *Heoruwulf*, genitive sing. *Here- Heoruwulfes*, + **tūn**. DEPN.

HARLESTON Suff TM 0160. 'Heoru- or Herewulf's estate'. *Heroluestuna* 1086, *-tun* c.1095, *Herleston(')* 1197–1344, *Harleston* 1610, 1661. If *Heorulfestune* 1015 S 1503 belongs here this is OE pers.n. *Heoruwulf*, genitive sing. *Heoruwulfes*, + **tūn**, but the identification is not certain. DEPN, Baron.

HARLESTONE Northants SP 7064. 'Herulf's farm or village'. *Erlestone, Herolvestone -tune* 1086, *Heruleston* 13th, *Herleston(e)* 1169–1372, *Harleston* 1367, *Harlston* 1482, *Halstone* 1675, 1712. OE pers.n. *Herulf* < *Here-* or *Heoruwulf*, genitive sing. *Herulfes*, + **tūn**. Nth 83 gives pr [hælsən].

HARLEY Shrops SJ 5091. 'Rock wood or clearing'. *Harlege* 1086, *Harleia* c.1200 etc. with variants *-leg(h)* and *-ley(e), Herleia* c.1090, 1121, 1138, *-le(g')* 1206, 1344–5, *Horle* 1271×2, *Hoorley* 1780. OE ***hær** + **lēah**. Sa i.144.

East HARLING Norf TL 9986. *Estherling'* 1242×3. ME adj. **est** + p.n. *(at) Herlinge* [1052×66]13th S 1519, *Herlinga* 1086, *(ad) Herl(l)inge* [1087×98]12th, *Herling'* 1194, 1197, 'the Herelingas, the people called after Herela', OE folk-n. **Herelingas* < pers.n. **Herela* + **ingas**. The earliest form is 'the homestead at or called Harling', *þet land at Herlingham* [1042×53]13th S 1535. 'East' for distinction from Middle and West Harling TL 9885, *Westherling'* 1242×3, *Midhirlyng* 1319. DEPN, ING 57–8.

HARLING ROAD STATION Norf TL 9788. P.n. Harling Road < Harling as in East HARLING TL 9986 + **road**, + ModE **station**.

HARLINGTON Beds TL 0330. 'Hill of the *Herlingas*, the people called after Herela'. *Herlingdon* 1086–14th cent., *(H)erlinge-dun(e)* [1181×90]13th, *Herlingedon* 13th cent., *Herlington* 1240, *Harlyngdon* 1489, *-ton* 1492. OE folk-n. *Herelingas* < pers.n. **Herel(a)* + **ingas**, genitive pl. *Herelinga*, + **dūn**. Bd 123, L 145.

HARLOW Essex TL 4711. Partly uncertain. *(at) Herlawe* [1043×5]13th S 1531, *Herlaua* 1086, *Herlaue -lawe* 1112–1303, *-lowe* 1345, *Harlawe* 1254. This name is taken to be the 'army mound', OE **here** + **hlāw**. The reference is probably to the small hill 250 yards NW of the railway station, the site of a Romano-Celtic temple, the base of which is surrounded by an earthwork probably marking the boundary of the temple precincts. It is tempting to suggest an alternative specific, OE **hearg** 'a heathen shrine', but the absence of any trace of the *-g* of *hearg* in the recorded spellings seems to rule this out. Ess 36, Pevsner 1965.221, Jnl 2.40.

HARLOW HILL Northum NZ 0768. P.n. *Hir- Hyrlawe* 1242 BF–1329, *Herlawe* 1346, *Harlawe* 1538, 'magpie hill', OE **hiġera, hiġre** + **hlāw**, + pleonastic ModE **hill**. Alternatively the specific might be as in HARBOTTLE NT 9304. NbDu 102, PNE i.246.

HARLOW HILL NYorks SE 2854. *Harlo Hill* 1597. P.n. *Harelaw* c.1190, *Herlawe* 1219, *Harlowe* 1544, 'the grey or stone hill or mound', OE **hār**, definite form **hāra**, or ***hær** + **hlāw**, + pleonastic ModE **hill**. YW v.117.

East HARLSEY NYorks SE 4299. *Esteharlsay* 1536. ModE **east** + p.n. *Herlesege* 1086, 'Herel's island', OE pers.n. **Herel*, genitive sing. **Her(e)les*, + **ēġ** in the sense 'hill jutting into flat land'. 'East' for distinction from West Harlsey, *West Harlesay* 1365, ModE **west** + p.n. *Herlesege, Herelsaie, Herselaige, Erleseie* 1086, *Herleseie -ey -ay* 1088–1316. YN 215, 212.

HARLTHORPE Humbs SE 7437. 'Herel's dependent settlement'. *Herlesthorp(ia)* 1150×60–1285, *Herlethorp(e)* 1199×1211–c.1400, *Harlethorp* early 13th, 1572, *-thropp* 1534. Pers.n. *Herel* + **thorp**. *Herel* may be a shortened form of an ON pers.n. such as *Herlaugr, Herleifr* etc. YE 241, SPN 139.

HARLTON Cambs TL 3852. 'Herela's estate'. *Herletona -e, Harle(s)tona* 1086, *Herleton(e)* 1185–1285, *Herlenton(a) -e* c.1150–1240, *Harltun* 1218, *Har(e)l(e)ton(e)* 1291–1549. OE pers.n. *Herela* , genitive sing. *Herelan*, + **tūn**. Ca 76 gives pr [hɔ:ltən].

HARMAN'S CROSS Dorset SY 9880. *Armons Cross* 1840. Surname *Harman* + ModE **cross** 'a crossroads'. Do i.67.

HARMBY NYorks SE 1289. 'Hjarni's village or farm'. *Ernebi* 1086, *Hernebi -by* 1086–1404, *Harneby* 16th cent. ON pers.n. *Hjarni*, genitive sing. *Hjarna*, + **bȳ**. YN 252, SSNY 29.

HARMER GREEN Herts TL 2516. *Harmer grene* 1641, *Harmore* 1672. Either p.n. *Harmore* or surname *Harmer* + ModE **green**. Hrt 125.

HARMER HILL Shrops SJ 4922. 'Hill at Harmer'. *Harmear Hill* 1702, *Armour Hill* 1833 OS. P.n. Harmer as in Harmer Moss Plantation SJ 4822, *Harmer Heath* 1611, *Haremore* 1636, *Harmer* 1833 OS, the name of former marshland drained in the 17th cent., possibly 'the grey marsh', OE **hār** + **mōr**, + ModE **hill**. Raven 88.

HARMSTON Lincs SK 9762. 'Hermothr's estate'. *Herm(od)estune, Hermodestone* 1086, *Hermodeston(e)* 13th cent., *Hermeston* 1180–1508, *Harmeston* 1327–1576, *Harmston* from 1524. ON pers.n. *Hermóthr*, genitive sing. *Hermóths*, + **tūn**. Perrott 227, SSNEM 192.

HARNHAM Wilts SU 1328. 'The hares' river-meadow'. *Harnham* from 1115, *Hareham* 1130–1211, *Harham, Har(r)am* 13th cent. OE **hara**, genitive pl. **harena**, + **hamm**. An East and West Harham are distinguished by the 13th cent. *Estharnham (atte Bryggesend juxta Sarum), Westharnham*. Wlt 222.

HARNHILL Glos SP 0700. 'Grey' or 'hares' hill'. *Harehille* 1086, *Harnhill(e)* from 1177 with variants *-hull(e) -helle*. OE **hār**, dative sing. definite form **hāran**, or **hara**, genitive pl. **harena**, + **hyll**. Gl i.75, L 170.

HAROLD HILL GLond TQ 5591. P.n. Harold as in HAROLD WOOD TQ 5590 + ModE **hill**.

HAROLD WOOD GLond TQ 5590. 'Harold's wood'. *Horalds Wood* c.1237, *Harold(e)swood(e)* c.1272, 1488, *Horoldswo(o)d(e)* 1281, 1321, *Horellyswod* 1480. OE pers.n. *Harold*, genitive sing. *Haroldes*, + **wudu**. The wood formed part of the Liberty of Havering-atte-Bower held by earl Harold in 1066. In 1062 it is referred to as *(andlang) ðare strate wald* '(along) the street-wood' S 1036 in the bounds of Upminster, from its proximity to the Roman road to Colchester, Margary no. 3a. Ess 118.

HAROME NYorks SE 6482. '(The settlement) amongst the stones'. *Harun -em* 1086, *Harum* 1086–1471, *Harom(e)* 1301, 1572. OE ***hær**, dative pl. ***harum**. YN 70.

HARPENDEN Herts TL 1314. 'Harp valley'. *Herpedene* c.1060, *-den(e)* 1272–1440, *Harpeden(e)* 1196–1292, *Harpden* 1487–1619, *Harden* 1558–1728, *Harding* 1637–1719, *Harpenden(a)* 1285, 1296, 1719. OE **hearpe** + **denu**. The sense of the name is unknown. Cf. HARPSDEN Oxon SU 7680. The modern form with *-en-* is not the survival of an OE genitival form *hearpan* but an 18th cent. revival. Cf ASPENDEN TL 3528 for a similar late development of *-en-*. Hrt 37 gives former prs [ha:dən -iŋ].

HARPER'S BROOK Northants SP 9286. 'Brook of the harper' or 'of a man named Harper'. *(le) Harperesbrok -is- -broc* 1135×54–c.1270. OE **hearpere**, genitive sing. **hearperes**, or surname *Harper* + **brōc**. Nth 2.

HARPFORD Probably 'ford on the (military) highway'. OE **herepæth** + **ford**.

(1) ~ Devon SY 0990. *Harpeford* 1167–1244, *Herpeford* 1212–89, *Herpford* 1292, *Hawford* 1675. The reference is to the highway from Exeter to Lyme Regis. The original ford was ½ mile N of the modern road. D 590.

(2) ~ Somer ST 1021. *Herpoðford* [899×909]16th S 380, *[1065]18th S 1042, *Herpeford* 1236. D 590.

HARPHAM Humbs TA 0961. 'The harper's homestead' or possibly 'homestead where the harp is played'. *Arpen, Harpein* 1086, *Harpam* [1100×15], 1265, *Harpham* from 1199. OE ***hearpa** or **hearpe** + **hām**. YE 89.

HARPLEY H&W SO 6861. 'Hearpa's wood or clearing' *Hoppeleia* 1222, *Harpele(y)* 1275–1611, *Happeleye* 1293. OE pers.n. **Hearpa* from ***hearpa** 'a harper' + **lēah**. Alternatively the specific might be OE **hearpe** 'a harp' in the sense 'riddle, sieve' and the p.n. might mean 'wood or clearing where sieves are made'. Wo 75, PNE i.240.

HARPLEY Norf TF 7826. 'Salt-harp wood or clearing'. *Herpelai* 1086, 1121, *Harpelai* 1086, *-le* 1206, 1254. OE **hearpe** + **lēah**. The sense is 'clearing where salt-harps, i.e. sieves or riddles, are made'. DEPN, PNE i.240.

HARPOLE Northants SP 6960. 'The muddy pool'. *Horpol* 1086–1580 with variants *-poll(e) -pull, Har(e)pol(e)* 1258–1556, *Happall* 1557. OE **horh** + **pōl**. Nth 84 gives pr [a:pəl], L 28.

HARPSDEN Oxon SP 7680. 'Harp valley'. *Harpendene* 1086–1517 with variants *-den(') -dun* and *-don', Harpeden(e) -don'* 1182–1501, *Harpden* 1441, 1647, *Harpesden'* 1275×6 etc., *Harpsden alias Harden* 1762. OE **hearpe**, genitive sing. **hearpan**, + **denu**. Near the head of the valley the bounds of Newnham Murren attached to S 738, an original charter of 961, run *andlang hearp dene* 'along harp valley'. Here *hearp(e)* must denote a topographical feature, though its exact significance is unclear. O 72, L 99.

HARPSWELL Lincs SK 9389. Partly uncertain. *Herpeswelle -uuelle* 1086, 13th, *Harpeswella* c.1115, 1185×7, *-well* 1212. The generic is OE **wella** 'a spring', the specific possibly **hearperes**, genitive sing. of **hearpere** 'a harper', with dissimilatory loss of the second *r*, or **herepæthes**, genitive sing. of **herepæth** 'a highway', referring to the prehistoric track along the top of the W scarp of the limestone ridge or Cliff from Lincoln to the Humber. DEPN.

East HARPTREE Avon ST 5655. *Est Herptre* c.1185. ME adj. **est** + p.n. *Harpetrev* 1086, *Harptre* 1172, OE **hearpe** for *herepœth* 'public road' + **trēow** 'a tree', as in West HARPTREE ST 5656. DEPN, Kitson.

West HARPTREE Avon ST 5656. *West Herpetre* c.1185. ME adj. **west** + p.n. *Herpetrev* 1086, *Herpethreu* 1086 Exon, OE **hearpe** for *herepœth* 'public road' + OE **trēow** 'a tree'. 'West' for distinction from East HARPTREE ST 5655. DEPN, Kitson.

HARPUR HILL Derby SK 0671. *Harper Hill* 1773. Probably surname *Harper* + ModE **hill**. Db 373.

HARPURHEY GMan SD 8601. 'Harpour's enclosure'. *Harpourhey* 1320, *Harpouresheie* 1327, *Le Harperheye* 1502×3, *Harperhey* 1509. Refers to 80 acres of land granted to one William Harpour before 1322. Surname *Harpour* + ME **hey** (OE *heġe*). La 36, Jnl 17.33.

HARRIETSHAM Kent TQ 8652. 'Heregeard's promontory'. *Herigetes ham* [973×87]12th S 1511, *Hergerdes hā* 1037×40 S 1467, *Hergeardes ham* 1044×8 S 1473, *Hariardesham* 1086, *Herietesham* 1215, *Haretsham* 1610. OE pers.n. *Hereġeard*, genitive sing. *Hereġeardes*, + **hamm** 2, 3 or 5. The specific has been variously modified under the influence of similar elements such as OE *hereġeatu* 'heriot' and the French feminine pers.n. *Harriette*. PNK 294, ASE 2.21.

HARRINGAY GLond TQ 3188. A revival of an early form of the name HORNSEY TQ 3089. Mx 123, GL 56.

HARRINGTON Cumbr NX 9925. 'Estate called after Hæfer'. *Halfringtuna* c.1160, *Haverington(a) -tun* [c.1160]n.d.–1279, *Hafrincton* c.1200, *Haueryngton* 1396, *Harington* 1292, *-yngton* 1428. OE pers.n. *Hœfer* + **ing**[4] + **tūn**. The first spelling is regarded as erratic. Cu 399.

HARRINGTON Lincs TF 3671. 'The estate called after Hearra'. *Haringtona, Harintun* 12th, *Harin- Harminton* 1202, *Harington* 1212. OE pers.n. *Hearra* + **ing**[2] + **tūn**. This seems more likely than the alternative suggestion 'the settlement at the stony place', OE ***hǣring** < **hǣr* + *ing*[4] + **tūn**. DEPN, Årsskrift 1974.45, Cameron 1998.

HARRINGTON Northants SP 7780. Probably the 'settlement by the heather-covered land'. *Arintone* 1086, *-tona* c.1090,

Harinton' 1166, *Hed- Hetherin(g)ton(e) -yng-* 1184–1379, *Hath-Harrington, Hethryngton* 16th. OE ***hǣddring** < ***hǣddre** + **ing**[4], + **tūn**. Alternatively the specific might be OE pers.n. **Heathu-* or *Hǣthhere* + **ing**[4], and so the 'estate called after Heathu- or Hæthhere'. Nth 113, Årsskrift 1974.48.

HARRINGWORTH Northants SP 9197. 'The enclosure at the stony place'. *Haringwrth* [c.1060]13th–1332×6 with variants *-yng-* and *-worth*, *Haringeworde* 1086, *-wurða -worth* 1166–1274, *Harryngworth* 1323. OE ***hæring** < ***hær** + **ing**[2], + **worth**. The soil here is light gravel overlying rock. Forms with medial *-e-* may represent the locative–dative inflected form ***hæringe** 'at the stony place' or genitive pl. **Hæringa* of folk-n. **Hæringas*, the 'dwellers at the stony place'. Cf. HARRINGTON Lincs TF 3671. Nth 167, Årsskrift 1974.45.

HARRISEAHEAD Staffs SJ 8656. *Harrishey-head* 1671, *Harrisea Head* 1836 OS. Unknown p.n. *Harrishey*, probably pers.n. *Harry* + ME **hey** (OE *ġehæġ*) 'an enclosure', + ModE **head** 'a headland'. The site is a spur of high ground extending from Mow Cop at a height of 714ft.

HARROGATE NYorks SE 3055. Possibly 'the road to the cairn'. *Harwegate* 1332–7, *Haru-* 1343–1460, *Harrowga(i)te -gayte* 1333–1695, *Harrogate* from 1512, *Harlo(w)-* 1518–73. ON **hǫrgr** 'a heap of stones, a cairn' < **hargu-* + **gata**. The *Harlo(w)-* spellings are by asociation with HARLOW HILL SE 2854. YW v.108.

HARROLD Beds SP 9456. 'The grey wold'. *Hareuuelle* 1086, *Harewold(a) -wode* 13th–14th cents., *Horewald* 1247, *Harrold* 1346. OE **hār**, definite form **hāra**, + **wald**. This was probably part of the great forest of Bromswold as in NEWTON BROMSWOLD Northants SP 9965. Bd 32, L 227.

HARROW GLond TQ 1586. 'Heathen temple(s)'. *(æt) Hearge* 825 S 1436, *Hergas* [830 for 833]13th S 1414, *Herges* 1086–1235, *Her(e)wes* 1234–72, *Hergh* 1243, *Har(e)we* 1278–1341, *Harowe* 1369. OE **hearg**. The earliest reference, *gumeninga hergae* 764 for 767 S 106, suggests it was the ritual centre of an otherwise unknown people called the Gumeningas. Also known as Harrow on the hill, *Harowe atte Hille* 1398, ~ *on the Hill* 1426. The isolated hill here rising above the Middlesex plain would have been a typical site for a holy place. Mx 51, GL 57, Settlement 106.

HARROW HILL WSusx TQ 0809. *Harrow Hill* 1813 OS. Probably 'the heathen temple hill'. There are no early forms but Harrow Hill is the site of a small rectangular Iron Age hill-fort overlying Neolithic flint mines. The place either was or was perceived as a place of heathen worship. OE **hearg** + **hyll**. Sx 165, Thomas 1980.208, 211.

HARROWBARROW Corn SX 4070. 'The grey wood'. *Harebere* c.1286, *-beare* 1313–99, *Harrobear* 1748, *Harrowburrow* 1813. OE **hār**, definite form **hāra**, + **bearu**. PNCo 94, Gover n.d. 181.

HARROWDEN Beds TL 0646. 'Hill of the heathen shrines'. *Herghe- Hergentone* 1086, *Har(e)wedon* 13th cent., *Harouden* 1286. OE **hearg**, genitive pl. **hearga**, + **dūn**. Bd 91, Evidence 106, Signposts 158, L 144, 151.

Great HARROWDEN Northants SP 8870. *Major Harwedon* 1227, *Magna Harudon* 1284. ModE adj. **great**, Latin **maior**, **magna**, + p.n. *Hargindone -tone, Hargedone* 1086, *Harhgeduna* [1155×8]1329, *Harudon* 12th–1284, *Har(e)wedon* 1203–1350, *Harowedon* 1381, *Harodown* 1526, 'the hill of the (pagan) temples', OE **hearg**, genitive pl. **hearga**, + **dūn**. 'Great' for distinction from Little HARROWDEN SP 8671. Nth 125, Signposts 158–61.

Little HARROWDEN Northants SP 8671. *Parva Harwedon* 1220. Adj. **little**, Latin **parva**, + p.n. Harrowden as in Great HARROWDEN SP 8870. Nth 125, Signposts 158–61.

HARSTON Cambs TL 4251. 'Herel's estate'. *Herlestona, -ton(e)* 1086–1499, *Harlestone -a* 1086–1480, *Hard(e)leston(e)* 1218–1495, *Harlston* 1389, *Harston* from 1390, *Harson, Haston* 1757. OE pers.n. **Herel*, genitive sing. **Herles*, + **tūn**. If the form *Heorulfestune* 1015 S 1503 belongs here this would be 'Heorowulf's estate' but the identification is uncertain. Ca 84.

HARSTON Leic SK 8431. 'The grey stone'. *Herstan* 1086, *Harestan'* c.1130–1276, *-ston'* 1180–1407, *Harstan(e)* 1195–1347, *-ston(e)* 1348–1610. OE **hār**, definite form **hāra**, + **stān**. Lei 141.

HART Cleve NZ 4634. Uncertain. *Hert* c.1135–1512, *Herte* 1235–1335, *Harte* 1321–1621, *Hart* from 1438×9. A difficult name. Ekwall believed that the name *Heruteu* 'stag island', first recorded c.730 in Bede's *Ecclesiastical History*, originally referred both to Hartlepool and to Hart. This would normally have become lOE *Heortē*, ME *Herté*, rather than *Hert*. Early loss of *-é* is evidenced in the name Cocken NZ 2847, 'Cocca's stream or fishery', OE **Coccan-ēa**, *C- Kokene, Cochena, C-Kokenne ee* c.1138–1243, *Coken, Cokin -yn* from 1228, but this does not really parallel the earliness of *Hert*. Ekwall explained the endingless spelling as a back-formation from the district name **He(o)rtē-hērnes* 'the jurisdiction or district of Hart' which must early have become **He(o)rt(h)ērnes*, cf. the spelling *Heorternesse* [c.1040]11th. If Ekwall is right the etymology is OE **heorot** 'a stag' + **ēġ** 'an island' presumably used in the sense 'an island, a spur of high ground'. Hart occupies an elevated site overlooking the coastal plain. Nevertheless, in *Beowulf* a hall is called simply *Heorot* 'stag' and in the new world simplex animal names such as Bear, Beaver, Buffalo, Eagle, Elk and Hart itself are not uncommon. Bede's analysis of *Heruteu* as *insula cerui* 'island of the stag' is undoubtedly right, but his translation may be wrong: *Heruteu* might be not 'stag island, hart island' but 'Hart island, island belonging to Hart.' If so, *Hert* and *Herté* (Bede's *Heruteu*) were distinct names and distinct places. See further HARTLEPOOL. Studies 1931.75.

River HART Hants SU 7659. A back-formation from HARTFORDBRIDGE SU 7757. Gover 1958.108.

HARTBURN Northum NZ 0986. 'Stag stream'. *Herteburne* [1198]1271, 1203, *Hertburgh* 1284, *-burn* 1507, *Harbourne* 1663. OE **heorot**, genitive pl. **heor(o)ta**, + **burna**. NbDu 103, L 18.

HARTER FELL 'Hart hill'. ON **hjǫrtr**, genitive sing. **hjartar**, + **fjall**.

(1) ~ Cumbr NY 4609. *Herter- Harterfell* 1578. We ii.168.

(2) ~ Cumbr SD 2199. *Herter fel* c.1210. Cu 437.

HARTEST Suff TL 8352. 'The stag wooded hill'. *Hertest* [1042×66]12th S 1051, *Herte(r)st* 1086, *Herthyrst* c.1095, *Hertherst* 1200, *Hartyste* 1610. OE **heorot** + **hyrst**. DEPN.

HARTFIELD ESusx TQ 4735. 'Open land where stags are seen'. *Hertevel* 1086, *-feld* 1135×54–1428, *Hertisfeld* 1179, *Hertfeld* 1309. OE **heorot**, genitive pl. **heorta**, + **feld**. Cf. HARVEL Kent TQ 6563. Sx 365, SS 70.

Upper HARTFIELD ESusx TQ 4634. ModE adj. **upper** + p.n. HARTFIELD TQ 4735.

HARTFORD Cambs TL 2572. 'Army ford'. *Hereforde* 1086–1535, *Herford* 1147–1444, *Harford* 1410–1558, *Hertford* 1428–1558. OE **here** + **ford**, probably short for *herepæth-ford* in which *herepæth* means simply 'highway'. The term *here-ford* is not used of a Roman road – the ford on Ermine Street was 2 miles W between Huntingdon and Godmanchester – but there is no obvious 'highway' here unless a route going N from Godmanchester to Whittlesey, Thorney and Crowland is postulated. The road from Huntingdon to Old Hurst does not cross the Ouse. Perhaps the sense is 'ford beside the highway'. Hu 208, Jnl 24.46.

HARTFORD 'Hart ford'. OE **heorot** + **ford**. L 71.

(1) ~ Ches SJ 6472. *Herford* 1086, 1349, 1557, *Harford* 1646, *Hertford* 1291–1694, *Hartford* from c.1188. Che iii.188.

(2) ~ END Essex TL 6817. *Hert(forde)hende* 1367 sc. of Felsted. P.n. Hartford, *hereford* 12th, *Herteford* 1329 + ME **ende**. Ess 422.

(3) East ~ Northum NZ 2679. A modern colliery settlement

named after East Hartford Farm, *East Hartford* 1867 OS, ModE adj. **east** + p.n. *Hertford* [1198]1271, 1296 SR, *Hertfeud* (sic) 1219 BF, *Herford'* 1242 BF, *Hertford super Blitham* 1203, *Harford* 1663. 'East' for distinction from *West Hartford Hall* (lost) 1867 OS and West Hartford Farm NZ 2578 formerly *Red House* 1867 OS. NbDu 103 gives pr [harfəd], DEPN 71.

HARTFORDBRIDGE Hants SU 7757. *Hertfordbrigge* 1327. P.n. *Hertford* 1327 'hart ford', ME **hert** (OE *heorot*) + **ford**, + **brigge** (OE *brycg*). Cf. HARTLEY WINTNEY SU 7656. Ha 87, Gover 1958.114.

HARTFORTH NYorks NZ 1706. 'Hart ford'. *Hertfort* 1086, *-ford(e)* 1086–1328, *Hert- Hartforth* 1539, 1577. OE **heorot** + **ford**. YN 289.

HARTHILL 'Hart hill'. OE **heorot** + **hyll**. L 171.

(1) ~ Ches SJ 5055. *Hert'* [1208–29]1580, *Herthil, Harthil* 1259, *Hert(e)hull(e)* 1297–1517, *Harthill* [1325]1619, [1461×83]1574 and from 1589. Che iv.92.

(2) ~ SYorks SK 4980. *Hertil(l)* 1086–1200, *-hill(e)* 1190–1535, *-hull* 1291–1428, *Harthill* from 1198. YW i.153, L 171.

HARTHOPE BURN Northum NT 9623. P.n *Herthop* 1305, 'stag valley', OE **heorot** + **hop**, + ModE **burn** 'a stream'. NbDu 103.

East HARTING WSusx SU 7919. *Estherting'* 1195–6, *Esthertlinge* 1248. ME adj. **est** + p.n. *Heartingas* [970]c.1100 S 779, *Hertingas* [970]13th S 776, c.1200, *Hertinges -is -ynges* 1086–1469, *Herting(e) -ynge* 1130–1289 etc., 'the Heortingas, the people called after Heort', OE folk-n. *Heortingas < pers.n. *Heort* + **ingas**. 'East' for distinction from South HARTING SU 7819, North Harting (now ROGATE SU 8023) and West Harting, *Westhertlinge* 1248, *Westhertinges* 1296. Sps here and for West and South HARTING with *-rtl-* imply the diminutive form of the pers.n. **Heortla* as in HARTLEBURY H&W SO 8470. Sx 35, ING 35, MA 10.23, SS 64.

South HARTING WSusx SU 7819. *Suthhertlinges* 1248, *Suhthurtinge* 1279, *Suthertyng* 1281, *Suthhertynges* 1296. ME adj. **sūth** + p.n. Harting as in East HARTING SU 7919. Sx 35, ING 35.

HARTINGTON Derby SK 1260. 'Stags' hill'. *Hortedun* 1086, *Hurtendun* [1203]1272, *Hertendon(e) -in- -yn-* 1231–1372, *Hertingdon(e) -doun -yng-* 1244–1431, *-ton* 1306–1559, *Hartington -yng-* from 1541. OE **heorot**, genitive pl. **heor(o)ta**, + **dūn**. Forms with *-en- -ing-* seem to show a non-etymological intrusion. Db 364, L 144, 155.

HARTLAND Devon SS 2624. Probably 'newly cultivated land of Harton'. *Herti- Hirtilanda* 1130–1169, *Hertiland* 1198–1285 with variants *-la(u)nd, Hertland* 1196. P.n. Harton + OE **land**. Harton, *Heortigtunes* (genitive sing.) [873×88]11th S 1507, *Hertitone* 1086, *(H)ertinton* 1175, 1179, *Harton* 1565 etc., 'farm, village, estate called after Heort', OE pers.n. *Heort* + **ing**[4] + **tūn**, is the proper and original name of the settlement, later superseded by Hartland as a more appropriate name for this extremely large parish (17,000 acres). Cf. DONYLAND Essex TM 0222. D 71.

HARTLAND POINT Devon SS 2227. *Hartland Point* 1577. P.n. HARTLAND SS 2624 + ModE **point**. D 78.

HARTLAND QUAY Devon SS 2224. *Hartland Quay* 1809 OS. P.n. HARTLAND SS 2624 + ModE **key**.

HARTLEBURY H&W SO 8470. 'Heortla's fortified place or manor'. *Heortlabyrig* [817]17th S 1597, [980]11th S 1342, *Heortlanbyrig* [852×74]12th B 1320, [985]11th S 1351, *Hverteberie* 1086, *He- Hurtlebery* 12th, *Hurkelebery* 1271. OE pers.n. *Heortla*, genitive sing. *Heortlan*, + **byriġ**, dative case sing. of **burh**. Wo 242.

HARTLEPOOL Cleve NZ 5132. 'Hart Island pool or inlet'. This is a difficult name; its forms fall into three main groups:

I. *Herterpol* c.1160–1335, *Hiartar poll* [c.1170]c.1220, *Hertrepol* 1200–1521 with variants *-poll(e) -pole*.

II. *Hertepol* 1211–1420 with variants *-i-, -y-, -pole -poll' -pull* (1420), *Hertpol(e)* 1310×11, 1382.

III. *Hertelpol* 1200–1433, *Hertilpol(e)* c.1230–1511 with variant spellings *-yl-, -poel* (1260–c.1325), *-pull(e)* (1342–1473), *Hartlepoul* 1344, *Hartlepool* from 1531.

There is also a strange 17th cent. spelling *Hartinpoole* 'Hart in pool'.

Group I is consonant with ON **hjartar pollr** 'stag's pool' as in the ON *Morkinskinna* c.1170, but this cannot be the origin of the name since St Hild's 7th century monastery on the headland at Hartlepool is named by Bede c.730 as *Heruteu, id est Insula Cervi*, 'Heruteu, that is, Island of the Stag'. There are no grounds for doubting Bede's analysis (as opposed to his translation, as argued in the entry for HART) of *Heruteu* which is a compound of OE **heorot** 'a stag' and **ēġ** 'an island' (with re-inflected feminine nominative sing. ending *-u*). This would become ME *Herté* which, with the addition of OE **pol** 'a pool, a creek', produces the group II spellings. It seems unlikely that ME *Hertépol* could have been subject to folk-etymological reformation under ON influence since there is no evidence of other than sporadic Scandinavian settlement in this area: the *Morkinskinna* form is best regarded as a literary translation.

A possible explanation was suggested by Ekwall. In late OE times the district dependent on Hart NZ 4634 was known as *Heorternesse* [c.1040]11th, 'Hart *hērnes*, Hart jurisdiction', exactly as the district dependent on Berkeley Glos ST 8149 was known as *Beorclea hyrnesse* 1121, with OE **hērnes** (WSax **hȳrnes**) 'obedience, jurisdiction, a district subject to a single jurisdiction'. With loss of medial *-h- Herternes* would have been subject to false analysis as *Herter-ness* 'the headland called *Herter*' and hence have led to reformation of *Herté-pol* as *Herter-pol*. This form was subject to dissimilatory processes whereby the sequence *r - r* became *r - l* to produce the Group III spellings. The final form of the name is probably influenced by the 15th/16th century vogue for names containing the French definite article *le*: *Hart-le-pool* nicely fits the pattern of Dalton-le-Dale, Chester-le-Street etc.

Hartlepool occupies a rocky peninsula (the **ēġ** or island of the name) connected to the mainland by a narrow neck of blown sand only 500 yards wide; there is no evidence that it was ever a tidal island like Holy Island but in ancient times the connecting neck of land was probably narrower than now. The parallel between the monastic site at Hartlepool and that at Lindisfarne is striking. The *pool* referred not just to the harbour south of the peninsula but also to the *Slake* (Northern dial. **slake** 'a mud-flat, waste land bordering the sea-shore and covered with water at high tide'), a large shallow bay now drained, which formerly extended inland in a NW direction, emphasising the insular character of the headland. Studies 1931.75, VCH Durham iii.263, NbDu 104, TC 104, L 28, 36, 39, Jnl 8.21, Nomina 12.35.

HARTLEPOOL BAY Cleve NZ 5231. P.n. HARTLEPOOL NZ 5132 + ME **bay**.

HARTLEY 'Hart wood or clearing'. OE **heorot** + **lēah**.

(1) ~ Kent TQ 7634. *heoratleag* 843 S 293, *Hertle* c.1200–53. Described as a woodland-pasture belonging to Little Chart. KPN 187.

(2) ~ Kent TQ 6166. *Erclei* 1086, *Ercleie, Herclei* c.1100, all for *(H)ert-, Hertle(gh') -ley(e)* 1226–78, *Hartlegh'* 1254, *Hartley* 1610. K 42.

(3) ~ Northum NZ 3375. *Hertelawa* 1166, *-lawe* 1242 BF, *Hertlaw* 1296 SR, *Hartley* 1573. NbDu 103.

(4) ~ WESPALL Hants SU 6958. 'H held by the Waspail family'. *Hertlegh Waspayl* c.1270. Earlier simply *Her- Harlei* 1086, *(apud) Heortlegam* (Latin accusative sing.) 1156, *Hertlegh* 1236. Waspail is an AN surname meaning 'waster', literally 'waste-straw', *wast-* + *paille*. Ha 87, Gover 1958.120.

(5) ~ WINTNEY Hants SU 7656. 'H held by (the prioress of) Wintney'. *Wynteneyes Hertlegh* 1303, *Hertele Wynteneye* 1416. Also known as 'nuns' H', *Minechenehertelеye* 1236, ME

minchene (OE *mynċen*, genitive pl. *mynċena*) and 'monks' H', *Hurtle Monialum* 1270, Latin **monialis**. genitive pl. **monialum**. Earlier simply *Hertlega* 1218, *Hurtlegh* 1228, *-leye* 1327. Wintney is 'Winta's island', *Winteneia* [1139×61]1337, *Winteneya* 1218, OE pers.n. *Winta*, genitive sing. *Wintan*, + **ēġ** in the sense 'raised ground in marsh'. Ha 87, Gover 1958.115, DEPN.

(6) New ~ Northum NZ 3076. *New Hartley* 1925 OS. ModE adj. **new** + p.n. HARTLEY NZ 3375. A modern development that grew up around *Hartley Colliery* 1865 OS.

HARTLEY Cumbr NY 7808. 'Hard ridge of land'. *Hardclay* 12th–1286, *Hard(e)cla* 1280×90, 1292, *Hartecla* 1280–1415, *Harcla* 1189×99–1368, *Harklowe -klaue*, *Harkeleye* 1278, *Hertlay -ley* 1283–1552, *Hartlay -ley* 1370–1714. OE **h(e)ard** + ***clā**, a side-form of OE *clēa*, *clawu* 'a claw, a tongue of land between streams'. The reference is to the curving ridge between the river Eden and Hartley Beck. The name was subsequently assimilated to the common p.n. type *Hartley*. We ii.2.

HARTLIP Kent TQ 8364. 'Hart leap'. *Heordlyp* c.1100, *Hertlep(e)* 1218–54, *Hartlep* 1219, *Hartlyppe* 1610. OE **heorot** + ***hlīep** 'a leap, a jumping place' and often, as here, of a steep place or sudden drop in the ground. PNK 249.

HARTON NYorks SE 7061. Possibly 'the marauders' settlement'. *Heretun(e)* 1086, *Harton(a)* from 1293. OE **here** + **tūn**. Alternatively this could be 'Herra's settlement', with OE pers.n. **Her(r)a* as the specific or 'the settlement by the rocks', OE ***hær** + **tūn**. YN 37, DEPN.

HARTON Shrops SO 4888. *Harton* 1833 OS.

HARTON T&W NZ 3764. 'The stags' hill'. *Heortedun* c.1107, *Hertedunā -dun'* 1154×66, [1199×1212]1300, *Hertendun* c.1190×5, *Herterdun* (sic) [c.1190×5]1418, *Hertedone* [1189×99]1336, 1203×4, *Herton* [1196×1208]c.1400, 1296–1345, *Harton* from 1356. OE **heorot**, genitive pl. **heor(o)ta**, + **dūn**. NbDu 104, L 144, 158.

HARTPURY Glos SO 8025. 'The hard pear-tree'. *Hardepiry(e) -pyry(e) -ia* 12th–1380, *Hard(e)purye* 1316, 1387, *Hartpirie* 1324, *-pury* from 1535. OE **heard** + **piriġe**. The reference is to a tree with hard fruit like perry-pears. Gl iii.155.

HARTSHILL Warw SP 3294. 'Heardred's hill'. *Ardreshille* 1086, *Hardredeshella* 1151, *-hull* c.1170–1276, *-hille* 1207, *Hardeshull -is-* 1287–1549, *Hartishill* 1535, *Hartshull* 1541, *-hill* 1688. OE pers.n. *Heardrēd*, genitive sing. *Heardrēdes*, + **hyll**. Wa 84, L 170.

HARTSHORNE Derby SK 3221. 'Hart's horn'. *Heorteshorne* 1086, *Herteshorn(e) -is-* c.1141–1495, *Hartsorn* 1348, *Hart(e)shorn(e)* 1577, 1610 etc. OE **heorot**, genitive sing. **heor(o)tes**, + **horn**. The reference is to Horn Hill, SE of the village, and its supposed resemblance to a hart or stag's horn. Db 637.

HARTSOP Cumbr NY 4013. 'Hart's valley'. *Herteshop(e) -op(p)* before 1184–1481, *Hertsop(p)* 1393–1475×80, *Hartsop(p) -ope* 13th–1777. OE **heorot**, genitive sing. **heor(o)tes**, + **hop**. The reference is to a side-valley opening out of the valley of Brothers Water. We ii.223, L 115, 117.

HARTWELL Northants SP 7850. 'The harts' spring'. *Hertewell(e)* 1086–1293, *Hertwell(a)* 1148×66–1498, *Harwell* 1675. OE **heorot**, genitive pl. **heor(o)ta** + **wella**. L 31. Nth 100 gives pr [ha:təl].

Isle of HARTY Kent TR 0267. *Harty Island* 1819 OS. Earlier simply *Herte* 1086, *Hertei* [1087]13th, *He(o)rtege* c.1100, *Herteye* 13th cent., 'stag island', OE **heorot** + **ēġ**. PNK 250.

HARVEL Kent TQ 6563. 'Open land frequented by stags'. *(to) heorot felda (geate)* 'to Harvel gate' 939 S 447. OE **heorot** + **feld**. KPN 244.

HARVINGTON H&W SP 0549. 'Settlement at or called Hereford'. *heruerton* *[709]12th S 80, *herefordtun* [964]12th S 731, *Hereferthvn* 1086, *Herfortun* 1249–1535, *Hervington* 1508, *Harvington* 16th. P.n. *hereford* [?799 for 802]11th S 154, [?814]11th S 1261, OE **here-ford** 'army ford' possibly short for **here-pæth ford** 'highway ford', + **tūn**. The village is a mile NW of the ford across the river Avon. Wo 134, L 71, Jnl 24.42–8.

HARWELL Oxon SU 4989. 'The stream or spring by the hill called *Hara*, the grey one'. *(æt) haranwylle* [956]12th S 672, *Harawille* [973]12th S 790, *(æt) Harewillan* [985]12th S 856, *Harvvelle*, *Harowelle* 1086, *Harewella -well(e) -wille* c.1130–1401×2, *Harwell* from 1337. OE hill name **Hāra* < **hār** 'grey', definite form **hāra**, genitive sing. **Hāran*, + **wylle**. Brk 521, L 32.

HARWICH Essex TM 2431. 'Army camp'. *Herdwic* 1229, *Herewic -wyk -wik- wyz -wiz -wych(e) -wich* 1248–1342, *Herewycum* 1308, *Harewych* 1470. OE **here** + **wīċ**. The remains of a camp were still visible in the 18th cent. Ess 339 gives prs [hæridʒ] and [ha:dʒ].

HARWICH HARBOUR Essex TM 2632. P.n. HARWICH TM 2431 + ModE **harbour**.

HARWOOD Durham NY 8233. 'The grey, hoary wood'. *Harewude* 1235×6. OE **hār**, definite form **hāra**, + **wudu**. Part of the Forest of Teesdale where John Baliol enclosed a park in the early 13th century. The scenery must have been like that traversed by the hero of *Sir Gawain and the Green Knight*:

Hiȝe hillez on vche a halue, and holtwodes vnder,
Of hore okez ful hoge a hundreth togeder (742–3).

HARWOOD GMan SD 7411. 'The grey wood'. *Harewode* 1212–1332 etc., *Harwude* 1227, *Harwode* 1327, *Horewood* 1521. OE **hār**, definite form **hāra**, + **wudu**. La 46, Jnl 17.38.

Great HARWOOD Lancs SD 7332. *majori Harewuda* before 1123, *Magna Harwod(e)* 1303, 1327. Adj. **great**, Latin **maior**, **magna**, + p.n. *Harewude* 13th, 'grey wood', OE **hār** + **wudu**. 'Great' for distinction from Little Harwood SD 6929, *Little Harewud* 1246, *Parua Har(e)wode* 1327, *Little Harewode* 1493. La 72, 73, L 228, Jnl 17.46.

HARWOOD BECK Durham NY 8321. *Harwood Beck* 1857 Fordyce. P.n. HARWOOD NY 8233 + **bekkr**.

HARWOOD DALE NYorks SE 9695. *Harwoddale* 1577. P.n. *Har(e)wod(e)* 1301–95, 'the grey wood', OE **hār**, definite form **hāra**, + **wudu**, + ModE **dale**. YN 113 gives pr ['ærəddil].

HARWOOD FOREST Northum NY 9894. P.n. *Harewud* c.1155, *-wod(e)* 1278, 1356, *-wood* 1421, 'the grey wood', OE **hār**, definite form **hāra**, + **wudu**, + ModE **forest**. NbDu 105, L 228.

HARWORTH Notts SK 6191. Partly uncertain. *Hareworde* 1086, *-wrthe -worth -wurth -wrde* 1192–1344. The generic is OE **worth**, the specific either adj. **hār** 'grey', definite form **hāra**, ***hær** 'stony ground' or **hara** 'a hare' used as a nick-n. Nt 80, DBS 166 s.n. Hare first recorded 1166.

HASBURY WMids SO 9583. 'Hazel fortified place'. *Haselburi* 1270, *Halesbury* 1544. OE **hæsel** + **byriġ**, dative sing. of **burh**. The *Hales-* spelling is due to the influence of nearby Halesowen. Wo 294 gives pr [heizbəri].

HASCOMBE Surrey SU 9940. 'The witch's valley'. *Has(e)cumbe*, *Hasco(u)mbe* 1241–1382, *Hassecumbe* 1266, *Hescumb(e)*, *Escumbe* 1243–c.1270, *Haskham* 1541, *Hascom* 1635, *Hascomb or Askum* 1728. OE **hæġtesse**, **hætse** + **cumb**. Identical with Hestercombe Somer ST 2428. Sr 243 gives prs. [ha:s- a:skəm], EPN i.216, Studies 1936.176–7, L 90, 93.

HASELBECH Northants SP 7177. Either 'the hazel-tree ridge' or 'stream-valley'. *Esbece* 1086, *Haselbech(e) -bek -bec(che)* 12th–1314, *Haselbeech* 1608, *-bitch(e)* 1639, 1730. OE **hæsel** + **bæċ** or **bæċe**. Haselbech is situated at the end of a short ridge above the 600ft. contour line but runs down to a typical small *bæċe* valley on its N side. Nth 115 gives pr [heizəlbitʃ], L 12, 220.

HASELBURY PLUCKNETT Somer ST 4710. 'H held by the Plugenet family'. *Haselbare Ploukenet* 1431. Earlier simply *Halberge* (sic) 1086, *Heselberge* 1176, *Haselbere* 1327, *Hasylbere* 1610, 'hazel-tree hill, barrow or grove', OE **hæsel** + **beorg** varying with **bearu**. The manor was held by Alan de Plugenet in 1268. DEPN.

HASELEY Warw SP 2367. 'The hazel-tree wood or clearing'. *Haseleia* 1086, *-leye* 1275–1535, *Heseley* 1272. OE **hæsel** + **lēah**. Wa 210, L 204, 220.

Great HASELEY Oxon SP 6401. *Magna Hassele* 1284×5, *Great Haselee* 1301. ME adj. **grete**, Latin **magna**, + p.n. *Haselie* from 1086 with variants *-le(e) -le(g)a -legh(e) -ley*, 'the hazel-tree wood or clearing', OE **hæsel** + **lēah**. 'Great' for distinction from Little HASELEY SP 6401. O 128, L 204, 220.

Little HASELEY Oxon SP 6401. *Parva Hasel, ~ Hassell* 1278×9, *Little Hasel(l)ee* 1300–49. ME adj. **litel**, Latin **parva**, + p.n. *(œt) Hœsellea, Hœseleia* [1002]13th S 902, *Haselie* from 1086 with variants *-le(ge) -ley(e)*, 'the hazel-tree wood', OE **hæsel** + **lēah**. 'Little' for distinction from Great HASELEY SP 6401. O 128.

HASELOR Warw SP 1257. 'The hazel-tree ridge'. *Haselove* (for *-overe*) 1086, *Haselovere* 1153×89–1397, *Haselore* 1291–1535, *Hasler* 1633. OE **hæsel** + **ofer**. Wa 211 gives pr [hæzlər], L 176–7, 220, Gelling 1998.84.

HASFIELD Glos SO 8227. 'Hazel-tree open land'. *Hasfeld(e)* 1086–1492, *-feud* 1278, *-fild* 1535. OE **hæsel** + **feld**. Gl iii.148.

HASKAYNE Lancs SD 3508. Uncertain. *Hasken* 1329, 1530, *Haskeen* 1366, *Haskyn* c.1540, *Haskeyne* 1598. This is probably OW **hescenn** 'a marsh'. British *e* may have been a more open sound than OE *e* and hence may occasionally have been borrowed as *æ* (>ME *a*), cf. *poll hæscen* [977]11th S 832, a lost place in Cornwall, and HESKIN Lancs SD 5315. Cf. W *hesg* 'sedges, rushes', Ir *seiscean* 'marsh, fen'. La 120, LHEB 281, CPNE 130–1, 188, GPC 1861.

HASKETON Suff TM 2550. 'Haseca's estate'. *Hasce- Has(c)hetuna* 1086, *Hasketun* 1253, *-ton(e)* from 1219, *Haston* 1504, *Hasonn* 1578. OE pers.n. **Haseca* + **tūn**. DEPN, Arnott 9, Baron.

HASLAND Derby SK 3936. 'Hazel grove'. *Haselont* 1129×38, *-lond* 1269, 1373, 1430, *-lund* c.1200–69, *Has(e)land* 1310–52 etc., *Heselunt* c.1263. ON **hasl** occasionally replaced by **hesli**, + **lundr**. Db 260, SSNEM 154, L 207.

HASLEMERE Surrey SU 9032. 'The pool by the hazel-trees'. *Heselmere* 13th cent., *Haselmere* from 1255. OE **hæsel** + **mere**. Sr 204, L 26, 220.

HASLINGDEN GRANE Lancs SD 7522. P.n. Haslingden SD 7823 + dial. **grain** 'a small valley forking off from another' (ON *grein* 'branch, fork'). Simply called *Grayne* 1566, *ye Grane* 1681. La 91.

HASLINGDEN Lancs SD 7823. 'Valley where hazels grow'. *Heselingedon* 1242, *Haselendene -in-* 1246, *Haselingden* 1251, *Haselinden(e)* [1258]13th–1332, *Haslyngdene* 1341, *Haselden* 1577. OE ***hæsling** varying with **hæslen** + **denu**. La 91, Jnl 17.54, L 98, 220.

HASLINGFIELD Cambs TL 4052. Probably 'open land of the Hæselingas, the people called after Hæsela'. *Haslingefeld(e)* 1086, *Hes(e)lingefeld(a) -inga-* 1086–[1189×99]1308, *-ingfeld(e) -yng-* 1154×89–1509×47, *Haseling(e)feld(e) -yng(e)-* 1190–1558. OE folk-n. **Hœselingas* < pers.n. **Hœsel(a)* + **ingas**, genitive pl. **Hœselinga*, + **feld**. This seems the preferable explanation in view of the frequency of the *-inge- -ynge-* spellings which seem to rule out OE **hœseling* 'a hazel thicket'. The pers.n. *Hœsel(a)* is not recorded but there is a parallel in CG *Hasilo* which avoids having to posit a lost p.n. *Haseley* to give 'the people who live at Haseley'. Ca 77 gives pr [heizliŋfi:ld], L 244.

HASLINGTON Ches SJ 7356. 'Hazel farm or village' or 'farm, village called or at *Hœsling*'. *Hasillinton* early 13th, *Haselin(g)ton(e) -yn(g)-* 1216×72–1516, *Haslington* from 1425. OE **hæslen** adj. 'growing with hazels' varying with ***hæsling** 'place growing with hazels'. Che iii.12.

HASSALL Ches SJ 7657. 'The witch's nook'. *Eteshale* 1086, *Hatishale* 13th, *Hattesale -hall* late 13th, 1318, *Hatsale* 1425, *Hassale* late 13th–1472 with variants *Has(s)(h)ale, Hassall* from 1408. OE **hætse** (a reduced form of *hæġtesse*) + **halh**, dative sing. **hale**. Che iii.21 gives pr ['(h)æsɔ:l].

HASSALL GREEN Ches SJ 7858. *Hassall Green* 1831. P.n. HASSALL SJ 7657 + **green**. Che iii.19.

HASSELL STREET Kent TR 0946. *Hessole-street, Hassell Street* 1798, P.n. *Hertesole* 1272×1307(p), *Hersole* c.1453, the 'stags' muddy pool', OE **heorot**, genitive pl. **heor(o)ta**, + **sol**, + Mod dial. **street** 'a village'. Cullen.

HASSINGHAM Norf TG 3705. 'Homestead of the Hasingas, the people called after Hasu'. *Hasingeham* 1086, *Hasingham* 1254. OE folk-n. **Hasingas* < pers.n. **Hasu* from OE adj. *hasu* 'grey', genitive pl. **Hasinga*, + **hām**. DEPN, ING 135.

HASSOCKS WSusx TQ 3015. A modern development which grew up after the opening of the Brighton railway in 1846, named from a field called *Hassocks* on account of the rough tussocks of grass growing there. ModE **hassock** (OE *hassuc*). Sx 276.

HASSOP Derby SK 2272. 'Hætt valley'. *Hetesope* 1086, *Hatsop(e)* 1236–60, *Hassop(e) -opp(e)* 1281–1306 etc. OE p.n. **Hœtt* 'the hat' which is suggested as a lost name of one of the heights around Hassop, genitive sing. **Hœttes*, + **hop**. An alternative suggestion for the specific is OE nick-name *Hœtt* or **hætse** 'a witch'. Db 109, L 115–6.

HASTINGFORD ESusx TQ 5225 → HASTINGS TQ 8009.

HASTINGLEIGH Kent TR 0944. 'Wood or clearing of the *Hœstingas*, the people called after Hæst(a)'. *(of) Hœstinga lege* [993]14th S 877, *Hastingelai -lie* 1086, *Haestingelege, Hastingeleg, Hestingelaege* c.1100, *Hastingle(ye) -leghe* 1195–c.1400 with variants *Hest-* and *-yng-*. OE folk-n. *Hǣstingas* < pers.n. *Hǣsta* + **ingas**, genitive pl. *Hǣstinga*, + **lēah**. The Hæstingas of Hastingleigh are probably quite separate from the Hæstingas of Hastings ESusx TQ 8009. It seems unlikely that the name means the 'wood or clearing belonging to Hastings' marking one of its woodland-pastures although it could be a name marking the boundary of its territory. PNK 424, KPN 340, L 206.

HASTINGS ESusx TQ 8009. 'The Hæstingas, the people called after Hæsta'. *Hastingas* *[790]12th S 133, *Hastengas* *[960]13th S 686, *Hastinges* 1086–1428, *Hœstinges* c.1100, *Hasting* 12th–16th, *Hastyngs* 1342. OE folk-n. *Hǣstingas* < pers.n. *Hǣsta* + **ingas**. The Hæstingas were an important tribal group referred to in an 8th cent. Northumbrian chronicle as the *gens Hestingorum* which seems to have kept a separate identity as late as the early 11th cent. Their territory probably extended as far as Hastingford TQ 5225 one mile N of Hadlow Down, *Hastingeford* 1279, the 'ford (leading to the territory) of the Hæstingas', and perhaps to HASTINGLEIGH Kent TR 0945. The town is also referred to as *Hestingport* 1042×66 coins and *Hœstingaport* c.1100 ASC(D) under year 1066, 'the town (OE **port**) of the Hæstingas', and also as *(to) hœstingaceastre* [11th]16th BH, *(at) hastingecestre* 1204×15 ibid., 'the city (OE **ċeaster**) of the Hæstingas'. Sx 534, xxiv, TC 105, SS 65, BH 99.

HASTINGWOOD Essex TL 4807. 'Wold held by the Hasting family'. *Waude Hastinge* 1235, *Hastyngeswode* 1316, *Haslingwood* 1836. OE **wald** 'high forest' + family name *Hasting*. The 1836 form shows substitution of *hasling* 'a hazel-tree wood' for the family name. Related forms are *Northwalde Hastinge* 1280, *Northweld Hastyng* 1325 and *Hazelwood Common* 1777, 1805. Ess 87.

HASTOE Herts SP 9209. 'Hall site'. *Halstowe* 1275, 1296, *Hawstowe* 1603×25, *Haster* 1711, *Harstow* 1750. OE **hall** + **stōw**. Hrt 52.

HASWELL Durham NZ 3743. Uncertain. *Hedsowell* 1153×95, *Heswell(e)* 1153×95, 1233×44, 1360–1517, *Hesswell(e)* 1180–1483, *Haswell* from 1483, *Hasilwell* 1560. Previously explained as 'hazel-tree spring', OE **hæsel** + **wella**. But most p.ns. containing *hœsel* show some early forms with *l*, e.g. HASBURY W. Mids SO 9583, HESSAY NYorks SE 5253, HESWALL Mers SJ 2782, the only other exception being HASFIELD SO 8227. An alternative explanation is possibly 'witch's well', OE **hæġtesse, hætse**, as in HASCOMBE Surrey SU 9940, HESCOMBE Somer ST 5116 and HESSENFORD SX 3057, or perhaps a pers.n. cognate with G pers.n. *Hasso, Hesso*. NbDu 105, L 31, 200.

HATCH 'Fence, gate, grating, fish-trap, sluice'. OE **hæċċ**.

(1) ~ Beds TL 1547. *la Hache* 1232, *Hattche* 1539. Hatch lies on a stream. Bd 95.

(2) ~ Hants SU 6752. *Hache* c.1227, *The Hatch* 1817 OS. Possibly a grating set in the stream here. Gover 1958.124.

(3) ~ Wilts ST 9228. *Hascia* 1100×1135, *Hache* 1200, 1202. OE **hæċċ** referring either to a gate leading to a wood or forest or to a wooden trap for catching fish or eels. An East and West Hatch are distinguished, *Westhacha* 1225, *Esthache* 1242. Wlt 196.

(4) ~ BEAUCHAMP Somer ST 3020. 'H held by the Beauchamp family'. *Hache Beauchampe* 1243, *Hatche Beacham* 1610. Earlier simply *Hache* 1086, *Hach* 1212, *Hacch'* [1329] Buck. Probably a gate into forest. The manor was held before 1212 by Robert de Bello Campo (*beau-champ*). DEPN.

(5) ~ END GLond TQ 1391. The 'gate end' sc. of Pinner parish. *le Hacchehend* 1448. ME **hacche** (OE *hæċċ*) + **ende**. Probably referred to a gate of Pinner Park. Mx 64.

(6) ~ WARREN Hants SU 6148. *Heche* 1086, *Hac(c)he* 1212–1316. Ha 88, Gover 1958.126.

(7) West ~ Somer ST 2820. *Westhache* 1243, *West Hatch* 1610. ME adj. **west** + p.n. *Hache* 1201 as in HATCH BEAUCHAMP ST 3020. DEPN.

HATCHING GREEN Herts TL 1313. *Hatchin Green* 1619. Possibly originally Hatch End Green, cf. *Hatchen End* 1610. P.n. *Hatch End, ME **hacche** + **end**, + ModE **green**. Hrt 38.

HATCHMERE Ches SJ 5571. 'Lake with a hatch or fish-trap'. *Hached(e)- Hachemer(e)hurst* 1357, *Hatchmere* 1780. ME ***hachede*** 'with a hatch' (OE *hæċċe* + *-ede*) + **mere**. Che iii.215, 250.

HATCLIFFE Humbs TA 2100. 'Headda's slope'. *Hadeclive* 1086–1252, *-cliva* c.1115, *Haddecliva -e -clif -clyf(f)* 1219–1399, *Hatteclif(f) -clyf(f)* 1364–1632, *Hatclyf(e) -clif(f)* etc. from 1338. OE pers.n. *Headda* + **clif**. Li iv.104, L 134–6.

HATFIELD 'Heath-covered open land'. OE **hǣth** + **feld**.

(1) ~ Herts TL 2309. *Haethfelth(a)* [c.731]8th BHE, *(on) Hæþfelda* c.900 ASC(A) under year 680, [966×75]12th S 1484, *(æt, into) Hæðfelda* [c.890]c.1000 OEBede, [11th]12th K 1354, *Hathfeld* [1077]14th, 1327, *Hetfelle* 1086, *Hatfeld(e)* 1086–1500 including [1077]14th. Also known as Bishop's H, *Hadfeld Episcopi* 1279, for distinction from other Hatfields. The manor belonged to the bishop of Ely in 1086. Hrt 126.

(2) ~ H&W SO 5959. *Hetfelde* 1086, *Hethfeld(a)* 1123, 1131×48, *Hatfeld* 1291. He 97, Bannister 89.

(3) ~SYorks SE 6509. *Haethfeld* [c.731]8th BHE, *Hæðfeld* [c.890]c.1000 OEBede, *Hedfeld* 1086, 12th, *Heth(e)feld* 12th cent., *Hait(e)- Hayt(e)- Heit- Heytfeld* c.1180–1589, *Hatfeld* 1336–1582. The *Hait-* spellings are due to substitution of ON **heithr** for **hǣth**. YW i.7, L 238, 242, SSNY 138.

(4) ~ AERODROME Herts TL 2009. P.n. HATFIELD TL 2309 + ModE **aerodrome**.

(5) ~ BROAD OAK Essex TL 5516. *Hatfeld Brodehoke* 1121×36, *Hatfeld Regis atte Brodeok* the 'king's H at the broad oak' 1303, OE **brād**, definite form **brāda**, + **āc**. Earlier simply *Hadfelda, Hatfelde* 1086, *Hadfeld -feud* 1155–1241. Ess 39.

(6) ~ CHASE SYorks SE 7100. *The king's chace of Hatfeld* 1464, *Hatfeild -field chace* 1590, 17th cent. P.n. HATFIELD SE 6509 + OFr **chace** 'a chase, a tract of ground for breeding and hunting wild animals'. YW i.8.

(7) ~ HEATH Essex TL 5315. *Hathefeld Heth* 1442. P.n. Hatfield as in HATFIELD BROAD OAK TL 5516 + ME **heth**. Ess 41.

(8) ~ HOUSE Herts TL 2308. P.n. HATFIELD TL 2309 + ModE **house**.

(9) ~ MOOR SYorks SE 7006. P.n. HATFIELD SE 6509 + ModE **moor**.

(10) ~ PEVEREL Essex TL 7911. 'H held by the Peverel family'. *Hadfeld Peurell* 1166. P.n. Hatfield + family n. of Ralph Peverel, the 1086 tenant. Hatfield is *Hæþfelda* 1000×2 S 1486†, *Hafelda, Hadfeldam* 1086, *Hedfeld* 1155, *Hethfeld* 1263. Ess 287.

(11) Great ~ Humbs TA 1842. *Magna Hai- Haytfeld* c.1205, *Est Hattfeld* 1226, *Gt. Hattfeild* 1650. ModE **great**, Latin **magna**, + p.n. *Haifeld, Hai(e)felt* 1086, *Hetfeldia* [12th]c.1536, *Haitefeld(e) -ay-* [1145×66]1293–1342, *Haitfeld -ay-* 1199×1215–1402. The forms have been influenced by ON *heithr* 'uncultivated land'. 'Great' and *Est* for distinction from Little HATFIELD TA 1743. YE 67.

(12) Little ~ Humbs TA 1743. *Parva Ha(i)tfeld* c.1265, 1285, *Little Hatfeild* 1650. ModE **little**, Latin **parva**, + p.n. *Hei(e)feld* 1086. 'Little' for distinction from Great HATFIELD TA 1842. Also known as *Westhaitfeld* [13th]15th. YE 67.

HATFORD Oxon SU 3394. 'The ford by the projecting piece of land'. *Hevaford* 1086, *Havetford* 1176, *Havedford* 1220, *Hautford(e)* 1294–1428, *Hatford* 1573. OE **hēafod** + **ford**. The reference is to a slight elevation which raises the village above the marshy course of Frogmore Brook. Brk 390,L 67, 70, 159.

HATHERDEN Hants SU 3450. Possibly 'hawthorn-tree valley'. *Heðerden* 1193, *Hether- Haytherden(e)* 1256–80. This cannot be ME *hethere* 'heather' on the chalk soil here. Possibly, therefore, OE **hæġthorn** + **denu**. Ha 88, Gover 1958.164.

HATHERLEIGH Devon SS 5404. Partly uncertain. *Adrelie* 1086, *Hadreleia* 1086 Exon, *Hatherlegam* 1193 etc. with variants *-legh(e), Hatherley al. Hetherley* 1675. This is probably 'the clearing where heather grows', OE ***hǣddre** + **lēah**. It is possible, however, that the specific is a reduced form of OE *haguthorn* 'hawthorn' or a derivative of *hǣth* 'heath, heather' such as **hǣthra* or **hǣthor*. D 142, lix, PNE i.214.

Down HATHERLEY Glos SO 8622. *D(o)un- Down(e)hatherle(a) -leg' -lay(e)* 1263–1666. ME **doun** 'down' + p.n. *Athelai* 1086, *Hai- Heiderleia -lega* 1150, c.1200, *Heiþer- Hay- Heytherlei(a) -leg' -ley(a)* c.1210–1287, *Haþ- Hað- Hatherle(a) -leg' -lay(e)* 1200–1598, 'the hawthorn wood or clearing', OE **hagu- hæġthorn** + **lēah**. 'Down' for contrast with Up HATHERLEY SO 9120 further up Hatherley Brook. Gl ii.145.

Up HATHERLEY Glos SO 9120. *Uphatherley(e)* from 1287. ME **up** + p.n. *Hegberleo* (sic) [c.1022]15th S 1424, *Hetherlegh -leye* 1221 as in Down HATHERLEY SO 8622. Gl ii.146.

HATHERN Leic SK 5022. 'The hawthorn-tree'. *Avederne* 1086, *Hau(e)- Haw(e)thirn(e) -thorne -thurn(e)* before 1150–1340, *Hathern(e)* from 1277, *Hag(h)ethirne, Hacthurne* early 13th. OE ***hagu-thyrne**. Lei 379.

HATHEROP Glos SP 1505. Probably 'the lofty dependent farmstead'. *Etherop(e)* 1086, 1284, *Hadrop* 1086, *Het(h)rop* 1154×89, 1221–91, *Heythrop* 1164×79, 1211×13, *Hatherop(e)* 1221–1783. OE **hēah**, definite form **hēa**, + **throp**. The village stands on the top of a hill overlooking the r. Coln. Gl i.36 gives pr ['heiðrap].

HATHERSAGE Derby SK 2381. Either 'he-goat ridge' or 'Hæfer's ridge'. *Hereseige* (sic) 1086, *Hauerseg(g)(e) -v-* c.1220–1398, *-heg(g)(e)* 1230–1327, *Hatherseg(g)(e) -ir- -yr-* 1264–17th, *Hathersage* from 1512. Either OE **hæfer**, or OE pers.n. **Hæfer*, genitive sing. **hæferes**, *Hæferes*, + **ecg**. The reference is to Millstone Edge SK 2480. Db 111 gives pr [(h)aðəsidʒ], L 158.

HATHERTON Ches SJ 6847. 'Heath settlement'. *Haretone* 1086, *Hatherton* from 1262. OE ***hǣthor** + **tūn**. Ch iii.62, v.1.ii.208.

HATHERTON Staffs SJ 9510. 'Hawthorn-tree hill'. *Hagenþorndun* *[996 for 994]19th S 1380, *Hargedone* 1086, *Hatherdon* 1203–1565, *-ton* from 1227. OE **haguthorn** + **dūn**. St i.123.

HATLEY ST GEORGE Cambs TL 2851. 'St George's H'. *Hattele de S'c'o Georgio* 1279. P.n. Hatley as in East HATLEY TL 2850 + saint's name *George* either from the dedication of the church or from the family name of William de Sancto Georgio 1236,

†Probable identification.

and also for distinction from East HATLEY TL 2850. Also known as *Hung' hatele* 1218, *Hungryattele* 1316, 'hungry Hatley', alluding to poor soil. Ca 54.

Cockayne HATLEY Beds TL 2549. 'H held by the Cockayne family'. *Cocking Hatley*. Earlier simply *Hettenleia* [10th]14th S 1807, *Hættanlea* 975×1016 S 1487, *Hætlea* [c.1053]13th S 1517, *Hatelai* 1086, *Hetten- Hatten- Hateleia* [13th]c.1350, 'Hætta's wood or clearing' or 'wood or clearing of *Hætte*, the elevated ground', OE pers.n. **Hætta* or p.n. **Hætte* from OE *hætt* 'a hat, a piece of elevated ground', genitive sing. **Hættan*, + **lēa**, dative sing. of **lēah**. The manor passed to the Cockayne family in 1417. The addition is for distinction from HATLEY ST GEORGE and East HATLEY Cambs TL 2851, 2850. Bd 105, Ca 55, Wa xlii.

East HATLEY Cambs TL 2850. *Esthatteleia* 1199. OE adj. **ēast** + p.n. *(H)atelai, Eateleia* 1086, *Attelee -leye* 1198, 1250, *Hatteleia -le* 1164×96–1405, 'Hætte wood or clearing', OE hill-name **Hætte* 'the hat-like hill' < OE *hætt* as in the lost p.n. *Hattelawa* 'Hætte hill' in neighbouring Gamlingay TL 2452, + **lēah**. The reference is to the isolated 250ft. oval hill due W of Cockayne HATLEY TL 2549. Ca 54.

HATT Corn SX 3961. 'The hat'. *La Hatte* 1305. Fr definite article **la** + ME **hatte** (OE *hætt*). A fanciful name for a village standing on an isolated hill. Cf. HETT Durham NZ 2836. PNCo 94, Gover n.d. 216.

HATT HILL Hants SU 3126 → HATTINGLEY SU 6437.

HATTINGLEY Hants SU 6437. 'Wood or clearing at or called *Hætting*, the hat-like hill'. *Hattingele* 1204–1327 with variants *-ynge-* and *-lege -li(g)e -legh*, *Hattonly* 1817 OS. OE p.n. **Hætting* < **hætt** + **ing**[2], locative–dative sing. **Hættinge*, + **lēah**. Hattingley is situated on a hill; for *hætt* as a hill-n., cf. HETT Durham NZ 2836 and Hatt Hill Hants SU 3126, *Hatte* 1206, 1379. An alternative view would be to take *Hattinge-* as representing OE genitive pl. *Hættinga* of a folk-n. **Hættingas*, 'the people living at *Hætt*, the hat-like hill'. Gover 1958.79, 190, DEPN.

HATTON 'Heath settlement'. OE **hǣth-tūn**.

(1) ~ Ches SJ 5982. *Hattun* 13th., *Hatton* from c.1230. Described in the earliest reference as a *grangia* 'an outlying farm'. Che ii.149.

(2) ~ Derby SK 2130. *Hatun(e)* 1086, *Hatton(e)* c.1200–1281 etc. Db 563.

(3) ~ GLond TQ 1075. *Ha(i)tone* 1086, *Hatton* from 1211, *Haddo(u)n juxta Huneslowe* 11373. It lay on the western edge of Hounslow Heath. Cf. HEATHROW TQ 0875. Mx 14.

(4) ~ Lincs TF 1776. *Hatune* 1086, *Hattuna* c.1115, *Hatton'* 1212. DEPN, Cameron 1998.

(5) ~ Shrops SO 4690. *Hetton'* 1167, *Hatton'* 1227. Sa i.145.

(6) ~ Warw SP 2467. *Altone* 1086, 1187, *Hectona* (for *Hett-*) 1150, *Haiton* c.1154, *Hatton* from 1163. Wa 212.

(7) ~ HEATH Ches SJ 4561. *Hatton Heath* 1690. P.n. HATTON SJ 5982 + ModE **heath**. Referred to in 1281 as *bruera iacens inter Halton et Salhton* 'a heath lying between Halton and Saighton' (SJ 4462). Cf. BRUERA SJ 4360. Che iv.122.

(8) High ~ Shrops SJ 6124. *Heghe Hatton'* 1291×2, *~ sur Hyneheth* 1333, *Hatton Hine Heath alias High Hatton* 1692. ME adj. **hegh** (OE *hēah*) + p.n. *Hetune* 1086, *Hatton(')* 1242, 1255. For Hine Heath see STANTON UPON HINE HEATH SJ 5724. Sa i.146.

HATTONS LODGE Wilts SU 0688. Formerly either *Red Lodge* or *White Lodge* 1773. Wlt 41.

HAUGH HEAD Northum NU 0026. 'The head of the haugh, the nook of low lying land by the river'. *Haugh Head* 1866 OS. N dial. **haugh** (OE *halh*) + ModE **head**. The place lies at the S extreme of a triangle of land beside Harthope Burn.

Nether HAUGH SYorks SK 4196. 'Lower Haugh, the nook of land in the hill-side'. *Netherhalgh* 1396, 1408, *Ne(y)therhaugh(e)* 1540–1822. ME **nether** + OE **halh** used as a p.n. 'Nether' for distinction from Upper Haugh SK 4495, *Over Halk* 1406–9, *-halgh(e)* 1408–1595, *(Over) Haugh* 1534–86, *Haugh Upper* 1822. Both are *Haleges* 1206. YW i.182, 175.

HAUGHAM Lincs TF 3381. 'The high village or homestead'. *Hecham* 1086, 1188, *Hac- Hecham* 12th, *Hagham* 1191, 1212. OE **hēah** + **hām**. The development must have been lOE *Hǣhham* > *Hăhham* > *Hagham* > *Haugham* [hoːəm] beside *Hagham* > *Haffam* [hæfəm], the present pr. The road between Burwell and Tathwell rises to 336ft. near this hamlet. DEPN, Cameron 1998.

HAUGHLEY GREEN Suff TM 0364. *Haughley Green* 1837 OS. P.n. HAUGHLEY TM 0262 + ModE **green**.

HAUGHLEY Suff TM 0262. 'The hawthorn-tree or the enclosure wood or clearing'. *Hagele* [1035×44]13th S 1521, *Hagala* 1086, *Hagh- Haggele* 1251–1414, *Haghle* 1327, 1401×2, *Haule(gh) -leg -ley* 1254–1400, *Hawley* 1388 etc. OE **haga** 'a hawthorn-tree, a hawthorn fence, an enclosure' + **lēah**. Two unusual sps, *Haga- Heg(h)enet* 1165–1470, seem to show the Fr diminutive suffix-**ette** added to *hagan*. Haughley Castle was the centre of an important lordship with French influence. DEPN, Baron.

HAUGHTON 'Nook settlement'. OE **halh** in various topographical senses or in the administrative sense 'projecting part of an estate' + **tūn**. Sa i.146.

(1) ~ Shrops SJ 5516. *Haustone* 1086, *Halghton(')* c.1220–1346 with variants *Hal(c)h- Hali(h)- Halech-* and *Halugh-*, *Halton* 1271×2, *Haughton* from 1535. The reference is to the situation of the farm in a knob projecting from the N boundary of Upton Magna. Sa i.147.

(2) ~ Shrops SJ 3727. No early forms. The sense of *halh* is probably 'firm ground in marsh'. Sa i.147.

(3) ~Shrops SO 6896. *Halchtona* [c.1090]15th. The sense of *halh* is probably administrative since the parish boundary makes a detour to the N at this point. Sa i.147, Gelling.

(4) ~ Staffs SJ 8620. *Hal(s)tone* 1086, *Halec(h)tun -tone* 12th–14th, *Hal(u)ghton* 12th–1435, *Haughton* from 1559. In this name *halh* probably means 'patch of dry ground in marsh'. St i.164, L 105.

(5) ~ COMMON Northum NY 8072. *Haughton Common* 1866 OS. P.n. Haughton NY 9172, *Haluton* 1177, 1279, *Haluch- Halghton* 13th cent., *Haughton* 1611, + ModE **common**. Haughton lies in a nook or bend of the North Tyne river. NbDu 105.

(6) ~ GREEN GMan SJ 9393. *Haughton Green* 1843 OS. P.n. Haughton, *Halghton* 1307–22, *Halcton* 1322, + ModE **green**. The village stands in a bend of the river Tame. La 30.

(7) ~ MOSS Ches SJ 5756. *Haughton Mosse* 1574. P.n. Haughton SJ 5856, *Halec(h)ton* c.1180×1220, 1216×72, *Halc(h)ton(ia)* c.1240, *Hal(g)hton* 1282–1578, *Hauton'* 1259, 1394, *Haughton* from 1508, + dial. **moss** (OE *mos*) 'a bog, a swamp'. Haughton is situated in a small valley. Che iii.309.

HAUGHTON Notts SK 6772. 'The settlement by a hill-spur'. *Hoctun* 1086, *-ton'* 1192, 1201, *Hoghton* 1258–1387, *Houghton* 1327–1418, *Haughton* 1316, *Hawghton* 1536. OE **hōh** + **tūn**. Nt 81, L 167.

HAUGHURST Hants SU 5762 → BAUGHURST SU 5861.

HAUNTON Staffs SK 2310. 'Hagona's estate'. *Hagnatun* [941]14th S 479, *Hagheneton* 1249, *Hauneton* 1271. OE pers.n. *Hagona* + **tūn**. DEPN.

Little HAUTBOIS Norf TG 2521. *Hauboys Minor* 1254. ModE adj. **little**, Latin **minor** 'lesser', + p.n. *Hobbesse* *[1044×7]13th S 1055, [c.1140]14th, *Hobuisse, Ohbouuessa* 1086, *Hobbossa* 1183, *Hobissa* 1191, *-wiss -wise* 1200, *Hau(t)boys* 1242, *(de) Haltobosco* 1200, 'hummock meadow', OE ***hobb(e)** + ***wisse**. 'Little' for distinction from from Great Hautbois TG 2620, *Hauboys Major* 1254, which occupies a promotory beside the Bure which must be the *hobb(e)* in question. DEPN gives pr [hɔbis], PNE i.255, ii.270.

HAUXLEY Northum NU 2703. 'Hafoc's or the hawk's hill or tumulus'. *Hauekeslaw* 1204, *Hauekislawe* 1242 BF, *Hawkeslawe* 1428, *Hauxley* 1638, *Haxlee* 1697. OE **hafoc** or pers.n. *Hafoc*, genitive sing. **hafoces**, *Hafoces*, + **hlāw**. NbDu gives pr [ha:ksli], L 162.

HAUXLEY HAVEN Northum NU 2802. *Hauxley Haven* 1866 OS. P.n. HAUXLEY NU 2703 + ModE **haven**.

HAUXTON Cambs TL 4351. 'Heafoc's estate'. *(æt) Hafucestune* c.975, *Hauochestune* [1042×66]12th S 1051, *-tun -tone* 1086, *Hauekestune -ton(a)* 1086–1285 including [1042×66]13th S 1051, *Haukeston(e) -a* c.1060–1445, *Hau(e)ston(e), Haw(e)ston* 1272–1553, *Hawson* 1589. OE pers.n. **Heafoc*, genitive sing. **Heafoces*, + **tūn**. Ca 85 gives former pr [hɔ:sən].

HAUXWELL NYorks SE 1693. 'Hawk's spring'. *Hauoc(h)(e)swelle* 1086, *Hou- Haukeswella -e* 1177–1362. It is impossible to say whether the specific is a pers.n., OE *Heafoc*, ON *Haukr*, or appellative **heafoc** 'a hawk', genitive sing. *Heafoces, Hauks* or **heafoces**, + OE **wella**. YN 269.

HAVANT Hants SU 7106. 'Hama's fountain'. *(æt) hamanfuntan* [935]12th S 430, [980]12th S 837, *Havehunte* 1086–1291, *Hauunta* c.1170, *Havonte* 1315–41, *Havant* 1579. OE pers.n. *Hāma*, genitive sing. *Hāman*, + **funta**. OE *funta* is possibly a direct borrowing of Latin *fontana* and the reference here may be to a Roman spring on the ridge of Portsdown where the chalk is overlain by Eocene Beds and artesian fountains may occur. Gover 1958.14 gives pr [havənt], Ha 88, DEPN, Signposts 84, 86, Nomina 9.6.

HAVEN H&W SO 4054. *Lower Haven, Upper Haven* 1832 OS. Possibly the family name *Hevyn* recorded as holding land in neighbouring Dilwyn 14th–17th cents. Bannister 90.

The HAVEN Lincs TF 3541. 'The harbour'. *The Haven* from 1495. ON **hafn**, lOE **hæfen**. Earlier referred to as the *Port of Boston* 1307–1415. Formerly an arm of the sea reaching inland as far as the town of Boston. Payling 1940.117, Wheeler 446.

HAVENGORE ISLAND Essex TQ 9788. P.n. *Havengor(e), Havingor* 1493, 1578, 'the haven mud or triangle', OE **hæfen** + **gor** or **gāra**, + ModE **island**. The reference is to Wakering Haven, *Wakryng Haven* 1537, earlier simply *(atte) Havene* 1274, *the Havene* 1419×21. MM 16.82 notes that Havengore Island has a triangular shaped point of sand projecting into the haven from which it probably takes its name. The earliest reference is to the *Havenemersche* 1230. Ess 185.

HAVENSTREET IoW SZ 5690. 'The street of houses or hamlet belonging to the Hethen family'. *Thenestrete* 1248, *(la) Hethenestrete* 1255–1419, *Haven Street* 1769. Surname Hethen as in Richard le *Hethene* c.1240 + ME **strete**. Wt 32 gives prs ['hebm- 'hevnstri:t], 281, Mills 1996.58.

HAVERAH PARK NYorks SE 2353. *parco de Hai- Hey- Haywra* 1227–1518, *Averaie Park* 1563, *Havera Parke* 1632. P.n. *Haiwra* 'hunting enclosure nook', OE **(ġe)hæġ** + ON **vrá**, + ME **park**. One of the parks of the Forest of Knaresborough. The modern form is due to confusion with Mod dial. **haver** (ON *hafri*) 'oats'. YW v.119.

HAVERHILL Suff TL 6745. 'Oats hill'. *Hauerhella -hol* 1086, *-hell* 1190, *Haverhell* 1158, *Hauerill* 1610. OE ***hæfera** + **hyll**. An alternative possibility is OE **hæfer** 'a he-goat'. DEPN.

HAVERIGG Cumbr SD 1578. 'Oat' or 'goat ridge'. *Haverich* c.1170, *Haverig(g)* from c.1180. ON **hafri** 'oats' or **hafr** 'he-goat' + **hryggr**. Cu 415, SSNNW 131.

HAVERING-ATTE-BOWER GLond TQ 5193. P.n. Havering + ME **atte bower** 1272 referring to the nearby royal residence called *the Bowre* 1418 'the chamber', ME **bour** (OE *būr*). Havering, *Hauer- Hauelingas* 1086, *Haveringes* 1166 *-as* 1179×89, *Hauering(e)* from 1140, is 'the Hæferingas, the people called after Hæfer', OE folk-n. **Hæferingas* < pers.n. **Hæfer* + **ingas**. The same folk-n. occurs again in Haveringland Norf TG 1521, *Heueringalanda* 1086, 'open land of the Hæferingas', genitive pl. **Hæferinga* + **land**. Ess 111, ING 20, LPN 106.

HAVERINGLAND Norf TG 1521 → HAVERING-ATTE-BOWER GLond TQ 5193.

HAVERSHAM Bucks SP 8242. 'Hæfer's homestead or enclosure'. *(æt) Hæfæresham* [966×75]12th S 1484, *Havresham* 1086, *Haver(e)sham* from 1450, *Harsham* 1542, *Haversham vulgo Hasome* 1675. OE pers.n. **Hæfer*, genitive sing. **Hæferes*, + **hām** or **hamm**. Bk 8 gives pr [ha:ʃəm], L 47.

HAVERTHWAITE Cumbr SD 3483. 'Clearing where oats are grown'. *Haverthwayt* 1336. ON **hafri** + **thveit**. La 217 gives pr [havəθət], SSNNW 131, L 211.

Great HAW NYorks SE 0779. *Great & Little Haugh* 1817. ModE **great** + p.n. Haugh 'the enclosure', ultimately from OE **haga**. YW v.219.

HAWES NYorks SD 8789. 'The neck, the pass between mountains'. *Hawes* 1614, *the Hawes* 1666. N dial. **hause** (OE *hals*). YN 266 gives pr [tɔ:z] 'the hause'.

HAWESWATER RESERVOIR Cumbr NY 4814. P.n. Haweswater + ModE **reservoir**. A modern reservoir created by flooding Mardale, *Merdale* 1278, *Mardale* c.1540, 'lake valley' referring to the original Hawes Water, *Havereswater* 1199, *Hawse-water* 1671, *Hall's Water* 1726, *Hawes Water* 1770, 'Hæfer's lake', OE pers.n. **Hæfer*, genitive sing. **Hæf(e)res*, or ON pers.n. *Hafr*, secondary genitive sing. *Hafres*, + **wæter**. We i.16, ii.169.

HAWFORD H&W SO 8460. 'Ford by the enclosure'. *Hageford* [1182]18th, *Haweford* 1262, *Hawford* [1352]18th. OE **haga** + **ford**. A ford across the Salwarpe river. Wo 270.

HAWKCHURCH Devon ST 3400. 'Heafoc's church'. *Hauekescherich* 1201, *-cherch(e)* 1214, 1218, *Hauekecherche* 1292, *-churche* 1311. OE pers.n. *Heafoc*, genitive sing. *Heafoces*, + **ċiriċe**. D 655.

HAWKEDON Suff TL 7952. 'Hafoca's hill'. *Hauochen- (h)auokeduna* 1086, *Hafkindun* 1195, *Hauekedon* 1242, *Haukden* 1610. OE pers.n. **Hafoca*, genitive sing **Hafocan*, + **dūn**. The forms with medial *-n-* point to the pers.n. rather than *hafoca*, genitive pl. of *hafoc* 'a hawk'. DEPN.

HAWKERIDGE Wilts ST 8653. 'Hawk ridge'. *Hauekerigge* 1249, *Hauecrugge* 1279. OE **hafoc**, genitive pl. **hafoca**, + **hrycg**. Wlt 149.

HAWKERLAND Devon SY 0588. 'Hawker's land'. *Haueker(e)sland* 1227, 1316, *Hauekerland* 1238. OE **heafocere** or better the ME surname *Hawker* derived from it, genitive sing. **heafoceres**, *Hawkeres*, + **land**. D 587.

HAWKESBURY Avon ST 7687. 'Hafoc or hawk's fortified place' *Havochesberie* 1086, *Hauekesbir(i) -byr -bur(y)* 1167–1327, *Haukes-* 1248–1498, *Hawkesbury(e)* from 1534. OE pers.n. *Hafoc* or OE **hafoc**, genitive sing. *Hafoces*, **hafoces**, + **byriġ**, dative sing. of **burh**. Gl iii.29.

HAWKESBURY UPTON Avon ST 7887 → Hawkesbury UPTON.

HAWKES END WMids SP 2982. 'The Hawksty end' sc. of Allesley. *Hawxstye end* 1533, *Hawkestend* 1724, *Hawkes End* 1835 OS. P.n. Hawksty as in *Hauokestiestrete* 'Hawksty street' 1389, *Hawkestye* 1433, '*Hafoc* or the hawk's path', OE pers.n. *Hafoc* or **hafoc**, genitive sing. *Hafoces*, **hafoces**, + **stīġ**, + ModE **end**. Wa 153.

HAWKHILL Northum NU 2212. 'Hawk hill'. *Hauechil* 1177, *Hauekehall', Haukehill'* 1242 BF, *Hauckill* 1288, *Hawkill* 1346. OE **hafoc** + **hyll**. NbDu 106 gives pr [hɔ:kəl], L 171.

HAWKHOPE Northum NY 7189. 'Hawk side-valley'. *Haucop* 1325, *Hauckup* 1603, *Hawcup* n.d. Map. OE **hafoc** + **hop**. NbDu gives pr [hɔ:kəp].

HAWKHURST Kent TQ 7630. 'Hawk wooded hill'. *Hauekehurst* 1254, *Hauckherste* 13th cent. OE **hafoc**, genitive pl. **hafoca**, + **hyrst**, Kentish **herst**. PNK 335, L 197.

HAWKINGE Kent TR 2140. 'Hawk place' or 'place called after Hafoc or Hafeca'. *Hauekinge* 1204, *Hawekingg'* 1218, *Halkenge* 1278. OE **hafoc** or pers.n. **Hafoc* or **Hafeca*, + **ing**[2]. PNK 447, ING 205.

HAWKLEY Hants SU 7429. 'Hawk clearing or pasture'. *Hauecle* 1207, *Hauek(e)le(gh) -lye* 1248–1327, *Haukele* 1325. OE **hafoc** + **lēah**. Ha 88, Gover 1958.92.

HAWKRIDGE RESERVOIR Somer ST 2036. P.n. Hawkridge (no early forms) + ModE **reservoir**.

HAWKRIDGE Somer SS 8630. 'Hawk ridge'. *Hauekerega* 1194, *-regg* 1195, *Haweckrig* 1225, *Haukridge* 1610. OE **hafoc**, genitive pl. **hafoeca**, + **hrycg**. The village is perched on top of a ridge of high ground separating the Barle from its tributary stream the Danes Brook. DEPN.

HAWKSHEAD Cumbr SD 3598. 'Hauk's shieling'. *Hovkesete* 1198×1208, *Haukes(s)et, Hoxeta* 13th, *Haukesheved* 1336. ON pers.n. *Haukr*, genitive sing. *Hauks*, + **sǽtr**. The generic, which might alternatively be either OE **set** 'a fold' or ON **setr** 'a settlement', was subsequently substituted by ME **heved** 'head' (<OE *hēafod*), a hypercorrection due to the reduction of *heved* to *hed* in p.ns. like Gateshead. La 218 gives pr [hɔ:ksed, hɔ:ksəd], SSNNW 64.

HAWKSWICK NYorks SD 9570. 'Hauk's dairy-farm'. *Hochesuuic* 1086, *Ho(u)keswic -wik -wyk* 12th, 13th, *Haukeswic -wik(e) -wyk(e)* 1176–1398, *Hackyswyke* 16th cent. ON pers.n. *Haukr*, genitive sing. *Hauks*, possibly replacing OE **hafoc** 'a hawk', + **wīċ**. YW vi.124.

HAWKSWORTH Notts SK 7543. 'Hoc's enclosure'. *Hochesuorde* 1086, *Hokiswurth -wurd -word -worth -es-* 1203–1346, *Houkes-* 1179–1300, *Howkesworth* 1240, 1428, *Haukesworth* 1390, *Hauxworth* 1600, 1663. OE pers.n. *Hōc*, genitive sing. *Hōces*, + **worth**. The modern form is influenced by the word *hawk*, possibly consequent upon the ME development *ou* > *au*, Jordan §105 Anm. Nt 226.

HAWKSWORTH WYorks SE 1641. 'Heafoc's enclosure'. *on Hafeces-weorðe* c.1030 YCh 7, *Hauoc(h)esord(e), Heunochesuurde* 1086, *Heuekeswurth* 1257, *Heukesw(o)rd(e) -wrth, Haukisuuorth, Haukeswurth(e) -worth(e)* 13th–1428. OE pers.n. *Heafoc* from *heafoc* 'a hawk', genitive sing. *Heafoces*, + **worth**. YW iv.199.

HAWKWELL Essex TQ 8691. 'Hook spring'. *Hac(he)uuella, Hechuellā* 1086, *Hakewell(e)* 1202–1594, *Hakwell* 1444, 1552, *Hau- Hawk(e)well(e)* from 1389. OE **haca** perhaps referring to a tongue or hook of land between streams, later confused with ME *hawke*, + **wella**. Ess 186.

HAWLEY Hants SU 8558. Perhaps 'nook wood or clearing'. *Halely, Hallee* 1248, *Hallegh -leye* 1250–1454, *Hawelee* 1380, *Hawley* 1590. OE **h(e)alh**, genitive pl. **h(e)ala**, + **lēah**. Hawley lies on the county boundary beside the Blackwater River marshes; the sense of *healh* might be 'dry ground in marsh' or 'projecting corner of an administrative area'. Ha 88, Gover 1958.111, L 101–2.

HAWLEY Kent TQ 5571. 'The holy wood or clearing'. *Hagelei* 1086, *Hageli -leg* c.1100, *Halgelei -le(g)* c.1100–1203, *Hal(e)ghele* 1253×4–1303, *Haluelegh* 1241, *Halwele* 1303. OE **hāliġ**, definite form **hālga**, + **lēah**. PNK 50.

HAWLING Glos SP 0623. Probably 'the folk of or dependent on HALLOW H&W SO 8258'. *Hallinga -ing(e) -yng(e)* c.1050, 1086, 1202–1633, *-inges -ynges* c.1162–1424, *Hawling(e) -yng(e)* 1535–1688. The earliest occurrence is in an 11th cent. recension of a purported charter of 816 (S 179) which really belongs to Hallow in H&W. In the bounds occur *into Hallinga homme* 'into the enclosed pasture of the Hallingas' and *to Healling weallan* 'to the boundary walls of the Hallingas'. OE folk-n. *Hallingas*, either 'the folk dependent on Hallow' or 'the dwellers in the *halh*', the nook of land formed by the valley stretching E at the head of which Hawling stands. Gl ii.17, ECWM 184–96, ING 48.

HAWNBY NYorks SE 5489. Either 'Halmi's farm or village' or 'straw-thatch farm'. *Halm(e)bi -by* 1086–1399, *Haunneby* 1285, *Hawnbye, Haunby* 1538. ON pers.n. *Halmi*, genitive sing. *Halma*, or **halmr** + **bȳ**. The persistence of *-e-* spellings favours the pers.n. explanation. YN 203, SSNY 29.

HAWORTH WYorks SE 0337. 'Enclosure made with a hedge'. *Hauewrth'* 1209, *Haw(u)rth(e) -worth* 13th–1641. OE **haga** + **worth**. YW iii.261 gives prs ['auwəθ] and ['ɔuwəθ].

HAWORTH MOOR WYorks SE 0035. *Haworth Moor* 1858. P.n. HAWORTH SE 0337 + ModE **moor** (OE *mōr*).

HAWSKER NYorks NZ 9207. 'Hauk's enclosure'. *Houkesgart(h) -gard* c.1100–1227, *Haukesgard -garth -garth(e)* c.1115–1351, *Hakisgarth* 1330, *Haskerth, Horskarse, Harrsker* 17th cent. ON pers.n. *Haukr*, genitive sing. *Hauks*, + **garth**. YN 121.

HAWSTEAD Suff TL 8659. 'The shelter place'. *hersteda* 1086, *Haldsteda* 1086–1193, *-sted(e)* 1087×98, 1121×48, 1209, *Halsted(e)* 1087×98–1578, *Hausted(e)* 1148×53–1610. OE ***hald-stede**. The sense is perhaps 'dairy- or cattle-farm with a shelter'. Studies 1936.53, Stead 189.

HAWTHORN Durham NZ 4145. 'The hawthorn-tree'. *Hagathorn(e)* 1114×6, c.1116, 1214, *Hagethorn* 1180, *hauthorn(e)* c.1195×1200–1580, *Hawthorn* from 1315. OE **haga-thorn**. NbDu 106.

HAWTHORN HILL Berks SU 8774. *Hawthorn Hill* 1800. P.n. Hawthorn + ModE **hill**. Hawthorn, *Horethorn(e)* 1327–1495, *Hothorne* 1573, *Hoe-thorne* 1607, *Hawthorne* 1790, is 'the grey thorn-tree', OE **hār**, definite form **hāra**, + **thorn**. Brk 45, 877.

HAWTHORNTHWAITE FELL Lancs SD 5751. P.n. Hawthornthwaite + ModE **fell** (ON *fjall*). Hawthornthwaite, *Haghthornthayt* 1323, *Haghethornthwait* 1324, is the 'hawthorn-tree clearing', OE **haguthorn** or ON **hagthorn** + **thveit**. La 172.

HAWTON Notts SK 7851. 'The settlement in the hollow'. *Holtone* 1086, *Houtune* 1086, *-tone* 1154×89–1351, *Hautone* 1270, *Hawton* 1291, *How(e)ton* 1346, 1484. OE **hol** + **tūn**. Hawton lies on low ground by the river Devon. The name shows the ME dial. development *ou* > *au*, Jordan §105 Anm. Nt 215.

HAXBY NYorks SE 6057. 'Hakr's farm or village'. *Haxebi -by* 1086–1319, *Haxby* from 1317. ON pers.n. *Hákr*, genitive sing. *Háks*, + **bȳ**. YN 14, SSNY 29.

HAXEY Humbs SK 7699. 'Hakr's island'. *Acheseia* 1086, *Haxei(a)* 12th, 1212, *Haxay* c.1220, *Hakesay* 1372. ON pers.n. *Hákr*, genitive sing. *Háks*, + **ey**. Alternatively the specific might be ON *hak* 'a little hook, a barb', referring to the shape of the promontory on which Haxey stands. DEPN, SSNEM 154.

HAXTON DOWN Wilts SU 2050. P.n. Haxton SU 1449, *Hakenestan* 1172, *-ston(e)* 1212–1382, *Hakeleston* 1268–1403, *Hackeston* 1576, 'Hacun's stone', lOE pers.n. *Hacun*, genitive sing. *Hacunes*, + **stān**, + ModE **down**. Wlt 330.

Little HAY Staffs SK 1202. 'The small enclosure'. *Luttelhay* 13th, *Littlehay* 1327. ME **lutel** (OE *lytel*) + **hey** (OE *(ġe)hæġ*). Horovitz.

HAY-A-PARK NYorks SE 3758. *parca del, de, de la Hay(e), Hey* 1291–1597, *La Haye park* 1355, *Hay Park* 1817. OE **(ġe)hæġ** 'a hunting enclosure' + ME **park**. One of the parks of the Forest of Knaresborough. The medial *-a-* in the modern form preserves the ending of the Latin form of the name, *Haia* 1518. YW v.113.

HAYCOCK Cumbr NY 1410. *Hay Cocks* 1783. A descriptive name from the shape of the hill. Cu 442.

HAYDOCK Mers SJ 5696. Uncertain. *Hedoc* 1169, *Heddoch* 1170–1, *Heidoc* 12th, *Haidoc* [1190×99]1268, 1212, *Haydok* 1286, 1332, *Hadock* 1292, *Haydock* 1322, *Haddock* 1513–16. It has been suggested that this might be PrW **Heᵢð'ịog* 'place growing with barley', identical with the second part of the W name for Kentchurch, Llanheiddog H&W SO 4125, *Lenheydok* 1326. Cf. W *haidd* 'barley' + p.n. forming suffix PrW ***-ǭg**. Jnl 17.57 gives local pr [hædək]. La 99, LHEB 612, GPC 1814.

HAYDON BRIDGE Northum NY 8464. *Haydon Bridge* 1866 OS. P.n. Haydon NY 8465, *Hayden* 1236 BF–1479, *Hei- Heyden* 13th,

Haydon 1346, 'the hay valley', i.e. the valley where hay is made, OE **hēġ** + **denu**, + ModE **bridge**. NbDu 106.

HAYDON Dorset ST 6715. Probably 'hay hill, hill where hay is made'. *Heydone* c.1163†, *-don* 1253–88, *Haydon(')* from 1258. OE **hēġ** + **dūn**. Alternatively the specific could be either OE **heġe** 'a hedge' or **(ġe)hæġ** 'an enclosure'. Do iii.334.

HAYDON DEAN Northum NT 9844. P.n. Haydon + ModE **dean** (OE *denu*). Haydon, no early forms, is probably identical with Haydon in HAYDON BRIDGE NY 8464.

HAYDON HILL Wilts SU 3156. *Haydon Hill Castle* 1773. P.n. Haydon + ModE **hill**. There are no early forms for Haydon, a hill of 860ft, but the name is identical with Haydon in Haydon WICK SU 1387, the element *(ġe)hæġ* here referring to Fosbury Camp SU 3256, a bivallate Iron Age hill-fort. Wlt 357, Pevsner 1975.250.

HAYDON WICK Wilts SU 1387 → Haydon WICK SU 1387.

HAYE Corn SX 3469. 'Enclosure'. *(la) Haye* 1269 Gover n.d., 1610, *Heye* 1327, 1385. ME **hay** (OE *(ġe)hæġ*). PNCo 94, Gover n.d. 177.

HAYES 'Brushwood'. OE ***hǣs**, ***hēs**. PNE i.218.

(1) ~ GLond TQ 0980. *linga hæse* [790 for 795]10th S 132, *Hæse* 831 S 188, *Hesa* 1086, *Hese* 1232–1516, *Heys(e)* 1229–1682, *Hayes* 1524. OE **hese**. *linga* in the earliest citation is unexplained. At the time of DB the manor contained woodland for 400 swine. Mx 39.

(2) ~ GLond TQ 4165. *Hesa* 1177, *Hese* 13th. DEPN.

HAYFIELD Derby SK 0387. 'Open land where hay is obtained'. *Heyfeld* 1285–1584, *Hayfeld* 1307–43 etc. OE **hēġ** + **feld**. Db 114.

HAYLE Corn SW 5637. *Heyl* 1205, *Heyll* 1318, *Hayle* 1816. Primarily a 19th cent. industrial development named from the river HAYLE SW 5632 on which it stands. PNCo 95, Gover n.d. 611, RN 192, DEPN.

River HAYLE Corn SW 5632. 'Estuary'. (River called) *Heyl-penword'* 1260, *Heyl* [1205]15th, *Heyle, Haile, Hayle* c.1540–84, *Hale* 1610. Co ***heyl** 'estuary, salt river', probably identical with the Gaulish r.n. *Salia* and Scots Shiel, *Sale* 7th, from IE **sal-* 'stream, flowing water, current', and a whole series of Old European r.ns. in *Sal-* from Hungary to Norway. The first form has affixed a bad spelling of PENWITH, 'Hayle in Penwith'. PNCo 94, Gover n.d. 611, RN 192, ScotPN 189.

North HAYLING Hants SU 7303. *North Hayling* 1810 OS. ModE adj. **north** + p.n. *Hailinges* 1139×42, *Hallinges* 1215, *Haylinges* after 1284, *Heling, Heyling'* 1242, *Hayling* 1316, 'the *Hægelingas*, the people called after Hægel' as in HAYLING ISLAND SU 7201. 'North' for distinction from South HAYLING SZ 7299. Also known in the 13th and 14th cents. as *Northwode*. Ha 89, Gover 1958.15–6, ING 43.

South HAYLING Hants SZ 7299. *South Hayling* 1810 OS. ModE adj. **south** + p.n. Hayling as in North HAYLING SU 7303. Also known in the 13th and 14th cents. as *Southwode*. Ha 89, Gover 1958.16.

HAYLING BAY Hants SZ 7198. P.n. Hayling as in HAYLING ISLAND SU 7201 + ModE **bay**.

HAYLING ISLAND Hants SU 7201. *island of Helinghey* 1261, *Haylingg Island* 1346. P.n. Hayling + ModE **island**. Hayling, *(to) heglingaigae, (to) hæglingaiggæ* [956]12th S 604, *(ad) Heilincigæ* [c.1053]12th S 1476, *Helinghei, Halingei, Helingey* 1086, *Helinghey* 1261, is the 'island of the *Hæġelingas*, the people called after Hægel', OE folk-n. **Hæġelingas* < OE pers.n. **Hæġel* + **ingas**, genitive pl. **Hæġelinga*, + **īeġ**. The obscure form *Helingey* was subsequently replaced by the more transparent *Hayling Island*. Ha 89, Gover 1958.9, ING 43.

HAYLOT FELL Lancs SD 5961. P.n. Haylot + ModE **fell** (ON *fjall*). Haylot, *Hailett* 1584, *Hayloth* 1624 is the 'hay allotment', ModE **hay** (OE *hēġ*) + **lot** (OE *hlot*). La 181.

†The form *(æt) Hægdune* [1046]12th S 1474 may belong here or under Haydon Somer.

HAYNES Beds TL 1042. 'The enclosures'. *Haganes* 1086, *Haw(e)nes, Haunes* 13th–19th cents., *Haynes* c.1560. OE ***hagan**, pl. ***haganas**. OE **hagan* would account for the spelling *Hawnes*, but the present-day form is an example of lME *au* > *ā* perhaps also influence by ME *hain* 'walled enclosure, park' from OE *hæġen*. This place is called *Silver End* in 1834 OS, the original site having been at HAYNES CHURCH END TL 0841. Bd 151 gives prs [heinz] and [hɔːnz].

HAYNES CHURCH END Beds TL 0841. 'Church end of HAYNES TL 1042'. This was the original site of the settlement.

HAY STACKS Cumbr NY 1913. A descriptive name from the shape of the peaks. Cf. T. West, *A Guide to the Lakes in Cumberland, Westmorland and Lancashire*, 3rd ed. London 1784; 'The more southern is by the dalesmen, from its form called *Hay-rick*'. Cu 387.

HAYTON 'Hay farm'. OE **hēġ** + **tūn**.

(1) ~ Cumbr NY 1041. *Hayton* from 1278, *Heyton'* 1292. Cu 288.

(2) ~ Cumbr NY 5057. *Hayton* from c.1170, *Heyton* 1230–79. Cu 88.

(3) ~ Humbs SE 8245. *Haiton(e)* 1086, *-tun -ton(e) Hay-* [12th]1203–1542. YE 233.

(4) ~ Notts SK 7284. *Hayton* from 1154×89, *Heiton* 1175, *Heyton* 1249–1428. Nt 32.

(5) ~ 'S BENT Shrops SO 4280. 'Grassland belonging to H'. *Heytons Bent* 1718, 1729, *Hayton Bent* 1832 OS. P.n. *Heyton* 1233, *Upper- Lower Hayton* 1832, ModE **bent** 'heathland, open pasture-land overgrown with bent' (OE **beonet**). DEPN, Gelling.

HAYTOR VALE Devon SX 7777. P.n. Haytor as in Haytor Rocks, *Haytor Rocks* 1809 OS, and Haytor Down SX 7577, + ModE **vale**. Haytor Down is *Idetordoune* 1506, *Ittor Doune* 1687, *Eator Down* 1762, *Itterdown* 1789, Haytor *Idetor* 1737, probably identical with the *eofede tor* 'the ivy-grown rock' [739]10th of S 255, OE ***īfede**, ***ēofede** + **torr**. The modern form has been influenced by the name Haytor Hundred called after a lost Haytor somewhere between Totnes and Newton Abbot, *Heitor(r)* 1187, 1219, *Heytore* 1238, 1249, *Haytor* from 1242, 'enclosure rock', OE **(ġe)hæġ** + **torr**. D 476, 504, PNE i. 279, Jnl 1.27.

HAYWARDS HEATH WSusx TQ 3324. 'The heath belonging to Hayworth'. *Haywards Hoth* 1544, 1608, *Hewards Hethe* 1603, *Hayworths Hethe* 1607. P.n. or surname *Hayworth* + ME **hoth** (OE *hāth*), later replaced by StE **heath**. Sussex *hoath* survives in sps until *Heywards Hoath* 1675 and Sx 268 records pr [hjuːərds hɔːð]. It is unclear whether *Haywards* is a surname or a lost p.n. *Heyworth* 1261–1306, *Hayworth(e)* 1276–1594, 'the hedge enclosure', OE **heġe** + **worth**. Sx 268.

Great HAYWOOD Staffs SK 0022. *Magna Heywode* 1311, *Great Haywood* 1836 OS. ModE adj. **great**, Latin **magna**, + p.n. *Haiwode -a -uuode* 1086, *Heywode* 1279, 'the enclosed wood', OE **(ġe)hæġ** or **heġe** + **wudu**. 'Great' for distinction from Little HAYWOOD SK 0021. Horovitz.

Little HAYWOOD Staffs SK 0021. *Little Haywode* 1432, *Little Haywood* 1836 OS. ME adj. **litel** + p.n. Haywood as in Great HAYWOOD SK 0022. Duignan 77.

HAYWOOD OAKS Notts SK 6055. 'The wood with an enclosure, the enclosed wood'. P.n. *bosco de Heywud'* 1237, *Heywod(e)* 1333, 1335, *Haywod(e)* 1287–1390, OE **(ġe)hæġ** + **wudu**, + ModE **oak(s)**. A small parish in Sherwood Forest, one-time extra-parochial. Nt 118.

HAZEL GROVE GMan SJ 9286. 'Hazel grove'. *Hesselgrove* 1690, *-grave* 1775, *Hazel Grove formerly Bullock Smithy* 1860, cf. *domus vocatus Bullock-Smythy* 'house called ~' 1619. This was a boundary point of the Forest of Macclesfield where drovers had their cattle shod with leather shoes. Che i.256.

HAZELEIGH Essec TL 8203 → DENGIE TL 9801.

HAZELEY Hants SU 7459. Partly uncertain. *Heishulla* 1167, *Hei-Heyshull* 1203, 1280, *Hayshull* 1317–1423, *Haysull* 1324, *Hesill*

1586, *Hazle* 1694. The generic is OE **hyll** 'a hill', the specific possibly OE **hǣs, hēs** 'brushwood' although the *ei, ey, ay* spellings point rather to OE **(ġe)hæġ**, genitive sing. **(ġe)hæġes**, 'a fence, an enclosure'. The modern form in *-ley* is due to misinterpretation of the 17th cent. spellings in *-le* for *-ul*. Ha 89, Gover 1958.121.

HAZELRIGG Northum NU 0533. 'Hazel-tree ridge'. *Hesilrig'* 1242 BF, *Heselrig* 1288, *Hessilrig* 1296 SR, *Hesilryge* 1428, *Heslerig* 1663. OE **hæsel** + **hrycg** with N Country [g] for [dʒ]. NbDu 107.

HAZELSLADE Staffs SK 0212. 'Hazel-tree valley'. *Hazelslade* 1880 transferred from *Hazell slade* 1682 in Brereton in Rugeley. ModE **hazel** + **slade** (OE *slæd*). St i.59, 107.

HAZELWOOD Derby SK 3246. 'Hazel wood'. *Haselwod(e) -il-* 1306–45 etc., *-wood* 1586, *Heselwod(e) -il-* 1330, 1482. OE **hæsel** + **wudu**. Db 564.

HAZLEBURY BRYAN Dorset ST 1408. 'H held by the de Brienne family'. *Hasilber(e) Bryan'* 1547, *Haselbury Bryant* 1811. Earlier simply *Hasebere* 1201, 1280, 1311, *Haselber(e)* 1237–1489, 'hazel wood', OE **hæsel** + **bearu**. Guy de Brienne held land here in 1361: his family came from one of the Briennes in France. Do ii.100, Dauzat-Rostaing 117–8 s.n. Brive.

HAZLEMERE Bucks SU 8995. 'Hazel-tree lake'. *Heselmere* 13th. OE **hæsel** + **mere**. Bk 202, L 26, 220.

HAZLERIGG T&W NZ 2372. A modern industrial settlement. No early forms but probably identical with HAZELRIGG Northum NU 0533, 'the ridge where hazel-trees grow'.

HAZLETON Glos SP 0818. 'Hazel valley', later 'hazel farm'.
I. *Hasedene* 1086, *Heseldene* early 12th, *Haselden* 1587.
II. *Haselton(a)* 12th–1404, *Hasilton(a) -yl-* 1313–1733.
OE **hæsel** + **denu**, later **tūn**. Gl i.174.

HEACHAM Norf TF 6737. 'The homestead with a hedge'. *Hecham* 1086, 1203, *Hecgham* 1191, *Hec(c)ham, Hekham* 1254. OE **hecg** + **hām**. DEPN gives pr [hetʃəm].

HEADBOURNE WORTHY Hants SU 4932 → Headbourne WORTHY SU 4932.

HEADCORN Kent TQ 8344. Partly uncertain. *Hedekaruna* c.1100, *Hedecrone -cr(o)une* 1226–80. Possibly 'Hedeca's fallen tree(s)'. OE pers.n. **Hydeca*, Kentish **Hedeca*, + ***hruna**. PNK 213, DEPN, Ritter 153.

HEADINGLEY WYorks SE 2886. 'Forest clearing belonging to the Headdingas, the people called after Headda'. *Hedingelei(a)* 1086 *-l(e)'* 1236, 1237, *Haddingleia* 1155×62, *Haddingeleia -laie* before 1170–c.1200, *Heddingle(y) -l(a)y -yng-* 1240–1641. OE folk-n. **Head(d)ingas* < pers.n. *Head(d)a* + **ingas**, genitive pl. **Head(d)inga*, + **lēah**. YW iv.140.

HEADINGTON Oxon SU 5407. 'Hedena's hill'. *Hedenandun' -don' -dun* [1004]14th, 15th, *Hedenedonœ -e* [1004]14th S 909, *Hed(d)endona -dun' -don(e) -dene* 1140–1281, *Hedingdona -don(e) -yng-* 13th–1380, *Hedyngton* 1330–1573, *Heddington* 1510. OE pers.n. **He(o)dena*, genitive sing. **He(o)denan*, + **dūn**. O 30.

HEADLAM Durham NZ 1818. '(At the) clearings overgrown with heather'. *Hedlum* c.1190, 1242×3, *Hetlum* 1235, *Hedlom* 1306, *Hedlam* 1381–1629, *Heuedlam* 1398×1421, *Headlam* from 1551. OE **hǣth** + **lēah**, dative pl. **lēum**. From the 1398×1421 form *Heuedlam* it seems that the specific was early misunderstood as 'head'. NbDu 107, SN 10.110.

HEADLESS CROSS H&W SP 0365. A modern reinterpretation of *Headleys Cross* 1789. Earlier *Hedley Cross* 1464, *Hedles Crosse* 1549, 'Headley crossroads', p.n. *Hedley* 1275–1327, 'heath clearing', OE **hǣth** + **lēah**, + ModE **cross**. Wo 365, L 205.

HEADLEY 'Heath clearing'. OE **hǣth** + **lēah**. L 205.
(1) ~ Hants SU 8236. *Hallege* 1086, *Hetliga* c.1190, *-le(gh)* 1242, 1272, *Hedlegh(e)* 1248–1458 with variants *-ligh -ley*. Ha 89, Gover 1958.103.
(2) ~ Hants SU 5162. Short for *Headley Common* 1759, 1817 OS. Ha 89, Gover 1958.148.
(3) ~ Surrey TQ 2054. *Hallega* 1086, *Hadlega -e* 1186–1212, *-lee* 1218, *Hetleg(a) -lege* late 12th, 1279, *Hedlega -lege -legh -ley(e)* 1190–1601, *Hethlee -leg(he) -leye -ligh* 1247–1313, *Hedelegh* 1535, *Headley* from 1569. Sr 77, L 205.

HEADON Notts SK 7477. Partly uncertain. *Hedun(e)* 1086, c.1170, 1208×13, *-don(e)* 1234–1509, *Headon* 1566, *Heddon'* 1176–1254, *Haddona* 1178, *-done* 1211. Forms with *-dd-* suggest OE **hǣth** + **dūn** 'heath hill' but the clayey soil conditions make this doubtful. Possibly, therefore, 'the high hill', OE **hēah**, definite form **hēa**, + **dūn**, referring to nearby Mill Hill SK 7577 which rises to 215ft. NW of the village. Nt 52, L 155.

HEADS NOOK Cumbr NY 4955. *Hedesnewke* 1570, *Heddesnucke* 1572, *Headesnook* 1590. Possibly surname *Head* + ModE **nook**. Cu 89.

HEAGE Derby SK 3750. 'High edge'. *Hey(h)eg(e) -egge* 1257–1387, *Hei(g)heg(e) -egg(e)* 1327–1417, *Heeg(g)e* 1330–1485, *Highegg* 1354, *-ege* 1539, *-edge* 1541, *Headg(e)* 1566, 1577, 1610. OE **hēah** + **ecg**. Db 565 gives pr [i:dʒ], L 159.

HEALAUGH 'The high wood or clearing'. OE **hēah**, definite form **hēa**, + Anglian **lǣh** for *lēah*.
(1) ~ NYorks SE 5047. *Hailage -a* 1086, *Helag(e) -a* 1086–1285, *Helagh* 1250–1589, *Helaw(e)* 1259–1405, *-laugh* 1394–1609. Healaugh lies on a slight ridge rising from the surrounding lowlands. YW iv.240.
(2) ~ NYorks SE 0199. *Hale* 1086, *Helagh* 1200–1402, *Helawe* 13th cent., *Healaughe* 1531. YN 273 gives pr [i:lə].

HEALD GREEN GMan SJ 8585. *Heald Green* 1841. If this is an ancient name it would be p.n. *Heald*, ME **helde** 'the slope' (cf. David son of Leuk *del Helde* 1289), + ModE **green**. Che i.244.

HEALD MOOR Lancs SD 8826. Possibly dial. **hield** 'a slope' (OE *helde*). Heald Moor is bordered by steep slopes to N and S.

HEALE Devon SS 6446. 'At the nook'. *Hele* 1333, 1481. OE **hēale**, dative sing. of **healh**. Heale is situated at the head of a short side-valley. D 67.

HEALEY 'The high clearing or pasture'. OE **hēah**, definite form **hēa** (<**hēaha*) + **lēah**. L 206–7.
(1) ~ Lancs SD 8815. *Hayleg* 1260, *Helay -leye* 13th, *Heghlegh* 1332. La 60.
(2) ~ NYorks SE 1880. *Helagh* c.1280, 1327, *-laugh* 1406, *Healey* 1561. YN 232 gives pr [i:lə].
(3) ~ Northum NZ 0158. *Heley* 1268. Also kown as *Temple Helay* 1570, an estate once held by the Knights Templar. NbDu 107.

HEALEYFIELD Durham NZ 0648. Probably 'new arable land at *Healey*'. *Helayfeld* 1430, *heleyfeld* 1446×8, *Healeyfield* 1777×82. *Healey* is *Heleia* [1183]c.1382, *Heleya* c.1200, *Heley* 1284–1447×8, *Helegh* 1393, 'the high clearing', OE **hēah**, definite declensional form **hēa**, + **lēah**. NbDu 108.

HEALING Humbs TA 2110. 'The Hæglingas, the people called after Hægel'. *Heg(h)e- Hechelinge* 1086, *Heghelinga* c.1115, *Heiling(a)* 1166–1218, *Hailinges* 1180, *Hegeling'* 1202–18, *Hehling* 1212, *Heling(e) -yng* 1195–1576, *Heyling(e) -yng* 1242–1563, *Healing* from 1590. OE folk-n. **Hæġelingas* < pers.n. *Hæġel* + **ingas**. Li v.103, ING 66.

HEAMOOR Corn SW 4631. 'Hea marsh'. *Hay Moor* 1663. P.n. Hea, *la Hae* 1277, *Haye alias Anhaye* 1619, 'the enclosure', OE **(ġe)hæġ** with definite article Fr **la**, Co **an**, + ModE **moor**. PNCo 95 gives pr ['hei-].

HEANOR Derby SK 4346. '(At the) high ridge'. *Hainoure* 1086, *Hanovre* 1212, *Henouer(e) -oure -ower -owre* 1232–1393, *Henor(e)* 1355–1610, *Heynor(e)* 1474–1576. OE **hēah(a)** + **ofer**, locative-dative sing. definite declension **hēan-ofre**. Db 469 gives pr [heinə], L 175–6.

HEANTON PUNCHARDON Devon SS 5035. 'H held by the Punchardon family'. *Heanton Punchardun* 1297. Earlier simply *Hantone* 1086, *Hainton* 1213, *Hyaunton'* 1242, 1302, *Heanton* 1292, *Yantone* 1291, *Yeanton* 1675, '(at) the high or chief settlement', locative-dative sing. definite form **hēan-tūne** from OE **hēah** + **tūn**. The place was held by Robert de Ponte Cardonis ('thistle-bridge') in 1086, the French form of the surname Punchardon occurring in 1215. D 45 gives pr [heintən].

HEAPHAM Lincs SK 8788. 'Rosehip village or homestead'. *Iopeham* 1086, *Iopheim* c.1115, *Hepham* 1196–1212. OE **hēope** + **hām**. The present form of the name shows the expected development of ME *Hēpham* < OE *Héopham*, but the earliest spellings seem to show an alternative development with accent shift due to Scand influence, viz. late OE *Hēópham* > ME *Hjopham* as in SHIPTONTHORPE Humbs SE 8543. This is paralleled by the development of ON *Hjaltland* to *Shetland* and by that of OE *hēo* > Northern ME *scho* 'she'. DEPN, Cameron 1998.

North HEASLEY Devon SS 7333. *N*th *Heasley* 1809 OS. ModE adj. **north** + p.n. *Haseleg* 1249, *Hes(e)le* 1330, 1333, 'hazel wood or clearing', OE **hæsel** + **lēah**. 'North' for distinction from *South Heasley* 1809 OS (SS 7332). D 345.

HEASLEY MILL Devon SS 7332. *Heasley Mill* 1809 OS. P.n. Heasley as in North HEASLEY SS 7333 + ModE **mill**. D 345.

HEATH '(The) heath'. OE **hǣth**. L 245.

(1) ~ Derby SK 4466. *(le) Heth(e)* 1257–16th, *Heath(e)* from 1574. Earlier called *Lunt* 1086, *Lund* 1246–91, *Lound(e)* 13th cent.–1878, 'the grove or small wood', ON **lundr**. Db 261.

(2) ~ AND REACH Beds SP 9228. *Hetheredge* c.1750, *Heathanreach* 1785. Originally two places, *la Hethe* 1276, 'the heathland', and Reach, *Reche* 1276, 'raised strip', OE ***rǣċ**. The reference is to Watling Street, Margary no. 1e, which passes a little to the N of here, much of it with a markedly raised profile. Bd 124–5, L 245, 183–4.

(3) ~ END Hants SU 5962. Short for Heath End of Baughurst. Cf. *Heath side F.* 1817 in Berks SU 5764. ModE **heath** + **end**.

(4) ~ END Hants SU 4162 → EAST END SU 4162, NORTHEND SU 4163.

(5) ~ HAYES Staffs SK 0110. 'Enclosures on the heath'. *hethhey* 1570, *Heathy Hays* 1834 OS. ModE **heath** + **hay** (OE *(ġe)hæġ*). Cf. nearby New Hayes SK 0312, *(le) New(e)- Neuhey -hay(e)* 1348–1528, Boney Hay SK 0610 and CHESLYN HAY SJ 9707. Heath Hayes is situated in Cannock Chase which has many heath names such as Brindley Heath SJ 9914, Green Heath SJ 9913 and *Big Heath* 1834 OS. St i. 59, 60.

(6) ~ HILL Shrops SJ 7614. *Hethulle* 1261×2, 1294×5, *Heath Hill* 1833 OS. Gelling.

(7) ~ HOUSE Somer ST 4146. *Heath House* 1817 OS. ModE **heath** used here of level wetland. Heath House stands on the edge of higher ground overlooking the wetlands of the river Brue. Some of these lands are called moors, others heaths, cf. nearby Ashcott Heath ST 4439, *Burtle Heath* 1817 OS, Calcott Heath ST 4041, Edington Heath ST 3941, Shapwick Heath ST 4340, Walton Heath ST 4539.

(8) Haywards ~ WSusx TQ 3324 > HAYWARDS HEATH TQ 3324.

(9) The ~ Norf TG 1821. *Buxton Heath* 1828 OS.

(10) Upper ~ Shrops SO 5685. ModE **upper** + p.n. *La Hethe* 1255–95, *Hethe* 1272–1563, *Heath* from 1559×60, *The Heath(e)* 1566–1833 OS. Sa i.150.

Old HEATH Essex TM 0122. 'Old Heath, the old landing-place'. *Ealdehethe* before 1272, *(La) Eldeheth(e)* 1310×11, *(la) Oldheth* 1401, *(Le) Old(e)hith(e)* 1401×6, 1495. Earlier simply *Hetha(m)* 1158–1237 'the landing place'. OE adj. **eald**, definite form **ealde**, + **hȳth** contrasting with *La New(e)heth(e)* 1311, 'the new landing place' (modern Hythe). *Heath* is the normal development of the Essex variant *hēth* for *hȳth*. Ess 376.

HEATHCOTE Derby SK 1460. 'Heath cottage(s)'. *Hedcote* late 12th, 1263, *(grangia de) Hethcote* 1244–1610, *Heathcoate Grange* 1566. OE **hǣth** + **cot**, pl. **cotu**. A grange of Garendon Abbey. Db 369.

HEATHER Leic SK 3810. 'The heathland'. *Hadre* 1086, *Hedreia* 1199, *Hethre* 1221–1345, *Hether* 1222–1553, *Hei- Heyther* 16th cent., *Heather* from 1576. OE ***hǣther** < *hǣth* + *-or*. Lei gives pr ['hi:ðə:], SSNEM 377.

HEATHFIELD Devon SX 8376. 'The waste land'. *la Hethfeld* 1228. OE **hǣth** + **feld**. Also known as *Boviheʒfelde* c.1327, *Bovyhethfeld* early 15th, 'Bovey waste land', from p.n. BOVEY SX 8178, for distinction from Heathfield SX 6850, *Hetfeld* 1086. D 469.

HEATHFIELD ESusx TQ 5821. 'Uncultivated open land'. *Hadfeld'* 12th cent., *Hatfeld* 13th, *Het(h)feld(e)* 13th–16th, *Heffeud* 1255, *Heffeld* 1312. OE **hǣth** + **feld**. Sx 463, L 242.

HEATHFIELD Somer ST 1626. 'Heathland'. *Hafella, Herfeld* 1086, *Haðfeld* 1159, *Hethfeld* [1174×92, before 1258] Buck, 1610, *Het- Hedfeld(e)* [12th] Buck. OE **hǣth** + **feld**. DEPN gives pr [hefl], L 242.

HEATHFIELD MOOR NYorks SE 1167. P.n. *Higrefeld -t* 1086, *Hikerfeldebec* c.1175, c.1200, *Hi- Hyrefeld(e)* 1198–1260, *Hearf(i)eld* 17th cent., *Heathfield* 1770, 'open land frequented by magpies', OE **higera** + **feld**, + ModE **moor**. The obscure first part of this name has been subject to popular etymology. YW v.215.

HEATHROW AIRPORT GLond TQ 0875. P.n. Heathrow + ModE **airport**. Heathrow, *la Hetherewe* 1413×22, *Hitherowe* 16th cent., is the 'row (of houses) at the heath', ME **heth** (OE *hǣth*) + **rew** (OE *rǣw*). The reference is to Hounslow Heath, *brueria de Hundeslaue* 1275, *Houndesloweheth* 1382. Mx 38.

HEATHTON Shrops SO 8192. 'The heath settlement'. *Hethton'* 1356×7, *Hethtoun* 1360×1, *Heathton* 1833 OS. OE **hǣth** + **tūn**. Gelling.

HEATLEY Ches SJ 7088. 'Heath clearing'. *Hey- Hayteley* 1525–59, *Hae- Haitley* 1531, 1542, *Heatley* from 1666. OE **hǣth** + **lēah**. Che ii.37, L 204.

HEATON 'The high settlement'. OE **hēah**, definite form **hēa** (<**hēaha*) + **tūn**. See also CLECKHEATON SE 1825, EARLSHEATON SE 2521, KIRKHEATON SE 1818.

(1) ~ Lancs SD 4460. *Hietune* 1086, *Hetun* c.1160, *Heton* c.1170–1332. Situated on a tract of ground rising to 50ft. above the Lune flats and contrasting with MIDDLETON SD 4258 and OVERTON SD 4358. La 176.

(2) ~ Staffs SJ 9562. *Heton* 1230×32–1539, *Heyton* 13th–1619, *Heaton* from 16th. Oakden.

(3) ~ T&W NZ 2766. *Hactonam* c.1200, *Heton(')* 1242, 1296, 1366×7 32.280. NbDu 108, BF 1114, SR 62.

(4) ~ MOOR GMan SJ 8791. P.n. Heaton as in HEATON NORRIS + ModE **moor**.

(5) ~ NORRIS GMan SJ 8890. 'H held by the Norris family'. *Heton Norays* 1282, *~ Norreis* 1322, 1332. Earlier simply *Hetton* 1196, *Heton* 1212, 1276. Heaton occupies a piece of land that rises to over 200ft. and slopes steeply S and W. It was held from the 12th cent. onwards by the Norreys family. La 30, Jnl 17.30.

(6) ~ PARK GMan SD 8204. P.n. Heaton, *Heton* c.1200–92 etc., *Hetone* 1212, *Heiton* 1222, *Haton* 1246, *Heaton hill* 1577, *Yetton* 1872. Also known as *Ouirheton super Faghefeld* 'upper H on Fallowfield' 1276, *Heton Fawfeld* 1437, OE **fāg** 'variegated' + **feld** 'open land', or Great Heaton for distinction from Little Heaton SD 8305. Heaton Hall stands at an elevation of over 300ft. La 49, Jnl 17.39.

(7) Castle ~ Northum NT 9041. ModE **castle** + p.n. *Heton'* [1183]c.1320. The place belonged to Norham Castle. NbDu 108.

(8) Hanging ~ WYorks SE 2523. *Hanging Heaton* from 1557. ME prefix **hengande** 'steep' (*Hingand(e)* 1266 etc.), ModE **hanging** (*Hyngyng* 1490), + p.n. *Etun* 1086, *He(a)ton* 1315. YW ii.196.

(9) Upper ~ WYorks SE 1819. *Upperheaton* 1549. ModE **upper** + p.n. *Heton* 1240, *West ~* 1284. YW ii.225.

HEAVERHAM Kent TQ 5758. Possibly 'boar enclosure'. *Euerham* 1313. OE **eofor** + **hamm** 5. PNK 57, ASE 2.36.

HEAVILEY GMan SJ 9088. 'Heathy clearing'. *Hethyleg(h) -i-* 1283–1360, *Hevileye* 1285, *(le) Heuylegh* 1354, 1372, *Heaveley* 1577. OE **hǣthiġ** + **lēah**. Che i.296.

HEBBURN T&W NZ 3164. 'The high tumulus'. The forms fall into four types:

I. *Heabyrm* c.1107, *-byrme* c.1174, *Hebirme* c.1220.
II. *Heberin(e)* 1154×66–1340, *Heberm(e)* 1189×1212–1396.
III. *Hebern(e)* [1199×1216]1300, 1279–1423, *Hebarn(e)* 1410–1483.
IV. *Heburn(e)* 1256, 1345, 1375, *Hebburn* from 1381, *Hebborn(e)* 1411–1763 Kitchin, *Hibburn* 1411×7–1478.
There are also occasional metathesised forms *Hebrom -bron'* 1370, 1371, *Hebron* 1768 Armstrong.
Type I represents OE **hēa**, the definite declensional form of OE **hēah**, + **byrġen**. OE *byrgen* would normally give ME *birin(e)* which was frequently misread as *birm(e)*. Type II represents reduction of weak stressed *-ir-* in the second element of the compound to *-er-*. Type III shows confusion with ME *bern*, late ME *barn*, 'a barn', IV with ME *burn* 'a stream', the *-or-* spellings reflecting Northumbrian and Tyneside dial. [ɔ:] for [ɛ:] before *r* as in the lexical set NURSE. NbDu gives pr [hebərən], Wells 374–5.

HEBDEN NYorks SE 0263. Probably 'rosehip valley'. *Hebeden(e)* 1086, 13th, *Hebbeden(e)* 12th–1325, *Hebden* from 1198. OE **hēope** + **denu** with early voicing of *p* to *b* before *d*. YW vi.101.

HEBDEN BRIDGE WYorks SD 9927. 'The bridge across HEBDEN WATER'. *Hepdenbryge -brigg(e), ~ Brig* 1399–1596, *Heptenbridge* 1508, *Heptonbridg(e)* 1578–1775, *(le) Hebdenbrigg(e)* 17th cent., *~ Bridge* 1822. R.n. HEBDEN (WATER) SD 9631 + **brycg**. YW iii.188 gives prs ['ebdin 'brig] and [t'brig].

HEBDEN GREEN Ches SJ 6365. *Ebden Green* 1831, *Hebden Green* 1842. P.n. *Heppedene* 1225, 'rosehip valley', OE **hēope** + **denu**, + ModE **green**. Che iii.169.

HEBDEN MOOR NYorks SE 0466. *moram de Hebbeden* 1471, *Hebden more* 1631. P.n. HEBDEN SE 0263 + ME **mōr**. YW vi.101.

HEBDEN WATER WYorks SD 9631. 'The river of Hebden'. *aqua de Heppedene* 1279, *aqua voc' Hepden* 1314, *the Hedden* 1577, *the water of Hebden* 1582. P.n. Hebden, 'bramble or wild rose valley', OE **hēopa** + **denu**, + **wæter**. YW vii.128, L 98.

HEBRON Northum NZ 1989. 'The high tumulus'. *Heburn* 1242 BF–1296 SR, *Heborin* 1264, *Heburnne* 1346, *Hebbourn* 1663. OE **hēa**, definite declensional form of **hēah**, + **byrġen**. NbDu 108.

Great HECK NYorks SE 5921. *Great Heck* 1571. ModE **great** + p.n. *Hech* 1153–1193, *Hecca* 1156–96, *Heck(e)* from 13th, 'the hatch, the gate', OE **hæċċ**, N dial. **heck**. 'Great' for distinction from *Lytel Heck* 'little H' 14th. YW ii.18.

HECKFIELD Hants SU 7260. 'High open land'.
I. *Hechfeld* 1207, *Heghfeud -feld* 1272, 1380, *He(i)ghfeild* 1579, 1618, possibly surviving in the name Highfield House in the centre of the village.
II. *Hec- Hekfeld -feud* 1208–1487.
III. *Heggefeld* 1280.
OE **hēah** + **feld**. The normal development of *hēah* would be ME *hegh, heigh*, ModE **high**, but substitution of [k] for the peripheral phoneme /x/ ([ç]) occurs frequently in p.ns. producing *heck*. Type III is a false etymological association of the name with ME *hegge* 'a hedge'. This is one of a series of names in *feld* marking open heathland along the Berkshire-Hampshire boundary†. Ha 90, Gover 1958.120, L 239.

HECKINGTON Lincs TF 1444. 'The estate called after Heca'. *Hechintone, (H)echintune* 1086, *Hekyngton'* [1115]13th, *-ington(a)* [after 1125]13th, *Heckingtuna* after 1184. OE pers.n. *Heca* + **ing**[4], + **tūn**. Perrott 70, Cameron 1998.

East HECKINGTON Lincs TF 1944. ModE **east** + p.n. HECKINGTON TF 1444.

HECKMONDWIKE WYorks SE 2123. 'Heahmund's dairy-farm'. *Hedmūdewic* 1166, *Edmundewich* 1167, *Hec- Hekmund(e)wy(c)k(e) -wik(e)* 13th–1551 with variants *Heck(e)-* and *-mond(e)-*, *Hekynwik -inwyk* 1370, *Heccundwek -wyk(e) -wike* 1382–1415. OE pers.n. *Hēahmund* + **wīċ**. For the treatment of *h* [x] before a liquid consonant cf. KEIGHLEY SE 0540 and HEPMANGROVE Cambs TL 2883. YW iii.24 gives prs ['hekmənwaik], earlier ['(h)ekənwɑ:k].

HEDDINGTON Wilts ST 9966. 'The estate called after Heddi'. *Edintone* 1086, *Hedinton(a) -yn-* 1203–1332, *Hedington -yng-* 1222–1403. OE pers.n. *Heddi* + **ing**[4] + **tūn**. Wlt 263.

HEDDON-ON-THE-WALL Northum NZ 1366. 'Heather hill'. *Hedon super murum* 1242 BF, 1296 SR. P.n. *Hedun* 1175, *Heddun* 1262, *Heddon* 1291, OE **hǣth** + **dūn** + Latin **super murum**, ModE **on the wall** referring to Hadrian's Wall. NbDu 109.

HEDDON'S MOUTH Devon SS 6549. 'Mouth of the river Heddon'. *Heddon's Mouth* 1809 OS. The relationship of the name Heddon's Mouth to the river name Heddon is uncertain. D 7 suggests that the r.n. is a late back-formation from Heddon's Mouth, the name of the cove where the stream enters the sea, and that the cove name derives from the family name Heddon found in N Devon in the 17th and 18th cents.

HEDENHAM Norf TM 3193. 'Hedena's homestead or enclosure'. *Hedenaham* 1086, *Hedenham* 1180, *Heddenham* 1296. OE pers.n. *Hedena* + **hām** or **hamm** 2b 'land on a hill-spur'. DEPN.

HEDGE END Hants SU 4912. Short for the Hedge End of Botley. *Cutt Hedge End* 1759, *Hedge end* 1810 OS. ModE **hedge** + **end**. Ha 90, Gover 1958.38.

HEDGEHOPE HILL Northum NT 9419. *Hedgehope Hill* 1869. P.n. Hedgehope + ModE **hill**, a hill rising to 2348ft. Hedgehope, no early forms, may be the 'side-valley with a hedge', ModE **hedge** + dial. **hope** (OE *hop*).

HEDGERLEY Bucks SU 9787. 'Hycga's wood or clearing'. *Huggeleie* 1189×99, *-legh -le(ye)* 1237–1393, *He(d)geley* 1526–40, *Hugely al Hedgerley* c.1560. OE pers.n. **Hycga* + **lēah**. Bk 238, Signposts 106.

HEDGING Somer ST 3029. *la Heghyng* [late 13th] Buck, *Hedging Green* 1782, 1822 maps. Possibly 'high ground', OE ***hēahing** < *hēah* 'high' + *ing*[2], as in HEIGHINGTON Durham NZ 2422. The modern form would then be a late folk-etymological reinterpretation. Hedging overlooks the flat land of North Moor.

Castle HEDINGHAM Essex TL 7835. 'H with the castle'. ME **castel**, *Chastell'* 1248, *Castel(l)* 1304, 1513, *Castle* from 1388, + p.n. Hedingham, the 'homestead or estate called or at *Hethning*, the place called after Hethin', or the 'homestead of the Hethningas, the people called after Hethin'. There are three different types:
I (with dissimilatory loss of *n*): *Hed- Hidingham* 1086, *Hi- Hyding(e)ham -yng-* 12th–1358, *Hething(e)ham -yng-* 1100×35, 1314–1475, *Hy- Hithingham* 1349–1484.
II (with loss of *th*): *Haingheham* 1086–1122, *Heyng- Heing(e)ham* 1166–1235, *Hen(g)eham* 1212–1330, *Henyge-* 1248.
III (with simplification of *-thn-* to *n*): *Henyngham* 1373–1549.
OE p.n. **Hethning* < pers.n. **Hethīn* + **ing**[2], locative-dative sing. **Hethninge*, or folk-n. **Hethningas* < pers.n. **Hethīn* + **ingas**, genitive pl. **Hethninga*, + **hām**. Pers.n. *Hethīn* or *Heddīn* is not recorded but seems to occur in the lost hundred name HINKFORD Essex. 'Castle', referring to the Norman fortress built by the de Vere earls of Oxford, is for distinction from Sible HEDINGHAM TL 7834. The two settlements lie on the r. Colne and an alternative suggestion is OE **Hȳthingahām*, the 'homestead of the Hythingas, the people of the hithe or landing-place or called after Hytha', OE **hȳth** or pers.n. **Hȳtha* + **ingas**. Ess 438, ING 121, Pevsner 1965.110.

Sible HEDINGHAM Essex TL 7834. 'Sibil's H'. Pers.n. *Sibill(e)* 1230, 1285, *Sible* from 1238 with variants *Sybille, Symple* 1527, *Steple* 1594, *Sibly* 1713, + p.n. Hedingham as in Castle HEDINGHAM TL 7835. The reference is possibly to Sibil, widow of Geoffrey de Laventon, who held land here in 1237. Ess 438–9.

HEDLEY ON THE HILL Northum NZ 0759. P.n. *Hedley* from 1242

†DEPN cites a form *Hizfeld* 1194 Selborne which may or may not belong here. The DB form *Effelle* 1086 is eccentric.

BF, *Heddeley* [1307]14th, 'Clearing where heather grows' OE **hǣth-lēah**. + ModE **on the hill**. Also known as *Karlhedley* 1275, 'H held by the carle(s)' or 'Karli's H', ON pers.n. *Karli*, ODan *Karle* or ME *carle* (ON *karl* 'a freeman'), a N equivalent of ME *churl* (OE *ċeorl*). NbDu 109.

East HEDLEYHOPE Durham NZ 1440. ModE adj. **east** + p.n. *Hedleehope* 1440×1, *Hedleyhope* 1498, 'Hedley valley', p.n. Hedley + OE **hop**. Hedley, *Hedley* 1349–67, ~ *on the Hill* 1392, is the 'clearing where heather grows', OE **hǣth** + **lēah**.

HEDNESFORD Staffs SK 0012. 'Heddin's ford'. *Hedenedford* (for *Hedenes-*) c.1153, *Hed(d)enesford* 1307–1570, *Hednesford* from 1362, *Hedg(e)ford* 1653–1793. OE pers.n. **Heddīn*, genitive sing. *Heddīnes*, + **ford**. St i.57.

HEDON Humbs TA 1828. 'Heath hill'. *Hedon(a)* [1138×42], [1190]1301–1828, *Haduna* 1138×42, *Heddun(a), Heddon(a)* 1162×97–1252, *Haddon* 1535. OE **hǣth** + **dūn**. YE 39 gives pr [edn].

HEDSOR Bucks SU 9086. 'Heddi's ridge'. *Heddesore* 1195–1241, *-overe* 1235, *Hedsore* 1316, 1489, *Hedyssoer* 1552. OE pers.n. *Heddi*, genitive sing. *Heddes*, + **ofer**. Two early spellings, *Heddenesora* 1223, *Hedleshore* 1195, point to alternative diminutive forms of the pers.n., *Heddīn* and **Heddel*. Further, it is uncertain whether the generic is really *ofer* or *ōra*. Bk 181, L 179, 181, Jnl 2.31, 22.35.

HEGDON HILL H&W SO 5854. *Egdon Hill* 1832 OS. P.n. Egdon + ModE **hill**. There are no early forms for Egdon but it appears to be a name ending in OE **dūn** 'a hill'. It forms an eminence rising to 829ft. NW of Pencombe SO 6052.

HEGGERSCALES Cumbr NY 8210. Partly obscure. *Hegelstale* (for *-scale*) 1380, *-scale(s)* 1402, 1422, *Haggilskale* 1597, *Higgelskales* 1608, *Heggerscale* 1777, 1782. Unknown element + ON **skáli** 'a shieling'. The specific might be a pers.n. related to OSwed *Hæghulf*, OG *Hagiwolf*, or an otherwise unrecorded p.n. **Heg-gill* 'bird-cherry ravine', ON **heggr** + **gil**, and so '**Heggill* shielings, shielings at bird-cherry ravine'. We ii.5, xi.

Potter HEIGHAM Norf TG 4119. 'The potter's H'. *Hegham Potter(e)* 1182–1357, *Hecham Potter(e), Potter(e)hecham* 13th cent., *Potter Heitham* (for *Heicham*) 1250, *Potteres Hecham* 1254. OE, ME **pottere**, genitive sing. **potteres**, genitive pl. **pottera**, + p.n. *Echam* 1086, *Hecham* 1101–1275, *Hegham* 1257, 1286, *Heyham* 1268–1379, *Heigham* 1413, 'the homestead with a fence', OE **heċċ** + **hām**. Large heaps of pot-sherds from the RBrit period have been found here. Nf ii.121–2.

HEIGHINGTON Durham NZ 2422. Probably the 'settlement at the high ground'. *Hekington* 1215×18, *He(h)ington* 1228, *hekenton(a)* 1256, late 13th, [1189×99]1336, *Hey(i)ncton, Hey(n)gton(a)* late 13th, *Heghyn(g)ton(a) -ing-* [1183]c.1382, 1311–1589, *Heighington* from 1313. OE ***hēahing** 'high ground', OE *hēah* + *ing*[2], + **tūn**. Heighington, an important administrative centre and *caput* of Heighingtonshire, is the first village on the high ground overlooking the Tees lowlands. Cf. HECKINGTON Lincs TF 1444. NbDu 110, Årsskrift 1974.56.

HEIGHINGTON Lincs TF 0369. 'The estate called after Hyht'. *Hicting- Hyctingetun, Hictinton* [1154×89]13th, *Hi- Hyghtington -yng-* 1320–1412, *Heghington* 1547×51, *Heighington* from 1564. OE pers.n. *Hyht* + *ing*[4] + **tūn**. The modern form of the name seems to point to an alternative formation, OE ***hēahing** 'the high ground' + **tūn** as in HEIGHINGTON Durham NZ 2422. The village lies on elevated ground above the r. Witham fens. Perrott 326, Cameron 1998.

South HEIGHTON ESusx TQ 4503. *Sutheghton* 1327. ME adj. **south** + p.n. *Hectone* 1086, *Hayton al. Heighton* 1735, 'high settlement', OE **hēah** + **tūn**. 'South' for distinction from Heighton Street TQ 4703, *H(i)ecton(a)* 12th cent., *Heghton (juxta Westferles)* (West Firle) 13th–15th cents. Sx 363, SS 76.

HEIGHTON STREET ESusx TQ 4703 → South HEIGHTON TQ 4503.

HELBECK Cumbr NY 7915 → HILLBECK NY 7915.

HELE 'Nook, valley, recess, land in a river bend'. OE **hēale**, dative sing. of **healh**. PNE i.223–4, L 100–111.

(1) ~ Devon SS 5347. *Hela* 1086, *Hele* from 1311. The site lies in a short deep valley opening from the sea. D 46.

(2) ~ Devon SS 9902. *Hele* from 1242[†]. The reference is to a small recess in the hills W of the valley of the river Culm. D 556.

(3) Giffords ~ Devon SS 5206 → HELE SS 9902 footnote.

HELFORD Corn SW 7526. Partly uncertain. *Helleford* 1230, *Hayleford* 1318, *Hailleford* 1347, *Hail- Heilford* c.1540, *Hayl- Heylford* 1564, *Helford* from 1576. Co **heyl** 'estuary, salt river' + **ford**. The difficulty is that all the medieval forms refer to the harbour or estuary implying the sense 'a passage for ships, an arm of the sea' for *ford*, although this is only recorded in Westmorland dial. in 1891 and must be due to Norse *fjord* which cannot have influenced usage in Cornwall. Possibly the assumption that *Heyl* was the original name of the Helford River is wrong and it referred only to the side inlet at Helford itself in which case *ford* can have its usual meaning. The village is first specifically mentioned in 1564. PNCo 95, Gover n.d. 506, RN 192.

HELFORD RIVER Corn SW 7626. P.n. HELFORD SW 7526 + ModE **river**. It is usually assumed that the original name of the river was *Heyl*, Co **heyl** 'estuary, salt river', but this may be a mistaken assumption and Carew's form *Haill* 1602 may be a back-formation from the name of the village. PNCo 95, RN 192.

HELHOUGHTON Norf TF 8626. 'Helgi's estate'. *Helge- Helga- Hælga- Halgetuna* 1086, *Helgetun* c.1150, *Helewetone* 1157. ON pers.n. *Helgi*, genitive sing. *Helga*, + **tūn**. The development of the name must have been *Helgetun* > *Helweton* > *Helowton* which was assimilated in spelling to nearby Houghton TF 2928. DEPN, SPNN 201.

HELLAND Corn SX 0770. 'Ancient church site'. *Hellaunde* 1284–1369, *Hellond(e)* 1285–1342, *Helland* from 1296. Co ***hen-lann** (*hen* + **lann*) reinterpreted as if an English name in *land*. Identical with Helland SW 9049, *Henland* 1086, and Helland SW 7531. PNCo 96, Gover n.d. 115, DEPN.

HELLESDON Norf TG 7928. 'Hægel's hill'. *Hægelis- Haglesdun* [c.985]12th Abbo, *Hailesduna* 1086, *-don* 1180, 1196, *Heilesdon* 1199. OE pers.n. *Hæġel*, genitive sing. *Hæġeles*, + **dūn**. Hellesdon is the place where St Edmund suffered martyrdom. DEPN.

HELLIDON Northants SP 5158. Partly uncertain.

I. *El(l)iden* 12th.

II. *Helidon* 1193–1242, *Heleydon* 1424, *Heliden(e) -y-* 1220–1452, *Hellyden* 1538.

III. *Hey- Heiliden* c.1210, *Heylydon* 1549.

IV. *Haliden'* 1246, 1537.

Perhaps 'Hægla's valley', OE pers.n. **Hæġla* + **ing**[4] + **denu**. This would account for all forms on the assumption that form IV shows folk-etymological association with OE **hāliġ** 'holy'. Alternative suggestions are **hǣliġ**, a Nb mutated side form of **hāliġ** 'holy' and therefore unlikely here, or **hǣliġ** 'unstable'. It is also uncertain whether the generic is **denu** or **dūn**: the village lies on a hill overlooking a *denu*-type valley. Nth 24.

HELLIFIELD NYorks SD 8556. Either 'Helgi's open land' or 'the holy tract of open land'. *Hælgefeld, Helgefeld -t -flet* 1086, *Helg(h)efeld* 12th–1234, *Helg(h)feld(e)* 1198–1346, *Helehefeld* late 12th, *Heleg(h)efeld* 1200–1246, *Helyfeld* 1373, *Hellifield* 1586. ON pers.n. *Helgi*, genitive sing. *Helga*, or OE **hǣliġ**, definite form **hǣlga**, + **feld**. YW vi.158, SSNY 157, L 243.

HELLINGLY ESusx TQ 5812. 'Wood or clearing of the hill dwellers'. *Hellingeleghe* 1216×72–1329, *Hellinglegh* 14th–16th,

[†]The identification of *Hiele* 1086 with this place is wrong. DB Devon 39.8 shows that it belongs to Giffords Hele SS 5206.

Hellingleight 1591. OE folk-n. **Hyllingas* < **hyll** + **ingas**, SE dial. form **Hellingas*, genitive pl. **Hellinga*, + **lēah**. For the 1591 form cf. FAIRLIGHT TQ 8612. Sx 438 gives prs [helin'lai] and [heriŋ'lai], SS 66.

HELLINGTON Norf TG 3103. 'Helgi's estate'. *Halgatuna -tona* 1086, *Hel(e)geton, Helegheton* 1254. ON pers.n. *Helgi*, genitive sing. *Helga*, + **tūn**. DEPN, SPNN 201.

HELMDON Northants SP 5843. Probably 'Helma's valley'. *Elmedene* 1086, 1235, *Helmesden* 1166, *Helme(n)den* 12th–1428, *Helmedon'* 1220, *Halme- Hamel- Hemelden* 1162, 1222. OE pers.n. **Helma*, genitive sing. **Helman*, + **denu**. Some early forms show confusion with ***hamol** 'mutilated'. It is not impossible that *Helma* is the name of a nearby hill < OE **helm** 'a helmet' used of a summit. Helmdon lies on a sloping side of a ridge of high ground overlooking the valley of the Tove. Nth 54.

HELMINGHAM Suff TM 1957. 'The homestead of the Helmingas, the people called after Helm'. *helming(h)e- Helminc- Elmingheham, Helmingh(e)am* 1086, *Helmingueham* [1158×62]1331, *Helmingeham* 1173, *Helmingham* from 1086. OE folk-n. **Helmingas* < pers.n. *Helm* + **ingas**, genitive pl. **Helminga*, + **hām**. DEPN, ING 130, Baron.

HELMSHORE Lancs SD 7821. Uncertain. *Hellshour* 1510. Probably OE **helm** + ***scor(a)**; the sense would be either 'steep slope with a cattle shed', or 'steep slope of the hill called *Helm*, the helmet'. La 91, PNE i.242.

HELMSLEY NYorks SE 6186. 'Helm's forest clearing'. *Elmeslac, Almeslai* 1086, *Helmeslac(h)* 1155–1252, *-leia -ley -legh -lay* from c.1170, *Hem(e)sley* 16th cent., 17th cent. OE pers.n. *Helm*, genitive sing. *Helmes*, + **lēah**, Anglian **lǣh**. YN 71 gives pr [emzlə].

HELMSLEY MOOR NYorks SE 5991. P.n. HELMSLEY SE 6186 + ModE **moor** (OE *mōr*).

Gate HELMSLEY NYorks SE 6955. *Gatehemelsay* 1438, *Gethemsley* 1574. ME **gate** 'road' (ON *gata*) + p.n. *Hamelsec* 1086, *Hemel(e)say -ey* 1177–1300, 'Hemele's island', OE pers.n. *Hemele*, genitive sing. *Hemeles*, + **ēġ** in the sense 'hill-spur'. 'Gate' is for distinction from Upper HELMSLEY SE 6956 and refers to the Roman road from York to Malton, Margary no.81a, on which Gate Helmsley stands. YN 10 gives pr [ge:t emzlə].

Upper HELMSLEY NYorks SE 6956. ModE adj. **upper**, ME *Over* 1301, + p.n. *Hamelsec(h)* 1086, as in Gate HELMSLEY SE 6955 which stands on lower ground. YN 11.

HELPERBY NYorks SE 4369. 'Hjalp's farm or village'. *Helperby* from [972]11th S 1453, *heolperbi* [972]11th ibid., *(H)ilprebi* 1086, *Helprebi -by* 1086–1202. ON feminine pers.n. *Hjalp*, genitive sing. *Hjalpar*, + **bȳ**. YN 23, SSNY 30.

HELPERTHORPE NYorks SE 9570. 'Hjalp's outlying farm'. *Elpetorp* 1086, *Help(e)torp* 1196, c.1300, *Helpert(h)orp(e)* from c.1165. ON feminine pers.n. *Hjalp*, genitive sing. *Hjalpar*, + **thorp**. YE 123, SSNY 60.

HELPRINGHAM Lincs TF 1340. Probably 'the homestead of the Helpricingas, the people called after Helpric'. *Helperi(n)cham* 1086, *Helprincheham* 1180, *-ingeham* 1213–1327, *Helpringham -yng-* 1138–1556. OE folk-n. **Helprīcingas* < pers.n. *Helprīċ* + **ingas**, genitive pl. **Helprīċinga*, + **hām**. Formally the specific could be a p.n. *Helpering*, an *ing*[2] derivative of OE **helpere* 'a helper' which occurs in the name of a fishing weir on the r. Tyne, *Helperyare* 1344. Formations in *ing*[2] often refer to water-courses. Possibly the reference here might be to Helperingham Eau in the sense 'ancillary drain'. Perrott 74, ING 141, DAJ 2.57.

HELPSTON Cambs TF 1205. 'Help's estate'. *hylpestun (ge mære)* 'Helpston boundary' [948]12th S 533, *Hylpestun* [c.980] c.1200 B 1130, *Helpestun(e) -ton* 1125×8 etc., *Helpston* 1390, *Helpson* 1571. OE pers.n. **Help*, genitive sing. **Helpes* + **tūn**. Nth 236.

HELSBY Ches SJ 4875. 'Village on a ledge'. *Helesbe* 1086, *-is- -ys- -by* 13th–1540, *Hellesbi -is- -ys- -by(e)* 1200–1656, *Helsby* from 1553. ON **hjallr** 'a hut, a ledge on a mountainside', secondary genitive sing. **hjalles**, + **bȳ**. The village occupies a narrow shelf between marshland and the foot of the precipitous Helsby Hill 462ft., SJ 4975. Che iii.235 gives pr ['helzbi], SSNNW 32.

HELSINGTON Cumbr SD 4988. Partly uncertain. *Helsingetune* 1086, *Helsington -yng-* 1187–1823, *Hesel- Haselington'* 1195–1256, *Henston* 1519, *Hel(l)ston* 1577, 1653. If the *Hesel- Hasel-* forms really do belong to this name they point to OE **hæsling** + **tūn** 'settlement at the hazel copse'. If not, we probably have to do with an OE **Helsinga-tūn* 'settlement of the folk who dwell on the *hals*', OE folk-n. **Helsingas* < OE *hals* + *-ingas*, genitive pl. **Helsinga*, + **tūn**. OE *hals* meant 'a neck of land, a col, a pass' and here would have to refer to the long ridge of Underbarrow Scar between Brigsteer and Holeslack, SD 4889–4988. The folk-ns. *Hœlsingas* in the OE poem *Widsith* and *Hœlsingar* in the Danish and Swedish p.ns. Hälsingör, Hälsingland and Hälsinge may be compared. However, some doubt has been cast on whether Underbarrow Scar could really be referred to as a *hals*. We i.108, Addenda xv.

HELSTON Corn SW 6527. 'Farm, village, manorial centre called or at *Henlis*'. *Henlistone* 1086, *Henleston* 1284, *(burgus de) Helleston* 1187, 1310, *Helston* from [1365]late 14th. P.n. *Henlis*, Co ***hen-lys** 'ancient court, ruins' (Co *hen* + **lys*), *Hellys* 1396, *Helles* late 17th, + OE **tūn**. Helston was probably an old administrative centre later replaced by the Domesday manor of Winnianton SW 6620, *Winetone* 1086. PNCo 96, DEPN, CPNE 130.

HELSTONE Corn SX 0881. 'Manorial centre, land called or at *Henlis*'. *Henliston* 1086, *Henleston* 1201–1302, *Helleslond* 1284, *Helleston in Trigg* 1297, 1310. P.n. **Henlis*, Co ***hen-lys** 'ancient court, ruins' as in HELSTON SW 6527, + OE **tūn** 'manorial centre' varying with **land** 'estate'. Probably an old administrative centre replaced by LESNEWTH SX 1390 'new court'. PNCo 96, Gover n.d. 533, DEPN, CPNE 130.

HELTON Cumbr NY 5122. 'Settlement on the slope'. *Helton* from c.1160, ~ *Moruyll*, ~ *Flechan* 1278 etc. OE **helde** + **tūn**. The manorial additions referring to possession by the *de Moreville* and *Flechan* families are for distinction from HILTON Cumbr NY 7320. We ii.200, L 162.

HELVELLYN Cumbr NY 3415. Possibly 'yellow upland moor'. *Helvillon* 1574, 1577, *Lauuellin* 1600, *Helvellin* 1786, 1799 ff, *Helvellyn* from 1800. Possibly Cumbric ***hal** (corresponding to Co *hal* 'moor, marsh', W *hâl* 'moor') + ***melyn** 'yellow' (Co *melyn*, W *melyn*) with *v* for *v̄*; an OW *halmelen* 'yellow moor' is on record. Coates 1988.31 suggests the reference is to white bent-grass on the lower slopes of the mountain which may have led Coleridge to enthuse over starting to 'ascend a hill bright yellow-green' as he set out for the summit of Helvellyn in August 1800. Jnl 3.50, TT 30, GPC 1815.

HELWATH BECK NYorks SE 9599. P.n. *Helewath* 1231, *Helwath* 1369, 'the ford made with flat stones', ON **hella** + **vath**, + ModE **beck** (ON *bekkr*) 'a stream'. YN 117.

HELWITH NYorks NZ 0703. 'The ford made with flat stones'. *Helwath(e)* 1280, 1283, *Helwith* 1577. ON **hella** + **vath**. YN 291.

HELWITH BRIDGE NYorks SD 8169. *Helworthe Brigge* 1590, *Hellith Bridge* 1771. P.n. Helwith 'ford made with flat stones', ON **hella** + **vath**, + ModE **bridge**. The reference is to a ford across the river Ribble. Cf. HELWATH BECK SE 9599, HELWITH NZ 0703. YW vi.222.

HEMBLINGTON Norf TG 3411. 'The estate called after Hemele'. *Hemelingetun* 1086, *Hemelington* 1252. OE pers.n. *Hemele* + **ing**[4] + **tūn**. The 1086 form may point to OE folk-n. *Hemelingas* 'the people called after Hemele', genitive pl. *Hemelinga*, + **tūn**. DEPN.

HEMBURY Devon ST 1103. Otherwise known as Hembury Fort, *Hembury Ford* 1670, 1700, ~ *Fort* 1672. The reference is to the largest earthwork in Devon which gave its name to BROAD-

HEMBURY ST 1004 under which earlier forms of the name are recorded. D 558, Thomas 1976.95.

HEMINGBROUGH NYorks SE 6730. 'The fortified place of the Hemingas, the people called after Heming'. *borg heminga* [1026]17th Knytlinga Saga with variant *helminga* [1026]13th, *Hemingburg(h) -yng-* [1080×6]17th–1488, *-browghe* 1571, *Hamiburg* 1086, *Hemming(e)burc(h) -yng-* 1121×8–1408. ON pers.n. *Hemingr* + OE **burh**. Hemingbrough may take its name from the Jómsborg Viking, Jarl Hemingr, who captained the northern detachment of Swein's housecarls at a place near York, now lost, called *Slésvík*. Possible alternative explanations would be **Hemingaburh* 'the fortified place of the Hemingas, the people called after Hema', OE folk-n. **Hemingas* < pers.n. *Hema* + **ingas**, genitive pl. **Heminga*, + **burh**, or **Hemmingeburh* 'the fortified place by the fish-weir', OE ***hemming**, locative–dative sing. ***hemminge**, + **burh**. Hemingbrough is situated on the old course of the river Ouse. YE 260, SSNY 147, Townend 34.

HEMINGBY Lincs TF 2374. Partly uncertain. *Hamingebi* 1086, *Heninghebi* c.1115, *Hemmingebi* 1173, *Hemmingkebi* 1212. This is usually explained as 'Heming's village or farm', ODan pers.n. *Heming* + **bȳ**, but the regularity of the medial *-e-* sp points to the weak variant **Hemingi*, genitive sing. **Heminga*. Alernatively we might suggest OE ***hemming**, locative–dative sing. ***hemminge** '(the place) at the enclosure' referring to a dam in the r. Bain as in *le Hemmyngyar* 1438×9, a fishery in the r. Tyne. DEPN, SSNEM 53, DAJ 2.57.

HEMINGFORD ABBOTS Cambs TL 2870. 'H held by the abbot' sc. of Ramsey. *Hemmingford Abbatis* 1276. P.n. Hemingford + ModE **abbot**, Latin **abbas**, genitive sing. **abbatis**. Hemingford, *Hemmingeford* [974]13th S 798, [1040×2]14th S 997, *Hemminggeford* c.1000, *Hemmingaford* [1012]12th, *Hemmingfordia* [1043×9]14th S 1106, *Emingeford, alia Emingeford* 1086, *Hem(m)ingeford* 1150–1286, *Hemmingford -yng-* 1218–1316, *Hemingford -yng-* from 1248, appears to be OE **Hemminga-ford* 'ford of the Hemmingas, the people called after Hemma or Hemmi', OE folk-n. **Hemmingas* < pers.n. *Hemma* or *Hemmi* + **ingas**, genitive pl. **Hemminga*, + **ford**. The Hemmingfords, however, lie on the Great Ouse and the specific may well be OE ***hemming** 'a weir, an enclosure for fish', locative–dative ***hemminge**, or folk-n. **Hemmingas* 'the people who live at Hemming, the fish weir, the people who work the *hemming*' and so 'ford at Hemming, ford of the Hemmingas, the weir-people'. 'Abbot's' for distinction from HEMINGFORD GREY TL 2970. Hu 260.

HEMINGFORD GREY Cambs TL 2970. 'H held by the Grey family'. *Hemmingford Grey* 1316. P.n. Hemingford as in HEMINGFORD ABBOTS TL 2870 + family name *Grey*. Hu 260.

HEMINGSTONE Suff TM 1553. 'Heming's estate'. *Ha- Hemingestuna* 1086, *Hemingeston(e)* 1212–50, *Hemmingestune -ton(e)* 1205–1316, *Hemyngston* 1346, 1524, *-ing-* 1568, 1610. ON pers.n. *Hem(m)ingr*, ME genitive sing. *Hem(m)inges*, + **tūn**. DEPN, Baron.

HEMINGTON Northants TL 0985. Partly uncertain. *Hemmingtune* [1077]13th, *-ton* 1219–1346 with variant *-yng-*, *Hemmingetonam -thon, Hemmi(n)cton* 12th, *Heminton(e)* 1086–1428 with variant *-yng-*. Usually explained as 'the farm or village of Hemma or of his people', OE pers.n. *Hemma* + **ing**[4] + **tūn** or **Hemminga*, genitive pl. of folk-n. **Hemmingas* 'the Hemmingas, the people called after Hemma' < *Hemma* + **ingas**, + **tūn**. But an OE ***hemming** 'that which is like a hem, a border, an enclosure' < OE **hemm** + **ing**[2] may have existed. Hemington lies close to the county border with Cambridgeshire. Nth 212, Nomina 7.40.

HEMINGTON Somer ST 7253. 'Estate called after Hemma or Hemmi'. *Hami(n)tone* 1086, *Hammingtona* 1086 Exon, *Heminton* 1176, *Hemmington* 1212, *Hemington* 1610. OE pers.n. *Hemma* or **Hemmi* + **ing**[4], + **tūn**. DEPN.

HEMLEY Suff TM 2842. 'Helma's wood or clearing'. *Helmele(a), Halmeleia, (H)almelega* 1086, *Helmele* 1219, 1254, 1327, *Hemley* from 1524. OE pers.n. *Helma* + **lēah**. DEPN, Arnott 30, Baron.

HEMPHOLME Humbs TA 0850. 'Hemp island'. *Hempholm(e)* 1175×90, 1664, *-home* 1577, *Henepeholm* 1312. OE **hænep**, ME **hemp** + **holmr** in the sense 'raised ground in marsh'. YE 72.

HEMPNALL GREEN Norf TM 2493. *Hempnall Great Green* 1838 OS. P.n. HEMPNALL TM 2494 + ModE **green**. Contrasts with nearby *Silver Green* 1838 OS.

HEMPNALL Norf TM 2494. 'Hemma's nook'. *Hemen- Hamehala* 1086, *Hemehal* 1199–1242. OE pers.n. *Hemma*, genitive sing. *Hemman*, + **halh**. DEPN.

HEMPSHILL Notts SK 5244 → HENBURY Ches SJ 8873.

HEMEL HEMPSTEAD Herts TL 0506 → Hemel HEMPSTEAD.

HEMPSTEAD 'A homestead'. OE **hāmstede**, side-form **hǣmstede**.

(1) ~ Essex TL 6338. *Hamstedā, Hamesteda* 1086, *Hamsted(e)* 1203–1428, *Hampsted(e)* 1291–1428, *Hemsted(e)* 1319–1418, *Hempsted(e)* 1331–1519. Ess 511, Stead 201.

(2) ~ Glos SO 8117. '(The) high homestead'. *Hechanestede* 1086, *Heccamstuda* 1137, *Heghamstede* 1236, *Heyham(p)stede -stude* 1282–1391, *Hempsted(e)* 1535–1681. OE **hēah** + **hām-stede** with SW variant *stude* from OE **styde**. The village is on top of a low hill that rises some 45ft. above the flat low-lying land beside the Severn. Gl ii.165.

(3) ~ Norf TG 4028. *Hemsteda* 1086, *Hemsted(e)* 1212–1371, *Hempsted(e)* 1291–1535. Nf ii.96, *Stead* 182.

(4) ~ Norf TG 1037. *Henepsteda, Hemesteda* 1086, *Hemsted(e)* 1201–1455, *Hempsted(e)* 1121×35–1441. Stead 181.

(5) Hemel ~ Herts TL 0506. 'Homestead in the district called Hæmele'. *Hamelamestede* 1086, *Hamelhamsted(e)* 1173–1293, *Hemel* 1184–1428, *Hemlamsted(e)* 1384–1423, *Helmpsted* 1509×47, *Hemlamsted al. Hempsted* 1541, *Hemsted* 1598, 1675. OE district name *Hæmele* [c.705]17th S 1784 'the broken country', OE **hamol**, + **hāmstede**. The original name was reduced to *Hemsted* by the 16th cent.: *Hemel* is a modern addition to this, not an ancient survival. The reference is to the broken and uneven nature of the country round about with its steep hills and deep valleys. Hrt 40 gives former pr [hæməl hæmstid], Stead 240.

HEMPTON Norf TF 9129. 'Hemma's estate'. *Hamatuna* 1086, *Hemton* 1242, *Hempton* 1254. OE pers.n. *Hemma* + **tūn**. DEPN.

HEMPTON Oxon SU 4431. '(The place) at the high settlement'. *Hentone* 1086–1346 with variants *-ton(')* and *-thone, Hempton(e)* from 1285. OE **hēah**, definite form dative case **hēan**, + **tūn**. O 257.

HEMSBY Norf TG 4917. 'Heimir's farm'. *He(i)mes- Haimesbei* 1086, *Heimesbi* c.1200, *Hem(m)esby* 1103×6–1551. Pers.n. ON *Heimir* or ODan **Hēmer*, genitive sing. *Heimis*, + **bȳ**. DEPN, SPNN 201, Nf ii.54.

HEMSWELL Lincs SK 9390. Partly uncertain. *Helmeswelle* 1086, 1185–7, *He- Halmeswella* c.1115, *Helmeswell* 1202. Either 'Helm's spring', OE pers.n. *Helm*, genitive sing. *Helmes*, + **wella**, or the 'spring of the summit or shelter', OE **helm** 'a helmet, the summit of a hill, a shelter'. One of a series of streams rising along the spring-line of the Cliff which may at this point have been known as the *Helm*. Cf. Elmswell Humbs SE 9958, *Helmesuuelle -well(e) -wella* 1086–1371, *Elmesuuelle -well(e)* 1086–1598. DEPN, YE 154.

HEMSWORTH Dorset ST 9605 → HENBURY Ches SJ 8873.

HEMSWORTH WYorks SE 4213. 'Hymel's enclosure'. *Hamelesuurde, Hilmeuuord* 1086, *Hi- Hymelesw(o)rd(e) -wurd(a)* 12th–1428 with variants *-wurthe -w(u)rth -worth(e), Hi- Hymlesvrtha* 1188×1202, *-w(o)rd(e)* c.1190×1220–1245, *Hi- Hymmesworth* 1334–1504, *Hemmesworth* 1431, *Hemsworth* from 1556. OE pers.n. *Hymel*, genitive sing. *Hymeles*, + **worth**. YW i.264.

HEMYOCK Devon ST 1313. Originally a stream name, probably

'summer stream'. *Hamihoc* 1086, 1197, *Hamioc* 1204, *Hemioch* 1194, *-yok(e) -ioc -yoc* 13th cent. Probably a stream name going back to Brit **Samiāco*, a derivative of **samo-* 'summer'. The meaning would be a stream that never dries up. D 616 gives pr [hemik].

HENBURY Avon ST 5678. '(At the) high or chief fortification.' *Heanburg* [691×99]17th S 77, *-b'u* [757×75]11th S 1411, *(æt) Heanbyri(g)* [793×6]11th S 146, *Henberie* 1086, *Hembir -byr(y) -bur(y)* 1167–1535, *Henbury* from 1551. OE **hēah**, definite form dative case **hēan**, + **byriġ**, dative case of **burh**, referring either to Blaise Castle hill-fort, half a mile SW of the village, or to a fortified manor or 'chief place' of the Bishop of Worcester's estates here. Gl iii.130.

HENBURY Ches SJ 8873. 'Hemeda's fortified place'. *Hameteberie* 1086, *Hamedeberie* 1086, *-bury* c.1301, *Hendebiry -bury* 1288–1478, *Henbury* from 1383. The generic is OE **byriġ**, dative sing. of **burh** 'a fortified place, a manor house'. Che i.78 took the specific to be OE **hǣmed** 'cohabitation, marriage, adultery, fornication' and explained the name as 'fortification where a community lives, seemingly with indelicate connotations'. However, OE *hǣmed* is not otherwise attested in p.ns. Pers.n. **Hemeda* would be a weak parallel to *Hemedi* found in Hempshill Notts SK 5244, *Hemdeshyll'* 1275, and Hemsworth Dorset ST 9605, *Hemedesw(o)rde* 1086. It seems unlikely that there is any connection with *humet/hemed(e)* in KINSHAM H&W SO 3664, *Kingeshemede -a* 1216, 1287, and Presteigne SO 3164, *Humet* 1086, *Prestehemed* 1137–9. Do ii.260, Nt 150, He 117–8.

HENDON GLond TQ 2389. '(At) the high down'. *Hendun* [959]12th S 1293, *Handone* 1086, *Hendon* from 1199, *Hyndon* 14th cent. OE **hēah**, definite form dative case **hēan**, + **dūn**. The boundary of Hendon is referred to as *heandunes gemære* 972×8 S 1451. The old village clustered round the church of St Mary on a prominent hill visible for many miles. Mx 56.

HENDON T&W NZ 4055. Uncertain. *Hendon(')* 1197, 1379, *Hynden'* 1382, *hyndyn'* 1418, *Hendon* 1768. If the earlier forms, which occur in surnames, belong here, this name is identical with HENDON GLond TQ 2389. The later forms, however, point to either OE **hind** + **denu** 'hind valley' or **hīġna** + **denu** 'the monks' valley', OE genitive pl. **hīġna** of **hīwan** 'a household, a community'. The topography is undecisive since there is both a dene running down to the sea and a marked hill at NZ 394554. NbDu 110.

East HENDRED Oxon SU 5688. *Esthenred'* 1200, 1284, *Esthanreth* 1284, *East or Great Hendred* 1761. ME adj. **est** + p.n. *Hennarið, (æt) Hennariþe* [956]13th S 581,*(æt) Hennenriðe* [962]12th, *(æt) Henne riðe* [962]13th S 700, *Henneriðe* [964]12th, *(to) henna riðe* [964]13th S 724, *Henret, Enrede* 1086, *He- Hanred(a) -e -reth(a) -rith* 13th cent., *Hendreda* 12th, 'the wildfowl stream', OE **henn**, genitive pl. **henna**, + **rīth**. 'East' for distinction from West HENDRED SU 4488. Brk 479, L 29.

West HENDRED Oxon SU 4488. *Westhenred* c.1180 etc. with variants *-rede* and *-reth(e), Weshanred* 1219 etc., *Little or West Hendred* 1761. ME **west** + p.n. Hendred as in East HENDRED SU 5688. Brk 479.

HENFIELD WSusx TQ 2115. Partly uncertain. *Hanefeld* [770]14th S 49, *Hamfeld(e)* 1086, 1227×33, *Hanefeld(e)* 1166–1438, *Hanfeld(e)* 1169–1316, *Henfeld* 1272, 1324, *Heanvell* 1654, *Heffeild* 1706, *Henvill* 1721. The site is higher than the surrounding country and this is probably, therefore, '(the settlement) at the high open land', OE **hēah**, definite form dative case **hēan**, + **feld**. But OE **hana** 'a cock', genitive pl. **hanena**, pers.n. *Hana* or **hān** 'a stone, a rock', genitive pl. **hāna**, all later replaced by **henn**, genitive pl. **henna**, are possible alternatives. Sx 215, SS 70.

HENGISTBURY HEAD Dorset SZ 1790. 'H headland'. *Hengestbury heade* 1610. P.n. Hengistbury + ModE **head(land)**. Hengistbury, *Hedenesburia* 12th, *-bury* 14th, *Hensbury* 1540, is 'Heddin's fort', OE pers.n. *Heddīn*, genitive sing. *Heddīnes*, + **byriġ**, dative sing. of **burh**. The reference is to an Iron Age hill-fort. In the 16th cent. antiquarians associated the reduced form *Hensbury* with the quasi-historical Hengist who according to Bede and Nennius brought three ships of German exiles to Kent in the time of Vortigern and so began the Anglo-Saxon invasion. Do 84, Ha 90.

HENGOED Shrops SJ 2933. A common W p.n. meaning 'old wood'. *Hengoed* 1760. W **hen** + **coed**. Gelling.

HENGRAVE Suff TL 8268. 'Hemma's pasture'. *Hemegretham* (Latin accusative sing.) 1086, *Hemegrede* c.1095, 1198, *Hemmegredhe* 1157×80, *-grave* 1242, 1264, *Hengraue* 1610. Pers.n. *Hemma* + **grǣd**, ***grǣth**. The form **grǣth* is necessary to account for the change [ð] > [v]. DEPN.

HENHAM Essex TL 5428. Probably '(at) the high homestead'. *Henham* from 1086 including [1043×5]13th S 1531, *Anham* 1202, *Hanham* 1328, 1378. OE **hēah**, definite form oblique case **hēan**, + **hām**. Ess 528.

HENLEY '(The settlement) at or by the high wood'. OE **hēan-lēa(ge)**, dative sing. definite form of **hēah** + **lēah**. The Henleys are typically in low sites overlooked by high woodland ground. L 207.

(1) ~ Somer ST 4332. *Henleighe* [973]14th S 791, *Henlegh* 1243. DEPN.

(2) ~ Suff TM 1551. *Henleia -le(ie)* 1086, *Henley(e)* 1242–1661, *Hanley* 1219. DEPN, Baron.

(3) ~ -IN-ARDEN Warw SP 1566. *Henleye in Arderne* 1343–1541. P.n. *Henle* c.1180 + forest name ARDEN WMids SP 1859. 'But in truth it ought to have been written *Heanly*, as the ordinary sort of people doe still pronounce it' (Dugdale 1756). Wa 244.

(4) ~ –ON–THAMES Oxon SU 7682. *Henle super Tamisia* 1275×6, *Henle upon Thames* 14th cent. P.n. Henley + r.n. THAMES. Henley, *Henleiam* [1136×45]13th etc. with variants *-le(y)a -leg(h)' -le(e) -ley(e)*. O 74, L 207.

(5) ~ PARK Surrey SU 9352. *Henly Park* 1816 OS. P.n. *Henlei* 1086, *Hanlei -legee* 1229, *Henlea* [727, 1062]13th S 1035, 1181, *-lye* 1216×72, *-le(e)* 14th, *Hendeley in le Hethe* 1458, + ModE **park**. Sr 136, L 207.

HENLEY Shrops SO 5476. 'Hens' wood or clearing'. *Haneleu* 1086, *Hennele(g') -leye* 1242–1428, *Henelegh' -ley(e)* 1271–1318, *Henley* from 1374. OE **henn**, genitive pl. **henna**, + **lēah**. The sense may be 'clearing where hens are kept' rather than 'wild birds' wood'. Sa i.150.

HENLEY WSusx SU 8925. Uncertain. *Henly Common* 1813 OS. Possibly a descendant of *Hundligh*, a topographical surname recorded in Easebourne in 1296 and appearing as *Hounly* in 1640. The meaning would be 'the hound clearing', OE *hund* + *lēah*. Sx 19, SS 66.

HENLOW Beds TL 1738. '(At) the high hill'. *heanhlæwe* 980×90 S 1497, *Han(e)slau(e)* 1086, *Han(e)lawe* [1154×89]1261, 1220, *Henlawe* 13th cent., *Henlowe* 1302. OE **hēan**, dative sing. definite form of **hēah**, + **hlǣw**. Bd 170.

HENNOCK Devon SX 8380. '(At) the high oak-tree'. *Hanoch* 1086, *Hainoc* 1086 Exon, *Hanok* [c.1200]13th–1450, *Henoc* 1234, *Heanok(e)* 1292–1434, *Henyk* c.1540. OE **hēah** + **āc**, dative sing. definite form **(æt thǣre) hēan āce**. D 471 gives pr [henək].

Great HENNY Essex TL 8738. *Henie Magna* 1291, *Hely Magna* 1412. ModE adj. **great**, Latin **magna**, + p.n. *Ha- Henies, Heni* 1086, *Henie -(e)y(e)* 1291–1346, '(at the) high island', OE **hēah**, dative sing. definite form **hēan**, + **ēġ** in the phrase *æt thǣm hēan ēġe*. The reference is to the well-marked hill on which the village stands. 'Great' for distinction from Little Henny (lost) mentioned in 1229 (*Parva*). Ess 443.

HENSALL NYorks SE 5923. 'Hethin's nook of land'. *Edeshale* 1086, *Hethensale* 12th, *Hethens-* 13th–1379, *Heuens-* 13th, 1364, *le Henssall* 1406, *Hensall* from 1444. ON pers.n. *Hethinn* or OE **Hethīn*, genitive sing. *Hethinns* or **Hethīnes*, + **halh** referring to an area of raised ground in the Aire marshes. Some spellings show dial. substitution of [v] for [ð]. YW ii.19, SSNY 157, L 107.

HENSBARROW DOWNS Corn SW 9957. P.n. Hensbarrow + ModE **downs(s)**. Hensbarrow, *Hyndesbergh* 1284, *Hainborough* 1602, *Hensbery* 1610, is 'hind's barrow', OE **hind** 'a hind, a female deer', secondary ME genitive sing. **hindes**, + **beorg**, referring to a hill-top of 1034ft. crowned by a tumulus. Later misinterpreted as 'hen's barrow' which gave rise to the humorously named 19th cent. village of Cocksbarrow SW 9755, *Cox Barrow* 1813. PNCo 96.

HENSHAW Northum NY 7664. 'Hethin's nook'. *Hedeneshalch* 12th, *Hethingishalc* [before 1153]1262 RRS, *Hetheneshalgh* 1298, *Heinzhalu* 1279, *Henneshalgh* 1326, *Henshaw* 1597. Pers.n. ON *Hethinn* or OE **Hethīn*, genitive sing. **Hethīnes*, + **halh**. The reference is to a small side-valley of the river S Tyne. NbDu 111, L 108.

HENSTEAD Suff TM 4986. 'The wild hen place'. *Henestede* 1086, 1275, 1293, *Hensted(e)* 1254–1610. OE **henn**, genitive pl. **henna**, + **stede**. Stead 190.

HENSTRIDGE Somer ST 7219. 'Stallion ridge'. *Hengesteshrege* [924×39]13th S 1712, [946×55]13th S 1736, *Hengstesrig* [956 for 953×5]14th S 570, *Hesterige, Hengesterich* 1086, *Heynstrugge* 1243, *Henkestridge* 1610. OE **hengest**, genitive sing. **hengestes**, + **hrycg**. The ridge overlooks Henstridge Marsh, a place where wild horses must have run. Cf. nearby HORSINGTON ST 7023 and the name *horspol* 'horse pool' in the bounds of S 570. DEPN, Charters V.75–6.

HENTON Oxon SP 7602. '(The place) at the high settlement'. *Hentone* 1086 etc. with variants *-tona -ton(')*. OE **hēan-tūne**, dative sing. definite form of **hēah** + **tūn**. O 107.

HENTON Somer ST 4208. 'Hen farm'. *Hentun* *[1065]18th S 1042. OE **henn** + **tūn**. DEPN.

HENWOOD Corn SX 2673. 'Hens' wood'. *Hennawode* 1327, *Hennewode* 1338, 1386. OE **henn**, genitive pl. **henna**, + **wudu**. PNCo 97, Gover n.d. 162.

HEPBURN Northum NU 0624. 'The high tumulus'. *Hi- Hyb(b)urn(e)* 1242 BF–1428, *Hebburne* 1542. OE **hēah**, definite form **hēa**, + **byrġen**. Hepburn Bell NU 0825 is referred to [c.1040]12th HSC as *montem Hybberndune*, 'mount H'. The forms with *Hi- Hy-* can be explained as due to smoothing of *hēah* to *hēh-* and late OE palatal umlaut to *hī(h-)*. There are cairns on Hepburn Moor. NbDu 111, Jordan §69.

HEPMANGROVE lost near BURY Cambs TL 2883. 'Heahmund's grove'. *Height- Heyt- Heyghtmond(e)grove* 1297–1437, *Eydmundgrave* c.1300, *Heighmondegrove* 1359, *Hetmingrove* 1387, *Hekmangrove* 1498, *Highmangrove* 1517, *Hepmangrove* 1552. OE pers.n. *Hēahmund* + **grāfa**. For the forms cf. HECKMONDWIKE WYorks SE 2123. Hu 207.

HEPPLE Northum NT 9800. 'Land in the river bend where rosehips grow, rosehip nook'. *Hephal(e)* 1205–1346, *Hyephale* 1229, *Heppal* 1236 BF, *Heppell* 1428. OE **hēope** + **halh**. NbDu 111, L 108.

HEPSCOTT Northum NZ 2284. 'Hebbi's cottage'. *Hebscot* 1242 BF, *Hebbescotes* 1288, *Heppescotes* 1257, 1313, *Heppscot* 1310. OE pers.n. **Hebbi*, genitive sing. **Hebbes*, + **cot**. NbDu 111.

HEPTONSTALL WYorks SD 9828. Either 'rosehip farmstead' or 'stable belonging to or beside Hebden'. *Heptonstall* from 1253, *Hemsall* 1695. Either OE **hēope** 'hip, the fruit of the wild rose' or **hēopa** 'dog-rose, bramble' + **tūn-stall**, or r.n. Hebden as in nearby HEBDEN WATER SD 9631 + **stall**. YW iii.191 gives prs ['epms(t)əl, 'epənstəl].

HEPTONSTALL MOOR WYorks SD 9430. *Heptonstall Moor* 1596. P.n. HEPTONSTALL SD 9828 + ModE **moor** (OE *mōr*). YW iii.193.

HEPWORTH Suff TL 9874. 'The rosehip enclosure'. *hepworda* 1086, *Hepewurde* 1193, *Hepwrthe* 1196, *Hepworth* 1610. OE **hēope** + **worth**. DEPN.

HEPWORTH WYorks SE 1606. 'Heppa's enclosure'. *Heppeuuord* 1086, *Heppeworth* 14th cent., *Hepworthyn'* 15th. OE pers.n. *Heppa* + **worth**. YW ii.242.

HEREFORD H&W SO 5140. 'Army ford'. *Hereford* from 958 S 677 including 1086 and *[811]12th S 167. OE **here-ford** probably short for **here-pæth ford** 'highway ford'. The reference is to a N–S crossing of the river Wye. Herefordshire is *Herefordscir* 1016×35 S 1462, 1043×6 S 1469, 12th ASC(E) under year 1048. The forged charter of 811 identifies Hereford with *Hecana* and hence with the tribe called the *Westan-Hecani* by Florence of Worcester, a name that has not been explained. The earliest certain reference to Hereford is the presence of Wulfheard *Herefordensis ecclesiæ episcopus* 'bishop of the Herefordian church' at a council of the province of Canterbury in 803. For the early history of Hereford see WM 159–64. The W name is *Henffordd* 'old ford'. DEPN, L 67,71, ECWM 217.

Little HEREFORD H&W SO 5568. *Lvtelonhereford* 1086, *Herefordia parva* 1122. OE **lȳtel**, dative sing. definite form **lȳt(e)lan**, + p.n. HEREFORD SO 5140. DEPN, L 67,71.

HERGEST H&W SO 2755. Uncertain. *Hergesth* 1086, *Heregast* 1251, *Hergest* 1340. A possible explanation is 'beautiful basket', W *hardd* + *cest*, referring to the shape of nearby Hergest Ridge, although this does not account for the regular sp with *Her-*. He 116, Celtic Voices 193.

HERGEST RIDGE H&W SO 2556. *Hergest Ridge* 1833 OS. P.n. HERGEST SO 2755 + ModE **ridge**. Also known by its W name *Cefn Hergest*, W **cefn** 'back, ridge'. The ridge rises to 1389ft.

HERMITAGE 'A hermitage'. ME **ermitage**, ModE **hermitage**.

(1) ~ Berks SU 5173. *Le Eremytage* 1550, *Hermitage* 18th cent. Brk 252.

(2) ~ Dorset ST 6407. *The hermitage of Blakemor* 1390, *Ermytage* 1389, *le Hermitage* 1469, *Harmitage* 1650. Named from the priory or hermitage founded here in Blackmoor Forest in the 13th cent. Do 85.

(3) ~ WSusx SU 7505. *Armetage* 1635, 1664. There was formerly a chapel here attached to the manor of Prinsted SU 7605 connected with the Hermits of the Causeway. One Simon Coates of Westbourne SU 7505 is described as *ermyt* in his will of 1527. An earlier reference is *Ermeteslandes* 1513. Sx 56.

(4) The ~ Surrey TQ 2253. *The Hermitage* 1816 OS.

HERNE Kent TR 1865. 'The angle'. *(æt) Hyrnan* c.1100, *Herne* from 1269. OE **hyrne**, Kentish **herne**, probably in the sense 'recess in the hills, curving valley' referring to the valley that bends round S of the village. PNK 509, TC 108.

The HERNE Cambs TL 2590. 'The corner'. *la Hern* 1219, *þe Hirne* [1275]14th. OE **hyrne**. The place is in the extreme NW corner of Ramsey parish.

HERNE BAY Kent TR 1768. *Herne Bay* 1819 OS. A modern coastal settlement in Herne. P.n. HERNE TR 1865 + ModE **bay**.

HERNER Devon SS 5826. Unexplained. *Yarner* 1809 OS. D 353 associates this p.n. with a *Richard atte Hurne* whose name occurs in 1333 (from ME *hurne*, OE *hyrne* 'an angle, a corner'). But elsewhere in Devon Yarner seems to mean 'eagle bank', OE **earn** + **ōra**, Yarner Beacon SX 7762, *Yornere* 1333, *Yerner* 1399, *Yearner* 16th, or with OE **ofer**, Yarner SX 7778, *Yarnour -ouere* 1344, *Yerner* 1399, *Yarner* 1553. D 297, 353, 468.

HERNHILL Kent TR 0660. 'The grey hill'. *Haranhylle* c.1100, *Harehell' -hull(e) -hell'* 1226–78, *Harnhulle* 1250. OE **hār**, definite form oblique case **hāran**, + **hyll**, dative sing. **hylle**. PNK 303, L 170.

HERODSFOOT Corn SX 2160. 'The foot of *Heriard*'. *Heriott foote, Herryottes foote* 1613. P.n. *Heriard* 1490, *Heriod* 1617, *Herod Wood* 1758, Co ***hyr-yarth** 'long ridge' (Co *hyr* + **garth*), + English **foot** 'lower end'. Herodsfoot is situated at the bottom of a long steep ridge and contrasts with Herodshead, *Bronheriard -hiriard* 1284, *Brounhirgard* 1342, Co **bron** 'breast, hill'. PNCo 97, Gover n.d. 259.

HERONGATE Essex TQ 6391. *Herongate* 1678. P.n. Heron as in

Heron House and Heron Hall, *Hern(e)* 1376–1593, *Heron* 1514, 1595, 'the nook, the corner', ME **herne, hurne** (OE *herne, hyrne*), + **gate**. The earliest reference is *Fi-Fyndegode(s)hurne* 1232 (*Fingods Heron* 1588) from the ME nickname *Findegod*; a John Fyndegod was regarder of the Forest of Essex in 1365 and a John Vyndegode was vicar of West Ham in 1388. Ess 159, 161.

HERONSGATE Herts TQ 0294. Unexplained. *Heryngarste* 1599, *Herringat* 1656. Fergus O'Connor, the Chartist leader, bought an estate here to be divided into small-holdings for letting to subscribers to the National Land Company. After him it became known as *O'Connorville*. Hrt 81.

HERRIARD Hants SU 6645. 'Army yard or enclosure'. *Henert* (for *Heriert*) 1086, *Herierd* c.1160–1341, *Herey(h)erd* 1236–1334, *Heryott* 1559. OE **here** + **ġeard**. The significance of this name is uncertain. If Æthelwulf's victory over the Danes at *Acleah* in 851 is located at nearby Oakley SU 5650 this might have been the site of a Danish encampment before or after the battle since *here* is the normal word in ASC for the Danish host. An alternative occasion could be the campaign of 871 when the Danes defeated the West-Saxons at Basing SU 6652. Ha 91, Gover 1958.132.

HERRINGFLEET Suff TM 4797. 'The inlet of the Herelingas, the people called after Herela'. *Herlingaflet* 1086, *Herlingeflet(h)* 1202–70, *Herlingflet(e)* 1254–1327, *Horningflet* 1610. OE folk-n. **Herelingas* < pers.n. *Herela* + **ingas**, genitive pl. **Herelinga*, + **flēot**. For the folk-n. cf. East HARLING Norf TL 9986. DEPN, Baron.

HERRINGTON T&W NZ 3653. Partly uncertain. *Herintune* 1114×6, *-ton()* 1249×56, 1308, *Harintune* c.1116, *Herington() -yng-* c.1200–1545 including [1183]c.1320, *Har(r)yngton -ing-* 1512–80, *Herrington* from 1614. The 16th cent. spellings with *Har(r)-* show late ME *ar* < *er*, but the form *Harintune* c.1116 is probably too early for this. Possibly the specific is OE ***hæren** 'rocky' with reference to Herrington Hill NZ 346527 which was subsequently quarried, or ***hæring**, an **ing**[2] derivative of OE **hær** 'a rock, a height, a ridge'. Some late forms do in fact show suffix confusion with OE **dūn** 'a hill', *Heryngdon* 1382, *heryng- haryngdon* 1418. Cf. HARRINGTON Lincs TF 3671. NbDu 112, PNE i.217, 218, Årsskrift 1974.45.

HERRISON Dorset SY 6794 → FORSTON SY 6695.

HERSDEN Kent TR 2062. Hasted (1790, iii. 608) refers to this place as *Hersing now usually called Haseden*. The original form Hersing, *Hersing'* from 1270 with varants *Hersingghe, Hersing(h)e, Hersenge, Heresinge* 13th, probably represents OE ***hyrsing**, Kentish ***hersing** 'horse place or pasture', < OE **hors** or ***hyrse** 'mare' + **ing**[2]. This and two other similarly named places, Hersing Marsh on the lower Medway near Iwade TQ 9067, *Hersing Marsh* 1272×1307, and the lost Harsyng Great Marsh in Cliffe TQ 7376, *Mochelhersinge* 1397, all refer to marshes where wild horses ran, as Chaucer reminds us in the *Reeve's Tale* where the untethered clerk's horse set out with 'wehee'.

Toward the fen ther wilde mares renne.

The alternative suggestion, 'place called after Heresa or Heresige' is less likely in view of the triple occurrence of the name. K 515, ING 185.

HERSHAM Surrey TQ 1164. 'Hæferic's river-bend land'. *Hauerichesham -y-* 1174, *Hav-* 1294–1370, *Haurecheues- Hauerkes- Heueriches-* 13th cent., *Haverycham* 1336, *-icham* 1516, *Harsham* 1535, *Hauersham* 1595, *Hersham* from 1599. OE pers.n. **Hæferīċ*, genitive sing. **Hæferīċes*, + **hamm** 1 or 3. Sr 97, ASE 2.40.

HERSTMONCEAUX ESusx TQ 6312. 'Herst held by the Munceus family' who are first associated with the place in the late 12th cent. *Herst Munceus, Monceus -ceux* 1304–17th, *Harsmounceux* 1492, *Horsemo(u)nsex* 16th. Earlier simply *Herst(e)* 1086–13th, *Hurst(e)* 1166–13th, 'the wooded hill', OE **hyrst**, SE dial. form **herst**. Also known as *(H)esthurst(e)* 'east Hurst'. The additions 'east' and 'Monceaux' are for distinction from HURSTPIERPOINT TQ 2816. Sx 479, L 197, SS 68.

HERTFORD HEATH Herts TL 3511. *Hertfordhethe* 1546. P.n. HERTFORD TL 3212 + ME **hethe**. Hrt 213.

HERTFORD Herts TL 3212. 'Hart ford'. *Herutford, ad Herutforda* [c.731]8th BHE, *(æt) Heorotforda* c.900 ASC(A) under year 673, c.925 ibid. under year 913, *(æt) Hertforda* c.1000 ASC(D) under year 913, *(æt) Heortforde* c.1100 ASC(E) under year 673, *Heort* 11th cent. coins, *Heortford* 1130, *Hertforde Burg'* 1086 etc., *Hartford* 1621, 1675, *Harforde* 1622. OE **heorot** + **ford**. Hrt 225 gives pr [ha:fəd].

HERTFORD NYorks SE 9980. 'Goat ford'. *Haverford* 1230, *Harford* 1299, cf. *Hertford Bridge* 1750. OE **hæfer** + **ford**. YE 118.

River HERTFORD NYorks TA 0780. A very early back-formation from HERTFORD SE 9980. *Haver(e)ford* [1172]c.1300–1395, *Harford* 14th, 1577. YE 6, RN 195.

HERTINGFORDBURY Herts TL 3112. 'Fortified place of the Heortfordingas, the people of Hertford'. *Herefordingberie* 1086, *Hertfordingber' -yng- -biry -bury* 1212–1511 with occasional variant *-inge-*, *Hertefordingbury* 1279, *Hartyngfordbury* 1507. OE folk-n. **Heortfordingas* < OE p.n. *Heorotford* + **ingas**, genitive pl. **Heortfordinga*, + **byriġ**, dative sing. of **burh**. Hrt 227.

HESCOMBE or HISCOMBE Somer ST 5116. 'Witch valley'. *Hascv̄be, Hasce- Hassecomba, Hetsecome -a* 1086, *Hececumb* c.1100, *Hetsecumb* c.1150, *Hatsecumbe* c.1155. OE **hæġtesse, hætse** + **cumb**. DEPN, L 93.

High HESKET Cumbr NY 4744. *Hesketh superior* 1541, *Over Hesketh* 1619. Adj. **high**, Latin **superior**, ModE **over** 'upper', + p.n. Hesket. 'High' for distinction from Low Hesket NY 4646. The two places were also known as *Better* and *Warre* 'worse' *Hesketh* 1485. Hesketh, *Hescayth* 1285–1346 with variants *-skaith -skeyth* etc., *Eskeyth in foresta* 1292, *Hesket* from 1380, is 'race course', ON **hestr** + **skeith**. This explanation has been challenged because of the frequency with which *skeith* is compounded in Swedish p.ns. with names of animals not used in racing. There *skeith* probably refers to a stretch of land along a boundary used for grazing or left uncultivated. Hesket lies on the boundary between Leath Ward and Cumberland Ward. In English names, however, the absence of compounds of *skeith* with any other animal suggests that the traditional explanation can stand, cf. HESKETH BANK Lancs SD 4423. Cu 199, NoB 1950.4–9, Jnl 10.26–39, SSNNW 133.

Low HESKET Cumbr NY 4646. *Netherheskett* 1530, *Heskett inferior* 1541, *Low Heskett* 1623. Adj. **low** (Latin **inferior**, ME **nether**) + p.n. Hesket as in High HESKET NY 4744.

HESKET NEWMARKET Cumbr NY 3438. *Hesket New Market* 1751. P.n. Hesket + **new market**. Earlier simply *Eskhevid* c.1230, *-heved -hevid* c.1250–1340, *Hesket* from 1523, 'head of the ash-tree copse', ON **eski** + ME **heved** (<OE *hēafod*). The reduction of ME *heved* > *hed* led to assimilation of the name to that of High HESKET Cumbr NY 4744. Cu 277, SSNNW 232, L 161.

HESKETH BANK Lancs SD 4423. P.n. Hesketh SD 4322 + ModE **bank** 'a hill'. Hesketh, *Eskehagh* 1259, *Eschayt* 1285, *Heschath* 1288, *Heskayt(h) -skeyt(h) -skaith -skeith* 1292–1332, *Hesketh* 1323, is the 'race-course', ON **hestr** + **skeith**. La 138, Jnl 10.29, SSNNW 133 (which prefers the sense 'strip of land along a boundary used for grazing' for *skeith* cf. High HESKET Cumbr NY 4744); Jnl 17.78 gives local pr 'Heskett'.

HESKETH LANE Lancs SD 6141. P.n. Hesketh + ModE **lane**. No early forms but presumably identical with Hesketh in HESKETH BANK SD 4423 and High HESKET HESKET Cumb NY 4744. Jnl 10.30.

HESKIN GREEN Lancs SD 5315. P.n. Heskin SD 5115 + ModE **green**. Heskin, *Heskyn* 1257–1332, *Heskin* 1497 is 'the marsh, the bog', OW **hescenn**. La 130, Jnl 17.74.

HESLEDEN 'Hazel valley'. OE **hæsel** + **denu**. L 98.

(1) ~ Durham NZ 4438. A modern mining settlement originally known as *Castle Eden Colliery* 1863 OS. Named from Monk HESLEDEN NZ 4537.

(2) Cold ~ Durham NZ 4146. *Cald(e) Hesil(le)den* 14th cent., *Coldhessylden* 1551×2, *Cold Hesleden* 1649. Northern ME adj. **cald** + p.n. *Heseldene* [c.1040]12th, c.1104, 1242×3, *Hesil(le)den* c.1200–1510. 'Cold' for distinction from Monk HESLEDEN NZ 4537. The settlement is situated on an exposed hill.

(3) Monk ~ Durham NZ 4537. *Munkheselden(e)* 1324, *Monk Hesleden* from 1666×7, *(Monk) Hesleton, Hazelden, Haseldon* 18th cent. ME **munk** + p.n. *Hœseldene* c.1123, *-den* 1172×95, *Haselden* c.1190×5, *Heselden(e) -ll-* 1154×66–1561. An estate of Durham Priory distinguished from Cold HESLEDEN NZ 4146 by the addition 'Monk'. NbDu 112.

East HESLERTON NYorks SE 9276. ME adj. *Est* 1259 + p.n. *Esrelton, Haslintonis* (sic) 1086, *Haslintunœ* [c.1160]15th, *Heslertun(e) -ton(e)* 1086–1531, the 'settlement at the place where hazels grow'. OE ***hæsling** varying with ***hæsler** + **tūn**. 'East' for distinction from West HESLERTON SE 9175. See also HESLINGTON SE 6250. YE 121, PNE i.219.

West HESLERTON NYorks SE 9175. *(West) Haslerton'* 1194×6, 1276, *Westhaselton'* 1196. ME **west** + p.n. *Heslerton(e)* 1086–1385, *Haslintune* [c.1160]15th, as in East HESLERTON SE 9276. YE 122, PNE i.219.

HESLEY HALL SCHOOL SYorks SK 6195. P.n. Hesley Hall + ModE **school**. Hesley Hall is p.n. *Heselay* 1217, 1318, *Heseley(e)* 1252–1415, *Hesley* from 1236, 'hazel-tree clearing or wood'. OE **hæsel** + **lēah**, + ModE **hall**. Nt 80.

Low HESLEYHURST Northum NZ 0897. ModE adj. **low** + p.n. *Heselyhyrst* 1268, *Hesilhurst* 1296 SR(p), 'the wooded hill where hazel-trees grow', OE ***hæsli(n)g** + **hyrst**. The reference is to the ridge of Wards Hill NZ 0796; Low Hesleyhurst is situated on lower ground to the NE. NbDu 112.

HESLEYSIDE Northum NY 8183. Either 'Hesley hill-side' or 'hill-side where hazel-trees grow'. *Hesleyside* 1279. The specific is uncertain, either p.n. **Hesley* 'hazel-tree clearing' identical with Hesley in HESLEY HALL SCHOOL SYorks SK 6195, OE **hæsel** + **lēah**, or, OE ***hæsling** 'place where hazels grow', + **sīde**. The house lies in the lee of a long hill-side. NbDu 112, PNE i.219.

HESLINGTON NYorks SE 6250. 'Settlement by the hazel-wood'. *Haslin- Eslinton* 1086, *Heseligtuna* c.1150, *Heselin(g)-* 1156–1349, *Heslington -yng-* from 1348. OE ***hæsling** + **tūn**. YE 273.

HESSAY NYorks SE 5253. 'Hazel-tree island'. *(H)esdesai* 1086, *Hessey(e) -ai -ay* 1100–1535, *Heslesaia -hai, Heselseia* 12th. OE **hæsel**, genitive sing. **hæsles**, influenced by ON **hesli**, + **ēġ**. The reference is to an area of dry ground in Marston Moor. YW iv.256, SSNY 157, L 35, 39.

HESSENFORD Corn SX 3057. 'Witches' ford'. *Heceneford'* c.1286–1338, *Hes(s)enford* 1310, 1384, *Hestonford* 1610. OE **hæġtesse**, genitive pl. **hæġtsena**, + **ford**. PNCo 97, Gover n.d. 220.

HESSETT Suff TL 9361. 'The hedge fold'. *heteseta, Eteseca* (for *Hece-*) 1086, *Hecesete* c.1095, *Hecheset* 1203, *Heggeset* 1225, *Hagge- Hegessete* 1254, *Hesset* 1610. OE **hecg(e)** + **(ġe)set**. DEPN.

HESSLE Humbs TA 0326. 'The hazel-tree'. *Hase* 1086, *Hesel(l)* 1156×7–1488, *Hessell -yll* 1285–1479, *Hezell* 1502. OE **hæsel** influenced by ON *hesli*. YE 215 gives pr [ezl], SSNY 139.

HEST BANK Lancs SD 4766. P.n. Hest + ModE **bank** 'a hill'. Hest, *Hest* 1177–1332 etc., *Heest* 1246, *Heast* 1557, is OE ***hǣst** 'undergrowth, brushwood'. La 185, PNE i.219.

HESTON GLond TQ 1277. 'Brushwood settlement'. *Heston(e)* from 1123×33, *Histon(e)* 1392–1592, *Heeston* 16th cent., *Heason, Hessen, Hesson* 17th cent. **hǣs** + **tūn**. Heston is about 4 miles from HAYES TQ 0980 'the brushwood'. Mx 25 gives local pr [hesən], Encyc 386.

HESWALL Mers SJ 2782. 'Hazel-tree spring'. *Eswelle* 1086, *Haselwell* 1190×1200–1535 with variants *-il-* and *-welle*, *Heselwall* c.1200–17th with variants *Hessel- Hesele-*, *Heswall* from 1520. OE **hæsel**, Mercian *hesel*, ON *hesli*, + **wella**, Mercian *wælla*. Che iv.276 gives prs ['hezwɔ:l], ['hezwəl] and dial. ['hesəl], L 31.

HESWORTH WSusx TQ 0019. 'Here's enclosure'. *Hereswerth* 1296 (p). Cf. *Hesworth meade* 1640, *Heshworth Street* 'H village' 1724. OE pers.n. *Here*, genitive sing. *Heres*, + **worth**. Sx 127, SS 79.

HETHE Oxon SP 5929. '(At) the uncultivated land'. *Hedham* 1086, *Heða* 1176, *Heth(e)* from 1201. OE **hǣth**. The 1086 form reflects dative pl. **hǣthum**. O 217, L 245.

HETHERSETT Norf TG 1505. 'The dwellers among the heather'. *Hederseta* 1086, 1254, *-sete* 1252, 1276. OE ***hǣddre** + **sǣte**. DEPN, PNE i.214.

HETHERSGILL Cumbr NY 4767. The forms are too late for certainty. *Hedris- Hederesgill* 1509, *Hethersgill* 1583. Possibly 'ravine of the tall deer', ME **he(h)der** (<OE *hēahdēor*), genitive sing. **he(h)deres**, + ON **gil**. But this is only a guess. Cu 91.

HETHPOOL or HEATHPOOL Northum NT 8928. 'Pool beside Hetha'. *Hetpol* 1242 BF, *Het(h)pol(')* 1249–96 SR, *Hethepol* 1542, *Hethpoole* 1806 Tomlinson 507. This appears to be p.n. Hetha as in Great Hetha, a hill rising to 1129ft. overlooking Hethpool at NT 8827, with a hill fort, + OE **pōl**. The meaning and origin of Hetha is unknown. Otherwise this might be the 'heath pool', OE **hǣth** + **pōl**. The pool at Hethpool still exists fed by Elsdon Burn. NbDu 108.

HETT Durham NZ 2836. 'Hat'. *Hett* from 1185, *Het* c.1200–1383, *Hette* mid 13th–1580. OE **hætt**. The village sits hat-like on top of a prominent hill. NbDu 113.

HETTON-LE-HOLE T&W NZ 3547. 'H in the hollow'. *Hetton in le Hole* 1507–1638. P.n. *heppedun* 1180×96–after 1250, *hepedun(a)* early 13th, *Eppedun* c.1195×1200, *Hep(p)edon(a)* 1187–1336, *Hepdon* 1360–1431, *Hepton(e)* 1323–1524, *hettún* c.1170–80, *Hetton(a)* c.1200×20–1675, 'dog-rose or rosehip hill', OE **hēope** + **dūn**, + Fr definite article **le** for *en le* + ModE **hole**. Hetton lies in a hollow at the edge of the East Durham Plateau. NbDu 113, L 157.

HETTON NYorks SD 9658. Probably 'settlement on the heath-land'. *Hetun(e) -a -ton(a)* 1086–1355, *Hetton* from c.1220. OE **hǣth** + **tūn**. YW vi.92 gives pr ['etn].

HETTY PEGLER'S TUMP Glos SO 7900. An important long barrow known as *the Barrow* 1683. Pegler is a local family name and one *Edith Pegler* is a party to the deed of 1683. The same surname occurs associated with a round barrow, *Pegler's Knob*, in Donnington. Western dial. *tump*, of unknown origin, is a regular local word for a barrow, cf. Norn's Tump, Avening, West Tump, Brimpsfield, Whitefield's Tump, Minchinhampton etc. Gl ii.254, Trans BGAS 79.47–8, 93.

HEUGH Northum NZ 0873. 'The hill-spur'. *Hough* 1276, *le Hogh* 1298, *le Hugh* 1346, *Heugh* from 1628. Mod dial. **heugh**, OE **hōh**. NbDu 113, L 167.

HEVENINGHAM Suff TM 3372. 'The homestead of the Hefeningas, the people called after Hefa or Hefin'. *Heuenigge- Euelincham* 1086, *Eueningeham* 1193×4, *Heueningeham* 1199×1216, 1225, *Heveningham* from 1200. OE folk-n. **Hefeningas* < pers.n. *Hefa(n)* or **Hefīn* + **ingas**, genitive pl. **Hefeninga*, + **hām**. Cf. EXENING TL 6165. DEPN, ING 130, Baron.

HEVER Kent TQ 4744. '(Settlement) at the high ridge'. *heanyfre* [814]?10th S 175, *Heu(e)re* 13th cent. OE **hēah**, definite form oblique case **hēan**, + ***ȳfre**, Kentish ***ēfre**. Originally a woodland pasture belonging to Bexley. The site is the tip of a hill-spur overlooking the valley of the river Eden. KPN 133, L 178.

HEVERSHAM Cumbr SD 4983. 'Heahfrith's homestead'. *Hefresham* c.1050, *Eureshaim* 1086, *Heversham* [1090×7]1308, 1160×70–1398, *Everesheim* late 12th–1429, *Heversham* from 1256, *Hersham* 1453, 1481, *Hearsham* 16th cent. OE pers.n. *Hēahfrith*, genitive sing. *Hēahfrithes*, + **hām** partially replaced by ON **heim**. We i.87, SSNNW 193.

HEVINGHAM Norf TG 1921. 'The homestead of the Hefingas, the people called after Hefa'. *(H)euincham* 1086, *Hebingham* 1200×30, *Hevingham* from 1242×3. OE folk-n. **Hefingas* < pers.n. *Hefa* + **ingas**, genitive pl. **Hefinga*, + **hām**. DEPN, ING 136.

HEWELSFIELD Glos SO 5602. 'Hygewald's open land'. *Hualdesfeld(e)* c.1145, 1276, 1355, *Huwaldesfeld(e)* 1227–1307, *Hewallesfeld* 1445, *Huelsfield* 1591. OE pers.n. *Hyġewald*, genitive sing. *Hyġewaldes*, + **feld**. The earliest forms are compounds in OE **tūn**, *Hiwoldestone* 1086, *Hiwaldestun* 12th, 'Hygewald's estate'. Gl iii.233, L 240, 243.

Copt HEWICK NYorks SE 3471. 'H with the peak'. *Coppedehaiwic* 1208. OE adj. **coppede** 'having a peak' referring to the hill on which the village stands + p.n. *(on) Heawic* [972×92]11th S 1453, c.1030 YCh 7, *Hewich -wych -wyk(e) -wik(e)* 1165–1685, the 'high or chief dairy farm', OE **hēah**, definite form **hēa**, + **wīċ**. 'Copt', *Cop(p)ed -id -yd* 1297–1523, *Copt* from 1546, for distinction from Bridge Hewick SE 3370, *oþer Heawic* 'the other H' [972×92]11th S 1453, *Hauui(n)c* 1086, *Hewik -wyk -wic(k)* 12th–1517, ~ *ad pontem* 1316, ~ *atte brig* 1309. YW v.156, 155.

HEWISH 'Estate of a size to support a family'. OE **hīwisc**. PNE i.248. Cf also HUISH.

(1) ~ Avon ST 4046. *Hiwis* 1198, 1223. DEPN.

(2) ~ Somer ST 4208. *Hywys* 1327. DEPN.

HEXHAM Northum NY 9364. 'The *hagustald* or Hagustald's homestead'. *Hagustaldes ham* c.1121 under year 685, *Hestoldes- Hastaldes- Hesteldesham* 12th cent., *Hextildes- -eldes-, Hextildisham* 13th cent., *Hextelsam* 1314, *Hexham* from 1362, *Hexam* 16th, 17th. The earliest reference in Bede is to the church of Hexham, *Hagustaldensis ecclesia* [c.731]8th. Other early spellings refer to the site of Hexham as an island of good land in moorland, *Inhagustaldaesae -e(n)sae* [c.710×20]11th Eddi (C), *Inhaegustaldesei, Inaegustaldesae, (H)agustaldesiae, Agustaldaesei -aesiae, Inhegustaldesiae* [c.710×20]c.1100 Eddi (F), *(He)agost(e)aldes ea* c.1000, *Hagustaldesea* c.1200. OE appelative **hagustald**, ONb **hehstald**, or pers.n. *Hagustald*, **Hehstald*, genitive sing. **hehstaldes**, **Hehstaldes*, + **hām**, earlier **ēġ** 'island, patch of good land in moorland, hill-spur'. The modern form of the name cannot derive directly from *Hagustald* but from its Old Northumbrian dial. by-form *hehstald* which became ME *hestold* and seems to have been influenced by the feminine pers.n. *Hextild*. It is uncertain whether we should regard the specific as a pers.n. or as an appellative, OE **hagustald**, 'the occupier of a *haga*, an enclosure or messuage, a young warrior'. *Hagustald*, which was used as a pers.n., was also later a legal technical term for a younger son of a noble house who under Germanic law was prevented from inheriting more than a small-holding or a *haga* or *hæġ*. A similar variation between the generics *ēġ* and *hām* occurs in the name of another early monastic site at LASTINGHAM NYorks SE 7290. NbDu 114, Jnl 8.22, Ilkow 165, Nomina 17.119–36, NQ 244.422ff.

HEXHAMSHIRE COMMON Northum NY 8853. *Hexhamshire Common* 1867 OS. P.n. Hexhamshire + ModE **common**. Hexhamshire is the name of the Liberty or Regality of Hexham, an area of jurisdiction which belonged to the Archbishop of York until its suppression by Act of Parliament in 1572. Its ultimate origins lay in the *regio* which Queen Aethilthryth gave to St Wilfrid c.671×3. By 1295 the Liberty comprised the modern parishes of Hexham, Allendale, Whitley and St John Lee with the chapelry of St Oswald, an area of approximately 92 square miles. SWAH 169–70.

HEXTABLE Kent TQ 5170. 'High post'. *Hagestapel(e)* 1203, *Heghstaple* 1327, *Exstapul* 1471, *Hackstaple* 1778. OE **hēah** + **stapol**. Probably a (guide-)post situated on high ground. PNK 50.

HEXTON Herts TL 1030. 'Heahstan's estate'. *Hege- Hegœstanestone* 1086, *Heh- Hecstanestun(e)* 12th cent., *Hextanstone* 1290–1301, *Hextoneston* 1303, 1317, *Hexton(e)* from c.1380, *Hexon* 1663. OE pers.n. *Hēahstān*, genitive sing. *Hēahstānes*, + **tūn**. Hrt 112.

HEXWORTHY Devon SX 6572. 'Hexten's enclosure'. *Hextenesworth'* 1317, *Hextenworth* 1344, *Hextworthy* 1379, *Hextesworthy* 1417, *Haxary* 1809. ME pers.n. *Hexten* (OE *Hēahstān*), genitive sing. *Hextenes*, + **worthy**. D 194 gives pr [hæksəri].

HEYBRIDGE Essex TL 8505. 'The high, i.e. chief bridge'. *Heaghbregge* c.1200, *He(e)bregg(e) -brigg(e) -brugg'* 1222–1428, *Heybrigge -brugg* c.1300, *Highbridge* 1594. OE **hēah** + **brycg**, Essex dial. form **brecg**. This name superseded earlier *Tidwolditone, Tidwoldintune* [c.940]13th S 453, *Tidwoldingtone -weldington* [946×c.951]14th S 1483, *Tidwoldingtune* 1000×2 S 1486, *Tydwoldyngton* 1316, 'estate called after Tidwald', OE pers.n. *Tīdw(e)ald* + **ing**[4] + **tūn**, cf. *Tydwaldinton Heybrug* 1236. Ess 303, Jnl 2.45.

HEYBRIDGE Essex TQ 6498. 'Bridge by the enclosure'. *Heybrigge* 1323, OE **(ġe)hæġ** + **brycg**. Ess 255.

HEYBRIDGE BASIN Essex TL 8707. P.n. HEYBRIDGE TL 8505 + ModE **basin** 'part of a river or canal widened for the loading and unloading of barges'.

HEYBROOK BAY Devon SX 4949. No early forms.

HEYDON Cambs TL 4340. Probably 'valley with an enclosure'. *Haindena, Haidenam* 1086, *Haiden(e)* 1086, 1222, *Haidenn* 1194, *Heidenñ* 1199, *Hei- Heyden(e)* 1202–1339, *-don(e)* from 1246×54. Probably OE **(ġe)hæġ** + **denu** but the specific could be OE **hēġ** 'hay'. OE **denn** 'swine pasture' does not seem to occur in Cambs. Heydon lies at the head of a well marked valley. The *hæġ* may well be a reference to Heydon Dyke or Bran Ditch, a defensive earthwork 3½ miles long from Fowlmire to Heydon lying across the line of the old Icknield Way, possibly of Anglo-Saxon date marking the boundary betwen the East and Middle Angles or the Angles and the Mercians. Pevsner 1954.201, 311, 326, Ca 374, Ess 529.

HEYDON Norf TG 1127. 'Hay hill'. *Heidon* 1196, *Heydon* 1242, 1253. OE **hēġ** + **dūn**. DEPN.

HEYDON HILL Somer ST 0327. Cf. *Heydon Down, Haydown Barrow House* 1809 OS. Heydon is probably 'hay hill', ultimately OE **hēġ** + **dūn**. But earlier forms are needed.

HEYDOUR Lincs TF 0039. 'The high pass'. *Hei- Haidure* 1086–14th, *Heydor(e)* 1202–1602, *Heydour* from 1431. OE **hēah** + **duru** 'a door'. The reference is to a gap in the ridge W of the village. Perrott 495.

Lower HEYFORD Oxon SP 4824. *Nether Heyford* 1246×7. ModE adj. **lower**, ME **nether**, + p.n. *(H)egford* 1086, *Hei- Heyford(') -fordia -e* 1172–1308, 'the hay ford', OE **hēġ** + **ford**. 'Lower' for distinction from Upper HEYFORD SP 4926. O 218, L 71.

Nether HEYFORD Northants SP 6558. *Nether Heiford* 1240. ME adj. **nether** 'lower' + p.n. *Haiford, Heiforde* 1086, *Hei- Heyford* from 1178, 'the hay ford', OE **hēġ** + **ford**, i.e. a ford across the Nene used particularly at hay-time. 'Nether' for distinction from Upper HEYFORD SP 6659 on the opposite side of the river. Nth 85 gives pr [hefəd], L 71.

Upper HEYFORD Northants SP 6659. *Heiford Superiore* 1220, *Overheiford* 1253. ModE adj. **upper**, ME **over**, Latin **superior**, + p.n. Heyford, as in Nether HEYFORD SP 6558. 'Upper' for distinction from Nether HEYFORD SP 6558 on the opposite side of the river. Nth 85.

Upper HEYFORD Oxon SP 4926. *Heyford Superiore* 1390, *Upper ~* early 18th. ModE **upper**, Latin **superior** + p.n. *(to) Hegforda*

[995]13th S 883, *Haiforde* 1086, c.1173×89, *Heyford* from 1220 as in Lower ~ SP 4824. Also known as *Heifort Waren* 1254×5 etc. with manorial affix from the count of *Warenne*. O 220.

Upper HEYFORD AIRFIELD Oxon SP 5126. P.n. Upper HEYFORD SP 4926 + ModE **airfield**.

HEYSHAM Lancs SD 4161. 'Homestead at or by the brushwood' or 'where brushwood is obtained'. *Hessam* 1086, *Hesseim, Heseym* 1094, *Hesheim* 1180×99, *Hesam* 13th cent., *Hesham* c.1190–1332 etc., *Heghsham* 1323, *Heisham* 1325–1667, *Hyseham* 1557, *Heyshame* 1627. OE ***hǣs*** 'brushwood' or ***hǣse**, ***hēse** 'land overgrown with brushwood', + **hām**. La 178, PNE i.218, Jnl 17.101, 18.17.

HEYSHOTT WSusx SU 8918. 'The corner of land by the heath, the heath-covered corner of land'. *Hethsete* c.1100, *-shete* 1332, *Hetschite, Hesshite* 12th, *Heys(c)hete* 1279–1442, *Heashot* 1675. OE **hǣth** + **sċīete**. Sx 22 gives pr [hiːʃət].

HEYTESBURY Wilts ST 9242. 'Heahthryth's fortified place'. *Hestrebe* (sic) 1086, *Hec- Heg- Heh- He(i)ch- Heitredeberi(a) -biri* 12th cent. *Heȝtredebir'* 1242, *Heghdresbury* 1299, *Heghtresbury* 1320, 1428, *Haitesbury* 1445, 1487, *Haitsbury* 1570, *Heytesbury* 1618. OE feminine pers.n. **Hēahthrȳth*, genitive sing. **Hēahthrȳthe*, + **burh**, dative sing. **byriġ**. The pers.n. occurs again in *Hegtredebrug* 1232, the name of a bridge on the boundary of Sherwood Forest. Wlt 168.

HEYTHROP Oxon SP 3527. '(At) the high village'. *Edrope* 1086, *Heðrop* 11th, *Hethrope -throp(p) -thorpe* 1223–1381, *Heythrop* from 1259. OE **hēah**, dative sing. **hēa**, + **throp**. O 271.

HEYWOOD GMan SD 8510. '(The) high wood'. *Hewude, Heghwode* 1246, *(del) Hewod(e)* 14th cent., *Yewood* 1865. OE **hēah**, definite form **hēa**, + **wudu**. La 62, L 228.

HEYWOOD Wilts ST 8753. 'The wood with or by an enclosure'. *Heiwode* 1225, *Heywode* 1289. OE **(ġe)hæġ** + **wudu**. Wlt 149, L 229.

HIBALDSTOW Humbs SE 9702. 'Holy place of Hygebald'. *Hiboldestou* 1086, *Hibaldestowa* 1088, *-stoua* c.1115. Saint's name *Hygebald*, genitive sing. *Hygebaldes*, + **stōw**. The original name of this place, *Cecesege -i* [c.1030]11th Secgan, 'Cec's island', was gradually replaced in usage after the burial here of St Hygebald in the late 7th cent. DEPN, *Stōw* 189.

HICKLETON SYorks SE 4805. 'Woodpecker farm' or possibly, 'Hicel's settlement'. *(Ch)icheltone* 1086, *Hikalton* c.1175, *Hi- Hykil- -ylton -tun* 12th–1519, *Hi- Hykelton(a)* 1200–1503, *Hickleton* 1587. OE **hicol** or pers.n. **Hicel* + **tūn**. YW i.85.

HICKLING Norf TG 4124. 'The Hicelingas, the people called after Hicel(a)'. *Hikelinga* 1086, *Hikeling(e)* c.1150–1361, *-linges* 1191–5, *Hiclyng* 1327. OE folk-n. **Hicelingas* < pers.n. **Hicel(a)* + **ingas**. Nf ii.100, ING 58.

HICKLING Notts SK 6928. 'The place called after Hicel(a)'. *Hikelinge* [978×1016]14th, *Hege- Heche- Echelinge* 1086, *Hickeling'* [c.1155]14th–*Hykkelyng'* 1340 with variants *Hikkel-* and *-ing(g)e*, *Hikelinga* 1184–*Hykelyng* 1325, *Hiclingg* 1235, *-ing'* 1242, 1251. OE pers.n. **Hicel(a)* + **ing**². Nt 235, ING 70.

HICKLING GREEN Norf TG 4023. P.n. HICKLING TG 4124 + ModE **green**. Earlier simply *Hickling* 1838 OS, *Green* 1845. Nf ii.102.

HICKLING HEATH Norf TG 4022. P.n. HICKLING TG 4124 + ModE **heath**, *Heath* 1845. Nf ii.102.

HIDCOTE BOYCE Glos SP 1742. 'H held by the Boyce family'. *Hudicote Boys* 1327, *Hidcott Boys* 1601, *Hitcote Boyce* 1688. P.n. Hidcote + family name Boyce. The feudal tenants were Ernolf de Bosco 1200 and Ernald de Bosco 1212, whose surname is from Latin *boscus*, the OFr equivalent of which was *bois* 'a wood'. Hidcote, *Hudicota* *[712]12th S 83, *Hedecote* 1086, 1404, *Hidecot(e)* 1200, 'Huda or Hydeca's cottage(s)', OE pers.n *Huda* or its derivative **Hydeca* + **cot**, pl. **cotu**. The manorial affix is for distinction from Hidcote Bartrim, 'H held by Philip Bertram (1221), *Hudicote Bertram* 1274–1302, ~ *Bartram* 1327. Earlier *Hidicote* 1086. Gl i.243,244.

East HIDE Beds TL 1317. *Esthide* 1247. Earlier simply *la Hide* 1197, 'the hide of land'. OE **ēast** + **hīd**. There was locally also a *Westhide* 1276. Bd 153.

South HIENDLY WYorks SE 3912. 'The hinds' glade'. *Hindeleia, Indelie* 1086, *Hi- Hyndelei -leg -ley(a) -lay* 13th–1536, *Suthindelai -hynde-* 12th, 13th cents. OE **hind**, genitive pl. **hinda**, + **lēah**. 'South' for distinction from Cold Hiendly SE 3714, *Hindelei(a)* 1086, *Cold(e)hind(e)lay -ley* 13th cent., so called from its bleak, exposed position, and Upper Hiendly SE 3913, *Parua Hyndelay* late 12th, *Overhindley* 1623. YW i.270, 271, L 205.

HIGHAM 'High homestead'. OE **hēah** + **hām**. PNE i.238.

(1) ~ Derby SK 3959. *Hehham* 1155, *Heg(h)am* 1199–1410, *Hygham* 1323, *Higham* 1577. Db 299.

(2) ~ Kent TQ 7171. A modern development on the main road from Gravesend to Strood at a place called *Gadshill* 1598 Henry IV pt.i, *Gad shill* 1623 ibid., 1819 OS, and *(on) Godes hylle* [10th]12th S 1457 ' god's hill'. OE **god** 'a pagan god', genitive sing. **godes**, + **hyll**. It is named from Higham at TQ 7174, still so called in 1805 OS, but subsequently replaced by the name CHURCH STREET. The original Higham is *heahhaam* c.767 S 31, *Heh ham* [774]10th S 110 *Hecham* 1086. Also known as Higham Upshire, *Upschire* 1240, *Upshere* 1435, 'the upper district'. OE **upp** + **sċīr**. PNK 116, KPN 51, 53.

(3) ~ Lancs SD 8036. *Hegham* 1296, 1324, *Heigham* 15th, 16th cent. This was a vaccary also known as *Highamboth* 1464. La 80. Jnl 18.17 suggests that the generic might be OE **hamm** in the sense 'cultivated plot in marginal land'.

(4) ~ Suff TM 0335. *Hecham* 1086–1378 including [1042×66]12th S 1051, *Heih(h)am* 1086, *Hegham* 12th–1303, *Heigham* 1384, *Hy-Higham* 1449, 1568. DEPN, Baron.

(5) ~ Suff TL 7465. *Heyham* 1275, *Hegham* 1303. DEPN.

(6) ~ DYKES Northum NZ 1375. *Higham Dykes* 1663. P.n. *Hey-Heiham* [13th]14th, *Hecham* 1289, + ModE **dyke(s)**. Higham is situated on a marked hill. NbDu 115.

(7) ~ FERRERS Northants SP 9668. *Heccham Ferrar'* 1279, *Higham Ferrers* 1675. Earlier simply *Hecham* 1086 and forms similar to those of Cold HIGHAM SP 6653. The town stands high above the Nene Valley. The manor was held by William de Ferrières, earl of Derby, in 1166. Nth 191.

(8) ~ GOBION Beds TL 1033. *Heygham Gobyon* 1291, *Higham Gubyns* c.1520. Earlier simply *Echam* 1086, *Heham* 1166. The Gobion family held the manor from the 12th cent. It stands on a marked spur of land overlooking a valley. Bd 153.

(9) ~ ON THE HILL Leic SP 3895. *Higham-on-the-hill* 1583. Earlier simply *Hec(c)ham* before 1173–1316, *Hei- Hey(g)ham* 1254–1428, *Hyham* 1269–1400, *Higham* from 1390. Lei 499.

(10) ~ WOOD Kent TQ 6048. P.n. *Hegham* 1327–38, *Heghham* 1332, + ModE **wood**. PNK 178.

(11) Cold ~ Northants SP 6653. *Colehigham* 1541, *Coldhigham* 1616. Earlier simply *Hecham* 1086–1205, *Hegham* 1316–38, *Heygham* 1350, *(Little) Hygham* 1401. It stands high and exposed. 'Cold' and 'Little' for distinction from HIGHAM FERRERS SP 9668. Nth 91.

(12) Lower ~ Kent TQ 7172. A railway halt at a place confusingly called *Upper Higham* 1805 OS. ModE **lower** + p.n. HIGHAM TQ 7171. It was *upper* in 1805 in relation to the original Higham at TQ 7174 subsequently known as CHURCH STREET, but *lower* in relation to the modern development at TQ 7171.

HIGHAMPTON Devon SS 4804. 'High Hampton'. *Heghanton* 1303, *Hehampton* 1318. ME adj. **hegh** (OE *hēah*) + p.n. *Hanitone* 1086, *Hantona* 1086 Exon, *Hanton* 1285, '(at) the high settlement', OE **hēah**, dative sing. definite form **hēan**, + **tūn(e)**. 'High' for distinction from OKEHAMPTON SX 5895, *Ochanton* 1284, probably 'settlement on the river Okement' misunderstood as if 'oak-tree settlement', ME *oke* 'an oak-tree' + p.n. Hampton. D 145, 202.

HIGH BEACH Essex TQ 4097. 'The high beech-tree'. *High Beech*

1734. Cf. *Highbeach-green* 1670, *High Beach Gr.* 1805 OS. Nearby is Beech Hill, *Bochel -hille* 1108×18, 1216×72, *Bokenhull, Bokynhelle* 1270, *Beak Hill* 1805 OS, OE **bōc** ' beech-tree', ***bōcen** 'growing with beech-trees'. Ess 28.

HIGHBRIDGE Somer ST 3147. 'The high, i.e. most important bridge'. *Highbridge* from 1324. ME **high** + **bridge**. Contrasts with BASON BRIDGE ST 3445. DEPN.

HIGHBROOK WSusx TQ 3430. *Highbrookes* 1613. ModE **high** + **brook**. Sx 273.

HIGHBURTON WYorks SE1913. 'High Burton'. *Highbyrton -birton* c.1442, 1504, *Highburton* from 1528. ME adjective **high** + p.n. *Burton(e)* 1086, OE **byrh-tūn** 'fortified enclosure'. The village stands on higher ground than KIRKBURTON SE 1912. YW ii.245, WW 151.

HIGHBURY Somer ST 6949. A modern mining settlement: no early forms.

HIGHCLERE Hants SU 4360. 'High Clere'. *Hegclere* 1269, *Hegheclere* 1280, *Hyeclere* 1380, Latin and French forms *Alta Clera* 1208, c.1270, *Hauteclere* 1284. Latin **altus**, feminine form **alta**, OFr **haut**, feminine **haute**, ME **hegh** + p.n. Clere as in BURGHCLERE SU 4761. The highest and most westerly of the three Cleres. Also known as *(to, æt) west clearan* [931]12th S 412, [955]12th S 565, and *Bisshopes Clere* 1320. It was one of the manors of the bishops of Winchester. Ha 56, Gover 1958.153, DEPN.

HIGHCLERE CASTLE Hants SU 4458. An 18th cent. mansion remodelled by Barry in 1839–42. P.n. HIGHCLERE SU 4360 + ModE **castle**. Pevsner-Lloyd 288.

HIGHCLIFFE Dorset SZ 2093. *High Clift* 1759, earlier *Black Cliffe* 1610. ModE **high** + **cliff** (dial. form *clift*). A modern parish carved out of Christchurch. Do 85, Ha 92, Gover 1958.223.

HIGH CROSS Hants SU 7126. No early forms.

HIGH CROSS Herts TL 3618. *Heyecrouch* 1360, *Hyecrosse* 1418. ME **heh** + **crouch, cross**. Hrt 199.

HIGH CROSS BANK Derby SK 2817. P.n. High Cross + ModE **bank** 'a hill-slope'. High Cross is probably a reference to *Lyntoncrosse* 1410, p.n. LINTON SK 2716 + ModE **cross**. Db 635.

HIGHER END GMan SD 5303 sc. of Orrell SD 5305. No early forms.

HIGHER TOWN Scilly SV 9315. The 'higher farm' of St Martin's. *Higher Town* 1748. ModE adj. **higher** + dial. **town** 'a farm'. 'Higher' for contrast with Middle Town SV 9216 and Lower Town SV 9116. PNCo 97.

HIGH FORCE Durham NY 8727. *(The) High Force* 1857 Fordyce, 1863 OS, a spectacular water-fall on the upper Tees. ModE **high** + N country dial. **force** 'a water-fall' (< ON *fors*). 'High' for distinction from the Low Force at NY 9027.

HIGHFIELD Northum NY 7391. 'The high open land or field'. *The Heefeld* 1695 Map, *Highfield* n.d. Map. ME **he** (OE definite form *hēa*) + **feld**.

HIGHFIELD T&W NZ 1458. No early forms.

HIGHFIELDS Cambs TL 3559. A modern development on the site of *High Field Wood* 1836 OS. The reference is to the high field of Caldecot. Ca 157.

HIGH GREEN H&W SO 8745. *High Green* c.1811 OS. ModE **high** + **green**. So called for distinction from Birch Green SO 8545, *Birch Green* 1831 OS, and *Kinnersley Green*, c.1811 OS, p.n. KINNERSLEY SO 8743 + ModE **green**.

HIGH GREEN Norf TG 1305. *High Common* 1838 OS. The name seems to have been changed under the influence of nearby *Whipple Green* 1838 OS.

HIGH GREEN SYorks SK 3397. *High Green* 1726. ModE **high** + **green**. YW i.250.

HIGHGREEN MANOR Northum NY 8091. P.n. *Highgreen* 1868 OS + ModE **manor**. Highgreen is ModE **high** + **green**.

HIGH HURSTWOOD ESusx TQ 4926. This is a misdivision due to a false etymology of *Hayhurst Wood* 1607, p.n. Hayhurst + ModE **wood**. Hayhurst, *Hayhurst* 1602, *Hyhayhurst* 'high Hayhurst' 1659, is 'wooded hill with or by the enclosure', ModE **hay** (OE *hæġ*) + **hurst** (OE *hyrst*). Sx 390.

HIGH KNOWES Northum NT 9612. 'The high hills'. *High Knowes* 1869 OS. ModE adj. **high** + N dial. **knowe** (OE *cnoll* 'a hill-top'). The reference is to a summit of 1294ft.

HIGH LANE GMan SJ 9585. Originally 'lane leading to the hill-spur'. *Ho Lane* 1690, *High Lane* 1842. ModE **hoe** (OE *hōh*) + **lane**. The *hōh* is the hill on which Disley stands at SJ 9685. Che i.284.

HIGH LANE H&W SO 6760. No early forms.

HIGHLEADON Glos SO 7723. 'Leadon held by the monks' sc. of St Peter's Gloucester. *Hi- Hyneledene* 13th–1541, *High Leaden als. Hyneleaden* 1629, *Hileadon* 1631. OE **hīġna**, genitive pl. of **hīwan**, 'a household or religious community' + r.n. LEADON. The affix is for distinction from UPLEADON SO 7527. Cf. HIGHNAM SO 7919. Gl iii.158.

HIGHLEIGH WSusx SZ 8498. Possibly 'the hillock-island'. *Hilegh* [c.700]17th S 1173, *Hyligh(e)* 1327, 1332, *Hylegh* 1428. OE ***hyġel** + **ēġ** subsequently misdivided and understood as if a name in *leigh* (OE *lēah*). The village occupies a slight rise in marshland. Sx 86, SS 65.

HIGHLEY Shrops SO 7483. 'Hugga's clearing'. *Hugelei* 1086, *Hug(g)el(eg)' -ley(e) -legh'* 1233–1428, *Hugley* 1399, *Higley* 1551–1833, *Highley* from 1535. OE pers.n. *Hugga* + **lēah**. The p.n. was reformed under the influence of popular etymology owing to its position near the end of a long ridge. Sa i.151.

HIGHMOOR CROSS Oxon SU 7084. *Highmoor Cross* 1830 OS. P.n. *Highmoor* 1830 OS + ModE **cross**. A dry upland moor unless *moor* obscures an original *mere*–name. O 78.

HIGHNAM Glos SO 7919. '*Hamm* held by the monks' sc. of St Peter's Gloucester. *Hi- Hynehamme* 12th–1326, *Hynehomme* c.1230–1355, *Hynham* 15201287, *Hinam* 1675. OE **hīġna**, genitive pl. of **hīwan** as in HIGHLEADON SO 7723 + p.n. *Hamme* 1086, OE **hamm** 'the water-meadow'. The parish includes low-lying meadows between the Severn and the Leadon. Gl iii.159.

HIGH NEBB SYorks SK 2285. ModE **high** + **neb** (OE *nebb*) 'a nose, a hill'.

HIGH PIKE Cumbr NY 3135. *High Pike* 1869 OS. A peak on Caldbeck Fells rising to 2152ft overlooking Low Pike at NY 3235. ModE **high** + dial. **pike** 'a mountain peak'.

HIGHSTED Kent TQ 9061. 'The high place'. *Hecsted(e)* 1197–9, *Hec(h)stede* c.1220, *Heg(h)- Heystede* 1254–1313. OE **hēah** + **stede**. PNK 265, Stead 90, 218.

HIGH STILE Cumbr NY 1614. 'High ascent'. *High Steel* 1783, *High Stile* 1784. Adj. **high** + **stile** <OE *stiġel* 'a steep slope'. The reference is to a mountain rising to 2644ft. Cu 356.

HIGH STREET Corn SW 9653. *High Street* 1748. ModE **street** 'street, row of houses'. An 18th cent. village high up on former downland. PNCo 97, Gover n.d. 424.

HIGH STREET Cumbr NY 4411, NY 4515. The Roman road from *Galava* in Ambleside to *Brocavum* (Brougham), which crosses the mountains of west Westmorland, Margary no.74. *High Street* 1793, 1823, presumably 'high Roman road'. Also simply *the Streete* 1650. Earlier *Brethstrette, Brestrett, Brethstrede* 1220×47, *Bredestrete* 1256, probably 'paved way of the Britons', OE *Brettas*, genitive pl. *Bretta*, + **strēt**, possibly influenced by ON **breithr** 'broad'. We i.21.

HIGH STREET Suff TM 4355. No early forms.

HIGH STREET GREEN Suff TM 0055. *High Street Green* 1837 OS.

HIGHTOWN Ches SJ 8761. The high part of the town of Congleton. Che ii.298.

HIGHTOWN Mers SD 3003. *High Town* 1702, *Hightown* 1842 OS. A single farmstead in 1842, the modern development began in 1965. Pevsner 1963.123, Room 1983.52.

HIGHWAY Corn SX 1453. *Highway* 1699. A hamlet on the main road to Boddinnick ferry. PNCo 97, Gover n.d. 270.

HIGHWAY Wilts SU 0474. 'The hay road'. *Hiw(e)i* 1086, *Hy- Hiwey*

-weie 1219–1347, *Heiewei* 1153, *Heywey(e)* 1156, 1321, 1346, *Hegh(e)weye* 1289–1345, *Hygheweye* 1332. OE **hēġ**, WS **hīeġ** + **weġ**. W 269.

HIGHWORTH Wilts SU 2092. 'The high enclosure'. *Hegworth* 1232, *Heworth* 1249, *Heyghewrthe* 1281, *Hyghworth* 1296. ME adj. **hegh** (OE *hēah*) + p.n. *Wrde* 1086, *Wortha* 1094, OE **worth**. The settlement lies on a prominent hill. Also known as *Hauteworth* 1231–1316, OFr *haut* 'high', and *Altaworth* 1305, Latin *alta* 'high'. Wlt 25.

HILBOROUGH Norf TF 8200. Short for Hilborough Well or Worth, 'Hildeburg's spring or enclosure'. *Hildeburhwella* 1086, *Hildeburwrthe* 1242, *Hilburgwrth* 1254, *Hilborough* 1824 OS. OE feminine pers.n. *Hildeburh*, genitive sing. *Hildeburge*, + **wella** and **worth**. DEPN.

HILDENBOROUGH Kent TQ 5648. 'Borough of Hilden'. *Hildenborough* 1389. P.n. Hilden + ME **burgh**. Hilden, *Hy-Hildenn'* 1240, *Hi- Hylden(e) -denn(e)* 1254–1458, is 'hill pasture'. OE **hyll** + **denn**. Expected Kentish dial. **hell** seems to occur once in *Heledenne* 1347. PNK 178.

HILDERSHAM Cambs TL 5448. 'Hildric's homestead'. *Hildricesham* 1086, *Hildrichesham -rikes-* 1086–1313, *Hi-Hyldresham -is-* 1239–1388, *Hy- Hildersham* from 1285, *Hilderson* 1698. OE pers.n. **Hildrīċ*, genitive sing. **Hildrīċes*, + **hām**. Ca 107, O li.

HILDERSTONE Staffs SJ 9434. 'Hildwulf's estate'. *Hildvlvestvne, Heldvlves tone* 1086, *Hildulveston* 1227, *Hildeleston* c.1250, *Hyldreston* 13th, *Hilderston* 1577. OE pers.n. *Hildwulf*, genitive sing. *Hildwulfes*, + **tūn**. DEPN, Horovitz.

HILDERTHORPE Humbs TA 1765. 'Hildiger or Hildigerth's outlying settlement'. *Hilgertorp -re-* 1086, *Hildredtorpe* [1100×35]19th, *Hildert(h)orp(e)* [1100×35]1246–1446, *Hilderthropp* 1600. ON masculine pers.n. *Hildiger* or feminine *Hildigerth* + **thorp**. Alternatively the pers.n. might be CG *Hildigard* or *Hildigar*. YE 102 gives pr [ildəθrəp], SSNY 60.

HILGAY Norf TL 6298. 'Island of the Hydlingas, the people called after Hydla'. *Hillingeiœ* [974]14th S 798, *(æt) Hyllingyge* 11th, *Huling- Hidlingheia* 1086, *Helingeia* 1103×6, *Helegeye* 1254. OE folk-n. **Hydlingas* < pers.n. **Hydla* + **ingas**, genitive pl. **Hydlinga*, + **ēġ**. Alternatively the folk-n. might derive from pers.n. **Hȳthla*. The reference is to a patch of dry ground in the fens beside the river Wissey. Cf. HILLINGTON TF 7125. DEPN.

HILGAY FEN Norf TL 5895. Cf. *Hilgay Sedge Fen, Hilgay West Fens* 1824 OS. P.n. HILGAY TL 6298 + ModE **fen**.

HILL 'hill'. OE **hyll**.

(1) ~ Avon ST 6495. 'The hill'. *Hilla* 1086, *Hill(e), Hyll(e)* from 1248, *Hulla -(e)* 1154×89–1690. OE **hyll**. The church stands on a prominent hill rising to 140ft. from the plain. Gl ii.233, L 170.

(2) ~ H&W SO 9848 → Lower MOOR SO 9847.

(3) The ~ Cumbr SD 1783. *the Hill* 1597, 1657. Cu 418.

(4) Upper ~ H&W SO 4753. *Upper Hill* 1832 OS. Earlier simply *(La) Hull(e)* 1200×13, 1332, 'the hill', Fr definite article **la** + ME WMidl dial. form **hull** (OE *hyll*). 'Upper' for distinction from Lower Hill SO 4654, *Lower Hill* 1832 OS. He 106.

HILLAM NYorks SE 5028. '(Settlement) at the hills'. *(on) hillum* [963]14th S 712, c.1050 YCh 9, *Hillum* 1070–1476, *-om(e)* 1320–1559, *Hillam* 1822. OE **hyll**, locative-dative pl. **hyllum**. The terrain here begins to rise in small hillocks towards the hillier region in the W. YW iv.43, L 170.

HILLBECK Cumbr NY 7915. 'Stream that flows from a cave-like recess or dark ravine'. *Hellebec -be(c)k* 12th–1352, *Helbek(e) -beck(e)* 1363–1777, *Hil(l)beck(e)* from 1700. Modern dial. **hell-beck** (<ON *hellir* 'cave' + *bekkr*). The reference is to Swindale Beck already said in Harrison's *Description of Britain* (1577) to have been 'called Helbecke bycause it commeth from the derne and elinge mountaines' – from the hidden and remote mountains. We ii.67, L 14.

HILL BROW WSusx SU 7926. No early forms. ModE **hill** + **brow**. A summit on the road from Petersfield to Haslemere S of Rake.

HILLDYKE Lincs TF 3447. 'The hill ditch'. *Hylldyk'* [before 1155]14th, *Hilledic* 1179, *-dich* 1215, *-dik* 1271, *Hyl(le)dyk(e)* 1282, *Hill Dyke* 1824 OS. OE **hyll** + **dīċ**. An ancient drain running from the SW corner of the East Fen at TF 3650 to the Frith Bank Drain at TF 3246, formerly navigable for boats. The reference is to the slightly raised ground on which Sibsey stands at TF 3551. Payling 1940.116, Wheeler 77 and Appendix I.20.

HILLEND Durham NZ 0135. *Hill End* 1862 OS. ModE **hill** + **end**.

HILLERTON Devon SX 7298. Partly uncertain. *(on) healre dune* [739]11th, *Helleredun* 1238, *Esterhellerdon, Hildredon* 1322, *Helliton* 1809. Unknown element + OE **dūn** 'a hill'. The unknown element may be an unrecorded OE **hēalor* 'a swelling, a hill', genitive sing. **hēalre*, related to OE *hēala* 'hernia, rupture' and *hēalede* 'hydrocelous'. It occurs again in 972 in S 786, *healre mere*. Alternatively *healre* could be the feminine dative sing. of an adjective **hēal* agreeing with *dūn* which varies between masculine and feminine gender: no such adjective, however, is known. D 360, PNE i.239.

HILLESDEN Bucks SP 6828. 'Hildi's hill'. *(æt) hildes dune* [949]16th S 544, *U- Ilesdone* 1086, *Hildesdon* 1184–1315, *Hillesdun* 1179, *-don* 1242, 1506, 1826, *Hillsdon* 1553, *Hillersden* 1755. OE pers.n. *Hildi*, genitive sing. *Hildes*, + **dūn**. Bk 62 gives pr [hilzdən], Nt addenda, L 150, Jnl 2.23.

HILLESLEY Avon ST 7689. 'Hild's wood or clearing'. *(on) Hil(l)eah* 972 S 786, *Hildeslei* 1086, *Hi- Hyldesleg' -le -le(i)gh -ley(e)* 1220–1492, *Hillesley* from 1439. OE feminine pers.n. *Hild*, genitive sing. *Hilde*, or *Hildi* masculine, genitive sing. *Hildes*, + **lēah**. Gl iii.30.

HILLFARRANCE Somer ST 1624. 'Furon's (estate called) Hill'. *Hull Ferun* 1253, *Hill farrence* 1610. Earlier simply *Hylle* [1066×86]n.d. ECW†, *Hilla* 1086, *Hulla* 1182, 'the hill', OE **hyll**. The estate was held in 1182 by Robert Furon whose surname is a byname from OFr *furon* 'pilferer, ferret'. DEPN.

HILLHEAD Devon SX 9053. 'Head of the hill'. *Hillhead* 1809 OS, ModE **hill** + **head**.

HILL HEAD Hants SU 5402. 'The head of the hill'. *Hill Head* 1810 OS. Cf. *North- Suthhulle* 1189×99. ModE **hill** + **head**. Gover 1958.30.

HILLIARD'S CROSS Staffs SK 1511. No early forms.

HILLINGDON GLond TQ 0882. 'Hilda's hill'. *Hildendune* 1078×85–1206, *Hillendon(e) -in-* 1086–1404, *Hillingdon* 1201. OE (masculine) pers.n. *Hilda*, genitive sing. *Hildan*, + **dūn**. Mx 41.

HILLINGTON Norf TF 7125. 'Settlement of the Hydlingas, the people called after Hydla'. *Helinge- (N)idlinghetuna* 1086, *Hillingeton* 1177, 1181, *Hellingeton* 1185. OE folk-n. **Hydlingas* < pers.n. **Hydla* + **ingas**, genitive pl. **Hydlinga*, + **tūn**. Alternatively the folk-n. may derive from pers.n. **Hȳthla*. Cf. HILGAY TL 6298. DEPN.

HILLMORTON Warw SP 5373. *Hullemortone* 1265–1357, *Hillemorton* 1309. Originally two separate places, *Hulle and Morton* 1247; Hill, *chapel of Hulle* 1247, is OE **hyll**, 'that part (sc. of Morton) standing on the Bank' (Dugdale 1756), and Morton, *Moritona* c.1080, *Mortone* 1086, *Morton* 1235, OE **mōr** + **tūn**, 'that below in a moorish ground' (idem). Wa 131.

HILL ROW Cambs TL 4475. 'Row of houses at Hill'. *Hilrowe* 1638. P.n. Hill + ModE **row**. Hill, 'the hill', is *(on) hylle* [970]13th S 780, *Hy- Hille* 1086–1298, *Helle* 1086, 1170, *Heilla, Heille, Hœlle* 1086, *Hulle* 1380, OE **hyll**. The reference is to the North Hill TL 4476, a ridge of higher ground (121ft.) which projects into the fens. Ca 233.

HILL ROW DOLES Cambs TL 4267. Originally Hill Doles, *Hulledoles* 1396, *Hilledoles* 1417, p.n. Hill as in HILL ROW TL

†The identification is tentative.

4475 + ME **dole** (OE *dāl*) 'a share' sc. in common meadow land in the fens. Ca 235.

HILLTOP Hants SU 4003. *Hill Top* 1810 OS. ModE **hill** + **top**.

HILMARTON Wilts SU 0275. 'The estate called after Helmheard'. *Helmerdingtun* [962]14th S 707, *Helmerintone, Helmertune, Adhelmertone* 1086, *Helmerton* 1198–1428, *Hilmerton* 1349, *Hilmarton* 1558×1601, *Hill Martyn* 1699. OE pers.n. *Helmheard* + **ing**[4] + **tūn**. Wlt 268 gives pr [hil'mɑirtən].

HILPERTON Wilts ST 8759. 'The estate called after Helpric'. *Help(e)rinton, Helperitune* 1086, *Helpring(e)ton* 1205, 1282, *Hulprincton(a) -ing- -yn(g)-* 1232–1416, *Hylperton -ir-* 1268, 1359, 1571, *Helperton* 1675. OE pers.n. *Helprīc*, side-form **Hylprīċ*, + **ing**[4] + **tūn**. Wlt 127.

HILSEA Hants SU 6603. Possibly 'holly island' *Helesye* 1236, *-eye* 1259–93, *Hulleseie* 1248, *Hul(e)sey(e)* 1281–1388, *Hullsea* 1620, *Hulsey alias Hilsey* 1702. OE ***hyles** + **īeġ** probably in the sense 'raised ground in marsh'. Ha 92, Gover 1958.25.

HILTON 'Hill settlement'. OE **hyll** + **tūn**.

(1) ~ Cambs TL 2966. *Hiltone* 1196 etc., *Hulton* 1227. Hu 261.

(2) ~ Cleve NZ 4611. *Hiltun(e) -tona* 1086, *Hilton* from 12th. YN 172.

(3) ~ Derby SK 2430. *Hiltune* 1086, *-ton(e)* from 1208, *Hulton(e)* 1208–1466, *Hylton(e)* 1250–1495. Db 568.

(4) ~ Shrops SO 7795. *Hulton'* 1255×6, late 13th., *Hilton* 1833 OS. The situation is a slope below high ground.

HILTON 'Settlement on a slope'. OE **helde** + **tūn**.

(1) ~ Cumbr NY 7320. *Hil- Hylton'* 1256, 1539–1664, *Helton* 1289–1777. Hilton is on the lower slope of Roman Fell. Also known as *Helton under Lyth* 1279, ON **hlíth** 'slope, hill-side', and *Helton Baco(u)n* 1292 'Hilton held by the Bacon family'. We ii.103, L 162, 166.

(2) ~ Dorset ST 7803. *Eltone* 1086, *Heltona* 1086 Exon, *Haltone* c.1086, *Halcton'* (possibly for *Haltton'*) 1210, *Helton(')* 1212–1774, *Hilton(')* 1280, 1795. Probably OE **helde** + **tūn**. The village lies at the foot of a road over high ground. Alternative possibilities are OE *h(e)alh* 'a nook' referring to the upper end of the valley in which the village lies, or *helde* 'tansy'. The modern form is due to association with ME *hill*. Do iii.208 gives pr ['hiltn].

(3) ~ Durham NZ 1621. *Helton(e)* 1180–1421, *heltun* 1185, *Hilton* from early 13th.

HIMBLETON H&W SO 9458. 'Hop farm'. *hymeltun* [816]11th S 179, [884]18th S 219, [896]17th ECWM 107, *(æt) Hymeltune* c.977 S 1373, *hymel tune* 11th S 1593, *Himeltvn* 1086, *Humelton(e)* 1240–1389, *Himbulton -ble-* 1564, 1570. OE **hymele** + **tūn**. In the [884]18th and [956]11th bounds of Himbleton (S 219, 633) occurs the name *hymel broc*, now the Bow Brook. Wo 135, Hooke 129, 133, 168.

HIMLEY Staffs SO 8791. 'The wood or clearing where *hymele* grows'. *Himelei* 1086, *Humelilega* 1185, *Humelele* 1242, *Himelegh* 1286. OE **hymele**, a plant-n., usually the hop-plant, + **lēah**. DEPN, Horovitz.

HINCASTER Cumbr SD 5184. 'Wild bird fortification'. *Hennecastr(e)'* 1086–1301, *Henkastre -castre -er* 1210×20–1609, *Hane-* 1260–1585, *Hinkaster* 1301, *Hin- Hyncastre -er* 1530–1777. OE **henn**, genitive pl. **henna**, + **ceaster**. No Roman remains have been noted here. For the name type, cf. CRASTER Northum NU 2519. We i.89.

HINCHINGBROOKE HOUSE Cambs TL 2271. *Hinchingbrooke House* 1835 OS. P.n. Hinchingbrooke + ModE **house**. Hinchingbrooke is *Hynchelingbrok* 1260, *Inchinbrok* 1378, *Hynchingbroke* 1535. The forms are too late for certainty but this might be OE *Hynċelinga-brōc* 'the stream of the Hyncelingas, the people called after Hyncela', OE folk-n. **Hynċelingas* < pers.n. *Hynċela* + **ingas**, genitive pl. **Hynċelinga*, + **brōc**. An unusual substitution of *f* for *h* occurs in the forms *Fynchyn(g)broke* 1402–78. Hu 261.

HINCKLEY Leic SP 4293. 'Hynca's wood or clearing'. *Hinchelie* 1086, *-lai -leia -lay* c.1130–1327, *Hy- Hinkele(e) -ley* 1207–1517, *Hy- Hinkle -c-* 1246–1395, *Hinckley* from 1578. OE pers.n. *Hynca* + **lēah**. Lei 501.

River HINDBURN Lancs SD 6366. 'Hind stream'. *Hyndborn, Hinburne* 1577. OE **hind** + **burna**. La 169, RN 197.

HINDERCLAY Suff TM 0276. 'The tongue of land growing with elder-trees'. *Hilderíclea* [978×1016]13th S 1219, *Hildercle* [978×1016]14th ibid., 1254, *-clea* 1086, *Hyldreclea* c.1095, *Hindercley* 1610. OE ***hyldre** + **clēa**. DEPN.

HINDERWELL NYorks NZ 7916. Probably 'spring where elder grows'. *(H)ildrewelle* 1086, *Hilderwell(e) -y-* 1139–1475, *Hinderwell -y-* from 1468. OE ***hyldre** + **wella**. An alternative possibility is 'Hild's well'. In this case the OE pers.n. *Hild*, genitive sing. *Hilde*, has been replaced by ON pers.n. *Hildr*, genitive sing. *Hildar*. The allusion would be to abbess Hild; there is a St Hilda's Well at Hinderwell but this may be a folk-etymology for OE *hyldre*. YN 138, SSNY 256.

HINDFORD Shrops SJ 3333. *Hy- Hinford* 1544–1602, *Hindford* from 1723. Gelling.

HINDHEAD Surrey SU 8835. 'The hind headland'. *Hyndehed* 1571, *Hynde head* 1610. ME **hind** + **hed** 'a head', probably referring to the high ground to the S of the Devil's Punch Bowl. Sr 214, L 160.

HINDHOPE LAW Northum NY 7697. *Hindhope Law* 1869 OS. P.n. Hindhope NY 7799 as in *Hindhope Burn* 1869 OS + N dial. **law** 'a hill' (OE *hlāw*). The reference is to a peak of 1394ft. Hindhope is the 'hind side-valley', ModE **hind** + N dial. **hope** (OE *hop*). The reference is to a side-valley of the river Rede.

HINDLETHWAITE NYorks SE 0851. '*Hindley* clearing'. *Hyndeletheyt* 1269, *Hyndelaythwayt* 1388, 1405. OE p.n. **Hindalea* 'the forest clearing of the hinds', OE **hind**, genitive pl. **hinda**, + **lēah**, + ON **thveit**. YN 254.

HINDLETHWAITE MOOR NYorks SE 0580. P.n. HINDLETHWAITE SE 0581 + ModE **moor** (OE *mōr*).

HINDLEY 'Hinds' wood or clearing'. OE **hind**, genitive pl. **hinda**, + **lēah**. L 205.

(1) ~ GMan SD 6204. *Hindele* 1212–46, *-leye -legh(e)* 1259–1346 with variant *Hynde-*, *Hindley* 1479. La 102, Jnl 17.58.

(2) ~ GREEN GMan SD 6303. P.n. HINDLEY SD 6204 + ModE **green**. A modern colliery settlement.

HINDLIP H&W SO 8758. 'Hinds' leap-gate'. *Hindehlyp, hindehlep* [966]11th S 1309, *Hindelep* 1086, *Hyndelepe* 12th–1327. OE **hind**, genitive pl. **hinda**, + ***hlēp**, **hlȳp**. Wo 139.

HINDOLVESTON Norf TG 6298. 'Hildwulf's estate'. *Hildolueston* [1047×70]13th S 1499, *Hidolfestuna* 1086, *Hildoveston(e)* 1206, 1254. OE pers.n. *Hildwulf*, genitive sing. *Hildwulfes*, + **tūn**. DEPN.

HINDON Wilts ST 9132. 'The religious or secular community's hill'. *Hendon* 1249, 1675, *Hynedon(e)* 1268–1356, *Hyndon* 1281. OE **hīga** 'a member of a household or family, a monk or nun', genitive pl. **hiġna**, + **dūn**. In nearby Fonthill a certain Æthelm Higa is mentioned 899×924 S 1445, possibly a member of the royal household at Wardour. Wlt 191.

HINDRINGHAM Norf TF 9836. 'The homestead of the Hindringas'. *Hindringham* [1047×70]13th S 1499, *Hindringa- Hindring(e)- Hidring- Indregeham* 1086, *Hindringeham* 1203, *Hindringham* 1204, 1254. OE folk-n. **Hindringas*, genitive pl. **Hindringa*, + **hām**. The folk-n. is unexplained unless it is possibly a compound formation of an OE *hinder* 'behind' in the sense 'the people who live behind' sc. the hills. But this is very uncertain. DEPN, ING 136.

HINGHAM Norf TG 0202. 'The homestead of the Heingas'. *Hengeham* [c.1000]12th, *Hinicham, Haincham, Hincham Regis, Ahincham* 1086, *Hengham* 1158–1242×3, *Heingeham* 1173–1200 etc. OE folk-n. **Hēingas* < **Hēahingas*, genitive pl. **Hēinga*, +

hām. The folk-n. **Hēahingas* is either 'the people called after Hēaha or Hega', OE pers.n. *Hēaha* or *Hega* + **ingas** or possibly 'the people who live at or by the high place', OE adj. **hēah** + **ingas**. The village stands on high ground and contrasts with nearby DEOPHAM TG 0400 'the homestead at Deop, the deep place'. DEPN, ING 136.

HINKFORD Essex. A lost hundred name. The 'ford of the Hethningas, the people called after Hethin'. *Hidingfort -forda, Hiding(a)forda -fort, Hiding(h)a- Hedingfort, Hidingh(e)- Hidingeforda* 1086, *Hainge- He(y)ing(e)ford* 1167–1235, *Hynkford* after 1420. OE folk-n. **Hethningas* < pers.n. *Hethīn* or *Heddīn* + **ingas**, genitive pl. **Hethninga*, + **ford**. The hundred meeting-place was at Castle HEDINGHAM TL 7835, the 'homestead of the Hethningas'. An alternative suggestion is folk-n. **Hȳthingas* 'the people of the hithe or landing-place' or 'called after Hytha', OE **hȳth** or pers.n. **Hȳtha* + **ingas**. Ess 405, ING 121.

HINKLEY POINT POWER STATION Somer ST 2146. P.n. Hinkley Point + ModE **power station**. Hinkley Point, *Inkley Point* 1809 OS, is unexplained p.n. + ModE **point**.

North HINKSEY Oxon SP 4905. *Northengestes'* 1241, *Northengesteseye* 1284, *Northchenxtesey* 1316, *Northenxeye* 1327, *Northenxy* 1509×10. ME adj. **north** + p.n. *Hengestesie* [821]12th S 183, *-ieg* [821]13th ibid. *(ad) Hengestesige* [955×7]12th S 663, *-eia* 1205×21, *Hengseya* 1233×4, *Henxeye* c.1250, 'Hengest's or the stallion's island', OE pers.n. *Hengest* or **hengest**, genitive sing. *Hengestes*, **hengestes**, + **īeġ**. Brk 450, L 39.

HINSTOCK Shrops SJ 6926. 'The dependent settlement of the domestic servants'. *Hi- Hynestok'* 1242, *Hinestock* 1255, *Hynestok* 1281, *Hinstok* from 1282 with variants *-stoke -stock*. ME **hine** (OE *hīġna* genitive pl. of *hīwan* 'a household, a family and its dependants') + p.n. *Stoche* 1086, OE **stoc**. The reference of *hine* is unknown; it may be that the profits of the manor were assigned to the upkeep of the household servants of the barony of Wem to which the manor belonged. Another sense of *hine* is a religious community but there is no record of any ecclesiastical ownership of Hinstock. Sa i.152, Studies 1936.29.

HINTLESHAM Suff TM 0843. 'Hyntel's homestead'. *Hintlesham* from 1086 including [1035×44]13th S 1521, *Huntlesham* 1168–1364. OE pers.n. **Hyntel*, genitive sing. **Hyntles*, + **hām**. DEPN, Baron.

HINTON '(At the) high settlement'. OE **hēah**, definite form dative case **hēan**, + **tūn**. PNE i.238.

(1) ~ Avon ST 7376. *Heanton* 1216×72, *Henton(e)* 1256–1639, *Hi- Hynton* from 1572. The reference may have been to the 'enclosure' of the Iron Age hill-fort on Hinton Hill at ST 7476. Gl iii.49.

(2) ~ Hants SZ 2195. *Hentune* 1086, *Henton(e)* 1242–1327. Also known as *Henton Aumarle* 1379, ~ *Damarle* 1412, *Hynton Admyral* 1592, 'H held by the Albemarle family'. Reginald de Albemara held the manor in 1242. Ha 93, Gover 1958.222.

(3) ~ AMPNER Hants SU 5927. 'H held by the almoner' sc. of St Swithun's priory, Winchester. *Hinton Amner* 13th. P.n. *(to) heantune* [1045]12th S 1007, *Hentune* 1086, *Hyentona* c.1170, 1180, + OFr **aumoner**. Ha 93, Gover 1958.71, DEPN.

(4) ~ BLEWETT Avon ST 5956. 'H held by the Blewett family'. *Hentun Bluet* 1246, 1288. Earlier simply *Hantone* 1086. Blewett is a French family name from the Fr nickname *bleuet, blouet* 'bluish'. DEPN.

(5) ~ CHARTERHOUSE Avon ST 7758. 'H with the charterhouse'. *Henton Charterus* 1273. Earlier simply *Hantone* 1086, *Henton* 1212. The charterhouse or Carthusian priory of Locus Dei ('the place of God') was founded here in 1232 by the widow of William Longespee, earl of Salisbury, to whose family the estate belonged in 1086. DEPN, Pevsner 1958N.204.

(6) ~ ST GEORGE Somer ST 4212. 'St George's H'. *Hentun Sancti Georgii* 1246, *George Henton* 1610, *Hinton St George* 1811 OS. P.n. Hinton + saint's name *George* from the dedication of the church. Earlier simply *Hantone* 1086, *Heanton* 1219, *Hyneton* [late 12th, 1279×80]14th. DEPN.

(7) ~ ST MARY Dorset ST 7816. 'St Mary's H'. *Henton' abbatisse Sancti Edwardi* 'the abbess of St Edward's H' 1212, *Henton Mary* 1535, *Hynton Mare* 1509×47, ~ *Marye* 1627. Earlier simply *(at, to) Hamtun(e)* (sic ? for *Hain-*) [944]15th S 502, [958]15th S 656, *Haintone* 1086, *Henton(e)* 1244–1547. The village occupies a 300ft. hill overlooking the r. Stour. Do iii.159.

(8) ~ WALDRIST Oxon SU 3899. 'H held by the Valery family'. *Henton Walery* 1362, ~ *Walrushe* 1591, *Hinton Walrush* 1676, ~ *Waldridge* 1761–1897. P.n. *Hentone* 1086–1517 with variant *-ton(')* + manorial affix from the *de Sancto Waleric* family from St-Valery-sur-Somme who held land here from the 12th cent. The bounds of the *Heantuningas* 'the people of Hinton' are mentioned in [958]13th S 654, *(on) heantunninga gemære*. Brk 391, Tengvik 113.

(9) Broad ~ Wilts SU 1076. 'Broad H'. *Brod(e)henton* 1319–48, *Brodhinton* 1339. ME adj. **brode** (OE *brād*) + p.n. *Han- Hentone* 1086, *Henton(e)* 12th–1316. 'Broad' for distinction from HINTON PARVA SU 2383 and Great HINTON ST 9059. The village stands on a swell of higher ground. Wlt 296.

(10) Great ~ Wilts ST 9059. *Greate Henton* 1547. ModE **great** + p.n. *Henton'* 1216. Situated on a spur of high ground. 'Great' for disinction from Broad HINTON SU 1076 and HINTON PARVA SU 2383. Wlt 142.

HINTON 'The Community's estate'. OE **hīna**, genitive pl. of **hīwan** 'a household, the members of a family', + **tūn**. Such a manor was set aside for the support of the domestic servants of a religious or other household. PNE i.247.

(1) ~ Northants SP 5352. *Hinton(e)* from 1086 with variant *Hyn-*. Nth 37.

(2) ~ Shrops SJ 4108. *Hen- Hunton* 1272, *Hynton* from 1555 with variant *Hin-*. Sa ii.38 says 'probably the estate of the religious community' with Mercian *hīona, hēona* for *hīna*, but the site is at the end of high ground and this could well be OE **hēah** 'high', definite form dative case **hean**. The name contrasts with nearby Halston SJ 4107, *Alston'* 1255×6, *Au(l)ston* 1552–1808, *Haulston(')* 1575, 1672, *Hallston* 1672–1724, 'Ealh's estate', OE pers.n. **Ealh*, genitive sing. **Eales*, + **tūn**, Sa ii.37.

(3) ~ IN-THE-HEDGES Northants SP 5536. *Hynton in the edge* 1549, *Hinton in the hedges* 1754. Earlier simply *Hintone* from 1086 with variant *Hyn-*. Nth 54.

(4) ~ MARTELL Dorset SU 0106. 'H held by the Martel family'. *Hineton Martel* 1226, *Hynton Martel* 1303. Earlier simply *Hinetone* 1086. The reference is probably to the former monastery of Wimborne Minster which held land here. The earliest reference is *to þære hina gemære* 'to the boundary of the community' sc. of Hinton [946]14th S 519. In 1212 the manor was held by Eudo Martel. Do ii.146.

(5) ~ ON THE GREEN H&W SP 0240. *Hinton on the Green* 1537. Earlier simply *Hinetvne* 1086, *Hyne- Hini- Hynyton(e)* 1154×89–1327. The estate belonged to St Peter's Abbey, Gloucester. The boundary of Hinton is referred to as *hinhæma gemæru* 'the boundary of the inhabitants of H' [1042]18th S 1396. Gl ii.45.

(6) ~ PARVA Wilts SU 2383. 'Little H'. *Little Hynyngton* 1311, *Lettell Hentune* 1663, *Little Hinton* 1828. ME adj. **litel** + p.n. *Hynyton* 12th S 312, *Hi- Hyneton(e)* 1242–1332, *Hynyn(g)ton* 1279–1311. Latin **parva** is a modern antiquarian sustitution for English 'Little', for distinction from Great HINTON ST 9059. Wlt 286.

(7) Cherry ~ Cambs TL 4857. *Cheryhynton* 1576. ModE **cherry** 'from the abundance of cherry trees formerly growing there' (Camden 1798) + p.n. *Hintone -a* 1086, *Hi- Hynton(e)* from 1218, *Hingtone* 13th, *Hynkton* 1420, *H(e)yneton* 1269, 1299. Ca 141.

HINTS Shrops SO 6175. 'The roads'. *Hintes* 1242, *Hyntes* 1292. PrW ***hïnt** 'a way, a path' + secondary ME pl. **-es**. DEPN.

HINTS Staffs SK 1503. 'The road'. *Hintes* 1086–1220 with variant

Hyntes, Hynce 1577. PrW ***hïnt** with AN **-s** or pl. **- es**. The village lies on Watling Street. DEPN, Duignan 80, Horovitz.

HINWICK Beds SP 9361. 'Hen farm'. *Hene- Haneuuic(h)* 1086, *Hanewich* 1204, *Henewic(h) -wik* 12th cent., *Hinewic -wike* 12th–14th cents., *Hinwick* 1235. OE **hænn**, genitive pl. **hænna**, + **wīċ**. Bd 39 gives pr [hinik], Ca lii.

HINXHILL Kent TR 0442. 'Stallion hut or shelter'. *Hengestesselle* [c.864]18th Hasted iii. 271, [?871×99]13th S 1652, *Haenostesyle* 11th, *Hangsel* c. 1140×44, *Heng(e)sell(e)* 1254, 1275, *Henxhulle -hille -helle* 1253×4, 1288, 1291. OE **hengest** + **ġesell** later replaced by **hyll**, Kentish **hell** 'a hill'. For another example of confusion between *ġesell* and *hyll* see WORMSHILL TQ 8857. PNK 412, DEPN, L 171, Cullen.

HINXTON Cambs TL 4945. Either 'stallion farm' or 'estate called after Hengest'. *Hestitone, Histetune -tone* 1086, *Hincstitona* 1086 InqEl, *Hengstiton* 1202, *Hinx(s)ton(e)* 1218–1548, *Henx- Heng(e)ston(e)* 1220–1384. Either OE **hengest** or pers.n. *Hengest* + **ing**[4], + **tūn**. Ca 94.

HINXWORTH Herts TL 2340. 'Stallion farm'. *Hain(ge)ste- Hamsteuuorde* 1086, *Heingstewurðe* 1176, *Hengst(e)worth(e)* 1196–1276, *Henxworth* 1516, *Hingstewurd* 1195, *Hinxt(e)worth* 1314, *Hinxworth* 1594. OE **hengest**, genitive pl. **hengesta**, + **worth**. Hrt 158.

HIPPERHOLME WYorks SE 1225. '(The place amongst the) osiers'. *Huperun* 1086, *Huperum* 13th cent., *Hi- Hyperum* c.1200–1419, *-om* 1321, *-ome* 1371, *Hi- Hyprum* 14th cent., *Hipperholme* 1562. OE ***hyper**, dative pl. ***hyperum**. YW iii.79 gives pr ['(h)iprəm].

HIPSWELL NYorks SE 1897. Possibly the 'stepping-stones spring or stream'. *Hiplewelle* 1086, *Hipp(e)leswell(e)* 12th, 13th cent., *Hippeswell(e)* 1184–1561. OE ***hyppels** + **wella**. YN 244, PNE i.276.

HIPSWELL MOOR NYorks SE 1497. P.n. HIPSWELL SE 1897 + ModE **moor** (OE *mōr*).

HIRST 'A wooded hill'. OE **hyrst**. L 197.

(1) ~ Northum NZ 2887. *Hirst* from 1242 BF, *Hyrst* 1268. NbDu 115, L 197.

(2) ~ COURTNEY NYorks SE 6124. 'H held by the Courtney family'. *Hirst(e) Court(e)nay* from 1303. P.n. *Hyrst* c.1030 YCh 7–1623, *Hurst(e)* 1279–1608, + manorial addition *Courtney* from the Courtney family in 1235 and 1303 and for distinction from Temple HIRST SE 6025. YW iv.21, L 197.

(3) Temple ~ NYorks SE 6025. ME manorial prefix *Temple* from 1310 + p.n. *Hyrst* c.1030 YCh 7–1411, *Hirst(e)* 1175×7–1609. 'Temple' referring to tenure of the estate by the Knights Templar is for distinction from HIRST COURTNEY SE 6124. YW iv.22.

HISCOTT Devon SS 5426. 'Stallion farm'. *Hengestecoth'* 1242, *-cote, Henxtecote* 1326, *Heyscote* 1346, *Eastacote* 1809 OS. OE **hengest** + **cot**, pl. **cotu**. D 121.

HISEHOPE RESERVOIR Durham NZ 0246. P.n. Hisehope + ModE **reservoir**. Hisehope, *Histeshope* c.1170, *histilhope* [1153×95]late 15th, *hyshope* late 15th., is a difficult name. There is also a Hisehope Burn, *Hystleyhop(e)burn'* 1260. The specific might be ON **hestr** 'a horse, a stallion' and so 'valley of the (wild) horses, valley where horses were coralled' or 'valley called *Histley*, the clearing of the horses'. NbDu 115, Nomina 12.72.

HISTON Cambs TL 4363. Unexplained. *Histuna -e -tone* 1086, *Hi- Hyston(e)* 1157×63–1434, *Heston'* 1165–1363, *Huston(a)* 1188×99–1251. Really we cannot say what this name is except a compound in OE **tūn** 'farm, village, estate'. The specific might be OE **hys(s)e**, genitive pl. **hyssa**, 'a son, a youth', also 'a shoot, a tendril'. Ca 153 gives pr [hisən].

HITCHAM Suff TL 9851. 'The fence homestead'. *He(t)cham* 1086, *Hec(c)ham* 1198, 1254, *Hitcham* 1610. **OE hæċċ, heċċ** or **hæċċe** + **hām**. DEPN.

HITCHIN Herts TL 1829. 'The Hicce'. *Hicca* (genitive pl.) [7th]c.1000 Tribal Hidage, *(ad) Hiccam* (accusative case for ? *Hiccum*) [944×6]13th S 1497, *Hiz* 1086, *Hicche* 12th–1428 including *[1062]13th S 1036 with variants *Hi- Hyche, Hiches, Hichene* 1147, *Hicchen* 1316–50, *Hytchyn* 1521. OE folk-n. *Hiċċe* of unknown origin, dative pl. *Hiċċum*. RN 197 suggests that this was an old name of the river HIZ (itself a late antiquarian revival of the DB spelling) possibly derived from W *sych* 'dry' with normal *h* for *s*. Norden in his *Discription of Hartfordshire* 1598 says the name derives from a famous wood called *Hitch*, but he may have confused the name with Wychwood, OE *Hwiccanwudu*. Hrt 8.

HITHER GREEN GLond TQ 3874. 'Nearer green' sc. to Lewisham. *He(a)ther, Hither Green* 18th cent. *Further Green* is recorded 19th cent. and preserved in Further Green Road. ModE **hither** + **green**. Encyc 395.

HITTISLEIGH Devon SX 7395. 'Hyttin's wood or clearing'. *Hiteneslei* 1086, *Hittenessleghe* 1285, *Hutteneslegh'* 1238–1391, *Hyttyngeslegh* 1433, *Hittesleigh* 1809 OS, *Hitsley oth. Hittisleigh* 1820. OE pers.n. *Hyttīn*, genitive sing. *Hyttīnes*, + **lēah**. D 441.

HIXON Staffs SK 0026. 'Hyht's hill'. *Hvstedone* 1086, *Huchtesdona* 1130, *Huhtesdon* 1239, *Huytesdon* 1289, *Hykstone* 1477, *Hickson* 1686, *Hixton* 1778. OE pers.n. *Hyht*, genitive sing. *Hyhtes*, + **dūn**. DEPN, Duignan 80, L 155, Horovitz.

River HIZ Herts TL 1833. *Hiz* c.1750. This is a late antiquarian adoption of the DB form of the name HITCHIN TL 1829, *Hiz*. Hrt 8.

HOADEN Kent TR 2659. 'Heath hill'. *Haddon'* 1240, *Hodon(e) -doun* 1254–1348. OE ***hāth** + **dūn**. PNK 530.

HOAR CROSS Staffs SK 1223. 'The grey cross'. *Horcros* 1230, *Hare- Horecros* 1236–68, *Whorecrose* 1513. OE **hār**, definite form **hāre**, + **cros**. DEPN, Duignan 80, Horovitz.

HOARWITHY H&W SO 5429. 'The white-beam tree'. *La Horewythy* 1272×1307, *Horewethye* 1519. ME **hore-withi**. He 100.

HOATH Kent TR 2064. 'The heath'. *Hoth(e)* 1278–1332, *La Hothe* 1289. OE ***hāth**. PNK 512, L 245.

East HOATHLY ESusx TQ 5216. *Esthothelegh* 1438, *Est Horthlyght* 1527. ME adj. **est** + p.n. *Hodlegh(e)* 1287–1332, 'heath wood or clearing', OE ***hāth** + **lēah**. 'East for distinction from West HOATHLY TQ 3632. Sx 400 gives pr [hu:əd'lai], L 205.

West HOATHLY WSusx TQ 3632. *Westhodleghe* 1288, 1361, *Westhothleg* 1347. ME adj. **west** + p.n. *Hadlega* 1121, *Hodlega* 1155, 'the clearing growing with heather', OE ***hāth** + **lēah**. 'West' for distinction from East HOATHLY TQ 5216. Sx 270, SS 66.

HOBARRIS Shrops SO 3178. Uncertain. *Owaris, Obaris* 1272, *Obbarrys* 1284, *Hobarrys* 1381, *Howbarris, Hobarrough* 17th, *Hopharris* 1847. It is suggested that Hobarris was a constituent of an area originally known as *Hop*, OE **hop**, 'the secluded valley', which was subsequently divided into Hobendrid, *Edretehop* 1086, 'Edric's (portion of) Hop', and Hobarris 'Hop belonging to Harris'. Sa i.153.

HOBHOLE DRAIN Lincs TF 3748. *Hobhole Drain* 1824 OS. P.n. *Hob Hole* 1799, 'goblin hole', ModE **hob** (ME *hob(be)*) + **hole**. This is the main drain of the East Fen running from Tynton TF 3962 to the r. Whitham at Hobhole TF 3639 constructed in 1801. Payling 1940.120, Wheeler 225 and Appendix I.20.

HOBLAND HALL Norf TG 5001. *Hopland Hall* 1837 OS. P.n. Hobland + ModE **hall**. The reference is to a late 18th cent. mansion destroyed by fire in 1961. No early forms for Hobland are recorded and it is uncertain whether this or *Hopland* is the earler form of the name. Hobland lies in the parish of HOPTON ON SEA TG 5300; the 1837 spelling may have been influenced by the parish name.

HOBSON Durham NZ 1755. No early forms.

HOBY Leic SK 6617. 'The farm or village at the spur of land'. *Hobie* 1086, *Hoby* from 1212, *Houbi -by* after 1150–1454,

How(e)by c.1291–1629. OE **hōh** + **bȳ**. Lei gives pr ['hu:bi], SSNEM 53.

HOCKERING Norf TG 0713. 'The hill people' or 'the hill place'. *Hokelinka* 1086, *Hokeringhes* 12th, *Hogring* 1205, *Hokering* 1206×7, *Okeringe* 1210×12, *Hokeryngges* 1336. OE folk-n. **Hoceringas* < ***hocer** + **ingas**, or p.n. **Hocering* < ***hocer** + **ing**[2]. The village lies close to the high ground of Hockering Heath at an elevation of 180ft. DEPN, Studies 1936.141, PNE i.255, ING 58.

HOCKERTON Notts SK 7156. 'The hill settlement'. *(H)ocretone* 1086, *-tun -ton(a)* c.1150–13th, *Hokerton* 1252–1327. OE ***hocer** + **tūn**. The village lies at an elevation of 122ft. SE of Hockerton Moor which rises to 221ft. at SK 7067. Nt 186.

Great HOCKHAM Norf TL 9592. *Great Hockham* 1838 OS. ModE adj. **great** + p.n. *Hocham* 1086–1254, *Hougham* 1204, 'homestead or enclosure where hock or mallow grows', OE **hocc** + **hām** or **hamm**. Alternatively the specific could be pers.n. *Hocca* and so 'Hocca's homestead'. Various species of mallow or hollyhock ('the holy hock') were cultivated for their medicinal properties. DEPN, Studies 1936.108, PNE i.255.

HOCKLEY Essex TQ 8293. 'Mallow wood or clearing'. *Hocheleia(m), Hacheleiā* 1086, *Hoccheleia* c.1150, *Hockele(g) -leye* 1198–1315. OE **hocc** 'hock, mallow', genitive pl. **hocca**, + **lēah**. *Hock* survives in the flower name *hollyhock* the 'holy hock'. Alternatively the specific might be the OE pers.n. *Hocca*. Ess 187, Studies 1936.108.

HOCKLEY HEATH WMids SP 1573. *Huckeloweheth* c.1280, *Hokeley heth* 1542. P.n. *Hockelouwe* 13th, *Ho(c)kelowe* 14th, 'Hocca's tumulus', OE pers.n. *Hocca* + **hlāw**, + ME **heth**. Wa 294.

HOCKLIFFE Beds SP 9726. 'Hocga's cliff'. *Hocganclife* 1015 S 1503, *Hocheleia* 1086, *Hoccliue* 1185 etc., *Hockley* 1576. OE pers.n. **Hocga*, genitive sing. **Hocgan*, + **clif**. The church stands on a marked rise NW of the present village. Bd 126, L 133, 136.

HOCKWOLD CUM WILTON Norf TL 7288. 'Hockwold with Wilton' a parish name. See WILTON TL 7288. Hockwold is 'the open upland ground where hock or mallow grows'. *Hocuuella* 1086, *Hocwood* 1198, *-wolde* 1242. OE **hocc** + **wald**. DEPN, Studies 1936.108, PNE i.255.

HOCKWOLD FENS Norf TL 7087. *Hockwold cum Wilton Fens* 1824 OS. P.n. Hockwold as in HOCKWOLD CUM WILTON TL 7288 + ModE **fen**.

HOCKWORTHY Devon ST 0319. 'Hocca's enclosure'. *Hocoorde, Hocheorde* 1086, *Hockeworthy* 12th–1378 with variant *Hokke-*. OE pers.n. *Hocca* + **worthiġ**. D 533.

HODCOTT Berks SU 4781 > WHITCOTT KEYSETT Shrops SO 2782.

River HODDER Lancs SD 7050. Uncertain. *Hodder* [930]14th S 407, 1577, *Hoder* 1226–1483, *Od(d)er* 16th. Possibly PrW ***hōth-** (W *hawdd* 'easy, prosperous, happy, pleasant') + ***duβr**, the 'pleasant stream'. La 139, RN 198, YW vii.129, GPC 1827.

HODDESDON Herts TL 3709. 'Hodd's hill'. *Hodesduna, Odes- Dodesdone* 1086, *Hodesdone* 1086–1322, *Hodgesdon* 1554, *Hodson* 1649, 1675. OE pers.n. or hill name **Hod*, genitive sing. **Hoddes*, + **dūn**. Hrt 228 gives pr [hɔdzdən], formerly [hɔdsən].

HODDLESDEN Lancs SD 7122. 'Hodde's valley'. *Hoddesden(e)* 1296–1311, *Hodelesdon* 1324, *Hodlesden* 1507. ME pers.n. *Hodde* later replaced by diminutive *Hoddel*, genitive sing. *Hoddes, Hoddeles*, + **denu**. La 75, L 98.

HODGE BECK NYorks SE 6294. *Hodge Beck* 1861 OS.

HOD HILL Dorset ST 8510. *Hod-hill* 1774. Hill name *Hod* + ModE **hill**. Hod, *Hod* c.1270, is either OE **hōd** 'a hood' in allusion to the shape of the 470ft. hill, or **hōd** 'a shelter' referring to the Iron Age hill-fort which crowns it. Do ii.117.

HODNET Shrops SJ 6128. 'The pleasant valley'. *Odenet* 1086, 1232, 1267, *Hodenet(h)* c.1090–1398, *Hodnet* from 1230. If the loss of the second *n* due to dissimilation is acceptable this is identical with the p.n. type seen in W Hoddnant S Glam SS 9768, Co Huthnance, Old Breton *Hudnant*, PrW ***hǭd** (W *hawdd*) + **nant**. Sa i.154, DEPN.

HODTHORPE Derby SK 5476. No early forms but cf. *Hob Wood* 1840 OS.

HOE Norf TF 9916. '(The settlement at the) hill-spur'. *Hou* 1086, *Ho* 1165, *Hó* 1166, *La Hoge* 1200. OE **hōh**, dative sing. **hō**. DEPN.

HOE GATE Hants SU 6213. *Hoe Gate* 1810. P.n. Hoe + ModE **gate**. Hoe, *Hov* 1086, *Ho* 1236, 1242, is 'the hill-spur', OE **hōh**, dative sing. **hō**. Ha 94, Gover 1958.51, DEPN.

HOFF Cumbr NY 6717. 'Heathen temple'. *Hofes* 1158×66, *Hoff(e)* 1251–1704. ON **hof**. The plural form *Hofes* probably refers to the two settlements at Hoff and Nether Hoff NY 6620. We ii.97.

HOG'S BACK Surrey SU 9248. A long ridge named from its profile. *Hogs Back* from 1823. Its earlier name was *Geldedon* 1195–1255, *Gilde- Gulde-* 13th cent., *montem de Geldeford* 1225, *monte de Guldedoune* 1300, *Gill Down* 1744, apparently from OE ***gylde** + **dūn**. The first element is the same as that in GUILDFORD SU 9949 nearby; its appearance in the hill name may be by ellipsis for **Geldeford-dūn* 'Guildford down'. Sr 8–10.

HOGGESTON Bucks SP 8025. 'Hog farm'. *Hochestone* 1086, *Hoggeston* from 1200. OE **hogg**, genitive sing. **hogges**, + **tūn**. Three and a half miles W lies Hogshaw SP 7322, *Hocsaga* 1086, *Hoggeshag* 1199, *Hogeshawe* c.1240, 'hog wood', OE **hogg**, genitive pl. **hogga**, + **sceaga**. Bk 67 gives pr [hogstən], 135, L 209.

HOGHTON Lancs SD 6126. 'Settlement at the hill-spur'. *Hoctonam* c.1160, *Hoc- Hogh- Houton* 13th cent., *Howghton* 1577. OE **hōh** + **tūn**. The reference is to the steep-sided ridge on which Hoghton Tower stands, a perfect example of a *hōh* or *heugh*. La 132, L 167, Jnl 17.75.

HOGNASTON Derby SK 2350. 'Settlement called or at *Hoccanæfesn*'. *Ochenavestun* 1086, *Hokenaston(e)* 1241–1446, *Hognaston* from 1433. OE p.n. **Hoccanæfesn* + **tūn**. *Hoccanæfesn* is 'Hocca's swine-pasture', OE pers.n. *Hocca*, genitive sing. *Hoccan*, + **æfesn**. Db 376.

HOGSHAW Bucks SP 7322 → HOGGESTON SP 8025.

HOGSTHORPE Lincs TF 5372. 'Hogg or hog's outlying farm'. *Hocgestorp* 1173×83, *Hogges-* 1180–98, *Hoggesthorp* 1242. OE pers.n. *Hogg* or **hogg** 'a hog', genitive sing. *Hogges*, **hogges**, + **thorp**. SSNEM 128.

HOLBEACH Lincs TF 3624. 'The hollow or concave ridge'. *Holebech -ben, Holobec(h)* 1086, *Holbecha* 1133×9, *-beche* 1240–1505 and [833]16th S 189, *-beth* [851]16th S 200, *Holebech(e)* 1169–1339 and [810]17th S 1189, *-bec -be(c)k(e)* 1206–87, *Holbeach* from 1554, *-bedge* 1652 *-bidge* 1674. OE **hol**, definite declensional form **hola(n)**, + **bæċ** partly replaced by ON **bekkr**. Holbeach has a raised site from which small streams flow away. Payling 1940.28, SSNEM 218, L 126.

HOLBEACH BANK Lincs TF 3527. P.n. HOLBEACH TF 3624 + ModE **bank**. The reference is to the ancient Sea Bank, *the sea dyke* 1471, *the Sea bank(e)* 1572, *Roman Bank* 1812. The bank is probably not in fact of Roman origin. Payling 1940.59, vi.

HOLBEACH DROVE Lincs TF 3212. *Magna drava de Holbech* 1331, c.1360, *Holbich Drove* 1652. P.n. HOLBEACH TF 3624 + ME **drave, drove** 'a road on which cattle are driven'. Payling 1940.32, PNE i.136.

HOLBEACH HURN Lincs TF 3927. 'H corner'. *Holebechehyrne* 1328, 1408, *Holbech Hiern* 1352, *Holbech(e)hyrn(e)* 1408–1525, *Holbeach Hirne* 1824 OS. P.n. HOLBEACH TF 3624 + ME **hirne** (OE *hyrne*) 'an angle, a corner, a spit of land', here referring to a corner in the ancient Sea Bank described in 1408 as 'a sea-girt island'. Payling 1940.32, xviii, Cameron 1998.

HOLBEACH MARSH Lincs TF 3829. *Marsh of Holebeche* 1313, *Holbeach Marsh* 1824 OS. P.n. HOLBEACH TF 3624 + ME **marsh** (OE *mersc*). The marsh was enclosed from the sea in the 17th cent. Payling 1940.32, Wheeler 123f.

HOLBEACH ST JOHNS Lincs TF 3418. P.n. HOLBEACH TF 3624 + saint's name *John*. An ecclesiastical parish was established here on what was once Holbeach Fen in 1867 with a church dedicated to St John the Baptist built in 1840. Pevsner 1964.574, Wheeler Appendix I.21.

HOLBEACH ST MARKS Lincs TF 3731. P.n. HOLBEACH TF 3624 + saint's name *Mark*. An ecclesiatical parish established on Holbeach Marsh in 1869 with a church dedicated to St Mark. Pevsner 1964.575, Wheeler Appendix I.21.

HOLBEACH ST MATTHEW Lincs TF 4132. P.n. HOLBEACH TF 3524 + saint's name *Matthew*. An ecclesiastical parish established on Holbeach Marsh in 1869 with a church dedicated to St Matthew. Pevsner 1964.575, Wheeler Appendix I.21.

HOLBECK Notts SK 5473. 'Stream in a hollow'. *Hollebec* 1227, *-beck* 1280, *Holebec -be(c)k* 1227–1315, *Holbek* 1280–1329, *How(e)be(c)ke* 16th cent. OE **hol** or **hola** + ON **bekkr**. Nt 83 gives pr [houbek], SSNEM 226.

HOLBERROW GREEN H&W SP 0259. *Holborough Green* 1831 OS. P.n. Holberrow + ModE **green**. Holberrow, *Holbarewe* 1275, *Holberwe* 1357, *Holborough* 1628, appears to be the 'hollow hill or barrow', OE **hol** + **beorg**, although a single spelling *Hulleberewe* 1327 may point to OE **hulu** 'a hovel' as the specific. The reference seems to be to the hill rising to 321ft. due S of the village. The specific is for distinction from INKBERROW SP 0157. Wo 327.

HOLBETON Devon SX 6150. 'Settlement in the hollow bend'. *Holbouton* 1229, *Houbouton* 1238, *Holbogatone* 1245×57, *-bogh(e)* 1289–1346, *Holboweton* 1278–9, *Holbeton* 1444. OE **hol** + **boga** + **tūn**. The village lies at the upper end of a small valley that makes a right-angle bend. D 275 gives prs [hɔb]- and [houbətən].

HOLBORN GLond TQ 3181. 'Stream in the hollow'. *Holeburne* 1086–1335 including *[951×9]13th S 1450, *Howeborne* 1551, *Holbourne by London* 1567. OE **holh** + **burna**. The hollow is still discernible in part by the course of Farringdon Road. The short lower navigable part of the stream was known as the Fleet (as in Fleet Street), *Fleta* 1110×13, *Flete* 1199, OE **flēot** 'a creek, an inlet'. The modern pronunciation of Holborn is seen in the pseudo-French form *Hautborne* 1703. Mx 113 gives pr [houbən], Encyc 395.

HOLBROOK '(The) brook in a hollow'. OE **hol**, definite form **hola**, + **brōc**. L 15.

(1) ~ Derby SK 3644. *Hol(e)broc -brok(e)* 1086–1306 etc. The stream is now called Bottle Brook. Db 570.

(2) ~ Suff TM 1636. *Holebroc* 1086–1327, *Holbrok* 1610. DEPN.

(3) ~ BAY Suff TM 1733. *Holbrook Bay* 1805 OS. P.n. HOLBROOK TM 1636 + ModE **bay**.

HOLBURN Northum NU 0436. 'The stream by the hill-spur'. *Ho(u)burn'* 1242 BF, *Houburn(e)* 13th cent., *Hulbourne* 1361, *Holbo(u)rne* 1539. OE **hōh** + **burna**. The reference is to an arm of Greensheen Hill on which the village stands. The spelling with *l* is due to association with OE *hol* 'a hollow' and the p.n. HOLBORN TQ 3181. Both *hōh* and *hol* become ME *hou* in compounds. NbDu 115.

HOLBURY Hants SU 4303. 'The hollow fortification'. *Holebiri* 1186, *Holebury* 1316. OE **hol**, definite form oblique case **holan**, + **byriġ**, dative sing. of **burh**. The reference is unknown. Ha 94, Gover 1958.198, DEPN.

HOLBURY PURLIEU Hants SU 4204 → DIBDEN PURLIEU SU 4106.

HOLCOMBE 'The deep or hollow coomb'. OE **hol**, definite form **hola**, + **cumb**. PNE i.257, L 92–3.

(1) ~ Devon SX 9574. *(æt) holacumba* 1050×73, *Holacumbe* 1069, *Holecome* 1086. The settlement lies at the mouth of a deep narrow coomb running down to the sea. D 493.

(2) ~ Somer ST 6749. *(la) Holecumbe* 1243, 1276, [n.d.] Buck, *Holcombe* 1610. DEPN.

(3) ~ ROGUS Devon ST 0518. 'H held by Rogo'. *Holecombe Roges* 1281. Earlier simply *Holecoma* 1086, *-cumbe* 1238. The earliest occurence is S 653, *(on) holancumbes land scare* [958]12th, 'to the boundary of the hollow coomb', and the following charter forms are all said to belong here, *Holancumb* [998]18th S 895, *æt Holancumbe* [1012]12th S 1422 and *æt Holacumbe* [1046]12th S 1474. The tenant in 1086 was Rogo or Rogus. D 535.

(4) Long ~ Somer SS 7636. No early forms. Seems to contrast with Shortacombe SS 7634.

HOLCOT Northants SP 7969. 'The cottage(s) in the hollows'. *Holecote* 1086–1305, *Halecote* 1086, *-a* 1183, *Holcot(e)* from 1286, *Hocott* 1590. OE **holh**, genitive pl. **hola**, + **cot**, pl. **cotu**. Nth 139 gives pr [houkət].

HOLDEN Lancs SD 7749. 'Hollow valley'. *Holdene* 1305, *Holden* 1332. OE **hol** + **denu**. La 91.

HOLDENBY Northants SP 6967. 'Haldan's village or farm'. *Aldenesbi* 1086, *Haldenebi -by* 1169–1317, *Haudenebi -esby* 1200–1, *Haldenby* 1298–1428, *Holmby al. Holdenbye* 1568. Pers.n. *Haldan* < ON *Halfdan* + **bȳ**. Nth 85 gives pr [houmbi], SSNEM 54.

HOLDERNESS DRAIN Humbs TA 1135. P.n. Holderness + ModE **drain**. Holderness is the name of the low-lying marshy peninsula between the North Sea and the Humber. *Heldernes(se) -re-* 1086–1188, *Holdernes(s -e)* from 1178, 'the hold's headland'. ON **hǫldr** 'an owner of allodial land, a higher yeoman', genitive sing. **haldar**, + **nes**. Cf. the Danish p.n. Holdnæs in Schlewig on the side of Flensborg Fjord, *Haldenes* 1209. Holderness Drain is all that is left of a former creek or channel extending north from the Humber called the SWINE TA 1335. YE 14, SSNY 96.

HOLDGATE Shrops SO 5689. Short for Castle Holdgate, 'Helgot's castle'. *Castrum Helgoti* 1109×18, *Castellum Helgod* c.1200, *Castelholegode* 1253, *Hologodecastel* 1199, *Holgate Castel* c.1540, *Howgate Castle* 1577, *Houldgate Cast* 1695, *Hol(e)god* 1294–1428, *Holgot(e) -cote* 1346–1431, *Holgate* 1549–1812, *Holdgate* 1732. Holdgate is simply the name of the DB tenant, *Helgot*, which replaced the original p.n. *Stantune* 1086, 'the stone farm'. The tendency to use the Norman pers.n. alone as the p.n. began in the late 13th cent. Surprisingly it underwent little folk-etymological reformation apart from the occasional *Hol(e)cote* spellings until the appearance of the form *Holgate* in the 16th cent. For a similar use of a pers.n. as a p.n. cf. GOODRICH H&W SO 5719. Sa i.155.

HOLDINGHAM Lincs TF 0547. 'The homestead of the Haldingas, the people called after Halda'. *Holing- Haldingeham* 1202, *Haldingham -yng-* 1202–1791, *Had(d)- Hawdingham* 17th cent. OE folk-n. **Haldingas* < pers.n. **Halda* + **ingas**, genitive pl. **Haldinga*, + **hām**. Perrott 286, ING 141.

South HOLE Devon SS 2219. *Southhole* 1364. ME adj. **south** + p.n. *Hola* 1086, *Hole* 1310, 'the hollow', OE **hol**. 'South' for distinction from *Northehoole* 1566 in the far N of Hartland parish. D 74.

HOLE IN THE WALL H&W SO 6128. *Hole in the Wall* 1831 OS. Possibly a translation of the DB place called *Turlestane* 1086 not otherwise identified, OE **thyrel-stān** 'pierced stone' as in THURLSTONE Devon SX 6742 referring to a coastal rock with a hole in it. The reference here is unknown – perhaps to a lost megalithic tomb. He 87.

HOLEMOOR Devon SS 4205. 'Hollow marshy spot'. *Holemore* 1623, *Holemoor* 1809 OS. OE **hol** + **mōr**. D 132.

HOLE PARK Kent TQ 8332. P.n. Hole + ModE **park**. Hole, *Hole* 1278–1347, *The Hole* 1813 OS, is OE **hol** 'a hole, a hollow'. PNK 353.

HOLFORD 'Ford in a hollow'. OE **hol**, definite form **hola**, + **ford**. PNE i.257.

(1) ~ Somer ST 1541. *Holeford* 1086, [1295] Buck, *Houlford* 1610. This is properly Holford St Mary (saint's name *Mary* from the

dedication of the church) for distinction from Rich's Holford ST 1433 and Treble's Holford ST 1533, *æt twam Holaforda* 'at two Holfords' 1066×86 ECW, *Holeford(e)* 1086. All three places are fords in deep valleys. DEPN.

(2) Rich's ~ Somer ST 1433. See under HOLFORD ST 1541.

(3) Treble's ~ Somer ST 1533. See under HOLFORD ST 1541.

HOLKER Cumbr SD 3677. 'Fen in a hollow'. *Holecher* 1086, *Holkerre* 1276, *-ker* from 1332, *Howker* 1577. OE, ON **hol** or OE **hola** 'a hollow' + ON **kjarr**. La 197 gives pr [hɔ:kə], SSNNW 135.

HOLKHAM Norf TF 8943. Possibly 'the homestead at the hollow'. *Holcham* 1086–1203. OE **holc** + **hām**. The reference may be to a lake or hollow in what later became Holkham Park, or possibly to Holkham Gap, a break in the coastal sand dunes. DEPN.

HOLLACOMBE 'Hollow or deep coomb'. OE **hol** + **cumb**. Identical with HOLCOMBE.

(1) ~ Devon SS 6311. *Holecumb(e)* 1235–61, *Hollowcombe* 1809 OS. D 374.

(2) ~ Devon SS 3703. *Holecome* 1086, *-cumbe* 1238, *-combe* 1299, *Hollacomb* 1809 OS. D 146.

HOLLAND 'Land on or by a hill-spur'. OE **hōh** + **land**. For *oi* spellings in this name see BzN 1981.304–7.

(1) ~ Lincs TF 2445. The name of the district between the Nen and the Welland as far S as Crowland and Tydd St Mary's, and between the Welland and the Witham W to South Forty Foot. *Hoylandia* 1060, *Hoilant -d* 1086–1190, *Hoy- Hoiland(e)* 1093–1280, *Holanda -(e)* c.1155–1418, *Holland* from 1266. One of the three ancient divisions of the county. The reference may be to the low inland hill-spurs around Donington TF 2035 and Kirton TF 3038 or to the particular hill-spur of Elloe between Moulton TF 3024 and Whaplode TF 3224, *Elleho* 1086, *Elneho* 1158, *Elho* 1160, 1165, *Ello(e)* from 1275, 'Ella's hill-spur', pers.n. *Ella*, genitive sing. *Ellan*, + **hōh**, which gave its name to Elloe Wapentake. Sps with *Hoy- Hoi-* probably represent a new analogical stem *hōȝ-* developed in the paradigm of *hōh* on the model of adj. *wōh - wōȝes - wōȝe*. Payling 1940.1, 9, SNP 2.35, L 167, 246, LNQ i.141–4, xiii.129, 170–4, Wheeler Appendix I.13, 28, BzN 1981.304ff, Brunner §242 A.1.

(2) ~ FEN Lincs TF 2346. *marisco de hoyland* [late 12th]13th, ~ *de Hoiland* 1202, *Holand(e)fen(ne)* from 1331, *mariscum inter Holandiam et Kesteven* 1331. P.n. HOLLAND TF 2445 + ME **fenne**. Payling 1940.7, Cameron 1998.

(3) Great ~ Essex TM 2119. *Hoylaund Magna* 1248–91, *My- Mikyll Holand* 1462×3, *Great Heylond* 1362, *moche holland* 1552. ME **great**, **michel**, **much**, Latin **magna** + p.n. *Holande* [c.1000]c.1125, *Holandā, Hoilandam* 1086, *Hoyland(e)* 1218–1328. Spellings with *oi, oy* are indications of vowel length. The modern spelling is probably due to the influence of the Dutch Holland and HOLLAND Lincs TF 2445. Ess 340, L 167, 248–9.

(4) New ~ Humbs TA 0823. A new settlement which developed in the 1830s at the terminus for passengers crossing the Humber by ferry to Hull at the end of the Manchester, Sheffield and Lincolnshire railway line. Named from the similarity of the low-lying surrounding countryside to the district of HOLLAND TF 2445 Lincs. Room 1983.78.

(5) Up ~ Lancs SD 5105. 'Upper Holland'. *Upholand* 1226, 1298. Adj. **up** + p.n. *Hoiland* 1086, *Holond* [1190]1268–1348, *Hollande* 1202, *Holand(e)* 1224–1430, *Holland* 1530. The *oi* of the Domesday Book spelling is an AN spelling for *ō*. 'Up' for distinction from Downholland as in DOWNHOLLAND CROSS SD 3607. Up Holland stands on the slope of a ridge and the reference is probably to land taken in at the limits of cultivable land on the edge of the high moor. La 105, L 247–8, Jnl 17.59.

HOLLAND-ON-SEA Essex TM 2016. A modern resort development taking its name from *Holland Hall* 1805 and Little Holland (lost in Clacton), *Parva Hoilande* 1212, Latin **parva** + p.n. Holland as in Great HOLLAND TM 2119. Ess 340.

HOLLESLEY Suff TM 3544. 'The wood or clearing of the hollow'. *Holeslea* 1086, 1177–91, *-leia -le(ga)* 1156–1336, *Hollesleye* 1568. OE **hol**, genitive sing. **holes**, + **lēah**. The reference may have been to the deep inlet of Hollesley Bay. DEPN, Baron.

HOLLESLEY BAY Suff TM 3944. *Hollesly Bay* 1805 OS. P.n. HOLLESLEY TM 3544 + ModE **bay**.

HOLLINFARE Ches SJ 6991. 'Ferry at Hollin(s), the holly-trees'. *Le Fery del Holyns* 1352, *(the) holynfeyr'* 1504, *Hollen Ferry* 1565. P.n. Hollins, OE **holeġn** 'a holly-tree', pl. **holeġnas**, + ME **ferry** (ON *ferja*) 'a ferry'. La 95.

HOLLINGBOURNE Kent TQ 8455. 'Stream of the Holingas, the people called after Hola or the people of the hollow'. *(to) holinganburnan* c.975 B 1322, *(æt) Holungaburnan* 1015 S 1503, *Holingaburnan* 1042×66 S 1047, *Holingeborne* 1086[†], *Holingaburna, Holingeburne* c.1100, *Holing(e)b(o)urn(e)* 13th cent. OE folk-n. **Hōlingas* < pers.n. *Hōla* or **hol** + **ingas**, genitive pl. **Hōlinga*, + **burna**. K 217, KPN 306, L 16.

HOLLINGTON Derby SK 2339. 'Holly farm'. *Holintune* 1086, *-ton(e) -yn-* 1141–1610, *Hollynton, Holyngton* 1364–1519. OE **holeġn** + **tūn**. Db 572.

HOLLINGTON ESusx TQ 7911. 'Holly settlement'. *Holintune, Horintone* 1086, *Holinton* 12th–1340, *Holyngton* 1390, *Hollyngton* 1535. OE **holeġn** + **tūn**. Sx 503, SS 76.

HOLLINGTON Staffs SK 0639. 'The holly enclosure'. *Holyngton* 13th, 1408, *Hollington* 1580. OE **holeġn** + **tūn**. DEPN, Horovitz.

HOLLINGWORTH GMan SK 0196. 'Holly enclosure'. *Holisurde* 1086, *Holinewurth'* early 13th, *Holinw(o)rth(e)* 13th–1860 with variant *-yn-*, *Holingworth* 1286–1359 with variant *-yng-*, *-ll-* c.1621. OE **holeġn** + **worth**. Che i.309.

HOLLINGWORTH LAKE GMan SD 9314. P.n. Hollingworth SD 9315 + ModE **lake**. The reference is to a modern reservoir. Hollingworth, *Holyenworth* 1278, *Hollinworth* 1582, *Hollingworth* 1843 OS, is 'holly enclosure', OE **holeġn** + **worth**. La 56.

HOLLINS GMan SD 8108. 'The holly-trees'. *Hollins* 1843 OS. Ultimately from OE **holeġn**.

HOLLINSCLOUGH Staffs SK 0666. 'Howel's ravine'. *Howelsclough* c.1395, 1574, *Howesclogh* 1472, *Hoarse Clough* 1583, *Hollesclough(e)* 1596–1755, *Hollins Clough* 1775. ME pers.n. *Howel* < W *Hywel*, genitive sing. *Howels*, + **clough** (OE *clōh*). The modern form with dial. **hollin** (OE *holeġn*) has been reformed under the influence of nearby *Hollins* 1614, 1840 in Derbys, 'the holly-tree'. Oakden, Db 373, Horovitz.

HOLLINWOOD Shrops SJ 5236. *Hollins Wood* 1833 OS. Mod dial. **hollin** (OE *holeġn*) 'holly-tree' + **wood**.

HOLLIWELL POINT Essex TR 0396. P.n. Holliwell + ModE **point**. Holliwell, short for *Halywell Marshe* 1540, *Halliwell mershe* 1546, *Hollowel* 1805 OS, belonged to Holywell priory in Shoreditch, Middlesex, which was named from a *fons qui dicitur Haliwelle* 'a spring called Haliwelle', the holy well or spring, OE **hāliġ** + **wella**. Ess 212, Mx 146.

HOLLOWAY Derby SK 3256. '(The) hollow way'. *Hol(e)way - wey(e)* 1200×10–1357, *Holewey(e)s* 1350–7, *-o-* 1357, *Holloway* from 1519, *-s* 1789, 1805. OE **hol**, definite form **hola** + **weġ**. L 83.Db 360.

HOLLOWELL Northants SP 6972. 'The spring in a hollow'. *Holewell(e)* 1086–1348, *Holowell* 1317, *Hollowell* 1701. OE **holh** + **wella**. Nth 72.

HOLLOWELL RESERVOIR Northants SP 6873. P.n. HOLLOWELL SP 6972 + ModE **reservoir**.

HOLLYBUSH H&W SO 7637. *Holly Bush* 1831 OS. ModE **holly** + **bush**.

HOLLY END Norf TF 4906. No early forms. Presumably short for 'holly end of Emneth township'. ModE **holly** + **end**.

[†]Another corrupt spelling occurs at DB 4d, *Hoilingeborde*.

HOLLY GREEN Bucks SP 7703 → PITCH GREEN SP 7703.

HOLLYM Humbs TA 3425. '(The settlement) at the hollows'. *Holam* 1086, 14th cent., *Holun* 1086, *Holume* 1260, *Holaim -eym -aym* 1235×9–1578, *Holem* 14th cent., *Holym* 1339–1534, *Hollym* from 1519. OE **hol**, dative pl. **holum**. The *-aim* spellings are due to false association with ON *heim*. YE 26.

HOLLYWOOD H&W SP 0877. *Holly Wood* 1831 OS. ModE **holly** + **wood**. Wo 354 cites the surname *del Holies, atte Holyes* c.1250.

HOLMBURY ST MARY Surrey TQ 1144.'St Mary's H'. P.n. *Hom(e)bery, Hombur'* 15th cent., *Holmbury* from 1610, unknown element + OE **burh**, dative sing. **byriġ**, 'fortification' referring to the Iron Age fort on Holmbury Hill + saint's name *Mary* from the church dedication. Sr 250, Thomas 1976.202.

HOLME ON **holmr** has various senses in p.ns. derived from the basic sense 'island', viz. 'raised ground in marsh, enclosure in marginal land, land in a river-bend, river meadow, promontory'. L 50–2.

(1) ~ Cambs TL 1987. *(Glatton cum) Hulmo* (ablative case) 1167, *Hulm(e)* 1217, 1244, *Holme* from 1252. The site is water-surrounded. Hu 188.

(2) ~ Cumbr SD 5279. *Holme* from 1086, *Hulme* 1608. We i.60 gives pr [hɔum], SSNNW 135, L 52.

(3) ~ Notts SK 8059. *Holm(e)* from 1203, *Howme* 1570. The sense is 'the raised ground in the marsh'. Nt 187.

(4) ~ CHAPEL Lancs SD 8728. P.n. Holme + ModE **chapel**. Holme, *Holme* from 1305, is ON **holmr** here in the sense 'river meadow'. La 84.

(5) ~ CULTRAM Cumbr NY 1750. 'Holme belonging to or called or at Cultram, the isthmus homestead'. *Holmcultram* [c.1180]1307, *Home Cultram* 1588. Earlier simply *Culterham* 1100×35. P.n. **Culdir* + OE **hām** with later ME **holme** 'water meadow' prefixed in Celtic word order. **Culdir* means 'narrow piece of land, isthmus', PrW ***cūl** 'narrow' + **tir** 'land'. Cu 288, GPC 629.

(6) ~ HALE Norf TF 8807. *Holmhel* 1267, *Holm Hale* 1838 OS. Apparently short for Holm and Hale. Holm, *Holm* 1086, 1254. The reference is to an island of dry ground in the marsh. The battle of the men of Kent and the Danes fought *æt þam Holme* 11th ASC(C) under year 902 has sometimes been identified with Holme Hale. For Hale → Holme HALE TF 8807. DEPN.

(7) ~ ISLAND Cumbr SD 4278. A peninsula in the Kent estuary.

(8) ~ NEXT THE SEA Norf TF 7043. *Holme next the Sea* 1824 OS. P.n. *Holm* 1035×40 S 1489, 1086, *Hulmum* 1216×72, + ModE **next the sea**. DEPN.

(9) ~ -ON-SPALDING-MOOR Humbs SE 8038. 'Raised ground in Spalding marsh'. *Holm(e) in Spaldingmore -yng-* from 1280. Earlier *Spaldigge- Spaldingholm(e)* 12th cent., and simply *Holm(e)* 1086. ON **holmr** + p.n. SPALDINGMOOR. YE 234 gives pr [ɔum], SSNY 96.

HOLME NYorks SE 1006. 'The holly-tree'. *Holne* 1086–1591, *Holm(e)* from 1316. OE **holeġn**. The form *Holme* is due to the assimilated *Holm-* of HOLMFIRTH SE 1508. YW ii.269.

HOLME WYorks SE 1006. 'The holly-tree'. *Holne* 1086–1591, *Holme* from 1316. OE **holeġn**. *Holne* has been assimilated to the *Holm-* of HOLMFIRTH SE 1508 two miles to the NE. YW ii.269 gives prs [oun, oum].

HOLME LACY H&W SO 5535. 'Ham held by the Lacy family'. *Homme Lacy* 1221, *Hamme Lacy* 1243, *Hum' Lacy* 1291, *Holme Lacy* 1831 OS. P.n. *Hammæ* 1085, *Hamme* 1086–1320, OE **hamm** 1 'land in a bend' sc. of the river Wye, + family name Lacy from Lassy, Calvados, Normandy. He 104, L 43, 50.

HOLME MARSH H&W SO 3454. *Holmes Marsh* 1833 OS. P.n. Holme, *Hulmo* 1229, *Hom* 1547×53, *Holmes* 1833 OS, OE **hamm** 2a 'promontory of dry land into marsh' + ModE **marsh**. He 138.

HOLME ON THE WOLDS Humbs SE 9646. *Holme super Wolde* 1578. This is not by origin a *holmr* name as the earlier forms make clear, *Hougon* 1086, *Hogum* 1100, *Haum* 1130×8, *Houm(e) -w-* c.1150–1371, *Houhum* 1194×1214, *-w-* 1332, '(the settlement) at the hills' or 'at the tumuli', ON **haugr**, dative pl. **(í) haugum** or OE **hōh**, dative pl. **hōgum**. Holme is situated on a spur of the Wolds commanding extensive views over low ground. *Haugr* could, however, refer to tumuli. YE 163 gives pr [oum], SSNY 97, L 52.

West HOLME Dorset SY 8885. *Westholn(e)* 1288, 15th cent., *Westholm(e)* from 1288. ME adj. **west** + p.n. Holme as in East Holme SY 8986, *Estholn(e)* 1288, 15th cent., *Estholm(e)* 1288, 16th cent., *East Holm* c.1586, earlier simply *Holne* 1086–1431, 'the holly-tree', OE **holeġn**. Do i.147, 82.

HOLMER H&W SO 5042. 'Pond in a hollow'. *Holemere* 1086, *-mer(e)* 1272, 1291, *Hol(e)mar(e)* 1309–34. OE **hol** + **mere**. He 105, L 27.

HOLMER GREEN Bucks SU 9096. *Homer Green* 1766. P.n. Holmer, *Holemer* 1154×89, *Holemerefeld* [c.1250]14th, *Holmere* 1339, 'pond in the hollow', OE **hol**, definite form **hola**, + **mere**, + ModE **green**. Bk 155 gives pr [houmə], L 27, Jnl 2.28.

HOLMES CHAPEL Ches SJ 7667. 'The chapel of Hulme'. *Holme chapell* 1400×5, 1463, *Hulmesschapell* 1483, *Holmes Chapel* from 1591. P.n. Hulme, genitive sing. Hulmes, varying with Holmes, + ME **chapel**. Hulme, *Hulm* 12th, *Hulme* 1216×72–1621, *Holm(e)* 1278, 14th, 1671, is the 'water meadow', ME dial. **hulm** (from late OE *holm*, ON *holmr*, rather than ODan **hulm*). Che ii.278, v.1.ii.237. Addenda gives 19th cent. local pr ['ouəmz].

HOLMESFIELD Derby SK 3277. 'Open land by *Holm*'. *Holmesfelt* 1086, *-feld* 1154×89–1275 etc., *Homesfeld* 1394, 1402, *Howne-Houn(s)feld* 16th cent. P.n. **Holm*, genitive sing. **Holmes*, + OE **feld**. **Holm* is taken to be OE **holm** in the sense 'mound, hill' as OSax **holm**. Holmesfield lies on a hill rising to 595ft. at Lidgate SK 3177. Db 264, AB 25.28, PNE i.258.

HOLMES WOOD Lancs SD 4317. *Holmes wood* 1571. P.n. Holmes SD 4318 + ModE **wood**. Holmes is the plural of ON **holmr** in the sense 'islands' in marshland. The surrounding land was formerly waterlogged, cf. the p.ns. Mere Brow SD 4218, Mere Farm SD 4217, Mere Side SD 4316 and Martin Mere SD 4214. Holmeswood rises to 50ft. above the flatts. La 137.

HOLMEWOOD Derby SK 4265. Uncertain. *Holm Wood* 1829. ModE **holm** (? OE *holeġn* 'holly', ON *holmr* 'island, enclosure in marginal land') + **wood**. Db 271.

HOLMFIRTH NYorks SE 1508. 'Holme wood'. *(le) Holnefrith* 1274–1324, *foresta de Holne* 1274, *Holm(e) Frithe(s) -fryth* 1302–1647, *-furthe* 1576, *-firth(e)* from 1598. P.n. HOLME SE 1006 + ME **frith** (OE *fyrhthe*). YW ii.289.

HOLMPTON Humbs TA 3623. 'Settlement at the shore-meadows'. *Ulmetun* 1086, *Holmeton(e) -tune -tona* 1086–1547, *Hompton, Humpton* 16th cent. ON **holmr** + **tūn**. YE 21, SSNY 113.

HOLMROOK Cumbr SD 0799. Possibly 'water meadow by a bend'. *Holmronk* (sic for *-crouk*) 1332, *Holmcrooke* 1569, *Holmrook* 1647. ME **holme** (ON *holmr*) + **kroke** (ON *krókr*). The evidence is somewhat scanty, but Holmrook lies by a bend in the river Irt. Cu 377.

North HOLMWOOD Surrey TQ 1647. ModE **north** + p.n. *Homwude* 1241, 1243, *(in bosco de) la Homwode* 1307, *Homewod* 1538, *Holm(e)wood(e)* 1597, 1604, *Homewood* 1765, either 'the home wood', ME **home** (OE *hām*) + **wode** (OE *wudu*), distinguishing the wood nearer the settlement of Dorking from another in the Weald; or 'the wood by an enclosure in marginal land', ME **homme** (OE *hamm 5a*) + **wode**. The modern spelling reflects reinterpretation of the specific as dial. *holm* 'holly' (OE *holēġn*). 'North' for distinction from South HOLMWOOD TQ 1745. Sr 271, L 228.

South HOLMWOOD Surrey TQ 1745. ModE **south** + p.n. Holmwood as in North HOLMWOOD TQ 1647.

HOLMWOOD STATION Surrey TQ 1743. P.n. Holmwood as in South HOLMWOOD TQ 1745 + ModE **station**.

HOLNE Devon SX 7069. 'Holly-tree'. *Holle* 1086 and 15th cent., *Holl* 1535, *Hole al. Holne* 1714, *Holne* from 1166. OE **holeġn**. D 301 gives pr [houl].

HOLNEST Dorset ST 6059. 'Holly wood'. *Holeherst* 1185–96, *Holenhurst(e)* 1268–88, *Holnest(e)* from 1278. OE **holeġn** + **hyrst**. Do iii.336 gives pr ['houlnest].

HOLSWORTHY Devon SS 3403. Probably 'Heald or Healda's enclosure'. *Haldeword* 1086, *Haldeurdi* 1086 Exon, *Haldewurth* 1228, *Halleswrthia -worth(e) -wordi* late 12th–1291, *Haldeswrthy -wrthi -worth* 1277–1389, *Holdesworthe* 1308, *Houlsworthy* 1675, *Healdesworthe* c.1320, *Hyalleswort hi* 1326. OE pers.n. *Healda* or *Heald*, genitive sing. *Healdes*, + **worthiġ**. An alternative possibility is OE ***heald** 'a slope', an unmutated form of OE *hielde*; Holsworthy occupies a hill-site 430ft. above sea level. D 146 gives pr [hɔlzəri].

HOLSWORTHY BEACON Devon SS 3508. *Holsworthy Beacon* 1809 OS. The site is a prominent hill 2 miles N of Holsworthy SS 3403 rising to 635ft.

HOLT 'A wood'. OE **holt**. L 196.

(1) ~ Dorset SU 0203. *Holt(e)* [1286]1313, 1307 etc., *le Holte* 1427. Short for *Winburneholt* 1185, 1225×6, *Wi- Wymburn(e)holt(e)* 1230–88, 'wood near Wimborne'. A royal chace and forest first mentioned in DB in the entry for Horton as *foresta de Winburne* 1086. Do ii.150.

(2) ~ H&W SO 8362. *holte* 1086. Wo 141.

(3) ~ Norf TG 0738. *Holt* from 1086. DEPN.

(4) ~ Wilts ST 8661. *Holt(e)* 1154×89, 1242 etc., *la Holte* 1257. Wlt 120, L 196.

(5) ~ END H&W SP 0769. 'The holt, i.e. wood end' sc. of Beoley. *Holt End* 1831 OS.

(6) ~ HEATH H&W SO 8163. *Holt Heath* 1832 OS. P.n. HOLT SO 8362 + ModE **heath**.

HOLTBY NYorks SE 6754. 'Holti's' or 'wood farm or village'. *Ho-Boltebi* (sic) 1086, *Holteby* 12th–1316. ON pers.n. *Holti*, genitive sing. *Holta*, or OE, ON **holt**, genitive pl. **holta**, + **bȳ**. YN 9, SSNY 30.

HOLTON 'Settlement on or by a hill-spur'. OE **hōh** + **tūn**. *Hol-* is a back-spelling for *Hou-* < *Ho^{u}h-* < *Hoh-* as a result of the vocalisation of pre-consonantal *-l-*.

(1) ~ CUM BECKERING Lincs TF 1181. P.n. *Houtune -tone* 1086 *-tuna* c.1115, *-ton'* 1198–1202 + p.n. BECKERING TF 1181. DEPN, Cameron 1998.

(2) ~ LE CLAY Lincs TA 2802. 'H in the clay land'. *Howlton in le Claie* 1615, *Holton le Clay* 1671. P.n. *Holtun -tone* 1086, *-ton* from 1565, *Houton(a)* c.1115–1602, *Hocton'* 1202, *Houcton* 1276, *Houghton* 1352–1779 + Fr definite article **le** short for *en le* and ModE **clay**. Li v. 109, Cameron 1998.

(3) ~ LE MOOR Lincs TF 0898. 'H on the moor'. *Houton in Mora* 1327, ~ *in le More* 1375. P.n. *Hoctvn(e)* 1086, *-ton'* 1168–77, *Houtuna -ton'* c.1115, 1181–1461, *Holton* from 1556 + Fr definite article **le** short for *en le* + ME **mōr**. Li iii.31.

HOLTON Oxon SP 6006. 'The settlement in the nook'. *(ondlong) Healhtunes (gemæres)* 'along the boundary of H' 956 S 587, *Halc- Halghton(')* 1192–1401, *Halton(a)* 1167–1510, *Haulton(')* 1297–1675, *Holton* 1822. OE **healh** + **tūn**. O 176.

HOLTON Suff TM 4077. 'Hola's estate'. *Holetuna* 1086, *-tun* 12th, *-ton* 1254, *Holton* from 1327. OE pers.n. *Hōla* + **tūn**. DEPN, Baron.

HOLTON Somer ST 6826. 'Nook settlement'. *(æt) Healtone* [959×75]13th S 1765, *(æt) Healhtune* [984×1016]12th S 1538, *Altone* 1086, *Halton* 1219, *Holton* 1610. OE **healh** + **tūn**. The village is situated in a valley in the upper reaches of the river Cam. DEPN, L 101.

HOLTON HEATH Dorset SY 9591. *Holton Heath* 1811. P.n. Holton + ModE **heath**. Holton, *Holtone* 1086, *Holton* from 1211 is the 'settlement in or near the hollow' or 'in or near the wood', OE **hol** or **holt** + **tūn**. Do i.163, 159.

HOLTON ST MARY Suff TM 0636. 'St Mary's H'. P.n. *Holetuna* 1086, *-ton* 1254, 1258, *-ton* 1270, *Holton* from 1336, 'Hola's estate', OE pers.n. *Hola* + **tūn**, + saint's name *Mary* from the dedication of the church. DEPN, Baron.

HOLWELL Dorset ST 7010. 'Ridge or bank in a hollow'. *Holewala* 1188, *-wal(e)* 1194–1431, *Holewell(a)* from 1206. OE **hol** + **walu** later confused with *wella*. The sense may have been 'sunk hedge, ha-ha'. Do iii.383 gives pr ['houlwel].

HOLWELL Herts TL 1663. 'Spring in the hollow'. *Holewelle* 1086–1303 including [969]12th S 774 and *[1066]12th S 1043. OE **hol** + **wella**. Hrt 12.

HOLWELL Leic SK 7323. 'The spring in the hollow'. *Holewell(e)* 1086–1428, *Holwell* from 1267. OE **hol** + **wella**. Lei 172, L 31.

HOLWELL Oxon SP 2309. 'The holy well'. *Haliwell(') -y-* 1222–1354, *Holewell(a)* [1189]1372, 1335, 1675. OE **hāliġ** + **wella**. O 325, L 31.

HOLWICK Durham NY 9027. 'Dairy-farm in the hollow'. *Holewyk -wic* 1235–1279×81, *Holwick* 1577. OE **hol** + **wīc**. Holwick lies in the valley of the Tees tucked into the foot of a dramatic cliff called Holwick Scar (ON **sker** 'a rock'). YN 307, DEPN.

HOLWORTH Dorset SY 7683. 'Enclosure in a hollow'. *(at) Holewertþe, (apud) Holewourthe* [843 for 943]17th S 391, *Holverde* 1086, *-worth(e)* 1280–1450, *Holworth* from 1275, *Helwarde* 1575. OE **hol** + **w(e)orth**. Do i.140 gives prs ['hɔləd] and ['houlwə:θ].

HOLY CROSS H&W SO 9279. *Holy Cross* 1831 OS. A cross-roads presumably once marked with a cross.

HOLY ISLAND Northum NU 1241, 1243. The alternative name for LINDISFARNE through association with St Cuthbert. *Halialand* [1093]c.1190, *Haliela(u)nd(e)* c.1150 RRS–1421, *Halyeland* 1296, 1305, *(the) Holy(e) Iland(e), Holly Elande* 16th., *Holy Island* from 1548, *Holly Ileand* 1715. Also known simply as *Eilande* [1084]c.1170 DEC. OE **hāliġ** + **ēġland**. The mainland estates belonging to the monastery on Holy Island were known collectively as Islandshire, *Ealondscire* 1099×c.1122 DEC, *Elandscira* c.1122 DEC, *Aelandscire* c.1124×8 DEC. NbDu 116.

HOLY ISLAND SANDS Northum NU 1042. *Holy Island Sands* 1866 OS. P.n. HOLY ISLAND NU 1241 + ModE **sand(s)**.

HOLYBOURNE Hants SU 7340. 'Holy stream'. *Haliborne* 1086, *-burna* 1167, *Hali- Halyburne -bo(u)rne* 1236–1551, *Hollyborne* 1554. OE **hāliġ** + **burna**. Gover 1958.104 give pr [hɔli-], DEPN, L 19.

HOLYMOORSIDE Derby SK 3369. '(Place) beside Holy Moor'. *Holy Moor-side* 1756. P.n. Holy Moor SK 3268 + ModE **side**. Holy Moor, *Howley More* 1584, 1634, *Hawley More* 1585, *Holly Moore* 1599, is possibly 'hill-clearing', ON **haugr** or OE **hōh** + **lēah**, to which **mōr** was added. There is a well-defined hill-spur rising to 994ft. at SK 3168. Db 223.

HOLYPORT Berks SU 8977. Probably 'filthy, muddy market town'. *Horipord'* 1220, *Holliporte, Hollyport* 1586–1790, *Holyport* from 1395. OE **horiġ** + **port**. Brk 45.

HOLYSTONE Northum NT 9502. 'The holy stone'. *Halistane* [1240]14th., *Hali- Halystan(e)* 1242 BF–1426, *Helistan* 1242 BF, *Halystone* 1539, *Holystone* 1724. OE **hāliġ** + **stān**. The reference is unknown, but may be to a huge overhanging rock in the ravine of nearby Dove Crag burn. Holystone has a tradition of religious association. Five barrows overlook the village and at the nearby St Ninian's or Lady's Well St Paulinus is reputed to have baptised 3000 converts. NbDu 116, Tomlinson 345, Dixon 274, 280.

HOLYWELL 'Holy spring or stream'. OE **hāliġ** + **wella**. PNE i.225, L 31.

(1) ~ Cambs TL 3370. *Haliewelle* 1086, *Hal(l)iwell -y-* 1231–1601, *Hollywell* 1600, c.1750. Hu 209 gives pr [hɔliwəl].

(2) ~ Corn SW 7638. *The Holy Well* 1663. A well in a cave on the

shore widely known for cures. The village is a 20th cent. holiday village. PNCo 98, Gover n.d. 369.

(3) ~ BAY Corn SW 7559. *Holywell Bay* 1813. P.n. HOLYWELL SW 7658 + ModE **bay**. Earlier *Holy Well Porth* 1795, dial. **porth** 'a cove', and *Porth Island* 1630, *Porth Reylen* 1637, *Porraylan* 1686, probably 'cove of the little estuary', Co **porth** + **heyllyn** (Co *heyl* + diminutive suffix *-ynn*) reinterpreted as if English *island*. PNCo 98, Gover n.d. 369.

(4) ~ GREEN WYorks SE 0819. P.n. *(de) Sacro fonte* before 1284, *Heli- Helywell(e)* 1285, 1298, Latin **sacer** + **fons**, OE **hǣliġ**, a mutated variant of **hāliġ**, + **wella**, + suffix *green* 1583. The reference is to a nearby spring dedicated to St. Helen. YW iii.50.

(5) ~ LAKE Somer ST 1020. 'Holywell stream'. *Holywell Lake* 1809 OS. P.n. Holywell + Mod dial. **lake** 'a stream' (OE *lacu*).

(6) ~ ROW Suff TL 7077. *Holywell Row* 1837 OS. P.n. Holywell + ModE **row**. Contrasts with nearby BECK ROW TL 6977 and WEST ROW TL 6775.

HOLYWELL Dorset ST 5904. Unexplained. *Ailwell* 1811 OS.

HOM GREEN H&W SO 5822. *Hom Green* 1831 OS. P.n. Hom, OE **hamm** 2a 'promontory of dry land into marsh', + ModE **green**.

HOMER Shrops SJ 6101. 'The pond in a hollow'. *Holmere* 1550 etc., cf. *Homer Common* 1833 OS. ME **hole** (OE *holh*) + **mere**. Gelling.

HOMERSFIELD Suff TM 2885. 'The open land of the Humber'. *Humbresfelda* 1086, *Humre- Humbresfeld* c.1130, *Humeresfeud -feld* 1227, 1254, *Humersfeld* 1254, *Homersfeld* 1524, 1610. R.n. Humber as in River HUMBER TA 2316 + OE **feld**. Homersfield is on the r. Waveney, an English r.n. Possibly *Humber*, an unexplained r.n., was the original name. Alternatively, 'Hunbeorht's open land', OE pers.n. *Hūnbeorht*, genitive sing. *Hūnbeorhtes*, + **feld**. DEPN and Baron give pr [hʌməzfi:ld].

HOMINGTON Wilts SU 1226. 'The estate called after Humma'. *Humming tun* [956]12th S 635, *Humitone* 1086, *Humintona -e -yn-* 1130–1310, *Hominton* 1244, *Humington -yng-* 1236–99. OE pers.n. **Humma* + **ing**[4] + **tūn**. Wlt 223 gives pr [hʌmiŋtən].

HONEY HILL Kent TR 1161. By origin probably a complimentary name for good land ('flowing with milk and honey') although it might allude to a place where bee-hives stand or, confusingly, to land with sticky soil. Field 1972.107.

HONEYBOURNE H&W SP 1144. 'Honey stream'. *huniburna -e* *[709]12th S 80, *Hvnibvrne* 1086, *Honiburne* 1275, 1291. OE **huniġ** + **burna**. Originally a stream name referring to sweet water. Wo 264 gives pr [hʌnibʌn].

HONEYCHURCH Devon SS 6202. Probably 'honey church'. *Honecherde* (sic for *-cherche*) 1086, *-chercha* 1086 Exon, *Hunichirche* 1238, *Honycherche* 1261–1378, *-churche* 1316–1535. OE **huniġ** + **ċyriċe**. The reference is unknown but it may have been a church where bees swarmed under the eaves or honey was blessed. D 165.

HONILEY Warw SP 2472. 'The honey clearing'. *Hunilege* 1208, *-le(ye)* 1276–1588, *Hunnely* 1667. OE **huniġ** + **lēah**. Cf. nearby HONINGTON SP 2642. Wa 213 gives pr [hʌnili:], L 205.

HONING Norf TG 3227. 'The Haningas, the people who live by Han, the hill' or 'Haning, the hill-like place'. *Hanninge* *[1044×7]13th S 1055, *Haninga* 1086, *Hanig(g)es* 1121–1243, *Haning(g)es* [1147×9]14th–1273, *Haninghe -yng(e)* 1286–1306, *Honyng(e)* 1286–1548. OE folk-n. **Hāningas* < OE **hān** 'stone, rock' possibly used as a p.n. + **ingas** or OE **hān** + **ing**[2]. Honing stands on a marked eminence called Howard's Hill crowned by the parish church. Nf ii.159, ING 58, DEPN gives pr [houniŋ].

HONINGHAM Norf TG 1011. 'The homestead of the Huningas, the people called after Hun or Huna'. *Huni(n)cham* 1086, *Huningeham* c.1184, *Huningham* 1202, 1254, *Honingham* 1205. OE folk-n. **Hūningas* < pers.n. *Hūn* or *Hūna* + **ingas**, genitive pl. **Hūninga*, + **hām**. DEPN gives pr [hʌniŋəm], ING 136.

HONINGTON Lincs SK 9443. *Hundinton(e) -tune, Hondintone* 1086, *Hundingtun'* 1135×54–1200, *Hundin(g)tun(e) -ton* 1175–1338, *Unnigtona* 1150×60, *Hunington* 1219–1332, *Honington* from 1261. Usually taken to be the 'estate called after Hund', OE pers.n. *Hund* + **ing**[4] + **tūn**. Alternatively the 'settlement at *Hunding*, the dog place', OE ***hunding** < *hund* + *ing*[2] perhaps referring to a place where hunting hounds were kept as in the Sw and Dan p.ns. *Hynding*. Honington lies beneath a prominent limestone plateau which is the site of the only major Iron Age fort in Lincs. Perhaps it was the plateau which was known as the *Hunding* 'that which is like or has the characteristic of a hound', cf. Hound Hill Do ii.169, Houndall D i.255, Hound Tor ibid.198. Perrott 498, Årsskrift 1974.50, Cameron 1998.

HONINGTON Suff TL 9174. Possibly 'the settlement of the Huningas, the people called after Hun(a)'. *hunegetuna* 1086, *-ton* 1254, *-tune* 1216×72, *Honeweton* 1305, *Hynnyton* 1610. OE folk-n. **Hūningas* < pers.n. *Hūn(a)* + **ingas**, genitive pl. **Hūninga*, + **tūn**. For the development *-inga-* > *-iga-* > *-ega-* > *-ewe-* cf. CANEWDON Essex TQ 8994 < *Caningadūn*, CONEY WESTON Suff TL 9578 < *Cunegestuna* 1086 (ON *konungr*), DANBURY Essex TL 7805 < *Deningabyrig*, MANUDEN Suff TL 4926 < *Manningadenu*, MONEWDEN Suff TM 2358 < *Mundingadenu*. DEPN, Ess 179.

HONINGTON Warw SP 2642. 'The honey farm'. *Hunitona* [1043]17th S 1000, *Huniton(e)* 1086–1445 with variants *Honi-* and *Hony-*, *Honington* [c.1043]17th S 1226, 1267–1466 with variants *Honin- Honyn- Hunin-*. OE **huniġ** + **tūn**. Cf. nearby HONILEY SP 2472. Forms with medial *-ing-* are probably by analogy with names in *-ingtun* rather than evidence for OE pers.n. *Hūna* + **ing**[4] + **tūn** 'the estate called cafter Huna'; the existence of an OE **huning* 'honey place' is also possible. Wa 281 gives pr [hʌniŋtən].

HONISTER PASS Cumbr NY 2213. P.n. Honister as in Honister Crag NY 2113 + ModE **pass**. Honister, *mountain called Unnisterre* 1751, *Honister Crag* 1784, is recorded too late for certainty; it may be ON *Húna stathir* 'Huni's place'. Cu 356.

HONITON Devon ST 1600. 'Huna's farm or village'. *Honeton(e)* 1086–1289, *Huneton -tune* 1210–58, *Hunyton -i-* 1238–75, *Honiton* 1251, *Honninton* 1270, *Honyngton* 1293, 1297. OE pers.n. *Hūna* + **tūn** varying with *Hūna* + **ing**[4] + **tūn**. D 639 gives pr [hʌnitən].

HONLEY WYorks SE 1311. 'Cock wood or clearing'. *Hanelei(a) -lay* 1086, 1243, *Honeley -lay* 1252–1482, *Honley* from 1274, *Hanley* 1323–1571. OE **hana** + **lēah**. The forms show WMidl *on* for *an*. The specific could alternatively be OE pers.n. *Hana*. YW ii.271.

HOO Suff TM 2558. '(The settlement) at the hill-spur'. *Ho* 1164–1275 including [1042×66]13th S 1051, *Hou, Hoi* 1086, *Hoe* 1254–86, *Hoo* from 1286. OE **hōh**, locative–dative sing. **hō(e)**. DEPN, Baron.

HOO ST WERBURGH Kent TQ 7872. 'St Werburgh's Hoo'. *Hoo Scē Wereburge* 1245. P.n. Hoo + feminine saint's name *Werburg* from the dedication of the church and for distinction from ST MARY HOO TQ 8076. Cf. *Sancta Wereburh de Hoo* c.1100. Hoo, *Hogh, (in) Hoge* [c.687 and 691]12th S 233, *(Æt) Hoe* [696×716]11th S 22, *Hohg* [738]12th S 27, *Hou* c.975 B 1321, *Hov, How* 1086, *Hoge* 12th ASC(E) under year 686, is OE **hōh** 'a hill-spur', applied to the ridge of high ground running from Cliffe Woods TQ 7373 to St Mary Hoo across what was described as the island of *Hebureagh* S 233 and *Heabureagh* ASC, later *Euery* 1518, and surviving in *Aviary Fm* 1801, 1805 OS, NW of Allhallows TQ 8377, probably 'Heahburg's island'. OE feminine pers.n. *Hēahburg*, genitive sing. *Hēahburge*, + **ēġ**. PNK 120, KPN 20,36, L 167, Jnl 8.22,23.

St Mary HOO Kent TQ 8076. 'St Mary's Hoo'. *(p'ochia) Scē Marie de Ho* 1240, *Ho St. Mary*, ~ *Scē Marie* 13th, *Seynte Maryho* 1292. Saint's name *Mary* from the dedication of the church + p.n.

Hoo as in HOO ST WERBURGH TQ 7872. According to DB there were six churches in the lordship of Hoo (*Hov*) which must be Hoo St Werburgh, High Halstow, Stoke, St Mary Hoo, Hoo All Hallows and the lost St Margaret in Hoo, *eccl. Scē Margarete* 1291, *St Margaret in Hoo* 1349. PNK 121.

HOOD NYorks SE 5082. 'The hood-shaped hill'. *Hod(e)* 12th–1376. OE **hōd**. Hood Grange is named from nearby Hood Hill, a detached conical peak of the Hambleton Hills. YN 195 gives pr [ud].

HOOE Devon SX 5052. '(At) the hill-spur'. *Ho* 1086, 1201, *Hoo* 1200, 1246. OE **hōh**, dative sing. **hō**. The site is a steep slope dropping down from Staddon Heights SX 4951. D 256.

HOOE ESusx TQ 6809. 'The hill-spur'. *Hov* 1086, *Hou* [1154×89]15th–1332, *Hoo* 1261–16th. OE **hōh**. The village stands on a hill above marshes. Sx 487.

HOOK GLond TQ 1764. 'The hook-shaped spur of land'. *Hoke* 1227 etc., *Hook* 1680. OE **hōc**. The shape of the parish is also that of a hook. Sr 58, LPN 114.

HOOK Hants SU 7254. 'The angle or nook'. *Hoc* 1223, *(la) Hoke* 1236–1327×77, *Houke* 1312. OE **hōc**. Originally applied to a wood forming an angle between streams. The present development is due to the LSWR station built here for Odiham. Ha 95, Gover 1958.128.

HOOK Humbs SE 7625. 'The bend'. *Hooc* [1154×89]copy, *Huuc* before 1227, *Huk(e)* 13th cent., *Houk(e)* 1314–1492, *Hook(e)* 1331, 1535–1822. OE ***hūc** varying with **hōc**. Hook lies in a sharp bend of the r. Ouse. YW ii.20.

HOOK Wilts SU 0784. 'The hook'. *la Hok(e)* 1238–75, *Houke juxta Lydeyerd* 1327. OE **hōc** 'a hook or spit of land' referring to a hill-spur of 465ft. NW of the hamlet. Wlt 275.

HOOKE Dorset ST 5300. 'The hook'. *Lahoc* 1086, *La Hoke* 1244, *Hok* 1209, *Houc* 1268. OE **hōc**. The reference is to a bend in the r. Hooke which takes its name from the place, its earlier name having been *Toller* as in TOLLER FRATRUM SY 5797. Do 89.

HOOKGATE Staffs SJ 7435. *Hook Gate* 1731. Cf. Hookgate Farm Somer ST 7235, Hook-a-gate Shrops SJ 4609, *Hooka Gate* 1833 OS, and Hooker Gate T&W NZ 1357, *Huckergaite* 1596, *Hookegate* 1602, 'the huckster's road', late ME **hukker** 'a petty dealer' + **gate**. Horovitz.

HOOKWAY Devon SX 8598. 'Hook road, road with a bend'. *Hocweie* 1205, *Hokeweye* 1210×12, *Hockeweye* 1262, *Hocweye* 1285, 1290. OE **hōc** + **weġ**. The road bends to negotiate a dip in the ground. D 407.

HOOKWOOD Surrey TQ 2642. 'The wood at *The Hook*'. *Hokewode* c.1450, *-wood* 1534. P.n. *la Hoke* 1235, *Hok'* 1332, 'the bend' sc. in the r. Mole, ME **hok** (OE *hōc*), + ME **wode** (OE *wudu*). Sr 288.

HOOLE Ches SJ 4267. 'At the hollow'. *Hole* [1119]1150–1665, *Houle* 1345, 1347, 1646, *Hoole* from 1522. OE **hol**, **holh**, dative sing. **hole**. Referred to in c.1195 as *Vallis Demonum* 'the valley of the devils'; perhaps a haunt of thieves. Che iv.129.

Much HOOLE Lancs SD 4723. *Magna Hole* [c.1235]1268–1327, *Much Hole* 1260, *Great Hoole* 1320, *Hole Magna* 1332, *Grett Wholle* 1551. Adjs. **much**, **great**, Latin **magna**, + p.n. *Hull(e)* 1204–46, *Hole(s)* 1212–46, *Hoole* 1508, *How(le)* 1577, 'the shed(s), the hovel(s)', OE **hulu**, pl. **hule**. 'Much' for distinction from Little Hoole, *Litlehola* [c.1200]1268, *parva Hola* [before 1220]1268, *Parva Hole* [before 1251]1268–1332, *Little Hoole* 1423. La 137, Jnl 17.78.

HOOTON 'Settlement by a spur of high ground'. OE **hōh** + **tūn**.

(1) ~ LEVITT SYorks SK 5291. 'H held by the Levitt family'. *Hoton Livet* 1243, *Houton Lyvet* 1246, *Hutton Levet* 1588, *Hootonlevitt* 1641. Earlier simply *Hoton(e)* 1086–1533. *Livet* was the name of a family which held land here in the 13th and 14th cents. YW i.136.

(2) ~ PAGNELL SYorks SE 4808. 'H held by the Painel family'. *Houton' Painelli* 1194, *Hooton Pannell* 1539. Earlier simply *Hotone, -tun* 1086, *Hoton(a) -tun(a)* c.1095–1546, *Hocton* 1196. A fee of the *Painel, Pagnel* family from the later 11th cent. YW i.87 gives pr ['hutn 'panal].

(3) ~ ROBERTS SYorks SK 4897. 'Robert's H'. *Hoton(e) Robert(i)* 1280–1446, *Hooton Roberte* 1535–1609, *~ Robberts* 1683. Earlier simply *Hotun* 1086, *Houeton* 1226. The affix is either from Robert son of William mentioned in 1226, or from *Robertus filius Willelmi de Hoton* who held the vill in 1285. YW i.124.

HOPE 'A valley, a side-valley opening off a main valley, a secluded valley, a remote enclosed place'. OE **hop**. L 111–21.

(1) ~ Derby SK 1783. *(æt) Hope* [926]14th S 397, 1086–1210 etc., *Hoppa* 1169–1199×1216, *Hoppe* 1188–1266, *Hawpe* 1541. Db 120.

(2) ~ Shrops SJ 3401. *Hope* 1242. DEPN.

(3) ~ BAGOT Shrops SO 5974. 'H held by the Bagard family'. *Hope Baghod* 1255, *Hope Bagar(d)* 1291–1546, *~ Bagotte* 1595, *Baggoteshope* 1577. P.n. *Hop', Hope* 1242–1301 + manorial addition from the tenant in 1242, Robert Bagard, whose surname was subsequently corrupted to Bagot. Sa i.156.

(4) ~ BOWDLER Shrops SO 4792. 'H held by the Bullers family'. *Hope Bothelers, Hup Budlers* 1255, *Hopebul(l)ers, ~ Budlers* 1255–1376 with variants *Boillers* and *Bowelers*, *~ Bowdler* 1535. P.n.*Hop', Hope* 1201–85 + manorial addition from Robert de Bullers, one of the tenants in 1201. The DB form is *Fordritishope* 1086, probably the OE pers.n. *Forthrǣd*, genitive sing. *Forthrǣdes*, + **hop**. Sa i.157.

(5) ~ MANSEL H&W SO 6219. 'Hope held by the Maloisel family'. *Hoppe Maloisel* 1160×80, 1214, *Hope Mal Oysel* 1243, *~ Maynessel* 1538. Earlier *Hope Gingenei* c.1140×5 and simply *Hope* 1086. Henry Maloisel 'bad bird' succeeded Guingené as lord of the manor c.1160. Guingené is a Breton pers.n., OBret *Uuin* (British **u̯indo-* 'white'). A remote side-valley off Castle Brook, a tributary of the Wye. He 105, Bannister 97.

(6) HOPESAY Shrops SO 3983. 'H held by the de Say family'. *Hope de Say* 1255, *Hope Say(e)* from 1302. P.n. *Hope* 1086–1272 + manorial addition from Picot de Say, the holder in 1086. Sa i.157.

(7) ~ UNDER DINMORE H&W SO 5052. *Hope sub' Dinnemor* 1291. Earlier simply *Hope* 1086–1203. A side-valley off the river Lugg. Dinmore SO 4850, *Dunemor(e)* 1189–1272, *Dinnemor* 1291, *Dynemore* 1338, is either W **din** + **mawr** 'great fort' referring to the hill-fort on Dinmore Hill SO 5151 or English 'Dynna's marsh', OE pers.n. *Dynna* + **mōr**. He 106, 71.

(8) Sollers ~ H&W SO 6033. 'H held by the Sollers family'. *Hope Solers* 1242. Earlier simply *Hope* 1086. The manor was held by Walter de Solariis, son of James de Solariis in 1242, cf. Sollers DILWYN SO 4255, BRIDGE SOLLERS SO 4142. A remote valley hemmed in on all sides by high ground. He 181.

HOPE Devon SX 6740. Uncertain. *la Hope* 1281, *atte Hope* 1330(p), 1394(p), *Hope* 1809. This is either OE **hop** 'enclosure in marshland, small valley' or **hōp**, ON **hóp** 'a small bay or inlet'. The reference is either to the bay here or to a long narrow valley stretching inland from the coast. Cf. HOPE'S NOSE SX 9563. D 308.

HOPE'S NOSE Devon SX 9563. 'The nose or point of Hope'. *Hope's Nose* 1765. P.n. Hope + ModE **nose**. The reference is to a projecting headland near Hope Farm (lost SX 9463), *Hope* 1765, OE **hop** 'a valley'. However, there is hardly a valley to speak of and it is tempting to see this as an instance of ON **hóp** 'a bay, an inlet' referring to the curving bay between Black Head SX 9464 and Hope's Nose. D 519.

HOPEHOUSE Northum NY 6780. 'The valley house'. *Hope Ho.* 1866 OS. The reference is to *Chirdon Hope* 1866 OS, the valley of Chirdon Burn. N dial. **hope** (OE *hop*).

HOPESAY Shrops SO 3983 → HOPE.

HOPTON 'The fen or valley settlement'. OE **hop** 'a side-valley, an enclosure in fen' + **tūn**.

(1) ~ Staffs SJ 9426. *Hotone* 1086, *Hoptuna* 1167, *Hopton* from 1204. H lies in a bend of a narrow side-valley of the r. Sow, but

the sense could also be 'enclosure in heathland'. DEPN, L 115.

(2) ~ Suff TL 9979. *Hoppe- Ho- Opituna* 1086, *Hopeton(e)* 1156×60–1344, *Hopton* from 1316. Hopton is situated on flat low-lying ground near the coast but is nearly surrounded by high ground. DEPN, Baron.

(3) ~ CANGEFORD Shrops SO 5480. 'H held by the Cangefort family'. *Hopton Cangefot* 1271–1316, ~ *K- Candivant* 1318, 1336, ~ *Cangeford* from 1356. P.n. *Hopton(')* 1255×6 etc., + manorial addition from the surname of Herbert Cangefort mentioned in 1199 in connection with nearby Clee St Margaret. Also known as *Hopton in the Hole* 1647–1803. The village lies near the head of a deep valley. Sa i.158.

(4) ~ CASTLE Shrops SO 3678. *Hopton Castle* 1577. P.n. *Opetune* 1086, *Oppetune* [c.1155]1348, *Hopton(')* 1242 etc., + ModE **castle**. The reference is to the 14th cent. stone keep of the 12th cent. Norman castle. The village stands at the meeting of three valleys. Sa i.159, Raven 94.

(5) ~ ON SEA Norf TG 5300. P.n. *Hoppe- Opituna* 1086, *Hopeton* 1242, + ModE **on sea**. Hopton lies in a valley in the sand dunes. DEPN.

(6) ~ TITTERHILL Shrops SO 3577. P.n. Hopton as in HOPTON CASTLE SO 3678 + p.n. Titterhill.

(7) ~ WAFERS Shrops SO 6376. 'H held by the Wafer family'. *Hopton' Wafr'* 1271×2, ~ *Wafers* 1535. P.n. *Hopton* 1086 + manorial addition from Robertus le Wafre 1255. The village lies in a deep valley. Sa i.159.

(8) Upper ~ WYorks SE 1918. ModE **upper** + p.n. *Hopton(e) -tun* 1086–1592. 'Upper' for distinction from Lower Hopton SE 1919. Upper H is the old village and lies in the valley of Valance Beck which runs into the Calder. YW ii.198, L 113–6.

HOPWAS Staffs SK 1805. Probably 'alluvial land by an enclosure' rather than 'the side-valley subject to flooding'. *Opewas* 1086, *Hopewaes* 11th, *-was* 1256, *Hopwas* from 1166. OE **hop** + **wæsse**. H is in a small side-valley of the Trent but one not typical of the normal *hop*. DEPN, L 59–60, 116, Horovitz.

HOPWOOD H&W SP 0375. 'Valley wood'. *(in) hoppuda* [849]11th S 1272, *[934]11th S 428, *Hopwod* 1208–75. OE **hop** + **wudu**. It has been suggested that *hop* in this name means 'enclosure' rather than 'valley' ('wood near an enclosure in waste'). The reference may, however, be to the remote valley NE of the hamlet called *Hopwood Dingle* 1831 OS, now Grovely Dingle. A farm belonging to Hopwood is recorded in [849]11th S 1272, *(ut pið) hoppudes pica*. Wo 333, L 113, 116, 229, Hooke 135, 144.

HORAM ESusx TQ 5717. *Horeham* 1813 OS. Earlier forms are needed.

HORBLING Lincs TF 1135. 'The muddy (part of) *billing*, the hill'. *(H)orbelinge* 1086, *Horbeling* 1185–1375, *-billing(e) -bulling -belling* 1190–1212, *Horbling* from before 1187. OE **horu** 'mud' + ***billing**, locative-dative sing. **billinge**, as in BILLINGBOROUGH TF 1134. Horbling lies a m. N of Billingborough, both settlements standing at the edge of high ground to the W, the *billing*, and marshes to the E. Perrott 122, V i.100.

HORBURY NYorks SE 2918. 'Fortification on dirty land'. *(H)orberie* 1086, *Horebir' -biri -y-* 12th–1254, *Horbir' -byr' -biri -y -byry* 1147–1577, *Horbury* from 1329. OE **horu** + **burh**, dative sing. **byriġ**. The reference is probably not to the site of the present town on a hill, but to a fortification nearer the river Calder on lower ground near the old ford. YW ii.150.

HORDEN Durham NZ 4441. 'Dirty valley'. *Horedene* [c.1040]12th, *Hordene* early 13th, *Horden* from 1260. OE **horu** + **denu**. In or near Horden was *Horetun* [c.1040]12th, 'dirty farm'. NbDu 117.

HORDERLEY Shrops SO 4187. Possibly 'the storehouse wood or clearing'. *Orderleg, Hordesle* 1255×6, *Horderley* from 1611. OE **hord-ærn** + **lēah**. Gelling.

HORDLE Dorset SZ 2795. 'Hoard hill'. *Hordella* 1100×35, *Hordhulle* 1135×54, *Hordehull, Hurdhill* 1331. OE **hord** 'treasure' + **hyll**. The reference is unknown. Ha 95.

HORDLEY Shrops SJ 3830. Partly uncertain. *Hordelei* 1086–1535 with variants *-leia -leg(h) -lee* and *-ley(e), Hordileg'* 1197×1203, *Hordylee -leye* 1271×2, 1306×7, *Horderle, Hordesleg'* 1271×2, *Hordley* 1577. Possibly OE ***horda** + **lēah** 'the wood or clearing of the hoards' possibly referring to the discovery of Roman coin hoards such as that found in 1950. But ME forms with *-i-, -y-* are difficult to explain from OE genitive pl. **horda**. Sa i.159.

Lower HORDLEY Shrops SJ 3929. Adj. **lower** + p.n. HORDLEY SJ 3830.

HORHAM Suff TM 2172. 'The dirty or muddy homestead or enclosure'. *Horham* from c.1095 including [942×c.951]13th S 1526, *Horam, Horan(t)* 1086. OE **horh, horu**, + **hām** or **hamm**. DEPN and Baron give pr [hɔrəm].

Great HORKESLEY Essex TL 9731. *Magna Horkesleg* 1254×5, *Meche Orkesley* 1501, *Great Harsley* 1546–8. ModE adj. **great**, Latin **magna**, ME **much, mech**, + p.n. *Horchesleia* c.1130, *Horkesle(ye)* 1212–1346, *Horskeleg'* 1198–9, *Horscley* 1546, possibly 'foul, dirty wood or clearing', OE **horsc, horx** + **lēah**. 'Great' for distinction from Little HORKESLEY TL 9531. Ess 392 gives pr [hɔsli], PNE i.262.

Little HORKESLEY Essex TL 9531. *Parua Horkesleg* 1254×5. ModE **little**, Latin **parva**, + p.n. Horkesley as in Great HORKESLEY TL 9731.

HORKSTOWE Humbs SE 9818. 'Place of shelter'. *Horchetou* 1086, *Horchestou* c.1115, *Horkestoue* c.1140, *-stow(e)* 1175–1640, *Horkstowe* from 1278. OE ***horc** 'a shelter' + **stōw**. OE **horc* is thought to be a cognate of MLG *hurken* 'to squat', dial. *hurk* 'crouch, cower'. Horkstowe was a place where people squatted together for some purpose. Li ii.158, PNE i.260.

HORLEY Oxon SP 4143. 'The wood or clearing at Horn, the horn or tongue of land'. *Hornelie* 1086–1344 with variants *-le(ia) -l', Hornlege -le(e) -ley(e) -legh* 1222–1428. OE **horn**, dative sing. **horne**, + **lēah**. Horley lies in a 'horn' of land between two streams. O 399.

HORLEY Surrey TQ 2842. 'The wood or woodland clearing belonging to Horne'. *Horle(e)* late 12th, *-leg' -l(i)e* 1203 etc., *Hornle -ly* 1230–70, *Horneleye* 1279, *-le* 15th, *Horelee* 1296, *-leye* 1305. P.n. HORNE TQ 3444 + OE **lēah**. Sr 292.

Great HORMEAD Herts TL 4030. *Magna Horemade* 1240, *Grete Hornemeade* 1583. ModE adj. **great**, Latin **magna**, + p.n. *Horemede* 1086–1297, *-made* 1197–1277, 'dirty meadow', OE **horu** + **mǣd**. 'Great' for distinction from Little Hormead TL 4029, *Parva Horemede* 1248. Hrt 179.

HORNBLOTTON GREEN Somer ST 5833. *Hornblotton Green* 1817 OS. P.n. Hornblotton + ModE **green**. Hornblotton, *Hornblawerton juxta Dichesgate* [855×60]13th S 1699, *Hornblaweton* 1086, *Hornblauton* 1236, *Hornbloutone* 1327, *Horn Blowton* 1610, is the 'estate of the hornblower(s)', OE **hornblāwere**, genitive pl. **hornblāwera**, or ***hornblāwa**, + **tūn**. Costen suggests that this was the estate granted to the man who led the king's hunt in the royal forest of Selwood. DEPN, Origins 90.

HORNBY 'Horni's village or farm'. ON pers.n. **Horni*, genitive sing. **Horna*, + **bȳ**.

(1) ~ NYorks NZ 3605. *Hornebia* 1086, *Hornebi -by* 1199–1367, *Hornby* from 1421. The topography is against derivation from **horn** 'a ridge of land'. YN 280 (which cites the wrong DB forms), SSNY 30.

(2) ~ NYorks SE 2293. *Hornebi -by* 1086–1361, *Hornby* from 1469. YN 240, SSNY 30.

HORNBY Lancs SD 5868. 'Village, farm on a horn of land'. *Hornebi* 1086, 1212, *-by* 1227–1332 etc., *Hornby* from 1297. ON **horn** or ***horni** + **bȳ**. Hornby Castle stands on a small hill projecting into the river Wenning; but the reference may rather be to the larger promontory of Baines Park Wood SD 5969 between the Lune and the Wenning. La 180, SSNNW 33, Jnl 17.101.

HORNCASTLE Lincs TF 2669. 'The Roman fort on the promontory'. *Hornecastre* 1086, *-castra -e* 1130, 12th, *Horncastell* c.1360,

1386. OE **horn** + **cæster** later replaced by ME **castel**. The horn or promontory refers to the site of the Roman town on a tongue of land at the confluence of the Bain and the Waring which encompass it on three sides. Possibly, therefore, to be identified with *Bannovalium* the 'strong spur', Brit ***banno- *banna** 'peak, horn, promontory' + ***u̯al- *u̯ali̯o-** 'strong'. There is, however, no independent evidence for this identification and Caistor TA 1101 on an elevated spur of the Wolds remains an alternative candidate. DEPN, RBrit 265, Cameron 1998.

HORNCHURCH GLond TQ 5487. 'Horn church'. *Horn(e)church* from 1233, *Hornedechirche* 1291, *Hornyngecherche* 1383. OE **horn**, varying with **hornede** 'provided with horns' + **ċyriċe**. OE *horn* could be used of animal horns or of hornlike projecting gables, but its significance in this name is unknown. The figure of a bull's head with horns on the eastern gable is not ancient; the horned bull's head on the 14th cent. priory seal is probably a rebus. The name was Latinised as *Monasterium Cornutum* 1222, and before that it was simply *ecclesia de Haweringis* 1163, the *church of Haveryng(e)* 1391. Ess 112, Encyc 404.

South HORNCHURCH GLond TQ 5283. A recent development on former farmland and market gardens. ModE **south** + p.n. HORNCHURCH TQ 5487. LPN 114.

HORNCLIFFE Northum NT 9249. 'The steep slope at the projecting horn of land'. *Horneclyff -clyf* [1183]c.1320, *Hornecliff(e)* 1208×11 BF–1539, *Horclife -kliffe* 1560, *Harkley* 1580. OE **horn**, ***horna** + **clif**. NbDu 117, L 134, 136.

HORNDEAN Hants SU 7013. 'Shrew or weasel valley'. *Harmedene* 1199, 1287, 1379, *Harnden* 1593, *Horn Dean* 1810 OS. OE **hearma** + **denu**. The precise meaning of *hearma* is unknown. Cf. OHG *harmo*. It is possible that *hearma* was used as a pers.n. Ha 95, Gover 1958.57.

West HORNDON Essex TQ 6288. *Westorin(n)don* 1274. ME adj. **west** + p.n. *Tornindunam, Torinduna* 1086. *Torindune* 1185, *Thorendon -indun' -undun* 1185–1251, *Thorn(e)don* 1254–91, 'thorn-tree hill', OE **thorn** + **dūn**. The earliest spelling, if not a mistaken duplication of *n*, and the persistent occurence of syllabic *-en*, suggest that the original specific may have been OE **thornen* 'growing with thorn-trees', a reformed version of OE *thyrnen*. AN *T-* for *Th-* in the combination *Est- West-Thorndon* together with the influence of the name Horndon in HORNDON ON THE HILL TQ 6683 account for the development *West-Torndon* > *West-Orndon* > *West Horndon*. 'West' for distinction from East Horndon TQ 6389, *Esthorendun'* 1240. Ess 158.

HORNDON ON THE HILL Essex TQ 6683. *Horn(e)don at the Hill* 1445, *~ on le Hyll* 1508. Earlier simply *Hornidvne* 1041×66 coins, *Hornindunа* 1086, *Hornin(g)don -inge- -ig-* 1200–86, *Horndun* 13th, *-don* from 1320, 'hill called or at Horning, the horn-like place', OE p.n. **Horning* < **horn** + **ing**^{2}, + **dūn**. The reference is to the isolated hill on which the village stands and for distinction from the quite different p.n. West HORNDON TQ 6288. Ess 157.

HORNE Surrey TQ 3444. 'The horn-shaped piece of land'. *Horne* from 12th, *Hornne* 1241, *Hoorne* 1397–1613, *Hourne* 1496, 1719, *Whorne* 1635, 1658. OE **horn**, perhaps referring to a projection of land to the SE of the church. Sr 322.

HORN HILL Bucks TQ 0192. 'Hill called the Horn'. *Horn Hill* 1822 OS. However, if the surname of Richard ate Hurne 1377×99 refers to this place, the specific would be OE *hyrne* 'an angle, a recess' referring to a recess in the hills here. Hrt 82.

HORNING Norf TG 3417. Short for *Horning Lower Street* 1838 OS. The church site is at TG 3516, the Hall at TG 3716, *Horning Upper Street* at UPPER STREET 3517. Horning is 'the Horningas, the dwellers at the horn or bend' sc. in the river Bure. *Horningga, (at) Horninggen* [1020×22]13th S 984, *Horninghe* *[1044×7]13th S 1055, *Horningam* (Latin accusative case) 1086, *Horning(ge) -inge* 1186–1337, *Hornyngges* 1316. OE folk-n. **Horningas* < **horn** + **ingas**. Nf ii.163, ING 58.

HORNINGHOLD Leic SP 8097. 'Wood at or called *Horning*, the horn-like hill', or 'of the Horningas, the people of the horn'. *Horniwale* 1086, *Horningewald* 1163, 1167, *Horni(n)gwald' -yn-* 1174×82–[1343]15th, *Hornin(n)gwold(e) -yn-* [1233]14th–1606, *Horningold(e) -yn-* 1495–1578, *-hold(e)* from 1557. OE p.n. **Horning* 'that which is like a horn, a horn-like hill' < OE **horn** 'a horn' + **ing**^{2}, locative–dative sing. **Horninge*, or folk-n. **Horningas* 'the people of the horn' < **horn** + **ingas**, genitive pl. **Horninga*, + **wald**. The former explanation is probably the better, referring to one of the hill-spurs close to the village. Lei 230, BzN 1967.375–9.

HORNINGLOW Staffs SK 2425. 'Horning hill'. *Horninglow* from 1100×35, *-lowe* 1327. Hill-n. **Horning* 'the horn-shaped hill', OE **horning** < *horn* + *ing*^{2}, + **hlāw**. The village lies at the E end of a narrow curving hill-spur overlooking the Trent. DEPN.

HORNINGSEA Cambs TL 4962. 'Island of Horning'. *(æt) Horninges ige* c.975 ASCh, *Horningesie* 1086, *-ei(e) -eia -ey(e) -yng-* 1086–1505 including [870–1170]12th, *Hornsey* 1719, 1808. OE p.n. **Horning* 'the horn-shaped hill' < OE **horn** + **ing**^{2}, genitive sing. **Horninges*, + **ēġ**. The site is a long narrow hill beside the Granta. Ca 145 gives former pr [hɔ:nsi], L 37–9.

HORNINGSHAM Wilts ST 8141. 'The homestead of the horn-shaped hill'. *Horningeshā* 1086, *-ham* c.1155–1332 with variant *-yng-*, *Horningham* 1086, *Horningeham* 12th, *Hornesham* 1405, *Hornison* 1585. OE ***horning** 'that which is shaped like a horn, a spit of land' < **horn** + **ing**^{2} perhaps used as a p.n., genitive sing. ***horninges**, + **hām**. The reference is to the headland below which the church stands. However, an OE pers.n. **Horning* from *hornung* 'a bastard' parallel to OG *Horning* is also possible. Wlt 168.

HORNINGTOFT Norf TF 9323. 'House site of the Horningas, the people who live at the horn'. *Horninghetoft* 1086, *Horningetoft* 1203, 1270. OE folk-n. **Horningas* < **horn** + **ingas**, genitive pl. *Horninga*, + **toft**. The reference of *Horningas* is unknown unless they are the same people as at HORNING TG 3417. DEPN.

HORNSBY Cumbr NY 5150. 'Orm's village or farm'. *Ormesby* c.1210–1533, *Hormesby* c.1230, *Hornesby* c.1241, *-bye* 1589. ON pers.n. *Ormr*, genitive sing. *Orms*, + **bȳ**. Cu 79, SSNNW 37.

HORNSEA Humbs TA 2047. 'Lake at *Hornness*, the projecting horn of land'. *Hornesse* 1086, 1301, 1358, *Hornessei* 1086, *Hornseie* [1087×95]copy, *Hornsey* 1529, *Hornsea* 1828. P.n. **Hornness*, ON **horn** + **nes** or OE **horn** + **næs**, + ON **sǽr** or OE **sǣ**. Originally the name of Hornsea Mere into which a long narrow peninsula projects. Cf. the Danish lake name *Hornisseu* 1157×82. YE 63, SSNY 97.

HORNSEA MERE Humbs TA 1947. *mara(m) de Hornseia* 1112×22, *~ de Hornesse* 1302, *lacum de Hornse* 1197×1200, *marra(m) de Hornse* 13th cent., *Hornsye Marre* 1595. P.n. HORNSEA TA 2047 + **mere** varying with ON **marr** 'a pool'. YE 65.

HORNSEY GLond TQ 3089. 'Woodland enclosure at or of Hæring'.

I. *Haringeie -ey(e)* 1201–1428, *-heye* 1243, *Heringeye -heie* 13th cent., *Haryngeay* 1465, *Haringaue* 1535.

II. *Haringes(h)eye* 1243–1471, *Haryngsey* 1401, *Harnesey* 16th cent., *Hornsey* 1564.

Two forms of this name existed, OE **Hæring-hæġ* 'the enclosure called or at H' and the genitival compound **Hæringes-hæġ* 'the enclosure of H'. The latter developed to become modern Hornsey and the former was revived for the name of Harringay House which then gave its name after 1880 to the district of Harringay in Hornsey. OE **hæring* is possibly an *ing*^{2} derivative of OE **hær* 'a rock, a heap of stones' cognate with Swed *har* 'stony ground' and so meaning 'stony place'. Mx 121, GL 55–6.

HORNTON Oxon SP 3945. 'The settlement at the horn-shaped spur of land'. *Hornigeton'* 1194, *Hornin(g)tun' -ton(e) -yn(g)-* c.1195–1385, *Hornton* 1317. OE ***horning** 'a horn-shaped spur of land', here between two streams on either side of the village, + **tūn**. O 399.

HORRABRIDGE Devon SX 5169. Probably 'the grey bridge'. *Horebrigge* 1345, 1377, *la Horabrigḡ* 1327×77, *Hollowbridge* 1675, *Harrowbridge* 1809 OS. The generic is ME **brigge** (OE *brycg*), the specific probably ME **hore** 'grey' from OE *hār* although **hore** from OE *horu* 'filth, dirt' is also possible. D 247.

HORRELL Corn SX 3088 → WRELTON NYorks SE 7686.

HORRINGER Suff TL 8261. Probably a doublet of the name Horningsheath, 'the heath of Horning', for which there is a variety of forms, *Horning(g)esh'de -hœð* [942×c.951]13th S 1526, *Horningasearðe, Horningeseorðe* 'the ploughed land of Horning' 11th, *Horningeserda -worda* 'the ploughed land' and 'the enclosure of H' 1086, *-eorda* c.1095, *-hearde* 1166, *Horning(g)esherth Magna, ~ Parva* 'Great' and 'Little H' 1254, *Herninghertһe* 1610. The common generic is OE p.n. **Horning*, OE ***horning** 'the bend place, the horn-like place' < *horn* + *ing*[2], the various specifics OE **hǣth**, **erth** and **worth**. DEPN.

East HORRINGTON Somer ST 5846. *Esthorningedon* 1268. ME adj. **est** + p.n. *Hornningdune, oðer Hornningdune* *[1065]18th S 1042, *Horningdon* 1243, 'hill called Horning, the horn-shaped hill', OE ***horning** (< *horn* + *ing*[2]) used as a p.n. + **dūn**. 'East' for distinction from West HORRINGTON ST 5747. DEPN, PNE i.262.

West HORRINGTON Somer ST 5747. *West Horrington* 1917 OS. ModE adj. **west** + p.n. Horrington as in East HORRINGTON ST 5846.

HORSEA ISLAND Hants SU 6304. *Horsea Island* 1810 OS, cf. *insula de Easthorsey* 1698. P.n. *Horsy* 1236, *Horsye* c.1300, 'horse island', OE **hors**, genitive pl. **horsa**, + **īeġ** in the sense 'dry ground in marshland', + pleonastic ModE **island**. The reference is probably to wild horses; when the Miller in the *Reeve's Tale* tries to cheat the students Alan and John of their corn, he distracts them by untying their horse which immediately makes off with a joyful *wehee* for the fen *ther wilde meres renne* (CT A.4065). Gover 1958.9.

HORSEBRIDGE ESusx TQ 5911. A modern popular etymology for Hurst Bridge the 'bridge by the wooded hill'. *Hors-Herstbregge* 1279, *-brugg'* 1327, *Hurstbregg'* 1285, *Horstbridge* 1553. OE **hyrst**, SE dial. form **herst**, + **brycg**, SE dial. form **brecg**. Sx 440.

HORSEBRIDGE Hants SU 3430. '*Hose* bridge'. This name has nothing to do with horses as the forms demonstrate: *Hosebrig'* 1236, *-brigg'* 1257–1331, *Husebrugge* 1280. *Horsbrydg* 1591 is a late folk-etymology. The original p.n. *Hosebrigge* is probably OE **hos(s)** 'a shoot, a tendril' used as a p.n. meaning 'reed-bed, thicket' or the like, + **brycg**. Cf. the minor ns. *la Hose* 14th in Toppesfield, *the Hose* 1534 in Colchester, *le Hose* 1487 in Felsted, all in Essex. Alternatives are OE *hōs* 'a company, a band', *hosa -e* 'a sheath, a hose' transferred to some long thin feature like a stocking, possibly a stream name, or **hās* beside *hǣs* 'brushwood'. Gover 1958.185, Ess 464, Brk 205–6.

HORSE BRIDGE Staffs SJ 9653. *Horse Bridge* 1815. Short for *Horseyate Bridge* 1604, 'horse-gate bridge'. Oakden.

HORSEBROOK Staffs SJ 8810. 'The brook frequented by horses'. *Horsebroc* 1262×72, *Horse(e)brok(e)* 1306–1627. OE **hors**, genitive pl. **horsa**, + **brōc**. St i.38.

HORSEHAY Shrops SJ 6707. 'The horse enclosure'. *Horsehay* 1759, *The Horsehays* 1833 OS. ModE **horse** + dial. **hay** 'a woodland enclosure'. Probably a reference to horse rearing to meet the demand for draught animals in the collieries and ironworks. VCH, Gelling.

HORSEHEATH Cambs TL 6147. Probably 'horse heath'.
I. *Horseda -e* 1086–1433, *Hors(e)hed(e)* 1411–1511.
II. *Horsei* 1086, *Horsey(e) -eie -eia -eya* 1195–1378.
III. *Horesathe* 1198, *-eth* 1285, *Horseth(e)* 1218–1416, *Horsheth(e)* 1272–98, 1426, *-heath* 1311.
Types I and III point to 'horse heath', OE **hors** + **hǣth** although the early loss of *-h-* is unusual. Type II points to an alternative 'horse island', OE **hors** + **ēġ**. Ca 108, L 245.

HORSEHOUSE NYorks SE 0481. *Horse House* 1861 OS.

HORSELEY FEN Cambs TL 4083. *Horsley or Horselode Fen* 1750. This name must be taken with Horslode Fen Farm which is *Hornigslade* [1240]14th, *Horningslade -yng-* c.1250, 1402, *Horneslade* 1430, *le Horslade* 1457, *Horselodde -loud* 1489, 1539, *Horslods Fens* 1829, 'Horning watercourse', OE p.n. **Horning* < **horning** 'bend, corner, spit of land', or pers.n. **Horning* < *hornung* 'bastard', genitive sing. *Horninges*, + **lād**. A good example of a name which seems to be transparent but actually has nothing to do by origin with the horses that ran wild in the fens. Ca 251, 249, PNE i.262.

HORSELL Surrey SU 9959. 'The dirty shed' rather than 'dirty shelf or slope'. *Horsele, Horisell -sselle -shull -(s)hill -sull(e), Horshell', Hors(h)ull, Horeshull(e)* 13th cent., *Horeswell* 1348, *Horsell* from 1487. OE **horu** or **horiġ** + **(ġe)sell** with reinterpretation of the generic as **hyll** and **wella**. Although the village is situated on shelving ground, the absence of spellings in *-lf, -lv* is against derivation from OE *scylf*; contrast GOMSHALL TQ 0847, OXHILL Warw SP 3145, BASHALL EAVES SD 7043, and BRAMSHALL Staffs SK 0633. Sr 128, PNE ii.118.

HORSEY Norf TG 4522. 'Horse island'. *Hors(h)eia* 1086, *Horseye* 1135×54–1398, *Horsey* 1392. OE **hors** + **ēġ**. An area of higher ground in the coastal marshes east of Hidding Broad and Horsey Meer where wild horses must once have grazed. Nf ii.107.

HORSEY ISLAND Essex TM 2324. *Hors(s)ey(e) (island)* 1304, 1540–1. P.n. *Horsehey(e)* 1212, 1218, *Horsheia* 13th, + pleonastic ModE **island**. Horsey could be either 'horse enclosure', OE **hors** + **(ġe)hæġ**, or more likely 'horse island', **hors** + **ēġ** with inorganic *h*-spellings. Originally a haunt of wild horses. Ess 341.

HORSFORD Norf TG 1916. 'Horse ford'. *Hosforda* 1086, *Horsford* 1254. OE **hors** + **ford**. The ford must have led to HORSHAM Norf TG 2115 and may perhaps better be interpreted as short for 'Horsham ford'. DEPN.

HORSFORTH WYorks SE 2438. 'Horse ford, ford that must be ridden'. *Ho(r)seford(e)* 1086, *Horseford(e)* 12th–1316, *Horsford* 1154×79–1327, *-forth* from 1378. OE **hors**, genitive pl. **horsa**, + **ford**. Late 9th cent. coins with the legend ORSNA FORD, OHSNA FORD, assigned to Horsforth, imply OE genitive pl. *horsna* from a weak noun **horsa* beside strong *hors*, recorded only as the pers.n. *Horsa*. It was an ancient crossing of the r. Aire. YW iv.148, xi.

HORSHAM H&W SO 7358. 'Horse farm or enclosure'. *Horsham* from 1271. OE **hors** + **hām** or **hamm**. Wo gives pr [ho:səm].

HORSHAM WSusx TQ 1730. 'The horse enclosure'. *Horsham* from 1233 including [947]13th S 525 and [963]13th S 714, *Horsam* 1301, *Hosham* 1601. OE **hors** + **hamm** 3 'a river meadow' or **hām** 'a homestead'. Horse-breeding and keeping is evidenced elsewhere in this area, cf. nearby WARNHAM TQ 1533 and STEDHAM SU 8622 on the r. Rother. Horsham lies beside the Arun but as an early-recorded parish may be an example of *hām*; *hamm* and *hām* seem to have fallen together at an early date in Sussex. Sx 225, ASE 2.50, SS 82, 81.

HORSHAM ST FAITH Norf TG 2115. 'St Faith's H'. *Ecclesia S. Fidis de Horsham* 'St Faith's church of H' 1163. P.n. *Horsham* 1086, 'the homestead or enclosure where horses are kept or collected', OE **hors** + **hām** or **hamm**, + saint's name *Faith* from the dedication of the church. Horsham is close to HORSFORD TG 1916. DEPN.

HORSINGTON Lincs TF 1968. 'The stud farm'. *Horsi(n)tone* 1086, *Horsintun* c.1140, *-ton* 1202, *Horsington* from 1254. OE ***horsing** 'horse place' < *hors* + *ing*[2] + **tūn**. An alternative possibility is the 'estate called after Horsa', pers.n. *Horsa* + **ing**[4] + **tūn**. DEPN, Årsskrift 1974.50.

HORSINGTON Somer ST 7023. 'The horsekeepers' estate'. *Horstenetone* 1086, *Horsinton* 1179, 1225, *Horsington* [c.1239]B. OE **horsthegn**, genitive pl. **horsthegna**, + **tūn**. The reference is probably to horse farming along the river Cale; cf. the nearby p.n. HENSTRIDGE ST 7219. DEPN.

HORSLEY 'The horse-pasture'. OE **hors**, genitive pl. **horsa**, + **lēah**. L 206. Frequently refers to a place where wild horses were corralled. L 206.

(1) ~ Derby SK 3744. *Horselei* 1086, *Hors(e)ley(e) -a -leg(e) -leg(h) -leia -lie* [1155×8]1316, 1157–late 12th etc. Db 471.

(2) ~ Glos ST 8398. *Horslei(a)* 1086, 1103×12, *-ley* 1261–1614, *Horsle* 1221, *-ley* from 1274. Gl i.91.

(3) ~ Northum NZ 0966. *Horseley* 1242 BF, *Horsley iuxta Tynam* 1296 SR. NbDu 117.

(4) ~ Northum NY 8496. *Horsley* 1869 OS.

(5) ~ CROSS Essex TM 1227. *Horssey Cross* 1768, *Horsley Cross* 1777. ModE **cross** 'a cross-roads'. Ess 345.

(6) East ~ Surrey TQ 0954. *Esthorselegh* 1235. ME **est** + p.n. *Horslei* 1086, *-lee* 1200, *Horseleia* 1203. 'East' for distinction from West HORSLEY TQ 0752. Sr 139, L 206.

(7) West ~ Surrey TQ 0752. *Westhorsleg(h)* 13th cent. ME **west** + p.n. *(on) Horsalæge* 871×89 S 1508, *Horslege* [1042×66]11th S 1047, *Orselei* 1086, as in East HORSLEY TQ 0954. Sr 141, L 206.

HORSLEYCROSS STREET Essex TM 1228. P.n. HORSLEY CROSS TM 1227 + Mod dial. **street** 'a hamlet'.

HORSMONDEN Kent TQ 7040. There are two types;

I. *Horsbundenne* c.1100, *Horsburden(n)e* 1147×82–1210×12, *Horsenden* 1226;

II. *Horsmindenn'* 1232, *-munden(n)e* 1235–59, *-monden* from 1259.

Type I is either 'Horseburn (woodland-)pasture', p.n. **Horsburn* < OE **hors** + **burna**, + OE **denn**, or 'reed woodland-pasture', OE ***horsbune** < **hors** + **bune**, + **denn**. Type II shows assimilation to other Kentish names in *-monden*, Delmonden TQ 7330, Spelmonden TQ 7037 etc., OE ***mund-denn** 'a protected woodland-pasture'. PNK 196, Jnl 18.43, 47.

HORSPATH Oxon SP 5704. '(The settlement) at the horse paths'. *Horspadan* 1086 etc. with variants *-pade -paða -path(e) -peth -pathes*. OE **hors** + **pæth**, dative pl. **pathum**. O 178.

HORSTEAD Norf TG 2619. 'Horse place'. *Ho(r)steda* 1086, *Horsted(e)* c.1148–1484. Horstead is near HORSFORD TG 1916 and HORSHAM TG 2115. The open country N and W of these places must have been an important area where horses ran wild. DEPN, Stead 182.

HORSTED KEYNES WSusx TQ 3828. 'H held by the Keynes family'. *Horsted(e) Kaynes, ~ Keynes* 1261, 1295, *Hosted Caines* 1721. P.n. *Horsted(e)* from 1086, 'the place where horses are kept', OE **hors** + **stede**, + manorial addition from the family of William de Cahainges, the DB tenant, who came from Cahagnes between Vire and Bayeux. Sx 336, Stead 257, SS 87.

HORTON 'Settlement on muddy soil'. OE **horu** + **tūn**. Sa i.160.

(1) ~ Berks TQ 0175. *Hortune* 1086, *Horton* from 1376. Bk 239.

(2) ~ Bucks SP 9219. *Hortone* 1086, *Horton* 1464. Bk 97.

(3) ~ Dorset SU 0307. *Hortun* [1033]12th S 969, *(æt) Hortune* [1061]12th S 1031, 1086, *Horton(e)* from 1212. The earliest reference to this settlement is *op hore tuninge gemære* 'to the boundary of the *Horetuningas*, the people of Horton' [946]14th S 519. Do ii.159.

(4) ~ Lancs SD 8550. *Hortun -ton* from 1086. YW vi.170.

(5) ~ Northants SP 8254. *Horton(e)* from 1086. Nth 148.

(6) ~ Northum NU 0230. *Horton' (Turbervill)* 'H held by the Turbervill family' 1242 BF. The manor was held by William Turbervill in 1242. NbDu 118, DEPN.

(7) ~ Somer ST 3214. *Horton* 1242. DEPN.

(8) ~ Staffs SJ 9457. *Hortun(e)* 1226, c.1227×30, *Horton* from 1239. Oakden.

(9) ~ Wilts SU 0463. *Horton(a)* from 1158. Wlt 251.

(10) ~ –CUM–STUDLEY Oxon SP 5912. P.n. Horton + Latin preposition **cum** + p.n. Studley. Horton is *Hortun* [1105×12]14th S 942 etc. with variants *-tuna -ton(e) -thona*, Studley, *Stodleya* [c.1180×90]c.1425 etc. with variants *-leia -le(e) -legh(e)* etc., *Studeleyge* 1526, 'the stud pasture', OE **stōd** + **lēah**. O 179, L 206.

(11) ~ GREEN Ches SJ 4549. *Horton Green* 1831. P.n. *Horton* from 1240, + ModE **green**. Che iv.55.

(12) ~ HEATH Hants SU 4917. *Horton Heath* 1810 OS. P.n. *Hortun* [963×75]12th Gover 1958.75.

(13) ~ IN RIBBLESDALE NYorks SD 8072. *Horton in Ribb(b)(e)lesdale* from 13th. P.n. *Hortune* 1086–1679 + *in Ribblesdale* for distinction from HORTON SD 8550. YW vi.218.

(14) ~ KIRBY Kent TQ 5668. 'H held by the Kirby family'. *Horton Kirkeby* 1379. Earlier simply *Hortune, hortone* 1086, *Hortune* [1087]13th, c.1100, *Horton(e)* from 1198. Gilbert de Kirkeby was tenant in 1253×4 whose family originated from Kirkby Hall in Lancs. PNK 43.

(15) ~ MOOR NYorks SD 8274. P.n. HORTON IN RIBBLESDALE SD 8072 + ModE **moor** (OE *mōr*).

(16) New ~ GRANGE T&W NZ 1975. ModE adj. **new** + p.n. Horton Grange as in *Horton Grange Farm* 1863 OS. Horton Grange, *grangiam de Horton* 1242 BF, was an outlying grange belonging to Newminster Abbey, p.n. *Horton* 1346 + ME **graunge**. NbDu 118.

HORTON Avon ST 7584. Partly uncertain. There are three types:

I. *Horedone* 1086.

II. *Hortune -ton(a)* from 1167 including [c.1075]1367 and [c.1115]1333.

III. *Heorton* 1291, *Herton(e)* 1303–74.

Probably 'dirty village', OE **horh, horu** + **tūn**. The DB form may be an eccentric form but type III may point to OE *heort* 'a stag' as an alternative specific. Gl iii.35.

HORTON COURT Avon ST 7685. A Norman manor house of c.1140. P.n. HORTON ST 7584 + ModE **court**. Verey i.279.

HORWICH GMan SD 6411. Uncertain. *Horewych(e)* 1254–1331 with variant *-wich(e), Horwyge* 1539, *Horridge* 1641. The earliest forms refer to the forest of the lords of Manchester here and accordingly the name has been taken to be *(æt) hāran wićan* '(at the) grey wych-elms', OE **hār** + **wiće**. However, formally the name could equally well be 'dirty farm', OE **horu** + **wīć**. La 44 gives pr [ɔridʒ].

HORWOOD Devon SS 5027. 'The grey wood'. *Horewod(e)* 1086, *Hareode, Horew(o)da* 1086 Exon, *Harewde* 1219, *Horwude -wode* 1196 etc., *Horwood al. Worewood* 1692. OE **hār**, definite form **hāra**, + **wudu**. D 115.

Great HORWOOD Bucks SP 7731. *Great Horwood* 1834 OS. ModE **great** + p.n. *Horwudu* *[795 for 792]13th S 138, *Hereworde* (sic) 1086, *Horewode* 1228, *Horwode* 1301, 'dirty wood', OE **horu** + **wudu**. 'Great' for distinction from Little HORWOOD SP 7930. Bk 68 gives pr [hɔrud].

Little HORWOOD Bucks SP 7930. *Little Horwood* 1834 OS. ModE **little** + p.n. Horwood as in Great HORWOOD SP 7731.

HOSE Leic SK 7329. 'The hill-spurs'. *Hoches* 1086, *Howes* 1086–1539, *Hoose* 1535–1610, *Hose* from 1544. OE **hōh**, pl. **hōhas, hōas, hōs**. The village is situated at the foot of the Wolds which here form a series of spurs. Lei 157, L 167.

HOT POINT Corn SW 7112. P.n. Hot + ModE **point**. Hot, *Hot* 1813, and nearby *Hot Cove* 1888, are obscure. Possibly OCo **hoet** 'a duck' used of a rock, although this would normally become *hos*. PNCo 98.

HOTHAM Humbs SE 8934. '(The settlement) at the shelters'. *Ho(l)de, Hodhu'* 1086, *Hodum* 1166, *Hothum* 1153×66–1401, *Hotham* 1444. OE ***hōd**, dative pl. ***hōdum**. The earliest reference is to a place 'east of *Hod*', *be fastan hode* for *be eastan Hode* [963]13th. The reference is to huts or shelters for seasonal use as in herding or woodcutting, here perhaps seed-gathering. OE medial [d] has been replaced by Scandinavian [ð]. YE 225 gives pr [uðəm], SSNY 147.

HOTHFIELD Kent TQ 9644. 'Heathland'. *Hathfelde* c.1100, *Hatfeld(e)* 1210×12–1278, *Hothfeld* 1254. OE ***hāth** + **feld**. PNK 413, L 242.

HOTON Leic SK 5722. 'Settlement at the hill-spur'. *Hoh- Holetone* (sic) 1086, *Houton(e)* late 12th–1399, *Hooton'* 1258, *Hot(t)on(e)* from 1277, *Ho(u)ghton* 1370, 1485, 1630. OE **hōh** + **tūn**. Lei 302 gives pr ['hu:tən].

HOUGH. 'A heel, a hill-spur'. OE **hōh**. L 167–9.

(1) ~ Ches SJ 7151. *le Hothg* 13th, *(le) Ho* 1241–78, *(le) Hogh* 1287–1517, *Hough* from 1287. Hough is situated on high ground N of Wybunbury. Che iii.64 gives prs [hʌf] and [ʌf], Addenda ii 19th cent. local [þ'uf], L 167.

(2) ~ GREEN Ches SJ 4887. No early forms.

HOUGH-ON-THE-HILL Lincs SK 9246. *Hough on the Hill* from 1576, *Hoffe* ~ ~ ~ 1666. Earlier simply *Hach(e), Hag* 1086, *Hach(e)* 13th cent., *Hagh'* 1208–1428, *Hauh* 1278, *Hoch* 1188, *Hogh(e)* 1262–1505, *Hough(e)* 1389–1597, 'the enclosure', OE **haga**. The village is situated on a sharply rising hill-slope. There was early confusion with OE **hōh** 'a heel, a hill-spur'. Perrott 375, L 156.

HOUGHAM Lincs SK 8844. 'Homestead or river-meadow at Hough'. *Hac(h)am* 1086, *Hagham* 1210–1517, *Hauham* 1212, *Hocham* 1325, 1385, *Hogham(e)* 1373–1554, *Hougham* from 1505. P.n. Hough as in HOUGH-ON-THE-HILL SK 9246 + **hām** or **hamm**. The spellings are indecisive but the site is a typical *hamm* 3. On the other hand in the Middle Ages Hougham was a major manorial centre with a church possibly of Anglo-Saxon origin. Perrott 379, Pevsner 1964.579.

West HOUGHAM Kent TR 2640. *West Hougham* 1819 OS. ModE **west** + p.n. Hougham as in Church Hougham TR 2740, ModE **church** + p.n. *Hichā, Huchā* 1086†, *Hucham* [1087]13th, *Huhcham* c.1100, *Hugham* 1177×7–1278, *Huam* 1211×2, *Hougham* 1271, either 'Huhha's homestead or estate' or 'homestead or estate on the hill-spur'. OE pers.n. **Huhha* or **hōh** + **hām**. The Houghams occupy a long narrow ridge which would suggest that the specific is *hōh*, but OE *ō* does not normally appear as *u* so early. Perhaps we have here a hitherto unrecorded element with zero-grade *huh-* of Gmc *hauh-* as in OE **hugol*, G *Hügel* 'a hillock'. The site of the church here is a small hillock. PNK 562, KPN 363, ASE 2.30, Cullen. DEPN.

HOUGHTON 'Settlement at or by a hill-spur'. OE **hōh** + **tūn**. L 167. Cf. HUTTON.

(1) ~ Cambs TL 2871. *Hoctune* 1086, *Hou(c)hton* 1240, 1303, *Houton* 1279, 1295, *Houghton* 1309. The reference is to the headland of Houghton Hill TL 2972 where the terrain rises steeply from the Great Ouse to 120ft. Hu 210.

(2) ~ Cumbr NY 4059. *Hotton* 1246, *Houghton* from 1279. Houghton and Houghton Hall stand on a low hill-spur. Cu 110.

(3) ~ Hants SU 3432. *Hovstvn* 1×, *Holstvne* 4× 1086, *Hoc(h)-Hog(h)tone* 1171–1329, *Houton* 1240, *Houghton(e)* 1316, 1428. The DB spellings show <s> for [ç] and back-spelling <l> for [u]. The boundary of the people of Houghton is referred to as *hohtuninga mearcel* [982]12th S 840. Also known as Houghton Drayton, *Drayton* 1267, 'the settlement where loads are dragged, portage village', referring to a crossing of the Test and Test marshes at this point, or to a point where boats could be dragged up out of the water. Ha 95, Gover 1958.180.

(4) ~ WSusx TQ 0111. *Hohtun* *[683]14th S 232, *Hocton* [957]14th S 1291, c.1230, 1271, *Hout(t)on, Hogton* 1263–79, *Hoghton* 1327–1428. The village stands on a spur of land projecting into the Arun valley. Sx 128 gives pr [houtən], SS 75.

(5) ~ CONQUEST Beds TL 0441. 'H held by the Conquest family'. *Houghton Conquest* 1316. Earlier simply *(H)oustone* 1086, *Hocton* 1202, *Hoghton* 1220, *Houghton* 1287. The Conquest family are associated with this place in 1223. Bd 74.

(6) ~ HOUSE Beds TL 0339. P.n. HOUGHTON + ModE **house**.

(7) ~-LE-SPRING T&W NZ 3449. Probably 'H in the copse'. *Ho(u)ghton in le Spryng* 1410–1556, *Houghton le Spring* 1647. P.n *Hoht'* 1147, *Hoctun* c.1149×52, c.1170, *Hoctona -(e)* [1183]c.1320, 1197–1242×3, *Hovthun* [c.1149×52]c.1220×30, *Houaton* 1304, *Houeton'* [1312]15th, *Houghton* from 1349, + Fr definite article **le** for *en le* + ME *spring* 'a young shoot, a young plantation, a copse'. Alternatively the addition might be manorial from the family name *Le Spring* later misunderstood. A *John del Spryng* of Houghton occurs in an IPM of 1420. Houghton lies at the foot of a prominent spur of the East Durham Plateau. NbDu 118, Surtees I.i.145.

(8) ~ ON THE HILL Leic SK 6703. *Houghton on the Hill* 1624, ~ *super Montem* 1723. P.n. *Hohtone* 1086, *Hoctona -(e)* 1130–1280×92, *Houcton(e)* 1242–1360, *Houton(e)* 1273–1370, *Houghton* from 1306, + ModE **on the hill**, Latin **super montem**. Lei 231.

(9) ~ REGIS Beds TL 0224. 'The king's H'. *Kyngeshouton* 1287, *Houghton Regis* 1353. ME **king**, genitive sing. **kinges**, Latin **rex**, genitive sing. **regis**, + p.n. *(ad) Hohtune* [975]12th, *Houstone* 1086, *Hohtun -ton* 1156, 1227. It was already a royal manor at the time of DB. Bd 128, YE xlvii.

(10) ~ ST GILES Norf TF 9235. 'St Giles's H'. Earlier simply *Hohttune, Houtuna* 1086, *Hocton* 1212. The land rises to over 200ft. on either side of the valley and to 259ft. at TF 9036. Saint's name *Giles* from the dedication of the church. Also known as *Houghton in the Dale* 1838 OS. DEPN.

(11) Great ~ Northants SP 7958. *Magna Houhtone* 1234, *Michelhoughton* 1342. Latin **magna**, ME **micel**, + p.n. *Hohton(e)* 1086, 1200, *Hoctona* 1131–1251, *Houton(a)* 1197. 'Great' for distinction from Little HOUGHTON SP 8059. Nth 149 gives pr [houtən].

(12) Little ~ Northants SP 8059. *Hoctona Parva* c.1220, *Parva Houghton* 1316. ModE adj. **little**, Latin **parva**, + p.n. Houghton as in Great HOUGHTON SP 7958. Nth 149.

(13) New ~ Derby SK 4965. ModE adj. **new** + p.n. Houghton as in Stony HOUGHTON SK 4966. A modern development.

(14) New ~ Norf TF 7927. ModE adj. **new** + p.n. Houghton as in Houghton Hall, *Houton* 1254, *-tone* 1291. New Houghton is *New Town* 1824 OS, referring to the 'village' begun in 1729 consisting of two rows of five houses S of the south gates of the park. There are several hill-spurs around the hall, the most prominent being S of the hall where the village of Harpley lies at TF 7826. DEPN, Pevsner 1962NW.210.

(15) Stony ~ Derby SK 4966. ModE adj. **stony** + p.n. *Holtune* 1086, *Hoc(h)ton(')* c.1275–1315, *Hoghton(e)* 1289–1560. 'Stony' from the nature of the soil. Db 292.

HOULSYKE NYorks NZ 7308. *Howlsike* 1861 OS. ModE dial. **howl** (OE *hol*) 'a hollow' + **sike** (OE *sīc*) 'a stream'.

HOUND GREEN Hants SU 7359. *Hound Green* 1817 OS. ModE **hound** + **green**.

HOUNSLOW GLond TQ 1276. 'Hund' or 'hound's tumulus'. *Hundeslawe* 1217–75, *-lowe* 1275–95, *Houndeslowe* 14th cent., *Hownslo(we)* 16th cent. OE pers.n. *Hund* or **hund**, genitive sing. *Hundes*, **hundes**, + **hlāw**. Hounslow was originally part of the parish of Heston but gave its name to a hundred in DB, *Honeslauu hundredum* 1086 (later Isleworth hundred). The hundred meeting place must have been the mound or barrow that once existed here. Mx 26, 24.

HOUSEDON HILL Northum NT 9032. No early forms. P.n. Housedon, unexplained, as in *Housedonhaugh* 1865 OS, + ModE **hill**. A hill rising to 877ft.

HOVE ESusx TQ 2805. 'The hood'. *la Houne in vill. Preston* (for *Houve* 'in the vill of Preston') 1288, *Huve* 14th, *Hoove* 1675. OE **hūfe** in a transferred sense applied to some sort of small building. Sx 293.

†The DB forms have not previously been accurately reported.

HOVERINGHAM Notts SK 6946. 'The homestead on the hump of ground' or 'of the dwellers on the hump'. *Horingehā* 1086, *Horingham -yng-* 1316, 1343, *Horryngham -ing-* 1525–1669, *Houringham* c.1160, *(H)oueringeham* 1167, 1235, *Hoveringham* from 1218. Either OE ***hofering** < ***hofer** + **ing**², locative–dative sing. ***hoferinġe**, or folk-n. **Hoferingas* < ***hofer** + **ingas**, genitive pl. **Hoferinga*, + **hām**. The village is situated on a slight ridge between two streams. Nt 169, ING 148.

HOVETON Norf TG 3018. 'Hofa's' or 'ground-ivy farm or estate'. *Houetonne* *[1044×7]13th S 1055, *Houetuna* 1086, *Houeton(e) -v-* 1127–1428, *Houtun* 1270. OE pers.n. *Hofa* or **hōfe** + **tūn**. Ground-ivy or alehoof was much used in medicine. Nf ii.167, PNE i.256.

HOVINGHAM NYorks SE 6675. Partly uncertain. *Hovingham -yng-* from 1086, *-ingeham* 1157, 1228, 1308. Possibly the 'homestead at Hofing, the *hof* place', OE p.n. **Hofing* < **hof** + **ing**², locative–dative **Hofinge*, + **hām**. *Hofing* would be a derivative of OE *hof* 'an enclosure, a dwelling, a house, a temple'. Hovingham is the site of a Roman villa which may have been called **Hofing* by the Anglo-Saxons, the 'place called after the house, the house-site'. But another sense of *hof* may have been 'hill' as in cognate Norw *hov* 'a small hill' and the derivatives OS *huvil*, MDu *hovel, huevel*, OHG *hubal* 'a hill', OE *hofer* 'a lump, a swelling'. In this case Hovingham would be 'village called or at **Hofing*, that which is like a swelling', referring to one of the promontories of the Howardian hills at the foot of which the village lies. Alternatively it might be OE p.n. **Hofinga-hām* 'the homestead of the Hofingas, the people called after Hofa', OE folk-n. **Hofingas* < pers.n. *Hofa* + **ingas**, genitive pl. **Hofinga*, + **hām**. But the evidence for the genitive pl. form *-inga-* is slender. YN 51 gives prs [ɔvinəm, ɔuiŋəm].

HOW Cumbr NY 5056. '(The) hill'. *le howe* 1610. Fr definite article **le** + ME **howe** < ON *haugr*. The reference is to nearby How Hill, *le Hill* 1611. Cu 89.

HOWARDIAN HILLS NYorks SE 6472. *Howardian Hills* 1860 OS. Named from the Howard family of CASTLE HOWARD SE 7170.

HOW CAPLE H&W SO 6130 → How CAPLE SO 6130.

HOWDEN Humbs SE 7528. 'The head valley'. *Heafud-Hœafuddene* [959]c.1200 S 681, *Hovedene -u-* 1086–1576, *Houden* 1231–1424, *How-* from 1403. OE **hēafod** 'head' replaced by ON **hǫfuth** + **denu**. It has been argued that *heafod* is used here in the sense 'headland, spit of land' with reference to the raised ground around Knedlington. This once formed a headland between the Ouse and the ancient course of the Derwent which is held to be the valley in question although there is no clear topographical feature characteristic of the element *denu*. Dr Gelling has very tentatively suggested that this may be an extremely ancient name going back to the time of Germanic mercenaries in the last century of Roman Britain at a time when *denu* still retained its basic sense 'flat space' referring to the Vale of York as a whole. YE 250, SSNY 158.

HOWDEN MOORS SYorks SK 1695. P.n. Howden + ModE **moor**. Howden, *Holden(e)* 1332, 1441, is 'the hollow valley', OE **hol** + **denu**. YW i.235, L 98.

HOWDEN MOORS RESERVOIR SYorks SE 1693. P.n. HOWDEN MOORS SK 1695 + ModE **reservoir**.

HOWDEN-LE-WEAR Durham NZ 1533. Probably 'hollow valley'. *Howden* 1382. OE **hol** + **denu**. *Le Wear* 'on the Wear' is a modern addition as in nearby WITTON-LE-WEAR NZ 1431.

HOWE Norf TM 2799. 'The hill or barrow'. *Hou, Howa* 1086, *Howe* 1254. ON **haugr**. DEPN.

The HOWE Cumbr SD 4588. 'The hill'. *How(e)* [1535]17th, 1669. ME **howe** < ON *haugr*. We i.84.

HOWE GREEN Essex TL 7403. P.n. Howe + ModE **green**. Howe, *How Farm* 1805 OS, is ultimately from OE **hōh** 'a hill-spur', cf. *Westho* 1291. How Farm occupies the end of a ridge of high ground. Ess 267.

HOWE STREET Essex TL 6914. 'Howe hamlet'. *hoostret* 1395×6, *Howstrete* 1594. P.n. *le Hoo* 1395×6 'the hill-spur', OE **hōh**, dative case **hō**, + ME **strete** 'a hamlet'. Ess 273.

HOWE STREET Essex TL 6934. 'Howe hamlet'. P.n. Howe as in *How F.*ᵐ 1805 OS + dial. **street**. The name Howe recurs in *Obourns or Howhall* 1486–93 which was the home of Richard atte Hoo 1327. It represents OE **hōh** 'a hill-spur'. Ess 427.

HOWELL Lincs TF 1346. Partially uncertain. *Welle, Huuelle* 1086, *Huwell(a)* 1165–6, *Howell(e)* from 1169. The generic is OE **wella** 'a spring', the specific uncertain, possibly OE **hūfe** 'a cap' used of some covering over the spring. Perrott 48, L 32.

HOWGILL FELLS Cumbr SD 6799. P.n. Howgill + ModE **fell(s)**. Howgill, *Holegile* 1235–68, *Holgil* 1467, *Howgill* 1550, is 'hollow, deep ravine', OE **hol**, ON **holr** + **gil**. YW vi.264, L 99.

HOWGRAVE NYorks SE 3179. 'The grove in the hollow'. *Hograve* 1086, 1088, *Hogram -em* 1086, *Holgrave* 12th cent., *Hougrave* 1184–1362, *Howegrave* 1403, 1536. OE **hol** + **grāf**. YN 220.

HOWICK Northum NU 2517. 'The high or chief dairy-farm'. There are three types:

I. *Hewic* c.1124×8, 14th, *Hewick* 13th DEC.

II. *Hawic(k) -yk* 1230–1374.

III. *Howyc* 1242 BF, *Ho(u)wy(c)k(e)* 1281–15th cent., *Howike* 1296 SR.

Type I appears to be OE **hēah**, definite declensional form **hēa**, + **wīċ**. Form II can be explained from the smoothed form ***hǣh** of **hēah** with shortening in the compound name, possibly influenced by ON **há(r)**. Form III seems to show influence of OE **hōh** 'a hill-spur'. It is not absolutely certain, however, that the form I spellings belong to this name. NbDu 117, PNE i.238, DEC 95–6.

HOWICK HAVEN Northum NU 2616. *Howick Haven* 1868 OS. P.n. HOWICK NU 2517 + ModE **haven**.

HOWLE Shrops SJ 6923. 'The hillock'. *Hugle* 1086, *Huggel'* 1255×6, *Hull(e)* 1221×2, c.1291, *Howle* from 1250, *Whole* 1655. OE ***hugol**, dative sing. ***hug(o)le**. The reference is to the small hill-spur on which Howle Manor stands. Sa i.161.

HOWLETT END Essex TL 5843. 'Howlett end' sc. of Wimbush. No early forms. A family name *Hulot* is recorded here in 1270. Ess 547.

HOWSHAM '(The settlement) at the houses'. ON **hús**, dative pl. **húsum**. A name-type common in areas of Scand settlement, although formally it could represent OE **hūs, hūsum**.

(1) ~ Humbs TA 0404. *Vsun* 1086, *Husum* c.1115–1343, *Housom* 1281–1438, *Howsham* from 1566. Li ii.76, SSNEM 149.

(2) ~ NYorks SE 7632. *Huson* 1086, *Husum* 12th–1285, *Housum* 13th cent., *Hou- Howsom* 1297–1542, *Howseham* 1549. YE 145 gives pr [u:zəm].

HOW STEAN BECK NYorks SE 0572. P.n. How Stean, *Holstan* 1307, 'the hollow stone', OE **hol** + **stān**, + ModE **beck**. In local dial. *stān* became [stiən]. YW v.216.

HOWTON H&W SO 4129. Probably 'Hugh's estate' after Hugh de Lacy who held the estate before c.1150. *Hug(g)ton(e)* 1163×7–1243, *Huetune* c.1182×86, *Houton* 1324–1407. ME pers.n. *Hugh* + **tūn**. He 109.

HOWTOWN Cumbr NY 4419. 'Hill village'. *How(e)town(e)* 1676–1770. ME **howe** (< ON *haugr*) + **toun** (<OE *tūn*). The village lies at the foot of a steep ridge rising to 1191ft. at NY 4418. We ii.219.

HOXNE Suff TM 1877. 'The hough'. *Hoxne* from c.1035 including [942×c.951]13th S 1526, *Hox(a)na, (H)oxa* 1086, *Hoxe* 1121×4, *Hoxna* [1154×89]1232, *Hoxon* 1524, 1610, 1674. OE **hōhsinu** 'the heel sinew, a hough'. The reference is to a spur of land shaped like a horse's hough. Skeat's suggestion of a folk-n. **Hoxan* is not, however, impossible. DEPN and Baron give pr [hɔks(ə)n].

HOYLAKE Mers SJ 2187. *Highlake* 1766, *High Lake* 1794, *Hoylake* 1806. Originally a hamlet which grew up around a hotel built for sea-bathers in 1792. The name was transferred from *Hyle*

Lake 1687–1809, *the Hyle-Lake* 1751, *High Lake* 1689–1793, *Hoyle Lake* 1796, *Hoylake* 1813, 'Hile Lake', p.n. Hile as in East HOYLE BANK SJ 2291 + ModE **lake**, an anchorage or roadstead, now silted up, off the NW coast of Wirral; inside East Hoyle Bank. Che iv.299.

HOYLAND 'Newly cultivated land by a hill-spur'. OE **hōh** + **land**. Cf. HOLLAND. Forms with <o> and <u> spellings represent OE *hō-land*, those with <oi> and <oy> OE *hōȝ-land* with *hōȝ* from analogical inflected forms **hōȝes*, **hōȝe* beside expected *hōs, hō(e)*. BzN 16.304–6.

(1) Upper ~ SYorks SE 3501. ModE **upper** (*Uper-* 1672) + p.n. *Hoiland* 1199, *Holand(e)* 1240, 1316, *Overhoyland* 1583×5. YW i.112.

(2) NETHER ~ SYorks SE 3600. 'Lower H'. P.n. *Ho(i)land* 1086, *Holand(e) -landa* early 13th–1548, (*Nether-* 1390), *Huland* 1406, *Hoyland(e)* 1580–1822, + ModE **nether** (*Netherhooland* 16th cent.). In the 1086 *oi* spellings *i* is probably a diacritic of length for *ō*. Later *oy* spellings represent local dial. *oi* for *ō*. YW i.111, L 246.

(3) ~ SWAINE SYorks SE 2604. 'Sveinn's H'. *Holand(e) Swayn(e)* 13th, *Holanswayne* 14th cent., *Hulandswayne* 1451. P.n. *Holan(de), Holant* 1086, *Holand(e)* 12th–1468, *Hoi- Hoyland* c.1195–1822, + ON pers.n. *Sveinn* borne by several 12th cent. tenants. YW i.308.

East HOYLE BANK Mers SJ 2291. *Hoyle Bank* 1806. Earlier *Hyle Sand* 1687, *Hoyle Sand* 1757. ModE dial. **hile** 'a hillock' (OE **hygel*, ME *huyle*) referring to a sand-bank off the NW coast of Wirral. 'East' for distinction from West Hoyle Bank SJ 1487. Che iv.299.

HUBBERHOLME NYorks SD 9278. 'Hunburg's homestead'. *Huburgheha'* 1086, *Hulber- Hobre- Huburham* 13th, *Hubberham* 1577, *-holme* 1686. OE fem. pers.n. *Hūnburg*, genitive sing. *Hūnburge*, + **hām**. YW vi.116.

HUBBERT'S BRIDGE Lincs TF 2643. *The Bridge over the great dreane called Hubards bridge* 1653, *Hubbard Bridge* 1824 OS. Surname *Hubbard* as in Allen Hubbat 1551, John Hubbert 1552, + ModE **bridge**. Cf. *Hobarts Hurne* 1529, *Hubbert Syke* 1547. The bridge carries the B 1192 over the South Forty Foot or Land Eau in Holland Fen W of Boston. Payling 1940.84, Cameron 1998.

HUBY NYorks SE 2747. 'Hugh's farm or village'. *Huby* from 1198, *Hug- Hugheby* 13th, *Hewby(e) -bie* 1536–1748. ME pers.n. *Huwe, Hew* < CG *Hugo*, + ON **bȳ**. YW v.52 gives pr ['iubi].

HUBY NYorks SE 5665. 'Farm or village at the hill-spur'. *Hobi* 1086–1179, *-by* 1135–1399, *Houby* 1326, *Huby* from 1398, *Hewby* 1538, 1614. OE **hōh** or ON **haugr**, + **bȳ**. Huby lies at the end of a low spur of land. YN 18.

HUCCLECOTE Glos SO 8717. 'Hucela's cottage(s)'. *Hochilicote* 1086, *Hu(c)kli- Hukelin(g)cot(e)* 13th cent., *Hokelecot(e)* c.1260–1329, *Huc(c)lecote* c.1560, 1590. OE pers.n. **Hucel(a)* + **ing**[4] + **cot**, pl. **cotu**. Gl ii.147.

HUCKING Kent TQ 8458. 'The Huccingas, the people called after Hucca'. *Hocking'* 1225, 1252, *Hukinges, Hucking', Hoking'* 1226, *Huckinges* 1270, *Houkyng(ge)* 1311–92, *Huckyng* 1610. OE folk-n. **Huccingas* < OE pers.n. **Hucca* + **ingas**. PNK 219, ING 12.

Great HUCKLOW Derby SK 1777. *Magna Hockelawe* 1251, ~ *Hokelowe* 1323–82, *Great Hucklow* 1610. Adj. **great**, Latin **magna**, + p.n. *Hochelai* 1086, *Hoccalawe* 1101×8, *Hokelow(e)* 1262–1390, *Hu(c)klow(e)* [1199] 1265, 1285 etc., 'Hucca's burial-mound' or 'hill' referring to Hucklow Edge and the high ground N of the village which rises to 1365ft. at SK 1879. OE pers.n. **Hucca* + **hlāw**. 'Great' for distinction from Little Hucklow SK 1678. Db 131.

Little HUCKLOW Derby SK 1678. *Parva Hokelawe* 1216×72, ~ *Hokelowe* 1285, 1382, *Litle Hokelowe* 1454, *Little Hucklow* 1526. Adj. **little**, Latin *parva*, + p.n. Hucklow as in Great HUCKLOW SK 1777.

HUCKNALL Notts SK 5349. 'Hucca's nook of land'. *Hochenale, Hochehale* 1086, *Hockenhala* 1160, 1161, *Huccenhal'* 1163, 1165, *Hukenhal(e)* c.1190–1442, *Hucknall* 1590. OE pers.n. **Hucca*, genitive sing. **Huccan*, + **halh**. Also known as *Hokenhale Torkard* 1287, *Huckney Torquet* 1700 from the family of Geoffrey Torchard who held the manor in 1195 and John Torcard in 1235. Nt 118 gives pr [uknə tɔ:kəd].

HUDDERSFIELD WYorks SE 1416. Possibly 'Hudræd's piece of open country' or 'open land of the shelter'. *Oderesfelt, Odresfeld* 1086, *Huderesfeld* 1114×31–1421, *Hudresfeld* 12th–1362, *Huddredisfeld* 1241, *Huthersfe(i)ld(e)* 1545–1623. Either OE pers.n. **Hudrǣd*, genitive sing. **Hudrǣdes* or OE ***hūder**, genitive sing. ***hūderes**, + **feld**. YW ii.295 gives pr ['uðəzfild], LSE xviii.136.

HUDDINGTON H&W SO 9457. 'Estate called after Huda or Hod'. *hudigtuna* (genitive pl.) 11th S 1591, *Hvdintvne* 1086, *Ho- Hudinton -yng-* 1232–16th, *Huddington* 16th. OE pers.n. *Hūda* or *Hod* + **ing**[4] + **tūn**. The first form occurs in an unattached boundary clause of Crowle in the form *into hudigtuna gemære* 'to the boundary of the *Huding tunas*, the Huding enclosures'. A *hodes ac* 'Hod's oak-tree' occurs in the bounds of neighbouring Dormston 11th S 786 at or near Hollow Court SO 9758. Wo 142, Hooke 384, 194.

HUDSWELL NYorks NZ 1400. Partially uncertain. *Hudres- Undreswelle* 1086, *Hud(e)leswell* early 12th, *Hud(d)es-* 12th–1519, *Hutles-* 1204. The generic is OE **wella** 'a spring', the specific uncertain, possibly OE ***hūd(e)**, ***hūder**, ***hūdel(s)** 'hut, shelter, cover', used like *helm* of a hut-shaped hill. Older explanations saw the specific as unattested pers.n. **Hūdel*. YN 245, LSE 1987.136.

HUGGATE Humbs SE 8855. 'The road to the mound or mounds'. *Hughete* 1086, *Hugat* 12th–1347, *-gate* 12th–1493, *Hugget(t)* 16th cent. ON ***hugr** + **gata**. The reference is to the ancient road to York and to the tumuli on Huggate Wold. YE 174 gives pr [ugit], SSNY 158.

HUGHENDEN VALLEY Bucks SU 8696. P.n. Hughenden + ModE **valley**. Hughenden is a 19th cent. reformation influenced by the pers.n. Hugh. The early forms are:

I. *Huchedene* 1086, *Huc(c)henden* 1199–1539, *Hutchenden* 1607, 1633.

II. *Hichenden(a)* [c.1145]c1230, 1241, 1535, *Hidgingdon* 1700, *Hitchenden* 1703, 1806.

III. *Hug(g)eden(e)* 1154–1241.

Type III could be 'Hycga's valley', OE pers.n. **Hycga*, genitive sing. **Hycgan*, + **denu**, types I and II possibly 'Huhha's valley', OE pers.n. **Huhha*. The precise relationship of these forms and their phonological development is unclear: neither type could produce the pr [u:əndən] given at Bk 182, L 106.

HUGHLEY Shrops SO 5698. 'Hugh's clearing'. *Huleye, Hewelege* c.1291, *Hughleye* 1334. Earlier simply *Leg'* 1203×4, *(La) Legh'* 1231, OE **lēah**. The ME pers.n. prefixed to *lēah* is that of *Hug' de Leg'* 1203×4, for distinction from other neighbouring *lēah* names. Sa i.161.

HUGH TOWN Scilly SV 9010. *The Hugh or New Towne* 1652. From the hill and promontory called *The Hew* c.1585, *the Hew Hill* 1593, *the Hugh Hill* 1652, ME **heugh** (OE *hōh*) 'a heel, a spur of land'. 'New' by contrast with Old Town SV 9110. PNCo 98.

HUISH 'A household, a family, the measure of land that would support one household or family'. OE **hīwisc**. Identical with HEWISH. Found only in D, Do, So and W. PNE i.248.

(1) ~ Devon SS 5311. *(H)iwis* 1086 etc. with variants *Hywis(c)h, Hywys(s)h*. D 93.

(2) ~ Wilts SU 1463. *Iwis* 1086, *Hi- Hywis* 1162–1242, *Huwes* 1291, *Hewishe* 1629. Wlt 319.

(3) ~ CHAMPFLOWER Somer ST 0429. 'H held by the Champflower family'. *Hywys Champflur* 1274, *Hewishe Cam̃flowre* 1610. Earlier simply *Hiwis* 1086, [1187] Buck. Thomas de Champflower (Chanflur, Campo Florido) was the

terre-tenant in 1166. The name comes from Champfleury, Aube or Marne, Latin *Campus floridus* 1154 'flowery field'. DEPN, Dauzat-Rostaing s.n. Camps, VCH.

(4) ~ EPISCOPI Somer ST 4326. 'The bishop's H'. P.n. Huish + Latin **episcopi**, genitive sing. of **episcopus**. The manor was held by the bishop of Wells. Earlier simply *Hiwissh* [973]14th S 791, *Hiwisc* *[1065]18th S 1041, *Hewish* 1610. DEPN.

(5) North ~ Devon SX 7156. *Northywys* 1285. ME adj **north** + p.n. *Hewis* 1086, *Hiwiss'* 1242, *Hywysh* 1291. 'North' for distinction from South Huish SX 6941, *Suthhywish* 1302, earlier simply *Heuis* 1086, *Hywis* 1242. D 303, 304.

HULCOTE Bucks SP 8516. Possibly 'Hucca's cottage(s)'. *Hoccote* 1200, *Huccote -ck-* 1227–1390, *Hulecote* 1228, *Hulcote* 1535. OE pers.n. *Hucca* + **cot**, pl. **cotu**. The only early form with *-l-* is *Hulecote* 1228: it is likely that from the 16th cent. the name has been influenced by HOLCOT Northants SP 7969 or Beds SP 9540. Bk 151.

River HULL Humbs TA 0646. Uncertain. *(neah) þare eá Húl, éá Hull* [c.1030]11th Secgan, *amnem Hul(l)* [c.1085]12th, *Hul* c.1200–1312, *Hull* from 1279. This has been taken to be Dan **hul** 'deep' and so 'the deep river, the river which flows through a cut channel', presumably a fenland drain. However, the section of *Secgan* in which the name first appears is thought to be of pre-Viking origin and a Celtic source is preferable, e.g. from the root *seul-* 'mud' as in Lithuanian *sulà* 'birch sap', Latvian *sula* 'fluid, sap', OPrussian *sulo* 'curdled milk', with Brit lenition of *s* to *h*. The sense would be 'muddy river, fen river'. This root, an *-l-* extension of IE *sew-* 'to flow, to rain', is represented in OE *sol* 'mud, wet sand', *solu* 'filth, mire', and other cognates; but no parallel r.n. seems to survive in Wales. Possible continental parallels, however, are the Sill at Innsbruck, *Sulle* 1187, and the *Sulmana*. YE 6, RN 201, IEW 913.

HULLAND Derby SK 2446. 'Newly-cultivated land by a hill-spur'. *Hoilant* 1086, *Hoyland* 1230, *Holand(e)* 1257–1545. OE **hōh** + **land**. Db 574, L 167, 248–9.

HULLAND WARD Derby SK 2547. *ward of Holand* 1262, *Holande Warde* 1330. P.n. HULLAND SK 2446 + **weard**. One of the administrative divisions of Duffield Frith, an area formerly more extensive than the modern parish. Db 576.

HULLAVINGTON Wilts ST 8982. 'The estate called after Hunlaf or Hundlaf'. *Hunlavintone* 1086–1343 with variant *-yn-*, *Hundlavinton -yn-* 1203–1310, *Hunlavyngton* 1347, 1415, *Hullavington, Hullyngton* 1583. OE pers.n. *Hūn-* or *Hundlāf* + **ing**[4] + **tūn**. Wlt 71 gives pr [hʌliŋtən].

HULLBRIDGE Essex TQ 8194. 'Bridge over the Wholve'. *pontem de la Wlwne* (for *Wlwue*) Latin 'bridge of the W' 1207, *Whoulnebregg'* for *Whoulue-* 1375, *Wolvebrigge* 1377, *Whulbridge* 1492, *Hul(l)brigge* 1480. R.n. *Wholve*, ME **wholve** 'a passage for water' < OE *hwealf* 'arch, vault, concave, hollow', an old name of the CROUCH, + **brigge, bregge**. Ess 187, YE lvii.

HULME 'Island' in various senses. ON **holmr**. L 50.

(1) ~ GMan SJ 8396. *Hulm* 1246, 1310, *Hulme* from 1440, *Holme* 1577. Earlier *Ouerholm, Noranholm* 1226, *Norholm* 1227, OE **uferra** 'upper' and **north**. La 33 gives prs [hu:m] and [hju:m], SSNNW 135 no.8.

(2) ~ END Staffs SK 1059. *Holme End* 1840 OS. P.n. Hulme as in Hulme House SK 0961, *Hulme* from 1227, *Hulm, Holm under Kevermund* 1293. The sense here is probably 'enclosure in marginal land'. DEPN, L 52.

(3) ~ WALFIELD Ches SJ 8465. Originally two distinct manors, *Hulle et Walmefeld* 1295, *Hulm et Wallefeld* 1319, *Hulm-Wallefeld* 1338, *Hulmewalfield* 1413. Hulme, *Hulm'* c.1262–1307, *Hulme* 1296, is 'the island, the water-meadow', ME dial. **hulm** (<ON *holmr*) and Walfield, *Wallefeld* c.1262–1361, *Walfield* from 1361, *Walmefeld* 1295, is 'open land with a spring', OE **wælla** varying with **wælm**, + **feld**. Che ii.302.

(4) Cheadle ~ GMan SJ 8786. 'Hulme, the water-meadow belonging to Cheadle'. *Chedle Hulm* 1345. Earlier simply *Hulm* late 12th–1496, *Holme* 1337–1585. Che i.247, SSNNW 136 no.12.

(5) Upper ~ Staffs SK 0161. *Overhulm'* c.1240, *-hulm(e)* c.1553×39–1653 etc., *Uver(e)hulm(e)* 13th, *Upper or Over Hulme* 1775, 'the upper water-meadow'. ME **over**, ModE **upper** + p.n. Hulme as in *Middelhulm'* 13th SJ 9962 and in *Nether Hulm(e)* 1240. Oakden.

HULNE PARK Northum NU 1615. *Hulne Park* 1868 OS. P.n. *Hol* 1271, *Holne* 1283–1334, *Holen* 1296 SR, *Hull* 1290, 'the holly-tree (wood)', OE **holeġn**, + ModE **park**. The loss of final *-n* in some 13th cent. forms may be compared with ModE *mill* < OE *mylen* and dial. *kill* < OE *cylen*. NbDu gives pr [hul].

HULNE PRIORY Northum NU 1615. *Hulne Abbey* 1868 OS. P.n. Hulne as in HULNE PARK NU 1615 + ModE **priory**. The reference is to the Carmelite friary founded here in 1242.

Abbey HULTON Staffs SJ 9148. 'H belonging to the abbey'. *Abbehilton* 1587, *Abby Hulton* 1678. ModE **abbey** + p.n. *Heltone* 1086, *Hiltona* 1166, *Hulton* from 1235, 'the hill settlement', OE **hyll** + **tūn**. Hulton was a Cistercian abbey founded in 1219. DEPN, L 170.

Little HULTON GMan SD 7203. *Little Hulton* 1843 OS. Adj. **little** + p.n. *Hilton* 1200–17th, *Hylton* 1219, 1256, *Hulton* from 1212, 'hill settlement', OE **hyll** + **tūn**. 'Little' for distinction from Middle Hulton SD 7006, *Medyll Hilton* 1552, and Over Hulton SD 6706, *Overhilton* 1521. All three places occupy sites along the same ridge of high ground. La 43, Jnl 17.36.

HULVER STREET Suff TM 4786. 'Hulver village'. ModE **hulver** 'holly' + dial. **street**.

HUMBER BRIDGE Humbs TA 0224. The largest single span suspension bridge in Europe. R.n. HUMBER SE 9523 etc. + ModE **bridge**.

HUMBER COURT H&W SO 5356. P.n. Humber + ModE **court**. Humber, *Hūbre* 1086, *Humbre* 1200×16–1341, is named from Humber Brook, an unexplained pre-English r.n. identical with River HUMBER TA 2316. RN 202, Bannister 99.

River HUMBER Humbs SE 9523, TA 1326, TA 2415. Unexplained. *Humbri fluminis* [c.720]10th BHA, [725]9th with variants *Umbri, Humbris*, [c.731]8th BHE with variants *Hymbri, Umbri*, 1160×2, *Humbrae, Humbre, Hymbrae fluminis* with variant *Hymbre* [c.731]8th BHE, *Humbre* c.895 ASC(A)–1475, *Humber* from 1279. RN 204 draws attention to a lost W r.n. *Humir, Humur, Humri riuuli* c.1150, *Humer* 15th, which might represent OW **Humbr-* from a base **Sumbr-* with Celtic *su-* 'good, well'. The sense would be 'the good river', used either literally (e.g. rich in fish) or magically (as a spell against the river's dangerous currents etc.). However, the etymology and meaning of Humber and the ten other English examples of this name are really quite unknown. The name of this important river is probably pre-Celtic. YE 8, RN 201ff, LHEB 510–1, 519.

HUMBERSIDE AIRPORT Humbs TA 0910. Modern p.n. (1974) *Humberside* < r.n. HUMBER SE 9523 etc. + ModE **side**, + ModE **airport**.

HUMBERSTON Humbs TA 3005. 'The Humber stone'. *Humbrestone* 1086, 15th, *Humberstein* c.1115–1240, *-stain -stayn* 1228–1423, *Humbrestan -er-* 1154×89–1450, *Humberston* from 1242. R.n. HUMBER SE 9523 etc. + OE **stān** varying with ON **steinn**. It referred to a great boundary stone recorded in 1634 as lying just at the place 'where Humber looseth himself in the German ocean'. Li v.116, SSNEM 202.

HUMBLETON Humbs TA 2234. Probably 'farm or village by the rounded hillock'. *Humeltone, Umelton* 1086, *Humbletun -ton* 1154×80–1828. OE ***humol** or Scand **humul** + **tūn**. There are two low glacial mounds in the neighbourhood. But other possibilities are ON *humli* 'hop' perhaps replacing OE *hymele*, or ON pers.n. *Humli*. YE 54, SSNY 114.

HUMBLETON Northum NT 9728. 'The mutilated hill'. *Hameldun* 1169, *-don -il-* 13th cent., *Homeldon(e) -yl-* 1296–1538, *Humbledon* 1403, *-ton* 1580. OE ***hamol** + **dūn**. There is a well marked cleft in the hill which has also been terraced for crop growing. NbDu 120, Tomlinson 502.

HUMPHREY HEAD POINT Cumbr SD 3973. P.n. Humphrey Head + ModE **point**. Humphrey Head, *Hunfrid'heved -frides-* 1199, *Umfrayhede* 1537, *Oumfray head* 1577, is 'Hunfrith's headland', OE pers.n. *Hūnfrith*, genitive sing. *Hūnfrithes*, + **hēafod**. La 196.

HUMSHAUGH Northum NY 9271. 'Hun's nook'. *Hounshale* 1279, *-halgh* 14th cent., *Homeshalk* 1318, *Humshaugh* 1663. OE pers.n. *Hūn*, genitive sing. *Hūnes*, + **halh**. The reference is to land in a bend of the N Tyne river. NbDu 121, L 108.

HUNCOAT Lancs SD 7730. 'Huna's cottage(s)' or 'honey cottages, place where honey is made'. *Hunnicot* 1086, *Huncote* 1167×8, 1332 etc., *Hun(n)ecotes* 1241, *Huncotes* 1246. OE pers.n. *Hūna* or OE **hunig**, + **cot**. La 91, Jnl 17.54.

HUNCOTE Leic SP 5197. 'Huna's cottage(s)'. *Hunecot(e)* 1086–1329, *Huncot(e)* from 1259. OE pers.n. *Hūna* + **cot**, pl. **cotu**. Lei 508.

HUNDERTHWAITE Durham NY 9821. 'Clearing of the hundred' or 'of the hundred court'. *Hundredestoit(h)* 1086, *Hundresthuait* [1184]15th, 1302, *Hunderthuait -thwayt* 1208–1352. OE **hundred**, genitive sing. **hundredes**, + ON **thveit**. The land division names of Yorkshire are found only in the form of Scandinavian wapentakes. But hundreds must have existed before the Scandinavian settlements and this may be a survival. Alternatively *hundred* may be used here to denote a subdivision of a wapentake or the wapentake court. YN 307, SSNY 97.

HUNDERTHWAITE MOOR Durham NY 9319. *Hunderthwaite Moor* 1863 OS. P.n. HUNDERTHWAITE NY 9821 + **mōr**.

HUNDLE DALE Humbs SE 8158. 'Hundolf's valley'. *Hundoluesdale* 1156–7. ON pers.n. *Hundulfr* or OG *Hundulf*, genitive sing. *Hundulfes*, + **dalr**. YE 131, SPN 145.

HUNDLEBY Lincs TF 3866. 'Hundulf's village or farm'. *Hundelbi* 1086, 1188, *Hundelbia* c.1145, 1141×54, *Hundelby* 1209–53. Pers.n. ON *Hundulfr* or OG *Hundulf* + **bȳ**. DEPN, SSNEM 54, Cameron 1998.

HUNDON Suff TL 7348. 'Huna's valley'. *Hvne(n)dana* 1086, *Hundeden(e)* 1219, 1263, *Honeden* 1610. OE pers.n. *Hūna*, genitive sing. *Hūnan*, + **denu**. DEPN.

The HUNDRED H&W SO 5264. *Hundred* 1832 OS. The hundred history of Herefordshire is complex. If ancient, this name may refer to land belonging to the hundred of Leominster which included nearby Ashton SO 5164 and Middleton on the Hill SO 5464. *DB Herefordshire* Note 7.

HUNDRED ACRES Hants SU 5911. No early forms. ModE **hundred** + **acre(s)**.

HUNDRED END Lancs SD 4122. Refers to the N end of Leyland Hundred, the district S of the Ribble.

The HUNDRED FOOT WASHES Cambs TL 4988. *The Washes* 1824 OS. ModE **wash** (OE *(ġe)wæsc*) 'land liable to flood' applied locally to land on the banks of the Ouse or Old West River, the early name of which seems to have been *Wasselode* 1251, *(S)quasselode* 13th, *Wasschlood* 1549, 'watercourse through lands that flood'. The Old West River once took the west branch of the Ouse from Earith Bridge to Benwick. In 1651 the Hundred Foot or New Bedford River was constructed parallel to the Old Bedford River (a new channel for the Ouse made in 1631) to take the water from Earith to Downham. Ca 17, 210, 212, L 59.

HUNDRED STREAM Norf TG 4521. *Hundred Stream* 1837 OS. A name of the upper Thurne which formed the boundary between West Flegg and Happing hundreds. ModE **hundred** + **stream**. RN 205.

HUNGARTON Leic SK 6907. 'Hunger farm or village'. *Hungreton(e)* 1086–1351, *Hungerton(e)* c.1130–1610, *Hungarton(e)* from 1440. OE **hungor** + **tūn**. The reference is to the barrenness of the soil which is here clay and gravel overlying clay. Lei 304, Gaz.

Emneth HUNGATE Norf TF 5107. 'Emneth hamlet called Hungate'. No early forms, but for **gate** in this sense see EASTGATE TG 1423.

HUNGERFORD Berks SU 3368. 'Ford leading to poor land'. *Hungreford* 1101–99, *Hungerford* from [1147]c.1315. OE **hungor** + **ford**. Brk 301, PNE i.269, L 69. 2.

HUNGERFORD NEWTOWN Berks SU 3571 → Hungerford NEWTOWN.

HUNGRY LAW Northum NT 7406. *Hungry Law* 1869 OS. This type of name is a common derogatory term meaning 'barren, unfertile hill'. Mod dial. **law** 'a hill' (OE *hlāw*).

HUNMANBY NYorks TA 0977. 'Farmstead or village of the houndsmen'. *Hundemanebi* 1086, *Hundeman(n)ebi -by* c.1135–1331, *Hundmaneby* 1241, *Hunmanby* from c.1130, *Humanby* 1286, *Hummon- Hunnonby(e)* 16th. ON ***hunda-mann**, genitive pl. ***hunda-manna**, + **bȳ**. YE 108, SSNY 30.

HUNNINGHAM Warw SP 3768. Probably 'the homestead of the Huningas, the people called after Hun or Huna'. *Hvningehā* 1086, c.1165, *-ingenham* 1291, *Huningham* c.1203–36, *Honyngham -ing-* 1288–1535, *Hunningham* 1242. OE folk-n. **Hūningas* < pers.n. *Hūn* or *Hūna* + **ingas**, genitive pl. **Hūninga*, + **hām**. But the existence of a p.n. **Hūning* 'the place called after Huna' < pers.n. *Hūna* + **ing**[2], locative–dative sing. **Hūninġe*, or of an OE **huning* 'honey-place' is also possible. Cf. HONINGTON SP 2642. Wa 133, ING 128.

HUNSDON Herts TL 4114. 'Hun's hill'. *Hones- Hodesdone* 1086, *Hunesduna -done* 12th–1324, *Hunsdon* from 1368. OE pers.n. *Hūn*, genitive sing. *Hūnes*, + **dūn**. Hrt 193.

HUNSINGORE NYorks SE 4253. 'Ridge at *Hunsiging*, the place called after Hunsige'. *Holsingoure, Ulsingouere* 1086, *Hulsingour(e) -ure* 1086–1227, *Hunsing(h)ouer(e) -(h)our(e) -yng-* 1194–1397, *Hunsingor(e) -yng-* from 1303. OE p.n. **Hūnsīġing* < pers.n. *Hūnsīġe* + **ing**[4], + **ofer**. The reference is to a low ridge overlooking the river Nidd. YW v.18.

High HUNSLEY Humbs SE 9535. ModE **high** + p.n. *Hundeslege* 2×, *Hvndreslege* 1× 1086, *Hundeslai -lei(e) -le(y)* 1100–1410, *Hunsley* from 1570. Either 'Hund's clearing' or 'the hundred clearing', pers.n. *Hund*, genitive sing. *Hundes*, or **hundred**, genitive sing. **hundredes**, + **lēah**. The choice depends on the weight to be given to the unique spelling with *-r-*. YE 204.

HUNSONBY Cumbr NY 5835. 'Village or farm of the dog-keepers'. *Huns(h)wanby* 1292, *Hundswa(y)nby* 1345, 1348, *Hunsanby* 1339, *Honsonby* 1393. ON **hunda-sveinn** + **bȳ**. ON *hunda-sveinn*, which was probably originally genitive pl. *hunda-sveinna*, has been influenced by OE *swān*. Cf. HUNMANBY TA 0977. Cu 207, SSNNW 33.

HUNSTANTON Norf TF 6741. 'Hunstan's estate'. *Hunstanestun* 1035×40 S 1489, c.1150, *Hunestanestuna* 1086. OE pers.n. *Hūnstān*, genitive sing. *Hūnstānes*, + **tūn**. DEPN gives pr [hʌnstn].

HUNSTANWORTH Durham NY 9449. 'Hunstan's enclosure'. *Hunstanworth* from [1183]c.1382, *Hunstanesworde* 1195×8, *-wrthe* 1242×3, *Hunsonworth, Hunsenwood* 17th. OE pers.n. *Hūnstān* + **worth**. NbDu 121.

HUNSTON Suff TL 9768. 'The huntsman's estate'. *Hunterstuna* 1086, *-ton* 1197, *Hunterestun* 1095, *Hunston* 1610. OE **huntere**, genitive sing. **hunteres**, + **tūn**. DEPN.

HUNSTON WSusx SU 8601. 'Huna's stone'. *Hunestan* 1086–1235, *Hunstan(e)* 1271–1329, *Hunston* from 1332, *Hunston al. Hunson al. Huston* 1756. OE pers.n. *Hūna* + **stān** probably referring to a boundary stone. Sx 71.

HUNSTRETE Avon ST 6462. 'Hund's tail of land'. *Hundesteorte* [1132]14th Secretum, *Hundestert(e)* 1189 Glast-1277, *Houndesterte* 1327, *Hundestret* 1316, *Hound Street* 1817 OS. OE pers.n. *Hund*, genitive sing. *Hundes*, + **steort** 'a tail, a projecting piece of land'. DEPN s.n. Houndstreet, PNE ii.151.

HUNT END H&W SP 0364. 'Hunt end' sc. of Feckenham. *Hunt End* 1831 OS. Surname Hunt, *le Hunte, Hounte* 1275, 1327, ultimately from OE *hunta* 'a huntsman'. Wo 319.

HUNT HOUSE NYorks SE 8198. 'Houses of the huntsman or -men'. *Huntereshuses* 1252. ME **huntere**, genitive sing. or pl. **hunteres**, + **hūs**. YN 83.

HUNT'S CROSS Mers SJ 4385. *Hunt's Cross* 1842 OS. Surname *Hunt* + ModE **cross** 'a crossroads'.

HUNTINGDON Cambs TL 2371. 'The huntsman's hill'. *Huntandun* [973]14th S 792, c.950 ASC(A) under year 921, c.1000 HistEl, *Huntedun* 1086, *Huntendon(ia) -in-* 12th–1259, *Huntyngdon* 1286. The shire is mentioned as *Huntandunscir* c.1200 ASC(E) under years 1011 and 1016, *Hunta- Huntedunscir* 11th ASC(C and D) under same years. OE **hunta**, genitive sing. **huntan**, + **dūn**. Alternatively the specific might be the OE pers.n. *Hunta*. Hu 261, TC 112, L 145, 153.

HUNTINGFIELD Suff TM 3473. 'The open land of the Huntingas, the people called after Hunta'. *Huntingafelda -e* 1086, *Huntingefeld -feud* 1156–1254, *Huntingfeld(')* 1185–1336, *-fild* 1568. OE folk-n. **Huntingas* < pers.n. *Hunta* + **ingas**, genitive pl. **Huntinga*, + **feld**. DEPN, Baron, L 244.

HUNTINGTON H&W SO 2553. 'The huntsmen's estate'. *Hantinetune* 1086, *Huntintone* 1137×9, *Huntington* 1275. OE **hunta**, genitive pl. **huntena**, + **tūn**. He 108.

HUNTINGTON NYorks SE 6256. Partly uncertain. *Huntindune* 1086, *Hunten-* c.1160, *Huntin(g)don -yn(g)-* 1159–1536, *-ingedon'* 1187, *-ingtuna -ton(a) -yng-* from c.1150, *-inton(a)* c.1180–1294. The evidence is given in full to illustrate that there are alternative forms of this name, viz. OE **Huntan* or **huntan**, + **dūn**, 'Hunta's or the huntsman's hill' or possibly genitive pl. **huntena**, 'the huntsmen's hill'; or OE p.n. **Hunting(e)* + **dūn**, 'the hill called or at **Hunting*' in which *Hunting* might be the 'place called after Hunta', OE pers.n. *Hunta* + **ing**[2], locative-dative **Huntinge*, or OE **hunting** 'that which has to do with the hunt'. YN 12, BzN 1968.173, L 145.

HUNTINGTON Staffs SJ 9713. 'The huntsmen's hill'. *Huntendon(')* 1167–1236, *-in(g)- -yn(g)-* c.1255–1456, *Huntyngton -ing-* from 1444. OE **hunta**, genitive pl. **huntena**, + **dūn**. The place is referred to in DB as *Estendone* 1086, apparently '(the place) east of the hill', OE **ēastan** + **dūn**, which is topographically difficult; the modern settlement lies W of the high ground of Huntington Belt. St i.68, L 145.

HUNTLEY Glos SO 7219. 'The huntsman's wood or clearing'. *Hvntelei* 1086, *Huntele(i) -leg(h) -ley(e)* 1211×3–1535, *Huntleia -ley(e)* c.1145–1592. OE **hunta** + **lēah**. Gl iii.191.

HUNTON Kent TQ 7149. 'The huntsmen's estate'. *Huntindune -tune* c.1100, *Huntinton(e)* 1210×12–1264, *Huntingtun' -ton(')* 1226–56, *Hunton* 1610. OE **hunta**, genitive pl. **huntena**, + **tūn**. PNK 160.

HUNTON NYorks SE 1892. 'Huna's settlement'. *Hunton(e)* from 1086. OE pers.n. *Hūna* + **tūn**. YN 241.

HUNTSHAM Devon ST 0020. 'Hun's homestead or enclosure'. *Honesham* 1086–1316, *Hun(n)esham* 1238, 1297, *Hunsam* 1543. OE pers.n. *Hūn*, genitive sing. *Hūnes*, + **hām** or **hamm**. D 540 gives pr [hʌnsəm].

HUNTSPILL Somer ST 3045. 'Hun's creek'. *Hunespulle* [793×800]13th S 1692, *-pille* 1084, [n.d.] Buck, *Hons- Hvnespil* 1086, *Hunespil(le)* 1170, 1177, *Honespull* [n.d.] Buck, *Huntspill* 1610. OE pers.n. *Hūn*, genitive sing. *Hūnes*, + **pyll**. DEPN.

HUNTSPILL LEVEL Somer ST 3245. *Huntspill Level* 1809 OS. P.n. HUNTSPILL ST 3045 + ModE **level** 'flat land, fen'.

East HUNTSPILL Somer ST 3134. ModE adj. **east** + p.n. HUNTSPILL ST 3045.

HUNTWORTH Somer ST 3134. 'The huntsman's enclosure'. *Hvnteworde* 1086, *-worth* 1225, *Huntworth* 1610. OE **hunta** (or pers.n. *Hunta*) + **worth**. This seems to have been the estate of the huntsman in the royal forest of North Petherton. DEPN.

HUNWICK Durham NZ 1832. 'Huna's farm'. *Hunewic* [c.1040]12th, [1072×83]15th, c.1104, 1242×3, *-wyc, -wyk* [1183]c.1382, 1242, 14th cent., *Hunwyk* [1183]c.1320–1542, *Hunwick* from 1603. OE pers.n. *Hūna* + **wīċ**. NbDu 121.

HUNWORTH Norf TG 0735. 'Huna's enclosure'. *Huneuurda, Hunaworda* 1086, *Hunewrth, Hunesworth* 1211–1254. OE pers.n. *Hūna* + **worth**. DEPN.

HURDSFIELD Ches SJ 9274. 'Open land of the hurdle'. *Hirdelesfeld* 13th, *Hyrdes- Hirdes- Herdes- Hurdesfeld -is-* mid 13th–1369, *Hursfield* 1724. OE **hyrdel**, genitive sing. **hyrdeles**, + **feld**. Che i.106, L 240, 244. Addenda gives pr ['hərdz-'hərtsfi:ld], 19th cent. local ['u:tsfılt].

HURGIN Shrops SO 2379. *Hirgan* 1833. Possibly a W p.n., perhaps **hir** 'long' + **gên** 'cheeks'.

The HURLERS Corn SX 2571. *The Hurlers* 1584 Gover n.d., 1602. Three Bronze Age stone circles named from the Cornish ball-game of *hurling* because the stones were felt to resemble the players at a hurling match. PNCo 98, Gover n.d. 162, Pevsner 1951.89, Thomas 1976.59.

HURLEY Berks SU 8283. 'Wood or clearing at a recess in the hills'. *Herlei* 1086, *-leia* 1130 etc., *Hurleya -leia* 1086 etc. with variants *-lea -leg -le(e), Hurnleia* 1106×21, *Hurnl(e)y -le* 13th cent. OE **hyrne** + **lēah**. Brk 60, L 206.

HURLEY Warw SP 2496. 'The corner wood or clearing'. *Hurle* 1154×89, *Hurlega -ley(e) -le(ge)* c.1180–1658, *Hurnlee -lega* c.1180–1262. OE **hyrne** + **lēah**. So called because of its location in the furthest corner of Kingsbury Lordship. Wa 18, L 206.

HURN Dorset SZ 1297. 'Angle or corner of land'. *Herne* 1086, *(la) Hurne* 1242, 1367, *Hern* 1811 OS. OE **hyrne**. The reference is either to the piece of land between the Moors river and the r. Stour, or to the position of the place in the NW of old Christchurch parish. Do 90, Ha 96.

HURSLEY Hants SU 4225. 'Mares' clearing or pasture'. *Herselega Episcopi* 'bishop's Hursley' 1167, *Hurseleye* 1171–1379 with variants *-le(gh), Hurseleghe* 1263. OE ***hyrse**, genitive pl. ***hyrsa**, + **lēah**. It was part of the bishop of Winchester's manor of Merdon SU 4226, *Merdona* 1138, *Merendon(a)* 1167, *Meredon* 1184, itself possibly the 'mares' hill', OE **miere**, genitive pl. **mierena**, + **dūn**. Ha 96–7, Gover 1958.175, Studies 1936.65, PNE i.276, L 206.

HURST 'Wooded hill'. OE **hyrst**, SE dial. form **herst**. PNE i.276, L 197.

(1) ~ Berks SU 7973. *Herst* 13th cent., *Hurst(e)* from 1224×5, *La Hurst(e)* 1242–1376. The parish name is St Nicholas Hurst from the dedication of the church. Brk 99.

(2) ~ GMan SD 9400. *Hurst* 1843 OS.

(3) ~ NYorks NZ 0402. *Hirst* 1539. YN 294, L 197.

(4) ~ CASTLE Hants SZ 3189. *Castle or fortress of Hurst* 1543. One of a chain of maritime castles built by Henry VIII 1541×4. It was built on a spit of land called *Hurst* 1434, *an herd sand called the Hurst* 1539, ME **hurste** 'a sand-bank' < OE *hyrst*. Ha 97, Gover 1958.227, Pevsner-Lloyd 302.

(5) ~ GREEN ESusx TQ 7327. *Herst grene* 1574. P.n. Hurst + ModE **green**. Hurst, *Hurst* 1342, *Herst* 1428, is 'the wooded hill', ME **hurst** (OE *hyrst*). Sx 460.

(6) ~ GREEN Lancs SD 6838. P.n. Hurst as in STONYHURST SD 6939 + ModE **green**. Hurst, *Hurst* from [c.1200]1268, *del Hirst* 1335, is OE **hyrst** 'a wooded hill'. La 141.

(7) ~ GREEN Surrey TQ 3951. 'The green at H'. *Herste grene* 1577. ME **herst, hirst** or surname *Hirst* + ME **grene**. Cf. William *ate Hurst* 1332. Sr 334.

(8) ~ MOOR NYorks NZ 0403. P.n. HURST NZ 0402 + ModE **moor** (OE *mōr*).

HURSTBOURNE PRIORS Hants SU 4346. 'H held by the prior' of St Swithin's, Winchester. *Hesseburna Prioris* 1167,

Husseburne Prior' 1291, *Husborn Pryors* 1564. P.n. Hurstbourne + ModE, Latin **prior**, Latin genitive sing. **prioris**. Also known as Nether H, *(on ðam) neoðeran hyssebur-nan* [875×88]11th S 1507, and *Down Hursborne* 1602, *Down Husband* 1707, for distinction from Up Hurstbourne or HURSTBOURNE TARRANT SU 3853 which lies further up the Bourne Rivulet. Hurstbourne, *(juxta) Hissaburnam* (Latin) [764×93]13th S 268, *(æt, by, to) Hysseburnan* [873×88]11th S1507, [900]11th S 359, *Esseborne* 1086, *Husseburn(e)* 1168–1277, is the 'tendril stream', OE **hysse** + **burna**. OE *hysse*, used in poetry to mean 'son, youth, young warrior', is also recorded in the sense 'shoot, scion' glossing Latin *pampinus*. This is the sense in which it is used in p.ns. for wet places such as *hyssa pol* [901]12th in the boundary clause of S 362 (Stockton, Wilts), HYCEMOOR Cumbr SD 0989 and HUSBORNE CRAWLEY Beds SP 9535. It is an *i*-mutation variant of OE *hos(s)* used to gloss Latin *botrus, rhamnus, vimen* 'grape, thorny shrub, withy' and seems to refer to some kind of aquatic plant, in this case to the trailing tendrils of reed canary grass, *Phalaris arundinacea*, which fills the burn. 'Priors' for distinction from H TARRANT SU 3853. Ha 97, Gover 1958.154, 159, BT Supplement s.v. *hyse*, PNE i.278, Jnl 23.48.

HURSTBOURNE TARRANT Hants SU 3853. 'H belonging to Tarrant' sc. abbey. P.n. Hurstbourne as in HURSTBOURNE PRIORS SU 4346 + p.n. Tarrant as in Tarrant Abbey, Dorset, to which it was granted by Henry III whence the names *Husseburne Regis* 1291 and *Kings Hursborne* 1628. Also called *Huphusseburn'* 1242, *Up Husband* 1719, 1822, because of its situation further up the Bourne Rivulet from Down H or H PRIORS SU 4346. Ha 97, Gover 1958.159.

HURSTPIERPOINT WSusx TQ 2816. 'H held by the Pierpoint family'. *Herst Perepunt* 1279, *Hirste Perpund* 1288, *Pyrpountesherst* 1359, *Parpoynthurst* 1439, *Harstpount* 1518×29, *Hurst upon ye point* 1722. P.n. *Herst* 1086, 'the wooded hill', OE **hyrst**, + manorial addition from the family of Robert de Pierpoint (*Petroponte* Latin 'stone bridge'), the DB tenant from Pierrepont near Falaise. Also known as *Westherst* 1261 in contrast to *Esthurst* 1302, 1340 referring to HERSTMONCEAUX TQ 6312. Sx 274, 479, SS 67.

High HURSTWOOD ESusx TQ 4926 → HIGH HURSTWOOD TQ 4926.

HURSTWOOD RESERVOIR Lancs SD 8931. P.n. Hurstwood + ModE **reservoir**. Hurstwood, *Hurstwode* 1285, *Hirst(e)wod(e)* 1323–97, is 'Hurst wood', OE **hyrst** 'wooded hill' probably used as a p.n., + **wudu**. La 84, Jnl 17.52.

HURT WOOD Surrey TQ 0843. *le Hurtwood* 1713. Possibly the same as *(in bosco voc.) le Churte* 1546, *Chart Wood* 1603×25; if so, this is 'the wood at the rough ground', OE **ċert** + **wudu**. Otherwise 'stag wood', OE **heorot**. Sr 249, 250.

HURWORTH 'Enclosure made of hurdles'. OE **hurth**, genitive pl. **hurtha**, + **worth**.

(1) ~ BURN RESERVOIR Durham NZ 4033. R.n. Hurworth Burn + ModE **reservoir**. Hurworth is *hurdewurd'* 1197, *-wrth(e)* 1242×3, *Hurthwrth* 13th, *Hurth(e)worth(e)* 1338–1500, *Hurworth Bryan* from 1378, from *Brian* son of *Alan de Hurdewrth* 1242×3. Also known as *Hurworth on the More* 1481–1624 and *Blackehurworth* 1624. NbDu 122.

(2) ~ -ON-TEES Durham NZ 3010. *Hurth(e)worth(e) on Teise* 1355, *Hurworth super Teyse* [1417]15th, ~ *on Tees* 1558. Earlier simply *hurdewurda* 1155×89, *Hurthew(u)rthe* c.1190×1200, 1242×3, *Hurth(e)worth(e)* 1235–1380×1 *Hurdworth* 1442–1539, *Hurworth* from 1314. NbDu 122.

HURY Durham NY 9519. Unexplained. *Hury* 1863 OS.

HURY RESERVOIR Durham NY 9618. P.n. HURY NY 9519 + ModE **reservoir**.

HUSBORNE CRAWLEY Beds SP 9535. Originally two separate places, *Husseburn et Crawele* 1276, *Husbond Crawley* 1535. Husborne, *(of) Hyssebumnan* [960]11th S 772, *Husseburn(e)* 13th cent., *Hisse- Hesseburn* 1220, [1220]13th, is the 'stream with tendril-like plants', OE **hysse** + **burna**; Crawley, *Crawelai* 1086, the 'wood or clearing frequented by crows', OE **crāwe**, genitive pl. **crāwa**, + **lēah**. Bd 118, PNE i.278, L 18, 205.

HUSTHWAITE NYorks SE 5174. 'Clearing with a house or houses'. *Hustwait -twayt(e)* 1167–1581, *Husthweyt -ayt* c.1180, 1283. ON **hús** + **thveit**. The p.n. *Hustveit* occurs in Norway. YN 191 gives pr [ustwit].

HUTCHERLEIGH Devon SX 7850. 'Hycci's wood or clearing'. *Hucheslegh* 1428, *Huchley* 1563, *Hitchesley* 1732, *Hutcherly* 1809 OS. OE pers.n. **Hyċċi*, genitive sing. **Hyċċes*, + **lēah**. **Hyċċi* would be a derivative of the pers.n. *Hucc*. It or the variant **Hyċċa* occurs again in Hutcherton SS 6332, *Hiccheton* 1330. D 317, 352.

HUTCHERTON Devon SS 6332 → HUTCHERLEIGH SX 7850.

HUTHWAITE Notts SK 4659. 'The clearing on or by the hill-spur'. *Hodweit* 1199, *Hothweit -thweyt* 1208–1302, *Houghthweyt* 1317, *Houthwayt -thwait* 1330, 1335, *Huthwaite* 1693, *Heught Weyt* 1537. OE **hōh** + ON **thveit**. Nt 119 gives pr [huθweit].

HUTTOFT Lincs TF 5176. 'The hill-spur curtilage'. *Hoto(f)t* 1086, *Hotoft* c.1115–75, *Hottoft* 1202. OE **hōh** + **toft**. Huttoft is situated on low ground which neverthless forms a marked elevation rising above the surrounding marshes. DEPN, SSNEM 214, Cameron 1998.

HUTTON 'Settlement on or by the hill-spur'. OE **hōh** + **tūn**.

(1) ~ Avon ST 3558. *Hotvne, Hutone* 1086, *Hocton* 1243, *Hutton* 1291. DEPN.

(2) ~ Cumbr NY 4326. Short for Hutton John, 'John's H', *Hoton John* 1279, ~ *Johan(nis)* 1292–1362, *Hoton Jon* 1323, *Huttonion(e)* 1584, 1609. The identity of the *John* whose name became attached to this place is unknown. Cu 210.

(3) ~ Essex TQ 6394. *Houton* 1200–55, *Hoton(e)* 1235–1428. Although this is a compound of *hōh* and *tūn*, in this name *hōh* was already being used as a p.n. in its own right, *Atahov* 1086, 'at Ho', OE **æt** + **hōh**, *Hou* [1086–1157]12th, *Hoo* 1323, *Ho ... now called Hoton* 1347. Ess 160.

(4) ~ Humbs TA 0253. *Hottune* 1086, *Hoton(e) -tona -tun* 1086–1526, *Hutton* from 1521, *Huton, Hewton* 16th cent. The village lies on a low shelf of land overlooking Skerne Beck. It was frequently linked with p.n. Cranswick, *(cum) Cra(u)ncewyke* 1210×20, *Crauncewyk -wik* 1197×1210–1419, *(by) Crauncewyk* 1310 for distinction from other Huttons. YE 156.

(5) ~ Lancs SD 4926. *Hotun* before 1180, *Hoton* [c.1200]1268–1332 etc., *Hotton* 1272×7, 1461, *Hooton* 1539, *Hutton* 16th. The village stands on a low spur of higher ground jutting out into the Ribble flats. La 136, Jnl 17.77.

(6) ~ BONVILLE NYorks NZ 3300. 'H held by the Boneville family'. *Hutton Bonevill* 1316. P.n. *Hotune -ton* 1086 + manorial addition *Bonville* from Robert de Boneville who held the estate in the early 13th cent. YN 211.

(7) ~ BUSCEL NYorks SE 9784. 'H held by the Buscel family'. *Hutton Bussalle, Bussel, Buscel(l)* 13th. cent. P.n. *Hotun(e)* 1086 + manorial addition *Buscel* from the Bushell family who held the estate in the 12th and 13th cents. YN 101.

(8) ~ CONYERS NYorks SE 3273. 'H held by the Conyers family'. *Hotonconyers* 1198, *Howton Coniers* 1530. P.n. *Hot(t)on(e), Hotune* 1086 + manorial addition *Conyers* from the Conyers family who held the estate from the early 12th cent. YN 281.

(9) ~ CRANSWICK Humbs TA 0252. This place is properly simply CRANSWICK TA 0252 to which Hutton has been prefixed from nearby HUTTON Humbs TA 0253.

(10) ~ END Cumbr NY 4538. *Hutton End* 1704. P.n. Hutton as in HUTTON IN THE FOREST Cumbr NY 4535 + ModE **end**. So named for distinction from Unthank End NY 4535. Cu 209.

(11) ~ HENRY Durham NZ 4236. 'H held by Henry (de Hotone)'. *Huton Henrie, Henry* 1611–37, *Hutton Henrie* 1619, ~

Hen(d)ry 1717. Earlier simply *Hotun* [c.1040]12th–1233×4, *Hoton(a)* [1183]c.1382, 1196×1215–1564, *Hottone* [1183]c.1320–c.1339, *Huton* 1360–1637. The addition *Henry* commemorates the grant of 30 acres to Finchale Priory by *Henricus de Hotone* in 1196×1215. The village extends along a ridge of high ground. NbDu 122.

(12) ~ IN THE FOREST Cumbr NY 4535. *Hoton' in the Forest* 1246–1402, *Hutton in the forest of Inglewood* 1572, *Hutton Forest* 1580. Earlier simply *Hoton'* 1181–1487. See also INGLEWOOD FOREST Cumbr NY 4639. Cu 208.

(13) ~ -LE-HOLE NYorks SE 7090. 'H in the hollow'. Earlier known as *(Hege)hoton* 1204 and *Hoton Underheg* 1285, *Hewton under Heighe* 1579. P.n. *Hotun* 1086 + uncertain addition. *Hege* is usually said to represent ME *hay* (OE *(ġe)hæġ*) 'land enclosed for hunting' which is certainly possible; but it might alternatively represent OE **hēah* 'a high place, a height'. The village lies in the deep valley of Hutton Beck (whence the modern addition *(en) le Hole* 'in the hollow') at the foot of Hutton Ridge. YN 60.

(14) ~ MAGNA Durham NZ 1212. 'Great H'. *(Magna) Hoton* [1157]15th, 1252. P.n. *Hottun, Hotune* 1086 + Latin **magna**. YN 300.

(15) ~ ROOF Cumbr SD 5678. 'H held by the Roof family'. *Hotun -ton* 1086–1461, *Hutton* from 1461, ~ *riof* [1090×7]1308, ~ *Rophe, Roffe, Ruffe, Rouf* 1161×84–1544, ~ *Roof(e)* 1376 and from 1612. The feudal affix *Roof* has not been identified; the forms point to pers.n. *Riulf, Riolf* <OG *Ricwulf*. We i.35.

(16) ~ ROOF Cumbr NY 3734. *Hotune- Hotonerof* 1278, *Hotonrof(e)* ~ *Roff(e), Rouf, Ruff* 1279–1465, *Hutton Roof* from 1671. The suffix has not been explained. Cu 210.

(17) ~ RUDBY NYorks NZ 4606. Short for *Hottona juxta Rodeby* 'H by Rudby' 1204, *Hooton Rudbye* 1582. P.n. *Hotun -ton* 1086–1412, + p.n. RUDBY NZ 4607. Rudby is on the opposite bank of the river Leven. YN 174.

(18) ~ SESSAY NYorks SE 4776. Short for *Hotton et Cessay* 'H and Sessay' 1316. P.n. *Hotun* 1252 + p.n. SESSAY SE 4575. Also known as *Hotun juxta Tresk* 'H by Thirsk' 1252, p.n. HUTTON + p.n. THIRSK SE 4382. High ground lies ½ a mile to the E of the village. YN 187.

(19) ~ WANDESLEY NYorks SE 5050. 'H and W'. *Hoton Wandeslay* from 1250, ~ *Wannesley* 1535, *Hewton, Hooton Wansley* 16th. P.n. *Hoton* 1190–1535 + p.n. *Wandeslage* 1086, *-ley* 1294, 'Wandel's clearing', OE pers.n. **Wandel* or **Wand*, genitive sing. **Wand(el)es*, + **lēah**. A compound p.n. representing two originally separate manors. YW iv.253 gives pr ['(h)utn 'wanzlə].

(20) High and Low ~ NYorks SE 7568. ModE **high, low** + p.n. *Hotun* 1086, *(Bardolf) Hoton* 1186–1202, *Hotton Bardouf* 1226, *Hoton upon Derwent* 1316. Both Huttons lie on or beside a steep ridge jutting into a bend of the river Derwent. High Hutton was held by the Bardolf family, Low Hutton or Hutton upon Derwent by a tenant called *Colsuain -sweyn -swayn* 13th from ON *Kolsveinn*. YN 40.

(21) New ~ Cumbr SD 5691. *Nova Hoton* 1337, 1340, *New Hutton* from 1556. Earlier simply *Hoton* 1274–1376, *Hutton* 1475. 'New' for distinction from Old HUTTON SD 5688. Also known as *Hoton in Laya*, ~ *en le Haye, in the Haye* 1283–1676, OE **(ġe)hæġ** 'an enclosure, a hunting enclosure', referring to the extensive hunting park on the hills to the E above Kendal Castle, cf. the names Kendal Park SD 5391, High Park SD 5390, Hayfellside and Hayclose SD 5490. We i.131.

(22) Old ~ Cumbr SD 5688. *Vetus Hoton* 1332, *Ald-* 1406–1518, *Old Hutton* from 1547. Earlier simply *Hotun -ton* 1086–1461, *Hutton* 1558. 'Old' for distinction from New HUTTON SD 5678. We i.127.

(23) Priest ~ Lancs SD 5373. 'H held by the priest(s)'. *Presthotone* 1307, *Prist- Presthoton* 1401–1443. Earlier simply *Hotune* 1086, *Hoton* 1327–82. 'Priest' for distinction from HUTTON ROOF Cumbria SD 5678. La 188, Jnl 17.104.

(24) Sand ~ NYorks SE 6958. *Sandhoton'* 1231–1438, *-huton'* 1577. ME, ModE **sand** + p.n. *Hot(t)un(e)* 1086, *Hoton(e)* 1086–1399. YN 38.

(25) Sheriff ~ NYorks SE 6566. *Shi- Shyref- Schirefhoton* 1199×1213, 1244 etc. Manorial prefix *Sheriff* + p.n. *Hotun(e) -ton(e)* 1086, 1154×60. The estate was held originally by Bertram de Bulmer, sheriff of York, d.1166, and subsequently by the Nevilles who became sheriffs in the 13th cent. The village occupies a low ridge of hill with the church at one end and the castle at the other and commands an extensive view to the S. YN 31.

HUXLEY Ches SJ 5161. Partly uncertain. *Huslehe* 1185, *Huxelehe -leg(h) -ley(e)* 1202–1363, *Huxley* from 1271 with variants *-le(i)gh(e) -lay* etc. The generic is OE **lēah** 'a wood, a clearing', the specific uncertain; perhaps OE **husc, hucs, hux**, genitive pl. **huxa**, 'insult, scorn, mockery' and so 'woodland or glade where ignominy is offered or suffered'; or OE pers.n. **Hucc*, genitive sing. **Hucces*. Che iv.101.

HUYTON-WITH-ROBY Mers SJ 4491. A combination of two p.ns., Huyton, the 'landing-place settlement', *Hitune* 1086, *Huton(a)* 1189×96, 1243, *Hutton* 1268, *Huyton(')* from [1191]late 14th, *Hyton* 1341, 1423, OE **hȳth** + **tūn**, and ROBY SJ 4390. La 113 gives pr [haitn], TC 112, Jnl 17.63.

HYCEMOOR Cumbr SD 0989. Partly obscure. *Hysemore* 1391, 1537, *Haysmore, His(e)more* 16th cent. The forms are too late for certainty, but this might be OE **hysse**, the name of some sort of water-plant, + **mōr**. Cu 346.

HYDE Glos SO 8801. 'The hide'. *Hida* 1234, *la, le Hide, Hyde* 1248–1438. OE **hīd** 'a hide, an estate of one hide extent'. Gl i.97.

HYDE GMan SJ 9494. 'The hide of land'. *Hida* early 13th, *Hide* 1240–1596, *Hyde* from early 13th. OE **hīd**. Che i.279.

HYDE Hants SU 1612. 'The hide'. *la Hyde* 1414. Earlier *la Suthyde* 'the south hide' 1301. ME **hide** 'a hide of land' (OE *hīd*). Gover 1958.215.

HYDE HEATH Bucks SP 9300. *Hyde Heath* 1822 OS. P.n. Hyde + ModE **heath**. Hyde, *Hida* [1150–68 etc.]14th, *Missenden Hyde* 1550, is 'the hide' of land held by Missenden abbey, OE **hīd**. So called for distinction from SOUTH HEATH SP 9102. Bk 155, O, YN addenda.

HYDE PARK GLond TQ 2780. *Hide Park* 1543. The park was created by Henry VIII as a royal preserve after the dissolution of the monasteries. P.n. Hyde + ModE **park**. Hyde, *Hida* 1204, *la Hyde* 1257, *la Hyde by the town of Westminster* 1353, was a hide of land, OE **hīd**, about 100–120 acres, which was originally a parcel of the manor of Ebury, *Eyebury* 14th cent., *Ebery* 1535, 'Eye manor', p.n. *Eia* 1086, *Eye* 1087×97–1406 'the island', OE **ēġ**, + ME **bury** (OE *byriġ*, dative sing. of *burh*). Ebury formed the W part of the present city of Westminster; the name survives in Ebury Bridge, Square and Street. Mx 168, Encyc 259, 412.

HYDESTILE Surrey SU 9740. Partly uncertain. *Hide Style* 1710. Uncertain element + ModE **stile** (OE *stiġel* 'a stile or steep ascent'). Sr 204.

North HYKEHAM Lincs SK 9466. *Northniche, Nort(h)hicam* 1086, *Northic(h)am -hycam* 1212–1539, *North(h)ik- -(h)ykam(e)* 1331–1464, *North Hyckham* 1808. OE **north** + p.n. Hykeham as in South HYKEHAM SK 9364. Also known as *Little Hickam* 1246, *Est Hykam* 1303, and *Chapel Hickam* 1723. Perrot 255.

South HYKEHAM Lincs SK 9364. *Suthhic(h)am -hik- hyk(h)am* 1212–1516, *South Hykeham* 1737. OE **sūth** + p.n. *Hicham* 1086–1327, *Hicam(e)* late 12th, 1583–1607, *Hi- Hyk(h)ame* 1195–1604, 'Hica's homestead', OE pers.n. **Hīca* + **hām**. OE **Hīca* corresponds to G *Heiko*, but an alternative specific might be OE **hīċe** 'a titmouse' with Scand [k] for [ʃ]. 'South' for distinction from North HYKEHAM SK 9466. Perrott 242.

South HYLTON T&W NZ 3556. *South Hylton* 1863 OS. ModE **south** + p.n. *Heltun* 1153×95, c.1170, 1174×89, *Helton(')* 1153×95, c.1162×74–1321, *Hylton, Hilton* from early 13th., 'the settlement on the slope', OE **helde** + **tūn**. The original manor house of Hylton, Hylton Castle and chapel, occupy sloping ground above Hilton Dene at NZ 3558. North Hylton, *North Hylton* 1863 OS, and South Hylton grew up as settlements on the N and S bank of the river Wear where Hylton ferry formed an important river crossing. The later form of the name shows assimilation to ME *hill* 'a hill'. NbDu 122.

HYTHE Hants SU 4208. 'The landing-place'. *la Huthe* 1248–1318, *Heothe* 1422, *Heith* 1681. OE **hȳth**. The reference is to a landing place on Southampton Water. Ha 98, Gover 1958.198.

HYTHE Kent TR 1634. 'The landing place'. *(on) Hyþe* 12th ASC(E) under year 1052, *Hede -æ* 1086, *Hethe, Hede, Heða* c.1100–1176×7, *Hee, Heia, Heya* 1199–1225, *Heth* 1215, 1228 etc. OE **hȳth**, Kentish **hēth**. Although Hythe is by the sea the original reference must have been to a landing-place on the r. Lympne long before the construction of the Royal Military Canal. The element *hithe* is normally used of a landing-place on a river, an inland port. PNK 459, TC 112, Cullen.

HYTHE END Berks TQ 0172. *Hythe End* 1822 OS. P.n. Hythe + ModE **end**. Hythe, *huth* [672×4]13th S 1165, *Huthe* [1261×72]15th, *la Huthe* 1436, *Hythe* 1574, is OE **hȳth** 'the landing place'. The reference is to an ancient landing place on the Thames. Sr 122.

I

IBBERTON Dorset ST 7807. 'Eadbriht's estate'. *Abristeton* 1086, *Abristentona* 1086 Exon, *Edbrich(c)testun* 1199×1216, *Edbrytyngton als. Edbrichton* 1283, *Eb(b)richton* 1245–99, *Ebyrton* 1424, *Ibrigton* 1291, 1423–8, *Yberton(')* [1384]16th, 1431, *Ibberton* 1575. OE pers.n. *Ēadbeorht -briht* + **ing**[4] varying with genitive sing. *Ēadbrihtes* + **tūn**. Do iii.214 gives pr ['ebəRtn].

IBLE Derby SK 2557. 'Ibba's hollow'. *Ibeholon* 1086, *I- Ybol(e)* 1154×9–1330, *I- Ybul* 1159×66–1516, *Ibeol, Ible* from 1307. OE pers.n. *Ibba* + **hol**. The 1086 form is from the dative pl. **holum**. The reference is to Griffe Grange Valley SK 2556, *le Griff* 1260–94, *the Gryffe* 1489, ON **gryfja** 'a deep narrow valley'. Db 380 gives pr [ibəl].

IBSLEY Hants SU 1509. 'Ibbi or Tibbi's wood or clearing'. *Tibeslei* 1086, *-leia* 1166, *Tybesle* 1282, *Ibeslehe* 13th, *Ibbesleg' -lig' -le(ya) -lye* from 1236. OE pers.n. **Ibbi* or **Tibbi*, genitive sing. **Ibbes, *Tibbes*, + **lēah**. It is impossible to say whether the original form of the name was *æt Ibbeslea* or *æt Tibbeslea*. In the first case the *T-* forms are due to transference of the *t* of *æt* to the pers.n., in the second the *I-* forms are due to misdivision of the sequence *æt T-*. Ha 99, Gover 1958.216.

IBSTOCK Leic SK 4010. 'Ibba's dairy-farm'. *Y- Ibestoche* 1086, late 12th, *-stok(e)* 12th–1434, *Ibbestok(e)* 1222–1463, *Y- Ibstok(e)* 1356–1576, *Ibstock(e)* from 1551, *Ipstok(e)* 1325–1486. OE pers.n. *Ibba* + **stōc**. Lei 509, Studies 1936.28.

IBSTONE Bucks SU 7593. 'Ibba's stone'. *Hibestanes, Ebestan* 1086, *Ibbastana* c.1150, *Ibbestane* 1184, *-ston* 1262, *Ipston* 1417. OE pers.n. *Ibba* + **stān**. The reference is unknown. It may have been a boundary stone on the old county boundary with Oxfordshire. Bk 185 gives pr [ipstən].

IBTHORPE Hants SU 3753. 'Ybba or Ibbas's secondary settlement'. *Ebbedrope* 1236, *Ybe- Ebethrop* 1269, *Ibthorp* 1457. OE pers.n. *Ybba* or *Ibba* + **throp**. An outlying settlement of Hurstbourne Tarrant. Ha 99, Gover 1958.160.

IBWORTH Hants SU 5654. 'Ibba's enclosure'. *Hibbewrth* 1233, *Ibbew(o)rth(e)* 1245–1351. OE pers.n. *Ibba* + **worth**. Ha 99, Gover 1958.140.

ICKBURGH Norf TL 8194. 'Ica's fortified place'. *Iccheburna, Ic(c)heburc* 1086, *Ykeburc* 1193, *Ikeburc* 1199. OE pers.n. **Ica* + **burh**. DEPN.

ICKENHAM GLond TQ 0786. 'Tica's homestead'. *Ticheham* 1086, *Tike(n)ham* 1176–1305, *Ikeham* 1203, *(H)ikenham* 1235–1387, *Icknam* 1737. OE pers.n. *Tica*, genitive sing. *Tican*, + **hām**. Initial *T-* was lost in the phrase *at Tickenham* misinterpreted as if *at Ickenham*. Mx 43.

ICKFORD Bucks SP 6407. 'Icca's ford'. *Iforde* 1086, *Y- Icford* 1175–1489, *Ikeforde* 1226–1300. OE pers.n. **Ic(c)a* + **ford**. Bk 124.

ICKHAM Kent TR 2258. 'Estate comprising a yoke of land'. *Ieccaham* [724]15th S 1180, *Iocc ham* [785]?9th S 123, *Iocham* [786]?12th S 125, *Jocham* [791]12th S 1614, *Geocham* 958 S 1506, *Ycham* [958]13th S 1506, *Ioccham* [1006 for 1001]15th S 914, *Ieoccham* 1042×66 S 1047, *Gecham* 1086, *L- Ieoc- Ie(a)cham* c.1100, *Icham* 1233. OE **ġeoc** 'a yoke, a measure of land (about 50 to 60 acres), a small estate' + **hām**. PNK 521, Jnl 8.34.

ICKLEFORD Herts TL 1831. 'Ford associated with Icel'. *Ikelineford* 1154×89, *Hiclingford* 1216×72, *I(e)klingford -yng-, I- Ykeleford* 13th cent., *Icleford* 1220. OE pers.n. *Iċel* + **ing**[4] + **ford**. Ickleford lies on the line of Icknield way though the two names do not seems to be connected. Hrt 12.

ICKLESHAM ESusx TQ 8816. 'Icel's promontory'. *Ikelesham, Icoleshamme* [772]13th S 108, *Ycles- Iklesham* 12th–14th, *Y- Ikelesham* c.1197–15th, *-hamme* 1379. OE pers.n. *Iċel*, genitive sing. *Ic(e)les*, + **hamm** 2, 3 or 7. There is also a unique form *Hykelingesham* 1288, 'the *hamm* of Hykeling', showing the genitive sing. of an *-ing* suffixed variant form of the p.n., **Iceling* 'place called after Icel'. Sx 510, ASE 2.26, SS 83, BzN 1967.149.

ICKLETON Cambs TL 4943. 'Estate called after Icel'. *(æt) Icelingtune* 975×1016 S 1487, *Inchel- Hichelintone* 1086, *Ik(e)lington(e) -yng-* 1122–1547, *Ic- Iklyngton* 1272–1547, *Y- Iclinton -yn-* 1188×94–1309, *Ikel(e)ton(e)* 1251, 1272, 1549×53, *Ickleton* 1574. OE pers.n. *Iċel* + **ing**[4] + **tūn**. Ca 94.

ICKLINGHAM Suff TL 7772. 'The homestead of the Ycclingas, the people called after Yccel'. *ecclinga- etclingaham* 1086, *Echelincgham* [1086]c.1180, *Ikelingeham* 1242×3, *Ikelingham* 1254, *Icklingham* 1610. OE folk-n. **Ycclingas* < pers.n. **Yċċel* + **ingas**, genitive pl. **Ycclinga*, + **hām**. DEPN, ING 130–1, NM 96.113–22.

Lower ICKNIELD WAY Herts SP 8912. Lies at the base of the Chiltern Hills. ModE **lower** + p.n. ICKNIELD WAY TL 2836.

Upper ICKNIELD WAY Herts SP 9212. Crosses the summit of the NW scarp of the Chiltern Hills. ModE **upper** + p.n. ICKNIELD WAY TL 2836.

ICKNIELD WAY Beds TL 2836 TL 0322, Bucks SP 9818, Cambs TL 4041, Herts TL 1630, 2634, Oxon SU 6285, 6892, Suff TL 7770, TL 8176. *Ic(c)enhilde weg* [903]12th S 369, [944 ? for 942]13th S 496, *Icenhylte* [903]10th S 367, *Ikenhild via* 1209, *Hykenhilte* 1274. The name of this important ancient British track-way from Dorset to Norfolk has not been satisfactorily explained. *Icenhylte* could be taken as 'Ica's wood', OE pers.n. **Ic(c)a*, genitive sing. **Ic(c)an*, + **hylte**, *Iccenhilde* as 'Ica's slope', OE **hield**. Another possibilty is connection with the Romano-British tribe settled in Norfolk, the *Iceni*, and so perhaps 'the way leading to the Iceni'. Bd 4, xl. Hrt 6, FLH 14.52. From its being one of the four privileged roads of the Laws, an early extension of the use of this name took place to roads in Worcestershire and elsewhere.

ICKWELL GREEN Beds TL 1545. *Ickwell Green* 1834 OS. P.n. Ickwell + ModE **green**. Ickwell, *Ikewelle* [c.1170]13th, *Gikewelle* [c.1180]13th, *Yi- Yekewell* 1287, *Ikwell* 1552, is probably the 'beneficial stream', OE **ġēoc**, definite form **ġēoca**, + **wella** rather than 'Gica's stream' with unrecorded pers.n. **Ġic(c)a*. Bd 95.

ICKWORTH HOUSE Suff TL 8161. *Ickworth House* 1836 OS. An 18th cent. mansion of the 4th earl of Bristol begun in 1795. P.n. *Ikewrth* [942×c.951]13th S 1526, *(æt) Iccawurðe* *[1042×66]12th S 1124, *KKeworthā* (for *Icke-*) 1086, *Icceuuorde* c.1095, *Ikewrthe* 1254, *Ikesworth* 1610, 'Icca's enclosure', OE pers.n. **Icca* + **worth**, + ModE **house**. DEPN, Pevsner 1974.285.

ICOMB Glos SP 2122. 'Icca's coomb'. *Ican- Iccacumb* *[781]11th S 121, *Iccacumb* [964]12th S 731, *Iccecumb* 11th, *Iccvmbe -cumb(e) -combe* 1086–1675, *Icvmbe -combe* 1086, 1220, 1611. OE pers.n. **Ic(c)a* + **cumb**. There is also a DB form *Iacumbe* which may be a mistake for *la cumbe* 'the coomb' referring to the lost Combe Baskerville in the same parish. Gl i.220 gives pr ['ikəm].

IDBURY Oxon SP 2320. '(At) Ida's fortified place'. *Ideberie* 1086–1428 with variants *Yde- -byre -bire* and *-bur(y), I- Yddebur(y) -bir'* 1241–1428, *Edbury* 1675. OE pers.n. *Ida* + **byriġ**, dative sing. of **burh**. O 357.

IDDESLEIGH Devon SS 5608. 'Eadwig's wood or clearing'. *Edeslege, Iweslei* 1086, *Edwislega* [1107]1300, *-leg(he), Edus-(Y)edolvesleg(h), Edulvesly* 13th cent., *Yd(d)usleghe* 1296, 1308, *Eddisleigh* 1339, *Yiddesle* 1330, *Yeddeslegh* 1428. OE pers.n. *Ēadwīġ*, genitive sing. *Ēadwīġes*, + **lēah**. Some forms point to a different pers.n., *Ēadwulf*. D 93 gives pr [idʒli].

IDE Devon SX 8990. Unexplained. *(æt) Ide* 1050×73 and from 1086, *Ede* 1511, 1577, 1634, *Eide* 1628. Possibly originally a stream-name, but no explanation is known. The present stream-name is Alphin, *the Alphin* 1797, a back-formation from ALPHINGTON SX 9190. The dedication of the church to St Itha, an Irish saint, may derive from the p.n. D 497 gives pr [i:d], RN 208.

IDE HILL Kent TQ 4851. 'Edith's hill'. *Edythehelle* 13th, *Ide Hill* 1610. OE feminine pers.n. *Ēadgȳth*, genitive sing. *Ēadgȳthe*, + **hyll**, Kentish **hell**. PNK 70.

IDEFORD Devon SX 8977. 'Giedda's or song ford'. *Lvdeford* (for *Ivde-*) 1086, *Ivdaforda* 1086 Exon, *Yudde(s)ford* 1275–1440, *Hiddeford* 1285, *Idyforde* 1291, *Ideford* 1440, *Yed(d)eford(e)* 1353–1441, *Edeford* 1440, 1577, *Edford* 1571. OE pers.n. **Ġiedda* or **ġiedd**, genitive pl. **ġiedda**, + **ford**. In the latter case the sense would be 'ford where people gather for speeches or singing'. D 474 gives pr [idifəd], DEPN.

IDEN ESusx TQ 9123. 'Yew-tree swine pasture'. *Iden(e)* from 1086. OE **īw** + **denn**. Sx 530, SS 88.

IDEN GREEN Kent TQ 8031. 'Iden common'. *Iden Green* 1813 OS. P.n. Iden + ModE **green**. Earlier simply *Idunne* 1194, *Ydenne* c.1200, *Idenne* 1293, 1327, *Iden* 1473, 'yew-tree (swine-)pasture'. OE ***īġ, īw** + **denn**. PNK 348.

River IDLE Notts SK 7497. Possibly 'the idle, slow, lazy river'. *Idlae* [731]8th BHE, *Idle* [c.890]c.1000 OE Bede with variants *iddlé, idle*, 1297 & from 1576, *Iddil, Iddel* [958]14th S 679, 1349, *Yddil* 1200×3, *(H)i- Yddel* 1247–1342, *I- Ydel(l) -ill* 1280–1440, *E(e)dle* 1602. OE **Īdle* from **īdel**. Vowel length seems to vary in this name, forms with *-dd-* showing shortening before *-dl-* in the inflected form *Idle*, forms with *-d-* maintaining the long vowel of the uninflected form of the adjective **īdel**. The name may, however, be pre-English, cf. the Breton r.n. Isole, *Idol(a)* 11th. Nt 5, RN 207–8.

IDLICOTE Warw SP 2844. 'Yttela's cottage(s)'. *Etelincote* 1086, *Utilicot(e)* 1154×89–1324 with variants *Uteli- Udeli-* and *-cota, Itelicota* 1154×89, *Itlicote* 1232, *Ydelicote* 1441, *Idlycote al. Utlycote* 1568, *Utlyngcote* 1285. Either OE pers.n. **Yttela*, genitive sing. **Yttelan*, or **Yttela* + **ing**[4] + **cot**, pl. **cotu**. Wa 281.

IDMISTON Wilts SU 2037. 'Idhelm or Idmær's estate'. *(at) Idemestone* [947]14th S 530, [948]14th S 541, [970]14th S 775, *Idemestona, Idemistonam* 1189, *I- Ydemeston(e) -is-* 1230–1315, *Ydmeston* 1428, *Idemereston* 1268, *Idmerston* 1675. OE pers.n. **Idhelm* or **Idmǣr*, genitive sing. **Idhelmes*, **Idmǣres*, + **tūn**. Wlt 380.

IDRIDGEHAY Derby SK 2849. 'Eadric's forest enclosure'. *Edrichesei* 1230, *-hay(e)* 1298, 1327, *Edrich(e)hay(e)* 1330–1436, *Idrychay* 1376, *Idrichhay* 1387, *Idersey, Ideressay, Ydershey* 16th cent. OE pers.n. *Ēadrīċ*, genitive sing. *Ēadrīċes*, + **(ġe)hæġ**. Db 577.

IDSTONE Oxon SU 2584. 'Eadwine's estate'. *Edwineston(') -wyne-* 1199–1428, *Edwynston(e)* 1274, 1342, *Ediston* 1363, *Idston* 1661. OE pers.n. *Ēadwine* or ME *Edwin*, genitive sing. *Ēadwines, Edwines*, + **tūn**. Brk 345.

IFIELD WSusx TQ 2537. 'The yew-tree open land'. *Ifelt* 1086, *Yfeld* 1210–12, *Eyefeild* 16th, *Ivel* 1726. OE ***īġ, īw** + **feld**. Sx 207 gives pr [aivəl], SS 70.

IFOLD WSusx TQ 0231. 'The fold in or by the well-watered land'. *Ifold* from 1296, *Ivold* 1592. OE **īeġ** + **falod**. Sx 106, SS 80.

IFORD ESusx TQ 4007. 'Island ford'. *Niworde* 1086, *Yford* 1087×1100, *Iford(e)* c.1100–15th, *Iver* 1624. OE **īeġ** + **ford**. Most likely this is the 'ford to the island'. It lies on the edge of the former Ouse flood plain. The 1813 OS shows two islands of higher ground, Upper and Lower Rise, with a track from the former to Iford, *insula juxta Lewes voc. Sutheye, Northeye* 'island by Lewes called South Eye, North Eye' 1279, *Southrye* 1363, *the South Rye* 1567, later *the Hither and Further Rhies*, OE **sūth, north** + **ēġ**. The later forms with *R-* arise from misdivision of ME *atter ye* < OE *æt thǣre īeġe*, the place 'at the island'.The DB form shows prosthetic *N-* from the inflected definite article, ME **atten I-ford** < OE *æt þǣm I-forde*. Sx 317 gives pr [aivərd], 322, Locus 1(2).

IFTON HEATH Shrops SJ 3236. P.n. *Iftone* 1272, *I- Yfton* late 13th–early 14th. Unidentified element + OE **tūn**, + ModE **heath**. DEPN, Gelling.

IGHTFIELD Shrops SJ 5938. 'Open land by the river *Giht*'. *Istefelt* 1086, *-feld* 1230, 1308, *Hic(h)tefeld' -feud* 1175–1256, *Y(c)hte- Ichtifeld'* 1210×12–1228, *Ightfelde* 1376 etc., *-field* 1703, *Eigh(t)fi(e)ld, Eitfield* 1675–1764. R.n. **Ġiht* as also in ISLIP Oxon SP 5214 + OE **feld**. Sa i.162 gives pr [aitfi:ld].

IGHTHAM Kent TQ 5956. 'Ehta's estate'. *Ehteham* c.1100, *Eytheham* 1226, *Ei- Eytham* 1232, 1284, *Igte- Yhytte- Egtham* 1270, *Heghtham, Itham* 1278 etc. OE pers.n. **Ēhta* + **hām**. PNK 153, KPN 362, Studies 1931.24, ASE 2.29.

IGHTHAM MOTE Kent TQ 5853. A small moated medieval manor house known simply as *(La, Le) Mote* 1327×77–1690, *The Mote* 1819 OS. P.n. IGHTHAM TQ 5956 + OFr, ME **mote** 'a moat'. K 153, Newman 1969W.330.

IKEN Suff TM 4155. Possibly 'Ica's river'. *Ykene* 1212, *Ikene* 1236–1346, *Ikano* 1254, *Iken* from 1286. OE pers.n. **Ican*, genitive sing. **Ican*, + **ēa**. Very likely this was an earlier name of the r. Alde, itself a late back-formation from the p.n. ALDEBURGH TM 4656. The alternative suggestion that the name of this large and important estuary is identical with the ancient unexplained r.n. ITCHEN Hants SU 4616, Warw 4062, is attractive except for the difficulty of explaining the stop [k] for the expected affricate [tʃ]. DEPN and Baron give pr [aikən].

ILAM Staffs SK 1350. Possibly '(the settlement) at the hills'. *Hilum* [1002×4]11th S 1536, [1004]11th S 906, 1227, *Hylum* 1255, *Ylum* 1116×33–1546, *Ilum* 11227–1514, *I- Ylom(e)* 1327–1581, *Ilam* 1293 and from 1547. Oakden following Wrander 34, 64, 122, 133 suggests ON **hylr** 'a deep place in a river', dative pl. **hylum**; but OE ***hyġel** 'a small hill', dative pl. ***hyġlum**, referring to the nearby ridges of Ilam Tops is equally possible.

ILCHESTER Somer ST 5222. 'Roman fort on the river Ivel'. *Givel- Giuelecester* 1086, *Iuelcestr'* 1157–1212, *Yvelchester* [1174×80]14th, *-cestre -cester* [1206×7, 1209×12]B, [c.1350] Buck, *Gyvelcestre* [1236] Buck, *Jevelchestre* [1366] Buck, *Ilchester* from 1586. R.n. Ivel + OE **ċeaster**. Ivel is the old name of the Yeo, *(on) Gifle* [963 for 946×51]12th S 516, *Yevel* [852 for 878]17th S 343, *Givell* 1243, *Ivel(le)* c.1540, identical with the folk-n. *Gifle* recorded in the Tribal Hidage from an unrecorded OE cognate of Gothic **gibla* 'gable', OHG *gibil*, G *Giebel* probably used to denote a fork in the course of a river. DEPN, RN 221.

ILDERTON Northum NU 0121. 'Elder-tree farm or village'. *Ildretona* c.1125, *Ilderton(e) -ir-* 1228–1428, *Hilderton -ir-* 1189–1291, *Hillerton* 1346. OE ***hyldre**, ME **hilder**, + **tūn**. NbDu 122, PNE i.274.

ILFORD GLond TQ 4586. 'Ford across the river Hyle'. *Ilefort* 1086, *-ford* 1232–93, *Illeford* 1226–1330, *Hil(l)e- Hyleford* 1234–1373, *Ilford* 1458. R.n. Hyle + **ford**. Hyle, *Hile* [958]12th S 676, *Hyle* [c.1250]15th, is the old name of the Roding. The OE form, **Hīl* represents a Celtic root **sil-* 'trickling stream', cognate with the r.ns. Sihl in Germany, OHG *Sil-aha*, and Sile in Italy, *Silis* [77]5th–9th Pliny; and the root of Latin *sileo*, Gothic *anasilan* 'be silent', ON *sil* 'still place in a river'. The modern r.n. Roding is a back-formation from the Rodings in Essex. The ford was where the Roman road to Chelmsford, Margary no.3a, crossed the river. Ess 97, 11, RN 206, TC 208.

ILFRACOMBE Devon SS 5147. 'Ælfred's coomb'. *Alfreincome* 1086, *Alferðingcoma* 1167, *Aufredyncomb* 1208, *Alfrede(s)cumbe* 1234, 1249, *Al- Aufrithecumb(e) -fride-* 13th cent., *Ilfredecumb(e)*

1262–1309, *Ilvercombe* 1302, *Ylfrycomb* 1346, *Ilfarcomb vulg. Ilfracomb* 1675. OE pers.n. *Ælfrǣd*, genitive sing. *Ælfrǣdes*, varying with *Ælfrǣd* + **ing**[4] + **cumb**. Some forms, however, suggest pers.n. *Ælfrith* or *Ælferhth*, and the *Il-* spellings a variant form **Ielfrǣd*. D 46.

ILKESTON Derby SK 4642. 'Ealac's hill'. *Tilchestune* (with *T-* from *(æ)t Ilchestune*) 1086, *(H)elkesdone* 1100×35–1330, *(H)ilkesdon(e) -is-* 1236–1330, *-ton(e)* 1276–1306 etc., *Ilston* 1554. OE pers.n. *Ēalāc*, genitive sing. *Ēalāces*, + **dūn** subsequently replaced by **tūn**. Db 473, L 155.

ILKETSHALL ST ANDREW Suff TM 3787. 'St Andrew's I'. *ecclesia Sancti Andree ... de Iketeshale* 'St Andrew's church of I' 1176, *St Andros* 1610, *S^t. Andrew's Ilketsal* 1838 OS. P.n. *Ilchet(el)es- Elcheteshala, Ulkesala* 1086, *Ilketeshal(e)* 1176–1327, *Ilketeshalla -hal(e)* 1188–1357, *Hulketeleshal* 1228, *Ilketsale* 1523, *-hall* 1568, 'Ylfketill's nook', ON pers.n. **Ylfketill* varying with *Ulfketill*, secondary genitive sing. **Ylfketilles*, + **halh**, + saint's name *Andrew* from the dedication of the church. The reference is probably to a side-valley in the hills here. DEPN, Baron.

ILKETSHALL ST LAWRENCE Suff TM 3883. *ecclesia Sancti Laurentii ... de Ilketeleshale* 1176, *S^t. Lawrence Ilketsal* 1837 OS. P.n. Ilketshall as in ILKETSHALL ST ANDREW TM 3787 + saint's name *Lawrence* from the dedication of the church. Baron.

ILKETSHALL ST MARGARET Suff TM 3585. *ecclesia Sancte Margarete ... de Ilketeleshale* 1176, *S^t. Margarets Ilketsal* 1837 OS. P.n. Ilketshall as in ILKETSHALL ST ANDREW TM 3787 + saint's name *Margaret* from the dedication of the church. Baron.

ILKLEY WYorks SE 1147. 'Yllica's wood or clearing'. *hillicleg* [972×92]11th S 1453, *on yllic-leage* c.1030 YCh 7, *Il(l)ecliue, Illiclei(a)* 1086, *(H)illec- Ylleclai -lay -ley* 12th cent., *I- Ylkelay -le -ley(e)* 1176–1604, *Ilkley* from 1198, *Ighlay -ley* 17th cent. OE pers.n. *Yllic(a)* + **lēah**. It has been widely asserted that the name is identical with the RBrit p.n. *Olicana* [c.150]13th Ptol. The reference of this form, however, is in doubt and the name of the Roman fort at Ilkley was probably *Verbeia* related to the r.n. WHARFE. YW iv.210, xii, RBrit 430. YW gives prs ['ilklə], earlier ['i:flə].

ILKLEY MOOR WYorks SE 1146. *Ylkeley More* c.1540. P.n. ILKLEY SE 1147 + ModE **moor** (OE *mōr*). YW iv.214.

ILLEY WMids SO 9881. Partly uncertain:
I. *Hilley* 1199×1216, *-leye* 1216×72.
II. *Illeg* 1255, *Illeleya* 1271.
The generic is OE **lēah** 'a wood or clearing', the specific an OE pers.n. either *Hilla* or **Illa*. Wo 297.

ILLINGWORTH WYorks SE 0728. 'Enclosure called after Ylla or Illa'. *Ill- Yllingw(o)rth(e) -yng-* 1276–1616. OE pers.n. *Ylla* or *Illa* + **ing**[4] + **worth**. YW iii.114, xii.

ILLOGAN Corn SW 6643. 'St Illogan's (church)'. *(Ecclesia) Sancti Illogany* 1291–1354, *(ecclesia) Sancti Elugani apud Egloshal'* 1302, *rector Sancti Elugani, Yllugani* 1308, *Seyntlugan alias Seyntlocan* 1436, *Illuggan* 1563, *Luggan* 1610. Saint's name *Illogan* from the dedication of the church. Nothing is known of him. The name of the churchtown *Egloshal'* is found as *Eglossalau* 1235, surviving as *Eglish Hallow* 1820, possibly 'church of the marshes', Co **eglos** + **halow**, pl. of **hal**, although this hardly fits the topography. PNCo 98, Gover n.d. 605.

ILLSTON ON THE HILL Leic SP 7099. Partly uncertain. *(N)elvestone* 1086, *Elueston'* 1176, 1185, *Jelverston* 1318, *Ilueston(e) -is-* c.1130–1443, *Ilston(')* from 1410, *Y- Ilson* 1537, 1572. Possibly 'Iolfr's village or farm'. ON pers.n. *Iólfr*, ME genitive sing. *Iólfes*, + **tūn**. However, doubt has been cast on ME *e* as a reflex of ON initial *jó*. Possibly, therefore, OE pers.n. *Ælfhere*, genitive sing. *Ælfheres*, + **tūn**. Lei 232, SSNEM 377.

ILMER Bucks SP 7605. 'Leech pond'. *Imere* 1086, *Ilmere* 1161–1346. OE **īl** + **mere**. 'Leech' here seems the sense preferable to usual 'hedgehog'. Bk 125, PNE i.280, L 26.

ILMINGTON Warw SP 2143. 'The elm-tree hill'. *Ylman dunnes gemære* 'the boundary of I' [978]11th S 1337, *(æt) Ylmandune* c.1000 Ælfric, *Edelmitone* 1086, *Ilmedon(e)* 1086, 1247–1428, *-myn-* 1269–1428, *-myng-* 1306, *Ilmyngton* 1535. OE ***ylme** + **dūn**. Wa 304, L 154.

ILMINSTER Somer ST 3514. 'Minster on the river Isle'. *Yleminister* [725]13th S 249, *Illemynister* 995 S 884, *Ileminstre* 1086, [1275×92]Bruton, *Ilminster* 1610. R.n. ISLE ST 4023 + OE **mynster**. DEPN, RN 216.

ILSINGTON Devon SX 7876. 'Estate called after Ielfstan'. *Lestintone* 1086, *Ilestintona* 1086 Exon, *Elstington* 1186×91, *Ilstingtun* c.1200–1342 with variants *Yl-, -yng- -in-*, and *-ton*. OE pers.n. **Ielfstān*, a variant of *Ælfstān*, + **ing**[4] + **tūn**. D 475.

East ILSLEY Berks SU 4981. *Esthildesleg'* 1189, *Estildesle -yldesl' -le* 1242×3 etc., *Esteldesley* 1396×7, *Est Ilsley or Market Ilsley* 1761. ME adj. **est** + p.n. *Hildes- Eldes- Hislelei* 1086, *Hildeslegam* [1146]15th etc. with variants *Hildesl(ea)* and *Hyldesle*, *Illesley* 1442, 'Hildi's wood or clearing', OE masc. pers.n. *Hildi*, genitive sing. *Hildes*, + **lēah**. The same pers.n. also forms the specific of a lost Berks hundred name, *Hilleslau(e), Hillislau* 1086, OE *Hildes hlæw*, 'Hildi's tumulus', in the bounds of Compton Beauchamp [955]13th S 564. 'East' for distinction from West ILSLEY SU 4782. Brk 502, 343.

West ILSLEY Berks SU 4782. *Westyldesl'* 1220, *Westhildesley* 1224×5, *Westildesle -yl-* 1230 etc., *Westyllesley -ys-* 1517. ME adj. **west** + p.n. Ilsley as in East ILSLEY SU 4981. Brk 502.

ILTON NYorks SE 1978. 'Ylca's settlement'. *Hilchetun, Ilcheton* 1086, *I- Ylketon* 1184–1535, *Ilkton* 1581, *Ilton* 1558. OE pers.n. *Ylca* + **tūn**. YN 233.

ILTON Somer ST 3517. 'Settlement on the river Isle'. *Atiltone* (i.e. *(at) Iltone*) 1086, *Ilton* from 1243. R.n. ISLE ST 4023 + OE **tūn**. DEPN, RN 216.

IMBER Wilts ST 9648. 'Imma's pond'. *Imemerie* 1086, *Immemera -mer(e)* 1146–1221, *Im(m)ere* 1291–1347, *Immer al. Hymbemer* 1435, *Immer al. Imber* 1540. OE pers.n. *Imma* + **mere**. Lost names in this parish include *Imendone* 1161 'Imma's hill', OE **dūn**, and *Ymmanedene* [968]14th S 765 'Imma's valley', OE **denu**. There are wells at Imber which is situated in the middle of a waterless part of Salisbury Plain. Wlt 170, Grundy 1920.83.

IMMINGHAM Humbs TA 1814. 'Homestead of the Immingas, the people called after Imma'. *Imungeham, In Mingeham* 1086, *Immung(h)eham* c.1115, 1205, *Emmingham* [1090×6]13th, *Imingeham -ynge-* 1220, 1305, *Immingham* from 1233 including [1100×35]c.1240. OE folk-n. *Immingas* < pers.n. *Imma* + **ingas**, genitive pl. *Imminga*, + **hām**. Li ii.163, ING 145.

IMPINGTON Cambs TL 4463. 'Estate called after Impa or Empa'.
I. *Impintune -a* [1042×66]13th S 1051, *Impintune -ton(e)* 1086–1574, *Impitune -tone -y- -e-* 1086–1482, *Impington -yng-* from 1271.
II. *Empintuna -tona -ton(e) -yn-* 1086–1298, *Epintone* 1086, *Empington* 1201, 13th, 1315.
OE pers.n. **Impa* or **Empa* + **ing**[4] + **tūn**. Ca 178.

INCE 'The island'. PrW ***ïnïs**.
(1) ~ Ches SJ 4576. *Inise* 1086, 1398, *Ynes* c.1150–1446, *Yns, Ins* 1284–1446, *Ince* from 1415. Ince occupies a low ridge amidst the marshland of the Gowy and Mersey estuaries. Che iii.251.
(2) ~ BANKS Ches SJ 4578. *Ince Banks* 1842. P.n. INCE SJ 4576 + ModE **bank(s)**. A sand-bank in the Mersey estuary. Che iii.252.
(3) ~ BLUNDELL Mers SD 3203. 'Ince belonging to the Blundell family'. *Ins Blundell* 1332, *Ince Blundell* 1842. P.n. *Hinne* 1086, *Ines, Ynes* 1212–1375. + family name *Blundell* into whose possession the manor passed c.1200. Ince Blundell is

situated on slightly higher ground in flat fen country. 'Blundell' for distinction from INCE-IN-MAKERFIELD SD 5094. La 118, Jnl 1.48 corrigenda, 17.65.

(4) ~ -IN-MAKERFIELD GMan SD 5904. *Ins in Makerfeld* 1332. P.n. Ince + district name MAKERFIELD, for distinction from INCE BLUNDELL SD 3203. Ince, *Ines* 1202–1327 etc., *Ynes* 1206–1261, *In(e)s* 1262–1341, *Ince* 16th, refers to a raised area of dry land amid mossland. La 103, Jnl 17.58.

INDIAN QUEENS Corn SW 9159. *Indian Queens* 1802, *Indian Queen* c.1870, earlier *The Queens's Head* 1780. A 19th cent. village which grew up around an inn described in 1802 as 'a single house which is rather a post-house than an inn'. This popular 18th cent. inn name probably refers to the famous Red Indian queen Pocahontas who landed in Plymouth in 1616, travelled to London and died at Gravesend and was celebrated in popular ballads. PNCo 99.

INGATESTONE Essex TQ 6499. 'Ing by the stone'. *Gi- Gynge Attestone* 1289–1428, *ȝingatestone* 1320, *Ingerston(e)* 1481, *Ingatston(e)* 1482, *Yng at Stone* 1542. P.n. *Ingā* 1086, *ginga* c.1090, *Gi- Gynge(s)* 1241–1543 as in INGRAVE TQ 6292, + ME **atte stone** 'at the stone' probably referring to a mile-stone near the church on the Roman road to Colchester, Margary no.3a. Ess 253 gives pr [iŋgətstən], Jnl 2.44.

INGBIRCHWORTH RESERVOIR SYorks SE 2206. P.n. INGBIRCHWORTH SE 2206 + ModE **reservoir**.

INGBIRCHWORTH SYorks SE 2206. 'Meadow Birchworth'. *Y- Ingbirchworth(e) -byrch-* 1433–1641. Prefix *Ing* (ON **eng**, *Hing-* 1326) + p.n. *Berceuuorde* 1086, *Bi- Byrchewrd -w(o)rth* 13th–1435, the 'birch-tree enclosure', OE **birce** + **worth**. *Ing* for distinction from Roughbirchworth SE 2601, *Bercewrde -uuorde* 1086, *Bi- Byrchewurth -worth(e)* 1246–1547, *Roygh-* 1299. YW i.330, 335.

INGESTRE Staffs SJ 9724. Uncertain. *in Gestreon* (for *Ingestreon*) 1086, *Ingkestrent* c.1200, *Ingestre(n)(t)* 1236, 1242, 1371, *Higestront* 1242, *Ingestraund* 1250. The n. has been explained as a compound of OE ***ing** 'a hill, a peak' and **ġestrēon(d)** 'property, treasure', but this is very uncertain. Alternatively the generic could be nominative or dative pl. of OE **trēo(w)** 'a tree'. DEPN, PNE i.282, ii.163, 186, Brunner §250.2 A.4.

INGHAM Lincs SK 9483. Partly uncertain. *Ingeham* 1086, *Ing(h)eham* c.1115–1610, *Yngeham* 1086–1325, *Ingaham* 1163, *Ingham* from 12th. This has usually been taken to be 'Inga's homestead', OE pers.n. **Inga* + **hām**. It has recently been suggested, however, that the specific is not a pers.n. but a term used of an early royal centre from Gmc **Ingwia* 'Inguian' as in the tribal name *Inguiones*. The name would then mean something like 'the homestead of the Ing people, the devotees of the god Ing'. See following entries. DEPN, SSNEM 378, LSE 18.233, 236–7, Cameron 1998.

INGHAM Norf TG 3925. Partly uncertain. *Hincham* 1086, *Ingham* from [1127×34]14th, *Ingeham* 1189×99–1248. This is usually held to be 'Inga's homestead', OE pers.n. **Inga* + **hām**, but the rarity of this pers.n., the relative scarcity of forms with medial *-e-* and the complete absence of forms with *-en-* representing OE genitive sing *-an* led K. I. Sandred to suggest that *Ing* in this name and in the Lincs and Suffolk Inghams, SK 9483, TL 8570, and in INGWORTH Norf TG 1929, might have been a designation of the Anglian king as a member of the Germanic Ingwionic dynasty of the *Ingaevones* or *Ingvaeones* mentioned by Tacitus and Pliny. If so this would be a very early royal estate. Nf ii.110, LSE 18.231–40.

INGHAM Suff TL 8570. Possibly the 'homestead of the devotees of Ing'. *In cham, ing- Ingham* 1086, *Ingham* 1610. Traditionally the specific of this name has been explained as the unrecorded OE pers.n. **Inga* + **hām**. Recently it has been proposed that here and at INGHAM Norf TG 3825 the specific is a very early element ***ing** < Gmc **ingwia* used to mark places which were royal or state property. Dr Insley suggests that this name alludes to a *Kultverband* of devotees of the Gmc demigod Ing, the eponymous ancestor of the Inguiones, and that the p.n. denotes a homestead occupied by devotees of the cult. DEPN, LSE 18.231ff.

INGLEBOROUGH NYorks SD 7474. '*Ingle* fort'. *Ingelburc(h)* 1165–1293, *Ingilburg(h)* 1401, *Inge-* 1177, 1251. OE p.n. **Ingle* + **burh**. **Ingle* probably represents either OE **ing* 'a peak' + *hyll*, or a derivative **ingel*. Ingleborough is one of the highest peaks in Yorkshire with a striking outline. There are remains of an extensive hill-fort on the summit. YW vi.242.

INGLEBY 'The Englishmen's village'. OE *Engle*, genitive pl. *Engla*, + ON **bȳ**.

(1) ~ ARNCLIFFE NYorks NZ 4400. Short for *Engelby juxta Erneclyf* 'I by Arncliffe' 1303. P.n. *Englebi -by* 1086, 1231, + p.n. ARNCLIFFE SE 4599. 'Arncliffe' for distinction from INGLEBY GREENHOW NZ 5806. YN 178, SSNY 30.

(2) ~ GREENHOW NYorks NZ 5806. Short for *Engilby juxta Grenehoue* 'I by Greenhow' 1285. P.n. *Englebi -by* 1086, 1203–7, *Ingolby* 1291, *Ingleby* 1301, + p.n. *Grenehou* c.1180, 'the green mound', OE **grēne** + ON **haugr**, as in Greenhow Plantation, Moor and Bank, NZ 5902, 6002–3, and for distinction from INGLEBY ARNCLIFFE NZ 4400. YN 167, SSNY 31.

INGLESBATCH Avon ST 7061. 'Engel's stream'. OE pers.n. **Engel*, genitive sing. **Engeles*, + **beċe, bæċe**. The same pers.n. occurs in nearby Englishcombe ST 7162. PNE i.24.

INGLESHAM Wilts SU 2098. 'Ingen's promontory'. *(æt) Inggeneshamme* c.950 S 1539, *Incgenæsham* [c.968×71]12th S 1485, *Ingeneshamme* [971×83]14th S 1498, *Inglesham* from 1160 with variant *Yng-*. OE pers.n. **Ingen* or **Ingīn*, genitive sing. **Ingenes*, + **hamm** 1 or 2. The situation is at the edge of a patch of higher ground in a tongue of land between the river Cole and the Thames. Wlt 28 gives pr [iŋgəlsəm].

Upper INGLESHAM Wilts SU 2096. ModE **upper** + p.n. INGLESHAM SU 2098.

INGLETON Durham NZ 1720. 'Ingjald's farm or village'. *Ingeltun* c.1104, *-ton* 1376–1523, *Ingilton* 1406–1532, *Ingleton* from 1498. ON pers.n. *Ingjaldr* possibly replacing OE *Inġeld* + **tūn**. OE *Inġeld* would normally yield ME **Inyeld*; the hard [g] is due either to substitution of the Scandinavian for the English personal name (or for an entirely different pers.n.) or to the replacement in pronunciation of English [j] by Scandinavian [g]. NbDu 122.

INGLETON NYorks SD 6973. 'Settlement beside *Ingle*'. *Inglestune* 1086, *Ingelton* 1202–1346, *Ingleton* from 1303. OE p.n. **Ingle* as in INGLEBOROUGH SD 7474 + **tūn**. YW vi.242.

INGLEWHITE Lancs SD 5440. Uncertain. *Inglewhite* 1662. The evidence is too late for certainty, but this might be a compound of ON pers.n. *Ingólfr* or *Ingjaldr* + OE ***thwīt**, ON ***thvít** cognate with **thveit** 'a clearing'. La 19, 149, Jnl 1.39 referring to ENEP 92–5 where this interpretation is defended against the alternative view of PNE ii.221, 265 identifying the generic as OE ***wiht** 'a bend, a curving recess, a bend in a river or valley'. Inglewhite is situated in a shallow valley beside a brook which goes through a right-angled bend SW of the village.

INGLEWOOD FOREST Cumbr NY 4639. P.n. Inglewood + ModE **forest**. Inglewood, *Englewod'* c.1150, *-wode -wud(e)* c.1158–1315, *Inglewod' -wod(e) -wood(e)* from 1267, is the 'forest of the Angles', OE *Engle*, genitive pl. *Engla*, + **wudu**, to distinguish English territory from that of the Britons of Strathclyde. Cu 38, L 229.

INGMIRE HALL Cumbr SD 6391. *Ingmer Hall* 1666. P.n. Ingmire + ModE **hall**. Ingmire, *Yngmyre* 1506, *Ingmire* 1626, 1648, is 'meadow swamp', ME **ing** (ON *eng*) 'meadow, pasture' + **mire** (ON *mýrr*) 'mire, bog'. YW vi.268.

INGOE Northum NZ 0374. Partly uncertain. *Hinghou* 1229, 1244,

Ingou 1242 BF, *Inghow* 1324, *Yengew, Ingowe* 1346, *Yngoo* 1524. The generic is OE **hōh** 'a hill-spur', the specific uncertain, possibly an OE **ing* 'a hill, a peak' and so 'hill-spur called *Ing*'. NbDu 122, L 167, 169.

INGOLDISTHORPE Norf TF 6832. Originally 'Ingolf's', later 'Ingald's outlying farm'. *Inguluesthorp* [1101×7]14th, *-torp* [c.1190]13th, *Ingaldesthorp* [c.1160×70]14th, *Ingaldestorp -galt-* 1203–54, *Ingoldestorp'* 1204, 1220, *-thorp* 1328–36. ODan pers.n. *Ingulf* later replaced by Anglo-Scand *Ingald* (OE *Ingæld*), genitive sing. *Ingulfes, Ingaldes*, + **thorp**. Substitution of *-ald -old* for *-ulf* occurs in other names in the ME period. The earliest reference is simply *Torp* 1086†. SPNN 223, 229.

INGOLDMELLS Lincs TF 5668. 'Ingiald's sand-banks'. *in Guldelsmere* 1086, *Ingoluesmera* 1095×1100, *in Golves-Ingolvesmeles* [1144×55, 1147×55]13th, *Ingoldes-* 1180, *Ingaldemoles* 1212. ODan pers.n. *Ingiald*, genitive sing. *Ingialds*, + **melr**. In several forms initial *In-* has been mistaken for the Latin preposition *in* and the generic confused with *mere* 'a pond'. Some sps point to alternative pers.n. *Ingólfr*. The soil is rich loam overlying clay. DEPN, SSNEM 155, Cameron 1998.

INGOLDMELLS POINT Lincs TF 5768. *Ingoldmells Point* 1824 OS. P.n. INGLODMELLS TF 5668 + ModE **point** 'a promontory'. This is the most easterly location in Lincolnshire.

INGOLDSBY Lincs TF 0130. 'Ingiald's farm or village'. *In Goldesbi, Ingoldesbi* 1086, *Ingoldesby* before 1184, *Yngoldebi* 1202, *Ingaldeby* 1237. ODan pers.n. *Ingiald*, secondary genitive sing. *Ingialdes*, + **bȳ**. The DB form *Goldesbi* is due to mistaking *In-* as the preposition *in*. There seems to have been association with the element *gold*. DEPN, SSNEM 55, Cameron 1998.

INGRAM Northum NU 0116. 'The homestead at the pastureland'. *Angerham* 1242 BF, *Angreham* 1244–1333, *Angran* 1255–1346, *Yngram* 1507. OE ***anger** + **hām**. The situation of the village in the river Breamish water-meadows suggests that the generic might be OE **hamm** rather than **hām**. This element has not, however, been certainly identified in Northern p.ns. NbDu 123, PNE i.11.

INGRAVE Essex TQ 6292. 'Ralf's Ing'. *Gynge Rad'i* (i.e. *Radulphi*) 1248, 1322, *Jynge Rauf* 1332, *Ingrave* 1476. P.n. *Ingam* 1086, *Gi- Gynge* 1248, + pers.n. *Ralf* from Ralf son of Turold of Rochester, the DB tenant, and for distinction from MARGARETTING TL 6601 'Margaret's Ing', FRYERNING TL 6400 'the brothers' Ing', MOUNTNESSING TQ 6297 'Robert de Mounteney's Ing' and INGATESTONE TQ 6499 'Ing by the stone'. The *Ing* in all these names is the folk-n. **Ġīġingas* or **Ġēġingas* 'the inhabitants of the *ġē* or district', a large area of land SW of Chelmsford, OE **ġē** + **ingas**. A possible parallel in Germany is Gauingen near Zwiefalten, Baden-Württemberg, 'the inhabitants of the *gau* or district' although this is now taken rather to be 'the people called after Gawo'. Ess 161, ING 20, Reichardt 1983.49.

INGS Cumbr SD 4498. 'Meadows'. *the Inges* 1546, *Chappleton Inges* 1577, *Ings* 1625. ME **ing** (ON *eng*). The 1577 form derives from the *chapel town* or 'chapel village' and the *Chapel on the Inges* 1675. We i.171.

INGST Avon ST 5887. Uncertain. *Inste* 1231–1630, *Yenst* 1547, *Ingst* 1779. Possibly PrW **ïnïs** 'island' with excrescent *-t* as in *against* (*aȝenest* late 13th cent., Somerset dial. for *aȝenes*). The place stands on a shallow hillock in low-lying marshland. Excresent *-t* is met with especially in non-standard varieties of English in a number of words of similar phonetic shape (against, amongst, once etc.) and this name may have been influenced by the nearby p.n. AUST ST 5789. Gl iii.120, Jordan 199, Dobson 437.

INGWORTH Norf TG 1929. Partly uncertain. *Ingewrda, Inghewurda* 1086, *Ingewrde -w(o)rthe* [1140×53]13th, 1207–1436, *Ingworth(e)* 1275–1548. This is probably 'Inga's enclosure', OE pers.n. **Inga* + **worth**, but an alternative possibility is that *Ing* in this name has the same significance as suggested for INGHAM TG 3925. LSE 18.230–40, esp. 237.

INHURST Hants SU 5761 → BAUGHURST SU 5861.

INKBERROW H&W SP 0157. 'Inta's hills'. *IN tanbeorgas* (sic for *Intan-*) [789]11th, *(æt) intebeorgas, Intan beorgan, INTANBEORGE* [803]11th S 1260, *(æt) intanbergum* [803]17th ibid., *(æt) intanbeorgan* [822×3]17th S 1432, [977]11th S 1331, [984]11th S 1349, *(æt) intebeorgan, intanbeorgum* [822×3]11th S 1432, *(æt) intebyrgan* c.1010×1023 S 1460, *Inteberge -a* 1086, *Inteberg(h)* 1230–1315, *Inteberwe* 1327, *Inkbarowe* 1275. OE pers.n. *Inta*, genitive sing. *Intan*, + **beorg**, dative sing. **beorge**, pl. **beorgas**, dative pl. **beorgum, beorgan**. 'Inta's' for distinction from Holberrow in HOLBERROW GREEN SP 0259. The village is surrounded by several small hills referred to also in the p.n. Bouts SP 0358, *Boltes* 1271, 1357, OE ***bultas**, pl. of ***bult** 'a small hill'. The territory and boundary of the Inkberrow people is referred to as *incsetena land* and *inc setena gemære* [963]11th S 1305, OE folk-n. *Incsǣte* 'the people of Inkberrow', genitive pl. *Incsǣtena*. Wo 324–5, L 128, Hooke 33, 91, 96, 151–2, 261, 330.

INKPEN Berks SU 3764. Possibly 'peak enclosure'. *(æt) Ingepenne* 931×9 S 1533, *Hingepene* 1086, *Ingepenne* 1167, *Y-Ingelpenn(e)* 13th cent., *Ynkepenne* 1241. OE ***ing** 'hill, peak' or ***ing-hyll** + **penn**. Stenton's interpretation of the name as 'Inga's hill', OE pers.n. *Inga* + PrW ***penn** (Brit **penno-*) is a possible alternative. Brk 309, 885, L 182.

INNER SOUND Northum NU 2035. 'The inner channel'. *Inner Sound* 1866 OS. ModE **inner** + **sound** 'a channel' (OE *sund* 'the sea'). The reference is to the passage between Bamburgh and the Inner Farne Islands.

INNSWORTH Glos SO 8621. 'Ine's enclosure'. *Ineswurth* [794]13th, *Ines- Ynesworde -worth* c.1230–1327, *Innesworth* 1266–1598, *Insworth* 1557–1775. OE pers.n. *Ine*, genitive sing. *Ines*, + **worth**. Gl ii.159.

River INNY Corn SX 2383. Probably 'ash-tree river'. *Æni* 1044 S 1005, *Eny* [c.1160]15th, 1229, *Innye* early 17th. Co **enn**, double plural of **onn** 'an ash-tree', + name suffix ***-i**. PNCo 99, Gover n.d. 681, RN 211.

INSKIP Lancs SD 4638. Uncertain. *Inscip* 1086, *Hinskipe, Inscype* 1246, *Inskip* from 1330. Inskip is situated on a small plateau surrounded by low marshy country. The first element might be PrW **ïnïs** 'an island' compounded with OE **cȳpe** 'an osier-basket for catching fish, a kipe', and so 'fishtrap(s) at the island'. La 161, DEPN, PNE i.124, 303, Jnl 1.48 Corrigenda, 17.94.

INSTOW Devon SS 4730. 'St John's holy place'. *Iohannestov* 1086, *Jone- Y(e)onestowe* 13th. Saint's name *John* from the dedication of the church to St John the Baptist, + **stōw**. Cf JACOBSTOWE SS 5801. D 117.

INWARDLEIGH Devon SX 5699. 'Inwar's wood or clearing'. *Inwardlgh'* (sic) 1242, *In(d)warlegh* 1285, 1303, *Inworley* 1538×44, *Inwardleigh vulgo Ingerleigh* 1765. Earlier simply *Lege* 1086, OE **lēah** 'a wood or clearing'. Inwar was the tenant at the time of Edward the Confessor. D 149 gives pr [inəli] and [iŋli].

INWORTH Essex TL 8717. 'Ina's enclosure'. *Inew(o)rth* 1206–71, *Inneworth* 1321–46, *Innorth* 1439, *Inforth(e) -ford(e)* 1543–1604. OE pers.n. **Ina* + **worth**. Ess 394.

IPING WSusx SU 8522. 'The Ipingas, the people called after Ipa'. *Epinges* 1086, *I- Ypinges* 12th–1288, *Ipping* 1291, 1334, *Ipyng(e)* 1296–1332 etc. OE folk-n. **Ipingas* < pers.n. **Ipa* + **ingas**. Sx 22 gives pr [aipiŋ], probably a sp pronunciation as the vowel of *Ipa* was probably short, ING 36, SS 64.

IPPLEPEN Devon SX 8366. 'Ipela's fold or enclosure'. *(to) Ipelanpænne, Iplanpen(ne)* [956]12th S 601, *Iplepene* 1086,

†There is also a corrupt DB form *In evlvesthorp*.

Ip(p)el(e)pen(ne) 1172–1274. OE pers.n. **Ipela*, genitive sing. **Ipelan*, + **penn**. D 513.

IPSDEN Oxon SU 6385. 'The valley of the upland'. *Yppesdene* 1086–1428 with variants *I-* and *-den -don, Y- Ipesden(')* 1194–1235×6, *Ipseden'* 1469. OE **yppe**, genitive sing. **yppes**, + **denu**. There is no need to posit an unrecorded pers.n. **Ippe*. O 56, L 99.

IPSTONES Staffs SK 0250. 'Ippa's stone(s)'. *Yppestan* 1175, *I- Yp(p)estanes* c.1190–1350, *Ipstone* 1220, 1242, 1560, *Uppestan* 1261, *Ipstones* from 1310 with variants *Y-*. OE pers.n. *Ippa* + **stān**, pl. **stānas**. An alternative possibilty is 'the hill with stones', OE **yppe** possibly referring to a look-out place. The village occupies a hill-top site and the parish abounds in outcrop stones. Oakden.

IPSWICH Suff TM 1644. 'Gip or Gipi's trading place or port'. *(to) Gipes wíc* 993 ASC(A), *(æt) Gipeswic* c.1120 ASC(E) under year 1010, *Gypeswic* 12th ASC(C, D) under year 991, *Gypes- Gepes- Gipe(s)wiz, Gipeswic* 1086, *Gip(p)eswic(h)* 1170–1265, *Ipswich* from 1255. OE pers.n. **Ġip* or **Ġipi*, genitive sing **Ġipes*, + **wīċ**. An alternative suggestion is OE ***ġip(s)** on the base of *ġipian* 'to yawn' in the sense 'gap' referring to the broad estuary of the Orwell. Cf. GIPSEY RACE Humbs TA 0967. DEPN, TC 113.

IRBY 'Village or farm of the Irishmen'. ON **Íra-bȳ**. Used of isolated settlements of Norwegian Vikings from Ireland.

(1) ~ Mers SJ 2584. *Erberia* [1096×1101]1280, c.1150, *Irreby* [1096×1101]1280, 1181×1232–1579, *Ireby* [1181×1232]1300, 1288–1660, *Irby* from 1288. ON **bȳ** is confused in the earliest spelling with OE **byriġ**, dative sing. of **burh**. Che iv.264, SSNNW 32.

(2) ~ IN THE MARSH Lincs TF 4763. P.n. *Irebi* c.1115, 1154×89, 1212, *Yreby* 1257. DEPN, SSNEM 55.

(3) ~ UPON HUMBER Humbs TA 1904. *Irby super Humbre* 1401×2. P.n *Iribi* 1086, *Irebi* 1086–1202, *-by* 1242–1502, *Irby* from 1303 + r.n. HUMBER SE 9523. Li v.124, SSNEM 55.

IRCHESTER Northants SP 9265. 'Yra's Roman fort'. *Yranceaster* [973]15th S 792, *Irencestre* 1086–1428 with variants *Hire-* (1086), *Yren-* and *Yrin-*, *Irn(e)cestre* 1261–1327, *Irecestre* 1167–1339, *Irchestre* 1291, *Irchester al. Irenchester* 1565, *Archester* 1510. OE pers.n. **Yra*, genitive sing. **Yran*, + **ċeaster**. The walls of the (unidentified) Roman settlement were still standing in the 18th cent. Nth 192 gives pr [a:tʃestə], Pevsner 1973.266.

IREBY Cumbr NY 2339. 'Village, farm of the Irishmen or man'. *Irebi* c.1160, *Ireby* from c.1150. ON *Íri*, genitive sing. and pl. *Íra*, + **bȳ**. Also known as Low Ireby, *Nether Irby* 1292, *Market Ireby* 1305, *Lawe Irby* 1333, *Bassa Irby* 1344, for distinction from High Ireby NY 2337, *Heghireby* 1279, *Magna Ireby* 1291, *Ireby Alta* 1399. Cu 299, SSNNW 33.

IREBY Lancs SD 6575. 'Village, farm of the Irishmen'. *Irebi* 1086, *Yreby -bi* 1212, 1215, *Ireby* from 1241, *(H)Yrby* 1292, 1297. ON *Íri*, genitive pl. *Íra*, + **bȳ**. La 183, SSNNW 33, Jnl 17.102.

IRELETH Cumbr SD 2277. 'Slope of the Irishmen or man'. *Irlid* 1190, *Ireleyth* c.1200, *Irelith* 1292. ON *Íri*, genitive pl. *Íra*, + **hlíth**. A grange of Furness Abbey, also known as *Yerlethcote* 1539 'Ireleth huts or cottages'. La 205, SSNNW 138, L 166.

IRESHOPEBURN Durham NY 8638. P.n Ireshope + OE **burna**. *Ishoppburn* 1647, *Ireshopeburne* 1685. Ireshope, no early forms, is apparently ON **Íri**, secondary genitive sing. **Íres**, + **hop**, 'the Irishman's side-valley'. ON *Íri* would be a reference to someone of Norwegian origin from Ireland who had crossed over the Pennine pass into a side-valley in upper Weardale. Pronounced [aizəp]. Nomina 12.35.

IRESHOPE MOOR Durham NY 8436. *Ireshope Moor* 1863 OS. P.n. Ireshope as in IRESHOPEBURN NY 8638 + **mōr**.

Kirk IRETON Derby SK 2650. 'Ireton with the church'. *Kirk(e)irton* 1370, *Kirkyrton* 1577. ME **kirk** + p.n. *Hiretune* 1086, *I- Yrton(e)* 1243–1485, *Ireton* 1261, 1326, 1451, 'farm, village of the Irishmen', ON **Íri**, genitive pl. **Íra**, + **tūn**. The reference is probably to Norwegians from Ireland. 'Kirk' for distinction from Little Ireton SK 3141, *Parva Ir(e)ton(e)* 1313–30 etc., *Lytle Ireton* 1543, *Little Yerton* 16th cent. Db 381, 508, SSNEM 264.

IRLAM GMan SJ 7294. 'Village, homestead or promontory by the river Irwell'. *Urwilham -wel- -uuel-* c.1190, *Yre- Irwelham* 1259, *Irwilham* 1250–1451, *Irelame* 1574. R.n. IRWELL SJ 7913 + OE **hām** or possibly **hamm**. La 39, Jnl 17.34.

IRNHAM Lincs TF 0226. Possibly the 'homestead on the (stream called) *Yerne*'. *Grene- Gerneham* 1086, *Erneham* 1090×1100, 1108, *I- Yrneham* 1166–1553, *I- Yrnham(e)* 1166–1642. OE r.n. **Ġerne* + **hām**. A difficult name. The suggested specific would be an otherwise unrecorded name of the stream by which the village is situated, identical with the r.n. YARE TG 1108, RBrit *Gariennos*, OE **Ġerne*, with possible sense 'bubbling stream'. Alternatively 'Georna's homestead', OE pers.n. **Ġeorna*. Perrott 189, Payling 94, RBrit 366, LHEB 434, Cameron 1998.

IRON ACTON Avon ST 6883 → Iron ACTON.

IRON CROSS Warw SP 0652. *Iron Cross* 1831 OS. ModE **iron** + **cross** 'a crossroads'.

IRON-BRIDGE Shrops SJ 6703. An 18th century town named from the bridge over the river Severn designed by Thomas Farnolls Pritchard and cast by Abraham Darby III in 1778 at his foundry in nearby Coalbrookdale, the first cold cast iron bridge in the world, opened in 1781. Room 1983.56.

IRONVILLE Derby SK 4351. *Ironville* 1837. A small market town which developed in the second half of the 19th cent. around the works of the Butterley Iron Company. ModE **iron** + **ville**. Cf. COALVILLE SK 4214, EASTVILLE TF 4057, FRITHVILLE TF 3250. Db 189, Room 1983.56.

IRSTEAD Norf TG 3620. Probably 'the mud or marsh place'. *Irsted(e)* c.1140–1535, *Yrstede* [1153×68]13th. Probably OE **gyr** + **stede**. The place is situated in fenland. Nf ii.171, PNE i.212Stead 183.

River IRT Cumbr NY 0900. Uncertain. *Irt* from 1279, *Erte, Urt* 1560, 1570. A pre-English river n. possibly from an OW **Iret* < W *ir* 'fresh, green' + suffix *-et-*. Cu 17, RN 211, GPC 2025.

River IRTHING Cumbr Northum NY 6670. Uncertain. *Irthin(am)* [1169]18th, 1256, *Irthing* [c.1195]18th, 1292, 1612, *Yrthinam* [1154×89]18th, *Irthyn(e)* 1383–9, *Hirthenam* [1183]1402, *Hirtinam* [1184, c.1210]18th, *Erthin(am)* [c.1210, 1275]18th, *Erthingg'* 1278, *Irdin* 1561, *Irding(e), Erding* 1603–4. Possibly an OW r.n. **Irtīn*, a derivative of IRT with suffix *-īn-*. NbDu 123, Cu 18, RN 212.

IRTHINGTON Cumbr NY 4961. 'Settlement on the river Irthing'. *I- Yrthinton* 1169, c.1220, *Irthington -yng-* from c.1170, *Erthington* c.1194–1295. R.n. IRTHING NY 6670 + OE **tūn**. Cu 91.

IRTHLINGBOROUGH Northants SP 9470. Partly uncertain. *Yrtlingaburg* 780 S 1184, *Erdi(n)burne* 1086, *Yrtling Burch* 1125×8, *Irtlingburg* 12th–1428 with variants *Irtel- -lyng-* and *-burgh, Yrthlingburc* 1212, *Irthyngborough* 1428, *Ertlinburg'* 1205, *Artilburgh* 1469, *Artleborowe* 1583. This is either the 'fortification of the Yrtlingas, the people called afterf Yrtla', OE folk-n. **Yrtlingas* < pers.n. **Yrtla* + **ingas**, genitive pl. **Yrtlinga*, + **burh**, or, if the *-th-* forms are original, the 'fortification of the ploughmen', OE **yrthling** 'a husbandman, a farmer, a ploughman', genitive pl. **yrthlinga**, + **burh**. DEPN suggests that the reference was to an old fort used for stalling oxen. OE **yrthling** is also the name of a bird, possibly the wagtail. Nth 182 gives prs [a:təl- ja:təlbərə].

IRTON NYorks TA 0084. 'Settlement of the Irishman or -men'. *Iretun(e)* 1086, 1170, *I- Yrton(a)* from c.1223, *Urton* 1572. ON *Íri*, genitive sing. and pl. *Íra*, + OE **tūn**. YN 101.

River IRWELL GMan SJ 7913. Possibly 'wandering stream'. *Urwil* 1184–1222, *-will* [1242×50]1268, *Irewel* [c.1200]14th, *Yrewel* 1292, *Ir(r)ewelle* 1277–8, *Irwell* from 14th. OE **irre**, **eorre** 'straying, wandering' + **wylle, wella**. La 27, RN 213.

River ISBOURNE Glos SP 0334. 'Esa's stream'. *in Esenburnen* *[709]12th S 80, *Biesingburnan* [862]14th S 1782, *Esegburna* *[778]17th S 113, *on Esigburnan* [n.d.]12th S 1559, *Biesingburnan, on is esingeburna, of, in esingburnan* *[930]12th S 404, *(of, on, in) Esingburnan* [1002]13th S 901, *Eseburn(e)* 13th cent., *E(a)seburn(e)* c.1540, *Isbourne* 1779. OE pers.n. *Ēsa* + **ing**[4] + **burna**. Gl i.9, RN 214.

ISCA Devon SX 9292. Short for ISCA DUMNONIORUM 'Isca of the Dumnonii', the Romano-British name of Exeter, originally a river name, the river EXE q.v. *Ἴσκα* [c.150]13th Ptol, *Isca Dumnoniorum* 4th Peut. *Isca* was a common Celtic r.n. from the IE root **is-* 'move swiftly' + suffix *-*kā* to which *Dumnoniorum* 'of the Dumnonii' was added for distinction. See EXETER SX 9292 for further details. RBrit 378, BzN 1957.241.

River ISE Norhants SP 8875. Uncertain. *(andlang) ysan* [956]12th S 592, *Yse* c.1270, 1292, *Ise* from 1247, *Use* c.1540. If the *y* of the charter form *Ysan* represents OE *y*, as seems probable, *Yse* might be a mutated form of the r.n. OUSE, OE *Use*, < **Ūsiōn*; if not, the zero grade **is-* of the Old European root **eis-* 'move swiftly', found in the r.ns. ESK, EXE, may be in question. Nth 3, RN 214, BzN 1957.238–42.

ISFIELD ESusx TQ 4417. 'Isa's open land'. *Isefeld -feud* 1215–1426, *Ysfeld* 1320. *Isvill* 1711. OE pers.n. *Īsa* + **feld**. Sx 396, SS 70, L 243.

ISHAM Northants SP 8874. 'The homestead on the river Ise'. *Isham* [974, 1077]13th, 1086 etc., *Ysham* [1060]14th S 1030–1314, *Isam* 1235, 1428. R.n. ISE SP 8875 + OE **hām** Nth 127 gives pr [aisəm].

River ISIS Glos ST 1796. An alternative name for the Thames above its junction with the Thame used particularly in Oxford. *Isa* [c.1350]c.1400, *Ise* 1347, 1387, c.1540, *The Isis* c.1540. *Isis or Ouse* 1577. Artificially reduced from the form *Tamise* for the r.THAMES, as if *Thame* + *Ise*. Modern *Isis* is due to the common Latin form *Tamesis* again falsely analysed as if *Thame* + *Isis*, not, alas, to association with goddess name Isis. RN 215, Chronique d'Egypts 39.67.

River ISLE Somer ST 4023. A pre-English r.n. of unknown origin and meaning. *Yle* [c.1200]c.1260–14th including [693, 725, 762, 966]13th S 240, 261, *Ile* 1280, c.1540, 1811 OS. RN 215.

ISLE ABBOTTS Somer ST 3520. 'Estate called Isle belonging to the abbot' sc. of Muchelney Abbey. *Ile Abbatis* 1291, *Abbotlelye* (sic) 1610. P.n. Isle + Latin **abbas**, genitive sing. **abbatis**. Isle, *Yli* [966]13th S 740, *Ile* 1086, is the r.n. ISLE ST 4023. 'Abbotts' for distinction from ISLE BREWERS ST 3621.

ISLE BREWERS Somer ST 3621. 'Estate called Isle held by the Brewer family'. *Ile Brywer* 1275, *Ilbruers* 1610. Earlier simply *I(s)le* 1086 from the r.n. ISLE ST 4023. The estate was held by Richard Briwer in 1212. DEPN.

ISLEHAM Cambs TL 6474. 'Gisla's homestead'. *(in) Yselham* *[895]12th S 349, 1167×8, 1236, *Iselham* 12th–1548, *Isleham* from 1284, *Gisleham* 1086, *Gy- Giselham* 1086–c.1250, *Hy- Hiselham* 1218–1433, *Yselham -il-* 1285–1416. OE pers.n. *Ġīsla* + **hām**. Ca 192 gives pr [izləm].

ISLEHAM FEN Cambs TL 6276. *Isleham Fen* 1836 OS. P.n. ISLEHAM TL 6474 + ModE **fen**.

ISLE OF DOGS GLond TQ 3778. So named from *Isle of doges ferm* 1593. Possibly a nickname of contempt for an area earlier known as Stepney Marsh, *(in) landmarisco de Stebeh'* 13th cent., *marsh of Stebenhithe* 1365, *Stepheneth mershe* 1432. Mx 135, Encyc 423.

ISLE OF WALNEY Cumbr SD 1768 → WALNEY ISLAND.

ISLE OF WIGHT S2 4985 → Isle of WIGHT.

ISLEWORTH GLond TQ 1675. 'Gislhere's enclosure'. *Gislheresuuyrth* *[677]16th S 1246, *Gisteleswworde* 1086, *Istlesworde -wurtha* 12th cent., *Istleworth* 1231–1675, *Yistel(es)worth -il-* 1275–1333, *Isleworth* 1675. OE pers.n. *Ġislhere*, genitive sing *Ġislheres*, + **worth**. The phonetic development of the name shows intrusive *t* between *s* and *l* and assimilation of initial [j] to following [i] which may account for the long vowel ([ji] > [i:]). A curious form with initial *Th-* had considerable currency showing association with the word *thistle*, *Thistelworth -le-* 1416–1641, *Thistlewood* 1650. Mx 27.

ISLEY WALTON Leic SK 4225 → Isley WALTON.

ISLINGTON GLond TQ 3085. 'Gisla's hill'. *(of) Gislandune* c.1000, *Iseldon(e)* 1086–1554, *Isendon(e)* 1086, 1235, *Islyngton* 1464. OE pers.n. *Ġisla*, genitive sing. *Ġislan*, + **dūn**. Mx 124.

ISLIP Northants SP 9878. 'Slippery place by the river Ise'. *Is slepe, Hyslepe* [c.980]c.1200, *Islep, Slepe* 1086, *I- Yslep(e)* 1175–1316, *Islip* from 1275. R.n. ISE SP 8875 + OE ***slǣp** 'a slippery place'. The reference may have been to the hill-side on which the village stands, or to a slippery place on the banks of the river Nene. The main problem is that Islip lies close to the Nene and to Harper's Brook. For this explanation to hold one of these rivers must once have borne the same name as the r. Ise. Nth 183 gives pr [aizlip], Studies 1936.186.

ISLIP Oxon SP 5214. 'The slippery place by the r. *Giht*'. *Giðslepe* [1065×6]13th S 1148, *Giht-* [1065×6]14th S 1147, *Gyhte-* 1204, *Letelape* 1086, *Githeslape* c.1215, *I- Ysteslape -(a) -slepe* 1165–1279, *Ystelapa, Istelep(e)* 1188–1331, *Yslape* 1192, *Islape -slep(e) -slepp* 1285–1428, *Ycte- Ighte- Ichteslepe* 13th cent. R.n. *Giht*, the original name of the r. Ray SP 5917 + OE **slǣp**. The reference is to the steep descent on either or both sides of the river to the crossing point. Another reference to this place may be *Slæpi* *[681]12th S 1168. O 221 gives pr [aislip].

ISTEAD RISE Kent TQ 6369. No early forms.

ISURIUM NYorks SE 4066. The Roman city of Aldborough, capital of the Brigantes. *Ἰσούριον* (Isurium) [c.150]13th Ptol, *Isurium -iam, Isubrigantum* [c.300]8th AI, *Coguveusuron* (for *Cocuve(da), Isur(i)on*) [c.700]13th Rav. R.n. URE SE 2085, 4662, **Isurā*, + suffix *-ion*. RBrit 379.

River ITCHEN Hants SU 4616. Uncertain. *icene, (on, of) icenan* *[701]12th S 242, [c.830]12th S 284, *[854]12th S 309, [868]12th S 340, 909 S 376–[1026]12th S 962, *Iccene* [c.830]12th S 284, *(on) Ycœnan, Ycenan, Icœnan* [825]12th S 273–[909]12th S 376, *Ichen(e)* 1205×22–1304, *Icchyn* 1425. OE r.n. *Icene* identical with River ITCHEN Warw SP 4062 and Itchen in ITCHINGTON Avon ST 6587. At the base may be an IE root **iak-/*ik-* surviving in W *iach* 'healthy' and posited for the continental r.ns. Jekker, the Dutch name of the Belgian Geer, Jaca, Huesca, Spain, < *Iacca*, Yonne, France, < *Icauna*, Aygues, Vaucluse, France, < *Icarus*, and *Iceavus*, Cote-d'Or, France. Ha 100, Gover 1958.3, RN 217, RIO 1956.98–9, GPC 1994.

River ITCHEN Warw SP 4062. Uncertain. *(on) Ycenan, Ycœnan* 998 S 892, *(in) Ycenan* 1001 S 898, *Ichene, Huchene* 1262, *Ichene* 1336. The name has been related to the RBrit tribal name *Iceni* perhaps from a Celtic root **iak-* **ik-* 'healthy' which seems to occur in the French r.ns. Aigues/Eygues, a tributary of the Rhône, Vaucluse, *Icarus*, and Yonne, *Icauna*. However, this is extremely uncertain and the r.n. may be pre-Celtic. Wa 3, RN 217–9, RBrit 374.

ITCHEN ABBAS Hants SU 5333. 'I held by the abbess' sc. of St Mary's, Winchester. *Ichene Monialum* 1167, *Ichyn alias Abbesse Ichyn* 1539. P.n. Itchen + Latin **monialis**, genitive pl. **monialum** and ME **abbesse**. Itchen, *Icene* 1086, *Ichene* 1205–1327, is transferred from r.n. ITCHEN SU 4616. Ha 100, Gover 1958.80.

ITCHEN STOKE Hants SU 5632 → Itchen STOKE.

West ITCHENOR WSusx SU 7901. *Westigenore* 1243, *Westichenor(e)* 1243–91. ME adj. **west** + p.n. *Iccannore* *[683]14th S 232, *Icenore* 1086, *Ikenora* 12th, 'Icca's flat-topped hill', OE pers.n. **Iċċa* or *Yċċa*, genitive sing. **Iċċan, Yċċan*, + **ōra**. 'West' for distinction from the extinct parish of *Estychenore* 1268, ME **est** + p.n. *Ichnore* 1263. Sx 81–2, Jnl 21.18.

ITCHINGFIELD WSusx TQ 1328. 'The open land of the Eccingas, the people called after Ecci'. *Ec(c)hingefeld* 1222, 1256, *Hecchingefeld* c.1225, *Hechingfeld* c.1230, 1278, *Ec(c)hyngfeld(e)* 1386–1442, *Eachyngfeild* 1483, *Hedge and Fylde* 1541, *Itchingfeeld* 1581. OE folk-n. **Eċċingas* < pers.n. *Eċċi* + **ingas**, genitive pl. **Eċċinga*, + **feld**. It is uncertain whether the *H*-sps are examples of inorganic *h* or evidence of pers.n. *Heċċi* and folk-n. **Heċċingas*. Sx 176, MA 10.24, SS 70.

ITCHINGTON Avon ST 6587. 'Settlement, estate on the river *Itchen*'. *(æt) Icenantune* [967]11th S 1312, 1316, [991]11th S 1364, *Icetune* 1086, *I- Ychinton(e) -yn-* 1220–1525, *-yng- -ing-* 1554, c.1560. Lost r.n. *Itchen* identical with River ITCHEN Hants SU 4616 and Warw SP 4062 + **tūn**. The reference is either to Ladden Brook (*Loden, Laden* c.1540), a back-formation from LATTERIDGE ST 6684 or to an unnamed brook which flows from Itchington to join Ladden Brook. Gl iii.20.

Bishop's ITCHINGTON Warw SP 3957. 'I belonging to the bishop' sc. of Coventry and Lichfield. *Uchenton Ēpi* 1247, *Ichinton Episcopi* 1291, *Bisshopesychengton* 1384. ME **bishop**, genitive sing. **bishopes**, Latin **episcopi**, + p.n. *Icetone* 1086, *Ichenton(a)* [1135]1348, 1259, 'the settlement on the river Itchen', r.n. ITCHEN SP 4062 + **tūn**. Wa 171.

Long ITCHINGTON Warw SP 4165. *Longa Hichenton* c.1185, *Long Ychenton, Igenton* 13th, ~ *Igynton* 1449. ME adj. **long** + p.n. *(æt) Yceantune* 1001, *Icentone* 1086, *Ichenton* 1202, *Ichyngton* 1293, 'the settlement on the river Itchen', r.n. ITCHEN SP 4062 + **tūn**. 'Long' for distinction from Bishop's ITCHINGTON SP 3957. Wa 133.

ITTERINGHAM Norf TG 1430. 'Homestead of the Ytringas, the people called after Ytra or Ytri'. *Vtrincham, Ultringham* 1086, *Itringham* 1203, 1242×3, *Iteringham* 1202–54. OE folk-n. **Ytringas* < pers.n. **Ytra* or **Ytri* + **ingas**, genitive pl. **Ytringa*, + **hām**. DEPN, ING 136.

ITTON Devon SX 6899. 'Giedda's farm or village'. *Hedeton* 1244, *Ȝydeton* c.1262, *Yetton* 1475, *Yedeton* 1526. OE pers.n. **Ġiedda* + **tūn**. D 449.

River IVE Cumbr NY 4143. Uncertain. *Yue* 1285, *Ive* from 1307. This is usually explained as ON *Ífá* 'yew stream', ON ***īwa-** 'yew' + **á** 'a river', comparable with the ON r.n. *Ífing*. But an attractive alternative explanation 'impetuous one' has been suggested from the root seen in OHG *eibar, eifir* (ModG *Eifer* 'enthusiasm, eagerness') and OE *āfor* 'bitter, sharp, hard, rough'. Cu 18, RN 219, SSNNW 138.

IVEGILL Cumbr NY 4143. 'Valley, ravine of the river Ive'. *Yuegill* 1361, *Iuegill* c.1369, *Ivegill* 1412, *Igyll* 1446, *Hivegill* 1664–1704. R.n. IVE NY 4143 + ON **gil**. Cu 134, SSNNW 138.

River IVEL Beds TL 1938. *Gifla* c.1150, *Givle* [c.1180]13th, *Yivele* 1294. The 1150 form is actually the genitive pl. of folk-n. *Gifle* 'the people who live by the river Ivel'. The name is identical with YEOVIL Somer ST 5575. Cf. NORTHILL TL 1446, SOUTHILL TL 1442. Bd 8, RN 220.

IVELET NYorks SD 9398. 'Ifa's slope'. *Ivelishe* (sic) 1298, *Iflythe* 1301. OE pers.n. *Ifa* + **hlith**. YN 272.

IVER Bucks TQ 0381. 'End of the promontory'. *Evreham* 1086, *Eure, Evre* 1189–1346, *Iver* from 1382. OE ***yfre**. The DB form is compounded with OE **hām** 'a homestead'. Bk 239, L 178.

IVER HEATH Bucks TQ 0282. *Everheth* 1365. P.n. IVER TQ 0381 + ME **heth**. Bk 241.

IVESLEY Durham NZ 1741. 'Ifa's clearing'. *Yuesleia* 1174×93, 1185×95, *Ivesley* from 1319, ~ *burdon'* 1382, *Islaye -leye* 1582. OE pers.n. *Ifa*, secondary genitive sing. *Ifes*, + **lēah**. The manorial addition refers to tenure by the Burdon family.

IVESTON Durham NZ 1350. 'Ifa's stone'. *Yuestan* 1153×95–1456, *Iuestan(e)* [1183]c.1320–1535, *I- Yvestone* 1472×3, 1580, *Iveston* from 1498, *Iseton* 1576 Saxton. OE pers.n. *Ifa* + **stān**. NbDu 123.

IVINGHOE ASTON Bucks SP 9518 → Ivinghoe ASTON.

IVINGHOE Bucks SP 9416. 'Hill-spur of the Ifingas, the people called after Ifa'. *Evingehou* 1086, *Iuingeho* 1195, 1199, *Ivingho* 1227, *Ivanhoe* 1665. OE folk-n. **Ifingas* < pers.n. *Ifa* + **ingas**, genitive pl. **Ifinga*, + **hōh**. Bk 96, L 169.

IVINGTON H&W SO 4756. 'Estate called after Ifa'. *Iuintune* 1086, *-ton(e)* 1160×70, c.1250, *Ivynton* 1328. OE pers.n. *Ifa* + **ing**[4] + **tūn**. He 125.

IVINGTON GREEN H&W SO 4656. *Ivington Green* 1832 OS. P.n. IVINGTON SO 4756 + ModE **green**.

IVY HATCH Kent TQ 5854. 'Heavy hatch'. *Heuyhatche* 1325(p), *Ivy Hatch* 1700, 1819 OS, *Heavy Hatch* 1714. ME **hevy** (OE *hefiġ*) + **hacche** (OE *hæċċ* 'a hatch, a grating, a gate'). PNK 153.

IVYBRIDGE Devon SX 6356. 'Ivy-covered bridge'. *Ivebrugge* 1292, *Ivybrygg' -brigg'* 1313–16, *Ivy Bridge* c.1550. The earliest reference is the Latin ablative case form *Ponte Ederoso* 1280. ME **ivy** + **brigge**. D 278.

IVYCHURCH Kent TR 0227. 'Ivy-church'. *Iuecirce* c.1100, *Ivichurch, Ivechirch* 1242 etc. OE **īfiġ** + **ċiriċe**. PNK 479.

IWADE Kent TQ 8967. 'The ford'. *Ywada* 1177×8, *Iwade* from 1235, *Wade* 1323, 1331, *Iwede* 1313. OE **ġewæd**, Kentish **ġewed**. The reference must have been to an extensive crossing over marshland and the tidal streams between Iwade and Sheppey. KPN 257 fn.1.

IWERNE 'Yew-tree river'. Brit **Iu̯ernos* from **iu̯o-* 'yew'.

(1) ~ COURTNEY or SHROTON Dorset ST 8512. 'I held by the Courtenay family', also known as the 'sheriff's estate'.

I. *Yuern' Curtenay* 1244, *Iwern Curtenay* 1273, earlier simply *Werne* 1086, *Iwern(e)* from 1219. The r.n. is recorded as *Iwern broc* 'Iwerne brook' [958]15th S 656.

II. *Schyreuetone* 1337, *Shro(w)ton* 1633, 1679, 1774, *Iwerne Courtnay als. Shroton* 1795. ME **shireve** + **toun**.

There is also an isolated *Iwerton'* 1242 'the Iwerne estate'. At the time of DB the manor was held by Baldwin of Exeter, sheriff of Devon. In the 13th cent. the manor belonged to the Courtenay earls of Devon. Do iii.37 gives prs ['ju:ən] and [ʃrɔ:tən].

(2) ~ MINSTER Dorset ST 8614. The 'Iwerne estate held by the minster', sc. Shaftesbury abbey. *Evneminstre* 1086, *I- Ywerne Menstre, Munstre, Min(i)str(e)* 1280–1428. Earlier simply *(at) ywern, (in) hywerna* [871×7]15th S 357, *(at) iwern* [956]14th, *Ywern(e)* 1227–c.1500. Do iii.123 gives pr ['ju:ən].

IXWORTH Suff TL 9370. 'Gicsa or Gycsa's enclosure'. *(æt) Gyxeweorde* c.1040 S 1225, *(of) Ixewyrðe* 11th, *Adixewrð, Giswortham, Icsewrda* 1086, *Ixewurða* 1168, *Ikesworthe* 1610. OE pers.n. **Ġicsa* (**Ġixa*, **Ġycsa* **Ġyxa*) + **worth**. DEPN.

IXWORTH THORPE Suff TL 9172 → Ixworth THORPE.

J

JACK HILL NYorks SE 2051. *Jack Hill* 1616. ME pers.n. *Jakke, Jack*, + OE **hyll**. YW v.125.

JACOBSTOW Corn SX 1995. 'Holy place of St James'. *Jacobestowe* 1270×2, *Jacobstow* 1291 Gover n.d., *Yacobestowe* 1302 Gover n.d., *Yapstowe* 1414, 1440. Saint's name Latin *Jacobus* (James) from the dedication of the church + OE **stōw**. PNCo 100, Gover n.d. 3.

JACOBSTOWE Devon SS 5801. 'Holy place of St James'. *Jacopstoue* 1331, *Jacobstouwe* 1334, *Stowe Sancti Jacobi* 1353, *Stawe Seint Jake* 1387. Saint's name *Jacob* (*James*) from the dedication of the church + ME **stowe**. An earlier reference, *Jacobestow* 1297, may belong here or under Jacobstow Corn SX 1995. D 151.

JARROW T&W NZ 3265. '(The settlement) amongst the Gyrwe, the fen people'. The name is recorded in four forms:

I. The prepositional case, *(in) Gyruum* [c.716]12th, [c.731]8th BHE–late 12th, *(in) Giruum, Gyrwum* early 12th cent., *Girwuū* 1203×4.

II. The nominative/accusative case, *Gyruæ* 1080×2, *Gi- Gyrwe -ue* c.1107, 1335.

III. *iaro* 1180, *Jaru(e)* 1189×1212, 1364, *Jarw(e)* 1228–1371, *Jarow(e)* 1300–1465, *Jarrow(e)* from 1627, *Gerra* 1587.

IV. *Yarow* 16th, *Iarro* 1723, *Yarrow* 1728.

OE folk-n. *Gyrwe*, dative pl. *Gyrwum*, from the root **gerw-* seen in OE *gyr, gyru, gyrwefenn* 'marsh'. An identical folk-n. is known from the Cambridgeshire fens; Peterborough Abbey lay *in regione Gyruiorum* 'in the territory of the Gyrwe' (BHE 4.6) and Crowland Abbey *on middan Gyrwan* 'among the Gyrwe' (*Guthlac* 172). It is not necessary, however, to posit a migration from Cambridgeshire to the Tyne: the reference is to Jarrow Slake, a large area of mud at the mouth of the r. Don left uncovered by the outgoing tide. The modern form of the name derives from form II which would normally develop with late palatalisation of initial *g* as **Yirwe* > **Yeru* > *Yaru* > *Yarrow*. The development *ir* > *er* is remarkable and the stage *ar* is reached in this name a century or more earlier than normally expected. The AN scribes of Durham Priory substituted OFr [dʒ] for English [j] to produce ModE Jarrow [dʒærə, dʒærou]. ME <J> sps are ambiguous as between [j] and [dʒ], cf. JESMOND NZ 2566 and OUSE BURN NZ 2369, JERVAULX and r.n. URE. NbDu 123, IntroSurvey 104, Reallexikon 16.37.

JAYES PARK Surrey TQ 1440. 'The enclosure called after the Jayes family'. *Jayes* c.1680. Family name *Jayes* + ModE **park**. Cf. William *Joye* 1332, Matilda *Joyes* 1381. Sr 277.

JAYWICK Essex TM 1513. *Gey wyck* 1584, *Jewick* 1768–1805. Earlier *Clakyngeywyk, Clakenjaywyk(k)e* 1438–41, *Clakynjay* 1441, *Clackyngia Wyke* 1459–1546, *Clac- Claketon Jaewyke, Jaywick, Jawyke* 1513–53. The origin of this name in all probability is OE *Claccinġe-wīc*, 'dairy farm at or called Claccing', OE p.n. **Claccing* 'place called after Clacc' < pers.n. **Clacc* + **ing**[2], or 'the hill-like place', OE ***claccing** < ***clacc** + **ing**[2] as in nearby Great CLACTON TM 1716, locative–dative sing. **Claccinġe*, + **wīċ**. The exceptional development must have been *Claccinġe-wic* > *Claccen-ġewic* in which *Claccen* was taken as a form of the name Clacton > *Clacton Jaywick*. Ess 335.

JERVAULX ABBEY NYorks SE 1785. P.n. Jervaulx + OFr **abbeie**. Jervaulx, *Jor(e)vall(e)* [c.1140]copy–1361, *Yorevall* 1312, *Gereuall(e)* 1196, 13th cent., *Gervaus* 13th cent., *J- Gervax* 15th cent., *Gerv(e)is* 16th cent., *Jarvaux* 1539, is 'the valley of the river Ure', r.n. URE SE 2085 + OFr **vals**. *Jor(e)- Yore-* forms are regular, cf. river URE SE 2085. *Ger- Jar-* forms presuppose an alternative unrecorded form of the r.n., **Yere* with OE *eo* > ME *ye* as in YEARSLEY SE 5874, EVERLEY SE 9789, *Yereley* 1577, and ME *Yerk* for YORK SE 6052, and AN sound substitution of [dʒ] for [j]. The reference is to the Cistercian monastery moved here in 1156. Cf. RIEVAULX SE 5785. YN 250 gives prs [dʒa:vis, dʒɔ:vou], Anglia 48.291.

JESMOND T&W NZ 2566. 'The mouth of the *Yese*'. *Gesemue* 1204–79, *Jesemuthie* [c.1200]copy, *-muth(e)* 1242–1378, *Jeze- Jese- Yesmue* late 13th, *Gesmond* 1414, *Jesmuth alias Jesmund* 1428, *Jesmound* 1514, *Jazment, Jasemond* 18th. R.n. *Yese* + OE **mūth**. *Yese* is the old form of the r.n. OUSE BURN. The pronunciation with [dʒ] instead of [j] is due to AN sound substitution. The generic *-muth* has been replaced by OFr **mont, mond** 'a hill'. NbDu 124, IntroSurvey 104, RN 318.

JEVINGTON ESusx TQ 5601. 'Settlement called or at *Geofing*, the place called after Geofa, or of the Geofingas, the people called after Geofa'. *Lovinge- Lovringetone* (sic) 1086, *Govingetona* [1189]14th, *Gevingeton* 14th, *Ge- Jevington* 1280–15th. Either OE p.n. **Ġeofing* < pers.n. **Ġe(o)fa* + **ing**[2], locative-dative sing. **Ġeofinge*, or, OE folk-n. **Ġeofingas* < pers.n. **Ġeofa* + **ingas**, genitive pl. **Ġeofinga*, + **tūn**. Sx 421, SS 76.

JOHNBY Cumbr NY 4333. 'John's farm'. *Johanbi* c.1208, *Johan(n)eby* 1212–1303, *Jo(e)nebi -by* 1209–1368, *Jonesby -bi* c.1210–14, *Joune- Jeune- Jauneby* 13th cent., *Jonby* 1332–70, *Johnby* from 1348. ME pers.n. *Joh(a)n* + **bȳ**. A 12th cent. name formation: the particular John in question is unknown. Cu 197 gives prs [dʒounbi, dʒwouənbi], SSNNW 33.

JORDANS Bucks SU 9791. *Jurdens* 1766. Probably a reduction of *Jourdemayns*, a possessive form from the family name Jourdemayn found in the district in 1301. Bk 220.

JUMBLES RESERVOIR GMan SD 7314. P.n. Jumbles SD 7037 + ModE **reservoir**. Cf. *Jumbers Wood* 1843 OS. ModE **jombyll** 'jumble' used in a topographical sense of some muddled or disorderly feature, e.g. vegetation, or of a stream that flows in a disordered way. Cf. f.n. Jumble Db 67, *Jumbles* 1547–51, Jumble Db 116, *Jumbles* 1549, and other examples collected in YW iii.115. The original application here was to a small wooded side-valley off Bradshaw Brook at SD 7315.

JUMP SYorks SE 3801. 'The abrupt descent'. *Jump* 1841. EModE **jumpe**. YW i.103, iii.173.

KABER Cumbr NY 7911. 'Jackdaw hill'. *Ca- Kaberg(h)* 12th–15th, *-ber* 1257, 1377 and from 1546, *Cay- Kay- Keyber* 1569, 1657, 1710. ON ***ká** 'jackdaw' also recorded as a by-name, + **berg**, possibly replacing OE ***cā** + **beorg**. We ii.5 gives pr ['ke:bə], SSNNW 112.

KEA Corn SW 8142. *Kea* 1813. Transferred from Old KEA SW 8441 in 1802. PNCo 130.

Old KEA Corn SW 8441. Adj. **old** + p.n. Kea for distinction from KEA SW 8142 whither the church of St Kea was moved to a more convenient location in 1802. Kea, *Landighe, Sanctus Che* 1086, *ecclesia Sancte Kee* 1451, *Kee* 1535 Gover n.d., *Key(e)* 1595, 1634 Gover n.d., is saint's name *Ke* (Kea) from the dedication of the church. He was believed to have sailed to Cornwall from Ireland in a granite trough and to have gone on to Brittany where he died at Cleder. *Landighe*, the name of the churchtown, is also recorded as *Landege(i) -eye -ai -ay* 1185–1311 Gover n.d., *Landekey* 1420 Gover n.d., and survives in the farm name Landegea, Co **lann** 'church site' + ***to-**, the honorific prefix in pet forms of saints' names, + saint's name *Ke* with lenition of *K-* to *G-*. Identical with Landkey Devon SS 5931, *Landechei* 1166, Llandygai Gwyn SH 5971, and *Lantokai*, the old name for Leigh-on-Street Somerset SS 9032, and of St-Quay in Brittany. PNCo 130, D 341.

KEADBY Humbs SE 8311. 'Kæti's village or farm'. *Ketebi* 1185, 1199, *-by* 1275. Dan. pers.n. *Kǣti*, genitive sing. *Kǣta*, + **bȳ**. Cf. Dan. p.n. Kædeby. DEPN, SSNEM 81.

KEAL Lincs TF 3763. Short for East Keal, *Estrecale* 1086, *Oustcal'* (corrected from *Ousteal*) c.1115, *Austercales* 1142, *Estrekales* [1199]1330, *Ostkele* 1281. OE comparative adj. **ēasterra** or ON **eystri** 'more easterly' partly replaced by ON positive **austr** 'east' + p.n. *Cale* as in West Keal TF 3663, *Cale, Westrecale* 1086, *Cal'* c.1115. The two settlements together are *Cales* c.1135, *Keles* 12th, 'the ridges, the Keals'. ON **kjǫlr** (PrON **kialuR*) 'a keel' used in the topographical sense 'ridge'. E and W Keal stand on neighbouring ridges resembling upturned boats. Sps with *a* are probably due to the interchange of *e* and *a* in AN. DEPN, SSNEM 155.

KEAL COTES Lincs TF 3661. 'The cottages belonging to Keal'. *Kelecotes* 1396, *Keal Cotes* 1824 OS. P.n. KEAL TF 3763 + ModE **cote** (OE *cot*). Cameron 1998.

KEARSLEY GMan SD 7505. 'Cress hill'. *Cherselawe* 1187, *-lawa* 1188, *Kersleie* c.1220, *Keresley* 1501. OE **cærse, cerse** + **hlāw** later replaced by **lēah** 'a clearing'. La 43, Jnl 17.36.

KEARSLEY Northum NZ 0275 → KERESLEY WMids SP 3284.

KEARSTWICK Cumbr SD 6080. 'Valley clearing'. *Kestw(h)ayte -ait* 1546–76, *-th(w)aite -thwayte* 1547–61, *Keisthwaite* 1640, *Keastwyck -wick* 1589, 1832, *Keis(t)wicke* 1644, *Kears(t)wick* 1777, 1848. ON **kjóss** 'valley, recess' + **thveit**. The place lies in a small valley N of Kirkby Lonsdale. The spelling with *-ear-* is a modern attempt to represent the diphthong [iə]. We i.43 gives pr ['kiəstwik].

KEARTON NYorks SD 9999. Possibly 'Kæri's estate'. *Kirton* 1298, *Kerton* 1301, 1646. This may be ON pers.n. *Kǣri(r)* + OE **tūn**. YN 271.

KEASDEN NYorks SD 7266. Probably 'valley where cheese is made'. *Kesedene* 1165–1240, *Kesen-* 1165×77, *Kesden(e)* 1165–1401, 1608, *Keasden* from 1617. OE **ċēse** with ON substitution of [k] for [tʃ], + **denu**. YW vi.233.

KEDDINGTON Lincs TF 3488. 'Estate called after Cedd(a)'. *Cadi(n)ton, Cadintone* 1086, *Chedingtuna* c.1115, *Kedyngton(a)* [c.1150]1409, 1180. OE pers.n. *Ċedd(a)* with ON [k] for English [tʃ] + **ing**[4] + **tūn**. DEPN, Cameron 1998.

KEDINGTON Suff TL 7046. 'The estate called after Cyda'. *Kydington* [1043×5]13th S 1531, *Kidituna* 1086, *Kedintune* 1154×89, *-ton* 1200, *Kediton* 1610. OE pers.n. *Cyda* + **ing**[4] + **tūn**. DEPN.

KEDLESTON Derby SK 3041. 'Ketill's estate'. *Chetelestune* 1086, *Ket(t)el(l)estun -ton(e) -is-* c.1199–1458, *Ketilston(e) -yl-* 1372–c.1600, *Ked(e)leston* from 1385. ON pers.n. *Ketill*, secondary genitive sing. *Ketilles*, + **tūn**. Db 580, SSNEM 192.

River KEEKLE Cumbr NY 0120. 'Winding stream'. *Chechel(a)* c.1125, *Kekel* 1295, *Kikil* c.1450, *Kekyll* 1500. Identical with Norwegian r.n. *Kykla* < ON adj. ***kikall** (as in Norwegian Kikallvaagen, the name of a winding inlet) + **á**. Cu 18, SSNNW 224 s.n. Corkickle.

KEELBY Lincs TA 1609. 'Village, farm on a keel-like ridge'. *Chele- Chilebi* 1086, *Chelebi* c.1115, 1157×81, *Kelebi -by* 1143×7–1642, *Kelesby* 1181×5, *Key- Keilby* 1500–1666, *Keelby* from 1380. ON **kjǫlr** + **bȳ**. Li ii.174, SSNEM 55.

KEELE Staffs SJ 8045. 'The cow's hill'. *Kiel* 1169–1203, *Kyel* 1173, 1230, *Kel* 1211. OE **cū**, genitive sing. **cȳ**, + **hyll**. DEPN, L 171, Horovitz.

KEELEY GREEN Beds TL 0046. This is *Wootton Keeley* 1834 OS, p.n. WOOTTON TL 0045 + Keeley, probably to be identified with *Gyldewode* 1446, 'wood where golden flowers grow', OE ***gylde** + **wudu**. Bd 86, PNE i.211.

KEEPWICK Northum NY 9571. Possibly 'the place where trade takes place'. *Kepwike* 1279, *Kep(e)wyk* 1298, 1479, *Keepicke* 1653. OE **ċēap** + **wīċ** with ON [k] for [tʃ]. Keepwick lies on the same fells as Stagshaw Bank, some 3 miles distant, the site of a famous cattle, sheep and horse fair from the late 13th cent. However, substitution of [k] for [tʃ] is not expected in Northum. NbDu 125.

River KEER Lancs SD 5472. Uncertain. *Kere* [1262]1268, 1292, c.1350, *Keere* c.1350, *Keri* c.1540, *Kery* 1577. Possibly the 'dark, shadowy stream', British ***kēr-** < IE **keiro-*. RN 224 compares the Gaelic r.n. Ciarán Water, Scotland, NN 2682, 'the little dusky one'. But this root is not otherwise recorded in British and remains doubtful. La 169, RN 223.

KEEVIL Wilts ST 9258. 'The wood or clearing in or at the hollow'. *(on) Kefle wirtrim* 'to Keevil woodbank' (in the bounds of Steeple Ashton) [964]14th S 727, *Chivele* 1086, c.1210, *Ki- Kyvele* 1211–1425, *Ki- Kyvel* 1231, 1249, *C-, kuvel(e)* 1242–1397, *Keveleigh al. Kevell* 1560–1. OE **cȳf** or **cȳfel** 'a tub, a bucket' used topographically of a depresion in the ground, + **lēah**. Keevil lies in a small valley. Alternatively the specific might be an OE pers.n. **Cȳfa*. Wlt 142, Grundy 1920.71.

KEGWORTH Leic SK 4826. Possibly 'Cægga's enclosure'. *Cache- Cogeworde* 1086, *Caggworth'* c.1130, *Cagewrdhe* 1199, *Kagworthe -word* 1209×19, 1216×72, *Keggeworth(e)* 1244–1428, *Kegworth(e)* from 1209×35, *Geggewðe* 1209–11. OE pers.n. **Ċægga* + **worth**. An OE pers.n. *Ċeagga* is recorded; initial [k] for [tʃ] would be due to Scandinavian influence. Lei 370.

KEHELLAND Corn SW 6241. 'Grove of the ancient church site'. *Kellyhellan -i-* 1284–1335, *Kyllyhel(l)an* 1403, 1464, *Kahellan* 1707. Co **kelli** + ***hen-lann**. The reference is unknown unless it was to a destroyed 'round' here which may have been a Dark-age Cemetery. PNCo 101, Gover n.d. 584.

KEIGHLEY WYorks SE 0540. 'Cyhha's wood or clearing'. *Chichelai* 1086, *Ki- Kykelei(e) -le(y) -lay* 12th–1504, *Ki- Kyghele(y) -lay(e)* 1246–1545, *Kytheleg* 1263, *Keighley* from 1571. OE pers.n. **Cyhha* + **lēah**. OE *h* [x] in various positions was either eliminated or substituted by phonetically similar sounds, in this case both by [k] and by [θ], the latter being the form that prevailed in pronunciation in spite of the sp. The same pers.n. and sound substitution also appear in the early name of nearby North Beck SE 0440, *Kicheburn* 13th, *Kitheburna* 1333, 'Cyhha's stream'. YW vi.2 gives prs ['ki:θlə], ['kɛiθlə] and obsolete ['kɛilə].

KEIGHLEY MOOR WYorks SE 0039. P.n. KEIGHLEY SE 0540 + ModE **moor**.

KEINTON MANDEVILLE Somer ST 5430. 'K held by the Mandeville family'. *Kyngton Maundevill* 1280, *Keinton Mandefield* 1811 OS. Earlier simply *Chintvne -tone* 1086, *Kynton* 1243, 'royal estate', OE **cyne** + **tūn**. The manor was held by William de Mandevill and subsequently by Geoffrey de Maundeville in 1243. DEPN.

KEISBY Lincs TF 0328. 'Kisi's village or farm'. *Chisebi* 1086, 1154×89, *Ky- Kiseby* c.1150–1642, *Keisby* from 1406. ON pers.n. *Kisi*, genitive sing. *Kisa*, + **bȳ**. Perrott 197, SSNEM 55.

KELBROOK Lancs SD 9044. 'Stream flowing through a ravine'. *Chel- Cheuebroc* 1086, *Kelbroc* 1221, *-brok(e)* 1285–1542, *-brooke* 1524. OE **ċeole** 'throat, gorge, ravine' with initial [k] due to ON influence + **brōc**. The stream descends through a deep high-sided valley E of the village. YW vi.33.

KELBY Lincs TF 0041. 'Village or farm on the wedge-shaped piece of land'. *Chele- Chillebi* 1086, *Kellebi -by* 1199–1332, *Keleby* 1242–1451, *Kelby* from 1330. ODan ***kæl** + **bȳ**. Identical with the Dan p.ns. Kelby and Källby. The village stands on a ridge. Perrott, SSNEM 56.

KELD Cumbr NY 5514. 'Spring'. *Keld(e)* from 1540, *Keilde, Keeld* 1612–13. ON **kelda**. The reference is to a spring on the W side of the hamlet half a mile up the Lowther from Shap Abbey, whence its name *th'Abbey Keld* 1732. We ii.165, L 22–3.

KELD NYorks NY 8901. *Keld(e)* 1538, 1577. Short for *Appeltrekelde* 1301, 'spring by the apple-tree', OE **æppeltrēow** + ON **kelda**. YN 272.

KELDHOLME NYorks SE 7086. 'Water-meadow by the spring'. *Keld(e)holm* 1170×86, 1201 etc. ON **keldr** + **holmr**. YN 64.

KELDY CASTLE NYorks SE 7791. *Keldy Castle* 1861 OS; cf. *Keldy Grain* ibid.

KELFIELD NYorks SE 5938. Partly uncertain. *Chelchefelt -d* 1086, *Kelk(e)feld(e)* 12th–1363, *Chalche- Calcefeld* c.1150, 1154, *Kellef(f)eld* 1219, 1299, *Kelfeld* 1342–1546. The generic is OE **feld** 'open land', the specific apparently OE ***ċelċe**, ***ċælċe** 'a chalky place, chalky ground'. But the soil here is not chalky. The site is flat alluvial land and the reference is possibly to a place where chalk was brought by river for spreading on the fields. YE 266, SSNY 159, Jnl 19.54.

KELHAM Notts SK 7755. Uncertain. *Calun* 1086, *Kelum* 1156–1335, *-hom* 1280, *Kellum* 1194–1268, 1646, *-om(e)* 1428, *Kellam* 1556, 1578, *Kelham* 1556. Either **kjǫlum** '(the settlement) at the keels or ridges', locative–dative pl. of ON **kjǫlr** referring to hill-ridges resembling upturned boats W and NW of the village, or ***kælum**, dative pl. of ODan ***kæl** used topographically of a wedge-shaped piece of land, again probably referring to the Kelham Hills. The latter suggestion would best suit the short vowel in this name. ODan ***kæl** has also been suggested for the lost Yorks p.n. Kelsit, *Chelestuit* 1086, 'the clearing on the wedge-shaped gravel patch' and occurs in the Danish p.ns. Keldby (*Nudansk Ordbog* I-II, 6 ed. Copenhagen 1969) and *Källby* (Ingemar Ingers, 'Ortnamn i Lund II', *Gamla Lund* Årsskrift 52, 1971, 43–5). Nt 187, SSNEM 156, SSNY 98.

Great KELK Humbs TA 1058. *Mekyl Kelk* 1490, *Great ~* 1520. ME **mekil**, ModE **great**, + p.n. *Chelc(he)* 1086, *Kelc -k* [before 1080]13th–1506, 'the chalky place', OE **ċelċe**, an *i*-mutated form of *calc* 'chalk'. Great and Little Kelk lie either side of a belt of chalky glacial gravel forming a fine dark soil with a multitude of chalk fragments. YE 92, lix, Jnl 19.49.

Little KELK Humbs TA 1060. *Parva Kelk(e)* 13th–1342, *Littelkelk* 13th. ME **litel**, Latin **parva**, + p.n. *Chelch(e)* 1086 as in Great KELK TA 1058. YE 92.

KELK BECK Humbs TA 0957. P.n. Kelk as in Great KELK TA 1058 + ModE **beck** (ON *bekkr*).

KELLATON Devon SX 8039. 'Cylla's farm or village'. *Kyleton* 1377×99, *Kyllaton* 1600. OE pers.n. **Cylla* + **tūn**. D 333.

KELLAWAYS Wilts ST 9795 → East TYTHERTON ST 9675.

Nether KELLET Lancs SD 5068. *Kellettam inferiorem* [n.d.]15th, *Netherkellet(t)* 1299–1343. Adj. **nether**, Latin **inferior**, + p.n. *Chellet* 1086, 13th, *Kellet* 1194–1297 etc., 'the slope of the spring', ON **kelda** + **hlíth** 'a slope, a concave slope, a hollow'. 'Nether' for distinction from Over KELLET Lancs SD 5269. Nether Kellet lies on the slope of a hill. The spring is probably the one referred to as the 'earl's spring', *Yerleskelde* [1235×68]1268. La 186, SSNNW 138–9, Jnl 17.103.

Over KELLET Lancs SD 5269. *Ovre- Overkellet* 1277, 1278, 1332, *Kellet superiori* c.1275. Adj. **over** 'upper' (Latin **superior**) + p.n. Kellet as in Nether KELLET Lancs SD 5068. La 187, SSNNW 138–9, Jnl 17.103.

KELLETH Cumbr NY 6605. 'Slope with a spring'. *Keldelith(e) -lyth* 1154×89–1337, *-lit* 1336, *-lett* 1588, *Kellet(t)* 1286, 1564, *Kelleth* from 1615. ON **kelda** + **hlíth**. We ii.43, SSNNW 139, L 22, 166.

KELLING Norf TG 0942. 'The Cyllingas, the people called after Cylla'. *Chi- Chyllinge, Killinge* [c.975]12th, *Kellinga, Challinga* 1086, *Kellinges* 1177–1202, *Kelling(')* 1242×3, 1275, *Kylling* 1275. OE folk-n. **Cyllingas*, Norf dial. **Cellingas*, < pers.n. **Cylla*, **Cella*, + **ingas**. DEPN, ING 58–9.

KELLINGLEY NYorks SE 5224. Either 'woodland clearing at or called Ceolling, the place called after Ceolla' or 'of the Ceollingas, the people of *Ceollingtun* (Kellington)'. *Kellinglaia(m) -ley(am)* 1144–60, *Kelinglai(am) -ley -yng-* 1147–54, 1467, *Kellingeleia* c.1190. OE p.n. **Ċeolling* < pers.n. *Ċeolla* with ON /k/ for /tʃ/ + **ing**[2], or folk-n. **Ċeollingas* 'the people of Kellington', genitive pl. **Ċeollinga*, + **lēah**. In either case it seems likely that both Kellingley and KELLINGTON SE 5524 are named after the same person. YW ii.56.

KELLINGTON NYorks SE 5524. 'The estate called after Ceolla'. *Chelin(c)- Chellinc- Ghelintune -tone* 1086, *Kel(l)ington(a) -yng-* 12th–1428. OE pers.n. *Ċeolla* with ON [k] for [tʃ] + **ing**[4] + **tūn**. See also KELLINGLEY SE 5224. YW ii.59.

KELLOE Durham NZ 3436. 'Calf hill'. *Kelflaw* [1133×40]1338, *-lawe -lau(ia)* 12th cent., *Chillau* 1135×9, *Kelvelaw* 1254, *Kellaw(e)* c.1230–1476, 1622, *-low(e)* 1256–1739, *Kello* 1561, *Kelloe* from 1605×6. OE **celf** + **hlāw**. NbDu 125, L 163.

KELLY Devon SX 3981. 'Wood'. *Chenleie* 1086, *Chelli* 1166, *Kelli -y* 1219–1391, *Kellegh* 1428. Co **kelli**. The DB form is probably to be disregarded. D 184.

KELLY BRAY Corn SX 3571. 'Grove of the hill'. *Kellibregh* c.1286, *Kellibre* 1306, *Kellebrey, Kellybray* 1337. Co **kelli** + ***bre**. However, the earliest spelling may rather point to **kelli** + adj. ***brygh** 'variegated, speckled' and so 'variegated grove'. PNCo 101, Gover n.d. 208.

KELMARSH Northants SP 7379. 'Pole marsh'. *Cailmare, Keilmerse* 1086, *Keyl- Keilmers(e) -mers(c)h(e)* 12th–1428, *Kelmers(e)* 1199, 1223, *-mersh* 1322. OE ***ċeġel** with [k] for [tʃ] due to Scandinavian influence, + **mersc**. The allusion is probably to a guide-post in the marsh. Nth 116, Studies 1936.164, PNE i.87, L 53.

KELMSCOT Oxon SU 2599. 'Cenhelm's cottage(s)'. *Kelmescote* 1234–1340, *Kelmscott* 1761. OE pers.n. *Cēnhelm*, genitive sing. *Cēnhelmes*, + **cot**, pl. **cotu**. O 325.

KELSALE Suff TM 3865. Possibly 'Celi's nook'. *Chylesheala, Keleshala, Kereshalla, Cheressala* 1086, *Kelleshalle* 1228,

Keleshale 1254–1336, *Kelshale* 1344, *Kelsale* from 1399. OE pers.n. **Cēl(i)* or **Cǣl(i)* or **Ċēol* with Scandinavian [k] for [tʃ], genitive sing. **Cēles*, **Cǣles* or **Ċēoles*, + **halh** referring to a small valley. DEPN, Baron.

KELSALL Ches SJ 5268. 'Kell's nook'. *Kelsale* 1260–1581, *-hale* 1303–1444, *Kelsall* from 1358. ME pers.n. *Kel(le)*, genitive sing. *Kel(le)s*, + **hale**, dative sing. of OE **halh**. Che iii.277, L 106.

North KELSEY Lincs TA 0401. *Norchelsei, Nortchelesei* 1086, *Nordchelesia* c.1115, *Northkel(l)esey(e) -eie* before 1218–1504, *Northkelsey* from 1318. OE **north** + p.n. *Chelsi, Colesi* (for *Cel-*) 1086, *Chaleseia* c.1115, *Kelesey(e) -eie* 1123–1214, *Kellesey(e)* 1264–1422, unidentified element + OE **ēġ** 'an island' in the sense 'a spur of higher dry ground in marsh'. The specific may be OE *ċēol* 'a keel' with Scand [k] for [tʃ] in the topographic sense 'ridge' – both N and S Kelsey occupy ridges which reach into low ground and may have been seen as upturned keels; or OE pers.n. **Ċēol*. Li ii.178.

South KELSEY Lincs TF 0498. *Sudkeleseia* 1177, 1204, *Suthkelesey(e)* before 1218–1461, *South Kelsey* from 1303. OE **sūth** + p.n. Kelsey as in North Kelsey TA 0401. Li iii.36.

North KELSEY BECK Humbs TA 0302. *North Kelsey Beck* 1824. P.n. North KELSEY TA 0401 + ModE **beck** (ON *bekkr*).

KELSEY HEAD Corn SW 7660. *Kelsey Head* c.1870. P.n. Kelsey SW 7659 + ModE **head** 'a headland'. Kelsey, *Kelse* 1349, *Kelsey* 1539 is unexplained. PNCo 102.

KELSHALL Herts TL 3236. 'Celli's hill'. *Keleshelle* [11th]c.1250 S 1051, [c.1060]14th, *Cheleselle* 1086, *Keleshell' -hulle -hill(e)* 1200–1428, *Kelshill* 1311, *Kelsell* 1545, *Kelshoe* 1621, *Kelsey* 1627, *Kelshall* 1638. OE pers.n. **Cylli*, Herts dial. **Celli*, genitive sing. **Celles*, + **hyll**. Hrt 159 gives prs [kelʃɔ:l, kelsi:].

KELSTON Avon ST 7067. 'Calf's estate'. *Calvestona* 1100×35, *-ton* 1178, *Kelveston* 1260. Anglo-Scand pers.n. **Calf* (ON *Kalfr*), genitive sing. **Calfes*, + **tūn**. DEPN.

KELTON Durham NY 9220. Possibly 'calf farm', OE **celf** + **tūn**, but there are no early forms.

KELVEDON Essex TL 8616. 'Cynelaf's valley'. *(æt) Cynlaue dyne* 998 S 1522, *Cynlæuedene* 11th, *Kynleuedene* [1066]13th S 1043, *Kinleof(e)- Ki- Kenleueden(e)* 1066–1272, *Chelleuadanā*, *Chelleuedana* 1086, *K- Celleveden(e)* 1157–1346, *-don* 1251, *Kelvedon* from 1307×27, *Kelden* 1476–1538, *Kewydon* 1552. OE pers.n. *Cynelāf*, Essex dial. form *Cenelāf*, perhaps varying with feminine pers.n. *Cyne- Cenelēofu*, genitive sing. *Cyne- Cenelēofe*, + **denu**. There is considerable difficulty in distinguishing the forms for Kelvedon from those for KELVEDON HATCH TQ 5698 both of which were Westminmster estates. Ess 290 gives pr [keldən], 58.

KELVEDON HATCH Essex TQ 5698. 'Kelvedon gate'. *Kel(e)wedon(e) Hac(c)h(e)* 1276, 1344, *Kellowdon Hatch* 1544. P.n. Kelvedon + ME **hacche** (OE *hæċċ*). Kelvedon, *Kylwendun(e)* *[1066]13th S 1043, *Kilwendune, Killeuenduna, Keleuuendun(e)* 1066, *Kel(u)endunā, Kalendunam* 1086, *Kel(e)wedon(e)* 1219–1344, *Kelden -don* 1285, 1566, *Kelv(e)don* from 1291, is '(the place at) the speckled hill', OE **cylu**, Essex dial. form ***celu**, dative sing. definite form **cylwan**, ***celwan**, + **dūn**. Ess 58, PNE i.123.

KELYNACK Corn SW 3729. 'Holly grove'. *Chelenoch* 1086, *Kellenyek* 1284, *Kellen(y)nek* 1300–30. Co **kelin** 'holly' + adjectival suffix **-ek**, 'place of holly, where holly grows'. PNCo 102, Gover n.d. 632.

KEMACOTT Devon SS 6647. 'Cyma's cottage(s)'. *Chymecote* 1330, *Kemecotte* 1537. OE pers.n. *Cyma* + **cot**, pl. **cotu**. D 66.

KEMBERTON Shrops SJ 7304. 'Cenbriht's estate'. *Chenbritone* 1086, *Kembricton'* 1242–1387 with variants *-brytton -brighton(')* etc., *Kembirton* 1284×5, *-berton* from 1397. OE pers.n. *Cēnbriht* (*Cēnbeorht*) + **tūn**. Sa i.164.

KEMBLE Glos ST 9897. 'The border'. *C- Kemele* *[682 for 688]13th S 231, *[688]13th S 234, *[854]14th S 305, [1065]14th S 1038, 1180, 1327, *Chemele* 1086, *Kembyll* 1535, *Kemble* 1613. PrW ***cuμil** < Brit **comel-*, IE **com-pel*, as in W *cyfyl*. Kemble lies in an area where control of territory seems to have fluctuated. The reference is to an ancient boundary, perhaps of the Dobuni or between Saxon held territory around Cirencester and a British one around Malmesbury. Gl i.75, iv.24, LHEB 487, 663, GPC 725, *TransBGAS* 114.34.

KEMERTON H&W SO 9437. 'Estate called after Cyneburg'. *KINEBURI, kineburhgingtun, cyneburgingctun* [840]11th S 192, *Caneberton, Chine- Chenemertvne* 1086, *Kenemerton(e) -tune* 1190–1411, *Kenmerton* 1447–1612, *Kemmerton* from 1478. OE feminine pers.n. *Cyneburg* + **ing**[4] + **tūn**. It appears that the original pers.n. was subsequently assimilated to OE pers.n. *Cynemǣr*. The reference is possibly to the Mercian princess Cyneburg, daughter of Penda, rather than to the late 7th cent. Kyneburg, sister of Osric and abbess of Gloucester, since this was a Worcester estate. Her land is referred to in the bounds of *Caldinccotan* in Bredon, *cyne burge lond gemœre* 'Cyneburg's land boundary' 984 S 1347. Gl ii.59 gives pr ['kemətən], Hooke 98, 307.

KEMP TOWN ESusx TQ 3303. A district of Brighton laid out by Thomas Read Kemp c.1830. Sx 292.

KEMPLEY Glos SO 6729. Possibly 'Cenepa's wood or clearing'. *Chenepelei* 1086, *Kenepele(ge) -leg(h) -lai -leye* 1215–70, *Kempele(a) -leg(a) -ley(e) -leigh* 1195–1529, *Kempley* from 1316. OE pers.n. **Cenepa* + **lēah**. **Cenepa* would be a byname from OE *cenep* 'moustache, horse's bit' which may also have been used as a topographical term (? 'peg, stump') or as a plant name. Gl iii.172.

KEMPSEY H&W SO 8549. 'Cemmi's island'. *(monasterium que dicitur) kemesei* 'the monastery called K' [799 ?for 820]11th S 154, *(monasterium) Kemeseg* [798×822]17th Hooke 104, [847]17th ibid., *(into) cymesige* [977]11th S 1332, *Kymesei, Chemeshege, Kemesige* 11th, *Chemesege* 1086, *Kemeseia -eye* 1208–91, *Kemsey* 1615. OE pers.n. **Cemmi*, genitive sing. **Cemmes*, + **ēġ** in the sense 'dry ground in marsh'. Kempsey lies close to the Severn in land liable to flood. The alternative pers.n. suggested, **Cymi*, a derivative of *Cymen* is less likely: *e*-spellings for OE *y* are less likely than *i (y)* spellings showing the raising of *e* > *i* before *m*. Wo 144, L 38.

KEMPSFORD Glos SU 1597. 'Cynemær's ford'. *(æt) Cynemæres forda* 9th ASC(A) under year 800, *Chenemeresforde* 1086, *Ki- Kynemere(s)ford(e)* 1086–1303, *-mer(s)-* 1221–1354, *Kymmysford* 1455, *Kemysford(e)* 1535, 1542, *Kemsford* 1577–1690, *Kempsford* 1610. OE pers.n. *Cynemǣr*, genitive sing. *Cynemǣres*, + **ford**. Gl i.38.

KEMPSTON Beds TL 0347. 'Settlement at or called *Cembes*'. *Kemestan* *[1060]14th S 1030, *Cœmbestun* [1042×66]14th, *Cœmbestune* [1077]17th, *Camestone* 1086, *Kembeston* 13th cent., *Kem(e)ston* 13th–14th cents., *Kempston* 1247. PrW p.n. **Cembes* identical with CAMBOIS Northum NZ 3083, Cemaes Gwyn SH 3693 and Cemmaes Powys SH 8306, ultimately from Brit **cambo-* 'crooked, bent', + **tūn**. The place is situated at a bend of the river Ouse. Bd 75.

KEMPTON Shrops SO 3683. 'Cempa's farm or village'. *Chenpitune* 1086, *Kempeton* 1255–1327, *Kempton(')* from 1291×2. OE pers.n. **Cempa* + **tūn** or **Cempa* + **ing**[4] + **tūn**. Sa i.165.

KEMSING Kent TQ 5458. 'Place called after Cymesa'. *cymesinc, cyme singes (cert)* 'rough ground belonging to Kemsing' [822]11th S 186, *Cymesing* [944]13th S 501, [955×59]14th S 662, *Cimisinga* c.1100, *Kemesinge(s) -ynge* 1153×6–1253×4. OE pers.n. **Cymesa*, Kentish **Cemesa*, + **ing**[2]. PNK 56, KPN 142, ING 186, Charters IV.104.

KENARDINGTON Kent TQ 9732. 'Estate called after Cyneheard'. *Kynardingtune* 11th, *Kenardintona -ton(e) -yn-* 1175–1253×4. OE pers.n. *Cyneheard*, Kentish *Ceneheard*, + **ing**[4] + **tūn**. KPN 156, DEPN.

KENCHESTER H&W SO 4442. 'Cena's Roman town'. *Chenecestre* 1086, *Kenecestr(e)* 1166, 1243, *Kenchestre* 1397, 1428. OE pers.n.

Cēna + **ċeaster**. The reference is to the Roman town of Magnis, dative pl. of Brit **magno-* 'stone, rock' and so 'At the rocks'. He 108, RBrit 407.

KENCOT Oxon SP 2504. 'The cottage(s) called after Cena'. *Chenetone* 1086, *Chenicota -e* c.1130, 1163, *Kenicot(') -cot(h)e -y-* c.1150–1361, *Kenningcote* c.1195, *Kencot* from c.1190×1200. OE pers.n. *Cēna* + **ing**[4] + **cot**, pl. **cotu**, replacing earlier **tūn**. O 326.

KENDAL Cumbr SD 5192. Short for Kirkby Kendal, 'church village in Kentdale'. *Cherchebi* 1086, *Kircabikendala* [1090×7]1308, 1157, *Ki- Kyrkebi(e) -by* [1154×89]1294, 1196–1609 with addition *(in) Kendal(e) -dall* from 1120×30, *Ki- Kyrkby in Kendal(e)* 1295–1748, *Ki- Kyrby Kendal(l)* 15th cent. Known as *(burgus de) Kendale -dall* from 1452, *Kendal(l) town(e)* 1519–1622. ON **kirkju-bȳ** + affix *Kendal* from the barony of Kendal, *baronia de Kendal(l) -dala -dale* 1150×5–1761, or directly from the valley (ON *dalr*) of the river KENT, for distinction from KIRKBY LONSDALE Cumbr SD 6178 and KIRKBY STEPHEN Cumbr NY 7708. We i.114, SSNNW 34, L 96.

KENDERCHURCH H&W SO 4028. 'St Cynidr's church'. *Ecclesia Sancti Kenedri* 1291, *Kend(er)church(e)* from 1397. Earlier simply *Sancti Kenedri* 'St Cynidr's' c.1200, 1291. Earlier W forms are *Lann Cruc* [1045×1104]c.1130, 'church site at *Cruc*, the hill shaped like a tumulus' referring to Mount Hill, OW **lann** + **cruc** possibly already used as a p.n., and *Lanncinitir lann icruc* [1066×87]c.1130. OW **lann** + saint's name *Cynidir* later Anglicised as saint's name *Cynidir* + ME **church**. He 109, Bannister 103.

KENILWORTH Warw SP 2972. 'Cynehild's enclosure'. *Chinewrde* 1086, *Chenille- Cheningeuurda, Kinilde- Kiningewrda, Chenildeworda* 1100×35, *Kenelleworth* 1195–1495 with variants *Kenel(e)- -yl(l)-* and *-il(l)e-*, *Chenelingwurthe* 1194–1315 with variants *Kineling- Kenelyng-* and *-worth*, *Kelingworth* 1316, *Kyllingworth* 1447–1624. OE feminine pers.n. *Cynehild*, genitive sing. *Cynehilde*, + **worth**, varying with *Cynehild* + **ing**[4] + **tūn**. Wa 172.

KENLEY GLond TQ 3259. 'Cena's wood or clearing'. *Kenele(e)* 1255–1348, *Kenle* 1403. OE pers.n. *Cǣna* + **lēah**. Sr 45.

KENLEY Shrops SJ 5600. 'Cena's wood or clearing'. *Chenelie* 1086, *Kenele(gh)* 13th, *Kenleg'* from 1203×4 with variants *-legh' -lye* and *-ley(e)*, *Kendl(e)y* 18th cent. OE pers.n. *Cēna* + **lēah**. Sa i.165.

KENN Avon ST 4169. Originally a r.n. transferred to the settlement. *Chen* 1086, *Chent* 1086 Exon, *Kenne* 1200. PrW **Cent* < Brit **Cantiā*, identical with the continental r.ns. Cance, a tributary of the Rhone, Chanza, a tributary of the Guadiana, Spain, Kinz near Aachen, Kanzach, a tributary of the Danube, formerly *Cantaha*, and others. Ultimately from the root **kantho-* 'corner, angle, bend' seen in OCo, W *cant* 'enclosure, ring, circle'. Cf. QUANTOCK HILLS Somer ST 1537. RN 224, RBrit 297–9, Pokorny 526, Bahlow 250, GPC 418.

KENN Devon SX 9285. *Chent* 1086, *Ken* 1167 etc. The place is named from the River KENN SX 9385. D 498.

River KENN Avon ST 4269. Probably '(stream of many) bends'. *Chen* 1086, *Chent* 1086 Exon, 1161, *Ken* 1220. The lower course of the Kenn is full of meanders. See KENN Avon ST 4169.

River KENN Devon SX 9385. *Ken* 1577, 1610, *Kenton brooke* 1577 (from KENTON SX 9583). The earliest forms are as under KENN SX 9285. Ultimately a Brit r.n. ***cunētiu** of unknown meaning and origin identical with KENNET etc. D 7, RN 224.

KENNERLEIGH Devon SS 8107. 'Cyneweard's wood or clearing'. *Kenewarlegh'* 1219, *Kinwardelegh* 1244, *Kyn(e)warde(s)ley(e) -legh* 1281, 1336, 1481, *Kynnerley* 1577, *Kenworleighe* 1609. OE pers.n. *Cyneweard* + **lēah**. D 408.

KENNET Cambs TL 6968. Transferred from the river KENNET TL 6969. *Kenet* 1086–1488, *Chenet* 1086, *Kent* 1316, *Kenett* 1570. Ca 193, RN 226.

East KENNET Wilts SU 1167. *Estkenette* 1267, *East Kynnet* 1558×1601. ME adj. **est** (OE *ēast*) + r.n. KENNET SU 2369, *(æt) Cynetan* 939 S 449, *Chenete* 1086, *Kenet(e)* 13th cent. 'East' for distinction from West KENNET SU 1168. Wlt 297.

West KENNET Wilts SU 1168. *Westkenete* 1288, 1327, *Westkynet* 1332. ME adj. **west** (OE *west*) + r.n. KENNET SU 2369, *Chenete* 1086. 'West' for distinction from East KENNET SU 1167. Wlt 294.

River KENNET. Unexplained. OE **Cynete* from Brit ***cunētjū** of unknown meaning. RN 225, RBrit 328.

(1) ~ Berks SU 5355. Wilts SU 2369. *Cunetione* [c.300]8th AI, *Cunetzone* [c.700]14th Rav. *(inter Tamesen et) Cynetan* [893]11th Asser, *(on) Cynetan* [944]13th S 500, [956]12th S 591, [984]13th S 855, [1050]13th S 1020, *Cynete* [984]13th S 855, [1050]13th S 1020, *Kenete* [1180×1200]c.1200–[1276×7]13th, *Kennet* from 1574. The two earliest forms are the name of the Roman town at Mildenhall on the Kennet, SU 2169. Brk 11, RN 225, RBrit 328.

(2) ~ Cambs TL 6969. *Kenet* 1249. See also the forms of KENNETT TL 6968. Ca 7, RN 226 ff., RBrit 328, Britannia i.71.

KENNFORD Devon SX 9186. 'Ford across the river Kenn'. *Keneford* 1300. R.n. KENN SX 9385 + OE **ford**. D 499.

KENNET AND AVON CANAL Wilts ST 8761, SU 2363. A canal designed by John Rennie and Robert Whitworth, begun in 1786, completed in 1810, connecting Newbury with Bath. R.ns. KENNET SU 2369, AVON SU 1301 + ModE **canal**. Pevsner 1975.52.

KENNICK RESERVOIR Devon SX 8084. P.n. Kennick + ModE **reservoir**. Kennick, *Kaynek -ok* 1294, *Kennok* 1504, previously explained as an old stream name related to W *cain* 'clear, bright, fair' + suffix *-oc*, is more likely to be identical with the second element of the Cornish names Castle Canyk, Trekennick and Treskinnick from Co ***keynek** 'ridged place', Co *keyn* 'back, ridge' + p.n. suffix *-ek*. Kennick is on a ridge between two valleys one of which contains the reservoir. D 431, CPNE 45–6.

KENNINGHALL Norf TM 0386. 'The nook of land called or at *Cening*, the place called after Cena'. *Keninchala, Cheninkehala* 1086, *Keninghale* 1212, 1254. OE p.n. **Cēning* < pers.n. *Cēna* + **ing**[2]. DEPN.

KENNINGTON Kent TR 0245. 'Royal manor'. *Chintun* 1072, *Kynigtune* 11th, *Chenetone* 1086, *Keningtune* [1087]13th, *Kynig- Chenetune* c.1100, *Kenin(g)ton(e)* 13th cent. OE **cyne**, Kentish **cene** varying with **cyning, cening**, + **tūn**. PNK 414, DEPN.

KENNINGTON Oxon SP 5202. 'The estate called after Cena'. *Chenitun* *[821]12th S 183, *Chenigtun* [821]13th ibid., *Cenigtun* [956]12th S 614, *(æt) Cenintune* [956×7]13th S 1292, *Chenitun, Genetune* 1086, *Kenitune, Kenintone* [1066×87]13th, *Kenington'* 1229×30. OE pers.n. *Cēna* + **ing**[4] + **tūn**. Brk 453.

KENNY HILL Suff TL 6679. No early forms.

KENNYTHORPE NYorks SE 7865. 'Cenhere or Cynhere's outlying farm'. *Cheretorp* 1086, *Kenert(h)orp* 12th–1336, *Kinnerthorp* [c.1180]15th, *Kenythorpe* 1546–1600, *-thrope* 1609. OE pers.n. *Cēn- Cynhere*, + **thorp**. YE 141 gives pr [keniθrəp], SSNY 61.

KENSINGTON GLond TQ 2579. 'Estate called after Cynsige'. *Chenesit'* 1086, *Kensin(g)ton* from 1235, *Kinsentona -tun, Kinsne- Chinsentuna* 12th cent. OE pers.n. *Cynsiġe* + **ing**[4] + **tūn**. Mx 128.

KENSWORTH Beds TL 0316. 'Cægin's enclosure'. *Ceagnesworthe* [975]13th, *Canesworde* 1086, *Keneswurth* [1189×99]1227, 13th cent. etc., *Kensworth* 1375. OE pers.n. **Cǣġīn*, genitive sing. **Cǣġīnes*, + **worth**. Bd 178, DEPN.

KENSWORTH COMMON Beds TL 0317. *Kensworth Common* 1834 OS. P.n. KENSWORTH TL 0316 + ModE **common**.

River KENT Cumbr NY 4502. Obscure. *Kent(e)* from c.1170, *Kient* 1205×16, *Kenet* 1246, 1256, *Keent* 1278–1340, *Ken(ne)* c.1540–1754, *Can(um)* 1586–1622. British ***cunētjū** of unknown meaning. We i.8, RN 226.

KENT The county name. *Cantium* c.50 BC Caesar *Gallic War* V.13, *Κάντιον* (Cantium) 20 BC×AD 20 Strabo, *Cantia* (accusative *Cantiam*, genitive *Cantie, Cancie)* *[605]14th S 2–[1041×5]13th S 1091, [c.720]10th BHAbb, [c.731]8th BHE, *Cent lond* 9th ASC(A) under year 457, *Cent* 9th ASC passim with variant *Cænt* 12th ASC(E) under year 616, [1051×66]13th S 1092, *Chenth* 1086. The inhabitants of Kent are variously *Κάντιοι* (*Cantii*) [c.150 BC]c.1300 Ptol with variant *Καντικοί* (*Cantici*), *Cantiaci* 8th Rav, *Cantuarii* (genitive pl. *Cantuariorum, Cantþariorum* *[675]15th S 7–[861]13th S 330, and *Cantware* 9th ASC passim. British **Cantịon* meaning probably 'corner land, land on the edge'. PNK 1, Jnl 8.40, RBrit 297–300, Charters IV passim.

KENTCHURCH H&W SO 4125. 'St Cein's church'. *ecclesia de Kein* 1194, *Ecclesia Sancte Keyne* 1205, *Kenschirch* 1300, *Keyn(e)churche* 1339, c.1380, *Kenchurch* 1535. Earlier simply *Sancta Kaenœ* ?1100, *Sancta Keynœ, Keina* 1195, 1217×29. The earliest form is W *Lan Cein* [1045×1104]c.1130 which survived as late as *Llankeyne* 1469. OW **lann** + saint's name *Cein* later Anglicised as saint's name *Cein* + ME **churche**. He 110, Bannister 103, O'Donnell 92.

KENTCHURCH COURT H&W SO 4225. *Kentchurch Court* 1831 OS. P.n. KENTCHURCH SO 4125 + ModE **court**. The reference is to a 14th cent. castle rebuilt by Nash c.1800. Pevsner 1963.200.

KENTFORD Suff TL 7166. 'The ford across the r. Kennett'. *Cheneteforde* [1066×87]1318, [1114×30]c.1350, *-fort* 1109, *Keneteford* 1203–1324, *Kenforde* 1610. R.n. *Kenet* [c.1080]12th, 1161, 1276, *Chenet* 1086, 1160, identical with r.ns. KENNET Berks SU 5366 and KENT Cumbr NY 4502, SD 5086, + OE **ford**. DEPN, RN 226.

KENTISBEARE Devon ST 0608. Partly uncertain. *Chentesbere* 1086, *Kentelesbar -be(a)re -bire* 1212×23–1291, *Kanteleber* 1219, *Kaentlesbyar* 1297, *Kentesbere* 1577. The generic is OE **bearu** 'a wood, a grove'. The specific must be taken with that of Kentis Moor ST 0506, *Kentelesmore* c.1200, *Kentysmore* 1445, and KENTISBURY SS 6243, possibly from a Co ***cantel**, **cant* probably meaning 'district, region' or 'edge, border' + adjective suffix *-*el*, cf. Cant Corn and possibly Cargentle Corn, **cant* + *-*yel* (if not from *cuntell* 'an assembly'). The meanings could be 'district wood', 'district moor' and 'district fort'. D 564, 566, CPNE 37–8.

KENTISBURY Devon SS 6243. Possibly 'district fort'. *Chentesberie* 1086, *Kentesbir' -byri -bury* 1242, 1275, 1316, *Kentelesberi -bere -bur'* 1260–1281, *Kantelesberi* 1285, *Kynsbury* 1523. Possibly Co ***cantel**, secondary genitive sing. ***canteles**, + OE **byriġ**, dative sing. of *burh*, referring to enclosures on Kentisbury Down SS 6343. See the discussion under KENTISBEARE ST 0608. D 49.

KENTIS MOOR Devon ST 0506 → KENTISBEARE ST 0608.

KENTMERE Cumbr NY 4504. 'Pool on the river Kent'. *Kent(e)mer(e)* from 1247×60, *Kenetemere* 1274. R.n. KENT Cumbr NY 4502 + **mere**. The reference is to a lake drained about a century ago. We i.165, L 26–7.

KENTMERE RESERVOIR Cumbr NY 4408. P.n. KENTMERE Cumbr NY 4504 + ModE **reservoir**.

KENTON Devon SX 9583. 'Settlement on the river Kenn'. *Chentone* 1086, *Kenton* from 1167, *Ky- Kinton* 1267, 1276. R.n. KENN SX 9385 + **tūn**. D 499.

KENTON GLond TQ 1888. 'Cena's estate'. *Keninton* 1232, *Kenyngton in parochia de Harghes* 'K in the parish of Harrow' 1307, *Kenton al. Kynyton* 1548. OE pers.n. *Cǣna* + **ing**^{4} + **tūn**. Mx 53, 23.

KENTON Suff TM 1966. Partly uncertain. *Chene- Kenetuna* 1086, *Keneton(a)* 1179–1286, *Kenton(e)* from 1286, *Kingeston* 1252. As the 1252 form shows this was a royal manor; possibly, therefore, OE **cyne-tūn** 'royal estate'. But pers.ns. *Cēna* or *Cyna* are also possible, 'Cena or Cyna's estate'. DEPN, Baron.

KENTS BANK Cumbr SD 3976. 'Bank of the river Kent'. *Kentsbanke* 1491, *Kentisbanke* 1537. R.n. KENT Cumbr NY 4502 + ME **banke**. La 196.

KENT'S GREEN Glos SO 7423. *Kent's Green* 1746. Surname *Kent* + ModE **green**. Gl iii.188.

KENT'S OAK Hants SU 3224. No early forms. Surname *Kent* + ModE **oak** 'an oak-tree'.

KENTWELL HALL Suff TL 8647. *Kentwell Hall* 1836 OS. A 16th cent. manor house. P.n. *Kanewella* 1086, *Kenetwelle* 1162, *Kenetewell* 1168, 1176, *Kentewelle* 1156×80, 'Kent spring or stream', r.n. Kennet as in the r.ns. KENNET Berks SU 5366 and KENT Cumbr NY 4502, SD 5086, + OE **wella**. The reference is either to the stream which rises $\frac{1}{4}$ m N of the hall, or to the r. Glem, an OE name possibly superseding pre-English Kennet. DEPN.

KENWICK Shrops SJ 4230. 'Cena's farm'. *Kenewic* 1203–80, *-wike* 1205×10, *Kenwick* 1610. OE pers.n. *Cēna* + **wīċ**. DEPN.

KENWYN Corn SW 8245. 'White ridge'. *Keynwen* 1259–1311, *-wyn* 1316–76, *Kenwen* 1265–84, c.1400, *Kenuen* 1610, *(capella) Sancti Kenwyni* 1342. Co **keyn** 'back, ridge' + **guyn** 'white' with lenition of *gw-* to *w-*. Variation between *e* and *y* spellings is common with Co *guyn*. Later misunderstood as a saint's name and wrongly associated with St Keyne. PNCo 102, Gover n.d. 461, DEPN, CPNE 120.

KENYON GMan SJ 6295. Possibly 'Enion's mound'. *Kenien* 1212–69, *Kenion* 1236, *Kenian -yan* 1243–1332 etc. PrW ***crüg** + W pers.n. *Eniōn*. It is suggested that *Crug Enion* was misunderstood as *Cru Cenion*. The reference, if this suggestion is right, would have probably been to the former Bronze Age barrow at this place. La 98, DEPN citing ProcLCAS 21(1903), Jnl 17.56.

KEPWICK NYorks SE 4690. Probably 'place where trade takes place'. *Cap- Chipuic* 1086, *Chepewic* 1166, *Kepwic -uuic -wyche* 1202–34, *Kep(p)ewyk -wick* 1224–1348, *Kepyk -ec* 1451, 1505. OE **ċēap** + **wīċ** with ON [k] for [tʃ]. Cf. KEEPWICK NY 9571. YN 201 gives pr [kepik], SSNY 147.

KERESLEY WMids SP 3284. Partly uncertain. *Keresleia* [c.1144]1348, *-le(ga) -leye* 1179–1451, *Kerusley* c.1170–1368, *Caresley* 1553–1656. The generic is OE **lēah** 'a wood or clearing', the specific probably a pers.n., perhaps a reduced form of OE *Cēnhere*, genitive sing. *Cēnheres*. It is unlikely that we have an instance of ON pers.n. *Kærir*, genitive sing. *Kæris*, so far W. Neither Kearsley Northum NZ 0275, *Kerneslawe* 1244, *Kareslaw* 1361, 'Kjarni's hill', or KERSALL Notts SK 7162, 'the nook of the cutting', OE *cyrfes-halh*, are comparable. Wa 174 gives pr [kɛ:əzli], NbDu 124.

KERNE BRIDGE H&W SO 5819. Partly unexplained. *Kerne Bridge* 1831 OS. Perhaps a topographical use of W **cern** 'cheekbone, side of the head, side of a hill, an exposed slope' used as a p.n. + ModE **bridge**. GPC 468.

KERRIDGE Ches SJ 9377. 'Boulder ridge'. *Caryge* [13th]1611, *Carigge* 1467, 1471, *Kay- Cairug(ge)* [1270]17th, 1341–57, *Kayryche* 1363, *-ridge* 1611, *Kerich* 1686, *Kerridge* 1842. OE ***cǣġ** 'a stone, a boulder' + **hrycg**. Che i.188.

KERRIS Corn SW 4427. Uncertain. *Kerres* 1302, 1310, *Kerrys* 1439 Gover n.d., *Kirthies* 1610. Also known as *Kerismoer* 1301–2 'Great Kerris', Co **meur**. There is an ancient enclosure nearby called Roundago and it seems likely that Kerris is connected with Co ***ker** 'a fort, a round', possibly as the past participle of a verb meaning 'to fortify' equivalent to W *caeru* or as a noun equivalent to W *caered* 'wall'. PNCo 102, Gover n.d. 655, CPNE 54.

KERRY'S GATE H&W SO 3933. *Careys Gate* 1831 OS. Surname *Carey* (W *Ceri, Kerry*) + ModE **gate**. WSurn 68.

KERSALL Notts SK 7162. Partly uncertain. *Cherueshale* 1086, *Ki- Kyrueshal(e)* 1195–1332, *Kyrneshale* 1196, 1305, *Ky- Kirnesal(e)* 1280–1335, *Kirsale* 1361, 1537. There has been confusion of *u* and *n* in this name, *u*-forms probably being primary. The specific might, therefore, be OE **cyrf** 'a cutting' and the name mean 'the nook of land of the cutting', OE **cyrfes-halh**, referring to a short side-valley opening off the deep valley of The Beck which runs between high banks at this point. Nt 188, L 103, 111.

KERSEY Suff TM 0044. 'Cress island'. *Cæresige* 1000×2 S 1486, *Careseia* 1086, *Karsee* 1220, *Kerseye* 1235, *Carsey* 1610. OE **cærse** + **ēġ**. DEPN, L 39.

KERSHOPE BURN Cumbr NY 5285. P.n. Kershope + eModE **burn**. *Carsoppburne* 1580, *water of Cressop*, *Kyrsopp* 1583, *Kirksop R.* 1695, *Kersope* 1740. Kershope, *Cresope* [before 1165]1307, *Cresop'* 1200, *Creshop(') -hoppe* 13th cent., *Cryssop* 1552, *Carsopp* 1580 and *Gresopa* [c.1165]1307, *Greshop(e)* 1201–1307, *Gressop* 1251–79, is either 'side-valley where cress grows', OE **cærse** + **hop**, or 'grassy side-valley', OE **gærs**, **græs** + **hop**. Cu 18, 61, L 115–6.

KERSHOPEFOOT Cumbr NY 4782. *foot of Cryssop* 1547×53, *the foote of Carsopp* 1580, *Kirksop Foot*, *Kirsopfoote*, *Cressopp fote* 17th. P.n. Kershope as in KERSHOPE BURN Cumbr NY 5285 + ME **fōt**. Cu 61.

KERSHOPE FOREST Cumbr NY 5181. P.n. Kershope as in KERSHOPE BURN NY 5285 + ModE **forest**.

KERSOE H&W SO 9940. 'Criddi's hill-spur'. *(æt) criddesho* *[780]11th S 118, *Crideshoth* (sic for *-hoch*) [1182]18th, *Crydesho(o)*, *Crideshoe* 1275–1445, *Crisso* 1548, *Kersowe*, *Kirsoe* 1588, 1635. OE pers.n. **Criddi*, genitive sing. **Criddes*, + **hōh**. The village stands on a well-marked spur of land. Wo 122, L 168.

KERSWELL Devon ST 0706. 'Cress spring'. *Carsewelle* 1086, *Carseuilla* 1086 Exon, *Karswille* 1201. OE **cærse** + **wylle**. D 558.

KERSWELL GREEN H&W SO 8646. *Kerswell Green* 1831 OS. P.n. Kerswell + ModE **green**. Kerswell, *Chirswell* [1182]18th, *Keres- Kers(e)well* 1208–1649, *Caswell* 1772, 1789, is 'cress spring', OE **cærse** + **wella**. Wo 146, L 31.

KESGRAVE Suff TM 2145. Apparently 'the cress ditch or grove'. *Gressegraua* 1086, *Kersigrave* 1231, *Ke(r)ssegrave* 1254, *Kessegrave* 1286–1336, *Kesgrave* from 1524. OE **cærse**, **cresse** possibly varying with ***cærsing** 'place growing with cress, a cress-bed' + ***grafa** or **grāf**. But the place lies in heathland where cress seems unlikely. DEPN, Baron.

KESSINGLAND Suff TM 5386. Possibly 'the newly cultivated land called or at *Cærsing*, the place growing with cress'. *Kessingalanda -inge-* 1086, *Kersingeland'* 1183×4, *Kessingeland* 1219, *Cassingeland* 1225, 1251, *Kersinglond'* 1286, *Kessingland(e)* from 1242, *Keslande* 1610. OE ***cærsing** < *cærse* + *ing*[2], locative–dative sing. ***cærsinge**, + **land**. Alternatively the 'newly cultivated land of the Cyssingas, the people called after Cyssi', OE folk-n. **Cyssingas* < pers.n. **Cyssi*, SE form **Cessi* as in KESTON GLond TQ 4164, + **ingas**, genitive pl. **Cyssinga* with occasional intrusive *r*. DEPN, Baron, L 247.

KESTEVEN Lincs. 'The meeting place or district of *Ced*, the wood'. *condenso sylucæ quæ uulgo Ceoftefn* (for *Ceos-*) *nuncupatur* 'the thickets of the wood commonly called C' late 10th Æthelweard, *Chetsteven* 1086, *Ketsteuen(e)* 1185–1295, *Kesteven(e)* c.1140–1348, *Kestene* 1329, *Keston* 1509×47. PrW ***cęd** (Brit **ceto-*) + ON **stefna**. One of the ancient divisions of the county of Lincs lying S of Lincoln between Northants and Leics on the W and the Witham and the S Forty Foot Drain on the E. The W parts were formerly thickly wooded. Cf. KETTON SK 9804. DEPN, Perrott 37.

KESTLE MILL Corn SW 8459. *Kestell Mill* 1659, *Kestle Mill* 1717. P.n. Kestle SW 8559 + ModE **mill**. Kestle, *Castel* 1194, *Kestell* 1345 is Co **castell** 'castle, village, tor', varying with pl. form **kestell** although not necessarily with pl. meaning. The sense is probably 'hamlet'; there are seven examples of this name throughout Cornwall. PNCo 102, Gover n.d. 327, CPNE 42.

KESTON GLond TQ 4164. 'Cyssi's stone'. *Cysse stanes (gemæro)* 'the bounds of K' 973 S 671, *Chestan* 1086, *Keston'* 1240. OE pers.n. **Cyssi*, Kentish dial. form **Cessi*, + **stān**. The boundary of the Cystaningas, the people of Keston, is mentioned, *cystaninga mearc* 862 S 331, 987 S 864. The reference is probably to a boundary stone of Cyssi's estate, or possibly to a stone-built dwelling. KPN 211, TC 208.

KESWICK 'Cheese farm'. OE **ċēse-wīċ** with substitution of ON [k] for OE [tʃ].

(1) ~ Cumbr NY 2623. *Kesewic* c.1240, *-wyk -wik* 1266–1403, *Keswyk* 1383, *Keswick* from 1589. Pronounced [kezik]. Cu 301, SSNNW 203.

(2) ~ Norf TG 2004. *Ch- Kesewic* 1086, *Kesewic* 1216×72. DEPN.

(3) ~ Norf TG 3533. *Casewic* c.1150–1275, *-wyk(e)* 1254–1531, *Caswycke* 1548, *Cesewyk* 1291, *Kesewike -wyk* 1316–1428, *Keswyk* 1461. The puzzling *Case-* spellings are probably due to substitution of OScand *ā* for OE *ē*. Nf ii.137, SN 2.33.

(4) East ~ WYorks SE 3644. *Est(e)keswyc(k) -wyk(e) -wi(c)k(e)* 12th–1615. ME **est** + p.n. *Chesinc -ing* 1086, *Keswic(h)* 12th cent. The DB spellings are due to misreading of *-uic* as *-inc*. 'East' in relation to DUNKESWICK SE 3046 on the other branch of the Wharfe. YW iv.184 gives pr ['kezik], SSNY 147.

KETTERING Northants SP 8678. 'The Cytringas'. *(to, æt) Cytringan* [956]12th S 592, *Kyteringas* [972]12th S 787, *Keteiringan* [?963]12th S 1448, *Ketering* [972]12th S 787(p.252)–1347 with variants *-inge -yng(e) -ingge*, *Ketterynge* 1557. Unexplained OE folk-n. **Cytringas*, dative pl. **Cytringum*. Nth 184.

KETTERINGHAM Norf TG 1602. 'Homestead of the Ceteringas'. *(at) Keteringham* [1052×66]13th S 1519, *Ke- Kitrincham*, *Keterincham* 1086, *Keteringeham* 1194, *Keteringham* 1242×3, 1254. OE folk-n. **Ceteringas* of unknown origin, genitive pl. **Ceteringa*, + **hām**. *Ceteringas* is probably to be associated with the *Cyteringas* at KETTERING Northants SP 8778, and possibly with the G p.n. Kettermann near Lüdinghausen, *Ketteringe* c.1150. DEPN, ING 137, Piroth 84.

KETTLEBASTON Suff TL 9650. 'Ketilbjǫrn's estate'. *kitelbeornastuna* 1086†, *Kytel- Chethelbernestun* 1095, *Ketelberneston* 1208, *Kettlebaston* 1610. ON pers.n. *Ketilbjǫrn* partly Anglicised as *Cytelbeorn*, ME genitive sing. *Ketilbernes*, *Cytelbeornes*, + **tūn**. DEPN, PNDB 304.

KETTLEBROOK Staffs SK 2103. Short for *Kettlebrook Colliery* 1834 OS. A brook runs through a wide curving valley here to join the Tame. ModE **kettle** (OE *ċetel* 'a deep valley surrounded by hills') + **brook**. St i.12.

KETTLEBURGH Suff TM 2660. 'The hill by the valley'. *Chetelberia*, *Chettlebiriga*, *Cetel- Ketdesbirig*, *Kettleberga*, *Ketelbiria* 1086, *Keteleberga* 1188, *Ketelberg(a) -e -bergh* 1190–1292, *Ketelberwe* 1235, *Kettlebergh* 1346, *-burgh* 1327, *Ketleborogh* 1610. OE **ċetel** with Scandinavian [k] for [tʃ] + **beorg**. The church is on a hill overlooking a small deep valley. DEPN, Baron.

Ab KETTLEBY Leic SK 7223. 'Abba's Kettleby'. *Abbeket(t)elby* 1236–1518 with variant *Abe-*, *Ap(p)e-* 1327–1610, *Abby-* 1333, *Abbey-* 1524. OE pers.n. *Abba* + p.n. *Chetelbi* 1086, 12th, *Ketelbi(a)* late 12th, *Ket(t)elby* c.1130–1610, 'Ketill's village or farm', ON pers.n. *Ketill* + **bȳ**. 'Ab' for distinction from Eye Kettelby SK 7316, 'Kettleby on the river Eye', r.n. EYE SK 8018 + p.n. *Chitebie* 1086, *Ket(t)elby -il- -yl-* 1200–1610, 'Ketill's village or farm'. Lei 170, 176, SSNEM 56 nos. 2 and 3.

KETTLENESS NYorks NZ 8315. 'Cauldron headland'. *Kettleness* 1861 OS. ON **ketill** + **nes**. *Cauldron* is used in coastal and other names of a place where water churns. Cf. CAULDRON SNOUT NY 8128, KETTLEWELL SD 9772.

KETTLESHULME Ches SJ 9879. 'Ketil's water-meadow'. *Ketelisholm* 1285, *-es-* 14th, *Ketel(l)eshulm(e)* 1285–1358, *Kettleshulm(e)* from 1347. ODan pers.n. *Ketil(l)*, genitive sing. *Ketil(l)es*, + **hulm** (ON *holmr*). Che i.110, SSNNW 139, L 52.

†Farley wrongly prints *bitel-*. In the MS the first letter is *k* altered from *l*.

KETTLESING NYorks SE 2256. 'Ketil's meadow'. *Ketylsyng'* 1379, *Ket(t)lesing(e)* 1590, 17th cent. ON pers.n. *Ketill*, genitive sing. *Ketills*, + **eng**. YW v.132.

KETTLESING BOTTOM NYorks SE 2257. P.n. KETTLESING SE 2256 + OE **botm**.

KETTLESTONE Norf TF 9631. 'Ketil's estate'. *Ketlestuna* 1086, *-ton'* 1199, 1209, *Keteleston'* 1209. ODan pers.n. *Ketil*, secondary genitive sing. *Ketiles*, + **tūn**. DEPN, SPNN 256.

KETTLETHORPE Lincs SK 8475. 'Ketil's outlying farm'. *Ketel(s)torp* 1220, *Ketelestorp* 1249. ODan pers.n. *Ketil*, genitive sing. *Ketils*, + **thorp**. DEPN, SSNEM 128.

KETTLEWELL NYorks SD 9772. 'Bubbling spring or stream'. *Cheteleuuelle* 1086, *Ketelwell(a) -uella -welle* 12th–1468, *Ket(t)lewell* 1597, 1608. OE **ċetel** + **wella**. Initial [k] is due to the influence of ON *ketill*. OE *ċetel* is also used in the sense 'a deep valley surrounded by hills' which would also be appropriate for this place. YW vi.107.

KETTON Leic SK 9804. Parlty uncertain. *Chetene* 1086, *Chetena* 1146, *Chetenea* 1163, *Ketene* 1174–1384, *Keten* 1209–1383, *Keittunia* late 12th, *Keton(e)* 1322–1505, *Ketton* from 1519. Possibly OE pers.n. **Cēta*, genitive sing. **Cētan*, + **ēa** 'river'. With the loss of the final syllable the name was remodelled as if an OE name in **tūn**, 'farm or village'. No such OE pers.n. is recorded, but no satisfactory alternative suggestion is known. Lei 667.

KEVERAL Corn SX 2955 → CHICKERELL Dorset SY 6480.

KEW GLond TQ 1877. 'Hill-spur called the Key'. *C- Kayho* 1327–1529, *Keyhow* 1439, *Kayo(we)* 15th cent., *Keyo(we)* 16th cent., *Kewe* 1535, *Kew al. Kyo* 1648. OE p.n. **Cæġ* from **cæġ** + **hōh**. Sr 58, L 168.

KEWSTOKE Avon ST 3363. 'Kew's Stoke'. *Kiustok* 1274. Earlier *Stoke super mare* 'S on sea' 1265. Saint's name *Kew* as in ST KEW Corn SX 0276 + OE **stoc** 'dependent farm'. DEPN, Studies 1936.19.

KEX BECK NYorks SE 0953. 'Teasel stream'. *Kexebec* 12th, 13th, *Kexbec* 1203, 1244, *-bek* 1381, 1386. ME **kex** 'teasel' + **beck** (ON *bekkr*). YW vii.130.

KEXBROUGH SYorks SE 3009. Possibly 'Kept's fortification'. *Ceze- Chizeburg* 1086, *Kesceburg(h)* c.1170, 1325, *Keseburc* 1194, *Kes(s)eburgh* 1316–1434, *Kesburgh* 1376, 1415, 1545, *Kexburgh(e)* 1402–1581, *-brough* 1540. ON pers.n. *Keptr*, genitive sing. *Kepts*, + OE **burh**. YW i.318, SSNY 147.

KEXBY Lincs SK 8785. 'Keptr's village or farm'. *Cheftesbi* 1086, *Chezbi, Chetesbi* (changed to *Chetlesbi* in a contemporary hand) c.1115, *Kigges- Keftesbi* 1194, *Keftesby* 1202, *Kestesbi* (for *Keftes-*) 1212. ON pers.n. *Keptr*, genitive sing. *Kepts*, + **bȳ**. DEPN, SSNEM 56.

KEXBY NYorks SE 7051. Either 'Keikr's' or 'teasel farm or village'. *Kexebi -by* 1175–1316, *Kexby* from 1285. Either ON pers.n. *Keikr*, genitive sing. *Keiks*, or ME **kex**, + **bȳ**. YE 272, SSNY 7.

KEXWITH NYorks NZ 0505. 'Clearing overgrown with teasels'. *Kexthwayt* 1280, 1283, *Kextwayte* 1301. ME **kex** + ON **thveit**. YN 291.

KEXWITH MOOR NYorks NZ 0305. P.n. KEXWITH NZ 0505 + ModE **moor**.

KEY GREEN Ches SJ 8963. 'Cattle pasture'. *Key Green* 1831. Dial. **kye** (plural of *cow*) + **green**. Che ii.292.

KEYHAM Leic SK 6706. Possibly 'homestead on the (hill called the) Key'. *C- Kaiham -ay-* 1086–1570, *Cahiham* c.1130, *Caham* 1247, *Kayam* 1252, *Ka(y)me* 1502–41, *Keyham* from 1535. OE **cǣġ** + **hām**. The c.1130 form may represent **cǣġ** + **hēah-hām**. Keyham lies on a long narrow hill. Lei 307, L 168.

KEYHAVEN Hants SZ 3090. 'The cow's haven'. *Ki- Kyhavene* c.1170–1316, *Kayhaven* 1532, *Key Haven* 1610. OE **cū**, genitive sing. **cȳ**, + **hæfen**. The reference is probably to a harbour where cows were shipped to and from the Isle of Wight. Ha 103, Gover 1958.227.

KEYINGHAM Humbs TA 2425. 'Village, homestead of the Cæingas, the people who live on the *Cæg*, or the settlement at or called *Cæging*'. *C(h)aingeha'* 1086, *Caingeham* 1115–1275, *K- Caing- Keing- Cayngham* 1190–1542, *Ken(n)ingham* 17th cent. OE **Cæġingas*, genitive pl. **Cæġinga*, or **Cæġing*, locative-dative **Cæġinge*, + **hām**. At the base of this name is OE **cǣġ** 'a key, a projecting piece of land, a key-shaped hill'. OE **Cæging* would be a common *-ing* derivative in the sense 'that which is like a key'. Keyingham lies on a small prominent hill in the Holderness marshes. Earlier explanations took the specific to be an unattested OE pers.n. **Cǣġa*. Cf. KEW TQ 1877, KEYSOE TL 0762. YE 32 gives pr [kenigəm, keniŋəm], ING 153, BzN 148.

KEYMER WSusx TQ 3115. 'The cow's pond'. *Chemere, Chimele* 1086, *Ki- Kymera -mer(e)* c.1090–1672, *Keymer* from 1598. OE **cū**, genitive sing. **cȳ**, + **mere**. Keymer lies low in well-watered land; there must have been a mere here at one time. Sx 276, SS 74.

KEYNSHAM Avon ST 6568. Probably 'Cægin's river-bend land or promontory'. *Cægineshamme* c.1000, *Cainesham* 1086, *Keinesham* 1170. OE pers.n. **Cǣġīn*, genitive sing. **Cǣġīnes*, + **hamm** 1 or 2a. The river Avon makes a large loop here called Keynsham Hams into which a spur of higher ground projects, the site of a Roman villa. OE pers.n. **Cǣġīn* would be a diminutive of the names **Cǣga*, **Cǣġi* formerely posited for CASSIOBURY, CASHIO Herts TQ 0897, CAYNHAM Shrops SO 5473, Cainhoe Beds TL 1036 and KEYSOE Beds TL 0763. *Cæg-* in these names, however, probably represents either OE **cæġ** 'a key' used in a topographical sense of a key-shaped hill or ***cǣġ** 'a stone, a boulder'. Possibly there was a diminutive **cǣġīn* used in a similar way here. DEPN, TC 118, ESt 43.41.

KEYSLEY DOWN Wilts ST 8634. *Keesley Down* 1817 OS. P.n. Keysley + ModE **down**. Keysley as in Keysley Farm, a sheep farm, ST 8635, is *Keasley* 1648, *Keesley* 1812, but the evidence is too late for explanation; it may be a compound in OE **lēah** 'a wood, a clearing'. Wlt 174.

KEYSOE Beds TL 0763. Partly uncertain. *Chaisot, Caissot* 1086, *Kaiesho* [1154×89]14th, *Kais(h)o* 13th cent., *Keysho* 1227, *Casoe* 1767. Probably 'spur, heel of the hill called *Cæg*, the Key', OE **cǣġ** 'a key', probably used as a p.n. in the topographical sense 'something shaped like a key', genitive sing. **cǣġes**, + **hōh**, dative sing **hō**. This interpretation is preferable to positing an unrecorded pers.n. **Cǣġ*. A different unrecorded OE **cæġ* cognate with MDu *kei* 'stone' has also been suggested for this and other names. Cf. CAINHOE TL 1036 and CASSIOBURY Herts TQ 6897. Bd 14 gives pr [keisou], ESt 43.41, L 168.

KEYSOE ROW Beds TL 0861. *Keysoe Row* 1835 OS. P.n. KEYSOE TL 0763 + ModE **row**, 'a row of buildings'.

KEYSTON Cambs TL 0475. 'Ketill's stone'. *Chetelestan* 1086–1166, *Ketelestan* 1172, 1209, *Ketelston* 1248, *Ketstan* 1227, *Keston -stan* 1255–1442, *Keyston* from 1553. AScand pers.n. *Ketel* < ON *Ketill*, genitive sing. *Keteles*, + **stān**. The reference is probably to a boundary stone perhaps marking the western edge of the old county of Huntingdonshire. Hu 243 gives pr [kestən].

KEYWORTH Notts SK 6130. 'Jackdaw enclosure'. *Cauorde -w-* 1086, *Kaword(i)a* [1178]14th, *-word -wurth'* 13th, *Caworth* 1499, *Kewurda -worð -wurth -worth* 1154×89–1359, *Keyworth* from 1482. OE ***cā** + **worth**. Nt 248.

KIBBLESWORTH T&W NZ 2456. 'Cybbel's enclosure'. *Kibleswrðe* c.1170×80–1185, *Kibbleswurd'* 1185, *Kibleswrth(e)* 1189×95–mid 13th, *Kybbelesuurth'* before 1290, *Ky- Kiblesworth* [1286]1332–1627, *Kibblesworth* from 1365×6. OE pers.n. **Cybbel*, a diminutive of **Cybba*, genitive sing. **Cybb(e)les*, + **worth**. NbDu 126.

KIBWORTH BEAUCHAMP Leic SP 5893. 'K held by the Beauchamp family'. P.n. *Chiburd(e), Chi- Cliborne* 1086, *Chiburd(e), Kibwrd'* c.1130, *Ki- Kybewrd(a) -w(o)rth* 1160–1309, *Ki- Kybbew(o)rth(e)* late 12th–1443, *Ki- Kybw(o)rth(e)* from c.1130, 'Cybba's enclosure', OE pers.n. **Cybba* + **worth**, + manorial affix *Beaucham* 1306, *Be(a)uchamp(e)* 1326–1607,

Beacham 1604. 'Beauchamp' refers to the tenure of the manor by Walter de Bellocampo c.1130 and is for distinction from the adjoining manor of KIBWORTH HARCOURT SP 6894. Also known as *Nether Kibworth* 1722. Lei 233.

KIBWORTH HARCOURT Leic SP 6894. 'K held by the Harcourt family'. P.n. Kibworth as in KIBWORTH BEAUCHAMP SP 6893, the adjoining manor, + manorial affix *Har(e)curt* 1242–1350, *Har(e)court(e)* 1308–1604. The manor was held by Ivo de Haruecurt in the late 12th cent. Also known as *Over Harcourt* 1705, *Vuer-* 1209×35, OE **uferra**. The village lies on higher ground than Kibworth Beauchamp. Lei 234.

KIDBROOKE GLond TQ 4076. 'Kite brook'. *Chitebroc* c.1100, *Katte-* 1191, *Ketebroc -brok'* 13th cent., *Cuttebroc* 1226, *Ketel-* 1238. OE **cȳta** + **brōc**. The 1191 form suggests an alternative name was also extant with OE *catt* 'a wildcat' unless this is a mistake for *Kutte-*. OE *ȳ* would produce ME *i, e* and *u* spellings. KPN 1.

KIDDEMORE GREEN WMids SJ 8508. *Kid(d)imoor(e) Green* 1686, 1755, 1775, *Kerrimore Green -y-* 1688, *Kiddermore Green* 1834. P.n. *Kudimor -y-* 1308, *Kyrre- Kyr(r)y- Kerrymore* 1383–1723, *Kadimore* 1659, *Kiddemore* 1874, possibly 'the brushwood marshland', ME ***kerry** < *ker* (ON *kjarr*) + *-y* (OE *-iġ*) + **more** (OE *mōr*), + ModE **green**. St i.38.

KIDDERMINSTER H&W SO 8376. 'Cydder's monastery'. *Chideminstre* 1086, *Kedeleministre* [1154]13th, *Ki- Kedemenistra* 1168, 1190, *Kyderemunstre* c.1200, *Kyddermynster* 1550. OE pers.n. **Cydder* varying with *Cydda* and *Cydela* + **mynster**. Wo 247, TC 118.

KIDDINGTON Oxon SP 4122. 'The estate called after Cydda'. *Chidintone* 1086, *C- Kudinton(a) -yn- -tone* 1148–1360, *C- Kudington(e) -yng-* 1230–1428, *Kedyngton* 1517. OE pers.n. *Cydda* + **ing**[4] + **tūn**. O 358.

Over KIDDINGTON Oxon SP 4122. 'Upper K'. *superiori Cudintone* 1241, *Ouerkudingtona* 1232. ME **over**, Latin **superior** + p.n. KIDDINGTON SR 4122. Also known as *Soukudintona* 'south K' 1241. O 358.

KIDLINGTON Oxon SP 4913. 'The estate called after Cyddela'. *Chedelintone* 1086, *Kedelintona -ton(e)* 1183×5–1263, *C- Kudelintona -tun -ton(e) -yn-* c.1130–1380, *C- Kudelington(e) -yng-* 1206–1388, *Ki- Kydlyngton* 1428. OE pers.n. **Cyd(d)el(a)* + **ing**[4] + **tūn**. O 272.

KIDMORE END Oxon SU 6979. *Kydmer end* 1551×2, *Kidmere End* 1704, *Kidmore End* 1713. P.n. Kidmore + ModE **end**. One of the Chiltern *mere* 'pond' names probably with ModE *kid* 'a young goat'. O 76.

KIDSGROVE Staffs SJ 8354. *Kydcrowe* c.1596, *Kidcrow(e)* 1656, 1686, *Kidcrew* 1695, c.1733, 1747, *Kidsgrove* 1807. A late rationalisation of a name in dial. **crew** < W *creu* 'a hut, a hovel, a pen, a sty'. In Chesh dial. *kidcrew* was a name for a calf-crib. Horovitz.

KIDSTONES NYorks SD 9581. 'Kid stones'. *Kydestanes* 1270, *Kidderstanes* 1301, *Kydestancauce* 'Kidstones causeway' 1331, *Kidston(s)* 1613, 1680. OE ? ME **kide** 'kid, a young goat' + OE **stān**. The reference is probably to the rocks of Kidstones Scar where wild goats were to be seen. The 1331 form refers to the road called *Causeway* 1844, now the B 6190, linking Bishopdale and Wharfedale, OFr **caucie**, ME **cauce**. YN 265, YW vi.119.

KIELDER Northum NY 6293. By origin a r.n., probably identical with CALDER, 'the rapid, violent stream'. *Keldre* 1309, 1370, *Keilder* 1325, 1663, *Kailder* 1330, *Keylder* 1542. PrW ***caled-duβr** (Brit **Caletodubro-*). Ekwall reconciles the two forms Kielder and Calder by suggesting early weakening of the first syllable to give OE **Celdér* which with subsequent restressing and lengthening before *ld* would give lOE **Cḗlder*, ME **Kēlder*, ModE *Kielder*. NbDu 127, RN 231, 62, LHEB 563.

KIELDER BURN Northum NY 6596. *Kylder water* 1583, *Kielder Burn* 1868 OS. P.n. KIELDER NY 6293 + ModE **water** and **burn**. RN 231.

KIELDER CASTLE Northum NY 6393. *Kielder Castle* 1868 OS. P.n. KIELDER NY 6293 + ModE **castle**. An 18th cent. castellated shooting box built for the Dukes of Northumberland. Tomlinson 226.

KIELDER FOREST Northum NY 6691. P.n. KIELDER NY 6293 + ModE **forest**. A man-made forest created by the Forestry Commission in the 1920's.

KIELDERHEAD MOOR Northum NT 6800. *Kielderhead Moor* 1868 OS. P.n. Kielderhead + ModE **moor**. Kielderhead is the 'source of the Kielder', p.n. KIELDER NY 6293 + ModE **head**.

KIELDER WATER Northum NY 6788. P.n. KIELDER NY 6293 + ModE **water**. The largest man-made lake in Europe, created by the Northumbrian Water Authority 1975–82.

KILBURN Partly uncertain. The generic is OE **burna**, the specific perhaps OE pers.n. *Cylla*. But the occurrence of several examples of this name in different counties strongly suggests that the specific is an as yet unidentified appellative. L 19.

(1) ~ Derby SK 3845. *Ki- Kyleburn(e)* 1179–1424, *Ki- Kylle-* 13th cent., *Kelle-* 1234, *Ki- Kylburn(e) -bo(u)rn(e)* 1275–1330 etc. Db 475.

(2) ~ NYorks SE 5179. *Chileburne* 1086, *Killebrun(na) -brunne* 12th–1231, *Ki- Kylebrunn(e)* 13th, *-burne* 1293, *Ki- Kylburne* 1249–1399. OE pers.n. **Cylla* + **burna** alternating with ON **brunnr**. YN 195, SSNY 139.

KILBY Leic SP 6295. Uncertain. *Cilebi* 1086, *Kildebi -by* 1195–1209×19, 1514, *Ki- Kyl(l)ebi -by* 1165–1353, *Ky- Kilby* from 1305. Possibly a Scandinavianisation of OE *ċilda-tūn* 'estate of the young men of noble birth', OE **ċild**, genitive pl. **ċilda**, + **tūn** replaced by ON **bȳ** and with substitution of ON [k] for [tʃ]. Lei 444, SSNEM 56.

KILCOT Glos SO 6925. 'Cylla's cottage(s)'. *Chilecot* 1086, *Ki- Kyldecota -e* 1167, 1385, *-coton'* 1278, *Ki- Kyllicote -e- -kote* 1221–1361, *Kyl(l)cot(e)* 1339–1580. OE pers.n. *Cylla* + **cot**, pl. **cotu**, dative pl. **cotum**. Gl iii.176.

KILDALE NYorks NZ 6009. Partly uncertain. *Childale* 1086, *Kildalam, Ki- Kyldale* from 1119. The generic is ON **dalr**. The specific is possibly ON **kíll** 'a narrow bay' in the sense 'narrow valley' or 'wedge-shaped piece of land' as in Danish and Swedish p.ns. YN 166, SSNY 99, L 96.

KILDWICK Lancs SE 0146. 'Dairy-farm of the young men'. *Childeuuic* 1086–1135×54, *-wic* 1120×47, 1135×54, *Ki- Kyldewik(e) -wyk(e) -wick(e)* 1135×40–1587, *Kilderwyk* 1293, *Kyldwyk -i-* 1293–1441. OE **ċild**, genitive pl. **ċild(r)a** with initial [k] due to ON influence, + **wīċ**. YW vi.18.

KILHAM Humbs TA 0664. Perhaps the place at 'the springs'. *Chillun -on* 1086, *-um* 1179–95, *Ki- Kyllum* 1100×8–1404, *Kyllam* 16th cent., *Kilham* from 1678. OE **cyle** 'coolness, cold, chill' used concretely in the sense 'that which is cold, a cold spring', dative pl. ***cylum**. Cf. the *fontem qui vocatur Chill* 'the spring called C' in Durham. The reference could be to the springs which burst out in the village street in wet winters. An alternative explanation is from OE *cyln* 'a kiln', dative pl. *cylnum*, as in KILHAM Northum NT 8832. YE 97.

KILHAM Northum NT 8832. '(The settlement) at the kilns'. *Killum* 1176–1442, *Kylnom* 1323, *Kilholme* 1480, *Kylham* 1542. OE **cylnum**, locative–dative pl. of **cyl(e)n**. NbDu 127.

KILKHAMPTON Corn SS 2511. 'Farm, estate called or at *Kylgh*'. *Chilchetone* 1086, *Kilcton* [c.1175]13th, *Kilkamton* 1194, 1330, *Kilcaneston* 1200 Gover n.d., *Kilcanton* 1202, *Kylchampton* 1238. P.n. **Kylgh*, Co ***kylgh** 'circle', + OE **tūn** later assimilated to the pattern of English p.ns. in *-hampton*. The unidentified estate called *Kelk* [c.839]late 14th given to the bishop of Sherborne by king Egbert may belong here. The reference is presumably to some lost archaeological feature. The parish is called *Kilcham- Kilkamlond* 1303–80 Gover n.d. PNCo 103, Gover n.d. 7, Charters III.81, DEPN.

KILLAMARSH Derby SK 4580. 'Cynewald's marsh'. *Chinewoldemaresc* 1086, *Ki- Kyn(e)wald(es)- mers(s)h(e) -mer-*

sch(e) 1212–1547, *-marshe, Kinoldemers, Kinaudemers* 1204, 1221×30, *Ky- Kinwal(l)mers(c)he* 1282–1475, *-mars(c)h(e)* 1462–1603, *Ky- Kilwalmerch* 1457, *Killomershe* 1577, *Killamarsh* 1676. OE pers.n. *Cynewald* + **mersc**. Db 273, L 53.

KILLERBY Durham NZ 1919. 'Kilvert's village or farm'. *culuerdebi* [1091×2]12th, 1197, *Kyllerby* 1310–1538, *Kylwerby* [1183]c.1320, *Killerby* from 1350, *-bie* 1623. ON pers.n. *Kilvert* < **Ketilfrithr* apparently influenced by OFr *culvert* 'a freedman', + **bȳ**. NbDu 127, Nomina 12.24.

KILLINGHALL NYorks SE 2858. 'Nook of land associated with Cylla, or called or at *Cylling*'. *Chene- Chenihalle, Chilingale, Kilingala* 1086, *Chilingehal'* 1165, *Kil(l)ingehal'* 1205, 1206, *Ki- Kyllinghal(a) -hall(e)* 12th–1675. OE pers.n. **Cylla* + **ing**[4] + **halh**, or p.n. **Cylling* 'place called after Cylla' < pers.n. **Cylla* + **ing**[2], + **halh**. For a similar compound see KILNWICK SE 9949. YW v.100, L 108, 110.

KILLINGHOLME Humbs TA 1416. '*Chelvinge* island'. *Chelvingeholm -hov* 1086, *Chiluingheholm* c.1115, *Kiluingeholm* c.1141–1219, *Kiluingholm* 1143×7–1264, *Killingeholm* 1194–1236, *Killingholm(e)* from 1197. OE **Celfing*, locative–dative **Celfinge*, + ON **holmr**. OE **Celfing* would be an **ing**[2] derivative of OE **celf** 'a calf', meaning perhaps 'calf-place, place where calves are raised or pastured'. *Holmr* is used here in the sense 'island of dry ground in a low-lying marshy area'. Alternatively the specific might be OE folk-n. **Ċeolfingas*, 'the people called after or Ceolfa', < pers.n. *Ċeolfa*, a known short form of *Ċēolwulf*, + **ingas**, genitive pl. **Ceolfinga*, + **hām** later replaced by ON *haugr* and *holmr* and with Scandinavian [k] for [tʃ]. Li ii.193, SSNEM 218.

KILLINGTON Cumbr SD 6188. 'Farm, village, estate called after Cylla'. *Killinton(a)* 1175–97, *Ki- Kyllington* 1193–1743. OE pers.n. **Cylla* + **ing**[4] + **tūn**. We i.39.

KILLINGTON RESERVOIR Cumbr SD 5991. P.n. KILLINGTON Cumbr SD 6188 + ModE **reservoir**.

KILMAR TOR Corn SX 2574. Unexplained.

KILMERSDON Somer ST 6952. 'Cynemær's hill'. *Ku- Kynemersdon* [951]14th S 555, *Chenemeresdone* 1086, *Kinemeresdon* 1176, *Kyn(e)mer(e)s- Kynneres- Kyngmersdon* [12th] Buck, *Kenemeresdon* [c.1350] Buck, *Kenersdon* [1413] Buck, *Kilmarston* 1610. OE pers.n. *Cynemǣr*, genitive sing. *Cynemǣres*, + **dūn**. DEPN.

KILMESTON Hants SU 5962. 'Kenelm's estate'. *(to) cenelmestune, (æt) kenelmestune* [961]12th S 693, *Cylmestuna* [984×1000]12th S 1820, *Chelmestvne* 1086, *Kelmeston* 1257–1333, *Kylmyston alias Kymston* 1597, *Kimpston* 1586. OE pers.n. *Cēnhelm*, genitive sing. *Cēnhelmes*, + **tūn**. Ha 103 gives pr [kimpstən], Gover 1958.71 [kimstən].

KILMINGTON Devon SY 2798. 'Farm, village, estate called after Cen- or Cynehelm'. *Chenemetone* 1086, *Chienemetona* Exon, *Culminton* 1194, *Kelmeton(e)* 1219–38, *Kelminton -myn-* 1271, 1281, *Kilminton* 1296, *Kulmyngton* 1346, 1359. OE pers.n. *Cēnhelm* or *Cynehelm* + **ing**[4] + **tūn**. D 640.

KILMINGTON Wilts ST 7736. 'The estate called after Cynehelm'. *Chelme- Cilemetone* 1086, *Chilmaton* 1086 Exon, *Culminton -yn(g)-* 1251–1544, *Kylmyngton* 1403. OE pers.n. *Cynehelm* + **ing**[4] + **tūn**. Wlt 174.

KILNDOWN Kent TQ 7035. Probably 'kiln hill'. *Kelwedoune* (? for *Kelne-*) 1391. ME **kiln**, Kentish **keln** (OE *cyln*) + **doun** (OE *dūn*). PNK 310.

KILNHURST SYorks SK 4597. 'Wooded hill with a kiln'. *Kil- Kyln(e)hirst(e) -hyrst(e)* 12th–1558, *-hurst* 1420. OE **cyln** + **hyrst**. YW i.115, L 198.

KILN PIT HILL Northum NZ 0355. No early forms.

KILNSEA Humbs TA 4015. 'The kiln lake'. *Chilnesse* 1086, *Ki- Kylnese* 1228–1359, *Kilnesey* 1273–1519, *Ki- Kylnse* 14th cent., *Kelsay* 1725. OE **cyln** + **sǣ**. YE 15 gives pr [kilsi].

KILNSEY NYorks SD 9767. Uncertain. *Chileseie* 1086, *Ki- Kylnese(y)ia -ei -ay -ey* 1146–1641, *Kynsaye, Kylsaye in Craven* 16th. Possibly OE **cyln** + ***sǣge**, 'marsh near the kiln'. YW vi.86.

KILNWICK Humbs SE 9949. 'Cylla's dairy farm', or 'dairy farm called or at *Cylling*'. *Chileuuit -d* 1086, *Killewic* 1200, 1201, *Killingwic -wik(e) -y-* 1163–1414, *Ki- Kylnewic -wyk* 13th cent. OE pers.n. **Cylla* + **wīċ**, alternating with OE p.n. **Cylling* 'place called after Cylla' < pers.n. **Cylla* + **ing**[2], locative–dative **Cyllinge*, + **wīċ**. YE 160 gives pr [kilik], BzN 1968.174.

KILNWICK PERCY Humbs SE 8250. 'K held by the Percy family'. *Kylnewyke Parcy* 1519, *Kyl(l)wike Perseye* 1539, 1545. P.n. *Chelingewic, Chileuuic, Chilleuuinc* 1086, *Kyllyngwyc(h) -wyk(e) -i-* 12th–1385, *Killingewic -wik* 13th cent., as in KILNWICK SE 9949, + manorial addition *Percy*. YE 179 gives pr [kilik piəsi], BzN 1968.174.

KILPECK H&W SO 4430. Partly unexplained. *Chipeete* 1086, *Cilpedec* c.1150, *Kilpeec -pedet -pech* 1167–93. OW **cil** 'corner, back, nape, nook' + unexplained element. The reference may be to one of the small valleys opening from Worm Brook or to the ridge on which Kilpeck stands. The second element may be compared with the lost Monmouth p.n. *(nant) pedecou* c.1150, the pl. of *pedec*. It must be W. The church here is dedicated to St David, *Lann Degui Cilpedec* c.1150. The DB tenant was Cadiand or Catgendu, an OW pers.n. DEPN, DB Herefordshire 1.53, O'Donnell 90, Bannister 105.

KILPIN Humbs SE 7727. Probably 'the calf-enclosure'. *Celpene* [959]c.1200 S 681, *Chelpin* 1086, *Kilpin* from 1199. OE **celf** + **penn**. YE 252.

KILSBY Northants SP 5671. 'The child's village or farm'. *Kildesbig* [1043]17th S 1000, *Chidesbi* (sic) 1086, *Kylesbia* c.1156, *Chilesbei* c.1225, *Ki- Kyldesby* 1223–1459. OE **ċild**, genitive sing. **ċildes** with [k] for [tʃ] due to Scandinavian influence, + **bȳ**. The sense of *cild* here is unknown. Nth 24, SSNEM 57.

KILTON Somer ST 1643. 'Settlement of or at Kilve'. *Cylfantune* 873×88 S 1507, *Chilvetvne* 1086, *Kilveton* [1295] Buck, *Kilton* 1610. P.n. KILVE ST 1442, OE **Cylfe*, genitive sing. *Cylfan*, + **tūn**. DEPN.

KILVE Somer ST 1442. Uncertain. *Clive* 1086, *Kelua* 1186, *Kylve* 1243, *Culue* 1329, *Kylue* 1610. This appears to be an OE ***cylfe** which might be cognate with ON *kylfa* 'a club, the prow of a ship' used as a hill name. DEPN, PNE i.123.

KILVINGTON Notts SK 8043. 'The estate called after Cylfa'. *Cheluintun, Chelvintone, Chelvinctune* 1086, *Ki- Kylvinton -tun* c.1180–1258, *Kilvington(a)* from c.1190 with variants *Ky-* and *-yng-*. OE pers.n. **Cylfa* + **ing**[4] + **tūn**. Nt 215, SN 4.120–40.

North KILVINGTON NYorks SE 4285. ME prefix *North* from 1292 + p.n. *Chelvin(c)tun(e)* 1086, *Keluintune* 1088, *Ki- Kyluinton* 1185×95, 1257, *Ki- Kylvington* from 13th, 'estate called after Cylfa', OE pers.n. **Cylfa* + **ing**[4] + **tūn**. YN 200, 205.

South KILVINGTON NYorks SE 4284. ME prefix *Suth* from 1216×72 + p.n. Kilvington as in North KILVINGTON SE 4285. YN 200, 205.

North KILWORTH Leic SP 6183. Prefix *Norþ-* c.1265, *North(e)-* from 1288 + p.n. *Chiveleswordé, Cleveliorde* 1086, *Ky- Kiuelingeworth' -v- -w(u)rth(e)* 1207–98, *Ky- Kiuelingworth(e) -v- -yng-* 1206–1386, *Kiuelewurd -wurða -worth'* 1177–1208, *Kiueleswurð* 1208, *Ky- Kil(l)ingworth(e) -yng- ´*c.1291–1585, *Ky- Kil(l)eworth'* 1203–1406, *Ky- Kil(l)worth(e)* 1412–1610, 'Cyfel's enclosure', OE pers.n. **Cyfel*, genitive sing. **Cyfeles*, + *worth*, varying with 'enclosure at or called Cyfeling', OE **Cyfeling* 'place called after Cyfel', dative sing. **Cyfelinge* or 'enclosure of Cyfel's people', OE **Cyfelingas*, genitive pl. **Cyfelinga*. 'North' for distinction from South KILWORTH SP 6081. Lei 446.

South KILWORTH Leic SP 6081. Prefix *Alia* 'the other' 1209×35, *Suth-* 1237–1370, *South(e)* 1309–1610, + p.n. Kilworth as in North KILWORTH SP 6183. Lei 448.

KIMBERLEY HOUSE Norf TG 0904. A new house built in 1712. P.n. KIMBERLEY TG 0704 + ModE **house**. Pevsner 1962NW 220.

KIMBERLEY Norf TG 0704. 'Cyneburg's wood or clearing'. *Chineburlai* 1086, 1161, *Cheneburlai* 1162, *Kineburle* 1254. OE feminine pers.n. *Cyneburh* + **lēah**. DEPN.

KIMBERLEY Notts SK 5044. 'Cynemær's wood or clearing'. *Chinemarel(e)ie* 1086, *Kinemarle* c.1200, *Kynmarley(e) -mer-* 1299–1404, *Kymmerleye* 1383, *Kemberley* 1405, *Kymberley* 1580, 1640. OE pers.n. *Cynemǣr* + **lēah**. Nt 148.

KIMBLE Bucks SP 8206. 'Royal Bell, the bell-shaped hill'. *Chenebelle (Parva)* 1086, *Ky- Kinebelle* 12th–13th cents., *Kymbell* 1369. The boundary of the people of Kimble is *Cynebellinga gemære* [903]10th S 367. OE **cyne** + **belle** probably used as a p.n. The reference is to the prominent Pulpit Hill 813ft., three-quarters of a mile SE of Great Kimble, crowned with Bulpit Wood Hill-fort SP 8305. 'Royal' refers to Great Kimble for distinction from Little Kimble. Bk 163.

KIMBLESWORTH Durham NZ 2547. 'Cymel's enclosure'. *Kimelwrthe* 1242×3, *Ki- Kymelesw(o)rth(e)* 1242×3–1479, *Kymblesworth* [1214×33]1370–1530, *Ki-* from 1611. OE pers.n. **Cymel*, genitive sing. **Cym(e)les*, + **worth**. NbDu 127.

KIMBOLTON Cambs TL 0967. 'Cynebald's estate'. *Chenebaltone* 1086, *Ki- Kynebalton* 1232–93, *-bauton* 1284–1428, *Kymbalton* 1329–1585, *Kimbolton* from 1330. OE pers.n. *Cynebald* + **tūn**. Hu 243 gives pr [kimәltәn].

KIMBOLTON H&W SO 5261. 'Cynebald's farm or estate'. *Kinebalt'* 1186×99, *Kimbalton* 1216×72, *-bolton* from 1200×50. OE pers.n. *Cynebald* + **tūn**. He 111.

KIMCOTE Leic SP 5586. 'Cynemund's cottage(s)'. *Chenemundescote* 1086, *Ky- Kinemunde(s)cot(e)* 1167–1379, *Ky- Kinemundcote* 1216×72, 1384, *Kynnemouncote* 1405, *Ky- Kilmundecot(e)* 1209×35–1470, *Ky- Kilmundecot(e)* 1209×35–1470, *Kylmyncote* 1424–1524, *Ky- Kilmecote* 1453–1610, *Ky- Kimcote* 1507–1720. OE pers.n. *Cynemund*, genitive sing. *Cynemundes*, + **cot** (pl. **cotu**). Lei 449.

KIMMERIDGE Dorset SY 9179. Uncertain. *Cameric, Cuneliz* (sic) 1086, *Kimerich* 1212–[1344]14th, *Kymeryg(g)e* 1489, 1553, *Kimbridge* 1575, *Kimmeridge* 1795. Probably OE **cȳme** 'comely, lovely, glorious' + ***riċ**? 'narrow strip of land, stream or ditch, narrow road'. The precise senses of *riċ* are unknown. It is related to the word 'reach' and words in cognate languages meaning 'hair-parting, stripe, furrow'. A possible reference here is to the straight ridge of high ground running a mile SE to Swyre Head for which the commoner term 'ridge' was later substituted. Do i.84.

KIMMERSTON Northum NT 9535. 'Cynemær's farm or village'. *Kynemerston', Lynemerstoc* (sic) 1242 BF, *Kynemereston* 1244, *Kymmerston* 1296 SR, *Kin- Kylmerston* 1346. OE pers.n. *Cynemǣr*, genitive sing. *Cynemǣres*, + **tūn**. NbDu 127.

KIMPTON Hants SU 2846. 'Cyma's estate'. *Chementune* 1086, *Keminton* 1167, *Cyminton* 1256. OE pers.n. **Cyma*, genitive sing. **Cyman*, + **tūn**. Ha 103.

KIMPTON Herts TL 1718. 'Estate called after Cyma'. *Kamintone* 1086, *Camentone* 1235, *Cumi(n)tone -y- -en-* 1199–1281, *Ki- Kymiton(e)* 1211–1303, *Kyminton(e) -yng-* 1271, 1291, *Kemi(n)ton(e) -y(n)-* 1216×72–1303, *Kympton* 1369, 1428. OE pers.n. *Cyma* + **ing**[4] + **tūn**. Hrt 15.

KINDER RESERVOIR Derby SK 0588. R.n. Kinder + ModE **reservoir**. Kinder, *Kinderwodebroc* c.1250, *Kinder broocke* 1640, is hill-name Kinder as in KINDER SCOUT SK 0988 + ME **wode, broke**. Db 10.

KINDER SCOUT Derby SK 0988. 'Kinder cliff'. *Kinder Scout* 1767. P.n. Kinder + ON **skúti**. Kinder, *Chendre* 1086, *Kender* 1275, *Kunder* 1299, *Kinder, Kynder -yr* from 1285, is of unknown origin. It is the name of the chief summit of the Peak District, 2088ft. high, with precipitous sides cloven by rocky chasms. The name is pre-English, possibly pre-Celtic. Db 116, 114.

KINETON Glos SP 0926. 'The royal manor'. *Ki- Kynton(e) -toniam -tune* 12th–1681, *Ki- Kyneton* 1287, 1292 and from 1653, *Kyngton(e)* 1327–15th. OE **cyne-tūn**. It was held of the king by his thane Brictric at the time of Edward the Confessor. Gl ii.14.

KINETON Warw SP 3351. 'The royal manor'. *Cyngtun* [968]18th S 773, *Quintone* 1086, *Kinton(a)* 12th–1316, *Kington(e)* 1211–1535, c.1830, *Kyneton* 1285, *Kineton* from 1656. lOE **cing** (OE *cyning*) varying with **cyne** 'royal' + **tūn**. The manor was held by the king in 1066. Wa 282.

KINGHAM Oxon SP 2624. 'The homestead of the Cægingas, the people called after Cæga'. *Caningeham* 1086, *Canyngesham* 1285, *Keingaham* 11th, *Kei- Keyingham* 1235–1397, *Kyngham* 1285–1428. OE folk–n. **Cǣġingas* < pers.n. **Cǣġ(a)*, genitive pl. **Cǣġinga*, + **hām**. O 360, ING 127.

KING'S ACRE H&W SO 4741. *Kingsacre* 1831 OS. According to Bannister 105 this name goes back to c.1281 (Richard Johnson, *Customs of Hereford*).

KINGS CROSS STATION GLond TQ 3083. Named in 1852 after the district known as *Kings Cross* from a statue of George IV which was erected in 1836 and taken down in 1845. It stood at the cross-roads at the junction of Gray's Inn Road, Euston Road and Pentonville Road. The original name of the district, Battlebridge (mentioned in *Oliver Twist* 1837–8), survives in Battle Bridge Road. It was *Bradeford* 1207, 'the broad ford' sc. over the Holborn, OE **brād**, definite form **brāda**, + **ford**, later *Bradefordebrigge* c.1387, which was eventually corrupted to *Battle Bridge al. Batford Bridge* 1622. Mx 140, GL 64, Encyc 447–8.

KINGSAND Corn SX 4350. 'King's beach'. *Kings sand* 1602. Surname *King* + ModE **sand**. PNCo 104, Gover n.d. 233.

KINGSBRIDGE Devon SX 7344. 'The king's bridge'. *(to) cinges bricge* 962, *Kingesbrig'* 1230, 1244. lOE **cing**, genitive sing. **cinges**, + **brycg**. An important crossing of the upper end of Kingsbridge Estuary under royal protection like the king's street or highway. D 305.

KINGSBRIDGE Somer SS 9837. No early forms.

KINGSBURY GLond TQ 1989. 'The king's manor'. *(to) Kyngesbyrig* [1044×51]12th S 1121, *Chingesberie* 1086, *Kingesbir' -bury* 1199–1316. lOE **cing**, genitive sing. **cinges**, + **byriġ**, dative sing. of **burh**. This appellation commemorates the grant of the manor to Westminster Abbey by Edward the Confessor. Its earlier name was *(æt) Tuneweorðe* [957]12th S 645, *Toneworth* 1399×1413, 'Tuna's farm', OE pers.n. *Tuna* + **worth**. Mx 61–3.

KINGSBURY Warw SP 2196. 'Cyne's fortified place'. *Chinesberie* 1086, *Ki- Kyn(n)esburi -beri(a) -bir(e)* 12th–1549, *Kingesbery* 1213, *Kins- Kynnes- Kyngesbury* 17th cent., *Kingsbury* from 1642. OE pers.n. *Cyne*, genitive sing. *Cynes*, + **byriġ**, dative sing. of **burh**. The modern form is a false etymology. Wa 16.

KINGSBURY EPISCOPI Somer ST 4321. 'K held by the bishop' sc. of Bath. Earlier simply *Cyncgesbyrig* *[1065]18th S 1042, *Chingesberie* 1086, *Kingsbury* 1610, 'the king's fortified place or manor', lOE **cing**, genitive sing. **cinges**, + **byriġ**, dative sing. of **burh**. DEPN.

KINGSCLERE Hants SU 5258. 'The king's (parcel of) Clere'. *Kyngesclere* 1100×35. lOE **cing**, genitive sing. **cinges**, + p.n *Clere* 1086, 1154×8, as in BURGHCLERE SU 4761. Kingsclere was a royal demesne manor before and after 1066. Ha 56, Gover 1958.146.

KINGSCOTE Glos ST 8190. 'The king's cottage(s)'. *Chingescote* 1086, *Kyngescota -cot(e) -cott(e)* 1191–1584, *Kingscote* from 1398. OE **cyning**, **cing**, genitive sing. **cinges**, + **cot** pl. **cotu**. Gl ii.237.

KINGSCOTT Devon SS 5318. 'Cyni's cottage(s)'. *Kynscot* 1281, *Kynescote* 1330, *Kynnescote* 1501. OE pers.n. *Cyni*, genitive sing. *Cynes*, + **cot**, pl. **cotu**. D 120.

KING'S DELPH Cambs TL 2595. 'The king's channel'. *(to) Cynges dælf* c.1150 ASC(E) under year 963, *Kyngesdelf* [1020×23]12th S 1463, *(in) Kinges delfe* *[c.1050]14th S 1110, *Ki- Kyngesdelf* c.1223–1390. lOE **cing**, genitive sing. **cinges**, + **(ġe)delf**. Now the name of a marsh, originally of an artificial watercourse, probably Roman, reputedly made by king Canute whence the reference *Cnoutes delf Kynges* 'king Canute's channel' [1052]14th. Also known as Cnut's Dyke and Oakley Dike. Ca 207, 208, 260.

KINGSDON Somer ST 5126. 'The king's hill'. *Kingesdon* 1194, 1201, *Kingsdowne* 1610. lOE **cing**, genitive sing. **cinges**, + **dūn**. DEPN.

KINGSDOWN Kent TR 3743. 'The king's hill'. *Kingesd.* [1176×7]n.d, *Kyngesdoune* 14th cent. ME **king**, genitive sing. **kinges**, + **doun**. PNK 577.

West KINGSDOWN Kent TQ 5762. ModE **west** + p.n. *C- Kingesdoñ* 1199, *-dun(a) -don(e)* 1154×89–1226 etc., 'the king's hill', lOE **cing**, genitive sing. **cinges**, + **dūn**. PNK 44.

KINGSEY Bucks SP 7406. 'The king's island'. *Kingesie* 1197 etc. with variants *Ky-* and *-eya -(h)eye*, *Kingsey* from 1284×6. lOE **cyng**, genitive sing. **cynges**, + **īeġ**. Also known earlier as 'great island', *Eia Magna* 1196, *mangna Eya* 1241, in contrast to Litney 'little island', *Lytteleye* 1361, and simply *Eye* c.1200. O 111–2.

KINGSFOLD WSusx TQ 1636. 'The king's fold'. *Kyngesfold* 1296, *Kingesfeld* 1482. ME **king** (OE *cyning*), genitive sing. **kinges**, + **fold** (OE *falod*). Sx 239.

KINGSFORD H&W SO 8281. 'The ford of the Ceningas'. *cenunga ford* [964]17th S 726, *Keningeford* [1189×96]c.1250–1306, *Keingford* 1240, *Kynges- Kingsford* from 1346. OE folk-n. **Cēningas* 'the people named after Cen or Cena', OE pers.n. *Cēn(a)* + **ingas**, genitive pl. **Cēninga*, + **ford**. Wo 259.

KINGSFORTH Humbs TA 0319. *Kingsforth* from 1658. This might be the 'king's ford' but the evidence is too late to offer a secure explanation. L ii.34.

KINGSHALL STREET Suff TL 9161. 'K hamlet'. P.n. Kingshall + Mod dial. **street**. Cf. *King'shall Green, King's hall Farm, Nether Street* all 1837 OS.

KING'S HEATH WMids SP 0781. *Kyngesheth* 1511, *King's Heath* 1650. ME **king**, genitive sing. **kinges**, + **heth**. Part of the manor of King's NORTON, a royal manor in 1086. Wo 354.

Great KINGSHILL Bucks SU 8898. *Great Kinghill* 1822 OS. ModE adj. **great** + p.n. *Kingeshulle* 1196, *Kingsell* 1675, 'the king's hill', lOE **cyng**, genitive sing. **cynges**, + **hyll**. The manor was held of the crown after it was forfeited by Odo of Bayeux. 'Great' for distinction from nearby Little KINGSHILL SU 8998. Bk 184. Jnl 2.33 gives pr [kiŋzul].

Little KINGSHILL Bucks SU 8998. *Lit Kingshill* 1822 OS. ModE adj. **little** + p.n. Kingshill as in Great KINGSHILL SU 8898.

KINGSKERSWELL Devon SX 8868. 'Kerswell held by the king'. *Kyngescharsewell -karswelle* 1270, *Kings Keswell* 1675, *Kingscraswell* 1736. Earlier simply *Carsewelle* 1086, *-willa* 1086 Exon, *Carsuill'* 1194, *Karswell* 1212. OE **cærse** + **wylle** 'spring growing with cress'. 'King's' for distinction from ABBOTSKERSWELL SX 8569. The manor was held by the king in 1086. D 514.

KINGSLAND H&W SO 4461. 'The king's Leen estate'. *Kingeslen(a)* 1137×9, 1195, *Kingeslone -lane* early 13th–c.1433 with variants *Kynges-* and *-leone*, *Kyngeland* 1539. lOE **cing**, genitive sing. **cinges**, + district name *Lene* 1086, *Lenes* 1278, as in The LEEN SO 3859. 'King's' referring to tenure by the crown in the time of Edward the Confessor and for distinction from EARDISLAND SO 4158 'the earl's Leen estate' and MONKLAND SO 4557. 'Leen' has been assimilated to ModE **land** reflecting these different holdings. He 113, Bannister 106.

KINGSLEY Ches SJ 5574. 'The king's wood or clearing'. *Chingeslie* 1086, *-le(y)e* 1154–81, *Kyngesley(a) -leg(h)* etc. 1157–1613, *Kingsley* from 1192. lOE **cing**, genitive sing. **cinges**, + **lēah**. Che iii.239.

KINGSLEY Hants SU 7838. 'The king's wood or clearing'. *Kyngesly* c.1210, 1233, *Kyngesle -lye* 1256–1352. lOE **cing**, genitive sing. **cinges**, + **lēah**. Ha 104, Gover 1958.105.

KINGSLEY Staffs SK 0147. 'The king's wood or clearing'. *Chingeslei(a)* 1086, *Kingeslegh* 1227. lOE **cing**, genitive sing. **cinges**, + **lēah**. DEPN, L 206.

KINGSLEY GREEN WSusx SU 8930. Cf. *Kingsley Marsh* 1813 OS.

KING'S LYNN Norf TF 6119 → King's LYNN.

KINGSNORTH Kent TQ 8072. No early forms. Identical with *North Street* 1805 OS. In the same parish occur Broad Street TQ 7671, *Brodestrete* 1478, and *Hoo Street* 1805 and in High Halstow *Fen Street* and *Hog Street* 1805, all with Mod dial. **street** 'a village, a hamlet'. Cf. BROAD STREET TQ 8356. PNK 120.

KINGSNORTH Kent TR 0039. 'The king's detached piece of land or woodland'. *Kingesnade* 1226, *Ki- Kynge(s)snade -snode* 13th cent. lOE **cing**, genitive sing. **cinges**, + **snād**. PNK 416.

KING'S STAG Dorset ST 7210. 'The king's boundary stake'. *Kyngeslake* (sic for *-stake*) c.1325, *Kingestake* 1337. ME **king**, genitive sing. **kinges**, + **stake** (OE *staca*). The boundaries of Lydlinch, Pulham and Hazelbury Bryan meet at Kingstag Bridge here. The boundary stake is referred to as *truncum qui stat in tribus divisis* 'the trunk which stands at three boundaries' [1216×72]14th. Do iii.349.

KINGSTANDING WMids SP 0794. *Kings Standing* 1840 OS. A royal hunting station. ModE **king**, genitive sing. **king's**, + **standing** 'a hunter's station from which to shoot game'. Wa 200.

KINGSTEIGNTON Devon SX 8773. 'Teignton held by the king'. *Teintone Regis* 1259, *Kingestentone, Kings Teyntune* 1274, *Kingstenton* 1628, *Kings Staynton* 1675. Earlier simply *Tegntun* c.1050 ASC(A) under year 1001, *Teintone* 1086, *Teynton* 1224×44, *Teignton* 1224×44, 'settlement on the river Teign', r.n. TEIGN SX 7689 + **tūn**. The manor was held by the king in 1086. D 478.

KINGSTHORNE H&W SO 4932. No early forms.

KINGSTHORPE Northants SP 7653. 'The king's Thorpe'. *Kingestorp* 1190, *Kyngesthorp* 1305–1442. ME **king**, genitive sing. **kinges**, + p.n. *Torp* 1086–1175, *Trop* 1195, *Thorp* 12th, 1232, *Throp* 1217, 'the outlying settlement', ON **thorp**. It was already a royal manor at the time of DB. Nth 133, SSNEM 119 no.26.

KINGSTON 'Royal estate'. OE **cyninges-tūn**. PNE i.124, Carole Hough, 'The place-name Kingston and the Laws of Æthelbert', *Studia Neophilologica* 69, 1997, 55–7.

(1) ~ Cambs TL 3455. *Chingestone* 1086, *Ky- Kingestona -tuna -ton(e)* 1086–1443, *Chinston* c.1265. A royal manor in the time of king Edward 1042×66. Ca 163.

(2) ~ Devon SX 6347. *Kingeston(e)* 1242–1394 with variant *Kyng-*. D 279.

(3) ~ Devon SX 9051 → KINGSWEAR SX 8851.

(4) ~ Dorset ST 7509. *Kingeston* 1580, *Kinston* 1605, *Kinson* 1774, *Kingston* 1811. The manor of Hazelbury Bryan in which Kingston lies was formerly held of the king. Do ii.101.

(5) ~ Dorset 9579. *Chingestone* 1086, *Ki- Kyngeston* 1212–1575, *Kingston* c.1586. Kingston belonged to Shaftesbury abbey; the name records its gift by king Eadred in 948. Do i.15.

(6) ~ Hants SU 1402. *Kyngeston* 1280. In Ringwood which was a royal demesne in 1086. Ha 105, Gover 1958.220.

(7) ~ IoW SZ 4781. *Chingestune* 1086, *Ki- Kyngeston(e)* 1250–1583, *Kingston* from 1375. Wt 163, Mills 1996.64.

(8) ~ Kent TR 1951. *Kingestun* c.1090, *-tun'* c.1227×40(p), *Kingeston'* 1172×3–1242×3 with variant *Kynges-*. PNK 315, Cullen.

(9) ~ BAGPUIZE Oxon SU 4098. 'K held by the Bagpuize family'. *Kingeston' Bagepuze* 1284. P.n. *(æt) Cyngestun* [956 for 975×8]13th S 828, *Chingestune* 1086, + manorial affix from the family n. of Ralph de Bagpuize who held it in 1086. The bounds of K are mentioned as *cing hæma gemære* [958]13th S 654 and as *kingtuninge gemære* [958 for 959]12th S 673. Brk 412.

(10) ~ BLOUNT Oxon SU 7399. 'K held by the Blount family'.

Kyngestone Blont 1379. P.n. *Chingestone* 1086 + manorial affix from the family of Hugo le Blund who occurs here in 1274×9. O 103.

(11) ~ BY SEA WSusx TQ 2305. *Kingston by Sea* 1730. P.n. *Chingestune* 1086, *Kingeston* 1199, + ModE **by sea**. Also known as *Kingeston Bouci* 1315, *Kinson Bousye* 1641, 'K held by the Boucy family' from Robert de Busci who held the manor in 1199. His family is said to have come from Boucé in Normandy. Sx 245, SS 75.

(12) ~ DEVERILL Wilts ST 8437. 'That part of D held by the king, the king's D estate'. *Deverel Kyneston* 1249, *Kingiston Deverel* 1260, earlier *Devrel* 1086, *Kingesdeverell* 1206. See r.n. DEVERILL. Also known simply as *Kyngeston(e)* 1291, 1428, *Kingson* 1675. Wlt 173.

(13) ~ LISLE Oxon SU 3287. 'K held by the L'Isle family'. *Kyngeston Isle* 1372, ~ *Lisle* 1412. P.n. *Kingeston'* 1220, + manorial affix from the family of Gerardus de Insula de Kingeston who occurs here 1275×6. Brk 372.

(14) ~ NEAR LEWES ESusx TQ 3908. *Kyngeston juxta Lewes* 1340. Earlier simply *Kyngestona* 1087×1100. 'Near Lewes' for distinction from KINGSTON BY SEA TQ 2305. Sx 310, SS 75.

(15) ~ ST MARY Somer ST 2229. P.n. Kingston + saint's name *Mary* from the dedication of the church. Earlier simply *Kyngestona* [1155×8]1334, *Kyngestone* 1327, *Kingston* 1610. DEPN.

(16) ~ SEYMOUR Avon ST 4067. 'K held by the St Maur family'. *Kingeston Milonis de Sancto Mauro* 'Milo of St Maur's K' 1196, *Kyngeston Saymor* 1327. Earlier simply *Chingestone* 1086. There are various places in France called St Maur in the departments of Manche, Oise and Seine. DEPN.

(17) ~ UPON HULL Humbs TA 0928. The town was known at various times as Wick, Hull and Kingston.

I. *Wyk'* 1160×80–1279, *(la, le) Wyke (super, juxta Hull')* 'Wick on, beside Hull', 13th cent., *(le) S(o)uthwyk'* 14th cent. OE **wīc** 'dairy-farm, trading place' or ON *vík* 'a creek, an inlet'.

II. *(portum, villa de) Hul'* 'port, vill of H' 1228, *(portum, villa de) Hull* from 1261. Transferred from r.n. HULL.

III. *Kyngeston -i- (super, on, upon Hul(l -e))* 1275–1548, *Kingston super Hull(e)* 1306 etc. In 1292 Edward I exchanged certain lands here with the monks of Meaux in order to secure the port of *Wike* after which the place became known as 'the king's town', ME **king**, genitive sing. **kinges**, + **toun**. YE 209 gives pr [ul].

(18) ~ UPON THAMES GLond TQ 1869. *Kyngeston super Tamisiam* 1321, *Kingestoun upon Thames* 1589. Earlier simply *Cyninges tun* 838 S 1438, *Cingestun* [838]12th S 281, 11th ASC(C, D) under year 924, *Chingestune* 1086, *Kingeston* 1164–1506. Kingston was a royal residence in Saxon times: at least two kings were crowned there, Athelstan in 925 and Ethelred Unræd in 978/9. Sr 59, TC 209.

(19) Collingbourne ~ Wilts SU 2355 → COLINGBOURNE KINGSTON.

(20) Winterborne ~ Dorset SY 8697 → WINTERBORNE KINGSTON.

KINGSTON ON SOAR Notts SK 5027. P.n. *Cynestan* 1082, *Chinestan* 1086, *Ky- Kinestan* 1158, 1189×99, *Ky- Kineston* 1198–1343, *Kyngeston* 1433, *Kingston* 1514, 'the royal stone', OE **cyne** + **stān**, + r.n. SOAR, a modern addition for distinction from other KINGSTONS. Nt 252.

KINGSTONE H&W SO 4235. 'Royal estate'. *Chingestone* 1086, *Kingestuna -e -tone* 1148×55–c.1218. 'King's' referring to tenure by the crown at the time of Edward the Confessor and for distinction from the episcopal estate at Bishop EATON SO 4439. lOE **cing**, genitive sing. **cinges**, + **tūn**. He 113.

KINGSTONE Somer ST 3713. 'The king's stone'. *Stane*[†], *Chingestone* 1086, *Chingestana* 1086 Exon, *Kingestan* 1194, 1212, *Kingston* 1610. lOE **cing**, genitive sing. **cinges**, + **stān**. DEPN.

KINGSTONE Staffs SK 0629. 'The king's estate'. *Kingeston* 1166–1275, *Kingston* 1663. lOE **cing**, genitive sing. **cinges**, + **tūn**. But cf. SN 1997.55–7. DEPN, Horovitz.

KINGSTOWN Cumbr NY 3959. A modern development, possibly named from nearby Kingmoor NY 3858, *Kinge moore, the Kyngmoore* 1589–90, *the king(e)s moor(e)* 1619–58, *a great Waste called Kingmoor* 1764, a hamlet onetime vested in the crown. Cu 94.

KING STREET Cambs TF 1108. *King Street* 1824 OS. The Roman road from Ailsworth to Ancaster via Bourne, Margary no. 26.

KING STREET Ches SJ 6969. *Kingstreete* 1609. Earlier *regalis via super Ruddehet* 'royal road across Rudheath', part of the royal demesne of DRAKELOW SJ 7070. The reference is to the Roman road from Northwich to Middlewich, Margary no. 70a. Che i.44.

KING STREET Lincs TF 1108. 'The king's highway or Roman road'. *le kingestrete* 1507×9, *lez kingstrete* 1550×2. ME **king**, genitive sing. **kinges**, + **strete** (OE *strǣt*). The second major Roman road running from the S into Lincs, Margary no. 260. Where it crosses Deeping fens also known as *Stonestrete* 1507×9, *Stonystrette* 1530×2. The notion of the 'king's highway' arose in the Anglo-Saxon period; persons or things in any way specially connected with the king were considered as falling under his protection, and deeds done against the king's servant, in his house or on the highway, were subject to severer penalties than those done against ordinary people or in places not under his protection. Perrott 44.

KINGSWEAR Devon SX 8851. 'The king's weir'. *Kingeswere* 1170×96–1493 with variant *Kyng-*. lOE **cing**, genitive sing. **cinges**, + **wer**. Kingswear was originally a part of Brixham parish which contains a Kingston SX 9051, *Kyngeston* 1292–1306, 'the king's estate'. The weir is recorded because of the importance of its revenues from the catching of fish. D 515.

KINGSWINFORD WMids SO 8888. 'Swinford held by the king'. *Kyngesswynford* 1322. ME **king**, genitive sing. **kinges**, + p.n. *Svinesford* 1086, 'the pig's ford', OE **swīn**, genitive sing. **swīnes**, + **ford**. 'King's' for distinction from Old Swinford SO 9083, *Old Swynford* 1291, ME adj. **old** + p.n. *Swinford* [951×5]c.1500 S 579, *Svineforde* 1086, OE **swīn**, genitive pl. **swīna**, + **ford**. DEPN, Wo 309, L 5–6, 71.

KINGSWOOD 'Royal wood'. OE **cyninges** + **wudu**. PNE i.124

(1) ~ Avon ST 6573. *Ki- Kyngeswod(e)* 1231–1512. A royal forest. Gl iii.80.

(2) ~ Bucks SP 6819. No early forms but in a perambulation dated 1298 of Bernwood, the wood referred to in GRENDON UNDERWOOD SP 6720 and WOOTTON UNDERWOOD SP 6815, there is a mention of 'the lord king's wood'. Bk 104.

(3) ~ Glos ST 7492. *Ki- Kyngeswod(a) -w(u)d(a) -w(o)de -wood(e)* 1166–1593, *Kyngswood* 1509×47. OE **cyning**, **cing**, genitive sing. **cinges**, + **wudu**. Gl iii.38.

(4) ~ H&W SO 2954. 'The king's wood'. *Kynges Wode* 1268, *Kingwode* 1335. He 116.

(5) ~ Kent TQ 8450. *le kingeswode* 1468. Associated with *le kingeswode* in 1468 is *le Frithe* 'the wood', a survival from *Cyninges firhðe* [850]13th S 300 'the king's wood. OE **cyning**, genitive sing. **cyninges**, + **fyrhthe**, **wudu**. KPN 191.

(6) ~ Surrey TQ 2456. *Kingeswode -wude* c.1182–1225. Granted to Merton Priory by Henry II. Sr 78, L 229.

(7) ~ Warw SP 1871. *Kyngeswood* 1334, 1488. Wa 289.

KINGTON 'A royal estate'. OE **cyne** + **tūn** replaced by **cyng** + **tūn**.

(1) H&W SO 9955. ~ *Cyngtun* 972 S 786, *Chintvne* 1086, *Ki- Kyn(e)ton(e)* [c.1086]1190–1730, *Kyngton* 1235–1327. The earlier name of this estate was *piduuella* [930]16th S 404, [1002]13th S 901 from Piddle Brook SO 9648; it was granted by king Athelstan in 904 to abbot Cynath and by king Æthelred to

[†]The identification of this form is not certain.

archbishop Ælfric in 1002 as a result of which the descriptive term 'royal manor' became current. Wo 330, Hooke 158–9, 351, Jnl 20.25.

(2) ~ LANGLEY Wilts ST 9277 → Kington LANGLEY ST 9277.

(3) ~ MAGNA Dorset ST 7623. 'Great K'. *Magna Kington'* 1242×3, *Grant Kington* 1272, *Michelekenton'* 1280, *Kyngton-Magna* 1367, *Machil Kyngton* 1412. Latin adj. **magna**, OFr **grand**, ME **michel** + p.n. *Chintone* 1086, *Kinton(')* 1203–52. The smaller of two DB manors held by a king's thegn at the time of king Edward and in 1086. 'Magna' for distinction from Little Kington ST 7723, *Chintone* 1086, *Parva Kynton* 1238, *Parva Ki- Kyngton(e)* 1242–1431, *Little Kington* from 1290. Do iii.41, 71.

(4) ~ ST MICHAEL Wilts ST 9077. 'St Michael's K'. *Kyngton Michel* 1279, *Kyngton Michaelis* 1428, *Michells Kynton* 1503, *Kynton Sci Michaelis al. Miles Keynton* 1672. P.n. *Chintone* 1086, *Chinctuna* 1174×91, *Kinton(e)* 1186–1221, *Kingtun* 1244, *(at) Kingtone* [934]14th S 426, + saint's name *Michael* from the dedication of the church. Wlt 100 gives pr [kaintən].

(5) Little ~ Dorset ST 7723 → KINGTON MAGNA ST 7623.

(6) West ~ Wilts ST 8177. *Weskinton* 1195, *Westkinton* 1196–1281, *Westkington(e) -yng-* 1211–44, *West Cyngton* 1322, *Westkineton* 1242. ME **west** for distinction from KINGTON ST MICHAEL ST 9077. Wlt 80 gives pr [kaintən].

KING WATER Cumbr NY 5466. Uncertain. *King* from 1169, *aqua de Keeng* 1292. Cu 19.

KINGWESTON Somer ST 5231. 'Cyneweard's estate'. *Chinwardestvne* 1086, *Kynewardeston* 1243, *Kyneweston* 1610. OE pers.n. *Cyneweard*, genitive sing. *Cyneweardes*, + **tūn**. DEPN.

KINLET Shrops SO 7230. 'The royal portion'. *Chinlete* 1086, *Kinlet* from 1242 with variants *Ky-* and *Cy-*, *Killet* 1193–5, *Kindlet* 1666, *Kinglett* 1730. OE **cȳne** + **hlēt** contrasting with Shirlett SO 6697, the 'shire portion'. The significance of these names is unknown; in DB Kinlet is said to have belonged before 1066 to Edith, presumably queen Edith, wife of Edward the Confessor. Sa i.166.

KINNERLEY Shrops SJ 3321. 'Cyneheard's wood or clearing'. *Chenardelei* 1086, *Kynardel' -ley(e) -legh* 1230–1427, *Kynardley* 1412, *Kynnerley* 1575–1726, *Ki-* from 1577. OE pers.n. *Cyneheard* + **lēah**. Some forms with genitival *es* are also found, *Ki- Kynardesl(e)' -ley(e) -lee* 1223–1398. Sa i.166.

KINNERSLEY 'Cyneheard's wood or clearing'. OE pers.n. *Cyneheard*, genitive sing. *Cyneheardes*, + **lēah**.

(1) ~ H&W SO 3449. *Chinardeslege* 1123, *Kynard(e)sl(eye)* c.1200–1311†. He 116.

(2) ~ H&W SO 8743. *Ki- Kynardesle(ge)* 1221–75, *Kynerslege* 1314, *Kynnersley* 1650. Wo 228.

KINNERSLEY Shrops SJ 6716. 'Cyneheard's island'.

I. *Chinardeseie* 1086, *Chi- Kinardeseia* 1138–1223, *Kynardeseye* 1255–late 14th, *Kynnardsey* 1697, *Kynnersey* c.1500.

II. *Kinardesheye -heia* c.1175, 1185, *Kynardesheye* 1256, [1256]14th.

III. *Kynardesleye* 1291×2, *Kynesley* 1577, 1672, *Kinnersley* 1687.

OE pers.n. *Cyneheard*, genitive sing. *Cyneheardes*, + **ēġ**. The village is situated at the S edge of a large oval-shaped island in the Weald Moors, an area of lake and marsh until drained by Telford in the late 18th and early 19th cents., made by the 200ft. contour. Types II and III show alternative forms of the name in OE **(ġe)hæġ** 'an enclosure in woodland' and **lēah** 'a woodland clearing'. Sa i.167, Raven 106.

KINNINVIE Durham NZ 0521. A single farmstead. No early forms. This is a name transferred from a Gaelic p.n. area. Cf. Kininvie House Grampian NJ 3144, Gaelic *Cinn Fhionn Mhuighe* '(at the) headland of the white plain'. A-A 27.

KINOULTON Notts SK 6730. 'Cynehild's farm or village'.

I. *Kinildetune* [978×1016]14th S 1493, *Chineltune -tone* 1086, *Ki- Kynalton'* [c.1155] 14th, c.1230–1314.

II. *Cheneldestona* 1148×53, *Kineldestowe* c.1175, *-ald-* 1216, *Kynewaldestouwe* 1235, *Kynoldeston* 1330.

III. *Kynolton* 1302.

OE feminine pers.n. *Cynehild*, genitive sing. *Cynehilde*, + **tūn**. Form II shows confusion with pers.n. *Cyneweald* + OE **stōw**. Nt 237.

KINSHAM H&W SO 3664. 'The king's (estate called) *Hemede*'. *Kingesmede* 1210×12, *Kingeshemede -a* 1216–1272×1307, *Kinsam* 1522. ME **king** (lOE *cing*), genitive sing. **kinges**, + p.n. **Hemede* of uncertain origin. Presteigne, Powys SO 3164, is *Prestehemed* 1137×9, 'the priests' estate called *Hemede*', OE **prēost**, genitive pl. **prēosta**, + p.n. **Hemede* for which DEPN suggests OE **hǣmed** 'a household, a society' although the word is only recorded in the sense 'marriage, sexual intercourse'. A better suggestion, therefore, is OE **hemm-mǣd** 'boundary meadow' referring to the tract of meadowland surrounded by hills between Presteigne and Kinsham and lying across the England–Wales boundary. This is an attractive suggestion though it does not well accord with *Humet* 1086, the DB form for Presteigne. He 117, PNE i.217.

KINSLEY WYorks SE 4114. 'Cyne's wood or clearing'. *Chineslai -lei* 1086, *Ki- Kynneslay -ley* 13th–1556, *Kinsley* from 1535. OE pers.n. *Cyne*, genitive sing. *Cynes*, + **lēah**. YW i.265.

KINTBURY Berks SU 3866. '(At the) fortified place on the river Kennet'. *(æt) Cynetan byrig* 931×9 S 1533, *Cheneteberie* 1086, *Keneteberi* 1188–1284 with variants *-bir(ia) -bur' -byr'*, *Kentebur* [c.1260]c.1280, *Kentbury* 1517, *Kintbury* from 1591. R.n. KENNET SU 5355 + OE **byriġ**, dative sing. of **burh**. Brk 313.

KINVER Staffs SO 8484. Partly uncertain. *Cynibre* 736 S 89, *Cynefaresstan* [964]17th S 726, *Chenevare* 1086, *-fara* 1130, *Ki- Kenefara* 1177, 1183, *Kyngfare* 1596, *Kinfare* 1834. This is probably a Brit n., **Cunobriga*, PrW **Cənobrëȝ*, assimilated by the Anglo-Saxons to OE p.ns. in *cyne-* 'royal'. The generic is PrW ***breȝ** 'a hill', the specific possibly related to the Celtic root **cuno-* 'dog'. DEPN, LHEB 455, 647.

KIPPAX WYorks SE 4130. 'Cippa's ash-tree'. *Chipesch* 1086, *Ki- Kyp(p)eis -ais -eys* c.1090–1270, *Kypask* 14th cent., *Ki- Kypax* 1246–1579, *Kippax* from 1448. OE pers.n. *Ċippa* with ON [k] for /tʃ/ or **Cyppa* + **esc, æsc** later replaced by ON **askr** and metathesised to *ax*. YW iv.90, SSNY 160.

KIRBY 'Village with a church'. ON **kirkju-bȳ**. In many cases this ON descriptive term must have replaced an earlier English name. See also KIRKBY.

(1) ~ BEDON Norf TG 2805. 'K held by the Bidun Family'. *Kirkeby Bydon* 1291. Earlier simply *Kerkebei* 1086. The estate was given to Hadenald de Bidun 1100×1135 whose family came from one of the French Bidons. DEPN.

(2) ~ BELLARS Leic SK 7117. 'K held by the Bellars family'. P.n. *Che- Chirchebi* 1086, *Ki- Kyrkebi -by* 1166–1551, *Ki- Kirkby* 1370–1576, *Kirby* from 1502, + manorial suffix *Bel(l)er(s) -ar-* 1361–1630. Also known as Kirby *iuxta Maltun*, ~ *Meltun* 12th cent., and Kirby *super Werc* 'on the river Wreake' 12th. Hamo Beler held the manor in 1166. Lei 288, SSNEM 57 no.1.

(3) ~ CANE Norf TM 3794. 'K held by the Cane family' from Caen, France. *Kyrkeby Cam* 1282, *Kirkebycaam* 1375. Earlier simply *Kerkeby* 1086, *Kyrkeby* [1087×98]12th. The estate was held by Walter de Cadamo in 1205, and by Maria de Cham in 1242. Early spellings of Caen are *Cadon* 1021×5, *Cadomo* 1032, from Gaulish **catu-magos* 'field of battle'. DEPN, Dauzat-Rostaing 129.

(4) ~ GRINDALYTHE NYorks SE 9067. *Kirkby in Crendalith* 1367. P.n. *Chirchebi* 1086, *Kirkebi -by* 12th–1623, *Ki- Kyrkby* 1187–1601 + p.n. GRINDALYTHE SE 8967–9971 for distinction from other Kirbys. YE 124, DEPN.

(5) ~ HALL Northants SP 9292. P.n. *Chercheberie* 1086,

†The DB from *Curdeslege* probably belongs to a lost p.n. in Brilley, H&W SO 2649.

Chirchebi 1162, *Ki- Kyrkeby* 12th–1324, *Kirby* 1568, + ModE **hall**, a mansion begun in 1570. There is no church here now. Nth 167.

(6) ~ HILL NYorks NZ 1306. *Kirkby on the Hill* 1534. P.n. *Kirkebi -by* c.1160–1379 + ModE **hill**. YN 290.

(7) ~ HILL NYorks SE 3908. Short for *Kirby under hill* 1665. P.n. *Chirchebi* 1086 + ModE **hill**. Also known as *Ki- Kyrkeby in Mora, ~ super Moram* 'on the moor' 1224–30. YN 180.

(8) ~ KNOWLE NYorks SE 4787. Short for *Kirby sub Knol* 'K under hill' 13th. P.n. *Chirchebi* 1086 + *Knowle*, the name of *Knowle Hill*, a high round-topped hill, from OE **cnoll**. YN 200.

(9) ~ -LE-SOKEN Essex TM 2222. 'K in the soke' sc. of The NAZE TM 2624. *Kirby-le-Soke* 1645. P.n. Kirby + Fr definite article **le**, short for *en le*, + OE **sōcn**, *in the Sokne* 1385. Earlier simply *Kirkby* c.1127, *Ky- Korkebi -by* 1181–1558×1603. Ess 340, Jnl 2.45.

(10) ~ MISPERTON NYorks SE 7779. P.n. *Chirchebi* 1086, *Ki- Kyrkeby -bi* 1094–1408, *Kirby* 1665 + p.n. *Mispeton* 1086, *Mi- Mysperton(a)* from 1137×61, possibly 'the settlement by the medlar-tree', OE ***mispeler** + **tūn**. Misperton was a separate manor in DB. Also called *Kirkebye Overkare* 'K above the marsh', ON **kjárr**. YN 75.

(11) ~ ROW Norf TM 3792. 'Row of houses at Kirby'. *Kirby Row* 1838 OS. P.n. Kirby as in KIRBY CANE TM 3794 + ModE **row**.

(12) ~ SIGSTON NYorks SE 4194. ~ *Si- Sygeston* 1244. Originally two separate names, *Kirchebi* 1088, *Kirkeby* 13th cent. and SIGSTON SE 4195. YN 211.

(13) ~ UNDERDALE Humbs SE 8058. *Ki- Kyrkeby in Hundoluesdale -dala* 1156×7, ~ *(in) Hundolfdale* 14th cent., ~ *(in) Hundoldale* 1254, ~ *Hundale* 1286, ~ *Underdale* 1542. P.n. *Cherche- Chirchebi* 1086, *Ki- Kyrkeby -bi* 1088×93–1572, + p.n. HUNDLE DALE. YE 129, SSNY 31.

(14) ~ WISKE NYorks SE 3784. P.n. *Chi(r)che- Cherchebi* 1086, *Kirkebi -by* 1086, c.1180 etc. + r.n. WISKE NZ 4300. YN 274.

(15) Cold ~ NYorks SE 5384 → Cold KIRBY NYorks SE 5384, 'Kæri's farm'.

(16) Monks ~ Warw SP 4683. *Kirkebi Monachorum* 1199, *Monkenkirkeby* 1416, *Monks Kyrkeby* 1427, *Monkes Kirbye* 1546. ME **munk**, Latin **monachus**, genitive pl. **munkes**, **monachorum**, + p.n. *Kirkeberia -biria* 1077, *Chircheberie* 1086, *Kirke- Kyrkebi* c.1160–1535. Land here was given to the monks of St Nicholas of Angers in 1077. The earlierst forms point to OE **ċiriċe** + **byriġ** 'the church fortification or manor', OE **ċiriċe** + **byriġ**, dative sing. of **burh**, later replaced by ON **kirkju-bȳ**. Wa 112.

(17) West ~ Mers SJ 2186. *Westkirkby* 1287–1633, *Westkirby* from 1287. Earlier simply *Cherchebia* [1081]12th, *Ki- Kyrkeby* [1137×40]1271–1535. 'West' for distinction from *Kirkby* in WALLASEY SJ 2992. Che iv.294, SSNNW 34 no.1.

KIRBY MUXLOE Leic SK 5204. P.n. *Carbi* 1086, *C- Karebi -by* late 12th–1277, *Karebi -by* 1208–1511, *Kerbi -by* c.1282–1518, *Ky- Kirby* from 1432, 'Kærir's village or farm', ON pers.n. *Kærir* + **bȳ**, + suffix *Muxlo(w)e* 1606–1729. The name has been assimilated to the frequent ON p.n. type Kirby from *Kirkju -bȳ*, 'village with a church'. Muxloe is the local surname *Mucksloe* 1623 adopted for distinction from neighbouring Kirby names after the assimilation had occured. The earliest form of the suffix, *Muckelby(e) -le-* is probably a late nick-name, 'great *bȳ*', ME *mickle, muckle*. Lei 512, SSNEM 58.

Cold KIRBY NYorks SE 5384. ModE adj. **cold** + p.n. *Carebi* 1086, *Kerebi -by* 1170–1343, *Kerby* 1143–1393, *Kyerby* 1541. 'Kæri's farm or village', OEScand pers.n. **Kæri*, genitive sing. **Kæra*, + **bȳ**. YN 197, SSNY 31.

KIRDFORD WSusx TQ 0126. 'Cynethryth's ford'. *Kinredeford'* 1228, 1238, *Ken- Ke(r)- Kunredeford* 1240–1379, *Kerdeforth* 1374, *Kir(d)ford* 1640. OE femine pers.n. *Cynethrȳth*, genitive sing. *Cynethrȳthe*, + **ford**. Sx 102 gives prs [kɑ:dfu:rd] and [keifu:rd], SS 72.

KIRKANDREWS-ON-EDEN Cumbr NY 3558. 'St Andrews church on Eden'. *Kirk Andrews upon Eden* 1777. Earlier simply *Kirkandres, eccl. sci' Andres* c.1200, *Kirk(e)- Kircandres* 1261–1485, *Kirkanders* 1576. ON **kirkja** + saint's name *Andrew* from the dedication of the church. The elements are in Celtic word order with the specific following the generic. Cu 141, SSNNW 200.

KIRKBAMPTON Cumbr NY 3056. 'Church Bampton'. *Kyrkebampton'* 1292, *Kirk(e)banton* 1314, 1589, *Kyrk-* 1333, *Kirkbampton* 1372. ON **kirkja** + p.n. *Banton(a)* c.1185–1547, *Bamton'* 13th cent., *Bampton* 1227–1352, 'tree farm or village', OE **bēam** + **tūn** perhaps referring to a prominent tree near the settlement or to a place where beams were made. Also known as *Magna Bampton* 1352, 'great Bampton'. 'Church, great' for distinction from Little Bampton, *Parua Bampton* 1227, 1278, *litilbantton* 1584. Cu 142.

KIRKBRIDE Cumbr NY 2356. 'St Bride's church'. *Chirchebrid* 1163, *Kirkebrid(e)* c.1185–1385 with variants *Kyrke-* and *-bryd(e), Kirkbrid(e)* from 1190. ON **kirkja** + Irish saint's name *Bride*. The elements are in Celtic word order with the specific following the generic. Cu 144, SSNNW 53.

KIRKBURN Humbs SE 9855. This place has been variously called Burn-house, Westburn, Burn, Burns, Kirkburn and 'at the kirk-burns'.

I. *Burnous* 1086, *Burn(n)us* 12th cent.

II. *Westburn(e)* 1086, 1316.

III. *Brun(ne), Bronne, Brunnas* 13th cent.

IV. *Kirkebrunnon* 1272, *Kirkebrun(e)* 1274, 1362, *Kirkburn(e)* from 13th.

Variously OE **burna** + **hūs**, **west** + **burna**, **burna** (ME pl. **brunnes**), and **kirkja** + **burna**, dative pl. **burnum**, **brunnum**. OE *burna* has been influenced by ON *brunnr* 'a spring'. Kirkburn, along with Eastburn SE 9955, *Au(gu)stburne* 1086, *Estbrunne* 1274–1367, *-burn(e)* 1316–1559, OE **ēast** + **burna** influenced by ON *austr* and *brunnr*, and Battleburn, *Bordelbrun(ne)* 1227, 1228, *Bottilbourne* 1539, *Batle Burn* 1650, 'Bordel's stream', stands on Eastburn Beck alongside which there are several springs, especially in Kirkburn itself. YE 166 gives pr [kɔkbɔn] beside [kərkbən].

KIRKBURTON WYorks SE 1912. 'Burton with the church'. *Ki- Kyrk(e)burton(e)* 1535–1605. ME **kirk** (ON *kirkju*) + p.n. *Bertone* 1086, *Burgtun'* 1164×81, *Bir- Byrtun -ton(a)* 1091×7–1457, *Burton(e)* 12th, OE **byrh-tūn** in which **byrh** is the genitive sing. of **burh**, varying with **burh-tūn** 'a fortified enclosure'. 'Kirk' for distinction from HIGHBURTON SE 1913. The church is first mentioned in 1147. YW ii.245, WW 151.

KIRKBY 'The village with a church'. ON **kirkju-bȳ**. Many such names, originally descriptive terms, must have replaced earlier English village names. See also KIRBY.

(1) ~ Lincs TF 0692. *Kyrchebeia* [1146]13th, *Ki- Kyrkeby* 1209–1559, *Kirkby* from 1576. Li iii.52, SSNEM 57 no.11.

(2) ~ Mers SJ 4198. *Cherchebi* 1086, *Kyrkeby* 1228–1341, *Kirkeby* 1311, 1332, *Kirkby* 1843 OS. La 116, SSNNW 34 no.2.

(3) ~ NYorks NZ 5305. *Cherchebi* 1086, *Kircha- Kirkabi, Ki- Kyrkeby (in Cliveland)* c.1150. P.n. Kirkby + p.n. Cleveland as in CLEVELAND HILLS SE 5899. YN 168.

(4) ~ FELL NYorks SD 8763. *Kirkby Fell* 1771. P.n. Kirkby as in KIRKBY MALHAM SD 8961 + ModE **fell** (ON *fjall*). YW vi.132.

(5) ~ FLEETHAM NYorks SE 2894. Short for *Ki- Kyrkeby cum Fletham* 13th cent. P.n. *Chir- Cherchebi* 1086 + p.n. FLEETHAM SE 2894. The village of Kirkby (SE 2895) has disappeared. It was originally a distinct place from Fleetham. YN 239.

(6) ~ GREEN Lincs TF 9857. *Kirkeby super le gren(e), on le Grene* 1409–1541, *Kirkeby Grene* 1473. Earlier simply *Cherchebi* 1086, *Ki- Kyrkeby* 1185–1428. Perrott 338, SSNEM 57 no.3.

(7) ~ IN ASHFIELD Notts SK 5056. *Chirchebi* 1086–1170, *Ki- Kyrkebi -by* 1174–1347. Ashfield is an old district name, *Esfeld*

1216, *Esse- Essh- Assefeld* 13th cent., *Asshefeld* 1305, 1316, 'the open land with ash-trees', OE **æsc** + **feld**, cf. SUTTON IN ASHFIELD SK 4959. Nt 120, 12.

(8) ~ -IN-FURNESS Cumbr SD 2282. *Kirkebi* 1191×8, *Kirchabi* 1175×1200, *Kirkeby* 1227, 1292. Also known as Kirkby Ireleth, *Kirkebi Irlid* 1180×99, *Kirkeby Ir(e)lith* 1278, 1332, for distinction from KIRKBY LONSDALE etc., from IRELETH Cumbr SD 2277, that part of Dalton parish adjoining Kirkby to the S. La 220 gives pr [kə:bi], SSNNW 34.

(9) ~ LA THORPE Lincs TF 0945. 'K by Laythorpe, Leithulfr's outlying farm'. *Kirkeby by Laythorpe* 1263, *Kirkeby Laythorp(e), Leythorp* 14th cent., *Kirkby Lathorp* 1634×42, *Kirkby-la-Thorpe* 1896 Gaz. Earlier simply *Che- Chirchebi* 1086, *Kirkeby* 1195–1370. Laythorpe refers to a lost village in the parish, *Ledulue- Ledulftorp* 1086, *Lei- Layltorp -thorpe* 1185–1298, *Lei- Ley- Laythorp* 13th cent., ON pers.n. *Leithúlfr* + **thorp**. The false analysis of Laythorpe as Fr definite article *la* + *thorp* appears to be a piece of 19th cent. antiquarianism. Perrott 78–81, SSNEM 57 no.113.

(10) ~ LONSDALE Cumbr SD 6178. *Cherkebilonesdale* 1176, *Kircabi Lauenesdale* [1090×7]1308, *Ki- Kyrk(e)bi -by* 1120×30–1707 with affix *(in) Lonesdale -dall* etc., 'valley of the river Lune'. Earlier simply *Cherchbi* 1086. We i.42, SSNNW 34.

(11) ~ MALHAM NYorks SD 8961. Originally two separate names, *Chirchebi* 1086, *Ki- Kyrk(e)by -bi* 1154–1682, and *Malghum* 1154–91, *Malgam* 1250, *Malhom* 1496, *in Malg(h)edale* 12th cent., *Malledal'* 1246, *in Malhamdale* 1597 as in p.n. MALHAM SD 9063. YW vi.132.

(12) ~ MALLORY Leic SK 4500. 'K held by the Mallory family'. *Ky- Kirkeby Mal(l)or(r)e -ur-* 1269–1376, ~ *Mal(l)ory(e)* 1361–1550. Earlier simply *Cherchebi* 1086†, 1109×1204, *Ky- Kirkebi -by* 1202–1550, *Ky- Kirkby* 16th cent., *Ky- Kirby* 1578, 1539, 1610. Ricardus Malore held the manor in 1202. Lei 528, SSNEM 57 no.6.

(13) ~ MALZEARD NYorks SE 2374. P.n. *Chirchebi* 1086, *-aby* 1154–91, *Kyrkebi -by* 1101–1493, *Ki- Kyrkby* from 1285, + suffix *Malesard* 1101–9, *Malas(s)art* 1154–1376, *Malserd* 1284, *Malzarde* 1468, apparently a lost p.n. OFr **mal-assart** 'the bad clearing'. YW v.209.

(14) ~ MILLS NYorks SE 7085. P.n. Kirkby as in KIRKBY MOORSIDE SE 6986 + ModE **mills**.

(15) ~ MOORSIDE NYorks SE 6986. 'K at the head of the moor'. *Ki- Kyrkebi -by (Moresheved)* c.1170–1391, *Kirkebymoreshede* 1399, *Kirkeby Moresyd* 1489. P.n. *Chirchebi* 1086 + ME *moresheved*, OE **mōr**, genitive sing. **mōres**, + **hēafod**. YN 64.

(16) ~ ON BAIN Lincs TF 2462. *Kyrkeby super Bein* 1226. P.n. *Che- Chirchebi* 1086, *Chirchebi* c.1115, + r.n. BAIN TF 2473. DEPN, SSNEM 57 no.7.

(17) ~ OVERBLOW NYorks SE 3249. 'The iron-smelters' K'. P.n. *Che- Chirchebi* 1086, *Ki- Kyrkebi -by* 12th–1560, *Ki- Kyrkby* from 1355, + suffix *Oreblowere* 1211, *Oreblawer(s) -blowere* 1242–1391, *Oure- Over- Ouerblowers -blawers* 1355–1465, *Overblowers* 1512–1617, *-blowe* 1759, OE **ōra-blāwere**. Iron-smelters were once common in the area. The development of *Over-* is due to the falling together in the local dial. of *over* and *ore* as [ouə]. YW v.42 gives pr ['kə:bi 'ouəblo:], DEPN, Settlement 1979.33–4.

(18) ~ STEPHEN Cumbr NY 7708. 'Stephen's K'. *Cherkaby Stephan* 1090×7, *Cherkebi Stephan* 1176, *Kircabistephan* 1157, *Ki- Kyrk(e)bi -by* 1200–1652 with affix *Stephan(i), Steffan, Stephen, Kirby steven* 17th cent., *Corkby Steven* 1573, probably from abbot Stephen of St Mary's Abbey, York, to which the church was given by Ivo Taillebois. But the church is also dedicated to St Stephen. We ii.8 gives prs [kə:kbi-] and ['kə:bi'stevn], SSNNW 34.

(19) ~ THORE Cumbr NY 6325. *Ki- Kyrk(e)bi -by thore* 1179–1752 with variants *-thor -thur(e), Kirbethure* 1630, *Kirby Thure* 1668. The affix is ON pers.n. *Thórir*, the name of an early unidentified owner. We ii.117, SSNNW 34.

(20) ~ UNDERWOOD Lincs TF 0727. 'K in the forest'. *Kyrkby Underwode* 1535. Earlier simply *Cherchebi* 1086, *Kirkeby* c.1160–1597. Patches of woodland still surround the village on the NW, W and S and again to the E. Perrott 124, SSNEM 57.

(21) South ~ WYorks SE 4510. *Sudkirkebi* 12th cent., *Suthkirkebi -by, -kyrkeby* c.1130×40–1314. OE **sūth** + p.n. *Che- Chirchebi* 1086. YW ii.40, SSNY 32.

KIRKCAMBECK Cumbr NY 5368. 'Church on Cam Beck'. *Kirkecamboc* c.1280, *-kambok' -cambok'* 1284, 1289, *Kircambok* 1292, *Kirkambek* 1363, *Kirkcamocke* 1589. ON **kirkja** + river n. Cam Beck, *Camboc* 1169–1386 with variants *Kam-, -bo(c)k, Camock(e)* 1603, *Crooked Cambeck* 1622, PrW ***cammọ̄g** (< Brit **cambo-* + suffix **-āco-*, W *camog*) 'crooked one'. Cu 56 gives pr [kərka:mək], 7, Jnl 1.43, 44, SSNNW 194, GPC 403.

KIRKDALE Mers SJ 3493. 'Church valley'. *Chirchedele* 1086, *Kirk(e)dale* from 1185. ON **kirkja** + **dalr**. The reference is to a valley leading to Walton church SJ 3594. La 116, Mills 101, SSNNW 140.

KIRKFELL Cumbr NY 1910. 'Church fell'. *Kerkefell, Kirkefelle* 1338. ON **kirkja** + **fjall**. Cu 392.

KIRKHAM Lancs SD 4232. 'Church homestead'. *Chicheham* [sic] 1086, *Chercheham* 1094, *Chircheham-* c.1130, *Kyrkham, Kircheham, Kyrcham* 1094, *Kirkeham* 1190–1257, *-heim* 1196, *Kyrk(h)eym -haym* 13th cent., *Kirkham* from [1212]1268. Probably OE **ċiriċe** + **hām** Scandinavianised to *Kirkeheim*, ON **kirkja** + **heimr**. La 152, Jnl 17.88, 18.17.

KIRKHAM NYorks SE 7365. 'Church village'. *Chercha', Cher- Chircham* 1086, *Cherc(he)ham* [1123×8]1336, *Ki- Kyrk(e)ham* from 12th, *-heim* 1238, *Kirkaham* c.1175, *-haim* c.1200, *Kirkam* 1288. OE **cirice** + **hām** influenced by ON **kirkja** + **heim**. YE 143 gives pr [kɔkəm].

KIRKHAMGATE WYorks SE 2922. Probably 'the road to Kirkham'. *Kirkhamgate* 1709, 1817, *Kirkham Gate* 1843 OS. P.n. *Kirkham* 'the church estate', ON **kirkju** + OE **hām**, is probably an unrecorded alternative name of WOODKIRK SE 2724; *gate* is ON **gata** 'a road'. YW ii.145.

KIRKHARLE Northum NZ 0182. 'Church Harle'. *Kyrkeherle* 1242 BF. ME **kirk** + p.n. *Herle* 1170–1430, *Herlee* 1196, possibly 'Herela's wood or clearing', OE pers.n. **Herela* + **lēah**. 'Kirk' for distinction from Littleharle NZ 0183, *Parva Herle* 1279. Also known as East Harle, *Hestherll'* 1296 SR. NbDu 128.

KIRKHEATON Northum NZ 0177. 'Church Heaton'. *Kyrke Heton* 1296 SR. ME **kirk** + p.n. HEATON. 'Kirk' for distinction from CAPHEATON NZ 0380. Also known as *Parva Heton* 1242 BF, *Little Heton* 1232 Ch. NbDu 128, DEPN.

KIRKHEATON WYorks SE 1818. 'Heaton with the church'. *Kirk(e)heton(e)* from late 13th, *Kirk(e)heaton* from 1577. ON **kirkja** + p.n. HEATON, 'the high settlement', *Heptone* 1086, *Hetun -ton(e)* 1170–1457. DB's *Heptone* is probably an eye-slip error from *Leptone* which immediately preceeds it. 'Kirk' for distinction from Upper HEATON SE 1819, EARLESHEATON SE 2521, and Hanging HEATON SE 2523. YW ii.225.

KIRKLAND 'Church land'. ON **kirkjuland**. SSNNW 140, L 246.

(1) ~ Cumbr NY 6432. *Kyrkeland* c.1140–1332 with variants *Kirke-, -launde - lond(e), Kyrkland* 1485. The origin of the name is unknown. Cu 214.

(2) ~ Cumbr NY 0717. *Kirkland* 1586. Cu 407.

KIRKLEATHAM Cleve NZ 5921. 'Leatham with the church'. *Kyrkelidun* 1181, *Kirkledom* 1491. ME **kirke** (< ON *kirkja*) + p.n. Leatham. Also known as *Weslide, Westlid(um) -un, Westude* (sic for *-lide*) 1086, 'west Leatham'. Leatham means '(at) the slopes', ON **hlíth**, dative pl. **hlíthum**, probably replacing OE **hlith**, dative pl. **hleothum**, and referring to the foothills of the Cleveland hills by which the village is situated. The name is in systemic contrast with other dative pl.

†There is also an erroneous DB form *Cherebi*.

names in the district, e.g. UPLEATHAM NZ 6319 (from which it is distinguished as 'west' or 'church' Leatham), MARKSKE NZ 6322, COATHAM NZ 5925 and ACKLAM NZ 4817. YN 155, SSNY 99.

KIRKLEVINGTON Cleve NZ 4209. 'Levington with the church'. *Kirkelevingtona* [1230×50]15th. ME **kirke** (< ON *kirkja*) + p.n. *Levetona, Lentune* 1086, 'the settlement on the r. Leven', r.n. LEVEN NZ 4906 + **tūn**. *Kirk* for distinction from *Castel Levinton* 1219, a medieval motte represented by Castle Hill NZ 4610. YN 173.

KIRKLEY Suff TM 5391. 'The church wood or clearing'. *Kirkelea* 1086, *-le(e)* 1200–1372, *Kirkley* 1523, *Kirtlow* 1610. OE **ċiriċe** replaced by ON **kirkja** + **lēah**. DEPN, Baron.

White KIRKLEY Durham NZ 0235. Unexplained. *White Kirtles* 1768 Armstrong, *White Kirkley* 1863 OS. The forms *Whitekirtilland* 1382, 1418, *Whitekirketilfeld* 1382, *Whitekirtilfeld* 1418 which have been associated with this name probably do not belong here. NbDu 214.

KIRKLINGTON Notts SK 6757. 'The estate called after Cyrtla'. *Cyrlingtune* [958]14th S 659, *Cherlinton* 1086, *Kirtlingtun* c.1175, *Ki- Kyrtelington -yn(g)-* 1286–1428, *Kirklyngton -ing-* from 1342. OE pers.n. **Cyrtla* + **ing**[4] + **tūn**. Nt 170.

KIRKLINGTON NYorks SE 3181. 'The estate called after Cyrtla'. *Cherdinton* 1086, *Chirtlintuna* c.1150, *Kirtlyngton -ing-* 12th–1449, *Kirk(e)lin(g)ton -yn-* 1276–1575. OE pers.n. *Cyrtla* + **ing** + **tūn**. YN 220.

KIRKLINTON Cumbr NY 4367. 'Kirk or Church Linton, the settlement by the river Lyne'. *Kirkeleuin(g)ton* 1278–1498 with variants *Kyrk(e)-, -yn(g)-, Kirkelinnton* 1279, *Kirklinton* from 1583. ME **kirk** (<ON *kirkja*) + p.n. *Leuendon* 1167, *Leuenton* c.1170–1249, *Leuinton'* 1177–1293, *Levington -yng-* 1231–1343, *Leineton* 1310, 'settlement by the river Lyne', r.n. LYNE Cumbr NY 4972 + **tūn**. 'Kirk' for distinction from Randalinton NY 4065, *Randolflevington* 1275–1365, *Randollyngton* 1493, 'Randolf's Linton' referring to an unidentified early owner, and WESTLINTON Cumbr NY 3964. Cu 101, 53.

KIRKNEWTON Northum NT 9130. 'Newton with the church'. ME **kirk** + p.n. *Niwetona* [1123×8]1336 Ch, *Neuton(')* 1242 BF, 1296 SR, *Niwetona in Glendale* 1336, 'the new settlement', OE **nīwe**, definite form **nīwa**, + **tūn**. 'Kirk' for distinction from Westnewton NT 9030, *alteram Neuton* 'the other N' 1242 BF. NbDu 129, DEPN.

KIRKOSWALD Cumbr NY 5541. 'St Oswald's church'. *Karcoswald* (sic) 1167, *Kirkos(e)wald* 1212 etc. with variants *Kyrk(e)- Kirc- Kyrc-*, *Kirkcouswould* 1659. ON **kirkja** + saint's name *Oswald*, king of Northumbria, from the dedication of the church. The elements are in Celtic word order with the specific following the generic. Cu 215 gives prs [kə:ku:zl, kə:ki:zl], pronounced locally [kɛkuzld], SSNNW 200.

KIRKSANTON Cumbr SD 1380. 'St Sanctan's church'. *Santacherche* 1086, *Kirkesantan* before 1152–1333, *Kirksanton* from 1175×90. ON **kirkja** + Irish saint's name *Sanctán*. Identical with Kirk Santon Isle of Man, *ecclesia Sancti Santani* 13th. The elements are in Celtic word order with the specific following the generic in all forms apart from Domesday Book 1086. Cu 415, SSNNW 54.

KIRKSTONE PASS Cumbr NY 4009. *A pass called Kirkstone* 1671. P.n. Kirkstone + ModE **pass**. Kirkstone, *Kirkestain* before 1184, *Kyrkestone* 16th, *Kirkstone* 1793, is 'church stone', ON **kirkja** + **steinn**, referring to some stones near the gorge of the pass, called *High-trough* 1778, *High-cross* 1793, also known as *the Rayse of Kyrkestone* 16th, from ON **hreysi** 'a cairn'. 'Kirk' is often used of cairns and ancient stone remains, being thought of as the debris of former churches. In Iceland names like *Alfakirkja* and *Tröllakirkja* reflect folklore beliefs that elves and trolls had their own churches. We ii.223.

KIRKWHELPINGTON Northum NY 9984. 'Church Whelpington'. ME **kirk** + p.n. *Welpinton* 1182, *Whelpington* 1267, *Welpington Est* 1296 SR, the 'estate called after Hwelp', OE pers.n. **Hwelp* + **ing**[4] + **tūn**. 'Kirk' for distinction from a lost *West Whelpington*. NbDu 129.

KIRMINGTON Humbs TA 1011. Partly uncertain. *Chernitone* 1086, *Chirnig- Cherli(ng)tuna* c.1115, *Kirningtun(a) -e -yng-* 1143–1696, *Kirmington(')* from 1219. DEPN's explanation of this name as the 'settlement of the Cynemæringas, the people called after Cynemær' hardly fits the forms. The form *(æt) Coringatune* [1066×8]12th ASWills may belong here but clearly does not fit the run of forms for the name. Probably, therefore, unidentifed pers.n. + **ing**[4] + **tūn**, 'estate called after X'. Li ii.214.

KIRMOND LE MIRE Lincs TF 1892. *Kirmond in the myre* 1607. P.n. *Chevremont* 1086, *Chesfremund* c.1115, *Keuermunt* c.1152, *Kuermunt* c.1150, 'goat hill, OFr **chevre** + **munt**, + Fr definite article **le** for *en le* + ModE **mire** (ON *mýrr*). The name *Chévremont* is common in France whence it has been transferred here. DEPN, Cameron 1998.

KIRSTEAD GREEN Norf TM 2997. *Kirstead Green* 1838 OS. P.n. Kirstead TM 2998 + ModE **green**. Kirstead Green is the settlement on the main road from Norwich a mile S of Kirstead which is *Kerkestede* [1087×9]12th, *Kirkested(e)* 1206–1411, *Kirstede* 1428, 'the church site', OE **ċiriċ-stede** partly Scandinavianised with [k] for initial [tʃ]. There is a 12th cent. church here. The earliest form is *Kerchestuna* 1086, 'the church estate'. DEPN, Stead 183.

KIRTLING Cambs TL 6857. 'Place associated with Cyrtla'. *Chertelinge, Curtelinga -e* 1086, *Chertelinges* 1166, *Kert(e)ling(a) -(e) -yng(g)(e)* 1167–1509, *Ki- Kyrt(e)ling(g)(e) -yng(e)* 1268–1489, *K- Cart(e)lyng(e)* 1450–1553, *Cartelage* 1553, *Cateledge* 1616, *Kertlidge* 1640, *Kirtling, Catlige or Catlage* 1808. OE p.n. **Cyrtling* < *Cyrtla* + **ing**[2], locative–dative sing. **Cyrtlinġe*. Ca 126 gives former pr [kætlidʒ], ING 198.

KIRTLING GREEN Cambs TL 6855. *Kirtling Green* c.1825. P.n. KIRTLING TL 6857 + ModE **green**. Ca 127.

KIRTLINGTON Oxon SP 5019. 'The estate called after Cyrtela'. *(æt) Kyrtlingtune* c.1000 ASC (B, C) under year 977, *Kirtlingtun' -ton(e) -yng-* early 13th–1398, *Cherie- C(h)erte- Cortelintone* 1086, *C- Kurtlintun(a) -yn- -ton* c.1216–1316. OE pers.n. **Cyrt(e)la* + **ing**[4] + **tūn**. O 223.

KIRTON 'The church settlement'. ON **kirkja** + **tūn** probably replacing earlier OE *ċiriċe-tūn*.

(1) ~ Lincs TF 3038. *Che- Chirchetune* 1086–1161, *Kirchetun* 1154×89, *Ki- Kyrketon* 1169–1535, *Ki- Kyrton* from 1391. OE *ċiriċe* has been replaced by ON **kirkja** and some forms show substitution by the full Scand equivalent **kirkju-bȳ**, *Kirkebi in Hoiland(a), Kyrkebi* 1188–94. Payling 1940.91, SSNEM 184.

(2) ~ Notts SK 6969. *Kirchetona* [1154×89]1316, *Ki- Kyrketon* 1192–1428, *Kirton* from 1581. This seems to have replaced the earlier name *Schidrin(c)tune, Schitrintone, Schidrictune* 1086, 'the settlement on the **Scitering* or dung stream', OE ***sciter-ing** < ***scitere** + **ing**[2], + **tūn**. Nt 52, SSNEM 185.

(3) ~ Suff TM 2739. *K- kirketuna* 1086, *Ki- Kyrketon(e)* 1254–1554, *Kyrton* 1289, 1524, 1539, *Kirton* 1610. DEPN, Arnott 31, Baron.

(4) ~ END Lincs TF 2840. *Kerton End* 1653, *Kirton End* 1824 OS. Short for the *Townys end off Kirton* 'K town end' 1526. P.n. KIRTON TF 3038 + ModE **end**. Payling 1940.94.

(5) ~ HOLME Lincs TF 2642. *Ky- Kirk(e)tonholm(e)* 1316–1549, *Kyrton home, Howme* 1524, 1528, *Kirton Holme* from 1707. P.n. KIRTON TF 3038 + ME **holm** 'flat land by a river, island of dry ground in marsh'. Payling 1940.91.

(6) ~ IN LINDSEY Humbs SK 9398. P.n. *Chirchetone* 1070×87, 1086, *C(h)irceton* 1086, *Kirketune* 1155×60 + p.n. LINDSEY. DEPN, SSNEM 184.

KISLINGBURY Northants SP 6959. 'The fortification at the gravelly place'. *Cifel- Ceselingeberie* 1086, *Ki- Kyseling(e)byr' -bury -biry* 12th–1360, *Kyslyngbury* 1308. OE ***ċiseling** < ***ċisel** +

ing[2], locative-dative sing. ***ċiselinge**, with [k] instead of English [tʃ] due to Scandinavian influence, + **byriġ**, dative sing. of **burh**. The specific might also represent OE **Ċiselinga*, genitive pl. of **Ċiselingas* 'the people who live at the gravelly place' < ***ċisel** + **ingas**. The soil is partly clay, partly gravel. DEPN suggests alternatively OE folk-n. **Cȳselingas*, 'the people called after Cysel(a)', OE pers.n. **Cȳsel(a)* + **ingas**, genitive pl. **Cȳselinga*. Nth 86.

KIT HILL Corn SX 3771. *Kit Hill* 1809 OS. A hill rising to 1096ft.

KIT'S COTY HOUSE Kent TQ 7460. *Kits Cotty House* 1819 OS. A megalithic burial chamber formerly covered by a long barrow, now consisting of three uprights and a massive capstone. Newman 1969W.137, Thomas 1976.158.

KITHURST HILL WSusx TQ 0712. P.n. *Kiteherst* 14th, 'the kite wooded hill', ME **kite** (OE *cȳta*) + **hurst** (OE *hyrst*), + ModE **hill**. Sx 162.

KITLEY Devon SX 5651 → KITWOOD Hants SU 6633.

KITWOOD Hants SU 6633. *Kitwood* 1817. Possibly 'kite wood'. Cf. Kitley Devon SX 5651, *Kitelhey* 1310, *Kytelegh* 1333, OE **cȳta** + **lēah**. D 263.

KIVETON PARK SYorks SK 4983. Possibly 'the settlement on the hill called *Cyf*'. *Ciuetone* 1086, *Ky- Kive- Kiuetun -ton* 13th–1439, *Keueton* 1297–1532, *Keton* 1410–1544. OE **cȳf** 'a tub' used topographically as a p.n. + **tūn**. An alternative possibility is 'Cifa's farmstead', OE pers.n. **Ċifa* + **tūn**. This would normally give *Ch-* as in CHEVELEY, CHEVINGTON, CHIEVELEY etc. *K-* would have to be explained as due to Scand sound substitution. YW i.156.

KNAITH Lincs SK 8384. 'The landing-place at the bend'. *Cheneide* 1086, *Kneia* 1199, *Keneya, Cneie* 1225, *Cnaythes* 1254, *Knayth* [c.1225]14th. OE **cnēo** 'a knee, a knee-shaped bend' + **hȳth**. Knaith is situated at a sharp bend in the r. Trent. DEPN, Cameron 1998.

KNAPHILL Surrey SU 9658. '*Knap* hill'. *Knephull* 1440, *Kneppehill* 1474, *Knep-* 16th, *Knob Hill* 1680, P.n. *la Cnappe* 13th cent., 'the hillock', OE **cnæpp** used as a p.n., + **hill**. Sr 158, cf. PNE i.101.

KNAPP Somer ST 3025. 'Hill-top'. Cf. *High Knap, Lower Knap* 1809 OS. Ultimately OE **cnæpp**. (High) Knapp is situated on a spur of high ground overlooking the river Tone flood-plain.

KNAPS LONGPEAK Devon SS 2081. 'The long peak of Knap'. P.n. Knap + ModE **long peak**. Knap, cf. *Knap F., Knap Head* 1809 OS, which probably occurs in the surname of Thomas atte Knappe 1330, is OE **cnæpp**, 'a hill-top'. D 79.

KNAPTON Either 'Cnapa's farm or village' or the 'estate of the young man'. OE pers.n. *Cnapa* or **cnapa**, + **tūn**. *Cnapa* is either a by-name from OE *cnapa* 'child, youth, servant' with a significance like OE *cild* or *cniht*, or an Anglicisation of ON pers.n. *Knappi* (genitive sing. *Knappa*).

(1) ~ NYorks SE 5652. *C- Knapeton(e)* 1086–1282, *C- Knapton* from [1119]15th, 1244. YW iv.230.

(2) ~ NYorks SE 8875. *C- Knapeton(a)* 1086–1345, *C- Knapton(a)* from [1190]13th. Alternatively the specific of this n. might be OE **cnæpp** 'the top of a hill, a summit' referring to the steep headland due S of the village rising to 600ft. YE 136.

(3) ~ Norf TG 3034. *Kanapatone* 1086, *Gnapenton* 1193, *Cnapeton* 1254. DEPN.

KNAPWELL Cambs TL 3362. 'Cnapa's spring'. *(at) Cnapwelle* [1043×5]13th S 1531, [1060]13th S 1030–1329, *Cnapanwelle* [1077]17th, *Cnapen(e)welle* c.1350, *Chenepewella* 1086, *Knapwell(e)* from 1208. OE pers.n. *Cnapa*, genitive sing. *Cnapan*, + **wella**. Ca 168.

KNARESBOROUGH NYorks SE 3557. 'Cenheard's fortification'. *Chenaresburg* 1086, *Chen- Canardesburg* 12th cent., *C- Knaresburg(h) -burc(h)* 1120–1491, *Knaisbrowgh* 1573, *Knairsb(o)rough* 17th cent. OE pers.n. *Cēnheard*, genitive sing. *Cēnheardes*, + **burh**. YW v.110 gives prs ['neizbrə, 'nɛəzbrə], Settlement 30.

KNARSDALE Northum NY 6753. 'Valley of Knar, the rugged rock'. *Cnaresdale* c.1240, *Knaresdal* 1254, *-dale* 1291, *Gnaresdale* 1255. P.n. *Knar* c.1275, *Knarre* 13th, 1325, OE ***cnearr**, ME *knarre*, dial. *gnar*, genitive sing. *Knares*, + **dæl**. NbDu 129, DEPN, PNE i.102, L 96.

The KNAVOCKS Corn SW 5943 → NAVAX POINT SW 5944.

KNAYTON NYorks SE 4387. 'Cengifu's estate'. *Keneuetun, Cheneue- Cheniue- Chennieton* 1086, *Cneve- Knayveton(e)* 13th cent., *Knayton* from 1354. OE pers.n. *Cēnġifu*, genitive sing. *Cēnġife*, + **tūn**. YN 206 gives prs [ne:tən, ni:ətən].

KNEBWORTH Herts TL 2520. 'Cnebba's enclosure'. *Chenepeworde* 1086, *Knebwurth -worth* from 1215 with variants *Kn- Cnebbe-*. OE pers.n. *Cnebba* + **worth**. The original site is at Old KNEBWORTH TL 2320. Hrt 130.

Old KNEBWORTH Herts TL 2320. ModE **old** + p.n. KNEBWORTH TL 2520. Old Knebworth is the original site of the settlement.

KNEESALL Notts SK 7064. Partly uncertain. *Cheneshale* 1086, *Keneshale* 1230, *C- Kneshala -hal(e)* 1175–1302, *Cneeshala* 1187, *Knees(h)all* 1684. The specific is an OE pers.n. in *Cēn-*, possibly *Cēnsiġe* or **Cēnhēah*, reduced to *C(e)nes-* in the genitive sing., + **halh** 'a nook'. The reference is to a small valley W of the village. Nt 53, L 103, 110.

KNEESWORTH Cambs TL 3444. 'Cnen's enclosure'. *Kneneswrde* c.1216, *Knenisword* 1272, *Knensworth* 1363, *Cnesworth, Gneswrth* c.1218, *Kenesw(o)rth* 1235–1456, *Kne(e)sw(o)rth(e)* 1236–1457, *Nesworth* 1696. The early spellings point to OE pers.n. **Cnēn* < *Cēnthegn*, genitive sing. **Cnēnes*, + **worth**. Ca 56.

KNEETON Notts SK 7146. 'Cengifu's farm or village'. *Cheniveton(e)* 1086, *Kenyveton* 1291, *Kni- Knyveton(a)* c.1145–1454, *Kneveton* c.1250–1346, *Kneton* 1535, *Kneeton al. Kneeveton* 1682. OE feminine pers.n. *Cēnġifu*, genitive sing. *Cēnġife*, + **tūn**. Nt 226.

KNIGHTACOTT Devon SS 6439. 'The knights' cottage(s)'. *Nightacot* 1809 OS. Cf. the surname of Richard de Knyghtecote 1330. OE **cniht**, genitive pl. **cnihta**, + **cot**, pl. **cotu**. The exact sense of OE *cniht* here is uncertain. The original meaning was 'youth, servant, soldier' later 'young retainer, knight'. D 31.

KNIGHTCOTE Warw SP 3954. 'The cottage(s) of the servants' or 'of the knights'. *Kynttecote* (sic) 1232, *Knittecote* 1242, *Knychtecote* 1279, *Knyghtcote* 1303–1502, *Knightcourt* 1762. OE **cniht**, genitive pl. **cnihta**, + **cot**, pl. **cotu**. Possibly a reference to the Knights Templar who held an estate at nearby Temple Herdewyke SP 3752, *Templeherdewyke* 1285. Wa 269.

KNIGHTON 'Knights' estate'. OE **cniht**, genitive pl. **cnihta**, + **tūn**. The exact sense of *cniht* is uncertain: its original sense was 'youth, servant, soldier', later 'young retainer, knight'.

(1) ~ Devon SX 5249. *Knytheteton* (sic) 1281, *Knyghton* 1548. D 261.

(2) ~ Leic SK 6001. *Cnihteton(e)* 1086–1218, *Cnichtingtunam* 1146, *Cnihtinton(e)* 1205, *Knicht(t)on'* 1269, 1293, *K- Criton' -y-* 1195–1285, *Knyghton(e)* 1297–1551. Lei 452.

(3) ~ Staffs SJ 7240. *Chenistetone* 1086 with AN [s] for [ç], *Knihteton* c.1205. DEPN, Horovitz.

(4) ~ Staffs SJ 7427. *Chnitestone* 1086 probably for *Chnistetone* with AN [s] for [ç], *Knichton* 1222. DEPN.

(5) ~ DOWN Wilts SU 1144. *Knighton Down* 1817 OS. P.n. Knighton as in Knighton Farm SU 1545, *Wenistetone* (sic for *Ken-*) 1086, *Knytheton* [1154×89]1270, *Knyghteton* 1332. Wlt 366.

(6) Chudleigh ~ Devon SX 8477. 'Knighton near Chudleigh'. *Knyghton juxta Chudlegh* 1464. Earlier simply *Chenistetone* 1086, *Knicheton* 1285, *Knyghteton(e)* 1299, 1382. D 471.

(7) East ~ Dorset SY 8185. *East Knighton* 1774. ModE adj. **east** + p.n. *Knyt(t)eton* 1250–1421, *Kni(h)tteton, Kni- Knysteton* 13th cent., *Kny(ȝ)gh(t)(t)(e)ton(e)* [1294]1313–1498, *Knighton juxta Bynedon* 1313. 'East' for distinction from *West Knighton* 1774. Do i.177.

(8) West ~ Dorset SY 7387. *Westknyghton* 1452. ME adj. **west** + p.n. *Chenistetone* 1086, *Cniht(t)eton* 1214, *Kny(g)ht(et)on(e)* 1288–1481, *Knighton* from 1222. 'West' for distinction from East KNIGHTON SY 8185. Do i.207.

KNIGHTWICK H&W SO 7355. 'The young men's farm'. *Cnihtapice* [964]12th S 731, *(æt) cnihte pican* [1014×16]18th S 1459, *Cnihtewic* 1086. OE **cniht**, genitive pl. **cnihta**, + **wīċ**. According to DB the estate was held by Robert the bursar and its revenues were used for the household supplies of the monks (sc. of Worcester), *et est de dominico victu monachorum*. 'Young man, schoolboy, novice' is one of the senses of OE *cniht* which therefore well fits this name. Wo 147, Hooke 145, 329, DB Worcestershire 2.67.

KNILL H&W SO 2960. 'Little hill'. *Chenille* 1086, *Cnulla* 1158×64, 1243, *Cnille* 1220. OE ***cnyll(e)**. The reference is to the small hill on which the church stands. He 118, PNE i.103.

KNIPTON Leic SK 8231. 'Settlement by the steep hill'. *Cnipeton(e)* 1086, 1235×53, *Gniptone* 1086, *Gnip(p)eton'* 13th, *Gni- Gnypton(')* c.1130–1511, *Kni- Knypton(')* 1208–1610. ON **gnípa** + OE **tūn**. Knipton lies in a valley with hill-sides rising steeply on either side. Lei 142, SSNEM 184.

KNIPTON RESERVOIR Leics SK 8130. *Knipton Reservoir* 1824 OS. P.n. KNIPTON SK 8231 + ModE **reservoir**.

KNITSLEY Durham NZ 1148. 'Clearing where faggots are obtained'. *Kni- Knycheley* 1279–1547, *Knychley* 1461–1627, *Knitsley* 1561. OE **cnyċċ**, genitive pl. **cnyċċa**, + **lēah**. NbDu 130.

KNIVETON Derby SK 2150. 'Cengifu's farm or village'. *Chenivetun* 1086, *Cni- Kni- Knyveton(e)* 1169–1241 etc., *Kneveton* 1290–1384, *Kneton* 1572, *Kneighton* 1729. OE feminine pers.n. *Cēnġifu*, genitive sing. *Cēnġife*, + **tūn**. Db 383.

KNOCK Cumbr NY 6627. 'Hillock'. *Chonoc-salchild* 1150×62, *C- Knok(e)*, *Knoc* 1256–1398, 1580, *Knock(e)* from 1634. OIr **cnocc** referring to the prominent conical eminence of Knock Pike which rises to 1306ft. half a mile N of the village at NY 6828. The affix in 1150×62 appears to be a family name from p.n. SALKELD NY 5536 and is recorded in the neighbouring parish. We ii.114, xiii, Jnl 2.71.

KNOCK FELL Cumbr NY 7230. *Knock Fell* 1720. P.n. KNOCK Cumbr NY 6627 + **fell**. We ii.115.

KNOCKHOLT Kent TQ 4759. 'At the oak-wood'. *Nocholt* 1353, *le Nokelte* 1374. ME **atten okholt** < OE *æt tham āc holte*. In the modern spelling of this name the *-n* of *atten* has attached itself to the proper form of the name, **Ockholt*, which is the earliest form recorded, *Ocholt* 1197, *Ok- Ocholt(e)* 1270–92. ME **ok** (OE *āc*) + **holt**. PNK 27.

KNOCKHOLT POUND Kent TQ 4859. *Knockholt Pound* 1819 OS. P.n. KNOCKHOLT TQ 4759 + ModE **pound** (OE, ME *pund*).

KNOCKIN Shrops SJ 3322. 'The small hill'. *Cnukin* 1195×6, 1196×7, *Knukin -yn* 1198, 1291×2, *(The) Knuckin* 1622–1799, *(Le) Knokin* 1221×2–1276, *Knokyn* 1255–1470, *-yng* 1351, *Kno(c)king* c.1540–1739, *(Ye) Knucking(e)* 1609–1744, *Nucking* 1777. W **cnycyn** 'a little mound' partly associated with OE **cnocc** 'a hillock'. The reference is to small oval hills at the E and W ends of the village. Sa i.168.

KNODISHALL Suff TM 4261. 'Cnott's nook'. *Chenotessala*, *Cnotesheala -e* 1086, *Knodeshal(e)* 1234–1348, *Knoteshal* 1275, *Cnoteshal(e)* 1286–91, *Knatshall* 1523, 1610. OE pers.n. *Cnott*, genitive sing. *Cnottes*, + **halh** in the sense 'valley'. Alternatively the specific might be OE **cnott*, ME *knot* 'a hill' although this is normally a northern word. DEPN, Baron.

KNOLE Kent TQ 5354. 'Hill-top, knoll'. *Cnolle* 1327, *Knoll(e)* 1346–1535. OE, ME **cnoll**. PNK. K 65.

KNOLL'S GREEN Ches SJ 8079. 'Hill-top green'. *Knowl Green* 1672, *Knolls Green* 1842. Dial. **knowle** (OE *cnoll*) + **green**. Che ii.68.

KNOOK Wilts ST 9341. 'The hillock'. *Cunuche* 1086, *C- Knuk(e)* 1211–1262, *Knouk* 1316, *Knooke* 1581. PrW ***cnucc**. Possibly a reference to Horse Hill ST 9522. Wlt 171 gives pr [nuk].

KNOSSINGTON Leic SK 8008. Partly uncertain. *Nossitone*, *Closintone* 1086, *Cnos(s)inton(i)a -ton(e)* before 1160–1216×72, *Cnossington'* 1254–1324, *Knos(s)ington(e) -yng-* 1203–1604, *Knoston(')* 1405–1727. Possibly 'settlement at or called *Cnossing*, the hill place', OE ***cnoss** + **ing**[2] + **tūn**, or 'settlement called after Cnoss(a)', OE pers.n. **Cnoss(a)* + **ing**[4] + **tūn**. The village is situated on a prominent hill. An OE **cnoss* would be related to *cnossian* 'to strike' and ON *knauss* 'a knoll' a crag', Sw dial. *knös*, Dan dial. *knøs* 'a sand-hill'. Lei 234.

KNOTT Cumbr NY 2933. 'Rocky hill'. *Knot* 1794. ON **knott**. Refers to a summit of 2329ft. Cu 279.

KNOTT END-ON-SEA Lancs SD 3548. 'Hill-end'. P.n. *Knott* + ModE **end**. The reference is to a slight hill at the mouth of the river Wyre referred to [c.1265]1268 as *Hacunshou Cnote*, p.n. HACKINSALL SD 3447 + ME **knot** 'a hill'. La 160.

KNOTTING Beds TL 0063. '*Cnottingas*, the people of the knot or hill' or *Cnotting*, 'the knot-like hill'. *Chenotinga* 1086, *C- Knotting(e)* 1163, 13th cent., *Cnotinges* 1224, *Knottyngges* 1331. Either OE folk-n. **Cnottingas* < **cnotta** + **ingas** or p.n. **Cnotting* < **cnotta** + **ing**[2]. The place lies on a hill of 311ft. Bd 15, ING 67.

KNOTTINGLEY WYorks SE 4923. 'The wood or clearing of the Cnottingas, the people called after Cnotta'. *Notingelai -leia* 1086, *Nottingle(ya) -laia* 1147×54–c.1200, *Cnottingele(g)* 12th–1254, *Cnottingaleia* 1155×70, *Knottingley* from 1119×21. OE folk-n. **Cnottingas* < OE pers.n. *Cnotta* + **ingas**, genitive pl. **Cnottinga*, + **lēah**. The alternative suggestion 'people who live at or by the hillock', OE **cnotta** + **ingas**, is unlikely on topographical grounds. YW ii.73.

KNOTTY ASH Mers SJ 4091. *Knotty Ash* 1842. A modern name referring to a gnarled ash-tree which grew at the top of Thomas Lane in an open area of ground. Originally simply *Ash* c.1700. Mills 102, Room 1983.61.

KNOTTY GREEN Bucks SU 9391. *Knocklocks Green* 1766, *Knattocks or Knotty Green* 1806, *Knotty or Knocklock Green* 1826. The evidence is too late to offer a clear explanation: possibly surname + ModE **green**. Bk 230.

KNOWBURY Shrops SO 5775. *Knowbury Hall Farm* 1832 OS. Probably ultimately < OE **cnoll** + **byriġ**, dative sing. of **burh**.

KNOWESGATE Northum NY 9885. 'Gate at Knowes, the hills'. P.n. *Knowes* 1867 OS, dial. **knowe** 'a hill' (OE *cnoll*) + **gate**. A gate on the turnpike road from Newcastle to Jedburgh; also a station on the North British railway line from Morpeth to Bellingham. The reference is to the high ground of Whitehill NY 9985.

KNOWL HILL Berks SU 8279. *Knoll Hyll* 1536, *Knowl Hill* 1761. Earlier simply *La Cn- Knolle* 1299 etc. OE **cnoll** 'cnoll, hillock' taken as a p.n. to which pleonastic ModE **hill** was later added. Brk 62.

KNOWLE 'A hill-top, summit of a large hill, hillock, knoll', especially one shaped like a truncated cone. OE **cnoll**. PNE i.103, L 136–7.

(1) ~ Avon ST 6070. *Canole* 1086, *Cnolle* 1196. The reference is to a half-mile ridge of high ground rising to 200ft. DEPN, L 136.

(2) ~ Devon SS 4938. *(la) Knolle* 1298, 1412. The reference is possibly to the hill and earthworks called The Castle on the opposite side of the valley. D 34.

(3) ~ Devon SS 7801. *Knolle* 1281. Knowle lies on the side of a ridge of higher ground. D 408.

(4) ~ Shrops SO 5973. *Knowl* 1684. Cf. *Knowl Gate, Knowl Hill* 1832 OS. Gelling.

(5) ~ WMids SP 1877. *Gnolla* c.1200, *(la) C- Knoll(e)* 1221–1589, *Knowl* 1540. Wa 62, L 136–7.

(6) ~ GREEN Lancs SD 6338. P.n. Knowl, *Cnolle, Knolle* 1246, *Knol* 1262, 1274, + ModE **green**. *Green* names are frequent in this district, possibly for squatter settlements on waste ground as around Birmingham, cf. Seed Green SD 6437, Ward Green SD 6337, Frances Green SD 6236, HURST GREEN SD 6838. La 145.

(7) Church ~ Dorset SY 9381. 'Knowle with the church'. *Churic(h)cnoll(a)* 14th, *Churchecnolle* c.1350, *-knowle* 1658. ME **chirche** (OE *ċiriċe*) + p.n. *Cnolle, Chenolle, Glole* (sic) 1086, *Canolla* 1086 Exon, *Cnoll(e)* 1181–1340, *Knoll(e)* 1202–c.1586, *Knowle* 1621, 'Church' for distinction from Bucknowle SY 9481, *Bubecnolle* 1285, *Bubbeknolle* 1306, 1452, *Bouknolle* 1412, *Bucknoll* 1584, 'Bubba's Knowle', OE pers.n. *Bubba* + **cnoll** used as a p.n. Do i.87, 89.

KNOWLTON Kent TR 2853. 'The hill settlement'. *Cnoltun* 1070×82, *Chenoltone* 1086, *Cnoltune* c.1100, *Cnolton(e)* 1226–53×4, *Knolton(e)* 1254. OE **cnoll** + **tūn**. The site is a hill-top between valleys. PNK 584, Cullen.

KNOWSLEY Mers SJ 4395. 'Cenwulf's wood or clearing'. *Chenulueslei* 1086, *Cnusleu* 1189×96, *Knuvesle* 1199, *K- Cnusleia -e* 1199–[1220]1268, *C- Knous(e)ley -legh* 1229–1376, *Knowesley* 1228, *Knowsley* 1346. OE pers.n. *Cēnwulf*, genitive sing. *Cēnwulfes*, + **lēah**. La 113, Jnl 17.63.

KNOWSLEY HALL Mers SJ 4493. P.n. KNOWSLEY SJ 4395 + ModE **hall**. The 16th cent. and later mansion of the Stanleys, Earls of Derby. Pevsner 1969S.132.

KNOWSLEY INDUSTRIAL ESTATE Mers SJ 4398. P.n. Knowsley as in KNOWSLEY HALL SJ 4493 + ModE **industrial estate**.

KNOWSTONE Devon SS 8223. 'Cnut's stone'. *Chenudestane, Chenvestan* 1086, *Cnud- Cnutstan* 1199–1230, *Cnuston'* 1242, *Knowston* 1489. AScand pers.n. *Cnut* (ON *Knútr*), genitive sing. *Cnutes*, + OE **stān**. The 'stone' may have been at the place called Rock just east of the village. D 340 gives pr [naustən].

KNOX KNOWE Northum NT 6502. 'Knox hill'. *Knox Knowe* 1868 OS. Possibly surname *Knox* + dial. **knowe** (OE *cnoll*). A hill rising to 1636ft.

East KNOYLE Wilts ST 8830. *Esteknoyle* 1467. ME adj. **est** + p.n. *(æt) Cnugel* [948]14th S 531, [955×7]14th S 666, *(on) Cnugellege* [983]15th S 850, *Chenvel* 1086, *C- Knoel* 1187–1332, *Cnohill* 1236, *Knoyhull* 1451, *Noyle* 1693, OE ***cnuwel** 'a knuckle' used topographically of the broken ground between East and West Knoyle. The charter sps with *-g-*, all preserved in late copies, are probably back-sps for *-w-* rather than evidence for an inexplicable variant **cnugel*. 'East' for distinction from West KNOYLE ST 8532. Also known as *Bisshopes Knoyle* 1570, *Knowell Episcopi* 1346, *Childecnowell* 1204, 'children's Knoyle', OE genitive pl. **ċilda**, and *Magna Knowel* 'great Knoyle' 1285. The manor was once held by the bishop of Winchester. Wlt 175 gives pr [nɔil], Studies 1936.190, BzN 1981.287–8.

West KNOYLE Wilts ST 8532. *Westeknoyle* 1467. ME adj. **west** + p.n. Knoyle as in East KNOYLE ST 8830. Also known as *Knowell Odierne* 1347 and *Knoel Parva* 'little Knoyle' 1428. The manor was once held by Hodierna, the mother of Alexander Neckham and foster mother of Richard I. Wlt 175.

KNUTSFORD Ches SJ 7578. 'Knut's ford'. *Cunetesford* 1086, *Knut(e)sford(e)* from 1294 with variants *Knot(t)es- Knuttis- Nut(t)s-* etc. to 1672. ODan pers.n. *Knūt*, genitive sing. *Knūt(e)s*, + **ford**. There is a tradition going back to the 17th cent. associating the name with King Canute (d.1035). However, the frequency of forms in *Knot(t)-* may well point to a possible OE **cnottan-ford* 'ford at a hillock' (OE **cnotta**) in which *cnotta* was later replaced by ODan *knūt* with the same meaning, subsequently reinterpreted as if a pers.n. Che ii.72, SSNNW 235, L 67,69. Ch ii. Addenda gives 19th cent. local pr ['nutsfərt].

KNYPERSLEY Staffs SJ 8756. Partly uncertain. *Kniperslee* c.1247, *Knypersleye, Knniprislega* 1272×3, *Knipersley Hall* 1836 OS. Unidentified element or pers.n. + OE **lēah** 'a wood or clearing'. Duignan 89, Horovitz.

KOKOARRAH Cumbr SD 0496. Unexplained. *Kokoarrah* 1864 OS. The generic is probably ON *ǽrgi* 'a shieling'.

KUGGAR Corn SW 7216. Possibly an old stream name meaning 'winding stream'. *Coger, Cogar* 1324, *Cuger* 1336. Stream name **Cu- Cocrā* (Brit **kukro-* 'crooked') as in River COCKER NY 1228. PNCo 104.

KYLOE Northum NU 0540. 'The cow pasture'. *Culei* [c.1170]n.d., *Culeia* [1195]1335, *Kylei -ley* 1208×10 BF–1366×7, *Ky- Kilay* 1344, 1460, *Kylo(w), Kylhowe* 16th cent., *Kilo, Killey* 17th cent., *Keiloe, Keiley, Kylo* 18th cent. OE **cū** varying with pl. **cȳ**, + **lēah**. The modern form is a hypercorrection due to the development of OE *hlāw* to *ley* in weak stress in N p.ns. NbDu 130.

KYLOE HILLS Northum NU 0439. P.n. KYLOE NU 0540 + ModE **hills**.

River KYM Beds TL 1066. A back-formation from KIMBOLTON Cambs TL 0967, the chief town on its banks, cf. *aqua de Kenebauton'* 1228. The original name was *Haile* [c.1180]13th as in Hail WESTON Cambs TL 1662. RN 232, 188.

North KYME Lincs TF 1552. *Nortchime* 1086, *Northkime* 1219–1316, *North Kyme* from 1226. OE **north** + p.n. *Kime, Kyme* c.1115–1552, *Chimba, Ki- Kymba -e* 12th cent., 'the depression, the hollow', OE ***cymbe**, a derivative of *cumb*. 'North' for distinction from South KYME TF 1749. Perrott 328.

South KYME Lincs TF 1749. *Suthkyme* 1316, *Southkyme* from 1323. OE, ME **sūth** + p.n. *Chime* 1086, *Chimbe* 1174×84, *Kimbes* 1201, *Kime* c.1180–1280, *Kyme* 1228–1428 as in North KYME TF 1552. The village is in a low-lying position on Kyme Eau with higher ground rising to the N. Perrott 82.

KYNANCE COVE Corn SW 6813. *Kinance Cove* 1813. P.n. Kynance + ModE **cove**. Kynance, *Penkeunans* 1325 ('head of Kynance', Co **penn**), *(Pen)kynans* 1613, is Co ***kew-nans** 'ravine' (Co **kew* 'a hollow' + *nans* 'valley') later reinterpreted as if the first element were Co **ky** 'a dog'. PNCo 104, Gover n.d. 575, CPNE 57, 58.

KYRE PARK H&W SO 6263. *Kyre Park* 1832 OS. A medieval and later house. P.n. Kyre + ModE **park**. Kyre, *Cyr* 11th, *Cver, Chvre* 1086, *Cura -e* [c.1086]1190–1431, *Cuyre* 1308, 1415, *Keeres Common* 1667, is by origin a stream-n. *Cura* 13th, *Kyer Brook* 1618, 1638, of unknown origin. Wo 55 gives pr [ki:ər], RN 233.

L

LACEBY Humbs TA 2016. 'Leif's village or farm'. *Leuesbi* 1086–[1132]1403, *Leyseby* c.1115, *Laifse- Leuse- Leissebi* 12th cent. ON pers.n. *Leifr*, genitive sing. *Leifs* + **bȳ**. SSNEM 58, Cameron 1998.

LACEY GREEN Bucks SP 8200. 'Lacey's green'. *Lacyes Green* 1693, *Leasey Green* 1766. Surname *Lacey* + ModE **green**. Bk 173 gives pr [li:əsi], Jnl 2.30.

LACH DENNIS Ches SJ 7072. '(The part of) Lach held by the Dane'. *Lache Deneys* 1260, ~ *Dennis* 1312. P.n. *Lece* 1086, *Lache* 13th cent., *Lach* from 1287, OE **leċe**, **læċċ** 'a boggy stream', + OFr **daneis** 'Danishman' probably alluding to the TRE tenant Colben (ON *Kolbeinn*, ODan *Kolben*) and distinguishing this part of Lach from that held by William fitz Nigel in 1086. Che ii.186 gives pr ['latʃ], L 25.

LACHE Ches SJ 3964. 'The boggy stream'. *Leche* 1086–1150, *Leeche, Layche* 18th, *Lache* from 1285. OE **læċċ**, **leċe**. Che iv.162 gives pr ['latʃ], L 25.

LACKFORD Suff TL 7970. 'Leek ford'. *Lec- Lacford* 11th, *Le(a)cfordā* 1086, *Leacforde* c.1095, *Lacford* 1253, *Lackforde* 1610. The sps point to OE **lēac** + **ford**, but a derivation from OE **lāc** 'play, frolic' would parallel PLAYFORD TM 2148 and explain the r.n. LARK, a back-formation from this name. DEPN.

White LACKINGTON Dorset SY 7198. Properly Whitelackington, 'estate called after Wihtlac'. *Wyghtlakynton* 1354, *White Lackington* 1811 OS. OE pers.n. *Wihtlāc* + **ing**[4] + **tūn**. ADM.

LACOCK Wilts ST 9168. 'The little stream'. *Lacok* [854]c.1400 S 305, 1230, 1232, *Lacoc(h)* 1086, *Lak- Lacoc* 1189–1306. OE ***lacuc** referring to the little stream which flows into the Avon just east of the village. Wlt 102.

LADBROKE Warw SP 4158. Partly uncertain. *Hlodbroc* 998 S 892, *Lodbroc* 1086–1535 with variants *-brok(e)* and *-brooke, Ladbrok* 1320. The generic is OE **brōc** 'a brook', the specific possibly OE **hlot** 'a lot', referring to a stream used for the purpose of divining the future or of drawing lots. Wa 135, DEPN, L 15–16.

LADDINGFORD Kent TQ 6948. 'Ford over the river Loden'. *Lodeneford* 1201–78 with variant *Lodenes-, Lodingford* 1782, *Latingford* 1819 OS. R.n. **Lodene* + OE **ford**. The present name of the river here is Teise, but this is a back formation from TICEHURST ESusx TQ 6930 first recorded in Harrison's *Description, Theise* ('it ariseth about Theise Hurst') 1577, *Teise* 1620. **Lodene* is identical with the r.ns. LODON, LODDON, British **Lutna* 'muddy river', a suitable appellation for Teise. PNK 170, RN 398.

LADDUS FENS Cambs TF 4701. *Fenne called Lodwas* 1553, *Ladus fenns* 1632, *Lodas Fenne* 1687. P.n. Laddus + ModE **fen(s)**. Laddus, *(of) ladwere* c.975, *Lodwer(e)* 1221–1445, *Lod(e)weres* 1251, *Ladwers* 1580, is 'the lode weirs, weirs in the watercourse called Lode', OE **lād** + **wer**, pl. **weras**. Lodwere was a fishery (*piscaria*). Ca 268.

LADE BANK Lincs TF 3954. 'The water-channel embankment'. *Lade Bank in Leake* 1631, *Newlade* 1653, *Lade Bank* 1824 OS. Mod dial. **lade** (OE *lād*) 'a fenland drainage channel' + **bank**. Payling 1940.126, 134.

LADOCK Corn SW 8950. 'St Latoc's church'. *Ecclesia Sancte Ladoce* 1268–1374, *Sancta Ladoca* 1291, *Sent Ladek* 1358, *Ladocke* or *Lazacke* 1596, *Lassick* 1653, *Ladock* 1610. Saint's name *Latoc* from the dedication of the church. A saint unique to Cornwall of whom nothing is known. The form *Egloslagek* 1354, Co **eglos** 'a church', shows Co *d* > *dʒ* whence the later spellings *Lazacke, Lassick*. PNCo 105 gives pr ['ladak], older ['lazik], Gover n.d. 467, DEPN, CMCS 12.52.

LADYBOWER RESERVOIR Derby SK 1986. P.n. Ladybower SK 2086 + ModE **reservoir**. Ladybower, *Lady Bower* 1584, 1640, is 'lady's bower', ModE **lady** (OE *hlǣfdiġe*) + **bower** (OE *būr*). Db 85.

LAINDON Essex TQ 6889. 'Hill of (the river) *Lyge*'. *(of) Ligeandune* [c.1000]c.1125, *Legenduna, Leiendunā* 1086, *Lei- Leyndon(e)* 1226–1344, *Lang(e)don* 1365, 1373, 1768, *Laundon* or *Langdon* 1594. Lost OE r.n. **Lyġe* identical in form with river LEA GLond TQ 3695, genitive sing. **Lyġan*, + **dūn**. This must have been the name of one of the head waters of the Crouch which rises in the hill on which Laindon stands. For the genitival composition cf. the *Leynton* forms for LEYTON GLond TQ 3886 and Limbury Beds TL 0624, *Lygeanburg* 9th ASC(A) under year 571. Some late confusion with nearby LANGDON HILLS TQ 6786 took place. Ess 161, BdHu 155.

LAKE Wilts SU 1339. 'The stream'. *Lak(e)* from 1289. OE **lacu** probably referring to a side-channel of the Avon. Wlt 372.

LAKENHAM Norf TG 2307. 'Laca's homestead or meadow'. *Lakemham* (sic) 1086, *Lakenham* 1211, 1247, *Lakeham* 1212. OE pers.n. **Lāca*, genitive sing. **Lācan*, + **hām** or **hamm** 3. The village lies alongside the river Yare. DEPN.

LAKENHEATH Suff TL 7182. 'The landing place of the Lacingas, the people of the stream or called after Laca'. *(æt) Lacingahið* [1015×6]14th S 948, *Lakingheðe* [1021×3]15th S 980, *Laringahetha, Laringeheta, Lakingahetha -heda* 1086, *Lachingahutha* c.1120, c.1150, *Lakenheath* 1610. OE folk-n. **Lacingas* < **lacu** + **ingas** or **Lācingas* < pers.n. **Lāca* + **ingas**, genitive pl. **Lac- Lācinga*, + **hȳth**. DEPN, L 62.

LAKENHEATH STATION Suff TL 7186. P.n. LAKENHEATH TL 7182 + ModE **station**.

LAKENHEATH WARREN Suff TL 6267. *Lakenheath Rabbit Warren* 1836 OS. P.n. LAKENHEATH TL 7182 + ModE **warren**.

LAKESIDE Cumbr SD 3787. No early forms. A modern resort on Windermere, terminus of the Lakeside and Haverthwaite railway and the summer steamers to Bowness and Waterhead.

LALEHAM Surrey TQ 0568. Probably 'the river-meadow where withies grow'. *Læleham* [1042×66]13th, *Lælham* [1062]13th S 1035, *Lelehā* 1086, *Lelham* 1134–1275, *Lalham* 1207–1467, *Laleham* from 1274. OE **lǣl** + **hamm** 3. The place is by the Thames. The generic might, however, be **OE hām** 'a homestead, a village, an estate'. Mx 16.

LAMALOAD RESERVOIR Ches SJ 9775. P.n. Lamaload + ModE **reservoir**. Lamaload, *Lo(w)melowe* 1479, *Lomelode* 1503, *Lamelode* 1519, 1611, is 'loamy track or watercourse', OE **lām** + **lād**. Che i.144.

LAMARSH Essex TL 8935. 'Loam stubble-field or marsh'. *Lamers* 1086, *Lam(m)ers(se)* 1217–1346, *Lammersch -ersh(e)* 1221–1428. OE **lām** + **ersc** or **mersc**. The soil varies from 'a rich sandy loam, excellent for turnips, to a strong soil on which hops have been grown'. Ess 444 gives pr [læmiʃ], DEPN, L 53.

LAMAS Norf TG 2422. 'Loam marsh'. *Lammesse* *[1044×7]13th S 1055, c.1150, *Lamers* 1086, *Lammasse* 1186. OE **lām** + **mersc**. DEPN gives pr [læma:ʃ], Forsberg 56.

LAMBERHURST Kent TQ 6736. Probably 'wooded hill of the lambs'. *Lamburherste* c.1100, *-herst -hurst* 1199, c.1200, *Lamberhurst* from 1226 with varants *-hurste -herst(e) -hirst*. OE **lamb**, genitive pl. **lambra**, + **hyrst**, Kentish **herst**. One spelling *Lamburneherst* 1270 suggests an alternative etymology 'wooded hill of Lamburn, the lamb or loam stream', OE **lamb** or **lām** + **burna**, but it is impossible to know whether this is the original form or an example of 13th century folk-etymology. PNK 200, L 197.

LAMBETH GLond TQ 3074. 'Lamb landing-place'. *Lambehiðe -hyðe* *[1062]13th S 1036, c.1150 ASC(E) under year 1041–1608 with variants *-hethe* 1240–1323 and *-huthe* 1263–1410, *Lamhyth(a) -heth(a) -huth(e)* 1088×9–1365, *Lambeth* from 1255. OE **lamb** + **hȳth**. A place where lambs were embarked or landed, referring like Rotherhithe TQ 3680, *Retherhith -heth(e) -huth(e)* 1127–1408, *Rotherhuthe -hethe* 1238–1358, 'cattle landing-place', OE **hryther**, to traffic in animals on the Thames. Before the embankment of the river it was largely marshland and polders. The DB form *Lanchei* is either anomalous or represents a variant form of the name 'lamb quay' with OFr **kay**. Sr 22, 28, Encyc 454.

LAMBLEY Northum NY 6858. 'The lamb pasture'. *Lambeley(a)* 1201, 1242 BF, *Lambeleye* 1256, *Lamley* 1296 SR, 1542. OE **lamb**, genitive pl. **lamba**, + **lēah**. NbDu 131, DEPN, L 206.

LAMBLEY Notts SK 6245. 'The lambs' pasture'. *Lābeleia* 1086, *Lameleya -leia -leg(h) -ley(e)* 1192–1333, *Lamleye* 1280. OE **lamb**, genitive pl. **lamba**, + **lēah**. Nt 171, L 206.

LAMBOURN Berks SU 3379. 'Stream where lambs are washed'. *(æt) Lambburnan* [873×88]11th S 1507, *(æt) Lamburnan* [962×91]11th S 1494, *Lam- Lanborne* 1086, *Lamburna -(e) -born'* 1156 etc. OE **lamb** + **burna**. Lambourn was distinguished from Upper LAMBOURN SU 3180 from the 13th cent. by the prefix **cheping** (OE *ċēping*) 'market', *Chepinglamburn(e)* 1227 etc., *Chipping Lamborne* 1754. Brk 333, L 18.

River LAMBOURN Berks SU 4370. 'Lamb stream'. *Lamburnam* [943]13th S 491, [955×9]12th S 665, *(on) lámburnan* 949 S 552, *(ut on) lamburnan* [956]13th S 622, *(on) lam burnan* [958]12th S 577, *(on) lamburnam* [968]13th S 761, *Lamborn'* 1376×7, *Lambourn* 1441. OE **lamb** + **burna**. The alternative explanation of the specific as OE *lām* 'loam' is based on the form in S 552. Against it must be set the form *Lambburnan* S 1507 cited under LAMBOURN SU 3379. Brk 12, RN 236.

Upper LAMBOURN Berks SU 3180. *Vplamburn'* 1182 etc. with variants *Up- Hup-* and *-born' -burn(e)*, *Upper Lamborn* 1761. Adj. **upper** + p.n. LAMBOURN SU 3379. Upper Lambourn lies higher up the valley than the main settlement at Lambourn or Chipping Lambourn as it was called for distinction. Brk 333.

LAMBOURN DOWNS Berks SU 3392. *Lamborn Downs* 1761. P.n. LAMBOURN SU 3379 + ModE **down**, pl. **downs**. Brk 339.

LAMBOURNE END Essex TQ 4894. P.n. Lambourne + ModE **end**. Lambourne at TQ 4896, *Lamburnā* 1086, *-bourne* from 1291, is the 'lamb or loam stream', OE **lamb** or **lām**, + **burna**. Ess 60.

LAMBRIGG FELL Cumbr SD 5894. *Lambrigg Fell* 1709, p.n. Lambrigg + **fell**. Lambrigg, *Lambrig -rigg(e) -ryge* [1154×89]1294, 1190×1200–1689, is 'lamb ridge', ON **lamb** + **hryggr** possibly replacing OE **lamb** + **hrycg**. We i.134, SSNNW 141, L 169.

East LAMBROOK Somer ST 4318. *Estlambrok* 1268. ME adj. **est** + p.n. *Landbroc* *[1065]18th S 1042, 1227, *Lambrok* 1201, 1610, originally a r.n. probably meaning 'boundary brook', OE **land** + **brōc**. 'East' for distinction from West and Mid Lambrook ST 4418, 4318, *West Lambrook, Mid. Lambrook* 1811 OS. DEPN.

LAMBS GREEN WSusx TQ 2136. No early forms. Surname *Lamb* + ModE **green**.

LAMERTON Devon SX 4576. 'Settlement on the Lumburn'. *Lambretone* 1086, *Lamberton(e)* 1232–84, *Lamerton(')* from 1242. Earlier simply *(of) lamburnan* [c.970]n.d. B 1247. R.n. Lumburn + OE **tūn**. Originally the settlement was simply called after the stream, the Lumburn, 'the loam-stream', OE **lām** + **burna**, with subsequent development of *ā > ọ > ọ > ŭ* before <m> > ʌ. The pasturage here is bad for sheep but the soil is loamy so that derivation from OE *lamb* is unlikely. The forms *Lambre* [1154×89]15th and *Lamber* 1750 are back-formations from *Lamberton*. D 185 gives pr [læmətən], 9, Forsberg 1950.55.

LAMESLEY T&W NZ 1558. Probably 'the lamb's pasture'. *Lamesleie* c.1200, *-leya* early 13th, *-legh'* 1248, *Lambesley* 1290×1318, *Lamesley* from 1485, *Laymsle* 1485, *Lammesley* 1287. OE **lamb** or pers. n. *Lambi*, genitive sing. **lambes**, *Lambes*, + **lēah**. The final *b* of *lamb* was lost early as in some forms of LAMBETH GLond TQ 3074. NbDu 131, L 206.

LAMONBY Cumbr NY 4135. 'Lambin's village or farm'. *Lambeneby* 1257, *Lamben- Lambineby* 1276–1367, *Lamenby* 1301, *Lamonbye* 1565. ME pers.n. *Lambin* (a diminutive of CG *Lambert*) + **bȳ**. The name may have been introduced from Flanders where St Lambert of Maastricht was venerated. There is an exact parallel at Lamonbie, a lost vill in Applegarth, D & G NY 1084, *Lammynby* 1488, *Lamanby* 1505, *Lamynby* 1535. Cu 240, SSNNW 34.

LAMORNA Corn SW 8950. 'Valley of the *Mornow*'. *Nansmorno* 1302, *-ou* 1319, *Lamorna alias Nansmorna* 1387, *Nansmorna* 1460 Gover n.d., *Lamorney* 1584 Gover n.d. Co **nans** + stream name *Mornow* of unknown meaning. PNCo 105, Gover n.d. 618.

LAMORRAN Corn SW 8741. 'Church site of Moren'. *Lannmoren* [969]11th S 770, *Lan-* 1268, *Lam(m)oren* 1194, 1291, 1342, *(Seint) Moren* 1525, *Moran* 1610. Co ***lann** + saint's name *Moren* of whom nothing is known. There is a St Moran honoured in Brittany. PNCo 105, Gover n.d. 471.

LAMPLUGH Cumbr NY 0820. Partly obscure. *Lamplou* c.1150–1279, *Lam- Landplo(gh) -ploh -plou* c.1160–1419, *Lamplew* 1419, *Lamplugh* 1580. Apparently a compound of PrW ***lann** 'enclosure, church' (< Brit **landā* as in *Vindolanda*) + the same element as in Nanplough in Cury Corn SW 6721, *Lamploigh* 1334, *Nansblogh* 1334, 1356, *Lamprow* 1757, 'bare valley', OCo **nant** + **blogh** 'bare, bald' (W *blwch*). For other examples of *nant* replaced by *lann* see LAMORNA Corn SW 4424, LANGORE Corn SX 2986 and CPNE 143. Cu 404 gives pr [lampla], Jnl 1.48, Ogam 7.81, CPNE 23, L 121–2.

LAMPORT Northants SP 7574. 'The long town, the long market-place'. *Langeport* 1086–1428, *Langport al. Lamport* 1559. OE **lang**, definite form **langa**, + **port**. Nth 127.

LAMYATT Somer ST 6535. 'Lamb gate'. *Lambageate* [955×9]13th S 1756, *Lamieta* 1086, *Lamiet(e) -iet(t)e* 1185–1249, *Lamyat* 1610. OE **lamb**, genitive pl. **lamba**, + **ġeat**. The reference is to the 'gate' or gap in the horseshoe of hills surrounding Evercreech on the opposite side of which lies DITCHEAT ST 6236. DEPN.

LANA Devon SX 3396. This is thought to be the place from which Lucas atter Lane was named 1330; *atter lane* represents OE *æt thǣre lane* 'at the lane'. OE **lanu**. D 169.

LANCASTER Lancs SD 4761. 'The Roman fort on the river Lune'. *(Cherca)loncastre* 1086, *Lanecastrum* [1094]15th, *Loncastra* 1127, *Lancastre* c.1140–1246 etc., *Lancaster* from 1262, *Langcastre -kastre* 13th. R.n. LUNE SD 6075 + OE **cæster**. La 174, Jnl 17.99.

LANCASTER CANAL Lancs SD 5041. P.n. LANCASTER + ModE **canal**. The Preston and Lancaster canal was begun in 1793 and carried on to Kendal by 1819. Pevsner 1969.26.

LANCASTER SOUND Cumbr SD 3366. P.n. LANCASTER SD 4761 + ModE **sound** 'a channel'.

LANCHESTER Durham NZ 1647. '(The) long Roman fort'. *Langescestre* c.1149×52, *Langecestr(e)* 1197–1304, *Lang-* [1183]c.1320, 1281–1408, *Langchestr(e)* [1183]c.1382–1509, *-chester* 1549, 1569, *Lanchester* from 1380. OE **lang**, definite declensional form **lange**, + **ċeaster**. This was RBrit

Longovicium, Longouicio [c.420]15th ND. Possibly the first element of this name, *longo-*, lies behind OE *lang, langa*. *Longovicium* is thought to mean 'place of the *Longovices*, the ship-fighters'. NbDu 131, RBrit 398, Nomina 4.32.

North LANCING WSusx TQ 1804. *Northlauncynges* 1296. ME adj. **north** + p.n. *Lancinges* 1086–1265, *Lanzinges, Launching* 13th, *Launcyng(e)* 1288–1633 with variants *Launs-* and *-ing*, 'the Hlancingas, the people called after Hlanc(a)', OE folk-n. **Hlanċingas* < pers.n. *Hlanc(a)* + **ingas**. The OE pers.n. Wlencing, one of the sons of Ælle, the supposed founder of the South Saxon kingdom, has been used as ground for proposing folk-n. **Wlanċingas*, **Wlænċingas*, < pers.n. **Wlanca*, **Wlænċa*, but the absence of any sps *Wl-* makes this unlikely. Cf. LINCHMERE SU 8631. Sx 199, ING 36, MA 10.23, SS 64.

South LANCING WSusx TQ 1803. *Suthlauncynges* 1235. ME adj. **south** + p.n. Lancing as in North LANCING TQ 1804. Sx 199.

LANDBEACH Cambs TL 4765. Properly Land Beach, 'Beach by the land'. *Land(e)bech(e)* 1218–1480, *Lan- Lang(e)- Lambech(e)* 1268–1547. ME **land** + p.n. *Bece, Bech* 1086, *Bech(e)* 1242–1434, '(at) the ridge', OE **bæċ**, locative–dative sing. **beċe**. 'Land' for distinction from WATERBEACH TL 4965: both places are situated on a ridge of higher ground N of Cambridge, Landbeach inland from the Granta. Also known as *Inbeche* c.1250–c.1330, 'the inland ridge'. Cf. the form *Vtbech* for WATERBEACH. Ca 179, L 12, 126.

LANDCROSS Devon SS 4523. Uncertain. *Lanchers* 1086, *Lancars(e) -kars -chars* 1242–1348, *Lancras* 1318, *-crasse* 1624, *Landcross* 1822. This is either a late rationalisation of OE **hlanc** 'long, lank' + **ears** referring to a prominent rounded hill here, or perhaps rather an unknown name in Co ***lann** 'church-site'. D 94.

LANDFORD Wilts SU 2518. Either 'the long ford' or 'the lane ford'. *Langeford* 1086–1332, *Laneford* 1242–1439, *Landeford* 1295, *Lanford* 1525, *Landford* from 1594. Either OE **langa**, definite form of **lang**, or **lanu** + **ford**. The stream is a small one so that 'ford crossed by a lane' seems more likely than *lanu* in the sense 'the slow-moving part of a river'. Wlt 386.

LANDFORD MANOR Wilts SU 2620. Modern LANDFORD SU 2518 has developed at what was originally *Landford Bridge* 1811 OS and along the road across *Landford Common*. Landford manor is now used for the original village at the church site. Landford manor itself is a 16th century house with a front of 1712. Pevsner 1975.291.

LANDKEY Devon SS 5931. 'Church-site of St Ke'. *Landechei* 1166, *Lan- Londekey(e)* 1225–1428, *Lankey* 1604. Co ***lann** + honorific prefix ***to-** + saint's name *Ke*. This saint's name occurs also in Old Kea Corn SW 8441, *Sanctus Che* 1086, *Landighe* 1086, *Kee* 1440, LEIGH-ON-STREET Somer SS 9032, *Lantokai* BBCS 14, 1950–2, 113, and Llandygai, Caernarvonshire. D 341 gives pr [læŋki], CPNE 219.

LANDRAKE Corn SX 3760. 'Clearing'. *Landerhtun* 1018 S 951, *Landrei* 1086, *Lanrak -rach* 1291–1501, *Larrake* c.1605. *Landrac* 1262, *Landrake* 1547. Co **lanherch**. Cf. LANERCOST Cumbr NY 5563, LANNER SW 7139, MUCHLARNIK SX 2156. PNCo 105 gives local pr ['larik] as in Larrick SX 3078 and 3280. The first form is compounded with OE **tūn**. Gover n.d. 226.

LANDSCOVE Devon SX 7766. No early forms.

LAND'S END Corn SW 3425. 'End of the land'. *Landeseynde* 1337, *the Londesende of Engelond* 1427, *le Londys-ende* 1478 Gover n.d. ME **land**, genitive sing. **landes**, + **ende**. Translated into Co as *Pen an gluas, Pen an ulays* 1504, Co **pen** 'end' + definite article **an** + **gulas** 'land' with lenition of *gw-* to *w-*. Also known as *Inglendesende* 1311 and earlier as *Penwihtsteort* c.1121 ASC(E), *Penwæðsteort* 11th 997 ASC(D) both under year 997, *Penwiðsteort* 1052 ASC(C,D), 'the tail of Penwith', p.n. PENWITH + OE **steort**. PNCo 105, Gover n.d. 662, DEPN.

LANDULPH Corn SX 4361. 'Church-site of Dylyk or Delek'. *Landelech* 1086, *Landylp -dilp(e)* 1280–1610, *-dulp(e)* 1333, 1432, 1511, *Landhilp* or *Landylik* c.1485. Co ***lann** + pers.n. *Delek* or *Dilic*. The church at Landulph is dedicated to St Leonard. It is uncertain whether the pers.n. should be linked to St Dilic, one of the children of Broccan, or St Illick or the Breton St Dilecq, or why the final consonant developed from *k* to *p* and *f*. PNCo 106, Gover n.d. 229.

LANDWADE Cambs TL 6268. 'Estate or district ford'.

I. *Langwaðe* [1060]14th, *Lang(e)worth* 1235, *-wath -uad(e) -wad(e)* 1246–1476.

II. *Landuuade* 1066×87, *Landwad(e)* 1236–1475, *wath(e)* 1195–1366, *Lanwad(e)* 1236–1682, *Landewade* 1298–1419.

It is impossible to be sure whether the orgininal form was OE *langa-wæd* (with later substitution of ON *vath*) 'the long ford' or *land-wæd* 'the land ford', but the consensus of opinion seems to be towards the latter. The sense of *land-wæd* is also unknown; the site is on the Cambridge-Suffolk border and the sense may be 'ford crossing from one county or district to another'. The county boundary here makes a very odd projection into Cambridgeshire round Newmarket. Ca 194.

LANDYWOOD Staffs SJ 9906. 'The clearing in the wood'. *Laund i' th' Wood* 16th, *Londewood* 1670, *Landywood* from 1695. ME **launde in the wode**. St i.71.

LANE END Bucks SU 8091. *Lane End* 1822 OS. Short for 'lane end of Fingest' and contrasting with Bolter End SU 7992, CADMORE END SU 7892, ROCKWELL END SU 7988 and Wheeler End SU 0893.

LANEAST Corn SX 2283. Partly obscure. *Lanast* [1076]15th, c.1200, 1373, *-ayst(e) -eyst* 1311–1464, *Lanerst* 1291, *Lannest(e)* 1548, 1554 Gover n.d. Co ***lann** 'church site' + unknown element, possibly a pers.n. The church is dedicated to SS Sidwell and Gulval. PNCo 106, Gover n.d. 65, DEPN.

LANEHAM Notts SK 8076. '(The settlement) at the lanes' or 'at the streams'. *Lanun* 1086–1253, *-um* 1185–1324, *-om* 1267, 1428, 1504, *Laneham* 1543. OE **lanu, lane**, locative-dative pl. **lanum**. It has been suggested that here and in the r.n. ASLAND Lancs SD 4524 OE *lanu* referred like Scots dial. **lane** to a brook whose movement is scarcely perceptible, the smooth, slow-moving part of a river. Laneham is situated in marshy land by the Trent. Nt 53, L 78.

LANERCOST PRIORY Cumbr NY 5563. P.n. Lanercost + ModE **priory**. Also known as *Abbay Leonardcoast* 1671, *Abbey Lander coast* 1712. A house of Augustinian canons was founded here in c.1166. Lanercost, *Lanrecost* 1169–1402, *Lanercost(e)* from 12th, *Lanercoaste* 1603, is partially obscure. The first element is thought to be PrW, W **lannerch** 'open space, open space in a wood, a clearing, a glade', the second possibly the pers.n. *Awst* (< Latin *Augustus*). The name is in Celtic word order. Cu 71, CW 1953.67, Jnl 1.48, CPNE 142, GPC 2095.

LANESHAW BRIDGE Lancs SD 9240. R.n. Laneshaw + ModE **bridge**. There are no early forms for the river Laneshaw which must remain unexplained.

LANESHAW RESERVOIR Lancs SD 9441. R.n. Laneshaw as in LANESHAW BRIDGE SD 9240 + ModE **reservoir**.

LANESHAW RESERVOIR NYorks SD 9441. P.n. Laneshaw as in *River Laneshaw* and *Laneshaw Bridge* 1857 OS, a combination of *Lane* and *Shaw* shown separately on 1857 OS, ModE **lane** + **shaw**, + ModE **reservoir**.

LANEYGREEN Staffs SJ 9607. *Loany Green* 1704, *Lowney Green* 1834, *Lanes Green* 1775, *Laney Green* 1817. Possibly dial. **loning** 'a lane' + **green**; but the forms are too late for certainty. St i.113.

LANGAR Notts SK 7234. 'The long gore or triangle of land'. *Langare* 1086–1422, *Langar* from 1162. OE **lang** + **gāra**. The reference is to a long low hill-ridge that stretches E from SK 7133. Nt 227.

LANGBAR NYorks SE 0951. 'The long hill'. *Lang(e)berh(e)* 1199, 1344, *-berg(h)e* 1203–4, 1616, *-bargh(e) -barre* 16th cent., *-ber*

1694. OE **lang**, definite form **langa**, + **beorg**. Identical with LANGBAURGH NYorks NZ 5511. YW v.69.

LANGBAURGH NYorks NZ 5511. 'The long hill'. *Langebergh(e)* 1231–[1335]15th, *Langberg* [13th]15th, *Langbarge* 1572, *Langbarff* 1599. OE **lang**, definite form **langa**, + **beorg**. The long, narrow ridge was the meeting point of the wapentake to which it gave its name, *Langeberg(e) Wapentac* 1086–[1339]15th, and which was subsequently adopted for the administrative district created in 1974. Pronounced [læŋgba:f]. YN 165, 128, Room 1983.62.

LANGCLIFFE NYorks SD 8265. 'The long bank'. *Lanclif(f)* 1086, 13th cent., *Lang(e)clif(f) -clyf(f)* c.1160–1560. OE **lang** + **clif**. A long steep bank above the river Ribble terminating at its N end in a great limestone cliff. YW vi.147.

Little LANGDALE Cumbr NY 3103. *Langedenelit(t)le* 1157×63, *Lyttyllangedene* 1390, *Little Langdale* 1704. Adj. **little** + p.n. Langdale as in LANGDALE PIKES NY 2707. 'Little' for distinction from Great Langdale NY 3006, *Great Langdale* 1706. We i.203.

LANGDALE END NYorks SE 9391. P.n. Langdale + ModE **end**. Langdale is *Langadale* [c.1200]15th, *Lang(e)dale* [1335]14th 'the long valley', OE **lang**, definite form **lange**, + **dæl**. YN 112.

LANGDALE FELL Cumbr NY 6500. *Langdale fells* 1615. P.n. Langdale NY 6405 + ModE **fell** (ON *fjall*). Langdale, *Langedal(a) -dale* 1178–1337, *Langdale* from 1310, is 'the long valley', OE **lang**, definite form **langa**, or ON **langr**, + ON **dalr**, presumably referring to the long side-valley opening from the Lune at NY 6405 rather than the valley of the Lune itself. The form *Langedena* 1179 may suggest that the original form of the name was purely English, **lang(e)** + **denu** but it is also possible that there has been confusion with Langdale as in LANGDALE PIKES NY 2707. We ii.47, 43, SSNNW 236.

LANGDALE PIKES Cumbr NY 2707. *Langdale Pikes* 1787. P.n. Langdale NY 3006 + ModE **pike** 'a pointed summit, a peak' (OE *pīc*). Langdale, *Lang(e)den(e)* 1179–1630, *Langdal(e)* from 1578, is 'the long valley', OE adj. **lang**, definite form **lange**, + **denu** later replaced by **dale** (<ON *dalr*). We i.203, 206, SSNNW 194.

East LANGDON Kent TR 3346. *Estlangedoun* 1291. ME adj. **est** + p.n. *(to) langandune* [861]13th S 330, *Langedone* 1201×2, 'the long hill', OE **lang**, definite form oblique case **langan**, + **dūn**, dative sing. **dūne**. 'East' for distinction from West LANGDON TR 3247. KPN 159, DEPN.

West LANGDON Kent TR 3247. *Westlangedone* 1291. ME adj. **west** + p.n. Langdon as in East LANGDON TR 3346. DEPN.

LANGDON BECK Durham NY 8531. 'Langdon stream'. *Langdon flu:* 1576 Saxton, *Langdon Beck* 1866 OS. P.n. Langdon + ON **bekkr**. Langdon, *Langdon* 1768 Armstrong, is the 'long hill', OE **lang** + **dūn**, referring to Chapel Fell which stretches from Noon Hill NY 8535 to Swinhope Head NY 8933.

LANGDON COMMON Durham NY 8533. *Langdon Common* 1866 OS. P.n. *Langdon* as in LANGDON BECK NY 8531 + ModE **common**.

LANGDON HILLS Essex TQ 6786. *Langdon Hill(e)s* 1485, 1537. P.n. Langdon + pleonastic ModE **hill(s)**. Langdon, *Langenduna* 1086, *Langedon(e)* 1203–1344, is '(place) at the long hill', OE **lang**, definite form oblique case **langan**, + **dūn**. Some forms, e.g. *Layndon hill* 1594, show confusion with nearby LAINDON TQ 6889. Ess 162.

LANGENHOE Essex TM 0018. '(Place at the) long hill-spur'. *Langhou* 1086, *-(e)ho* 1168–1274, *Langenho(o) -howe* 1252–1365. OE adj. **lang**, definite form dative case **langan**, + **hō**, dative sing. of **hōh**. The reference is to a hill-spur which thrusts out between Geeton creek and Pyefleet channel. Cf. nearby FINGRINGHOE TM 0220. Ess 316, L 168.

LANGFORD 'The long ford'. OE **lang**, definite form **langa**, + **ford**. PNE i.181, L 68–9.

(1) ~ Beds TL 1841. *langa(n)forda* 980×90 S 1497, *Longaford* [944×6]13th S 1497, *Langeford* 1086–14th, *Langford* 1428. Bd 106.

(2) ~ Devon ST 0203. *Langeford* 1086, 1299. D 561.

(3) ~ Essex TL 8408. *Langefort, Langheforda* 1086, *Langford* from 1252. Ess 304.

(4) ~ Oxon SP 2402. *Langefort* 1086 *-ford(ia)* 1155×8–1393, *Langford* from 1316. O 327.

(5) ~ BUDVILLE Somer ST 1122. 'L held by the Budville family'. *Langford Budevill* 1305. Earlier simply *Langeford* 1086, 1212, *Langford* 1610. The manor was held by Richard de Buddevill before 1212. His surname probably represents Boudeville, Seine-Maritime, *Bodiville* c.1160. DEPN, Beaurepaire & Laporte, *Dict. topographique du Dép. de Seine-Maritime*, Paris 1982–4, s.n. Boudeville.

(6) ~ END Beds TL 1654. *Langford End* 1835 OS. 'End of Langford'.

(7) Hanging ~ Wilts SU 0337. 'The part of L at the steep slope'. *Hangindelangeford* 1242. ME adj. **hanginde** + p.n. *(æt) langanforda* [963 for 943]13th S 1811, *Langeford* 1086, 1224, a ford across the river Wylye. 'Hanging' for distinction from Little LANGFORD SU 0436 and Steeple LANGFORD SU 0437. Wlt 227, L 68–9.

(8) Little ~ Wilts SU 0436. *Parva Langeford* 1211, *Lutelanggeford* 1307×26, *Litelangeford* 1332. ME adj. **litel** (OE *lȳtel*), Latin **parva**, + p.n. Langford as in Hanging LANGFORD SU 0337 and Steeple LANGFORD SU 0437. Wlt 227.

(9) Lower ~ Avon ST 4560. ModE adj. **lower** + p.n. *Langeford* 1086. The reference is to a crossing of a tributary of the r. Yeo. 'Lower' for distinction from Upper Langford ST 4659, *Upper Langford* 1817 OS. DEPN.

(10) Steeple ~ Wilts SU 0437. 'The part of L with the church steeple'. *Stupel- Stepellangeford* 1294, *Steple Langford* 1305, 1558×1601, 1605. ME **stepel** (OWS *stīepel, stȳpel*) + p.n. Langford as in Hanging LANGFORD SU 0337 and Little LANGFORD SU 0436. Wlt 227.

LANGFORD Notts SK 8258. 'Landa's ford'. *Landeford(e)* 1086–1399, *Lan(d)ford(e) -forth* 1327–1735, *Langford* from 1630. OE pers.n. **Landa* + **ford**. Nt 205.

LANGHAM Essex TM 0233. 'Homestead of the Lahhingas, the people called after Lahha'. *Laingaham* 1086, *-(e)ham* 12th, 1239, *Laingheham* 1146×8, *Lehyngeham* 1153, *Leingeham* 1190×6, *La Wingeham* 1130, *Lawingeham* 1197, *lauhinge- Lavinga- Lahingeham* 1139–89, *Laghenham* 1254, *Leng(e)ham* 1238–91, *Langham* 1291×2. ME *Langham* is a folk-etymological form 'long homestead' for original OE **Lahhingahām*, the 'homestead of the Lahhingas', OE folk-n. **Lahhingas* < pers.n. **L(e)ahha* + **ingas**, genitive pl. **Lahhinga*, + **hām**. Ess 395, ING 122.

LANGHAM Leic SK 8411 'The long homestead'. *Langeham* 1154×89–1334, *Langham* from 1202. OE **lang**, definitive declensional form **langa**, + **hām**. Lei 692.

LANGHAM Norf TG 0041. 'The long homestead or village'. *Langham* from [1047×70]13th S 1499, *Langaham* 1086. OE **lang**, definite form **langa**, + **hām**. DEPN.

LANGHAM Suff TL 9769. 'The long village or homestead'. *Langham* from 1086, *Langeham* 1205. OE **lang**, definite form **langa**, + **hām**. DEPN.

LANGHO Lancs SD 7034. A modern development named from Old Langho SD 7035, *Langeshag'* 1203, *Langale* [13th]14th, 'long haugh', OE **lang** + **halh**, dative sing. **hale**. The earliest spelling, however, points to OE *langa sceaga*, 'the long copse'. La 72, Jnl 17.46.

LANGLEE CRAGS Northum NT 9622. *Langlee Crags* 1869 OS. P.n. Langlee as in LANGLEEFORD NT 9421 + ModE **crag**. The reference is to a hill rising to 1390ft.

LANGLEEFORD Northum NT 9421. *Langleeford* 1869 OS. P.n. Langlee NT 9623 + **ford**. Langlee is 'the long clearing or pasture', OE **lang(a)** + **lēah**. It is situated in the long narrow valley of Harthope Burn.

LANGLEY 'The long wood or clearing'. OE **lang**, definite form **langa**, + **lēah**. PNE ii.15, 18, L 205.

(1) ~ Berks TQ 0079. *Langeley* 1208, *Langley Maries* 1546, *Langley Marsh* 1826, *Langley Marish* 1925. The late added suffix comes from the *Mareys* family who once held the manor. Bk 241.
(2) ~ Ches SJ 9471. *Langel'* 1286, *le longlegh* 1384, *Longley* 1467, 1611, *Langley* from 1508. Che i.150.
(3) ~ Essex TL 4435. *Lang(e)leg(a)* 1166, 1225, *-le(ie)* 1205, 1212. Ess 551.
(4) ~ Hants SU 4401. *Langlie* 1086, *(boscus de) Langele* 1201–98. Ha 106, Gover 1958.198.
(5) ~ Herts TL 2122. *Langeleye* 13th cent. Hrt 17.
(6) ~ Kent TQ 8051. *(bituihn) longanleag* 814 S 173, *Langvelei* 1086, *Langele(g)* 1226–1242×3. KPN 132.
(7) ~ Warw SP 1963. *Longelei* 1086, c.1168, *-le* 1316, *Langeleia -lege -le(ye)* c.1140–1535. Wa 215.
(8) ~ WSusx SU 8029. *Langele* 1454. Sx 42.
(9) ~ BURRELL Wilts ST 9375. 'L held by the Burel family'. *Langelega -le(y)e Burel, Borel* 1289, 1327. P.n. *Langelegh* [940]14th S 473, *Langhelei, Langefel* 1086, + manorial addition from the family name of Peter Burel who held the estate in 1242, apparently a descendant of the undertenant Borel in 1066. Wlt 105.
(10) ~ HILL Glos SP 0029. *Longley hull* 1482. P.n. Langley + ME **hull** (OE *hyll*). Langley is *Langeleia* 1216×72, *Lang(e)ley* 1547, 1605. Gl ii.34.
(11) ~ MARSH Somer ST 0729. *Langley Marsh* 1809 OS. P.n Langley + ModE **marsh**. Earlier simply *Langle* *[1065]18th S 1042. DEPN.
(12) ~ PARK Durham NZ 2144. A modern colliery village. P.n. Langley + ModE **park**, possibly with reference to Bearpark. Langley is *Langleiam* [1183]c.1382, *Langeley* c.1200–1361, *Langley* from late 13th. NbDu 131.
(13) ~ STREET Norf TG 3701. *Langley Street* 1838 OS. P.n. Langley as in Langley Priory TG 3500 + ModE **street** 'a hamlet'. Langley Street lies ½ mile south of Langley, *Langale* 1086, *Langeleg* 1201, *-le* 1254. DEPN.
(14) Abbots ~ Herts TL 0902. *Abbotes Langele* 1263. ME **abbot** sc. of St Alban's, genitive sing. **abbotes**, + p.n. Langley as in Kings LANGLEY TL 0702. Hrt 44.
(15) Kings ~ Herts TL 0702. *Kyngeslangeley* 1346, *Lengele Regis* 1428. ME **king**, genitive sing. **kinges**, Latin **rex, regis**, + p.n. *(æt) Langalege* [1042×9]13th S 1228, *Langelai -lei* 1086, *-leie* [1177×99]1301. Also known as *Childe Langeleya* [c.1090]14th, *Childer Langele* 1338, *Chilternlang(e)ley(e)* 1346–1556, 'the children's L', OE **ċild** 'young nobleman, young monk', genitive pl. **ċildra** (later confused with p.n. Chiltern). The revenue of the estate was used for support of the monks of St Alban's. The manor belonged to the Honour of Berkhamstead which was annexed to the crown in the 12th cent. 'Kings' for distinction from Abbots LANGLEY TL 0902. Hrt 44.
(16) Kington ~ Wilts ST 9277. 'That part of L belonging to Kington'. *Langley Kington* 1636, *Kington Langley* 1699. P.n. Kington as in KINGTON ST MICHAEL + p.n. *Langhelei* 1086, *Langeleghe* 1243, *Longley* 1547. 'Kington' for distinction from LANGLEY BURRELL ST 9375. Also known as *Northlangele(ye)* 1289–1307, *North Langlay* 1544. Wlt 101.
(17) Kirk ~ Derby SK 2838. 'L with the church'. *Ky- Kirk(e)lang(e)le(ye)* 1281–1337 etc., *-long(e)le(ye)* 1269–1462. ME **kirk** + p.n. *Langelei* 1086, *-leya -le(y)e -leg'* 1230–1502, *Lang(e)le(ye) -legh* 1264–1452. 'Kirk' for distinction from Meynell LANGLEY SK 3039. Db 476.
(18) Meynell ~ Derby SK 3039. 'Langley belonging to the *Meynell* family'. *Longeleg' Meynill'* 1273, *Longelemeygnell* 1304, *Langelle Meynill* 1284×6, *Langeleye Meynill* 1376, *Meignyl Longeley* 1326. Earlier forms as under Kirk LANGLEY SK 2838. 'Meynell' for distinction from Kirk LANGLEY SK 2838. Db 476.

LANGNEY ESusx TQ 6302. '(At) the long island'. *Langelie* (sic) 1086, *Langley* 1672–1716, *Langania* 1121, *Langane* c.1121–14th. OE **lang**, definite form dative case **langan**, + **īeġ**. The reference is to a ridge of ground between Bourn Level and Pevensey Level. Sx 447 gives pr [laŋli], SS 71, L 39.

LANGNEY POINT ESusx TQ 6401. *Langney Point* 1813 OS. P.n. LANGNEY TQ 6302 + ModE **point**. The earlier name was *fallesia de Langneia* 'Langley cliff' 13th, *the Cliff* 1564. Latin **falesia**, ModE **cliff**. Sx 448.

LANGOLD Notts SK 5887. 'The long shelter'. *Langald(e)* 1246–1572, *Langold* from 1318. OE **lang** + **hald**. YW i.144.

LANGORE Corn SX 3068. Uncertain. *Langover* 1431–74. The evidence is too late for certainty but this might be 'valley with a stream', Co **nans** + **gover** with change of *Nans-* to *Lan-* as in LAMORNA SW 4424. An alternative possibility is 'long ridge', OE **lang** + **ofer**, referring to the ridge of high ground running EW from SX 2885 to 3286. PNCo 106, Gover n.d. 148.

LANGPORT Somer ST 4226. 'The long town or market'. *(to) langport* [914×9]16th BH1, *(to) langeport* [914×9]13th BH3, 4, *Longport* c.930 coins, *Lanporth* 1086, *Langeport* 1086, 1225, 1358, *Langport* 1251, [1308, 1358] Buck, *Lamport* [1228]B. OE **lang, long**, definite form **langa, longa**, + **port**. The town extends along a ridge rising above the surrounding marshes. DEPN, BH 114.

LANGRICK Lincs TF 2648. 'The long stretch of river'. *langraca* [1162]13th, *Langerak' both'* ('booth') 1243, *Langrake* 1260, *Langrick* 1824 OS. OE **lang** + **racu** 'a hollow, a stream, the bed of a stream', here in the sense 'reach, straight stretch of water' referring to the course of the r. Witham at this point. Identical with LONG DRAX NYorks SE 6828, *Langrak(e)* 13th–1616. Cameron 1998.

LANGRIDGE Avon ST 7469. 'The long ridge'. *Lancheris* 1086, *Langerig(ge)* 1225, 1276. OE **lang**, definite form **langa**, + **hrycg**. The reference is to a steeply descending ridge of land between streams at the top of which the medieval church stands. DEPN.

LANGRIGG Cumbr NY 1645. 'Long ridge'. *Langrug(ge)* 1189, 1191, *Langrig(g)* from c.1230. OE or ON **lang** + ON **hryggr**. The reference is to a long narrow ridge of higher ground stretching from Westnewton NY 1344 NE to NY 1745. Cu 304, SSNNW 142.

LANGRISH Hants SU 7023. 'The tall or long rush-bed'. *Langerische* 1236, *-risshe -ryshe -risch -rysshe* 1280–1490. OE **lang**, definite form **lange**, + **risc**. Ha 106, Gover 1958.62.

LANGSETT SYorks SE 2100. 'The long hill-side'. *Langeside -syde* 12th–1375, *Lang-* 1208×22–1608, *Langsett* from 1550. OE **lang**, definite form *lange*, + **sīde**. The reference is to the long steep slope on the N side of the Little Don valley. YW i.331, L 187.

LANGSETT RESERVOIR SYorks SE 2000. P.n. LANGSETT SE 2100 + ModE **reservoir**.

LANGSTONE Hants SU 7105. 'The tall or long stone'. *Langeston* 1289, 1810 OS, *Langestone* 1307. OE **lang**, definite form **langa**, + **stān**. The reference is unknown but may have been to a lost menhir. Ha 106, Gover 1958.15, DEPN.

LANGSTONE HARBOUR Hants SU 6802. *Langston Harbour* 1810 OS. Earlier *port of Langeston* 1364, *Langston Haven* 1687. P.n. LANGSTONE SU 7105 + ModE **harbour**. Gover 1958.9.

LANGSTROTHDALE CHASE NYorks SD 8979. *(foresta de) Lang(e)strod(e)* c.1190–1316, *(foresta, chacea de) Lang(e)stroth(e)* 1198–1598. P.n. Langstrothdale + **chase** 'a hunting forest', OFr **chace**, Medieval Latin **foresta**. Langstrothdale, *Lang(e)strother* 1266–1434, *Langstroth(e) Dale* from 1462, is the 'valley of the long stretch of marshy ground overgrown with brushwood', ME **lang** + p.n. *Strothdale*, OE **strōd** varying with **strōther** + **dæl**. The reference is to the long deep valley of the river Wharfe between Buckden and Beckermonds which gave its name to the medieval chase or hunting forest. YW vi.117 gives pr ['laŋstədil].

LANGTHORNE NYorks SE 2491. 'The tall thorn-tree'. *Langetorp* 1086, *Langethorn(e)* 1086–1350, *Langthorne* 1285, 16th cent. OE

langa + **thorn**. The DB form with ON **thorp** may be a mistake for **thorn**. YN 238.

LANGTHORPE NYorks SE 3867. 'Long Thorpe'. *Langthorp(e)* from [1300]15th, 1576. ME **long** + p.n. *Thorpe*. Short for *Langlivetorp* 12th, *Langle(i)thorp* [1157]15th, 1300–1, 'Langlif's Thorp', ON feminine pers.n. *Langlifa*, genitive sing. *Langlifu*, + p.n. *Torp* 1086, ON **thorp**. YN 180.

LANGTHWAITE NYorks NZ 0002. 'The long clearing'. *Langethwait* 1167–1341. ON **langa** + **thveit**. YN 296.

LANGTOFT Lincs TF 1212. 'The long curtilage'. *Langetof* (sic) 1086, *-toft(e)* 1167–1535, *Langtoft* from 1291. ON **langr** or OE **lang**, definite form **lange**, + **toft**. The village consists of two rows of tofts extending in a narrow line along Langtoft Outgang Road which gave access to Langtoft Common and Deeping Fen. Perrott 410, SSNEM 149.

LANGTON '(The) long village'. OE **lang**, definite form **langa**, + **tūn**.
(1) ~ Lincs TF 3970. *Langetune* 1086, *Langhetuna* c.1115. DEPN.
(2) ~ Lincs TF 2368. *Langetone* 1086, *Langhetuna* c.1115. DEPN.
(3) ~ NYorks SE 7967. *Lanton* 1086, *Langatuna* [1177×99]1308, *Langetun -ton(a)* [1169]13th, 1190–1353, *Langton* from 1303. YE 141.
(4) ~ BY WRAGBY Lincs TF 1476. P.n. *Langetone* 1086, *-tuna* c.1115 + p.n. WRAGBY TF 1378. DEPN.
(5) ~ HERRING Dorset SY 6182. 'L held by the Harang family'. *Langeton Heryng* 1336, *Langton Herynge* 1406. Earlier simply *Langetone* 1086, *Langetun* 1221. The Harang family was here from the 13th cent. Do 97.
(6) ~ MATRAVERS Dorset SZ 0078. 'L held by the Mautravers family'. *Lang(e)ton Maw- Mautravers* 1428–36, ~ *Matrevers* [1497]17th. Earlier simply *Langeton(e)* c.1165–1546. Also known as *Langeton Walisch* 1376, ~ *Wallis* c.1586 from the *le Waleys* family which was here 1276–85 etc. The *Mautravers* family was here 1281–1333 etc. Do i.33.
(7) Church ~ Leic SP 7293. 'L with the church'. *Ki- Kyrk(e)* 1315–1523, *Chi- Church(e)* 1316 and from 1509 + p.n. Langton as in East LANGTON SP 7292. Lei 237.
(8) East ~ Leic SP 7292. Latin prefix *alia* 'the other' c.1130, c.1299, *Est-* 1211–1536 + p.n. *Langhœtonœ* [1081]c.1131, *Lagintone* 1086, *Langeton(e)* 1086–1465, *Langton* from 1086. 'East' for distinction from West Langton SP 7192, *West Lang(e)ton(e)* 1211–1717. Lei 236.
(9) Great ~ NYorks SE 2996. P.n. *Langetun -ton* 1086 etc. with prefixes Latin *magna* 1285, ME *Mekyl* 1536, ModE **great**. YN 277.
(10) Little ~ NYorks SE 3094. *parua Langeton* 1292. Latin **parva**, ModE **little** + p.n. Langton as in Great LANGTON SE 2996. YN 277.
(11) Tur ~ Leic SP 7194. *Tur Langton* 1835 OS. Incorrectly for Turlangton on account of the proximity of East, West and Church LANGTON. See TUR LANGTON SP 7194.

LANGTON Durham NZ 1619. 'The long hill'. *Langadun* [c.1040]12th, [1072×83]15th, c.1107, *Langeton(e)* 1242×3–1351, *Langton* from 1336, *Lankton* 1585. OE **lang**, definite declensional form **langa**, + **dūn** later replaced by **tūn**. Langton is situated at the E end of a ridge of higher ground which extends for a mile and a half from West Side House NZ 1420. NbDu 131.

LANGTON GREEN Kent TQ 5439. *Lengthington Green* 1819 OS.

LANGTREE Devon SS 4515. 'The tall tree'. *Langtrewa* 1086, *Langetre(we) -triwe* 1228 etc. OE **lang**, definite form **langa**, + **trēow**. D 95.

LANGWATHBY Cumbr NY 5633. 'Village at *Langwath*'. *Langwadebi -by* 1159–1242, *-wathebi -by* 1227–79, *Langwathby* from 1242, *Langhumby* 1490, *Langanby(e)* 1576, 1777. P.n. **Langwath* (<ON **lang-vath** 'long ford') + **bȳ**. The reference is to a ford across the river Eden at which the settlement developed. Cu 218, SSNNW 35, L 82–3.

LANGWITH 'Long ford'. OE **lang** or ON **langr** + ON **vath**, L 52.
(1) ~ Derby SK 5269. A modern industrial settlement that grew up around *Whaley Furnace* 1840 OS, midway between Upper Langwith SK 5169, *Lang(e)wath(e)* 1208, *Langwad -wat* 1229–14th cent., *Langueth* 1245, *-weth* 1540, *Overlangwith* 1404 (also known as *Bassetlang(e)wath(e) -langwith* 1330–1613) and Nether LANGWITH SK 5370. Db 294, SSNEM 171.
(2) ~ JUNCTION Derby SK 5768. P.n. LANGWITH SK 5269 + ModE **junction**. A railway junction.
(3) Nether ~ Notts SK 5370. *Netherlangwat* 1252, 1328, *Netherlangwithe* 1540. ME **nether** + p.n. *Languath -wath(e)* [c.1169]1291–1394. 'Nether' for distinction from Upper LANGWITH SK 5169 further up the river Poulter. Nt 84, SSNEM 171.

LANGWOOD FEN Cambs TL 4385. *Langwodefen* 1251. P.n. *Langwode* 'the long wood', OE **lang**, definite form **langa**, + **wudu**, + **fenn**. Ca 251.

LANGWORTH Lincs TF 0676. 'The long ford'. *Longwathe* [1043×66]late 13th, *Langwat* c.1115, 1175×6, *Langwath* [1170]1291, 1202. ON **langr** or OE **long** + ON **vath**. Identical with the common Dan p.n. Langvad. Another Langworth occurs at TF 2456. DEPN, SSNEM 156, L 82.

LANHERNE or Vale of MAWGAN Corn SW 8964. Partly obscure. *Lanherwev* 1086, *-ueu* (for *-neu*) 1086 Exon, *Lanhern* c.1187, *-herno(n)* c.1250–1302, *La Herne* 1279, *Lanhorn* 1285, 1306, 1448. Co ***lann** 'church site' + unknown element, possibly a pers.n. **Hernow*. The old name of the church-town of ST MAWGAN SW 8765. The valley in which these two places are situated is known as *Vale of Lanherne* 1836, *Vale of Mawgan or Lanherne* 1906. PNCo 175, Gover n.d. 348.

LANHYDROCK HOUSE Corn SX 0863. P.n. Lanhydrock + ModE **house** referring to the 17th cent. mansion of the Robartes family, begun by Sir Richard Robartes, a Truro tin and wool merchant. Lanhydrock, *Lanhideroc* 1201, *Lanhidrok -hydrek* 1299–1342, *Lanhetheroock* 1610, is 'church-site of Hydrek', Co ***lann** + saint's name *Hydrek* of whom nothing is known. PNCo 106, Gover n.d. 342, Pevsner 1951.74.

LANIVET Corn SX 0364. 'Church-site at *Neved*'. *Lannived* 1268, *Lan(n)yvet* 1283–1443, *Lanivet* 1276, *Laneuet* 1610. Co ***lann** + p.n. **Neved*, OCo ***neved** 'pagan sacred place, sacred grove' < Brit **nemeto-*. PNCo 160, Gover n.d. 343, CPNE 172, Jnl 1.50.

LANK RIGG Cumbr NY 0912. *Lank Rigg* 1865 OS. A peak rising to 1795ft.

LANLIVERY Corn SX 0758. Probably 'church-site of Livri'. *Lanliveri -iri* c.1170–1291, *Lanlyvri* 1235, *Lanlivery* 1428. Co ***lann** + pers.n. **Livri*. There is a place in Brittany called Lanlivry, but the church here is dedicated to the female St Bryvyth, of whom nothing is known. PNCo 106, Gover n.d. 402.

LANNER Corn SW 7139. 'Clearing'. *Lan(n)ergh* 1311, 1327, *Lannarth* 1542. Co **lanherch**. PNCo 106, Gover n.d. 515.

LANREATH Corn SX 1856. 'Church-site of Reydhogh'. *Lavredoch* (for *Lan-*) 1086, *Lanreydhou* 1260, *-reythou* 1266, 1364, *-reitho* 1283, *Landreyth* 1377, *Lanreth* 1610, 1656. Co ***lann** + pers.n. **Reydhogh*. The church is dedicated to St Dunstan and an unknown St Manakneu. PNCo 107 gives pr [lan'reθ], Gover n.d. 263, DEPN.

LANSALLOS Corn SX 1750. 'Church-site of Salwys'. *Lansaluus -salhvs* 1086, *Lancelewys* 1283, 1312, *-selewys -is* 1291–1320, *-salwys* 1326, *Lansaloes* 1566, *-solas* 1563, *-sallos* 1610. Co ***lann** + pers.n. *Salwys* or *Selwys*. The church is dedicated to the otherwise unknown female saint Ildiern. PNCo 107, Gover n.d. 267, DEPN.

LANTON Northum NT 9231. 'The long settlement'. *Langeton* 1242 BF, 1255, *Langton* 1296 SR, *Lanton* 1638. OE **lang**, definite form **langa**, + **tūn**. A form *Langecestre* is also recorded 1296 SR 118. NbDu 132.

LAPFORD Devon SS 7308. Probably 'Hlappa's ford'. *Slapeford* 1086, *Eslapaforda* 1086 Exon, *Lap(p)eford* [1107]1300, 13th cent., *Lapford* 1272. The DB spellings with *Sl-* and *Esl-* are graphies for [çl], and point to OE pers.n. **Hlappa* or ***hlæpe**/***hlēapa** 'a lapwing' + **ford**. An alternative possibility, if the DB forms were discounted, would be OE **læppa** + **ford**, 'ford at the edge' sc. of the parish. D 369, Forsberg 61, Jnl 19.45, 50.

LAPLEY Staffs SJ 8712. 'The wood or clearing at the end' sc. of the estate or township. *Lappeley(e)* [c.1061]13th S 1237, 1291–1526, *Lepelie* 1086, *Lapley* from 1263. OE **læppa** + **lēah**. St i.167.

LAPWORTH Warw SP 1671. 'Hlappa's enclosure'. *Hlappawurthin* [816]11th S 179, *Lappawurthin* 11th, *Lapeforde* 1086, *Lappewrthe -worth* late 12th–1325, *Lapworth* from 1275. OE pers.n. **Hlappa* + **worthiġn**, later replaced by **worth**. Alternatively this might be 'lap enclosure', OE **læppe** in the sense 'skirt, outlying portion of a district'. Wa 288.

LARK STOKE Warw SP 1943 → LAVERSTOKE Hants SU 4948.

River LARK Suff TL 6476. *Lark* 1735, 1764. A back-formation from the p.n. LACKFORD TL 7970. The sp *Lark* for *Lac(k)* is due to fanciful association either with the bird-n. *lark* or with *lark* 'frolic' especially if the p.n. represents OE *lāc-ford*. RN 236–7.

LARKHILL Wilts SU 1244. A modern military camp begun in 1920×24 named from Lark Hill, *Larkehull* 1360, *-hill* 1617, 'lark hill', ME **larke** (OE *lāwerce*) + **hill** (OE *hyll*). Wlt 365.

LARLING Norf TL 9889. 'The Lyrlingas, the people called after Lyrel or Lyrla'. *Lur(l)inga* 1086, [1086]c.1180, *Lurlinges* 1180, *Lerlinges* 1223, *Lirlinge* 1242×3, 1254, *Lyrlinge* 1275, *Lyrlyngges* 1346. OE folk-n. **Lyrlingas* < pers.n.**Lyrel* or **Lyrla* + **ingas**. The present form is due to the late ME change *er* > *ar*. DEPN, ING 59.

LARRICK Corn SX 3078, 3280 → LANDRAKE SX 3760.

LARTINGTON Durham NZ 0117. 'Settlement on the stream called *Lerting*'. *Lyrtingtun* [c.1040]12th, *Lertinton* 1086, *Lyrtington -yng-* c.1130–1327, *Lertinctona* 1154×69, *Lirt(h)ington* 1252–1301, *Lertington* 1403. R.n. **Lerting* varying with **Lyrting* + **tūn**. **Lerting* or **Lyrting*, an *ing*[2] derivative of OE **lort(e)*, **lurt(e)* 'dirt, mud, a muddy place, a swamp', must have been the earlier name of Scur Beck on which Lartington stands. YN 308, RN 259, PNE ii.27, Årsskrift 1974.45.

LASHAM Hants SU 7105. 'The lesser estate'. *Lessam* c.1110, *Les(s)- Less(e)ham* 1195–1210, *Lasham* from 1174[†]. OE **lǣss**, definite form **lǣssa**, + **hām**. Lasham, a pre-conquest royal estate, was a detached part of Odiham hundred. Ha 107, Gover 1958.133, DEPN, WD 56.157.

LASKILL NYorks SE 5690. 'Low shielings'. *Lauesc(h)ales* 1170, 1200–1, *Laygskales* 1301. ON **lagr** + **skali**. Laskill is in the bottom of the deep valley of the river Rye. YN 72.

LASTINGHAM NYorks SE 7290. The 'island', later the '(monastic) estate of the Lastingas, the people called after Last'. *læstinga- lestinga eu, laestinga- laestenga eu (ig), Læstingæ, læstinga æi, lestinga ei* [c.731]8th BHE, *Læstinga ea* [c.890]c.1000 OE Bede, *Lestingaheu, Lestingaea* c.1130, *Lesting(e)ham* 1086, *Lestingham -yng-* [1086–9]15th–1665, *Lastingham* from 1393. OE folk-n. **Læstingas* < pers.n. *Læst* + **ingas**, genitive pl. **Læstinga*, + **ēġ** and later **hām**. The site of the Anglo-Saxon monastery here is a piece of land round which the Beck flows. The sense of **ēġ** is probably a 'patch of good land in moorland'. YN 60, BzN 1968.175, Nomina 17.134–6.

LATCHINGDON Essex TL 8800. Partly uncertain. *Læce(n)dune* *[1065]14th S 1040, *Lacedune* *[1065]13th S 1043, *Lachenduna -tunam, Lacen- Lessenduna* 1086, *Le- Lachedon* 1216–58, *Lachindon -yn-* 1246–1367, *-ing- -yng-* from 1297, *Lashin(g)don* 1594–1610. The generic is OE **dūn** 'a hill', the specific perhaps OE ***læċċe** 'a trap, a snare' as ME *latch*, genitive sing. ***læċċan**. Alternatively, since the early forms show only *-c-*, OE ***læċen** 'abounding in water channels' < **lǣċe* + *en*. This would be an irregular formation, but neither *lǣċe* 'a leech, a physician' nor *lǣċe* 'a leech' blood-sucking worm, fit the preponderance of ME *Lach-* spellings. Ess 216, Forsberg 33, PNE ii.10.

LATCHLEY Corn SX 4073. 'Grove of the wetland'. *Lachesleigh* [mid 13th]1619, *Lacchislegh* 1318, *Lacchelegh* 1337, *Lacheley* 1534 Gover n.d. OE ***læċ(ċ)**, ***læċe**, genitive sing **læċ(ċ)es**, + **lēah**. Below the village is flat, marshy ground, lying in a bend of the river Tamar. PNCo 107, Gover n.d. 181.

LATELY COMMON GMan SJ 6698. *Lately Common* 1843 OS.

LATHBURY Bucks SP 8745. 'Fortified place, manor house made of beams'. *Late(s)berie* 1086, *Lateberia -biry -bur(e)* 1167–1247, *Lathbury* from 1241, *Lat(t)hebury* 1276–1373. OE **lætt**, genitive pl. **latta**, ME **laththe**, + **byriġ**, dative sing. of **burh**. Bk 8.

LATIMER Bucks TQ 0099. *Lattimers* 1526. Short for *Isenhampstede Latymer* 1389, *Is- Eystramsted(e) Latemers* 1552, 'Isenhampstede held by the Latimer family', p.n. *Isenhampstede* + family name Latimer from William Latimer who held the manor in 1330. Isenhampstede, *Yselhamstede* 1220, *Isenhampstede* 1346, is 'Isa or Isela's homestead' as in CHENIES TQ 0198, the other division of the original estate of Isenhampstede. An alternative possibility is that the name shows *n/l* confusion and that the specific is an old r.n. **Isene* from OE *īsen* 'iron' for the Chess, itself a late back-formation from Chesham. Bk 225, D l, Stead 263, Jnl 2.32.

LATRIGG Cumbr NY 2724. Partly uncertain. *Latterigg* 1666, *Latrig(g)* 1769, 1784. Unidentified element + ON **hryggr** referring to a short ridge of hill rising to 1200ft. This element occurs again in the forms *Laterhayheved* 1220, *-hefed* 1256, *Laterayheved* c.1260, which belong here, p.n. *Laterhay* + OE **hēafod** 'head'. *Later* may represent either ON **látr** 'a lair' or OIr **lettir** (Gaelic *leitir*) 'a hill, a slope', and *hay* OE **(ġe)hæġ** 'an enclosure'. If this is right the original meaning of the name was 'head of the lair or hill enclosure'. Cu 321.

LATTERIDGE Avon ST 6684. 'Ladda's ridge'. *Laderugg(e)* 1176–1449 with variant *Ladderug'* 1221, *Latridgg* 1557. OE **ladda** 'lad, servant' or the by-name derived from this, *Ladda*, + **hrycg**. Gl iii.2.

LATTIFORD Somer ST 6926. 'Beggars' ford'. *Lodreford* 1086, *Lodereforda* 1086 Exon, *Loderford* 1243, *Lotterford* 1811 OS. OE **loddere**, genitive pl. **loddera**, + **ford**. DEPN.

LATTON Wilts SU 0995. 'The herb or vegetable garden'. *Latone* 1086, *Latton(a)* from 1133, *Lacton* 13th cent. OE **lēac-tūn**. Wlt 45.

LAUGHTERTON Lincs SK 8375. 'The vegetable garden'. *Lahtreton'* 1213×23, *Lactertun* 1227, *La(c)hterton'* 1253, 1272, *Laghterton* 1316. OE **leahtric** + **tūn**. Cf. LEIGHTERTON Glos ST 8291. Identification with *Leugttricdun* [675×92]12th S 1806 is very uncertain as this form contains *dūn* and Laughterton does not lie on a hill. DEPN, Cameron 1998.

LAUGHTON 'The leek enclosure or vegetable garden'. OE **lēac-tūn**.

(1) ~ ESusx TQ 5013. *Lestun -tone* 1086, *Lectone* 13th, *Lacton(e)* 1229–14th, *Laughton* from 1338. The DB forms show AN *s* for [ç]. Sx 402, SS 77.

(2) ~ Leic SP 6689. *Lachestone* 1086, *Lacton(')* 1200–54, *Laghton(')* 1273–1510, *Laughton(')* from 1402, *Lectone* 1219, *Leicton* 1223, *Lei- Leyton(e)* 1204–77. The name shows two developments; (1) *lēac-tūn* > *lǣctūn* > *læctun* > *lahton* > *lauhton* > *lauton* and (2) *lēac-tūn* > *lǣctūn* > *lēctun* > *lehton* > *leihton* > *leiton*. Lei 238.

(3) ~ Lincs SK 8497. *La(ce)stone, Lac- Loctone* 1086, *Lactuna* c.1115, *Lactun* 1212, *Lecton* 1209×35. DEPN.

(4) ~ -EN-LE-MORTHEN SYorks SK 5287. P.n. *Lastone* 1086, *Laghton(a)* c.1201–1522, *Lacton(a)* 13th, 14th cents., *Laughton* from 1546, + suffix *in le Morthing* 1574, *Lawton le Morthen* 1761, from the district name MORTHEN. YW i.141, 168, YAJ 58.23.

[†]DB *Esseham* 1086 is probably due to mistaken analysis of *Lesseham* as if *L'Esseham* with prefixed Fr definite article.

LAUNCELLS Corn SS 2405. Partly obscure. *Landsev* 1086, *Lanceles* 1204–61, *Launcel(e)s -celles* 1238–1435. Co ***lann** 'church-site' + unknown element. PNCo 107, Gover n.d. 11.

LAUNCESTON Corn SX 3384. '*Lan-Stefan* estate or manor'. *Lanscavetone* [for *Lanstave-*] 1086, *Lanstavaton* [c.1125]14th, c.1180, 1261, *Lanzave(n)ton* 1176, 1184, *Lanceuetona*, *Lanstaueton* early 13th, *Lancaveton* 1228, *Lanceton* 1303, *Launston* 1342 Gover n.d., 1561, 1610, *La(w)nson* 1478. P.n. **Lan-stefan* 'church-site of St Stephen' (Co ***lann** + saint's name) + OE **tūn**. The DB form actually refers to St Stephens, the original site across the river at SX 3285. In 1155 the canons of St Stephen moved to the present site which was up to then called *Dunhevet* 1086 'hill-end', OE **dūn** + **hēafod**. The Co form of the name survived until *Lesteevan* 1602. Pronounced ['la:ns(t)ən]. PNCo 107, Gover n.d. 145.

LAUNDE ABBEY Leic SK 7904. *Lawnde Abbey* 1617. P.n. Launde + ModE **abbey**, the name of a house occupying the site of a priory of Augustinian canons founded in 1125. Launde, *(la) Landa* 1155×60–1528, *(la) Launda* 1333, *(la) Laund(e)* from 1202, is OFr, ME **launde** 'an open space in woodland, a forest glade, a woodland pasture'. Lei 174.

LAUNTON Oxon SP 6022. 'The long settlement'. *Langtune* [1057×66]13th S 1139, *Lanton* 1086, *Langeton(e) -tona -tun(a)* 1206–1384, *Langton* 1428. OE **lang** + **tūn**. O 228 gives prs [lɔ:ntən, la:ntən].

LAVANT WSusx SU 8608. Transferred from the r. Lavant. *Lavent* 1227, *Lovente* 14th cent. Also known as *Estlavant* 1498 for distinction from Mid Lavant SU 8508, *Midlouente* 1288, *Medil Lavant* 1542. The forms of the r.n. are *Louente -v-* 12th, 1225, *Lauant -v-* 1610, a formation of uncertain origin. An alternative name for the place is *Loventone, Lovintune* 1086, *Loventon* 1121, *Mydloventon* 1264, *Midlave(n)ton* 1250, 'the settlement or estate on the r. Lavant'. RN 264 compares r.ns. Lavant, a tributary of the Drau in Carinthia, *Labanta* 860, *Laventa* 888, and Lafnitz, a tributary of the Raab in the Steiermark, *Labenza* 864, *Lavenza* 1126, both in Austria, but Flussnamen 53, 76 derives these names from Old European **Albant(i)a* < **albh-* 'white' with Slavic metathesis. Better, therefore, the root seen in Latin *lābor* 'glide', Ir *lobh* 'rot, decay', OIr *lobat* 'putrescent' with Old European r.n. suffix *-nta/-nto-*. Bahlow 307 compares the Fr r.ns. *Lova* (Louve), *Lovria, Loverna, Lovissa* and the Belgian p.n. Louvain, *Lovene*. Sx 50, SS 74.

LAVENDON Bucks SP 9153. 'Lafa's valley'. *Laven- Lawe(n)dene* 1086, *Lavenden(e)* 1201–15th, *Lavendon(e)* from 1232, *Laundone -den(e)* 1232–1639. OE pers.n. *Lāfa*, genitive sing *Lāfan*, + **denu**. Bk 9 gives pr [la:ndən], L 98.

LAVENHAM Suff TL 9149. 'Lafa's homestead'. *Lauanham* 962×92 S 1494, *Lauen(ham)* 1086, *Lave(n)ham* 1254. OE pers.n. *Lāfa*, genitive sing. *Lāfan*, + **hām**. DEPN.

High LAVER Essex TL 5208. *Heye Lafare -laffar'* 1285, *Hyghlaver* 1422. ME adj. **hegh** + p.n. *(æt) Lagefare* [1004×14]13th S 1495, [1042×66]12th, *Laghe- Lagafarā* 1086, *Lage- Lahefar(e)* 1170–1272, *Lau- Lawefar(e)* 1197–1301, *Lauvar(e)* 1291, 1325, 'stream passage', OE **lagu** + **fær**. The reference is to a stream-crossing by the Roman road from Great Dunmow to London, Margary no. 30. Also known as Great Laver, *Magna Lagefare* 1212, for distinction from Little LAVER TL 5409. Ess 61 gives pr [leivə], L 67, Jnl 2.40.

Little LAVER Essex TL 5409. *Little Laffour* 1555. Earlier *Parva Lagefare* 1212. ModE adj. **little**, Latin **parva**, + p.n. Laver as in High LAVER TL 5208. Ess 61.

Magdalen LAVER Essex TL 5108. '(Mary) Magdalen's L'. *Laufar(e) Magdalen(e)* 1256, 1269, *Mawdlen Laver* 1549. Also simply *Mawdelyn* 1460. Saint's name *Magdalen* from the dedication of the church + p.n. Laver as in High LAVER TL 5208. Ess 62.

River LAVER NYorks SE 2072. 'Babbling or noisy stream'. *Laver* from 1180. PrW **Labar* < Brit **Labarā* (W *llafar* 'vocal, resounding'). Identical with Laber, the name of four small streams in Bavaria, the Lièvre in Alsace, *Lebraha* 9th, the W *Llafar*, and *Labarus*, Silius Italicus 4.232. YW vii.130, RN 238, LHEB 272, 558, Holder ii.113–4, GPC 2084.

LAVERSTOCK Dorset ST 4200 → LAVERSTOKE Hants SU 4948.

LAVERSTOCK Wilts SU 1530. 'The lark outlying farm'. *Lavvrecestoches, Lavertestoche* 1086, *Laverkestok* 1221–50, *Laverstok* 1255, 1316, *Larkestok* 1310–11, *Larstocke* 1634. OE **lāwerce** + **stoc**. Wlt 381 gives former pr [la:rstɔk], Studies 1936.23.

LAVERSTOKE Hants SU 4948. 'Lark dependent farm'. *Lavrochestoche* 1086, *Laverchestoch* 1155×8–75, *Laverkestok(e)* 1236–1341, *Laverstok* 1219, *Larkstoke* 1378. OE **lāwerce** + **stoc**. Identical with LAVERSTOCK Wilts SU 1530, Laverstock Farm Dorset ST 4200, *Larkestok* 1244, *Laverstoke* 1285, Laverstoke (lost) Bucks TQ 0184, *Laverkestoke* 1138–68, and Lark Stoke Warw SP 1943, *Lavirkestoch'* 1236, earlier simply *Stoch* 1086. Ha 107, Gover 1958.137, Studies 1936.21–6.

LAVERTON Glos SP 0735. 'Lark estate'. *Lawertune* c.1160, *Laberton* 1329, 1505, 1687, *Lau- Laverton* from 1532. OE **lāwerce** + **tūn**. Gl ii.4.

LAVERTON NYorks SE 2273. 'Settlement on the river Laver'. *Laure- Lavreton(e)* 1086, *Lau- Laverton -ton* 12th–1294, *Lawreton* 1606, *La(y)o- Lareton* 15th cent. R.n. LAVER SE 2072 + **tūn**. YW v.211 gives prs ['la:tən, 'lɛətən, 'lavətən].

LAVERTON Somer ST 6926. Partly uncertain. *Lavretone* 1086, *Laurton* 12th, 1196, *La(u)werton* 1238, 1243, *Lauerton* 1610. The generic is OE **tūn** 'settlement, farm, village, estate', the specific possibly OE **lāwerce, lāferce** 'a lark', **lǣfer** 'a rush, a reed, a yellow iris, a reed bed' or a r.n. identical with LAVER SE 2072. The forms of the name may have been influenced by nearby WOOLVERTON ST 7853. DEPN.

East LAVINGTON WSusx SU 9416. *Estleuyngton* 1288. ME adj. **est** + p.n. *Levitone* 1086, *Le- Lovinton* 1212×14, apparently a shortened form of an alternative name for this place, *Wellauenton* 1209, *Wullauinton -en-* 1230–59, *Wollavyngton* 1330. 'The estate called after Wulflaf', OE pers.n. *Wulflāf* + **ing**[4] + **tūn**. 'East' for distinction from BARLAVINGTON SU 9716 and West LAVINGTON SU 8920. Sx 109, SS 75.

Market LAVINGTON Wilts SU 0154. 'L with the market'. *Markett Lavington* 1681, earlier *Chepynglavynton* 1397. ModE **market**, ME **cheping** (OE *ċeping*) + p.n. *Laventone* 1086, *-ton(a)* 1175, 1255, *Lavinton(e) -(a)* 1091–1225, 'Lafa's estate', OE pers.n. *Lāfa*, genitive sing. *Lāfan*, or *Lāfa* + **ing**[4], + **tūn**. Also known as Steeple Lavington, *Stupellavintona* 1242, *Stepel Lavynton* 1255, referring to the church tower. 'Market' for distinction from West LAVINGTON SU 0053. Wlt 240 gives pr [læviŋtən].

West LAVINGTON Wilts SU 0053. 'The west part of L'. *West Lavington* 1628. ModE adj. **west** + p.n. Lavington as in Market LAVINGTON SU 0154. Wlt 240.

West LAVINGTON WSusx SU 8920. A detached part of East or Woolavington parish. See East LAVINGTON SU 9416.

West LAVINGTON DOWN Wilts ST 9949. *West Lavington Down* 1817 OS. P.n. West LAVINGTON SU 0053 + ModE **down**.

Church LAWFORD Warw SP 4476. 'L with the church'. *Kirkelalleford* c.1204, *Chirche Lalleford* 1235–1406, *Chyrcelalford* 1308, *Church Lawford* 1535. ME **chirche** + p.n. *Lellevorc* [1077]17th, *Leileforde* 1086, *Ledleford* c.1090, *Lelle-* 1143, early 13th, *Lelesforde* 1161×70, *Lalleford* 1154×89–1281, *Lall-* 1445, 'Lealla's ford', OE pers.n. **Lealla* + **ford**. 'Church' for distinction from Little Lawford SP 4677, *Parva Lalleford* [1155]1235, *Litellalford* 1453, *Little Lawford* 1548, and Long LAWFORD Warw SP 4775. Wa 136, DEPN.

Long LAWFORD Warw SP 4775. *Long(a) Lalleford* c.1204, *Lungelelleford* c.1310, *Longlalford* 1542, *Long Lauford* 1546. ME **long** + p.n. *Lelle- Lilleford* 1086, *Lalleford* 1189, as in

Church LAWFORD SP 4476. Most of the houses stand in one street ¾ mile long. Wa 138.

LAWHITTON Corn SX 3582. '*Landwithan* estate or manor'. *Langvitetone* 1086, *Lanwitinton* 1188, *Lawittetone* 1260, *Lawhitton* 1610. P.n. *Landwithan* + OE **tūn**. *Landuuithan* 905 B 614, *Landwiþan* 980×8, appears to be 'tree chuch-site', Co ***lann** + **gwythen** 'a tree'. The original site, however, was at Oldwit SX 322817, *Yoldelanwyta* 1302, *Olde Lawhytta* 1348 'Old Lawhitton', at the head of a valley and away from the church. This is probably, therefore, an instance of *lann* replacing *nans* as in LAMORNA SW 4424 and LANGORE SX 3086. If so the original name was 'valley of Gwethen', Co **nans** (OCo *nant*) + pers.n. **Gwethen*. PNCo 108, Gover n.d. 151, DEPN.

LAWKLAND NYorks SD 7766. 'Land where leeks are grown'. *Lauk(e)land* 12th, 14th cent., *Lau(e)k(e)land(e)* 1383, 1589, 1614, *Lakeland* 17th cent., 1789. ON **laukr** + **land**. An exact parallel is ON *Laukeland*. Development of lME *au* to *ā* as in the 17th cent. spelling is a widespread development. YW vi.226, Nomina 13.109.

LAWLEY Shrops SJ 6708. 'Lafa's clearing'. *Lauelei* 1086, *-le(ye) -leg(h)'* 1177–1302, *Lauley* 1392, 1396, 1517, *Laweleg(h)'* 1271×2, 1381, *Lawley* from 1494, *Laley* 1577–1735. OE pers.n. *Lāfa* + **lēah**. Sa i.170.

LAWNHEAD Staffs SJ 8324. 'The head of the clearing'. *Lawn Head* 1833 OS. P.n. *the Launde* 1585, ME **launde** 'a forest glade', + **hede**. Contrasts with nearby *Ashwood Head* 1833 OS at the head of Ash Wood SJ 8224. This was originally woodland subsequently cleared, cf. the neighbouring names KNIGHTLEY SJ 8125, Walton's Rough SJ 8325, Anne's Well Wood SJ 8424, *Humphreys Wood* 1833 OS, Big Wood SJ 8323.

LAWSHALL Suff TL 8654. 'The hill-dwelling or -shelter'. *Lawesselam* 1086, *Laueshel* c.1095, *Laweshell* 1194, 1196, *Lausell', Lausill', Laweshill'* 1204, *Laugesale, Laugetsille* 1253, *Lawshill* 1610. OE **hlāw** + **sele** or **(ġe)sell**. DEPN, Studies 1936.48, 49.

LAWTON H&W SO 4459. 'Hill settlement'. *Lavtone -tvne* 1086, *Lauton* 1160×70, *Lawtone* 1397. OE **hlāw** + **tūn**. Lawton lies at the end of a ridge of raised ground. He 113.

Church LAWTON Ches SJ 8255. 'Lawton with the church'. *Chirche-lauton* 1333–1490, *Kirke-* 1356. ME **chirche** + p.n. *Lautune* 1086, *-ton* early 13th–1527, *Laughton* 1288–c.1540, *Lawton* from 1311, 'settlement by a mound', OE **hlāw** + **tūn**. 'Church' for distinction from BUGLAWTON SJ 8863. Che ii.320.

LAXFIELD Suff TM 2972. 'Leaxa's open land'. *Lessafeld, Lesse- Laxefella, Laxefelda* 1086, *Lexfelde* c.1095, *Lexefeld(e)* 1168–95, *Laxefeld(')* 1156×62, 1168, [1189×99]1396, *Laxfeud -feld(e)* 1208–1610. OE pers.n. *Leaxa* + **feld**. DEPN, Baron.

LAXTON Humbs SE 7925. 'Estate called after Laxa'. *Laxinton* 1086–1310, *-ing-* 1199, 1252, 1282, *-inge-* 1199, *Lexin(g)ton* 1230–51, *Laxton* from 1285. OE pers.n. **Laxa* + **ing**⁴ + **tūn**. YE 254.

LAXTON Northants SP 9596. 'Leaxa's farm or village'. *Lastone* 1086, *Laxton(a)* from 1166, *Laxeton(a)* 1131–1275, *Laxinton(a)* 12th, 1222. OE pers.n. *Leaxa*, genitive sing. *Leaxan*, + **tūn**. Nth 168, SSNEM 378.

LAXTON Notts SK 7266. 'Leaxa's farm or village'. *Laxintune* 1086, *-ton(a)* c.1200–1235, *-en-* 1218, *-ing-* 1281, *Laxton* from c.1190, *Les(s)in(g)ton(a) -tun* c.1190–1275, *Lexinton* 1204–98, *Lexing- Laxington* 1608, 1637. OE pers.n. *Leaxa*, genitive sing. *Leaxan*, possibly varying with *Leaxa* + **ing**⁴, + **tūn**. Nt 54.

LAYCOCK WYorks SE 0341. Uncertain. *Lacoc* 1086, *Lacock(e)* 1273, 1545, *Laycock* from 1411. Previously explained as 'the little stream', OE ***lacuc** as in LAYCOCK Wilts SE 0341, but this is a mountain stream for which *lacuc* seems an unlikely term. YW vi.7.

LAYER BRETON Essex TL 9417. 'L held by the Breton family'. *Leyre Bretones* 1254. P.n. Layer as in LAYER DE LA HAYE TL 9620 + family name of Lewis Brito 'the Breton' 12th. Ess 317.

LAYER DE LA HAYE Essex TL 9620. 'L held by the de la Haye family'. *Leyr(e), Leire de la Haye* 1272 etc. Earlier simply *Legrā* 1086, *Leg(e)ra -e, Leira -e, Leyre, Lehere, Lehegra, Legehere* 1119–1554, *Layer, Leyer* 17th, probably identical with the lost r.n. *Leire* seen in LEIRE Leics SP 5290 and in LEICESTER SK 5904 and the r.ns. Loire, *Ligeris*, and Ligoire, *Ligorius*. 'De la Haye' from the family of Maurice de Haia 12th and for distinction from Layer Marney TL 9217, *Leyre Marinnye*, 'L held by the Marney family', and LAYER BRETON TL 9417. All three places lie along Layer Brook now partly submerged in Abberton Reservoir. Ess 316.

LAYER MARNEY Essex TL 9217 → LAYER DE LA HAYE TL 9620.

LAYHAM Suff TM 0340. 'The shelter homestead'. *Hligham* 1000×2 S 1486, *Las- Lei- Latham* 1086, *Laiham* 1207, *Layham* 1610. OE ***hlīġ**, ***hlēġ** + **hām**. DEPN, PNE i.252.

LAYTHAM Humbs SE 7439. '(The settlement at) the barns'. *Ladon(e)* 1086, *Lathum* 1226–1350, *-om* 1199×1212–1539, *Laytham* from 1562. ON **hlatha**, dative pl. **hlathum**. YE 238, SSNY 87.

East LAYTON NYorks NZ 1609. *(Est) Laton* 1184–1530, *Est Leiton* 1256. ME adj. **est** + p.n. *Latton, Latone* 1086, *Laghton* 1346, the 'leek enclosure, kitchen garden'. OE **lēac-tūn**. 'East' for distinction from West LAYTON NZ 1409. YN 300.

West LAYTON NYorks NZ 1409. ModE **west** + p.n. *Lastun -ton* 1086 as in East LAYTON NZ 1609.

LAZENBY Cleve NZ 5719. 'The settlement of the freedmen'. *Le(i)singe- Lesighebi, Laisinbia* 1086, *Lei- Leysing(e)bi -by* 12th–1279×81, *Lesingby* 1300. ON **leysingi**, genitive pl. **leysinga**, + **bӯ**. Leising is the name of a DB tenant in nearby Kirkleatham and Normanby, but the genitive sing. of by-n. *Leising* should give **Leisingesbi* not *Leisingebi*. Cf. LAZONBY Cumbr NY 5439. YN 160 gives pr [lεəzənbi], SSNY 32, SN 70.16.

LAZONBY Cumbr NY 5439. 'The village or farm of the freedmen'. *Leisingebi* 1165–94, *Leysingby* 1154×89–1427 with variants *Lei- Lai-* and *-yng-*, *Laysonby* 1293, *Lazenbye* 1588. ON **leysingi**, genitive pl. **leysinga**, + **bӯ**. Cf. LAZENBY Cleve NZ 5719. Cu 219, SSNNW 35.

LEA '(The settlement) at the wood, clearing or pasture'. OE **lēah**, locative–dative sing. **lēa**. L 203.

(1) ~ Derby SK 3257. *Lede* 1086, *Lea* from 1154×9, *(la, le) Lee* 1237–1350. If the 1086 Db form is significant it seems to point to OE **lǣth** 'lathe, ? landed property, ? meadowland'. Db 360.

(2) ~ H&W SO 6621. *Lecce* 1086, *Leche* 1160×70, *Lacu* 1201, *La Le(e)* 1219–75. He 119.

(3) ~ Lincs SK 8286. *Lea* 1086, *Le* c.1115, *Lee* 1135×54, 1212. There is still a stretch of ancient woodland in the parish. DEPN, Cameron 1998.

(4) ~ Shrops SJ 4108. *Lya* 1271×2, *(La) Lee* 1276–1601 etc. Sa ii.39.

(5) ~ Shrops SO 3589. *(La) Lee* 1255–1334 etc., *Legh'* 1271×2, *Lea* 1836 OS. Gelling.

(6) ~ Wilts ST 9586. *Lia* 1190, *la Le(e)* 1242, 1248, 1346, 1370, *Lea al. Lee* 1581, *La Lye al. Le* 1600. Wlt 61.

(7) ~ TOWN Lancs SD 4731. *Town* is a modern addition to the original name of the township, *Lea* 1086, *Lehe* [before 1190]1268, *Le(e), Legh* 13th cent. Lea Town N of Savick Brook was also known as English Lea, *Engleshel'* 1201, *Le Engleis* 1207, *Englesshelee* 1385, *Lee Anglicana* 1422, for distinction from French Lea S of Savick Brook (now Lea SD 4930), *Le Franceis* before 1194, *Lee Francia, Fraunceis* 1259, *Lee Gallica* 1377, *Frenkyssele* 1277, *Le Frensshe Lee* 1356, which was given to a Norman, Warin de Lancaster, before 1189. La 146, Jnl 17.84.

River LEA or LEE GLond TQ 3695, Herts TL 2609. 'Bright river' or 'river dedicated to Lug'.

Type I. *(on, be, amnem) Lig(e)an* [880]c.1125 B 856–7, c.1100 ASC(D) under year 895, [1062]13th S 1036, [before 118]12th, [c.1130]12th., *(on) Lygan* 895 ASC(A), *(betweox) Lygean* 913 ASC(A), c.1100 ASC(D) under years 896, 913, *(on) Liggean* c.1000 ASC(B) under year 895.

Type II. *Luye* [c.1130]12th–c.1540 with variants *Luia, Luy(a)*.

Type III. *(le, la) Leye* 1274–1440, *(le, la) Ley* 1416–1586.

Type IV. *Lea* from 1576.

OE *Lyġe*, oblique case *Lyġan*, cognate with the W r.n. Lleu, ultimately from IE **leug-* 'bright, light'. Three different regular developments of OE *y* produced ME forms **Lye, Luye* and *Leye*, the ancestor of the modern form Lea. Mx 4, RN 239, LHEB 310, GPC 2166.

River LEACH Glos SP 1707. 'The boggy stream'. *Lec* [718×45]11th S 1254, *Leche* 1570, 1690, *Leech* 1612, *Lachebrok* 12th. OE **læċċ, leċe** (Mod dial. *letch*). Gl i.9, RN 241, L 25.

LEADENHAM Lincs SK 9552. 'Leoda's homestead or estate'. *Ledeneham* 1086, 1201, *Ledenham* 1178–85, *Langeledenham* [1042×55]c.1150, *Long(e)-* 1301–1559, *Long Leadenham* 1687. OE pers.n. **Lēoda*, genitive sing. **Lēodan*, + **hām**. The village extends along the line of the Lincoln to Grantham road. Perrott 381.

LEADGATE Cumbr NY 7043. 'Swing-gate'. *Lid(d) Gate* 1703, *Leadgate* 1761. Ultimately OE **hlid-ġeat**, influenced by ModE **lead** and **gate**. Cu 178.

LEADGATE Durham NZ 1251. 'Swing-gate'. *Lydyate* 1404, *Lidgate* 1590, *The Lide Yate, Lidge yeat, Lidyat* 17th cent., *Leadgait* 1617. OE **hlid-ġeat**. The modern form of the name has been wrongly associated with lead mining in the Durham dales. NbDu 132.

River LEADON Glos SO 7628. 'The broad river'. *Ledene* [972]10th S 786, [978]11th S 1338–1287, *Leden* c.1235, 1248, *Ledon* 1542, *Leddon* 1565, *Leadon* from 1619. PrW **Lıdan* < Brit **litano-* as in OW *litan*, ModW *llydan*, Irish *leathan*, < IE **pltano-* cognate with Latin *platanus*, Greek *πλάτανος*. This element occurs also in the unidentified RBrit p.n. *Litanomagus* 'broad place', in *Litana Silva*, a forest in Cisalpine Gaul, *Litanobriga* 'broad hill' near Chantilly, Oise, France, and Ledesma, Salamanca, Spain, originally *Bletisama* 'very broad'. Gl i.10 gives pr ['ledn], RN 241–2, LHEB 478, 534, 672, RBrit 394, GPC 2251.

LEAFIELD Oxon SP 3215. 'The field, the open stretch of land'. *la Feld(e)* 1213–1298, *le Feild* 1641. OFr definite article **la, le** + ME **feld** (OE *feld*). O 361.

LEAGRAVE Beds TL 0523. Probably 'the light grove'. *Littegraue* 1224, *Little- Lihtegraue* 1227, *Lightgrave* 14th cent., *Lygrave* 1504. OE **lēoht, līht**, definite form **līhta -a**, + **grāf**. The name has been assimilated successively to OE *lȳtel* and the r.n. LEA TL 2609. This explanation is preferable to positing an unrecorded pers.n. **Lihtla*. Bd 154, PNE ii.23, L 193–4.

LEAKE 'The stream'. ON **lœkr** possibly replacing OE ***leċe**. L 25.

(1) ~ COMMONSIDE Lincs TF 3952. '(The settlement) beside Leake common'. *Leake Common Side* 1824 OS. P.n. Leake Common, p.n. Leak as in Old LEAKE TF 4050 + ModE **common**, + **side**.

(2) ~ HURN'S END Lincs TF 4249. *Leake Hurn End* 1824 OS. P.n. Leake as in Old LEAKE TF 4050 + p.n. *Hurn's End* 1812, 1813, dial. **hurn** (OE *hyrne*) 'a portion of a village situated in an angle or corner' + ModE **end**. Payling 1940.126.

(3) East ~ Notts SK 5526. *Ester Leke* 1154×89, *Est(er)lek* 1278, 1280, *Esterleyk* 1305, *Este Leake* 1567. OE **ēast**, comparative **ēasterra**, + p.n. *Lec(c)he* 1086, *Lek(e)* before 1163–1297, *Leyc -k, Leic* c.1225–39. Also known as *Magna Leyke* 1390 and *Greit Leke* 1531. 'East' and 'Great' for distinction from West LEAKE SK 5226. Both places lie on the banks of a small stream called *Lekbroc* in 1289. Nt 252–3.

(4) New ~ Lincs TF 4057. *New Leake* 1868 Wheeler, 1896 Gaz. A modern hamlet on reclaimed fenland in the parish of Eastville created in 1812. ModE adj. **new** + p.n. Leake as in Old LEAKE TF 4050. Wheeler Appendix i.25, Pevsner 1964.231.

(5) Old ~ Lincs TF 4050. ModE adj. **old** + p.n. *Leche* 1086, *Lech* 1191×2, *Lec(a)* 1158–1219, *Leke* late 12th–1535, *Leik* 1221, *Leek(e)* 1316–1535, *Leake* 1824 OS. Leake is surrounded by drainage channels. 'Old' for distinction from New LEAKE TF 4057. Payling 1940.125, SSNEM 157, Cameron 1998.

(6) West ~ Notts SK 5226. *West(er)lek* 1240, 1242, *Westleke* 1276, *-leyk* 1280, ~ *Leake* 1568. OE **west**, comparative **westerra** + p.n. Leake as in East LEAKE SK 5526. Also known as *Lytel Leyk* 1330, *Parva Leek* 1358. Nt 252–3.

LEALHOLM NYorks NZ 7607. '(The settlement) amongst the twigs'. *Lelun, Laclum* 1086, *Lelum(e)* 1273, 1301, *Lelom* 1301, 1349, *Lelhom(e)* 1273, 1410, *Leleholme* 1579. OE **lǣl**, dative pl. **lǣlum**. For parallel formations cf. RYSOME Humbs TA 3623, SNAIZEHOLME SD 8386. YN 133 gives pr [li:ləm], L 52.

LEALHOLM MOOR NYorks NZ 7509. P.n. LEALHOLM NZ 7607 + ModE **moor** (OE *mōr*).

River LEAM Warw SP 4568. 'Elm-tree river'. *(on) Limenan* [956]11th S 623, *(on) leomene, leomenan, (of) leomanan* [1033]c.1200 S 967, *Lemine -ene* 1232–85, *Leeme* 1411, *Leame* 1576. OE r.n. *Leomene* < **Limane* and *Limene* < PrW **Liμan, *Liμon* from Brit. **Lemanā* or **Lemonā* and **Lemenā*, formations on the root **lem*, cognate with Latin *ulmus* < **lmo-* and English *elm* < **elmo-*. The same root occurs in the r.ns. LYMPNE, LYMN, LEM BROOK, LEMON, LOMAN and in the continental r.ns. Limagne, Puy-de-Dôme and Allier, France, *Lemane, Limane*, and Lac Leman (Lake Geneva), *Lacus Lemannus*. Wa 4, RN 243 gives pr [lem], LHEB 486, 673, RBrit 385.

LEAMINGTON HASTINGS Warw SP 4467. 'L held by the Hastings family'. *Lemyngton Hasting* 1285, *-ings* 1488. P.n. *Lvnintone* 1086, *Lementon(a) -in- -yn-* 1174–1304, *-yng-* 1289, 'the settlement on the river Leam', r.n. LEAM SP 4568 + OE **tūn**, + manorial addition from the family of Aytropius son of Humfrey Hastinges who held the estate in 1280. Also known as *Estlemyn(g)ton* 1221–1327, *East Lemington* 1603, for distinction from Leamington Spa SP 3265. Wa 138, DEPN.

Royal LEAMINGTON SPA Warw SP 3265. The title *Royal Leamington Spa* was granted after a visit by Queen Victoria in 1838 to *Lemynton Prioris* 1325, 1416, *Lemyngton Prioris* 1535, 'Leamington held by the prior' sc. of Kenilworth to whom it was granted by Henry I and for distinction from LEAMINGTON HASTINGS SP 4467. Earlier simply *Lamintone* 1086–[1154×89]15th, *Leminton(a) -yn-* 1198–1316, 'the settlement on the river Leam', r.n. LEAM, *Lemene* 13th cent., + **tūn**. Wa 175.

LEAP HILL Northum NT 7207. *Leap Hill* 1869 OS rises to 1540ft.

LEARMOUTH Northum NT 8637. 'The mouth of the river *Lever*'. *Leuremue* 1176, *Liver- Levermu(w)e* 13th cent., *Levermuth* 1346, 1461, *Leremouthe* 1542. R.n. Lever + OE **mūtha**. Lever, *riuulo de Leu', Leuer* 1296, is the 'stream where rushes grow', OE **Lǣfre*, a derivative of **lǣfer** 'a rush', the lost name of Willow or Pressen Burn NT 8336, 8234. NbDu 132, RN 246.

LEASGILL Cumbr SD 4984. 'Bright, bare ravine' or 'Ljosa's ravine'. *Lesegill* 1458, *Leasgill -gyll* from 1504. ON **ljóss** 'light, bright, bare' or f byname *Ljósa*, + **gil**. We i.88, SSNNW 143.

LEASINGHAM Lincs TF 0548. 'The homestead called after Leofsige'. *Less-* (for *Lefs-*), *Leuesingham* 1086, *Levesingham -yng-* c.1160–1355, *Levesingeham* 1196–1227, *Lefsing-* 1202, *Lesing- -yng-* 1303–1562×7, *Lessing -yng-* 16th cent., *Lezingham* 1672, 1685, *Leasingham* from 1790. OE pers.n. *Lēofsiġe* + **ing**[4] + **hām**. Perrott 287, ING 142.

LEATHERHEAD Surrey TQ 1656. Probably 'the public ford' *(æt) Leodridan* [873×88]11th S 1507, *Leret* 1086, *Ledred(e)* 1155–1504, *Ledd-* c.1190–1420, *Ledered(e)* 1241–1428, *Lethered* 1470–1603, *-hed(d)(e)* 16th cent., *Leatherhead* from 1630. OE **lēode** 'people' + OE **rida -e** 'a riding path, a ford which can be ridden'. Cf. THETFORD Norf TL 8783 for examples of a similar name-type. The alternative suggestion, 'grey or brown ford', PrW ***lēdrïd** (Brit **lētorito-*) is unnecessary. Sr 78–9, Jnl 12.70–4, DEPN 292, L 62, 79–80.

LEATHLEY NYorks SE 2347. 'Woodland clearing on the slopes'. *Ledelai* 1086, *Lethelei(a) -lay -ley* c.1190–1540, *Leeleia* 1196, *Lelay -ai -ey* 13th cent., *Leath(e)ley* from 1548. OE **hlith**, genitive pl. **hleotha**, + **lēah**. Leathley is situated on the steeply sloping edge of an expanse of hill-country. YW v.54 gives prs ['li:þlə, 'leiþlə], L 165, 166.

LEATON Shrops SJ 4618. Partly uncertain. *Letone* 1086, *-ton(a) -tone* 1243×8–1404, *Leeton(')* 1250, 1255, 1397, *Leytun'* 1250, *-ton* 1609, 1611, *Leaton* from 1577, *Leighton* 1613, 1772. Unknown element + OE **tūn**. None of the previously suggested etymologies, OE *lēac* 'a leek', *lēah* 'a wood', *(ġe)lǣt* 'a road junction', *hlēo* 'a shelter', happily fits the run of spellings for this name or for Leaton SJ 6111. One possibility is OE ***hlēġ** 'a shelter' postulated for LEYBURN NYorks SE 1190, an *i*-mutation side-form of *hlēo* with analogical *ġ* as in *nīġe* beside *nīwe* 'new'. However, the occurrence of two Leatons in areas in which Lea names are common suggests that there may have been a compound **lēa(h)-tūn* roughly equivalent to the p.n. type Wootton. Sa i.171 gives pr [leitən], Campbell 120.1, 2, 411, PNE i.252, ii.50, Gelling.

LEAVELAND Kent TR 0054. 'Leofa's newly-cultivated land'. *Levelant* 1086, *Liofe- Liveland* c.1100, *Leveland(e)* c.1175–1254. OE pers.n. *Lēofa* + **land**. PNK 286, L 247, 249.

LEAVENHEATH Suff TL 9537. 'Levin's heath'. (heath of) *Levynhey* 1292, *Levenesheth* 1351. ME pers.n. *Levin* (OE *Lēofwine*) + **hethe** (OE *hǣth*). DEPN.

LEAVENING NYorks SE 7862. Possibly the 'place called after Lethen'. *Ledling(h)e et alia Ledlinge* 1086, *Legh(en)- Ley- Lev- Lethenyng -ing(g)* 13th cent., *Leu- Levening(e) -yng* 1284–1549. OE p.n. **Lethening* < pers.n. **Lethen* (probably a shortened form of *Lēoftheġn* or the like) + **ing**². Medial [ð] was replaced by other fricative sounds, [γ] and finally [v]. YE 148.

LEAVES GREEN GLond TQ 4162. *Lese Green* 1500, *Leves Green* c.1762, *Leaves Green* 1819 OS. Family name *Leigh*, genitive *Leigh's* + ModE **green**. LPN 134.

LEAVESDEN AIRPORT Herts TL 0900. P.n. Leavesden + ModE **airport**. Leavesden, *Levysdene* 1398, *Levesden* 1504, 1518, cf. *Levesdene wode* 'Leavesden wood' 1333, is 'Leve's valley', ME pers.m. or surname *Leve* (OE *Lēof(a)*), genitive sing. *Leves*, + **dene** (OE *denu*). Nearby must have been *Levescroft* 1348. Hrt 77.

LEBBERSTON NYorks TA 0782. 'Leodbriht's settlement'. *Ledbestun -ez-* 1086, *Ledbrithun* 1181, *-breston(a)* [1190–1251]16th, *Lebreston* 1285–1408, *Lyberston* 1550. OE pers.n. *Lēodbeorht -briht*, genitive sing. *Lēodbrihtes*, + **tūn**. YN 105.

LECHLADE Glos SU 2199. Probably 'channel, water-course of the river Leach'. *Lecelade* 1086, *Lec(c)helad(e)* 12th–1508 with variant *Li- Ly-* 1211–1494, *Lechlad(e)* from 1213. R.n. LEACH SP 1707 + OE **(ġe)lād**. The original course of the r. Leach from Little Farrington SP 2201 to the Thames followed the county boundary near Kelmscot SU 2498. It is suggested the *ġelād* in this name refers to the present channel which joins the Thames at St John's Bridge SU 2299 and that this was originally a mill-stream driving Lechlade Mill SU 2299. An alternative possibility, however, is 'river-crossing near the river Leach'. Gl i.40 gives pr ['letʃleid], L 25,73–4.

LECK BECK Lancs SD 6376. Identical with the name of the hamlet of Leck SD 6476 which stands on the river with ON **beck** added later. *Lech* 1086, *Lec* [1184–1202]1268, *Leec* [1196]1268, *Leck(e)* from 1212, *Leek* 1332, 1370, 'the brook', ON **lœ́kr**, **lǽkr**, Cf. LEAKE, LEEK. Very likely it replaces the original British name of the stream, **Calācon*. See under Nether BURROW Lancs SD 6175. La 184, L 25, SSNNW 143, Jnl 17.102.

LECK FELL Lancs SD 6678. P.n. Leck as in LECK BECK + ModE **fell** (ON *fjall*).

LECKFORD Hants SU 3737. '(Side-)channel ford'. *(to) Leahtforda, Leg(h)ford* [947]15th S 526, 1419, *Lec(ht)ford* 1086, *Lec- Lekford* 1218–1329. OE ***lēaht** 'irrigation channel, side-channel' cognate with *lecht* in Belgian Anderlecht, + **ford**. The Test flows in multiple streams here. Ha 107, Gover 1958.180, DEPN, Forsberg 1950.68.

LECKHAMPSTEAD Berks SU 4375. 'Homestead where leeks are grown'. *Lechamstede* *[811 for 815]12th S 166, *[821]12th S 183, [955×9]12th S 665, [1045×8]12th S 1404, *Læhham stede* [821]13th S 183, *(æt) Leachamstede, (æt) lecham stede* [943]13th S 491, *Lecanestede* 1086, *Lechhamesteda* 1167, *Lec- Lekhamsted(e)* 1195–1459, *Leckhampsted(e)* 1316–49. OE **lēac** + **hāmstede**. Brk 254, Forsberg 65, Stead 267.

LECKHAMPSTEAD Bucks SP 7237. 'Leek homestead'. *Lechamstede* 1086–1262, *Leke-* 1307–8, 1517, 1766, *L(e)yk(e)-* 1323–1512, *Lekhampstede* 1316, 1360. OE **lēac** + **hāmstede**. Bk 43, Stead 264.

LECKHAMPTON Glos SO 9419. 'Garlick or leek homestead'. *Lechame- Lechantone* 1086, *Lec- Lekhamton* 1211–21, *-hampton(e)* 1226–1327, *Lekehampton* 1424–1538, *Leck(e)-* from 1535, *Leckington* 1691. OE **lēac** + **hām-tūn**. Gl ii.109.

LECONFIELD Humbs TA 0143. 'Open land beside the stream'. *Lachinfeld -t* 1086, *Lec- Lekingfeld -yng-* 1130×8–1530, *Lekenfeld* 1199–1504. OE ***leċing** with Scand [k] for [tʃ] + **feld**. YE 189.

LEDBURY H&W SO 7137. 'Fortified place or manor on the river Leadon'. *Liedeberge* 1086, *Ledeb(ury)* 1150×4–1364, *Ledeberi(a) -birie, Lydebury* 1162–1280, *Ledbury* 1568. R.n. LEADON SO 7628 + OE **byriġ**, dative sing. of **burh**. He 119.

LEDGEMOOR H&W SO 4150. Unexplained. *Lidgmoor* 1832 OS.

LEDICOT H&W 4162. Partly uncertain. *L(e)idicote* 1086, *Ledicote* 1210, *Lydecote* 1317. Unidentified element + OE **cot**, pl. **cotu** 'cottages'. He 181.

LEDSHAM Ches SJ 3574. 'Leofede's homestead or village'. *Levetesham* 1086, *Leuedesham* [1096×1101]1280, *Ledesham* 1287–1524, *Ledsham* from 1387, *Ledsam -om* 17th. OE pers.n. *Lēofede*, genitive sing. *Lēofedes*, + **hām**. Che iv.217 gives pr ['ledʃəm] and ['ledsəm].

LEDSHAM WYorks SE 4529. 'Homestead near Leeds'. *Ledes-ham* c.1030 YCh, *Ledesham(a)* 1086–1428, *Ledsham, Ledsam* 1559, 1612. P.n. LEEDS SE 3034 + **hām**. YW iv.49 gives pr [ledsəm].

LEDSTON WYorks SE 4328. 'Settlement, estate belonging to Leeds'. *Ledestun(e) -tona* 1086–1379, *Ledston* from 1545. P.n. LEEDS SE 3034 + **tūn**. YW iv.50.

LEDWELL Oxon SP 4228. 'The stream or spring called *Hlyde*, the loud one'. *Ledewelle* 1086, *Lydewell* 1270, *Ludewell'* 1273×4, *Ledwell* from 1186×91. OE r.n. **Hlȳde* 'the loud one' + **wella**. O 280, RN 273, Mills 207.

LEE 'Wood, clearing, pasture'. OE **lēah**, dative case **lēa**. PNE ii.20, L 198–207.

(1) ~ Devon SS 4846. *Legh* 1416. D 48.

(2) ~ Hants SU 3617. *Ly* 1236, *la Legh* 1256, *Lee* 1316. Ha 108, Gover 1958.185, DEPN.

(3) ~ Lancs SD 5655. *Mikelegh, Litelegh* 1323, *Mikel- Litellegh* 1324. Preceded by the adjs. **micel** 'great' and **lytel** 'little'. La 172.

(4) ~ Shrops SJ 4032. *Lega* [c.1090]15th, *Legh'* 1291×2 etc., *Lee* 1833 OS. Gelling.

(5) LEEBOTWOOD Shrops SO 4798. 'Lee in Botwood'. *Lega in Bottewode* c.1170×6, 1253, late 13th, *Leg' de Bottewud'* 1212, *Leg' Bottewode, Leybotwood, Leye Bottewode, Legh' Bottewode* 13th cent., *Lebotwood* 1577, *Leebotwood* 1713. P.n. *Lega* 1204–1273, + p.n. *Botewde* 1086, 1189, *Bottewode, Botewd* 12th, 'Botta's wood', for distinction from other settlements called *Lege* in Botwood, one called *Bot(t)elegee* 1185 belonging to the Knights Templar, probably a shortened form of **Bottewodelege*, and others due to assarting by Haughmond Abbey which was granted *duas landas et terram* 'two clearings and land' here in 1163×6. Sa i.172.

(6) ~ BROCKHURST Shrops SJ 5427. 'Lee by Brockhurst'. *Legh' sub Brokhurst* 1271×2, *Leye subtus Brockhurst* 1316,

Leebrockhurst 1608. P.n. *Lege* 1086–1392 with variants *Legh(')* and *Ley(e)*, OE **lēah**, locative–dative **leaġe**, + p.n. Brockhurst, 'the wooded hill frequented by badgers', OE **brocc** + **hyrst**, as Brockhurst Warw SP 4683, *Brochurst(e)* 1221, and elsewhere, for distinction from LEEBOTWOOD SO 4798, Leegomery SJ 6612, *Lege Cumbray -Cumbrei* 1249, earlier simply *Lega* 1086–1202, with variants *Lege* and *Legh'*, 'Lee held by the Cumbray family', and other Lea and Lee names in Shrops. Sa i.174.

(7) ~ CLUMP Bucks SP 9004. P.n. Lee as in The LEE SP 9004 + ModE **clump** 'cluster of trees'.

(8) ~ MILL Devon SX 5955. *Legh Mille* 1437. P.n. LEE + ME **mill**. D 255.

(9) ~ MOOR Devon SX 5761. *Leemore* 1649, *Leigh Moore* 1695. P.n. LEE + + ModE **moor**. D 260.

(10) ~ -ON-THE-SOLENT Hants SU 5600. *Lie* 1212, *(la) Lye* 1242, 1280, *Lee* from 1281. 'On the Solent' was added in the late 19th cent. when the village became a watering place. Also known as *Lee Britain* 1695 from the 13th cent. Breton manorial tenants John le Bret 1242, John de Britannia 1280. Ha 108, Gover 1958.30, DEPN.

(11) Nash ~ Bucks SP 8408. *Nash Lee* 1834 OS, cf. *Nashleyfield* 1706. The name is probably a redivision of ME **atten ash ley** 'at the ash-tree wood or clearing' (OE *æt thǣm æsc-lēa*). Lee, *(la) Lega* [1146]14th, 1181, 1302, *Leia* [1186×95, 1302, 1320]14th, *La Legh* 1241, *(la) Leye* [1183]14th, 1284, *(the) Lye* 1537, 1552, *The Lee* locally, originally denoted a woodland area including Nash Lee. Known as [ði: li:] to-day as a result of a County Council ruling. Cf. nearby North LEE SP 8309. Bk 150, 152, O addenda, Jnl 2.26.

(12) North ~ Bucks SP 8309. *Northleye* [n.d.]14th, *Northlee* 1834 OS. ME **north** + p.n. Lee as in Nash LEE SP 8408. Jnl 2.26.

(13) The ~ Bucks SP 9004. *Lega* 1181, *La Legh* 1241, *La Leye* 1284, *Lye* 1537. Bk 152.

LEEBOTWOOD Shrops SO 4798 → LEE.

LEECE Cumbr SD 2469. 'Woodland clearings'. *Lies* 1086, *Lees* 1269–1332, *Leghis* 1341, *Lece* 1577. OE **lēah**, pl. **lēas**. La 209, L 203.

LEEDS & LIVERPOOL CANAL Lancs SD 5209, SD 7832. Constructed 1770–4. Pevsner 1969.26.

LEEDS Kent TQ 8253. Uncertain. *Esledes* 1086, *Hlyda, Hledes* c.1100, *Ledes* 1185×6–13th. At least three explanations are possible, none of them without difficulty: OE **hlēda,hlȳda** 'a seat, a shelf', OE **hlid** 'a door, a gate, an opening', OE stream name **Hlȳde*, Kentish *Hlēde* 'the loud one'. In the case of the first and second suggestions secondary ME pl. *-es* must be assumed (and with *hlid* ME *e* from OE *eo*, **lede(s)* < *hleodu*) and in the case of *Hlyde* an unparalleled genitive sing. *-es* 'belonging to the noisy one'. The village lies on a stream which flows down a narrow valley forming a gap in higher ground with shelving terrain on either side. PNK 221.

LEEDS NYorks SE 3034. '(The district of) the people living beside the river *Lat*'. *Loidis* [c.731]8th BHE, [c.890]c.1000 OEBede, 12th, *Ledes* 1086–1470, *Ledis -ys* 1175–1548, *Leeds* from 1518. Brit **Lāt- Lādenses* 'the people of the r. **Lāt-*. This must have been an earlier name for the r. AIRE related to W *llawd*, Ir *láth* 'heat (in animals), ardour, passion', meaning something like 'the boiling or violent river'. The development was **Lādenses* > late Brit **Lōdeses* borrowed as PrOE **Lōdis* which underwent *i*-mutation through the stage *Loidis* to *Lēdes*. YW iv.124, Antiquity 20.209.

LEEDSTOWN Corn SW 6034. *Leedstown* 1867. A 19th cent. village created by the Duke of Leeds whose family inherited estates here as early as 1785. PNCo 108, Gover n.d. 590, Room 1983.63.

LEEGOMERY Shrops SJ 6612 → LEE BROCKHURST SJ 5427.

LEEK Staffs SJ 9856. 'The brook'. *Lec* 1086, 1240, 1317×28, *Lech* c.1100–1230, *Lek* 1232–1340, *Le(e)ke* 1226–1635, *Leek* from c.1291. OE ***leċe** or ON **lœ́kr**. Oakden, PNE ii.26, L 25.

LEEMING NYorks SE 2989. Originally a stream name of uncertain meaning, possibly the 'shining stream'. *(aquam de) Lemyng* 13th, *Leming* [1154–89]1348, 1231, 1280, 1577, *-yng(e)* 13th–1576, *Leminges be(c)k(e)* 16th cent. Brit ***lemo-** + **ing**[2]. YN 227, RN 247.

LEEMING BAR NYorks SE 2890. P.n. LEEMING SE 2989 + ModE **bar** 'a barrier'.

River LEEN Notts SK 5226. *Liene* c.1200, *Lene* 1218–1445, *Leen* from 1232, *Line* c.1540, *Lyne* 1675. Possibly identical with the specific of LEOMINSTER H&W SO 4959, *Leomynster* [c.1000]11th, W *Llanllieni*, LYONSHALL H&W SO 3355, *Lenehalle* 1086, and the second element of EARDISLAND, KINGSLAND and MONKLAND H&W, SO 4258, 4461 and 4657, *Le(i)ne* 1086, *Erleslen* 1250, *Kingeslen* 1230, *Munkelen* c.1180. All of these contain ME *Lēn* probably representing OE **Lēon*, **Lion* < OW ***lion**, ***lian**, pl. **llieni** as in *Llanllieni*, a r.n. formed on the IE root **lei-* 'to flow'. Nt 5, RN 247, DEPN.

The LEEN H&W SO 3859. *Lene* late 12th, 1310, *Leone* 1334. The farm-name preserves the district name Leen seen in EARDISLAND SO 4158, KINGSLAND SO 4461, LEOMINSTER SO 4959, LYONSHALL SO 3355, MONKLAND SO 4557 and *Leonis (monasterium)* 1080×90, *Lene* 1086. The earliest form occurs in the phrase *Lionhina gemæres* 958 S 677 'the boundary of the Lion community', probably a reference to the monks of Leominster abbey. OE *Lēon*, *Līon* is an Anglicisation of OW *Lien* < **Lion*, **Lian* 'the stream district' referring to the rivers Arrow and Lugg W of Leominster. He 6–9, 156, ECWM 142, DB Herefordshire 1.5, DEPN.

LEES 'Woods, clearings, meadows'. OE **leāh**, pl. **lēas**. L 203.

(1) ~ Derby SK 2637. *Leghes* 1302. Later *Dalburyleghes* 1402, *Dalbye Lees* 1547, 'Lees belonging to Dalbury'. Db 548.

(2) ~ GMan SD 9504. *The Leese* 1604. La 29, L 203.

(3) ~ SCAR LIGHTHOUSE Cumbr NY 0953. P.n. Lees Scar + ModE **lighthouse**. The reference is to a skerry in Silloth Bay off Blitterlees NY 1052. Blitterlees is known locally as The Lees, *The Leys*, *Lies* 16th, *The Lees* 1672. Other skerries along the coast are Beck Scar NY 0849 (off Beckfort), Catherinehole Scar NY 0850 and Lowhagstock Scar NY 0749. Blitterlees, *Blatterleese(e)* 1538, *Blaterleese* 1605, 1635, *Blitter-leese* [1538]1603, *Bletter Lees* 1575, 1657, may contain OE **blæcthorn** 'blackthorn', but the evidence is too late for certainty. Cu 293, L 203.

LEEZ LODGE LAKES Essex TL 7118. P.n. Leez Lodge, *Leighs Lodge* 1777, 1805 OS, + ModE **lake(s)**. *Leighs* as in Great and Little LEIGHS TL 7317, 7116. Ess 424.

LEGBOURNE Lincs TF 3684. 'The tricking stream'. *Lecheburne* 1086, *-burna* c.1115, *Lekeburna* 12th cent., *-burne* [c.1150]1409, *Lecceburne -a* 1158, c.1160. OE ***leċe** replaced by ON **lœ́kr** + **burna**. SSNEM 219, L 19, Jnl 23.45.

LEGBURTHWAITE Cumbr NY 3119. Possibly '*Legborg* clearing'. *Legberthwait* 1303, *Legburgthwayte -bo(u)rtwhat -berthwayte* 16th. P.n. **Legborg* 'Legg's fortified place', ON pers.n. *Leggr* + **borg**, + ON **thveit**. Cu 313, SSNNW 237.

High LEGH Ches SJ 7084. 'High clearing'. *Alta Lega* 1318, *High Legh* from 1488. Adj. **high** + p.n. *Lege* 1086, *Lega* 1209, *Legh* from [1153×81]1285. OE **lēah**. Che ii.45 gives pr [li:].

LEGSBY Lincs TF 1385. 'Legg's farmstead or village'. *Lagesbi* 1086, *Leggesbi* 1202, 1212. ON pers.n. *Leggr*, genitive sing. **Leggs**, + **bȳ**. There is a f.n. *Leggeshou -howe* [12th]1409, 'Legg's burial mound', ON **haugr**, in the adjacent township of Lindwood. For a similar pairing see GRIMESTHORPE SYorks SE 3587. DEPN, SPN 184, SSNEM 58.

LEICESTER Leic SK 5804. 'Roman fort of the *Ligore*, the dwellers by the river *Ligor*'. *(of) Ligera ceastre* c.924 ASC(A) under years 917, 921, *(æt) Ligra- Lig(e)receastre* 11th ASC (B,C,D) under year 914, *Legra- Ligra(n)ceaster* 11th ASC (B,C,D) under year 918, *Ligora- Ligera- Ligrece(a)ster* c.960, 11th ASC (A,B,C,D) under year 942, *Leogereceastre* before 1118,

Legeceastre c.1000, *Ledecestre* 1086, *Legecestria -cestr(e)* 12th cent., *Lei- Leyrcestr(e)* 1189–1544. Also recorded in Latin as *Legorensis civitatis* 803 B 312, *Legoracensis civitatis* [839×44]12th B 440. OE folk-n. **Ligore*, genitive pl. **Ligora*, + **ċeaster**. The reference is to *Ratae Coritanorum* 'Ratæ of the Coritani', the tribal capital of the Coritani or Corieltauvi. The name is best regarded as a new descriptive term for a deserted site since neither the memory of the Corieltauvi or of Ratæ (Brit ***rātis** 'the ramparts' cognate with W *rhawd* as in *beddrawd* 'a grave-mound' and Irish *ráth* 'the earthen rampart surrounding the residence of a cheiftain, a fort') survived†. The r.n., however, from which the folk-n. is derived is preserved in nearby LEIRE SP 5290. Lei 113, RN xlii, Forsberg 112, RBrit 443.

LEICESTER FOREST EAST Leic SK 5303. A motorway service station named from the civil parish of Leicester Forest East. Leicester Forest is *foresta de Leycestrie* c.1150, *Laycytwr foryste* 1524, *Leycester Forest* c.1545, *Lecester Forest* 1610. Earlier *Hereswode* 1086, 'wood of the host, the common woodland', OE **here**, genitive sing. **heres**, + **wudu**. OE *here* 'host, army' may bear its special sense 'a Danish army' here referring to the period when Leicester was recognised as a specifically Danish stronghold; or it may bear the sense 'host, people' as ON *herr*. The wood is described in 1086 as *silva totius vicecomitatus* 'the wood of the whole sheriffdom', cf. Edda 108, *herr er hundrað* 'a hundred makes a *herr*'. Lei 103.

LEIGH 'Wood or clearing'. OE **lēah**. PNE ii.18–22, L 198–207.

(1) ~ Dorset ST 6108. *Lega* 1228, *Legh* 1244, *Leyghe* 1327. Do 97.

(2) ~ GMan SD 6500. *Lecthe* c.1265, *Leeche* 1276, *Legh* 1276, 1292, *Leegh* 1341, *Leth* 1276, 1451. Probably originally a district name embracing both ASTLEY SD 6505 and TYLDESLEY SD 7001. *Leth* forms show substitution of [θ] for the peripheral phoneme [ç] as in Tanfield Leith Durham NZ 1854 and KEIGHLEY WYorks SE 0540. La 100.

(3) ~ H&W SO 7853. *Lege* 1086, *Leya* 1251, 1275, *Lye* 1389, 1675. The original reference was to an extensive area of woodland commemorated in a number of different names. This part was originally called *beornothes -noðes leah* 'Beorhtnoth's wood or clearing' 972 S 786 in the list of Pershore abbey lands for distinction from other parts called *eadpoldincg lēah* 'Ealdwold's wood or clearing' ibid., Braces Leigh SO 7950, *Lega Ricardi* c.1150, 'Richard's Leigh' subsequently held by the Bracy family, and Leigh SINTON SO 7850. Wo gives pr [lai], Wo 190, 204, Hooke 179, 215–9.

(4) ~ Kent TQ 5446. *Lega* c.1100, *(La) Legh(e)* 1239–79. PNK 84.

(5) ~ Shrops SJ 3303. *Lege* 1199. Bowcock 140.

(6) ~ Surrey TQ 2246. *Leghe* c.1184, *(La) Legh* 1255–98, *Leye* 1313, *Lye* 1434, 1541, *Lee alias Leighe* 1569, *Leight* 1558×1603, 1680. Sr 297 gives pr [lai], L 203.

(7) ~ Wilts SU 0692. *Lia* 1242, *la Leye* 1249, *la Legh* 1279, *Lighe* 1561. Wlt 46 gives pr [lai].

(8) ~ BECK Essex TQ 8182. *Leighbeck*. 1563. This is a spit or 'beak' (OFr *bec*) of marshland at the end of Canvey Island opposite Leigh-on-Sea TQ 8385. P.n. Leigh + ModE **beak** (ME *bek*).

(9) ~ DELAMERE Wilts ST 8879. 'L held by the Delamere family'. *Lye Dallamer* 1558×1601, *Lydallimore* 1637, *Ligh Dallimor* 1722. P.n. *Leye* 1236, *la Legh*, *Lega* 1242, *Lye* 1412, + manorial addition from the family name of Adam de la Mare who held the manor in 1236 and 1242. Wlt 106 gives pr [lai dælimɔ:r].

(10) ~ GREEN Kent TQ 9032. P.n. Leigh, *la Legh'* 1254, *(atte) Leghe* 1327, *(ate) Lee* 1352, + ModE **green** 'common'. PNK 358.

(11) ~ -ON-SEA Essex TQ 8385. A modern seaside resort, earlier simply *Legrā* (sic) 1086, *(La) Leghe* 1227–47 etc., *Leigh(e)* from 1419. Ess 188 gives pr [li:].

(12) ~ SINTON H&W SO 7850 → Leigh SINTON.

(13) ~ UPON MENDIP Somer ST 6947. *Lye vnder Mendip* 1610, *Leigh upon Mendip* 1817 OS. Earlier simply *(æt) Leage* [984×1016]12th S 1538, *Legh* 1243. The identification of both of these forms is probable rather than certain. DEPN.

(14) ~ WOODS Avon ST 5673. 'Wood(s) at L'. *Leigh Wood* 1830 OS. P.n. Leigh as in Abbots LEIGH ST 5474 + ModE **wood(s)**.

(15) Abbots ~ Avon ST 5474. *Legh of the Abbot of St Augustin* 1243. Earlier simply *Lege* 1086. Held by the abbot of St Augustine's abbey, Bristol. DEPN.

(16) Asthall ~ Oxon SP 3012. 'The woodland of the people of Asthall'. *Esthallingel'* before 1270, *Estallingeleye* 1272, *Asthallingeleye* 1270×90, *Asthawyngleye* 1406, *Astally* 1797. OE **Ēasthalingas* < p.n. ASTHALL SP 2811 + **ingas**, genitive pl. **Ēasthalinga*, + **lēah**, dative sing. **lēaġe**. O 300.

(17) Bessels ~ Oxon SP 4501. 'L held by the Bessels family'. *Besilles Lee* 1538. Family name of Petrus Besyles 1428 + p.n. *Leie* 1086, *Leye* 1284, *Legh(e)* 1229 etc. Possibly identical with the lost *Ærmundes lea* [942]13th S 777 'Earnmund's wood or clearing'. Brk 443.

(18) Braces ~ H&W SO 7950 → LEIGH SO 7853.

(19) Bradford ~ Wilts ST 8362. 'L by Bradford'. *Bradfordslye* 1571, *Bradforde Leighe* 1637. P.n. Bradford as in BRADFORD-ON-AVON ST 8260 + p.n. *Lighe* 1412. W 119.

(20) Church ~ Staffs SK 0235. 'L with the church'. ModE **church** + p.n. *Lege* [1002×4]11th S 1536, [1004]11th S 906, 1086. 'Church' for distinction from Upper and Lower Leigh SK 0136. DEPN.

(21) East ~ Devon SS 6905. Adj. **east** + p.n. *Legh'* 1242. 'East' for distinction from West Leigh SS 6805. D 366.

(22) Little ~ Ches SJ 6176. *Little Legh* 1666, ~ *Leigh* 1831. Earlier simply *Lege* 1086, *Leg(a)*, *Leye*, *Legh* 1295–1724, *Leigh* from 1486. Also known as *Leigh juxta Bartington* 1860, *iuxta Berterton* 1344 etc. Che ii.115.

(23) North ~ Oxon SP 3812. *Nordleg alias Nordelegh* 1233, *Nort(h)leg(h) -leghe* [1237]c.1300–1314, *Northley(a) -leye -le(e)* c.1250–1526. ME adj. **north** + p.n. LEIGH for distinction from South LEIGH SP 3908. O 274, L 203.

(24) South ~ Oxon SP 3908. *So- Suthleye* early 13th, *Suthlege -legh(e)* 1285–1327. Earlier known as *Stanton(')lega* 1190, 1197, from its proximity to Stanton Harcourt and for distinction from North LEIGH SP 3812. Part of a district probably called simply Leigh including SANDLEIGH SP 4501. O 276.

(25) The ~ Glos SO 8726. *Lalege* 1086, *(la) Leia*, *Leya*, *Leye* 13th cent., 1503, *(la) Legh* 1248–1478, *(ye, the) Leygh*, *Leigh(e)* 1377–1830. French definite article **la**, **le** + OE **lēah**. Gl ii.83 gives pr [li:], L 203.

(26) Westbury ~ Wilts ST 8650. 'L beside Westbury'. *Lye juxta Westbir'* 1302, *Westbury Leyghe* 1581. P.n. WESTBURY ST 8751 + p.n. *Lia* 1242, *Lye* 1249–75. Wlt 150.

Great LEIGHS Essex TL 7317. *Magna Legh(e)* 1230 etc., *Great Leyges* 1338, *Much Leighes*, ~ *Leeze* 1588, 1662. ME **grete**, **much**, Latin **magna**, + p.n. *Legā*, *Legram* 1086, *Legh(e)* 1230–1303, *Ley(e)s* 1271–1319, 'the wood or clearing', OE **lēah**. The pl. form probably refers to Great and Little Leighs. Ess 256 gives pr [li:z].

Little LEIGHS Essex TL 7116. *Parva Legh(e)* 1303, *Little Lyghes* 1393. ME **litel**, Latin **parva** + p.n. Leighs as in Great LEIGHS TL 7317. Ess 256.

LEIGHTERTON Glos ST 8291. Probably 'lettuce farm'. *Lettrint(t)one -en-* 12th, 1205×24, after 1412, *Lechtrinton'* 1221, c.1230, *Leghtryngton -in-* 1287, 1313, *Lei- Leyghterton* 1289–1600. OE **leahtric** + **tūn** with analogical sustitution of ME *-in(g)-* for *-i(c)-*. Gl iii.25.

LEIGHTON 'Leek enclosure, herb garden'. Later (ME) 'kitchen garden, vegetable garden'. OE **lēac-tūn**. PNE ii.18.

(1) ~ NYorks SE 1678. *Suthleghton* 14th, *Lighton* 1540. YN 232 gives pr [li:tən].

†The forms for RATBY SK 5105 do not support the tentative suggestion in Signposts 47 that Brit **rātis* is the specific of this name referring to the hill-fort of Ratby Burroughs SK 4906 or Bury Wood.

(2) ~ Shrops SJ 6105. *Lestone* 1086, *Lecton'* 1198–1272, *Le(c)hton, Leghton* c.1200–1577, *Leihton* 1212, 1255, *Leighton* from 1318, *Leyton* 1255×6, *Layton* 1683–1761. Sa i.176.
(3) ~ Somer ST 7043. *Lecton* [1185] Buck, *Leighton* 1817 OS.
(4) ~ BROMSWOLD Cambs TL 1175. *Leghton super Bromeswolde* 1287, *Leyghton Bromeswold* 1549. P.n. *Lectone* 1086–1237, *Lehtone* 1227, *Leghton* 1253, + district n. *Brouneswald -wold* 1286–7, 1347, 'Brun's wold', OE pers.n. *Brūn*, genitive sing. *Brūnes*, + **wald**. Bromswold was the name of all the undulating clay country between the Nene valley and the E boundary of Northants. Hu 245, 246, Fox 1989.78–9.
(5) ~ BUZZARD Bucks SP 9225. 'L belonging to the Busard family'. *Leghton Busard* 1287. Earlier simply *Lestone* 1086, *Le(c)htone* 1164–1240, *Lei- Leyton* 1206, 1227. There is no record of the Busard family having property here; the 17th cent. forms of the name, *Budeserte* 1553×1603, *Be(a)udesert, Budezard* 1643, 1646, are antiquarian inventions. Bd 129.
(6) Low ~ Derby SK 0085. *Loe Laughton* 1714, cf. *Lawe Laughen Field* 1565, *Loughton Mills* 1643. Db 154.

LEINTHALL EARLS H&W SO 4467. 'The earl's Leinthall'. *Leintall Comites* 1275. P.n. Leinthall + ModE **earl**, Latin **comes**, genitive sing. **comitis**. The reference is uncertain; it cannot refer to the Mortimer Earls of March whose title was created only in 1328. Leinthall, *Le(n)te- Len- Lintehale* 1086, is the 'nook of land by the river Lent', r.n. Lent + OE **hall**. The r.n. is well documented elsewhere: it is the old name of the Cole in Berks, *(innan, of, on) Lentan* *[854]12th S 312, [931]13th S 413, and of a brook near King's Norton H&W, *(of, on) Leontan, Liontan* [699×709]11th S 64, OE **Leonte* corresponding to W *lliant* 'flood, torrent, stream' from the root **lei-* 'to flow'. 'Earls' for distinction from LEINTHALL STARKES SO 4369. DEPN, RN 249–50, GPC 2174.

LEINTHALL STARKES H&W SO 4369. 'Starker's L'. *Leinth. Sterk.* 1216×72, *Leinhale Starkare* 13th. P.n. Leinhall as in LEINTHALL EARLS SO 4467 + pers.n. *Starker* < CG *Starcher*. DEPN.

LEINTWARDINE H&W SO 4074. 'Enclosure by the river Lent'. *Lenteurde* 1086, *Leintwordyn* 1289. R.n. Lent as in LEINTHALL EARLS SO 4467 + OE **worth** varying with **worthiġn**. DEPN, Bannister 114.

LEIRE Leic SP 5290. *Legre* 1086, *Leghere* 1176, *Leir(e)* from c.1130, *Lear(e), Layr(e), Layer* 16th cent. Probably a British r.n. which also forms the specific of the p.n. LEICESTER SK 5804 which is said to have been named *a Legra fluvio* 'from the river *Legra*' by William of Malmesbury writing in the 12th cent. The exact form and the origin of the r.n. are unknown. The OE form seems to have been **Ligor*, **Legor*, which may be compared with the Gaulish r.ns. *Liger* or *Ligeris*, the modern Loire, *ad flumen Ligerim* [c.50 BC]8th–9th BG 7.5.4, *Ligorius*, the Ligoire, and the tribal name *Ligures* behind which Bahlow sees a Celto-Ligurian root **lig* 'marsh'. Lei 454, RN xlii, Forsberg 113, Bahlow 300 s.n. Lieg.

LEISTON ABBEY Suff TM 4464. *Leiston Abbey* 1837 OS. P.n. LEISTON TM 4462 + ModE **abbey**. A Premonstratensian house founded in 1182.

LEISTON Suff TM 4462. 'Leif's estate'. *Ledes- Leh- Lees- Leistuna* 1086, *Leestuna -ton(a)* 1156–[1189×99]1396, *Legestona* 1168, *Leystona* [1189]1235, *Leiston(e)* from 1205. The forms are difficult but would best suit ON pers.n. *Leifr*, ME genitive sing. *Leifes*, + OE **tūn** 'settlement, village, estate'. An alternative might be OE **hlīġ*, **hlēġ* 'protection, shelter', genitive sing. **hlīġes*, **hlēġes*, but this seems less likely. DEPN, Baron, PNE i.252.

LEITH HILL Surrey TQ 1343. *Lyth Hill* 1626, *Leith Hill* 1816 OS. P.n. *la Lida* 1166–7, *La Lythe, Lithe* 1263–6 (p), *Leth* 1610, *Leethe* 1628, *Leith* 1781, 'the hill-side with a hollow', OE **hlīth**, + ModE **hill**. Sr 279–80, PNE i.252–3, L 165.

LELANT Corn SW 5437. 'Church-site of Anta'. *Lananta* c.1170–1418, *Lalant* 1478, *Lalante* c.1540. Co ***lann** + saint's name *Anta*. The church, however, is dedicated to St Euny, as in the 1610 form *Vny Lalant*. PNCo 108, Gover n.d. 635, DEPN.

LELLEY Humbs TA 2032. 'Wood or clearing where brushwood is collected'. *Lelle* 1246–1435, *Lelley* from 1275. OE **lǣla** + **lēah**. YE 40.

LEM HILL H&W SO 7275. Stream-n. *Lemp(e)* 1577, a Celtic r.n. 'elm-tree stream', cf. MIr *lem*, + ModE **hill**. RN 243.

Lower LEMINGTON Glos SP 2234. *Nether Lymyngton* c.1560, *Nether Lemington* 1587. ModE adjs. **nether, lower** + p.n. *Leminingtvne* 1086, *Lemel- Lemeninton'* 1220, 1221, *Limentone* 1086, *Lyminton* 1303, *Lemynton(e)* 1287–1481, *Lemmyngton'* 1287, *Lemyngton -ing-* 1426, 1540, 'settlement on the stream *Limen*', PrW r.n. **Λιμαη* < Brit **Lemanā* + **ing**[2] + **tūn**. For the r.n. see River LEAM Warw SP 4568. 'Lower' for distinction from Lemington Manor SP 2233, formerly Over or Upper Lemington, *Ouerlemyntone* 1493, *Over Lemington* 1587. Gl i.246.

LEMMINGTON HALL Northum NU 1211. *Lemmington Hall* 1868 OS. A mid-18th cent. country house. P.n. *Lemetun* 1157, *Lemechton* 1185, *Lemouton* 1236 BF, *Lemocton'* 1242 BF, *Lemot(h)on* 1289–1308, *Lematon* 1395, 1428, *Lemadon, Lemontone* 1247, *Leman(g)ton* 1278, 1402, *Lemmanton* 1589, *Leamondon -endon* 17th cent., *Lemingdon -ton* 1724, *Leamockdon* 16th cent., 'the settlement where brook-lime grows', OE **hleomoc** + **tūn**, + ModE **hall**. ME *Lemocton* became *Lematon*; the development of forms with *-n-* may have been spontaneous in the unstressed syllable as in WILLIMONTSWICK NY 7763, or it may have been due to the influence of ME *leman* 'a lover, a sweetheart'. NbDu 133, Pevsner 1992.374.

River LEN Kent TQ 8054. A back-formation from LENHAM TQ 8952. *Leno* (Latin ablative case) 1607, *Len* 1612. RN 249.

Abbotes LENCH H&W SP 0151, Atch LENCH H&W SP 0350 → Church LENCH SP 0251.

Church LENCH H&W SP 0251. 'Lench with the church'. *Chirichlench, Ciricleinc* [1054, 1070]13th. Earlier simply *(æt) lench* *[860×5]12th S 226, [1042×6]17th S 1856, [1044×51]12th S 1058. OE **ċiriċe** + district name Lench, 'the hill-side' < OE **hlenċ**. The hill-side is referred to as *lencdune* *[709]12th S 80 and the various Lench estates as *terræ de lench* ibid. 'Church' for distinction from I. Ab or Abbots Lench SP 0151 *Abeleng* 1086, *Abbelench* 1227, *Ab(e)lench* 1316–1544, *Abs Lench* 1704, 'Æbba's L', earlier *(æt) lenc* [983]11th S 1345; II. Atch Lench SP 0350 *Lench, Achelenz* 1086, *Aches Lenche* 1262–1300, *Acch(e)lench* 1495, 1535, 'Æcci's L'; III. Lenchwick SP 0347 *lenchwic* *[709]12th S 80, *Lencuuike, Lencuueke* [714]16th S 1250, *Lenchewic* 1086, 'L dairy farm', OE **wīċ**; IV. Rous LENCH SP 0153; and V. Sheriff's Lench SP 0149 *Schyruelench* 1271, 'sheriff's L', given by Odo, Bishop of Bayeux to Urse the Sheriff, earlier *Lench Alnod* *[716]12th S 83, *Lench Alnoth* 14th, 'Ælnōth's L', and *Lenz Bernardi* [c.1086]1190, 'Bernard's L'. Wo 148, 264, 330, PNE i.250, Hooke 24, 46, 47.

Rous LENCH H&W SP 0153. 'L held by the Rous family'. *Rous Lench* from 1445. Earlier known as *Biscopesleng* 1086, and *Lench Randolf* 1230–1431. P.n. Lench as in Church LENCH SP 0251 + family name *Rous*. The manor was held by the bishop of Worcester in 1086, by one Randolf in the late 12th cent. and by the Rous family in the 14th cent. Wo 149.

LENHAM Kent TQ 8952. Partly uncertain. *Westrelenham, Se þestraleanham* 'the western Lenham' [804]15th S 159, *Leanham* [?838]13th S 1649, *Leanham, East Leanaham* [850]13th S 300, 858 S 328, [961]13th S 1212, *Lenham* from [1087]13th. Possibly 'Leana's estate', OE pers.n. **Lēana* + **hām**. PNK 223, KPN 94, 359, Studies 1931.24, Forsberg 1950.71, ASE 2.29, Charters IV.64, 77.

LENHAM HEATH Kent TQ 9149. *Lenham Heath* 1819 OS. P.n. LENHAM TQ 8952 + ModE **heath**.

LENTON Lincs TF 0230. 'The estate called after Lafa'. *(æt) Lofintune* [1066×8]c.1200, *Lavintone, Parva Lavintune* 1086, *Lauingtun'* after 1167, *Lavington* 1230–1796, *Launton(a)* 1093×1100, 1219, *Laington* 1316, *Lainton* 1318, *Lenton* 1665. OE pers.n. *Lāfa* + **ing**^{4} + **tūn**. Perrott 195, Cameron 1998.

LENWADE Norf TG 0918. Partly uncertain. *Langewade* 1198×9, *Lon(d)ewade* 1257, *Lonwade* c.1330. The earliest form points to 'the long ford', OE **lang**, definite from **lange**, + **(ġe)wæd**, the later to 'the land ford', OE **land** + **(ġe)wæd**. The ford takes the road from Norwich to Fakenham across the Wensum; 'long' would suit the topography. DEPN prefers 'the lane ford', OE **lanu** + **(ġe)wæd**, but *lanu* might better be used here in the sense 'slow-moving stream'.

LEOMINSTER H&W SO 4959. 'Monastery in the district called Leen'. *Leomynster* c.1000 S 1534 and Saints, 1046 ASC(C), *Leofminstre* 1086, *Lemenestre* 1275, *Lemestre* 1428. District name Leen as in The LEEN SO 3859 + OE **mynster**. Possibly a straight translation, as Leland suggested, of the W version of the name *Llianllieny* 'church on the streams' or 'of *Lion*, the stream district'. DEPN, Bannister 114.

LEPE Hants SZ 4498. 'The leap'. *Lepe* from 1277, *L(h)upe* 1280–1350, *Leope* 1324–1494, *Leape* 1579. OE(WS) **hlīep(e), hlȳp** 'a leap, a jump'. The reference is normally to a stream that can be leapt or to a fence that some animals can leap while others are restrained. Ha 108, Gover 1958.197.

LEPPINGTON NYorks SE 7661. 'Estate called after Leppa'. *Lepinton(e)* 1086–1335, *Lepenton'* 1196, *Lepington' -yng-* 1246–1525, *Leppington* from 1279×81. OE pers.n. *Leppa* + **ing**^{4}, or genitive sing. *Leppan*, + **tūn**. YE 146.

LEPTON WYorks SE 2015. Probably the 'settlement on the hill-slope'. *Lepton(e) -tun(a)* 1086–1552. OE **hlēp** + **tūn**. The exact significance of **hlēp** in this name is uncertain; it normally means 'a leap, a thing to leap from or over, a waterfall'. YW ii.229.

LERRYN Corn SX 1457. Unexplained. *Leryon* 1284, 1289, *Lerion* 1289, *Leryan* 1325 Gover n.d., 1470, *Lerine* c.1540. Comparison may be made with OW *Cair Lerion* 9th HB, one of the 28 Cities of Britain, which also has not been explained (or identified), and Middle Breton *Coit-Lerien*. PNCo 109, Gover n.d. 302.

LESBURY Northum NU 2311. 'The leech's fortified place or manor'. *Lechesbiri* c.1190, *Le(s)cebyr(y) -bir'* 13th cent., *Lessebury* 1296 SR, *Lescebiri* 14th cent., *Lesbery* 1507. OE **lǣċe**, genitive sing. **lǣċes**, + **byriġ**, locative–dative sing. of **burh**. NbDu 133.

LESNEAGE Corn SW 7722 → MENEAGE.

LESNES ABBEY WOODS GLond TQ 4778 → ABBEY WOOD TQ 4778.

LESNEWTH Corn SX 1390. 'New court'. *Lisniwen* 1086, *Lisniwet -neweth* 1201, *-newic* 1233, *Lysnewyth* 1238, 1302, *Lesnewythe* 1421. Co ***lys** + adj. **nowyth, newyth** 'new'. 'New' for contrast with an 'old court' elsewhere, possibly HELSTON SX 0881 (*Hen-lys* 'old court' + **tūn**). PNCo 109, Gover n.d. 70, DEPN.

LESSINGHAM Norf TG 3928. Probably 'homestead of the Leofsigingas, the people called after Leofsige'. *Losincham* 1086, *Lesingham* 1254–1455. OE folk-n. **Lēofsīġingas* < pers.n. *Lēofsīġe* + **ingas**, genitive pl. **Lēofsīġinga*, + **hām**. Nf ii.112, ING 137.

LESSONHALL Cumbr NY 2250. Uncertain. *Lassenhall* 1507, 1539, *Lasselhall, Lacy Hall* 1518×29, *Lesson Hall* 1627. Probably a surname + **hall**, but this place was earlier known as 'Little Warton', *parua Wauertoun'* 1278–1325, and *Lassenhall* might represent OE **(æt thām) lǣssan hale** '(at the) lesser haugh'. Cu 159.

LETCHMORE HEATH Herts TQ 1597. *Lachemeresheth* 1299, *Lechmore Heath* 1544. P.n. Letchmore as in *Westlechemere* 'the muddy pond' 1341, OE ***lǣċe**, ***leċe** + **mere**, + ME **heth** (OE *hǣth*). There was once a large pond on the common here on the London clay liable to be muddy. Hrt 61.

LETCHWORTH Herts TL 2132. 'Enclosure farm' or 'lockable enclosure'. *Leceworde* 1086, *Lec(c)hew(o)rth* c.1190–1346, *Luchewrth(ia)* c.1190, 1216×72, *Letchworth* 1638. OE ***lyċċe**, Herts dial. ***leċċe**, + **worth**. Hrt 132, Studies 1936.182, 184, PNE ii.30.

LETCOMBE BASSET Oxon SU 3785. 'L held by the Bassett family'. *Ledecumbe Basset* 1247, *Letecumbe Basset* 1320. P.n. *Ledecumbe* 1086, *-cumba -cumb(e) -co(u)mbe* c.1130–1327, *Hledecumba* [1154×89]c.1200, *Lhedecumbe* 1385, *Li- Lo- Ludecumb(e) -combe* 1205–1338, *Letcombe* from 1383, 'Leoda's coomb', OE pers.n. **Lēoda* + **cumb**, + manorial affix from the Bassett family who held here in the 12th cent. The *Hle- Lhe-* spellings might suggest OE *hlēda* 'a seat, a bench' used in the topographical sense 'ledge' but the *-o-* and *-u-* spellings point to OE *-eo-*. Also known as *Upledecumbe* 'upper L' 1220 etc. (OE, ME **up(p)**) for distinction from LETCOMBE REGIS SU 3886. Brk 323, L 90.

LETCOMBE REGIS Oxon SU 3886. 'The king's L'. *Ledecombe Regis* 1356–1463, *Lodecombe Regis* 1400. P.n. *Ledencumbe* 1086, as in LETCOMBE BASSETT showing the genitival form of the pers.n. **Lēoda, *Lēodan*, + Latin **rex**, genitive sing. **regis**. The manor was already a royal one in 1066. Also known as *Dunledcombe* 'lower L' 1316 (OE, ME **dūn(e)**) from its position lower down the valley from Letcombe Bassett. Brk 323.

LETHERINGHAM Suff TM 2758. 'The homestead of the Letheringas'. *Ledringa- Letheringaham* 1086, *Ledrincgeham* [1086]c.1180, *Letheringham* from 1235. OE folk-n. **Letheringas* < unknown element or pers.n. + **ingas**, genitive pl. **Letheringa*, + **hām**. Late OE *Letheringas* might represent earlier **Hlēothringas* 'the dwellers on the *Hlēothre* the sounding stream' from OE *hlēothor* 'sound, melody'. This could have been an earlier name of the Deben, itself a back-formation from Debenham. DEPN, ING 131.

LETHERINGSETT Norf TG 0638. Partly uncertain. *Le- Laringaseta* 1086, *Letheringsete* 1254. Both this place and LETHERINGHAM Suffolk TM 2757 lie on streams which may have been called **Hlēothre* 'the noisy one' < OE *hlēothor* 'noise, melody'. Possibly therefore, 'the dwelling or fold of the Hleothringas, the people who live by the Hleothre', OE folk-n. **Hlēothringas* < **Hlēothre* + **ingas**, genitive pl. **Hlēothringa*, + **(ġe)set**. DEPN.

LETOCETUM Staffs SK 1006. 'The grey wood'. The RBrit name of the Roman town at Wall SK 0906 also used as the basis of the p.n. LICHFIELD $2\frac{1}{2}$ miles N at SK 1209. *Etoceto* (sic for *Leto-*) [4th]8th AI, *Letoceto* [c.700]13th Rav. Brit **Lētocēton* < **leito-* 'grey' + *caito-* 'wood', the equivalent of English Harwood 'hoar wood', a name type applied to parcels of the ancient wildwood which here must have extended from Wall to Lichfield. RBrit 387, NQ 1997.454.

LETTAFORD Devon SX 7084. 'Clear, bright ford'. *Lottreford* 1244, *Lutterforde* 1390, *Litterford* 1505. OE **hlūttor** + **ford**. It is possible that the stream itself was called *Hlūtre* 'the bright one', a name which has not survived independently here but is found elsewhere. D 470.

LETTON 'Kitchen garden'. OE **lēac-tūn**.

(1) ~ H&W SO 3346. *Letvne* 1086, *Lecten -tun -ton(e)* c.1130–1464, *Leitun* 1159×64, *Lettun'* 1243, *Letton* from 1431. He 126.

(2) ~ H&W SO 3870. *Lectune* 1086, *Lec(t)hon(e)* late 12th–1535, *Letton* from 1292. He 127.

(3) ~ LAKE H&W SO 3547. *Letton Lake* 1833 OS. P.n. LETTON SO 3346 + dial. **lake** 'a stream'.

LETTON HALL Norf TF 9705. *Letton Hall* 1838 OS. An 18th cent. mansion house. P.n. *Let(e)tuna* 1086, *Lecetuna* [1086]c.1180, *Lecton* 1200, either 'the vegetable garden', OE **lēac-tūn**, or perhaps rather 'the stream settlement', OE ***leċe** + **tūn**, + ModE **hall**. DEPN.

LETTY GREEN Herts TL 2810. *Letty Green* 1545. Surname Letty, cf. Richard *Lety* 1296, + ModE **green**. Hrt 228.

LETWELL SYorks SK 5687. 'Impeded stream'. *Lettewell(e)* c.1150–1460, *Letwell* from 1505. ME **lette** 'an obstruction, a stoppage' + **welle**. The term describes a spring whose flow is in some way impeded. There is now no well or spring near the village. YW i.143, L 32.

LEVEDALE Staffs SJ 9016. 'Leofgyth's nook'. *Levehale* 1086, *Levedehal(e)* 1198–1242, *Levedale* from 1198, *Ley- Le(a)dall* 16th cent. OE feminine pers.n. *Lēofgȳth*, genitive sing. *Lēofgȳthe*, + **halh**. The 1086 form may point to a shortened form of the pers.n., *Lēofe*. The sense of *halh* here is 'a projecting corner of land' referring to its position relative to the parish centre at Penkridge. St i.89.

LEVEN Humbs TA 1045. Uncertain. *Leuen(e), Leven(e)* 1086–1840. Originally a stream name identical with r. LEVEN NZ 4906. YE 72.

River LEVEN NYorks NZ 4906. Uncertain. *Leuen(e)* [1218×31]15th, 1268–1586, *Leven* from 1293, *Leaven* 17th. Various explanations have been proposed:

(1) PrW **Livon, Liμon* from Brit **Lemonā* usually explained as 'the elm-wood river' from Brit **lemo-*, late Brit **levo-* 'an elm-tree';

(2) PrW **Livon, Liμon* from **lim-/*lem-* 'a marsh' as in Gk *λιμήν* 'a harbour', *Lemannus lacus* (Lake Geneva), the Fr r.ns. Limours, Limagne and p.ns. Limeux (Somme), Limous, *Lemos* 1250×63, Limoux, Limousin, *de Lemosinis* 1260, and Latin *limosus* 'muddy';

(3) PrW **Libni* from Brit **Libnio-* or **Limnio-* from the root **lib-/*leib-* 'drop, pour' as in Gk *λείβω*, Lat *lībo* 'pour out' or **sli-m-no-* 'slippery, fluent' as in W *llyfn* 'smooth, flowing calmly without ripples'. YN 4, RN 251, LHEB 488, 672, GPC 2254.

LEVENS Cumbr SD 4886. 'Leofa's headland'. *Lefuenes* 1086, *Leu- Levenes* 1170×84–1439, *Levens* from 1352. OE pers.n. *Lēofa* + **næss** referring to the ridge of high ground between the rivers Kent and Gilpin which unite at the S end of the parish. We i.90, L 173.

LEVENSHULME GMan SJ 8794. 'Leofwine's island'. *Lewyneshulm* 1246, *Levensholme* 1322, *Lensom* 1587. ME pers.n. *Lewin* (OE *Lēofwine*), genitive sing. *Lewines*, + **hulme** (ON *holmr*) in the sense 'dry ground in marsh'. La 31 gives pr [levenzu:m], SSNNW 237, L 50, 52, Jnl 17.30.

Little LEVER GMan SD 7507. *Parva Lefre* 1212, *Little Lethre* 1221, *Little Levere* 1331. ME adj. **litel**, Latin **parva**, + p.n. *Leoure* 1227, *Lever* 1246, of uncertain origin; it could be OE **lǣfere** 'rushes', pl. of **lǣfer**, or more likely an earlier name of the Croal, **Lǣfre* 'rush stream'. (There are no early forms recorded for the r.n. Croal). Cf. LEARMOUTH Northum NT 8637 and the Alsace p.n. Liépvre on the r. Liepvrette, a tributary of the Ill near Sélestat, in German the Leber-bach, *Lebraha* 781 (OHG *aha* 'river'). 'Little' for distinction from Great Lever SD 7207, *Magna Leure* 1285, *Great Leure* 1326 and Darcy Lever SD 7308, *Darcye Lever* 1590, held by Sir Thomas D'Arcy c.1500. La 45, Jnl 17.37.

LEVERINGTON Cambs TF 4411. 'Estate called after Leofhere'. *Leverington(e)* from c.1130 with variants *-in- -yn(g)-*, *Luringtune* [c.1250]14th, *Lewryngton* 1494, 1541. OE pers.n. *Lēofhere* + **ing**[4] + **tūn**. Ca 271.

LEVERS WATER Cumbr SD 2799. Uncertain. *Lever Water* 1774, *Levers Tarn* 1786, *Levers Water* 1830. The evidence is really too late for certainty, but this might be OE **lǣfer**, **lēfer** 'a rush, a reed, a yellow iris, levers, a reed bed'. La 192 f.n.2 gives pr [li:vəz wɔ:tə], PNE ii.11.

LEVERTON Lincs TF 3947. 'Farm, village where reeds or rushes ('levers') grow'. *Leuretune* 1086, *Leuerton* 1166–1531, *Leirton* 1200, 1202, *Leverton* from 1212. OE **lǣfer, lēfer** + **tūn**. Payling 1940.128.

South LEVERTON Notts SK 7881. *Sudleg'ton* 1275, *Suthlegertone* 1279, *Suthleerton* 1280, *Southleverton* from 1334, *Suthleyrton* 1236. ME adj. **sūth** + p.n. Leverton as in North LEVERTON SK 7882. Nt 33.

North LEVERTON WITH HABBLESTHORPE Notts SK 7882. *Nordleg'ton* 1275, *Northlegherton* 1280, *Northleverton* from 1302, *Northleyrton* 1276. ME adj. **north** + p.n. *Cledretone* 1086, *Legretone* 1086, *Leg(h)erton* 1200–1244, *Leertona* [1163]13th, *Leir- Leyrton(e)* 1166–1329, *Leverton* from 1174, *Letherton* 1302, the 'settlement on the river *Legre*'. The various spellings with *-g(h)- -v- -th-* point to original guttural spirant [γ] which was replaced by the more productive neighbouring spirant phonemes [v] or [ð] and under the influence of the existing OE appellatives **læfer** 'a rush, a reed, an iris' and **lether** 'leather'. The r.n. **Legre* (or **Hlegre* if the DB form *Cledre-* is not simply a mistake), of unknown meaning and origin, appears again in LEIRE Leics SP 5290, in LEICESTER itself SK 5804, and in LAYER BRETON and LAYER DE LA HAYE Essex TL 9418 and TL 9620, both of which lie on *Layer Brook*. 'North' for distinction from South LEVERTON SK 7881. See also HABBLESTHORPE. Nt 33.

LEVINGTON Suff TM 2339. 'Leofa's estate'. *Leuetuna, Leuentona* 1086, *Leventon* 1273–86, *Levington* from 1254. OE pers.n. *Lēofa*, genitive sing. *Lēofan*, + **tūn**. DEPN, Baron.

LEVISHAM NYorks SE 8390. 'Leofgeat or Leofede's village or homestead'. *Leu- Lewecen* 1086, *Leuezham* 13th cent., *Leu- Levesham* 1242–1619, *Lewsam, Leas(h)am* 1577. OE pers.n. *Lēofġēat* or *Lēofede*, genitive sing. *Lēofġēates, Lēofedes*, + **hām**. Cf. LEDSHAM Chesh SJ 3574. YN 92 gives prs [liusəm, levisəm].

LEVORRICK Corn SX 0144 → MEVAGISSEY SX 0144.

LEW Oxon SP 3206. '(At) the tumulus or burial mound'. *(æt) Hlæwe* [984]13th S 853, *Lewa* 1086, *Lewe(s)* 1185–1346. OE **hlǣwe**. One of the most conspicuous landmarks in W Oxfordshire (Blair 1994.45). O 327, L 162.

River LEW Devon SS 5301. Possibly 'bright stream'. *Lyu* 1281, 1370. Co **lyw** 'colour', W *lliw* as in the W r.n. Lliw. A tributary of the Torridge. D 8, RN 253, CPNE 151.

LEWANNICK Corn SX 2780. 'Church-site of Gwenek'. *Lanwenuc* [c.1125]14th, *-wennoc* 1201 Gover n.d., *-wenech* 1261, *Lawanek* 1334, *Lewannick* 1610. Co ***lann** + pers.n. *Gwenek*. This is the equivalent name to St Guenoc in Brittany, cf. also Llanwenog Dyfed SN 4945. But the church here is dedicated to St Martin. PNCo 109, Gover n.d. 154.

LEWDOWN Devon SX 4487. 'Hill by the river Lew'. *Leywedon* 1270. R.n. Lew + OE **dūn**. Lew, a tributary of the Lyd, *Lywe* 1565, is identical in meaning with the River LEW SS 5301. For earlier forms see LEWTRENCHARD SX 4586. D 188, 8.

LEWES ESusx TQ 4110. 'The gash'. *(to) Lœwe* [911×19]1562, [925×c.935]12th, *(wiþ) Lœwe* c.960 S 1211, *(juxta) Laewes* [961]13th S 1212, *Lœw(e), Lœ* 10th coins, *Lewes* from 1086; Latin forms *Lewias, Lewiarum, Lewiis* accusative, genitive, dative pl., c.1089–1169, *(de) Laquis, (pro) castellatione aquarum* 1086. OE **lǣw** 'injury, gash, mutilation' referring to the great gap in the South Downs through which the river Ouse flows to the sea and where the town of Lewes is situated. There is also an extensive area of low marshy ground subject to flooding along the tidal Ouse which led to the popular etymology of *Lewes* as if *L'ewes* 'the waters' with OFr *ewe* and reflected in the DB Latin forms. Forsberg 1997 *passim*.

LEWISHAM GLond TQ 3975. 'Leofsa's homestead'. *Lievesham* 918 B 506, 661, *Liofesham* 11th, *Levesham* 1086, *Leuesham* 1081, 1203, 1275, *Leusham* c.1762. OE pers.n. **Lēofsa*, a short form of *Lēofsiġe*, + **hām**. The boundary of Lewisham and the people of Lewisham are recorded as *Liofshema mearc* 862 S 331 and *Leofsnhœme mearc* 987. DEPN, LPN 135.

LEWKNOR Oxon SU 7179. 'Leofeca's flat-topped ridge'. *(æt) Leofecanoran* 990×2 S 1454, *Leovechenoran* [1146]c.1225, *Levec(h)anole* 1086, *Leuekenor(e) -v- -c(h)- -a* 12th–1285, *Leukenor(e)* 1207–1428, *Lukenor* 1675. OE pers.n. *Lēofeca*, genitive sing. *Lēofecan*, + **ōra**. O 112, L 179–80, Jnl 21.17, 22.35.

LEWORTHY Devon SS 6738. 'Leofa's enclosure'. *Loveworthi* 1330, *Lyvaworthi* 1333. OE pers.n. *Lēofa* + **worthiġ**. D 31.

LEWTRENCHARD Devon SX 4586. 'Lew held by the Trenchard family.' *Lyu Trencharde* 1261, *Trenchardeslieu* 1316, *Lewtrancharde* 1431. Earlier simply *Lewe* 1086, *Lyu, Lyw(e), Liw* 13th cent., *Lewe* 1298. Originally a r.n. identical in sense with LEW SS 5301. The Trenchard family first occurs here in 1238, but a Richard Trencard occurs in Devon as early as c.1100. D 187 gives pr [træn∫əd].

East LEXHAM Norf TF 8516. *Estlechesham* 1242. ME **est** + p.n. *Lecesham* 1086, *Lechesham* 1158–97, 'the leech's estate'. OE **lǣċe**, genitive sing. **lǣċes**, + **hām**. 'East' for distinction from West LEXHAM TF 8417. DEPN.

West LEXHAM Norf TF 8417. *Westlechesham* 1242. ME adj. **west** + p.n. Lexham as in East LEXHAM TF 8516. DEPN.

LEY Corn SX 1776. 'Clearing'. *Leye* 1436 Gover n.d. Cf. cross called *Lay Crosse* 1515, probably a stone monument. Ultimately from OE **lēah**, dative sing. **lēaġe**. PNCo 109, Gover n.d. 288.

LEYBOURNE Kent TQ 6858. 'Stream called *Lille*, the sounding one'. *(ofer) lylle burnan* [942×6]12th S 514, *Lillanburnan, Lilleburna* c.975 B 1321–2, *Lelebvrne* 1086, 1181, *Leeburn'* 1194, *Leiburn' -burne* 1203 etc. OE stream n. **Lille* + **burna**. Cf. LILBOURNE Northants SP 5677, LILBURN Northum NU 0224. KPN 255, Forsberg 1950.197.

LEYBURN NYorks SE 1190. 'Shelter stream'. *Leborne* 1086, *Layburn(e)* from 12th, *Laibrunn, Lei- Leybroun* 13th. OE ***hlēġ** + **burna** varying with ON **brunnr**. Cf. LEATON Shrops SJ 4618. YN 257, PNE i.252.

LEYCETT Staffs SJ 7846. 'The fold or folds called or at *Levering*, the place called after Leofhere'. *Leveringsete* 1275, *Lover(e)sete* 1278, 1307, *Levereshevеd* 1327, *Levershede* 1474, *Leversete* 1334, *Lysot* 1475, *Leycett* 1548, *Leasit(t)* 1733. P.n. *Levering* < OE pers.n. *Lēofhere* + **ing**², + **set**, pl. **setu**. Horovitz.

LEYLAND Lancs SD 5421. 'Estate with a high proportion of non-arable ground; untilled land'. *Lailand* 1086, *Leiland(ia)* c.1160, 1212, *Leyland* from 1243. OE ***lǣġe** + **land**. The reference is to the low-lying land along the river Lostock used as pasture and meadow rather than arable. OE **lǣġe* is cognate with ON *lágr* 'low', MHG *læge* 'low, flat, poor'. La 133, TC 124, L 248–9.

LEYSDOWN ON SEA Kent TR 0370. Partly uncertain. *Legesdun* c.1100, *Leesdon(a)* 1174×5, 1175×6, *Lesdon(e)* 1208–62, *Leysdon(')* 1247–59. Possibly 'beacon hill', OE **līeġ** 'flame', Kentish ***lēġ**, genitive sing. ***lēġes**, + **dūn**. Alternatively the specific might be **lēaġes**, genitive sing. of **lēah** 'a wood or clearing' although there is no firm evidence for the use of *lēah* as the first element of p.ns. PNK 252, L 146.

LEYSTERS H&W SO 5563. Unexplained. *Last* 1086, *Lastes* 1160×70–1243, *Lastr(es)* 1220–1354, *Laysters* 1575, *Leysters* 1832 OS. He 118.

LEYTON GLond TQ 3886. 'Settlement on the r. Lea'. *Lugetune* 1042×66, *Lygetun(e)* [1066]13th S 1043, *Lei(n)tuna, Leintun* 1086, *Lu(i)- Luyton* 1200–1266, *Leyton* from 1224. R.n. LEA TG 3695 + OE **tūn**. Ess 101.

LEZANT Corn SX 3379. Probably 'church-site of Sant'. *Lansant* [c.1125]14th, 1276–1352, *Lassant* 1471, *Lazant* 1544, *Lesante* 1610. Co ***lann** + pers.n. *Sant*. PNCo 109, Gover n.d. 156.

LICHFIELD Staffs SK 1209. 'The open land at or called *Lyccid*'. *Anliccitfelda* [c.715]11th Eddi, *Onlicitfelda* [c.715]c.1100 ibid., *Lyccidfelth* [c.731]8th BHE(M), *licidfelth* 8th ibid.(B), *liccidfeld* 8th ibid.(C), *licidfelt* 9th ibid.(N), *Liccedfeld -t* [c.890]11th OE Bede, *(on) Licet felda* 11th ASC(C) under year 1039, c.1050 ibid.(D) under year 1053, c.1120 ibid.(E) under years 716 and 731, *(on) Licedfelda* 11th ibid.(C) under year 1053, *Li- Lecefelle* 1086, *Lichesfeld* 1130, 1164, *Lichefeld* c.1145. Bede also has the adj. form *Lyccitfeldensis* (M), *liccitfeldensis* (B, C). P.n. *Lyccid* or *Liccid* < PrW **Luitgēd* or *Lētgēd* < LETOCETUM 'the grey wood', the name of an ancient parcel of wildwood and of the Roman town at Wall SK 0906 ½ m. to the S, + **feld**. This is the place where St Chad established his episcopal see as bishop of the Mercians in 669. DEPN, LHEB 327, 332–4, 647, Medieval Settlement 213–4, L 190, 238–43, Horovitz.

LICKEY H&W SO 9975. Partly unexplained. *la Lec- Lekheye -hay(e)* 1255–1408, *Lykheye -ehay* 1271, 1280, *Ly- Lickhay* 1473×5, 1675, *Bromsgrove Lickey* 1831 OS. As the regular occurrence of the French definite article shows the name was early regarded as an appellative, probably in ME **haie** (OE *(ġe)hæġ*) 'a forest enclosure' and referring to a tract of land of some extent. Through it runs a small brook; the specific may, therefore, be OE ***leċe** 'a stream'. Wo 342, PNE ii.22.

LICKEY END H&W SO 9772. *Lickey End* 1831 OS. Probably the 'Lickey end of Bromsgrove' in contrast to *Rock End* ibid. P.n. LICKEY SO 9975 + ModE **end**.

LICKFOLD WSusx SU 9226. 'The leek enclosure'. *Lykfold* 1332. ME **lek** (OE *lēac*) + **fold** (OE *falod*). Sx 27.

LIDDINGTON Wilts SU 2081. 'Settlement on the *Hlyde*, the noisy one'. *(at) Lidentune* [940]15th S 459, *Ledentone* 1086, 1377, *Lidetona -e* 1156, 1166, *Lidinton* 1205, *Ludinton -yn(g)-* 1242–1394, *Lydington* 1279. OE stream name *Hlȳde* < OE adj. *hlūd* 'loud', genitive sing. *Hlȳdan*, + **tūn**. The stream itself Lid Brook, *Lydbroke* 1558×1603, is referred to in boundary clauses as *(on, anlang) lyden, liden* [940]14th S 459 and *(andlang) Hlydan* and *(innan) hlydan æwylmas* 'to the springs of Hlyde' [854]12th S 1588, and its valley as *Lyde Cumb* 'the coombe of Hlyde' [940]15th S 459. Wlt 8, 282, RN 272, Grundy 1919.177, 1920.12.

LIDDINGTON CASTLE Wilts SU 2179. *Liddington Castle* 1828 OS. P.n. LIDDINGTON SU 2081 + ModE **castle** referring to the Iron Age hill-fort of the 5th–4th cent. BC known also as *Battlebury* 1670. Wlt 283.

LIDGATE Suff TL 7258. 'The swing-gate'. *Litgata* 1086, *Lidgate* 1254. OE **hlidġeat**. DEPN.

LIDLINGTON Beds SP 9939. 'Settlement of the Lytlingas, the people called after Lytel(a)'. *Litincleton* 1086, *Lit(t)lingeton* 1180–1227, *Lit(t)lington* 1204 etc., *Lidlington* 1780. OE folk-n. **Lȳtlingas* < pers.n. **Lȳtel(a)* + **ingas**, genitive pl. **Lȳtlinga*, + **tūn**. Bd 77.

LIFTON Devon SX 3885. 'Settlement, estate on the river Lew'. *(æt) Liwtune* [873×88]11th S 1507, *(fram) liwtune* [c.970]n.d. B 1247, *Listone* (for *Lif-*) 1086, *Liftuna, Li- Lyfton(e)* 1158–1331†. R.n. Lew as in LEWDOWN SX 4487 and LEWTRENCHARD SX 4586 + OE **tūn**. D 188, Forsberg 1950. 145.

LIGGARS Corn SX 2622 → LYDEARD ST LAWRENCE ST 1232.

LIGGER or PERRAN BAY Corn SW 7256. *Ligger or Piran Bay* 1813, cf. Ligger Point SW 7558, *Ligger Point* 1813, for which Gover n.d. 377 cites a form *Legoe* 1673. The evidence is too late for explanation, but one possibility is 'grey promontory', Co **los** (earlier *lot*) + **garth** as in Liggars SZ 2262. The alternative name Perran Bay is taken from the saint's name preserved in PERRANPORTH SW 7554. PNCo 110.

LIGHTHORNE Warw SP 3455. 'Light or bright thorn-bush'. *Listecorne* (sic) 1086, *Lith(e)- Ly(ch)tethurn(e)* 1242–1340, *Lightethurn(e)* 1370, *Lighthorne* 1545, *Lighterne* 1481–1545. OE **liht** + **thyrne**. Wa 255 gives pr [laithɔ:n], L 221.

LIGHTWATER Surrey SU 9262. Originally the name of a pool. *Light Water Moor* 1719, *Lightwater Pond* 1765. ModE **light** + **water**. Sr 153.

LIGHTWOOD Staffs SJ 9241. 'The bright wood'. *Lyhtwude* c.1230, *Lyghtwode* 1306. OE **leoht** + **wudu**. Horovitz.

LILBOURNE Northants SP 5676. 'Lilla's or the chattering stream'. *Lineburne* 1086, *Lilleburne* 1086–1437 with variants *Lylle-* and *-bourne*, *Lilburne* 1393. OE pers.n. *Lilla* or stream-name **Lille* + **burna**. Nth 72, L 17, cf. Forsberg 1950. 115, SN 6.142f.

East LILBURN Northum NU 0423. ModE adj. **east** + r.n. Lilburn as in LILBURN TOWER NU 0224.

†Probably here also belongs *Leowtun* 931 S 416.

LILBURN TOWER Northum NU 0224. *Lilburn Tower* 1868 OS. A ruined medieval tower later replaced by a 19th cent. house. P.n. *Lilleburn(e)* 1177–1334, *Lilburn* 1428, originally a stream name, 'the chattering stream', OE **Lille* + **burna**, + ModE **tower**. Also known as *Parva Lilleburn* 1201 Cur, *West Lilleburn* 1256 Ass, for distinction from East LILBURN NU 0423. NbDu 134.

LILFORD HALL Northants TL 0384. P.n. Lilford + ModE **hall**, a 17th cent. mansion. Lilford, *Lilleford(e)* 1086–1428, *Lyllford* 1430, *Lillingford* 1205, *Lillesford* 1284, is either 'Lilla's ford', OE pers.n. *Lilla* + **ford**, or the 'chattering ford, ford of the chattering stream', OE **Lille*. A ford of the river Nene. Nth 185, cf. Forsberg 1950. 114–6, SN 6.142f.

LILLESHALL Shrops SJ 7315. 'Lill's hill'. *Linleshelle* 1086, *Lileshelle* 1167, *-hull' -hill* 1200–1737, *Lilleshul(l') -hil(l') -hell'* 1198–1660, *Lincel* c.1540, *Lilsell, Linshull* 17th, *Linshall* 1704, *Lilleshall alias Linsell* 1708. OE pers.n. *Lill*, genitive sing. *Lilles*, + **hyll**. Cf. *Lilsætna gemære* [963]12th S 723, 'the boundary of the Lilleshall people'. Sa i.176.

LILLEY Herts TL 1226. 'Wood or clearing where flax or lime-trees grow'. *Linleia* 1086, *Li- Lynlege -legh -le(y)e* 1204–1445, *Lilley* from 1275. OE **līn** or **lind** + **lēah**. Hrt 18.

West LILLING NYorks SE 6465. *West Lillinge* 1282. ME **west** + p.n. *Lilinge -a* 1086, *Lilling(a)* 1167–1295, probably a singular **ing**[2] formation either on OE pers.n. *Lil(l)a* and so 'place called after Lila' or on the word ***lill(e)** 'slope, valley with a river' which has been postulated for LILLINGTON Dorset ST 6212. Either sense would be possible for **Lilling* in this name; the Lillings lie on sloping ground N of the r. Foss. Alternatively **Lilling* might have been the original name of the river itself. Like the pers.n. *Lil(l)a* derived from it, OE **lill(e)* is taken to be a lall-word so that a formation **Lilling* would mean the 'chattering, babbling stream'. East Lilling, *Estlillyng* 1317, is a deserted medieval village site at SE 6664. YN 32, SN 6.142, BzN 1967.349.

LILLINGSTONE DAYRELL Bucks SP 7039. 'L held by the Dayrell family'. *Li- Lutlingestan Daireli*, 1166. Otherwise simply *Lelinchestune* 1086, *Lil(l)ing(e)stan* 1086–1249, *Lilling(e)ston* 1194–1242, 'boundary stone of the Lytlingas, the people called after Lytel or Lytla', OE folk-n. **Lȳtlingas* < pers.n. **Lȳtel* or *Lȳtla* + **ingas**, genitive pl. **Lȳtlinga*, + **stān**. 'Dayrell' (from *de* + *Airelle*, a place between Bayeux and Caen) for distinction from LILLINGSTONE LOVELL SP 7140 which lay in Oxfordshire prior to 1844. The stone may, therefore, have marked the county boundary. Bk 44.

LILLINGSTONE LOVELL Bucks SP 7140. 'L held by the Lovell family'. P.n. Lillingstone as in LILLINGSTONE DAYRELL SP 7039 + family name Lovell. Bk 44.

LILLINGTON Dorset ST 6212. Either 'estate called after Lylla' or 'on the stream called **Lylle*'. *Lilletone* 1166, *Lillinton* 1180, 1232, *Lillington(')* from 1244 with variants *Lyll- Lull-* and *-yng-*. OE pers.n. **Lylla* + **ing**[4] or stream name **Lylle* + **ing**[2], + **tūn**. Do iii.343.

LILSTOCK Somer ST 1644. 'Lylla or Lulla's outlying farm'. *Lvlestoch* 1086, *Lullinstoke* 1204, *Lillingstok* 1285, *Lystoke* 1610. OE pers.n. **Lylla* or *Lulla*, genitive sing. **Lyllan, Lullan*, varying with **Lylla, Lulla* + **ing**[4], + **stoc**. DEPN, Studies 1936.19.

Great LIMBER Lincs TA 1308. *Magna Linberga* c.1115, *Moche Lymbergh* 1493–1500, *Gret Lymber* 1529. ModE **great**, ME **much**, Latin **magna** + p.n. *Lindbeorhge* [1066×8]c.1200 Wills, *-berge -bergh* 1200, *Lin- Limberge, Linberghā* 1086, *Linberga -(e)* c.1115–1246, *Limberga -(e)* 1157–1291, *Lymber* 1430–1568, *Limber* 1576, 'the lime-tree hill', OE **lind** + **beorg**. 'Great' for distinction from Little Limber TA 1210, *Parva Linberge* c.1115. Li ii.219.

LIMBRICK Lancs SD 6016. No early forms. Possibly ON **lind** + **brekka** 'lime-tree slope'. La 129.

LIMBURY Beds TL 0624 → LAINDON Essex TQ 6889.

LIMEBROOK H&W SO 3766 → LINGEN SO 3667.

LIMEFIELD GMan SD 8012. *Limefield* 1843 OS. Either 'the lime-tree field' or the 'field spread with lime'.

LIMINGTON Somer ST 5422. Partly uncertain. *Limin(g)tone* 1086, *Liminton(e)* [1189×1215]14th, 1235, *Li- Lemington* 1243, *Lymmyngton* 1610. The generic is OE **tūn** 'settlement, farm, village, estate', the specific possibly a lost r.n. identical with LEAM SP 4568. DEPN.

LIMPENHOE Norf TG 3903. 'Limpa's hill-spur'. *Lim- Linpeho* 1086, *Limpenho* 1193 etc. OE pers.n. **Limpa*, genitive sing. **Limpan*, + **hōh**. The hill-spur overlooks the Yare marshes. DEPN.

LIMPLEY STOKE Wilts ST 7860 → Limpley STOKE ST 7860.

LIMPSFIELD Surrey TQ 3953. 'The open land of or called *Limen*'. *Limenesfeld* 1086, *-feld(e)* c.1121–1398, *Linesfeld* 1087, *Limenefeld* and *Lemene(s)- Lymene- Lyuenes- Lymnes- Lemne-* with *-feld(e)* or *-feud* 13th cent., *Lymmesfeld* 1325, *Lempnes-* 1426, *Lympesfeild* 1636. P.n. **Limen* from Brit ***Leman-** 'growing with elms', genitive sing. **Limenes*, + OE **feld**. Sr 323, RN 245, LHEB 287, L 239, 243.

LINACRE RESERVOIR Derby SK 3372. P.n. Linacre 'Newly-cultivated land where flax is grown', OE **līn** + **æcer**, ON **lín-akr**. + ModE **reservoir**. Linacre is *Lin(n)- Lynacra -acer -aker -akir* 1189–1252 etc. Db 221.L 232–3.

LINBY Notts SK 5350. 'Lime-tree village or farm'. *Lidebi* (sic for *Līde-* ie. *Linde-*) 1086, *Lindeby -bi* 1163–1282, *Lindby* 1232×51, *Linneby* 1233, *Lynby* 1392, *Lymby* 1474. ON or OE **lind**, genitive pl. **linda**, + **bȳ**. Nt 122, SSNEM 58.

LINCHMERE WSusx SU 8631. 'Wlenca's pool'. *Wlenchemere* 1186–1314, *Wel(l)enchemere -ynche-* 1428–1535, *Lenchmere* 1332, *Lynchmere* 1582. OE pers.n. **Wlenċa* + **mere**. The mere probably survives as Linchmere Marsh S of the church. Sx 24.

LINCOLN Lincs SK 9771. 'The *colonia* by the pool'. *Λίνδον* (*Lindon*) [c.150]13th Ptol, *Lindo* [4th]8th AI, *Lindum colonia* [late 7th]13th Rav, *(in) Lindocolino, Lindocolinæ ciuitatis* [c.731]8th BHE, *(to) lindcolne* [c.890]c.1000 OEBede, *(to) Lindcyl(en) (ceast(re))* [c.890]c.1000 OEBede–c.1050, *Lindcoln* 975–8 Coins, *Lincolnia* 1139–1322×9, *Lincoln* from 1176, *Nicol(e)* 1195–1390. Brit ***lindo-** 'water, pool', PrW **līnn*, + Latin **colonia** 'a town'. The town or *colonia* was established for the settlement of retired veterans of the Ninth Legion which garrisoned the fortress at Lincoln. It takes its name from its situation on the high ground overlooking a widening of the r. Witham later called Brayford Pool. Three forms of the name had some currency, (1) RBrit **Lindocolōnia* > PrW **Lindogolūnia*, **Lindgolūn* > OE **Lindcolun*, the ancestor of the present name; (2) PrW **Lindogolūnia*, **Lindogolūn* > **Lindgolün* > OE *Lind(o)colīn* re-Latinised by Bede as *Lindocolina* and **Lindculīnu* > lOE *Lindcylene* with i-mutation; and (3) *Nicol(e)* with dissimilatory AN interchange of *n* and *l*. L i.1, Jnl 8.34.

LINCOMB H&W SO 8268. 'Flax or lime-tree coomb'. *(on, of) lincumbe* *[706]12th S 54, *Li- Lyncumbe -combe* 1275–1676. OE **līn** or **lind** + **cumb**. Wo 244, Hooke 36, 38.

LINCOMBE Devon SX 7458. Partly uncertain. *Lyncombe* 1306, 1478. It is impossible to say whether this is 'flax valley', OE **līn** + **cumb**, 'lime-tree valley', OE **lind**, 'lime-tree', OE **hlinċ** in the sense 'ledge of ploughland on a hill-side', or whether the specific is a r.n. identical with the river Lyn in LYNTON SS 7149. D 292, lviii.

LINDAL IN FURNESS Cumbr SD 2575. 'Lime-tree valley'. *Lindale* c.1220, *Lindal* from c.1225. ON or OE **lind** + **dalr**. 'In Furness' for distinction from Lindale SD 4180. La 207, SSNNW 144, L 96, 222.

LINDALE Cumbr SD 4180. 'Lime-tree valley'. *Lindale* from 1246. ON or OE **lind** + **dalr**. La 199, SSNNW 144, L 96, 222.

LINDFIELD WSusx TQ 3425. 'The open land of the lime-trees'.

Lindefeldia, Lendenfelda [c.765]13th S 50, *Li- Lyndefeld* 12th–1483, *Linfeld* 1590. OE **lind**, genitive pl. **linda**, + **feld**. Sx 340 gives pr [linvəl], SS 70.

LINDFORD Hants SU 8036. Short for Lindford Bridge. *(pontem de) Linford* 'L bridge' 1269, *Lyngfordesbrygge* 1298, *Lymfordesbrigge* 1300. The generic is OE **ford** 'a ford', the specific either OE **līn** 'flax', **lind** 'a lime-tree', or **hlin** 'a maple-tree'. Ha 108, Gover 1958.103.

LINDISFARNE Northum NU 1243. Uncertain. *Lindisfarnae* [699×705]c.900, *Lindisfaronaeae* [699×705]c.1200, *Lindisfarne, Lyndisfaronaee* [699×705]13th, *Lindisfaronee* [699×705]14th, *Lindisfarna* [c.710×20]11th Eddi, *Lindisfarne insula* [c.710×20]c.1100, *Lindisfarena ea* [c.890]c.1000, *Lindisfar(e)na* 1121 ASC(E) under years 779, 793, *Lindisfarne* c.1107. The forms dated 699×705 are AC, those dated c.710×20 Eddi, and that dated c.890 from OE Bede. Bede himself regularly refers to Lindisfarne using the Latinised adjective construction *Lindisfarnensis insula* or *Lindisfarnensium insula*, 'island of the Lindisfarners'. The name has traditionally been explained as the 'island of the travellers to and from Lindsey', OE ***Lindisfara** 'a Lindsey traveller', genitive pl. ***Lindisfarena**, + **ēġ**. Bede regularly calls the people of Lindsey the *Lindisfari*. Cf. LINDSEY Humbs SK 9398. Symeon of Durham states that the island was named from a stream called *Lindis* which flowed into the sea, a mere 2 feet in breadth and only visible at low tide. This is clearly a reference to the tidal waters which twice daily cut Lindisfarne off from the mainland but is incredible as an explanation of the name of the island. Cox suggests that both Lindisfarne and the Farne Islands share the same name (of unknown origin) and that the tidal island was distinguished from the others by prefixing PrW ***lïnn** 'water, a pool' (Brit **lindo-*). In either case the adoption of the name precedes the assimilation of *nd* > *nn* which was probably complete by the end of the sixth century. NbDu 135, AC 94, Eddi 6, SD 87, LHEB 512–3, Jnl 8.24. See further Celtic Voices 241–59.

LINDLEY WOOD RESERVOIR WYorks SE 2149. P.n. Lindley Wood + ModE **reservoir**. Lindley Wood is named from Lindley SE 2250, the 'lime-tree wood or clearing', *lindeleh* [972×92]11th S 1453, *on Linde-leage* c.1030 YCh 7, *Lillai(a)* 1086, *Li- Lyndele(a) -ley -lay* 1167–1524, OE **lind** + **lēah**. YW v.60, L 222.

LINDRIDGE H&W SO 6769. 'Lime-tree ridge'. *Lynderycge* 11th, *Ly- Lindrug(g)e* 1175–1360. OE **lind**, genitive pl. **linda**, + **hrycg**. Wo 57, L 169, 222.

LINDSELL Essex TL 6427. 'Lime-tree hall'. *Lindeselā -seles* 1086, *Lindsel(e)* from 1123, *Lynsell, llynssyll, lynzel* 1533–41. OE **lind** + **sele**. Ess 486.

LINDSEY Humbs Lincs SK 9398. 'The island of or called *Lindes*'. Type I: *(in) lindissi* [c.704×14]9th Gregory, *prouincia Lindissi, lindissae prouinciae* (genitive sing.), *prouincia Lindisfarorum* [c.731]8th BHE, *Lindissa* 8th Alcuin, *Lindisse* c.892 ASC(A) under years 838, 873, 874, [c.890]11th OEBede, c.1122 ASC(E) under year 874, *in urbe Lindesse, iuxta Lindisse urbe* late 9th Æthelweard under year 841, *Lindisse* [997]12th S 891, c.1122 ASC(E) under year 678, *-isse* ibid. under year 627;
Type II: *Lindesige* c.1050 ASC(D) under year 838, 11th ASC(C) under year 1016, c.1122 ASC(E) under years 993, 1013, 1014, *Lindesig, Lindissig* [c.894]11th Asser, *Lindesege* 1066 ASC(C), *Lindesi* 1086, *Lindeseye* c.1140.
The two forms of this name correspond to (1) PrW **Lindes* < **Lindenses* 'the people of *Lindon* (Lincoln)' < Brit ***lindo-** 'water, pool' referring to Brayford Pool in Lincoln, + unknown element seen in Bede's *Lindissi*, OE *Lindesse*; (2) PrW **Lindes* + OE **ēġ, īeġ** 'an island' seen in the spellings *Lindisige -ege*, probably a popular etymology reformation of *Lindissi*.
The reference is to the district on the spine of higher ground S and E of the river Witham and bounded on the W by Sincil Dyke, known from a ford across the Witham as Wigford, *Wich(e)ford* c.1107–1202, *Wick(e)ford(a)* later 12th–1553, *Wigeford* 1196, *Wigford* from 1555, 'the ford leading to the *wic*'. This was the trading district of Lincoln (as opposed to the administrative centre in the *colonia* on the hill), accessible by boat from the sea. This may explain Bede's *prouincia Lindisfarorum* 'province of the people who travel to and fro to *Lindissi*' and references to the bishop of Lincoln as *Lidensis Faronensis episcopus* in Alcuin and later. The trading post gave its name to the Anglo-Saxon kingdom of Lindsey and subsequently to that part of the country of Lincoln between the Humber and the Witham known until 1974 as the parts of Lindsey. DEPN, L i.45, Jnl 8.41, ASE 18.1–32.

LINDSEY Suff TL 9744. 'Lelli's island'. *Lealeseia* c.1095, *Leleseia* 1191, *Lelleseye* 1233, *Lynsey* 1610. OE pers.n. **Lelli*, genitive sing. **Lelles*, + **ēġ**. The DB forms for this name are *blalsega* and *balesheia* (wrongly printed by Farley as *haleseia*). *L/h* and *h/b* are confusibilia in this text so that behind these sps probably lies a form **Laleseia*. DEPN, Domesday Studies 136.

LINFORD Essex TQ 6779. No early forms.

LINFORD Hants SU 1806. 'Lime-tree ford'. *(on) lind ford* 961 S 690, *Lyndford* 1365, 1432, *Lymford* 1300, *Lynford* 1331. OE **lind** + **ford**. Gover 1958.220.

LINGDALE Cleve NZ 6716. Cf. *Lingdale Head* 1861 OS. A modern colliery village. N dial. **ling** 'heather' + **dale**.

LINGEN H&W SO 3667. Probably a W r.n. by origin, 'fair water'. *Lingen* from c.1150 including [704×9]13th S 1801, with variants *Lingein(e) – ayn(e)* 1190–1535 and *Lingham* 1086. A mile SE of Lingen on the same stream is Limebrook SO 3766, *Li- Lyngebrok(e)* 1221–1527, *Lymebroke* 1535, which suggests that Lingen is by origin a stream name. DEPN suggests W *llyn-gain* from W **llyn** 'lake, pool, pond' (PrW **lïnn*, OW *linn*) + **cain** 'fine, fair, beautiful'. He 127, 128, DEPN, CPNE 149, GPC 390, 2272.

LINGFIELD Surrey TQ 3843. Probably 'the open country of the *Leah*-people'. *(on) Leangafelda* 871×89 S 1508, *Lingedefelda* [11th]copy, *Lingefeld, Lynge- -feud* 12th–13th cents., *Linchesfeud* 1270, *Lengefelde* 1279, *Li- Lyngge- Lygnefeld* 14th cent., *Lynke-* 1507, *Linvil* 1675. OE folk-n. **Lēangas* < **Lēahingas* 'the inhabitants of a *lēah* or forest clearing', genitive pl. **Lēanga*, + **feld**. The earlier explanation from OE ***hlenċ**, genitive pl. ***hlenca**, + **feld**, hardly fits the forms. Sr 327, DEPN, L 239, 244.

LINGMELL Cumbr NY 2008. Unexplained. *Lingmale* 1578. The specific might be ON **lyng** 'heather' but the evidence is too late. Cu 392.

LINGWOOD Norf TG 3608. 'Heather wood'. *Lingewode* 1199, *Lingwude* 1254. ME **ling** (ON *lyng*) + **wode** (OE *wudu*). DEPN.

LINKENHOLT Hants SU 3658. Partly uncertain. *Linchehov* 1086, *Ly- Linkeholt* c.1145–1256, *Li- Lynkenholt(e)* from 1209, *Nyncknoll, Ninkenholt* 16th. This is a difficult name. The earlier forms point to OE **hlinċ**, genitive pl. **hlinca**, + **holt** 'wood of the terraces', referring to the magnificent terraced roads here. But the forms with *-en-* are difficult. They can hardly represent OE **hlinċen* for which pr [tʃ] would be expected. Possibly in this name the ME adjectival-genitive ending *-ene* has replaced *-e* from OE *-a* as *monkene* occasionally replaces *monke* < OE *muneca*. Alternatively there may have been another quite different name for the wood, 'linnets' wood', OE **lineċe** 'a linnet', genitive pl. ***linecna** with depalatalised [k] in the syncopated ending *-ecna* < *-eċena*. For the 16th cent. forms cf. *Nicole* 1195–1390 for Lincoln. Ha 108, Gover 1958.160.

LINKINHORNE Corn SX 3173. 'Church-site of Kenhoarn'. *Lankinhorn* [c.1175]1378, 1229, 1259, *-kine-* 1235, *Lankynhorn* 1342–1548. Co ***lann** + pers.n. **Kenhoarn*. The patron saint of the church, however, is St Mylor. PNCo 110 gives pr ['liŋkinhɔ:n], locally [liŋkin'hɔ:n], Gover n.d. 160, DEPN.

LINLEY Shrops SO 3592. Probably 'the lime-tree wood or clearing'. *Lindele* 1208×9, *Linlega* c.1150, *Lingl', Linleg* 1255. OE **lind**, genitive pl. **linda**, + **lēah**. Sa i.178, Bowcock 142.

LINLEY GREEN H&W SO 6953. No early forms. The hamlet is situated in LINTON SO 6853 and may share the same specific, 'flax' or 'lime-tree wood or clearing' (OE *līn* or *lind* + *lēah*).

LINLEY HILL Shrops SO 3695. *Linley Hill* 1836 OS. P.n. LINLEY SO 3592 + ModE **hill**. Rises to 1349ft.

LINSHIELS Northum NT 8906. 'The shielings by the torrent'. *Linsel'* 1242 BF, *Lynsheles* 1292–1346, *Lynshields* 1618, *Lyndesele* 1314, *Lindsheildes* 1632 Dixon 32. OE **hlynn** + ***scēla**. Dixon 32 describes the Coquet here as rushing 'with great impetuosity between rugged and precipitous cliffs of porphyry, forming a succession of waterfalls, deep dark pools and rippling streams'. Some spellings show confusion with **lind** 'a lime-tree' which became *linn* in N dialects. NbDu 135.

LINSTEAD Suff TM 3377. Partly uncertain. *Linesteda -e* 1086, *Li- Lynsted(e)* 1199–1610. Either OE **līn** 'flax' or **hlin** 'a maple-tree' + **stede** 'a place'. DEPN, Stead 191.

LINSTOCK Cumbr NY 4258. 'Place where flax is grown'. *Linstoc* 1212, *-stok* 1278–1535 with variant *Lyn-, Linstock* from 1279. OE **līn** + **stoc**. Cu 110.

LINTHWAITE WYorks SE 1014. 'Flax or lime-tree clearing'. *Lindthait* 1185×1202, *Lin- Lynthwait -thwayt(e) -thweyt* 1208–1599. OE **līn** or **lind** + ON **thveit**. YW ii.273, L 211.

LINTON 'Flax-farm'. OE **līn** + **tūn**. It is not always possible to distinguish between *līn* 'flax' and *lind* 'a lime-tree' in this compound.

(1) ~ Cambs TL 5646. *(æt twam) Lintunum* 'at the two Lintons' [1008]12th S 919, *(duo) Lintunum, (in) Lintune* [1008]12th, *Lintona* [1008]14th, *Lintone, alia Lintone* 1086, *Linton Magna, ~ Parva* 'Great, Little Linton' from 1218. Ca 109.111

(2) ~ Derby SK 2716. *(at) Lintone* [942]13th S 483, 1086, *-ton(a)* c.1125–1281 etc., *Lintone, Linctune* 1086. If the 1086 form is significant it points to 'heather farm' with ON **lyng** or 'ledge farm' with OE **hlinċ**. Db 640, Forsberg 138.

(3) ~ H&W SO 6853. *Lintvne -tone* 1086, *-tun* 1156. DEPN.

(4) ~ -ON-OUSE NYorks SE 4960. *Linton super Usam* 1336. P.n. *Lvctone, lucton* (sic) 1086, *Lyn- Linton* from 1176 + r.n. OUSE SE 4959. YN 20.

LINTON Kent TQ 7549. 'Estate called after Lilla'. *Lilintuna* c.1100, *Li- Lyllin(g)ton' -yng-* 1226–1316, *Lynton(e)* 1315–1535. OE pers.n. *Lill* or *Lilla* + **ing**[4] + **tūn**. PNK 138.

LINTON NYorks SD 9962. Partly uncertain; probably 'settlement by the rushing stream'. *Lipton* (sic) 1086, *Lyn- Linton* from c.1150. OE **hlynn** + **tūn**. The village stands on Linton Beck, a swift-flowing stream; but 'flax farm', OE **līn** + **tūn**, is also a possible explanation. YW vi.103.

LINWOOD Hants SU 1809. 'Lime-tree wood'. *Lyndwode* c.1170–1327, *Lynwode* 1503. OE **lind**, genitive pl. **linda**, + **wudu**. Gover 1958.212.

LINWOOD Lincs TF 1086. 'The lime-tree wood'. *Lindvde* 1086, *Lindwda* c.1115, *Li- Lynd(e)wod(e)* after 1182–1487, *Lynwod(e)* 1305–1562, *-wood* 1514–84. OE **lind** + **wudu**. There is another Linwood at TF 1260. Li iii.58.

LIPHOOK Hants SU 8431. Partly uncertain. *Leophok* 1364, *la Leephook* 14th, *(atte) Lepok* 1404, *Liephok, Lepook, Lippuck* 15th, *Liphooke* 1690. Apparently a compound of OE (WSax) **hlīep** 'a leap' + **hōc** 'a hook', possibly 'angle or nook of ground at the leap', but the name is not fully understood. Ha 109, Gover 1958.99, DEPN.

LISCARD Mers SJ 3092. 'Hall at the rock'. *Lisnekarke -carke* 13th–1307 with variants *-caryc -ic -cark* and *Lys(e)ne-, Liscark* 1260–1468 with variants *Lys-* and *-kark, Liskard* 1350–1511, *Liscard* from 1558. PrW **lis ən garreg*, ***lïs** 'a hall, a court' + definite article ***en** + ***carreg**, referring either to one of the reefs at Black Rocks or Red Noses SJ 3095 or to the whole of the rocky hill on which Liscard lies. Not the same name as LISKEARD Corn SX 2654. Che iv.324.

LISCOMBE Somer SS 8732. 'Pigsty coomb'. *Loscumb* 1251, *Spire Liscomb* 1809. OE **hlōse** + **cumb**. DEPN.

LISKEARD Corn SX 2564. Probably 'court of Kerwyd'. *Lys Cerruyt* c.1010, *Liscarret* 1086, *-caret -karet* c.1150–1284, *Leskered* 1229, *Liskyrres* 1298, *Lyskerde* 1378 Gover n.d. Co ***lys** + pers.n. **Kerwyd*. PNCo 110 gives pr [lis'ka:rd], Gover n.d. 272.

LISS Hants SU 7727. 'The palace, estate centre'. *Lis* 1086–1306 with variant *Lys, Lissa -e, Lysse* 1174–1413, *Lyss* 1810 OS. PrW ***lïs**. Ha 109, Gover 1958.59.

East LISS Hants SU 7827. ModE adj. **east** + p.n. LISS SU 7727.

LISS FOREST Hants SU 7829. P.n. LISS SU 7727 + ModE **forest**. Cf. *Forest Field Farm* 1810 OS.

LISSETT Humbs TA 1458. 'Dwelling or fold at the pasture'. *Lessete* 1086, *Leset* c.1180–1494, *Lesset* 1194×1214–1316, *Licit, Lyssett* 16th cent. OE **lǣs** + **(ġe)set**. YE 77.

LISSINGTON Lincs TF 1083. 'The estate called after Leofsige'. *Lessintone* 1086, *Lissigtuna* c.1115, *Lissingtona* c.1200, *Leusinton* 1202, *Lissinton* 1203, *Linsinton* 1242. OE pers.n. *Lēofsiġe* + **ing**[4] + **tūn**. DEPN.

LITCHAM Norf TF 8817. Possibly 'the enclosure homestead or meadow'. *Licham, Lec(c)ham* 1086, *Lucham* 1197–1302, *Litcham* 1254, *Luycheham* 1547. OE ***lyċċe** + **hām** or **hamm**. DEPN, Studies 1936.183, Forsberg 194.

LITCHBOROUGH Northants SP 6354. Partly uncertain. *Liceberge* 1086, *Lichesberga* 1184–1366 with variants *Lyches-* and *-berg -ber(e)we -bar(e)we, Lichebarue* 12th–1366 with similar variants. The generic is OE **beorg** 'a hill', the specific uncertain; either OE ***liċċ** 'a stream', **līċ** 'a body, a corpse' referring to an ancient burial place, or ***lyċċe** 'an enclosure'. Of these the last is perhaps most likely. Nth 25, Studies 1931.59, Studies 1936.182–4, L 128.

LITCHFIELD Hants SU 4653. *Lychefeld* 1539. Apparently transferred from LICHFIELD Staffs SK 1209 and replacing earlier *Liveselle* 1086, *-schella -sulve -shull* 1167–1349, *Lidesull'* 1212, *Lidescelve* 1219, *-s(c)hulve* 1228, 1272, *Ludes(h)ulve -shelf* c.1270–1428, *Leveshulle*, unidentified element + OE **scylf(e)** 'shelf, a terrace'. Ha 109, Gover 1958.149, DEPN.

LITHERLAND Mers SJ 3498. 'Newly-cultivated land of the slope'. *Liderlant* 1086, 1114×16, *Litherland(e)* from 1202, *Liverlande* 1221. ON **hlíth** 'a slope', genitive sing. **hlíthar**, + **land**. Also known as *Dunlytherlond* 1298, *Dounelithirlond* 1392, 'down Litherland'. The ancient hamlet is situated at the foot of a small hill. La 117, SSNNW 145, L 166, 246, 249, Jnl 17.65.

LITLINGTON Cambs TL 3142. 'Settlement of the Lytelingas, the people called after Lytel(a)'. *Lit(e)lingetona -ton(e)* 1086, 1183×4, 1306, *Lit(t)lington(e) -yng-* 1187–1468, *Lyllington* 1570, *Lylyngton* 1610. OE folk-n. **Lȳtlingas* < pers.n. **Lȳtel(a)* + **ingas**, genitive pl. **Lȳtlinga*, + **tūn**. Ca 56.

LITLINGTON ESusx TQ 5201. 'Lytela's estate'. *Litlinton* 1191, 13th, *Litleton'* 1199, *Lit(y)(e)lington* 1201–16. OE pers.n. *Lȳtela*, genitive sing. *Lȳtelan*, or *Lȳtela* + **ing**[4], + **tūn**. Sx 412.

LITTLEBOROUGH GMan SD 9316. 'Little fort'. *Littlebrough, Lyttlebrugh* 1577. OE **lȳtel** + **burh**. The reference is unknown. Possibly a station on the Roman road over Blackstone Edge, Margary no.720a. La 58.

LITTLEBOROUGH Notts SK 8282. 'The little fortified place'. *Litelburg* 1086, *Littilburg(h)* c.1190, 1291, *Lyttleburg'* 1276, *-burgh* 1342. OE **lȳtel** + **burh**. This is the Roman settlement of *Segelocum* on the left bank of the r. Trent where the Roman road from Lincoln to Doncaster (Margary No.28a) crossed the river on a paved ford. Identical with Bede's *Tiouulfingacaestir* [c.731]8th HE, near which Paulinus conducted a mass baptism in the r. Trent. *Tiouulfingacaestir* means 'Roman fort of the people called afer Tiowulf'. Nt 35, RBrit 453.

LITTLEBOURNE Kent TR 2057. 'The little Bourne estate'. *(terram que appellatur) Litleburne* [696 or 711]13th S 16 (with 15th cent. variant *Littelb'ne*), *Liteburne* 1086, *Litleburne* 1197 etc. OE **lȳtel** + p.n. *Bourne* as in BEKESBOURNE TR 2055, BISHOPSBOURNE TR 1852 and PATRIXBOURNE TR 2055, all divisions of an original *Bourne* estate < OE **burna**. The reference is to the Little Stour TR 2462 on which the village stands. KPN 21, Charters IV.36, Cullen.

LITTLEBREDY Dorset SY 5889. *Litelbride* 1086, *Litlebridie* 1204. OE **lȳtel** + p.n. Bredy as in Long BREDY SY 5690. Do 42.

LITTLEBURY Essex TL 5139. '(At the) little fort'. *(æt, into) Lyt(t)lanbyrig* [1004]12th S 907, 975×1016 S 1487, *Littelbirig* [1042×66]12th S 1051, *Litelbyriā* 1086, *Lit(t)leberi -bir(e) -bury* 1202–1324. OE **lȳtel**, definite form oblique case **lȳtlan**, + **byriġ**, dative sing. of **burh**. The reference is to Ring Hill Iron Age hill-fort. An alternative early name of the fort is *Stirchberry, Sterchebury -bery* 1387, later *Strawberry*, r.n. *Styric* [1004]12th S 907 + **byriġ**. *Styric* 'little Stour', r.n. *Stūr* + suffix **iċ**, is another name for the Cam or Granta. Ess 530, 12, Jnl 2.47, Thomas 1976.123.

LITTLEBURY GREEN Essex TL 4938. *Littlebury Green* 1805 OS. P.n. LITTLEBURY TL 5139 + ModE **green**. The original name was *Stretley Green* 1768, earlier simply *(to) stræleage* (sic) [1004]12th S 709, *Strethlai* [1008]12th, *Stratlai* [1042×66]12th S 1051, *Strat(te)leya -leia* 1387, 'wood or clearing by the Roman road', OE **strǣt** + **lēah**. The reference is to the Roman road from Braughing to Great Chesterford, Margary no. 21b. Ess 530.

LITTLE COMMON ESusx TQ 7107. *Little Common* 1813 OS.

LITTLECOTE Wilts SU 3070. 'The little cottage(s)'. *Litlecota* 1187, *-e* 1329. OE **lȳtel** + **cot** (pl. **cotu**). Wlt 288.

LITTLEDALE Lancs SD 5561. 'Little valley'. *Luteldale* 1226, *Liteldale* 1251. OE **lȳtel** + lOE **dæl** or ON **dalr**. La 177, SSNNW 237, Nomina 10.175.

LITTLEDEAN Glos SO 6713. 'Little Dean'. *Parva -ua Dene* 1220–1437, *Li- Lyt(t)le, Ly- Littel(l) Dene* 1328–1594. ME adj. **litel**, Latin **parva**, + p.n. *Dene* 1086, 'the valley', OE **denu**, referring to the valley of the forest of Dean. 'Little' for distinction from MITCHELDEAN SO 6618, the other main division of the forest. Gl iii.225.

LITTLE END Essex TL 5400. *Little End* 1777 sc. of Stanford Rivers. ModE **little** + **end**. Ess 79.

LITTLE FEN Cambs TL 5868. *Little Fen* c.1840. ModE **little** + **fen**. Ca 189.

LITTLEHAM 'The little enclosure'. OE **lȳtel**, definite form **lȳtla** + **hamm**.

(1) ~ Devon SY 0281. *(to) Lytlanhamme* [1042]12th S 998, *Liteham* 1086, *Litt(e)leham* from 1242. D 590.

(2) ~ Devon SS 4323. *Liteham* 1086, *Lit(t)leham* from 1177. Alternatively the generic might here be OE **hām**. 'Little' perhaps for distinction from ABBOTSHAM SS 4226. D 96.

LITTLEHAMPTON WSusx TQ 0202. *Lyttelhampton* 1482. ME adj. **litel** + p.n. *Hantone* 1086, *Hamton(e)* 1229, 1296, *Hampton* 1230–1428, 'the homestead enclosure', OE **hām-tūn**. The reason for the prefix *Little* is unknown. Sx 169, ASE 2.34, SS 77.

LITTLEHEMPSTON Devon SX 8162. 'Little Hempston'. *Parva Hæmestone* 1176, *Hemestone minor* 1264, *Littlehempstone* 1608. Earlier simply *Hamistone -es-* 1086, *Hemmeston'* 1242–1313, 'Hæmi's estate', OE pers.n. *Hǣmi*, genitive sing. *Hǣmes*, + **tūn**. 'Little' for distinction from neighbouring BROADHEMPSTON SX 8066. D 514.

LITTLEHOUGHTON Northum NU 2316. *Parva Houcton* 1242 BF, ~ *Hotton* 1296 SR. ModE **little**, Latin **parva**, + p.n. HOUGHTON. 'Little' for distinction from LONGHOUGHTON NU 2415.

LITTLEMORE Oxon SP 5302. 'The small marsh'. *Luthlemoria* c.1130, *Lu- Letlemore* 1220, *Lit(t)lemor(e) -il- -a* 1191–1517. OE **lȳtel**, definite form **lȳtla**, + **mōr**. O 180, L 55.

LITTLEOVER Derby SK 3334. 'Little Ridge'. *Parva Ufre* 1086, ~ *Oura, Oure, Ovre* 1177×82–1318, ~ *Over(e), Overa* 1243–1577, *Lyttleover* 1577. Latin **parva**, ModE **little** + OE **ofer** used as a p.n. 'Little' for distinction from MICKLEOVER SK 3135. Db 478.

LITTLEPORT Cambs TL 5686. 'Small town'. *Litelport* 1086, *Lit(t)elport* c.1250 etc., *Lit(t)leport(e)* from 1170. OE **lȳtel** + **port**. Ca 225.

LITTLESTONE-ON-SEA Kent TR 0824. A modern seaside resort first developed in 1886 as an extension of New Romney TR 0624. It is named after the former rocky headland here called *Little Stone* 1816 OS in contrast to GREATSTONE-ON-SEA TR 0822. Room 1983.65.

LITTLETHORPE NYorks SE 3269. *Littlethorpe* from 1653. ModE **little** + p.n. *Torp'* 1086, 1190, 1234, *Thorp(p)* 1201–1584, ~ *juxta Ripon* 14th, ON **thorp** 'an outlying farm'. This was one of the *berewics* of Ripon in DB. YW v.173, SSNY 69.

LITTLETON 'Little settlement, farm, village, estate'. OE **lȳtel** + **tūn**, locative–dative form 'at the little settlement', **lȳtlantūne**. PNE ii.30.

(1) ~ Ches SJ 4466. *Litelton* 1435, *Littleton* 1724. Local name for *Parva Cristentona* 1150, *Littil Cristelton* 1311, 'the smaller part of CHRISTLETON' SJ 4465. Che iv.113.

(2) ~ Hants SU 4532. *Littletone* 1171, *-ton* 1205. Ha 110, Gover 1958.177.

(3) ~ Somer ST 4930. *Liteltone* 1086. 'Little' in comparison with nearby Somerton. DEPN.

(4) ~ Surrey TQ 0768. *Littleton* from [1042×66]13th with variants *Li- Lytle-* 1184–1291 and *Li- Lyt(t)el-* 1282–1469, *Litlinton* 1206, 1242, *Lutelington -yng-, Littelyngton* 1274–1356. Mx 16.

(5) ~ DOWN Wilts ST 9751. 'Hill belonging to L'. P.n. Littleton as in LITTLETON PANELL SU 0054 + ModE **down**.

(6) ~ DREW Wilts ST 8380. 'L held by the Drew family'. *Littleton Dru* 1311. P.n. *Litletun* [1065]14th S 1038, *Liteltone* 1086, + manorial addition from the family name of Walter Driwe or Dreu who held here in 1220; Drew is an OFr pers.n. of Gmc origin (OHG *Drogo*). 'Little' in relation to one of the nearby *tūns*, perhaps Grittleton ST 8580. W 76.

(7) ~ -ON-Severn Avon ST 5890. *Littleton upon Severn* 1728 (~ *super Sabrinam* 1586). Earlier simply *Lytletun, Lutletone* [986]c.1400, *Litletun* [1065]14th S 1038, *Liteltone* 1086. Gl iii.118.

(8) ~ PANELL Wilts SU 0054. 'L held by the Panell family'. *Lutleton Paynel* 1317. P.n. *Liteltone* 1086 + manorial addition from the family name of William Paynel who held here in 1249. 'Little' in relation to West Lavington SU 0053. Wlt 241.

(9) High ~ Avon ST 6458. *Heghelitleton* 1324. Earlier simply *Liteltone* 1086. 'High' for distinction from Stoney Littleton ST 7356, *Liteltone* 1086, and Littleton ST 4030, *Liteltone* 1086. DEPN.

(10) Middle ~ H&W SP 0747. *Middle Littelton* 1831 OS. ModE adj. **middle** + p.n. *litletona* *[709]12th S 80, *Liteltvne* 1086, *Lutleton -lin-* 1227–1327. Three Littletons are early mentioned, *þry lytlentunes* [n.d.]12th S 1591a, two, *lytltun et alia litletun* [714]16th S 1250. Simply called *Middleton* 'the middle village' 1251. Wo 265, Hooke 23, 46, 377.

(11) North ~ H&W SP 0847. *Northlitleton* 1251. ME adj. **north** + p.n. Littleton as in Middle LITTLETON SP 0747. Wo 265.

(12) South ~ H&W SP 0746. *Suthlitinton* (sic) 1251. ME adj. **suth** + p.n. Littleton as in Middle LITTLETON SP 0747. Wo 265.

(13) West ~ Avon ST 7675. *Westlytelton* 1458, *West Littleton* 1707. Earlier simply *Lit(t)leton'* 1221, 1340–1687. 'Little' in relation to Tormarton ST 7678 in which parish it lies, and 'west' in relation to Littleton Drew ST 8380. Gl iii.51.

LITTLE TOWN Cumbr NY 2319. *Litleton* 1578, *Litletowne* 1595. ME **litel** + **toun** (OE *tūn*). Cu 373.

LITTLETOWN Durham NZ 3443. *Little towne* 1613 is exactly what it appears to be, 'little settlement'. It is probably identical with the *Litle Pittington* recorded in a will of 1581, just as South Pittington was called *Suthtun* in 1366 and West

Merrington *Westerton* in 1647. See Kirk MERRINGTON NZ 2631, PITTINGTON NZ 3244 and WESTERTON NZ 2331.

LITTLEWICK GREEN Berks SU 8380. *Little Wick Green* 1761, *Littlewick Green* 1790. P.n. Littlewick + ModE **green**. Littlewick, *Lidlegewik* [1052×66]13th S 1477, *Lidleuuike -wyk* 1256–1340, *Lityl- Littilwike* 1497, seems to be composed of a p.n. **Littley* and OE **wīċ** 'a dairy-farm' and hence 'Littley dairy-farm'. *Lidle(g)-* derives from *(to) hild leage* [940]13th S 461 which is usually taken to be a mistake for *(to) hlid-leage*, OE **hlid** 'slope' or **hlid** 'gate' + **lēah** 'wood, clearing', dative sing. **lēaġe**, and so 'wood or clearing at a slope' or 'with a gate'. But OE **hild-lēah** 'battle clearing' is also possible. Brk 72, Forsberg 1950.111.

LITTLEWINDSOR Dorset ST 4404 → BROADWINDSOR ST 4302.

LITTLEWORTH H&W SO 8850. *Littleworth* 1831 OS. No early forms cf. LITTLEWORTH SK0112.

LITTLEWORTH Oxon SU 3197. *P'ua Worth'* 1241, *Petyte Wrth, Lytleworth, Parva Worth* 1284. ME **litel**, Latin **parva**, Fr **petite** + p.n. *(æt) Wyrðæ* [c.968×71]12th S 1485, *Ordia* 1086, *Worda* c.1187–1363, 'the enclosure', OE **worth, wyrth**. 'Little' for distinction from LONGWORTH SU 3999. Brk 367.

LITTLEWORTH Staffs SK 0112. *Littleworth* 1834 OS. A derogatory name for poor land. ModE **little** + **worth**. St i.59.

LITTON Derby SK 1675. 'Hill-side farm or village'. *Litun* 1086, *Lit- Lytton(e)* from early 13th. OE **hlith** + **tūn**. The village is situated in the hollow of a slope. Db 137.

LITTON NYorks SD 9074. Possibly the 'settlement at the slope'. *Litone* 1086, *-tuna* 1173, *Li- Lytton(a) -tun(a)* c.1150–1671. OE **hlith** + **tūn**. The moorland rises steeply behind the village. YW vi.125, L 166.

LITTON Somer ST 5954. 'The settlement on the river *Hlyde*'. *(æt) Hlytton'* [1061×6]13th S 1116, *Hlittun* *[1065]18th S 1042, *Litvne* 1086, *Lutton* 1157–1388, *Lidtone* 1176, *Lytton* 1610, *Litton* 1742. Lost r.n. **Hlȳde* 'the noisy one' (< OE *hlūd*) + **tūn**. A small stream flows strongly through the village. DEPN, Turner 1951.156.

LITTON CHENEY Dorset SY 5590. 'L held by the Cheyne family' who were here from the 14th cent. Earlier simply *Lideton* 1204, *Lidinton* 1212, *Ludeton* 1232, *Lutton* 1258, 'settlement on the stream called *Hlyde*, the noisy one', OE r.n. **Hlȳde* (from OE *hlūd* 'loud') or **Hlȳding* (from **Hlȳde* + **ing**[2]), + **tūn**. Do 100.

LITTONDALE NYorks SD 9172. 'Litton valley'. *Littun(e)dale, Li- Lytton(e)dale* 12th–1632. P.n. LITTON SD 9074 + ON **dalr**. Although the river that runs through the dale is the Skirfare, the dale is named from the village at its head. YW vi.126.

Great LIVERMERE Suff TL 8871. *Leuermere mag* 1610, *Great Livermere* 1836 OS. Earlier *Maius Liuremere* c.1095. ModE **great**, Latin **magna, maius** 'greater', + p.n. *Leuuremer* [1042×66]12th S 1051, *Liuer- Liuelmera* 1086, *Liuermere* c.1095, *Liuremere* 12th, 'levers lake, the lake growing with reeds or irises', OE **lǣfer, lēfer** later replaced by **lifer** 'liver' referring either to the curving shape of the lake or to its thick clotted water, + **mere**. There is still a lake at this place. 'Great' for distinction from Little Livermere, *litla liuermera* 1086. DEPN.

LIVERPOOL Mers SJ 3490. 'Creek with thick water'. *Liuerpul* before 1194–before 1240, *-pol(e)* 1211–1416 with variants *Liver- Lyver-, Lyverpull* 1348, 1359 etc., *Lyverpool(e)* 1515–1897, *Litherpol(e)* 1222×26, 1586, *-poole* 1752, *Lir- Lyr(e)pole, Lirepoole* 16th cent. OE **lifer** 'liver' used in the sense 'thick, clotted water' + **pōl**. The original reference was to a pool or tidal creek now filled up into which two streams drained. La 116, TC 125, L 28, Jnl 17.65.

LIVERPOOL AIRPORT Mers SJ 4183. P.n. LIVERPOOL SJ 3490 + ModE **airport**.

LIVERPOOL STREET STATION GLond TQ 3381. Liverpool Street was built in 1829 over a winding street known as Old Bethlehem from the hospital which once stood to the north. It was renamed after Lord Liverpool, Prime Minister 1812–27. The station was opened in 1874 as the new city station of the Great Eastern Railway whose original terminus was at Shoreditch. Encyc 477.

LIVERSEDGE WYorks SE 2023. Uncertain. *Liuresech, Livresec* 1086, *Liu- Liv- Lyverseg(g) -segg(e)* 1154×89–1543, *Liversedge* from 1597. Three alternatives have been suggested: (1) 'Leofhere's edge or scarp', OE pers.n. *Lēofhere*, genitive sing. *Lēofheres*, + **ecg**; (2) 'Liver-sedge' with OE **secg**, 'sedge, reeds, rush': *liversedge* could be an otherwise unrecorded plant name like *liverwort*; (3) the first element might be OE **lifer** 'liver' in the sense 'thick clotted water' as in LIVERPOOL. The topography is not decisive. YW iii.27, L 158–9.

LIVERTON Cleve NZ 7115. Probably 'settlement on the (stream called) *Lifer*'. *Liuretun* 1086, *Livertun* 1165×75, [after 1180]15th, *Liverton* from 12th, *Leverton* c.1200–1577. OE **lifer** + **tūn**. OE *lifer* derives from a root which meant 'fat, sticky' as in Gk λιπαρός; it is suggested that the word was used of streams with thick muddy water as in the Norw r.n. *Levra*, earlier *Lifr-*. Some forms, however, suggest the alternative OE **lǣfer, lēfer** 'a rush, a reed, a yellow iris' probably used collectively in the sense 'a reed bed'. Liverton stands on Liverton Beck. YN 141, PNE ii.11.

River LIZA Cumbr NY 1514. 'Light or bright river' *Lesar* 1292, *Lesagh* 1294, 1322. ON **ljóss** + **á**. Identical with the Icelandic *Ljósá* and Norwegian *Ljøsa*. Cu 20.

LIZARD Corn SW 7012. 'Court on a height'. *Lv- Lisart* 1086, *Lesard* 1250–1348, *Lezarth* 1323, *Lezard* 1451, *Lezarde* 1610, *Lysard* 1284–1427, *Lyzart* c.1540. Co ***lys** + **ardh**. PNCo 110, Gover n.d. 564.

LIZARD POINT Corn SW 6911. *le Forlond de Lysard* 1427, *Lyzart Poynt* c.1540, *Lesard poynt* 1610. P.n. LIZARD SW 7012 + ModE **point**. An important navigational landmark. PNCo 110.

LLANDINABO H&W SO 5128. 'Iunabwy's church'. *Lann Hunapui* [1045×1104]c.1130 LL, *Landinabou -dynabo* 1214–1334. OW **lann** + hypocoristic **ti** + pers.n. *Iunabury* (*Iunapeius*). According to LL, Llandinabo, in Latin *podium Iunabui* 'Iunabwy's estate', was given to St Dyfrig, one of whose disciples, *Iunabui presbiter*, 'Iunabwy the priest', was one of the witnesses. He 131, Bannister 119, CPNE 144, O'Donnell 89.

LLANFAIR WATERDINE Shrops SO 2476. 'Mary-church at Waterdine'. *Lanweyrwaterden* 1376, *Llanvaier-* 1577, *Llanfair Water Dene* 1614, *Llanvair Waterdine* 1808. Originally a purely English name, *Watredene* 1086, *Waterdene* [1155]1348, *Watirden* [1331]1348, 'the water valley', i.e. the wet valley of the river Teme, OE **wæter** + **denu**, to which W *Llanfair* 'St Mary's church', a common W name, W **llan** 'a church' + saint's name *Mair* 'Mary' with lenition of *M-* to *F-*, was subsequently prefixed. The process of replacing an older name by a church name was common in Wales, rare in England; the present instance is an interesting compromise between the two practices. Sa i.178.

LLANGARRON H&W SO 5321. 'Church on the river Garren'. *Lann Garan* [8th]c.1130, *Langara(n)* 1148×63–1356. OW **lann** + r.n. *Garran* 1558, 1577, *Garan, Garon* 1582, OW **garan** 'a crane'. RN 169 suggests the original may have been *Dufr Garan* or the like, 'crane water'. He 131, O'Donnell 92.

LLANGOLLEN CANAL Ches SJ 5747. P.n. Llangollen SJ 2142 + ModE **canal**. Another name for part of the Ellesmere Canal.

LLANGROVE H&W SO 5219. 'Long grove'. *Longegroue* 1372, *Langrove* 1831. ME **longe** + **grove**. An interesting example of Welsh spelling mistakenly applied to an English name. He 132.

LLANROTHAL H&W SO 4718. 'Ridol's church'. *Lann Ridol* [1045×1104]c.1130, *Lanruat'* c.1163, *Lanrothal -el* 1278, 1341, *Llanrithalle* 1397. OW **lann** + pers.n. *Ridol*. In the cartulary of Monmouth Priory this appears as St Ruald or Roald, *ecclesia(m) Beati Rualdi* 1131×44, ~ *Sancti Roaldi* 1160×80, due to confusion with *Roald, Rual*d, a Breton form of Frankish

(H)rôdwald popular among the Breton knightly class. The W pers.n. may have been that of the original incumbent rather than the dedication. He 134.

LLANVEYNOE H&W SO 3031. 'Church of St Beuno'. *Llanveyno* 1832 OS for W *Llanfeuno*. OW **lann** + saint's name *Beuno*. Although the evidence is late there is ancient sculpture at the church and this must be an early unrecorded church site. Bannister 120, Pevsner 1963.240, Richards 119, Bowen 1956.82.

LLANWARNE H&W SO 5028. 'Church by the alder swamp'. *Lann Guern* [8th]c.1130 LL, *(una æcclesia uocata) Ladgvern* 'a church called L' 1086, *Lanwara(n)* 1131–1341, *Landwar(a)* 1158, 1163, *Llanwarne* 1535. OW **lann** + **guern**. Later W forms from LL *Lann Guern Teliau ha Dibric* [1045×1104]c.1130, *Lann Guern Aperhumur* [1066×87]c.1130, suggest that two churches, St Teilo's and St Dyfrig's, were combined on one site 'at the confluence of the Gamber', W **aber** + r.n. Gamber, *Am(h)yr* c.1130, *Gamer* 1612, identical with River AMBER Derby SK 3463 (initial *G-* was mistakenly added as if *Amyr* was the lenited form of a word beginning with *g*). He 136, Bannister 121, RN 12, O'Donnell.

LLANYBLODWELL Shrops SJ 2423. 'The church in Blodwell, the blood spring'. *Llanblodwell* 1535, 1666, *Llanyblodwell* 1728 etc., *Llany(ne)mblodwell* 1576, 1602×3, *Llanymlodwell* 1707. Late prefix W **llan** 'a church' + p.n. *Blodwelle* c.1200, *-wall* 1577, *-well* 1508, one of the 'sweet waters' of Shropshire, OE **blōd** + **wella**. The explanation of the name is unknown. Cf. LLANFAIR WATERDINE SO 2476. The grammatically correct form is *Llanymlodwell* with nasal mutation *B-* to *M-* after the preposition *yn (ym)* 'in'. Sa i.178, Morgan 1997.36.

LLANYMYNECH Shrops SJ 2620. 'The church of the monks'. *Llanemeneych'* 1254, 1272, *Llanymenych* 1310, *-mynech'* 1486, 1614 etc., *-minich* 1628×9, *Lanamynech, Llanaminecke* 17th, *Lanamunough* 1714, *Thlanmenygh* 1307. W **llan** 'a church' + **mynach**, pl. **myneich**. The reference is unknown. Sa i.179, Morgan 36.

LLAWNT Shrops SJ 2531. *Lawnt yr afon goch* 'red river L' 1604, *Llawnt* 1837 OS. W **lawnt** (ME *launde*) 'a woodland glade'. GPC 2055.

LLWYN Shrops SO 2880. 'The grove, the copse'. *Llwyn* 1833 OS. W **llwyn**.

LLYNCLYS Shrops SJ 2824. 'The quagmire court'. *Thenclis* [12th]late 13th, *Fenches* 1272, *(the two) Lenhokys, Lenkolrys* 1302, *Llynklis* 1295, *Llynclys* 1393, 1602–62. W **llwnc** + **llys**. O'Donnell 101, Morgan 36.

Long LOAD Somer ST 4623. *Longlode* 1610, *Long Load* 1811 OS. ModE adj. **long** + p.n. *La Lade* [13th]B, 1285, 1292, OE **(ġe)lād** 'an artificial channel, a canalised river, a canal'. The reference is to a canalisation of the river Yeo for transport in the Somerset Levels. 'Long', for distinction from Little Load ST 4624 on the opposite bank, refers not to the length of the canal but to the shape of the village which straggles along the road from Long Sutton to Martock. DEPN.

LOADPOT HILL Cumbr NY 4518. Unexplained. *Loadpot* 1823. We ii.202.

LOANEND Northum NT 9450. 'The lane end'. *Loan End* 1865 OS. N dial. **loan** + ModE **end**. *Loan* is a variant form of *lane* < OE *lane, lone, lanu*.

LOCKENGATE Corn SW 6425. 'Lockable' or 'rocking gate'. *Locken Gate* 1748, *Locking Gate Toll Gate* 1839. Cf. ME present participle adj. **loggande** 'rocking, loose' from *loggen, lokken* 'to rock' as in LOGAN ROCK SW 3922. The reference is to a turnpike toll-gate. PNCo 111.

LOCKERIDGE Wilts SU 1467. 'The ridge of the folds or enclosures'. *Locherige* 1086, *Lokerugge -krigge -rygge* 1155×69–1332. OE **loc**, genitive pl. **loca**, + **hrycg**. Wlt 306.

LOCKERLEY Hants SU 3026. 'Looker's wood or clearing'. *Locher(s)lei, Locherlega* 1086, *Lokerlay* 1194–1352 with variants *-leg(h) -le(y)e -lig'*, *Lockerleye* 1271. OE ***lōcere** (ME *lokere*) 'looker, keeper, shepherd', genitive sing. ***lōceres**, pl. ***lōcera**, + **lēah**. Ha 110, Gover 1958.189, DEPN, PNE ii.26.

LOCKERLEY HALL Hants SU 2978. A mansion built in 1868. P.n. LOCKERLEY SU 3026 + ModE **hall**. Pevsner-Lloyd 322.

LOCKING Avon ST 3659. Probably like Lockinge Oxon SU 4186 originally a stream name, 'the playful one'. *Lockin* 1212, *Lokkinges* 1249, 1264. OE r.n. **Lācing* < **lāc** + **ing**[2]. The forms for Lockinge are *Lacinge* [868]12th S 1201, *Lakinge* [868]13th ibid., *Lachinge(s)* 1086, *Lockinge* 1227×8, *Lockinges* 1235×6, and for Locking Brook on which the village lies, *Lackincg* [868]12th S 1201, *lacing* 965 S 594. DEPN, Brk 13, 486, Forsberg 1950.1ff.

LOCKINGTON Humbs SE 9947. Either 'estate called after Loca' or 'farm or village with or at the enclosure'. *Loche- Lecheton* 1086, *Loketon* 1305, *Lokintun -ton(a) -yn-* 1154×60–1314, *-en-* 13th cent., *Lockington(a) -yng-* [1178]1238–1504. OE pers.n. **Loca*, genitive sing. **Locan* varying with **Loca* + **ing**[4], or OE **loca**, genitive sing. **locan**, varying with ***locing**, + **tūn**. YE 160 gives pr [lɔkitən], Årsskrift 1974.52.

LOCKINGTON Leic SK 4627. Probably the 'estate called after Loc'. *Lochentun* 12th, *Lokinton(e) -yn-* c.1130–1518, *Lokington' -yng-* 1240–1540, *Lockington* from 1548. OE pers.n. *Loc(a)* + **ing**[4] or genitive sing. *Locan*, + **tūn**. Lei 372.

LOCKLEYWOOD Shrops SJ 6928. *Lockleywood* 1833 OS. P.n. Lockley as in *(bosco de) Lockele* 1291×2, *(a wood called) Lokkeleye*, 'the wood or clearing with an enclosure', OE **loca** + **lēah**, + ModE **wood**. Gelling.

LOCKS HEATH Hants SU 5107. *Locks Heath* 1810 OS. Surname *Lock* + ModE **heath**. Cf. pers.n. Rober Lok 1327 and the ns. *Lockesgrene, Locks Brydge* 1545. Gover 1958.32.

LOCKTON NYorks SE 8489. Probably 'Loca's estate'. *Lochetun* 1086, *Loketon* 1167–1303, *Lokinton(e) -yn-* 1198–1322, *Lok- Locton(e)* 1285–1577. OE pers.n. **Loca* + **tūn** varying with genitive sing. **Locan* or **Loca* + **ing**[4] + **tūn**. But the specific might alternatively be OE **loca** 'an enclosure'. YN 91.

LOCKWOOD BECK RESERVOIR Cleve NZ 6713. R.n. Lockwood Beck + ModE **reservoir**. Lockwood Beck is p.n. Lockwood, *Locwyt* 1273, 'enclosed wood, or wood by the enclosure', OE **loc** + **wudu** partially replaced by ON **vithr**, + **bekkr**. YN 147.

LODDINGTON Leic SK 7802. 'Estate called after Luda or Lude'. *Ludinton(e)* 1086–1265, 1483, *Lodinton(e)* c.1130–1350, *-ing-* c.1130–1535. OE pers.n. *Luda* or feminine *Lude* + **ing**[4] + **tūn**. Lei 308.

LODDINGTON Northants SP 8178. 'The estate called after Loda'. *Lodintone* 1086–1428 with variants *-ing- -yng-*, *Ludington* 1305. OE pers.n. **Loda* + **ing**[4] + **tūn**. Nth 116, Forsberg 150–1.

LODDISWELL Devon SX 7248. 'Lodd's spring'. *Lodeswille* 1086–1345 with variant *-well(e)*, *Lodgewyll* 1552, *Lodswell* 1684. OE pers.n. **Lodd*, genitive sing. **Loddes*, + **wylle**. D 306 gives pr [lɔdzwəl].

River LODDON Berks SU 7568, Hants SU 6758. 'Muddy river'. *Lodena* [1190] n.d., after 1212, *Lodene* 1227–1400, *Lodon* 1543. British r.n. **Lutnā*, a possible adjectival formation on the base seen in Latin *lutum*, OIr *loth* 'mud'. Gaelic *lón* 'marsh, mud, meadow' is thought to be a parallel formation < **lut-no-*. RN 258 compares the Gaulish r.ns. Lodéve, Hérault, *Lutēvā*, and Loze, Côte-d'Or, *Lutōsā*. Brk 13, Gover 1958.3, Holder ii.351–4.

LODDON Norf TM 3698. The settlement name is probably an earlier name of the river Chet, itself a late back-formation from Chedgrave. *(into, at) Lodne* [1042 or 3]13th S 1490, [1051×2 or 1053×7]13th S 1082, *Lot(h)na, Lodna -nes* 1086, *Lodne* [1087×98]12th, 1198, *Lodnes* 1208, 1209, *Lodene* 1270, *Loden* 1275. The hundred of Loddon is *Lod(d)inga, Loth(n)inga, Lotninga* 1086, *Lodinge* [1086] c.1180, *Lodenynges, Lodningges* 1275 representing OE **Lodningas* 'the dwellers on the river Loddon', r.n.

Lodne + **ingas**. The base is probably British **Lutnā* 'muddy river'. DEPN, RN 258, LHEB 578, ING 59.

LODE Cambs TL 5362. 'The watercourse'. *Lada* 1154×89, c.1251, *la Lade* c.1260, 1279, *(le) Lode* 1345, 1462. Also known as *Botesham Lode* 1396, 1521, *Bosham Load* 1712. OE **lād**. This is a drainage channel of Bottisham Fen leading from Bottisham to the Granta. Ca 131.

LODERS Dorset SY 4994. Uncertain. *Lodre(s)* 1086, *Loddre(s)* 1244, *Loderes* 1291. Probably a Celtic r.n. for the stream now called the Asker, a back-formation from ASKERSWELL SY 5292. Possibly Co ***loch** 'pool' used as a r.n. + **dour** 'water'. For *loch* used as a r.n. cf the Northumberland r.n. Low as in LOWICK NU 0139. Do 100, CPNE 87, 152.

River LODON H&W SO 6152. *Loden* 1577, 1586. Identical with River LODDON Hants SU 7568 and LODDON Norfolk TM 3698 from an old name of the Chet. RN 258.

LODSWORTH WSusx SU 9223. 'Lodd's enclosure'. *Lodesorde* 1086, *-wurða* 1165, *Loddesw(o)rth* 1272–1428, *Lud(e)sworth* 1288, 1627. OE pers.n. **Lodd*, genitive sing. **Loddes*, + **worth**. The river at Lodsworth, the Lud, is a back-formation from the p.n. Sx 26 gives pr [lʌdzwəθ], SS 78.

The LOE Corn SW 6425. 'The pool'. *La Loo* 1337, *The Lowe* 1610. OCo ***loch**. PNCo 111.

LOFT HILL Northum NT 8513. 'Tall hill'. *Loft Hill* 1869 OS. Probably a reflex of ON **loft** 'a hill' + ModE **hill**. It rises to 1508ft. and is called *Toft Hill* by Dixon 14. PNE ii.26–7 s.v. *lopt*.

LOFTHOUSE NYorks SE 1073. '(The settlement) at the two-storied houses'. *Lofthusum* 1175–1314, *Loft(e)hous(es)* 1198–1535. ON **lopt-hús**, dative pl. **-húsum**. YW v.203.

LOFTHOUSE GATE WYorks SE 3324. P.n. Lofthouse SE 3325 + ModE **gate**. The reference is to a former gate on the turnpike road from Wakefield to Leeds. Lofthouse, *Locthuse*, *Loftose* 1086, *Lofthus(e)* 13th cent., *Lofthouse* from 1303 is ON **lopt-hús** 'a house with a loft or upper chamber'. YW ii.136.

LOFTUS Cleve NZ 7118. '(At) the loft-houses'. *Loctus(h)um*, *Loctehusum* 1086, *Lofthus(es)* [12th]15th–1301, *Loftus* from 1160×75. ON **lopt-hús** 'a house with a loft', dative pl. **lopt-húsum**. The DB scribe was unfamiliar with the consonant cluster *-ft-* which he represented as *-ct-*, a regular spelling for OE [xt], a combination of fricative and stop similar to [ft]. YN 140, SSNY 87.

LOGAN ROCK Corn SW 3922. 'Rocking stone'. *The Logan Stone* 1745. ME present partciple adj. **loggande** 'rocking' from *loggen* 'to rock'. PNCo 111.

LOGGERHEADS Staffs SJ 7336. *Loghead* 1657, *Logerheads* 1775, *The Logger Heads* 1833 OS. Probably an inn-name from ModE **loggerhead** 'a stupid person'. *Loggerheads* was a common sign in the 17th cent. alluded to by Shakespeare in *Twelfth Night* 2.iii.17. It consisted properly of two (but sometimes of three) wooden heads with the inscription 'We three loggerheads'; the innocent spectator upon asking the whereabouts of the third was himself pointed to. Pub Names 157, Horovitz.

LOLWORTH Cambs TL 3664. 'Loll or Lull's enclosure'. *Lollesw(o)rth(e)* 1202–1378 including [1034]17th, *Lolesuuorde* 1086, *Lulleswrðe* 1199, *Lollew(o)rth(e)* 1242–1553, *Lolworth(e)* from 1346, *Lolor* 1617, *Lowlow* 1702, 1788. OE pers.n. *Loll* or *Lull*, genitive sing. *Lolles, Lulles,* + **worth**. Ca 180 gives pr [loulə].

LONDESBOROUGH Humbs SE 8645. 'Lothen's fortification'. *Lodenesburg* 1086, *Lonesburgh* 1275–1343, *Lounesburg(h)* 1285–1341, *Lownesburgh* 1306, *Loandesburgh* 1608. ON pers.n. *Lothinn*, lOE *Lothen*, genitive sing. *Lothenes*, + **burh**. YE 231 gives pr [lounzbrə], SSNY 148.

LONDINIUM GLond TQ 3181. The RBrit name of LONDON q.v.

LONDON (GATWICK) AIRPORT WSusx TQ 2640. P.n. Gatwick, no early forms, but possibly an ancient name, 'the goat farm', OE **gāt** + **wīc**, appearing in the surnames of Richard de Gatewik 1329 and Margery de Gatewyk 1332, + ModE **airport**. Sr 291.

LONDON BRIDGE STATION GLond TQ 3280. *lunduna bryggjur* [c.1015]13th Townend 52, *(æt) Lundene brigce* [963×75]12th S 1377. A bridge across the Thames has stood here from Roman times suffering many destructions by storm, fire or rebuilding (1014, 1091, 1136, 1176, 1212–13, 1282, 1633, 1823–31 and 1967–72). The first London Bridge Station, a temporary wooden building, was in use by the London and Greenwich Railway Co. in 1836 and a proper station built in 1840–44. Encyc 483.

LONDON GLond TQ 3079. Unexplained. *Londinium* [c.115]11th Tacitus *Annals* 14.33, *Λονδίνιον* (Londinium) (a *polis* of the Cantii) [c.150] 13th Ptol, *oppidum Londiniense* 3rd Panegyric, *civitate Londiniense* 314 Council of Arles, *Lo- Lundinio* [4th]8th AI, *Londinio* inscriptions, *Lundin(i)um* late 4th Ammianus Marcellinus, *Λινδόνιον* (Lindonium) [6th]15th Stephanus of Byzantium, *Londinium, Landini* Rav, *Lundonia* *[597×604]12th S 1244, [c.731]8th BHE, [748]15th S 91, 12th ASC(F) under year 886, *(on, æt, to, fram) Lundene* 9th–12th ASC passim, *(on) Lunden* 11th ASC(C) under year 1050, 1121 ASC(E) under years 886, 1094, *(in) Lundenne* [c.890]c.1000 OEBede, *(of) Lundone* 1121 ASC(E) under year 656, *Lundres* (French form) 12th, *Lundin, Lunden(e), Lu- Londenne, Londene* [c.1200]13th Layamon. There are also forms with *burh* and *ceaster*, the settlement within the Roman wall, and with *wic*, the later Aldwych: *(to, of, binnan) Lundenbyr(i)g* 9th–11th ASC passim[†], *Lunden burg, burh* 9th ASC(A) under years 851, 886, 1121 ASC(E) under year 1077, *þa burh Lundene* 12th ASC(F) under year 886; *Lundenceaster* [c.890]c.1000 OEBede; *Lunden wíc* 1121 ASC(E) under year 604.

The former explanation, 'town of Londinos', British pers.n. **Londinos* < **londo-* 'fierce, merry, active' as in OIr *lond* 'wild, angry' has been abandoned; the pers.n. is not on record and the vowel length of **lŏndo-* conflicts with that of *Lōndinium* > *Lūndinium*. The same difficulty confronts Jensen's derivation from IE **lendh-/*londh-* 'water, pool' which would otherwise account for the variant form *Lindonium*.

London's mythical history is told by Layamon in the *Brut* 1012ff: the city was founded by Brutus who called it *Troye the Newe* to remind his people of their original homeland. Later it was called *Trinouant* and when king Lud ruled he named it *Kaerlud*. In Layamon's own day it was called *Lundin* by the English and *Lundres* by the French. DEPN, LHEB 308 n.1, *Studia Onomastica Monacensia* III.427, RBrit s.n., Biddle 1984.23–4. For a recent attempt to explain the name see TPS 1998. 203–29.

Little LONDON An ironic name.

(1) ~ ~ Hants SU 6259. *Little London* 1695. Gover 1958.141.

(2) ~ ~ Hants SU 3749. *Little London* 1695. Gover 1958.165.

(3) ~ ~ Lincs TF 2421. *Little London* 1789. Payling 1940.49.

(4) ~ ~ ESusx TQ 5620. A jocular name for the house built here in the 19th century. Sx 11, 255 gives two other examples, Little London in Chichester, *Litellondon* 1454, and Little London in Ardingly, apparently alluding to growth of housing here in the 19th cent. There are many other instances from Buck to WYorks.

LONDONDERRY NYorks SE 3087. *Londonderry* c.1855 OS.

LONDONTHORPE Lincs SK 9538. 'Copse outlying farm'. *Lunde(r)torp* 1086, *Lunderthorp(e)* late 12th–1553, *Lundentorp'* 1237, *-thorp* 1562×7, *Londynthorpe* 1374, *Londonthorp(e)* from 1438×9. ON **lundr**, genitive sing. **lundar**, + **thorp**. The specific has been influenced by the p.n. London. Perrott 500, SSNEM 113.

[†]The form *Lunden byrig* is also used once as the subject of a sentence, 11th ASC(C) under year 982.

The LONG MAN ESusx TQ 5403.

The LONG MYND Shrops SO 4193. 'The long hill'. *Longameneda* [12th]copy, *Longa Muneta* 1212, *Longemynede* 1275. OE adj. **lang** + PrW ***mōnïth** (W *mynydd*) 'a mountain' referring to the long range of hills stretching NE from Plowden SO 3887 to Picklescott SO 4399. DEPN.

LONGBENTON T&W NZ 2668. ModE adj. **long** + p.n. *Bentun* c.1190, *Benton'* 1242, either 'the bean farm', OE **bēan** + **tūn**, or 'the farm or village where bent-grass grows', OE **beonet** + **tūn**. Also known as *Magna Ben(e)ton* 1256, 1296, 'great Benton', for distinction from Little Benton NZ 2867, *Parvam Benton(am)* 1236, 1242 BF, 1296. NbDu 17.

LONGBOROUGH Glos SP 1829. 'The long barrow'. *Langeberg* 1086–1296, *Lang(e)ber(u)we* 1236–1392, *Long(e)berg(h)* 1301–61, *Longborowe* 1571, 1592, *-borough* 1731. OE **lang**, definite declensional form **langa**, + **beorg**. A long barrow stands ½ mile above the village at SP 1729. Gl i.246, xiii.

LONGBRIDGE 'Long bridge or causeway'. OE **lang** + **brycg**.

(1) ~ Warw SP 2762. *Longebrige* 1250, *-brigge -brugge* 14th. Also recorded with Latin forms *longum pontem* (accusative case) 1123, *Longo ponte* (ablative case) 1190, and French *Longpont, Longepunde, Langpond, Langpound* 1262. Wa 263.

(2) ~ WMids SP 0177. *Long Bridge* 1831 OS. ModE **long** + **bridge**. A crossing of the r. Rea.

(3) ~ DEVERILL Wilts ST 8640. 'The long bridge across the r. Wylye or Deverill'. *Deverill Langebrige* 1316, *Deverall Longe Bridge* 1546. Known earlier by its French name *Peverel* (sic for *D-*) *Lungpont* 1239 or simply *Longepont* 'the long bridge' 1275. The earliest occurrence is simply *Devrel* 1086, for which see River DEVERILL. Wlt 166.

LONGBURTON Dorset ST 6412. *Langebourton'* 1460, *Longburton* 1575. ME adj. **lang, long** + p.n. *Burt(h)on'* 1244–1552, 'fortified enclosure', OE **burh-tūn**. 'Long' for distinction from Little Burton ST 6412, *parva Borton'* 1450, *Little Burton* 1569×74. The reference is to the length of the village. These two places were also known respectively as *Esterborton'* 'eastern B' 1460 and *West(er)burton'* 1484–1531. Do iii.298, WW 152.

LONGCLIFFE Derby SK 2255. 'The long steep slope'. *Longcliffe Bancke* 1620. ME **long** + **cliff**. Db 353.

LONG COMMON Hants SU 5014. No early forms. ModE **long** + **common**.

LONGCOT Oxon SU 2790. 'The long cottages'. *Longcote* 1332, *-cotes* 1388. ME adj. **long** (with variant form *Lang-* 1412, 1547) + p.n. *Cotes* 1233–1327, *Cote* 1284, 1307, ME **cote**, pl. **cotes**. Longcot was a township of the royal demesne of *Kyngescote* 1270 near Shrivenham. Brk 374.

LONG CRAG Northum NU 0606. *Long Crag* 1868 OS. ModE adj. **long** + **crag**. An elongated hill-top rising to 1047ft.

LONGDEN Shrops SJ 4406. 'The long hill'. *Langedune* 1086, *-don(e)* 1251–1385, *Longe-* 1235–1421, *Longdon* 1294, 1546–1764, *Longden* from 1615. OE **lang**, definite declensional form **lange**, + **dūn**. The reference is to a smooth whale-back 400ft. high ridge which contrasts with the 'sugar-loaf' shape of many of the other nearby hills. Sa i.180.

LONGDENDALE Derby SK 0498. '*Langedene* valley'. *Langedenedele* 1086, *-dale* 1158–1310, *Long(e)den(e)dal(e)* 1245–85 etc. P.n. **Langedene* + **dæl**. *Langedene* is 'the long valley', OE **lang**, definite form **lange**, + **denu**. Db 69, L 95, 98.

LONGDON H&W SO 8336. 'The long hill'. *(into) langan dune* 972 S 786, *Longedvne, Langedune* 1086, *Longdon* from 1378. OE **lang**, definite form oblique case **langan**, + **dūne**, dative sing. of **dūn**. Wo 208.

LONGDON Staffs SK 0814. 'The long hill'. *(æt) Langandune* [1002×4]11th S 1536, *Langedun* 1158, *-don* 1195. OE **lang**, definite form dative case **langan**, + **dūn**. DEPN, L 143, 152–5.

LONGDON ON TERN Shrops SJ 6115. *Longedon supra Tyren* 1491, *Longdon uppen Tyren* 1516, ~ *upon Tern* 1767. P.n. *Languedune* 1086, *Lange- Longedon(')* 13th cent., *Longdon* 1516, 'the long hill', OE **lang**, definite declensional form **lange**, + **dūn**, + r.n. TERN. Sometimes unusually shortened to *Longe* 1610, *Long super Tearne* 1658, *Long* 1755–1808. Longdon is raised above the Weald Moors by a ridge and two islands made by the 175ft. contour. Sa i.180.

Upper LONGDON Staffs SK 0614. Cf. *Longdon Upper End* 1834 OS. ModE **upper** + p.n. LONGDON SK 0814, originally in contrast to *Brook End* 1834 OS from How Brook in Longdon.

LONGDOWN Devon SX 8691. 'Long hill'. Cf. *Longdown End, Longdown Heath* 1675. ModE **long** + **down**. D 442.

LONGDOWNS Corn SW 7343. *Long Downs* 1841. A 19th cent. hamlet on former downland. PNCo 111.

LONGFIELD Kent TQ 6068. 'The long stretch of open ground'. *Langafelda* [973×871]12th S 1511, *Langafel* 1086, *Langefeu(l)d -feld(e)* 1226–54. OE **lang**, definite form **langa**, + **feld**. KPN 292, 294.

LONGFORD 'The long ford'. OE **lang**, definite form **langa** + **ford**. L 68–9.

(1) ~ Derby SK 2137. *Lang(e)ford(e)* 1197–1676, *Long(e)-* 1199×1216–1275 etc., *-forth* 1527, 1536. Db 581.

(2) ~ Glos SO 8421. *Lang(e)ford(e)* 1100×35–c.1560, *Long(e)ford(e)* 1137–1702. Carries the Roman road from Gloucester to Worcester, Margary no.180, across Horsbere Brook. Gl ii.149.

(3) ~ GLond TQ 0576. *Longeford(e)* 1294, 15th cent., *Langeford(e)* 14th cent., 1410, *Lang(e)- Le Langeforth* 1430. An oblique crossing of the river Colne. Mx 39, GL 67.

(4) ~Shrops SJ 6434. *Langeford'* 1232, *Le Longford* 1331. The reference is to a causeway and crossing of the river Tern a mile to the W on the course of the Roman road from Bletchley towards Newport. Margary no.19, p.294. Sa i.181.

(5) ~ Shrops SJ 7218. *Laganford* (sic) 1002×4 Charters II, *Langan-* 1004 ibid., *Langeford* 1086–1316, *Longe-* c.1182–1428 with variants *-forth* and *-forde, Longford* from 1199. The reference is probably to a crossing of Strine Brook by means of a causeway. Sa i.181.

(6) ~ WMids SP 3583. *Longford* 1835 OS. Cf. *Longfordwey* 1411. Wa 111.

LONGFRAMLINGTON Northum NU 1301. 'Long or great F'. *Framlington Magna* 1296 SR, *Longframlington* 1868 OS. Latin **magna** 'great' varying with ModE **long** + p.n. *Fremelintun* 1166, *Franglingtune* (sic) 1166, *Framelinton* 1170, *Framlincton* 1196, *Framelington* 1242 BF, *Framlyngton* 1346, 'the estate called after Framela', OE pers.n. **Framela* + **ing**[4] + **tūn**. Also known locally as *High Town*. 'Long' for distinction from Low Town NU 1300, formerly Low or Little Framlington, *Framlington Parva* 1296 SR, *Lowframelington* 1868 OS. NbDu 89 gives pr [framptən], DEPN.

LONG GILL NYorks SD 7858. 'The long ravine'. *Longegill* 1557, 1592. ME **long** + **gill** (OWScand *gil*). YW vi.163.

LONGHAM Dorset SZ 0698. Probably the 'long part of Ham(preston)'. *Longeham* 1541, *Longham* from 1575. ModE **long** + p.n. Ham as in HAMPRESTON SZ 0598. This place is probably to be identified with *Hamme (D)aumarle* 1298–1381 and *Esthamme* 1280, 1288, 1333 or *Estehame Preston'* 1288, parts of Hampreston. Do ii.226.

LONGHAM Norf TF 9415. 'The homestead of the Lawingas, the people called after Lawa'. *Lawingham* 1086, 1200, *Lowingeham* 1199×1200, *Laingeham* 1208, *Laungham* 1268. OE folk-n. **Lāwingas* < pers.n. **Lāwa* + **ingas**, genitive pl. **Lāwinga*, + **hām**.The place is in Launditch Hundred, *Lave(n)- Lawendic* 1086, *Lawendich, Lowedich* 1202, 'Lawa's dyke', OE pers.n. **Lāwa*, genitive sing. **Lāwan*, + **dīċ**. The pers.n. **Lāwa* is not on record. Perhaps, therefore, it should be *Lāfa* with early vocalisation of medial [v] > [w]. DEPN, ING 137.

LONGHIRST Northum NZ 2289. 'The long wooded hill'.

Langherst 1200, *-hirst* 1242, 1296 SR, *-hurst* 1242 BF, 1297. OE **lang** + **hyrst**. NbDu 136, DEPN, L 198.

LONGHOPE Glos SO 6818. 'Long Hope or valley'. *Lang(e)hop(e)* 1248–1466, *Long(e)hop(e)* 13th–1675. OE **lang**, definite form **lange**, + p.n. *Hope, Hop(a)* 1086–c.1340, OE **hop**. 'Long' for distinction from HOPE MANSELL H&W SO 6219. Gl iii.191, L 113, 116.

LONGHORSLEY Northum NZ 1494. 'Long H'. *Longhorsley* 1868 OS. ModE adj. **long** + p.n. *Horselega* 1197, *-ley* 1236, 1242 BF, 1296 SR, 'the horse pasture, the clearing where (wild) horses are corralled'. OE **hors**, genitive pl. **horsa**, + **lēah**. 'Long' for distinction from HORSLEY NZ 0966. NbDu 117, L 206.

LONGHOUGHTON Northum NU 2415. 'Long or great H'. *Houcton Magna* 1242 BF, *Magna Hotton* 1296 SR, *Long Houghton* 1868 OS. Latin **magna**, ModE **long**, + p.n. *Howton* 1281, *Hoghton* c.1325, 'the farm or village beside the hill-spur', OE **hōh** + **tūn**. The reference is to Howlet Hill NU 2315. 'Long' for distinction from LITTLEHOUGHTON NU 2216. NbDu 118.

LONGLEAT HOUSE Wilts ST 8043. The great Elizabethan mansion of the Marquess of Bath, named from the long leat or channel which brought water from Horningsham to the mill of the convent on the site of which the house stands. *Langelete* 1235, 1315, *le* ~ 1257, *Longleyte* 1546. OE **lang**, definite form neuter **lange**, + **ġelǣt**. Wlt 169.

LONGLEY GREEN H&W SO 7350. *Longley Green* 1832 OS.

LONGMOOR CAMP Hants SU 7930. P.n. Longmoor + ModE **camp**. A modern military camp. Longmoor, *Longemore* 1298, is 'the long moor or marsh', ME **long**, definite form **longe** (OE *longa*) + **mor** (OE *mōr*). Ha 111.

LONGNEWTON Cleve NZ 3816. 'Long Newton'. *(Longa) Neutona, (Lange) Newetone* 1235, *Long Neuton* 1281, *Lang Newton* 1367–1586, *Longnewton* from 1508. OE **lang** + p.n. NEWTON.

LONGNEY Glos SO 7612. '(In, on the) long island'. *in Longanege* [972]10th S 786, *Langenei -ey(e)* 1086–1291, *Longaneia* c.1100, *Longeney(e)* 13th–1407, *Longney* from 1316. OE **lang**, dative sing. definite form **langan**, + **ēġ**. The reference is either to a low stretch of ground enclosed on three sides by the Severn or to a long mud island exposed when the river is at low tide. Gl ii.185, L 39.

LONGNOR Shrops SJ 4800. Probably 'the long alder-copse'. *Langanara* 1121, *Longenalra -alr(e)' -olr(e)* 1155–1338, *Longenour' -ore* 1255×6, *Longnor* from 1587. OE **lang**, locative-dative sing. definite declensional form **langan**, + **alor** 'an alder-tree' probably used collectively, dative sing. **alre**. An alternative translation of *langan alre* is 'at the tall alder-tree'. Sa i.182.

LONGNOR Staffs SK 0965. 'The long flat-topped hill'. *Langenour(e)* 1227, *Longenov(e)re* 1277, *Long(e)nor(e)* from 1332. OE **lang**, definite form dative case **langan**, + **ofer**. Oakden.

LONG NOSE SPIT Kent TR 3872. P.n. *Long Nose* 1819 OS + ModE **spit** 'a tongue of land, a long narrow shoal'.

LONGPARISH Hants SU 4344. 'The long parish'. *Langeparisshe* 1389, *Long Parish* 1558×1603. ME **lang**, definite form **lange** (OE *langa*), + **parishe** (OFr *paroche*). This name has supplanted MIDDLETON SU 4244 as the parish name. Ha 111, Gover 1958.171.

LONGRIDGE Lancs SD 6037. Originally the name of the long ridge stretching from SD 6137 to 6840, now LONGRIDGE FELL. *Langrig* 1246, *Longerige* 1409, *Longridge hill* 1577. OE **lang** + **hrycg** varying with ON **hryggr**. The ridge gave its name to the town and *Chapel of Langgrige* 1521, *Longerydche chap* 1554. La 140.

LONGRIDGE FELL Lancs SD 6540. P.n. Longridge as in LONGRIDGE SD 6037 + ModE **fell** (ON *fjall*).

LONGRIDGE TOWERS Northum NT 9549. *Longridge Ho.* 1865 OS. P.n. Longridge, *Lungridg* 1695 Map, 'the long ridge', ModE **long** + **ridge**, + ModE **house**, **tower**. The reference is to a ridge of high ground stretching NE from West Longridge NT 9549 to East ORD NT 9851. Longridge Towers is a 19th cent. neo-Tudor mansion. Pevsner 1957.209.

LONGSDON Staffs SJ 9654. 'The hill of Long or Lang'. *Longesdune* 1216×72, *Longesdon'* 1242–1632, *Langesdun'* c.1246×61, *-don'* 1275, 1292, *Longsdon als. Longston* 1612. OE pers.n. **Long, *Lang* < *lang* 'tall', or hill-n. **Long, *Lang* 'long one', genitive sing. **Longes, *Langes,* + **dūn**. The village stands on a long hill. Identical with Great and Little LONGSTONE Derbys SK 2071. DEPN.

LONGSHIPS Corn SW 3225. 'Long ships'. *Langeshipes* 1347, *Longshippys* 1512 Gover n.d., *Longships rock* 1576. ME **lang** (OE *lang*) + **ship**, pl. **shipes**. A fanciful name for rocks in the sea. PNCo 111, Gover n.d. 662.

LONGSIGHT GMan SJ 8696. *Longsight* 1843 OS. Presumably a spot commanding a long view.

LONGSLEDDALE Cumbr NY 5002. *Langsleiddall* 1466, *Longsleddell* 1492, *Langsleddale* 1578, *Long Sleddle* 1738. Adj. **long** + p.n. *Sleddal(e) -all* 1229–1622, a tautological name compounded of OE **slæd** and ON **dalr**, both of which mean 'valley'. Possibly the Viking settlers regarded **slæd** as the name of the valley. The name occurs twice in Yorkshire, at NZ 6112 and SD 8586. We i.160 gives pr [laŋ'sledəl], SSNNW 250.

LONGSLOW Shrops SJ 6535. 'Wlanc's tumulus'. *Walanceslau* 1086, *Wlankeslaw(e) -lauwe -lowe* 1200–1314, *Wlonkesla(u)we -lowe* 1212–1431, *Lonkeslowe* 1489, *Longeslowe* 1672, *-low* 1712. OE pers.n. *Wlanc*, genitive sing. *Wlances,* + **hlāw**. Although no burial mound is known here or at the other Shrops *hlāw* names, it is suggested that Wlanc may have been the name of a late 6th–early 7th cent. Mercian aristocrat of the last generation still to practise pagan burial. Sa i.183.

LONGSTANTON Cambs TL 3966. *Longa Staunton(e)* 1272, *Long Stantone* 1281. ME adj. **long, lang** + p.n. *Stantone* 1086, *Stanton(e)* from 1161, 'stone farm, village'. 'Long' for distinction from FENSTANTON TL 3168. Ca 183.

LONGSTOCK Hants SU 3547. 'The long Stoke'. *Langestok* 1233–1428 with variant *-stoke*. ME **lang**, definite form **lange** (OE *langa*), + p.n. *(œt) Stoce* [982]12th S 840, *Stoches* 1086, 'dependent farm', OE **stoc**. The village straggles along the road. It is not known what place it was depedent on, possibly Leckford SU 3737 which, like Longstock, was an estate of St Mary's abbey, Winchester. Ha 111, Gover 1958.181, Studies 1936.22.

Great LONGSTONE Derby SK 2071. *Magna Longesdon(e) -is-* 1275–1539, *Mikell Longesdone* 1365, *Mykell Longeston* 1546, *Great Longstone* 1840 OS. Adj. **great**, Latin **magna**, ME **micel**, + p.n. *Lang(e)sdune* 1086, 1225, *Langesdon(e) -is-* 1199×1216–1365, *Langedon* 1200–1280, *Longesdune* 1086, *-don(e) -is-* 1215–16th, *Longston* 1339, 1577, 1610, possibly '*Lang*'s hill' in which **Lang* is OE hill-name 'the long one', referring to the long escarpment of Longstone Edge, High Rake and Deep Rake SK 1872–2273 (ME **rake** 'a narrow path'). 'Great' for distinction from Little LONGSTONE SK 1871. Db 138.

Little LONGSTONE Derby SK 1871. *Little Langesdon(e)* 1252, *Parva* ~ 1300, 1365, *Parva Longesdon(e) -is-* 1216×72–1430, *Little Longstone* 1840 OS. Adj. **little**, Latin **parva**, + p.n. Longstone as in Great LONGSTONE SK 2071. Db 141.

LONG STREET Bucks SP 7947. 'Long village'. *Long Street* 1835 OS. ModE **long** + **street**.

LONGTHORPE Cambs TL 1698. *Lange-, Longethorp(e)* 1285. ME adj. **long**, definite form **longe**, + p.n. *Torp, þorp* [972]c.1200 B 1130, *Torp* 1086, *Thorp(e)* [1189]14th, 'outlying farm', OE, ON **thorp**. Nth 227.

LONGTON 'The long village'. OE **lang(a)** + **tūn**.

(1) ~ Lancs SD 4725. *Longe- Langetuna, Langetona* 1153×60,

Langeton 1178–1212, *Longeton* 1243–1332 etc., *Longton* from 1374. The township is long and narrow (4 miles long, 1 mile wide) and the village 'straggles' 2.5 miles along a road. La 136 citing VCH Lancs vi.69, Jnl 17.77.

(2) ~ Staffs SJ 9143. *Longeton* 1212, 1216×72, *Langetun'* 1242. The village extends a mile or more along the Roman road from Derby to Stoke on Trent, Margary no. 181. DEPN, Horovitz.

(3) New ~ Lancs SD 5025. Adj. **new** + p.n. LONGTON SD 4725.

LONGTON LONG BLANDFORD Dorset ST 8905 → BLANDFORD FORUM ST 8806.

LONGTOWN Cumbr NY 3868. '(The) long village'. *Longeton* 1267, *Langetoune* 1584, *Longtowne* 1590. ME **longe** (OE *lang*, definite form *langa*) + **toun** (OE *tūn*). Cu 53 gives pr [lɔŋtu:n].

LONGTOWN H&W SO 3229. 'Long town'. *Longa Villa* 1540, *Longton* c.1540. ME **long**, Latin **longa**, + **toun**, Latin **villa**. The settlement straggles along the Hay to Abergavenny road. Earlier references to this place are *Novi Castelli* (genitive case) 1187 and *Nova Villa* 1232, 'new castle' and 'new town'. The reference is to the new castle built here as the centre of the marcher lordship of Ewyas Lacy, superseding earlier centres at CLODOCK SO 3227 and WALTERSTONE SO 3425. He 57.

LONG VALLEY Hants SU 8352. No early forms. ModE **long** + **valley**.

LONGVILLE IN THE DALE Shrops SO 5493. P.n. *Longefewd* 1255, *Longfeld* 1291 as in Cheney LONGVILLE SO 7387 + ModE **in the Dale** for distinction. DEPN.

Cheney LONGVILLE Shrops SO 7387. 'L held by the Cheyney family'. *Longefelde Cheyne* 1421, *Chenes Longvill* 1577, *Cheyny Longveld* 1641. Family name Cheyny as in Roger de Cheyny 1315 + p.n. *Languefelle* 1086, *Lange- Langafeld(a)* 1087×94–c.1200, *Longafelda -(e) -feud(e)* 1142×6–1421, *Longfeld* 1592, *Longville* 1674×5, 'the long stretch of open land', OE **lang**, definite declensional form **langa**, + **feld**. Originally a district name including ACTON SCOTT SO 4589, *Acton' in longefeld'* 1261×2 and LONGVILLE IN THE DALE SO 5493. Sa i.75.

LONGWICK Bucks SP 7905. 'Long Wick'. *Long Wyke* 1320. ME adj. **long** + p.n. Wick from ME **wyke** (OE *wīc*) 'the dairy farm'. The village that grew up at Wick is long and straggling. Bk 173.

LONGWITTON Northum NZ 0788. *Langwotton* [1340]14th cent. ME adj. **lang** + p.n. *Wittun* 1236 BF, *W(o)otton* 1296 SR, 1340, 'the wood-farm', OE **wudu** + **tūn**. 'Long' for distinction from NETHERWITTON NZ 1090. NbDu 218, L 227.

Upper LONGWOOD Shrops SJ 6007. ModE adj. **upper** + p.n. *The Longwood* 1671, *Long Wood* 1833 OS, 'the long wood'. Gelling.

LONGWOOD HOUSE Hants SU 5424. P.n. *Langwode* 1272 *la Longewode* 1307, 'the long wood', ME, OE **lang**, definite form **langa**, + **wudu**, + ModE **house**. Gover 1958.73.

LONGWOOD WARREN Hants SZ 3295. *Longwode Warenna* 1453. P.n. Longwood as in LONGWOOD HOUSE SU 5424 + ME **warenne**. Gover 1958.73.

LONGWORTH Oxon SU 3999. *Langewrth'* 1284, *Longwrth* 1294. ME adj. **lang, long** + p.n. *Wyrðe* [811]12th S 166, *Weorþe* [955]12th S 567, *(æt) Wurðe* [958]13th S 654, *Weorthe* [959]12th, *(to) wyrþe* [959]13th S 673, *Wurtha* [1066×87]13th, *Worthe* 1401×2, 'the enclosure', OE **w(e)orth**. 'Long' for distinction from LITTLEWORTH SU 3197. Brk 393.

LONTON Durham NY 9524 → River LUNE NY 9524.

LOOE BAY Corn SX 2753. *Looe Bay* 1813. P.n. LOOE SX 2553 + ModE **bay**. PNCo 111.

LOOE Corn SX 2553. 'Pool, inlet'. *Loo* 1284–1365 including [c.1220]1320, *Lo* 1237–1312. East Looe is *Estloo* 1364, 1442, West Looe *Westlow* c.1540, *West Loo* or *Portbyhen* 1622, earlier *Porthbighan* c.1286 'little harbour' (Co **porth** + **byghan** 'small'). Co ***loch**. PNCo 111, Gover n.d. 279, 298.

LOOE or ST GEORGE'S ISLAND Corn SX 2551. *St George's Island* 1584, c.1605, *Looe Island* 1699. P.n. LOOE SX 2553 + ModE **island**. Earlier, *insulam Sancti Michelis de Lammana* 'St Michaels's island of L' [13th]1727, *Lamayne* c.1547, *Lemain* 1839, *S. Michaell's Isle* 1610, 'church site of a monk', Co ***lann** + **managh** 'monk' (OCo *manach*). PNCo 86, 111, Gover n.d. 298.

LOOSE Kent TQ 7652. 'The shed, shelter, or pigsty'. *Lose* [c.832]17th, 1189×99, 1204, *Hlose* c.1100, *Loose* 1610. OE **hlōse**. PNK 139.

LOOSLEY ROW Bucks SP 8100. 'Row of houses at L'. *Louslerowe* 1464, *Lesley Row* 1553×1601, *Loosley Row* 1694. P.n. Loosley + ME **rowe**. Loosley, *Losle* 1241, is 'wood or clearing with a pigsty', OE **hlōse** + **lēah**. It has been claimed that a form *Lulleslede* 1351 belongs here. If so this can only be 'Lull's valley', OE pers.n. *Lull*, genitive sing. *Lulles*, + **slæd**. The spelling *Losle* 1241 is from the pers.n. Galfridus de Losle, whose surname may not belong here at all. Bk 173, D 1, Sr xxxix, Jnl 2.30.

LOPCOMBE CORNER Wilts SU 2535. *Lobcombe Corner* 1772, 1817 OS, *Lapcombe Corner or Coniger Hill* 1773. P.n. *Lobecombe* 1342, possibly a compound of OE ***lobb** 'something heavy or clumsy' used in p.ns. in a topographical sense not determined or **lobbe** 'a spider', + **cumb**, + ModE **corner** 'a place where two roads meet' referring to the junction of the roads from Andover and Winchester to Salisbury. Lopcombe refers to the valley stretching N from Lopcombe Corner where *Lobcombe Barn* is found 1817 OS. Wlt 386.

LOPEN Somer ST 4214. Uncertain. *Lopen(e)* 1086, *Lopena* 1166, *Luppena* 13th, *Lopen* 1244. This has been tentatively explained as 'Lufa's pen or fold', OE pers.n. *Lufa* + **penn**. Lopen stands on hilly ground overlooking fenland; it is just possible that it might be the dative pl. ***lopum** 'at the hills' of the element suggested for LOPPINGTON Shrops SJ 4729 and LOPHAM Norfolk TM 0683. DEPN.

LOPHAM Norfolk TM 0683. Normally taken to be 'Loppa's homestead or village'. *Lopham* from 1086, *Lopp-* 1198. OE pers.n. *Loppa* + **hām**. Lopham stands on hilly ground, however, and the specific might be the element proposed for LOPPINGTON Shrops SJ 4729 and LOPEN Somer ST 4214. DEPN.

North LOPHAM Norf TM 0383. *North Lopham* 1836 OS. ModE adj. **north** + p.n. *Lopham* 1086, 1177, *Loppham* 1198, 'Spider' or 'Loppa's homestead', OE **loppe** or pers.n. **Loppa*, + **hām**. 'North' for distinction from South LOPHAM TM 0481. DEPN.

South LOPHAM Norf TM 0481. *South Lopham* 1836 OS. ModE adj. **south** + p.n. Lopham as in North LOPHAM TM 0383.

LOPPINGTON Shrops SJ 4729. Partly uncertain. *Lopitone* 1086–1452 with variants *-tun'* and *-ton('), Lopynton(') -in-* 1199–1535, *-i(n)gton* 1230, *Lopyngton* 1395, *Loppington* from 1535, *Lapp-* 1656–1770. Possibly OE pers.n. *Loppa* + **ing**[4] + **tūn** 'the estate called after Loppa'. The pers.n. is thought to occur also in LOPHAM Norfolk TM 0683 and in two OE charters, *Loppancomb'* [947]14th S 524 (Berks iii.694) and *Loppandyne* 1000×2 S 1486. It derives, however, from OE **loppe** 'a spider', so that an alternative might be OE ***lopping** 'place infested with spiders' + **tūn**. Both Loppington and Lopham occupy hill sites; possibly at the root of both names is a hitherto unidentified element ***lop** 'a hill' cognate with ModE **lump** and the element *lob* seen in the Devon names Lobb SS 4737, *Loba* 1086, Lobb Farm SX 5658, *Lobbe* 1356, Lopwell SX 4764, *La Lobbapilla* [1291]1408, and Labdon SS 6611, *Lobbeton* 1330, and thought to mean 'a hill, a hillock'. Cf also LOPEN Somer ST 4214. Sa i.184, Årsskrift 1974.51, DEPN, D 33 and refs.

LOPWELL Devon SX 4764. Uncertain. *La Lobbapilla* [1291]1408, *Lophill* 1809 OS. This might be an OE **lobb** 'a lump' + **pyll** 'a tidal creek, a pool in a river' but the 1408 form may be a bad spelling for original *Lobbpilla* for *Lobb-wylle*, 'Lobb well'. The situation is a prominent hill overlooking the river Tavy just where it begins to widen out to the estuary. Such an element would be related to OE *lobbe* 'a spider' on a root meaning 'thick' as in Swedish *lubbig* 'fat', ON, Swedish *lubba* 'fat woman'. D 226, Forsberg 1950.149f.

LORBOTTLE Northum NU 0306. 'Leofwaru's homestead'. *Luuerbotle -v-* 1178, 1219 BF, *-batte* 1236, *Leuerboda* 1176, *-botle* 1178, *Loverbothill'* 1236, 1242 BF, *Liver-* 1244, *Liuuerboth* 1253, *L(o)urbotil(l) -botel(l)* 1280–1428, *Lurebodil* 1296 SR, *Leyrbotel* 1368, *Lowrebotell* 1532, *Lorbottle* [1650]17th, *Lurbottle* [1648]1888, [1663]17th. OE feminine pers.n. *Lēofwaru*, genitive sing. *Lēofware*, + **bōtl**. NbDu 136, Tomlinson 355–6.

LORBOTTLE HALL Northum NU 0408. P.n. LORBOTTLE NU 0306 + ModE **hall**. Lorbottle Hall is an early 19th cent. house. Pevsner-Grundy etc. 1992.210.

LORD'S SEAT Cumbr NY 2026. 'The lord's mountain pasture'. *Lauerdesate* 1247, *the Lordseatt* 16th. OE **hlāford**, genitive sing. **hlāfordes**, + ON **sǽtr**. Cu 409.

LORTON Cumbr NY 1525. 'Settlement on the *Hlora*'. *Loretona* [c.1150]n.d., *-tuna* [1171×5]1333, *Lorton(e)* from 1197. ON r.n. **Hlóra* 'roaring stream' + **tūn**. The stream-name *Lora* occurs in Norway. Cu 408, SSNNW 187.

LORTON VALE Cumbr NY 1526. P.n. LORTON NY 1525 + ModE **vale**, referring to the valley of the river Cocker.

LOSCOE Derby SK 4247. 'Wood with a lofthouse'. *Loftskou* 1281, *Loscowe -skowe* 1277–1564, *Losco(e)* c.1325 and from 1500. ON **loft** + **skógr**. Db 434, SSNEM 171.

LOSTOCK GRALAM Ches SJ 6975. '(That part of) Lostock held by Gralam'. *Lostoke Graliam* from 1288 with variants *Gralan, Gral(y)am, Lostock Gralam* 1505. P.n. *Lostoch(e)* 1150–early 14th, *Lostock(e)* from 1154×89 with variants *-stoc -stok(e)* to 1532, is 'pig-sty hamlet', OE **hlōse** + **stoc**, + pers.n. *Gralam, Graland*, grandson of Hugh de Runchamp to whom this moiety of the manor was granted c.1070×1101. For the other moiety see ALLOSTOCK SJ 7571. Che ii.189 gives pr ['lostok 'greiləm], Notes 29.

LOSTOCK JUNCTION GMan SD 6708. P.n. Lostock + ModE **junction**. A railway junction. Lostock, *Lostok* 1205–1332, *Lostoc* 1212, 1220 etc., *Lostoke* 1451, is 'pig-sty hamlet', OE **hlōse** + **stoc**. La 45, Jnl 17.37.

LOSTWITHIEL Corn SX 1059. 'Tail of the forest'. *Lostwetell* 1194, *Lostuuidiel* c.1194, *Lostudiel* 1195×8, *Lostwithiel* from 1234, *Listhyell* 1610. Co **lost** 'tail' + ***gwthyel** 'woody place' (Co *gwyth* 'trees' + adjective suffix **-yel*). **Gwythyel* occurs again in WITHIEL SW 9965 8 miles to the W. It is suggested that Withiel may have been a name of the whole intervening upland region. PNCo 111, Gover n.d. 402, DEPN.

LOTHERSDALE NYorks SD 9646. 'The beggar's valley'. *Lodresdene* 1086, *Lodderesden'* 1202, 1314, *Letheresden* (sic for *Loth-*) 1279–81, *Lothersden -daill* 16th cent., *-dale* 1675. OE **lod-dere**, genitive sing. **lodderes**, + **denu** later replaced by ON **dalr**. An alternative suggestion for the specific has been OE pers.n. *Hlōthhere*, but the forms seem to support the explanation given. The change of /d/ to /ð/ before *r* and *er* is well attested in English from about 1400, Jordan §298, Dobson §384. YW vi.31, L 99.

LOUDWATER Bucks SU 9090. Originally a descriptive name of the river also called Wooburn as in WOOBURN SU 9087 and Wye as in High WYCOMBE SU 8593. *la Ludwatere* 1241, *Lowdewater* 1485. ME **loud** + **water**. Bk 203 quotes Lipscomb's *History and antiquities of the county of Buckingham* 1847, iii.652: 'Its name is probably derived from the noise incessantly made by the rapidity of the stream which rushes with great impetuosity towards Wooburn and its junction with the Thames'.

LOUGHBOROUGH Leic SK 5319. 'Luhhede's fortified place'. *Lu-Locteburne* 1086, *Lucteburg(h)* 1225–c.1284, *L(o)ughteburg(h)* 1239–1436, *Lut(t)eburc -burg(h)* 1180–late 13th, *L(o)ught-burgh(e)* 1295–1517, *Lug(h)burgh* 1382–1509, *Loughburgh* 1416–1525. OE pers.n. *Luhhede* + **burh**. Lei 373.

LOUGHTON Bucks SP 8337. 'Luha's estate'. *Lochintone* 1086, *Lughton(a)* 1219–1376, *Loughton(e)* from 1300, *Louton* 1284, 1302. OE pers.n. *Luha* + **tūn**. The 1086 form has connective **ing**[4]. Other early spellings show different treatments of OE *h*, [x] > [f] in *Lufton* 1152×8, [x] > [k] in *Luctune* 1199, [x] > [γ] > [w] in *Luwetune* 1235. Bk 20 gives pr [lautən].

LOUGHTON Essex TQ 4296. 'Luhha's estate'. *Lochetuna -ā, Lochintunam* 1086, *Luke(n)tun', Luchentuna* 1177, *Loke- Luche-Luketon(')* 1200–1338, *Lukintone* *[1062]13th S 1036, *Lutton(e)* 1270–1558×1603, *Lughton* 1331–79, *Loughton* from 1338, *Lufton, Lowton, Laughton* 16th cent. OE pers.n. *Luhha*, genitive sing. *Luhhan*, + **tūn** possibly varying with *Luhha* + **ing**[4] + **tūn**. Ess 64 gives pr [lautn].

LOUGHTON Shrops SO 6183. 'The lake settlement'. *Loketona* 1121, *Loche-* 1138, 1155, *Luchton(')* c.1143–1255, *Luhtune* c.1230, *Lughton(')* 1291–1327, 1803, *Louhton'* late 14th, *Loughton* from 1405, *Lowton* 1696. OE **luh** + **tūn**. No lake or pool is recorded here in modern times. Sa i.184 gives pr [loutən].

LOUND 'The wood'. ON **lundr**. Identical with Lund in Denmark and Sweden.

(1) ~ Lincs TF 0618. *Toftlund, Totf et Lund* 1086, *Lund* late 12th–[1344]15th, *Lound(e)* from 1298. Two distinct settlements have been erroneously run together in some of the DB forms. Perrott 207. SSNEM 157.

(2) ~ Notts SK 6986. *Lund* 1086, *Lund(e)* 1171–1280, *(le) Lound* from 1296. Nt 85, SSNEM 157 no.1.

(3) ~ Suff TM 5099. *Lunda* 1086, *Lund* 1254–91, *Lound(e)* from 1316. DEPN, Baron.

(4) East ~ Humbs SK 7899. ModE adj. **east** + p.n. *Lvnd* 1086. 'East' for distinction from Graize LOUND SK 7798. SSNEM 157.

(5) Graize ~ Humbs SK 7798 → GRAIZELOUND SK 7789.

LOUNT Leic SK 3819. 'The wood'. *Lunda* c.1150, *le Lounte* 1347, *the Lount* 1446, *Lount* 1806. ON **lundr**. Lei 401, SSNEM 171 no 3.

LOUTH Lincs TF 3387. R.n. LUD TF 3387 used as a p.n. *Ludes, Lvde* 1086, *Luda* 1093–1271, *Lude* 1160×75–c.1540, *Luth(e)* c.1155–1245, *Louth(e)* from 1234. The monastery at Louth is *Hludensis monasterii* [c.1130]12th SD under year 791. The substitution of *th* for *d* is due to Scand influence. Louth was the market centre for the densely populated wolds from early times. RN 261, Cameron 1998.

River LOVAT or OUZEL Bucks SP 8831. *Lovente* [c.1200]c.1260, 1307, *Lavente, Louette, Luuente* 1276, *Lovatt* 1847. An ancient r.n. of uncertain origin and meaning apparently identical with LAVANT WSusx SU 8608 and possibly Lavant, Carinthia, *Labanta* 888, and Lafnitz in Styria, *Labenza* 864, which may belong to the IE root **lēb-/ləb-* seen in Latin *labō* 'waver', *lābor* 'glide'.†

LOVE CLOUGH Lancs SD 8127. 'Luha's ravine'. *Lugheclogh, Lufclough* 1324, *Lufclogh* 1325, 1425, *Luffecloch* 1464. OE pers.n. *Luh(h)a* + **clōh**. The OE guttural spirant [x] has been substituted with [f] as, e.g., in the 1152×8 spelling *Lufton* for LOUGHTON Bucks SP 8337. La 92, Bk 20.

LOVER Wilts SU 2120. No early forms. Perhaps a version of Low Ford.

LOVERSALL SYorks SK 5798. 'Leofhere's nook of land or little valley'. *Loures- Luures- Geuershale* 1086, *Luuresale* c.1200, *Luu- Luver(e)s(h)al(e)* c.1205–1246, *Liversal* 1234, *Lou-Loversal(e)* 1276–1546. OE pers.n. *Lēofhere*, genitive sing. *Lēofheres*, + **halh**. YW i.34, L 107, 110.

LOVES GREEN Essex TL 6404. No early forms. Surname *Love* as Robert Love 1306 + ModE **green**. Ess 280–1.

LOVINGTON Somer ST 5930. 'Estate called after Lufa'. *Lovintvne* 1086, *Louinton* 1187, *Louington* 1610. OE pers.n. *Lufa* + **ing**[4] + **tūn**. DEPN.

†Otherwise, however, Krahe 1964.53. See also River OUZEL. RN 263.

LOW GATE Northum NY 9063. *Low Gate* 1867 OS. A turnpike gate.

LOW GRANGE Cleve NZ 4625 → COWPEN BEWLEY NZ 4824.

LOW MILL NYorks SE 6795. *Low Mill* 1860 OS. ModE **low** + **mill**.

LOW MOOR Lancs SD 7341. No early forms. Contrasts with Little Moor SD 7440.

LOW ROW Cumbr NY 5863. *Low Row* 1867 OS. 'Low' for distinction from *Middle Row* 1867 OS.

LOW ROW NYorks SD 9897. *Low Row* 1860 OS. ModE **low** + **row**.

LOWBRIDGE HOUSE Cumbr NY 5401. *Lowbridge House* 1857. Earlier *Bryghouse in Banisdale* 1688, *Bridge Ho* 1836. The reference is to a bridge on the A6 across Bannisdale Beck, *Banendes- Banandesdala -e* 1175×84–1539, 'Bannand's valley', ON pers.n. **Bannandr*, secondary genitive sing. **Bannandes*, + **dalr**. We i.150, 137, SSNNW 102.

LOWCA Cumbr NX 9821. Unexplained. *Loweowe* (for *-cowe*) [1327×77]16th, *Lowkay, Lowcoe, Lowker* 1655×73. Lowca is situated on a headland, suggesting that the generic may be OE **hōh** or ON **haugr**. Cu 400 gives pr [laukə].

LOWDHAM Notts SK 6646. Probably 'Hluda's homestead'. *Ludehā* 1086, *-ham* 1149×53–1276, *Ludhā* 1086, *-ham* 1088–1242, *Loud(h)am* 1257–1319, *Lowdam* 1497, *Lewdham al. Lowdham* 1726. OE pers.n. **Hlūda* + **hām**. Lowdham is situated on Cocker Beck, a slow-moving clay stream unlikely to have been called **Hlūde* 'the loud one'. Nt 171.

LOWER DOWN Shrops SO 3384 → Lower DOWN SO 3384.

LOWER GREEN Norf TF 9837. *Lower Green* 1838 OS. Short for Hindringham Lower Green and contrasting with Hindringham *Summer Green* 1838 OS at TF 9836.

The LOWER HOPE Essex TQ 7077. *The Hope* 1787, 1805 OS. Earlier *Tilbery Hope* 1486. ME **hope** 'an inlet, a reach of a river' as in STANFORD-LE-HOPE TQ 6882. Ess 15.

LOWESBY Leic SK 7207. Partly uncertain. *Glowesbi* 1086, *Lo(u)sebi -by* c.1160–1576, *Lowesby(e) -bie* 1241–1535. Probably 'Lauss or Lausi's village or farm', ON by-name *Lauss* or *Lausi*, genitive sing. *Lausa*, + **bȳ**. Lei 309, SSNEM 58.

LOWESTOFT Suff TM 5493. 'Hlothver's toft'. *Lothu Wistoft* 1086, *Lothewistoft* 1212, *Lowistoft* 1219–1419, *Lowestoft* 1524, *Lestoft* 1610. ON pers.n. *Hlothvér*, genitive sing. *Hlothvis*, + **toft**. DEPN, Baron.

LOWESWATER (the lake) Cumbr NY 1221. *Lawes- Lausewatre* c.1203, *Loweswatre* 1230, *Lawsewater* 1539, *Loosewater* 1578, is possibly ON lake name **Laufsœr* 'leafy lake' + **wæter**. **Laufsœr* is found in Sweden, with suffixed article, as *Lövsjön*; **wæter** was added when *Laufsœr* was no longer understood. Cu 34.

LOWESWATER (the village) Cumbr NY 1420. *Lousewater* c.1160–1385, *Lowswater* 1186, *Laweswater -watre* 1188–1397, *Loweswater* from 1367. Transferred from LOWESWATER, the lake, NY 1221. Cu 410 gives pr [lauzwɔtə], SSNNW 238.

LOWESWATER FELL Cumbr NY 1319. P.n. LOWESWATER NY 1221 + ModE **fell**.

LOWGILL Cumbr SD 6297. 'Low ravine'. *Lowgill* 1823. ModE **low** + **gill**. 'Low' for distinction from Howgill SD 6396, *Highgill* 1823. We i.31.

LOWGILL Lancs SD 6564. 'Low ravine'. *Lawgill* 1520, 1528. ON **lágr** + **gil**. La 182.

LOWICK Cumbr SD 2986. 'Leafy nook'. *Lofwic* 1202, *Lowyk* 1246, *Laufwik -wic -wyk* 15th. ON **lauf** + **vík** 'creek, bend in a river, nook in hills'. Probably refers to bends in the river Crayke at Lowick Bridge SD 2986 and Lowick Green SD 2985. La 213, SSNNW 145.

LOWICK Northants SP 9780. Partly uncertain. *Lude- Luhwic* 1086, *Luf- Lofwyc -wik -wyk* 12th–1376, *Luff(e)wyk -wich -wi(c)k* [c.1148]1348–1823, *Louf- Lughwyk* 1305, *Lowicke* 1262, *Lowick* 1823. Possibly 'Luffa's farm', OE pers.n. **Luffa* + **wīċ**, but the occasional form with *-(g)h-* suggests OE pers.n. *Luhha*. Substitution of [f] for [x] is regular in ME, e.g. *tough* spelled *tuf* in the 14th cent. < OE **tōh**. Nth 185 gives pr [louik, lʌfik].

LOWICK Northum NU 0139. 'Settlement, farm beside the river Low'. *Lowich* 1180, *Lowic -wyc -wik -wyk(e)* 1228–1539. R.n. Low + OE **wīċ**. Low, *Low* c.1540, is the 'tidal creek or stream', OE **luh**, a Celtic loan-word cognate with Welsh *llwch* 'lake, loch' and Ir, Gael *loch* 'a lake or arm of the sea', which survives in N dial. *low* 'a shallow pool left in sand by the retiring tide'. NbDu 137, RN 264, PNE ii.27, GPC 2234. Cf. however Celtic Voices 242.

LOWSONFORD Warw SP 1868. Partly obscure. *Lowston ford* 1496, *Lowson ford* 1682. The specific, which also occurs in *Lowston end(e)* 1496, 1561, *Los(t)on, Lowson end* 1608, looks like an unidentified p.n., and so 'ford leading to or at *Lowston*', OE **ford**, and 'Lowston end' sc. of Rowington SP 2069. Wa 218.

LOWTHER Cumbr NY 5323. *Lauuedra* 1174, *Lauder* 1170×8, 1184, *Lauther* c.1225, *Lawther* 1370, *Loudr(e) -er* 1195–1247, *Louthre -er -ir* 1190×1200–1476, *Lowther* from 15th. Transferred from the r.n. LOWTHER NY 5120. We ii.182 gives pr ['lauðə].

LOWTHER CASTLE Cumbr NY 5223. *Lowther Castle* 1811. Earlier *Lowther Hall* 1656. P.n. LOWTHER NY 5323 + ModE **castle**. The hall was rebuilt as a castle in 1806–11. We ii.184, Pevsner 1967.272.

River LOWTHER Cumbr NY 5120. A British r.n. formed on the root *lou̯-* 'to wash'. *Lauther* 1157×86, 1204×14, *Louther* early 13th–1408, *Lowther* from 1249, *Lowdyr* 1473, *Lowder* 1577, *Loder* c.1540–1750. As the forms for Lowther NY 5323 show, *d* may be original in the name, linking it with the Scottish r.n. Lauder. The alternative derivation from ON *lauthr-á* 'foam, lather river' has phonological difficulties since ON *au* normally becomes [ɔ:] and not [au]. We i.9 gives pr [lauðə], RN 266, SSNNW 423.

LOWTHORPE Humbs TA 0860. 'Logi or Lági's outlying settlement'. *Log(h)e- Langetorp* 1086, *Lout(h)orp(e)* 12th–1376, *Lowthorpe(e)* from 1285, *Laut(h)orp* 13th cent., *Lawthorp(e)* 16th cent. ON pers.n. *Logi*, genitive sing. *Loga*, or *Lági*, genitive sing. *Lága*, + **thorp**. DB *Lange-* is probably an error for *Lauge-*. YE 93 gives pr [louθrəp], SSNY 62.

LOWTON GMan SJ 6297. 'Hill settlement'. *Lauton* 1202–1332 etc., *Law(e)ton* 1432, 1500, *Laitton* 1201. OE **hlāw** + **tūn**. La 99.

LOWTON COMMON GMan SJ 6397. P.n. LOWTON SJ 6297 + ModE **common**.

LOXBEARE Devon SS 9116. 'Locc's wood'. *Lochesbere* 1086, *Lockesbere* 1205–1326 with variants *Lokkes-* and *-bear(e)*. OE pers.n. *Locc*, genitive sing. *Locces*, + **bearu**. D 540.

LOXHILL Surrey TQ 0038. 'The hill called after the Lock family'. *Locks Hill* 1842. Cf. Henry *Lok* 1332. Family name *Lock* + ModE **hill**. Sr 244.

LOXHORE Devon SS 6138. 'Locc's ridge'. *Loches(h)ore* 1086, *Lokkesore* 1244–1324 with variants *Lockes-*, *Lockis-* and *-hore*. OE pers.n. *Locc*, genitive sing. *Locces*, + **ōra**. D 63, L 180.

LOXLEY SYorks SK 3089. 'Locc's wood'. (A wood called) *Lockeslay -ley* 13th, 1332, *Lokeslay -ley* 1332–85, *Loxley Firth* 1637. OE pers.n. *Locc*, genitive sing. *Locces* + **lēah**. For an alternative possibility see r. LOXLEY SK 2989. YW i.225.

LOXLEY Warw SP 2453. 'Locc's wood or clearing'. *Lockeslea, Loccheslega* 11th, *Locheshā -lei* 1086, *Lochesle* 1100×35–1428 with variants *Lockes- Lokkes-* and *-legh -ley(e), Loxley* 1535. OE pers.n. *Locc*, genitive sing. *Locces*, + **lēah**. The boundary of the inhabitants of Loxley is referred to in S 1350 [985]11th as *Locsetna gemære*. Wa 235.

River LOXLEY SYorks SK 2989. *Loxley Water* 1637. P.n. LOXLEY SE 3089 + ModE **water**. The specific of Loxley might, however, be itself an original r.n. identical with the r. Lossie in Morayshire, Scotland, *Λόξα* (Loxa) [c.150]13th Ptol, [c.670]13th Rav 'crooked or winding one', Brit ***losk-** cognate with OIr *losc* 'crippled' and possibly Gk *λοξός* 'oblique' < ***lok-s-**. Cf also the r.n. LOX YEO Somerset. Another known earlier name was Steyne, *Stene* 13th cent., *Steyne* 16th cent., OE **stǣne** 'stony place, stony river'. RBrit 399, LHEB 536, YW vii.131, 138.

LOXTON Avon ST 3755. 'Settlement on the river Lox'. *Lochestone* 1086, *Lockeston* 1189×99 *Berkeley*, *Lokestone* 1212, *Loxton* 1259. Loxton lies beside the Lox Yeo River *(on, & lang) Loxan, (into) Locxs, (of) Loxs* [1086]15th. A tributary of the Avon at ST 7165 was also formerly called Lox, *(innan, andlang) Loxan* [931]12th S 414, *(into, be, of) Loxan* [946]12th S 508, with which must be compared the Scots river Lossie Grampn NJ 2059, *Loxa* [c.150]13th Ptol, [c.700]13th Rav. This probably represents a Brit r.n. **Losca* < Celtic **lok-sko-* possibly cognate with Gk λοξός 'oblique' and meaning 'winding one'. Whether or no metathesis of *sk* > *ks* can be assumed for the Scots r.n., there is no difficulty in English as the r.n. Exe from *Isca* shows. DEPN, RN 267, LHEB 536 f.n.2, 539 f.nn.1, 2, Pokorny 308, RBrit 399.

LOXWOOD WSusx TQ 0331. 'Locc's wood'. *Logeswod* 1263, *Lo(c)keswod(e)* 1288–1385. OE pers.n. *Locc*, genitive sing. *Locces*, + **wudu**. Sx 134.

LUBENHAM Leic SP 7087. '(The settlement) at the hill-spurs of Luba or Lubba'. *Lobenho* 1086, 1087×110, after 1150, *Lubbenho(u)* 1203–76, *Luban- Lubeham* 1086, *Lubbenham* 1323–1610, *Lubenham* from c.1291. OE pers.n. *Luba* or *Lubba*, genitive sing. *Lub(b)an*, + **hōh**, dative sing. **hō**, dative pl. **hōum, hōm**. The village is situated on the southern prong of a forked hill-spur. The dative pl. *hōm* was subsequently confused with OE *hām*. Lei 239, Notes 20.

LUCCOMBE Somer SS 9144. 'Lufa's coomb'. *Lucum* [959×75]13th S 1769, *Locūbe* 1086, *Locumba* [c.1150]Bruton, *Loucumba* 1183, *Lo- Luc(c)umb(e)* [13th, c.1350] Buck, *Luuecumbe* c.1271, *Luckum* 1610, *Luckham* 1809 OS. OE pers.n. *Lufa* + **cumb**. An attractive alternative suggestion is OE **lufu** 'love' + **cumb**, 'coomb where courtship takes place'. DEPN, L 90, 92.

LUCCOMBE CHINE IoW SZ 5879. *Luccombe Chine* 1781. P.n. Luccombe + Hants/IoW dial. **chine** (OE *ċinu*) referring to the bowl-shaped valley descending to the sea at the head of which Luccombe stands. The chine is actually called *Bowl Hoop* 'bowl-shaped bay' in 1790 and 1851. Luccombe, *Lovecvmbe* 1086, *Louecumb(e) -v-* before 1155–1536, *Luvecumbœ -e* 1135×54, 1258, *Lowcumbe -combe* 1537, c,1540, *Luckome* 1611, is probably 'Lufa's valley', OE pers.n. *Lufa* + **cumb** referring to the same feature. An alternative possibility, however, is OE **lufu** + **cumb** 'love valley'. The place was a grange of Quarr abbey in the middle ages. Wt 47, Mills 1996.68.

LUCCOMBE VILLAGE IoW SZ 5879. A modern holiday village named after Luccombe as in LUCCOMBE CHINE SZ 5979.

LUCKER Northum NU 1530. Uncertain. *Lucre* 1169, 1242 BF, 1255, *Luker* 1242 BF, *Lucker* 1296 SR. This could be ON nominative pl. **lúkur** 'the hollows' from **lúka** as in the Norwegian p.n. *Luker* (Norske Gaardnavne iii.195), or ON **ló-kjarr** 'sandpiper marsh'. Lucker is situated in low-lying ground on Waren Burn and either suggestion would suit the topography. Scandinavian names are rare, however, in Northumberland: perhaps OE **luh** 'a pool' + ME **ker** 'marsh' is preferable. NbDu 137, DEPN, PNE ii.27, Jnl 1.31.

LUCKETT Corn SX 3873. Probably 'Leofa's cottage(s)'. *Lovecott* 1557, *Lucot* 1813. OE pers.n. *Leofa* + **cot**, pl. **cotu**. However, the evidence is too late for certainty. PNCo 112, Gover n.d. 209.

LUCKINGTON Wilts ST 8384. 'The estate called after Lucca'. *Lochintone* 1086, *Lokinton(e) -yn-* 1201–1341, *Lokyngton* 1296–1360, *Luckington* from 1598. OE pers.n. **Lucca* + **ing**[4], + **tūn**. The same pers.n. occurs in Luckley Farm ST 8285 in the same township, *Lokelee* 1332, *Luckley* 1648, 'Lucca's wood or clearing'. Wlt 77.

LUCKWELL BRIDGE Somer SS 9038. *Luckwell Br.* 1809 OS. Earlier forms are needed to know whether Luckwell means 'lucky well' or perhaps 'Luca's well or spring'.

LUCTON H&W SO 4364. 'Settlement on the river Lugg'. *Lugton* 1185, 1193, *Lucton* 1255. R.n. LUGG SO 5251 + OE **tūn**. DEPN, RN 268[†].

River LUD Lincs TF 3387. 'The loud one'. *Ludena* [c.1163, late 12th]13th, *Ludhena, Ludeney* [12th]1314, *Lude* c.1540, *Lud fluuiolus* 1586, *River Ludd* 1824 OS. OE r.n. **Hlūde* < **hlūd** 'loud', genitive sing. **Hlūdan*, + **ēa** 'a stream'. RN 261, Cameron 1998.

LUDBOROUGH Lincs TF 2995. Possibly the 'fortified place of the Louth district'. *Ludeburg* 1086, 1253, *-burc* c.1115, [1154×89]1409, *Ludburc -burg(h)(')* 1177–1610, *Lo(u)theburgh* 1297–1374. P.n. LOUTH TF 3387. Louth is by origin the r.n. LUD TF 3387; Ludborough stands on an affluent of the Lud; either the whole district drained by the river is referred to or the stream at Ludborough had the same name as the r. Lud. Li iv.25, RN 261.

LUDDENDEN WYorks SE 0426. 'The *Ludding* valley'. *Luddingden(e)* 1284–1645, *Ludingden(e)* 1296, 1297, *Loudingedene* 1462, *Luddinden* 1563, *Luddenden* 1805. R.n. **Ludding*, OE **Hlūding* 'the loud stream', OE **hlūd** + **ing**[2], + **denu**. YW iii.132.

LUDDENHAM COURT Kent TQ 9963. P.n. Luddenham + ModE **court**. Luddenham, *Dodeham* (sic) 1086, *Ludeham* 1211×2, *Lo- Ludenham* 1242×3, 1253×4 etc., lies at the edge of Luddenham Marsh and is probably, therefore, on topographical grounds, 'Luda's river-meadow or dry ground', OE pers.n. *Luda*, genitive sing. *Ludan*, + **hamm** 3 or 4. KPN 245, ASE 2.37.

LUDDESDOWN Kent TQ 6766. 'Hlud's hill'. *Hludesduna* c.975 B 1321, *Hludes dune* c. 975 B 1322, c.1100, *Ledesdvne* 1086, *Ludesdon'* 1185×6, 1203, 1229 etc. OE pers.n. **Hlūd*, genitive sing. **Hlūdes*, + **dūn**. An early alternative form of the name seems to have been *hludes beorh* [939]13th S 447, 'Hlud's hill or barrow'. PNK 101, KPN 241, 244, L 146.

LUDDINGTON Humbs SE 8216. The 'estate called after Luda'. *Ludintone* 1086, *Ludingeton* 1229. OE pers.n. *Luda* + **ing**[4] + **tūn**. Luddington lies on the old course of the r. Don but the very flat topography does not support derivation from OE r.n. **Hlūde* or **Hlūding* 'the loud one', as suggested by Årsskrift 1974.30. DEPN.

LUDFORD Lincs TF 1989. 'The ford leading to Louth'. *Lude(s)forde* 1086, *Ludesfort, Ludeforda* c.1115, *Ludeford(e)* c.1115–1271. P.n. LOUTH TF 3387 + OE **ford**. This is Ekwall's explanation, since Ludford lies on the r. Bain. This, however, is a Scandinavian name and the river may originally have been called OE *Hlūde* 'the loud one' like the Lud. Alernatively 'Luda's ford', OE pers.n. *Luda*. RN 261, DEPN, Cameron 1998.

LUDGERSHALL Bucks SP 6617. 'Nook of the trapping-spear'. *(æt) Lutegaresheale* 1015 S 1503, *-hale* 1164–1331, *Lotegarser* 1086, *Ludegarshall* 1241, *Lurgosall* 1536, *Lurgysall* 1552, *Ludgarsell* 1766. OE ***lūtegār**, genitive sing. ***lūtegāres**, + **healh**. The reference is to a spear set as a trap for impaling wild animals in a nook or indentation of the 250ft. contour. It is not certain that the first spelling belongs here rather than under LUDGERSHALL Wilts SU 2650 or LURGASHALL Sussex SU 9326. Bk 104 gives pr [lə:gəsəl], Tengstrand 222ff, StN 14.92ff, Signposts 171, L 109, ECTV no.179, Jnl 2.23.

LUDGERSHALL Wilts SU 2650. 'The nook of land with a spear-trap'. *Litlegarsele* 1086, *Lu- Lotegareshal(e)* 1166–1318, *Ludgarshale* 1422, *Luggershaull* c.1540, *Lurgshall* 1675. OE ***lūtegār** 'a trapping spear, a spear set as a trap for impaling wild animals', genitive sing. ***lūtegāres**, + **healh**, dative sing. **heale**. Wlt 367 gives pr [lʌgərʃ ɔ:l], PNE ii. 28f, Tengstrand 222ff.

LUDGVAN Corn SW 5033. Probably 'place of ashes'. *Luduhā* 1086, *Lvdhā* 1086 Exon, *Luduhanum* 1087×91, c.1150, *Luduan -wan* c.1170, 1201, 1602, *Lud(e)won -uon -von* c.1180–1420, *Ludgwon* 1291, *Lusuon* 1366, *Lugevan* 1438, *Lusewan* 1454,

[†]The DB form *Lvchetvne* 1086 does not belong here.

Ludgion 1610, *Lydgan* 1750. Co **lusew** (< **ludw*, cf. Breton *ludu*) + p.n. forming suffix ***-an**. PNCo 112 gives pr [lʌdʒən] and [lidʒən], Gover n.d. 642–3, DEPN.

LUDHAM Norf TG 3818. Either 'Luda's homestead' or 'homestead on the Hlude, the loud stream'. *Ludham* from 1086 including [1022×22]13th S 984 and *[1044×7]13th S 1055. OE pers.n. *Luda* or stream name *Hlūde* 'the loud one' + **hām** or **hamm** 'a meadow'. If the specific is a stream name the reference is to Womack Water. Nf ii.114, Forsberg 166.

LUDLOW Shrops SO 5175. 'The tumulus by the loud one'. *Ludelaue* 1138–1347 with variants *-lawa* and *-lawe*, *Lodelawe* 1242–1347, *-loue -low(e)* 1272–1438, *Ludlowe* 1372, *Ludlo* c.1540. R.n. **Hlūde* 'the loud one' < OE *hlūd* referring to the river Teme + **hlāw**. A large tumulus was demolished in 1199 when the parish church was enlarged. Sa i.186.

LUDWELL Wilts ST 9122. 'The loud spring'. *Luddewell'* 1100×35, *Ludewell(e)* 1195–1450, *Ludwell* from 1498. OE **hlūd**, definite form **hlūda**, + **wella**. Wlt 189.

LUDWORTH Durham NZ 3641. 'Luda's enclosure' or 'enclosure by the *Hlūde*'. *Ludew(o)rth(e)* 1156×62–1428, *Lode-* c.1260×90–1321, *Ludwrthe* 1215×40, *-worth(e)* from c.1261, *Ludde-* 1260–1360. OE pers.n. *Lūda* or stream name **Hlūde* + **worth**. Both names derive from the OE adjective *hlūd* 'loud, noisy'. NbDu 137.

North LUFFENHAM Leic SK 9303. OE prefix **north**, *Nor-* 1179–1250, *Nord-* 1185–8, *Nort* 1210–1302, *North(e)* from 1197, + p.n. *Luffenham* from 1086, *Luffe-* 1105×7–1206, *Luf(f)ham* 1179–1250, with variant *Loffen-* 1205–1342, 'Luffa's homestead', OE pers.n. *Luffa*, genitive sing. *Luffan*, + **hām**. 'North' for distinction from South LUFFENHAM SK 9401. An Anglo-Saxon cemetery here has produced late military metalwork of the *foederati* and early pottery. R 256, Jnl 5.30.

South LUFFENHAM Leic SK 9401. ME prefix **sūth**, *Sut-* 1210–20, *Sud-* 1219–50, *Suth-* 1209×19–1480, *South-* from 1356 + p.n. Luffenham as in North LUFFENHAM SK 9303. R 267.

LUFFINCOTT Devon SX 3394. 'Cottage(s) called or at *Luhhiṅg*, or called after Luhha'. *Lughyngecot'* 1242, *Loghingecote -ynge* 1285, 1303, *Logghyngcote* 1350, *Luffingecote* 1275, *Luffyncote* 1428. OE p.n. **Luhhing* 'place called after Luhha', OE pers.n. *Luhha* + **ing**[2], or OE pers.n. *Luhha* + **ing**[4] + **cot**, pl. **cotu**. Also known as *Loghynton* 1346. D 152.

River LUGG H&W SO 5251. 'Bright river'.

I. English forms: *(neah þære éá) Lucge* c.1025 Saints–[before 1118]12th, *Lugge* 1219–17th.

II. Welsh forms: *Lhyguvy* 1572, *Lugw(y)* 1831.

A W r.n. borrowed into OE as **Lugge* from the IE base **lewk-* 'gleam, light' seen in W *llug* 'radiance, light', Gk *λευκός* 'brilliant, white'. RN 268, He 10, GPC 2221.

LUGUVALIUM Cumbr NY 4055. The Romano-British name of CARLISLE. *Luguvallo -valio* [4th]6th AI, *Lugubalium* [late 7th]14th Rav, 731 BHE, *Lugubalia, id est Luel, nunc dicitur Carleil* 'Lugubalia, that is, Luel, now called Carlisle' [1129]12th. The name means 'belonging to *Luguvalos', a pers.n. meaning 'strong as Lugus': Lugus was a Celtic divinity whose name occurs again in the name type *Lug(u)dunum* from which are descended Leiden, Holland, and Lyon, France. Cu 40–42, RBrit 402, O'Donnell 80.

LUGWARDINE H&W SO 5541. 'Enclosure by the river Lugg'. *Lagordin* 1067×71, *Lvcvordine* 1086, *Lugwardin -yn(e)* 1143×8–1412. R.n. LUGG SO 5251 + OE **worthiġn**. He 137.

LULHAM H&W SO 4041. 'Lulla's river-bend land or homestead'. *Lvllehā* 1086, *Lulham* from 1318. OE pers.n. *Lulla* + **hamm** or **hām**. Bannister 124.

LULLINGSTONE CASTLE Kent TQ 5364. *Lullingstone Castle* 1819 OS. P.n. Lullingstone + ModE **castle**. A mainly 18th century house incorporating earlier work. Lullingstone is *Lolingeston(e)* 1086, *Lolinge(s)tune, Lullingestuna -stana* c.1100, *Lullingestan* 1200, *Lu- Lollingeston(e) -stan(e) -ynges-* 1208–1380. The variation between *-tūn -ton* and *-stan -stone* indicates that there were felt to be two variant forms of the name throughout the Middle ages, the 'settlement or estate of Lulling' and the 'stone of Lulling', OE **tūn** and **stān** respectively. *Lulling* is either an OE pers.n. or an alternative name of the River Darent at this place, 'the prattling stream', from the root of ME *lullen* 'soothe with sound', MDu *lollen* 'mutter', Sw *lulla*, Da *lulle* 'hum'. PNK 46.

LULLINGTON Derby SK 2513. 'Estate called after Lulla'. *Lullitune* 1086, *Lullington(e) -yng-* 1245–1304 ec., *-in- -yn-* 1291–1386, *Loll-* 1269–1552. OE pers.n. *Lulla* + **ing**[4] + **tūn**. Db 640.

LULLINGTON Somer ST 7851. 'Estate called after Lulla'. *Loligtone* 1086, *Lullyngton* 1272. OE pers.n. *Lulla* + **ing**[4] + **tūn**. DEPN.

LULSGATE BOTTOM Avon ST 5065. P.n. Lulsgate + ModE **bottom**. Lulsgate, *Lollesyate* 1281, *Lollusyat'* 1327 LS (p), is 'Lull's gate', OE pers.n. *Lull*, genitive sing. *Lulles*, + **ġeat**. The reference may be to the col or 'gate' between Downside ST 4965 and Felton ST 5165, to the N and S of which the land rises to 663ft. and 641ft. respectively.

LULSLEY H&W SO 7455. 'Lull's island'. *Lolleseie, Lulleseia* 12th, *Lolles(s)eye* 1316–1656, *Lulsey* 1535–1763, *Lulsley* 1832 OS. OE pers.n. *Lull*, genitive sing. *Lulles*, + **ēġ** probably in the sense 'hill jutting into flat land' beside the river Teme. Wo 59, L 38.

LULWORTH CAMP Dorset SY 8381. A modern military camp, home of the R.A.C. Gunnery School, begun in 1940. P.n. Lulworth as in East LULWORTH SY 8682 + ModE **camp**. Newman-Pevsner 446.

LULWORTH COVE Dorset SY 8279. *Lulworth Cove* 1774. Earlier referred to as *the Creke of Lulworthe* 1539 and *Lulworth hauen*. P.n. Lulworth as in East LULWORTH SY 8682 + ModE **cove**. Do i.133.

East LULWORTH Dorset SY 8682. *Estlolleworth(e)* 1268–1439, *-lulle-* 1285–1531. Earlier simply *Lulvorde, Loloworde* 1086, *Lu- Lolleworth(e)* 1199–1439, *Lulleswrða -e -wrde* 1202–5, 'Lulla's enclosure', OE pers.n. *Lulla* + **worth**. 'East' for distinction from West LULWORTH SY 8280. Do i.122 gives prs ['lʌlwə:θ, 'lʌləθ, 'lɔlwə:θ] and ['lauwə:θ].

West LULWORTH Dorset SY 8280. *Westlullewrth* 1258, *-worth(e)* 1268–1442, *West Lel(l)is, Lilles Worth* 1288, *Westlulworth* from 1442. ME adj. **west** + p.n. Lulworth as in East LULWORTH SY 8682. Do i.129.

LUMB WYorks SE 0321. 'The pool'. *Lom* 1307, 1309, *(the) Lum(me)* 14th cent., *The Loome* 1569, *Lower, Upper Lumbe* 1709. OE ***lumm**, referring to a pool or mill-dam in Lumb Brook. YW iii.148.

LUMBY NYorks SE 4830. 'Farmstead or village at the wood'. *Lundby* [963]14th S 712, c.1030 YCh 7, *Lumbi -by(e)* from 1200. ON **lundr** + **bȳ**. YW iv.54, SSNY 32.

Great LUMLEY Durham NZ 2949. *Magna Lomley* c.1200–1635, *Great Lomley* 1633. ModE **great**, Latin **magna**, + p.n. *Lummalea* [c.1040]12th, *Lummelei -ley(e)* c.1190–1316, *Lumelei(a) -ley(e)* c.1200–1406, *Lome-* c.1200–1399×1400, *Lumley* from 1277, 'wood or clearing of or by the pools', OE ***lumm**, genitive pl. ***lumma**, + **lēah**. OE **lumm* survives as modern dial. *lum* 'a deep pool in the bed of a river'. The reference is most likely to deep places in the river Wear where fish could be trapped. 'Great' for distinction from Little Lumley, i.e. Lumley Castle, NZ 2851, *Lumley Parva* 1326–1609. NbDu 137, Forsberg 1950.174.

LUMLEY MOOR RESERVOIR NYorks SE 2270. No early forms. P.n. Lumley as in Lumley Moor SE 2271 and Lumley Farm SE 2270 possibly identical with Lumley in Great LUMLEY Durham NZ 2949 'wood or clearing of the pools', OE ***lumm**, genitive pl. ***lumma**, + **lēah**, + ModE **moor**, + **reservoir**.

LUND ON **lundr** 'a grove'. See also LOUND.

(1) ~ Humbs SE 9748. *Lont* 1086, *Lond* 13th, *Lund(e)*

1177×9–1583, *Lownde (upon the Wo(u)lde)* 16th cent. There was *silva minuta* 'underwood' here in DB. YE 163 gives pr [lund], SSNY 100.

(2) ~ NYorks SE 6532. *Lund(e)* from [1066–9]14th, *Lont* 1086, *Lownde* 1488. This name is not in SSNY, which on p.100 wrongly identifies Lund on the Wolds in Banton Beacon division of Harthill Wapentake (YE 163) with this Lund in Ouse and Derwent Wapentake. YE 259.

LUNDY Devon SS 1345. 'Puffin island'. *Lundey, til Lundeyar* [1139×48]14th, *Lundeia* 1199, *Lundeth* (for *-ey*) c.1250, *Lunday -ey* 1281–1353. ON **lundi** + **ey**. Puffins are still frequent visitors to the island. D 19.

LUNE FOREST Durham NY 8323. *Lune forest* 1576 Saxton. R.n. LUNE NY 8820 + ME **forest**.

LUNE MOOR Durham NY 8923. *Lune Moor* 1863 OS. R.n. LUNE NY 8820 + ModE **moor**.

River LUNE Possibly the 'health-giving river'. The forms point to an OE **Lōn* (with later Northern raising of ō to ü, Jordan §54) which would have been borrowed from a PrW **lōn* < British **lān-* (cf. Don < **Dānum*) identical with ModW *llawn* 'full', *llanw* '(flood-)tide, flowing, flood'. This has been compared with the Norwegian r.n. *Fulla* 'full stream' and Irish Slane, Meath, '(homestead of) fullness'. Alternatively a second Brit **lān-* has been posited, from the same root as Gaelic, Irish, Old Irish *slán* 'healthy, whole', cognate with Latin *salvus*. Although such a root is unknown in P-Celtic, in which it could have suffered early loss through homophonic clash with the aforementioned **lān-*, it is appropriate to river naming in the sense 'cleansing, health-giving', as probably in the Irish names River Slaney, Toberslane ('health-giving well'). RN 270–1, IPN 254, GPC 2112.

(1) ~ Durham NY 8820. *Loon* 1201, *Lune* from 1561. Beside the Lune is Lonton NY 9524, pronounced [luntən], *Lontun(e) -ton(e)* 1086–1301, *Lunton(e)* [1184]15th, 1577, 'settlement, village by the r. Lune'. YN4.

(2) ~ Lancs SD 6075. *Loin* [1156×60]14th, *Lon* [1150×70]15th, [1170×84]1268–1292, *Loon* [1186×90]1268, [1342]1412, *Loune* 1344, *Lune* from c.1540. La 167, RN 270, Jnl 17.97.

LUNSFORD'S CROSS ESusx TQ 7210. *Lunsford's Cross* 1547, *Luntsford Cross* 1813 OS. Surname *Lundsford* + ModE **cross**. Lundsford near Etchingham is *Lundresford* 1176, 'Lundrǣd's ford'. A John de Lunnesford is recorded in nearby Ninfield in 1340. Sx 489, 514.

LUNT Mers SD 3401. 'The wood or grove'. *Lund* [1251]1268, c.1275, *(del) Lunt* 1344. ON **lundr**. La 118, SSNNW 146, L 207.

LUNTLEY H&W SO 3955. 'Lunta's wood or clearing'. *Lvtelei* (for *Lunte-*) 1086, *Luntelee* 1216×68, *-ley(e)* 1361–92, *Luntlenlena, Lutlenene* 1158×64, *Luntlena* c.1170. OE pers.n. **Lunta* + **lēah** varying with the district name Leen as in the LEEN SO 3859. He 70.

LUPPITT Devon ST 1606. 'Lufa's hollow or pit'. *Lovapit* 1086, *Luuepuet* 1175, *Loveputte -pytte -pitte -u-* 1267–1377, *Loweputte* 1257, *Loveputt nowe called Luppitt* c.1630, *Lippitt* 1767. OE pers.n. *Lufa* + **pytt**. This might refer to the deep valley below the village or to a pit for trapping animals. D 641.

LUPTON Cumbr SD 6205. 'Hluppa's farm or village'. *Lupetun -ton* 1086, 1202, *Luppe- Loppeton(a)* 1189–99, *Lupton* from 1190. OE pers.n. **Hluppa* + **tūn**. We i.46.

LUPTON BECK Cumbr SD 5580. *Lupton bek* 1292. Earlier *rivulum de Lupton* 1190×5. P.n. LUPTON SD 5581 + ON **bekkr**. We i.47.

LURGASHALL WSusx SU 9327. 'The nook of the trapping spear'. *Lutesgareshal(e)* 1135×54, *Lu- Lotegareshal(e) -hall* 1224–1331, *Ludgareshale* 1471, *Lurga(r)sale* 1529, 1535, *Lurshall* 1604. OE ***lūtegār**, genitive sing. ***lūtegāres**, + **healh**. Sx 111 gives prs [lə:gəsəl] and [lə:gə 'seil] as in the local rhyme

Bake good bread, brew good ale, Say the bells of Lurgasale.
Sx 111, PNE ii.28.

LUSBY Lincs TF 3367, 'Lutr's village or farm'. *Luzebi* 1086, *Luceby* [1115]14th, 1125, [1147]13th, *-bi* [1125]1269, *Lusceby* [c.1150]1269. ON pers.n. *Lútr*, genitive sing. *Lúts*, + **bȳ**. SSNEM 58, Cameron 1998.

LUSTLEIGH Devon SX 7881. Partly uncertain. *Leuestelegh'* 1242, 1327, *Luuesteleg(h)* 1276, 1277, *Lustelegh(e)* 1276, 1282, *Lisleigh* 1672, *Listleigh al. Lustleigh* 1712, *Lisley* 1749. Either OE pers.n. **Leofġiest* or ME surname *Luvesta* 'dearest one' + **legh** (OE *lēah*) 'wood or clearing'. D 480.

LUSTON H&W SO 4863. 'Lussa's estate'. *Lustone* 1086, *-tun(')* *-ton* 1123–1291, *Lussetone* c.1285. OE pers.n. **Lussa* + **tūn**. He 138.

LUTON Beds TL 0921. 'Settlement, estate on the river Lea'. *Lygeton* [795 for 792]13th S 138, *Loitone* 1086, *Luitun -ton* 1156–1415, *Luton* from 1195. R.n. LEA TL 2609 + OE **tūn**. Bd 156.

LUTON Devon SX 9076. Partly uncertain. *Leueton* 1238, *Lynynton* (for *Lyuyn-*) 1292, *Luneneton* (for *Luuene-*) 1303, *Lunaton* (for *Luua-*) 1346, *Luton al. Louton al. Loveton al. Levaton* 1811. Probably 'Leofa's estate', OE pers.n. *Lēofa*, genitive sing. *Lēofan*, or *Lēofa* + **ing**[4] + **tūn**. D 488 suggests the OE feminine pers.n. *Lēofwynn* or masculine *Lēofwine* without genitival *-s*.

LUTON Kent TQ 7666. 'Leofa's estate'. *Leueton(e)* 1240–85, *Lyeueton'* 1313, *Luton* c.1600. OE pers.n. *Lēofa* + **tūn**.

LUTON HOO Beds TL 1018. *Luton How* 1782, a mansion built by Robert Adam for the earl of Bute and subsequently much altered. Earlier *Hoo in Luton* 1480 and simply *le Hoo* 1276, OFr definite article *le* + OE **hōh**, dative sing. **hō**, 'the hill-spur'. Bd 159.

LUTON INTERNATIONAL AIRPORT Beds TL 1220.

LUTTERWORTH Leic SP 5484. 'Enclosure by the river *Hlutre*'. *Lutresurde* 1086, *Lut(t)reworth(e)* 1202–1318, *Luterworth(e)* 1203–1524, *Lutterworth(e)* from 1204. R.n. **Hlūtre* 'the bright one' < OE **hlūt(t)or** 'clear, bright' + **worth**. *Hlutre* must have been a name of the river Swift on which Lutterworth stands. Lei 456.

LUTTON Devon SX 5959. 'Lutta's estate'. *Luttat(h)on* 1242, 1216×72, *Ludeton* 1242, *Lutteton* 1216×72. OE pers.n. **Lutta* + **tūn**. The 1242 form points to 'estate on the r. *Hlude*, the loud stream' from OE *hlūd*, but this appears to be an isolated spelling. It would have to refer to the river Piall, an unexplained r.n. *Pial(l)* 13th. D 270.

LUTTON Lincs TF 4325. 'Settlement by the pool or inlet'. *Lvctone* 1086, *Lochtona* c.1175, *Lutton* from 1177, *Lughton* 1303. OE **luh** + **tūn**. The reference is to a former inlet represented by Lutton Leam, dial. **leam** 'a water-course or drain', also called Lutton Gowts, *Lutton Gote* 1509×47, ME **gote** 'a water-course', or *Lutton Eye* 1508×57, OE **ēa**. Payling 1940.34, Wheeler 129.

LUTTON Northants TL 1187. 'The farm or village on the stream called *Luding*'. *(æt) Lundingtune* [963×92]c.1200, *Ludi- Lidin- Lidentone* 1086, *Lud- Lodyn(g)ton(e) -in(g)-* [1052×65]13th K 904–1430, *Loudington* 1329, *Lutton* 1428. OE r.n. **Hlūding* 'the loud stream' < OE **hlūd** 'loud' + **ing**[2], + **tūn**. Lutton overlooks a stream now called Billing Brook. On formal grounds this could also be 'the estate called after Luda', OE pers.n. *Luda* + **ing**[4] + **tūn**. Nth 204, Forsberg 167, Årsskrift 1974.30.

East and West LUTTON NYorks SE 92–9369. *Estlutton* 1234–1533, *Westlutton* from 1285. ME **est, west** + p.n. *Ludton* 1086, *duabus Luttunis* 'the two Ls' [1108×14]c.1300, *Luttons two* 1650, *Luttons-Ambo* 1828, *Lutton(a)* 1166 etc., 'Luda's settlement', OE pers.n. *Luda* + **tūn**. *Ambo* is Latin 'both'. YE 122.

LUTWORTHY Devon SS 7616. Possibly 'Lutta's enclosure'. *Loteworthy* 1281. OE pers.n. **Lutta* + **worthiġ**. D 401.

LUXBOROUGH Somer SS 9737. 'Lulluc's fortified place or hill'. *Lolochesberie* 1086, *Lollokesbourgh* c.1240, *Lochberg* [1142×66]Bruton, *Loches- Lokesberg(e)* [1139×61–1192×6] Bruton, 1150×61, *Loge- Lokeberge -bergh* [late 12th, 1204×13]Bruton, *Lusebrege* [n.d.] Buck, *Luxborough* 1610. OE pers.n. *Lulluc*, genitive sing. *Lulluces*, + **byriġ**, dative sing. of **burh**, varying with **beorg**. DEPN.

LUXHAY RESERVOIR Somer ST 2017. P.n. Luxhay + ModE **reservoir**. There are no early forms for Luxhay. It may be a formation in OE **(ġe)hæġ** 'a fence, an enclosure' with an OE pers.n. as specific (cf. LUXBOROUGH SS 9737) but more evidence is needed for this name.

LUXULYAN Corn SX 0558. 'Chapel of Sulian'. (Chapel of) *Luxulian* 1282, *Lussulian* 1319 Gover n.d., *Luxulien* 1333 Gover n.d., *Luxilyan* 1343, *Luxulyan* 1356 Gover n.d. Co ***lok** + saint's name *Sulian*, as in the Breton saint Sulien. The church, however, is dedicated to SS Ciricius and Julitta. PNCo 112 gives pr [lʌk'siliən], Gover n.d. 407.

Upper LYDBROOK Glos SO 6015. ModE **upper** + p.n. *Lu(i)debroc* 1224, *Ludebrok(e)* c.1275–1306, *Ludbro(o)k(e)* 1436–1642, *Li- Lydbrooke* 1616, 1641, 'stream called *Hlyde* or the loud one', OE **hlȳde** + **broc**. Upper and Lower Lydbrook are 19th century developments in the stream-valley distinguished on the 1831 OS map as *Lidbrook* and *Lydbrook* respectively. Gl iii.218.

LYDBURY NORTH Shrops SO 3586. *Northlidebiry* 1254, *Lidebur' Nor(h)t* 1255, *Lydburg North* 1641. P.n. *Lideberie* 1086, *-bir' -bery -bury* c.1150×60, 1179×80, *Ledebir' -bur' -byr -bury* 1208–1307, *Liddebury* 1271×2, *Ludbury* 1747, 'the fortified place or manor house by *Hlȳde*, the loud stream', OE r.n. **Hlȳde* < **hlūd** 'loud', + **burh**, dative sing. **byriġ**, + pseudo-manorial adj. **north**. The reference is to the r. Kemp. 'North' for distinction from LEDBURY SO 7137, a different name but with similar ME spellings. Sa i.187.

LYDCOTT Devon SS 6936. '(The) little cottage'. *Litlecote* 1261, *Litecote* 1330, ME **litel** + **cote** (OE *lȳtel*, definite form neuter *lȳtle*, + *cot*). D 58.

LYDD Kent TR 0421. Possibly '(the settlement) at the gates'. *(ad) Hlidum* [774]10th S 111, endorsed *To Hlidum, (on) hlidum* and *Lyden in marisco de Romenal, Hlide* 11th., *Lyde, Lide(s)* 13th., *Lydde* 1610. OE **hlid**, dative pl. **hlidum**. The usual sense of OE *hlid* is 'lid, covering', but there is some evidence that it was also used of the closure of an aperture in a wall or fence, cf. *on Lullan hlyd on ða hegestowe* B iii. 213, 8, the compound *ceasterhlid* 'city gate' and ON *hlith* 'gateway, gap, door', ultimately < IE **kel-* 'something woven, a fence'. The same difficulty occurs also in LYDHAM Shrops SO 3391. KPN 55, DEPN.

LYDD AIRPORT Kent TR 0621. P.n. LYDD TR 1421 + ModE **airport**.

LYDD-ON-SEA Kent TR 0820. A modern development named after LYDD TR 0421.

River LYDDEN Dorset ST 7208. 'Broad stream'. *Lidenne* 1244, *Ludene* 1288, *Lydden* 1575.[†] OE r.n. **Lydene* < PrW **Lidan*, Brit **Litano-* 'broad' cognate with Gael *leathan* 'broad', OIr *lethan*, OW *litan*, W *llydan* and *Litana*, the name of a forest in Roman Gaul. Do 102, RN 242, LHEB 286, 554, 673, GPC 2251.

LYDDEN Kent TR 2645. 'Sheltered valley'. *Hleodaena* c.1090, *Liedenne* 1176, *Leden(n)e* 1205×6–1261×2, *Liden, Lyden(e)* 13th., *Lyeden(e)* 1304, 1313. OE **hlēo** + **denu**. A long narrow valley heads SW across the parish. PNK 452, DEPN, Cullen.

LYDDINGTON Leic SP 8797. 'Settlement on the *Hlyding*'. *Lidentone* 1086, *Ly- Lidinton(e) -yn-* 1167–1354, *-ing- -yng-* 1190–1547, *Ly- Liddington(') -yng-* from 1444. OE r.n. **Hlȳding* 'the loud stream' < OE **hlūd, hlȳde** + **ing**[2], + **tūn**. A small tributary of the r. Welland drops here some 200ft. in two miles. R 274 prefers derivation from OE pers.n. *Hlyda* + **ing**[4] + **tūn**, 'the estate called after Hlyda'.

LYDE ARUNDEL H&W SO 4943 → PIPE AND LYDE SO 5044.

Lower LYDE H&W SO 4944 → PIPE AND LYDE SO 5044.

Upper LYDE H&W SO 5144 → PIPE AND LYDE SO 5044.

LYDE SAUCEY H&W SO 5143 → PIPE AND LYDE SO 5044.

Bishop's LYDEARD Somer ST 1629. 'L held by the bishop' sc. of Wells. *Lydiard Episcopi* 1291, *Bishops Lediard* 1610. ModE **bishop**, Latin **episcopus**, genitive sing. **episcopi**, + p.n. Lydeard as in LYDEARD ST LAWRENCE ST 1232. DEPN.

LYDEARD ST LAWRENCE Somer ST 1232. 'St Lawrence's L'. *Lydiard Sancti Laurencii* 1291, *Lawrencliddeard* 1610. P.n. Lydeard + saint's name *Lawrence* from the dedication of the church. Lydeard, *Lid(e)geard* *[854]12th S 311, [899×909]16th S 380, *(of) Lidigerde* [1066×86]12th ASCharters, *Lidegeard* *[1065]18th S 1042, *Lidegar, Lediart, Lidiard* 1086, *Lidiard, Lideyard* [c.1250] Buck, is the 'grey ridge', PrW ***lēd** + ***garth**, identical with Liggars Corn SX 2622, W Llwydiarth Clwyd SJ 2237, Powys SN 9980, SH 7710, SJ 0616, Gwyn SH 4285 and Llwydarth MGlam SS 8590 and Dyfed SN 0923, *litgarth* 12th. Three Lydeards are distinguished in DB, Lydeard St Lawrence, Bishop's LYDEARD and East Lydeard ST 1729. The OE forms have been Anglicised through association of the generic with OE **ġeard** 'a fence, an enclosure'. DEPN, LHEB 332, CPNE 102, 153, Pembs 414.

LYDFORD Devon SX 5185. 'Ford over the river Lyd, the loud one'. *Hlydanforda* c.1000 ASC(C,D) under year 997, *Lydanford* 979×1016 coins, *Hlidaford* [1018]11th, c.1120 ASC(E) under year 997, *Lideforde -fort* 1086[†], *Lyde- Lideford* from 1232×9, *Ludford* 1379. R.n. Lyd, *(to) lidan* [914×9]16th, 12th BH1, 2, *(to) hlydan* [914×9]13th BH3, *Lide* 1249, *Lyd* 1575, 'the noisy stream', OE **Hlȳde* from **hlūd** 'loud', + **ford**. D 191, 9, RN 272, BH 110.

LYDFORD Somer ST 5631. 'Ford across the *Leden*'. *Ledenford* [744]14th S 1410, *Lideford* 1086, 1194, *Ludeford(e)* 1194–5, 1312, *Lydford* 1610. Lost OE r.n. **Leden* < Brit **litano-* 'broad' as in the r.ns. LEADON SO 7628, LODDON SU 7568 and LYDDEN ST 7208, + **ford**. *Leden* must have been an alternative name for the river Brue. The *Lude* forms show that it was subsequently taken to be the common OE r.n. *Hlȳde* 'noisy one' (< OE *hlūd*). DEPN, RN 273.

LYDGATE WYorks SD 9225. 'The swing-gate'. *Lydgate -yate* 1596. OE **hlid-ġeat**. YW iii.183.

LYDHAM Shrops SO 3391. Uncertain. *Lidum* 1086, *-am* 1271×2, *Lydom* 1290–1687, *Lideham* 1250–56, *Lydham* from 1258, *Ledom -on* 1447–1659, *Lyddum -om* 1577–1691, *Leedham* 1728. OE **hlid** 'a lid, a door, a window-shutter', dative pl. **hlidum**. The significance is unknown. Identical with LYDD Kent TR 0421. The vowel is normally short *i*, but open syllable lengthening to *ē* seems also to have occurred. Sa i.188.

LYDIARD MILLICENT Wilts SU 0986. 'L formerly held by Millisent'. *Lidgard Milisent* 1257, *Li- Lydiard -yard Milisent, Milicente* 1275, 1289. P.n. *(æt) Lidge(a)rd, Lidegæard* [900]12th S 1284, *Lediar, Lidiarde* 1086, 'the grey ridge', PrW ***lēd** + ***garth**, + manorial addition from Millisent, the mother of Richard who held the village in 1199, and for distinction from neighbouring Lydiard Tregoze SU 1084, *Lidiard Tregoz* 1196, held by Robert Traigoz in 1242, by John Tregos in 1275, a surname of French origin from Tregoz near St Lô, La Manche. Identical with LYDEARD ST LAWRENCE Somer ST 1232, Liggars Corn SX 2622 and Welsh Llwydiarth and Llwydarth, *Litgarth* c.1150. Wlt 35 gives pr [lidjəd], xxxix, CPNE 102 s.v. **garth*.

LYDIARD TREGOZE Wilts SU 1084 → LYDIARD MILLICENT Wilts SU 0986.

LYDIATE Mers SD 3704. 'Swing-gate'. *Leiate* 1086, *Lichet* 12th, *Lid(d)iate, Lideyate* [1190–c.1220]1268, *Liddigate* 1202, *Lydiate* from [c.1225]1268 with variant *-yate*. OE **hlid-ġeat**. La 120, Jnl 17.66.

LYDLINCH Dorset ST 7413. 'Ridge beside or bank of the r. Lydden'. *Litelinge* 1166, *Lidelinz -ling -lins* 1182–1279, *-linch*

[†]The OE forms *(on, of) Ludenhame* [956 for 953×5]14th S 570 probably do not belong here; the forms *(on, bi) Lideman, Lidenan* [968]14th S 764 are corrupt.

[†]A bad form *Tideford* occurs at *DB Devon* 34.3, a misreading of Exon's *lidefort* which has a smudge running from the top of the *l*.

-lynch(e) 1285–1486, *Ludelynch(')* 1303, 1406, 1462, *Lydlynch(e)* 1491–1575, *-linche* 1533. R.n. LYDDEN ST 7208 + OE **hlinċ** perhaps in the sense 'river-terrace used as a road, a terrace-way'. Do iii.346, L 164.

LYDNEY Glos SO 6303. 'Lida's hill-spur'. *Lideneg* [c.853]12th ECWM no 75, *on Lidanege* [972]10th S 786, *Lidenei* 1086, *Li- Lydenai -ay(e) -ei(a) -ey(e) -e(e)* 1167–1478, *Li- Lydney* 1473–1662. OE pers.n. *Lida* or **lida** 'a sailor' from which the pers.n. derives, genitive sing. *Lidan*, **lidan**, + **ēġ** 'island' in the sense 'hill-spur'. The reference is to the temple, fort and castle site in Lydney Park at SO 6102. Gl iii.257, L 38, Verey 1970V.294–5, Thomas 1976.130.

LYDNEY SAND Glos ST 6399. *Lydney Sand* 1830 OS. P.n. LYDNEY SO 6303 + ModE **sand**. A sand-bank in the Severn.

LYE WMids SO 9284. 'The clearing'. *The Lye* 1550, 1834 OS. Earlier forms occur in the surnames *de Lega* 1275, *atte Leye* 1327. OE **lēah**. One of a sequence of *lēah* names in this area, BRIERLEY SO 9287, HAGLEY H&W SO 9181, ROWLEY SO 9687, WORDSLEY SO 8987. Wo 310, L 203.

Lower LYE H&W SO 4066. *Leye Inferior* 1292, *Netherleye* 1303, 1305, *Nether Lighe* 1560. ModE adj. **lower**, ME **nether**, Latin **inferior**, + p.n. *Liya* [704×9]13th S 1801, *Lege* (4×), *Lecwe*, *Lega* 1086, *Leye* 1243–1527, 'the wood', OE **lēah**. 'Lower' for distinction from nearby Upper Lye SO 3965, *Overley(e)* 1263, *Ov'lee* 1535. He 30, Bannister 125.

LYE GREEN Bucks SP 9703. *Ley Green* 1689, *Leigh Green* 1822 OS. P.n. Lye + ModE **green**. Cf. nearby ASHLEY GREEN SP 9705. Lye, *Lye* 1550, *Leigh* 1766, is ultimately from OE **lēah** 'a wood or clearing'. Bk 213, Jnl 2.31.

LYFORD Oxon SU 3994. 'Flax ford'. *(to) Linforda* [944]12th S 494, *Linford* [1032]12th S 964, 1086, 13th., *Li- Lyford(')* from 1224. OE **līn** + **ford**. A ford used for transporting flax. The boundary of the Linford people is mentioned as *(oð) linfordinga gemere* [940]12th S 471. Brk 413.

LYMBRIDGE GREEN Kent TR 1243. 'Lymbridge common'. *Limbridge Green* 1819 OS. P.n. Lymbridge + ModE **green**. Lymbridge is *Lemering'* 1254, *Limering* 1257, *Lymeryng'* 14th, possibly representing OE pers.n. *Lēofmǣr* + **ing**^{2}, 'place called after Leofmær'. The modern form suggests that the OE name was in the locative–dative form **Lēofmǣringe* with assimilation of the suffix [ɪndʒ]. PNK 429.

LYME PARK Ches SJ 9682. *parcum de Lyme* 1466. P.n. *Lyme* from 1312, + ME **park**. Lyme is the ancient name for the uplands of the Lyme or the Pennine massif on the SE border of Lancashire, the E and SE borders of Cheshire, and the N borders of Staffordshire and Shropshire, *Lima* 1121–1297 (mostly referred to as a wood), *Lime* 1161–1724, *Lyme* from 1243, *Lyne* 1319–1534, 'the district of the Elm, the country where the elm-tree grows', Brit ***lēmono-**, ***lēmano-**, PrW **llïṽon -an*, **llēṽon -an*. Che i.198, 2–6.

LYME REGIS Dorset SY 3392. 'The king's Lyme'. *Lyme Regis* from 1285. P.n. Lyme + Latin **regis**, genitive sing. of **rex**. Lyme, *Lim* [774]12th S 263, *Lym* [938]14th S 442, [957]14th S 644, 1086, *Lime* 1086, is originally a Celtic r.n. occurring also in nearby UPLYME Devon SY 3293, from the root **līm-* seen in λειμών 'wet place', Latin *līmus* 'mud, slime' < IE **lei-* 'flow, water'. Do 102, RN 274, 1950.Forsberg 117.

LYMINGE Kent TR 1641. 'The Limen district'. *Lymingae* 697 S 19, 700 or 715 S 21, *Liminge* [689]13th S 12, 798 S 153, *Liminiae(a)e* 741 for ?750 S 24, *Limmincge* c.1000, *Leminges* 1086, *Lim(m)inges* 11th, *Lymynge* 1610. R.n. *Limen* as in LYMPNE TR 1235 + OE ***ġē**. *Limen* is the name of the old East Rother which originally flowed into the sea not at Rye but at Lympne before the construction of the Royal Military Canal which took over its course. Although Lyminge lies on an arm of the Stour and not of the *Limen* it preserves the name of the ancient *ġe* or district, later the lathe, of the *Limenware* 'the *Limen* dwellers', cf. *on limenweara wealde* 'in the common wood of the *Limenware*' [724]15th S 1180, *in limen wero wealdo* (Latin) [786]12th S 125, *Limowart Lest, Limwarlet, Linuuarlest* 1086. It extended along the coast from the Rother Levels to the bounds of Eastry, 'the eastern *ġe*'. The significance of the name Lyminge is probably that it lay near the edge of the *ġe* and was the first place of importance across the boundary of Eastry. ['lɪmɪndʒ]. KPN 25, 32, 61, RN 243–4, ING 227, Forsberg 1950.120–1, Origins 69–73, Charters IV.33.

LYMINGE FOREST Kent TR 1545. P.n. LYMINGE TR 1641 + ModE **forest**.

LYMINGTON Hants SZ 3295. 'Settlement on the r. Limen'. *Lentvne* 1086, *Lemynton(e), Limneton, Limenton(a)* 1100×35–1299, *Li- Lymin(g)ton* from 1185, *Lemeton* c1210. R.n. **Limene* + **tūn**. The name is identical with Leamington in LEAMINGTON HASTINGS and Royal LEAMINGTON SPA Warw SP 4467, 3265, which lie on the r. LEAM SP 4568, *(on) Limenan* [956]11th S 623. Lymington lies on the r. Boldre, *Bolre* 1199×1216, *aqua de* ~ 1253, 1298, *Balderwater* 1671, a name transferred from BOLDRE SZ 3198 which must have replaced *Limene* at an early stage. Ha 112, Gover 1958.210, DEPN.

LYMINGTON RIVER Hants SU 3102. A back-formation from LYMINGTON SZ 3295.

LYMINSTER WSusx TQ 0204. 'The monastery church associated with Lulla'. *Lullyngmynster* [873×88]11th S 1507, *Lolinminstre* 1086, *Lileministre, Lillemenstre* 1242, c.1270, *Limenistr(e) -minstere* 1201–42, *Lymmynstre* 1438, *Lymyster* 1511, 1584, 1641, *Limster* 1617. OE pers.n. *Lulla* + **ing**^{4} + **mynster**. Sx 169 gives pr [limstə].

LYMM GMan SJ 6887. 'The torrent'. *Lime* 1086, *Lyme* c.1200 etc., *Li- Lymme* 1194 etc., *Lymm* 1635. OE **hlimme** referring to the noise of Slitten Brook in its course through a ravine in the middle of the village. Che ii.36.

LYMORE Hants SZ 2993. Uncertain. *Leymore* 13th, *Lymoor* 1680. Uncertain element + OE, ME **mōr**. Gover 1958.228.

LYMPNE CASTLE Kent TR 1134. A late medieval manor house with decorative battlements. P.n. LYMPNE TR 1235 + ModE **castle**. Newman 1969W.378.

LYMPNE Kent TR 1235.

I. Latin forms: *Ad Portum Lemanis* [4th]8th AI, *Lemanis* [7th]13th Rav, *Lemanio* (possibly *Lemauio*) 4th Peutinger Table, *Lemmanis* [5th]10th ND, *(in) Limneo (portu)* c.1000.

II. English forms: *(of, on) Liminum* 805×10 S 1188, 811 S 1264, 812 S 169, *Limes* 1086, *Limene* 1291, *Lymen* 1396, *Limne* 1475, *Lymme* 1480.

Type II implies an OE plural **Liminas* from Latin *Lemanis*, type I the dative case of a plural name **Lemanae*. If this is so a probable sense is 'marshes, flooded areas' from the root **lim-* 'mud, marsh' seen in W *llif* 'stream, flood', Greek λίμνη 'stagnant water, lake'. *Limanis*, however, may have been a sing. i-stem formation wrongly taken as a plural by bilingual Anglo-Saxons. The settlement name, which originally referred to the Roman fort of Stutfall Castle TR 1134, is transferred from the r.n. *Limene*, the old name of the East Rother river, which used to flow to Lympne along the course of the Royal Military Canal constructed as a defence work during the Napoleonic wars in 1806. Early forms (selection only) are *Lemana* [7th]13th Rav, *Liminaea* 700 or 715 S 21, *Limenœ* [724]15th S 1180 (with 13th cent. variants *Liminae, Lymene*), *Liminœa* 741 for ?750 S 24, *-ea* 798 S 153, *-aee* 732 S 23, 773 for 833 S 270, *Limenaea* [c.848]13th S 1193, *Limene muþa* 893, 894, 896 ASC(A), *Li- Lymene* 13th cent., *Limene* c.1540, a British r.n. *Lemana* which also lies behind the r.ns. LEAM Northants, Lemon Devon SX 7972, 8071, *Lymenstream* 11th S 1547, Lymn Lincs TF 3868, *Lim(in)e* [12th]1331, 1276, *Leman* 16th cent., LEVEN NYorks and also Fife NO 2601 and Highland NN 2160, and *Lemannus Lacus* (Lake Geneva), *Limonum* (Poitiers, Vienne, France), Limagne (Puy de Dôme and Allier, France), and many other continental water names. These have mostly been referred to the root **lem-* **lim-* 'elm-tree' (W *llwyf*, Irish *leamh*, OIr *lem*, cognate with Latin *ulmus*) so that the r.n. *Lemana* 'elm-tree river, river in an elm-wood' would parallel British *Derventio* 'river in an oak-wood'. There seems no good reason, however, to exclude the root cited above from the explanation of at least some of these rivers, and it would admirably suit the topography of the present instance. The pr is [lim]. RN 244, RBrit 385–7 with references, esp. LHEB 282, 486, 630, 672–3 and Britannica 1.78, Charters IV.92, 162.

LYMPSHAM Somer ST 3354. Possibly 'Limpel's homestead'. *Lympelesham* 1189, *Linpelesham* 1225, *Limpelesham* 1254, *Limpsham* 1610. OE pers.n. **Limpel*, genitive sing. **Limpeles*, + **hām**. Alternatively the specific might be genitive sing. of p.n. such as **Līn-pōl* 'flax pool' or **Lind-pyll* 'lime-tree pool'. DEPN.

LYMPSTONE Devon SX 9984. Partly uncertain.
I. *Levestone* 1086, *-ton* 1249, 1285.
II. *Lemineston* 1219, *Ly- Limeneston(e)* 1254–1377.
III. *Leveneston* 1238–1392, *Luveneston(e)* 1275–1334.
IV. *Limestone* 1434, *Limpston* c.1630, *Lympstone al. Lymson* 1726.
Variation between *v* and *m* might point to a PrW origin with late Brit *v* < Brit *m* in the r.n. **Limen*, 'elm-tree river'. Alternatively types I and III can be explained as 'Leofwine's estate', OE pers.n. *Lēofwine*, genitive sing. *Lēofwines*, + **tūn**. Type II probably reflects OE *mn* from *fn* (as in *hremn* from *hræfn* 'raven'), and type IV is its normal development. D 591, RN 244–5.

LYNDHURST Hants SU 3008. 'Lime-tree hill'. *Linhest* (sic) 1086, *Li- Lyndeherst -hirst -hurst* 1165–1331, *Lindhurst* from 1196, *Lynhurst* 1382, *Lyndust* 1758. OE **lind** + **hyrst**. Ha 113, Gover 1958.207.

LYNDHURST ROAD STATION Hants SU 3310. A railway station on the road between Tatton and Lyndhurst SU 3008.

LYNDON Leic SK 9004. Either the 'lime-tree' or the 'flax hill'. *Ly- Lindon(e)* 1167–1553. OE **lind** or **līn** + **dūn**. R 189, L 156.

LYNE Surrey TQ 0166. 'The lime-tree'. *Lyne* 1624, cf. *la Linde* 1208(p), *la Lynde* 1272, 1274 and *Lyndehacche* 1446. OE **lind**. Sr 110, L 222.

Black LYNE Cumbr NY 5481. *Blakelevyn, Black Leven* 16th cent. A branch of the river LYNE Cumbr NY 4972. See also White LYNE NY 5473. Cu 21.

River LYNE Cumbr NY 4972. Uncertain. *Leuen* 1259–1671, *Levin, Livin, Leaven* 16th, *Leving or Line* 1703. The river is formed of the junction of two rivers, the Black Lyne, *Blakelevyn* early 16th, and the White Lyne, *Whit Leven* 1576. Cu 21, RN 251.

White LYNE Cumbr NY 5473. *Whit Leven* 1576, cf. John de *Wyteleuenwra* 1288 'White Lyne nook' (ON **vrá**). ME adj. **white** + river n. LYNE NY 4972. One of two branches of the river Lyne distinguished as Black and White. See Black LYNE NY 5481. Cu 21.

LYNEAL Shrops SJ 4433. 'The maple-tree nook'. *Luny- Lunehal* 1221, *Lunyal* 1280, *Lenyall* 1610, *Lineal* 1833 OS. OE **hlynn** + **halh**. Bowcock 141, DEPN, PNE i.254.

LYNEHAM Oxon SP 2720. 'Flax water-meadow or homestead'. *Lin(e)ham* 1086–1216 with variants *Lyn(e)- Leynham* to 1285. OE **līn** + **hamm** 1 or **hām**. Lyneham is situated on the N bank of the Evenlode in a *hamm*-type site. O 362.

LYNEHAM Wilts SU 0279. 'The dry-ground in the marsh or cultivated plot where flax is grown'. *Li- Lynham* 1224–1349, *Lyneham* 1571, *Linum* 1649. OE **līn** + **hamm**. Wlt 270.

LYNE HOUSE Surrey TQ 1938. *Lyne Ho.* 1816 OS. P.n. (wood of) *Lynde* 1289, *Lynd* 1535, *Lyne farm* 1603×25, 'the lime-tree', OE **lind**, + ModE **house**. Cf. William *atte Lynde* 1372. Sr 85.

LYNEMOUTH Northum NZ 2991. 'The mouth of the river Lyne'. *Lynemuwe* 1242 BF, *Linemue* 1296 SR, *Lynmuth* 1342. R.n. *Lina* [c.1040]12th HSC–14th, *Lyne* from 13th, + OE **mūtha**. Lyne is a Brit r.n., **Līnā* cognate with the r.ns. LYNHER Cornwall SX 3269, *Linar* 11th, LYNOR Devon ST 0210 , *Linor* [958]12th S 653, and the W r.n. Afon Llynor, Merionethshire ST 0538, from an IE root **lei-*/**li-* 'to pour, flow, drip' or **lei-*/**li-* 'slimy, slippery, to glide'. NbDu 138, RN 275, BzN 1957.220, D 9, Co 112.

LYNG Norf TG 0617. 'The terrace or terrace-way'. *Ling* 1086–1254, *Lins* 1157. OE **hlinċ**. DEPN, L 164.

LYNG Somer ST 3228. 'The length'. *(to) lengen* [914×9]13th BH1–5–16th, *Lenga* (rubric), *Relengen* [937]18th (for ? *æt thære lengen*) S 432, *Lege* (for *Lēge* i.e. *Lenge* or *Lengen*) 1086, *Leng(a)* c.1180–1225, *-e* [1295] Buck, *Lyng* 1610. This is usually explained as OE **hlenċ** 'a terrace' with voicing of [tʃ] to [dʒ]. But Lyng, properly East Lyng, lies at the E end of a low ridge of dry ground extending into the marshland of Curry and Stan Moors, probably called 'the extent, the length', OE ***lengen** < *lang* + *-en* noun suffix (Gmc *-innjo-*). DEPN, VCH vi.53, BH 114.

River LYNHER Corn SX 3663. A pre-English r.n. of uncertain origin. *(to) linar* 1018 S 951, *Liner* c.1125, *Lyner* 1284–1478, *Lynher* 1797. Cf. r. Lynor in Leonard Devon ST 0009, 0351 (now called Spratford Stream), *(to, andlang) linor* [958]12th S 653, 1408, *Li- Lynor* from [c.1160]14th. (Leonard ST 0210 is *Limor, Lannor* 1086, *Linor, Lannar* 1086 Exon, *Ly- Linor* 1196–1330). A possible source is the IE root **ley-* 'flow' as in W *lli* 'stream, flood, sea'. PNCo 112 gives prs [lainə], formerly [linə], Gover n.d. 682, RN 275, D 9, 549, GED 231, L 34, GPC s.v. llif2, lli2.

LYNMOUTH Devon SS 7249. 'Mouth of the river Lyn'. *Lymmouth* 1330. R.n. Lyn, *Lyn* 1281, 'the torrent', OE **hlynn**, + ME **muth**. The West and East Lyn rivers and their affluents sweep down narrow gorges from Exmoor to form a dangerous and sometimes destructive torrent. D 64, 9, RN 274.

King's LYNN Norf TF 6119. ModE **king**, genitive sing. **king's** + p.n. *Lena, Lun* 1086, *Lynna* c.1105, *Linna* c.1140, *Lenn* [1087×98]12th, 1196, 'the lake', PrW ***lïnn**, referring to a pool at the mouth of the Ouse. The hundred of Lynn is mentioned, *(on) Lynware hundred* 'the hundred of the people of Lynn' 11th, OE **ware**. 'King's' for distinction from North Lynn, *Nordlen* 1199, and South and West Lynn. DEPN, Forsberg 199.

LYNSTED Kent TQ 9460. 'Lime-tree place'. *Li- Lyndestede* 1212×20–1270, *Li- Lynstede* 1254–1270. OE **lind**, genitive pl. **linda**, + **stede**. PNK 276, Stead 220.

LYNT BRIDGE Wilts SU 2198 → River COLE SU 2294.

LYNTON Devon SS 7149. 'Settlement on the river Lyn'. *Linton(e) -a* from 1086 with variant spelling *Lyn-*. R.n. Lyn as in LYNMOUTH + OE **tūn**. D 64, RN 274.

LYON'S GATE Dorset ST 6505. *Lyons Gate* 1817 OS.

LYONSHALL H&W SO 3355. 'Nook of land belonging to the Leen district'. *Lenehalle* 1086, *Lineshalla* 1096, *Lenhal(es)* 1173×4–1291, *Leonhales* 1271–1535, *Lynhales or Lionshall* 1833 OS. District name Leen as in The LEEN SO 3859 + OE **halh**, pl. **halas**. Lyonshall is at the extreme W of the district centred on Leominster. He 138, L 104, 110.

LYPE HILL Somer SS 9537. *Lype Hill* 1809 OS. Earlier forms are needed, but this probably represents OE **hlēp**, WS ***hlīep, hlȳp** 'a steep place, a sudden drop in the ground'. PNE i.251.

LYTCHETT MATRAVERS Dorset SY 9495. 'L held by the Maltravers family'. *Li- Lyc(c)het(t) Mautravers(e)* 1280–1484, *~ Matravers* 1508. Earlier simply *Lichet* 1086, 1297, 1357, 'the grey wood', PrW ***lēd** (W *llwyd* < Celtic **lēto-*) + ***cę̄d**, identical with the first part of LICHFIELD Staffs SK 1209. 'Matravers', for distinction from LYTCHETT MINSTER SY 9693, refers to tenure by Hugh Maltravers in 1086. Do ii.29 gives pr ['litʃət mə 'trævəz], GPC 2239.

LYTCHETT MINSTER Dorset SY 9693. 'L belonging to the minster' sc. of Sturminster Marshall of which it was a chapelry. *Licheminster, Lyccemynistr'* 1244, *Ly- Lichet(t) Minystre, Min(i)stre, Mynstr(e), Mynster* 1280–1539. P.n. Lychett as in LYCHETT MATRAVERS SY 9495 + ME **minstre**. Do ii.33.

Great LYTH Shrops SJ 4507. *Magna Lya* 1236, ~ *Lega, Leye* 1255×6, ~ *Lyth'* 1291×2, *Greatt and Litle Leathe* 1586. ModE adj. **great**, Latin **magna**, + p.n. *(La) Lithe* 1160–72, c.1175 etc., 'the concave hill-side', OE **hlith** sometimes confused with *lēah*. The reference is to the concave W edge of the Lyth Hill escarpment SJ 4606, a prominent edge overlooking the river Cound which also has a hollow on its E side. 'Great' for distinction from Little Lyth SJ 4706, *Parva Lega, Leye* 1255×6, *Parva Leyth(o)* 1271×2, 1291×2, named from the same feature. Sa ii.114.

LYTHAM Lancs SD 3627. 'At the concave hills'. *Lidun* 1086, *Lythum* [1189×94]1336, 1300, 1332 etc., *Lithum* 1201, 1212, *Lethum* 1341, 1506, 1577. OE **hlith** or ON **hlíth**, dative pl. **hlithum**, referring either to the slight ground rise above the moorland of the Ribble estuary or perhaps to coastal sand dunes worn into concave shapes. Spellings with *-e-* point to lengthening of open-syllable *ĭ* > *ē*. Cf. KIRKLEATHAM NZ 5921, UPLEATHAM NZ 6319. La 155, PNE i.252, Jnl 17.90, L 165, LLH 6.11.

LYTHAM ST ANNE'S Lancs SD 3228. St Anne's parish in LYTHAM SD 3627. A holiday place which began to develop in the 1870s and took its name from the new church dedicated in 1873. The original name was *Kilgrimol* mentioned as a cemetery [c.1190]1336, *Kylgremosse* 1531, AScand pers.n. **Cyll-Grimr* + **hol**. The cemetery was washed into the sea by 1532, and the coast re-established as a result of changed tidal patterns consequent on the dredging of the Ribble. La 155, Mills 129, SSNNW 139, Nomina 10.174.

LYTHE NYorks NZ 8413. 'The slope'. *Lid* 1086–1210, *Li- Lyth(e)* from 1201, *Leth* 1401, *Lieth* 1623. ON **hlíth** varying with OE **hlith**. The 17th cent. form suggests a pronunciation with [i:] which must come from the OE word with ME lengthening of short i > *ē* in open syllables as in UPLEATHAM NZ 6319. The modern pr with [ai] must come from the ON form with *ī*. The reference is to the steep slope which borders on the sea-coast NW of Whitby. YN 137, SSNY 100.

LYVEDEN NEW BUILDING Northants SP 9885. P.n. *Louenden* 1175, *Luue- L(i)eueden(e)* 1178–1287, *Li- Lyveden(e)* from 1220, 'Leofa's valley', OE pers.n. *Lēofa*, genitive sing. *Lēofan*, + **denu**, + ModE **new building**, a large late 16th century summer house built for Sir Thomas Tresham, otherwise known as New Bield, in replacement of an earlier house known as the Old Build or Bield, ModE **bield** 'a lodging'. Nth 178, L 98, Pevsner 1973.299–300.

River LYVENNET Cumbr NY 6120. Uncertain. *aquam de Lyvennc* (for *-ut*) 1232×5, *Lyuened* 1292, *Leveneth -yd* 13th, *Liuennat* 1576, *Lyvennet* 1777, *Little-Vennat* 1704, *Little Vennet* 1808. Possibly to be connected with the late Brit ***levo-** (Brit **lemo-*) 'elm-tree' or the root ***(s)leib-** 'slippery'. The Welsh bard Taliesin locates one of the abodes of Urien, the 6th cent. ruler of the British kingdom of Rheged, at a place called *Lluyuenyd* or *Llwyfenydd*, probably to be identified with this name. The Anglo-Saxons called the river 'meadow-stream', *mǣd-burna* as in King's and Maulds MEABURN. We i.10 gives pr [lai'venit], xxxv, Antiquity xx.210–11.

MABE Corn SW 7533. 'Church-site of Mab'. *(parochia de) la Vabe* 1524, *Lavape* 1535, *Mape* 1549, *St Mabe* 1559, *Mabe* 1610. Co ***lann** + pers.n. *Mab* with lenition of *M-* to *V-*, and later restoration of the unmutated *M-*. The church, however, is dedicated to St Laud or Lo, a 6th cent. bishop of Coutances in Normandy, *(capella) Sancti Laudi* [1309]late 14th, *(parochia) Sancti Laudi* 1327–1405. PNCo 113, Gover n.d. 517.

MABLETHORPE Lincs TF 5084. 'Malbert's outlying farm'. *Malber- Mal(te)torp* 1086, *Maltorp* c.1115, *Malmertorp* 1154×89, *Malberthorpa* [c.1180]1409, *Malbretorp* 12th, *Melbertorp* 1209×10. CG pers.n. *Malbert* < *Madabert* + **thorp**. Probably a late colonisation of the dunes. DEPN, SSNEM 114, Jnl 7.54.

MACCLESFIELD Ches SJ 9173. 'Maccel's open land'. *Maclesfeld* 1086, [c.1096]1150, *Macclesfeld* [c.1096]1280–1656 with variants *Ma(c)kles- Makkles- -is-* etc., *Macclesfield* 1548. OE pers.n. **Maccel*, genitive sing. **Macc(e)les*, + **feld**. Che i.113, ii.vii which gives 19th cent. prs ['maksfılt, 'maksılt].

MACCLESFIELD CANAL Ches SJ 9066. P.n. MACCLESFIELD SJ 9173 + ModE **canal**. A canal some 30 miles long linking the Peak Forest Canal at Marple to the Grand Trunk Canal at Church Lawton.

MACCLESFIELD FOREST Ches SJ 9772. *(la) Forest(a) de Macclesfeld -felt* 1337–47, *Macclesfeld Forest* 1439, *Maxfield Forest* 1656. P.n. MACCLESFIELD SJ 9173 + ME **forest** 'a hunting preserve'. Che i.125.

MACKWORTH Derby SK 3137. 'Macca's enclosure'. *Macheuorde* 1086, *Mac- Makwurth(e) -worth(e) -wrthe -wrde* 1157–1540, *Macke- Makkeworth(e) -wurth* 1211–1455, *Mackworth(e)* from 1313. OE pers.n. **Macca* + **worth**. Db 479.

MAD WHARF Mers SD 2608. *Mad Wharf* 1842 OS. Apparently ModE adj. **mad** 'demented' + **wharf** 'a sand-bank'. Perhaps an allusion to unstable shifting sands.

MADEHURST WSusx SU 9810. 'The speech wooded hill'.
I. *Meslirs* 1135×54, *Medliers -leris -l(y)ers* 1188–1535, *Melleriis* 12th, *Melers* 1248.
II. *Medhurst* 1255, 1262, *Madhurst* 1279–1625.
III. *Madehurst* 1327, 1423, *Maydhurst* 1610.
OE **mæthel** + **hyrst**. The reference is to No Man's Land near Up Waltham SU 9512, *Nomanneslond* 1279, 1493, the site of the meeting-place of the rapes of Arundel and Chichester and where Forest courts were also held. Type I indicates AN difficulties with the sound [ð] in *mæthel* and Type III folk-etymological association with *maid* 'a maiden'. Sx 129 gives pr [mædəst], 77, SS 67.

MADELEY Staffs SJ 7744. 'Mada's wood or clearing'. *Madanlieg* [975]11th S 801, *Madelie* 1086, *Mad(d)eley(e)* 13th. OE pers.n. **Māda*, genitive sing. **Mādan*, + **lēah**. There is another Madeley at SK 0638, *Madelie* 1086, *Madeleye* 1176. DEPN, Duignan 99, L 203.

MADELEY HEATH Staffs SJ 7845. *Madeley Heath* 1833 OS. P.n. MADELEY SJ 7744 + ModE **heath**.

MADINGLEY Cambs TL 3960. 'Wood or clearing of the Madingas, the people called after Mada'. *Mading(e)lei* 1086, *Mad(d)ingele(e) -ley -ynge-* 1086–1480, *Maddingle(e) -leye -yng-* 1203–1470. OE folk-n. **Mādingas* < pers.n. **Māda* + **ingas**, genitive pl. **Mādinga*, + **lēah**. Ca 181 gives pr [mædiŋli].

MADLEY H&W SO 4238. 'Madda's wood or clearing'. *Medelagie* 1086, *Maddele(ia) -ley(e)* c.1200–1360, *Madele(ye)* c.1200–1327, *Madley* 1428. OE pers.n. *Madda* + **lēah**. The earlier W name seems to have been *Lann Ebrdil* [8th]c.1130 LL 'St Ebrdil's church', OW **lann** + saint's name *Ebrdil*, the mother of St Dyfrig. In the *Life of St Dyfrig* in LL c.1130 the p.n. is given as *Matle*, an interesting piece of medieval etymologising to provide St Dyfrig's birth place with a suitable name, W *mad* 'fortunate, auspicious, lucky' (MW *mat*) + *lle* 'a place'. He 139, GPC 2121, 2300.

MADRESFIELD H&W SO 8047. 'Mower's' or 'Mæthhere's open land'. *Madresfeld* [c.1086]1190–1327, *Matysfyld* 1522, *Matchfield* 1597. OE **mǣthere** or pers.n. *Mǣthhere*, genitive sing. **mǣtheres**, *Mǣthheres*, + **feld**. Wo 209 gives pr [mædəzfi:ld], L 240, 244, Anglia 110.156.

MADRON Corn SW 4531. 'St Madern's (church)'. *(Ecclesia) Sancti Maderni* 1203–91, *Madern* 1310, *Eglsomaddarn* 1370, *Maddron* 1524, *Eglos Maderne* 1616. Co **eglos** 'church' + saint's name *Madern* from the dedication of the church. Madern is possibly identical with the W female saint Madrun. The churchtown name, *Landythy* 1616, Co ***lann** + unknown element, survives as Landithy SW 452317. PNCo 113 records a pr [mædərn], Gover n.d. 646.

MAER Staffs SJ 7938. 'The lake'. *Mere* 1086, 1242, *Mare* 1198, *Meire* 1586. OE **mere** A large natural lake still exists here. DEPN, Duignan 99, Horovitz.

Lower MAES-COED H&W SO 3530. *Lower Maescoed* 1832 OS. ModE adj. **lower** + p.n. *Maischoit* 1137×9, *Mascoy(t)* 1232, 1327, *The Mescotts* 1812, 'the plain by the wood', PrW ***maes** + ***coid**. The reference is to the meadowland alongside Dulas Brook. 'Lower' for distinction from Middle and Upper MAES-COED SO 3333 and 3334. He 58, Bannister 127.

Middle MAES-COED H&W SO 3333. *Middle Maescoed* 1832 OS. ModE adj. **middle** + p.n. Maescoed as in Lower MAES-COED SO 3530.

Upper MAES-COED H&W SO 3334. *Upper Maescoed* 1832 OS. ModE adj. **upper** + p.n. Maescoed as in Lower MAES-COED SO 3530.

MAESBROOK Shrops SJ 3021. 'The boundary brook'. There are two types:
I. *Meresbroc* 1086.
II. *Maysbroc'* 1271×2, 1292×5, *-brook* 1397, *Masbro(o)ke* 1455–1738, *Maesbroc* 1427, *-brooke* 1578, *Macebrook* 1762–3.
The translation given is that of type I, the DB form, OE **(ġe)mǣre**, genitive sing. **(ġe)mǣres**, + **brōc**. Type II is either a reinterpretation of this with W **maes** 'an open field, a plain' or the original form, in which case type I is an error. Additional forms from the 12th cent. are needed for a decision on this point. Cf. MAESBURY MARSH SJ 3125. Sa i.191.

MAESBURY MARSH Shrops SJ 3125. P.n. Maesbury SJ 3025 + ModE **marsh**. For Maesbury there are three types;
I. *Meresberie* 1086, *-bury* 1294–1337.
II. *Mersbir' -buri* 1272, *Mersshbury* 1306×7.
III. *Mesbury* 1392×3–1706, *Measburie -y* 1582–1805, *Maesbury* from 1602, *Masbury* 1674. The name exhibits a similar development to (but spellings different from) MAESBROOK SJ 3021, type I being 'the fortified place of the boundary', OE **(ġe)mǣre**, genitive sing. **(ġe)mǣres**, + **burh**, dative sing. **byriġ**, susequently partly reinterpreted as 'the marsh fort or manor' with OE **mersc** (type II) and later still with W **maes** 'a plain' (type III). The boundary is that of Wales, conceived of in general rather than in precise terms. Maesbury is the precur-

sor of Oswestry, a post-conquest creation on its N edge. The W name was Llysfeisir, *Llesveyser* 1397, 'the court of Meisyr', PrW **lïs** + W feminine pers.n. *Meisyr'*. Sa i.192, O'Donnell 106.

MAGHAM DOWN ESusx TQ 6111. 'Magham high ground'. *Magham Downe* 1686. P.n. Magham + ModE **down**. Magham, *Megham* 13th, 14th, *Magham* 1261, is probably 'Mæcga's homestead', OE pers.n. **Mæcga* + **hām**. The same name is thought to occur in Magreed near Heathfield, *Magerede* 1409, 'Mæcga's cleared land'. Sx 437, 465.

MAGHULL Mers SD 3702. Partly uncertain. *Magele* 1086, *Maghele* [before 1190]1268, 1322, *Mah(h)al(e)* c.1200–83, *Magh(h)al(e)* 1219–1332, *Maele* 1323, *Male* 1501. Possibly 'Maga's nook', OE pers.n. **Maga* + **hale**, dative sing. of **halh** in the sense 'dry ground in marsh'. Another suggestion is that the specific is PrW ***maȝ** 'plain, open country' < Brit **mago-*, W *ma*. La 119 gives prs [mə'gul, mə'gʌl], older [me:l], L 107, Jnl 17.66, GPC 2293.

MAIDEN CASTLE Dorset SY 6688. An Iron Age hill-fort overlying a Neolithic causewayed camp of c. 3000 BC *Mayden Castell* 1607. A traditional name for a prehistoric or Roman fortified site, meaning perhaps 'fortification so strong it could be defended by girls' or 'fortification never captured' or even 'place frequented by girls, lovers' haunt'. Another example, Maiden Castle Roman Fortlet at NY 8713 in Cumbria, is *Maydencastel* 1292, and the earliest recorded in England is a hill-side called *Maidanecastell* 1175×86 referring to entrenchments in Saxton NYorks at SE 4437. As a traditional name for such features it was not confined to the British Isles where the OS Gazetteer gives 14 instances, and examples are known from Byzantine Macedonia (*Γυναικοκαστρο*), between Antakya and Haleb (Arabic *Qaṣr-el-Banāt* 'castle of the maidens') and from medieval romance (Chrétien de Troyes, *Le Conte du Graal* 7371ff). The Dorset Maiden Castle is sometimes identified with the city *Δούνιον* (i.e. *Dunium* erroneously for *Dunum*) < Brit ***dūno-** 'fort' mentioned by Ptolemy as the capital of the Durotriges, but *Dunium* is more likely to be the fort at Hod Hill ST 8510. The British victims of the attack by the Roman second legion Augusta under Vespasian in c. AD 44 were uncovered in the famous excavation by Sir Mortimer Wheeler in 1934–8. Do i.377, YW iv.70, Cu 255, PNE ii.31, RBrit 344, Newman-Pevsner 483.

MAIDEN LAW Durham NZ 1749. *Maiden Law* 1768 Armstrong, 1820 Surtees, 1843 *TA*. The name must be taken with the old name of the Manor House of Lanchester a mile and a half to the South (NZ 1747), *Maydenstanhall* 1382–1479, 'maiden-stone hall'. The significance of *maiden* in place-names is uncertain. It is often associated with earthworks or old fortifications, possibly in the sense 'fortification so strong that it could be defended successfully by girls' or 'that has never been taken', or meaning 'place frequented by maidens, a lovers' retreat'. Surtees (ii.305) refers to 'a very remarkable tumulus at Maiden Law, which I believe has never been opened'. The phrase 'maiden-stone' suggests an ancient monolith accredited with fertility or matrimonial powers. See the discussion of MAIDEN CASTLE, SY 6688.

MAIDEN WAY Cumbr NY 6537. Name of the Roman road from Kirkland over Melmerby Fell to Alston, Margary no.84. *Maydengat(h)e* c.1179, *le Maidingate* 1294, *the Maiden-way* 1777. OE **mæġden** + ON **gata** later replaced by ModE **way**. To be associated with this name is the lost *Maiden Castle* 1823 at Low Abbey, Kirkby Thore, NY 6527. For *mæġden* in p.ns. see MAIDEN CASTLE Dorset SY 6688. We i.19.

MAIDENCOMBE Devon SX 9268. 'The maidens' valley'. *Medenecoma* 1086, *-cumbe* 1219–99, *Medencumb* 1303, *Maiden(e)cumbe* 1228, 1238. OE **mæġden**, genitive pl. **mæġdena** + **cumb**. D 460.

MAIDENHEAD Berks SU 8881. 'Landing place of the maidens'. *Maideheg'*, *Maidenhee* 1202, *Maydehuth'* *-heth'* 1241, *Maidenhith(')* *-hach* *-eth* 1262–1402, *Maidenheued, antiq. nom. South Ailington* 1542, *Southealington hodie Maidenhead* 1600. OE **mæġden**, genitive pl. **mæġdena**, +**hȳth**. An ancient landing place on the Thames. The earlier name, **South Elington*, is a compound of ModE **south** and the p.n. *Elentone* 1086, *Elinton'* *-ynton(')* *-intona* 1167–1336, OE plant-name **e(o)lene** 'elecampane, *Inula helenium*' or ***ǣling** 'eel fishery' + **tūn**. The reference of *maiden* is unknown. Brk 53, 54, 849, TC 132.

MAIDENWELL Corn SX 1470. 'Maidens' spring'. *Medenawille* 1347, *Medenewille* c.1450. OE **mæġden**, genitive pl. **mæġdena**, + **wylle**. Cf. MAIDWELL Northants SP 7476. PNCo 113, Gover n.d. 250.

MAIDFORD Northants SP 6052. 'The maidens' ford'. *Merdeford* (sic) 1086, *Maideneford* 1166, *Maide-* *Maydeford* 1175–1453. OE **mæġden**, genitive pl. **mæġdena**, + **ford**. Possibly a place where maidens gathered at times for some purpose. Nth 41, L 70.

MAIDSTONE Kent TQ 7655. Possibly 'the people's stone'. *(to) mægþan stane, (de) mæides stana* [c.975]12th B 1321, 1322, *Med(d)estan(e)* 1086, *Maegdestane* 11th, *Maidestan* 1159×60, 1218, *Maydenestan'* 1205, *Maidstone* 1610. Apparently OE **mǣġth**, genitive sing. **mǣġthe**, + **stān**.Cf. FOLKESTONE TR 2336. The name was nevertheless early understood as the 'maiden's or maidens' stone', OE **mæġth**, genitive sing. **mæġthe**, genitive pl. **mæġtha**, and **mæġden**, genitive sing. **mæġdenes**. The reference is not known. K 140, KPN 306, TC 132.

MAIDWELL Northants SP 7476. 'The maidens' spring'. *Medewelle* 1086, *Mayde-* *Meidewell(e)* 1154×89–1475, *Maydenwell* 13th cent. OE **mæġd**, a shortened form of **mæġden**, genitive pl. **mæġda**, + **welle**. Cf. MAIDFORD SP 6052. Nth 117, L 32.

MAINSTONE Shrops SO 2787. 'The strength stone'. *Meyneston* 1284, *Meynston* 1333, *Maynstan* 1336, *Maynston(')* 1344–1672, *The Mainstone* 1640. OE **mæġen** + **stān**. Although this p.n. type is widespread in England for a boundary marker in the sense 'great stone' the reference here seems to be different. Preserved in the church is a small smooth boulder of about 230 lbs weight which was used as a test of strength: young men would try to lift it to face height and cast it over their left shoulder as a demonstration of strength. Sa i.193 referring to TSAS 1st series 7.127, Raven 126.

MAISEMORE Glos SO 8121. 'Great field'. *Maies-* *Mayesmore* *-mor(a)* 12th–1387, *Mais-* *Maysmor(e)* 1221–1542, *Mai-* *Maysemor(e)* 1248–1701. PrW ***ma(ȝ)es** + **mọr**. Gl iii.161.

MAKERFIELD An extensive district partly in Mers partly in GMan stretching from Ince-in-Makerfield SD 5904 to Newton-le-Willows (formerly Newton-in-Makerfield) SJ 5894. 'Open land at *Maker*'. *Macrefeld* 1121, *Maker(e)feld* 1213–61, *Machesfeld(a)* 1123, 1169, *Makeresfeld* 1204–15, *Macresfeld* 1280, 1291. P.n. **Maker*, PrW ***maguïr** 'walls, ruins, remains', W *magwyr*, Co **magoer*, genitive sing. **Makeres*, + OE **feld**. Cf. Maker Corn SX 4351, *Makere* 1346, *Magre* 1428, Magor Gwent ST 4287, *Magor* 13th. The allusion is unknown but it has been suggested that it might be to the ruins of the Roman settlement at Wigan. La 93, TC 41, Jnl 18.25, GPC 2320.

MALBOROUGH Devon SX 7039. 'Mærla's hill'. *Malleberg(e)* *-bergh* *-burgh(e)* 1249–1356, *Marleberg(e)* 1270–89, *Marlburrough* 1561, *Maulborrough* 1751. OE pers.n. **Mǣrla* + **beorg**. D 307 gives pr [mɔ:lbərə].

MALDON Essex TL 8506. 'Hill with a cross'. *(to, æt, at) Mældune* c.925 ASC(A) under years 913, 920, 921, c.1000 ASC(A) under year 993, c.1200 ASC(E) under year 991, *Mæld(v), Mæl(dvn), Mal(d), Mel, Mæli, Mld* 959–1100 coins, *Meldunam* [c.990]12th, *Mealduna* before 1066, *Mealdon* *-dun(e)* [1068]1309, *Melduna* *-donam* 1086, 12th, *Malduna* 1098, *-don* from 1212, *M(e)audon(a)* 1219–54, *Mawdon* 1567. OE **mǣl** + **dūn**. Ess 218 gives pr [mo:ldən], Jnl 2.43.

New MALDEN GLond TQ 2168. *New Malden* 1874. ModE **new** + p.n. *Meldon(e)* 1086–1188, *Meaudon(a)* 1215–64, *Maldon* 1225–1325, *Maldene* 1241, *Mauden* 1249, *Maulden* 1602, 'the hill with a cross or crucifix', OE **mæl** + **dūn**. A superior 19th-cent. development due largely to the coming of the railway and known colloquially as 'The Montpelier of Surrey'. Originally in Kingston parish it became a separate authority in 1866 and achieved borough status in 1936. Sr 64, Encyc 557, LPN 145.

MALHAM NYorks SD 9063. Uncertain. *Malgun-on* 1086, *Malgum* 1176–1312, *Mal(g)hum* 12th–1461, *Mal(l)um* 1285, 1311, *Mallom(e)* 1417–1548, *Malgham* 1303, 1597, *Malham* 1535–1623, *Mallam* from 1457, *Maum* 18th cent. There are two views about the origin of this name. It may mean (the settlement) 'at the hollows' from ON **malr** 'a sack, a bag' used in a transferred sense of the two great gorges of Malham Cove and Gordale. ON *malr* is from earlier Gmc **malhaz* and it is this unusually archaic form with the retained *-h-* in the dative pl. **malhum* which would have to lie behind the various spellings with *g, gh* and *h*. ON *-h-*, however, was normally lost in this position by about 700, i.e. before the Scandinavian Settlement in England. Alternatively, therefore, it is (the settlement) 'at the gravelly places or lakes' from ON ***malgi**, ***malugr**, again in the dative pl. form ***malgum** referring either to gravelly lakes like Malham Tarn or to the gravelly sites along the river Aire. ODan **malugh* is cited to explain the Dan p.n. Malle, *Malugh* 1322 and the Swed lake Maljen **Malghe*. YW vi.133, SSNY 100.

MALHAM TARN NYorks SD 8966. *Malham Terne* 1650. P.n. MALHAM SD 9063 + ME **tern** (ON *tjǫrn*). Short for *Malhom Water Tern* 1540, p.n. *Malhe- Malgh- Malgewater, Mallewatre* 12th cent., *Mawater* 1593, p.n. MALHAM SD 9063 + OE **wæter**, + ME **tern**. YW vi.139.

MALLERSTANG COMMON Cumbr SD 7798. *Mallerstang Common* 1776. P.n. Mallerstang + ModE **common**. Mallerstang, *Malrestang* 1223–1322, *Malvestang* 1228, *-verstang* 1462, *Mallerstang(e)* from 1272, *Mol(l)erstang* 1687–1716, is probably 'pole on the hill called *Moel-fre*', ON **stǫng** + p.n. **Moel-fre* 'bare hill' < PrW ***mēl** + ***breȝ**. The reference is probably to a landmark or guide post. In the Middle Ages Mallerstang Forest, *Mallerstangchace* 1315, was a hunting preserve (ME **chace**, **forest**). We ii.17, 11 which gives pr ['malə 'stæŋ], formerly ['mɔ:stən], SSNNW 239, L 129, GPC 2474, 313 s.v.*bre*.

South MALLING ESusx TQ 4211. *Suth Mallinges* [838]c.1300 S 1438 rubric, *Suthmelling* 1232, *Su(t)malling(g)* 13th cent., *South Mawling* 1628. ME adj. **south** + p.n. *(æt) Mallingum* (dative case plural) 838 S 1438, *Meallinges* [838]c.1300 ibid., *Mellinges* 1086, *Melling* 1121, *Malling(e)s* 13th cent., 15th, 'the Meallingas, the people called after Mealla', OE folk-n. **Meallingas* < pers.n. **Mealla* + **ingas**. 'South' alluding to the settlement's position in the far south of the original hundred of Malling. Malling, the territory of the *Meallingas*, is a well studied example of an ancient multiple estate possibly of pre-English origin. Sx 354, ING 36, SS 65, Settlement.Eng 20–9.

East MALLING Kent TQ 6957. *Estmalling(es)* 1206, *Estma(u)lling'* 1226. East Malling is already distinguished from West MALLING TQ 6757 in the 10th cent. when the boundary of the people of East Malling is referred to as *east meallinga gemære* [942×6]12th S 514. OE **ēast** + p.n. *Meallingas* [942×6]12th ibid., *(de) Meallingis* (Latin), *(of) Meallingum* (Old English) [c.975]12th B 1321–2, *(æt) Meallingan* [995×1005]12th S 1456, *Metlinges* (probably East M), *Mellingetes* (probably West M) 1086, *Me(a)llinget(t)es, Melling(a)etes* 11th, *Mallinges* 1182–1227×35, *Malling'* 1226, 'the Meallingas, the people called after Mealla', OE folk-n. **Meallingas* < pers.n. **Mealla* + **ingas**. The spellings *Me(a)llinget(t)es* seem to show diminutive forms with the French suffix *-ette*. PNK 148, KPN 252, ING 13.

West MALLING Kent TQ 6757. *West Malling* 1819 OS. ModE adj. **west** + p.n. Malling as in East MALLING TQ 6957.

MALLOWDALE FELL Lancs SD 6259. P.n. Mallowdale SD 6061 + ModE **fell** (ON *fjall*). Mallowdale, *Malydall* 1574, *Malladale* 1640 is possibly just 'mallow dale, dale where mallow grows', OE, ME **malwe** (Latin *malva silvestris*) + **dale** (ON *dalr*). La 181.

MALMESBURY Wilts ST 9387. 'Maildub's fortified place'.

I. *Maildufi urbs* [c.731]8th BHE, *Maldubia civitate* [c.770]9th *Life of Boniface*, *Maelduburi -beri* [871×99]14th S 356, *Mailduberi* [901]12th S 1797, *Maldulfes burgh* [c.890]c.1000 OEBede, *Meldulfesbirg* *[681]13th S 73, *Maeldubesburg* [685]13th S 1169.

II. *Maldumesburuh* [701]10th B 106, *-burg, Mealdumes- Mel-Maeldunesburg* 14th cent. versions of puported 7th–10th cent. charters (S 71, 73, 322, 356, 629, 796, 841, 1169, 1245), *Maldemesburuh* [974]13th S 796, *(into) Mealdœlmesburuh* [10th]c.1150 Wills, c.1120 ASC(E) under year 1015, *Maldmesbyrig* c.1050.

III. *Ealdlemesbyrig* c.1050 ASC(C,D) under year 1015.

IV. *Malmesberie -biry* 1086 etc. with variants *-bury* and *Mamesberie* 1086, *-bur'* 1198, *Maumesberi* 1200, *Mamsbury* 1652, 1737.

The monastery here is also referred to as *(in) Maldubiensi monasterio* [758]13th S 260 and *(ad) Maldunense monasterium* [745]13th S 256, the church as *Maildubiensis œcclesia* [871×99]12th S 356, and the city as *Meldufuensis Burgi* *[934]15th S 454.

OIr pers.n. *Maildub* 'the black prince' (*maglo-* 'prince' + *dubo-* 'black'), secondary genitive sing. *Maildubes*, + OE **burh**, dative sing. **byriġ**. The later forms, types II and IV, are due to confusion of this name and that of St Aldhelm (preserved in the unique type III), the famous 7th cent. abbot of Malmesbury. There is no reason to question William of Malmesbury's ascription of the foundation of the monastery to the Irishman *Mailduf*. The current pr is [mɑ:mzb(ə)ri]. Wlt 47.

MALPAS Ches SJ 4847. 'The difficult passage'. *Malpas* from 1150 including [1121×9]1285 with variants *-p(h)ase -pace -paas* etc., *Maupas(se)* 1208–1653. OFr **mal** + **pas**. The reference is to the crossing of Bradley Brook by the Roman road from Chester to Wroxeter, Margary no. 6a, just S of Hough Farm at SJ 4945. The French name replaced the earlier name for the valley of Bradley Brook at this point, which was *Depenbech* 1086, *Depenbache -in-* 1344–1523, 'at the deep stream valley', OE **dēopan**, dative sing. definite form of **dēop** 'deep', + **beċe**, **bæċe**. Che iv.17 gives prs ['mælpəs] and ['mɔ:lpəs], and older local ['mɔ:pəs], L 67.

MALPAS Corn SW 8442. 'Difficult passage'. *Le Mal Pas* [late 12th]late 13th, *Malpas* 1383, *Molpus alias Mopus* 1674. OFr **mal** + **pas** referring to a crossing of the Tresillian river. The first citation comes from Béroul's *Tristan* romance and refers to a place where Tristan has a rendezvous with Yseut on a mound in marshy ground 'at the end of the plank bridge at the Malpas' (lines 3295ff). PNCo 114, Gover n.d. 435.

MALTBY Either 'Malti's farm or village' or 'malt farm'. ON pers.n. *Malti*, genitive sing. *Malta*, + **bȳ**, or **malt** (perhaps replacing OE *mealt*) + **bȳ**. The pers.n. *Malti* is rare in Scandinavia before the 13th cent. and malt was one of the commonest exports from England to Scandinavia in the Middle Ages. On the other hand, the regular occurrence of medial *-e-* points to the pers.n. See also MAUTBY. SSNEM 59.

(1) ~ Cleve NZ 4613. *Maltebi* 1086, *-bi -by* 1176–1406, *Mauteby* 1222–1310, *Maltby* from 1616. YN gives pr [mɔ:tbi], SSNY 33.

(2) ~ SYorks SK 5392. *Maltebi -by* 1086–1464, *Mauteby* 1221, 1276, *Maltby* from 1377. YW i.137, SSNY 33.

(3) ~ LE MARSH Lincs TF 4681. 'M in the marsh'. P.n. *Malte(s)bi -by* 1086, *Maltebi* c.1115, *Mautesbi* 1176, *-by* 1219, + Fr definite article **le** (for *en le*) + ModE **marsh**. There was

another Lincs Malby now lost at TF 3184, *Maltebi* 1086, c.1115. SSNEM 59, Cameron. 1998

MALTMAN'S HILL Kent TQ 9043. *Maltman's Hill* 1841 TA. Surname Maltman meaning 'maltster' + ModE **hill**. Cullen.

MALTON NYorks SE 7871. *Maltune* 1086. Identical with Old MALTON SE 7972; also known as New Malton, *Nova Malton* 1301. YN 45.

Old MALTON NYorks SE 7972. Latin prefix *Veteri* 1173 (dative case), ME *Ald* 1399, + p.n. *Maaltun* c.1130, *Maltune* 1086, [12th cent.]16th cent., *Malton(e)* from 1173, the 'moot village', OE **mæthel** 'speech' + **tūn**. Loss of medial *-th-* was due to AN influence. SSNY 116 suggests the same etymology for Malton as for High Melton, viz. ON **methal** 'middle' replacing OE **middel**. But the sequence of forms for each place is quite different. YN 43 gives pr [ɔ:d mɔ:tən].

Great MALVERN H&W SO 7846. *Magna Malverna* 1228, *Moche Malv'ne* 1521. ModE adjs. **much, great** + p.n. *Mælfern* c.1030, *Malferna* 1086, *Maluern(i)a* 1156–1535, 'the bare hill', PrW **mēl, moil** + ***brïn**. The reference is to the range of hills rising to 1395ft. due W of the town at Worcester Beacon. William of Malmesbury offers a typical medieval etymology: *Malvernense monasterium quod mihi per antifrasin videtur sortitum esse vocabulum. Non enim ibi* male *sed bene et pulcherrime religio vernat*, 'Malvern monastery which seems to me to have got its name antiphrastically. For not *ill* (*male*) but well and beautifully does religion flourish there (*vernat*)' (*Gesta Pontificum* 296). 'Great' for distinction from Little MALVERN SO 7740. DEPN, Wo 210, 129.

Little MALVERN H&W SO 7740. *parve, minor Malvernie, La petite Maluerne* 1232, 1275. ModE adj. **little**, Latin **parva, minor** 'lesser', Fr **petite**, + p.n. Malvern as in Great MALVERN SO 7846. Wo 150.

West MALVERN H&W SO 7646. ModE adj. **west** + p.n. Malvern as in Great MALVERN. A modern development.

MALVERN HILLS H&W SO 7742. *Maluerne hilles* [c.1367×70]c.1400 Piers Plowman. P.n. Malvern as in Great MALVERN SO 7846 + ModE **hills**. A range of hills extending in a straight line N and S for nine miles.

MALVERN LINK H&W SO 7847. *Malvern Link* 1608. P.n. Malvern as in Great MALVERN SO 7847 + ME **link** (OE *hlinċ*) 'a ridge or bank', used as a p.n. *Link, la Lynke, (atte) Lynkes* 1215–1327, *Linche, la Lynche* 1215, 1275, *The Link* 1776, 1831 OS referring to an outlying ridge of the Malvern Hills. Wo 207.

MALVERN WELLS H&W SO 7742. 'The springs at Malvern'. *Malvern Wells* 1831 OS. P.n. Malvern as in Great MALVERN SO 7846 + ModE **wells**. Medicinal waters were discovered here in the 17th cent. and became famous late in the following century. Pevsner 1968.219.

MAM TOR Derby SK 1283. *Manhill* 1577, 1610, *Mantaur* 1630, *Mame Torre* 1640, *Mam Tor* 1668. Earlier forms of the name are preserved in *Mamsichefeld* 1378, *le Mamsechefylde* 1455, *Mamsitchfield* 1688, 'new arable land by Mam sitch', OE **sīċ** 'a small stream'. Mam Tor is a summit of 1709ft. in the Peak District with an Iron Age hill-fort and Bronze-Age round barrows. It has been suggested that Mam is Irish **mam** 'a breast' used in a transferred topographical sense and so a p.n. related to other supposedly Irish-Scandinavian names in S Derby, e.g. MAMMERTON, MERCASTON and Kirk IRETON. More likely, however, is OE **mamme** 'a teat' as in MAMHEAD Devon. Db 55–6, xxxiii, Addenda vi.

MAMBLE H&W SO 6971. Uncertain. *momela gemære* 'the bounds of the *Momele*' [?781×96]11th S 1185, *Momele* 1275–1428, *Mamele* 1086, 1232, 1255, *Mamull, Momyll, Maumble, Momble, Momehill* 16th cent., *Mamble* 1604. Mamble occupies a hill site (480ft.). The form *momela* of S 1185 seems to be genitive pl. of an otherwise unrecorded folk-n. **Momele*, perhaps a derivative of OE *mamme* 'teat' used topographically as in MAM TOR Derby SK 1283 or MAMHEAD Devon SX 9381. If, however, the name was originally applied to the whole Clows Top massif, a Celtic origin is more likely, cf. MIr *mamm* 'breast', Ir *mám* 'a large round hill'. Wo 60.

MAMMERTON Derby SK 2136. 'Gravel farm'. *Malmerton(e) -tona* 1197–1281, *Malmarton* 1272–1477, *Mamerton* 1415. The persistent *a* spellings point to ON **malmar**, genitive sing. of **malmr** 'sand, gravel' + **tūn** rather than OWScand pers.n. *Melmor* < OIr *Máelmuire* 'servant of Mary' + **tūn** as previously suggested. Db 582, SSNEM 197.

MANACCAN Corn SW 7625. Uncertain. *(Ecclesia) Sancte Manace in Menestre* 1309, *Managhan* 1395, *Manacca* 1535, *Manachan* 1623, *Manacken* 1657. It is unknown whether the dedication to St Manaca is genuine or derived from an older p.n. *Managhan* 'place of monks', Co **manach** 'monk' used as plural, + p.n. forming suffix ***-an**. *Menestre*, earlier *Ministre* 1259, represents OE **mynster** 'a monastery, a minster church'. There is, however, no trace or record of a former monastic church here. PNCo 114, Gover n.d. 565, DEPN.

MANACLE POINT Corn SW 8121. *The Manacles poynte* 1576. Named from the offshore rocks called the MANACLES SW 8120. PNCo 115.

The MANACLES Corn SW 8120. *The Manacles* 1576, 1605, 1610, *rock called Mannahackles alias Manacles* 1619, (pronounced) *Meanácles* 1808. This could be ModE **manacle(s)** used of the trap the rocks offer to shipping; or if the accentuation given in 1808 is reliable Co **men** (or plural **meyn**) + **eglos** 'church stone(s)' referring either to their fanciful resemblance to a church, or to the sighting of the spire of St Keverne's church at SW 7921 viewed as a landmark from the sea. According to A. L. Salman (1903) another of the rocks which stand on a direct line with Trenoweth on the mainland is called Mantrenoweth 'Trenoweth Rock' (now Maen Chynoweth SW 8121). PNCo 115, Gover n.d. 555.

MANATON Devon SX 7581. 'Estate held in common'. *Manitone* 1086, *Magnetona, Manitona* 1086 Exon, *Maneton(e)* 1234–1443. OE **(ġe)mǣne** + **tūn**. D 481 gives pr [mænətən].

MANBY Lincs TF 3986. 'Manni or the men's village or farm'. *Mannebi* 1086–c.1200, *Magnebi* 1212. ON pers.n. *Manni*, genitive sing. *Manna*, or OE **mann**, ON **mathr**, genitive pl. **manna**, + **bȳ**. DEPN, SSNEM 59, Cameron 1998.

MANCETTER Warw SP 3296. '*Manduessedum* Roman fort'.

I. *Manduessedo* [4th]8th AI ablative case.

II. *Manecestre* 1195–1313, *Mam(m)e- Mamne- Maumecestre* 1196–1296, *Mancestre -ter* 1251–1610, *Mancett(o)ur -cettre -setur, Manchester* 16th cent.

Brit p.n. *Manduessedum* 'horse-chariot' (Brit **mandu-* 'small horse, pony' + **essedo-* 'war-chariot') + OE **ċeaster**. Wa 86, RBrit 411.

MANCHESTER GMan SJ 8397. 'The Roman fort *Mamucium*'.

I. *Mamucio, Mam- Mancunio* [3rd]8th AI, *Mantio, Mautio, Mancio* [c.700]13th Rav.

II. *Mameceaster* 923 ASC(A), *Mamecestre* 1086–1227 etc., *-mm-* 1184–94, *Mamchestre* 1385, 1441, *Mancestre* 1310, 1384, *-chestre* 1330, *Maunchestre* 1348, *Mainchester* 1421×2, *Manchester* from 1480.

PrW ***mamm** 'breast' as in *Mamucium* + OE **ċeaster**. The RBrit name is ***mamma** 'breast, breast-like hill' referring to the site of the fort, + suffixes **-ūc-** and **-io-**. La 33, TC 133, RBrit 409, Nomina 4.32, Jnl 17.32.

MANCHESTER INTERNATIONAL AIRPORT GMan SJ 8184.

MANCHESTER SHIP CANAL Ches SJ 6990. P.n. MANCHESTER SJ 8397 + **ship canal**. This famous canal constructed in 1887–94 extends some 35 miles from Eastham to Manchester Docks.

MANEA Cambs TL 4989. Probably 'the common island'. *Moneia -eya -eye* 1177–1326, *Maneia -ey(e) -(a)ye* 1178–1575, *Mawna* 1499, *Mane* 1570, *Monea* 1576, 1695. OE **(ġe)mǣne**, oblique case definite declension **mānan**, + **ēġ** 'dry land in marsh'. It is recorded in 1251 that the whole vill of Littleport had rights of common on *Moneyeslode*, Manea Lode. However, OE *manu* 'a mane' would actually fit the forms better, although it is difficult to see what the sense might be. Cf. MANEY WMidl SP 1295. Ca 235, L 37, 39.

MANEY WMidl SP 1295. Possibly 'the common island'. *Maney(e)* from 1285. OE **(ġe)mǣne**, definite form oblique case **(ġe)mānan**, + **ēġ**. Cf. however, MANEA Cambs TL 4989.

MANFIELD NYorks NZ 2213. 'Manna's open land'. *Mannefelt* 1086, *-feld* 1202, *Manefeld(e)* 1086–1280, *Manfeld* from 1310. OE pers.n. *Man(n)a* + **feld**. YN 284, L 243.

MANGOTSFIELD Avon ST 6577. 'Mangod's stretch of open country'. *Man(e)godesfelle -feld(e)* 1086–1339, *Mangotesfeld* 1327–1492, *Maggottes- Magger(y)sfeld* 16th cent. CG pers.n. *Mangod*, genitive sing. *Mangodes*, + **feld**. Gl iii.98, L 240, 243.

River MANIFOLD Staffs SK 1152. 'The river with many turns'. *Water of Manifould* [1434]17th, *Manyfold(e)* from 1573. ME **manifold** (OE *maniġfald*). An appropriate descriptive name for this river, which must have replaced an earlier name now lost. St i.13, RN 277.

MANKINHOLES WYorks SD 9623. 'Mancan's hollows'. *Mankanholes -can-* 1275–1602, *Mankinholes* from 1309. OIr pers.n. *Manc(h)án* + OE **hol**, ME pl. **holes**. YW iii.176.

MANLEY Ches SJ 5071. 'Common wood or clearing'. *Menlie* 1086, *Manl(ey) -leye* from 1265 with variants *-le(e) -legh*. OE **(ġe)mǣne** + **lēah**. Che iii.245.

MANNINGFORD ABBOTS Wilts SU 1458 → ~ BOHUNE SU 1357.

MANNINGFORD BOHUNE Wilts SU 1357. 'M held by the Bohun family'. *Manyngeford Bon* 1279, *Maningford Boun* 1316, ~ *Bohun* 1321. P.n. *Maning(a)ford* (Manningford Abbots) [987]16th S 865 and 1505, *Mane- Maniford* 1086, *Man(n)i(n)geford* 13th cent., 'the ford of the Manningas, the people called after Manna', OE folk-n. **Manningas* < pers.n. *Manna* + *ingas*, genitive pl. **Manninga*, + **ford**, + manorial addition from the family name of John de Boun who held here in 1263 and for distinction from the other divisions of the manor, Manningford Abbots SU 1458, *Manyngford Abbatis* 1289, held by the abbot of Hyde Abbey, Winchester and MANNINGFORD BRUCE SU 1358. Wlt 320, L 70.

MANNINGFORD BRUCE Wilts SU 1358. 'That part of M held by the Bruce family'. *Manyngeford Breuse* 1279, *Manyngford Breouse*, ~ *Brehuse* 1289. P.n. Manningford as in MANNINGFORD BOHUNE SU 1357 + manorial addition from the family name Bruce (*Breuse, Breuose, Brewosa* 13th cent.). Wlt 320.

MANNINGS HEATH WSusx TQ 2028. No early forms. Surname *Manning* + ModE **heath**.

MANNINGTON Dorset SU 0605. 'Estate called after Mann or Manna'. *Manitone* 1086, *Manitun' -ton(')* 1100–1390, *Maninton(')* 1244, 1278, *Manington(e) -yng-* 1279–1387, *Mannington* 1774, *Manaton* 1811. OE pers.n. *Mann* or *Manna* + **ing**[4] + **tūn**. Do ii.151.

MANNINGTREE Essex TM 1031. 'Mann's tree'. *Manitre -y-* 1248–1428, *Mei- Mai- Meyntre(e)* 1248–1535, *Maningtre -yng-* 1308–1454. OE pers.n. **Mann(a)* + **ing**[4] + **trēow**. Alternatively this might be 'many trees' with OE **maniġ**. Ess 343.

MANSELL GAMAGE H&W SO 3944. 'M held by the Gamages family'. *Maumeshull' Gamages* 1243, *Malmeshul(l)(e) Gamag(e)* 1291–1318. P.n. Mansell + family name *Gamages* from Gamaches-en-Vexin, Eure, France. Mansell, *Mælueshylle* 1043×6 S 1469, *Malveselle* 1086, *Mau- Maweshull(a)* 1160×70, 1199, *Malmeshull(e)* 1292, *Mauneshulla* 1195, is 'gravel hill', OE ***malu** 'gravel ridge', possibly used as a p.n. **Malwe* (cf. Warw 339), genitive sing. ***malwes**, later replaced by **malm** 'sand, sandy soil', genitive sing. **malmes**, + **hyll**. The soil here is stiff loam on gravel. 'Gamage' for distinction from MANSELL LACY SO 4245. He 141, L 170.

MANSELL LACY H&W SO 4245. 'M held by the Lacy family'. *Maumeshull' Lacy* 1243, *Malmeshull(e) Lacy* 1291–c.1433. P.n. Mansell as in MANSELL GAMAGE SO 3944 + family name *Lacy* from Lassy, Calvados, France. He 141.

MANSERGH Cumbr SD 6082. 'Man's shieling'. *Manzerge* 1086, *Manserg* 1187–1349, *Mansergh(e)* from 1301 with variant *-(h)argh*. ON pers.n. (OEScand) *Man*, genitive sing. *Mans*, + **ǽrgi**. The original village was at OLD TOWN SD 5983. We i.49 gives pr ['manzə], SSNNW 65.

MANSFIELD Notts SK 5361. 'The open land by the river Maun'. *Mamesfeld(e)* 1086–1332, *Māmesfed* (sic for *-feld*) 1086, *-feld* 1153–1334, *Manesfeld* 1202–55, *Maunsfeld* c.1250–1307, *Mannesfeld* 1291–1399, *Mawnnes- Mauncefeld* 15th. R.n. MAUN SK 6166, *Mam(m)*, genitive sing. *Mames*, + OE **feld**. The change *m* > *n* is an assimilatory one before dental-alveolar *s*. Nt xli, 123.

MANSIONVILLE Ches SJ 9077 → PRESTBURY SJ 9077.

MANSTON Dorset ST 8115. 'Mann's estate'. *Manestone* 1086, *Man(n)eston'* 1196–1543, *Manston(')* from 1268. OE pers.n. *Mann*, genitive sing. *Mannes*, + **tūn**. Do iii.45.

MANSTON AIRPORT Kent TR 3366. A famous RAF base during World War II. P.n. Manston + ModE **airport**. Manston, *Man(n)eston(e)* 13th cent., is 'Mann's estate'. OE pers.n. *Mann*, genitive sing. *Mannes*, + **tūn**. KPN 17.

MANSWOOD Dorset ST 9707. *Mangewood* 1774. Perhaps 'mange wood' from ModE *mange* 'a parasite skin disease of animals' used of a wood thought to be diseased or to harbour parasites. Do ii.142.

MANTHORPE Lincs TF 0715. 'Manni or the men's outlying settlement'. *(æt) Mannethorp* [c.1067]12th Wills, *Mannetor(p)* 1086, *-torp* 1167–1210, *-thorp* 1227, [1216×72]15th, *Manthorp(e)* from 1298. ON pers.n. *Manni*, genitive sing. *Manna*, or OE **mann**, ON **mathr**, genitive pl. **manna**, + **bȳ**. There is another Lincs Manthorpe at SK 9137, *Mannetorp* 1185, 1212, *Manthorp(e)* from 1276. Perrott 207, 489, SSNEM 114, 129.

MANTON Humbs SE 9302. 'Farm or village on sandy soil'. *Malmetun* [1061×6]13th, *-tune*, *Mameltune* 1086, *Malmetuna* c.1115, *-tona* c.1145. OE **malm** + **tūn**. DEPN.

MANTON Leic SK 8804. Probably 'Manna's farm or village'. *Mannatonam* 1120×9, *Manneton* 1123×9, 1209×19, *Maneton(e)* 1219–78, *Manton(e)* from 1244. OE pers.n. *Manna* + **tūn**. R 195.

MANTON Wilts SU 1768. 'Manna's estate'. *Manetune* 1086, *Mani(n)ton -yn(g)-* 1229–81, *Manton* from 1259. OE pers.n. *Manna*, genitive sing. *Mannan* or *Manna* + **ing**[4], + **tūn**. Wlt 308.

MANUDEN Essex TL 4926. 'Valley of the Manningas, the people called after Manna'. *Magghedenna, Menghedanā, Magellana* 1086, *Manegedan'* 1123–33. *Manegadenna* c.1150, *Maneg(h)eden* 1225–55, *Manuedene* c.1150, *Manew(e)den(e) -don* 1274–1614, *Manuden(e)* from 1274, *Mal(l)eden -dyne* 1475–1627. OE folk-n. **Man(n)ingas* < pers.n. *Man(n)a* + **ingas**, genitive pl. **Man(n)inga*, + **denu**. The phonological development was *Maninga-* > *Maniga-* > *Manega-* > *Manewe-* as in CANEWDON TQ 8994. There was considerable variation in the local pr, [ma:ndən, mæləndain, mælətain] and [mæliŋdən] all being recorded 19th cent. prs. Ess 551.

MAPLEBECK Notts SK 7160. 'The maple-tree stream'. *Mapelbec -berg, Mapleberg* 1086, *Mapelbec* 1165, *-bek* 1291–1325, *Malebe(c)ke, Ma(y)lbeke, Malebecke al. Mapplebecke* 16th. OE ***mapel** + ON **bekkr**. Nt 189, SSNEM 219, L 14, 222.

MAPLE CROSS Herts TQ 0392. 'Maple-tree cross-roads'. *Mapull Crosse* 1485×1509. ME **maple** + **cross**. Hrt 83.

MAPLEDURHAM Hants SU 7321 → BURITON SU 7320.

MAPLEDURHAM Oxon SU 6776. 'The maple-tree homestead'. *Mapeldreham* 1086–1285, *-der-* 1209–86, *Mapuldurham* 1274–1395. OE **mapuldor** + **hām**. O 59.

MAPLEDURWELL Hants SU 6951. 'Maple-tree spring'. *Mapledrewelle* 1086, *Mappedreuuella* c.1125, *Mapeldurwelle* 1183–1223. OE **mapuldor** + **wella**. Ha 114, Gover 1958.127, DEPN.

MAPLEHURST WSusx TQ 1824. 'The maple-tree wooded hill'.

Cf. *Maplehurst Wood* 1485 and NUTHURST TQ 1926. ME **mapel** + **hurst**. Sx 232.

Great MAPLESTEAD Essex TL 8034. *Great Mapeltrested* 1242–56. ME adj. **gret** + p.n. *Mapulderstede* [1042]14th, *[1066]13th S 1043, *Mapoldrestede* [1083]14th, *Mapelduresstede -dere- -ter(e)- -tre-* c.1174–1357, *Mapledestedam, Mappesteda* 1086, *Mapelstede -il-* 1303–1465, *Maplested* 1654, 'maple-tree place', OE **mapuldor** + **stede**. 'Great' for distinction from Little MAPLESTEAD TL 8233. Ess 445, Stead 203.

Little MAPLESTEAD Essex TL 8233. *Little Mapeltrested* 1242–56. ME adj. **litel** + p.n. Maplestead as in Great MAPLESTEAD TL 8034. Ess 445.

MAPLETON Derby SK 1648. 'Maple-tree farm'. *Mapletune* 1086, *-ton* from 1573, *Mapelton(e) -tona -il-* 1159×60–1439, *Mapulton* 1229–1577. The local spelling is now *Mappleton*. OE ***mapel** + **tūn**. Db 387.

MAPLIN SANDS Essex TR 0388. P.n. Maplin + ModE **sand** 'a sand-bank'. Maplin occurs in *Mablynge Swene* 'Maplin channel' 1421, as opposed to East Swin in the Blackwater estuary, *Atteswenne* 1310, *le Suyn* 1362, *The Swyn* 1365, OE ***swīn** '(at) the channel'. *Mablynge* is unexplained. Ess 18 gives pr [meiplin], 16.

MAPPERLEY Derby SK 4343. 'Maple-tree wood'. *Maperlie* 1086, *-leya, -ley(e)* etc. early 13th–1269 etc. OE ***mapel** or **mapuldor** + **lēah**. Db 583, L 222.

MAPPERTON Dorset SY 5099. 'Maple-tree settlement'. *Ma(l)peretone* 1086, *Mapeldoreton* 1236, *Maperton* 1244. OE **mapuldor** + **tūn**. Do 104.

MAPPLEBOROUGH GREEN Warw SP 0866. *Mapleborough Green* 1656. P.n. *Mepelesbarwe* [840×52]12th S 191, *Mapelberge* 1086, *-berga -berewe* 13th, 'maple-tree hill', OE ***mapel** + **beorg**, + ModE **green**. The earliest form was OE **mapuldor** + **beorg** as is shown by the form *Mapoldreu geat* 'maple-tree gate' in the bounds of Beoley H&W SP 0669 972 S 786 and the form *Mapeldreborch* c.1200. A further form, *Mapeldosbeordi* [714]17th S 1250, is probably a spurious charter and certainly in a late unreliable copy. Wa 226.

MAPPLETON Humbs TA 2243. 'Maple-tree farm or village'. *Mapleton(e)* 1086–1599, *Mapeltun -ton(a)* 1160×2–1359, *Mapiltun -ton* 1230–1494. OE ***mapel** + **tūn**. YE 63.

MAPPOWDER Dorset ST 7306. 'The maple-tree'. *Mapledre* 1086, *-dra* 1088–1332, *Mapodr(e) -udr(e)* 1227–1428, *Mapowdre* 1438–47, *Mapowder* 1446–1567. OE **mapuldor**. Do iii.259.

MARAZION Corn SW 5130. 'Little market'. *Marghasbigan* c.1200, [c.1265]14th, *-byghan -bi(gh)an* 1302–89, *Marhasbean* 1311, 1532, *Marhasvean* 1591, *Marhasion* 1601. Co **marghas** 'market' + **byghan** 'small', Latinised and also translated into French in the surname of John de *Parvo Foro*, John de *Petitmarche* [c.1220]14th. 'Little' for distinction from Market Jew, the alternative name of the town but originally a separate place, *Marghasyou* [c.1210]14th, *Marchadyou* [c.1265]14th, *Marcasiou* 1310, *Marghasiewe* 1601, *Market Jew* 1610, 'Thursday Market', Co **marghas** + **yow**, referred to in [c.1080]12th as *marcatum die quinte ferie*. PNCo 115, Gover n.d. 601, CPNE 140, DEPN gives pr [ma'razi:n].

MARBURY Ches SJ 5645. 'Fortified place near the lake'. *Merberie* 1086, *-bury* 1260–1656 with variants *Mere-* and *-buri*, *Marburia* [c.1130]1479, *-bury* from 1289. OE **mere** + **byriġ**, dative sing. of **burh**. Che iii.106.

MARCH Cambs TL 4197. Possibly 'the boundary'. *Mercha -e* 1086, *Merch(e)* 1170–1355, *March(e)* from 1286. OE **mearc**, locative-dative sing. **merċe** < **marki-*. Alternatively this could be OE folk-n. *Merċe* 'the Mercians'. Ca 253.

MARCH GHYLL RESERVOIR NYorks SE 1251. P.n. March Ghyll, *Marches* 1847, probably the 'boundary ravine', OFr **marche** + ON **gil**, + ModE **reservoir**. The valley is crossed by the Roman road from Ilkley to Aldbrough, Margary no.720b, close to the boundary of Ilkley parish. YW v.67.

MARCHAM Oxon SU 4596. 'Smallage river-meadow'. *(æt) Merchamme* [900]12th S 358, *Mercham, (to) Merchamme* [965]13th S 734, *Mercham* 12th–1383×4 including *[835]13th S 278 and [1066×87]13th. OE **mereċe** 'smallage, wild celery, *apium graveolens*' + **hamm** 1. This plant grows near salt water and in brackish marshes. There is a salt-spring here and the plant still thrives. Brk 414, L 43, 49.

MARCHAMLEY Shrops SJ 5929. Partly uncertain. *Marcemeslei* 1086, *Merchemeslege -lega* 1185–c.1190×94, *Marc(h)emeleg(')* 1206–72, *Marchamlegh* 1229, *-ley(e)* from 1316, *Marchumleg(h) -ley(e) -leie -om-* 1242–1796, *Marshomley* 1695. This is usually taken to be the 'wood of the dwellers at *Mercham*', OE **Merċhæma-lēah* in which **Merċhæma* is the genitive pl. of **Merċhæme* 'the dwellers at **Merċhām*', the homestead or village of the Mercians, OE **Merċa-hām*. Doubt has been cast on this explanation because of the primarily southern distribution of *hǣme* and the preponderance of *-hum-* and *-hom-* sps from the second half of the 13th cent. until 1796. In spite of this doubt, however, no entirely satisfactory alternative has been proposed which takes account of the early sps in *-(h)eme(s)-*. Sa i.196 suggests *Merċum-lēah* '*Merċum* wood' in which *Merċum* is the dative pl. of *Merċe* 'the Mercians' used as a p.n. 'at the Mercians'. DEPN takes the specific to be OE pers.n. *Merchelm*, genitive sing. *Merchelmes*. It still seems tenable to view the ME sps in *-um- -om-* as reflecting loss of stress in the middle syllable of the name, OE **Mérchǽma-léah* > *Márchamléy* > *Márchumley*, although the question remains why this seems to be the case only here and in BARTHOMLEY Ches SJ 7652. Sa i.194–7.

MARCHINGTON Staffs SK 1330. 'Settlement of the dwellers at *Mercham*'. *Mærchamtun* 1002, *Merchametone* 1086, *Mercinton* 1154×89, *Mercington* 1230. OE **Merċhǣme*, genitive pl. **Merċhǣma*, + **tūn**. The lost p.n. Mercham, *æt Mærcham* [951]13th S 557, may be identical with MARCHAM Oxon SU 4596, or represent OE *Merċa-hām* 'the homestead of the Mercians'. DEPN.

MARCHINGTON WOODLANDS Staffs SK 1128. *Marchington Woodlands* 1836 OS. P.n. MARCHINGTON SK 1330 + ModE **woodlands**.

MARCHWOOD Hants SU 3810. 'Wild celery wood'. *Merceode* 1086, *Merchewude -wode* 1254–1490, *Marchewude* 1280. OE **mereċe** 'wild celery, smallage' + **wudu**. This plant thrives in wet ditches near the sea, appropriate to the situation of Marchwood. Ha 114, Gover 1958.199.

Little MARCLE H&W SO 6736. *Parva Marcle(y)* 1208, 1334, ~ *Markel* 1243. ModE adj. **little**, Latin **parva**, + p.n. *Merchelai* 1086, *Marchelai* 1148×9, *Markelai(a) -le(ia) -leye* 1160×70–1317, *Markehill* 1652, the 'boundary wood', referring to the boundary between the sub-kingdom of the Magonsæte and the Hwicce. 'Little' for distinction from Much MARCLE SO 6633. He 141, Bannister 129.

Much MARCLE H&W SO 6633. *Magna Markel(e)(y)* 1243–1358. ModE adj. **much**, Latin **magna**, + p.n. Marcle as in Little MARCLE SO 6763. He 141.

MARDALE COMMON Cumbr NY 4811. *Mardale Common* 1865. P.n. Mardale + ModE **common**. Mardale, *Merdale* 1278, *Mardale* from c.1540, is 'lake valley', OE **mere** + ON **dalr**. The lake is now Hawes Water. We ii.176, 169, SSNNW 239, L 26, 96.

MARDEN H&W SO 5147. 'Enclosure in the Maund'. *Mav-Maurdine* 1086, *Mawerdine -(a) -wurdin -wordin -worthin* 1125–1275, *Maurden(e)* 1252, 1291, *-dyn* 1286–1419. District name Maund as in MAUND BRYAN SO 5650 + OE **worthiġn**. He 143, Bannister 129.

MARDEN Kent TQ 7444. Possibly 'pasture for mares'. *Meredæn* [765×85]12th S 37, *Maeredaen* c.1100, *Mereden -denn(e)* 1218–40, *Merenden', Merdenn'* 1254. OE **miere**, Kt **mere**, + **denn**. But the specific could equally well be OE **(ġe)mǣre** 'boundary', **mere** 'pool' or pers.n. **Mǣra*. KPN 75, PNK 314.

MARDEN Wilts SU 0857. 'The fertile valley'. *Mercdene -a* [941]15th S 478, 1170, *Merh dœne* [963]12th S 715, *Meresdene* 1086, *Mergdena* 1167, 1169, *Meregh- Mergh(e)dene* 1280–1341, *Merewedene* 1298, *Merden(e)* 1242, 1275, 1406. OE **mearh, mearg** 'marrow, fat' + **denu**. The forms are consonant with this etymology showing the normal development of OE fricative *g* > *w*. The two earliest sps are therefore best seen as examples of sound substitution of stop [k] for voiceless [x] rather than as evidence for the alternative explanation, 'boundary valley', OE *mearc* + *denu*. Cf. Merrow Surrey TQ 0250, *Marewe* 1185–1556, *Merewe* 1187–1488, *Marrowe* 1583, 'fertile ground'. Wlt 321, Sr 142.

East MARDEN WSusx SU 8014. *Estmeredun* 12th. ME adj. **est** + p.n. *Meredone* 1086, *-dune* 1135×54, 'the boundary hill', OE **(ġe)mǣre** + **dūn** referring to the boundary between Hants and Sussex. 'East' for distinction from North MARDEN SU 8016, Upmarden SU 7914, *Upmerdon* 'upper M' [918×39]14th S 1206, and West MARDEN SU 7713. Sx 51 gives pr [mɑ:rn].

North MARDEN WSusx SU 8016. *Northm'den* 1288. ME adj. **north** + p.n. Marden as in East MARDEN SU 8014. Sx 51.

West MARDEN WSusx SU 7713. *Westmerdon* 1279. ME adj. **west** + p.n. Marden as in East MARDEN SU 8014.

MARE FEN Cambs TL 5488. Possibly the 'boundary fen' rather than 'mare fen' (a perfectly possible sense). *Marffen* 1636, *the Mare Fen* 1637. Ultimately from OE **(ġe)mǣre** and **fenn**. It lies on the parish boundary. Cf. Mare Fen in Swavesey, *Marefenne* 1677, also on a boundary. Ca 228, 173. O li gives pr [ma:fn].

MARE TAIL Lincs TF 4437. *Mare Tail* 1824 OS. ModE **mare** + **tail**. The reference is to a sand-bank in The Wash.

MAREFIELD Leic SK 7407. 'Martens' open land'. *Merdefeld(e)* 1086–1333, *Merðefeld'* 1177, *Mardefeld'* 1199–1426, *Merfeld* 1443–1502, *Marefeld* 1540, *-field* 1608. OE **mearth**, genitive pl. **meartha**, + **feld**. Lei 240, L 241, 244.

MAREHAM LE FEN Lincs TF 2761. *Mareham in the fenne* 1644, *Mareham le Fen* 1824 OS. P.n. *Marun* 1086, *Marum* c.1180–1341, '(the settlement) at the pools', OE ***marum**, dative pl. of **mere** replacing ***marim**, + Fr definite article **le** for *en le* + ModE **fen** for distinction from Mareham on the Hill TF 2868. OE *mere* is an *i*-declension noun < Gmc **mariz*. In this declension the dative pl. ending *-im* was early replaced by *-um* of the *a*-declension nouns. If Mareham is correctly explained as above with unmutated root vowels it must be a very ancient name. Possible alternative explanations are *marrum*, dative pl. of ON *marr* 'fen, marsh' or *marum*, dative pl. of OE *mare* 'silver-weed' and so 'among the silver-weed'. DEPN, ING 66, Campbell §600, Brunner §262 A.2, Grieve 449 s.n. Jewelweed.

MAREHAM ON THE HILL Lincs TF 2868. *Maring of the hill* 1517, *Mareham on the Hill* 1824 OS. P.n. *Mœringe* [c.1060]c.1359, *Meringhe* 1086, *Maringa* [1143×53]14th, 1154×89, *Maring(es)* c.1180–1342, *Marringes* 1190×3, *Meringa* 1154×89, *Meringes* 1219, + ModE **on the hill** for distinction from Mareham le Fen TF 2761 by the forms of which this p.n. has been influenced. The *œ* and *a* spellings cast doubt on the explanation from OE folk-n. **Meringas* 'the lake people' < **mere** + **ingas**. Possibly this is OE **maring* 'place growing with silver-weed' < *mare* + *ing*². DEPN, ING, Cameron 1998.

MARESFIELD ESusx TQ 4624. 'Open land by the marsh'. *Mersfeld(e)* 1234, 14th, *Mars(e)feld* 1240–1327, *Maresfeld* 1340. OE **mersc** + **feld**. Sx 349, SS 70.

MARFLEET Humbs TA 1429. 'Pool stream or creek'. *Mereflet -flot* 1086, *Merflet(e)* 12th–1395, *Marflet(e)* 1285–1539, *-fleete* 1598. OE **mere** + **flēot**. There is a small stream flowing from the mere, now a bog, into the r. Hull. YE 213 gives pr [ma:flit], Jnl 29.81.

MARGARETTING Essex TL 6601. 'Margaret's Ing'. *Gynge Margarete* 1328×30, *Margaretesyngge* 1345, *Margarettyng* 1500, *Margett-Inn* 1657. Saint's name *Margaret* from the dedication of the church + p.n. *Gingā* 1086, *Gi- Gynge(s)* 1238–1428, *Yenge* 1331, *Yng(ge)* 1386–7, as in INGRAVE TQ 6292. Ess 258 gives pr [mə 'gritən].

MARGATE Kent TR 3571. 'Gate(s) by the pool'. *Meregate* 1254–92, *Mergate* 1258. OE **mere** + **ġeat**, pl. **gatu**, dative pl. **gatum**. The reference is to a dip in the hills here. KPN 259, TC 134.

MARHAM Norf TF 7009. Partly uncertain. *Merham* [1042×66]12th S 1051, 1292, *Marham* from 1086. Usually said to be 'the mere homestead', OE **mere** + **hām**, with possible replacement of OE *mere* by ON *marr* 'fen, marsh'. The village is situated on a ridge of high ground overlooking Turf Fen so that the specific might rather be **hamm** 2a 'a promontory of high ground into marshland'. The specific could also alternatively be OE *mere* 'a mare' and so 'the horse farm'. Mares ran wild in the fens as we know from the *Reeve's Tale*. DEPN.

MARHAM CHURCH Corn SS 2203. 'St Marwen's church'. *Maronecirche* 1086, *Marwenecherche* 1275, (church of) *Sancta Merwenna alias Marwenchurch* 1400, *Marwynchurche alias Marhamchurch* 1417, *Morkam* 1610. Saint's name *Marwen*, one of the 24 sons and daughters of King Brochan, or English St Merwenn, a 10th cent. abbess of Romsey, from the dedication of the church, + OE **ċiriċe**. PNCo 115.

MARHOLM Cambs TF 1402. 'Enclosure or homestead by the mere'. *Marham* [1052×65]14th–1525, *Marram* 1428, *Marome* 1526, *Marholme* 1739. OE **mær(e)** + **hamm** or **hām**. *Marholme* is a 'learned' spelling. Nth 237 gives pr [mærəm].

MARIANSLEIGH Devon SS 7422. 'St Marina's Leigh'. *Marinelegh(e)* 1238–1337, *Seyntemarilegh'* 1242, *Maryon-Marne- Maryneslegh* c.1400, *Marle al. Marynalegh* 1465, *Marley al. Mariansleigh* 1747, *Mary Annesleigh* 1765. Saint's name *Marina* + p.n. *Lege* 1086, 'wood or clearing', OE **lēah**. The church here is dedicated to St Mary, but this may be an alteration of earlier Marina. D 382 gives prs [ma:li] and [mɛ:ri ænzli:].

MARK Somer ST 3847. 'Boundary house'. *Mercern* *[1065]18th S 1042, *(at) Merkerun* (OE), *Merke* (Latin) [1066×75]13th S 1241, *Merker* 1164, *Merk(e)* 1201, 1225, *Marcke* 1610. OE **mearc** + **ærn**. The boundary of Mark is *Merkemere* [973]14th S 793 (OE *ġemǣre*). DEPN.

MARKBEECH Kent TQ 4742. No early forms. Possibly 'boundary beech-tree', ModE **mark** + **beech** (OE **mearc bēce**).

MARKBY Lincs TF 4878. Either 'Marki's village or farm' or 'the village or farm at the frontier or wilderness'. *Marche(s)bi* 1086, *Marchebi* c.1115–c.1200, *Markebi -by* 1177–93. ODan pers.n. *Marki*, an early loan from LG, or ON **mǫrk** < **marku* + **bȳ**. Markby lies out in the fenland and there is no other ancient settlement site between it and the sea. DEPN, SSNEM 59

MARK CAUSEWAY Somer ST 3637. *Mark Causeway* 1817. P.n. MARK ST 3847 + ModE **causeway** referring to a raised path or road across the Somerset Levels to Mark.

MARK CROSS ESusx TQ 5831. 'Boundary cross'. *Markecross* 1509. ME **mark** (OE *mearc*) + **cross**. The cross marked the boundary between Rotherfield, Mayfield and Wadhurst. Sx 378.

MARKENFIELD NYorks SE 2967. 'Open land at the boundary'. *Merchefeld* 1086, *Merchingfeld* 1135×53, *Merkingfeld(e) -yng-* c.1190–1431, *Merkyngefelde* 1198, *Merkenfeld* 1229, 1297, *Markenfe(i)ld* from 1461. OE **(ġe)merċe**, ***merċing**, + **feld**. The *ing* and *en* forms may also point to an alternative form of the specific, viz. OE folk-.n. *Merċingas* or *Merċe* 'the boundary dwellers', genitive pl. *Merċinga, Mercna*. The medial /k/ of Markenfield points either to OE *Mercna-feld* or to ON substitution of /k/ for /tʃ/ in one of the other suggested etymologies. A lost 12th cent. p.n. in the township, *Ripestic*, is very probably OE *Hrypa* + **sticca**, 'stick, boundary mark of the *Hrype*, the people of RIPON' SE 3171. See also MARKINGTON SE 2965. YW v.177.

MARKFIELD Leic SK 4810. 'The open land of the Mercians'.

Merchenefeld 1086, *Merken(e)feld* after 1204–1346, *-ing- -yng-* 1258–1325, *Merkefeld'*, 1265–1535, *Marke-* 1404–1574, *Merkfeld* 1131–1484, *Markfeld(e)* 1425, 1576. OE *Merċe*, genitive pl. *Mercna*, + **feld**. Lei 576.

East MARKHAM Notts SK 7473. *Estmarcha' -ham* 1192–1240, *Estmarcam* c.1245, *Est Markham* 1292. ME adj. **est** + p.n. *Marchā* 1086–1175, *Markeham* 1211, the 'boundary village', OE **mearc** + **hām**. However, the boundary in question is not known. An alternative possibility might be OE **mearh** 'a horse', an element not previously identified in English p.ns. although the OG equivalent, *marah, march* is found in continental p.ns. as also is ON **marr** and OE **mere** 'a mare'. Also known as *Myche Markham* 1529×32 and *Grett Markham* 1541, 'Much' and 'Great M'. 'East' for distinction from West MARKHAM SK 7272. Nt 55.

West MARKHAM Notts SK 7272. *Westmarchā* 1086, *West Markham* from 1154×89. OE adj. **west** + p.n. Markham as in East MARKHAM SK 7473. Also known as *Little Markham* 1327 and *Parva Markham* 1588. Nt 56.

MARKINGTON NYorks SE 2965. 'Boundary settlement, settlement of the boundary people' or 'settlement of the people of Markenfield'. *Mercinga tun, (on) Mercingtune* c.1030 YCh 7, *Merchinton(e) -tona(m)* 1086–1166, *Merkinton -yn-* c.1190–1298, *-ing- -yng-* 1198–1488, *Markinton(e)* 1210×12, *Markington -yng-* from 1428. OE folk-n. **Merċingas* 'the people of the *(ġe)merċe* or boundary', genitive pl. **Merċinga*, varying with ***merċing** 'a boundary, a boundary place', + **tūn**. A lost p.n. in the parish is *Scharstykes* 1335, *Sherestikes* 1358, 'the boundary sticks', OE **scearu** + **sticca**. **Merċingas* might, however, be an elliptical folk-n. formed from the specific of MARKENFIELD SE 2967. The pronunciation with /k/ instead of /tʃ/ is due to ON influence. YW v.179, 184, xii, Årsskrift 1974.53.

MARKSBURY Avon ST 6662. 'Mæruc or Mæric's fortified place'. *(at) Merkesburi* [936]14th S 431, *Mercesburh* [941]12th S 476, *Merkesbiry* [959×75]13th ECW, *Mercesberie* 1086. OE pers.n. **Mǣruc* or **Mǣriċ*, genitive sing. **Mǣr(u)ces*, **Mǣr(i)ces*, + **byriġ**, dative sing. of **burh**. Wansdyke runs a mile N of the village and it is tempting to suggest *mearc* as the specific here referring to the dyke, but ME *Mark-* spellings would then be expected, not *Merk-* DEPN.

MARKWELL Corn SX 3658. 'Ælmarch's spring'. *ælmarches wylle* 1018 S 951, *Markewelle* 1199, *Markwill'* 1290. OE pers.n. *Ælfmearh*, genitive sing. *Ælfmearges*, + **wylle**. PNCo 115, Gover n.d. 227.

MARKYATE Herts TL 0616. 'Boundary gate'. *Markʒate* 1119–46, *Markegate -yate* 1248–1310, *Merc- Merk(e)yate* 1247–1352, *Market(t) Street(e)* 1660, 1675, *Marget* 1750. OE **mearc** + **ġeat**. The reference is to the Herts-Beds boundary. Hrt 47 gives pr [ma:kit].

Peters MARLAND Devon SS 4713. *Petermerland* 1242, *Merland(e) Sancti Petri* 1244, 1316, 1431, *Petrysmerland* 1303. Saint's name *Peter*, Latin genitive sing. *Petri*, from the dedication of the church, + p.n. *Mirland* 1086, *Merlanda* 1184, 'newly cultivated land by the mere', OE **mere** + **land**. There is much marshland in the valley of the river Mere, *Meer* 1630, a late back-formation from this name and MERTON SS 5212. 'Peters' for distinction from Little Marland SS 4911 *L^t* Marland 1809 OS, *Merland* 1086, also known as *Pye Merland* 1250, *Merland Pye* 1303, with unexplained manorial addition. D 97, 106, L 26, 248–9.

MARLBOROUGH Wilts SU 1969. Probably 'Mærla's barrow'. *Merleberge -berg(a)* 1086–1218, *Marleberg(e)* 1091–1302, *Mærle beorg* c.1150 ASC(E) under year 1110, *-burgh* 1246, *Marborowe* 1485. OE pers.n. **Mǣrla* + **beorg**. The reference is to an ancient mound which later formed the nucleus of the castle. An alternative suggestion for the specific is OE *mearġealla* 'gentian'. Identical with MALBOROUGH Devon SX 7039. Wlt 297 gives pr [mɔ:lbərə].

MARLBOROUGH DOWNS Wilts SU 1575. P.n. MARLBOROUGH SU 1969 + ModE **downs**. Cf. *le Downe* 1540, *le Downe in Marleborough* 1553. Wlt 17.

MARLCLIFF Warw SP 0950. 'Mearna's hill'. *Marnan clive, mearnan clyfe* [883×911]11th S 222, *(to) Marana cliue* [1005]16th S 911, *Marleclive -clyve* 1275, 1289, 1501, *Marlcleve* 1524, *-clive* 1656. OE pers.n. **Mearna*, genitive sing. **Mearnan*, + **clif**. Wa 202.

MARLDON Devon SX 8663. Probably 'gentian hill'. *Mergheldon(e)* 1307–8, *Merledone* 1307, 1370, *Mareldon* 1524. OE **meargealle** + **dūn**. D 516.

MARLESFORD Suff TM 3258. 'Mærel's ford'. *Mar- Merlesforda* 1086, *Merlesford* 1264, *Marle(s)ford* from 1235. OE pers.n. **Mǣrel*, genitive sing. **Mǣrles*, + **ford**. The same pers.n. has been posited for Marlston Chesh SJ 3864. DEPN, Baron, Che iv.163.

MARLEY GREEN Ches SJ 5845. *Marley Green* 1831. P.n. *Marley* [1621]1656, the 'boundary wood or clearing', OE **(ġe)mǣre** + **lēah**, + ModE **green**. Che iii.107, L 206.

MARLINGFORD Norf TG 1309. Partly uncertain. *Marding- Marþingforð* [c.1000]13th S 1525, *Merlinge- Marthingeforda* 1086, *Mearthingforde* [1087×98]12th, *Merlingeford* 1197. This might be 'ford at Marthing, the marten place', OE p.n. **Mearthing* < **mearth** + **ing**², locative–dative **Mearthinge*, + **ford**; or 'ford of the Mearthingas or Mearthlingas, the people called after Mearth or Mearthel', OE **Mearth(l)ingas* < pers.n. **Mearth* or **Mearthel* + **ingas**, genitive **Mearth(l)inga*, + **ford**. DEPN.

MARLOW Bucks SU 8586. 'Land left after the draining of a pond'. *Merelafan* (dative pl.) 1015 S 1503, *Merlaue* 1086–1240, *Merlaw(a)* 1189–1204. OE **mere** + **lāfum** (lOE *lāfan*), dative pl. of **lāf**. Bk 186.

Little MARLOW Bucks SU 8788. *Parva Merlauuia* 1209×19. ModE adj. **little**, Latin **parva**, + p.n. MARLOW SU 8586. Bk 186.

MARLPIT HILL Kent TQ 4447. *Marl pit Hill* 1819 OS. ModE **marl-pit** + **hill**.

MARNHULL Dorset ST 7818. Partly uncertain. *Marnhull(e)* from [1267]14th. Possibly 'Mearna's hill', OE pers.n. **Mearna* + **hyll** or 'soft-stone hill', OE ***mearn** + **hyll**. A soft limestone is found here. Do iii.166 gives pr ['ma:nəl].

MARPLE GMan SJ 9588. 'Boundary stream'. *Merpille* [early 13th]1287, *-pil(l)* 1288–90, *-pull* 1286–1492, *-pel* 1248–1398, *Merhull* 1283–1351, *Marple* from 1355. OE **(ġe)mǣre** + **pyll**. Marple lies on the r. Goyt on the historical boundary between Cheshire and Derbyshire. Che i.281.

MARR SYorks SE 5105. 'Fen, marsh'. *Marra, Marle* 1086, *Mar(r)a* c.1110–1250, *Mar* 1196–1525, *Marre* 1316–1733. ON **marr**. The reference is probably to a marshy patch on the hillside. YW i.74, SSNY 100, L 34.

MARRICK NYorks SE 0798. Uncertain. *Marige* 1086, 1252, *Marrich* c.1150, *Marrig(g)* 1157–1400, *Marrik(e) -yk* from 1301. The etymology of this name is very uncertain. Suggestions have included 'the horse ridge', ON **marr** + **hryggr** referring to a place where horses ran wild, although *marr* is only recorded in poetry, or 'the boundary ridge' or 'road', OE **(ġe)mǣre** + **hrycg** with Scandinavian [k] for [dʒ] or OE ***riċ** 'a narrow strip, a narrow road'. The actual boundary, however – that between the wapentakes of Gilling West and Hang West – is the river Swale, not the prominent ridge on which Marrick stands, and *ġemǣre* in p.ns. usually gives *Mear- Meer-* forms. YN 294, SSNY 257, L 169.

MARSDEN WYorks SE 0411. 'The boundary valley'. *Marchesden(e)* 1177×93–1349, *Marcheden(e)* 1275–1487, *Marsden* since 1540. OE **merċels** + **denu**. The reference is to the boundary between Yorkshire and Lancashire. YW ii.276.

MARSDEN BAY T&W NZ 3965. *Marsden Bay* 1863 OS. P.n. *Marston* 1768, *Marsdon* 1863, forms which are too late for certain explanation, + ModE **bay**.

Little MARSDEN Lancs SD 8536. *Little Merkelstene* 1242, *Little Merlesden* 1251, *parua Merclesden* 1296, *Little Mersden* 1496. ME adj. **litel** + p.n. *Merkesden* 1195ff., *Merkelesden* 1246, *Merclisden -es-* 1258–1332, *Marclesden* 1363, *Merseden* 1478, OE **merċels** 'a mark, a mark to shoot at, a boundary mark' + **denu**, and so possibly 'boundary valley'. The reference is either to the valley of Catlow Brook which flows through Nelson to join Pendle Water, or to the valley of Pendle Water itself. 'Little' for distinction from Great Marsden, *Majori Merkedenna* 1180×93, *Merclesden major* 1242, *Gret Merclesden* 1251. Marsden has been engulfed by the later settlements of Nelson and Brierfield. La 86, Mills 111, L 99, Jnl 17.53.

MARSETT NYorks SD 9086. 'Maur's shieling'. *Mouressate* 1283, *Moursette* 1285. ON pers.n. **Maurr*, secondary genitive sing. *Maures*, + **sǣtr**. **Maurr* would be a nickname from ON *maurr* 'an ant'. For the development ON *au* > *a*, cf. ADDLEBROUGH SD 9487, LASKILL SE 5690, SCRATBY Norfolk TG 5115. YN 264.

MARSH 'Marsh, marshland'. OE **mersc**, ME **mersh, marsh**.

(1) ~ Devon ST 2510. *Le Mersch in Ertecomb* 1307, 'the marsh in Yarcombe'. D 653.

(2) ~ GIBBON Bucks SP 6423. 'M held by the Gibwen family'. *Gibbemers* 1272, *Mersh Gibwyne* 1292, *Marsh Gibbyon* 1553×1601. Earlier simply *Merse* 1086–1284. The connection with the Gibwen family dates back to the 12th cent. Bk 54, L 53.

(3) ~ GREEN Devon SY 0493. *Marsh Green* 1809 OS. P.n. *Mershe juxta Huntebeare* 1420, 'marsh beside Houndbeare', *Mershe Bawdyn* 1491 'marsh held by Baldwin the sheriff' in 1086, + ModE **green**. D 595.

(4) ~ GREEN Kent TQ 4444. *Marsh Green* 1819 OS. ModE **marsh** + **green**.

(5) ~ GREEN Shrops SJ 6014. *The Marsh Green* 1788ff. ModE **marsh** as in *The Marsh* 1833 OS + **green**. Gelling.

(6) ~ MARGARET Dorset ST 8218. 'Margaret's marsh'. *Margaret(t) Marsh(e)* from 1575. Short for earlier *Margaretysmerschchurche* 1395, 'the church of Margaret Marsh'. The reference is probably to Margaret Auch(i)er or Margaret de Leukenore, abbesses of Shaftesbury Abbey 1315–29 and 1350–62 respectively, to which the marsh belonged, rather than to St Margaret to whom the church is dedicated. 'Margaret' for distinction from GUY'S MARSH ST 8420, another division of the same marsh. Do iii.163.

MARSHALLS HEATH Herts TL 1515. *Marshalls Heath* 1834 OS, *Martials Heath* c.1840. Surname Marshall, cf. Adam *le Mareschal* 1272×7, + ModE **heath**. Cf. *bosc. voc. grava Marescalle* 'wood called Marshal Grove' 1279, *Marshaleswode* 1390. Hrt 58.

MARSHAM Norf TG 1923. 'The marsh homestead'. *Marsam* 1086, *Marsham* 1252. OE **mersc** + **hām**.

MARSHAW Lancs SD 5953. Partially uncertain. *Marthesagh* 1323, *Marcheshawe* 1324, *Marchshagh*, *Marschasheued* c.1350. The generic is OE **sceaga** 'a copse, a small wood', the specific probably ME **marche** 'a boundary' (OFr *marche*) or, if the 1323 spelling is reliable, OE **mearth** 'a marten, a weasel'. In either case the name has subsequently been reformed with ME **mershe**, **marshe**, OE *mersc*. Marshaw lies beside the Marshaw Wyre a mile from the old boundary between Lancashire and the West Riding of Yorkshire. La 172, Mills 111.

MARSHBOROUGH Kent TR 3057. 'Mæssa's hill or barrow'. *Masseberge* 1086, *Messeberg(he) -bergh* 1246–1348, *Mersebergh'* 1348. OE pers.n. **Mæssa* + **beorg**. PNK 588.

MARSHBROOK Shrops SO 4489. The name of a halt on the railway line from Church Stretton to Ludlow. No early forms of the name are recorded but cf. *Marsh Wood, Marsh Farm, Marsh Hill* 1833 OS.

MARSHCHAPEL Lincs TF 3599. *Mersch Chapel* 1347, *Marshchapell* 1457. ME **mersh** as in *Fullestowe merske* 'Fulstow marsh' early 13th + **chapel**. Marshchapel is a daughter settlement from Fulstow TF 3297 on land reclaimed from the sea. Its prosperity is due to salt manufacture. Jnl 7.54, 56, LAASR n.s. iv, 1.38, Cameron 1998.

MARSHFIELD Avon ST 7873. 'Open land on the boundary' sc. between the counties of Gloucestershire and Wiltshire. *Meresfeld(e)* 1086–1248, 1492, *Mersfeld* 1100×35–1535, *Marsfeld(e)* 1221–1675, *Mers(s)h(e)feld(e)* 1397–1493, *Mars(s)h(e)fe(i)ld(e)* 1414–1619, *Marshfield* 1712. OE ***mǣrse** or **(ġe)mǣres**, genitive sing. of **(ġe)mǣre**, + **feld**. *Marsh* is a late false etymology. Gl iii.59, L 53, 240, 243.

MARSHGATE Corn SX 1591. 'Gate at or leading to a marsh'. *Marsh Gate* 1748. ModE **marsh** + **gate**. PNCo 115, Gover n.d. 64.

MARSHLAND Humbs SE 7920. *Merskland(ia)* 1302–1412, *Mersshland* 15th cent., *Mars(s)hland* from 1473. OE **mersc** + **land**. The reference is to the large area of marsh along the S side of the Ouse. YW ii.2.

MARSHLAND Norf TF 5312. *Marshland* 1824 OS. ModE **marshland**.

MARSHLAND FEN Norf TF 5508. *Marshland Fen* 1824 OS. P.n. MARSHLAND TF 5312 + ModE **fen**.

MARSHLAND ST JAMES Norf TF 5209. A modern settlement name from MARSHLAND Norf TF 5312.

MARSHSIDE Kent TR 2265. This is *Marsh Row* 1814, 1819 OS. ModE **marsh** + **side**. The place lies at the edge of Chislet Marshes TQ 2365. Cullen.

MARSHSIDE Mers SD 3619. *Marsh Side* 1842 OS. A modern development in Southport beside former marshland.

MARSHSIDE SANDS Mers SD 3421. P.n. MARSHSIDE SD 3619 + ModE **sand(s)**. Called *The Wharf* 1842 OS, ModE **wharf** 'a sand-bank'.

MARSHWOOD Dorset SY 3899. 'Marshy wood'. *Merswude* 1188, *Mershwod* 1288, *Marshwode* 1358. OE **mersc** + **wudu**. Do 106, L 53, 228.

MARSHWOOD VALE Dorset SY 4097. *Merswodeuaal* 1319. P.n. MARSHWOOD SY 3889 + ME **vale** 'a wide valley'. Do 106.

MARSKE NYorks NZ 1000. 'The marsh'. *Mersche* 1086. OE **mersc** with [sk] for OE [ʃ] either due to Scand influence or from the dative pl. **merscum** '(the settlement) at the marshes'. YN 293 gives pr [mask].

MARSKE-BY-THE-SEA Cleve NZ 6322. 'By the sea' is a modern addition to p.n. Marske, *Merscum* [1043×69]12th, *Mersch(e)* 1086, *Mersc* 1086–1223, *Mersk(e)* [1180]copy-1401, *Marske* from 1442, *Mask(e)* 1577–1714, 'at the marshes', **OE mersc**, d.pl. **merscum**. This name is in systemic contrast with other dative pl. names in the district, e.g. ACKLAM NZ 4817, COATHAM NZ 5925, KIRKLEATHAM NZ 5921, UPLEATHAM NZ 6319. YN 154 gives pr [mask] which is still used locally today.

New MARSKE Cleve NZ 6220. A 19th cent. town built by Pease and Partners for workmen at their Upleatham ironstone mine from 1857. Pevsner 1966.239.

MARSTON 'Marsh settlement'. OE **mersc** + **tūn**. L 34, 53.

(1) ~ Ches SJ 6775. *Merstona* 1188, *-ton* 13th–1581, *Mers(s)h(e)ton* 1337–1584, *Marston* from 1330. Che ii.118.

(2) ~ H&W SO 3657. *Merstone* 1086–1250, *Mershton* 1326, *Marshton* 1334. He 156.

(3) ~ Lincs SK 8943. *Merestune -ton(e)* 1086, *Merstun(e) -tuna* 1160–1212, *Merston(e)* 1171–1723, *Mers(s)hton* 1292–1396, *Marston* from 1501. Perrott 383.

(4) ~ Oxon SP 5208. *Mersttune* [c.1069]13th, *Merston' -tona -ton(e)* 13th cent., *Mershton(e)* 1285, 1316, *Marston* from 1531. Earlier simply *(æt) Mersce* [1065×6]13th S 1148, [1065×6]14th S 1147, '(the settlement) at the marsh'. O 181, L 53.

(5) ~ Staffs SJ 8314. *Mersetone* 1086, *Merston* 1203–1428, *Marston* from 1567. St i.140, Horovitz.

(6) ~ Staffs SJ 9227. *Mertone* (sic) 1086, *Mershton* 1316. DEPN.

(7) ~ Warw SP 2095. *Merston(e)* 1086, *Merston* 1290 (*juxta La Lee* 1428). Wa 84, DEPN.
(8) ~ Wilts ST 9656. *Merston* 1198–1370, *Mers(c)hton(e)* 1289, 1291, *Marston* 1559. Wlt 244.
(9) ~ GREEN WMids SP 1785. *Marston Green al. Marston Culy* 1830. P.n. *Merstone* 1086, + ModE **green**. Also known as *Merston de Cu(y)ly, Quilly, Culey* 1262–1661, from Ralph de Cuyly who held the manor in 1272 and 1287. The addition is for distinction from another manor in the same place, *Waure Merston* 1257, *Wavers Marston* 1725, held by William de Waver in 1189×99. Wa 59.
(10) ~ MAGNA Somer ST 5922. 'Great M'. *Great Merston* 1248. Earlier simply *Merstone* 1086, *Merscetun* [1091×1106, 1135×7]14th, *Mers(c)etone* [1100×18–1135×66]14th, *Merston* [1174×80, 1211×52]14th, *Marston* 1610. DEPN.
(11) ~ MEYSEY Wilts SU 1297. 'M held by the Meysey family'. *Merston Meysi* 1259, *Marshtone Meysi* 1302, *Marston Measey* 1773. P.n. *Merston(e)* 1199–1239, *Mason* 1738, + manorial addition from the family name of Roger de Meysi who held here in 1211. Wlt 29 gives popular pr [mɑ:sən].
(12) ~ MONTGOMERY Derby SK 1338. *Marston Mountegomery* c.1350. Earlier simply *Merston(e)* 1243–1472. 'Montgomery' for distinction from MARSTON ON DOVE SK 2329. Held by William de Mungumerie in 1243. Db 585.
(13) ~ MOOR NYorks SE 4953. *mora de Merston* 1281, *Marston More* 1499, ~ *Moore* 1655. P.n. Marston as in Long MARSTON SE 5051 + ME **more** (OE *mōr*). YW iv.255.
(14) ~ MORETAINE Beds SP 9941. 'M held by the Morteyn family'. *Merston Morteyn* 1383, *Marston Morton* 1666. Earlier simply *Mer(e)stone* 1086–c.1335, *Mersh(e)ton* 14th cent., *Marson* 1558×1603. The bounds of the Marston people are referred to as *gemæru mersctuninga* [969]11th S 772. Bd 79, BHRS 9.5ff.
(15) ~ ON DOVE Derby SK 2329. *Merstun* 1086, *-tona -ton(e)* late 11th–1610, *Marston* 1453. 'On Dove' for distinction from MARSTON MONTGOMERY SK 1338. Db 586,
(16) ~ ST LAWRENCE Northants SP 5342. 'St Lawrence's M'. *Mersshton Scī Laurencii* 1330, *Lawrence, Larrens Marston* 16th. P.n. *Merestone -a* 1086–1336, *Merstone(e)* 1181–1396, + saint's name *Lawrence* from the dedication of the church and for distinction from MARSTON TRUSSELL SP 6985. Nth 54.
(17) ~ STANNETT H&W SO 5755. *Marston Stannett* 1896 Gaz. P.n. *Mersetone* (1×), *Merstvne* (3×) 1086, *Merston(a)* 1160×70–1333, *Maristun* 1304, + unidentified element, possibly a family name of manorial origin, or perhaps representing OE *stāniht* 'stony, rocky'. Also known as *Marston Chapel* 1832 OS and *Little Marston* Pevsner 1963.270. He 160.
(18) ~ TRUSSELL SP 6985. 'M held by the Trussell family'. *Merston Trussel* 1235, 1298. P.n. *Mersitone* 1086, *Merston(e)* 1220–1361 + family name of Richard Trussel who held the manor in 1233. Nth 117.
(19) Butler's ~ Warw SP 3250. 'M held by the Butler family'. *Merston le Botiler* 1176, *Mershton Boteler* 1291, *Bot(e)ler(i)smerston* 1387, *Butler Merston* 1544. P.n. *Mersetone* 1086, *-a* 1140, + manorial addition from the family of Ralph Boteler who held the estate c.1135×54. Wa 256.
(20) Broad ~ H&W SP 1446. *Brademerston'* 1224, 1241, *Brod(e)mers(h)ton* 1306–1504, ~ *Marston* 1571. Earlier simply *Merestvne* 1086. 'Broad' for distinction form Long MARSTON Warw SP 1548. Gl i.253.
(21) Fleet ~ Bucks SP 7715 → North ~ SP 7722.
(22) Lea ~ Warw SP 2093. *Leemerston* 1535, *Lea Marston* 1651. A combination of p.n. Lea, *Leth* 1086, *La Lee* 1221–1545, ~ *in Merston* 1306, 'the clearing, the pasture', OE **lēah**, and p.n. MARSTON SP 2095. Wa 84.
(23) Long ~ Herts SP 8915. *Longe Merston* 1333, *Longmarsson* 1603×25. ME adj. **long** + p.n. *Mers(s)hton* 1287, 1307, *Merston(e)* 1288, 1294. Hrt 52.
(24) Long ~ NYorks SE 5051. *Long Marston* from 1607. ModE **long** + p.n. *Mersetone* 1086, *Mersctona* 12th, *Merston(a) -tun* 1086–1524, *Marston* 1250–1546. 'Long' in allusion to the length of the village. YW iv.254.
(25) Long ~ Warw SP 1548. *Longa Merston(e)* 1285–1535. OE, ME adj. **long** + p.n. *Merstuna* [1043]15th S 1000, *Merston(e)* [c.1043]15th S 1226, 1198–1340, *Merestone* 1086. Also known as *Drey(e)- Dri(e)- Dry(e)merston(e)* 1248–1540, *Marston Sicca* (Latin) 1535, *Dry Marston* 1610. Gl i.248.
(26) North ~ Bucks SP 7722. *Normerstone* 1233. Earlier simply *Merstone* 1086. 'North' in relation to the lost village of Fleet Marston SP 7715, 'M by the fleet or stream', *Fletemerstone* 1223–1302, *Fletemarston* 1509, *Fleet Masson* 1690, OE **flēot** + p.n. *Merstone* 1086. Bk 107, 136.
(27) Priors ~ Warw SP 4857. 'M held by the prior' sc. of Coventry in 1242. *Prior(i)s Merston* 1316. ME **priour**, Latin **prior**, genitive sing. **prioris**, + p.n. *Merston* 1236, 1242. Wa 271.
(28) South ~ Wilts SU 1988. *Suthmershton* 1331, *Southemaston, Southmarson* 16th. ME **suth** + p.n. *Merston* 1204. 'South' for distinction from MARSTON MEYSEY. Wlt 29.

MARSTOW H&W SO 5519. 'Martin's holy place'. *Martinestowe* 1277, 1334. Saint's name *Martin* + OE **stōw**. Apparently a translation of *Lann Martin* [1045×1104]c.1130 LL, PrW ***lann** 'church-yard, church-site' + saint's name *Martin*. He 145.

MARSWORTH Bucks SP 9214. 'Mæssa's enclosure'. *(æt) Mæssanwyrðæ* [966×75]12th S 1484, *Missevorde* 1086, *Messeworth* 1200–1377, *Massewurth* 1201, *Mosseworth* 1338, *Mersworth* 1766. OE pers.n. **Mæssa*, genitive sing. **Mæssan*, + **worth**. **Mæssa* is the hypocoristic form of a name such as **Mǣrsīġe* or **Mǣrstān*. Bk gives pr [ma:zəθ].

MARTEN Wilts SU 2860. 'The pond settlement'. *Mar- Mertone* 1086, *Merton(e)* 1187–1428. OE **mere** + **tūn**. There is a small pond beside the road, the last source of water before the long climb of Chute Causeway Roman road over the Downs. Wlt 347, Jnl 24.38.

MARTHALL Ches SJ 8075. 'Weasel nook'. *Marthel(e)* late 13th–1616, *Marthall* from 1342. OE **mearth** + **hale**, dative sing. of **halh**. Che ii.82 gives pr [ˈmartɔ:l].

MARTHAM Norf TG 4518. 'The homestead or meadow where martens are seen'. *Martham* from 1086. OE **mearth** + **hām** or **hamm**. Nf ii.59.

MARTIN Hants SU 0619. 'Boundary settlement'. *Merton(e)* [944×6]14th S 513–1428, *Meretun* 1225, *Merten* 1316, *Mertyn* 1491, *Martin* 1756. OE **(ġe)mǣre** + **tūn**. Martin lies close to the boundary with Wiltshire to which it formerly belonged and also close to the great earthwork called Bokerly Ditch SU 0220–0616 which has been interpreted as the 5th–6th cent. boundary between the Saxons and the Celts and may be the feature referred to here. Wlt 402, Ha 114, Jnl 24.39, Newman-Pevsner 69, 104.

MARTIN Lincs TF 1259. Probably 'the boundary settlement'. *Martuna* [1135×54]13th, *Marton* 1185–1580, *Merton* 1388–1821, *Marten* 1512, *Martin* from 1625. OE **(ġe)mǣre** + **tūn**. Martin lies near the boundary between Kesteven and the S Riding of Lincs. Jnl 24.33 and Cameron 1998, however, take it as an example of OE *mere-tūn* 'a pond settlement'. Perrott 329.

MARTINDALE Cumbr NY 4319. 'Martin's valley'. *Martindale -yn-* from 1220×47. ME pers.n. *Martin* + ON **dalr**. The 17th cent. chapel dedicated to St Martin takes its name from this unidentified pers.n. We ii.215, SSNNW 239, L 96.

MARTINDALE COMMON Cumbr NY 4317. P.n. MARTINDALE NY 4319 + ModE **common**. In the Middle Ages Martindale Forest, *forestam de Martyndale* 1220×47, was a hunting preserve (ME **forest**). We ii.219.

MARTIN DALES Lincs TF 1761. *Martin Dales* 1824 OS. P.n. MARTIN TF 1259 + ModE **dale** (OE *dāl*, ON *deill*) 'a share' sc. of the common open field or, as here, of the meadow.

MARTIN DROVE END Hants SU 0520. Cf. *Droveende,*

Drovefurlong 1518, *Drove End* 1811 OS. P.n. MARTIN SU 0619 + ModE **drove** (OE *drāf*) 'a road along which cattle or sheep are driven'. W 403.

MARTINHOE Devon SS 6648. 'Hill-spur of the Mættingas'. *Matingeho* 1086–1219, *Mattynho* 1303, *Martenhoo al. Mattinghooe* 1702. OE folk.n. **Mættingas* 'the people called after Mætta', pers.n. *Mætta* + **ingas**, genitive pl. *Mættinga*, + **hōh**. There is a high spur of land here running out to the coast. The modern form may have been influenced by the nearby name Combe Martin. D 65.

MARTIN HUSSINGTREE H&W SO 8860. *Marten Hosentre* 1535, *Martin Hosyngtre* 1545–1600. Two originally distinct place names, Hussingtree, *(in, to) husantreo* 972 S 786, *Hvsentre* 1086, *Hosintre -yn-* 1255–1525, 'Husa's tree', OE pers.n. *Hūsa*, genitive sing. *Hūsan*, + **trēow**, and Martin, *(on) meretune* 972 S 786, *Merton* 1256–1428, 'mere settlement', OE **mere** + **tūn**, a settlement offering water to travellers on the Roman road from Birmingham to Gloucester, Margary no. 180. There were or are small ponds in the village. Wo 213 gives pr [hasentri:], Jnl 24.38.

MARTINSCROFT Ches SJ 6589. 'Martin's enclosure'. *Martinescroft(e)* 1332. ME pers.n. *Martin*, genitive sing. *Martines*, + **croft**. La 96.

MARTINSTOWN Dorset SY 6488. 'St Martin's town'. *Martyn towne* 1494, *Martynston* c.1500, *Martyn(e)s Towne* 1569, 1615. Originally called *Wintreburne* 1086 and *Wynterburn Sancti Martini* 'St Martin's Winterburne' 1244, 1268 etc., from the dedication of the church, one of five different estates in the WINTERBO(U)RNE valley south of Dorchester. Do i.375.

MARTLESHAM Suff TM 2446. Partly uncertain. *Merlesham* 1086, 1289, 1533, *Marlesham* 1216–91, *Martelsham* 1254, 1316, *Martlesham* from 1286, *Martyllesham* 1533, *Mertlesham* 1610. The earliest sps point to 'Mærel's homestead', OE pers.n. **Mǣrel*, genitive sing. **Mǣrles* as in MARLESFORD TM 3258, + **hām**. In this case the *t*-forms show intrusive *t* between *r* and *l* parallel to the dial. development of *d* in words like *parlour* and *curls*. DEPN, Arnott 12, Baron, EDG §298.

MARTLESHAM HEATH Suff TM 2445. *Martlesham Heath* 1805 OS. P.n. MARTLESHAM TM 2446 + ModE **heath**.

MARTLEY H&W SO 7560. 'Marten wood or clearing'. *Mærtleages* genitive sing. c.1030, *Mertlega* 11th, 1178, *Merlie* (3×), *Mertelai* (1×) 1086, *Merlega* 1155–92, *Martelea* 1184, *Martleye* 1327. OE **mearth** + **lēah**. Wo 62, L 205.

MARTOCK Somer ST 4619. Uncertain. *Mertoch* 1086, *Mærtoc* 1086 Exon, *Mertok* [1142×66]Bruton, 1176, 1230, *Meretok -oc* 1204, 1209, *Merestok* 1265, *Martoc* c.1350, *Martock* 1610. Possibly a compound name in OE **stoc** 'outlying farm' with loss of *-s-* as in the DB form for Edstock ST 2340, *Ichetoche* 1086, *Ichestoke* 13th, 'Icca or Ycca's outlying farm'. If this is so the specific is probably OE **mere** 'lake' rather than **mearc** 'boundary'. Martock would then be the 'lake *Stoc*' in contrast with STOKE SUB HAMDON ST 4717. DEPN, Studies 1936.19.

MARTON 'Marsh settlement'. ON **marr** + **tūn**. Formally difficult to separate from MARTON 'pond settlement'.

(1) ~ Cleve NZ 5216. *Martun(e) -ton'* 1086 etc. YN 162, HistCleve 357.

(2) ~ NYorks SE 7383. *Martun -tone* 1086, *Marton'* from 1167. YN 76.

(3) ~ ABBEY NYorks SE 5869. P.n. MARTON-IN-THE-FOREST SE 5968 + ModE **abbey** referring to the lost Marton Priory, a house of Augustinian canons founded c.1150. Pevsner 1966.240.

(4) ~ -IN-THE-FOREST NYorks SE 5968. *Marton(a) in Galtres* 1278. P.n. Marton + ModE **in the forest**, referring to the forest of GALTRES SE 5663 (lost). Marton is *Martun* 1086, *Marton(a)* from c.1170. YN 28.

MARTON 'Pond settlement'. OE **mere** + **tūn** (ONb **mær-tūn**). Formally difficult to separate from MARTON 'marsh settlement'. Jnl 24.31–41.

(1) ~ Ches SJ 8568. *Merutune, Mereton* 1086, *Merton* mid.12th–1612, *Marton* from late 13th. The former existence of a lake here is commemorated in the name of Mere Farm and Mere Cottages, *The Mere* 1819, *Marton Mere* 1848. The mere, which covered about 14 acres, was drained in 1848–9. Che i.80, 82, Jnl 24.32.

(2) ~ Lincs SK 8481. *Martone* 1086, *Martuna* c.1115. DEPN, Jnl 24.37.

(3) ~ NYorks SE 4162. *Marton(e)* from 1086, ~ *in Burg(e)scire* 1203. There were two pools by the village. For Burghshire see ALDBOROUGH SE 4066. YW v.88, Jnl 24.38 no. C9.

(4) ~ Shrops SJ 2802. *Mertune* 1086, *-ton(')* 1228–72, *Marton* from 1255. Near the village is a lake called Marton Pool SJ 2902, *Martine Poole, meately large and plentiful of Fische* Leland c.1540. Sa i.199.

(5) ~ Warw SP 4068. *Merton* 1151–1618, *Marton* from 1535. DB *Mortone* 1086 belongs not here but under HILLMORTON SP 5373. Wa 140, DB Warw 16.35 note, Jnl 24.36–7.

(6) East ~ NYorks SD 9050. *Est Marton* early 13th. ME adj. **est** + p.n. *Martun -ton* 1086–1542. 'East' for distinction from West Marton SD 8950, *West* before 1217. YW vi.39, Jnl 24.38 no. C8.

(7) Great ~ Lancs SD 3335. *Great Marton* 1297 etc., *Merton Magna* 1327, ME adj. **grēt**, Latin **magna**, + p.n. *Meretun* 1086, *Merton(a)* 1176–1332 etc., *Mareton* 1183–4, *Marton* from 1249. The reference is to Marton Mere SD 3435, formerly a lake of considerable size reaching closer to the village. 'Great' for distinction from Little Marton SD 3434, *Little Marton* 1297. La 156.

(8) Long ~ Cumbr NY 6624. *Long(e) Marton* from 1616, *Langmartton* 1634, *Long Merton* 1698. ModE adj. **long** + p.n. *Meretun* c.1170, *Merton(e)* 1214–1478, *Marton* 1332–1617. Large and small fishponds are referred to here in 1272, *magnum, parvum vivarium de Merton*. 'Long' from the shape of the parish which is long and narrow, not the village. We ii.113, L 26.

(9) New ~ Shrops SJ 3434. *New Marton* 1837 OS. 'New' for distinction from Old Marton SJ 3434, *Old Marton* 1837 OS.

(10) West ~ NYorks SD 8950 → East MARTON NYorks SD 9050.

MARWOOD Devon SS 5437. 'Boundary wood'. *Merode, Merehode, Merevde* 1086, *Merew(o)de* 1219–1343, *Merwode* 1263, 1291. OE **(ġe)mǣre** + **wudu**. This place is on the boundary between Braunton and Shirwell Hundreds. D 50.

MARYLEBONE GLond TQ 2881. 'Mary's stream'. *Maryburne* 1453, *-bourne* 1492, 1509, *Marybon* 1542, *Marylebone* 1626. Saint's name *Mary* from the dedication of the church + ME **burn**. Reduction of *burn* to *bone* is common in local speech. The intrusive *le* is probably influenced by St Mary le Bow and other names in which the addition of the French definite article became fashionable. Probably it was wrongly understood to mean 'Mary the good'. The original name of the stream in question is Tyburn, *(andlang) teoburnan* 951 for ?959 S 670 (S 1450), *(into, andlang) t(h)eoburnan* [979×1016]14th, *Tiburne* 1086, *Tyburn'* 1216×72, *Tyborne otherwise called Maryborne* 1490, 1504, 'boundary stream', OE ***tēo** + **burna**. It rose in Hampstead, flowed through Marylebone, across Oxford Street and formed the boundary of the manor of Ebury which belonged to Westminster Abbey and Westminster itself. Mx 137, 6–7, RN 424, LMxAS n.s. 4.75–84, Encyc 923.

MARYLEBONE GMan SD 5807. No early forms. Presumably transferred from MARYLEBONE GLond TQ 2881.

MARYPORT Cumbr NY 0336. *Mary-port* 1762. A harbour was created here between 1750 and 1760 by Humphrey Senhouse who named it after his wife Mary. The original name was *Alynfoote* 1656, *Elmefoote* 1567, *Ellom-foot* 1722, *Elnefoot* c.1745, 'foot, i.e. mouth, of the river ELLEN'. Cu 305, Pevsner 1967.159.

MARYSTOW Devon SX 4382. 'Holy place of St Mary'. *Ecclesia Sancte Marie Stou, Stowe* 1266×78, *Stowe Beate Marie* 1282,

Marystowe 1344, *Marstow* 1790. Saint's name *Mary* from the dedication of the church + OE **stōw**. D 199.

MASHAM NYorks SE 2280. 'Mæssa's homestead or village'. *Massan* 1086, *Masham* from [1153]14th, *Massam* 1286–1530. OE pers.n. **Mæssa* + **hām**. Masham gave its name to a medieval *shire, Mashamshire -shyre, Massamshire* 12th–14th cents., p.n. Masham + OE **scīr**. YN 234 gives pr [mæsəm], YN 230.

MASHAM MOOR NYorks SE 1079. P.n. MASHAM SE 2280 + ModE **moor** (OE *mōr*).

MASHBURY Essex TL 6511. 'Mæcca's fortified place'. *Maisseberia, Mæisbyrig* [1068]1309, *Masceberiam* 1086, *Ma- Messebir(ig) -bury* 1203–1324, *Mas(sc)hebyri -bery -bury* 1287–1421, *Maysbury* 1540, 1544. OE pers.n. *Mæċċa* + **byriġ**, dative sing. of **burh**. Ess 488.

MASONGILL NYorks SD 6675. 'Titmouse ravine'. *Maisinggile* c.1200, *Maisingill -gyll* 1467, 1609, *Meysongill* 1719. ON **meis-ingr** + **gil**. YW vi.250.

Great MASSINGHAM Norf TF 7922. *Magna Massingham* 1254. ModE adj. **great**, Latin **magna**, + p.n. *Masincham* (3×), *Massingeham, Marsinghara* (sic), *Marsincham, Masingheham, Masinicham* 1086, *Massingham* from 1198, 'homestead of the Mæssingas, the people called after Mæssa', OE folk-n. **Mæssingas* < pers.n. **Mæssa* + **ingas**, genitive pl. **Mæssinga*, + **hām**. DEPN, ING 137.

Little MASSINGHAM Norf TF 7924. *Massingham Parva* 1254. ModE adj. **little**, Latin **parva**, + p.n. Massingham as in Great MASSINGHAM Norf TF 7722. DEPN, ING 137.

MASSINGHAM HEATH Norf TF 7721. *Massingham Heath* 1824 OS. P.n. Massingham as in Great MASSINGHAM TF 7922 + ModE **heath**.

MATCHING Essex TL 5212. 'The Mæccingas, the people called after Mæcca'. *Matcingam -e, Metcingã* 1086, *Macinga* 12th, *-es* 1123–1236, *Machinges* 1163, 1286, *Mac(c)hing(ge)* 1232–1377, *Mescing* 1228, *Mecchinge -yngg* 1276, 1324, *Massing(g) -inges* 1254–1303, *Macyn -in* 1509, 1665–1709. OE folk-n. **Mæċċingas* < pers.n. **Mæċċa* + **ingas**. Ess 45, ING 37.

MATCHING GREEN Essex TL 5311. *Machynggrene* 1359. P.n. MATCHING TL 5212 + ME **grene**. Ess 46.

MATCHING TYE Essex TL 5111. Either 'Matching enclosure' or 'Matching held by the Tye family'. *Matchyngetye* 1281. P.n. MATCHING TL 5212 + either OE **tēag** or family name from Alexander de la Tye. Ess 46.

MATFEN Northum NZ 0371. 'Mata's fen'. *Matefen* 1159, *Mate(n)fen* 1182, *Matesfen* c.1190, *Mat(t)efen* 1200–1327, *Matfen* from 1220 BF. OE pers.n. *Mata*, genitive sing. *Matan*, + **fenn**. NbDu 140, DEPN.

MATFIELD Kent TQ 6541. Probably 'Mætta's open land'. *Mattefeld* 1227×38, 1258, *Mattfeld* 1254, *Mettefeld* 1275, 1346, *Macchfeld*1278. OE pers.n. **Mætta* + **feld**. PNK 193.

MATHON H&W SO 7345. 'The treasure'. *Matham* [1014]14th S 932, *Matma -e* 1086, *Madma -e* 1137×9, 1160×70, *Mathm(a)* 1163×86. OE **māthm** in the sense 'gift'. He 146.

MATLASKE Norf TG 1535. 'Speech ash-tree'. *Matelasc -esc* 1086, *-aske* 1198. OE **mæthel** + **æsc** with [k] for OE [ʃ] from ON **askr**. This must have been the hundred meeting place. DEPN.

MATLOCK Derby SK 3060. 'Speech oak-tree'. *Meslach* 1086, *Matlac -k* 1196–1330, *-oc -ok(e)* 1204–1597, *Matlock(e)* from 1323. OE **mæthel** + **āc**. The 1086 form shows typical AN confusion between [θ] and [ç] spelled *s*. Hundred or wapentake meeting-places were frequently held at a spot marked by a prominent tree. Db 388, L 219.

MATLOCK BATH Derby SK 2958. *Matlock Bath* 1840 OS. P.n. MATLOCK + ModE **bath**. Medicinal springs here came into repute about 1698.

MATSON Glos SO 8515. 'Mæthhere or Mattere's hill'. *Mat(t)res-don(e) -dunn* 1121–1331, *Mattesdon(e) -dun* 13th–1542, *Matson* from 1535. OE pers.n. *Mǣthhere*, **mǣthere** 'a mower' or **Mattere* or ***mattere** 'a mat maker', genitive sing. **Mǣthheres*, **Matteres*, **mǣtheres**, ***matteres**, + **dūn**. Gl ii.167.

MATTERDALE END Cumbr NY 3923. *Matterdail End* 1663, 'the end of Matterdale', p.n. Matterdale + ModE **end**. Matterdale, *Mayerdale* (for *Maþer-*) c.1250, *Matherdal(e)* 1323–1419, *Madredale* 1380, *Matterdall* 1559, *-dale* 1615, is 'madder valley', ON **mathra** + **dalr**. The plant in question was probably a species of Galium or Asperula. Cu 221, SSNNW 146, L 96.

MATTERSEY Notts SK 6989. 'Mæthhere's island'. *Madressei, Madreisseig* 1086, *Mathersai -ey(e) -ay* 13th–1343, *Mareseia -eie* 1176–1330 with variants *-eye -aie*, *Marsey(e)* 1267–1321, *Mattersey al. Marsey* 1539, 1690. OE pers.n. *Mǣthhere*, genitive sing. *Mæthheres*, + **ēġ**. Nt 86 records an obsolete pr [ma:si].

MATTINGLEY Hants SU 7357. 'Wood or clearing of the *Mattingas*, the people called after Matta'. *Matingelege* 1086, *-leia -le(g)a* 1100×35–1248, *Mattingele(g) -ly* 1205–50. OE folk-n. **Mattingas* < pers.n. **Matta* + **ingas**, genitive pl. **Mattinga*, + **lēah**. Ha 115, Gover 1958.121.

MATTISHALL Norf TG 0501. Probably 'Mætt's nook'. *Mateshala* 1086, *-hal* 1200, *Matteshala* 1193×4. OE pers.n. **Mætt* genitive sing. **Mættes*, + **halh**. A possible alternative is to regard the specific as a unique instance of the rare OE *m(e)att* 'a mat' perhaps used of a patch of high ground here if it was masculine rather than feminine gender[†]. The nook is the small valley in which the village stands. DEPN.

MATTISHALL BURGH Norf TG 0511. 'Mattishal Berg'. P.n. MATTISHALL TG 0511 + p.n *Berk* 1204, *Parva Berg* 'little Berg' 1254, OE **beorg** 'a hill'. 'Mattishall' for distinction from SOUTHBURGH TG 9904, *Berg Maior* 'greater Berg' 1254. DEPN.

MAUGERSBURY Glos SP 2025. 'Mæthelgar's fortified place'. *Meilgaresbyri*' *[714]16th S 1250, *(æt) Mæþelgares byrig* [949]c.1600 S 550, *into Mædelgares byrig* [n.d.]12th S 1553, *Maðelgæresbyri* *[1016]12th S 935, *Malgeresberiæ* 1086, *Malgaresburi -biria -byr' -bury(e)* early 13th–1590, *Malgarsbury* 1374, 16th, *Maw- Maugersbury(e)* 1557, 1572, 1682. OE pers.n. **Mæthelgār* (possibly from OG *Madalger*), genitive sing. *Mæthelgāres*, + **byriġ**, dative sing. of **burh**. The reference is to the Iron Age hill-fort on Icomb Hill SP 2023. Gl i.222 gives pr ['mɔ:gəzberi].

MAULDEN Beds TL 0538. 'Cross hill'. *Meldone* 1086, *Meudon* 1152×8, *Mald- Maud- Meaudon* 12th cent., *Maulden* c.1550. OE **mǣl** + **dūn**. For another example of a hill marked by a cross cf. TRIMDON Durham NZ 3634. Bd 80, L 145, 151.

River MAUN Notts SK 6166. *Mome* [1216×72]16th, c.1300, 1330, *Mone* 1335, *Maun* 1327×77, 1650, *Man(n)* c.1450–1663, *Mam* c.1490. With these forms must be taken the name of the source of the river, *Mammesheved* 1232, 1339, *Manshead al. Manswell* 1663. It seems likely that the element **mamme** was not originally a river name but a hill name as in MAM TOR Derby SK 1283, referring to one of the rounded sand-stone hills near Mansfield. The r.n. is then a back-formation from *Mammesheved*, properly 'head of the hill called *Mamme*' misunderstood as the 'source of the river *Mamme*'. The modern form of the name has been influenced by the assimilatory change *m* > *n* in the p.n. MANSFIELD. Nt 5, RN 280.

MAUNBY NYorks SE 3586. 'Magni's farm or village'. *Manne(s)bi* 1086, *Magnebi -by* 1157–1301, *Maun(e)by* from 1310. ON pers.n. *Magni*, genitive sing. *Magna*, + **bȳ**. OFr pers.n. *Magne* would also be possible. YN 274, SSNY 33.

MAUND AUBIN H&W SO 5549 → MAUND BRYAN SO 5650.

MAUND BRYAN H&W SO 5650. 'Maund held by Brian' son of Nicholas de Machena from whom this portion of the Maund district was earlier known as Maund Nicholas. *Magne*

[†]The gender is unknown but as a loan from late Latin *matta* it might be thought to have been feminine, with either strong genitive **matte* or weak genitive **mattan*, not **mattes*.

Nicholai 1195, *Mawen(e) Nichol* 1303–46, *Mawne Nicholl* 1431, *Magene Brian'* 1243, 1292. District name Maund + pers.n. *Nicholas* from Nicholas de Magen' fl. 1143×55 and his son *Brian*. Maund, as in Maund Maurice, Maund Aubin SO 5549, *Magena Maur'* 1160×70, *Mawene Aubin* 1240 from Maurice de Machena fl. 1148×62 and Robert de Sancto Albino fl.1148×54, later Rowberry, *Ruberh* 1148×55, 'rough hill', and Whitechurch Maund SO 5649, *Manneswyteschirche* 1317, *Witechurche Magne* 1428, Rosemaund SO 5648, *Mag(g)e, Magene, Magga* 1086, *Rous Maune* 1373, 'Thomas le Rus's Maund', and MARDEN SO 5147, is an ancient district name, *Magana* [675×90]13th S 1798, whose inhabitants were known as the *Magonsǣte* 'the dwellers beside Maund', *on Magonsetum* 811 S 1264, *in pago Magesætna* 'in the district of the Magonsæte' 958 S 677, *mid Mage sæton* 'among the Magesæte' c.1120 ASC(E) under year 1016. Two explanations of this name have been offered, neither of them satisfactory. One relates it to the RBrit p.n. *Magnis*, Kenchester SO 4442, from Brit **magno-* 'stone, rock'. Kenchester, however, is far to the W of Maund. Furthermore Brit internal *gn* had become *in* by c.550 and **magno-* was normally borrowed into OE as PrW **main* as in BROADMAYNE Dorset SY 7286, Mainstone D 228, Redmain Cu 267, Triermain Cu 116. The ME forms of Maund with *-w-* can only derive from an OE *Magan*. The other explanation very hesitantly suggests OE *maga* 'stomach', dative pl. *magum*, late OE *magan*, 'at the stomachs' referring to the flood-plain of the r. Lugg. This would, however, be the sole instance of OE *maga* used in a topographical sense. The origin of the name Maund remains in fact obscure. He 13, 38–9, Archeologia 93.39, Britannia 1.76, LHEB 466 n.3, Jnl 1.49, RBrit 407, Signposts 102-5, DB Herefordshire 7.5, WM 82.

MAUND MAURICE H&W → MAUND BRYAN SO 5650

MAUTBY Norf TG 4812. 'Malti's farm'. *Malteb(e)y -bei* 1086, *Maltebi -by* 1168–1342, *Mautebi -by* 1198–1482, *Mautby* from 1448. ON pers.n. *Malti*, genitive sing. *Malta*, + **bȳ**. Nf ii.10, SPNN 296.

MAWBRAY Cumbr NY 0846. 'The maidens' fort'. *Mayburg'* 1175, c.1187, *-burch -burgh -broughe* 1262–1581×97, *Mawbray* from 1552, *Mawbroughe* 1605, *Mawburgh or Malbray* 1816. OE **mǣġe** varying with genitive pl. **māga**, + **burh**. The reference is to the Roman fort at this place. For *maiden* in p.ns. see MAIDEN CASTLE Dorset SY 6688. Cu 296.

MAWDESLEY Lancs SD 4914. 'Maud's clearing or pasture'. *Madesle* 1219, *Moudesley -legh* 1269–1332, *Maldislei* 1295, *Maudesley* 1398, 1500 etc., *Mawdesley* 1485×6. ME pers.n. *Maud* < OFr *Mahaut*, genitive sing. *Maudes*, + **lēah**. La 136, Jnl 17.77.

MAWGAN Corn SW 7025. 'St Mawgan's (church)'. *Sanctus Mawan* 1086, *Santmauuant* 1086 Exon, *(in) Sancto Maugan, (villa) Sancti Ma(l)gani* 1206, *(Ecclesia) Sancti Maugani* 1291, 1342, *Seynt Mawgan in Maneke* 1470, *Maugon* 1610. Saint's name *Maucan* from the dedication of the church. He was venerated in Wales as St Meugan and in Brittany as St Maugan or Mogan. For *Maneke* see MENEAGE. PNCo 117, Gover n.d. 570, DEPN.

MAWLA Corn SW 7045. Unexplained. *Maula* 1302–91, *Mowla* 1490, *Mawla* 1610. PNCo 118, Gover n.d. 364.

MAWNAN Corn SW 7827. From the patron saint of the church. *(Ecclesia) Sancti Maunani* 1281–1361, *Seynt Maunan* 1398, *Mawnan* 1535 Gover n.d. Nothing is known of this saint. PNCo 118, Gover n.d. 519, DEPN.

MAWNAN SMITH Corn SW 7728. *Mawnan smith* 1699. Probably from a smithy at the crossroads here. P.n. MAWNAN SW 7827 + ModE **smith** 'a smithy' < OE *smiððe*. PNCo 118, Gover n.d. 520.

MAXEY Lincs TF 1208. 'Maccus' island'. *Macuseige* [?963]12th S 1448, *Macusi(g)e, Macesige* [963×92]12th B 1130, *Makesia -eya -eia -e(y)* 1184–1428, *Maxey* 1515. Irish pers.n. *Maccus* + **ēġ**. Nth 237.

MAXSTOKE Warw SP 2386. 'Maca's outlying farm'. *Makestoka -e* 1169–1538, *Macstok* 1262, *Maxstoke* 1538. OE pers.n. **Mac(c)a* + **stoc**. DB *Machitone* 1086 belongs not here but to Mackadown SP 1686. Wa 87.

MAYFIELD ESusx TQ 5827. 'Open land where mayweed grows'. *Magavelda* 12th, *Mawefeld* 1253, *Mathe- Megthefeud* 13th, *Maighfelde al. Mawghfelde* 1547, *Mavill, Mayfyld* 16th. OE **magethe, mæġthe** + **feld**. OE *magethe* develops to *maw-*, *mæġthe* to *may-*, cf. the forms of this word cited in OED, *mawthen, maythe*: *th* was lost as in the word *mayweed* itself. *Mesewelle* 1086 may be an error for *Megevelle*: if so it may be added to the above forms. Sx 381 gives prs [meivul] and [mefəl], SS 70.

MAYFIELD Staffs SK 1545. 'Open land growing with madder'. *Medevelde* 1086, *Matherfeld* c.1180–1327, *Mathelfeld* 1309, *Mathe- Mayfeld* 14th. OE **mæddre** + **feld**. DEPN, Duignan 100, Horovitz.

MAYFORD Surrey SU 9956. Partly uncertain. *May(n)- Mai(n)- Mein- Meyford(e)* 13th cent. Possibly 'Mæga's ford', OE pers.n. *Mǣġa*, genitive sing. *Mǣġan*, + **ford**. Sr 158.

MAY HILL Glos SO 6921. *May Hill* 1703. The name is said to refer to a folk custom of adjacent parishes contending for possession of the hill on May day. Gl iii.193.

MAYLAND Essex TL 9101. Partly uncertain. *la Mailanda* 1188×92, *Mayla(u)nde -lond(e)* 1202×3–1323×9. Possibly OE **mæthe** 'mayweed' + **land** 'newly cultivated land' or **æt thǣm ēġlande** 'at the island' with initial *M-* carried over from the definite article *thǣm*. Ess 219, DEPN, Jnl 3.41.

Fryer MAYNE Dorset SY 7386, Little MAYNE SY 7287 → BROADMAYNE SY 7286.

MAYPOLE GREEN Norf TM 4195. *May Pole Green* 1838 OS. ModE **maypole** + **green**.

King's MEABURN Cumbr NY 6221. *Meburn(e) Regis* 1279–1429, *Ki- Kynges Meburn(e)* 1292–1368, *Kings Meaburn* from 1617. Earlier simply *Mebrun(n)* 1158–1241, *Meburn(e)* 1200–1601. The earliest form is *Meabruna Gerardi* 12th, 'Gerard's Meaburn'. Originally a stream name or descriptive term of the river Lyvennet, it probably means 'meadow stream', OE **mǣd, mēd** + **burna**: cf. the forms for Maulds MEABURN NY 6216 referring to the same stream. The manor belonged to the crown. We ii.141, L 18, 250.

Maulds MEABURN Cumbr NY 6216. Stream name Meaburn (referring to the river Lyvennet), *Mai- Mayburne* 1154×89, c.1200, *Mebrun(e) -born(e)* 1153×82–1210, *Meburn(e)* 13th–1602, *Medbrunne* 13th, *Meu- Mew(e)b(o)urn* 1282–1440, *Meaburn* from 1575, + suffix *Mauld(is), Maulds* c.1210–1407, *Matildis* 1244, *Maud(e)* 1278–1422, referring to Maud, wife of William de Veteriponte who granted her the estate here c.1174. The occasional spellings *Med-* and *Mew(e)-* point to OE **mǣd, mēd** and **mǣwe**, all meaning 'meadow' as the specific of the stream name. We ii.156 gives pr ['mɔːdz'miːbən], L 18, 250.

MEADLE Bucks SP 8005. 'Meadow hill'. *Ma- Medhulle* 1227, *Medell* 1541. OE **mǣd** + **hyll**. Bk 172.

MEADOWTOWN Shrops SJ 3101. *Medowtowne* 1577, *Meadow Town* 1836 OS. ModE **meadow** + **town**. Gelling.

MEAL BANK Cumbr SD 5495. Uncertain. *Meal(e)ban(c)k(e)* 1578–1612. *Meal-* in Cumbrian p.ns. is ambiguous, derivable variously from OE **mǣle** 'coloured, variegated', ON **melr** 'sand-hill, bent-grass', ON **methal** 'middle', or even PrW ***mę̄l** 'bare hill'. Cu 136, 37.

Raven MEALS Mers SD 2705. *Ravenesmeles* 1190×4, 1246, 1269, *-muels -mo(e)les* 1232–84, *Ravenmeles* 1468, *Ravens Meols* 1842 OS. ON pers.n. *Hrafn*, secondary genitive sing. *Hrafnes*, + p.n. *Mele* 1086, *Molas* 1094, ON **melr** 'a sand-bank', secondary ME pl. **meles**. 'Raven' for distinction from North MEOLS SD 3518. The old manor has been partly washed away by the sea. La 125, SSNNW 147.

MEALSGATE Cumbr NY 2042. Unexplained element + ModE **gate**. *Meals Gate, Meal's Gate* 1777. Cu 270.

MEARBECK NYorks SD 8160. 'The boundary stream'. *Mearbeck(e)* 1554, 1637. OE **(ġe)mǣre** + ON **bekkr**. The stream from which the place is named forms part of the boundary between Long Preston and Settle. YW vi.152.

MEARE Somer ST 4541. 'The lake'. *Mere* 1086, 1225, *Meare* 1610. OE **mere**. Short for *Ferramere* [670]14th S 227, [680]14th S 1249, *[971]14th S 783, [c.1125]12th, 1154×89, *Ferremere* *[725]12th S 250, [c.1125]12th, *Ferlingemere* [725]14th, *Fearningamere* [1016]12th S 935, *Feringemere* [1129×39]14th, *Farningmere* 1168, *Faryngmere* [1230]14th, *Ferningemere* [1218×9]c.1500, *Faringmere* 1307×27, *Ferlyngmere* 1327×8, 'mere of the Fearningas, the fern-dwellers', OE folk-n. **Fearningas*, genitive pl. **Fearninga*, + **mere**. Described in DB as an island with 10 fishermen and three fisheries. It lay by a lake of five miles circumference which was drained late in the 18th cent. An Iron Age lake village once existed here. DEPN, Turner 1950.119, Pevsner 1958S.233.

MEARE GREEN Somer ST 3326. This is *Bear Green* 1809 OS, *Mare Green* 1968 Bartholomew. Earlier forms are needed.

River MEASE Derby SK 2711. Probably 'the bog, the swamp'. *Meis* 13th cent., *Meys* 1247–1347, *Mays* 1330, *Mese* 1577, 1586, 1610, *Meese* [c.1600] 1717. OE **mēos**. 'bog, swamp' transferred to the river which flowed through the swamp. In the area of MEASHAM SK 3312 the bog is still evident. The *ei, ey* spellings are assumed to be AN graphies for ME *ē* (< OE *ēo*) as in the case of the river TEES. If, however, the diphthong is original, the name would be unexplained. Db 12, St i.14, RN 280.

MEASHAM Leic SK 3312. 'The homestead on the r. Mease'. *Messeham* 1086, *Mei- Meysham* c.1160–1535, *Mai- Mays(h)am* 1189×99–1541, *Measam* 1599, *Measham* 1722. R.n. MEASE SK 2711 + OE **hām**. Lei 562 gives pr [mi:ʃəm].

MEATHOP Cumbr SD 4380. 'Middle plot of enclosed land'. *Midhop(e) -hopp* 1184×90–1256, *Mithehop* 1190×1210, *Methop(e) -hopp* 1278–1650, *Meathop* 1700. OE **midd** replaced with ON **mithr** + **hop** 'piece of raised ground in marshland'. We i.76, SSNNW 240.

MEAUX Humbs TA 0939. 'Sand-bank lake'. *Melse* 1086–[1238]15th, *Melsa* 1149×50–1465, *Mealsa, Meausa* 12th cent., *Meaus* 13th cent., *Meaux* from 1291. ON **melr** + **sǽr** or OE **sǣ**. The name is exactly parallel to the Norwegian lake-name *Mœlsjø*, but in 1150 it became the site of an important Cistercian monastery and this resulted in the French influence on the name which was also associated with the name of the French abbey of Meaux, S.-et-M. The abbey chronicle offers an etymology of Mel-sa as *sapor mellis* 'flavour of honey' because of the 'amenities of the place' and the 'sweetness of religion'. YE 43 gives pr [mius].

MEAVY Devon SX 5467. Originally a river-name. *Mœwi* 1031 S 963, *Mewi* 1086[†], *Meuui* 1086 Exon, *Mewy* 1257–1372, *Mevie* 1571, *Meavie* 1589. The place is named from the river Meavy, *mœwi* 1031 S 963, *Mewyz* 1154×89, *Mewy* 1291, *Meve* 1589, a pre-English r.n. of unknown origin. Rav *Maina, Mavia* [8th]13th has been cited here; if *Mavia* is the correct reading it may be related to a root meaning 'lively, quick'. It is unlikely, however, that this stream was of sufficient importance to feature in Rav and *Maina* may, in any case, be the correct reading and irrelevant here. [v] for [w] is a late SW dial. feature. D 229 gives pr [meivi], 9, RN 282, RBrit 419.

MEDBOURNE Leic SP 8093. 'The meadow stream'. *Medburna* c.1076–[late 13th]1404, *-burn(e)* 1086[†]–1550, *-bourn(e)* from 1293, *Medeburn(e)* 1286–1402, *Meadbourn* 1700. OE **mǣd** + **burna**. A broad stream divides the length of the village which is situated in rich meadowland. Lei 241, L 18, 250.

MEDBURN Northum NZ 1370. No early forms. Cf. MEDBOURNE Lei SP 7993.

MEDDON Devon SS 2717. 'Meadow hill'. *Madone* 1086, *Meddon(')* from 1234, *Mededon* 1242, 1303. OE **mǣd** + **dūn**. D 75 gives pr [meidən].

River MEDEN Notts SK 5565. 'The middle river'.
I. *Medine* 1227–1300, *Medyne* c.1300, *Medene* 1232, 1287, *Meden* from 1327×77, *Mayden, Meyden* 16th.
II. *Medme* c.1250–1335, *Mod(o)me, Modeme* 1338–9, *Medom* 1393.
OE **meodume**. Type I is due to the difficulty of distinguishing *in* and *m* in manuscript. The Meden is the middle of the three parallel rivers Poulter, Meden and Maun. Cf. River MEDINA IoW SZ 5094. Nt 6 records an obsolete pr [meidən], RN 283 gives pr [mi:dn].

MEDEN VALE Notts SK 5769. R.n. MEDEN + ModE **vale**.

River MEDINA IoW SZ 5094. 'The middle river'. *Med(e)me* 1279–1307, *Medine* c.1200–1461×83, *Medina* 1781. OE **medume**. The modern form is a Latinisation of *Medine*, itself a misreading of *Medme*. The Medina flows S to N across the middle of the island. Wt 5, RN 283, Mills 1996.71.

River MEDLOCK GMan SJ 9099. 'Meadow stream'. *Medlak'* 1292, *Medeloke -lake* [1322]15th, *Medlakbroke* 1391–2, *Medlok* c.1540, 1590, *Medlocke* 1577. OE **mǣd** + **lacu**. La 27, RN 285.

MEDMENHAM Bucks SU 8084. 'The medium sized homestead or river-bend land'. *Medmeham* 1086–1240, *Medmenham* from 1204, *Mednam* 1643. OE **medume**, dative case definite form **meduman**, + **hām** or **hamm** 1. If this name contains *hamm* it contrasts with the much larger *hamm* at REMENHAM Berks SU 7784. Alternatively *medume* may have meant 'situated in the middle' between the *hamms* of Remenham and Bisham. Bk 190, D, Sx addenda.

MEDSTEAD Hants SU 6537. 'Meadow or Mede's place'. *Medested(e)* 1202–1431, *Mayd(e)stede* 1271, 1579, *Medsted alias Mensted* 1530. OE **mǣd** or pers.n. **Mēde* + **stede**. Ha 115, Gover 1958.79, Stead 270.

River MEDWAY Kent TQ 6446, 8673. A pre-English r.n. probably meaning 'middlewater'. *Meduuuœian* (genitive sing.) [764]12th S 105, *Medeuuœge* [765×85]12th S 37, *Meduwege* 811 S 165, *Medwœg* [894]c.1000 Asser, *Medeway(e) -waie -wey(e)* 1184–1431, *Medway* from 1281. IE ***medhu** + r.n. **Waisā* on the root **wis-/*weis-* 'water' seen also in the r.ns. WEY, WYE etc. The Medway is the main dividing river of Kent and opens a marked gap between the W and E stretches of the North Downs. 11th and 12th century spellings with diphthongs *eo, io, ea, ia, Meodowœge* [790 for 860]11th S 327, *Meadowege, Miadowegan, Miadawegan* [868]12th S 339, *Miodowœge* [880]12th S 321, are due to back-mutation of *medu-* > *meodu-* or to association with OE *me(o)du* 'mead' (or to both). RN 285, BzN 1957.238.

MEECHING ESusx TQ 4400. 'The Mecingas, the sword-people or the people called after Meaca'. *Mechinges -as -ynges* 1087×1100–1295, *Mechyng* 1296–1493, *Meeching(e)* 1656, 1729. OE folk-n. **Mēċingas* < **mēċe** or pers.n. **Mēaca* + **ingas**. This is the old name for NEWHAVEN ESusx 4401. Sx 323, ING 37.

MEER END WMids SP 2472. 'The boundary end' sc. of Balsall. *Meyre End, More End* 1540. ModE **mere**. (OE *(ġe)mǣre*) + **end**. The place lies on the parish boundary. Wa 55.

MEERBROOK Staffs SJ 9960. 'The boundary brook'. *Merebroc*

[†]There is also an erroneous spelling *Metwi* at 29.9 wanting in Exon. It probably arose through misreading of *œ* in *Mœwi* as &, Latin *et*.

[†]DB also has the erroneous form *Metorne*. A scattering of other forms with *-t-*, *Metburna* [1100–35]1333, [1154×89]early 15th, *Metburn(e)* [c.1250]1404, 13th, 1451, 1519, may point to an alternative explanation, 'may-weed stream', OE **mæġth**.

1294, 1330, *-broke* 1338, *Meerbrook* 1775. OE **(ġe)mǣre** + **brōc**. The forms refer to the brook from which the settlement took its name. St i.14, Horovitz.

MEERING Notts → MORNINGTHORPE Norf TM 2129.

MEESDEN Herts TL 4432. 'Moss or bog hill'. *Mesdone* 1086, *-don* 1213–1361, *-dene* 1291, 1428, *Mesendune -don* 1207, 1218. OE **mēos** varying with adj. ***mēosen** 'mossy, boggy' + **dūn**. Hrt 183, L 56, 152.

MEETH Devon SS 5408. 'Hayland'. *Meda* 1086, *Meðe* 1175, *(la) Methe* 1209–1535, *Mathe* 1275, *Meythe* 1585. OE **mǣth**. D 98.

MEIR Staffs SJ 9342. Probably 'the boundary'. *Mere* 1242–1564, *La Mere* 1250, *Meyre* 1564, *Meir* 1656, *Meare* 1695. OE **(ġe)mǣre**. The village lies by the parish boundary but this could also be OE **mere** 'a pond'. Oakden, Horovitz.

MEIR HEATH Staffs SJ 9340. Cf. *Mear Heath Mill* and *Mear Heath Plant*[n]. 1836 OS. P.n. MEIR SJ 9342 + ModE **heath**.

MELBECKS MOOR NYorks NY 9400. P.n. *Melbecks* 1676, the 'sand-bank streams', ON **melr** + **bekkr**, + ModE **moor**. YN 271.

MELBOURN Cambs TL 3844. Partly uncertain. *Meldeburna* [970]c.110 S 779, [970]18th S 776, 1086, *(æt) (M)eldeburnan* c.975, *Meldeburne* [1042×66]13th S 1051–1434, *Melleburne -borne* 1086, *Mel(e)burne -bo(u)rn(e)* 1204–1530. The generic is OE **burna** 'a stream' but the specific uncertain. OE **melde** 'orach' is possible; although it is denied that this plant could be found in this neighbourhood, the exact botanical species implied by the word in OE is uncertain. Another possibility is the OE pers.n. *Melda* although forms in the genitive sing. *Meldan* would be expected. Ca 58.

MELBOURNE Derby SK 3825. 'Mill stream'. *Mileburne* 1086, *Mel(e)burna -(e) -born' -bourn(e)* from 1163, *Melleburna -(e) -born(a)* c.1190–1330. OE **myln** + **burna**. Db 641.

MELBOURNE Humbs SE 7544. 'Middle stream'. *Middelburne, Midelborne* 1086, *Mileburn* 1200, *Methel(e)burn* 1230–1342, *Mechel-* 1276–1415, *Melburn(e)* 1305–1540. OE **middel** later replaced by ON **methal** + **burna**. The *Mechel-* spellings are errors for *Methel*. The stream, now known as The Beck, runs parallel to the later Pocklington canal through the middle of Thornton parish, which extends from Storwood SE 7144 to Allerthorpe SE 7847. YE 236, SSNY 139, L 18.

MELBURY Devon SS 3719. Partly uncertain. *Melebyre* 1244. The specific may be OE **mǣle** 'variegated' or a pers.n. *Mǣla*; the generic is OE **byriġ**, dative pl. of **burh**. D 105, 130, 204.

MELBURY Dorset. 'The variegated fort'. OE **mǣle** + **burh**, genitive sing. **burge**, dative sing. **byriġ**.

(1) ~ ABBAS Dorset ST 8820. 'M held by the abbess' sc. of Shaftesbury. *Melebury Abb(at)isse* [1304]c.1407, 1346, *Melbury Abb'isse, Abbesse* 1291, 1326 etc. P.n. MELBURY + **abbas**, a reduced form of Latin **abbatissa**, genitive sing. **abbatissae**. Earlier simply *(at) Meleburge (imare)* 'at the boundary of M', *(on) meleberig (dune)* 'to Melbury down' [956]14th S 630, *Meleberie* 1086. The reference is to the undated cross-dyke on Melbury hill. Do iii.129.

(2) ~ BUBB Dorset ST 5906. 'M held by the Bubbe family'. *Melebir Bubbe* 1244, *Maleburi Bobbe* 1290. P.n. MELBURY *Meleberie* 1086 + family name *Bubbe*. Do 107.

(3) ~ OSMOND Dorset ST 5707. 'M held by William son of Osmond'. *Meleberi Willelmi filii Osmundi* 1202, *Melebur Osmund* 1283, *Melbury Osmond* 1316. P.n. MELBURY *Melesberie* (sic for *Mele-*) 1086 + pers.n. *Osmund*. Do 107.

(4) ~ SAMPFORD Dorset ST 5706. 'M held by the Saunford family'. *Melebury Saunford* 1312, *Melbury Sandford* 1361, ~ *Samford* 1386. P.n. MELBURY + family name *Saunford*: this family was here from the late 13th cent. Do 107.

MELCHBOURNE Beds TL 0265. 'Milk stream' i.e. one with turbid water or by pastures yielding good milk. *Melceburne* 1086, *Melcheburn(e)* 1163–1438, *-bourn(e)* 14th cent. OE ***melċe** + **burna**. Bd 16, L 18.

MELCHET COURT Hants SU 2722. A 19th cent. country mansion. P.n. Melchet + ModE **court** 'a manor house'. Melchet, *Milchet* 1231–1300, *Melchet* from 1275, is 'bare-hill wood', PrW ***mę̄l** + ***cę̄d**, the ancient name of Melchet Forest, *Milchet(e) silva* 1086, *Melchetwode* 1255. Wlt 14, 383, Ha 115, LHEB 326.

MELCOMBE BINGHAM Dorset St 7602. 'M held by the de Byngham family'. *Melcombe Byngham* 1431. Earlier simply *Mel(e)come* 1086, *Melecumb(e)* 1206–91, *-combe* 1316, 'milk valley', OE **meoluc** + **cumb**. The sense is 'valley where milk is produced, a fertile valley'. The family name affix *Bingham* is for distinction from Melcombe Horsey ST 7502 'M held by the de Horsey family', *Melcombe Horsey* 1535, earlier simply *Malecom -cumb* 1151×7, *Melecum(be)* 1152–1428, and Melcombe Regis SY 6879, 'the king's M', *Melcoumbe Regis* 1336, *Melcombe ~* from 1371, earlier simply *Melecumb* [1100×7]13th, 1223–1340, *Melcoumbe* 1280–1375. The affixes refer to the families of Robert de Byngham 1265 and Adam de Preston of Horsey, Somer, 1374. The two manors were also distinguished as 'lower' and 'upper M', *Nether Melecumb(e) -combe* 1265–1433 and *Upmelecumb(e)* 1280–89. Do iii.216 gives pr ['milkəm 'ha:Rsi], 217, L 93.

Bingham's MELCOMBE Dorset ST 7702. *Bynghammes Melcombe* 1412, *Binghams Melcombe* 1644. Surname *Byngham* + p.n. Melcombe as in MELCOMBE BINGHAM ST 7602. This name is now used for the early 16th cent. manor house built for Sir John Horsey as distinct from the village. Do iii.217, Newman-Pevsner 278.

MELDON Devon SX 5592. 'Variegated hill'. *Meledon(e)* 1175–1292, *Melledon* 1244. OE **mǣle** + **dūn**. D 203, L 143, 157.

MELDON Northum NZ 1183. 'Crucifix hill'. *Meldon* from 1242 BF. OE **mǣl** + **dūn**. Cf. TRIMDON Durham NZ 3633. NbDu 140, L 145, 158.

MELDON RESERVOIR Devon SX 5691. P.n. MELDON SX 5592 + ModE **reservoir**.

MELDRETH Cambs TL 3746. Probably 'mill-stream'. *Melrede* 1086, *-a* 1086–1423, *Melreðe -reth(e)* 1086–1550, *Meldreth -dred(e)* 1086–1643. OE **myln**, SE dial. form **meln**, + **rīth**. Ca 60.

Long MELFORD Suff TL 8645. *Long Melford* 1805 OS. ModE adj. **long** + p.n. *Melafordā* 1086, *Meleforde* c.1095, *-ford* 1235, *Melforde* 1610, 'the mill ford', OE **myln** + **ford**. DEPN.

MELKINTHORPE Cumbr NY 5525. 'Melkan's outlying settlement'. *Melcanetorp* c.1150, *Melkan- Melcanthorp(e)* c.1199–1777, *Melkenthrop(p) -in-* 1278–1716. Pers.n. *Melkan* (<OIr *Māelcian* or OW *Mailcun*) + **thorp**. We ii.183, SSNNW 58.

MELKRIDGE Northum NY 7363. 'Milk ridge'. *Melkrige* 1279, *Melkerigg* 1292, *Milkrigg* 1479, *-ridge* 1663. OE **meoluc** + **hrycg**. The reference is to rich milk producing pasturage. NbDu 140.

MELKSHAM Wilts ST 9063. Uncertain. *Melcheshā* 1086, *Melkesham* 1155–1254, *Milkesham* 1229–1553, *Mylsham* 1558×79, *Milsham, Melsam* 1675. Understood in the Middle Ages as OE **meoluc** 'milk', genitive sing. **meoluces**, + **hām** or **hamm**. If this unusual genitival compound is original the sense is 'homestead or meadow where good milk is produced'. Probably, however, a folk-etymology of an earlier name now lost. Wlt 128 gives prs [melksəm] and [melsəm].

MELLING 'The Mellingas, the people called after Molli', or p.n. Melling, 'the place called after Molli'. OE folk-n. **Mellingas* < pers.n. *Molli* + **ingas** or p.n. **Melling* < pers.n. *Molli* + **ing**[2]. The vowel of the folk-n. must be due to *i*-mutation. ING 71.

(1) ~ Lancs SD 6071. *Mellinge* 1086, *Mellynges -ing(u)es* 1094–1271, *Melling* from c.1190, *Malling* 1206, 1229. La 180, Jnl 17.101, 18.20.

(2) ~ Mersey SD 3800. *Melinge* 1086, *Mellinges* 1194, 1256, *Melling* [from 1190]1268. La 119, Jnl 17.66, 18.20.

MELLIS Suff TM 1074. 'The mills'. *Melle(l)s* 1086, *Melles* 1158–1610, *Mellys* 1291, 1523. OE **myln**, SE dial. form **meln**, pl. **melnas**. DEPN, Baron.

MELLOR 'The bare hill'. PrW ***mę̄l-βre**, Brit **mailo-* + **brigā*.

(1) ~ GMan SJ 9888. *Melver -uer* 1216×72–1371, *Mellour* 1424, 1428, *-er* 1432, *-or* from 1569. Db 144.

(2) ~ Lancs SD 6530. *Malver* c.1130, *Meluer(e) -ir -yr* 1200×8–1322, *Melure* 1274–1327, *Mellour* 1428, 1508. The village stands on the slope of Mellor Moor, a hill of 733ft. La 73.

(3) ~ BROOK Lancs SD 6430. P.n. MELLOR + ModE **brook**.

MELLS Somer ST 7249. 'The mills'. *(at) Milne* [942]14th S 481, *Mvlle* 1086, *Melnes* 1196, *Melles* 1225, 1610. OE **myln**, pl. **mylnas**, dative pl. **mylnum**. The reference must have been to water-mills on the river Frome which runs through the village. A single mill is mentioned in DB. DEPN.

MELLS Suff TM 4076. 'The mills'. *Mealla* 1086, *Melne* c.1160, *Melnes* 12th, *Melles -ys -is* 1254–1568. OE **myln**, SE dial. form **meln**, pl. **melnas**. DEPN, Baron.

MELMERBY Cumbr NY 6137. 'Melmor's village or farm'. *Malmerbi* 1201, *Melmerbi -by(e)* 1240–1552, *Melmorby* c.1210–1552, *Mellorby -er-* c.1580–1750. Pers.n. *Melmor* (<OIr *Māelmuire* 'servant of Mary' or *Māelmaire*) + **bȳ**. Cu 223 gives pr [melərbi], SSNNW 35.

MELMERBY NYorks SE 3376. 'Farm or village on sandy ground'. *Malmerbi -by* 1086–1243, *Melmerby* from 12th. ON **malmr** 'a sandy field', genitive sing. **malmar**, + **bȳ**. The form of the name has been influenced by MELMERBY SE 0785 which is of different origin. YN 219, SSNY 33.

MELMERBY NYorks SE 0785. 'Melmor's farmstead or village'. *Melmerbi* 1086, 1202, *Melmor(e)by* from 1184. OIr pers.n. *Máelmuire* [me:lmurjə] borrowed in OWScand as *Melmor* + ON **bȳ**. YN 255, SSNY 33.

MELMERBY FELL Cumbr NY 6538. *Melmerby Fell* 1589. P.n. MELMERBY NY 6137 + ModE **fell** (ON *fjall*). Cu 224.

MELPLASH Dorset SY 4898. 'Multicoloured pool'. *Meleplays* 1288, *Melepleisch* 1312, *Meleplash* 1337. ME **mele** (OE *mǣle*) + **plaish** (OE *plæsc*). Do 107.

MELSONBY NYorks NZ 1908. Partly uncertain. *Malsenebi* 1086, *-sambi* 1182, *Melsanebi* 1154–1310, *-sambi -by* 1189–1400, *Melsonbye* 1540. The earlier explanation, 'Máelsuthain's village or farm', has been questioned, since the medial *th* of the OIr pers.n. *Máelsuthain*, pronounced [me:l'suthən], was retained in England as the recorded forms *Mælsuthan*, the name of a moneyer to Eadgar and Eadward II, 959×78, and *Mæglsothen* in a charter dated 956 (S 628) demonstrate. No alternative explanation has been proposed. YN 297, SSNY 297.

MELTHAM NYorks SE 0910. Partly uncertain. *Meltha'* 1086, *Meltham* from 1255. Possibly OE ***melt** 'melting, smelting, malting' or ***melta** 'a smelter', + **hām**. YW ii.282.

MELTON 'Middle farm, estate or village'. OE **middel** + **tūn** with Scandinavian *methal* for *middel*.

(1) ~ MOWBRAY Leic SK 7519. 'M held by the Mowbray family'. P.n. *Medeltone* 1086, *Meal- Meauton'* 1174–1282, *Melton(e)* c.1130–1610, + manorial affix *Mowbray* from 1284. Willelmus de Mobray, whose family came from Montbrai, Manche, held the manor in 1200. Lei 175, SSNEM 184 no.1.

(2) ~ ROSS Humbs TA 0710. 'M held by the Ros family'. *Melton Roos* 1375. P.n. *Medeltone* 1086, *Meltuna -ton(a)* c.1115–1526, + family name of Willaim de Ros 1303 from Rots in Calvados, Normandy, near Caen. Li ii.232.

(3) Little ~ Norf TG 1606. *Lithle Meddeltone* [c.1050]13th S 1516, *Parva Meltuna* 1086, OE **lȳtel**, Latin **parva** + p.n. *Middil- Methelton* [c.1050]13th S 1516, *Meltuna* 1086. 'Little' for distinction from Great Melton TG 1306, *Magna Melton* 1242, earlier simply *Middil- Methelton*. DEPN.

(4) West ~ SYorks SK 4201. ME prefix *West(e)-* from 1401 + p.n. *Middeltun, Medel- Merelton(e)* 12th cent., *Metheltona* c.1225, *Melton(a)* from 13th. YW i.114, SSNY 116.

MELTON Suff TM 2850. Probably 'the mill settlement'. *Melton* from [1042×66]12th S 1051, *Meltuna* 1086. OE **myln**, SE dial. form **meln**, + **tūn**. DEPN.

MELTON CONSTABLE Norf TG 0433. 'M held by the constable' sc. of the bishop of Norwich. *Melton Constable* 1320. Earlier simply *Maeltuna* 1086, *Meutone* 1212, possibly 'settlement with a crucifix', OE **mǣl** + **tūn**, or 'speech settlement', OE **mǣl** + **tūn**. DEPN.

MELTONBY Humbs SE 7952. Partly uncertain. *Melte- Metelbi* 1086, *Meltenebi -by* 1170–1281, *Melt(h)aneby* 1203–28, *Meltonby* from 1297. Possibly an original ON **methal-tūn** 'middle farm or village' + **bȳ**. YE 183, SSNY 33.

MELVERLEY Shrops SJ 3316. Partly uncertain. *Meleurlei* 1086, *Melewerleye* 1291×2, *Melwardleye, Mulverdeleg, Maluerdelegh'* 13th(p), *Meluerdeley* 1394×5, *Meluerlegh -leye* 1291×2, *Melverly* 1357, *Mil- Melverley* 1635. The generic is OE **lēah** 'a wood, a clearing', the specific possibly an earlier English p.n. **Milford* 'the mill ford' referring to an otherwise unknown crossing of the Severn, or a W p.n. *Melwerne* 1385, *Melwern* 1400–18th although this is perhaps more likely to be a W adaptation of the English n. through association with W *mêl* 'honey' and *gwern* 'an alder-tree, a marsh where alder-trees grow'. Sa i.200, Morgan 1997.39.

MEMBURY Devon ST 2703. 'Fortification of or at the stone(s) or called *Main*'. *Mane- Maiberie* 1086, *Maa- Manberia* 1086 Exon, *Manbire* 1238, *Menbure -byri* 1238, 1279, *Membiri -byr(i) -bery* 1154×89–1428, *Meymbury* 1333. PrW ***main**, Co **men**, pl. **meyn**, possibly used as a p.n., + OE **byriġ**, dative sing. of **burh**. Alternatively the specific might be OE adj. **mǣne** 'common'. The reference is to Membury Castle, an undated hill-fort at ST 2802 overlooking the village. D 644, CPNE 161, Thomas 1976.96.

MENABILLY Corn SX 0951. Unexplained. *Menabilyou* 1354, *Mynabile* 1525, *Mynnybelly alias Mennebile* 1573. PNCo 118.

MENDHAM Suff TM 2782. 'Mynda's homestead'. *My- Mi- Mendham* [942×c.951]13th S 1526, *Men(d)- Menneham* 1086, *Mendham* from 1168. OE pers.n. **Mynda*, SE form **Menda*, + **hām**. DEPN, Baron.

MENDIP FOREST Somer ST 5054. P.n. Mendip as in MENDIP HILLS ST 5255 + ModE **forest**. Formerly one of the five royal forests of Somerset and a favourite summer hunting ground of Saxon kings. Whitlock 110.

MENDIP HILLS Somer ST 5255. *Mendip Hills* 1610. P.n. Mendip + ModE **hills**. Mendip, *Munedup* [705×12]13th S 1670, [late 10th]c.1247, *(aput) menedip* [740×56]1395 Charters III, *Munidop* [c.1135]13th, *Menedep(e)* 1185, [13th]Bruton, *Mendep, Menedup* 1225, *Munedep* 1235, *Minedepe* 1236, is unexplained. It seems unlikely to be PrW ***mönïth** 'a hill' + OE **hop** since the later element is not certainly evidenced in SW p.ns. and is never so early reduced to *-ep*. An alternative might be OE **yppe** 'a raised place, a platform, a hill'. DEPN, L 172, Charters III.81.

MENDLESHAM Suff TM 1065. 'Myndel's homestead'. *Mun(d)lesham, Melne- Menlessam* 1086, *Menlesham* 1163–1228, *Mendlesham* from 1165. OE pers.n. **Myndel*, SE form **Mendel*, genitive sing. **Mendles*, + **hām**. DEPN.

MENDLESHAM GREEN Suff TM 0963. *Mendlesham Green* 1837 OS. P.n. MENDLESHAM TM 1065 + ModE **green**.

MENEAGE Corn. 'Monk land'. *Manech* [c.1070]14th, *Manahec* 1269, *Manaek* 1291, *Mane(e)ke* 1343, 1461, *Menek* early 16th. Co **managhek** (Co *manach* 'monk' + adjectival ending *-ek*). A district name which included the parishes of Mawgan SW 7025, St Martin SW 7323, St Anthony-in-Meneage SW 7825, Manaccan SW 7625 and St Keverne SW 7921. The court of the district was at Lesneage SW 7722 recorded as *Les-manaoc* [967]11th S 755, Co **lys** 'court' + p.n. Meneage. PNCo 118, Gover n.d. 541.

MENHENIOT Corn SX 2862. 'Land of Hunyad or Hynyed'. *Mahiniet* 1260, *Mahenyet* 1318, *Mahynyet* 1362, *Mehenett* 1553, *Manhunghet* 1291, *Menhynyheth* 1342, *Minheneth* c.1540, *Minhinet* 1610. Co ***ma** 'plain, open country, place' later replaced with **men** 'stone', + pers.n. **Hynyed* or *Hunyad*. PNCo 119, Gover n.d. 192, DEPN.

MENSTON WYorks SE 1744. 'Estate called after Mensa'. *mensinctun* [972×92]11th S 1453, *on Mensingtune* c.1030 YCh 7, *Mersintone* 1086, *Mensintune -ton(a)* 1172–1333, *Mensington(e) -yng-* c.1230–1617, *Menston* from 13th. OE pers.n. **Mensa* + **ing**[4] + **tūn**. YW iv.201.

MENTMORE Bucks SP 9019. 'Mænta's marsh or moor'. *Mentemore* 1086 etc. OE pers.n. **Mænta* + **mōr**. Bk 81, L 55.

MEOLE BRACE Shrops SJ 4810. '(The share of) Meole held by the de Braci family'. *Melesbracy* 1274, *Melebracy* 1301, *Moles Bracy* 1276, *Molebracy* c.1291, *Meole Bracy* 1334, 1346, *Milbrace* 1535, *Meole Brace* from 1586, *Brace Meale* 17th. P.n. *Mela(m)* 1086, *Mele(s), Mole(s), Meola, Meoles, Mueles, Meules, Meeles, Moeles* 13th cent., *Meele* 1597, 1615, + family name *de Braci*. Meole is probably the OE name for Meole Brook, *Mola(m), Meolam* c.1130, *Me(o)le, Meola* 13th) from OE **me(o)lu** 'meal, flour' referring to cloudy water. The pl. forms probably arise from the existence of several manors, one of which was granted to the Buildwas Abbey, *Monkemeol* 1334 'monks' Meole' later called Crow Meol, and another held by Adolf de Braci in the early 13th cent. Sa 1.201–3.

North MEOLS (lost) Mers SD 3518. *Normalas* c.1190, *Nor Muelis* 1229, *Nortmelis* 1243, *Northmeles* 1229–1332, *Northmeales -meeles* 1417–16th cent., *North Meols* 1842 OS. Adj. **north** + p.n. *Moles* before 1149, *Moeles* 1153×60, *Mels* 1311, ON **melr** 'sandbank', secondary ME pl. **meles**. In 1086 this was *Otegrimele, Otringemele* with ON pers.n. **Oddgrímr* or *Authgrímr* for distinction from the lost vill of Argarmeols which was washed away before 1503, *Erengermeles* 1086, *Argarmelis* 1243, *Argarmeles* 1254, with ON pers.n. *Arngeirr* or **Aringeirr*. 'North' perhaps for distinction from Raven MEALS SD 2705. La 125, SSNNW 99, 150, Jnl 17.70.

East MEON Hants SU 6822. *Estmunes* c.1270–1327, *Estmoene(s) -meones* 1280–1317, *East Mean* 1610, *East Meon* 1810 OS. ME adj. **est** + p.n. *(æt) Meone* [873×88]11th S 1507, 12th cent. including S 417, 619, 718, 754, 811, *Mene(s)* 1086, *Meonis, M(v)enes, Mienes -a* 12th cent., transferred from the r.n. MEON. 'East' for distinction from West MEON SU 6423. Ha 69.

River MEON Hants SU 5407. Uncertain. *Meonea* [786×93]13th S 269, *Meone* 12th including S 276, 283, 417, 446, 463, 600, 754, 811. OE r.n. **Mēone* and *Mēon-ēa* 'the river Meon' of uncertain meaning and origin, perhaps ultimately IE **mew-* 'wet, slime' with *n*-suffix **meun-* > OE **mēon-* and so cognate with Armenian *-moyn* 'immersed', Ir *mūn* 'brain' and with different suffix OE *mēos* 'moss' < **meus-*. The people of the Meon district are referred to in early sources as the *Meonware*, r.n. Meon + OE **ware**, genitive pl. **wara**, e.g. *Meanuarorum prouincia* 'the province of the *Meonware*' c.731 BHE, *Mean- Meonware mægð* [c.890]c.1000 OEBede in which spellings with <ea> are due to the N change *ēo* > *ēa*, *Meonwara snað* 'the detached ground of the M' [826]12th S 276, ~ *snad* [939]12th S 446. Gover 1958.4, 63, RN 228, Pokorny 741.

West MEON Hants SU 6423. *Westmunes* c.1270, *-me(o)nes -mune -meone* 1291–1383, *Westmean alias Westmeon* 1701. ME adj. **west** + p.n. Meon as in East MEON SU 6822. Gover 1958.72.

MEON HILL Warw SP 1745. Unexplained. *Mene -a* 1086–1405, *Muna -e, Mona* 1158×9–15th, *Meone* 1221–1594, *Meene, Mean(e)* 17th cent. The forms point to an OE **Mēon-* which may be compared with the Hants. r.n. MEON, *Meonea* [786×93]13th, 'the river Meon'. The reference here is, however, to an isolated hill rising to 635ft. with a hill-fort on top; Meon may have been a name of one of the nearby brooks but comparison should be made with the hill-n. Myne in MINEHEAD Somerset SS 9746. Gl i.254, Thomas 1980.214.

MEONSTOKE Hants SU 6119. 'Outlying farm on the r. Meon or dependent on Meon'. *Menestoch(e) -es* 1086, *Mienstohc* 1156. *Meonestok* c.1200, *Mean Stoke* 1577. P.n. Meon as in E and W MEON or River MEON + OE **stoc**. Ha 116, Gover 1958.50, RN 288, Studies 1936.22–3.

MEON VALLEY Hants SU 6016. R.n MEON SU 5407 + ModE **valley**.

MEOPHAM Kent TQ 6466. Possibly 'homestead of Meapa or of the Meapas'. *Meapaham* [788]12th S 129, [973×87]12th S 1511, 1042×66 S 1047, *Meaphám* 939 S 447, [1006 for 1001]11th S 914, *Me(a)peham* [940]12th S 1210, [961]12th S 1212, 1270–3, *Mepeham* 1086–1240, *Meppe-* 1226, 1220, *Map(p)e-* 1231–78. OE pers.n. **Mēapa*, or folk-n. **Mēapas*, genitive pl. **Mēapa*, + **hām**. But neither folk-n. nor pers.n. are otherwise known. According to the 939 charter the name was regarded as comic, *ludibundisque vocabulis nomen indiderunt æt Meaphám*, which must be due to association with the rare, possibly colloquial, ME word *mepen* < OE **mēapian* 'wander, stray' related to ModE *mope*. KPN 71, 359, PNK 103, Studies 1931.24, ASE 2.29, EMEVP 409–10, Zettersten 32–4.

MEOPHAM STATION Kent TQ 6467. A railway station on the line from London to Rochester. P.n. MEOPHAM TQ 6466 + ModE **station**.

MEPAL Cambs TL 4481. Possibly 'Meapa's nook'. *Mepahala* 12th, *Mepehale* 1362, 1379×87, *Mephal(e) -hall* 1199–1605, *Mep(p)le* 1636, 1675, *Meypole, Maple* 1695. OE pers.n. **Mēapa* + **halh**. Alternatively we may have a folk-n. **Mēapas*, genitive pl. **Mēapa*, as perhaps in MEOPHAM Kent TQ 6466, 'nook of the Meapas'. Ca 237 gives pr [mepəl], Studies 1931.25.

MEPPERSHALL Beds TL 1336. Probably 'Malpert's nook of land'. *Malpertesselle, Maperteshale* 1086, *Maperteshala -e* 12th–13th cents., *Meperteshale* 13th–15th cents, *Mepereshale* 1316, *Mepsall* 1610. Pers.n. *Malpert* < CG *Mathalperht*, genitive sing. *Malpertes*, + **halh**, dative sing. **hale**. An alternative specific might be OE **mapultrēow**, **mapuldor** 'a maple-tree': but the forms do not well support this suggestion. Bd 170 gives pr [mepʃul], L 102.

MERBACH H&W SO 3045. 'Boundary stream'. *Merbache* 1216×41, *Merebache* 1292. OE **(ġe)mǣre** + **bæċe**. A parish boundary runs along a stream-valley on the east side of Merbach Hill. He 55.

MERCASTON Derby SK 2643. 'Merchiaun's farm'. *Merchenestune* 1086, *Murc- Murkan(e)ston(e) -is-* c.1141–1298, *Murc- Murkaston(e)* 1252–1391, *My- Mirc- Mirkaston(e)* 1330–1675. PrW pers.n. **Merchiōn* (Latin *Marciānus*), secondary genitive sing. **Merchiōnes*, + **tūn**. Db 587.

Mynydd MERDDIN H&W SO 3428. 'Mount Merddin'. *minid ferdun* [early 8th]c.1130 LL, *Mynydd Ferddyn* 1832 OS. PrW **mönïð** + p.n. Merddin as in Merddyn Clwyd SH 8759 and Marddyn-y-bit Gwyn SH 3470 from W **merddyn**, **murddun** 'a ruin'. He 56–7, GPC 2503.

MERDON Hants SU 4226 → HURSLEY SU 4225.

MERE 'Lake' OE **mere**. Jnl 25.38–50.

(1) ~ Ches SJ 7281. *Mara, Mera* 1086, *(la) Mara* [c.1176]17th, c.1262, *Mere* from c.1230, *Meyr(e), Meir(e), Meer(e)* 1331–1661. Che ii.51.

(2) ~ Wilts ST 8132. Probably 'the lake'. *Mera* 1086–1156, *Mere* from 1086. OE **mere**. This was a watery area, but OE **(ġe)mǣre** 'boundary' referring to the point where Dorset, Somerset and Wiltshire meet is an alernative possibility. Wlt 178, L 26.

(3) ~ BROW Lancs SD 4118. ModE **mere** + **brow** 'an eyebrow, the brow of a hill'. The reference is to the one-time extensive lake which stretched from there to Martin Mere SD 4214 and whose existence is remembered in other nearby names such as HOLMESWOOD SD 4317, Mere Side SD 4316, Meer Sands SD 4315, Mere Hall SD 4016. It is now drained by The Sluice SD 4116.

(4) ~ GREEN WMids SP 1299. *Mere Green* 1821. Cf. *Mere Pool* 1834 OS. P.n. Mere + ModE **green**. Wa 51.

(5) The ~ Shrops SJ 4035. *Ellesmere Mere* 1833 OS. The lake from which ELLESMERE SJ 3934 takes its name.

MERECLOUGH Lancs SD 8730. 'Boundary ravine'. *Meerclogh* 1311. OE **(ġe)mǣre** + **clōh**. La 84.

MEREWORTH Kent TQ 6553. 'Mæra's enclosure'. *Meran worð* 843 S 293, *Mœreweorðe* 960×88 S 1458, *Mœreuurtha* c.975 B 1321, *(to) Mœranwyrþe* c.975 B 1322, *Marovrde* 1086, *Merew(u)rth(e) -worthe* 1179×80–1253×4, *Marewr(th') -wurthe* c.1180–1233. OE pers.n. *Mǣra* + **worth**. The 843 charter refers to this place as *illa famosa loco* 'that famous place' which appears to be due to popular identification of the specific of the name with OE adj. *mǣre* 'famous'. KPN 188, PNK 158.

MEREWORTH CASTLE Kent TQ 6653. A Palladian villa built in 1723. P.n. MEREWORTH TQ 6553 + ModE **castle**. Newman 1969W.403.

MERIDEN WMids SP 2482. 'The pleasant valley'. *Mereden* 1230–2, *Mu- My- Mir(r)yden(e) -i-* 1285–1456, *Meriden(e)* from 1443. OE **myriġ** + **denu**. By origin probably a nick-n. since the original name of the parish was *Ailespede* 1086, *Ellespethe* c.1155, *Allespathe* 1235–1403, *Al(l)spath(e)* from 1325, 'Ælli's path', or 'path to Allesley', OE pers.n. *Ælli* as in neighbouring ALLESLEY SP 2981, genitive sing. *Ælles*, + **pæth**. No doubt the same man is commemorated in both names. Wa 64.

MERRINGTON Shrops SJ 4720. 'The pleasant hill'. *Muridon* c.1220×30–1478, *Mu- Meryden* 16th, *Meriton* 1577, *Mer(r)ington* from 1655. OE **myriġ** + **dūn**. *Muridon* replaces an earlier name, *Gellidone* 1086, *Gulidon(a) -dun(e)* 1155–1277, *Gilydon'* 1291×2, unidentified element + **dūn**. If an OE equivalent of ON *gol, gul* 'wind' existed there might have been an OE adj. **gyliġ* 'windy'. Sa i.204.

Kirk MERRINGTON Durham NZ 2631. 'Merrington with the church'. *Kirke Merington* 1296, *Kirk Merington* 1565. Also known as *Est Merington -y-* 1296–1580, *East Merrington* 1644×5 and *Great Merrington* 1675. One of three settlements on a prominent ridge of high ground south of Durham City. The whole estate or ridge was called *Mœrintun* c.1123, *-tune* c.1170×4, the subsequent forms of which are *Merington(') -y-* c.1270–1564, *Merrington* 1644×5, 1872, 'estate called after Mæra', OE pers.n. *Mǣra, Mēra*, + **ing**[4] + **tūn**: however, the name might be topographical, 'village, estate called or at **Mǣring*, the conspicuous hill or place', OE adjective **mǣre** 'famous, great, splendid' + **ing**[2] + **tūn**. The tower of Kirk Merrington church is a prominent landmark for many miles in all directions. Two Merringtons are already distinguished in the 12th cent., *duas merinctonas* [1189×99]1336, *duas Meringtonas* c.1190×5, one of which was the later Kirk Merrington, cf. *ecclesia sancti Johannis cum villa sua Merinctona at alia Merinctona* 1154×66. No known boundary exists here so that derivation from OE *(ġe)mǣre* 'a boundary' is not in question. See further MIDDLESTON MOOR NZ 2535 (for Middle Merrington) and WESTERTON NZ 2331 (for West Merrington). NbDu 129.

MERRIOTT Somer ST 4412. Possibly 'mare gate'. *Meriet* 1086–[1275×92]Bruton, *Merriet* [13th]Bruton, *Muriet* 1329, *Meryot* 1610. OE **mere**, WS **miere**, + **ġeat**. Perhaps a gate leading to the marshland along Broad river where mares ran wild. Alternatively the specific might be OE **mere** 'a pool' or **(ġe)mǣre** 'a boundary'. It was subsequently associated with ME *muri, meri* 'merry'. DEPN.

MERRIVALE Devon SX 5475. 'Pleasant open land'. *Miryfeld* 1307–1474 with variants *Meri- Mere-, Merivill* 1612, *Merrifield -ville* 18th. OE **myriġ** + **feld**. A common name type, cf. Merrifield in Inwardleigh, *Murifelde* 1353, *Merefylde* 1562, Mearfield in Stoke Climsland Corn, *Merifeld* 1338, Merrifield in Anthony Corn, *Merifeld* 1414 etc. D 247, 150, L 239.

MERROW Surrey TQ 0250 → MARDEN Wilts SU 0857.

MERRYMEET Corn SX 2766. 'Pleasant meeting place'. *Merrymeet* 1699. ModE **merry** + **meet**. A common name for places where several roads meet. PNCo 119, Gover n.d. 194.

East MERSEA Essex TM 0414. First distinguished in 1219, ME adj. **est** + p.n. Mersea as in MERSEA ISLAND TM 0314. Ess 319.

West MERSEA Essex TM 0112. First distinguished in 1281, ME adj. **west** + p.n. Mersea as in MERSEA ISLAND TM 0314. Ess 320.

MERSEA FLATS Essex TM 0513. P.n. Mersea as in MERSEA ISLAND TM 0314 + ModE **flat**. The reference is to a sand-bank off the coast of Mersea Island.

MERSEA ISLAND Essex TM 0314. *Mersey Island* 1805 OS. P.n. Mersey + pleonastic ModE **island**. Mersey, *Merese(ge)* [959×9]12th, *(on) Meres ig(e)* c.900 ASC(A) under year 895, *Mereseg -igge* c.1000 ASC(B) under same year, *(at) Mereseye* [946×c.951]14th S 1483, *(æt) Myresigœ, (et) Myresige* [962×91]11th S 1494, *Myrœsegœ, (into) Myresiœ, (into) Myresie* 1000×2 S 1486, *Meresai(am)* 1086, *-eia -ee -eya -ie- eiee* 1163–1303, *Mersay -ey(e) -eia* 1202–1362, *Martsey* 1267, *Marza* 1545, is 'pool island', OE **mere**, genitive sing. **meres**, + **ēġ**. *My-* spellings are probably inverted spellings resulting from the common use of *e* for OE *y* in the Essex dial. The pool probably refers to the marshland N of the island between Langenhoe and the Colne. Ess 319, Jnl 2.45.

River MERSEY Mers SJ 4081. Probably the 'boundary river'. *Mœrse* [1002]11th, *Mersham* (Latin accusative) 1086, *Merse* [1141×2]n.d., 1200–1577, *Mersee* 1209–1577, *Mersey* from c.1540. OE **(ġe)mǣre**, genitive sing. **(ġe)mǣres**, + **ēa**. From about 600 the Mersey formed the boundary between Northumbria and Mercia. La 26, RN 289, Jnl 17.29.

MERSHAM Kent TR 0539. Probably 'Mærsa's homestead or estate'. *mersaham* 858 S 328, 863 S 332, *Mœrseham* [995×1005]12th S 1456, *Merseham* 1053×61 S 1090, 1042×1066 S 1047, 1086–1278, *Mersham* 1191, 1253×4. OE pers.n. **Mǣrsa* + **hām**. It is just possible, however, that the specific is an ancient folk-n. **Mǣrse*, genitive pl. *Mǣrsa* corresponding to the German tribal name *Marsi*. KPN 199, 359, PNK 417, Studies 1931.25. ASE 2.29.

MERSTHAM Surrey TQ 2953. 'The horse-paddock settlement'. *(æt) Mearsæt ham* 947 S 528, *Mersetham* [1042×66]11th S 1047, *Merstan* 1086, *Mersteham* 12th–1286, *Merst(h)am* 13th cent. etc., *Mestham* 1207–1757 including [727, 933, 967, 1062]13th S 1181, 420, 752, 1035, *Merstham* from 1568, *Mearstham* 1680. OE ***mearh-sæt** + **hām**. The alternative explanation 'weasel-lair settlement' from OE ***mearth-sæt** + **hām**, is less likely in view of the names of the adjacent estates CHALDON TQ 3255, OE **ċealfa-dūn* 'the upland pasture for cows', and Gatton TQ 2752, OE **gāta-tūn* 'goat-farm', which imply a specialisation of animal husbandry in the Banstead-Coulsdon region in the Middle-Saxon period. Sr 300 gives prs [mə:stəm] and [mə:stəhæm], the former reflecting the common medieval and early modern spelling. Sr 291, 300–1, PNE ii.94, ASE 2.48, Rumble 1976.174–5.

MERSTON WSusx SU 8903. 'The marsh settlement'. *Mersitone* 1086, *Mers(h)ton(e)* 1274–1414, *Marston* 1569, *Merson* 1584. OE **mersc** + **tūn**. Sx 72.

MERSTONE IoW SZ 5284. 'The marsh settlement'. *Messe-Merestone* 1086, *Merston* c.1200 etc., *Mershton(e)* 1349–1608, *Marshtone* 15th, *Marston* 1559. OE **mersc** + **tūn**. Wt 16 gives prs ['mesn, 'mərstn], Mills 1996.71.

MERTHER 'Saint's grave'. Co ***merther**. Cf. MERTHYR.

(1) ~ Corn SW 6329 → SITHNEY SW 6328.

(2) ~ Corn SW 8644. *Merther* from 1549. Earlier *Eglosmerthe* (for *-merther*) 1201, *Eglosmerther* 1302, 1327 'church at Merther', Co **eglos** + ***merther** misunderstood as if 'church of Merther' and hence the form *St Murther* 1569. The patron saint, unique to this church, is named as St Coan otherwise Comgan, traditionally an Irish chieftain exiled to Skye where he founded a monastery; cf. the Scottish p.ns. Kilchoan NM 4963, 7799, 7913 and Kirkcowan NX 3260. PNCo 120, Gover n.d. 471.

MERTHYR Corn SW 4035 → MORVAH SW 4035. Cf. MERTHER.

MERTON 'Pond settlement'. OE **mere** + **tūn**. Jnl 24.30–41.

MERTON Devon SS 5212. 'Settlement, estate on the river Mere'. *Merton(e)* from 1086, *Marton* 1428, 1675, *Martyn -in* 1535, 1765. R.n. Mere, *Meer* 1630, + **tūn**. The r.n. means 'the boundary', OE **(ġe)mǣre**: but its late appearance suggests that it is in fact a back-formation from the p.ns. MARLAND SS 4713 and MERTON. D 99, 10, RN 288.

(1) ~ GLond TQ 2569. *Merton(e)* [967]14th S 747 etc., *Mereton(e)* 1086–1196, *Mareton(a)* 1176, *Marten* 1500, *Martin* 1605, *Mirton al. Marten* 1679. Sr 25, Mills 1998.150.

(2) ~ Norf TL 9098. 'The lake settlement'. *Meretuna* 1086, *Mertuna* 1121. OE **mere-tūn**. Merton is on the northern edge of the Breckland where there are natural meres, e.g. West Mere near Tottinton, Mickle Mere, Fowlmere. The mere at Merton is no longer extant, but the village was probably so named as being a place of water for travellers on nearby Peddars Way. DEPN, Jnl 24.35.

(3) ~ Oxon SP 5717. *Meretone* 1086, *-ton(')* 1199, 1390, *Merton* from 1316. O 182, L 26, Jnl 24.35.

MESHAW Devon SS 7519. 'Bad, unfertile clearing'. *Mauessart* 1086, *Malessart* 1175, *Mausard(e)* 1242–1346, *Mausawe* 1249, *Meus(h)awe -shagh* 1291–1340. Fr **mal** + **essart**, later replaced by ME **shawe** (OE *sceaga*). DEPN, however, suggests that the *sceaga, shawe* forms are original and were later reinterpreted by the Normans. D 382.

MESSING Essex TL 8918. Uncertain folk-n. *Metcinges* 1086, *Mec- Meds- Metz- Mecz- Mettings* 12th, *Mescinges -ynges* 1206–90, *Mescing(e)* 1199–54, *Messing(e)* from 1235. It is uncertain whether the folk-n. here is OE **Mæċċingas* as in MATCHING TL 5212 or **Mēċingas* as in MEECHING ESusx TQ 4400. Ess 396, ING 21, Nt xxxvii.

MESSINGHAM Humbs SE 8904. 'Homestead of the Mæssingas, the people called after Mæssa'. *Mœssingaham* [c.1067]12th, *Messingeham* 1086, 1181, *Massing(e)ham* c.1115, *Massingham* from 1265. OE folk-n. **Mœssingas* < pers.n. **Mœssa* + **ingas**, genitive pl. **Mœssinga*, + **hām**. Identical with MASSINGHAM Norf TF 7620. DEPN, ING 145.

METFIELD Suff TM 2980. 'The open land with meadow'. *Medefeld* 1214–1344, *-feud* 1286, *Metfeld* 1524, 1568, 1610. OE **mǣd** + **feld**. DEPN, L 244, 250.

METHERINGHAM Lincs TF 0761. 'Medric's homestead or estate'. *Medric(h)esham* 1086, *Meding- Mericham* [13th]copy, *Metrinc- -ing-, Metryng- Medringham -yng-* 1162–1526, *Mederinge-* 1193, 1202, *Metheringe-* 1219–20, *Med- Methering- -yng-* 1202–1563. OE pers.n. **Mēdrīċ*, genitive sing. **Mēdrīċes*, varying with p.n. **Mēdrīċing* 'the place called after Medric' < **Mēdr(īċ)* + **ing**[2], + **hām**. Perrott 331, ING 142, BzN 1968.176.

METHLEY WYorks SE 3926. Partly uncertain. *Medelai -lei(a) -lay -ley* 1086–1427, *Methelei(a) -le(y) -lay* 1155×62–1486, *Meydlay* 1379, *Medlay -ley* 1439–1608, *Meith- Meethley* 1522, *Methley* from 1541. Either ON **methal** 'middle' + OE **lēah** 'forest clearing' or **ēġ** 'an island', or OE **mǣth** + **lēah** 'forest clearing where grass is mowed'. Topography suggests **methal** + **ēġ** in the sense 'land between two rivers', viz. the Calder and the Aire. YW ii.125, L 35.

METHWOLD Norf TL 7394. 'Middle wold' sc. between HOCKWOLD TL 7288 and NORTHWOLD TL 7597. *Medelwolde* [1042×66]12th S 1051, *Methelwalde* 1086, *-wolda* 1171. OE **middel** replaced by ON **methal** + **wald**. The chalk wolds form the first high ground E of the fens. DEPN.

METHWOLD FENS Norf TL 6593. *Methwold Fen* 1824 OS. P.n. METHWOLD TL 7394 + ModE **fen**.

METHWOLD HYTHE Norf TL 7194. *Methwold Hithe* 1824 OS. P.n. METHWOLD TL 7394 + ModE **hithe**, **hythe** 'a landing place'. The place is at the W end of the wolds where they dip to meet the fens.

METTINGHAM Suff TM 3689. 'The homestead of the Mettingas, the people called after Metti'. *Metingaham* 1086, *Metingham* [1176]1331, 1230–1610 with variant *-yng-*. OE folk-n. **Mettingas* < pers.n. **Metti* + **ingas**, genitive pl. **Mettinga*, + **hām**. The form *Mydicaham* [942×c.951]13th S 1521 has been identified with Mettingham but this is a corrupt sp and useless for etymology; it may represent an OE *Myndicanham* 'Myndica's homestead'. DEPN, Baron, ING 131.

MEVAGISSEY BAY Corn SX 0246. *Mevagissey Bay* 1813. P.n. MEVAGISSEY SX 0144 + ModE **bay**. PNCo 120.

MEVAGISSEY Corn SX 0144. '(Church of) Meva and Ita'. *Meffagesy* c.1400, *Mavagisi* 1410, *(ecclesie) Sanctae Meva et Ida* 1427, *Mevegysy* 1440, *Mevagesi* 1546, *Mevagizy* 1580, *Meuegisie, Meva and Issey alias Mevagissey* 1639, 1737. Saint's names *Memai* (pronounced *Mevai*) and *Iti* (pronounced *Idi*) from the dedication of the church, with Co copula **hag** 'and', **Mevai hag Idi*, with assibilation of Co medial *d* to *z*. For Idi (or Iti) cf. ST ISSEY SW 9271. The original name is preserved as Levorrick, a tenement near the church, 'church-site of Moroc', *Lammorech* c.1210, *Lanmorek* 1259–1366, *Lamorek* 1302–42, *Laverocke* c.1570, Co ***lann** + saint's name *Moroc*. PNCo 120, Gover n.d. 410, CMCS 12.62.

Great MEW STONE Devon SX 9149. ModE adj. **great** + p.n. *Meweston* 1383, *Mew Stone* 1813 OS, 'seagull rock', OE **mǣw** + **stān**. D 261.

MEXBOROUGH SYorks SE 4700. 'Meoc's fortification'. *Mechesburg* 1086, c.1160, *Mekesburc(h) -burg(h)* c.1135–1535, *Mexburgh(e)* from 16th. Pers.n. OE **Mēoc*, genitive sing. *Mēoces*, or ODan *Mīuk*, genitive sing. *Mīuks*, + **burh**. YW i.77, SSNY 148.

MICHAELCHURCH Saint's name *Michael* + ME **chirche**.

(1) ~ H&W SO 5225. *Mycheleschyrch* 1334, *Michelchirche* 1488, a translation of the earlier W name, *Lann mihacgel cil luch* 'St M's church at *cil luch*' [1045×1104]c.1130 LL, PrW ***lann** + saint's name *Michael*, + p.n. Gillow SO 5328, *Gilhou* 1228, 1350, *Kilho* 1280, 'lake-nook', OW **cil** + **luch**. He 191, 99.

(2) ~ ESCLEY H&W SO 3134. 'St M's church on Escley Brook'. *Michaelchurch Escley* 1832 OS. Earlier simply *Michaeleschirche* c.1275. Escley Brook is *Eskle* 1577, 1586, *Eskill, Hesgill* 1577, the village *Eskelyn* 1232, 1327, possibly derivatives of a r.n. identical with river ESK Cumbr NY 3289, SD 1297, NYorks NZ 7407. W forms of the name on maps are *Llanyhangleskle* 1577, *Llanihangleskle* 1611, OW **lann** + W form of Michael, *Mihangel*, + p.n. *Eskle*. RN 151, Bannister 134, DEPN.

MICHAELSTOW Corn SX 0878. 'St Michael's holy place'. *(ecclesiam) Sancti Michaelis de Hellesbiri* 1279, *Mighelestowe* 1302, *Mihelstou* 1311, *Stouwe Sancti Michaelis juxta Hellesbiri* 1315, *Mychelstowe* 1351. Saint's name *Michael* from the dedication of the church + OE **stōw**. *Hellesbiri* is p.n. *Hen-lys* 'old court' as in Helstone SX 0881, + OE **byriġ**, dative sing. of **burh** 'a fort', referring to the hill-fort of Helsbury Castle SX 0879. PNCo 122, Gover n.d. 72.

MICHELDEVER Hants SU 5139. Possibly 'marsh waters'. *Mycendefr* [862]12th S 335, *Myceldefer* [900]11th S 360, [904]13th S 374, [984×1000]12th, *Miceldevre* 1086, *Michelderura, Muchel(e)diure -deure -doura -dever* 1167–1297, *Michell Daver* c.1496. If the earliest form is reliable this is a purely W name, PrW ***mïgn** < Brit **mūkinā*, IE **mewg-*, **mewk-* 'slimy, slippery', + ***dïþr** < Brit **dubrī*, later reinterpreted as containing OE **miċel** and meaning 'Great Dever'. The river flows through the marshes of the Test. Ha 116, Gover 1958.82, DEPN, LHEB 285, GPC 2454b.

MICHELDEVER STATION Hants SU 5142. P.n. MICHELDEVER SU 5139 + ModE **station**.

MICHELDEVER WOOD Hants SU 5337. *Micheldever Wood* 1817 OS. P.n. MICHELDEVER SU 5139 + ModE **wood**.

MICHELMERSH Hants SU 3426. 'The great marsh'. *(æt) Miclamersce* [985]12th S 857, *Muchel(e)mareis -mareys* 1167–1280, *Muchelmersh(e)* 13th–1332, *-marshe* 1407. OE **miċel**, definite form **micla**, + **mersc**. The reference is to the wetland at the confluence of the Dun and the Test. Ha 116, Gover 1958.181.

MICKFIELD Suff TM 1361. 'The large open land'. *Mucelfel(da)* 1086, *Miclefeld* c.1095, *Mikel(e)feld -feud* 1242–1346, *Mickfield* 1610. OE **myċel**, definite form **mycla**, + **feld**.

MICKLE FELL Durham NY 8124. 'Great fell'. *Mickle Fell* 1863 OS. Mod dial. **mickle** (OE *miċel*) + **fell** (ON *fjall*).

MICKLEBY NYorks NZ 8013. 'The large farmstead or village'. *Michelbi* 1086, *Mikelby* 1247 etc. ON **mikill** or OE **miċel** with Norse [k] for [tʃ], + **bȳ**. YN 137, SSNY 34.

MICKLEFIELD WYorks SE 4433. 'The great stretch of open country'. *on miclanfelda* [936]13th S 712, *Miclafeld* c.1030 YCh 7, *Mikelfeld(a)* 1160×75–1404, *Micklefeild* 1636. OE **miċel** + **feld**. YW iv.57, L 242.

MICKLEHAM Surrey TQ 1753. 'The big river-bend land'. *Micle- Michel(e)ham* 1086–1294, *Mukelam, Mikeleham* 12th, *Mi(c)kel- Mikle- Mikkelham* 13th cent., *Micklam* 1569–1622. OE **miċel** + **hamm** 1 and 3 perhaps reinterpreted as **hām**. Sr 81 gives pr [mikləm], ASE 2.48, NoB 48.158.

MICKLEOVER Derby SK 3135. 'Great Ridge'. *Ufram majorem* 1087, *Oure Majoris* 1215, *Magna Oura, Owra, Ovre* 1223×9–1318, ~ *Overa -(e)* c.1232–1577, *Mycheloure* 1386, *Myckellouer* 1601, *Great Over* 1575–1608. OE **miċel**, Latin **magna**, + ***ofer** used as a p.n. Earlier simply *(æt) Vfre, Vfra* [1011]13th, *Ufre* 1086, 1226, *Over(e)* 1086–1330. 'Mickle' for distinction from LITTLEOVER SK 3334. Db 483, L 175–6.

MICKLETON Durham NY 9623. The 'large farm or village'. *Micleton* 1086. OE **miċel** + **tūn**. YN 309.

MICKLETON Glos SP 1643. 'The great manor'. *Micclantun, (to) Mycclantune* [1005]12th S 911, *Mucletv[n]e* 1086, *Mic- Mik(e)letun(a) -ton* 1091, 1221, c.1560. OE **miċel**, dative sing. definite form **miclan**, + **tūn**. It was by far the largest manor in this area. Gl i.249.

MICKLEY NYorks SE 2577. 'The great enclosure'. *Michelhach* 12th, *Mikelhae -hagh -haye* 12th–1293, *Miklay* 1379, *Mickley* from 1481. OE **miċel** + **haga**. There was also a (lost) Littley, *Litlehae* c.1180. YW v.200.

MICKLEY SQUARE Northum NZ 0762. *Mickley Square* 1864 OS. P.n. *Michelleie* c.1190, *My-, Mi(c)keley* 13th cent., *Mykley* 1428, 'the great clearing or pasture', OE **miċel** + **lēah**, + ModE **square**. NbDu 141, BF 1120, SR 14.

MIDDLE FELL Cumbr NY 7444. *Middle Fell* 1754. Middle fell between the waters of the river Nent and the South Tyne. Cu 178.

MIDDLE FEN Cambs TL 5679. *Middelfen* 1251. ME **midel** + **fen**. Ca 223.

MIDDLEHAM 'Middle homestead or estate'. OE **middel** + **hām**.

(1) ~ NYorks SE 1287. *Medelei -lai* 1086, *Mid(d)elham* 1184 etc., *Midelam* 1530. YN 252.

(2) Bishop ~ Durham NZ 3331. 'M belonging to the bishop' sc. of Durham. *Bisshopmydelham* 1412, *-medlam* 1514, *Myddelham Episcopi* 1544, *Bushop Midlom* 1580, *Bishop Middleham* from 1647. P.n. *Middelham* 1133×41–c.1700, *Midelham* 1146–1516, *Mid(d)ilham* 1259×60–1418, *Middle'm* 1561, *Midlam(e)* 1580, 1675, 1717, + ME **bishop**, Latin **episcopus**, genitive sing. **episcopi**. The manor was a chief residence of the Bishop of Durham, the earthworks of whose castle still survive. The name refers either to the manor's central position in an early multiple estate whose original *caput* was at Sedgefield, or to its location between two other important episcopal residences at Durham and Stockton. NbDu 141, Nomina 2.30.

MIDDLEHOPE 'Middle valley or side-valley'. OE **middel** + **hop**.

(1) ~ Shrops SO 4988. *Mildehope* 1086, *Middelhop(e)* 1204–1360, *Midlinghope* 1291, *Mydlop* 1577. Sa i.204.

(2) ~ MOOR Durham NY 8841. *Middlehope Moor* 1866 OS. P.n. Middlehope + ModE **moor**. Middlehope, *Midelhope* 1418, is the 'middle side-valley' of Weardale. NbDu 141.

MIDDLE LEVEL Cambs TL 3393. The middle level of Bedford Level. *Middle Levell* 1652. ModE **middle** + **level**. Ca 211.

MIDDLE LEVEL MAIN DRAIN Norf TF 5405. In his *Discourse* of 1642 Vermuyden divided Bedford Level or the Great Level of the Fens into three areas *North, Middle* and *South Levell* 1652. Ca 211.

MIDDLEMARSH Dorset ST 6707. 'Middle marsh'. *Middelmersh* 1227, *-mers* 1244, *Midelmersh* 1318. OE **middel** + **mersc**. Do 108.

MIDDLE MOOR Cambs TL 2589. *Middelmor* 1286. ME **midel** + **moor**. Hu 215.

MIDDLE MOOR Northum NU 1423. *Middlemoor* 1867 OS. ModE **middle** + **moor**. The moor lies midway between North Charlton Moor NU 1422 and Brockdam Moor NU 1524.

MIDDLESBROUGH Cleve NZ 4920. Possibly 'middlemost fortification'. *Mid(e)lesburc(h) -burgh* [1114×40]15th–1314, *Middelburg(h)* 1273, c.1291, 1613, 1846, *Middlesbrough* from 1407. OE **midleste** + **burh**. The significance of the name is not understood, but the older explanation 'Midele's fortification' depends on an OE pers.n. not independently recorded. YN 160 gives a pr [midəlzbruf], YW v.218.

MIDDLE SHIELD PARK Cumbr NY 6070. P.n. Middle Shield, *the Midle Shealds* 1589, + ModE **park**. Shield is ME **schele** 'a shieling'. Cu 98.

MIDDLESMOOR NYorks SE 0974. '*Middles* moor'. *Middelismor(e) -es-* c.1190–1296, *Mildesmore* 1198, *Mi- Myd(e)lesmore* 1249–1609, *My- Middlesmor(e)* from 1345. P.n. **Middles* + OE **mōr**. **Middles* is a common name for land between two streams or valleys. Middlesmoor lies between the valleys of Armathwaite Gill, How Stean Beck and the river Nidd. YW v.217.

MIDDLESTON MOOR Durham NZ 2535. P.n. Middlestone + **mōr**. Middlestone NZ 2531, *Midelton* 1367, *Middleton* 1576, *Middleston* 1584–1647, is the 'middlest village' of the three Merringtons, OE **middel**, later replaced by **midlest**, + **tūn**. Earlier known as Middle Merington, *Media merigtone* 1242×3, *Mid Meryngton*, ~ *Mer(r)ington* 1296–1580, *Middel- Midilmeryngton -ing-* 1311–1424. It is probably to be identified with the *Middeltun* of Cnut's grant (1031) to St Cuthbert of Staindrop and its dependent vills which included nearby Eldon NZ 2527, *Middeltun* [c.1040]12th, c.1107, *Midilton* [1072×83]15th. See Kirk MERRINGTON NZ 2631 and WESTERTON NZ 2331. NbDu 141.

MIDDLESTOWN WYorks SE 2617. *Midlestowne* 1556, *Middles Town* 1817. ModE **middles**, a term frequently used of land between two rivers or river valleys, in this case Smithy Brook and the (unnamed) stream running NW into the Calder at SE 2817, + **town**. Short for *Middle- Mid(d)els(c)hitelington -yng-*, ME **midel** + p.n. SHITLINGTON SE 2615. YW ii.206.

MIDDLETON 'Middle settlement'. OE **middel** + **tūn**. This p.n. type may sometimes denote a settlement which performed some central function for a group of communities, e.g. a market, rather than one equidistant from two others.

(1) ~ Cumbr SD 6286. *Middeltun -ton(a)* 1086–1376, *Midelton* 1180–1452, *Middleton in Lonesdale* 1401. We i.53.

(2) ~ Derby SK 1963. *Middeltune* 1086, *Mi- Myd(d)elton(e) -il- -ul-* c.1225–43 etc. Db 394.

(3) ~ Derby SK 2756. *Middeltune* 1086, *Mid(d)elton(e) -il- -ul-* c.1220–97 etc. Db 396.

(4) ~ GMan SD 8606. *Middelton* 1194–1332, *Midel-* 1212, 1317, *Mid(d)il-* 1327, 1381. La 53, Jnl 17.41.

(5) ~ Hants SU 4244. *Middeltvne* 1086, *-ton* 1228–1360, *Middleton alias Longparish* 1636. Probably the middle settlement of LONGPARISH SU 4344 in contrast to East Aston SU SU 4345, *Estetun* 1256, 'the eastern settlement' and Forton SU 4143, *Forton* 1228, 'the ford settlement'. The people of Middleton are referred to as the *Middelhæme* in S 956, *be middel hæma mearc* 'by the boundary of the *Middelhæme*' 1019. Ha 111, Gover 1958.171.

(6) ~ H&W SO 5469. *Middeltone* 1086. Also called *Dirty Middleton* 1832 OS for distinction from MIDDLETON ON THE HILL SO 5464. Not an obvious 'middle place'.
(7) ~ Lancs SD 4258. *Middeltun* 1086, *-ton* c.1190, 1199ff., *Midelton -il-* 1212, 1337 etc. The middle farm or village of the Heysham peninsula. Contrasts with HEATON SD 4460 and OVERTON SD 4358. La 176, Jnl 17.100.
(8) ~ Norf TF 6615. *Mideltuna* 1086. DEPN.
(9) ~ Northants SP 8490. *Middelton* 1197, *-il-* 1285. Between East Carlton SP 8389 and Cottingham SP 8490. Nth 169.
(10) ~ Northum NU 0023. There are three settlements here, *Tres Middleton* 1201, 1242 BF, viz. MIDDLETON HALL NT 9825 q.v., North Middleton NU 0024, *North Middiltun* 1236 BF, *Midilton North* 1296 SR, also known as *Middelton cum Rodum membro suo* 'M with its member Roddam' 1242 BF, and as *Midilest Midilton* 1296 SR, and South Middleton NT 9923, *Suth Middiltun* 1236 BF, *Middelton Suth* 1296 SR, also known as *Middilton Thome* 'Thomas's Middleton' 1242 BF. Middleton is situated between Ilderton and Wooler. NbDu 141.
(11) ~ Northum NZ 0685. *Middilton* 1296 SR. Also known as *Midd(e)leton Morel* 'M held by the Morel family' 1242 BF, 1346, for distinction from South Middleton NZ 0583, *Middelton' de Suth* 1242 BF. NbDu 141.
(12) ~ Northum NU 1035. *Middelton* 1242 BF, 1346, *Medelton* 1250, *Middilton* 1296 SR. NbDu 141.
(13) ~ NYorks SE 1249. *Middeltun(e)* [c.972]11th BCS 1278, 1086, *Meðeltune* c.1030 YCh 7, *Mithelton* 1205×19, *Mi- Myd(d)eltun -ton(a)* 1086–1379, *Mid(d)leton(a)* from c.1175. The forms have been influenced by ON **methal**. YW v.65.
(14) ~ NYorks SE 7885. *Mid(d)eltun(e)* 1086. YN 80.
(15) ~ Shrops SO 2999. *Mildetune* 1086, *Middelton* 1228, *-le-* 1361. Half-way between Rorrington SJ 3000 and Priest Weston SO 2997. Sa i.205.
(16) ~ Shrops SO 5477. *Middeltone* 1086, *-ton* 1255 etc. Not an obvious 'half-way' place. Sa i.205.
(17) ~ Shrops SJ 3129. *Middleton* 1272. DEPN, Sa i.205.
(18) ~ Suff TM 4367. *Mid(d)el- Mildetuna* 1086, *Middelton(')* 1191–1523, *Midleton* 1610. Baron.
(19) ~ Warw SP 1798. *Midel- Mildentone* 1086, *Middleton* from 1175. Half-way between Tamworth and Coleshill. Wa 20.
(20) ~ WYorks SE 3028. *Milde(n)tone* 1086, *Middelton(e)* 1155×8–1329, *Mideltun -ton'* 1185–1434, *Middleton juxta Rothwell* 1621. YW ii.139.
(21) ~ CHENEY Northants SP 4942. 'M held by the Chendut family'. *Middelton Cheyndu(i)t* 1342–3, *Middleton Cheynie* 1558. Earlier simply *Mideltone* 1086, *Middelton* 12th. The manor was held by Simon de Chendut in the 12th cent. Cf. Kings SUTTON SP 4936. Nth 55.
(22) ~ COMMON Durham NY 9531. *Middleton Common* 1866 OS. P.n. MIDDLETON-IN-TEESDALE NY 9425 + ModE **common**.
(23) ~ GREEN Staffs SJ 9935. *Middleton Greene* 1607×8. P.n. *Middleton* 1272, *Midelton* 1375, + ModE **green**. Horovitz.
(24) ~ HALL Northum NT 9825. One of the *Tres Middelton* of 1201 → MIDDLETON Northum NU 0023. *Mydleton Hall* 1542 Tomlinson. P.n. MIDDLETON + ModE **hall**. Also known as *le Midlest Middiltun* 1236 BF, *Middilmost Middelton* 1344, as *Middelton Nicholai* 'Nicholas' Middleton' 1242 BF, and as *North Midilton* 1296 SR. NbDu 141.
(25) ~ -IN-TEESDALE Durham NY 9425. *Mideltone-in-Tesedale* 1235×6 etc. with many variant spellings. Earlier simply *Middeltona -tun* [c.1160×83]16th, *My- Middilton* 1418, 1540, *Midelton(a) -tune* [c.1160×83]16th–1640. NbDu 141.
(26) ~ ONE ROW Durham NZ 3512. 'M with a single row of houses'. *Midilton Eraw* 1485, *Mid(d)elton Araw(e) -le-* 1506–1632×3, *Midleton one Rawe* 1620. So called from having only one row of cottages fronting the street and for distinction from MIDDLETON ST GEORGE NZ 3611. Also known as *Ouermyddelton'* 1382–1624 with variant spellings.
(27) ~ -ON-SEA WSusx SZ 9799. P.n. *Middelton(e)* 1086–1493 + ModE **on sea**. Sx 142.
(28) ~ ON THE HILL H&W SO 5464. *Middleton on the Hill* 1832 OS. Earlier simply *Mittleton* 1200×50, *Middletone* 1269×74, *Mydetone* 1397. Earlier, however, this was *Miceltvne* 1086, *Miclatun'* 1123, *Micleton'* c.1250, 'the large farm or village', OE **miċel**, definite form **micla**, + **tūn**. 'On the Hill' for distinction from MIDDLETON SO 5469. He 147.
(29) ~ ON THE WOLDS Humbs SE 9449. P.n. *Middeltun(e)* 1086, *-ton(a)* 1190×1210–1372 + ModE **on the wolds**. The place lies midway on the road from Market Weighton to Driffield. YE 164.
(30) ~ PRIORS Shrops SO 6290 → MIDDLETON.
(31) ~ ST GEORGE Durham NZ 3611. 'St George's M'. *Midelton' Sancti Georgii* 1313, 1314, *Middelton' Seint George* 1382, *Medylton George* 1473–1620 with many variant spellings. P.n. *Middilton* 1259, 1406, + saint's name *George* from the dedication of the parish church situated here and for distinction from the other settlement in the parish, MIDDLETON ONE ROW NZ 3512. Middleton St George, a deserted village, was also known as Low Middleton, *Nethermiddilton* 1320–1619 with variant spellings, *Low Middleton* 1717, and Middleton One Row as Over Middleton, *Ouermyddelton'* 1382–1624 with variant spellings. NbDu 141.
(32) ~ SCRIVEN Shrops SO 6887. 'M held by the Scriven family'. *Skrevensmyddleton* 1577, *Midelton Screeven* 1737. P.n. *Middeltone* 1086, *Middleton* 1284×5, + manorial addition Scriven, probably a family name from OFr *escrivein* 'a writer' added for distinction from other Middletons. Not an obvious 'half-way' place. Sa i.206.
(33) ~ STONEY Oxon SP 5323. 'Stony M'. *Myddylton Stony* 1552. P.n. *Mideltone* 1086 etc. with variants *-intune -intone- -in- -yn(g)ton* 12th–14th, + adj. **stony** referring to the soil. O 228.
(34) ~ TYAS NYorks NZ 2205. 'M held by the Tyas family'. P.n. *Midelton* 1086, *Middeltun -ton* 1086 etc., + manorial addition *Tyas* 14th representing OFr *(le) Tyeis* from CG *diutisc* 'German'. YN 285.
(35) Stoney ~ Derby SK 2375. *Stony Middelton (by Eyom)* 1332, 1577, 1610. 'Stony' with reference to the nature of Middleton Dale and its cliffs. Earlier simply *Middeltun(e)* 1086, 1236, *-il- -ton(e)* 1200–1490. Db 147.

MIDDLETON PRIORS Shrops SO 6290. '(That part of) M belonging to the prior', sc. of Wenlock Priory. *Middleton Priors* 1833 OS. P.n. *Mittilton* c.1200, *Mittelington* 1222, *Miteltone* 1291, *Muttulton* 1349, of uncertain origin, possibly **(ġe)mȳthlēah** + **tūn** 'the settlement at *(ġe)mȳth-lēah*, the wood at the junction of streams', + ModE **prior** for distinction from Middleton Baggot SO 6290, 'the part of M held by the Bagard family', cf. HOPE BAGOT SO 5874. DEPN.

MIDDLE TONGUE NYorks SD 9181. *Middle Tongue* 1860 OS. ModE **middle** + **tongue**. ON *tunga*, ModE **tongue**, was used in the topographical sense of 'a tongue of land', especially 'a tongue of land formed at the confluence of two rivers'. Here it is applied to land between two grains of Cragdale Water.

MIDDLEWICH Ches SJ 7066. 'The middlemost Wich or salt-works'. *Mildestuich* (for *Midlest-*) 1086, *le Mydlestwych* 1394, *the* ~ 1517, *Mideluuiche* [1070×1101]1302, *Middlewich* from 1273. Earlier simply *seo wic* [1002×4]c.1100, *Wich* 1086–1260. OE **middel** + **wīċ** 'salt-works'. Site of the Roman town of Salinae 'the salt-works', *Salinis* [c.700]13th Rav. 'Middle' for distinction from NORTHWICH SJ 6573 and NANTWICH SJ 6552. Che ii.240, RBrit 451. Addenda gives 19th cent. local pr [-witʃ, -wɛitʃ, -waitʃ].

MIDDLEWOOD GREEN Suff TM 0961. *Middlewood Green* 1837 OS. P.n. Middlewood, ModE **middle** + **wood**, + **green**.

MIDDLEWOOD STATION GMan SJ 9484. P.n. Middlewood + ModE **station**. A halt on the railway line from Stockport to Buxton. Middlewood, *Midlewood* 1611, is so-called because it is in or adjoins three townships, Poynton SJ 9283, Norbury SJ 5647 and Marple SJ 9588. Che i.283.

MIDDLEZOY Somer ST 3732. 'Middle Sowy'. *Middlesowy* 1227, *Middlesoy* 1610. ME adj. **midel** + p.n. *Se- Soweie* *[725]12th S 250, *Sowy* [725]14th S 251, [854]14th S 303, *[971]13th S 783, [c.1250, c.1350] Buck, *Sowi* 1086, [1205, 1246] Buck, 'Sow island', r.n. Sow identical with River SOW Staffs SJ 8628 and River SOWE Warw SP 3777 + OE **ēġ**, WS **īeġ** 'island'. 'Middle ' for distinction from WESTONZOYLAND ST 6124. DEPN, RN 375.

MIDDRIDGE Durham NZ 2526. 'Middle ridge'. *Midderigg'* [1183]c.1320, *Miderige* 1242×3, *Midrige -rigg(e) -y-* 1259×60–1624, *Midrich* 1382, 1418, *-ridge* 1466×7, *Middridge* from 1613. OE **midd** + **hrycg**. The exact topographical sense is unclear: PNE ii.40 suggests **mid** preposition and the meaning 'between ridges'. There are very slight rises N and S of the village, but the more likely sense is 'middle ridge' of the three prongs of outlying high ground with Merrington and Ferryhill to the N and the high ground at Royal Oak to the S. NbDu 142, L 117.

MIDGE HALL Lancs SD 5123. 'Nook of land infested by midges'. *Miggehalgh* 1390. OE **mycg** + **halh** in the sense 'dry ground in marsh'. La 133, L 107.

MIDGEHOLME Cumbr NY 6358. Possibly 'marshy hollow infested by midges'. *Midgwham* 1677, *Midgham* 1698. The forms are too late for certainty but this could be ModE **midge** + dial. **wham** 'marshy hollow' (<OE *hvamm*). But the specific could also be OE **micg** 'liquid drainings from manure'. Cu 103.

MIDGHAM Berks SU 5567. 'Midge infested riverside meadow'. *Migeham* 1086–1547, *Miggeham* 1260–1341, *Mi- Mygham* 1199–1378, *Midgham* 1761. OE **mycg** genitive pl. **mycga** + **hamm** 1. On formal grounds the generic might be *hām* but the semantics of the name and its situation on the Kennet favour *hamm*. Brk 255.

MIDGLEY WYorks SE 0326. 'Midge-infested wood or clearing' or one 'with a manure heap'. *Micleie* 1086, *Miggelay -ley* 13th–1452, *Midg(e)ley* from 1536. OE **mycg** or **micge** + **lēah**. YW iii.132, L 205.

MIDHOPE MOORS SYorks SK 1998. *Middop Moor* 1841. P.n. Midhope + **moor**. Midhope, *Middehop* 1255, 1358, *Myd- Midhop(e)* before 1279–1424, is the 'land in the middle of the valley, or, between the valleys', OE **mid** + **hop**. Midhope lies in the middle of the Little Don valley, but the name may well denote the high ground on which Upper Midhope stands between two small valleys. YW i.225, 237, L 115, 117.

MIDHOPE STONES SYorks SK 2399. P.n. Midhope as in MIDHOPE MOORS + ModE **stones**.

MIDHURST WSusx SU 8821. 'The middle wooded hill' or 'the place in the middle of the *hurst*'. *Mid(d)eherst* 1185, 1190. OE **midd**, definite form **midda**, or preposition **mid**, + **hyrst**. Sx 27.

MIDLOE Cambs TL 1664 → SOUTHOE TL 1864.

MIDVILLE Lincs TF 3857. *Midville* 1824 OS. ModE **mid** 'middle' + **ville**, an element briefly in vogue for new 19th cent. p.n. formations. A modern creation for one of several new townships formed in 1812 on newly drained fenland. It lies midway between Stickney TF 3456 and EASTVILLE TF 4057. A *West Ville* occurs 1824 OS 4 m. SW of Stickney at TF 3052. Cf. also FRITHVILLE TF 3150. Wheeler 228, Room 1983.71.

MILBORNE PORT Somer ST 6718. 'Milborne town or borough'. *Milleburnport* 1249, *Milborn port* 1610. P.n. Milborne + ME **port** 'a town with market rights'. Milborne Port became a borough in 1225. Earlier simply *(æt) Mylenburnan* 873×88 S 1507†, c.950 S 1539, *Mi- Melebvrne -burne* 1086, *Meleborn* [1156×7]14th, 'the mill-stream', OE **myln** + **burna**. 'Port' for distinction from Milborne WICK ST 6620. DEPN, PNE ii.70.

MILBORNE ST ANDREW Dorset SY 8097. 'St Andrew's M'. *Mileburne Sancti Andree* 1272×1307, *Muleburne(e) St Andrew* 1294, *Mi- Myl(e)bo(u)rn(e) Seynt Andrew(s)* 1391–1512. P.n. *(æt) Muleburne* [843 for 934] n.d. S 391, *Muleborn* [843 for 934]17th ibid., *-burn(e)* 1288–1334, *Milbo(u)rne* 1412, 1626, 'mill stream', OE **myln** + **burna**, + saint's name *Andrew*, Latin *Andreas*, genitive sing. *Andree*, from the dedication of the church. Do i.306, L 1, 17, 97.

†The identification is not certain: ECW 126–7 suggests this may refer to the great royal manor of Silverton, Devon, *Borna* 1086.

MILBOURNE Northum NZ 1175. Probably the 'mill stream'. *Meleburna* 1158, 1202, *Melleburn* 13th cent., *Mulneburn* 1201, *Miln(e)burn* 1255–1479, *Milleburn* 1242 BF, *Milburn* 1296 SR, *-bourn* 1346. OE **myln** + **burna**. The persistent spellings with *mel(l)e*, however, suggest that the specific may originally have been something other than OE **myln**, possibly OE **mǣle**, **(ġe)mǣl** 'dyed, stained, multicoloured', or **mǣle**, **mēle** 'a cup, a bowl, a basin' used in a topographical sense. The stream flows here through a deep narrow glen. NbDu 142, DEPN, Tomlinson 78.

MILBURN Cumbr NY 6529. 'Mill stream'. *Milleburn(e)* 1178–1278, *Miln(e)brunn' -burn(e)* 1202–1679, *Milburn(e) -bourn(e)* from 1370. OE **myln** + **burna**. We ii.120, L 1, 17, 97.

MILBURN FOREST Cumbr NY 7232. *Milburn For.* 1727, *the Forest of Milburne* 1777. Earlier *Milburnfelchace* 1357, 'the chase or hunting preserve of Milburn Fell', p.n. MILBURN NY 6529 + ME **chace, forest**. We ii.122.

MILBURY HEATH Avon ST 6690. *Mi- Mylburgheth(e) -bury- -borow-* 1497, *Milberrowe Heath* 1622. This is thought to be p.n. Milbury 'the fortified place near the mill', ME **mill** + **burgh, bury**, + ModE **heath**. A better etymology, however, might be 'Mildburg's heath', OE feminine pers.n. *Mildburh*, genitive sing. *Mildburge*, + **hǣth**. Gl iii.15.

MILCOMBE Oxon SP 4134. 'The middle valley'. *Midelcvmbe* 1086–1268 with variants *-cumb(a), Middelcumbe -combe* 13th cent. *Milecumbe* 1206, *Mylcombe* 1530, *Millcome* 1586. OE **middel** + **cumb**. O 400.

MILDEN Suff TL 9546. Possibly 'the place where milds grows'. *Mellinga* 1086, *Meldinges* 1121×48, 1319, *Melling'* 1160, *Mauling* 1228, *Meulinges* 1235, *Medling(g')* 1254, 1287, *Mildinge* 1610. OE ***melding** < **melde** + **ing**². *Melde* 'milds' is the name of various kinds of arrach or orache, all of which were used in medicine. However, the occasional *-es* form may point to an OE folk-n. **Meldingas* 'the people called after Melda', OE pers.n. **Melda* + **ingas**. DEPN, ING 54.

MILDENHALL Suff TL 7174. Probably 'Milda's nook'. *(æt) Mildenhale* [1043×4]14th S 1069, *Mitdene- mudenehalla* 1086, *Middelhala* 1130, 1162, *Mildehal* 1158, *Mildenhale* c.1200, *Mildnall* 1610. Probably OE pers.n. **Milda*, genitive sing. **Mildan*, + **halh** identical with MILDENHALL Wilts SU 2169. The specific caused the DB scribes some difficulty and was later partly replaced by ME *middel*. DEPN.

MILDENHALL Wilts SU 2169. 'Milda's nook of land'. *Mildanhald* (sic) [801×5]12th S 1263, *Mildenhalle* 1086, *Milde(n)hal(e) -hall* 13th cent., *Midnall* 1539, *Mylenoll* 1558×1601, *Minal* 1760. OE pers.n. **Milda*, genitive sing. **Mildan*, + **healh** in the sense 'valley'. Wlt 301 gives pr [mainəl], L 101.

MILDENHALL FEN Suff TL 6678. *Mildenhall Fen* 1836 OS. P.n. MILDENHALL TL 7174 + ModE **fen**.

MILE ELM Wilts ST 9969. *Mile Elm* 1773. A mile from Calne along the Melksham road. Wlt 260.

MILE END Essex TL 9827. *Milend(e)* 1156×80, 1257, 1539, *La Milhende iuxta Colecestr'* 'the mile end by Colchester' 1285, *La Mile ende, Mi- Mylende* 1287, 1291, *Myle End alias Myland* 1561. ME **mile** + **ende**. About one mile from Colchester. Ess 376, Jnl 2.46.

MILE END Glos SO 5911. *Mile End* 1830. One mile from Coleford SO 5710. Gl iii.230.

MILEBUSH Kent TQ 7545. *Mile Bush* 1819 OS. One mile on the road N from Marden. ModE **mile** + **bush**.

MILEHAM Norf TF 9119. 'Mill village'. *Mele- Muleham* 1086, *Meleham* before 1122, 1160. OE **mylen** + **hām**. DEPN.

MILES PLATTING GMan SJ 8699. *Miles Platting* 1843 OS. The second element is dial. **platting** 'a small bridge', and the first presumably the genitive sing. of **mile** 'a mile'. Situated approximately a mile from the centre of Manchester.

MILFIELD Northum NT 9333. *Melfelde* late medieval Hope-Taylor 15–16, *Melfeld* 1637 Camden, *-field* c.1715 AA 13.11. Since the time of Camden, Milfield has been identified with Bede's *Maelmin* [c.731]8th HE ii.14, the royal residence of the kings of Northumbria after their abandonment of Yeavering. If this is so, the specific of Milfield preserves the first element of *Maelmin*, a British name from PrW ***mę̄l**, ***moil** 'bare' (Brit **mailo-*) + probably ***mönïth** 'hill' (Brit **monïio-*, cf. Latin *mons*). The reference is to Coldside Hill NT 9132. Camden 815(d), Jnl 8.24.

MILFORD 'Mill ford'. OE **mylen** + **ford**. L 72.

(1) ~ Derby SK 3545. *Muleford* 1086, *Milford* 1836 OS. This place was unimportant in the Middle Ages and only gained prominence with the development of industry here in the 18th cent. Db 589.

(2) ~ Devon SS 2322. 'Mill ford'. *Meleforde* 1086, *Milleford* 1238, 1310, *Muleford* 1316. D 75 gives pr [milvəd].

(3) ~ Surrey SU 9442. *Muleford(e)* 1235–1452, *Meleford* 1344, *Mulford(e)* 1384, 1502, *Mylford* 1529, *Millforde* 1603. Sr 216, L 72.

(4) ~ Staffs SJ 9721. *Millford* 1798.

(5) ~ ON SEA Hants SZ 2892. *Melleford* 1086, *Melne- Mulneford* 1100×35, *Milneford* 1152. A mill is mentioned here in DB. 'On Sea' is a late 19th cent. addition to enhance the holiday associations of the place. Ha 117, Pevsner-Lloyd 335.

(6) South ~ NYorks SE 4931. *Suth Muleforth* 1285. ME **suth** + p.n. *(on) niy senforda* (sic for *mylenforda*) [963]14th S 712, *Myleford* c.1030 YCh 7, *Milford* from 1109×12. 'South' for distinction from North Milford SE 5039. The ford carried the London road from Sherburn S over Mell Dike. YW iv.58, 69.

MILK HILL Wilts SU 1064. *Melkhulle, Melkehylle* 1425, *Mylkehill* 1570. OE **meoluc** + **hyll**. A hill with pasturage producing good milk. Wlt 324.

MILKWALL Glos SO 5809. 'Milk spring'. *Milkwall* 1844. Ultimately OE **meoluc** + **wælla**. Either a spring with milky water or one used by cattle or where milk was cooled. Gl iii.230.

MILL BANK WYorks SE 0321. *(the) Milne Banke* 1624, *Millbanck* 1709, earlier *Solandmylnebanke* 1492. The mill, *Solandemyle* 1462 'Soyland mill' (p.n. Soyland as in SOYLAND MOOR SD 9817), was the corn mill for the graveship of SOWERBY SE 0423. YW iii.148.

MILL CORNER ESusx TQ 8223. An Andrew atte Melne is mentioned in the 1296 SR. Sx 523.

MILL END Bucks SU 7885. *Mill End* 1822 OS, sc. of Hambleden SU 7886.

MILL END Herts TL 3332. The 'mill end' sc. of Sandon. *le Melnende* 1277. ME **meln** (OE *myln*, Herts dial. *meln*) + **ende**. Hrt 166.

MILL GREEN Essex TL 6400. *Mill Green* 1777. ModE **mill** + **green**. Ess 255.

MILL HILL GLond TQ 2292. '(Wind)mill hill'. *Myllehill* 1547. ModE **mill** + **hill**. Mx 59.

MILL LANE Hants SU 7850. The lane leading to Itchel Mill, *Itchel Mill* 1816 OS. ModE **mill** + **lane**.

MILLAND WSusx SU 8327. 'The mill land'. *Mylland* 1556. ME **mill** + **land**. Sx 45.

MILLBRIDGE Surrey SU 8442. 'The mill bridge'. *Milbridge* 1613. ModE **mill** + **bridge**. Sr 180.

MILLBROOK Beds TL 0138. 'Mill stream'. *Melebroc* 1086, *Mele- Mulebrok* 13th cent., *Millbrook* 1227. OE **myln** + **brōc**. Bd 82.

MILLBROOK Corn SX 2671. 'Mill stream'. *Milbrok* 1342, *Millbroke* 1393. OE **myln** + **brōc**. PNCo 123, Gover n.d. 232.

MILLBROOK Hants SU 3813. 'Mill brook or marsh'. *(æt) melebroce* [956]10th S 636, *Melebroc* 1086–1205, *Mulebrok(e)* c.1124–1327. OE **myln** + **brōc**. A ford is mentioned here in 1045 S 1008, *(on) mylebroces ford*. Ha 117, Gover 1958.40, DEPN.

MILLBROOK STATION Beds TL 0040. P.n. MILLBROOK TL 0138 + ModE **station**.

MILLER'S DALE Derby SK 1473. *Millers-dale* 1767. This name superseded *Milndale* 1630, *Mil(l)hous(e)dale* 1633, 1650, *Milnhouse Dale* 1840, 'mill dale' and 'mill-house dale', OE **myln**, **myln-hūs** + **dæl**. Named either from Tideswell mill-house, *Tiddeswal Mylnehowse* 1438, 1442, or Wormhill mill-house, *Worm-hille Millehowses* c.1460. Db 180.

North MILLFORD NYorks SE 5039. ME prefix *North(e)* 13th + p.n. *Mileford(e) -ia* 1086, *Milford* c.1190×6, as in South MILFORD SE 4931. 'North' for distiction from South MILFORD SE 4931. YW iv.58, 69.

MILLGREEN Shrops SJ 6828. *Mill Green* 1833 OS.

MILLHOLME Cumbr SD 5690. 'Mill water-meadow'. *Miln(e)holme* 1530, 1680, *Millholme* 1707. ME **milne** (OE *myln*) + **holm** (ON *holmr*). We i.133.

MILLHOUSE Cumbr NY 3637. 'Mill house'. *del Miln(e)hus* 1292, *the Milnehous of Castle Soureby* 1340, *Milhouse* 1608. OE **myln** + **hūs**. Cu 245.

MILLINGTON Humbs SE 8351. Partly uncertain. *Mil(l)eton(a)* 1086, *Milington(a)* [1150×61]1206–1360, *-in-* 1254, 1294, 1363, *Millington(a)* 1156×7–1527, *Midlington* 1227. Either OE **myl(e)n** 'a mill' or pers.n. *Mil(l)a*, gentive sing. *Mil(l)an* varying with *Milla* + **ing**^{4}, + **tūn**. YE 178, DEPN.

MILLMEECE Staffs SJ 3333. 'Meece with the mill'. *Mulnemes* 1289. ME **muln** (OE *myln*) + p.n. *Mess* 1086, *Mes* 1163, 1208, *Meis* 1218, originally a r.n. from OE **mēos** sb. 'a moss, a marsh, a bog' or adj. 'mossy'. 'Mill' referring to a mill mentioned in 1163 for distinction from Coldmeece SJ 8532, *Coldemes* 1272, 'cold M'. St i.14, RN 281.

MILLOM Cumbr SD 1780. '(Settlement at) the mills'. *Millum* c.1180–1550 with variants *Mil- Myllum*, *Milnum* c.1205–1338, *Millom* from 1303. OE **myln**, dative pl. **mylnum**. The name is a local one by origin and implies membership of a larger unit. In fact it seems to have supplanted an earlier name, *Hougenai*, p.n. *Hougun* 1086 + OE **ēġ** or ON **ey** 'an island'. *Hougun*, ON **haugr** 'a hill', dative pl. **haugum** 'at the hills', was the name of a manor held by Earl Tostig in 1065 comprising the vills of *Hougun*, Bootle, Whicham, *Hougenai*, and Kirksanton. Cu 414, SSNNW 326, TLCAS 18.94, 97.

MILL SIDE Cumbr SD 4484. *Mill Side* 1865 OS. Cf. *Milne Beck* 1741. We i.80.

MILL STREET Norf TG 0118. *Mill Street* 1838 OS. ModE **mill** referring to the Swanton Paper Mill (1838) on the river Wensum + **street** probably in the sense 'hamlet'.

MILLTHROP Cumbr SD 6691. 'Outlying farmstead with or near a mill'. *Milnethorpe* 1535, *-throppe* 1608, 1621, *Milthropp(e)* 1606, 1625. OE **myln** + **thorp**. YW vi.269.

MILLTOWN Derby SK 3561. 'Mill settlement'. *Milnetown* (sic) 1316, *Mylneton(e)* 1454, 1519, *Mylnetowyne* 1543 *-towne* 1547–74. ME **miln** + **toun**. The mill settlement in Ashover parish. Db 191.

MILLTOWN Devon SS 5538. 'The mill settlement' of Marwood parish. *Miltoune* 1609. ME **mill** + **toun**. D 52.

MILNROW GMan SD 9212. 'Mill row'. *Mylnerowe* 1554, *Mylneraw* 1577, *Milnrow* 1843 OS. OE **myln** + **rāw** 'a row of houses'. Earlier known as *Milnehouses* 13th, *Milnehus* 1292. The reference is to a mill on the river Beal. La 56.

MILNTHORPE Cumbr SD 4981. 'Outlying settlement with a mill'. *Milntorp* 1272, *Mi- Myln(e)thorp(e)* 1282–1682, *Mi- Myl(l)throp(pe)* 1579–1754. OE **myln** + ON **thorp**. We i.95 gives pr ['milθrəp], SSNNW 204, 355–6.

MILSON Shrops SO 6473. Uncertain. *Mulstone* 1086, *Mulston(')* 1221–1431, *Muleston(e)* 1210×12–92, *-tūn* 1255, *Mulnestone* 1291×2, *Milston(')* 1291×2, 1535, 1683×4, *Millson* 1672–1743.

Clearly understood in 1291×2 as if from OE **mylen-stān** 'a millstone', but this would be an unusual p.n. Millstones were normally concealed in the mill-building; Sa i.208 suggests that there may have been some peculiarity of construction which caused the grindstones to be visible. This was probably originally either 'Mula's stone', OE pers.n. *Mūla*, genitive sing. *Mūlan*, + **stān**, or rather 'Mul's estate', OE pers.n. *Mūl*, genitive sing. *Mūles*, + **tūn**, early reinterpreted as if from *mylen-stān*. It is noteworthy that while there are no *-stan* spellings there is one in *-tūn*. Sa i.208.

MILSTED Kent TQ 9058. Partly uncertain. *Milstede* c.1100–1313 with variants *Myl- Milne-* and *-sted, Mild(e)sted(e)* 1213–1270, *Middelstede* 1226, *Milcstede* 1278. It is clear that this name was predominantly taken to be 'mill-place', OE **myln** + **stede**, in the Middle Ages, but this may have been, like the isolated *Middelstede* 1226, a folk-etymology of the difficult *Mildestede*, or *Milde-* may be for *Middel-* as in MILTON REGIS TQ 9064. PNK 253, Stead 221.

MILSTON Wilts SU 1645. Possibly 'the middlemost settlement'. *Mildestone -ton* 1086–1428, *Midleston* 13th cent. *Milleston* 1309, *Mylston* 1468, *Millson* 1603×25. OE **midlest** + **tūn**. Wlt 369.

MILTON 'Middle settlement or estate'. OE **middel** + **tūn**.

(1) ~ Cambs TL 4762. *(fram) Middeltune* c.975, [975]12th, [1042×66]14th S 1051, *-ton(e)* 1086–1399, *Milton* from 1275. Between Impington and Fen Ditton. Ca 182.

(2) ~ Oxon SU 4891. *(æt) Middeltune* 956 S 594, *Middeltune* 1086, *Milton* from 1369. M lies between Steventon and Sutton Courtney. Brk 416.

(3) ~ Oxon SP 4535. *Middelton'* 1199–1316 with variants *-ton(e)*, *Milton* from 1386. O 401.

(4) ~ ABBAS Dorset ST 8001. 'M belonging to the abbot' sc. of Milton Abbey. *Middelton' Abbatis* 1268–1392, earlier simply *Middeltone* [843 for 934]17th S 391, *(of, to) Middel tune* late 10th ASC(A) under year 964, *Middeltune* c.1000, 1086, *Mid(d)elton(e)* 1162–1558, *Milton(')* from 1268. Do iii.219.

(5) ~ ABBOT Devon SX 4079. 'M held by the abbot' sc. of Tavistock Abbey in 1086. *Middeltone* 1086, *Middelton Abbots* 1297, *Myddeltone al. Myltone* 1391. D 215.

(6) ~ BRYAN Beds SP 9730. 'M held by the Brian family'. *Midelton Brian* 1272×1307. Earlier simply *Milden- Middelton(e)* 1086. Bd 130.

(7) ~ CLEVEDON Somer ST 6637. 'M held by the Clevedon family'. *Milton Clyvedon* 1408. Earlier simply *Mideltvne* 1086, *Middleton* 1201, *Milton* 1610. Midway between Stoney Stratton ST 6539 and Bruton ST 6834. Held by William de Clyvedon c.1200. DEPN.

(8) ~ DAMAREL Devon SS 3810. 'M held by the Albemarl family'. *Middelton Aubemarle* 1301, *Milton Daumarle* 1339. Earlier simply *Mideltone* 1086. The estate was held by Robert de Albamarla in 1086. D 152, DEPN.

(9) ~ ERNEST Beds TL 0156. 'M held by Ernis'. *Mildentone* 1086, *Middetone* 1086–1372, *Milton* 1372. The manorial addition, *Erneis* 1291, *Ernys* 1334, *Harnes* 1526, has been assimilated to the pers.n. Ernest. Robert son of Erneis is associated with Milton in 1227. The name goes back to Frankish *Arnegis*. Bd 24 gives pr [a:nist].

(10) ~ HILL Oxon SU 4790. *Milton Hill* 1830. P.n. Milton + ModE **hill**. Brk 416.

(11) ~ KEYNES Bucks SP 8733. 'M held by the Cahaignes family'. *Middeltone Keynes* n.d. FF, *Milton Keynes* 1422. Earlier simply *Mid(d)eltone* 1086. Hugh de Cahaignes held land in Bucks in 1166. His surname derives from either Cahaignes, Eure, or Cahagnes, Calvados. Milton Keynes is now the name of a new town founded in 1967. Bk 36, DBS s.n. Caines.

(12) ~ KEYNES VILLAGE Bucks SP 8839. This name was coined only after the foundation of the new town of MILTON KEYNES in 1967. Properly it is the original settlement of Milton Keynes.

(13) ~ LILBORNE Wilts SU 1960. 'M held by the Lilborne family'. *Mydilton Lillebon* 1249, *Milton Lilborn* 1412. P.n. *Mideltone* 1086 + manorial addition from the family name of William de Lilebone who held here in 1236 and whose family came from Lillebone, Seine Inférieure, France. Midway between Pewsey and Easton and S of Wootan Rivers. Wlt 349.

(14) ~ MALSOR Northants SP 7355. 'M held by the Malsor family'. *Middelton Malsores* 1395, *Middleton Mallesworth* 1542, *Milton al. Middleton Malsor* 1781. Earlier simply *Mideltone* 1086–1428 with variants *Middel- -le- -il-*. The place lies between Gayton SP 7054 and Wooton SP 7656. *Mallesworth* is a hypercorrect spelling due to belief that final *-or* was a reduction of *-worth*. Nth 150.

(15) ~ ON STOUR Dorset ST 8028. *Milton upon Stoure* 1397 etc. P.n. *Midel- Miltetone* 1086, *Mid(d)elton(e)* 1235–1431, *Milton(')* from 1275, + r.n. STOUR. The place lies half-way between Gillingham and Silton. Do iii.16.

(16) ~ REGIS Kent TQ 9064. 'The king's M'. *(to) Middeltune þæs cynges* 1052 ASC(E). Otherwise simply *(æt) Middeltune* 839 ASC(A), 1086, 1087, *-ton(e)* 1086–1200, *Milde(n)tone* 1086, *Mildetune* c.1100. OE pers.n. *Middeltūn* + Latin **rex**, genitive sing. **regis**, translating OE *þæs cynges* 'of the king'. K 254.

(17) ~ -UNDER-WYCHWOOD Oxon SP 2618. P.n. *Mideltone* 1086–1393 with variants *Mid(d)elton(e) -il-* and *-le-*, *Milton* from 1391×2, + preposition **under** + district n. WYCHWOOD. O 363.

(18) Great ~ Oxon SP 6302. *Magna Middeltone* 1233, *Magna Mi- Myltone* 1285, *Great Milton* 1360. ME **great**, Latin **magna**, + p.n. *Mid(d)eltone* 1086–1274 with variants *-tune -ton(e) -t(h)ona* and *-il-*, *Mi- Mylton(e)* from 1285. O 141.

(19) Little ~ Oxon SP 6100. *Parva Middeltone* 1233, *parva Miltone* 1285, *Little Milton* from 1301. ME adj. **litel**, Latin **parva**, + p.n. *Mideltone* 1086, *Milton(e)* 1285. 'Little' for distinction from Great MILTON SP 6302. O 143.

(20) New ~ Hants SZ 2495. The name of the new development around the London South Western Railway's branch station opened here in 1888. Confusion with Milton SZ 6699 near Portsmouth is said to have led the postmistress of the new sub-post-office near the station to have suggested the name after the 'old' Milton ½ a mile away at SZ 2394, *Mideltvne* 1086, *Middelton(e)* 1242–1416, *Milton or Medilton* 1474, the 'middle settlement' of the area bounded by Barton, Chewton, Hinton, Bashley, Hordle and Downton. Ha 117, Gover 1958.228.

(21) South ~ Devon SX 6942. *South Milton* 1809 OS. ModE adj. **south** + p.n. *Mideltone* 1086, *Middelton* 1285, *Milton* 1428. Milton is the central place of the peninsula formed by the Avon, the Kingsbridge Estuary and the sea. 'South' for distinction perhaps from MILTON ABBOT SX 4079. D 309.

(22) West ~ Dorset SY 5096. ModE adj. **west** + p.n. *Mideltone* 1086, *Midelton* 1212, *Middleton* 1288, *Mylton* 1303. 'West' for distinction from Milton Abbas and 'middle' in relation to the other *Power* estates, POORTON SY 5197 and POWERSTOCK SY 5196. Do 109.

MILTON 'Mill settlement'. OE **myln** + **tūn**.

(1) ~ Cumbr NY 5560. *Milneton* 1285–1387, *Milton* from 1399. The mill settlement of Farlam NY 5558. Cu 85.

(2) ~ Staffs SJ 9050. *Mulneton* c.1287, *Mylton* 1613×4. Horovitz.

(3) ~ COMBE Devon SX 4865. *Milton* 1546, 1809 OS. D 226.

(4) ~ GREEN Ches SJ 4658. *Milton Green* 1671. P.n. *Milneton* 1257, *Mulneton* 13th–1471, *Mi- Mylton* 1510–1607, + ModE **green**. Che iv.90.

MILVERTON Somer ST 1225. 'Settlement at Milford, the mill ford'. *Milferton* [1061×6]13th S 1240, *Milvertone -tvne* 1086, *Melverton* 1235, *Milverton* 1610. OE **myln** + **ford** probably used as a p.n. *Mylnford + **tūn**. DEPN.

Old MILVERTON Warw SP 3067. ModE adj. **old** + p.n. *Malvertone* 1086, *Mulverton(a)* 1123–1408, *Milverton* from 1210, probably 'the settlement at *Milford*', OE p.n. *Myln-ford* 'the mill ford', + **tūn**. A mill is mentioned here in DB. 'Old' for distinction from Milverton SP 3066, a modern suburb of Leamington. Wa 176.

MILWICH Staffs SJ 9732. 'Mill farm or salt-working place'. *Melewich, Mvlewiche* 1086, *Mulewich* 1166, *Millewyz* 1236. OE **myln** + **wīċ**. The place lies in an area of salt springs. DEPN, Horovitz.

South MIMMS Herts TL 2200. *Suthmimes* 1253, *-mimmes -mymmes* 1256–1312. ME **south** + p.n. *Mimes* 1086, 1274, *Mi- Mymmes* 1211–1316, of unknown origin and meaning. An otherwise unknown folk-n. **Mimmas* has been suggested, but this is pure guesswork. 'South' for distinction from North Mimms, 3 miles N, *Northmimmes -myn-* 1236–1301, earlier *Mimmine* 1086, *Mimmis* 1138, 1142. Hrt 65, Mx 76.

MINCHINHAMPTON Glos SO 8700. 'Hampton held by the nuns' sc. of the Trinity at Caen, Normandy, to whom the manor was granted c.1080. *Hampton(e) Monialium* 1220×30–1553, *Mun(e)chenehampton(e)* 1221–1416, *Min(e)chen- Mynchen-* 1221–1756. OE **mynċen**, genitive pl. **mynċena**, Latin **moniales**, genitive pl. **monialium**, + p.n. *Hantonia -tone* 1082, 1086–1358, *Hampton(e)* 1215–1631, 'the high settlement', OE **hēah**, dative sing. definite form **hēan**, + **tūn**. Gl i.95.

MINDRUM Northum NT 8432. 'Ridge mountain'. *Minethrum* [c.1040]12th HSC, 1176, *Mi- Myndrum* 13th cent., PrW ***mönïth** + ***drum**. Cf. f.n. Menadrum Cornwall CPNE 163. NbDu 143, PNE ii.41, Jnl 1.49.

MINEHEAD Somer SS 9746. '(Place at) the headland(s) called Myne'. *(æt) Mynheafdon* [1046]12th S 1474, *Manehefue* 1084, *Maneheve* 1086, *Maneheafe* [1090×1100]12th, *Man- Menhaved* [1174×91–1252]Buck, *Menehewed* 1225, *Menheve* [1192×6, 1237×42]Buck, *-haved -heved* 1198–[1287]Buck, *Manhed* [13th]Buck, *Menhed* [1237]Buck, *Myn(e)heved* 1284–1352, *Mynehed* 1400, *Minehed* 1459, *Mynhead* 1610. Hill-name Myne SS 9248, *Mene* 1086, + OE **hēafod** referring to North Hill, the prominent rounded headland N of the town. The earliest form represents OE dative pl. *hēafdum*. Myne is probably an ancient hill-name originally applied to the high ground along the coast from Selworthy, cf. the Buckland Cartulary form *Menelond* for land on Myne there. Myne may be identical with the prominent but unexplained hill-name MEON HILL Warw SP 1745, *Mene -a* 1086–1405, *Muna* 1158–15th, *Meone* 1221–1594, rather than PrW *mönïth* with early loss of final *-th* before *hēafod* or a back-formation from *Menedun* 1225 which Ekwall explained as lost hill-name *Myned* < PrW ***mönïth** 'hill, mountain' + OE **dūn**, 'the hill called Myned'. Turner suggests Co **myn**, W **min** 'edge, brink, lip'. The first part of *Menheved* seems early to have been understood as the plural of the word *man* which frequently replaced it. DEPN, Gl i.254–5, Turner 1950.119, 1952–4.17, L 172, CPNE 167.

MINETY Wilts SU 0390. 'Mint island'. *Mintih -g* [880]c.1400 S 320, *Minty* [884]c.1400 S 322, *Mi- Mynti -ty* 1156–1314. OE **minte** + **īeġ, īġ** in the sense 'dry ground in marsh'. Wlt 61 gives pr [minti], L 39.

Upper MINETY Wilts SU 0090. ModE **upper** + p.n. MINETY SU 0390. Upper Minety is now used of the old village around the church; Minety of the new village by the station.

MININGSBY Lincs TF 3264. Partly uncertain. *Melingesbi* 1086, *Mithinges- Minigesbia* 1142, *Mithingesbi* 12th cent., *Miniggesby* [c.1150]late 13th, *Midhungesbi* [1154×89]1330, *Mynyngesby* 1346. Forms with *-th-* predominate until 1346. This could, therefore, be ON pers.n. *Mithjungr*, genitive sing. *Mithjungs*, + **bȳ**. But *Mithjungr* is only known as the name of a mythical giant. Sound substitution of *n* for *th* is also seen in Castle HEDINGHAM Essex TL 7835. SSNEM 59.

MINIONS Corn SX 2671. Unexplained. *Minions* 1897. A 19th cent. mining village named from the nearby tumulus called Minions Mound, *Mimiens borrough* 1613 'Minions barrow'. *Mimiens* is presumably for *Minniens* but its origin remains obscure. PNCo 123.

MINLEY MANOR Hants SU 8258. A French brick château built in 1858–60. P.n. Minley + ModE **manor**. Minley, *Mindeslei* 1086, *Mundeleya* 1189×99, *-le* 1236, *Mendeley* 1280, *Mynley* 1516, is 'Mynda's wood or clearing', OE pers.n. **Mynda* + **lēah**. Ha 117, Gover 1958.112, Pevsner-Lloyd 338.

Church MINSHULL Ches SJ 6660. '(That part of) Minshull with the church'. *(le) Chirch(e)munschull* 1289–1416, *Churche-* 1342, *Church(e) Mynshul(l)* [1428]1521, [1460]1471, 1498. ME **chirche** + p.n. *Maneshale, Manessele* 1086, *Mun(s)chulf(e)* [c.1130]1230–1344 with variants, *Muneshull* c.1200–1420 with many minor variants, *Minshull* from 1316, 'Mann's shelf', OE pers.n. *Man(n)*, genitive sing. *Mannes*, + **scylfe**. Church Minshull is situated on a shelf of land on the W bank of the river Weaver. 'Church' for distinction from Minshull Vernon SJ 6760, *Minshull Vernon* 1394, earlier simply *Manessele* 1086, *Muneshull'* c.1200 etc., '(That part of) Minshull held by Warin de Vernon'. Che iii.154, ii.247.

MINSKIP NYorks SE 3864. 'Land held in common'. *Minescip -skip* 1086, 13th cent., *Men(e)- Mœn- Mansc(h)ipe* 12th cent., *Menskip* 1444, *Minskip* from early 13th with variants *Myn-*, *-skippe* and *-skyp*. OE **(ġe)mǣnscipe** with ON [sk] for [ʃ]. YW v.85.

MINSTEAD Hants SU 2811. 'Mint place'. *Mintestede* 1086, *Minstede* 1248–1445 with variant *Myn-*. OE **minte** + **stede**. The reference may be to cultivated mint. Ha 118, Gover 1958.208, Stead 270.

MINSTER OE **mynster** 'monastery', ME **minster** 'a large church'.

(1) ~ Kent TR 3164. Originally called South Minster, *Suðmynster* [696×716]11th S 22, for distinction from *(æt) North Mynstre* [943]13th S 489 (lost in Thanet). An abbey was founded here in 669. In 1086 the estate is referred to as *Tanet*. See Isle of THANET TR 3267. KPN 22, Newman 1969E.378, Charters IV.139–82.

(2) ~ Kent TQ 9573. Originally 'Seaxburg's monastery', *Sexburgamynster* c.1100. Otherwise simply *Menstr(e)* 1253–92. The monastery was founded c.670 by Seaxburg, widow of king Eorconbeorht of Kent (640×664). PNK 256.

(3) ~ LOVELL Oxon SP 3111. 'M held by the Lovell family'. *Ministr' Lovel* 1278×9, *Mu- Minstre Lovel* 1341–7, *Mynster Lovell* 1517. P.n. *Minstre* 1086–1390 + manorial affix from the family of Maud, widdow of William Luvel or Lupellus 'the wolf-cub' (OFr *lovel*), who founded a priory here 1200×6. There must, however, have been a minster and a community of priests serving the district before 1086. For the family name see also LILLINGSTONE LOVELL SP 7140. O 364.

MINSTERACRES Northum NZ 0255. 'Newly cultivated land where mill-stones are found'. *Mynstanesacres* 1268, *Mi- Mynstanaker* 13th cent., *Milnstoneacres* 1347, *Mynstracres* 1506. OE **myln-stān** + **æcer**, pl. **æcras**. NbDu 143, L 233.

MINSTERLEY Shrops SJ 3705. 'The wood or clearing belonging to the minster church'. *Menistrelie* 1086, *Munstreleg* 1246, *Munsterleg(h)' -ley(e) -le(i)e* 1255–1296, *Minstreley* 1333, *Mynsterley* 1403. OE **mynster** + **lēah**. The reference is probably to Westbury where the mention of two priests in DB suggests the status of a minster church. Sa i. 209.

MINSTERWORTH Glos SO 7717. 'The enclosure held by the monastery' sc. of St Peter's, Gloucester. *Mynsterworðig* c.1030, *Munstreworth(e) -worþe -w(u)rth(e)* 12th–1404, *Mi- Mynstrew(u)rth -worth(e) -er-* 1221–1680. OE **mynster** + **worth(iġ)**. Gl iii.162.

MINTERNE MAGNA Dorset ST 6504. 'Great M'. *Great Mynterne* 1363, *Mynterne Magna* 1596. P.n. Minterne + Latin adj. **magna**. Minterne, *Minterne* [987]13th S 1217, *Mynterne* 1268, is 'mint house', i.e. either 'house at the place where mint grows' or 'building with a mint', OE **minte** or **mynet** + **ærn**. 'Great' for distinction from Minterne Parva ST 6603, 'little M', *Minterne Parva* from 1314. Do 109.

MINTING Lincs TF 1873. Uncertain. *(in duobus) Menting(h)es* 1086 referring to Minting and Little Minting TF 1673, *(in) Mintingis* c.1115, *Mentinges* [c.1125]1336, 1240, *Minthinges* c.1150, *Mintinges, Mintingke, Mineting'* 1212, *Muntinges* 1218×9, *Menting'* 1235×6, *Minting* 1824 OS. Possibly OE folk-n. **Myntingas* 'the people called after Mynta' < OE pers.n. **Myn(i)ta* + **ingas**, but the pl. may refer to the two settlements of Great and Little Minting. An alternative might, therefore, be an OE **minting* 'place where mint grows' < *minte* + *ing*[2] although this is difficult to equate with the *e* and *u* spellings. ING 66.

MINTON Shrops SO 4390. 'The settlement by (the Long) Mynd'. *Munetune* 1086, *-ton(e)* 1198–1329, *Muni- Miniton(')* 1198×9–1276, *Moneton(e)* 1268–1376, *Mynton* 1535–1665. P.n. Mynd as in The LONG MYND SP 4193, + OE **tūn**. Sa i.209.

MINWORTH WMids SP 1592. 'Mynna's enclosure'. *Meneworde* 1086, *Mun(n)ew(o)rth(e)* 1154–1354, *Minnewurth* 1232, *Mynworth* 1372. OE pers.n. **Mynna* + **worth**. Wa 46.

MIRFIELD WYorks SE 2019. 'The pleasant stretch of open country' or 'the open land where games or sports were held'. *Mirefeld -felt* 1086, *Mi- Myr(e)feld* 1170×85–1391, *Mi- Myrfeld(e)* 1154×89–1537, *Mi- Myrifeud, Mirifeld* 13th cent. OE **myriġ** 'pleasant' or **myrġen** 'joy, pleasure' + **feld**. If the latter suggestion is correct the name is comparable in significance with WAKEFIELD SE 3320. YW ii.197, xi, TC 136, L 242.

Great MIS TOR Devon SX 5676. 'Mist tor'. *Mystorre* 1240, *Mistorr* [1291]1408. OE **mist** + **torr**. Probably so called because this tor was more frequently enshrouded in mist or fog than others in the neighbourhood. Misdon in Inwardleigh, *Myston* 1345, *Miston* 1609, may for the same reason be 'mist hill'. D 195.

River MISBOURNE Bucks SU 9696. Possibly 'river called Mysse, the mossy one'. *Misseburne* 1407, 1427, *Mysse-* 1412, *Messeborne* 1475, *the Miss or Mease* 1847. Possibly an old r.n. **Mysse* < OE **mos** 'moss' + suffix **-jōn**. Cf. River MEASE Derbys SK 2711. RN 294.

MISERDEN Glos SO 9308. 'Place of the Musard family'. *Musardera* 1186, 1190, *(la) Musarder(e)* 1211–1461, *Musarden(e) -erd-* 1291–1628, *My- Miserden* from 1488. Family name *Musard* + OFr suffix **-ere -erie**, Latin *-aria*. The original name of the manor was *Gren(e)hā- -hamstede* 1086–1310 'green homestead', OE **grēne** + **hām-stede**. It was held in 1086 by Hascoit Musard in whose family it remained until the 14th cent. and became known as 'the Musardery'. The form in *-den* is an adaptation to English p.ns. in *denu* 'a valley'. Gl i.129.

MISLINGFORD Hants SU 5814 → Forest of BERE SU 6711.

Great MISSENDEN Bucks SP 8901. *Magna Messenden* 1237×40. ModE adj. **great**, Latin **magna**, + p.n. *Missedene* 1086, *Messendena -den(e)* from [1133 etc.]14th, *Mussend'* 1154×89, *Messenden* 1181–1262, 'valley of the river *Mysse* or where *mysse* grows', r.n. **Mysse* as in MISBOURNE SU 9696 or OE ***myssen** 'growing with *mysse*' + **denu**. The exact sense of **mysse* is unknown; the word has to be a derivative of *mos* 'moss, lichen, bog' and cognate with Dan *mysse* 'water arum or marsh marigold (*Caltha palustris*)', Swed *missne* 'water arum' or 'buckbean (*Menynanthes trifoliata*)'. According to Jnl 2.27 water arum (*Caltha palustris*) is an alien species introduced from the continent and the most likely plant is the buckbean or bogbean which grows in spongy bogs and marshes and is recorded among the flora of Bucks. This plant was formerly held to be of great value as a remedy against scurvy and other conditions. 'Great' for distinction from Little MISSENDEN SU 9298. Also known as 'St Peter's M', *Messenden Sancti Petri* 1262. Bk 152, DEPN, PNE ii.47, Jnl 2.27, Grieve 117f.

Little MISSENDEN Bucks SU 9298. *Parva Messenden* 1232. ModE adj. **little**, Latin **parva**, + p.n. Missenden as in Great MISSENDEN SP 8901. Also known as *Messenden Attewhytechirch* 1262, 'M at the white church'. Bk 152, Jnl 2.27.

MISSON Notts SK 6894. 'The marsh, the fen'. *Misna* 1086, *Mi- Mysne* 1086–1358, *Mi- Mysena -e* 1179–c.1300, *My- Misen -yn(e)* 1280–1428, *Messen* 1524. OE ***mysen**. Nt 87 gives pr [mizən], Studies 1936.114.

MISTERTON Leic SP 5583. 'Minster village'. *My- Minstreton(e)* 1086†–1335, *My- Minsterton(e) -ir-* 1189–1409, *My- Misterton' -ir-* 1264–1610. OE **mynster** + **tūn**. Lei 457.

MISTERTON Notts SK 7694. 'The estate belonging to the minster'. *Ministretone* 1086, *Minsterton(a) -ir-* 1130–1209, *Misterton* from 1205 with variants *-tre-* and *-tir-*. OE **mynster** + **tūn**. The reference is unknown. Nt 36.

MISTERTON Somer ST 4508. 'Minster estate'. *Mintreston* 1199, *Musterton* 1316. OE **mynster** + **tūn**. Misterton was an estate belonging to Crewkerne which must have been the minster in question. DEPN, Costen 145, 148.

MISTLEY Essex TM 1231. 'Mistletoe wood or clearing'. *Mittesleam* 1086, *Misteleg(h) -ley(e)* 1225–1546, *Mis(s)ley* 1594, 1768. OE **mistel** + **lēah**. Ess 343 gives pr [misli].

MITCHAM GLond TQ 2868. 'Large homestead or village'. *Michelham* 1086, 1195, *Mic(c)ham* [1140×54]14th etc. including *[727, 933]13th S 1181, 420. OE **miċel** + **hām**. The boundary of the Mitcham people is given as *(bi) michamingemerke* [967]14th S 747. Sr 51.

MITCHELDEANE Glos SO 6618.'Great Dean'. *Muckeldine* 1224, *Magna Dene* 1259–1478, *Mucheldene* 1337–1409, *Mi- Mychel(l)de(a)ne -il(l)-* 1317–1675, *Mitcheldean* 1680. ME adj. **michel, muchel** (OE *myċel*), Latin **magna**, + p.n. *Dena, Dene* 1220, 1250, 'the valley', OE **denu**, as in Forest of DEAN SO 6311. 'Great' for distinction from LITTLEDEAN SO 6713. Gl iii.234, L 103.

MITCHELL Corn SW 8654. 'The maid's hollow'. *Meideshol* 1239, *La Medissole* 1277, *La Medeshole* 1284–1333, *Medschole* 1401, *Metsholle* 1435, *Medishole alias Michell* 1596, *Michael* 1610. Apparently OE **mæġd**, genitive sing. **mæġdes**, + **hol**, referring to a dip in the road where the village stands. PNCo 123, Gover n.d. 330.

River MITE Cumbr SD 0998. *Mighet* 1209, *Mite* from 1292. A British r.n. on the root **meigh-* 'to urinate' + suffix *-ētion* or *-et*. Comparable to Icelandic r.n. *Mígandi* and Norwegian *Migande* from ON *míga*. Cu 22.

MITFORD Northum NZ 1768. Partly uncertain. *Midford(')* 1195–1267, *Middeford* 1296, *Mydford -forth* 1489, *Mitford* 1242 BF, *Mytfourth* 1560, *Mithford* 1315. Either 'the middle ford', OE **midd** + **ford**, or 'the rivers' meet ford', OE **(ġe)mȳthe** + **ford**. The village lies between the rivers Font and Wansbeck at their junction. NbDu 143, L 11, 70.

MITHIAN Corn SW 7450. Unexplained. *Mithien* 1201, *Mithian -yan* 1288–1329. PNCo 124, Gover n.d. 364.

MITTON Staffs SJ 8815. 'The river-junction settlement'. *Mutone* 1086, *Muton'* 1194, *Mutton'* 1200×3–1506, *Mitton(')* from 1203. OE **(ġe)mȳthe** + **tūn**. The reference is to the junction of Whiston Brook with Church Eaton Brook at SJ 8914. St i.86, L 11.

Great MITTON Lancs SD 7139. *Magnam Mitton* 1103. Adj. **great**, Latin **magna**, + p.n. *Mitune* 1086, *Mi- Mytton* from [1206×35]1268, 'settlement at the confluence', OE **mȳthe** + **tūn**. Great Mitton lies between the Hodder and the Ribble at their conjunction. 'Great' for distinction from Little Mitton SD 7138 on the opposite bank of the Ribble, *Little Mitton* 1242–1322 etc., *parua Mitton* 1296, adj. **little**, Latin **parva**, + p.n. Mitton. YW vi.198, La 77, Jnl 17.48.

MIXBURY Oxon SP 6033. '(At) the fortified place by or with the dunghill'. *Mixberia* c.1130–1428 with variants *-bir(e) -bur(ia)* and *-bury, Mi- Myxbury* from 1265×7. OE **mixen** + **byriġ**, dative sing. of **burh**. O 230.

MIXON Staffs SK 0457. 'The dunghill'. *Mixen, Mixne* 1199–c.1291,

†There is also a bad DB spelling *Ministone*.

Mixenn -ene 1203–1332, *Mi- Myxton* 1516–1614. OE **mixen**. Oakden.

MOBBERLEY Ches SJ 7879. 'Clearing at or belonging to *Motburh*'. *Motburlege* 1086, *Modburle -berleg(a) -birleye* late 12th–1514 with variants, *Moberlehe* c.1170, *Mobburley -erleia -le(y)* from early 13th. P.n. **Motburh* 'fortification where meetings are held', OE **mōt** + **burh**, + **lēah**. Che ii.65.

MOCCAS H&W SO 3542. 'Pigs' rough grazing'. *Mochros* [6th, 7th, early 8th]c.1130 LL, *Moches* 1086, *Mocres*, *Mocke(r)s* 1115×20–1303, *Mokkas* 1316, 1341. W **moch** + **rhos**. LL translates the name as *locus porcorum* 'place of pigs'. The exact English equivalent occurs four miles further down the Wye at Swinmoor SO 4240, *Swynemor* 1348. He 147, Bannister 136, 182, DEPN, LHEB 569, GPC 2468.

MOCKBEGGAR WHARF Mers SJ 2591. *Mockbeggar Wharf* 1757. A sand-bank, dial. **wharf**, off the Wallasey shore named after Mockbeggar Hall (Leasowe Castle) SJ 2691, *Mock Beggar Hall* 1690, *a manor house formerly called New Hall, afterwards Mock Beggar, and now Leasowe Castle* 1819, a sporting lodge built in 1593 at the earl of Derby's race-course here and enlarged in 1818. ModE **mock-beggar** 'an uncharitable or deserted house which disappoints the expectations of a beggar' + **hall**. Che iv.333, v.1.ii 283, Pevsner-Hubbard 1971.372.

MOCKERKIN Cumbr NY 0923. 'Corcan's head or hill-top'. *Moldcorkyn -kin* 1208, *Molcorkilne -kork-* c.1225, *-korkyn* 1370, *Mokerkyn* 1505, *Mockerkin* 1578. ON ***moldi** + OIr pers.n. *Corccān*. The reference is to the hill called Mockerkin How (814ft). The elements are in Celtic word order. Cu 410, SSNNW 147.

MODBURY Devon SX 6551. 'Meeting-place fortification'. *Mortberie*, *Motbilie* 1086, *Motberia -bilia* 1086 Exon, *Modberi(a)* 1174×83, 1181, *-bury* from 1238. OE **(ġe)mōt** + **byriġ**, dative sing. of **burh**. At one time this must have been the meeting place of Ermington hundred. Cf. Modbury in Buckland Filleigh and Modbury Village, Dorset SY 5189. D 279, 91.

MODDERSHALL Staffs SJ 9236. 'Modred's nook'. *Modredeshale* 1086, *Modreshalle* 1305, *Mothersall* 1551, *Moddershall* 1836. OE pers.n. *Mōdrēd*, genitive sing. *Mōdrēdes*, + **halh** in the sense 'small valley'. DEPN, L 105, 110, Horovitz.

MOGGERHANGER Beds TL 1349. Partly unexplained. *Mogarhangre* 1216, *Moger- Moke(r)hanger* 1240–1428, *Morhanger* 1675. Unexplained ME element **moker** + **hanger** (OE *hangra*). A possible solution might be sought in a Gmc *-az-* formation on the root **muka-* 'soft' (IE **meuk-* 'slip, slimy'), OE ***mocor** 'soft, slippery place', as perhaps in the r.n. *Mucra* cited by Bahlow 337 s.n. Mockstadt and the p.n. Möckern. Cf. Latin *mūcor*, *mūcus*. Bd 91, L 195–6.

MOIRA Leic SK 3115. *Moira* 1831. A modern industrial development due to the discovery of fire-clay there at the end of the 18th cent. on land belonging to the earl of Moira. The first pits were sunk in 1804, and miners cottages were erected in 1811. The title is Irish and comes from Moira, County Down, Northern Ireland, from OIr **mag** 'field, plain' + **ráth** 'ring-fort, rampart' and means 'plain of the ring-fort'. Lei 343, Room 1983.72, IPN 239.

MOLASH Kent TR 0251. 'Speech or assembly ash-tree'. *Molesse* 1226, 1316, *-eysse -essh(e)* 1270–1315. lOE **māl** + **æsc**. PNK 378, PNE ii.34.

River MOLE Devon SS 7327. Probably a back-formation from MOLLAND SS 8028 and North and South MOLTON SS 7329, 7125. *Moll* 1553, *Moule* 1577, *Moul* 1610. D 10, RN 295.

River MOLE Surrey TQ 1263, 2347. A back-formation from the p.n. Molesey as in East MOLESEY TQ 1568, near which the river enters the Thames. *Moule* 1577, *Molis* 1586, *Mole* from 1612. Earlier *aqua de Mulesia* 'Molesey water' 1214, *the back river of Moulsey, Moulsey river* 1595. The river was formerly called the Emel, *(on, andlang) Emenan* [983]c.1325 S 847, *(on) Æmenan, (andlang) Emenan* [1005]13th S 911, *Emele* 1307, *Emel* 1429, *Emlynstreame* 1534, OE **ǣmen* 'misty, causing mists', parallel to the Swedish r.n. Emån, the Norwegian Eima, and possibly the Dutch Eem, *Ema* 8th. The change from *-n-* to *-l-* is due to AN influence. In 1629 it was called *Letherhead River* from p.n. LEATHERHEAD TQ 1656. It may also have been called the **Dorce*, cf. DORKING TQ 1648. Sr 4–5, RN 295, 146–7.

MOLEHILL GREEN Essex TL 5624. *Mole Hill Green* 1777. This seems to be a continuation of *(le) Morynges(grene)*, *Moryesgrene* 1348, *Moryes(ffeild)* 1383, *Moringes Green* 1768, p.n. *le Morynges* + ME **grene**. The meaning of *Morynges* is unclear; possibly an *ing*[2] derivative of OE *mōr* meaning 'the marsh places'. Ess 535.

MOLESCROFT Humbs TA 0240. Probably 'Mul's enclosure'. *Molecroft* 1086, 1316, *Molescroft* from 1086, *Mule(s)-* 12th cent., *Mol-* 1292, *Molles-* 13th–15th, *Mowsecroft* 1579. OE pers.n. *Mūl*, genitive sing. *Mūles*, + **croft**. Alternatively the specific might be OE **mūl**, ON **múll** 'a mule' and so the 'mule's field'. Cf. MOLESEY Surrey TQ 1568. YE 200.

East MOLESEY Surrey TQ 1568. ModE **east** + p.n. *Muleseg* [672×4]13th S 1165, *Muleseige* [933]13th S 420, *Molesham* 1086, *Moleseya -ia -ey(e) -ei(e) -heye, Mulesia -ey(e) -ei(e)* 1129–1244, *Muleseye* [967]13th S 752, *Mulleseya* 1201–79, *Moulsey* 1569–1609, 'Mul's island', OE pers.n. *Mūl*, genitive sing. *Mūles*, + **ēġ**. The DB form is corrupt. 'East' for distinction from West Molesey TQ 1368, *Westmoleseie* 1200. Sr 94, L 38.

MOLESWORTH Cambs TL 0075. 'Mul's enclosure'. *Molesworde* 1086, *Mulesw(u)rth -worth* 1208–1316, *Moll- Mullesw(o)rth* 1253–1467, *Molesworth* from 1303. OE pers.n. *Mūl*, genitive sing. *Mūles*, + **worth**. Hu 246 gives former pr [mʌlzwəθ].

MOLLAND Devon SS 8028. 'Newly cultivated land at or called *Mol*'. *Mollande* 1086–1204, *Mouland* 1202–49, *Mollond(e)* 1238–86. Unknown element *Mol* as in River MOLE SS 7327 and North and South MOLTON SS 7329, 7125 + OE **land**. DEPN suggests that *Mol* is a late adoption of W *moel* 'bare' used as a name for part of the Exmoor hills between Barcombe Down SS 7531 and Round Hill. D 342, DEPN, L 248–9.

MOLLAND COMMON Devon SS 8130. Cf. *Molland Down* 1809. P.n. MOLLAND SS 8028 + ModE **common**.

MOLLINGTON Ches SJ 3870. 'Estate called after Moll'. *Molintone* 1086, *Molin- Molynton* early 13th–1581, *Mollington* from 1287. Also known as *Great Mollyngton* 1582, *Magna Molinton* 1271, and Mollington Torold, *Molynton Thorot* 1286, from the family owning land here from 1271. OE pers.n. *Moll* + **ing**[4] + **tūn**. Che iv.177.

MOLLINGTON Oxon SP 4447. 'The estate called after Moll'. *(æt) Mollintune* 1015 S 1503, *Mollitone* 1086, *Mollinton* 1235–1332, *Mollyngton* 1285–1428. OE pers.n. *Moll* + **ing**[4] + **tūn**. Wa 271.

MOLTON 'Estate at or called *Mol*'. Unknown element *Mol* as in MOLLAND SS 8028 + OE **tūn**.

(1) North ~ Devon SS 7329. *Nortmoltone* 1086, *Northmolton* from 1266. OE adj. **north** + p.n. *Moleton* 1283. D 344.

(2) South ~ Devon SS 7125. *Sudmoltone* 1086, *Suthmolton* 1238, *Sumouton* 1246. OE **sūth** + p.n. Molton as in North MOLTON SS 7329. D 346 gives pr [zaumoultən].

MONEWDEN Suff TM 2458. Possibly 'the valley of the Mundingas, the people called after Munda'. *Mun(e)ga- Mange- Mungedena* 1086, *Munegeden'* 1194, *Mungeden' -don'* 1166–1243, *Moneweden(e)* 1254–1336, *Moneden* 1568, 1610. OE folk-n. **Mundingas* < pers.n. *Munda* + **ingas**, genitive pl. **Mundinga*, + **denu**. For the development *-inga-* > *-iga-* > *-ega-* > *-ewe-* cf. HONINGTON TL 9174, Coney WESTON TL 9577. DEPN, Baron.

Great MONGEHAM Kent TR 3451. *Mongeham the Great* 1508, ~ *Magna* 1570, *Great Mongeham* 1819 OS. ModE adj. **great**, Latin **magna**, + p.n. *Mundelingeham* [761]13th S 28, *Mundlingham* 843×63 S 1197, *Mundingeham* 1086, c.1100,

Mouingham (for *Mon-*), *Est monigham* 1087, *Muningeham* 1195, *Monigeham* 1251, 'homestead, estate called *Mundeling*, or at *Mundelinge*'. OE p.n. **Mundeling* 'place called after Mundel' < OE pers.n. **Mundel* + **ing**², locative-dative **Mundelinġe*, possibly varying with folk-n. **Mundelingas* 'the people called after Mundel' < pers.n. *Mundel* + **ingas**, genitive pl. **Mundelinga*, + **hām**. 'Great' and *Est* 'east' for distinction from Little Mongeham TR 3350, *Mongeham Parva* 1570, *Little Mongeham* 1819, to which also the 761 form refers when the vill is described as *vicus antiquus* 'the old village'. Also known as Up ~, *Op- Vpmonighm̄* c.1295, *Hupmonynghm̄* c.1380. PNK 569, ING 120, BzN 1967.382, Cullen.

MONKERTON Devon SY 9693 → SOWTON SX 9792.

MONKESBURY COURT H&W SO 6143 → WESTHIDE SO 5844.

MONKHIDE H&W SO 6143 → WESTHIDE SO 5844.

MONKHOPTON Shrops SO 6293. 'H belonging to the Monks' sc. of Wenlock Priory. *Hop'ton' Monachorum* 1291×2, *Hopton Priors* 1302, *Munkehopton* 1577, *Monk Hopton* 1711. ModE **monk**, Latin **monachus**, genitive pl. **monachorum**, + p.n. *Hopton* 1255, 'the valley settlement', OE **hop** + **tūn**. Situated at the mouth of a valley running SE. 'Monk' for distinction from Hopton SJ 5926, HOPTON CANGEFORD SO 5480, ~ CASTLE SO 3678, ~ TITTERHILL SO 3577 and ~ WAFERS SO 6376. Sa i.209.

MONKLAND H&W SO 4657. 'The monks' land', originally 'the part of Leen held by the monks' sc. of the abbey of St Peter de Castellion, Conches, France. *Moneclond* 1316. Earlier *Leine* 1086, *Monecheslene* 1160×70, *Monekesleona -lane* 1137, 1193, *Monekenelane* 1277×92, *Monkelene* 1291, 1328. OE **munuc**, genitive pl. **muneca** (ME *monekes, monekene*) + p.n. LEEN SO 3859. The estate was already held by the monks of Conches in 1086. He 148.

MONKLEIGH Devon SS 4520. 'Leigh belonging to the monks' sc. of Montacute, Somerset. *Munckenelegh* 1244, *Monckeleghe* 1265, *Monkelegh* 1286. ME **munkene** + p.n. *Lege* 1086, *Legh* 1100×35, 'the wood or clearing', OE **lēah**. D 100.

MONKOKEHAMPTON Devon SS 5805. 'Okehampton belonging to the monks' sc. of Glastonbury. *Monuchementone* 1086, *Monacochamentona* 1086 Exon, *Munekeokementon* 1242, *Monk Oakinton* 1675, *Monke Ockhampton oth. Monk Ockington* 1759. OE **munuca**, genitive pl. of **munuc**, + p.n. Okehampton, 'settlement, estate on the river Okement', identical with OKEHAMPTON SX 5895. D 153.

MONK'S HEATH Ches SJ 8474. *Munkesheth* c.1389, *Monk(e)she(a)th(e)* 1565–c.1620. ME **munk** (OE *munuc*), genitive sing. **munkes**, + **heth** (OE *hǣth*). Lands in this district were owned by Dieulacres and Poulton abbeys. Che i.96.

MONKS HOUSE ROCKS Northum NU 2033. P.n. Monks House NU 2033, *le Monkeshouse ex parte boreali rivuli Broxmouth* 'M on the N side of Broxmouth stream' 1495, a grange of Lindisfarne Priory. ME **monk**, genitive pl. **monkes**, + **house**, + ModE **rock**. NbDu 144.

MONKSIDE Northum NY 6894. 'The monks' hill-side'. *Monkside* 1868 OS. ModE **monk** + **side**. A hill that rises to 1683ft.

MONKSILVER Somer ST 0737. 'Silver held by the monks' sc. of Goldcliff in Monmouthshire. *Monkesilver* 1249, *Monksiluer* 1610. ME **monke** possibly representing OE genitive pl. *muneca* + p.n. *(æt) Sulfhere* 11th K 897, *Selv(er)e*, *Selvre* 1086, *Siluria* [1154×89]1290 of uncertain meaning and origin. Robert de Chandos gave Silver to his new priory at Goldcliff in 1113. DEPN.

MONKTON 'Monks' estate'. OE **muneca**, genitive pl. of **munuc**, + **tūn**. PNE ii.45.

(1) ~ Devon ST 1803. *Muneketon* 1244, *Monketon* 1284–1455, *Munkton* 1272×1326. The reference to monks is unexplained. D 627.

(2) ~ Kent TR 2865. *Munccetun* (for *Munece-*) [961]13th S 1212, *Monocstvne* (for *Monoce-*) 1086, *Munec- Muneketune*, *Mvnechetun* c.1090, *Monke- Mun(e)keton -tune* 13th cent. KPN 288.

(3) ~ T&W NZ 3263. *Munecatun* c.1107, c.1170×4, *Muneketon* 1189×1212, 1314, *Munchetun(ā)* 1154×66, c.1190×5, *Munke- Monketon'* [1153×95]14th, 1196×1208–1536, *Munkes- Monkeston* 1303, 1366, *Monkton* from 1365. A holding of the monks of Jarrow. NbDu 144.

(4) ~ COMBE Avon ST 7762. *Combe Monkton* 1817 OS.

(5) ~ DEVERILL Wilts ST 8537. 'The monks' D estate'. *Moneketon Deverel* 1270, *Mounton Deverall* 1558. P.n. *Munketon* 1189, 1249, ME **monkton** (OE *munuc*, genitive pl. *muneca*, + *tūn*) + r.n. *Deverel* [924×39]18th S 1714, *Devrel* 1086. The manor already belonged to Glastonbury Abbey by 1086. Also known as *Vuerdeverel* 'over D' [924×39]18th S 1713. Wlt 174.

(6) ~ FARLEIGH Wilts ST 8065 → Monkton FARLEIGH ST 8065.

(7) ~ HEATHFIELD Somer ST 2526. *Monkton Heathfield* 1809 OS. P.n. Monkton as in West MONKTON ST 2628 + p.n. Heathfield as in Creech Heathfield ST 2726, *Creech Heathfield* 1809.

(8) ~ UP WIMBORNE Dorset SU 0113. 'The monks' estate of or called Up Wimborne'. *Wymborne Monkton* 1504, *Mounckton Up Wimborne* 1617. ModE **monkton** 'monks estate' (ultimately from OE *muneca-tūn*) + p.n. *Up Wimburn*, ~ *Wymburne* [1154×89]1496, 1213, 1232 etc. Earlier simply *Winburne* 1086, by origin a r.n. meaning 'meadow stream', OE **winn**, **wynn**, + **burna**. 'Up' means 'higher up the river' compared with WIMBORNE MINSTER SU 0100. Also known as *(Up)wymborne Abbatis* 'abbot's W' 1288–1430. Both affixes, *Monkton* and *Abbots*, refer to possession of the estate by Cranborne priory in 1086 or later by Tewkesbury abbey. Do ii.265.

(9) Bishop ~ NYorks SE 3266. 'Monkton held by the bishop' sc. of York. ME prefix *Bysshop* 1402 + p.n. *Muneca tun* c.1030 YCh 7, *Monucheton(e)* 1086, *Munketon(a)* 1141×4–1356. Possibly part of the land originally granted by king Alchfrith to St Wilfrid in 661 for the foundation of a monastery at Ripon. Later called *Bishop* for distinction from Moor and Nun MONKTON SE 5056, 5057. It formed part of the Ripon estates of the archbishops of York. YW v.176.

(10) Moor ~ NYorks SE 5056. *Monkton super Moram* 1198, *Moremonketon* 1402. OE **mōr** + p.n. *Monechetune -ton(a)* 1086–1100×8, *Munechatun* 1175×86, *Muneketon* 13th cent., *Mu- Monketon* 12th–1428, *Monkton* from 1198. The reference is probably to possession of the estate by Christ's Church priory in York. 'Moor' refers to Marston Moor and was added for distinction from nearby Nun MONKTON SE 5057. YW iv.257.

(11) Nun ~ NYorks SE 5057. *Nunmonkton* 1397. ME **nunne** + p.n. *Moneche- Monuchetone* 1086, *Mo- Munketon* 1147–1403. 'Nun' for distinction from Moor MONKTON SE 5056. A priory of nuns was founded here in the 12th cent. YW v.2.

(12) West ~ Somer ST 2628. *Westmonketon* 1397. Earlier simply *Monechetone* 1086, *Moneketon* [1380] Buck, *Monkton* 1610. The estate belonged to Glastonbury abbey. 'West' for distinction from MONKTON HEATHFIELD ST 2526.

(13) Winterborne ~ Dorset SY 6787 → WINTERBORNE MONKTON.

MONKWEARMOUTH T&W NZ 3951 → River WEAR Durham NZ 1134.

MONKWOOD Hants SU 6731. *Munckwodd* 1548. ModE **monk** + **wood**. The reference is unknown but nearby names include *Moncton* 1810 OS and *Abbots Croft* 1840. Ha 118, Gover 1958.88.

MONNINGTON ON WYE H&W SO 3743. *Mointon' super Weyam* 1292, *Monynton supra Wyam* 1418. P.n. Monnington + r.n. WYE. Monnington, *Manitvne* 1086, *Manintona* 1160×70, *Moni(n)ton(')* -*y*- 1207–1334, is the 'estate held in common', **(ġe)mǣne**, oblique case definite form **mānan**, + **tūn**. He 148.

River MONNOW H&W SO 4717. Possibly the 'little Wye'.

I. Welsh forms: *Mi- Myngui, Mynugui, Mynui* c.1130, *Mynwy, Monnwy* 15th.
II. English forms: *Munuwi muða* 11th, *Mune* [c.1200]c.1260, *Monowe* 1564, 1567.

The Monnow is a tributary of the Wye and might be Brit ***minu-** 'lesser' + r.n. WYE. Alternatively Brit **monou̯ĭā** has been suggested < **mono-* 'neck' + suffix *-u̯ĭā*, OW *-ui*, ModW *-wy*. RN 296, LHEB 379, 672 fn., 676.

MONTACUTE Somer ST 4916. 'Pointed hill'. *munt acuht* [1066×87]13th, *Montagud* 1086, *(de) Monteacuto* (Latin) 1156–[1363×86]14th Buck, *Montis Acuti* (Latin genitive case) [1186] Buck, *Mu- Montagu* [1129×39, 1192]14th, 1309–[1361×2]14th M, *Montacu* [1408]Buck, *Montacute* 1610. OFr **mont** + **agu** as in the Fr ns. Montagut, Montaigu, Montégut. The modern form comes from the Latin *Mons Acutus* [c.1125]12th, dative *Monte Acuto* as in *in monte accuto* 1084. The reference is to St Michael's hill, a remarkable conical hill formerly crowned by the castle of the count of Mortain to which the name originally applied. It replaced the earlier name of the estate, *Biscopestone* 1086, *-tun -ton(e)* [1091×1106–1152×8]14th, *Bisshop(e)stone* [1361×2]14th M, 'the bishop's estate' apparently referring to a bishop called Tumbeord (i.e. Tunbeorht of Winchester) in a 13th cent. list of the contents of the ancient *Liber Terrarum Glastoniae* 'book of the Glasonbury lands' (MS Trin.Coll.Cantab.724). He is there described as *de Logderesdone .i. Montagu Glastoniae*, 'of *Logderesdone*, that is, the Glastonbury Montacute', a name which is elsewhere recorded as *Logworesbeorh* [?676×85]13th ECW, *Logderesbeorgum* [851]13th ECW, *Lodegaresbergh* [854]14th S 303, *Logderesdone* [871×9]13th S 1703, *Legperesberh* [c.1125]12th, *Loggaresbeorg* 13th, unidentified element (sometimes assimilated to OE **lūtegār* 'trap-spear', genitive sing. **lūtegāres*), + **beorg**, dative pl. **beorgum**. DEPN, Turner 1950.120.

MONTACUTE HOUSE Somer ST 4917. *Montacute Ho.* 1811 OS. P.n. MONTACUTE ST 4916 + ModE **house** referring to the Elizabethan mansion built on the lands of the former Cluniac priory here. Pevsner 1958S.245.

MONTFORD Shrops SJ 4114. Partly uncertain. *Maneford* 1086–1272, *Moneford* 1203×4–1695, *Muntford* 1255×6, *Mountfort* 1412, *Montford* from 1601. The later bridge here, *pons de Muneford* 1255×6, *Monfordbrigg* 1374, was a traditional meeting place for English and Welsh potentates at times of negotiation. The name is probably either OE *(æt thǣm) ġemǣnan forde* '(at) the common ford', OE **(ġe)mǣne** 'common, shared' + **ford**, or *ġemāna-ford* 'the meeting ford', OE **(ġe)māna** 'fellowship, association, intercourse, commerce, conjunction' also 'what is held in common, common property', + **ford**. The modern form of the name may have arisen through false association with Simon de Montfort in the 13th cent. It became more or less official in 1741 when Henry Bromley, lord of the manor, was created Lord Montfort. Sa i.210.

Old MONTSALE Essex TR 0097. *Old, New Mountsale* 1643. ModE adj. **old** + p.n. Montsale, short for Montsale Wick, *Momsale(wyk)* 1438, *Monnsalewyk* 1441, *Momlynshale* 1459, unexplained p.n. probably ending in OE **halh** 'a nook' + **wīċ**. This is one of several *wicks* or dairy farms on Southminster Marsh including *Middlewyk* and *Southwyk* 1438, and *Burnham Wick* (cf. BURNHAM-ON-CROUCH TQ 9496), *Bridge Wick, Dunmore Wick, Little Wick, North Wick, Ray Wick* (cf. RAY SAND TM 0500), and *West Wick*, all 1805 OS. Ess 226.

MONXTON Hants SU 3144. 'The monks' estate'. *Monekeston* 1307×27, *Monkestone* 15th, *Monkston* 1817 OS. ME **muneke**, genitive pl. **munekes**, + **toun** (OE *tūn*). Originally simply a ME descriptive phrase it eventually replaced earlier *Anne de Bek* 1269, *Anna de Becco* c.1270, 'Ann held by the abbey of Bec' in Normandy. Cf. AMPORT SU 3044, ANDOVER SU 3645 and Abbots ANN SU 3243. Ha 118, Gover 1958.167, DEPN.

MONYASH Derby SK 1566. 'Many ash-trees'. *Maneis* 1086, *Manias(s)* 1254×7, 1279, *Moni- Monyasch(e) -ass(c)h(e) -ash(e)* 1198×1208–1285 etc. OE **maniġ** + **æsc**. 'Many' for distinction from nearby One Ash SK 1665, *Aneise* 1086, *Anaix* 1101×8, *Oneasshe* 1301, *Onyashe* 1345. Db 148–9.

Lower MOOR H&W SO 9847. 'Lower' is a modern addition for distinction from Upper Moor SO 9747, parts of the ancient parish of *More et Hylle* 1086, *Mora juxta Fladebure* 'M beside Fladbury' 1306, *Hulle and More* 1375, 1431. The two names go together: Moor, OE **mōr**, is the marshland settlement close to the river Avon, Hill SO 9848, *hylle* [1046×53]17th S 1406, OE **hyll**, is the upland settlement on high ground to the N. Wo 135, Hooke 343.

The MOOR Kent TQ 7629. 'The waste land'. *(de) Mora* 1240, c.1270, *(de) la More* 1258–78, *atte More* 1292–1327, *Hawkhurst Moor* 1813 OS. OE **mōr**. PNK 340.

MOORBY Lincs TF 2964. 'The fen village or farm'. *Morebi* 1086, 1170×89, *-by* c.1200, 1254. ON **mór** + **bȳ**. DEPN, SSNEM 60, Cameron 1998.

MOORCOT H&W SO 3555. 'Marsh cottage(s)'. *Mor(e)cote* 1160×70, 1334, 1591, *Mor(e)cott* 1619. OE **mōr** + **cot**, pl. **cotu**. He 157.

MOORDOWN Dorset SZ 0994. Probably 'hill in marshy ground'. *Mourdon* late 13th, *Mourden* 1300, *Moredown* 1759. Probably OE **mōr** + **dūn**. But OE *more* 'root, tree-stump' is also possible. Do 110, Ha 119.

MOORE Ches SJ 5884. 'The marsh'. *Mora* [early 12th]17th, *More* early 13th–1560, *Moore* 1517. OE **mōr**. Che ii.153.

MOORENDS SYorks SE 6915. *Moorend* 1771, ~ *Ends* 1822. ModE **moor** + **end**. Moorends lies at the edge of Thorne Moor. YW i.5.

MOORHALL Derby SK 3174. 'Hall, house on the moor'. *Morehall* 1583. Earlier *Morehowsse* 1563, *Moorehowse* 1579. ME **mōr** + **hous, hall**. Db 205.

MOORHAMPTON H&W SO 3947. 'Moor homestead'. *Moramptone* 1341. ME **more** (OE *mōr*) + **hampton** (OE *hām-tūn*). Bannister 139.

MOORHOUSE 'The house(s) on or at the moor'. OE **mōr** + **hūs**.
(1) ~ Cumbr NY 3356. *The Murhous* 1340, *(le) Morehous* 1485, *Moorhouse* 1574. A settlement on the moor of Burgh by Sands NY 3259. Cu 128.
(2) ~ Notts SK 7566. *Morhus* 1231, *-huses* 1232, *Laxton Morehouse* 1324. As the 1324 form makes clear, these were the moor-houses of Laxton parish. Nt 55.

MOORLAND OR NORTHMOOR GREEN Somer ST 3332. *Moorland* 1809 OS. A settlement on the edge of *Northmore* [before 1276] Buck, 1610, the north marsh of Lyng, ME **north** + **more** (OE *mōr*).

MOORLINCH Somer ST 3936. 'Pleasant terrace'. *Merlinch* *[725]12th S 250, *Mirieling* [971]13th S 783, *Merielinz* 1196, *Murieling* 1202, *Mirielinch* 1256, *Morlynch* 1610. OE **myriġ** + **hlinċ**. The village lies on the S side of Polden Hill overlooking Kings Sedge Moor; the reference is to the terraced road cut into the side of the hill W of the village. DEPN.

MOORS RIVER Dorset SZ 1099. A recent name derived from East and WEST MOORS SU 0802. Do 110.

MOORSHOLM Cleve NZ 6814. '(At) the houses on the moor'. *Morehusum* 1086, *Mores(h)um* [12th]15th, 1328, *(Magna, parva) Morsum* 'Great, Little M' 1257–1404, *-som* 1285–1404, *Moresham* 1610. OE **mōr** + **hūs**, dative pl. **mōr-hūsum**. YN 143.

MOORSIDE GMan SD 9507. *Moor Side* 1843 OS.

MOORTOWN IoW SZ 4283. 'The marsh settlement'. *Mortone* 1320, *Mourton* 1383. ME **mor** + **toun**. The place is referred to as the *(township of) More* 1248. Wt 69, Mills 1996.72.

MOORTOWN Lincs TF 0799. *Moor Town otherwise Moreton* 1808, *the hamlet of Moorton or Riverhead* 1842. ModE **moor** + **town**. The moor is recorded much earlier as *moram de Suth Kelsey* 1220×40, *Kelsey Moor* 1667–83. Li iii.38.

MOOTA HILL Cumbr NY 1436. 'Hill called Moota'. *Mewtey Hill*

1537. Moota, *Mewthow* 1537, *Mewto(o)* 1576, 1656, *Muta* 1675, is 'assembly hill', OE **(ġe)mōt** + **hōh**. Cu 267.

MORBORNE Cambs TL 1391. 'Marsh stream'. *Morburn(e)* 1086–1428, *-borne* from 1255, *Marborn* 1610, 1675. OE **mōr** + **burna**. Hu 192, L 54.

MORCHARD 'Great wood'. PrW ***mōr** + ***cę̄d**.

(1) ~ BISHOP Devon SS 7607. 'M belonging to the Bishop' sc. of Exeter. *Morchet Episcopi* 1207, 1256, *Bisshopes Morchestre* 1289, *Bisschoppesmorechard* 1311. Earlier simply *Morchet* 1086, 1165, *Morceta* 1086, *Morcherd* 1226, *Mor- Marchut* 1675. P.n Morchard + Latin **episcopus**, genitive sing. **episcopi**, varying with ME **bishop**. 'Bishop' refers to possession by the bishop of Exeter already in 1086 and for distinction from Cruwys MORCHARD SS 8712. D 408, CPNE 67.

(2) ~ ROAD STATION Devon SS 7505. A railway halt on the line from Exeter to Barnstaple at the turning to Morchard Bishop SS 7607.

(3) Cruwys ~ Devon SS 8712. 'M held by the Crues family'. *Cru(w)ys Morchard* from 1257×80, *Morcestr(e) Crues* 1279–91. Earlier simply *Morc(h)et* 1086, *Morceth* 1086, 1242. 'Cruwys' refers to possession by Alexander Crues in 1242 and for distinction from MORCHARD BISHOP SS 7607. D 380 gives pr [kru:z], CPNE 67.

MORCOMBELAKE Dorset SY 4094. 'Morcombe stream'. *Morecomblake* 1558. P.n. *Morcombe* + ME **lake** (OE *lacu*). *Morcombe* is 'marshy valley', OE **mōr** + **cumb**. Do 111.

MORCOTT Leic SK 9200. 'The cottage(s) at the marshland'. *Morcot(e)* 1086–1543, *Morecot(e)* 1263–1555, *Morcott* from 1495. OE **mōr** + **cot**, pl. **cotu**. The topography points to the sense 'marshland' rather than 'moor' in this name. R 282.

MORDA Shrops SJ 2827. 'Great Taf'. *Mordaf* 1295. PrW ***mọ̄r** + r.n. *Taf* with lenition of *t-* to *d-*. R.n. *Taf* is identical with Afon Taf Dyfed SN 1622–3209, *Tam, Taf* c.1140, *Thaph* 1191, and with River Taff Glam SO 0701–ST 1380, *Taam, Taaph* c.1075, *Tamius* c.1100, *Tám, Taf* c.1140, *Tauus* c.1150, from Brit **Tamos* from the IE root **tā-/*tə-* 'flow' as in river THAMES. DEPN, BzN 1957.256, FT 374–92, O'Donnell 100.

MORDEN Dorset SY 9195. 'Hill in marshy ground'. *Mordune* 1086, *-don(e)* 1086–1568, *Morden* from 1196. OE **mōr** + **dūn**. Do ii.58.

East MORDEN Dorset SY 9194. *Est Mordun*, ~ *Morden* 1250. ME adj. **est** + p.n. MORDEN SY 9195. Do ii.58.

Guilden MORDEN Cambs TL 2744. 'Golden Morden'. *Gildene Mordone* 1204. OE adj. **gylden** used of a place which is rich and productive + p.n. *(æt) Mórdune* 1015 S 1053, *Mordune* 1086, *-don(e)* from 1198, 'marsh hill', OE **mōr** + **dūn**. Ca 61, L 145, 153.

Steeple MORDEN Cambs TL 2842. 'Morden with the tower'. *Stepel Mordun* 1242. ME **stepel** + p.n. Morden as in Guilden MORDEN TL 2744. Ca 61, L 153.

MORDIFORD H&W SO 5737. Partly uncertain. *Mordford'* 1148×63, 1186×98, *Mordeford(e)* 1143×8–1535, *-i- -y-* c.1160–1489. Unknown element + OE **ford**. He 148.

MORDON Durham NZ 3226. 'Marsh hill'. *Mordun* [c.1040]12th, 1104×7, late 13th, 1371, *Mordon* from 1294, *Morden* 1559–1636. OE **mōr** + **dūn**. Mordon is situated at the point where the elevated ground to the E dips down to the marshlands of the river Skerne. Surtees described it as 'surrounded with rich low grounds verging to the marsh'. NbDu 145, Surtees iii.44, L 54, 145, 158.

MORE Shrops SO 3491. 'The marsh'. *La More* 1199–1401, *The Mo(o)re* 1575–1601×2, *Mora* 1226×8–72, *More* from 1291×2. OE **mōr**. The place overlooks the wide valley of the river Onny. Sa i.211.

MOREBATH Devon SS 9524. 'Bathing place in marshy ground'. *Morbade* 1086, *Morbath(a)* 1086 Exon–1316 with variant *-bathe*. OE **mōr** + **bæth**. There are chalybeate springs here and marshy land close at hand. D 536.

MORECAMBE Lancs SD 4364. A modern holiday resort which owes its development and name largely to The Morecambe Bay Harbour and Railway Company established here in 1848. The name Morecambe, which replaced the old name POULTON-LE-SANDS in 1870, is itself an antiquarian revival of Ptolemy's *Morikambe* first identified with Morecambe Bay (formerly Kent Sands) in Whitaker's *History of Manchester* 1771 and generally accepted since. *Morikambē* 'curved sea, curve of the sea' is a compound of Brit ***mori-** 'sea' and ***cambo-** 'curved, crooked'. La 176 fn., RBrit 420, Pevsner 1969.178, TC 137, Mills 113, Room 1983.73.

MORECAMBE BAY Cumbr SD 3567. *Morcambe Bay* 1852 OS. P.n. MORECAMBE Lancs SD 4364 + ModE **bay**.

MORELEIGH Devon SX 7652. 'Moor wood or clearing'. *Morlei* from 1086 with variant *-legh(e)*. OE **mōr** + **lēah**. D 309.

MORESBY Cumbr NX 9921. 'Maurice's village or farm'. *Moresceby* c.1160, *Moricebi* 1195–1428 with variants *Moryce- Morise-* and *-by*, *Moresby* 1498. OFr pers.n. *Maurice* + **bȳ**. Cu 421 gives pr [morizbi], SSNNW 36.

MORESBY PARKS Cumbr NX 9919. A modern development. P.n. MORESBY NX 9921 + ModE **parks**.

MORESTEAD Hants SU 5025. 'Moor place'. *Morsted(e)* 1158–1321, *M(o)urstede* 1306–41. OE **mōr** in the sense 'barren upland' + **stede**. Ha 119, Gover 1958.78, Stead 271.

MORETON 'The moor settlement'. OE **mōr** + **tūn**. This name-type often implies that the vill was a constituent member of a larger territorial or organisational unit. PNE ii.42. Cf. MORTON, MURTON.

(1) ~ Dorset SY 8089. *Mortune* 1086, *-ton(e)* from 1194, *Mourton(e)* 1332–1462. Do i.135.

(2) ~ Essex TL 5307. *Mortunā* 1086, *Morton(e)* from 1156. Ess 69.

(3) ~ Mers SJ 2690. *Mortona* 1272×1307, *Morton* 1291–1547, *Moreton* from [1278]14th. Che iv.319.

(4) ~ Oxon SP 6904. *Mortune* [before 1152]c.1200–1365, *Moreton* from 1346. O 146, L 55.

(5) ~ Staffs SJ 7917. *Mortone* 1086, *Morton* 1242–1652, *Moreton* from 1428. St i.157.

(6) ~ CORBET Shrops SJ 5623. 'M held by the Corbet family'. *Morton' Corbet* 1255–1577, *Moreton Corbett* 1734. P.n. *Mortone* 1086–1719 with variants *-ton(a) -tuna* + manorial addition from the Corbet family in the 13th cent. The reference is to marshland by the river Roden. Sa i.212.

(7) ~ -IN-MARSH Glos SP 2032. *Morton(e) & Henemers'* 1248, ~ *in Hennemersh* 1253, ~ *Inmarsh* 1648. Earlier simply *Mortun* *[714]16th S 1250, *Mortvne* 1086. *In marsh* is an adaptation of the affix *Hennemerse* 1235, 'marsh haunted by wild hen-birds', OE **henn**, genitive pl. **henna**, + **mersc**, an area of marshland extending from Bourton-on-the-hill SP 1732 and Chipping Camden SP 1539 on the W to Lemington SP 2234, Barton-on-the-Heath SP 2532, Little and Long Compton SP 2630, and Sutton-under-Brailes SP 3037 on the E. Gl i.251, 230.

(8) ~ JEFFRIES H&W SO 6048. 'Geoffrey's M'. *Morton Jeffrey* 1273, ~ *Geffrey* 1316. Earlier simply *Mortvne* 1086. He 150.

(9) ~ MORRELL Warw SP 3156. *Morton Merehill* 1285, *Morton Merell* 1514, ~ *Morrell* 1604. Not a manorial name but short for *Morton(e) et Merhulle* 1316, 1327, two distinct vills later united. Moreton is *Mortone* 1086–1535 with variant *-ton*, Morrell *Merehull* 1279, 1316, *Merhull* 1332, 'the boundary hill', OE **(ġe)mǣre** + **hyll**. Wa 256.

(10) ~ ON LUGG H&W SO 5045. *Morton juxta Logge* 1291. Earlier simply *Mortvne* 1086. See River LUGG SO 5251. Bannister 139.

(11) ~ PINKNEY Northants SP 5749. 'M held by the Pinkeni family'. *Morton Pynkenye* 1346, *Moreton Pynckney* 1590. Earlier simply *Morton(e)* from 1086. Also known as *Geldene- Guldenemortone, Gilden Moreton* 13th, 'golden Moreton', OE **gylden**. The manor was held by the family of *Pinkeni* from Picquigny in Picardy from 1199 and was parcel of the Honour

of Pinkney the *caput* of which was at Weedon Lois SP 6047. 'Golden' presumably because a specially wealthy manor. Nth 41.

(12) ~ SAY Shrops SJ 6334. 'M held by the de Say family'. *Morton' de Say* 1255, *Morton Say* 1284–1389, ~ *Se(a), Sew, Sey* 17th, *Moreton Say* 1394. P.n. *Mortune* 1086–1719 with variant *-ton* + manorial addition from Robert de Say in 1255. The reference is to marshland at the headwaters of the r. Duckow. The *Sea* spellings are regular in the 17th and 18th cents. when the vowels of *say* and *sea* rhymed as in Pope's famous lines on Hampton Court:
Where thou, great Anna, whom three realms obey,
Dost sometimes council take, and sometimes tea.
Sa i.212.

(13) ~ VALENCE Glos SO 7809. 'M held by the Valence family'. *Morton Valence* from 1276. Earlier simply *Mortvne* 1086. The manor was held by William de Valence in the mid 13th cent. Gl ii.186.

(14) Maids ~ Bucks SP 7035. 'The maidens' M'. *Maidenes Morton* 1480×96, *Maydes Morton* 1553×1601. ME **maiden**, ModE **maid**, genitive pl. **maidens, maids**, + p.n. *Mortone* 1086, *Murton, Moorton* 14th. According to tradition the church here was built in the 15th cent. by two maiden ladies of the Peyvre family. Bk 45.

(15) North ~ Oxon SU 5689. *Northmorton'* 1220. ME adj. **north** + p.n. *Mortun* [879×99]12th S 354, *Mortune* 1086 etc. 'North' for distinction from South MORETON SU 5688. Brk 524.

(16) South ~ Oxon SU 5688. *Sudmorton'* 1220, *Suthmorton'* 1242×3, *Southmorton* 1332, 1517. ME **south** + p.n. *Mortun* *[879×99]12th S 354, *Mortune* 1086. 'South' for distinction from North MORETON SU 5689. Brk 524.

MORETONHAMPSTEAD Devon SX 7586. *Morton Hampstead* 1493, *Moreton Hampstead* 1692. Earlier simply *Morton(e)* 1086–1291, 'moor settlement or estate', OE **mōr** + **tūn**. The reason for the later addition of *Hampstead* is unknown. D 483.

MORETON'S LEAM Cambs TL 2999. A cut made in 1478 by John Morton, Bishop of Ely, from the Old Nene near Peterborough to Guyhirn. Called *the New Leame* 1529 and *river of Neane called the New Leame al Moretons Leame* 1616. ModE dial. **leam** 'a drain or watercourse in fen districts'. The earliest citation in the OED is 1601, *Ye new leame that he* [sc. Bishop Morton] *caused to be made for more convenient cariage to his towne...*, cf. 1646 ibid., *Doctor Morton for his private commodity... brought certain Leames or bigger ditches to his owne grounds about Wisbitch*. Ca 210.

MORETON'S LEAM Suff TL 3099. Identical with MORETON'S LEAM Cambs.

MORGAN'S HILL Wilts SU 0367. *Morgans Hill* 1773. Surname *Morgan* (otherwise found here first in 1804) + ModE **hill**. Wlt 251.

MORGAN'S VALE Wilts SU 1921. A modern residential district in a place called *Morgans Bottom* 1820, 'Morgan's valley'. Surname *Morgan* (found in adjacent parishes from 16th cent.) + ModE **vale**. Wlt 398.

MORLAND Cumbr NY 5922. 'Wood at the moor or marsh'. *Morlund(ia)* 1133×47–1405, *Morland* from 1154×89. ON **mór** (or OE **mōr**) + **lundr** later replaced by **land**. We ii.143, SSNNW 148, L 56, 208.

MORLEY 'Moor wood, clearing or pasture'. OE **mōr** + **lēah**. L 54, 206.

(1) ~ Derby SK 3941. *(æt) Morlege* [1002]c.1100 S 1536, [1004]14th S 906, *(in) Morleage* 1009 S 922, *-leia* 1086, *-ley(e), -legh(e)* etc. 1186×94–c.1250 etc., *Morelei* 1086. Db 486.

(2) ~ Durham NZ 1227. *Morley* from 1295. *Morleypitte*, a coal mine, is already mentioned in 1440×1. NbDu 145, L 54, 206.

(3) ~ WYorks SE 2627. *Morelei* 1086, *Morlai -lay -lei -ley -leg(a)* 1121–1504. Cf. *Morelei Wapentac* 1086, which probably met at TINGLEY SE 2826. YW ii.182, iii.1.

(4) ~ GREEN Ches SJ 8282. P.n. *Morlegh* c.1200, *Morley(e)* from 1330, + ModE **green**. Che i.229.

(5) ~ ST BOTOLPTH Norf TM 0799. 'St Botolph's M'. *Morley St. Botolph* 1838 OS. P.n. *Morlea* 1086, *Morleg* 1201 + saint's name *Botulf* from the dedication of the church and for distinction from Morley St Peter TG 0600 a mile S, *Morley St. Peter* 1838 OS. DEPN.

MORNINGTHORPE Norf TM 2129. Properly Morning Thorpe. *Maringatorp* 1086, *Meringetorp* 1198, *Mourning Thorpe* 1838 OS, and simply *Torp* 1086. Formally this is 'Thorp belonging to the Mæring- or Meringas', OE folk-n. **Mǣringas* 'the people called after Mæra' < pers.n. *Mǣra* + **ingas**, or folk-n. **Meringas* 'the people who live at the mere' < **mere** + **ingas**, genitive pl. **Mǣr-* **Meringa*, + ON **thorp**, 'an outlying settlement'. The folk-n. **Mǣr-* **Meringas* probably became a p.n. **Mer-* or **Maring* in its own right to which *thorp* was later added as in nearby Saxlingham Thorpe TM 2197. See SAXLINGHAM NETHERGATE TM 2297 and for the lost p.n. MAREHAM LE FEN Lincs TF 2761 and Meering Notts, *Meringe* 1086–1278, *Meringes* c.1155–1294, 'the dwellers by the mere' referring to a pool formed by the alternative course of the Trent known as the Fleet. DEPN, Nt 191.

MORPETH Northum NZ 1986. 'The murder path, the path where a murder took place'. *Morthpath* c.1200, *Morpath* 13th, 14th cents., *Mor(e)peth* 1199, 1346. OE **morth** + **pæth**. OE *pæth, peth* in N country names is often used of Roman roads or for a path that descends a hill at an angle. No Roman road is known at Morpeth and the latter sense must apply here; the Great North road sweeps to the right to descend the hill obliquely to cross the river Wansbeck. The town itself is of post-Conquest origin. NbDu 145, TPS 1961.109, L 63, 78.

MORREY Staffs SK 1318. *Morrey* from 1553, *Murry* 1686. Possibly ME **murrey** 'a mulberry, the colour mulberry'. Horovitz.

MORRIS FEN Cambs TF 2906. 'Marsh fen' or possibly 'St Mary's fen'. *Marrys ffenne* 1540, *Marys* 1550, *Morris Fen* 1836. OFr **mareis** or saint's name *Mary*, genitive *Mary's*, + **fen**. Ca 283.

MORSTON Norf TG 0048. 'Marsh settlement'. *Merstona* 1086, *Merston* 1252, *Marston* 1185. OE **mersc** + **tūn**. DEPN.

MORT BANK Cumbr SD 3067. A sand-bank in Morecambe Bay. Possibly ModE *mort* 'a salmon in its third year'.

MORTE BAY Devon SS 4343. *Mortbay* 1577. P.n. *Morte* as in MORTHOE SS 4545 + ModE **bay**. D 54.

MORTE POINT Devon SS 4445. *Mort Point* 1809 OS. P.n. *Morte* as in MORTHOE SS 4545.

MORTHEN SYorks SK4789. 'The moorland assembly'. *Mordinges* 1164×81, *Morthyng -ing* c.1205–1588. ON **mór** + **thing**. The site of the meeting-place, a hill in Whiston, gave its name to the whole district. YW i.141, 168, YAJ 58.23.

MORTHOE Devon SS 4545. Either 'hill-spur called *Morte*' or 'Morta's hill-spur'. *Morteho* 1086–1450, *Marto* 1628, *Moorthoe* 1739. OE hill-name **Morte* or pers.n. **Morta* + **hōh**. Both hill-name and pers.n. could be related to Norw *murt* 'small fish', ModE dial. *murt* 'small person', Icel *murtr* 'short, stumpy', MHG *murz* 'stump'. If correct, **Morte* was probably the original name of Morte Point. D 52.

MORTIMER Berks SU 6564. This is the local name shortened from STRATFIELD MORTIMER SU 6664. It has achieved general currency.

MORTIMER CROSS H&W SO 4263. *Mortimers Cross* 1832 OS. Cf. nearby Mortimer's Rock SO 4163, *Mortimers Rock* 1832 OS.

MORTIMER WEST END Hants SU 6363. Properly the 'west end of STRATFIELD MORTIMER' Berks SU 6764 known locally simply as Mortimer. It was held in 1086 by Ralph Mortimer. Ha 119, Brk 216.

MORTON 'Moor or marsh settlement'. OE **mōr** + **tūn**. PNE ii.42.

(1) ~ Avon ST 6490. *Morton(e)* from 1307. Gl iii.15.
(2) ~ Derby SK 4060. *(æt, into, mid) Mortun(e)* [956]14th, [c.1002]c.1100 S 1536, [1004]14th S 906, 1086, *-ton(e)* from 1186. Db 275.
(3) ~ Lincs TF 0924. *Mortun(e)* 1086, late 12th, *Morton* from 1185. The reference of *mōr* here is to marshland. Perrott 126.
(4) ~ Lincs SK 8091. *Morton* from 1242×3. The reference of *mōr* here is to marshland. Perrott 268.
(5) ~ Norf TG 1217. *Morton* 1196. DEPN.
(6) ~ Shrops SJ 2924. *Mortune* 1086, *Morton* from c.1230. Probably the main source of hay for the manor of Oswestry. Some spellings show occasional 'Welshification', e.g. *Mortyn* 1600×1, *Mortin* 1627. Sa i.213.
(7) ~ BAGOT Warw SP 1164. 'M held by the Bagot family'. *Morton Bagod* 1262, *~ Bagot* 1282. P.n. *Mortone* 1086 + manorial addition from the family of William Bagod who was associated with the manor 1154×89. Wa 215.
(8) ~ -ON-SWALE NYorks SE 3291. P.n. *Mortun(e)* 1086 + suffix *on Swale* 1281. Cf. River SWALE SE 2796. YN 275.
(9) Abbots ~ H&W SP 0355. 'Morton held by the abbot' sc. of Evesham. *Abbotes Morton* 1418. Earlier simply *mortun* *[708]12th S 78, *mortona* *[709]12th S 80, *Mortun* *[714]16th S 1250, *Mortvne* 1086. 'Abbots' for distinction from Morton Underhill SP 0159, *Morthone Underhull* 1280. Wo 331, 328.
(10) East & West ~ WYorks SE 1042. *Mortun(e) -ton(a)* 1086–1605. YW iv.172.

MORVAH Corn SW 4035. 'Sea grave'. *Morveth* 1327–77, 1500, *(capella) Sanctae Brigide et Morvethe* 1390, *Sancta Morpha* 1435, *Morva* 1565. Co **mor** 'sea' + **beth** 'grave' with lenition of *b* > *v*. The nearby farm-name Merthyr SW 401354, *Merder* 1302 Gover n.d., *Merthyr* 1523, Co **merther** 'place with saint's relics' confirms this interpretation. In 1478 William of Worcester wrote that St Morianus was buried in the parish of *Sanctus Morianus* above the sea-shore W of Penzance which may be a reference to Morvah. The church is dedicated to St Bridget of Sweden (died 1373, canonised 1391) and St Morveth is due to a mistaken reinterpretation of the p.n. PNCo 124, Gover n.d. 651.

MORVAL Corn SX 2656. Unexplained. *Morval* 1238, *Moruale* 1610. Possibly Co **mor** 'sea' + unidentified element. PNCo 124, Gover n.d. 281.

MORVILLE Shrops SO 6794. 'Open land at *Mamer*'. *Membrefelde* 1086, *Ma- Momerfeld(')* c.1138–1382, *Momeresfeud* 1255×6, *Momersfeld(')* 1291×2, 1317, *Morfelde* 1464, *Morveld* 1554–1614, *-vill* 1577, *-ville* 1704, *Marvil -vel* 18th. Unknown element or name **Mamer*, genitive sing. **Mameres*, + OE **feld**. *Mamer* might be an ancient name of Mor Brook, itself a back-formation from Morville, or of Aston Hill SO 6693 which rises to 726ft.; but in neither case is a satisfactory etymology to hand. The forms forbid association with the r.n. Mimram Herts TL 2116, OE **Memere* of unknown meaning, *(betweox) Memeran* 913 ASC(A), [before 1118]12th, *Mimeram* [c.1130]12th, or with the hill-name **Mam* in MAMBLE, MAMHEAD, MAM TOR etc. whether of British or English origin. Sa i.213.

MORWENSTOW Corn SS 2015. 'St Morwenna's holy place'. *Morwestowa* 1201, *Morwennestohe* 1273, *Morwenestowe* 1284, 1291, *Stow* 1610. Saint's name *Morwenna* from the dedication of the church + OE **stōw**. In the 15th cent. she was believed to be buried here. PNCo 124, Gover n.d. 18, *Stōw* 190.

MORWICK HALL Northum NU 2303. P.n. *Morewic* 1171–1242 BF, *-wyc* 1242 BF, *Morwic(k) -wyc -wike* 1212 BF(p)–1296 SR, *Morrick* 1628, *Morweek* 1682, the 'moor farm', OE **mōr** + **wīċ**, + ModE **hall**, a Georgian house c.1740–50. NbDu 145 gives pr [mɔrɪk], Pevsner-Grundy etc. 1992.401.

MOSBOROUGH SYorks SK 4280. 'Fortified place of the moor'. *Moresburh* [1002×4]11th S 1536, *-burg* 1086, *Morisbur' -burge -burgh(e)* 13th–1435, *Morsburgh* 1448, *Mosburgh(t)* 1417–1575, *-broughe* 1570. OE **mōr**, genitive sing. **mōres** + **burh**. Db 248.

MOSEDALE Cumbr NY 3532. 'Valley with a moss or bog'. *Mosedale* from 1285, *Mosdale* 1305–1610, *Mosdall* 1620. ON **mosi** (or OE **mos**) + **dalr**. Cu 304 gives pr [mouzdəl], SSNNW 148, L 57, 96.

MOSELEY WMids SP 0883. 'Mouse wood or clearing'. *Museleie* 1086, *Moseley* from 1221, *Mousley* 1577. OE **mūs**, genitive pl. **mūsa**, + **lēah**. Alternatively the specific could be OE pers.n. *Mūsa*, 'Musa's wood or clearing'. Wo 356, L 205.

MOSS SYorks SE 5914. 'The swamp'. *del Mose* 1416(p), *Moss(e)* from 1446, *The Moss* 1636. OE and ME **mos**. The reference is to the great tract of marshland between the rivers Aire and Don. YW ii.48.

MOSS BANK Mers SJ 5197. 'Hill by a bog'. *Moss Bank* 1843 OS. ModE **moss** + **bank**. Cf. Kings Moss SJ 5001 in the same river valley. Now a suburb of St Helens.

MOSS BAY Cumbr NX 9826. No early forms. ON **mosi** or OE **mos** 'a moss, a bog' + ModE **bay**.

MOSSDALE NYorks SD 8391. 'Bog valley'. *Mussedale* 1280, 1283, *Mos(e)dale* 1285–1607. ON **mosi** + **dalr**. Still called *Mosedale* in J.E. Morris's 'Little Guide' to *The North Riding of Yorkshire* published in 1906. YN 267.

MOSSDALE MOOR NYorks SD 8090. P.n. MOSSDALE SD 8391 + ModE **moor** (OE *mōr*).

MOSSLEY GMan SD 9701. 'Wood, clearing with or by a bog'. *Moselegh* 1319, *Mossley* 1422. OE **mos** + **lēah**. La 29.

MOSSLEY HILL Mers SJ 3887. *Mosley* 1842 OS. Probably 'wood, clearing by a bog', OE **mos** + **lēah**.

MOSS MOOR WYorks SE 0014. P.n. *Moss(e)* 1641, 1817, ModE **moss** (OE *mos*) 'bog, swamp, moss', + ModE **moor**. YW iii.74.

MOSS SIDE GMan SJ 8496. 'Beside the bog'. *Mossyde* 1530, *Mosside* 1564, *Moss Side* 1594. OE **mos** + **sīde**. La 32.

MOSS SIDE Lancs SD 3830. No early forms. Dial. **moss** 'a bog, a swamp' + **side**. The name of a railway halt on the edge of Lytham Moss, cf. Moss Hall Farm SD 3529.

MOSTERTON Dorset ST 4505. 'Mort's thorn-tree'. *Mortestorne* 1086, *Mortesthorn* 1209, *Mostreton* 1354, *Musterton* 1431. OE pers.n. **Mort*, genitive sing. **Mortes*, + **thorn** later replaced by ME **ton** (OE *tūn*). Do 111.

MOSTON GMan SD 8701. 'Bog settlement'. *Moston* from 1195. OE **mos** + **tūn**. The reference is probably to the area called *White Moss* 1843 OS preserved in the name Moss Farm SD 8704. La 36, Jnl 17.33.

MOTCOMBE Dorset ST 8425. 'Assembly coomb'. *Motcumbe* 1244–1310, *Motecumb(e)* 1244–70, *Motcombe* from 1502. OE **(ġe)mōt** + **cumb**. Probably the meeting place of Gillingham hundred. Do iii.47, L 93.

MOTE HOUSE Kent TQ 7855. A mansion, now a Cheshire Home, built for Lord Romney in 1793–1801, presumably replacing a moated house. *La Mote* 1266, 1278, *Mot(t)e* 1484, *The-moat* 1690. OFr, ME **mote**. PNK 142, Newman 1969W.397.

MOTTISFONT Hants SU 3226. 'Fountain of the waters' meet'. *Mortesfvnde, Mortelhvnte* 1086, *Motesfont(e) -funt* 1167–1306, *Mottesfunte* 1243. OE **(ġe)mōt**, genitive sing. **(ġe)mōtes**, + **funta**. The reference is to the meeting of the waters of the Dun and the Test and the spring at Mottisfont priory 'full of water as clear as crystal, ever flowing and yielding perhaps 2 million gallons of water daily'. Ha 119, Gover 1958.189, Signposts 84, 86, Nomina 9.5.

MOTTISTONE IoW SZ 4083. 'The speakers' stone'. *Modrestan* 1086, *Motereston(e)* 1291, 1325, *Motestan* 1175–87, *-stone* 1255–1408, *Mottestone* 1318–1535, *-is-* 1374, *Motston* 1398, 16th, *Matston* 1451, *Motson* 1595. OE **mōtere**, genitive pl. **mōtera**, + **stān**. The reference is to a large menhir called Long Stone on the hill above the village. Wt 164 gives prs ['mat(i)stn, mɔtsn] and ['matəstn], Mills 1996.72, Thomas 1976.156.

MOTTRAM IN LONGDENDALE GMan SJ 9995. *Mottrum in Longedonedale* 1332, *Mottram in Longedenedale* 1308. P.n.

Mottrum 1211×25, *Mottram* from 1308, identical with Mottram in MOTTRAM ST ANDREW SJ 8778, + p.n. *Langedenedele* 1086, *-dale* [c.1153×81]1318, c.1251–1475, *Long(e)den(e)dale* [1181×1232]1318 and from 1246, 'the dale of *Longdene*, the long valley', p.n. *Longdene*, OE **lang** + **denu**, + OE **dæl**, ON **dalr**, referring to the valley of the river Etherow. Che i.313, 2.

MOTTRAM ST ANDREW Ches SJ 8778. 'St Andrew's M'. *Mottram St Andrew* is a 19th cent. modification of *Mottram Andrew(e)* 1408, 'M held by Andrew', an unidentified feudal tenant. Earlier simply *Motre* 1086, *Motterom* 1240, 1349, *-um* c.1280, *Mottrom* mid 13th–1379, *-am* from mid 13th, possibly '(at) the assembly trees', OE ***mōt-trēum** (OE *(ġe)mōt* + *trēow*, dative plural *trēum*), or 'at the speakers', OE **mōterum**, dative pl. of **mōtere**. The additions *Andrew*, later *St Andrew*, are for distinction from MOTTRAM IN LONGDENDALE SJ 9995. Che i.202.

MOUDY MEA Cumbr NY 8711. 'Pasture infested with moles'. *Moudy Mere* 1843, ~ *Mea* 1863. Lakeland dial. **moudie** 'a mole' (shortened form of *moudy-warp*) + Mod dial. **mea** 'pasture'. Refers to a summit of 1701ft. on Stainmore. We ii.78.

MOUGHTON NYorks SD 7971. Possibly '*Mow* hill'. *Moughton* 1687. The first element is thought to be a derivation of OE **mūga** 'a heap, a heap of stones, a cairn', ModE *mow* 'a rick of corn or hay', possibly used as a p.n. in the sense 'lofty hill'. The reference is to a mountainous area of exposed limestone with great scars and cliffs. The second element is probably **dūn** rather than *tūn*. YW vi.230.

MOULDSWORTH Ches SJ 5171. 'Hill enclosure'. *Moldeworthe* 1153×81, *-wurth -worth* 1230–1562 with variants, *Moldesworth* [1257]17th, 1565, *-wrth* 1260, *Moldsworth* from 1372. OE **molda** 'crown of the head' in a topographical sense + **worth**. Pseudo-genitival *-s* was added in this name on the mistaken assumption that the specific was a pers.n. derived from ON *Moldi* or **Moldr* or the surname *Mold*. Che iii.279 gives pr ['moulzwɔːθ] and ['mouldzwəθ], SPN 197, DBS 235 s.n. Maud.

MOULSECOOMB ESusx TQ 3307. 'Mul's coomb'. *Mu- Molescumba* 1087×1100, *Molscumb* 1296, *Mousecombe* 1615. OE pers.n. *Mūl*, genitive sing. *Mūles*, + **cumb**. Sx 294 gives pr [mauskəm], SS 69.

MOULSFORD Oxon SU 5983. 'Mul's ford'. *Muleforda* c.1110, *-ford'* 13th cent., *Mulleford* c.1250, *Mulford* 1517, *Mullesford* 1130–1536, *Mulesford* 1271, 1324, *Molesford(e)* 1194×5–1327, *Molsford* 1395, *Moulsford* 1600. OE pers.n. *Mūl* < **mūl** 'a mule', genitive sing. *Mūles*, + **ford**. The *Mul(l)e-* forms suggest a weak variant **Mūla*, genitive sing. **Mūlan*. Moulsford must have been a ford on the Thames. Brk 527, L 69.

MOULSOE Bucks SP 9041. 'Mul's hill-spur'. *Moleshou* 1086–1176, *Mo- Mulesho* 1152–1284, *Mulso* 1435–1806, *Moulsoe* c.1550. OE pers.n. *Mūl*, genitive sing. *Mūles*, + **hōh**. The village stands on a prominent eminence falling away to the E. Bk 36 gives pr [mʌlsou], L 168.

MOULTON Ches SJ 6569. 'Mul(a) or Muli's farm or village' or 'mule farm'. *Moletune* 1086, *Multon* 1260–1579, *Molton* 1375, 1544, *Moulton* from 1519. Pers.n. OE *Mūla* or ON *Múli*, or OE **mūl** + **tūn**. There was also an OE by-name *Mūl* 'half-breed' which might perhaps here refer to a man or men of mixed Welsh-English parentage. Che ii.207 gives pr ['moultən], ['mou(l)tn] and ['mou?n], SSNNW 190, Jnl 12.6.

MOULTON Lincs TF 3024. Probably 'Mula's farm or village'. *Multune* 1086, *-ton(a)* c.1200–1535, *Muleton(a)* 1169–1330. OE pers.n. *Mūla* + **tūn**. An alternative possibility is 'mule farm', OE **mūl**, gentive pl. **mūla**, + **tūn**. Payling 1940.36.

MOULTON Northants SP 7868. Probably 'the mule farm or village'. *Multon* before 1076, 1086–1359, *Mol- Muleton(e)* 1086–1275, *Molenton* 1205, *Multon* 1266, *Moulton* from 1576. OE **mūl** + **tūn**. Forms with medial *-e-* may point to genitive pl. **mūla** or possibly to the OE pers.n. *Mūla*. Nth 134.

MOULTON NYorks NZ 2303. 'Mula or Muli's settlement'. *Moltun* 1086, *Muleton'* 1176–1220, *Multon* 13th–1441, *Mowton* 1577, 1613. Pers.n. OE **Mūla* or ON *Múli*, genitive sing. *Múla*, + **tūn**. YN 286 gives pr [moutən], SSNY 128.

MOULTON Suff TL 6964. Probably 'Mula's estate'. *Muletunā* 1086, *-ton* 1198, 1235, *-tun* 1241, *Moulton* 1610. OE pers.n. *Mūla* + **tūn**. An alternative possibility is OE **mūl**, genitive pl. **mūla**, + **tūn**, 'the mule farm'. DEPN.

Great MOULTON Norf TM 1690. *Muleton Maior* 1254. ModE adj. **great**, Latin **major** 'greater', + p.n. *Mulantun* 1035×40 S 1489, *Muletuna* 1086, [1183]14th, 'Mula's estate', OE pers.n. **Mūla*, genitive sing. **Mūlan*, + **tūn**. Also known as Moulton St Michael from the dedication of the church and for distinction from Little Moulton, *Muleton Minor* 1254. DEPN.

MOULTON CHAPEL Lincs TF 2918. *Multon Chapell* 1488, *Chapel of St James in Mulketon* (sic) 1549. P.n. MOULTON TF 3024 + ME **chapel**. Payling 1940.40.

MOULTON SEAS END Lincs TF 3227. 'The sea end of M'. *Multon Sea-end called sea lands* 1590, *Moulton Sea End* 1824 OS. Earlier simply *The see ynd* 1514. The village stands on the ancient Sea Bank and marks the original inland extent of The Wash. Payling 1940.40.

MOUNT Corn SW 7856. *Mount* 1841. A 19th cent. hamlet situated on a hill. PNCo 124.

MOUNT Corn SX 1467. *Mount* 1839. A 19th cent. hamlet situated on former downland, probably a shortened form of the nearby farm-name Mount Pleasant SX 1468, *Mount Pleasant* 1612. PNCo 125.

MOUNT CABURN ESusx TQ 4409. Partly unexplained. *Mount Carbone* 1770, *Mt Caburn* 1784. An Iron Age hill-fort for which unfortunately there are no earlier forms. Sx 353 gives prs [kɔː-] and [kaːbəːn], Thomas 1976.206f.

MOUNT EDGECOMBE Corn SX 4353. *Monte Edgcomb*. A house built by Sir Richard Edgcumbe c.1547. OFr, ME **mont** 'a hill' + family name Edgcumbe (from Edgcumbe Devon SX 398791). PNCo 125, Gover n.d. 232.

MOUNTFIELD ESusx TQ 7320. 'Munda's open land'. *Montifelle* 1086, *Montefelde* 12th, *Mundifeld* [12th]15th–c.1215, *Munde-* 1265–16th, *Mountefylde*, *Munfilde* 16th. OE pers.n. *Munda* + **feld**. Spellings with *-i-* may show reduction of OE medial **ing**[4]. Sx 474, Ritter 111.

MOUNT GRACE PRIORY NYorks SE 4498. 'The hill of grace'. A French name for the Carthusian priory founded here in 1396. *Monte Grace, Mountgrace, Montem Graciœ* 1413–67. The original name of the site was *Bordlebi, Bordelbia, Bordalebi* 1086, *Bordelesby* 13th, *Bordelby* 1243–1405, a name in ON **bȳ** the specific of which is either pers.n. **Bordel* or an earlier p.n. **Bordal* 'ridge valley', OE ***bor** + ON **dalr**. YN 214, PNE i.41–2, SSNY 22.

MOUNT HARRY ESusx TQ 3812. Unexplained. *Mountharry* 1610. Possibly to be identified with unlocated *Haregedon* 1203, *Harewedon* 1332 'hill with a heathen temple', OE **hearg** + **dūn**. Sx 316.

MOUNT HAWKE Corn SW 7147. *Mount Hawk* 1813. An 18th/19th cent. village named from a local family *Hawke* first appearing in the parish in 1758. PNCo 125, Gover n.d. 364.

MOUNTJOY Corn SW 8760. 'Stone house'. *Meyndi* 1277, *Meyndy* [1284]1602, *Mayndi* 1334, *Munjoy* 1735. Co ***meyn-dy** with assibilation of medial *d* > *dʒ* permitting subsequent reinterpretation as the common minor name *Mountjoy*. PNCo 125, Gover n.d. 318.

MOUNTNESSING Essex TQ 6297. 'Mounteney's Ing'. *Gi- Gynges Willelmi de Munteni* 'William de Munteni's Ing' 1198, *Mounteneye Giginge* 1363, *Mountnes(s)ing* 1535, 1628, *Monesyng* 1562, *Munnings-end* 1663. Surname *Mounteney* from Robert de Mounteney 1141×51 + p.n. *Gingā* 1086, *Gi- Gynges* 1198–1261, *Geinge* 1282, as in INGRAVE TQ 6292. Ess 260 gives prs [mʌnt'nisən,'mʌnəsiŋ].

MOUNT PLEASANT Corn SX 1468 → MOUNT SX 1467.

MOUNTSORREL Leic SK 5814. 'Sorrel-coloured hill'. *Munt Sorel* 1152, *Muntsorel* late 12th–1294, *Mountsor(r)el(l)* from 1276, *Mun- Monsorel* late 12th–1323. OFr **mont** + **sorel**. The reference is to Castle Hill formerly occupied by a Norman stronghold commanding the narrow pass between the Soar and the wild granite country to the W and beneath which the farm grew up. The hills are famous as the source of a hard red-grey granite. Lei 385, Gaz.

MOUSEBERRY Devon SS 7616 → MUSBURY SY 2794.

MOUSEHOLE Corn SW 4626. 'Mouse hole'. *Pertusum muris* 1242, *Musehole* 1284–1347, *Mousehole* from 1302, *Mosehole* 1300–71, *Mosal* 1400, *Mowssel -al* [1580]18th. OE **mūs** + **hol** (latinised as **pertusum muris** 'hole of a mouse') referring to a large cave in nearby cliffs called the Mousehole SW 468257. The Cornish name was *Portheness* 1267, *Porthenys juxta Mosehole* 1310, 'harbour of the island', Co **porth** + **enys** referring to St Clement's Isle SW 4726. PNCo 125 gives pr [mauzl], Gover n.d. 655.

MOUSEN Northum NU 1232. 'Mul's fen'. *Mulefen* 1166, *Mulesfen* 1186–1267, *Mullesfen* 1219 BF, *-is-* 1236 BF, *Mulsfen* 1296 SR, *Mulssen* 1428, *Mowssen* 1538, *M(o)ulsfen* 1628. OE pers.n. *Mūl* (by origin a by-name meaning 'half-breed'), genitive sing. *Mūles*, + **fenn**. The place is situated where the high ground of Chatton Moor dips down to the low ground of Newlands Burn. NbDu 146.

MOW COP Staffs SJ 8557. 'Mow hill'. *Mowle-coppe* 1621, *Mowcop Hill* 1656, *The Mole Cop or Mow Cop* 1842 OS. P.n. *Mowl* 13th, *(Rocha de) Mowel* c.1270, 1280, *Mouhul* 1317, *Moul'* 1313, *Mole* 1437–68, 1657–1720, OE **mūga** 'a heap', + **hyll**, + ModE **cop** (OE *copp*) 'a summit'. The reference must have been either to the heap-like shape of the hill here rising to 1100ft. or to a cairn on the Chesh-Staffs boundary. DEPN, Che ii.308.

MOWSLEY Leic SP 6489. 'Wood or clearing infested with (field-)mice, mouse wood'. *Muselai* 1086, *-le(e) -leia -leye* 1199–1331, *Mousele(e) -ley(e)* 1277–1551, *Mowsley* from 1605. OE **mūs**, genitive pl. **mūsa**, + **lēah**. Lei 243 gives pr [mouzli], L 205.

MOXBY NYorks SE 5966. 'Moldr's farm or village'. *Molz- Molscebi* 1086, *Molesby -bi* 1158–1318, *Molsby* 1538. ON pers.n. **Moldr*, genitive sing. **Molds*, + **bȳ**. For the spelling see ROXBY NZ 7616. YN 29 gives pr [mouzbi], SSNY 35.

MOYLES COURT Hants SU 1608. *Moyles Court* 1470, earlier *Meoles maner'* 1395. Surname *Moyle* as in Richard de Molis 1242, Roger de Moeles 1280 and John de Mules 1330, + ME **manere, court** 'a manor house'. A mid-17th cent. brick manor house. Also known as *Rokford Meolys alias Elyngham* 1400, *Rokford Moyles alias Moyles Court* 1470, from p.ns. ROCKFORD SU 1508, ELLINGHAM SU 1408. Gover 1958.213, Pevsner-Lloyd 210.

MOZIE LAW Northum NT 8215. *Mozie Law* 1869 OS, *Hozie Law* (sic) 1878 Newm (map). Unidentified element + N dial. **law** (OE *hlāw*). A hill rising to 1812ft. on the England-Scotland border.

MUCHELNEY Somer ST 4224. '(At the) great island'. *Myclaneya* [725]14th S 249, *Michelnie* [762]14th S 261, *Miclani* [964]14th S 729, *(of) Miclanige* 995 S 884, *Micelenye, Michelenie* 1086, *(on) Myclan ȳge* 12th ASC(H) under year 1114, *Muchelene(ia)* 1160, [1174×80]14th, *Muchelnay -ey* [c.1221, 1228]Buck, *Muchenay* 1610. OE **miċel**, definite form dative case **miclan**, + **īeġ**. An island of dry ground in the marshes between the rivers Isle and Yeo. Contrasts with the Little Island (lost) ST 4126, *Lytlenige* *[1065]18th S 1042, *Litelande* 1086, *Litelaneia* 1086 Exon, OE **lȳtel**, definite form dative case **lȳtlan**, + **īeġ**. DEPN, L 39.

MUCHLARNIK Corn SX 2156. 'Great Larnick'. *Muchele Lanrack* 1307, *Muchellanrek* 1382, *Michellarniak* 1415. Adj. ME **muchel** (OE *myċel*) + p.n. Larnick, *Lanher*, *Lanner* 1086, Co **lanherch** 'clearing'. 'Much' for distinction from the other part of the divided manor, Little Larnick SX 2255, *Lyttyllavracke* (with *v* for *n*) 1525. PNCo 126, Gover n.d. 294.

MUCKING Essex TQ 6881. Usually taken to be 'the Muccingas, the people called after Mucca'. *Muc(h)inga* 1086, *Mucc- Muckinges* 1199–1200, *Mucking* from 1219 with variant *Mock-* 1178, 1285, *Mockynges* 1270. OE folk-n. **Muccingas* < pers.n. *Mucca* + **ingas**. It is not entirely clear, however, that this is a pl. formation rather than an *ing*[2] sing. formation on the root seen in ModE *muck*, a 13th cent. word probably of Scandinavian origin on the Gmc root **meuk-/*muk-* 'soft'. An OE **Mucing* 'mud place' is suggested by Dr Gelling referring to Mucking Creek which is characterised by great silver mudbanks on either side: cf. the p.n. *Hlōsmoc* cited AEW 397 s.n. *mykr*. Ess 163, ING 22, Excavations, 96.

MUCKLESTONE Staffs SJ 7237. 'Mucel's estate'. *Moclestone* 1086, *Mukleston* 1221, *Muxton* 1833. OE pers.n. *Mucel*, genitive sing. *Muceles*, + **tūn**. Horovitz.

MUCKLETON Shrops SJ 5921. 'The estate called after Mucel'. *Moclyton* 1255×60, 1271, *-i-* c.1270, 1313, *-in-* 1268, *Muckleton* 1833 OS. OE pers.n. *Mucel* + **ing**[4] + **tūn**. Bowcock 167.

MUCKTON Lincs TF 3781. 'Muca's farm or village'. *Machetone* 1086, *Munchetone* c.1115, *Muketun* c.1110, 1154×89. OE pers.n. *Muca* + **tūn**. DEPN.

MUDDIFORD Devon SS 5638. 'Moda's enclosure'. *Modeworthi* 1303, *Modworthy* 1400, *Mud-* 1515×18, *Mudford or Mudworthie* 1558, *Muddiford* 1809 OS. OE pers.n. **Mōda* + **worthiġ**. D 51.

MUDEFORD Dorset SZ 1892. 'Muddy ford'. *Modeford* 13th, *Muddiford* 1826. ME **mudde** + **ford**. Do 111, Ha 120.

MUDFORD Somer ST 5719. 'Muddy ford'. *Mv(n)diford* 1086, *Modiford(e) -y-* [1091×1106–1279×8]14th, *Mudiford(e)* [1135×7–1189×99]14th, 1176, 1201, [13th] Buck, *Mudeford* 1201, *Muttford* 1610. OE adj. ***muddiġ** + **ford**. DEPN, PNE ii.44, L 68.

MUDGLEY Somer ST 4445. Partly uncertain. *Mudesle* 1157, *-liegh* 1164, *Modeslega* 1176. Unidentified element or pers.n. apparently in the genitive sing. case + **lēah** 'wood or clearing'. DEPN.

MUGGINTON Derby SK 2843. 'Estate called after Mogga or Mugga'. *Mogintona -yn- -ton(e)* 1086–1577, *Mug-* c.1141–1330, *Mugginton* from 1243, *Mog- Muggyngton* 1380–1556. OE pers.n. **Mogga* or **Mugga* + **ing**[4] + **tūn**. Db 509.

MUGGLESWICK Durham NZ 0450. 'Dairy farm at or called **Muceling*'. The spellings fall into four types:

I. *Mucelingwic* c.1170–1291 with variant sp *-wyk*.

II. *Mucelyngeswyk(e)* [1183]14th, c.1260–1340 with variants *Muck- Moc- -ing-*.

III. *Muclyswik* late 13th, *Mokeleswyk'* 1304, *mukleswik* 1372×3.

IV *Mugleswyk(e) -wi(c)k(e)* 1367–1717, *Muggleswick* from 1565. OE pers.n. *Mucel* + **ing**[4] + **wīċ**.

I is 'the *wīc* (called or at) **Muceling*, the place called after Mucel'; II 'the *wīc* of **Muceling*'; III 'Mucel's *wīc*; IV 'Mucel's *wīc*' with voicing of intervocalic [k] to [g].

Muggleswick was a vaccary belonging to the Prior of Durham, later emparked and used as a hunting domain, *parcum de Muklingwyc* [1199×1216]1300, *the payrk of Mugleswyke* 1550. NbDu 146, BzN 1968.149.

MUGGLESWICK COMMON Durham NZ 0146. *The Comon* 17th, *ye, The Common* 1695–1809, *Muggleswick Common* 1801. P.n. MUGGLESWICK NZ 0450 + ModE **common**. Mellard 9.

MUKER NYorks SD 9097. 'The narrow newly cultivated field'. *Meuhaker* 1274, *Muaker* 1577, *Mewker* 1606. ON **mjór** + **akr**. YN 272, L 231.

MULBARTON Norf TG 1900. 'Milk Barton or outlying dairy-farm'. *Molkeber(tes)tuna* 1086, *Mulkebertun* 1250, *-ton* 1254. OE **meoluc** 'milk' + **beretūn**. One of the DB spellings shows a false analysis of the name as if 'Molkebert's estate'. No such name exists. DEPN.

MULFRA Corn SW 4534 → NEW MILL SW 4534.

MULGRAVE CASTLE NYorks NZ 8412. P.n. Mulgrave + ModE **castle**. A Norman and later castle with marvellous 18th cent. grounds. Mulgrave, *Mulegrif -grive* [1155–1227]15th,

Mul(e)greve 1224–1415, *Mulgrave* from 1335, *Mowgrave* 1577, 1613, is 'Muli's valley', ON pers.n. *Múli*, genitive sing. *Múla*, + **gryfja**. First recorded simply as *Grif* 'the valley' 1086. This element was especially used of the steep-sided narrow valleys running down to the coast in this part of Yorkshire as at Corngrave Cleve NZ 6420, GRIFF SP 3588, HAZELGROVE SJ 9286, Falsgrave TA 0287 and SKINNINGROVE NZ 7119. YN 137 gives pr [mougriv], SSNY 101, Pevsner 1966.260.

MULLION Corn SW 6719. 'St Mellion's (church)'. (Peter of) *Sanctus Melan(i)us* [c.1225, 1241]14th, *(ecclesia) Sancti Melani* 1262, 1309, *Seynt Melan* 1284, *Eglosmeylyon* 1342, *Seynt Mylyan*, ~ *Melyon* 1420–79, *Melion* 1535, *Mullyan* 1580, *Mullion* 1708. Saint's name *Melanius* (Mellion, Melaine) from the dedication of the church. He was a Breton or Frankish bishop of Rennes in the 6th cent. Cf. ST MELLION SX 3865 and St Mellons Gwent ST 2287, *(ecclesia de) sancto Melano* 14th. PNCo 126, Gover n.d. 574, DEPN.

MULLION COVE Corn SW 6617. *Mullion Cove* c.1870. P.n. MULLION SW 6719 + ModE **cove**. Also called by its Cornish name *Porth Mellin* 1840, 'cove of the mill', Co **porth** + **melin**. PNCo 126, Gover n.d. 577.

MULLION ISLAND Corn SW 6719. *Mullion Island* 1813. P.n. MULLION SW 6719 + ModE **island**. Earlier known as *Inispriven* c.1540, 'reptile island'. Co **enys** + **prif** 'worm, reptile', either pl. **prevyou** or diminutive **prevan**, probably a fanciful name from its shape. Also called *Gull Rock* 1699, 1748. PNCo 126, Gover n.d. 576, CPNE 194.

MUMBY Lincs TF 5147. 'Mundi's farm or village'. *Mund(e)bi* 1086, *Mon- Munbi* c.1115, *Mumbi* c.1115–c.1200. ON pers.n. *Mundi*, genitive sing. *Munda*, + **bȳ**. This explanation has been felt to be unsatsifactory on the grounds that the pers.n. is not independently recorded in Lincs (or Yorks). OE, ON **mund** 'a hand' also meant 'protection'. Perhaps the name meant 'protected village' in some unknown sense; in MUNDON Essex the reference may be to protection from floods. DEPN, SSNEM 60, Jnl 8.51, L 126.

MUNCASTER CASTLE Cumbr SD 1096. P.n. Muncaster + ModE **castle**. Muncaster, *Mulcastre* [1100–1135]1412–1509 with variant *-caster, Mole- Mulecastre* 1185×1201–1278, *Mulncastre* 1278, *Mulcastr' et non Moncastel* 1292, *Moncastre* 1389–1505, *Munkaster* 1397, is 'Muli's fort', ON pers.n. *Múli* + OE **cæster** 'a Roman fort' possibly referring to the Roman fort of *Glannoventa* at Ravenglass SD 0895. Sensitivity to the Anglo-Norman change *Mul-* to *Mun-* is evidenced by the 1292 form. Cu 423.

MUNDERFIELD ROW H&W SO 6551. 'Row of houses at Munderfield'. *Munderfield Row* 1832 OS. P.n. Munderfield + ModE **row**. Munderfield, *Mandefeld* 1194, *Manderfeld* 1224, 1268×75, *Mundrefeld* 1224, is 'Munda's open land', OE pers.n. *Munda*, genitive sing. *Mundan*, + **feld**, with ME substitution of *-re-* for *-ne-* < OE *-an*. He 29.

MUNDERFIELD STOCKS H&W SO 6550. 'Stocks at Munderfield'. *Stocks* 1832 OS. ModE **stocks**.

MUNDESLEY Norf TG 3136. Probably 'Mul's wood or clearing'.
I. *Muleslai* 1086, *-le* c.1150.
II. *Munesle* 1208, 1254.
Type I is clearly 'Mul's wood', OE pers.n. *Mūl*, genitive sing. *Mūles*, + **lēah**. Type II may have developed from Type I by the dissimilatory process *l - l* > *n - l*, cf. Lindsey Suff TL 9744, *Linsey* 14th < earlier *Lealeseia* [1087×98]12th, *Lelesia* 1191, 'Lelli's island'. Alternatively both types might go back to an OE *Mundles-lēah* 'Mundel's wood' < OE pers.n. **Mundel* in which genitive sing. *Mundles* became both *Mules-* and *Munes-*. DEPN.

MUNDFORD Norf TL 8093. 'Munda's ford'. *Mundefort* [1042×66]12th S 1051, *-forda* 1086, *-ford* 1242. OE pers.n. **Munda* + **ford**. DEPN.

MUNDHAM Norf TM 3397. 'Munda's homestead'. *Mundaham* 1086, *Mundham* from 1158, *Mundeham* 1197, 1264. OE pers.n. **Munda* + **hām**. DEPN.

South MUNDHAM WSusx SU 8700. *Suth Mundham* 1410. ME adj. **suth** + p.n. *Mundham(e)* *[683]14th S 232, [688×705]14th S 45, *Mundreham* 1086, 1162, 1332, 'Munda's homestead or enclosed land', OE pers.n. *Munda*, genitive sing. *Mundan* (source of the *Mundre-* sps by AN sound substitution), + **hām** or **hamm** in any of senses 1–5, especially 3 and 4, 'river meadow' or 'dry ground in marsh'. 'South' for distinction from North Mundham SU 8702, *se northra Mundan ham* 'the more northerly M' [680]10th S 230 (in which South M is the *other Mundan ham*), *North Munham* 1278. Sx 72, ASE 2.33, SS 81.

MUNDON Essex TL 8702. 'Security hill'. *Mundunā* 1086, *Mundon(ia) -e* from 1119. OE **mund** + **dūn**. OE *mund* is probably used in a concrete sense as in *mund-beorg* 'protecting hill' (unless this is an error of *munt-beorg*) and ModE *mound* 'hedge, fence, tumulus'. The best suggestion is that this low hill afforded security in times of flood: the place lies between Purleigh Wash TL 8502 and Mundon Wash TL 8703, ModE **wash** 'low-lying land subject to flooding'. Ess 220, L 152, Mills 236.

MUNGRISDALE Cumbr NY 3630. Probably 'St Mungo's *Grisedale*'. *Mounge Grieesdell* 1600, *Mungrisedale* 17th cent., *Mungrisdale* 1781. Saint's name *Mungo* (alias St Kentigern) from the dedication of the church + p.n. *Grisedale* 1285–1487 with variant *Gryse-, Grysdale* 1292, *Grysdell* 1559, 'valley of the pigs', ON **gríss**, genitive pl. **grísa**, + **dalr**. Cu 226 gives pr [mʌŋraizdəl], SSNNW 128, L 94.

MUNSLEY H&W SO 6640. 'Mul's wood or clearing'. *Mvnes- Moneslai, Mvleslage* (2×) 1086, *Mulesle* 1201, *Mo- Munesl(e)(ia) -leye* 1160×70–1256. OE pers.n. *Mūl*, genitive sing. *Mūles* with *l/n* confusion, + **lēah**. He 150.

MUNSLOW Shrops SO 5287. Partly uncertain. *Mosselawa* 1167, *Musselawa -e* 1187–1272, *Munselowe* 1252–1334, *Mounselawe -low(e)* 1331–1571, *Mounslowe* 1549–1750, *Munslow* from 1577. Unknown element or name **Munse* or **Munsel* + OE **hlāw** 'a tumulus'. The forms *Mulslaye -leie* [c.1110×15]1348 are unreliable. Sa i.215.

MURCOTT Oxon SP 5815. 'The marsh cottage(s)'. *Morcot'* [c.1191]c.1280–1340 with variant *-cote, Morecote* 1345–1452. OE **mōr** + **cot**, pl. **cotu**. O 207, L 54–5.

MURROW Cambs TF 3707. 'Marsh row (sc. of cottages)'. *Morrowe* 1376–84, *(le) Murrow(e)* 1420–1515, *Murroughe, Mooroe* 16th. ME **moor** (OE *mōr*) + **rowe** (OE *rāw*). Ca 293.

MURSLEY Bucks SP 8128. 'Myri's wood or clearing'. *Muselai* 1086, *Mures- Murse- Me(r)selai(e)* 12th, *Muresle* 1200–1350, *Mursele(e)* 1252, 1392, *Morrysley* 1552, *Muresley* 1766. Near Mursley was a boundary called *myres gemære*. OE pers.n. **Myri*, genitive sing. **Myres*, + **lēah**. Bk 70 gives pr [mə:zli], Jnl 2.23.

MURTON 'Marsh or moor settlement'. OE **mōr** + **tūn**. L 55.
(1) ~ Cumbr NY 7221. *Morton* 1235–1547, *Murton* from 1376. We ii.103.
(2) ~ Durham NZ 3947. *Morton* c.1200–1647. Also known as ~ *Daudre* 1277, *Est Murton* 1364, *Murton alias East Murton alias Murton on the Moor* 1611, *Murton-le-Whins* 1624. John Shaclock's grant of this manor to William Shipperdson in 1624 included 50 acres of furze ('whin'). Furze was a valuable crop: crushed it was used for cattle food and horse fodder, it was burned as fuel, and its ashes used as a substitute for soap or as a fertilizer. NbDu 146, Surtees I.ii.8, Grieve s.v. *gorse*.
(3) ~ Northum NT 9648. *Morton* 1312, *Murton* 1384. The moor-settlement of Ord township. NbDu 146.
(4) ~ NYorks SE 6552. *Mortun -ton* 1086–1295, *Murton* from 1391. Situated at the S edge of Murton Moor. YN 10.
(5) ~ FELL Cumbr NY 0918. P.n. MURTON NY 0720 + ModE **fell** (ON *fjall*). Murton is *Morton'* 1203–1585, *Murton juxta Lamplogh* 1335. The moor-settlement of Lamplugh. Cu 406.

(6) ~ FELL Cumbr NY 7524. *Murton Fell* 1863. P.n. MURTON NY 7221 + ModE **fell** (ON *fjall*). We ii.105.

MUSBURY Devon SY 2794. 'Mouse fortification', i.e. infested with field-mice. *Mvsberie* 1086, *-biri(a) -bery* 1166–1260×6, *Mu- Moseburi -biri(a) -bury* 1204–89, *Mousebur(gh)* 1281, 1356. OE **mūs**, genitive pl. **mūsa**, + **byriġ**, dative sing. of **burh**. The reference is to Musbury Castle Iron Age hill-fort at SY 2894. The name type recurs at Mouseberry SS 7616, *Mousbiri* 1330, referring to a tumulus, and there are Cornish parallels at Carloggas SW 8765, 9063, 9554, Co **ker** 'fort' + **logaz** 'mice'. But a pers.n. *Mūsa*, Co *Logot*, is also possible. D 646, 402, CPNE 152, Thomas 1976.96.

MUSCOATES NYorks SE 6880. 'Musi's' or 'the mouse-infested cottages'. *Musecote(s)* [1154–1333]16th, *Mouscotes* 1293. ON pers.n. *Músi*, genitive sing. *Músa*, or OE **mūs**, genitive pl. **mūsa**, + **cot**, ME pl. **cotes**. YN 65, SSNY 128.

Great MUSGRAVE Cumbr NY 7613. *Magna Musegrave* 1289, *Great Musgrave* from 1577. ModE adj. **great**, Latin **magna**, + p.n. *Musegrave* 12th–1335, *Musgrave* from 1203×27, 'mouse-infested wood', OE **mūs**, genitive pl. **mūsa**, + **grāf**. 'Great' for distinction from Little Musgrave NY 7513, *Musgrave parva* 1369, *Little Musgrave* 1712. We ii.60, L 194.

North MUSKHAM Notts SK 7959. *Nordmuschā* 1086, *-muscham* 1242, *Northmuscham -muskham* 1281. ME adj. **north** + p.n. *Muschā* 1086, *-ham* 1159×81–1241, *Muskeham* 1210, 1335, *Muskam* 1312, *Muscom* 1618, *Musc(h)amp* c.1150–1428 mostly (p), 'Musca's homestead', OE pers.n. *Musca* + **hām**, cf. the G p.n. Muschenheim, Kr Lich, *Muscanheim* 8th. 'North' for distinction from South MUSKHAM Notts SK 7957. Nt 191.

South MUSKHAM Notts SK 7957. *Sudmuscham* 1242, *Suthmusc(h)am* 1290. OE adj. **sūth** + p.n. Muskham as in North MUSKHAM SK 7959. Nt 191.

MUSTON Leic SK 8237. 'Mouse farm, settlement infested with mice'. *Mustun(e)* [1100×35]1333, 1154×89–1216×72, *-ton(e)* 1191–1610. OE **mūs** + **tūn**. Lei 146.

MUSTOW GREEN H&W SO 8774. *Muster Green* 1831 OS.

MUTFORD Suff TM 4888. Partly uncertain. *Mutford(a)* from 1086, *Muthford* 1264. The absence of *-o-* sps and the early date of the *- u-* sps tell against OE **(ġe)mōt** 'a moot, a legal assembly' + **ford**. Possibly an early example of ME *mūt(e)* 'a hunting pack' < OFr *muete* although this sense is not recorded for the OFr word before 13th cent. Possibly, therefore, the 'ford frequently used by huntsmen and their hounds'. DEPN.

MUTTERTON Devon ST 0205. Uncertain. *Motterton* 1571, *Mutterton* 1578. The evidence is too late for certainty but this might be the 'speakers' enclosure or estate', OE **mōtere**, genitive pl. **mōtera**, + **tūn**. D 562.

MYDDLE Shrops SJ 4623. Uncertain. *Mulleht* 1086, *Muthla* 1121, *Mudla* 1155, *Mudle* c.1220–1450, *-ley* 1291×2, *Mudele* 1234–1431, *Mudell* 1379, *Middell* 1272, *Middle* 1535 etc., *Myddle* from 1554. The forms are untypical for a compound in OE **lēah** 'a wood, a clearing' so that OE **(ġe)mӯthe-lēah** 'the wood or clearing at the junction of streams' is unlikely. Possibly OE ***(ġe)mӯthel** 'the little stream-junction': but there is no such landscape feature at the site today. Sa i.216.

MYLOR BRIDGE Corn SW 8036. *Mylor Bridge* 1745. Parish name Mylor, *(Ecclesia) Sancti Melori* 1258, 1308, *Mylore* 1562, *Myler* 1568, from the patron saint of the church, Mylor (Meler), venerated there, at Merther Mylo, Linkinhorne SX 3173, at Amesbury Wilts SU 1541, and in Brittany, + ModE **bridge**. Earlier *Ponsnowythe* 1562, *Penowith* 1699, Co **pons** + **nowyth**, which occurs translated as *New Bridge* 1597. PNCo 126, Gover n.d. 521, 523, DEPN.

MYNDTOWN Shrops SO 3989. 'The town of (Long) Mynd'. *Myntowne* 1577, 1641, *Mintown* 1695–1808, *Mindtown* 1803, 1808. P.n. *Munete* 1086, *(La) Munede* 1255–1431, *Mynde* 1431, PrW **mönïth** 'a mountain' as in The LONG MYND + ModE **town**. Sa i.217.

MYNE Somer SS 9248 → MINEHEAD SS 9746.

Afon MYNWY H&W SO 4717 → River MONNOW.

MYNYDDBRYDD H&W SO 2841 → WALTERSTONE SO 3425.

MYTCHETT Surrey SU 8855. 'The big corner of land'. *le Muchelesshete* 1340, 1461, *Medechowte*, *Mydchotelese* 1548, *Mitchet* 1765, *Mitchett* from 1842. OE **miċel** + **scēat**, with ModE **leaze** 'a meadow' in 1548. Situated in the extreme SW angle of Frimley parish. Sr 127.

MYTHOLM WYorks SD 9827. '(The place at) the river mouths'. *the Mythome, My- Mithom* 17th cent., *Mytholm* 1771. OE **(ġe)mӯthe**, dative pl. **(ġe)mӯthum**. The reference is to the confluence of Colden Water and the r. Calder. YW iii.189.

MYTHOLMROYD WYorks SE 0126. 'The clearing at or near or belonging to *Mithum*'. *(le) Mithomrode* 1286–1323, *Mi- Mythom(e) Royd(e) -roid* 1478–1709, *Mytholmroyd* from 1622. OE ***rod, *rodu** which frequently appears as *roid* in the dial. of the southern half of WYorks, + p.n. *Mithum* identical with MYTHOLM SD 9827 referring to the confluence of Cragg Brook and the r. Calder. YW iii.159 gives prs ['maːðəm-'maiðəmroid].

MYTON-ON-SWALE NYorks SE 4366. P.n. Myton + r.n. SWALE SE 2796. Myton, *(æt) nyðtune* (for *myðtune*) [972]11th S 1453, *mytun* [972]16th ibid., *Mitune* 1086, *Myton(e)* from c.1100, *Mitton* 1247–1406, is the 'settlement at the rivers' meet', OE **mӯthe** + **tūn**. Situated ½ mile from the confluence of the Swale and the Ure. The White Battle of *Mytoun* mentioned by Barbour 1319, was fought on land between the two rivers. YN 23 gives pr [mitən] with the shortened vowel reflected in the 1247–1406 spellings.

NABURN NYorks SE 6045. Partly uncertain. *Naborne* 1086–1538, *Naburn(e)* from 1167, *Neiburn* 1200, *Naubourne* 1370. The generic is OE **burna** 'a stream' referring to a small tributary of the Ouse at this place. Four alternatives have been suggested for the specific: ON **ná-** 'near', ON **nár** 'a corpse', OE **nafu** 'nave, hub, bend', and ONb ***naru** 'narrow'. YE 274, StN 10.103, SSNY 258.

NACKINGTON Kent TR 1554. 'The hill called after Næta or called or at *Nating*, the wet place'. *Natindone* [c.850]13th S 1239, *(æt) Natincgdune, (to) Natyngdune -ing-, (suþ) Natingþune* [993 for 996]14th S 877, *Natinduna* c.1100, *-don* 1200–40, *Natingdon* 1240, 1247, *Nakynton* 1415. OE pers.n. **Næta* + **ing**^{4} or p.n. **Nating* < OE **næt** 'wet' + **ing**^{2}, + **dūn**. PNK 544, KPN 348, Charters IV.97.

NACTON Suff TM 2240. 'Hnaki's estate'. *Nachetuna* 1086, *-ton* 1165, *Naketon(')* 1232–1346, *Nacton* from 1524. ON pers.n. *Hnaki*, genitive sing. *Hnaka*, + **tūn**. DEPN, Baron.

River NADDER Wilts SU 0130. 'The flowing'. *Noodr* [705]12th B 114, *(be, on, to) Nodre* [860]15th S 326, [901]14th S 354, [956]14th S 630, *(æt, of, andlang) Noddre* [937]14th S 438, [956]14th S 631, 1255–1327×77, *Nadder* 1540. A Brit r.n. identical with Gk *νάτωρ, ναέτωρ* 'flowing' on the root of *νάω* 'to flow', IE **(s)nā-* . Cf. Skr *nadī* 'river'. Wlt 9, RN 297.

NADEN RESERVOIR GMan SD 8516. R.n. Naden + ModE **reservoir**. Naden, *Naueden* c.1300, *Naueden(e)* 1323–4, *Neuedene* 1325, is 'peak valley', OE **nafu** 'nave, hub (of a wheel)' + **denu**. OE *nafu* must have been used like cognate ON *nǫf*, genitive sing. *nafar*, in the sense 'projecting peak', here with reference to the 1378ft. conical peak of Knoll Hill SD 8416. Norn *nov* <ON *nǫf* and *niv* < ON *nef* 'nose' <**nafja-*, are used in exactly the same way in p.ns. of a high point of land. Cf. also the cognates ON *nebbi, nabbr*, dial. *nab* and OE *nebb* 'beak'. La 60.

NAFFERTON Humbs TA 0559. 'Nattfari's farm or village'. *Nadfartone* 1086, *Natferton* 1180×90, *Nafretune -ton* [before 1080]15th, 1202–1363, *Nafferton* from 1202. ON pers.n. *Náttfari*, genitive sing. *Náttfara*, + **tūn**. The topography does not support the suggestion that the specific is a p.n. **Nadford* as in NAFFORD H&W SO 9441 from OE **nēatford* 'a cattle ford'. YE 94, SSNY 128.

NAFFORD H&W SO 9441. Probably the 'cattle ford'. *Nadford* 1086, *Nasford* [c.1220]15th, *Nafford* 1290. OE **nēat** + **ford**. An alternative possibility might be 'Nata's ford', OE pers.n. *Nata*, genitive sing. *Natan*, + **ford**. Wo 188.

NAILSEA Avon ST 4770. 'Nægel's island'. *Nailsi* 1196, *Naylesye* c.1300. OE pers.n. **Næġel*, genitive sing. **Næġles*, + **īeġ**. DEPN, L 38.

NAILSTONE Leic SK 4107. 'Nægl's farm or village'. *Nay- Nail(l)eston(e) -is-* c.1200–1531, *Nay- Nailston(e)* 1392–1576, *Ney- Neileston'* 1224–1362, *Ney- Neilson* 1513–1617, *Nel(l)son* 1509–1714. OE pers.n. **Næġl*, genitive sing. **Næġles*, + **tūn**. Lei 520.

NAILSWORTH Glos ST 8499. 'Nægl's enclosure'. *Neilesw'nda* (sic) c.1170, *Nai- Nay- Nei- Neylesw(o)rth(e)* 1220–1434. OE pers.n. *Næġl*, genitive sing. *Næġles*, + **worth**. A *Negles leag* [716×45]11th S 103, 'Nægl's wood or clearing', occurs in the bounds of adjacent Woodchester. Gl i.102.

NANCEGOLLAN Corn SW 6332. 'Whetstone valley'. *Nansecolon -en* 1327, 1350, *Nannsygollen* 1356, *Nansegollan* 1620. Co **nans** 'valley' + ***ygolen** 'whetstone'. PNCo 126 gives pr [nansi'gɔlən], Gover n.d. 590.

NANCLEDRA Corn SW 4936. Probably 'valley of Clodri'. *Nanscludri* 1302, 1343, *-cludry* 1327, *Nicledera* 1699, *Nancledry* 1748. Co **nans** + pers.n. *Clodri* or *Cludri*. PNCo 127 gives prs. [nan'kledri] and ['kledri], Gover n.d. 665.

NANPANTAN Leic SK 5017. Short for 'Nan Pantain's house'. *Nan Pantain's* 1754, *Nanpantam* 1831. The original reference was to a single building. Lei 380.

NANPEAN Corn SW 9656. 'Little valley'. *Nanspian* 1332, *-pyan* 1380. Co **nans** + **byghan** 'small' with unvoicing of *b* > *p* after *s*. PNCo 127, Gover n.d. 424.

NANSTALLON Corn SX 0366. Probably 'valley of the river Alan'. *Lantalan* 1201, 1412, *Lantallen* [1284]1602, *Nanstalen* 1392, *-an* 1513. PrCo ***nant** (Co *nans*) + r.n. *Alan* (the original name of the river Camel SX 0168), later reinterpreted as **nans** + pers.n. *Talan*. Interchange between original **nant, nans* and **lann* is common to all the Brittonic countries. PNCo 127, Gover n.d. 100, Jnl 1.50, CPNE 143.

NANTMAWR Shrops SJ 2524. 'Great valley'. *Nant mawr* 1837 OS. W **nant** 'a glen, a brook' (PrW **nant* 'a valley') + **mawr** (PrW **mōr*).

NANTWICH Ches SJ 6552. 'The famous (salt-)works'. *Nametwihc* 1194, *-wich -wych(e) -witch -wick* 12th–1775 often preceded by the definite article *le, la, Nantwich* from 1257†. ME **named** 'well-known, renowned' + p.n. *Wich* 1086–1527 with variants *Wic(he), Wyke* etc., OE **wīċ** 'a collection of buildings for a particular purpose', in this case the manufacture of salt. The W name for Nantwich is *Nant yr Heledd Wen* 'stream of the white salt-pit' alluding to the famous whiteness of its salt. Also known as *Wicus Maubanc,* ~ *Malbanc* 12th–16th with many variants, from the surname *Malbanc* of the post-Conquest barons of Nantwich. Not identical with the Roman *Salinis* for which see MIDDLEWICH SJ 7066. In DB there were eight salthouses, Leland speaks of over 300. By 1624 there were still about a hundred, but thereafter the industry declined and ceased by the mid 19th cent. Che iii.30 gives prs ['næn?witʃ], ['næntwitʃ] and older local [-weitʃ, -waitʃ], Pevsner-Hubbard 1971.285.

NAPHILL Bucks SU 8596. 'Hill called Knap or with a summit'. *Knaphill* 1599, *Napple* c.1825. ModE **knap** (OE *cnæp*). The land here rises to a summit of 587ft. Bk 184 gives pr [næpəl].

NAPPA NYorks SD 8553. Partly uncertain. *Napars* 1086, *Nappai -ay(e) -ey* 1182×5–1649. Possibly OE **hnæpp** 'a bowl' in the topographical sense 'something bowl-shaped' + **(ġe)hæġ** 'an enclosure'. YW vi.172.

NAPPA NYorks SD 9690. Partly uncertain. *Nappay -ey* 1251–1610, *Nappa* 1665. Identical with NAPPA SD 8553. YN 262.

NAPTON ON THE HILL Warw SP 4661. P.n. *(N)eptone* 1086, *Neptone* c.1154–76, *Napton(e)* from 1173, *Cnapton* 1235, 'the settlement at *Cnap*, the hill', OE p.n. **Cnæp* (OE **cnæpp** 'a prominent hill') + **tūn**, + pleonastic ModE **on the hill**. Napton is situated by a prominent isolated hill. The old name for the inhabitants, OE **Cnæp-hǣme* 'the people of *Cnæp*' survives in the forms *Nepemmeford, Nephembroc* 1206 refering to the brook that devides Southam and Napton from Myer

† This form comes from printed calendars and is suspect in view of the propensity of editors to transcribe ms *Wich Malbank* as *Nantwich*. The first incontestable instance cited is *Nantewich'* 1340.

Bridge southwards and a ford across it. The specific might be OE **hnæpp** 'a bowl' in a topographical sense 'inverted bowl', but *cnæpp* seems better. Wa 140, PNE i.254.

River NAR Norfolk TF 2812. *Nar, Narra Fluvius* [before 1641]1723. A back-formation from NARBOROUGH TF 4712 probably invented by the antiquarian Sir Henry Spelmanm who first records it. RN 299.

NARBOROUGH Leic SP 5497. 'The northern fort'. *Norburg(h)* c.1200–1411, *North-* 1211–1444, *Narborow(e)* 1518–1610, *-borough* 1586. For *ŏ* > *ă* cf. RATBY SK 5105. Lei 522.

NARBOROUGH Norf TF 7412. Partly uncertain. *Nereburh* 1086, *-burg* c.1150–1254, *Narburgh', Nerburg'* 1242×3. Uncertain element + OE **burh** 'a fort' referring to the Iron Age fort in Camphill Plantation overlooking the river Nar. R.n. Nar, first occurring in 1641, is a back-formation from this name and from nearby Narford TF 7613, *Nereforda* 1086, *-ford* 1190–1250, *Nerford(ia)* 1176, 1195, possibly invented by Sir John Spelman in his *Icenia* published in 1723. Ekwall suggested OE **neru* 'a narrow place' for the specific < **narwīn-* on the base of adj. *nearu* 'narrow' like *lengu* 'length', *strengu* 'strength' beside *lang, strang*. Narborough is thus 'the fort at *Neru*, the narrow place', or perhaps better 'the fort beside the *Nere*, the narrow river', OE **Nere* < **narwjōn*. In either case the reference is to the narrowness of the river valley directly upstream from here. DEPN, RN 299.

NARE HEAD Corn SW 9136. *The Nare* 1690, *Nare Head* c.1870, earlier *Penare Point* c.1540. P.n. Pennare SW 9238 with loss of unstressed first syllable + ModE **head** 'a headland'. Pennare, *Pennarde* [c.1100]14th, *Penhart* 1238, *-hard* 1303, *-arth* 1309, is 'headland', Co ***pen-arth**. PNCo 127, Gover n.d. 486, DEPN s.n. Penare.

NARE POINT Corn SW 7925. *Penare Point* c.1540, *Nare Poynt* c.1605. P.n. Pennare SW 7924 with loss of unstressed first syllable + ModE **point** 'a promontory'. Penare, *penarð* [967]11th S 755, is 'headland', Co ***pen-arth**. PNCo 127, Gover n.d. 556, DEPN.

NARFORD Norf TF 7613 → NARBOROUGH TF 7412.

NASEBY Northants SP 6877. 'Hnæf's fortified place'. *Navesberie -u-* 1086, *Nauesbi -by* 1166–1475 with variant *Naves, Naseby* 1447. OE pers.n. *Hnæf*, genitive sing. *Hnæfes*, + **byriġ**, dative sing. of **burh**, later replaced by ON **bȳ**. Nth 73, SSNEM 14.

NASEBY FIELD Northants SP 6879. P.n. NASEBY SP 6877 + ModE **field** in the sense 'battle-field'. The site of a civil war battle between Charles I and Cromwell, 14th June 1645.

NASEBY RESERVOIR Northants SP 6677. P.n. NASEBY SP 6877 + ModE **reservoir**.

NASH Bucks SP 7834. '(At) the ash-tree'. *Esse* 1231, 1302, *Asshe* 1389, *Nassche* 1520. OE **æsc**. The modern form arose by misdivision of ME *atten ashe* (< OE *æt thǣm æsce*). Bk 71, L 219.

NASH H&W SO 3062. 'At the ash-tree'. *Essis* 1244, *La Ayse, (La) Asshe* 1287, 1307, *Nasche* 1239. ME **atten ashe**. He 173.

NASH Shrops SO 6071. '(The settlement at) the ash-tree'. *Fraxinus* (Latin) 1210×2, 1274(p), *(La) Esse* 1242–1346, *Esses, Asses, Frene* (French) 1255×6, *Ass(c)he* 1308–91, *Nazsche* 1391, *Nasshe* 1395, *The Nassh(e)* 1575–1654, *Nash* 1701. OE **æsc**. Initial *N-* is due to false analysis of ME *atten ashe* 'at the ash-tree' (OE *æt thǣm æsce*) as if *atte nash*. Sa i.218.

NASH LEE Bucks SP 8408 → Nash LEE.

The NASS Essex TM 0011. There are no early forms but this is presumably ultimately from OE **næss** 'a headland'.

NASSINGTON Northants TL 0696. 'The settlement on the promontory'. *Nassingtona* [1017×34]13th etc. with variants *-in- -yn(g)-* and *-ton, Nassintone* 1086, *Nessinton(e)* c.1152–1241. OE ***næssing** < **næss** + **ing**², + **tūn**. Nassington stands on a broad headland above the Nene. The complete absence of forms with medial *-e-* casts doubt on the earlier explanation from OE folk-n. **Næssingas*, genitive pl. **Næssinga* + **tūn**, the 'settlement of the Næssingas, the dwellers on the ness', whether this refers to the site of Nassington or the ness of Nassaborough Hundred on which Peterborough is situated. Nth 204, Årsskrift 1974.42.

NASTY Herts TL 3624. '(Place at) the east enclosure'. *Nastey* 1406. The clue to this name is given by the forms *Nasteyfeld* 1422×61, *Astyefilde* 1556, which show that the initial *N* is actually inorganic and transferred from the definite article in the ME phrase *atten Astey* for *atten ast hey*, OE *æt thǣm ēast hæġe*, **ēast** + **(ġe)hæġ**. Hrt 134.

NATEBY Cumbr NY 7706. 'Nettle farm or village'. *Nateby* from 1242. ON **nata** + **bȳ**. We ii.20 gives pr ['ne:tbi], SSNNW 36.

NATEBY Lancs SD 4644. 'Nettle village or farm'. *Nateby -bi* 1203, 1204 etc. ON **nata** + **bȳ**. Nettles were used for food, fodder, and medicine and in textile production. La 164, SSNNW 36, Jnl 17.96.

Up NATELEY Hants SU 6951. 'Upper Nately'. *Upnatelye* 1272, *-le(y) -lygh* 1274–1327. ME **uppe** + p.n. *Nataleie* 1086, *Nattelega* 1100×35, *Natelega -leg' -le(ye)* c.1195–1330, the 'wet wood or clearing', OE ***næt**, definite form ***nata**, + **lēah**. The soil and subsoil are clay. 'Up' for distinction from Nately Scures SU 7053, 'N held by the de Scures family' from Escures, Calvados, Normandy, *Neteleye Skures* 1272, *Nately Scures* 1327, *Skewers* 1817 OS. Ha 121, Gover 1958.127.

NATIONAL EXHIBITION CENTRE WMids SP 1983.

NATLAND Cumbr SD 5289. Probably 'nettle wood'. *Natalund* 1170×80, *Natelund* 1246–92, *-lond -land -laund* 1279–1446, *Natland* from 1170×80. ON **nata** + **lundr**. It has been suggested that the specific might be the rare ON pers.n. *Nati*, the name of a mythological giant, compounded wih **lundr** in the sense 'sacred grove', but this is very uncertain. We i.112, SSNNW 149.

NAUGHTON Suff TM 0249. Possibly 'Nagli's estate'. *Nawelton* c.1150, *Nauelton* 1191, *Navelton* 1254, *Naughton* 1610. ON pers.n. *Nagli*, genitive sing. *Nagla*, + **tūn**. The evidence is not unambiguously in support of this explanation and an alternative might be OE **Nafel-tun* < **nafela** 'a navel' + **tūn** on the assumption that *nafela* was used like *nafu* and *nafetha* to mean 'the nave of a wheel' as occasionally in late ME. Naughton might then be the settlement where naves were made. DEPN.

NAUNTON 'At the new settlement'. OE **nīwe**, definite form dative case **nīwan**, + **tūn**.

(1) ~ Glos SP 1123. *Niwetone -tune* 1086, 1185, *Newenton(e) -in- -yn-* 1235–1501, *Nawinton -en-* 1484–1625, *Nau- Nawnton* 1476–1683. Gl i.199.

(2) ~ H&W SO 8739. *Newentone* [c. 1120]17th, [1299]18th, *-yn-* 1275, 1375, *Nounton* [1182]18th, *Nawnton* 1584. A new settlement in the parish of Ripple. Wo 158.

(3) ~ BEAUCHAMP H&W SO 9652. 'N held by the Beauchamp family'. *Naunton Becham* 1563. Earlier simply *(in) niuuantune, nipan tun* 972 S 786, *Newentvne* 1086, *Niwenton, Newinton(e) -yn- -en-* 1166–1428, *Na- Newington* 1666, 1667. The Beauchamps who came from Beauchamps, Manche, held the manor from the end of the 11th cent. Wo 215.

NAVAX POINT Corn SW 5944. *Navax Point* c.1870. P.n. The Knavocks SW 5943 + ModE **point** 'a promontory'. Knavocks, *Navax or Knavocks* 1582 Gover n.d., *Knavocks* 1838, is possibly 'the summerland', Co definite article **an** + ***havek** (Co *haf* 'summer' + adjective suffix *ek*), with English *-s* added later. PNCo 128, Gover n.d. 585, CPNE 124.

NAVENBY Lincs SK 9857. 'Nafni's village or farm'. *Navenebi -u-* 1086–1328, *Navenbi -by* from 1185, *Navenesbi -by* c.1160–1246, *Naunebi* c.1200, *Naneby* 1332–1666, *Nain- Nawnby* 16th cent., *Namby* 1396. ON pers.n. *Nafni*, genitive sing. *Nafna*, + **bȳ**. Perrott 229, SSNEM 60.

NAVESTOCK HEATH Essex TQ 5397. *Navestock Heath* 1805 OS. P.n. Navestock + ModE **heath**. Navestock, *Nasingestok* *[867 ?

for 967]13th S 337, *Næsingstoc* [c.970]17th[†], [c.1000]c.1125, *Nasestocā, aliam Nessetocham* 1086, *Nastok(e)* 1234–c.1300, *Navestok(e)* 1283–1498, *Knavestock* 1749, is the 'outlying farm at or called *Nasing*, the headland place', OE p.n. **Næsing* < **næss** + **ing**², + **stoc**. Alternatively the specific could be NAZEING TL 4106 and the name mean 'outlying farm belonging to Nazeing'. Ess 69, ING 22, Studies 1936.27, YE lvi.

NAVESTOCK SIDE Essex TQ 5697. *Navestock Side* 1805 OS. P.n. Navestock as in NAVESTOCK HEATH TQ 5397 + ModE **side**.

NAWORTH CASTLE Cumbr NY 5662. *Castrum de Naward* 1365, *Naward, Naworth Castle* 1589, p.n. Naworth + ME **castel** (Latin **castrum**). Naworth, *Naward(e)* early 14th–1777, *Naworthe* 1323, 1534, *Naworth* from 1589, is 'narrow stronghold', OE **nearu** + **w(e)ard**. The reference is to the narrow ridge between deep valleys occupied by the castle. Cu 67 gives pr [na:wθ].

NAWTON NYorks SE 6584. 'Nagli's estate'. *Naghelton, Nageltone -tune, Nagletune* 1086, *Nagelton* 12th cent., *Nau-Nawelton* 1202–1298, *Nalton(a)* 1301–[1333]16th, *Nawton* from 1665. ON pers.n. *Nagli*, genitive sing. *Nagla*, + **tūn**. YN 65, SSNY 129.

NAYLAND Suff TL 9734. '(The settlement) at the island'. *Eilanda* 1086, *Eiland* 1167, *Leiland* 1234, *Neiland* 1227, *Neyland* 1610. ME *atte Neilande* < *atten Eiland*, OE *æt thǣm* **ēġlande**, dative sing. of **ēġland**. The change may particularly have occurred in the p.n. STOKE-BY-NAYLAND TL 9836, *Stoke atte Neylaunde* 1303. DEPN.

The NAZE Essex TM 2624. 'The promontory'. *the Na(i)sse* 15th, *the Na(y)se, Nesse* 1514–1671. Earlier *Eduluesnæsa* *[924×39]13th S 453, *(to) Eadolfesnæsse -ulf-* 11th ASC(C) under year 1049, ASC(D) under year 1050, *(to) Eadulfes næse* c.1150 ASC(E) under year 1052, *Ælduluesnasā* 1086, *Edulvesnasa -e -olv-, Adulves nasa* 12th cent., 'Eadwulf's promontory', OE pers.n. *Ēadwulf*, genitive sing. *Ēadwulfes*, + **næss**. The original reference was to the whole manor or soke which included Kirby-le-Soken TM 2222, Thorpe-le-Soken TM 1822 and Walton-on-the-Nase TM 2521. Ess 354, Jnl 2.45.

NAZEING Essex TL 4106. 'The Næsingas, the people who live at the *næss* or headland'. *Nesingan, (into) Nassingan* *[1062]13th, *Nasinga -ā* 1086, *Nas(s)inges, Nesing(es)* 1225–52, *Nasing* 1228–70, 1805 OS, *Naysinge* 1593. OE folk-n. **Næsingas* < **næss** + **ingas**. The situation is a hill-spur. Ess 25, ING 22.

Lower NAZEING Essex TL 3906. A modern name for the place called *Nassingbury* 1550, *Nasing Bury* 1805 OS. Earlier simply *le Bery* 1403, '(at) the manor house', OE **byriġ**, dative sing. of **burh**. Ess 26.

NEACROFT Hants SZ 1896. 'The new enclosure'. *la Newecroft* 1271×1307, *Necrofts* 1759. OE **nīwe, nīġe** + **croft**. An assart in the New Forest. Gover 1958.222.

NEAL'S GREEN Warw SP 3384. Probably named after the family of John Neel of Walsgrave 1327, 1332. Wa 109.

NEASHAM Durham NZ 3310. 'Homestead or village at the nose-shaped bend' of the river Tees. *Neshann* (sic) 1157, *Nesham* 1155×89–1803, *Neseham* 1382–1535, *Neceham* 1418–1629, *Neesam* 1549, *Neasam* 1598, *Neasham* from 1535. OE ***neosu** + **hām**. The first spelling is a mistake for *-ham*: it occurs in a Papal bull issued in Rome where English names would have been unfamiliar and is spelled *Nesham* in Henry II's confirmation. Pronounced [ni:səm]. NbDu 147.

NEATISHEAD Norf TG 3421. Partly uncertain. *Netheshird* [1020×22]13th S 984, *-hirda* *[1044×7]13th S 1055, *Snateshirda* 1086, *Neteshird(e) -herd(e) -hyrde* [1127×34]13th, 1150–1410, *Nettisheade* 1591. On the strength of the unique DB form this is usually explained as 'Snæt's household', OE pers.n. *Snǣt*, genitive sing. *Snǣtes*, + **hīred**. The loss of initial *S-* would be parallel to the loss of initial *S-* in the p.n. NOTTINGHAM < *Snotingaham*. However, if the DB spelling is ignored as an inverted spelling, this would be 'the retainer's household', OE **(ġe)nēat**, genitive sing. **(ġe)nēates**, + **hīred**. Nf ii.173, Forsberg 61n., Nomina 9.21, 27.

[†]Studies 1936 and DEPN give this form from *Bodl(James 23)* as *Naesingstoc*.

NECTON Norf TF 8709. 'The settlement at the neck (of land)'. *Neche- Neketuna* 1086, *Neketona* 1168. OE **hnecca** + **tūn**. Necton lies beside a ridge or 'neck' of land. DEPN.

NEDDERTON Northum NZ 2381. 'Adder farm or village'. *Nedertun* [c.1040]12th HSC, *Nedderton'* [1183]c.1320, *Nedirtona* [1183]c.1320, *Neterton* 1207, *Nodirton* 1307, *Netherton* 1865 OS. OE **nǣddre** + **tūn**. The alternative suggestion that the specific is OE **neothera** 'lower' is impossible since Nedderton stands on a hill. DEPN.

NEDGING TYE Suff TM 0149. 'Nedging common pasture'. *Nedging Tye* 1837 OS. P.n. *(æt) Hnyddinge* 1000×2 S 1486, *Neddinge* [1042×66]12th S 1051, *Niedingā* 1086, *Neoddinge* [1086]c.1180, *Nedding(g)e* 1121×3, *Nedding'* 1235, *Nedging* 1610, 'the place called after Hnydda or Hnyddi', OE **Hnydding*, SE form **Hnedding* < pers.n. **Hnydda*, **Hnedda*, or **Hnyddi*, **Hneddi*. + **ing**², + Mod dial. **tye** (OE **tīġ*). Cf. CHARLES TYE TM 0252. The modern form may be due to metathesis of [dʒ] in the locative–dative sing. form **Hnyddinġe*. DEPN, ING 198f.

NEEDHAM Norf TM 2281. 'Needy homestead'. *Nedham* 1352×3, 1428. ME **nede** + **hām**. DEPN.

NEEDHAM MARKET Suff TM 0954. *Nedham Markatt* 1388, *Nedeham Markett* 1511. P.n. *Nedham* 1286–1344, *Neidham* 1331, *Nedeham* 1610, possibly 'the needy, poor homestead or village', OE **nīed, nēod** 'need' + **hām**, + ME **market**. DEPN, Baron.

NEEDINGWORTH Cambs TL 3472. 'Enclosure of the Hnyddingas, the people called after Hnydda', or 'enclosure called or at Hnydding, the place called after Hnydda'. *Neddingewurda* 1161, 1163, *Nedingewrht* 1234, *Niddingeworth* 13th, *Nidding- Ny- Nidingworth* 1260–1417, *Nedyngworth -ing-* 1452–1535. OE folk-n. **Hnyddingas* < pers.n. **Hnydda* + **ingas**, genitive pl. **Hnyddinga*, or p.n. **Hnydding* < pers.n. **Hnydda* + **ing**², locative-dative **Hnyddinge*, + **worth**. Hu 209, ING 199.

The NEEDLES IoW SZ 2984. *Nedlen* 1333, *les Nedeles* 1409, *the Needles* 1525. ME **nedle** (OE *nǣdl*), pl. **nedlen**, later **nedles**. Three famous pointed rocks off the W point of the island. Wt 228, Mills 1996.73.

NEEDS ORE POINT Hants SZ 4397. *Needes Or Poynte* 1616, *Needs Oar Point* 1810. P.n. *Needes ore* 1585 + ModE **point**. Needs Ore is probably surname *Need* + ModE **ore** (OE *ōra*) 'shore, foreshore'. Gover 1958.202.

NEEDWOOD FOREST Staffs SK 1624. P.n. *Nedwode* 1198–1248, *Neydwode* c.1265, *Needwodde* 1540, OE **nēd** 'need, distress, poverty' + **wudu**, possibly a wood resorted to in need as a refuge by outlaws, + ModE **forest**. A hunting preserve, cf. *Chacia nostra de Nedwoode* 1248, between the rivers Trent, Dove and Blithe. DEPN, Horovitz.

NEEN SAVAGE Shrops SO 6777. '(The portion of) N held by the Savage family'. *Nene Sa(u)vage* 1255 etc., *Neen Savage* from 1305. P.n. *Nen* [?781×96]11th S 1185, *Nene* 1086 etc., the ancient name of the river Rea, of unknown origin and meaning identical with r. NENE Lincs-Cambs-Northants TF 4719–SP 6959, + manorial addition from the Savage family. Sa i.218.

NEEN SOLLARS Shrops SO 6672. '(The portion of) N held by the Solers family'. *Solers Nene* 1271×2, *Nene Solers* 1284–1320, *Neen Solars* 1305, *Nenesalers* 1577. P.n. *Nene* 1086–1242 as in NEEN SAVAGE SO 6777 + manorial addition from Roger de Solers who acquired the manor 1190×96. Sa i.219.

NEENTON Shrops SO 6388. 'The river Neen estate'. *Newentone* 1086, *-ton* 1221×2, *La Nienton(a)* c.1090, c.1200, *Nenton(')* 1242–1346, *Neenton* from 1362. R.n. Neen as in NEEN SAVAGE SO 6777 + OE **tūn**. There is confusion of the r.n. Neen with the dative sing. definite inflectional form *nīwan* of OE *nīwe* 'new'. Sa i.220.

NELSON Lancs SD 8637. A part of MARSDEN SD 8536 renamed from The Lord Nelson Inn. The name first appears on a map dated 1818, but in Cassell's *Gazetteer of Great Britain and Ireland* published in 1896 it is still referred to as *Nelson or Nelson-in-Marsden*. La 86 fn., Room 1983.75.

NELSON VILLAGE Northum NZ 2577. A modern development commemorating Lord Nelson.

NEMPNETT THRUBWELL Avon ST 5360. A combination of two separate names, Nempnet, *Emnet* c.1200, *Empnete* 1242, OE **emnet** 'a plain, a plateau, level ground' with prefixed *N-* from the ME dative form of the definite article **atten** (< OE *æt thǣm* 'at the'), and Thrubwell, *Trubewel* 1201, *Tribnelle* 1227, *Trubewelle* 1239, *Threbwell* 1299, probably 'gushing spring', OE ***thrybb** connected with ModE *throb*, + **wella**. DEPN

River NENE Northants SP 5959, TL 0385. Norf TF 4617. A pre-English r.n. *(of, on) Nyn* [948]12th S 533, [964]12th S 1566, *(on þa éa æt) Nýn* [964]12th ibid., *Nén* [972]12th S 787, *Nen* [1020×23]12th S 1463, [before 1085]12th, *(neah þære éá) Nén* [c.1025]11th, *Nene* from [c.1200]c.1260, *Neene, Nyne* 16th. The OE form was *Nēn* of unknown origin. *Nyn* has been taken to represent PrW **Nuin* < Brit **Nēn*, but the etymology of this form is unexplained. Furthermore the PrW sound change *ē* > *ui* is held to be too late to have occurred by the time of English settlement in this area unless an enclave of British speech is posited in the fens. Attempts to relate *Nēn* to IE **sneigʷh-* 'snow', cf. OIr *snigid* 'it is raining', Skr *sníhyati* 'is moist', *snéha* 'slime, grease', or **neigʷ-* 'wash', cf. OIr *nigid* 'washes', *necht* 'pure', νίζω 'wash', are guesswork. The name is certainly pre-English, possibly pre-Celtic. Nth 3 gives pr [nen], RN 299 [ni:n] and [nen], Chantraine 745.

NENTHEAD Cumbr NY 7843. 'Source of the river Nent'. *Nentheade* 1631, *Ninthead* 1705. R.n. Nent + ModE **head**. Nent, *Nent* 1314, is probably derived from PrW ***nant** 'a glen, a brook'. Cu 22.

NESBIT Northum NT 9833. 'The nose-shaped bend'. *Nesebit* 1242 BF, *-byt -bite* 1255, *Nesbyt(e)* 1255, 1296 SR. OE ***neosu** + **byht**. The reference is to a projecting ridge of hill. NbDu 147.

NESS 'Promontory, headland'. OE **ness, næss**. L 173.

(1) ~ Ches SJ 3076. *Nesse* 1086–1656, *Ness* from 1392. The reference is to the promontory at Burton Point SJ 3073. Che iv.220.

(2) ~ NYorks SE 6978. *Ne(i)sse* 1086 etc. The reference is to a prominent ridge between Holbeck and the river Rye. YN 51.

(3) Great ~ Shrops SJ 3919. *Ness Magna* 1311, *Greatenesse* 1553. ModE **great**, Latin **magna**, + p.n. *Nessham* (for *Nessam*, accusative case of Latin form *Nessa*) 1086, *Nessa* c.1090–1175, *Nesse* 1154–1341, 'the headland', OE **ness** referring to the ridge of Nesscliffe thought of as a headland or promontory jutting into marshland. 'Great' for distinction from Little NESS SJ 4119. Also known as *Nesse Extraneus* 1271×2, *Ness le Estraunge* 1311, *Straungeness(e)* 14th cent., *Ness(e) Strange* 1657 etc. from its possession by the *Lestrange* family in the mid 12th cent. Sa i.220.

(4) Little ~ Shrops SJ 4119. *Nesse Parva* 1284×5, *Little Nesse* 1635 etc. ModE **little**, Latin **parva**, + p.n. *Nesse* 1086 and *Nesse Alayn* 1271×2 from its possession by the *Fitz Alan* family. OE **ness** as in Great NESS SJ 3919. Sa i.221.

(5) White ~ Kent TR 3971. 'White headland'. *Whyte nasse* 1596, *White Ness* 1819 OS. ModE **white** + **ness**. Nearby is FORENESS TR 3871. Cullen.

(6) ~ POINT NYorks NZ 9606. Cf. *North Cheek or Bay Ness* 1857 OS referring to Robin Hood's Bay. ModE **ness** (OE *næss* or ON *nes*) 'a promontory' + **point**. Also called NORTH CHEEK NZ 9606.

NESSCLIFFE Shrops SJ 3819. 'Ness cliff'. *Nesscliff* 1833 OS. OE **ness** as in Great NESS SJ 3919 + **clif**.

NESTON Ches SJ 2977. 'Settlement at the ness or belonging to Ness'. *Nestone* 1086, *Nestuna* [1096×1101]1150, *Neston* from [1130×50]1285, *Nesson* 1596–1723. OE **ness, næss** 'a promontory or headland' or p.n. Ness + **tūn**. In the first explanation the reference is to Burton Point headland at SJ 3073, in the second to the nearby village of NESS SJ 3076. Che iv.222 gives pr ['nestən] and ['nessn].

NESTON Wilts ST 8668. 'The headland settlement'. *Neston* from 1282. OE **næss** + **tūn**. The settlement stands on an isolated hill which must be the *næss* in question. Wlt 97.

NETHERAVON Wilts SU 1449. 'The lower part of Avon'. *Nederauena* 1149×53, *Netheraven* 1212†. OE **neothera** + r.n. AVON. 'Nether' for distinction from UPAVON SU 1355. Wlt 331.

NETHERBURY Dorset SY 4799. 'Lower fortified place'. *Niderberie* 1086, *Nutherbir* 1226, *Nitherbury* 1285, *Netherbury* 1288. OE **neothera** + **byriġ**, dative sing. of **burh**. Do 112.

NETHERBY Cumbr NY 3971. 'Lower village or farm'. *Netherby* from 1279. ON **nethri** (or OE **neothera**) + **bȳ**. Netherby is the site of the Roman *Castra Exploratorum* 'camp of the Scouts' and it has been suggested that 'nether' refers to its location in contrast to *Blatobulgium* near Middlebie, Dumfries NY 2176, *Middeby* 1291, *Middelby* 1296, 'middle village or farm', OE **middel** + **bȳ**, and the Roman Camps at Burnswark, Dumfries NY 1878, supposedly *Overby*, but this is fanciful. Cu 53, SSNNW 36, 35.

NETHEREND Glos SO 5900. 'Lower end' sc. of Alvington. Contrasts with Nupend SO 6001, *Newpend* 1813, *Nupend* 1830, ultimately from ME **atten upp ende** 'at the upper end'. ME **nether** + **end**. Gl iii.250.

NETHERFIELD ESusx TQ 7018. 'Open land where snakes are seen'. *Nedrefelle, Nirefeld* 1086, *Ned(d)refeld -er-* c.1123–1332, *Neddres-* 1176, 14th, *Nadder- Nadrefeld* 1187–14th cent., *Nethirfeld* 1469. OE **nǣd(d)re** + **feld**. Sx 498.

NETHERHAMPTON Wilts SU 1129. 'Lower Hampton'. *Otherhampton* 1209, *Nother-* 1242, *Netherhampton* from 1249, *Netheryng(e)ton* c.1570, *Netherington* 1675, 1777. OE **neothera** + p.n. Hampton, OE **hāmtūn** 'farm'. 'Nether' possibly because S of the river with reference to Wilton. Wlt 217 gives former pr ['neðəriŋtən].

NETHER ROW Cumbr NY 3237. 'Lower row of houses'. *Netherraw* 1658. ME **nether** (OE *neothera*) + **raw** (OE *rāw*). 'Nether' in contrast to High Row NY 3535, *High Row* 1723. Cu 279.

NETHERSEAL Derby SK 2813. 'Lower *Scheyle*'. *Nether Scheyl(e)* 1216×72, ~ *Shayle* 1454. OE **neothera** + p.n. *Scel(l)a, Sela* 1086, *Scegla* c.1125, *Sceyle* c.1141, *S(c)heile -eyle* 1225–1396, *Seal* 1272×1307, 'the little copse', OE **sceġel**. 'Nether' for distinction from OVERSEAL SK 2915. Db 645.

NETHERTHONG WYorks SE 1309. 'The lower part of Thong'. *Netherthonge* 1448. ME adj. **nether** + p.n. Thong, *Thoying* 13th, *Thwonge(s)* 1313–1610, 'the narrow strip of land', OE **thwang**. 'Lower' for contrast with UPPERTHONG SE 1308. The reference is to a narrow strip of land varying in width from 50 to 100 yards and over 2 miles long linking the two parts of Thong township. YW ii.286.

NETHERTON 'The lower settlement'. OE **neothera** + **tūn**.

(1) ~ Devon SX 8971. *Nitherton* 1242, *Nythereton* 1269. The settlement lies at the bottom of a steep slope beneath Combeteignhead SX 9071. D 460.

(2) ~ H&W SO 9941. *(ad) neoþeretune* *[780]11th S 118, *Neotheretvne* 1086, *Netherton* 1256, *Noþerton* 1256. The boundary of the Netherton people is *neoðere hæma gemære* [1042]18th S 1396. N was parcel of the royal estate of Cropthorne, but it is 'lower' probably in respect of another member of the estate, Elmley which lies higher up on the lower slopes of Bredon Hill. Wo 150, Hooke 30, 362.

†The DB forms *Nigravre, Nigra avra* 1086 are aberrant.

(3) ~ Mers SD 3500. *Netherton* 1576. Possibly the 'nether town' in contrast to Sephton Town SD 3400 which is on a hill. La 118.

(4) ~ WYorks SE 2716. *Netherton* 1523. 'Lower' for contrast with MIDDLESTOWN SE 2617, both being parts of SHITLINGTON SE 2615, as the earlier forms show, *Schitelington inferior* 'lower S' 13th, *Nether Shut- Shet- Shit(e)lington -yng- -tun* 1326–1687. YW ii.206.

NETHERTON Northum NT 9807. Probably the 'adder farm or village'. *Netterton* 1207, *Ned(d)erton -ir-* 1242 BF–1428, *Nethreton* 1479. OE **nǣddre** + **tūn**. ME *d* before a vowel or syllabic *r* regularly became [ð] from 1400, Jordan §298, cf. NETHERFIELD Sussex TQ 7118. Alternatively the specific could be OE **neothera** 'lower'; Netherton lies downstream from Biddlestone and downhill from Burradon. But ME spellings with *-tt-*, *-d(d)-* are rare with this element. Cf. NEDDERTON NZ 2381. NbDu 147.

NETHERTOWN Cumbr NX 9907. 'Lower town'. *Nethertowne* 1720. 'Nether' in contrast to Middletown NX 9908. Both are probably named in relation to Egremont NY 0110. Cu 413.

NETHERWITTON Northum NZ 0990. 'Lower Witton'. ModE adj. **nether** + p.n. *Wittun* 1236 BF, *Witton* 1242 BF, the 'wood farm or village', OE **widu-tūn** (*widu* is the earlier form of *wudu*). Also known as *Witton cum le Schell* 'W with the shieling' 1296 SR, *Witton by the Water* 1379. 'Nether' and the other additions are for distinction from LONGWITTON NZ 0788. The shieling mentioned in 1296 is Witton Shields NZ 1290. Part of the 'wood farm' of Hartburn parish. NbDu 218.

NETLEY Hants SU 4708. 'Wood or clearing where laths are obtained'. *(æt) lætanlia* [955×8]14th S 1491, *Latelie* 1086, *-leghe* 1284, *Lettelege -le(gh) -ley(e)* 1239–1338, *Letele(ye) -lye -legh* 1241–1341, *Latteleye* 1256, *Netteley, Nettley* 1496. OE **lætt**, genitive pl. **latta**, + **lēah**. The *N-* forms may be due to the influence of the name NETLEY MARSH SU 3313. Ha 121, Gover 1958.39.

NETLEY MARSH Hants SU 3313. *Netley Marsh* 1759. P.n. Netley + ModE **marsh**. Netley, *Nateleg'* 1248, *-le* 1256–1411, is usually identified with *Natan leaga* 9th ASC(A), *Nazanleog* 1121 ASC(E), both under year 508. Although the Chronicle claims that the Netley district is named after the Welsh king Natanleod killed here by the West-Saxon leaders Cerdic and Cynric, the name can only mean the 'wet wood or clearing', OE ***næt**, definite form oblique case ***natan**, + **lēah**, dative sing. **lēaġe**. Ha 122, Gover 1958.200, DEPN.

NETTLEBED Oxon SU 7086. 'The nettle bed'. *Nettlebed* from 1246×7 with variants *Net(t)el-* and *-bedd(e) -byd*. OE **netel** + **bedd**. O 131.

NETTLEBRIDGE Somer ST 6448. *Nettle Br.* 1817 OS.

NETTLECOMBE Dorset SY 5195. 'Coomb where nettles grow'. *Netelcome* 1086, *Netlecumb* 1206, *Nettelcumbe* 1244. OE **netel** + **cumb**. This is probably not just descriptive but a reference to nettles as a crop. Do 112, L 90, 93.

NETTLEDEN Herts TL 0210. 'Netley valley'. *Net(t)eleydene* c.1200, *Netteleyden* 1309, *Netleden* 1432. P.n. **Netley* 'nettle clearing' < OE **netel** + **lēah**, + **denu**. Hrt 48.

NETTLEHAM Lincs TF 0075. 'Homestead, village where nettles grow'. *Netelhā, in Etelehā* 1086, *Netelham* c.1100, 1166. OE **netel(e)** + **hām**. DEPN.

NETTLE HILL Cumbr NY 7107. *Nettle Hill* 1858. ModE **nettle** + **hill**. We ii.40.

NETTLESTEAD Kent TQ 6852. 'Nettle homestead'. *Netelamstyde* 871×89 S 1508, *Netlasteda, Netlestede* [c.975]copy B 1321–2, *Nedestede* 1086, *Netlested(e)* 1212–48. OE **netel** + **hāmstede**. KPN 227, Stead 221, ASE 2.31.

NETTLESTEAD GREEN Kent TQ 6850. 'Nettlestead common'. *Nettlested green* 1819 OS. P.n. NETTLESTEAD TQ 6852 + ModE **green**.

NETTLESTON IoW SZ 6290. 'Farm at or called *Nuteles*'. *Hoteleston(e)* (sic) 1086, *Nu- Notel(e)ston(e)* 1269–1413, *Netleston* 1352. P.n. **Nuteles*, OE **hnutu** 'nut-tree' + **lǣs** 'pasture' or **lēah** 'wood or clearing', genitive sing. **lēas**, + **tūn**. Wt 198, Mills 1996.74.

NETTLESTONE POINT IoW SZ 6291. *Nettles Heath Point* 1720. P.n. *Nettles Heath* < *Nettles*, a back-formation from NETTLESTON SZ 6290, + ModE **heath** + **point**. The form *Nettles Hythe* 1611, 'Nettles landing-place' shows that *heath* in this name is SE dial. for *hythe*. Wt 199, Mills 1996.74.

NETTLETON Lincs TA 1000. 'Farm, village where nettles grow'. *Neteltone* 1086, *-tuna* c.1115, *Netleton(')* 1175. OE **netel(e)** + **tūn**. DEPN, Cameron 1998.

NETTLETON Wilts ST 8278. 'The nettle settlement'. *Netelin(g)tone* [944]14th S 504, *(at) Netelingtone -yng-* [956]14th S 625, 1289, 1305, *Netelinctona* 1189, *Niteletone* 1086, *Netleton -el-* 1197–1331. OE ***neteling** 'a nettle place, a place growing with nettles' < OE **netele** + **ing²**, + **tūn**. Nettles may have been cultivated here for food or medicinal use or mark a deserted settlement with enriched soil. Wlt 80, Årsskrift 1974.49.

NEVENDON Essex TQ 7390. Possibly 'Hnefa's valley'. *Nezendenā* 1086, *Nevenden(e)* 1218–1322, *Neweddon(a)* 1222, 1255, 1412. OE pers.n. **Hnefa*, genitive sing. **Hnefan*, + **denu**. This pers.n. would correspond to OHG *Nebi, Nefi*, ON *Hnefi* and be related to OE *Hnæf*, all from the root of ON *hnefi* 'fist'. An alternative suggestion is that the initial *N-* comes from ME *atten* in the phrase *atten efen dene* (OE *æt thǣm efnan dena*) 'at the level valley' which fits the topography here. Ess 164 gives pr [nɛːndən], DEPN.

NEWARK 'The new building', sometimes referring to new fortifications. OE **nīwe** + **(ġe)weorc** or ON **nȳ** + **virki**.

(1) ~ Cambs TF 2100. *Nieuyrk* [1189]1332, *Newerc -k* 1227, 1308, *Newewerk* 1330. Nth 227, lii.

(2) ~ -ON-TRENT Notts SK 7953. P.n. Newark + r.n. TRENT. Newark, *Newarcha* [1054×7]12th S 1233, *Niweweorce* [c.1080]c.1195, *Neuuerche, Newerc(h)e* 1086, *Newerk* 1242, is the 'new fortification'. The date of the 'new work' is unknown but pre-Conquest. 'New' is probably merely descriptive rather than for distinction from *Aldewerch* c.1230, *Aldewerke* 1289, *Old-work* 1722, the site of the RBrit fort of Margidunum at Castle Hill SK 7041. Both settlements lay on the Foss Way. Nt 199, 222.

North NEWBALD Humbs SE 9136. *North New(e)bald* from 1348. ME adj. **north** + p.n. *(to) msebotle* (for *niwebotle*) [963]14th S 716, *(æt) neowe boldan* [972×92]11th S 1453, *Niuuebold, Niwebolt* 1086, *Neu- New(e)bald* from 1285, 'the new building', OE **nīwe** + **botl, bold**. 'North' for distinction from South Newbald SE 9135, *(Suth)neubald* [1201×4]c.1300. The sequence *-old-* was a rare sound-group in Northern ME and the unusual *-bald* spelling is probably by analogy with Northern *ald* 'old' for Southern *old*. YE 226 gives pr [niuboːd], 227.

NEWBIGGIN 'New building'. ME **new** + **bigging**.

(1) ~ Cumbr NY 4729. *Neubighyng super Stainton* c.1200, *Neubiggyng -inge* 1278, 1288, *Newbiggin* 1610, *Newbiken* 1670. Cu 188 gives pr [nibikən], SSNNW 204 no.2.

(2) ~ Cumbr NY 5549. *Niewebigginge* 1202, *Newbegin* 1603. Cu 184, SSNNW 204 no.1.

(3) ~ Cumbr NY 6228. *Neu- Newbig(g)ing(e) -y-* 1179–1543, *Newbiggen* 1365, *-biggin* 1704. We ii.126, SSNNW 204 no.3.

(4) ~ Cumbr SD 2669. *Neubygging, Newebigginge* 1269. La 208, SSNNW 205 no.8.

(5) ~ Durham NY 9127. *Neubinghinge* (sic) [1333]16th., *-bigging -yng* 1372–1461, *Newbigging in Teasdale* 1512, *Newbyggyn* 1564. NbDu 148.

(6) ~ NYorks SD 9985. *Neu- Newbigging(e)* c.1230–50. YN 268.

(7) ~ NYorks SD 9591. *Neubigging* 1228 etc. YN 262.

(8) ~-BY-THE-SEA Northum NZ 3087. P.n. *Niwebiginga* 1187, *Neubigging* 1242 BF, 1268, *-bighing* 1296 SR, + ModE **by the sea**. NbDu 148, DEPN.

(9) ~ COMMON Durham NY 9131. P.n. NEWBIGGIN NY 9127 + ModE **common**.

(10) ~ -ON-LUNE Cumbr NY 7005. *Newbig(g)in -byg(g)in* from [1154×89]1645, *Nubegin(e)* 1647. We ii.32, SSNNW 204 no.6.

NEWBOLD 'New building'. OE **nīwe** + **bold**.

(1) ~ Derby SK 3773. *Newebold* 1086, *Newbold* from 1230 with variants *-bald* 1234–1351, *-bolt* 1243–1310, *-baud* 1226–69. Db 277.

(2) ~ Leic SK 4019. *Neubold'* 1212–1495, *Newbold(e)* 1254–1572. Lei 422.

(3) ~ ON AVON Warw SP 4976. *Neubold on Avene* 1349. P.n. *Neo- Neabaldo* 1077, *Newebold* 1086–1604 with variants *Niwe- Neu- New-*, + r.n. AVON. Wa 115.

(4) ~ -ON-STOUR Warw SP 2446. *Newebold super Stoure* 1364, 1392. P.n. *Nioweboldan* (dative pl. form) [991]11th S 1366, *Neubold* 1208, *Nobold* 1695, + r.n. STOUR. The *-an* of the earlier form shows the original sense was 'at the new buildings'. Wo 173 gives pr [noubəld].

(5) ~ PACEY Warw SP 2957. 'N held by the Pacey family'. *Neubold Pacy* 1235. P.n. *Niwebold* 1086, *Nevbald* 1185, *Neubold* 1221, *Newbold* 1247, + manorial addition from the family of Robert de Pasci who held the manor in 1221, and Adam de Pasci in 1199×1216. Wa 257.

(6) ~ VERDON Leic SK 4403. 'N held by the Verdun family'. P.n. *Ni- Newebold* 1086, *Neubold* 1274–1462, *Newbold(e)* 1226–1610, + affix *Verdo(u)n* 1318–1610 from Nicholas de Verdun who held the manor in 1226. Lei 523.

NEWBOROUGH Cambs TF 2006. *Newborough* 1823. ModE **new** + p.n. Borough as in Borough Fen, *Burgfen* 1307, *Boroughe Fenne* 1557, *Peterborowe Fen* 1614. Borough or Peterborough Fen was an extraparochial district of fenland N of Peterborough. The parish of Newborough was formed out of it in 1822, the remainder becoming known as Oldborough. Nth 239, 231.

NEWBOROUGH Staffs SK 1325. 'The new borough'. *Neuboreg* 1280, *Neuburgh* 1327. ME **new** + **burgh**. A new borough created by Robert de Ferrers in 1263 in a place previously called *eadgares lege* [1008]13th S 920, *Edgareslege* 'Eadgar's wood or clearing' 1086, OE pers.n. *Ēadgār*, genitive sing. *Ēadgāres*, + **lēah**. The name survives in Agardsley Park SK 1327. DEPN, Duignan 106, Charters II.59.

NEWBOTTLE Northants SP 5237. 'New building'. *Neubote, Niwebotle* 1086, *Neubotha* 1121×9, *Neubotl(e)* 1148×66–1331 with variants *New(e)- Niwe-* and *-bottle*. OE **nīwe** + **bōtl**. Nth 56.

NEWBOURN Suff TM 2742. 'The new stream'. *Neubrunna* 1086, *Neubrounia* [c.1160]1331, *Neubrunne -broun(e)* 1254–1316, *Neuburne* 1291, 1332, *New(e)bourn(e)* from 1327. OE **nīwe** + **burna** influenced by ON **brunnr**. DEPN, Arnott 15, Baron.

NEWBRIDGE Corn SW 4231. *New Bridge* 1839. A 19th cent. hamlet at a bridge. PNCo 128.

NEWBRIDGE Hants SU 2915. No early forms. ModE **new** + **bridge**.

NEWBRIDGE IoW SZ 4187. 'The new bridge'. *Newbryge* 1378. ME **newe** + **brigge**. The road to Calbourne crosses the Caul Bourne here. Wt 211 gives pr ['nu:bridʒ], Mills 1996.74.

NEWBROUGH Northum NY 8767. 'The new borough'. *Nieweburc* 1203, *Neweburgh* 1256, *Neuburgh* 1329, *Newbrough* 1542. ME **newe** + **burgh**. A new borough founded by the Cumin family, the grant for a market at Thornton being made in 1221. NbDu 148, DEPN, Tomlinson 150.

NEWBUILDINGS Devon SS 7903. *New Buildings* 1809 OS.

NEWBURGH Lancs SD 4810. 'The new borough'. *Neweburgh* 1431, *Newburgh* 1529, *Newborow* c.1540. ME **newe** + **burgh**. La 123.

NEWBURGH PRIORY NYorks SE 5476. P.n. Newburgh + ModE **priory**. Newburgh, Latin *Nouo Burgo* 1199, *Newburg* c.1250, *Neuburgh* [13th]15th, is the 'new fort', OE **nīwe** + **burh**. YN 192.

NEWBURN T&W NZ 1965. 'The new fortification or borough'. *Nieweburc* 1204, *Neuburgum* 1281, *Nieweton* 1206, *Nyweburne* c.1175, *Neuburn(e)* 1203–96. OE **nīwe** + **burh**. There is interesting suffix variation between *burh* and *tūn* in this name before it became fixed, revealing its original status as an appellative before it became a name proper. There are said to be traces of ancient fortification here, but the place was refounded as a royal borough in the 12th cent. For the subsequent change *burh* > *burn*, cf. SOCKBURN NZ 3407. NbDu 149, Tomlinson 92.

NEWBURY Berks SU 4767. 'The new market-town'. *Neuberie* [c.1080]n.d., *Niweberiam* 1094×1100 etc. with variants *Newe-* and *-beri -bir' -bur(i) -bury -byr'*, *Nubre* 1401. OE **nīwe** + **byriġ**, dative sing. of **burh**. Newbury was a new borough founded by the Domesday tenant-in-chief Arnulf de Hesdin at the point where the Oxford-Winchester road crosses the river Kennet. The earlier name of the manor was *Ulvritune* 1086 'estate called after Wulfhere'. Brk 257, 259, TC 140.

NEWBY 'New village or farm'. OE **nīwe** + ONY **bȳ**. SSNY 7.

(1) ~ Cumbr NY 5921. *Neubi -by* 1158–1631, *Newby(e)* from 1256. We ii.146, SSNNW 37 no.3.

(2) ~ NYorks NZ 5012. *Neubie -by* [c.1236]15th–1463. YN 169.

(3) ~ NYorks SD 7270. *Neubi -by* 1154×89–[1267]copy, *Newby* from [1165×77]copy, *Nuby -bie* [1210×30]copy, 1642. YW vi.233.

(4) ~ BRIDGE Cumbr SD 3786. *New bridge* 1577, *Newbybridge* 1659. Probably not a genuine *newby*. Refers to the bridge over the river Leven and possibly the specific is a local family name *Newby*. La 217.

(5) ~ EAST Cumbr NY 4758. *Neubi* c.1175, *Neuby* c.1190–1363, *Newby* 1399. E of Carlisle and for distinction from NEWBY WEST NY 3653. Cu 92, DEPN, SSNNW 37 no.2.

(6) ~ HALL NYorks SE 3467. P.n. NEWBY + ModE **hall**. An early 18th cent. brick mansion. Newby is *Neuby* 1170×80–1444, *Neweby* 1166, *Newby* from 1452. YW v.152, Pevsner-Radcliffe 1974.375.

(7) ~ MOSS NYorks SD 7472. P.n. NEWBY + **moss** 'a bog'. YW vi.235.

(8) ~ WEST Cumbr NY 3653. *Neuby* before 1211, *Newby* from 1424. W of Carlisle and for distinction from NEWBY EAST NY 4758. Cu 130, SSNNW 37 no.1.

(9) ~ WISKE NYorks SE 3687. P.n. *Neuby* [1157]15th + r.n. WISKE NZ 4300, SE 3497, ~ *super Wisk* 1285. YN 275.

NEWCASTLE ME **newe** + **castel**.

(1) ~ Shrops SO 2482. *Novum Castrum* 1284. The reference is to a Norman motte SE of the village. DEPN, Raven 146.

(2) ~ AIRPORT T&W NZ 1971. P.n. Newcastle as in NEWCASTLE UPON TYNE NZ 2464 + ModE **airport**.

(3) ~ -UNDER-LYME Staffs SJ 8546. *Nouum Oppidum sub Lima* 'new town under L' 1168, *Novum castellum subtus Lymam* 1173. P.n. Newcastle + p.n. Lyme as in LYME PARK Chesh SJ 9682. The castle is first referred to as *novum castellum de Staffordshira* 1149. Also known as *Novum Castrum super Are* 'N on Are' (Latin), *Nef Chastel sus Are* (French) and *Newcastle super Are* 14th in which *Are* is probably the old name of Lyme Brook, *Lyme Brook* 1686, itself a back-formation from LYME SJ 9682. *Are* is probably identical with the r.ns. ORE Suffolk TM 3845, OARE Somer SS 8047, *Are* 1086, and Ayr, Scotland NS 5826, *Ar* 1177, *Are* 1197, *Air* c.1300, from Old European **Ara* 'water' as in the continental r.ns. Ahr, a tributary of the Rhine near Remagen, *Ara* 975, Ahre or Aar, a tributary of the Orke near Frankenberg, Aare, a tributary of the Rhine in Switzerland near Berne, *Area* 1115, Ara, a tributary of the Cinca in NE Spain etc. DEPN, BzN 1957.228, Berger 31, TC 141.

(4) ~ UPON TYNE T&W NZ 2464. The earliest forms are Latin, *Novum Castellum super Tinam* 1168, *Novum Castrum super Tynam* 1296, and simply *Novum Castellum* 1130–1226×8. The castle was begun by Robert Curthose, son of William the

Conqueror, the present keep belonging to the period 1172–7. Earlier names for the city or its site are *Munecaceastre* c.1107, 'the monks' Roman fort', OE **munuc**, genitive pl. **munuca**, + **ċeaster**, referring to the Roman camp where the three monks from Winchcombe tried to revive monastic life in the North East in the 11th cent. and specifically identified by Symeon of Durham with the later *Novum Castellum*; and *Pons Aelii* or *Aelius*, 'Hadrian's bridge', *Aelius* being the gentile name of the Emperor. NbDu 149, DEPN, RBrit 441.

NEWCHAPEL Staffs SJ 8654. *Newe Chappell* 1611. The name replaced earlier *Turvoldesfeld* 1086, *Thurfredsfeld* 1212, *Thuredesfeld* 1227, *Thuresfeld* 1253, *Thursfield als. New Chap.* 1747, 'Thorvaldr or Thorfrethr's open land', ON pers.n. *Thorvaldr* varying with *Thorfrethr* + OE, ME **feld**, as in Thursfield Lodge SJ 8655. Horovitz, SPN 396, 402.

NEWCHAPEL Surrey TQ 3642. *the new chapel* 1534. ModE **new** + **chapel**. A chapel attached to the manor of Hedgecourt (Sr 322) and apparently replacing the one recorded there in 1365. Sr 319.

NEWCHURCH IoW SZ 5686. 'The new church'. *Niechirche* 1135×54, *Niucherche* 1171, *Ni- Ny- Newecherche -chirche -churche* 1228–1414, *Neuchirche* 1255–1476, *Newchurch* 1510. OE ***nīġe, nīwe** + **ċiriċe**. The church was founded in 1087 or slightly earlier. Wt gives pr ['niutʃərtʃ], Mills 1996.74.

NEWCHURCH Kent TR 0531. 'The new church'. *Nevcerce* 1086, *Niwancirce* c.1100, *Newechirch(e)* 1198, 1235. OE **nīwe**, definite declensional form oblique case **nīwan**, + **ċiriċe**. A new church related to the drainage of Romney Marsh. The earliest fabric of the present building is 13th cent. There are ruins of an earlier church at Eastbridge and a chapel at Orgarswick. PNK 470, Gelling 1981.4–9.

NEWCHURCH IN PENDLE Lancs SD 8239. No early forms, but the chapel was created in 1529 and the church has 16th cent. fabric. The township was originally a vaccary known as Goldshaw Booth, *Goldianebothis* [for *Goldiaue-*] 1324, *Goldiaue* 1325, *Goldeshagh* 1459, *Nethir- Overgoldshagh* 1464, *Nether- Overgouldshey* 1502, 'Goldgeofu's booths or temporary shelters', OE pers.n. *Goldġeofu*, genitive sing. *Goldġeofe*, + ODan, ME **bōth(e)**. Alternatively the vaccary itself might have been called *Goldġeofu* 'the gold-giver' like the complimentary field names *Land of Promise, Fill Barns, Make Me Rich* etc. The modern form *Goldshaw*, a hypercorrection from *Goldshay* as if from OE **sceaga**, probably developed as a result of the assimilation of *dj* in the form Goldiaue to [dʒ]. For Pendle see PENDLE HILL Lancs SD 7941. Pevsner 1969.182, La 80, Mills 115, Jnl 17.50.

NEWCOTT Devon ST 2308. 'New cottage(s)'. No early forms.

NEWDIGATE Surrey TQ 1942. '*New wood* gate'. *Niudegate* c.1167, *Ne-* c.1167–1428, *Niwudegate* 1241, *Niwode-* 1263, 14th cent., *New(e)de- Niuede- Niwede-* 13th cent., *Nudygate alias Nydgate* 1552, *Newdigate* from 1723. ME ***niwude* 'new wood' (OE *nīġe* + *wudu*) + **gate**. The wood is probably that recorded c.1350 as *Nywode, Newode*. Sr 83–4, PNE ii.50.

NEWELL GREEN Berks SU 8871. *Newwell Green* 1761, *Newell Green* 1800. Probably surname Newell + ModE **green**. Brk 118.

NEW END H&W SP 0530. *New Inn* 1831 OS.

NEWENDEN Kent TQ 8327. '(Place) at the new pasture'. *Neuuenden* [1072]copy, *Newedene* 1086, *Ni- Newe(n)den(ne)* c.1100–1240. OE **nīwe**, definite declensional form oblique case **nīwan**, + **denn**. PNK 342.

NEWENT Glos SO 7226. 'New settlement'. *Noent* 1086, 1240, *No(u)went* 1167, 1199, *Newent* from 1221. A British name, identical with the common French p.n. Nogent < *Novientum*, Gaulish **novios* + suffix **-entum*. Gl iii.173.

NEW FERRY Mers SJ 3485. *New Ferry* 1840 OS. 'New' for distinction from ROCK FERRY SJ 3386.

NEWFIELD Durham NZ 2033. 'The newly cultivated land'. *le Newfeld* 1382, *le Newefeld* 1418, *a parcell of land called Newfield or Hermeth Hugh lying upon Byers Moore* 1647. OE **nīwe** + **feld**. *Hermeth Hugh* 1647 is 'hermit hill-spur', ME **ermite** + **hōgh**, N dial. **heugh** (< OE *hōh*). NbDu 148.

NEW FOREST Hants SU 2806. *Nova Foresta* 1086, c.1115, 1231, *Noveforest* 1154, *The New Forest* 1608. Latin **nova**, ModE **new** + Latin **foresta** 'a royal hunting reserve', ModE **forest**. A wasteland extended by William I by destroying a number of villages so that it was new in his times as a legally expanded area under forest law. Ha 122, Gover 1958.8.

NEWGATE Norf TG 0543. *Newgate* 1838 OS. ModE **gate** probably in the sense 'hamlet' as in EASTGATE TG 1423.

NEWGATE STREET Herts TL 3005. 'N village'. *Neugatestret* 1177×99, *le Newgatestrete* 1468. P.n. *le Neugate* 'the new gate' 1371, ME **new** + **gate**, + **strete** in the sense 'village, hamlet'. The *gate* may have been an entrance into Hatfield Chase. Hrt 128.

NEW GROUNDS Glos SO 7205. *New Grounds* 1779. Earlier *lez Newegayned groundes* 1609 referring to land recovered from the marshy waste along the r. Severn. ModE **new** + **ground(s)**. Gl ii.249.

NEWHALL 'New hall'. OE **nīwe** + **hall**.

(1) ~ Ches SJ 6145. *Nova Aula* 1227–1430, *La Nouehall* 1252, 1378, *Newhall* from 1256×7 with variants *(la, le, the) Newe- Neu- -halle -all -haule*. Latin **nova** + **aula**, ME **newe** + **halle**. A new house built by the lords Audley c.1227, but already dismantled c.1540. Che iii.101.

(2) ~ Derby SK 2921. *Novahalle* 1150×9, *Nova Aula* 1185–1431, *Niewehal'* 1197, *(la, le) Neu- New(e)hall(e)* 1285–1330 etc. The two earliest forms are partially and totally Latinised respectively. Db 660.

NEWHAM 'New homestead'. OE **nīwe** + **hām**.

(1) ~ Northum NU 1728. *Neuham* 1242 BF, 1288, *Newham* 1296 SR. Also known as *Neuham Comyn* 1242 BF when it was held by David Comyn. NbDu 148.

(2) ~ HALL Northum NU 1729. P.n. NEWHAM NU 1728 + ModE **hall**.

NEWHAVEN ESusx TQ 4401. 'New harbour'. *Newhaven* 1586. ModE **new** + **haven**. The harbour was formed when the river Ouse was diverted from Seaford to flow into the sea at a place formerly called Meeching, *Mechinges* 1087×1100–13th, *Mechyng* 1296–15th, OE folk-n. **Mēċingas* the 'people of the sword' or the 'people called after Meaca', OE **mēċe** or pers.n. **Mēaca* corresponding to OHG *Mauco*, + **ingas**. 'New' for contrast with the 'Old Haven' at SEAFORD TQ 4899. Sx 323, ING 37, SS 65.

NEW HAYES Staffs SK 0312 → HEATH HAYES SK 0110.

NEWHEY GMan SD 9311. 'New enclosure'. *New Hey* 1843 OS. ModE **new** + **hay** 'fence' (OE *(ġe)hæġ*).

NEWHOLM NYorks NZ 8610. 'New homestead'. *Neu(e)ham* 1086, *Neweham* [c.1100–1125]15th. OE **nīwe** + **hām**. YN 124.

NEW HOUSES NYorks SD 8073. 'New houses'. *Newhouse* 1560, *-howses* 1635. ModE **new** + **house** (OE, ME pl. *hūs*). YW vi.223.

NEW HYTHE Kent TQ 7159. 'The new landing-place'. *La Newehethe* 1254, 1292, 1320, *Neuheth* 1323, *New hyth* 1610. French definite article **la** + ME **niwe**, **newe** + **hethe** (OE *hȳth*). 'New' for distinction from the lost Old Hythe, *attenhaldehithe* 1254, *Ealdehethe* 1292. A landing-place on the Medway. PNK 149.

NEWICK ESusx TQ 4121. 'New farm'. *Niewica* [1154×89]14th, *Newik(e)* 1218, 13th. OE **nīġe** + **wīċ**. Sx 316, SS 78.

NEWINGTON 'At the new farm or settlement' OE **(æt thǣm) nīwan tūne**.

(1) ~ Kent TR 1837. *Neventone* 1086, *Niwan tune* c.1100, *Neweton(')* 1201–13. K 453.

(2) ~ Kent TQ 8564. *Newe- Neu(ue)tone* 1086, *Newentune -tone* [1087]13th, *Niwan- Niuuentune* c.1100, *Newe(n)ton(')* 1172×3–1223. PNK 259.

(3) ~ Oxon SU 6096. *Niwantun* 1042×66 S 1047, *(æt) Niwantune*

1042×52 S 1229, *Niwetune* 1086, *Ni- Newenton(e)* 1275–1364, *Newington alias Newenton* 1707. O 132.

(4) North ~ Oxon SP 4239. *Northneweton'* 1268, *Northnewynton* 1324, *North Newton* 1675. ME adj. **north** + p.n. *Neweton'* 1200, *Newin- Neuin- Newenton'* 13th. O 401.

(5) South ~ Oxon SP 4033. *Suthnewentone* 1285, 1320, *Southnewington* 1537. ME **south** + p.n. *Niwetone* 1086, 1208, *Nevtone* 1086, *Neuton(e)* 1204–c.1224, *Neuuentone* 1163–94, *Newenton(e)* 1163–1428. O 277 gives pr [nu:tən].

(6) Stoke ~ GLond TQ 3286. 'N made of stocks or at the place of stumps'. *Neweton Stocking, ~ Stoken, Stokneweton* 1274, *Stokene Neuton, Stoke Newenton* 1294, *Stokenewington* 1535. ME **stocken** (OE *stoccen*) or **stocking** (OE *stoccing*) + p.n. *Neutone* 1086, *Newtun, Newenton* 13th cent. Also known as *Newtun Canonicorum* 1254 from its possession by the canons of St Paul's, and *Newtun juxta Clerekennewelle* 1274, for distinction from Newington Barrow TQ 3185, '(that part of) Newington held by (Thomas de) Barewe' in 1271, *Neweton Barrewe* 1274, a manor held by the de Barewe family, now called Highbury, 'the high manor', *Heybury* 1349×96, *Highbury* 1535 (ME **hih** (OE *hēah*) + **bury** (OE *byriġ*, dative sing. of *burh*)), from its position in relation to other neighbouring manors, Canonbury, *Canonesbury* 1373, which was held by the canons of St Bartholomew's, Smithfield, and Barnsbury, *Iseldon Berners* 1274, *Bernersbury* 1406, that part of Islington held by William de Berners in 1235. Mx 159, 125.

NEW INVENTION Shrops SO 2976. Said to be derived from the name of an inn so called, a semi-punning name for a public-house ('inn-vention'). Room 1983.79.

NEWLAND 'New arable area'. OE **nīwe** + **land** L 246–7, 249.

(1) ~ Glos SO 5509. *Neweland(e)* 1248–1461, *(la) Neu- Newlond(e)* 1248–1536. Latin forms are *Noua -v- terra* 1221–1327, *Nova Landa* 1256. The reference is to land taken out of the Forest of Dean and cleared for cultivation in the 12th and 13th cents. Gl iii.236.

(2) ~ H&W SO 7948. *Nova Terra* [1127]1321, *Nova Landa* 1232–55, *Newelond* 1327, *Newland* 1831 OS. Wo 215.

NEWLAND NYorks SE 6924. 'The newly recovered land'. *New(e)land* 1234–1587, *Newelaund', la Neouelaund* 1246. OE **nīwe** + **land**. The reference is to low-lying marshland originally cut off from the W by an old course of the river Aire. YW iv.14.

NEWLAND(S) 'New arable area'. OE **nīwe** + **land**, L 246–7, 249.

(1) ~ Northum NZ 0955. *Novalanda* 1268, *Neulond* 1345, *Newlands* 1866 OS. NbDu 148.

(2) ~ HAUSE Cumbr NY 1917. 'Newlands pass'. *Newland Hose* 1783, *~ hawse* 1784. P.n. Newlands NY 2320 + **hawse** 'a neck, a col, a pass' (OE *hals*). Newlands, *Neulandes* 1318, *Neweland* 1369, is 'land newly taken into cultivation'. Newlands Hause is a pass on the road from Buttermere to Newlands. Cu 356, 371.

NEW LANE Lancs SD 4213. A halt on the railway line from Manchester to Southport.

NEWLYN Corn SW 4628. 'Army pool'. *Nulyn* 1279, *Lulyn* 1290–mid 14th, *Nywelyn* 1337. *Neulin* 1370. Co **lu** 'host, army' in the sense 'fleet' + **lyn** 'pool', with dissimilation of *l - l* to *n l*. PNCo 128, Gover n.d. 653, CPNE 155.

NEWLYN DOWNS Corn SW 8354. *St Newlin downe* early 17th. P.n. Newlyn as in ST NEWLYN EAST SW 8256 + ModE **down(s)**. PNCo 129.

NEWMARKET Suff TL 6463. *la Newmarket* 1418. The new market is earlier referred to in Latin as *Novum Forum* 1200 and *Novum Mercatum* 1219. ME **new** + OFr **mercat**. DEPN.

NEWMARKET HEATH Suff TL 6163. P.n. NEWMARKET TL 6463 + ModE **heath** as in *The Heath* 1836 OS.

NEW MILL ModE **new** + **mill** (OE *myln*).

(1) ~ Corn SW 4534. *Newmill* 1748. Short for *Mulfra Newe Mill* 1621. Along with *Mulfra Mill* and *Mulfra Stampinge Mill* it belonged to the hamlet of Mulfra SW 454347, *Moelvre* c.1250–1327, 'bare hill', Co ***moyl** + ***bre** with lenition of *b* to *v*. Co 129, Gover n.d.649.

(2) ~ Herts SP 9212. *le new milne* 1603×25. Hrt 54.

(3) ~ WYorks SE 1609. *Newmylle* 1462. YW ii.240.

NEW MILLS 'New mills'. OE **nīwe** + **myln**.

(1) ~ Ches SJ 7782. *New (Corn) Mill* 1831. Che ii.68.

(2) ~ Corn W 9052. A translated name. *Melynewyth* 1364, *Melenowith alias New Myll* 1596. Co **melin** 'mill' + **nowyth** 'new'. PNCo 129.

(3) ~ Derby SJ 9985. *The Queen's Mill, called Berde Mill or New Mill* 1565, *(the) New Miln(e)* 1625–41. The plural is a late development. The reference is to a mill on the right bank of the river Goyt. The form *Berde Mill* contains ME **berde** 'a beard' in the sense 'fish-weir, fish-trap' as in the fishery name *Berde yar'* [1128]14th on the river Tyne. Db 150, DAJ 2.55, L.C.Wright, *Technical Vocabulary to do with Life on the River Thames in London c. AD 1270–1500*, Oxford University D.Phil thesis, 92.

NEWNHAM 'At the new estate'. OE **nīwe** + **hām** in the phrase *æt thǣm nīwan hām*.

(1) ~ Glos SO 6911. *Nevneham* 1086, *Niweham* 12th cent., *New(e)- Neuham* 1218–1316, *Newenham* 1217–1517, *Neunham* 1221, *Newnham* from 1492. Gl iii.197.

(2) ~ Hants SU 7153. *Neoham* 1100×35, *Niweham* 1167, *Niwenham* 1212, 1291, *Newenham* 1316–54, *Newnam* 1468. The site is a patch of sand amidst the surrounding clay, probably a secondary settlement in more marginal land as the result of population growth in mid Saxon times. Ha 122.

(3) ~ H&W SO 6469. *neowanham* [?781×96]11th S 1185, *Neoweham* [c.1160]c.1240, *Noweham* [1206]c.1250, *Newenham* 1240, 1392, 1535. Wo 55, Hooke 82.

(4) ~ Herts TL 2437. *Neuhā* 1086, 1119–46, *Newenham* 12th–1217, *Neunham* 1291. Hrt 113.

(5) ~ Kent TQ 9557. *Newenham* 1177–1254, *Neuham* 1247. PNK 287, ASE 2.37, 47.

(6) ~ Northants SP 5759. *æt niwanham* [1021×3]11th S 977, *Newæham* [1020]12th S 957, *Neuenham* 1166–1476 with variants *Newn- Niwen*. Nth 26.

Long NEWNTON Glos ST 9192. *Lange Newenton(e)* 1287, *Longnewnton* 1585. ME adj. **lang** + p.n. *Niuentun* [681]13th S 73, *Newenton(e) -tuna* 1065–1258, '(at) the new settlement', OE **nīwe**, dative sing. definite form **nīwan**, + **tūn**. The reference is either to the length of the parish or to the lane along which the houses are dispersed. A new settlement from Tetbury ST 8993. Gl i.103, Wlt 63.

NEW PARK Hants SU 2905. *New Parke* 1650. ModE **new** + **park**. An enclosure in the New Forest. Gover 1958.207.

NEWPORT 'New town'. OE **nīwe** + **port**.

(1) ~ Devon SS 5632. *Neuport by Barnstaple* 1295, *Nyweport* 1311, *Burgus de Nyuport* 1330. 'New' in relation to Barnstaple. D 27.

(2) ~ Essex TL 5234. *Neuport* 1086–1321, *New(e)port* 1100–1510. Ess 531, Jnl 2.47.

(3) ~ Glos ST 7097. *Neu- New(e)port(e)* 1287–1652. A new town was founded here with a chantry and two fairs in 1343×8. Gl ii.208.

(4) ~ Humbs SE 8530. *Neu- Newport* 1368, 1828. A village developed on reclaimed fenland. YE 248.

(5) ~ IoW SZ 5089. *Neweport* 1202–1591, *Neuport* 1227–1514, *Newport* from 1255. The Latin form *Novo Burgo* is found 1189×1204. Wt 175 gives local pr ['nipərt], Mills 1996.75.

(6) ~ Shrops SJ 7419. 'The new market-town'. *Neuport* 1221–1501, *Le ~* late 13th, *Neweport* 1319–1494, *Newport* from 1443. ME **newe** + **port**. Also known in a Latinised form as *Novus Burgus* 1136–1322. Believed to have been founded by Henry I within the royal manor of Edgmond. Sa i.222.

(7) ~ PAGNELL Bucks SP 8743. 'N held by the Paynel family'.

Neuport Paynelle 1220. Earlier simply *Neuport* 1086. Bk 21 gives pr [nju:pət pænel], a pr recorded since the spelling *Panell* 1367.

NEWPORT Norf TG 5016. 'New port'. ModE **new** + **port**. A modern development with no early forms.

NEWPOUND COMMON WSusx TQ 0627. *Newpound Common* 1813 OS. P.n. Newpound 'the new pound', ModE **new** + **pound**, + ModE **common**. A Richard *atte Punde* 'at the pound' is recorded 1296. Sx 135.

NEWQUAY Corn SW 8161. 'New quay'. *Newe Kaye* 1602, *New Key* 1748, *New Quay alias Towan* 1813. ME **new** + **key** 'a quay, a wharf'. The construction of a quay is mentioned in 1440 on the sea-shore at *Towan blustry*, the original Cornish name, *Tewenplustri* 1308, *Towan Blistra* 1906, MCo ***tewyn** 'sand-dune(s)' + unidentified element. PNCo 129, Gover n.d. 325.

NEWQUAY BAY Corn SW 8162. *New Quay Bay* 1813. P.n. NEWQUAY SW 8161 + ModE **bay**. PNCo 130.

NEW RIVER Cambs TL 5869. A cut made in 1610 and so named in 1634. Ca 9.

NEW RIVER Lincs TF 2518. *New River* 1790. A new channel of the river Welland. Payling 1940.15.

NEW ROW Lancs SD 6438. No early forms. Adj. **new** + **row**.

NEWSHAM '(Settlement) at the new houses'. OE (**æt thǣm**) **nīwe** + **hūsum**.

(1) ~ Northum NZ 3079. *Ne(h)usum* 1200, *Neuhusum* 1207, *Neusum* 1207, 1242, *Neusome* 1296 SR, *Newsam* 1461, *Newsome* 1728. NbDu 150, DEPN.

(2) ~ NYorks NZ 1010. *Neuhuson* 1086, *Neusom(e)* 1336.

(3) Little ~ Durham NZ 1217. ModE adj. **little** + p.n. *newehusa* [1091×2]early 12th, *Neusom* [1183]c.1320–1355, *Neusome* 1388, *Neusum* 1432, *New(e)sham* from 1484.

(4) Temple ~ WYorks SE 3532. 'N held by the Temple'. *Temple Newsom(e)* 1323. ME *Temple* from possession of the estate by the Knights Templar, + p.n. *Neuhusu'* 1086, *-husum* 1197, *Newsom(e) -um* 1276–1654. YW iv.116.

NEWSHOLME '(The settlement) at the new houses'. **nīwe** + **hūs**, dative pl. **nīwe-hūsum**, or **nīwan** + **husum**.

(1) ~ Humbs SE 7229. *Neuhusa'* 1086, *Neu- Niwehus* 12th cent., *Newe- Neusum -om* 1202–1432, *Newsome* 1523. YE 243, SSNY 87.

(2) ~ Lancs SD 8451. *Neuhuse* 1086, *Neusum* 1226, 1381, *Neusom* 1285, 1303, *Newsholme* from 1567. YW vi.173.

NEWSTEAD 'The new farmstead'. OE **nīwe** + **stede**. Stead 111.

(1) ~ Northum NU 1527. *(apud) Novum locum qui dicitur Neubigginge* 'at the new place called *Newbigginge*' c.1230, *Novo Loco, Novum Locum* 1279, *New(e)stede* 1339, *Newstede* 1377, 1424. In the early 13th cent. the owners of the barony of Ellingham left their old residence at *Osberwic -wyc -wick -wyk* 1242 BF–1296 SR, *Osburwick* 1278, *Osborwyk* 1346, 'Osburg's farm', OE feminine pers.n. *Ōsburg*, genitive sing. *Ōsburge*, + **wīċ**, and built a new one subsequently known as the *Newbigging* or Newstead. NbDu 148, 168 s.n. Rosebrough, Stead 294.

(2) ~ NYorks SE 5478. *Newestede* 1541. ME **newe** + **stede**. The new site of Byland Abbey as opposed to what may have been the original site at OLDSTEAD SE 5380. YN 194, Stead 295.

(3) ~ Notts SK 5252. Latin *de Novo Loco, Novi Loci, Novum Locum* (dative, genitive and accusative cases) 1169–1441 (*in Syrewde, in Shyrewud* 13th), *New- Neustede* 1302–68, 'the new place in Sherwood'. This was the new site chosen for the Augustinian Priory founded by Henry II. Nt 128, Stead 288.

NEWTHORPE NYorks SE 4632. 'The new outlying farm'. *(on) Niwan-þorp* c.1030 YCh 7, *Newthorp* 13th cent., *-thorpe* from 1446, *-throp(e) -thropp* 1558–1641. OE **nīwe**, definite form dative case **nīwan**, + **thorp**. YW iv.60, SSNY 63.

NEWTIMBER PLACE WSusx TQ 2613. P.n. *Nitimbre* 960 S 687, 1291, *Niuembre* (sic) 1086, *Newetymber* 1270, 'the new building', OE **nīġe**, **nīwe** + **timber**, + ModE **place** 'a mansion house', in this case of the 17th cent. Sx 286.

NEWTON '(The) new settlement'. OE **nīwe**, definite form **nīwa**, + **tūn**. Some spellings show OE variant **nīġe** for **nīwe**.

(1) ~ Cambs TF 4314. *(to) niwantune* c.972, *Newen- Nuenton* 1285, *Neuton(e)* c.1213–1457. Ca 274.

(2) ~ Cambs TL 4349. *Neutune* [1042×66]13th S 1051, *Neuton(e) -(a)* c.1060–1445, *Newton(a)* from 1139. Ca 86.

(3) ~ Ches SJ 5375. *Neuton* 1260, *Newton* from 1293. A new settlement in Delamere Forest. Che iii.248.

(4) ~ Ches SJ 5059. *Newton iuxta Tatenhall* 1298. Che iv.96.

(5) ~ Cumbr SD 2371. *Neuton* 1190, 1336, *Newtona* 1191×8. La 202.

(6) ~ H&W SO 3433. *Neuton(e)* 1247×72, 1287, 14th. He 151.

(7) ~ H&W SO 5054. *Newentone, Niwetvne* 1086, *Niweton'* 1123, *Neutona* 1243. He 150.

(8) ~ Lancs SD 4431. *Neutune* 1086, *Neuton* [before 1242]1268–1332 etc. A new settlement in marshland beside the Ribble associated with nearby Scales SD 4530, *Skalys* 1501, *Skales, Scalis* 1537, 'the huts', ON **skáli**. The marshes were exploited for grazing. La 150, SSNNW 360, Jnl 17.87.

(9) ~ Lancs SD 5974. *Neutune* 1086, *Neuton super Lon* [before 1219]1268. A new settlement presumably colonised from Whittington. La 185.

(10) ~ Lancs SD 6950. *Neutone -ton(a)* 1086–1316, *~ in Bogheland* 1285, *Newton (in Bowland)* from 1367. YW vi.206.

(11) ~ Lincs TF 0436. *Nev- Neutone* 1086. DEPN.

(12) ~ Norf TF 8315. *Newton* 1824 OS.

(13) ~ Northants SP 8883. *Newe- Neutone* from 1086 with variants *New- Niwe-, Niwen- Ne(u)wenton* 1162–1378. Nth 170.

(14) ~ Northum NZ 0364. *Neuton* 1296 SR, 1346. Also known as *Neuton del West* 1242 BF for distinction from Newton Hall NZ 0465, *Neuton del Est* 1242 BF. NbDu 148.

(15) ~ Staffs SK 0325. *Niwetone* 1086, *Neuton* 1252, 1306. Horovitz.

(16) ~ Suff TL 9140. *Niwetunā, Neuuetona* 1086, *Newton* 1610. DEPN.

(17) ~ Warw SP 5378. *Niwetone* 1086, *Neuton(e) juxta Clifton(e)* 1310–44, *Newton upon Dounesmore* (DUNSMORE) 1424. Wa 117.

(18) ~ WYorks SE 4427. *Niuueton -tun* 1086, *Newton(am)* c.1206, *Neuton Waleys* 13th. Held along with BURGHWALLIS by the *Waleys* family. YW ii.49.

(19) ~ Wilts SU 2322. *Neuton juxta Whitcherch* 1289. 'N by Whiteparish'. Wlt 391.

(20) ~ ABBOT Devon SX 8671. 'N belonging to the abbot' sc. of Torre. *Nieweton' abbatis* (Latin 'of the abbot') 1201, *Nyweton Abbatis* 1270, 1289, *~ Abbots* 1338. Earlier recorded in Latin form *Nova Villa* late 12th and French form *la Novelevile in parte pertin' ad manerium de Teyngewyk* 'the new vill belonging to the manor of *Teignwick*' (i.e. Highweek SX 8472) 1275. Granted to Torre Abbey by William de Briwere in 1196. D 473.

(21) ~ ARLOSH Cumbr NY 1955. *Neutonarlosk'* 1345, *Newton Arloche* 1538, *~ Arlosh* 1649. *Arlosh -sk* 1185–1411, *Arloss(c)h(e)* 1332–1418, is W **ar llosg** '(upon the) place cleared by burning'. A new town established by Holm Cultram Abbey to take the place of the town of Skinburness destroyed by the sea. Also known as *Langnewton* 1576 for distinction from WESTNEWTON NY 1344 and *Kirkebi Johannis* 'St John's church settlement' 1305 from the dedication of the church. Cu 291, SSNNW 34 no.8, GPC 2210.

(22) ~ AYCLIFFE Durham NZ 2724. 'New town by Aycliffe'. A new town created in 1947 for workers on the nearby Aycliffe Industrial Estate which originated in 1940 as a Royal Ordnance factory. The new town preserves the name of the ancient village of AYCLIFFE NZ 2882. Room 1983.81.

(23) ~ BERGOLAND Leic SK 3709. 'N held by the Burgoland family'. P.n. *Neuton(e)* 1086, 1417, *Newton* + affix *Burgilon -el- -ul-* 1391–1561, *Burkelande* 1563, *Burgeland* 1626. Also known as *Neuton(e) Botiler -el-* 1242–1513. The manor was held by

Roger de Burgelym c.1225 and by William de Butiller in 1317. Lei 405.
(24) ~ BEWLEY Cleve NZ 4626. 'The new settlement in the manor of Bewley'. *Neuton de Beaulu* 1334, *Neuton Beaulu* 14th cent. with variants *beauli(e)u, Beulie, Beulewe* etc., *Newton beaulieu* 1387, ~ *Bewley* from 1580. Earlier simply *(le) Neuton* 1296–1381. ME **newe-toun** + p.n. Bewley as in COWPEN BEWLEY NZ 4824.
(25) ~ BLOSSOMVILLE Bucks SP 9251. 'N held by the Blosseville family'. P.n. *Niwetone* 1175, *Newentone* 1202, + family name *Blosseville* 1202, 1316, *Blostmevill(e)* 1311, 1340, from Blosseville-sur-Mer, Seine-Maritime. Bk 11.
(26) ~ BROMSWOLD Northants SP 9965. *Newenton al. Newton Bromswolde* 1639. Earlier simply *Niwetone, Neuuenton(e)* 1086 etc. Bromswold, *Bruneswald* 12th, *Brouneswold* 1339, 1430, 'Brun's forest', was an area of woodland on the borders of Huntingdonshire and Northamptonshire. Nth 193.
(27) ~ BY TOFT Lincs TF 0587. *Neuton iuxta Tofte* 1272×1307, *Toft Neuton* 1324–97, *Toftnewton(')* 1335–1762. P.n. *Nev- Neutone* 1086, *-tun -ton(')* after 1169–1363, *Newton(')* 1210–1576, + p.n. TOFT TF 0488. Li iii.63, SSNEM 150 no.1.
(28) ~ DALE NYorks SE 8290. *Neuton(e)dale* 1322. P.n. NEWTON-ON-RAWCLIFFE SE 8190 + ME **dale** (ON *dalr*). YN 85.
(29) ~ FERRERS Devon SX 5448. 'N held by the Ferrers family'. *Niweton Fereirs* 1306, *Newenton Ferers* 1328. Earlier simply *Niwetone* 1086, *Newenton* 1249. D 282.
(30) ~ FLOTMAN Norf TM 2098. 'Floteman's Newton'. *Neuton Floteman* 1291. Earlier simply *Niwetuna* 1086. Pers.n. *Floteman* occurs in DB representing OE *flotmann* 'a sailor, a pirate'. DEPN, PNDB 251.
(31) ~ HARCOURT Leic SP 6397. 'N held by the Harcourt family'. *Neuton Harecurt* 1275, *Neweton Harcourt* 1437. Earlier simply *Niwetone* 1086, *Neuton(e)* 1086–1439, *Newton(')* from 1393. The manor was held by Richard de Harcurt in 1236. Lei 266.
(32) ~ HEATH Dorset SZ 0084. *Newton Heath* 1811. P.n. *Nyweton* 1404, *Newton* 1575, + ModE **heath**. This is the 'new settlement' (ME **niwe** + **toun**) of Edward I's proposal to lay out a new town and harbour in this place in 1286; the proposal was never fully implemented. Do i.48, 45, Newman-Pevsner 405.
(33) ~ HEATH GMan SD 8800. 'Heath beside Newton'. P.n. *Newton* from 1322, + *Heath* 1843 OS. La 36.
(34) ~ KYME NYorks SE 4645. 'N held by the Kyme family'. P.n. *Ne- Niuueton* 1086, *Neuton(e)* 1086–1428, *Neweton* 1190–1269, *Newton* from 1461, + manorial addition *Kyme* from 1275. The *Kyme* family held land here from 1240. YW iv.79.
(35) ~ -LE-WILLOWS Mers SJ 5894. 'N in the willows'. *Newton-in-Makerfield otherwise Newton-Le-Willows* 1895. Earlier simply *Neweton* 1086, 1201, *Ni(e)weton* 1177, 1202, *Neuton* 1212–1332, and ~ *Macreffeld* 1257, ~ *in Makerfeld* 1332, *Makerfield* 1843 OS. 'Le-Willows', short for *en le willows*, seems to be a modern addition for distinction from other Newtons. For the original addition, see MAKERFIELD. La 98, Jnl 17.56.
(36) ~ -LE-WILLOWS NYorks SE 2189. 'N among the willow-trees'. *Neuton in le Wilughes* 1300. P.n. *Neuton* 1086 + Fr definite article **le** short for *en le* + ME **willowes**. YN 241.
(37) ~ LONGVILLE Bucks SP 8431. 'N held by (the church of) Longueville'. P.n. *Neutone* 1086, *Newenton(a)* 1152×8, 1402, + p.n. *Longe- Lungeville* 1241, *Longfylde* 1526, 1607. Newton was granted c.1152×8 to the church of St Faith of Longueville by Walter Giffard, earl of Buckingham, who was lord of Newton and of Longueville-sur-Scie, Seine-Maritime, the *caput* in Normandy of the honour of the Giffard earls. Bk 22.
(38) ~ -ON-OUSE NYorks SE 5159. *Niweton' super Vsam* 1176 etc. P.n. *Neuton* 1086, *Neweton(e)* 1086–1330, *Niwenton'* 1167, + r.n. OUSE SE 4959. The 1167 form retains traces of an OE locative–dative form *(æt thǣm) nīwan tūne*. YN 20.
(39) ~ -ON-RAWCLIFFE NYorks SE 8190. P.n. NEWTON + p.n. Rawcliffe. Newton is *Neuton, Newetone, Newetun(e)* 1086, 1242. Rawcliffe is recorded separately as *Rouclif -clyff* 1334, 1408, *Rocliffe* 1619, 'the red bank', ON **rauthr** + **klif**. YN 84, 87.
(40) ~ ON THE MOOR Northum NU 1705. *Neuton supra moram* 1242 BF, 1296 SR, *Newton super moram* 1346, *Newton on the Moor* 1868 OS. NbDu 148.
(41) ~ ON TRENT Lincs SK 8374. P.n. *Neutone* 1086, *Neotune* c.1115, *Newetun'* 1154×66, *Newentone* 1155×62, [1158×62]14th, + r.n. TRENT. DEPN, Cameron 1998.
(42) ~ or ST MARY'S HAVEN Northum NU 2424. *St Mary's or Newton Haven* 1868 OS. P.n. Newton as in High NEWTON-BY-THE-SEA NU 2325 + ModE **haven**.
(43) ~ POPPLEFORD Devon SY 0889. 'N or new settlement called or at Poppleford'. *Neweton Popilford* 1305, *Newton Popler* 1675. Earlier simply *Poplesford* 1226, 1274, *Popelford -le-* 1257–1306, 'pebble ford', OE **popel** + **ford**. The original name, Poppleford, refers to the round stones known as 'Budleigh pobbles' found here, to which Newton was added at a later date. It is not clear in relation to what place it was new. D 592.
(44) ~ PURCELL Oxon SP 6230. 'N held by the Purcell family'. *Newentone Purcel* 1285–1474 with variants *-ton* and *Purcell(e)*. P.n. *Niweton'* [1198]c.1300, *Newent(h)on' -tona* 1213–1320, *Neu- Newton'* 13th cent. + manorial affix from the Purcell family who held the manor in 1198. O 231.
(45) ~ REGIS Warw SK 2707. 'N held by the king'. *Kyngesneuton* 1259, *Kings Newton al. Newton in the Thistles* 1654, *Neuton Regis* 1291. P.n. *Newintone -en* 1155–75, *Niwetone* 1189, + Latin **rex**, genitive sing. **regis**. The estate was royal demesne until 1158×9. The earliest forms are properly dative sing. 'at the new *tūn*', OE *(æt thǣm) nīwan tūne*. Wa 20.
(46) ~ REIGNY Cumbr NY 4731. 'N held by the Reigny family'. *Neutonrey(g)ny(e)* 1275×6–1332 with variants *Reny, Reyner, Newton Reynye* from 1316 with variants *Renye, Ranney* etc. Earlier simply *Niweton'* 1185–1203, *Neweton'* 1250–1354. Held by William de Reigny in 1185 and said to have been given to Turstan de Reigny by Henry I. Two forms *Newinton'* 1201 and *Neuinton* 1202 may show traces of the OE dative inflexional form **(æt thǣm) nīwan** + **tūne**. Cu 227.
(47) ~ ST CYRES Devon SX 8797. 'St Ciricius' N'. *Nyweton Sancti Ciricii* 1330, *Seynt Serys Newton* 1525, *Newton St Cires* 1605. Earlier simply *(æt) niwan tune* 1050×73, *Niwe(n)tone* 1086, *Nywetone* 1264. The church is dedicated to St Ciricius. D 410.
(48) ~ ST FAITH Norf TG 2217. *Newton St. Faith* 1838 OS lies in the parish of HORSHAM ST FAITH TG 2215.
(49) ~ ST LOE Avon ST 7064. 'N held by the St Lo family'. *Nywetonseyntlou* 1336. Earlier simply *Niwetone* 1086. The manor was held in 1122 by Roger de Sancto Laudo from St-Lô, Manche, France, *S. Laudi* 1056, named after Lauto, a 6th cent. bishop of Coutances. DEPN, Dauzat-Rostaing 611.
(50) ~ ST PETROCK Devon SS 4112. 'St Petrock's N'. *Nyweton Sancti Petroci* 1416, *Nuton Petrocke* 1557. Earlier simply *(æt) Nywantune* 938 S 388, *Niwetone* 1086, 1252, *Nietona* 1086 Exon, *Neweton* 1292. It was granted to St Petrock's monastery at Bodmin by Athelstan in 938. D 101.
(51) ~ SOLNEY Derby SK 2825. 'N held by the Solney family'. *Neuton(e) Sulney, Solney* from c.1300. Earlier simply *(æt) Niwantune* [956]14th, [c.1002]c.1100 S 1536, *Nywantun* [1004]14th S 906, *Newetun* 1086, *New(e)ton(e)* 1244–1342 etc., *Neuton(e)* 1135×9–1428. The manor was conveyed from Ralph de Argosis to his brother Alfred de Solenneio in 1205. Db 647.
(52) ~ STACEY Hants SU 4140. 'N held by the Sacy family'. *Nyweton Sacy* 14th, *Newton Stacey* 1817 OS. For the change *Sacy* > *Stacey* cf. BARTON STACEY SU 4341. Earlier simply *Niuuetone* [903]16th S 370, *Niwetone* 1086, *Nywentone, Niwinton* 13th, *Newton juxta Wherwell* 1389. Gover 1958 173.
(53) ~ TONY Wilts SU 2140. 'N held by the Tony family'. *Newenton Tony* 1338, *Nywetone Tony* 1332. P.n. *Newentone*

1086, *Niwetona* -e [1154×89]1270 etc., + manorial addition from the family name of Ralph de Toenye who held here in 1254 and whose family came from Tosny, Eure, France. The sps show that the original form of the name was in the locative-dative case, *(æt thǣm) nīwan tūne* 'at the new settlement'. W 370.

(54) ~ TORS Northum NT 9127. *Newton Tors* 1865 OS. P.n. Newton as in KIRKNEWTON NT 9130 + ModE **tor** (OE *torr* 'a rock, a rocky outcrop or peak'). The reference is to a peak that rises to 1761ft.

(55) ~ TRACY Devon SS 5226. 'N held by the Tracy family'. *Newton Tracy* 1402. Earlier simply *Newentone* 1086, *Nywethon'* 1242. Henry de Tracy held the manor in 1242. D 118.

(56) ~ UNDER ROSEBERRY Cleve NZ 5613. 'N beneath *Othenesberg*'. *Newetunie sub Ohtnebercg, Neuton sub Otneberch* [12th]15th. Earlier simply *Newetun, Nietona* 1086. For *Othenesberg* see ROSEBERRY TOPPING NZ 5712. YN 163.

(57) ~ UPON DERWENT Humbs SE 7249. *New(e)- Neuton on, super Derwent* 1371. P.n. *Niweton'* 1190, 1194, *New(e)- Neuton* 1246–1504 + r.n. DERWENT SE 7035. YE 188.

(58) ~ VALENCE Hants SU 7232. 'N held by the Valence family'. *Nyweton Valence* 1346. Earlier simply *Newentone* 1086 from the prepositional form *æt thǣm nīwan tūne, Ne- Niwenton(e)* 1167–1339, *Neuton* 1251. The estate was held by William de Valencia whose family came from Valence, Seine-et-Marne, Normandy. A late foundation on marginal land on a hill capped with clay-with-flint. Ha 123, Gover 1958.92.

(59) Bank ~ NYorks SD 9153. 'N held by the Bank family'. *Ban(c)k(e) Newton* 1546 etc. Earlier simply *Neuton(e) -tune* 1086–1350, *New(e)ton* 1271, 1428. Also known as *Cauld Newton* 1399, and *Newton in Craven* 1401. *Bank* refers to the local family of that name which held the manor for three centuries. YW vi.55.

(60) Bircham ~ Norf TF 7633. 'Newton belonging to Bircham'. *Bircham Newton* 1824 OS. P.n. Bircham as in Great BIRCHAM TF 7632 prefixed to p.n. *Niwetuna* 1086. DEPN.

(61) Cold ~ Leic SK 7106. ME prefix **cold**, *Cold(e)* 1279–1623, + p.n. *Niwetone* 1086, *Neuton(e)* c.1130–1521, *Newton* from 1405. Also known as *Neuton(e)- Newton Burdet(t) -ytt* 1242–1610 and *Newton Marmion* 1563–1616. 'Cold' on account of its exposed, bleak situation. William Burdet held land in the village from 1271. The late appearance of the Marmion affix may indicate that it is an antiquarian addition. Lei 311.

(62) East ~ Humbs TA 2637. *Aust, Est Neuton* [12th]c.1536, *East Newton* from 1512. ON **austr**, ME **est**, + p.n. *Niuuetone* 1086, *Neuton(a)* 1285–1341, *Newton iuxta Aldeburgh* 'N by Aldbrough' 12th–13th. YE 60.

(63) High ~ Cumbr SD 4082. *Neutun* 1086, *Newton* 1537. 'High' for distinction from Low Newton SD 4082. The two Newtons are situated at the top and bottom of the road ascending the flank of Newton Fell. La 199.

(64) High ~-BY-THE-SEA Northum NU 2325. ModE **high** + p.n. *Newton by the Sea* 1866 OS, *Neuton super mare* 1242 BF, *~ juxta mare* 1296 SR, 1346. 'High' for distinction from Newton Seahouses NU 2424. NbDu 148.

(65) Maiden ~ Dorset SY 5997. 'The maidens' N'. *Maydene Neweton* 1288, *Mayden Nywton* 1303, *Maydene Newenton* 1325. ME **maidene** (OE *mægdena*, genitive pl. of *mægden*) + p.n. *Newetone* 1086. The reference is possibly to possession of the manor by nuns but the precise allusion is unknown. Do 103.

(66) North ~ Somer ST 2930. *North Neuton* [n.d.] Buck, *Northneweton* [1397] Buck. ModE adj. **north** + p.n. Newton ST 3031, *Newe(n)- Niwetone, Newetvne* 1086. 'North' for distinction from *Neuton Commitis* and *~ Regis* [1186] Buck and *Tukereniweton* [n.d.] Buck. DEPN.

(67) North ~ Wilts SU 1257. *norþ niwetune* [892]14th S 348, *Norþneuton* 1242, otherwise simply *Nywantun* [933]14th S 424, *Newetone* 1086, *Newenton* 1212, 1268, *newington* 1571. Wlt 322

(68) Old ~ Suff TM 0562. *Vetus Neuton* 1271×1307, *Eldneuton* 1418. ModE **old**, Latin **vetus**, ME **eld** (OE *eald*) + p.n. *Niuetuna, newetuna -tona, neutūn* 1086, *Neweton* 1196, *Newton* 1610. DEPN.

(69) Out ~ Humbs TA 3821. *Outneuton -newton* from c.1265. OE **ūt(e)** 'out, remote' + p.n. *Nieuueton(e)* 1086, *Neu- Newton(a)* 1145×60 etc. YE 19.

(70) Place ~ NYorks SE 8872. 'N held by the Place family'. *Place Newton* 1560. Earlier simply *Neu(ue)ton* 1086, *Neu- Newton(a)* 1219–1543. A William Playce was tenant here in 1285. The surname represents either ME *place* '(market-)place', OFr *pleix, plais* 'an enclosure or coppice surrounded with an interwoven fence of living wood' or a p.n. derived from it, or OFr *plaise* 'a plaice' used as a nick-n. for a fish merchant. YE 137.

(71) South ~ Wilts SU 0834. *Sutneuton* 1242, *Sutnewenton* 1281. ME **suth** + p.n. *(in, æt) niwantune* [943]14th S 492, [946×7]12th S 1504, *Nyweton* c.1190. 'South' for distinction from North NEWTON SU 1257. Wlt 228.

(72) Trowse ~ Norf TG 2406. *Trowse Newton* 1838 OS. Short for two separate names, *Trowse cum Newtone* 1316, 'Trowes with Newton'. P.n. Trowse, 'the tree house' as in TROWSE NEWTON TG 2406 + p.n. *Newotona* 1086. DEPN.

(73) Water ~ Cambs TL 1097. *Waterneuton* c.1300. Earlier simply *Niwantune* [937]14th S 437, [973]14th S 792, c.1000, *Newetone* 1086. The place lies on the Nen. Hu 193.

(74) Welsh ~ H&W SO 5018. *Welshnewton* 1505, *Newton Wallia* 1535. ModE adj. **Welsh** (Latin *Wallia*) + p.n. *Newtone* 1313–97. 'Welsh' for distinction from English Newton or Newton Huntly at Newton Court, Monmouthshire, Gwent, on the other side of the Monnow, SO 5214.

(75) West ~ Humbs TA 1937. *West Newton* 1512. ModE adj. **west** + p.n. *Niuuetun -tone, Neutone* 1086, *Newton* from c.1265. Also known as *Newton Constable* 1285. 'West' for distinction from East NEWTON TA 2637. Formerly held by the family of *Constable* who held the 'new town' as well as the parent village of BURTON CONSTABLE. YE 61.

(76) West ~ Norf TF 6927. *West Newton* 1824. W of Appleton.

(77) Wold ~ Humbs TA 0473. *Wald(e) Newton* 1367–1556, *Wawd ~* 1604, *Would ~* 1617. ME **wald** + p.n. *Neuton(e)* 1086, *Neu- Newton(e) -tun* 12th–1828, *Neu- Newton(e) super Wald(am)* [1141×54]1464, *~ in Waldo* [c.1230]. Situated on the Yorkshire Wolds. YE 114.

(78) Wold ~ Humbs TF 2498. *Wald neuton* [1154×89]13th–1431, *Woldneuton* 1248, 1297. ME **wald** + p.n. *Neutone* 1086, *-tun -ton(a)* 1170–1431. Situated on the Lincolnshire Wolds. Li iv.134.

NEWTOWN 'New farm'. ME, ModE **new** + **town** (OE *nīwe* + *tūn*).

(1) ~ Ches SJ 6248. No early forms. Che iii.122.

(2) ~ Cumbr NY 5062. *Neuton* 1278, *Newton* 1485, *Irthington Newetowne* 1568. Cu 92.

(3) ~ Derby SJ 9984. *Newtown* 1860. A modern name for part of DISLEY Ches transferred to Derby in 1934. Db 155, Che i.271.

(4) ~ Dorset SZ 0493. A new village built in the late 18th cent. to rehouse the inhabitants of Moor Crichel displaced by Humphrey Sturt beyond the borders of his park. ModE **new** + **town**. Do ii.262, Newman-Pevsner 300.

(5) ~ Hants SU 6113. *Newtown* 1859. In the Forest of Bere. Gover 1958.51.

(6) ~ Hants SU 2710. No early forms. In the New Forest.

(7) ~ Hants SU 4763. *Neweton, la Nywetone* 1316, *Newtoen* 1579. Earlier Latin *Novus Burgus* 1218, *Nova Villa* 1284. A new town founded by the bishop of Winchester in the second decade of the 13th cent. as a commercial venture on the lands of his Clere manors in imitation of and competition with NEWBURY Berks SU 4666. It was not successful. Gover 1958.155.

(8) ~ Hants SU 3023. No early forms. In the New Forest.
(9) ~ H&W SO 6145. *Towns End* 1831 OS, sc. of Stretton Grandison.
(10) ~ IoW SZ 4290. Properly the 'new town of Francheville'. *Niwetune* 1189×1204, *nouum burgum de Francheuile, Frankevile* 1254, *(le) New(e)tone* 1255–1441, *Newetowne* 1415, *Newton* 1591 etc. Francheville, *Francheuile, Fraunchevile -vyle* 1257–1592, is the 'free town', OFr **franc**, feminine form **franche**, + **ville**. Wt 82 gives pr ['nju:tæun], Mills 1996.75.
(11) ~ Northum NT 9731. *Newtown* 1866 OS.
(12) ~ Northum NU 0300. Latin *Nova Villa* 1242 BF, *Newtown* 1248, *Le Neuton* 1309. Also known as *Neuton' in Roubir'* 'N in Rothbury' 1296 SR. NbDu 149.
(13) ~ Northum NU 0425. *Newtown* 1695 Map.
(14) ~ Shrops SJ 4731. *Neuton by Wemme* 1373, *Newton vill* 1586×7, *Newtowne* 1558×9. Gelling.
(15) ~ Staffs SJ 9060. A modern development not occurring on the 1842 OS where the site is called *Horton Hay House*, cf. *the haye of Horton* 1282 etc. Horovitz.
(16) ~ Wilts ST 9128. *Newtown* 1820 OS. Wlt 196.
(17) ~ BAY IoW SZ 4192. *New Town Bay* 1769, cf. *Newto(w)ne haven* 1583. P.n. NEWTOWN SZ 4290 + ModE **bay**. Wt 83, Mills 1996.76.
(18) ~ -IN-ST MARTIN Corn SW 7423. *Neweton* 1620, *Newtown* 1870. A 19th cent. hamlet. See ST MARTIN SW 7323. PNCo 130, Gover n.d. 568.
(19) ~ LINFORD SK 5110. 'The new settlement at Linford'. *Neuton' Lynforthe* 1446, *Newtown(e) Lyndeford* 1512, *~ Li- Lynford* from 1513. Earlier simply *Neuton* 1325–71, *(La) Neweton* 1331, 1474, *le Newetowne* 1405. ME **newe-toun** + p.n. *Lyndeford* 1293, *Lyndenforth'* 1327–71, *Lynford* 1474, 'the ford where lime-trees grow', OE **linden** + **ford**. Lei 386.
(20) Hungerford ~ Berks SU 3571. *New Town* 1761. P.n. HUNGERFORD SU 3368 + ModE **new town**. Brk 305.

NEW YATT Oxon SP 3713. 'The new gate'. *Newgate* 1797, *New-gate* 1833 OS. ModE **new** + dial. **yatt** (OE **ġeat**). O 447.

NEW YORK Lincs TF 2455. *New York* 1824 OS. The name New York occurs frequently in f.ns. for a remote piece of land. In this instance its use may have been prompted by the proximity of Boston. It probably dates to the reclaiming of Wildmore Fen in the 18th cent. One report claims that the builder who developed the settlement was a native of York. Field 148, Room 1983.83, Stokes.

NEW YORK T&W NZ 3270. *New York* 1863 OS. Names of this type, unless there is a special American connection, most often originate as field names, often transferred to farm-houses. They were often given as ironic allusions to the remoteness of the field from the farm-house.

NIBLEY Avon ST 6982. 'Peak wood or clearing'. *Nubbelee* 1189, *Nibelee* 1287, *Ni- Nybley* from 1444. OE ***hnybba** or ***hnybbe** + **lēah**. Gl iii.68, ii.240.

North NIBLEY Glos ST 7495. *North Nibley* 1522. ModE adj. **north** + p.n. *(to) Hnibban lege* [940]12th S 467, *Nub(b)elee -lei(a) -leg(e) -legh -ley(e)* 1189–1444, *Nibbeleg(h) -ley* 1248, 1441, 1492, *Ni- Nyblega -ley* c.1250–16th, 'wood or clearing of the peak', OE ***hnybba**, ***hnybbe**, genitive sing. ***hnybban**, + **lēah**. The reference is to Nibley Knoll (on which the Tyndale Monument was built in 1866) referred to in the bounds of S 467 as *nybban beorh* 'hill of the peak'. 'North' for distinction from NIBLEY Avon ST 6982. Gl ii.240.

NICHOLASHAYNE Devon ST 1016. 'Nicholas's (woodland) enclosure'. *Nycolesheghe* 1305, *Nicholeshaine* 1612, *Nicholson* 1814. ME pers.n. *Nicholas* + **hain** (OE ***hæġn**). D 613.

River NIDD NYorks SE 3357. 'Flowing water'. *fluuium Nidd* [c.731]8th BHE, *Nide stream* [c.890]c.1000 OEBede, *Nid* [c.715]11th Eddi–1734, *Nidd* from 1181×90. PrW **Nīd*, Old European **Nidā, *Nidi-, *Nido-* from IE **neid-/*nid-* 'flow'. This is a widespread r.n. formation occurring e.g. in Neath W Glamorgan, *Nidum* [c.300]8th, Nethe Belgium, *Hnita* 726, *Nita* 1008, Nethe, Höxter, Germany, *Nithega* 935, Nidda, a tributary of the Main, *Nida* [c.300], *Nidda* 9th, Nied, a tributary of the Saar, *Nida* [c.300], *Nita* 1018, *Neda* 1121, Neide, East Prussia, *Nyda* 1343, Nitja Norway. See also STRATTON Corn SS 2206. RN 302, YW vii.132, LHEB 558, BzN 1956.5–8, 1957.251, Krahe 1964.48, Berger 198.

NIDD NYorks SE 3060. Transferred from the r.n. NIDD SE 3357. *Nit(h)* 1086, *Nid, Nyd* 1165–1428, *Ni- Nydd(e)* 1167–1574. YW v.97.

NIDDERDALE NYorks SE 0976. 'The valley of the river Nidd'. *Nidderdale* from c.1150, *Netherdale* 1249–1595. R.n. NIDD SE 3357 with ON genitive sing. inflectional ending **-ar**, + **dalr**. YW v.76 gives pr ['niðədil].

NINE ASHES Essex TL 5902. 'The nine ash-trees'. *Nine Ash* 1768, *Nine Ashes* 1805 OS. ModE **nine** + **ash**. Ess 75.

NINE BARROW DOWN Dorset SZ 0081. *Nine Barrow Down* 1774. Part of the main Purbeck ridge so called from the many tumuli there, originally 19 according to local tradition. Do i.39.

NINE STANDARDS RIGG Cumbr NY 8206. *Nine Standards Rigg* 1861. P.n. Nine Standards + dial. **rig** 'a ridge'. The Nine Standards, *the Nine standares* 1636, *ye Nine standers* 1687, were a set of boundary marks on the hill between Winton NY 7810 and Hartley NY 7808, ModE **stander** 'an upright pillar'. We ii.29.

NINEBANKS Northum NY 7853. 'The nine benches'. *Ninebenkes* 1228, *Nen-* 1230, *Nynbenkys* 1479, *Nyne Benkes* 1542, *Ninebanks* c.1715 AA 13.9. OE **nigon** + **benċ**. The reference is possibly to 'steps' on the road up the West Allen valley. NbDu 150.

NINFIELD ESusx TQ 7012. 'Open land taken in'. *Nerewelle* (sic) 1086, *Nanefeld* 12th cent., *Ni- Nemene(s)feld -feud* 1340–1535, *Nindfeild* 1672, *Ninvill al. Ninfeild* 1707. OE past participle adj. **(ġe)numen** or **nīwe-numen**, definite declension dative form **-numenan**, + **feld** in the phrase *æt þǣm nīwe-numenan felde* 'at the newly taken-in open land'. The reference is probably to land newly taken in for use as pasture by the coastal communities to the S. Sx 487, SS 70.

NINGWOOD IoW SZ 4088. Probably the 'enclosed wood', i.e. taken into cultivation or by the land newly taken into cultivation. *Lenimcode* (sic) 1086, *Ni- Nyng(e)wod(e)* 1189×1204–1583, *Nyngwood* 1635. OE ***niming** + **wudu**. The DB form represents Fr definite article + *nimc* < *niming* + *ode* for *wode*. Wt 211, Mills 1996.76.

NITON IoW SZ 5076. 'The new settlement'. *Neeton* 1086, *Neuton* 1155×8, *Neweton(e)* 1181–14th, *Nitona* 1189×1204, *Ny- Niton* from 1305, *Ny- Nighton* 1488×9–1611. OE ***nīġe, nīwe** + **tūn**. Wt 180 gives prs [nitn, naitn] and [noitn], Mills 1996.76.

NOAK HILL GLond TQ 5493. *Noak Hill* 1490, *Nookhill* 1523, *Noke Hill* 1570. The home of Richard *ate Noke* c.1290 from misdivision of ME *atten oke* (OE *æt thæm āce*) 'at the oak-tree'. Ess 116, LPN 162.

NOBOTTLE Northants SP 6763. 'The new building'. *Neubote* 1086, *Neubottle* 12th–1511 with variants *Niwe- Newe-* and *-botle -botel*. OE **nīwa-bōtl**. Identical with NEWBOTTLE. Nth 80.

NOCTON Lincs TF 0664. 'The wether-sheep farm'. *Nochetune* 1086, *-tun(i)a* 1155, *Nocton(')* from 1177. OE **hnoc**, genitive pl. **hnoca**, + **tūn**. Perrott 334.

River NOE Derby SK 1385. Pre-English r.n. of uncertain origin and meaning. *Noue* before 1300, *Now(e)* 1330, 1577–1622, *Noe* 1636. Comparison with the G r.ns. Nahe, a tributary of the Rhine, *Nava* 1st-4th cents., *Nauua* 8th, *Nah(a), Naa* 12th–13th cents., Nau, a tributary of the Danube at Ulm, *Nawa* 1150, Naaf, a tributary of the Agger, *Nafe* 13th, the Spanish r.n. Navia, and with different suffix, the r. NADDER Wilts SU 1003, suggests IE **nāu̯* 'ship, trough, valley, groove' as a possible source of this name. An alternative suggestion is Brit ***nāv-** < IE **(s)nāv-* 'flow', as in W *nawf, nofio* 'swim', Latin *no, nare*, Gk *νάω*, Skr *snāuti* 'drips'. From the same source comes the Romano-British name of the fort at BROUGH on Noe SK 1882, *(a) Navione* RIB 2243, [c.700]14th Rav, Brit ***nav-io-**. Db 13, RN 303–4, Flussnamen 100, Chantraine 748f, RBrit 423–4, Berger 192.

NOKE Oxon SP 5413. '(The settlement) at the oak-trees'. *Ac(h)am* 1086, *Ok(e)* 1213–1476, *Noke* from 1382 with variants *Nocke* 1590, *Noake* 18th. OE **āc**. The name arose through misdivision of ME *atten oke* 'at the oak tree' (OE *æt þǣm ācum*). O 232, L 219.

NO MAN'S HEATH Ches SJ 5148. 'Heath nobody owns'. *Nomonheth* 1483, *No Man Heath* 1671. Che iv.9.

NO MAN'S HEATH Leic SK 2909. *No mans Heath* 1835 OS. A place on the county boundary between Warw and Leic. The sense is boundary land claimed by more than one parish or owner. Field 150.

NOMANSLAND Devon SS 8313. *Nomansland* 1809 OS. Situated on Witheridge Moor on the boundary between Witheridge and Templeton.

NOMANSLAND Wilts SU 2517. *No Mans Land* 1811 OS. An extra-parochial place on the Hampshire border between the parishes of Landford and Branshaw Hunts. Wlt 391.

NONELEY Shrops SJ 4828. 'Nunna's wood or clearing'. *Nuneleg* 1221, *Nunnileg, Noni- Nonelegh, Noneleye* 13th. OE pers.n. *Nunna* + **lēah**. Bowcock 173.

NONINGTON Kent TR 2552. 'Estate called after Nunna'. *Nunningitun* c.1100, *Noninton' -yn -ig-* 1240–91, *Nonyngton* 1282. OE pers.n. *Nunna*+ **ing**[4] + **tūn**. PNK 534.

NOOK Cumbr NY 4679. *The Nuke* 1587. ModE *nook* occurs frequently in field and minor names in Cumbria. Cu 63.

NOOKTON FELL Durham NY 9148. *Knucton Fell* 1866 OS. P.n. Nookton + ModE **fell**. Nookton NY 9247 is *Knokeden(e)* c.1190–1382, *Nuketon* [1596]18th, *Knuck- Knowckden, Knewkdon* 17th cent., *Nuckton* 1768. The generic is OE **denu** 'a valley', and the specific probably a word for a hill related to ON *knjūkr* 'a mountain peak', OE **cnocc*, Dan dial. *knok*. This remote valley reaching up into mountainous country over 1800ft. high is aptly named 'mountain valley'. NbDu 150.

The NOOSE Glos SO 7207. A bar of sand-bank in the river Severn. *The Noose* 1779, but not connected with ModE *noose*. Rather this represents ME **atten ose** misdivided, earlier *atte(n) wose*, OE *æt thǣm wāse*, 'at the mud'. Cf. Wapping GLond TQ 3408, known in 1345 as *Wappinge atte Wose*, 'Wapping at the mud'. Gl iii.254, Mx 152.

NORBITON GLond TQ 1969 → SURBITON TQ 1867.

NORBURY 'North fortification or manor house'. OE **north** + **byriġ**, locative-dative sing. of **burh**.

(1) ~ Ches SJ 5647. *Norberie* 1086, *-bury* from 1305, *Northburi -bury* [12th]17th, 1250–1492. The most northerly part of the parish of Whitchurch Shrops SJ 5441. Che iii.109.

(2) ~ Derby SK 1242. *Nortberie* 1086, *Northbury* 1269–1448, *Norburie* late 11th cent., *-bury* from 1252. 'North' for distinction from SUDBURY SK 1632. Db 591.

(3) ~ Glos SO 9915. *Norberry* 1839. The reference is to Norbury Camp, an Iron Age hill-fort. 'North' for distinction from Southbury SO 9913, *Soncheberi* (sic for *Southe-*) 1306, probably 'land south of the fortification'. Gl i.155.

(4) ~ Shrops SO 3693. 'The north manor-house'. *Norbir'* 1236, *Northbur'* 1242, *Norbury* from 1316. Sa i.223.

(5) ~ Staffs SJ 7825. *Nortberie* 1086, *Nordbiri* 1198, *Northbury* c.1291–1428 with variant *-burgh, Norbury* from 1316. N of SUTTON 'the south *tūn*' SJ 7622 and WESTON JONES SJ 7524. St i.173.

NORDELPH Norf TF 5500. *Nordelph* 1824 OS. ModE **north** + **delf, delph** 'a fenland drainage channel'.

NORDEN Dorset SY 9483. 'North hill'. *Northdon* 1291, *Nordon(e)* 1381, c.1586–1861, *Norden* [1381]16th, 1795. OE **north** + **dūn**. The reference is to the ridge called Norden Hill $\frac{1}{4}$ mile to the S which rises to 600ft. Do i.16.

NORDEN GMan SD 8514. This is a modern settlement and appears to be a corruption of p.n. Naden in NADEN RESERVOIR SD 8516.

NORDLEY Shrops SO 6996. 'The north wood or clearing'. *Nordleia* [c.1090]15th, *Nortleia* c.1200, *Northleg'* c.1225. OE **north** + **lēah**. 'North' for distinction from ASTLEY ABBOTS SO 7096, 'the east wood or clearing'. Gelling.

NORFOLK '[The territory of] the northern people'. *(on) Norfolke* [1043×5]13th S 1531 *(on) Norðfolce* l.11th ASC(D) under year 1076, *Nordfolc, Norfulc* 1086. OE **north** 'north' + **folc** 'folk, people'. Named in contrast to SUFFOLK, the southern people of the East Anglian region.

NORHAM Northum NT 9047. 'The north estate' sc. of the community of St. Cuthbert. *Northham* [c.1040]12th HSC, *Northam* [1097]copy, *Norham* from c.1107×23, c.1160×5 DEC, *Norram* 1584. OE **north** + **hām**. This name probably began life as an appellative description of the place originally called *Ubbanford* c.1030 Secgan, c.1107 SD, 'Ubba's ford', which it subsequently displaced. Norham was the administrative centre of Durham abbey's estates in N Northumberland known collectively as Norhamshire, *Norhamscire -a* 1099×c.1122, c.1122 DEC, *-syr'* 1208×10 BF, *Northamscire* c.1124×8 DEC, *-shire* 1464 FPD. NbDu 150, DEPN.

NORLEY Ches SJ 5772. 'North clearing'. *Norl(oȝ')* 1239, *Norley* from 1316 including [1239]17th, *Northleg'* [1239]1580, *Nortlegh, North(e)ley(e) -le(gh)* 1259–1359. OE **north** + **lēah**. N of Eddisbury SJ 5569 and Oakmere SJ 5769 in Delamere Forest. Che iii.249.

NORLEYWOOD Hants SZ 3597. P.n. Norley + ModE **wood**. Norley, *Norley F.* 1810 OS, occurs as the specific of *Northlyghesdych* 'the ditch of Norley' 1298, *Northleghe by Baddesleye* 1306, 'the north wood or clearing', ME (OE) **north** + **legh** (OE *lēah*), on the edge of Beaulieu Heath. Ha 123, Gover 1958.205.

NORMANBY 'Village, farm of the Norwegians'. ON *Northmenn*, genitive pl. *Northmanna*, + **bȳ**. A name-type indicating isolated settlements of Norwegians.

(1) ~ Humbs SE 8816. *Normanebi* 1086, c.1115, *Nordmanabi* c.1115. DEPN, SSNEM 61.

(2) ~ NYorks SE 7381. *Normanebi -by* 1086–1308, *Northmannabi* c.1130, *Normanby* from c.1150. YN 57, SSNY 34.

(3) ~ -BY-SPITTAL Lincs TF 0088. *Normanby next Spittel* 1545. P.n. *Normanestov, Normanebi* 1086, *Nordmanabi* c.1115, + ModE **spittle**, 'a hospital, a hostel' referring to Spital in the Street SK 9590, a medieval hostel for the poor on the Roman road from Lincoln to the Humber, Margary no.2d. SSNEM 60, Pevsner-Harris 1964.375.

(4) ~ LE WOLD Lincs TF 1295. 'N on the wold'. *Normanby on the Hill* 1563, ~ *super montem* 1558–1762, ~ *super Waldam* 1294, ~ *on the Wolds* 1824. P.n. *Normane(s)bi* 1086, *Nordmanabi* c.1115, *Norman(n)ebi -by* 1150–1338, *Northmannebi* c.1200, *Normanby* from 1192, + Fr definite article **le** short for *en le* + ModE **wold** for distinction from NORMANBY-BY-SPITTAL. Li iii.71, SSNEM 61, Cameron 1998.

NORMANDY Surrey SU 9351. *Normandie -y, Normanby* 17th cent. Named from an inn called 'The Duke of Normandy'. Sr 136.

NORMAN'S GREEN Devon ST 0503. *Normans Green* 1809 OS. Pers.n. or surname *Norman* + ModE **green**. Adjacent to *Pusham Green* 1809 OS. A John Norman is recorded here in 1244. D 568.

NORMANTON 'Settlement of the Northman or -men'. OE *Northman*, genitive sing. *Northmannes*, genitive pl. *Northmanna*, + **tūn**. SSNEM 262, 266.

(1) ~ Derby SK 3433. *Normanestune* 1086, *Normantun(e) -ton(e)* 1086–1585, *-menton'* 1195–1242. *Northman* in this name appears to be singular. Db 649.

(2) ~ Lincs SK 9446. *Normenton* 1086, *-tun* 1154×89, *Normanton(a)* from 1185. Perrott 385.

(3) ~ Notts SK 7054. *Nor(d)mantune* [958]14th S 659, *Normantun* 1086, *-ton* from 1252, ~ *juxta Suthwell* 1280 etc. Nt 176.

(4) ~ WYorks SE 3822. *Norme- Normatune, Normantone* 1086, *Normanton* from 1177×86, *Northmanton(e)* 13th cent. YW ii.121.

(5) ~ LE HEATH Leic SK 3712. 'N on the heath'. *Normanton othe heth(e)* 1299–1376, *on Heth* 1376, *super le Heth* 1327, 1356. Earlier simply *Normanton* 1247–1576. Probably an English village taken over by a group of Norwegians at the time of the original Viking settlement. It lies on the great heath that once stretched W from the edge of Charnwood Forest. Lei 524, SSNEM 262.

(6) ~ ON SOAR Notts SK 5123. P.n. *Normantun(e)* 1086, *-ton(e)* from 1086, + r.n. *(super) Sore* 1224. Nt 254.

(7) ~ -ON-THE-WOLDS Notts SK 6233. P.n. *Normantun -tone* 1086, *-ton* from 1303 + affix *super le Woulds* 1682, Latin **super**, ModE **on**, Fr definite article **le**, ModE **the** and **wolds**. Nt 238.

(8) ~ ON TRENT Notts SK 7968. P.n. *Normentune -tone* 1086, *Normanton* from c.1130, + r.n. *(super) Trentam* 1272 etc. Nt 193.

(9) South ~ Derby SK 4456. *South Normanton* 1563, 1577, for distinction from Temple NORMANTON SK 4167. Earlier forms for South Normanton are *Normentune* 1086, *Normantun -ton(a) -tone* before 1166–1216×72 etc. Db 280.

(10) Temple ~ Derby SK 4167. *Normanton Templer* 1330. Held by the Knights Templar 1185. Also known as *North Normanton* 1460 for distinction from South NORMANTON SK 4456. Earlier simply *Normantune* 1086, 1185, *-ton(a)* early 13th etc. Db 282.

NORMOSS Lancs SD 3337. 'North bog'. Adj. **north** + dial. **moss** (OE *mos*) 'a swamp, a bog'. Situated at the northern edge of Marton Mere SD 3435, formerly a lake of considerable size. See Great MARTON Lancs SD 3335.

NORRINGTON Wilts ST 9623 → NORRINGTON COMMON ST 8864.

NORRINGTON COMMON Wilts ST 8864. P.n. Norrington as in *Norrington Green* c.1840 + ModE **common**. It is in the N part of Broughton Gifford parish. Early forms of Norrington ST 9623 in Alvediston parish, *Northinton(e) -yn(g)-* 1227–1327, *Norrington al. Northington* 1568, suggest that this name type derives from OE *north in tūne* '(the place) in the north of the township'. Cf. EASTINGTON Glos SP 1313 and SIDDINGTON Glos SU 0399. Wlt 119, 199.

NORRIS GREEN Mers SJ 3994. *Norris Green* 1842 OS. Surname *Norris* + ModE **green**. The name of the family that lived at Speke Hall. Mills 116.

NORRIS HILL Leic SK 3216. Surname Norris + ModE **hill**. Cf. *Norris Hill Road* 1807. Lei 342.

NORSEBURY RING Hants SU 4940 → DANEBURY Hants SU 3237.

NORTHALLERTON NYorks SE 3794. *Northallerton* from 1371. ME adj. **north** + p.n. *Aluretune, Aluertun(e), Alverton(e)* 1086–1444, 'Alfhere's farm, village, estate', OE pers.n. *Ælfhere* + **tūn**. 'North' for distinction from other Yorkshire Allertons. YN 210.

NORTHAM Devon SS 4429. 'North homestead or estate'. *Northam* from 1086, *Norham* 1219, 1238. OE **north** + **hām**. N of Bideford. D 102.

NORTHAM Hants SU 4312. 'The north river-bend land'. *Norham* 1151 and simply *Homme* 1291. Possibly the *Northam* in Redbridge Hundred held by Ansketel in 1086. OE **north** + **hamm** 1. N of Southampton. Ha 124.

NORTHAMPTON Northants SP 7560. 'The northern *Hamtun*'. *Norðhamtun* 12th ASC(C) under the year 1065, *Northantone* 1086, *Norhthamtun* 1122 ASC(E), *Norhamptoun* c.1300. OE **north** + p.n. *Hamtun* 10th ASC(A) under the year 917, 12th ASC(E) under the year 1140, *Hamton* 972×92 B 1130, *Hantone* 1086, *Hamtona* 1219, OE **hām-tūn** 'a homestead enclosure'. 'North' for distinction from SOUTHAMPTON SU 4112. Nth 6, xix, DEPN.

NORTHAW Herts TL 2802. 'North enclosure, north enclosed wood'. *sylvam quae dicitur North Haga* 'wood called N' [1077–93]14th, *Northawe* [1189×99]1301, *(pro) bosco de Norhal'* 1161, *boscum de la Northaga* 1201, *Northaw al. Northall* 1603×25. OE **north** + **haga**. 'North' for distinction from a lost *Suthhag wood* [1189×99]1301, *Southawe* 1432 in East Barnet. All this area was part of a dense forest on the Middlesex-Herts border. Hrt 113.

NORTHBOROUGH Cambs TF 1508. 'Northern fortified place'. *Norðburh* 12th ASC(E) under year 656, *Northburg* 12th–1338, *Northborough al. Norborough* 1541, *Narborow* 1549, 1702. OE **north** + **burh**. 'North' probably with reference to Peterborough. Nth 239.

NORTHBOURNE Kent TR 3352. 'The north stream'. *Northburne* *[618]13th S 6 (with 15th cent. variant *Nortburne*), 1240, *Norborne* 1086. OE **north** + **burna**. The reference is to the stream called North Stream TR 3556. KPN 8, Charters IV.23.

NORTHCHAPEL WSusx SU 9529. *North-Chapell* 1514. ME **north** + **chapel**. A chapelry in and N of Petworth. Sx 113.

NORTH CHEEK NYorks NZ 9606. *North Cheek or Bay Ness* 1857 OS. ModE **north** + **cheek**. Contrasts with SOUTH CHEEK NZ 9802. *Bay Ness* refers to Robin Hood's Bay.

NORTHCHURCH Herts SP 9708. Short for North Berkhamstead Church. *le Northcherche* 1347, *Northchurche* 1357–1556. ME **north** + **cherch**. The earliest references are to *ecclesia Sancte Marie Berchamsteda* 1209×35 and *ecclesia de North Berchmastede* 1254, 1291, 'St Mary's church of North Berkhamstead'. Hrt 48.

NORTHCOTE MANOR Devon SS 6218. P.n. Northcote + ModE **manor**. Northcote, *Northecoth'* 1242, *Northcote* 1303, *Narracot(t)* 1759, 1809 OS, is 'north cottage(s)', ME **north** + **cot**, pl. **cote** (OE , *cot, cotu*). 'North' from Burrington SS 6416 and in contrast to Aylescott SS 6116, *Aylescote* 1333 'Æġel's cottage(s)', and Upcott SS 6118 in the same parish. D 363.

NORTHCOTT Devon SX 3392. 'North cottage(s)'. *Northcote* 1263, 1345, *Northcot Hamlet* 1809 OS. ME **north** + **cot**, pl. **cote** (OE *cot, cotu*). N of St Giles on the Heath SX 3590. D 154.

Upper NORTH DEAN Bucks SU 8498. *Uppr North Dean* 1822 OS. ModE adj. **upper** + p.n. *Northdene* 1325, 'the northern part of the valley' sc. of HUGHENDEN VALLEY SU 8695, ME **north** + **dene** (OE *denu*) as in Hughen*den*. 'Upper' for distinction from Lower North Dean SU 8498, *Low North Dean* 1822 OS. Bk 184.

NORTH DOWNS Kent TQ 6762. A range of high ground in N Kent stretching from Biggin Hill TQ 4159 to Charing TQ 9549 and contrasting with the SOUTH DOWNS ESusx TQ 3707.

NORTHEASE ESusx TQ 4106 → SOUTHEASE TQ 4205.

NORTH END ModE **north** + **end**.

(1) ~ Avon ST 4167. North end of Yatton ST 4365. *North End* 1817 OS.

(2) ~ Hants SU 6502. No early forms. Originally probably the N end of Copnor SU 6602.

(3) ~ Hants SU 4163. *Northende* 1441 sc. of East Woodhay SU 4061 and in contrast with Heath End SU 4162 and East End SU 4161, *Estende* 1447. Gover 1958.158.

(4) ~ WSusx TQ 1209. *North End* 1813 OS, sc. of Findon TQ 1208. Sx 199.

NORTHEND Avon ST 7768. North end of Batheaston: no early forms.

NORTHEND Bucks SU 7392. Northend 1830 OS sc. of Turville Park.

NORTHEND Warw SP 3952. 'The north end' sc. of Burton Dassett parish. *Northende* 1285–1332, *Nurthen al Northend* 1649. Contrasts with a lost Southend, *Suthende Derset* c. 1315, *Southend* 1332. Wa 269.

NORTHENDEN GMan SJ 8290. 'North enclosure'. *Norwordine*

1086, *-wurding -warding* c.1270, *Northwrthinam* (Latin accusative) *-w(o)rthin -yn* [1119]1280–c.1310, *Northurthin -yne -urdene -erden(e)* c.1310–1831, *Northenden* from 1439, *Northyn* 1379, *Northen* 1527, 1656, *Northern* 1841. OE **north** + **worthiġn**. Che i.234 gives prs ['no:rðəndən], older ['nourðin -ən].

NORTHEY ISLAND Essex TL 8706. *Northey Island* 1805 OS. P.n. Northey + pleonastic ModE **island**. Northey alone, *Northeye* 1413, is 'north island', OE **north** + **ēġ**, a name which superseded earlier *Siriches(h)eie* 1141×51, 1438, *Sydriches Hey* 1163×87, 'Sigeric's island or enclosure', OE pers.n. *Sīgeriċ*, genitive sing. *Sīgeriċes*, + either **(ġe)hæġ** or **ēġ**. Ess 218.

NORTH FEN ModE **north** + **fen**.
(1) ~ Cambs TL 2909. *North fen* 1697. N of Morris Fen. Ca 283.
(2) ~ Cambs TL 5169. *The Northffenne* 1331. The north fen of Waterbeach. Ca 187.

NORTHFIELD WMids SP 0279. 'The north open land'. *Nordfeld* 1086, *Northfeld* [c.1086]1190, *Norfeld* 1577. OE **north** + **feld**. 'North' like King's Norton SP 0579 with reference to Bromsgrove SO 9670. Wo 348, L 240, 244.

NORTHFLEET Kent TQ 6274. 'North Fleet'. *Norflvet* 1086, *Nor(th)flet* 1201, 1214. OE **north** + stream name *(to) flyote* [c.975]12th B 1322, OE **flēot** 'estuary, inlet, stream'. 'North' for distinction from SOUTHFLEET TQ 6171. KPN 303.

NORTH GREEN Norf TM 2288. Short for *Pulham North Green* 1838 OS. Contrasts with nearby BUSH GREEN TM 2187, SHELTON GREEN TM 2390, CROW GREEN etc. and *Pulham South Green* 1838 OS.

NORTH HEATH WSusx TQ 0621. ModE **north** + **heath**.

NORTH HILL Corn SX 2776. *Northulle* 1275–1311, *-hille* 1342, 1420. Earlier simply *Hulle* 1276. This appears to be ME **hull** (OE *hyll*) 'a hill' but is in fact a reinterpretation of *Northindle* 1260, *Northinle* 1270, earlier *Henle* 1238 of obscure origin, possibly OE **hind** + **lēah** 'hind wood or clearing', or Co **hen** + **le** 'ancient place'. 'North' for distinction from SOUTH HILL SX 3372. PNCo 130, Gover n.d. 166.

NORTH HILL Somer SS 9447. *North Hill* 1809 OS. ModE **north** + **hill** referring to the hill N of Minehead.

NORTH HINKSEY VILLAGE Oxon SP 4905. P.n. North HINKSEY SP 4905 + ModE **village**.

NORTHIAM ESusx TQ 8324. 'North Hiam'. *Nordhyam* c.1200, *Nordiam* 1675, *Northyham -hamme* c.1210–16th. OE adj. **north** + p.n. *Hiham* 1086, 'high' or 'yew-tree promontory or cultivated plot', OE **hēah** or more likely **īw, īġ** + **hamm** 2b or 7. 'North' for distinction from Higham or South Higham, the site of New Winchelsea built c.1280, *Iham* 1200–1509, *Suthyhomme* 1339. Sx 522, SS 83.

NORTHILL Beds TL 1446. 'The northern Gifle'. *Nortgiuele -gible* 1086, *Northgille* 1185, *Northgiuel(e) -yeuil -iuel* 13th cent., *-yevel(e)* 14th cent., *Noryevele* 1346, *Norrell* 1443. OE **north** + folk-n. *Ġifle* 'the people who live by the r. Ivel'. Cf. River IVEL TL 1938. Contrasts with and balances SOUTHILL TL 1442. Bd 93. YE xlvi mentions a pr [norəl] and an old rhyme *Green grass and sorrel, Five bells at Norrell.*

NORTHINGTON Hants SU 5737. 'Settlement of the *North-hǣme*, the people who live to the north or at Northam, the northern estate'. *(æt) Northametone* *[903]16th S 370, *Norhameton'* 1166×7, *Nor(t)hampton(e)* c.1300–1551, *Northyngton* 1533. Either OE **north** + **hǣma-tūn** or folk-n. **North-hǣme*, genitive pl. **North-hǣma*, + **tūn**. The reference is unknown but might be to Alresford, Swarraton or even Winchester since the estate belonged to Hyde abbey, Winchester, from the 10th cent. The name is in systemic relationship with other nearby *hæma-tun* names e.g. QUIDHAMPTON SU 5150. Ha 124, Gover 1958.83, DEPN.

NORTHINGTOWN H&W SO 8161 → SINTON GREEN SO 8160.

NORTHLANDS Lincs TF 3453. 'Newly taken in land to the north' sc. of Sibsey. *Sibsey Norland* 1824. ModE **north** + **land**.

NORTHLEACH Glos SP 1114. 'North Leach'. *Northlecch(e)* 1200–1330, *-lech(e)* 12th–1514, *-lacche* 1429, *-leach(e)* 1627, 1641. ME **north** + p.n. *Lecce* 1086, *Lec(c)h(e)* 1086–1220, from the r.n. LEACH SP 1707. 'North' for distinction from EASTLEACH SP 2005. Gl i.175.

NORTHLEIGH Devon SY 1995. 'North Leigh'. *Northleghe* 1291, *Norley* 1555. ME adj. **north** + p.n. *Lege* 1086, *Legh* 1228, 'the wood or clearing', OE **lēah**. 'North' for distinction from SOUTHLEIGH SY 2093. D 628 gives pr [nɔ:rli:].

NORTH LEVEL Cambs TF 2404. The North Level of Bedford Level, *North Levell* 1652. ModE **north** + **level**. Ca 211.

NORTHLEW Devon SX 5099. 'North (estate called) Lew'. *Northlyu* 1282, *Northleu* 1327×77, *Northlieu* 1353. ME adj. **north** + p.n. *Levia* 1086, *Lewe* 1274, *Liw* 1228, *Lyu* 1242, 1258, identical with r.n. LEW SS 5301. 'North' for distinction from LEWTRENCHARD SX 4586 on a different r. Lew. D 154, DEPN.

NORTHMOOR GREEN OR MOORLAND Somer ST 3332. See MOORLAND OR NORTHMOOR GREEN.

NORTHMOOR Oxon SP 4203. *Northmore* 1367. ME adj. **north** + p.n. *(to þam) more* [1059]c.1100 S 1028, *(la) More* 1200–1361, 'the marsh', OE **mōr**. O 366, L 54.

NORTHOLT GLond TQ 1285. 'The northern nooks of land'. *(æt) norð healum* c.950×68 S 1447, *Northala* 1086, *-hal(l)e* 1214–1341, *-hold* 1593, *-holt* 1610. OE **north** + **healh**, dative pl. **healum**. Contrasts with SOUTHALL TQ 1380. Mx 44.

NORTHORPE 'Northern outlying farm'. ON **north** + **thorp**.
(1) ~ Lincs TF 0917. *Nortorp* 1202, *-thorp* 1300, 1547, *Nordthorp* early 13th. The place lies ¼ m. N of Thurlby. Perrott 436.
(2) ~ Lincs SK 8997. *Nortorp* 1202. Earlier simply *Torp* 1086–1198 including *Þorp* *[1061×66]12th S 1059. 'North' for distinction from *Southorpe*, a lost vill marked by Southorpe Farm SK 8895, *altero Torp* 'the other T' 1086. SSNEM 119.

NORTHOVER Somer ST 5223. 'North ridge'. *Nordoure* 1180, *Northovere* 1242, *North Ouer* 1610. OE **north** + **ofer**. The reference is probably to a ridge on the river bank as seen from the river rather than simply the river bank. DEPN.

NORTHOWRAM WYorks SE 1126. 'North Owram'. *Northuuerum* 13th cent., *Northowram* from 1358. ME adj. **north** + p.n. *Ufrun* 1086, *Huuerū* 1166, *Ourom(e)* 1402, 1594, '(place) on the ridges', dative pl. **uferum** of OE **ufer**. The village stands on a flat-topped ridge overlooking Shibden Dale. Cf. SOUTHOWRAM SE 1123. YW iii.97, L 176–7.

NORTHREPPS Norf TG 2439. 'North Repples'. *Norrepes* 1086, *Nordrepples* 1185, *Northreppes* 1254. OE adj. **north** + p.n. Repples 'the strips of land', OE **reopul**, ***ripul** 'a strip of land'. The reference may be to cultivated strips of land in the fens. 'North' for distinction from SOUTHREPPS TG 2536. Cf. also REPPS TG 4117. DEPN.

NORTH SCALE Cumbr SD 1870. 'North shieling-hut'. *Northscale* from 1247. ON **northr** (or OE **north**) + **skáli**. La 205, SSNNW 66.

NORTH SHORE Lancs SD 3038. No early forms, adj. **north** + **shore**, the northern part of Blackpool shore. 'North' for distinction from South Shore SD 3033.

NORTH SIDE Cambs TL 2799. A modern name: the north side of the river Nene from Whittlesey in which parish it lies. In 1787 it was known by the name of its public house, the *Dog in Doublet*, in 1824 *Dog and Doublet* OS. ModE **north** + **side**. Ca 264.

NORTH STREET Hants SU 6433. 'North hamlet'. *Northstrete* 1548. ME **north** + **strete**. Identical with *Ropley Street* 1817 OS, p.n. ROPLEY SU 6431 + ModE **street** 'a hamlet' and for distinction from Gilbert Street SU 6532, *Gylbertstrete* 1548, family name *Gilberd* 1294 + ME **strete**. Gover 1958.88.

River NORTH TYNE Northum NY 8974. *North Tyne* 1866 OS. Adj. **north** + r.n. TYNE NZ 0261.

NORTHUMBERLAND 'The land of the *Northhymbre*'. *Norðhymbra lond* e.10th ASC(A) under year 895, *Norðhymbralande* 11th ASC(D) under year 1065, *Norðan hymbre* 12th ASC(E) under year 867, *Norhumberland* 1130. Group-name

Northhymbre 'dwellers north of the Humber', used of both the people and the land, OE **north** 'north' + r.n. HUMBER Humbs SE 9523, + OE **land** 'land'. The application of the name, originally that of a pre-Conquest kingdom extending from the Humber into southern Scotland, became progressively more restricted. NbDu xiii. DEPN.

NORTHWAY Glos SO 9233. 'Northern enclosure'. *Northihaia* [1100×35]1496, *Nordihaia* [1107]1300, *Northeia -eye* 1100×35–1639, *North(e)wey* 1568, 1570, *-way* 1702. OE **north** + **ġehæġ**. Gl ii.54.

NORTH WEST PASSAGE Scilly SV 8411. *The North West Channell* 1689. PNCo 130.

NORTH WHARF Lancs SD 3249. No early forms, adj. **north** + **wharf** (OE *hwearf*) 'a sand-bank'.

NORTH WEST POINT Devon SS 1348. The North West Point of Lundy Island.

NORTHWICH Ches SJ 6573. 'The northern Wich or salt-workings'. *Norwich* 1086–1646 with variant *Nore-*, *Northwich* from [1119]1150 with variants *Nort(he)- Nord-* and *-wic(he) -wyk -wi(c)ke* etc. OE **north** + **wīċ** 'a collection of buildings for a particular purpose', in this case the manufacture of salt. Northwich, Middlewich and Nantwich are collectively known as 'The Wickes', *les Wyches* c.1200. Northwich, originally only a few acres in extent, was an enclave in Witton township SJ 6673, known simply as *Wich* 1086, *le Wiche* 1299, 'the industrial estate' OE **wīċ**, while Witton, *Witune* 1086, *Witton* from c.1200, 'settlement at the **wīċ**', OE **wīċ** + **tūn**, was the residential quarter. Cf. the parallel in DROITWICH H& W SO 8963 and Witton SO 8962. Che ii.192. Addenda gives 19th cent. local pr [-witʃ, -wɛitʃ, -waitʃ].

NORTHWICK Avon ST 5686. 'North dairy-farm'. *(to) norþwican* [955×9]12th S 664, *Northwik(e) -wyk(e) -wick(e)* from 1284. OE **north** + **wīċ**, dative pl. **wīcum**. N of Redwick. Gl iii.137.

NORTHWOLD Norf TL 7597. 'North wold'. *Northuuold* [970]c.1100 S 779, *Nortwalde* 1086. OE **north** + **wald**. N of METHWOLD TL 7394 'the middle wold' and HOCKWOLD TL 7288. DEPN.

NORTHWOOD 'North wood'. OE **north** + **wudu**. L 228.

(1) ~ Derby SK 2664. *Northwod(e)* 1301–81, *-wood* 1625. N of Darley. Db 82.

(2) ~ GLond TQ 1090. *Northwode* 1435. ME **north** + **wode**. Originally the name of a wood and farm lying N of Ruislip. The present town dates from the building of the railway c.1880. Mx 47.

(3) ~ IoW SZ 4894. *Nortwuda* 1181×5, *North(e)w(o)de* 1199×1216–1552, *Northwood* 1364 and from 1544. N of Parkhurst Forest. Wt 185 gives pr ['nɔrθwud] and ['nɔrðud], Mills 1996.78.

(4) ~ Shrops SJ 4633. *Northwood* 1833 OS.

(5) ~ GREEN Glos SO 7216. *Northwood Green* 1830. P.n. Northwood + ModE **green**. Northwood, *Northw(o)da -wode -wood(e)* 1154×89–1698. Gl iii.203.

NORTH YORK MOORS NYorks SE 7398. No early forms.

NORTON 'North settlement'. OE **north** + **tūn**.

(1) ~ Ches SJ 5582. *Nortune* 1086, *-tun'* c.1200–1285, *-ton* from [1137]17th. N in relation to SUTTON WEAVER SJ 5479, cf. also ASTON SJ 5578 and WESTON SJ 508, all members of the ancient parish of Runcorn. Che ii.173.

(2) ~ Cleve NZ 4421. *Norðtun* [late 10th]11th S 1661, *Nortuna* 1108×14, *Norton(')* from 1228. Norton was part of an episcopal estate with its *caput* at Stockton 2 miles to the south.

(3) ~ Glos SO 8524. *Nortvne - tune* 1086, *-tuna -tuun -ton* 1221–1672. Four miles N of Gloucester. Gl ii.150.

(4) ~ H&W SO 8851. *Nortona* 1208–1346†. Wo 151.

(5) ~ H&W SP 0447. *nortona* *[709]12th S 80, *(æt) norðtune* 1016×23 S 1423, *Nortvne* 1086. 'North' in relation either to Lenchwick SP 0347 or to Evesham SP 0344. The boundary of the people of Norton is *norðamere* (for *norðhæma gemære*) *[709]12th S 80. *Mortun* [714]16th S 1250 is probably a mistake for *Nortun*. Wo 264, Hooke 23, 46, 47.

(6) ~ Herts TL 2234. *(æt) norðtune* [1007]11th S 916, *Norton(e)* from 1086. N of Letchworth. Hrt 114.

(7) ~ IoW SZ 3489. *Northtone* 1248, *Nortone* 1271, *Norton* 1608. N of Freshwater. Wt 130, Mills 1996.78.

(8) ~ Northants SP 6063. *Norton(e)* from 1086 with variant *-tune* 1242. N in relation to Dodford SP 6160. Nth 27.

(9) ~ NYorks SE 7971. *Norton(e) -tun(a)* 1086 etc. N in relation to SUTTON GRANGE SE 7970. YE 140.

(10) ~ Notts SK 5772. *Nortun(a)* 1159–81, c.1179 etc. with variant spelling *-ton(e)*, *Norton Cuckney* 1546, *Northtowne Cuckney* 1585. 'North' in relation to CUCKNEY SK 5671 and once actually recorded as *Norkukeneia* 'north C'. Nt 88.

(11) ~ Shrops SJ 7200. N of Stockton SO 7399. An exact parallel to the Durham Norton-Stockton pairing.

(12) ~ Shrops SJ 5609. N of Wroxeter SJ 5608.

(13) ~ Shrops SO 4681. *Nortune* 1086. N of Onibury. Sa i.224.

(14) ~ SYorks SE 5415. *Norton(e) -tun(a)* 1086–1596. The place is N in the parish. There is a Sutton at SE 5512, *Sutone* 1086, *Sutton* from 1170×80, 'south farmstead', OE **sūth** + **tūn**. YW ii.49, 50.

(15) ~ Suff TL 9565. *Nocturna, nortuna -e -turna* 1086, *Norton* 1610. DEPN.

(16) ~ WSusx SU 9206. *Norton* 1279. N of Aldingbourne SU 9205. Sx 65.

(17) ~ Wilts ST 8884. *Nort(h)un* [931]c.1400 S 415, [937]12th S 436, *Nortone* 1086. N with reference to Hullavington. W 72.

(18) ~ BAVANT Wilts ST 9043. 'N held by the Bavant family'. *Nortonbavent* 1381. P.n. *Nortone* 1086 + manorial addition from the family name of Alesia de Bavent who held here in 1301. N with reference to Sutton Veny ST 8941. W 154.

(19) ~ BRIDGE Staffs SJ 8730. *Norton Bridge* 1836 OS. P.n. N as in Cold Norton SJ 8732, *Calde, Colde Norton* 1227, + ModE **bridge**. DEPN, Horovitz.

(20) ~ CANES Staffs SK 0108. 'N in Cannock Chase'. *Norton-super-le-Canok* 1289, *Norton Kaims, Kaynes* 1566, *Cank Norton, Norton Caynes, Norton on Canuch* 1579. P.n. *(æt) Norððtune* [951]14th S 554, *Nortone* 1087, + pseudo-manorial Canes ? < p.n. CANNOCK SJ 9710. DEPN, Duignan 107, Horovitz.

(21) ~ CANON H&W SO 3847. 'Norton held by the canons' sc. of Hereford cathedral. *Norton Canons* 1327. Earlier simply *Nortvne* 1086. The canons of Hereford held this estate already by 1086. DEPN.

(22) ~ DISNEY Lincs SK 8959. 'N held by the d'Isney family'. *Norton de Iseny* 1299. Earlier simply *Nortune* 1086. The vill was held by a family from Isigny in Normandy; a William de Ysini is mentioned c.1150. DEPN, Cameron.

(23) ~ EAST Staffs SK 0208. P.n. Norton as in NORTON CANES + ModE **east**.

(24) ~ FERRIS Wilts ST 7936. 'N held by the Ferrers family'. *Nortone* 1086 + manorial addition from the family name of John de Ferrers who held here in 1368. Also known as *Norton Mussegros* 1368 from Robert de Muscegros who had a holding here in 1253. N of Stourton ST 7734. Wlt 175.

(25) ~ FITZWARREN Somer ST 1925. 'N held by the Fitzwarren family'. Earlier simply *Nortone* 1086, *Nortun(e) -tone* [1135×7, 1174×80]14th. Also know as *Nortun by Tantone* (Taunton) [1091×1106, 1100×18]14th. DEPN.

(26) ~ GREEN IoW SZ 3488. P.n. NORTON SZ 3489 + ModE **green**. Earlier *(le) Norton Com(on)* 1608, 1781, and *More Green* 1775–1852 from a lost p.n. *le More* 1299 'the moor or marshy ground'. Wt 130, Mills 1996.78.

(27) ~ HAWKFIELD Avon ST 5964. '(That part of) Norton held by the Hauteville family'. *Norton Hautevill* 1238. The Domesday manor of *Nortone* was divided between the Hautevilles and the Malrewards. Reginal de Alta Villa

†The form *norðtun* [989]11th S 1359 is wrongly attributed to Norton SO 8851 instead of Bredon's Norton SO 9339. ECWM no.327.

(Hauteville) held this portion before 1219. N of Chew Magna ST 5763. DEPN.

(28) ~ HEATH Essex TL 6004. P.n. Norton as in NORTON MANDEVILLE TL 5804 + ModE **heath**.

(29) ~ IN HALES Shrops SJ 7038. P.n. *Nortune* 1086 + p.n. Hales as in HALES Shrops. Sa i.223.

(30) ~ -IN-THE-MOORS Staffs SJ 8951. *Norton super le Mores* 1285. P.n. *Nortone* 1086 + ModE **in**, Latin **super**, Fr definite article **le**, and ME **more**. Also known as *Norton under Keuremunt* 1227. DEPN.

(31) ~ -JUXTA-TWYCROSS Leic SK 3207. 'N beside Twycross'. P.n. *Norton(e)* 1086–1576 + affix *iuxta Twycros(se)* 1327–1611 from p.n. TWYCROSS SK 3305. Also known as *Hogge Norton* 1333, *Hoggis -es Norton* 16th cent. alluding to pig farming and for distinction from East NORTON SK 7800 and Kings NORTON SK 6800. Lei 552.

(32) ~ -LE-CLAY NYorks SE 4071. *Norton in the Cley* 1536. P.n. *Norton(e)* 1086–1578 + OFr definite article **le** short for *en le* + ME **cley**. N in relation to KIRBY HILL SE 3968. Also known as *Norton in le Drit* 'N in the dirt' 1301. YN 182.

(33) ~ LINDSEY Warw SP 2263. 'N held by the Limsey family'. *Nortonelyndeseye* 1288, ~ *Lymsey -say* 1316, 1535. P.n. *Norton(e)* c.1080 etc., *Mortone* (sic) 1086, + manorial addition from the Limesi family from Limésy, N of Rouen, holders of the manor in the 12th cent. Wa 216, DEPN.

(34) ~ MALREWARD Avon ST 6065. '(That part of) Norton held by the Malreward family'. *Norton Malreward* 1238. Cf. NORTON HAWKFIELD ST 5964. William Malreward held this part of the manor in 1238. DEPN.

(35) ~ MANDEVILLE Essex TL 5804. 'N held by the Mandeville family'. *Norton' Maundeuil'* 1262. Earlier simply *Nortuna -ā* 1086, *Norton(a)* [1068]1309. N with reference to its location in the hundred of Ongar or because N of *Plumtunā* 1086 if this is to be identified with Plunker's Green in Doddinghurst. The manorial affix comes from the family of Ernulf de Maundevill 1251. Ess 71, 153.

(36) ~ MARSHES Norf TG 4100. *Norton Marsh* 1838 OS. P.n. Norton as in NORTON SUBCOURSE TM 4198 + ModE **marsh**.

(37) ~ ST PHILIP Somer ST 7755. 'St Philip's N'. *Norton Sancti Phillipi* 1316. P.n. Norton + saint's name *Philip* from the dedication of the church. Earlier simply *Nortvne* 1086. DEPN.

(38) ~ SUBCOURSE Norf TM 4198. 'N held by the Subcourse family'. *Norton Supecors, Subcors* 1282, ~ *Soupecors* 1326. Earlier simply *Nortuna* *[1044×7]13th S 1055, 1086. The correct form of the family name may have been *Surlecors* 'on the river'; a Herman Sorlecors occurs 1177. DEPN.

(39) ~ SUB HAMDON Somer ST 4615. *Norton under Hamedon* 1246. P.n. *Nortone* 1086 + Hamedon as in HAM HILL ST 4816. *Sub* is Latin for *under*. DEPN.

(40) Bishop ~ Lincs SK 9892. 'N held by the bishop' sc. of Lincoln. *Bischopnorton* 1402. Earlier simply *Nortune* 1086, *Nordtuna* c.1115. The manor was held by the bishop at the time of DB. DEPN.

(41) Blo ~ Norf TM 0179. 'Blue, i.e. bleak Norton'. *Blonorton* 1291. ME adj. **blo** (ON *blá(r)*) + p.n. *Nortuna* 1086. DEPN.

(42) Bredon's ~ H&W SO 9339. *Northton in Bredon* 1320. Hill-n. BREDON SO 9236 + p.n. *norðtun* [780]11th S 116‡, [989]11th S 1359, *NORÐTUN, (æt) norðtune* 1058 S 1405, *Nortvne* 1086. Wo 151, Hooke 31, 152, 368.

(43) Brize ~ Oxon SP 3007. 'St Brice's N'. *Brice Norton* early 18th. Saint's name *Brice* from the dedication of the church + p.n. *Nortone* 1086 etc. However, the manor was also known as *Northone -ton(') Brun* 1284×5–1316, *Nort(h)on(e) Bruyn* 1320–1517, *Brimesnorton* (for *Brunes-*) 1388, *Bresenorton alias Norton Bruyn* 1517 from the family of William le Brun mentioned in 1200 in connection with land here. It is possible that *Brize* is a corruption of *Brunes* and the source of the dedication rather than vice versa. O 306.

(44) Burnham ~ Norf TF 8243. *Burnham Norton* 1457. P.n. Burnham as in BURNHAM DEEPDALE TF 8043 + p.n. *Norton* 1300. DEPN.

(45) Chipping ~ Oxon SP 3127. 'N with the market'. *Chepingnorthona* 1224 etc. ME **cheping** (OE *cēping*) + p.n. *Nortone* 1086 for distinction from Over NORTON SP 3128. O 368.

(46) Church ~ WSusx SZ 8795. ModE **church** + p.n. *Norton* c.1300. N of Selsey SZ 8593 which is referred to as *Suthton* in the same document. Sx 83.

(47) Cold ~ Essex TL 8500. *Coldenorton* 1350, *Coldnorton alias Gold Norton* 1605. ME adj. **cold** + p.n. *Nortunā* 1086. N of the river Crouch. Ess 221.

(48) East ~ Leic SK 7800. *Est(e)norton(e)* 1271–1557, *East Noorton* 1604. ME **est** + p.n. *Norton(e)* 1086–1610. 'East' for distinction from King's NORTON SK 6800. Lei 312.

(49) Greens ~ Northants SP 6649. *Grenesnorton* 1465, 1580. Earlier simply *Norton(e)* from 1086. Also known as *Norton Davy* 1329. N in relation to Towcester. Held by Henry Grene, knight, in 1369, and by David, son of Griffin, in the 13th cent. Nth 42.

(50) Hook ~ Oxon SP 3533. → HOOK NORTON SP 3533.

(51) King's ~ Leic SK 6800. ME prefix **king**, genitive sing. **kinges**, *Ki- Kynges* [1189×99]1253, 1237–1306, *King's* 1798, + p.n. *Norton(e)* 1086–1610. Also known as *West Norton(e)* 1284–1511 for distinction from East NORTON SK 7800. Norton is described in DB as parcel of the royal demesne appendent to the manor of Great Bowden, but the affix *King's* fell into disuse until the late 18th cent. when it was reintroduced by local antiquarians. Lei 243.

(52) King's ~ WMids SP 0579. *Kinges Norton* 1221, *Northone Regis* (Latin) 1286. ME **kyng**, Latin **rex**, genitive sing. **regis**, + p.n. *Nortune* 1086. The manor was held by the king in 1086. The most northerly *berewic* of the great composite estate of Bromsgrove SO 9670. Wo 350, DEPN.

(53) Over ~ Oxon SP 3128. 'Upper N'. *Ou' Norton* 1268, *Over Norton* from 1316. ME **over** + p.n. *Norton(')* from 1188. Also known as *Parva Norton'* 'little N' 1213, *Spitulnorton'* 'N with the hosptal' [before 1217]c.1400, and *Coldenorton'* c.1217×8 for distinction from Chipping NORTON SP 3127. The priory of the church of St John the Evangelist at Cold Norton, the hospital house (ME **spitel**) and the church were 'canonically instituted' in the period 1148×58. O 369.

(54) Pudding ~ Norf TF 9228 → Wood NORTON TG 0127.

(55) Wood ~ Norf TG 0127. *Wudnorton* 1199. ME **wode** (OE *wudu*) + p.n. *Nortuna* 1086. 'Wood' for distinction from Pudding Norton TF 9228, *Pudding Norton* 1276, earlier simply *Nortuna* 1086. DEPN.

NORWELL Notts SK 7761. 'The north spring'. *Nortwelle* 1086–1275, *Norwell* from 1166, *North(e)well(e)* 1259–1316. OE **north** + **wella**. 'North' for distinction from SOUTHWELL SK 7054. Nt 193 gives pr [nɔrəl], L 32.

NORWICH Norf TG 2308. 'The north hamlet'. NORDVICO, NOR(Ð)WIC c.900–939 Coins, *Norðwic* c.1122 ASC(E) under year 1004, [1035×40]11th S 1489, 1038, *Northwic* [c.1000]13th S 1525, *Norwic, Noruic* 1086, *Northwich(e)* c.1200, 1389, *Norwic(')* 1130–1401, *(de) Norwico, (apud, iuxta) Norwicum* 1176–1664, *Norwiz* 1194–1361, *Norwich(')-e* from 1173. OE **north** + **wīċ**. The modern city is a fusion of at least four separate settlements along the river Wensum, Conesford, *Cunesford(e)* 1147×9–1307, 'the king's ford', ODan **kunung**, secondary genitive sing. **kununges**, + **ford**, Coslany, *Coslania -(i)e -y(e)* 1146×9–1428, 'reed river island' or 'Cosa's river island', PrW ***cors** or OE pers.n. **Cosa*, + OE **lane, lanu** 'a lane, the course of a river in

‡The identification is not entirely certain and the reference may be to an estate in Gloucestershire, cf. D. Hooke, *The Anglo-Saxon Landscape: the Kingdom of the Hwicce*, Manchester 1985.84.

meadowland' + **ēġ**, Westwick, *Westwik -wic -wyc -wyk* 1153×60–1340, 'the western hamlet or district', and Northwick 'the northern hamlet' E and N of Westwick at the crossing of E-W and N-S routes. The crossing developed into a lively market centre and because of its importance gave its name to the whole city with *wīċ* very probably in the Middle Saxon sense 'trading place'. Norwich may thus be N in relation to this group of nuclear hamlets, not N of Dunwich or Ipswich as usually argued. Nf i.1ff, 98, 114, 152, Williamson 79–80.

NORWICH AIRPORT Norf TG 2113. P.n. NORWICH TG 2308 + ModE **airport**.

West NORWOOD GLond TQ 3171. ModE **west** + p.n. *Norwude* 1176, *Northewode* 1272, *Norwood* 1509×47, OE **north** + **wudu**. An extensive area of forest N of Croydon not entirely cut down until the 19th cent. Sr 50.

NORWOOD GREEN GLond TQ 1378. *Norwood Green* 1724. P.n. Norwood, *(apud) Northuuda* [830 for 833]12th S 1414, *Northwode* 1254, 1353, *Norwode* 1235, 1453, 'the north wood', OE **north** + **wudu**, + ModE **green**. Norwood was a chapel of ease to Hayes, but in relation to which place it was regarded as N is uncertain. Mx 45.

NORWOOD HILL Surrey TQ 2443. *Norwood Hill* 1816 OS. P.n. *Northewode* 1272, 1359, *(boscus voc.) Norwood* 1509×47, 'the north wood', OE **north** + **wudu**, + ModE **hill**. N of Croydon. Sr 50.

NOSELY Leic SP 7389. 'Nothwulf's wood or clearing'. *Noveslei* 1086, *-lai* c.1131, *Noues- Nousele(e) -ley(e)* 1209×35–1438, *Noweselee -ley(e)* 1334–1572, *Noseley* from 1509. OE pers.n. *Nōthwulf*, genitive sing. *Nōthwulfes*, + **lēah**. Lei 244.

NOSS MAYO Devon SX 5447. 'Maheu's Noss'. *Nesse Matheu* 1286, *Nasse Mayow* 1418. P.n. Noss, + OFr pers.n. *Maheu* 'Matthew'. Noss, *(La) Nasse* 1309–1400, is OE **næss** 'a headland'. Matheu son of John held the manor 1284–1309. D 257.

NOSTERFIELD NYorks SE 2780. Either 'the sheepfold' or 'the hillock field'. *Nostrefeld* 1204, *-er-* 1257–1521. OE **ēowestre** or ***ōster** + **feld**. Initial *N-* is due to misdivision of ME *atten ostrefelde* (OE *æt thǣm eowestre- osterfelda*) as *atte Nostrefelde*. YN 223, PNE ii.56, L 243.

NOTGROVE Glos SP 1120. Either '(at) the wet groves' or 'Nata's groves'. *ad Natangrafum* [737×40]11th S 99, *Nategraua -v- -grave* 1086–1407, *Notgrove* 1535. Either OE **næt**, dative pl. definite form **natan** for **natum**, or pers.n. **Nata*, genitive sing. **Natan*, + **grāf**, dative pl. **grāfum**. Gl i.175, L 193–4.

Black NOTLEY Essex TL 7620. ME adj. **blake**, *Blake* 1252, OFr **neir**, *Neir-* 1254, *Neyre* 1274, + p.n. *Hnutlea* 998 S 1522, *Nuthleam, Nutlea(m)* 1086, *Nut(t)el(eg) -leye* 1203–1300, *Not(t)(e)le(ye) -ley* from 1212, 'nut-tree wood or clearing', OE **hnutu** + **lēah**. Also known as Great or Magna Notley for distinction from White or Little NOTLEY TL 7818. Ess 293 gives pr [nɔ:tli].

White NOTLEY Essex TL 7818. *White* 1235, 1240, Latin *Alba* 1236, *Parva* 'little' 1212 + p.n. Notley as in Black NOTLEY TL 7620. Ess 293.

NOTTINGHAM Notts SK 5741. 'The homestead of the Snotingas, the people called after Snot'. *Snotengaham* late 9th ASC(A) under year 868, c.1000, c.1050 ASC(C,D) under same year, c.894 Asser, *Snotengeham* 924×939 coins, *-ingaham* c.925 ASC(A) under years 922 and 924–c.1120, *-ingeham* 1086, *Snotingham* c.1275, *Notingeham* 1130–1230, *Nottingham* from 1172×4. Folk name **Snotingas* < pers.n. *Snot* + **ingas**, genitive pl. **Snotinga*, + **hām**. The spirantisation and loss of initial [s] in the combination [sn] is due to AN influence, as in some forms of SNEINTON with the same OE pers.n. Nt 13 records local pr [nɔtiŋgəm] beside [nɔtiŋəm], ING 148, Cu lxxv.

NOTTINGTON Dorset SY 6682. 'Estate called after Hnotta'. *Nouington* (sic for *Notting-*) 1199, *Notinton* 1212, *Not(t)yngton* 1388–1483, *Nottington* 1594, *Nettington* (sic) 1811 OS. OE pers.n. *Hnott(a)* + **ing**[4] or genitive sing. *Hnottan*, + **tūn**. The specific might, however, contain the OE adj. *hnott* 'bare, bald' + *ing*[2], **hnotting* 'the bare, bald place'. Do i.200.

NOTTON Wilts ST 9169. 'The cattle farm'. *Natton* 1232–86, 1536, *Notton* from 13th, *Netton* 1284. OE **nēat** + **tūn**. Wlt 103.

NOTTON WYorks SE 3413. 'Wether-sheep farm'. *Notone, Nortone* 1086, *Nottun', Notton(a)* 1170–1610, *Nocton* 1186–1259. OE **hnoc** + **tūn**. YW i.283.

NOUNSLEY Essex TL 7910. Cf. *Nownsley Bridge* 1589, *Nounsley Green* 1777, 1805 OS, *Nounceley River* 1795, *Ouncelley End or Nouncels Green* 1890. The evidence is too late to give an etymology. Ess 290.

NOUTARD'S GREEN H&W SO 7966. *Noutard Green* 1832 OS. ME surname *Nout(h)ard* < AScand *noutherd* 'neatherd' + ModE **green**.

NOX Shrops SJ 4110. Short for *Nock's House* or the like. *Nocks* 1768, *Nox* 1783, 1833 OS. Richard Nock built an alehouse at this squatter settlement on the S fringe of Ford Heath in 1653. It was called the *Star and Ball*, now Nox House. The surname *Noc* occurs as early as 1255 < ME *atten oc* 'at the oak-tree'. Sa ii.44.

NUFFIELD Oxon SU 6787. Partly uncertain. *Tocfeld* [1180–1200]c.1200, *Thocfeld* 1241, *Toh- Togfeld* early 13th, *Tuffeld -feild(e) -filde -fyld(e)* 1349–1723, *Tuffield alias Nuffield* 1580, *Nuffield* from 1350. This has been explained as '(the settlement at) the open land or piece of land at the hill-spur', OE **hōh** + **feld**. The *T-* and *N-* forms would then derive from ME *atte* and *atten (H)oghfeld* respectively by misdivision. Many of the early forms, however, point rather to 'the tough open land', OE **tōh**, definite form **tō**, **tōga**, + **feld**. Nuffield is near the crest of the Chilterns. O 133, L 241.

NUNBURNHOLME Humbs SE 8548. 'The nuns' Burnholme'. *Nonneburnholm* 1543. ME **nunne** probably representing OE genitive pl. *nunna* + p.n. *Brunha'* 1086, *-ham* 1285, *Brunum* 13th cent., *Brunnum -om* [1189×99]1308–1418, *Brunhum -hom* 1286–1375, *Burn(e)holm(e)* 1521, '(the settlement) at the streams', ON **brunnr**, dative pl. **brunnum**. There are many springs and little streams round about. There was a Benedictine nunnery here. YE 180 gives pr [nun'bɔnəm], SSNY 101.

NUNEATON Warw SP 3691. 'Eaton with the nunnery'. *Nonne Eton* 1247, *Nunne Eton(e)* 1272–1309, *Nonneaton al. Nuns Eaton* 1654. ME **nunne** + p.n. *Etone* 1086–1376 'the river settlement', OE **ēa-tun**. The reference is to the river Anker. A Benedictine nunnery was founded here 1135×54. Wa 88.

NUNEHAM COURTENAY Oxon SU 5599. 'N held by the Courtenay family'. *Newenham Courteneye* 1320, *Newnham Courteney* 1476. P.n. *Neu- Neweham* 1086, *Newenham* [1236]c.1300–1428, '(at) the new homestead, OE **nīwe**, definite form dative case **nīwan** + **hām**, + manorial affix from the family name of John de Curtenay 1242×3 whose family came from Courtenay, Loiret. O 183.

NUNNEY Somer ST 7345. Partly uncertain. *Nuni* [946×55]13th S 1742, *Noiun* (for *Nonni*) 1086, *Nony* 1219, *Nuni* 1243, *Nuñye* 1610. Either 'Nunna's island', OE pers.n. *Nunna* + **īeġ** or 'the nuns' island', OE **nunne**, genitive pl. **nunnena**, + **īeġ**. DEPN, L 38.

NUNNINGTON NYorks SE 6679. 'Estate called after Nunna'. *Nonnin(c)- Noning- Nunnigetun(e)* 1086, *Nunintun -ton'* 1167, 1257, 1308, *Nunnington* from 1169. OE pers.n. *Nunna* + **ing**[4] + **tūn**. YN 54.

NUNNYKIRK Northum NZ 0892. Possibly 'Nunna's church'. *graungia vocata Noonekirke* 1536 Newm, *Graunge called Nonnykyrke* 1547 ibid. OE pers.n. *Nunna* + **ċiriċe** influenced by ON **kirkja**, but the forms are rather late for certainty. It was a grange of Newminster Abbey.

NUNTHORPE Cleve NZ 5314. 'Outlying farm held by the nuns'. *Nunnethorpe* 1301, *Nunthorpe* from 1328. Earlier simply *Torp* 1086–c.1200. ME **nunne** prefixed to ON, ME **thorp**. Nunthorpe was an outlying farm of Great Ayton; for a while there was a small Cistercian nunnery there. YN 166 gives pr [nunθrəp], SSNY 69, HistCleve 419.

NUNTON Wilts SU 1526. 'Nunna's estate'. *Nunton(a)* from 1209, *Nonyhampton, Nonyton* 1281. OE pers.n. *Nunna*, genitive sing. *Nunnan*, or *Nunna*, + **ing**[4] + **tūn**. Wlt 394.

NUNWICK Northum NY 8744. 'Nunna or the nuns' estate'. *Nun(n)ewic* 1165. OE pers.n. *Nunna* or OE **nunne**, genitive pl. **nunnena**, + **wīċ**. NbDu 151, DEPN.

NUPEND Glos SO 6001 → NETHEREND SO 5900.

NURSLING Hants SU 3716. 'The nutshell estate'.
I. *Nhutscelle* [8th]c.800 *Life of Boniface*, *Nutscille* c.1127;
II. *(æt) Nu(h)tscillingæ* [877]12th S 1277, *Hnut Scillingc* *[909]11th S 376, *(to) hnutscillingæ* [956]12th, *Notesselinge* 1086, *Nutselling(es) -sill-* c.1124–1204, *Nutscelling(e) -schull- -shill- -shell- -yng* c.1230–1330, *Nusselyng* 1269, *Nurselynges* 1483, *Nurslyng* 1536.
OE ***hnutusciell** 'the nutshell' was the name of the 8th cent. Benedictine monastery here of which St Boniface (Winfrith) became a monk in 702. It is after him that the Nursling nuclear power station was called Winfrith in 1960 (echoing Winscales in Cumbria). 'Nutshelling' with later addition of suffix **ing**[2] was the name of the monastic estate. Its boundary is *(on) Hnut scyllinga mearce* 'the mark of the *Hnutscyllingas*, the monks of the Nutshell'. Ha 124, Gover 1958.182.

NURSTEAD Kent TQ 6468 → NURSTED WSusx SU 7621.

NURSTEAD FARM Suffolk TL 9783 → NURSTED WSusx SU 7621.

NURSTED WSusx SU 7621. Probably 'the nut-tree place'. *Nursted(e)* 1236, 1316, *Nurstead* 1813 OS. OE **hnutu** + **stede**. Cf. Nurstead Kent TQ 6468, *Notestede* 1086, *(H)nutsted(e)* c.1100–1377, *Norstede* 1413, and Nurstead Farm, Suffolk TL 9783, *hnutstede* 1000×2 S 1486, *Nut(e)sted(e)* c.1190–1327, *Nustede* c.1190, *Nussted* 1369. Stead 192, 222, 271.

NUTBOURNE WSusx SU 7705 → SOUTHBOURNE SU 7705.

NUTBOURNE WSusx TQ 0718. 'The north stream'. *Nordborne* 1086, *-burne* 1195, *Nudburn* 1262, *Nubburne* 1263, *Nut(e)burn(e)* 1274–1327. OE **north** + **burna**. The reference is to the r. Chilt which forms a northern branch of the Stor at TQ 0617. Sx 154.

NUTFIELD Surrey TQ 3049. 'The open country with nut trees'. *Notfelle* 1086, *-feud* 13th cent., *-feld* 1376, *Nutfeld(a)* from 1169, *Notefeld* 1255, *Nu(t)te-* 1307, *Nhut-* 1313. OE **hnutu** + **feld**. Sr 302, L 239, 243.

South NUTFIELD Surrey TQ 3049. ModE **south** + p.n. NUTFIELD TQ 3049.

NUTHALL Notts SK 5144. 'The nook of land where nuts grow'. *Nutehale* 1086–1235, *Note-* 1211–1302, *Nutehall'* 1242, *Nuthall* 1442, *Nottale* 1391, *Nottoll* 1520. OE **hnutu** + **halh**. The reference is to a shallow valley. Nt 149 gives local pr [nʌtl], L 103.

NUTHAMPSTEAD Herts TL 4234. 'Nut-tree homestead'. *Nuthamsted(e)* 1141–1240, *Not-* 1255–1355, *Nutsted* 1676, 1730. OE **hnutu** + **hām-stede**. Hrt 183 gives common pr [nʌtstid], Stead 241.

NUTHURST WSusx TQ 1926. 'The nut-tree wooded hill'. *Nothurst* 1228–1361, *Nutehurst* 1230, 1592, *Nuthurst* from 1371. OE **hnutu** + **hyrst**. Sx 231, SS 67.

NUTLEY ESusx TQ 4427. 'Nut-tree wood or clearing'. *Nutlye* 1249, *Nutteley, Notley* 1291. OE **hnutu** + **lēah**. Sx 350, SS 67.

NUTWELL SYorks SE 6304. 'Nut-tree spring'. *Nutwell* 1841. ModE **nut** + **well**. YW i.38.

NYETIMBER WSusx SZ 8998. 'The new building'. *Neuetunbra* 1135×54, *Niwetimb'* 1279, *Ni- Nytimbre -tymbre* 1275–1508. OE **nīwe, nīġe** + **timber**. Sx 95.

NYEWOOD WSusx SU 8021. 'The new wood'. *la Nywode* 1337. ME **new, nye** (OE *nīwe, nīġe*) + **wode** (OE *wudu*). An earlier form *Iwode* 1332 shows loss of initial *N-* through misdivision of the ME phrase *at then nye-wode* as if *at then ye-wode*. Sx 37.

NYLAND HILL Somer ST 4550. *Nyland hill* 1610. P.n. Nyland + ModE **hill** referring to a hill rising to 251ft. from the marshes beside the river Axe. Nyland, *Nyland* 1817 OS, is probably a misdivision of ME **atten iland** 'at the island' as if *atte niland*. The original name to which 'the island' refers was *Andreyesie* [? 670×72]13th ECW, *Ederedeseie* *[725]12th S 250, *Ætheredesie, Ædredesye, Æthersig* [959×75]13th ECW, S 1761, *Adredeseya* [971]13th S 783, *Adredesia* 1344, 'Eadred's island', OE pers.n. *Ēadrǣd*, genitive sing. *Ēadrǣdes*, + **īeġ**. DEPN.

NYMET 'Pagan sacred place, sacred grove'. PrW ***nīved**, British **nemeto-* related to OIr *nemed* 'sacred place', Old Frankish *nimid*, Gaul *nemeton*, Latin *nemus* 'sacred wood', Gk *νέμος* 'wood'. LHEB 286, 555, 673, RN 304–5.

(1) ~ ROWLAND Devon SS 7108. 'Roland's Nymet estate'. *Nimet Rollandi* 1242, *Nymet Rolaund* 1292, *Nemytroland* 1558. P.n. Nymet + pers.n. *Roland*. Earlier simply *Limet* (sic) 1086 used originally as a r.n. for the r. later known as the Yeo SS 7306 which means simply 'the river', OE **ēa**. Forms for the r.n. are *(on, oþ) Ni- Nymed* [739]11th S 255, *(on, oþ, fram) Nymed* [739]10th, 15th ibid., [974]10th S 795, *(on) Nímed* [739]15th S 255, *Nymet* 1282. 'Rowland' refers to Rolandus de Nimet who held the manor in 1166 and is for distinction from NYMET TRACEY SS 7200. D 370, RN 304, PNE ii.50, Jnl 1.50.

(2) ~ TRACEY Devon SS 7200. 'Nymet estate held by the Tracy family'. *Nemethe Tracy* 1270×6, *Nymet Tracy* 1274. P.n. Nymet + family name Tracy. Nymet, *Limet* (sic) 1086, is the original name of the river Yeo as in NYMET ROWLAND SS 7108. D 360.

NYMPSFIELD Glos SO 8000. 'Open land of Nymed'. *(in) Nymdesfelda* [862]c.1400 S 209, *Ni- Nymdesfeld(e)* 1086–1535, *Ni- Nymesfeld* 1434, 1457, *Nymsf(i)eld* 1508–1700, *Nympsfyld* 1562. PrW p.n. **Nιμed* <Brit **nemeto-*, secondary genitive sing. **Nιμdes*, + **feld**. Celtic *nemeton*, cognate with Latin *nemus* 'a grove', is a widespread word meaning 'sacred place, sacred grove'. Cf. *Nemetobala*, possibly the RBrit name of the sacred site at Lydney, Glos SO 6102, *Nemeto Statio*, probably the Roman fort at North Tawton, Devon SX 6699, *Nemetacum Atrebatum*, Arras, France, *Nemetodurum*, Nantère, France, *Nemetobriga* near Pueblo de Tribes, Spain. Gl ii.243, RBrit 254, 424, LHEB 278, 286, 487, 555, L 240, 243.

NYMPTON 'Estate on the r. *Nymet*'. R.n. NYMET + OE **tūn**.

(1) Bishop's ~ Devon SS 7523. *Nemetone Episcopi* 1269, *Bysshopes Nymet* 1334, *Bishops Nymeton* 1377. Earlier simply *Nimeton(e)* 1086, 1238. Nymet, *Nimet* 1238, 1249, must have been the earlier name of the r. Mole. 'Bishop's' referring to possession of the manor by the bishop of Exeter in 1086 and for distinction from George and King's NYMPTON SS 7022 and 6819. D 383.

(2) George ~ Devon SS 7022. 'St George's N'. *Nymet(h) Scī Georgii* 1281–1308, *Nymet Georgii* 1345, *Nymeton Sancti Georgij* 1291, *George Nemyton* 1523. Saint's name *George* from the dedication of the church + p.n. *Limet, Nimet* 1086, *Nimet* 1249 as in Bishop's NYMPTON SS 7523. D 348, DEPN.

(3) King's ~ Devon SS 6819. *Kyngesnemeton* 1254, *Kyngesnymyngton* 1341, *Kingsnimpton* 1649. ME **king**, genitive sing. **kinges**, p.n. *Nimetone* 1086–1242 as in Bishop's NYMPTON SS 7523. For 'king's' cf. Queen's Nympton, a name coined in 1900 in honour of Queen Victoria. D 385.

(4) King's ~ STATION Devon SS 6616. A halt on the railway line from Exeter to Barnstaple.

NYNEHEAD Somer ST 1422. '(Estate consisting of) nine hides'. *Nigon Hidum* [? c.890]lost ECW, *(of) Nigonhidon* 11th ECW, *Nichehede* 1086, *Nigenid* [1091×1106]14th, *Nyghud* [12th]14th, *Nigheyd* [1100×18]14th, *Nigahyde* [1135×7]14th, *Nyghehid, Nighehyde* [1152×8]14th, *Nynhead* 1610. OE **nigon** + **hīd**, pl. **hīde**, dative pl. **hīdum**. DEPN.

NYTON WSusx SU 9305. Probably 'the new settlement or farm'. *Nyton* 1327 and from 1593, *Neiton* 1593. ME **nye** (OE *nīġe*) + **toun** (OE *tūn*). Sx 64 prefers a different solution, misdivision of the ME phrase *atten eyton* 'at the islands', OE **ēġeth, īġeth**, dative pl. **ēġthum, īġthum**, as if *atte neyton*. Nyton is on low-lying marginal land. PNE ii.50.

OADBY Leic SK 6200. 'Authi's farm or village'. *Oldebi* 1086, *Outhebi -by* 1199–1399, *Oudebi -by* c.1125–1535, *Odeby(e)* 1245–1684, *Oadby* 1629. Scand pers.n. ON *Authi*, ODan *Øthi*, genitive sing. *Autha*, *Øtha*, + **bӯ**. Lei 460. For a different explanation see SSNEM 61.

OAD STREET Kent TQ 8662. 'Old street'. *Holdestrete* 1254, *Oldestrete* 1313, *Oat street* 1819 OS. ME **old** + **strete** probably in the dial. sense 'hamlet'. The reference is unknown. Contrasts with nearby Key Street TQ 8864, *Kaystrete* 1254, Chestnut Street TQ 8763, *Chestnut Street* 1819 OS, *Bobbing Street* 1819 etc. PNK 244, 245.

OAKAMOOR Staffs SK 0544. '*Ocwall* moor'. *Ocuallemor* 1327, *Okwallemore* 1328, *Okemore* 1602, *Oakallmore* 1636, *Oackamore* 1680, *Oakeymore* 1686, *Oakway Moor* 1693. P.n. **Ocwalle*, 'the oak-tree spring', OE **āc** + Mercian **wælla**, + **mōr**. Horovitz.

OAKE Somer ST 1525. '(At) the oak-trees'. *(æt) Acon* 11th ECW, *Acha -e* 1086, *Oke* 1610. OE **āc**, dative pl. **ācum**. DEPN.

OAKEN Staffs SJ 8502. '(The settlement) at the oak-trees'. *Ache* 1086, *Ak'* 1242, *Oca* 1253, *O(c)ke* 1293, 1327, *Hoken* 1292, *Oken* 1327, *Woken* 1577. OE **āc**, dative pl. **ācum**. DEPN, Horovitz.

OAKENCLOUGH Lancs SD 5447. 'Ravine where oak-trees grow'. No early forms, **oaken** (OE *ācen*) + dial. **clough** (OE **clōh*).

OAKENGATES Shrops SJ 7010. 'The gap where oak-trees grow'. *Okenȝate* 1414, *Okynyate* 1535, *Oakenyates -yeates* 17th cent., *(the) Oaken Gates* from 1707. ME **oken** (OE *ācen*) + **yate** (OE *ġeat*). The reference is to a narrow valley crossed by Watling Street and forming a gap in hilly ground. Sa i.224.

OAKENSHAW 'Oak-copse'. OE **ācen** + **sceaga**. L 209.

(1) ~ Durham NZ 1937. *Akynshawe* 1440×1, *Akinshaw* 1598, *Yeatonshaw* 1768. OE **ācen** + **sceaga**. The 1768 form reflects a local pronunciation (which has not survived) with N dial. [ja] for ME *ā*. Orton 355–66.

(2) ~ WYorks SE 1727. *Akescahe* 1246, *Okens(c)haw(e)* 13th–1597, *Okenshey -shay* 16th cent., *Wockingsha* 1576. Development of ME *aw* through *ā* to *ay* is characteristic of the SW of WYorks. In the same area OE *ā* was rounded to *o* as in the Midlands and ultimately raised to [u:], [wu] in initial position as in the spelling *Wocking-*. YW iii.17, vii.78, 80–87, cf. Nomina 13.109–14.

OAKFORD Devon SS 9021. 'Oak-tree ford'. *Alforde* (sic) 1086, *Acford(e)* 1166–1260, *Okeford* 1249, *Wocford* 1559. OE **āc** + **ford**. D 387.

OAKGROVE Ches SJ 9169. No early forms but cf. *G^t. Oak*, *Broad Oak* 1842 OS. Che i.155.

OAKHAM Leic SK 8508. Probably 'Oca's river-meadow'. *Ocheham* 1086, *Oc- Okham(e)* 1067–1471, *Okeham'* 1190–1610, *Okam* 1208–1349, *Oakeham* 1685. OE pers.n. *Oc(c)a* + **hamm**. The specific was popularly associated with ME *ōk* 'an oak-tree'. In view of the topography of the town which lies on a tongue of land between two streams the generic is probably *hamm* 'land hemmed in by water or marsh' rather than *hām* 'a homested'. Formerly the county-town of Rutland. R 102, L 47, Jnl 5.30.

OAKHANGER Hants SU 7735. 'Wooded slope of oak-trees'. *Acangre* 1086, *Achangra -e* 1174–1250, *Oc- Okhangre -er* from 1250. OE **āc** + **hangra**. Ha 126, Gover 1958.94, L 195.

OAKHILL Somer ST 6347. *Oakhill* 1817 OS. Early forms are needed but this is presumably ModE **oak** (OE *āc*) + **hill** (OE *hyll*).

OAKINGTON Cambs TL 4164. 'Estate called after Hoc(c)a'. *Hochinton(e)* 1086, *Hokinton(a) -e -yn-* 1086–1529, *Hogintune* 1086, *-yng- -ington* 1415–1539×41, *Okin(g)ton' -yn(g)-* 1176–1570, *Oak(e)ington* 1660, 1685. OE pers.n. *Hoc(c)a* + **ing**[4] + **tūn**. Ca 182.

OAKLE STREET Glos SO 7517. *Oakle Street* 1830. A hamlet near the Roman road to Wales, Margary no.61. P.n. Oakle + ModE **street**. Oakle, *Acle* 1263×84, *Ocle* 1320, 1327, is 'oak wood or clearing', OE **āc** + **lēah**. Gl iii.197.

OAKLEY 'Oak-tree wood or clearing'. OE **āc** + **lāeh**. L 203.

(1) ~ Beds TL 0153. *Acleia* [c.1043×65]14th S 1231, *Achelai* 1086, *Akle* 13th cent., *Oc(k)le(y)* 1247–18th, *Oakley* c.1750. The identification of the first spelling is not firm. Bd 24 gives pr [ɔkli], L 203.

(2) ~ Bucks SP 6312. *Achelei* 1086, *Acle* 1222–1355, *Ocle* 1242–1485, *Okele(y)* 1341–1619, *Okeley al. Whokeley* 1607. A clearing in the former great forest of Bernwood. The 17th cent. form shows a dial. pr with excrescent *W-*, cf. OVING SP 6821. Bk 126, 132.

(3) ~ Hants SU 5650. *(to) Aclea* [10th]12th S 1821, *Aclei* 1086, *Akele* 1201. Also known as *Chircheocle* 1206, *Church Oakley* 1817 OS. Possibly the site of the battle between king Æthelwulf and the Danish host mentioned ASC(A, E) under year 851 as *æt Aclea*. The whole district included East and North OAKLEY SU 5750, 5354, in the parish of WOOTTON ST LAWRENCE SU 5953. Another neighbouring woodland name is North WALTHAM SU 5646. A church is mentioned here in 1086. Ha 126, Gover 1958.140, DEPN.

(4) ~ Suff TM 1677. *Acle* 1086–1291, *Akle* 1286, *Ocle(y)* 1260–1610, *Oakley* 1674. DEPN, Baron.

(5) Little ~ Essex TM 2128. *Parva Acle* 1330. ModE adj. **little**, Latin **parva**, + p.n. *Adem* (sic for *Acle(a)m*) 1086, etc. as in Great OAKLEY TM 2028. Ess 345.

(6) ~ GREEN Berks SU 9276. *Okeley Greene* 1607, *Oakley Green* 1790, 1800. P.n. *Aukelay -ley* 1220, *Ac(c)le* 13th cent., *Oc(c)le*, *Okele* 14th cent., + ModE **green**. Oakley is a possible location for the battle *æt Aclea* described in ASC under year 851 when Æthelwulf defeated the Danes. Brk 46, L 203.

(7) East ~ Hants SU 5750. *Estacle* 1236, *Est Ocle* 1295. ME adj. **est** + p.n. OAKLEY SU 5650. Gover 1958.140.

(8) Great ~ Essex TM 2028. *Magna Acle* 1256, *Moche Okeley* 1552. ModE adj. **great**, Latin **magna**, ME **much**, + p.n. *Accleiam* 1086, *Acle(e)* 1219–1330, *Ocle* 1294. 'Great' for distinction from Little OAKLEY TM 2128. Ess 345.

(9) Great ~ Northants SP 8785. *Magna Acle* 1246, *Ockele Magna* 1316, *Mykyll Okeley* 1478, *Great Ockley* 1675. ModE **great**, ME **micel**, Latin **magna**, + p.n. *Aclea* [11th]13th–1303 with variants *-le(y) -lei*, *Achelau* 1086, *Westokle* 1275, *Oc(k)le* 14th. 'Great' for distinction from Little OAKLEY SP 8985. Nth 170.

(10) Little ~ Northants SP 8985. *Little Acle* 12th, *Parva Acle* 1220, *Ockele Parva* 1316, *Little Ockley* 1675. ModE adj. **little**, Latin **parva**, + p.n. Oakley as in Great OAKLEY SP 8785. Nth 170.

(11) North ~ Hants SU 5354. *Northacle* 1278, *North Oclye* 1298. ME adj. **north** + p.n. *Acleah* [824]12th, *Acle* 1206, *Ocle* 1256 as OAKLEY SU 5650. Ha 126, Gover 1958.147, DEPN.

OAKRIDGE Glos SO 9103. 'Oak-tree ridge'. *Okerigge* 1459, *-ru(d)ge* 1535, 1540, *Oakridge* 1775. ME **oke** (OE *āc*) + **rigge**, **rugge** (OE *hrycg*). Gl i.119.

OAKS Shrops SJ 4204. 'The oak-tree(s)'. *Hach* 1086, *Akes*

1199–1256, *Okes* 1271–1713, *Oakes* 1565–1702. OE **āc**, probably originally singular, with inorganic -*s* which was frequently added to monosyllabic names after the Norman Conquest. Sa i.225.

OAKSEY Wilts ST 9993. 'Wocc's island'. *Wochesie* 1086, *Wokesai -eia -ey(e)* 1196–1428, *Woxey* 1452, *Okyssey* 1535, *Oxey* 1585, *Wokesey alias Oxhey* 1638, *Oaksey* 1828 OS. OE pers.n. **Wocc*, genitive sing. **Wocces*, + **īeġ** possibly in the sense 'hill jutting into flat land' or 'patch of good land in moors'. Wlt 63, L 38.

OAKTHORPE Leic SK 3213. 'Áki's outlying farm'. *Achetorp* 1086, *Actorp* c.1130, *Okethorp(e)* late 13th–1539, *Oc(k)t(h)orp(e)* early 13th–1426. OE pers.n. *Áki*, genitive sing. *Áka* + *thorp*. The specific was early identified with OE *āc*, ME *ōk* 'an oak-tree'. Lei 566, SSNEM 114.

OAKWOODHILL Surrey TQ 1337. *Oakewood hill* 1664. P.n. *Oc- Okwode* 1263, 1272×1307, *Ockwod* 1422, 1575, 'the oak-tree wood', OE **āc** + **wudu**, + ModE **hill**. Sr 263.

OAKWORTH WYorks SE 0338. 'The oak-tree enclosure'. *Acurde* 1086, *Akew(u)rth(e) -worth(e)* 1154×89–1379, *Okeworth* 1368–1688. OE **āc** + **worth**. YW vi.7.

OARE Kent TR 0062. 'The flat-topped ridge'. *Ore -a* 1086, *Oran, Ores* c.1100, *Ore(s)* [1087]13th, 1203–51. OE **ōra**. PNK 289, Jnl 21.21, 22.27.

OARE Somer SS 8047 → CULBONE HILL SS 8247.

OARE Wilts SU 1563. '(The settlement) at the flat-topped hill'. *(æt þam) oran* [933]14th S 424, *Ore* 1229–1332, *Oare* 1428. OE **ōra**, dative case **ōran**. The settlement lies on one side of Oare Hill. The S 424 form occurs in a boundary clause of a piece of land called *Motenesora* 'Moten's hill', OE pers.n. *Moten*, genitive sing. *Motenes*, + **ōra**. The N end of Oare Hill is called Martinsell Hill, *Mattelesore* 1257–1300, *Martinshall hill* 1549, 'Mæthelhelm's hill', OE pers.n. *Mæthelhelm*, genitive sing. *Mæthelhelmes*, as in the name of the Iron Age hill-fort on the summit, *Mætelmesburg* 940 S 470. These seem to have been two separate Anglo-Saxon land-holdings centred on Oare Hill. Wlt 325, 351, Grundy 1919.190–1, 248, L 179, 180–1, Jnl 21.17, 22.32–5.

OASBY Lincs TF 0039. Possibly 'Asvithr's village or farm'. *Asedebi* 1086–1202, *-by* 1242×3, *Asdebi* 12th, *Assedeby* 1200–11, *Os(e)by* 1298–1567, *Oasby* 1640. ON pers.n. *Ásvithr* + **bȳ**. A Dan pers.n. *Asede* recorded in 1408 is taken to be a form of *Ásvithr*. SSNEM 61, Perrott 496.

OBORNE Dorset ST 6518. 'The crooked stream'. *(æt) womburnan* [970×5]12th S 813, *Waburnham* [970×5]14th, *(in) Wonburna* [998]12th S 895, *Wocburne* 1086, *Wogburne* 12th, *Woborn(e)* [1145]12th–1585, *Oburne* 1479–1575. OE **wōh** + **burna** and dative sing. definite form **wōgan, wōn**, + **burnan**. The reference is to the winding course of the r. Yeo. Do iii.354.

OCCLESTONE GREEN Ches SJ 6962. *Occleston Green* 1831. P.n. *Aculuestune* 1086, *Ac(c)ulueston -is-* early 13th, *Ac(c)leston -is-* c.1165–[1244]17th, *Oc(c)liston -es-* 1233–1831, 'Acwulf's farm or village', OE pers.n. *Ācwulf*, genitive sing. *Ācwulfes*, + **tūn**, + **green**. Che ii.252 gives pr ['ɔkəlstən].

OCCOLD Suff TM 1570. 'The oak-tree copse'. *Acholt* [1042×66]12th S 1051, *Acolt* 1086–1378, *Ac- Akholt* 1156–1286, *Ocolt* 1254–1384, *Ockold* 1610. OE **āc** + **holt**. DEPN, Baron.

River OCK Oxon SU 4095. 'The salmon'. *(on) æoccænen* [856]12th S 317, *(on) æoccæn* [944]12th S 503, *(on, and lang, to, innan, of) eoccen* [931]c.1200 S 1546, [955]12th S 567, [965]12th S 734, [940]12th S 471, [956]12th S 605, [958 for 959]12th S 673, [1032]12th S 964, *(ut on, on) eoccan* [940]12th S 471, [1032]12th S 964, *(Be tweox, on) eoccene* [944]12th S 494, [956]13th S 603, [958]12th S 650, [960]12th S 682, [965 for 975]12th S 829, [968]12th S 759, [970]c.1200 S 1544, *(betweox, on) Eccene* [944]12th S 494, [958]13th S 654, *(on) eccen* [958]13th ibid., [958 for 959]12th S 673, *Eoche* 1066×87, [1100×35]c.1240, *Ocke* [1100×35]c.1240, 1241×2, c.1540, *Ock* 1335. OE **Eoccen*, **Ēocen*, < PrW **Iogan* < Brit **Esocona*, a formation of the word for a salmon, W *eog*, MW *ehawc*, Co *ehog*, Gaelic *iach* < **esāco-* whence Gallo-Latin *esox, isox*. Brk 14, RN 306, LHEB 363, GPC 1225.

OCKBROOK Derby SK 4236. 'Occa's brook'. *Ochebroc* 1086, *Okebroc(h) -brok(e)* 1185–1584, *-brooke, Oc- Okbrok(e) -broc* c.1250–1500, *Ockbr(o)ok(e)* 1577, 1586, 1610. OE pers.n. *Oc(c)a* + **brōc**. Db 487.

North OCKENDON GLond TQ 5984. ME *North* 1397 + p.n. *Wokendune* 1066×87, 1067, 1335, *Wochenduna -ā* 1086, *Wokendon(a)* 1085–1392, *Wokindun -don(e) -y-* 1066–1365, *Ok(k)endune -(a) -ck-* 1100–1261, *Okyngton* 1448–54, 'Wocca's hill', OE pers.n. **Wocca*, genitive sing. **Woccan*, + **dūn**. 'North' for distinction from South OCKENDON Essex TQ 5982 and CRANHAM GLond TQ 5787, *Wochendunā* 1086. Ess 125.

South OCKENDON Essex TL 5982. *South Wokynden* 1291, *Southokyngdon* 1505. ME adj. **suth** + p.n. *Wokendune* 1066×87, 1067, 1335, *-don(a)* 1085–1397 with variant *-yn-*, *Wochenduna, Wochadunā* 1086, *Ok(k)- Ockendune -(a)* 1100×35, 12th, 1261, 'Wocca's hill', OE pers.n. *Wocca*, genitive sing. *Woccan*, + **dūn**. 'South' for distinction from North OCKENDON TQ 5984. Ess 126.

OCKHAM Surrey TQ 0756. 'Occa's homestead'. *Bochehā* (sic) 1086, *Oc- Okham* 13th cent., *Ockham* from 1314, *Ockam* 1610. OE pers.n. *Occa* + **hām**. The DB form is corrupt. Sr 143, ASE 2.32.

OCKLEY Surrey TQ 1440. 'Occa's wood or woodland clearing'. *Hoclei* 1086, *Oc-* 1200, *Ho(c)keleye, Occhel', Ockel(e)(e), Okeleg(e)* 13th cent., *Okkele(gh)* 1255–1428. OE pers.n. *Occa* + **lēah**. Sr 276.

OCLE PYCHARD H&W SO 5946. 'O held by the Picard family'. *Acle Pichard(i)* c.1179×86, 1243, *Oclepichard* 1336. P.n. Ocle + family name Picard 'the Picardian'. Ocle, *Aclea* [1017×41]17th, *Acle* 1086–1148×63, *Ocle* 1232, is the 'oak-tree wood or clearing', OE **āc** + **lēah**. Roger Picard held the manor in 1243. 'Pychard' for distinction from Lyvers Ocle SO 5746, the portion of 'Ocle held by the monks of Lyre', Normandy, *Acla monachorum de Lira* 1160×70, *Acle Lyre* 1243, 1299, *Ocle Lyre* 1334, *Lyres Oc(c)le* 1346, 1520. Hugh Donkey gave one hide of land in Ocle Pychard to this abbey in 1100. He 154.

ODCOMBE Somer ST 5015. 'Uda's coomb'. *Vdecome* 1086, *Odecumba -e* [1091×1106–1303]14th, *Odcombe* 1610. OE pers.n. *Uda* + **cumb**. DEPN, L 92.

ODDA'S CHAPEL Glos SO 8729. A Saxon chapel called after Odda whose name appears in an 11th cent. inscription there.

ODD DOWN Avon ST 7362. *Odd Down* 1817 OS.

ODDENDALE Cumbr NY 5913. 'Odelin's valley'. *Odehenedale* 1262, *Odelmedal* (for *Odeline-*), *(H)odelingdale* 1279, *Oldyngdale* 1515, *Od(d)leda(i)le* 1538–1688, *Odding- Oldedale* 17th, *Oddenda(i)le* from 1634. CG pers.n. *Odelin* (or feminine *Odelina*) + ON **dalr**. We ii.157, SSNNW 242.

ODDINGLEY H&W SO 9159. 'Wood or clearing of the Oddingas, the people named after Odda'. *odDINGALEA* [816]11th S 179, *oddungalea, ODDUNCALEA, (to) oddungga lea* [943 for 963]11th S 1297, *Oddvnclei* 1086, *Oddingelega -leye* [c.1086]1190, 1275, *Oddyngley* 1327, 1535. OE folk-n. **Oddingas* < pers.n. *Odda* + **ingas**, genitive pl. **Oddinga*, + **lēah**. Wo 152, Hooke 107, 251.

ODDINGTON Glos SP 2225. 'Estate called after Otta'. *Otintone -ton(a) -tune -yn-* [862]c.1400 S 209, 1066×87, 1086–1340, *-ing-* 1138, *Otindon(a) -yn- -en-* 1086–1306, *Odynton* after 1412, *Odyngton -ing-* 1535 etc. OE pers.n. **Ot(t)a* + **ing**[4], possibly varying with genitive sing. **Ot(t)an*, + **tūn**, partly confused with **dūn** 'hill'. Gl i.224.

ODDINGTON Oxon SP 5515. 'Otta's hill'. *Otendone* 1086, *Ottendun(e) -don(e)* [1137]c.1200–1428 with variants *Ot- -in- -yn-* and *-doun*, *Odingtone* 1526. OE pers.n. **Otta*, genitive sing. **Ottan*, + **dūn**. **Otta* is a hypocoristic form of *Ordstān* occurring again in OT MOOR SP 5614.

ODELL Beds SP 9658. 'Woad hill'. *Wade(he)lle* 1086, *Wahella* 1162, *Wahull* 1222–1428, *Wodhull -hill* 13th–15th cents., *Wodell* 1485, *Odyll* 1492, *Odell* 1526. OE **wād** + **hyll**. Odell was the head of an important barony and its name is frequently recorded in its Normanised form with loss of medial *d*. For subsequent loss of *W-* cf. Dobson 419–21. Bd 34 gives prs [wʌdel, oudəl], DEPN, L 171.

ODIHAM Hants SU 7451. 'Wooded estate'. *Odiham* from 1086 with variant *Ody-*, *Wudiham* c.1121 ASC(E) under year 1116, *Odyam* 1559. OE **wudiġ** + **hām**. Ha 126, Gover 1958.115 which gives pr [oudiəm].

ODSTOCK Wilts SU 1426. 'Odda's outlying farm'. *Odestoche* 1086, *Od(d)estok(e)* 1199–1319, *Odstocke al. Adstocke* 1732. OE pers.n. *Odda* + **stoc**. Wlt 223, Studies 1936.23.

ODSTOCK DOWN Wilts SU 1324. *Odestocke Downe* 1594. P.n. ODSTOCK SU 1426 + ModE **down**. Wlt 223.

ODSTONE Leic SK 3907. Either 'Odd's farm or village' or 'settlement on the protruding piece of land'. *Odeston(e) -is-* 1086–1553, *Oddeston(e) -is-* early 13th–1610, *Odston* 15th cent., *Odson* 1597. ON pers.n. *Oddr* or ON **oddr** 'point', ME genitive sing. *Oddes* or **oddes**, + **tūn**. Odstone stands on a marked promontory of high ground between two river valleys. Lei 536, SSNEM 193.

OFFA'S DYKE H&W SO 2959. Glos SO 5407. Shrops SO 2678. *Offedich* 1184. OE pers.n. *Offa*, genitive sing. *Offan*, + **dīċ**. The old boundary between England and Wales said by Asser to have been built by King Offa (*Life of Alfred* 14). It is referred to in S 1555, an 11th cent. survey of Tidenham ST 5596, *bufan dic* 'above the dyke', and in W as *claud offa* in the Red Book of Hergest, later as *le Dikes in Mayleskoyt* 1279 (Mailscot SO 5513), OE **dīċ**, MW **claud**. The modern form seems to be a translation of W *Clawdd Offa* rather than a direct descendant of ME *Offedich* (OE *Offan dīċ*). *Offan dic* [854]12th S 310 refers to a different earthwork in Somerset. DEPN, Gl iii.210.

OFFCHURCH Warw SP 3565. 'Offa's church'. *Offechirch* 1139–1291, *-churche* 1221, *Ofchirche* 1262–1386. OE pers.n. *Offa* + **ċiriċe**. Wa 177.

OFFENHAM H&W SP 0546. 'Uffa's river-bend land'. *offeham* *[709]12th S 80, *uffaham* [714]16th S 1250, *uffenham* [n.d.]12th S 1591a, 1392, *Offenham* 1086, 1535. OE pers.n. *Uffa*, genitive sing. *Uffan*, + **hamm** 1. The settlement lies in a bend of the Avon. Wo 266, Hooke 23, 46, 377.

OFFHAM ESusx TQ 4012. 'Crooked valley'. *Wocham* c.1092, *Wogham* 1296–17th, *Ofham* 1302, *Offam* 1332. OE **wōh** + **hamm** 6a, 7. The name shows ME substitution of [f] for the neighbouring peripheral fricative phoneme [x]. Sx 316, SS 82.

OFFHAM Kent TQ 6557. 'Offa's homestead or estate'. *(on) offahames (gemære)* 'to the boundary of Offham' [942×6]12th S 514, *Offe- Offaham* [c.975]12th B 1321–2, *Offaham* 1044×8 S 1473, *Ofeham* 1086, *Offe(n)ham* 1202–78, *Offam* 1313. OE pers.n. *Offa*, genitive sing. *Offan*, + **hām**. KPN 254, ASE 2.29.

Great OFFLEY Herts TL 1427. *Gret Offle* 1467. ME adj. **gret** + p.n. *Offanlege* [944×6]13th S 1497, *Offelei, Offelei altera* 1086, *Offelege -legh -le(ye)* 1200–1303, *Affley* 1608, 'Offa's wood or clearing', OE pers.n. *Offa*, genitive sing. *Offan*, + **lēah**. 'Great' for distinction from Little Offley TL 1328, *Parva* 1307. Hrt 19.

Bishop's OFFLEY Staffs SJ 7829. 'O held by the bishop' sc. of Lichfield. *Bissopstoffeleg* 1285. ME **bishop** + p.n. *Offeleia* 1086, 'Offa's wood or clearing', OE pers.n. *Offa* + **lēah**. 'Bishop's' for distinction from High OFFLEY SJ 7826. Also known as *Cyprian's Offley* 1203 from Sir Cyprian de Offley who held here. DEPN, L 203.

High OFFLEY Staffs SJ 7826. *Alta Offyleye* 1316. ModE **high**, Latin **alta**, + p.n. *Offelie* 1086 as in Bishop's OFFLEY SJ 7829. High O is situated on a marked hill rising to 433ft. DEPN, L 203.

OFFORD CLUNY Cambs TL 2267. 'O held by (the monks of) Cluny'. *Offord Clunye* 1257. Earlier simply *V- Opeford* 1086, *U- Oppeford* 1195–1202, *Offord* from 1200, 'upperford', OE **up**, **uppe** + **ford**. Further upstream on the Great Ouse from the crossing of the Roman road at Godmanchester. 'Cluny' for distinction from OFFORD D'ARCY TL 2267. Hu 262.

OFFORD D'ARCY Cambs TL 2267. 'O held by the Dacy family'. *Offord Willelmi Daci* 'William Daci's O' 1220, *Offord Darcye al. Dacie* 1596. P.n. Offord as in OFFORD CLUNY TL 2297 + family name *Dacy*. Hu 263.

OFFTON Suff TM 0649. 'Offa's estate'. *Offetuna* 1086, *Offin(g)ton(e)* 1166–1281, *Offigetun'* 1198, *Ofton* 1270–1610. OE pers.n. *Offa*, genitive sing. *Offan*, varying with *Offa* + **ing**[4], + **tūn**. DEPN, Baron.

OFFWELL Devon SY 1999. Probably 'Offa's spring'. *Offewille* 1086, *Offewell(e)* 1219–84, *-wille* 1315, *Uffewell -wille* 1238–1303. OE pers.n. *Offa* + **wylle**. D 628.

OGBOURNE Wilts. The original name of the river Og SU 1973 which is a back formation from Ogbourne. It means 'Occa's stream', OE pers.n. *Occa* + **burna**. It survived as the name of the estate centred on the river subsequently subdivided into OGBOURNE MAIZEY, OGBOURNE ST ANDREW and OGBOURNE ST GEORGE. The forms are *Oceburnan* [967×7]12th S 1504, *Ocheburne -borne* 1086, *Ocche- Ockeburne* 1133–1257, *Auquebourne* 1390, *Ockborn* 1558×1603, *Oakeborne* 1609. Wlt 303, L 18.

(1) ~ MAIZEY Wilts SU 1871. 'O held by the Meysey family'. *Ockeburn' Meysey* 1242. Robert de Meysey held a manor there in 1242. Wlt 303.

(2) ~ ST ANDREW Wilts SU 1972. 'St Andrew's O' from the dedication of the church. *Okeborne Sc̄i Andr'* 1289, *Ogborne Sent Androes* 1544. Also known as *Ocheburna parva* c.1143, *Little Okeburne* 1296, Wlt 303.

(3) ~ ST GEORGE Wilts SU 2074. 'St George's O' from the dedication of the church. *Okeburne Sci Georgii* 1332, *Oggeburn St George* 1449. Also known as *Ocheburna magna* c.1143 'great O', and *North Okeburne* 1275. Wlt 303.

OGDEN RESERVOIR Lancs SD 8039. P.n. Ogden + ModE **reservoir**. Ogden, no early forms, is probably 'oak-tree valley', ME **ōk** (OE *āc*) + **dene** (OE *denu*). Cf. Ogden GMan SD 9512, *Akeden* 1246, *Aggeden* 1276, *Okeden(e)* 1276, 1324, and Ogden in Chadderton, Oldham, *Okeden* 1332. La 50, 56, Jnl 17.42.

OGDEN'S PURLIEU Hants SU 1811 → DIBDEN PURLIEU SU 4106.

OGLE Northum NZ 1378. 'Ocga's hill'. *Hoggel* 1169, *Ogle, Ogg(e)l(e)* 1180–1346, *Oghyll* 1255, *Oggill* 1296 SR. OE pers.n. *Ocga* + **hyll**. *Ocga* is recorded as the name of a son of king Ida of Bernicia. NbDu 151.

OGSTON RESERVOIR Derby SK 3760. P.n. Ogston SK 3859 + ModE **reservoir**. Ogston, *Oggodestun* [c.1002]c.1100 S 1536, *(æt) Oggodestune* [1004]14th S 906, *-ton* 1252, *Oug(h)edestun(e)* 1086, *(H)oggodeston(e)* 1154×9, *Oggetiston* 14th cent., *Oggeston* 1272×1307–1532, *Ogston* from 1577. CG pers.n. *Oggod*, genitive sing. *Oggodes*, + **tūn**. Db 217.

OGWELL 'Wocga's spring'. OE pers.n. **Wocga*, genitive sing. **Wocgan*, + **wylle**.

(1) East ~ Devon SX 8370. *Estwoggewill* 1275, *East Ogwell* 1809 OS, ME adj. **ēst** (OE *ēast*) + p.n. *(to) woggan wylle* [956]12th S 601, *(oð) wocga willes (hafod), (of) wogga will (lacu), (of) wocgga willes (heafde)* 'to, from the head/stream of *Wocgan wielle*' 11th S 1547, *Wogwel* 1086, *Oghawillæ -wille, Woguwel, Wogewil* 1086 Exon, *Wogewill'* 1242. 'East' for distinction from West OGWELL SX 8270. D 461 gives pr [ougwel], RN 308.

(2) West ~ Devon SX 8270. *Westwogewelle* 1278, *W. Ogwell* 1809 OS. ME adj. **west** + p.n. Ogwell as in East OGWELL SX 8370. D 461.

OH ME EDGE Northum NY 7099. *Oh Me Edge* 1869 OS. This is the name of the steep southern face of Wool Heath which rises to 1809ft. in East Kielder Moor.

Childe OKEFORD Dorset ST 8312. Partly uncertain. *Chiltacford*

1212, *Childacford(e)* 1227–1346, *Childocford(e)* 1236–1501 with variants *Chyld-* and *-ok-*. Earlier simply *Acford* 1086, *-ford(e)* 1155–1245, 'oak-tree ford', OE **āc** + **ford**. The origin of the prefix *Child* is uncertain: it might represent OE *ċielde* 'a spring' ('Okeford with or by the spring') or *ċilda*, genitive pl. of *ċild* 'a young nobleman' ('O held by the young noblemen') although the only spellings with medial *-e-* from genitive pl. *-a* are late and insignificant (*Childe(h)okeford* 1397–1431). One of the two manors of Childe Okeford was held in 1086 by Harold Godwinsson, grandson of Wulfnoth *cild*. It is unlikely, however, that the p.n. can refer to possession by this family as the genitive form *cildes* would be expected. 'Child' is for distinction from Shilling Okeford, now SHILLINGSTONE ST 8210. Do iii.59.

OKEFORD FITZPAINE Dorset ST 8010. 'O held by the Fitz Payn family'. *Acford Fitz Paen* 1316, *Ocford Fitz Payn* 1321, *Ok(e)ford(e) Fitz Payn* 1332, 1411. Earlier simply *Acford* [939×46]1726 S 1719, *-ford(e)* 1100–1356, *Okeford(')* from 1315, 'oak-tree ford', OE **āc** + **ford**. 'Fitzpaine' from Robert Fitz Payn who inherited the manor in 1264 and for distinction from Childe OKEFORD ST 8312. Do iii.178.

OKEHAMPTON Devon SX 5895. 'Settlement, estate on the river Okement'. *(æt) ocmund -mond tune* c.970 B 1245, *Ochementone* 1086, *Okementon(a)* 1167, 1275, *-maton* 1219, *Ocumptona* 1222, *Okhamptone* 1316, *Okington, Okenton* 16th. R.n. OKEMENT SS 5901 + OE **tūn**, later associated with the common p.n. element **hāmtūn**. D 202 gives local prs [ɔkiŋtən, ɔkəntən].

OKEHAMPTON CAMP Devon SX 5893. The reference is to the artillery camp there: in the late 19th cent. the Royal Artillery encamped in Okehampton Park there for three months in the year.

OKEHAMPTON COMMON Devon SX 5790. P.n. Okehamptom + ModE **common**.

River OKEMENT Devon SS 5901. Uncertain. *Okem* 1244, *Okemund* 1281, *Ockment* 1577. D 11, RN 308.

OLANTIGH Kent TR 0548. 'Holly enclosure'. *Olenteye* 1270, *Holmthege, Holitege* 1272×1307, *Olentegh'* 1313, *Olantigh* 1607. OE **holeġn** + **tēag**. PNK 386.

OLD Northants SP 7873. '(The settlement) at the forest'. *Walda -e* 1086–1249, *Waud(e)* 1205–51, *Wolde* 1316–1439, *Olde, Owlde, Wo(o)ld* 16th, *Old* from 1597. OE **wald**. The reference is to the situation of the village at the junction of wood to the N and open land to the S. Nth 128, L 225–6.

The OLD HALL Humbs TA 2717. *Old Hall* 1824 OS.

The OLD MAN OF CONISTON Cumbr SD 2797. A conspicuous summit on the Furness Fells rising to 2635ft. *Old Man* 1800, cf. *Old Man Quarry* 1786. An 18th cent. name from an ancient cairn said to have existed with two others on the summit called 'The Old Man, his Wife and Son'. TCWAAS 1918.93, La 193 f.n.2.

OLDBERROW Warw SP 1266. 'Owl hill'. *Ulenbeorge* [709]12th S 79, *Ulbeorge* [714]16th S 1250, *Oleberge* 1086, *Ulleberga* 1190, *-berwe* 13th, 1307, *-burwe* 1280, *Oule-, Ulleberewe* 14th cent., *Ulbarewe* 1346, *Olbarwe -berewe* 1321, 1327, *Owlbarrow* 1545, *Owlburrow* 1622, *Oldberrow* 1831 OS. OE **ūle**, genitive sing. **ūlan** + **beorg**. Cf. nearby ULLENHALL SP 1267. Wo 267, L 128.

OLDBOROUGH Devon SS 7706. Uncertain. *Oldburrough* 1701, *Oldburrow* 1809 OS. The evidence is too late to say whether this is OE **burh** 'a fortified place' or **beorg** 'a hill, a barrow'. D 410.

OLDBURY 'The old fortified place' or 'Alda's fort or manor'. OE **ald**, definite form dative case **aldan**, or pers.n. *Alda*, genitive sing. *Aldan*, + **byriġ**, dative sing. of **burh**.

(1) ~ H&W SO 6322 → CAPLER CAMP SO 5933.

(2) ~ Shrops SO 7192. *Aldeberie* 1086–1292 with variants *-byr'* and *-bur(y)'*, *Oldebur(y)'* 1261–1535, *Oldbury* 1535, *Owdburye* 1575×6, *Wo(w)bury* 16th. OE **ald**, definite form dative case **aldan**, + **burh**, dative sing. **byriġ**. The reference is probably to the undated earthworks on Panpudding Hill SO 7192 referred to in 1571 as *le Tirrett* ('turret') *alias le Old Castle*. Sa i.226.

(3) ~ Warw SP 3194. *Aldburia* 12th, *Aldeborowe* 1235, *-bury* 1262, *Oldebury* 1278. The reference is to Oldbury Iron Age hill-fort. Wa 92, Thomas 1980.214.

(4) ~ WMids SO 9889. *Aldeberia* 1174, *Oldebure* 1270. Wo 299.

(5) ~ CASTLE Wilts SU 0469. *Oldebyry* 1265, *Oldborough Castle* 1828 OS. A triangular bivallate Iron Age hill-fort of the 3–2nd cents. BC. Wlt 262, Thomas 1976.238.

(6) ~ -ON-SEVERN Avon ST 6192. *Oldbury on Sev'n* 1662. Earlier simply *Aldeburhe* 1185, *-beri* 1208, *Oldebur(i) -bury* 1287–1533, *Wolbery* 1556. The reference is to an ancient circular encampment just N of the village. Gl iii.8.

(7) ~ ON THE HILL Glos ST 8188. *Old(e)bury sup. montem* 1696, *~ on the hill* 1756. Earlier simply *(on) Ealdanbyri* [972]10th S 786, *Aldeberie* 1086, *Audibure* 1216, *Old(e)bur(y) -bure -byr(y)* 1273–1479. Possibly 'old' when the centre of population moved to Didmarton ST 8287. 'On the hill' for distinction from OLDBURY-ON-SEVERN ST 6192. Gl iii.28.

(8) ~ SANDS Avon 5893. *Oldbury Sand* 1830. A sand-bank in the course of the river Severn off Oldbury. P.n. OLDBURY + ModE **sand(s)**.

OLDCLIMS Corn SX 3775 → STOKE CLIMSLAND SX 3674.

OLDCOTES Notts SK 5988. 'Owl cottages'. *Ulecotes* c.1135–1309, *-cote* 1201–44, *Oulecotes* 13th–1492, *Ol- Ulkotes -cotes* c.1280–1457, *Ouldcots* 1598, *Oldcotes* 1603×25. OE **ūle** later replaced by ModE **old** + **cot**, secondary pl. **cotes**. Nt 99.

OLDFIELD H&W SO 8465. *Oldfield* 1831 OS.

OLDFORD Somer ST 7850. *Oldford* 1817. Early forms are needed but this is presumably ModE **old** (OE *eald*) + **ford** (OE *ford*).

OLD HALL Lancs SD 4616. Short for Rufford Old Hall, a late 15th cent. half-timbered house built by the Hesketh family. 'Old' for distinction from Rufford New Hall built in 1760. Pevsner 1969.213–4.

OLDHAM GMan SD 9204. 'Old island'. *Aldholm* 1222×6, 1246, *-hulm* 1227, *Oldelum* 1276, *Oldum -om* 1292–1537, *Oldham* from 16th, *Owdam* 1546, *Owdham* 19th. OE **ald** + ON **holm** in the sense 'island of higher dry ground in a bog'. The sense may be 'promontory which is the site of ancient settlement'. La 50, TC 145, SSNNW 242, Jnl 17.40.

OLD HEATH Essex TM 0122 → Old HEATH Essex TM 0122.

OLD HOWE Humbs TA 1156. A stream-name. Cf. *Emartland Hows, the Old Hoo* 1569. NCy dial. **howe** 'a hollow, a valley, a ditch' (OE *hol*). YE 76.

OLDHURST Cambs TL 3077. 'Wold Hurst, Hurst on the wold'. *Waldhirst -hurst* 1227, *Woldhurst -hyrst* 1258–1318, *Woldhurst* 1350, 1546. OE **wald** + p.n. *Hirst* 1228, 1285, 'wooded hill', OE **hyrst**. 'Old' for 'Wold' (with dial. loss of initial *w-*, Dobson §421) for distinction from WOODHURST TL 3176. Hu 211.

OLDLAND Avon ST 6771. 'Ancient cultivated land'. *Aldelande* 1086, *Oldelond(e) -land(e)* 1287–1533, *Oldland* 1482, *Wol(l)and* 1577, 1639. Gl iii.81, L 248–9.

OLD PARK Shrops SJ 6909. Cf. *Old Park Furnaces* 1833 OS. An estate owned by Isaac Hawkins Browne where deposits of iron and coal were found and an ironworks built in 1790. By 1806 the Old Park Company was the second largest ironworks in Britain. It closed in 1880. The name refers to a park recorded here from 1506. Raven 152, Gelling.

OLD PEAK NYorks NZ 9802. *Old Peak or South Cheek* 1857 OS. The name of the headland at the south end of Robin Hood's Bay, also known as SOUTH CHEEK NZ 9802. ModE **old** + **peak** 'a headland'.

OLD SOAR MANOR Kent TQ 6154. An 18th cent. house with the medieval solar end of an earlier house. *Old Soar* 1819 OS. Earlier *Sore(lands)* 1480×90. Probably a manorial name with the surname of John *le Suur* 1254, Roger *le Soure* 1292, from OFr *sor* 'reddish brown'. PNK 156, Newman 1969W.445, DBS 326.

OLD SOUTH Lincs TF 4735. No early forms. A sand-bank in The Wash.

OLDSTEAD NYorks SE 5380. 'Old site'. Latin *veterem locum* 1247, ModE *Oldsteade* 1541. ME **old** + **stede**. The reference is uncertain, but since one of the senses of *stede* is 'a religious house' possibly this is a reference to an earlier site of Byland Abbey which is said to have been temporarily established in the neighbourhood of Hood SE 5082. Otherewise *stede* in the late northern sense 'farmstead'. Cf. NEWSTEAD SE 5478, Notts SK 5252. YN 196, Stead 295.

OLD SWAN Mers SJ 3991. *Old Swan* 1842 OS. Transferred from an inn called *The Three Swans* until 1824 and re-named in that year *The Old Swan* for distinction from a newly-opened rival called *The Swan*. Mills 117.

OLD TOWN. A term often used for the site of a deserted mediaeval village. GV 20.

(1) ~ Cumbr SD 5983. *Aldton* 1451, 1461, *Holdtown* 1501, *Oldtowne* 1630. Situated half a mile W of MANSERGH and the parish church at SD 6082. We i.51.

(2) ~ Northum NY 8891. *Old Town* 1695 Map.

OLDWAYS END Somer SS 8624. No early forms. The modern trunk-road (A 361) runs a little S of this point which lies on a parallel track from South Molton to Exebridge.

OLD WINCHESTER HILL Hants SU 6420. *Old Winchester Hill* 1810 OS. P.n. Old Wichester + ModE **hill**. A prominent hill with Bronze-Age round barrows and an Iron Age hill-fort. Thomas 1976.142.

OLD WIVES LEES Kent TR 0754. *Oldwives Lease* 1690. A popular re-formation of earlier *Old Woodes Lease* 1569, *Oldwood Leeze, ~ lease* 1575. P.n. *Ealdewode* 1278–1347, *(H)oldewode* 1292, 'the old wood', OE **eald**, definite form **ealda**, + **wudu**, + Kentish dial. **lees** 'a common or open space of pasture ground' (OE *lǣs*). PNK 375, Cullen.

OLDWIT Corn SX 3281 → LAWHITTON SX 3582.

OLENACUM ROMAN FORT Cumbr NY 2646. The identification of *Olenacum* with the Roman fort at Old Carlisle is disputed. RBrit believe it to be rather the Roman fort at Elslack. The forms are *Ὀλίκανα (Olicana)* [c.150]13th Ptol, *Olerica* [c.700]13th Rav, *Olenaco* [c.408]15th ND. If the last form (in the dative case) is correct, the name means something like 'property or estate of a man called **Olen-*', pers.n. **Olen-* (the pers.n. suffix is unknown) + p.n. suffix **ācum**. For Old Carlisle RBrit suggest instead *Maglona* [c.408]15th ND, 'high, outstanding, noble place', Brit ***maglo-** + suffix **-ona**, identical with Maguelonne, Hérault, France, *Magalona*, and Moulons Char.-Maritime, *Magalonum*, in which, however, the specific may rather be the personal name *Magalos* of identical origin (and ultimately the source of OIr *mál*, W, Breton *mael* 'prince'). RBrit 430, 405, Dauzat-Rostaing 484.

OLIVER'S BATTERY Hants SU 4527. *Cromwell's Battery* 1810, 1810 OS, *Oliver Cromwell's Battery* [1866×9]1908. Pers.n. *Oliver* as in Oliver Cromwell who besieged Winchester in 1645 + ModE **battery**.

OLLERTON Ches SJ 7776. 'Alder-tree settlement'. *Alretune* 1086, *(H)Ol(l)reton(e)* late 13th–1334, *Ol(l)erton* from 1319. OE **alor** + **tūn**. Che ii.79. Addenda gives 19th cent. local pr ['u:lər-'oulərtn -tən] now ['ɔlərtən].

OLLERTON Notts SK 6567. 'The alder-tree settlement'. *Alretun* 1086, before 1190, *-ton(a)* 1176–1335, *Allerton* 13th–1362, *Ollerton* from 1316. OE **alor**, genitive pl. **alra**, + **tūn**. Nt 89.

OLLERTON Shrops SJ 6525. Possibly 'Ælfhere's estate'. *Allerton* 1284×5, *Alverton(e)* 1292–1317, *Ollerton* 1833 OS. OE pers.n. *Ælfhere* (or *Ælfweard* or *Ælfrǣd*) + **tūn**. The form of the name has been influenced by dial. *oller* 'an alder-tree'. Bowcock 176.

New OLLERTON Notts SK 6668. A modern colliery development. ModE adj. **new** + p.n. OLLERTON SK 6567.

OLNEY Bucks SP 8851. 'Olla's island'. *(æt) Ollanege* [979]12th S 834, *Olnei* 1086, *Ol(e)neye* 1233, 1237×40, 1452, *Ouneya -eia* 1207×8. OE pers.n. *Olla*, genitive sing. *Ollan*, + **ēġ**. Bk 12 gives pr [ouni], L 38.

OLTON WMids SP 1382. 'The old settlement'. *Oltun* 12th, *Olton* from 1295, *Alton* 1221, *Oulton* 1466, *Owlton* 1610. OE **ald** + **tūn**. The original settlement of Solihull. There are a number of unusual sps for this place, *Hulton* 1374, *Elton* 1401 and *Oken end* 1656 for *Olton ende* 1581. If the form *Oudelton* 1198 belongs here a different explanation is needed. Wa 71.

OLVESTON Avon ST 6087. 'Ælf's estate'. *(æt) Ælves- Alfestune* [955×7]12th S 664, *Alvestone* 1086, *Olveston(')* from 1167, *Olston* 1422, 1562, *Wolston, Owlson* 16th. OE pers.n. *Ælf*, genitive sing. *Ælfes*, + **tūn**. Gl iii.119.

OMBERSLEY H&W SO 8464. 'Bunting wood or clearing'. *AMBReslege* *[706]12th S 54, *Ambresl(e)ie* [714]16th S 1250, *Ambreslege* [n.d.]12th S 1594, *Aumbresleg -ley* 1229×51, 1280, *Ombresleye* c.1300. OE **amer**, genitive sing. **am(e)res**, ***ambres**, + **lēah**. A spring called *Ombreswelle* is mentioned in S 54 and the bounds of the estate, *ombersetena gemære*, occur [980]11th in S 1342, and as *Ombresetene gemæres* [n.d.]18th in S 1597. Wo 268, Hooke 23, 36, 304, 371, 402, V 14.

OMPTON Notts SK 6865. 'Ealhmund's farm or village'. *Almuntone, Almentune* 1086, *Elmeton(a)* 1156, 1203, *Almeton(')* 1181–16th, *Owmton, A(u)mpton* 16th, 17th, *Ompton* 1699. OE pers.n. *Ealhmund* + **tūn**. Nt 56.

ONE ASH Derby SK 1665 → MONYASH SK 1556.

ONECOTE Staffs SK 0555. 'The lonely cottages'. *Anecote* 1199–1240, 1578, *Onecote* from c.1265, *Oncot(t)* 1414, 1535–1695, *Uncote -cott* 1479–1607. OE **āna**, neuter gender **āne**, + **cot**, ME pl. **cote** (OE *cotu*). Oakden gives prs [ɔnkət] and [wɔnkout]. There is another Onecote, *Onecote* 1381, at SJ 8626. Horovitz gives pr *On-cot*, sometimes *Won-cot*.

Chipping ONGAR Essex TL 5502. 'O with the maket'. *Chepyng Aungre* 1328. ME **cheping** (OE *ċēping*) + p.n. *Angra* 1086, *Angr(e)* 1158–1239, *Aungre* 1277–1378 including [1043×5]13th S 1531, *Ongre* 1416, 'grassland', OE **anger**. 'Chipping' for distinction from High ONGAR TL 5603. Ess 71 gives pr [ɔŋgə].

High ONGAR Essex TL 5603. *Alta Aungre* 1245, *Hecch Aungre* 1328. ME **hegh**, Latin **alta**, + p.n. Ongar as in Chipping ONGAR TL 5502. Ess 71.

ONGAR HILL Norf TF 5824. *Hungry Hill* 1824 OS. If the 1824 form is reliable this is a derogatory name applied to poor soil.

ONGAR STREET H&W SO 3967. *Hunger Street* 1963 Bartholomew. A Hunger Street occurs in Hereford, *Hungerstrete* 1375, *Hongery Strete* 1610. Bannister 100.

ONIBURY Shrops SO 4579. 'The fortified place or manor house by the river Onny'. *Aneberie* 1086, 1121, *Onebur' -byr' -buri -bury* 1247–1769, *Onybury* 1334–1766. R.n. ONNY So 3987 + OE **burh**, dative sing. **byriġ**. Sa i.227.

High ONN Staffs SJ 8216. *Hyonne* 1430, *Hyghon* 1577. ME **high** + p.n. *Othnam* 1081, *Otne* 1086, *Onna, Othna* c.1130, *Onne* 1221, 'the kiln', OW **otyn** (W *odyn*). 'High' for distinction from Little Onn SJ 8416, *Little onne* 1293, ME **litel** + p.n. *Anne* 1086. DEPN, GPC 2618, TSAHS 37.139, Horovitz.

ONNELEY Staffs SJ 7543. 'The lonely wood or clearing'. *Anelege* 1086, *Oneleia* 1185, *Onilegh* 1211, *Onyleye* 1293. OE **āna** + **lēah**. DEPN.

River ONNY Shrops SO 3987. 'The single river'. *Onye* 1236–1301, *Onie* c.1250, *Oneye* 1301, *Oney* c.1540, *On(n)y* 1577, *On(e)y* 1675. OE **āna** + **ēa**, dative sing. **ānan** + **ē**, **īe**. The Onny has two headwaters of equal size known as the East and West Onny and which unite to form a single river at Eaton SO 3789. The earlier derivation from Brit **Onnīo-* 'the ash-tree river' and comparison with the supposed Hants r.n. Ann in ABBOTTS

ANN SU 3243, AMPORT SU 3044 and Little Ann SU 3343 cannot be right in view of the persistent early sps with single *-n-*. Sa i.227–8, RN 310.

ONSLOW VILLAGE Surrey SU 9949. A modern development named after the landowner, the earl of Onslow.

OPENSHAW GMan SJ 8897. 'Open, unenclosed wood'. *Openshawe* 1272, *Oponshaghe, Openshagh* 1322. OE **open** + **sceaga**. La 35, Jnl 17.32.

ORBY Lincs TF 4967. Probably 'Orri's village or farm'. *Heresbi* (sic) 1086, *Orreby* c.1115, *Orrebia* 1148×56, *Orrebi* 1202. ON pers.n. *Orri* or **orri** 'a black-cock' from which the pers.n. is derived, genitive sing. *Orra*, + **bȳ**. DEPN, SSNEM 61.

ORBY MARSH Lincs TF 5167. P.n. ORBY TF 4967 + ModE **marsh**. This is simply *Salt Marsh* 1824 OS.

ORCHARD Dorset ST 8215. '(Place) beside the wood'. *(ad, at, to) Archet* [939]15th S 445, [963]15th S 710, 1176, *Orchet* 1317–1576, *Orchard(')* from 1427. As the spellings show the modern form is a 15th cent. rationalisation of PrW ***ar** + ***cę̄d**. Identical with the common W p.n. Argoed. Do iii.133.

ORCHARD PORTMAN Somer ST 2421. '(Estate called) Orchard held by the Portman family'. P.n. *(æt) Orceard* *[854]12th S 310, *Orchyard* 1225, *Orchard* 1610, 'the orchard', OE **orċeard**, + family name Portman from Walter Portman who held the estate in the 15th cent. Orchard ford is mentioned in S 345, *Orcerdford* [882]12th. 'Portman' for distinction from Orchard Wyndham ST 0740, *Orchard* 1424, which was held by Johannes de Wyndeham c.1619. DEPN.

ORCHESTON Wilts SU 0545. 'Ordric's estate'. *Orc(h)estone* 1086, *Orcheston(e)* from 1167, *Orkeston* 1227, 1316, *Ordeston* 1296, *Ordrycheston* 1315, *Orston* 1524. OE pers.n. *Ordrīċ*, genitive sing. *Ordrīċes*, + **tūn**. The village is divided between Orcheston St Mary and Orcheston St George from the dedication of the two churches, *Orcheston Seynt Jorge* 1279, *Orcheston Marie* 1404. Wlt 234.

ORCHESTON DOWN Wilts SU 0748. *Orcheston Down* 1817 OS. P.n. ORCHESTON SU 0545 + ModE **down**.

ORCOP H&W SO 4726. 'Ridge summit'. *Orcop(pe)* 1137–1341, *Orkhope* 1280, *Arcoppe* 1535. OE **ōra** + **copp**. Orcop is named from Orcop Hill SO 4828, a summit of 958ft. on the long ridge N of the village. One of the most Northerly examples of WS *ōra* in p.ns. Cf. PERSHORE SO 9446. He 155, L 137, 179, 181, Jnl 22.37.

ORCOP HILL H&W SO 4828. *Orcop Hill* 1831 OS. P.n. ORCOP SO 4726 + ModE **hill**.

East ORD Northum NT 9951. ModE adj. **east** + p.n. *Horde* 1196, *Orde* 1208, 1228, *Owrde* 1539, the 'projecting ridge of land', OE **ord**. 'East' for distinction from West, Middle and South Ord NT 9551, 9650 and 9850. As the scatter of Ord names suggests, the reference is to a long ridge of high ground running parallel to the river Tweed, on which LONGRIDGE TOWERS is situated at NT 9549. NbDu 152.

ORE ESusx TQ 8311. 'Flat-topped hill'. *Ore, Ora* c.1123–15th, *Oure* 1610. OE **ōra**. Sx 504, Jnl 21.22.

River ORE Suff TM 3845. *Orus* (Latin) 1577, *Ore* 1735. A back-formation from p.n. ORFORD TM 4250. RN 310.

ORETON Shrops SO 6580. Probably 'the ridge settlement'. *Oreton* 1649, 1833 OS, *Orton* 1695. OE **ofer** + **tūn** referring to a ridge of high ground SW of the village. Gelling.

ORFORD Ches SJ 6190. 'The upper ford'. *Orford* from 1332, *Overforthe* 1465. OE **uferra** + **ford**. Site of the crossing of Orford Brook by the Roman Road from Warrington to Wigan, Margary no. 70b. La 96.

ORFORD Suff TM 4250. 'The ford by the flat-topped hill'. *Oreford* 1164–1327, *Orford* from 1202. OE **ōra** + **ford**. DEPN, RN 310, Baron.

ORFORD NESS Suff TM 4549. *Orford Ness* 1805 OS. P.n. ORFORD TM 4250 + ModE **ness** 'a headland'.

ORGREAVE Staffs SK 1516. 'The grove at the corner or spit of land'. *Ordgraue* 1195, *Ordegrave* 13th, *Orgrave* 1203, 1262. OE **ord** + **grāfa**. DEPN, PNE ii.56, Duignan 112, Horovitz.

ORLETON 'Alder-tree settlement'. OE **alor**, genitive pl. **alra** + **tūn**.

(1) ~ H&W SO 4967. *Arletvne* 1086, *Olreton* 1291, *Orleton(e)* 1292–1357. One spelling, *Erleton* 1431, shows confusion with OE *eorl*, owing to the late survival in this area of [ø] for OE *eo*. He 155.

(2) ~ H&W SO 6967. *(æt) ealre tune* [1014×16]18th S 1459, *Alretvne* 1086, *Olreton* 1275, *Orleton* 1357. Wo 67, Hooke 329.

ORLINGBURY Northants SP 8672. Partly uncertain. *Ordinbaro* 1086, *Orlinberg(a)* 1131, *Ordlingbœre* 1066×75, *(H)or(d)lingber(e) -berg(h) -berg -burgh* 13th cent., *Orlibergh* 1388, *Orlingbury, Orlebere* 1614. The generic shows confusion between **bǣr** 'a woodland pasture', **bearu** 'a wood, a grove', **beorg** 'a hill' and **burh** 'a fortified place'. The specific is probably OE pers.n. **Ordla* + **ing**[4], and so 'the pasture or wood or hill or fort called after Ordla'. Nth 128 records a one-time pr [ɔ:libi:ə], DEPN.

ORMERSFIELD FARM Hants SU 7852 → DOGMERSFIELD SU 7852.

ORMESBY 'Orm's village or homestead'. ODan pers.n. *Orm*, secondary genitive sing. *Ormes*, + **bȳ**.

(1) ~ Cleve NZ 5317. *Ormesbi(a)* 1086, *Ormesby* from 12th cent. YN 157, SSNY 34.

(2) ~ ST MARGARET Norf TG 4915. 'St Margaret's O'. *Ormebi Sancte Margarete* 1213. P.n. *Ormesby* [c.1020]14th S 1528, *Ormisby* [c.1020]late 13th ibid., *Ormesbei -by -bey, O(s)mesbei, Orbeslei* (sic) 1086, *Ormesby* from 1209, + saint's name *Margaret* from the dedication of the church. 'St Margaret' for distinction from ORMESBY ST MICHAEL TG 4814. Nf ii.13.

(3) ~ ST MICHAEL Norf TG 4814. *Ormysby St Mighell* 1538. P.n. Ormesby as in ORMESBY ST MARGARET TG 4915 + saint's name *Michael* from the dedication of the church. In the Middle Ages there were two other churches in Ormesby in addition to St Margaret's and St Michael's dedicated to St Andrew and St Peter. Both have disappeared. Nf ii.13.

North ORMSBY Lincs TF 2993. *Northormesby* 1355–1671. ME **north** + p.n. *(æt) Vrmesbyg* [1066×8]12th Wills, *Ormesbi* 1086–1218, *-by* from 1147, 'Ormr's village or farm', ODan pers.n. *Orm*, genitive sing. *Orms*, + **bȳ**. 'North' for distinction from South ORMSBY TF 3775. The earliest spelling shows contamination by OE **Wyrm*. Also known as *Nunormesby* 1386–1554 from the 12th cent. Gilbertine priory here. Li iv.32, SSNEM 62.

South ORMSBY Lincs TF 3775. ModE **south** + p.n. *Ormesbi* 1086, c.1115, *Ormeresbi* c.1115, *Hormeresbi* late 12th, *Hormesby* 1242×3, 'Ormarr's village or farm', ON pers.n. *Ormarr*, ODan *Ormær*, ME genitive sing. *Ormares*, + **bȳ**. 'South' for distinction from North ORMSBY TF 2993. DEPN, SSNEM 62.

Great ORMSIDE Cumbr NY 7017. *Magna Ormesheved* 1278–1415, *Great als. Myckle Ormeshead* 1587, *Great Ormside* from 1687. Adj. **great**, Latin **magna**, + p.n. *Ormeshouad* 1145×61, *Ormesheued -v-* 12th–1415, *-hed(e)* 1360–1547, *Ormside* 1625, usually explained as 'Orm's headland', ON pers.n. *Ormr*, secondary genitive sing. *Ormes*, + ON **hǫfuth** later replaced with OE **hēafod**. However, pers.n. *Ormr* derives from ON *ormr* 'a snake, a worm' and the recurrence of the type Ormes Head and Worms Head (SS 3887) in Wales suggests that this name might rather be 'snake's head' referring to some distinctive feature of the hills. 'Great' for distinction from Little Ormside NY 7016, *Ormesheved parva* 1415, *Little Ormside* 1687. We ii.89 gives prs ['ɔ:msaid] and ['ɔ:msit], SSNNW 151, L 161–2.

ORMSKIRK Lancs SD 4108. 'Orm's church'. *Ormeschirche* [1189×91]14th, 1286, *Ormeskierk* 1203, *Ormiskirke -kyrke, -es-* [1232]14th–1552, *Ormeschurch(e)* c.1300, 1317. ON pers.n. *Ormr*, genitive sing. *Orm(e)s*, + OE **ċiriċe** varying with ON **kirkja**. The *Orm* in question is not known, nor whether he was founder of the church or an early owner. The fact that Ormskirk was held by a man called Orm de Ormeskierk in 1203 may be no more than coincidence since Orm was the most common masculine pers.n. of Scandinavian origin in mediaeval Lancs. La 121, SSNNW 54, Nomina 3.56, Jnl 17.67.

ORPINGTON GLond TQ 4565. 'Estate called after Orped'. *Orpedingtun* [1032]n.d. S 1465, *1044×66 S 1047, *Orpinton* 1086, *Orpington* 1226, *Arpington* c.1762. OE pers.n. **Orped* from OE **orped** 'active, valiant' + **ing**[4] + **tūn**. DEPN, LPN 170.

ORRELL GMan SD 5305. 'Ore hill'. *Horhill* 1202, *-hull* 1204–5, 1294, *Orhille* 1206, *Orul* before 1220, *Oril* 1272, *Orel(l)* 1292, 1332. OE **ōra** + **hyll**. Originally referred to the high ground at SD 537039 where presumably bog-ore or bog-iron was at one time found. The identical name occurs in Sefton Mers at SD 3496, *Orhul* 1299, *Orell* 1347, 1385, *Orrell* 1547. La 105, 117, Jnl 17.59.

ORSETT Essex TQ 6481. 'Ore pits'. *Aetorseaþan* [957]17th, *(of) Orseaþun, (of þam westrum) Orseaþum* 'from the western ridge pits' [c.1000]c.1125, *Dorseda, Orsedā* 1086, *Orset(e)* 1230–1361. OE **ōra** + **sēath**, dative pl. **sēathum**. Orsett stands on a ridge above marshes where bog-ore must have been dug. Ess 165, PNE ii.55.

ORSLOW Staffs SJ 8015. 'Horse stream or pond'. *Horslage* 1195, *Horselawe -lowe* 1208–1468, *Ors(e)lowe* 1242–1604, *Orslow* from 1755. OE **hors**, genitive pl. **horsa**, + the rare element **lagu**, ME *lawe* identical with the reflex of OE *hlāw* 'hill, mound'. St i.142.

ORSTON Notts SK 7741. 'Osica's farm or village'. *Oschintone* 1086, *-ton(a)* [1093]13th, 1162×5, *Oskinton(a)* 1167–1247, *Orskinton* 1197–1261×5, *Horston* c.1275–1428, *Orston* from 1272, *Orson* 17th. OE pers.n. **Ōsica*, genitive sing. **Ōs(i)can*, or **Ōsica*, + **ing**[4] + **tūn**. The intrusive *-r-* in this name appears to be due to confusion with OE **hors** 'a horse' and the need for distinction from OSSINGTON SK 7564. Nt 227 records obsolete pr [ɔ:s(ə)n].

ORTON Cumbr NY 6208. Either 'higher settlement' or 'settlement by the ridge'. *Overton -u-* 1239–1674, *Orton* from 1263. OE **uferra** or **ofer** + **tūn**. Overton lies at the upper end of Chapel Beck under the steep slope called Orton Scar NY 6209. We ii.42, L 173 ff.

ORTON Northants SP 8079. Partly uncertain. *Overton(e)* 1086–1316, *Oreton(e)* 1283–1306. The specific could be either OE **uferra** 'upper, higher' or OE ***ofer**, ***ufer** 'a flat-topped ridge'; the generic is OE **tūn** 'a setlement'. Orton is situated on the edge of level ground overlooking the steep slope down to Slade Brook (OE **slæd** 'a valley'). Nth 118, L 178.

ORTON LONGUEVILLE Cambs TL 1696. 'O held by the Longueville family'. *Overton Henrici de Longa Villa* 'Henry de Longueville's O' 1220, *Overton Lungheuille* 1227. Earlier simply *(æt) Ofertune* [958]13th S 674, *Ovretune -tone* 1086, *Ouerton* 1200, *Orton* from 1546, 'settlement at the flat-topped ridge', OE **ofer** + **tūn**. 'Longueville' for distinction from ORTON WATERVILLE TL 1596. The family came from Longueville-sur-Scie, Seine-Maritime. Hu 193, Jnl 22.39.

ORTON-ON-THE-HILL Leic SK 3003. P.n. *Wortone* 1086, *Ou- Overton(e) -ir-* 1087×1100–1575, *Orton* from 1518, + affix *le Hyll* 1570, *-on-the-hill* from 1575, for distinction from the lower settlement at the now lost *Westone* 1086, *Weston'* 1211–1354. It is uncertain whether this is 'the ridge settlement', OE **ofer** + **tūn**, or 'the upper settlement', OE **uferra** + **tūn**. Lei 553, L 178.

Great ORTON Cumbr NY 3254. *Great Orreton'* 1267, *magna Orton* 1485. Adj. **great**, Latin **magna**, + p.n. *Orreton'* 1210–1363, *Ortun* c.1225, *Orton'* 1232–1424, *Overton* 1270–1580, *Warton* 1586, *Worton* 1800, 'Orri's farm or village', ON pers.n. *Orri* or possibly ON **orri** 'black grouse' from which the by-name derives, genitive sing. *Orra*, + **tūn**. The *Over-* spellings are due to confusion with ORTON NY 6208. 'Great' for distinction from Little ORTON NY 3555. Cu 144, SSNNW 190, L 178.

Little ORTON Cumbr NY 3555. *Little Orreton'* 1267, *litil Orton* 1424. Adj. **little** + p.n. Orton as in Great ORTON NY 3254. Cu 144.

Water ORTON Warw SP 1891. *Water Ouerton* 1546, ~ *Orton* 1605. ModE **water** + p.n. *Overton(e)* 1262–1346, *Oreton* 1431, 'the ridge settlement', OE **ofer** + **tūn**. The village is situated by a ridge of higher ground overlooking the r. Tame. Wa 47, PNE ii. 53–4.

ORWELL Cambs TL 3650. 'Spring by the point'. *Or(e)d- Or(d)euuelle* 1086, *Orewell(e)* 1173–1428, *Orwell(e)* from 1201. OE **ord** + **wella**. The 'faire springe' is mentioned c.1640; the village lies below the tip of a hill which thrusts into the valley there. Ca 79 gives former pr [ɔrəl].

River ORWELL Suff TM 2138. There are three types:
I. *(into) Arewan* c.1100 ASC(D) under year 1016, *(into) Arwan* 1121 ASC(E) under same year, *Arewe* [before 1118]12th, c.1130.
II. *Orewell* 1341, *Orwell* from 1575.
III. *Ure, Urus* (Latin) 1577, 1586.
Orwell Haven is *portus de Orewell'* 1216, 1223, *Orewell(e)* 1275–c.1386, *Orwell* 1370, *Orwell hauen* 1610.
Orwell is OE **Orwan wella* 'the stream of the *Orwe*' from *Arwe* identical with r. ARROW Warw SP 0861. *Orwe* for *Arwe* may be an example of Scand *w*-umlaut whereby **Arṷa* became **Ǫrṷa* as Dan *hør* 'flax' < **harṷa*. RN 311.

OSBALDESTON Lancs SD 6431. 'Osbald's farm or village'. *Osbaldeston* from 1246, *Osbaston* 1577. OE pers.n. *Osbald*, genitive sing. *Osbaldes*, + **tūn**. La 70 gives prs [ɔ:bistn] and [ɔzbəldestn], Jnl 17.45 local pr 'Osboston'.

OSBASTON Leic SK 4204. 'Osbern's farm or village'. *Sbernestun* (sic) 1086, *Osberneston'* 1194–1251, *Osberston'* 1253–1443, *Osbarston'* 1428–30, *Osbaston(')* from 1329. ON pers.n. *Ásbjǫrn*, ME *Osbern*, ME genitive sing. *Osbernes*, + OE **tūn**. The DB spelling *Sbernes-* probably stands for *Esbernes-* with omission of the first vowel as if it were the prosthetic *E-* frequently found in Anglo-Norman spellings of words beginning in *S-*. If so this would be evidence for the Danish *i*-mutation variant *Esbjǫrn*. Lei 524, SPN 18, SSNEM 193.

OSBORNE BAY IoW SZ 5395. P.n. Osborne as in OSBORNE HOUSE SZ 5194 + ModE **bay**. Formerly called *the Medehole* 1514, *Meade hole* 1591, 1600, ModE **mead** 'a meadow' + **hole** 'a hollow'. Wt 245, Mills 1996.79

OSBORNE HOUSE IoW SZ 5194. P.n. Osborne + ModE **house** originally referring to an 18th cent. house preceeding that built for Queen Victoria 1845×51. Osborne, *Austeburn* 1316, *-bourne* 1327, *Austerborn* 1339, *Auseborne* 1514, *Osborne* 1559, is probably the 'sheepfold stream', OE **eowstre** + **burna**. Mills 1996.79, Pevsner-Lloyd 756.

OSBOURNBY Lincs TF 0638. 'Osbeorn's village or farm'. *Esb'ne- Osbernede- Osb'nebi* 1086, *Osbernebi -by* 1220×35–1689, *Osbernedebi -by* 1204–1412, *Osbernby* 1257–1393, *Osbornbe -by* 1491–1602, *Osburnby* 1495–1607, *Osbournby* 1634×42. Anglo-Scand pers.n. *Āsbeorn* (ODan *Æsbiorn*) + **bȳ**. The form *Esb-* is the EScand mutated variant; the others show substitution of cognate OE *Ōs-* for *Ās-*. Sps with medial *-e-* look like ODan genitive sing. *Æsbiorna(r)*; sps in *-edebi* are probably corrupt. Perrott 131, SSNEM 62, Cameron 1998.

OSCROFT Ches SJ 5067. 'Sheep croft'. *Ouuescroft* 1272–90, *Ouescroft* 1288, 1393, *Owes-* 1314, 1347, *Oscroft* from 1503. OE **eowe**, secondary genitive sing. **eowes**, + **croft** 'a small enclosed field'. Che iii.281.

OSEA ISLAND Essex TL 9106. *Osey Island* 1805 OS. P.n. Osea + pleonastic ModE **island**. Osea, *Uueseiam* 1086, *Oveseye* 1303–1408, *Osey* 1610, is 'Ufi's island', OE pers.n. *Ufi*, genitive sing. *Ufes*, + **ēġ**. Part of the island was called *Ovesland* 1412 from the same man. Ess 311.

OSGATHORPE Leic SK 4319. 'Osgod's outlying farm'. *Osgodtorp* 1086, *Osgodest(h)orp* c.1130, *Osgotthorp* 13th, *Osga(re)sthorp* 14th cent., *Osgarthorp -er-* 1253–1431, *Osgathorp(e)* from 1412,

Angodestorp(e) 1199–1255. ON pers.n. *Ásgautr*, ME *Osgod* partly replaced by Norman *Angod*, + **thorp**. Lei 389, SSNEM 115.

OSGODBY 'Asgaut's farmstead or village'. ON pers.n. *Ásgautr*, genitive sing. *Ásgauts*, secondary genitive sing. *Ásgautes*, + **bȳ**.

(1) ~ NYorks SE 6433. *Ansgote(s)bi* 1086, *Angotebi -by* [1184]17th, 1204–1311, *Osegotebi -by* 13th cent., *Osgotebi -by* 1200–1316, *Osgodby* from 13th. YE 261 gives pr [ɔzgəbi], SSNY 34.

(2) ~ NYorks TA 0584. *Asgozbi* 1086, *Angotby* [c.1160]15th cent., *Osgotby* [c.1160]15th cent., 1285, 1408, *Osgodebi* c.1170, 1252, [1333]16th cent., *Osgodby* from 1301. YN 104, SSNY 34.

OSGODBY Lincs TF 0792. 'Osgot's village or farm'. *Osgote(s)bi* 1086, *Osgotebi -by* c.1115–1379, *Osgotby* 1316–1527, *Ansgotebi* 1168–9, *Angoteby -bi* 1153–1236, *Osgodby* from 1322. AScand pers.n. *Ōsgot* < ON *Ásgautr* + **bȳ**. Li iii.52, SSNEM 62.

OSGOODBY NYorks SE 4980. 'Asgaut's farmstead or village'. *Ansgotebi* 1086. ON pers.n. *Ásgautr*, + **bȳ**. A shrunken village. YN 190, SSNY 34.

OSMASTON Derby SK 2043. 'Osmund's farm or village'. *Osmundestune* 1086, *-mond-*, *-is-*, *-ton(e)* c.1141–1452, *Osmunston* 1264, *Osmoston* 1451, *Osmaston* from 1515. OE pers.n. *Ōsmund*, genitive sing. *Ōsmundes*, + **tūn**. Db 595.

OSMINGTON Dorset SY 7283. 'Estate called after Osmund'. *(at) Osmyntone* [843 for 934]c.1200, *(apud) Osmingtone* [843 for 934]17th S 391, *Osmentone* 1086, *Osmin(g)ton(e) -yn(g)-* 1212–1341. OE pers.n. *Ōsmund* + **ing**[4] + **tūn**. The dedication of the church there to St Osmund (d.1099) derives from the p.n. rather than vice-versa. Do i.211.

OSMINGTON MILLS Dorset SY 7381. P.n. OSMINGTON SY 7283 + ModE **mill**. A mill was recorded here in 1086. Do i.214.

OSMOTHERLEY NYorks SE 4597. 'Asmund's wood or clearing'. *Asmundrelac* 1086, *Osmunderle -ley(e) -lai(e)* 1088–1418, *Osmoderl(a)y* 1536, 1558, *Osmothʳly* 1577. ON pers.n. *Ásmundr* + Anglian **lǣh** (OE *lēah*). YN 213, SSNY 161.

OSPRINGE Kent TR 0060. 'Spring'. *Ospringes* 1086–1242×3, *Ospringe* 11th, *Ospring(')* 1163×4, 1200, *Ospringa* 1194, *Osprinc* 1199, *O(ff)sprenge* 1205, 1211×2. OE ***ofspring**. KPN 148, ING 233.

OSSETT WYorks SE 2820. 'The fold frequented by thrushes' or, possibly 'Osla's fold'. *Osleset* 1086, *Osel(e)set(e)* 13th cent., *Osset(e)* 1274–1641. OE **ōsle** + **(ġe)set**. An alternative view is that the specific is an OE pers.n. **Ōsla*. YW ii.188.

OSSINGTON Notts SK 7564. 'Osica's farm or village'. *Oschintone* 1086, *Oskintun'* [1174×81] *-ton(e)* 1154×89, 1201, *Oscington* 13th cent., *Ossinton* 1280, *Ossington* 1312. OE pers.n. **Ōsica*, genitive sing. **Ōs(i)can*, or **Ōsica* + **ing**[4] + **tūn**. Nt 195.

OSTEND Essex TQ 9397. *Ostend* 1777, 1805 OS. Ess 212.

OSTERLEY PARK GLond TQ 1577. *Osterley Parke House* 1576. The present Osterley Park is an 18th cent. reconstruction of an earlier mansion. P.n. *Ostrele -er-* 1274, *Esterlee* 1294, *Oysterle(y)* 1302, 1342, 'sheepfold wood or clearing', OE **eowestre** + **lēah**, + ModE **park**. Mx 25, Encyc 585.

OSWALDKIRK NYorks SE 6279. 'St Oswald's church'. *Oswaldescherca, Osuualdescherce* 1086, *villa de Scō Oswaldo* 'St O's vill' 1167, *Oswaldkirke* [c.1170]16th, *Chirch-Ki(e)rkos(e)wald* 13th cent. OE saint's name *Ōswald* + OE **ċiriċe** replaced by ON **kirkja**. The 13th cent. inversion of the elements is due to Irish-Norwegian influence. It is suggested that the St Oswald in question is the 10th cent. Archbishop of York rather than the 7th cent. Northumbrian king since land was held there by the Archbishops of York in the Middle Ages. YN 55, SSNY 149.

OSWALDTWISTLE Lancs SD 7327. 'Oswald's tongue of land'. *Oswaldthuisel* 1208×25, *Osowoldestuisil* 1229×38, *Oswalde(s)twisel* 1246, *Os(e)waldestwisel -twysell* 14th. OE pers.n. *Oswald* + **twisla** 'fork of a river, junction of two streams'. The original site lies between two streams. The local pronunciation is [ɔzəltwisəl]. La 90, Jnl 17.54.

OSWESTRY Shrops SJ 2929. 'Oswald's tree'. *Oswaldestr'* c.1180, *Oswaldestroe, id est Oswaldi arborem* 1191, *Oswaldestr(eo) -tre(e)* early 13th–1500, *Osewoldestry* 1394×5, *Osewastre* 1324, 1398, *-westre* 1399, *Oeswestre, Oasewestre* 1427, *Oswestry* 1701. OE pers.n. *Ōswald*, genitive sing. *Ōswaldes*, + **trēow**. Trees were often cited as boundary-markers; the original *Oswaldestrēow* was probably a boundary marker on the N edge of MAESBURY SJ 3125.

The first reference to Oswestry in c.1180 is to the *Castellum de Osewaldestr'* (otherwise known as *Luvre* 'the work' 1086) the site of which is confirmed in DB as being in the manor of Maesbury. When the centre of administration shifted from Maesbury to Oswestry and the collegiate church was built there it was probably dedicated to St Oswald because the place name suggested him as the appropriate saint. Although type II sps *Mes-* for Maesbury (originally *Meresberie*) are not recorded before 1392×3, this type may have been current earlier and the superficial resemblence between Maesbury and Bede's name *Maserfelth*, the site of the battle in which the pagan king Penda of Mercia defeated and killed king Oswald of Northumbria in 642, may have helped the growth in the 12th and 13th cents. of the tradition that the meaning of Oswestry was 'the cross of St Oswald' which in turn gave rise to the W name for the place, *Croesoswald* 1254, *Croes Oswallt* late 13th. In reality, however, the site and the identification of *Maserfelth* remain quite unknown. Oswestry, politically part of England, nevertheless remained in the diocese of St Asaph until the disestablishment and to that extent was recognised as part of Wales. Sa i.231 gives local prs [ɔzəztri] and [ɔdʒeztri], O'Donnell 108.

Old OSWESTRY Shrops SJ 2931. An Iron Age hill-fort 1 mile N of Oswestry. Thomas 1976.184.

OT MOOR Oxon SP 5614. 'Otta's marsh'. *Ottemor(e) -mour* [c.1191]c.1280–1363. OE pers.n. **Otta*, genitive sing. **Ottan*, + **mōr**. O 208 takes *mere* in *(to, of) ottanmere* [1005×12]14th S 943, to be a miscopied form rather than OE *mere* 'a pond', and suggests that the pers.n. is the same as in ODDINGTON SP 5515. *Ottanmere* is, however, a point feature referring to a pond in the bounds of Ot Moor. L 54–5.

OTFORD Kent TQ 5259. 'Otta's ford'. *Ottan forda* 9th ASC(A) under year 773, 12th ASC(E) under year 774, *Ottafordam* 1016, *Otteford* 1160×1–1232 including [830 for 833]13th S 1414. OE pers.n. **Otta*, genitive sing. **Ottan*, + **ford**. KPN 90, PNK 58, L 69.

OTHAM Kent TQ 7953. 'Otta's homestead or estate'. *Otehā* 1086, *Otteham* 1222–78. OE pers.n. **Otta* + **hām**. KPN 90, ASE 2.29, Sx 437.

OTHERY Somer ST 3831. 'The other island'. *Othery* [12th]14th, *Othri* 1225, *Otheri* 1263, *Audre* 1610. OE **ōther** + **īeġ**. Othery is the other (or second) island with reference to *Sowi*, MIDDLEZOY 3733 or ATHELNEY ST 3428. DEPN, Gelling 1998.93.

OTHONA Essex TM 0308. The Romano-British name of a coastal fort now partly eroded, of unknown origin and meaning. *Othona, Othonae* [5th]15th ND with variant *Othana*, *Ythancaestir* c.731 BHE, *Yðþanceaster* [c.890]c.1000 OEBede, *Ithancestre* c.1250, *Effecestre -cestrā* 1086. *Othona* appears to be a Latin name related to or assimilated to the Roman pers.n. *Otho*. Since British did not possess the sound [θ] this name, if British, must have been **Ottōna* of unknown origin. British **Ottōna*, Latin *Othōna* would have given PrW **Othün*, OE **Uthīn* and by i-mutation **Ythin* whence Bede's form *Ythan* + **cæster**. Alternatively *Ythan* may have been, or may have been understood as the genitive sing. of an unknown OE pers.n. **Ytha* or **Yththa*. Ess 210, RBrit 434, LHEB 568, 570, 662, FT 597, Zachrisson 1927.83.

OTLEY Suff TM 2055. 'Otta's wood or clearing'. *Otelega -leia* 1086,

Ot(t)eleia -leg' -le(ya) -leye 1198–1568, *Otley* from 1523. OE pers.n. *Otta* + **lēah**. DEPN, Baron.

OTLEY WYorks SE 2045. 'Otta's wood or clearing'. *of Ottan lege* [972×92]11th S 1453, *on Ottanleage* c.1030 YCh 7, *Othelai -lei(e)* 1086–c.1130, *Ot(t)elai -lei -lay -ley* 1086–1546, *Otley* from 1434. Pers.n. *Otta*, genitive sing. *Ottan*, + **lēah**, dative sing. **lēaġe**. YW iv.203, L 203.

River OTTER Devon SY 0996. 'Otter river'. *Othery* [963]14th S721, *(on) otrig* [1061]1227, *Ot(t)ery* 1244, 1249, *Oter* 1540, *Autri* 1577. OE **oter** + **īe, ī(ġ)**, dative sing. of **ēa**. D 11, RN 313.

OTTERBOURNE Hants SU 4523. 'Otter stream'. *Oterburna* [963×75]12th S 827, *Otrebvrne* 1086, *Oterburne -bourn(e)* 1212–1428. OE **oter** + **burna**. Ha 127, Gover 1958.177, L 18.

OTTERBURN 'Otter stream'. OE **oter** + **burna**.

(1) ~ Northum NY 8893. *Oterburn* 1217, 1242 BF, 1336 SR. NbDu 152.

(2) ~ NYorks SD 8857. *Otreburne* 1086, 1304, *Oter-* 1170–1358, *Otterburn(e)* from 1285. YW vi.142.

(3) ~ CAMP Northum NY 8895. P.n. OTTERBURN + ModE **camp**. A modern military training camp.

OTTERDEN PLACE Kent TQ 9454. A castellated house, part 16th cent., part early 17th cent. P.n. Otterden + ModE **place** 'a mansion'. Otterden, *Otringedene* 1086, *Ottrindaenne, Ottringedene* c.1100, *Ot(e)ringeden'* 1181×2–1283, *Otringden'* 1254, is '*Otering* pasture', OE p.n. **Otering* 'place called after *Oter* or otter place' < OE pers.n. *Oter* or **oter** + **ing**² + **denn**. PNK 228, BzN 1967.354, Newman 1969E.397.

OTTERHAM Corn SX 1690. Either 'river meadow or enclosed land on the river Ottery' or 'river meadow where otters are common'. *Otrham* 1086, *Ottram* 1086 Exon, *Oterham* 1231–1428, *Otterham* from 1234. R.n. OTTERY SX 2788 or OE **oter** + **hamm**. PNCo 131, Gover n.d. 76, RN 313.

North OTTERINGTON NYorks SE 3689. ME prefix *North* 1292 + p.n. *Otrin(c)tun(e) -tona* 1086–12th, *Otheringeton'* 1208, *Ot(e)rington'* 1208–92, 'settlement at Otering, the otter stream', OE r.n. **Otering* < **oter** + **ing**², + **tūn**. Both North and South Otterington lie on the river Wiske which it is suggested was known here as the 'otter place' or 'otter stream'. Older explanations now abandoned include OE folk-n. **Ot(e)ringas*, genitive **Ot(e)ringa*, + **tūn**, and OE pers.n. *Ohtere* or **Oter* + **ing**⁴ + **tūn**. Cf. OTTERTON Devon SY 0885, OTTRINGHAM Humbs TA 2624. YN 207–8, BzN 1967.354, Årsskrift 1974.50.

South OTTERINGTON NYorks SE 3787. *Sonotrinctune* 1088. OWN **sunnr**, ModE **south** + p.n. *Ostrinctune, Otrintona* 1086, as in North OTTERINGTON SE 3689. YN 207–8, BzN 1967.354, Årsskrift 1974.50.

OTTERSHAW Surrey TQ 0264. 'Otter wood'. *Otershaghe* [871×99]13th S 353, *Otershawe* with variants *Ottere-* and *-shagh* 13th cent., *Otreshaghe* 1301, *Oterschaghe* 14th. OE **oter** + **sceaga**. Sr 111, L 209.

OTTERTON Devon SY 0885. 'Settlement, estate on the river Otter'. *Otritone* 1086, *Oterinton* 1235, *-ingthon'* 1242, *Oterytone* 1261, *Otriton* 1325, *Auterton* 1577. R.n. OTTER SY 0996 + OE **tūn**. D 593.

River OTTERY Corn SX 2788. 'Otter river'. *Otery* 1284, *Atery* 1522, *Atrye, Aterey* 16th. OE **oter** 'an otter' + **ēa** 'a river', dative sing. **īe**. The 16th cent. spellings show dial. *ŏ* > *ă*. PNCo 131, Gover n.d. 76, 683, RN 312, Morsbach §272.

OTTERY Original form of the r.n. OTTER SY 0996.

(1) ~ ST MARY Devon SY 1095. 'St Mary's Ottery or river Otter estate'. *Otery Sancte Marie* 1242, *Otreg St Mary* 1275, *Autre Sct Maries* 1577. Earlier simply *(æt) Oteri* before 1100, 1190, *Otri* 1086. R.n. OTTER SY 0996 + saint's name *Mary*. The earliest reference to this place is the *Otrig land* granted to the church of St Mary at Rouen in a 13th cent. facsimile copy of a charter of 1061. D 603.

(2) Mohun's ~ Devon ST 1905 → UPOTTERY ST 2007.

(3) Venn ~ Devon SY 0791. 'Marsh Ottery'. *Fenotri* 1158–1269, *Fennotery* 1346, *Venawtrie* 1606. OE **fenn** with S [v] for [f] + r.n. OTTER SY 0996. 'Venn' for distinction from the other river Otter estates, OTTERY St Mary SY 1095 and OTTERTON SY 0885.

OTTRINGHAM Humbs TA 2624. 'Homestead, estate at or called *Otering*, the otter place or otter stream'. *Otringe- Otrengha', Otrege* 1086, *Otringham -yng-* 1198–1365, *Ot(e)ringeham* [1155×77]14th, 1186–1273, *Ottrinkeha'* 1166, *Ottringham* from 1230. OE p.n. **Otering* < **oter** + **ing**², locative–dative **Oteringe*, + **hām**. The DB form *Otrege* may point to OE *oter-ēġ* 'otter island'. YE 31, ING 154, BzN 353.

OUGHTERSHAW NYorks SD 8681. 'Uhtred's shieling', later 'Uhtred's copse'. *Uhtrede- Huctredescal(e)* [1241]copy, *Ughtershaw(e)* 1540, 1695, *Owghtershawe* 1551–6, *Outersha* 1657, *-shaw* 1702. Apparently OE pers.n. *Ūhtrēd* + ON **skáli** later replaced by ME **shawe** (OE *sceaga*) after the vocalisation of *-áli* and *-aga* to *-aw*. The pers.n. *Ūhtrēd* continued in use in Yorkshire as late as 17th cent. YW vi.117.

OUGHTIBRIDGE SYorks SK 3093. 'Uhtgifu's bridge'. *Uhtinabrigga* (for *Uhtiua-*) [1161]16th, *Doutybrigg* 'of O' 1279(p), *V- Ughti- -y- -brig(g) -bryg* 1359–1633, *Oughty Bridge* 1822. OE feminine pers.n. *Ūhtġifu*, genitive sing. *Ūhtġife*, + **brycg** influenced by ON **bryggja**. YW i.226.

OULSTON NYorks SE 5474. 'Ulf's estate'. *U- V- Wluestun -ton* 1086–1440, *Ulston* 1498, 1613, *Owlston* 1572, *Owston* 1577. ON pers.n. *Úlfr* possibly replacing OE *Wulf*, ME genitive sing. *Ulfes*, + **tūn**. YN 192 gives prs [oustən] and [oulstən], SSNY 129.

OULTON Cumbr NY 2450. 'Ulfi's farm or village'. *Ulveton* c.1200–1367, *Ulton* 1370–1581, *Owton, Owlton* 1570. ON pers.n. *Ulfi* (possibly replacing OE **Wulfa*), genitive sing. *Ulfa*, + **tūn**. Cu 307 gives pr [u:tən].

OULTON Norf TG 1328. 'Othulf's estate'. *Oulstuna* 1086, *Oueltun(e)* 1198, *Owelton(a)* 1225, 1253, *Oulton(')* from 1219. Anglo-Scand pers.n. *Ōthulf* < ON *Authúlfr*, genitive sing. *Ōthulfs*, + **tūn**. DEPN, SPNN 83.

OULTON Staffs SJ 9135. Partly uncertain. *Oldeton* 1280, *Oldington(e)* 1268, 1371, *O(u)lton* 1412–1615, *Over, Nether Wolton als Over, Nether Olton* 1572, 1612. The evidence is insufficient to decide whether this is 'the estate called after Alda', OE pers.n. *Alda* + **ing**⁴ + **tūn**, or 'the old settlement', OE **ald**, definite form **alda**, dative case **aldan**, + **tūn**. St i.174, Duignan 112.

OULTON Suff TM 5294. 'Ali's estate'. *Aleton* 1203, *Alton* 1275, *Olton* 1220–1327, *Oulton* from 1286. ON pers.n. *Áli*, genitive sing. *Ála*, + **tūn**. DEPN, Baron.

OULTON WYorks SE 3628. 'The old settlement'. *Aleton* 1180, *Olton* 1251–1594, *Old(e)ton* 1297–1586, *Owlton* 1598, *Oulton* from 1651. OE **ald** + **tūn**. YW ii.141.

OULTON BROAD Suff TM 5292. *Oulton Broad* 1837 OS. P.n. OULTON TM 5294 + Mod dial. **broad** 'an extensive piece of fresh water'.

OULTON STREET Norf TG 1527. 'Oulton hamlet'. *Oulton Street* 1838. P.n. OULTON TG 1328 + ModE **street**. The hamlet lies a mile SE of Oulton (which has no houses now) near *Oulton Green*.

OUNDLE Northants TL 0488. Unexplained. *(in) Undolum* [c.720]11th Eddi, *Undola* c.1000, *(in prouincia quae uocatur) Inundalum, (in prouincia) Undalum* [c.731]8th BHE, *Undalum* 12th ASC(E) under year 709, *(to) Undelan, Vndelum* [?963]12th S 1448, 1566, *Undele* 1086–1316, *Undle* 1183–1375, *Oundel(l)* 1301, 1381, *Owndele* 1382 *-dale* 1675. An ancient district name in the dative-plural inflectional form **-um** suggesting a tribal name **Undalas* or *Undalan*, cf. *on Undalana mægþe* c.1000 OE Bede, in which *-ana* (for *-ena*) is the expected genitive pl. of an OE **Undalan*. This has been derived from OE *un-dāl* 'without share, undivided' (DEPN) but this is very uncertain. Nth 213 gives pr [aundəl], Jnl 8.41.

OUSBY Cumbr NY 6135. 'Ulf's village or farm'. *Vlmesbi* 1190–1, *Ulmsby* c.1205, 1291, *Vlvesbi -by* 1195–1375, *Ullesby(e) Ullis- Ullys-* 1279–1582, *Owsbie* 1573. ON pers.n. *Ulfr*, genitive sing. *Ulfs*, + **bȳ**. The first two spellings are insignificant against the weight of the majority of forms and are assumed to be due to scribal error. Cu 228 gives pr [u:zbi], SSNNW 37.

OUSDEN Suff TL 7459. 'The owl's valley'. *Vuesdana* 1086, *-dene* 1198, *Ouesden* 1610. OE **ūf**, genitive sing. **ūfes**, + **denu**. DEPN.

OUSE BURN T&W NZ 2369. *Useburn* 1671, *Ewes Burn* 1732. R.n. *(in) Jhesam* [early 13th]copy, *Eyse* 1275, *Yese* 1292, 'the gushing stream', + ME **burn** (OE *burna*). The underlying form is OE **Ġēose*, an agent noun formed from an OE verb **ġēosan* cognate with ON *gjósa* 'to gush', Gmc **geusan*. OE **Ġēose* would normally yield ME *Yese*, but a form **Ġeóse* with accent shift to the second element of the diphthong *ēo* would yield ME **Yose*, the probable source of the 17th and 18th cent. sps. The modern name has been influenced by the r.n. OUSE NYorks SE 4959. NbDu 153, RN 318.

River OUSE ESusx TQ 4208. 'Mud river'. *Wos* 1288, *Ouse* 1612. OE **wāse**. This seems to have been the original name of the river at Lewes where it was also called *aqua de Lewes* 'Lewes water'. There is no credence to the theory that the name *Ouse* derives from misunderstanding of this as if *aqua del Ewes*. Its upper reaches were formerly known as the Midwen or Modwind, *Midewinde* 'the middle winding river' that divides Sussex into two halves, OE **midd** + **winde**. Sx 6, RN 317.

River OUSE NYorks SE 4959. Unexplained. *Usa* [c.780]9th–1280, *usan* [959]12th (accusative, dative case) S 681, [963]14th (genitive) S 712, *Use* 1066–c.1540, *Ouse* from 1268. Attempts to explain this name from PrW **Ûs* < Brit **Usso-* < IE **udso-* from the zero grade of the PIE root **wed-* 'water' are guesswork. The name is probably pre-Celtic, possibly pre-IE. The pr [u:z] is the normal development of ME *Use, Ouse* in N England. YN 5, YE 9, RN 314, LHEB 342.

Little OUSE RIVER Norf TL 8387. *Owsa parva* 1575, *Little Ouse* 1735. ModE **little**, Latin **parva**, + r.n. Ouse as in River Great OUSE TF 5915, TL 5990, 4671, 2065, SP 8646. RN 314.

River Great OUSE Bucks SP 5936, 8646 Norf TF 5915, Cambs TL 5990, 4671. A pre-English r.n. of uncertain antiquity. *Owse magna* 1575. ModE adj. **great**, Latin **magna** + r.n. *(on, neah þære éá, amnem, andlang) Usan* [880]c.1125 B 856, c.1025, [before 1085]12th both Saints, 1121 ASC(E) under year 1010, *Use* 12th–c.1400 including [937]14th S 437, and [979]12th S 834, *Ouse* from 1279. The OE form was *Ūse*, probably representing PrW **Ûs* < **Usso-* < IE **udso-* formed on the root **ued-/ud-* 'water' as in OE *wæter*, Gk *ὕδωρ*, Skr *udán-* 'water, wave'. Against this are the forms of a lost p.n. meaning 'mouth of the river Ouse' in the bounds of Whittlesey on the old Cambridge-Huntingdon border, *Wyshamm(o)uth(e)* 1244–1341, *Wy- Wisem(o)uth(e)* c.1250–1603, which seem to show that the r.n. was orginally *Wise* from the root **vis-* 'to be wet'. The oblique case of *Wise, Wisan*, could have become *Wusan* by combined back-umlaut. Loss of initial *w-* before *u* would then give *Usan* and new nominative *Use*. See further WISBECH TF 4609 and the r.n. WISSEY TF 8401. Ca 11–13, RN 313, 315–7.

OUSEFLEET Humbs SE 8223. 'Tidal inlet or side channel of the r. Ouse'. *Use- Vseflet(e)* 1100×8–1336, *Ouseflet* 1304, 1331, *Uslytte, Uslet, Usflett* 16th cent. R.n. OUSE + OE **flēot**. Such channels were often used for fish-traps. YW ii.7, Jnl 29.81.

OUSTON Durham NZ 2554. 'Ulkill's stone'. *Vlkilstan* 1244×9, *Ulkestan* 1310, *Ulstan* [1286]1332, *Vlston* 1558, *Owsten* 1571, *Ouston alias Owston alias Ewston* 1620×1. Pers.n. *Ulkill* (from ON *Ulfketill*) + OE **stān**. The reference is probably to a boundary stone. NbDu 153, Nomina 12.35.

OUTGATE Cumbr SD 3599. *Out Gate* 1865 OS. The word means 'road leading out of the town or village'.

OUTHGILL Cumbr NY 7801. Possibly 'desolate ravine'. *Hothegill* 1324, *Outhgill* 1800. ON **authr** + **gil**. We ii.17.

OUTLANE WYorks SE 0818. 'The public road called *the Outlane*'. *le Outelone* 1326, *Outlane* 1593. OE **ūt** + **lāne**, the 'exit' sc. from the village to the open fields or pastureland. YW ii.303.

OUTWELL Norf TF 5103. 'Wood' or 'outer Well'. *Utuuella* 1086, *Ude(wu)welle, Wodewelle* [c.1130]17th, *Witewelle* 12th, *Wythewella* 1260, *Vt- Utwell(e)* 1221–1304, *Outwell(e)* from 1324. OE **wudu** partly substituted by ON **vithr** + p.n. Well as in UPWELL TF 5002. Outwell is further downstream than UPWELL TF 5002 which accounts for the later substitution of OE **ūt(e)**. The two places originally formed a single unit called Well. Ca 276.

OUTWOOD Surrey TQ 3245. 'The outlying wood'. *Outwood* from 1640. OE **ūt** + **wudu**. In the Surrey Weald. The contrastive form Inwood 'home wood' is also found in Surrey. Sr 47, 286, 138.

OUTWOOD WYorks SE 3223. 'The outlying wood'. *bosco forinseco* 1305, *Outewode* 15th cent., *the Outwood* 1587. Latin **boscus** + **forinsecus**, OE **ūt** + **wudu**. The great demesne wood of Wakefield Manor mentioned in DB. YW ii.156.

River OUZEL or LOVAT Bucks SP 8831. Ouzel is a late alternative name of the LOVAT. *Ousel* 1847. The original form is uncertain but said to be *Whizzle Brook* of unknown origin: if this is correct the name has been partially assimilated to the r.n. Ouse as in River Great OUSE SP 5936 to which it is a tributary. RN 318.

OVENDEN WYorks SE 0827. Either 'above the valley, the upper valley' or 'Ofa's valley'. *Ouen- Ovenden(e)* early 13th–1822. OE **ofan** + **denu**. Alternatively the specific could be OE pers.n. *Ofa*, genitive sing. *Ofan*. YW iii.113, L 98.

OVER Avon ST 5882. 'Flat-topped ridge'. *Ofre* 1005 S 913, *Ov(e)re* 1247–1427, *Woober* 1535. OE **ofer**. Gl iii.107, L 174, 177.

OVER Cambs TL 3770. 'Flat-topped hill'. *Ouer* *[1060]14th S 1030, *Over(e)* from 1247, *Oure, Ovre* 1086–1337, *Owre* 1576. OE **ofer**. Ca 168. O li gives pr [u:və].

OVERBURY H&W SO 9537. 'Upper fortification'. *Uuerabyrig* [?757]18th ECWM 215, *DE UFERABYRIE, uferesbreodum uel uferebiri, UFERA byrig* [875]11th S 216, *Ovreberie* 1086. OE **uferra** + **byriġ**, dative sing. of **burh** probably referring to Bredon Hill Iron Age promontory fort at SO 9540 which situated at 961ft. lies higher than nearby Conderton Camp hill-fort at SO 9738. Also called 'Upper Bredon', *Uuera breodun* *[964]12th S 731. Wo 153, Hooke 28, 125, 145, Thomas 1976.145, 146.

OVERCOMBE Dorset SY 6982. No early forms but the name probably means 'higher, upper coomb'. It lies at the opposite end of Weymouth Bay from Melcombe Regis SY 6879. Do i.234.

OVERSEAL Derby SK 2915. 'Upper *Scheyle*'. *Vuerseila* 1186×1205, *Overe Scheyle* 13th cent., *Overseale* 1577. OE adj. **uferra** + p.n. *Scheyle* as in NETHERSEAL SK 2813. Db 651.

OVERSTONE Northants SP 8066. 'Ofi's farm or village'. *Oveston* 12th–1468, *Ouston* 1275, *Oveston* 1284–1468, *Overstone* 1610. OE pers.n. *Ofi*, genitive sing. *Ofes*, + **tūn**. Nth 134.

OVERSTRAND Norf TG 2440. Partly uncertain. *Othestranda* 1086, *Overstrand* from 1231, *Ove(r)stonde -ir-* 1254, 1275. If the 1086 form is reliable this might be 'the other shore', OE **ōther** + **strand**, contrasting with nearby SIDESTRAND TG 2539. Otherwise this is 'shore with a ridge', OE **ofer** + **strand**, referring to the Cromer Ridge. DEPN.

OVERTON 'The ridge settlement'. OE **ofer** + **tūn**. L 178.

(1) ~ Lancs SD 4358. *Ouretun* 1086, *Ouretonam* 1094, *Ouerton* 1177–1332 etc., *Overton* from 1201, *Orton* 1577. The reference is to the situation on the bank of the Lune, but the name participates in a little contrastive system of names on the Heysham peninsula with HEATON SD 4460 and MIDDLETON SD 4258. La 175, Jnl 17.100.

(2) ~ Shrops SO 5072. *Huverton* 1174×85, *Overton* from 1199. Overton occupies a flat-topped hill overlooking a small valley running down to the r. Teme. DEPN, Gelling.

(3) Cold ~ Leic SK 8110. *Caleverton* c.1130, *Cald(e)ouirton* 1201, *Caud-* 1209×35, *Cold(e)-* 1212–1714. OE **cald** 'bleak, exposed' + p.n. *(æt) Ofertune* 1066×8, *Ovretone* 1086, *Overton(e)* 1198–1714. 'Cold' describes the site on a high ridge. Lei 236.

(4) Market ~ Leic SK 8816. ME prefix **market**, *Market(t)es- -is-* 1200–1372, *Market-* from 1267 + p.n. *Overtune* 1086, *Overton(e) -v-* 1200–1610. Known for its market as early as 1200. R 656.

OVERTON Hants SU 5149. 'The upper settlement'. *Uferantun, (to) Uferatune* [909]12th S 377, *Ov(e)retvne* 1086, *Overton(a)* from 1167. OE **uferra**, inflected form **uferran**, + **tūn**, dative sing. **tūne**. The reference is to the position of Overton higher up the river Test than Laverstoke. Ha 127, Gover 1958.137.

West OVERTON Wilts SU 1367. *Westovertone* 1275. ME adj. **west** + p.n. *Uferan tune* (dative case), *Uferan tunes* (genitive) 939 S 449, *Ovretone* 1086, *Overtun' -tone* from 1165, *Uverton* 1242, 1263, 'the upper settlement', OE **uferra** + **tūn**, dative case **uferran** + **tūne**[†]. Also known as *Uverton Abbisse* 1263 'abbess's Overton' because held by the abbess of Wilton. 'West for distinction from East Overton (lost). Wlt 305, L 178.

OVER WATER Cumbr NY 2535. 'Blackcock, or, Orri's water'. *Orre Water* 1687, *Orr water* 1777, *Over Water* c.1784. ON **orri** or pers.n. *Orri* derived from it, genitive sing. *Orra*, **orra**, + ME **water**. The modern form is a hypercorrection on the assumption that *Orr* stands for *o'er* from *over*. Cu 35.

OVING Bucks SP 6821. 'The Ufingas, the people called after Ufa'. *Olvonge* 1086, *Huuinga Mainfelini* 'Mainfelin's O' 1167, *Vuinges* 1189×99, 1237×40, 1320, *Ouinges* 1241, 1262, *Ovinge* 1240, 1242, *Wooven* 1687. OE folk-n. **Ufingas* < pers.n. *Ufa* + **ingas**. The 17th cent. form shows a dial. pr with excrescent *W-*, cf. OAKLEY SP 6312. Bk 107 gives pr [u:viŋ], Jnl 2.33 [u:viŋ], ING 46, Jnl 2.24.

OVING WSusx SU 9005. 'The Ufingas, the people called after Ufi'. *Vuinges* *[956]13th S 616, 1183, *Ouing(g)es* 1230–96, *Ovyng(h)e* 1272×1307, 1282, 1316, *Ouvinge* 1282, *Ooving* 1579. OE folk-n. **Ūfingas* < pers.n. *Ūfi* + **ingas**. An *ufes ford* 'Ufi's ford' *[680 for ?685]10th S 230 is recorded in the bounds of nearby Pagham SZ 8897 probably referring to the same man. Sx 75 gives pr [u:viŋ], 93, ING 37, MA 10.23, BzN 2.366, 384, SS 64.

OVINGDEAN ESusx TQ 3503. 'The valley at or of Ofing, the place called after Ofa, or the valley of the Ofingas, the people called after Ofa'. *Hovinge- Hoingesdene* 1086, *Ovingedena -e* c.1094–14th, *Vuinge-* 1198. OE p.n. **Ōfing* < pers.n. *Ōfa* + **ing**^{2}, locative-dative sing. **Ōfinge*, or folk-n. **Ōfingas* < pers.n. *Ōfa* + **ingas**, genitive pl. **Ōfinga*, + **denu**. Cf. OVING WSusx SU 9005 of which Ovingdean may have been a colony. Sx 311, BzN 1967 2.384, SS 71.

OVINGHAM Northum NZ 0863. 'Homestead called or at *Ofing*'. *Ovingeham* c.1200, *Ovingham* 1244, *Owyncham* 1296 SR, *Ovyngeham* 14th cent., *Owingham* 1610 Map. OE **Ōfing* 'the place called after *Ōfa*', locative-dative sing. **Ōfinğe*, + **hām**. The pers.n. *Ōfa* also occurs in nearby OVINGTON NZ 0663. Pronounced [ɔvindʒəm]. NbDu 154, BzN 1967.383.

OVINGTON Durham NZ 1314. 'Wulfa's farm or village'. *Ulfeton* 1086, *Olueton* [1184]15th, *Ulvington* 1251–1343, *Ouinton* 1576, *Ovington* 1665. OE pers.n. *Wulfa*, genitive sing. *Wulfan*, + **tūn** varying with *Wulfa* + **ing**^{4} + **tūn**. The loss of the initial *W-* is due to the influence of the ON pers.n. *Úlfr*. YN 299 gives pr [ouvintən].

OVINGTON Essex TL 7742. 'Estate called after Ufa or Ufe'. *Oluintuna, Ouitunam* 1086, *Uvinton* 1227, *Ovin(g)ton(e) -yn(g)-* 1227–1432, *Oviton -y-* 1227–1428. OE pers.n. masculine *Ufa* or feminine *Ufe*, + **ing**^{4} + **tūn**. Ess 448.

OVINGTON Hants SU 5631. 'Estate called after Ufa'. *(æt) Ufinctune* [963×75]12th S 826, *(æt) Ufintuna* [c.965]12th S 1820, *Ovyngton -in-* from 1185. OE pers.n. *Ufa* + **ing**^{4} + **tūn**. Ha 128, Gover 1958.73 which gives pr [ɔv-].

OVINGTON Norf TF 5103. 'The estate called after Ufa'. *Uvinton* 1202, *Uvington* 1254, *Oviton* 1263. OE pers.n. *Úfa* + **ing**^{4} + **tūn**. DEPN.

OVINGTON Northum NZ 0663. 'Ofa's farm or village' or 'settlement called or at *Ofing*'. *Oventhuna* [1155×89]1271, *Ovintun -ton* c.1200, *-i(n)gton* 1242 BF, 1255, *Owynton* 1296. Either this is OE pers.n. *Ōfa*, genitive sing. *Ōfan*, + **tūn**, or OE **Ōfing* 'place associated with or called after Ofa' + **tūn**. If the first suggestion is correct, as the spellings suggest, the name has been influenced by nearby OVINGHAM NZ 0863. The pers.n. *Ōfa* occurs in both, and it is possible that they should be understood as the **tūn** and the **hām** respectively of the estate called *Ōfing*. Possibly connected with these two places is *Osingadun* [699×705]c.900 AC 126, if this is in fact a mistake for *Ofingadun* 'the hill of the people of Ofa', where St Cuthbert had a vision of the death of Hadwald. NbDu 154, DEPN.

OWER Hants SU 3216. There are two Owers in Hants, this one and Ower SU 4701. The latter is *Hore* 1086, probably representing OE **ōra** 'shore'. Unfortunately the forms for these two places, *Ure* 1202, *Ore* 1284, *Oure* 1256, are confused in the sources. This one may represent OE **ofer** 'flat-topped ridge' referring to its site on a narrow ridge between streams. Ha 128, Gover 1958.198, DEPN, Jnl 21.21, 22.29.

OWERMOIGNE Dorset SY 7685. 'Ower held by the Moigne family'. *Our(e) Moyngne, ~ Moigne* 1314–1459, *Ovre Moigne* 1375, *Owre Moygne* 1575. Ower, *Ogre* 1086, *Ogre(s)* 1210–1348, *Oghre* 1244, *Oweres, Hore* 1212, *Ore* 1288–1430, *Our(e)* 1219, 1313, *Owre* 1431, is obscure. The 1375 spelling *Ovre* is a false etymology as if the forms *Ore, Oure* were derived from OE *ofer* 'a ridge'. The earliest forms, however, clearly point to OE ***ogre** with subsequent vocalisation of [ɣ] to [w], an element of unknown origin, possibly related to the RBrit name for the Lizard peninsula, *Ocrinum* or *Ocrum*, and *Ocra*, a peak in the Julian Alps, Greek *ὄκρις* 'point, corner', Latin *ocris* 'a rugged mountain' (Umbrian *ukar, ocar* 'hill citadel'), Irish *ochair* 'corner, edge'. Most recently Professor Coates has suggested British **ogro-drust-* (PrW **oirthrus*, W *oerddrws*) 'wind-gap(s)' referring to gaps in the hills between the village and the sea. The phonological development is complex but possible. The manor was held by Radulphus Monachus ('the monk') in 1212, by William le Moyne in 1244. Do i.138 gives pr ['ɔ:rmoin], RBrit 429, Chantraine s.n. ὄκρις, IF 100.244–51, GPC 2625.

North OWERSBY Lincs TF 0694. ModE **north** (1824) + p.n. *Ares- Oresbi* 1086, *Ouresbi(a) -by* c.1115–1529, *Oresbi* 1182–1203, *Owresbi -by* 1198–1634, *Owersby* from 1463, probably 'Avarr's farm or village', ON pers.n. *Ávarr*, secondary genitive sing. *Ávares*, + **bȳ**, rather than the 'sand or gravel farm or village', ON **aurr**, secondary genitive sing. **aures**, + **bȳ**. North Owersby is *ad finem borealem de Owresby* 'at the N end of O' 1445. 'North' for distinction from South Owersby TF 0694, *South Owersby otherwise Owersby South End* 1822. Li iii.78–81, SSNEM 63, AEW 20.

OWLSWICK Bucks SP 7906. 'Wulf's farm'. *Wulueswik, Ulueswike* [c.1200]14th, *Wulueswyk* 1242, *Ulfysyke, Ulueswyk* 1262, *Holswyck* 1552, *Owleswicke* 1617. lOE pers.n. *Ulf* (< ON *Úlfr*) influenced by OE *Wulf*, genitive sing. *Ulfes*, + **wīċ**. Possibly identical with 'the herdsman's buildings', *þa herde wic* 903 of S 367. Bk 172 gives pr [elsik] (sic), ECTV p.180, Jnl 2.30.

OWMBY-BY-SPITTAL Lincs TF 0087. P.n. *Ovne- Oune(s)- Dunebi* 1086, *Ounebi* c.1115, 1166, *Ouenebi* 1176, *Auneby* 1230×40, *Oudnesby* 1242×3, *Othenby* 1303, *Oumbi* 1202, 'the deserted farm or village', ON **authn** + **bȳ**, + ModE **spittle**, 'a hospital, a hostel'. The reference is probably to a deserted site taken over by Viking settlers rather than to a deserted Viking settlement, but the specific might alternatively be Anglo-Scand pers.n. *Outhen* corresponding to ON *Authunn*, ME genitive

[†]The form *æt Ofærtune* [949]12th S 547 in a charter the boundary clause of which bears no relation to that of S 449 possibly belongs to Orton Waterville Cambs TL 1596.

sing. *Outhunes*. For Spital see NORMANBY-BY-SPITAL TF 0088. Li ii.260–1, SSNEM 33, Cameron.

OWSLEBURY Hants SU 5123. 'Blackbird fort'. *Oselbyrig* [963×75]12th S 827, *Oselbury* 1185, *Oslebery -bury* 1245, 1301, *Ouselbery* 1395, *Owselbury* 1579. OE **ōsle** + **byriġ**, dative sing. of **burh**. The reference is unknown. Ha 128 gives prs [ʌsl-] and [ʌzlbəi], Gover 1958.73.

OWSTON 'Eastern farm or village'. ON **austr**, perhaps replacing OE **ēast**, + **tūn**.

(1) ~ Humbs SE 8000. *Ostone* 1086, *Oustuna, Houstun* 1154×89, *Ouston* 12th. The village lies at the E edge of the Isle of Axholme. DEPN, SSNEM 185.

(2) ~ FERRY Humbs SE 8000. P.n. OWSTON SE 8000 + ModE **ferry**. An ancient crossing of the Trent. Also known as *West Kinnards Ferry* 1824 OS.

OWSTON Leic SK 7707. 'Oswulf's farm or village'. *Osulvestone* 1086, *Os(s)ulueston(e)* 1148×66–1349, *Osel(l)eston(e)* c.1291–1428, *Olueston* 1196–1551, *Ouston* 1629, *Ouson* 1710, *Ow(e)ston* 1603–1719. OE pers.n. *Ōs(w)ulf*, genitive sing. *Os(w)ulfes*, + **tūn**. Lei 246 gives pr ['oustən].

OWSTWICK Humbs TA 2732. 'East dairy-farm'. *Hostewic -uuic* 1086, *Oustwic -wik(e) -wyk(e)* 1199×1215–1385. ON **austr** probably replacing OE **ēast** + **wīċ**. YE 58, SSNY 142.

OWTHORPE Notts SK 6733. 'Ufi or Ufa's outlying farm'. *Ovetorp* 1086, *-thorp* 1276, *Uuetorp* c.1190–1337, *Outhorp(')* c.1200–1370, *Owethorp* 1433, *Outhroppe* c.1570, *Oughthrope* 1681. Pers.n. ON *Úfi*, genitive sing. *Úfa*, or OE *Ufa* + **thorp**. Nt 238 gives prs [ouθɔrp] and [ouθrəp], SSNEM 115.

OXBOROUGH Norf TF 7401. 'Fort, manor house where oxen are kept'. *Oxenburch* 1086, *Oxeburg* 1194. OE **oxa**, genitive pl. **oxna**, + **burh**. DEPN.

OXENHOLME Cumbr SD 5290. 'Ox pasture'. *Oxinholme* 1274, *Oxenholme* from 1568. ON **oxi** or OE **oxa**, genitive pl. **oxna**, + **holmr**. We i.121, SSNNW 121.

OXENHOPE WYorks SE 0335. 'Valley of the oxen'. *Hoxnehop' -en-, Oxenehope, Oxope* 12th, *Oxenhop(e)* early 13th–1605. OE **oxa**, genitive pl. **oxna**, + **hop**. A deep side-valley of the r. Worth. YW iii.263.

OXEN PARK Cumbr SD 3187. *Oxen Park* 1864 OS.

OXENTON Glos SO 9531. 'Ox hill'. *Oxendon(e) -dun(a)* 1086–1779, *Oxynton* 1287, *Oxenton* from 1573. The earliest reference is in the bounds of Teddington S 1554, *on Oxna dunes cnol* [n.d.]c.1100 'to the knoll of Oxenton, the hill of the oxen', OE **oxa**, genitive pl. **oxna**, + **dūn**. Gl ii.61.

OXENWOOD Wilts SU 3059. 'The wood of the oxen', *Oxenwod(e)* 1257, 1275. OE **oxa**, genitive pl. **oxna**, + **wudu**. Wlt 355.

OXFORD Oxon SP 5106. 'The ox ford'. *Oxnaforda* c.925 ASC(A) under year 912, *Oxenaforda* [1059]c.1100 S 1028, *Oxeneford* 1086 etc., Latin *Oxonia* from 1185. OE **oxa**, genitive pl. **oxna**, + **ford**. O 19, L 71.

OXFORD CANAL Oxon SP 4837.

South OXHEY Herts TQ 1193. ModE **south** + p.n. *(æt) Oxangehæge* 1007 S 916, *Oxehai -y* 1165, 1272×1307, *Oxhey* 1648, the 'ox enclosure', OE **oxa**, genitive sing. or pl. **oxan**, + **ġehæġ**. Cf. nearby BUSHEY TQ 1395. Hrt 106.

OXHILL Warw SP 3145. 'Ohta's shelf'.

I. *Octeselve* 1086, 12th, *O(c)hteselue* 1135×54, 1221, *Ohteshulue* 1221.

II. *Oxeschelfa -schelua -self -sulve -schulf* 1184–1485, *Oxshelve -s(h)ulve -shulf* c.1350–1507.

III. *(H)ostes(s)hulle* 1176–1267.

IV. *Okschull* 1276, *Oxhulle* 1298, 1330, *Oxhille* 1298×9, *Oxhill* 1535.

OE pers.n. *Ōhta* + **scylf** with subsequent folk-etymological reshaping of both elements, *Ox* for *Ōhta* by 1184 (type II), *hyll* for *scylf* by 1176 (type III). Oxhill lies at the edge of a plateau between two hills. Wa 283, L 187.

OXLEY WMids SJ 9102. 'The ox pasture'. *Oxelie* 1086, *Oxeleg* 1236. OE **oxa** + **lēah**. DEPN, L 206.

OXLEY'S GREEN ESusx TQ 6921. Surname Oxley + ModE **green**.

Isle of OXNEY Kent TQ 9127. Pleonastic ModE **isle** + **of** + p.n. *Oxnaiea* [724]13th S 1180 (with variant *Oxenaeya*), *Oxenai* 1086, *Oxeneia* 1165×6, *Oxney* 1610, 'isle of oxen', OE **oxa**, genitive pl. **oxna**, + **īeġ**. KPN 34, Charters IV.163.

OXSHOTT Surrey TQ 1460. 'Occa or Ocga's corner of land'. *Hocchessata* [1135×54]1318, *Hokeset'* 1202, *Okesseta* 1179, *Oggeschate -sethe, Oggasahte, Hoggeset(e), Ogsath -sethe, Oksshete, Oxsete* 13th cent. OE pers.n. *Occa* or *Ocga* + **scēat**. In the NE angle of Stoke D'Abernon parish. Sr 95–6.

OXSPRING SYorks SE 2606. 'Ox spring' or 'copse'. *Ospring -inc* 1086, *Oxspring -yng* from 1154×9. OE **oxa** + **spring**. YW i.334.

OXTED Surrey TQ 3952. 'The stand of oak-trees'. *Acstede* 1086, *Acsted(e)* 1197–1270, *Axsted' -stede -stude, Hocstede, Ocsted(e), Ockstede, Oxsted* 13th cent. OE **āc** + **stede**. Sr 332–3, Stead 247–8, L 218.

OXTON Notts SK 6351. 'The ox farm'. *Oxetune* 1086, *-ton* 1240, 1287, *Oxton* from 1212, *Oxen* 1566, *Oxon* 1638. OE **oxa** + **tūn**. Nt 172 records local pr [ɔks(ə)n].

OXWICK Norf TF 9124. 'Ox farm'. *Ossuic* 1086, *Oxewic* 1242. OE **oxa** + **wīċ**. DEPN.

P

PACKINGTON Leic SK 3614. The 'estate called after Pacca'. *Pakinton(e) -yn-* 1043–1539, *Pachinton(e)* 1086, 1221, *Pakenton'* 1200–65, *Pakyngton'* c.1291–1580 including [c.1043]15th S 1226, *Packington* from 1535. OE pers.n. **Pac(c)a* + **ing**[4] + **tūn**. Lei 568.

PADBURY Bucks SP 7230. 'Pad(d)a's fortified place'. *Pateberie* 1086, *Paddeberi* 1167, *-bury* 1363×98. OE pers.n. *Pad(d)a* + **byriġ**, dative sing. **burh**. Bk 55.

PADDABURN MOOR Northum NY 6578. *Paddaburn Moor* 1866 OS. R.n. Padda Burn NY 6478 + **moor**. Padda Burn, *Padda Burn* 1866 OS, is probably 'the toad stream', OE ***padde** + **burna**.

PADDINGTON GLond TQ 2482. 'Padda's estate'. *(in) Padintune* [959]12th S 1293, *Paddingtone* *[998]14th S 894, *Padington* c.1045, *Padinton(e)* 12th–13th cents., *Pad(d)inton* from 13th. OE pers.n. *Pad(d)a* + **ing**[4] + **tūn**. Mx 132, GL 76.

PADDINGTON STATION GLond TQ 2681. P.n. PADDINGTON TQ 2482 + ModE **station**, built originally in 1838, relocated and rebuilt 1850–4. Encyc 591.

PADDLESWORTH Kent TR 1939. 'Pædel's enclosure'. *Peadleswurthe* 11th, *Pathelesworth* 1307, *Padelesworth* 1341, 1535. OE pers.n. **Pædel*, genitive sing. **Pædeles*, + **worth**. KPN 307, PNK 436, DEPN.

PADDOCK WOOD Kent TQ 6745. *Paddock Wood* 1819 OS. P.n. Paddock + ModE **wood**. Paddock, *Parrok* 1279, 1346, *Parrocks* 1782, is OE **pearroc** 'a small enclosure, a paddock'. PNK 194.

PADDOLGREEN Shrops SJ 5032. *Para Green* 1687, 1689, *Pady Green* 1737, *(The) Padergreen, Pader Green* 1762–72, *Paddoe Green* 1766, *Padall Green* 1775, *Paddock Green* 1750, 1794, 1805, *Pad(d)a Green* 1807, 1833 OS. Probably ME **parrok**, ModE **paddock** + **green**. About a mile from Edstaston Park.

PADGHAM ESusx TQ 8025 → PADIHAM Lancs SD 8033.

PADIHAM Lancs SD 8033. Uncertain. *Padiham* from 1251, *Padingham* 1292–1305, *Padynngeham* 1311. Usually explained as 'the homestead of the Padingas, the people called after Pada', OE folk-n. **Pad(d)ingas* < pers.n. *Pada* + **ingas**, genitive pl. **Pad(d)inga*, + **hām**. But the early reduction of *ing* to *i* points rather to an OE **Pad(d)inghām* 'homestead associated with Padda', OE pers.n. *Pad(d)a* + **ing**[4] + **hām**. In addition Padiham lies by a bend in the river Calder so that the generic might well be OE **hamm** 'land in a river-bend' like Padgham ESusx TQ 8025, *Padiham* c.1200–1327, *-hamme* c.1245, *Padinham* 1261. In this case an alternative specific, OE ***pad(d)ing** 'toad place' < OE **padde*, ON *padda* + *ing*[2], suggests itself, cf. PADDINGTON GLond TQ 2482. La 79, ING 151, Sx 520, Årsskrift 1974.52, Jnl 17.49, 18.19.

Nether PADLEY Derby SK 2578. 'Lower Padley'. *Nether Padeley* 1330, ~ *Padley* from 1451. OE adj. **neothera** + p.n. *Paddeley(e) -le(ghe)* c.1230–1361, *Padeley(e) -lay* 1332–1496, 'Padda's or toad clearing', OE pers.n. *Padda* or **padde** + **lēah**. 'Nether' for distinction from Upper or Little Padley SK 2579, *Parva Padley* 1415, 16th, *Over Padley* 1620. Db 158, 113.

PADON HILL Northum NY 8192. *Pedon* before 1769 map, *Padon's* or *Peden's Pike* 1888. Said to be named after Alexander Peden, 'one of the most noted of the ousted Scotch ministers in the reign of Charles II, who held conventicles on it among the wild Borderers'. Tomlinson 322.

PADSTOW Corn SW 9175. 'St Petroc's holy place'. *Sancte Petroces stow* 11th ASC(C) under year 981, *(at) Petrocys stowe* c.995, *Petrokestowe* 1297, *Patri(ke)stowe* 1318–43, *Pade(r)stowe* 1347, *Patisto* 1418, *Padstowe* 1525, ~ *otherwise Petherickstowe* 1686. Saint's name *Petroc* from the dedication of the church (Co *Pedrek* < *Peder* (Latin *Petrus*) + *-ek*), genitive sing. *Petroces*, + OE **stōw**. The modern form is due to the 14th cent. confusion of *Petroc* with St Patrick, while locally *Petroc*, Co *Pedrek*, survived as *Petherick*. Also known as *Eldestowe* 1201, 1284, *Aldestowe* 1249, 1297, *Oldestowe* 1337, the old 'stow' or holy place, OE **eald**, definite form feminine gender **ealde**, + **stōw**, used as a p.n. in contrast to Bodmin, the new 'stow' to which St Petroc's monastery was transferred, possibly as a result of the Viking raid of 981; and also as *Langwehenoc* 1086, *-uihenoc* 1086 Exon, *Lanwethenek* 1350, *Lodenek* c.1540, *Laffenake* c.1600, 'church-site of Gwethenek', Co ***lann** + saint's name *Gwethenek* with lenition of *Gw-* to *w-*. PNCo 131, Gover n.d. 354–5, *Stōw* 190.

PADSTOW BAY Corn SW 9179. P.n. PADSTOW SW 9165 + ModE **bay**. Earlier *Patystoo havyn* 1478, *Padstow haven* 1576, ME **haven** 'a harbour' (lOE *hæfen*). Also known as *Hægelmutha* 11th 'mouth of the river *Heyl*', r.n. *Heyl* as in EGLOSHAYLE SX 0072 + OE **mūtha**. PNCo 131.

PADWORTH Berks SU 6266. 'Peada's enclosure'. *(æt) Peadanwurðe, (to) peadanwyrðe* [956]13th S 620, *Peteorde* 1086, *Pedewrtha* c.1160–13th cent., with variants *-wurtha -worth, Padeworth'* 1220–1395 with variants *Padde-* and *-w(u)rth(') -w(o)rth(e), Padworth* 1761. OE pers.n. *Peada*, genitive sing. *Peadan*, + **worth**. Brk 214.

PAGHAM WSusx SZ 8897. 'Pæcga's promontory'. *Pecgan ham, Pacganham* *[680]10th S 230, *Pageham* 1086–1377, *Pagan-Pagnaham* 1297, 1300, *Paggam* 1564. OE pers.n. **Pæcga*, genitive sing. **Pæcgan*, + **hamm** 2–5 referring to the situation between Pagam Harbour and the sea. Sx 92, ASE 2.27, SS 81.

PAGLESHAM CHURCHEND Essex TQ 9292. P.n. Paglesham + ModE **church end** from the location of the parish church and for distinction from PAGLESHAM EASTEND TQ 9492. Paglesham, *Paclesham* [1065]14th S 1040, *[1066]14th S 1043,1085×9–1216×72, *Pakelesham* 1117–1461 with variants *-is- -ys-*, *Pachesham* 1086–1475×80 with variant *-kes-*, *Pagesham* 1361, *Paglesham* from 1504×15, is 'Pacol's homestead', OE pers.n. **Pacol*, genitive sing. **Pacoles*, + **hām**. Ess 189 gives prs [pækəlsəm, pægəlʃəm] and [peglʃəm].

PAGLESHAM EASTEND Essex TQ 9492. P.n. Paglesham as in PAGLESHAM CHURCHEND TQ 9292 + ModE **east end**, *East End* 1768. Ess 190.

PAIGNTON Devon SX 8960. 'Estate called after Pæga'. *Peinton(e)* 1086–1238, *Peinctun* early 13th, *Peington(e) -yng-* 1267–1316, *Painton'* 1215, 1230, *Paington* 1809 OS, 1837. OE pers.n. *Pæġa* + **ing**[4] + **tūn**. The spelling *Painton* was substituted for usual *Paington* by the railway company since 1850. D 517.

PAILTON Warw SP 4781. 'The estate called after Pæġla'.
I. *Pailintona* 12th, *Payl(l)inton(a) -yn-* 1222–1332, *-ing- -yng-* 1302–84;
II. *Pallinton(a)* 1217, *Palyn(g)ton -ing- -en-* 1304–1413, *Pallentuna* [1077]17th;
III. *Paylton* 1461.
Either OE pers.n. **Pæġla* + **ing**[4] or genitive sing. **Pæġlan*, + **tūn**. Wa 118.

PAINSHAWFIELD Northum NZ 0660. *Painshaw Field* 1864 OS.

PAINSWICK Glos SO 8609. 'Pain's Wick'. *Painswik(e) -wyk(e)*

-*wick(e)* 1237–1602, *Painsik* 1708. Earlier simply *Wiche* 1086, *Wik(a)*, *Wyk(e)* 13th cent., 'the dairy-farm', OE **wīċ**. The manor was held by Pain Fitzjohn who died in 1137. Gl i.132.

PAKEFIELD Suff TM 5390. 'Paga's open land'. *Pag(g)efella* 1086, *Pagefeld(e)* 1198, 1327, 1401×2, *Pakefeld(')* 1228–1523, *-feud* 1254, *-feild* 1524, *-felde* 1610. OE pers.n. *Paga* + **feld**. DEPN, Baron.

PAKENHAM Suff TL 9267. 'Pacca's homestead'. *Pakenham* [942×c.951]13th S 1526, 1610, *Paccenham* 11th, *Pachenhā* 1086. OE pers.n. *Pacca*, genitive sing. *Paccan*, + **hām**. DEPN.

PALESTINE Hants SU 2640. *Palestine* 1840. A modern development associated with a halt on the railway line between Andover and Salisbury. Transferred from the former Palestine, modern Israel, a name meaning 'land of the Philistines'. Gover 1958.193.

PALEY STREET Berks SU 8676. *Payley-Street* 1711, *Paly Street* 1761. P.n. Paley + ModE **street**. Paley occurs in the lost nearby *Pailehirst* 15th and seems to represent OE *_Pǣganlēah_ 'Pæga's wood or clearing', OE pers.n. *Pǣġa*, genitive sing. *Pǣġan*, + **lēah**. *Street* normally indicates a Roman road but here it is probably Mod dial. **street** 'a hamlet'. Cf. PANGBOURNE SU 6370. Brk 73.

PALGRAVE Suff TM 1178. Probably 'the grove where stakes are obtained'. *(at) Palegrave* [962]13th S 1213, *Pallegrafe* 11th, *Palegraua* 1086, *Palegraue* 1199–1330 including [11th, probably before 1038]13th S 1527, *Palgrave* from 1291. OE **pāl**, genitive pl. **pāla**, + **grāf(a)**. If, however, the 11th cent. sp with *-ll-* is significant, the specific could be OE pers.n. *_Palla_, 'Palla's grove', or even __*pall__ 'a ledge'. The 'grove of the ledges' would refer to the ridge on which the village stands. DEPN, Studies 1936.145, Baron, PNE ii.59.

Great PALGRAVE Norf TF 8312. *Great Pograve* 1278. ModE adj. **great** + p.n. *Pag(g)raua* 1086, *Paggrave* 1202, of uncertain origin. The specific might be a pers.n. such as OE *_Pæcga_ or *_Pacca_,† the generic **græf** 'a grave' or **grāf(a)** 'a grove'. If, however, *pag-* in this name can be related to Gmc *_pagila-_ (whence MLDu *pēghel, peil* 'a stake', OE *pægel* and ultimately ModE *peg*) the meaning would be 'the grove where pegs or stakes are made'. DEPN, PNE ii.58 under __*pagol__.

Sea PALLING Norf TG 4226. 'P by the sea'. ModE **sea** + p.n. *Pal(l)inga* 1086, *Pallenges -inges* 1199–1224, *Palling(e)* 1202–1428, *Pawling(e)* 1637, 1686, either 'the terrace, the plateau', OE __*palling__ < __*p(e)all__ + **ing**^{2}, or 'the Pallingas, the people called after Pælli', OE folk-n. *_Pallingas_ < pers.n. *Pælli* + *ingas*. The village stands on a ledge of higher ground a little back from the sea. Nf ii.119, ING 59, Studies 1936.145, PNE ii.59.

PALLINGHAM WSusx TQ 0422 → POLING TQ 0404.

PALLINSBURN HOUSE Northum NT 8939. P.n. Pallinsburn NT 9138, *Pallinsburn* 1865 OS + ModE **house**. A 20th cent. house containing an 18th cent. core. Pevsner 1992.540.

PALTERTON Derby SK 4768. Partly uncertain. *(æt) Paltertune* [c.1002]c.1100 S 1536, *-ton(e)* 1214×33–1287 etc. with variants *-ir-* 1330–36, *-ur-* 1380, *Paltretune* 1086, *Pautirton -er-* 1268, 1286. The generic is OE **tūn** 'estate, village, farm', the specific identical with the early forms of the r.n. POULTER near the source of which the village lies. The r.n. may, however, be a back-formation from the p.n. which would then remain unexplained. See further r. POULTER SK 6575. Db 295.

PAMBER END Hants SU 6158. 'The end of Pamber'. *Pamber End* 1759. P.n. Pamber + ModE **end**. Pamber, *Penberga* 1166, *Penberg* 1204, *Pembergh* 1234, *Pem- Penber(e)* 1199–1292, *Panbere* 1236–1314, *Pambear* 1306, is the 'hill or barrow with an enclosure', OE **penn** + **beorg**. The generic has been reshaped as if OE *bǣr* 'a woodland swine-pasture' because this was a woodland area with common rights of pannage. Another possibility is W hill-n. *_Penn_ from PrW __*penn__ and so 'Penn hill'. Ha 129, Gover 1958.140–1.

†Searle's *Paga* is a misreading of Ƿaga AC8.

PAMBER GREEN Hants SU 6059. *Pamber Green* 1817 OS. P.n. Pamber as in PAMBER END SU 6158 + ModE **green**.

PAMBER HEATH Hants SU 6162. No early forms. In 1817 OS the area is marked *Pamber Forest*. P.n. Pamber as in PAMBER END SU 6158 + ModE **heath**.

PAMPHILL Dorset ST 9900. 'Pempe's hill' or 'hill called Pempe'.
I. *Pamphilla* 1168, *Pamphill* from 1524.
II. *Pemphull(e)* 1323, 1407, *Peympehull(e)* 1332–1459, *Py- Pimphill(')* 1496–1653.
Explanation must begin from *_Pempe_ possibly from an OE pers.n. *_Pempa_, an i-mutation by-form of *_Pampa_ corresponding to the ON pers.n. *Pampi*, and so 'Pempe's hill'. *Pampi*, however, is a by-name from a word meaning 'small, thick, compact' (ModSwed *pamp*, ModDan *pamper*, ModG dialedct *pfampf*): the alternative explanation is, therefore, an OE hill-name *_Pamp_ or *_Pempe_. Do ii.163, Studies 1926.145.

PAMPISFORD Cambs TL 4948. 'Pampi's enclosure'. *Pampesuuorde* 1086, *-w(o)rda -uuorð(a) -wurda -w(o)rth(e)* 1086–1532, *-forde* 1489, *Pamsforth(e) -ford* 1535–1627, *Panser* 18th. OE pers.n. *_Pampi_, genitive sing. *_Pampes_, + **worth**. Ca 111.

PANCRASWEEK Devon SS 2905. Probably 'St Pancras's Week'. *Pancradeswike* 1197, *Wyke Sci Pancratii* 13th, *Weeke St. Pancars* 1570. Saint's name *Pancras* (Latin *Pancratius*) + p.n. Week, OE **wīċ** 'a dairy farm'. The church is dedicated to St Pancras, but some early spellings, *Wykepranhard* 1285, *Prankardiswyk* 1303, *Prangardeswik* 1316, suggest that the dedication may derive from the p.n., which may have had a different origin, possibly 'Pancoard's Week' from the CG pers.n. Pancoard. D 156.

PANFIELD Essex TL 7325. 'Open country by the river Pant'. *Penfeldā* 1086, *Pan(t)feld -feud* 1233–1428, *Pa(u)mfeld* 12564–1303. R.n. PANT TL 6631 + OE **feld**. Ess 448 gives pr [pænfl].

River PANG Berks SU 5971. A back-formation from Pangbourne, *(on) panganburnan* [956]12th S 607, (water called) *Pangeburne* 1271. See PANGBOURNE SU 6376. Brk 16, RN 319.

PANGBOURNE Berks SU 6376. 'Stream of the Pægingas, the people called after Pæga'. *(at) Peginga burnan, Pægeinge burnan* [844]13th S 1271, *Pange- Pandeborne* 1086, *Pangeburna* c.1160 etc. with variants *-burne -burnia -born'*, *Pangbourn* 1517. OE folk-n. *_Pǣġingas_ < pers.n. *Pǣġa* + **ingas**, genitive pl. *_Pǣġinga_, + **burna**. Cf. PALEY STREET SU 8676. Brk 167, RN 319, PNE i.63, 301.

PANNAL NYorks SE 3051. Possibly 'nook of land in the broad, shallow pan-shaped valley'. *Panhal(e)* 1170–1457, *Panehal(e)* 13th cent., *Panall* 1301, 1377, *Pannall* 1409–1590. OE **panne** + **halh**. The exact sense of *panne* is uncertain; it might alternatively here be an early instance of the sense 'depression in the ground in which water stands', recorded from 1594: hence possibly 'nook of land with a hollow where water stands'. YW v.116, L 108.

PANSHANGER AERODROME Herts TL 2712. P.n. Panshanger + ModE **aerodrome**. Panshanger, *Paleshangre* 1198–1225, *Pe- Paneshangre* 1206–1302, *Panshanger al. Passager* 1567, is the 'slope of the enclosure', OE **penn**, E Saxon dial. form **pænn**, genitive sing. **pennes, pænnes**, + **hangra**. For the *l* forms cf. BENGEO TL 3213 and BENINGTON TL 3023. Hrt 226.

East PANSON Devon SX 3692. *East Panson* 1809 OS. ModE adj. **east** + p.n. *Panestan* 1086, *Panneston* 1238, 1242, 1303, 'pan stone', i.e. stone shaped like a pan, OE **panne** + **stān**. 'East' for distinction from West Panson, SX 3491, *West Panson* 1809, and Hollow Panson SX 3692, *Hollow Panson* 1809. D 164.

PANT Shrops SJ 2722. Short for *Pant Trystan* 1837, 'Tristan's valley', W **pant**.

River PANT Essex TL 6631. *(in ripa) Pentae amnis* 'on the bank of the r. Pant' [c.731]8th BHE with variants *Paente, Pante, Pente* in later mss, *(in) Pente stæðe* [c.890]c.1000 OEBede with variant *Pante, (ripa) Pente amnis* 1104×8, *Pentæ amnis* [c.1350]c.1400, *Pantan (stream), (ofer) Pantan* [c.1000]18th *Battle of Maldon*, *Paunte* 1258, *Po(o)nte* 1285–1638, *Pant(e)* from 1586. OE r.n. *Pante* representing PrW ***pant**, Brit **panto-* 'hollow, valley' taken as a r.n. The *Pente* spellings may represent a side-form < **Pantiōn-* as in the r.n. Lyd, OE *Hlȳde* < **hlūdiōn-*, in LYDFORD Devon SX 5084. The passage in BHE (iii.22) refers to the city of *Ythancæstir* (Roman *Othona*) near Bradwell-on-Sea which shows that the whole length of the river was formerly known as the Pant. Subsequently the reaches below Witham were renamed Blackwater. According to Ess 9 which gives local pr [pɔnt], the estuary was still called Pont or Pant in the 1930s. RN 319.

PANTON Lincs TF 1778. Partly uncertain. *Pantone* 1086, *Pantuna* c.1115, *Pam- Pantun* 12th. OE **panne** or ***pamp** + **tūn**. OE **pamp* is thought to mean something like 'lump' or 'hillock', *panne* 'a pan' in the sense 'depression'. There are several hill formations in the neighbourhood. DEPN, PNE ii.59, Cameron 1998.

PANXWORTH Norf TG 3112. Possibly 'Panc's ford'. *Pankesford* 1086–1202, *Pancforda* 1086, *Pangeford* 1254. Pers.n. **Panc* of unknown origin, genitive sing. **Pances*, + **ford** later replaced in weak-stress position by *worth*. *Panc* appears as the surname of Robert Panke of Mancetter 1221 and is thought to be cognate with OSwed *panka* 'a kind of fish' used as a byname (cf. also Sw *panka* 'a young bream', Sw dial. *panke* 'a young pig', *panker* 'a young boy' etc.). Possibly the reading *Pantha* (for *Pancha*, variant *Pendae)* in the Welsh Annals for 657 is related. Under the lost p.n. Pankford in Dilham TG 3325, *Pancforda* 1086, *Pangford* 1272, 1308, Nf ii.151 prefers the fish-n. DEPN, Redin 69.

PAPCASTLE Cumbr NY 1131. 'Roman fort of the hermit'. *Pabecastr'* 1260–79, *Papecastr(e)* 1266–1399, *Papecastel* 1285, *Papcastle al. Papcaster* 1675, *Pope's Castle* 1737. ON **papi**, genitive sing. **papa**, + OE **cæster** later replaced with ME **castel**. The reference is to the Roman fort of *Derventio* (from the river n. Derwent). Cu 308, SSNNW 205.

PAPLEY GROVE Cambs TL 2761 → PAPWORTH EVERARD TL 2862.

PAPPLEWICK Notts SK 5451. 'The dairy farm at the pebbly place'. *Papleuuic* 1086, 1154×89, *Papilwik* [1154×89] 1356, 1242, 1286, *-wyk* 1332, *Papelwic(ke) -wyc -wik -wyk* 1199–1298, *Paplweeke, Paplique* 17th. OE **papol** + **wīċ**. Some of the fields to the E of the valley are very pebbly. Nt 130 records local pr [papəlwig].

PAPWORTH EVERARD Cambs TL 2862. 'P held by Evrard' sc. de Beche (1155×75). *Pappeworthe Everard* 1254. Earlier simply *Papeuuorde* 1086, *Pappew(o)rd(a) -w(o)rth(e)* 1086–1553, *Papw(o)rth(e)* from 1240, 'Pap(p)a's enclosure', OE pers.n. *Pap(p)a* + **worth**. The same pers.n. occurs in nearby Papley Grove TL 2761, *Pappele* 1279–1385, 'Pappa's wood or clearing'. 'Everard' was wrongly taken as a saint's name, *Papworth S^t. Everard* 1835, because of and for distinction from neighbouring PAPWORTH ST AGNES TL 2664. Ca 171.

PAPWORTH ST AGNES Cambs TL 2664. Properly 'Agnes's P' rather than 'St Agnes's P'. *Papw(o)rth(e) Agnetis, ~ Anneys, ~ Annes* 1218, 1240, 1539, *Papworth S^t. Agnes* 1835 OS. P.n. Papworth as in PAPWORTH EVERARD TL 2862 + pers.n. *Agnes* from *Agnes de Papewurda* 1160 later mistaken as a saint's name. Cf. PAPWORTH ST EVERARD TL 2862. Ca 171.

PAR Corn SX 0753. Possibly '(the) harbour'. *Le Pare* 1573, *the Parre* 1665, *Par* 1732. Fr definite article **le** + Co **porth** with normal later Cornish loss of *th* and SW dial. Cf. Porth SX 074530, which preserves the original form *Porth* 1327. PNCo 132, Gover n.d. 389, Morsbach §272.

PARADISE GREEN H&W SO 5247 → WALKER'S GREEN SO 5247.

PARBOLD Lancs SD 4911. 'Pear-tree homestead'. *Iperbolt* [sic] 1195, *Perbold(e)* [1199–1233]1268, *-bolt -balt* 13th cent., *Perebold* 1202, *Parbold* from 1243, *Parbaud* 1302×3. OE **peru** + **bold**. La 130, Jnl 17.73 gives local pr 'Perbot, Parbot'.

PARBROOK Somer ST 5636. *Parbrook* 1817 OS. Possibly 'pear-tree stream', ultimately from OE **peru** + **brōc**. Early forms are needed.

PARDSHAW Cumbr NY 0924. 'Perd's hill'. *Perdishaw -hau* [c.1205]n.d., *-hou -howe* c.1300–1397, *Pardishou* c.1260, *Paradyshowe* 1382, *Pardsey* 1578, *Pargay* 1660, *Pardshawe* 1601. OE pers.n. **Perd*, genitive sing. **Perdes*, + ON **haugr** referring to the hill at NY 1025 where Pardshaw Hall stands. Cu 367 gives pr [pa:rdzə], SSNNW 243.

PARHAM HOUSE WSusx TQ 0614. *Perham* [959]12th S 1293, 1086–1438, *Parham* 1086, 1121×5, 1200, *Parrum* 1575, 'the pear orchard', OE **peru** + **hamm** 5b, + ModE **house**, an Elizabethan mansion dated 1583 incorporating an older and smaller fortified house. Sx 152 gives pr [pærəm], ASE 2.33, SS 82.

PARHAM Suff TM 3060. 'The pear-tree homestead'. *Per(re)ham* 1086, *Pereham* 1206, *Perham* 1254–1416, *Parham* from 1280. OE **peru** + **hām**. DEPN gives pr [pærəm], Baron.

PARK CORNER Oxon SU 6988. *Park Corner* early 18th. ModE **park** + **corner**. O 137.

PARK END Northum NY 8775. *Park End* 1769 Armstrong. The reference is to the medieval deer-park of Wark.

PARK GATE Hants SU 5208. Cf. *Little Park* and *Park Farm* 1810 OS. A gate on the turnpike from Southampton to Titchfield. ModE **park** + **gate**.

PARK HEAD Corn SW 8470. *Park Head* c.1870. P.n. Park SW 849709, *Park* 1813, + ModE **head** 'a headland'. No park is recorded here but *park* is also used in Cornwall in the sense 'field'. The Co name of the headland is preserved in the farm-name Pentire SW 8570, *Pentir* c.1210, Co ***pen-tyr** 'headland'. Also recorded as *Pencarne Point* 1699. PNCo 132, Gover n.d. 337.

PARKEND Glos SO 6108. Short for Park End Iron Works, *the iron works at the Parke end* 1661, named from *le (the) Parke* 1635, 1669. A Forest of Dean iron settlement. Gl iii.230.

PARKESTON Essex TM 2332. A new station and quay at Harwich opened February 1883 replacing the old pier, named after Charles Henry Parkes, chairman at the time of the Great Eastern Railway. Ess 346.

PARKGATE 'The park gate'. ME **park** + **gate**.

(1) ~ Ches SJ 2878. *Parkgate* 1707. A settlement which grew up to cater for the passenger traffic to Ireland and the 18th cent. novelty of sea-bathing. In 1724 described as 'some houses upon the Water-side in Great Neston...called *Park Gates*'. The reference is to *(the) Parkgate* 1690, an entrance to what was once Neston Park, *parcum de Neston(a)* c.1258, 1350, *Neston Park* 1569. Che iv.223, Room 1983.88.

(2) ~ Ches SJ 7874. *Parkgate* 1831. Named from an entrance to the park at Peover Hall SJ 7773. Che ii.88.

(3) ~ Surrey TQ 2044. *atte Parkgate* 1307(p), *Parke gate* 1636. Sr 86.

PARKHAM Devon SS 3821. 'Paddock homestead'. *Percheham* 1086, *Parkeham* 1242, 1269, *Parcham* 1254, 1292. OE **pearroc** + **hām**. D 104.

PARKHAM ASH Devon SS 3620. P.n. PARKHAM SS 3821 + p.n. *Orshe* 1607, *Ash* 1648, 1809 OS, 'ash-tree', ModE **ash**. Other examples of intrusive *r* are Arscott SS 3505 and 3800, *Ess(h)ecote* 1238, 1277, *Asshecote* 1292, *Arscott* 1563, and *Ayshcote* 1330, *Arscote* 1333, 'ash-tree cottage(s)', ME **ash** + **cote** (OE *cot*, pl. *cotu*), all three showing Devon dial. [a:ʃ] (unless Arscott SS 3800 is really ME **ersh** (OE *ersc*) 'earsh-land, ploughed-land' + **cote**). D 105, 147, 127.

PARKHURST IoW SZ 4991. 'Enclosure or park wood'. *Perkehurst* c.1200, *Parkhurst* from 1255. OE **pearroc** + **hyrst**. A royal forest also referred to simply as *(le) Parke* 1271–1608. Wt 104–5.

PARKHURST FOREST IoW SZ 4790. *King's chace of Parkehurste* 1364, *the kyng's forest...called Parkehurst* 1533, *Parkhurst Forest als. Avington Forest als. Parkehurst Chace* 1600. P.n. PARKHURST SZ 4991 + ModE **forest**. Wt 105, Mills 1996.81.

PARKSTONE Dorset SZ 0391. 'The park (boundary-)stone'. *Parkeston(e)* 1326–1543, *Parkson* 1774. ME **park** + **stone**. The reference is to an early park in the manor of Canford. Do ii.38.

PARLEY COMMON Dorset SZ 0999. *Parley Common* 1811. P.n. Parley as in West PARLEY SZ 0898 + ModE **common**. Do ii.233.

PARLEY CROSS Dorset SZ 0897. 'Parley cross-roads'. P.n. Parley as in PARLEY COMMON SZ 0999.

West PARLEY Dorset SZ 0898. *Westperele* 1305, *Westperle* 1331–1618, *West Parley* from 1618. Earlier simply *Perlai* 1086, *-le(a) -lee -legh -ley* 1187–1575, *Parlea* 1194, *-le* 1399, *-ley* 1558 etc., 'pear-tree wood or clearing', OE **peru** + **lēah**. 'West' for distinction from East Parley SZ 1097. Do ii.231.

Great PARNDON Essex TL 4308. *Magna Pyrindon'* 1291, *Great Parnedon* 1436, *Much Pardington* 1486–93. ModE **great**, Latin **magna**, ME **moche** + p.n. *Perendunā, -in-* 1086, *Pe- Parendon(e)* 1202–87, *Perndon* 1325–46, 'pear-tree hill', OE ***peren** 'growing with pear-trees' + **dūn**. 'Great' for distinction from Little Parndon TL 4310, *Little Permdon* 1359, *Petyparyndon* 1412. Ess 48.

PARRACOMBE Devon SS 6644. 'Valley with an enclosure'. *Pedrecũbe* 1086, *Parrecumb(e)* 1238, 1285, *Parracombe* from 1291, *Pearecumbe -combe* 1297, 1307. OE **pearroc** + **cumb**. A similar DB spelling for *pearroc* occurs in *Apedroc* 1086 for Parrock Sussex. The reference is probably to the prehistoric remains at Holwell Castle SS 6744. D 66, Sx 368.

River PARRETT Somer ST 3927. A pre-English r.n. of uncertain meaning and origin. *Pedride* [?676×85]13th S 1665, [794]18th S 267, *Pedrit* [702]13th S 244, *Pedrete* [708]13th S 1176, *Pedredistrem* [725]14th S 251, *Pedridun* [802×39]17th ECW, *(oþ) Pedridan* 9th ASC(A) under year 658, *Pedridan muþa* 'the mouth of Parrett' ibid. under year 845, *(be eastan, os, oð) Pedredan* 9th ASC(A), c.1100 ASC(F), 1121 ASC(E) under years 658, 845, *Pedryd* [852 for 878]18th S 343, *Pedret* [c.1200]c.1260, c.1540, *Peret* 1233–c.1350 including [973]14th S 791, *Paret* 1243, *Parrett* 1575–86, 1809 OS. This is clearly OE *Pedride* corresponding to OCo **pedreda* possibly representing Brit **petru-rit-* which might mean 'fourfold stream', 'fourfold ford' or 'stream with four fords', cf. the W r.n. Pumryd Gwyn SH 8719 '(river with) five fords'. RN 320, CPNE 176.

PARSON CROSS SYorks SK 3592. No early forms but a *Parsonfeld* 1451 and *Parson greene* are recorded. YW i.251.

PARTINGTON GMan SJ 7191. 'Farm, village called after Pearta'. *Partinton* 1260, *-yn-* 1321, *-ing- -yng-* from 1417. OE pers.n. **Pearta* + **ing**^{4} + **tūn**. The *-ing* forms are so late that the original form may have been 'Pearta's farm', genitive sing. **Peartan* + **tūn**. Che ii.27.

PARTNEY Lincs TF 4168. 'Pearta's island'. *Peartaneu* [c.731]8th BHE, *Peortanea* [c.890]11th OEBede, *Partene -ai* 1086, *Partenay* 1208. OE pers.n. *Pearta*, genitive sing. *Peartan*, + **ēġ** in the sense 'island of dry ground in marsh'. DEPN, Jnl 8.25, L 38.

PARTON Cumbr NX 9720. A modern name. *Parton* 1794. Cu 426.

PARTRIDGE GREEN WSusx TQ 1919. *Partridge Green* 1813 OS. Surname *Partrych* recorded here from 1332 + ModE **green**. Sx 188.

PARWICH Derby SK 1854. Partly uncertain. *pioperpic* [963]17th, *Peuerwick* [966]14th S 739, *-wic -wik(e)* late 12th–1340, *Pevrewic* 1086, *Peverwick -wych(e)* 1241–1338, *Per(e)wych(e) -wich(e)* 1305–1577, *Parwych(e) -wich(e)* 1382, 1452, 1500 etc. The old explanation, r.n. **Pever* (PrW **peβr*, W *pefr* 'bright, radiant' as in PEOVER Chesh) + OE **wīċ**, has had to be abandoned in the light of the recently (1983) discoverd charter of 963. The etymology of this name is now an open question. Db 403, RN 322–3, LHEB 287, ASE 13.141, GPC 2714.

PASSENHAM Northants SP 7939. 'Passa's river-bend land'. *(to) Passan hamme* 921 ASC(A), *Passon- Pas(s)eham* 1086, *Passeham* [1154×89]1383, 1242–62, *Passenham* from 1284, *Pasnam* 1568, *Parsnam* 1675. OE pers.n. *Passa*, genitive sing. *Passan*, + **hamm** 1. Nth 101 gives pr [pa:znəm], L 43, 47, 49.

PASTON Norf TG 3234. Partly uncertain. *Pastuna* 1086, *-tun(e)* c.1150–1216×72, *Paxton'* 1193–9, *Paston* from 1190. The generic is OE **tūn** 'farm, village, estate', the specific possibly a pers.n. such as **Pæċċi*, genitive sing. **Pæċċes*, or OE ***pæsc(e)**, the source of ModE *pash* 'puddle' and cognate with MDu *pasch* 'pasture-land'. Nf ii.184.

PATCHAM ESusx TQ 3009. 'Pecca's homestead or promontory'. *Piceham* 1086, *Peccham* 1087×1100–15th, *Peck- Pec- Patcham* 1664. OE pers.n. **Peċċa* + **hām** or **hamm** 6. Sx 293, ASE 2.33, SS 86.

PATCHING WSusx TQ 0806. 'The Pæccingas, the people called after Pæcc or Pæcci'. *Pettinges* (sic for *Pecc-*) [947]12th S 1631, *Pæccingas* 960 S 687, *[1006 for 1001]13th S 914, 1042×66 S 1047, *Petchinges* 1086, *Pacchyng* 1298, 1327. OE folk-n. **Pæċċingas* < pers.n. **Pæċċ(i)* + **ingas**. Sx 248, ING 37.

PATCHOLE Devon SS 6142. 'Pætti's hollow'. *Patsole* 1086, *Pacheshole* 1326, 1333. OE pers.n. **Pætti*, genitive sing. **Pættes*, + **hol**. The village lies in a small side-valley. D 49.

PATCHWAY Avon ST 6081. 'Peot's enclosure'. *Petsage* 1216×72, *Petshagh* 1276, *Petshawe* 1287–1606, *Padchewaye* 1542, *Patches als. Patshawe* 1624, *Patchway als. Pathchawe* 1629. OE pers.n. *Pēot*, genitive sing. *Pēotes*, + **haga**. The modern form is due to popular association with the words *patch* and *way*. Gl iii.108.

PATELEY BRIDGE NYorks SE 1565. *Patheleybrig(ge)* 1320–1533, *Pai- Paythlaybrig(ges)* 1474–1588, *Padlabrig* 1475, *Padely Bridge* 1486, *Paytley Brigges* 1503. Earlier *Patleiagate* [1175]copy. P.n. Pateley + N dial. **brig**. Pateley, *Pathlay -le* 1202, 1405–1598, is usually taken to be 'the woodland clearing of the paths', OE **pæth**, genitive pl. **patha**, + **lēah** referring to the routes from Knaresborough up Nidderdale and from Ripon to Wharfedale which intersect here at an important crossing of the river Nidd. The first mention of a bridge is in 1320. An alternative explanation is ME ***padil** 'a shallow place in water' referring to the ford that must have preceded the bridge, cf. the lost p.n. *Padlewath* 1227, *Patheleswathe, Patheslayewathe* [1210–30]copy, *Pathelaie- Paithelay- Patylaywath(e)* [1269]copy, with ON **vath** 'a ford', and the lost ford-n. *le Padylford -ul-* 1466, *Padelford* 1467 in Bodington, Glos. Cf. ModE *paddyll*, Scots *paidle* 'to walk in shallow water'. YW v.149 gives pr ['pɛətlə brig] which implies ME *ā̆* by open-syllable lengthening of *păthe* from OE genitive pl. *patha*; xii, vi.xi, 48, Gl ii.76.

PATELEY MOOR NYorks SE 1967. Short for Pateley Bridge Moor, *moram de Padeley Brigge* 1481. P.n. Pateley as in PATELEY BRIDGE SE 1565 + ME **mōr**. YW v.151.

PATHADA Corn SX 3262 → River PARRETT.

PATHFINDER VILLAGE Devon SX 8493.

PATIENT END Herts TL 4227 → EAST END TL 4527.

PATMORE HEATH Herts TL 4526. *Patmore Heath* 1805 OS. P.n. *Patemere* 1086–1320, *Patte- Patesmere* 1203–1324, *Pattmore* 1607, 'P(e)atta's pond', OE pers.n. *P(e)atta* + **mere**, + ModE **heath**. Hrt 169.

PATNEY Wilts SU 0758. 'Peatta's island'. *(æt) Peatanige, (to) Peattan- Pittanige* [963]12th S 715, *(æt) Peattaniggi* [1047×52]12th B 1403, *Pateney(e)* 1171–1300, *Patney* 1284. OE pers.n. **Peatta*, genitive sing. **Peattan*, + **īeġ** in the sense 'raised ground in marsh'. Wlt 314, L 38.

PATRIEDA Corn SX 3073 → River PARRETT.

PATRINGTON Humbs TA 3122. Obscure. 'Farm, village, estate of the Pateringas or called or at Patering'. *(æt, to) yaterinsa-paterings- paterins(a)tune* (for *Patering(a)tune*) [1033]14th S 968, *Patrictone* 1086, *Patrington -yng-* 1194×5–1786, *Patringeton* 1283. OE folk-n. *Pateringas*, genitive pl. *Pateringa* varying with p.n. *Patering 'that which is called *Pater', genitive sing. Pateringes, locative–dative *Pateringe, + tūn. Pater is unexplained; in DB the p.n. is associated with St Patrick to whom the church is dedicated. But this cannot explain the OE forms corruptly preserved in the 14th cent. *Magnum Registrum Album* of York Minster or the later form of the name. YE 25 gives pr [patθrintən], BzN 1968.151.

PATRIXBOURNE Kent TR 2055. 'Bourne held by (William) Patricius'. *Pat'keburn'* 1207, *Patrickesburne* 1242×3. Earlier simply *Borne* 1086, *Burna* 1172×3, 1174×5, 'the stream.' OE **burna**. The manor was held by William Patricius in 1174×5. PNK 545.

PATSHULL Staffs SJ 8001. 'Pættel's hill'. *Pecleshella* 1086, *Pecdeshull* 1166, *Patleshul(l)* 1200–55, 1373, *Patteshille* 1427, *Patsell* c.1565. OE pers.n. *Pættel*, genitive sing. *Pætteles*, + **hyll**. DEPN, Horovitz.

PATTERDALE Cumbr NY 3915. 'Patrick's valley'. *Patrichesdale* before 1184, *-k-* 1292, *Pat(e)ri(c)kedale -ryk-* c.1189–1348, *Patricdale -i(c)k- -yk-* 1247–1629, *Patredale* 1363, *Patyrdale* 1375, *Patterda(i)le -dall* 1578–1823. ME pers.n. *Patric* (<OIr pers.n. *Patraic*), genitive sing. *Patrices*, + ON **dalr**. The dedication of the church and of the name of St Patrick's Well (1777) was probably suggested by the name of the valley. *Patric* continued in use in Westmorland throughout the Middle Ages but the identity of this individual is unknown. We ii.220, SSNNW 152, L 97.

PATTINGHAM Staffs SO 8299. 'The homestead called or at *Peatting*'. *Patinghā* 1086, *Pat(t)ingeham* 1158–1235, *Pattingcham* 12th, *Patyncham* 1448. OE p.n. **P(e)atting* 'the place called after P(e)atta', pers.n. *P(e)atta* + **ing**[2], locative-dative sing. **P(e)atinge*, + **hām**. DEPN, Duignan 114 gives pr 'Pattinjam', ING 150, 170, BzN 1967.384.

PATTISALL Northants SP 6754. 'Pætti's hill'. *Pascelle* 1086, *Pateshull(e) -hill(e)* 1176×81–1675, *Patsell* 1542, *Patchall* 1657, *-ell* 1702. OE pers.n. **Pætti*, genitive sing. **Pættes*, + **hyll**. Nth 92 gives pr [pætʃ ɔ:l], L 170.

PATTON BRIDGE Cumbr SD 5597. *Patton Bridge* 1668, p.n. Patton + ModE **bridge**. A bridge over the river Mint. Patton, *Patun* 1086, *Pattun -ton(e)* 1154×89–1662, is 'settlement by the Roman road', OE **pæth** + **tūn** referring to the Roman road from Natland to Low Borrow Bridge, Margary No.7b. We i.145.

PAUL Corn SW 4627. 'St Paul's (church)'. *(ecclesia) Sancti Paul(in)i* 1259, *St Paulinus* 1266, *(ecclesia) Sancti Pauli de Breweny* 1323, 1329, *Pawle* 1437, *Paul alias Paulyn* 1717. Saint's name *Paul* or *Paulinus* from the dedication of the church. He was a Welsh saint Paulinus Aurelian (otherwise Paulinus or Pol) who founded monasteries in Brittany at Ploudalmézeau, St Pol-de-Léon and on the Ile de Batz. The alternative name of the churchtown survives as Brewinney SW 456270, Co ***bre** 'a hill' + unidentified element. PNCo 132, Gover n.d. 653.

PAULERSPURY Northants SP 7245. 'Pury held by the Pavelli family'. *Pirye Pavely* c.1280, *Pauleyespirye* 1319, *Pawlerspyrry* 1535, *Paulsperry* 1712. Earlier simply *Pirie* 1086, c.1220 or *West Pyria* 12th, 'the pear-tree', OE **pyriġe**. The manor was held by Robert de Pavelli in 1086. 'West' in 12th for distinction from POTTERSPURY SP 7643 which may have influenced the form of Paulerspury with the intrusive *-er-*. Nth 103.

PAULL Humbs TA 1626. Uncertain. *Pagele* 1086, *-a* 12th cent., *Paghel* 1086–1348, *-ell* 1285–1354, *Paule* 1329–1549, *Paull* from 1504. The forms are consistent with an OE ***pagol** cognate with ODu **pagil* 'a litle peg', MDu *pegel* 'a little knob', LG *pegel* 'a stake'. The reference might be either to a stake on the bank of the Humber used as a landmark and boundary stake, or to the small glacial mound on which the church stands. Alternatively the name may be shortened from **pagol-lēah* 'wood or clearing where pegs are made'. YE 36, lix.

PAULL HOLME Humbs TA 1824. *Paghelholm* 1297, 1354, *Pauleholme* 1416. P.n. PAULL TA 1626 + p.n. *Holm(e)* 1086–1401, *(in, de) Hulmo* 1190–1285, ON **holmr** here in the sense 'island, promontory, raised ground in marsh'. YE 37.

PAULL HOLME SANDS Humbs TA 1823. *Holme Sand* 1678, *Paull Sand* 1824. P.n. PAULL HOLME TA 1824 + ModE **sand(s)**. YE 38.

PAULTON Avon ST 6556 'Settlement by the ledge'. *Palton* 1171, 1201, *Pealton* 1194, *Peanton* (sic) 1225ff. OE ***peall** + **tūn**. The 13th–14th cent. surname *de la Palle* recorded in the nearby village of Cameley suggests that *peall* was used as a p.n. *Pall* 'The Ledge' referring either to the 510ft. summit SW of the village or the plateau on which the village stands. DEPN, PNE ii.59, Studies 1936.145.

PAUPERHAUGH Northum NZ 1099. 'Nook of land at *Papworth*'. *Papwirthhalgh* [c.1120]copy, *Pap(w)ur(th)halgh* [c.1250]copy, *Pappeworthhalugh* 1309, *Pepperhaugh* 1798, *Pauperhaugh sometimes called Pepperhaugh* 1888 Tomlinson 36. Lost p.n. Papworth 'Pappa's enclosure', OE pers.n. *Pappa* + **worth**, + OE **halh**. NbDu 155 gives pr [pepəha:f].

PAVENHAM Beds SP 9955. 'Papa's river-bend land'. *Pabeneham* 1086, *Papenham* 1195–1406, *Pabenham* 1240–1535, *Pakenham* 1276, 16th, *Patenham* 1576, *Pavenham* 1491, *Pavingham* 1766. OE pers.n. **Papa*, genitive sing. **Papan*, + **hamm** 1. The village lies beside a great bend in the river Great Ouse. Two alternative developments took place, medial *p* > *b* > *v*, cf. BAVINGTON Northum NY 9880, BAVERSTOCK Wilts SU 0232, and *p* > *k* > *t*, cf. Hepmangrove Beds formerly *Heytmongrove, Hekmangrave*. Bd 36 gives prs [pe:ətnəm, pævənəm], formerly [peikənəm], L 47.

PAWLAW PIKE Durham NZ 0032. *Parlo Pike* 1768. P.n. Pawlaw + ModE **pike** 'a pointed summit of a hill or mountain'. Pawlaw, *Little Pawly, the Easter Pawlawe* 1647 is possibly 'Paga's hill', OE pers.n. *Paga* + **hlāw**, but the forms are too late for certainty. The same specific occurs in *Pawfeld* 1382, a lost place in the ancient waste of the adjoining parish of Wolsingham. NbDu 156.

PAWSTON Northum NT 8532. 'Palloc's farm or village'. *Pachestenam* [c.1130]copy, *Palestun* 1175, *Paloxtun* 1227, *Palwiston'* 1242 BF, *Palx(s)ton* 1255–1441, *Palston* 1292, *Palkeston* 1335, *Pawston* 1542. OE pers.n. **Palloc*, genitive sing. **Pall(o)ces*, + **tūn**. NbDu 155.

PAXFORD Glos SP 1837. 'Pæcc's ford'. *Paxford* from 1208. OE pers.n. *Pæċċ*, genitive sing. *Pæċċes*, + **ford**. Gl i.235.

PAXHILL PARK WSusx TQ 3526. P.n. *Packshill, Paxhill* 18th + ModE **park**. Paxhill is a late reformation of p.n. *bacanscylfes* *[c.765]14th S 50, *la Backeshelue* 1279, *Bacshelve* 1339, *Backshells* 1636, 'Bacca's shelf', OE pers.n. *Bac(c)a*, genitive sing. *Bac(c)an*, + **scylf**. Sx 342, ii.v.

Great PAXTON Cambs TL 2164. *Magna Pacstonia* 1164, *Magna Paxton al. Muche Paxton* 1588. ModE adj. **great**, Latin **magna**, ME **muche**, + p.n. *Parchestune, Pachstone* 1086, *Pacstonia* 1154×89, *Paxton* from 1164, 'Pæcc's estate', OE pers.n. **Pæċċ*, genitive sing. **Pæċċes*, + **tūn**. The same pers.n. occurs in the lost stream n. *Paxebroc* n.d. in Paxton. 'Great' for distinction from Little PAXTON TL 1862. Hu 263, xlii.

Little PAXTON Cambs TL 1862. *Parua Paxton* 1227. ModE adj. **little**, Latin **parva**, + p.n. Paxton as in Great PAXTON TL 2164. Hu 264.

PAYHEMBURY Devon ST 0801. 'Paie's Hembury'. *Paihember* 1236–1422 with variants *Pay(e)-* and *-bury -biri -byre, Payhaumbir'* 1242, *Pahembury* 1272. ME pers.n. *Paie* (OE *Pǣġa*) + p.n. *Han- Henberie* 1086, *Hamberia* 1086 Exon, *Haumbire* 1238, '(at) the high fortification', OE adj. **hēah**, dative sing. definite form **hēan**, + **byriġ**, dative sing. of **burh**, referring to Hembury Iron Age hill-fort at ST 1103 also alluded to in the name BROADHEMBURY ST 1004. D 566, Thomas 1976.95.

PAYTHORNE Lancs SD 8352. Uncertain. *Pathorme, Pathorp*

[sic] 1086, *Paththorn* late 12th, *Pathorn(e)* 1196–1590, *Paythorne* from 1558. Possibly 'Pai's thorn-tree', ON pers.n. *Pái* + **thorn**. DB varies between *thorn* and *thorp* and alternatives have been suggested for the specific, OE ***patha** 'a wayfarer', ON **pá** 'a peacock'. OE **pæth** is impossible in view of the long vowel required to explain the modern form. YW vi.174, SSNY 258.

PEACEHAVEN ESusx TQ 4101. New Zealand and Australian troops were stationed here at the beginning of World War II and the place became known as *Anzac-on-Sea*. The present name was chosen by competition shortly after the war, both to symbolise peace and to echo the name of Newhaven. Land for the new resort was bought by a local businessman in 1914. Room 1983.89.

High PEAK Derby SK 1188. *Alto Pech* 1196–7, *Altus Peccus* 1197–1200 etc., *Hye Peeke* 1490, *High Peak* 1610, and as a wapentake name, *wapentachium de Alto Pecco* 1219. ModE **high**, Latin **altus**, + p.n. *Peak, Peac (lond)* 924 ASC(A), *Pec* 1086–1307, *Pek', Pecco, Pecko, Pekko* 1104×14–late 14th, *Peyk* 1310, *Peek* 1347, 1357 etc., OE **pēac** 'hill, peak', Latinised as *Peccus*, dative case *Pecco*, nowadays applied to the hilly district of N Derbyshire but originally referring to a much larger area. The inhabitants of the area were called the *Pecsæte* in the Tribal Hidage, the earliest reference to the Peak, *Pecsætna lond* [7th]c.1000 'the land of the settlers of the Peak'. Cf. *in pago Pecset* [963]17th referring to Ballidon SK 2054. Which particular hill gave its name to the area is uncertain. Db 24, 1, ASE 13.141.

PEAK DALE Derby SK 0976. A modern name. P.n. Peak as in High PEAK + **dale**.

PEAK FOREST Derby SK 1179. 'Peak hunting-forest'. *foresta de Pecco, Pecko* 1223 etc., ~ *de Peck* 1277, *foreste del Peek* 1383, *Peake Forest* 1577. P.n. Peak as in High PEAK + ME **forest**. Db 159.

PEAKIRK Cambs TF 1606. 'St Pega's church'. *Pegekyrk* [871]c.1200, *(æt) Pegecyrcan* [1016]12th S 947, *Pei- Peychirche* 1140–1247, *Pei- Peykirke* 1198–1390, *Paykyrk* 1526. Saint's name *Pega* + OE **ċiriċe** replaced by ON **kirkja**. Pega was the sister of St Guthlac. She lived as an anchoress here, not far from her brother's hermitage at Crowland. She died on pilgimage at Rome c.719. Nth 241.

PEAMORE Devon SX 9188 → POUNDON Bucks SP 6425.

PEASEDOWN ST JOHN Avon ST 7057. A nineteenth cent. colliery village. No early forms. The church (1892×4) is dedicated to St John. Pevsner 1958N.242.

PEASEMORE Berks SU 4577. 'Pond where peas grow'. *Praxemere* (sic) 1086, *Pesemer(e)* [1166]13th etc., *Pesmere* 1385, *Peysmer* 1535. OE **pise, peosu** + **mere**. Brk 261.

PEASENHALL Suff TM 3569. 'The nook of land growing with peas'. *Pese(n)hala, Pisehalla -healle* 1086, *Pesehal(e) -a* 1163–1291, *Pesenhal(e)* 1228–1524, *Peasynhale* 1568×9, *Pesnall* 1610, *Peasnall* 1674. OE **pise, peosu** and adj. ***pisen**, ***peosen** + **halh** in the sense 'a small valley'. DEPN, Baron, PNE ii.66.

PEASLAKE Surrey TQ 0844. 'The stream by which peas grow'. *Pys(s)he- Pyschelake* 15th cent., *Pis-* 17th cent., *Peaslick* 1749. OE **pise** + **lacu**. Sr 251.

PEASMARSH ESusx TQ 8822. 'Marsh where peas grow'. *Pisemers(h)e* 12th cent., *Pesemers(h)e -mersse* 1248–16th, *-marshe* 1420. OE **pise** + **mersc**. Sx 531.

PEATLING MAGNA Leic SP 5992. 'Great Peatling'. P.n. Peatling + Latin affix **magna** 'great' from 1254 (as prefix 1224–1467), ME **micel**, *Mekyll- Much(e)-* 15th cent. for distinction from PEATLING PARVA SP 5989. Peatling, *Petlinge* 1086–1300, *Petling' -yng'* c.1130–1535, *Pelling* 1166, *Peteling -yng* 1247–1564, *Peatling* from 1480, is probably 'the place called after Peotla', OE pers.n. **Pēotla* + **ing**[2]. The plural forms *Pellingis* c.1131, *Pell- Pedlinges* c.1160–1237, *Petlinges* 1203–47, may refer to the two settlements of Peatling Magna and Peatling Parva, or alternatively represent OE **Pēotlingas*, 'the people called after Peotla', OE pers.n. **Pēotla* + **ingas**. Lei 462, ING 70.

PEATLING PARVA Leic SP 5989. 'Little Peatling'. P.n. *alia Petlinge* 'the other P' 1086 etc. as in PEATLING MAGNA SP 5992 + Latin affix **parva** from 1343 (prefix 1225–1535), ME **litel**, *Litle- Lytell-* 16th cent. Lei 463, ING 70.

PEATON Shrops SO 5385. 'The peak settlement'. *Pet(t)on'* 1255×6, *Petun, Petton'* 1271×2, *Peton* 1274, *Peeton* 1349. OE **pēac** + **tūn**. There is a small hill by the village. Identical with PETTON SJ 4326. Gelling.

PEBMARSH Essex TL 8533. 'Pebba's stubble-field'. *B- Pebenhers, Pebeners* 1086, *Peb(b)eners(e) -(h)ersh* 1205–1324, *Pubemers(a)* 1141–63, 13th, *Peb(e)marsh* 1295, 1371, *Pedmersh(e)* 1320, 1535, 1555. OE pers.n. *Pybba*, Essex dial. form **Pebba*, genitive sing. **Pebban*, + **ersc** later confused with ME *mersh*. Ess 449 gives pr [pedmiʃ].

PEBWORTH H&W SP 1346. 'Peobba's enclosure'. *Pebewrthe* [840 for 844×52]12th S 191, *(æt) Pebbewurðy* c.1010×23 S 1460, *Pebeworde, pebeuuorde* 1086, *Peb(b)ew(o)rth(a) -(e) -wurth* 12th–1533, *Pebworthine* 1278, *-worth(e)* from 1286. OE pers.n. **Peobba* + **worth**. Gl i.252, Hooke 100.

PECKET WELL WYorks SD 9929. *Pecket Well* 1808, cf. *Pecket-Clough* 1769, *the Pecket* 1771. Local tradition associates the spring with St. Thomas á Becket. In reality the name derives from the surname *Peckett*. YW iii.207, xiii, DBS 267.

PECKFORTON Ches SJ 5356. 'Settlement at *Peckford*'. *Pevreton* 1086, *Pecfortuna* [1096×1101]1280, *Pe(c)forton* from 1284. P.n. *Peckford* 'ford by the peak', OE **pēac** + **ford**, + **tūn**. The reference is to a crossing of the River Gowy near Peckforton Point, a prominence on Peckforton Hills SJ 5256. Che iii.331, L 69.

East PECKHAM Kent TQ 6648. *Est Pecham* c.1180. OE adj. **ēast** + p.n. *Pettham* (for *Pecc-*) [961]13th S 1212, *Peccham* [c.975]12th B 1321–2, *Pechehā* 1086, *Pecham* 1189×99, 1210×12, 'homestead, estate by the peak'. OE ***pēac** + **hām**. The reference is to the 300ft. summit at TQ 6652. 'East' for distinction from West PECKHAM TQ 6452. KPN 289.

West PECKHAM Kent TQ 6452. *Westpecham* 1226. OE adj. **west** + p.n. Peckham as in East PECKHAM TQ 6648. KPN 289.

PECKLETON Leic SK 4701. 'Peohtla's farm or village'. *Pechintone* 1086, *Pequintona* 1190×1204, *Pecclinton* early 13th, *Pech(e)lington* 13th, *Peghtlyngton* 1401, *Peyhtelton* 1265, *Peccilton' -(c)k- -yl-* 1209×35–1518, *Peckleton* 1576. OE pers.n. **Peohtla*, genitive sing. **Peohtlan*, + **tūn** varying with **Peohtla* + **ing**[4] + **tūn**. Lei 527.

PEDDARSWAY Norf TF 7724. The Roman road from Ixworth to Holme next the Sea, Margary no. 33b. *Peddars Road, ~ Way* 1824 OS, *Peddar Way* 1838 OS. ME **peddere** 'a pedlar' + **way**. According to Margary 263 the parallel road, Margary no. 333, N of Cockley Cley TF 7904, was called *Pedderysty alias Saltersty* 1399×1413 (OE **stīġ**), which he takes to be generic names for an ancient trackway. This may, however, have been a mistake for Peddars Way.

PEDLEY BARTON Devon SS 7712 → THELBRIDGE BARTON SS 7812.

PEDMORE WMids SO 9182. 'Pybba's marsh'. *Pevemore* 1086, *Pubemora* 1176, *Pebb(e)more* 1291, 1346, *Pebmore* 1291–1327, *Pedmore* from 1291. OE pers.n. *Pybba* + **mōr**. Wo 305, L 55.

PEDNGWINIAN POINT Corn SW 6521 → GUNWALLOE FISHING COVE SW 6522.

PEDWELL Somer ST 4236. 'Peda or Peoda's spring'. *Pedewelle* 1086, *-well* 1201, 1243. OE pers.n. *Peda* or *Pēoda* + **wella**. DEPN.

PEEL FELL Northum NY 6299. *Peel Fell* 1868 OS. P.n. Peel Borders NY 6099 + ModE **fell**. The most westerly spur of the Cheviot range rising to 1975ft. A peel (tower) is marked at NY 6099.

PEGSWOOD Northum NZ 2287. 'Pegg's enclosure'. *Peggis-*

Piggesworth 1242 BF, *Peggeswurthe* 1258, *Peggisworth* 1296 SR, *Pegsworth* 1663–1800. Pers.n. or surname *Pegg*, genitive sing. *Pegges*, + **worth**. NbDu 156, DBS 268.

PEGWELL BAY Kent TR 3563. *Pegwell Bay* 1819 OS. P.n. *Pegwell* 1799, + ModE **bay**. The meaning of Pegwell is uncertain. PNK 602.

PELAW Durham NZ 2752. Uncertain. *Pelhou* [1183]14th, *Pellowe* 1242, *Pel(l)awe* 1279, 1313. Possibly 'Peola's hill-spur', OE pers.n. **Pēola* + **hōh**. This personal name is not, however, independently recorded. Pelaw is situated on a pointed spur of land. Possibly the specific is ME *pēl* 'a pole, a palisade' with reference to the hill's pole-like shape, or ME *pele* 'a triangular shaped shovel' used in the topographical sense 'peel-shaped, shovel-shaped hill'. Cf. PELAW WOOD NZ 2842 and PELTON NZ 2553. NbDu 156.

PELAW T&W NZ 2962. Unexplained. *Pelaw* 1863 OS. Unidentified specific + dial. **law** 'a hill' (OE *hlāw*). Early sps are needed.

PELAW WOOD Durham NZ 2842. P.n. Pelaw + ModE **wood**. Pelaw, *Pellow* 1420, is of uncertain meaning; the site is a pointed spur of land and the generic is either OE *hōh* 'a hill-spur' or *hlāw* 'a hill', but the specific presents the same problem as PELAW NZ 2752 and PELTON NZ 2553.

PELDON Essex TL 9816. Partly uncertain. *(at) Piltendone* [946×c.951]13th S 1483, *(æt) Peltandune* [962×91]11th S 1494, 1000×2 S 1486, *Peltendunā* 1086, *-done* 1237–1303 with variants *-in- -yn-*, *Peltedon'* 1205, *Pel(e)don* 1285–1457. Either OE pers.n. **Pylta* or ***pylta** 'that which thrusts forward, a thrusting hill', genitive sing. **Pyltan* or ***pyltan**, + **dūn**. The village occupies a modest hill rising to 125ft. from marshland. Also known as *Pelden in le Plaish* 1553, ModE **plash** (OE *plæsc*) 'a marshy pool'. Ess 321.

Brent PELHAM Herts TL 4330. 'Burnt Pelham'. *(la) Arse Pelham* 1210, 1254, *Barndepelham* 1230, *Brende-* 1241–1334, *Brentpelleham* 1399, *Burnet Pelham* 1619. OFr **arse**, ME **brende** + p.n. *Pelehā* 1086–1221, *Pelham* from 1207, 'Peola's homestead', OE pers.n. **Pēola* + **hām**. 'Brent' for distinction from Furneux and Stocking PELHAM TL 4327, 4529. Hrt 184 gives pr [bɛ:nt].

Furneux PELHAM Herts TL 4327. 'P held by the Furnelle family'. *Pelham Furnelle* 1240, ~ *Forney* 1291, ~ *Furneus* 1293, ~ *Fourneaux* 1324, *Furnysshe Pelham* 1541. Family name *Furnelle* of Simon de Fornell' 1232 + p.n. Pelham as in Brent PELHAM TL 4330. Hrt 184 gives pr [fɛ:nix] (sic for [fɛ:niks]).

Stocking PELHAM Herts TL 4529. 'P made of logs'. *Stokenepelham* 1235–94, *Stocken Pelham* 1255, *Stokking* ~ 1428. ME adj. **stoken** (OE *stoccen*) + p.n. Pelham as in Brent PELHAM TL 4330. Hrt 184.

PELSALL WMids SK 0203. 'Peol's nook'. *Peoleshale* [996 for 994]17th S 1380, *Peleshale* 1086, *-hala* 1167. OE pers.n. *Pēol*, genitive sing. *Pēoles*, + **halh** in the sense 'a small valley', dative sing. **hale**. DEPN, L 105.

PELTON Durham NZ 2553. Uncertain. *Pelton* from 1242×3. Possibly 'Peola's farm or village', OE pers.n. **Pēola* + **tūn**. This personal name is not, however, independently recorded. Cf. PELAW NZ 2752 and PELAW WOOD NZ 2842. NbDu 156.

PELUTHO Cumbr NY 1249. Possibly 'pilloats hill'. *Pellethowe* 1538, *Pellathow* 1532, *Pelutho* 1575. ModE **pilloats** 'oats whose husks peel off' + dial. **how** (<ON *haugr*). Cu 297 gives prs [peləte] and [pelidə], Studies 1936.105–6.

PELYNT Corn SX 2055. 'Parish of Nennyd'. *Plvnent* 1086, *Plounent* c.1200 Gover n.d., *Plenint* 1229, 1236 Gover n.d., *Plynt* 1478, *Plenynt alias Pelynt* 1577, *Plinte* 1610. Co **plu** + saint's name *Nunet* 1442 as in Eglwys Nunydd W Glam SS 7985. PNCo 132.

PEMBRIDGE H&W SO 3958. 'Pena's bridge'. *Penebrvge* 1086, *-brug(g)e* 1100–1328, *-brigg(e)* 1275–1346, *Penbrugge* [c.1148]14th, *Pembrug'* 1317. OE pers.n. **Pena* + **brycg**. The pers.n. is unknown but alternative suggestions, pers.n. **Pæġna* as in PAINTON Devon SX 8960, or OE *penna*, genitive pl. of *penn* 'a pen', fit the forms less well. He 156.

PEMBURY Kent TQ 6241. Partly uncertain. *Peppingeberia -b'i -byr'* c.1100, 1218, 1250, *Peping(e)bir' -byr' -bur(i) -ber(y)* 1205–70, *Papingbyr* 1257, *Peapyngeberi* 1309, *Pemburye* 1575. The persistence of early spellings with *-pp-* casts doubt on the explanation of K 185, 'look-out fort or manor', OE ***pēping**, locative–dative ***pēpinge**, + **byriġ**, dative sing. of **burh**, although Pembury is high up (499ft.) and commands views of the surrounding ground. Possibly, therefore, the 'fortified place of the Peppingas, the people called after Peppa', OE folk-n. **Peppingas* < pers.n. **Peppa*, Kentish for *Pyppa*, + **ingas**.

PENARE 'Headland'. Co ***pen-arth**. Cf. PENNARE.

(1) ~ Corn SX 1550. *Pennarth* [1284]1602, *Penhard* 1303, *Penarth* 1306, *Pennare* 1547. The reference is to the promontory now called Dodman Point. PNCo 132.

(2) ~ Corn SW 9724 → NARE POINT SW 7925.

PENCARROW HEAD Corn SX 1550. P.n. Pencarrow + ModE **head** 'a headland'. Pencarrow, *Pencarowe* c.1482, *Pencarrow* 1748, is 'stag's head', Co **pen** 'head' + **carow** 'stag', a fanciful name taken from the headland's two-horned appearance from the sea. PNCo 132, Gover n.d. 271.

PENCOMBE H&W SO 6052. 'Coomb with a pen'. *Pencumb(a)* c.1115×25, 1160×70, *Penc(o)umbe* 1243–1342. OE **penn** + **cumb**. He 158, L 92.

PENCOYD H&W SO 5126. 'Wood's end'. *Pencoyt* 1291, 1334, *Pencoid* 1535. OW **pen** (PrW **penn*) + **coit** (PrW **cę̄d*, **coid*). The Welsh Annals (Harleian 3859) record a battle under year 722 at a place called *Pencon* which is identified with Pencoyd, but this is very uncertain. An earlier name is probably *Cil Hal* [6th]c.1130 LL. Identical with PENGE GLond TQ 3570 and PENKETH Lancs SJ 5687. He 161.

PENCRAIG H&W SO 5621. 'Top of the crag'. *Penncreic* [c.860×66]c.1130 LL, *Pencreid* 1334. PrW ***penn** + ***creig**. He 145, O'Donnell 94, L 183.

PENDEEN Corn SW 3734. 'Headland of the fort'. *Pendyn* [1284]1446, 1306, 1326, *-dyne* 1416, 1452, *Pendeen* 1588. Co **pen** + ***dyn**. Originally the name of the headland called Pendeen Watch SW 3735, although no promontory fort is known there, it was transferred to the medieval settlement at Pendeen House SW 3835 and then to the 19th cent. mining village. PNCo 133, Gover n.d. 633.

PENDEEN WATCH Corn SW 3735. 'Pendeen look-out'. P.n. PENDEEN SW 3734 + ModE **watch** 'a look-out place'. PNCo 133.

PENDENNIS POINT Corn SW 8231. P.n. Pendennis + ModE **point** 'a promontory'. Pendennis, *Pendinas* c.1540, *Pendynas, Pendennys* 1546, is 'fort headland', Co **pen** 'headland' + ***dynas** 'fort', probably referring to a lost Iron Age promontory fort rather than Henry VIII's castle of 1540×3. PNCo 133, Gover n.d. 499.

Forest of PENDLE Lancs SD 8338. *foresta de Penhull* n.d. ME **forest** 'woodland area devoted to the preservation and hunting of game' + p.n. Pendle as in PENDLE HILL SD 7941. La 68, PNE i.184.

PENDLE HILL Lancs SD 7941. P.n. Pendle + ModE **hill**. Pendle, *Pennul* 1258, *Pennehille* 1296, *Penhul(l), Penhill* 14th cent., is the 'hill called Penn', PrW ***penn** 'a head, end, top, height' + OE **hyll**. The name refers to a distinctive hill rising to 1831ft. It shows the Anglo-Saxon failure to perceive PrW ***penn** as an appellative rather than a name, and late pleonastic addition of ModE **hill** when the original compound became obscured by the reduction of its second element to [əl]. Cf. PENDLETON SD 7639. La 68, Mills 120, Jnl 17.44.

PENDLEBURY GMan SD 7801. '(At the) fortified place on Penhill'. *Penelbiri, Pinnelberia* 1202, *Pennelbiry* 1246, *Penhil-*

Pennyl(les)byry, *Penhulbury* 13th, *Penlebyri* 1218×40, 1235×60, *Pendilsbury* 1537, *Pendulbury* 1561, *Pendlebury* from 1567. OE p.n. **Penn-hyll*, PrW ***penn** 'a head, a hill' + pleonastic OE **hyll**, + **byriġ**, dative sing. of *burh*. The reference of *burh* in this name is unknown. An early alternative form of the name is also found, *Penesbire* 1206, *Pennebire* 1208, 1226, *Penisburia* 1212, *Pennesbyry* 1278, '(at the) fortified place of or at *Penn*'. *Penn* and *Pen-hill* refer to the long ridge of high ground stretching from Wardley SD 7602 where it reaches a height of 303ft. to SD 7901. La 42, Jnl 17.35.

PENDLETON Lancs SD 7639. 'Settlement beside Pendle'. *Peniltune* 1086, *-ton* 1241, *Pen(n)ulton* 1242, 1262, *Penhulton* 1272, *Penhil(l)ton(e)* 1305, 1311. P.n. Pendle as in PENDLE HILL SD 7941 + OE **tūn**. La 77, Jnl 17.48.

PENDOCK H&W SO 7832. Possibly 'head of the barley field'. *(æt) Pendoc* [888]17th S 1839, *(et)peonedoc* [875]11th S 216, [964]12th S 731, [967]11th S 1314 with variant *penedoc*, *Penedoc* [1052×62]17th S 1857, *Pe(o)nedoc* 1086, *Pendoke* 1327. PrW ***penn** + ***heiddioc** < **heidd* + *iauc* 'of barley, barley place' (ModW *haidd*). For OE spellings of **penn* with *eo* cf. PENSELWOOD Somerset ST 7531, PINHOE Devon SX 9694. Wo 154, DEPN, Hooke 102, 125, 145, 264, 337, GPC 1814.

PENDOGGETT Corn SX 0279. 'The top of two woods'. *Pendeugod* 1289, 1297, *Pendewegoys, Pendouket* 1302, *Pendoget* 1467. Co **pen** + **dew** 'two' + OC **cuit** (Co *cos*) with lenition of *c-* > *g*. The village is situated on a watershed at the head of two wooded valleys. PNCo 133, Gover n.d. 119.

PENDOMER Somer ST 5210 → PENSELWOOD ST 7531.

PENGE GLond TQ 3570. 'The head of the wood'. *(se wudu þe hatte) pænge* 'the wood called P' [957]12th S 645, *Penceat* 1067, *Pange* 1204, c.1230, *Penge* from 1206. PrW ***penn** + ***cę̄d**. This name, identical in origin with Pencoyd H & W SO 5126, Penketh Lancs SJ 5687 and the Welsh Pencoeds, should have become *Penchet*, but *-et* seems to have been taken as a Fr diminutive suffix which was dropped, giving **Pench* and subsequently *Penge*. Penge was originally a wood or swine-pasture 'seven miles, seven furlongs and seven feet in circumference' (S 645) attached to the manor of Battersea of which it remained part until 1888. Sr 14.

PENHALE POINT Corn SW 7559. *Penhale Point* 1813. Lost farm-name *Penhale* (as in Penhale Camp SW 7658) + ModE **point** 'a promontory'. Penhale, *Penhal* 1327, is 'head of the marsh', Co **pen** + **hal**. The farm was situated at the upper end of a marshy depression. PNCo 133, Gover n.d. 378.

PENHALE SANDS Corn SW 7656. *Penhale Sands* 1888. P.n. Penhale as in PENHALE POINT SW 7559 + ModE **sands** 'dunes'. Earlier *Piran Sands* 1813 from the patron saint of the parish, St Piran, as in PERRANZABULOE SW 7752. PNCo 133.

PENHALURICK Corn SW 7037. '*Penhal* rampart'. *Penhaleluryk* 1485, *-louryk* 1512. P.n. **Penhal* 'head of the marsh' (Co **pen** + **hal**) + Co ***luryk** 'coat of mail, breastplate' in the sense 'breastwork, rampart' as in Latin *lorica* from which it derives. **Penhal* occurs again in nearby PENHALVEAN SW 7137. CPNE 155, Gover n.d. 531.

PENHALVEAN Corn SW 7137. 'Little *Penhal*'. *Penhalvyhan* 1319, *Penhalvian* 1404. P.n. **Penhal* 'head of the marsh' (Co **pen** + **hal**) + **byghan** 'small' with lenition of *b-* to *v-*. 'Little' for contrast with Penhalveor SW 7037, *Penhalmur* 1319, *Penhalvoer* 1404, p.n. **Penhal* + Co **meur** 'great' with lenition of *m-* to *v-*. PNCo 133, Gover n.d. 531.

PENHALVEOR Corn SW 7037 → PENHALVEAN SW 7137.

PEN HILL Dorset ST 8517 → Sixpenny HANDLEY ST 9917.

PENHILL NYorks SE 0486. '*Pen* hill'. *Penle* 1202, *Penhill* 1577. PrW ***penn** 'head, end, top, height, a hill' probably uswed as a p.n. + OE **hyll**. A pleonastic formation identical with PENDLE HILL SD 7941. YN 256.

PENHURST ESusx TQ 6916. 'Pena's wooded hill'. *Penehest* (sic) 1086, *-herste -hurste* 12th cent.-13th, *Penherst -hurst* c.1210–1414. OE pers.n. **Pēna*, a SE dial. form of **Pǣġna*, or **Pena*, + **hyrst**, SE dial. **herst**. Sx 476, SS 68.

PENISTON SYorks SE 2403. 'The settlement, estate of or by the hill called *Penning*'. *Pen- Pangeston(e)* 1086, *Peningestun -ton -ynges-* 12th cent., *Penig(g)estun -ton* 13th cent., *Penyeston* 1283, *Penyston -is-* from 1295. OE p.n. **Penning* < PrW ***penn** 'a hill' + OE **ing**[2], genitive sing. **Penninges*, + **tūn**. The reference is to the great ridge lying between the Don and Little Don. YW i.336, BzN 1968.152, L 183.

PENJERRICK Corn SW 7730. Partly uncertain. *Penhegerick* 1327, *Pennanseyryk* 1333, *Penseryk* 1517. Co ***pen nans** 'valley's head' + unidentified element, probably a stream name in **-yk**, e.g. **eyrik* corresponding to W *eirig* 'warlike, fierce'. PNCo 133, Gover n.d. 499.

River PENK Staffs SJ 9005. *Penck(e)* 1568–1610 etc., *Penk* from 1577. A back-formation from PENKRIDGE SJ 9213. Earlier the whole p.n. was transferred to the river, *(on lang) Penchrich* [996]17th S 1380, *Pencriz* c.1250–1300. St i.15.

PENKETH Ches SJ 5687. 'End of the wood'. *Penket* 1243 etc., *Penketh* from 1259, *Penkith* 1259. PrW ***penn** + ***cę̄d**. Cf. Pencoed M Glam SS 9581 and many other examples in Wales, Penquite Corn (4 examples), PENCOYD H& W SO 5126. La 106, L 171, 183, 190.

PENKRIDGE Staffs SJ 9213. 'The headland tumulus'.

I. Latin form: *Pennocrucio* [4th]8th AI.

II. English forms: *Pencric* [958]14th S 667, [c.1000]11th S 1534–1251, *Pancriz* 1086–1259, *Pencriz'* 1136–1328, *Pencrich(e) -crych(e) -krich(e) -krych(e)* 1156–1538, *Penkridge* from 1755.

PrW p.n. **Penngrüg* < RBrit *Pennocrucium*, the name of the Roman settlement at Water Eaton SJ 9010, Brit ***penno-** 'a head, a hill, an end' + ***crouco-** 'a hill, a tumulus' with suffix *-*i̯o-*. The reference is to a tumulus at Rowley Hill SJ 9011, the discovery of which has rendered previous discussion in RBrit 436 and Signposts 41, 58–9 unnecessary. The modern sp suggests the name has been reinterpreted as if containing the element *ridge*. St i.87, LHEB 228, 309, 315, 557, L 138, 182.

PENLEE POINT Corn SX 4449. *Penlee poynte* 1584 Gover n.d. P.n. Penlee + ModE **point**. Penlee, *Penleigh* 1337, is 'headland of the slab-stone', Co **pen** + ***legh**. Cf. Penlee Point SW 4726. PNCo 134, Gover n.d.283.

PENN Bucks SU 9193. 'The hill-top'. *Penna de Tapeslawa* 1188, *Lapenne* 1197–1222. PrW ***penn** sometimes preceded by Fr definite article **la**. The village stands on a well marked hill. The Taplow referred to in the 1188 form is probably an unrecorded place of that name rather than TAPLOW SU 9182 8 miles to the S. Bk 229.

Lower PENN Staffs SO 8696. *Netherpenne* 1271. ModE adj. **lower**, ME **nether**, + p.n. *Penne* 1086 etc. 'the hill', PrW ***penn** (Brit *penno-*). 'Lower' for distinction from Upper Penn SO 8995, *Overpenne* 1318, earlier *Overtone* 1086, 'the upper settlement', OE **ofer** + **tūn**. DEPN.

PENN STREET Bucks SU 9295. 'Penn hamlet'. *Penn Street*. P.n. PENN SU 9193 + Mod dial. **street**.

East PENNARD Somer ST 5937. *East Pennard* 1243. ME adj. **est** + p.n. *Pengerd* [681]?10th S 236, *Penger .i. Pennard* [681]13th S 236, *Pennard* [705]14th S 705, *[725]12th S 250, [854]14th S 303, 'end of the hill', PrW ***penn** + **garth** < **garto-* as in Penyard in WESTON UNDER PENYARD SO 6323, *Penyerd* 1227, *Peniard -ierd* 1228, and the common W name Peniarth. The minster at Pennard is *Pengeard mynster* 955 S 563. The reference is to Pennard Hill (395ft. at ST 5738) under the end of which the village lies. From Pennard came *Eanulf Penearding* 'Eanulf of Pennard' 899×924 S 1445. 'East' for distinction from West PENNARD ST 5438. Cf. LYDEARD ST 1232. DEPN, Turner 1950–2.115.

West PENNARD Somer ST 5438. *West Penard* 1610. ME adj. **west** + p.n. Pennard as in East PENNARD Somer ST 5937.

PENNARE Corn SW 9238 → NARE HEAD SW 9136. Cf. PENARE.

PENNERLEY Shrops SO 3599. *Penalley* 1836 OS. Possibly a made-up name for the lead-mining community that grew up here in the 1830s and 1840s, possibly from W **pen** 'a head, a height, a hill'.

The PENNINES NYorks SE 0077. A name invented c.1747 by Charles Julius Bertram, the compiler of the forged treatise *De Situ Britanniae* attributed to Richard of Cirencester and published in Copenhagen in 1757. In his *Britannia* published in 1586 Camden, commenting on the names *Pen-mon maur, Pendle* and *Pennigent*, related them to the Pennine Alps in Switzerland containing the element *pen* 'the summit of a mountain', and to the Appenines in Italy. In a later passage he alluded to *Ingleborrow, Pendle* and *Penigent* as 'our Appennine' (*in Appennino nostro sunt eminentissimi*). These passages provided the source for Bertram's invention in I.6.33 of *De Situ Britanniae*. The whole province, he says, is divided into two equal parts by mountains called *Alpes Penini*. And in his 7th *Iter* he invents a place called *Ad Alpes Peninos* between *Rerigonio* (Ptolemy's *Rerigonium*, a town and bay near Loch Ryan, Wigtownshire) and *Alicana* (Ptolemy's *Olicana* formerly thought to be Ilkley and now the Roman fort at Elslack SD 9249). The forgery was generally accepted as genuine by 18th century antiquarians, and although shown up for what it was in the 19th century the name has stuck. The genuine *Alpes Penninae* in Switzerland do indeed derive from Celtic **penno-* 'a mountain summit'. See further J.E.B. Mayor's introduction to vol.II of *Ricardi de Cirencestria Speculum Historiale de Gestis Regum Angliae*, Rolls Series 1869 and Antiquity vii, 1933, 49–60.

PENNINGTON Cumbr SD 2677. 'Farm, village paying a penny rent'. *Pennigetun* 1086, *Peni(g)ton(a)* 1157×63–1202, *Peninton* 1187, *Penyngton* 1327. OE **pening** + **tūn**. La 210 gives pr [penitn].

Lower PENNINGTON Hants SZ 3139. ModE **lower** + p.n. Pennington as in (Upper) Pennington SZ 3095, *Penyton* 12th–1578, *Penintune, Penigtone* 13th, *Penington* 1272, *Pennington alias Pennyton* 1682, 'estate on which a penny geld is payable', OE **pening** + **tūn**. Cf. PENTON MEWSEY Hants SU 3347. Ha 130, Gover 1958.227.

PENNY BRIDGE Cumbr SD 3083. A bridge over the river Crake in the late 16th cent. and named after the Penny family of Crake Side. La 213, Mills 1976.120.

PENNYMOOR Devon SS 8611. *Pennymoor* 1765. Presumably a moor the rental of which was one penny. D 381.

PENPILLICK Corn SX 0756. 'Little Penpell'. *Penpelic* 1302, *Penpillyk* 1389. P.n. Penpell SX 0756 + Co **-yk**. Penpell, *Penpel* c.1194, is 'far end', Co **pen** + **pel** 'far, distant', referring to its situation at the southernmost tip of Lanlivery parish. PNCo 134, Gover n.d. 428.

PENPOL Corn SW 8139. 'Head of the creek'. *Penpol* from 1275. Co ***pen** + **pol**. Penpol is situated at the head of Restronguet Creek. PNCo 134, Gover n.d. 449.

PENPOLL Corn SX 1454. 'Head of the creek'. *Penpol* c.1250, 1327. Co ***pen** + **pol**. Penpoll is situated at the head of Penpoll Creek off the river Fowey. PNCo 134, Gover n.d. 302.

PENRITH Cumbr NY 5130. 'Ford headland'. *Penred* 1167–1290, *Penreth* 1197–1597, *Penrith* from 1242, *Peareth* 1397, 1585, *Perith* 1511. PrW ***penn** (< Brit **penno-*) + ***rïd** (< Brit **ritu-*). The reference is probably to Beacon Hill (NY 5231, 937ft.) and the ford over the river Eamont at Eamont Bridge NY 5228. An exact parallel is found in Perret (Brittany), *Penret* 9th. Cu 229 gives prs ['penriθ], [pi:əriθ], [pi:riθ] and [pi:rəθ], O'Donnell 80, L 79–80, 183.

PENROSE Corn SW 8770. 'End of the hill-spur'. *Penros* 1286, 1549. Co **pen** + **ros**. Penrose is situated at the end of high ground between two streams. PNCo 134, Gover n.d. 334.

PENRUDDOCK Cumbr NY 4227. Perhaps 'little-ford headland'. *Pendredoch* 1276, *Penreddok* 1278–1348, *Penruddok'* 1339–1540, *Penruddock* 1576. PrW ***penn** (< Brit **penno-*) + **rïdǭg** (< Brit **ritu-* + **-āco-*), a diminutive of ***rïd** 'a ford' corresponding to OBret *Ritoch, Redoc*. Cu 213.

PENRYN Corn SW 7834. 'Promontory'. *Penryn* from [1236]1275, *Penrin* 1259, 1307, *Peryn* 1337. Co ***pen-ryn**. The town lies on a ridge between two valleys. A local pronunciation [pe'rin] survived at least until the 17th cent. For PENRYN RIVER, a back-formation from Penryn, → STRATTON SS 2206. PNCo 134, Gover n.d. 510.

PENSAX H&W SO 7269. 'Hill of the Saxons'. *(to) Pensaxan* [n.d.]12th S 1595, *Pensex* 1240, 1355, *Pensax* 1327. PrW ***penn** + folk-n. **Sachson*. Wo 67, Hooke 392, Jnl.1.51, LHEB 539, L 183.

PENSBY Mers SJ 2683. 'Farm, village at or of *Penn*'. *Penisby* c.1229–1329, *Pennisby -es-* 1270×80–1535, *Pensby* from c.1574. P.n. **Penn*, PrW ***penn** 'top, end, hill' taken as a p.n., genitive sing. **Pennes*, + **bȳ**. The reference is to Heswall Hill SJ 2682 which overlooks the village. Che iv.271, SSNNW 38.

PENSELWOOD Somer ST 7531. *Penzlewood* 1811 OS. Short for *Penne in Selewode* 1345, p.n. *(æt) Peonnum* 9th ASC(A) under year 658, 12th ASC(E) under same year, *(æt) Peonnan* 12th ASC(E) under year 1016, *Penne* 1086, *(boscus de) la Penne* 'wood of the pen' 1274, *Pen* 1610, dative pl. ***peonnum** of OE ***penn**, ***peonn** < PrW ***penn** 'a hill' or possibly OE **penn** 'a pen', + p.n. *(be eastan) Seal wydu* 'east of Selwood' 9th ASC(A) under year 878, *(be westan) Seal wuda* 'west of Selwood' ibid. under year 894, *Seluudu* [c.894]c.1000 Asser, *Selewode* [1135×66–1252]Buck, *-wuda* 1168, *Selwood* 1610, 'sallow-tree wood', OE **sealh** + **wudu**. The reference is probably to a series of summits reaching to between 721ft. and 945ft. marking the edge of Salisbury Plain. According to Asser, the W name of the wood was *Coit maur* 'great wood', which Simeon of Durham translated as *Mucelwudu* 12th, OE **myċel** + **wudu**. At one time 15 miles long and six miles broad it constituted an almost impassable barrier between Somerset and Wiltshire which held up the Anglo-Saxon advance westwards for almost a century. The addition *in Selewode* is for distinction from Pendomer ST 5210, 'Penne held by the Domer family', *Penne dommere* 1311, earlier simply *Penne* 1086, *Penna* 1180. DEPN, Whitlock 112.

PENSFOLD FARM WSusx → PEVENSEY TQ 6405.

PENSFORD Avon ST 6263. Partly uncertain. *Pensford* from 1400. Uncertain element + **ford**. Pensford lies at the edge of a spur of high ground which may have been called *Pen* from PrW ***penn** 'a hill'. DEPN.

PENSHAW T&W NZ 3253. Uncertain. *Pench', Pencher* [1183]c.1320–1763, *Pyncher* 1510, *Penshire* 1582, *Peyncher* 1622, *Panshaw, Pa(i)nsher, Pancher* 18th cent., *Old* and *New Painshaw* 1863 OS. This has usually been taken to be a compound of PrW ***penn** 'a hill, a headland' + an element ***cerr** related to ONb *carr* 'a rock' assumed to be of W origin (cf. W *carreg*, Ir *carraig* 'a rock'). However that may be, the name has undergone several folk-etymological reinterpretations by association with (1) OE **ċerr**, ***ċeare** 'a turn, a bend', N dial. *chare* 'a narrow winding lane', and (2) ModE **shaw** 'a copse, a wood'. NbDu 154, GPC 431.

PENSHURST Kent TQ 5243. Partly uncertain. *Penesherst* [1072]copy, 1203–64, *Pennes herst* c.1100, *Penecestr(e)* 1263–1346, *Peneshurst* 1270. The generic is OE **hyrst**, Kt **herst** 'a wooded hill', the specific apparently either the genitive sing. of **pen** 'a hill', PrW **penn* used as a p.n. referring to one of the headlands here, or possibly the genitive sing. of an OE pers.n. **Pefen*. PNK 86.

PENSHURST PLACE Kent TQ 5244. 'Penshurst mansion'. P.n. PENSHURST TQ 5243 + ModE **place**. The reference is to the famous 14th cent. manor house here.

PENSHURST STATION Kent TQ 5146. A halt on the railway line between Tonbridge and Reigate. P.n. PENSHURST TQ 5243 + **station**.

PENSILVA Corn SX 2969. *Pensivillowe* 1592 Gover n.d., *Pensilva* 1868. Co **pen** + p.n. Silva of unknown origin as in *Silva Down* 1840. A 19th cent. mining village. PNCo 134, Gover n.d. 189.

PENTEWAN Corn SX 0147. Possibly 'foot of the river Tewan'. *Bentewoin* 1086, *Bentewyn* 1297, *Pentewen -towyn* 1327, 1478. Co **ben** + r.n. **Tewan* as in Towan SX 0149, *Bewintone* 1086, *Deuuintona* 1086 Exon, *Tewinton* 1181, *-yn(g)ton* 1275–1386, *Tewyn* 1337, *Towyn* 1543, 'the manor of Tewan', r.n. **Tewan* + OE **tūn**. **Tewan* is suggested as the old name of the St Austell river at the foot of which Pentewan is situated. Cf. the r.n. Tywynni Powys SN 8618, possibly 'bright river' (cf. W *tywyn* 'radiance', *tewyn* 'ember, torch'). PNCo 135, Gover n.d. 385.

PENTIRE 'headland' Co ***pen-tyr*** (*pen* 'head' + *tyr* 'land'). CPNE 183.

(1) ~ Corn SW 7961. *Pentirbighan* 1239 Gover n.d., c.1270, *Pentyrvaen* 1317, *-bian* 1373, 'little Pentire', p.n. Pentire as in West PENTIRE SW 7760 + Co **byghan**. Co 135, Gover n.d. 367.

(2) ~ Corn SW 8570 → PARK HEAD SW 8470.

(3) ~ POINT Corn SW 9280. *Pentire Point* 1576, p.n. Pentire SW 9380, *Pentir* c.1230, 1297, *Pentyr* 1284–1320 (the headland) + ModE **point**. PNCo 135, Gover n.d. 128.

(4) West ~ Corn SW 7760. *Pentir* 1202 Gover n.d., c.1270, *West Pentire* 1813. PNCo 135, Gover n.d. 367.

PENTLOW Essex TL 8146 → PENTNEY Norf TF 7212.

PENTNEY Norf TF 7212. Perhaps 'Penta's island'. *Penteleiet* 1086, *Pateneia* 1195, *Pentenay* 1200, *Penteneya* 1254, *Paunteneye* 1293. OE pers.n. **Penta*, genitive sing. **Pentan*, + **ēġ**. The DB form shows AN *-l-* for *-n-* and diminutive **ēġeth** for **ēġ**. The pers.n. *Penta* seems to be required for the p.n. Pentlow Essex TL 8146, *(at) Pentelawe* [1043×45]13th S 1531, [11th]12th–1255, *-lauuā* 1086, 'Penta's tumulus'. On the other hand, Pentlaw and Pentney both stand on or by rivers (the Stow in the former, the Nar in the latter) and the question arises whether both at one time may not have been known by the name *Pœnte, Pente* cognate with the river PANT Essex Tl 6631. DEPN, Ess 451, RN 319.

PENTON MEWSEY Hants SU 3347. '(The part of) *Peningtun* held by the Maisy family'. *Penyton Meysi* 1255–1364 with variants *Me(y)sy, Meisy, Pennington Mewsey* 1606. Earlier simply *Penitone* 1086, 'estate on which a penny geld is payable', OE **pening** + **tūn**. The manor was held in 1212 by Robert de Meisy from Maisy, Calvados, France, who is probably the man referred to in the form *Penintona Roberti* 'Robert's P' 1167. The additions are for distinction from the other parts of the manor, Penton Grafton SU 3247, *Penitone* 1086, known also as *Peninton Abbatis* 'the abbot's (part of) P' 1167 and later as *Penitone Grefteyn* 13th, *Penytun Gresteyn* 1269, *Penyton Croftyn* 1316, '(the part of) P held by Grestein abbey', Eure, France, as already in 1086. Modern <f> must be due to misreading of 17th cent. long *s* (<ʃ>), aided by the name of the prominent landowner, the earl of Grafton. Ha 130, Gover 1958.167, 170, DEPN.

PENTRE Shrops SJ 3617. 'The village'. *(Ye) Pentrey* 1677, 1684, *The Pentr(e)y* 1694, 1699, *Pentre* 1836. W **pentref**.

PENTRICH Derby SK 3852. Possibly 'boar's hill'. *Pentric* 1086, 1171×5, *Pentriz* 1162×82–1348, *Pentrich(e) -trych(e)* 1229 etc., *Pentridge* 1577–1767. OW **penn** (Brit **penno-*) + ***tirch**, pl. of **twrch**. Cf. Pentyrch Wales ST 1082, PENTRIDGE Dorset SU 0317. Db 490, L 139, 182.

PENTRIDGE Dorset SU 0317. 'Boar hill'. *Pentric* 1086, [1107]1300, [1100×35]1496†, *-riche* 1264–1428, *-rig(')* 1288, 1382, *-rygge* 1494, 1535, *-ridge* 1618. PrW or PrCo ***penn** + PrW ***tïrk**, PrCo ***tirk**, genitive sing. of W *twrch*, Co *torch*. The form *Pentringtone* [944×6]14th S 513 means 'estate of the men of Pentridge'. Do ii.235, CPNE 222.

PENWITH The ancient name of the whole Land's End peninsula. Probably 'end district', cf. *Penwihtsteort* c.1121 ASC(E), *Penwæð steort* c.1050 ASC(D) both under year 997, *Penwið steort* 11th ASC(C,D) under year 1052, p.n. Penwith, *Penwid* 1186, *-wed* 1194, Co **pen** + ***weth** as in *fynweth* 'end, limit', + OE **steort** 'tail' referring to Land's End. Co 136, DEPN.

PENWITHICK Corn SX 0256. Probably 'top of the wood'. *Penwythyck* 1357, *Penwythek* c.1550. Co **pen** + ***gwythek** 'woody place' (Co *gwyth* 'trees' + adjectival suffix *-ek*). PNCo 136, Gover n.d. 385.

Higher PENWORTHAM Lancs SD 5728. Adj. **higher** + p.n. *Peneverdant* 1086, *Penuertham -uerdham* before 1149–1212, *Penford(i)ham* c.1190, *Pendrecham* 1200, *Penw(o)rtham* 1201–1332, *Pentfortham* 1204, *Penwortham* from 1210 with variants *-wirtham -w(o)rham* 1242, *-w(u)rtham* 1246, 1255, *-wardine* c.1540, probably the 'enclosed homestead at *Penn*', PrW ***penn** 'a head, top, end, height' taken as a p.n. + OE **worth-hām**. The reference is to the headland overlooking the Ribble where the Norman motte and bailey castle was subsequently built. 'Higher' for distinction from Lower Penwortham SD 5327. La 135, Jnl 17.76, 18.17.

PENYARD H&W SO 6122 → WESTON UNDER PENYARD SO 6323.

PEN-Y-GHENT NYorks SD 8473. Partly unexplained. *Penegent* 1307, *Penaygent* 1377×99, *(montem voc') Penygent -i-* 'hill called P' 1598, 1676. PrW ***penn** 'head, end, top, height, hill' + definite article **y** + unknown element or name. YW vi.219, xi, Antiquity 12.46, JRS 38.55.

PENZANCE Corn SW 4730. 'Holy headland'. *Pensans* 1284–1367, *Pensaunce* 1552, *Penzaunce* 1582, *Penzans* 1698. Co **pen** + **sans** referring to the chapel of St Mary on the headland first mentioned in 1327. PNCo 136, Gover n.d. 646.

PEOPLETON H&W SO 9350. 'Settlement at the gravelly place'. *(in) piplincgtune* 972 S 786, *Piplintvne* 1086, *Piplintona* 1166–1291, *Pupplinton -yn-* 1240–1428, *Pipulton* 1401, *Pippleton* 1577, *Peopleton* from 17th. OE ***pyppeling** perhaps used as a p.n. < ***pyppel** + **ing**², + **tūn**. The soil there is clay, loam, gravel and sand. The modern spelling is an example of popular etymolgy. Wo 216 gives pr [pipelton], Hooke 179, ODEE s.v. pipple.

Lower PEOVER Ches SJ 7474. 'The lower part of Peover'. *Lower Peover* 1724, earlier *Parua Pefria* c.1200, *Little Pever* [1465]17th, *Little Peover* 1715. Latin **parva**, ME **litel**, ModE **lower**, + p.n. Peover. 'Lower' for distinction from Peover Superior SJ 7773, *Over(e)pevre, Ovre-, ~ Superior* 1287–14th, *Over Peover* 1559, *Higher Peover* 1724, Latin **superior**, ME **over**, ModE **higher**, + p.n. *Pevre* 1086, *Peu- Pev(e)re -er* c.1232–14th, *Pevur* 1382, *Peover* 1414. This is the r.n. Peover Eye, *Peuerhe(e)* 13th–1348, *Peever Eye* 1619, *Peover-Eye* c.1620, 'the river Peover, the Bright One', PrW ***Peβr** < Brit **Pebro-* (ModW *pefr* 'bright, shining, sparkling') + OE **ēa** 'a river'. Cf Polpever Corn, River Peffery. Che ii.90, ii.85 which gives pr ['piːvər], i.33, GPC 2714.

PEOVER HEATH Ches SJ 7973. *Over Peover Heath* 1831. P.n. Peover as in Lower PEOVER SJ 7474 + ModE **heath**. Che ii.88.

PEOVER SUPERIOR Ches SJ 7773 → Lower PEOVER SJ 7474.

PEPER HARROW Surrey SU 9344. Partly uncertain. *Pipereherge* 1086, *Piperherge* 1166, *-hargh(e) -hergh, Pyperharewe* 13th cent., *Peperhargh(e), Pepurharowe* 14th cent., *Pepperharowe* 1593, 1719. The generic is OE **hearg** 'a heathen temple', the specific either OE pers.n. *Pip(e)ra* or **pīpera**, genitive pl. of **pīpere** 'a piper'. The same specific occurs in nearby Pepperhams SU 9033, *Piperham* c.1180, *Pyperhammes* 1453 with OE **hamm** 5 or 6, 'an enclosed piece of land'. If the sense is 'heathen temple of the pipers' this may be a reference to the use of musical instruments in pagan worship. Sr 206, 207, ASE 2.23, SelPap 77, Evidence 107.

PEPLOW Shrops SJ 6324. 'Pebble tumulus'. *Papelau* 1086, *-lawe* c.1155, *Pepelawe* c.1138, 1255×6, *Pep(p)elowe* 1284–1520, *Peplow* from 1562, *Pippelawe* 1255×6, *Pyppelowe* 1294×9. OE **papol**

†The Muchelney Cartulary form *Pentric* [762]13th may also belong here.

varying with ***pyp(p)el** 'a pebble' + **hlāw**. A possible mound still exists by fields with pebbly soil. Sa i.233–4.

PEPPERHAMS Surrey SU 9033 → PEPER HARROW SU 9344.

PERDREDDA Corn SX 3456 → River PARRETT.

PERIVALE GLond TQ 1682. 'Pear-tree valley'. *Pyryvale* 1508, *Peryvale* 1524. ME **pirie** (OE *piriġe*) + **vale** (OFr *val*). This replaces the former name Little Greenford, *Greneford parva* 13th cent., *Little Greneford* 1386, earlier simply *Greneford* 1086, the 'green ford', OE **grēne** + **ford**. 'Little' for distinction from Great GREENFORD TQ 1382. Mx 46.

PERLETHORPE Notts SK 6571. 'Thorpe held by the Peverel family'. *Peureltorp* 1159, *Peverelt(h)orp* 1212–1316, *Perlethorp(e)* from 1376, *Per(i)l- Parlethorp, Pellethropp* 16th, *Pal(e)- Pailthorp(e) -throp* 17th. Family name *Peverel* + p.n. *Torp* 1086, 'the outlying settlement', ON **thorp**. The main difficulty with this name is that there is no independent evidence that Perlethorpe ever belonged to the Peverel Honour. Nt 91 records obsolete pr [peilθɔ:p], SSNEM 119.

PERRAN or LIGGER BAY Corn SW 7256. *Ligger or Piran Bay* 1813. Saint's name *Piran* as in Perranzabuloe SW 7752, in which parish the bay is situated, + ModE **bay**. Ligger is from Ligger Point SW 7558, probably 'grey promontory', Co **los** (OCo *lot*) + **garth**, as in Liggars SX 2262. PNCo 110, Gover n.d. 377.

PERRANARWORTHAL Corn SW 7738. 'Peran in Arworthel'. (Church of) *Sanctus Pieranus in Arwothel* 1388, *St Peran Arwothal* 1543, *Peran Arwothel* 1612, *Perranarworthall* 1674. Saint's name *Piran* (Perran) from the dedication of the church, + p.n. *Arewethel* 1181, *Harewithel* 1187, *Arwr(u)thel* 1196–8, *Arwo(y)thel* 1284–1342, *Arwothal* 1610, 'beside the marsh', Co ***ar** + ***gothel** 'watery ground' with lenition of *g* to *w* and confusion with ***gwythel** 'thicket'. The addition to the saint's name is for distinction from PERRANUTHNOE SW 5329 and PERRANZABULOE SW 7752. St Piran was a monk from Ireland or Wales, the patron of tinners. His cult flourished in Cornwall, Wales and Brittany. PNCo 136, Gover n.d. 524, DEPN.

PERRANPORTH Corn SW 7534. *Perranporth* 1720 Gover n.d., cf. *St Perins creeke* 1577. Saint's name *Piran* as in Perranzabuloe SW 7752, in which bay the place is situated, + ModE dial. **porth** 'a harbour, a cove'. The accentuation is on *-porth*. PNCo 136, Gover n.d. 378.

PERRANUTHNOE Corn SW 5329. 'Peran in Uthnoe'. *Peran Uthnoe* 1202, *(ecclesia) Sancti Pierani de Udno* 1373. Otherwise simply *(ecclesia) Sancti Pierani* 1348. Saint's name *Piran* + p.n. Uthnoe as in Edno-Vean SW 5429 'little Uthnoe' (Co **byghan**), *Udno parva* 1308, earlier simply *Odeno(l)* 1086, *Hithenho* 1214, *Hutheno* 1229, 1235, of unknown origin. The suffix is for distinction from PERRANARWORTHAL SW 7738 and PERRANZABULOE SW 7752. PNCo 136, Gover n.d. 609, DEPN.

PERRANZABULOE Corn SW 7752. 'Peran in the sand'. *Peran in Zabulo* 1535, *St Piranus in the sandes* c.1540, *Peran in the Sāde* 1610. Earlier *Lanpiran* (held by the canons of *Sanctus Pieranus*) 1086, *Lanberan* 1303, and simply *(in) Sancto Pirano* 1195, *Seynt Peran* 1336, Co ***lann** 'church-site' + saint's name *Piran*. 'Zabuloe' (Latin *in sabulo* 'in the sand') for distinction from PERRANARWORTHAL SW 7738 and PERRANUTHNOE SW 5329. PNCo 137, Gover n.d. 375.

South PERROT Dorset ST 4706. *Sutpetret* 1084, *Sutpe(d)ret -petret -perret* 1086 Exon, *Suthperet* 1268, *Southperot* 1428. Earlier simply *Pedret* 1086. OE adj. **sūth** + r.n. PARRETT ST 4706. 'South' for distinction from North PERROTT Somer ST 4709. Do 119, RN 320.

North PERROTT Somer ST 4709. *North parret* 1610. ME adj. **north** + p.n. *(æt) Peddredan, (apud) Perret* [1061×65]13th S 1116, *Peret* 1086, *Perrete* 1219, transferred from the r.n. PARRETT ST 3927. 'North' for distinction from South PERROTT Dorset ST 4706. RN 320.

PERRY Cambs TL 1466. East Perry, *Est Perye* 1323, is *Peri* 1147, *Pirie* [c.1180]c.1230, West Pery *Pirie* 1086–1286 etc., *Pery* 1463, *Pury* 1478, *Pyrry* 1484, 'the pear-tree', OE **piriġe**. Hu 242, 271.

River PERRY Shrops SJ 3926. 'The river Pefr, the bright one'. *Peueree* [1154×89]19th, [early 13th]19th, *Pevereye* c.1250, 1268×71, *Peverey* 1335, 1577. PrW ***peβr** (W *pefr*) 'radiant, bright, beautiful' used as a r.n. + OE **ēa**. Identical with PEOVER Ches SJ 7474 and the Scottish r.n. occuring in Peffer Burn Lothn NT 5080, River Peffery Highld NH 5259, Strath Peffer Highld NH 5159, *Pe- Paforyn* 1247, Peffermill Lothn NT 2871 and Paphrie Burn Tays NO 5166. RN 322, Scot PN 164.

PERRY BARR WMids SP 0692. 'P beside Barr'. P.n. *Pirio* (Latin dative case) 1086, *Piri(e)* 1176, 1242, 'the pear-tree', OE **piriġe**, + p.n. Barr as in Great BARR SP 0495. DEPN, L 219.

PERRY GREEN Herts TL 4317. 'Pear-tree green'. *Perry Grene* 1543. ME **perry** (OE *piriġe*) + **grene**. Hrt 177.

PERSHORE H&W SO 9446. 'Ridge by the osier beds'. *(at, et) Pershoram* (Latin accusative) [671 for 674×9]14th S 70, *(in) perscoran* 972 S 786, *Persceoran* (dative case) [1051×5]18th S 1409, *Perscora* *[1065]18th, *Perscoran* *[1066]12th S 1043 (accusative case), [1042×66]14th S 1143, S 1145, [1062×6]14th S 1146, *Periscoran* (accusative case) [1042×66]14th S 1144, *Persore, P'sore* 1086, *Pershore* 1542, *Parshior, Parshore* 1464–1675. OE ***persc** + **ōra**. The town lies on the riverward side of the ridge which overlooks the r. Avon. This is the most northerly example of *ora* in p.ns., showing the extent of West Saxon settlement in territory which subsequently formed the Mercian sub-kingdom of the Hwicce. Wo 217 gives pr [pa:ʃər], Hooke 19, 177, 338–41, Ritter 133, PNE ii.62, L 179, 181, Jnl 22.36.

PERTENHALL Beds TL 0865. 'Pearta's nook of land'. *Partenhale* 1086, 13th, *Pertenhal(e)* 1202–1504, *Partnale* 16th, *Pertenall* 1526. OE pers.n. **Pearta*, genitive sing. **Peartan*, + **halh** probably in the sense 'dry ground in marsh', dative sing. **hale**. For the pers.n. cf. PARTINGTON GMan SJ 7191, PARTNEY Lincs TF 4168. Bd 16 gives pr [pa:tnəl].

PERTHY Shrops SJ 3633. 'The bushes'. *Perthy* 1837 OS. W **perth**, pl. **perthi**.

PERTON Staffs SO 8598. 'The pear-tree farm'. *Pertone* 1086, *Periton* 1193, *Perton'* 1198, *Purton* 1686. OE **pere-tūn**. Perton lies in Tettenhall parish of which Pitt's Agricultural Survey of 1791 says, 'This parish has one singularity in the fruit way: it has produced a peculiar kind of pear called Tettenhall pear and known by no other name; many hundreds of trees grow in this parish though not, or scarcely, to be found at all at any considerable distance. The tree is large and a plentiful bearer, the fruit well flavoured and bakes and boils well, but will not keep long enough even for carriage to any considerable distance unless some time before it is ripe; the average annual produce of this parish is many thousand bushels more than its own consumption... in plentiful seasons, and it is seldom otherwise, large quantities are eaten by hogs which are suffered to pick them up as they fall'. DEPN, PNE ii.63, Horovitz.

PETER BLACK SAND Norf TF 6231. *Peter Black* 1824 OS. A sandbank in The Wash.

PETER'S GREEN Herts TL 1419. *Peters Green* 1834 OS.

PETERBOROUGH Cambs TL 1999. 'St Peter's *Burg*'. *Petreburgh* 1333, 1385, *Peterburgh* 1397, *Peterborough al Borough Saynt Peter*. Earlier simply *Burg* 1086, *Burh* c.1115, *Burc* 1155. Saint's name *Peter* from the dedication of the abbey + OE **burh** 'fortified place, town'. Peterborough is a renaming of the original monastery of Medeshamstede which was ravaged by the Danes in 870 (ASC(E) under that year), restored by Bishop Athelwold in 963 (ibid.) and fortified by abbot Kenulf sometime before his translation to Winchester in 1006. According to the Laud MS of the Chronicle he first built the wall around the minster and called it *Burch* (ibid.). Medeshamstede, *Medeshamstedi* c.731 BHE, *Medeshamstede* 12th cent. passim, is probably 'Medi's homestead', OE pers.n.

*Mēdi, genitive sing. *Mēdes, + **hām-stede**. According to the Laud Chronicle under the year 654 Peada king of Mercia and Oswiu king of Northumbria founded the monastery here and called it *Medeshamstede* because there was a spring here called *Medeswæl* 'Medi's spring' (OE **wella**) or 'pool' (OE **wǣl** 'well, pool', probably a pool in the river Nene where fish were caught). Nth 224, Stead 260, TC 50.

PETERCHURCH H&W SO 3438. 'St Peter's church'. *Sancti Petri* 'St Peter's' 1160×70, *Pat' church* 1291, *Peterescherche* 1302. Saint's name *Peter* + ME **chirche**. A marginal note in the *Herefordshire Domesday* shows that this name replaced *Almvndestvne* 1086 'Alhmund's estate', OE pers.n. *Alhmund*, genitive sing. *Alhmundes*, + **tūn**. He 162, DB Herefordshire 29.9, Nomina 10.73.

River PETERILL Cumbr NY 4352 → CHICKERELL Dorset SY 6480.

PETERLEE Durham NZ 4240. A new town founded in 1948. The name is happily coined by a simple combination of the forename and surname of a former Durham miners' trade union leader, Peter Lee (1864–1935). Room 1983.91.

PETERSFIELD Hants SU 7423. 'Peter's field'. *Peteresfeld* 1181–9, *Petresfeud -feld* 1248–1327. Apparently saint's name *Peter* from the dedication of the church + ME **feld**. This was a twelfth cent. new town and the name therefore unlikely to conceal an earlier name such as 'Peohtere's open land', OE pers.n. *Peohtere*, genitive sing. *Peohteres*, + **feld**. Ha 130, Gover 1958.60.

PETERSTOW H&W SO 5624. 'St Peter's holy place'. *Peterestow'* 1207. A translation of W *Lannpetyr* 'Peter's church' [1045×1104]c.1130 LL. Saint's name *Peter* + OE **stōw** translating PrW ***lann**. He 166, Stōw 192.

PETHAM Kent TR 1251. 'Pit valley'. *Piteham* 1086, *Pyt- Peteham* c.1090 DM, *Pethom* 1203, *Petham* from c.1090 DM. OE **pytt** possibly used as a p.n. + **hamm** 6 'valley-bottom hemmed in by higher ground'. KPN 288, ASE 2.22, Cullen.

Little PETHERICK Corn SW 9172. *(parocia) Sancti Petroci minoris* 1327, 1334, *Pedrock minor alias Nasenton* 1535, *Litle Pedrock* 1549, *Petherocke alias Nansenton* 1563, *Little Pethericke* 1634. Saint's name *Pedrek* + Latin **minor** 'the lesser' for distinction from the major place of veneration of the saint at PADSTOW SW 9175. The older name of the parish was *Nansfonteyn* 1281 'valley of the fountain', Co **nans** + **fenten** influenced by OFr **fontein**. PNCo 137, Gover n.d. 346.

North PETHERTON Somer ST 2933. *Nortpetret* (the hundred) 1084, *Nort- Nordperet, Nordpereth, Peretvne, Peret* 1086, *Norperton* 1243, *Northperton'* (hundred) 1276, *Northpederton -per(e)ton(e)* [1269–95] Buck, *North petherto͞* 1610. OE adj. **north** + p.n. *Per(r)eton* [1165–1388] Buck, *Perton* [before 1199–13th] ibid., *Pederton* [1279, 1408] ibid., as in South PETHERTON ST 4316. DEPN, RN 320.

South PETHERTON Somer ST 4316. *Sudperetonne* (hundred) 1084, *Sv- Sudperet, Svdperetone* 1086, *Suthpederton* [1386]Buck, *S. Pedderton* 1610. OE adj. **sūth** + p.n. *Perreton* [1174×91–1289]Buck, the 'settlement on the r. Parrett', r.n. PARRETT ST 3927 + OE **tūn**. RN 320.

North PETHERWIN Corn SX 2889. *North Piderwine* 1259, *Nordpydrewyn* 1269, *North Petherwyn* 1524. ME adj. **north** + p.n. *Pidrewina* [1171]13th, of uncertain origin. The church is dedicated to St Paternus, whose title in Cornish might be *Padern wynn* 'Padern the white, blessed Padern' (Co **guyn** with lenition of *gu* to *w*). But the change of *a* to *i* in the first syllable is not explained. 'North' for distinction from South PETHERWIN SX 3182. PNCo 137, DEPN.

South PETHERWIN Corn SX 3182. *Suthpydrewin* 1269, *Southpiderwyne* 1321–3, *South Petherwin* 1518. ME adj. **suth** + p.n. *Pidrewyn* [c.1147]n.d., *Pitherwyne* 1275, *Piderwyne* c.1220, [c.1293]15th, of uncertain origin. Cf. however the old name of the Hundred of Pyder, *Piderscire* 1130, 1186, p.n. *Pider* of unknown origin + OE **scīr** 'a shire, a district'. 'South' for distinction from North PETHERWIN SX 2889. Gover n.d. 170, DEPN.

PETROCKSTOW Devon SS 5109. 'St Petroc's holy place'. *Petrochestov* 1086, *Patricstouwe* 1268, *Pad(e)rokestowe* 1281, 1286, *Stowe Sancti Petroci* 1342, *Pad(e)stowe* 1356, 1450, 1618, *Petrockstowe* 1618. Saint's name *Petroc* from the dedication of the church, genitive sing. *Petroces* (Latin *Petroci*), + OE **stōw**. Identical with PADSTOW Corn SW 9175. D 105.

PETT ESusx TQ 8714. 'The pit, hole'. *Ivet* (sic) 1086, *Pett(e)* 1195–16th, *Putt(e)* 13th–14th cents., *Potte* 1386. OE **pytt**, SE dial. form **pett**. Pett lies on a hill and the feature referred to in the name is unknown. Cf. PITT DOWN Hants SU 4128. Sx 513.

PETTAUGH Suff TM 1659. 'Peota's enclosure'. *Pet(t)ehaga* 1086, *Pethag(he) -hagh* [1189×99]1396, 1242–1346, *Pethawe* 1286, 1327, *Pettawe* 1524, *Petaugh* 1610, *Pettaugh* 1674. OE pers.n. *Peota* + **haga**. An alternative possibility is 'the pit enclosure', OE **pytt**, SE dial. form **pett** + **haga** referring to the deep valley of the stream here. Cf. Peat's Corner TM 1960 on the same stream. DEPN, Baron.

River PETTERIL Cumbr NY 4543. An obscure pre-English river name. *Petrel(l)* 1268–1385, *Peterel(l)* 1292–1636, *Petrille* 1389, *Petterell* 1576, *Peterill* 1586. RN 323 gives pr [petərəl], Cu 23.

PETTISTREE Suff TM 2954. 'Peter's tree'. *Peterestrie* 1202, *Petrestre* 1253–1327, *Petristre* 1291, *Peterstre* 1316, 1344, *Petistre* 1336, *Petestre* 1568, *Pystre* 1524, *Petistrie* 1610. Pers.n. *Peter*, genitive sing. *Peteres*, + ME **tre** (OE *trēow*). Baron. DEPN, however, prefers an OE pers.n. and gives 'Peohtred's tree'.

PETTON Devon ST 0024. 'Peatta's estate'. *Peteton* 12th, *Petton* from 1368, *Pat(t)eton* 1219, 1238, *Pyaton* 1303. OE pers.n. **Peatta* + **tūn**. D 531.

PETTON Shrops SJ 4326. 'The peak settlement'. *Pectone* 1086–1307 with variant *-ton(a), Peton(e)* 1203–1317, *Petton(')* from 1255. OE **pēac** + **tūn**. The church and motte stand on a conical hill commanding extensive views. Identical with PEATON SO 5385. Sa i.234.

PETWORTH WSusx SU 9721. 'Pytta's enclosure'. *Peteorde* 1086, *Petew(u)rda -wurða* c.1136–93, *Puetewurth* 1181, *Puettewurda* 1182, *Putteworth al. Petworth* 1399. OE pers.n. **Pytta* + **worth**. Sx 114 gives pr [petəθ], Ritter 96. DEPN prefers pers.n. *Pēota*.

PETWORTH HOUSE WSusx SU 9721. P.n. PETWORTH SU 9721 + ModE **house**. The reference is to a 17th cent. mansion house.

PEVENSEY ESusx TQ 6405. 'Pefen's river'.

I. *Pævenisel* [795]17th S 1186, *Peuen-* *[790]12th S 133, *[960]13th S 686, *Pevenesel* 1086–14th with variants *-ess-, -ell*.

II. *Pefenesea* [947]16th S 527, *Peuenesea* 12th ASC(E) under year 1042, *Pefena sæ* 11th ASC(D) under year 1050, *Peuenesee -ey(e)* 1198–1407, *Pemse(y)* c.1375–17th, *Pensay* 1537.

Type II is OE pers.n. **Pefen*, genitive sing. **Pefenes*, + **ēa**. Type I with *-el* is said to be due to French influence; the generic might, however, represent OE ***ēgel** 'a small island'. The pers.n. occurs again at Pensfold Farm in Slinfold WSusx TQ 1131, *Peuenesfeld* 1301. The Romano-British name of Pevensey, *Anderidos, Anderitos*, is discussed under the WEALD TQ 6035. Sx 443 gives prs [pemzi] and [pimzi], 160, Zachrisson 1910. 27, SS 71.

PEVENSEY BAY ESusx TQ 6603. *Pemsey Bay* 1545. P.n. PEVENSEY TQ 6405 + ModE **bay**. Sx 445.

PEVENSEY LEVELS ESusx TQ 6307. P.n. PEVENSEY TQ 6405 + ModE **level**. Earlier referred to as *marisc. int' Peuenes' et Heilesham* 'the marsh between P and Hailsham', *mora de Peuenese* 'the marsh of P' 13th cent., *the marsh of Peveneseye* 1317. Sx 445.

PEVERIL CASTLE Derby SK 1482. *Terri* (sic) *castelli Willi peurel* 'the land of the castle of William Peverell' 1086, *Castelli de*

Pech' 1173 etc. Described in an interlineation in DB as *in Pechefers* for *Pechesers* 'Peak's arse', an old name for Peak Cavern. Surname *Peverell* + ME **castel**, Latin **castellum**. Db 56, BdHu xli.

PEVERIL POINT Dorset SZ 0478. *Peverell Poynt* 1539. Surname *Peverell* + ModE **point**. Do i.58.

PEWSEY Wilts SU 1660. 'Pefi's island'. *(æt) Pefesigge* [873×88]11th S 1507, *Pevesige* 940 S 470, *Pevesie -ei* 1086, *-eye* 1261–1316, *Pusye* 1378, *Pewse* 1524. OE pers.n. **Pefi*, genitive sing. **Pefes*, + **īeġ** in the sense 'well-watered land'. Wlt 350, L 38.

PEWSEY DOWN Wilts SU 1757. *Peusedowne* 1332, *Downe Pewsey* 1585, Latin form *montem de Peusey* 1279. P.n. PEWSEY SU 1660 + ME **doun** (OE *dūn*). Wlt 352.

PHILHAM Devon SS 2522. Uncertain. *Fyleham* 1262, *Fileham* 1321, *Philham* 1809 OS. This might be the 'enclosure for fillies', OE ***fylle** cognate with ON *fylja*, OHG *fulī(n)* + **hamm** or possibly 'homestead or enclosure at the dirty place', OE ***fȳle** + **hām** or **hamm**. D 79.

PHILLACK Corn SW 5638. 'St Felec's (church)'. *(Ecclesia) sancte Felicitatis* 1259–15th, *Felok* 1388, *Seynt Felleck* c.1530, *Felak* 1537, *Phillacke* 1613. Saint's name *Felec* from the dedication of the church. Nothing is known of this saint who was identified in Latin sources with Felicity. The name occurs again in Bephillick SX 2259, *Bodfilek* 1319, 'Felec's dwelling'. PNCo 138, Gover n.d. 610, DEPN, CMCS 12.48.

PHILLEIGH Corn SW 8739. 'St Fili's (church)'. *(Ecclesia) Sancti Filii de Eglosros'* 1312, *Fili* 1450, *Phillie* 1613. Saint's name *Fili* from the dedication of the church. Nothing is known of this saint whose name also occurs in Carphilly Glamorgan ST 1587 and in the Breton p.ns. Kerfili and Lamphily. The alternative name of the churchtown was *Egloshos* 1086, *Eglossos* 1086 Exon, *Eg(g)losros* 1201–1342, 'church of the moorland', Co **eglos** + ***ros** probably referring to Roseland as in ST JUST IN ROSELAND SW 8435 although there is no evidence that Philleigh was ever the mother church of Roseland. PNCo 138, Gover n.d. 474, DEPN, CMCS 12.45.

PHOENIX GREEN Hants SU 7555. *Phoenix Green* 1817 OS contrasting with *Bear's Green*, Murrell Green SU 7455 and West Green SU 7456. 18th cent. inn-name *Phoenix* + ModE **green**. Gover 1958.115, Inn Names 22, 90.

PICA Cumbr NY 0222. No early forms. The site is a pointed hill rising to 597ft. Possibly, therefore, 'pointed hill', dial. **pike** (OE *pīc*, ON *pík*) + ON **haugr**.

PICCOTTS END Herts TL 0509. *Pigots end* 1599 sc. of Hemel Hempstead. Surname Piccott, cf. William *Pycot* 1287, + ModE **end**. Contrasts with *Counters End* c.1840 at TL 0407, *Green End* 1588, Hillend, *Hyllende* 1545, *Moor End* 1623, Pouchen End, *Punchin End* 1598, *Powchen end* 1617 at TL 0206, and *Wardsend* 1623. Hrt 42–3.

North PICKENHAM Norf TF 8606. *Nortpykenham* 1291. ME adj. **north** + p.n. *Pichen- Pi(n)kenham* 1086, *Pikenham* 1198, 'Pica's homestead', OE pers.n. **Pīca*, genitive sing. **Pīcan*, + **hām**. 'North' for distinction from South PICKENHAM TF 8504. DEPN.

South PICKENHAM Norf TF 8504. *Sutpikeham* 1242. ME adj. **sūth** + p.n. Pikenham as in North PICKENHAM TF 8606.

PICKERING NYorks SE 7984. Partly uncertain. *Pichering(a)* 1086, 1165, *Pikeringes* 1109–38, 1234, *Pikering(a) -e* from 1157. OE p.n. **Picering* of unknown meaning and origin varying with folk-n. **Piceringas* 'the people of Pickering'. The suggestion that **Picering* might be a singular **ing**² formation on OE *pīc-ōra* 'the place by the slope of the *pīc* or hill' is unacceptable in the light of recent work on the meaning and distribution of *ōra*, a word of southern provenance and early obsolete in place naming. YN 85, ING 75–6, DEPN, Jnl 22.26–42.

PICKET PIECE Hants SU 3947. No early forms. Nearby is Picket Twenty Farm SU 3845, *Pickiltrenthey* 1817 OS, which seems to have been influenced by this name. Picket Piece is a f.n. for an angular piece of land, ModE **picked** (ME *piked*) + **piece**. Cf. PICKET POST SU 1906. Ha 130, Gover 1958.165, O 389, 461, Field 1972.165.

PICKET POST Hants SU 1906. 'The pointed post'. *Picked Post* 1759, *Picked Post* 1759–1811 OS. ModE **picked** (ME *piked*) + **post**. Ha 131, Gover 1958.218.

PICKHILL NYorks SE 3483. 'Pica's nook of land'. *Picala* 1086, 1301, *-e* [1184]15th cent., *Pi- Pykehal(e)* 1207–1328, *Pykel* 1327, *Pickhill* from 1718. OE pers.n. **Pīca* + **halh**. YN 224. For a different view, 'the nook of the peaks', cf. L 110, 304.

PICKLESCOTT Shrops SO 4399. 'Picel's cottage(s)'. *Pikelescote* before 1231, *Piclescote* 1255. OE pers.n. *Pīcel*, genitive sing. *Pīc(e)les*, + **cot** (pl. **cotu**). Bowcock 182.

PICKMERE Ches SJ 6977. 'Pike lake'. *Pichemere* 1154×89, *Pikemere* 13th–1512 with variants *Pyke-* and *-mer*, *-meyre*, *Pickmere* from 1209 with similar variants. OE **pīc** 'a pike (fish)' + **mere**. Che ii.120.

PICKWELL Devon SS 4540. Possibly 'Piddoc or Piddic's spring'. *Wedicheswelle* (for *Ped-*) 1086, *Pediccheswella* 1086 Exon, *Pido(c)keswell -ekes* 1207–30, *Pydekewill'* 1303, *Pyckewyll al. Pydyckswell* 1560. OE pers.n. **Piddoc* or **Piddic*, genitive sing. **Piddeces*, + **wylle**. D 44.

PICKWELL Leic SK 7811. 'Peak spring'. *Pichewell(e)* 1086–[mid 13th]1404, *Pi- Pykewell(e)* 1202–1518, *Pi- Pykwell -c-* c.1130–1536, *Pick(e)well* from 1295. OE **pīc** + **wella**. The reference is to a conical hill overlooking the village. Lei 189, L 32.

PICKWORTH Leic SK 9913. 'Pica's enclosure'. *Pichewurða* 1170–1, *Py- Pikewurda -wurða -wurthe -worth(e) -word(a)* 1174–1316 with variants *-kk- -ck-* 1213–1301, *Pic- Pik- Pykworthe* 1293–1503, *Py- Pickworth* from 1286. OE pers.n. *Pīca* + **worth**. R 157.

PICKWORTH Lincs TF 0433. 'Pica's enclosure'. *Pichev(o)rde, Picheu(u)orde* 1086, *Pikewurda* 1174×6, *-word* late 12th, 1275, *-wurth* 1242×3, *Py- Pikworth* 1291–1576, *Pickworth* from 1548. OE pers.n. *Pica* + **worth**. Perrott 133.

PICTON Ches SJ 4371. 'Pica's farm or village'. *Picheton* 1086, *Picton* from c.1200 with variants *Pyc(k)- Pick-*. OE pers.n. *Pīca* + **tūn**. Che iv.132.

PICTON NYorks NZ 4107. 'Pica's farm or estate'. *Pi- Pyketon(a) -tun* c.1200–1310, *Pickton* 1285. OE pers.n. **Pīca* + **tūn**. YN 173.

PIDDINGHOE ESusx TQ 4303. 'Hill-spur of the Pyddingas, the people called after Pydda'. *Pidingeho* 1199×1216–13th cent., *Pi- Pydingho(we)* 13th–14th cents., *Pedyngho(o)* 1288, 15th cent., *Peddinghooe* 1674. OE folk-n. **Pyddingas* < pers.n. **Pydda* + **ingas**, genitive pl. **Pyddinga*, + **hōh**. Sx 324, L 169.

PIDDINGTON Northants SP 8054. Possibly 'the settlement by the stream called *Piding*'. *Pidenton(e)* 1086, 12th, *Py- Pidin(g)ton -yn(g)-* 1204–1405. OE r.n. **Piding* < ***pide -u** + **ing**² + **tūn**. The alternative explanation is 'the estate called after Pydda', OE pers.n. **Pydda* + **ing**⁴ + **tūn**. Nth 150.

PIDDINGTON Oxon SP 6417. 'Pyda's estate'. *Petintone* 1086, *Pedyngton(a)* [1151×2–1383]c.1444, *-in(g)ton -ynton* 1220–[1349]c.1444 including [c.1175]c.1425, *Pidentuna* [1152×3]c.1444, *Pydentona* [c.1160]c.1425, *Pi- Pydington(e) -yng-* 1242–[1360]c.1425, *Pyddington* 1285, *Pudyn(g)ton* 1299, 1311. OE pers.n. **Pyda*, genitive sing. **Pydan*, + **tūn**. The *-ing-* forms are probably secondary. O 184.

PIDDLE OE ***pidele** 'marsh, marshy stream', a diminutive of ***pide**.

(1) ~ BROOK H&W SO 9648 → KINGTON SO 9955.

(2) North ~ H&W SO 9654. *Northpidele* 1271. ME adj. **north** + p.n. *Pidelet* 'little P' 1086–1175, *Pydele* 1234, from the stream name *pidele* *[708]12th S 78, *(in, on) pidelan* [963]11th S 1305, 972 S 786. OE ***pide** appears in the alternative form of this name + OE **wella**, *piduuella, (æt, of) pidpellan* [930]16th S 404 with 13th cent. variant *pid puella, piduella, (on, æt, of) pid-*

wellan -wyllan [1002]13th S 901. 'North' for distinction from Wyre PIDDLE SO 9647. The form *Pidelet* is a French form with diminutive suffix **-et**. Wo 222, RN 324, PNE ii.64, Hooke 43, 159, 182, 262, 351.

(3) River ~ or TRENT Dorset SY 8591. *Pidelen stre(a)m* [966]15th S 744, *(iuxta) Pydelan* *[987]13th S 1217, *Pi- Pydele* 1229–88, *Pudele* 1325, *Piddle* from 1586. The alternative name Trent, *Terente* before 1118, *Trent(e)* c.1540, is a back-formation due to misunderstanding of the name PIDDLETRENTHIDE SY 7099. The name Piddle occurs in a series of estates along the course of the stream. Do 119, RN 324, 416.

(4) Wyre ~ H&W SO 9647. *Wyre Pidele* 1208. Earlier simply *Pidele* 1086. P.n. Wyre as in WYRE FOREST SO 7576 + p.n. Piddle as in North PIDDLE SO 9654. Wo 155.

PIDDLEHINTON Dorset SY 7197. 'The monks' Piddle estate'. *Pidele* called *Hinctune* (sic for *Hine-*) 1082×4, *Pidel Hineton* from 1244 with variants *Pydel- Pudel(l)e-* and *-hynton*. R.n. PIDDLE SY 8591 + p.n. *Hineton* 'monks' estate', OE **hīġna**, genitive pl. of **hīwan** 'a household, a religious community', + **tūn**. Also called simply *Pidele* 1086 and *Hi- Hynepid(d)el(l)(e) -pydel(l)e* 1244–1423, 'monks' Piddle', OE **hīġna** + r.n. PIDDLE. The estate belonged to the abbey of Marmoutier in the late 11th cent. Do i.309 gives pr ['pidl'hɛ:ntən].

PIDDLETRENTHIDE Dorset SY 7099. 'The 30 hide Piddle estate'. *Pidele Trentehydes* 1212, *Pydle Trenthide* 1288. R.n. PIDDLE SY 8591 + p.n. Trenthide '30 hides', OFr **trente** + ME **hide** (OE *hīd*, pl. *hīde*). This estate was assessed at a value of 30 hides in DB where it is simply called *Pidrie* 1086. Identical with *Uppidelen* [966]15th S 744 'P higher up-stream', OE **upp** + r.n. PIDDLE. Do 120.

PIDLEY Cambs TL 3377. 'Fen'. *Py- Pidele* 1228–1387, *Pud(d)ele* 1319, 1327, *Pydley* 1535. OE ***pidele** rather than pers.n. *Pyd(d)a* + **lēah** as Hu 211.

PIDLEY FEN Cambs TL 3480. *Pidley Fen* 1836 OS. P.n. PIDLEY TL 3377 + ModE **fen**.

PIEL BAR Cumbr SD 2361. *Peel Bar* 1852 OS. A sand-bar across the channel leading to Barrow-in-Furness harbour named after PIEL ISLAND SD 2363.

PIEL ISLAND Cumbr SD 2363. Named from a peel built by the monks of Furness Abbey to protect their harbour, *the pyle of foudray* 1577, ME **pele** 'a small castle or tower'. The original name of the island was Fouldrey, *Fotherey* c.1327, *-ay* c.1400, *Foderaye* 1537, *Fouldrey* 1586, discussed under FURNESS SD 2087 which is named after it. La 200.

PIERCEBRIDGE Durham NZ 2015. 'Osier causeway'. *Persebricge* [c.1040]12th, *Persebrig(g)e -bryg* 1104×7–1418, *Percebrig(g')* 1306–1441, *Pearsbridge* 1592, *Peircebridge* 1717. OE **persc** + **brycg**. The reference can hardly be to a crossing of the Tees; more likely it referred to a causeway of faggots laid across marshy ground. NbDu 157.

PIGDON Northum NZ 1588. Probably the 'pointed hill'. *Pikedenn* 1205, *Pi- Pykeden* 1226–1311, *Pyk(e)don* 1346, *Pykden* 1428, *Pykton* 1465. OE ***pīc**, adj. ***pīcede**, + **dūn**. The village stands on a hill and can hardly be an example of OE **denu**. Furthermore the alternative OE **denn**, 'a swine pasture', had little currency outside Kent and Sussex. An alternative possibility for the specific is OE pers.n. *Pīca*. NbDu 157, PNE ii.63.

PIKE HILL MOSS NYorks NZ 7701. P.n. Pike Hill, ModE **pike** (OE *pīc*) 'a peak' + **hill**, + **moss** 'a bog, a swamp'.

PIKEHALL Derby SK 1959. Possibly 'Pica's hall'. *Pikenhall* 1313, *Pikehall* 1701. OE pers.n. *Pīca*, genitive sing. *Pīcan*, + **hall**. The generic might, however, be OE **healh** 'a nook'. More early forms are needed. Db 369.

PIKESTONE FELL Durham NZ 0332. P.n. Pikestone + ModE **fell**. A large track of high moor named from *le Pykedstan* 1408, *Pikestone* 1647, 1768, 'the pointed stone', ME **piked** + **stān**. It was probably a boundary stone.

PILGRIMS HATCH Essex TQ 5895. The 'pilgrims' hatch or gate'. *Pylgremeshacch* 1483. ME **pelegrim** + **hacche**. The name refers to the visits of pilgrims to St Thomas's chapel dedicated in 1221. Ess 137.

PILGRIM'S WAY Kent TQ 6561. A modern, long-distance pathway. Called *Kentish Drove* 1819 OS, from ModE **drove** 'a road along which horses or cattle are driven'.

PILHAM Lincs SK 8693. Possibly 'Pila's homestead or village'. *Pileham* 1086, 1202, *Phileham* 1139. OE pers.n. **Pīla* + **hām**. An alternative possibility is OE **pīl** 'a spike, an arrow, a shaft, a pile' in some unknown sense, genitive pl. **pīla**: perhaps the reference was to a village where piles or arrows were made. DEPN, PNE ii.64.

PILL Avon ST 5275. 'The creek'. *Pill or Crockern Pill* 1830 OS. SW dial. **pill** 'a small stream' < OE *pyll* 'a tidal creek'. Crockern is either 'pottery' (ultimately < OE *crocc-ærn*) or the genitive pl. of ME *crokkere* 'a potter' (< OE *croccere*) and so 'potters' Pill'. PNE ii.75

PILLAR Cumbr NY 1712. A mountain peak rising to 2927ft. so named in 1783. Cu 387.

PILLATON Corn SX 3664. 'Farm, village of stakes, or where shafts or stakes or pill-oats are obtained'. *Piletone* 1086, *Pilatona* 1086 Exon, *Pi- Pyletone* 1259–1349, *Py- Pilaton(e)* 1291–1425. OE **pīl**, genitive pl. **pīla**, or **pil-āte** 'pill-oats' + **tūn**. PNCo 139, Gover n.d. 198, DEPN, PNE ii.64, Studies 1936.105.

PILLERTON HERSEY Warw SP 3048. 'P held by the Hersey family'. *Pilardinton Hersy* 1247, *Pilarton Herce* 1487. P.n. *Pilardetone* 1086, *Pilardeston* 1169, 1230, and other forms as under PILLERTON PRIORS SP 2947, 'Pilheard's estate', OE pers.n. *Pīlheard* + **ing**[4] + **tūn**, + manorial addition from the family of John de Hersy 1235 who held the manor and for distinction from PILLERTON PRIORS SP 2947. Wa 257.

PILLERTON PRIORS Warw SP 2947. 'P held by the prior' sc. of Sheen, who held the manor after the abbot of St Evroult 1086 followed by the monks of Ware. *Pilardinton Prior* 1247. P.n. *Pilardintone, Pilardetvne* 1086, *Pil(l)ardinton(e) -(a)* 12th–1438, *Pillardyngton* 1375, 1538, *Pylhardington* 1232, 'the estate called after Pilheard', OE pers.n. *Pilheard* + **ing**[4] + **tūn**. Wa 257.

PILLEY SYorks SE 3300. 'Wood or glade where shafts or stakes are obtained'. *Pillei(a)* 1086–1196, *Pi- Pyllay -ley* 1194–1593. OE **pīl** + **lēah**. YW i.297, L 205.

PILLING Lancs SD 4048. Partly obscure. *Pylin* 1194×99, [1201]1268, 1270, *Pelyn* 1320, *Pillyn(g)* 16th cent. A grange of Cockersand Abbey, but originally a river name, Pilling Water SD 4346, recorded as *Pylin* 13th cent., *Pylyn* 1364. Pilling Moss SD 4046 is *Mussam de Pilyn* c.1280. At the base of this name is OE **pyll** 'a tidal creek, a pool in a river', Mod dial. *pill*, but the precise formation is unclear. The consistent early spellings with *-in -yn* tell against an OE derivative in *ing*. OE *pull, pyll* are probably loan-words from PrW; the suffix may, therefore, be PrW *-īn*. La 140, 165, RN 326, Jnl 17.97.

PILLING LANE Lancs SD 3749. 'Lane to Pilling'. P.n. PILLING SD 4048 + ModE **lane**.

PILNING Avon ST 5585. '*Pyllyn* end' sc. of Redwick and Northwick township. *Pilnen* 1579, 1639, *Pylnynge als. Pylyn* 1624, *Pilnend* 1654×9. P.n. *Pyllyn* 1529×32 + ModE **end** later reduced to *-in* and *-ing*. *Pyllyn* is SW dial. **pill** 'a stream' + ME weak declensional ending **-en**, 'the streams'. *Pyllyn* end contrasts with Chittening ST 5382, *Chitnend* 1658, and Rednend ST 5784, *Riddenende* 1564 'clearing end', OE ***ryden** + **ende**, in the same township. Gl iii.138–9.

PILSBURY Derby SK 1163. Either 'Pil's fortified place' or 'fortified place of the spike'. *Pilesberie* 1086, *Pillesbury* 1262–1543, *Pilsbury* 1577. OE pers.n. **Pīl* or **pīl** 'a spike, a stake', genitive sing. **Pīles*, **pīles**, + **byriġ**, locative–dative sing. of **burh**. Db 370.

PILSDON Dorset SY 4199. Partly uncertain. *Pilesdone* 1086, *-don* 1196, *Pillesdun -don* 1168, 1236. The generic is OE **dūn** 'a hill', the specific uncertain. The reference is probably to the hill on which the village stands. The specific seems to be OE **pīl** 'a spike, a shaft, a pile', either of a hill marked by a stake ('hill of the stake') or of the hill itself. This seems more likely than an unattested pers.n. **Pīl* as proposed in PILSBURY Derby SK 1163. Do 120.

PILSDON PEN Dorset ST 4101. *Pilsdon Pen* 1811 OS. P.n. PILSDON SY 4199 + either PrW ***penn*** 'a top, a height, a hill' which seems to have survived as a local term, or OE **penn** 'a small enclosure, an enclosure of animals, a fold' referring to the bivallate Iron Age hill-fort which crowns the hill. Pilsdon Pen is the highest point in Dorset rising to 908ft. Do 120, Newman-Pevsner 315.

PILSLEY Derby SK 4262. Possibly 'Pinnel's clearing'. *(æt) Pilleslege* [c.1002]c.1100 S 1536, *-leage* [1004]14th S 906, *-leia* c.1200, c.1220, *Pyllesley(e)* 1272–1535, *Pinneslei(g)* 1086, *Pileslea* 1170, *-lege* 1253, *Pilsley* 1577. OE pers.n. *Pinnel*, genitive sing. *Pin(ne)les* with assimilation of *nl* to *ll*, + **lēah**. The forms seem to support this pers.n. rather than OE **Pīl* or **pīl** as in PILSLEY SK 2471. Db 289.

PILSLEY Derby SK 2471. Either 'Pil's clearing' or 'clearing of the spike'. *Pirelaie* (sic) 1086, *Pilisleg'* 1205, *Pillesleya -ley(e) -leg'* 1224–1376, *Pyllesley* 1431–1547, *Pilsley* from 1577. OE pers.n. **Pīl* or **pīl** 'a spike, a stake', genitive sing. **Pīles*, **pīles**, + **lēah**. Db 161.

PILTON Leic SK 9102. Partly uncertain. *Pilton(')* from 1202 with variant *Pyl-* 1248–1610. The generic is OE **tūn** 'farm, village', the specific either OE **pyll** 'a pool in a river' or 'a small stream' referring to the river Chater which the village overlooks, OE **pīl** 'a shaft' and so 'settlement where shafts or piles are obtained', or ON **píll** 'a willow'. R 289.

PILTON Northants TL 0284. 'Pileca's farm or village'. *Pilchetone* 1086, *Py- Pilke(n)ton* 1225–1582, *Pilton* from 1313. OE pers.n. **Pīleca*, genitive sing. **Pīlecan*, + **tūn**. Nth 217.

PILTON Somer ST 5940. 'Creek settlement'. *Piltune* *[725]12th S 250, *Piltone* 1086, *Pulton* 1243, [1330]Buck, 1610. OE **pyll** + **tūn**. Pilton lies N of the same water-course as PYLLE ST 6038. DEPN.

PIMPERNE Dorset ST 9009. Uncertain.
I. *(to, of) pimpern(welle)* [935]15th S 429, *Pi- Pympr(e)* 1178–1288, *Pi- Pympern(e)* from 1210×12.
II. *Pinpre* 1086–1230, *Pi- Pynperne* 1234–1406.
It seems likely that the generic is, or was thought to be, OE **ærn** 'a house'. The form of the specific is uncertain: *pimp* can be explained from *pinp* by assimilation of *n* > *m* before a following bilabial but not vice-versa. The forms with *n*, however, might be AN spellings. No Gmc element **pimp* is securely evidenced although ModE dial. *pimp* 'a pimple, a bundle of brushwood' first recorded in 1742 and ModE *pimple* itself (recorded from c.1400) point to the existence of a word with the transferred sense 'small hill' ultimately cognate with Gk *πομφός* 'pimple', Latin *papilla* 'nipple' and Balto-Slavic words on the base *pampa-* 'navel, bud, swelling'. Possibly, therefore, 'house at the small hill'. Alternatively the name might be of British origin, PrW ***pïmp***, PrCo ***pimp*** 'five' + **prenn** 'tree' and so 'five trees'. The name remains elusive. Do ii.110, Trautman BSW 205, SNP 6.96.

PIMPERNE DOWN Dorset ST 8910. *Pimperne Down* 1811. P.n. PIMPERNE ST 9009 + ModE **down**.

PINCHBECK Lincs TF 2325. Possibly the 'minnow stream'. *Picebech* 1086, *Pincebec* 1086–1303, *-beck(e)* 13th cent., *-bek* 1263–1399, *Pinchbeck* from 1346. OE ***pinċ*** + ON **bekkr** probably replacing OE **beċe**. Alternatively this could be 'finch ridge', OE ***pinca*** + **bæċ**. Payling 1940.42, PNE ii.65, SSNEM 219, L 126.

PINCHBECK WEST Lincs TF 2024. P.n. PINCHBECK TF 2325 + ModE **west**. Formerly called *Pinchbeck Bars* and *Dovehirne* 1824 OS, this was made into an ecclesiastical parish in 1815. In 1587 jurors in the Court of Sewers said 'there shall be a pair of *barres* with a gate at *Dowhirne*'; cf. *barehedrove* 'drove with a bar' 1331. *Hyrne* is frequently used in fenland names of a corner of land bounded by water-courses. Payling 1940.46, Wheeler Appendix i.12, 30.

PINFOLD Lancs SD 3811. No early forms, but this must be ME **pynfold** (OE **pynd-fald*) 'a place for confining stray animals, a pound'.

PINHOE Devon SX 9694. Uncertain. There are three types:
I. *(æt) Peon hó* c.1050(A) ASC under year 1001.
II. *Pinnoch* 1086, *(on) Pynnoc* c.1120.
III. *Pinho* 1216×27, 1238, *Pynnow* 1390.
I is unidentified element, possibly a bad spelling for PrW **penn* 'a hill', + OE **hōh** 'a hill-spur'; II could either represent OE **pennuc**, ME **pinnok** 'a small pen' or be spellings with stop [k] for the final voiceless aspirate of *hōh* in III, OE pers.n. *Pinna* or **pin** 'a pin, a peg' + **hōh**, possibly 'hill-spur called the Pin'. Pinhoe stands on a prominent hill-spur. The later forms do not support association with PrW **penn* 'top, height, hill'. D 443 gives pr ['pinhou].

PINKWORTHY POND Somer ST 7242. P.n. Pinkworthy ST 7241 + ModE **pond**. There are no early forms for Pinkworthy or its doublet Pinkery Farm ST 7241. The generic is ultimately from OE **worthiġ** 'an enclosure'. The specific might be any of *pink* 'plant of the genus *Dianthus*', *pink* 'a finch', ME *pinnok* (OE **pennuc*) 'a small pen' or an element cognate with Co *pynnoc*, MW *penawc*, < PrW *penn* + *-*ōg* 'a hill'. Cf. Pinnock Glos SP 0728 *Pignoscire* 1086, *Pi- Pynnokes(c)ir' -s(c)hir(e)* 1195–1543 'Pinnock shire', *Pi- Pynno(c)k(e)* 1327–1715. Cf. Turner 1950–2.116 s.n. Pinksmoor.

PINNER GLond TQ 1289. 'Pin-shaped ridge'. *Pin(n)ora* 1232, *Pinnore* 13th–14th cents., *Pinnere* 1332. OE **pinna** + **ōra**. The reference is to the humped ridge traversing Pinner Park. The river Pinn is a late back-formation from this name. Mx 63, 5, Jnl 21.17, GL 78.

Great PINSEAT NYorks NY 9602. *Great Pinseat* 1860 OS. A summit of 1914ft. NW dial. **seat** 'the summit of a hill'.

PINVIN H&W SO 9549. 'Penda's fen'. *Pendefen* 1275, *Pyndeven* 1493, *Pennefynne* 1542, *Pynvin* from 16th cent. OE pers.n. *Penda* + **fenn**. This is probably not a reference to Penda, king of Mercia, but an instance of Penda as a short form of a pers.n. such as Pendrǣd, Pendwald or Pendwulf. Wo 223, L 41.

PINXTON Derby SK 4554. 'Penec's farm or village'. *Penekestone* 1212, *Pen(c)keston -is-* 1208–1422, *Pynkeston* 1283–17th cent., *Pinxton* from 1593. OE pers.n. **Penec*, genitive sing. **Pen(e)ces*, + **tūn**. Db 291.

PIPE AND LYDE H&W SO 5044. A combination of two separate names: I. Pipe SO 5044, *Pipe* 1086, *(La) Pipe* c.1179×86–1535, OE **pīpe** 'a pipe, a conduit' apparently referring to a small tributary of the Lugg, *Pipe brook* 17th; and II. Lyde as in Lyde Arundel SO 4943, Lyde Saucey SO 5143, Lower and Upper Lyde SO 5144 and 4944, *Leode, Lvde* (3×) 1086, representing OE stream-n. **Hlȳde* 'the loud one' from OE ***hlȳde*** 'noisy stream', referring to the same stream formerly called *Ludebroc* 1173, c.1175. It still chuckles noisily as it flows over rocks or stones beneath the bridge N of the church. He 166–7, RN 272, 327.

PIPE GATE Shrops SJ 7340. *The Pipe Gate* 1724, *Pipe Gate* 1833 OS. A gate on the turnpike between Nantwich and Stone. The reference is unknown.

PIPE HILL Staffs SK 0909. *Pypehill* 1562. P.n. *Pipe* c.1140, 1361, *Magna, Parva Pipa* 'great, little Pipe' 1166, OE **pīpe** 'a pipe, a conduit' + ModE **hill**. Lichfield was for many centuries supplied with water from springs rising in Pipe manor and conveyed by pipes to the city. Also kown as *Herdewykepipe* 1349, *herdewyk, Pypeherdewyk* 1374. DEPN, Duignan 119, PNE ii.65, Horovitz.

PIPE RIDWARE Staffs SK 0917 → Pipe RIDWARE.

PIPEWELL Northants SP 8485. 'The piped spring, the spring

with the conduit'. *Pipewelle* 1086–1551, *Pyp(p)- Pipwell(e) -(a)* 1166–1428. OE **pīpe** + **wella**. Nth 175 gives pr [pipwəl], L 32.

PIPPACOTT Devon SS 5237. 'Pyppa's cottage(s)'. *Pippecote* 1311–56 with variants *Puppe- Pyppe-*. OE pers.n. **Pyppa* + ME **cote** (OE *cot*, pl. *cotu*). D 34.

PIRBRIGHT Surrey SU 9455. 'The pear-tree scrubland'. *Perifrith -frid, Pierrefrid -t* 12th cent.(p), *Perefricth -friht, Pirifricht -friȝt, Pyrybrith, Purifricht, Purefright* etc. 13th cent., *Piribryht -bright* etc. 14th cent., *Pur- Pyrbright* 15th cent. OE **piriġe** + **(ġe)fyrhth**. Sr 144–5, L 191, 219.

PIRTON H&W SO 8847. 'Pear orchard'. *Piritun* [760]17th, *(in) pyritune* 972 S 786, *Peritvne* 1086, *Pi- Pyriton* 14th cent., *Pyrton* 1523. OE **pyriġe** + **tūn**. Wo 223, Hooke 28, 179.

PIRTON Herts TL 1431. 'Pear orchard'. *Peritone* 1086–1467 with variants *Pery-* and *-tun*, *Piriton(e)* 1185–1570 with variants *Piry- Pyri- Pyry-*, *Pyrtone* 1303. OE **piriġe** + **tūn**. Hrt 21.

PISHILL Oxon SO 7289. 'Hill where peas grow'. *Pesehull'* 1195, *Pushill(e) -hull(e)* 1204–1777, *Pishulla* 1205, *Pishill* from 1310. OE **peosu, pise** + **hyll**. O 84, L 171.

PITCH GREEN Bucks SP 7703. *Pitch Green* 1822 OS. Contrasts with nearby Forty Green SP 7603, *Forty Green* 1822 OS, Holly Green Sp 7703, *Holly Green* 1822 OS, and Skittle Green SP 7702. The last example suggests that the sense is 'green where pitch is played'.

PITCH PLACE Surrey SU 9852. Uncertain. *Pitch Place* 1680. Sr 164.

PITCHCOMBE Glos SO 8508. 'Pincen's coomb' or 'coomb where pitch is produced'. *Pi- Pychenecumb(e) -co(u)mb(e)* 1211×13–1361, *Pinchecomb'* 1246, *Pinchenecumbe* 1267, *Pitchcomb(e)* 1584, 1627. OE feminine pers.n. **Pinċen*, genitive sing. **Pinċene*, or adj. **piċen**, + **cumb**. Gl ii.169, L 93.

PITCHCOTT Bucks SP 7720. 'Pitch cottage(s)'. *Pichecote* 1176–1328, *Pychcote* 1400. OE **piċ** + **cot**, pl. **cotu**. Possibly a place where pitch or resin or mineral pitch was stored or traded. Bk 136.

PITCHFORD Shrops SJ 5303. 'Pitch ford'. *Piceford* 1086, *Pi- Pycheford(e)* 1198–1583, *Pi- Pycheford* 1189–1577, *Pitchford* 1593. OE **piċ** + **ford**. So named because pitch or bitumen oozes out of a well ¼ mile N of the village, where the road to Shrewsbury used to cross Row Brook (diverted eastwards in 1833). Sa i.236.

PITCOMBE Somer ST 6733. 'Marsh coomb'. *Pidecome* 1086, 12th, *-combe -cumbe* [1139×61–13th]Buck, *Pydecombe* [c.1220]Buck. OE ***pide** + **cumb**. DEPN, PNE ii.64.

PITMINSTER Somer ST 2219. 'Pippa or Pyppa's minster'. *(to) Pipingmynstre* [938]12th S 938, *(æt) Pippingmynstre* [941]12th S 475, *Pipeminstre* 1086, *Pupmunstre* 1330, *Putmynstre* 1327, *Pitmyster* 1610. OE pers.n. *Pippa* or *Pyppa* + **ing**[4] + **mynster**. DEPN.

PITNEY Somer ST 4528. 'Pytta or Peota's island'. *Petenie* 1086, *Puttenaya* 1225, *Petteney* 1230, *Pitney* 1610. OE pers.n. **Pytta* or *Pēota*, genitive sing. **Pyttan, Pēotan*, + **īeġ**. DEPN.

PITSEA Essex TQ 7488. 'Pic's island'. *Piceseia* 1086, *Pich- Pikesey(e) -e(i)a -eya* 1119–1423, *Pe(t)chesey(e)* 1285, 1329, 1423, *Pisseye* 1401, *Pytsay* 1488, *Pitsey* 1602. OE pers.n. *Pīċ*, genitive sing. *Pīċes*, + **ēġ**. For the change [tʃ] > [ts] cf. BLETSOE TL 0258 and PITSTONE SP 9415. Ess 167.

PITSFORD Northants SP 7568. Probably 'Piht's ford'. *Pidesford* 1086, *Pitesford* 1086–1242, with variant *Pittes- Pictes- Picces- Picheford* 13th, *Pisseford* 13th–1341, *Pi- Pysford* 1337–1779. OE pers.n. **Piht* (**Peoht*), genitive sing. **Pihtes*, + **ford**. The variety of forms point to ME *-ts-* with derived variants *-tʃ-* and *-s-*. Nth 135.

PITSFORD RESERVOIR Northants SP 7669. P.n. PITSFORD SP 7568 + ModE **reservoir**.

PITSTONE GREEN Bucks SP 9315. *Pitstone Green* 1834 OS. P.n. Pitstone + ModE **green**. The green lies a mile N of Pitstone itself, *Pincenes- Pincelestorne* 1086, *Pichelest(h)orne -yorne -þorne* 1220–1316, *Pi- Pychestorn* 1227, 1235, *Pittleshorn* 1526, *Pittlestone -thorne* 1734, 1766, *Pightlesthorne or Pilstone* c.1825, 'Picel's thorn-tree', OE pers.n. *Pīċel*, genitive sing. *Pīċeles*, later confused with ME **pightel, pichel** 'a small enclosure', + **thorn**. Bk 99, L 221.

PITT DOWN Hants SU 4228. *Pit Down* 1810 OS. P.n. Pitt + ModE **down**. Pitt, *Putta* 1167, *la Putte* 1256, 1316, *Pette* 1260, *Putt alias Pitt* 1639, is OE **pytt** 'a pit, a hollow' referring to a valley N of the hamlet. Ha 131, Gover 1958.176.

PITTINGTON Durham NZ 3244. 'Hill called after Pytta'. *duo Pittindunas* c.1123, *Pitindun -en-* c.1150–1218×34, *-don(e)* c.1162–14th, *Pi- Pyt(t)ingdon -yng-* c.1220–1446, *Pitingtun* 1189×1212, *Pi- Pyt(t)ington -yng-* from c.1220, *Petyngton* 1351–1522, *Pydington -yng-* 1364, 1382, *Piddington* 1675. OE pers.n. **Pytta* + **ing**[4] + **dūn**, possibly varying with **Pytting-dūn* 'hill called **Pytting*, the place called after *Pytta*'. The two settlements mentioned in c.1123 are North Pittington NZ 3244 and Pittington Hallgarth NZ 3243, the manor house of the Prior of Durham. NbDu 157, L 157.

PITTON Wilts SU 2131. 'Putta's or the hawk's estate'. *Putenton'* 1165, *Putteton* 1267, *Putton(e)* 1198–1402, *Putton alias Pitton* 1726. OE pers.n. *Putta* or ***putta** 'a hawk, a kite', genitive sing. *Puttan*, ***puttan**, + **tūn**. Wlt 378.

PITY ME Durham NZ 2645. *Pity Me* 1862 OS. Examples of this common name type are (or were) to be found also at NZ 1929, NZ 3028 (near Coldsides, Standalone, and Linger and Die), NY 9640 etc. Despite the speculation the Durham name has given rise to, it is no more than an example of wry name giving for an exposed locality on difficult land, short for something like 'Pity me for having to live or work in such an unpropitious place!' Such names have little history, having mostly arisen in post-enclosure times when farm-houses or pit settlements began to be built away from old village centres in the former open fields or on the moors.

PIXEY GREEN Suff TM 2475. *Pixey Green* 1837 OS. Contrasts with nearby *Ashfield Green, Barley Green* TM 2474, *Buttlesea Green* and *Caterpole Green* all 1837 OS.

PLAISH Shrops SO 5396. 'The shallow pool'. *Plesc, Plæsc* 963 S 723†, *Plesham* (Latin accusative sing. in *-am*) 1086, *Ples(s)e, Plessh* 13th, *Playsse* 1255, 1508, *Plaishe* 1590, *Plas(s)he* 1304–1602. OE ***plæsc** (Mod dial. *plash*). The reference is either to a pool associated with a spring near Holt Farm which formerly supplied the village with water, or to an area of wet ground at Plaish Hall which sometimes floods the roadway. Sa i.236.

PLAISTOW WSusx TQ 0030. 'Sport or games place'. *La Pley(e)stowe* 1271, 1331. OE **pleġ-stōw**. Sx 106.

PLAITFORD Hants SU 2819. 'Playing ford'. *Pleiteford* 1086, 1391, *Pleitesford* 1234, *Pleyte- Playteford* 1263–1495, *Platford* 1610, 1695. OE ***pleġet** + **ford**. A ford beside which games were held. Cf. PLAYFORD Suff TM 2148. W 383 gives pr [plætfəd], Ha 132, PNE ii.67.

PLASTOW GREEN Hants SU 5361. *Low*[r]., *Up*[r]. *Plastow Green* 1817 OS contrasting with nearby *Summer-hurst Green* and *Catts Green*. P.n. Plastow + ModE **green**. Plastow, *Pleystowe* 1509×47, *Plaistowe* 1575, is ultimately from OE **pleġ-stōw** 'a sport place, a place where people gather for play'. Gover 1958.147.

PLATT Kent TQ 6257. No early form. ModE **plat** 'a plot of ground'.

PLAWSWORTH Durham NZ 2647. 'Enclosure for sports'. *Plauswrda* 1135×9, *-word* [1133×40]1338, [1183]c.1382, 1338, *-worth* 1360–1635, *Plawesworth(a)* before 1174–1501, *Plawsworth* from 1481. OAngl **plaga**, ME **plawe**, pl. **plawes**, + **worth**. NbDu 157.

PLAXTOL Kent TQ 6053. 'Play place'. *Plextole* 1386, *Plaxtol* 1650. OE ***pleġ-stall**. PNK 157, PNE ii.67.

†Identification uncertain.

PLAYDEN ESusx TQ 9121. 'Play pasture'. *Pleidenā -den(ne)* 1086–14th, *Plaidenam* c.1195, *Playdenne* 1340. OE **plega** + **denn**. The sense is probably 'hollow where deer (or some other wild animals) play'. Alternatively the specific might be an OE pers.n. **Plega*. Sx 533, SS 88, 138–59, Nomina 1979.103.

PLAYFORD Suff TM 2147. 'The ford where sports are held'. *Playford* from 1269 including [c.1040]13th S 1224, *Plegeforda* 1086, *Pleiford(a)* 1130–1254, *Pleyford(e)* 1246–1568. OE **pleġford** < **pleġa** + **ford**. Cf. GLEMSFORD TL 8348. DEPN, Baron.

PLAY HATCH Oxon SU 7475. 'Gate at a playing field'. *Playhatch* 1603×4. ModE **play** + **hatch**. O 69.

PLAYING PLACE Corn SW 8141. *Playing Place* 1884, earlier *Kea Playing Place* 1813, from its situation in Kea parish. A 20th cent. village that grew up near an arena used for performances of the Cornish dramas or games. Slight traces of the arena survive at SW 814419. PNCo 139.

PLEALEY Shrops SJ 4206. 'The play clearing'. *Pleyle* 1255–1756 with variants *-ley(e)* and *-leg(h')*, *Plealey* from 1615. OE **pleġa** + **lēah**. A glade where human sports or animal mating rituals took place. Cf the p.ns. Playley Glos SO 7631 and Deer Play Northum NY 8490, Lancs SO 8627 and Durpley Devon SS 4213. Sa ii. 47 gives pr [pleili], DEPN, PNE ii.67.

PLEASINGTON Lancs SD 6426. 'Estate called after Plesa'. *Plesigtuna* 1196, *Plesinton* 1208–1332 etc., *-ing- -yng-* 1228–1497. OE pers.n. **Plēsa* + **ing**[4] + **tūn**. In the same township occurs *Plesley* 1284, 'Plesa's wood or clearing'. La 74, Jnl 17.47, 18.22.

PLEASLEY Derby SK 5064. 'Plesa's clearing or wood'. *Pleselei(a) -ley(e) -lay(e)* c.1165–16th cent., *Plesseley* 1506, *Plesley(e) -lay(e)* 1310–18th cent., *Pleasley* from 1577. OE pers.n. *Plēsa* + **lēah**. Db 292 gives pr [plesli].

PLENMELLER Northum NY 7163. 'Mellor end'. *Plenmenewre* 1256, *Playn- Pleinmelor(e)* 1279, 1307, *Plenmeller* 1663, *Plainmeller* 1866 OS. PrW ***blain** + p.n. *Mellor*. Mellor is identical with MELLOR GMan SJ 9888, Lancs SD 6530, W *Moelfre* 'bare hill' <PrW ***męl** + ***breȝ**. The reference is to the lofty plateau of Plenmeller Common NY 7162–7762. NbDu 158, Jnl 1.44, 49.

PLENMELLOR COMMON Northum NY 7361. *Plainmeller Common* 1866 OS. P.n. PLENMELLOR NY 7163 + ModE **common**.

PLESHEY Essex TL 6614. 'The enclosure'. *Plaseiz* c.1150, *Plaisseiz* c.1230, *Pless(e)y -i(z)* 1302–76, *Pleshe* 1298, *Pleshey* from 1402. The Latin forms are *Plassetum, (de, in) Plasseto* 1212–1428. OFr **plessis**. Pleshey was originally part of Good Easter as the *Liber Eliensis* makes clear, *famosa uilla de Estre alio nunc nomine Plassiz uocitata* 'the famous vill of Easter now alternatively called Pleshey' 12th. A *plessis* was part of a forest enclosed by a fence of living wood with branches interlaced (*plessé*). Ess 488 gives pr [plʌʃə].

PLOUGHFIELD H&W SO 3841. 'Open land where sports are held'. *Ploufeld* 1273,1415, *Plowfeld* 1345. OE **plaga** with ME vocalisation of [ɣ] > [w] + **feld**. He 170.

PLOVER HILL NYorks SD 8475. ModE **plover** + **hill**. A summit of 2231ft.

PLOWDEN Shrops SO 3887. 'Play valley'. *Plau- Pleweden* 1252, *Plou- Ploweden(e)* 1282–1316, *Playden* 1610, *Plowden or Ployden* 1700. OE **plaga** + **denu**. A valley where sports were held or wild animals played, probably deer. Cf. PLEALEY SJ 4206, PLAYDEN ESusx TQ 9221. Bowcock 185, DEPN, PNE ii.67.

PLOXGREEN Shrops SJ 3603. *Plox Green* 1836 OS. ModE **plock** 'a small piece of ground', pl. **plocks** 'the plots'. Houses were built here by 1766. Sa ii.26.

PLUCKLEY Kent TQ 9245. 'Plucca's wood or clearing'. *Plvchelei* 1086, *Plukele* 1086×7, *Plucele(a), Plukele -lai* c.1090, *Plukeleya* 1142×8, 1198×1205, *Plo- Plukkele(ga)* 1210×12–1292. OE pers.n. *Plucca* + **lēah**. In 1086 Pluckley still had woodland offering pannage for 140 pigs. PNK 393.

PLUCKLEY STATION Kent TQ 9243. A halt on the railway line from Ashford to Tonbridge. P.n. PLUCKLEY TQ 9245 + ModE **station**.

PLUMBLAND Cumbr NY 1539. 'Plum-tree grove'. *Plumbelund* [c.1150]c.1225, *Plumlund* c.1175–1328, *-land* 1278–1612, *Plumbland* from 1443. OE **plūme** + ON **lundr**. Named, according to Reginald of Durham (c.1150), from the very dense woods with which it is surrounded, *lund* meaning *nemus paci donatum* 'grove devoted to peace'. Cu 309 gives pr [plimlən], SSNNW 243, L 207.

PLUMLEY Ches SJ 7175. 'Plum-tree wood or clearing'. *Plumleya -leia* [1119]1150, *-leg(h) -ley* from 1259×60, *Plomle(y)* c.1290–1452. OE **plūme** + **lēah**. Che ii.90.

PLUMPTON 'Plum-tree farm'. OE **plūme** + **tūn**.

(1) ~ Cumbr NY 4937. *Plumton'* 1212, c.1460×80, *Plumpton* from 1256. Cu 234 gives pr [pluntən], cf. the spelling *Pluntonhead* 1580 for PLUMPTON HEAD NY 5035.

(2) ~ ESusx TQ 3613. *Plumtone* 1086, *Plumptune -ton(e)* c.1110 etc., *Plimpton*. Sx 303, SS 75.

(3) ~ Lancs SD 3833. Short for Great Plumpton, formerly Great Fieldplumpton, *Le Graunte Fildeplumpton* 1323, adj. **great**, OFr **grand**, + p.n. Fieldplumpton. 'Great' for distinction from Little Plumpton SD 3832, *Little Fildeplumpton* 1323. Fieldplumpton is really p.n. FYLDE + p.n. *Pluntun* 1086, *Plumton* [1190–1217]1268, 1226, 1257 etc., *Plumpton* from 1327, for distinction from WOODPLUMPTON Lancs SD 5034. A specialised settlement within the estate of Amounderness. La 151, Jnl 17.88, 21.34.

(4) ~ GREEN ESusx TQ 3616. P.n PLUMPTON TQ 3613 + ModE **green**.

(5) ~ HEAD Cumbr NY 5035. 'Head of Plumpton'. *Pluntonhead* 1580, *Plumpton Head* 1597, p.n. PLUMPTON NY 4937 + ModE **head** in the sense 'top end' in contrast with the bottom end at Plumpton Foot NY 1597, *the foote of Plumpton* 1578, *Plumbton foote* 1597. Plumpton is a scattered village lying along the Roman road from Penrith to Carlisle. Cu 234.

PLUMSTEAD 'Plum-tree farm, plum orchard'. OE **plūm** + **stede**.

(1) ~ Norf TG 1334. *Plumestede* 1086, *Plumsted(e)* 1179–1417. *Stead* 183.

(2) Great ~ Norf TG 3010. *Magna Plumstede* 1275. ModE **great**, Latin **magna**, + p.n. *Plum(m)esteda -stede* 1086, *Plumsted(e)* 1121–1370. Also known as *Gri- Grumene Plumstede* 1254 and *Grim(m)ere, Greymmer(e) Plumsted(e)* 1275–1370 possibly representing ME *grome* 'groom, boy, servant', genitive pl. *gromene* and ON pers.n. *Grimarr*. 'Great' for distinction from Little ~ TG 3513. *Stead* 184.

(3) Little ~ Norf TG 3513. *Parva Plumsted(e)* 1257–1370, *Little Plumstede* 1265. ME adj. **little**, Latin **parva** + p.n. Plumstead as in Great PLUMSTEAD TG 3010. *Stead* 184.

PLUMTREE Notts SK 6133. 'The plum-tree'. *Pluntre* 1086, *Plumtre* c.1180–1352, *Plumptre, Plumbtre* 13th, *Plomptre* 1419, *Plumtree* 1574. OE **plūm-trēow**. The plum-tree was an early meeting-place, cf. the references to a *Wapentak de Plumptre* 1266, *Plumptre wapentake* 1310. Nt 239, L 212, 214.

PLUNGAR Leic SK 7633. 'Plum-tree point of land'. *Plumgar* 1243, 1254, 1525, *Plungar(e)* from c.1130, *Plumgard'* c.1130–1242, *-garth'* c.1291–1518, *Plungard(e)* 1155×68–1308. OE **plūme** + **gāra**. Many forms show the influence of ON **garthr** 'an enclosure'. Lei 181, SSNEM 214.

PLUSH Dorset ST 7102. 'The pool'. *Plyssche, Plisshe* [891]14th S 347, *(ad) Plussh'* [941]14th S 474, *Plys(se)* 1268, 1288, *Plussh* 1412. OE ***plysc**. Do iii.245, DEPN.

River PLYM SX 5464. A back-formation from PLYMSTOCK SX 5152 and PLYMPTON SX 5356. *Plyme* 1238, *Plym* 1359. D 12, RN 328.

PLYMOUTH Devon SX 4755. 'Mouth of the river Plym'. *Plymmue* 1230, 1254, *Plimmuth* 1234, *Plummuth* 1281, 1318, *Plemmuth* 1297, 1350. R.n. PLYM SX 5464 + ME **mouthe** (OE **mūtha**). The original name of the place was Sutton Prior 'the prior's

Sutton', *Sutton Prior vulgariter Plymmouth nuncupatur* 'S P commonly called P' c.1450. Earlier simply *Sutona* 1086, *Sutton(a)* 1201, 1275, 'the south settlement' relative to Stoke in Devonport SX 4655. D 233, TC 151.

PLYMOUTH BREAKWATER Devon SX 4751. P.n. PLYMOUTH SX 4755 + ModE **breakwater**.

PLYMPTON Devon SX 5356. 'Plum-tree farm'. *Plymentun* [899×909]c.1500 S 380, *Plintone* 1086, 1131, *Plinton* 1179–1263, *Plimtun Plimton* 1225, *Plympton* from 1244×62, *Plum(p)ton* 1166–1275. OE **plȳme** + **tūn**. The earliest form may point to an OE adj. ***plȳmen** 'growing with plum-trees'. D 251, TC 151.

PLYMSTOCK Devon SX 5152. 'Outlying farm with a plum-tree'. *Plemestocha* 1086, *Plumstok* 1228, *Plimstok* 1244, 1281. OE **plȳme** + **stoc**. The name might, however, be short for *Plympton Stock*, 'outlying farm belonging to Plympton'. D 256.

PLYMTREE Devon ST 0502. 'Plum-tree'. *Plvmtrei* 1086, *-tre(oue) -trewe -trowe* c.1200–1364, *Plimtre* 1219, *Plymtre* 1291. OE ***plȳm-trēow**. D 567.

POCKLEY NYorks SE 6386. Possibly 'Poca's forest clearing'. *Pochelaf -lac* 1086, *Pokelai -lay* [1184–98]16th cent., 1279×81, *Pockeley(a)* 1252, 1259, *Poklee* 1285. OE pers.n. **Poca* + Anglian **lǣh** (OE *lēah*). The specific might, however, be OE appellative *pocca* 'a pouch, a bag', used in some topographical sense, perhaps of the valley below the village. YN 72.

POCKLINGTON CANAL Humbs 7445. P.n. POCKLINGTON SE 8049 + ModE **canal**.

POCKLINGTON Humbs SE 8049. 'Estate called after Pocela'. *Poclinton* 1086–1276, *-ing- -yng-* 13th–1524 with variants *Pochel- Pokel-*. OE pers.n. **Pocela* + **ing**[4] + **tūn**. YE 182.

POCK STONES MOOR NYorks SE 1160. *Poxstones Moor* 1771. Named from Great Pock Stones SE 1060, *Poxtonnes* 1613. In Staffs *pox-stone* was 'a hard stone of greyish colour' (EDD). In Ripon Minster in the 16th cent. there was a stone called the *Pokstane* of St Wilfrid, probably used as a charm against the pox. The reference was probably to pitted stones. YW v.127, xii.

PODE HOLE Lincs TF 2122. 'Frog hole or hollow'. *Pode Hole* 1824 OS. ME **pode** + **hole**. Wheeler Appendix i.30.

PODIMORE Somer ST 5424. *Puddimore* 1811 OS. Earlier spellings are needed to decide whether this is 'marsh with a ditch', OE **pudd** + **ing**[4] + **mōr**, 'Puda's marsh', OE pers.n. *Puda* + **ing**[4] + **mōr** or a doublet of PODMORE Staffs SJ 7835 'frog marsh', ME **pode** + **mōr**.

PODINGTON Beds SP 9462. 'Estate called after Puda or Poda'. *Podinton, Potintone* 1086, *Podinton -yn-* 13th–14th cents., *Pudin(g)ton* 1163–1227. OE pers.n. *Puda* or *Poda* + **ing**[4] + **tūn**. Bd 37.

PODMORE Staffs SJ 7835. 'Frog moor'. *Podemore* 1086, *-mor* c.1235, 1288, 14th, *Podmore* 1559. OE ***pāde** (ME *pode*) + **mōr**. DEPN, PNE ii.68, Horovitz.

PODS BROOK Essex TL 7126. *Pods Brook* 1805 OS. Probably surname *Pod* + ModE **brook**. This is the name of the r. Brain above Braintree. For another possible name see RAYNE TL 7222. Ess 4.

POINT CLEAR Essex TM 0915. No early forms. It seems to be the same as St Osyth Stone Point, *St Osyth Stone* 1777. It is probably named from St Cleres Hall TM 1214, *Seynclers Hall* 1555, *Seyntcleres Hall* 1558×79, *Sincklers* 1624, from the surname *St Clere* recorded there in 1273. Ess 350–1.

POINTON Lincs TF 1131. Perhaps the 'estate called after Pohha'. *Pochinton(e), Poclintone* 1086, *Pointon(')* from 1150×60, *Puinton* 1186–1210, *Poyngtun' -ton'* 13th cent. OE pers.n. *Pohha* + **ing**[4] + **tūn**. *Pohha* is a by-n. from *pohha* 'a pouch, a bag, a swelling'. The specific of the name might, therefore, be an OE **pohhing* 'that which is like a pouch, bag or swelling' < *pohha* + *ing*[2] referring to some natural feature, and so the 'settlement at **Pohhing*' referring to rising ground W of the village. Perrott 134, BzN 1981.308.

POKESDOWN Dorset SZ 1292. Probably 'goblin hill'. *Pokesdoune* 1300, *-don* 1333, *-downe* 1540. ME **poke** (OE *pūca*), genitive sing. **pokes**, + **doun** (OE *dūn*). Alternatively the specific might be the OE pers.n. *Poc*. Do 121, Ha 132.

POLAPIT TAMAR Corn SX 3389. *Poolapit Tamer* 1625. P.n. Poolapit + r.n. TAMAR. Poolapit is probably the same as Bullapit SX 3289, *Bulapit* [1149]15th 'bull-pit', OE **bula** + **pytt**. This could be a reference to a pit for bull-baiting or to a natural hollow. PNCo 139.

POLBATHICK Corn SX 3456. Partly uncertain. *Polbarthek* 1365, *Polvethick* 1699, *Polvathick* 1748. Co **pol** 'a creek' + unidentified element, possibly a r.n. *Barthek* comparable with the second element of TREBARTHA SX 2677 and the W stream-name Barthen. Polbathick is situated at the head of a long creek. PNCo 139, Gover n.d. 222.

POLDEN HILLS Somer ST 4435 → CHILTON POLDEN ST 3739.

POLDHU POINT Corn SW 6619. *Poljew Point* 1888. P.n. Poldhu (as in Poldhu Cove) + ModE **point**. Poldhu, *Polsewe* 1533, *Poljew* 1678, 1813, is 'dark cove', Co **pol** + **du** 'black'. PNCo 140 gives pr [poldʒu:], Gover n.d. 576.

POLEBROOK Northants TL 0787. Probably the 'pouch brook'. *Pochebroc* 1086, *Pokebroc -broke* 1207–1608, *Polebroc -brok(e)* 1254–1608. OE **pohha**, **pocca** later replaced with **pōl** 'a pool' + **brōc**. The reference would be to the shape of the valley down which the stream runs; other less likely suggestions are OE **puca** 'a goblin', OE ***poc(c)e** 'a frog'. Nth 215, DEPN, L 15.

POLEGATE ESusx TQ 5805. 'Gate by a pool'. *Powlegate Corner* 1563, *Poolgate* 1579. ModE **pool** + **gate**. A modern settlement developed since the opening of the railway. Sx 437.

POLESHILL Somer ST 0823 → POULSHOT Wilts ST 9759.

POLESWORTH Warw SK 2602. 'Poll's enclosure'. *Polleswyrð* c.1000, *-worth* 1247–1627, *Polesw(o)rda -worða -w(o)rth(e)* c.1155–1316, *Poulesworth* 1477, *Powl(e)sworth(e)* 1599–1615. OE pers.n. **Poll*, genitive sing. **Polles*, + **worth**. Wa 21.

POLGOOTH Corn SW 9950. Either 'pool of a goose' or 'pool or pit of a water-course'. *Polgoyth* 1500×1. Either Co **poll** + **goth** 'a goose' or + **goth** 'a water-course'. Since the earliest reference is the name of a tin-mine, it is suggested that the latter is more likely and that **poll** may here have the sense 'pit'. PNCo 140, Gover n.d. 394.

POLHAMPTON Hants SU 5250 → QUIDHAMPTON SU 5150.

POLING WSusx TQ 0404. 'The pole place or the pole people'. *Paling'* 1156×8, 1275, *Paling(e) -yng* 1341–1580 including [1154×89]1361, *Polynge* 1305 etc., *Powling* 1610, 1669, *Palinges -ynges* 1199–1347. OE p.n. **Pāling* < **pāl** + **ing**[2] or folk-n. **Pālingas* < **pāl** + **ingas**. It is impossible to know whether this was originally a sing. p.n. or a pl. folk-n. or what the significance of *pāl* may have been, a palisaded enclosure or a place where stakes were obtained, or a nick-n. derived from *pāl*. The Palingas seem to occur again in *Palinga schittas* 'the sheds (OE ***scydd**) of the Palingas' S 562, now Limbo Farm SU 9624, and in p.n. Pallingham Farm TQ 0422, *Palingham* 1233–1436, *Pallinge- Palyngeham* 1276–1342, 'the river-bend land belonging to the Palingas or at *Paling*, the stake place', OE **hamm** 1. Sx 118, 134, 171, ING 39, 124, SS 64, 84, MA 10.23.

POLKERRIS Corn SX 0952. Partly uncertain. *Polkerys* 1585, *Polkeryes, -kerries* c.1605. Co **poll** 'a cove' + unidentified element which could formally be *kerys* 'fortified', though in what sense is unclear. There is a whirlpool in the Menai Straits called Pwll Ceris, *Polkerist* [10th]c.1100 which may contain a pers.n. *Cerist* but it is unclear whether the two names can be connected. PNCo 140, Gover n.d. 428.

POLLINGTON Humbs SE 6119. Probably 'the farm or village at or called (the) *Pofeling*'. *Pouilgleton* 1154×60, *Pouelington(a)* 1180×90–c.1262, *Polington(a) -yng-* 1154×89–1449, *Pollington* from c.1262. OE ***pofeling** + **tūn**. Formally **pofeling* is a diminutive of OE **pofel*, the probable origin of Scots *poffle* 'a small piece of land', but its precise meaning in OE is unknown. Cf POOL WYorks SE 2445. YW ii.21, BzN 1968.184.

POLMASSICK Corn SW 9745. 'Bridge of Madek'. *Ponsmadek* [1301]14th, *Polmasek* 1469. Co **pons** 'a bridge' later replaced with **pol** 'a pool, a stream' + pers.n. *Madek* with Cornish *d* > *z*. PNCo 140, Gover n.d. 394.

POLPERRO Corn SX 2050. Partly uncertain. *Portpira* 1303, *Port(h)pera* 1321–1535, *Porthpire* 1361, *-pyre* 1391, *Polpera* 1522, *Poulpirrhe* c.1540, *Polparrow* 1748. Co **porth** 'a harbour' later replaced with **pol** 'a pool, a stream, a cove' + pers.n. **Pera* or **Pyra*. PNCo 140, Gover n.d. 300.

POLRUAN Corn SX 1250. 'Harbour of Ruveun'. *Porruwan* early 13th, *Porthruan* 1284, *Polruan* from 1292. Co **porth** 'a harbour' later replaced with **pol** 'a pool, a stream, a cove' + pers.n. *Ruveun*. PNCo 141, Gover n.d. 271.

POLSHAM Somer ST 5142. 'Pawol's promontory into marsh'. *Paulesham* *[1065]18th S 1042, *Pauleshamesmede* 'Polsham meadow' 1361, *Polsham* 1610. OE pers.n. **Pāwol*, genitive sing. **Pāwoles*, + **hamm** 2a. **Pāwol*, a by-n. from *pāwa* 'a peacock', seems preferable to *Paul*. DEPN.

POLSTEAD Suff TL 9938. 'The pool place'. *Polstede* [962×91, probably after 975]11th S 1494, 1000×2 S 1486, *Polesteda* 1086, *Polsted(e)* c.1190–1610. OE **pōl** + **stede**. Pools are marked here on the 1805 OS. Stead 105, 192.

POLTIMORE Devon SX 9696. Uncertain. *Ponti- Pvltimore* 1086, *Pultemor(e) -ti- -ty-* 1219–1378, *Poltimor(e)* from 1263. Cf. Poultney Leics, *Pontenei* 1086, *Pultenei* 1292, and the lost p.n. *Pultimor* 13th, 14th, near Landimore Wales SS 4693. Unknown element + OE **mōr**. D 444, ii.xiii.

POLYPHANT Corn Sx 2681. 'Toad's pool'. *Polefand* 1086, *Polofant* 1086 Exon, *Polefant* [c.1170]1476, *-faunt -fond* 1296–1432, *Pollyfont* 1578. Co **pol** + ***lefant**. PNCo 141, Gover n.d. 155.

POLZEATH Corn SW 9378. Probably 'dry cove'. *Polsegh* 1311 Gover n.d., *Bulsath Bay* 1584, *Pulsath baye* c.1605, *Polzeath* 1748. Co **pol** + **segh**. The sense is 'a cove dry at low tide'. PNCo 141, Gover n.d. 129.

New POLZEATH Corn SW 9379. ModE adj. **new** + p.n. POLZEATH SW 9378. A 20th cent. holiday village so named in 1972. PNCo 141.

PONDERS END GLond TQ 3695. *Ponders ende* 1593 sc. of the parish of Enfield. Surname *Ponder* + **end**. A ponder may have been a dweller by or keeper of the fish- or mill-pond of the parish. Mx 74.

PONDERSBRIDGE Cambs TL 2691. *Ponds bridge* 1688, 1834 OS. Ca 264.

PONSANOOTH Corn SW 7537. Probably 'bridge of the watercourse'. (Bridge called) *Pons an Oeth* 1521, *Ponsewoth* 1555, *Ponsonwoth Mills* 1582, *Ponsanwoth* 1613, 1620. Co **pons** + definite article **an** + **goth** with lenition of *g* to *w*. Alternatively the final element could be **goth** 'a goose' and so 'goose bridge'. PNCo 141, Gover n.d. 152.

PONSFORD Devon ST 0007 → PONSWORTHY SX 7073.

PONSWORTHY Devon SX 7073. Probably 'ford of the river *Pont*'. *Pauntesford* 1281, *Pontesford(e) -is-* 1347, 1374, *Ponceford* 1600, *Pounsford al. Pounsworthy* 1674. R.n. **Pont* from PrW ***pant** 'a hollow, a valley' as in River PONT Northum NZ 1676 + OE **ford**. Another Ponsford occurs at ST 0007, *Pantesfort* 1086, *Pauntes- Pontesford(e)* 13th cent. D 528, 561.

River PONT Northum NZ 1676. 'Valley river'. *Ponte* 1268, *Pont* from 1479. PrW ***pant** + OE **ēa**. NbDu 159, RN 332.

PONTEFRACT WYorks SE 4521.'The broken bridge'.

I Latin forms: nominative *Ponsfractus*, accusative *Pontemfractum*, dative, ablative *Pontefracto* 1120×2–1452.

II French and English forms: *Pontefratch -freyt -fract(e)* 1100–1817, *Pontifract(e)* 1415, *Pumfrate* c.1190, *Pomfret(t)(e)* 1472–1638.

The site is that of the present Buswith Bridge which carries the road from Pontefract to Ferrybridge over a stream called Wash Dike. The original name of the town survives in TANSHELF SE 4422. It was superseded on the foundation of the priory of Cluniac monks and the establishment of the village as the seat and castle of Ilbert de Lacy, a great Norman landowner in the area: cf. for similar developments BARNARD CASTLE Durham NZ 0516 and RICHMOND NYorks NZ 1701. The modern pr ['pʌntifrækt] is a spelling pronunciation based on the Latin form of the name, the older pr [pʌmfri] and the local forms ['pɔm- 'poumfrit] represent OFr, AN *pontfreit* with assimilation of *-n-* to *-m-* before *-f-*. YW ii.75.

PONTELAND Northum NZ 1673. 'Pont island'. *Ponteland* from 1256, *Punteyland, Pount Eland, Pont Eyland* late 13th cent. R.n. PONT + p.n. *Eland* 1242 BF–1428, 'the island', OE **ēġland**, here used of land surrounded by marshes. L 248–9 tentatively offers an alternative analysis, r.n. *Ponte* + **land**, 'newly cultivated land by Pont river'. NbDu 159, RN 332.

PONTESBURY Shrops SJ 4006. Possibly 'Pant's fortified place'. *Pantesberie* 1086, *-buri* 1217, *Pontesbir(i)' -bur(y) -buri(e) -byr'* 1203×4 etc., *Ponsepre* 1454, *Pomspry* 1752, *Ponsbury(e)* 1495–1746, *Ponsbrey* 1643. OE pers.n. *Pant*, genitive sing. *Pantes*, + **burh**, dative sing. **byriġ**. A mile to the east is Pontesford Bridge, which might have borne the W r.n. *Pant* as in River PONT Northum NY 1676; but the distance of Pontesbury from the stream tilts the balance in favour of the pers.n. explanation. It is suggested that Pontesbury itself was an early fortified place; but there are Iron Age hill-forts on Pontesbury Hill SJ 3905. There are no grounds for identifying *Posetnes byrig* 9th ASC under year 661 with this place. Sa i.238.

Great PONTON Lincs SK 9230. *Magna Pa'ptvne, magna Pa'tone* 1086, *Magna Pantone -tun* c.1160–1241, ~ *Paunton* 1237–1723. ModE adj. **great**, Latin **magna**, + p.n. *Pamtone, Pamptune* 1086, the 'hill settlement', OE ***pamp** + **tūn**. 'Great' for distinction from Little PONTON SK 9232. Both villages lie in the Witham valley beside elevated ground. Alternatively the specific might be OE **panne** 'a pan' and topographically of a depression. Perrott 471, Cameron 1998.

Little PONTON Lincs SK 9232. *Alia Pātone, parua Pantone* 1086, *Parua Paunton* 1235–1402, *Little Paunton* 1339–1821. ME adj. **litel**, Latin **parva**, + p.n. Ponton as in Great PONTON SK 9230. Perrott 473.

PONTRILAS H&W SO 3927. 'Bridge of Elys'. *Pontrilas* 1750, 1786. W **pont** + **yr** + p.n. *Elys*. This is a 16th cent. translation of *Elstones Bridge* 1577, 1611, 1670, p.n. *Elstone* + ModE **bridge**. Elstone, *Elwistone* 1086, *Heliston(a)* c.1180×96, c.1206, *Elyston, Elston* 1300, is 'Ælfwig's estate', OE pers.n. *Ælfwīġ*, genitive sing. *Ælfwīġes*, + **tūn**. He 110, Bannister 155, DB Herefordshire 1.56.

PONTS GREEN ESusx TQ 6716. Surname *Pont* + ModE **green**. Sx 479.

PONTSHILL H&W SO 6422. 'Hill of the Pant'. *Panchille* 1086, *Paunteshull* 1219×34, *Penshull* 1300, *Pants Hill* 1831 OS. Stream-n. *Pant* < PrW ***pant** 'a hollow, a valley', genitive sing. *Pantes*, + OE **hyll**. But the OE pers.n. *Pant* is an alternative possibility. He 203.

POOKS GREEN Hants SU 3710. No early forms. Possibly from the surname Pook, cf. Richard le Puke 1327 (OE *pūca* 'goblin'). Gover 1958.199.

POOL WYorks SE 2445. Uncertain. *(on) Pofle* c.1030, *Pouele* 1086–c.1290, *Pouel* 1166–1327, *Poule* 1172–1583, *Pole* before 1266–1591, *Pool* from 1686. OE ***pofel** of uncertain meaning, possibly 'a small piece of (? sandy or gravelly) soil'. Later, as a result of the vocalisation of medial *-f-* before *-l-*, confused with OE *pōl* 'a pool'. YW iv.208.

POOL Corn SW 6741. *Pool* 1813 OS.

South POOL Devon SX 7740. *Suthpole* 1284, 1300. Earlier simply *Pole* 1086, *Pole* 1242, OE **pōl** 'a pool, a creek' referring to Southpool Creek, an arm of Kingsbridge Estuary. 'South' for distinction from North Pool SX 7741, *Pola* 1086, *Northpole* 1238. D 327.

POOLE Dorset SZ 0190. 'The pool, the creek'. *Pole* 1183–1484 with definite article *la* 1220–1428, *Pool(e)* from 1392. OE **pōl**. The reference is to Poole Harbour. Do ii.41.

POOLE HARBOUR Dorset SZ 0088. *Poole Harbour* 1811, earlier *portus de Pola* 1268 and *la Hauene de la Pole* 1364, *Poole Haven* 1535×43. P.n. POOLE SZ 0190 + ModE **harbour**. Do ii.44.

POOLE KEYNES Glos SU 0095. 'Poole held by the Keynes family'. *Pole Canes* 1610. Earlier simply *Pole* 1086–1315 including *[10th]13th S 1552, 'the pool', OE **pōl**. The manor passed by marriage to Sir John Keynes in the 14th cent. Gl i.79 gives pr ['pu:l 'keinz], L 28.

POOLEY BRIDGE Cumbr NY 4724. *Powley Bridge* 1671, *Pooley Bridge* 1793. P.n. Pooley + ModE **bridge**. Pooley, *Pulhou(e)* 1252–1308, *Poulhou* 1284, *Pullou -au* 1292, *-o* 1493×1500, *Pulley* 1549, *Powley* 1573–1714, *Pooley* from 1672, is 'hill by a pool', OE **pōl** + ON **haugr**. The reference is to a pool or deep place in the river Eamont and the conical hill which rises to 775ft. at NY 4624. We ii.211, SSNNW 244.

POOLHILL Glos SO 7329. *Poolehill* 1619. Probably named from John Poole of Pauntley SO 7329. A local surname from *la Pulle* c.1280, 'the pool', OE **pōl**, **pull**, + ModE **hill**. Gl iii.183.

POORTON Dorset SY 5179. Partly uncertain. *Powr- Pourtone* 1086, *Poertona* 1168, *Pourton* 1212, *North- Suthporton* 1288. The generic is OE **tūn** 'farm, village, estate', the specific an unidentified element, possibly an old r.n. found also in POWERSTOCK SY 5196 and formed on the o-grade of PrW **pefr* 'radiant' seen in PEOVER. Do 121, DEPN, Studies 1936.21.

POPHAM Hants SU 5643. Possibly 'pebble estate'. *Popham* from 1167 including *[903]16th S 370, *Popehā* 1086. Uncertain element + OE **hām**. The site is a patch of clay-with-flints surrounded by loamy soil; perhaps the specific is OE ***pop(p)**, a short form of *popel* 'a pebble'. Ha 133, Gover 1958.83.

POPHAM'S EAU Norf TF 5300. *Pophams Eau* 1824 OS. Surname *Popham* + ModE **eau** 'a drainage canal'.

POPLAR GLond TQ 3781. 'The poplar-tree'. *Popler* 1327, *(le) Pop(e)ler(e)* 14th cent. ME **popler**. Poplar was a hamlet of Stepney until 1817. Mx 133.

Nether POPPLETON NYorks SE 5655. 'Lower P'. Prefix *Nether* 1538, earlier Latin postfix *Inferior* 1316, + p.n. *Popeltune* [972×92]11th S 1453, *-ton* 1190–1317, *Popleton(e) -tuna -e* 1086–1605, *Poppleton* from 1590, the 'settlement on pebbly soil', OE **popel** + **tūn**. 'Nether' from its low-lying position on the bank of the river Ouse. YW iv.231.

Upper POPPLETON NYorks SE 5554. Prefix *Over* 1409, *Upper or Land* 1822, earlier Latin postfix *Superior* 1316, + p.n. Poppleton as in Nether POPPLETON SE 5655. 'Upper' and 'Land' from its situation away from the river Ouse, above Nether Poppleton. YW iv.231.

PORCHFIELD IoW SZ 4491. 'Land belonging to the Port family'. *Portsfildes* 1559, *Porchfield* 1769. Family name Port + ModE **field(s)**. A Hugh and Thomas de Porta occur in 1232 and 1299 and a John le Port in 1326. Wt 83, Mills 1996.83.

PORINGLAND Norf TG 2701. 'Newly cultivated land of the Porringas'. *Porringa- Porringhelanda* 1086, *Porringeland* [1087×98]12th, *Poringland Maior, ~ Minor*. OE folk-n. **Porringas* of unknown origin, genitive pl. **Porringa*, + **land**. DEPN, L247, Dutt 183 gives pr *Por'land*.

PORKELLIS Corn SW 6933. 'Hidden entrance'. *Porthkelles* 1286, 1337. Co **porth** in the sense 'gateway, entrance' + **kellys** 'lost'. The reference is unknown. PNCo 141, Gover n.d. 539.

PORLOCK Somer SS 8846. 'Harbour enclosure'. *(æt) Port locan* 10th ASC(A) under year 918, *(æt) Por locan* 12th ASC(D) under year 915, *Portloc* 1086, [1252]Buck, *Porlock* 1610. OE **port** + **loca**. The reference may be to the Bronze Age megalithic stone circle just S of the church. DEPN, Pevsner 1958S.276.

PORTBURY Avon ST 5075. 'The harbour fort'. *Portbrig* [899×925]13th S 1707, [979×1016]13th S 1781, *Porberie* 1086, *Portberi -bury* 1159, 1196. OE **port** + **byriġ**, dative sing. of **burh**. The reference is to the port of PORTISHEAD ST 4576 and the small prehistoric fort on Conygar Hill. DEPN.

PORTCHESTER Hants SU 6105. 'Roman fortification called or at *Port*'. *Porceastra* [904]12th S 372, *Porteceaster* [963×75]12th S 816, *Por(t)cestre* 1086, *Porchester* 1412. OE **port** 'a haven, a harbour' probably used as a p.n. + **ċeaster**. The Roman fort survives and was transformed into a medieval castle in the 12th cent. If, as seems likely, it is to be identified with the RBrit *Portus Ardaoni* or *Ardaonium*, *Ardaoneon* [c.700]13th Rav, *Portum Adurni* [c.408]15th ND, the 'port of the height' referring to Ports Down behind and above the site, the RBrit name survives in the specific either as the OE loan-word **port** or as a p.n. **Port*. Ha 133, Gover 1958.22, RBrit 441–2.

The PORTER OR LITTLE DON RIVER SYorks SK 2399. Possibly from the surname *Porter* which occurs in the local name *Porter Field* 1770×9. Another Porter Brook, *Porter watter* 1587, flows into the r. Sheaf in Eccleshall, Sheffield. YW vii.135, i.242.

PORTESHAM Dorset SY 6085. 'Enclosure belonging to the manor' sc. of Abbotsbury. *Porteshamme* 1024 S 961, *-ham* from 1086, *Portsham* 1653. OE **port**, genitive sing. **portes**, + **hamm** 4 'dry ground in marsh', cf. nearby Marsh Farm. Do 122.

PORTGATE Devon SX 4185. 'Gate on the Portway'. *Portcot oth. Portgate* 1774. But the name occurs in *Porteyetepark* 15th, 'Portgate park'. OE **port** from **port-weġ** 'road leading to a town' + **ġeat**. The reference is to the road from Launceston to Okehampton. D 208.

PORTGAVERNE Corn SX 0080. Partly uncertain. *Porcaveran* 1337, *Portkaveran, Porte Kaverne* 1343, *Portkeveren* 1610. Co **porth** 'cove, harbour' + unknown element as in Tregaverne SX 0180, *Tregaveren* 1284, Co **tre** 'estate, farmstead'. Possibly Co **gavern** 'young goat' used of the stream that runs into the sea here; cf. Welsh stream-name Gafran. PNCo 142 gives pr [po:rt'geivərn], Gover n.d. 112, 114.

PORTH NAVAS Corn SW 7527. 'Cove of the sheep'. *Porhanavas* 1649, *Porranavas* 1729. Co **porth** + definite article **an** + **daves** 'sheep'. PNCo 143, Gover n.d. 506.

PORTH RESERVOIR Corn SW 8662. Built in 1962. P.n. Porth SW 8362, *Porthe* [1284]1602, 'cove, harbour', Co **porth** + ModE **reservoir**. PNCo 143, Gover n.d. 327.

PORTHALLOW Corn SW 7923. 'Cove of the river Alaw'. *porð alaw, perd alau* [967]11th S 755, *Porthaleu* 1333, *-alou* 1376, *Porthalow* 1466, *Porthallow* 1529. Co **porth** + r.n. Alaw as in Llyn Alaw, Anglesey SH 3986, *Alaw, Alow* 14th, and the continental r.n. *Alava* which represent an Old European *u*-extension of the widespread IE root **al-* 'to flow'. PNCo 142 gives pr [pralə], Gover n.d. 558, RN 4, BzN 1957.226.

PORTHCOTHAN BAY Corn SW 8572. *Porthcothan Bay* c.1870. P.n. Porthcothan SW 8672 + ModE **bay**. Porthcothan, *Porkehuuson* c.1242, *Porthcuhuthon* 1251, *Pordgohoython* [1295]14th, *Porthkehothon* 1298, *Porthcothyan* 1423, is Co **porth** 'harbour, cove' + unidentified element. PNCo 142, Gover n.d. 352.

PORTHCURNO Corn SW 3822. 'Cove of horns'. *Porth Cornowe* [1580]18th, *Port-Curnoe* c.1605. Co **porth** + **cornou**, pl. of **corn** 'a horn, a corner'. The reference is to the two headlands or horns which enclose the bay. PNCo 142, Gover n.d. 640.

PORTHLEVEN Corn SW 6225. Probably 'harbour of the Leven'. *Porthleven* 1529 Gover n.d., *Port-levan* c.1605. Co **porth** + r.n. **Leven* 'the smooth one', Co **leven**. Cf. the r.n. LEVEN Cumbr SD 3483, *Levena* 12th, NYorks NZ 4906, 6009, *Leven* 1293. PNCo 146, Gover n.d. 494, RN 250–2.

PORTHMEOR Corn SW 4337. 'Great cove or harbour'. *Pordmur* 1304, *Porthmur* 1313, *-meor* 1720. Co **porth** + **meur**. PNCo 142, Gover n.d. 667.

PORTHOLLAND Corn SW 9641. Partly uncertain. *Portallan*

[1288]c.1300, *Porthallan* 1339, 1379, *Porthsalen* 1576, *Porthollon* 1699. Co **porth** 'cove, harbour' + unidentified element, possibly **Alan* as in r.n. ALLEN SX 0678. In recent times the name was anglicised as if Port Holland. PNCo 43 gives pr [pɔ:t'holənd], Gover n.d. 415.

PORTHOUSTOCK Corn SW 8021. Possibly 'nightingale's cove'. *Portheustech* c.1255, *-ek* 1360, *Pordeustek* c.1290, *Porthustok* 1500, *Porthowstock* 1543, *Proustock* 1699. Co **porth** + **eustek** or pers.n. **Eustek*. PNCo 143 gives pr [praustok], Gover n.d. 558.

PORTHPEAN Corn SX 0350. 'Little harbour'. *Porbichan* c.1270 Gover n.d., *Porthbyhan* 1297, *Portbian* 1307, *Por(t)pighan* 1337, 1492. Co **porth** + **byghan**. 'Little' for contrast with Polmear 'great harbour', the Cornish name for CHARLESTOWN SX 0351. PNCo 143, Gover n.d. 386.

PORTHTOWAN Corn SW 6847. 'Towan cove'. *Porthtowan* 1628. Co **porth** + p.n. Towan SW 6948, *Tewyn* 1316, *Towan* 1699, 'sand-dunes', MCo ***tewyn**. The village is 19th cent. PNCo 143, Gover n.d. 607.

PORTINGTON Humbs SE 7830. Either 'farm, village belonging to the town' or 'estate called after Porta'. *Portiton* 1086, *Portinton* 1086–1281, *-ing-* from 1276. OE **port** + or pers.n. **Port(a)* + **ing**[4] + **tūn**. In the former case the reference would be to Howden. YE 247.

PORT ISAAC Corn SW 9980. Partly uncertain. *Portusek* 1337, *Port(h)issek* c.1540, *Porthtyseke* 1576. Co **porth** + unidentified element. PNCo 144 gives local pr [port'izik], Gover n.d. 113.

PORT ISAAC BAY Corn SX 0181. *Port Isaac Bay* c.1870. P.n. PORT ISAAC SW 9980 + ModE **bay**. PNCo 144.

PORTISHEAD Avon ST 4576. 'The harbour headland'. *Porteshe.* 1086, *-heved* 1200, 1225. OE **port**, genitive sing. **portes**, + **hēafod**. The village lies beneath a headland of Portishead Down which formed a shelter for ships. DEPN, L 161.

PORTLAND HARBOUR Dorset SY 6876. Formed in 1861 with the building of Portland Breakwater. The area of water enclosed was previously known as *Portland Road* 1710, 1811 OS. P.n. Portland as in Isle of PORTLAND SY 6971 + ModE **harbour** and **road** 'sheltered water where ships may ride'. Do i.223.

Bill of PORTLAND Dorset SY 6768. ModE **bill**, 'a narrow promontory' + p.n. Portland as in Isle of PORTLAND SY 6971.

Isle of PORTLAND Dorset SY 6971. *insulam de Portland(e)* [978×84]14th S 938. ModE **isle**, Latin **insula**, + p.n. *(on) Portland(e)* [862]14th S 1782, c.1050 ASC(C) under year 982 etc., *-laund(e)* 1244–90, *Portlanda* 1086–1275, 'land or estate attached to Port', p.n. *Port* late 9th ASC(A), 12th ASC(E) both under year 837, OE **port** 'port, harbour', + **land**. Do i.217, L 246, 249.

East PORTLEMOUTH Devon SX 7438. ModE adj. **east** + p.n. *Porlemvte* 1086, *Por(t)lemue* 1242–1268×77, *Porthelemuthe* 1308, *Portelemouth* 1346, *Portlemouth al. Pottlemouth* 1708, unidentified element + OE **mūtha** 'mouth, estuary'. 'East' for distinction from West Portlemouth SX 7039, *Westportelemuth* 1292, 1311, *Portelmuth juxta Malleburgh* 'beside Malborough' 1313. The specific must refer to the Kingsbridge Estuary SX 7441 and probably contains PrCorn PrW **porth** 'a harbour, a gate', perhaps with Co ***heyl** 'estuary'. D 328, 308, DEPN, Jnl i.50.

PORTLOE Corn SW 9339. 'Inlet harbour'. *Portlo* 1253 Gover n.d., *Porthlowe* c.1653, *Por(t)low* c.1690. Co **porth** + ***loch** 'creek, tidal pool'. PNCo 144, Gover n.d. 487.

PORTMELLON Corn SX 0143. 'Mill cove'. *Portmelyn* 1338 Gover n.d., *Porth-* 1359 G, *Porthmelen* 1469. Co **porth** + **melin**. PNCo 144, Gover n.d. 411.

PORTMORE Hants SZ 3397. *Portmore* 1759. Possibly 'moor belonging to the port' sc. of Lymington. Gover 1958.205.

PORT MULGRAVE NYorks NZ 7917. ModE **port** + surname Mulgrave. The harbour was begun in 1856 as an outlet for the Staithes Ironstone Co., later the Mulgrave Ironstone Co., to ship ironstone to the blast-furnaces at Jarrow. The harbour finally closed in 1930 and the breakwater was blown up to prevent possible use by the Germans in the second World War.

PORTON Wilts SU 1936. Partly uncertain. *Po(e)rtone* 1086, *Porton(e)* from 1232, *Pourton(e)* 1160–1316, *Power- Pouerton(e)* 1211–1409. The generic is **tūn** 'settlement, farm, estate', the specific identical with that of POORTON and POWERSTOCK Dorset SY 5197 and SY 5196, which is recorded as *Power* in MS Cotton Faustina Aii, possibly a river name of an unknown pre-English origin. The river at Porton is the Bourne 'the stream', OE **burna**. Wlt 381, Studies 1936.21.

PORTQUIN Corn SW 9780. 'White harbour'. *Porquin* 1201, *Porthguyn* 1297, *-quyn* 1298–1346, *Portwyn* 1303. Co **porth** + **guyn**. PNCo 144, Gover n.d. 113.

PORTQUIN BAY Corn SW 9850. *Portquin Bay* c.1870, cf. *Portguin haven*, *-cove* c.1605. P.n. PORTQUIN SW 9780 + ModE **bay**. PNCo 144.

PORTREATH Corn SW 6545. 'Beach cove or cove at Treath'. *Portreath* [1495]17th, *Porth treyth* [1495]1866, *Portreath* 1582. Co **porth** + **trait** (with OCo *t* = [θ]) or p.n. *Treath* 1699 representing OCo **trait** 'a beach'. PNCo 144, Gover n.d. 607, CPNE 223.

PORTS DOWN Hants SU 6406. 'The hill of *Port*'. *Portesdon(e)* 1086, *Portesdon* 1167, *Portesdune* 1175. P.n. **Port* 'the port' as in PORTCHESTER SU 6105, genitive sing. **Portes*, + OE **dūn**. The reference is to the long ridge of high ground overlooking Portchester and Porstmouth Harbour. Ha 133.

PORTSCATHO Corn SW 8735. 'Harbour of boats'. *Porthskathowe* 1592. Co **porth** + **skathow**, pl. of **skath**, a special kind of large rowing boat. PNCo 145 gives pr [(pə)'skaðə], Gover n.d. 452.

PORTSEA Hants SU 6300. 'The island of *Port*'. *Portesig* [982]14th S 842, *Porteseia* c.1125, 1165, 1218, *Portsea* 1638. P.n. **Port* 'the port' as in PORTCHESTER SU 6105, genitive sing. **Portes*, + **īeġ**. Ha 134, Gover 1958.10, DEPN.

PORTSEA ISLAND Hants SU 6501. *Isle of Portsea* 1638, *Portsea Island* 1810 OS. P.n. PORTSEA SU 6300 + pleonastic ModE **island**. Gover 1958.10.

PORTSLADE-BY-SEA ESusx TQ 2604. 'Port's road'. *Porteslage -lamhe* (sic) 1086, *Portes Ladda* c.1100, *Porteslad(e)* 1179–15th. OE pers.n. *Port*, genitive sing. **Portes**, + (ġe)**lād**. Portslade lies on the line of an ancient highway dating back to Roman times or earlier. Sx 289, BHArch 3(1926).

PORTSMOUTH Hants SU 6501. 'The mouth of *Port*, the harbour'. *Portes muþa* c.890 ASC(A) under year 501, *Portes muða* 12th ibid. (E), *(on) Portesmuðe* 1114, *Portismouth* 1204, *Porch(e)mouth(e)* 16th. P.n. **Port* 'the port' (OE **port**) as in PORTCHESTER SU 6105, genitive sing. **Portes*, + OE **mūtha**. The chronicle account of the arrival in 501 of Port with his two sons and two ships, their seizure of land and defeat of a young British noble, is a typical ætiological foundation myth due to the workings of folk-etymology. The reference is simply to a geographical location (like Bournemouth before 1810) and settlement here seems first to have occurred in the 12th cent. Ha 134, Gover 1958.23, Origins 84–5 and n.12, Pevsner-Lloyd 389.

PORTSMOUTH ARMS STATION Devon SX 6319. A halt on the Barnstaple to Exeter railway line named from the local public house.

PORTSMOUTH HARBOUR Hants SU 6203. *Portsmouth Harbour* 1810 OS. P.n. PORTSMOUTH SU 6501 + ModE **harbour**. Earlier *Portsmouth Haven* 1522, *Porte haven* 1593. Gover 1958.9.

PORT SUNLIGHT Mers SJ 3484. A model village and soap factory of international repute begun in 1888 and named after the brand of soap called Sunlight soap. Che iv.249, Pevsner-Hubbard 1971.304.

PORTSWOOD Hants SU 4214. 'The town wood'. *(on) portes wuda* [1045]12th S 1012, *Porteswuda* 1167, *-wude* 1197. OE **port**, genitive sing. **portes**, + **wudu**. The reference is to Southampton. Ha 134, Gover 1958.45, DEPN.

PORTWAY Hants SU 4048. 'The road leading to the town'. *Portweye* 1298, 1395, *Port Way* 1817 OS. ME **portwey** (OE *portweġ*). The Roman road from Silchester to Old Sarum, Margary no.4b, referred to in charters as *on ðære, andlang stræte* [956]12th S 613, [901]14th S 365, and *to þere ealdan stræt* 'to the old road' [931]12th S 412, OE **strǣt**. Wlt 16, Gover 1958.6.

PORTWAY Warw SP 0872. *Portway* 1831 OS. Ultimately from OE **port-weġ** 'a road leading to a town'.

The PORTWAY Shrops So 4294. *The Port Way* 1833 OS. The name of an ancient track along the top of the Long Mynd. OE **port-weġ** 'a highway, a Roman road', literally a 'road leading to a town'. Cf. Portway H&W SO 3844 on Margary No. 63a, H&W SO 4845 on Margary 6c, Portway Farm Wo 3, Portway D 501, an ancient non-Roman track, and Portway Roman road Wilts SU 2138, Hants SU 3848–6160, Margary no.46, *Portway* 1364. Wlt 16.

PORTWRINKLE Corn SX 3553. Partly uncertain. *Port Wrickel* 1605, *Porth Wrinkle* 1699, *Portwrincle* 1748. Co **porth** + *Wrikel* as in Trewickle SX 3554, *Trewikkel* [c.1190]13th, Co **tre** 'farmstead' + unidentified element. PNCo 145, Gover n.d. 235–6.

POSLINGFORD Suff TL 7748. 'The enclosure of the Poslingas, the people called after Posel'. *Poslingeorda -wrda, PoslindeWRda* 1086, *Poslingeuuorde* c.1095, *Poselingwrtha* 1195, *Poslingforde* 1610. OE folk-n. **Poslingas* < pers.n. **Posel* + **ingas**, genitive pl. **Poslinga*, + **worth**. DEPN, ING 14.

POSSINGWORTH PARK ESusx TQ 5420. P.n. Possingworth + ModE **park**. Possingworth, *Posingeworde* 12th, *-worth* 1281, *Possingwrthe* 1268, is the 'enclosure of the Posingas, the people called after Posa'. OE folk-n. **Posingas* < pers.n. **Posa* + **ingas**, genitive pl. **Posinga*, + **worth**. DEPN cites a form *Poselingewurth* 1238 which seems to show a diminutive form of the pers.n. **Posel* suggested also for in POSTLING Kent TR 1438 and the lost Sussex p.n. Puslingham, *Posselingehamme*. Sx 407, 432, SS 79.

POSTBRIDGE Devon SX 6579. 'Bridge over which the post is carried'. *Post bridg* 1675. Also called *Poststone Bridge* 1720. A bridge on the important post-road across Dartmoor from Exeter to Yelverton. D 196.

POSTCOMBE Oxon SU 7099. 'Valley at a gate–post or postern'. *Postlecumbe* 1246×7, 1278×9, *Postelcombe* 1327×77–1415, *Postercumbe* 1246×7, *Postcombe* from early 18th. ME **postel** 'a door-post, a gate-post', + **cumb**. O 113.

Great POSTLAND Lincs TF 2612. *Great Pursant* 1652, *Great Por(t)sand* 1680, 1824 OS, *Great Parsand* 1789. ModE **great** + p.n. *(island called) le Purceynt* 1415, *le Purceinte* 1416, *Postland* 1775, 'the enclosed land, the purcinct', AN **purceynte(e)**. The word is related to *precinct* and refers to an island of drained land in Deeping Fen belonging to Crowland Abbey. 'Great' for distinction from Little Postland TF 3113, *Little Porsand* 1824 OS, a small detached portion of the purcinct at Whaplode Drove. Payling 1940.15.

POSTLING Kent TR 1438. Possibly 'the pill, the small hill'. *pi-Postinges* 1086, *Pistinges, Postling(e) -yng* c.1100–1316, *-es* 1211×12–1261×2. The most likely explanation of this difficult name is that it is OE **posling** 'a pill, a pastil' < **posl** + **ing**[2] used in a transferred topographical sense. The base of this word is a formation found in other Gmc languages meaning variously 'small bag, pocket, something swollen, a blister' as ON *posi, púss*, OE *posa*, OHG *pfoso*, Mod Norw *pūs*, OSwed *pusin* 'swollen', < Gmc **pusan*. The reference here is to the small round hill S of the village at Postling Lees. PNK 456, ING 14.

POSTWICK Norf TG 2907. 'Poss or Possa's farm'. *Possuic* 1086, *Posswyc, Poswyk* [c.1147]14th, *Possewik* [1175×86]14th, 1254. OE pers.n. **Poss(a)* + **wīċ**. DEPN gives pr [pɔsik].

POTSGROVE Beds SP 0529. 'Pit or pot-hole grove'. *Potesgraue* 1086–1428, *Pottesgrave* 13th–15th cents. OE **pott**, genitive sing. **pottes**, + **grāfa**. Bd 131, L 194.

POTTEN END Herts TL 0108. *Potten End* 1834 OS sc. of Great Berkhamstead. Surname Potten, cf. John *Potyn* 1556, + ModE **end**. Hrt 28.

POTTERHANWORTH Lincs TF 0566. *Potterhaneworth* 1327. Earlier simply *Haneuuorde -worde* 1086, 'Hana's enclosure', OE pers.n. *Hana* + **worth**. 'Potter' for distinction from Old HANWORTH TF 0383. According to Gaz (1896) 'great quantities of Roman pottery were found on the site of the present schoolhouse at the time of its erection'. DEPN, Cameron.

The POTTERIES Staffs SJ 8745. 'The district called "the Potteries" is an extensive tract of country in the hundred of North Pyrehill and County of Stafford, comprehending an area of about eight miles long and six miles broad' 1825 OED. Although this is the first recorded occurrence of the district name, the pottery industry existed there since the 17th cent. and underwent great expansion in the later 18th cent. In 1839 there were at least 150 kilns and it was the chief seat of English pottery manufacture. The main centres were Burslem, Fenton, Hanley, Longton, Stoke-on-Trent and Tunstall. St i.3.

POTTERNE WICK Wilts SU 0057 → Potterne WICK SU 0057.

POTTERNE Wilts ST 9958. 'The house where pots are made, the pottery'. *Poterne* 1086–1291 with variant *-ern(a), Pottern* 1281. OE **pott** + **ærn**. Wlt 244.

POTTERS BAR Herts TL 2501. *Potterys Barre* 1509, *Potters Barre* 1548. Surname Potter, cf. Geoffrey *le Pottere* 1294, William *Pottere* 1327×77, + ME **barre** 'a bar, a gate' referring to one of the forest gates of Enfield Chase. Mx 77.

POTTER'S CROSS Staffs SO 8484. No early forms.

POTTERSPURY Northants SP 7643. 'Potter's Pury'. *Potterispirye* 1287, *Potterespury* 1326, *Potterspury* from 1498. OE occupational term or surname **pottere** 'a potter', genitive sing. **potteres**, + p.n. *Perie* 1086–1324 with variants *Pirie, Pery*, the 'pear-tree', OE **pyriġe**. The village was the site of a pottery from ancient times. Also known as *Estpirie* 1229 for distinction from West or PAULERSPURY SP 7245. Nth 105.

POTTER STREET Essex TL 4608. *Potterestrate* 1382. Surname *Potter* from Maurice le Pottere who held land here before 1382 + ME **strete** 'a hamlet'. Ess 37, YE lvi.

POTT HALL Ches SJ 9479. *le Halle of Pott* 1432, *Pot Hall* 1528, *Pott Hall* 1737. P.n. *Potte* [1270]17th, *Pott* [13th]1611, 1703, + ME **halle**. Pott is OE **potte** 'a deep hole' referring to the dingle or deep clough at the head of which Pott lies. Che i.130.

POTT ROW Norf TF 7021. *Roydon Row* 1824 OS. Originally p.n. ROYDON TF 7022 + ModE **row**. The reason for the change is unknown.

POTT SHRIGLEY Ches SJ 9479. *Shriggele pot* 1348, *Potte Shryggelegh* 1354, *-shrigley* from 1420. P.n. Pott as in POTT HALL SJ 9479 + p.n. Shrigley as in SHRIGLEY PARK SJ 9479. Che i.130. Addenda records 19th cent. local pr ['pɔt'sigli].

POTTO NYorks NZ 4703. Possibly the 'hill near the deep hole'. *Pothow(e)* 1202–1385, *Pottowe* 1285, *Potto* from 1548. ME **potte** + **howe** (ON *haugr*). ME *potte* usually describes 'a deep hole in a river-bed, a deep natural hole, esp. in the mountain limestone'. Since this hardly applies to the valley of Potto Beck, the specific may alternatively be lOE *pott* 'a pot', and so 'pot hill' with reference to its shape, or to the manufacture or even the discovery of ancient pots at this place. YN 176.

POTTON Beds TL 2249. Probably 'pottery (settlement)'. *Pottune* [c.980×90]n.d., [986]n.d. ECTV, [10th]14th S 1807, 975×1016 S 1487, *Potton* from 1203. OE **pott** + **tūn**. There is no obvious topographical feature to which *pott* might have been applied: probably, therefore, the sense is a place where pots are made. Bd 106.

POTTON ISLAND Essex TQ 9591. *Potten Island* 1805 OS, 1930 MM. P.n. Potton + ModE **island**. Potton, *Magna Potting(e)* 1244, 1259×62, *Pottyng(e)* 1419, 1578, *Pottent* 1740, is probably a singular **ing**[2] formation on OE **pott** 'a pot, a deep hole' referring to the fens here, cf. *mariscus de magna Potting* 1259×62. Ess 205, ING 190, MM 16.76.

POUGHILL Corn SS 2207. Uncertain. *Poche(he)lle* 1086, *Poccahetilla* 1086 Exon, *Pochewell* 1227, *Pocwell* 1247, *Pohewille* 1269, *Poghawell* 1284, *Poghewille* 1314, *Poffill* c.1605, *Poffyl* 1610. The specific is either OE **pohha** 'pouch, bag' in a topographical sense, or the pers.n. *Pohha* derived from it, the generic either OE **hyll** 'a hill' or **wylle** 'a spring'. Cf. POUGHILL Devon SS 8508. PNCo 145 gives pr [pofil], Gover n.d. 25, DEPN.

POUGHILL Devon SS 8508. 'Pouch' or 'Pohha's hill'. *Pocheelle, Pochehille* 1086, *Pokehill* 1219, *Poghahille* 1238–1381 with variants *Pog(he)-* and *-hull(e), Powhill* 1602. OE **pohha** or pers.n. *Pohha* + **hyll**. If 'pouch' is right the reference must be to one of the deep valleys either side of the village. But 'pouch' was also used as a nickname. Cf. POUGHILL Corn SS 2207 pronounced [pofəl]. D 415.

POULSHOT Wilts ST 9759. 'Paul's wood'. *Paveshou* (sic) 1086, *Paulesholt* 1187–1274, *Pollesholde* 1585, *Polshed* 1558×1603, *Poulsall al. Polshead* 1675, *Polshott* 1632. Pers.n. *Paul*, genitive sing. *Paules*, + **holt**. The pers.n. *Paul* was early current in the West Country. Cf. POLSHAM Somer ST 5142 and Poleshill Somer ST 0823, *(æt) Pauleshele* [c.1070]12th K 897, *Pouselle* 1086, *Pauleshulle, Pawleshele* 1278×84. Wlt 130 gives pr [poulʃət], L 196.

River POULTER Derby SK 6575. *Paltr'* 1422×71, *Palter* 1589, 1663, *Paulter* 1609, *Poulter* 1775. The forms are late and this is probably a back-formation from unexplained p.n. PALTERTON SK 4768. The earlier name was *Clun* as in CLOWNE Derby SK 4975. Nt 7.

POULTON 'Pool settlement'. OE **pōl, pull** + **tūn**.

(1) ~ Glos SP 1001. *æt Pultune* [855]11th S 206, *Pultun -ton(e)* 1100–1577, *Poulton* from 1221. Gl i.79.

(2) ~ -LE-FYLDE Lancs SD 3439. *Poltun* 1086, *Pulton(a)* 1094–1332 etc. Also known as *Magna* or *Kirkepulton* 'church Poulton' 1330, for distinction from Little Poulton SD 3539, *Parva Pulton* 15th. The Poultons are situated beside SKIPPOOL, the 'ship pool'. *Le Fylde* 'in the FYLDE' for distinction from POULTON-LE-SANDS, the old name of Morecambe. La 157.

(3) ~ -LE-SANDS Lancs SD 4264 (lost). The original name of Morecambe. *Poltune* 1086, *Pulton* 1201–1332 etc., *Poulton* 1226. The reference is unknown. *Le Sands* 'on the sands' for distinction from POULTON-LE-FYLDE. La 176.

POUND BANK H&W SO 7374. ModE **pound** 'a pound, an enclosure into which stray cattle are put' as in *Hortons Pound* 1832 OS + **bank**.

POUNDON Bucks SP 6425. 'Peafowl hill'. *Paundon* 1255, *Pondon* 1291, *Powendone* 1316. OE **pāwa, pāwe**, genitive sing. **pāwan**, + OE **dūn**. This rare specific seems to occur in Peamore Devon SX 9188, *Peumera* 1086, *Paumera -a* 1194, 1250, 'peafowl pond'. Bk 56, D 496, L 150.

POUND HILL WSusx TQ 2937. *Pound Hill* 1816 OS. ModE **pound** + **hill**.

POUNDSTOCK Corn SX 2099. 'Dairy farm with an animal pound'. *Podestot, Pondestoch* 1086, *Pund- Puntestok(e)* 1201–97, *Poundestoke* 1261–1357, *Ponstok vel Poundstok* 1442. OE **pund** + **stoc**. PNCo 145, Gover n.d. 77, DEPN.

POWBURN Northum NU 0616. *Powburn* 1868 OS. Probably the 'stream called Pow' like the three examples of Pow Beck in Cumbria, NX 9712, *Pol* c.1175, NY 2424, *Watter of Powe* 1589, *becke called Powe* 1605, NY 3949, *Powbeck* 1618. At the root of all these is PrCumb **poll** 'a pit, a pool, a mire'. RN 329–30, Cu 24, PNE ii.69, Jnl 1.50.

POWDERHAM Devon SX 9684. 'Promontory at the polder'. *(a)poldraham* c.1100, *Poldreham* 1086, *Pouderham* 1219–1303, *Poudram* 1257×71, 1291. OE ***polra** 'marshy land' + **hamm** 2a. Powderham lies at the tip of a promontory of high ground reaching into marshland: *poldra* probably refers to reclaimed marshland. D 502, PNE ii.69.

POWERSTOCK Dorset SY 5196. Partly uncertain. *Povrestoch* 1086, *Pourstoke* 1195, *Porstok* 1224. Uncertain element + OE **stoc** 'outlying farm, secondary settlement'. The specific may be an old r.n. found also in POORTON SY 5197. Either 'outlying settlement on the Power' or 'outlying settlement belonging to *Power*': a place called *Power* is mentioned in reference to a grant by king Egbert 800×36 in MS BM Cotton Faustina A ii and may be the old name of Poorton. Do 122, Studies 1936.21.

POWICK H&W SO 8351. 'Farm(s) called after Poha'. *(in) poincguuic, (into) poincg pican* 972 S 786, *Poiwic* 1086, *Poywy(c)k -wik* 1212–91 etc., *Powyck* 1535. OE pers.n. **Poha* + **ing**^{4} + **wīċ**, dative pl. **wīcum**. Wo 223.

POXWELL Dorset SY 7484. Uncertain. *Poceswylle* [987]13th S 1217, *Pocheswelle* 1086, *Pokeswel(l)(e)* 1188–1575, *Poxwell* from 1535. Probably 'Poc's spring', OE pers.n. *Poc*, genitive sing. *Poces*, + **wylle**, referring to a spring just S of the village. But the forms are not incompatible with 'Poca's' or 'frog rise', OE pers.n. *Poca* or ***poc(c)e** 'a frog' + **swelle** or **(ġe)swell** 'steeply rising ground' referring to the steep hill ascended by the old Osmington road. Do i.143–4.

POYLE Surrey TQ 0376. *Poyle* 1580. Held by a family called *(de) la Poille* who perhaps originated in Poilly-sur-le-Homme, La Manche, or Poiley, Eure, in Normandy. Sr 182.

POYNINGS WSusx TQ 2612. 'The Puningas, the people called after Pun or Puna'. *Puningas* 960 S 687, *Poninges* 1086–1489 with variant *-ynges, Punninges* 1203–48, *Poinynges* 1369. OE folk-n. **Pūningas* < pers.n. **Pūn(a)* + **ingas**. Apparently identical with Püning in Munster, Germany, *Puningum* 890 etc. Sx 286 gives prs [pʌninz] and [pʌnənz], ING 39, 124, MA 10.23, SS 65, Piroth 105.

POYNTINGTON Dorset ST 6420. 'Estate called after Punt'. *Ponditone* 1086, *Pond- Pundintone -tun(e)* 1091–1254, *Puntintun(a) -ton(')* 1135–1249, *Punt- Pontington* 1250–65, *Poyntington* from 1326. OE pers.n. *Punt* + **ing**^{4} + **tūn**. Do iii.387.

POYNTON Ches SJ 9283. 'Estate called after Pofa'. *Poninton* (sic for *Pou-*) 1248–1349, *-yn(g)-* 1270–1501, *Povinton* 1249, 1292, *Pouindon -ton* c.1280, 1286, *-yng-* 1289, *Poynton* from 1325. OE pers.n. **Pofa* + **ing**^{4} + **tūn**. Che i.207. Addenda gives 19th cent. local pr ['pɛ:əntn, 'pɛintn].

Higher POYNTON Ches SJ 9483. ModE adj. **higher** + p.n. POYNTON SJ 9283.

POYNTON GREEN Shrops SJ 5618. *Poynton Green* 1833 OS. P.n. Poynton SJ 5717 + ModE **green**. Poynton, *Peventone* 1086, *Peuinton(e)* 1221–1272×84, *Peuyngton'* 1271×2, *-yncton* 1276, *-ington* 1376, *Peinton'* 1291×2, *Peynton* 1379–1672, *Paynton* 1585–1751, *Pointon* 1737, *Poynton* 1790, is 'Peofa's farm or village', OE pers.n. *Pēofa*, genitive sing. *Pēofan* + *tūn*, or *Pēofa* + **ing**^{4} + **tūn**. For loss of intervocalic [v] cf. BOYNTON Humbs TA 1368, Loynton Staffs SJ 7724, BzN 16.281. The sound development *ei* > *oi* is a remarkable parallel in English to OFr *ei* > *oi*. Sa i.240.

POYSTREET GREEN Suff TL 9858. *Poy Street or Posford Green* 1837 OS.

PRAA SANDS Corn SW 5828. Uncertain. *Parah or Prah* 1714, *Pra Sand* 1813, *Prah Sands* 1888. The evidence is too late but the name may be identical with lost *Polwragh* 1331, 'hag's pool', Co **pol** + **gruah** 'hag, old woman, witch', with later replacement of *pol* by *porth*. PNCo 145 gives pr [prei], Gover n.d. 494.

PRATT'S BOTTOM GLond TQ 4762. *Sprat(t)s Bottom* 1791, 1799, 1819, *Pratts Bottom* 1779, 1801, 1821. Surname *Pratt* + ModE **bottom** 'a valley bottom'. Prat(t) is a local surname from the 14th cent., cf. nearby Pratt's Grove. Encyc 636, LPN 182.

East PRAWLE Devon SX 7836. ModE adj. **east** + p.n. *Pral(l)'* 1169–1340, *Praule* 1203–1308, *Prahulle* 1204, possibly 'look-out hill', OE ***prā(w)** + **hyll**. The reference is to the prominent headland of Prawle Point SX 7735. 'East' for distinction from

West Prawle SX 7637, *West Prawle* 1300. Earlier *Prenla* (for *Preula* i.e. *Pre-hill*) 1086. D 319, Studies 1931.79–80, PNE ii.72.

PRAWLE POINT Devon SX 7735. *Prawle P^t^* 1809 OS. P.n. Prawle as in East PRAWLE SX 7836 + ModE **point**.

PRAZE-AN-BEEBLE Corn SW 6335. 'Meadow of the conduit'. *Praze-an-beble* 1697, *Praze* 1699. Co **pras** + definite article **an** + ***pybell** 'pipe' with lenition of *p* to *b*. PNCo 146, Gover n.d. 590.

PREDANNACK WOLLAS Corn SW 6616. 'Lower Predannack'. *Predennek Wooles* 1339. P.n. Predannack + **goles** 'lower' for distinction from Predannack Wartha SW 668167 'Upper Predannack', *Predannekwartha* 1289 Gover n.d., Co **guartha**. Predannack, *Bridanoc* 1196, *Bro- Brethanec* 1201 Gover n.d., *Predannek* 1284, 1302, *Pradnek* c.1740, is probably '(headland) of Britain', p.n. *Predenn* 'Britain' (< *Britannia*) + adjectival suffix **-ek** meaning 'place of', no doubt referring originally to the more important Lizard Point SW 6911 rather than Predannack Head SW 6516, which is probably a coinage from the p.ns. Predannack Wollas and Wartha. PNCo 146 gives pr [pradnik], Gover n.d. 577.

Church PREEN Shrops SO 5498. 'Preen with the church'. *Chircheprene* 1255×6. ME **chirche** + p.n. *Prene* 1086–1557 with variant *Prena, Prune* 1208–1462 with variant *Pruna, Preone* c.1291, 1334, *Preene* 1316, *Preen* 1552, OE **prēon** 'a pin, a brooch', probably used as the name of the ridge on whose slope the two Preens are situated. Also known as *Pruna Magna* 1255 'great Preen'. 'Church' for distinction from Holt Preen, now simply Holt SO 5396, *Parva Prona* 1255 'little Preen', *Holtprene* 1255×6, OE **holt** 'a wood'. Sa i. 240, Studies 1936.195.

PREES Shrops SJ 5533. 'The brushwood, the copse, the grove'. *Pres* 1086–1327, *Prees* from 1316, *Preese, Preece* 1608–1766. PrW ***prês** (W *prys*). An English adaptation from W in the 7th cent. after W monosyllable vowel lenghtening. Sa i.241, LHEB 343.

PREESE Lancs SD 3736 → PREES Shrops ST 5533.

PREESE HALL Lancs SD 3736 → PREESSALL SD 3647.

PREES GREEN Shrops SJ 5631. P.n. PREES SJ 5533 + ModE **green**.

PREESGWEENE Shrops SJ 2936. 'Gwen's copse or grove'. *Prys Gwen* 1602, *Prysgwaen* 1837. W **prys** + pers.n. *Gwen*. Morgan 1997.

PREES HIGHER HEATH Shrops SJ 5636. Cf. Prees Heath SJ 5537, *Prees Heath* 1833 OS and Prees Lower Heath SJ 5732, *Prees Lower Heath* 1833 OS. P.n. PREES SJ 5533 + ModE **heath**. Prees Heath and Prees Higher Heath are modern developments which grew up to service travellers and long distance lorry drivers on the busy A49 and A41.

PREESSALL Lancs SD 3647. There are three forms of this name:-

I. *Pressouede* 1086, *Preshoued* [c.1190]14th, *-(h)oueth -hout -hefd* 13th.

II. *Pressoure* 1094, c.1190, *-houere -o(u)ra* 12th, *Presoure* 1202.

III. *Presho(u)* c.1190–1332 etc., *Preeshowe -hou* 14th, *Priso* 1590.

This is PrW ***prês** 'brushwood, copse' taken as a p.n. + (1) ON **hǫfuth** 'head, headland', (2) OE **ofer** 'tip of a ridge, promontory, flat-topped ridge', (3) ON **haugr** 'mound, hill', and so 'Preese headland, ridge and hill'. Preessall is situated on a low ridge near the mouth of the river Wyre. The name was subsequently remodelled as if a compound in OE **halh**. PrW **prês* occurs again in the simplex name Preese Hall Lancs SD 3736, *Pres* 1086, *Prees* [c.1200]1268–1332 etc., *Prese* [c.1200]1268, and PREES Shrops SJ 5533. La 159, 153, LHEB 343, CPNE s.v. ***prys**, ***prysk**, L 161–2, Jnl 17.92, 89, Sa i.241.

PRENDWICK Northum NU 0012. 'Prænda's farm'. *Prendewic, Prendwyc* 1242 BF, *Pride- Prenderwyk, Prandewick* 1256, *Prendewyk* [1275]copy, *Prandewich* 1279, *Prendwyke* 1428, *Prendyke* 1542. OE pers.n. **Prænda* + **wīċ**. NbDu 160, DEPN.

PRENTON Mers SJ 3086. 'Pren's farm or village'. *Prestune* 1086, *-tona* [1096×1101]1280, 1150, *Prenton* from 1260, *Printon* 17th. OE pers.n. *Præn* + **tūn**. The earliest spellings probably represent genitive sing. **Prenes* + **tūn**. Che iv.272.

PRESCOT Mers SJ 4792. The 'priests' cottage(s)'. *Prestecota* 1178, *-e* 1189×6, 1254, 1329, *Prestcot(e)* 1246, 1341, *Prescote* 1462×3, *-cot(t)* 1562. OE **prēost**, genitive pl. **prēosta**, + **cot**, pl. **cotu**. This was the rector's manor of Whiston SJ 4791. La 108, Jnl 17.61.

PRESCOTT Shrops SJ 4221. 'The priests' cottages'. *Prestecote* 1250–c.1280. OE **prēost**, genitive pl. **prēosta**, + **cot**, pl. **cotu** (ME *cote*). Bowcock 189.

PRESHAW HOUSE Hants SU 5723. A 17th cent. house with later additions. P.n. Preshaw + ModE **house**. Preshaw, *Presagh'* 1207, *Pre(i)shaghe* 1236, 1353, *Pre(i)shawe* 1266, 1388, *Presthawe* 1272, 1360, *Presher* 1759, is of uncertain origin, possibly PrW ***prês** 'brushwood' used as a p.n. + OE **sceaga** 'copse, scrub'. The name has been influenced by OE *prēost* 'a priest' and perhaps also by ME *pre(y)* 'a meadow'. Gover 1958.70.

PRESSEN Northum NT 8335. 'Priest fen'. *Prestfen* 1176–1255, *Pressefen* 1278, *Presfen* 1296, 1300, *Pressen* from 1251. OE **prēost** + **fenn**. NbDu 160.

PRESTBURY Ches SJ 9077. 'Priests' manor'. *Prest(e)bur'* c.1170×82, *-bury* 1215–c.1540, *Prest(e)bire* [late 12th]1287, *Presbury* 1428. OE **prēost**, genitive pl. **prēosta**, + **byriġ**, dative sing. of **burh**. Che i.212 reports that people in nearby Bollington sometimes refer to Prestbury as *Mansionville* in allusion to the more pretentious style of the place. Ch ii.vii gives pr ['pres(t)bəri], 19th cent. local ['presbəri -buri -beri].

PRESTBURY Glos SO 9723. 'The priests' fortified place or manor'. *into Preosdabyrig* [899×904]16th S 1283, *Presteberie -bir(e) -byr' -bur(y)* 1086–1557, *Prestbiri* 1189×99, *-bury(e)* from 1305. OE **preost**, genitive pl. **preosta**, + **byriġ**, dative sing. of **burh**. The reference is to the priests of Cheltenham minster whose lands were leased to the Bishop of Hereford in the 8th cent., to whom it belonged by 1066. Gl ii.110.

PRESTHOPE Shrops SO 5897. 'The priests' valley'. *Presthope* 1167, *Prestehop* 1221, *Presthop* 1245, 1534×5. OE **prēost**, genitive pl. **prēosta**, + **hop**. Bowcock 190, DEPN.

PRESTLEIGH Somer ST 6340. 'Priest wood or clearing'. *Preostle* *[1065]18th S 1042, *Presteleg -leye* [13th]Buck, *Priestleigh* 1807 OS. OE **prēost**, genitive pl. **prēosta**, + **lēah**.

PRESTON 'The priest(s') estate'. OE **prēost**, genitive pl. **prēosta**, + **tūn**. PNE ii.73. See also PURSTON JAGLIN WYorks SE 4319.

(1) ~ Devon SX 8574. *Prustaton* 1335, *Preston* 1356, *Pruston* 1620. D 480.

(2) ~ Dorset SY 7083. *Prestun* 1228, *Preston(e)* from 1285. This estate was a prebend of Salisbury cathedral. Do i.231.

(3) ~ ESusx TQ 3107. *Prestetone* 1086. Aso known as *Bisshopespreston* 1319, *Preston Ep'i* 1340, 'the bishop's P'. It was held in 1086 by the bishop of Chichester. Sx 295, SS 75.

(4) ~ Glos SP 0400. *Prestitvne* 1086, *Preston* from 1221. Gl i.80.

(5) ~ Glos SO 6834. *Prestetvne* 1086, *Preston(e)* from 12th. Gl iii.184.

(6) ~ Herts TL 1724. *Prestun(e)* 1185, 1416, *Preston(e)* from 1225. Hrt 22.

(7) ~ Humbs TA 1830. *Prestun(e) -tone* 1086, *-tun(a) -ton(a)* 1115–1828, *Pri- Pryston* 1316, 1570, *Purston* 1574. YE 40.

(8) ~ Kent TR 2561. *Prestetvn(e)* 1086, c.1100, *Prestune* [1087]13th, 1154×89, *Preston(')* from 1200. The abbot and convent of St Augustine, Canterbury, held the manor in 1086. PNK 518.

(9) ~ Lancs SD 5429. *Prestvne* 1086, *Preston(a)* from 1094, *Presteton* 1180ff. La 146, Jnl 17.83.

(10) ~ Leic SK 8702. *Prestetona* 1130, *Preston(e)* from 1208. Lei 680.

(11) ~ Northum NU 1825. *Preston* from 1242 BF. NbDu 160.

(12) ~ Wilts SU 0378. *Preston* 1289. Wlt 271.

(13) ~ BAGOT Warw SP 1765. 'P held by the Bagot family'. *Preston Bagot* 1263. P.n. *Prestetone* 1086, *Preston* from c.1185 + manorial addition from the family of Ingeram Bagot who held the estate during the reign of Henry II (1154×89). Wa 217.

(14) ~ BISSETT Bucks SP 6529. 'P held by the Biset family'. *Preston Byset* 1327. Earlier simply *Prestone* 1086, *Prestinton* 1162. The manor was held in 1167 by Manasser Biset, steward to Henry II. Bk 63.

(15) ~ BROCKHURST Shrops SJ 5324. 'P by Brockhurst'. *Preston' subtus Brokhurst* 1291×2, *Preston Brokhurst* 1385, 1494, ~ *Brockhurst* from 1669. P.n. *Preston(e)* 1086 + p.n. Brockhurst as in LEE BROCKHURST SJ 5427 for distinction from other Prestons. Also known as *Preston T(h)oret*, ~ *Turet* 1255 and ~ *Corbet* 1291 from manorial owners *Toret* and the Corbet family. Cf. MORETON CORBET SJ 5623 in which parish Preston lies. Sa i.242.

(16) ~ BROOK Ches SJ 5680. *Preston-on-the-Hill, usually called Preston Brook* 1860. Preston Brook, so named first in 1819, either 'the brook of Preston' or 'the part of Preston at the brook' is the more important part of the township of Preston on the Hill SJ 5780, *Prestun* [12th]17th, *Preston(a)* from 1157×94, *Preston super le Hill* 1373. Che ii.156.

(17) ~ CANDOVER Hants SU 6041. *Preston Candovere* 1255. P.n. Preston + p.n. *Candovre -devre* 1086. Also known as *Prestecandevere* 'priests's Candover' 13th cent., OE **prēost**, genitive pl. **prēosta**, + p.n. *Cendefer* *[701]12th S 242, [873×88]14th S 1507, [956]12th S 589, *Kendefer* [900]11th S 360, 'beautiful waters', PrW ***cen** (Brit **kanịo-*) + **dïβr**. The priests' Candover is only one of three estates of the same name, the others being Brown Candover SU 5739, *Brune Candevere* 1256, earlier simply *Candevre* 1086, held by the Brune family in the 15th and 16th cents., and Chilton Candover SU 5940, *Chilton Candevere* 1286, earlier simply *Candevre* 1086, *Chiltecandevere* 13th, 'the Candover at *Chilte*', from Brit ***cilta** 'a steep slope'. Priests' Candover is already held by clerks (*clerici*) in 1086. The name Candover may be compared with the forms *Cendefrion* [967]11th S 755 and *cendeurion* [977]11th S 832 for the stream at Condurrow in St Keverne parish, Cornwall. Ha 48, Gover 1958.84–5, RN 68, LHEB 285, 602, CPNE 49.

(18) ~ CAPES Northants SP 5754. 'P held by the Capes family'. *Preston Capes* from 1300. Earlier simply *Prestetone* 1086, *Preston(a)* 1174. Also known as *Preston othe Hull* 'on the hill' 1421 and *Magna Preston* 'great Preston' 1428. The manor was held by Hugo son of Nicholas de Capes in 1234. 'Great' for distinction from Little Preston SP 5854, *Parva Preston* 1220, *Little Preston* 1557. The village stands on a hill-top. Nth 28.

(19) ~ GUBBALS Shrops SJ 4919. 'Godebald's P'. *Preston Gubald* 1255, ~ *Gobaldes* 1292×5, ~ *Gubbales* 1654, ~ *Gubballs* 1694, *Preeston Gubbals* 1654, *Presson Gobolls* 1751. P.n. *Prestone* 1086 + manorial addition from *Godebold* 1086, a priest who was a tenant of St Alhmund's church in Shrewsbury. Sa i.243.

(20) ~ HILL Northum NT 9223. *Preston Hill* 1869 OS. A remote peak in the Cheviots whose name is unexplained.

(21) ~ ON STOUR Warw SP 2050. *Preston super Steuram, Sturham* 1291, ~ *upon Sture* 1504. P.n. *Preston* 1086 + r.n. STOUR. Gl i.253.

(22) ~ -ON-THE-HILL Ches SJ 5780 → PRESTON BROOK SJ 5680.

(23) ~ ON WYE H&W SO 3842. *Prestone super Weye* 1221. Earlier simply *Prestretvne* (sic) 1086, *Prestetune -ton'* c.1175, 1176, *Preston(a) -(e)* 1252, 1375. Also known as *Prestone Superior* 'Upper P' for distinction from PRESTON WYNNE SO 5647. He 169.

(24) ~ ST MARY Suff TL 9450. 'St Mary's P'. P.n. *Preston* [1052×66]13th S 1519, 1610, *Prestetona, p'stetune* 1086, + saint's name *Mary* from the dedication of the church. DEPN.

(25) ~ -UNDER-SCAR NYorks SE 0691. *Preston undescar* 1568. P.n. Preston + ModE **under** + dial. **scar** 'a bare place on a hillside, a precipice, a cliff' (ON *sker* 'a rock'). Earlier simply *Prestun -ton* 1086 etc. YN 257.

(26) ~ UPON THE WEALD MOORS Shrops SJ 6815. *Preston' in Wyldemor* 1261×2, ~ *super Le Wyldemor* 1305×6, *Preston on the Wild-Moores* 1709. P.n. *Prestune* 1086 + ModE *Weald Moors* as in EYTON UPON THE WEALD MOORS SJ 6514. Sa i.243.

(27) ~ WYNNE H&W SO 5647. 'P held by the Wynne family'. *Preston Wynn* 1832 OS. Earlier simply *Prestetvne* 1086, *Preston(e)* 1210–1334. The manor was held by Dionisia la Wyne in 1303. Also known as *Preston Inferior* 'lower P' for distinction from PRESTON ON WYE SO 3842. He 170.

(28) East ~ WSusx TQ 0602. *Est Preston* 1327. ME adj. **est** + p.n. *Prestetune* 1086, *Preston(a)* from 1179. 'East' for distinction from West Preston in Rustington, TQ 0401, *West Prestone* c.1230. Also known as *Preston Milers* 1288 from the family of Humfridus de Millieres who held the manor in 1179. Sx 172–3.

(29) Long ~ NYorks SD 8358. *Long(e) Preston* from 1537. ModE **long** + p.n. *Prestune -tuna -ton(a)* 1086–1785. Also known as Preston in Craven, *in Cravana -en* 1233–1480. YW vi.160.

(30) White ~ Cumbr NY 5977. *two hills called the Black and White Prestons* 1794. White Preston rises to 1384ft., Black Preston a little to the SW to 1152ft. Cu 64.

PRESTWICH GMan SD 8103. 'Priest farm or village'. *Prestwich* from 1194 with variants *-(e)wic -wi(t)che -wych(e) -wyk -wik(e)*. OE **prēost** + **wīċ**. La 49, Jnl 17.39, 21.45.

PRESTWICK Northum NZ 1872. 'The priest or priests' farm'. *Prestwic* 1242 BF, *Prestewike* 1296 SR, *Prestick* 1428. OE **prēost**, genitive pl. **prēosta**, + **wīċ**. The priest's farm within Ponteland parish. NbDu 160 gives pr [prestik].

PRESTWOOD Bucks SP 8700. 'The priest wood'. *Prestwude* [1184×91]14th, *Prestwood* 1535. OE **prēost** + **wudu**. The foundation charter of Missenden abbey (1133) to which this place belonged grants the abbey one virgate of the land of Arnulf the priest and two bits of wood later exchanged for other land including land formerly belonging to Ralph the priest. The name probably commemorates the original donor Arnulf. Bk 156, Jnl 2.28.

PRICKWILLOW Cambs TL 5982. *Pri(c)kewylev* 1251, 1277, *Prikewiluh* 1279, *Prickwillowe* 1570. ME **prick-willow**. A prick-willow is a tree from which 'pricks' or wooden skewers and the like were made. Ca 222.

PRIDDY Somer ST 5251. 'Earth houses'. *prediau* [9th]1395 Charters, *Pridi* c.1180, *Pridia -ie* c.1185, 1219, *Prydie* [13th]Buck, *Priddy* 1610. PrW ***prith** + ***tïȝ**, MW pl. **-tyeu**. DEPN, Jnl 1.50–1, Charters III.81.

PRIMETHORPE Leic SP 5293. 'Prim's Thorpe'. *Prymesthorp* 1316, *Pry- Primethorp(e)* from 1575. OE pers.n. *Prim* + p.n. *Torp'* 1086, 1203, *Thorp(e)* 1260–1610. Lei 433, SSNEM 119 no.31.

PRIMROSE GREEN Norf TG 0616. *Primrose Green* 1838 OS. Contrasts with nearby *Colends Green, Stephens Green* and *Paseling Green* all 1838 OS.

PRIMROSE HILL Cambs TL 3889. No early forms.

PRINCE'S PARK Mers SJ 3688. A public park laid out for Richard Vaughan Yates in 1842 on the pattern of Regent's Park, London, to which the name alludes. Pevsner 1969.233.

PRINCETHORPE Warw SP 4070. 'Præn's outlying farm'. *Prandestorpe* 1221, *Prenestorp -thorpe* 1232–1332, *Prensthorp* 1289–1393, *Prinsthorp* 1300, *Prince-* 1461, *Prinstrop* 1600, OE pers.n. *Præn, Pren, Prend*, genitive sing. *Pren(d)es*, + **thorp**. An outlying hamlet of Stretton on Dunsmore. Wa 141.

PRINCETOWN Devon SX 5873. A modern settlement which grew up around the prison originally built in 1808 for French and American prisoners of war and named after the Prince Regent. D 196.

PRINKNASH PARK Glos SO 8813. *Prynkenasshe Parke* 1542,

Prinknash Park 1798. A country house dating from c.1300 to 1898. P.n. Prinknash + ModE **park**. Prinknash, *Pri- Prynkenesse* 1121, *-esche* after 1412, *-asshe* 1544, is 'Princa's ash-tree', OE pers.n. **Princa*, genitive sing. **Princan*, + **æsc**. The same pers.n. may occur in Prinkham Farm, Lingfield, Surrey, *Prinkeham* 1207–1362. Gl ii.170 gives prs ['priŋknæʃ] and ['priniʃ], Sr 330.

PRIOR PARK Cumbr SD 1490. *Prior Park* 1840 OS.

PRIORS FROME H&W SO 5739 → River FROME SO 6544.

The PRIORY IoW SZ 6390. A house dating from c.1790, now a Dominican priory. The site of the 12th cent. alien Benedictine priory of St Mary, founded c.1156 as a cell of the abbey of Lyre in Normandy. A certain Emmanuel Badd, 'a verie poor man's sonn ... by God's blessing and ye losse of 5 wyfes ... grew very rich, purchasing ye priory, and mutch other landes in ye Island' (*The Oglander Memoirs*, ed. W. H. Long, London 1888.87). Wt 200, Mills 1996.83, Pevsner-Lloyd 749.

PRIORY WOOD H&W SO 2645. *Priory Wood* 1833 OS. So named from the Cluniac priory founded on the site of Priory Farm in 1129×30. Pevsner 1963.102.

PRISKE Corn SW 6720 → PRISTON Avon ST 6960.

PRISTON Avon ST 6960. 'Settlement or estate called or at *Prisc*'. *Pris(c)tun* *[931]12th S 414, *Prisctone* 1086, *Prisctun* before 1087, *Pristona -e* 12th, *Prisshtone* 1327. This is probably PrW ***prisc** 'brushwood' taken as a p.n. + OE **tūn**. Cf. Priske Corn SW 6720 and related ***pris** in PREES Shrops ST 5533, Preese Lancs SD 3736, *Pres* 1086, and Preeze Corn. DEPN, PNE ii.73, DB Somer 384, 368, CPNE 194, GPC 2925 s.v. *prysg, prys*.

PRITTLEWELL Essex TQ 8787. 'Prattling spring'. *Pritteuuellā* 1086, *Prit(t)e- Prytewell(e)* 1191–1305, *Pri- Pryt(e)l(e)well(e)* 1165–1309, *Peperell* 1544. OE ***pritol**, definite form ***pritola**, + **wella**. The spring at Prittlewell Priory was celebrated as 'the strongest in the hundred' (Benton, *History of Rochford Hundred*, 1878–8, ii.446). This was the original name of the parish later known as Southend-on-Sea. OE **pritol* is an ideophonic variant of *prattle*. Ess 190, YE lvii, PNE ii.73.

PRIVETT Hants SU 6726. 'Privet copse'. *Prevet* 1207–1364, *Pruuet* c.1245, *Privet* from 1248. The earliest reference is in ASC to a place where Sigebriht, the deposed king of Wessex, was killed in 755 by a herdsman *æt Pryfetes flodan* 9th (A), *æt Pryftes flodan* 12th (E), 'at the water channel or flooding place of P'. OE ***pryfet**. Ha 135, Gover 1958.74.

PRIXFORD Devon SS 5436. Possibly 'Piroc's enclosure'. *Pirkewurth* 1238, *Pirkesworthi* 1330, 1333, *Prikesworthy or Pyrkesworthy* 1499, *Prexford* 1809 OS. PrW pers.n. **Piroc*, secondary genitive sing. **Pir(e)ces*, + **worthiġ**. D 51.

PROBUS Corn SW 8947. 'St Probus's (church)'. *Canonici S' Probi ten' Lanbrebois -bra-* 'the canons of Probus hold Probus' 1086, *(ecclesia) Sancti Probi* [1123]1379, 1269, 1361, *Lanbrobes* 1302, *Seynt Probus* 1466, *Lambrobus* c.1500, 1621, *Lamprobus* 1759. Saint's name *Propus* (Probus) from the dedication of the church. Nothing is known of him. The name is sometimes preceded by Co ***lann** 'church-site'. PNCo 146, Gover n.d. 475, Do iii.357–8, CMCS 12.51, DEPN.

PROSPECT Cumbr NY 1140. *Prospect* 1869 OS. Occupies an elevated position with views of Allonby Bay.

PRUDHOE Northum NZ 0962. 'Pruda's hill-spur'. *Prudho* 1173, *Prudehou* 1212 BF 201, *Pr(o)ud(e)hou -houe* 1217–1471, *Prodow* 1296 SR, *Pruddo* 1610 Map, *Priddowe* 1642. OE pers.n. *Prūda* + **hōh**. The name was probably thought of as 'proud height' after the great Umfraville castle was built c.1161, ME **proud** (OFr *proutz*). NbDu 161 gives pr [prudə], L 168, Pevsner-Grundy etc. 1992.545.

PUCKERIDGE Herts TL 3823. 'Goblin ridge'. *Pucherugge* 1294, *Pokerigge* 1310, *Puckeridge* 1680. OE **pūca** + **hrycg**. Hrt 198.

PUCKINGTON Somer ST 3718. 'Estate called after Puca'. *Pochintvne* 1086, *Pokintuna* 1086 Exon, *Pukinton* 1201, 1244, *Pokinton* 1610. OE pers.n. **Pūca* + **ing**[4] + **tūn**. DEPN.

PUCKLECHURCH Avon ST 7076. 'Pucela's church'. *(æt) Puclancyrcan* c.1016 ASC(D) under year 946, *Pucelancyrcan* [950]19th S 553, *Pukelescircean* [939×46]13th S 1724, *Pukelescirce* [979×1016]13th S 1777, *Pvlcrecerce* 1086, *Poc- Pok(e)lechirch(e)* 1167–c.1560, *Pukeleschurch -cherche* 1168–1324, *Pekelecherche -chirch(a)* 1185–1259, *Pucklechurch(e)* from 1564. OE pers.n. **Pūcela*, genitive sing. **Pūcelan*, + **ċiriċe**. Gl iii.64.

PUCKPOOL POINT IoW SZ 6192. *Puckpool or Popful Pt* 1829. P.n. Puckpool + ModE **point**. Puckpool, *Chockepon* (sic) 1086, *Cukepole* 1255, 14th, *Cokepo(u)le* 1255–1431, *Poupoll* 1611, *Puck Pool* 1769, is of uncertain origin, either 'Cuca's pool, creek or stream', OE pers.n. **Cuca* + **pōl**, or the 'lively stream', OE **cucu** (*cwicu*) or r.n. *Cuce* 'the lively one' + **pōl**. The modern form seems to be due to popular association with ModE **puck**, **pook** 'a goblin'. Wt 194, Mills 1996.84.

PUDDINGTON Ches SJ 3373. 'Estate called after Put(t)a'. *Potitone* 1086, *Potinton(a) -yn-* [12th]1311–1579, *Pudington* c.1100–1660, *Puddyngton* 1353, *-ing-* from 1388, *Poddin(g)ton -yn(g)-* 1260–1601. OE pers.n. *Put(t)a* + **ing**[4] + **tūn**. Che iv.214.

PUDDINGTON Devon SS 8310. 'Estate called after Put(t)a'. *Potitone* 1086, *Potin(g)ton* 1219–1334, *Puti(n)ton* 1238, 1242, *Pod(d)yngton* 1281, 1310. OE pers.n. *Put(t)a* + **ing**[4] + **tūn**. D 389.

Turners PUDDLE Dorset SY 8393. *Turnerepidel* 1242, *Toners(s)pydele -pudel(l)(e)* 1268–1458, *Tournerspedyll* 1428, *Turnerspudell* 1535, earlier simply *Pidele* 1086, *Pudele* 1303–84. Also known as *Py- Pidel(e) Tunere, Tonere* 1264–1316, *Pudel Toners* 1361. Surname *Tonere* + r.n. PIDDLE SY 8591. The manor was held by Walter Tonitruus in 1086 and by Henry Tonere in 1280 etc. Do i.294.

PUDDLEDOCK Norf TM 0592. *Puddle Dock* 1838 OS. The place is situated in Old Buckenham fen. A small inlet on the N bank of the Thames near Blackfriars was also called *Puddle Dock* 1633–1720, according to Stow 'after one Puddle that kept a wharf on the west side thereof'. OED s.v. puddle sb.6.b, EncycLond 643.

PUDDLETOWN Dorset SY 7594. 'Settlement on the r. Piddle'. *Pi(t)retone* 1086, *Pi- Pydel(e)ton(e)* 1212–1379, 1495, *-towne* 1539, *Pudel(e)to(u)n* 1280–1539. R.n. PIDDLE + OE **tūn**. In 1956 there was controversy over the proper form of the name of the village. Dorset County Council wanted to change it to Piddletown to conform with the other villages of the Piddle valley, but local protest against the change won the day, mainly on the grounds of the expense involved, but also because Puddletown sounded 'nicer'. Do i.314 gives prs ['pʌdltaun] and ['pidltaun].

PUDLESTON H&W SO 5659. Probably 'Hawk's hill'. *Pillesdvne* 1086, *Putlesdun(a) -don(e)* 1123–1291, *Pud(e)lesdon'* 1222–1334, *Pudleston* 1287. OE **pyttel** 'a hawk, a mousehawk', genitive sing. **pytteles**, + **dūn**. But the specific might be the pers.n. *Pyttel* derived from *pyttel* or possibly even OE **pidele* 'a fen, a marshland' often used of streams. Puddleston overlooks the upper reach of Stretford Brook. He 170, L 144, Studies 1936.91.

PUDSEY WYorks SE 2233. 'Pudoc's island'. *Podechesai(e)* 1086, *Pudekessey(a)* 12th, *-ssai* 13th, *Pudekesei(a) -aie -aye -ey* 12th–13th cent., *-heye -hee -hay* 13th, *Pudsey* from 1394. OE pers.n. **Pudoc*, genitive sing. **Pudoces*, + **ēġ**. If the generic is correctly identified as **ēġ** 'island', it must be used here and in some other northern names of places on high ground in the sense 'island of good land in moorland' or 'hill-spur'. For this reason alternative suggestions have been sought for the specific, e.g. OE **(ġe)hæġ** 'enclosure' or **hēah** 'high ground, height'. YW iii.236 gives pr ['pudsə], L 35.

PUESDOWN Glos SP 0717 → PULESTON Shrops SJ 7322.

PULBOROUGH WSusx TQ 0518. Either 'the hill of the pools' or 'Pulla's hill'. *Poleberge* 1086, *-berga -beregh* 1166, 1285, *Pul(l)eberg(a) -ber(h)e* 1167–1279, *Pulberg(e)* 1247–1510, *-berwe* 1304, *-burgh* 1309, 1496, *-borough* 1479. Either OE **pōl**, genitive pl. **pōla** referring to pools or deep places in the r. Arun, or OE pers.n. **Pulla*, + **beorg**. Sx 152 gives pr [pulbər].

PULESTON Shrops SJ 7322. Probably 'Peofel's hill'. *Plivesdone* (sic) 1086, *Pi- Pyvelesdon(') -dun' -u-* 1200–1334, *Piulesdone* c.1271, *Pulesdune* c.1275, *-don* 1436, *Pulston* 1586, *Pilston* 1587–1605, *Puleston* 1601, *Peilsone* 1639. Probably OE pers.n. **Peofel*, genitive sing. **Peof(e)les*, + **dūn**. Cf. Puesdown Glos SP 0717, *Peulesdon'* 1236, with the same pers.n., either a diminutive of OE *Peuf(a)* or a form **Pefel* of *Pefi* in PEWSEY Wilts SU 1660. Sa i.244, Gl i.174.

PULFORD Ches SJ 3758. 'Ford of the Pull'. *Pulford* from 1086. OE ***pull** 'a pool, a creek' taken as a river name, + **ford**. The ford on Pulford Brook was an important passage on the road from Chester to Wrexham. In Wales it was regarded as one of the limits of the kingdom of Powys; *The Dream of Rhonabwy* c.1200 begins with the statement that Madawg son of Maredudd ruled Powys from one end to the other, that is, from Porffordd (i.e. Pulford) as far as Gwavan in the highlands of Arwystli (Pen Pumlumon-Arwystli SO 8188). Che iv.155, L 28, *The Mabinogion* tr. Jeffrey Gantz, Harmondsworth 1976.178.

PULHAM Dorset ST 7008. 'Homestead or enclosure by the pools'. *Poleham* 1086, 1279, *Puleham* 1130–1291, 1428, *Pullam* 1212–44, 1440–66, *Pulham* from 1237. OE **pōl**, genitive pl. **pōla**, + **hām** or **hamm** 2a or 4, 'promontory or dry ground in marshland'. Do iii.264.

PULHAM MARKET Norf TM 1986. P.n. *Polleham* [1042×66]12th S 1051, *Pul(la)ham* 1086, *Pulham* 1251, 'the homestead or meadow of or by the pools', OE **pōl**, **pull**, genitive pl. **pōla**, **pulla**, + **hām** or **hamm**, + ModE **market**. Pulham Market was earlier *Pulham St Mary Magdalen* 1837 OS for distinction from PULHAM ST MARY TM 2185. The church at Pulham Market is dedicated to St Mary Magdalen. According to Gaz 1897, Pulham St Mary Magdalen was commonly called Pulham Market from the ancient market subsequently removed to Harleston. DEPN.

PULHAM ST MARY Norf TM 2185. *Pulham St Mary the Virgin* 1837 OS. P.n. Pulham as in PULHAM MARKET Norf TM 1986 + saint's name *Mary* from the dedication of the church.

PULLOXHILL Beds TL 0634. 'Pulloc's hill'. *Polochessele* 1086, *Pollokeshill* 1205, *Pullokeshull* 13th–14th cents., *-hill* 1287. OE pers.n. **Pulloc*, genitive sing. **Pulloces*, + **hyll**. For the pers.n. cf. Poulston Devon SX 7754, *Polochestona* 1086. Bd 160, L 170.

PULSTON Dorset SY 6695 → FORSTON SY 6695.

PULVERBATCH Shrops SJ 4202. Partly uncertain. *Polrebec* 1086, *-beche -bach(e)* 1242–1346, *Pulrebech(e) -bache* 1180×9–1361, *Puluerbach'* 1261×2, *Pulverbech(e) -bach(e)* 1284–1561, *Pulverbatch* from 1605, *Poulderbech -bach(e) -batch* 1557–1672, *Powderbach(e)* 1590–1739. Unknown element ***pulfre** + OE **bæće** 'a valley with a stream'. Also known as *Castell Polerebech* 1284×5 etc. for distinction from Church PULVERBATCH ST 4302. **Pulfre* might be an otherwise unrecorded OE loan from Latin *pulvis, pulverem* 'dust'. Sa i.245.

Church PULVERBATCH Shrops SJ 4302. '(The part of) P with the church'. *Chyrche Pulrebach* 1271×2, *Churchpolrebache* 1301. ME **chirche** + p.n. PULVERBATCH SJ 4202 which was also called Castle Pulverbatch. Known locally as Churton, 'the church town', *Chirchetona de Pulrebeche* 1221, *Cherton* 1535, *Chorton* 1667, *Churton* 1697, 1742. Sa i.246.

PUNCKNOWLE Dorset SY 5388. Probably 'Puma's hill-top'. *Pomacanole* 1086, *Pomecnolle* 1268, *Pomcnolle* 1303, *Pomknolle* 1391. OE pers.n. *Puma* + **cnoll**. Do 124.

PUNNETT'S TOWN ESusx 6220. *Pannets Fm* 1823. Surname *Pannet* found at HERSTMONCEAUX in 1645 + ModE *town*. Sx 467.

Isle of PURBECK Dorset SY 9581. The isle of *Purbek* 1380, *the Ile of Purbyke* c.1500. ME **ile** + p.n. *Porbi(che)* 1086, *Purbik(e) -byk(e)* c.1170–1482, *Purbek* 1380–1512, 'bittern or snipe ridge', OE **pūr** + ***bic(a)** 'a bill, a beak, a beak-like projection' referring to the prominent central chalk ridge running across the district. The sense is 'ridge resembling a bittern's beak' or 'ridge frequented by bittern (or snipe)'. For *pūr* see PURLEIGH TL 8301. The earliest reference is S 534(1), an Anglo-Saxon charter describing eight hides of land in Corfe Castle parish as *pars telluris Purbicinga* 'part of the land of the men of Purbeck' [948]15th. This land was called the Isle of Purbeck because it is a peninsula almost surrounded by water. Do i.1.

PURBECK HILLS Dorset SY 9181. *Purbeck Hill* 1811. P.n. Purbeck as in Isle of PURBECK SY 9581 + ModE **hill(s)**. Do i.2.

PURBROOK Hants SU 6707. 'Goblin or water-sprite's brook'. *Pokebroc* 13th–1542 with variants *-brok(e) -brouk*, *Pukebrok* 1248–1312, *Putbrooke* 1611, *Purbeck* 1675. OE **pūca** + **brōc**. Ha 135, Gover 1958.21.

PURFLEET Essex TQ 5578. Partly uncertain. *Purteflyete* 1285, *Pourt(e)flet(e)* 1312–58, *Purflete* 1313. Unidentified element + OE **flēot** 'a stream'. There seems to have been an element **purte* of unknown meaning, possibly associated with water in this name and in Portpool in Holborn, *Purtepole* 1203, *Purtewelle* 13th in Manuden, *Purtemere* 1259 in Farnham, *Purtemere* in Walthamstow and *purtan ige* [962]12th S 705 in Wiltshire. Ess 130, 649, 108.

PURITON Somer ST 3241. 'Pear-tree farm'. *Piriton* [854]14th S 303, 1212, *Peritone* 1086, *Peryton* 1610. OE **piriġe, pyriġe** + **tūn**. DEPN.

PURLEIGH Essex TL 8301. 'Bittern or snipe wood or clearing'. *Purlea* 998 S 1522, *-lai* 1086, *-le(ye) -leg(h) -leigh* from 1212. OE **pūr** + **lēah**. The precise species designated by *pūr* is uncertain; the word glosses both *bicoca* 'snipe' as an alternative to *hæferblæte* and *onogratulus* as an alternative to *raradumbla* 'bittern' cf. G *Rohrdommel*. Either bird could have been common in the marshland here. Ess 222, Studies 1931.80.

PURLEY Berks SU 6676. 'Wood or clearing frequented by bitterns'. *Porlei -laa* 1086, *Purleia* [c.1180]c.1200 etc. with variants *-l' -l(y)e -legh*, *Pourle(g')* 1297. OE **pūr** + **lēah**. Brk 215, L 205.

PURLEY GLond TQ 3161. 'Pear-tree wood or clearing'. *Pirlee* 1200, *-le(a)* 1208, 1227, *Purle(e)* 1220–1346. OE **pyriġe** + **lēah**. Sr 54.

PURLIEU Hants SZ 1899 → DIBDEN PURLIEU SU 4106.

PURLOGUE Shrops SO 2877. Unexplained. *Porteloke* 1272, *Portlok(e) -loge -loc* 1272, 1284, *Porthlok* 1283, *Parlogue* 1641, *Perlogue* 1833 OS. Formally identical with PORLOCK Somer SS 8846, OE **port** 'a gate, a harbour, a town' + **loca** 'an enclosure'; *loca* might refer to Offa's Dyke there, but there is no town here and no gap or gate in the Dyke. Gelling.

PURLS BRIDGE Cambs TL 4787. *Purls Bridge* 1720, *Purl's Bridge* 1824 OS. Ca 236.

PURSLOW Shrops SO 3681. 'Pussa's tumulus'. *Posselau* 1086, *-lawe* 1242, *-low(e)* 1327, 1397, *Pusselawa* c.1200×10, *-lawe* 1255×6, *-lowe* 1271–1431, *Puslowe* 1449, *Purslaw* 1577, *-low* 1641. OE pers.n. *Pussa* + **hlāw**. The reference is to a vanished burial mound, probably the marker for the meeting-place of Purslaw Hundred. For intrusive *-r-* cf. MARSWORTH Bucks SP 9214, *Masseworth* 1201–1526, DRAYTON PARSLOW Bucks SP 8428, *Drayton Passelowe* 1421–1501 from the Fr surname *Pass l'ewe*. Sa i.247, Bk 66, 97.

PURSTON JAGLIN WYorks SE 4319. 'P held by Jakelyn'. *Preston(e) Jakelin -yn* from 1269, *Pruston Jackling(e) -lynge* 1581–1641, *Purston Jackling* 1658. Earlier simply *Preston(e) -tona* 1086–1605, 'the priests' estate', OE **prēosta** + **tūn**. *Jakelyn* is the name of an as yet unidentified feudal tenant. YW ii.87.

PURTON 'Pear-tree farm, pear orchard'. OE **piriġe** + **tūn**. L 219.

(1) ~ Glos SO 6904. *Piriton(e) -y-* 1297–1439, *Puryton* 14th, *Pyrton* 1607. Gl ii.235.

(2) ~ Glos SO 6704. *Periton(e) -tune* 1086–1270, *Piritun -ton(e) -y-*

1200–1419, *Puriton -y-* 1268, 1330, *Perton'* 1202, *Pi- Pyrton* c.1275–1642, *Purton* 1662. Gl iii.259.

(3) ~ Wilts SU 0987. *Perytun* [796]13th S 149, 1156–1282, *(æt) Peritune* [854]12th S 305, *Puriton(e) -a* [796]13th S 1586, 1242, 1332, *Purtun* 1247, *(at) Pirituna* [1065]14th S 1038, *-tone* 1086–1304. The earliest form of the name seems to have been *æt pirigean* 'at the pear-tree', the form that lies behind the corrupt readings *æt Piertean* [796]13th, *æt Piergean, et Pirigean* [796]c.1400 S 149. Wlt 37, L 219.

(4) ~ STOKE Wilts SU 0990 → Purton STOKE SU 0990.

PURY END Northants SP 7045. 'Pury end' sc. of Paulerspury. *Perry End* 1672. P.n. Pury as in PAULERSPURY SP 7245 + ModE **end**. Contrasts with Tew's End SP 7245, *Tewesend* c.1825, from the family name Tew evidenced from 14th cent., and Plumpton End SP 7245 'the Plumpton end of Paulerspury' from p.n. *Plumton'* 1275, *Plumpton Pyrye* 1315, *Plompton End* 16th, 'the plum farm', OE **plūm** + **tūn**. Nth 104.

PUSEY Oxon SU 3596. 'The island where peas grow'. *Pesei, Peise* 1086, *Peseia -ia -i(e) -y(e) -ee -ey(e)* [1163]13th & 13th cent., *Pusie -eia -ye -ey(e)* from [1066×87]13th. OE **pise, peosu** + **īeġ**. Brk 395, L 37, 39.

East PUTFORD Devon SS 3616. *Est Putteford* 1316. ME adj. **est** + p.n. *Potiforde* 1086, *Potteford* 1234, *Putteford* 1242, 'Putta's ford', OE pers.n. *Putta* + **ford**. 'East' for distinction from West PUTFORD SS 3515. D 106.

West PUTFORD Devon SS 3515. *Westerputford* 1270, *Westpotteford* 1316. ME adj. **west(er)** + p.n. *Pote- Pvdeforde, Pote- Podiford* 1086, *Putteford(e)* 1238–1311 as in East PUTFORD SS 3616. D 160.

PUTLEY H&W SO 6437. 'Hawk or Putta's wood or clearing'.

I. *Poteslepe* 1086, *Poteslepa* 1160×70.

II. *Putteleghe* 1158×65, *-ley(e)* 1327-62.

Form II represents OE **putta** or the pers.n. *Putta* derived from it + **lēah**. Form I is probably **putta** or *Putta* + ***slǣp** 'a slippery muddy place'. He 171, DB Herefordshire 10.4, Studies 1936.92, PNE ii.127.

PUTNEY GLond TQ 2274. 'Putta's landing place'. *Putelei* 1086, *Puttenhuth(e)* 1279–1332, *-hethe(e)* 1332, 1502, *-hith* 1461, *Putneth* 1474, *Putney* 1503. OE pers.n. *Putta*, genitive sing. *Puttan*, + **hȳth**. Sr 27.

PUTTENHAM Herts SP 8814. 'Putta's homstead'. *Puteh*ā 1086, *Put(t)enham* from 1155. OE pers.n. *Putta*, genitive sing. *Puttan*, + **hām**. Hrt 50.

PUTTENHAM Surrey SU 9348. 'Putta's settlement'. *Puteham* 1199–1286, *Pote(n)- Putte(n)-* 13th cent., *Potten-* 1342, *Putne-* 1535, *Putnam* 1618. OE pers.n. *Putta*, genitive sing. *Puttan*, + **hām**. Sr 209, ASE 2.32.

PUXTON Avon ST 4063. 'Pukerel's estate'. *Pukereleston* 1212, 1227. Family name *Pukerel*, genitive sing. *Pukereles*, + OE, ME **tūn**. A Robert Pukerel is mentioned in Somerset in 1158×9 and 1176. The name perhaps derives from OFr *poacre* 'gout'. DEPN.

PYECOMBE WSusx TQ 2812. Either 'the valley of the pointed hill' or 'the valley infested with insects'. *Picumba* 1087×1100, 1180×1204, *Piccumbe* c.1100, *Pikomb* 1284, *Pycombe* 1272–1439. Either OE **pīc** or **pēo, pīe** + **cumb**. In the first interpretation the reference is to the conical shape of Wolstonbury Hill TQ 2813 (675ft.) which may have been called *Pic* 'the pike' in Anglo-Saxon times. Sx 287, PNE ii.62.

PYEMORE Cambs TL 4986. 'Insect' or 'magpie marsh'. *Pymoor* 1497. OE **pīe** + **mōr**. Ca 224.

PYLE IoW SZ 4778. Uncertain. *Pyle* 1769, *Pile* 1810 OS. Walter *atte Pyle* 1320 and Richard de *La Pile* may take their surname from here; if so, this is OE **pīl** 'a post, a stake' in some unknown application, possibly a boundary marker. Wt 116, Mills 1996.84.

PYLLE Somer ST 6038. 'The creek'. *Pil* [705]14th S 247, *þæt pyl* 955 S 563, *Pille* 1086, *Pulle* 1276, *Pull* 1610. OE **pyll** referring to the stream called WHITELAKE ST 5240. DEPN.

PYMMES BROOK Middlesex → SOHAM Cambs TL 5973.

PYNE FARM Devon SS 8010 → WASHFORD PYNE SS 8111.

Canon PYON H&W SO 4649. 'Pyon held by the canons' sc. of Hereford. *Piona Can'* 1160×70, *Pyone Canonicorum* 1221. ModE **canon**, Latin **canonici**, genitive pl. **canonicorum**, + p.n. *Pionia* before 1066–1294, *Pevne* 1086, *Pyone* 1292, 1320, 'gnats' island', OE **pēo**, genitive pl. **pēona**, + **ēġ** probably in the sense 'dry ground surrounded by marsh'. 'Canon' for distinction from King's PYON SO 4450. He 171.

King's PYON H&W SO 4450. 'Pyon held by the king'. *Piona Regis* 1160×70, *Kyngespewne* 1535. ME **king**, genitive sing. **kinges**, Latin **rex, regis**, + p.n. *Pionie* 1086, as in Canon PYON H&W SO 4649. A royal estate up to 1066. He 172.

PYRFORD Surrey TQ 0359. 'The pear-tree ford'. *(æt) Pyrianforda* [956]13th S 621, *Piriford* 1067, c.1104, *Peliforde* 1086, *Pyry- Pire- Puri- Puryford* 13th cent., *Pyrford* from 1573. OE **piriġe**, genitive sing. **piriġan**, + **ford**. The ford rosses the river Wey. Sr 132, L 219.

PYRTON Oxon SU 6895. 'The pear orchard'. *(to) Pirigtune* [987]11th S 1354, *Peritone, Pir(i)tune* 1086, *Perton(')* 1208, 1517, *Purton* 1300, *Pyreton(')* 1266, 1361. OE **piriġe** 'a pear-tree' + **tūn**. The site of Pyrton is earlier *(æt) Readanoran* 'at the red ridge' [887]11th S 217, [late 10th]11th S 104. O 86, L 219.

PYTCHLEY Northants SP 8574. 'Piht's wood or clearing'. *pihtleslea* [956]12th S 592, 1086–1302 with variants *Pyht-* and *-le(gh)*, *Picteslei* 1086–1249 with variants *-le(a)*, *Piteslea* 1086–1398 with variants *Pyt-* and *-le(y)*, *Py- Pightesle(e)* 14th cent., *Pitchesle(ye)* 1323, *Pycheley* 1544. OE pers.n. **Piht* (*Peoht*), genitive sing. **Pihtes*, + **lēah**. Nth 130 gives pr [paitʃli].

PYWORTHY Devon SS 3103. 'Insect enclosure'. *Paorde* 1086, *Peworthy -wrthe -wrthi* 1239–87, *Pyworthe* 1291, *-worthy* from 1310, *Pury* 1547, 1786. OE **pēo, pīe** + **worthiġ**. *Pēo, pīe* may have been used as a nickname here. The next hamlet is Hopworthy SS 3002, *Hoppeworthe* 1318, *-thi* 1333, *Hopworthy* 1464, 'grasshopper enclosure', OE ***hoppa** + **worthiġ** where *hoppa* may also have been used as a nickname. D 162, 163.

Q

QUABBS Shrops SO 2180. 'The marshy place'. *Quabs* 1833 OS. Mod dial. **quab** (OE *cwabba*).

QUADRING Lincs TF 2233. 'Mud Havering'. *Guedhauringe, Quedhafring* [11th]12th, *Quadheueringe, Quedhaveringe -u-* 1086, *Quad- Quedhauering(e)* 1170–1254, *Quadauring* 1253, *Quadring* from 1271. OE **cwēad** + p.n. *Hæfering*, an *ing*[2] derivative of OE ***hæfer** 'higher ground'. Quadring is an island of higher ground in the fens; cf. Dan Hevring, G Haferungen, OG **haver*. An alternative possibility would be OE *hæfer* 'a he-goat' or pers.n. **Hæfer*, 'goat place' or 'place called after Hæfer'. Bahlow, with his dilection for *Sumpfnamen*, posits a fen word *hav-r* for G names such as Haverbeck. Payling 1940.95 gives pr [kweidriŋ], FmO i.1322, ING 62, Bahlow 192, 204, Steenstrup *De dænske Stednavne* 334.

QUAINTON Bucks SP 7420. Probably 'the queen's manor'. *Chentone* 1086, *Quenton* 1175–1371, *Quei- Queynton* 1235–1524. OE **cwēn**, genitive sing. **cwēne**, + **tūn**. Bk 108 gives pr [kweintən], PNE i.121.

QUANTOCK HILLS Somer ST 1537. *Cantucdun* [678]13th, *mons de Cantok* 1314, *Quantoke hills* 1610. Hill name *Cantuc* + OE **dūn**, Latin **mons**, ModE **hill**. *Cantuc* as in *Cantucdun* above and *Cantucuudu* (OE **wudu** 'wood') [672 for 682]16th S 237, [late 10th]c.1247, later *Cantoche* 1086, *Canthoc* before 1107, *Cantoc* [c.1136]18th, [1156×7]14th, *Kantoc* before 1187, *Cantok(')* 1274–1327, is Brit **Cantāco-*, a formation on the root **gantho-* 'corner, angle, bend' seen in OCo, W *cant* 'enclosure, circle, rim'. The meaning seems to be 'full of windings, abounding in corners or bends'. Cf. Camel in Queen CAMEL ST 5924 and KENN Avon ST 4169. DEPN, Turner 1952–4.18, CPNE 37, GPC 418.

East QUANTOXHEAD Somer ST 1343. *Estcantokeshende* 1327, *East Quantokehead* 1610. ME adj. **est** + p.n. *Cantocheve* (for *-heved*) 1086, *Cantokesheued -heved* 1185, 1212, 'the head or end of Quantock', hill name *Cantuc* as in QUANTOCK HILLS ST 1537, genitive sing. *Cantuces*, + OE **hēafod** sometimes confused with **ende**. 'East' for distinction from West QUANTOXHEAD ST 1141. DEPN.

West QUANTOXHEAD Somer ST 1141. *Westcantokeshende* 1327, *West Quantokhead* 1610. ME adj. **west** + p.n. Quantoxhead as in East QUANTOXHEAD ST 1343.

QUARLEY Hants SU 2743. 'Millstone wood or clearing'. *Cornelea* 1167, *Corn(e)le* 1248–1331, *Querly -li(e) -le(e)* 1269–1327, *Corle* 1272, *Qwharley* 1438. OE **cweorn**, genitive pl. **cweorna**, or **cweorne** 'a hand-mill', + **lēah**. Ha 136, Gover 1958.168.

QUARNDON Derby SK 3341. 'Quern hill, hill called *Quern*, the quern-stone'. *Querendunie -don(e)* 1154×89–1535, *Qwerendon, Quern(e)don(e)* 1252–1547×51, *Quarndon* from 1391. Earlier simply *Cornun* 1086, *Querne* 1544×7, *Quarne* 1622–1767. OE **cweorn** used as a p.n. + **dūn**. Cf. QUORNDON (QUORN) Leic. Db 492, L 145.

QUARRINGTON Lincs TF 0544. 'Quern village or farm'. *Querentonam* [1051]late 13th, *-tone* [1149]late 13th, *Cornintone, Corninctun(e) -tone* 1086, *Querington(e)* 1219–1346 including [1078]late 14th, *Querendon -in(g)-* 1224–1346, *Quarington* 1496–1642, *Quarrington* from 1526. OE **cweorn** 'a quern, a hand-mill, a mill' + **tūn**. The *ing-* forms are probably a secondary development from *cweorn* with parasitic vowel *cweoren, cweorin*, cf. QUARRINGTON HILL Durham NZ 3337. An alternative OE specific *cweorning* is possible but that would properly mean 'stream where querns are found' as in WHORLTON Durham NZ 1014. There is no stream here; the sense is probably 'settlement where querns are obtained'. Perrott 84.

QUARRINGTON HILL Durham NZ 3337. *Quarrington Hill* 1863 OS. P.n. Quarrington + ModE **hill**. Quarrington, *Querindune -don* [1183]14th, *Queringdon(e) -yng-* [1183]c.1382, 1242–1563, *-ton* 1242–1339, *Quarrington* 1647, is 'quern hill, hill where stones for querns are found or prepared', OE **cweorn** + **dūn**. NbDu 161.

QUARRY BANK WMids SO 9386. *Quarry Bank* 1834 OS. ModE **quarry** + **bank**.

The QUARRY Glos ST 7399. *Quarr* 1728. Mod dial. **quar**, short for *quarry*. Gl ii.217.

QUATFORD Shrops SO 7490. 'The ford leading to Quatt'. *Quatford* from 1086. P.n. QUATT SO 7588 + OE **ford**. Sa i.247.

QUATT Shrops SO 7588. Uncertain. *Quatte* 1208–1582, *Quat* 1320–1833 OS, *Qwatt* 1550×1, *Quatt* 1661, *Cutt* 1604. OE ***cwætt** of uncertain meaning and origin, possibly related to *quat* 'a pimple, a pustule, a small boil'. DB has *Quatone* 1086 which is either aberrant or a nonce compound with OE **tūn**. Sa i.247.

QUEBEC Durham NZ 1743. *Quebec* 1862 OS. Originally the name of a farmhouse, transferred from the name of the Canadian city settled in 1608.

QUEDGELEY Glos SO 8013. Possibly 'Cweod's wood or clearing'. *Quedesley(a) -leia -leg' -leye* 1135×54–c.1560, *Cudesleya* 1199×1216, *Quodgley* 1577, *Quedg(e)ley* 1610, 1648. OE pers.n. **Cweod*, genitive sing. **Cweodes*, + **lēah**. Gl ii.xii, 187.

QUEEN ADELAIDE Cambs TL 5681. So named from a public-house. Queen Adelaide was the consort of William IV 1830×37. Ca 223.

QUEEN CAMEL Somer ST 9524 → Queen CAMEL.

QUEEN'S GROUND Norf TL 6893. *Queens Ground* 1824 OS.

QUEENBOROUGH Kent TQ 9172. 'The queen's borough'. *Quenesburgh* 1367–70, *Queneburgh'* 1376, 1436. ME **quene** + **burgh**. Edward III erected a castle here in 1361–77 as a defence against French raids (demolished in 1650). The accompanying town was made a borough by charter in 1368 and named in honour of queen Philippa of Hainault. PNK 261, Newman 1969E.403.

QUEENSBURY WYorks SE 1030. This place was so named at a public meeting held on 8th May 1863. Before that the village was known by the name of its public house, the *Queen's Head* 1821. YW iii.87, Room 1983.100.

QUENBY HALL Leic SK 7006. *Quenby Hall* 1720. P.n. Quenby + ModE **hall**. The reference is to the Jacobean mansion built for George Ashby c. 1620. Quenby, *Queneberie* 1086, *Quenebi(a) -by* c.1130–1402, *Quenby* from 1334, is originally 'the Queen's manor', OE **cwēn**, genitive sing. **cwēne**, + **byriġ**, dative sing. of **burh**. OE *byriġ* was subsequently replaced by ON *bý*. Lei 305, SSNEM 14, 208.

QUENDON Essex TL 5130. 'The women's valley'. *Kuenadanam* 1086, *Quen(e)den(e)* 1248–1399, *Quendon* 1323, *Quenendon -den, Quenynden* 1281. OE **cwene**, genitive pl. **cwenena**, + **denu**. Ess 532.

QUENIBOROUGH Leic SK 6412. 'The queen's manor'. *Cuinburg* 1086, *Quenburg(h)* 1154×89–1409, *Queniburg(h) -bur(c) -y-* c.1130–1510, *Queneburch -burg -borou(ghe)* 1154×89–1536, *Queningburc -burg(h) -yng-* 1236–1394. OE **cwēn** + **ing**[4] + **burh**. Lei 314.

QUENINGTON Glos SP 1404. Partly uncertain. *Qv- Queninton(e) -tona -yn-* 1086–1434, *-ing- -yng-* 1282–1587. It is impossible to be sure whether this is 'the women's estate', OE **cwene**, genitive pl. **cwenena**, + **tūn**, or 'estate called after Cwen or Cwena', OE feminine pers.n. *Cwēn(e)*, + **ing**[4] + **tūn**. Gl i.44 gives pr ['kwenintən].

QUERNMORE Lancs SD 5160. 'Moor where mill-stones are found or prepared'. *Quernemor(e)* 1228–1323, *Wernelmor* 1281, *Whernemor* 1322, *Quermore* 1342, *Whermore* c.1500, *Quarmore* 1684×5. OE **cweorn**, genitive pl. **cweorna**, + **mōr**. La 173 gives pr [wɔ:mə], Jnl 17.98.

QUETHIOCK Corn SX 3164. Probably 'wooded place'. *Quedoc* 1201, *Quedic -yk -ik* 1207–1342, *Quethek* 1535, *Quethyock* 1620. OCo **cuidoc** (OCo *cuit*, Co *cos* 'wood' + adjectival suffix OCo *-oc*, Co *-ek* 'place of'). Identical with CHIDEOCK Dorset SY 4292. PNCo 146 gives pr [kwiðik], Gover n.d. 200, DEPN.

QUIDENHAM Norf TM 0287. 'Cwida's homestead'. *Cuidenham* 1086, *Quideham* 1177. OE pers.n. **Cwida* corresponding to OHG *Quito*, genitive sing. **Cwidan*, + **hām**. DEPN.

QUIDHAMPTON Hants SU 5150. 'Settlement of the *Cweadhœme*, the inhabitants of *Cweadham*'. *Quedementune* 1086[†], *Qued- Quidhamtun* 1245, *Quedhampton* 1280, *Quidhampton* from 1302. OE folk-n. **Cwēadhǣme*, genitive pl. **Cwēadhǣma*, + **tūn**. The *Cweadhœme* must have been the inhabitants of a place called *Cweadham*, the 'dirty homestead', OE **cwēad** 'dung' + **hām**, and are distinguished from the *Polhœme* of nearby Polhampton SU 5250, *(œt) Polehametune* [956]12th S 613 who lived at **Polham* 'the pool estate', OE **pōl** + **hām**, the *Northhœme* of NORTHINGTON SU 5637 and the *Suthhœme* of Southington SU 5049, *Suthampton* 1346–1428, *Sothyngton* 1412, who lived at **Suthham* 'the south estate', OE **sūth** + **hām**. These four places are in systemic relationship within Overton parish. Cf. QUIDHAMPTON Wilts SU 1030. Ha 136, Gover 1958.138, Wlt 226, DEPN.

[†]This form is usually equated with CORHAMPTON SU 6120 which fits this position in the DB text order and hundred. It does not, however, fit the other known forms for Corhampton.

QUIDHAMPTON Wilts SU 1103. 'The dirt or dung farm'. *Quedhampton* 1249–1333, *Quid-* 1289, *Quyddamton* 1558×1603, *Quidington* 1675. OE **cwēad** + **hāmtūn**. Wlt 226 gives pr [kwid'hæmtən], formerly [kwidiŋtən].

QUIES Corn SW 8376. Probably 'the sow'. *Quies* 1813. Co **guis**. The reference is to a rock off the coast of Trevose Head SW 8576 called in English *Cow and Calf alias So. Rock* 1699 in which *So.* is probably a mistake for *Sow* (*South* would make no sense). Cow and Calf occur again a few miles to the south off Park Head at SW 8370 and the two locations may have been confused. PNCo 146, Gover n.d. 352.

QUINTON Northants SP 7754. Probably 'the queen's manor'. *Quintone* 1086, 1200, *Quynton* 1337–1511, *Quenton(a)* 1148×66–1428, *Queenton* 1342. OE **cwēn** + **tūn**. OE **cwēn** in p.ns. is difficult to distinguish from **cwene** 'a woman' and pers.n. **Cwēna*. The single spelling *Quenynton* 1287 could point to OE *Cwēningtūn* 'estate called after Cwena' or to OE *cwenena-tūn* 'the women's estate'. Nth 151.

Lower QUINTON Warw SP 1847. *Nether Quenton* 1292, 1325. ME adj. *nether* + p.n. *Quenton(e) -ton(a) -thon* [840×52]12th S 191, 1168–15th, *Qvenintvne* 1086, *Queninton* 1221, *Quy- Quinton* from 1416, 'the queen's estate', OE **cwēn** + **tūn**. The *Quenin-* forms are due to confusion with QUENINGTON Glos SP 1404. 'Lower' for distinction from Upper QUINTON SP 1746. Gl i.254.

Upper QUINTON Warw SP 1746. *Uuer Quenton* 1248, *Over(e) Quenton(e)* 1274–92. ME adj. **over** + p.n. Quinton as in Lower QUINTON SP 1847. Gl i.254.

QUODITCH Devon SX 4097. 'Dirty or well-manured land'. *Quidhiwis* 1249, *-hiwische* 1318, *Quydiche* 1609, *Cowditch* 1765, 1809 OS. OE **cwēad** 'dung, filth' + **hiwisc** 'land to support a family'. D 127.

QUORNDON or QUORN Leic SK 5016. 'Hill where quern-stones are obtained'. *Querendon(e)* 1154×89–1535, *Querindon' -yn(g)-* 1258–1412, *Quer(n)don'* 1271–1539, *Quarndon'* 1393–1561, *Quornedon'* 1210, 1327. OE **cweorn** + **dūn**. The short form occurs as *Querne* 1549–54, *Quarn(e)* 1506–1708. Lei 391.

RABY Mers SJ 3180. 'Boundary village'. *Rabie* 1086, *Raby* from [1096×1101]1280, *Robi -y* 1208×11, 1321. ON **rá** + **bȳ**. Possibly marked the boundary of Scandinavian settlement in the Wirral. Che iv.228, SSNNW 38.

RABY CASTLE Durham NZ 1221. *(in) castro de Raby* 1440. P.n. Raby + ONFr **castel**. Raby is *Raby* from [c.1040]12th. On formal grounds this could be either ON **rá** 'a roe-deer' or ON **rá** 'a boundary' + **bȳ**. Swedish *Raby*, Danish *Raaby*, the Rabies in Cheshire and Cumbria and Roby in Lancs have usually been taken to be 'boundary village' for local reasons, but the Durham Raby lies on no known boundary. In the 13th cent. the customary rent for the estate was the annual offering of a stag at the shrine of St Cuthbert in Durham. This suggests that the name may then have been understood to mean 'deer farm, deer estate'. Possibly Raby originated as the hunting lodge of the great multiple estate of Staindropshire parallel to the hunting lodge of the Bishop of Durham in the forest of Weardale in Aucklandshire. Whereas these hunting lodges were not envisaged as permanent structures, in Staindropshire the new secular lord, Dolfin son of Uhtred, made Raby the seat of his lordship. The present castle was licensed in 1378. NbDu 161, Settlement 24, Boyle 710, Nomina 12.24.

RACKENFORD Devon SS 8518. 'Ford suitable for riding at *Rack*, the throat'. *Rachenefode* 1086, *-forda* 1086 Exon, *Rac- Rakarneford(e) -erne-* 1147×61–1387, *Rach- Rakeneford* 1204–1303, *Rakereford'* 1246, *Ratten- Rackingford* 1628. OE **hraca** 'throat, pass, narrow track' referring to the depression SW of the village and probably used as a p.n. + **ærne-ford** 'ford for riding' as in ARNFORD NYorks SD 8356. D 389, YW vi.158, DEPN.

RACKHAM WSusx TQ 0513. 'Rick hill' or 'rick-yard'. *Recham* 1166–1447 with variants *Rek-, Rak- Racham* 1295, 1307. OE **hrēac** + **hamm** 2a 'a promontory into marshland' or 5b 'an enclosed plot, a close'. If this was not originally a rick-yard, OE **hrēac-hamm*, the reference may have been to Rackham Hill TQ 0512 once possibly called *Hrēac* 'the hay-rick'. Sx 156, ASE 2.50, SS 82.

RACKHEATH Norf TG 2715. Partly uncertain. *Rachei(th)a* 1086, *Racheia* [1153×68]14th, 1197, *Racheth* 1252. Uncertain element + OE **hȳth** 'a landing place'. The specific might have been OE **hraca** 'a throat' or **racu** 'a hollow, a stream, a reach of river' or ON **rák** 'a rough path over a hill'. The first two would refer to the valley of The Springs which must have provided navigable water into the Bure broads; the latter possibly to the path leading up from Rackheath church. DEPN.

New RACKHEATH Norf TG 2812. ModE adj. **new** + p.n. RACKHEATH. A modern development later than 1838.

RADCLIFFE 'The red cliff', OE **rēad**, definite form **rēada** with early shortening of OE *ēa* > *ĕa* > *æ*> ME *a*, + **clif**. L 133–5.

(1) ~ GMan SD 7807. *Radeclive* 1086–1202 etc., *-cliv(a) -clif -clyve* 1194–1332, *Radclif -clyf* 1346–1442, *Radcliffe* 1500, *Ratcliffe* 1577, *Redeclif* c.1200. The form of c.1200 shows remodelling with ME *red*; Radcliffe stands on a cliff of red sandstone on the side of the Irwell. La 48, Jnl 17.39.

(2) ~ Northum NU 2602. *Radcliffe Colliery, ~ Terrace* 1868 OS. The site is a low hill rising to 23ft.

(3) ~ ON TRENT Notts SK 6439. P.n. *Radeclive* 1086–1327 + r.n. *(super) Trentam* 1269. There is a steep slope of red clay here by the Trent. Nt 240.

RADCLIVE Bucks SP 6733. 'The red cliff'. *Radeclive* 1086–1379, *Redeclive* 1314, *Raclyff* 1526, *Ratcliff* 1537, *Ratcley* 1524, *Ratley* 1675. OE **rēad**, dative sing. definite form **rēadan**, + **clife**, dative sing. of **clif**. Bk 46 gives pr [rætli], O xlv, L 133.

RADCOT Oxon SU 2899. 'The red cottages' varying with 'the reed cottages'.

I. *Raðcota* 1163, *Rad(e)cote* 13th cent., *Radcott* 1761.

II. *Redcot(e)* 1176–1272, *Rotcot(e)* 1220–1428.

Form I is OE **rēad** + **cot**, pl. **cotu**, with shortening of *ēa* before the double consonant group; form II is OE **hrēod** + **cotu**, in which the spellings <e> and <o> represent monphthongisation of *ēo* to [e:] and [ø] respectively, the latter a characteristic WMidl and SW development. O 320.

RADFORD 'The red ford'. OE **rēad**, definite form **rēada**, + **ford**.

(1) ~ Avon St 6757. *Reddeford* 1249, *Radeford* 1275. D 257.

(2) ~ Notts SK 5540. *Redeford* 1086, *Radeford* 1212. Nt 150.

(3) ~ Warw SP 3280. *Raddeford* 1354, *Radford* 1411. Wa 167.

(4) ~ SEMELE Warw SP 3464. 'R held by the Simely family'. *Raddefford Simily* 1279, *Radford Seemley* 1619. P.n. *Redeford* 1086, *Radeford* c.1154, manorial addition from by Henry de Simely who held the estate 1100×1135. Wa 178 gives pr [simili].

RADLETT Herts TL 1600. Uncertain. *Radelett* 1453. Possibly from OE **rād** + **ġelǣte** 'a road junction'. The evidence, however, is insufficient for certainty. Hrt 61.

RADLEY Oxon SU 5299. '(The settlement) at the red wood or clearing'. *Radelege* [c.1180]13th, *Radeley* 1242×3. OE **rēad**, definite form dative sing. **rēadan**, + **lēah**, dative sing. **lēaġe**. Brk 455, L 205.

RADNAGE Bucks SU 7897. '(At) the red oak-tree'. *Radenhech* 1162, 1237×40, *Radenach(e)* 1175–1228 etc., *Redenhache* 1185, *Radenage* 1440, *Radenedge* 1552, *Radnidge* 1643. OE **rēad**, dative sing. definite form **rēadan**, + **ǣċ**, dative sing. of **āc**. Bk 191, Wa xli, L 219, Jnl 2.31.

RADSTOCK Avon ST 6854. 'Road Stoke'. *Radestok* 1221, 1243, *Rodestoke* 1276. OE **rād** + p.n. *Stoche* 1086, OE **stoc** 'an outlying farm'. The place lies on the Roman road called Fosse Way. Also known as *Stokes Elie de Clifton* 1198, 'Elias of Clifton's Stoke', and possibly identical with *(æt) Welewe stoce* [984]12th S 854, 'Wellow Stoke', i.e. 'Stoke, outlying farm on the river Wellow' or 'belonging to WELLOW' ST 7358. Cf. however ECW 118. DEPN, Studies 1936.19.

RADSTONE Northants SP 5840. Probably the 'rood-stone', a stone used as the socket of a rood or cross. *Rodeston(e)* 1086–1415, *-tun* 1166–1223, *-stan* 1201, 1227, *Rudstan* 1198, *Roddeston(e)* 1200–1381, *Radeston* 1288, *Radston* 1629. The forms are ambiguous as between OE **rōd** + **stān** and ***rod(u)** 'a clearing' + **stān**, 'the stone in the clearing'. Nth 56, PNE ii.87.

RADWAY Warw SP 3748. 'The red way'. *Rad- Rodewei, Radvveia* 1086, *Radewey* 1198–1492, *Radwey* from 1497. The reference is to an ancient trackway running from Brailes SP 3139 below Edge Hill to Knightcote SP 3954, Bishops Itchington SP 3957, Ufton SP 3762 and Stoneleigh SP 3372 passing Radford Semele SP 3464, and to the red colour of the earth. Wa 272.

RADWAY GREEN Ches SJ 7754. *Radway Green* 1831. Apparently 'green at the roadway', OE **rād** + **weg** + ModE **green**. Che iii.7 gives pr ['rædə].

RADWELL Herts TL 2335. 'The red spring'. *(to, of) readan wylles heafdon* 'the heads of the red spring' [1007]11th S 916, *Radeuuelle* 1086, *-welle* 1185–1347. OE **rēad**, definite form oblique case **rēadan**, + **wylle**. Hrt 160.

RADWINTER Essex TL 6037. *Redeuuintram* 1086, *Redewinte -went' -wynt'* 1212–85, *Radewinter -wynter* 1166–1344, *Rodwinter* 1539, *Roadwinter* 1605. Possibly the 'red vineyard', OE **rēad**, definite form **rēade**, + ***winter**, or 'Rædwynn's tree', OE feminine pers.n. **Rǣdwynn*, genitive sing. **Rǣdwynne*, + **trēow**: but *tre* spellings are rare. Ess 512, PNE ii.269.

North RADWORTHY Devon SS 7534. *Northradeworthy* 1330. ME adj. **north** + p.n. *Raordin* 1086, *Radewrthi, Redewrth* 1199, *Rade- Redewurth* 1234, 'Reada or Ræda's enclosure', OE pers.n. *Rēada* or *Rǣda* + **worthiġ**. 'North' for distinction from South RADWORTHY SS 7432. D 344.

South RADWORTHY Devon SS 7432. *Suthradewurthi* 1262. ME adj. **south** + p.n. Radworthy as in North RADWORTHY SS 7534. D 344.

RAGDALE Leic SK 6619. Possibly 'gulley valley'. *Ragendel(e)* 1086, *Rachen- Rakendale* c.1130, 1166, *Rache- Rakedal(e)* 1516, *Raggedal(e)* 1243, 1262, *Ragdale* from 1428. OE **hraca** 'throat, gulley', genitive sing. **hracan**, + ON **dalr**. The dale rises northwards from the Wreake Valley. Lei 299, L 95–6.

RAGNALL Notts SK 8073. 'Ragni's hill'. *Ragenehil* 1086, 1205, *-hull'* 1230–1471, *Ragenel* 1223, *Ragnell* 1433–1581, *Ragnall* 1704. ON pers.n. *Ragni*, genitive sing. *Ragna*, + **hyll**. Nt 57.

RAINFORD Mers SD 4801. Probably 'boundary strip ford'. *Raineford* before 1198, *Reineford* 1202, *Rayn(e)ford* 1246–1354, *-forth* 1454, 1568, *Rayn(e)sford* 1262–1503. ON **rein** + OE **ford**. Rainford is on the boundary between the old hundreds of Newton and West Derby. Nearby are *Rain Hey* 1842 OS 'boundary strip enclosure' and *Rayndene* c.1300 'boundary strip valley'. This explanation (Mills 124) supersedes 'Regna's ford', OE pers.n. *Reġna* + **ford**, and 'rowan-tree ford', ON **roynir** 'a rowan-tree' or **royni** 'a place growing with rowan-trees' + **ford**, although on formal grounds both are possible. Cf. RAINHILL Mers SJ 4991. La 110, SSNNW 245, Jnl 17.61.

RAINHAM GLond TQ 5282. Possibly 'Regna's homestead'. *Rena- Rene- Raineham* 1086, *Renham* 1205–1309, *Rein- Reynham* 1234–1346, *Rayn(h)am* 1254, 1432. OE pers.n. **Reġna* + **hām**. But comparison has been made with RAINHAM Kent TQ 8165, *Roeginga hám* 811 S 168, *Ren(e)- Raynham* from 1130, 'the homestead of the Roegingas'. Ess 127, ING 120, 122.

RAINHAM Kent TQ 8165. 'Homestead, estate of the Rœgingas, the people called after Rœga'. *Roeginga hám* 811 S 168, *Raenham* c.1100, *Ren(e)ham* 1129×30–1270, *Reynham* 1240–70. OE folk-n. **Rǣġ- Rēġingas* < pers.n. **Rǣġa*, **Rēġa* + **ingas**, genitive pl. **Rǣġ- Rēġinga*, + **hām**. PNK 261, KPN 115, ING 20, ASE 2.30.

RAINHILL Mers SJ 4991. Probably 'boundary strip hill'. *Reynhull -hill* 1246, *Rayn(h)ull* c.1190–1354, *Raynhill* 1400, *Raynyl* 1258, *Reynel* 1292. ON **reinn** + **hyll**. The hill forms the boundary against Eccleston SJ 4895. On formal grounds, however, the specific could equally well be OE pers.n. *Reġna* or ON **roynir** 'a rowan-tree'. Cf. RAINFORD SD 4801. La 107, Mills 124, SSNNW 245, Jnl 17.60.

RAINOW Ches SJ 9576. 'Raven hill-spur'. *Rauenok'* [c.1270]17th, 1283, before 1303, *-ock(e)* [13th]1611, 1325, *Raven(h)oh* 1287, *Ravenow(e)* 1320–57, *Ran(e)ow(e)* 1329–1445, *Rainow* 1724. OE **hræfn** + **hōh**. Spellings in *-ok, -ock(e)* probably show substitution of stop [k] for spirant [x] in *hōh* rather than ME *oke* 'an oak-tree'. The village lies on the shoulder of a hill. Che i.137.

RAINTON 'Estate called after Rægna'. OE pers.n. *Rœgna*+ **ing**[4] + **tūn**. *Rœgna* is a shortened form of *Rœġenwald*.

(1) ~ NYorks SE 3775. *Reineton, Rainincton* 1086, *Renyng- Renington* [1157]15th–1319, *Raynton* 1535. YN 184.

(2) ~ West Durham NZ 3146. *West Reinigtone* 1242×3, *~ reyningtona* c.1260, *~ rayngton'* 1348, *~ Raynton* 1345–1592, *West Rainton* from 1597. ME **west** + p.n. Rainton for distinction from East Rainton NZ 3477, *Est Reynton* 1296, *Estraynton(')* 1313–1675, *Estrayngton'* 1317, 1348, *Estraynington'* 1322, *East Rainton* from 1581. The two Raintons are mentioned 1154×66 *Raintonā 7 aliā Raintonā*, c.1190×5 *duas Renigtonas*. The earliest form is *Reiningtun* c.1107 'estate called after Rægna' in Symeon of Durham's *History of the Church of Durham*. Symeon reports that the estate was named after one *Reingualdus* son of Franco, one of the four faithful companions of St Cuthbert's coffin during its wanderings after the community left Lindisfarne in 875. According to Symeon, Reinguald's son Riggulf lived to be 210 years old and his great great great grandson, Symeon's informant, was still alive in Symeon's own day. The date at which Rainton received its present name may reasonably be computed to have been c.900, but it is likely that the settlement existed long before its association with *Rœġna*. NbDu 162.

(3) East ~ T&W NZ 3347. *Est Reynton* 1296–1384, *~ Raynton* 1313–1675, *East Rainton* from 1581. ME adjective **est** + p.n. Rainton as in West RAINTON Durham NZ 3146.

RAINWORTH Notts SK 5958. Probably the 'boundary ford'. *Reynwath* 1268, 1278, *Raynwath* 1278, 1287, *Raynwathford* 1300, *Reinwarth Forthe* c.1500. ON **rein** 'a boundary strip' + **vath**. The ford is on the boundary between two wapentakes. Alternatively the specific might be ON **hreinn** 'clean', 'the clean ford'. Nt 116 gives prs [reinɔːθ, renɔːθ], SSNEM 172.

RAISBECK Cumbr NY 6407. *Rayes- Reise- Reysebek* 1310, *Raisbek* 1342, *Raisbeck* from 1631. Transferred from the r.n. Rais Beck, *Rayssebeck* c.1270, 'cairn stream', ON **hreysi** + **bekkr**, referring to a tumulus at NY 6306. We ii.44, i.12, SSNNW 152.

RAITHBY Lincs TF 3184. Probably 'Hreithi's village or farm'. *Radresbi* 1086, *Reithebi* 12th, 1202, *Raitheb'* c.1180, *Red(d)ebi* 1191, 1198, *Re(i)theby* 1200. ON pers.n. *Hreithi*, genitive sing. *Hreitha*, + **bȳ**. This explanation fits all the recorded forms except that of DB, which appears to be 'Hreitharr's village or farm', ON pers.n. *Hreitharr*, secondary genitive sing. *Hreithares*, + **bȳ**. DEPN, SSNEM 63.

RAITHBY Lincs TF 3767. 'Hrathi's village or farm'. *Radebi* 1086, 1154×89, *Radabi* 1142×53, *Rathebi* 1170×8, *Ra(d)thebi* 1202. ON pers.n. *Hrathi*, genitive sing. *Hratha*, + **bȳ**. DEPN, SSNEM 63.

RAKE WSusx SU 8027. Probably 'the narrow path'. *Rake* 1549. ME **rake** (OE *hraca* 'a throat'). The reference is probably to the long narrow ridge followed by the Portsmouth-Petersfield road at this point. Alternatively the reference may have been to the deep hollow on the E side of the ridge. Sx 41, PNE ii.263.

RALFLAND FOREST Cumbr NY 5413. *forest of Rauphland* 1612, *Ralfland Forest* 1823. P.n. Ralfland + ME **forest**. Ralfland, *Rafland* c.1200, *Ralphland* 1726, is 'land belonging to Ralf', ME pers.n. *Ralf* (OG *Radulf*) + **land**. Also occurs as *Rowland Park* 1748. We ii.169.

RAM LANE Kent TQ 9646. No early forms. Perhaps ModE **ram** + **lane** or possibly the surname Ram as in Richard Ram of Westwell 1370. PNK 403.

RAME Corn SW 7233. Unexplained. *Raan* 1556, *Rame* 1650. The reading of the first form is uncertain. PNCo 147.

RAME Corn SX 4249. Uncertain. *Rame* from 1086. Possibly OE ***hrama** 'post, barrier' corresponding to OHG *(h)rama* 'post, frame' and dial. *rame* 'the base framework of anything, dried sticks', or **hramu* from the root seen in *hremman* 'hinder', *hremming* 'hindrance, obstacle'. Perhaps applied to Rame Head as a barrier to navigation. PNCo 147, Gover n.d. 232, DEPN.

RAME HEAD Corn SX 4148. 'Rame headland'. *Ramyshead* 1405, *Ramehed* c.1540. Cf. *le forland de Raume* c.1490 Gover n.d. P.n. RAME SX 4249 + ME **hēd** 'a headland'. PNCo 147, Gover n.d. 233.

RAMPISHAM Dorset ST 5602. Partly uncertain. *Ramesham* 1086, 1238, *Rammysham* 1288, *Rampsham* 1401. The specific might be OE **ramm** 'a ram', genitive sing. **rammes**, pers.n. *Ram*, genitive sing. *Rammes*, or **hramsa** 'wild garlic'; the generic is probably **hamm** 3 'an enclosure, a river meadow'. Do 125.

RAMPSIDE Cumbr SD 2466. 'Ram's head'. *Rameshede* 1292,

-heved(e) 1336, 1400, *Ramsyde* 1539. OE **ramm**, genitive sing. **rammes**, + **hēafod**. Refers to the shape of the headland. La 203 gives pr [ramsaid], L 160.

RAMPTON Cambs TL 4268. 'Ram farm'. *Ramtona -ton(e)* 1086–14th, *Rantone* 1086, *Ram(p)tune* 1086, *Rampton(e)* from 1247. OE **ramm** + **tūn**. Ca 183.

RAMPTON Notts SK 7978. Probably the 'ram farm'. *Rametone* 1086, *Ramentone* 1130, *Rampton* from 12th, *Raveneston'* 1194, *Ranton* 1200, *Raunton* c.1540. OE **ramm**, genitive pl. **ramma**, + **tūn**. The isolated form *Raveneston'* 1194 has given rise to speculation that the specific is OE **hræfn** 'a raven' or ON pers.n. *Hrafn*. It is probably best regarded as an erroneous back-spelling as a result of the late OE assimilatory sound change [vn] > [mn, nn]. Nt 58, SSNEM 379.

RAMSBOTTOM GMan SD 7816. 'Wild garlic valley-bottom'. *Romesbothum* 1324, *Romsbothum* 1509, *Ramysbothom* 1540. OE **hramsa, hramse** + ***bothm**. Alternatively the specific could be OE **ramm**, genitive sing. **rammes**, 'ram's valley-bottom'. Known locally as *Rammy*. La 64, L 86.

RAMSBURY Wilts SU 2771. 'Hræfn or the raven's fort'. *Rammesburi* [947]14th S 524, *-biri -bury* 1091, 1240, *Hremnesbyrig* c.990 Crawford, *Remmesbiri(a) -bury* 1227–1310, *Ramesberie* 1086, *Remsbury* 1558×1603. OE pers.n. *Hræfn* or **hræfn**, genitive sing. *Hræfnes*, **hræfnes**, + **byriġ**, dative sing. of **burh**. This name-type occurs three times in Berks minor names, in each case in association with a hill-fort or a feature which may have been taken to be so. No such feature, however, has yet been discovered here. Wlt 287, Signposts 144–5.

RAMSDEAN Hants SU 7022. Uncertain. *Ramesdene* 1233, *Rammesdene* 1248, *Ramesdune* 1257, *Rammesdon* 1316, 1414, *Ramsden* 1810. Perhaps 'wild garlic valley', OE **hramsa** + **denu**. But the specific could be any of **hræfn** 'a raven' or pers.n. **Hræfn** derived from it, genitive sing. **hræfnes**, **hræmnes**, or **ramm** 'a ram', genitive sing. **rammes**. Ha 137, Gover 1958.63.

RAMSDELL Hants SU 5857. Probably 'wild garlic dell'. *Ramesdela* 1170, *Ram(m)esdelle* 1248–1379. OE **hramsa** + **dell**. There is a tiny valley just N of this place. Ha 137, Gover 1958.143.

RAMSDEN BELLHOUSE Essex TQ 7194. 'R held by the Belhus family'. *Ramesden(e) Belhus* 1261. P.n. Ramsden + family name of Richard de Belhus of Cambridgeshire 1200. Ramsden, *Ramesdanam -dunā* 1086, *Ram(m)esden(e)* 1197–1390, *Ramnesdene* 1303, *Ravensden* 1564–73, appears to be 'Hræfn's' or 'raven's valley, OE pers.n. *Hræfn* or **hræfn**, genitive sing. *Hræfnes*, **hræfnes**, with regular early assimilation *fn > mn > mm*. However, the evidence for original *-fn-* is late and an alternative suggestion is OE **ramm** 'a ram' or **hramsa** 'wild garlic'. 'Bellhouse' for distinction from Ramsden Cray as in CRAYS HILL TQ 7192. Ess 168.

RAMSDEN Oxon SP 3515. 'Wild garlic valley'. *Rammesden'* 1246×7, 1278×9, *Ramesden* 1307, 1341. OE **hramsa** + **denu**. O 370, L 98.

RAMSDEN HEATH Essex TQ 7195. P.n. Ramsden as in RAMSDEN BELLHOUSE TQ 7192 + ModE **heath**.

RAMSEY Cambs TL 2885. 'Wild garlic island'. *Ramesige* 11th ASC(C) under year 1034, *(æt) Hramesige, Ramesie* c.1000, *(on) Ramesige* 1011, *Rammesege* c.1050 ASC(D) under year 1050, *Ramesege* c.1050 ASC(D) under year 1045, *[c.1060]14th S 1109, *Ramesie* [1040×42, 1053×7]14th S 997, 1108, *Ramesia* [1043×9, 1050×2]14th S 1106/1107, 1189×99, *Hramesege* 975×1016 S 1487, *Rames(eie)* 1200, *Rammes(eye)* 1227. OE **hramsa** + **ēġ**. Hu 212, DEPN, Writs 57–61.

RAMSEY Essex TM 2130. 'Hræfn, raven, ram's or wild garlic island'. *Rameseiam* 1086, *-eye* 1234–1303, *Rammesye* 1223. OE pers.n. *Hræfn*, **hræfn** or **ramm**, genitive sing. *Hræfnes*, **hræfnes** or **rammes**, or **hramsa**, + **ēġ**. Ess 346.

RAMSEY FORTY FOOT Cambs TL 3187. The reference is to *Vermu(y)dens Eau* 1654, 1655, the *Adventurers' Fortyfoot Dreyne* (1664) made in 1651 by Sir Cornelius Vermuyden. Cf. p.n. RAMSEY TL 2885 and ModE **adventurer, drain, eau**. Ca 209.

RAMSEY HOLLOW Cambs TL 3186. *Ramsey Hollow* 1824 OS. P.n. RAMSEY TL 2885 + ModE **hollow** 'a valley, a basin'.

RAMSEY ISLAND Essex TL 9605. P.n. Ramsey + pleonastic ModE **island**. Ramsey, *Ramingeseye* c.1210, *Ramsey* 1509×47 is uncertain element + OE **ēġ**. The specific, as in *Ram(m)yng(e)smersh(e)* 1483, 1485, is an **ing**[2] formation on either **hræfn** 'a raven', perhaps a pers.n. **Hræfning*, genitive sing. **Hræfninges*, ***hrama** 'a post, a barrier, a framework', 'island of the *hraming* or barrier', perhaps a fish-trap, or **ramm** 'a ram'. Ess 227.

RAMSEY MERESIDE Cambs TL 2889. The reference is to nearby Ramsey Mere, *Ramesmere* [13th]c.1350. P.n. RAMSEY TL 2885 + ME **mere**. Ca 215.

RAMSEY ST MARY'S Cambs TL 2588. A modern development, no early forms. P.n. RAMSEY TL 2885 + saint's name *Mary*.

RAMSGATE Kent TR 3765. 'Raven's gate'. *Ramis- Ram(m)esgate* 13th cent.–1357, *Remmesgate* 13th cent. OE **hræfn, hrefn, hremm**, genitive sing. **hræfnes, hremmes**, + **ġeat**. The reference is to a gap in the cliffs. Alternatively the specific may pers.n. *Hræfn* derived from *hræfn*. PNK 602.

RAMSGILL NYorks SE 1171. Probably 'the ravine where wild garlic grows'. *Ramesgile* [1198]15th, *-gill -gyll* 1363, 1591, *Rammesgille* 1343, 1345, *Ramsgill* from 1499. OE **hramsa** 'wild garlic', + ON **gil**. Alternative possibilities are either 'Hrafn's' or 'ram's ravine', ON pers.n. *Hrafn*, genitive sing. *Hrafnes*, or OE **ramm**, genitive sing. **rammes**, + **gil**. YW v.215.

RAMSHORN Staffs SK 0845. Possibly 'the wild garlic promontory'. *Rumesoura* 1197, *Romesovere* 13th, *Romesor* 1309, *Ramsore* 1538 etc. Uncertain element + OE **ofer** 'a ridge' referring to the narrow ridge on the tip of which the settlement lies. The specific could be either OE **hramsa** 'wild garlic', or **ramm** 'a ram', genitive sing. **rammes**, while a single form, *Ramnesovere* 1327, suggests OE **hræfn, hramn** 'raven' as a another possibility. Horovitz gives pr *Ramser*. DEPN, Duignan 124, PNE ii.80, L 175–6.

RANBY Notts SK 6581. 'Hrani's village or farm'. *Rane(s)bi* 1086, *Ranesby* 1330, *Raneby* 1280–1340, *Ranby* from 1236. ON pers.n. *Hrani*, genitive sing. *Hrana*, secondary *Hranes*, + **bȳ**. In the same parish is a lost *Ranel(o)und* 13th, 'Hrani's grove'. Nt 66, SSNEM 63.

RAND Lincs TF 1078. 'The bank'. *Rande* 1086–1202, *Randa* c.1115. OE **rand**. The village stands at the edge of high ground overlooking a stream. DEPN.

RANDALINTON Cumbr NY 4065 → KIRKLINTON NY 4367.

RANDWICK Glos SO 8306. Possibly 'edge-farm'. *Rendewiche* 1121, *-wik(e) -wyke* 1135×50, 1248, *Ri- Ryndewice -wyk(e)* 12th–1308, *Randewyk'* 1248, *Randwicke* 1635, *Ronwi(c)ke, Ran- Ren(d)wyke, Renwike* 16th cent., *Runnick* 1713, 1715. OE ***rend** + **wīċ**. The reference may be to the edge of Standish Wood – in 1086 Randwick was an outlying part of Standish – or to the border between the hundreds of Whitstone and Bisley. The rare OE **rend* < **rand-jō-* was later replaced by the commoner *rand*. Gl ii.189.

RANGEMORE Staffs SK 1823. *Rauenwolmesmor, Rauenesmor* 1337. Cf. *Rangemore Farm, House, Wood* 1836 OS. The early sps seem to point to 'raven-spring moor' and 'raven's moor', OE **hræfn** + Mercian **wælm**, genitive sing. **hræfn-wælmes** and **hræfnes**, + **mōr**. Horovitz.

RANGEWORTHY Avon ST 6986. Partly uncertain. *Rengeswurda* 1167, *Rengew(o)rth(e)* 1248–c.1603, *-worthye* 1539, *Rungewurthy -worth(i)* 1247, 1346, 1497, *Ri- Ryngew(o)rth* 1291–1479, *Rangeworthy* 1646, *Rangery* 1609. Uncertain element + OE **worth(iġ)** 'an enclosure'. None of the suggested etyma of the specific are really satisfactory: OE **hrynge** 'a pole, a stake' would not normally give ME *Renge-* spellings; OFr **renge** 'rank, row, felled wood, brushwood' would be an unusual compositional element with OE *worthiġ*; and an OE pers.n. **Hrencga* is unknown and unparalleled. Gl iii.11.

RANSKILL Notts SK 6587. 'Raven shelf'. *Raveschel* 1086, *Rauenschel* c.1280, *Raven(e)skelf* 1275–1359, *Ranskelf* 1364, *Ranskyll* 1513. ON **hrafn** + **skjalf**. The specific might be ON pers.n. *Hrafn*, and the whole name could be a Scandinavianised version of OE **hræfn** + **scelf**. The village lies on a slight slope above the level ground along the r. Idle. Nt 93, SSNEM 158, L 187.

RANSON MOOR Cambs TL 3893. 'Ravenesho marsh'. *Ramyshomore* 1345, *Rames- Raunshammor(e)* 1389, *Ransome moor(e)* 1618, *Ransonmo(o)re* 1636. P.n. *Ravenesho* 1227 'raven's hill-spur' (OE **hræfn**, genitive sing. **hræfnes**, + **hōh**) + ME **more** (OE *mōr*). There is a low ridge of land at Doddington S of the moor. Ca 252.

RANTON Staffs SJ 8524. Partly uncertain. *Rantone* 1086, *Ranton(ie)* c.1182, 1209, *Ram(p)ton* 1208, *Rontun* 1236, *Ramton* c.1540. Possibly 'the ram farm', OE **ramm** + **tūn**, or 'the bank settlement', OE **rand** + **tūn**. The village stands beside Clanford Brook and also near the hundred boundary. DEPN, Horovitz.

RANWORTH Norf TG 3514. 'Shore enclosure'. *Randwrðe* *[1044×7]13th S 1055, *Randuorda* 1086, *Randewrtha* c.1158, *-worth* 1203, *Randeswrth* 1242. OE **rand** + **worth**. Ranworth lies on the shore of Ranworth Broad. DEPN.

Market RASEN Lincs TF 1089. *Marketrasyn* 1358, *Market Rasen* 1619. ME **market** + p.n. *(æt) ræsnan* [973]13th S 792, *Resne* 1086, *Rase -a* 1086–1223, *Rasne* 1185–1279, *Rasen(')* from 1202, '(the settlement) at the planks', OE **ræsn**, dative pl. **ræsnum**, probably referring to a plank bridge over the R. Rase (itself a back-formation from the p.n.) at Middle or West Rasen or over marshy ground here. 'Market', *altera* 1086 and 'east' *Estrasne -en* 1153–1545 for distinction from Middle and West Rasen TF 0889 and 0689. Li iii.94.

Middle RASEN Lincs TF 0889. *Media Rasa* c.1115, ~ *Rasum* 1212, ~ *Rasne -en* 1209–1535, *Middelrasen(')* 1201–1374, *Middle Rasen* from 1282. ME **middle**, Latin **media**, + p.n. Rasen as in Market RASEN TF 1089. Li iii.102.

West RASEN Lincs TF 0689. *Westrasen(')* from 1202, *-rasun -rason -rasyn('), Rasyn, Rasin* 1236–1526. ME **west** + p.n. Rasen 1086 as in Market RASEN TF 1089. Li iii.114.

RASKELF NYorks SE 4971. 'Roe-deer ledge'. *Raschel* 1086, *Raskel(l)* 1169–1613, *Raskelf* 13th cent. ON **rá** + **skjalf** 'a shelf' referring to a low ridge of land overlooking the river Kyle on which the village stands. The name may be a Scandinavianisation of earlier OE **rā** + **scelf**. YN 26 gives pr [ræskil], SSNY 102.

RASTRICK WYorks SE 1321. 'The long narrow ridge with a resting-place, or that requires a rest'. *Rastric -rik -ryk* 1086–1500, *Rastrick(e)* 1180×1202–1641. ON **rǫst** or OE **ræst** + ***ric**. The village lies on the Roman road to Lancashire, Margary no.712, which here climbs the steep hill-side from the r. Calder. YW iii.38, SSNY 162, L 184.

RAT ISLAND Devon SS 1443. A small island at the S tip of Lundy Island, shaped like a rat.

RATAE Leic SK 5804. The name of the Roman city of Leicester, the capital of the Corieltauvi. *Ratis* 2nd inscrs, *Ῥάτε* (Rate) [c.150]13th Ptol with variants *Ῥάχε* and *Ῥάγε* (*Rache, Rage*), *Ratas, Ratis* [c.300]8th AI, *Rate Corion* [c.700]13th Rav. Probably a Latinised form of British **rātis* 'an earthern rampart, a fortification, a fort', cognate with Irish *rá(i)th* the 'earthern rampart surrounding a chief's residence, a ring-fort'. Cf. the Gaulish p.ns. *Argentorate* (Strasbourg), *Corterate* (Coutras, Gironde, France). RBrit 443, Britannia 1.78, Holder II.1075, Dauzat-Rostaing 1963.224, 665, IPN 132.

RATBY Leic SK 5101. Possibly 'Rota's village or farm'. *Rotebi(e)* 1086–1214, *-by* after 1204–1512, *Routebi* c.1206, *-by* 1326, *Rootby* 1484–1513, *Rotby* 1464–1540, *Ratby(e)* from 1549. OE pers.n. **Rōta* + ON **bȳ**. Lei 530, SSNEM 63.

RATCLIFFE '(The) red cliff'. OE **rēad**, definite form **rēade**, + **clif**. L 132–5.

(1) ~ CULEY Leic SP 3299. 'R held by the Culey family'. P.n. *Redeclive* 1086, *Radecliue -y- -v-* 1226–1375, *-clif -clyf(f)* 1358–1426, *Radcliue -clif -clyf(fe)* c.1130, 1357–1428, *Ratclif(f) -y-* 1493–1554, *-cliffe* from 1576, + manorial affix *Cule* 1392, *Culey* from 1506. The manor was held by Johannes de Cuylly in 1228. Lei 557.

(2) ~ ON THE WREAKE Leic SK 6314. P.n. *Radecliue -v-* 1086–1361, *-clif(f)' -y-* 1332–1486, *Radclif(f)(e) -y- -clyve* 1285–1547, *Ratcliff' -y-* 1456–1610, + r.n. WREAKE SK 6616, *super (le) Wrethek, Wre(y)k(e)* etc. 1259–1699, *opon Wrethek, Wreyk* 1456, 1486. Lei 316.

RATHMELL NYorks SD 8059. 'The red sand-bank'. *Rodemele* 1086, *Routh(e)mel(e) -mell* [12th]copy–1335, *Rau- Rawth(e)mell* [1394]copy–1706, *Rathemell* [1266]copy, *Rathmell* from 1474. ON **rauthr** + **melr**. Identical with *Rauthamelr* Iceland. YW vi.148, SSNY 102.

RATLEY Warw SP 3847. 'The clearing with tree-roots'. *Rotelei* 1086, *-le(gh) -ley* 1155–1302, *Rotteleia -le(g) -leye* 1154×89–1535, *Rotleye* 1387, *Ratley* from 1538. OE **rōt**, genitive pl. **rōta**, + **lēah**. The form *Ratclyffe* 1549, 1609 is a hypercorrection due to the frequent reduction of *-clif* to *-cley* in p.ns. This explanation supersedes that in Wa 272 (pers.n. *Rōta* or adj. **rōt** 'cheerful'). NQ 241.128–9.

RATLINGHOPE Shrops SO 4097. 'The secluded valley at or called *Roteling*, the place called after Rotel'. *Rotelingehope* 1086–1271×2, *Rotelinghop(e)* 1255–1301, *Rol(l)ynghope* 1271×2, 1535, *Ratlinghope* from 1641, *Rotchchop* 1255, *Ratchop* 1737, *Rotelynch'hope* 1306×7. OE p.n. **Rōteling* < pers.n. **Rōtel* + **ing**[2], locative-dative sing. **Rōtelinġe*, + **hop**. The alternative explanation, 'the valley called after Rotel', OE pers.n. **Rōtel* + **ing**[4] + **hop**, does not account for the assibilated pronunciation evidenced in the last three sps cited. Sa i.248; BzN 1967.385 gives a local pr [rætʃəp] and offers an alternative pers.n., OE *Hrōthulf*, Nomina 6.34, L 114–21.

RATTEN RAWE Lancs SD 4241. No early forms. *Raton-raw* 'rat-infested row of houses' (OFr **raton** + ME **rawe**) is a regular pejorative name for poor houses.

RATTERY Devon SX 7461. 'The red tree'. *Ratrev* 1086, *Radetre* c.1240, *-trewe* 1316, *Rattre* 1242, 1620, *Rattrew -tru* 1270, 1284, *Ratterie* 1675. OE **rēad**, definite form **rēade**, + **trēow**. The spelling *Ratworthie* 1620 is an inverted form due to the frequent reduction of names in *-worthy* to *-ery*. D 310, L 214.

RATTLESDEN Suff TL 9759. Possibly 'the valley where rattle grows'. *Rattesdene* [1042×66]12th S 1051, *Ratesdana, rachest- raste- ratlesdena* 1086, *Retlesden* 1198, *Ratlesden* 1198, 1200, 1610. This has been taken to be 'Ræt or Rætel's valley', OE pers.n. **Ræt* or **Rætel*, genitive sing. **Ræt(l)es*, + **denu**. Such a pers n. would belong to the OE appellative *ræt* 'a rat' with the meaning 'rat' or 'little rat' which seems unlikely. An alternative and preferable suggestion is OE **hrate**, **hratele**, genitive sing **hrates**, **hrateles**, a plant name found in the glosses *hratele: bobonica, hrate: bobonaca* and *hrætelwyrt: hierobotanum*, the latter glossed as 'vervain'. Connection of the Anglo-Saxon name with ModE *rattle* as in Red and Yellow Rattle is uncertain. DEPN, DML s.n. *bubonica*.

North RAUCEBY Lincs TF 0246. *Nord- Northrouceby* 1242–1731, *-rauceby* 1552, 1808. ME **north** + p.n. *Ros(ce)bi* 1086, *Rosceby* 1197–1327, *Rou- Raucebi -by* 1146–1607, 'Rauthr's village or farm', ON pers.n. *Rauthr*, genitive sing. *Rauths*, + **bȳ**. 'North' for distinction from South RAUCEBY TF 0245. Pr with [s] from original genitive sing. *-s* shows this name to be a formation by Danish speakers. Perrott 289, SSNEM 64.

South RAUCEBY Lincs TF 0245. *South Rouceby* 1242–1731, *S. Rauceby* 1778. ME **south** + p.n. Rauceby as in *Altera -o Rosbi* 'the other R' 1086 and North RAUCEBY TF 0246. REF.

RAUCEBY STATION Lincs TF 0344. P.n. Rauceby as in North RAUCEBY TF 0246 + ModE **station**. The railway link from Grantham to Sleaford was built in 1857.

RAUGHTON Cumbr NY 3947. 'Settlement by Roe Beck'. *Ragton'* 1182, *Rach- Ra(g)hton* 1186–1470, *Rauthon* 1288, *Raughton* from 1292. Stream name Roe as in ROE BECK NY 3941, *Rawe* 1272, + **tūn**. Cu 134, SSNNW 423.

RAUGHTON HEAD Cumbr NY 3845. 'Headland belonging to Raughton'. *Raughtonheved* 1367, *Raughton Hede* 1476. P.n. RAUGHTON NY 3947 + OE **hēafod**. Probably a stretch of hill pasture used by the men of Raughton. Cu 245.

RAUNDS Northants SP 9972. 'The borders'. *(æt) randan* [983×5]1200 S 1448a, *Rande* 1086–1316 with variants *Randa* and *Raunde, Raundes* from 12th with variants *Randes* and *Raunds, Rawns* 1442. OE **rand** 'a border, an edge', pl. **randas**. The first spelling represents dative pl. **randum** 'at the borders'. The place lies on the eastern edge of the county. Nth 194 gives pr [ra:ns].

Great RAVELEY Cambs TL 2581. *Magna, Graunt Rauele* 1297, 1364. ModE **great**, Latin **magna**, Fr **grand**, + p.n. *Ræflea* *[1062]14th S 1030, [1077]17th, *Rauelai* 1163, *Ravele(ye)* 1227 etc., unknown element + OE **lēah** 'wood, clearing'. The absence of any trace of forms with *-n-* renders Ekwall's suggestion OE *hræfn-lēah* 'raven wood' unlikely. Hu 217.

Little RAVELEY Cambs TL 2579. ModE **little** + p.n. Ravely as in Great RAVELEY TL 2581. Cf. *Rauelea et altera R.* 1167. Hu 217.

RAVEN'S KNOWE Northum NT 7806. 'Raven's hill'. N dial. **knowe** (OE *cnoll*). A Cheviot peak rising to 1729ft.

East RAVENDALE Humbs TF 2399. *Estravendale* 1238–1428, *E(a)st Randall* 1519–1679. ME adj. **est** + p.n. *Raven(e)- Rauenedal(e), altera Rauenedale* 1086, *Rauenedala* c.1115, *Randall* 1535, 1539×40, 'raven valley', ON **hrafn**, genitive pl. **hrafna** + **dalr**, or OE **hræfn**, genitive pl. **hrafna** + **dæl**. 'East' for distinction from West RAVENDALE TF 2299. Li iv.151, SSNEM 159.

West RAVENDALE Humbs TF 2299. *Westravendale* 1202–1428, *Westrandall* 1553–1842. ME adj. **west** + p.n. Ravendale as in East RAVENDALE TF 2399. Li iv.152.

RAVENFIELD SYorks SK 4895. 'Raven's open land'. *Rau- Ravenesfeld* 1086–1276, *Rau- Ravenfeld* 1234–1431, *Ranfeld -feild -field* 1338–1608. OE **hræfn** or ON **hrafn** or a pers.n. derived from these words, genitive sing. **hræfnes**, **hrafns** + **feld**. YW i.172, SSNY 162.

RAVENGLASS Cumbr SD 0896. 'Glas's lot or share'. *Rengles* c.1180, *Renglas* 1208–92, *Ranglas(s)* c.1240–1323, *Ravenglas(se)* 1297–1540. OIr **rann** + Ir pers.n. *Glas*. The spelling with *Raven-* is a popular hypercorrection due to the regular ME change *raven* > *rane*, e.g. in *Ranestandale* 1291 for RAVENSTONEDALE NY 7204 and in RENWICK NY 5943. Cu 425.

RAVENINGHAM Norf TM 3996. 'Homestead of the Hræfningas, the people called after Hræfn'. *Rauenicham,Raueringham -i(n)cham, Rauelincham, Rauingeham -incham* 1086, *Rafningeham* 1177, *Raueningham* 1202–3, *Ravenigham* 1236, 1242×3. OE folk-n. **Hræfningas* < pers.n. **Hræfn* + **ingas**, genitive pl. **Hræfninga*, + **hām**. DEPN, ING 137.

RAVENSCAR NYorks NZ 9801. Probably 'raven cliff or rock'. *Rauenesere* [1312]late 14th. ON **hrafn** + **sker**. The reference is to the lofty headland known as The Peak. Cf. OLD PEAK NZ 9802. YN 111.

RAVENSDEN Beds TL 0754. 'Hræfn or raven's valley'. *Rauenesden(e)* [c.1150]13th–1493, *Raunston* 1528. OE pers.n. **Hræfn* or **hræfn**, genitive sing. *Hræfnes*, **hræfnes**, + **denu**. Bd 61 gives pr [ra:nzdən], L 98.

RAVENSEAT NYorks NY 8603. *Raven Seat* 1860 OS. The hamlet must have been named from the 1881ft. summit on *Ravenseat Moor* 1860 OS. ModE **raven** + NW dial. **seat** 'the summit of a hill'.

RAVENSER lost, near Easington Humbs TA 3919 → SPURN HEAD TA 3910.

RAVENSHEAD Notts SK 5654. 'Raven's head'. *Raveneshevéd* 1205, 1287, *Ravensheved* 1327×77. OE **hræfn**, genitive sing. **hræfnes**, + **hēafod** in the sense 'headland'. This is the highest ground in the neighbourhood and is probably named from the bird rather than a pers.n. Nt 129, L 161.

RAVENSMOOR Ches SJ 6250. 'Raven's waste or marsh'. *Raven(e)smor(e)* [early 13th]1244–1621, *-moor* 1831, *Raynmore* 1460, *Ranmore* 1514, 17th cent. OE **hræfn**, genitive sing. **hræfnes**, + **mōr**. Ancient common described in 1656 as 'a very sweet and fruitful piece of ground... hitherto preserved for the relief of the poor neighbours to it'. Che iii.135 gives pr ['reivnzmɔ:r], older local ['ranmər] and ['ramnər].

RAVENSTHORPE Northants SP 6670. 'Hrafn or Hræfn's outlying farm'. *Ravenstorp* 1086–15th with variant *-thorp(e)*, *Raunest(h)orp* 1247, 1346, *Rawns-* 1477, *Ransthorpe* 1539, *Rawnstruppe* 1631. Pers.n. ON *Hrafn* or OE **Hræfn*, genitive sing. **Hrafnes*, + **thorp**. Nth 87 gives pr [rɔ:nstrəp], SSNEM 115.

RAVENSTHORPE WYorks SE 2220. A modern name modelled on the old name for Camber Dyke, *Ravenesbroc* 1296, *Ravensbrooke* 1525, 'Hrafn's or raven's brook', OE **hræfn** or ON **hrafn**, genitive sing. **hræfnes**, + **brōc**. YW ii.192.

RAVENSTHORPE RESERVOIR Northants SP 6770. P.n. RAVENSTHORPE SP 6670 + ModE **reservoir**.

RAVENSTONE Bucks SP 8550. 'Hræfn, Hrafn or the raven's estate'. *Raveneston* 1086, *Raueneston(e) -tun* 1167–c.1260, *Rauenstone* 1304, *Raunston* 1379–1526. OE pers.n. **Hræfn*, ON *Hrafn*, or OE **hræfn**, genitive sing. **Hræfenes, Hrafn(e)s*, or **hræfnes**, + **tūn**. Bk 12 gives pr [rɔ:nstən].

RAVENSTONE Leic SK 4013. 'Hræfn or Hrafn's farm or village'. *Ravenestvn* 1086, *-ton(e)* c.1130–1429, *Rauenston' -v-* 1262–1534, *Raunston'* 1462–1641, *Ranstone* 1522, *Raunson* 1641. OE pers.n. **Hræfn* or ON pers.n. *Hrafn*, genitive sing. **Hræfnes*, + **tūn**. The alternative DB form *Ravenstorp* is either a scribal error or an attempt to replace a native by a Scandinavian element. Lei 569, SSNEM 193, Db 652.

RAVENSTONEDALE Cumbr NY 7204. 'Valley of the raven stone'. *Rau- Ravenstandal(e)* 1154×89–1577 with variant spellings *-stain- -steyn-* 1223–92, *Ravenstondale -dall* 1593–1704, *Ranestandale* 1291, *Ri- Rystnedall* 1578–86, *Rissendaill* 1605, 1625. OE ***hræfn** + **stān** 'the raven stone' referring to some characteristic rock or stone not now known, + ON **dalr**. 13th cent. spellings show the influence of ON **steinn**. We ii.30, SSNNW 246, L 94.

RAVENSTONEDALE COMMON Cumbr NY 6900. *Ravenstonedale Common* 1861. P.n. RAVENSTONEDALE NY 7204 + ModE **common**. We ii.37.

RAVENSTOWN Cumbr SD 3675. Cf. *Ravynse Windor* 1491, *Raven Winder* and *Chanon Winder, Winder Moor* 1864 OS. ME **raven** (OE *hræfn*) or pers.n. *Raven*, genitive sing. **ravenes**, *Ravenes*, with **wind** + **erg** 'windy shieling'. Mills 150.

RAVENSWORTH NYorks NZ 1407. 'Hrafn's ford'. *Ravenesu(u)et* 1086, *Raven(e)swath* [1184]15th–1427. ON pers.n. *Hrafn* or OE **Hræfn*, secondary genitive sing. *Hrafnes*, + **vath**. The specific might, however, be the appellative **hrafn** 'a raven'. YN 292, SSNY 102, L 82.

RAW NYorks NZ 9405. 'The row of houses'. *Raw* 1857 OS. Short for *Fyling Rawe* 16th, the 'row of houses at Fyling'. P.n. Fyling as in FYLINGDALES + ME **raw** (OE *rāw*). YN 118.

RAWCLIFFE Humbs SE 6822. 'The red bank'. *Rouþeclif* [1079×85]14th, *Routheclif(f) -clyf* 13th cent., *Ro(u)decliue -a* 12th cent., *Rouclif(f) -clyf(f)* 1189–1334, *Rawclyff(e)* 1379, 1583. Probably OE p.n. ***rēade-clif** with substitution of ON **rauthr** 'red' for OE **rēad**. YW ii.22, SSNY 162.

RAWCLIFFE NYorks SE 5855. 'The red cliff'. *Roud(e)clif(e)* 1086, *Rout(h)ecliua, Routhecliue* 1170–1294, *-clyff* 1301, 1323, *Roucliff -clyf(f)* 14th cent. ON **rauthr** + **klif**, referring to the high reddish bank of the river Ouse. YN 15, SSNY 162.

RAWCLIFFE BRIDGE Humbs SE 7021. *Rawcliffe Bridge* 1817. P.n. RAWCLIFFE SE 6822 + ModE **bridge**. YW ii.23.

Out RAWCLIFFE Lancs SD 4042. *Outerouthclif* 1327, *Outrotheclife* 1332, *Outrauclif* 1443. ME **out** (OE *ūt*) + p.n. *Rodeclif* 1086, *Boutecliue* [for *Route-*] 1206, *Routheclive -clyve -clif* 1219–1323, *Raucliff* 1537, the 'red cliff', ON **rauthr** + **klif** possibly replacing OE **rēad** and **clif**. 'Out' for distinction from Middle and Upper Rawcliffe, *Middle Routheclive* 1249, *Uproucheclive* 1246, *-routheclive* c.1275, *Uprauclyf* 1369. Upper Rawcliffe is higher up the Wyre rather than on higher ground and not, therefore, the original settlement site which was probably at Rawcliffe Hall SD 4140. The reference is to slightly elevated ground overlooking the river Wyre where the soil colour is exposed. La 160, 161, SSNNW 246, L 132, 125, Jnl 17.93.

RAWDON WYorks SE 2139. 'The red hill'. *Roudun, Rodum* 1086, *Roudun -don* late 12th–1374, *Raudon* 13th–1367, *Rawdon* from 1246. ON **rauthr** possibly replacing OE **rēad** + OE **dūn**. YW iv.152, SSNY 162, L 157.

RAWMARSH SYorks SK 4496. 'The red marsh'. *Rodemesc* (sic) 1086, *Routhemersk* 1293, *-mers(he)* 1298–1334, *Rowmareis* 1190×1210, *-mers(c)h(e)* 14th, 15th cent., *Raumarays* 1297, *-mersche* 14th cent., *Rawmarsh* from 1502. ON **rauthr** possibly replacing OE **rēad** + **mersc**. From the red earth everywhere about this place. YW i.175, SSNY 162, L 53.

RAWRETH Essex TQ 7794. 'Heron stream'. *Raggerea* 1177, *Regriue -rug(g)e* 1181–4, *Rag(h)er(ee)* 1189–1303, *Ragheride* 1201, *-rith* 1302–5, *Rawe- Raurey(e)* 1238–61, *Raureth(e)* 1267–1414. OE **hrāgra** + **rīth**. Ess 192.

RAWRIDGE Devon ST 2006. Partly uncertain. *Rou(e)rige* 1086, *Roveruge* 1191–1421 with variants *-rigge -rygge -righ -regge, Raurygge* 1360. The generic is OE **hrycg**, the specific identical with *hrofan* in *of hrofan hricge* [986]12th S 861. This might be pers.n. **Hrofa*, genitive sing. **Hrofan*, or the dative sing. definite form of an adjective ***hrof** 'scabby, crusty' beside *hrēof* 'rough, scabby, leprous'. The form with zero-grade ablaut, however, should be a noun, cf. ON *hrufa* 'scaly skin, crust' OHG *ruf* (ModG *Rufe*). D 650.

RAWTENSTALL Lancs SD 8123. 'Farmstead or vaccary on the rough ground'. *Rowtonstall* 1323, *Routonstall* 1324–5, *R(o)unstall, Rotenstall* 1507. OE **rūh** probably used as a noun 'rough ground' + ***tūn-stall**. Described as a vaccary in 1324, the reference being to rough pasture. Identical with RAWTONSTALL WYorks SD 9729. La 92, SSNNW 424, Jnl 17.54.

River RAWTHEY Cumbr SD 7595. 'Red river'. *Rautha* [1224]1651, *Roudha, Routha* 1235×55, *Rowthey* 1777, *Rawthey* 1861. ON **rauthr** + **á**. So called from an exposure of the Old Red Sandstone in its course. Immortalised for its pebbly madrigal accompanying Bunting's bragging tenor bull in *Brigg Flats*. We i.12, L 11.

RAWTONSTALL WYorks SD 9729. 'Farmstead on the rough ground'. *Ructunstal* 1238, *Routunstal(e)* 1265, *Routon-* 1265, 1275, *-stall* 1274–1342, *Rattonstall* 1583, 1817, *Rawtonstall* from 1606. OE **rūh** probably used as a noun 'rough ground' + ***tūn-stall**. The reference is to rough pasture. The modern form with *Raw-* is due to a late ME confusion of *rū* and *rau* in dial. YW iii.197, SSNNW 423.

River RAY Oxon SP 5917. 'The river'. *le Ree of Ettemore* (Ot Moor) 1363, *the Reie* 1586. The name arose through misdivision of ME *atter ee* 'at the river' (OE *æt thǣre ēa*). Its earlier name was *Geht* 844×5 S 204, *Giht* [983]13th S 843, *Ychte* 1185, *Yhyst* [1298]c.1444, of uncertain origin and meaning. O 9, RN 209, 336.

River RAY Wilts SU 1191. *The Rey* 1576, *the Rhe* 1577, 1586, *(the) Rea* 1630, 1733, *Rye* 1754. Not really a r.n. at all, but due to misdivision of ME *atter ee* 'at the river' < OE *æt thǣre īe* as if *atte ree*. The true name, preserved in the p.n. WROUGHTON SU 1480, was *uuorf* [962]13th S 705, *Worfe* [796]13th S 1586, [1008]13th S 918, *Wurf* [943]12th S 486, [956]12th S 638, *Werf(f)e* 1228, OE **Werf*, **Weorf*, **Worf* 'the winding river' < Brit **u̯erb-* 'turn, twist'. Wlt 9, RN 336, 469, L 20.

RAYBURN LAKE Northum NZ 1192. P.n. Rayburn + ModE **lake**. Rayburn is probably the 'roe-deer stream', ultimately from OE **rā** + **burna**.

RAYDON Suff TM 0538. 'Rye hill'. *Reinduna -e, Rienduna* 1086, *Reidunia* 12th, *Reindun* c.1200, *Reydon* 1254–1523, *Raydon* from 1568. OE **ryġen** + **dūn**. Identical with REYDON TM 3440. DEPN, Baron.

RAYLEES Northum NY 9291. 'Roe-deer clearings or pastures'. *Raleys* 1377, *Ralees* 1409, *Reelees* 17th cent. OE, ME **rā** + **lēah**. Either a place frequented by roe-deer, or where they were shot. NbDu 163.

RAYLEIGH Essex TQ 8090. 'Wood or clearing of the wild she-goats or female roe-deer'. *Rageneiā, Ragheleiam* 1086, *Ra(e)lega -legh -leg(he)* 1173–1337, *Reilee, Reyleg(h) -lee -ley(e)* 1181–1303, *Rayley -le(i)gh* from 1216×72. OE **rǣġe**, genitive pl. **rǣġena**, + **lēah**. Ess 194.

RAYNE Essex TL 7222. Uncertain. *(æt) Rægene* [961×95]11th S 1501, *(æt) Hrægenan* 975×1016 S 1487, *Ræine* *[1066]13th S 1043, *[1065]14th S 1040, 1085, *Raines* 1086, 1166, *Rei- Reynes* 1195–1294, *Rayne* from 1121. An OE ***hrægene** seems to lie behind these forms in the sense 'hut, shelter' related to OE *ofer-hrǣġan* which is thought to mean 'tower above', cf. G *ragen* 'tower, loom' of unknown origin. This is very uncertain, however, and, since the *Hr-* and *R-* spellings are of more or less equal authority, Ess 452 may be right in ignoring the former and pointing to a pre-English r.n. for Pods Brook, cognate with G r.n. Regen, a tributary of the Danube at Regensburg, *Reganum* [c.700]14th Rav, *Regan* [819]9th, 1107, *(iuxta) Reganam* 882, *Régino* 1003, *Regin* 1029, *Regen* 1140, an *-ina/-ana* formation on the IE root *reg-* 'wet, moisten, water'. An exact parallel would be the G town Regen, *Regen* 1239, *Regn* 1148, on the Black Regen N of Passau. The r. RYE NYorks SE 6382 is formed on the same root. Krahe 1964.104, Berger 218, LbO 313.

East RAYNHAM Norf TF 8825. *East Raynham* 1838 OS. ModE adj. **east** + p.n. *Reineham* 1086, *Reinham* 1199, 'Regna's homestead', OE pers.n. *Reġna* + **hām**. 'East' for distinction from South and West RAINHAM TF 9724, 8725. DEPN.

South RAYNHAM Norf TF 8724. *Sutreineham* 1086. OE adj. **sūth** + p.n. Raynham as in East RAYNHAM TF 8825. DEPN.

West RAYNHAM Norf TF 8725. *West Raynham* 1838 OS. ModE adj. **west** + p.n. Raynham as in East RAYNHAM TF 8825.

RAY SAND Essex TM 0500. *Reysand, le Reysant* 1367. P.n. Ray + ME **sand** 'a sand-bank'. Ray might represent ME *atte Ray* < *atter Ay* < OE *æt thǣre ēġe* 'at the island' or the fish-name **raye**. Ess 18, 321.

RAYTON SYorks Notts SK 6719. 'The reeve farm'. *Rouuetone, Reneton* (for *Reue-*) 1086, *Reueton* 1287–1333, *Reton* 16th, *Rayton* from 1616, *Ryton* 1677. OE **(ġe)rēfa** + **tūn**. Nt 108.

River REA H&W SO 6773. *(le) Ree* 1310, 1523–1619, *Rea* 1577, 1619. This is simply the fossilised form *rea* due to misdivision of OE *in þære ea* 'to the river' [?781×96]11th S 1185, OE feminine definite article **sēo**, dative sing. **thǣre**, + **ēa**, with original r. name *Nen*, British **Nēn-* of unknown origin. RN 336, Hooke 82.

River REA Shrops SO 6773. This refers to the lower reaches of REA BROOK SO 6586.

REA BROOK Shrops SO 6586. 'The river'. *in þære éa nen* *[?781×96]11th S 1185, *Ree* 1310, 16th, *la, le Ree* 1579, 1610. The original name of the river was *Nēn* as in NEEN SAVAGE SO 6777 etc. The later form derives from the OE phrase *in thǣre ēa* 'to the river' misdivided in ME as if *in the Rea*. RN 336, Hooke 82.

REACH Cambs TL 5666. 'The strip'. *Rete, Reche* 1086, *Reche* c.1122–1552, *La Reche* 1230. This is OE **rǣċ**, ModE 'reach', used here in the sense 'raised ridge' referring to Devil's Dyke. The place was often described as a *litus* 'a shore'. It lies at the edge of higher ground reaching into Burwell Fen at the end of Devil's Dyke, a post-Roman linear earthwork aligned on Reach Lode, itself an artificial cut probably of Roman origin which served as a channel for sea-going ships loading and unloading here. Ca 136, lix, L 183–4, Pevsner 1954.210, 367.

READ Lancs SD 7634. 'Roe-deer headland'. *Revet* 1202, *Reved* 1246–1311, *Revid* 1258–1332, *Rieheved* 1418, *Rede* 1554×8. OE **ræġe** + **hēafod**. The reference is to the hill-ridge on which Read Hall stands overlooking the confluence of Sabden brook with the Calder. La 79, Jnl 17.49.

READING Berks SU 7173. '(At, among) the Readingas, the people called after Reada'. *(to) Readingum* c.900 ASC(A) under year 871, *(æt) Readingan* [962×91]11th S 1494, *Red(d)inges* 1086, *Radinges -as -um -ingges* 1152, 1153 etc., *Redyng(g)es* 1295 etc., *Reding'* 1152, *Redyng -inge* 1330 etc. OE folk-n. **Rēadingas* < pers.n. **Rēada*, an original by-name cognate with ON *Rauthr* 'the red-haired one' + **ingas**, locative-dative pl. **Rēadingum*. Brk 170, ING 45, TC 158.

READING STREET Kent TQ 9230. 'Reading hamlet'. *Reeding Street* 1710, *Reading-street* 1798. P.n. *Reedinges* 1216×72, *Reding(') -yng* 1240–1619, *la Ryding* 1254, 'the clearing', OE ***ryding**, Kentish ***reding**, + ModE **street**. PNK 359, ING 180, Cullen.

READ'S ISLAND Humbs SE 9622. Surname *Read* + ModE **island**. Earlier called *Ferriby Sand or Old Warp* 1824, ModE dial. **warp** 'sediment deposited by a river'. PNE ii.248.

REAGILL Cumbr NY 6017. 'Ravine haunted by foxes'. *Reue- Revegil(e) -gill(e)* 1176–1526, *Regill -gyll* 1374–1617, *Reagill* from 1592. ON **refr**, genitive pl. **refa**, + **gil**. We ii.157, SSNNW 153.

REARSBY Leic SK 6514. 'Reitharr or Rethar's village or farm'. *Re(d)resbi* 1086, *Reresbi -by -is- -ys-* early 13th–1535, *Rearesby(e)* 1576, 1610. Pers.n. ON *(H)reitharr*, ODan *Rethar*, ME genitive sing. *Rethares*, + **bȳ**. Lei 317, SSNEM 64 no.1.

REASE HEATH Ches SJ 6454. Possibly 'heath where races were held'. *Rease Heath (Hall)* 1772, *Raseheath (Hall)* 1789, *Rees Heath* 1831. ModE **race**, dial. [riəs, ri:s], + **heath**. Nearby Mile House, *The Mile House* 1831, may have been the distance mark for competitors. Che iii.153 gives pr [ri:s-].

RECULVER Kent TR 2269. Partly uncertain. *Racolvr'* 1241, *Racolv(r)e* 1243, *Rakulvre* 1247, *Raculu(e)re* 1270. The earlier form of this name was *Raculf(e)*, the direct descendant of the RBrit name REGULBIUM TR 2169. The present form is probably due to misunderstanding of the abbreviation mark in spellings like *Raculf'* 1226 etc. as if standing for *-re*, perhaps aided by association with ME *culver* < OE *culfre* 'pigeon, dove'. For a similar instance cf. SEVENOAKS TQ 5255. KPN 12.

REDBOURN Herts TL 1012. 'Reed stream'. *(æt) Reodburne* [1041×9]13th S 1228, *Redborne* 1086 etc. with variants *-b(o)urne*. OE **hrēod** + **burna**. This etymology was known in the 11th cent., *Redburna quod tale nomen trahit ab arundineto* 'R so called from a reed-bed' (*Gesta Abbatum Sancti Albani* Rolls Series i.54). The original name of the river Ver. Hrt 78 gives pr [redbən].

REDBOURNE Humbs SE 9700. 'The reedy or the red stream'. *Rad- Redburne* 1086, *(in) Ratburno* 1090, *Redburna* c.1115, *Rodburn* 1224. This type of name is usually taken to be a compound of OE **hrēod** 'reed' + **burna**, but the *a-* spellings rather suggest OE **rēad** 'red' as the specific. DEPN.

REDBROOK Glos SO 5310. 'Reed stream'. *Redebroc* 1216, *-broke* 1490, *Redbro(o)ke* 1536–1642. OE **hrēod** + **brōc**. Gl iii.237.

REDCAR Cleve NZ 6024. 'The red or reed marshland'. *Red(e)ker(re)* c.1170–1422, *Rideker* 1271, *Readcar* 1653. OE **rēad**, definite form **rēada**, or **hrēod**, genitive pl. **hrēoda**, + ON **kjarr**. The land is low-lying and the rocks of reddish hue. YN 156.

REDCLIFF BAY Avon ST 4475. No early forms, but probably self-explanatory.

RED DIAL Cumbr NY 2546. *Red-Dyal* 1685, *Red Deal* 1816. Also in Cumbr is Blue Dial NY 0740, *Blue Dial* 1774. The reference is to sun-dials carved out of turf, cf. 3 Henry VI, II, v, 21–5: 'Oh God! methinks it were a happy life/ To be no better than a homely swain;/ To sit upon a hill, as I do now,/ To carve out dials, quaintly, point by point,/ Thereby to see the minutes how they run'... Cu 331, 307, 469, NQ 185.266–7.

REDDISH GMan SJ 8998. Probably 'reed ditch'. *Rediche* 1212–84, *-dich* 1212–1332 etc., *Reddich* 1262, *Reedyche* 1325, *Redissh* 1404, *Reddish* 1577, *Radich(e)* 1226, 1324, *Radyshe* 1550. OE **hrēod** + **dīċ**. *Rad-* forms point to an alternative possibility, OE **rēad** + **dīċ**, 'red ditch'. The reference is probably to Nico Ditch SJ 87–8894, *Mykeldiche* 1190–1212 'great ditch', which formed the N boundary of the township. La 30, Jnl 17.30.

REDDITCH H&W SP 0467. 'The red ditch'. *la Rededich, le Rededych* 1247–1348, *The Redde Dych* 1536, *Reddich, Reddyche* 1394–1535. The earliest mention is the Latin form *(de) Rubeo Fossato* c.1200. ME **rede** (OE *rēad*) + **diche** (OE *dīc*). So named from the colour of the soil. Wo 364.

RED DOWN Corn SX 2685. Partly uncertain. *Richadune* [c.1233]15th, *Rydun* 1536, *Ryghdon* 1550, *Reghdon* 1551, *Reddowne* 1582, *Rightadown* 1633 Gover n.d. Unidentified element + OE **dūn** 'a hill'. The reference is to a long straight ridge of hill so this name might be identical with Rightadown Devon SS 4203, OE **riht** + **dūn** 'straight hill', although here the spellings rather point to OE ***ric** 'a narrow strip'. PNCo 147, Gover n.d. 144, D 132.

REDE Suff TL 8055. Uncertain. *Re(o)da, Riete* 1086, *Reode* c.1095, *Rede* 1254, 1610, *Wrede* 1269. This might be OE **hrēod** in the sense 'reed-bed', ***rēod** 'a clearing' or ***wrēode** 'a shelter, a protection'. DEPN, PNE ii.82, 279.

River REDE Northum NY 8298. 'The red one'. *Rede* from c.1200, *Reade* 1577, *Ridde* 1577, 1586. OE **Rēade* < adj. **rēad**. Red from the presence of ironstone. RN 337.

REDENHALL Norf TM 2684. Pehaps 'reedy nook'. *Rada(na)- Redanahalla* 1086, *Redehal* 1166, *Redhala* 1186, *Redenhal* 1199. OE ***hrēoden** + **halh**. DEPN gives pr [red]-, PNE i.264.

REDESDALE Northum NY 8396. 'The valley of the r. Rede'. *Redesdale* from 1075, *Reddesdal'* 1279, *Riddesdale* 1203–1446. R.n. REDE NY 8298, genitive sing. *Redes*, + OE **dæl**. NbDu 163 gives pr [ridzdəl], RN 337.

REDESDALE CAMP Northum NY 8298. P.n. REDESDALE NY 8396 + ModE **camp**. A modern military training camp.

REDESDALE FOREST Northum NT 7501. P.n. REDESDALE NY 8396 + ModE **forest**. A modern Forestry Commission plantation.

REDESMOUTH Northum NY 8682. 'The mouth of the r. Rede'. *Riddesmouth* 1577, *Rhedes mouth* 1586, *Readsmouth* 1695 Map. R.n. REDE NY 8298, genitive sing. *Redes*, + ModE **mouth**. RN 338.

REDFORD Durham NZ 0730. Possibly 'dirty ford'. *le Roteford* 1314, *Rotiford* 1339, *Rette- Retyford* 1341, 1377, *Rutynford* 1369, 1422, *(Ouer)ridforth* 1382, 1418, *Ritforth* 1418, *(High, Low) Redford* 1768 Armstrong. OE ***hrotiġ** + **ford**. This OE adjective also occurs in the name of a medieval fishing weir or yair on the Tyne, *Roti* 1279, *Rutyare* 1344. Nevertheless, the variety of spellings is puzzling: a mutated variant **hrytig* would be possible, or connection with the obscure OE adj. *hrūt* and its mutated variant *hrȳte*, glossed *balidinus* (? 'dark brown') by Ælfric, might be sought. This would then be a reference to the peat stained water of the stream at the ford. NbDu 165, DAJ 2.56.

REDGRAVE Suff TM 0477. Probably 'the reed ditch'. *(on) Redgrafe* 11th, *-graue -v-* from c.1095, *Redegraue* 1179. OE **hrēod** + **græf**. But 'the red grove' would also be possible, OE **rēad** + **grāf(a)**. DEPN.

REDHILL Avon ST 4963. *Red Hill* 1817 OS.

REDHILL Surrey TQ 2850. 'The red slope or hill'. *Redehelde*

1301(p), *Redd hyll* 1588, *red hill* 1601. OE **rēad** + **helde** later replaced by **hill**. The colour refers to the sandstone here. Sr 306, L 162.

REDISHAM Suff TM 4084. Probably 'Read's homestead'. *Redesham* 1086–1524, *Redisham* 1291 and from 1524, *Redsham* 1610. OE pers.n. **Rēad*, genitive sing. **Rēades*, + **hām**. DEPN, Baron.

REDLAND Avon ST 5875. 'Third part of an estate'.
I. *(la) Redelond(e)* c.1200–1386, *Rydelond* 1266, *Rydland* 16th cent.
II. *Thriddeland* 1208×13–1348 with variant *Thrydde-, Thyr(de)land* 16th cent.
OE **thridda** + **land**. The reference is unknown. The name was early subject to false analysis as if it were *the Riddeland* and to wrong explanation as *Rubea terra* 'red land' 1230. Gl iii.142.

REDLINGFIELD Suff TM 1870. 'The open land of the Rædlingas, the people called after Rædel or Rædla'. *Radinghefelda* 1086, *Rad(d)ingefeld(a) -yngefelde -ingefeld* [1120×35]1285–[1189×99]1396, *Radlingefeld* 1166, *Redlingefeld(')* 1182–1203, *Redelyngfeud* 1286, *Redlingfelde* 1568. Folk-n. **Rǣdlingas* < pers.n. *Rǣdel* or **Rǣdla* + **ingas**, genitive pl. **Rǣdlinga*, + **feld**. DEPN, Baron.

REDLYNCH Somer ST 7033. 'Reed marsh'. *Roliz* 1086, *Retlis, Redlisc* 1086 Exon, *Redlis -lys* 1219–43, *Rodlis* 1243, *Redelich* 1294, *Redlysh* 1327, *Redelyche* [1373]B, *Redlinch* 1225, *Ridlinch* 1610. OE **hrēod** + ***lisc** later replaced by ME **linch** (OE *hlinċ*) 'a terrace'. The replacement is due to the fact that Redlynch lies on the slope of a hill rising to 528ft. There is, however, a lake in Redlynch Park S of the village in what was no doubt originally marshy ground. DEPN, Studies 1936.109.

REDLYNCH Wilts SU 2021. 'The red ledge'. *Radelynch(e)* 1282–1402, *Redlynche* 1567. OE **rēad**, definite form **rēada**, + **hlinċ**. The settlement occupies a ledge of higher ground at the edge of a slope. Wlt 395.

REDMARLEY D'ABITOT Glos SO 7531. 'R held by the D'Abitot family'. *Rydmareleye Dabytot* 1355, *Redmarley Dabitot* 1747. P.n. Redmarley + family name D'Abitot; the manor was held by Urse d'Abitot in 1086. Redmarley, *(æt) Reode-Rydemæreleage* [963]11th S 1306, [978]11th S 1338, *Hrydmærlean* 1035, *Rud- Rid- Ryd(e)merlege -mar- -ley(e)* 1086–1481, is '*Hreodamere* wood or clearing', OE p.n. **Hreodamere* 'reed pool', OE **hrēod**, genitive pl. **hrēoda**, + **mere**, + **lēah**. *Hreodamere* would be a pond where reeds and rushes were gathered for domestic use. Gl iii.184, L 26.

REDMARSHALL Cleve NZ 3821. 'Red-mere's hill'. *Redmereshil(l')* 1195×1221, 1272×85, *-is-* [c.1240]late 13th., *Redmershil(l')* c.1225–1559, *Redmershall* 1436–1596, *Redmarsell* 1460, *Readmarshall* 16th cent., *Redmarshall* from 1624. OE p.n. **rēad-mere**, genitive sing. **rēad-meres**, + **hyll**. Winter flood water used formerly to stand on the red clay soil here. Surtees iii.76.

REDMILE Leic SK 7935. 'Red earth'. *Redmeld(e)* 1086–1553, *-mell'* 1448–1582, *-mild(e) -mylde* 1208–1525, *-mill* 1239, *-mile* from 1526. OE **rēad** + **mylde**. Lei 178.

REDMIRE NYorks SE 0491. 'Reed pool'. *Rid(e)mare* 1086, [1173]16th, *Ridemere* 1166–1403, *Redmire* from 1665. OE **hrēod** + **mere**, probably referring to a place where reeds were cut for thatching etc. YN 257.

REDMIRE MOOR NYorks SE 0493. *Redmire Moor* 1860 OS. P.n. REDMIRE SE 0491 + ModE **moor**.

REDMIRES RESERVOIR SYorks SK 2685. P.n. Redmires + ModE **reservoir**. Redmires, *Redmires* 1660, *Rudmire* 1817, is 'the reed marsh, the marsh where reeds are gathered', OE **hrēod** + ON **mýrr**. YW i.200.

REDMOOR Corn SX 0861. 'Red marsh'. *Redemor* 1301, *Rademoure* 1427 Gover n.d., *Redmore* 1748. OE **rēad** + **mōr**. PNCo 148, Gover n.d. 405.

REDNAL Shrops SJ 3628. 'The red nook'. *Radenhale* 1269×75, 1343, 1468, *Rednall(')* 1478 etc., 1837 OS. OE **rēad**, dative case definite form **rēadan**, + **halh**, dative case **hale**. Gelling.

REDNEND Avon ST 5784 → PILNING ST 5585.

RED PIKE Cumbr NY 1615. 'The red peaks'. *le Rede Pike* 1322. Fr definite article **le** + ME **red** + **pike**. A mountain peak rising to 2479ft. Cu 387.

RED ROCK GMan SD 5809. *Red Rock* 1843 OS.

RED ROW Northum NZ 2599. *Red Row* 1868 OS. Probably a row of buildings with a red tile roof or painted red.

REDRUTH Corn SW 7042. 'Red ford'. *Ridruth* 1259, 1324, *Rudruth* 1283, *Redruth* from 1302, *Rysdruth* 1361, *Reesdruth* 1435, *Ridreth* 1333, *Rydryth* 1342, *Unyredreth* 1563 ('St Euny Redruth' from the patron saint of the church). Co **rid** 'ford' (later **res*) + **ruth** 'red' (later *reth*). 14th and 15th cent. spellings show the expected Co developments *rid* > *res* and *ruth* > *reth* but the older form of the p.n. pervailed under the influence of the traditional spelling. The reference is probably to mining waste. PNCo 148, Gover n.d. 613.

REDSHIN COVE Northum NU 0150. *Redshin Cove* 1865 OS. P.n. Redshin + ModE **cove** 'a small creek on the sea-shore'.

RED STREET Staffs SJ 8351. *Redstreete* 1608. The place is mentioned in Plot's *Natural History of Staffordshire* (1686) as a place where iron-ore was mined.

REDWICK Avon ST 5486. 'Reed farm'. *(to) hreodwican* [955×9]12th S 664, *Redeuuiche* 1086, *-wi(c)k(') -wyk(e)* 1248–1658 with variant *Rad(d)(e)-*. OE **hrēod** + **wīcum**, dative pl. of **wīċ**. The sense may have been 'dairy-farm among the reeds' or 'farm where reeds are obtained'. Gl iii.136.

REDWORTH Durham NZ 2423. 'Reda's enclosure'. *Redeworth* [1183]c.1320–1377, *Redworth* from 1337. OE pers.n. *Rēda* + **worth**. NbDu 164.

REED Herts TL 3636. 'Rough ground'. *Retth* 5×, *Rete* 1× 1086, *Ru(i)th* 1185–1230, *Reth* 1205, *Red(e)* 1204–1428, *Reed* from 1291. OE **rȳth**, a derivative of adj. *rūh* 'rough'. Hrt 161.

REEDHAM Norf TG 4101. 'Reedy homestead or promontory'. *Redham* *[1044×7]13th S 1055, 1158, *Redeham* 1086. OE **hrēod** + **hām** or **hamm** 1 or 3. DEPN.

REEDNESS Humbs SE 7923. 'Reedy headland'. *Rednesse* 1164×77–1456, *Rede-* 14th cent., *Reedness* from 1609. OE **hrēod** + **nes**. YW ii.9, L 172–3.

REEKER PIKE Northum NY 6682. *Reeker Pike* 1868 OS. N dial. **pike** 'a pointed hill'. Rises to 1209ft. in the wastes of Wark Forest.

REEPHAM Lincs TF 0373. 'The reeve's manor or estate'. *Refa(i)m, Refan* 1086, *Refham* c.1115, 1196, 1202. OE **(ġe)rēfa** + **hām**. The estate belonged to Peterborough abbey and the reference is to the reeve of the abbey. DEPN.

REEPHAM Norf TG 1023. 'The reeve's estate'. *Refham* 1086–1254. OE **(ġe)rēfa** + **hām**. DEPN gives pr [ri:fəm].

REETH NYorks SE 0399. 'The rough common'. *Rie* 1086, *Ri-Ryth(e)* [1184]15th–1575, *Reth(e)* 1401–1581, *Reeth* from 17th cent. OEScand **ryth**. Cf. the nearby p.ns. HEALAUGH SE 0199 'the high clearing' and Riddings SE 0299 'the clearings'. YN 273.

REGENTS PARK GLond TQ 2882. *The Regents Park* 1817. Named from the Prince Regent, afterwards George IV. The park was created as part of a planned development including Regent Street and corresponded roughly to the Tudor hunting *park of Maryborne* 1588, *Maryborne Park* 1574, *Marrowbone Park* (sic) 1649, p.n. Marylebone + ModE **park**. Mx 138, Encyc 662.

REGILBURY Avon ST 5262 → RIDGEHILL ST 5362.

REGULBIUM Kent TR 2169. 'The great headland'. *Regulbi(o)* [c.400]15th ND. RBrit p.n. **Ro-gulbion* < British ***ro-** + ***gulbio-**. The name of the Roman fort at RECULVER, which survived as one of the earliest British names borrowed into OE not later than AD. 450 as *recuulf* 679 S 8, *Reculf* 9th ASC(A) under year 669 and *Racuulfe* [c.731]8th BHE, *Raculf(e)* c.748–62 S 31–1254 including [696×716]11th S 22, [747]13th S 1612, 949 S 546 and 12th ASC(E) under year 669. Other forms include

Raculfcestre [784]13th S 38, 'Raculf Roman camp', *Raculfinga mearce* 944 S 497 'boundary of the people of Raculf', and with a false etymological interpretation *(in monasterio quod nominatur) Ricuulfi* c.767 S 31 'in the monastery called Ricwulf's'. RBrit 446, KPN 12, 51, LHEB 554, 559, 600.

REIGATE Surrey TQ 2750. 'The roe-deer gate'. *Reigata* c.1170, *Reigat(e)* from 1196, *Regata* 12th cent., *-e* 1203–1302, *Rigate* 12th–1561, *Ray-* 1235–1382, *Reigh-* 1604. OE **rǣġe** + **ġeat**. Probably a reference to hunting, the gate being a gap through which deer were driven towards the arrows of the huntsmen as in the description in *Sir Gawain and the Green Knight* 1151ff. The place was earlier called *Cherchefelle* 1086, *Crec(c)hesfeld* 12th cent., 'the open land marked by a barrow or an abrupt hill', OW **cruc** (PrW **crūg*, Brit **crouco-*) + OE **feld**. Sr 304–5, 281–2.

REIGHTON NYorks TA 1375. 'The settlement at or by the strip of land'. *Ri- Rycton(a) -tone* 1086–1331, *Ri- Rychton(a)* [1125×30]c.1300–1331, *Ri- Ryhtuna -ton(a)* [1154×56]16th–1304, *Ri- Ryghton* 1251–1597, *Reighton* 1828. OE ***ric** + **tūn**. The reference is to the long ridge of land on which the village lies. YE 107, lix, gives pr [ri:tən].

REIGHTON SANDS NYorks TA 1476. *Reighton Sands* 1856 OS. P.n. REIGHTON TA 1375 + **sand**.

REJERRAH Corn SW 8056. Probably 'ford of Gorvoy'. *hryd worwig* 960 S 684, *Roswyrou* 1327, *-worou* 1373, 1409, *-worra*, *Resorrow* 1630, *Rejorrow -jorre* c.1690. OCo **rid** + pers.n. **Gorvoy*. PNCo 148, Gover n.d. 373.

RELUBBUS Corn SW 5631. Partly uncertain. *Reyglubith* 1249, *(Pons-)releubes* 1298 ('bridge of Relubbus'), *Resleubes* 1302, *Rellehoubes* 1355, *Reloubys* 1357, *Relobus* 1522. OCo **rid** 'ford' + unidentified element occuring also in Trelubbas SW 6629, *Trelaheubes* 1319, *-lehoubes* 1337 (**tre** 'estate, farmstead'). PNCo 149.

REMENHAM Berks SU 7784. 'Homestead or water-meadow at the river-bank'. *Rame(n)ham* 1086–1502, *Remenham* from 1086, *Remnam* 1535. OE **reoma**, back-mutated variant of **rima** 'border, rim', genitive or dative sing. **reoman**, + **hām** or **hamm** 1. Brk 66.

REMENHAM HILL Berks SU 7882. No early forms. P.n. REMENHAM SU 7784 + ModE **hill**. Brk 67.

REMPSTONE Notts SK 5724. 'Hrempi's farm or village'. *Rampestune, Rāpestone* 1086, *Rampeston(a)* c.1200–1340, *Repestone* 1086, *Rempestunam -tune -ton(e)* c.1155–1340, *Rempston* 1228, *Remston* 1327–1578, *Remson* 1541. OE pers.n. **Hrempi*, genitive sing. **Hrempes*, + **tūn**. Nt 254.

RENDCOMB Glos SP 0209. 'Valley of the *Hrinde*'. *Rindecome -cvmbe* 1086, *Ri- Ryndecumb(e) -comb(e)* 12th–1372, *Rende-* 1261–1533, *Rencomb(e)* 1460, 1752. Stream name *Hrinde* + OE **cumb**. *Hrinde* occurs in the genitive case in the bounds of S 202 *[852]13th *on Hrindan broc* 'to the brook of *Hrinde*', possibly 'the thrusting stream, the pusher' from OE *hrindan* 'to push, thrust'. Gl i.160, L 93.

RENDHAM Suff TM 3564. 'Possibly 'the cleared homestead'. *Rimd- Rin(de)ham* 1086, *Rindham* 1086–1323, *Rendham* from 1254. OE **rȳmed** + **hām**. If, however, the *Rimd-* sp is to be ignored, the specific might be OE ***rind(e)**, ***rend** 'a hill, a ridge, an edge, a border'. DEPN, Baron, PNE ii.83.

RENDLESHAM Suff TM 3253. 'Rendel's homestead' according to Bede. *Rendlaesham* [c.731]8th BHE, *Rendlesham* [c.890]c.1000 OEBede and from 1086. OE pers.n. **Rendel*, genitive sing. **Rendeles*, + **hām**. This the place of the baptism of king Swithhelm of the East Saxons 653×64, described by Bede as a *uicus regius* 'a royal village'. Although he translates the name as *mansio Rendili*, 'Rendil's residence', it is not impossible that the specific is rather an appelative such as OE **rændel*, a diminutive of OE *rand* 'a shore', and later in E Anglia 'a marshy reed-covered strip of land between a river and its embankment'. It lies near the r. Deben a little over 2 m NE of the royal burial mounds at Sutton Hoo TM 2848. DEPN, PSI 24 (1948) iii.228–51.

RENDLESHAM FOREST Suff TM 3449. No early forms. P.n. RENDLESHAM TM 3253 + ModE **forest**.

RENHOLD Beds TL 0852. 'Roe-deer nook'. *Ramhale* (sic) [1189×99]13th, *Ranhale* 1220, *Ronhale* 1227–1428, *Ronnall, Rey- Raynold* 16th cent. OE **rā**, genitive sing. **rān**, genitive pl. **rāna**, + **halh**, dative sing. **hale**. The development of final *-d* is common in local pronunciations of words ending in *-l*. Bd 62 gives pr [renəld], L 111.

RENISHAW Derby SK 4477. 'Reynold's copse'. *Reynold schaie* c.1216, *Reynalddeschawe* 1281, *Raynoldeshaw* 1545, *Reynishaw* 1570, *Renishawe* 1595. Pers.n. *Reynold* (OG *Reginald*), genitive sing. *Reynoldes*, + **sceaga**. For the unusually early form *-schaie* beside *-schawe*, cf. Nomina 12.89–104. Db 248.

RENNINGTON Northum NU 2118. 'The estate called after Regna'. *Renninton* 1175, *Reni(n)gton* 1242 BF–1307, *Rynington* 1538. OE pers.n. *Reġna* + **ing**[4] + **tūn**. Cf. East and West RAINTON NZ 3347, 3146. NbDu 165.

RENWICK Cumbr NY 5943. 'Hræfn's farm'. *Raveneswic(k)* c.1177–c.1225, *Rauenwich' -wic -wyk(e) -wyc(h)* 1178–1479, *Raneswic* c.1180, *Ranewich -wyk* 1191, 1279, *Renwyck* 1568, *Rennok* 1576. OE pers.n. **Hræfn*, genitive sing. **Hræfnes*, + **wīc**. Cu 235 gives pr [renik].

REPPS Norf TG 4117. 'The strips of land'. *Repes* 1086, *Reppes* [1101×7]13th, 1152–1547, *Rep(p)les* 1191–1211. OE ***ripul**, ***reopel**. The reference is to strips of cultivated land in the fens. Nf ii.69.

REPTON Derby SK 3027. 'Hill of the people called *Hreope* or *Hrype*'.

I. *(on) Hreopandune, (to) Hreopendune* c.1100 ASC(D) under years 755 and 874, *Hreopa dune, Hreopedune* c.900 ASC(A) under years 755, 874, 875, c.1020 LVH, before 1118 FW.

II. *(æt) Hrypadune* [848]c.1200 S 197, *Hrypadun* [730×40]9th Felix, [10th]c.1050 Guthlac, *Hrypandun* before 1118 FW.

III. *Rep(p)enduna -(e) -don(e)* 1086–1313, *Rep(p)edon(a)* 1135×54–1252, *Rep(p)indon(e) -ing- -yng-* 1228–1556, *-ton(e)* 1330–1747.

IV. *Repton(e)* 1443, from 1513.

V. *Rap(p)endun(e) -don(e)* 1086–1248.

OE folk-n. *Hreope* varying with *Hrype*, genitive pl. *Hreopa*, *Hrypa*, + **dūn**. For the folk-n. cf. p.ns. RIPLEY SK 4050, RIPON SE 3171. Db 653, L 145, 155.

North RESTON Lincs TF 3883. *North Riston* 1274, ~ *Reston* 1824 OS. ME **north** + p.n. *Ristone* 1086, *-tuna* c.1115, 1175×81, *-ton'* 1170, 1202, *Rustun* 1193, 'the brushwood settlement', OE **hrīs** + **tūn**. The reference is either to local vegetation or to a place where brushwood was traded. DEPN, Cameron 1998.

South RESTON Lincs TF 4083. *Suthriston'* 1271. ModE **south** + p.n. Reston as in North RESTON TF 3883.

RESTRONGUET Corn SW 8316 → TRUNCH Norf TG 2834.

RETEW Corn SW 9257. Partly uncertain. *hryt ar þugan* 1049 S 1019, *Ritteu* 1302, *Retteu* 1320, *Retewe* c.1546. If, as seems likely, the S 1019 form belongs here, this must be 'ford on the Dugan', OCo **rid** 'ford' + stream n. **Dugan*, a formation unique in Cornwall but paralleled in Wales (Rhydardaeau Dyfed SN 4326 etc.). **Dugan* would be an alternative name of the Fal. Otherwise, OCo **rid** + uncertain element, possibly **du** ('black ford') or **tew** 'fat, dense' referring to the water at the ford and so 'thick, muddy ford'. PNCo 149, Gover n.d.332, CPNE 8–9.

East RETFORD Notts SK 7080. *Estredford* 1393, *Est Radford* 1449, *Estretford* 1280. ME adj. **est** + p.n. *Redford(e)* 1086–1675, *Rad(e)ford(')* c.1155–1212, *Retford* 1219–1227 etc., *Ratford* 1227–1280, 'the red ford', OE **rēad** + **ford**. 'East' for distinction from West Retford SK 6981, *Westredford(e)* 1267, 1535, *Westretford(e)* 1276, *Westratford* 1290. Nt 58, L 68.

RETTENDON Essex TQ 7698. Possibly 'rat-infested hill'. *(æt) Rettendune* 1000×2 S 1486, *Retendon(e) -in(g)- -yng-* 1227–1483,

Radendunam, Ratendunā 1086, *Ratendune* [1042×66]12th S 1051, *Ratin(g)don -yng-* 1248–1334. OE ***rætten** + **dūn**. This place, named *Gilberts* on *Rettenden Common* 1805 OS, is not the historic centre of the village, which was rather at RETTENDON PLACE TQ 7796. Ess 262, PNE ii.79.

RETTENDON PLACE Essex TQ 7796. P.n. RETTENDON + ModE **place**. This is simply *Rettenden* 1805 OS and seems to be the original centre of the village.

REVESBY Lincs TF 2961. 'Refr's village or farm'. *Resuesbi* 1086, *Reuesbi(a)* 1142, 1154, *-by* 1163. ON pers.n. *Refr*, secondary genitive sing. *Refes*, + **bȳ**. DEPN, SSNEM 64.

REWE Devon SX 9499. 'Row (of houses)'. *Revwe* 1086, *Rewe* from 1242. OE **rǣw**. D 445.

REYDON Suff TM 4977. 'Rye hill'. *Rienduna* 1086, *Reindon(e)* 1170–1236, *Reidon(')* 1173–[1189×99]1396, *Reydon(')* from 1242×3, *Roydon* 1610. OE **ryġen** + **dūn**. Identical with RAYDON TM 0538. DEPN, Baron.

REYMERSTON Norf TG 0206. 'Raimer's estate'. *Raimerestuna* 1086, *Reimerestona* 1168. OG pers.n. *Raimer* or OE **Reġenmǣr*, genitive sing. *Raimeres*, + **tūn**. DEPN gives pr ['remǝstǝn].

RHODES MINNIS Kent TR 1542. 'Rhodes common'. *Rhoads Minnis* 1819. A combination of earlier p.n. *Rode* 1327–57, OE ***rod**, ***rodu** 'a clearing' and **(ġe)mǣnnes** 'common land' as in the local surnames *atte Menesgate* 1327, *de Men(n)essegate* 1334, 1338. PNK 436.

RHODESIA Notts SK 5680. Transferred from Rhodesia, the former British colony in southern Africa (1889–1978) named after Cecil Rhodes.

RHYDD H&W SO 8345. 'The cleared land'. *la Ridde* 1241, 1276, *The Rhydd* 1832 OS. OE ***ryde**. This place lies on the Severn and the form of the name has been influenced by W *rhyd* 'a ford'. Wo 212, L 79.

RHYDYCROESAU Shrops SJ 2430. 'The ford of the crosses'. *Rhyd y Croese* 1680, *Rhydycroesau* 1837 OS. W **rhyd** + definite article **y** + **croes**, pl. **croesau**. The place lies on the Wales–England border at a cross-roads. Gelling.

RIBBESFORD H&W SO 7874. 'Ford with a bed of ribwort'. *(æt) ribbed forda* [1014×23]18th S 1459, *Ribetforda* 11th, *Ribeford* 1086–1275, *Ribbesford* 1232. OE **ribbe** + **bedd** with later genitival *-s* + **ford**. Wo 68.

River RIBBLE Lancs SD 6434. Unexplained.

I. *Rippel* [c.715]11th Eddi, [c.1350]c.1400, *Ripam* 1086, *Rypel* 1387.

II. *Ribbel* [930]14th, [1002]11th–14th, *Ribel(l)* [1189×90]1412, c.1229–[1338]1412, *Rib(b)am, Ribbe* 1130–1199×1216, *Ribble* [c.1130]12th, 1262×3, and from 1577.

This cannot be derived from the RBrit name of the Ribble estuary, *Belisama* [c.150]13th, nor from the other possible RBrit name of the river **Bremetona* discussed under RIBCHESTER Lancs SD 6535. No satisfactory solution has been proposed. La 65, Jnl 17.44, RN 340, RBrit 267.

RIBBLE HEAD NYorks SD 7779. *Riblehead* 1674. R.n. RIBBLE SD 6434 + ModE **head** 'source'. YW vi.247.

RIBBLESDALE NYorks SD 8158. 'The valley of the r. Ribble'. *Riblesdale* [1198]copy–1462, *Rybbesdale* [1258]n.d., c.1300×20. R.n. RIBBLE SD 6434, genitive sing. *Ribles*, + ON **dalr**. One of a handful of p.ns. found in medieval poetry -

Mosti ryden by Rybbesdale
wilde wymmen forte wale
and welde wuch ich wolde,
founde were the feyrest one....

'Could I ride by Ribblesdale to choose wanton women and possess whichever one I wanted, I would find the fairest one....' YW vi.112.

RIBBLETON Lancs SD 5630. 'Settlement on the river Ribble'. *Ribleton* 1201–1354, *Ri- Ryb(b)elton -il-* 1226–1332. R.n. RIBBLE Lancs SD 6434 + OE **tūn**. La 146, Jnl 17.83.

RIBCHESTER Lancs SD 6535. 'Roman fort on the river Ribble'. *Ribelcastre* 1086, *Ribbelcestre -il-* 1215–1332, *Rib(b)il- Riblechastre* 1335–62, *Ribbecestre* 1202, 1246 etc., *-chestre* 1246, *Rybchestre* 1292. R.n. RIBBLE Lancs SD 6434 + OE **ċeaster**. The RBrit name of the fort was *Bremetonacum* [284×305]8th < r.n. **Bremetona* 'roaring river' < British ***brem-** as in BREMENIUM Northum NY 8398, + suffix *-āco-*, 'settlement, place on the Bremetona'. The reference is either to the Ribble, a r.n. of unknown origin, or to the stream that flows past Ribchester from high ground to the NW to join the Ribble at SD 6535. La 144, RBrit 276, Jnl 17.81.

RIBDEN Staffs SK 0747 → WEAVER HILLS SK 0946.

Great RIBSTON WYorks SE 3954. Latin prefix *Magna* 1303, ModE **great** + p.n. *Ripesta(i)n* 1086, *Ri- Rybestan(n) -stain -stayn* 1173–1247, *Ribbestain -stein -steyn* 1196–1327, *Ri- Rybstain -stayn -stan(e)* 1217–1502, *Ri- Rybston* 1535, 1540, 'rock or stone where rib-wort or hound's-tongue grows', OE **ribbe** + **stān**. YW v.19.

Little RIBSTON WYorks SE 3853. *Parva Ribstayn* 13th, *Little Ribestayn* 1316, *Litil, Little Ribston(e)* 1459, 1570, 1622. Latin **parva**, ME **litel**, + p.n. *Ripestain -sten* 1086, *Ribestein -stain* 1214, 1217 as in Great RIBSTON SE 3954. YW v.31.

RIBY Lincs TA 1807. 'Rye farm'. *Ribi* 1086–c.1300, *-by* from c.1150, *Riebi -by* 1159–1461. OE **rȳġe** + ON **bȳ** possibly replacing earlier OE *rȳġe-tūn*. Li ii.249, SSNEM 64.

River RICCAL NYorks SE 6382. 'The calf of the river Rye'. *Rycaluegy'* (for *Rycaluegreynes* 'the streams of R') [1251]16th, *Ricolvegraines* [1333]16th, *Ricoll(e), Ricall* 16th cent. Nearby Riccal House SE 6780 is *Ricalf* 1086, 1293, *Rycal* 1316. R.n. RYE SE 5784 + OE **calf** or ON **kalfr**. The Riccal runs parallel to the Rye, like a calf following its dam. *Calf* is more usually used of a small rock or island beside a larger one like the famous Cow and Calf rocks at Ilkley, or the Calf of Man SC 1666. YN 5, RN 341.

RICCALL NYorks SE 6237. Probably 'Rica's nook of land'. *Richale* 1086–1311, *Richehal(e)* 1190–4, *Ry- Rikehal(e)* 1230–1306, *Rikinhal'* 1230, *Rical'* 1276, *Rickall* 1316. OE pers.n. **Rīca*, genitive sing. **Rīcan*, + **halh**. YE 265, L 108.

RICHARDS CASTLE H&W SO 4969. *Castellum Ricardi* c.1180–6, 1235×6, *Castrum Ricardi* 1212 etc., *Castel Richard* 1321, *Richardescastel* 1349 etc. ME pers.n. *Richard* < CG *Ricard* (W Frankish **Rikhard*), genitive sing. *Richardes*, + OFr, ME **castel**. The castle is said to have been built by and named after Richard fitz Scrob, one of Edward the Confessor's Norman favourites who held land in H&W and Shrops in 1066. Sa i.249, Pevsner 1963.276, PNDB 349.

RICHBOROUGH Kent TR 3260. 'The rich fort'. *Richborough* 1471–9, *Richborowe, Richeborough, Russhebourough* 16th cent. This is probably a respectable re-formation of earlier *Ratteburg'* 1197, *Ret(t)eburg'* 1270, *Retesbrough* 14th, *Ratb(o)urgh* 1462, 1464, understood as if 'rat's fort', OE **ræt**, genitive pl. **rata**, + **burh**, a term which, like Rat's Castle TQ 6253 and Ratsbury, is a common name for ancient ruins or defences, here referring to the Roman fort of RUTUPIAE. Cf. Ratsborough Essex TQ 9598, *Rattesbourgh -borowe* 1303, 1536. The original sense, however, must have been 'Roman fort called Ratte', to judge from the 11th OEBede form *Rœtte* representing *Repta* in Bede's *Reptacaestir* 'Roman fort called Repta' or 'of the Repte'. Bede himself believed the name to be a corruption of Latin *Rutubi portus* 'port of Rutubus' which he found in a text of Orosius; *ciuitas quae dicitur Rutubi portus, a gente Anglorum nunc corrupte Reptacaestir uocata* 'the city called the port of Rutubus, now corruptly called Reptacaestir by the English' [731]8th BHE I.1. This can only be explained by supposing that, contrary to the most likely explanation of *Rutupiae*, in late Roman times the name was wrongly taken to represent **Ro-tupīās* with British prefix *ro-* 'great' as in REG-

ULBIUM. This form, being unstressed, would have become **Rətúpīās*. With English accentuation, syncope, and substitution of *pt* for the unfamiliar group *tp* the subsequent development seems to have been **Rətúpias* > **Rétupias* > **Retpi* > **Repte*, genitive pl. *Repta*. But this is very uncertain. PNK 531, ESS 226, IPN 149, RBrit 450, LHEB 655, 661–2, Cullen.

RICHMOND GLond TQ 1874. There was a royal palace here destroyed by fire in 1499, rebuilt by Henry VII and renamed after his earldom of RICHMOND NYorks NZ 1701, *Shene otherwise called Richemount* 1502, *West Shene nowe called Rychemond* 1515×18. The earlier name, *(on) Sceon* [942× c.951]14th S 1526, *Schenes* 1225, *Shene* 1230–1601, meant 'sheds or shelters', OE ***scēo**, pl. ***scēon**, although later antiquaries associated the name with the noun *sheen* and held that it was 'called for the beauty thereof *Shine* or *Shene*'. Sheen was the favourite summer residence of Richard II and his queen where they daily fed 10,000 guests. Sr 65, Encyc 668.

RICHMOND NYorks NZ 1701. 'The strong hill'. *Richemund(e)* 1108×14 etc. Transferred from one of the numerous Richemonts in France by Earl Alan the Red, a second cousin of the Conqueror, who received the estates of the Saxon Earl Edwin of Mercia and built the great castle here as his new estate centre to replace Gilling. The site was originally called *Hindrelac* 1086, 'the hind's woodland glade'. YN 287, Jnl 15.3.

RICHMOND PARK GLond TQ 1972. *The New Park of Richmond al. Richmond Great Park* 1649. The present park was enclosed by Charles I in 1637 to provide a hunting ground close to Richmond Palace and Hampton Court. Earlier citations, *le Newe Parc de Shene* 1463, *park called Richmond Park* 1541, refer to an older park, now Old Deer Park and Kew Gardens. Sr 66, Encyc 576, 668.

RICKINGHALL Suff TM 0475. 'The nook of the Ricingas, the people called after Rica'. *Rikinghale* [978×1016]13th S 1219, *Ricynga- Rikingehale* 11th, *Rickingahala, Richingehal(l)a* 1086, *Rikingehal(e)* 1224, 1254, *Ri- Rykinghal(e) -yng-* 1257–1568, *Rickingale* 1610. OE folk-n. **Rīcingas* < pers.n. **Rīca* + **ingas**, genitive pl. **Rīcinga*, + **halh** referring to a small side-valley. DEPN, Baron.

RICKLING Essex TL 4931. 'The Riclingas, the people called after Ricola'. *Richelingā* 1086, *Riclinges, Rigeling* 1214, *Rikelinges* 1235, *Ry- Rikeling(e)* 1238–56. OE folk-n. **Rīcelingas* < feminine pers.n. *Rīcūla* + **ingas**. This may be a reference to Ricola, a member of the royal house of Kent who became queen to Sledda, king of Essex in the late 6th cent. Ess 532, ING 23.

RICKMANSWORTH Herts TQ 0594. 'Ricmær's enclosure'. *Prichemareworde* (sic) 1086, *Riche- Ri- Rykemaresworth -wurth* c.1180–1303, *Rikeman(n)esw(o)rth* 1248–94. OE pers.n. *Rīcmǣr*, genitive sing. *Rīcmǣres*, + **worth**. Hrt 80.

RICKNIELD STREET SYorks. *Rikenild stret* c.1400. A local form for ICKNIELD WAY q.v. ME *at the Rikenilde strete* by false analysis of *at there Ikenilde strete*, in which *there* is by origin the dative singular feminine form **thǣre** of the OE definite article **se, sēo, thæt**. Bd 5.

RIDDLECOMBE Devon SS 6113. Perhaps 'valley where riddles (sieves) are made'. *Ridelcome* 1086–1428 with variants *-cumbe -combe, Ridelecumbe* 1238, *Riddelcumbe* 1323. OE **hriddel** + **cumb**. Alternatively the specifc might be p.n. *Ridel* identical with Ruddle Corn SX 0055, *Rydel* 1296, *Ridel* 1327, *Ridell* 1420, of obscure origin, possibly an *-ell* diminutive of Corn *rid* 'ford' or Corn **rid** + **hal** 'muddy ford'. D 356, ii.xiii.

RIDDLESDEN WYorks SE 0842. 'Rethel's valley'. *Red(e)lesden(e)* 1086–1362, *Ri- Ryd(e)lesden(e)* after 1166–1577. OE pers.n. *Rēthel* or *Hrēthel*, genitive sing. *Rētheles*, + **denu**. YW iv.172, L 98.

RIDGE 'The ridge'. OE **hrycg**, ME **rigge**.

(1) ~ Dorset SY 9386. *Rygge* 1431, *Rydge* 1632. Do i.73.

(2) ~ Herts TL 2100. *la Rigge, Regge, Rugge* 1248–1487, *Ridge al. Rudge* 1638. The place lies on a narrow ridge of land. Hrt 83.

(3) ~ Wilts ST 9532. *Rigge* 1195, *Rydge, Ruygge, Rudge* 16th. Wlt 187.

RIDGE HILL H&W SO 5035. *Ridge Hill* 1831 OS. A long narrow hill S of Hereford probably originally called simply *Ridge*.

RIDGEHILL Avon ST 5362. 'Rocky ground'. *Ragiol* 1086, *Ragel* 1193, *Rachel* 1254, *Raggel* 1289, *Ridge Hill* 1817 OS. Cornish dial. **radgell** 'an excavated tunnel, a number of stones lying about'. This element occurs again in nearby Regilbury ST 5262, *Ragelbiri, Rachelburi* c.1200, and in Radjell Corn. DEPN, EDD, CPNE 195.

RIDGE LANE Warw SP 2995. *Ridge Lane* 1835 OS. The lane crosses a ridge of high ground at right angles.

Brown RIDGE NYorks SE 1077. No early forms. ModE **brown** + **ridge**.

RIDGE WAY 'A road along a ridge'. OE **hrycg** + **weġ**. Arch J 25.69–194.

(1) ~ Berks SU 5481. Oxon SU 3683. Wilts SU 1168, 1677. *(andlang) hric weges* 939 S 449, *hric weg* [956]12th S 585, *(on) hrycwæg* [856]12th S 317, *(on) hrucg weg, (of) hrucg wœgœ* [944]12th S 503 (all in Woolstone Oxon), *(oth thone) hricg weg* [944 ? for 942]13th S 496 (in Blewbury Oxon), *Ruggeweye* c.1225 (in Uffington Oxon), *la, le Rigweye* 1281, 1289, *The Ridge Way* 1761. An ancient track-way along the N edge of the Wilts and Berks downs extending from Wansdyke at SU 1160 to Streatley at SU 5581. Wlt 17.

(2) ~ Wilts SU 0151. *Rigwege* 1216×72, *-weie* 1224. An ancient track-way following high ground on Salisbury Plain from Imber ST 9648 to SU 1250. Wlt 241, 242.

RIDGEWAY Avon ST 6286. 'Ridge way'. *Rugewei* 1191, *Rigeweye* 1276, *Rudgeway(e)* 1587. OE **hrycg** + **weġ**. The reference is to the Roman road from Seamills to Gloucester, Margary no. 541, which here keeps along high ground bordering the Vale of Berkeley and the Severn estuary. Gl iii.112.

RIDGEWAY CROSS H&W SO 7147. *Ridgeway Cross* 1832 OS. A cross-roads on the ancient route or ridgeway running NW–SE W of Bromyard, near the B4220. The name occurs again in Ridgeway Oak SO 7046.

RIDGEWAY OAK H&W SO 7046 → RIDGEWAY CROSS SO 7147.

RIDGEWELL Essex TL 7340. Partly uncertain.

I. *Rideuuella* 1086, *Ri- Rydeswell* 1294, 1324, *Rygewell* 1483.

II. *Rad(d)eswell(a)* 1163×4–1443.

III. *Redeswell(e)* 1189×99–1346.

IV. *Rod(d)eswell(e)* 1272–1497.

V. *Redeleswell'*. 1254.

The isolated *Redeleswell'* of 1254 suggests 'Rædel's spring', OE pers.n. *Rǣdel*, genitive sing. *Rǣdeles*, + **wella**, but this hardly accounts for the full and puzzling range of types of this name. Ess 453 gives pr [redʒwəl].

RIDGEWOOD ESusx TQ 4719. 'Wood on a ridge'. *Regewod* 1488. ME **rigge**, SE dial. form **regge** (OE *hrych, hrecg*), + **wode** (OE *wudu*). Sx 397.

RIDGMONT Beds SP 9736. 'Red hill'. *Rugemund* 1227, *-munt* 1286, *Rougemont* 14th cent., *Ridgemond* 16th cent. OFr **rougemont**. The village stands on a reddish sandstone ridge: hence the change of *Rouge-* to *Ridge-*. Bd 82.

RIDING MILL Northum NZ 0161. *Rydinge mylne* 1575. P.n. Riding + ModE **mill**. Riding, *Ryding* 1262, 1296 SR, *le Ruddyng* 1298, *la Ridding* 1335, is 'the clearing', OE ***rydding**. NbDu 165, PNE ii.90, Jnl 1.33.

RIDLEES CAIRN Northum NT 8404. *Ridlees Cairn* 1869 OS. P.n. Ridlees NT 8405, *Reddeleys* c.1320, *Redlees* 1720, 'the cleared pastures', OE **(ġe)rydd** + **lēah**, pl. **lēas**, + ModE **cairn** (Gael *carn* 'a heap of stones'). The cairn marks a summit of 1357ft. It was once a prominent feature but has been much reduced by artillery fire (PSA 5.i.152). NbDu 166, PNE ii.90.

RIDLINGTON Leic SK 8402. The 'estate called after Redel or Red(w)ulf'. *Redlinctune* 1086, *Ry- Ridelinton' -yn-* 1167–1327,

Red(e)lington' -yng- 1220–1505, *Ry- Rid(e)lington' -yng-* 1209×19–1610. OE pers.n. *Rēd(w)ulf* or its short form **Rēdel* + **ing**[4] + **tūn**. R 206.

RIDLINGTON Norf TG 3430. Partly uncertain. *Ridlinketuna* 1086, *Red(e)lingtun -ton(') -yng-* 1199–1535, *Rid(e)lington* 1254–1346. Possibly OE pers.n. *Rēdel* + **ing**[4] + **tūn** as RIDLINGTON Leics SK 8402. Nf ii.188, R 206.

RIDSDALE Northum NY 9084. *Redesdale* 1868 OS. The *Rid-* form is a variant of REDESDALE NY 8396 now applied to the village at NY 9084.

Hamstall RIDWARE Staffs SK 1019. 'R with the homestead'. *Hamstede Ridewale* 1236, *Hamstal Ridewar* 1242. ME **hamstede** varying with **hamstall** 'a homestead, a demesne farm' + p.n. *Rideware* 1004, *Rideuuare -ware, Ridvare* 1086, *Ridwara* 1169, *Rydeware* 1281, by origin a folk-n. **Ridware* 'the ford-dwellers', PrW ***rïd** 'a ford' probably taken as a p.n. + OE **ware**. The Ridware occupied an extensive area between the rivers Blithe and Trent, which was divided into three manors by 1086, Hamstall, Mavesyn and Pipe RIDWARE SK 1019, 0717 and 0917. Hill RIDWARE SK 0818 seems to be a later division. DEPN, L 79, 80, Horovitz.

Hill RIDWARE Staffs SK 0818. 'R with the hill'. *Hill Ridware* 1834 OS. P.n. *le Hul* 1346, 'the hill', OE **hyll**, + p.n. Ridware as in Hamstall RIDWARE SK 1019. Horovitz.

Mavesyn RIDWARE Staffs SK 0717. 'R held by the Malveisin family'. *Ridewale Mauvaisin* 1236. Family n. Malveisin who held land here from at least 1100×35 + p.n. *Ridvare* 1086 as in Hamstall RIDWARE SK 1019. DEPN.

Pipe RIDWARE Staffs SK 0917. 'R held by the Pipe family'. *Pipe Ridware* 14th. Surname of Robert de Pipe from PIPE HILL SK 0909, who held the manor c.1285, + p.n. *Riduuare* 1086 as in Hamstall RIDWARE SK 1019. DEPN.

RIEVAULX NYorks SE 5785. 'The valley of the r. Rye'. *Rievalle, Ry- Rieualle* 1157–1202, *Riesuals* 1161, *Ryvaus* 1301–1491, *Ryvaux* 1390–1497. R.n. RYE SE 5784 + OFr **vals**. An AN name originally applied to the Cistercian monastery founded here in 1131. Cf. JERVAULX ABBEY SE 1785. <x> is an OFr graphy for [ts], a sound which developed from the OFr nominative sing. inflexional *-s* after vocalisation of preceding *l* and survived in the traditional prs ['rivis, 'rivəz]. Pronounced today ['ri:vo]. YN 73.

The RIGG Northum NY 6483. *The Rigg* 1868 OS. N dial. **rigg** for *ridge*, applied to the high ground between Bach Burn and Humble Burn in the wastes of Kielder Forest, all over 1000ft. high.

RIGGS MOOR NYorks SE 0373. *Riggs Moor* 1771. P.n. *Rigge(s)* 1607–1637, 'the ridges', N dial. **rigg** (ON *hryggr*), + ModE **moor**. YW v.219.

RIGHTADOWN Devon SS 4203 → RED DOWN Corn SX 2685.

RIGMADEN PARK Cumbr SD 6184. P.n. Rigmaden + ModE **park**. Rigmaden, *Rig(g)- Ryg(g)maiden* 1255–1738, *Rigmaden* 1647, is 'the maiden's ridge', ME **rigg** (ON *hryggr*) + **maiden**. The p.n. is in the Celtic word order introduced into NW England from Ireland by Norwegian Vikings, i.e. generic + specific rather than specific + generic. We i.51, SSNNW 246.

North RIGTON WYorks SE 2849. Adj. **north** + p.n. *Riston(e)* 1086, *Ri- Rygton* 12th–1640, *Ri- Ryghton* 1279–81, the 'ridge settlement', OE **hrycg** replaced with ON **hryggr** + **tūn**. Situated on the three mile ridge topped by Almscliff Crag. 'North' for distinction from East Rigton SE 3643, *Est Ryghton* 1530, earlier simply *Riston, Ritone -tun* 1086, *Rigton* 1172–1641. The AN spelling *Ris-* and late English spellings *Righ- Rygh-* show traces of a pronunciation with [ç] instead of [g]. YW v.44.

RILEY HILL Staffs SK 1115. *Riley Hill* 1834 OS. The evidence is too late for certainty but Riley looks like 'the clearing where rye grows', OE **rȳġe** + **lēah**.

RILLA MILL Corn SX 2973. 'Rillaton mill'. *Rillamulle* 1399. P.n. Rilla(ton), *Risleston* 1086, *Rillectona* 1130, *Ridlehtuna* [c.1155]15th, 'farm, estate at or called *Rid-legh*, ford of the flat stone' Co p.n. *Rid-legh* (< OCo **rid** + ***legh**) + OE **tūn**, + ME **mulne** (OE *myln*). PNCo 149, Gover n.d. 164.

RILLINGTON NYorks SE 8574. 'Estate called after Redla or Hrethel'. *Redlinton(e), Renliton* 1086, *Rillintun -ton(a) -y-* 1175×84–1268, *Ridlin(c)ton* 1188, 1229, *Rillington* from 1190. OE pers.n. **Rēdla* or *Hrēthel* + **ing**[4] + **tūn**. YE 138.

RIMINGTON Lancs SD 8045. 'Settlement on the Riming, the boundary stream'. *Renitone* 1086, *Rimington(a)* from 1147, *Remin(g)ton -yng-* c.1250–1646, *Rimmington* 1522, 1665. OE r.n. **Riming* < **rima** 'rim, edge, boundary' + **ing**[2], + **tūn**. The reference is to Ings Brook which once formed the boundary between Lancashire and Yorkshire at this point. YW vi.176 gives pr [rimintən].

RIMPTON Somer ST 6021. 'Rim or border settlement'. *(æt) Rimtune* [938]12th S 441, *Rimtun* [956]12th S 571, *Rintone* 1086. OE **rima** + **tūn**. The village lies on the boundary with Dorsetshire. DEPN.

RIMSWELL Humbs TA 3128. Possibly 'the boundary hill'. *Rimeswelle -uuelle* 1086, *Ri- Rymeswell(e)* late 12th–1430, *Ri- Rymswell* 1584–1828. OE **rima** + ***swelle, (ġe)swell**. Rimswell lies S of Roos or Keyingham Drain, the boundary here between Mid and South Holderness, on rising ground opposite the promontory of Roos. Alternatively 'Rymi's spring', OE pers.n. **Rȳmi*, genitive sing. *Rȳmes*, + **wella**. YE 28 gives pr [rimzil], SSNY 258.

RING'S END Cambs TF 3902. 'End of the ring'. *Rings End* c.1840. The reference is to *banckes called the Rynge of Waldersey and Coldham* 1607, the ring being the land within the bank, viz. WALDERSEY TF 4304 and Coldham Field TF 4402. Ca 296.

RINGLAND Norf TG 1313. Possibly 'newly cultivated land of the Rymingas, the people called after Rymi'. *Remingaland* 1086, *Ringeland* 1206, *-lond* 1219. OE folk-n. **Rȳmingas* < pers.n. **Rȳmi* + **ingas**, genitive pl. *Rȳminga*, + **land**. DEPN, L 247.

RINGMER ESusx TQ 4412. 'Circular pool'. *Ry- Ringemere* 1275–91, *Ry- Ringmere* 1290–16th, *Ringmer* 1564. OE **hring** + **mere**. There are other Ringmeres in Norfolk and Dorset. Sx 355, L 27, Jnl 25.40–1.

RINGMORE Devon SX 6545. 'Cleared moor'. *Reimore* 1086, *Ri- Ryd(e)mor(e)* 1242–1381, *Rydymore* 1281, *Ryn(n)more* 1434, 1438, *Reyn- Reymmore* 1435, *Ryngemore* 1600. OE **(ge)ryd(d)** + **mōr**. Forms with *n* suggest the alternative specific ***ryden** or ***ryd-ding** 'a clearing' subsequently assimilated to *ring*. The same name occurs at Ringmoor Down SX 5666, *Rydemore* [1291]1408, *Rynmore* 1535, 1547, *Ridemore Down al. Ringmore Downe* 1558×1601, and Ringmore SX 9271, *Rumor* 1086, *Redmor(e)* 1275–89, *Ryd- Rid(e)more* 1281–1318. D 283, 239, 460.

RINGSFIELD Suff TM 4088. 'Hring's open land'. *Ringesfelda -fella* 1086, *-feld(e)* 1267–1316, 1610, *Ringefeld* 1235, 1264, *Ringesfilde* 1568. OE pers.n. **Hring*, genitive sing. **Hringes*, + **feld**. DEPN, Baron, L 241, 245.

RINGSHALL Bucks SP 9814. 'Hring's nook'. *Ringeshale* 1235, 1262, *Ryngsole* 1448, *Ringsell* c.1660. OE pers.n. **Hring*, genitive sing. **Hringes*, + **halh**. Hrt 37, Bk 95.

RINGSHALL Suff TM 0452. 'Ring shelter'. *Ringhesehla, Ring(h)eshala* 1086, *Ri - Ryngeshal(e)* 1198–1286, *Ri- Ryngesell' -s(s)ele* c.1236–1344, *Ringsale* 1610, *Rinshall* 1674. OE **hring** + **(ġe)sell**. The reference is unknown. DEPN, Baron.

RINGSHALL STOCKS Suff TM 0551. *Ringshall Stocks* 1837 OS. P.n. RINGSHALL TM 0452 + ModE **stock** 'a tree-stump'.

RINGSTEAD 'Circle place, site with a ring', possibly a reference to a stone circle, cf. Scandinavian p.ns. Ringsted, Ringstad. OE **hring** + **stede**. Stead 103, 184, 261.

(1) ~ Norf TF 7040. 'Circle place' or 'enclosure'. *Ringstyde* *[1060]14th S 1030, *Ringstede* *[1060]14th S 1109, *Rinc(s)teda* 1086, *Ri- Ryngsted(e)* 1053–1474.

(2) ~ Northants SP 9875. *Ringsted(e)* 12th–1298, *Ringested(e)* 1203–1348. Nth 195.

(3) ~ BAY Dorset SY 7581. *Ringstead Bay* 1773. P.n. Ringstead + ModE **bay**. Ringstead, *Ringestede* 1086–1475, *Ring- Ryngsted(e)* 1227–1554. An alternative for the specific might be OE **hringe** 'a salt-pan'. Ringstead is a deserted village on the coast. Do i.142, 212.

RINGWOOD Hants SU 1405. 'Boundary wood'. *(to) rímucwuda* [955]14th S 582, *(æt) runcwuda* [955×8]14th S 1491, *rimecuda* 961 S 690, *Rincvede* 1086, *Ri- Rynk(e)wod(e) -wude* 1213–1367, *Ringwod(e)* 1199–1231. OE ***rimuc** 'edge, border, boundary, ridge' + **wudu**. The reference is either to situation on the edge of the New Forest or to the county boundary with Dorset to the W. Ha 138, Gover 1958.219.

RINGWOOD FOREST Hants SU 1108. No early forms. A modern plantation on former heathland.

RINGWOULD Kent TR 3648. Partly uncertain. *Roedligpealda* [861]13th S 330, *Rudelingewealde* 1216×72, *Ridelingwold' -we(a)ld(e)* 1240–1311, with variants *Ry- Ru- Re-*, *Redelingewalde* 1254, *Ringwald'* 1184×5. Uncertain element, apparently OE ***(h)rēdling**, + **weald** 'forest', and so '*(H)redling* forest'. The 861 spelling, if reliable, points to a derivative of **rōd** 'a pole, a rod' with i-mutation of *ō* > *œ̄* > *ē*: the sense would be 'forest where poles are obtained'; later forms point rather to OE **hrēodling* 'reedy place' from OE **hrēod** 'a reed', but this seems an unlikely element to be compounded with OE **weald**. Another possibility is OE *Hrœ̄thling*, 'the place called after Hrœ̄thel' with [d] for [ð] as in RODMERSHAM TQ 9261. Forms with *inge* may point to locative-dative case **(H)rēdlinge*, or possibly the genitive pl. of a folk-n. **(H)rēdlingas* 'the people of Hredling'. KPN 275, Cullen.

RINSEY Corn SW 5927. 'Point house'. *Renti, Rentin* 1086, *Rendy* 1244, 1284, *Rynsi* 1333, *Rensy* 1367, *Ringie* 1660. Co ***rynn** + **chy** (OCo). Rinsey is situated just inland from Rinsey Head. The second DB form may not belong here. PNCo 149, Gover n.d. 494.

RIPE ESusx TQ 5110. 'The strip or ridge'. *Ripe* 1086, *Rip(p), Ryp(pe)* 13th, 14th cents. OE ***ripp** cognate with LG *riep* 'shore, slope', EFris *ripe* 'edge' and *rippel* 'a strip, undergrowth', Norw *ripel* 'strip, thin stem'. Sx 404.

RIPLEY Derby SK 4050. Either 'strip-shaped clearing' or 'wood of the people called *Hrype*'. *Ripelei* 1086, *-le(ye)* 1224, 1291, *Rip- Rypley(e) -lay* 1177, 1284×6, 1323, 1501 etc., *Rippe- Ryppelegh(a) -le(y)a -ley(e) -le* 1154×9–1410. OE ***ripel** or folk-n. *Hrype*, genitive pl. *Hrypa*, + **lēah**. Db 493, L 205.

RIPLEY 'Strip-shaped wood or clearing'. OE ***ripel** + **lēah**.

(1) ~ Hants SZ 1698. *Riple* 1086, 1338, *Ripela* 1167, 1272, *Ry- Rippele(y)* 1280–1349. Gover 1958.230, DEPN, PNE ii.84, L 205.

(2) ~ Surrey TQ 0456. *Ripelia, Rippelle* [1204]1321, *Reppely, Rip(e)le, Ruppe- Rypele* 13th cent., *Rippe-* 13th–1393, *Ripley* from 1488. Sr 146–7, PNE ii.84, L 205.

RIPLEY NYorks SE 2860. 'Woodland clearing of the Hrype'. *Ripeleie -leia* etc. 1086–1501, *Ri- Ryppelei(a) -ley* etc. 1175–1501, *Ripley* from 1277. OE folk-n. *Hrype*, genitive pl. *Hrypa*, + **lēah**. The place probably marked the boundary of the territory of the *Hrype*, whose capital was at RIPON SE 3171. YW v.101.

RIPLINGHAM Humbs SE 9631. 'Homestead of the Riplingas, the people living on the ridge', or 'the homestead or estate called or at *Ripling*'. *Ripingha'* 1086, *Ripplingeham* 1180, *Rip(p)lingham -y-* c.1175×88–1399. OE folk-n. **Rip(p)lingas* < ***rip(p)el** 'a strip of land' + **ingas**, genitive pl. **Rip(p)linga*, or OE p.n. **Rip(p)ling* < ***rip(p)el** + **ing**[2], locative–dative **Rip(p)linge*, + **hām**. The underlying element is OE ***rip(p)el** 'a strip of land', Mod dial. *ripple* 'a strip of woodland', here used in the sense 'hill-ridge'. YE 204, ING 154.

RIPON NYorks SE 3171. '(The settlement amongst or in the territory of) the Hrype'. *Inhrypis, Inhripis, Hrypis* [c.715]11th BHA, *Inhrypum* [c.731]8th BHE, [c.890]c.1000 OEBede, *hreopum, hripum* ibid., *in Hripum* 11th, *Ri- Rypum -un* 1086, *Ripon* from c.1000. OE folk-n. *Hrype*, dative pl. *Hrypum*, Latinised *Hrypi*, dative pl. *Hrypis*. This was an important Anglian people whose territory (Riponshire, *Ripsire, Ripeshire* 1173, OE *Hrype*, genitive pl. *Hrypa*, + **scīr** 'a territory, a jurisdiction') reached as far south as RIPLEY SE 2860 and also had an offshoot at REPTON SK 3027. *Hrype*, like other OE folk-ns. such as *Engle, Dene, Mierce*, is an *i*-declension formation, implying a Gmc form **Hrup-iz*, but the ultimate etymology of this is unknown. OE spellings with *eo* by *u*-mutation of *i* unrounded from *y* in late ONb imply a short vowel. An OE *Hryp* is recorded among the ancestors of the kings of East Anglia. YW v.164.

RIPPINGALE Lincs TF 0928. 'The valley of the Hrepingas, the people called after Hrepa'. *Repinghale* 1086–1486 with variants *-yng-, -hal(l)* and *-ale* (from 1219), *Rep- Rapingehale, Ripingehal', Ripingahala* 12th cent., *Rippinghale -yng-* c.1160, 1400–1616, *-ale* from 1502. OE folk-n. *Hrepingas* < pers.n. *Hrepa* + **ingas**, genitive pl. *Hrepinga*, + **halh**, locative–dative sing. **hale**. The Hrepingas are mentioned ASC(E) under year 675 in a grant of lands by king Ethelred of Mercia to the newly founded monastery of *Medeshamstede* (Peterborough). Rippingale was one of the Peterborough estates featuring in late copies of various spurious AS charters (S 68, 70, 162, 189, 200, 213, 538, 741) forged to replace lost originals. Perrott 140, L 103, 110.

RIPPLE 'The strip (of land)'. OE ***rip(p)el**.

(1) ~ H&W SO 8737. *rippell* [680 for 678×93]11th S 52, *Rippel* [757×96]17th Hooke 33, [821×2]17th Hooke 95, [852×72]17th Hooke 101, 1086–1549 with variant *Ry-*. The reference is probably to a tongue of higher land along the Severn west of Ripple. Wo 158, Hooke 20, 33, 95, 101.

(2) ~ Kent TR 3450. *Ryple* [1087]13th, *Roppelega* 1211×12, *Ripple* from 1235 with variants *Ripplee, Riplegh* and *Ripp(el)lega* 13th. The reference is to a strip of high ground forming a spit or ridge of land. PNK 574, 403, 383.

RIPPONDEN WYorks SE 0319. 'Valley of the Ryburn'. *Ry- Riburneden(e)* 1307–1465, *Ri- Rybondeyn -den* 16th, *Riponden* 1555, *Ripponden* from 1653. R.n. RYBURN SE 0419–22 + **denu**. YW iii.65, L 99.

RIPPON TOR Devon SX 7475. *Rippentorr* 1726, *Rippen Tor* 1827. D 478.

Abbots RIPTON Cambs TL 2377. 'R held by the abbot' sc. of Ramsey. *Ripton Abbatis* 1163, *Ripton Abbottes al. Saynt John's Ripton* 1579. ModE **abbot**, Latin **abbas**, genitive sing. **abbatis**, + p.n. *Riptune* 1086, *Ripetona* c.1139, *Ripton* from 1163, 'settlement by the edge or slope', OE ***rip(p)** + **tūn**. 'Abbots' for distinction from nearby Kings RIPTON TL 2576. Both places lie in a shallow valley below gently shelving ground. The exact sense of ***ripp** here is unclear. Hu 218.

Kings RIPTON Cambs TL 2576. *Ripton Regis* 1163, *Kyngesripton* 1381. ME **king**, genitive sing. **kinges**, Latin **rex**, genitive sing. **regis**, + p.n. Ripton as in Abbots RIPTON TL 2377. Hu 218.

Monks RISBOROUGH Bucks SP 8004. 'R held by the monks' sc. of Christ Church, Canterbury. *Monks Ryseberg* 1290, *Monkenrisbourgh* 1346, *Monkyng Ryseberg* 1509. ModE **monk**, genitive pl. **monks**, ME **moneken**, + p.n. *at Risenburga* [944 or 5]14th S 882, *Risberghe* *[995]13th S 1378, *Risebergh* 1086–1349, *Ryseborowe* 1302 'brushwood-covered hill(s)', OE **hrīsen** + **beorg**. The earliest form, *æt þǣm ēasteran Hrisanbyrge* 'at the eastern R' 903 S 367, probably refers to Monks R, but several other early forms, *Risenbeorgas* [1005]17th S 911, *Hrisanbeorgen* [966×75]12th S 1484, *Hrisebyrgan be Ciltemes efese* *[1006 for 1001]11th S 914 (Latin *Hrisbeorgam*), *Hrysebyrgan* 1042×66 S 1047, *Hrisbeorgan*

1020×38 S 1464, 1038×50 S 1466, are nominative-accusative and dative pl. respectively referring either to the hills of the Chiltern eaves or to the two settlements of Monks and Princes Risborough. Bk 170, YN xliv, L 128.

Princes RISBOROUGH Bucks SP 8003. *King's Rysburgh* 1290, *Earl's Risebergh* 1337, *Rysburgh Principis* 1359, *Pryns Risburgh* 1433, *Pryncyn Ryseborough* 1509, ME **king, erl, prince**, Latin **princeps**, genitive sing. **principis**, + p.n. Risborough as in Monks RISBOROUGH SP 8004. One OE form, *Hrisanbeorgan* [966×75]12th S 1484, belongs here. Originally a royal manor, it came into the custody of the Black Prince in 1342. In 1337 it was held by the earl of Cornwall. Bk 170.

RISBURY H&W SO 5556. 'Brushwood fort'. *Riseberia* 1086, 1177, *Risebur(y)* 1212–1303. OE **hrīs**, genitive pl. **hrīsa**, + **byriġ**, dative sing. of **burh**. The reference is to Risbury Camp Iron Age hill-fort at SO 5455. He 108.

RISBY 'Village or farm in the brushwood, or where brushwood is collected'. ON **hrís** + **bȳ**. Cf. the Dan p.ns. Risby and Rejsby.

(1) ~ Humbs SE 9214. *Risebi* 1086–1196, *Risabi* c.1115, *Risby* from 1254. DEPN, SSNEM 65.

(2) ~ Lincs TF 1491. *Risebi* 1086–1219, *Risabi* c.1115, *Risby* from 1268. Li iii.173, SSNEM 65.

(3) ~ Suff TL 8066. *Rysebi* 11th, *Re- Risebi, Riseby* 1086, *Resebi* 1179, *Rissebi* 1166, *-by* c.1265, *Risbye* 1610. Sps with *Re-* may alternatively point to OEScand **ryth** 'a clearing' as the specific, genitive sing. **rytz**. DEPN, PNE ii.91

RISE Humbs TA 1542. '(The settlement) amidst the brushwood'. *Risun -on* 1086, *Rise* from c.1265 with variants *Rys, Ryse*. OE **hrīs**, dative pl. **hrīsum**. Cf. RYSOME GARTH TA 3622. YE 70.

RISEGATE Lincs TF 2129. 'Rye's pasturage, way or drainage channel'. *(manerium de) Ri- Rys(e)gate* 1275–1512, *sewera de gosberchirch quod vacatur Risgate* 'Gosberchurch sewer called R' 1331, *Risgate Eay* 1653. Family name *Rye* as in Ranulph de Ry 1250, John de Rye 1280, genitive sing. *Ryes*, + ME **gate** (ON *gata*) 'a way, a pasturage'. The forms for 1331 and 1653 with dial. **ea(u)** (OE *ēa*) 'a drainage channel' suggest that *gate* was also used in the fenlands for a water-course. Payling 1940.88.

RISE HILL Cumbr SD 7388. Perhaps 'brushwood hill'. *Rysell* 1771. Ultimately from OE **hrīs** + **hyll**. A hill-ridge rising to 1825ft. YW vi.262.

RISELEY 'Brushwood clearing' i.e. where brushwood can be obtained. OE **hrīs** or ***hrīsen** + **lēah**.

(1) ~ Beds TL 0463. *Riselai* 1086, *Rise(e)le(y)* from 1202. Bd 18, L 205.

(2) ~ Berks SU 7263. *Rysle* 1300, 1572, *Riseley* 1761. Brk 109, L 205.

RISHANGLES Suff TM 1668. 'The brushwood wooded slope'. *Risangra* 1086, *Rishang(r)'* 1171, 1203, *Rissangeles* 1254, *Ri- Ryshangles* from 1286. OE **hrīs** + **hangra**. DEPN, Baron.

RISHTON Lancs SD 7230. 'Settlement, farm where rushes are gathered'. *Riston* 1200×8, *Ruston* 1243, *Ri- Ryssheton* 1322, 1371, *Russhton* 1332. OE **risc** + **tūn**. La 72, Jnl 17.46.

RISHWORTH WYorks SE 0318. 'Enclosure where rushes grow'. *Ri- Rysseworde* 12th, *-wrth(e)* late 12th–1309, *-worth* 14th cent., *Ri- Rysche- Ry- Rissh(e)worth(e)* 1275–1540, *Rus(s)(c)h(e)worth* 1316–1724. OE **risc** + **worth**. ME *i* in the neighbourhood of *r* sometimes appears as *u* in Y dialects, but influence from StE *rush* is probable in this name. Rushes once grew in profusion in the river valley here. YW iii.71 gives pr ['ruʃəθ].

RISHWORTH MOOR WYorks SE 0017. P.n. RISHWORTH SE 0318 + ModE **moor** (OE *mōr*). YW iii.74.

RISING BRIDGE 'Brushwood causeway'. OE ***hrīsen** + **brycg**. L 66.

(1) ~ ~ Lancs SD 7825. No early forms.

(2) ~ ~ Northants near Geddington SP 8983. *Risenbrige* c.1220, *Rysingbridge* 1612. Nth 166.

RISLEY 'Brushwood clearing'. OE **hrīs** or adjective **hrīsen** + **lēah**. Probably refers to places where brushwood could be gathered. Identical with RISELEY.

(1) ~ Ches SJ 6592. *Ryselegh* 1284, *Risselley* 1285, *Riselegh* 1328, 1332. La 97.

(2) ~ Derby SK 4635. *Riselei(a)* 1086, *Ri- Ryseleg' -legh(e) -ley(e)* 1212–1440, *Risley* from 1263. Db 496.

RISPLITH NYorks SE 2468. 'Slope overgrown with brushwood'. *Respleth* 1490, 1549, *-lith(e)* 1608–38, *Risplethe* 1587. Dial. **risp** from OE **hrispe* + **hlith**. YW v.188.

Great RISSINGTON Glos SP 1917. *Magna Risinton* 1218, *Magna Ri- Rysindon(e) -yn-* 1291–1435, *Magna Rysyngdon* 1303, *Greate* ~ 1584. ModE **great**, Latin **magna**, + p.n. *Risendune -a -don(e)* c.1075, 1086–1333, *Ri- Rysindon(e) -yn- -dun* 12th–1252, 'hill overgrown with brushwood', OE **hrīsen** + **dūn**. 'Great' for distinction from Little RISSINGTON SP 1919. Gl i.201, L 149.

Little RISSINGTON Glos SP 1919. *Litell, Parua Ri- Rysindon(a) -yn-* c.1230–1571. ME **litel**, Latin **parva**, + p.n. *Risedvne* 1086 etc., as in Great RISSINGTON SP 1917. Gl i.202.

Wyck RISSINGTON Glos SP 1921. 'Outlying farm belonging to Rissington'. *Wy- Wik(e) Ri- Rysindon -yn- -en- -dun* c.1170–1492, *Wyke Risington -yng-* 1558, 1575. OE **wīc** 'dairy farm, outlying farm' + p.n. *Risendvne* 1086 etc., as in Great RISSINGTON SP 1917. A secondary settlement from Great Rissington. Gl i.203.

Long RISTON Humbs TA 1242. *Long Ruston* 1611. ModE adj. **long** + p.n. *Ristun(e)* 1086–1160×75, *Ri- Ryston(a)* 1199×1215–1583, *Reston(a)* 1150×60–1316, *Russton* 1465, *Rouston* 1524, 'brushwood farm or village', i.e. where brushwood is collected, OE **hrīs** + **tūn**. The village is two miles from RISE TA 1542, the settlement 'amidst the brushwood'. 'Long' for distinction from RUSTON PARVA TA 0661. YE 70.

RIVAR Wilts SU 3161. '(The settlement) at the promontory'. *(on tha) yfre* 931 S 416, *Ryver* 1609. OE **yfer**. The form with *R-* is due to misdivision of ME *atter iver* (OE *æt thære yfre*) as if *atte river*. The settlement lies just below a very steep hill-side. Wlt 355 gives pr [raivə], L 178–9.

RIVENHALL END Essex TL 8316. 'The end of R'. *Rivenhall End* 1805 OS. P.n. Rivenhall TL 8217 + ModE **end**. Contrasts with SILVER END TL 8019. Rivenhall, *Reuenhala* [1068]1309, *-hal(e)* 1275–1532, *Ruwenhala -e* [1068]1309, *Ruuuenhalā* 1086, *Ruenhalā -e* 1086–1314, *Riuuenhalā* 1086, *Ri- Rywe(n)hale* 1185–1313, *Rivenhale* 1195–1360, is the place 'at the rough nook or sheltered place', OE **hrēof**, definite form oblique case **hrēofan**, + **halh**, dative sing. **hale**. The reference is to a hollow near the church. The *u*-forms are probably due to confusion with ME *ruwe* < OE *rūh* 'rough'. Ess 295 gives pr [rivnɔ:l].

RIVER BANK Cambs TL 5386. A modern coinage, no early form. Cf. however *communem ripariam in Reche* 'common river-bank in Reach' 1453. Ca 137.

RIVERHEAD Kent TQ 5156. 'Head of the river'. *Riverhead* from 1656. The reference is to the head waters of the river Darent. This form, however, is a false etymology for earlier *Reddride* 1278, *Reydrythe* 1292, later *Re- Rotherdon -den* 1479–1560 and *Retherhė(a)d* 1619, 1656, *Rotherhith or Rethered now called Riverhead* 1778. Possibly OE **hrēod** + **rith** 'reed stream' later confused with ME *rether, rother* 'an ox, cattle' or identical with Rotherhithe GLond TQ 3579, *Retherhith* 1127, 'the landing place for cattle'. PNK 62, DEPN.

RIVINGTON Lancs SD 6214. 'Settlement beside Rivington Pike'. *Ro- Rawin- Revington* 1202, *Ru(h)winton* 1212, *Rou- Row- Rovington -yn(g)-* 1227–1448, *Ruyn- Ruwinton* 1246, *Roynton* 1332, *Reuynton* 1338, *Ryvington* 1346, *Riventon* c.1540. OE hill-name **Hrēofing* 'the rugged one' < **hrēof** + **ing**[2], + **tūn**. The reference is to Rivington Pike SD 6514, a hill of 1193ft., *Roinpik* before 1290, *Rovyng* 1325, *Rivenpike* c.1540, *Riuenpike* *hill, Rauenpike* 1577, OE **Hrēofing* + **pīc**. According to Leland c.1540, the alternative name of this rough mountain was *Faierlokke*. The old form *Roynton* survives in the name Roynton Cottage. La 48, 28, Jnl 17.38, 18.22.

ROA ISLAND Cumbr SD 2365. Possibly the 'red island'. *The Roa* 1577. Possibly ON **rauthr** + **ey**. But earlier forms are needed. La 205 gives pr [rɔ:ə].

River ROACH Essex TQ 9592. A late back-formation from ROCHFORD TQ 8790. In 1777–1848 it was known as *Broom Hill River* from Broomhills in Great Stambridge, *Bromehyll* 1449, and before that as *Walflestreme* 1374, *Wal(l)fle(e)t(e)* 1594–1676, the original name of the estuary between BARLING TQ 9289 and Wallasea Island TQ 9693, *Walfliet* 1216×72, *Wal(e)flet(e)* 1229–1578, *Wallet or rather and more trulie Walfleet Iland* 1594, 'wall estuary', OE **weall** + **flēot** referring to sea-walls, cf. 'between Crouch Creek and St Peter's Chapel ... upon the very shore was erected a wall for the preservation of the land ... and all the Sea shore which beateth on that wall is called the Walfleet and ... up in Crouche Creeke at the end of the wall ... is an island called Commonly and corruptly Wallet (but I take it Walfleete) Island' 1588. Ess 10, 205, MM 16.75.

ROACHILL Devon SS 8422. No early forms.

The ROAD Scilly SV 8912. 'The anchorage'. *Roade* 1689. ModE **road** in the sense 'a sheltered piece of water for ships, an anchorage, a roadstead'. PNCo 150.

ROADE Northants SP 7551. 'The clearing'. *Rode* 1086 etc., *Rhode* 1629, *Road* 1775. OE ***rod**, ***rodu**. Nth 106, PNE i.86, L 208.

ROADWATER Somer ST 0338. *Higher Roadwater* 1809 OS. Roadwater seems to be another name for Washford River. It could be a compound of OE ***rodu** 'a clearing' + **wæter** 'a stream', but earlier forms are needed.

ROBERTSBRIDGE ESusx TQ 7323. *(de) Ponte Roberti* '(of) Robert's bridge' 1176–1332, *Robartesbregge* 1445, *Roberdisbrigge* 1475. OFr pers.n. *Robert* + ME **brigge**, SE dial. form **bregge** (OE *brycg, brecg*). The reference is to Robert de St Martin, the founder of the abbey of Robertsbridge. The upkeep of bridges was a medieval act of charity. In the 17th cent. the name is occasionally recorded as *Rother(s)bridge* after the eastern river Rother, itself a 16th cent. back-formation from ROTHERFIELD TQ 5529. Sx 459.

ROBERTTOWN WYorks SE 1922. *Robertowne* 1658, *Robert Town* 1709. Originally *Robert Lyversegge* 1375 'that part of Liversedge held by Robert', an unidentified feudal tenant. YW iii.28.

ROBIN HOOD'S BAY NYorks NZ 9505. *Robin Hoode Baye* 1532. Rhymes of Robin Hood were so widely popular by the end of the 14th century as to attract disapproval – Sloth in *Piers Plowman* knew them more readily than his Paternoster (V.396). 'The many Robin Hood's Hills, Wells, Stones, Oaks or Butts, which are found in all parts of the country ... are a reflex of his legend, and occur as far from his haunts as Gloucestershire and Somerset. He had no more to do with them than the Devil, to whom similar natural features are also ascribed.' So E.K. Chambers, who believed the original outlaw to have been a Robin Hood who was imprisoned in 1354 for taking wood and deer in the Forest of Rockingham in Northants. Most of the stories are located in Barnsdale near South Elmsall SE 4801 or in Sherwood. YN 118.

ROBOROUGH 'The rough hill'. OE **rūh**, definite form **rū(h)a**, + **beorg**.

(1) ~ Devon SS 5717. *Raweberga* 1086, *Rauburga* 1193, *Ruaberga* 1166, *Rughe- Ro(g)heberg(e) -berhe, Rouberwe* 13th, *Rouhburgh* 1349, *Roweburghe* 1291, 1384. D 118.

(2) Plymouth (ROBOROUGH) Airport Devon SX 5060. *Rueberge* 1114×16, *Roweberwe* 1284, *Rouburgh* 1387. D 225.

ROBY Mers SJ 4390. 'Boundary village'. *Rabil* (sic) 1086, *Rabi* 1185, *Raby* 1238–1327, *Roby* from c.1199. ON **rá** + **bȳ**. Roby is on the border with Childwall SJ 4189. La 113, SSNNW 38, Jnl 17.63.

ROBY MILL Lancs SD 5107. P.n. Roby + ModE **mill**. There are no early forms for Roby; possibly identical with ROBY Mers SJ 4391.

ROCESTER Staffs SK 1139. Partly uncertain. *Rowcestre* 1086, *Roffecestre* 12th, *Rouecestre -v-* 1208, 1225, *Rocestre* 1246. DEPN suggests 'Hrothwulf's Roman camp', OE pers.n. *Hrōthwulf*, genitive sing. *Hrōthwulfes*, + **ċeaster**, comparing the early forms for ROUSHAM Oxon SP 4724. A better alternative is 'the rough camp', OE **rūh**, definite form **rūe** or **rūhe** with early [w] for [ɣ] as in ROBOROUGH Devon SS 5717 or Rowner Hants SU 5801, *(oþ) ruwan oringa gemære* '(to) the boundary of the people of Rowner, the rough hill' [948]12th S 532, + **ċeastre**. DEPN, Ha 141.

River ROCH GMan SD 8712. Possibly a back-formation from ROCHDALE SD 8913. There are two distinct types:

I. *Rached* [13th]14th, *Rachet* 1292.

II. *Rach* 12th–[c.1300]14th, *Rache* [c.1200]14th, 1292, 1577, *Reche* 1292, *Ritch* 1590, *Roch* 1843 OS, 1897, *Roach* 19th.

Type I may be a back-formation from *Rachedham*, the earlier form for Rochdale, type II a back-formation from later *Rachedale*. Alternatively the r.n. may be primary and both *Rachedham* and *Rachedale* take their name from it. No such Celtic r.n., however, is known, and in the current state of knowledge it is impossible to be sure which name is derived from which. See further ROCHDALE SD 8913. La 28 and RN 344 give pr [ro:tʃ].

ROCHDALE GMan SD 8913. 'Valley of the river Roach'. *Rachedal'* 1190×8, *-dal(e)* 1242–1341, *Rechedale* 1276, *Rochedale* 1246, 1292, *Rochdale* 1843 OS, *Rachdall* 1598, *Ratchda* 1865. R.n. ROCH SD 8712 + ON **dalr**. An alternative earlier type is *Recedham* 1086, *Rachetham* before 1193, *Rachedham* before 1193, 1296, *Rechedham* 1195×1211, *Racheham* 13th, either 'homestead, village by the r. Roch', r.n. ROCH + OE **hām**, or 'hall village', OE **reċed**, **ræċed**+ **hām**. The relationship between the r.n. Roch and the p.n. Rochdale is unclear. It is impossible to be sure whether the r.n. is primary in the types *Recedham* and *Rachedal'* or whether the sequence of development was *Recedham* 'hall village' misunderstood as 'village by the *Reced*' > r.n. type I *Rached* > new coinage *Rached-dale, Rachedale* > r.n. type II *Rache*. The variation between *Re-* and *Ra-* spellings is explicable by the co-existence in OE of a side-form **ræċed** as well as usual **reċed**, but this does not establish the etymology, it merely suggests that at some stage the name was at least understood to mean 'hall village'. In fact **reċed** otherwise only occurs in OE in poetry; it looks like a specifically poetic word and its appearance in a p.n. would be unique. La 54 gives 19th cent. prs [ratʃ] and [ratʃit], TC 161, Jnl 17.42, 18.18.

ROCHDALE CANAL GMan SD 9420. P.n. Rochdale SD 8913 + ModE **canal**. The canal begun in 1786 and completed in 1804 linked the Duke of Bridgewater's Canal at Manchester to the Calder and Hebble Navigation at Sowerby Bridge near Halifax.

ROCHE Corn SW 9860. 'The rock'. *(de, in) Rupe* 1201–1357, *La Roche* 1201–1428, *Rock* 1478, *Roche* 1610, *Roach* 1748. French definite article **la** + **roche** (varying with Latin *rupes* and English *rock*). The reference is to the prominent granite tor at SW 991596 called Roche Rock. PNCo 150 gives pr [routʃ], Gover n.d. 416.

ROCHE ABBEY SYorks SK 5489. 'Rock abbey'. Latin *Abbatia, monasterium de Rupe* 1147–1445, *Roch(e)* from 1235. ME **roche**, Latin **rupes** 'a rock', referring to the rocky bank now called Table Rock beside which the Cistercian abbey was founded c.1147. YW i.137.

ROCHESTER Kent TQ 7369. 'Durobrivæ Roman town'.

I. RBrit forms: *Durobrivis -brovis* [4th]8th AI, *Durobrabis* [c.700]13th Rav, *Dorubreui* [c.731]8th BHE, *Dorobreuia* [868]12th S 339.

II. Later Latin forms: *ciuitas Hrofi* [c.731]8th BHE, [762]12th S 32, [789]12th S 131, ~ *Hrobi* [823]12th S 271, [842]14th S 291, [860]10th Rochester 24, ~ *Hroui* [838]10th S 280, ~ *Roffi* [850]14th S 299, *castellum Hrobi* [855]12th S 315, and adjectival *Rof(f)ensis (ecclesia, episcopus)* *[597×604, 618, 695]12th–14th S 1244, 4, 6, *Hrofensis ciuitas* [c.731]8th BHE 'church, bishop, city of R'.

III. English forms: *Hrofaescaestrae -caestir* [c.731]8th BHE, *Hrofiscestri* [765]12th S 34, *Hrofescester* [788]12th S 129, *-ceaster* c.1122 ASC(E) under year 604, *-caester* [789]12th S 130, *-ceastre* 9th ASC(A) under years 633, 644 etc., c.995 S 1458, 894, 897 ASC(A), c.1050 ASC(D) under year 1023, [980×87]12th S 1457, *Hrofeceastre* 11th ASC(C) under year 986, [995]12th S 885, [995×1006]12th S 1456, c.1122 ASC(E) under years 616, 885 etc. to 1114, *-cestre* 12th ASC(D) under year 1058, *Rove- Rouecestre* 1086, *Rofescœstre -ceastre* c.1122 ASC(E) under years 643, 740, 839, *Rofe ce(a)stre -cœstre* c.1122 ibid. under years 655, 675 etc. to 802, 12th ASC(H) under year 1114, *Roue ce(a)stre* after 1154 ASC(E) under years 1123–40, *Rochester* 1610.

IV. *ciuitas Hrofibreui* [604]12th S 1, *Hrofesbreta* [765×85]12th S 37.

Type I is the Latin dative–ablative case form of Brit **Durobrīu̯ās*, a pl. name compounded from **duro-* 'fort, walled town' and **brīu̯ā* 'bridge', and so 'bridges fort'. Another example of this compound occurs for the Roman town at Chesterton, Water Newton, Hunts TL 1297. Although Bede knew the authentic RBrit form of this name, it must have been heard locally with an accentuation **Duróvrīw* or **D'rovrīw* which became OE **Rofi*. This permitted folk-etymological association with OE *hrōf* 'roof' and accounts for types II and III, presumably originally 'the Roman city *Rofi*'. The specific must early have been misunderstood as the genitive sing. of a pers.n. **Rofus* or **Hrofus* and hence Bede's explanation of *Hrofaescaestrae*, the English name for *civitas Dorubreui*, the city of *Dorubrevum*, viz. that it was named *a primario quondam illius (sc. gentis Anglorum) qui dicebatur Hrof* 'after one of their former chiefs called Hrof' (BHE 2.3), a similar explanation to that which invented another unattested pers.n. for PORTSMOUTH. Such etymological speculation seems to have been fashionable among upper-class commentators in the Anglo-Saxon period. Type IV shows a curious conflation of *Hrof* and *Dorubreui*. RBrit 346, Britannia 1971.xvi-xvii, LHEB 267, TC 162.

ROCHESTER Northum NY 8398. Partly uncertain. *Roff'* 1208, *Rouschestre* 1325, *Rochester* 1695 Map. Unknown element + OE **ċeaster**. The reference is to the Roman fort of BREMENIUM NY 8398. If *Roff'* stands for *Roffensis* it is identical with the mediaeval Latin adjective for ROCHESTER Kent. DEPN.

ROCHFORD Essex TQ 8790. Partly uncertain.

I. *Rochefort* 1086, *-ford* 1200–1306, *Rochford* from 1227.

II *Rochesford* 1180–1337.

III. *Rac(c)h(e)ford* 1200–1535.

IV. *Rech(e)ford* 1243–1484.

The hundred name is *Rokesford* 1085×9, *Rochesfort* 1086, *-ford* 1180, *Rochefort* 1086, *-ford* 1172–1274, *Rac(c)h(e)ford* 1216×72, 1452. The form *Rochefort* is due to assimilation to the common Fr p.n. type Rochefort 'strong rock' and its alternative form Ro(c)quefort. The original was probably *Hrōca-ford* 'rook ford', OE **hrōc**, genitive pl. **hrōca**, + **ford**, or *Hrōces-ford* 'Hroc's ford', OE pers.n. *Hrōc*, genitive sing. *Hrōces*, + **ford**. Types III and IV, however, point to *Ræċċa-ford* 'ford of the hunting dogs', OE **ræċċ**, genitive pl. **ræċċa**, + **ford** as in ROCHFORD H&W SO 6268. The precise relationship between these different types is unclear. Ess 196 gives pr [rɔtʃfəd].

ROCHFORD H&W SO 6368. 'Hunting-dog's ford'. *Recesford* 1086, *Ræccesford* 11th, *Rechesford* c.1240, *Rache(s)ford* 1255–1461, *Rochesford* 1249, *Rocheford* 1322. OE **ræċċ**, genitive sing. **ræċċes**, later assimilated to OFr, ME **roche** 'a rock, a cliff' or **roche** 'a roach'. Wo 69.

ROCK Corn SW 9475. Cf. *Rock House* 1813. Short for *Black Rock* 1748, itself a partial replacement of *Blaketore* 1337, 'black crag', OE **blæc** + **torr**. The 1337 form refers to a ferry across the channel estuary also known as *passag' de Penmayn* 1303, Co **maen** 'rock', *passage of Blaketorre* 1337. PNCo 150, Gover n.d. 129.

ROCK H&W SO 7371. 'At the oak-tree'. *del Ak* 1224, *Ake -a* 1253–1338, *(la) Rok(e)* 1259–1550, *Rock(e)* from 1309, *(la) Rook(e)* 1366–1545. OE **āc** in the phrase *æt thǣre āce* which developed to ME *atter oke* misdivided as if *atte roke* 'at Rock'. Wo 69.

ROCK Northum NU 2020. 'The rock'. *Roch* 1164, *Rok'* 1242 BF, *Rock'* 1296 SR. ME **rokke**, referring to outcrops of limestone here. NbDu 167.

ROCKBEARE Devon SY 0295. 'Rook wood'. *Rochebere* 1086, *Rokeber(e)* 1196–1261, *-bear(e)* 1261–1322, *-byar* 1316. OE **hrōc**, genitive pl. **hrōca**, + **bearu**. D 594.

ROCKBOURNE Hants SU 1118. 'Rook stream'. *Rocheborne* 1086, *Rokeb(o)urne* 1155–1341. OE **hrōc**, genitive pl. **hrōca**, + **burna**. Ha 139, Gover 1958.216.

ROCKBOURNE DOWN Hants SU 1021. *Rockborn Down* 1811 OS. P.n. ROCKBOURNE SU 1118 + ModE **down**.

ROCKCLIFFE Cumbr NY 3561. 'Red cliff'. *Rod- Roudecliua* 1185, *Routh(e)clive -clif(f)* c.1187–1348, *Rade- Redeclive* 1203, *Rouckecliffe* 1305, *Rouclef -clif Raw-* 1346–1485, *Rocley* 1552. ON **rauthr** probably replacing OE **rēad**, + OE **clif**. Cu 146, SSNNW 246, L 132, 135–6.

ROCKEN END IoW SZ 4975. *Rocken End* 1769. This may be the 'end of the rocks', ME **rokke** + adjectival or plural suffix **-ene** or a reduced form of **end**, + **end**. Wt 119, PNE i.152, Mills 1996.87.

ROCK FERRY Mers SJ 3386. 'Ferry at the rock'. *The Rock Ferry* 1757. Earlier simply *le Fere* 'the ferry' 1350, *passagium et batillagium de Bebynton* 'the toll and ferry-boat of Bebington' 1357. The rock was an outcrop on the Wirral shore of the Mersey which gave its name to the ferry and to the *house called the Rocke* 1644. Che iv.246.

ROCKFORD Hants SU 1508. 'Rook ford'. *Rocheford* 1086, 1195, *Rokeford* 1286, 1371, 1503, *Roc- Ro(c)kford* from 1295. OE **hrōc**, genitive pl. **hrōca**, + **ford**. Other spellings, *Reche- Rachesford* 1167, *Racfordesbroke* 1199×1215, if reliable, may point to different specifics, OE **ræċċ** 'hunting dog', **racu** 'bed of a stream'. For similar problems cf. ROCHFORD Essex TQ 8790, H&W SO 7371 and also MOYLES COURT Hants SU 1608. Gover 1958.213.

ROCKHAM BAY Devon SS 4546. *Rockham Bay* 1809 OS.

ROCKHAMPTON Avon ST 6593. Probably 'rock homestead'. *Rochm̃tvne* 1086, *Roc- Rokhamton'* 1211×13–1322, *-hampton* 1248–1489, *Rockington -yn-* 1566, 1631. OE **rocc** + **hām-tūn**. Alternatively the specific might be OE **hrōc** 'a rook'. Gl iii.13.

ROCKINGHAM Northants SP 8691. 'The homestead of the Hrocingas, the people called after Hroc'. *Rochingeham* 1086–1131, *Rokingeham -ynge-* 1174–1223, *Rokingham -yng-* 12th–1316. OE folk-n. **Hrōcingas* < pers.n. **Hrōc* + **ingas**, genitive pl. **Hrōcinga*, + **hām**. On formal grounds the specific could be identical with RUCKINGE Kent but a precise parallel seems to be preserved in ODu *Hrokingahem* 815×44. Nth 171, ING 147.

ROCKINGHAM FOREST Northants SP 9791. *foresta de Rochingeham* 1157. An ancient hunting preserve. P.n. ROCKINGHAM SP 8691 + ME **forest**. Nth 1.

ROCKLAND ALL SAINTS Norf TL 9996. 'All saints' Rockland'. *Roclund Omnium Sanctorum* 1254. P.n. Rockland + ModE **all saints**, Latin **omnes sancti**, genitive pl. **omnium sanctorum**, from the dedication of the church. Rockland, *Rokelund* 1086, *Roclund* 1254, is 'rook wood', ON **hrókr**, genitive pl. **hróka**, + **lundr**. 'All Saints' for distinction from ROCKLAND ST MARY TG 3104 and ROCKLAND ST PETER TL 9997. DEPN.

ROCKLAND ST MARY Norf TG 3104. 'St Mary's Rockland'. *Rockland St. Mary* 1838 OS. P.n. Rockland as in ROCKLAND ALL SAINTS TL 9996 + saint's name *Mary* from the dedication of the church and for distinction from ROCKLAND ALL SAINTS TL 9996 and ROCKLAND ST PETER TL 9997.

ROCKLAND ST PETER Norf TL 9997. 'St Peter's Rockland'. *Rokelund Sancti Petri* 1291. P.n. Rockland as in ROCKLAND ALL SAINTS TL 9996 + saint's name *Peter* from the dedication of the church. DEPN.

ROCKLEY Wilts SU 1671. 'The rooks' wood or clearing'. *Rochelie* 1086, *Rokele(ya) -ley* 1155–1310, *Roukle(y)* 1301, 1332, *Rookle(y)* 1335, 1591. OE **hrōc**, genitive pl. **hrōca**, + **lēah**. Wlt 304.

ROCKWELL END Bucks SU 7988. 'Rockwell end' sc. of Hambleden parish. *Rockall End* 1822 OS. P.n. *Rocolte* c.1307, *Rokholte* 1340, 'rook wood', OE **hrōc** + **holt**, + ModE **end**. The modern spelling is a piece of false etymology. Bk 180 gives pr [rɔkəl].

RODBOURNE Wilts ST 9383. Originally a stream name, 'the reedy stream'. *Reodburna* [701]13th S 243, [758]13th S 260, [982]13th S 841, *Rodburne* ibid., *Rod(e)borne* 16th, *Redburn(a) -borne* c.1125–1289. OE **hrēod** + **burna**. Wlt 51, L 18.

RODD H&W SO 3262. 'The clearing'. *(La) Rode* 1220–1355. OE ***rodu**. He 174.

RODDAM Northum NU 0220. '(The settlement) at the clearings'. *Roden* [1135×54]copy, *Rodun* 1203, 1230, *Rodon* 1222, *Rodun* 1236 BF–1289, *Rodom* 1307, 1308, *Roddome* 1542, *Rodham* 1663. OE ***rod**, ***rodu**, locative-dative pl. **rodum**. NbDu 167, PNE ii.86.

RODDEN Dorset SY 6184. 'Red hill'. *Raddun* 1221, *Raddon* 1244, *Roddon* 1420, *Rodden* 1637. OE **rēad** + **dūn**. The soil is a rich red clay here. Do 127.

RODE 'Clearing'. OE ***rod**, ***rodu**. PNE ii.86.

(1) ~ Somer ST 8053. *Rode* from 1086, *la Rode* 1230, *Road* 1817 OS.

(2) ~ HEATH Ches SJ 8057. *Rodeheze* 1280, *Rode Heath* 1741. P.n. Rode as in Odd RODE SJ 8157 + ME **heth** (OE *hǣth*). Che ii.306.

(3) North ~ Ches SJ 8966. *Northrode* from c.1284. Earlier simply *Rodo* 1086, *Rode* 1259–1460. 'North' for distinction from Odd RODE SJ 8157. Che i.59, L 208.

(4) Odd ~ Ches SJ 8157. 'Hod's Rode'. *Hoderod(e)* 1258, 1286, *Odd(e)rode* from 1286. *Hod* later confused with OE pers.n. *Odda* or ME **odde** 'odd, the odd one of three, the third, unique, standing alone', + p.n. *Rode* 1086, 1259 etc. 'Odd' for distinction from North RODE SJ 8966 and RODE HEATH SJ 8057. The tenant after whom the manor is named is recorded as *Adam dictus Hod de Odderorode* 1286, whose by-name is derived from ME **hod** 'a hood' (OE *hōd*) and is identical with the surname *Hood*. Che ii.306.

RODEHEATH Ches SJ 8766. *Rode Heath* 1831. P.n. Rode as in North RODE SJ 8966 + ModE **heath**. Che i.60.

River RODEN Shrops SJ 5915. 'The swift river'. *Rodene* 1249 etc., *Roden(')* from 1256, *Roder* c.1540, *Roddon* 1577. On the Roden was the RBrit settlement of Rutunium (probably at Harcourt Mill SJ 5524), *Rutunio* [c.300]8th AI, Brit **Rutunion*, a regular formation with *-i̯o- suffix on a r.n. which in this case must have been Brit **Rutūnā* from the IE root **reu-* 'move swifly' seen in Latin *ruo*. RN 344, Sa i.250, RBrit 448.

RODEN Shrops SJ 5716. Transferred from the River RODEN SJ 5915. *(æt) Hrodene* c.1000 S 1534, *Roden* late 12th, *Rodene* 1242ff. The initial *H-* of *Hrodene* is inorganic, cf. *hlafe* for *lafe* in the same document. RN 344, Sa i.250.

RODHUISH Somer ST 0139. 'Huish at *Rode*, the clearing'. *Radehewis* 1086, *Rodhuish* 1809 OS. In spite of the 1086 spelling this seems to be OE ***rodu** perhaps used as a p.n. like RODE ST 8053 + **hīwisc** 'estate of sufficient size to support a family'. DEPN.

The RODINGS Essex TL 5813. *duae Rotinges* 'two Rodings' [1042×66]12th, *duae Rodinges* [c.1050]14th K 907, *Ro(d)inges* 1086, *Rodingas* 1100×35, 1198, *Rothinges* [10th]12th, 1119, 1281, 1306, *Roinges -ynges* 1114–1254, *-ing(e) -yng(e)* 1119–1254, *Rothing(e) -yng* 1248–1374, *Rooding* 1535, 1713. The modern name *Rodings* refers to the separate settlements called Abbess, Aythorpe, Beauchamp, High, Leaden, Margaret and White RODING. The original plural *Rodinges*, however, is OE folk-n. **Hrōthingas* 'the people called after Hrotha' < OE pers.n. *Hrōtha* + **ingas**. The *Roinges -ynges* spellings show loss of intervocalic [ð] as in AN and certain other English p.ns. e.g. TAYNTON Glos SO 7321. Ess 490 gives pr [ru:ðiŋ], ING 23f, Pope §346–7, 1176.

Abbess RODING Essex TL 5711. 'R belonging to the abbess' sc. of Barking. *Roinges Abbatisse* 1237, *Abbeys Rothyng* 1518. ModE **abbess**, Latin **abbatissa**, genitive sing. **abbatissae**, + p.n. Roding as in The RODINGS TL 5813. Ess 75.

Aythorpe RODING Essex TL 5815. 'Aitrop's R'. *Rohinga Willelmi fil Ailtrop* 'William son of Ailtrop's R' c.1200, *Aythorp* 1482. Pers.n. *Aitrop* later taken as a p.n. *Aythorp* + p.n. Roding as in The RODINGS TL 5813. Earlier known as *Roinges -ynges Grimbaldi* 'Grimbald's R' 1141–63, and also as *Roynges St Mary* 'St Mary's R' 1235 from the dedication of the church. Ess 491 gives local pr [eiθrɔp].

Beauchamp RODING Essex TL 5809. 'R held by the Beauchamp family'. *Roynges Beuchamp* 1238, *Bechamp Roding* 1400, *Becheme* 1503. Family name of William de Beauchamp of Bedford + p.n. Roding as in The RODINGS TL 5813. Also known as *Rothyng Bothulf* 'St Botulph's R' 1254 from the dedication of the church. Ess 76.

High RODING Essex TL 6017. *High Roinges* 1224, *Alta Rothingg* 1303, *Hyrowthyng* 1462. ME adj. **hegh**, Latin **alta**, + p.n. Roding as in The RODINGS TL 5813. Also known as *Roinges Doun' Bard'* 'Doun Bardulf's R' 1194, *Roinges Bardulf* 1229. Doun or Dodo Bardulf was the donor of three acres in Roding to Dunmow priory. Ess 492.

Leaden RODING Essex TL 5913. *Ledene Roinges* 1219, *Roing(es) Plumbata, la Plumbee* 1238. ME adj. **lede**, Latin **plumbatus**, feminine gender **plumbata** agreeing with *ecclesia* unexpressed, Fr **plumbee**, + p.n. Roding as in The RODINGS TL 5813. The adjective refers to the fact that this was the first church in the district to be roofed with lead, *ecclesiam de Roinges quae vocatur Ledenechirche* 'the church of Roding called Leaden Church' [c.1100]13th. Ess 493 gives pr [li:dn].

Margaret RODING Essex TL 5912. 'St Margaret's R'. *Roinges Sce Margar'* 1235–55, *Margaret Rothyng, Rooden* 1593, 1697. Saint's name *Margaret* from the dedication of the church + p.n. Roding as The RODINGS TL 5813. Ess 494.

River RODING Essex TQ 4294. *Rodon* 1576–86, *Roding* from 1586. A back-formation from p.n. Roding as in the RODINGS TL 5813. Earlier names were *Angrices burne* [1062]12th K 813, 'the stream of *Angric*, Ongar stream', p.n. Ongar + OE **ric**, *(innan, andlang) Hile, (andlang) ealdan Hilæ* 'along the old Hile' [958]12th B 1037, *Hyle* c.1250, *Iuell* 1577, 'the trickling stream', cf. ILFORD TQ 4586, and Shellow in the neighbourhood of SHELLOW BOWELLS TL 6108. Ess 10, RN 206–7.

White RODING or ROOTHING Essex TL 5613. *Blauncherothinges* 1281, *Alba Rothinges* 1306, *(White)rouching(g)e* (for *-routh-*) 1329, *White Ruding* 1688. ME **white**, Fr **blanc**, feminine form **blanche**, Latin **albus**, feminine **alba**, + p.n. Roding as in The RODINGS TL 5813. The reference is to the whiteness of the church walls. This was the largest of the Rodings and so also known as *Magna Roynges* 'great R' 1235. Ess 494.

RODINGTON Shrops SJ 5814. 'The settlement on the river Roden'. *Rodintone* 1086–1751 with variants *-tun* and *-ton(')*, *Roddington* 1254, 1601–1809, *Rodyngton* 1359–1535, *Rodington* from 14th. R.n. RODEN SJ 5915 + OE **tūn**. The form of the name was early assimilated to the pattern of *-ingtūn* names. Sa i.250.

RODLEY Glos SO 7411. 'Wood or clearing where reeds grow or are gathered'. *Rodele -a* 1086–1326, *-ley(e) -legh* 1276–1527, *Rodleg(h) -lee -ley(e)* 1221–1662, *Rudelai -leya* 1086–1200, *Redleg(a) -ley(e) -lei(a)* 1100×35–1292, *Radleg(a) -ley* 1159–1279. OE **hrēod**, genitive pl. **hrēoda**, + **lēah**. Gl iii.203.

RODMARTON Glos ST 9498. Partly uncertain. *Redmertone* 1086, 1301, *Rodmarton(e)* 1220–1328, *Rodmerton -tun* 1227–1587. This could be 'reed-pond settlement', OE **hrēod** + **mere**, perhaps used as a local p.n. referring to a place where reeds could be gathered for domestic use, + **tūn**. However, since the place is on the Wiltshire border, 'reed border settlement' would be possible, OE **hrēod** + **mǣre** + **tūn**. This would explain the *mer/mar* variation in the spellings. Cf. DIDMARTON ST 8287 and TORMARTON Wilts ST 7778, also on the same border. Gl i.105, L 26.

RODMELL ESusx TQ 4106. 'Red soil'. *Ra- Redmelle* 1086, *Red(e)melde* 1087×1100–1432, *Radmelde* 13th, *Radmell* 1638. OE **rēad** + ***mylde**, SE dial. form ***melde** (a derivative of OE *molde*). Sx 325.

RODMERSHAM Kent TQ 9261. 'Hrothmær's homestead or estate'. *Rodmaeresham* c.1100, *-meres-* 1198–1235. OE pers.n. **Hrōthmǣr*, genitive sing. **Hrōthmǣres*, + **hām**. K 263, ASE 2.29.

RODNEY STOKE Somer ST 4850 → STOKE.

RODSLEY Derby SK 2040. Uncertain. *Redlesleie, Redeslei* 1086, *Redeslege -leye* 1244, 1260, *Reddeslye* 1272, *Roddeslea -ley(e) -is-* 1183–1470, *Raddesley* 1330, *Radsley* 1577, 1610. If the DB form *Redlesleie* is reliable and not an error, this might be '*Redley* clearing', p.n. **Redley*, OE **Hrēod-lēah*, 'clearing where reeds grow', OE **hrēod** + **lēah**, genitive sing. **Hrēodlēas*, + a second pleonastic **lēah**. Otherwise the name remains unexplained though subsequently associated with ME *rodde* 'a rod' as if 'clearing where rods are obtained'. Db 598, DEPN.

ROE BECK Cumbr NY 3941. 'Moss stream'. *Rauhe, Rawe* 1272, *Raw* 1331, 1623, *Raugh* c.1333, 1687. The best explanation is that this is an OE r.n. **Ragu* < OE **ragu** 'moss, lichen'. Cu 25, NoB 1926.153–161, RN 346, SSNNW 423.

ROECLIFFE NYorks SE 3765. 'The red bank'. *Routhecliue -cliva -clif(f)* 1170–1320, *Rouclif -clyf(f)* 1319–1412, *Roecliffe* from 1566. ON **rauthr** + **klif**. The reference is to the lofty bank of the river Ure above which the village stands, cf. RAWCLIFFE SE 5855. YW v.86.

ROEHAMPTON GLond TQ 2373. 'Rook Hampton'. *Rokehampton* 1350–1605, *Rowhampton* 1553, *Roehampton* 1645. ME **roke** (OE *hrōc*) + p.n. *Hampton(e)* 1332, 1382, 'home farm', OE **hām-tūn**. The earliest reference is to *Est Hampton* 1318. 'Rook' and 'east' for distinction from Hampton Court TQ 1668. Sr 28.

ROFFEY WSusx TQ 1932. 'The rough enclosure'. *(La) Rogheye* 1281–1432, *le Rugheye* 1342, *Roughey* 1446, ~ *or Roughway* 1830, *Ropheye* 1574. OE **rūh** + **(ġe)hæġ**. Sx 228.

ROGAN'S SEAT NYorks NY 9103. The highest summit (2204ft.) between Swaledale and Arkengarthdale. Surname Rogan + ModE **seat** 'a mountain top'.

ROGATE WSusx SU 8023. 'Roe gate'. *Ragata* 12th, *la Ragat'* 1229, *Rogate* from 1203. OE **rā** + **ġeat**. Some sps show the influence of feminine **rǣġe** (*La Regate* 1230, 1261, *Reygate* 1279) and others show confusion with adj. **rūh** 'rough' (*La Rugate* 1261, *Rough- Rowgate* 1412, 1453) when **rā** had become *ro* and **rūh** *row*. The original name was North Harting, cf. East HARTING SU 7919. Sx 38.

ROGER SAND Lincs TF 4841. *Roger Sand* 1824 OS. A sand-bank in The Wash.

ROGERLEY Durham NZ 0173 → FROSTERLEY NZ 0337.

ROKER T&W NZ 4059. Unexplained. *Roca Battery, Roca Point* 1768 Armstrong, *Roker Battery* 1863 OS. Possibly transferred from Cape and Fort Roca on the coast of Portugal.

ROLLESBY Norf TG 4415. 'Rolf's farm or vilage'. *Rollesby* [1044×7]13th S 1055, *Rotholfuesby -bei, Roluesb(e)i* 1086†, *Roluesby -bi* 1196–1242, [1127×34]14th, *Rollesby -bi* 1193–1535 including [1127×34]14th. ON pers.n. *Hrólfr*, secondary genitive sing. *Hrólfes*, + **bȳ**. *Hrólf* is a reduced form of **Hróthúlfr* which appears three times among the DB spellings. DEPN gives pr [roulzbi], Nf ii.73, SPNN 209, 212.†

ROLLESTON Leic SK 7300. 'Rolf's farm or village'. *Rouestone* 1086, *Rolueston(e)* 1170–1375, *Rolleston(e)* from 1195, *Rolston'* 14th cent., *Rolson* 1513, *Rowl(e)ston* 1539–1624, *Rowston* 1604. ON pers.n. *(H)rólfr*, ME genitive sing. *Rolfes*, + **tūn**. Lei 248, SSNEM 194.

ROLLESTON Notts SK 7452. 'Hroald's farm or village'. *Roldestun* 1086, [c.1200]14th, *-ton(e)* c.1200–1428, *Rollestone* 1086, *-ton(e)* 1180–1428 etc., *Rouliston* 1242, *Rowlston* 1603×25. ON pers.n. *Hróaldr*, secondary genitive sing. *Hróaldes*, + **tūn**. Nt 173 gives pr [roulstən], SSNEM 193.

ROLLESTON Staffs SK 2327. 'Hrothwulf's estate'. *Roðulfeston* [942]13th S 479, *Rólfestun* [1002×4]11th S 1536, *Rolvestvne* 1086, *Rolleston* 1291. OE pers.n. *Hrōthwulf*, genitive sing. *Hrōthwulfes*, + **tūn**. Horovitz gives pr *Roll-ston*. DEPN.

Great ROLLRIGHT Oxon SP 3231. *Rollandri maiore* 1086, *Magna Rolaunderit'* 1220, *Great Rollandryght* 1380, *Great Rollywright* 1346, *Great Rowlright*. ME adj. **grete**, Latin **magna**, + p.n. *Rollandri, Rollendri* 1086, *Rollendriz* 1090, *Rollendricht* [1091]12th–1363 with variants *-rit(h) -ric(h)t -right* etc., possibly 'the groove at Rodland', PrW **Rodlandrĭch* < Brit **Roto-landā* 'the wheel precinct' + W **rhych** 'a groove', refering to the Neolithic stone circle known as The King's Men and to the gorge known as Danes Bottom just SW of the original nucleated village of Great Rollright. The name, borrowed into OE as **Rollandrih*, seems to have been subject to refashioning as if meaning 'Roland's right' sc. of possession. O 371, DEPN, Celtic Voices 199.

ROLLRIGHT STONES Warw SP 2831. *Rolle-rich stones* 1607. P.n. Rollright as in Great ROLLRIGHT Oxon SP 3231 + ModE **stone(s)**. The reference is to the New Stone-Age stone circle of seventy-seven monoliths. O 372 f.n. 2, Thomas 1980.177–8.

ROLSTON Humbs TA 2145, sometimes spelled ROWLSTON. 'Rolf's estate, farm or village'. *Roolfestone, Roluestun* 1086, *Rolleston'* 1203–1512, *Rolston(e)* from c.1265, *Rowston, Roulston* 16th cent. ON pers.n. *Hrólfr*, genitive sing. *Hrólfs*, + **tūn**. YE 63, SSNY 129.

ROLVENDEN Kent TQ 8431. Partly uncertain. *Rovindene* 1086, *Ruluindaenne* c.1100, *-den(a) -denn(e)* 1184×5–1240, *Roluindenn'* 1242×3, *Rolvenden* 1610. Possibly the 'woodland-pasture called after Rolf', lOE pers.n. *Rolf* + **ing**[4] + **denn**. *Rolf* is probably a late form of OE pers.n. *Hrōthwulf*; but the absence of *-ing-* spellings casts some doubt on this explanation. K 350.

ROLVENDEN LAYNE Kent TQ 8530. P.n. ROLVENDEN TQ 8431 + p.n. *Leyne* 1215, *la Lene* 1254, ME **leyne, lain** 'a tract of arable land, arable land lying fallow'. PNK 345, PNE ii.24.

ROMALDKIRK Durham NY 9922. 'St Rumwold's church'. *Rumoldesc(h)erce* 1086, *ecclesia St Rum(b)aldi* 1244, *Rumbald(e)kirk(e)* [1184]15th, 1285, 1343, 1576, *Romerkirk -kyrke* 1576, 1606. OE saint's n. *Rumwald* + ON **kirkja** replacing OE **ċiriċe**. According to legend Rumwold was a prince of the royal house of Northumbria who 'whan he was borne cryed wt lowd voyce sayeng thre tymes togyder these wordes, "I am a chrystyan," and than required the Sacrament of baptym and after to haue masse and was communed and than he made a noble sermon wt meruaylous good eloquence and lyued thre dayes and so departid and lyeth in buckyngham ful of myracles'. YN 309, ODS 424.

ROMANBY NYorks SE 3693. 'Rothmund's farm or village'. *Romundrebi* 1086, *Romundebi -by* 1086–1316, *Romundby* 1348,

†Two other erroneous forms also appear in DB, *Rothbfuesbei* and *Thoroluesby* by confusion with ON pers.n. Thórólfr.

Romanby from 1398. ON pers.n. *Róthmundr*, genitive sing. *Róthmundar*, + **bȳ**. YN 210, SSNY 35.

ROMAN RIVER Essex TL 9920. Probably from the family name of John Romayn 1377, later taken as the adj. Roman. *in stangno de Romine, Romyne* 'in the pool of R' 1338, *bank of Remyne* c.1561, *Remyn Creek* 1565, *Romans Creek* 1686, *Roman River* 1777, 1805 OS. Near the river is Roman Hill TM 0021, *Roman Hill House* 1777, *Romynbridge* 1432, *Rammynsbregge* c.1561, now Manwood Bridge, and *Romaynemad* 'R meadow' 1396. Ess 11, 316, 388.

ROMANSLEIGH Devon SS 7220. 'St Rumon's Leigh'. *Reymundesle* 1228, *Romundesleg h(e)* 1267–1406 with variants *-mond-* and *-is-*, *Rumonsleigh vulgo Rumsleigh* 1765. Saint's name *Rumon* from the dedication of the church, + p.n. Leigh, *Liege* 1086, 'the wood or clearing', OE **lēah**. The identical name occurs at Rumonsleigh Cottages in Tavistock, *Rumonslegh* 1349, *Remmenlegh* 1371, *Romanesleghe* 1416, where St Rumon was the patron saint of the abbey. Early forms show assimilation to the names *Raymond* and *Romund* (OE *Hrōthmund*). Cf. also RUAN MINOR Corn SW 7115. D391.

The ROMAN WALL Cumbr NY 4661. *murum quo olim Romani Britanniam insulam praecinxere* 'the wall with which the Romans once surrounded the island of Britain' 731 BHE, *þone langan weall þe þa romaniscan worhtan* 'the long wall the Romans built' [c.996×7]1025×50 Ælfric, *murum antiquum* 1169, *the Pict wal* c.1540, *the Wall* 1553, *the picts wall* 1601–1777, *the Pight Wall* 1603, 1610. Cu 39.

ROMBALDS MOOR WYorks SE 0945. 'Rumbald's moor'. *Rumblesmor(a) -mo(o)re* 1150×60–1617. OG pers.n. *Rumbald*, genitive sing. *Rumbaldes*, + OE **mōr**. YW iv.165.

ROMFORD GLond TQ 5188. 'Wide ford'. *Romfort* 1177, *-ford* from 1306, *Rumford* 1199–1250. OE **rūm** + **ford**. Ess 117 gives pr [rʌmfəd].

ROMILEY GMan SJ 9491. 'The roomy clearing'. *Rumelie* 1086, *Rumilegh* 1321, *Romileg -lee* 13th, *-ley* from 1512. OE **rūm**, definite form **rūma**, or adj. ***rūmiġ**, + **lēah**, dative sing. **lēa(ġe)**. Che i.292.

New ROMNEY Kent TR 0624. *Newe Romney* before 1624, *New ~* 1710. ModE adj. **new** + p.n. Romney as in Old ROMNEY TR 0325. 'New' for distinction from Old ROMNEY TR 0325. New Romney is in fact an old settlement, one of the Cinque ports, reputedly a planned town of the Saxon period. Newman 1969W.416–7.

Old ROMNEY Kent TR 0325. *Vet' Rumenol* 1227, *Elderumenal* 1248, *-romenal* 1252, *Old Romynall* 1441, *Old Rumney* 1527, *Old Romney* 1610. ModE adj. **old**, ME **elde** (OE *eald*), Latin **vetus**, + p.n. Romney, originally a r.n., *flumen qui vocitatur rumenesea* 'river called r.' [924 for 890, 905 or 920]10th S 1288, *Rumenea* [895]12th S 1627, after 1154 ASC(E) under year 1052, c.1200 ASE(F) under year 1051, *Romenel* 1086, *Rumenel -al* 1130, 1247, possibly 'Rumen's river'. OE pers.n. **Rūmen*, genitive sing. **Rūmenes*, + **ēa**.† For this kind of compound of a pers.n. + *ēa*, and also for AN **-el** see PEVENSEY ESusx TQ 6405. RN 347, KPN 236, 327, PNK 485.

ROMNEY MARSH Kent TR 0529. *marisco de Romenal* [774]10th S 111, *Romney Marsh* 1710. P.n. Romney as in Old ROMNEY TR 0325 + ModE **marsh**, Latin **mariscus**. The marsh is mentioned in the p.n. ST MARY IN THE MARSH TR 0267 and is earlier simply *Mersc* 9th ASC(A) under year 796, OE **mersc**, its inhabitants *Mersc warum* (dative case) 9th ASC(A) under year 838, *Merscuuare* [774]10th S 111, *Mersc ware* c.1122 ASC(E) under year 796. OE **mersc** + **ware** 'dwellers', DEPN, Cullen.

ROMNEY SANDS Kent TR 0823. *Romney Sand* 1816 OS. P.n. Romney as in Old ROMNEY TR 0325 + ModE **sand**.

†The form *Ruminella* [1027×35]12th S 1061 suggested for Romney in KPN 327 probably does not belong here.

ROMSEY Hants SU 3521. 'Rum's island'. *rummæsig* [966×75]12th S 1484, *romes(e)ye* [967×75]14th S 812 *(æt) Rumesige* 971 ASC(A), *(to) rumesigge* [c.1000]14th Saints, *to Rumesege* 1026 ASC(E), *Romsey* 1086, *Rumeseia -eya* etc. 1167–1223, *Rumsey* 1578, 1610, *Romesia -eye* etc. from 1189. OE pers.n. **Rūm*, genitive sing. **Rūmes*, + **īeġ** in the sense 'raised ground in marshland'. Ha 140, Gover 1958.183 which gives pr [rʌmzi], DEPN, L 38.

ROMSLEY H&W SO 9680. 'Wild garlic clearing'. *Romesle(ye)* 1270–93, *Rummesleye* 1355, *Ramsselie* 1604. OE **hramsa** with WMid *ō* for *ā*, + **lēah**. Alternatively the specific might be OE **ramm** 'a ram', genitive sing. **rammes**. Wo 300.

ROMSLEY Shrops SO 7882. Probably 'the raven's wood'. *Hremesleage* 1002×4 S 1536, *Rameslege* 1086, *Ram(m)esleg'* 1203×4, *Rem(m)esleg' -ley(e)* 1255–1369, *Rommesleye -le(i)gh* 1287–1387, *Rumisley* 1287, *Roummesleye* 1294, *Rummesle* 1348, *Romsley* from 1672, *Rams(e)ley* 17th. OE **hræfn**, genitive sing. **hræfnes**, + **lēah**. The various OE forms of **hræfn** include *(h)refn, hremn, (h)remm* and *(h)ræm*; with WMid *o* for *a* before a nasal, this sound change accounts for the variety of spellings of the name. Sa i.251.

ROOKBY Cumbr NY 8010. 'Rook-haunted farmstead' or possibly 'Hroca's farmstead'. *Rochebi(a)* 1178–9, *Rokeby* 1201–15th cent., *Rukeby* 1323–1389, 1615, *Rook(e)by -bie* 1597–1777. ON **hrókr** or OE **hrōc**, genitive pl. **hróka**, **hrōca**, or pers.n. OE *Hrōca* or ON **Hróki*, genitive sing. *Hróka*, + **bȳ**. We ii.5, SSNNW 39.

ROOKEN EDGE Northum NY 7896. Cf. *Rooken Cairn* 1869 OS. P.n. Rooken NY 8096, *Rooken* 1869 OS, + ModE **edge**. A ridge that reaches to 1287ft.

ROOKHOPE Durham NY 9442. 'Rook valley'. *Rokehope* c.1190, *Roc- Rokhop* 1242×3, 1300, 1382, *Ruk(e)hop* 1323×4–1418, *Rokop* 1338, 1418, *Rokeup* 1580, *Rucupe* 1685, *Rookhope* [1569]18th. OE **hrōc**, genitive pl. **hrōca**, + **hop**. Alternatively the specific could be the OE pers.n. *Hrōca* derived from **hrōc**. NbDu 168, L 116–7.

ROOKLEY IoW SZ 5084. 'Rook wood or clearing'. *Roclee* 1202, *Rok(e)le(y)* 1235–1428, *Roukley* 1328, 1350, *Rookley* 1769. OE **hrōc**, genitive pl. **hrōca**, + **lēah**. Wt 20 gives pr ['rukli], Mills 1996.88.

ROOKS BRIDGE Somer ST 3652. *Rooks Bridge* 1817 OS. Probably surname *Rook* + ModE **bridge**.

ROOS Humbs TA 2930. 'The promontory'. *Rosse* 1086–1650, *Ros* 1190–1542, *Russe* 1202, 1208, *Rose* 1285, 1418, 1531, *Roos* from 1414. PrW ***ros** with ME lengthening in open syllable. The village stands on a low promontory overlooking the marshland of Roos or Keyingham Drain. YE 56 gives prs [ru:əz, rus, rɔ:z], Jnl.1.50.

ROOSEBECK Cumbr SD 2567. Originally a stream-name, 'Rose brook'. *Rosbech* 1227, *Rosebec -beke* 1269, *Rosebek* 1418. P.n. Rose + ON **bekkr**. The brook rises near Rose SD 2269, *Rosse* 1086, *Ros* 1155–1246, *Roos* 1336, *Rwse* 1537, PrW ***ros** 'marsh, moor', dial. *ross*, or possibly here 'promontory' referring to the Rampside promontory. La 208, 202.

ROPLEY Hants SU 6431. 'Hroppa's wood or clearing'. *Roppele* 1172–1414 with variants *-ley(e) -legh*, *Ropele(ia) -leye* 1198–1284, *Ropley* 1610. OE pers.n. **Hroppa* + **lēah**. Ha 140, Gover 1958.87, DEPN.

ROPLEY DEANE Hants SU 6331. *Ropley Dean* 1810 OS. P.n. ROPLEY SU 6431 + ModE **dean** 'a valley' (OE *denu*). Ropley Dean lies at the mouth of the valley running up to Ropley. It is referred to in the names *Westdene, Dene-furlong* 1548. Gover 1958.87.

ROPSLEY Lincs SK 9934. 'Hrop's clearing'. *Ri- Ropeslai* 1086, *Ropesle* 1185, 1212, 1354, *Roppesle* 1202–1432 with variants *-leia -le(e)* etc., *Ropsley* from 1603. OE pers.n. **Hrop(p)*, genitive sing. **Hroppes*, + **lēah**. Perrott 505.

RORRINGTON Shrops SJ 3000. 'The estate called after Hrōr'. *Roritune* 1086, *-ton(')* 1261–1316, *-inton'* 1255×6, *-ynton'* 1334,

-yngton 1535, *-ington* 1629, 1693. OE pers.n. **Hrōr* + **ing**[4] + **tūn**. Sa i.251.

ROSE ASH Devon SS 7821 → Rose ASH SS 7821.

ROSE Corn SW 7754. 'Moor'. *The Rose* 1697. Co ***ros**. PNCo 150.

ROSEACRE Lancs SD 4336. 'Newly-cultivated land with a cairn'. *Ra(y)sak'* 1249, *Raysacre* 1283, *Rosaker* 1569. ON **hreysi** + **akr** or OE **æcer**. Situated on a slight rise of ground amid former carrs. Half a mile SE lies WHARLES Lancs SD 4435, possibly 'the mounds' and the feature referred to here. La 152, SSNNW 154, Jnl 17.88.

ROSEBERRY TOPPING NYorks NZ 5712. '*Othenesberg* hill'. *Ounsbery or Rosebery Topping* 1610. Earlier simply *Othenes- Ohensberg* [1119, 1129]15th, *Outhensbergh* [1239]15th, *Ounesbergh* [c.1310]15th, *Ousebergh* 1404. P.n. *Othenesberg* + Y dial. **topping** 'a hill'. *Othenesberg* is probably 'Othin's rock', ON god-name *Óthinn*, genitive sing. *Óthins*, + **bjarg**, although the ON pers.n. *Authunn* would also be possible. Roseberry Topping rises to 1488ft. with a very distinctive cone-like profile which must have attracted attention and have led to association with divinity. There is an exact parallel in *Onsbjærg* in Samsø, Denmark, and in the English *Wodnesberge* [825]12th S 272 'Woden's tumulus', the ancient name of the neolithic long-barrow now known as Adam's Grave in Alton Priors Wilts SU 1163, and WOODNESBOROUGH Kent TR 3056. The development of the name shows (1) loss of *-th-* which is usually an OFr development but may have occurred in Scandinavian (cf. the modern Danish pronunciation of Odense), (2) false analysis of *Newton-under-Ouesberg* as *Newton-under-Rouesberg* and subsequent association with *rose* as in ROSEDALE SE 7296. YN 164, SPN 41, xciv, Wlt 318, Gelling 1998.80.

ROSEBROUGH Northum NU 1326. *Rosebrough* 1866 OS. Possibly a descendant of the name *Osberwic* discussed under NEWSTEAD Northum NU 1527.

ROSEDALE NYorks SE 7196. 'Rossi's' or the 'horse valley'. *Russedal(e)* [1130×58]1201, [1155×70]15th, *Rossedal(e)* 1186×94–1541, *Rosedale* from 1376. ON pers.n. *Rossi*, genitive sing. *Rossa*, or **hross**, genitive pl. **hrossa**, + **dalr**. Possibly a valley where wild horses were corralled. The pronunciation [ro:zdil] has been influenced by ME *rose*. YN 80, SPN 225.

ROSEDALE ABBEY NYorks SE 7295. P.n. ROSEDALE SE 7196 + ModE **abbey**. A priory for nuns was founded here by Robert de Stuteville in the reign of Henry II.

ROSEDALE MOOR NYorks SE 7299. P.n. ROSEDALE SE 7196 + ModE **moor**.

ROSEDEN Northum NU 0321. 'Rush valley'. *Russeden* 1242 BF-1428, *Ruscheden'* 1296 SR 179, *Russhden* 1428, *Rossedoun* 1580, *Rosden* 18th cent. OE ***rysc** + **denu**. NbDu 168, PNE ii.85.

ROSEHILL Shrops SJ 6630. *Rose Hill* 1833 OS.

ROSELAND Corn SW 8535. 'District of or called *Rose*, the promontory'. *Rolland* 1201, *Rosland* 1259, 1610, *Ros* 1261. Co ***ros** + OE **land**. Refers to the peninsula comprising the parishes of St Anthony in Roseland, Gerrans, St Just in Roseland and Philleigh which was also known as *Eglosrose*, 'the church of Rose(land)'. Cf. The Rhos peninsula in Dyfed SM 8209 which gave its name to Roose Hundred, *Ros* 1136–15th. PNCo 150, Pembs 570.

ROSEMARY LANE Devon ST 1514. 'Lane where rosemary grows'. *Rosemary Lane* 1809 OS.

ROSEMAUND H&W SO 5648 → MAUND BRYAN SO 5650.

ROSEMULLION HEAD Corn SW 7928. *Rosemullion Head* 1813, cf. *Rosmilion poynte* c.1605. P.n. Rosemullion SW 7827, *Rosmylian* 1318, *Rosemullian* 1562, 'headland of Milyan', Co ***ros** + OCo pers.n. *Milian*, or possibly 'clover headland', Co **melhyon**, + ModE **head** 'a headland'. PNCo 150, Gover n.d. 521.

ROSENANNON Corn SW 9566. 'Moorland of the ash-tree'. *Rosnonnen* 1326. Co ***ros** + definite article **an** + **onnen** 'ash-tree'. PNCo 150, Gover n.d. 359.

ROSEWORTHY Corn SW 6139. Probably 'ford of Gorhi'. *Ritwore* 1086, *Rituuori* 1086 Exon, *Red- Ridwuri* 1183, 1196 Gover n.d., *Redwrthi* 1201 Gover n.d., *Rydwory* 1302, *Red- Reswori* 1289, *Riswory* 1351, *Roswory* 1351, 1468, *Rosworthy* [1580]18th, *Rudgewery* c.1605. OCo **rid** + pers.n. **Gorhi*. The name was early assimilated to the English pattern of names in *-worthy* but the old pronounciation [rə 'zʌri] or ['zʌri] is still remembered. PNCo 151, Gover n.d. 598.

ROSGILL Cumbr NY 5316. 'Horse ravine'. *Rossegil(e) -gyl(e) -gill* late 12th–1372, *Rossgill* 1260×70, 1540, *Rosgill -gyll* from 1628. ON **hross**, genitive pl. **hrossa**, + **gil**. Probably a place where wild ponies from Shap Fells were corralled by the monks of Shap Abbey. We ii.170, SSNNW 154.

ROSLEY Cumbr NY 3245. 'Horse pasture'. *Rosseley(e)* 1272–1317, *Rosley* 1578. OE **hors** replaced by ON **hross**, genitive pl. **hrossa**, + **lēah**. Cu 330 gives pr [rɔsle], SSNNW 247.

ROSLISTON Derby SK 2416. 'Hrothlaf's farm or village'. *Redlauestun* 1086, *Rostlavestona -ton(e) -lawes-* 1189×99–1279, *Rosliston* from c.1232, *Rost(e)laston(e)* 1296–1546, *Roslaston* 1353, 1393, 1577. OE pers.n. *Hrōthlāf*, genitive sing. *Hrōthlāfes*, + **tūn**. Spellings in *Ros-* reflect AN sound-substitution of [ç] for [θ]. Db 657.

ROSS Northum NU 1337. 'The promontory or moor'. *Rosse* 1208×10 BF, *Ross* 1249. PrW ***ros**. Cf. dial. *ross* 'a marsh'. NbDu 169, Jnl 1.50–1.

ROSS-ON-WYE H&W SO 6024. 'The promontory'. *Rosse* 1086, *Ros* 1199, 1242, *Roos* 1291. PrW ***ros**. DEPN.

ROSSALL POINT Lancs SD 3147. P.n. Rossall + ModE **point**. Rossall, *Rushale* 1086, *Rossall* from 1212 with variant *-ale* 1216–92, *Roshal(e)* 1222, 1228, *Russal* 1292, *Rosso hall* 1577, is probably ON **hross** + **hali** 'horse-tail' referring to a tongue or tail of land used for pasturing horses, although the earliest spelling with *-u-* might rather point to OE ***rysc** 'a rush', genitive pl. ***rysca**. If so the suffix was OE **halh** 'a nook of land'. La 158, L 111, Jnl 17.92.

Forest of ROSSENDALE Lancs SD 8525. Partly uncertain. *Rocendal* 1242, *Rossendale* from 1292 with variants *Rosen- Roscin- Rossyng- Rostyndale* 14th–16th cent. The generic is probably ON **dalr** 'a valley' referring to the deep valley that runs SW from Lums SD 8324 to join the Irwell at Waterfoot SD 8321, the specific possibly a diminutive of PrW ***ros** 'a moor'. La 92, SSNNW 247, Jnl 17.55.

ROSSINGTON SYorks SK 6298. 'The settlement at or called *Rosing*, the promontory place'. *Rosington' -tun -yng-* c.1190–1657, *Rosinton(e)* 13th cent., *Rossinton* 1676. OE p.n. **Rosing* < PrW ***ros** 'a promontory' + OE **ing**[2], + **tūn**. The reference is to land in a bend of the r. Torne. YW i.49, Jnl i.87, BzN 1968.185, Årsskrift 1974.45.

ROSTHERNE Ches SJ 7483. 'Rauth's thorn-tree'. *Rodestorne* 1086, *-thorn(e)* 13th, *Rosthorne* [1154×89]17th–1581, *-thern(e)* from 1324, *Rothesterne* c.1188, *Rouestorn -thorn(e) -t(h)erne* 1214–1530, *Raston* 1546, 1554. ON pers.n. *Rauthr*, genitive sing. *Rauths*, + OE **thorn** 'a thorn-tree', varying with **thyrne** or ON **thyrnir** 'thorn-bush, thorn-thicket'. Che ii.56 gives pr ['rɔstə:n], SSNNW 154, L 221.

ROSTHWAITE Cumbr NY 2514. 'The cairn clearing'. *Rasethuate* 1503, *Ray- Raistwhat* 1564, *Rosthwait* 1786. ON **hreysi** + **thveit**. Cu 353.

ROSTON Derby SK 1341. Partly uncertain. *Roschintun -tone* 1086, *Ros(s)ington(e) -yng-* 1263–1652, *Rossinton(e) -yn-* 1269–1340, *-en-* 1281, 1329, *Rosson, Rawson* 16th cent., *Rawston* 1577, 1610. The generic is OE **tūn** 'farm, village, estate', the specific possibly pers.n. **Roschin*, **Roskin* < ON *Roskill*, *Rosketill*. But substitution of *-in* in shortened forms of *-ketill* is very rare. *Roskin* might, therefore, be a diminutive of OFr *Rosce*, OG *Rozzo*. Alternatively the name might be an **ing**[2] formation on an unidentified specific. Db 591 gives pr [rɔsn], SSNEM 379, OES 215.

River ROTHAY Cumbr NY 3308. 'Trout river'. *Routha* 13th–1681×4, *Rowthey* 1452–1614, *Rawthey* 1454, *Rothay* 1793. ON ***rauthi** 'the red one' used as an appellative for the trout, + **á**. In 1671 it was reported that the Rothay had 'a great plenty large trouts'. We i.12, RN 335–6, L 11.

ROTHBURY FOREST Northum NZ 0599. P.n. ROTHBURY NU 0501 + ModE **forest**.

ROTHBURY Northum NU 0501. Probably '(the settlement) at the red fortification'. *Routhebiria* c.1125, *Routhbiry* [1258]14th, *Rothbyry* 1278, *Roberi -y* 1176–1271, *Robire -bir' -bur'* 1212 BF, 1242 BF, *Rothebyr'* 1296 SR. ON **rauthr** + **burh**, locative-dative sing. **byriġ**. The bed-rock here is red. However a Norse-English hybrid name is unlikely here and the specific could well be OE pers.n. **Hrōtha* and so 'Hrotha's fort'. The same pers.n. may occur in nearby Rothley as in ROTHLEY LAKES NZ 0490. NbDu 169, Jnl 8.9–11.

River ROTHER ESusx TQ 6125, 9525. *Rother* 1575. A 16th cent. back-formation from ROTHERFIELD TQ 5529. The earlier name was *Liminel* c.1180, *Lym(m)ene* 1279 as in LYMPNE TR 1235, which survives in the r.n. Limden (+ **denu** 'valley'), a branch of the Rother at TQ 7027. Sx 7, RN 347.

River ROTHER Hants SU 7625, WSusx SU 9420. A late back-formation from Rotherbridge SU 9620, *Redrebrige -brvge* 1086, *Rutherebrygg'* 1275, *-brigge* 1302, *Rotherbridge* 1813 OS, 'ox bridge', OE **hrīther, hrȳther**, genitive pl. **hrīthra**, + **brycg**. The original name of the river was *(andlang) scíre* [956]12th S 619, *scyre* [959×63]12th S 811, *Schir(e)* [c.1160]13th, 'the bright one', OE *Scīre* from adj. **scīr**. Ha 140, Gover 1958. 5, RN 347, 362.

River ROTHER SYorks SK 4492. Probably 'the chief river'. *Roder* c.1170–1383, *Rother* from 1539. Brit ***ro-** + PrW ***duβr**. RN 348, YW vii.136.

ROTHERBRIDGE Hants SU 9620 → River ROTHER SU 7625.

ROTHERBY Leic SK 6716. 'Reitharr's village or farm'. *Redebi* 1086, *Rederbia* c.1130, *Reidebi* 1154×89, *Retherby -ir-* 1206–1536, *Ratherby -bie* 1487–1582, *Rotherby(e)* from 1344. ON pers.n. *(H)reitharr* + **bȳ**. The later forms have been reshaped as if the specific were dial. **rother** 'an ox'. Lei 301, SSNEM 64 no.3.

ROTHERFIELD ESusx TQ 5529. 'Open land of the cattle'. *Ridrefeld* [795]17th S 1186, [790]13th S 133, *(æt) Hryðeranfelda* [873×88]11th S 1507, *Ridrefelda* *[960]13th S 686, *Hryðerafeld* c.1030, *Reredfelle* 1086, *Retheresfeld* 1089. OE **hrȳther**, SE dial. form **hrēther**, genitive pl. **hrȳthera, hrēthera**, + **feld**. The reference is probably to detached pastureland used by cattle drovers from the coastal community of East Bletchington near Seaford. Sx 376 gives prs [rʌdəfəl] and -[vəl], L 239, 244, SS 70.

ROTHERFIELD GREYS Oxon SU 7282. 'R held by the Grey family'. *Retherfeld Grey* 1313×4, *Rotherfelde Grey* 1316. P.n. *Redrefeld* 1086, *Reðeresfeld(')* 1194–6, *Reð- Retherfeld(e)* 1195–1402, *Rutherefeld* 1240–1321, *Rotherfeld(e)* 1275–1401, 'the open land where cattle are grazed', OE **hrȳther** + **feld**, + manorial affix *Grey* early 13th. O 77.

ROTHERFIELD PEPPARD Oxon SU 7181. 'R held by the Pipard family'. *Retherfeld Pippard* 1289–1517 with variants *-feud* and *Pyp(p)ard, Rotherfeld Pyppard* 1411×12, 1474. P.n. *Redrefeld* 1086, *Retherfeld* [c.1110]15th etc. as in ROTHERFIELD GREYS SU 7282 + manorial affix *Pipard*, a family of important feudal tenants of the Honour of Wallingford. O 79.

ROTHERHAM SYorks SK 4492. 'Homestead, estate on the r. Rother'. *Rodreham* 1086, *-er-* c.1195–1528, *-en-* c.1185–1234, *Rotherham* from 1461. R.n. ROTHER SK 4492 + **hām**. RN 348, YW i.184, TC 162.

ROTHERHITHE GLond TQ 3680 → LAMBETH TQ 3074, GREENHITHE Kent TQ 5974.

ROTHER LEVELS ESusx TQ 8725. R.n. ROTHER TQ 6125 + ModE **level**.

ROTHERSTHORPE Northants SP 7156. 'Rethær's Thorpe or outlying farm'. *Retherest(h)orp -thro(u)p* 1231–1387, *Rotherestrop* 1333, *Rothersthrope* 1456. Earlier simply *Torp* 1086, 1194, *T(h)rop* 12th–1296 and locally *Thrupp* 1675. ODan pers.n. *Rethær* (ON *Hreitharr*), secondary genitive sing. *Rethæres*, + **thorp**. The specific was later confused with *rother* 'cattle' (OE *hryther*). Nth 151 gives pr [θrʌp], SSNNW 119 no.32.

ROTHERWICK Hants SU 7156. 'Cattle farm'. *Hrytheruuica* c.1100, *Retherwyk -wik* 1193–1294, *Rutherwyc -wik* 1235, 1262, *Rotherwyk* 1362. OE **hrīther, hrȳther** + **wīċ**. Ha 140, Gover 1958.117, DEPN.

ROTHLEY Leic SK 5812. 'Wood or woodland glade with clearings'. *Rodolei* 1086 *-leia* c.1130, *Rothele(e) -lei(e) -leg(h) -ley(e)* 1254–1553, *Rothley* from 1414, *Rowtheley* 1520, *Roelai -lay -le(gh)* 1155–1237, *Role(e) -ley(e) leg(h)'* 1206–1278. OE ***roth** + **lēah**. Lei 394 gives pr ['rouθli:].

ROTHLEY LAKES Northum NZ 0490. *Rothley Lakes* 1868 OS. P.n. Rothley NZ 0488, *Ruelea* 1195 P, *Rotheley* 1233–96 SR, possibly 'Hrotha's clearing', OE pers.n. **Hrōtha* + **lēah**, + ModE **lake(s)**. The same pers.n. may occur in nearby ROTHBURY NU 0501. But the specific might alternatively be OE ***roth** 'a clearing' and so 'the clearing pasture'. NbDu 169.

ROTHWELL Lincs TF 1599. 'The clearing spring'. *Rodo- Rodewelle -vvelle -uuelle* 1086, *Rothewelle* 1086–1386, *-well(')* 1272–1576, *Rothwell* from 1428, *Rowell(')* 1154×89–1261. OE ***rod**, OE, ODan ***roth**, + **wella**. Li iv.154, PNE ii.86, 88.

ROTHWELL Northants SP 8181. 'The spring by the clearing'. *Rodewelle* 1086, *Rothewell(e)* 1227–1439, *Rothwell* from 1314, *Roewella* 1152×73, *Rowell(e)* 1184–1730. OE ***roth** + **wella**. Nth 118 gives pr [rouəl].

ROTHWELL WYorks SE 3428. 'Spring by the clearings'. *Rodouuelle, Rodevvelle* 1086, *Rouell(a) -uuelle -well(a) -welle* 1130×40–1313, *Rothewell(a) -welle* 1189–1531, *Rothwell* from 1280, *Roythe- Roithwell* 16th cent. OE ***roth**, genitive pl. **rotha**, + **wella**. YW ii.143.

ROTSEA Humbs TA 0651. 'Rot or Rota's lake'. *Rotesse* 1086, 1285, 1509, *Rot(t)ese* 1204–c.1362, *Rotse* 1210×20, 1378, 1408, *Ratsey* 17th cent. OE pers.n. *Rōt*, genitive sing. *Rōtes* or *Rōta*, + **sǣ**. YE 157 gives pr [rotsə, rɔtsə], DEPN.

ROTTEN END Essex TL 7229 → BEAZLEY END TL 1428.

ROTTINGDEAN ESusx TQ 3702. 'The valley of the Rotingas, the people called after Rota' and 'the valley of Roting, the place called after Rota'. *Rotingeden(e) -yng-* 1086–1331, *Rottingedene -ynge-* 1087×1100–1327, *Rotingesdena* 1121. OE folk-n. **Rōtingas* < pers.n. *Rōta* + **ingas**, genitive pl. **Rōtinga* and p.n. **Rōting* < pers.n. *Rōta* + **ing**[2], genitive sing. **Rōtinges*, + **denu**. The original name was probably *Rōting* 'place called after Rota'. Rottingdean is variously 'the valley at or called Roting' and 'the valley of the people of Roting'. Sx 311, L 99, SS 71.

ROTTINGTON Cumbr NX 9613. 'Farm, village, estate called after Rota'. *Rotington* c.1125–1610, *Rodintona* c.1125, c.1130, *Rodington -yng-* 1279–1496, *Rottington* from 1295. OE pers.n. *Rōta* + **ing**[4] + **tūn**. Cu 428.

ROUD IoW SZ 5180. 'The reed-bed'. *Rode* 1086, 1522, *Rud(a)* 1248, 1305, *Roude* 1287×90–1462. OE **hrēod**. The situation is low-lying beside the Yar. OE **rūde** a 'rue-bed' is also possible. Wt 156 gives pr [ræud], PNE i.264, Mills 1996.88.

ROUGHAM Norf TF 8320. 'Settlement, homestead in rough ground'. *Ruhham* 1086, 1203, *Rugham* 1182, *Rucham* 1198. OE **rūh** 'rough' probably used here as a noun 'the rough' + **hām**. DEPN gives pr [rʌfəm].

ROUGHAM GREEN Suff TL 9061. P.n. *Rucham* [942×c.951]13th S 1526, [978×1016]13th S 1219, *Ruhham* 11th, *Ruh- Rudhā* 1086, *Rugham* 1254, *Rougham* 1610, 'the homestead on rough ground', OE **rūh** + **hām**, + ModE **green**. DEPN, Baron.

ROUGH CLOSE Staffs SJ 9239. *Rough Close* 1836 OS. ModE **rough** + **close**.

ROUGH COMMON Kent TR 1259. *Rough Common* 1819 OS. ModE **rough** + **common**.

ROUGHLEE Lancs SD 8440. 'Rough clearing or pasture'.

Rughley 1296, *Rugh(e)legh* 1324–5, *le Roughlee* 1515. OE **rūh** + **lēah**. La 81.

ROUGH PIKE Northum NY 6286. *Rough Pike* 1868 OS. ModE adj. **rough** + N dial. **pike** 'a pointed hill'. Rises to 1219ft. in the wastes of Kielder Forest.

ROUGHSIKE Cumbr NY 5275. *Roughsike* 1865. ModE adj. **rough** + dial. **sike** (OE *sīc*, ON *sík*) 'a small stream'.

ROUGH TOR 'The rough tor'. OE **rūh**, definite form **rū(h)a**, ME **rowe**, + **torr**.

(1) ~ ~ Corn SX 1480. *Roghe- Rowetorr* 1284. D 199. Co 151 gives pr ['rautə(r)], Gover n.d. 106.

(2) ~ ~ Devon SX 6079. *Rowetor* 1532. D 199 gives pr [rautər].

ROUGHTON 'Settlement in rough ground'. OE **rūh** + **tūn**, where OE **rūh** is used as a noun, cf. ROUGHAM TF 8320.

(1) ~ Lincs TF 2464. *Rocstune* 1086, *Ructuna* c.1115, *Ruchtuna* 1163, *Ructon* 1202, 1232. DEPN, SSNEM 379.

(2) ~ Norf TG 2136. *Rus- Rostuna, Rugutune* 1086, *Rocton* 1196, *Ruchton* 1254. DEPN gives pr [rautn].

(3) ~ Shrops SO 7594. *Ruwynton'* 1261×2, *Ru- Rouh- Rouchton'* 1271×2, *Rughton'* 1291×2, *Roughton* from 1316, *Rowton* 1318. The earliest form suggests specific ***rūhing** 'rough place'. DEPN, Gelling.

(4) ~ CASTLE Shrops SJ 3712. *Rowton Castle* 1577. P.n. *Rutune* 1086, *-ton(e)* c.1210×12–1374×7, *Routon(')* 1255–1715, *Roughton* 1346. Pn. Roughton + ModE **castle** referring to an early 19th cent. castellated mansion replacing a Queen Anne house on the site of a medieval castle destroyed by Llewellyn, Prince of Wales, in 1482. Sa i.252, Raven 169.

ROUNDHAY WYorks SE 3337. 'The round hunting-enclosure'. *(Le) Rund(e)heia -hai(a) -hey -haie -hay(e)* before 1153–1294, *(la) Round(e)hay(e) -hey(e)* 1287–1532. OFr **rond** + OE **(ġe)hæġ**. YW iv.113.

ROUND HILL ModE adj. **round** + **hill**.

(1) ~ Cumbr NY 7436. *Round Hill* 1568. Cu 179.

(2) ~ NYorks SE 1253. *Round hill* 1858. An apt description of this feature. YW v.67.

(3) ~ NYorks SE 1476.

ROUNDHILL RESERVOIR NYorks SE 1577. P.n. ROUND HILL SE 1476 + ModE **reservoir**.

ROUND ISLAND Scilly SV 9017. *Rownd Ylond* c.1540. Named from its shape. PNCo 151.

ROUNDSTREET COMMON WSusx TQ 0528. *Roundstreet Common* 1813 OS. P.n. *Rowner Street* 1738 + ModE **common**. P.n. Rowner as in Rowner Farm TQ 0726, *Ruwenore* 1261, *Rouenere* 1288, *Roughenore* 1327, '(the settlement) at the rough ridge', OE **rūh**, definite form dative case **rū(g)an**, + **ōra**, + ModE **street** 'a hamlet'. Contrasts with NEWPOUND COMMON TQ 0627. Sx 135, 154.

ROUNDWAY Wilts SU 0163. 'The cleared road'. *(apud) Rindweiam* 1149, *Ri- Ryndway -wey(e)* 1262–1340, *Ryng(e)wey(e)* 1289, 1428, *Rundewey* 1493, *Roundwaye* 1619. OE **rȳmed** with assimilation of *m* to *n* before *d*, + **weġ**. Wlt 253, L 83.

ROUNTON NYorks NZ 4103. 'Village or farmstead or marked by poles' or 'where poles are obtained'. *Runtune* 1086, *Rungtune -ton* 1128×35–1508, *Rungeton(e)* 1218–1330, *Westruncton* 1562, *Estrungeton* 1324. OE **hrung** 'a rung, a staff, a pole', genitive pl. **hrunga**, + **tūn**. If the reference is not to manufacture of rungs or poles it might be to a causeway made of poles; cf. RUNCTON Norf SU 8802. YN 217.

ROUSDON Devon SY 2991. 'Down held by the Ralph family'. *Rawesdon* 1285, *Rowston* 1529×32, 1670, *Rowsedown* 1739. Also known as *Doune Rauf(e)* 1334, 1340, *-Rafe* 1480. The family of Radulfus (Ralph, Rafe) held the manor of Down in 1156. Cf. ROSE ASH SS 7821. Down, *Done* 1086, *(la) Dune* 1156–1248, *Doune* 1303, is 'the hill', OE **dūn**. D 647 gives pr [rauzdən].

ROUTH Humbs TA 0942. Uncertain. *Rute* 1086, *Rutha* 1086–1205, *Rue* 13th, *Routh(e)* from 1293. Possibly ON **rúðr** 'scurf' in the sense 'rough ground', or ***rūth** or ***rutha** 'a clearing'. YE 71 gives pr [ru:θ], SSNY 102.

ROW Corn SX 0976. 'Row' of cottages. *Row* 1888. A 19th cent. hamlet. PNCo 151.

ROW Cumbr SD 4589. 'Row of houses'. *(the) Row* 1535, 1716, *Raw(e)* 1699, 1703. OE **rāw**. We i.85.

Stoke ROW Oxon SU 6884. 'The row of houses at or belonging to a place called Stoke'. *Stoke Rewe* 1435, *Stokerowe* 1659. ME p.n. *Stoke* + **rewe**, **rowe** (OE *rǣw, rāw*). O 58.

ROWDE Wilts ST 9862. 'The reed-bed'. *Rode* 1086, *(la) Rode* 1230–83, *Rude(s)* 1180–1268, *Roud(es)* 1263–1316, *Rowd(e)* 1558×1603. OE **hrēod** used in a collective sense. Wlt 246 gives pr [roud].

ROWE DITCH H&W SO 3859. 'The rough ditch'. *Roge- Rugedich* 1219. OE **rūh**, definite form **rūga**, + **dīċ**. A linear earthwork of uncertain date parallel to Offa's Dyke. He 157.

ROWFOOT Northum NY 6860. *Rowfoot* 1869 OS.

ROWHEDGE Essex TM 0221. 'Rough enclosure'. *Rouhegy* 1346, *le, la Roweheg(ge)* 1375–1509×47, *Le Roughhegge* 1435. OE **rūh** + **hecg**. Ess 388.

ROWHOOK WSusx TQ 1234. 'The rough hook or corner of land'. *Rowhooke Landes* 1601. The earliest occurrence is in the surname of Henry *atte Rowhook* 1327 of Warnham. ME **row** (OE *rūh*) + **hok** (OE *hōc*). Sx 157.

ROWINGTON Warw SP 2069. 'The estate called after Hroca'. *Rochintone* 1086, *Rokinton* c.1130–1291, *Ruhinton* 1221, 1227, *Ruginton* 1232, *Rowintone* 1275–91, *Rouen- Rowghyn- Roughynton* 1313–7, *Rowyngton* 1526, *Rounton* 1451, 1462, *Rownton* 1614, *Rowington vulgarly Rownton* 1656. OE pers.n. *Hrōca* + **ing**[4] or genitive sing. *Hrōcan* + **tūn**. Wa 217 gives pr [rauiŋtən].

ROWLAND Derby SK 2172. Probably 'boundary grove'. *Ralunt* 1086, *-lund* 1200, 1216×72, *Rolund* 1230, *-lound'* 1348, *Roland* 1236–1481, *Rou- Rowland* from 1332. ON **rá** + **lundr**. ON **rá** can mean 'boundary' or 'roe-deer'. Rowland is a small elongated parish separating Great Longstone and Hassop. Db 162, SSNEM 159, L 208.

ROWLAND'S CASTLE Hants SU 7310. Originally 'Rolok's castle', *Rolokescastel* 1307×27, *Roulakescastel* 1381, ME pers.n. **Rolok*, **Rolak* < OHG **Hrōdlaik*. Later 'Roland's castle', *Roulandes Castell* 1369, possibly influenced by the name of the hero of the 12th cent. Fr epic poem *The Song of Roland*. Ha 140, Gover 1958.17.

ROWLANDS GILL T&W NZ 1658. *Rowlands Gill* 1863 OS. Surname *Rowland* + ModE **gill** (ON *gil*).

ROWLEDGE Surrey SU 8243. Probably the same as *La Rowedich* [c.1200]1285, *Rowl ridge, Row ditch* 18th cent., 'the rough or overgrown ditch', OE **rūh** + **dīc**. Perhaps identical with the ditch referred to in *to dic geate* [909]c.1135 S 382. Sr 171.

ROWLEY REGIS WMids SO 9687. 'King's R'. *Rowley Regis* 1834 OS. P.n. *Roelea* 1173, *Ruelega* 1174, *Ruleye* 1272, 'the rough wood or clearing', OE **rūh**, definite form **rū(g)a**, + **lēah**, + Latin **regis**, genitive sing. of **rex**. DEPN.

ROWLEY Shrops SJ 3006. 'The rough hill with a cleft'. *Rowlyth* 16th, *Rowleth* 1705. OE **rūh** + **hlith**. Gelling.

ROWLSTONE H&W SO 3727. 'Rolf's estate'. *Rolueston* 1300, *Rouleston* 1317, 1383, *Rol(l)eston* 1338, 1540. lOE pers.n. *Rolf* (< ON *Hrólfr*), genitive sing. *Rolfes*, + **tūn**. He 175.

ROWLY Surrey TQ 0440. 'The rough woodland clearing'. *Rowley* 1559, *Row Lye* 1823. ME **row** (OE *rūh*) + **ley** (OE *lēah*). Cf. William *de Roweleg* 1294, John *atte Rouly* 1335. Sr 257.

ROWNER Hants SU 5801 → ROCESTER Staffs SK 1139.

ROWNEY GREEN H&W SP 0471. *Rowney Green* 1831 OS. P.n. Rowney + ModE **green**. Rowney, *(la) Ruen- Rowenheye* 1244, 1275–6, *Rowney* 1669 is 'the rough enclosure', OE **rūh**, definite form dative case **rūgan**, + **ġehæġ**. Wo 335.

ROWNHAMS Hants SU 3817. 'The rough enclosure'. *Rowenham*

1269, *Ro- R(o)ughenham* 1301, 1379, *Rownham* 1306, *Roundhams, Roundham Ho.* 1810 OS. OE **rūh**, dative sing. definite form **rūgan**, + **hamm** 2b 'land on a hill-spur' referring to the tongue of ground projecting N of Rownham and culminating in TOOTHILL SU 3818. The modern form has acquired a pseudo-manorial *-s* as if 'Rownham's manor'. Ha 141, Gover 1958.41.

ROWSHAM Bucks SP 8518. 'Hrothwulf's homestead'. *Rullesham* 1189×99, 1227, 1363 with variant *Rol(l)es-* 1198–1375, *Rolvesham* 1363, *Ruls(h)am* 1465, *Rowsham* 1518, *Rowsome* 1640. OE pers.n. *Hrōth(w)ulf*, genitive sing. *Hrōth(w)ulfes*, + **hām**. Bk 89 gives pr [rauʃəm].

ROWSLEY Derby SK 2566. 'Rolf's wood or clearing'. *Reuslege* 1086, *Rolvesle* 1204, *Roulesley(e) -leg(h) -lie -ly -is-* c.1240–1373, *Rollesley(e)* 1388–1578, *Rowsley* from 1550. Pers.n. *Rolf*, genitive sing. *Rolfes*, + **lēah**. Rolf probably represents ON *Hrólfr* rather than the rare English pers.n. *Hrōthwulf*. Db 163, SSNEM 220.

ROWSTON Lincs TF 0856. 'Hrolfr's estate'. *Rouestune* 1086, *-ton* 1209–1337, *Rolueston* 1202×23, before 1221, *Rouston* 1242–1594, *Rowston* from 1526. ON pers.n. *Hrólfr*, genitive sing. *Hrólfs*, + **tūn**. It has been suggested that because of the rarity of spellings with *-l-* the specific might be OE *hrōf* 'a roof' or ON *hróf* 'a boat-shed' used topographically of the rounded promontory on which Rowston stands at the edge of the fens. Perrott, SSNEM 380, Cameron 1998.

ROWTON Ches SJ 4464. 'Rough settlement'. *Roweton'* from 13th with variants *Rue- Rou- Row-* (from 1417). Short for Rough Christleton, *Rowa Christletona* 12th, *Ruhcristelton* c.1200, *Rowton alias Roughe Chrystleton* 1579, 'the rough part of Christleton', OE adj. **rūh** + p.n. CHRISTLETON SJ 4465. The reference is to rough ground. Che iv.114 gives pr ['rautən].

ROWTON Shrops SJ 6119. 'The settlement at Rough Hill'. *Routone* 1086, *Rowelton(')* 1195–late 14th, *Ru(gh)elton -tun'* 13th, *Roulton(')* 1233–1590, *Rowton* from 1569. The generic is OE **tūn** 'an estate', the specific probably an OE p.n. < **rūh** + **hyll** 'rough hill'. The alternative tentatively proposed in Sa i.253 of an OE diminutive ***rūhel** 'a small rough place' from OE adj. **rūh** 'rough' would depend on the use of the adj. as a noun, as possible in the p.n. type ROUGHTON SO 7594 and ROUGHTON CASTLE SJ 3712.

ROXBY 'Rauth's farm or village'. ON pers.n. *Rauthr*, genitive sing. *Rauths, Rauz*, + **bȳ**.

(1) ~ NYorks NZ 7616. *Roscebi, Rozebi* 1086, *Raucebi* [1154×8]15th, *Roucebi* 1285–1425, *Rokesby* 1575. Spellings with *z* and *c* represent [ts] which survives in the pronunciation given by YN 139 [rouzbi], but was substituted for by [ks] in the 16th cent., as also in FLAXBY SE 3958, KEXBROUGH SE 3009, MOXBY SE 5966, ROXBY (lost in Thornton Dale SE 8283) and THROXENBY TA 0189. These are inverted forms arising after the development of [ks] to [s, z] as in BARKISLAND SE 0519, AYSGARTH SE 0188 etc. YW vii.90, YN 139, SSNY 35.

(2) ~ NYorks lost in Thornton Dale SE 8283. *Roze- Rosebi* 1086, *Roucesby* 1250–1408, *Roxbie* 1577. For the phonology see ROXBY NZ 7616. YN 90 gives pr [rɔuzbi], SSNY 35.

(3) ~ HIGH MOOR NYorks NZ 7511. P.n. ROXBY NZ 7616 + ModE **high moor**.

ROXBY Humbs SE 9217. 'Hrok's village or farm'. *Roxebi, Roscebi* 1086, *Rochesbi* c.1115, *Rokesbi* 12th. ON pers.n. *Hrókr*, genitive sing. *Hróks*, + **bȳ**. DEPN, SSNEM 65, Cameron 1998.

ROXTON Beds TL 1554. 'Rook hill'. *Rochestone -done* 1086, *Rokesden* 13th cent., *-don* 13th–14th cents., *Roxton* 1449. OE **hrōc**, genitive sing. **hrōces**, + **dūn**. The village is situated on the top of a rounded hill. It is possible that the specific is a pers.n. **Hrōc* derived from the bird-name. Bd 64.

ROXWELL Essex TL 6408. 'Hroc's spring'. *Rokeswelle* 1291–1310. OE pers.n. *Hrōc*, genitive sing. *Hrōces*, + **wella**. Ess 264.

ROYAL BRITISH LEGION VILLAGE Kent TQ 7257. A village set up as a residential and rehabilitation centre for ex-servicemen after the founding of the British Legion in 1921. Originally known as *British Legion Village*, the 'Royal' prefix was added when royal patronage was granted to the Legion in 1971. Room 1983.105.

ROYAL MILITARY ACADEMY Berks SU 8661. *Military College* 1816 OS. Commonly referred to as Sandhurst, the college was originally established at High Wycombe in 1799, transferred to Marlow in 1802, and finally settled here in 1812. Gaz s.n. Sandhurst.

ROYAL MILITARY CANAL Kent TR 0133. A canal begun in 1807 as part of the land defences in the Napoleonic war, named after George III. Room 1983.105.

ROYAL SOVEREIGN ESusx TV 7393. The name of a lightship.

ROYDON Essex TL 4009. 'Rye hill'. *Ruindune* 1086, *Rei- Roi- Roy- Reyndon(e)* 1154×89–1354, *Rei- Reydon* 1248–1383, *Roydon(e)* from 1467. OE **rȳġen** 'growing with rye' + **dūn**. Ess 49.

ROYDON 'Rye hill'. OE **rȳġe** + **dūn**.

(1) ~ Norf TF 7022. *Reiduna* 1086, *Ridone* 1254. DEPN.

(2) ~ Norf TM 0980. *(et) Rygedune* 1035×40 S 1489, *Regadona, Ragheduna* 1086, *Reydon* 1242. DEPN.

ROYSTON Herts TL 3541. The 'settlement at Roys'. *Roiston* 1286, *Royston* 1440. P.n. *Roys* 1282 + ME **toun** (OE *tūn*). Earlier this was *Crux Roaisie* 'Roese's cross' 1184, *Cruce Rohaysie* 1192 and simply *Rohesie* 1229, Latin **crux** + feminine pers.n. *Ro(h)ese* < CG *Hrōhohaidis*, possibly to be identified with Rose, wife of Eudo Dapifer, a steward of William I. The place developed in the 12th cent. at the spot where Icknield Way and Ermine Street cross one another. The base of the cross still exists. The original reference may have been to the priory founded *apud Crucem Rohesie* 'at Roese's cross' in the reign of Henry II. Hrt 161 gives pr [raistən], ASE 631, BzN 16.314–5.

ROYSTON SYorks SE 3511. 'Roarr or Hror's farm or village'. *Rorestun(e) -ton(e)* 1086–1428, *Roston* 1268–1466, *Rus-* 1409–1531, *Rois- Royston(e)* from 1411. Pers.n. ON *Róarr*, genitive sing. *Róars*, or OE **Hrōr*, genitive sing. **Hrōres*, + **tūn**. The modern spelling reflects the SWYorks sound change *ō* > *ui* before dentals written *ui* or *oi*. YW i.285, SSNY 129, BzN 16.327–8.

ROYTON GMan SD 9107. 'Rye farm or village'. *Ritton* 1226, *Ryton* 1260–1323, *Ruyton* 1327, 1332, *Royton* 1577. OE **rȳġe** + **tūn**. La 52.

RUAN LANIHORNE Corn SW 8942. A combination of saint's name Ruan from the dedication of the church and the church-site name, *Lanryhorn* 1270, *-rihoerne* 1297, *Larihorne -y-* 1309–28 Gover n.d., *(parochia) Sancti Rumoni de Lanyhorn* 1327, *Ruon alias Laryhorin* 1535, *Ruan Lanhorne* 1569, *Ruanlanyhorne* 1614. Saint's name Ruan < OCo *Rumon* + p.n. Lanihorne 'church-site of Rihoarn', Co ***lann** + pers.n. **Rihoarn*. St Ruan is first found as patron saint of this parish in a 10th cent. list. His shrine was at Tavistock and he was also patron of Romansleigh Devon SS 7220, *Romundesleghh(e)* 1267–1406. The church-site name was added for distinction from RUAN MINOR and Major SW 7215 and 7016. PNCo 151, Gover n.d. 481, D 391.

RUAN MINOR Corn SW 7215. 'The lesser Ruan's (church)'. *Rumon le meinder* c.1250 Gover n.d., c.1395, *(ecclesia) Sancti Rumoni Parvi* 1277, c.1320, *Rewan Minor* 1543, *Ruan Vean* 1569, *Litle Ruan* 1610. Saint's name Ruan < OCo *Rumon* + Latin **minor** (OFr *meindre*, Co *byghan* with lenition of *b* > *v*, English *little*) for distinction from Ruan Major SW 7016, *Great Ruan* 1610, earlier simply *(ecclesia) Sancti Romoni* 1208, *Rumon le greinder* c.1250 Gover n.d., and RUAN LANIHORNE SW 8942. PNCo 151, Gover n.d. 580, 578.

RUARDEAN Glos SO 6217. Perhaps 'rye enclosure'. *Rwirdin* 1086, *Rowardyn* 12th–1509×47 with varaints *Re- Ru-, Ruardin -yn* 1255–1437, *Ruardeane* 17th cent. OE **ryġe** + **worthiġn**.

Alternatively the specific might be PrW *riu (W *rhiw*) 'a hill'. Gl iii.240 gives prs. ['ru:ədi:n] and ['ruəR'di:n].

RUARDEAN WOODSIDE Glos SO 6216. P.n. RUARDEAN SO 6217 + p.n. *Woodside* 1831 OS.

RUBERY WMids SO 9977. Short for *Robery Hills* 1650, *Rubery Hill* 1831 OS. Rubery probably represents OE **rūh** 'rough', definite form **rū(g)a**, + **beorg** 'a hill'. Wo 357.

RUCKCROFT Cumbr NY 5344. 'Rye croft'. *Rucroft* c.1233, *Ruc- Rukroft* c.1240, 1380, *Ru(c)hcroft(e)* c.1260, 1280, *R(o)ughcroft* 1339, 1344, *Rowecrofte* 1538, *Roocrofte* 1541, *Rook-Croft* 1649, *Ruckcroft* 1710, *Roe Croft* 1736. ON **rugr** + OE **croft**. So also Cu 169 and SSNNW 248. But the name may well be rather 'rough croft', OE **rūh**.

RUCKINGE Kent TR 0233. Probably 'the rookery' rather than the 'place called after Hroc'. *Hroching* [786]11th S 125, *Rocinga* [791]13th S 1614, *Hrocing* [805]17th S 39, *Rochinges* 1086, *Rocinge(s)* 11th, *Hrocinges* 1066×87, *Rokinges* 1203–1261×2, *Roking' -yng' -inge* 1202–1261×2, *Rukynges* 1242×3. OE ***hrōcing** < **hrōc** + **ing**[2], locative-dative sing. ***hrōcinġe** < earlier ***hrōcingi**. Cf. Ruckinge Grove in Herne and Rucking Hall in Capel-le-Ferne, *Ruckinghall* 1698, *Ruckinge Hill* 1727. The apparent plural is probably the AN nominative sing. *-s* ending sporadically occurring in p.ns. Cf. STOWTING TR 1242. PNK 471, KPN 68–9, ING 206, BzN 1967.343, Cullen.

RUCKLAND Lincs TF 3378. Possibly 'the ridge grove'. *Rocheland* 1086, *Rokelunda* 1153×69, *Roclund* c.1180, 1212, *Rokelaunde* 1212. ODan ***roki** + **lundr**. If this explanation is correct the name is identical with Dan Rågelund, *Rogelund* [1339]15th. Ruckland stands on a marked ridge. However, the specific could as well be OE **hrōc** or ON **hrókr** 'a rook', genitive pl. **hrōca**, **hróka**, and so 'the rooks' grove', as in the Norfolk ROCKLANDS TL 9896, TG 3104, TL 9897. SSNEM 159.

RUCKLEY Shrops SJ 5300. 'Rook wood or clearing'. *Ruclee* 1221×2, *Rokele(ye)* 1249–97, *Rocley -le(ye) -legh'* 1251–92, *Ruckley* from 1276. OE **hrōc**, genitive pl. **hrōca** + **lēah**. Sa i.254.

RUDBY NYorks NZ 4607. 'Farm or village at the clearing' or 'Ruthi or Rudda's farm'. *Rodebi* 1086, *Rudebi -by* c.1150–1228, *Ruddeby* c.1190–1402, *Rudby* from 1285, *Ruthby* 1489. ON **ruth** or pers.n. *Ruthi*, genitive sing. *Rutha*, or feminine pers.n. *Rudda*, + **bȳ**. YN 174, SSNY 35.

RUDDINGTON Notts SK 5733. 'Rudda's farm or village'. *Rod(d)intone -tun* 1086, *Rudinton(e) -yn-* 1182–1266, *Ru- Rotinton(e) -yn-* 1219–1302, *Rodington -yng-* 1293–1383, *Rudingtun* c.1190, *-ton(a) -yng-* 1205–1359, *Ruddington* 1280. OE pers.n. *Rudda*, genitive sing. *Ruddan* or *Rudda* + **ing**[4], + **tūn**. Nt 248.

RUDGE Somer ST 8251. *Ridge* 1817 OS. Ultimately from OE **hrycg**. The hamlet lies at the tip of a ridge of high ground.

RUDGWICK WSusx TQ 0833. 'The ridge farm'. *Regwic* 1210, *Reggewik* 1225, *Rugewik* 1240. OE **hrycg** + **wīc**. Sx 156 gives pr [ridʒik], SS 78.

RUDHALL H&W SO 6225. Partly uncertain. *Rudhale* 1364, 1396. The generic is OE **halh** 'a nook', the specific possibly the plant name **rūde** 'rue' and so the 'nook of land where rue grows' or ME **rud** 'marigold'. He 175.

East RUDHAM Norf TF 8228. *Est Rudham* 1254. ME adj. **est** + p.n. *Rudeham* 1086, 1147, *Ruddaham* 1163, 'Rudda's homestead', OE pers.n. *Rudda* + **hām**. 'East' for distinction from West RUDHAM TF 8127. DEPN.

West RUDHAM Norf TF 8127. *Westrudham* 1254. ME adj. **west** + p.n. Rudham as in East RUDHAM TF 8228. DEPN.

RUDLAND RIGG NYorks SE 6595. Unexplained p.n. Rudland (no early forms) + Mod dial. **rigg** (ON *hryggr*).

RUDLOE Wilts ST 8469. 'The ridge hill or barrow'. *Riglega* 1167, *Ry- Riggelawe* 1249–1398, *Ruggelewe* 1330, *Ridlawe* 1249, *Ridlowe* 1497, *Rudlow(e)*, *Ridgelow* 1767. OE **hrycg** + **hlāw**. Wlt 84.

RUDSTON Humbs TA 0967. 'The rood-stone, the stone used as a cross'. *Rodestan* 1086, 1276, *-stein* 1086, *-stain -y-* 1086–1297, *Rud(de)stan* [1156×75]c.1300–1500 with variants *-stane*, *-stain*, *Rudstone* 16th cent., *Rudston* 1625. OE **rōd** + **stān** influenced by ON **steinn**. The reference is to a great Bronze Age monolith in the churchyard. An alternative possibility is that the specific is OE **rudu** 'redness', since the stone is tinged with red from the weathering of iron in the sandstone. YE 98 gives pr [ruds(t)ən], SSNY 63, Thomas 1976.155.

RUDYARD Staffs SJ 9557. 'The rue garden'. *Rudegeard* [1002×4]11th S 1536, *Rvdierd* 1086, *Rudehard* 1199, *Rud(e)yerd(e) -yord(e)* 1240–1521, *Rod(i)ehierd* c.1255, *Rudeyard* 13th–1427, *Rudyard* from 1560. OE **rūde** + **ġeard**. Oakden. Horovitz suggests rather OE **rudiġ** + **ġeard** 'red yard' pointing to a prominent area of red earth at Red Earth Farm, *Redyerth* 1563, and the neighbouring names Redshaw Wood and Rad Brook (? < OE **rēad** 'red') which flows into Rudyard Lake.

RUDYARD RESERVOIR Staffs SJ 9459. P.n. RUDYARD SJ 9557 + ModE **reservoir**. Cf. *Reservoir Ho.* 1815. An artificial lake constructed under an Act of 1797 to feed the Trent, Mersey, Leek and Cauldon Canals. Oakden.

RUE HILL Staffs SK 0847 → WEAVER HILLS SK 0946.

RUFFORD Lancs SD 4515. 'Rough ford'. *Ruchford* 1212, *R(o)ughford* 1318–32, *Roghforth* 1411, *Rufford* from [c.1200]1268. OE **rūh** + **ford**. Refers to a ford of the river Douglas. La 137, Jnl 17.78.

RUFFORTH NYorks SE 5351. 'The rough ford'. *Ruford(e) -fort* 1086, *Ruefordam* [early 13th]16th, *Rufford(e)* 1165–1589, *Ruhford* c.1190, *Rughford* 1300–1487, *Ruffurth* 1535, *-forth* from 1589. OE **rūh** + **ford**. The reference is to a crossing of Moor Drain towards Hutton Wandesley. YW iv.232, L 68.

RUFUS STONE Hants SU 2712. *Rufus's Stone* 1789. A stone erected in 1745 marking the traditional site of the death of William II (Rufus) in 1099. Latin nickname *Rufus* 'the red' + ModE **stone**. Gover 1958.209

RUGBY Warw SP 5075. Probably 'Hroca's fortified place'. *Rocheberie* 1086, *Roche- Rokebi -by* 1154×89–1627, *Roukeby* 1373, *Rukby* 1484, *Rugby* from 1525. OE pers.n. *Hrōca* + **byriġ**, dative sing. of **burh**. The specific was replaced by ON **bȳ** under the influence of Danelaw p.ns. to the N and E. Wa 143.

RUGELEY Staffs SK 0417. 'Ridge wood or clearing'. *Rvgelie* 1086, *Rug(g)elega -leia* 1155–1291, *Rug(g)ele(e) -ley(e)* 1189–1551, *Rigeley* 1304–1562, *Ry- Ridg(e)ley* 1562–1643. OE **hrycg** + **lēah**, dative sing. **lēaġe**. The reference is to a ridge of high ground at SK 0317 W of Hagley Hall. St i.105, L 169. Duignan 129 records local pr [ridʒli].

RUISHTON Somer ST 2624. 'Rush settlement'. *(æt) Risctune* *[854]12th S 310, *Risctun* [979 for 879]12th S 352, [995×1002]12th S 1242, *Rysctune* [? c.890]n.d. ECW, *Riston* 1610, *Ruishton* 1809 OS. OE **risc**, ***rysc** + **tūn**. This could refer to the site of the settlement or, more likely, allude to a place where rushes were grown and gathered. DEPN.

RUISLIP GLond TQ 0987. 'Rush leap'. *Rislepe* 1086, *Ris(se)- Risshelep(e)* 1230–1483, *Res(s)- Rus(se)- Russhelep(e)* 1227–1365, *Risslip* 1246, *Ryselypp* 1530, *Ruislip(p)* 1597, 1621. ***rysc** + **hliep**. The reference is perhaps to a place where the river Pin could be jumped. The modern spelling with *ui* preserves a ME spelling for [y] as in *buy, build, bruise* etc. Mx 46, GL 82.

RUISLIP COMMON GLond TQ 0889. *Ruislip Common* 1904. P.n. RUISLIP TQ 0987 + ModE **common**. LPN 196.

RUMBURGH Suff TM 3481. Possibly 'the fort made of tree-trunks'. *Romburch* [1047×64]copy, *Rā- Rōhurc* 1086, *Rumburg(he)* c.1130–1316, *Romburg* c.1189, *Rumburgh* from 1327, *Rumboro* 1610. OE ***hrun-burh** < **hruna** + **burh**. OE **rūn** 'deliberation, council' would also be possible but not **rūm** 'broad' in the absence of *Rume-* sps < definite form **rūme**. DEPN, Baron.

RUMFORD Corn SW 8970. Uncertain. *Rumford* 1699. This could

be 'wide ford' (OE **rūm** + **ford**) or from the surname Rumford (itself from Romford, Essex). The evidence is too late for certainty. PNCo 152, Gover n.d. 335.

RUMPS POINT Corn SW 9381. *The Rumps* 1813. ModE **rump**. Alludes to the shape of the headland seen from the sea. PNCo 152, Gover n.d. 130.

RUNCORN Ches SJ 5182. '(At the) wide bay'. *æt Rūm cofan* [915]c.1000 ASC(B), *Runcofan* c.1118 *-coua* 1115–1211, *Runcorna -e* [c.1134]17th,14th–1549, *Runcorn* from 1262. OE **rūm** + **cofa**. The reference is to the lagoon above Runcorn Gap SJ 5083 where the Mersey opens out again after the restriction caused by the promontories of Widnes and the former Castle Rock N and S of the river respectively. Che ii.176.

North RUNCTON Norf TF 6415. *Northrungetone* 1276. ME adj. **north** + p.n. *Run(c)getun* 11th, *Runghetuna* 1086, 'rung or pole settlement', OE **hrung**, genitive pl. **hrunga**, + **tūn**. This might be a place where rungs were made, but since the twin village of South Runcton lies 4 miles S across the river Nar and its marshes it seems more likely that the reference is to a causeway of wooden poles laid lengthwise across the fenland linking the two places. Such plank roads are known both from the continent, for example at Stapel, Lower Saxony, < OHG *stapula-* 'a wooden post', and from the Somerset Levels, notably Sweet Track. DEPN, R. Hachmann, *The Germanic Peoples* London 1971, illus. 15, M. Jones *England before Domesday*, London 1986, 125ff.

South RUNCTON Norf TF 6308. *Suthrungetone* 1291. ME adj. **suth** + p.n. Runcton as in North RUNCTON TF 6415. DEPN.

RUNCTON WSusx SU 8802. Partly uncertain.

I. *Rochintone* 1086, *Rogenton(a)* 1110–1332;

II. *Rongenton* 1179, *Rung(u)eton* 13th cent., *Rongeton al. Rounton commonly called Romton* 1540.

Type I is comparable with ROWINGTON Warw SP 2069, *Rochintone* 1086, *Rouhwinton* 1291, either 'the estate called after Hroc' or 'the settlement called *Ruhing*, the rough place'. It is compatible with type II on the assumption that the abbreviation mark for *n* was regularly omitted in the latter; perhaps, therefore, 'the estate called after Runa or Rune', OE pers.n. *Rūna* or feminine *Rūne* + **ing**[4] + **tūn**. The same pers.n. occurs again in Rowley in Ticehurst, *runanleagesmearc* 'the boundary mark of Runa or Rune's wood or clearing' 1018 OSFacs iii.39, *Runle* 1296. Sx 74, 453, SS 74.

RUNCTON HOLME Norf TF 6109. 'Runcton island'. *Rungeton Holm* 1276. P.n. Runcton as in North RUNCTON TF 6415 + ME, ON **holm**, **holmr** 'a piece of dry land in fen'. DEPN.

RUNFOLD Surrey SU 8747. 'The tree felling'. *hrunig fealles wæt* 'the ford of the tree felling' (OE *wæd*) [973×4]c.1150 S 820, *Runifall(e), Runy- Ronifalle, Runfæll -falle, Run(n)efall, Ronefalle -welle* etc. 13th cent., *Runvale* 16th cent., *Renfold, Runfolde* 11607, *Runfull* 1816 OS. OE ***hruna** + ***(ġe)fall**. Sr 172.

RUNHALL Norf TG 0507. Partly uncertain. *Runhal* 1086, 1206, *-hale* 1254. The generic is OE **halh** 'a nook of land', the specific either pers.n. **Rūna* corresponding to ON *Rúni*, CG *Rūno*, or **hruna** 'a fallen tree, a log', or **rūn** 'council'. The largest category of elements compounded with **halh** is pers.ns. but 'nook where there are fallen trees' is equally possible, perhaps referring to a bridge formed from tree-trunks. DEPN L 110.

RUNHAM Norf TG 4611. Partly uncertain. *Rom- Ronham* 1086, *Runnaham* 1163, *Runham* from 1165. The generic is OE **hām** 'homestead, estate'; the specific offers the same problems as for RUNHALL TG 0507. Nf ii.19.

RUNNEL STONE Corn SW 3620. Possibly 'tidal current rock'. *Raynoldis stone* 1461×83, *a Rocke called Reynolde stone* 1528, *Renaldstowe* (for *-stone*) [1580]18th, *Rundleston* 1695. OE **rynel** + **stān**. OE **rynel** normally means 'water-channel'; here it is used of tidal currents. PNCo 152, Gover n.d. 461, Jnl 24.13.

RUNNINGTON Somer ST 1121. Probably 'the estate called after Runa'. *Rvnetone* 1086, *Ro- Runneton* 1202, 1233, *Runnyngton* 1306, *Runton* 1610. OE pers.n. *Rūna* + **ing**[4] + **tūn**. Alternative possibilities include OE **(ġe)rūna**, 'the councillor's estate', and ***hruna** 'estate where logs are cut and obtained'. DEPN.

RUNSWICK BAY NYorks NZ 8016. P.n. Runswick + ME **bay**. Runswick, *Reneswike -wyk* 1273, 1348, *Ri- Rynneswyk* 1293, 1407, *Runswick* from 1577, is 'Reinn's inlet', ON pers.n. *Reinn*, genitive sing. *Reins*, + **vík**. The pers.n. might alternatively be OE *Rœġen*. YN 139 gives pr [runzik], SPN 217.

East RUNTON Norf TG 1942. *East Runton* 1838 OS. ModE adj. **east** + p.n. *Runetune* 1086, *-ton(a)* 1175×86–1209, 'Runa or Runi's estate'. OE pers.n. **Rūna* or ON *Rúni*, genitive sing. *Rúna*, + **tūn**. 'East' for distinction from West RUNTON TG 1842. DEPN.

West RUNTON Norf TG 1842. ModE adj. **west** + p.n. Runton as in East RUNTON TG 1942. West Runton has the church and is the parent settlement of East and West Runton.

RUNWELL Essex TQ 7494. Partly uncertain. *Runweolla, Runewelle* *[c.940]12th, *Runewellā* 1086, *-welle* 1258, *Renewell* 1291, 1299, 1374, *Roundwell* 1547. Uncertain element or pers.n. + OE **wella**. The specific might be any of OE ***hruna** 'a tree-trunk, a log' used in forming a conduit, **rūn** 'a secret, a mystery, a council' referring to a wishing well, or perhaps, in view of the *Rene-* spellings OE ***ryne**, ***rene** 'a ditch, a channel' (ModE *reen, rhine*) varying with **run** 'a stream' (ModE *run* sb. II.9.). Ess 265.

RUSCOMBE Berks SU 7976. 'Rot's enclosed land'. *Rothescamp(e)* [1091–1223]13th, *Rotescamp* 1167, [1220]13th, *Rotescomb* [1226]13th, *Ruscomp(e)* 1284–1368, *Ruscombe* from 1535. OE pers.n. **Rōt*, genitive sing. **Rōtes*, + **camp**. Brk 127, Signposts 77, 78.

RUSHALL H&W SO 6435. No early forms.

RUSHALL Norf TM 1982. Partly uncertain. *Riuessala* 1086, *Riuishale* 1175, *Riveshale* 1242, 1254, *Reueshall* 1264. The generic is OE **halh** 'a nook of land' referring to the valley in which the hamlet lies. The specific might be OE **hrif**, 'belly, womb', genitive sing. **hrifes**, applied to the valley ('nook of the womb') although the feature seems too shallow for this. Alternatively an OE pers.n. **Rīf* might be posited from the adj. *rīf* 'violent, fierce, ravenous, noxious'. Neither solution is entirely satisfactory. DEPN.

RUSHALL Wilts SU 1255. 'Rust's nook'. *Rusteselve* (2×) 1086, *-hala -e* 1160–1332, *Rosteshale* 1281–9, *Rosshall* 1416, *Russall* 1535, *Rushall* 1627. OE pers.n. **Rust*, genitive sing. **Rustes*, + **hale**, dative sing. of **halh**. Wlt 323, L 101.

RUSHALL WMids SK 0301. 'Rush nook'. *Rischale* 1086, *Rushale* 1195, 1242. OE **risc** + **hale**, dative sing. of **halh** in the sense 'a small hollow'. DEPN, L 105.

RUSHBROOKE Suff TL 8961. 'The rushy brook'. *Ryssebroc* [964]13th S 1483, *ryscebroc* 1086, *Rushbrok* 1610. OE **risc**, ***rysc**, genitive pl. **risca**, ***rysca**, + **brōc**. DEPN.

RUSHBURY Shrops SO 5192. 'The rush fortified place or manor'. *Riseberie* 1086, *Ryssebury* c.1291, *Rusberia -bir' -bury(e)* etc. c.1158–1723, *Russhebur(y)* 1283–1549, *Rushbury* from 1577. OE ***rysc**, genitive pl. ***rysca**, + **burh**, dative sing. **byriġ**. Sa i.254.

RUSHDEN 'Valley where rushes grow'. OE ***ryscen** + **denu**.

(1) ~ Herts TL 3031. *Risendene* 1086, *Ressen- Ryshen- Riss(h)enden(e)* 1210–1308, *Ri- Ressedene* 1203–94, *Russ(h)endene(e)* 1255–1336, *Rusheden* 1603×25, 1713. Hrt 163 gives common pr [rizdən].

(2) ~ Northants SP 9566. *Ris(e)dene* 1086, *Rissenden* [1109×22]1356–1316, *Russen-* 1205, 1251, *Rishendon* 1288, *Russhenden* 1337, *Russheden* 1428. Nth 195 gives pr [ruʒden], L 98.

RUSHFORD Suff TL 9281. 'The rushy enclosure'. *Rissewrth* c.1060, *Riseurdā* 1086, *Rischewrthe* 1242, *Rushforth* 1610. OE **risc, *rysc**, genitive pl. **risca, *rysca**, + **worth**. DEPN.

RUSH GREEN GLond TQ 5187. *Rush Greene* 1651, *Rush Green* 1777. ModE **rush** + **green**. Ess 120, LPN 196.

RUSHLAKE GREEN ESusx TQ 6218. *Ruslake grene* 1567. P.n. Rushlake + ModE **green**. Rushlake, *Rysshelake* 1537, is 'rush streamlet', ME **rish** (OE *rysc*) + **lake** (OE *lacu*). Sx 470.

RUSHMERE 'The rushy mere'. OE **risc**, ***rysc**, genitive pl. **risca**, ***rysca**, + **mere**.

(1) ~ Suff TM 4987. *Rise- Ri- Ryscemara* 1086, *Russhemere* 1286, 1314, *Kushmere* (for *R-*) 1610. DEPN, Baron.

(2) ~ ST ANDREW Suff TM 2046. 'St Andrew's R'. *Ri- Ryscemara, Rissemera* 1086, *Ryshmere* 1316, *Russemere* 1193–5, *Russhemer* 1524, *Rushmere* 1610. Saint's name *Andrew* from the dedication of the church. DEPN, Baron.

RUSHMOOR Surrey SU 8740. A modern administrative name from Rushmoor Bottom, meaning literally 'waste ground overgrown with rushes'. Room 1983.105.

RUSHMORE HOUSE Wilts ST 9518. P.n. *Ros(se)mere* 1263–1310, *Rysshemere* 1321, *Rushemere* 1570, *Rushmore* 1618, probably 'the rush pool', + ModE **house**, a large Georgian/early Victorian mansion. The earliest sps could point to OE **hrossa-mere* 'the horses' pool', but the later forms probably indicate OE ***rysc**, genitive pl. ***rysca**, + **mere**. A ME variant *rosshe* of *rush* is otherwise evidenced only from the 15th cent. Wlt 202, Pevsner 1975.526.

RUSHOCK H&W SO 8871. 'Rushy place or brook'. *Rvssococ* (sic) 1086, *Russoc -ok(e)* 1200–1327, *Roshoke* 1210. OE ***riscuc**. It is uncertain whether the DB spelling is a genuine form, perhaps for *riscuc-hōc* 'hook of land at *Riscuc*', or an instance of dittography. Wo 255, xliv, PNE ii.85.

RUSHOLME GMan SJ 8695. '(Settlement at) the rushes'. *Russum* 1235, *Ryssham* 1316, *Rysum* 1320, *Resshum* 1417, *Russhum* 1420, *Rysshulme, Rysholme* 1551. OE ***ryscum**, dative pl. of **risc**, ***rysc** 'a rush'. La 31 gives pr [ruʃəm], Jnl 17.31.

RUSHTON 'Rush farm or village'. OE **risc**, ***rysc**+ **tūn**.

(1) ~ Ches SJ 5864. *Rusitone Riseton(e) -don* 1086, *Riston* 1237, *Ruston* 1154×89–1359 with variant *-tone, Rushton* from 1242 with variants *Rus(s)(c)h(e)-* and *-t(o)un, Rishton* 1308–1672 with variants *Ruys(s)h(e)ton* 1323–1416. Che iii.291.

(2) ~ Northants SP 8482. *Riseton, Ricsdone* 1086, *Riston* 1102–1329, *Ruston* 12th–1268, *Ri- Ryssh(e)ton* 1305–1401, *Rush- Russheton* from 1346. Nth 120.

(3) ~ Shrops SJ 6008. *Ruston* [1190]15th, *Ruch- Ruston* 1255×6, *Rushton* 1271×2, 1833 OS.

(4) ~ SPENCER Staffs SJ 9462. 'R held by the Spencer family'. *Ruyston Spencer* 1240, *Russheton Spencer* 1399. P.n. *Risetone* 1086, *Ruston* 1227–1544, *Rushton* 1334–1775. Manorial addition from Hugh le Despenser who held here c.1328×43 and for distinction from Rushton James SJ 9261, *Ruyston James* 1240, *Rushton ~* from 1306, p.n. *Risetone* 1086 + manorial addition from the 13th cent. FitzJames family. DEPN, Horovitz.

RUSHWICK H&W SO 8253. There are two forms:

I. *rixuc* [963]11th S 1303 with variant *riscyc.*

II. *Russewyk* 1275, *Rushwyke* 1348, 1510.

Form II represents OE **risc** + **wīc** 'dairy farm where rushes grow', but this is almost certainly a re-interpretation of form I representing OE ***riscuc** 'a rushy place or stream', the name of a small rivulet occurring in the bounds of Cotheridge also called *cyrces pulle* 'the church's stream' in an alternative set of 11th cent. bounds (B 1107). Wo 94, xliii, PNE ii. 85, Hooke 257, 259.

RUSHY KNOWE Northum NY 9299. 'Rushy hill'. *Rushy Knowe* 1869 OS. ModE adj. **rushy** + N dial. **knowe** 'a hill' (OE *cnoll*) rising to 1065ft.

RUSHYFORD Durham NZ 2828. 'Rushy ford'. *Risseforthe* 1242×3, *-ford* 1259×60, *Rushey ford* 1675, 1728×35. OE **risc**, genitive pl. **risca**, + **ford**. The ford carried the Great North Road across Rushyford Beck, a tributary of the river Skerne. The name was Latinised as *vadum cirporum* 1336 'ford of rushes', but the evidence is rather late and the specific may actually have been OE **riscen** 'growing with rushes' or an OE ***risciġ** 'rushy'. NbDu 170, PNE ii.85.

RUSKINGTON Lincs TF 0851. 'Rush farm or village'. *Rischinton(e)* 1086–1197, *Reschintone* 1086, *Ri- Ryskinton(e) -yn- -en-* 1185–1642, *Ri- Ryskington -yng-* 1236–1556, *Resk-* 1265–1467, *Ruskington* from 1399. OE **riscen** 'rushy' or ***riscing** 'rushy place' with ON [sk] for English [ʃ] + **tūn**. Ruskington is situated in low-lying fenland. Perrott 298, SSNEM 185.

RUSLAND Cumbr SD 3488. 'Rolf or Roald's newly cultivated land'. *Rolesland(e)* 1336, 1400, *Rwseland* 1537. ON pers.n. *Hrólfr* or *Hróaldr* + secondary genitive sing. **es**, + ON or OE **land**. La 217, SSNNW 155, L 248–9.

RUSPER WSusx TQ 2037. 'The rough enclosure'. *Rusparre* 1219, 1261, *La ~* 1304, *Rugespere -sparre* 1233–78, *Rou- Rowspar(re) -sper* 1247–1487, *Rusperre* 1309, *Roosper* 1529, *Rusper* 1597. OE **rūh** + ***spearr**, ***spær**, ME **sperre** 'a spar, a shaft, a rafter', possibly used in the sense 'enclosure'. Sx 232, PNE ii.135.

RUSPIDGE Glos SO 6512. Possibly 'brushwood hill-side'. Cf. *Ruspatch Meend* 1770, *Ruspedge Meend* 1831 OS. Dial. **risp** 'sedge; bush, twig, branch' (OE **rispe*) + **edge** (OE *ecg*) with **meend** (ME *munede*) 'forest waste'. Gl iii.223.

RUSSELL'S WATER Oxon SU 7189. *Russels Water* early 18th. O 85.

RUSTINGTON WSusx TQ 0401. 'The estate called after Rusta'. *Rustincton'* 1185, *Rustinton -yn-* 1190–1368, *Rustyton -i-* 14th cent., *Ruston* 1589, *Russen* 1723, *Rustington* from 1255. OE pers.n. **Rusta* + **ing**[4] + **tūn**. Sx 172, Ritter 108.

RUSTON 'Brushwood settlement'. OE **hrīs** + **tūn**.

(1) East ~ Norf TG 3427. *Estriston* 1405. ME adj. **est** + p.n. *Ristuna* 1086, *Riston(e)* 1129–1208. The village lies at the edge of the Mown Fen. 'East' for distinction from Sco RUSTON TG 2821. DEPN.

(2) Sco ~ Norf TG 2821. 'Wood R'. *Scouriston* [c.1280]14th, *Skouriston* 1425. ON **skógr** + p.n. *Ristuna* 1086, *Riston(e) Ry-* 1257–1402 including [1134×40]14th. 'Sco' for distinction from East RUSTON TG 3427. Nf ii.189.

RUSTON PARVA Humbs TA 0661. *Little Ruston* 1583, *~ Parva* 1828. P.n. *Roreston* 1086, *Ruston* from 1167, 'Roarr's farm or village', ON pers.n. *Róarr* or possibly OE *Hrór*, genitive sing. *Róars*, + **tūn**. Latin **parva**, ModE **little**, for distinction from Long RISTON TA 1242 when the name of that village was pronounced Ruston. YE 93 gives pr [la:tl riəstən], SSNY 129.

RUSWARP NYorks NZ 8809. 'Silted land overgrown with brushwood'. *Rise- Rysewarp(e)* 1145×8–[1351]15th, *Ruswarpe* 1665. OE **hrīs** + **w(e)arp**. The sense 'sediment deposited by a river' retained by Mod dial. *warp* and *warp-land* 'land formed by the silt of a river' derives from the root of OE *weorpan* 'to throw', cf. LG *werp* in the p.n. Antwerp. Ruswarp is situated on the river Esk. YN 125 gives pr [ruzəp], PNE ii.248.

RUTHERNBRIDGE Corn SX 0166. 'Ruthen bridge'. (Bridge of) *Ruthen'* 1412, *Rothyn brygge* 1518, *Ruthan Bridge* 1748. R.n. Ruthen, *Rodan* 1200, *Roethon* 1310, *Ruthen* 1412, 'liquid one' (Co ***roeth** cognate with W *rhwyth* 'juice, liquid' + suffix **-an** or **-en**), + English **bridge**. W Rhwyth occurs as a stream name in Dyfed. PNCo 152, Gover n.d. 101.

RUTHVOES Corn SW 9260. 'Red bank'. *Ruthfos* 1298, 1350, 1545, *Rud- Ridfos* 1302 Gover n.d., *Ruthas* c.1696. Co **ruth** + **fos**. PNCo 152 gives pr ['rʌðəz].

RUTLAND WATER Leic SK 9207. A modern man-made lake completed in 1977. P.n. Rutland + ModE **water**. Rutland, *Roteland* [1053×66]12th S 1138, 1086, *-land(e)* 1080×7–1554, *Rotland'* 1264–1462, *Rut(t)elande* 1396–1586, *Rutland(e)* from 1416, is 'Rota's land', OE pers.n. *Rōta* + **land**. R 1.

RUTUPIAE Kent TR 3260. The name of the Roman fort at RICHBOROUGH. *Ῥουτουπίαι (Rutupiae)* [c.150]13th Ptol, *Ritupis* AI, *Ritupium* IM both [c.300]8th, *Rutupias* [late 4th] Ammianus Marcellinus, *Rutupi, Rut(h)ubi Portus* [c.415]

Orosius, *Rutupis, Rittupis* [c400]15th ND, *Ru- Ratupis* [c.700]13th–14th Rav, *Rutubi portus* [c.731]8th BHE. Brit **Rutupi̯as*, probably a formation on ***rutu-** 'mud' meaning 'mud-flats, muddy creeks'. Forms with *Ritu-* have been assimilated to Brit ***ritu-** 'ford'. RBrit 448, BBCS 26.395–8, Britannia 1.78, LHEB 661.

RUYTON-XI-TOWNS Shrops SJ 3922. *Royton of the Eleven Towns* 1758, *Ryton* ~ 1784. P.n. *Ruitone* 1086, *Ruyton* from 1330, *Ruiton* 1611–1750, *Ruton(e)* 1242–1416, *Ry- Riton* 1497–1777 OE **ryġe-tūn** 'tye farm' + ModE **eleven towns** for distinction from other Ruytons and referring to the eleven townships which once constituted the parish. Sa i.255.

RYAL Northum NZ 0174. 'Rye hill'. *Ryhill'* 1242 BF, 1268, *-hull* 1255, *-hil'* 1296 SR, *Riell* 1346, *Ryall* 1663. OE **ryġe** + **hyll**. NbDu 170.

RYAL FOLD Lancs SD 6621. No early forms, but other instances of the p.n. Ryal often represent OE **ryġe** + **hyll** 'rye hill'.

RYALL Dorset SY 4094. 'Rye hill'. *Rihull* 1240, *Rioll* 1468, *Ryalle* 1494. OE **ryġe** + **hyll**. Do 127.

RYARSH Kent TQ 6759. 'Rye plough-land'. *Riesce* 1086, *Resce, Reiersce* c.1100, *Ryers(se), Ryhers(e), Ryershe* 13th cent., *Ryersh* 1610. OE **ryġe**, Kentish ***reġe**, + **ersc**. PNK 149.

Great RYBURGH Norf TF 9527. *Riburg Magna* 1291. ModE adj. **great**, Latin **magna**, + p.n. *Reieburh* 1086, *Rieburc* 1165, 'rye fort or manor', OE **ryġe** + **burh**. 'Great' for distinction from Little RYBURGH TF 9628. DEPN.

Little RYBURGH Norf TF 9628. *Parva Reienburh* 1086, *(in) Parvo Riburc* 1198. ModE adj. **little**, Latin **parva, parvus**, dative sing. **parvo**, + p.n. Ryburgh as in Great RYBURGH TF 9527, 'the fort or manor growing rye', OE **ryġen** + **burh**. DEPN.

River RYBURN WYorks SE 0419–0422. 'Reed stream'. *Riburn* 1308, *Rybo(u)rne* 1316, 1589, *Roburn* 1462, *Reddbourn, Ridbourne* 1492. OE **hrēod** + **burna**. This seems the best explanation in view of the *Ro- Redd- Rid-* spellings (some of which come from the forms for RIPPONDEN SE 0319), but OE **hrīfe** 'fierce' with early assimilation of *-f-* to the following *-b-* or **ryġe** 'rye' are not impossible. The lost p.n. Crummock Holme in Norland SE 0722, *Cromackholme* 1652, suggests that the earlier name of the river was **Crummōg* 'the crooked one', from PrW ***crumm** 'crooked' + suffix **-ǭg** (Brit *crumbāco-*). YW vii.136, iii.54.

RYDAL Cumbr NY 3606. 'Rye valley'. *Ri- Rydal(e) -dall* 1240–1777. OE **ryġe** + ON **dalr**. We i.208 gives pr ['raidəl], SSNNW 248.

RYDAL FELL Cumbr NY 3609. *Rydal Fell* 1865. P.n. RYDAL NY 3606 + ModE **fell** (ON *fjall*). Rydal was a former forest or hunting preserve, *foresta de Ridale* 1274. We i.211.

RYDAL WATER Cumbr NY 3505. *Rydal(l)-water* [1576]1681×4–1865. P.n. RYDAL NY 3606 + ModE **water** in the sense 'lake'. The old name was *Routh(e)mer(e)* 13th cent., *Rowthmere* 1452, *Routha-meer* 1671, 'lake of the river Rothay', r.n. ROTHAY NY 3308 + OE **mere**. The Rothay flows through the lake. We i.17.

RYDE IoW SZ 5992. 'The stream'. *(la) Ride* 1257–1353, *(la) Ryde* from 1274, *Rythe* 1420. OE **rīth**. Wt 193, Mills 1996.89.

RYDE ROADS IoW SZ 5893. P.n. RYDE SZ 5992 + ModE **road** 'an anchorage for ships'.

RYDER'S HILL Devon SX 6569. *Ryders Hill* 1809 OS. Probably from the surname Ryder. A prominent height on Dartmoor rising to 1691ft. The earlier name was *Gnatteshulle* or *Gnatesburghe* 1240, *Gnatteshill, Gnattleshill* or *Knattleburroughe* 1608, 'gnat or Gnat's hill', ME **gnat** (OE *gnæt*) possibly used as a pers.n., genitive sing. **gnattes** or *Gnattes*, + **hill** (OE *hyll*) and **burgh** (OE *beorg* or *burh*). Cf. Great Gnat's Head SX 6167. D 197.

RYE ESusx TQ 9220. '(At) the island'. *Ria, Rya* 1130–1232, *Rie, (la) Rye* 1135×1154–14th. OE **ēa** with misdivided definite article prefixed from the OE prepositional phrase **æt thǣre īeġe**, ME *atter ye* > *atte Rye*. The site, a rock promontory, was up to the 14th cent. a true island separated from the mainland by marsh subject to inundation by the sea. The town was laid out as a new town by the abbey of Fécamp which held the manor in the 12th cent. Its original name was *Rammesleah* [1028×35]18th S 982, *Rames- Hrammeslege* [1005]12th S 911, *Rameslie* 1086, 'clearing growing with wild garlic', OE **hramsa** + **lēah**. Sx 536, vi-vii.

RYE BAY ESusx TQ 9617. P.n. RYE TQ 9220 + ModE **bay**.

RYE DALE NYorks SE 5982. R.n. RYE SE 5784 + ModE **dale** (ON *dalr*).

RYE FOREIGN ESusx TQ 8822. 'Rye outside the boundary'. P.n. RYE TQ 0617 + ME **foreign**. Rye was originally a separate liberty within the larger liberty or area called Rye Foreign. Sx 537.

RYE HARBOUR ESusx TQ 9419. P.n RYE TQ 9220 + ModE **harbour**.

River RYE NYorks SE 5784. Uncertain. *Ria(m)* [1132]16th–1252, *Rye* from 1181×1216. Two alternative sources have been proposed;

(I) Brit **Rīu̯a* from a root **rī-* 'flow' seen also in the W r.n. Rhiw, *Rue* 1578, and cognate with OE *rīth* 'brook', r.n. Rhine < *Rein-no-*, and Lat *rīvus* < **rei-u̯-os*, all from PIE **er-/*or-/*r-* with various suffixes;

(II) Brit **Reg-iā* from the root **reĝ-* 'water, wet' probably seen in Lat *rigāre* and the continental r.ns. Regen, a tributary of the Danube, *Regana* 882, Rioncy, *Rioncy* 1326, alternative name of the Grand Eau of the Ormont Valley < **Reg-ontia*, and Rionzi, a tributary of the Petit Flon at Lausanne, also < **Reg-ontia*.

YN 5, RN 349, BzN 1957.253, Flussnamen 96, 104.

RYHALL Leic TF 0310. 'Nook where rye is grown'. *Rihala* c.1121 ASC(E) under year 963, *Righale* [1042×1055]13th S 1481, *Riehale* 1086, *Ry- Rihal(e)* 1107–1415, *Ryhall(e)* from 1426, *Ryall(e)* 1400–1539. OE **ryġe** + **halh**, dative sing. **hale**. The 'nook' is formed by a bend of the river Gwash. R 160, L 110.

RYHILL WYorks SE 3914. 'Hill where rye is grown'. *Rihella -helle* 1086, *Ri- Ryhil(l) -hyll* 12th–1400, *Ryle, Rile* 1382–1658. OE **ryġe** + **hyll**. YW i.261, L 171.

RYHOPE T&W NZ 4152. 'The rough enclosure or valley'. *duas Reof- Reophoppas* [c.1040]12th, *duas Refhopas* [1072×85]15th, *Riefhope* 1197, *Refhop(')* [1183]c.1382, c.1190–1385, *Revehop(pe)* 1327–c.1500, *Ryop(p)* 1582–1636, *Rive- Ryhopp* 1647, *Riop* 1763 Kitchin, *Ryhope* 1768 Armstrong. OE **hrēof** + **hop**. There are two valleys running down to the sea at NZ 4153 and 4151, but valleys opening to the sea are normally called *denes* in Co. Durham and the latter is indeed called *Ryhope Dene* (OE *denu*). L argues for the sense 'enclosure of or on the moor' for this name and may well be right. NbDu 170, L 112, 116.

RYKNILD STREET Derby SK 2930. *Ykenild* c.1200, 1216×72, *Hykenalestret* 1243, *Y- Ikenildestret(e)* 1275–1319, *Ikenelde-* 1314, 1327, 1546, *Hykenilde-* 1327×77, *Ikelynge-* 1340, *Ekeling-* 1535, *Ickle-Street* 1656, *Rikenildstret* c.1400. The Roman road from Bourton on the Water to Polterham, Margary no.18a-e. The name is transferred from ICKNIELD WAY with false analysis of the ME phrase *at there Ikenilde strete* as *at the Rikenilde strete* in which *there* represents the OE feminine dative sing. form **thǣre** of the definite article. Db 23, Wa 7, Wo 2.

RYKNILD STREET Warw SP 0862, WMids SP 0897. *Ick- Hykenildestrete* 1327×77, *Ickle-Street* 1656. Transferred from ICKNIELD WAY. Wa 7.

Great RYLE Northum NU 0212. *Mangnam Ryhil* (sic) 1236 BF, *Great Ryle* 1868 OS. ModE adj. **great**, Latin **magna**, + p.n. *Rihul* 1176, *Ryhill* 1242 BF, *Ryle* 1428, 'rye hill', OE **ryġe** + **hyll**. NbDu 170.

Little RYLE Northum NU 0211. *Parva Rihull'* 1212 BF, ~ *Ryhil* 1236 BF, *Little Ryle* 1868 OS. ModE adj. **little**, Latin **parva**, + p.n. Ryle as in Great RYLE NU 0212.

RYLSTONE NYorks SD 9758. Possibly the 'stream settlement'. *Rilestun(e) -is-* 1086, *Ri- Rylleston(a)* 1135×54–1574, *Rilston(e)* 1285–1524. Uncertain specific + OE **tūn**. EModE *rill* 'a small stream' is not recorded before 1538 and is thought to be a loan from LG *rille*. The first element could be OE **rynel** 'a brook' with regular assimilation of *nl* to *ll* in the genitive sing. *rylles* < *rynles*, or a r.n. identical with Norw *Rylla* in *Rynes* and *Rylledalen* from the root **hruzl-* seen in ON *hrollr* 'shivering, trembling', *hrolla* 'waver, totter, oscillate'. YW vi.93.

RYME INTRINSECA Dorset ST 5810. 'Ryme within'. *Ryme Intrinsica* 1611. P.n. Ryme + Latin **intrinseca** for distinction from Ryme Extrinseca, a former manor in Long Bredy SY 5690. Ryme, *Rima* 1160, *Ryme* 1229, is OE **rima** 'a rim, an edge, a border' referring to the situation either on the slope of a ridge or near the county boundary. Do 127.

RYSOME Humbs TA 3623. '(Settlement) amongst the brushwood'. *Rison, Utrisun* 1086, *Ri- Rysom* 1175×95 etc., *Rysome* from 1579. OE **hrīs**, dative pl. **hrīsum**. A place where brushwood was cut and collected. DB *Utrisun* 'outer Rysome' was for distinction from RISE TA 1452, cf. OUT NEWTON TA 3822. YE 22.

RYSOME GARTH Humbs TA 3622. *Risingarth* 17th cent., *Risom Garth* 1786. P.n. *Rison, Utrisun* 1086, *Ri- Rysom* from 1175×96, '(The settlement) amidst the brushwood, in the sticks', OE **hrīs**, dative pl. **hrīsum**. Cf. RISE TA 1542. YE 22.

RYTHER NYorks SE 5539. 'The clearing'. *Ridre, Rie* 1086, *Rider(a)* c.1150, 1205, *Ri- Ryther* from 1257, *Ria, Rie* 1166×79–1269, *Rider* 16th cent. OE ***ryther** 'a clearing', a derivative of ***roth** as in ROTHLEY SK 5812 and ROTHWELL Lincs TF 1599, Northants SP 8181, WYorks SE 3428. Early spellings with loss of *th* and *er* are AN, later spellings with *d* reflect the sound change *th* > *d* before *r*. YW iv.65.

RYTON 'Rye farm'. OE **ryġe** + **tūn**.

(1) ~ Glos SO 7332. *Ruyton* 1327, *Ri- Ryton* 1613–1784. Gl iii.169.

(2) ~ Shrops SJ 7602. *Ruitone* 1086, *Ruyton* 1316–1431, *Ruton(e)* 1203–1325, *Ryton* from 1535, *Royton* 1577, 1695, *Righton* 1610, 18th. According to DB the vill contained a mill paying *8 sesters of rye*. Sa i.256.

(3) ~ T&W NZ 1564. *Riton* 1138×59–1647, *Ritun* c.1190, *Ryton* from 1300. NbDu 170.

(4) ~ -ON-DUNSMORE Warw SP 3874. P.n. *Rietone* 1086, *-ton* 1221, *Rugintunia, Ruitonia, Rutunia* 13th, *Ruyton* 14th cent. and *[c.1043]17th S 1226, *Riton* 1598, + affix *super Donnesmor* 1327. Wa 178.

RYTON NYorks SE 7975. 'Settlement or estate beside the river Rye'. *Ritun -tone* 1086, *Rih- Rictona* [12th cent]16th, *Rigeton* c.1200, *Ryton* from 1282. R.n. RYE SE 5784 + OE **tūn**. YN 76.

River RYTON Notts SK 6482. *river Ryton* c.1825, *Rayton river* 1826. A back-formation from RAYTON Notts 6179. The original name of the river preserved in the p.n. BLYTH Notts SK 6286 was *(on, andlang) blidan* [958]14th S 679, 1349, *Blye* 1249, *Blida* 13th, 1330, *Blith(e)* 16th, 'the pleasant one', OE **blīthe**. Nt 8, RN 38.

S

SABDEN Lancs SD 7837. 'Spruce valley'. *Sapeden* c.1140, 1377, *Sappeden* 1377. OE **sæppe** 'a fir-tree' + **denu**. La 80, L 98.

SABDEN BROOK Lancs SD 7535. P.n. SABDEN Lancs SD 7837 + ModE **brook**.

SACOMBE Herts TL 3419. 'Swæfa's field, or the field of the Swæfe'. *Sueuechāp -cāpe, Seuechampe, Stuochampe* 1086, *Swavecaumpe* 1310, *Sauecampe -compe -caumpe -com(b)e* 1199–1436, *Sacam(e)* 1499, 1535, *Sacombe* 1505, *Sawcombe* 1598. OE pers.n. *Swǣfa* or folk-n. *Swǣfe*, genitive pl. *Swǣfa*, + **camp** possibly referring to settlement in an area formerly known to Latin speakers as the *campus*. Hrt 137 gives pr [seikəm], Signposts 76–8.

SACRISTON Durham NZ 2447. Short for Sacriston Heugh, *le Segrestayneheuh* 1311, *Sacristanhough* 1382, *Sacrestone Heughe* 1580, *Sacristonheughe* 1647×8, 'sacrist hill, hill where the sacrist (of Durham Abbey) has an estate'. OFr **secrestein, segrestein** + ME **hogh**, N dial. **heugh** (OE *hōh*). *Heugh* was dropped because *Sacriston* looked like a name in **tūn**. NbDu 171.

SADBERGE Durham NZ 3416. 'Flat-topped hill'. *Satberga* c.1150, *Sadberg* c.1220–1699, *-bergh(')* 14th cent., *-barghe* 1485, *Satberge* 1235, *Sadberge* from 1235×6, *-bardg* 1582, *-burge* 1406–1634, *Sadbiry* 1312, *-bery* 1387–1593, *-bury* 1408–1624, *Sedberg* 1615. ON **set** + **berg** as in SEDBERGH SD 6592 Cumbria. Sadberge was the centre of the only wapentake in Durham. NbDu 171, Nomina 12.30.

SADDINGTON Leic SP 6591. Possibly the 'estate called after Sada'. *Sad- Setintone* 1086, *Sadinton(e) -yn-* 1195–1350, *Sadingt(h)ona* late 12th, *Sadington(e) -yng-* 1231–1576, *Saddington* from 1536. OE pers.n. **Sada* + **ing**[4] + **tūn**. **Sada* is unrecorded but paralleled by the first element of Langobardic *Sadipertus*. Lei 249.

SADDINGTON RESERVOIR Leic SP 6691. P.n. Saddington + ModE **reservoir**.

SADDLE BOW Norf TF 6015. *Sadelboge* 1198, *Satelbowe* 1314. OE **sadulboga** 'the arched front part of a saddle'. Already in 1198 the reference was to an area of land; either it was similar in shape to a saddle-bow or perhaps there was once an arched bridge here. DEPN.

SADDLEBACK Cumbr NY 3127. Name of a mountain rising to 2847ft. *Saddle-Back or Blenk-Arthur* 1704, *Saddleback* from 1769, *Sattleback* 1789. See also BLENCATHRA. Cu 253, lxxix.

SADDLEWORTH MOOR GMan SE 0305. P.n. Saddleworth SE 0106 + ModE **moor**. Saddleworth, *Sadelwrth* late 12th, *-worth* c.1230–1379, *Saddleworth* 1572, is 'saddle enclosure', OE **sadol** used to describe a topographical feature resembling a saddle + **worth**. The exact application of 'saddle' in this name is not clear. YW ii.310.

SADGILL Cumbr NY 4805. Partially uncertain. *Sadgill -gyl(l)* 1238–1823, *Satgill* 1283, before 1307, *Saggill* 1546. Probably ON **sát** 'lurking-place (for hunters)' + **gil** 'cleft, ravine' rather than a compound with **sǣtr** 'shieling' which should retain its *-r* (as in SATTERTHWAITE SD 3392). The reference is to the medieval technique of secreting the hunters in an ambush in a valley where they could shoot down the quarry as it was driven towards them by the beaters, cf. the vivid description of the stag hunt in *Sir Gawain and the Green Knight* 1146ff. The narrow upper reaches of Longsleddale would be an ideal site for such an ambush. We i.161, SSNNW 155.

SAFFRON WALDEN Essex TL 5438 → Saffron WALDEN.

SAHAM TONEY Norf TF 9002. 'S held by (Roger de) Toni' in 1199. *Saham Tony* 1498. Earlier simply *Saham* 1086, 1168, *Seham* 1130, 'lake homestead or estate', OE **sǣ**+ **hām**. There is still a lake in the village. DEPN.

SAIGHTON Ches SJ 4462. 'Willow-tree farm or enclosure'. *Saltone* 1086, *-ton(a)* c.1150–1252, *Salghton* 1188–1696 with variants *Salh- Salig(h)- Salg-*, *Sauton* 1208×26, *Saughton* 1579–1749, *Saighton* from 1579. OE **salh** + **tūn**. Che iv.121 gives pr ['seitən] which develops from lME **Sāton* from *Sauton*, cf. Nomina 13.109.

ST AGNES Corn SW 7150. 'St Agnes's (parish, church)'. *(parochia) Sancte Agnetis* 1327–47, *Agnette alias St Tannes* 1586, *St Agnes* 1599. Saint's name *Agnes* from the dedication of the church, a 4th cent. martyr at Rome. PNCo 49, Gover n.d. 362.

ST AGNES Scilly SV 8807. 'The pasture headland'. (Island of) *Aganas* [1193]16th, *Hagenes(se)* 1194, *Agnas* 1244, *St Agnes* c.1540. ON **hagi** 'an enclosure' + **nes** 'a headland'. The name was subsequently transferred from the headland to the whole island. Identical with a lost 'meadow called *Hagenesse*' 1213 in Lincolnshire. The prefix *St* is a late addition not normally used locally. PNCo 49.

ST AGNES HEAD Corn SW 6951. *St Agnes Head* c.1870. P.n. ST AGNES SW 7150 + ModE **head** 'a headland'. PNCo 49.

ST ALBAN'S OR ST ALDHELM'S HEAD Dorset SY 9675. *St Alban's or St Aldhelm's Head* 1811. Earlier *the foreland of Seynt Aldem'* c.1500, *Aldelmes Point* 1535×43. Saint's name *Aldhelm* + ModE **head** 'a headland'. The reference is to St Aldhelm's chapel, a late 12th cent. building which stands on the promontory. Do i.65, Newman-Pevsner 358.

ST ALBANS Herts TL 1507. Short for 'St Alban's church or holy place and town'. *æcclesia Sancti Albani* [795 for 792]13th S 138, *(æt, into) Sancte Albane* [1020]12th S 957, [1002×5]13th S 1488, c.1150 ASC(E) under year 1116, *Sancte Albanes stow* 1007, *(æt) Sc̄e Albanes stowe* 12th ASC(H) under year 1114, *villa Sancti Albani* 1086, *la ville de Seint Alban* 1307×27, *Seint Auban* 1400, *Seynt Albones* 1421. Saint's name *Alban* from the shrine of the 3rd cent. British proto-martyr executed here and the abbey subsequently dedicated to him. For the RBrit name see VERULAMIUM TL 1307. Hrt 86, Stōw 191, ODS 11.

ST ALDHELM'S OR ST ALBAN'S HEAD Dorset SY 9675 → ST ALBAN'S OR ST ALDHELM'S HEAD.

ST ALLEN Corn SW 8250. 'St Alun's (church)'. *(ecclesia) Sancti Alluni* 1261, *Seynt Alun* 1270, *Sct Allan* 1610, *Eglosellan* 1840. Saint's name *Alun*, of whom nothing is known, from the dedication of the church. The 1840 form is a field name at the churchtown earlier recorded as *Eglossalan* 1235 Gover n.d., *Eglosalon* 1302, 1322, Co **eglos** 'church'. PNCo 50.

ST ANNE'S Lancs SD 3128. St Anne's parish in LYTHAM SD 3627. St Anne's church was built in 1873. Pevsner 1969.174.

ST ANN'S CHAPEL Corn SX 4170. (Chapel of) *Sancta Anna* 1500, *Sent Anne is Chapell* 1541. Saint's name *Anne*, the mother of the Virgin Mary, from the dedication of the chapel, + ModE **chapel**. PNCo 50.

ST ANN'S CHAPEL Devon SX 6647. *St Anns Chapel* 1809 OS.

ST ANTHONY-IN-MENEAGE Corn SW 7825. 'St Anthony's church in Meneage'. *(ecclesia) Sancti Antonini in Manahec'* 1269, *St Antony* 1522, *Sct Anton* 1610. Properly saint's name

Entenin, a Cornish king and martyr (Latinised as *Antoninus*) from the dedication of the church + district name MENEAGE for distinction from St Anthony in Roseland, *ecclesia Sancti Antonini regis et martyris* late 12th. The name Antoninus was popularly interpreted as Anthony. There is a well dedicated to him at Ventoninny in Probus (Co **fenten**). The site there was also known as *Lanyntenyn* 1344, 'church-site of Entenin', OCo ***lann**. PNCo 51, Gover n.d. 541, CMCS 12.45.

ST AUSTELL Corn SX 0152. 'St Austell's (church)'. *Ecclesiam de Austol'* [c.1150]13th, *(ecclesia) Sancti Austoli* [1169]1235, *Seint Austele* 1384 Gover n.d., *Austill* 1610, *St Tossell* 1654 Gover n.d. Saint's name *Austell* (Austol) from the dedication of the church. He is believed to have been a follower of St Mewan and is also honoured at St Méen-le-Grand in Brittany and at Llanawstl in Machen parish, Mid Glamorgan ST 2189. PNCo 51, Gover n.d. 380, CMCS 12.59.

ST AUSTELL BAY Corn SX 0650. *St Austell Bay* 1813. P.n. ST AUSTELL SX 0152 + ModE **bay**. PNCo 51.

ST BEES Cumbr NX 9711. 'St Bega's (church)'. *eccl. Sce bege* c.1135–1279, *Sancta Bega* 13th, *Seynt Beys* 1434, *St Bees* 1578. Saint's n. *Bega* or *Begu* from the dedication of the church. St Bega is a legendary 7th cent. nun, St Begu an Anglo-Saxon nun of Hackness mentioned by Bede: the two have been confused here. The ON name of the place was *Cherchebi* c.1125, *Kirkeby* c.1125–1404, ON **kirkjubýr** 'village with a church'. This was compounded with the Irish diminutive form of the saint's name, *Beghóc*, to give *Kirkebibeccoch* [1189×99]1308, *Kirkebybeghog* 1331, *Kirbye Beacock* 1593 and *Bechockirke* c.1210, *Beghokirk(e)* 14th, *Bughokirke* 1540 with many other variant spellings. Cu 430.

ST BEES HEAD Cumbr NX 9413. 'Headland by St Bees'. *saincte bees hede* 1523, *the Barugh or St Bees head* 1777. P.n. ST BEES NX 9711 + ModE **head**. The earlier name for the headland was simply ON **berg** 'the hill', *Berh* 1261, *lez berghe* 1496, surviving in the 1777 spelling *the Barugh*. Cu 430.

ST BLAZEY Corn SX 0654. 'St Blaise's (chapel)'. *(capella) Sancti Blasii* 1440, *Seynt Blasy* 1525, *S.Blais* 1610. Saint's name *Blasius* (Blaise) from the dedication of the church. He was a 4th cent. Armenian martyr invoked for afflictions of the throat and also the patron-saint of wool-combers, having been tortured with wool-combs. Also known as *Landrait* [1169]1235, *Landrayth* 1284, 'church-site on a strand', OCo ***lann** + **trait**. The estuary has since become silted up and built upon and the tide no longer reaches the church. PNCo 54, Gover n.d. 389.

ST BREOCK Corn SW 9771. 'St Brioc's (church)'. *(ecclesia) Sancti Brioci* 1259, 1310, ~ ~ *Breoci* 1362, *Breok* 1522, *Breage* 1610. Saint's name *Brioc* (Brieuc) from the dedication of the church. Born in Cardiganshire, educated by St German of Paris, he lived and worked in Brittany. There are dedications in S Wales and in Brittany including Saint-Brieuc. Also known as *Lansant* 1259, *Nansant* 1291, *Nanssent* 1335, *Nancent* 1841, 'valley of Sant or Sent', Co **nans** + pers.n. *Sant* or *Sent*, or 'holy valley' or 'valley of saints', Co **nans** + **sans** 'holy' or ***sens** 'saints'. PNCo 59, Gover n.d. 313.

ST BREOCK DOWNS Corn SW 9668. *St Breock Down* 1813. P.n. ST BREOCK SW 9971 + ModE **down(s)**. PNCo 59.

ST BREWARD Corn SX 0976. 'St Breward's (church)'. *(ecclesia) Sancti Brewveredi* c.1190, *Seynt Brewerd* 1380, *Semerwert* 1406, *Symon ward* c.1535, *Bruard* 1610. Saint's name *Branwalader* (Breward) from the dedication of the church. His cult was imported from Brittany to Jersey where he gave his name to St Brelade, and places in Wessex. PNCo 59, Gover n.d. 103.

ST BRIAVELS Glos SO 5604. *(castellum, villa de) Sco' Briau- Briavel(l) -ello -elli* 1130–1535, *St, Saynt, Seynt Brev- Breuel(l)* 1269–1490, ~ *Brevellys -is -(e)s* 1483–1642, *þe Castell' called Symme Revelles* 15th. Welsh saint's name *Briafel*, the baptismal name of St Brioc or Brieuc. Gl iii.242 Jnl 3.49.

SAINTBURY Glos SP 1139. 'Sæwine's fortified place'. *Svineberie* (sic) 1086, *Sei- Seynesberia -bur(ia) -bury* 1186–1501, *Sei- Seynebur(ia) -bir' -bury* 1220–1303, *Seynbury* 15th cent., *Seyntbury* 1570, *Sayn(t)bury* 1621, *St Berry* 1675, *Saintbury* 1707. OE pers.n. *Sǣwine*, genitive sing. *Sǣwines*, + **byriġ**, dative sing. of **burh**. Gl i.256.

ST BUDEAUX Devon SX 4558. 'St Budoc's (parish)'. *Seynt Bodokkys* 1520, *St Budox* 1624, *Saint Budeaux or Saint Buddox* 1796. Celtic saint's name *Budoc* as in BUDOC WATER Corn SW 7832. Earlier known as *Bvcheside* 1086, *Buddekeshyde -hide* 1242, *Bodekesid(e) -hid* 1263, 1284, *Butshead al. Boxhead al. Budocoshide al. St Budeox* 1671, 'the hide of land of St Budoc', saint's name *Budoc*, secondary genitive sing. *Budeces*, + OE **hīd**, a name that survives as Budsheal SX 4560. D 236–7 where pr [bʌdəks] is given.

ST BURYAN Corn SW 4025. 'St Berion's (church)'. *(ecclesia) Sancte Beriane* 1220–1342 including [c.939]14th, *(canonici) S' Berrione*, *Eglosberrie* 1086, *Seint Beryan* 1343, *Burian* c.1450, *Sct Burien* 1610. Saint's name *Berion* (Buryan) from the dedication of the church. He was believed to have been Irish. Also honoured at Berrien, Finistère. The DB form *Eglosberrie -a* contains Co **eglos** 'a church'. PNCo 61 gives pr ['beriən], Gover n.d. 616, CMCS 12.48.

ST CATHERINE Avon ST 7770. *St Catherine* 1830 OS. Saint's name *Catherine* from the dedication of the medieval church. An estate of Bath abbey.

ST CATHERINE'S POINT IoW SZ 4975. Saint's name *Katherine* + ModE **point**, from St Catherine's Down, Hill and Tower, a one-time sea-mark and lighthouse on the site of a chapel dedicated to St Catherine on Chale Down SZ 4977, *(land called) Seynt Kateryns* c.1440, *St Katherine on Chale-downe* 1520, *St Katherins hill* 1583. Wt 116, Mills 1996.89.

ST CLEER Corn SX 2468. 'St Cleer's (church)'. *(ecclesia de) Sancto Claro* 1212, *Seintclere* 1388, *S.Cler* 1610. Saint's name *Clarus* (Cleer) from the dedication of the church. His main cult is at St Clair, Normandy. He is said to have been English, to have lived in NW France in the 9th cent., and to have died a martyr's death for chastity. PNCo 69, Gover n.d. 252.

ST CLEMENT Corn SX 8543. 'St Clement's (church)'. *(ecclesia) Sancti Clementis* 1329, *Clemens* 1464, 1522, *St Clements* c.1530. Saint's name *Clemens*, a 1st cent. pope, from the dedication of the church. According to legend he was exiled to the Crimea where he was tied to an anchor and thrown into the sea. As a result maritime sites are often dedicated to him and he is the patron saint of Trinity House, the organisation in charge of lighthouses. Also known as Moresk, *Moireis* 1086, (church of) *Moreis* [1178]1523, *Moresc* 1261, *Moresk(e)* 1329–1422, of unknown meaning, possibly Fr *marais* 'a marsh'. PNCo 69, Gover n.d. 434.

ST CLEMENT'S ISLE Corn SW 4726. (An islet and a chapel of) *St Clementes* c.1540, *St Clement Isle* 1610. Saint's name *Clemens* (Clement) + ModE **isle**. For the significance of the saint's name see ST CLEMENT SW 8543. Also called *Moushole Ile* 1587 from MOUSEHOLE SW 4626. PNCo 69, Gover n.d. 654.

ST CLETHER Corn SX 2084. 'St Cleder's (church)'. *Seyncleder*, *(ecclesia) Sancti Clederi* 1249–1422, *Seyntclether* 1405, *Clethor* 1610. Saint's name *Cleder* from the dedication of the church. He was believed to have been one of the 24 offspring of King Broccan of Breconshire. He set up a monastic community in SW Wales and retired to Cornwall. His name also occurs at Cleder, Brittany, and at Ventonglidder SW 9049, 'holy well of St Cleder', Co **fenten** + pers.n. *Cleder* with lenition of *c* to *g*. PNCo 70, Gover n.d. 50.

ST COLUMB MAJOR Corn SW 9163. 'Great St Columba's (church)'. *(parochia, ecclesia) Sancte Columbe major'* 1335, 1342 Gover n.d., *Great Seynt Columbe* 1478, *St Cullumb Major* 1694. Also known as *Overaseyntcolumbe* 1427, *Saint Colombe the Over* c.1547. Earlier simply *(ecclesia) Sancte Columbe* c.1240. Saint's name (female) *Columba*, of whom nothing is known except her reputed martyrdom at Ruthvoes. 'Major' or 'Great' for distinction from the adjoining parish of St COLUMB MINOR SW 8462. PNCo 71, Gover n.d. 319.

ST COLUMB MINOR Corn SW 8462. 'Lesser St Columb's (chapel)'. *(capella) Sancte Columbe Minoris* 1284, 1314, *St Columbe the less* 1473, *St Cullum Minor* 1666. Also known as *Columbe the nether* 1549. Saint's name (female) *Columba*. 'Minor' or 'little' for distinction from neighbouring ST COLUMB MAJOR SW 9163. PNCo 71, Gover n.d. 325.

ST COLUMB ROAD Corn SW 9159. *St Columb Road Station* 1888. P.n. St Columb as in ST COLOMB MAJOR SW 9163 + ModE **road**. A 19th and 20th cent. village that grew up around a railway station at the road to Columb Major. PNCo 7.

ST DAY Corn SW 7342. 'St Day's'. *Seyntdeye* 1351, *Sendey* 1398, *Sent Day* c.1510. Saint's name *Day*, a saint widely honoured in Brittany, but not known to have been venerated here where the medieval church was dedicated to the Holy Trinity. PNCo 77 gives former pr [sn dai] (1949), Gover n.d. 513.

ST DENNIS Corn SW 9557. 'St Denys's (parish, church)'. *(parochia) Sancti Dionisii* 1327, *Seynt Denys* 1380, 1436, *Sct Denys* 1610. Saint's name *Denys* from the dedication of the church, a 3rd cent. martyr at Paris. It has been suggested, however, that this is really derived from Co ***dynas** 'a fort': the church occupies a hill-top and is surrounded by a fort-like structure. PNCo 78, Gover n.d. 390, CoArch 4.31–5, Pevsner 1951.148.

ST DOMINICK Corn SX 4067. 'St Dominica's (church)'. *(ecclesia) Sancte Dominice* 1263–1344, ~ *Sancti Dominici* 1355, *Seynt Domenike* 1375. Saint's name (female) *Dominica* from the dedication of the church. Nothing is known of her. PNCo 79, Gover n.d. 184.

ST ENDELLION Corn SW 9978. 'St Endilient's (church)'. *(ecclesia) Sancte Endeliente* 1260, ~ *de Sancto Endeliente* 1269, 1302, *Endelient* 1439, *Endelyn* 1522, *Delyn* 1543, *St Endellion* 1610. Saint's name *Endilient* from the dedication of the church. She was believed to be a daughter of the Welsh king Broccan, and according to legend a god-daughter of King Arthur who helped her when a local lord killed her cow, the site of her church being determined after her death as the place where the oxen drawing her body stopped. Another chapel dedicated to her existed on Lundy Island. PNCo 81, Gover n.d. 112.

ST ENODER Corn SW 8956. 'St Enuder's (church)'. *Heglosenvder* 1086, *(ecclesia) Sancti Enodori* 1270, *Enoder* 1522, *Sct Enedon* 1610. Saint's name *Enuder* from the dedication of the church. Nothing is known of him. There is said to have been a lost monastery of *Lan-Tinidor* at Landerneau, Finistère. PNCo 81, Gover n.d. 330.

ST ERME Corn SW 8449. 'St Hermes' (church)'. *(ecclesia) Sancti Hermetis* [1250]14th, *Egloserm* 1345, *Seynt Erme* 1456, *St Herme* 1592. Saint's name *Hermes* from the dedication of the church. Hermes was a 3rd cent. Roman martyr, also venerated at ST ERVAN SW 8970 and formerly at a chapel in Marazion. The 1345 form contains Co **eglos** 'a church' and survives in Egloserme, the name of the churchtown farm. PNCo 82.

ST ERTH Corn SW 5535. 'St Ergh's (church)'. (Vicarage of) *Sanctus Ercius* c.1270, *(ecclesia de) Sancto Ercho* 1307, *Seynterghe* 1332, *Saincte Erthes* 1566. Saint's name *Ergh* (Irish *Erc*) from the dedication of the church. He was believed to have been a brother of St Euny venerated at Lelant SW 5437 and Redruth SS 6941; alternatively identical with the Irish St Erc, bishop of Slane and pupil of St Patrick. The p.n. shows substitution of [θ] for [γ]. Also known as *Lanuthinoch* 1204, *Lannutheno* 1233, *Lanuthno* 1269, OCo ***lann** 'church-site' + unknown element as in PERRANUTHNOE SW 5329 probably taken as the saint's name *Euny*. PNCo 82, Gover n.d. 592.

ST ERTH PRAZE Corn SW 5736. P.n. ST ERTH SW 5535 + p.n. *Praze* 1748, Co **pras** 'a meadow, a pasture'. PNCo 82, Gover n.d. 594

ST ERVAN Corn SW 8970. 'St Hermes' (parish, church)'. (Richard) *de Sancto Hermete* c.1210, *Seint Erven* 1397, *St Eruan* 1584, 1610. Saint's name *Hermes* from the dedication of the church as in ST ERME SW 8449. However, there may have been a Celtic saint of this name, cf. Breton St Hervé, *Haerveu* 6th, and W St Erven recorded at Llangwm, Gwent, in the 12th cent., and another Breton St Erven. PNCo 82, CMCS 12.49.

ST EVAL Corn SW 8769. 'St Uval's (church)'. *(ecclesia de) Sancto Uvele* 1260, ~ *Uvelo* 1291, ~ *Sancti Uvely* 1342, 1348, *St Eval* 1525. Saint's name *Uvel* from the dedication of the church. Nothing is known of this saint who was also honoured in Brittany. PNCo 82 gives pr [evəl] instead of the expected (but unacceptable) [i:vəl], Gover n.d. 336.

ST EWE Corn SW 9746. 'St Euai's (church)'. *Sancta Ewa* 1282, *Saynthuwa* 1303, *Seynt Ewe* 1413, *St Tewe alias St Ewe* 1610. Saint's name *Euai* (feminine) from the dedication of the church. Nothing is known of this saint. Cf. Lanuah, the name of the church-town farm, *Lanewa* 1302, OCo ***lann** 'church-site' + saint's name *Euai*. PNCo 82, Gover n.d. 392, CMCS 12.61.

ST GENNYS Corn SX 1497. 'St Genesius's (church)'. *Sanguinas* (with *i* suprascript over the *a* of *San*), *Sanwinas* 1086, *(ecclesia) Sancti Genesii* [c.1160]15th, *St Ginnes* 1244, *Gennows* 1610, *St. Genny's* 1903. Saint's name *Genesius* from the dedication of the church. This could be one of several saints, possibly the 3rd cent. martyr of Arles, possibly an earlier local saint, cf. W St Gwynys, patron saint of Gwnnws, Dyfed SH 3441. PNCo 86 gives pr ['ginis], Gover n.d. 59.

ST GEORGE'S OR LOOE ISLAND Corn SX 2551 → LOOE OR ST GEORGE'S ISLAND.

ST GEORGES Avon ST 6273. A district of Bristol. *St Georges* 1817 OS.

ST GERMANS Corn SX 3657. 'St German's (church)'. *Sanctus Germanus* c.950, 994–1289, *(æccl'a* for *ecclesia) S' Germani* 1086, *Synt Germayn* 1328, *Seynt Germyn* 1440, *St Jermyn* 1539. Saint's name *Germanus* from the dedication of the church, either St Germanus of Auxerre or an unknown local saint. Formerly known also as Lannaled, *Lannaledensis*, *Lanaletensis* 10th, OCo ***lann** 'church-site' + unknown element, possibly an unexplained district name like Allett in Kenwyn SW 7948 or Aleth in Ile-et-Verlaine, Brittany, or a r.n. as in Afon Aled, Clwyd, SH 9570. PNCo 87, Gover n.d. 217.

ST GIDGEY Corn SW 9469 → ST ISSEY SW 9271.

ST GILES IN THE WOOD Devon SS 5318. *Seint Giles* 1501. Earlier *Stow St Giles* 1330 and *Ecclesia Sancti Egidii* 1379, 'holy place, ME **stow** < OE **stōw**, and church, Latin **ecclesia**, of St Giles'. Saint's name *Giles*, Latin *Egidius*, genitive sing. *Egidii*, from the dedication of the church. 'In the wood' is a modern addition for distinction from ST GILES ON THE HEATH SX 3690. D 119.

ST GILES ON THE HEATH Devon SX 3690. *St Gylses in le Hethe* 1585. Earlier simply *capella Sancti Egidii* [1202]14th, 1291, and *parochia Sancti Egidii* 1532, 'chapel and parish of St Giles'. Saint's name *Giles*, Latin *Egidius*, from the dedication of the church. 'On the heath' for distinction from ST GILES IN THE WOOD SS 5318. D 164.

ST HELEN'S Scilly SV 9017. Short for St Helen's island. *St Helene* 1564, *Sainte Ellens Isle* 1570. Saint's name *Helen*, possibly a corruption of the dedication of a ruined chapel of St Illid or Elidius on the island which may originally have been (the island of) *Sanctus Elidius* [c.1160]15th, (the island of) *Seynt Lyda* 1478, *Saynct Lides Isle* 1540. St Helen's may, however, have been a separate place. PNCo 95.

ST HELENS IoW SZ 6288. *S(ancta) Elena* (Latin) 1154×89–1462, *Sancte Elene* (English) 1248–1503, *St Helens* from 1278 with variants *Seyntelenes* 1418, *Seint Ellyns* 1544. Saint's name *Helena* from the dedication of the former Cluniac priory church only the tower of which remains, the rest having fallen into the sea. The original name of the place was *Edyneton* 1104, 1287×90, *Edeme- Ed(e)neton* 1346–1413×22, *Ed(d)ington* 1561, 'Eadwynn or Eadwine's estate', OE feminine pers.n. *Ēadwynn*, genitive sing. *Ēadwynne*, + **tūn**, or masc. *Ēadwine* + **ing**[4] + **tūn**. The DB spelling *Etharin* 1086 may represent OE *æt harum* 'at the rocks', OE **æt** + ***hær**, dative pl. ***harum**. Wt 195, 197.

ST HELENS Mers SJ 5095. Short for St Helen's chapel. *Sct Elyus chap.* (for *Elyns*) 1577. A medieval chapel of ease. The present industrial town began its development with coal-mining in the 17th cent., succeeded by glass-making in 1773. La 110, Room 1983.107.

ST HILARY Corn SW 5531. 'St Hilary's (church)'. *(ecclesia) Sancti Hilarii* [1178]1523, *Sanctus Hillarius* [c.1200]14th, *Hillary* 1524, *St Ilarye* 1581. Saint's name *Hilary*, Latin *Hilarius* from the dedication of the church, referring to St Hilary of Poitiers, a 4th century bishop and opponent of the Arian heresy. PNCo 97, Gover n.d. 601.

SAINT HILL WSusx TQ 3835. Possibly 'the singed or burnt hill'. *Saynt Hill* 1568. OE ***senġet** + **hyll**. Sx 333, PNE ii.118.

ST IPPOLLITTS Herts TL 1927. *Seynt Ipollitts* 1518–29, earlier *Polytes* 1412, *Polletts* 1475 and *Polledge* 1556. The earliest references are to the church of *S. Ypollitus* 1283 and the *villa de Sc̄o Ipolito* 1342. The church, dating from the 11th cent., is dedicated to St Hippolytus, a Roman priest, apologist and martyr of the 3rd cent. He was early confused with another martyr of the same name whose fate was to be torn apart by wild horses. This is one of only two such dedications in England. Sick horses were once brought to the shrine here through the N door of the church. The saint's name means 'loosed horse'. Hrt 13, Saints 232.

ST ISSEY Corn SW 9271. 'St Iti's (church)'. *(in) Sancto Ydi* 1195, *Seint Idde* 1189×99, *(ecclesia) Sancte Ide* 1257×80, *Seintysy* 1362, *St Issie* 1610. Saint's name *Ida* or *Ita* (OCo *Iti*) from the dedication of the church, said to have been one of the 24 sons and daughters of the W king Broccan of Breconshire. His (or her) name also appears in MEVAGISSEY SX 0145, in the Breton p.ns. Plouisy (Breton *ploue* 'parish'), Lannidy and Trévidy, and in the hamlet of St Jidjey or Gidgey SW 9469, *St. Jidjey* 1272×1316, a false analysis of Co *sans-Ysy* as if *San-Sysy*. Palatalisation of *d* in medial position to [dʒ] and eventually to [z] is a normal sound change in MCo. PNCo 99, DEPN, CMCS 12.62.

ST IVE Corn SX 3067. 'St Ivo's (church)'. *(eeclesia de) Sancto Ivone* 1201, 1256 Gover n.d., ~ *Sancti Ivonis* 1291, 1342, *Seynt Ive* 1390. Saint's name *Ivo* from the dedication of the church, believed to have been a Persian bishop who came to convert the pagan English. His cult centre was at St Ives Cambs TL 3171. PNCo 100 gives pr [iːv], Gover n.d. 186, DEPN.

ST IVES Cambs TL 3171. 'St Ivo's (resting place)'. *S. Yvo de Slepe* [1110]14th, 1130, *villa S. Yuonis* 1200, *St Ive* 1485. Saint's name *Ivo* or *Ives*. One of four bodies discovered here in 1001 was declared, following a peasant's dream, to be that of a Persian bishop Ivo or Ives, whose remains were taken to Ramsey but subsequently returned. The original name of the village was Slepe, *Slepe* from 1086†, 'slippery place, portage', OE ***slǣp**. Hu 221, 222, Studies 1936.186, PNE ii.127, Saints 247.

ST IVES Corn SW 5140. 'St Ya's (church)'. *(juxta) Sanctam Yam* 1284, *(parochia) Sancte Ye* 1327, 1333 Gover n.d., *Sent Ia* 1468, *Seint Ive* 1346, *Seint Ithe* 1347, *Seynt Yves* 1579, *St Ies* 1602, *Sct.Ithes* 1610. Saint's name *Ya* (Ia) from the dedication of the church. She was believed to have been an Irish saint, sister of SS Euny and Erth. She also occurs in the Breton p.n. Plouyé, *Ploehie* 1337, 'the parish of St Ie' (Breton *ploue*). The Co name of the town was *Porthye* 1284, 1331, *Porthia* 1291–1356, 'harbour of St Ya', Co **porth**. The form with [v] is late and may be due to the influence of St IVE SX 3067 and ST IVES Cambs TL 3171. PNCo 100, Gover n.d. 625, Dauzat-Rostaing 535.

ST IVES Dorset SU 1204. *St Ives* 1811 OS. This name has nothing to do with a saint, as the earlier forms show, *Iuez* 1167, *le Yuez* 1187, *(in) Yvetis* 1212, *Yvettis* 1250, representing OE ***īfet(t)** 'ivy-grown copse' from **īfiġ**. The 13th cent. spellings look like Latinised forms in the dative pl. *Saint* is in fact a late addition partly through association with nearby ST LEONARDS SU 1002 and partly through association with the Cambridge and Cornwall St Ives. Do 128, Ha 143.

ST IVES BAY Corn SW 5441. *roda* of *Sancta Ya* 1284, *Seint Ive Baye* 1346. P.n. ST IVES SW 5140 + ME **rode** 'a roadstead' and **baye**. PNCo 100, Gover n.d. 628.

ST JIDGEY Corn SW 9469 → ST ISSEY SW 9271.

ST JOHN Corn SX 4053. 'St John's (church)'. *(ecclesia) Sancti Johannis* [c.1160]13th, 1320–95, *Seynt Johan* 1372, *Sct. Johns* 1610. Saint's name *John* (the Baptist) from the dedication of the church. PNCo 100, Gover n.d. 225.

ST JOHNS H&W SO 8454. *St John's* 1832 OS. Saint's name *John* from the dedication of the 12th cent. church of St John in Bedwardine, 'Bede's enclosure', *Bedewordine* 1235, *Bed(e)wardyn* 1327–1501, OE pers.n. *Bēda* + **worthiġn**. Wo 89, Pevsner 1968.319.

ST JOHN'S BECK Cumbr NY 3121. *St John's Beck* 1868. The beck is named from the chapelry of St John's in the Vale through which it runs, *chappell of Seynte John* 1554. Cu 311.

ST JOHN'S CHAPEL Durham NY 8838. 'Chapel of St John'. *Sct Johns Chap.* 1576, *St John's Chappell in Werdale* 1678. The church of St John and its village are mentioned in 1335, *ecclesia S. Johannis cum villa sua*. NbDu 171.

ST JOHN'S FEN END Norf TF 5311. In 1824 OS this is *Terrington Fen End*. Both names are from TERRINGTON ST JOHN TF 5314.

ST JOHN'S HALL Durham NZ 0634. *St John's* 1862 OS. A detached portion of Stanhope parish presumably associated with ST JOHN'S CHAPEL NY 8838.

ST JOHN'S HIGHWAY Norf TF 5214. *St Johns Highway* 1824 OS. Saint's name *John* as in TERRINGTON ST JOHN TF 5314 + ModE **highway** as also in WALPOLE HIGHWAY TF 5113 and WALTON HIGHWAY TF 4912, all on the road from Wisbech to King's Lynn.

ST JUST Corn SW 3631. 'St Iust's (church)'. *(ecclesia) Sancti Justi* 1291, ~ ~ *in Penewith* 1297 Gover n.d., *Seint Just* 1342, *Yust(e)* 1342, 1524, *St Just alias St Towst* 1581. Saint's name *Iust* from the dedication of the church. Nothing is known of this saint. See also PENWITH, GORRAN HAVEN and ST JUST IN ROSELAND SW 8535. Salmon 138 gives pr [sən'tuːst]. PNCo 100, Gover n.d. 629, DEPN, CMCVS 12.9.

ST JUST IN ROSELAND Corn SW 8535. 'St Iust's church in Roseland'. *(ecclesia) Sancti Justi* [c.1070]17th, 1383, ~ *de Sancto Justo* 1202, *Sancti Justi in Ros* 1259, ~ ~ ~ *Roslonde* 1282, *Seynt Just en Rosland* c.1398, *St Ewest alias St Juste* 1578. Saint's name *Iust* from the dedication of the church as in ST JUST SW 3631. Still popularly pronounced *Yust* in 1887, a pronunciation which survived in *Portheast*, the Co name of GORRAN HAVEN. 'In Roseland' for distinction from ST JUST SW 3631. See ROSELAND. PNCo 101, Gover n.d. 453.

ST KEVERNE Corn SW 7921. 'St Achobran's (church)'. *(canonici) Sancti Achebranni* 1086, *in Sancto Akeverano* 1201, *St Kaveran* 1236, *Seynt Keveran* 1339, *Keyran* 1535, *Saint Kyerane* 1553, *Sct. Keuern* 1610. Saint's name *Achobran* (Keverne) from the dedication of the church. Nothing is known of this saint outside Cornwall. The Co name of the churchtown was *Lannachebran* 1086, *Lanheveryn* c.1500, *Lanhevran* 1504, Co ***lann** 'church-site' + saint's name *Achobran*. PNCo 102, Gover n.d. 551, DEPN, CMCS 12.47.

†The form *Slaepi* [681]12th S 1168 probably refers to Islip, Oxon, rather than Slepe. *Slepa* [974]14th S 798 does belong here but the charter is a forgery.

ST KEW Corn SX 0276. 'St Kew's (church)'. *Sancta Cypa* (with *p* for þ (*w*)) [961×3]14th S 810, *Sanctus Cheus* 1086, *Seint Kewe* 1373. Saint's name *Kew* (Ciwa) from the dedication of the church, as in the Breton p.n. Kew, *Caio* n.d. An alternative name for the churchtown survives in Lanow SX 0277, (monastery called) *Docco* [7th]11th, *Lanehoc* 1086, *Lantloho, Lannohoo* 1086 Exon, *Landeho* 1261, *Lanho* 1185–95, *Lannou* 1342, 1373, 'church-site of Docco', Co ***lann** + saint's name Docco (Doghow), also known as *Lanhoghow seynt* 1284, 'Lanow of the saints', OCo **sant**, pl. **seynt**. St Doghow and St Kew, joint patron saints of this place, were believed to be brother and sister who came to Cornwall from Gwent in SE Wales. PNCo 103, Gover n.d. 117, DEPN.

ST KEW HIGHWAY Corn SX 0375. *Highway* 1699. A hamlet in Kew parish on the main road from Camelford to Wadebridge, formerly a halt on the railway line. PNCo 103.

ST KEYNE Corn SX 2461. 'St Keyne's (church)'. *Ecclesia Sancte Keyne* 1291, *Seynt Kayn* 1525, *Cayne* 1610. Saint's name *Keyne* or *Kayna* from the dedication of the church. She was believed to be a daughter of King Broccan in Breconshire and is commemorated in Southey's ballad of *The Well of St Keyne* (1798):

If the husband of this gifted Well
Shall drink before his Wife,
A happy man henceforth is he,
For he shall be master of life.
But if the Wife should drink it first . . .

Another possible occurrence of her name is Llangeinor 'church of St Keyne', Mid Glamorgan SS 9187. PNCo 103 gives pr [kein].

ST LAWRENCE Corn SX 0466. 'St Lawrence's (hospital)'. *Hospitalis Sancti Laurencii extra Bodminiam* 1374, *Seyntlaurence* 1444. Saint's name *Lawrence*, a 3rd-cent. Roman martyr, from the dedication of the medieval leper hospital formerly at this place. PNCo 108, Gover n.d. 102.

ST LAWRENCE Essex TL 9604. 'St Lawrence's vill, St Lawrence's Dengie'. *villa Sancti Laurentii* 1235, *Seint Lorenz* 1254, *St Laurence Danesie* 1334. Saint's name *Lawrence* from the dedication of the church + Latin **villa** or p.n. DENGIE TL 9801. The original name was *Niuuelandā, Niwelant* 1086, *La Newe(n)lond(e)* 1303, 1346, 'the new (arable) land', OE **nīwe**, dative sing. definite form **nīwan**, + **land**. Ess 224, L 246.

ST LAWRENCE IoW SZ 5376. Short for 'St Lawrence's Wathe'. *Southwade Sancti Laurencii* 1287×90, *St Laurence* 1235, ~ ~ *de Wathe* 1340, *S. Lawrence* 1591. Saint's name *Lawrence* from the dedication of the church. The original name of the place was *Wathe* 1311–1632, 'the ford', OE **(ġe)wæd**, divided into *Underwathe* 'lower W' 1250×60, *Southwathe* 1287×90 and *Stoureswath* 'W belonging to the Estur family' of Gatcombe 1292. There was formerly a small stream flowing through the village. Wt 201, Mills 1996.90.

ST LEONARDS Bucks SP 9107. Short for 'St Leonard's chapel'. *capella Sancti Leonardi* [1187]14th, *capella Sancti Leonardi de Blakemere* 1250. Saint's name *Leonard* from the dedication of the chapel at a place originally called 'Black mere', OE **blæc** + **mere**. Bk 144, Jnl 2.25.

ST LEONARDS Dorset SU 1002. *St Leonards* 1811 OS. A chapel called *Sct. Leonarde* occurs in 1575. The origin lies in a medieval foundation called *domus Sancti Leonardi de Russeton* 1288, 'the house of St Leonard at *Russeton*', a lost p.n. meaning 'rush farm'. Do 128, Ha 143.

ST LEONARDS ESusx TQ 8009. Short for St Leonard's church or vill. *ecclesia Leonardi de Hastynges* 'church of Leonard of Hastings' 1279, *(villa de) Sco. Leonardo juxta Hasting* '(vill of) St Leonard's next Hastings' 1288, *Seynt Leonards* 1557. Saint's name *Leonard* from the dedication of the church. Sx 536.

ST LEONARD'S FOREST WSusx TQ 2231. *foresta S. Leonardi* 1213. Saint's name *Leonard* from the dedication of a chapel in the forest + OFr, ME **forest**. Sx 2.

ST LEVAN Corn SW 3822. 'St Salamun's (parish)'. *(Parochia) Sancti Silvani* 1327, *Seleven* 1523, *-an* 1545, *Sent Levane* 1569, *Sleuen* 1610, *Levan* 1657. Saint's name *Selevan* from the dedication of the church. This name represents OCo *Salamun* from Latin *Solomon* but was turned into *Silvanus* in official records. He also has dedications in Brittany at Seleven and Saloman. The local form *Sleven* has been wrongly analysed into S.Leven. Salmon 159 records a local pronounciation [slevən]. PNCo 109, Gover n.d. 638, CMCS 12.42.

ST MABYN Corn SX 0473. 'St Mabon's (church)'. *Sancta Mabena* c.1210, 1266, *(ecclesia de) Sancto Malbano* 1234, *(parochia) Sancti Maubani, Sent Maban* 1327, *Synt Mabyn* 1421. Saint's name *Mabon* from the dedication of the church. She was one of the 24 sons and daughters of King Broccan of Breconshire, usually regarded as female but sometimes treated as if masculine. PNCo 113, Gover n.d. 122.

ST MARGARETS H&W SO 3533. *St Margarets* 1825 OS. Saint's name *Margaret* from the dedication of the church, which dates back to the 12th cent. Pevsner 1963.284.

ST MARGARET'S AT CLIFFE Kent TR 3466. 'St Margaret's Atcliffe'. *Clyue scē Margarete* 1270, *St Margaret's at Clyffe* 1610. P.n. *Atcliffe* as in West CLIFFE TR 3444 + saint's name *Margaret* from the dedication of the church. Also referred to as either 'St Margaret', *S' Margarita* 1086, *Seinte Margerete* 1240, or simply *Cliue* c.1100. The original name *Atcliffe*, from OE preposition **æt** + **clife**, dative sing. of **clif**, has been misunderstood in the modern name as if a prepositional phrase qualifying the name St Margaret's; properly St Margaret's is added to *Atcliffe* as west is to West CLIFFE, originally *æt clife*, to distinguish the two vills. PNK 565.

ST MARGARET'S BAY Kent TR 3744. *St Margarets bay* 1710. Saint's name *Margaret* as in ST MARGARET'S AT CLIFFE TR 3644 + ModE **bay**. Cullen.

ST MARTIN. Saint's name from the dedication of the parish church to St *Martin* of Tours, a 4th cent. Gaulish bishop.

(1) ~ Corn SX 2655. *Martistowe* [c.1220]1320, *(ecclesia) Sancti Martini (de Lo)* 1282–1335, *St Marten by Loo* 1516, *S. Martyn* 1610. See also LOOE SX 2553. The earliest form is 'St Martin's holy place', saint's name + OE **stōw**. PNCo 116, Gover n.d. 279.

(2) ~ Corn SW 7323. *(capella) Sancti Martini* 1342, *Sent Martin yn Meneck* 1549, *S. Martyn* 1610. See also MENEAGE. There is also an alternative name for this place, *(parochia) Sancti Dydmini* 1327, *(ecclesia) Sancti Martini alias Dydemin* 1385. *Dydemin* is a completely unknown male saint. PNCo 116, Gover n.d. 567.

ST MARTIN'S Scilly SV 9215. Short for 'St Martin's island'. *Seynt Martyns, St Martines Isle* c.1540. Saint's name *Martin* from the dedication of the church. Earlier, probably, *Bechiek* (for *Bre-*) 1319, *Brethiek* 1336, *Brechiek* 1390, 'the place with arms, the island or district of promontories', Co **bregh** 'arm' + adjectival ending **-ek** referring to the shape of the island. PNCo 116.

ST MARTIN'S Shrops SJ 3236. Short for St Martin's Chapel or Church. *Capella de Sancto Martino* c.1222, c.1235, *Capella de Martineschirch'* early 13th, *Martinchirche* 1301, *St Martins* from 1562. Saint's name *Martin* from the dedication of the church + ME **chirche**. The W name is Llanfarthin, *Llanvarthin* 1495, W **llan** + saint's name *Martin* with lenition of *m* to *v*. Sa i.256, O'Donnell 106.

ST MARY IN THE MARSH Kent TR 0627. *Seyntemariecherche* 1240. Saint's name *Mary* from the dedication of the church, a Norman and later building in Romney Marsh. PNK 479, Newman 1969W.183.

ST MARY'S Scilly SV 9111. Short for St Mary's church, St Mary's island. (Church of) *Sancta Maria* of *Heumor'* (for *Hennor*) [c.1195]13th, (island of) *Sancta Maria* 1375, *Seynt Mary island* 1478. Saint's name *Mary* from the dedication of the church. *Hennor* represents the Co name of the single land-mass formed by northern Scilly before it sank and became subdivided, also recorded as (island of) *Ennore* [1193]16th, (castle of) *Enoer* 1306, (island called) *Enor* 1372, probably 'the ground, the land', OCo **en** + **doer** becoming *En-nor*. PNCo 116.

ST MARY'S BAY Kent TR 0927. Saint's name *Mary* as in ST MARY IN THE MARSH TR 0627 + ModE **bay**. A modern holiday resort.

ST MARY'S MARSHES Kent TQ 8078. *St Mary's Marsh* 1805 OS. Saint's name *Mary* as in St Mary HOO TQ 8076 + ModE **marsh**.

ST MARY'S or BAIT ISLAND T&W NZ 3575. *St Mary's or Bait Island* 1863 OS. Supposedly the site of a chantry chapel of St Mary. See BAIT ISLAND.

ST MARY'S or NEWTON HAVEN Northum NU 2424 → NEWTON OR ST MARY'S HAVEN NU 2424.

ST MARY'S SOUND Scilly SV 8909. *St Mary Sownd* c.1540. P.n. ST MARY'S SV 9111 + ModE **sound** 'a strait, a sailable channel'. PNCo 116.

ST MAWES Corn SW 8433. 'St Maudith's (village)'. *(villa) Sancti Maudeti* 1284, 1302, 1345, *Seint Mauduyt* 1342, *Seynt Maudys* 1318, *Seint Mausa* 1467, *St Maws* c.1540, *Sct. Moze* 1610. Saint's name *Maudith* (Maudez, Mawes) from the dedication of the church. An Irish saint widely venerated in Brittany, e.g. Ile Modez, St-Mandé, St-Mandé-sur-Brédoire, St-Maudez. The Co name of the town, *Lavousa, Lavausa* 1445, *La Vousa* c.1540, 'church-site of Mausa', Co ***lann** + saint's name, regularly shows MCo *s* for *d* and lenition of initial *m* to *v*. PNCo 116, Gover n.d. 453, Dauzat-Rostaing 613.

ST MAWGAN Corn SW 8765. 'St Maucan's (village, church)'. *(villa) Sancti Malgani* 1206 Gover n.d., *(ecclesia) Sancti Mauchani de Lanherno* 1257, *Seyntmogan* 1284, *(ecclesia) Sancti Maugani* 1291, *Mawgan* 1543, 1610. Saint's name *Maucan* as in MAWGAN SW 7025 from the dedication of the church. See also VALE OF MAWGAN OR LANHERNE SW 8765. PNCo 117, Gover n.d. 347, CMCS 12.47.

ST MELLION Corn SX 3865. 'St Mellion's (church)'. *(Robertus de) Sancto Melano* 1198–1280, *(ecclesia) Sancti Melani* 1291–1342, *St Melyn* 1544, 1610, *Millians* 1553. Saint's name *Melanius* (Melaine, Mellion) from the dedication of the church. He was a Breton or Frankish bishop of Rennes in the 6th cent., author of a letter threatening to excommunicate two British priests working in Brittany and trying to introduce insular Celtic practices. He was venerated also at St-Melaine, St-Melaine-sur-Aubance, St-Mélany and St-Meslin-du-Bosc. PNCo 118, Gover n.d. 191, DEPN, Dauzat-Rostaing 618.

ST MERRYN Corn SW 8873. 'St Merin's (church)'. *(Rector) Sancte Marine* 1259, *(ecclesia) Sancte Maryne* 1342, *Seynt Mer(r)yn* 1380–c.1520. Saint's name *Merin* or *Meren* from the dedication of the church, a Celtic saint honoured in Wales and Brittany, rather than *Marina*, an early saint of Asia Minor. Cf. the p.ns. Bodferin 'dwelling of Merin' (W *bod*) and Llanferin 'church site of Merin' (W *llan*). PNCo 120, DEPN.

ST MEWAN Corn SW 9951. 'St Mewin's (church)'. *Sancti Maweni* 1245, *(ecclesia) Sancti Mawani* 1291, ~ ~ *Mewani* 1297–1318, *Seynt Mewan* 1380, 1397, *Mowun* 1610. Saint's name *Mewinn* (OCo *Megunn*) from the dedication of the church, believed to have been a W aristocrat related to St Samson whom he accompanied to Brittany where he founded the monastery of St Méen Finistère. PNCo 121, Gover n.d. 351, CMCS 12.59.

ST MICHAEL. Saint's name *Michael* (the Archangel) from the dedication of the church.

(1) ~ CAERHAYS Corn SW 9642. *(capella) Sancti Michaelis de Karihaes* 1259, *S.Mich.Karyheys* 1400, *Michels* 1610. The Co name was *Lanvyhayll* 1473–8, Co ***lann** 'church-site' + saint's name with lenition of initial *m* to *v*. 'Caerhays', for distinction from ST MICHAEL PENKEVIL SW 8542, is the name of the manor house, Caerhayes Castle SW 9441, recorded in addition to the above forms as *Karieis, Caryheys* 1297, *Karihays* c.1300, *Kerihayes* 1313, *Caryhays* 1329, of unknown meaning and origin but possibly identical with the similarly unexplained Breton p.n. type Carhaix of which the most important is Carhaix-Plouger, Finistère, *caer Ahes* [1084]12th, *Karahes* 12th. PNCo 121, DEPN, Dauzat-Rostaing 148.

(2) ~ PENKEVIL Corn SW 8542. *(ecclesiam) Sancti Michaelis de Penkevel* 1261, 1264, *S.Michel* 1610. 'Penkevil', for distinction from ST MICHAEL CAERHAYS SW 9642, *Penkevel* c.1210–91 etc., is 'horse's head', Co **pen** + ***kevyl**. The reference is unknown, perhaps to a fancied resemblance of a piece of land to a horse's head, perhaps to the whole promontory between the Tresillan River and the River Fall. The name survives at Penkevel SW 8640. PNCo 122, Gover n.d. 473, DEPN.

ST MICHAELS H&W SO 5865. No early forms.

ST MICHAEL'S Kent TQ 8835. Short for 'St Michael's church', built 1862×3. This supersedes the earlier name Boars Isle, *Bu-Borwarsile* 1226–1313, *Burwardesile* 1253×4, *Borewardeshull'* 1292, *Bordes Isle* 1722, *Boresile* 1757, *Birds Isle* 1801, the 'shed' or 'miry place' of the citizens', OE **burhware**, genitive pl. **burhwara**, + either ***(ġe)sell** or **syle**. If this explanation is right the reference would be to the men of Canterbury lathe, the *Boruuar Lest* of DB, in direct contrast to the men of Thanet, the *Tenet-ware* who held nearby Tenterden. Newman 1969W.541, PNK 356, Cullen.

ST MICHAEL'S MOUNT Corn SW 5129. *Sanctus Michael juxta mare* [mid 11th]12th, *montem Sancti Michaelis de Cornubia* [c.1070]14th, *Mons Sancti Michaelis* 1169, *Mihœlesmunt* 1205, *Mount Mychell, le Mont Myghellmont* 1478, *Seynt Mychell Mount* 1479. Saint's name *Michael* (the Archangel) + OFr **mont, munt**. According to William of Worcester writing in 1478, St Michael was believed to have appeared here in AD 710. The Mount and its priory were given to Mont St Michel in France either by Edward the Confessor in c.1030, or by Robert, Count of Mortain, in c.1070. The Cornish names of the Mount, *Cora Clowse in Cowse* 1602, *Carrack Looes en Cooes* late 17th, 'grey rock in the wood', Co **karreck** + **los** + **an** + **cos**, are probably antiquarian coinages. PNCo 122, Gover n.d. 601, DEPN.

ST MICHAEL'S ON WYRE Lancs SD 4641. Short for 'St Michael's church'. *Michelescherche* 1086, *eccl. Sc̄i Mich' Sup' Wirū* c.1195, *Sancto Micaeli super Wiram* 1205, *Sainct Mihels* c.1540. Saint's name *Michael* from the dedication of the church + OE **ċiriċe**, + r.n. WYRE. La 160.

ST MINVER Corn SW 9677. 'St Menfre's (church)'. (Church of) *Sancta Menfreda* 1256, *Sancta Mynfreda* 1291, *Seynt Mynfre* 1374, *Minver* 1543, 1903, *Mynuer* 1610. Saint's name *Menfre* (Minver) from the dedication of the church, believed to have been one of the 24 children of King Broccan of Breconshire. Her name may also occur in Minwear Dyfed SN 0413. PNCo 123, DEPN.

ST NEOT Corn SX 1867. 'St Neot's (church)'. *(Clerici) S' Neoti* 1086, *Sanctus Neotus* 1201, *Senniet* [early 12th]14th, *Seynt Nyet* 1284, *Sct.Neot* 1610. Saint's name *Neot* from the dedication of the church. He was believed to have been a kinsman of King Alfred and to have been a monk at Glastonbury. His relics were later removed to St Neots in Cambs. There is also an alternative name for this place, *Neotes- Nietestov* 1086, *Neotestoce* [11th]12th, saint's name *Neot*, genitive sing. *Neotes*, + OE **stōw** and **stoc**, 'Neot's holy place'. PNCo 128, Gover n.d. 284, CMCS 12.50.

ST NEOT'S Cambs TL 1860. 'St Neot's (resting place)'. *S' Neod* 12th ASC(E) under year 1132, *villa S. Neoti* 1203, *St Nyot's* 1329, *Seynt Nedys* 1513. Saint's name *Neot*, a hermit monk d. c.877. He founded a small monastery at *Neotstoke*, St Neot, Cornwall, where he was buried. In 972×7 a monastery was founded at Eynesbury, Cambs, with monks from Thorney who 'by gift or theft' obtained his relics from Cornwall, all but an arm. Eynesbury was thereafter called St Neot's. Hu 265 gives prs [sənt ni:ts, snouts, sni:dz], Saints 351.

ST NEWLYN EAST Corn SW 8256. Properly simply Newlyn East, *Newlyn East* 1884, p.n. Newlyn + ModE **east** for distinction from NEWLYN SW 4628. Newlyn, 'St Niwelina's (church)', *(ecclesia) Sancte Niweline* 1259, ~ ~ *Neuline* 1264, *Seint Neulin* 1270, *Nulyn* 1543, *Newland* 1610, is from the patron saint of the church, St Niwelina, believed to have been martyred here by her father, a king. There is a Cornish form of the name with **eglos** 'a church', *Eglosnyulyn* 1415. PNCo 129, Gover n.d. 371, DEPN.

ST NICHOLAS AT WADE Kent TR 2666. 'St Nicholas's vill at Wade'. *Villa Sancti Nicolai* 1253×4–1287, *St Nicholas by Wade* 1456, ~ *at Wade* 1458. Saint's name *Nicholas* from the dedication of the church + p.n. Wade, OE **(ġe)wæd** 'a ford' referring to a crossing of the Wantsun at Sarre, very probably the ford mentioned as *on middel gewæd* 943 S 512 in a grant of land at an unidentified place called *Æt Miclan grafe* 'at the great grove' on Thanet. PNK 604.

ST OSYTH Essex TM 1215. *seynte Osithe* 1046, *Sancta Osida* 1187, *St Osith* 1280, 1302 etc., *Seint Osiez, Osyes, Oses, Osis* 1362–1538, *Toozy* 1532, *Sainte toosie -ei* 1602. Saint's name *Osith* (OE *Ōsgyth*) from the dedication of the priory church. Of obscure royal birth she is said to have married King Sighere of the East Saxons reigning in 664. She retired to a place called *Cicc* given her by Sighere, where she founded a nunnery, becoming first abbess and ultimately patron saint. The name St Osyth replaced earlier *Tit* (for *Cic*) 942 for 951, *Cicc* c.1000, c.1050, *Cice* 1086, 1123 ASC(E), *Chich(e)* 1044, 1198, 1218–60, *Chuch* 1216×72 which seems to be OE ***ċiċċ** 'the bend' referring to the shape of the creek here. Ess 347–8, PNE i.93, ODS 366, Jnl 2.45.

ST OSYTH MARSH Essex TM 1113. *St Osyth Marsh* 1805 OS. P.n. ST OSYTH TM 1215 + ModE **marsh**.

ST OWEN'S CROSS H&W SO 5424. The cross is marked as an antiquity OS 1831.

ST PANCRAS STATION GLond TQ 2982. P.n. St Pancras + ModE **station**, built in 1863–7. St Pancras, *(ad) Sanctum Pancratiū* 1086, *eccl. S. Pancratii* 'St Pancras's church' c.1183, *Parochia S. Pancratii extra Lond'* 'St Pancras's parish outside London' 1291, *St Pancras in the Fields* 1531, *Panrich, Pankeridge* 16th cent., takes its name from the dedication of the church. St Pancras was a martyr in the times of Diocletian. Mx 140 gives former pr [pæŋkridʒ], GL 84.

ST PETER'S Kent TR 3868. Short for 'St Peter's tithing or vill'. *borgha scī Petr'* 1254, *villa scī Petri* 1270. In 1292 a *Suthborgh'* and a *Northborgh' scī Petri* are distinguished. Saint's name *Peter* from the dedication of the church + ME **borgh** (OE *borg*), Latin **villa**. Now better known as BROADSTAIRS TR 3967. PNK 602, V i.129.

ST PETER'S FLAT Essex TM 0408. Cf. *S^t. Peters Sand* 1805. Saint's name *Peter* from nearby St *Peter's Chapel* + ModE **flat, sand** ' sand-bank'. *St Peter's Chapel* 1558, *S Peters chapell on the wall* 1594, and earlier without reference to the chapel *vill Sci Petri Attewalle* 'St Peter's vill at the wall' 1291, *Seynt Peters in Bradwell* 1571. This is the church built by St Cedd c. AD 654 upon the west wall of the Roman fort of OTHONA TM 0308, whence its earlier name *Chapel of la Vale* i.e. wall 1254 and *U-Wallam* 1123×33, 1163×83, *La Wall(e)* 1204–1428, OE **wall**. Ess 210, MM 16.75.

ST PINNOCK Corn SX 2063. 'St Pinnuh's (church)'. (Church of) *Sanctus Pynnocus* 1284, 1291, *Seint Pynnok* 1385, *Pinock* 1610. Saint's name *Pinnuh* from the dedication of the church, of whom nothing is known. PNCo 139.

ST RADEGUND'S ABBEY Kent TR 2741. *eccl. Scē Radegundis de Bradeshole* 1199, 'church of St Radegund at Bradsole'. The reference is to the ruins of a Premonstratensian abbey founded 1192×3 at a place called *Bradesole* 1204–34, 'the broad miry-pool'. OE **brād**, definite form **brāde**, + **sol**. PNK 456.

ST SAMPSON Corn SX 1254 → GOLANT.

ST STEPHEN Corn SW 9453. 'St Stephen's (church)'. *(ecclesia) Sancti Stephani* c.1166, ~ ~ *in Brannel* 1291, *Seynt Stevyns* 1478. Saint's name *Stephen*, the first martyr, from the dedication of the church. There is also a Cornish form of the name with **eglos** 'a church', *Eglostephen* 1578. The 1291 form has the addition of the unexplained manor name Brannel SW 9551, *Bernel* 1086, *Branel* 1201–1343, unidentified element ***bran** + adjective suffix **-el**. PNCo 158, Gover n.d. 421.

ST STEPHENS. Saint's name from the dedication of the church to *St Stephen*, the first martyr.

(1) ~ Corn SX 3285. *(Canonici) S' Stefani* 1086, *(ecclesia) Sancti Stephani de Lanstaveton* 1259, *Seint Stevenys* 1413, *Sct.Stephens* 1610. The original site of *Lann-Stefan*, the 'church-site of St Stephen', which gave rise to Launceston SX 3384 with the addition of OE **tūn** and a move of site across the river Kensey. Thereafter there were two churches of St Stephen, this the older one on the hill (the upper church) and the new one by the ford. Co 158, Gover n.d. 147.

(2) ~ Corn SX 4158. *(ecclesia) Sancti Stephani de Seint Estevene* 1270, *(vicarius) Sancti Stephani juxta Saltasshe* 1355, *S Stephens* 1610. PNCo 158, Gover n.d. 237.

ST TEATH Corn SX 0680. 'St Teth's (church)'. (Church of) *Sancta Tetha* [c.1190]17th, *Tethe* 1259, *(ecclesia) Sancte Thetthe* 1266, *Sancta Thetha* 1278, *Seynte Tetha* 1525, *St Etha* 1549, *Sct.Teath* 1610. Saint's name *Tedda* (Teth) from the dedication of the church. She was believed to have been one of the daughters of King Broccan of Breconshire; cf. Landéda in Brittany and a 9th cent. Breton priest called *Tedei*. There is also a Cornish form of the name with Co **eglos** 'a church', *Egglostetha* [c.1190]17th. PNCo 162, DEPN.

ST TUDY Corn SX 0676. 'St Tudic's (church)'. *Seintudi* 1201, *Sanctus Tudius* 1281, *Sanctus Tudicus* 1302, *Seynt Udy* 1522, *S Tudy* 1610. Saint's name *Tudi(c)* from the dedication of the church. He was also honoured in Brittany, where he was believed to have been a disciple of St Mawes (Maudez). *Tudi(c)* may be a pet form of Tugdual, an important Breton saint of the diocese of Quimper, cf. the p.ns. Ile-Tudy and Loctudy (Breton *loc* < Latin *locus* 'church, monastery'). The earliest recorded form is the Exeter Domesday Book Cornish form with Co **eglos** 'a church', *Hecglostudic* 1086. PNCo 174, Gover n.d. 138.

ST WENN Corn SW 9664. 'St Wenna's (church)'. *(ecclesia) Sancte Wenne* 1236–60, *Seynt Wenna* 1380, *Seint Wenne* 1439, *Sct Wenn* 1610. Saint's name *Wenna* from the dedication of the church, believed to have been one of the daughters of King Broccan of Breconshire. PNCo 178, Gover n.d. 358, DEPN.

ST WEONARDS H&W SO 4924. '(Church of) St Gwennarth'. *Lann Santguainerth* [1045×1104]c.1130, *Sancti Wenarch* 1143×55, *Sancti Waynard* 1291. Saint's name *Gewnnarth* from the dedication of the church. He 176.

ST WINNOW Corn SX 1156. 'St Winnoc's'. *Sanwinvec* 1086, *San Winnuc* 1086 Exon, *Sanctus Winnocus* 1166, *Sanctus Guennou* c.1300, *Seyntwynnowe* 1434, *St Gwinnowes* 1577, *S Winow* 1610. Saint's name *Winnoc* from the dedication of the church, probably a pet form of the Cornish and Breton St Winwaloe (as in Gunwalloe SW 6522 and Towednack SW 4838). PNCo 180, Gover n.d. 307.

SALCEY FOREST Northants SP 8051. P.n. Salcey + ME **forest**. Salcey, *Sasceya* 1206, *(bosco de) Salceto* 'the wood of S' 1212–1301, *Saucey* 1213–1391 with variants *Saucee, Sause(e), Saus(e)y* etc., *Salcey* from 1229, is OFr **salceie**, **sauceie**, late Latin **salicetum* 'a willow copse'. Nth 1 gives pr [sa:si].

SALCOMBE 'Salt coomb'. OE **sealt** + **cumb**.

(1) ~ Devon SX 7338. *Saltecumbe* 1244–1464 with variants *Salt-* and *-combe, Salcume* 1286. The reference is probably to the making of salt in salt-pans in the creek here. D 311.

(2) ~ REGIS Devon SY 1488. 'King's Salcombe'. *Salcombe Regis*

1717. P.n. Salcombe + **regis**, genitive sing. of Latin **rex** 'king'. Salcombe, *(æt) sealt cumbe* 1050×72, *Salt(e)cumbe -combe* 1219–1320, *Selcome* 1086, probably referring to the making of salt in salt-pans at Salcombe Mouth. D 595, ii.xiv.

SALCOTT Essex TL 9413. 'Salt cottage(s)'. *Salcotā* 1086, *Salt(e)cot(e)* 1199×1216–1326, *Salcot(e)* 1230, 1397, *Sawcott alias Salcott* 1606. OE **salt** + **cot**, pl. **cotu**. The reference is to salt-pans and salt making on the marshes here. Ess 322 gives prs [sɔkət] and [sɔ:kət].

SALDEN Bucks SP 8229 → SAMLESBURY Lancs SD 5930.

SALE GMan SJ 7891. '(Place at) the willow-tree'. *Sala* 1199×1216, 1305, *Sale* from 1199×1216. OE **sale**, dative sing. of **salh**. Che ii.5, L 221.

SALE GREEN H&W SO 9358. *Sale Greene* 1650. P.n. Sale + ModE **green**. Sale probably represents OE **salh**, dative sing. **sale** '(at) the willow-tree'. Wo 143.

SALEBY Lincs TF 4578. 'Salli's village or farm'. *Salebi -by* 1086, *Salesbi* 1166, *-by* 1209×35. ODan pers.n. *Salli*, genitive sing. *Salla*, + **bȳ**. DEPN, SSNEM 65.

SALEHURST ESusx TQ 7424. 'Sallow-willow wooded hill'. *Salhert* 1086–14th, *-hurst* 1226–14th, *Salehurst* 1253. OE **sealh** + **hyrst**, SE dial. form **herst**. Sx 457.

SALES POINT Essex TM 0209. No early forms.

SALESBURY Lancs SD 6732. Named from Salesbury Hall SD 6735, the 'fortified place of or by Sale'. *Sale(s)byry* 1246, *Salebiry -byry -bury -buri* 1258–15th, *Salesbyry* 1305. P.n. Sale as in Sale Wheel + OE **burh**, dative sing. **byriġ**. Sale Wheel, *Salewelle* 1296, 1305, *Salewell* 1311, is a wide deep pool in the river Ribble, OE **salh** 'a sallow, a willow', dative sing. **sale**, + **wǣl** 'a deep still part of a river'. Salesbury Hall lies on the line of the Roman road from Ribchester to Ilkley, Margary No.72a. La 70, Jnl 17.45.

SALFORD 'Salt ford' OE **salt** + **ford**. A ford on a salt-way. L 71.

(1) ~ Oxon SP 2828. *Saltford* 12th etc. including *[777]12th S 112, *Salword, Salford* 1086, *Salford* from [c.1100]late 12th. The village seems to have lain on a saltway from the Four Shires Stone to Chipping Norton. Nearby Rollright had salt–rights in Droitwich in 1086. O 373.

(2) ~ PRIORS Warw SP 0751. 'S held by the prior' sc. of Kenilworth in 1122. *Saltford prioris* 1221, *Salford Priors* 1505. P.n. *Saltfordia* [c.1086]1190, *Saltford* c.1200–1546, *Salford* 1086, *Sauford* 1218, *Sawford* 1576. 'Priors' for distinction from Abbots SALFORD SP 0650. Also known as *Saltford major* [714]17th S 1250 'greater S' and *Church Salford* 1697. Wa 220 gives pr [sɔ:lfəd].

(3) Abbots ~ Warw SP 0650. 'S held by the abbot' sc. of Evesham. *Abbot Saldford* 1314, *Saltford Abbot* 1327, *Salford Abbot* 1439, *Abbotts Sawforde* 1546. ME **abbot** + p.n. *Salford* 1086. Evesham abbey held land here already in 714. Also known as *Littleton Salford* 1651 from the adjacent parts of LITTLETON Glos. The affixes are for distinction from SALFORD PRIORS SP 0751. Wa 220.

SALFORD GMan SJ 8098. 'Willow ford'. *Salford* from 1086, *Sauford* 1169, 1201 etc., *Shel- Shal- Chelford* 13th. OE **salh** + **ford**. La 32, L 70, Jnl 17.31, 18.25.

SALFORD Beds SP 9339. 'Willow ford'. *Saleford* 1086–13th, *Salford* from 1247, *Sawford* 17th. OE **salh** + **ford**. Bd 131 gives pr [sa:fəd], L 70.

SALFORDS Surrey TQ 2846. 'The willow-tree ford'. *Saleford* 1355, *Sallforde* 1535, *Salvers Bridge* 1622. Cf. Stephen *de Salford* 1279, Philip *de Saleford* 1332 and *Salfordebrugg'* 1316. OE **salh**, genitive pl. **sala**, + **ford**. The modern form with genitival *-s* may be short for Salford's Station. Sr 295, Ess lx.

SALHOUSE Norf TG 2913. 'The willows'. *Salhus* 1291, *Sallowes* 1543. OE **salh**, pl. **salgas**. In later times the term 'sallow' was confined to species of the genus *Salix* such as *Salix cinerea* and *Salix caprea* as opposed to willow and osier. Here it may simply mean 'willow-trees'. DEPN.

Bardfield SALING Essex TL 6826. 'Saling by (Great) Bardfield'. *Berdeford(e) Saling* 1272–85, *Bradfield Salyng* 1535. P.n. Bardfield as in Great BARDFIELD TL 6730 + p.n. *Salinges* 1086–1282, *Saling(e) -yng(e)* from 1220, 'the Salingas, the people dwelling by the sallows or willow-trees', OE folk-n. **Salingas* < **s(e)alh** + **ingas**. The name may also represent OE ***saling** 'sallow copse' or 'sallow stream' < **s(e)alh** + **ing**[2]. The place is known for its cricket-bat willows. Ess 454, 504, ING 24.

Great SALING Essex TL 7025. ME **gret** (*Great alias Olde* 1489), **much** (*Moche, Myche* 1512) + p.n. Saling as in Bardfield SALING TL 6826. Ess 454, ING 24.

SALISBURY Wilts SU 1429. Spellings up to the end of the 12th cent. refer to the original site at Old SARUM SU 1322 before the new settlement on the plain beneath came into being. The forms for both places are presented together in order to give a complete sequence of types.

I. *Sorbiodoni, Sorvioduni* [4th]8th AI;

II. *(æt) Searobyrg* 9th ASC(A) under year 552, *(to, on, of) Sear(e)byrig, Sæ- Seares- Sere(s)byri(g)* 1085–1137 including ASC(E) under various years, *Sarisberie* 1086–1327×77 with variants *-beri(a) -biri(a) -bury*.

III. *Salesberia* before 1086, *Salesberia -e -bir'* 1131–1235, *Salesbury* from 1227, *Salsbery* 1575.

IV. *New Saresbury* 1427, *Neu Salesbery* 1450×3, *Cyty of Newe Sarum* 1586.

Brit **soru̯io-* of unknown meaning + **dūnon* 'a hill, a fort'. Another *Sorviō* p.n. is recorded from Raetia near Straubing on the Danube, Germany, *Soruiodoro* [4th]c.1200, for which Reitzenstein suggests pers.n. **Sorvius* + Gaulish **-duro* 'fort, walled town'. Brit **Sorvio-* became PrW **Ser'w'* through the process of *i*-affection and **Seoru-* by OE breaking. The OE forms of the name, type II, show folk-etymological remodelling of **Se(o)ru-* under the influence of the OE word **searu**, 'cunning device, cunning work, trick, armour', no doubt thought of as applicable to the massive earthworks of the Iron Age hill-fort at Old Sarum. Brit *dūnon* was translated by OE **burh**. The OE name *Searoburh* thus must have been thought of as meaning something like 'cunningly defended fort'. Forms with medial *-es-* constitute another piece of popular etymologising, as if the specific were a pers.n. requiring the genitive inflexional ending. Type III shows AN substitution of *l* for *r* and type IV the addition of *New* in contrast to Old SARUM, *Old Saresbury* 1429. Wlt 18 gives pr [sɔ:lzbəri], RBrit 461, JRS 38.58, Britannia 1.79, TC 167, LbO 370 s.n. Straubing.

SALISBURY PLAIN Wilts SU 0645. *planum Sar'* 1346, *Salesburge Playne* 1610, also simply *the Playne* c.1540. P.n. SALISBURY SU 1429 + ME **plain** 'a great open tract of land'. Wlt 17, PNE ii. 66.

Great SALKELD Cumbr NY 5536. *magna Salkild* 1285, *Myckelsalkeld* 1554, *Much Salkeld* 1564. ModE adj. **great** (Latin **magna**, ME **mikel**) + p.n. *Salchild* c.1110, c.1202, *Salkil* 1197–1233, *Salkild -kyld* 1229–79, *Salkeld* 1278, *Salehhild* 1164–7, 'sallow wood', OE **salh** + **hylte** later confused with ON **kelda** 'a spring, a stream'. 'Great' for distinction from Old or Little SALKELD NY 5636 on the opposite bank of the river Eden. Pronounced locally [sɔ:kld]. Cu 236.

Little SALKELD Cumbr NY 5636. *parva Salkyld, ~ Salkehild* 1279, *Lytle Salkeld* 1535. ModE adj. **little** (Latin **parva**) + p.n. Salkeld as in Great SALKELD NY 5536. Also known as Old Salkeld, *Veteri Salkil(l)* 1201–42, *~ Salkild* 1229, 1279, *Olde, Alde Salehhild* 1164–7 etc. with many variant spellings. Cu 236.

SALLE Norf TG 1015. 'The willow-tree wood or clearing'. *Salla* 1086, *Salle* from 1196, *Saulle* 1197. OE **salh** + **lēah**. DEPN gives pr [sɔ:l].

SALMONBY Lincs TF 3273. 'Salmundr's village or farm'. *Salmundebi* 1086–1206, *Salmonebi* c.1115. ODan pers.n. *Salmund*, + **bȳ**. SSNEM 65, DEPN.

SALMONSBURY Glos SP 1620 → BOURTON-ON-THE-WATER SP 1620.

SALPERTON Glos SP 0720. Possibly 'settlement on the salt-road'. *Salpretvne* 1086, *Salpertune -ton(e)* 1169–1535. OE **salt** + **herepæth** + **tūn**. The reference is to an ancient saltway from Droitwich to the *Saltewhich* or salt market in Cirencester which passes to the W of the village. Gl i.176, 19 no. 3, 67.

SALPH END Beds TL 0752. *Safe End* 1766. P.n. Salph + ModE **end**. Salph, *Salcho(u)* 1086–13th, *Salho* 13th, *Salpho* 1377 is 'willow-tree hill-spur', OE **salh** with substitution of [f] for the peripheral phoneme [x] + **hōh**, dative sing. **hō**. Bd 64 gives pr [sa:f end].

SALT Staffs SJ 9527. 'The salt pit'. *Selte* 1086, *Salt* 1167, *Saute* 1236. OE ***selte**, Mercian ***sælte** whence the modern form. There are saltworks in the area. DEPN, PNE ii.118.

SALTASH Corn SX 4259. '(The port of) Ash where salt is manufactured'. *Saltehasche* 1302, *Saltes(s)h* 1334–51, *Saltassh* 1338. ME **salte** + p.n. *(Burgh' de) Esse* 1201, 1316, *Aysh* 1284, *Asshe* 1297, 'the ash-tree', OE **æsc**. PNCo 153, Gover n.d. 237, DEPN.

SALTBURN-BY-THE-SEA Cleve NZ 6621. 'Salt stream'. *Salteburnam* 1180×90, 1293. OE **salt** + **burna**. The reference is to the alum which is found here. YN 143.

SALTBY Leic SK 8526. 'Salti's farm or village'. *Saltebi* 1086–1301, *-by* c.1150–1539, *Sautebi -by* 1185–1276, *Saltby(e)* 1328–1576. ON pers.n. *Salti*, genitive sing. *Salta*, + **bȳ**. Alternatively the specific might be OE or ON **salt** + **bȳ** referring to a nearby chalybeate spring. Lei 192, SSNEM 65.

SALTDEAN ESusx TQ 3802. *Saltdean Gap* 1740. ModE **salt** + **dean**. A dean that runs up from the sea. Sx 312.

SALTER Lancs SD 6062. 'Salt shieling'. *Salter* from 1612. ON **salt** + **ǽrgi**. La 181.

SALTER'S BANK Lancs SD 3028. No early forms. A sand-bank in the Ribble estuary off Lytham St Anne's.

SALTER'S BROOK BRIDGE Derby SE 1300. P.n. Salter's Brook + ModE **bridge**. Salter's Brook, *Salterbroke* 1509×47, *-Brook* 1695, 1753, OE **saltere** + **brōc**, is situated on the county boundary where the main Cheshire-Stocksbridge road crosses it: this must have been an old saltway. YW i.342.

SALTERFORTH Lancs SD 8845. 'Ford used by salt-merchants'. *Salterford* 1216×72, *Salterforth* 1626. OE **saltere** + **ford**. YW vi.37.

SALTERGATE NYorks SE 8594. 'Track used by salt-merchants'. *Saltergate* [1335]14th, 1619. OE **saltere** + ON **gata**. Possibly a track used by men carrying salt, i.e. alum, south from Cleveland where it was mined. YN 91.

SALTERSWALL Ches SJ 6267. 'Salter's well'. *Salterswall* from 1542. OE **saltere**, genitive sing. **salteres**, + **wælla**. The place is on a route from Middlewich to Chester called *Salteresway* in 1334. Che iii.173.

Budleigh SALTERTON Devon SY 0682. *Budley Salterton* 1765. P.n. Budleigh as in East BUDLEIGH SY 0684 + p.n. *Salterton* 1667. A modern town developed at the mouth of the river Otter where the salt-making houses of the manor of (East) Budleigh had existed for many years, *Saltre* [1210]1326, *Salterne in the manor of Buddeleghe* 1405, OE **salt-ærn** 'a building where salt is made or sold'. Salterton is 'settlement at the salt-house' or 'at Saltern', OE **salt-ærn** probably used as a p.n. *Salterne* + *ton* from OE **tūn**. D 583.

Woodbury SALTERTON Devon SY 0188. *Woodbury Salterton* 1809 OS. P.n. WOODBURY SY 0187 + p.n. *Salterton* 1306–1413, 'settlement of the salt workers', OE **saltere**, genitive pl. **saltera**, + **tūn**. 'Woodbury' for distinction from other Saltertons. D 602.

SALTFLEET Lincs TF 4593. 'Salt creek'. *Salfluet* 1086, *Saltflet* 1301, 1361. Cf. *Saltflethaven* 1346. OE **salt** + **flēot**. Originally a r.n. referring to the watercourse later called Long or Great Eau TF 4484 beside which stand the three settlements called SALTFLEETBY. DEPN, Jnl 29.80.

SALTFLEETBY Lincs. 'The village by Saltfleet'. *Salflatebi* 1086, *-fletebi -flede-* c.1115, *Saltfletebi* 1185. P.n. SALTFLEET TF 4593 + **bȳ**. There are three separate hamlets, S All Saints TF 4590, *Sauflet Omnium Sanctorum* 1254, S St Clement TF 4591, *Saltfletteby S. Clementis* 1254, and S St Peter, *Saltfletby Sancti Petri* 1254, distinguished from each other by their church dedications. Saltfleetby was a daughter settlement from Grimoldby TF 3988 and Manby TF 3986 on land reclaimed from the sea. DEPN, SSNEM 65, Jnl 7.56, Cameron 1998.

SALTFORD Avon ST 6867. 'Salt ford'. *Sanford* (sic) 1086, *Saltford* from 1291. OE **salt** + **ford**. The reference may be to the tidal nature of the river up to this point. The 'salt ford' would then contrast with the FRESHFORD further up-stream at ST 7860. An alternative suggestion is that this marked a saltway crossing of the river. DEPN, L 71, BAA 1983.25–34.

SALTHAUGH GRANGE Humbs TA 2321. *grangia de Salthagh(e)* [1349]15th. P.n. Salthaugh + ME **grange**, 'an outlying farm belonging to a religious house', in this case Meaux Abbey. Salthaugh, *Saltehache* 1150×3, *-hag(h)* 1172, [1275]15th, *Salthah* [1177]15th, *-hagh(e)* 1205–[1347]15th, is 'the salt enclosure, enclosure where salt is made', OE **salt** + **haga**. The site, which moved in the Middle Ages owing to the frequent inundations of the Humber, is near the old course of that river. YE 32.

SALTHOUSE Norf TG 0743. 'House where salt is made or stored'. *Salthus* 1086, 1242. OE **salt** + **hūs**. The reference is to the manufacture of salt by drying or boiling sea-water in pans in Salthouse Marsh. DEPN.

SALTMARSHE Humbs SE 7924. 'Brackish marsh' or 'marsh where salt is made or obtained'. *Saltemersc* 1086, *-mareis* 1194, *Saltmers* 1282, *Salt(e)mers(c)h* c.1348–1581, *-marsh(e)* 1316, 1546. OE **salt**, definite form **salta**, + **mersc** influenced by OFr *mareis*. Scandinavianised forms in *-mersk(e)* also occur 1285–c.1362. The site is on the Ouse which is tidal here; a field in the locality was called *le Ssalin* (sic) 1229, 'the salt pit'. YE 254, SSNY 138.

SALTOM BAY Cumbr NX 9575. Unexplained p.n. *Saltom* 1737 + ModE **bay**. Cu 453.

SALTON NYorks SE 7180. 'Enclosure of or by the willows'. *Saletun -ton* 1086, 1167×80, *Salton* from 1285, *Sauton* 1577. OE **salh**, genitive pl. **sala**, + **tūn**. YN 57 gives pr [sɔ:tən].

SALT SCAR Cleve NZ 6126. 'Salt reef'. ON **salt** + **sker**. The reference is to the outlying part of Redcar Rocks submerged at high tide, also known as *Crab Scar* 1846. In 1281 a wreck occurred at *Salcker in Clyvelond*. The form appears to be a compound of ON **salt** + **kjarr** 'marsh' and so 'salt-marsh', but it may be a bad spelling for Salt Scar, itself notorious for wrecks. YN 156, HistCleve 359.

SALTWICK BAY NYorks NZ 9211. P.n. Saltwick + ModE **bay**. Saltwick, *Saltewicke* 1540, is the 'salt creek', OE **salt** + ON **vík**. The reference is, as with SALTBURN-BY-THE-SEA NZ 6621, to alum. YN 122.

SALTWICK Northum NZ 1779. 'The salt-works'. *Sal(t)wic* 1242 BF, *Saltwyke* 1268, *Saltewicke* 1296 SR, *Saltik* 1676. OE **salt** + **wīc**. NbDu 171.

SALTWOOD Kent TR 1535. 'Salt wood'. *(æt) Sealtpuda* [c.985×1006]11th S 1455, *(æt) Sealtwuda* [993]14th S 877, *Saltuuda* [1026]12th S 1221, *Sealtwuda* [995×1005]12th S 1456, *Salteode* 1086, *Saltewda* 1161×2. OE **sealt** + **wudu**. The reference is possibly to the supply of wood for saltworks at the coast. KPN 317, L 229, Charters IV.118.

High SALVINGTON WSusx TQ 1206. A modern suburb of Worthing. ModE **high** + p.n. Salvington TQ 1305, *Saluinton* 1249, *Saluington -yng-* from 1250, 'the estate called after Sæwulf or Sælaf', OE pers.n. *Sǣwulf -lāf* + **ing**[4] + **tūn**. Sx 197, DEPN, SS 77.

SALWARPE H&W SO 8762. Originally a r.n. Forms for the settlement are *salouuarpe*, *SALOUUEARPAN* (dative case) [817]11th S 181, *Salewarpe* [c.1043]16th S 1226, *Salewarp*

[1043]16th S 1000, *SALOUUARPAN* (dative case), *sale þarpe, (into) sale þarpan* [n.d.]11th S 1596, *Salewarpe* 1086, and for the river *saluuerpe* [692]11th S 75, [716×7]11th S 102, 770 S 59, *(ondlang) saleþerþæn* *[706]12th S 54, *(ondlang) saleþerpan* [n.d.]12th S 1594, *Saluuarpe* [767]12th S 58, *Saleuuearpe, (in, of) salþarpan, (of) seal þarpan* *[770]17th S 60, *saloþeorþe -þearpe* [816]11th S 180, *Saloþearpe, (on) Saloþorpan, Saleþorp, (on) Saleþearpan* [n.d.]18th S 1597, *saloþorpe, (in) saloþorpan, saleþarpe, (in) saleþarpan* [n.d.]11th S 1596, *(on, ondlang) seale þeorpan* 972 S 786, *Salope brooke* c.1540. The meaning is 'alluvium thrower', OE **sealu** 'sallow stuff, dirty stuff' + **wearp(e)**. The river has a reputation for flooding and drowning nearby land; excavations have revealed 5–7th cent. salt-works here buried beneath a metre of silt. Wo 306 gives pr [solwʌp], formerly [sælep], RN 350, Hooke *passim*.

SALWAYASH Dorset SY 4596. *Salway Ash* 1811 OS.

SALWICK STATION Lancs SD 4632. P.n. Salwick + ModE **station**, a halt on the railway line from Preston to Blackpool. Salwick, *Saleuuic* 1086, *-wic* 1201, 1226, *-wyk* 1327, *Sawick* 1577, is probably the 'place where willow is grown and processed for basket-weaving', OE **salh** 'willow', genitive pl. **sala**, + **wīc** in the sense 'place associated with trade'. La 150, Jnl 17.86, 21.45.

SAMBOURNE Warw SP 0662. 'The sandy stream'. *Sandburne* 1086, *-burna* [c.1086]1190, *Saundburne* 1221–74, *Sombo(u)rne* 1316–27, 1535, *Samburne* [714]16th S 1250, 1547. OE **sand** + **burna**. Wa 221, L 18.

SAMBROOK Shrops SJ 7124. Probably 'sand brook'. *Semebre* 1086, *Sambroc* from 1255×6 with variants *-brock -brok(e) -brook(e)*. Probably OE **sand** + **brōc** though the absence of sps in *Sand-* is surprising. Sa i.257.

SAMLESBURY Lancs SD 5930. 'Fortified place of the shelf or ledge'. *Samerisberia* 1179, *Samelesbur(e) -biri -biry -buri -is-* 1188–1252, *Samlisbyri -buri -bury -es-* 1258–1311, *Sam(pn)esbiry -bury* 1276–8, *Samsbury* 1577, *S(c)hamelesbir(y)* 1225, 1246, *Scamelesbyry* 1277, *Shamp(e)lesbiry* 1246, 1277. OE **sceamol**, genitive sing. **sceam(e)les**, + **burh**, dative sing. **byriġ**. It is uncertain whether the site of the church on a ledge beside the Ribble at SD 5830 or of Samlesbury Hall at SD 6230 is in question, but OE **sceamol** 'a ledge, a bench' is used in this name in a topographical sense. The early spellings with *S-* and the pronunciation with [s] instead of [ʃ] may be due to AN influence and a native tendency to assimilation in the sequence *sc - s*, cf. Salden Bucks SP 8229, *Schaldene* 1175, OE **sceald**, SAPCOTE Leic SP 4893 OE **scēap**, and possibly Sidbury Hill Wilts SU 2150, *Shidbury* 1325, OE ***scydd**. La 69, Jnl 17.44.

SAMLESBURY BOTTOMS Lancs SD 6128. P.n. SAMLESBURY Lancs SD 5930 + **bottom** (OE *botm*) 'flat wet valley floor' referring to the situation in the Darwen valley.

SAMPFORD 'Sandy ford'. OE **sand** + **ford**.

(1) ~ ARUNDEL Somer ST 1018. 'S held by the Arundel family'. *Samford Arundel* 1240. Earlier simply *Sanford* 1086, *Samford* 1610. Roger Arundel ('swallow' from OFr *arondel*) held the manor in 1086. DEPN, DBS 12.

(2) ~ BRETT Somer ST 0840. 'S held by the Brett family'. *Saunford Bret* 1306. Earlier simply *Sanford* 1086. Simon Bret 'the Breton' held the manor c.1120. The soil here is described as 'rich sandy loam with some clay, overlying marl'. DEPN, Gaz.

(3) ~ COURTENAY Devon SS 6301. 'S held by the Courtenay family'. *Saundford Curtenay* 1262. Earlier simply *Sanford* 1086, *-ford* 1242, 1248, *Sandfort* 1093, *Sampford* 1274. The ford crosses the river Taw. D 165.

(4) ~ PEVERELL Devon ST 0314. 'S held by the Peverel family'. *Sanfordepeverel* 1339. The manor was held by Matilda Peverel in 1152. Earlier simply *Sanford(e)* 1086, 1212, *Saundford* 1237, 1300, *Sampforde* 1291. The ford crosses the river Lynor and contrasts with *stanford, anne stanihtne ford* and *fileþ leage ford* all in the bounds of *Æscforda* (Ashford ST 0415) [958]12th S 653. D 551.

(5) ~ SPINNEY Devon SX 5372. 'S held by the Spinney family'. *Saunford Spinee* 1281, *Saundford Spyneye* 1304. The manor was held by Gerard de Spineto in 1234 (Latin *spinetum* 'a spinney'). Earlier simply *Sanford* 1086, 1242, *Saunford* 1234, 1237, *Samford* 1262, *Sampford* 1291. The ford crosses the Walkham river. D 238.

(6) Great ~ Essex TL 6435. *Magna Saunford* 1230. ModE adj. **great**, Latin **magna**, + p.n. *Sanfordā -fort* 1086, *Sanford(e)* 1119–48, *Samford* 1232–1371, *Sampford* from 1303. Also known as *Auncien Samford* 'old S' 1371. A reference to sand beds here is found 1831×5. 'Great' for distinction from Little SAMPFORD TL 6533. Ess514.

(7) Little ~ Essex TL 6533. *Parva Samford* 1232. ModE adj. **little**, Latin **parva**, + p.n. Sampford as in Great SAMPFORD TL 6435. Also known as *Newesaumford* 1351. Ess 514.

SAMSON Scilly SV 8712. Short for St Sampson's island. (Island of) *Sanctus Sampson* [c.1160]15th, *Sampson* c.1585, *Sampson's Ile* 1652. Saint's name *Samson*, probably St Samson of Dol, but there is no record of such a chapel on the island. Cf. St Sampson on Guernsey. PNCo 153.

SANCREED Corn SW 4229. 'St Sancred's (church)'. *(ecclesia) Sancti Sancreti* 1235, ~ ~ *Sancredi* 1291, *Sanckras* [1580]18th, *Sancret* 1610. Saint's name *Sancred* from the dedication of the church. The 16th cent. form *Sanckras* shows the normal MCo development of *d* > *z*, a pronunciation which lasted into the 20th cent. The Co form of the name with **eglos** 'a church' is found in the forms *Eglossant* [c.1176]1300 (with Co **sant** 'saint' instead of the saint's name) and *Egglossanres* 1443. PNCo 153, Gover n.d. 657, DEPN.

SANCTON Humbs SE 9039. 'Farm or village on sand'. *Santun(e) -ton(a)* 1086–1505, *Sancton(a)* 1195×1211–1650, *Saunton* 1219–1589, *Sainton -y-* 1241, 1539. OE **sand** + **tūn**. Forms with *-c-* arose through popular association with OE *sanct* 'a saint'. The soil is loam overlying chalk and sand. However, this is the site of a huge pagan cemetery which marks Sancton and nearby GOODMANHAM SE 8843 as the main religious foci of the pagan kingdom of Deira. It is just possible, therefore, that the *sanct* forms are primary and indicate awareness of the site's 'sacred' nature from the beginning. YE 227 gives pr [santən], Myres 190.

SAND BAY Avon ST 3264. *Sand Bay* 1809 OS. ModE **sand** + **bay**.

SAND POINT Avon ST 3165. *Sand Point* 1809 OS. ModE **sand** as in SAND BAY 3264 + **point**.

Kirk SANDALL SYorks SE 6108. 'S with the church'. ME *Ki- Kyrke* from 1261, + p.n. *Sandalia -e* 1086, *Sandale* 1086–1822, *-hale* 1148, *Saundhal'* 1221, *Sandall(e)* from 1279, the 'sandy nook of land', OE **sand** + **halh**. The reference is to a piece of flat ground in a sharp bend of an old course of the r. Don. 'Kirk' for distinction from Long Sandall SE 6006, *Sandala -ela -al(i)e* 1086, *Parva Sand(h)al(e)* 1291–1330, ~ *Sandall* 14th, *Long Sandal* 1740. YW i.22.

SANDBACH Ches SJ 7560. 'Sandy valley-stream'. *Sanbec(d)* 1086, *-bache* 1308, *-bage* 1539, *-bitch* c.1703, *-batch -bich* etc. OE **sand** + **bæċe**. Che ii.269, L 12. Addenda gives 19th cent. local pr ['sɔnbitʃ].

SANDBANKS Dorset SZ 0487. *the Sand Bankes* c.1800, ~ *vulgarly called Cales* 1843, *ground called Cales* 1579. ModE **sand, bank** and surname *Cale*. Do ii.40.

SANDERSTEAD GLond TQ 3461. 'Sandy place'. *(an) sonden stede, (on) sondenstyde* 871×89 S 1508, *Sandestede* 1086–1220, *Sandersted(e)* 1221 etc. OE ***sanden** + **stede**. The earliest forms seem to support this rather than the alternative 'sandy homestead', OE **sand** + **hāmstede**. The soil here is sandy. Sr 53, LPN 203.

SANDFORD Identical with SAMPFORD 'Sandy ford'. OE **sand** + **ford**.
(1) ~ Avon ST 4259. *Sandford* 1817 OS.
(2) ~ Cumbr NY 7216. *Saunford* 1199–c.1250, *Sand(e)ford* from 1256, *-forth* 1380–1598. We ii.83, L 68.
(3) ~ Devon SS 8202. A 'sandy ford' over a tributary of the Greedy. *(æt) sand forda* [930]10th S 405, *(æt) Sandford(a)* [997]10th/11th S 890, [1008×12]11th S 1492, *Saunford* 1340, *Sanford* 1675. D 411.
(4) ~ Dorset SY 9389. *Sandford* 1811. Cf. *Sanford Ditche* 1606, ~ *Bridge* 1791, *Sampford Mill* 1671. The ford was where Sandford road crossed an unnamed tributary of the Piddle or Trent. Do i.164.
(5) ~ -ON-THAMES Oxon SP 5301. *Samford ultra Tamisiam* 'S the other side of the Thames' c.1225. P.n. *(æt) Sandforda* [1050]13th S 1022, *(æt) Sandfordan, (to) Sandforda* [1054]12th S 1025, *Sandford* from c.1225, *Sanford* 1086–1285, *Samford(')* 1154×89–1428. O 186.
(6) ~ ORCAS Dorset St 6220. 'S held by the Orescuilz family'. *Sandford Orescure* 1309, ~ *Orskuys* 1348, *Sampford Orkas* 1535. Earlier simply *Sanford* 1086, *Sandford* from 1243. The ford was where the road from Sherborne crossed the stream now called Mill Stream. The manor was held from the 12th cent. by a family whose name is variously spelled *Oriescuilz* 1177, *(D)orescuilz* 1195, 1210. Do iii.389.
(7) ~ ST MARTIN Oxon SP 4226. 'St Martin's S'. P.n. *Sanford* 1086–1273 with variant *-forde, Saunford(e)* 1225–79, *Sandford* from 1240, + saint's name *Martin* from the dedication of the church. O 279

SANDGATE Kent TR 2035. 'The sand gate'. *Sandgate* from 1256. OE **sand** + **ġeat**. The reference is to an opening in the cliffs to the beach. PNK 454.

SANDHOE Northum NY 9766. 'Sandy hill-spur'. *Sandho* 1225, 1232, *-hou* 1328, *-hoe* 1663, *Sandow* 1479, *Sandy* 1724. OE **sand** + **hōh**. NbDu 171 gives pr [sanda].

SANDHOLME Humbs SE 8230. 'Sandy meadow'. *Sandholm(e)* 1285–1548. OE **sand** + **holm**. YE 247.

SANDHOLME Lincs TF 3337. 'Sandy water-meadow or piece of raised ground in marsh'. *Sandholme* from 1531. ME **sand** + **holm** (ON *holmr*). Payling 1940.84.

SANDHURST 'Sandy wooded hill'. OE **sand** + **hyrst**.
(1) ~ Berks SU 8361. *San(d)herst* 1175, *Sandh'st -hurst -herst* 1185 etc. SE dial. form **herst**. Brk 128.
(2) ~ Glos SO 8323. *Sanher* 1086, *Sandhurst(e)* 12th–1616. Gl ii.152, L 197–8.
(3) ~ Kent TQ 7928. 'Sandy wooded-hill'. *Sandhyrste* c.1100, *Sandhurst(')* from 1229, *-herst* c. 1230, Kentish **herst**. PNK 343.

SANDHUTTON NYorks SE 3882. 'Hutton on sandy soil'. *Sandehoton* 12th, *Sand-Hutton* 1665. ME **sand** + p.n. *Hot(t)une* 1086, *Hoton* 1202, the 'settlement at a hill-spur', OE **hōh** + **tūn**. The village is situated on a low ridge projecting into the alluvial plain of the river Swale. YN 187.

SANDIACRE Derby SK 4736. 'Sandy newly-cultivated land'. *Sandiacre* from 1086 with variants *-y- -acr(a) -aker -akyr, Sandeacre* 1312, 1403, *Sendiacra -e* 1179–1201, *Seint, Saint Diacre* 1198–1210, *Saundiacre* 1234–1330. OE **sandiġ** + **æcer**. Db 497, L 232–3.

SANDILANDS Lincs TF 5280. 'Sandy land newly taken into cultivation'. ModE **sandy** + **land**.

SANDIWAY Ches SJ 6071. 'Sandy road'. *Sondeway* 1379, 1435, *Sondyway* 1503, *Sandyway* 1499, *-i-* from 1721. OE **sandiġ** + **weġ**. Che iii.208.

SANDLEHEATH Hants SU 1214. *Sandelheath* 1536, *Sandhill Heathe* 1590, *Sandel Heath* 1811 OS. P.n. Sandle + ModE **heath**. Sandle, *Sandehill'* 1201, is the 'sand hill', OE **sand** + **hyll**, referring to an outcrop of Bagshot Sand amid the London Clay. Ha 144, Gover 1958.215.

SANDLEIGH Oxon SP 4701. Possibly a modern formation from nearby Sandford and Leigh as in BESSEL'S LEIGH SP 4501. Brk 463.

SANDLING Kent TQ 7558. 'Sand hill'. *Sandlink(e)* 1293–5, *Sandlyng* 1466–1535. OE **sand** + **hlinċ**. PNK 135, ING 229.

SANDON 'Sand hill'. OE **sand** + **dūn**.
(1) ~ Essex TL 7404. *Sandun -don(a) -done* 1199–1303. The soil is varied, mostly a wet loam over clay, but the hill on which the village stands is light and sandy. Ess 266.
(2) ~ Herts TL 3234. *(of) Sandune* c.1000, *Sandone* 1086 etc. including *Sandonam* *[924×39]13th S 453. Hrt 164, L 144, 152, 155.
(3) ~ Staffs SJ 9429. *parua Sandone* 'little S' 1086, *Sandun* 1236. Sandon Hall SJ 9528 is *Scandone* 1086. The soil is sand and gravel. DEPN, L 144, 155.

SANDOWN IoW SZ 5984. 'Sandy water-meadow'. *Sande* 1086, *Sandome* 1204, *Sandham* 1271–1844, *Sandone* c.1300, *Sandam* 1432, 1591, *Sandown* 1759. OE **sand** + **hamm**. The 1086 spelling may represent OE ***sænde** 'a sandy place' possibly used as a p.n. In this case *Sandham* is better taken as the 'enclosure at Sande' and *Sandome* 1204 as OE **sandum** 'at the sands', dative pl. of **sand** or ***sænde**. Wt 203 gives pr ['zændæun], Mills 1996.92.

SANDOWN BAY IoW SZ 6183. *Sandhambaye* 1558×1603, *Sanden bay* 1600, *Sandown Bay* 1781. P.n. SANDOWN SZ 5984 + ModE **bay**. Wt 203, Mills 1996.92.

SANDPLACE Corn SX 2457. 'Place where sand is brought'. *Placeae sabulonis* [1326]15th, *Sandplace* 1667. The first form is a Latin translation of 'places of sand'. It is suggested that boats brought coastal sand here for spreading on the fields. PNCo 153, Gover n.d. 283.

SANDRIDGE Herts TL 1710. 'Sand ridge'. *Sandrige* 1086 etc. with variants *-rug(g)e -rigge*. OE **sand** + **hrycg**. The soil is reported to be 'light over gravelly'. Hrt 100, Gaz 1896.

SANDRINGHAM Norf TF 6928. Properly 'Sand Dersingham', that part of Dersingham which lies on sand. *Santdersingham* 1086, *Sandring(e)ham* 1195, 1254. OE **sand** + p.n. DERSINGHAM TF 6830. DEPN, ING 138.

SANDSEND NYorks NZ 8612. 'The end of the sand'. *Sandes(h)end(e)* 1254, 1279, 1301. OE **sand**, genitive sing. **sandes**, + **ende**. Situated at the W end of Whitby Sands. YN 137.

SANDTOFT Humbs SE 7408. 'The curtilage on the sand'. *Sandtofte* 1157. OE **sand** + **toft**. The village lies on blown sand. DEPN, SSNEM 164.

SANDWICH Kent TR 3358. 'Trading place on or at the sand, sandy trading place'. *Sanduuich* [963×71]14th S 808, *Sondwic* 9th ASC(A) under year 851, c.1122 ASC(E) under same year, *Sandwic* c.1100 Asser, 1023 S 959, 1037×40 S 1467, 1042×66 S 1047, *San(d)uuic, Sandwic(e)* 1086. OE **sand** + **wīc**. KPN 295.

SANDWICH BAY Kent TR 3759. P.n. SANDWICH TR 3358 + ModE **bay**.

SANDWICH FLATTS Kent TR 3561. *Sandwich Flats* 1819 OS. P.n. SANDWICH TR 3358 + ModE **flat** sb.6, 'a nearly level tract over which the tide flows'.

SANDWICK Cumbr NY 4219. 'Sandy creek'. *Sandwic -wik(e) -wyk(e) -wick(e)* from 1200. ON **sandr** + **vík**. The reference is to a bay in Ullswater. We ii.217, SSNNW 156.

SANDWITH Cumbr NX 9614. 'Sandy ford'. *Sandwath* 1260–1607, *Sandwith* from early 16th. ON **sandr** + **vath**. Cu 433 gives pr [sanəθ], SSNNW 156, L 82.

SANDY Beds TL 1649. 'Sand island'. *Sandeie* 1086 etc., *Saundeye* 13th–14th cents. OE **sand** + **ēġ**. A patch of greensand forms an island between two streams here. Bd 107, L 37, 39.

SANDY LANE Wilts ST 9668. *Sandy Lane* 1817 OS. ModE **sandy** + **lane**.

Great SANKEY Ches SJ 5788. *Great Sonky* 1325, 1332. Adj. **great** + p.n. *Sonchi* c.1180, *Sanki* 1212, *Sonky* 1243–1332, *Sanky* 1285, transferred from the stream name Sankey, *Sanki* 1202, *Sanky*

1228, 1257, *Sonky* 1228, possibly 'holy stream', British **Sanclo-*. 'Great' for distinction from Little Sankey on the opposite side of the stream. La 105,94, RN 351.

SANTON BRIDGE Cumbr NY 1101. *Santon Bridge* 1673. P.n. Santon + ModE **bridge**. Santon, *Santon* from c.1235, is 'sand settlement', OE **sand** + **tūn**. A place where silver-sand was obtained. Cu 402.

SANTON DOWNHAM Suff TL 8187 → Santon DOWNHAM.

SAPCOTE Leic SP 4893. 'Sheep cottage(s)'. *Scepe- Sapecote* 1086, *Scapecotes* 1230, *Sapecot(e)* 1225–1375, *Sapcot(e)* from 1285. OE **scēap** + **cot**, pl. **cotu**. Lei 532.

SAPEY COMMON H&W SO 7064. *Sapey Common* 1832 OS. P.n. Sapey as in Upper SAPEY SO 6863 + ModE **common**.

Upper SAPEY H&W SO 6863. ModE **upper** + p.n. *Sapina* 1180, *Sapy -i* 1210×12–1397, probably originally a stream-n. 'the sappy one', OE **sæpiġ**, definite form feminine **sæpiġe**, oblique case **sæpiġan**. 'Upper' for distinction from Lower Sapey SO 6960, *Nethersapi* 1275, earlier simply *(æt) Sapian, Sepian* *[781]11th S 121, *Sapie* 1086, 1221, *Sapy* 1212, 1235 and *Sapi Pichard* 1242 from Miles Pichard who held a knight's fee here in 1212. He 178, Wo 75.

SAPISTON Suff TL 9175. Partly uncertain. *Sapestuna* 1086, *-tune* c.1095, *-ton* 1204, 1254, *Sapston* 1610. Unidentified element or pers.n. + OE **tūn**. DEPN.

SAPPERTON 'The soap-makers' settlement'. OE **sāpere**, genitive pl. **sāpera**, + **tūn**.

(1) ~ Glos SO 9403. *Sapertun(e)* [c.1075]1367, *-ton(')* 1211×13–1602. *Sapere* is used as a by-n. by Adam le Sapere 1248. Gl i.137.

(2) ~ Lincs TF 0133. *Sapertone* 1086, *Saperton -tun* 1154–1597, *Sapperton* from 1327†. Perrott 493, SelPapers 76.

SARACEN'S HEAD Lincs TF 3427. Transferred from an inn name. Pub Names 232, Inn Names 8f, 82.

SARISBURY Hants SU 5008. Unexplained. *Sarebury* 1272×1307, *Sarisbury* 1538. There seems to have been some influence from the name SALISBURY Wilts SU 1429 but the origin of Sarisbury remains obscure. Gover 1958.31.

River SARK Cumbr NY 3369. *Serke* 1214, 1580, *Sarke water* 1552, *Sark* 1740. A pre-English river n. of unknown meaning identical with the lost Glos stream name *Sarke* [c.1340]16th. Cu 26, RN 352.

SARKFOOT POINT Cumbr NY 3265. P.n. Sark Foot + ModE **point**. Sark Foot is *the foot of Sarke* 1616, *Sark foot* 1740, river n. SARK + ModE **foot**. Cu 147.

SARNESFIELD H&W SO 3751. 'Open land of Sarn'. *Sarnesfelde* 1086, *Se- Sarne(s)feld(e)* 1123–1331. W **sarn** 'pavement, causeway' perhaps used as a p.n., English genitive sing. **sarnes**, + OE **feld**. He 178, L 240, 243.

SARRATT Herts TQ 0499. Uncertain. *Syreth* [1077×93]14th, *Si- Syret* 1166, 12th, *(La) Sareth* 1094–1296, *Seret(h)* 1154×89–1237. An OE ***sīeret**, ***sȳret** 'dry place' < **sēar** + **-et** has been suggested. Hrt 102, PNE ii.120.

SARRE Kent TR 2565. Uncertain. *ad Serrae, Seorre* [c.761]13th S 29 with variant *Serre, Syrran* c.1100, *S- Cer(r)e(s)* 1179×80–1262, *Serra* 1279. Possibly originally a r.n., cf. *Serres aqua salsa* 'Sarre saltwater' [c.1200]c.1260, *(water of, the river) Serre* 1392–3, *the little river Sarr* 1801, which Ekwall identifies with the Wantsum TR 2366, Wallenberg with the Sarre Penn TR 2364. If this is so, and if the form *Syrran* is a hypercorrect Kentish form for *Serran*, then the OE r.n. **Serre* is identical with French Serre, a tributary of the Oise, *Sera* 867, the Serio, a tributary of the Adda in Lombardy, *Sarius* 7th, and also of the SOAR SK 4925, SP 5599, the Scottish Sorn on Islay NR 3563, the German and French Sarre, *Sara, Saroa* 6th, the Latin *Saravius* 3rd–4th, and the Alsace Zorn, *Sorna* 724, all Old European formations on the IE root **ser-*/**sor-* 'flow'. If on the other hand the *e* spellings represent Kentish *e* for *y* and the correct form was OE (West Saxon) **Syrre* < Gmc **surrjōn*, the name may derive from the IE root **swer-* 'resound' with zero vocalisation as in Gk *ὕρον* 'swarm of bees' and with zero vocalisation and consonant gemination as in G *surren* 'hum, whirr', Latin *susurrus* 'whisper, rustling'. KPN 43, K 599, RN 353, Berger 230, Krahe 1964.40, Chantraine 1161 a, s.v. *ὕραξ*, Charters IV.179–80.

SARSDEN Oxon SP 2823. Partly unexplained. *Secendene* 1086, *Serchedene* c.1170, 1414–31, *Cercheden(e)* [c.1173]13th, 1251, 1368, *Cerches-* 1181–97, 1332, *Certes-* 1246×7, 1285, *Serseden* 1428, *Saresden* 1539. Unidentified element or name + **denu** 'a valley'. The forms do not support derivation from OE **ċiriċe** 'a church' with AN [s] for [tʃ], genitive sing. **ċirċan**, secondary genitive sing. **-es**. O 374.

SARSON Hants SU 3042 → THRUXTON SU 2945.

Old SARUM Wilts SU 1332. The forms quoted under SALISBURY before c.1200 normally refer to this site. Already, however, by 1187 the two sites were distinguished and the first mention of *Vetus Saresbir'* occurs. Later forms are *Old Saresbury* 1429, *the old castell of Sar'* 1524, *the Olde Castell of Sarum* 1540, *Olde Sarum* 1581. The form *Sarum* appears to be a late abbreviation *Sar'* for *Saresbury*. Wlt 372.

SATLEY Durham NZ 1143. Possibly 'clearing of the (? robbers') lairs'. *Sateley* 1228–1393, *Satteley* 1303, *Satley* from 1336. OE **sǣt**, genitive pl. **sǣta**, **sāta**, + **lēah**. Ekwall's suggestion that the specific is OE **seotu**, the pl. of **set**, 'stables, folds', requires the further assumption of the ONb sound change *eo* > *ea*. NbDu 172.

SATTERLEIGH Devon SS 6622. Possibly 'wood or clearing of the robbers'. *Saterleia* 1086–1372 with variants *-leye- -legh(e)*. OE **sǣtere** 'robber' < OE *sǣtian* 'ambush', genitive pl. **sǣtera**, + **lēah**. D 349, PNE ii.95.

SATTERTHWAITE Cumbr SD 3392. 'Shieling clearing'. *Saterthwayt* 1336. ON **sætr** + **thveit**. La 219 gives pr [satəθət], SSNNW 156, L 210.

SAUGHALL Ches SJ 3670. 'Willow nook'. *Salhare* 1086, *-hale* 1086–1278, *Salghale -hall(e)* [1096×1101]17th, 1286–1516, *Saughall* from 1397. OE **salh** + **halh**, dative sing. **hale**. Che iv.202 gives pr ['sɔːgl], L 106, 110.

SAUL Glos SO 7409. 'Willow-tree wood, clearing or pasture'. *Salle* 12th–1584, *Sal(l)eg'* 1221, *Sall* 1501–1675, *Sawle, Saul(l)* 1557–1744. OE **salh**, genitive pl. **sala**, + **lēah**. Gl ii.179, L 203.

SAUNDBY Notts SK 7986. 'Sandi's village or farm'. *Sandebi* 1086–1176 etc., *-by* 13th, *Saundeby* [1154×89]14th, 1230–88, *Saunby* 1398–1460, *Sawnby* 1527. ON pers.n. *Sandi*, genitive sing. *Sanda*, + **bȳ**. The soil is variously described as 'clayey, overlying red marl and indurated clay' (Gaz 1897) and 'keuper marl with skerry' (SSNEM) which seems to exclude OE **sand**, ON **sandr** 'sand' as the specific. Nt 38, SSNEM 66.

SAUNDERTON Bucks SP 7901. Perhaps '*Sandris* hill'. *Santesdone -dune* 1086, *Santres- Sandresdon* 1189×99–1200, *Sa(u)ntredon* 1196, *Saund'don* 1227. OE p.n. **Sand-hrīs* < **sand** 'sand' and **hrīs** 'brushwood', + **dūn**. For a name **Sandris* cf. ACRISE Kent TR 1942 'oak brushwood', Foulrice NYorks SE 6270, *Fulryse* 1301, 'foul brushwood', Galtres (lost) NYorks, *Galtris* 1155×89, 'boar brushwood' (PrON **galtuR*). The occurrence of forms with and without medial *-s-* usually suggests variation between genitival and non-genitival composition and a pers.n. OE **Sandhere*, genitive sing. **Sandheres*, has been suggested here; but the loss of *-s-* may have been analogical. Another possibility is OE ***sanden** 'sandy' + **dūn** with dissimilatory change *n - n* > *n - r*, but this does not account for the early forms with *-s-*. A sizeable patch of surface sand

†*æt Sapere tún* [969]11th S 1324 belongs not here but to Saberton in Beckford Glos SO 9736.

occurs here. Bk 192 gives prs [sɔ:ndətən] and [sa:ndətən], Problems 91, PNE i.265, YN 8, 28, Signposts 168, L 144, 150.

SAUNDERTON STATION Bucks SU 8198. P.n. SAUNDERTON SP 7901 + ModE **station**.

SAUNTON Devon SS 4537. 'Sand settlement'. *Santon(e)* 1086, 1242, *Saunton* from 1304, *Sampton* 1505. OE **sand** + **tūn**. Saunton lies at the edge of the sand dunes of Braunton Burrows. D 33 gives pr [sa:ntən].

SAUSTHORPE Lincs TF 3869. 'Sauthr's outlying settlement'. *Saustorp* 1175, *Sauztorp* 1189, *Saucetorp* 1195, 1222. ON pers.n. *Sauthr*, genitive sing. *Sauths*, + **thorp**. DEPN, SSNEM 131.

SAVERNAKE FOREST Wilts SU 2266. P.n. *silva quae appellatur Safernoc* 'the wood called S' [933]14th S 424, *Savernak -ac* 1155 etc., *-ack* 1675, *-ake* 1684, *Severnak* 1224–1355, + ModE **forest**. If, as is generally supposed, this is late Brit **Saβrena* as in the r.n. SEVERN + PrW suffix **-ǭg** < Brit **-āco-*, it must in origin be the name of one of the rivers in the forest, perhaps an earlier name of the Bedwyn. Wlt 15 gives pr [sævərnæk], L 189.

SAWBRIDGEWORTH Herts TL 4814. 'Sæbriht's enclosure'. *Sabrixteworde* 1086, *Sebrihteswrde* 1118, *Sebrichteworde* 1159, *Sebricte(s)w(o)rth* 1222, 1237, *Sabrihteswrda -bric(h)t- -bri(gh)t- -worth -wurth* 1159–1321, *Sabrigeworth* 1338, 1428, *Sabrysworth, Sabresford* 1469, *Sabbis- Sabys- Sabesford* 1489–1549, *Sapsworth* 1565, 1727, *-ford* 1646, 1662, *Sabridgeworth* 1646, *Sawbridgeworth* 1770. OE pers.n. *Sǣbriht*, genitive sing. *Sǣbrihtes*, + **worth**. Hrt 193 gives former pr [sæps(w)əθ].

SAWDON NYorks SE 9484. 'Willow-tree valley'. *Salden(e)* 13th–1562, *Sawdon* from 1570. OE **salh** + **denu**. YN 97, L 98.

SAWLEY Derby SK 4732. 'Willow hill'. *Salle* 1086, 1212, *Sallaw(a) -lawe* 1166–1330, *Sallou -low(e)* 1176–1660, *Sawla* 1535×43, *Sawley* from 1577. OE **salh** + **hlāw**. The reference is to the slight rise on which the church stands. The two forms *Salle* are errors for *Sallo*. Db 499.

SAWLEY 'Willow clearing or pasture'. OE **salh** + **lēah**.

(1) ~ Lancs SD 7746. *Sotleie* 1086, *Sallai(a) -lay(a) -lei(a) -ley* [1147]14th–1523, *Sawley* from 1546. YW vi.182, L 203.

(2) ~ NYorks SE 2467. *Sal-lege* c.1030 YCh 7, *Sallai(a) -lei(a) -lay -ley(a)* 1086–1527, *Sawley* from 1504. YW v.187, L 203.

SAWREY Cumbr SD 3795. 'Muddy places'. *Sourer* 1336, 1400, *Sawrayes* c.1535, *Sawrey* 1656. ON **saurar**, plural of **saurr** 'sour ground, mud, dirt, marshy ground'. Remodelled as if a compound of **saurr** and **ey** 'an island'. La 219 gives pr [sɔ:rə], SSNNW 156.

SAWSTON Cambs TL 4849. 'Salsi's estate' or 'estate of the Salsingas, the people called after Salsi'. *Salsinge- Delsingetune* (sic), *Selsingetona* [970]17th, *Salsintona* 1086, [1189×99]1315, *Salsiton(e) -a* 1086, *Salsetune* c.1280, *Salston(e)* 1285, 1312, *Sausetun* 1201, *Sauston(e)* 1272–1435, *Sawson* 1587, 1603×25. OE pers.n. **Salsi* + **ing**[4] + **tūn** or possibly folk-n. **Salsingas* < **Salsi* + **ingas**, genitive pl. **Salsinga*, + **tūn**. Ca 96 gives former pr [sɔ:sən].

SAWTRY Cambs TL 1683. 'Salt or salters' stream'. *Saltrede* 1086, *-a* 1183, *-retha* c.1350, *Saltreia, Saltre(y)* 1157–1363, *Saltereia -eye* 1147–1242, *Sauteria -eye* 1184–1228, *Sautre(ye)* 1235 etc. OE **salt** + **rīth**. The specific may have been OE **saltera**, genitive pl. of **saltere** 'a salter, a salt-worker'; if so the meaning is probably stream by which salt was transported. Hu 195.

SAXBY 'The village of the Saxons'. OE *S(e)axe*, genitive pl. *S(e)axa* + **bȳ**. It seems probable that the Danes did not distinguish between Angles and Saxons but used the two national terms synonymously. However, the specific might rather be ODan pers.n. *Saxi*, genitive sing. *Saxa*.

(1) ~ Leic SK 8220. *Saxebi* 1086–1270, *-by* c.1141–1380, *Sauceby* 1189, *Sawsby* 1577. Alternatively the specific could be ON pers.n. *Saksi*, genitive sing. *Saksa*. Lei 166.

(2) ~ ALL SAINTS Humbs SE 9916. *Saxebi* 1086–13th, *Saxeby* 1179–1441, *Saxby* from 1327. 'All Saints' from the dedication of the church and for distinction from SAXBY-BY-LINCOLN TF 0086. Li ii.254, SSNEM 6.

SAXBY Lincs TF 0086. 'Saksi's village or farm'. *Sassebi* 1086, *Saxsebi -abi* c.1115, *Saxeby* 1206. ODan pers.n. *Saxi*, genitive sing. *Saxa*, + **bȳ**. SSNEM 66.

SAXELBYE Leic SK 7020. Partly uncertain. *Saxelbie* 1086, *-by* from 1219. Uncertain element or name + ON **bȳ** 'farm' or 'village'. The spellings do not support derivation from a pers.n. ON *Sasúlfr*. The specific might be an OE r.n. **Saxel* derived from OE **seax** 'a knife'. *Sax* occurs as a r.n. in Danish Gladsakse.

Great SAXHAM Suff TL 7862. *Saxham Magna* 1254, 1610. ModE **great**, Latin **magna**, + p.n. *Saxham, soxa, Sexhā, sexham* 1086, 'the homestead of the Saxons', OE *Seax-hām* < *Seaxe* + **hām**. DEPN.

Little SAXHAM Suff TL 7963. *Saxham Parva* 1254, 1610. ModE **little**, Latin **parva**, + p.n. Saxham as in Great SAXHAM TL 7862. DEPN.

SAXILBY Lincs SK 8975. Partly uncertain. *Saxebi* 1086, *Saxlabi -le-* c.1115, *Saxelebi* 1143×7, 1160×6. Possibly 'Saxulfr's village or farm', ON pers.n. *Saxúlfr* + **bȳ** although the absence of forms with *-f-* is difficult. Possibly the specific is an unrecorded r.n. **Saxel* from OE **seax** 'a knife' referring to Foss Dyke, an artificial watercourse cut by the Romans to connect the Witham and the Trent. DEPN, SSNEM 66, Cameron 1998.

SAXLINGHAM Norf TG 0239. 'Homestead of the Seaxlingas, the people called after Seaxel'. *Saxeling(h)aham -ingham, Sexelingaham* 1086, *Saxlingham* from 1198. OE folk-n. **Seaxelingas* < pers.n. **Seaxel* + **ingas**, genitive pl. **Seaxelinga*, + **hām**. DEPN, ING 138.

SAXLINGHAM NETHERGATE Norf TM 2297. 'Saxlingham lower hamlet'. *Saxlyngham Neyergate* (sic for *Neþer-*) 1291. P.n. Saxlingham + ME p.n. *Nethergate* < **nether** + **gate** in the sense 'hamlet'. So called for distinction from Saxlingham Thorpe, *Saxlyngham Thorp* 1291, 'Saxlingham outlying farm', p.n. Saxlingham + **thorp**. Saxlingham, *Sexlingham* [1042×53]13th S 1535, *Sasil- Sa(i)selingaham, Saisilingeham, Sasselingham* 1086, *Saxlingaham* 1163, *Saxlingeham* 1202, *Saxlingham* 1242×3, is the 'homestead of the Saxlingas, the people called after Seaxel', OE folk-n. **Seaxlingas* < pers.n. **Seaxel* + **ingas**, genitive pl. **Seaxlinga*, + **hām**. ING 138.

SAXMUNDHAM Suff TM 3863. 'Seaxmund's homestead'. *Saxmonde- Sasmunde(s)ham* 1086, *Saxmundham* from 1213 with variants *-mond(e)-*. OE pers.n. **Seaxmund* + **hām**. DEPN, Baron.

SAXONDALE Notts SK 6839. 'The valley of the Saxons'. *Saxeden* 1086, *Saxenden* 1316, *Saxendala -e* c.1130–1358. OE folk n. *Seaxe*, genitive pl. *Seaxna*, + **denu** later replaced by ME **dale**. The name marks an isolated settlement of Saxons in predominantly Anglian territory. Nt 241, L 95, 99.

SAXON STREET Cambs TL 6859. *Saxon Street* 1836 OS. P.n. Saxon + Mod dial. **street** 'a hamlet' for distinction from Saxton Hall, *Saxon Hall* 1836 OS. Saxon, or Saxton, *Sextuna -tone* 1086, *-ton(e)* 1208–85, *Saxton(e)* 1236–1484, *Saxon* 1570, *Saxham* 1695, is probably the 'settlement of the Saxons', OE *Seaxe*, genitive pl. *Seaxa*, + **tūn**. The absence of any spellings of the form *Seaxetun* with medial *-e-*, however, is a problem; the name must therefore be a direct compound on the stem *Seax-* like *Swæf-* in SWAFFHAM TL 5562, TF 8109. Hypothetical OE ***seax** 'a stone, a rock' is unlikely. Ca 127.

SAXTEAD Suff TM 2565. 'Seaxa's place'. *Saxtedam* 1086, *Sasted(e)* 1254–1316, *Saccestede* 1269, *Sacstede* 1291, *Saxsted* 1327, 1524, 1568, 1610. OE pers.n. *Seaxa* + **stede**. Stead 193.

SAXTEAD GREEN Suff TM 2564. *Saxtead Green* 1837 OS. P.n. SAXTEAD TM 2565 + ModE **green**.

SAXTHORPE Norf TG 1130. 'Saks's outlying farm'. *Saxthorp, Sastorp, Saxiorp* (sic for *-torp*) 1086, *Saxthorp* 1254. ON pers.n. **Saks* + **thorp**. The name is identical with Saxtorp in Skåne, *Saxtorp* 1274, *Saxthorp* 1352. DEPN, SPNN 323–4.

SAXTON NYorks SE 4736. Probably 'Saxi's estate'. *Saxtun(a) -ton(a)* 1086–1593. ON pers.n. *Saxi*, genitive sing. *Saxa*, + OE **tūn**. Possible alternatives are 'settlement of the Saxons', OE *Seaxe*, genitive pl. *Seaxa*, + **tūn**, which would imply an early immigration of Saxons into WYorks, or, 'rock settlement', OE **seax** + **tūn**. YW iv.70, SSNY 129.

SAYERS COMMON WSusx TQ 2618. *Sayer's Common* 1665, *Sawyers* 1813. Surname *Sayer, Sawyer* + ModE **common**. A Walter *le Saghier* is recorded at Poynings in 1323. Sx 275.

SCABBACOMBE HEAD Devon SX 9251. P.n. Scabbacombe + ModE **head** 'a headland'. Scabbacombe, *Schobecumb* c.1250, is 'Sceobba's coomb', OE pers.n. *Sc(e)obba* + **cumb**. D 509.

SCACKLETON NYorks SE 6472. Uncertain, possibly 'quaking grass valley'. *Scachelden(e), Eschalchedene* 1086, *Skakilden* 1138–1408, *Sc- Skakelden(a)* 1231–1328. The generic is OE **denu** 'a valley', the specific a Scandinavianised form, with substitution of [sk] for [ʃ], of OE **sceacol** 'a shackle'. This word is a derivative of OE *sceacan* 'to shake' and may have also meant 'quaking grass, stubble' in p.ns., like Mod dial. *shackle*. Others have sought a link with ON **skekill** 'a corner, a point, a tongue of land' related to OHG *scahho* 'promontory'. YN 51, cf. YW iii.201, SSNY 163, L 99.

SCA FELL Cumbr NY 2006. 'Fell called Bald-head'. *Skallfeild* 1578, *Scottfield* 1749, *Scoffield* 1750, *the mountain Scofell or Scowfell* 1794, *Sca Fell* 1865 OS. ON **skalli** used as a nickname for the bare mountain + **fjall**. Rises to 3162ft. Cu 390 gives pr [scɔ:], *sic* for [skɔ:], Jnl 2.58.

SCAFELL PIKE Cumbr NY 2107. *Sca Fell Pike* 1865 OS. The peak of Scafell. P.n. SCA FELL NY 2006 + dial. **pike** 'a pointed summit' (OE *pīc*, ON *pík*).

SCAFTWORTH Notts SK 6691. 'Shaft enclosure'. *Scafteorde* 1086, *Scaftworð* 1202, *-worth* from 1323, *Sc- Skaftew(o)rth(')* 1226–1315. OE **sceaft**, genitive pl. **sceafta**, + **worth**. The reference is either to the structure of the enclosure from poles or shafts, or possibly to their manufacture, cf. HURWORTH. Alternatively the specific might be an OE pers.n. **Sceafta*. In either case English [ʃ] has been replaced by Scandinavian [sk]. Nt 38, SSNEM 215.

SCAGGLETHORPE NYorks SE 8372. Uncertain, possibly 'Skakel's outlying farm'. *Scachetorp* 1086, *Scaketorp* 1207, *Scakilt(h)orp -el-* [1154×9]15th, 1285, 1305, *Scagilthorpe* 1441. ON pers.n. **Skakull, Skakli* + **thorp**. Doubts have been raised about the occurrence of this pers.n. in England and the specific may therefore be topographical, related to that discussed under SCACKLETON SE 6472. SSNY 66, YE 139 gives pr [skaglθrəp].

SCALBY NYorks TA 0190. 'Skalli's village or farmstead' or possibly 'village by *Skalli*, the bald head'. *Sc- Skal(l)ebi -by* 1086–1280 (*-ll-* to 1400), *Scalby* from 1322, *Sc- Skawby* 16th cent. ON pers.n. *Skalli*, genitive sing. *Skalla*, + **bӯ**. *Skalli* might be the name of a bare hill, 'Bald-head', which could fit the topography here. SSNY 36, YN 108 gives pr [skɔ:bi].

SCALBY NESS ROCKS NYorks TA 0391. P.n. Scalby Ness + ModE **rock**. Scalby Ness is p.n. SCALBY TA 0190 + ModE **ness** (ON *nes*) 'a headland, a promontory'.

SCALDWELL Northants SP 7672. 'The shallow spring'. *Sca(l)de(s)welle* 1086, *Scalde- Schaude- Skaldewell(e)* 12th–1313, *Shaldewell* 1298. OE **sceald** with [sk] instead of [ʃ] due to Scandinavian influence + **wella**. Nth 131 gives pr [skɔ:ldwəl], SSNEM 220.

SCALEBY Cumbr NY 4463. 'Village by the shielings'. *Schaleb(er)y* c.1235, *Scaleby* from c.1245 with variants *Skale-* and *-bi*, *Skailbye* 16th. ON **skála**, genitive pl. of **skáli**, + **bӯ**. Also referred to as *villa de Scales* c.1180, *manerium de Scales* 1227, ON **skáli** + secondary ME pl. **-es**. Cu 106, SSNNW 39.

SCALEBY HILL Cumbr NY 4363. *Scalebyhill* 1787. P.n. SCALEBY NY 4463 + ModE **hill**. Cu 106.

SCALE HOUSES Cumbr NY 5845. 'Houses at Scale'. *Scailhouses* 1531, *Skalehowses* 1586. P.n. Scale + ModE **house(s)**. Scale, *le Schal* 1332, is short for Renwick Scale(s), *Rauenwykchales* 1278, *Ravenwik Scalez* 1485, 'Renwick shieling(s)', p.n. RENWICK NY 5943 + ON **skáli**, secondary ME pl. **skales**. Cu 236.

SCALES 'Huts, shielings'. ME **skales** < ON **skáli** + secondary ME pl. **-es**.

(1) ~ Cumbr NY 3426. *Skales* 1323. In Threlkeld township. Cu 253, SSNNW 68 no.4.

(2) ~ Cumbr SD 2672. *Scales* 1269, 1418. In Aldingham township. La 208, SSNNW 68 no.2.

(3) ~ Lancs SD 4530 → NEWTON SD 4431.

(4) ~ MOOR NYorks SD 7177. P.n. Scales *(the) Sc- Skales* 1379, 1674, *Skalls* 1619, + ModE **moor**. YW vi.247.

SCALFORD Leic SK 7624. 'Shallow ford'. *Scaldeford* 1086, *-ford(e)* 1107–1430, *Scaudeford* 1180–1340, *Scaldford* 1381, *Scalford* from 1392. OE **sc(e)ald** + **ford**. The modern pronunciation is due to ON substitution of [sk] for English [ʃ]. Lei 182, SSNEM 220.

SCALING Cleve NZ 7413. 'The shieling(s)'. *Skalynge -inge* [12th]15th–1577, *Scalingis* [1243×73]c.1498, *Estskaling* 1415. ME **skaling** < ON **skáli** + **ing**. YN 139, PNE ii.123.

SCALING RESERVOIR Cleve NZ 7412. P.n. SCALING NZ 7413 + ModE **reservoir**.

SCAMBLESBY Lincs TF 2778. Partly uncertain. *Scamelesbi* 1086, *Scamelbi* 1146, *Scam(e)lesbi -by* 1160–1202. It is unclear whether the specific is a pers.n. such as *Skammlaus* from ON adj. *skammlaus* 'shameless', *Skammel* or *Skammhals* 'short-neck' or an appellative such as OE **sceamol** 'a shamble, a bench, a stall' with ON [sk] for English [ʃ] used in a topographical sense such as 'shelf, ledge'. Scamblesby stands on a shelf in a broad but steep-sided river valley. DEPN, SSNEM 66.

SCAMMONDEN WATER WYorks SE 0516. P.n. Scammonden + ModE **water**. Scammonden, *Sc- Skambandene* 1275–1323, *Scammendene* 1316, *Scammonden* from 1581, is 'Skammbein's valley', ON pers.n. *Skammbein* 'short-leg' + OE **denu**. YW ii.304, L 98.

SCAMPSTON NYorks SE 8675. 'Skammr or Skammel's estate'. *Scameston(a)* 1086, *-tun -ton(a)* [a.1080]15th–1366, *Scamastuna* 1122×37, *Sc- Skamston* 1137×47–1596, *Scameliston'* 1202, *Skameleston'* 1244, *Skampston* 1351. ON pers.n. *Skammr* apparently varying with its diminutive form *Skammel*, genitive sing. *Skamm(e)s, Skammel(e)s*, + **tūn**. Alternatively the pers.n. might be the ON by-n. *Skammhals*. YE 138, SSNY 129, SPN 245, DBS 473 s.n. Scammel.

SCAMPTON Lincs SK 9479. Partly uncertain. *Scanton(e) -tune* 1086, *-tuna* c.1115, *Sc(h)amtona -tun(a)* 12th. This appears to be ON **skammr** 'short' possibly replacing a cognate OE ***scamm** + **tūn**, but the significance of such a name if correctly explained is unknown. DEPN, SSNEM 185.

SCAR HOUSE RESERVOIR NYorks SE 0577. P.n. Scar House + ModE **reservoir**. Scar House SE 0677, *Scarrehowse* 1607, is ModE **scar** (ON *sker* 'a rocky cliff') + **house** (OE *hūs*). YW v.219.

SCARBOROUGH NYorks TA 0488. 'Skarthi's fortified place'. *Escardeburg* 1155×63, 1256, *Scardeburc(h) -burg* 1159–1505, *Scarðeborc* c.1200, *Skarðabork* [c.1200]c.1390, *Scartheburg(h)* 1208 etc., *Scareburgh* 1414, *Scarbrowgh* 1573. ON pers.n. *Skarthi*, genitive sing. *Skartha*, + **borg**. The 13th cent. Icelandic *Kormákssaga* tells that Kormak and his brother Thorgils Skarthi, 'Thorgils Harelip', were the founders of the stronghold called *Skarthaborg* while raiding England. This was in 966–7. YN 105, APhS 1.320.

SCARCLIFFE Derby SK 4968. 'Steep slope with a gap'. *Sc- Skardeclif(f) -clive -clyve -clyf(e)* 1086–1413, *Sc- Skartheclif(e)* etc., c.1166–1336, *Sc- Skarclif(f)(e) -y-* from 1330. OE **sceard** influenced by ON **skarth** + **clif**. Refers to the valley which cuts into the limestone escarpment at SK 4968. Db 294, L 131, 135, SSNEM 220–1.

SCARCROFT WYorks SE 3641. 'Enclosure at the gap'. *Sc- Skardecroft(e)* 1160×75–1252, *Sc- Skarthecroft(e)* 1174–1348, *Scarecroft* 1285, *Scarcroft* from 1491. OE **sceard** replaced by ON **skarth** + **croft**. The reference could be to a gap in the hills, or a gap in a fence. YW iv.101.

SCARGILL Durham NZ 0510. 'Skakari's ravine'. *Seachregil, Scracreghil* (sic) 1086, *Scakregill* 1172, *Sca(c)kergill'* 1173–1294, *Skargill* from 1282. ON pers.n. *Skakari*, genitive sing. *Skakara*, + **gil**. The name has been reformed under the influence of dial. **scar** 'a rocky cliff' (< ON *sker*) with reference to nearby **scars** along the banks of the river Greta. YN 303, SPN 242.

SCARGILL HIGH MOOR Durham NY 9909. *Scargill High Moor* 1862 OS. P.n. SCARGILL NZ 0510 + ModE **high** + **moor**. Contrasts with *Scargill Low Moor*.

SCARGILL RESERVOIR NYorks SE 2353. P.n. Scargill + ModE **reservoir**. Scargill, *Skergill* 1323, is ON **sker** 'a rock, a scar, a rocky cliff, a bed of rough gravel' + **gil** 'a ravine'. YW v.120.

SCARHILL Devon SX 6794 → SCORRIER Corn SW 7244.

SCARISBRICK Lancs SD 3713. Possibly 'slope of the depression'. *Scharisbrec* c.1200, *Sc- Skaresbrec -k(e) -is-* 1200–1346. ON **skǫr**, ODan **skar**, genitive sing. **skares**, + **brekka**. Most of the area is low-lying carrs, but the village and Scarisbrick Hall lie on slightly higher ground. The depression might be a reference to the low valley of Sandy Brook SD 3813. La 124 gives pr [ske:zbrik], SSNNW 158, L 129, Jnl 17.69.

North SCARLE Lincs SK 8466. *Northscarle* from 1240, *North Skarle* 1332, 1602. ME adj. **north** + p.n. *Scarle* 1185–1482, *Scaruell' -nell* 1202, *Skarele* 1223×39 as in South SCARLE Notts SK 8463. Also known as *Parva Scarle* 1230 'little Scarle'. Probably a secondary settlement dependent on South Scarle. But the spellings *Sacaruell', Scarnell* (? erroneously for *-uell*) might point to an original OE **scearn** + **wella** 'dirty spring' subsequently modified under the influence of the forms of South Scarle. Perrott 258, SSNEM 227.

South SCARLE Notts SK 8463. *Suthscarl', Suth Scarle* 1240–1330, *South Skaryll* 1557. ME adj. **sūth** + p.n. *Scornelei* 1086, *Scarlai* 1147, 'dung wood or clearing', OE **scearn** with Scand [sk] for English [ʃ] + **lēah**. 'South' for distinction from North SCARLE Lincs SK 8466. Nt 206, SSNEM 221.

SCARNING Norf TF 9512. 'The dirty brook' and 'the dwellers on the dirty brook'. *Scernenga -inga, Scherninga, Scerniga* 1086, *Sc- Skerninges* 1199–1250, *Skerning(e)* 1223×4, 13th, *Schernigges* 1254. OE r.n. **Scearning* < **scearn** 'dung' + **ing**[2] varying with folk-n. **Scearningas* < OE **Scerne* 'dirty brook' + **ingas**. Initial [sk] for [ʃ] is due to Scandinavian influence, cf. the p.n. *Skjerninge* in Denmark. DEPN, ING 60.

SCARRINGTON Notts SK 7341. Probably the 'settlement on the dirty stream'. *Scarintone* 1086, *Scherninton', Shernintona* 1166, 1167, *Skeryngton -ing-* 14th cent., *Skarington* 1477. OE ***scearning** < **scearn** + **ing**[4] + **tūn**. The reference is to Carr Dyke SK 7443. Initial [sk] for [ʃ] is due to Scandinavian influence. Cf. the Danish p.n. Skerninge which is thought to be a derivative of **skarn**. Nt 228, Årsskrift 1974.45, SSNEM 185. Mx xxxii records local pr [skæritn].

The SCARS Northum NZ 2993. 'The reefs'. *The Scars* 1866 OS. ModE **scar** 'a low or sunken rock in the sea' (ON *sker* 'a rock, a scar, a reef, a skerry').

SCARTH GAP PASS Cumbr NY 1813. P.n. Scarth Gap + ModE **pass**. Scarth Gap, *Scarf Gapp* 1821, is ON **skarth** 'a notch, a cleft, a mountain pass' + **gap** 'a break or opening in a range of mountains'. A mountain pass linking Buttermere and Ennerdale. Cu 387.

SCARTH HILL Lancs SD 4206. 'Cleft hill'. P.n. Scarth, *Scarth* c.1190, ON **skarth** 'an opening, an open place in the edge of something, a gap, a pass', + ModE **hill**. La 123, SSNNW 158.

SCARTHO Humbs TA 2606. Probably the 'gap hill or mound'. *Scarhou* 1086, 1271, 1323, *Scarfho(u)* 1176–1214, *Scarho* 1177–9, *Scartho* from 1190. ON **skarth** + **haugr**. The reference is unknown as the area is now heavily built-up and the original topography irrecoverable. The alternative suggestions, ON *skarfr* 'a cormorant', *skarv* 'bare', depend upon *f*-spellings in a single unreliable source. Li v.135, SSNEM 160.

SCAWBY Humbs SE 9705. Possibly 'the village or farm on the bare slope'. *Scalebi* 1086, *Scallebi* 1086, c.1115, *Scallabi* c.1115. ON **skalli**, genitive sing. **skalla**, + **bȳ**. The village stands on the slopes of the Wolds. Alternatively the specific might be ON pers.n. *Skalli*, genitive sing. *Skalla*. DEPN, SSNEM 67.

SCAWTON NYorks SE 5483. 'Valley settlement'. *Scaltun* 1086–[1181]15th, *Sc- Skalton(a)* [1189]16th–1414, *Scaulton* 1328, *Scawton* 1575. ON **skál** 'a hollow' + OE **tūn**. This suits the topography, although on formal grounds ON **skáli** 'a shieling', perhaps replacing OE ***scēla**, is also possible. YN 56, SSNY 115.

SCAYNE'S HILL WSusx TQ 3623. *Skerns Hill* 1586, *Skaynes* 1588, *Scheanes, Scarmes, Skermes, Skaines (Hill)* 1819–49. Uncertain surname + ModE **hill**. The early 19th cent. forms show great uncertainty about this name. Sx 343.

The SCHIL Northum NT 8622. *The Schel* 1869 OS. A mountain peak on the England–Scotland border rising to 1985ft. Possibly identical with Shill in SHILLMOOR NT 9415 and SHILLMOOR NT 8807.

SCHOLAR GREEN Ches SJ 8357. 'Green at *Scolehalc*'. *Schollers Green* 1668, *Scol(l)ar Green* 1767, *Schollow Green* 1704, 1741, ~ *alias Schollar Green* 1813. P.n. *Scolehalc* + ModE **green**. *Scolehalc -haleth* 1272×1307, *Scol(e)halg(h)* 1286–1369 with variants *Skol(e)-* and *-halch* etc., *Scolale* 1280, *Sc(h)olehall* 1308–1410, is the 'nook with a hut', ON **skáli** + OE **halh**. Che ii.308, SSNNW 249.

SCHOLES 'The huts'. ME **skole** (ON *skáli*), pl. **skoles**. The reference is primarily to shielings.

(1) ~ WYorks SE 3737. *Sk- Scales* 1258–1543, *Scholes* from 1259. YW iv.109.

(2) ~ WYorks SE 1607. *Scoles* 1274–1392, *(le) Scholes* from 1284. YW ii.247.

Isles of SCILLY SV 8912. Unexplained. The ancient sources are Pliny NH 4.103 *Silumnus* [77]9th with variant readings *Silimnis*, Solinus 22.7, *Siluram insulam* [c.200]9th (with variant *Sillinas insulas* from 10th cent. mss) Sulpicius Severus, *Sylinancim* (accusative case) [c.400] (with variant *Sylinam*); the medieval and modern forms *Sulling* [c.1160]13th, *Sully* [1176]13th, *Syllingar*, ON plural presumably in the sense 'the people of Scilly' [c.1200]late 14th, *Sylly* 1460, *Iles of Scillye* [1568]1652, *Sillan* late 17th. The late 17th cent. form *Sillan* represents genuine oral tradition. Otherwise about the only certain thing that can be said is that the *c* of the modern spelling is not original but was added for distinction from ModE *silly* as this word developed in meaning from 'happy, blissful' to 'foolish'. PNCo 153.

SCOAT FELL Cumbr NY 1511. P.n. Scoat + ModE **fell** (ON *fjall*). Scoat, *le Scote in Bouthdale* 1338, *Leescote of Boulderdale* 1578 ('the Scoat of Bowderdale'), is ON **skot** 'projecting piece of land, especially high land; a rising hill; sometimes, a place where timber is shot down a hill'. Scoat Fell rises steeply to over 2700ft. Cu 441.

SCOLE Norf TM 1579. 'The sheds'. *Escales* 1191. ME **scale** < ON *skáli*, pl. **scales**. DEPN.

SCONCE POINT IoW SZ 3389. *Sconce Pt.* 1720. ModE **sconce** + **point**. The reference is to a *sconce* or small fort built here in the reign of Elizabeth I by Sir George Carey called *Sconce Fort* 1700 and *Carey's Sconce* 1781. The earlier name was *Scharpenorde* 1324 'the sharp point', OE **scearp**, definite form dative case **scearpan**, + **ord**. Wt 130, Mills 1996.92.

SCOPWICK Lincs TF 0758. 'The sheep farm'. *Scapuic,*

Scapeu(u)ic 1086, *Scap(e)wic -wyk* 1170–1250, *Scapwik -wyk* c.1150–1337, *Scaup(e)wik -wyk* 1242–1367, *Scopwick(e)* from 1558×9. OE **scēap** + **wīċ** with Scand [sk] for English [ʃ], [k] for [tʃ] and early shortening of OE *ēa, ǣ> ǣ*. Perrott 337, SSNEM 215.

SCORBOROUGH Humbs TA 0145. 'The temporary shelter in or at the wood'. *Scogerbud* 1086, *Sc- Skoureburg(h)* 13th–1323, *Scorburgh(e)* [13th]1304–1547. ON **skógr**, genitive sing. **skógar** + **búth**, as in the Dan p.n. Skovbo in Aalborg Amt, *Skoffuebod* 1487 from **skógar-bóth**. The final element **búth** was replaced by OE **burh** when a castle was built here. YE 162, SSNY 88.

SCORRIER Corn SW 7244. 'Mining waste'. *Scoria* 1330, *Scorya* 1350, 1404, *Scorrier* 1748. Latin **scoria**. Cf. the Devon p.n. Scarhill SX 6794, found as a surname *Scoriawell* 1333. PNCo 154, Gover n.d. 515, D 449, ii.xiii.

SCORTON Lancs SD 5048. 'Settlement by the ravine'. *Scourton* c.1550, *Skurton* 1563. ON **skor** + **tūn**. There is a deep ravine close to the church. La 164.

SCORTON NYorks NZ 2500. 'Settlement on the rift-like stream'. *Scorton(e) -tona* 1086–1665. ON **skor** possibly replacing OE ***scor(u)**, + **tūn**. The reference is to Scorton Beck. YN 278, SSNY 113.

SCOT'S GAP Northum NZ 0486. *Scot's Gap Station* 1866 OS. According to Tomlinson 257, the name commemorates an old Scottish raid.

SCOTBY Cumbr NY 4454. 'Village or farm of the Scots'. *Scoteby -bi* 1130–1291 with variant *Skote-*, *Scottebi -by* 1167–1339, *Scotby* from 1310. ON *Skotar* or OE *Scottas*, genitive pl. *Skota, Scotta*, + **bȳ**. Cu 163, SSNNW 39.

SCOTCH CORNER NYorks NZ 2105. *Scotch Corner* 1860 OS. A junction on the Great North Road where the main road to Scotland via Carlisle branches off.

SCOTFORTH Lancs SD 4859. 'Scots' ford'. *Scozforde* 1086, *Scotesford'* 1203, *Scoteford* 1204, *Scotford* 1203–1332 etc., *Scotforth* 1501. OE *Scot*, genitive sing. *Scottes*, genitive pl. *Scotta*, + **ford**. Possibly a reference to Scottish drovers, but the OE pers.n. *Scot(t)* is also possible. La 173, Jnl 17.99.

SCOTHERN Lincs TF 0377. 'The Scot's thorn-tree'. *Scot(st)orne* 1086, *Scotstorna* c.1115, *Scosthorne* 1203×6. OE *Scot*, genitive sing. *Scottes*, + **thorn**. DEPN, L 221.

SCOTLAND GATE Northum NZ 2584. A modern colliery settlement.

SCOTNEY CASTLE Kent TQ 6835. *Scotney Castle* 1819 OS. A 14th century moated castle with later developments. The name Scotney is a manorial name, i.e. a p.n. derived from the name of the family which held the manor as in Walter de *Scotenii* c.1180, later the barony of *Sc- Skoteneye* 1276–1383. The name is ultimately from Étocquigny, Seine Inférieure, France. PNK 201, Newman 1969W.486, DBS 309.

SCOTS HOLE Lincs TF 1264. Perhaps surname *Scott* + ModE **hole**.

SCOTTER Lincs SE 8800. Possibly the 'tree of the Scots'. *Scottere* [1061×6]12th S 1059, *Scot(e)re* 1086, *Scotra* c.1115. OE *Scot*, genitive pl. *Scotta* referring to the inhabitants of nearby SCOTTON SK 8899, + **trēow**. DEPN.

SCOTTERTHORPE Lincs SE 8702. Partly uncertain. *Scaltorp* 1086, 1212, *-thorpe* [1189]14th, *Salttorp* c.1115, *Scale- Scaldestorp* c.1128, *Scalkestorpe* [1067×9]c.1150, 12th, *Scalðorpe* 13th S 68 (a forged charter purporting to be of 664). The earliest form appears to be 'Skalli's outlying settlement' or 'outlying settlement on the bare slope', ON pers.n. *Skalli*, genitive sing. *Skalla*, or adj. **skalli** 'bald-head, bald hill' from which the pers.n. derives, + **thorp**. The name was subsequently influenced by the ON pers.ns. *Skáld* and *Skalk* and finally assimilated to nearby p.n. SCOTTER SE 8800. DEPN, SSNEM 116.

SCOTTON 'Settlement of the Scot or Scots'. OE *Scott* 'a Scot (or Irishman)', genitive pl. *Scotta* (in some instances), + **tūn**.

(1) ~ Lincs SK 8899. *Scottun* [1061×6]12th S 1059, 1212, *Scotone -tune* 1086, *Scottuna* c.1115. DEPN.

(2) ~ NYorks SE 1896. *Scot(t)une* 1086, *Scotton(a)* from [1199×1210]13th. YN 246.

(3) ~ NYorks SE 3259. *Scotone -tona* 1086, *Scotton* from 1167. YW v.92.

SCOTTOW Norf TG 2723. 'Scoto hill'. *Scoteho* *[1044×7]13th S 1055, *Scotohou* 1086, *Scothowe* 1117, *-houe* 1177. The DB form is taken to be OE p.n. *Scothōh* or *Scottahōh* 'hill-spur of the Scots' + pleonastic ON **haugr** 'a hill'. DEPN.

SCOULTON Norf TF 9800. 'Skuli's estate'. *Sculetuna* 1086, *-ton* 1198, 1212. ON pers.n. *Skúli*, genitive sing. *Skúla*, + **tūn**. DEPN, SPNN 337.

SCOUT HILL Cumbr SD 5682. *Scout Hill* 1865. ModE **scout** (ON *skúti*) 'a projecting rock' + **hill**. We i.48.

East SCRAFTON NYorks SE 0884. ModE **east** + p.n. *Sc(h)rafton* [1184]15th, 1270 etc., as in West SCRAFTON SE 0783. YN 253, SSNY 117.

West SCRAFTON NYorks SE 0783. ME *West* from 1285 + p.n. *Sc- Skraftun -ton* from 1086[†], the 'settlement at the hollow', OE **scræf** + **tūn**. YN 255.

SCRAINWOOD Northum NT 9909. 'The wood at the hollow place'. *Scravenwod'* 1242 BF, 1324, *Scrawenewude* 1255, *Scran(e)wod* 1288–1428. OE ***scræfen** + **wudu**. Scrainwood lies at the entrance to the deep valley of Spartley Burn. NbDu 172, PNE ii.114.

SCRANE END Lincs TF 3841. *Crane End* (sic) 1824 OS. P.n. Scrane + ModE **end**. Short for Scrane end of FREISTON TF 3743 in contrast to HALLTOFT END TF 3645. Scrane, *Scrainga* 1158, *(vetus) Screing(a)* 'old S' 1158–14th, *Scrainge(s)* 1195–1252, *Screhinges, Scrahing* 1202, *Escrahinghe* 1212, *Skraing(e)* 1396, 1485, is a singular **ing**[2] formation on an unknown element, possibly OE ***scraga** 'a trestle' and so OE ***scræġing** 'a structure of poles, a trestle-like contrivance' with Scand [sk] for English [ʃ]. Payling 1940.122, ING 62, ENEP 85, BzN 1967.344, SSNEM 224.

SCRAPTOFT Leic SK 6405. Partly uncertain. *Scraptofte* [1043]15th S 1226, *Scrapetoft* [1043]15th S 1000, *Scrapetoft(e)* 1205–1394, *Scraptoft(e)* from 1276, *Scrapentot* 1086. The generic is ON **toft** 'a curtilage', the specific either ON **skrap** 'scraps, scrapings' perhaps referring to arid, barren soil, or the rare ON pers.n. *Skrápi*, genitive sing. *Skrápa*. Scraptoft is situated on a small patch of sand and gravel overlying boulder clay. Lei 249, SSNEM 149.

SCRATBY Norf TG 5115. 'Skrauti's homestead or village'. *Scroutebi -by* [c.1020]14th and late 13th S 1528, *Scroutebei -bey, Scroteby* 1086, *Scrouteby -bi* 1202–1428, *Scrout- Scrowtby* c.1390–1547, *Scrotby(e)* 1535–76. ON pers.n. *Skrauti*, genitive sing. *Skrauta*, + **bȳ**. Nf ii.14, SPNN 336.

SCRATCHBURY Wilts ST 9144. 'The haunted fort'. *Scratchburie* 1609. *Scratch*, or *Old Scratch*, a name for the Devil (earlier *scrat*, OE **scratta*, cognate with OW *skratti* 'wizard, goblin, monster') + **bury** (OE *byriġ*, dative sing. of *burh* 'a fort'). The reference is to Scratchbury Camp, the most impressive earthwork in Wiltshire, a large univallate Iron Age hill-fort superbly sited on a spur overlooking the Wylye valley. Within it there are five Bronze Age round barrows, one very large, which account for the name. Wlt 154, Pevsner 1975.361, Thomas 1980.239.

SCRAYINGHAM NYorks SE 7360. Partly uncertain: possibly the 'village or homestead called or at *Scraying*'. *Screngha'*, *Escr(a)ingha'* 1086, *Scraingeham* 1157–1268, *-inges-* 1165×75, *Sc- Skrahingham* 1233–1351, *Sc- Skreingham -ey- -ai- -ay-* 1241–1416. P.n. **Scraying*, genitive sing. **Scrayinges*, locative-dative sing. **Scrayinge*, + **hām**. **Scraying* may represent OE ***scræġing** 'a structure of poles', an **ing**[2] derivative of OE

[†]Another DB form is *Scalftun* 1086.

*scraga 'a trestle' cognate with MHG *schrage*, ModG *Schragen*. But this is very uncertain. The specific has also been explained as an **ing**[2] formation on a pers.n., either OE **Scīrhēah* or ON *Skrái*. **Scraying* also occurs in SCRANE END TF 3841. YE 146, ING 154, BzN 1967.344–5, SSNY 149.

SCREDINGTON Lincs TF 0940. Partly uncertain. *Scredinctun -intune* 1086, *Scredinton(')* 1154×89–1327, *Scredington* from 1189×95 with variants *Sc- Skrid(el)- Skred(l)- Skret- Skre(c)k-* and *-yng*. Possibly an **ing**[2] formation on OE **scrēad** 'a shred, a strip cut off' and so OE ***scrēading** with Scand [sk] for English [ʃ] + **tūn**. The reference would be to the position of Scredington on a strip of land cut off by streams N and S; cf. ModE **screed** 'a strip of land' first recorded in 1615. Otherwise possibly the 'estate called after Scirheard', OE pers.n. *Scīrheard* + **ing**[4] + **tūn**. Perrott 86, Årsskrift 1974.42, SSNEM 186.

The SCREES Cumbr NY 1504. *Screes* 1783, *Eskdale Screes* 1794. ModE **scree** 'the precipitous stony slope upon a mountain side' (ON *skritha* 'a landslip'). The reference is to the precipitous south bank of Wast Water which rises over 1700ft. in less than half a mile. Cu 392.

SCREMBY Lincs TF 4467. 'Skræma's village or farm'. *Screnbi* 1086, 1170, *Scrembia* c.1160, *Scrembi -by* 1184×5, 1191ff, *Screinbi* 1202. ON pers.n. **Skrœma* + **bȳ**. DEPN, SSNEM 67, Cameron 1998.

SCREMERSTON Northum NU 0049. 'Skirmer's boundary stone or estate'. *Scrimestan* [c.1130]copy, *Scremerstun* 1208×10 BF, *Scremerestune* 1228, *Skremerston(e)* 1248, 14th cent., *Scrymmerstone* 1542. Surname *Skirmer*, genitive *Skirmeres*, + either OE **stān** or **tūn**. NbDu 172, DS s.n. Scrimgeour.

SCREVETON Notts SK 7343. 'The sheriff's estate'. *Screvin- Escreventone* 1086, *Screveton(e)* 1086–1211 etc. with variant spellings *Screue- Schreve-* and *Skreve-*, *Sc- Skriveton* 1200, 1276, *Sc- Skreton* 1323–1613, *Screyton* 1535. OE **scīr-ġerēfa**, genitive sing. **scīr-ġerefan**, + **tūn**. Initial [sk] for English [ʃ] is due to Scand influence. Nt 229 gives prs [skri:t(ə)n] and [skreitn], SSNEM 186.

SCRIVEN NYorks SE 3458. 'Hollow place with pits'. *Scrauing(h)e* 1086, *Scrauin* 1167, *Screu- Screvin -yn* 1173×85–1545, *Screving(e)* 1208, 1584, *Scriven* from 1301. OE ***screfen** 'hollow place, place with pits', a derivative of OE **scræf** 'a hole, a pit'. Spellings with *a* may represent a variant formation with the same meaning, OE ***scræfing**. ON [sk] has replaced OE [ʃ]. YW v.114, ING 77, SSNY 163.

SCROOBY Notts SK 6590. Partly uncertain. *Scrobi* 1086, 1185, *-by* 1225–1342, *Scruby* 1557, 1675, *Scrowbye* 1582, *Scrooby(e)* 1601, 1627. The site of Scrooby lies within the bounds of the lost *Scroppen þorpe* [958]14th S 679, which is probably 'Skroppa's outlying farm', ON feminine pers.n. *Skroppa*, English genitive sing. **Skroppan*, + **thorp**. The vowel of the specific of Scrooby, however, must have been long *ō* in ME to account for the modern form. Possibly, therefore, 'Skropi's village or farm', ON pers.n. *Skropi*, genitive sing. *Skropa*, + **bȳ**, or OE ***scrof** 'a hollow, a cutting' referring to the valley of the river Ryton, + **bȳ**, 'the village or farm by the cutting'. In the former case ME lengthening in open syllable would occur; in the latter vocalisation of [f] > [u]. Nt 96, SSNEM 67–8, PNE ii.107.

SCROPTON Derby SK 1930. 'Skropi's farm or village'. *Scrotun(e)* 1086, *Scropton(e)* from late 11th, *Scrapton* 1577–1767. ON pers.n. *Skropi* + **tūn**. Db 560, SSNEM 194.

SCRUB HILL Lincs TF 2355. *Scrubb Hill* 1896. ModE **scrub** + **hill**. A place in Wildmore Fen near Dogdyke which was not drained until the erection of a steam engine in 1841. Wheeler Appendix i.11, 33.

SCRUTON NYorks SE 3092. 'Scurfa's estate'. *Scurueton(e), Skurveton* 1086–1396, *Screwton* 1611. ON pers.n. *Skurfa* + OE **tūn**. Identical with SHERATON Durham NZ 4335. YN 238, SSNY 129.

SCULTHORPE Norf TF 8930. 'Skuli's outlying farm'. *Scula- Sculetorpa* 1086, *Sculetorp* 1174×80. ON pers.n. *Skúli*, genitive sing. *Skúla*, + **thorp**. DEPN, SPNN 337.

SCUNTHORPE Humbs SE 8910. 'Skuma's outlying settlement'. *Escumetorp* 1086, *Scumetorp* 1196, *Scummptorp* 1245, *Scumthorp* 1273×4. ON pers.n. *Skúma* + **thorp**. DEPN, SSNEM 117, Cameron 1998.

SEABOROUGH Dorset ST 4306. 'Seven hills or barrows'. *Seveberge* 1086, *Seuenbergh* 1306. OE **seofon** + **beorg**. Seven was a favourite number with barrows: cf. *Seven Barrows* 1811 in Wareham which actually refers to eight barrows. The reference here is unknown. Do 129, i.165, Newman-Pevsner 440.

SEACOMBE Mers SJ 3290. 'Valley by the sea'. *Secumbe* 13th, c.1227, *Secom(e) -um* 1303–1639, *Seacum* 1421–1663, *-combe* from 1659. OE **sǣ** + **cumb**. The reference is probably to the valley at Oakdale (Road) SJ 3190 that descends to Wallasey Pool (East Float). Che iv.329.

SEACOMBE CLIFF Dorset SY 9876. *Seacombe Cliff* 1811. P.n. Seacombe, *Secombe* [1306]1372 'valley opening to the sea', OE **sǣ**+ **cumb**, + ModE **cliff**. Do i.66

SEACROFT Lincs TF 5660. ModE **sea** + **croft**. A coastal development on Croft Marsh TF 5360, after which it is named.

SEAFIELD BAY Suff TM 1232. No early forms.

SEAFORD ESusx TQ 4899. 'Ford by the sea'. *Saforda* [795]17th S 1186, [790]12th S 133, *-forde* [1087×1100]15th–1231, *Seford* 1180–15th, *Sheford* 1230–15th. OE **sǣ**+ **ford**. Originally this was a ford across the old course of the river Ouse which, prior to the great storms in the 16th century, flowed into the sea at Seaford. After the diversion of the river to flow into the sea at Meeching, Meeching became the 'new haven' while Seaford was known until 1834 as the 'old haven'. See NEWHAVEN TQ 4401. Sx 363 gives pr [si:'fu:əd].

SEAFORTH Mers SJ 3297. *Seaforth* 1842 OS. After Seaforth House, itself so named by its owner Sir John Gladstone (the father of the Prime Minister) when he moved there in 1813 because his wife was a member of clan MacKenzie headed at the time by Lord Seaforth. Mills 132, Room 1983.109.

SEAGRAVE Leic SK 6117. Possibly 'grove with or near a pit'. *Setgraue, Sat- Segrave* 1086, *Sat- Set- Sed- Sadgrave* 1162×70–1281, *Sa- Segrave* c.1130–1610, *Seagrave* from 1207. OE **sēath** 'a pit, a pool, a spring' + either **grāf** 'a wood, a grove' or **græf** 'a ditch'. Lei 318.

Upper SEAGRY Wilts ST 9480. *Overe Segreve* 1420, *Over Segrith* 1465, *Over Seagry* 1773. ModE adj. **upper**, ME **over**, + p.n. *Segrete, Segrie* 1086, *Segreth* 1399, *Segre(a)* 1190–1363 with variant *-rey(e), Seggereye* 1225, *Segary* 1670 (*'vulgo') Seagrey* 1699, 'the sedge stream', OE **secg** + **rīth**. 'Upper' for distinction from Lower Seagrey ST 9580, *Nethere Segre* 1218. Wlt 72 records former pr [segəri], L 29.

SEAHAM Durham NZ 4249. 'Homestead, estate by the sea'. *Sæhā* [c.1040]12th, *Seham* c.1240×50–1628, *Seam* 1420, *Seaham* from 1635. OE **sǣ**+ **hām**. Pronounced [si:əm]. NbDu 173.

SEAHOUSES Northum NU 2131. 'The houses or buildings by the sea'. *North Sunderland Sea-Houses* 1888. A 19th cent. fishing settlement. Tomlinson 448.

SEAL Kent TQ 5556. 'The hall'. *Lasela* 1086, *Lausele, Sela* c.1100, *(la) Sele* 1233–92. French definite article **la** + OE **sele**. PNK 63.

SEAMER 'Lake, lake called *Sœ*'. OE **sǣ** 'lake' taken as a p.n. + **mere**.

(1) ~ NYorks NZ 4910. *Semer(s)* 1086, c.1150, *Samara -mare* [1133–c.1180]15th, 1218. In this name **mere** seems to vary with OE **mersc** and ON **marr**. The village lies on a hill and the original reference of the name must have been to a lake in the area of Seamer Carrs NZ 4809. YN 172.

(2) ~ NYorks TA 0183. *Semœr* 1086, *Semer(e)* 1086–1534, *Samare -mara* 1096×6–1224, *Semar(e) -(a)* [1155×65]15th–1529. In this name **mere** varies with ON **marr**. The reference was to a lake now drained in the area SW of the church at Seamer Carr TA

0281. There was once a Mesolithic settlement on a birch-wood platform on the edge of the lake. YN 102, Pevsner 1966.337.

SEARBY Lincs TA 0705. 'Sæfari's village or farm'. *Soure- Seurebi -v-* 1086, *Seure- Safrebi* c.1115, *Sauerbi* 1155×8, *Seuerbi -by* c.1155–1550, *Sereby* 1394–1671, *Seerby* 1526, *Searby* 1661. ON pers.n. *Sǽfari*, genitive sing. *Sǽfara*, + **bȳ**. Li ii.259, SSNEM 68.

SEASALTER Kent TR 0965. 'The sea salt-house'. *Seseltre* 1086, *Sæ- Sesealtre* c.1100, *Sesaltre -sautre* 1205–78, earlier *þ*ᵗ *sealt-tærn steal æt here pit* (sic for *here wic*) 'the salt-house site at Harwich' (Harwich Street in Whitstable) [946]13th S 518, *Sealterna steallas* [786]?11th S 125, *Sealtern, ðem sealtern et fefresham* 'the salt-house at Faversham' 858 S 328, *Sealternsteall* 863 S 332. OE **sǣ** + p.n. *Sealtærn*, OE **sealtærn**. PNK 494, KPN 308.

SEASCALE Cumbr NY 0301. 'Shieling(s) beside the sea'. *Sescales* c.1165, 1497, *Sescale* 1278, *Seaskaill* 1576. ON **sǽr** or OE **sǣ** + **skáli**. Already correctly interpreted in 1687–8 as derived from 'a Scale or Skeele, for cattle and sheepcot at ye sea'. Cu 433, SSNNW 69.

SEATALLAN Cumbr NY 1308. Probably 'Alan's shieling'. *Settallian* 1783. The evidence is very late but this appears to be ME pers.n. *Alein* + ON **sǽt** in Celtic word order. The name is now applied to a mountain rising to 2270ft. Cf. SEAT SANDAL NY 3411. Cu 442.

SEATHORNE Lincs TF 5765. A modern coastal development.

SEATHWAITE Cumbr NY 2312. 'Sedge clearing'. *Seuethwayt* 1292, *Sethwayt -thwait* 1542, *Seatwhait* 1566, *Seathwaite* from 1600. ON **sef** + **thveit**. Cu 351, SSNNW 159.

SEATHWAITE Cumbr SD 2296. 'Clearing by the lake'. *Seathwhot* 1592, *-what* 1598, *Seathwaite* 1786. ON **sǽr** + **thveit**. The reference is to Seathwaite Tarn SD 2598. La 223, Mills 132, L 29, 211.

SEATHWAITE TARN Cumbr SD 2598. P.n. SEATHWAITE + Mod dial. **tarn** 'a lake' (ON *tjǫrn*).

SEATON 'Farm, village, estate near the sea'. OE **sǣ** + **tūn**.

(1) ~ Corn SX 3054. Transferred from the River SEATON SX 3059. *Seythen* 1601, *Seaton* 1602, *Sythian* c.1605. Short for *Seton Bridge* c.1540. The forms *Seton* and *Seaton* are Anglicisations of the Cornish r.n. partly because the village is situated on the coast. PNCo 154, Gover n.d. 223.

(2) ~ Cumbr NY 0130. *Setona* c.1174, *Seton* 1230–1440, *Seaton* 1580. Cu 319.

(3) ~ Devon SY 2490. *Seton(e)* 1238–1310, *Seetone* 1445. Earlier known as *(æt) Fleote* [1005]12th S 910, *Flveta* 1086, *Fluta* 1086 Exon, from OE **flēot** 'estuary', referring to the mouth of the river Axe on which the town stands. D 629.

(4) ~ Durham NZ 3949. *Sætun* [c.1040]12th, *Seton* c.1180×95–1597, *Seaton* from 1575. NbDu 173.

(5) ~ Humbs TA 1646. *Settun* 1086, *Seton(a) -tone* [12th]c.1536–1409, *Seaton in Holdernes* 1562. In this instance the sense of **sǣ** is 'lake' rather than 'sea', with reference to Hornsea Mere about a mile from the village. YE 67.

(6) ~ Northum NZ 3276. *Seton* 1200, 1242 BF, ~ *de la Val(e)* 1270, 1296 SR. The manor was held by the Delaval family in the 13th cent. NbDu 173.

(7) ~ BAY Devon SY 2489. P.n. SEATON SY 2490 + ModE **bay**.

(8) ~ BURN T&W NZ 2373. *Seaton Burn* 1863 OS. P.n. SEATON Northum NZ 3276 + ModE **burn**.

(9) ~ CAREW Cleve NZ 5219. 'S held by the Carew family'. *Seton(e) Carrou* 1311×12, ~ *Carrowe* c.1325–1567, *Seaton Karewe* 1559, ~ *Carew* 1605. Earlier simply *Seton(e)* c.1200–1510. The vill was held by a Robert Carew in the reign of Henry I, whose family name is of Welsh origin. DS 94.

(10) ~ DELAVAL Northum NZ 3075. A 19th cent. colliery settlement named after *Seaton Delaval Colliery* 1865 OS, itself named after SEATON NZ 3276.

(11) ~ HALL NYorks NZ 7817. P.n. Seaton + ModE **hall**. Seaton is *Scetun(e)* 1086, *Seton* 1279–1412, *Seaton* 1571. YN 139.

(12) ~ ROSS Humbs SE 7841. 'S held by the Ross family'. *Sei- Seyton Ros(s)e* 16th cent. P.n. *Setton* 1086, 1285, *Seton(a)* 1086–1473, *Seaton* from 1523 + family name Roos or Ross, who held the village from the 12th to the 17th cent. The sense of **sǣ** is 'lake' rather than 'sea' here but the land has long been drained. YE 235.

(13) ~ SLUICE Northum NZ 3376. *Seaton Sluice* 1865 OS. P.n. SEATON NZ 3276 + ModE **sluice**. The reference is to the 17th cent. sluice built by Sir Ralph Delaval in Seaton Burn in order to provide a head of water to scour the easily silted bed of his harbour for coal exports. Tomlinson 60, Pevsner-Grundy etc. 1992.564.

(14) North ~ Northum NZ 2986. *North Seaton* 1866 OS. ModE **north** + p.n. *Seton(')* 1242–1296. 'North' for distinction from SEATON NZ 3276. NbDu 173.

SEATON Leic SP 9098. Partly uncertain. *Sei(e)- Segen- Segestone* 1086, *Sei- Seyton(e)* 1185–1610, *Seton'* 1205–1355, *Seaton* from 1546. The generic is OE **tūn** 'farm, village', the specific probably an OE pers.n. ***Sǣġa** (short for names such as *Sǣġeard, Sǣġeat* etc.) although a stream name **Sǣġe* 'the slow-moving one' from OE ***sǣġe** related to *sīgan* 'sink, set, move' has also been suggested. Seaton is situated on a hill overlooking meanders and a small tributary of the river Welland. R 293, DEPN.

River SEATON Corn SX 3059. Uncertain. *Sethul* [early 13th]15th, *Seychym* (for *Seythyn*) 1302, *Seythin* [1396]1593, *Seythen* 1441, *Setoun Ryver* c.1540, *Seton* 1577, 1586, *Seaton* 1602. The earliest spelling is corrupt and the original form is *Seythyn* of uncertain origin, possibly Co **seit** 'pot' (*t* = [θ]) + r.n. suffix **-en** 'river full of pot-holes', or Co **seth** 'arrow' + **-en**, perhaps meaning 'swift river'. Cf. the r.ns. Saith Dyfed SH 2729 and Saethon Gwynedd SH 2932. PNCo 154, Gover n.d. 223, RN 354.

SEATOWN Dorset SY 4291. 'Settlement by the sea'. *Setowne* 1469, *Seeton* 1470, *Seetowne* 1494, *Zetowne* 1508. ME **see** + **toun**. The 1508 form shows typical southern voicing of initial [s] > [z]. Do 129.

SEAT SANDAL Cumbr NY 3411. 'Sandulf's shieling'. *Satsondolf* 1274, *the Sate Sandall* 16th, *Seat Sandall* 1614. ON **sǽtr** + ON pers.n. *Sandulfr* in Celtic word order. It is now applied to a mountain rising to over 2400ft. Cf. SEATALLAN NY 1308. We i.199.

SEAVE GREEN NYorks NZ 5600. Mod dial. **seave** (ON *sef*) 'sedge, a rush', + ModE **green** 'a grassy spot, a village green'.

SEAVIEW IoW SZ 6291. *Sea View* 1839, cf. *Sea Grove* 1810 OS. The place commands a wide view over the Solent and the open sea. Wt 201, Mills 1996.93.

SEAVINGTON ST MARY Somer ST 3914. 'St Mary's S'. *Seuingto Maries* 1610. P.n. Seavington as in SEAVINGTON ST MICHAEL + saint's name *Mary* from the dedication of the church.

SEAVINGTON ST MICHAEL Somer ST 4015. 'St Michael's S'. *Sevenhampton Michaelis* 1291, *Seuington Michaell* 1610. P.n. Seavington + saint's name *Michael* from the dedication of the church. Earlier simply *Seofenempton* [1027×32]18th S 979, *Seovenamentone, Sevenehantvne, Sevenemetone* 1086, *Sevenamtone -tune* [late 12th]14th, *Sevenhampton* [13th–1389]Buck, 'village, estate of the *Seofonhǣme*, the people who live at a place called Seven Springs' or the like. OE folk-n. **Seofonhǣme*, genitive pl. **Seofonhǣma*, + **tūn**. In p.ns. compounded with *-hǣme* the first element is always a reduced form of another p.n. known or unknown. In this case the place must have been **Seofonwyllas* 'seven springs' or something similar. DEPN.

SEBERGHAM Cumbr NY 3541. Probably 'Sæburh's homestead'. *Setburg(e)ham* 1204, 1223, *Sedburgham* c.1292, *Saburgham* 1223–33, *Seburgham* 1223–1567, *Sebbram(e)* 1565, 1592, *Seabrougham* 1541. OE feminine pers.n. *Sǣburh*, genitive sing. *Sǣburge*, + **hām**. The early spellings in *Set- Sed-* are assumed to be due to the influence of ON **set-berg** 'seat-shaped hill' possibly with reference to Warnell Fell (993ft.) a mile to the SW at NY 3341. Cu 150 gives prs [sebrəm, sebərəm].

SECKINGTON Warw SK 2607. 'Secca's hill'.
I. *Seccandun* c.890 ASC(A) under year 755, *Secandune* c.1121 ASC(E) under same year, *Sekendune* 1194, *Se(c)kendon -in-* 1227–1401, *Seckingdun* 1232, c.1840, *Sekkyngdon* 1323, 1350.
II. *Sec(h)intone* 1086, *Sekyngton* 1327, *Segynton* 1535, *Seekington* 1834 OS.
OE pers.n. *Secca*, genitive sing. *Seccan*, + **dūn** apparently varying with *Secca* + **ing**[4] + **tūn**. Wa 24.

SEDBERGH Cumbr SD 6592. 'Seat-shaped, i.e. flat-topped, hill'. *Sedberg(e) -(a) -bergh(e)* 1086–1660, *-ba(a)rghe* 1535, 1601, *Sadberg(e) -bergh* 1231–1428, *Satberg* 1257. ON **set-berg**. Forms with *Sad- Sat-* are probably due to confusion with ON **sǽtr** 'a shieling', which appears in English as *set* and *sat*. YW vi.263.

SEDBURY Glos ST 5493. 'South fortification'. *Sothebur'* 1279, *Soncheberi* (sic for *Southe-*) 1307, *Sudburie -y* 1562–1639, *Sedbury* from 1638. OE **sūth** + **byriġ**, dative sing. of **burh**. The reference is to the site of a Roman camp overlooking the river Severn. Gl iii.265.

SEDBUSK NYorks SD 8891. 'Bush near the shieling'. *Setebu(s)kst(e)* 1280, 1283, *Sedbuske* 1611. ON **sǽtr** + **buskr**. YN 260.

SEDGEBERROW H&W SP 0238. 'Secg's grove'. *(DE) SEGGESBERUUE* [778]11th S 113, *(æt) segcesbearuue* [778]17th S 113 with variant *Segcgesbearuue* 11th, *Secgesbearuþe* [964]12th S 731, *Seggesbarve* 1086, *-berwe* 1275–1327, *Seggebarowe* 1420, *Sedgeberrowe* 1649. Many later spellings show confusion with OE **beorg** 'a hill, a barrow', but this is clearly by origin OE pers.n. *Secg*, genitive sing. *Secges*, + **bearu**. The boundary of the people of Sedgeberrow is *secg hæma gemæru* [1042]18th S 1396. Wo 164 gives pr [sedʒiberou], Hooke 78, 145, 362, L 190.

SEDGEBROOK Lincs SK 8537. 'Brook where sedge grows'. *Seckebroc* [c.1080]13th, *Sechebroc* 1086, [1152×62]1396, *Seg(g)ebroc -brok(e)* c.1160–1498, *Sedgebrook* 1558×79. OE **secg** + **brōc**. Perrott 475, L 15.

SEDGEFIELD Durham NZ 3528. 'Cedd's open land'. *Ceddesfeld* [c.1040]12th, *Seggesfeld* c.1174×89–1392, *Seg(g)efeld* [1183]14th, 1242–1564, *Seg-* 1408–1565, *Sedge(e)field -feild* from 1558. OE pers.n. *Ċedd*, genitive sing. *Ċeddes*, + **feld**. The pronunciation with [s] instead of [tʃ] is due to dissimilation in the awkward sequence [tʃ] - [dʒ]. NbDu 173, L 243.

SEDGEFORD Norf TF 7136. Partly uncertain. *Seces- Sexforda* 1086, *Sicheford* 1166, *Secheford* 1166, 1212, *Sechesforde* c.1140, *Sekeford* 1190. The generic is OE **ford** 'a ford', the specific either OE pers.n. **Seċċi*, genitive sing. **Seċċes*, or OE ***sǣċe**, a conjectural form related to and varying with **sīc** 'a drainage stream'. Sedgeford lies on a small stream draining into The Wash called Heacham River. The modern form shows popular association with ModE *sedge*. DEPN.

SEDGEHILL Wilts ST 8628. 'Sedge hill'. *Seghull(e)* 1100×35–1529, *Sege(s)hull* 1249–81, *Seggehull* 1289–1514, *Sedghyll* 1547, *Sedgell* 1716. OE **secg** + **hyll**. The parish is remarkable for its pools and streams, even in the higher ground in the south. Wlt 191, L 170.

King's SEDGE MOOR Somer ST 4133. *Sedege more* 1610, *Kings Sedge Moor* 1817 OS. Earlier *Kyngesmore* [before 1276] Buck, cf. *Quenes more* 1610, *Queens Sedge Moor* 1817 OS. A tract of marshland between Bridgewater and Somerton named from the abundant growth of common sedge.

West SEDGE MOOR Somer ST 3025. *West Sedgemoor* 1809 OS. ModE adj. **west** + p.n. Sedge Moor as in King's SEDGE MOOR ST 4133. Contrasts with *East Sedge Moor* 1817, ST 5241.

SEDGLEY WMids SO 9294. 'Secg's wood or clearing'. *Secgesleage* [985]12th S 860, *Segleslei* 1086, *Seggeslegh* before 1211. OE pers.n. *Secg*, genitive sing. *Secges*, + **lēah**. DEPN.

SEDGWICK Cumbr SD 5187. 'Sicg's dairy farm'. *Si- Sygeswic(k) -wyk(e)* 1180–1530, *-gg-* 1180–1777, *Sedgwicke* 1610. OE pers.n. **Sicg*, genitive sing. **Sicges*, + **wīc**. We i.97, SSNNW 424.

SEDLESCOMBE ESusx TQ 7818. 'Coomb of the house'. *Sales- Selescome* 1086, *-cumba* [1154×89]14th, *Setelescumbe* c.1180–13th, *Sedelescumbe -combe* 13th–14th cents., *Sedlescombe al. Sellescombe* 1584. OE **setl**, **sethl**, genitive sing. **setles**, **sethles**, + **cumb**. An alternative possibility might be 'coomb of the saddle', OE **sadel**, genitive sing. **sædles**, + **cumb**: the village stands on a prominent ridge. In Norwegian p.ns. 'saddle' is used of a prominent dip in hills. Sx 524 gives pr [selzkəm], L 89, 93.

SEEND Wilts ST 9461. 'The sandy place'. *Sinda* 1190, 1195, *Sei- Seynde* 1194–1249, 1553, *Sende* 1211–1322, *Seende* 1330, *Seene al. Seend* 1602, *Sean* 1675. OE ***sende**. The village occupies an isolated hill, an outcrop of iron-sand, and there is no reason to link it with the stream name *Semnet* discussed under SEMINGTON ST 8960 or to seek its origin from Brit **sento-* 'a way, a path' (PrW *hïnt*, W *hynt*). Wlt 131 records obsolete pr [si:n], PNE ii. 118.

SEEND CLEEVE Wilts ST 9360. *Seend Cleeve* 1773. P.n. SEEND ST 9461 + p.n. *Clyve* 1255, *les Cleves* 1341, 'the cliff, the steep slope', ME **clif**. Wlt 132.

SEER GREEN Bucks SU 9691. *Seare Greene* 1625, *Seer Green* 1822 OS. P.n. Seer + ModE **green**. Seer, *la Sere* 1223, 1258, *la Cere* 1274, *Sere* 1361, is probably 'the enclosure', OFr **serre**, rather than 'bare place', OE **sēar** or **sīere**. Bk 231, D l.

SEETHING Norf TM 3197. 'The Sithingas, the people called after Sitha'. *Sithinges -inga, Silinga* 1086, *Seinges* 1181–99, *Se(e)nges* 1201, 1254, *Sithing* 1323, *Sethyng* 1450. OE folk-n. **Sīthingas* < pers.n. **Sītha* + **ingas**. An alternative possibility is perhaps OE folk-n. **Sȳthingas* 'the people to the south' sc. of Norwich or S of the river Yare < **sūth** + **ingas**. Cf. SURLINGHAM TG 3106. DEPN, ING 61.

SEFTON Mers SD 3501. 'Farm, village where sedge is cut'. *Sextone* (sic) 1086, *Sefftun* before 1222, *Sefton* from 1242×3 with variants *Seff- Ceff-*. ON **sef** + OE **tūn**. La 118, SSNNW 188, Jnl 17.65.

SEFTON PARK Mers SJ 3787. A park laid out and completed in 1872. Pevsner-Hubbard 1971.148.

SEGHILL Northum NZ 2874. Partially uncertain. *Sihala, Syghal* 1271, *Seyhale* 1295, 1296, *Syhale, Sikhale* 1336, *S(e)ighale, Seghall* 14th cent., *Syghale* 1428, *Sy- Sighall* 16th cent., *Sighill* 1663, 1855, *Seghill* 1637 Camden, 1727, *Sedgehill* 1855. The generic is OE **halh** 'a nook of land, land enclosed in a river bend'. The forms of the specific seem to point to an OE element varying between **siġ** and **sigg**. Mawer suggested *Sigga* and **Siġa* as short forms for an OE pers.n. such as *Siġefrith* and so 'Sig(g)a's nook of land'. DEPN suggests a r.n. *Siġe* on the root of OE **sīgan** 'sink, descend, move' but this would not explain the preservation of [g] in this name. The final predominance of the *Seg-* form may be due to Camden's erroneous identification of Seghill with RBrit *Segedunum*, which is actually Wallsend. The village is a modern colliery development. NbDu 173, L 108, 111.

SEGSBURY Oxon SU 3884. *Segsbury* 1841. Probably an old name in OE **byriġ**, dative sing. of **burh** 'a fortified place' referring to the Iron Age hill-fort of Letcombe Castle or Segsbury Camp. O 324, Thomas 1976.181.

SEIFTON Shrops SO 4883. Possibly the 'settlement on the r. *Sifen*'. *Sireton* 1086, *Siditonia* 1177–86, *Ci- Cyneton* (for *Ciue-*) 1200–74, *Siveton* 1257, *Siui(ng)ton'* 1271×2, *Seueton* 1343, *Syffeton* 1549, *Syfton* 1583, *Seefton* 1695, r.n. **Sifen* + OE **tūn**. The early sps of the Yorkshire Seven include *Si- Syvene* forms. Sa i.272.

SEIGHFORD Staffs SJ 8825. 'Strife ford'. *Cesteforde* 1086–13th, *Sestesforde* c.1200, *Seteford* 1208, *Sesteford* 13th, 14th, *Cesterford* n.d. Ronton Cartulary, *Seythford* c.1565, *Cyford* 1644. OE **ċeast** + **ford**. Identical with Chesford Bridge Warw

SP 3069, *Chesford* 1279, *Chestford* 1291, 1370, *Chesterford* 1370. The isolated *Chester-* forms are due to folk-etymology. DEPN gives pr [sai-], Horovitz.

SEISDON Staffs SO 8495. 'The hill of the Saxons or of Seax'. *Sais- Seiesdon, Sais- Seisdone* 1086, *Saiesdona* 1130, *Seyxdun* 1236, *Seisdon* 1242, *Seasdon* 1590. OE folk-n. *Seaxe*, uninflected compositional form *Seax-*, or pers.n. *Seax*, genitive sing. *Seaxes*, + **dūn**. Seisdon is on the boundary between Staffs and Shrops: the reference is probably to Abbots Castle Hill and the presence of an isolated group of Saxons in Mercian territory. The sps *Sais- Seis-* are AN reflecting preconsonatal palatalisation of [k] as in *laissier, ais* < Latin *laxare, axem*. Horovitz gives pr *Sees-don*. DEPN, Pope §325.

SELATTYN Shrops SJ 2634. Probably 'the settlement of the gullies'. *Sulatun* 1254, *Sullatton* 1358, *Salatyne* 1532, *-tin* 18th cent., *Sylattin* 1564–1808, *Selattyn* from 1577. OE **sulh**, genitive pl. **sūla**, + **tūn**. Normally an OE p.n. *Sūla-tūn* would become ModE *Soul-* or *Sowton*; in this case the medial *-a-* has been preserved because the name was assimilated to the W stress pattern *Sūlá-tūn*. *Tyn* is the usual W development of **tūn**. The reference is to the broken nature of the ground here. Sa i.258, O'Donnell 107.

SELBORNE Hants SU 7433. 'Willow stream'. *seleborne* [903]14th S 370, *Selesburna* 1086[†], *Seleburn* 1197, *-b(o)urne* 1201–1341, *Shelborn -burn* 1579, 1610. OE **sealh**, genitive pl. **sēala**, + **burna**. Originally this was the name of the water-course now called Oakhanger Stream. Ha 145, Gover 1958.93, DEPN, L 18.

SELBY NYorks SE 6132. Partly uncertain. *Seleby -bi* c.1030 YCh 7, [c.1040]12th–1504, *Saleby(a) -bi(a)* c.1070–1154, *Selby* from 1221. Any explanation must take account of the *a/e* alternation in the 12th cent. spellings. The specific might therefore be OE ***sele**, ON **selja** 'a willow copse' varying with cognate OE **salh** 'a willow', genitive pl. **sala**. This is not to be taken as a Scandinavianised form of the *seletun* mentioned in ASC(E) under the year 779 (*recte* 782) where the ealdorman Beorn was burnt by the high reeves of Norhumbria, since this is most likely to be identified with a lost *Seletun* near Easington in Durham. YW iv.31, SSNY 36.

SELBY CANAL NYorks SE 5829. P.n. SELBY SE 6132 + ModE **canal**. A canal built 1775–8 to service the export of cloth. Hadfield i.34.

SELHAM WSusx SU 9320. 'The sallow copse homestead or estate'. *Seleham* 1086–1283, *Selham* from 1230, *Sellam* 1604, *Sul(e)ham* c.1200–1442, 1752. OE ***sele**, ***siele**, ***syle**, + **hām**. Sx 28, DEPN, PNE ii.117, ASE 2.33.

SELKER BAY Cumbr SD 0788. *Selkers Bay* 1794. P.n. Selker SD 0788 + ModE **bay**. Selker, *Seleker* c.1205, *Selker* 1390, *Seltar in Bootle* 1730, is 'willow marsh', ON **selja** + **kjarr**. Cu 347, SSNNW 159.

SELLACK H&W SO 5627. Short for 'church of St Suluc'. *Sellak* 1344, 1397, *Sellick* 1654. Earlier *Lann Suluc* [late 9th, 1045×1104]c.1130. PrW ***lann** + saint's name *Suluc*, a pet form of *Suliau* or *Tysilio*, OW *ty* 'thy' + *Suliau*, from the dedication of the church. Cf. Llandissilio Powys SN 1221. He 179, O'Donnell 89.

SELLAFIELD STATION Cumbr NY 0203. P.n. Sellafield NY 0204 + ModE **station**. A halt on the railway line from Barrow-in-Furness and Whitehaven. Sellafield, *Sellofeld* 1576, *Sellowfield* or *Sea-low-field* 1610, is 'open land at Sello', p.n. Sello, *Sellagh* 1278, unexplained, + ME **feld**. Cu 339.

SELLINDGE Kent TR 1038. Uncertain. *Sedlinges* 1086, 1226–75, *Sedling, Sellinge* 11th, *Silling'* 1204, *Selling'* 1219, 1232–3, *Sellinges* 1226, 1242×3, *Sellinge* 1242, *Shelynge* 1270. Apparently OE **Sedlingas* 'people of the dwelling' from OE **sedl** + **ingas**, varying with *Sedlingĕ*, locative–dative sing. of *Sedling* 'dwelling place' from OE **sedl** + **ing**[2]. Apparently identical with OFris *Sedlingi* and the first element of Zedelgem near Bruges, *Sedelingahem*. Cf. SELLING TR 0456. PNK 304, KPN 30, ING 206, BzN 1967.345.

SELLING Kent TR 0456. Uncertain. *Setlinges* 1086, *Sedling* 11th, *Sellinge* 1087, *Selling(e)* 1206–1316, *Selling(e)s* 13th, *Shellinge* 1248. Identical with SELLINDGE TR 1038. PNK 304.

SELLS GREEN Wilts ST 9562. *Sells Green* 1817 OS. Surname *Self* as in Selves Farm in Melksham parish and *Selfestreet* 1639, + ModE **green**. Wlt 133.

SELLY OAK WMids SP 0482. *Selley Oak* 1834 OS. P.n. *Escelie* 1086, *Selle(gh), Selley* 1221–1416, *Selvele* 1204, 1242, 'the shelf wood or clearing', OE **scelf** + **lēah**, + ModE **oak** 'an oak-tree'. The reference is to a plateau of high ground. Wo 349, DEPN, L 186, 206.

SELMESTON ESusx TQ 5007. 'Sigehelm's estate'. *S(i)elmestone* 1086, *Syelmestona* 1220, *Selmeston* from 1242, *Silmes-* 1376, *Symston* 1578, *Simpson* 1632, *Simson* 1765. OE pers.n. *Siġehelm*, genitive sing. *Siġehelmes*, + **tūn**. Sx 338, SS 75.

SELSDON GLond TQ 3562. Partly uncertain. *(on) Selesdune* 871×99 S 1508, *Sel(l)esdon(e)* 1225–1347, *Seldson* 1570. The generic is OE **dūn** 'a hill', the specific any of OE pers.n. **Seli* and so 'Seli's hill', or **sele** 'a hall' or **sele** 'a willow copse' and so 'hill of the hall or the willow copse'. Sr 54.

SELSET RESERVOIR Durham NY 9121. P.n. Selset + ModE **reservoir**. Selset, *High, Low Selset* 1865 OS, is probably identical with SELSIDE NYorks SD 7875.

SELSEY WSusx SZ 8593. 'Seal island', literally 'the seal's island'. *Seolesige* *[683]14th S 232, [714]14th S 42, *(in) Seolesiae, Selaesiae* [c.715]11th Eddi, *Selæs- Seles- Selaeseu* [c.731]8th BHE, *Siolesaei* 780 S 1184, *Selesie* [957]14th S 1291, *Se- Sylesea, Seolesige* [c.890]c.1000 OEBede, *Seleisie* 1086. OE **seolh** probably used as a collective, genitive sing. **seoles**, + **īeġ**. Translated by Bede as *insula vituli marini* 'the island of the marine calf'. For his form *eu* for Anglian **ēġ**, WS **īeġ** see Nomina 17.133. Sx 82, Jnl 8.26, SS 70.

SELSEY BILL WSusx SZ 8592. *Selsey Bill* 1740 P.n. SELSEY SZ 8593 + ModE **bill** (OE *bile*) 'a bill, a beak, a headland' as in BILL OF PORTLAND Dorset SY 6768.

SELSFIELD COMMON WSusx TQ 3434. P.n. *Selefeld* 1279–90, *Selesfeld, Schelefeld -feud* 1279, 'the open land of the hall', OE **sele**, genitive sing. **seles**, + **feld**, + ModE **common**. Sx 272, SS 70.

SELSIDE NYorks SD 7875. 'Willow-tree shieling'. *Selesat* 1086, *-seta -sete* 1190–1220, *-seth* 13th cent., *Selset* 1634, *Selside* from 1541. ON **selja** + **sǣtr**. SSNY 88 points out that the *r* of **sǣtr** is part of the stem, not the nominative sing. masculine ending, cf. Shetland *seter*, Mod Norw *sæter*. The second element might, therefore, rather be ODan ***sæta** f. 'a residence'. YW vi.220.

SELSTON Notts SK 4653. Partly obscure. *Salestune* 1086, *Salleston* 1258, *Sel(l)estun* 1223, 1226, *-ton* 1228–1346, *Selston* 1304, 1323, *Scellson, Selson, Celson* 17th. The specific is probably the reduced form of an OE pers.n. but no certainty is possible. The generic is OE **tūn** 'farm, village, settlement'. Nt 131 records local pr [selsən].

SELWORTHY Somer SS 9146. 'Sallow copse enclosure'. *Selevrde* 1086, *-worth* 1243, *Sel(e)worthi* [1255, 1259] Buck, *-worthy* [c.1350] Buck, *Syleworth* 1291. OE ***sele**, WSax ***syle**, + **worth**, later **worthy** (OE *worthiġ*). DEPN.

SELWORTHY BEACON Somer SS 9148. P.n. SELWORTHY SS 9146 + ModE **beacon**. An eminence rising to 1012ft. on the coast of N Somerset.

River SEM Wilts ST 9127 → SEMINGTON ST 8960.

SEMER Suff TM 0046. 'Sea mere'. *Seamerā* 1086, *Semere* c.1095–1254, 1610. OE **sǣ** 'a lake' used as a p.n. + pleonastic **mere**. DEPN.

SEMER WATER NYorks SD 9287. 'Semer lake'. *Semerwater* 1153, 1283, *Semmerwater* 1280. P.n. *Semer* identical with SEAMER +

[†]An erroneous form *Lesborne* occurs at 48c, d.

pleonastic OE **wæter**. The reference is to a natural lake approximately $\frac{3}{4}$ m. in length, said to contain a city drowned for its inhospitality to an angel. YN 264 gives pr [semǝwætǝ], Morris 1930.339–40.

SEMINGTON Wilts ST 8960. 'The settlement on the *Semnet*'. *Semelton* 1249–1344, *Semneton* 1216×72, 1289, *Sembleton* 1257–1555, *Semington -yng-* from 1470. R.n. **Semnet* + **tūn**. *Semnet*, the original name of Semington Brook, is *semnit* [964]14th S 727, *Sem(n)et(')* 1228, *Semelet, Semnet* 1216×72, *Semenet* 1279, possibly from Brit **Suminetā*, a diminutive of Brit **Suminā* suggested for the river Sem Wilts ST 9127, *(on) Semene* [983]15th S 850, 1244. *Sumina* occurs again as the Gallo-Roman name of the Somme in Gregory of Tours (6th cent.) and as *Sumena* in Rav [c.700]13th, the Sumène, Cantal, a tributary of the Dordogne, *Simina* 12th, *Sumena* 1585, the Sumène, Haute-Loire, *Sumera* 1254, *Sumena* 1305, Sumène on the Rieutard, Gard, originally a r.n. *Sumena* 1150, the diminutive in the Sumenat, a tributary of the Cantal Sumène, and *Sumeneta* 1513, the former name of part of the course of the Rieutard. W 143 gives pr [semiŋtən], 9, 10, RN 355–6, LHEB 674–5, Dauzat-Rostaing 665.

SEMLEY Wilts ST 8926. 'The wood or clearing by the river Sem'. *(on) Semeleage* [955]14th S 582, *Semele(g) -legh -ligh -lee* c.1190–1344, *Semlegh* 1329, *Sembly* 1572. R.n. Sem discussed under SEMINGTON ST 8960, + OE **lēah**. Wlt 209.

River SENCE 'Drink, draught, drinking cup'. OE **scenċ** applied to streams with a copious supply of good drinking water. RN 357.

(1) ~ Leic SP 6096. *Sence* from 1602. The river rises near Billesdon and runs past Great GLEN SP 6597 and GLEN PARVA SP 5798 to join the Soar at SP 5598. Its earlier name is recorded as *Glene* 1402, a British r.n. **glaniā* from ***glano-** 'clean, holy, beautiful'. Lei 97, RN 357.

(2) ~ Leic SK 3604. *Sheynch* 1307, *Sence* from 1602. On Sence Brook, a tributary of the r. Sence, lies SHENTON SK 3800, *Scenctune* 1002×24, 'the settlement on Sence Brook'. Lei 97, RN 356.

SEND Surrey TQ 0255. '(The settlement) at the sandy places'. *(æt) Sendan* c.950×68 S 1447, *Sande* 1086–1241, *Sandres* 1197–1241, *Sandes, Saundres, Saund(e), Sendes, Shende, Sonde(s)* 13th cent., *Send(e)* 1260–1485, *Seende* 12th–1428. OE ***sænde**, dative pl. ***sændum**. Situated on a patch of Bagshot sand. Identical with Scene Farm, Kent, *Seende* 1292, *Sende* 1304. Sr 146, PNK 461.

SENNEN Corn SW 3525. 'St Senan's (church)'. *(parochia) Sancte Senane* 1327–77, *Senan* 1524, *Sennan* 1610, *Zenning* 1697, *Zennen* 1702. Saint's name *Senan* of uncertain origin and gender from the dedication of the church. Possibly the 4th cent. Persian martyr Sennen. PNCo 155, Gover n.d. 660.

SENNEN COVE Corn SW 3526. *Sennen Cove* 1838. P.n. SENNEN SW 3525 + ModE **cove**. Earlier *Portsenen* 1370 Gover n.d., 1461, Co **porth**. PNCo 155, Gover n.d. 662.

River SEPH NYorks SE 5691. Uncertain. *Sef* [1170×85]13th, 1201, *Cepht* [13th]copy, *Cepth* [1260]16th. Possibly a Scand r.n. related to OSwed *sæver* 'calm, slow' or ON sef replacing, or replaced by OE **sēfte** 'gentle, mild'. YN 6, RN 357, PNE ii.117.

SESSAY NYorks SE 4575. 'Secg's' or 'sedge island'. *Sezai* 1086, *-ay* 1236, 1304, *Secey* 1182, *Ces(s)ay* 13th, *Sessay* 1483. OE pers.n. *Secg*, genitive sing. *Secges*, or OE **secg**, genitive sing. **secges**, + **ēġ** in the sense 'dry ground in marsh'. YN 187, L 39.

SETCHEY Norf TF 6313. Partly uncertain. *Seche, Siecche* 1202, *Sechiche* (for *-hithe*) 1242, *Sechithe -hyth* 1216×72, 1291. *Seche* appears to be OE ***sǣċe** 'a drainage stream' referring to the river Nar and *Sechithe* 'the landing place at *Seche*, p.n. *Seche* + OE **hȳth**. Another possibility is that *Seche* really represents OE *Sǣ-wīċ* 'lake farm' in the same way as nearby WINCH TF 6315 represents OE *Winn-wīċ* 'the meadow farm'. DEPN.

SETLEY Hants SU 3000. Partly uncertain. *Setle* 1331, *Westsetley* 1544. The generic is ME **ley** (OE *lēah*) 'wood or clearing', the specific any of **(ġe)set** 'a fold', **sǣt** 'a seat' or the antecedent of ModE *set* sb. in any of the senses 20 'an area marked out for a hunt', 23 'slip used for planting', 32 'the earth or burrow of a badger' (not recorded in this sense before 1898) or *set* ppl. 'planted, not self sown'. Ha 145, Gover 1958.207.

SETTLE NYorks SD 8163. 'The dwelling'. *Setel* 1086–1465, *-ell* 1218–1321, *-il(l) -yl(l)* c.1160–1592, *Settle* from 1553. OE **setl**. YW vi.150.

SETTRINGTON NYorks SE 8370. Partly uncertain. *Sendriton* 1086, *Seteringetune* [c.1090]12th, *Setteringtona* 1122×37, *Set(e)ringtun -yng- -ton* 1185×1208–1475. Possibly the 'estate at or called **Sǣtering*, the robber's place', OE **sǣtere** + **ing**^{2}. YE 139, BzN 1968.177.

River SEVEN NYorks SE 7380. Uncertain. *Sivena* 1100×13–1339, *Si- Syvene* [1155×8]15th–1328, *Seven* from c.1540. Possibly Brit **Suminā* of unknown meaning, the source also of the W r.n. Syfynwy SN 0324 and Gaulish *Sumina* which lies behind several Fr r.ns. Sumène, tributaries of the Dordogne, *Simina* 12th, *Sumena* 1585, and the Loire, *Sumera* 1254, *Sumena* 1305, and the r.n. Somme, *Sumenam fluvium* [c.576]7th Gregory of Tours, *Sumena* [c.670]13th Rav. YN 6, RN 358, LHEB 490, 519, 675.

SEVENHAMPTON 'Village of seven homesteads'. OE **seofon** + **hām-tūn**.

(1) ~ Glos SP 0321. *Sevenhamton(e)* 1086–1291, c.1560, *-hampton* from early 13th, *Senhampton* 1506–1614, *Sen(n)yngton* 1575. Gl i.177.

(2) ~ Wilts SU 2090. *Sevamentone* 1086, *Suvenhamtone* 1211, *Sevenhampton* from 1227, *Senhampton* 1330, *Sennington* 1608, 1616. Close by was *Sevenhamehulle* 1332, 'the hill of the people of Sevenhampton'. 27 gives pr [seniŋtən].

SEVEN KINGS GLond TQ 4586. Uncertain. *Sevekyngg(es)* 1285, *Sevyn Kynges* 1456. Early understood locally to refer to a meeting of seven Saxon kings. Possibly a popular reinterpretation of an ancient folk-n. **Seofecingas* 'the people called after Seofeca', OE pers.n. **Seofeca* + **ingas**. Ess 99, ING 24.

SEVENOAKS Kent TQ 5255. 'Seven oak-trees'. *Seouenaca* c.1100, *Sevenac(her)* 1200, *Seuenacre -ak'* 1240. OE **seofon** + **āc**, probably dative pl. **ācum** 'at the seven oak-trees'. The abbreviation <'> has occasionally been falsely taken for *er* in this name instead of *e* as if the name were 'seven acres'. PNK 64, KPN 13.

SEVENOAKS WEALD Kent TQ 5251. 'Sevenoaks forest'. *Seven Oaks Weald* 1819 OS, earlier simply *Walda* 1535. P.n. SEVENOAKS TQ 5251 + ModE **weald**. PNK 65

SEVEN STONES Corn SW 0424. *Seven steen* 1584, *Seven stones* 1588. A cluster of seven rocks in the sea between the Isles of Scilly and the coast of Cornwall, dry at low water spring tides. The earliest form is a Dutch translation. PNCo 155.

River SEVERN Shrops SJ 6902. There are 4 types:

I. Latin forms: *Sabrina* [c.115]9th Tacitus *Ann* 12.31, *Σαβρίνα* (Sabrina) [c.150]13th Ptol, [c.540]11th Gildas, [c.731]8th BHE, 11th–1300;

II. Middle Irish: *Sabrann* 12th.

III. Welsh: *Sabren* 1147, *Habrinum* [7th]c.1000 *Vita Samsoni*, c.1150, *Habren* [c.800]11th HB, [c.1350]c.1400, 1387, *Hafren* c.1150, *Haffren* c.1150, 1257, *Haveren* [1194×1215]13th, *Haveran* 15th;

IV. English: *Saberna, (on) Sæfyrne* [706]12th S 54, *Sæbrine* [816]11th S 180, *Sæfern(e)* [757×75]c.1000 S 142–1042 S 1395 including ASC, *Sauerna* 1086, *Saverne* [c.1130]12th–1393, *Seferne* [851]11th S 201, *Sefern* c.1000, *Seu- Severne* [c.1220]c.1260–1439, *Severn* from c.1540.

Sabrina is probably the name of the divinity of the river. Its origin is unknown but it seems to belong to a family of European r.ns. which may be of pre-IE origin, on a root **sab-* 'liquid', e.g. Sambre, Belgium, *Saba, Sabis*, Sèvre, Seine-et-Oise, France, *Savara* 6th, OIr *Sabrann*, the river of Cork (now the Lee) < **Sabrona*, and the lost r.n. Saferon in Beds,

Sauerne, Sauerna 13th. This root, whatever its origin, was taken into British with an *r*- extension and the regular Celtic suffix **inā*. One interesting point of issue is that British initial *s*- became *h*- by the middle half of the sixth century, while the English reached the Severn in 577. It seems unlikely that they would have heard other than type III represented by the spelling *Hafren*; the name of so important a river, however, may well have been known to the English long before this date. The development of *-b-* to *-v-* by lenition is regular. The normal OE form is *Sæfern* with <f> = [v], the regular development of which is ME *Savern*; ME *Severn* represents an OE Mercian variant *Sefern* with second fronting of OE *æ* > *e*. See also Savernake in SAVERNAKE FOREST Wilts SU 2266. RN 358, RBrit 450, LHEB 519–20, Gl i.10, Bd 9. RN 360 draws attention to Skr *sabar*- possibly meaning 'milk'.

SEVERN BEACH Avon ST 5484. No early forms. R.n. SEVERN + ModE **beach**.

SEVERN ROAD BRIDGE Avon ST 5590. A suspension bridge opened in 1966. R.n. SEVERN + ModE **road-bridge**. Verey ii.92.

SEVERN TUNNEL Avon ST 5286. A railway tunnel under the Severn, the longest in England, opened in 1886. R.n. SEVERN + ModE **tunnel**.

SEVERN VALLEY RAILWAY Shrops SO 7583.

SEVINGTON Kent TR 0340. Probably 'Sægifu's estate'. *Seivetone* 1086, *Siulelde- Siuledtune* c.1100, *Sa- Sey(e)ue- Seiueton'* 1221–92, *Sewenton' -in-* 1242×3, *Seventon* 1267. OE feminine pers.n. *Sǣġifu*, genitive sing. *Sǣġife* + **tūn**. PNK 419.

SEWARDS END Essex TL 5738. 'Sigeweard's end' sc. of Saffron Walden. *Si- Sywardeshond* 1285, *-(h)end* 1322–3, *Sewardesende* 1339–1453, *Sewers End* 1509×47, 1550. OE pers.n. *Siġeweard*, genitive sing. *Siġeweardes*, + **ende**. The name contrasts with AUDLEY END TL 5237. Ess 540.

SEWELL Beds SP 9922. 'The seven springs'. *Sewelle* 1086, 1260, *Seuewell* 1193–1286, *Sewell* from 1247. OE **seofen** + **wella**. A common p.n. type. Cf. SYWELL SP 8297. Bd 129.

SEWERBY Humbs TA 1968. 'Siward's village or farm'. *Siuuarbi* 1086, *Siuuardbi -ward-* 1086–1411, *Si- Sywardebi -by* 12th–1355, *Sewerby* from 1552, *Sureby* 16th cent., *Shourby* 1641. Pers.n. ON *Sigvarthr* or OE *Siġeward* + **bȳ**. YE 104 gives pr [siuwəbi], SSNY 37.

SEWORGAN Corn SW 7030. 'Ford of Goedhgen'. *Reswoethgen* 1302, *Roswoetgan -woeghan* 1361, 1448, *Sewothgan alias Seworgan* 1614, *Savorgan* 1661. Co ***res** (< OCo *rid*) + pers.n. *Goedhgen*. The unstressed first syllable has been lost. PNCo 155 gives pr 'Seᴜurgan', Gover n.d. 506.

SEWSTERN Leic SP 8821. Uncertain. *Sewesten* 1086, *-terna -tern(e)* 1166–1549, *Seustern(e)* c.1130–1446, *Sewstern(e)* 1298–1576, *Sheusterne* 1412, *Shewesterne* 1609. Possibly 'Sæwig's lake', OE pers.n. *Sǣwīġ*, genitive sing. *Sǣwīġes*, + ON **tjọrn**. However, neither element of this name is in reality clear. The suggested alternative, OE **seofon** 'seven' and ***sterne**, a supposed metathesised form of **(ge)streōn** 'property', is unconvincing. Lei 149 gives pr ['sju:stə:n], PNE ii.151.

SEZINCOTE Glos SP 1731. 'Gravelly cottage(s)'. *Ch(i)esne- Cheisnecot(e)* 1086, *Chesnecothe* 1236, *Sesnecot(e)* 1185, *S(c)hesnecote* 1196–1380, *Sesencote -yn-* 14th, 15th, *Sezingcott als. Seseencott als. Sesencott* 1610. OE **ċiesen** with later substitution of [s] for [tʃ] + **cot**, pl. **cotu**. The persistent *Ches- Ses-* spellings show analogical influence of OE **ċeosol** 'gravel' or **ċesolen** 'gravelly'. Gl i.257, xiii.

SHABBINGTON Bucks SP 6606. 'Estate called after Sceobba'. *Sobintone* 1086–1237, *S(c)hob(b)ington* 1241, 1278, *Shabbington or Shobbington* 1806. OE pers.n. **Sceobba* + **ing**[4] + **tūn**. Bk 128.

SHACKERSTONE Leic SK 3706. 'The robber's farm or village'. *Sacrestone* 1086–1268, *S(c)hakereston(e)* after 1250–1397, *S(c)hakerston(e)* 1283–1619, *S(c)hakeston* 1209×35–1553, *Shakston, Shaxton* 1401, 16th cent. OE **scēacere**, genitive sing. **scēaceres**, + **tūn**. Lei 533.

SHACKLEFORD Surrey SU 9445. 'The shackle ford'. *Sak(e)les- Sakel(s)- Shakeles- S(c)hakel(e)- Chakelford* 13th cent., *Shakul-* 1370, *Shaggle-* 1558×1603. OE **sceacol** + **ford**. The reference may be to a chain used to aid crossing the river. Alternatively it may be related to dial. *shackle* 'loose' in relation to the bed of the river. Sr 199, PNE ii.98–9.

SHADFORTH Durham NZ 3441. 'Shallow ford'. *Shaldeford* [1183]14th, 1313–1433, *-forth* [1183]15th, *Shaldford(e)* c.1308–1432, *Shauld- Shawdeforthe, Shaldfurth* 1580–1593, *Shadforth* from 1647. OE **sc(e)ald** + **ford**. NbDu 175, L 68.

SHADINGFIELD Suff TM 4384. 'The open land called or at *Scead-denu*, the boundary valley'. *Scadenafella* 1086, *Shadenefeld* [1154×89]1268, *Schadenesfeld* 1190, 1291, *S(c)hadenfeld* 1250, 1316, 1344, *Shan(e)feld* 1336, 1524, 1610, *Shadyngfeld* 1575 *-feylde* 1615. OE p.n. **Scēad-denu* < **scēad** + **denu**, + **feld**. Shadingfield parish lies on the boundary of Wangford hundred near Hundred river. DEPN, Baron.

SHADOXHURST Kent TQ 9737. 'Shaddock's wooded hill'. *Schettokesherst* 1239, *Scadockesherst* n.d., *S(c)hattokeshurst -herst -hirst* 13th cent., *Sadhokesherst* 1267, *Chaddekesherst* 1292. Uncertain family name + **hyrst**, Kentish **herst**. There is some similarity between the specific and the surnames Shaddock, Chaddock and Chattock usually explained as varants of Chadwick. They are not early recorded, however, and it may be preferable to assume an unrecorded *-uc* diminutive of OE **scēad** 'boundary' or **scēat** 'a corner of land', OE ***scēaduc**, genitive sing. **scēaduces** '(little) boundary wood', ***scēatuces** '(little) corner wood'; or a compound p.n. *Scēad-āc*, ME *Shed- Shadoke* 'boundary oak', or *Scēat-āc*, ME *Shet-Shatoke* 'corner oak'. PNK 363.

SHAFTESBURY Dorset ST 8622. Partly uncertain.

I. *(in to, on, æt, at) Sceaftesburi* [871×7]15th S 357(1), [956]15th S 630, *-byrig* [914×9]13th BH, [c.950]10th S 1539, [971×83]14th, c.1050 ASC(C) under year 982, 12th ASC(E) under year 1036, *Scæftesbyrig* c.1000, *Sheftesburi -bury -byri* 1269–c.1540, *Schaftesbir(y) -buri -bury* 1194–1356, *Shaftesbury* from 1242, *Shast(es)bury* 1391–1432, *Shasberrye -bur(r)y* 17th.

II. *(to) Sceafnesbirig* [951×5]14th B 912, *to sceaftenes byrig, to Scæftenesbyrig* [1015]11th S 1503(1, 2).

III. *Shafton(e)* [883]14th B 410, [860]15th B 499, 1258–1417.

IV. Latin forms: *monast' Septoniæ* [951×5]14th B 914, *(ad) S(c)eftonia(m)* early 12th ASC(F) under years 980, 1245, 1285.

V. *S(c)haston(')* 1260–1650.

Type I seems to be 'Sceaft's fortified place', OE pers.n. **Sceaft*. genitive sing. *Sceaftes*, + **byriġ**, dative sing. of **burh**. Type II suggests that beside **Sceaft* there was a pet form **Sceaften* < **Sceaftīn*, unless the spellings actually represent OE pers.n. *Sceaftwine* (of which *Sceaft* would be a short form). Alternatively the specific might be OE **sceaft** 'a shaft, a pole' used either literally or of the prominent steep-sided hill on which the town stands. In this case ***sceaften** would be an unrecorded derivative in the frequent noun suffix - **en** as in words such as *fæsten, hæġen, wēsten* etc. The sense would be 'place at the pole, place where poles stand', probably used as a p.n. and so the whole name would be 'the *burh* of **Sceaft* or **Sceaften*'. Type III is probably not a genuine form in **tūn** but an Anglicisation of Type IV, Latin *Shaftonia* (like *Wintonia* for Winchester). Type V is either a reduction of a form **Shafteston* or derives from III by misreading of *f* as long *s*. Holinshed's alternative name for Shaftesbury, *Paladour* 1577[†], is taken from Geoffrey of Monmouth's *oppidum Paladur* 'Paladur town' c.1138[††], i.e. 'shaft town', Latin *oppidum* + W *paladr*

[†]Chronicles I.446.

[††]Griscom ii.9.

'shaft', a nice early example of antiquarian etymologising and invention. Do iii.139–43, PNE i.151, BH 106.

SHAFTON SYorks SE 3911. 'Settlement marked by a pole, made of poles, or where poles are made'. *Sceptun -tone* 1086, *Sc(h)afton(a) -tun* c.1165–1379, *Shafton* from late 13th. OE **sceaft** + **tūn**. YW i.272.

SHALBOURNE Wilts SU 3162. 'The shallow stream'. *Scealdeburnan* [951×5]15th S 1515, *S(c)aldeburne* 1086, *S(c)haldebo(u)rne* 1202–1428, *Shawborne* 1547, 1599. OE **sceald**, definite form **scealda**, + **burna**. Wlt 354.

SHALCOMBE IoW SZ 3985. 'The shallow valley'. *Eseldecome* 1086, *Scaldecumbœ -a -e* 1135×54, *Shaldecumb(e) -comb(e)* 1284–1544, *Shalcome* 1526, *Shaucome* 1611, *Sharcomb* 1775. OE **sceald** + **cumb**. Wt 212 gives prs ['ʃa:kəm, 'ʃɔkəm] and ['ʃæɫ kəm], Mills 1996.93.

SHALDEN Hants SU 6941. 'The shallow valley'. *Seldene* 1086, *Scaldedene* 1167, 1175, *Shalden(e)* from c.1270. OE **sceald**, definite form feminine **scealde**, + **denu**. The earliest reference to this place is in the phrase *be scealdedeninga gemære* 'by the boundary of the people of Shalden' [1046]12th S 1013. Ha 145, Gover 1958.106.

SHALDON Devon SX 9372. *Shaldon* 1603×25. Unknown specific + ModE **don** representing OE **dūn** 'a hill'. The topography does not permit this to represent OE *sceald-denu* 'shallow valley' like SHALDEN Hants. Shaldon is a late-developed fishing village on the sands. D 460, Fox 64–5.

SHALFLEET IoW SZ 4189. 'The shallow stream or creek'. *(apud) Scaldeflotam* (Latin) [828]c.1300, *(æt) Scealdan fleote* [838]12th S 281, *(7lang) scealdan fleotes, (on) scealdan fleot* [949]14th S 543, *Sceldeflet* 1086, *Schaldeflete* 1262–1429, *Shalflete* 1397–1595, *Shalflett* 1583, 1704, *Shoflyt* 1600. OE **sceald**, definite form dative case **scealdan**, + **flēot**. The reference is to a narrow creek N of the village which flows into Caul Bourne. Wt 205 gives pr ['ʃæ(:)flət], Mills 1996.93, Jnl 29.80.

SHALFORD Essex TL 7229. The 'shallow ford'. *Celdefordā, Esceldeforde, Scaldefort* 1086, *Saldeford* 1200–93, *S(c)hald(e)ford* from 1275, *Sc(h)aude-* 1200, 1264, *Shawford* 1563–1678. OE **sceald**, definite form **scealda**, + **ford**. Ess 455 gives pr [ʃa:fəd].

SHALFORD Surrey TQ 0047. 'The shallow ford'. *Scaldefor* 1086, *-ford'* 1199–1294, *Scaude- Schalde- Sholdeford* 13th, *Shalde-* 1279–1428, *Schal-* 1462, *Shalforth* 1541, *Shalford* from 1550, *Shawl-* 1580. OE **sc(e)ald**, definite form **scealda**, + **ford**. Sr 246, PNE ii.100.

SHALFORD GREEN Essex TL 7127. *Shalfordegrene* 1580, *Shalford-pond Gr* 1805 OS. P.n. SHALFORD TL 7229 + ModE **green**. Ess 456.

SHALLOWFORD Devon SS 7144. No early forms. ModE **shallow** + **ford**.

SHALMSFORD STREET Kent TR 1054. *Shanford-Street* 1727, *Shalmsford* ~ 1819 OS. P.n. Shalmsford + Mod dial. **street** 'a hamlet'. Shalmsford, *Essamelesford* 1086, *S(c)(h)amelesford(e)* c.1100–1292, is 'Shamel's ford', OE **sc(e)amol** 'a shelf of land', probably used as a p.n. **Sc(e)amol*, genitive sing. **Sc(e)ameles*, + **ford**. The reference is to Chartham Downs which here form a ledge of high ground overlooking the ford site. PNK 371, PNE ii.100, Cullen.

SHALSTONE Bucks SP 6436. 'Settlement of the shallow'. *Celdestane -stone* 1086, *S(c)haldeston* 1227–1391, *Shaweston al. Shalston* 1522×47, *Shalson* 1584. OE **sceald**, genitive sing. **scealdes**, + **tūn**. The reference is to a crossing of the stream due N of the village. Bk 47 gives pr [ʃɔ:lstən].

SHAMLEY GREEN Surrey TQ 0343. *Shamley Green* 1816 OS. P.n. *Shambles* 1544, *Shamele(igh)* 1548, c.1550, *Shambley* 1607, 'the wood or woodland clearing with ridges or shelves of land', OE **sceamol** + **lēah**. Cf. also Thomas *ate Shamele* 1332. Sr 256.

SHANGTON Leic SP 7196. 'The settlement by the shank'. *San(c)tone* 1086, *Sanketon'* late 12th–1295, *Scanketon(e)* c.1130–late 13th, *S(c)hanketon(e)* 1206–1439, *S(c)hankton'* 1344–1610, *Schanton* c.1291, *Shanton'* 1426, *Saun(c)keton'* 1295, *Schangeton'* 1274–1491, *Shangton(')* from 1446. OE **scanca** + **tūn**. The reference is to a shank or narrow ridge projecting from high ground beside the village. Lei 250.

SHANKLIN IoW SZ 5881. 'Waterspout ledge'. *Sencliz, Selins* 1086, *S(c)hentling(') -lyng* 1287×90–1431, *Shenclyng' -ling* 1305, 1324, *Shenckling(e)* 1583, 1632, *Shenclyn* 1503, *Shanklin* from 1611. OE **scenċ** 'a drink, a draught, a cup' + **hlinċ**. The reference is to the cascade in Shanklin Chine. Wt 214, PNE ii.106, L 164, Mills 1996.94.

SHANKLIN CHINE IoW SZ 5881. *Chynklyng Chyne* 1550, *Chanklin Chine* 1769. P.n. SHANKLIN SZ 5881 + Hants/IoW dial. **chine** 'a ravine' (OE *ċinu*). Wt 215, Mills 1996.94.

SHAP Cumbr NY 5614. 'The heap'. *Hep* 12th–1398, *Hepp(e)* 1246–1468, *Yhep* 1241, *Yhap* 1292, *Shep* 1256, *S(c)hap* from 1279, *S(c)happ(e)* 1279–1773. OE **hēap** in the sense 'heap of stones' referring to the ruins of the megalithic stone circle a mile south of the town at NY 5613. The name underwent two parallel developments, the normal one without shift of stress in the OE diphthong *ēa*, viz. OE *héap* > ME *hēp*, and an unusual one with shift of stress and substitution of *sh* for *hj* (phonetically [ç]), viz. *héap* > *heáp* > **hjap* > *shap*. We ii.164, RES 1.437ff, Brno SE 1964.87 note 39, Thomas 1976.72.

SHAP FELLS Cumbr NY 5308. *Shap Fells* 1793. P.n. SHAP NY 5614 + ModE **fell(s)** (ON *fjall*). We ii.177.

SHAPWICK 'Sheep farm or trading place'. OE **scēap**, genitive pl. **scēapa**, + **wīċ**.

(1) ~ Dorset ST 9301. *Scapeuuic* 1086, *S(c)hapewyk(e)* 1176–1431, *S(c)hapwyk(e) -wik(e)* 1244 etc., *-wick* 1591, *Shepwyk(e)* 1238, 1431, 1593. Do ii.176.

(2) ~ Somer ST 4138. *Sapwic* *[725]12th S 250, *Schapwik* [729]14th S 253, *Schapewyke* *[971]14th S 783, *Sapeswich* 1086, *Schepwich* 1173, *Shapwi(c)ke* [13th]Buck. Possibly a Roman estate commemorated in the name *Abchester*. DEPN, M. A. Aston, *The Shapwick Project: a topographical and historical survey* 2nd Report, Bristol Extramural Dept. 1989.11, Origins 118.

SHARDLOW Derby SK 4330. 'Mound with a notch'. *Serdelau* 1086, *Sherdelawe* 1231, 1240, *S(c)hard(e)low(e)* from early 13th cent. OE **sceard** + **hlāw**. The mound has not been discovered. Db 501.

SHARESHILL Staffs SJ 9406. 'Scearf's hill'. *Servesed* (sic) 1086, *Sarueshul(l)* 1213, 1242, *Sarweshull'* 1225, *Sarneshull -hulf* (for *Sarues-*) 13th, *Shareweshulf* 1252, *S(c)har(e)shulfe* 1282–1347, *S(c)har(e)shul(l)(e)* 1271–1657, *S(c)har(e)shill -ys- -hyll* from 1370. OE pers.n. **Scearf*, genitive sing. **Scearfes*, + **hyll** varying with **scylf** 'a shelf' referring to the level area behind the church. The DB form may point to OE **hēafod** 'a head'. St i.115.

SHARKHAM POINT Devon SX 9354. *Sharkham Pt* 1809 OS. P.n. Sharkham of unknown origin + ModE **point**.

SHARLSTON WYorks SE 3918. Probably 'Scearf's estate'. *Scharuest(on)'* 12th–1296, *Scharneston(e)* c.1160–1286, *S(c)harweston* 1291×1312, 1297, *Sharston* 1379, *S(c)harleston(e)* 1428–1591, *Sharlston* from 1633. OE pers.n. **Scearf*, genitive sing. **Scearfes*, corresponding to ON *Skarfr*, + **tūn**. The *Scharles-* forms are late and probably developed through association with pers.n. *Charles* from *Scharnes-* by misreading of *u* as *n* and confusion of liquids *n* and *l*. YW ii.114, xi.

SHARNBROOK Beds SP 9959. 'Dung brook'. *S(c)ernebroc, Serneburg* 1086, *Shernebroc* 1163, *Sharnebroc* 1167, *Scharmbrook* 1785. OE **scearn** + **brōc**. Bd 39 gives pr [ʃa:mbruk], L 15.

SHARNFORD Leic SP 4891. 'The dirty ford'. *Scearnforda* 1002×4, *Scerneford(e), Sceneford* (1×) 1086, *S(c)harneford(e)* 1220–1539, *Sharnford(e)* 1276–1576, *-forth* 1516, *S(c)herneford(e)* 1284–1549. OE **scearn** + **ford**. Lei 537.

SHAROE GREEN Lancs SD 5333. P.n. Sharoe + ModE **green**.

Sharoe, *Sharoo, Shayrawe, Sharrow* 1502, *Sharoe* 1513, is the 'boundary ridge', OE **scaru** + **hōh**. Sharoe is situated on slightly rising ground between Sharoe Brook and a brook that formed the boundary between Broughton SD 5235 and Fulwood SD 5331. La 147.

SHAROW NYorks SE 3271. 'The boundary hill'. *Sharou* 1114×40, *S(c)harehow(e)* 1297–1328, *S(c)harow(e)* 1318–1676. OE **scearu** + **hōh**. The reference is probably to *The Mount*, a hill by the main road to Thirsk near the old boundary of the North Riding. YW v.157.

SHARPENHOE Beds TL 0630. '(Settlement at) the sharp or rugged hill-spur'. *Scarpe- Serpenho* 1197, *S(c)harpenho* 1232 etc. OE **scearp**, dative sing. definite form **scearpan**, + **hōh**, dative sing. **hō**. The side of the hill-spur has a series of folds in it which points to the sense 'rough, rugged' here. Bd 165, L 168.

SHARPERTON Northum NT 9503. 'Farm or village at or called *Sceard-* or *Scearp-beorg*'. *Scharberton'* 1242 BF, 1307, *Sharperton'* 1296. OE ***sceard** + **beorg** 'the notched hill' or ***scearp** + **beorg** 'the pointed hill' + **tūn**. The Roman road from High Rochester to Whittingham, Margary No.88, passes through a notch between hills at NT 9604. NbDu 175.

SHARPHAM HOUSE Devon SX 8257. P.n. Sharpham + ModE **house**. Sharpham, *Sharpeham* 1249, *Sherpham* 1340, 1377, is 'the pointed promontory', OE **scearp**, definite form **scearpa**, + **hamm** 2a, referring to the sharp bend of the Dart around the site of the house. D 315.

SHARPNESS Glos SO 6702. 'Sceobba's headland'. *Schobbenasse* 1368, *Shob(b)enas(s)he* 1444, 1575, *Shepnes(s)* 1516, *Sharpness (Point)* 1824. Earlier simply *Nesse* 1086. OE pers.n. *Sc(e)obba* + **næss**. Gl ii.236, L 172.

Higher SHARPNOSE POINT Corn SS 1914. *Upper Sharpnose Point* c.1870. ModE adj. **higher** + p.n. Sharpnose Point. 'Higher' for distinction from Lower SHARPNOSE POINT SS 1912 two miles to the S. *Sharpnose* is probably 'sharp nose' referring to the pointed character of the promontory. PNCo 155.

Lower SHARPNOSE POINT Corn SS 1912. *Lower Sharpnose point* c.1870. ModE adj. **lower** + p.n. Sharpnose Point as in Higher SHARPNOSE POINT SS 1914. PNCo 155.

SHARPTHORNE WSusx TQ 3732. *Sharpethorne* 1597. ME **sharp** + **thorn** 'a thorn-tree'. Sx 274.

SHARRINGTON Norf TG 0337. 'The dirty settlement'. *Scarnetuna* 1086, *-tune* 12th, *Sharnetone* 1254. OE **scearn** 'dung' + **tūn**. DEPN.

SHATTERFORD H&W SO 7981. 'Sewer ford'. *Sciteresforda* [996 for 994]17th S 1380, *Sheteresford* 1286. OE ***scitere**, genitive sing. ***sciteres**, + **ford**. Wo 32, Hooke 237.

SHAUGH PRIOR Devon SX 5463. 'S belonging to the prior' sc. of Plymouth. *Pryoursheagh* 1559. Shaugh, *Scage* 1086, *(La) S(c)hagh* 1303 etc., *Shea(gh), Shaff, Shaghe al. Shave, Shaye* 16th cent., *Shaue* 1281, is 'the wood', OE **sceaga**. D 258 gives pr [ʃei]. The name seems to have undergone several different developments, eME [ʃaγə] > (i) [ʃawə] (*Shaue* 1281); (ii) eME [ʃaγə] > [ʃavə, ʃaf] with substitution of the allied fricative [v] for [γ] and subsequent unvoicing to [f] in the final position; (iii) eME [ʃaγə] > ME [ʃawə] > [ʃau] > [ʃa:] whence modern [ʃei]. Cf. Jnl 13.109–14.

SHAVINGTON Ches SJ 6951. 'Estate called after Sceafa'. *Santune* 1086, *S(c)havinton* 1276–1845 with variants *S(c)hau- Shaw-* and *-yn- -en-*, *Shavington* from 1299, *Shenton* 1514–1802. OE pers.n. *Scēafa* + **ing**[4] + **tūn**, varying with genitive sing. *Scēafan*, + **tūn** 'Sceafa's farm or village'. Che iii.69 gives pr [ˈʃævintən] and local [ˈʃentən].

SHAVINGTON PARK Shrops SJ 6338. P.n. *Savintune* 1086, *-ton'* 1219, 1284×92, *Shauenton -in-* 1255×6, *-yn-* 1271–1419, *Schauyngton'* 1291×2, *Shavyngton* 1337, *-ing-* from 1618, *Shenton* 1695, 1731, 1760, 'Sceafa's estate', OE pers.n. *Scēafa*, genitive sing. *Scēafan*, varying with *Scēaf(a)* + **ing**[4] + **tūn**, + ModE **park**. Sa i.259.

SHAW 'Undergrowth, woodland, scrub'. OE **sceaga**. L 208–9, Nomina 12.89–104, 13.109–14 (for the *shay* form).

(1) ~ Berks SU 4868. *Sagas* c.1080, *Essages* 1086, *Shag(h)e, Shaga* 1167–1327, *Shawes* 1241, 1343, *Shaue* 1284, 1363. Brk 263, L 209.

(2) ~ GMan SD 9308. *Shaghe* 1555, *Shawe* 1600, *Shay* 1577, *Shaie, Saye* 1580. La 52.

(3) ~ Wilts ST 8865. *Schawe* 1256, *Shage* 1286, *Shaa* 1641, 1652. Wlt 129.

(4) ~ MILLS NYorks SE 2562. *Shaws Mill* 1817. P.n. *Shaw* 1594, + ModE **mill(s)**. YW v.185.

(5) High ~ NYorks SD 8792. ModE **high** + p.n. *S(c)hal(l)* 1218, 1301, either an Anglicised form of ON **skáli** 'a shieling' or an AN form of OE **sc(e)aga** 'a copse, a small wood'. YN 260.

SHAWBURY Shrops SJ 5621. 'The fortified place or manor house by the wood'. *Sawesberie* 1086, *Sagesbury, Scagesburie* late 12th, *Shaburia* 1130×47, *Schayebur'* 1242, *Schabury* 1306×7, *Sha-* 1535–1747 with variants *-biry -burie* and *-ber(r)y*, *Shawberia* from c.1165 with variants *Schau- S(c)haw-* and *-bir' -ber(r)y -burie, Shawbury* from 1376, *Schagebir'* 1201, *-bury* 1227, *Shagebyr'* 1237. OE **sceaga** + **burh**, dative sing. **byriġ**. Sa i.259.

SHAWDON HALL Northum NU 0914. *Shawdon Hall* 1868 OS. P.n. *Schaheden* 1232, *Schauden* 1242 BF, *Shawdon* 1542, 'the copse valley', OE **sceaga** + **denu**, + ModE **hall**, an 18th cent. Adam-style house. NbDu 175, L 209, Pevsner 1992.567.

SHAWELL Leic SP 5480. 'The boundary spring'. *Sawelle* 1086, *S(c)hadewell(e)* 1224–1338, *S(c)hathe-* 1270–1517, *Shathwell* 1432, *Sc(h)awell(e)* 1232, *Shawell* from 1507. OE **scēath** + **wella**. The village is close to but not on the county boundary which the stream crosses but does not form part of. The precise reference is unclear. Lei 464.

SHAWFORD Hants SU 4724. 'The shallow ford'. *Scaldeforda* 1208, *Scaudeford* 1233, *S(c)haldeford* 1263–1329. OE **sceald**, definite form **scealda**, + **ford**. Identical with SHADFORTH Durham NZ 3441, SCALFORD Leics SK 7624. Ha 146, Gover 1958.175.

SHAWFORTH Lancs SD 8920. There are no early forms, but this appears to be the 'copse ford', ME **schawe** (OE *sceaga*) + **forth** (OE *ford*).

River SHEAF SYorks SK 3484. 'The boundary river'. *Scheve, Sheue* 1183, 13th cent., *Scheth* 14th, *Sheath* 1637, *Sheaf(e)* from 1700. OE **scēath**. The river probably formed part of the boundary between Mercia and Northumbria. Later it marked the boundary between Yorkshire and Derbyshire. RN 360, YW vii.137, Db 16.

SHEARSBY Leic SP 6290. Partly uncertain. *Sv(ev)es- Sevesbi* 1086, *Seuesbi -by* 1195–13th, *S(c)heuesby -ys-* 1209×35–1455, *Shewesby* 1488, *Shethesby(e) -is-* 1436–1590, *Sheysby* 1517, *Sheasb(e)y* 1576–1691, *Shearsby* 1721. The DB forms possibly point to 'Swæf's village', OE pers.n. *Swǣf*, genitive sing. *Swǣfes*, + **bȳ**, but all the other forms point to OE **scēaf**, genitive sing. **scēafes**, + **bȳ**, 'the village of the sheaf'. Possibly OE **scēaf** is used of the steep hill on which the village stands. Lei 465, SSNEM 68.

SHEBBEAR Devon SS 4309. 'Shaft wood'. *(of) Sceft beara* 1050×73, *Sepesberie* 1086, *Scheaftberia* 1177, *Shaftebeare* 1319, *Syeftbere* 1262, *Sheftbeare* 1291, 1353, *Schebbeare* 1425. OE **sceaft** + **bearu**. Probably a wood where poles or shafts were obtained. The earliest spelling shows that **bearu** the dative case of which was historically **bearwe** was early assimilated in the Devon dial. to the declension of **wudu**, dative sing. **wuda**. D 107.

SHEBDON Staffs SJ 7625. 'Sceobba's hill'. *Schebbedon* 1267, *Shebdon Ley* 1572, *Shebben* 1686. OE pers.n. *Sceobba* + **dūn**. DEPN, Horovitz.

SHEDFIELD Hants SU 5513. Partly uncertain. *(to) scida felda* [956]12th S 600, *Side- Schide- S(c)hydefeld* 1256–1350.

Apparently 'open land of the shides', OE **scīd**, genitive pl. **scīda**, + **feld**. A shide is a piece of split wood, a board, a plank. Possibly this was a reference to a wooden footbridge on the line of the Roman road, Margary no.420, which here crosses a number of small streams. Ha 146, Gover 1958.47, DEPN, L 245.

SHEEN Staffs SK 1161. '(The settlement) at the sheds or shelters'. *(æt) Sceon* [1002×4]11th S 1536, [1004]11th S 906, 1086, *Shene* 1265. OE ***scēo**, dative pl. ***scēon**. Identical with Sheen, the ancient name of Richmond GLond TG 1874, *(on) Sceon* [942×c51]13th S 1526, *Shene* 1230–1601. DEPN, Sr 65, Studies 1936.55.

SHEEP ISLAND Cumbr SD 2163. *Sheep Island* 1852 OS.

SHEEPSCOMBE Glos SO 8910. 'Sheep coomb'. *Sepescumb'* 1248, *Shep(e)scombe* 1276–1740. OE **scēap**, genitive sing. **scēapes**, + **cumb**. Gl i.133 gives pr [ʃepskəm].

SHEEPSTOR Devon SX 5567. 'Tor on the steep hill'. *Sitelestorra* 1168, *Schitelestor(re)* 1219, *Scitestorr(e)* 1408, *Shistor* 1547, *Shepystorr* 1574, *Sheepstor al. Shittestor* 1695. OE **scyttels** + **torr**. Sheepstor lies at the foot of the steep rounded hill of Sheeps Tor (*la Torr apud Shitestorr* [1291] 1408). 'Sheep' is a late substitution to avoid the unpleasant sound of the older form of the name. D 238.

SHEEPWASH Devon SS 4806. 'Place for dipping sheep, a sheep-wash'. *Schepewast* 1166, *Schepways(s)e* 1253, 1276, *Schipwaysche* 1346, *Shipwash* 1675. OE **scēap** + **wæsce**. D 109, PNE ii.101.

SHEEPY MAGNA Leic SK 3201. 'Great S'. ME prefix **micel**, *Muchele* 1223, Latin **magna**, *Magna* 1276–1514 (postfix from 1277), ModE **great**, *Great(e)* 1538, 1610, + p.n. *Scepa* 1086, *Scepeie -ey* 1199, early 13th, *S(c)hepey(e)* 1209×35–1610, 'sheep island', OE **scēap** + **ēġ** in the sense 'dry ground in fenland'. Lei 538, L 39.

SHEEPY PARVA Leic SK 3301. 'Little S'. Latin prefix **parva**, *Parva* 1209×35–1535 (postfix 1277–1559), ModE **little**, *Lyttel* 1572, + p.n. *Scepehe* 1086 and other forms as in SHEEPY MAGNA SK 3201. Lei 539.

SHEERING Essex TL 5013. 'The Sceringas, the people called after Sceri'. *Sceringā* 1086, *Scheringes* 1248, *Ser(r)inges* 1212, 1244, *Shering(e) -yng* from 1240, *Sher(r)ing(es)* 1216×72, *Shir(r)ing(g) -yng(ge)* 1254–1391. Folk-n. **Sceringas* < pers.n. **Sceri* + **ingas**. The precise form of the pers.n. is unknown but it was cognate with OG *Scarius, Skerilo* and *Scering*. Ess 50, ING 24.

SHEERNESS Kent TQ 9174. 'Bright headland'. *Scerhnesse* 1203, *Shernesse* 1221, *Shirenasse* 1462, *Shiernas* 1579, *Sheerness* 1690. OE ***scīr** + **næss, ness**. PNK 264.

SHEET Hants SU 7524. 'The corner or nook'. *Syeta* c.1210, *Shete* 1236–1462, *la Syete* c.1255, *la Shyte* 1266, *Schute* 1316, *Sheete* 1613. OE ***scīete**. The reference is to Sheet's position in the NE corner of Petersfield parish in an angle between two headwaters of the western Rother. Ha 146, Gover 1958.60, DEPN, PNE ii.108.

SHEFFIELD SYorks SK 3587. 'Open countryside by the r. Sheaf'. *(E)scafeld, Sceuelt'* 1086, *Sefeld(ia)* 1161–1228, *S(c)hefeld(e)* 1202–1557. R.n. SHEAF SK 3484 + OE **feld**. Two spellings, *Sedfeld* 1184, *Sadfeld* 1185, show vestiges of the original OE form *Scēath-feld*, with AN *d* for *th*, but the majority show assimilation to the following *f*. RN 361, YW i.204, TC 169, L 237, 243.

SHEFFIELD BOTTOM Berks SU 6469. P.n. Sheffield + ModE **bottom**. Sheffield, *Sewelle* 1086, *Scheaffelda* 1167, *S(c)hefeld' -f(f)eld -feud* 1267–1350, *Sheff(i)eld* 1609, is 'open land with a shelter', OE **scēo** + **feld**. Brk 206, L 244.

SHEFFIELD PARK ESusx TQ 4023. P.n. Sheffield + ModE **park**. Sheffield, *Sifelle* 1086, *-feld'* 13th cent., *Shiffeld* 1316, *Shep(e)feld -feud* 13th–14th cents., is 'open land for sheeep', OE **scēap, scīp** + **feld**. Sx 347.

SHEFFORD Beds TL 1439. 'Sheep ford'. *Sepford* 1220, *Shipford* 1229, *Shepford* 1247, *S(ch)efford* 1247–71. OE **scēap** + **ford**. Bd 172, L 71.

Great SHEFFORD Berks SU 3875. *West Shifford als Shifford Magna* 1535, *West Shefford als Great Shefford* 1757. ModE **great**, Latin **magna**, + p.n. *Sif(f)ord* 1086–1204, *Shiford'* 13th cent., *Shipford* 1284, 'sheep ford', OE (Mercian) **scēp** + **ford**. 'Great' or 'West', *West Shifford(')* 13th cent., *West Schyp-Schipford'* 1284, *Westchifforde* 1428, for distinction from Little or East Shefford SU 3974, *Parva Siford'* 1222×3, *Ests(c)ifford'* 1242×3, *Shifford Parva* 1535. Brk 325–6, L 71.

SHEINTON Shrops SJ 6103. 'The beautiful farm, village or estate'. *Sc(h)entune* 1086, *Shenton(')* 1242–1765, *Seinton'* 1200, 1207, *Seynton* 1255, 1262, *Sheinton* from 1221×2 with variant *Sheyn- Shineton* 1687–1833. OE **scēne** + **tūn**. OE **scēne** gives ModE *sheen*, but with shortening in compound p.ns. *Shen-* as in SHENFIELD Essex TQ 6094, SHENINGTON Oxon SP 3742, SHENLEY Bucks TL 1900; the 17th cent. form *Shine-* presupposes ME *shīn* which may be due to a ME raising of *ē* > *ī* before *n*†; but the *Shein-* forms remain puzzling. Sa i.261 suggests that the name is in fact a contraction from OE **Scēningtūn* 'the farm, village, or estate called after *Scēna*'. Bowcock gives pr *Shine-ton*.

SHELDERTON Shrops SO 4077. Partly unexplained. *Sheldreton'* 1248, 1255, *Scelderton'* 1255×6, *Schelvertone* 1264–79, *Shelderton* 1832 OS. Unidentified element or pers.n. + OE **tūn**. Gelling.

SHELDON Derby SK 1768. Probably 'Shelf Haddon'. *Scelhadun* 1086, *Schelehaddon'* 1230, *S(c)hel- Shelladon(e)* c.1250–1476, *Shelledon* 1278, *Sheldon* from 1355. OE **scelf** + p.n. HADDON SK 2366. The village lies on the edge of a flat limestone hill, the whole of which must have been called *Haddon* 'heath hill' and perhaps also 'the shelf' at the 1000ft. contour at the head of a valley dropping steeply to the river Wye. On formal grounds the specific could be OE ***scēla** 'a summer hut' but this element appears to be confined to more northerly counties. Db 164, PNE ii.104.

SHELDON Devon ST 1108. 'Shelf valley'. *Sildene* 1086, *Schildene* 1185–1543 with variants *Shil- Schyl- Scil-, Shyeldone* 1269, *Shildone* 1334, 1336. OE **sci(e)lf** + **denu**. The village lies on a shelf of land overlooking a steep slope to the valley. D 569, L 97, 99, 187.

SHELDON WMids SP 1584. 'Shelf hill'. *Scheldon(e)* 1189–1403, *Sheldon* from c.1240. OE **scelf** + **dūn**. The village lies on a slope below a shelf of high ground. Wa 48, L 143, 154, 186–7.

SHELDWICH Kent TQ 0156. 'Farm, work-place with a shelter'. *Scildwich -uuic* [784]13th S 38, *Sceldwike* 1198, *Sheldwych(')* 1240–54, *Schaldwyk'* 1270, *Schaldwyche* 1291. OE **sceld** + **wīc**. PNK 239, KPN 59, PNE ii.104.

SHELF WYorks SE 1228. 'The shelf, the ledge'. *Scelf* 1086, 1268, *Shelf* from 1236, *Schelf(e)* 1275–1543. OE **scelf**. The site is a level area in hilly country. YW iii.85, L 186.

SHELFANGER Norf TM 1083. 'The sloping wood at the shelf'. *Sceluangra* 1086, *Scelfhanger* [1087×98]12th. OE **scelf** in the sense 'plateau' + **hangra**. DEPN, L 186, 195–6.

SHELFIELD WMids SK 0302. 'Shelf hill'. *Scelfeld* 1086, *S(c)helfhull* 1271, 1300. OE **scelf** + **hyll**. Probably this is correctly 'the hill called Shelf' referring to a low plateau. The DB form may be a genuine variant, 'the open land at or called Shelf'. DEPN, L 187.

SHELFORD Notts SK 6642. 'Shallow ford'. *Scelford(e)* 1086–1283 etc. with variant spellings *Schel-* and *Shel-*, *Sceldford* c.1160, *S(c)held-* 1251–90. OE ***sceldu** + **ford**. Nt 241.

Great SHELFORD Cambs TL 4652. *Magna Scelford* 1218. ModE **great**, Latin **magna**, + p.n. *Scelford* 1086–1279 including [1042×66]13th S 1051, *S(c)hel(e)ford(a)* c.1060–1473, *Escelford(e)* 1086, *Scelford(a) -fort* 1086–c.1120, *Scheldford* 1109×33–90, 1354,

†Cf. *quin* for *queen*, Dobson §132 Note (f)(viii), §136 Note 2.

'shallow ford', OE ***sceldu** + **ford**. 'Great' for distinction from Little SHELFORD TL 4551. Ca 87, PNE ii.104.

Little SHELFORD Cambs TL 4551. *Parva Scelford* 1218. ModE **little**, Latin **parva**, + p.n. Shelford as in Great SHELFORD TL 4652. Ca 87.

SHELLEY WYorks SE 2011. 'Wood or clearing on a shelf or ledge'. *Scelneleie, Sciuelei* 1086, *S(c)helveley -lay* 13th cent., *Schellai -lay -ley* 1220×30–1449, *Shelley* from 1344. OE **scelf** + **lēah**. The village stands on a shelf of level ground above a steep slope down to Shepley Brook. YW ii.24, L 187.

SHELL HAVEN Essex → THAMES HAVEN TQ 7581.

SHELLINGFORD Oxon SU 3193. Probably 'the ford of the Scearingas, the people called after Scear'. *Scaringaford* [931]12th S 409, *Serengeford* 1086, *Schalingef', Schalengeford* 1220, *Shal(l)ingford* 1241 etc. with variants *Schal-* and *-yng-*, *Shelingford'* 1284. OE folk-n. **Scearingas* < pers.n. **Scear* + **ingas**, genitive pl. *Scearinga*, + **ford**. Brk 396.

SHELL NESS Kent TR 0567. 'Shell headland'. *Shelnasse* 1558×1610, *Shellness Point* 1819 OS. ModE **shell** + **ness**. PNK 253.

SHELLOW BOWELLS Essex TL 6108. 'S held by the Bowels family'. P.n. Shellow + family n. *Boweles* 1297, *Bowels* 1428, from Lambert de Buella, the DB tenant, whose family came from Bouelles, Seine-Inférieure. Shellow, *Scelgam, Scelda* 1086, *Scel(e)ga* [1086]InqEl, *S(c)hel(l)eg(h)es* 1244–54, *S(h)el(e)we(s)* 1238–1374, is OE r.n. **Sceolge* by origin, the 'winding one' < OE **sceolh** 'squinting, awry'. The reference is to the Roding the course of which is here full of bends. Ess 495.

SHELL TOP Devon SX 5963. *Shell Top* 1809 OS. A hill on Dartmoor rising to 1557ft. so-called from its resemblance to a limpet shell. D 271.

SHELLY GREEN WMids SP 1476 → CHESWICK GREEN SP 1376.

SHELSLEY BEAUCHAMP H&W SO 7363. 'S held by the Beauchamp family'. *Sheldeslegh Beauchampe* 1255, *Shellysley Becham* 1535. Earlier simply *Celdeslai* 1086, *Sceldeslega* c.1150, 'wood or clearing of the shallow place', OE ***scelde**, secondary genitive sing. ***sceldes**, + **lēah**. 'Beauchamp' refers to possession by the Beauchamp family and is for distinction from SHELSLEY WALSH SO 7263 on the opposite side of the river Teme which divides the two places. Wo 77 gives pr [biːtʃəm].

SHELSLEY WALSH H&W SO 7263. 'S held by the Welshman'. *Seldeslege Waleys* 1275, *Shelsley welsh* 1577. Earlier simply *Sceldeslœhge* 11th, *Caldeslei* 1086, as in SHELSLEY BEAUCHAMP. The manor was held in 1211 by Johannes Walensis 'John the Welshman' called *le Waleys* in 1235. Wo 78.

SHELTON 'Settlement on or by a shelf of land'. OE **scelf** + **tūn**. PNE ii.104, L 168–9.

(1) ~ Beds TL 0368. *Eseltone* 1086, *Sheltune* 1197, *-ton* from 1242. Shelton lies beside the river Till on level ground at the bottom of a hill. Bd 19.

(2) ~ Norf TM 2290. *Sceltuna* 1086, *Scelton* 1203. The hamlet lies on a low plateau at the foot of a marked slope. DEPN, L 187.

(3) ~ Notts SK 7844. *Sceltun(e)* 1086, *-ton(e)* 1199×1216, 1204, 1208×13, *Shelton* from 1236, *Skelton* 1305, *Shilton* 1335. Situated on a slight elevation between the river Smite and Black Dyke. Nt 216, L 186.

(4) ~ GREEN Norf TM 2390. P.n. SHELTON TM 2290 + ModE **green**.

(5) Lower ~ Beds SP 9942. *Lower Shelton* 1834 OS. ModE adj. **lower** + p.n. *Es(s)eltone* 1086, *Sheltune* 1197, *-ton* from 1227. 'Lower' for distinction from Upper SHELTON SP 9943. Bd 80.

SHELVE Shrops SO 3399. 'The shelf'. *Selva* c.1225, *Selue* 1255, 1291×2, *Shelve* from 1261. OE **scelf**. The meaning is 'level area among hills'. Sa i.262.

SHELWICK H&W SO 5243. Either 'Dairy-farm on a shelf of land' or 'dairy-farm with a shelter'. *Scelwiche* 1086, *Sceldwica* early 12th, *Scelfwica* 1160×70, *Shelwick* from 1241 with variants *Schel-* and *-wyk(e), Sceldwyke* 1292. OE **scelf** or **sceld** + **wīc**. Shelwick occupies a shelf of higher ground overlooking the river Lugg, but the evidence for **sceld** is also persistent. Perhaps these were alternative names. He 105, PNE ii.104, 262.

SHENFIELD Essex TQ 6094. 'Bright open country'. *Scenefeldā* 1086, *-feld* 1165, *Shi- Shy- Shen(e)feld* 1224–1372. OE **sciene** + **feld**. Ess 169.

SHENINGTON Oxon SP 3742. '(The settlement at) the beautiful hill'. *Senendone* 1086, *Schenyndon* 1305, 1309, 1393, *Scheni'gton'* 1246×7, *Shenyngton* 1383. OE **scīene**, dative sing. definite form **scīenan**, + **dūn**. Alternatively the specific might be pers.n. **Scīena*, feminine **Scīene*, genitive sing. **Scīenan*, 'Sciena's hill'. Nearby Shenlow Hill SP 3542, *Shenlowe* c.1741, could be 'Sciena's burial mound'. O 402.

SHENLEY Herts TL 1900. 'Bright wood or clearing'. *Scenlai -lei, Senlai* 1086, *Scen(e)le(am) -le(y)e -legh* 12th cent. etc. OE **scēne** + **lēah**. Hrt 67.

SHENLEY BROOK END Bucks SP 8335. 'The brook end of Shenley'. *Shenly Brook End* 1683. P.n. Shenley + ModE **brook** + **end**. Shenley, *Senelai* 1086, *Shenle* 1198–1435, is 'bright clearing', OE **scīene** + **lēah**. Shenley was divided into a number of different parts, Great, Little, Maunsel, Nether and Over Shenley, *Schenle Magna* 1302, *Shenle Maunsel* 1303, *Nethere Senle* 1284, *Ouer(e)schenle* 1262. Shenley Brook End lay in Great Shenley and contrasts with SHENLEY CHURCH END SP 8336 which corresponds to Maunsel and Nether Shenley. Over Shenley is now lost and Little Shenley is now Westbury 'the western fortified place'. Bk 23–4, Jnl 2.22.

SHENLEYBURY Herts TL 1802. 'Shenley manor house'. *Shenley Bury* 1822 OS. P.n SHENLEY TL 1900 + Mod dial. **bury**.

SHENLEY CHURCH END Bucks SP 8336. 'The church end of Shenley'. *Shenly Church End* 1683. P.n. Shenley as in SHENLEY BROOK END SP 8335 + ModE **church** + **end**. Bk 24, Jnl 2.22.

SHENMORE H&W SO 3938. *Lower Shenmoor* 1831 OS.

SHENSTONE H&W SO 8673. 'Scene's estate'. *Senestone* 1221, *Schenestone, Scefneston* 1275, *Shenston* 1327, 1410. OE pers.n. *Scēne*, genitive sing. *Scēnes*, + **tūn**. Wo 256 suggests the less plausible pers.n. **Scēafen* on the basis of the 1275 form.

SHENSTONE Staffs SK 1104. 'The beautiful stone'. *Scenstan* 11th, *Seneste* 1086, *Scenestan* c.1130, *Shenestan* 1168, *Schenestone* 14th. OE **scēne** + **stān**. The reference is unknown. Alternatively the specific could be OE pers.n. *Scēne*. DEPN, Duignan 134.

SHENTON Leic SK 3800. 'Settlement on the river Sence'. *Scenctune* [1004]13th S 906, *Scentone* 1086, *Seinton'* 1195–1227, *S(c)heinton' -ey-* 1249–1541, *Shenton(e)* 1258–1576. R.n. SENCE SK 3604 + OE **tūn**. Lei 548.

SHEPHERD'S GREEN Oxon SU 7183. *Shepherd's Green* 1833 OS. ModE **shepherd** or surname *Shepherd* + ModE **green**. One of numerous squatter settlements in the Chilterns. Cf. TOKERS GREEN SU 6977.

SHEPHERDSWELL OR SIBERTSWOLD Kent TR 2647. A modern folk-etymological re-interpretation of original 'Swithbeorht's forest'. *(œt) Spyðbrihteswealde, Siberdeswalde* [944]13th S 501, *Sibrighteswealde* [944]14th ibid., *Swyðbeorteswald* [990]13th S 875, *Sibertesuuald -d -walt* 1086, *Siberdeswald'* 1231, *Si- Syperdesweld* 1477×8, *Shepardysweld* 1510×11, *Schepardswell* 1513, *Sheppards Wold* 1557, *Sybertswold* 1596, *Shepher's Well* 1665. OE pers.n. *Swīthbeorht -briht* later replaced by *Sigebeorht*, genitive sing. *Swīthbeorhtes*, + **weald**. KPN 251, L 223, 226–7, Charters IV.104, Cullen.

SHEPLEY WYorks SE 1909. 'Woodland clearing used for sheep'. *Scipelei, Seppeleie* 1086, *S(c)hepelay -le(y)* 13th cent. 1549, *S(c)heplay -ley* 1249–1540. OE **scēap** + **lēah**. Apart from DB *Scipe-* there are no traces of ONb **scīp**. YW ii.250.

SHEPPERDINE Avon ST 6295. 'Sheep enclosure'. *Shepewardin* 1215, *Shepardyn -ine* 1489, 1779. OE **scēap** + **worthiġn**. Occasionally reformed as if 'the shepherd's end' sc. of the parish, *Shepardynde* 1400, *Sheperdende* 1421. Gl iii.9.

SHEPPERTON Surrey TQ 0867. 'The shepherd farm or village'. *(in) Scepertune* [959]12th S 1293, [1066]13th S 1043, [1051×66]14th S 1130, *(in) Sceapertune* [1042×66]13th, *Scepirton* [1051×66]13th S 1130, [1066×87]12th, *-tune* [1042×66]13th, *Scepertone* 1086–1393 with variants *S(c)hep- Sep-*. OE **scēaphierde**, genitive pl. **scēaphierda**, + **tūn**. Mx 17.

Isle of SHEPPEY Kent TQ 9769. *Sheppey Isle* 1819 OS. P.n Sheppey + pleonastic ModE **isle**. Earlier simply *Scepeig* [699×716]11th S 22, *Scaepege* [850]13th S 300 with 14th cent. variant *Scaerege, Sceapige* 9th ASC(A) under years 832, 855, *Scepige* [894]c.1100 Asser, *Sceap ege* 12th ASC(E) under same years, ~ *ige* 12th ASC(E) under years 1016, 1052, *Scape* 1086, 'sheep island', OE **scēap** + **īeġ**. KPN 24, Charters IV.83.

SHEPRETH Cambs TL 3947. 'Sheep brook'. *Esceprid(e)* 1086, *Sepere(e) -eye -ethe* 12th–1299, *Shepreth(e)* from 1272 with variants *-ree -ryth -red(de)*. OE **scēap**, genitive pl. **scēapa**, + **rīth**. Perhaps a brook where sheep were washed. Ca 80.

SHEPSHED Leic SK 4719. 'The sheep's headland'. *Scepe(s)hefde* 1086, *Shepesheued* 1167, *Shepe-* 1191. OE **scēap**, genitive sing. **scēapes**, + **hēafod**. The meaning is probably a headland where sheep grazed. DEPN, NoB 14.131, Sr 405, L 160–1.

SHEPTON 'Sheep farm'. OE **scēap** + **tūn**, ***sclep** + **tūn**. Often a specialist unit in a multiple estate. Origins 93.

(1) ~ BEAUCHAMP Somer ST 4017. 'S held by the Beauchamp family'. *Septon Belli campi* (Latin) 1266, *Shepton Beauchamp* [1346×7]Buck. Earlier simply *Sceptone* 1086, *Shepton* 1610. The manor was held before 1212 by Robert de Bellocampo (Beauchamps, La Manche). DEPN.

(2) ~ MALLET Somer ST 6143. 'S held by the Mallet family'. *Scheopton Malet* 1226×8, *Shepton Malet* [1346×7]Buck. Earlier simply *Sepetone* 1086. The manor was held by Robert Malet 1100×35. DEPN.

(3) ~ MONTAGUE Somer ST 6731. 'S held by the Montague family'. *Shepton Montague* [1174×91]Buck, *Schuptone Montagu* 1285, *Shepton Montague* 1610. Earlier simply *Sceptone* 1086, *S(h)epton* [12th]Buck. In 1086 the manor was held by Drogo de Montacute (Montaigu or Montaigu-le-Bois, La Manche). DEPN.

SHEPWAY DISTRICT Kent is named after the ancient Lathe of Shipway, *Sippeweiam* 1205, *Shi- Shypweie -weye* 1220–54, *Shepweye* 1227, 1254, 'sheep path', OE **scēap**, **scīp** + **weġ**. PNK 368, Room 1983.112.

SHEPWAY Kent TQ 7753. No early forms. Cf. SHEPWAY DISTRICT.

SHERATON Durham NZ 4335. 'Skurfa's farm or village'. *Scurufatun* [c.1040]12th, *-ton* c.1275, *S(c)(h)uruetun -ton* 1172×95–1385, *Shoruton* 1372, *Shoro(w)ton -ou-* 1395–1510, *Sherowton* 1498, *Schereton* 1539, *Sheriftton* 1580, *Sheraton* from 1610. ON pers.n. *Skurfa* + **tūn**. NbDu 176, Nomina 12.20.

SHERBORNE 'Bright stream'. OE **scīr** + **burna**. L 18, Jnl 23.26–48. See also SHERBURN.

(1) ~ Dorset ST 6316. *(to þaere halgan stowe aet) scireburnan* 'to the holy place at S' [864]12th S 333, *(æt) Scire burnan* late 9th ASC(A) under years 860, 867, early 10th ibid. under year 910, 11th ASC(C, D) under years 978 and 910 respectively, *(in, æt) Scir(e)burna(n)* [893]early11th, [970×5]12th S 813, [998]12th S 895, [1007×14]12th S 1422 – [1145]12th, *Scireburne* 1086–1209, *Sherborn(e)* from 1415. The reference is to the upper course of the r. Yeo. Do iii.355 gives pr [ˈʃəːRbən].

(2) ~ Glos SP 1714. *Scireburne* 1086, *S(c)hi- Shyreburn -bo(u)rn(e)* 1193–1727. Gl i.203.

(3) ~ ST JOHN Hants SU 6255. 'S held by the St John family'. *Shireburna Johannis* 1167, *Schirberne Saint Johan* 1272. Earlier simply *Sireburne* 1086. The manor was held by Robert de Sancto Johanne in 1242. The addition is for distinction from Monk SHERBORNE SU 6056. Ha 147, Gover 1958.128.

(4) Monk ~ Hants SU 6056. 'The monks' S'. *Schireburne Monachorum* c.1270, *Monkeneschirbourne* 1306, *Shirebourn Moygnes* 1332. Earlier simply *Sireborne* 1086, *Shireburne* 1255. A priory was founded here in the early 12th cent. The three forms referring to the monks are Latin **monachus**, genitive pl. **monachorum**, ME **monk**, adjectival genitive pl. **monkene**, OFr **moine**. Ha 118, Gover 1958.141, DEPN.

SHERBOURNE Warw SP 2661. '(The) bright stream'. *Sciresburne* 1086, *Shireburne* 1221–1327 with variants *Schir-* and *-bourne*, *Sherborne* 1535. OE **scīr**, definite form **scīra**, + **burna**. L 18. Wa 222

SHERBURN 'The bright stream'. OE **scīr**, definite form **scīra** (in some instances), + **burna**.

(1) ~ Durham NZ 3142. *Sireburna* [1143×52]15th, *-burn(e), Syreburn(a)* c.1197×1217–c.1310, *S(c)hyrburn(e)* 1285–1500, *Shereborne* 1580, *Sherburne* 1647. NbDu 176, L 18.

(2) ~ NYorks SE 9577. *Schirebur', Schiresburne* 1086, *Scir(e)- Scyreburn(e) -borna* [1109×19]c.1300–13th, *S(c)hireburn(e) -y-* [1145×53]c.1300–1303, *S(c)hir- Shyr-* 1285–1531, *Scher(e)burne* 1286, 1302. YE 120.

(3) ~ IN ELMET NYorks SE 4933. *Sherburn in Elmet* 1528. P.n. Sherburn + **in** + p.n. ELMET. Sherburn is *(on) Scirburnan* [c.900]n.d. B 1324, *(to, of) Sciréburnan* [963]14th S 712, [972×92]11th S 1453, c.1030 YCh 7, *Scir(e)burn(e)* 1086–1299, *Schir(e)burna* 1156–1431, *Shir(e)- Shyreburn(e)* 1180×5–1495. YW iv.60.

SHERE Surrey TQ 0747. Originally a r.n., 'the bright one'. *Essira -e* 1086, *S(ch)ire, S(c)hyre* 13th–1535, *Shere* from 1462, *Shire alias Sheare* 1610. OE R.n. **Scīre* < **scīr**. The reference is to the river Tillingbourne. Sr 248.

SHEREFORD Norf TF 8829. 'The clear ford' or 'the ford of the clear stream'. *Sciraforda* 1086, *Shireford* 1206. OE **scīr**, definite form **scīra**, possibly used as a r.n. *Scīre* 'the bright one', + **ford**. Cf. the r.ns. SHERE Sr TQ 0747 and *Scir, Shire*, former name of the Rother. DEPN.

SHERFIELD ENGLISH Hants SU 2922. 'S held by the Lengleis family'. *Shyrfeld Engleys* 1373. Earlier simply *scirefelde* [11th]14th, *Sirefelle* 1086, *Schirefeld* 1291, the 'bright open land' or 'open land of the shires', OE **scīr**, definite form **scīra**, or **scīr**, genitive pl. **scīra**, + **feld**. Sherfield is close to the border between Wilts and Hants. The manor was held by Richard Lengleis 'the Englishman' in 1303. The addition is for distinction from SHERFIELD ON LODDON SU 6758. Ha 147, Gover 1958.190, DEPN, L 242.

SHERFIELD ON LODDON Hants SU 6758. *Sherfeld super Lodon* 1578. P.n. Sherfield + r.n. LODDON SU 6758. Sherfield, *Sirefelda* 1167, *S(c)hirefeld* 1179–1327, *Shervill* 1720, is probably the 'bright open land', OE **scīr**, definite form **scīra**, + **feld**. Like SHERFIELD ENGLISH, however, it is close to a county border and might be the 'open land of the shires', OE **scīr**, genitive pl. **scīra**, + **feld**. The original reference may have been to an extensive area of open country including the STRATFIELDS SU 6764, 6961, 6959, BURGHFIELD SU 6608, HECKFIELD SU 7260, WOKEFIELD SU 6765, and SWALLOWFIELD Berks SU 7264. Ha 147, Gover 1958.117, DEPN, L 242.

SHERFORD Devon SX 7744. 'The bright, clear ford'. *Scireford* [1057×65]14th S 1236, *Sireford* 1086, *Shireford(e)* 1281–1306, *Shyrford* 1312. OE **scīr**, definite form **scīra**, + **ford**. D 329.

SHERIFFHALES Shrops SJ 7612. 'Hales held by the sheriff'. *Schirrenghales, Schirrenchal'* 1271×2, *Schirreuehale(s)* 1291×2, *Shirevehalys* 1367, *Sherref Hales* 1398. ME **shireve** + p.n. *Halas* 1086, *Hales* c.1128×38 etc., 'the nooks', OE **halh**, pl. **halas**. The holder in 1086 was Reginald of Balliol, sheriff of Shropshire. Sa i.262.

SHERINGHAM Norf TG 1542. Properly *Lower Sheringham* 1830 OS for distinction from the original settlement at Upper SHERINGHAM TG 1441. Lower Sheringham, the sea-side resort, began to develop in the 1890s and early 20th cent. Pevsner 1962NE.314.

Upper SHERINGHAM Norf TG 1441. ModE **upper** + p.n. *Silingeham* 1086[†], *Siringeham* 1173×4, 1179, *Suringeham* c.1180, *Seringeham* 1207, *Sc(h)ringham* 1209×10, 1254, *Scheringham* 1242×3. The generic is OE **hām** 'homestead', the specific an unknown folk-n., possibly **Scīringas* 'the people called after Scira', OE pers.n. *Scīra* + **ingas**, genitive pl. **Scīringa*. This is actually the original Sheringham to which *upper* was added when Lower Sheringham, the sea-side resort, began to develop in the 1890s and early 20th cent. DEPN, ING 138.

SHERINGTON Bucks SP 8946. 'Estate called after Scira'. *Serintone* 1086, *S(c)herintone* 1185, 1227, *S(c)heryngton* 1355–1533, *Schirintone, Shiringtona* 1179–1378, *S(c)hrington(e)* 1278–1403, *Churton or Cherryngton* 1524. OE pers.n. *Scīra* + **ing**[4] + **tūn**. Bk 38 gives prs [ʃɛ:tən] and [tʃɛ:tən].

SHERNBORNE Norf TF 7132. 'The dirty stream'. *Scernebrune* 1086, *Scarnebrune* 1254. OE **scearn** 'dung' + **burna**. DEPN.

SHERRINGTON Wilts ST 9639. 'The mud or dung settlement'. *Scearntune* [968]14th S 766, *Scarentone* 1086, *Scharn(e)ton* 1268–1342, *Sherinton* 1166, *S(c)hernton* 1282–1332, *Shernton al. Sheryngton* 1560. OE **scearn** + **tūn**. Wlt 229.

SHERSTON Wilts ST 8586. 'The stone on a precipitous slope'. *Scorastan* 11th *Encomium Emmæ*, *Scorstan* 11th ASC(D) under year 1016, *Sceorstan* c.1121 ASC(E) under same year, *Sorstain, Sorestone* 1086, *Sorestan(e)* 1142×54–1204, *Shorestan(e)-ton* 1236–45, *S(c)herstan* 1258, 1305, *Sherston* from 1316, *Sharston* 1335, *Shaston, Sharson* 16th. OE ***scora** 'a shore, a steep slope' with WS *sceo-* for *sco-* + **stān**. Sherston stands on high ground overlooking the river Avon. The form *Scorranston* [896]11th S 1441 may belong here, but if so the sp is corrupt. A further form *skorsteini* [c.1025×30]17th is found in the Norse poem *Knútsdrápa* (dative case of **Skorsteinn*). Wlt 109 gives obsolete pr [ʃa:s(t)ən], Townend 1998.67.

SHERWOOD FOREST Notts SK 6060. P.n. Sherwood + ModE **forest**. Sherwood, *(of) scyryuda, scirwuda* [958]14th S 679, *Shirewuda* 1163–1322 with variant spellings *Sc(h)ire- -wude* and *-wode*, *Sherewode* 1325–1500, *Shearwood* 1576, is 'the shire wood, the wood belonging to the shire' sc. of Nottinghamshire in which the ancient shire villages once held woodland-pasture rights. After the conquest it became a royal forest. OE **scīr** + **wudu**. Nt 10.

SHERWOOD GREEN Devon SS 5520. P.n. Sherwood + ModE **green**. Sherwood, *Shirewode* 1238, is 'the bright wood or clearing', OE **scīr**, definite form **scīra**, + **wudu**. D 83.

SHEVINGTON GMan SD 5408. 'Settlement at *Chevin*'. *Shefinton* c.1225, *S(c)hevinton -yn-* c.1200–1332, *Shevyngton* 1312–1426. P.n. **Chevin*, PrW ***cevn** (Brit **cemno-*) 'back, ridge' + OE **tūn**. In or near Shevington were the lost *Schevynlegh* 1329 'wood or clearing at *Chevin*' and *Shevynhulldiche* 1362 'ditch at *Chevin* hill'. The reference is to a ridge of high ground stretching SE from Hunger Hill SD 5311 to Standish Hall SD 5609. La 128, 263, YW iv.204, Jnl 1.45, 17.71.

SHEVINGTON MOOR GMan SD 5410. *Shevington Moor* 1843. P.n. SHEVINGTON SD 5408 + ModE **moor**.

SHEVIOCK Corn SX 3755. Probably 'strawberry place'. *Savioch* 1086, *Sevioc* [12th]13th, 1226, 1286, *Sheviok(e) -y-* 1259–1428, *Shyuyoke* 1610. Co **sevi** 'strawberries' + adjectival ending **-ek** (OCo *-oc*). PNCo 156, Gover n.d. 234, DEPN.

North SHIELDS T&W NZ 3568. *Nortschelis* 1273, *le North Shels* 1379, *North Sheeles* 1663, ~ *Sheelds* 1723, ~ *Shields* from 1763. ME **north** + p.n. *Chelis* 1267, *the Sheles by Tinmouth Castle* 1562, *Sheales* 1576, *Sheilds* 1607, 'the (fishermen's) huts', ME **shele**, pl. **sheles**. These huts for sheltering fishermen and their gear had already become a permanent settlement by 1290 when the Corporation of Newcastle took out an action against the Prior of Tynemouth for 'building towns where no town ought to be'. It was claimed in reply that there were no more than three *sc(i)ales* there. 'North' for distinction from South Shields on the S bank of the Tyne. NbDu 176, Tomlinson 44.

South SHIELDS T&W NZ 3666. *Suthshelis* 1313, *(le) South Scheles* 1379–c.1400×15, *South(e) Sheles* 16th cent., *yͤ southshiylls* 1547, *South Shealds* 1675, ~ *Shields* 1744×5. ME **sūth** + p.n. *(le, les) Scheles* 1235–1382, *(le) Sheles* 1345–1587, *Sheils, the Sheals* late 16th, 'the (fishermen's) huts', ME **shele**, pl. **sheles**. 'South' for distinction from North SHIELDS on the N bank of the Tyne. NbDu 176.

SHIFNAL Shrops SJ 7508. 'Scuffa's valley'. *Shuffenhale* 1315–1459, *-ale* 1394, 1408, *Shuffnal* 1414–1657, *Scifnael* 1212, *Schyf(f)nall* 1547, *Shiffnall* 1600 etc., *Shifnal* 1713. OE pers.n. **Scuffa*, genitive sing. **Scuffan*, + **halh**, dative sing. **hale**, referring to the large shallow depression in which the town lies. There are a number of forged charter forms of which *Scuffanhalch'* S 72, in a charter purporting to date to 680 in a 12th cent. copy, is possibly an authentic form. An alternative name, *Iteshale* 1086, 1282, *Idesale* c.1175–1446, 'Idi's valley', OE pers.n. **Idi*, genitive sing. **Ides*, + **halh**, was in use for many centuries. Sa i.263.

SHILBOTTLE Northum NU 1908. 'Shipley house, *bōthl* belonging to Shipley'. *Siplibotle* 1228, *Schiplibotle* 1238, *Schepeling-Schip(p)lingbotil* 13th cent., *Shilbotill* 1336. P.n. SHIPLEY NU 1416 + OE **bōthl**. NbDu 177.

SHILDON Durham NZ 2325. 'Shelf hill'. *Seluedon'* 1211, *Sciluedon'* 1214, 1283, *Shiluedon* 1350, *Sheldon* 1306, 1549, *Shy-Shildon* from 1362×3. OE **scelf**, **scylfe** + **dūn**. NbDu 177, L 186.

SHILLINGBOTTOM Lancs → SKILLINGTON Lincs SK 8925.

SHILLINGFORD Formally the Shillingfords could be 'ford of the *Scillingas*, the people called after Sciella'. OE folk-n. **Sci(e)llingas* < pers.n. **Sci(e)lla* + ***ingas**, genitive pl. **Sci(e)llinga*, + **ford**. But three examples of this name type, together with the lost *scillinges broc* [862]12th S 335, suggest the presence of stream-name **Scielling* 'the noisy one' < **sciell** 'sounding, shrill' + **ing**[2]. A pers.n. *Scilling* is recorded. D 503, 532, O 139, Redin 23.

(1) ~ Devon SX 9824. *Sellingeford* 1179, *Shillyngford* 1333. D 532.

(2) ~ Oxon SU 5992. *Sillingeforda* [1156]13th, *Shillingford* from 1278 with variants *Shyll-* etc. O 139.

(3) ~ ST GEORGE Devon SX 9087. 'St George' from the dedication of the church is a recent addition for distinction from SHILLINGFORD SS 9824. Earlier simply *Esselingeforde, Selingeforde* 1086, *Sulling- Sillinge- Schyl- S(c)hillyngford(e)* 13th cent. D 503.

SHILLINGSTONE Dorset ST 8210. 'Eskelin's manor'. *Shillyngeston* 1444, *Shillingston* 1774. At the time of DB this manor was held by Schelin or Eschelinus and his descendants were here until the 13th cent. The original name of the manor was *Akeford(')* 1199–1303[†], *Acford(')* 1201–1428, 'oak-tree ford', OE **āc** + **ford**, to which the name Eskelin was appended as a manorial affix, *Acforde Eskelin* before 1155, *Acforde* of *Robert Eskylling* c.1155, *Acford(e) S(c)hilling(') -yng* 13th, 14th cents., *Ocford(') Sculling(')* 1280, *Shillyngokford(e)* 1418, for distinction from Child OKEFORD ST 8312 and OKEFORD FITZPAINE ST 8010. Do ii.238 gives pr [ˈʃilənstn].

SHILLINGTON Beds TL 1234. 'Hill called Scytteling'. *Scytlingedune* *[1060]14th S 1030, *Sethlindone* 1086,

[†]Another form, *Stringham* 1086, may belong here. If so, it is clearly erroneous.

[†]The DB form *Alford* 1086 is a mistake.

Scetling(e)don 1202–7, *S(c)hut- S(c)hitlingdon* 1236–1504, *Shitlington* 1287–1830, *Shilindon* 1780. OE hill-n. **Scytteling* < ***scyttel(s)** + **ing**[2], locative–dative sing. **Scyttelinġe*, + **tūn**. Additionally the specific might be OE folk-n. **Scyttelingas* 'the hill people' < ***scyttel(s)** + **ingas**, genitive pl. **Scyttelinga*. This explanation fits the topography and is preferable to positing an unrecorded pers.n. **Scytla* or **Scyttel*. Bd 173, L 145, 151.

SHILL MOOR Northum NT 9415. *Shill Moor* 1869 OS. *Shill* seems to be used in Northum of a steep pointed hill, which there rises to 1734ft. Cf. The SCHIL NT 8622 and SHILLMOOR Northum NT 8807.

SHILLMOOR Northum NT 8807. 'The shovel-shaped moor'. *Shouelmore* 1292, *Sholemore- Shelmerlaw* 1380, *Sholdmore* 1577, *Shumore* 1633 PSA 1.334–5, *Shillmore* 1642. OE **scofl** + **mōr**. The reference is to the narrow spit of land between the Coquet and Usway Burn. A mile to the NW rises Shillhope Law 1644ft., which is probably referred to in the 1380 forms. There is some evidence that *shill* is used in Northum for a steep pointed hill: this name suggests the derivation is from OE *scofl*. NbDu 177.

SHILTON 'Shelf settlement'. OE **scylf** + **tūn**.

(1) ~ Oxon SP 2608. *(into) Scylftune* [1044]12th S 1001, *Siltone* before 1180, *Schelton'* 1201, *Shulfton'* 1241–3, *Shilftun'* 1242×3, *Shulton(')* 1284–1359, *Shyltone* 1428. There is a magnificent flat 'shelf' here. O 328.

(2) ~ Warw SP 4084. *Scelftone* 1086, *Sceffetone* 1176, *Selton* 1169, 1199, *Shulton* 1221–1420, 1635, *Shilton* 1535. The site is a low plateau of level ground. Wa 119, L 187.

(3) Earl ~ Leic SP 4697. 'S held by the earl' sc. of Leicester. ModE prefix **earl**, *Erle* 1576, *Earle* 1603–12, + p.n. *Sceltone* 1086, *S(c)helton(e)* 1209–1429, *S(c)hulton(e)* 1263–1521, *Schilton(e) -y-* 1277, 1354, 1477, *Shilton* from 1437, The affix is late, alluding to Henry III's gift of Shilton to Edmund Plantagenet, earl of Leicester, in 1271. Lei 504.

SHIMPLING Norf TM 1583. 'The Scimplingas, the people called after Scimpel or Scimpla'. *Simplinga* (3×) 1086, *Scimplinge* (2×) [1087×98]12th, *Simplinges* 1185, 1187, *Scympling'*, *Simpling'*, *Scinpling*, *Symplinge* 1275. OE folk-n. **Scimplingas* < pers.n. **Scimpel* or **Scimpla* + **ingas**. Identical with SHIMPLING Suff TL 8551. There are also early forms in **hām**, *Simplingham* [c.1035]13th S 1527, *Simplingaham* (2×) 1086, *Scimplingeham* (2×) [1087×98]12th, 'homestead of the Scimplingas'. DEPN, ING 61.

SHIMPLING Suff TL 8651. 'The place called after Scimpel or Scimpla'. *Simplinga* 1086, *Simpeling* 12th, *Simplengges*, *Simpligges* 1230, *Simpling'* 1236, *Sci- Sympling'* 1275, *Shimplinge* 1610. This name looks more like OE sing. **Scimpling* < pers.n. **Scimpel* or **Scimpla* + **ing**[2] than pl. folk-n. **Scimplingas* 'the people called after Scimpel or Scimpla'. ING 55.

SHIMPLING STREET Suff TL 8752. 'Shimpling hamlet'. *Shimpling Street* 1836 OS. P.n. SHIMPLING TL 8651 + Mod dial. **street**.

SHINCLIFFE Durham NZ 2940. 'Demon hill'. *Scinneclif* 1107×23, [1107×23]c.1160, *-cliue* c.1190×95, *Schinecliue* 1242×3, *Sinclif* 1311, *Shincliff* 1342–1618 with variants *Shyn-* and *-clyff*, *-k(c)liff*, *S(c)hinkley* 1369–1675 with variants *S(c)hyn-* and *-ckley*. OE **scīn**, genitive pl. **scīna**, or **scinna** 'a phantom, an evil spirit, a spectre' + **clif**. NbDu 178 gives pr [ʃiŋkli].

SHINEY ROW T&W NZ 3252. *Shiney Row* 1863 OS. ModE **shiny** and **row**.

SHINFIELD Berks SU 7368. 'Open land of the Scieningas, the people called after Sciene'. *Selingefelle* 1086, *S(c)hiningefeld(')* 1167, 1254, *Shynyngfeld* 1324, [1349]c.1444, *Shynnyngfold* 1603×25, *Scenegefeld'* 1190–1, *S(c)henyngefeld* 1275×6, 1327, *Shenefeld* 1324, 1474. OE folk-n. **Scīeningas* < pers.n. **Scīene* + **ingas**, genitive pl. **Scīeninga*, + **feld**. Brk 103, L 239, 244.

SHINING TOR Ches SJ 9973. 'Bright rock'. *Shining Tor* 1831. ModE **shining** + **tor**. Che i.128.

SHIPBOURNE Kent TQ 5952. 'Sheep stream'. *Sciburna* c.1100, *Sibburn'* 1195 or 1196, 1226, *Scipb'ne* 1198, *Shirburn* 1240, *Schipburn'* 1278. OE **scēap, scīp** + **burna**. PNK 154, L 18.

SHIPDHAM Norf TF 9507. 'Sheep farm'. *Scipd- Scipedeham* 1086, *Sipedham* 1200, *Schipedham* 1254. OE ***scīpde** 'a flock of sheep' + **hām**. DEPN, PNE ii.101.

SHIPHAM Somer ST 4457. 'Sheep farm or enclosure'. *Sipeham* 1086, *Schepham* 1291, *Shepeham* 1610[†]. OE **scēap**, WS ***scīep** + **hām** or **hamm**. The gathering-point for sheep from the Mendips, possibly for the royal estate at Cheddar. DEPN, Origins 93.

SHIPHAY Devon SX 8965. 'Sheep enclosure'. *Shepehay* 1541. ME **shep** + **hay** (OE *(ge)hæġ*). D 512.

SHIPLAKE Oxon SU 7678. 'The stream where sheep are washed'. *Siplac, Sipelaca* 1086, *-lak(e)* 1209–1300, *Shy- Shiplak(e)* from 1237. OE **scēap** + **lacu**. O 81.

Lower SHIPLAKE Oxon SU 7779. ModE adj. **lower** + p.n. *Siplac, Sipelaca* [1163]c.1300, *Si- Syp(p)lak(e)* 1209–1300, *Sheplak'* 1238, *Schi- Shyplak(e)* 1204–1530, *Sheeplake* 1676, 'the sheep stream, the stream where sheep are washed'. OE **scēap, scīp** + **lacu**. 'Lower' for distinction from SHIPLAKE SU 7678. O 81, L 23.

SHIPLEY 'Sheep clearing or pasture'. OE **scēap** + **lēah**, (WS *scȳp*, *scīep*, Nb *scīp*, Angl *scēp*).

(1) ~ Northum NU 1416. *Scepley* 1247, *Scippele* 1252, *S(c)hipley* 1242–1428. NbDu 178.

(2) ~ Shrops SO 8195. *Sciplei* 1086, *Shiple(egh')*, *Schipleg'*, *Schyplegh'* 13th cent., *Shipley* from 1577, *Schepleg'*, *Shepple* 13th, *Shepley* 1599. *Shepley* is the expected form in Shrops, but the forms clearly point to OE **scīp** or **scīep**. Sa i.264.

(3) ~ WSusx TQ 1421. *Scapeleia* 1073, *Scapuleia* c.1080, before 1096, *Sepelei* 1086, *Shepele(ye)* 1235–1509, *Shipele* 1312. Sx 188, SS 66.

(4) ~ WYorks SE 1537. *Scipelei(a)* 1086, *Schipeleia -ley* 12th, 1288, *Shepele(y)* 1225, 1303, *Shipley* from 1468. ONb **scīp** + **lēah**. YW iii.267, L 206.

SHIPMEADOW Suff TM 3890. 'Sheep meadow'. *Scipmedu* 1086, *Shi- Shypmedwe* 1254–1344, *Shipmedowe* 1568, *Shepemedowe* 1610. OE **scēap, scīp** + **mǣd**, dative case **mǣdwe**. DEPN, Baron.

SHIPPON Oxon SU 4898. 'The cattle-shed'. *Sipene* 1086, *Schi- Shy- She- Shupen(e)* 1284–1377. OE **scypen**. Brk 439.

SHIPSTON-ON-STOUR Warw SP 2540. *Shipston on Stour* 1828 OS. P.n. *Scepuuæisctune, vadum nomine Scepesuuasce* 'a ford called Sheep's Wash' *[764×75]11th S 61, *in ripa Sturę fluminis Sceppæsctun* 'S on the bank of the river Stour' [964]12th S 731, *Scepwestun* 1086, *Sc(h)wastona* [c.1086]1190, *Sip(p)estone* 1275, *Shipston* 1542, 'the settlement at *Sceap(es)wæsce*, the sheep-wash', OE **scēap** varying with genitive sing. **scēapes** + **wæsce**, + **tūn**, + r.n. STOUR. Wo 164.

SHIPTON 'Sheep farm'. OE(WS) ***scīep** + **tūn**, (Nb) **scīp** + **tūn**. Cf. SKIPTON, SHEPTON.

(1) ~ Glos SP 0418. *Scip(e)tvne* 1086, *S(c)hi- Shypton(a)* 1221–1715. Originally two separate parishes known as Shipton Oliffe, ~ *Olyve* 1371, held by the Olive family (Thomas Olyve 1347) and Shipton Solers, ~ *Solars, Solace* 16th, ~ *Sollers* 1621, held by the Solers family (William Solers 1221, Robert de Solers 1236). Gl i.180.

(2) ~ Shrops SO 5692. *Scipetune* 1086, *Schipton'* 1255×6, *Shi- Shypton* from 1271×2, *Shipton* 1360, *Shepton* 1607, 614. Sa i.265.

(3) ~ BELLINGER Hants SU 2345. 'S held by the Berenger family'. *Sceptone -tun* 1086, *Shipeton* 1167, *Shupton* 1270. The manor was held by Ingelram Berenger in 1296. Ha 148.

(4) ~ GORGE Dorset SY 4991. 'S held by the Gorges family'.

[†]The form *Cympanhamme* [939×46]13th S 1733 is sometimes thought to belong here.

Shipton Gorges 1594. Earlier simply *Sepetone* 1086, *Sipton* 1214, *Scepton* 1268. The manor was held by Ralph de Gorges in the 13th cent. Do 131.
(5) ~ GREEN WSusx SZ 8099. No early forms.
(6) ~ MOYNE Glos ST 8989. *Shi- Shypton Moi- Moygne* 1287~1375, ~ *Moi- Moyn(e)* 1308–1728. Earlier simply *Scipetone, Scipton(e)* 1086. The manor was held by the family of Moygne from the 13th cent. (Radulfus le Moine 1221 etc.). Gl i.108.
(7) ~ -ON-CHERWELL Oxon SP 4816. *Shipton-upon-Charwelle* 1332. P.n. *Sceaptun* [1005]12th S 911, *Sciptone -tune* 1086, *Shipton(e)* from 1201 + r.n. CHERWELL SP 5212. O 280.
(8) ~ -UNDER-WYCHWOOD Oxon SP 2717. *Shipton under Whicchewood* 1322. P.n. *Sciptone* 1086, [1220×8]13th, *Schi- Schypton(e)* 1195–1346, *Shipton* from [1210]c.1425 + woodland n. WYCHWOOD. O 375.

SHIPTON 'Farm or village where briars grow'. OE **heope** + **tūn**.
(1) ~ Humbs SE 8543. *Epton* 1086, *Hyepton, Yheptona* 1176, *Yupton'* 1259, *Sipton* 1219, *S(c)hupton* 1234–1562, *Skip- Scipton* late 13th, *Shipton* 1532. The *Skip-* forms are Scandinavian as if from SHIPTON 'a sheep farm'. YE 228.
(2) ~ NYorks SE 5558. *Hipton* 1086, *Hepeton'* 1167, *Hieptuna, Hyepton, Yheptona, Yhupton* 12th, *S(c)hupton* [13th]14th–1541, *Shipton* from 1328. The history of this name demonstrates a well-known development due to change of stress in the diphthong *ēo*: *héo* > *heó* > *hye* > *shi*. YN 15, TYDS 1967.5–14, Brno SE 4.87.

SHIPTONTHORPE Humbs SE 8543. Originally two separate names, SHIPTON SE 8543 and THORPE LE STREET SE 8543.

SHIRBURN Oxon SU 6995. 'The bright stream'. *Scir(e)burn(e)* 1086, *Sireburne* 1086–1279 with variants *Syre-* and *-burn(a) -born, Shirb(o)urn(e)* from 1307. OE **scir** + **burna**. O 91.

SHIRDLEY HILL Lancs SD 3613. No early forms. The specific may represent the lost p.n. *Shirwall* 1476, the 'clear spring', OE **scīr** + **wælla**. La 120.

SHIREBROOK Derby SK 5267. Probably the 'shire brook'. *Scirebroc* 1202, *Schir(e)brok(e)* 1216×72–1405, *Shir(e)- Shyr(e)broc -brok(e)* 1236–1304 etc. OE **scīr** + **brōc**. The place is situated on the Derby–Notts border. The specific could, however, be **scīr** 'bright'. Db 298, L 15.

SHIREHAMPTON Avon ST 5377. 'Dirty Hampton'. *Shern(y)hampton* 1325–1420, *Sherehampton* 1486–1666, *Shirehampton* from 1551. Earlier simply *Hampton* 1285–1455. OE **scearniġ** + **hāmtūn** 'homestead' used as a p.n. The name has been reformed from pejorative 'dung Hampton' (ME *sherne* < OE *scearn*) to 'bright Hampton' (ME *shere*). *Scearamtone* [c.855]13th, usually cited here from William of Malmesbury's *De Antiquitate Glastonie Ecclesie* 98, probably does not belong: the name occurs among a list of estates in Somerset and Devon. Gl iii.132, ECW 201 fn.2.

SHIREMOOR T&W NZ 3171. *Shiremoor* 1863 OS. ModE **shire** + **moor**. The name commemorates the moor of Tynemouthshire, which was common to eight townships, Amble, Backworth, Bebside, Bewick, Elswick, Monkseaton, Preston and Wylam. EHR xli.12, Barrow 52.

SHIRE OAK WMids SK 0504. *Shire Oak* 1833 OS. A prominent oak-tree, possibly an old shire meeting-place.

SHIREOAKS Notts SK 5581. The name refers to oak-trees on the boundary between Nottinghamshire and Yorkshire. *Shirakes* 1100×1135, 1286, *S(c)hirokes* c.1160–1340, *Shire Okes* 1383. OE **scīr** + **āc**. Nt 108.

SHIRKOAK Kent TQ 9436. Unexplained. *Shik Oak* (sic) 1819 OS.

SHIRL HEATH H&W SO 4359. *Shirl Heath* 1832, cf. nearby *Shirl Wood* ibid.

SHIRLAND Derby SK 3958. Probably 'bright grove'. *Sirelunt* 1086, *Schir- Schyrlund -lound* 13th cent., *Schirlond* 1250, 1307, *-land(e)* 1291, 1316, *Shir(e)- Shyr(e)lond(e)* 1242–1415, *-land(e)* from 1291. OE **scīr** + ON **lundr** later replaced by OE **land**. The specific could, however, also be **scīr** in the sense 'grove where the shire meets'. Db 299, Forsberg 100, SSNEM 221, L 208.

SHIRLAW PIKE Northum NU 1003. *Shirlah Pike* 1868 OS. P.n. Shirlaw + N dial. **pike** 'a pointed summit'. Shirlaw is possibly the 'shire hill', OE **scīr** + **hlāw**. It rises to 1010ft. and marks the boundary between the royal manor of Rothbury and Longframlington.

SHIRLEY 'Bright wood or clearing'. OE **scīr** + **lēah**. Some of these compound names may contain **scīr** 'a shire' in the sense 'clearing where the shire moot is held'. Forsberg 1950.99, L 205.
(1) ~ Derby SK 2141. *Sirelei(e)* 1086, *-lai* 1230, *S(c)hir- S(c)hyrle(g) -legh(e) -ley(e)* 1159×66–1247 etc., *Sherley* 1577, 1610, *Skirleghe* 1281. Db 599, SSNEM 207.
(2) ~ Hants SU 4013. *Sirelei* 1086, *S(c)hirleg(h) -le* 1227–1341, *Sherley* from 1253. Ha 148, Gover 1958.40.
(3) ~ WMids SP 1279. *Syrley* c.1240, *Schirleye* 1307×27 etc., *Shirley* from 1403. Shirley lies close to the old Warwickshire-Worcestershire boundary. Wa 72, L 205.
(4) ~ MOOR Kent TQ 9332. *Shyrlemoor* 1535. P.n. Shirley + ModE **moor**. Shirley, *Sirle* 1199×1216, *Shirle, S(c)herle* 1240–1327, *Scharleghe* 1338, is either the 'shire' or 'the district wood', or 'the bright wood or clearing'. PNK 367, L 205.

SHIRRELL HEATH Hants SU 5714. *Sherrill heath* 1695. P.n. Shirrell + ModE **heath**. Shirrell, *(oð ðæt) scirhiltæ** [826]12th S 276, *(of) scyrhylte* [939]12th S 446, is either the 'bright wooded land' or the 'shire wood', OE **scīr** or + ***hylte**. Ha 148, Gover 1958.48, PNE i.276.

SHIRWELL Devon SS 5937. 'The bright or clear spring'. *Ascere- Sirewelle* 1086, *-willa* Exon, *Shirewelle* 1222–1346 with variants *Shyre- Schire-* and *-wille*. OE **scīr**, definite form **scīre**, + **wylle**. D 68.

SHITLINGTON WYorks SE 2615. Partly uncertain. *Scellin- Scelintone* 1086, *Schilenton* 1202×8, *Schilyngton* 1359, *S(c)hit- S(c)hyt(e)lington(a) -yng-* 1145×60–1822. This is usually taken to be the 'farm, village, estate called after Scyttel', OE pers.n. **Scyt(t)el* + **ing**[4] + **tūn**. OE **scyte** 'shooting', however, was probably used in p.ns. in the sense 'steep slope' and this may also have been the case with OE **scyt(t)el(s)** otherwise recorded in the sense 'bolt, bar, arrow, something that can be shot'. Shitlington lies at the summit of a steep hill above the r. Calder and may, therefore, be the 'settlement by the steep slope called **Scyt(t)eling*, that which has the characteristics of a *scyttel*'. This township name fell out of use on grounds of delicacy and was replaced by MIDDLESTOWN SE 2617 and NETHERTON SE 2716. YW ii.205.

SHOBDON H&W SO 4062. 'Sceobba's hill'. *Scepedvne* 1086, *Skopindona* 1143×8, *Sopedun'* 1200, *Sob(b)edon(a) -dune* 1160×70–1291, *S(c)hob(be)don(e)* from 1291. OE pers.n. *Sceobba* + **dūn**. He 180, L 154.

SHOBROOKE Devon SS 8601. 'Haunted brooke'. *(of) sceoca broces (forda)* 'from Shobrooke ford' 925×939 S 387[†], *Sotebroch* 1086, *Sokebroc, Schokebrocke* 1260, *Shog(ge)broke* 1329–1524, *Shabbroke* 1541, *Shawbrook* 1675, *Shobbrooke* 1694. OE **sceocca** 'evil spirit, demon' + **brōc**. The reference is to the stream now called Shobrooke Lake SS 8701. D 416.

SHOCKLACH Ches SJ 4449. 'Haunted stream'. *Socheliche* 1086, *Sokeleche -lach(e)* 1209–29, *Soklache* 1421, *Shokelach* from c.1240 with variants *Sc(h)ok(e)-* and *-lache -latch, Shocklach* from 1429. OE **scucca** 'evil spirit' + **læċċ, leċe** 'boggy stream'. Che iv.63.

North SHOEBURY Essex TQ 9286. *Nord Scobire* 1202, *North Showbery* 1607. OE **north** + p.n. *(to) Sceobyrig* c.900 ASC(A) under year 894, *Sceabyrig* c.1050 ASC(C), *Sceorebyrig* c.1000 ASC(D), *(Es)soberiam* 1086, *Soberia -biri(e) -beri* etc. 12th–1294,

[†]The identification is uncertain and this form might possibly refer to the lost Shutbrook Street in Exeter.

Shobir(e) -bery -byry 1229–1428, *Shooberry* 1724×6, 'the shoe fortification', OE **scēo** 'shoe, shoe-shaped piece of land', possibly used as a p.n., + **byriġ**, dative sing. of **burh**. The reference may have been to a sand-bank called *The Shoo* 1509×47, and to the remains of an encampment at South Shoebury supposed to be that raised by the Danes in 894 (cf. the ASC entry for that year). 'North' for distinction from South Shoebury TQ 9484, *S. Shoebury* 1805 OS, also known as *Great Shoreby* 1352, cf. *Shoebery alias Lytle Showbery alias North Showbery* 1607, *Parva Soberia* 1261. Ess 198, MM 16.82, Pevsner 1965.357.

SHOEBURY NESS Essex TQ 9383. *Shoberynes(se)* 1509×47, *Shooberry-ness* 1724×6. P.n. Shoebury as in North SHOEBURY + ModE **ness**. Ess 199, MM 16.82.

SHOEBURYNESS Essex TQ 9384. Formerly South Shoebury, now renamed after SHOEBURY NESS TQ 9383. For earlier forms see under North SHOEBURY TQ 9286.

SHOLDEN Kent TR 3552. 'Shovel hill'. *S(c)houeldun(e) -don(e)* 1176–1278, *Soldon'* 1242×3, *Sholdon* 1346, *Schoueldene* 1253×4. OE **scofl**, perhaps used as the name of the hill, 'The Shovel', + **dūn**. The reference is to a low hill-spur made by the 50ft. contour running out into marshland. PNK 575, KPN 204, L 143, 145.

SHOLING Hants SU 4511. Unexplained. *Sholling* 1251, *-yng* 1295, *S(c)hollinge* 1255–1304, *Sholling alias Showling alias Showland* 1766, cf. *Shoreland Common* 1810 OS. If the form *Sorlinga* 1167–70 belongs here this could be OE **scora** + **hlinċ** 'shore ledge' referring to its site on shelving ground near the mouth of the r. Itchen, but Ha 149 and Gover 1958.39 assign it to Shoreland in Hursley. ING 231, DEPN.

SHOLVER GMan SD 9507. Probably 'sloping shieling'. *Solhher* 1202, *Shollerg, Shollere* 1246, *S(c)holver* from 1246, *Sholgher* 1291, *-re* 1332, *Sholler* 1323. OE **sceolh** + ON **ǽrgi**. The form with *v* is due to substitution of a neighbouring fricative for the peripheral ME phoneme [ɣ]. Sholver stands on steeply sloping ground. La 52, SSNNW 211, Jnl 17.41.

SHOOTING HOUSE HILL Cumbr SD 2681. No early forms.

SHOP 'Workshop, smithy'. English **shop**.

(1) ~ Corn SS 2214. *Shop* 1840. Situated at a road junction, a common site for a blacksmith's workshop. PNCo 156.

(2) ~ Corn SW 8773. *Parkens Shop* 1748. Cf. the surname of Richard *Parkyn*, 1525. Situated at a road junction as SHOP SS 2214. PNCo 156.

SHOPLAND HALL Essex TQ 8988. P.n. Shopland + ModE **hall**. Shopland, *Scopingland* [946]17th S 1793, *Scopinglande* [c.1000]c.1125, *Scopelandā* 1086, *S(c)op(i)land(e)* 1195–1346, *Schopelaund(e) -londe -i-* 1219–1303, *Scopland* 1428, is the 'newly cultivated land at or called Shopping, the shed place', OE ***sc(e)opping** < **sc(e)oppa** + **ing**[2] perhaps used as a p.n., + **land**. A later folk-etymology is *Shepela(u)nde -londe* 'sheep land' 1254–1361. Ess 200, PNE ii.107, Jnl 2.43.

SHOREDITCH GLond TQ 3282. 'Shore ditch', the ditch leading to the shore sc. of the river Thames. *Soredich* c.1148, *Schoredich* 1236, *Soresdic(h)* 1183–1346, *S(c)horesdich(e)* 13th cent. OE ***scor(a)** + **dīċ**. Mx 145, GL 85.

SHOREHAM Kent TQ 5161. 'Homestead, estate by the steep bank'. *scorham* 822 S 186, *Sorham* 1210×12–1241, *S(c)horham* 1226–54. Shoreham lies in the Darent valley at the foot of steeply sloping land. OE ***scor(a)** + **hām**. KPN 144, PNE ii.113, ASE 2.30.

SHOREHAM AIRPORT WSusx TQ 2005. P.n. Shoreham as in SHOREHAM-BY-SEA TQ 2204 + ModE **airport**.

SHOREHAM-BY-SEA WSusx TQ 2204. A secondary development at Shoreham Beach known earlier as New Shoreham, *Noua Sorham* 1235, *Nywe Shorham* 1288, ME **niwe**, Latin **nova** + p.n. Shoreham as in Old Shoreham TQ 2006, *(de, in) Veteri S(ch)orham* 1235, 1261, *Eldesorham* 1279, 1300, ME **eld** (OE *eald*), Latin **vetus**, dative case **veteri**, + p.n. *Sorham* 1073, *Soraham* c.1075, *Sore(s)ham* 1086, *Shorham* from 1167, 'the homestead by the steep hill', OE ***scora** + **hām**. Old Shoreham lies on the bank of the Adur at the foot of the steep slope of the Downs. Sx 246–7, PNE ii.112, ASE 2.33, SS 86.

SHORESDEAN Northum NT 9546. This seems to be *Shoreswood Hall* 1865 OS. P.n. SHORESWOOD NT 9446 influenced by nearby Allerdean NT 9646.

SHORESWOOD Northum NT 9446. 'The enclosure of the slope'. *Scoreswurthin* c.1107×23 DEC, *-wurthe* c.1160×5 DEC, *-wirthe* c.1170×4 DEC, *Schoreswirtha* before 1195 DEC, *-worth* 1331, *Shoreswoode* 1530. OE ***scor**, genitive sing. ***scores**, + **worthiġn**, later **worth**, subsequently replaced in weak stress by **wood** (OE *wudu*). Shoreswood is situated on a hill that rises to 275ft. overlooking Norham. NbDu 179, PNE ii.112–3.

SHORNCOTE Glos SU 0296. 'Cottage(s) in a mucky spot, dunghill cottage(s)'. *S(c)hern(e)cote* 1086–1492, *Scornecote* 1221, *Sharnecote* 1375, *Sharcote* 1387, 1392. OE **scearn** + **cot**, pl. **cotu**. Gl i.83 gives pr [ʃaːŋkət].

SHORNE Kent TQ 6970. 'Steep place'. *Scorene* c.1100, *Sorne(s)* 1156×7–1232, *Shorna* 1158×9, 1159×60, *Shorne(s)* 1168×9–1240. The site is the slope of a hill-spur. OE ***scoren**. PNK 117, PNE ii.113.

SHORTACOMBE Somer SS 7634 → Long HOLCOMBE SS 7636.

SHORTGATE ESusx TQ 4915. Possibly 'gap by a cut off piece of land'. *Shertegate* 15th. OE ***scerte** + **ġeat**. Sx 403.

SHORT HEATH Leic SK 3115. 'Short heath'. *Shertheth* 1342, *Shortheath* 1625. ME **shert** (OE *sceort*) + **hethe**. Lei 568.

SHORT HEATH WMids SP 1093. *Short Heath* 1834 OS. ModE **short** + **heath**. Wa 34.

SHORTLANESEND Corn SW 8047. 'End of Short's Lane'. *Shortlane end* 1678, *Shorts Lanes End* 1716, *Shorts Lane End* 1748, 1813. Cf. the surname of William and Roger *Shorte* in this area, 1569. Probably short for the 'Short family's part of Penvounder' (*Penfounder* 1547, ~ *alias Penvounder* 1695, Co **penn** 'head' + **bounder** 'lane, pasture' with lenition of *b* to *v*, 'head, end of the lane'). PNCo 156, Gover n.d. 465.

SHORWELL IoW SZ 4583. 'Steep-hill spring'. *Sorewelle* 1086–1250×60, *Shorewell(')* 1220–1600, *S(c)horwell* from 1227. OE ***scor(a)** + **wella**. The reference is to a stream that rises here from the S slope of the chalk down. Wt 215 gives pr [ˈʃorəɫ], Mills 1996.95.

SHOSCOMBE Avon ST 7156. Partly uncertain. *Shenescomb* 1327 LS, *Shevescombe* 1380×1 FF, *Shoscomb* 1817 OS. Uncertain element + OE **cumb**. If the 1327 form were reliable this could be 'Sciene's coomb' but it is almost certainly a mistake for *Sheuescomb*.

SHOTESHAM Norf TM 2499. 'Scott's homestead'. *Shotesham* *[1044×7]13th S 1055, *Scotessam, Scotesham* 1086, *Schotesham* 1254. OE pers.n. *Scott*, genitive sing. *Scottes*, + **hām**. DEPN.

SHOTGATE Essex TQ 7692. *(Cabes or Caves now known as) Shotgate Farm* 1770, 1805. Earlier *Cames* 1494. Surname *Cam* (of various origins), genitive sing. *Cames* and ModE **shotgate** 'a gate that can be shot or slid to and fro'. Ess 194.

SHOTLEY BRIDGE Durham NZ 0852. *Shotleybrigg* 1613×4. P.n. Shotley Northumb NZ 0852 + ModE **bridge**. Shotley, *Shotley* 1242 BF, is probably 'clearing, pasture on a steep slope', OE ***scēot** + **lēah**. Shotley lies at the foot of steeply rising ground which excludes the other formal possibilities, OE **sceot** 'shooting', OE **(ġe)sceot** 'an inner room' or OE ***sceote** 'a pigeon'. NbDu 179, Studies 1936.147.

SHOTLEY GATE Suff TM 2433. *Shotley Gate* 1805 OS. P.n. SHOTLEY TM 2335 + ModE **gate**. Arnott xv notes a series of gate names in the Deben valley which seem to be connected with the *hards* or landing-places along the waterside.

SHOTLEY Suff TM 2335. Possibly 'pigeon wood or clearing'. *Scoteleia* 1086, *Soteleg* 1212, *Sc(h)ottele* 1242, 1303, *Shotley* 1524, 1610. OE ***sc(e)ote** + **lēah**. The place is on a slope and may alternatively be 'the wood or clearing on the steep slope', OE **scēote** + **lēah**. DEPN, Baron, PNE ii.108.

SHOTTENDEN Kent TR 0454. 'Hill called *Sceoting*, the steep place'. *Sotindona* c.1175, *Shotenden'* 1226, 1254, *S(c)hotin(g)don'* 1240. OE ***scēoting*** 'a steep place' < ***scēot*** 'a steep slope' + **ing**[2] probably already used as the name of the hill + **dūn**. Shottenden lies on the slope of high ground that rises to a peak of 501ft. at The Mount TR 0454. Another Shottenden (or Shottington or Shoddington) occurs at TQ 9946 in Westwell next to a conspicuous hill rising to 303ft. PNK 377, Cullen.

SHOTTERMILL Surrey SU 8832. 'The mill called after the Shottover family'. *Shottermill* from 1583, *Schotouermyll* 1607, 1641. Family name *Shottover* + ModE **mill**. Cf. Robert *Shottover* and a half-virgate called *Shottover* 1537. The family perhaps came from Shotover Oxon SP 5606. Sr 183.

SHOTTERY Warw SP 1855. Probably 'the Scots' stream'. *(æt) Scotta rith* [699×709]11th S 64, *Scotriðes gemære* 'the bounds of S' [1016]18th S 1388, *Scotrith* c.1208, *Shottrythe* 1272, *Shotrith* 1440, *Schotriue* 1272, *Shotryve* 1304, *Shotryff* 1488, *Shoteri* 1221, *Shottery* 1538, *Shatterey* 1554. The generic is OE **rīth** 'a stream', the specific probably OE *Scot* 'a Scot', genitive pl. *Scotta*, but **sceota** 'a trout' is also possible though fitting the recorded sps less well. Wa 239, L 29.

SHOTTESWELL Warw SP 4245. Probably 'Sceot's spring'. *Soteswalle* c.1135, *-well(e)* c.1140–1262, *Shoteswell* 1165–1393, *-tt-* from 1428, *Shatswell* 1705, 1756. OE pers.n. **Scēot*, genitive sing. **Scēotes* + **wella**. Wa 273, L 31.

SHOTTISHAM Suff TM 3244. 'Sceot's homestead'. *Scotesham* 1086, *Sotesham* 1186, *Schatesham* 1254, 1344, *Skatersham* 1279, *Shetesham* 1307, *Shettisham* 1313, *Shotisham* 1261, 1316, 1674, *Schatham* 1524, *Shatsham* 1610. OE pers.n., possibly pers.n. **Scēot*, genitive sing. **Scēotes*, + **hām**. DEPN, Arnott 66.

SHOTTLE Derby SK 3149. 'Hill with a steep slope'. *Sothelle* 1086, *S(c)hethell(e)* 1191–6, *S(c)hothull(e)* 1298–1564, *Shotell* 1372, 1541. OE **scēot** + **hyll**. Db 601, Studies 1936.147–51, L 170.

SHOTTON Durham NZ 4139. 'Farm, village at the steep slope'. *Sceottun* [c.1040]12th, *Scioton', Siottona* [1183]14th, *Shotton* from early 14th. OE ***sceot*** + **tūn**. NbDu 179.

SHOTTON Northum NT 8430. 'The hill of the Scots'. *Scotadun* [c.1040]12th, *Schotton* 1242 BF 1120, *-tone* 1284, *Scotton'* 1296 SR. OE **Scot**, genitive pl. **Scotta**, + **dūn**. The abbeys of Kelso and Melrose had interests in Shotton which lies just on the English side of the border with Scotland. The reference is to Shotton Hill which rises to 748ft. at NT 8429. NbDu 180.

SHOTTON COLLIERY Durham NZ 3940. P.n. SHOTTON NZ 4139 + ModE **colliery**.

SHOTWICK Ches SJ 3371. 'Hamlet at or called *Shoto*, the steep promontory'. *Sotowiche* 1086, *Shotowica* [1096×1101]17th, *Shotewic* 1240–1547 with variants *Schot(t)e-* and *-wyc -wyk(e) -wik(e)*, *S(c)hotwic -wik* from 1240 with variants, *Shotwick* from 1278. P.n. **Shoto*, OE **scēot** 'a steep slope' + **hōh** 'a promontory, a hill-spur', + **wīċ**. The ground rises gently rather than steeply from the Dee marshes and the church to over 100ft. Che iv.206 gives pr ['ʃ ɔtwik] and locally ['ʃ ɔtik].

SHOULDER OF LUNE Lancs SD 7734. The name of a sand-bank at the mouth of the river Lune from its shape like a shoulder.

SHOULDHAM Norf TF 6709. Partly uncertain. *Sculham* [1043×5]13th S 1531, *Sculdeham* 1086, *Schuldham* 1177, *Shuldham* 1251. The generic is OE **hām** 'homestead, estate', the specific perhaps an OE ***sculd*** 'obligation, debt, liability' related to **scyld**. The meaning would be 'estate for which rent is paid'. DEPN, PNE ii.115.

SHOULDHAM THORPE Norf TF 6508. 'Shouldham outlying farm'. *Shouldham Thorpe* 1824 OS. P.n. SHOULDHAM TF 6709 + **thorp**.

SHOULSBARROW COMMON Devon SS 7040. *Shoulsbury Common* 1907 Baring-Gould. P.n. Shoulsbarrow + ModE **common**. Shoulsbarrow, *Solsbury Castle* 1630, *Showlsborough (Castle)* 1809 OS, *Shrowlsbury Castle or Salusbury Castle* 1815, is an Iron Age hill-fort at SS 7039. The generic is ME **burgh**, **bury** (OE *burh*, dative sing. *byriġ*), the specific possibly an OE word related to MHG *schröuwel* 'a devil'. D 61, Thomas 1976.97.

SHOULTON H&W SO 8158. 'Settlement on a slope'. *Selge-Scolegeton* [c.1221]c.1240, c.1250, *Soutton* 1327, *Shelton* 1518, *Shoulton* 1649. OE **sceolh**, oblique case definite form **sceolgan**, + **tūn**. Cf. OHG *skelah*, MHG *schelch*, ModG *scheel* 'sloping, slanting'; Shoulton lies on sloping ground. Wo 133.

SHRAWARDINE Shrops SJ 4015. Partly uncertain. *Saleurdine* 1086, *Sera(w)ordina* 1121, 1155, *Scrawardin* 1166, 1195, *Shrawurthin* 1240, *S(c)hrawardyn* 1308–1443, *Shrawardine* from 1637, *Screwrdin* 1204–14th with variants *-wrðin -wardin(e) -worthin* etc., *Shradon* 1601, *-den* 1609–92, *Shredon* 1723. The generic is OE **worthiġn** 'an enclosure', the specific possibly OE **scræf**, **scref** 'a cave, a den, a hollow' or **scrēawa** 'a shrew' used as a pers.n. Sa i.265.

SHRAWLEY H&W SO 8064. 'Wood or clearing of the huts or hollows'. *Scref- scræfleh* [804]11th S 1187, *Escreueleia* c.1150, *Scrauele(ga)* 1220–35, *Shrauele(ye)* 1327–61, *Shrawley* 1431. OE **scræf**, genitive pl. **scrafa**, + **lēah**. Wo 78, Hooke 92.

SHREWLEY Warw SP 2267. 'The pit wood or clearing'. *Servelei* 1086, *Scraueleia* 1150, *S(c)hrauelee -leye* 1272×1307, 1300, 1320, *Shreveleg* 1198–1332 with variants *Sch-* and *-ley(e)*, *Shreweleg* 1289, *Shrewley* 1578. OE **scræf** 'a cave, a pit', genitive pl. **scrafa**, + **lēah**. Wa 222.

SHREWSBURY Shrops SJ 4912. 'The fortified place of the scrub-land'.

I. *(in) civitate Scrobbensis* 901 S 221.

II. *(at) Scropesbyrig* 12th ASC(F) under year 1006, *(into, on) Scrobbesbyrig* 1121 ASC(E) under years 1006, 1016, 1102, *Shrobbesbyri* c.1540, *Srovesbroc* 1271, 1282, *Shrovesbury -u-* 1331×5–1491, *Shrousbury* 1339–1507, *Shrowsbury* 1515–1755, *Shrewesbury* 1384–1599, *Shrewsbury* from c.1540, *Shroosbury* 1729.

III. *Saropesberia* 1066×87, [1100×35]1267, *Salopesberie* [1094×8]1332–1377 with variants *-biri -bury -bery* etc.

OE ***scrobb***, genitive sing. ***scrobbes***, + **byriġ**, dative sing. of **burh**. Coins minted at Shrewsbury 924–1066 are most frequently inscribed **scrob** which establishes ***scrobb*** as a variant of OE **scrybb** 'a scrub, brushwood'. The development of the name was *Shrobesbury* > *Shroves-* (cf. PAVENHAM SP 8955, *Pabenham* 1240–1535, *Paven-* 1492, BAVERSTOCK Wilts SU 0232, *Bab(b)estok* 13th cent., *Baverstoke* 1568) > *Shrousbury* (cf. BOWCOMBE IOW SZ 4686, *Bovecombe* 1086). Type III is AN with inorganic vowel due to the difficulty of initial *Sr-* (for English *Shr-*) and subsequent confusion of *r* and *l*. The W name is *Amwythig* from late 14th, *Moythike* c.1540, probably representing *am* 'about, around' + *gwŷdd* 'wood, trees' + adjectival suffix *-ig*, a rough translation of the English name. Sa gives pr [ʃrouzbəri], a pronunciation developed presumably on the analogy of [sou] for *sew, sow*, local pr [ʃru:zbəri]. Sa i.267–71.

SHREWTON Wilts SU 0643. 'The sheriff's estate. *Schyrreveton* 1255, *Sherritone* 1310, *Shryfton* 1529×32, *Shrewton alias Shreveton* 1682. OE **scīr-ġerēfa** + **tūn**. Edward of Salisbury, sheriff of Wiltshire, held the manor in 1086. It was one of the Winterbourne manors, originally called simply *Wintreburne* 1086, then *Winterbourne Syreveton* 1232 and finally Winterbourne was dropped. See WINTERBOURNE. Wlt 236.

SHRIGLEY PARK Ches SJ 9479. *Shrigley Park* 1842. P.n. Shrigley + ModE **park**, referring to the former Regency house *Shrigley Hall* 1611. Shrigley, *S(c)hriggeleg(h) -ley(e)* 1285–1438, *Shrigley* 1545, is 'missel-thrush glade', OE **scrīċ**, genitive pl. **scrīca**, + **lēah**. Che i.130.

SHRIPNEY WSusx SU 9302. Partly unexplained. *Scrippan eg* *[680 ? for 685]10th S 230, *Scrippeneye* 1229, *Shripeneye* 1288, *Shireppeny, Scharpeny* 1279 etc. Unidentified OE element or pers.n. ***scrippa***, genitive sing. ***scrippan***, + **ēġ** 'an island'. Sx 91, SS 70.

SHRIVENHAM Oxon SU 2489. 'The decreed or allotted river-meadow'. *Scriuenham* *[821]c.1200 S 183, *Scrivenanhom* *[821]c.1240 ibid., *(to) Scrifenanhamme* [c.950]11th Wills, *Shrivenham* from 1217 with variants *Scriv- Sriv-* and *S(c)hryv-*, *Seriveham* 1086, *S(c)hriueham* 1157–1284. OE past part. **scrifen** of *scrīfan* 'judge, give sentence, shrive', dative case definite form **scrifenan**, + **hamm** 1. The reference may be to an estate given to the church in satisfaction of an ecclesiastical claim or a penance. Brk 375, L 43, 50.

SHROPHAM Norf TL 9893. Partly unexplained. *Scer(e)p-Screpham* 1086, *Schrepham* 1166×7, *Shropham* from 1231 with variants *Scrop-* 1242, *Shorp-* 1283. The generic is OE **hām** 'a homestead, a village, an estate', the specific apparently either OE **sceorp** 'ornament, clothing, equipment' or **screopu** 'a strigil, a curry comb' neither of which carries conviction. The DB spelling *Scerep* with svarabhakti vowel suggests that the OE etymon was *scre(o)p-*, cf. DB *Siropesberie, Sciropescire* for Shrewsbury, Shropshire. We probably have to do with an unknown pers.n. here. DEPN.

SHROPSHIRE 'The county of Shrewsbury'. *(into) Scrobbesbyrigscire* 11th ASC(C) under year 1006, *(in) Scropscire* 11th Thorpe, *Sciropescire* 1086, *Salopescira* c.1095. OE **scīr** 'county, shire' + p.n. SHREWSBURY SJ 4912. After the earliest instance the p.n. is found in elliptical form, with *-byrig* dropped. *Salop-* is an Anglo-Norman spelling.

SHROPSHIRE UNION CANAL Staffs SJ 7526. Originally the *Birmingham and Liverpool Junction Canal* 1833 OS.

SHROTON OR IWERNE COURTNEY Dorset ST 8512 → IWERNE COURTNEY OR SHROTON.

SHRUB END Essex TL 9723. The 'shrub end' of Lexden Heath. *Shreb End* 1777. P.n. Shrub + ModE **end**. Earlier simply *Screb* before 1272, *Shrobben, Srobbe* 1272, *S(c)hrebbe* 1276–1538, 'place growing with shrubs', OE **scrybb**, Essex dial. form **screbb.** Contrasts with *Bottle End* 1805 OS. Ess 377.

Lower SHUCKBURGH Warw SP 4962. *Lower Sokeberw* c.1236, *Netheresockebergh* 1247, *Shukkeborgh Inferior* 1327. ME adjs. **lower**, **nether**, Latin **inferior**, + p.n. *Socheberge* 1086, *Suc(c)heberga* c.1154, 1163, *Schukeberg* 1236, *Schukesburgh* 1242, *Shukeborough* 1547, *Shugb(o)rough(e)* 1524, 1547, 'the goblin hill', OE **scucca** + **beorg**. 'Lower' for distinction from Upper Shuckburgh SP 4961, *Uvere Shukebergh* 1247. Wa 143.

SHUCKNALL H&W SO 5842. 'Demon's hill'. *Shokenhulle* 1377. OE **scucca**, genitive sing. **scuccan**, + **hyll**. Bannister 171.

Great SHUNNER FELL NYorks SD 8497. ModE **great** + p.n. Shunner Fell, probably the 'look-out fell', ON **sjón** 'sight, view', genitive sing. **sjónar**, + **fjall**. At 2357ft. this is the second highest hill in YN. Cf. SHUNNER HOWE SE 7399.

SHUNNER HOWE NYorks SE 7399. 'Look-out hill'. *Senerhou* 1223, *Shonerhom* (for *-how*) 1252, *Shonerhowes* 15th. ON **sjón** 'sight, eyesight, view', genitive sing. **sjónar**, + **haugr**. An exact parallel is the Norw p.n. Sjonhaug, ONorw *i Siónarhaugi* NG i.10. The spellings *Sener-* and *Shoner-* show two separate developments, earlier PrN **sēonaR* giving expected ME *sener*, late ON *sjónar* showing sound substitution of [ʃ] for [sj]. YN 130, SPN 241, NoB ix.162, PNE ii.123.

SHURDINGTON Glos SO 9218. Possibly 'estate called after Scyrda'. *Surditona* c.1150, *S(c)hurdentone -in- -yn-* 1148–1376, *-ing- -yng-* 1356–1605, *Shardington -yng-* 1327, *Sherrington, Shoryngdon -ington, Sharnton, Shernton, Shureton* 16th cent. OE pers.n. **Scyrda* + **ing**[4] + **tūn**. Gl ii.xii, 156.

SHURLOCK ROW Berks SU 8374. *Shurlock Row* 1761, *Southlake Row* 1790. P.n. Shurlock + ModE **row**. Shurlock, *Suthelak(')* 1242×3, 1260×1, *Suthlake* 1347, is 'south of the stream' sc. Ruscombe Lake, OE **sūthan** + **lacu**. Brk 113.

SHURTON Somer ST 2044. 'The sheriff's estate'. *Shur(r)eveton* 1219–20, *Shirevton* 1228. OE **scīr-ġerēfa** + **tūn**. DEPN.

SHUSTOKE Warw SP 2290. Possibly 'Sceot's outlying farm'. *Scotescote* (sic) 1086, *Schustoke* 1217–1548 with variants *Shu-* and *-stok, Shustock* 1625. OE pers.n. **Scēot*, genitive sing. **Scēotes*, + **stoc**. Alternatively the specific might be OE **scēot** 'a steep slope'. Shustoke is near a bold spur of the hill called Shustoke Hill. Wa 92 gives pr [ʃʌstək], Studies 1931.25.

SHUTE Devon SY 2597. 'Steep slope'. *Schieta* 1194×1206, 13th, *(La) S(c)hete* 1228–1465, *S(c)hute* from 1301, *Shoote* 1602. OE **scȳte**, dial. *shute* 'a steep slope'. Shute lies on the steep declivity of Shute Hill (520ft.). D 630.

SHUTFORD Oxon SP 3840. 'Scytta, Scyttel' or 'the archer's ford'. *Schiteford'* [c.1160]c.1225, *Su(i)telesford'* 1168, *Setelesford'* 1169, *Schutford* 1240×1, *Schutt(e)ford(e)* c.1250, 1278×9, 1483, *Shitford(e)* 1285, 1526. OE pers.n. **Scytta* varying with diminutive form **Scyttel*, + **ford**. **Scytta* represents OE appellative **scytta** 'a shooter, an archer' which may be the specific here. O 425.

SHUTHONGER Glos SO 8835. 'Wood on a steep slope'. *Shuthanger* 1779. Ultimately from OE **scȳte** + **hangra**. Gl ii.72.

SHUTLANGER Northants SP 7249. 'Sloping wood where shuttles are cut'. *Shitelhanger* 1162–1353 with variants *S(c)itel- S(c)hytel-* etc., *Schutelhanger* 13th–1329, *Shyttlanger* 1502. OE **scyttel** in various senses 'shuttle, dart, bolt, gate-bar' + **hangra**. Nth 106 gives pr [ʃʌtlæŋə], L 195–6.

SHUTTINGTON Warw SK 2505. 'The estate called after Scytta'. *Cetitone* 1086, *Schetintuna* c.1160, *Suttintona* 1154×89, *S(c)hutyngton* 1285, 1327, *Shitington* 1232, *Shittyngton* 1535, 1543. OE pers.n. **Scytta* + **ing**[4] + **tūn**. Wa 24.

SHUTTLEWOOD Derby SK 4672. 'Wood where bars are cut'. *Shittilwod* 1486, *Schittilwod* 1521, *Shuttlewood* 1540. OE **scyt(t)els** + **wudu**. OE **scyttels** meant 'shuttle, bolt, bar' and in dialects 'gate-bar'. Db 215.

SIBBERTOFT Northants SP 6882. 'Sibert or Sibern's curtilage'. *Sibertod* 1086, *-toft -y-* 12th–1317, *Sybbertofte al. Sibbtofte* 1564. Pers.n. OE *Siġebeorht* or ON *Sigbjorn* + **toft**. Nth 121, SSNEM 150.

SIBDON CARWOOD Shrops SO 4183. 'S by Carwood'. *Shepeton Corbet* c.1540, *Sipton Carswood* 1672, *Sibdon Carwood* 1705. The c.1540 form is corrupt. P.n. *Sibetune* 1086, *Sib(b)eton(e)* c.1155–1431, *Sibton* 1367–1647, 'Sibba's estate', OE pers.n. *Sibba* + **tūn**, + p.n. Carwood as in Lower and Upper Carwood SO 4086 and 4085, *Carwod'* 1306×7, *Carwood* 1672, 'the rock wood', OE **carr** + **wudu**. Sa i.271.

SIBERTSWOLD Kent TR 2647 → SHEPHERDSWELL.

SIBFORD FERRIS Oxon SP 3537. 'S held by the Ferrers family'. *Sibbard Ferreys* early 18th. P.n. *Scipford* 1086, *Si- Sybbeford(')* [c.1153]1280–1309, *Sibford* from c.1200, 'Sibba's ford', OE pers.n. *Sibba* + **ford** + manorial affix from the family n. of Robert de Ferrers, second earl of Derby. O 404.

SIBFORD GOWER Oxon SP 3537. 'S held by the Go(h)er family'. *Sibbeford' Goyer* 1220, *Sybeford Goher* 1251, *Sibbeford Gower* 1281–1314. P.n. *Sibford* as in SIBFORD FERRIS + manorial affix from the family n. *Goher* which occurs 1222–43. O 404.

SIBLYBACK LANE RESERVOIR Corn SX 2370. P.n. Siblyback Lane 'lane to Siblyback' + ModE **reservoir**. The reservoir was completed in 1969. Nearby Siblyback SX 2372, *Cibliback* 1567, *Sibliback* 1590, is 'ridge of land belonging to the Sibley family', family n. Sibley, *Sibyly* 1419, + ModE **back** (OE *bæc*). On the map the name has been wrongly inscribed Siblyback *Lake* Reservoir. PNCo 156, Gover n.d. 255.

SIBSEY Lincs TF 3551. 'Sigebald's island'. *Sibolci* 1086, *Ci- Sybeceia* 1151×3, *Sibeceie* before 1198, *-cey* [1199]14th. OE pers.n. *Siġebald*, genitive sing. *Siġebaldes*, + **ēġ**. DEPN, Cameron 1998.

SIBSON Cambs TL 0997. 'Sibbi's estate'. *Sibestune* 1086, [c.1150]14th, *Sibeston(e) -is-* 1279–1329, *Sibston* 1233, 1324, *Sybson* 1544, 1609. OE pers.n. *Sib(b)i*, genitive sing. *Sib(b)es*, + **tūn**. Hu 197.

SIBSON Leic SK 3500. Possibly 'Sigeberht's hill'. *Sibetesdone* 1086, *Sy- Sibedesdune -don* before 1173–1243, *Sy- Sib(b)esdon(e)*

-is- -ys- -ton 1229–1712, *Sy- Sibsdon* 1576–1627, *Sy- Sibson* 1541–1610. OE pers.n. *Siġeberht*, genitive sing. *Siġeberhtes*, + **dūn**. This pers.n. with weakly stressed second element offers less difficulty than unrecorded **Siġebed* suggested by DEPN and Lei 541.

SIBTHORPE Notts SK 7645. 'Sibba or Sibbi's outlying farm'. *Sibetorp* 1086–1214, *-thorp* 1230–80, *Sibbe- Sybbe-* 13th cent., *Sibthorp(e)* from 1245, *Sybthrope* 1582. Pers.n. OE *Sibba* or ON *Sibbi*, genitive sing. *Sibba*, + **thorp**. Nt 216, SSNEM 117.

SIBTON Suff TM 3669. 'Sibba's estate'. *Sib(b)etuna* 1086, *Si- Syb(b)eton* 1169–1571, *Sybton(')* 1424–1633, *Sy- Sipton* 1523–1674, *Sibton* 1610. OE pers.n. *Sibba* + **tūn**. DEPN, Baron.

SICKLESMERE Suff TL 8760. *Sicklesmere* 1836 OS. If this is an ancient name it represents 'Sicel's mere', OE pers.n. **Sicel*, genitive sing. **Siceles*, + **mere**.

SICKLINGHALL NYorks SE 3648. 'Nook of land at or called **Sic(e)ling*'. *Sidingal(e)* 1086, *Sieclinghale, Sicclinhala, Sicolinghal'* 12th, *Si- Syc- Siklinghale* 13th–1364, *Sicklinghall* from 1584, *Sikelingehal(l), Siclingehal* 13th. OE p.n. **Sic(e)ling* + **halh**. **Sic(e)ling* is probably an **ing**[2] derivative of OE pers.n. **Sicela*, 'the place called after Sicela'. YW v.47.

SIDBURY Devon SY 1391. 'Fortification by the river Sid'. *(æt) sydebirig* 1050×72, *Sideberi(e)* 1086–1303, *Sydebury* 1316. R.n. *Side* c.1250, *(le) Syde* 1284, 1420, *Sid river* c.1550, probably 'the wide one' from OE **sīd**, definite form **sīde**, + **byriġ**, dative sing. of **burh**, referring to the fortification of Sidbury Castle, an Iron Age hill-fort at SY 1291. D 596, Thomas 1976.97.

SIDBURY Shrops SO 6885. 'The south fortified place or manor house'. *Sudberie* 1086, *-bury* 1291×2–1788, *Suthbery* 1208×9, *Sidbury* from 1577. OE **sūth** + **byriġ**, dative sing. of **burh**. Directly S of Middleton Scriven. Sa i.272.

SIDBURY Wilts SU 2150. Partly uncertain. *Shidbury, Chydebur'* 1325, *Sydbury hyll* 1571, *Shudburie, Shudburrowe hill* 1591, *Chidbury* 1812. Possibly OE ***scydd** 'a hovel, a shed' + **byriġ**, dative sing. of **burh**. The reference is to a bivallate Iron Age hill-fort. Cf. SIDBURY Devon SY 1491. Wlt 343 gives popular pr [ʃedbəri], Pevsner 1975.359.

SIDBURY HILL Wilts SU 2150 → SAMLESBURY Lancs SD 5930.

SIDCUP GLond TQ 4672. 'Seat-shaped hill, flat-topped hill'. *Cetecopp* 1254, *Setecoppe* 1301, *Sedecoppe* 1332, *Sidycope* 1407. OE ***set** + **copp**. PNE ii.120, LPN.208.

SIDDINGTON Ches SJ 8471. '(Settlement) south of the hill'. *Sudendune* 1086, *Sudindun -don* mid 13th, *Sudingdon(e)* 1286–8, *Sidinton* 1269–1518 with variants *-yn-*, *Sidington* 1335, *-dd-* 1383 and from 1694. OE **sūthan** + **dūn** replaced by **tūn**. Che i.84. Che ii.vii gives former pr ['siðitn].

SIDDINGTON Glos SU 0399. '(Land) south in the township'. *Svdi(n)tone, Svintone* 1086, *Sueton, Suinthon, Swinton* 12th, *Suthintun(a) -ton(e) -yn-* 1146–1409, *Sudinton -yn-* 1201–1494, *Sutton juxta Cicestre* 1398, *Sydyngton -ing-* 16th cent. OE **sūth(an) in tūne**. Gl i.81.

SIDESTRAND Norf TG 2639. 'The wide shore'. *Sistran* 1086, *Sidestrande* 1189×99. OE **sīd** + **strand**. The name contrasts with nearby OVERSTRAND TG 2440. DEPN.

SIDFORD Devon SY 1390. 'Ford over the river Sid'. *Sydeford* 1283–1316. R.n. Sid as in SIDBURY SY 1391 + OE **ford**. D 598.

SIDLESHAM WSusx SZ 8598. 'Sidel's estate'. *Sidelesham* *[683]14th S 232, *Si- Sydelesham* 1227–1338, *Sid(d)lesham* from 1262. OE pers.n. **Sīdel*, genitive sing. **Sīdeles*, + **hām**. A form *Sideleshamstede* is also recorded [724]14th S 42 with OE variant **hām-stede**. Sx 85 gives pr [sidəlsəm], Stead 258, ASE 2.33, SS 86.

SIDLEY ESusx TQ 7409. 'The wide clearing'. *Si- Sydelegh(e)* 1279–14th, *Sidley Green* 1636. OE **sīd**, definite form **sīda**, + **lēah**. Sx 494.

SIDMOUTH Devon SY 1287. 'Mouth of the river Sid'. *(of) Sidemuða* 1072×1103, *Sedemuda* 1086. R.n. Sid as in SIDBURY SY 1391 + OE **mūtha**. D 598.

SIGFORD Devon SX 7773. 'Sicga's ford'. *Sigeforde* 1086, *Si- Syggeford* 1281, 1356. OE pers.n. *Sicga* + **ford**. A crossing of the river Lemon. D 477.

SIGGLESTHORNE Humbs TA 1545. 'Sigell's thorn-tree'. *Siglesto(r)ne* 1086, *Si- Syglest(h)orn(e)* [12th]15th, 1251–1468, *-gh-* 14th cent., *Si- Syst(h)orn* 1512, 1538, 1610. Pers.n. *Sigell*, genitive sing. *Sigels*, + **thorn**. *Sigell* is from ON *Sigulfr* probably replacing earlier OE *Siġewulf*. *Siġewulf* would normally become *Siulf* which seems to survive in the 16th cent. spellings and in the local pr [si:lsθrən]. YE 68, SPN 235, L 221.

SIGHTY CRAG Cumbr NY 6080. *Sighty Crag* 1865 OS. A mountain on the border between Cumbria and Northumberland rising to 1701ft.

SIGSTON NYorks SE 4195. 'Siggr's estate'. *Sig(h)estun* 1086, *Siggestune -ton* 1088–1474. ON pers.n. *Siggr*, secondary genitive sing. *Sigges*, + **tūn**. YN 212, SSNY 129.

SIKE MOOR NYorks SD 8078. *Sike Moor* 1858. ModE **sike** (OE *sīċ*) 'a small stream, a ditch' + **moor**. YW vi.223.

SILBURY HILL Wilts SU 0968. Uncertain. *Seleburgh* 1281, *Selbyri, Selburi hille* c.1540, *Selbarrowe hill* 1558×1603, *Selbury or Sibury Hill* 1663. The evidence is insufficient to ascertain whether the specific is OE **sele** 'a hall' or the generic OE **burh**, dative sing. **byriġ** 'a fort', or **beorg** 'a hill, a tumulus', although *bury* is a common 16th cent. corruption of *burgh* from *beorg*. The etymology thus remains as uncertain as the origin itself of this great enigmatic Bronze-Age man-made mound. Wlt 295, Pevsner 1975.99, Thomas 1976.228.

SILCHESTER Hants SU 6261. 'The Roman town *Cille*'. *Silcestre* 1086–1287, *Sele- Silechæstre* 1205, *Cilcestr(e)* 1228–1302, *Cylchestre* 1278–1327, *Sy- Silchestre* 1322–1431, *Chilchester* 1349, 1432. OE p.n. **Ċille* + **ċeaster**. This was the Roman city of Calleva Atrebatum, *Callevae* (genitive sing.) inscr., *Καλη̣οῦα* [c.150]13th Ptol, *Galleva Atrebatum, Calleva* [c.300]8th AI, *Caleba Arbatium* [c.700]13th Rav, 'the Calleva of the Atrebates', Brit **Callēu̯ā* '(the town in the) woods', from **calli-* (for **caldi-*) as in Co *kelli*, W *celli*, Ir, Gaelic *coill(e)*, OIr *caill* 'wood' + suffix *-evā*. Scholars have long sought to derive the specific of Silchester from the *Call-* of *Calleva*; a PrW form **Callēw* would have been taken into English as **Cællīw* which as the result of normal sound changes (breaking and loss of *-w* > **Ċeallī*, i-mutation > **Ċielli*) would eventually have produced WSax **Ċille*. The change OE *ċ* [tʃ] > [s] occurs sporadically in p.ns. and would have been encouraged here by the tendency to dissimilation in the sequence ċ – ċ, [tʃ] – [tʃ] to [s] – [tʃ]. The alternative suggestion, OE ***siele** 'a willow copse' is unlikely on topographical grounds. Ha 149, Gover 1958.121, DEPN, RBrit 291, Nomina 4.31, TT 39ff., GPC 459.

SILEBY Leic SK 6015. 'Sigulfr's village or farm'. *Siglesbie, Si- Seglebi* 1086, *Sygleby, Siglebi(a)* c.1130–1229, *Si- Sylebi -by* from 1207, *Si- Sylby* 1328–1518. ON pers.n. *Sigúlfr*, ME genitive sing. *Sigulfes*, + **bȳ**. The medial syllable has been reduced in weak stress and there is no need to posit an unrecorded r.n. **Sigol* from the root of OE *sīgan* 'sink, move'. Cf. SILSDEN SE 0446. Lei 319, SSNEM 68.

SILECROFT Cumbr SD 1381. 'Willow croft'. *Selecrotf* (sic for *-croft*) c.1205, *Selecroft* 1211–1358, *Selcroft* 1278–1579, *Sealcrofte* 1578, *Sillcroft* 1770. ON **selja** + OE **croft** 'a small enclosed field'. On formal grounds the specific could alternatively be OE **sele** 'a hall'. Cu 444, SSNNW 249.

SILKMORE Staffs SJ 9021. 'Marshland with a drain'. *Selchemore* 1086, *Selkemor(e)* 1224–1421, *Si- Sylkemore* 1408–1572, *Silkmore* 13th and from 1439. OE ***sīluc**, ***sīoluc**, ***sēoluc** + **mōr**. St 77.

SILKSTONE SYorks SE 2905. 'Sigelac's farm or village'. *Silches- Silcston(e)* 1086, *Si- Sylkestun(a) -ton(a)* c.1090–1572. OE pers.n. *Siġelāc*, genitive sing. *Siġelāces* + **tūn**. YW i.310.

SILKSTONE COMMON SYorks SE 2904. P.n. SILKSTONE SE 2905 + ME **commun**, 'common land'. YW i.312.

New SILKSWORTH T&W NZ 3753. A modern colliery settlement. ModE **new** + p.n. *sylcespurðe* [c.1040]12th, *Selceswurtha*

[1070×83]15th, *Sylkeswrtha* c.1162×74–*Silkesworth* 1675 with variants *Silches- Silckes- Selkes-, -wrðe -wrth(e) -uurth', -is-* and *-ys-, Silksworth* from 1632, probably 'Sigelac's enclosure', OE pers.n. *Sīgelac*, genitive sing. *Sīgelaces*, + **worth**. The occasional spellings in *Sel-* may, however, point to OE ***sēoluc**, ***sīoluc**, ***sīluc** 'a gully, a drain' as the specific. Silksworth is situated in a small but marked valley on the banks of a stream that drains into the sea at Hendon. NbDu 180, DEPN s.n. Silkmore.

SILLOTH Cumbr NY 1153. 'Barn(s) at the sea'. *Selathe* 1292, 1377, *-lathes* 1361, *Seelet, Selyth, Silluthe -eth(e)* 16th, *Silloth* 1718. ON **sǽr** or OE **sǣ** + ON **hlatha**. Silloth stands on the coast of the Solway Firth. Cu 293, SSNNW 59, L 29.

SILLS Northum NT 8200. *Sills* 1869 OS. In the absence of earlier forms the explanation of Sills, Silloans NT 8200, and Sills Burn NT 8201 must remain uncertain. OE **syle, sylu** 'a bog, a miry place' would be suitable and possibly also **syll** 'a threshhold, a foundation, a base' with reference to remains of one of the Roman camps here.

SILPHO NYorks SE 9692. 'Shelf-shaped hill-spur'. *Sifthou* (sic) [1145×8]15th, *Silfhou -how* [1155×65–early 14th]15th, *Silfow(e)* 1301, [1395]15th, *Silfey* 1577. OE **scylf** or **scylfe** with sound substitution of [s] for [ʃ] + **hōh** influenced by ON **haugr**. The sound substituton may be a Scandinavian one rather than Anglo-Norman. YN 115, SelPap 89, SPN 283, L 187.

SILSDEN WYorks SE 0446. 'Sigel's valley'. *Siglesden(e)* 1086–1312, *Sighelesden* 1131×40–1328, *Si- Sylesden(e)* c.1140–1323, *Silsden* from 1614. OE pers.n. **Sigel*, genitive sing. **Sigeles*, + **denu**. YW vi.19.

SILSOE Beds TL 0835. 'Sifel's hill-spur'. *Siuuile- Sewilessou* 1086, *Siuelesho* 1199–1330, *Silshoo* 1506. OE pers.n. **Sifel*, genitive sing. **Sifeles*, + **hōh**, dative sing. **hō**. For the pers.n. cf. *Sifeca* in the OE poem *Widsith*. Bd 161 gives pr [silsə], L 168.

SILTON Dorset ST 7829. 'Willow farm or village'. *Seltone* 1086–1415 with variant *-ton('), Salton'* 1268–1428, *Silton(')* from 1332. OE **sealh** + **tūn**. Cf. SALTON NYorks SE 7180. Do iii.64.

Nether SILTON NYorks SE 4592. ME *Nether* 1298 + p.n. *Silf- Siluetune* 1086, *Silton* from 1204, the 'settlement by a ledge of high ground', OE **scylf(e)** 'a shelf, a ledge' + **tūn**. *S-* instead of *Sh-* is an AN sound substitution. Nether Silton, with Over Silton, lies beneath a ledge-shaped spur of the Hambleton Hills. Also known as *Silton Paynill* 1285 from the tenure of land here in 1231 by Isabell Paynell. YN 201, 207, SSNY 259.

Over SILTON NYorks SE 4593. ME *Over* 1316 + p.n. Silton as in Nether SILTON SE 4592. Over Silton, with Nether Silton, lies beneath a ledge-shaped spur of the Hambleton Hills. Also known as *Parva Silton* 'little S' 1301. YN 201, 207, SSNY 259.

SILVER END Essex TL 8019. *Silver End* 1777. Silver End lies in Belchamp St Paul's parish and contrasts with RIVENHALL END TL 8316. Ess 411.

SILVERDALE Cumbr SD 4675. 'Silver valley'. *Selredale* 1199, 1246, *Sellerdal* 1246, 1341, *Celverdale* 1292, *Silverdale* 1320–46. OE **seolfor**, later replaced by ON **silfr** or ME **silver**, + ON **dalr**. The reference is to the silver-grey limestone rocks around the village. La 189, SSNNW 249.

SILVERDALE Staffs SJ 8146. *Silverdale* 1796, *Silverdale Iron Works* 1833 OS. Possibly transferred from SILVERDALE Lancs SD 4674. The ironworks were started there c. 1792. Horovitz.

SILVERLEY'S GREEN Suff TM 2975. *Silverleys Green* 1837 OS. Probably surname *Silverley* + ModE **green**. Contrasts with nearby *Hussey Green, Northwood Green, Sawn Green* and *Chaphall Green* all 1837 OS.

SILVERSTONE Northants SP 6644. 'Siulf's or Seulf's farm or village'. *Sulueston* [942]14th S 1606, *Si- Sylvestone* 1086–1823, *Selveston(e)* 1086–1314, *Silverstone* from 1260×90, *Sylson* 1484. OE pers.n. *Sigewulf* or *Sǣwulf*, genitive sing. *-wulfes*, + **tūn**. The forms with *-r-* are the result of early folk-etymological association of the name with ME **silver**. Nth 43 gives pr [silsən].

SILVERTON Devon SS 9502. 'Settlement at *Sulhford*, the gully ford'. *Sulfretona* 1086, *Suffertona* 1086 Exon, *Sulferton(e)* 1281–1318, *Seluerton* 1179, *Silverton(e)* from 1249. OE p.n. **Sulh-ford* literally ' furrow ford', OE **sulh** + **ford**, + **tūn**. The reference is to the deep valley or 'furrow' E of the village and the stream running through it. D 569.

SIMONBURN Northum NY 8773. 'Sigemund's stream'. *Simundeburn* 1230, *Symundesburn* 1291, *Simmonborn* 1596, *Simonburn* 1769 Armstrong. OE pers.n. *Sigemund* + **burna**. It is widely held that Simonburn was one of the castles built by Simon of Senlis, grandson of Waltheof, the last Saxon earl of Northumberland, that his is the name preserved in the p.n., and furthermore that it is really a name in **burh** rather than **burna**, like SOCKBURN Durham NZ 3407 and NEWBURN T&W NZ 1965. The forms for Simonburn, however, do not support this view. NbDu 180,

SIMON FELL NYorks SD 7574. *Simon Fell* 1817. Pers.n. *Simon* + ModE **fell** (ON *fjall*). YW vi.236.

SIMONSBATH Somer SS 7739. 'Simon's bathing place or pool'. *Simonsbath, Simon's Bath* c.1550, *Symmonsbath* 1657. Pers.n. *Simon* + ModE **bath**. D 530.

SIMONSTONE Lancs SD 7734. 'Simund's stone'. *Simundeston* 1246, *Symondestan* 1278, *Simundestan* 1292, *Simondestone* 1296, *Sy- Simoundeston* 1327, 1333. ME pers.n. *Simund*, genitive sing. *Simundes*, + **stān**. La 79, Jnl 17.49.

SIMPSON Bucks SP 8835. 'Sigewine's estate'. *Sevinstone* 1086, *Siwinestone* 1086–1237, *Sywynestone* 1241–1324, *Suenston* 1434, *Synston* 1485, *Symston* 1495, 1535, *Sympson* 1674. OE pers.n. *Sigewine*, genitive sing. *Sigewines*, + **tūn**. Bk 24.

SINDERBY NYorks SE 3481. Probably the 'southern farm or village'. *Senerebi* 1086, *Sindarbi* [1170×88]15th, *Sinderbi -by* from 12th. ODan **syndri** + **bȳ**. Sinderby lies S of Pickhill in Pickhill parish. Alternatively the specific might be ON pers.n. *Sindri*, genitive sing. *Sindra*, or OE **sinder** 'cinder, slag'. YN 225, SSNY 37.

SINDERHOPE Northum NY 8451. *Sindrop* 1695 OS, *Sinderhope* 1867 OS. Possibly OE **synder** 'asunder, apart' used as a prefix to denote 'land detached or separated from an estate', + **hop** and so 'the detached valley'.

SINDLESHAM Berks SU 7769. 'Sin(n)el's homestead or river-meadow'. *Si- Syndlesham* 1256–1397 and from 1620, *Sin(e)sham* 1240, 1761, *Silsham als Sindlesham* 1674. OE pers.n. **Sin(n)el*, genitive sing. **Sin(n)les*, + **hām** or **hamm** 1. Brk 137.

SINGLETON Lancs SD 3838. 'Settlement where shingles are manufactured' or 'with shingled roofs'. *Singletun* 1086, *Synglentona* 1094, *Si- Syngelton* 1177–1332, *Schingelton(a)* 1169–82, *Shingelton* 1246, 1362. OE ***scingol** + **tūn**. La 154, PNE ii.109, 121 s.v. **shingel**.

SINGLETON WSusx SU 8713. Possibly 'the brushwood farm or estate'. *Silletone* 1086, *Sengelton* 1135–1330, *Schingelton* 1181, *Si- Syngelton* 1189–1417, *Sangelton* 14th cent. OE ***sengel**, ***sængel** + **tūn**. The exact sense of **sengel** is unknown. Sx 53, Problems 16, PNE ii.118.

SINGLEWELL Kent TQ 6471. *Singlewell* 1819 OS. A reformation of earlier *la Chinglede Welle* 1240, *Shingledewell* 1316, 1333, 'spring with a shingle roof', ME **shinglede** + **welle**. This place has been identified with the lost *Shauecuntewelle* 1293 'spring called Shaved Cunt', but the two names are probably separate. PNK 100.

SINNINGTON NYorks SE 7485. 'Farm, village or estate on the river Seven'. *Siuenin- Sevenic- Siuerinctun* 1086, *Siviling- Si- Syvelington* 1183×93–1327, *Synnyngton* 1580. OE **Syfening*, an **ing**[2] formation on the r.n. SEVEN SE 7380, + **tūn**. YN 76, BzN 1968.180, Årsskrift 1974.31.

SINTON GREEN H&W SO 8160. *Sinton Green* 1832 OS. P.n. Sinton SO 8463 + ModE **green**. Sinton, *Suth- Sudintun* 1240–75, *Synton* 1483, is OE **sūth in tūne** 'place south in the estate' sc. of Grimley, in contrast to Northingtown Farm SO

8161, *Nor(th)inton* 1240, 1275, representing OE **north in tūne**. Wo 128, which wrongly cites S 1437 under this name instead of Leigh SINTON SO 7850.

Leigh SINTON H&W SO 7850. 'Sinton in Leigh'. *Sothynton in Lega* c.1275, *Sodyngton, Lye Sinton al. Syddington* 14th–16th. P.n. Sinton + p.n. LEIGH SO 7853. Sinton, *(to) suþtune* [825]18th S 1437, *Suthin(g)ton* 1275, 1316, is OE **sūth tūn** 'a southern settlement' varying with **sūth in tūne** as in SINTON GREEN SO 8160. Sinton lies S in the parish of Leigh. Wo 207.

SIPSON GLond TQ 0878. 'Sibwine's estate'. *Sibwineston, Sybwynston* early 13th cent., *Sibodeston'* 1214, *Sibbeston* 1318, *Sibston* 1391, *Shepiston al Sypson* 1593. The earliest forms point to 'Sibwine's estate' but later forms suggest an alternative pers.n. *Sigebald* or *Sigebod*. Mx 39.

SISSINGHURST Kent TQ 7937. Partly uncertain. *Saxingherste* c.1180, 1242, 1310, *Saxingeherste* 1206, *Syssingherst* 1610. Uncertain element + OE **hyrst**, OKt **herst** 'a wooded hill'. The specific could be either an OE p.n. **Seaxing* 'rock place' < OE ***seax** + **ing**[2] or the genitive pl. **Seaxinga* of a folk-n. **Seaxingas* 'the people called after Seaxa', OE pers.n. *Seaxa* + **ingas**. PNK 323.

SISTON Avon ST 6875. 'Sige's estate'. *Si- Syston(e) -tun(e)* 1086–1599, *Siston* 1599. OE pers.n. **Sige*, genitive sing. **Siges*, + **tūn**. Gl iii.67.

SITHNEY Corn SW 6328. 'St Sithney's (church)'. *(ecclesia) sancti Sythnini* 1230, ~ *de Sancto Sydnio* 1270, *Seynt Sitheny* 1379, *Seynt Synney* 1554, *Sithney* 1610. Saint's name *Sydhni* (Sithney) from the dedication of the church. Also venerated at Guisseny, Brittany ('village of Sydhni', OBret *guic* 'town part of a parish'), where he is the patron saint of mad dogs (his own choice instead of being patron saint of young women). An alternative name, *Merthersitheny* 1250, Co ***merther** 'saint's grave' + *Sydhni*, survives as Merther SW 6329. PNCo 157, Gover n.d. 526, DEPN, Dauzat-Rostaing 337 s.n. Guichen.

SITTAFORD TOR Devon SX 6383. Uncertain. *Siddaford Tor* 1809 OS. Possibly identical with *Chiteford* 1379. D 198.

SITTINGBOURNE Kent TQ 9063. Possibly 'stream called Siding, the spacious one'. *Si- Sydingeburn(')* 1200–53, *Si- Sydingburn(e)* c.1230–78, *Si- Sythingeburn(e)* 1258–62. OE stream name **Siding* < **sīd**+ **ing**[2] referring to Milton Creek + **burna**. For OE **sīd** applied to a river estuary, cf. Sid in SIDBURY and SIDMOUTH Devon SY 1391, 1390. An alternative explanation might be 'stream of the Sidingas, the dwellers on the slope', OE folk-n. **Sīdingas* < **sīde** + **ingas**, genitive pl. **Sīdinga*, + **burna**. PNK 264, RN 364.

SIX ASHES Shrops SO 7988. *Six Ashes* 1833 OS. Cf. Four Ashes Staffs SO 8087 referring to prominent trees.

SIXHILLS Lincs TF 1787. 'The six clearings or pastures'. *Sisse* 1086, *Sixla* c.1115, *-lei* 1196, *Sixele* 1212, *Sixil* c.1230, *Sixhill'* 1258. OE **sex** + **lēah**. DEPN, Cameron 1998.

SIX MILE BOTTOM Cambs TL 5756. *(the) Six Mile Bottom* 1801. A hamlet in a hollow six miles from Newmarket. Ca 139.

SIXPENNY Dorset ST 8416 → Sixpenny HANDLEY ST 9917.

SIXPENNY HANDLEY Dorset ST 9917 → Sixpenny HANDLEY.

SIZEWELL Suff TM 4762. Probably 'Sigehere's stream'. *Syreswell* 1240, *Syswell* 1280, *Cisewall'* 1336, 1344, *Sysewell* 1524, 1568, *Siswell* 1610. OE pers.n. *Sigehere*, genitive sing. *Sigeheres*, + **wella**. DEPN, Baron.

SKEEBY NYorks NZ 1902. 'Wood farm' or 'Skithi's farm or village'. *Schirebi* 1086, *Schittebi* 1187, *Sc- Skytheby* [12th]15th–1301, *Sketheby* 1421, *Skebye* 1565. ON **skíth** 'a billet of wood' or **skítha** 'a stick' or pers.n. *Skíthi*, genitive sing. *Skítha*, + **bȳ**. YN 288, SSNY 37.

SKEFFINGTON Leic SK 7402. Probably 'settlement on *Skefting*, the hill shaped like a shaft'. *Sciftitone* 1086, *Sc- Skef(f)tinton(e) -yn-* c.1130–1363 with variants *Sc(h)aft-, Scheft- Skeftington' -yng-* 1230–1369, *Skeffington* from 1254. OE hill-name **Sceafting* < OE **sceaft** 'a shaft' + **ing**[2] with ON [sk] for English [ʃ] + **tūn**. The village stands high on a hill. On formal grounds, however, the specific could be the OE pers.n. *Sceaft* and the name mean the 'estate called after Sceaft'. Lei 320, Årsskrift 1974.53, SSNEM 207.

SKEFFLING Humbs TA 3719. Uncertain. *Sk- Sc(k)eftling(e) -yng(e)* c.1150×76–1375, *Sheftling* [1154×60]14th, *Cheftling'* 1195, *Sceflinge* 1204, *Scaftling* 1227, *Sheftelyng* 1301, 1338, *Skef(f)lyng* 1486, 1531. Previous explanations include OE pers.n. **Sceftela*, a diminutive of *Sceaft* as in SHAFTESBURY ST 8622 with Scand *sk* for *sh*, or ON *Skapti*, + OE **ing**[2], 'the place called after or associated with Sceftela', or + **ingas** 'the people called after Sceftela'. Formally, however, the best explanation is a Scandinavianisation of either OE ***sceaft-ling** 'place marked by a shaft or shafts or where shafts are gathered', cf. SWAYTHLING Hants SU 4315, or of an **ing**[2] derivation of a p.n. **Sceaft-lēah* 'clearing where shafts are cut', as in the lost Kentish p.n. *Hedlinge*, a new sing. formation on the common p.n. **Hæth-lēah* or Hadley. It could, however, be purely Scand, as the lost Swedish p.n. *Skeftlingquærn* suggests. YE 20, ING 73, PNE i.285, 287.

SKEGBY Notts SK 4961. 'Skeggi's village or farm'. *Schegebi* 1086, *Sceg(g)ebi(a)* 1174–82, *Scheggeby* 1208, *Skeg(g)eby* 1221–96, *Skegby* from 1302. ON pers.n. *Skeggi*, genitive sing. *Skegga*, + **bȳ**. Nt 133, SSNEM 69.

SKEGNESS Lincs TF 5663. 'Beard headland', i.e. that juts out like a beard. *Sceggenesse, Schegenes* 12th, *Skegenes* 1256, *Skegnesse* [1198]14th. ON **skegg** + **nes**. In former times the sea came right up to the S side of Skegness. The vill on the dunes was probably not developed until after the construction of the 11th cent. sea banks. DEPN, SSNEM, Jnl 7.47–9.

SKELBROOKE SYorks SE 5012. 'The shieling stream'. *Scalebre -bro* 1086, *-broc* 1161–1221, *Skelbroc -brok(e)* 1220–86, *Skelbro(o)k(e)* 1298–1598. OE ***scēl** or ***scēol** with ON [sk] for [ʃ] + **brōc**. The stream n. *Skell* is probably a back-formation from Skelbrooke although it might represent an original Scand r.n. **Skjallr* 'the resounding one'. YW i(ii.43), SSNY 164, L 16.

SKELDYKE Lincs TF 3337. 'The boundary ditch'. *Skeldyk(e)* from 1281 with variant *-dik(e)*, *Sheldyke* 1331, 1485, *Kirton Skeldyke* 1529. ODan **skial** + OE **dīc**, ON **dík**. Payling 1940.92.

River SKELL NYorks SE 2269. 'Resounding stream'. *Scel* [12th]15th, *Schel* 12th, [c.1150]15th, *Skell(e)* c.1170–1680. ON **skjallr**. RN 367, YW vii.137.

SKELLINGTHORPE Lincs SK 9271. 'The outlying farm at or called *Skelding*, the shield-like place'. *Skeldin(g)torp -yng-* 1205–1402, *Skellingthorpe* from 1555. OE **scelding** 'that which is shaped like a shield' < **sceld** + **ing**[2] with Scand [sk] for English [ʃ] + **thorp**. The earliest forms of the name are *Sceldinhopa* [?1085×9]14th, *Scheldinchope* 1086, *S(c)helding-hop(e) -yng-* 1141–1318, *Schellinghop* 1086, *Schellyng(h)op(e)* 1350, 1535, *Skeldinghop(e) -yng-* 1237–1520, *Skellinghope* 1225–1464, 'the enclosure called or at *Skelding*', OE **scelding** as above + **hop** 'a patch of dry land, an enclosure in a marsh', the site on which the **thorp** later developed. On formal grounds the name could be the 'thorp or enclosure called after Sceld', OE pers.n. *Sceld* + **ing**[4]. Perrott 260, SSNEM 221.

SKELLOW SYorks SE 5210. 'The shieling nook of land'. *Scanhalle -a* 1086, *Scalehale* 1200, *Skel(e)hal(e)* 12th–1336, *-hall* 1371, 1473, *Skellawe* 1379, *Skellow(e)* from 1493. OE ***scēl** or ***scēol** with ON [sk] for [ʃ] varying with ON **skáli** + OE **halh**, probably refering to the hollow between the hills through which the Skell flows. The stream name is a back-formation from SKELBROOKE SE 5012. YW ii.34, SSNY 165.

SKELMANTHORPE WYorks SE 2310. 'Skelmer's outlying farmstead'. *Scelmer- Scemeltorp* 1086, *Scelmertorp* 1195, 13th, *Skelmerthorp* 1290–1479, *Skelmanthorp(e)* 1316–1596,

Skelmon(d)thorpe, Skelmotheroppe 16th. ODan *Shelmer*, ON pers.n. **Skjaldmarr* + **thorp**. YW ii.221, SSNY 67.

SKELMERSDALE Lancs SD 4605. 'Skelmer's valley'. *Schelmeresdele* [sic] 1086, *Skelmersdale* from 1202 with variants *-mares- -is-* and *-den*. ODan pers.n. *Skelmer*, ON **Skjaldmarr*, genitive sing. *Skelmeres*, + **dalr**. A 1960s new town preserving an ancient name. The reference is probably to the valley of the river Tawd SD 4806. La 122, SSNNW 160.

SKELTON 'Settlement on or by a shelf of land'. OE **scelf** with substitution of Scand [sk] for [ʃ], + **tūn**.

(1) ~ Cumbr NY 4335. *Sheltone* c.1160, *Schelton'* 1186–1279, *Skelton* from 1271. The village stands on a ledge projecting from a stretch of shelving land between the 700ft. and 800ft. contours. Cu 239, SSNNW 188.

(2) ~ NYorks NZ 0900. *Sc- Skelton'* from [12th]13th. YN 293.

(3) ~ NYorks SE 3668. *Scheltone -done* 1086, *Skelton* from 1198. YW v.153.

(4) ~ NYorks SE 5756. *Sc(h)eltun* 1086, *Skelton* from 1181×4. YN 16.

SKELTON Cleve NZ 6518. 'Settlement by the *Skell*'. *Sc(h)eltun* 1086, 1130×5. R.n. **Skell* + OE **tūn**. Skelton lies in a valley bottom beside a small stream. It is suggested that the specific is the name of this stream identical with the r.n. SKELL NYorks SE 2269. Alternatively it might be OE **scelf** with ON [sk] for English [ʃ], 'farm, village beneath the shelf'. YN 145, SSNY 116.

SKELWITH BRIDGE Cumbr NY 3403. *Skelleth-bridge* 1651, *Skelwith bridge* 1693. P.n. Skelwith NY 3403 + ModE **bridge**. Skelwith, *Schelwath* 1246, *Skelwath* 1332, *Skelwyth* 1537, is 'ford of the *Skell*', r.n. **Skell*, ON **Skjallr* 'the resounding one' referring to the waterfall called Skelwith Force, + **vath**. The waterfall is just above the bridge and can be heard from a great distance. We i.211, La 219, SSNNW 160.

SKENDLEBY Lincs TF 4369. Possibly the 'village or farm at or called *Scenehelde*, the beautiful slope'. *Scheueldebi* (for *Schen-*) 1086, *Schendelbi(a)* 1135–78, *Scendelbi* 1154×89. Possibly OE p.n. *Scēnehelde* < **scēne** + **helde** with Scand [sk] for [ʃ] + **bȳ**. Otherwise the specific appears to be OE **scendle** 'abuse, reproach, shame'. DEPN, SSNEM 69.

SKERNE Humbs TA 0455. A transferred n. from the r.n. SKERNE BECK TA 0453. *Schirne* 1086, *-y-* 1246, *Skir(e)n(a), Skirne -y-* 1197×1210–c.1400, *Skerne* from 1531. There are Scand parallels in the Norw p.n. *Skjern* and Dan *Skern*. YE 155.

River SKERNE Durham NZ 3026. 'Bright stream'. *Schyrnam* c.1190, *Skiren, Skyren* 1242–1417, *Skyrne* 1361–1647, *Skerne* from 1388. Identical with the Norwegian *Skirne* 'bright, clear river'. On the Skerne is Skerningham NZ 3018 *Skirning(h)eim* c.1107, *Sc(h)ir- Skirnig(e)ham* 1141–c.1280, *Schirnīgahā* 1154×66, *Sqir- Skirningham -yng-* c.1220–1623, 'homestead called or at *Skirning*', OE p.n. **Scīrning* 'place by the Skerne' or 'the Skerne stream' (r.n. Skerne + **ing**[2]) + **hām**, or 'homestead of the people who live by the Skerne', OE **Scīrningas*, genitive pl. **Scīrninga*, + **hām**, a type of name unknown in Scandinavia. Probably, therefore, *Skirna* is a Scandinavianisation of earlier OE **Scīran-ēa* 'the water of **Scīre*, the bright stream', with ON [sk] for English [ʃ]. NbDu 181, RN 367, ING 156.

SKERNE BECK Humbs TA 0453. Either 'bright stream' or 'cleansing, healing, baptism stream'. *aqua(m) de Ski- Skyren(a), Shyrne, Skyron* 13th cent., *Skyrna, Skirne* 14th cent. The sense depends on whether this is to be regarded as a Scandinavianisation with substitution of [sk] for [ʃ] of an OE **Scīran-ēa* 'water of the *Scire*, the bright one' (OE **scīr**) or an original ON formation on **skírn** 'cleansing, healing, baptism' as in the Norw r.ns. Skirna. YE 11.

SKERNINGHAM Durham NZ 3018 → River SKERNE NZ 3026.

SKEWSBY NYorks SE 6271. 'Wood farm or village'. *Scoxebi* 1086, *Scog(h)esby* early 14th cent., *Scousby* 1226, *Skuesby* 1408, *Skewsby* 1666. ON **skógr**, genitive sing. **skógs**, + **bȳ**. Identical with the Swed p.n. Skogsby and Skogby. The specific could, however, be ON pers.n. *Skógr*. YN 30, SSNY 37.

SKEYTON Norf TG 2425. 'Skeggi's estate'. *Scegu- Scedgetuna* (sic) 1086, *Sceketuna* c.1150, *Scegeton'* 1191, 1236. ON pers.n. *Skeggi*, genitive sing. *Skegga*, + **tūn**. The modern pronunciation of this name, [skaitn], and the loss of [g] are hard to account for; cf. SKEGBY Notts SK 4961, SKEGNESS Lincs TF 5663. One must presume that [g] became devoiced by contact with [t] and subsequently spirantised to give the sequence [skegton] > [skekton] > [skeçton] > [skeiçtən] > [ski:tən] > [skaitn]. DEPN, SPNN 335.

SKIDBROOKE Lincs TF 4492. 'Dirt brook'. *Sc(h)itebroc* 1086, c.1115, *Scitebroc* 1230, *Skydbrok* 1328. OE **scite** 'dung' with Scand [sk] for English [ʃ] + **brōc**. A daughter settlement from Cockerington TF 3790 on land reclaimed from the sea. DEPN, SSNEM 221, L 15, Jnl 7.56.

SKIDBY Humbs TA 0133. Either 'Skyti's village or farm' or 'dirty village or farm'. *Scyteby* [972×92]11th S 1453, *Schitebi -by* 1086–1303, *Skitebi -y-* 1200–1362, *Skitby* 1359, *Skidby* from 1566. ON pers.n. *Skyti*, genitive sing. *Skyta*, or **skítr** 'dung' perhaps replacing OE **scite** + **bȳ**. YE 208, SSNY 37.

SKIDDAW Cumbr NY 2529. Partially obscure. *Skithoc* 1230, *Schy- Skydhow* 1247, *Skiddehawe* 1256, *Skythou* 1260, *Skythow(e)* 1343–1450, *Skethowe* 1397, *Sketho(w), Skedowe* 1539, *Skedo* 1544, *Skyddowe* 1570, *Skiddaw* 1671, *Skiddey Topp* 1675, *Skidda* c.1690. The generic is ON **haugr** 'mound, hill' referring to the mountain top of 3054ft., the specific possibly ON ***skýti**, an *i*-mutation variant of ON **skúti** 'a projecting or overhanging crag'. If so the sense is 'craggy hill', a suitable description of the craggy summit, cf. Randal Crag NY 2529, Grey Crags NY 2627, Dead Crags NY 2630. Cu 319, TCWAAS 1918.101, SSNNW 160.

SKIDDAW FOREST Cumbr NY 2729. *forest of Skithoc* 1230, *foresta s[i]ue chacea de Skydhowe* 1519, *forest de Skethoo* 1541. P.n. SKIDDAW NY 2529 + OFr, ME **forest** 'a hunting preserve'. Cu 320.

SKILGATE Somer SS 9827. Possibly 'boundary gate'. *Scheligate* 1086, *Schiligata* 1086 Exon, *Schillegat* 1195, *Scilla- Scillegate* [13th] Buck, *Skilegate* 1243, *Skilgat* 1610. Late OE **scyl** (< ON *skil*) + **ġeat**. DEPN, PNE ii.124.

SKILLINGTON Lincs SK 8925. Partly uncertain. *(æt) Scillintune* [1066×8]c.1200 Wills, *Schillinton(e) -tun(e)* 1086–1212, *Schellintune* 1086, *Schillingetona* 1146, *Ski- Skylington -yng-* 1210–1428, *Skillington* from 1254×8. Uncertain element or name + **tūn**. The following suggestions have been made:

I. OE folk-n. *Scillingas* 'the people called after Sciella' < pers.n. **Sciell(a)* from OE **sciell** 'shrill' + **ingas**, genitive pl. *Scillinga*.

II. OE **scilling** 'a shilling'.

III. OE r.n. **Scilling* 'the resounding one' < **sciell** + **ing**[2].

IV. OE ***sciling** 'a boundary' from *scilian* 'to separate'.

The most plausible suggestions are (3) or (4); in the former case the reference would be to Cringle Brook which runs in a marked valley S of the village, in the latter to the county boundary with Leic which forms the W edge of the parish. In either case Scand [sk] has replaced English [ʃ]. Cf the p.n. type SHILLINGFORD and the p.n. Shillingbottom Lancs, *Shillingbothim* 1296, *-botham* 1305, the 'valley bottom of the *Scilling*'. Perrott 200, Årsskrift 31, SSNEM 186, La 64.

SKINBURNESS Cumbr NY 1255. '*Skinburgh* headland'. *Skynburneys* from 1298 with variants *Skin-* and *-ness -nees*, *Skymburnes(se)* 1305–17, *Skimburgh-Neese* [1538]1603. P.n. *Skinburgh* + OE **næss**, ON **nes**. *Skinburgh, Skyneburg'* 1175, *Schineburgh* 1185, *Skynburgh* 1301, is a Scandinavianised form with [sk] for [sh] of an OE **Scinn-burh* 'demon- or spectre-haunted fortification', OE **scinna** + **burh**. Cu 294 gives pr [skinbərni:z], SSNNW 249, L 172.

SKINNINGROVE Cleve NZ 7119. 'Skinnari's, or the tanner's ravine'. *Scinergreve* 1273, *Skynnergreve* 1301, 1404, *-gryf* 1348, *Skyn(n)ingrave* [1285]16th, 1579. ON by-n. *Skinnari* or **skinnari** 'a tanner', genitive sing. **skinnara**, + **gryfja**. Skinningrove is situated at the mouth of a deep ravine leading down to the sea; **gryfja** survives in Y dial. *griff* 'a narrow valley'. YN 142.

SKIPPOOL Lancs SD 3540. 'The ship pool'. *Skippoles* 1330, *Skippull* 1593. ON **skip**, or OE **scip** with ON [sk], + **pōl** or **pull**. Formerly an important harbour. La 140, Mills 134, SSNNW 250.

SKIPSEA Humbs TA 1655. 'The ship lake, the harbour'. *Ski- Skypse* 12th–c.1400, *Skipsea* from 1442. This could be OE **scip** with Scand [sk] for English [ʃ] + **sǣ** or ON **skip** + **sær**. The reference is to the lost Skipsea Mere; cf. HORNSEA TA 2047. YE 82.

SKIPTON 'Sheep-farm'. Northern dial. equivalent of SHIPTON. ONb **scīp** with initial [sk] for [ʃ] due to ON influence, + **tūn**.

(1) ~ NYorks SD 9951. *Scipton(e) -tun* 1086–1275, *Schipton* 1276, *Skipton* from 1120. Also known as Skipton-in-Craven, *in Cravene* 1241. See CRAVEN. YW vi.71, SSNY 138.

(2) ~ NYorks SE 3679. *Schipetune* 1086, *Skipton' super Swale* 1243. See River SWALE SE 2796. YN 186, SSNY 117.

SKIPWITH NYorks SE 6638. 'Sheep-farm'. *Schipewic -uuic* 1086, *Schipwic* 1203, *Scypwic, Skipwik(e) -y-* 1121×8–1282, *Scipe- Scypwiz* 1166, 1244, *-wis* 1225, *Schip(e)wit(h)* 12th, 13th cents., *Skipwith* from 1080×6, 1276. ONb **scīp** with initial [sk] for [ʃ] due to ON influence, + **wīċ** later replaced by ON **vithr** 'a wood'. YE 262 gives pr [skipiθ], SSNY 149, L 222.

River SKIRFARE NYorks SD 8875. 'The bright river'. *Chirphare* 1170×90, *Scirphare* [1170]15th, *Skyrfare* 1580. ON **skírr** + **far** 'a river course, a track'. YW vii.138.

North SKIRLAUGH Humbs TA 1439. ME adj. *North* c.1265 + p.n. *Schires- Scir- Schir(e)lai, Scherle* 1086, *Ski- Skyrlagh(e)* 1240–1402, *Skirlawe* 13th cent., *Skyrlaugh* 1568, the 'bright clearing'. OE **scīr** with Scand [sk] for [ʃ], definite form **scīra**, + **lēah**. 'North' for distinction from South SKIRLAUGH TA 1439. YE 49, 51.

South SKIRLAUGH Humbs TA 1439. ME adj. *South* 1240 + p.n. Skirlaugh as in North SKIRLAUGH TA 1439.

SKIRMETT Bucks SU 7790. 'The shire meeting place'. *la Skiremote* c.1307, *Skirmot* 1347, *Scermit* 1797, *Scirmet* 1826. OE **scīr** + **(ġe)mōt** with Scandinavian [sk] for OE [ʃ]. The reference is to the same meeting place as is alluded to in the nearby name FINGEST SU 7791. Bk 180.

SKIRPENBECK Humbs SE 7457. 'Stream or beck called *Skerpin(g)*'. *Scarpinberg* 1086, *Scarpenbec* 1086, *-in-* 12th cent., *Scerpinbec -en-* [before 1080], 1160×80–1247, *Skerpenbec -in(g)- -k* 1156×7–1293, *Sci- Skirpenbec(k) -y-* 1180×90–1579, *Scerkyngbech* 1251, *Ski- Skyrkenbec(k)* 1268–1546. ON ***skerpin(g)** 'a stream that sometimes dries-up' + **bekkr**. A derivative of ON skarpr 'dried-up' found also in Norw p.ns. such as *Skjerpen, Skjerpa* meaning 'dried-up, barren land'. YE 150 gives pr [skɔpmbek], SSNY 103, PNE ii.124.

SKIRWITH Cumbr NY 6132. 'Shire wood'. *Skirewit* 1205, *Skyrewyth* 1279–1338 with variants *Skire-* and *-wyt(h)*, *Skirwith* from 1215 with variants. OE **scīr** with substitution of Scandinavian [sk] for [sh] + ON **vithr** possibly replacing OE **wudu**. As with the name SHERWOOD the reference is to a wood used in common by the men of a whole district in contrast to a wood used by the inhabitants of a single village. Cu 242 gives prs [ski:riθ], [skəriθ] i locally [skɛrit], SSNNW 250, L 222, 229.

SKITTLE GREEN Bucks SP 7702 → PITCH GREEN SP 7703.

SLACK WYorks SD 9728. 'The hollow'. *Slack(e)* 1621, *Heptonstall Slack* 1771. ON **slakki**. YW iii.194.

SLACKHALL Derby SK 0781. 'Hall in the hollow'. *Slackhall* 1633. ME **slak** (OWScand *slakki*) + **hall**. The valley is referred to as *Slak* 1285, *(del) Sclak* 1335 (p). Db 66.

SLAD Glos SO 8707. 'The valley'. *The Slad* 1779. Mod dial. **slad** < OE **slæd**. Gl i.136, L 85, 122–3.

SLADE Devon SS 5146. 'The valley'. *Slade* 1809 OS. A Gilbert atte Slade is mentioned 1330 possibly from here. ModE **slade** (OE *slæd*) 'a valley'. D 48, L 122–3.

SLADE GREEN GLond TQ 5276. *Slade Green* 1561, 1905, *Slads Green* 1805. P.n. **Slade* from ME **slade** 'valley, low-lying marshy ground' (OE *slæd*) + **green** 'a village green'. LPN 209.

Gurney SLADE Somer ST 6249. 'S held by the Gurney family'. *Gurney Slade* 1817 OS. Family name Gurney as in BARROW GURNEY ST 5367 + p.n. Slade, ultimately from OE **slæd** 'a valley'.

SLAGGYFORD Northum NY 6752. 'The ford at *Slagging*, the muddy place'. *Slag(g)ingford* 1257, 1267, *Slaggiford* 1335, *Slaggyford* from 1350. OE, ME ***slagging** < OE, ME **slag** 'slippery, muddy' + **ing**[2], + **ford**. NbDu 181 cites another spelling, *Chaggeford* 1218, which, if genuinely belonging here, represents an alternative name with OE ***ċeacga** 'broom, gorse, brushwood', and so 'the brushwood ford'. PNE i.83, ii.128.

SLAIDBURN Lancs SD 7152. 'Sheep-pasture stream'. *Slateborne* 1086, *Slai- Slayteburn(a)* [1135×40]13th–1481, *S(c)lateburn(e) -borne* 1303–1638, *Slaydeburn* 1451, *Slaidburn* from 1500. OE **slæġet** partially confused with ME **s(c)late** 'slate' + **burna**. YW vi.203.

SLAITHWAITE NYorks SE 0814. Probably 'clearing where timber is felled'. *Sladweit -wait* 1178–93, *Slathwait -thweyt* 1191–1286, *Slagh(th)waite, Slaugh(th)wayte* 14th cent., *Slaythwayte* 1627, *Slawit* 1750. ON **slag** + **thveit**. The *Slaugh/Slay* variation is paralleled in *Haigh/Haugh* from OE **haga** or *Shay/Shaw* from **sceaga**. YW ii.307 gives pr ['slauwit], Nomina 13.109–14.

SLALEY Northum NY 9757. 'The muddy clearing'. *Slaveleia* 1166, *Slaueley(e)* 1242–1526, *Slaule* 1170, *Slauley* 1428, *Slalee* 1526. OE ***slæf** + **lēah**. NbDu 181, PNE ii.127, DEPN.

SLALEY FOREST Northum NY 9555. P.n. SLALEY NY 9757 + ModE **forest**. A modern Forestry Commission creation.

SLAPTON Bucks SP 9320. 'Settlement called or at Slæp, the slippery place'. *Slapetone* 1086, *Slaptone* 1223–84. OE ***slǣp** possibly used as a p.n. + **tūn**. Bk 100, Studies 1936.187.

SLAPTON Devon SX 8245. 'Settlement by the slippery place'. *Sladone* 1086, *Slapton(e)* from 1244. OE ***slǣp** + **tūn**. The reference might be either to the slipperiness of the hill-side on which Slapton lies or possibly to a slip-way or portage from the sea ½ m distant. D 330, PNE ii.127.

SLAPTON Northants SP 6446. 'The settlement at the slippery place'. *Slapton(e)* from 1086. OE ***slǣp** + **tūn**. Slapton is situated on ground sloping down to the river Tove. Nth 44, Studies 1936.186ff.

SLAUGHAM WSusx TQ 2527. 'The sloe promontory, the sloe garden'. *Slacham* 1087×1100, 1091×1125, *Slagham* 1248–1454, *Slaucham* 1272, *Slaweham* 1279, *Slaugham* 1324, *Slogham* 1590. OE **slāh** + **hamm** 2b or 5b. The church is situated on a hill-spur between streams, but this might simply be 'the sloe enclosure, the sloe garden'. Sx 277 gives pr [slæfəm], SS 82.

Lower SLAUGHTER Glos SP 1622. *Sloughtre -ter inferior'* 1327, 1535, *Nether ~* 1327–1504, *Lower Slater* 1698. ModE adj. **lower**, ME **nether**, Latin **inferior**, + p.n. *Sclostre* 1086, *Slochtre(s)* 12th, *Sloghters -tre -tir'* 1229, 1276, 1459, *Sloughtre -ter* 1328–1496, 'the miry place or stream', OE ***slōhtre**, a derivative of **slōh** 'slough, mire' cognate with the German *schluht* 'gorge, ravine', older *sluoht* 'water channel', and the p.ns. Schlüchtern on the Kinzig, *Sluohterin* 999, Schluchtern near Leingarten, *Sluhtra* 8th, Schlochtern near Melle, *Slohteren* 1186, and Dutch Slochteren near Groningen, *Slohtoron* 11th, and the r.ns. Schlücht, a tributary of the Wutach near Waldshut, and *Schlochterbach*. 'Lower' for distinction from

Upper SLAUGHTER SP 1523. The pl. forms refer to both places. Gl i.206 gives prs ['slɔ:tə] and ['sla:tə], L 58, Bahlow 425.

Upper SLAUGHTER Glos SP 1523. *Sloutre superiori* 1290, *Over* ~ 14th, *Uper Slater* 1698. ModE adj. **upper**, Latin **superior**, ME **over**, + p.n. Slaughter as in Lower SLAUGHTER SP 1622. Gl i.206.

SLAWSTON Leic SP 7794. Partly uncertain. *Slages- Slachestone* 1086, *Slag(h)estuna* [1100×35]1333, [12th]15th, *Sla(h)ston(e)* 1225–1447, *Slaueston(e) -w-* c.1130–1465, *Slauston'* c.1130–1610, *Slawston(e)* from 1232, *Slawson* 16th possibly 'Slagr's estate', ON pers.n. *Slagr*, genitive sing. *Slags*, + **tūn**. The pers.n., however, is extremely rare, and the specific might rather be ODan *slagh* 'depression in the ground, track' related to ON *slag* 'wet', *slakki* 'valley, hollow'. The village lies in a dip between peaks. Lei 252, SSNEM 194.

River SLEA Lincs TF 0948. *aquam Lafford* 'L water' [1154×89]17th, *aqua de Lafford* [12th]c1300, *aquam Slafordie* [1154×89]17th, *Sleford water* c.1540. A back-formation from SLEAFORD TF 0645, the specific of which is a feminine r.n. *Slīo* 'muddy, slimy water' < **Slīwō* cognate with G r.n. Schlei, *Slia*, at Schleswig, *Sliaswic* 12th, the Dutch *Slie* in the p.n. Sliedrecht 'the crossing of the Slie', and ON *slý* 'slimy water plants' < **slīwa*, Mod Norwegian *sly, sli* 'slime' and the fish name OE *slīw, slēo*, MLG *slī*, OHG *slio*, ModG *Schleie* 'a tench', all from IE **slei-* 'glide, slime, smooth'. Perrott 40, RN 371.

River SLEA SU 8038 → SLEAFORD SU 8038.

SLEAFORD Hants SU 8038. 'Slaughter ford'. *Sleyford* 1245, *Slayford* 1336–1424. ME **sley** (OE *sleġe*) + **ford**. The reason for this name is unknown. The r.n. Slea SU 8038 is a back-formation from the p.n. Ha 150, Gover 1958.104.

SLEAFORD Lincs TF 0645. 'The ford across the Slea'. *(of) Sliowaforde, (æt) Slioforda* [852]12th S 1440, *Sleifordam* [1051]late 13th, *Eslaforde* 1086, *(of) Sliowa forda* c.1121 ASC(E) under year 852, *Slaford(ia)* 1126–60, *Es(t)laf(f)orde* 1193–1255, *Sleford(e)* c.1170–1675, *Sleaford* from 1392, *Lafford* 1196–1723, *Slifford* 1356, *Slyforth* 15th cent. R.n. *Slio* as in r. SLEA TF 0948 + OE **ford**. The forms with prosthetic *E-* or loss of initial *S-* are AN. Perrott 88, RN 371, Cameron 1998.

SLEAGILL Cumbr NY 5919. Partially obscure. *Slegil(e) -gill(e) -y-* 1180–1777, *Sleuegile* 1294, *Slenegill* (for *Sleue-*) 13th, *Sleegill* 1612, 1704, *Sleagill* from 1636. ON **slefa** + **gil** 'a ravine'. ON **slefa** means 'saliva' and may have been used here of Sleagill Beck which is, however, not a 'trickling' but a vigorous stream; it may be the name of a stream or of a sort of snake as in the Norwegian names *Slævdal* and *Sleveland*; or it may be the by-name *Slefa*. We ii.xiv, 148 which gives pr ['sli:gil], SSNNW 161, L 99.

SLEAPFORD Shrops SJ 6315. 'The ford leading to Sleap'. *Sleford* 1582. P.n. *Sclepa* [1138]15th, *Slepe* 1255–1542, *Sleap* 1833 OS, 'the slippery place', OE **slǣp**, + ModE **ford**. Bowcock 217, Gelling.

SLEDDALE NYorks SD 8587. 'Sled valley'. *Seldale* 1280, *Sleddal* 1285. OE **slæd** perhaps already used as a p.n. + **dæl**. The dale is wide and flat in comparison to the ravine through which it is entered at Gayle. In modern dial. a *slade* is a grassy plain between hills, a flat-bottomed, often wet-bottomed valley. YN 267, EDD s.v.

Wet SLEDDALE RESERVOIR Cumbr NY 5411. P.n. Wet Sleddale + ModE **reservoir**. Wet Sleddale, *Weat-Sleddale* 1562, *Weet, Wate* ~ 17th, *Wet Sleddale* from 1693. ModE adj. **wet** (OE *wēt*) + p.n. *Sleddal(a) -dale -dall -dell* 12th–1692, probably 'dale with side-valleys', OE **slæd** 'valley' + ON **dalr**. 'Wet' for distinction from LONGSLEDDALE NY 4903. We ii.170, L 96, 122–3.

SLEDGE GREEN H&W SO 8134. *Sledge Green* 1831 OS. ModE **sledge** + **green**. The hamlet lies on the S side of the high ground no doubt used for sledging in winter.

SLEDMERE Humbs SE 9364. 'The pond in the valley'. *Slidmare* 1086, *Ledemare* 1086, 1197, *Led(d)emer(e)* 13th cent., *Sledmere* from 13th, *-dd-* 1219–1316. ONb **sled** + OE **mere** influenced by ON **marr**. YE 126.

East SLEEKBURN Northum NZ 2883. *East Sleekburn* 1865 OS. ModE adj. **east** + p.n. *Sliceburne* [c.1040]12th, *Slicke(s)burn(e)* 12th cent., *Sli- Slykeburne* 13th cent., originally a stream name (now Sleek Burn NZ 27–2883), either 'the muddy' or 'the smooth stream', OE ***slīc** or ***slicu** + **burna**. 'East' for distinction from West Sleekburn (lost) NZ 2784. NbDu 182, PNE ii.129.

SLEIGHTHOLME Durham NY 9510. 'Flat ground near water'. *Slethholm* 1254. ON **sletta** + **holmr**. YN 305 gives pr [sli:təm].

SLEIGHTHOLME MOOR Durham NY 9207. *Sleightholme Moor* 1860 OS. P.n. SLEIGHTHOLM NY 9510 + **mōr**.

SLEIGHTS NYorks NZ 8607. 'Level fields'. *Slechetes* [c.1223]15th, *Sleghtes* [c.1223]15th, 1347, 1429. ON **slétta** 'a smooth level field', earlier **sleht-*, dial. *sleet*. The reference is not to the modern village but to a flat expanse of ground at NZ 8507 in the otherwise steep-sided valley of the Esk where the medieval Eskdale chapel was situated. YN 120 gives pr [sli:ts].

SLEPE Dorset SY 9293. 'Slippery place'. *Slepe* from 1244, *Sleape* 1584, *Slape* 1811, *Sleppe* 1244, *Slep(p)* 1632. OE **slǣp**. The reference is to low-lying ground near Poole Harbour. Do i.73 gives pr [slip].

SLIMBRIDGE Glos SO 7403. 'Bridge or causeway across muddy ground'. *Heslinbruge* 1086, *Sli- Slymbrug(g)(e)* c.1153–1487, *-brig(ge) -bryg* 1166–1505, *-bridge* 1535–1727. OE **slīm** + **brycg**. Gl ii.247, L 66.

SLINDON Possibly 'the hill with a slope'. OE ***slinu** or ***slind** 'a slope' + **dūn**.

(1) ~ Staffs SJ 8232. *Slindone* 1086, *-don* 1242, *Sclindon* 1199. SLINFOLD TQ 1131 and SLYNE Lancs SD 4765. DEPN, Horovitz.

(2) ~ WSusx SU 9608. *Eslindone* 1086, *Slindon(e)* from 1188. Sx 96, PNE ii.129, SS 74.

SLINFOLD WSusx TQ 1131. 'The hill-side fold'. *Stindefald* (sic for *Slinde-*) 1165, *Slindefold* 1225–1409, *Slynfold* 1482. OE ***slind** + **falod**. Slinfold lies on the side of a hill. Sx 159, PNE ii.129, SS 79.

SLINGSBY NYorks SE 6974. 'Sleng's farm or village'. *Selunges- Eslingesbi* 1086, *Slengesbi -by* 1164×72–1403, *Sli- Slyngesby* 1202–1578. ON pers.n. *Slengr*, genitive sing. *Slengs*, + **bȳ**. YN 48, SSNY 37.

SLIP END Beds TL 0818. No early forms.

SLIPTON Northants SP 9579. 'The settlement at the slippery place'. *Slipton(e)* from 1086 with variant *Slyp-*. OE ***slip** + **tūn**. Slipton is situated on ground sloping down to a tributary stream of the river Nene. Nth 187.

SLOLEY Norf TG 2924. 'Wood or clearing where sloes grow'. *S(l)aleia* 1086, *Slalee -leg(h)' -leye* 1207–75, *Sloleg(e) -leye -leghe* 1251–1482. OE **slā** + **lēah**. Nf ii.191.

SLOOTHBY Lincs TF 4970. Partly uncertain. *Slodebi, Lodeby* 1086, *Slothebi* 1199, 1208, *Slodebi* 1200. Unidentified element or name + ON **bȳ** 'a farm, a village'. The specific might be ON **slóth** 'a track, a trail', cf. ModSwed dial. *slō* 'a winter track over water', or its derivative **slóthi** 'something trailed along' used as a by-name *Slóthi* 'lazy one' of either a person or a slow moving stream in the fens, cf. ModNorw *slōde* 'tall, gaunt fellow'. The form without initial *S-* is AN. DEPN, SSNEM 69.

SLOUGH Berks SU 9780. 'The slough, the muddy place'. *Slo* 1195, *le Slowe* 1437, *le Slough* 1443. OE **slōh**. Bk 243, TC 172.

SLYNE Lancs SD 4765. 'The slope'. *Sline* 1086–1332 etc. with variant *Slyne* from [1200×10]1412, *Slen(e)* c.1250–1332, *Sleen* 13th. OE ***slinu**. Near Slyne Hall SD 4866 is a small prominent hill. La 185 gives pr [slain], PNE ii.129, Jnl 17.103.

SMALLBRIDGE GMan SD 9115. *Smallbridge* 1843 OS.

SMALLBURGH Norf TG 3324. 'Hill by the river Smale'. *Smal(e)berga* 1086, *Smal(e)berg(e) -bergh* [1107×7]13th, 1177–1535, *-ber(e)we* 1221–1340, *-burghe* 1257, *Smalburgh*

-borough 1535, *Sma'boro* 1610. R.n. *Smale* + OE **beorg**. *Smale* is the old name of the river ANT TG 3617, meaning 'the narrow stream', OE **smæl**, definite form **smale**, + **ēa**. Nf ii.192.

SMALL DOLE WSusx TQ 2112. No early forms. Probably originally a f.n., 'the narrow allotment'. ModE **small** + **dole**.

SMALLEY Derby SK 4044. 'Narrow wood or clearing'. *(in) Smælleage* 1009, *Smalleg(a) -legh(e) -ley(e)* c.1200–1231 etc. OE **smæl** + **lēah**. Db 505.

SMALLFIELD Surrey TQ 3143. 'The narrow piece of open land'. *Smalefeld* 1522, *Smalfell* 1610. ME **smale** + **feld**. Sr 287.

SMALL HYTHE Kent TQ 8930. 'The narrow landing-place'. *la Smalelide al. Smelelide* 1252, *Smalhede* 1289, *-hide* 1290, *-hithe* 1377, 1380. OE **smæl**, definite form **smale**, + **hȳth**, Kentish **hēth**. The use of the French definite article in 1252 suggests that the name was still felt to be an appellative. PNK 360.

SMALLRIDGE Devon ST 3000. 'Narrow ridge'. *Smarige* 1086, *Esma(u)rige* Exon, *Smalrigg(e)* 1200, *-rugge* 1279. OE **smæl** + **hrycg**. The village occupies a narrow ridge between parallel streams. D 634.

SMALLWAYS NYorks NZ 1111. 'The narrow fords'. *Smalwathes* 1336. ME **smal** (OE *smæl*) + **wath** (ON *vath*), pl. **wathes**. Cf. WASS SE 5579. YN 292.

SMARDALE Cumbr NY 7308. Partially uncertain. *Smer(e)dal(e)* 1190–1657, *Smardale -dall* from 1422. OE **smeoru** or ON **smjór** 'grease, fat, butter' or ON **smǽra**, Lakeland dial. *smere*, 'clover', + OE **dæl**, ON **dalr**. Either, therefore, 'valley with rich (butter-producing) pasture' or 'clover valley'. We ii.25, SSNNW 162, L 96.

SMARDEN Kent TQ 8842. 'Rich pasture'. *Smeredaenne* c.1100, *-den(n)(e)* 1229–57. OE **smeoru** + **denn**. OE **smeoru** 'fat, grease' may here refer to wet, greasy land, since Smarden is situated low in what was described as watery unpleasant clay soil; but it could refer to rich pasture where butter is made. PNK 399.

SMEATHARPE Devon ST 1910. *Smith Harp* 1765, 1809 OS. *Harp* is probably used of a sieve-like trap used in a fishery, cf. *The Harp, Harp Close*. C 330, PNE i.240 and OED s.v. *harp* 5, D 651.

SMEATON 'The settlement or estate of the smiths'. OE **smith**, genitive pl. **smitha**, + **tūn**.

(1) Great ~ NYorks NZ 3404. ModE *Great* 1541 + p.n. *(on) smiþatune* 966×72 S 1660, *Smidetune -ton, Smet(t)on* 1086, *Smithetuna -ton* 1088–1231, *Smetheton* [1157]15th–[1366]18th, *Smeton* 1541. 'Great' for distinction from Little SMEATON NZ 3403. YN 281, 211.

(2) Kirk ~ NYorks SE 5116. 'S with the church'. ME prefix *Kyrke, Kirk* from 1311 (ON *kirkja*), + p.n. *Smedetone* 1086, *Smyde- Smidetona* 12th cent., *Smithetun, Smi- Smytheton(e)* 12th–1303, *Smetheton* 1276–1480, *Smeton* 1409–1535. Forms with *Smethe-* arise by ME open syllable lengthening of *ĭ* > *ē*. YW ii.51, 52.

(3) Little ~ NYorks NZ 3403. *litle Smithetune* 1088, *Litill Smeton* 1530. OE **lȳtel** + p.n. *Smidetun(e), Smitune* 1086, as in Great SMEATON NZ 3404. YN 281, 211.

(4) Little ~ NYorks SE 5126. Suffix *Minori* 12th, prefix *parva* 13th, *Lit(t)le* from 1315, Latin **minor** 'lesser', **parva**, ME **litel**, + p.n. Smeaton as in Kirk SMEATON SE 5116. YW ii.51, 52.

SMEETH Kent TR 0379. 'Smithy'. *Smiða* 1018, *Smeth(e)* 1245–1316. OE **smiththe**. K 422.

SMEETON WESTERBY Leic SP 6792. 'The *westerby* or west part of the village of Smeeton'. *Smetheton Westerby* 1316, 1327, *Smeton Westorby -ur-* 1402–21, *Smeeton Westerby* 1615. P.n. *Smi(te)tone, Esmeditone* 1086, *Smethe- Smy- Smitheton(')* 1203–1414, *Smeton(')* 1311–1516, *Smeaton, Smeeton* 1604–15, the 'settlement or estate of the smiths', OE **smith**, genitive pl. **smeotha** (whence the spelling *Smethe-*), + **tūn**, + p.n. *Westerby* from c.1130, the 'west village or farm', ON **vestr** + **bȳ**, probably a secondary settlement from Smeeton.

SMETHWICK WMids SP 0288. 'The smiths' dwelling or trading place'. *Smedeuuich* 1086, *Smethewic* 1221. OE **smith**, genitive pl. ***smeotha**, + **wīċ**. DEPN.

SMIDDY SHAW RESERVOIR Durham NZ 0446. P.n. Smiddy Shaw + ModE **reservoir**. There are no early forms recorded for Smiddy Shaw, but it appears to be N dial. **smiddy** 'a blacksmith's shop' + **shaw** 'a copse'.

SMISBY Derby SK 3419. 'The smith's village or farm'. *Smidesbi* 1086, *Smiðesbi* 1166, *Smi- Smythesby -bi -is-* c.1162–1758 *Smithsby, Smisby* from 1676. OE **smith**, genitive sing. **smithes**, or ON **smithr**, genitive sing. **smiths**, + **bȳ**. Db 658, SSNEM 69.

SMITH SOUND Scilly SV 8706. *The Smeth sownde, The Smythes sownd* c.1540. From the Smith islands, Great and Little Smith SV 8609, *The Smyth* c.1585, *Great, Little Smith* 1689, + ModE **sound** 'a strait, a sailable channel'. The origin of the name of Great and Little Smith is unknown. PNCo 157.

SMITHEY FEN Cambs TL 4570. Probably '*Smethey* marsh' rather than 'smithy fen'. *Smithyfen* 1343, *Smy- Smetheffen* 1483, 1493, *Smethy Fen* 1604. OE p.n. **Smēth-ēġ* 'smooth island' < OE **smēthe** + **ēġ**, + **fenn**, or even simply **smēthe** + **fenn** 'smooth fen'. Formally the specific could, however, be OE **smiththe**, ME **smythy** 'a smithy'. Ca 150.

SMITHFIELD Cumbr NY 4465. *Smithfield* 1866 OS.

SMITHINCOTT Devon ST 0611. 'The smiths' cottages'. *Smithenecota* 1223, *Smythenecote* 1306. OE **smith**, ME genitive pl. **smithene**, + **cot**, pl. **cotu**. D 538.

SNAILBEACH Shrops SJ 3702. 'Snail Valley', ModE **snail** + **beach** (OE bæċe) *Sneilbach* 1799. Contrasts with nearby Wagbeach SJ 3602, 'the stream-valley with shaking ground', *Waggebeche* 1321, OE ***wacga** + **bæċe**. Sa ii.25–6.

SNAILWELL Cambs TL 6467. 'Snail stream'. *Sneillewelle* [1042×66]14th S 1051, 1170, 1387, *Snelleuuelle* 1086, *Sneiluuelle -well(e)* [c.1050]12th–1354, *Snail(l)(e)well(e) -ay- -ey-* 1203–1552, *Sneil(l)eswelle -ey- -ay- -ai-* 1214–1436. OE **snæġel** + **wella**. The sense is either sluggish stream, stream that moves at a snail's pace, or stream where snails abound. Ca 196.

SNAINTON NYorks SE 9282. Partly uncertain. *Snechintun(e) -ton(e)* 1086, *-int'* 1166, *Sneington -y-, Snainton -y-* 13th cent. The forms are hardly compatible with an OE **Snocing-tūn* 'the settlement or estate called or at **Snocing*' in which the specific might be an OE pers.n. **Snoc* + **ing**[2], the 'place called after Snoc' as suggested by DEPN unless the formation is very early showing the results of i-mutation. Alternative suggestions are a topographical term related to OE **snōca** 'a point, a projection', or a derivative of an OE ***snæċċe**, ***sneċċe** 'a trap' (ME **snecche* Che iii.146) or ME **snekke** 'a catch, a latch' and in 19th cent. Craven dial. 'a small piece or tongue of land'. Snainton lies at the foot of high ground cut by narrow valleys, but there is no obviously projecting feature at this point. YN 97, OED s.v. *sneck* 1 and 2.b.

SNAITH Humbs SE 6422. 'The detached piece of land'. *Snaith* from 1086×85 with variants *Snayth, Snaid, Sneid, Sneith, Esneid, Esnoid -t* 1086. ON **sneith** perhaps replacing OE *snæd*. The significance of this name has usually been explained as referring to land cut off from its parish by the r. Aire, the original course of which may have differed from that of today. Snaith, however, was a peculiar, i.e. a parish exempt from the jurisdiction of the bishop, and it is tempting to associate the name with this status. YW ii.6422.

SNAIZEHOLME NYorks SD 8386. '(The settlement) among the twigs'. *Snaysum* 1280–5, *Snaysome* 1423. ON **sneis**, dative pl. **sneisum**. Cf. RYSOME TA 3623. YN 267.

SNAPE NYorks SE 2684. 'Poor pasture'. *Snape* from 1270. Icel **snap**. But OE ***snæp** 'a boggy piece of land' is also possible and topographically apt. YN 229, PNE ii.132.

SNAPE Suff TM 3958. 'Poor pasture'. *Snapes* 1086–1327 with variants *-ys -is*, *Snape* from 1275, *Snape bridge* 1610. ON **snap**, lOE ***snæp**. DEPN, PNE ii.132.

SNARESTONE Leic SK 3409. Probably 'Snaroc's farm or village'. *Snarchetone* (sic) 1086, *Snarkeston(e) -is-* 1188–1410, *Snar(e)ston* 1445–1576, *Snarson* 1502, 1523. OE pers.n. **Snaroc*, genitive sing. **Snar(o)ces*, + **tūn**. Lei 544, SSNEM 280.

SNARFORD Lincs TF 0582. 'Snǫrtr's ford'. *Sn- Svardes- Sner(t)eforde* 1086, *Snarteford(e)* 1115–1231, *Narte(s)ford(e)* 12th, *Snardeford* 1265. ON pers.n. *Snǫrtr*, genitive sing. *Snartar*, secondary genitive sing. *Snartes*, + OE **ford**. Forms without initial *S-* are AN. DEPN, SSNEM 221, L 69.

SNARGATE Kent TQ 9828. 'Gate(s) with a snare'. *Snergathe* c.1197, *Snargate* from c.1210 with varants *Sner(e)-* 1240–70. OE **sneare** + **ġeat**, pl. **gatu**. The reference is probably to fish traps. PNK 478.

SNAVE Kent TR 0129. 'The narrow strips or spits of land'. *Snaues* 1218–78, *Snathes* 1202, *Snaue* from 1240. OE ***snæf**, pl. ***snafas**, cognate with Norwegian *snav* 'a spit of land'. Snave stands on an old creek ridge. PNK 478, Cullen.

SNEACHILL H&W SO 9053 → SNETTISHAM Norf TF 6834.

SNEATON NYorks NZ 8907. 'Settlement on the slope'. *Snetune -ton* 1086–1665, *Snetton* 12th cent. OE **snǣd**+ **tūn**. 12th cent. spellings suggest that the expected shortening of *ǣ* (raised to *ē* before a dental consonant) before the cluster *-dt-* did occur. This etymology is preferable to the earlier one, the very rare ON pers.n. *Snjó, Snǽr*, + **tūn**. OE **snǣd** (related to *snīthan* 'to cut') has been variably interpreted as 'a detached piece of ground' or 'a piece of woodland'. Its Scand cognates, however, refer to something cut off slantwise, e.g. Mod Norw *sneid*, Dan dial *sned* 'a slope'. This sense would fit the topography of Sneaton admirably. YN 118, SSNY 259, PNE ii.133 s.v. *sneith*.

SNEATONTHORPE NYorks NZ 9006. 'Sneaton's dependent farm or village'. Short for *Sneton et Thorpe* [1349]15th. P.n. SNEATON + ON **thorp**. YN 119.

SNEINTON Notts SK 5839. 'The estate called after Snot(a)'. *Notintone* 1086, *-ington* 1263, *Snotinton(e)* 1165–97, *Snottinton* 1174, *Sneinton* from 1194, *Snainton -ay-* 1227–1486, *Snointun -ton* 1233. OE pers.n. *Snota* + **ing**[4] or genitive sing. *Snotan* + **tūn**. Initial *S-* was lost in the DB form through AN influence as in NOTTINGHAM SK 5741. Nt 174, ING 149.

SNELLAND Lincs TF 0780. 'Snell or Snjallr's grove'. *Esnelent, Sneleslunt* 1086, *Snelleslund* c.1115, 1202. Pers.n. OE *Snell*, genitive sing. *Snelles*, or ON *Snjallr*, genitive sing. *Snjalls*, + **lundr**. DEPN, SSNEM 161, L 207.

SNELSTON Derby SK 1543. 'Snell's farm or village'. *Snel(l)estune -tona -ton(e) -is-* 1086–1472, *Snelston(e)* from 1323, *Snelson* 16th cent. OE pers.n. *Snell*, genitive sing. *Snelles*, + **tūn**. Db 602, SSNEM 380.

SNETTERTON Norf TL 9991. Partly uncertain. *Snetretuna* 1086, *Sneterton* 1192 ff, *Sniterton(a)* 1195, 1254. Usually explained as 'Snytra's farm or village', OE pers.n. **Snytra* + **tūn**. But the pers.n. is not independently recorded and this may be an unidentified element occurring in SNITTERBY Lincs SK 9894 and other names there cited. DEPN

SNETTISHAM Norf TF 6834. Probaby 'Sneti's homestead'. *Snet(t)esham* 1086, *Snetesham* 1161–96. OE pers.n. **Sneti*, genitive sing. **Snetes*, + **hām**. The exact form of the pers.n. (which, or a variant of which, may also occur in NEATISHEAD TG 3421 and Sneachill H&W SO 9053, *(to) fnœtes wyllan* (for *snœtes-*) [n.d.]11th S 1601), is unknown. An alternative form **Snœt* is possible. DEPN.

SNILESWORTH MOOR NYorks SE 5296. P.n. Snilesworth + ModE **moor**. Snilesworth, *Snigleswath* 1150×70, [1243]15th, *Snygheleswath* [1186×9]copy, *Snileswath -y-* 1230–1376, *Snailesworthe* 1575, is 'Snigill's ford', ON pers.n. **Snigill*, genitive sing. **Snigils*, + **vath**. **Snigill*, by origin a by-name, is not on record. The specific might, therefore, be the ON appellative **snigill** 'a snail' from which the pers.n. would have to be derived. YN 204, SPN 256, L 82.

SNITTER Northum NU 0203. Uncertain. *Snitere* 1175, *Snittera* 1176, *Snitter* 1242 BF. Possibly the 'snipe shieling', OE **snīte** + ON **ǽrgi**. However, **ǽrgi** has not hitherto been identified as a p.n. element in Northum. DEPN suggests connnection with ME *sniteren* used of snow in *Sir Gawain and the Green Knight*,

the snawe snitered ful snart that snayped the wylde

'the snow came bitterly down pinching the wild animals cruelly', and dial. *snitter* 'a biting blast'. Snitter is situated on a ridge exposed to the NW wind; nearby is a farm called Snitter Windyside NU 0104. NbDu 183–4.

SNITTERBY Lincs SK 9894. Partly uncertain. *Esnetre- Snetrebi* 1086, *Snitrebi* c.1115, *Sniterbi -by* 1194–1212. Usually said to be 'Snytra's village or farm', OE pers.n. **Snytra* + **bȳ**. **Snytra* would be related to OE *snytre, snotor* 'wise' but is not independently recorded. It may be an unidentified element occurring also in SNITTERTON Derby SK 2760, *Snitterley*, the old name of BLAKENEY Norf TG 0243, SNETTERTON Norf TL 9991 and SNITTER Northum NU 0203. DEPN, Cameron 1998.

SNITTERFIELD Warw SP 2160. 'The open land of the snipe'. *Snitefeld* 1086, *Snitenefelt -feld* 1123–1267, *Snitenfel -feud* 1154×89–1282, *Sniterfelde* 1242, 1299, *Snit(er)feild* 17th. OE **snīte**, genitive pl. **snītena**, + **feld**. Wa 223, L 241, 244.

SNITTERTON Derby SK 2760. Partly uncertain. *Sinitretone* 1086, *Sniterton(e)* 1180–13th, *Snitterton* from late 12th. Usually explained as 'Snytra's farm or village', OE pers.n. **Snytra* + **tūn**. The pers.n. is not independently recorded and may be an unidentified element occurring in SNITTERBY Lincs SK 9894 and other examples there cited. Db 412.

SNITTON Shrops SO 5575. Partly uncertain. *Snitton* 1203×4, *Sni- Snytton* 14th cent. The generic is OE **tūn**, the specific perhaps the bird-n. **snīte** 'a snipe' or possibly OE **snid** 'a saw, a cut, a piece cut off', referring to a place where saws were made or to a topographical feature. Gelling.

SNODHILL H&W SO 3240. 'Snowy hill'. *Snauthill* 1196, *Snod(e)hull(e)* c.1230–1431, *Snowdell* 1397, *Snothill* 1540. OE ***snāwede** + **hyll**. S lies on the NE side of the Cefn Hill in the Black Mountains. He 164.

SNODLAND Kent TQ 7061. 'Newly cultivated land called after Snodd'. *Snodding land* [838]10th S 280, *Snodingcland* [973×87]12th S 1511, *Snodin(g)landes* [10th]12th S 1457, *(æt) Snoddinglande* [995×1005]12th S 1456, *Snodiland(e) -lond(e)* [c.975]12th B 1321–2, 1218–53×4, *Esnoiland* 1086, *Snodland* from 1242×3. OE pers.n. *Snodd(a)* + **ing**[4] + **land**. PNK 150, KPN 178, L 248–9.

SNOOK POINT Northum NU 2426. *Snook Point* 1866 OS. Dial. **snook** (OE **snōc(a)*) 'a point, a projection, a promontory' used of coastal features, + ModE **point**. Cf. BLYTH NZ 3181.

Great SNORING Norf TF 9434. *Great Snoring* 1838 OS. ModE adj. **great** + p.n. *(E)snaringa, Snaringes* 1086, *Snar(r)inges* 1166–1206, *Naringes* 1242×3, *Snoring* from 1314, 'the Snaringas', OE folk-n. **Snāringas* of uncertain origin. Long *ā* is required in this name to account for the subsequent change *a* > *ō*. A pers.n. **Snāra* has been proposed from an adj. **snear* 'swift' < Gmc **snarha-* from which derive also ON *snarr* 'brisk' and the associated ON pers.ns. *Snari, Snara*, OSwed *Snare*, OG *Snaring*, and the DB tenant *Snaring*. It is often held that the vowels of OE words ending in *lh* and *rh* underwent compensatory lengthening when *-h-* was lost to produce Anglian paradigms like *halh/hālas, Walh/Wālas, marh/māras*. Theoretically, therefore, a pers.n. **Snarha* could give OE *Snāra* and a folk-n. **Snāringas* 'the people called after Snara' < pers.n. **Snāra* + **ingas**. Unfortunately, however, there is no trace of the subsequent ME change *a* > *o* which presupposes lengthened OE *ā* in any of the words in which these conditions obtain. For the same reason the alternative suggestion of a lost r.n. **Snāre* 'the swift one' from the same adjective + **ingas**, 'the people who live by the Snare', is problematical. DEPN, ING 61, PNE ii.243, Campbell §241, Bülbring §529.

Little SNORING Norf TF 9532. *Parva Naringes* 1242×3, *Parva Snoring* 1343. ModE adj. **little**, Latin **parva** + p.n. Snoring, *Esnaringa, Snaringes* 1086, as in Great SNORING TF 9434. DEPN, ING 61.

SNOWHOPE HILL Durham NY 9434. P.n. Snowhope + ModE **hill**. Cf. Snowhope Close NY 9536, *Snawhopclos* 1382, *Snawehopclose, Snawpclose* 1418, 'enclosure in Snowhope, the side-valley where snow lies long'. OE **snāw** + **hop**. NbDu 185.

SNOWSHILL Glos SP 0933. 'Snow hill'. *Snawesille* 1086, *-hull(a)* 1171–1248, *Snoweshull(a)* 1175–15th, *-hill* 1287, 1535, *Snowshill* from 1535. OE **snāw**, genitive sing. **snāwes**, + **hyll**. The village is situated on a hill where the snow lies longest. Gl ii.21 gives pr ['snouzəl], L 170.

River SOAR Notts SK 5123. A pre-English r.n. meaning simply 'river, flow'. *Sora* [1147]12th–c.1350, *Sore* 1247–1683, *Soore* 1422×71, *Soare* 1683. PrW **Sār* < **Sarā*, identical with the r.ns. Sor, a tributary of the Usk, Saar, Germany, *Saravus* 4th, *Sara* 6th, Serre, France, a tributary of the Oise, *Sera* 867, Cère, France, a tributary of the Dordogne, Saire, France, a tributary of the Manche, *Sara* 1171, Sar, Spain, a tributary of the Ulla, *Sars*, the Zorn, Alsace, France, *Sorna* 724, etc., all derived from the IE root **ser-/*sor-* 'to flow' as in Latin *serum*, Greek *ὀρός* 'whey'. PrW **Sār* was borrowed into OE as **Sār* whence, by normal retraction of OE *ā*, ME *Sōr*. Nt 8, RN 375, BzN 1957.254, Krahe 1964.40, Berger 230.

SOBERTON Hants SU 6116. 'South grange'. *Svdbertvne* 1086, *Suberton(e)* 1167–1316, *Suthberton* 1280, *Soberton* from 1291. OE **sūth** + **beretūn**. S of Meonstoke and Corhampton; later (1205) a grange of Beaulieu abbey. Ha 151, Gover 1958.50 which gives pr [sʌbətən].

SOBERTON HEATH Hants SU 6014. *Soberton Heath* 1810 OS. P.n. SOBERTON SU 6114 + ModE **heath**.

SOCKBURN Durham NZ 3407. 'Socca's fortified place'. *æt Soccabyrig* 1121 ASC(E) under year 780, *Socceburg* [c.1040]12th, [1072×83]15th, c.1107, *So(c)keburn(e)* 1235–1313, *Sokburn(')* 1283–1520, *Sockburn* from 1431. OE pers.n. **Socca* + **burh**, dative sing. **byriġ**. For confusion of **burh** and **burn** in weak stress, cf. NEWBURN T&W NZ 1965. The Laud Chronicle locates the consecration of Hygebald as bishop of Lindisfarne at Sockburn in 780. NbDu 185.

Chipping SODBURY Avon ST 7382. 'Market Sodbury'. *Cheping -yng Sob(b)yr(i) -biri -bire -bury* 1269–1522. ME **cheping** (OE *ċēping*) + p.n. Sodbury as in Old SODBURY ST 5781. A market was granted here in 1218. Gl iii.51.

Old SODBURY Avon ST 7581. *Old(e)sobbur(y) -bir(e) -byry* 1385–1522. Earlier simply *Soppanbyr(i)g* [c.903]11th S 1446, *Sopeberia -bir(i)* 1086, [1100×35]1496, *Sopbire -byr' -buria* 1204–87, *Sobbur(y) -bir(e) -byry* 1222–1492, *Sodbury* from 1287, 'Soppa's fortified place', OE pers.n. **Soppa*, genitive sing. **Soppan*, + **byriġ**, dative sing. of **burh**. The reference is to Little Sodbury hill-fort at ST 7583. Gl iii.53.

SOHAM Cambs TL 5973. 'Marsh or lake homestead'. *(æt) Sægham* 1000×2 S 1486, *Sigham* 1198, *Saham* 1086–1463, *Seham* 1170–1327, *Soham* from 1294, *Some* 1427–1628. OE ***sǣġe** early confused with **sǣ**, ***sā**, + **hām**. The reference is to Soham Mere TL 5773. OE ***sǣġe** occurs again in *Medeseye* c.1200, the early name for Pymmes Brook Middlesex. Ca 196 gives former pr [soum], RN 284–5, Mx 5, PNE ii.93.

Earl SOHAM Suff TM 2362. 'The earl's S'. *Earl Saham* 1235, *Erlesoham* 1610, *Saham Comitis* 1254. ME **erl**, Latin **comes**, genitive sing. **comitis**, + p.n. *Saham* 11th–1195 including 1086, *Soham* from 1316, 'the lake homestead', OE **sā** + **hām**. The place was held by the earl of Norfolk. DEPN.

Monk SOHAM Suff TM 2165. 'S held by the monks' sc. of Bury St Edmunds. *Monks Saham* 1235, *Munekesoham* 1610, *Saham Monachorum* 1254. ME **monk**, genitive pl. **monkes**, Latin **monachus**, genitive pl. **monachorum**, + p.n. Soham as in Earl SOHAM TM 2362. DEPN.

SOHAM MERE Cambs TL 5773. *Soham Mere* 1836 OS. P.n. SOHAM TL 5973 + ModE **mere**. The mere is referred to as *mariscum de Ely*, c.1195, while Soham itself is regularly referred to as *Saham quæ est ad stagnum, Seham quae est villa juxta stagnum* 'Soham by the lake'. William of Malmesbury says Soham was a hill beside a lake once dangerous to boats wanting to reach Ely, but in his day a road had been made across the marsh so that it could be crossed on foot. Ca 196.

SOLDON CROSS Devon SS 3210. 'Soldon crossroads'. *Soldon Cross* 1809 OS. P.n. Soldon + ModE **cross**. Soldon, *Soldon* 1313 is probably 'mud hill' or 'hill with a muddy pond', OE **sol** + **dūn**. D 149, PNE ii.134.

SOLDRIDGE Hants SU 6535. 'Wallowing-place ridge'. *Solrigge* 1233–1327. OE **sol** 'mud, slough, wallowing-place for animals, dirty pond' + **hrycg**. Ha 151, Gover 1958.79, PNE ii.134.

SOLE STREET 'Sole hamlet'. P.n. *Sole* < OE **sol** 'a muddy pool', + Mod dial. **street**. OE **sol** appears as a simplex name in Soles Court TR 2550, *Soles* 1086.

(1) ~ ~ Kent TR 0949. *Sole Street* 1798. Earlier simply *on ðæt sol* 824 S 1434, *atte Sole* 1346. KPN 150, Cullen.

(2) ~ ~ Kent TQ 6567. *Solystrete* 1448, *Solstre(a)te* 1453, 1572. Earlier simply *atte Sole* 1327. K 112.

(3) ~ ~ Kent TR 0258. *Sole Street* 1819 OS. Earlier simply *(de) Solis* 1211×12, *(de) Soles* 1261×2, *at(t)e Sole* 1327, 1381, *La Sole* 1351. PNK 306.

The SOLENT Hants SZ 5098. Unexplained. *Soluente* c.731 BHE, [9th]c.1000 OEBede with variants *Sol(w)ente, (utt on) solentan* [948]12th S 532, *(of) Solentan* [980]12th S 836, *le Soland* 1395. OE **Sol(w)ente*, unexplained element *sol* + suffix *-wente* as in the Old European r.ns. Allan Water Borders NT 4707, *Alwente* 1153×65, Allen Northum NY 7961, *Alwent* 1275, Alwin Northum NT 9206 *Alewent* 1200, and Alwent Beck Durham NZ 1319, *Alewent* 1235 are all *-nt* extensions of Old European r.n. **Alaua*, itself a *-uā* formation on the IE root **el-/*ol-* 'to flow'. But no satisfactory IE root **sol-* is known. Speculative attemps to associate **sol-* with a non-IE language such as Phoenician can neither be accepted nor rejected in the present state of knowledge. Ha 151, Gover 1958.9, DEPN, ScotPN 187, TT 1ff.

SOLIHULL WMids SP 1579. 'The muddy hill'. *Solihull* from 12th, *Sul(l)ihull -y-* 1242–1514, *Sylhyl* 1340, *Sylill* 1650, *Silhill* 1708. OE ***syliġ** varying with ***soliġ** or ***suliġ** + **hyll**. Wa 67 gives former pr [silil].

SOLLOM Lancs SD 4518. 'Enclosure in the mire'. *Solayn-Salaynpul(l)* [c.1200]1268, *Solame* 1372, *Solem hey* 1451, *Sullam* 1539, *Solom* 1554. OE, ON **sol** 'mire, bog' + ON ***heġn** 'enclosure'. *Solaynpull* is a 'pool' or brook that falls into the Douglas; the modern form of the name is probably due to assimilation of *n* > *m* before the *p* of **pull**. La 138, SSNNW 162.

SOLWAY FIRTH Cumbr NX 8536–NY 0959. *Solway firth* 1695. P.n. Solway Dumfries NY 3165 + Scots **firth** 'estuary of a river'. Solway, *Sulewaht* c.1275, *Sol- Sulwath(e)* 1292–1461, *Sulway* 1429, *Sulweye* 1565 (all forms either *aqua de* or *water of* ~), is 'pillar ford', ON **súla** + **vath**, probably referring to the Lochmaben Stone, a large ice-borne granite boulder which marked the Scottish end of the ford across the firth and was a recognized meeting-place for the settling of border disputes. Cu 39, SSNNW 163, L 82.

SOLWAY MOSS Cumbr NY 3469. *Solome Mosse* 1543, 1552, *Sollome Mose* 1599, *Solom Moss* 1777, *Solway Moss* 1771. A morass NW of the river Esk as it bends W to enter Solway Firth from which the modern form of the name is taken. Originally, however, lost p.n. Solum + OE **mos**. Solum, *Solum*

1246, 1339, *Salom* 1280, *Solom* 1282, is OE **solum**, dative pl. of **sol** 'muddy place'. It meant '(place at) the mudflats'. Cu 40, L 58.

King's SOMBORNE Hants SU 3631. 'S held by the king'. *Kyngessumburne* 1265. lOE **cing**, genitive sing. **cinges**, for distinction from Little Somborne SU 3832, *Parva Sunburn'* 1204 and Up SOMBORNE SU 3932, + p.n. *Sūburne* 1086, *Sumburna -e -borne* 1155–1235, transferred from the stream-n. *(to) swinburnan* 'pig stream' *[909]12th S 381, OE **swīn** + **burna**. King's and Little S are distinguished in 1086 when the former is already in royal hands. Ha 104, Gover 1958.186–7, DEPN.

Up SOMBORNE Hants SU 3932. 'Upper S' i.e. higher up the valley. *Opsūburna* 1167, *Upsumburne* 1269. OE **upp** + p.n. Somborne as in King's SOMBORNE SU 3631. Ha 104, Gover 1958.186.

SOMERBY Leic SK 7710. Either 'Sumarlithi's village' or 'village of the summer travellers'. *Sumerlide- Sūmerdebie, Sumerdeberie* 1086, *Su- Somerdebi -by* 1169–1323, *Somerby* from 1313. Either ON pers.n. *Sumarlithi*, genitive sing. *Sumarlitha*, or ON **sumarlithar** 'summer travellers, sailors, Vikings', genitive pl. **sumarlitha**, + **bȳ**. Alternatively the specific might be OE ***sumor** + **hlith**, genitive pl. ***sumor-hleotha**, and so the 'village of the concave hill-slopes used in summer' referring to summer pastures. Somerby is situated between hills of over 600ft. Lei 186, SSNEM 70.

Old SOMERBY Lincs SK 9633. ModE adj. **old** + p.n. *Sū- Sūmerdebi* 1086, *Sumerdebi -by* 1086–1306, *Somer(e)deby* 1242–1412, *Sumerbi* 1185, *Somerby* from 1296. From the late 12th cent. the name was understood to be the 'summer village, the village with pasture used for summer grazing' but it is unclear whether its original form referred to summer grazing, ON **Sumar-hlíthar-bȳ*, the 'village of the summer slopes' or to possession, ON *Sumarlitha-bȳ*, the 'summer traveller's village', ON **sumarlithi**, genitive sing. **sumarlitha**, probably used as a by-name. There are hill slopes at Somerby. 'Old' for distinction from New Somerby which was formed in 1894 out of the W end of Somerby. Perrott 507, 466, SSNEM 70.

SOMERCOTES Derby SK 4253. 'Huts used in summer'. *Somercote* 1304–18, *-cotes* from 1330. OE **sumor** + **cot**, pl. **cotu**. On high ground S of Alfreton. Db 188.

North SOMERCOTES Lincs TF 4296. *Nort Sumercotes* 1281. ModE **north** + p.n. *Summercotes* 1086, late 12th, *Sumercotis* c.1115, the 'huts used in summer', OE **sumor** + **cot**, secondary pl. **cotes**. 'North' for distinction from South SOMERCOTES TF 4193. DEPN.

South SOMERCOTES Lincs TF 4193. *Suth Sumercotes* 1294. ModE **south** + p.n. Somercotes as in North SOMERCOTES TF 4296. The two Somercotes were daughter settlements from Cockerington TF 3790 on land reclaimed from the sea. Jnl 7.56.

SOMERFORD 'Ford used or usable in summer'. OE **sumor** + **ford**. L 69.

(1) ~ KEYNES Glos SU 0295. 'S held by the de Kaynes family'. *Somerford C- Kaynes* 1289–1547. Earlier simply *Sumerford* [685]13th S 1169, *[1065]13th S 1038, 1211, 1242, *Somerford* 1281. A ford across the r.Thames. Gl i.xi, 83.

(2) Great ~ Wilts ST 9682. *Brode Somerford* 1409, *Somerford Magna* 1588. ModE adj. **great**, ME **brode**, Latin **magna**, + p.n. *Sumerford* [937]12th S 436, *Somerford* [956]13th S 629, *So- Sumreford* 1086. 'Great' for distinction from Little SOMERFORD ST 9784. Also known as *Sumerford Mautravers* from the tenure of the manor by the family of Walter Maltravers 1196. Wlt 73, L 69.

(3) Little ~ Wilts ST 9784. *Little Somerford al. Somerford Maudítt* 1681. ModE adj. **little** + p.n. Somerford as in Great SOMERFORD ST 9682. The manor was held by John Maudith in 1275, cf. *Sumerford Mauduyt* 1268. Wlt 73.

SOMERLEY WSusx SZ 8198. 'The summer clearing or pasture'. *Sum(m)erlege* 1086, *Somerle* 1248. OE **sumor** + **lēah**. A pasture used in summer. Sx 89, SS 66.

SOMERLEYTON Suff TM 4796. 'Sumarlithi's estate'. *Sumerlede(s)tuna* 1086, *Sumerletun -ton(e)* c.1185–1327, *Sumerleyton* from 1316[†], *Somerley towne* 1610. ON pers.n. *Sumarlithi*, genitive sing. *Sumarlitha*, + OE **tūn**. DEPN, Baron.

SOMERSAL HERBERT Derby SK 1335. 'S belonging to the (Fitz)herbert family'. *Somersal(e) Herbert, ~ Herberd* c.1300–1472. Also known as *Kirke Somersale* 'S with the church' for distinction from Potter Somersal SK 1435, *alia Summersale* 1086, *Nethersomersale* 1318, *Potter Somersale* 1423, and Hill Somersal SK 1434 *Oversomersale* 1318, *Hyll Somersale* 1507. Somersal, *Summersale* 1086, *Sumer(e)shal(a) -al'* 1179–1267, is either 'summer nook of land, nook of land used in summer' or 'Sumor's nook of land', OE **sumor** or pers.n. **Sumor*, genitive sing. **sumores**, **Sumores*, + **halh**. Db 604, 610, L 106.

SOMERSBY Lincs TF 3472. Partly uncertain. *Summerdebi* 1086, *Sumerdebi* c.1115. Probably 'Sumarlithi's village or farm', ON pers.n. *Sumarlithi*, genitive sing. *Sumarlitha*, + **bȳ**; but the name presents the same problem as Old SOMERBY SK 9633. There are hill slopes at Somersby. DEPN, SSNEM 70.

SOMERSET The county name. *(mid) Sumor sæton* 9th ASC(A) under year 845, *Sumur sætna (dæl)* (genitive pl.) ibid. under year 878, *(on) Sumær sæton, (ofer) Sumer sæton* (dative pls.) 12th ASC(E) under years 1015 and 1048, *Sumersetescir* 1122. In Latin the Somerset region is *Summurtunensis paga* [c.894]11th Asser. The name represents OE *Sumorsǣte*, short for **Sumortūn-sǣte* 'the people who live at or who are dependent upon SOMERTON ST 4828'. DEPN.

SOMERSHAM Cambs TL 3677. 'Sumor's homestead'. *Summeresham* c.1000, *Sūmersham* 1086, *Sumeresham* 1163–1236 including [1042×66]14th S 1051, *Somersham* from 1303. OE pers.n. **Sumor*, genitive sing. **Sumores*, + **hām**. It is unlikely that this could be 'summer homestead, homestead used in summer' since in such names OE **sumor** is normally uninflected. The spelling with *-mm-* suggests, however, that the original name may have been *Sūthmeres-hām* 'the estate of South Mere' referring to the *Suðmere* mentioned in the bounds of the Ramsey banlieu in S 1563. Cf. Summerfield Norf TF 7438, *Sutmere* 1086, and a lost *Suðemeresfeld* [727]13th S 1181, [933]13th S 420, [967]13th S 752 near Banstead, Surrey. An alternative possibility might be pers.n. **Sunmǣr*. Hu 222 gives pr [sʌməsəm], Sr 69.

SOMERSHAM Suff TM 0848. 'Sumor's homestead'. *Sumers(h)am* 1086, *Sumeresham* 1242, *Someresham* 1270–1447, *Somersham* from 1280, *Somersam* 1524. OE pers.n. *Sumor*, genitive sing. *Sumores*, + **hām**. DEPN, Baron.

SOMERSHAM HIGH NORTH FEN Cambs TL 3581. *Somersham and High North Fen* 1836 OS. P.n. SOMERSHAM TL 3677 + ModE **fen**.

SOMERTON 'Summer settlement'. OE **sumor** + **tūn**.

(1) ~ Norf TG 4719. *Sumertonne* *[1044×7]13th S 1055, *Somertuna* 1086, *Sumerton(a)* 1153–1254, *Somerton(e)* from 1261. Somerton had a great expanse of grazing marsh where cattle were put during the summer. The name contrasts with nearby WINTERTON TG 4919 on the coast. Nf ii.76.

(2) ~ Oxon SP 4928. *Sumertone* 1086, *Somerton* from 1278. O 235.

(3) ~ Somer ST 4828. *Sumur tún* 9th ASC(A) under year 733, *Sumertun* [860]15th S 329, *Sumor tun* 1121 ASC(E) under year 733, 899×924 S 1445, *Sūmer- Sumertone* 1086, *Su- Somerton* [12th, 1335] Buck, *Somerton* [c.1252, 1348×9]Buck. The town, from which the shire of Somerset takes its name, occupies a marked hill-top site between the rivers Cary and Yeo. DEPN, Charters V.20.

[†]It is uncertain whether *Sumerledetune* [1042×53]13th S 1535 and *Somerledetone* [1052×66]13th S 1519 belong here or to Somerton TL 8153.

SOMPTING WSusx TQ 1704. 'The Suntingas, the marsh-people'. *Suntinges -yng-* 1186–1291, *S(t)ultinges* 1086, *Sumptinges* 1242, *S(o)untinge -ynge* 1267–1697, *Sumptyng* 1336–1637. OE folk-n. **Suntingas* < ***sumpt**, ***sunt** 'a marsh, a swamp' + **ingas**. The boundary of the Suntingas is *Suntinga gemære* 956 S 624. Sx 210 gives prs [sauntiŋ] and [sʌmtiŋ], PNE ii.168, ING 40, SS 64.

SONNING Berks SU 7575. 'The Sunningas, the people called after Sunna'. *Soninges* 1086, *Sunnings* [1146]n.d., *Sunningas* [1146]15th, *Sun(n)inges* 1167 etc., *Sunning* [c.1200]13th, *Sonning* from 1284. OE folk-n. **Sunningas* < OE pers.n. **Sunna* + **ingas**. The same name occurs in SUNNINGHILL SU 9567. Brk 132, ING 45.

SONNING COMMON Oxon SU 7080. *Sunnynge Commone* 1606. P.n. SONNING Berks SU 7575 + ModE **common**. O 70.

SOPLEY Hants SZ 1697. Probably 'Soppa's wood or clearing'. *Sopelie* 1086, *Soppeley(e) -le(gh) -lye* 1152–1341, *Sople* 1412. OE pers.n. **Soppa* + **lēah**. It is tempting to connect the specific in this name with OE *sopa* 'mouthful, sip' and *sopp* 'softmeal morsel' in some sense referring to the situation of the village beside marshland, but supporting evidence for such a usage is wanting. Ha 152, Gover 1958.229.

SOPWORTH Wilts ST 8286. 'Soppa's enclosure'. *Sopeworde* 1086, *Soppeworth* 1281–1315, *Sopw(o)rth(e)* 1251–1332, *Sapworth* 1665, 1773. OE pers.n. **Soppa* + **worth**. Wlt 112.

SOR BROOK Oxon SP 4437. No early forms. Possibly identical with r. SOAR Leic–Notts SK 4925, SP 5599. O 10, RN 374.

SOTBY Lincs TF 2078. 'Soti's village or farm'. *Sotebi* 1086–1178, *Soteby* 1154×89, 1225, *Sottebi* 1155, 1206. ODan pers.n. *Sōti*, genitive sing. *Sōta*, + **bȳ**. DEPN, SSNEM 71.

SOTHERTON Suff TM 4479 → WESTHALL TM 4181.

SOTS HOLE Lincs TF 1164. A place in Metheringham. No early forms.

SOTTERLEY Suff TM 4584. Partly unexplained. *Soterlega* 1086, *Soterle -ley(e)* 1188–1524, *Saterley* 1610. Unknown element or pers.n. + OE **lēah**. DEPN compares Sotterum, Frisia, *Sotrenheim* 10th, and Bahlow 455 Sottrum, Wümme-Moor, and Sottrum by Hölle, Hildesheim, all boggy places. Baron.

SOUDLEY Shrops SJ 7229. *Great Soudley* 1833 OS. Probably 'the south wood', OE **sūth** + **lēah**. S of Cheswardine.

Upper SOUDLEY Glos SO 6610. Upper and Lower Soudley are modern developments of *Soudley Green* 1830 OS, p.n. Soudley + ModE **green**. Soudley, *Suleie* 1221, *Suth(e)leg(e)* 1258, c.1275, *Sud(e)le(ye)* 1270, 1287, *Sowdeley* 1616, *Soudley* 1758, is the 'south wood or clearing', OE **sūth** + **lēah**. It lies south in the parish of East Dean SO 6520. Gl iii.219, L 206.

SOULBURY Bucks SP 8827. 'Fortified place by the furrow'. *Soleberie* 1086, *Sulheberi* 1151×4, *Sulebire* 1198–1241, *Sulbury* 1305–1628, *Soulbury* 1755. OE **sulh**, analogical genitive or dative sing. **sule**, + **byriġ**, dative sing. of **burh**. The reference is to the valley or 'furrow' due S of the village. Bk 82 gives pr [sʌlbəri].

SOULBY Cumbr NY 7410. Partly uncertain. *Sulebi -by* c.1160–1369, *Suleghby* 1247, 1256, *Souleby* 1278–1653, *Sowleby* 1581, 1662, *Soulby* from 1355, *Soolby* 1712. Probably ON **súla** 'a pillar, a post' + **bȳ**, i.e. 'farmstead constructed of posts' or 'where posts are obtained'. But formally the specific could be ON byname *Súla* or possibly OE **sulh** as the 1247, 1256 forms suggest if they are not examples of false etymology; the genitive sing. would be **sūle** with vowel-lengthening as required by the ME *Soule-* spellings although there is some doubt whether this vowel-lengthening through loss of *-h* did in fact occur. If the specific is **sulh** the sense is 'village, farm of the trench or furrow'. We ii.22 gives pr ['su:lbi], SSNNW 40.

SOULDERN Oxon SP 5231. 'The thorn-tree in the gully'. *Suleþorne* 1152–1200, *Sulthorn(a)* 1152–1576, *Solthorn(e)* 1197–1295, *Soulthorn* 1285, *-thern* 1316, *Saldern* 1389, *Sulderne* 1562, *Showldren* 1539, *Souldrone* 1526. OE **sulh** + **thorn**. O 235.

SOULDROP Beds SP 9861. 'Hamlet by the furrow or furrow-like valley'. *Sultrop* 1196, *Suldrop(e)* 13th–14th cents., *Souldrop* 1535. OE **sulh** + **throp**. There is a well-marked valley SW of the village. Bd 42.

The SOUND Devon SX 4752. *The Sounde* 1585. The reference is to Plymouth Sound, ModE **sound** 'a channel' (OE *sund* '(a place for) swimming'). D 20.

SOUNDWELL Avon ST 6575. *Soundwell* 1830. Gl iii.80.

SOURTON Devon SX 5390. 'Settlement of the neck or col'. *(of) swuran tune* c. 970 B 1247, *Svrintone* 1086, *Suret(h)on'* [c.1200]15th, *Surton* 1249, 1284, *Sourton(e)* from 1286, *Sworton*, *Sowerton* 1577. OE **sweora**, genitive sing. **sweoran**, + **tūn**. The reference is probably to the pass over Sourton Down traversed by the road from Tavistock to Okehampton. D 206 gives pr [su:rtən].

SOUTERGATE Cumbr SD 2281. 'Shoemaker road'. *Soutergate* 1332, *Sowtergate* c.1535. ON **sútari** or OE **sūtere** + ON **gata**. La 221 gives pr [sautəge:t], SSNNW 163.

SOUTHALL GLond TQ 1380. 'The south nook of land'. *Suhaull'* 1198, *Su(d)hal(e)* 13th cent., *Suthall(e)* 1261, *Southall(e)* from 1345, *Southold(e)* 1578, 1675, *Southolt* 1710. OE **sūth** + **healh**. 'South' for distinction from NORTHOLT TQ 1285. Mx 45.

SOUTHAM 'The South homestead'. OE **sūth** + **hām**.

(1) ~ Glos SO 9725. *Suth-ham* c.991 S 1308, *Surham* 1086, *Suham* 1209–76, *Sutham* 1269–1478, *Southam* from 1299. The S part of the ancient manor of Bishop's Cleeve SO 9527. Gl ii.89 gives pr ['sauðəm].

(2) ~ Warw SP 4161. South of Itchington SP 4165. *Suð- Suthham* [965 for 991]11th S 1308[†], *Suðham* 998 S 892, 1001 S 898, *Svchā* 1086, *Sutham* 13th–14th cents., *Southam* from 1316. Wa 144 gives pr [sauðəm].

SOUTHAMPTON Hants SU 4112. 'South Hampton' as opposed to NORTHAMPTON SP 7561. *(æt) suthamtunam* [962]12th S 701, *Suðhamtun* 11th ASC(C) under year 980, *Suthamtonia* 12th. OE adj. **sūth** + p.n. *(æt) Hamtune* c.900 ASC(A), 10th ASC(B), 11th ASC(C, D) all under year 837, *Hamtun* [900]11th S 360, 11th ASC(D) under year 981, *amtun* 924–39 coins, *(an) ham tune* [956]c.1000 S 636, *aamtun* c.973–c.1025 coins, *(on) hām tune* 1045 S 1008, *hantone -tvne* 1086, *(H)omtune* [825, 826]c.1130 S 272–3, 275–6, *Hamptone* [1087]12th, [840]c.1340 S 288, 'promontory settlement', OE **hamm** 2a + **tūn**, referring to the whole promontory of dry land between the mouths of the Itchen and the Test. A variant early name is *Ham-wih* [778]c.800 *Life of Willibald* referring to the year c.721, *Hamwig* [842]late 9th, *hamwic* c.973–1015 coins, 'the trading-place on the promontory', OE **hamm** + **wīċ**. There is no evidence that *Hamwic* referred to a different site from *Hamtun*; rather the two names allude to different aspects of this important Anglo-Saxon town, its role as a mercantile centre (**wic**) and as an administrative focus (**tun**). The need to distinguish the two Hamptons arose when they both acquired the status of shire-towns in the 10th cent. The distinction is first made in documents written at Abingdon abbey, a place midway between the two shires. For a full collection of forms and exhaustive discussion see Rumble 1980.7–20, Ha 152, Gover 1958.43.

SOUTHAMPTON (EASTLEIGH) AIRPORT Hants SU 4517. P.n. SOUTHAMPTON SU 4112 + p.n. EASTLEIGH SU 4519 + ModE **airport**.

SOUTHAMPTON WATER Hants SU 4506. *Southamton Water* 1695. Earlier *the water of Southampton* 1475, French *le eawe de Hanton* c.1300, *Hampton water* 1518. P.n. SOUTHAMPTON SU 4112 + ModE **water**.

SOUTH BANK Cleve NZ 5320. A modern industrial development on the south bank of the r. Tees.

SOUTHBOROUGH Kent TQ 5842. 'The southern manor'. *la South Burgh* 1450. Earlier *bo. de Suth'* 1270, and simply *Suth* 1306, 1314. ME **south** apparently as a p.n. + **burgh**. PNK 180.

[†]Idenitifcation not certain.

SOUTHBOURNE Dorset SZ 1491. A modern name for part of Bournemouth, originally applied to a terrace of shops near The Square, and subsequently applied to the seaside residential development east of the town centre c.1870. See WESTBOURNE SZ 0791. Do 133, Room 1983.114.

SOUTHBOURNE WSusx SU 7705. No early forms. So called for distinction from WESTBOURNE SU 7507 and Nutbourne SU 7705, *Notburn* [1154×89]14th, 1277, *Nut(h)burne* 1288, 1304×7, 'the nut-tree stream'. Sx 231, 433.

SOUTHBURGH Norf TF 9904. 'South Berg'. *Suthberg* 1291. ME adj. **sūth** + p.n. *Berc* 1086, *Berg Maior* 'greater Berg' 1254, OE **beorg** 'a hill'. 'South' and 'greater' for distinction from MATTISHALL BURGH TG 0511 which was known as *Parva Berg* 'little Berg' 1254. DEPN.

SOUTHBURY Glos SO 9913 → NORBURY SO 9915.

SOUTH CHEEK or OLD PEAK NYorks NZ 9802. *Old Peak or South Cheek* 1857 OS. The reference is to the headland at the south end of Robin Hood's Bay and the name contrasts with NORTH CHEEK NZ 9606. ModE **cheek** (OE *ċēoċe, ċēaċe*) applied to anything projecting at the side.

SOUTHCHURCH Essex TQ 9186. *Suðcyrcean* 1042×66 S 1047, *Sudcercā* 1086, *Suth(e)cherche -chirche -chyrch* 1239–1428, *Su(e)chirche* 1248, *Sycherche* 1520, *South Church alias Sea Church* 1699, 1746. OE **sūth** + **ċiriċe**. The south church of Rochford hundred. Ess 201.

SOUTH DOWNS ESusx TQ 3707. Cf. *The South Downs* 1787, a breed of sheep reared on the South Downs of Sussex and Hampshire. *Down* here has a different sense from OE **dūn** and means 'an open expanse of elevated land', especially in the pl., 'the treeless undulating chalk uplands of the south and south-east of England', Shakespeare's 'unshrubd down' (*Tempest*). Evelyn (1646) writes of 'Downs of fine grass like some places in the south of England'. OED s.vv. *down* sb., *southdown*.

SOUTHEASE ESusx TQ 4205. 'Southern brushwood-land'. *Sueise* [966]c.1400 S 746, *Suesse* 1086, 1279, *Suthese* 1268–14th, *Sowthees* 1590. OE **sūth** + ***hēse**. 'South' for distinction from Northease TQ 4106, *Northesia -hese* 12th, the 'northern brushwood-land'. The whole area was probably originally the *Hese* 'the brushwood, the swine-pasture'. Sx 326.

SOUTH END Cumbr SD 2063. *South End* 1852 OS, sc. of Walney Island.

SOUTHEND Berks SU 5970. 'The southern end' or district sc. of Bradfield SU 6072. *Southend* from 1648, *South End* 1973. ModE **south** + **end**. Brk 202.

SOUTHEND-ON-SEA Essex TQ 8885. Southend-on-Sea is a modern development at the south end of Prittlewell. There were no buildings on the shore here until the 18th cent. when oyster cultivation began and the first brick cottages for the oystermen were built in 1767. In 1791 a syndicate was formed to develop a resort at New Southend: Royal Terrace and the Royal Hotel commemorate the visit in 1804 of the Princess of Wales, but the real development of Cliff Town only began in 1859. The pier dates from 1889–95, 1923 and 1929. The name Southend derives from *venella vocat' Sowthende parochie Beate Marie de Pritwell* 'vennel (or lane) called *Sowthende* of the parish of the blessed Mary of Prittlewell' 1481. This was *Old Southend* 1805 OS, ½ m. E of New Southend. An earlier name seems to have been *Strat(esh)end* 1309, the 'end of the street' from Milton where grain from Prittlewell priory was ground before shipment from Southend to Canterbury. Ess 192, Pevsner 1965.350.

SOUTHEND MUNICIPAL AIRPORT Essex TQ 8789. P.n. SOUTHEND TQ 8885 + ModE **municipal airport**.

SOUTHERY Norf TL 6294. 'The southern island'. *Suthereye* [942]12th S 483, *Suðereye* [942×c.951]13th S 1526, *Sutreia* 1086, *Suthereie* [1087×98]12th. OE **sūther**, definite form feminine **sūthre**, or comparative form **sūtherre**, + **ēġ**. The southern of two islands of dry land in the fens S of Fordham, the other being HILGAY TL 6298. DEPN gives pr [sʌðəri].

SOUTHERY FENS Norf TL 6193. *Southrey Fen* 1824 OS. P.n. SOUTHERY TL 6294 + ModE **fen**.

SOUTHFIELD RESERVOIR Humbs SE 6519. Named from the south field of Snaith, *Sudcampun* 13th, *Sowthfeld* 1462, *(le) South(e)feld* 16th. OE **sūth** + **feld**. YW ii.27.

SOUTHFLEET Kent TQ 6171. 'South Fleet'. *suth fleotes* [c.975]12th B 1321, *Svdfleta* 1086, *Suthflete* 1218. OE adj. **sūth** + **flēot** 'stream' used as a p.n. as in NORTHFLEET TQ 6274. KPN 304.

SOUTHGATE 'South hamlet'. ModE **south** + dial. **gate** < ON **gata**.

(1) ~ Norf TF 6833. *Southgate* 1824 OS. A part of Snettisham.

(2) ~ Norf TG 1424. *Southgate* 1838 OS. The name is puzzling because it lies N of Cawston. It contrasts with EASTGATE TG 1423.

SOUTHGATE GLond TQ 3093. 'South gate'. *S(o)uthgate* 1370, 1372. ME **south** + **gate**. The hamlet grew up at the south gate of Enfield Chase. Mx 69.

SOUTH GREEN Essex TQ 6893. *South Green* 1777 sc. of Billericay. ModE **south** + **green**. Ess 147.

SOUTH HEATH Bucks SP 9102. *South Heath* 1822 OS. Contrasts with HYDE HEATH SP 9300.

SOUTH HILL Corn SX 3372. 'South Hindle'. *Suthhulle* 1270, *Suthhynle* 1306, *Southille* 1327. ME adj. **sūth** + p.n. Hindle (*Henle* 1238) as in NORTH HILL SX 2776. The connection between these two places is probably that North Hill, originally *Henle* of uncertain origin, is the earlier settlement and South Hill is a daughter settlement designated 'south' from the start for distinction. Subsequently the original *Henle* became known as North *Henle* (*Northindle* 1260) with later reinterpretation of both names as containing ME **hull** (OE *hyll*) 'a hill'. PNCo 157, Gover n.d. 203.

SOUTH HOLLAND MAIN DRAIN Lincs TF 3718. *Little South Holland Drain* 1812, *South Holland Drain* 1824 OS. A drain 14 miles long extending from Peak Hill in Crowland TF 2616 across the South Holland Fens to the Nene at Peter's Point TF 4720, originally constructed under the South Holland Drainage Act of 1793 and known as the New South Holland Drain for distinction from the Old South Holland or Shire Drain on the boundary between Lincs and Cambs. Its course was earlier called *Asgeresdich* [late 12th]13th, *Asgerdyke Bank* 1797, 'Asgeirr's dyke', ON pers.n. *Ásgeirr* + **dík**. Payling 1940.67, Wheeler 105ff, Cameron 1998.

SOUTHILL Beds TL 1442. 'The south Gifle'. *Sudgiuele -gible* 1086, *Suthyeu(e)le* 1227–1329, *Southyevel(l)* 1338–1451, *Southiell* 16th. OE **sūth** + folk-n. *Gifle* 'the folk who live by the river IVEL TL 1938'. Contrasts with NORTHILL TL 1446. Bd 96 give pr [sʌðil].

SOUTHINGTON Hants SU 5049 → QUIDHAMPTON SU 5150.

SOUTH KYME FEN Lincs TF 1848. *South Kyme Fen* 1799. Cf. North Kyme Fen TF 1653 and *(in) marisco de Kyme* 1154×89, *mariscum de Kima* 1276. P.n. South KYME TF 1749 + ModE **fen**. Together N and S Kyme Fens and Kyme Low Grounds formed a large tract of fenland between Billinghay Skirth TF 1755 (Mod dial. **skirth** 'a fen drain, a dike', ON *skurthr*, OSwed *skyrth*, ODan *skyrdh*) and Kyme Eau (*aq' de Kyme* [1315]15th, *aqua vocata le Ee de Kyme* 1342, *Kyme Rau* 1638×9, 'Kyme river', p.n. Kyme + ModE **eau** < OE *ēa*) and extending up to the Car Dyke. It was drained in the late 18th cent. and later. Perrott 40, 83, 329, Wheeler Appendix i.24.

SOUTHLEIGH Devon SY 2093. 'South Leigh'. *Suthlege* 1228, *S(o)uthlegh(e)* 1242, 1291, *Sowley* 1555. ME adj. **suth** + p.n. *Lege* 1086, 'the wood or clearing', OE **lēah**. 'South' for distinction from NORTHLEIGH SY 1995. D 631 gives pr [sauli].

SOUTH LEVEL Cambs TL 5985. 'South level' sc. of Bedford Level. *South Levell* 1652. ModE **south** + **level**. Ca 211.

SOUTHMINSTER Essex TQ 9599. 'The southern minster' sc. of Dengie. *Suðmynster* [c.1000]c.1125, *Sudmunstrā* 1086, *Suthmenstre -min(i)str(e)* 1221–1332, *Sum(m)enistr(e) -ministre -mynystre* 1212–1367, *Sydmynster* 1427†. OE **sūth** + **mynster**. Contrasts with the north minster of St Cedd at Bradwell-on-Sea (Bede's *Ythancœstir* at Othona). Ess 225.

SOUTH MOOR Durham NZ 1851. Short for *South Moor Colliery* 1865 OS, the colliery on the moor S of Oxhill. ModE **south** + **moor**.

SOUTHMOOR Oxon SU 4098. *Draycote More als Southemore* 1548×53, *South Moor als Draycott Moor* 1744. ModE **south** + **moor**. The original n. of the manor was *Draicote* 1086, *Draycot et Mora* 1220, 'the shed where drays are kept', OE **dræġ** + **cot**. Brk 404.

SOUTHOE Cambs TL 1864. 'Southern hill-spur'. *Sutham* (sic) 1086, *Sudho* 1187, 1220, *Southo(u)* 1255–76. OE **sūth** + **hōh**. 'South' in relation to Midloe Farm TL 1664, *Middelho* [1135×60]14th, the 'middle hill-spur' and the lost *Westho* 1286 and *Molesho* 1257. Hu 266, 262.

SOUTHOLT Suff TM 1968. 'The southern wood'. *Sudholda* 1086, *Sutholt* 1252–1327, *Southolt* from 1336. OE **sūth** + **holt**. DEPN, Baron.

SOUTHORPE Cambs TL 0803. 'Southern outlying settlement'. *Sudtorp* 1086–1200, *Suthorp(e)* 1227–1346, *Sowthorpe* 1541. OE **sūth** + **thorp**. S of Barnack. Nth 242.

SOUTHOWRAM WYorks SE 1123. 'South Owram'. *Suthuuerum* 13th, *Southourum* 14th cent., *Southowram* 1570. ME **south** + p.n. *Oure, Overe* 1086, *Vuerum, Huuerum* 13th, *Ourum* 14th cent., '(place at) the ridges', OE **uferum**, dative plural of **ufer**. 'South' for distinction from NORTHOWRAM SE 1126. YW iii.89 gives pr [sauθ 'aurəm], L 176–7.

SOUTHPORT Mers SD 3316. *Southport* 1842 OS. A modern development to the S of North MEOLS at a place originally called South Houses. In 1792 William Sutton of North Meols built a hotel here of driftwood. It was replaced with a stone building in 1798 and re-named *South Port* at a public dinner by a Dr Barton of Hoole. Whether the re-naming was a punning tribute to William Sutton (literally 'south-town'), an attempt to attach to the new development the favourable associations of other established resorts such as Southsea Hants, or for contrast with north 'ports' at Blackpool or Preston, the early names of the district are characteristically facetious. The hotel was subsequently called Duke's Folly (it stood at the junction of Duke Street and Lord Street) while a streamlet that used to drain into the sea at SD 3217 was known as *The Nile* 1842 OS. Mills 134, Room 1983.114.

SOUTHREPPS Norf TG 2536. 'South Repps'. *Sutrepes* 1086, *-repples* 1209, *Suthreppes* 1254. OE **sūth** + ***ripul**, ***reopul**, pl. ***reoplas** 'cultivation strips' perhaps used as a f.n. The reference is to strips of land in the fen. The name contrasts with NORTHREPPS TG 2439. DEPN.

SOUTHREY Lincs TF 1366. 'The southern island'. *Sutrei(e), Svdtrie* 1086, *Suderei* c.1115, *Surrea* 1163. OE **sūtherre** + **ēġ**. S with reference to Bardney 'Bearda's island' TF 1169, another island of dry ground in the Witham Fens. DEPN, L 39.

SOUTHROP Glos SP 2003. 'Southern dependent farmstead'. *Suthþrop* 12th, *Suthrop(e)* 1211×13–1327, *Southrop(e)* 1290–1822, *-thorp(e)* 1297–1675. OE **sūth** + **throp**. The most southerly of the *throps* in the district. Gl i.45 gives pr ['sauθrəp].

SOUTHROPE Hants SU 6744. 'South hamlet'. *Sudtrop(e)* 1164, 1221, *Suth(th)rop* 1236, 1256, *Southrupp* 1690. OE, ME **sūth** + **thorp, throp**. S relative to Herriard SU 6645. Gover 1958.133.

SOUTHSEA Hants SZ 6498. The name was originally that of Henry VIII's fort built c.1538×40, *le South Castell of Portesmouth* 1545, *Southsey Ca(stle)* 1579, *Southsea Castle* c.1600. ModE **south** + **castle**. The form of the name seems to have been influenced by the neighbouring names PORTSEA SU 6300, HILSEA SU 6603 and HORSEA ISLAND SU 6304. Ha 153, Gover 1958.26, Pevsner-Lloyd 426.

SOUTH SHORE Lancs SD 3033. The southern part of Blackpool shore, for distinction from NORTH SHORE SD 3038.

SOUTHSTOKE Avon ST 7461. 'South Stoke'. *Sudstoca* 1156. OE **sūth** + **stoc** 'outlying farm' sc. of St Peter's abbey, Bath. Earlier known as *Tottanstoc* [961]12th S 694, 'Totta's Stoke or outlying farm', OE pers.n. *Totta*, genitive sing. *Tottan*, + **stoc**. Cf North STOKE ST 7069. DEPN.

SOUTH STREET ESusx TQ 3918. 'South hamlet' sc. from Chailey. ModE **south** + dial. **street**. Cf. *South Common* 1813 OS.

SOUTH TOWN Hants SU 6536. *le Suthtowne* 1548. ME **suth** + **toun**. This is the 'south settlement' in Medstead parish SU 6537. Gover 1958.79.

SOUTHWAITE Cumbr NY 4545. *Southwaite* 1663, an alteration of earlier *Thouthweyt* 1272–1677 with variants *-thwayt(e) -thwait(e), Thoghtwayt* 1367–97, 'clay clearing', OE **thōhe** + ON **thveit**. Cu 203, SSNNW 251, L 211.

SOUTHWARK GLond TQ 3280. 'The southern fort or defence', i.e. south of the r. Thames at the southern end of London Bridge. *Suð, Suðg(g), Suðge(r), Suðie, Suðw, Siði, Sudew* 1014×1066 coins, *(to) Suð geweorke, (to) Suþ geweorce* 11th ASC(D) under years 1023 and 1052, *Sudwerca* 1086–1130 etc., *Suthingwerk* c.1100, *Suthwerc(a)* 1158–1335, *Suthewark* 1298, *Suthwark* 1413. The form in the Burghal Hidage is *Suðriganaweorc* 'the fort of the people of Surrey' [10th]11th B 1335, OE folk-n. *Suthrige*, genitive pl. *Suthrigena*. Sr 29 gives pr [suðək].

SOUTHWATER WSusx TQ 1525. *Suthwatre* 1346, *Southewater* 1570. ME **suth** + **water**. Sx 231.

SOUTHWAY Somer ST 5142. *Southway* 1817 OS.

SOUTHWEEK Devon SX 4393 → GERMANSWEEK SX 4394.

SOUTHWELL 'South spring'. OE **sūth** + **wella**.

(1) ~ Dorset SY 6870. *Southwell* from 1608. Situated in the S of the Isle of Portland. Do i.219.

(2) ~ Notts SK 7054. 'The south springs'. *(at) Suðwellan* [958]14th S 679, *(œt) Suðwillum* c.1000, *Sudwelle* 1086, c.1130, *(apud) Suðwaellam* 1137, *Suwell(a), Suelle* c.1150, *Suthwell(a)* 1166–1280, *Sowthewell* c.1540. OE **sūth** + **wella**, dative pl. **wellum**. 'South' in contrast to NORWELL SK 7761. Nt 175 gives pr [sʌðl].

SOUTH WEST POINT Devon SS 1343. The SW point of Lundy Island.

SOUTHWICK 'South farm or trading place'. OE **sūth** + **wīc**.

(1) ~ Hants SU 6208. 'South farm'. *Sudwic* c.1140, 1212, *Suwyca* [1199×1216]13th, *Suthwyk(e) -wik* 1291–1446. The place is S of the Roman road from Chichester to Bitterne, Margary no.421. Gover 1958.22 gives pr [sauðik].

(2) ~ Northants TL 0292. *(œt) Suthwycan* [c.980]c.1200 B 1130, *Sudwic* 1131, 1147, *Suwic'* 1204, 1209, *Suthwyk(e) -wic(k)* 1235–1326, *Sowick* 1657. The reference is to location at the southern edge of Rockingham Forest which is ringed by secondary settlements at Bulwick SP 9694, Blatherwycke SP 9795, Apethorpe TL 0295 and Woodnewton TL 0394. Nth 206 gives pr [sauðik].

(3) ~ T&W NZ 3758. *Suthewic* c.1107–1300, *-wik(') -wyk(e) -wyc* [1199×1216]1300, 1212×5–1371, *-wich* 1154×66, 1203×4, *Suthwic - wik -wyk* c.1107–1424, *Southwik' -wyk -wik* 1341–1480×1, *Sowthick* [1564]18th, *Suddick(e)* 1580, 1613, 1768, *Southwick* 1863. OE **wīc** can mean 'dairy-farm', 'work place' or 'trading place'. The position of Southwick on the river Wear, where in later times ship-building flourished, suggests that the latter senses may have been current here although elsewhere **wich** seems to be the expected form to designate a trading place, e.g FORDWICH Kent TR 1859. There is no *Northwick*: Southwick

†The form *œt Suðmynstre* 825 B 384 does not belong here, Jnl 2.43.

was the southernmost **wīc** of the estates of the joint Monkwearmouth-Jarrow Anglo-Saxon monastery. Still pronouned locally [sudik]. NbDu 185.

(4) ~ Wilts ST 5355. *Sudwich'* 1196–8, *Suthwick* 1249, *Sowthick* 1539, *Southweeke* 1558×1603, 1638. S of Trowbridge. Wlt 144 gives pr [sauðik].

(5) ~ WSusx TQ 2405. *Sudewic* 1073, *Suwic -wyk* 1231, 1242, 1438, *Southewick* 1279, *Southweeke* 1593, *Week* 1675, *Suthick* 1690. Probably the southern **wīc** of the manor of Kingston TQ 2205. Sx 248, SS 78.

SOUTHWOLD Suff TM 5076. 'The southern wood'. *Sudwolda -holda* 1086, *Sudwald* 1227, *Suthwaud(e)* 1259–86, *-wald(e)* 1275, 1286, *-wold(')* 1275–1568, *Sowewolde* 1610. OE **sūth** + **wald**. DEPN, Baron.

SOUTHWOOD Norf TG 3905. 'The southern wood'. *Suthuuide, Sudwda* 1086, *Suthwode* 1254. OE **sūth** + **widu, wudu**. DEPN.

SOUTHWOOD Somer ST 5533. *Southwood* 1817 OS.

River SOW Staffs SJ 8528. Unexplained. *Sowa* [1118]12th, *Sowe* c.1174–1699, *Sow* c.1540, 1577. There are several examples of an IE formation **soụos* meaning 'wetness' in European r.ns. such as Save, a tributary of the Garonne, *Sava*, the Croatian Sáva, *Σαυος* (Savos) before AD 21, the Savone in Campania, Italy, *Savo* 1st cent. AD etc., but there are difficult phonological problems in linking the Sow with this root. It remains, therefore, a pre-English r.n. of unknown origin and meaning. St i.18, RN 375, LHEB 372, 519, Krahe 1964. 50.

River SOWE WMids SP 3777. An Old European r.n. *Sowe* from 13th, *Souwe* 13th. IE **soụo-* 'wet, flow' as in OHG *sou* 'juice', *ὕει* 'rains' and r.ns. Sava (G Sau), Yugoslavia, *Σαυος* (Savos) before 21 BC Strabo VII.314, *Savos, Saus* [77]9th–10th Pliny NH III.61, *Savus* c.350–6th Justinian, Save, a tributary of the Garonne and a stream in the Isère district, the Savone in Campagna, Italy, *Savo* [77]9th–10th Pliny ibid., the Savite in Lithuania, the Sèvre, Seine-et-Oise, *Savara* 6th Venantius Fortunatus, the Sora, a tributary of the Save N of Laibach, *Sovra* 10th < **Savara* etc. Some sps for WALSGRAVE ON SOWE SP 3881, however, point to OE ***sōh, *sōg**, possibly a *k/g*-formation on the same root which is needed to explain eModE *sogge* 'a swamp, a bog' with affective consonant gemination, cf. Norw *søgg, soggjen*, ModIcel *söggr* 'damp', Norw *soggiast, sogna* 'become damp', and OE *sogetha* 'hiccough, eructation'. Cf. the r.n. *Sowi* in MIDDLEZOY and WESTONZOYLAND Somer ST 3733 and ST 3534. Wa 5, RN 375, LHEB 372, 519, Krahe 1964.50.

SOWERBY 'Farm, village on mud or sour ground'. ON **saurr bȳ**. A common name in England corresponding to OIcel *Saurbœr*, ONorw *Saurby*, Swedish *Sörby*. According to Landnámabók, Steinolf built a farm in Iceland and called it *Saurbœr* 'because the land there was very marshy'. It was not necessarily a derogative name since marsh-land could provide good pasture. SSNNW 40.

(1) ~ NYorks SE 4381. *Sorebi* 1086, *Sourebi -by* 1228 etc. YN 187, SSNY 37.

(2) ~ WYorks SE 0423. *Sorebi* 1086, *Sourby(e)* 13th–1566, *Sowerby* from 1475. YW iii.144 gives prs ['sɔə- 'sɔ:bi], SSNY 38 no.4.

(3) ~ BRIDGE WYorks SE 0623. A modern industrial town taking its name from a bridge across the r. Calder. *Sour(e)bybrig(g)e, ~ Brygge* 1424–1541, *Sowerby(e)-brigge* 1578, *-bridge* 1642. P.n. SOWERBY SE 0423 + ME **brigge**. YW iii.140 gives pr ['sɔ:bi 'brig], SSNY 38.

(4) ~ ROW Cumbr NY 3940. 'Row of houses at S'. *Sowerby Rawe* c.1662, *Sowerby Row* 1750. P.n. Sowerby + dial. **raw** (ModE *row*). Cu 244.

(5) Brough ~ Cumbr NY 7912. 'S by Brough'. *Brough Sowerby* 1670. P.n. BROUGH NY 7914 + p.n. *Sowreby* 1235, *Soureby* 1256–1422, *Sorby* 1596, 1682, *Sowerby(e)* from 1675, *~ iuxta Burgh* 1318–1617. 'Brough' for distinction from Temple SOWERBY NY 6127. We ii.69 gives prs ['bruf 'sauəbi, ~ 'sɔ:bi].

(6) Castle ~ Cumbr NY 3745. *Castle Sowerby* 1777. ModE **castle** + p.n. *Sourebi* 1185–1565 with variants *-by(e), Saurebi -by* 1190–1215. The castle is mentioned in 1186 but the exact site is unknown. Also known as *Kyrkesaurebi* 1191 and *Soureby Kyrkeby* 1367. Cu 244, SSNNW 40.

(7) Temple ~ Cumbr NY 6127. *Tempel Saureby* 1292, *Temple Sowerbie -by* 1605–1777. ME **tempel** + p.n. *Saureby* 1154×89–1230, *Sourebi(a) -by* 1177–1415. The estate was held by the knights Templar until their dissolution in 1312. We ii.124, SSNNW 40 no.7.

SOWTON Devon SX 9792. 'South estate'. *Southton* 1420, *Clyst S'ci Mich'is al. Sowton* 1535. ME **suth** + **toun** (OE *tūn*). As the 1535 form shows, 'St Michael's Clyst otherwise Sowton', is one of the estates on the river Clyst, also known simply as *Clis* 1086 and *Clist Fomicon* 1242, *Clyst Fomecon, ~ Fomechun* 1278 from the family of Fomicum which held the manor in 1242. See River CLYST SY 0098. 'St Michael's' from the dedication of the church. 'South' in relation either to Monkerton SY 9693, *Monketon* 1420 'the monks' estate' held by the abbey of Battle in 1086, West Clyst SY 9795, *West Clist al. Clist Moyes* 1776, earlier simply *Cliste* 1086, or CLYST HONITON SY 9893. D 445, 443, 574.

SOYLAND MOOR WYorks SD 9819. *Soyland Mo(o)r(e)* 1624. P.n. Soyland + ModE **moor** (OE *mōr*). Soyland, *Soland* 1274–1539, *Soyland(e)* 1542–1649, is the 'newly-cultivated swamp land', OE ***sōh** 'swamp' + **land**. In either case the reference is to the swampy moors which constitute much of the township. The sp <oy> probably represents ME analogical *sōȝ* parallel to *hōȝ* in the HOYLANDS. YW iii.62, L 246, 249, BzN 1981.330.

SPA COMMON Norf TG 2930. *Spa Common* 1838 OS. Nf ii.180.

SPADE MILL RESERVOIRS Lancs SD 6237. P.n. Spade Mill, no early forms, + ModE **reservoir(s)**.

SPADEADAM Cumbr NY 5870. Possibly 'hawthorn-tree'. *Spathe Adam* 1295, *Spadadam* 1399, 1589, *Spadeadam* 1576. Possibly W **ysbyddaden**. This unusual name underwent early remodelling as if a formation with pers.n. *Adam* in Celtic word order. Also recorded as *Spadeedholm, Spaidhead-holm* 1758, and *Spear Edom* in the ballad of *Hobie Noble* 1784. Cu 96 gives pr [spi:di:dəm], OBB 575.

SPADEADAM FOREST Cumbr NY 6373. Cf. *Spadeadam Waste* 1868 OS. P.n. SPADEADAM NY 5870 + ModE **forest**.

SPALDING Lincs TF 2422. 'The Spaldingas, the people of the district called *Spald*'. *Spallinge* 1086, *Spalling* 1190–1330, *Spaulling* 1202–30, *Spaldingis* [c.1074]14th, 1115–85, *-es, -ynges* 1199–1399 including [1135×54]14th, *Spalding* from 1158 with variants *Spaud-* 1212–75 and *-yng*. Charter forms purporting to date from 716 and later are either spurious or preserved in 16th and 17th cent. copies. OE folk-n. *Spaldingas* < district name **Spald* + **ingas**. The district name is inferred from a people called the *Spalde* or *Spaldas* in the 8th cent. Tribal Hidage who settled in the fenlands of Lincs with off-shoots in YE. The name is related either to OE **spald** 'spittle' < IE **spoi-* 'foam, froth' or ***spald** cognate with OHG *spalt* 'a ditch, a trench' referring perhaps to Car Dyke, the Roman canal or drain extending from Lincoln to an ancient course of the Nene near Eye TF 2202 or to similar works. Payling 1940.47, TC 173, Wheeler 10, Cameron 1998.

SPALDINGMOOR Humbs. *Spaldinghemore* 1172. R.n.*Spalding + **mōr**. A large tract of land between the r. Derwent and Market Weighton. See SPALDINGTON SE 7533. YE 13, DEPN.

SPALDINGTON Humbs SE 7533. 'The farm or village on the Spalding'. *Spellinton* 1086, *Spaldi(n)ggetun* 1100×35, *Spaldington(a) -yng-* 1154×89–1573. R.n. **Spalding*, an **-ing**[2] derivative of OE ***spald** 'a trench, a ditch, a fenland river', + **tūn**. The form *Spaldinggetun* may point to a variant specific *Spaldinga*, the genitive pl. of OE *Spaldingas* the 'folk living by the *Spald*'. If this explanation is correct, the reference is to the r. Foulness. YE 241, Årsskrift 1974.31.

SPALDWICK Cambs TL 1272. Partly uncertain. *Spalduice* 1086, *Spaldewic(k) -wyk* 1163–1428, *Spaldwycke* 1316, *Spaldingwik* 1286, *Spaldicke* 1583. Uncertain element + OE **wīc** 'farm, work-place'. Most likely the specific represents an OE ***spald** cognate with G *spalt* 'gap, opening, crevice' which occurs in the Bavarian p.n. Spalt, *Spalte* c.1135, in the sense 'cleft' (in hills), and in the probable sense 'trench, ditch' in the stream names *Spaltbeke* and *Spaltbächle*. Here it would be appropriate for the stream on which Spaldwick stands, which flows in a marked valley, especially W of Spaldwick, and further provides an etymon for the other fenland *Spald-* names, SPALDING Lincs TF 2422, SPALDINGTON Humbs SE 7533 and SPALFORD Notts SK 8369. The posited ***spald** would have been a homophone of OE **spāld** 'spit, spittle', possibly itself used to denote a certain kind of river or stream. Hu 247 gives pr [spɔ:ldik], ING 63–4, Bahlow 456, LbO 362, PNE ii.134.

SPALFORD Notts SK 8369. Possibly the 'ford of the *Spald*'. *Spaldesforde* 1086, *Spaldeford* 1182–1343, *Spawd- Spaudforth -ford* 1567–1783. The specific is probably a stream name, either OE **spāld** 'spittle, saliva, foam', or ***spald** 'a trench, a ditch' cognate with OG *spalt*. Alternatively, the 'ford of the Spaldas', OE folk-n. *Spaldas*, an Anglian tribe settled chiefly in the fenlands of Lincs and Humberside, genitive pl. *Spalda*. Nt 207, PNE ii.134, NoB xli.147.

SPARCELLS Wilts SU 0986 → SPARSHOLT Hants SU 4331.

SPARHAM Norf TG 0719. 'Spar homestead'. *Sparham* from 1086 including [mid 11th]13th S 1516, *Sperham* 1196, 1264. OE ***spearr**, ***spær**, + **hām**. The reference is probably to the manufacture of spars, shafts etc. DEPN gives pr [spærəm], PNE ii.135.

SPARK BRIDGE Cumbr SD 3084. There are no early forms but this probably contains OE ***spearca** 'brushwood' referring to a brushwood causeway as in Spark Bridge, Essex, *Sperke-Sparkebride* 1216×72, and SPARKFORD Somer ST 6026. Cf. also the p.n. type RISEBRIDGE. Ess 267.

SPARKFORD Somer ST 6026. 'Brushwood ford or causeway'. *Spercheford* 1086, *Sparkeforda* 1086 Exon, *Sperkeford* 1242, *Sparkford* 1610. OE ***spearca** + **ford**. DEPN, PNE ii.135.

SPARKWELL Devon SX 5857. 'Brushwood spring or stream'. *Sparchewelle* 1167, *Sperkwill* 1330. OE ***spearca** + **wylle**. The identical name occurs again at SX 7865, *spearcan wille* 1050×72, *Sperchewelle* 1086, *-willa* 1086 Exon, *Sparkewill'* 1242. D 255, 520, PNE ii.135.

SPARROWPIT Derby SK 0980. *Sparrowe pit(house)* 1617, *Sparrowe pit (hole)* 1620, *(the) Sparrowpitt(yatte)* 1640. The origin of the name is unknown unless it really is a compound with **sparrow** (OE *spearwa*) and **pit** (OE *pytt*) 'a pit, a natural hollow, a hole' here used of the deep valley in which the hamlet lies. Db 66.

SPARSHOLT Hants SU 4331. Perhaps 'wood where spars are obtained'. *(æt) spæresholte* [1047×57]12th S 1402, *Speresholt* 1167–1204, *Spersholt(e)* 1252–1327, *Sparesholt* 1257, *Sparshall alias Sparsholt* 1591. OE ***spearr**, ***spær**, genitive sing. ***spearres**, ***spares**, + **holt**. A further charter form, *(æt) sweoresholte* [900]12th S 359 with *w* due to misreading *p* as the runic letter wynn ƿ, may belong here, although the latest authority identifies it with Sparcells Wilts SU 0986, *Speresholt* 1263. The exact sense of ***spearr** in these names and SPORLE Norf TF 8411 is uncertain. Since the element occurs also as a generic as in *Wynburgespær -spear* 947 B 834, 963 B 1125, the sense 'enclosure in a wood' has been suggested. Ha 154, Gover 1958.177, Wlt 36, DEPN s.v. **spearr**, Lucerna 136 fn 1, L 196.

SPARSHOLT Oxon SU 3487. 'Spear wood'. *(æt) Speresholte* [963]12th S 713, *Spers(h)olt, Spersold* 1086, *Speresholt* 1156–1284, *Spersholt(')* 1212–1517, *Spars(e)holte* 1284. OE **spere**, genitive sing. **speres**, + **holt**. The obvious sense is 'wood where spear-shafts are obtained' but genitive sing. **speres** would not be expected; possibly, therefore, **spere** here has the sense 'spear-trap' and so 'the wood of the spear-trap'. Other examples of this name-type occur. Brk 489.

SPAUNTON NYorks SE 7289. 'Shingle farm'. *Spantun(e)* 1086, *Spaunton* from [1086×9]14th. ON **spánn** possibly replacing OE **spōn**, + **tūn**. The reference is either to buildings with shingle tiling, or to a place where shingles were manufactured. YN 61, SSNY 115.

SPAUNTON MOOR NYorks SE 7293. P.n. SPAUNTON SE 7289 + ModE **moor**.

SPAXTON Somer ST 2237. 'Spak's estate'. *Spachestone* 1086, *Spaxton* [1195–1295] Buck, 1227, 1610. Late OE pers.n. **Spac* (ON by-name *Spakr*), genitive sing. **Spaces*, + **tūn**. DEPN.

SPEEN Berks SU 4668. Uncertain. *Spene* [821]12th S 183, 1208 etc., *Spone* 1086, *Church(e)spene* 1460, 1555, *Speen als Church Speen* 1751. This seems to be OE ***spēne**, a *-ja-* stem derivative of OE **spōn** 'a chip, shaving'. Probably it meant something like 'place where wood chippings are left, or where shingles are made, or characterised by brushwood'. It was called Church Speen to distinguish it from its hamlet Woodspeen SU 4562, *Wodespene* 1275, *Wode Spene* 1325, *Woodyspene* 1547. Speen is the *Spinis* of the Antonine Itinerary but cannot derive directly from the Latin name which would have produced OE, ME **Spine*. The Latin name means 'at the thorn-bushes', being the ablative pl. of Latin **spīna**. It has been suggested that a PrW ***spīn**, equivalent to the Latin name, was adopted by the English and modified to ***spēne** 'chips, shingles' through popular etymology. This is possible but it may equally well be the case that the English perceived the countryside here in a similar way to their predecessors, this being reflected in the apparent coincidence of *Spinis* and *Spene* which need not be taken to represent onomastic continuity. Brk 266, RBrit 462, Britannia 1.79, Nomina 4.33, Signposts 34, 53, 58, Celtic Voices 40.

SPEEN Bucks SU 8499. *Speen* 1822 OS. Earlier forms are needed to ascertain what if any relationship this name has to SPEEN Berks SU 4568.

SPEETON NYorks TA 1574. 'Speech enclosure'. *Spretone* 1086, *Specton(e)* 1086–1139, *Spectune -ton(a)* 12th–1579. ONb **sp(r)ēč** + **tūn**. Probably a hundred meeting-place. YE 104.

SPEKE Mers SJ 4483. 'Brushwood'. *Spec* 1086, 1212, *Speke* from 1252 with variants *Spe(c)k(e)* and *Speek*. This appears to be OE **spēc** 'a small branch, a twig' presumably in the sense 'dry brushwood' but whether with reference to vegetation or to a brushwood causeway is unknown. A derivative OLG *speckia* is found in continental p.ns. meaning 'a causeway of fascines'. La 111.

SPELDHURST Kent TQ 5541. 'Wooded hill where split wood is obtained'. *Speldhirst* [765×85]12th S 37, *-herst(e)* c.1100–1347, *-hurst* from 1226. OE **speld** + **hyrst**, Kentish **herst**. The reference is to the manufacture of shingles or the like. PNK 94, KPN 76, L 198.

SPELLBROOK Herts TL 4817. 'Speech brook'. *Spelebrok* 1287, *-broc* 1387, *Spellbroke* 1640. OE **spell** + **brōc**. The reference is unknown but the implication is that this was a meeting-place at one time. Hrt 195.

SPELSBURY Oxon SP 3521. 'Speol's fortified place'. *(æt) Speoles byrig* early 11th B 1320, *Spelesberie* 1086–1428 with variants *Spelles-* and *-bury -beri -biri -bery, Spelsberi* 1230, *Spillesbury* 1343. Pers.n. **Spēol*, genitive sing. **Spēoles*, + **byrig**, dative sing. of **burh**. O 379.

High SPEN T&W NZ 1359. ModE **high** + p.n. *the Spen* [1228×37]1647, *(le) Spen* from c.1250, 'the (land enclosed by a) fence', OE ***spenn(e)**. The reference is probably to a boundary fence of Chopwell Wood, a medieval estate belonging to the Cistercian monks of Newminster Abbey. NbDu 186, PNE ii.136–7.

SPENCERS WOOD Berks SU 7166. *Spencers Wd* 1761, *Spencers Wood Common* 1790. Surname *Spencer* + ModE **wood**.

SPENNITHORNE NYorks SE 1388. 'Thorn-tree at the dyke'. *Speningetorp* 1086, *Spen(n)ingthorn(e) -yng-* 1184–1487, *Spinithorn* c.1150, *Spenythorn(e)* 1347–1410, *Spennithorne* 1614. OE ***spenning** + **thorn**. OE ***spenning** is an **ing**² derivative of OE **spenn** 'a hedge, a dyke'. YN 253.

SPENNYMOOR Durham NZ 2534. 'Moor with or by a fence or enclosure'. *Speñing mor'* 1305×6, *Spennyngmour* 1366–7, *Spen* 1345–69, *Spennymore* 1358–1580. OE ***spenning** + **mōr**. **Spenning* is an **ing**² derivative of OE ***spenn(e)** 'a fence, a hurdle, a piece of land enclosed with a fence'. NbDu 186, PNE ii.136–7.

SPETCHLEY H&W SO 8954. 'Speech clearing'. *spæcleahtun* *[816]11th S 179, [816]11th S 180 with variant *speacleahtun* 17th, *(æt) speclea* [967]18th S 1315, *(æt) spæclea* [988]11th S 1355, *Speclea* 1086, *Spechlega -leie* 1173–1246, *Spech(e)ley*, *Spetchley* 1561–1627. OE **spēċ** + **lēah**. The reference is to Low Hill, *Oswaldes hlaw* 'Oswald's hill or barrow' [977]11th S 1332, the meeting-place of Oswaldslow Hundred. It is the open space where the speeches at the Hundred meetings were made. The boundary of S is referred to as *spæc hæme gemære* 'boundary of the Spetchley people' [987]11th S 1369. Wo 165, 89, Hooke 107, 113, 150, 152, 311.

SPETTISBURY Dorset ST 9102. 'Woodpecker's fortification'. *Spehtes- Spesteberie* 1086, *Spectesb'i* 1162, *-bury* 1294–1386, *Spettesbury* 1349–1599 with variants *Speghtes- Sp(e)ytes- Spittes-*, *Spette- Spe(y)tebury* 1291–1460. OE ***speoht**, ***speht**, genitive sing. ***spe(o)htes**, + **byriġ**, dative sing. of **burh** 'a fort' referring to the Iron Age hill-fort of Crawford Castle or Spettisbury Rings ST 9101. Forms without medial *-s-* may represent OE genitive pl. ***spe(o)hta**. ***Spe(o)ht** corresponds to OHG *speht* and ME *specht* recorded from c.1450. It may have been used as a nick-name, as in the surname *Speight*. Do ii.63, DBS 328, Newman-Pevsner 395, Thomas 1976.120.

SPEXHALL Suff TM 3780. 'The woodpecker's nook'. *Specteshale* 1197–1316, *Spicteshal(e)* 1198, 1286, *Spec(c)eshal(e)* 1254, 1344, *Spetteshale* 1268–1346, *Speksall* 1610, *Spexall* 1674. OE ***spe(o)ht**, possibly used as a pers.n., genitive sing. ***spe(o)htes**, + **halh** in the sense 'side-valley'. Spexhall contrasts with nearby WESTHALL TM 4181. DEPN, Baron, PNE ii.137.

SPILSBY Lincs TF 4066. 'Spillir's village or farm'. *Spilesbi* 1086, 12th, *Spillesby* 1209×35. ON pers.n. **Spillir* from **spillir** 'a waster', genitive sing. **Spillis*, + **bȳ**. DEPN, SSNEM 71.

SPINDLESTONE Northum NU 1533. 'The spindle stone'. *Spilestan* 1165, *Spinestan* 1176, *Spindlestan* 1186, *Spinlestan* 1212 BF, *Spinilston* 1296 SR. OE **spinele** + **stān**. The reference is to a detached pillar of whinstone thought to resemble a spindle. NbDu 186, Tomlinson 440,

SPIRTHILL Wilts ST 9975. 'The hill with a spring'. *Speerful* (sic) *[1065]14th S 1038, *Spertella* 1153, *Sperthull* 1305, *Spirthille* 1535. OE ***spyrt** + **hyll**. A stream rises in a nick of the hill here. Wlt 88.

SPITHEAD Hants SZ 6395. 'The head of the spit or sand-bank'. *Spithead* 1629. ModE **spit** 'a narrow reef, shoal or sand-bank' + **head**. The reference is to the spit of Haslar, the heel of which is at Gilkicker Point and its underwater extension is Spit Sand, on which Spit Sand Fort stands at SZ 6397. Ha 154, Gover 1958.9.

SPITHURST ESusx TQ 4217. Possibly identical with the surnames *Splytherst* 1296, *Splidhurst* 1327, 'split wooded hill', i.e. one with a road through it, ME **split** 'a split, an opening, a gap' + **hirst** (OE *hyrst*), SE dial. form **herst**. Sx 314.

SPITTAL Northum NU 0051. 'The hospital'. *the Spitle* 1695 Map, *Spittle* 1769 Map. ME **spitel** (OFr *hospital*). A leper hospital was founded here by Edward I, supposedly where Spittal Hall Farmstead now stands. Tomlinson 540.

SPIXWORTH Norf TG 2415. Partly uncertain. *Spikesuurda* 1086, *Spicasurda* 1163, *Spicheswrtha* 12th. The generic is OE **worth** 'an enclosure', the specific either OE **spic** 'bacon' and so 'bacon farm', ***spic** 'brushwood', or perhaps a pers.n. **Spic*, a nick-name from one of these. DEPN.

SPOFFORTH NYorks SE 3651. 'Ford by the plot of land'. *Spoford(e)* 1086–1300, *Spotford* late 12th–1276, *Spofford* late 12th–1436, *-forth* from 1327. OE **spot** + **ford**. The ford crossed Crimple Beck. YW v.33.

SPONDON Derby SK 4035. 'Shingle hill, hill where shingles for tiling are obtained or made'. *Spondun(e) -don(a)* 1086–1219 etc., *Spoundon* 1306–1707, *Spoondon* 1614–1767. OE **spōn** + **dūn**. Db 605, L 145, 155.

SPOONER ROW Norf TM 0997. *Spooner Row* 1838 OS. Surname Spooner + ModE **row** sc. of cottages.

SPORLE Norf TF 8411. 'Clearing where spars are obtained'. *Sparle(a)*, *Esparlea* 1086, *Esparlaium* 1146, *Sperly* c.1195, *Sporle* 1254. OE ***spearr**, ***spær** + **lēah**. DEPN.

SPRATTON Northants SP 7170. 'Settlement where poles are obtained'. *Spreton(e)* 1086–1235, *Sprotone* 1086, *Sprotton* 12th–1484, *Spratton* 1613. OE **sprēot** + **tūn**. Nth 135.

SPREAKLEY Surrey SU 8341. 'The wood or woodland clearing with young shoots'. *Sprake- Sprachele* 1225, *Sparkeley* 1298, *Spraklye* 1596, *Spreaklie* 1600. OE **sprǣc**, genitive pl. **sprāca**, + **lēah**. Sr 179.

SPREYTON Devon SX 6996. 'Settlement in brushwood or where brushwood is obtained'. *Spreitone* 1086, *Spreyton(e)* from 1234. OE ***sprǣġ** + **tūn**. D 446, PNE ii.139.

SPRIDLINGTON Lincs TF 0084. 'The estate called after Spreotel'. *Spredelin- Sperlinctone* 1086, *Spridlinc- Spritlingtuna* c.1115, *Spridlinctun* 1212. OE pers.n. **Sprēotel* + **ing**⁴ + **tūn**. DEPN, Cameron 1998.

SPRINGFIELD WMids SP 0981. 'The field with a spring'. No early forms. ModE **spring** + **field**.

SPRING HILL RESERVOIR Lancs SD 8717. P.n. Spring Mill, no early forms, + ModE **reservoir**.

SPRINGTHORPE Lincs SK 8789. 'Outlying farm by the spring or copse'. *Springetorp* 1086, 1196, *Springthorp* c.1170×8. Either OE **spring** 'a spring, a well' or **spring** 'a young plantation, a copse' + **thorp**. DEPN, SSNEM 117.

River SPRINT Cumbr SD 5297. 'Bounding, leaping stream'. *Sprit(t) -y-* 1186–1579, *Sprytte*, *Spret(t)* 1186–1777. ON assibilated form of **sprint-* (whence ME *sprinten* 'to spring, to sprint', ON *sprettr* 'a run, a bound', *spretta* 'to jump, to spirt', Swed *spritta* 'to spirt'). Identical with the Norwegian r.n. *Spretta* and waterfall name *Spretten*. The modern form *Sprint* is probably a remodelled form to rhyme with the neighbouring r.s. Kent and Mint, since it is unlikely from the evidence that the original nasalised form survived. We i.xiii, 13, RN 376.

SPROATLEY Humbs TA 1934. 'Clearing where shoots grow or are gathered'. *Sprotel(i)e -lai* 1086–1294, *Sprottele(y) -lay* 1196–1546, *Sprotle(y) -lay* c.1265–1537. OE **sprota** or **sprott**, genitive pl. **sprotta**, + **lēah**. YE 52 gives pr [sprɔ:lə].

SPROSTON GREEN Ches SJ 7366. *Sproston Green* 1831. P.n. Sproston + ModE **green**. Sproston, *Sprostune* 1086, *-ton* from 1240×6, *Sprouston* 1230–1621, is 'Sprow's farm', OE pers.n. *Sprow*, genitive sing. *Sprowes*, + **tūn**. Che ii.254 gives pr ['sprɔstən] and ['sprɔssn], Addenda 19th cent. local ['sprɔussn].

SPROTBROUGH SYorks SE 5402. Either 'Sprotta's fortification' or 'fortification overgrown with sprouts and shoots'. *Sproteburg* 1086–1285, *Sprotteburg(h)* 1246–1303, *Sprotburg(h)* 1276–1590. OE pers.n. *Sprotta* or **sprotta** + **burh**. One of a line of fortifications W of Doncaster in the Don and Dearne valleys. YW i.64, SSNY 150.

SPROUGHTON Suff TM 1244. 'Sprow's estate'. *Sproeston* 1191–4, *Sprouton(')* 1181–1384, *Sprouston(e)* 1254, 1386, *Sprowghton* 1524, *Sproughton* 1610. OE pers.n. *Sprow*, genitive sing. *Sprowes*, + **tūn**. DEPN.

SPROWSTON Norf TG 2411. 'Sprow's estate'. *Sprowestuna, Sprotuna* 1086, *Sprouston* c.1125. OE pers.n. *Sprow*, genitive sing. *Sprowes* + **tūn**. DEPN.

SPROXTON Probably 'brushwood farm'. ODan ***sprogh**, genitive sing. **sproghes** + OE **tūn**. ODan ***sprogh** is related to ON *sprek* 'dry twig', MLG *sprok*, MDu *sproc, sprokkel* 'dry twig', OE *spræċ* 'twig, shoot', ME *spray* (OE **sprǣġ* or **sprēġ*). The alternative suggestions, ON pers.ns. *Sprógr* or **Sprok*, are unlikely – *Sprógr* is recorded only as the name of a horse, and **Sprok* not recorded at all.

(1) ~ Leic SK 8524. *Sprotone* (sic 3×) 1086, *Sproxcheston'* c.1130, *Sprokeston'* 1183–1236, *Sproxton(e)* 1166–1610, *Sprostona* c.1154, *Sprouston'* 1201, *Sprauston* 1549, *Sprawson* 1539. Lei 190 gives pr ['sprousən], SSNEM 187.

(2) ~ NYorks SE 6181. *Sprostune* 1086, *Sproxtun(a)* 1165×75–1202, *Sproxton* from 1226. YN 70, SSNY 114.

SPURN HEAD Humbs TA 3910. 'Headland called Spurn'. *Spun Head* 1610, *Spurn P* before 1678, *Spurn Head* 1786. ModE **spurn** sb. 'a sharp projection' + **head**. There were originally two forms of the name for this headland:

I. *Ravenserespourne* 1399, *Ravenser Spurne* 1406, p.n. Ravenser + *spurn*, Shakespeare's *Rauenspurgh* 1597 Richard II, 2.i.298.

II. *Odd juxta Ravenserre* [1235×49]copy, *Odrauenser, Rauenser Odd(e), Hodde* 13th cent., *Ravenserot* 1251, *Ravenserod(d -e)* [1273–1342]15th, lost p.n. *Rau- Raven(e)ser(e)* 1199×1216–[1361]15th, *Hrafnseyri* 13th Heimskringla, Orkneyingasaga (dat.sg.), + ON **oddr** 'a point of land'. Ravenser is 'Hrafn's sand-bank', ON pers.n. *Hrafn* or **hrafn** 'a raven', genitive sing. *Hrafns*, + ODan **ør**. The ON sources show ON *eyrr* for ODan *ør*. The place was washed away by the Humber in medieval times. YE 16, 17, 19.

SPURSTOW Ches SJ 5557. Possibly 'meeting-place on a spur'. *Spuretone* 1086, *Sporstow* 1180–1395 with variants *Spori- Sporu-* and *-stou(e) -stowe, Spurestou -is-* 1180×1220, *Spurstow(e)* from c.1280. OE **spura** 'a spur' or **spure** 'a heel' referring to the slight ridge on which the hamlet stands, + **stōw**. Che iii.315.

STACKHOUSE NYorks SD 8165. '(The settlement) at the stack houses, the houses near, or for, ricks'. *Stacuse* 1086, *-hus* 1285, *Stakhous* 1221, 1294, *Stac- Stakho(u)ssum -husum* 12th cent. ON **stakkr** + **hús**, dative pl. **húsum**. YW vi.144, SSNY 88.

STACKSTEADS Lancs SD 8421. No early forms.

STADDISCOMBE Devon SX 5151. 'Horse or stud valley'. *Stotescoma* 1086, *-c(o)umbe -camb* 1281–1334, *Stodescumbe* 1288, *Staddiscomb* 1809 OS. OE **stot** or **stōd**, genitive sing. **stot(t)es, stōdes** + **cumb**. In the same parish is Staddon, *Estotdona* 1086, *Stoddune* 1242, 'horse hill', OE **stot** + **dūn**. D 256, 257.

STADDLETHORPE Humbs SE 8428. 'Outlying farm with a wooden platform'. *Stadyethorpe* 1550, *Staddyl- Stertil- Saddlethorpe* 16th, *Staddlethropp* 1604. ModE dial. **staddle** 'a wooden platform on which hay-ricks are built' + **thorp**. YE 244.

STADHAMPTON Oxon SU 6098. P.n. Stadham + **ton**, a late addition due to the influence of Brook- and Chislehampton. Stadham, *Stodeham* 1130–47, *Stodham* 1246–1552, is 'the river-meadow where horses are kept', OE **stōd** + **hamm** 1. O 154.

STAFFIELD Cumbr NY 5442. 'Staff hill'. *Stafhole* c.1225–79, *Staffol(e)* c.1252–1777, *Staffold* 1270, *Staffeld* 1276–8, *Staffel(l)* 1307, 1508, *Stafful* 1348–1568, *Staffield oth. Staffell* 1806. ON **stafr** 'a post, a pole' + **hóll** 'an isolated hill'. The reference is to a small hill marked by a post or where posts were obtained. Cu 248 gives pr [stafl], SSNNW 164, L 169.

STAFFORD Devon SS 5811 → STOWFORD Devon SX 4386.

STAFFORD Staffs SJ 9223. 'The landing-place ford'. *STED, STEF, STAE(F)Ð, STÆTH, STÆÐ, STÆFORA* 10th Coins, *STAF, STÆF(FD), STAFORDE, STEFFOR* 11th Coins, *(æt) Stæf forda* 11th ASC(C) under year 913, *Stafforda* (accusative case) c.1050 ASC(D) under same year, *Stat- Stadford* 1086, *Statford* 1130. OE **stæth** + **ford**. The name points to early traffic on the r. Sow as well as on the roads which converge on the crossing place. DEPN, TC 174.

West STAFFORD Dorset ST 7289. *West(st)af(f)ord(e)* from 1285, *West Stanorde* (sic for *Stau-*) 1320, *Stavord West* 1340. ME adj. **west** + p.n. *Stanford* 1086, 1322, 1381, *Staford* 1086, *Staf(f)ord(e)* 1205–64, 'the stone ford', OE **stān** + **ford**, referring to a crossing of the r. Frome. 'West' for distinction from a lost *Est(st)aford* 1303–32, *East Stafford* 1569–1669. Do i.243, 209.

STAFFORDSHIRE 'The shire of Stafford'. *(into) Stæffordscire* c.1050 ASC(D) under year 1016, c.1120 ASC(E) under same year. P.n. STAFFORD SJ 9223 + OE **scīr**.

STAFFS & WORCS CANAL WMids SO 8788. A canal built in 1766–71 by James Brindley, designer of the first of the English canals, the Bridgwater canal of 1761. Pevsner 1968.271.

STAGSDEN Beds SP 9849. 'Stake valley'. *Stacheden(e)* 1086–15th, *Stachesden(e)* 1196–14th, *Staggeden* 1183, *Staggesden* 1228. OE **staca** + **denu**. Either a valley marked by a stake or more likely where stakes are manufactured or obtained. Bd 43, L 99.

STAINBURN NYorks SE 2448. 'The stony stream'. *Stanburn(e)* [972×92]11th S 1453, *Sta(i)nburne* 1086, *Stain- Stayn- Steinburn(e)* 12th–1520 with variant *-brun(e)* 12th, 13th. OE **stān** + **burna** replaced by ON **steinn** + **brunnr**. YW v.48.

STAINBY Lincs SK 9022. 'Stigandi's village or farm'. *Stigandebi* 1086, *Sty- Stiand(e)by* 1194–1366, *Steyenneby* 1291, *Steenby* 1353, 1808, *Stainby* 1651. ON pers.n. *Stígandi*, genitive sing. *Stíganda*, + **bȳ**. Perrott 187, SSNEM 71.

STAINCROSS SYorks SE 3310. 'The stone cross'. *Stainecrosse* 1589. Site of the meeting-place of Staincross Wapentake, *Stain- Stancros wapentac* 1086–1219, *Stain- Stayncros(s) -crosse* 1086–1620. ON **steinn** + **kross**. YW i.317,261, SSNY 104.

STAIN DALE NYorks SE 8690. 'The rocky valley'. *Staindal* 1185×95. ON **steinn** + **dalr**. YN 92.

STAINDROP Durham NZ 1220. 'Stony valley'. *Standropa(m)* [c.1040]12th, 1128, *Standrop(e) -dropp* c.1107–1612, *Stæindrop* 1128×9, *Steindrop(e)* 1128–1244×9, *Stayndrop(p)(e)* 1243–1563, *Stain-* from 1129, *Stain(e)- Stan(d)throp(p)* 1564–1647. OE **stǣner** (influenced by **stān** and ON **steinn**) + **hop**. Intrusive *d* developed early in this name as in late OE *gandra* 'a gander' < *ganra*. NbDu 187, Nomina 12.35.

STAINES Surrey TQ 0471. 'The stone'. *Stána* [969]13th S 774, [1053×66]11th S 1142, *(æt) Stane* c.1050 ASC(C, D), 1121 ASC(E) under year 1009, *(into) Stane* [1053×66]13th S 1142, *Stanes* 1086–1593, *Staines* 1578. OE **stān** + unexplained ME **-es**. Staines is the site of the Roman settlement called *Pontibus* 'at the bridges', and the reference is generally held to be to a Roman milestone. The development of the name is very irregular; it should have become **Stone* and the *-es* ending is equally peculiar whether representing the AN nominative sing. inflexional ending *-s* or ME pl. *-es*. Mx 18, TC 174.

STAINFIELD Lincs TF 0725. 'Stone field'. *Steinfeld* 1314×5, *Stainfield* from 1689. ME **stein** (ON *steinn*) + **feld**. This seems to be a reformation of earlier *Steintone, Stentvith* 1086, *Steynthveyt* 1274–1475 with variants *Stein-* and *-thweyt* etc., 'stone clearing', ON **steinn** + **thveit** varying with OE **tūn**. Doubts about the generic of this name are further illustrated by late spellings such as *Stainflete* 1562 and *Stainwheat* 1645×6. Perrott 120, SSNEM 161, L 211.

STAINFIELD Lincs TF 1073. 'Open land characterised by stones'. *Stain- Steinfelde* 1086, *Steinfelda* c.1115. ON **steinn** probably replacing OE **stān** + **feld**. DEPN, SSNEM 222.

STAINFORTH AND KEADBY CANAL SYorks SE 7311 → STAINFORTH SE 6411, KEADBY Lincs SE 8311.

STAINFORTH 'The stony ford'. ON **stein** probably replacing OE **stān** + **ford**.

(1) ~ NYorks SD 8267. *Stain- Stranforde* 1086, *Stain-*

Staynford(a) c.1190–1564, *-forth(e)* 1285–1659. The reference is either to a ford taking the road up Ribblesdale across Stainforth Beck, or to a ford across the Ribble itself to Little Stainforth. YW vi.154.

(2) ~ SYorks SE 6411. *Stenforde* 1086, *Steinford(e)* 1086, [1199]13th, *Stain- Staynford(e)* 1285–1589, *-forth* from 1405. Possibly the *Stanforda* [c.715]11th given to St Wilfrid c.660 Eddi 8. A crossing of the r. Don with a stone track. YW i.12, L 68–9, SSNY 165.

STAINING Lancs SD 3536. Uncertain. *Staininghe* 1086, *Stenig̃* 1208, *Staining* from 1246, *Stanynggas, Steyninges* 1211×40, *Stain- Stayninges* 1246. Probably OE ***stāning**, ***stǣning** 'stony place' varying with **Stāningas*, **Stǣningas* 'the people who live at *Staning*, the stony place'. The name has been influenced by ON **steinn**. Staining lies about a mile W of the Roman road from Ribchester to Poulton le Fylde, Margary No.703. La 156, SSNNW 206, Jnl 17.91.

STAINLAND WYorks SE 0719. 'Newly cultivated land with stones'. *Stanland* 1086, 1386, *Staineland* 13th cent., *Stainland* 13th–1604. OE **stān** replaced by ON **steinn** + **land**. YW iii.49, SSNY 138, L 246, 249.

North STAINLEY NYorks SE 2877. *(on) Norð Stanlege, (on) Nyrran Stanlege* c.1030 YCh 7, *Nordstanlai(a)* 1086. OE **north** + p.n. *Stanleh* [972×92]11th S 1453, *(E)stanlai* 1086, *Stain- Stayn- Steinlega -lei(a) -le(y) -lay* 1154–1576, the 'stony clearing', OE **stān** later replaced by ON **steinn**, + **lēah**. 'North' in relation to Ripon and in contrast to South STAINLEY SE 3063. *Nyrran Stanlege* is 'the nearer part of North Stainley', i.e. nearer to Ripon, OE ***nīerra**. YW v.159.

South STAINLEY NYorks SE 3063. ME prefix *South* from 1198 varying with *Kirke, Kyrk(e)* 13th–1609, ON **kirkja** 'a church', + p.n. *Stanlai -lay -lei(a) -ley* 1086–1436, *Stein- Stain- Staynlea -lei -le(ya) -lay* 12th–1482, the 'stony clearing', OE **stān** later replaced by ON **steinn**, + **lēah**. 'South' in relation to Ripon and in contrast to North STAINLEY SE 2877. YW v.95, L 205.

North STAINMORE Cumbr NY 8315. Adj. **north** + p.n. Stainmore as in STAINMORE COMMON NY 8517.

South STAINMORE Cumbr NY 8413. Adj. **south** + p.n. Stainmore as in STAINMORE COMMON NY 8517.

STAINMORE COMMON Cumbr NY 8517. P.n. Stainmore + ModE **common**. Stainmore, *Stanmoir* c.990, *-more* [1154×89]1348, 1622, *Stein- Steyn- Stainmor(e)* c.1230–1785, is 'rock moor', OE **stān** replaced by ON **steinn**, + **mōr**, referring to the boulders and many exposures of the carboniferous limestone that lies beneath the peat. We ii.71, SSNNW 193, L 54–5.

STAINMORE FOREST Durham NY 9410. *forestria de Stammore* (for *Stain-*) 1272, *forest of Staynemore* 1588. P.n. Stainmore + ME **forest** 'a hunting reserve'. Stainmore, *Stanmoir* c.990, etc. as in STAINMOOR COMMON NY 8517. YN 305, We ii.71, 80, SSNNW 193.

STAINSACRE NYorks NZ 9108. 'Stein's cultivated land'. *Stainsaker* 1090×6, *Steinsecher -eker* [12th]15th, *Stainsecre -echer, Staynseker* 12th–14th cent. ON pers.n. *Steinn*, genitive sing. *Steinns*, + **akr** varying with **ekra**. YN 123, SPN 263, L 233.

STAINTON 'Stone village or farm'. ON **steinn** frequently replacing OE **stān**, + **tūn**. The reference may be to stony ground or buildings in stone, but perhaps most frequently to a place where stone is quarried. PNE ii.144, 150.

(1) ~ Cleve NZ 4814. *Steintun -tvn* 1086. YN 171, SSNY 113.

(2) ~ Cumbr SD 5285. *Steintun* 1086, *Steynton* 1256, 1461, *Stain- Staynton* 1180–1657. We i.99, SSNNW 188 no.4.

(3) ~ Cumbr NY 4828. *Stainton'* from 1166, *Staynton'* 1292–1608. Cu 188 gives pr [stentən], SSNNW 188 no.3.

(4) ~ Durham NZ 0718. *Stantun* [c.1040]12th, [1072×83]15th, c.1107, *Stainton* from 1242×3, *Staynton in le Cragges* 1516. The quarry here has been in use for almost a thousand years and supplied stone for the building of Barnard Castle. It is still active.

(5) ~ NYorks SE 1096. *Steintun* 1086. The identification is uncertain. DG 534 refers this form (f.300a) to a lost Stainton in Stanghow Cleve. YN 270, SSNY 113, DBY IN6 note.

(6) ~ SYorks SK 5593. *Stantone* 1086, 13th, *Stainton(e), Staynton(a)* c.1130–1552. The occurrence of the affix *Stoney-* in 1739 suggests that **stān** refers to the exposed limestone rock here, rather than to a farmstead built of stone. YW i.130, SSNY 133.

(7) ~ BY LANGWORTH Lincs TF 0677. *Steintuna iusta Languat* c.1115. P.n. *Staintvne* 1086 + p.n. LANGWORTH TF 0676. DEPN, SSNEM 187 no.2.

(8) ~ LE VALE Lincs TF 1794. 'S in the valley'. *Stainton in Valle* c.1300, ~ *le Vale* 1828. P.n. *Stainton(')* from 1086 with variant *Stayn-* 1240–1535, *Steintuna -ton(')* c.1115–1559, *Staunton(')* late 12th, 13th, 1526, *Stanton* 1428, + Fr definite article **le** short for *en le* + ModE **vale**, Latin **vallis**, dative sing. **valle**. Also known as *Wald Staintone* 13th or *Stainton super Waldam* 'on the wold' 1296 and ~ *in le Hole* 1440. Li iii.124, SSNEM 187 no.3, Cameron.

(9) ~ WITH ADGARLEY Cumbr SD 2572. *Steintun* 1086, *Steynton* 1246, 1269, *Stayton(am)* 1269, 1276. Adgarley, *Eadgarlith* 1185, *Adgerlith* c.1300, is 'Edgar's slope', OE pers.n. *Ēadgār* + **hlith**. La 210 gives pr [stentn], SSNNW 188 no.1.

(10) Great ~ Durham NZ 3322. *Magna Staynton* 1366, *Great Stainton* 1592. ModE **great**, Latin **magna**, + p.n. *Stantona* c.1200, *Steinton* 1200, 1239, *Staynton* 1296–1588, *Stainton* 1560. Also known as Stainton-le-Street 'Stainton on the Roman road', *Staynton in strata* 1438×9–1588, ~ *in le Strete* 1461, *Stanton in yᵉ streat* 1567, *Stainton in the Streate* 1588. Some early forms for this name, *staninctona* [1091×2]12th, *Stanyntone* 1235×6, *Stanyngton* 1436, show that an early alternative form was OE **stāning** + **tūn** 'farm, village in the stony place'. The reference may have been to the Roman road from Thirsk to Durham, Margary no. 80a. 'Great' for distinction from Little STAINTON NZ 3420. NbDu 187, Nomina 12.35.

(11) Little ~ Durham NZ 3420. *parua steintun* 1153×75, ~ *staintū* 1233×44, *Staynton Parva* 1458–1623, *Litil Staynton* 1366, *Little Stainton* from 1621. ME **litel**, Latin **parva**, + p.n. STAINTON for distinction from Great STAINTON NZ 3322.

(12) Market ~ Lincs TF 2279. 'S with the market'. *Steynton Market* 1286. ME **market** + p.n. *Staintone* 1086, *Steintuna* c.1115. DEPN, SSNEM 187 no.4, Cameron 1998.

STAINTONDALE NYorks SE 9998. 'Stainton valley'. *Staynton Dale* 1562. P.n. *Steintun* 1086, 'stone enclosure', ON **steinn** probably replacing OE **stān**, + **tūn**, + ModE **dale** (ON *dalr*). YN 110, SSNY 113.

STAIR Cumbr NY 2321. 'Steep place' or 'step'. *Stayre* 1565, *Stare* 1566. ME **steir**, **stair** (OE *stæġer*). The reference is to a rise in the road. Cu 373.

STAITHES NYorks NZ 7818. 'The landing-places, the jetties'. *Stathes* 1577, 1665, *Stease* 1686. ME **stathe**, pl. **stathes** (OE *stæth*, or ON *stǫth*, earlier **stathwō*). The earliest form is *Setonstathes* 1451, 'Seaton landing place', from the p.n. now preserved in SEATON HALL NZ 7817. YN 139 gives pr [stiəz].

STAKEFORD Northum NZ 2785. 'The ford marked by stakes'. ModE **stake** (OE *staca*) + **ford**. Cf. the p.n. type STAPLEFORD.

STAKE PASS Cumbr NY 2608. P.n. Stake + ModE **pass**. Stake, *the Stake of Borrowdall* 1784, is possibly dial. **stake** 'a columnar rock'. The pass carries the Cumbria Way from Borrowdale into Great Langdale. Cu 353.

STAKE POOL Lancs SD 4148. No early forms; apparently 'creek, inlet marked by a stake', ModE **stake** (OE *staca*) + **pool** (OE *pōl, pull*).

High STAKESBY NYorks NZ 8810. ModE **high** + p.n. *Staxebi* 1086, *Stachesbi -by* 1090×6–1314, *Stakesbi -by* 1100×c.1115 etc., 'Stak's farm or village', ON pers.n. *Stakr*, genitive sing. *Staks*, + **bȳ**. YN 125, SSNY 38.

STALBRIDGE Dorset ST 7317. 'Post bridge'. *Stapulbrige* [860×6]14th, *-breicge* [998]12th S 895, *Staplebrige* 1086, *Stapelbrig(g)(e) -brug(g)(e) -brygg(e)* 1145–1402, *Stalbrig(g)(e) -brug(g)(e)* 1327–1552, *-bridge* from 1535. OE **stapol** + **brycg**. The sense is 'a bridge built on posts or piles'. Do iii.280.

STALBRIDGE WESTON Dorset ST 7216 → Stalbridge WESTON.

STALHAM Norf TG 3725. 'Stall homestead'. *Stalham* from 1086 including *[1044×7]13th S 1055. OE **stall** + **hām**. The various senses of **stall** include cattle or sheep stall, building site, fishing pool. The latter would be possible since Stalham Broad comes right up to the village. Nf ii.124.

STALHAM GREEN Norf TG 3824. *Stalham Green* 1838 OS. P.n. STALHAM TG 3725 + ModE **green**. Nf ii.125.

STALISFIELD GREEN Kent TQ 9552. *Stalisfield Green* 1819 OS. P.n. Stalisfield + ModE **green**. Stalisfield, *Stalisfeld* 1172, *Stalesfeld(e)* 1177–1253×4† is identical with the lost woodland-pasture name *stealles felde* [c.765]13th S 50, 'open land of the stall', OE **steall**, genitive sing. **stealles**, + **feld**. KPN 50, L 244.

STALLING BUSK NYorks SD 9185. 'The stallion's bush'. *Stalunesbusc* 1218, *Stalunbusk* 1283. ME **stalun** (OFr *estalon*) + **busk** (ON **buskr*). YN 264.

STALLINGBOROUGH Humbs TA 2011. 'The fortification called or at *Staling*'. *Stalingeburg* 1086, 1204, *Stalinburc* c.1115, *Stal(l)ingburgh* 1109–1344. The specific appears to be an OE ***stal(l)ing** or ***stæling** of uncertain origin: either

I. OE **st(e)all** 'a standing place, a stall for cattle, a fishing pool';

II. OE **stæl** 'a place, a spot'; or

III. OE **stalu** 'a stem, a post, the wooden support for the strings of a harp', whence ME *stale* 'the upright of a ladder'.

Stallingborough lies close to the Humber so that the sense 'fishing place' would be suitable. However the forms point to ***staling** or ***stæling** rather than ***stalling**, the sense of which is unclear. Both **stæl** and **stalu** derive from the root *stol-* seen in Greek *στόλος* 'beam, post, pole', Lat *stolo* 'a shoot, a sucker'. An OE ***stæling** might have had the sense 'support, post, structure of posts, landing stage' or the like. More recently a form ***staling** has been proposed as the source of ME *stalling* 'a young tree' so that the name would mean 'the fort growing with young trees'. An alternative explanation takes the specific as OE folk-n. **Stællingas*, genitive pl. **Stællinga*. The generic is OE **burh**. Li ii.268, NQ n.s. 45.286.

STALMINE Lancs SD 3745. Uncertain. *Stalmine* 1086, *Stalmin -myn(ne)* [1184]17th–1332, *Staylmyn* 1443, *Sto'min* 19th. The specific could be any of OE **stæll** 'a place for catching fish', **stall** 'a standing-place for cattle' or ON **stallr** 'stand, stable, crib', the generic apparently ON **mynni** 'mouth (of a river etc.)'. It is suggested that the course of the river Wyre may formerly have been further to the E and so closer to the village. The river-mouth was probably that of the stream N of the village as it flowed into the Wyre. La 159, SSNNW 251–2, Jnl 17.92.

STALYBRIDGE GMan SJ 9698. 'The bridge at Staley'. *Stalybridge* 1687. P.n. Stayley SD 9698 + ModE **bridge**. Stayley, *Staue- Staveleg(h)* early 13th–[1402]17th, *-ley* 1348–1560, *Staley(e) -legh* 1285–1559, *Stayley* from 1560, *Steal(e)y*, *Steeley* 17th, is the 'wood or clearing where staves are manufactured or obtained', OE **stæf**, genitive pl. **stafa**, + **lēah**. Che i.317.

STAMBOURNE Essex TL 7238. 'Stone stream'. *Stanburna* 1086, *-(e)* 1221–54 etc., *Stamburn* 1221–1317. OE **stān** + **burna**. Ess 456.

†The DB form *Stanefelle* 1086 is erroneous.

Great STAMBRIDGE Essex TQ 8991. *Great Staunbregg'* 1261, *muche Stambridge* 1578. Earlier *Magna* 1248, *Moche* 1503, *Mekill* 1493–1500, *Myche* 1548, 1768, Latin **magna**, ME **muche, mikel, grete** + p.n. *Stanbruge* 1086–1206, *-brige* 1087, the 'stone bridge', OE **stān** + **brycg**. 'Great' for distinction from Little Stambridge TQ 9892, *Parva* 1230, 1412. Ess 202 gives pr [sta:mbridʒ], Jnl 2.43.

STAMFORD 'Stony or stone-paved ford'. OE **stān-ford**.

(1) ~ Lincs TF 0307. *Stanford* [c.731]8th BHE 1727 including 1086 with variants *-forde* and *-fort(h)*, *Stean forda* 922, 942 ASC(A), *Stamford* from 1146. The town is situated at a point where the r. Welland was easily fordable throughout the year. The ford carried the Great North Road but the original site may have been where Ermine Street crossed the Welland at TF 0206. The identification of Bede's form is uncertain and depends on a 15th cent. Durham tradition. Stamford near York has also been suggested. Perrott 412, TC 174.

(2) ~ BRIDGE Humbs SE 7155. *Stanford brycg* c.1075, *~ brig(ge)* 1219–1475, *Stam-* from 1350. The ford, later replaced by a bridge, carried the Roman road from Barmby to Easingwold, Margary no. 80a, across the Derwent. In ASC the name is also found in the early Scandinavianised forms *Stemfordbrycg* (for *Stein-*) c.1100 and *Stængfordbrige* (for *Stægn-*) c.1150 while in Scand sources the OE form of the name is recorded in various forms, eg. *Stanforðabryggiur*. The alternative forms of the name, Latin *Pons belli* [before 1180]15th, 1130×5–1346, French *Punt de la Bataill(e)* 12th, *Ponte(r)bell* 13th, 'the bridge of the battle', refer to the battle of 1066 in which Harold Godwinson defeated Earl Tosti and Harold Sigurdson of Norway. YE 186.

STAMFORDHAM Northum NZ 0872. 'The homestead at *Stanford*, the stony ford'. *Stanfordham* 1187–1428, *Staun-* 1242 BF, 1246, *Stamfordeham* 1249, *Stanwardham* 1409, *Stanerden* 1559, *-ton* 1717. P.n. **Stanford* < OE **stān** + **ford**, + OE **hām**. NbDu 187 gives pr [stanətən].

STANBOROUGH Herts TL 2210. 'Stone hill'. *Stanberue* 1277, *-berwe* 1296, *Stanborowe (brydge)* 1461×83. OE **stān** + **beorg**. Hrt 128.

STANBRIDGE 'Stone bridge'. OE **stān** + **brycg**.

(1) ~ Beds 9623. *Stanbrug(g)(e)* 1165–1405, *-brig(g)e* 13th–15th cents., *Standbridge* 1785. The bridge across a tributary of the Ouzel lies a mile S of the village. Bd 132, L 65.

(2) ~ Dorset SU 0004. *Stanbrig(g)(')* 1230–1428, *-brug(g)(e)* 1254, 1428. The reference may have been to a causeway through marshy ground rather than simply to the bridge across the r. Allen. Do ii.149.

STANDEFORD Staffs SJ 9108. 'The stony ford'. *Stanieford* *[996 for 994]19th S 1380, *Staunford'* 1245×50, *Stoniford* 1300, *Stawntiford* 1506, *Staundefo(o)rde* 1595. A crossing of Deepmore Brook on the line of the Roman road running SE from Water Eaton, Maragry no.190. St i.39, Horovitz.

STANDEN Kent TQ 8540. 'Stony pasture'. *Stankyndenn'* 1334, 1347. OE **stāniht**, oblique case definite form **stānihtan**, + **denn**. Identical in form with Standen in Standen Street TQ 8030, *stanehtandenn* 858 S 328, *Stonekindenn'* 1240, *Stoneghyndenn* 1278. KPN 201.

STANDFORD Hants SU 8134. 'Stone ford'. *Stanford(a)* from 1208, *Sandford* (sic) 1816 OS. OE **stān** + **ford**. Gover 1958.104.

STANDISH GMan SD 5610. 'Stony pasture'. *Stanesdis* 1178, *Stanidis* c.1190, *-edis(s) -edissh* 1207–92, *Standische* 1288, *Standis(s)h* from 1304. OE **stān** + **edisc** 'enclosure, an enclosed park' for cattle, or deer. La 127, Jnl 17.71.

STANDLAKE Oxon SP 3903. 'The stony stream'. *Stanlache* c.1150–1636 with variants *-lac(h) -lak(e)* and *-layke*, *Standlac* [c.1200]c.1225, *Standlake* 1302, 1544. OE **stān** + **lacu** in the sense 'side-channel'. O 329.

STANDON Hants SU 4226. 'Stony valley'. *Standen(e)* 1167–1249,

Stonden 1257, *Standen T. Pike* 1811 OS. OE **stān** + **denu**. Ha 155, Gover 1958.176.

STANDON Herts TL 3922. 'Stone hill'. *Standune* [990×1001]13th S 1497, *Standone* 1086–1314 with variant *-don*. OE **stān** + **dūn**. Hrt 196.

STANDON Staffs SJ 8235. 'Stone hill'. *Stantone* 1086, *Standon'* 1190, 1248, *Staundon* 1277, 1597, *Standon vulg. Stawne* 1679. OE **stān** + **dūn**. DEPN, L 144, 155, Horovitz.

STANE STREET 'The Roman road paved with stone'. OE **stān** + **strǣt**.

(1) ~ Essex TL 5421. *(la) Stanstrete* 1181–1323, *-strat(e)* 1228–1328. The reference is to the Roman road from Boughing to Colchester, Margary no. 32. Ess 3.

(2) ~ Surrey TQ 1440. *Stanstrete* 1279(p), 1381, *Stenestret Causwaye* 1610, *Staens Street* from 1680, *Stone* - 1786. The reference is to the Roman road from London to Chichester, Margary no.15. Sr 8.

STANEGATE 'The stone road'. OE, ME **stān** + **gate** (ON *gata*).

(1) ~ Cumbr NY 4760. The Roman road from Carlisle to *Cilurnum*, the Roman fort at Chesters NY 9170, Margary no.85a. Also known as *Carelgate*, 'the road to Carlisle'. Cu III xvi.

(2) ~ Northum NY 7866. *Stanegate* 1867 OS. The name of the Roman road from Corbridge to Carlisle, Margary Nos.85a,b.

STANFIELD Norf TF 9320. 'Stony open land'. *Stanfelda* 1086. OE **stān** + **feld**. DEPN, L 242.

STANFORD 'Stone ford' i.e. one with a stony bottom or marked by a stone. OE **stān** + **ford**. PNE ii.144, L 67–9.

(1) ~ Beds TL 1641. *Stanford* from 1086, *Stamford* 1202, *Standford* 1535. Bd 97.

(2) ~ Kent TR 1238. *Stanford* from 1035. Carries the Roman road called Stone Street from Canterbury to Lympne, Margary no. 12, across the upper waters of the East Stour river. PNK 427.

(3) ~ BISHOP H&W SO 6851. 'S held by the bishop' sc. of Hereford. *Stanford(e) Episcopi* 1316–97. Earlier simply *Stanforde* 1174×86. The reference is to a crossing of the Frome at SO 6750. 'Bishop' for distinction from Stanford Regis, a deserted medieval village at SO 6650, *King(e)stanforde* 1243, *Stanford Regis* c.1273, earlier simply *Stanford* 1086, and Stanford Faucon (Court Farm) SO 6951, *Stannfordesfaucon* 1317, also known as *Parva Stanford* c.1267 and earlier simply *Stanford* 1086, another division of the Stanford estate held by the *ffaukun* or Falcon family of Hereford. He 89, 181.

(4) ~ BRIDGE H&W SO 7165. *Stanford Bridge* 1832 OS. P.n. Stanford as in STANFORD ON TEME SO 7066 + ModE **bridge**.

(5) ~ DINGLEY Berks SU 5771. 'S held by the Dingley family'. *Staneford Deanly* 1535, *Standford Dingley* 1744. Earlier simply *Stanworde* 1086, *Stanford(')* from 1235×6. Stanford was a crossing point on the River Pang. Robert Dyngley appears here in 1428. Brk 270.

(6) ~ IN THE VALE Oxon SU 3493. *Stanford in le Vale* 1496, *Standford alias Stanford in the Vale of the White Horse* 1599. Earlier simply *Stanford* from 1086. The boundary of the people of S occurs as *(œt) sanfordinga gemœre* (sic for *stan-*) [931]12th S 409. Brk 398.

(7) ~ -LE-HOPE Essex TQ 6882. 'S at the inlet'. *Stanford in the Hope* 1361, *in le Hope* 1475, ~ *Le Hope* 1535. P.n. Stanford + Fr definite artlicle **le** (short for *en le*) + ME **hope** 'small valley, bay, inlet' (OE *hop*, ON *hóp*). The bay or inlet here was known as *Tilberi Hope* in 1509, cf. LOWER HOPE TQ 7077. Stanford is *Stanford* from [1068]1309, *Standford* [1068]1377, *Stamford* 1348–9, *Stamvert -d* 1662, 1678. Ess 170 gives former pr [stæmvət].

(8) ~ ON AVON Northants SP 5878. *Stanford super Hauen* 12th, ~ *super Aven* 1367. Earlier simply *Stanford* from 1086. Nth 74.

(9) ~ ON SOAR Notts SK 5422. P.n. *Stanford* from 1086 + ModE **on** + r.n. SOAR SK 5123. Nt 255.

(10) ~ ON TEME SO H&W SO 7066. *Stanforde upon Temede* 1317. Earlier simply *Stanforde* 1086 + ModE **on** + r.n. + TEME. Also known as *Stanforde Esturmi* 1242 and ~ *Wassebourn* 1346 from the tenures of Johannes Sturmi 1242 and the Wassebourn family of Little Washbourne. Wo 79.

(11) ~ RESERVOIR Leic SP 6080. P.n. Stanford as in STANFORD ON AVON Northants SP 5878 + ModE **reservoir**.

(12) ~ RESERVOIR Northants SP 6080. P.n. Stanford as in STANFORD ON AVON SP 5878 + ModE **reservoir**.

(13) ~ RIVERS Essex TL 5301. 'S held by the Rivers family'. *Stan(e)ford ryueres* 1289, ~ *Ripar'* 1292, *Stanvord(e) Rithers* 1530. Earlier simply *Stanfort* 1086, *-ford* [1068]1309, *Stanverd* c.1640, and *villa de Ryvers* 1412. P.n. STANFORD + family name Rivers from Richard de Ripariis or Rivers 1213. Ess 77.

The STANG Durham NZ 0208. *Stang* 1860 OS. ModE **stang** (< ON *stǫng*, genitive sing. *stangar*) 'a pole'. Probably used as a guide-post or a cross although **stang** might have been the name of one of the hills here.

STANGHOW Cleve NZ 6715. 'Tumulus marked by a pole'. *Stanehou* 1273, *Stanghou(e)* 1280–1575. ON **stǫng** (< **stangu-*) + **haugr**. The earliest form has been influenced by OE **stān**. There are several barrows in the area. YN 146, HistCleve 263.

STANHOE Norf TF 8036. 'Stony hill-spur'. *Stanho(u)* 1086, *Stanho* 1173. OE **stān** + **hōh** possibly replaced with ON **haugr**. DEPN.

STANHOPE Durham NY 9939. 'Stone valley'. *Stanhope* from c.1160×70, *-hop(pe)* 1283–1685, *-op(pe)* 1408–1598, *Stane-Staynhop* 1368–1530. OE **stān** + **hop**. The reference is to a side-valley opening to the N from Weardale at the mouth of which the village stands. NbDu 187, L 116.

STANHOPE COMMON Durham NY 9642. *Stanhope Common* 1862 OS. P.n. STANHOPE NY 9939 + ModE **common**.

STANION Northants SP 9187. 'The stone house'. *Stanere* 1086, *Stanern(a) -e* 1162–1314, *Stanierne -(e)yerne* 1362–1486, *Stannyon* 1631. OE **stān**, later reformed as if from adj. **stāniġ**, + **ærn**. OE **ærn** is frequently used for a building with a special function, here probably a quarry-house or stone-working place. Cf. COLERNE Wilts ST 8171, POTTERNE Wilts ST 9958, *sealt-œrn*. Nth 171, DEPN.

STANLEY 'Stone clearing or pasture; clearing where stone is obtained'. OE **stān** + **lēah**. L 205.

(1) ~ Derby SK 4140. *Stanlei -leg' -leya -ley(e) -le(gh)* 1086–1229 etc. Db 607.

(2) ~ Durham NZ 1952. *Stanelay* c.1200, *-ley* 14th cent., *Stanley* from 1228. NbDu 187.

(3) ~ Mers SJ 3693. No early forms. It probably represents the family name *Stanley*, the surname of the earls of Derby.

(4) ~ Staffs SJ 9352. *Stanlega* 1130, *Stanleg' -le(gh) -le* c.1223–1317, *Stanlowe* c.1280, 1351, *Stanley(e)* from 1293, *Stanley otherwise Stanlow* 1587. Suffix confusion between **lēah** and **hlāw** 'a hill, a tumulus' is not unusual. Horovitz.

(5) ~ WYorks SE 3424. *Stanlei(e)* 1086, 1202, *Stanley* from 1274. YW ii.159.

(6) ~ FORCE Cumbr SD 1799. P.n. Stanley + dial. **force** (ON *fors*) 'a waterfall'. Stanley as in Stanley Hall, *Stanley hall, Standley heale* 1578, *Stanlehale* 1658, 'Stanley nook', OE **halh**, dative sing. **hale**. Cu 384.

(7) King's ~ Glos SO 8103. *Stanley Regis* 1216×72–1684, *Ki-Kyng(e)* ~ 1278–1705, *Kynges -ys -is* ~ 1327–1538. Earlier simply *Stanlea* 1160, *Stanlei(a) -le -leg(h) -ley(e)* 1164–1634.† 'King's' for distinction from Leonard STANLEY SO 8003; an ancient royal demesne held of the king by Tovi before and after the conquest. Gl ii.199.

†DB *Stantone* 1086 is erratic.

(8) Leonard ~ Glos SO 8003. 'St Leonard's S'. *Stanley Sancti Leonardi* 1261–1324, ~ *Leonards* 1349, *Leonard* ~ from 1535. Saint's name *Leonard* from the former dedication of the church + p.n. *Stanlege* 1086, *Stanlei(e) -ley(e)* 1100×35–1492. Gl ii.201.

STANLEY CROOK Durham NZ 1638. A recent combination of two older p.ns., Stanley NZ 1639 and CROOK NZ 1635, to distinguish the former from STANLEY NZ 1952. Stanley NZ 1639, *Stanlowe* 1364, *Stanley* 1484, *Stayn- Standley* 16th, is 'stone hill', OE **stān** + **hlāw**. Suffix confusion between **hlāw** and *-ley* < **lēah** is frequent in Northern p.ns., possibly helped by the existence of the by-form **hlǣw**. The surname *de Stanlawe* occurs in the county 1282–1418. Stanley is situated at the end of Broom Hill, a prominent ridge of high ground. NbDu 187.

STANLOW Ches SJ 4277 → Little STANNEY SJ 4174.

STANMER ESusx TQ 3309. 'Stone pool'. *Stanmere* [765]c.1300 B 197, 1086, 13th cent., *Stam(m)er(e)* 1086–18th. OE **stān** + **mere**. Sx 312, Jnl 25.44.

STANMORE Berks SU 4778. 'Stone pond'. *(æt) Stanmere* [948]13th S 542, 960 S 687, 1412, *Stanmore* 1761. OE **stān** + **mere**. The boundary of the Stanmore people is mentioned as *(of) stanmeringa gemere* [916]12th S 1542, ~ ~ *gemære* [1042]13th S 993. Brk 232.

STANMORE GLond TQ 1692. 'Stony pool'. *Stanmere* 1086–1411 including *[793]12th S 136, *Stand(e)more* 1562, 1605, *Stanmore* 1574. OE **stān** + **mere**. Also known as *Stanmere magna* 1235, *Great(e) Stanmare* 1392, 1565, *Much Stanmer* 16th and *Stanmer the more* 1563 for distinction from Whitchurch or Little Stanmore TQ 1890, *Stanmera* 1086, *alia Stanmera* 'the other S' 1106, *villa de Parvo Stanmere* 13th cent., *Whitchurch al. Little Stanmer* 1590. Mx 65.

Little STANNEY Ches SJ 4174. *Parua Staneya* 1278, *Little Stanney* 1583. Adj. **little**, Latin **parva**, + p.n. *Stanei* 1086, *Staney* 1163–1709 with variants *-eie -eia -eye*, *Stanney* from 1351, 'rock island', OE **stān** + **ēġ**. 'Little' for distinction from Great Stanney SJ 4075, *maior villa de Staneya* 1279, *Magna Stanney* 1499, *Greate Stanney* 1577. The district of Great and Little Stanney comprised the marshland south of the rocky promontory of Stanlow SJ 4277, *Stanlawa* 1172×8, *-law(e)* 1172–1819, *Stanlowe* from 1178×89 (*-low* from 1260) 'stone hill, rock hill', OE **stān** + **hlāw**. Che iv.180, 182, 185, L 39.

STANNINGTON 'The settlement at the stony place'. OE ***stāning** < **stān** + **ing**², + **tūn**.

(1) ~ Northum NZ 2179. *Stanigton'* 1242 BF 1116–1296 SR 59, *-ington* 1257, *Steynington'* 1256, *Stayngton* 1270, *Stainton* 1303. Stannington lies on the Great North Road which makes it possible that the first element is **stān** + **weġ**. NbDu 187, PNE ii.145,

(2) ~ SYorks SK 3088. *Stanygton* [13th]1310, *Stanington -yng-* [13th]1310–1609, *Stannyngton -ing-* from 1440. There is a steep rocky declivity to the N. YW i.227.

STANSBATCH H&W SO 3561. This is *Staunton bach* 1833 OS, p.n. Staunton as in STAUNTON ON ARROW SO 3760 + WMidl dial. **bache**, **batch** 'a stream, a valley', referring to a small stream South of the hamlet which rises near The Orls SO 3262 and flows into the river Arrow at Horseway Head. This is called *tanes bæc* 958 S 677, 'the Tan', a name which survives in Tan House SO 2448, *Tan House* 1833 OS, representing OE **tān** 'a branch, a valley or stream branching off the main valley or river', as in TANSLEY Derbyshire SK 3259 and TANSOR Northants TL 0590. The modern form seems to be due to a conflation of *Tan* and Staunton. He 183.

STANSFIELD Suff TL 7852. 'Stan's open land'. *Stanes- Stenesfelda* 1086, *Stanesfelde* c.1095, *Stanefeld* 1196–1253, *Stanfelde* 1610. OE pers.n. *Stān*, genitive sing. *Stānes*, + **feld**. DEPN.

STANSORE POINT Hants SZ 4698. *Stone Point* 1810 OS. P.n. Stone as in Stone Farm SZ 4599 + ModE **point**. The modern form of the name seems to go back to OE **stān**, genitive sing. **stānes**, as in Stanswood Farm SU 4600, *Stanevde* 1086, *Staneswode* 1283, the 'wood belonging to Stone', + ModE **ore** (OE *ōra*) 'shore, foreshore' as in nearby NEEDS ORE POINT SZ 4397. Stone, *Stanes* 1086–1327, *Stone juxta Falylee* (Fawley) 1294, seems to be first recorded as Bede's Latin *Ad Lapidem* [c.731]8th BHE, *æt Stane* [c.890]c.1000 OEBede, 'at Stone', OE **stān**, perhaps referring to a milestone on the Roman road that terminates at a one-time inlet on the coast here, Margary no.423. Ha 155, Gover 1958.199.

STANSTEAD Suff TL 8449. 'The stone place'. *Stanesteda* 1086, *-stede* c.1190, *Stansted(e)* 1197–1610. OE **stān** + **stede**. Stead 193.

STANSTEAD ABBOTTS Herts TL 3811. 'S held by the abbot' sc. of Waltham from 1212. *Stanstede Abbatis de Wautham* 1254, ~ *Abbots* 1318–25. P.n. *Stan(e)stede* 1086–1428, 'stone place', OE **stān** + **stede**, + ME **abbot**, Latin **abbas**, genitive sing. **abbatis**. Hrt 200, Stead 241.

STANSTED Kent TQ 6062. 'Stone-place'. *Stansted'* 1231, 1254. OE **stān** + **stede**. KPN 109, Stead 226.

STANSTED AIRPORT Essex TL 5422. P.n. Stansted as in STANSTED MOUNTFITCHET TL 5124 + ModE **airport**.

STANSTED HOUSE WSusx SU 7610. P.n. *Stanesteda* 1179–82, *Stansted(e)* from 1203, 'the stony place', OE **stān** + **stede**, + ModE **house**. Sx 55, Stead 258.

STANSTED MOUNTFITCHET Essex TL 5124. 'S held by the Mountfitchet family'. *Stan(e)sted(e) Mounfichet, Muffichet* 1288–1337. Earlier simply *Stanestedam* 1086, *Stan(e)sted(e)* 1201 etc., 'stone place', OE **stān** + **stede**. The 15th and 16th cent. forms are very confused (*stanford Mounfychet* 1472, *Stanfeld Monfichet* 1483, *Stapersted Mountfichet* 1485, *Stamford Schyched* 1503). Ess 533, Stead 205.

STANSWOOD Hants SU 4600 → STANSORE POINT SZ 4698.

STANTON 'Stone settlement'. OE **stān** + **tūn**. The reference may be to an ancient megalithic structure, more frequently to stony soil, or a quarry.

(1) ~ Derby SK 2719. *Stantun* 968, 1086, *-tona -ton(e)* c.1100–1292 etc. Db 659.

(2) ~ Glos SP 0734. *Stanton(e) -tona* 1086–1689, *Stawnton(a)* 1183, 1422×71, 1535, *Staunton(e)* 1248–1795. Gl ii.23 gives pr ['stɔːntən].

(3) ~ Northum NZ 1390. *Stantuna* 1200, *-tun* 1236 BF 598, *-ton'* from 1242 BF 1122. NbDu 187.

(4) ~ Staffs SK 1246. *Stantone* 1086, *Stanton* 1197. DEPN, Horovitz.

(5) ~ Suff TL 9673. *Stantun* 11th, *Sta(n)- Scantuna* 1086, *Stanton* 1610. DEPN.

(6) ~ BY BRIDGE Derby SK 3727. *Stantun* 1086, *-ton(e)* from 1325, ~ *atte Brigende* 1362, *Staunton* 1271–1593, ~ *juxta ponte Cordi vulgariter nuncupatum Swerkeston Bridge* 'S beside Cordy's bridge commonly called Swarkeston Bridge' 1532, referring to SWARKESTON Bridge across the river Trent at SK 3727 and to the family of Henry Cordy 13th cent. Also known as *Ston- Stonistanton -y-* 1243–1314 etc. There are good stone quarries in the area. Db 660–1.

(7) ~ -BY-DALE Derby SK 4638. *Stanton(e)* from 1086, *Staunton(e)* 1244–1335. Db 507.

(8) ~ DREW Avon ST 5963. 'Drogo's S'. *Stanton Drogonis* 1253, *Stantondru* 1219. Earlier simply *Stantone* 1086. *Drogo* is a Frankish pers.n. which became OFr *Dreu(s), Dr(i)u*. The place is named after three circles of standing stones at ST 6063 belonging to a major Neolithic religious site. DEPN, Pevsner 1958N.263, Thomas 1976.44.

(9) ~ FITZWARREN Wilts SU 1790. 'S held by the Fitzwarren family'. *Staunton Fitz Waryn* 1394, *Stanton Fitzwarren* 1535. P.n. *Stantone* 1086 + manorial addition from Fulco filius Warini 'son of Warinus' who held the manor in 1196. The soil is stonebrash overlying stonebrash and clay. W 30.

(10) ~ HARCOURT Oxon SP 4105. 'S held by the Harcourt family'. *Stantone Harecurt* 1268–81, *Stauntone* ~ 1285. P.n.

Stantone 1086 etc. + manorial affix from Robert de Harcourt, owner in 1193. The **stān** in this name refers to the stones known as The Devil's Quoits. O 282, Thomas 1976.000.

(11) ~ HILL Notts SK 4860. No early forms.

(12) ~ IN PEAK Derby SK 2464. *Staunton in Peake* 1272×1307, *Stanton(e) in le Peake* 1372. Earlier simply *Stantune* 1086, *Stanton(e)* from 1199×1216, *Staunton* 1202–1387. Db 165.

(13) ~ LACY Shrops SO 4978. 'S held by the Lacy family'. *Stantone* 1086 etc. with variants *-tun'* and *-ton('), Staunton(') -tun'* 1241–1790, + manorial addition *Lacy* first occurring in 1255. Roger de Lacy was the holder in 1086. Sa i.276.

(14) ~ LONG Shrops SO 5790. Originally 'long S', *Longa Stantona* c.1235, *Long(e)stanton* 1255–1833, subsequently *Staunton Longe* 1577, *Stanton Long* from 1698. ME adj. **long** + p.n. *Stantune* 901 S 221, 1086, *-ton'* 1167–95. The word order was remodelled so that *Long* became a pseudo-manorial addition in contrast to STANTON LACY at the opposite end of Corve Dale. The original reference was to the straggling nature of the village. Sa i.276.

(15) ~ -ON-THE-WOLDS Notts SK 6330. *Stanton super Wold* c.1240–1330, ~ *Othewold* 1324. P.n. *Stantun(e)* 1086, before 1163 etc. with variant spelling *-ton(e), Staunton* 1280–1318, *Staynton* 1325, + ModE **on the wolds**. Nt 249.

(16) ~ PRIOR Avon ST 6762. 'S held by the prior' sc. of St Peter's abbey, Bath. *Staunton Prior's* 1276. Earlier simply *(æt) Stantune* [963]12th S 711, *Stantun* [965]12th S 735, *Stantone* 1086. DEPN.

(17) ~ ST BERNARD Wilts SU 0962. 'St Bernard's S'. *Staunton Barnarde* 1553, *Stanton Barnard* 1572. P.n. *Stantun* [903]14th S 368, *(æt) Stantune* [957]14th S 647, *Stantone* 1086, *Stauntone* 1275, + saint's name *Bernard*. Also known as *Staunton Fits Herbard* 1402, 'S held by the Fitzherbert family' who must have descended from the Erebertus who held the manor in 1242. The reason for the later addition *Bernard* is unknown; to it an unauthorised *saint* has been prefixed wrongly inscribed on the 1817 OS as *Stanton Fitz Warren*†. The soil is marl overlying chalk. Wlt 323.

(18) ~ ST JOHN Oxon SP 5709. 'S held by the St John family'. *Stantona Johannis de Sancto Johanne* 'John of St John's S' 1155×61, *Stanton Seynt Johan* 1339. P.n. *Stantone* 1086 etc. + manorial affix from John de Sancto Johanne who gave the estate to Eynsham abbey 1135×49. O 187.

(19) ~ ST QUINTIN Wilts ST 9080. 'S held by the St Quintin family'. *Staunton Quyntyn* 1317, ~ *Seynt Quyntin* 1339. P.n. *Stantone* 1086 + manorial addition from the family name of Herbert de Sancto Quintino who held here in 1211. The soil is brash overlying clay. Wlt 112.

(20) ~ STREET Suff TL 9566. This is *Norton Street* 1837 OS, p.n. NORTON TL 9565 + dial. **street** 'a hamlet'.

(21) ~ UNDER BARDON Leic SK 4610. P.n. *Stanton(e)* from 1086, *Staunton* 1309–1612 + affix *de sub' monte Bardona* late 12th, ~ *undirberdon'* 1327, ~ *under Bardon* 1502–1610. Stanton lies under the granite hill of Bardon, *Berdon* 1240–1475, *Berghdone* 1438, *Bardon(')* from 13th, the 'barrow hill', OE **beorg** + **dūn**. Quarrying has removed any trace of barrows on this prominent hill. Lei 518, 483, L 142, 156.

(22) ~ UPON HINE HEATH Shrops SJ 5724. *Stanton upon Hyne Heth* 1327, ~ *upon Hindheath* 1720. P.n. *Stantune* 1086, *-ton(a)* from 1201, *Staunton(')* 1242–1720, + p.n. Hine Heath, *bruera de Hynehethe* 1291×2, *Hyne Heath* 1552, *Hind Heath* 1796, 'the hinds' heath', OE **hīwan**, genitive pl. **hīġna**, + **hǣth**. OE **hīwan** 'a household, members of a family' can refer in p.ns. to a monastic community or to domestic servants. The reference here is unknown. Sa i.277.

(23) ~ WICK Avon ST 6162. See Stanton WICK.

(24) Lower ~ ST QUINTIN Wilts 9180. *Lower Stanton* 1828 OS. ModE adj. **lower** + p.n. STANTON ST QUINTIN ST 9080.

(25) Stoney ~ Leic SP 49994. Prefix *Stony* 1312–1610 + p.n. Stanton, *Stantone* 1086, *Stantona -ton(e)* c.1160–1610, *Staunton(e)* 1261–1428. There are extensive stone and granite quarries here. Lei 545.

STANWARDINE IN THE FIELDS Shrops SJ 4124. *Stanwarthin in Le Felde* 1271×2, *Stanworthyng' in La Feld, Stanwordyn in Campo* 1291×2, *Stanworthin ythefeld* 1316. P.n. *Staurdine* 1086, *Stanwardin(e)* 1194 etc. with variants *-worthin -warthin -worthyn(g)* and *-wardyn, Stannarton -er-, Stannarden* 17th, 'the enclosure on stony ground', OE **stān** + **worthiġn**, + ModE **in the fields**, more correctly *in the field*, OE **feld** 'open country', sometimes with Fr definite article **le, la**, for distinction from Stanwardine in the Wood SJ 4227, *Stanward' in Bosco* 1231. Sa i.277.

STANWAY Essex TL 9324. '(At the) stone ways'. *(æt) Stanwægun* 1000×2 S 1486, *Stanewegā* 1086, *Stanuueie -wey(e)* 1119–1544. OE **stān** + **weġ**, dative pl. **wegum**. The reference is to the Roman road from London to Colchester, Margary no.3b, and to the originally separate villages of Great and Little Stanway with their separate parish churches of St Albright and All Saints (now in ruins). Ess 398, Pevsner 1965.367.

STANWAY Glos SP 0632. 'Stone road'. *Stan(e)wege -wei(e) -wey(a) -way* 1105–1729. OE **stān** + **weġ**. The reference is probably to the old salt-way through Hinton on the Green SP 0240 to Salter's Hill SP 0428 and beyond. Gl ii.24, L 83.

STANWELL Surrey TQ 0574. 'The stone spring'. *Stanwelle* 1086, *Stan(e)well(e)* 1199–1316, *Standwell* 1595. OE **stān** + **wella**. Mx 20, L 31.

STANWELL MOOR Surrey TQ 0474. 'Stanwell marsh'. *(super) moram de Stanewell* 1274, *Stanwellemore* 1363. P.n. STANWELL TQ 0574 + OE **mōr**. Cf. Walter de *la More* 1226. M 21.

STANWICK Northants SP 9771. 'The rocking stone'. *Stan wigga* [10th]c.1200, *Stanwige* 1086, *-wigga -(e) -wygg(e)* 1125×8–1428, *Stanewica* 1086, 1209×18, *-wyk(e)* 1285–1536, *Stanwik* 1232. OE **stān** + **wicga**. OE **wicga** normally means 'beetle', but the underlying sense is 'that which wags or wiggles'. The reference is probably, therefore, to a rocking stone like the Cornish logan-stones. Early remodelled as if a p.n. in **wīc** 'a farm, a work-place'. Nth 196 gives pr [stænik], DEPN.

STANWICK-ST-JOHN NYorks NZ 1811. 'St John's S'. P.n. *Stenweg(h)e, Steinueges -wege* 1086, *Stein- Stainwegges* 13th cent., *Stain- Staynwigges* 1285–1348, *Staynwyks* 1421, *Stanwyx* 1542, *Stanwick* 1665, the 'stone walls', ON **steinn** + **veggr**, + saint's name *John* from the dedication of the church. The reference is to constructions now represented by the extensive earthworks in Stanwick Park, also referred to in the p.n. ALDBROUGH ST JOHN NZ 2011.

STAPE NYorks SE 7993. *Stape* 1861 OS. Stape is on the line of Wade's Causeway, the Roman road from Malton to Whitby, Margary no.81b. The origin of the name might be OE **stæpe** 'a step' used of stepping stones, referring to the exposed stones of the road's foundation layer after the smoother surface had washed away. Since, however, this is a place where the road rises steeply from the valley bottom to the moors, a better suggestion is OE ***stēpe** 'a steep place, a declivity'. Margary 425.

STAPEHILL Dorset SU 0501. 'Steep hill'. *Staphill'* 1583, *Stapes-Hill* 1774, 1869, *Steep Hill* 1811, *Stape Hill* 1837, 1869. ModE **steep** + **hill**. The name must be ironical as there is only a slight rise here. Do ii.228.

STAPELEY Ches SJ 6749. 'Wood, clearing where posts are obtained'. *Steple* 1086, *Stapeleg'* from 1216×72 with variants *-le(e) -legh -ley* (from c.1240). OE **stapol** + **lēah**. Che iii.71, L 205.

STAPLE Kent TR 2756. 'The post'. *Stapl(es)* 1205, *Stapel'* 1240, 1254, *Staple* from 1240. OE **stapol**. PNK 525.

STAPLECROSS ESusx TQ 7822. 'Staple cross'. *Cruche de Staple*

†The church dedication is All Saints.

1296, *Staple Cross* 1768. P.n. *Stapele* c.1180, OE **stapol** 'pillar', + ME **crouche**, ModE **cross**. The pillar or cross marked the meeting-place of Staple Hundred, *Staple(hā)* 1086, *Staples* 1130, *Stapele* 12th–14th cents. Sx 520.

STAPLEFIELD WSusx TQ 2728. 'The open land with a post'. *Stapelfeld* 1315. OE, ME **stapol -el** + **feld**. Sx 268.

STAPLE FITZPAINE Somer ST 2618. 'S held by the Fitzpaine family'. P.n. *Staple* 1086, 1212, 1610, OE **stapol** 'a pillar' + family name Fitzpaine from Robert Fitzpaine, who held the estate in the 14th cent. The state may take its name from the large sarsen stone beside the church. DEPN.

STAPLEFORD 'A ford marked by posts'. OE **stapol** + **ford**.

(1) ~ Cambs TL 4751. *(ad) Stapelforda(m)* [956]12th S 572, *Stapelford(e)* c.1060–1345, *Stapleford(e)* from 1086. Ca 88.

(2) ~ Herts TL 3117. *Stapleford* from 1154×789 with variant *Stapel-*. Hrt 231.

(3) ~ Leic SK 8118. *Stapeford* 2× 1086, *Stapelford(e)* 1087×1100–1514, *Stapeleford'* 1199, *Stapleford* from 1159. The 1199 spelling may point to a form compounded with genitive pl. **stapola**. Lei 167, L 69.

(4) ~ Lincs SK 8757. *Stapleforde* 1086, *Stapleford(e)* from 1257, *Stapelford -il-* 1200–1361. A ford across the r. Witham. Perrott 264.

(5) ~ Notts SK 4837. *Stapleford* from 1086 with variant spellings *-el-* and *-il-*, *-forth* 1291–1432. Nt 151.

(6) ~ Wilts SU 0737. *Stapleford* from 1086 with variant *Stapel-* 1115–1219. Wlt 229, L 69.

(7) ~ ABBOTTS Essex TQ 5095. 'S held by the abbot' sc. of Bury St Edmonds. *Stapleford Abbottes* 1509×47, earlier ~ *Abbatis* 1235. P.n. Stapleford + ModE **abbot**, Latin **abbas**, genitive sing. **abbatis**. Stapleford is *Staplefordam -fort* 1086, *Stabelfoot* 1749. 'Abbotts' for distinction from STAPLEFORD TAWNEY TQ 5098. Ess 79 gives pr [steiplfut].

(8) ~ AERODROME Essex TQ 4996. P.n. Stapleford as in STAPLEFORD ABBOTTS TQ 5095 + ModE **aerodrome**.

(9) ~ TAWNEY Essex TQ 5098. 'S held by the Tany family'. P.n. Stapleford as in STAPLEFORD ABBOTTS TQ 5095 + family name *Taenny* 1254, *Tany* 1255, *Taweny* c.1480, from Richard de Tany, who inherited the manor by marriage c.1240. Ess 79.

STAPLEGROVE Somer ST 2126. 'Grove where posts are made'. *Stapilgrove* 1327, *Staplegroue* 1610. ME **stapel** (OE *stapol*) + **grove** (OE *grāfa*). DEPN.

STAPLEHAY Somer ST 2121. Apparently a compound of **staple** (OE *stapol*) 'a post, pillar' and **hay** (OE *(ġe)hæġ*) 'an enclosure'. Recorded simply as *Staple* 1809 OS.

STAPLE HILL Somer ST 2416. 'Hill marked by a post'. *Staple Hill* 1809 OS. ModE **staple** + **hill**.

STAPLEHURST Kent TQ 7843. 'Wooded hill where posts are obtained'. *Stapelherst* 1226, 1244, *-hurst(e)* 1242×3–1281. OE **stapol** + **hyrst**, Kentish **herst**. PNK 326, L 198.

STAPLERS IoW SZ 5189. 'Wooded hill where posts are obtained'. *Stapelhurst -herst* 1235–c.1300, *Staplers* from 1284. OE **stapol** + **hyrst**. Wt 247 gives pr ['stæplərz], Mills 1996.98.

STAPLE SOUND Northum NU 2337. *Staple* as in Staple Island NU 2337, *Staple Island* 1610, + ModE **sound** 'a narrow stretch of water between the mainland and an island' (OE **sund**). The cliffs at the S end of Staple Island are called *(The) Staples* 1610, 1798, from OE **stapol** 'a pillar'.

STAPLETON 'Settlement near or marked by a post' or 'where tree-trunks or posts are obtained'. OE **stapol** + **tūn**.

(1) ~ Avon ST 6176. *Stapleton* 1215, *Stapelton* 1221.

(2) ~ Cumbr NY 5071. *Stapelthein* (sic) 1188–9, *Stapelton -il-* 1190–1485, *Stapleton* from c.1210. Cu 112.

(3) ~ Leic SP 4398. *Stapletone* 1086, *Stapelton(e)* 1185–1530, *Stapleton* from 1313. Lei 529.

(4) ~ NYorks NZ 2612. *Staple(n)dun* 1086, *Stapeltun -ton* 1166 etc. YN 283.

(5) ~ Somer ST 4621. *Stapelton* 1212, *-le-* 1236. The reference may have been to the manufacture and supply of posts. DEPN.

STAPLETON 'The settlement at the steep place'. OE **stēpel** + **tūn**.

(1) ~ H&W SO 3265. *Stapleton'* from 1207, *Stepelton(e) -ul-* 1259–1473. Stapleton lies at the foot of a road climbing past Hell Peak (1089ft.). He 182.

(2) ~ Shrops SJ 4704. *Stepelton(e)* 1166–1399, *Stepleton* 1274–1785, *Stapelton' -le-* 1203×4, *Stapleton* from 1577. The reference is to Lyth Hill SJ 4606 which dominates the landscape and gave its name to the 1086 form of the name, *Hundeslit* 'Hund's part of Lyth', OE **hlith**. The forms with *Stapel-* and *Staple-* show early reinterpretation as if from OE **stapol**. Sa i.278.

STAPLEY Somer ST 1813. Possibly 'wood or clearing where posts are obtained'. *Staple* 1809, probably ultimately from OE **stapol** + **lēah**. Early forms are needed for certainty.

STAPLOE Beds TL 1460. 'Post hill-spur'. *Stapelho(u)* 1227–14th, *Staplo* 1512. OE **stapol** + **hōh**. A hill-spur marked by a post or where posts are manufactured and obtained or possibly 'post-shaped'. The place lies close to DULOE TL 1560. Bd 58, L 168.

STAR Somer ST 4358. *Star* 1817 OS. Possibly dial. **star** 'rough grass' although this is mainly a Northern and Eastern word (< ON *stǫrr*). Early forms are needed.

STARBOTTON NYorks SD 9574. 'Flat alluvial land where stakes are obtained'. *Stamphotne* (sic) 1086, *Stauerbot(t)en* [12th]15th–[1330]copy, *-bothem* [1334]copy, *-botell* [13th]copy, *Starbotene* [1268]15th, *-bot(t)en -botton -bottom* 16th. Probably OE ***stæfer** + **botm** replaced by ON **botn**; the DB spelling is probably corrupt but might point to ON **stafn** 'a pole' as the specific. YW vi.108, Names 5.109, SSNY 105, Jnl 20.40.

STARCROSS Devon SX 9781. Possibly 'stair cross'. *Sterrcrosse* 1578, *Star Crosse* 1689. The evidence is late but this might be ME **stair** + **cross** referring to a cross set up by fishermen by the steps giving access to the sea. Starcross is a late-developed fishing village; there were 'cellars' or store-houses here in 1578, only later converted for habitation. D 500, Fox 1996.66.

STARSTON Norf TM 2384. 'Styr's estate'. *Sterestuna* 1086, *-tun* [1087×98]12th, *Stirston* 1205. ON pers.n. *Styrr*, ODan *Styr*, secondary genitive sing. *Styres*, + **tūn**. DEPN, SPNN 349.

START BAY Devon SX 8444. *Start Bay* 1809 OS. P.n. Start as in START POINT SX 8237 + ModE **bay**.

START POINT 'A tail of land, promontory'. OE **steort**, ME **stert** + ModE **point**.

(1) ~ Corn SX 0486. *Start Point* c.1870. P.n. Start, *Start* 1841. Identical with START POINT Devon SX 8837 and the earlier name of TORPOINT SX 4355, *Stertpoynt* 1610. PNCo 157, Gover n.d. 84, 215.

(2) ~ Devon SX 8237. *Start Point* 1586. Start, *La Sterte* c.1550. Start Point is a prominent headland closing the S end of Start Bay. D 332.

STARTFORTH Durham NZ 0316. 'Roman-road ford'. *Stretford* [c.1040]12th, *Stradford* 1086, *Stred- Stretford(e)* c.1130–1316, *Stratford(e)* 1233–1576 Saxton, *Starforde* 1563, *Startforth* 1768 Armstrong. OE **strēt** + **ford**. The ford carried the Roman road from Bowes to Bishop Auckland, Margary no. 820, across the Tees. YN 304, L 62, 70.

STARTLEY Wilts ST 9482. 'The stiff wood'. *silvum q.v. Stercanlei* 'the wood called S' [688]13th S 1170, *Sterckle* 13th, *Sterkele* 1249, *Startley* 1558×79. OE **stearc** + **lēah**, dative sing. **(æt thǣm) stearcan lēaġe**. Wlt 73.

STATHE Somer ST 3728. 'The landing place'. *Stathe* from 1233. OE **stæth**. DEPN.

STATHERN Leic SK 7731. 'The stake thorn-tree'. *Stachedirne -tone* 1086, *Stakethern(e) -thirne -thyrne -thorne* 1235–1472, *Stac- Stakthern(e)* 1226–1362, *Stathern(e)* 1337–1610. OE **staca** + **thyrne**. The exact sense of the compound is uncertain, but

staca may be used in the sense 'boundary post' and so the 'thorn tree used as a boundary post'. Lei 194.

STATION TOWN Durham NZ 4036. A modern colliery settlement named after Wingate Station on the Hartlepool Branch of the North Eastern Railway opened in 1846.

Little STAUGHTON Beds TL 1062. *Parva Stokton* 1241, ~ *Stoutone* 1346. Adj. **little**, Latin **parva**. + p.n. *Estone* (sic) 1086, *Stoctuna -ton* 1167–1240, *Stowgh- Stoughton* 1390×2–1490, 'the outlying farmstead', OE **stoc** + **tūn**. 'Little' for distinction from Great STAUGHTON Cambs TL 1264. Bd 20 gives pr [stɔ:tən], Studies 1936.33.

Great STAUGHTON Cambs TL 1264. *Stocton Magna* 1232, *Moche Staughton* 1566. ModE **great**, Latin **magna**, ME **much** + p.n. *Stoctun* [975×1016]11th S 1487, *Tochestone* (sic) 1086, *Stocton(a)* 1163–1592, *Stokton* 1287–1502, *Stoghton* 1358, 1366, *Stoughton* 1504, OE **stoc** + **tūn** 'stock-farm' or possibly **stocc-tūn** 'enclosure made of stocks'. 'Great' for distinction from Little STAUGHTON Beds TL 1062. Hu 267.

STAUGHTON HIGHWAY Cambs TL 1364. *Staughton Highway* 1835 OS. P.n. Staughton as in Great STAUGHTON TL 1264 + ModE **highway** for distinction from Staughton Green TL 1365, *Staughton Green* 1835 OS. Staughton Highway is on the main road from Kimbolton to St Neots.

STAUNTON 'Stone settlement, settlement where stone is obtained'. OE **stān** + **tūn**.

(1) ~ Glos SO 5512. *Stantun(e) -ton(e)* 12th–1662, *Staunton(e)* 1221–1642. The reference here is to a former large standing stone, originally a rocking stone. Gl iii.247 gives pr [sta:ntən].

(2) ~ Glos SO 7929. *Stantun* [972]10th S 786, 1086, *Staunton* from 1221. Gl iii.186.

(3) ~ HAROLD HALL Leic SK 3721. P.n. Staunton Harold + ModE **hall**. Staunton Harold is *Stantone* 1086, *Stanton(e)* c.1130–1610, *Staunton(')* from [c.1218]14th + affix *Herald(e)* after 1250–1401, *Harold(e)* from after 1250. The manor was held by Harold de Stantona c.1160, otherwise called *Harold de Leec* 1185. Lei 399.

(4) ~ HAROLD RESERVOIR Derby SK 3723. P.n. Staunton Harold as in STAUNTON HAROLD HALL Leic SK 3721 + ModE **reservoir**.

(5) ~ IN THE VALE Notts SK 8043. *Stanton in le Vale* 1535, ~ *'ith Vale of Beaver* 1657, i.e. of Belvoir for distinction from STANTON-ON-THE-WOLDS SK 6330. P.n. *Stantone* 1086, *-ton(a) -(e)* before 1140–1284, *Staunton* 1269–1330, + ModE **in the vale**. Nt 217.

(6) ~ ON ARROW H&W SO 3760. *Staunton on Arrow* 1833 OS. Earlier simply *Stantun(e)* 958 S 677, 1086, 1243, *Stauntone juxta Penbrugge* 1397. He 182.

STAUNTON ON WYE H&W SO 3645. 'Stone hill'. *Staunton on Wye* 1833 OS. Earlier simply *Standvne* 1086, *Standon(a)* 1160×70, 1243, *Staundon(e)* 1273–1315. OE **stān** + **dūn**. He 184.

STAVELEY 'Wood or clearing where staves are obtained'. OE **stæf**, genitive pl. **stafa**, + **lēah**. L 200, 205.

(1) ~ Cumbr SD 4698. *Staue- Staveley(e) -lay(e) -ly* from 1189. We i.172.

(2) ~ Derby SK 4374. *Stavelie* 1086, *Stauele(ia) -leie -lega -leg(h) -l(e)y(e) -lay* 1154×89–1212 etc., *Staley -lee* 1383–1435 etc. Db 301.

(3) ~ NYorks SE 3662. *Stanlei(a)* (for *Stau-*) 1086, *Stave- Stauelai -lay -lei -ley(e)* 1202×8–1661. YW v.89.

(4) ~ -IN-CARTMEL Cumbr SD 3886. *Stavelay* 1282, *Staveley* 1491. La 199.

STAVERTON 'Settlement where stakes are made or obtained'. OE ***stæfer** + **tūn**.

(1) ~ Devon SX 7964. 'Settlement at *Stoford*'. *(æt) Stofordtune* 1050×72, *Stovretone* 1086, *Stovretona* 1086 Exon, *Stavert(h)on* from 1242. OE p.n. **Stoford* + **tūn**. *Stoford* is probably 'post ford', OE **stafa** + **ford** as in STOWFORD SX 4386. D 520.

(2) ~ Glos SO 8923. 'Stake settlement'. *Staruenton* 1086, *Stauerton(a)* early 13th–1560, *Star(e)ton* 1555, 1610–11. Gl ii.84 gives prs ['stævətən] and ['sta::vətən].

(3) ~ Northants SP 5361. *stæfertun* 944 S 495, *Staverton(e)* from 1086, *Star(e)ton* 1460–1702. Cf. *staver* in Netherlands p.ns. NGN vii.42. Nth 28 gives pr [stɛ:ətən], PNE ii.141.

(4) ~ Wilts ST 8560. *Stavretone* 1086, *Staverton* from 1212, *Stafferton* 1509–1733. Wlt 133 gives pr [stævətən].

STAWELL Somer ST 3638. 'Stony stream or spring'. *Stawelle* 1086, *Stanwelle* 1276†. OE **stān** + **wella**. DEPN.

STAXTON NYorks TA 0179. 'Stak's estate'. *Stac(s)tone, Staxtun* 1086, *Staxton(a)* [1180×95]c.1300–1583. ON pers.n. *Stakkr*, genitive sing. *Stakks*, + OE **tūn**. YE 118.

STAYTHORPE Notts SK 7553. The 'outlying farm where sedge grows'. *Startorp* 1086, 1197, *-thorp(e)* 1205–1392, *Staret(h)orp* 1196, 1244, 1288, *Stathropp* 1594, *Statrep* 1676. ODan **star** + **thorp**. Nt 196 records obsolete pr [statrəp], SSNEM 117, PNE ii.158 s.v. **storr**.

STEAN NYorks SE 0873. 'The rock'. *Steane* 1609–1817, *Stone* 1695, 1771, 1861 OS. OE **stān** with local pr [stiən]. YW v.215.

STEAN MOOR NYorks SE 0671. *Stone Moor* 1861 OS. P.n. STEAN SE 0873 + ModE **moor**.

STEARSBY NYorks SE 6171. 'Styr's farm or village'. *Estires- Stirsbi* 1086, *Stiresbi* c.1110×25–1308, *Steresbi -by* 1167–1399. ON pers.n. *Styrr*, genitive sing. *Styrs*, + **bӯ**. YN 28, SSNY 38.

STEART Somer ST 2745. 'Tail, projecting piece of land'. Cf. *Stert I., Stert Marsh* 1809 OS. Ultimately from OE **steort**. The pr is said to be dissyllabic by VCH.

STEBBING Essex TL 6624. 'The Stebbingas, the people called after Stybba' or possibly 'the people dwelling at the tree-stumps'. *Sti- Stabingā* 1086, *Steb(b)ingis -es* 1139–1256, *Stebbing(e)* from 1173, *Stubbing(es)* 1183–1291. OE folk-n. **Stebbingas*, Essex dial. for **Stybbingas* < pers.n. **Stybba*, **Stebba*, or possibly OE **stybb, stebb** 'a tree-stump', + **ingas**. Cf. Bardfield SALING TL 6826 for a similar formation. Ess 457, ING 24.

STEDHAM WSusx SU 8622. 'The stallion's river-bend land'. *Steddanham* 960 S 687, *Stedeham* 1086–1334, *Stedham* from 1308. OE **stēda**, genitive sing. **stēdan**, + **hamm** 1. There are, however, alternative forms of the name, *Stod(d)eham* 1187, 1279, *Sto- Studham* 1428–30, which point to OE **stōd** 'a stud' as the specific, while the 960 form, if reliable, might be 'Stedda's *hamm*', OE pers.n. **Stedda*, genitive sing. **Steddan*, a hypocoristic form of *Stēda*. The site is at the base of a meander loop of the r. Rother. Sx 29, ASE 2.46.

STEEN'S BRIDGE H&W SO 5457. *Steens Bridge* 1832 OS. A bridge across Humber Brook between Leominster and Bromyard. Perhaps ultimately from OE ***stǣne** 'a stony place'.

STEEP Hants SU 7425. 'The steep place'. *Stepe* 12th–1341, *(la) Stupe* c.1200–1316, *la Stiepe* c.1230, 1305, *Steepe* 1600. OE ***stīepe**. The reference is to the steep slope from the Downland towards Petersfield. Ha 155, Gover 1958.65, PNE ii.151.

STEEP HOLM Avon ST 2260. 'Steep island'. *Stepholm* 1189×99, *Stupeholm* 1216×72, *Stepelholme* 1331. OE **stēap** used as a p.n. **Stēape* 'the steep one' + lOE **holm** (< ON *holmr*). 'Steep' is here probably taken as the name of the island 'the island (called) Steep' (cf. *insula Stepen* Henry of Huntingdon 156) from the original OE name *(æt) Steapan Re(o)lice* 11th ASC(B, C, D) under year 915, 'the steep Relic', OE **stēap**, dative case definite form **stēapan** + p.n. *Relic* from OIr **re(i)lic(c)** 'graveyard' (ultimately < Latin *reliquiae* 'relics'), where a Viking fleet was starved out in 915. Steep Holm contrasts with nearby FLAT HOLM ST 2265, originally called 'broad Relic'. DEPN.

†*Stapelwille* [946×55]13th S 1740 is sometimes held to belong here, but this seems unlikely.

Great STEEPING Lincs TF 4364. *Magna Stepinge* 1231, *Great Steeping* 1824 OS. ModE **great**, Latin **magna**, + p.n. *Stepinge* 1086, 1154×89, *Stepinges* 1142×51, 1225, *Estepingues, Stepinga* 1154×89, *Stepping* 1205, *Steppinges* 1209×35, *Steping'* 1212, of uncertain origin. On formal grounds this could be OE ***stēap-ing** 'a steep place' < **stēap** + **ing**^{2} or folk-n. **Stēapingas* 'the people called after Steapa' < pers.n. *Stēapa* + **ingas**. 'Great' for distinction from Little Steeping TF 4362, *paruo Stepinge* 1154×89, earlier simply *Stepi* 1086 which may point to sing. p.n. *Stēaping* 'the steep place' < **stēap** + **ing**^{2} or 'the place called after Steapa'. DEPN, ING 67, BzN 1967.346.

STEEPING RIVER Lincs TF 4661. *The River of Steepinge alias the Limbe* 1661. P.n. Steeping as in Great STEEPING TF 4364 + ModE **river**. The original name of the river was *Lim(in)e* [12th]1311, *Lyme* [c.1195]14th, 1281, *Limine* (or *Limme*) 1276, *Lymm* or *Limb* 1889, identical with r. LEAM Warw SP 4568. RN 243, Cameron 1998.

STEEPLE 'Steep place'. OE **stīepel**.
(1) ~ Dorset SY 9181. *Stiple* 1086, 1464, *Stuple, Stupel(l)(e)* 1204–1509, *Stepel -le* 1222–c.1586, *Steeple* 1614. The village lies beneath a steep hill. Do i.95. The *Stup-* forms are from WS **stȳpel**.
(2) ~ Essex TL 9303. *Stepla(m)* 1086, *Steple* 1194 etc., *Lestepl', le Stepl', la Stepel'* 1232, 1239. Steeple lies beside a hill. Also known as *Stepeltun(e)* 1230–54, *-ton* 1227–1317, *Sti- Stupelton* 1255, 1282–3, 'estate called or at Steeple'. Ess 226.

STEER RIG Northum NT 8524. *Steer Rig* 1869. ModE **steer** + dial. **rig** 'a ridge'.

STEETON WYorks SE 0344. 'Farmstead built of or amongst tree-stumps'. *Stiuetune* 1086, *-ton* 12th, 13th, *Steueton* 1379, *Steton* 1572, *Steeton* 1594. OE **styfic** + **tūn**. YW vi.24.

STELLING MINNIS Kent TR 1446. 'Stelling common'. *Stelling Minnis* 1819 OS. P.n. Stelling + ModE **minnis** (OE *(ġe)mǣnnes*). The *minnis* is the area S of the settlement proper at TR 1447. It became attached to the settlement name on the pattern of nearby RHODES MINNIS TR 1542. Stelling, *Stellinges* 1086, c.1100, *Steallinge* c.1100, *Stelling* from 1240, is probably OE ***stelling** 'stall place, cattle fold', rather than an OE folk-n. **Steallingas* 'the people called after Stealla' < pers.n. **Stealla* + **ingas**. PNK 437, ING 15.

STENALEES Corn SX 0157. Possibly 'green Stennack'. *Stenaglease alias Stenylease* 1621, *Stenales alias Stone Rise* 1757. P.n. Stennack, Co **stenek** 'tin-place' (Co *stean* + adjectival suffix *-ek*) + **glas** for distinction from Stennagwyn SW 9654 'white Stennack', p.n. Stennack + **guyn**. This seems to be the most satisfactory explanation although the recorded forms for Stenalees actually point to Co ***lys** 'court' as the second element. This, however, would not make sense as Stenalees was never of administrative importance. PNCo 157, Gover n.d. 387.

STENBURY DOWN IoW SZ 5379. *Stenbury Down* 1781. P.n. Stenbury SZ 5279 + ModE **down**. Stenbury, *Staneberie* 1086, *la Steniberi* 1189×1204, *Stenebire -byre -beri* 1235–04, *Stenbyri* 1255, *-bury* 1351–1632, *la Steveneberi* 1294, *(la) Steuen(e)bury* 1305–83, is 'the stone fort', OE **stǣnen** + **byriġ**, dative sing. of **burh**. Forms with *-v-, -u-* are probably misreadings of *-n-*. Wt 161 gives pr ['stembəri], Mills 1996.98.

STEPNEY GLond TQ 3581. 'Stybba's landing-place'. *(of) Stybbanhyþe* c.1000, *Stibenhed(e)* 1086, *Stebbenheda* early 12th, *Stebenee* 1198, *Stepenhithe* 1370, *Stepney* 1466. OE pers.n. *Stybba*, SE dial. form *Stebba*, genitive sing. *Stebban*, + **hȳth**, SE form **hēth**. The early spellings show great variation between *Stiben-*, *Steb(b)en-* and *Stub(b)e-* and *-heth(e)*, *-hithe* and *-huth(e)*. The medieval parish of Stepney included Bethnal Green, Bow, the Isle of Dogs and Poplar. Mx 149.

STEPPINGLEY Beds TL 0135. 'Wood or clearing at Stepping, the steep place'. *Stepigelai* 1086, *Step(p)ingele(a) -legh* 1167–13th, *Step(p)ingle* 13th–1433. OE p.n. **Stēaping* or **Stēping* < **stēap** or ***stēpe** + **ing**^{2}, locative-dative sing. **Stē(a)pinġe*, with shortening of *ēa* in the four syllable compound, or OE folk-n. **Stēapingas* 'the people called after Steapa', OE pers.n. **Stēapa* + **ingas**, genitive pl. **Stēapinga*, + **lēah**, locative-dative sing. **lēa(ġe)**. Bd 84, BzN 1967.346.

Earl STERNDALE Derby SK 0967. 'S held by the earl'. *Erlisstenerdale* 1330, 1377, *Earlesterndalle* 16th cent. ME **erle**, genitive sing. **erles**, + p.n. *Stenredile* (sic) 1244, *Sternedale* 1251, *Stenerdale -ar-* 1415, *Sterndall* 1531, the 'valley with rocky ground', OE **stǣner** + **dæl**. 'Earl' for distinction from King STERNDALE SK 0972. Held by William de Ferrers, Earl of Derby in 1244. Db 366.

King STERNDALE Derby SK 0972. 'S held by the king'. *Kyngus Stenerdale* 1419, 1437, *Ky- King(e)stern(e)dale* from 1498. ME **king**, genitive sing. **kinges**, + p.n. *Stauredal(e) -dala* (for *Stanre-*) 1101×8–1210, *Stever- Steuerdale* (for *Stener-*) 1301–1431, *Stenirdale -er-* late 12th cent.–1472, *-re-* 1263, 1285, *Stern(e)dale* 1461×1483, *Stearndale* 1509×47, 'valley with rocky ground' as in Earl STERNDALE SK 0967 from which it was so distinguished. It was part of the royal manor and forest of the Peak. Db 134.

STERNFIELD Suff TM 3961. Possibly 'Sterni's open land'. *Sternesfelda, Sterne(s)fella* 1086, *Sternesfeud* 1254, 1286, *Sternefeld* 1235–1344, *Sternffeld* 1524, *Starnefilde* 1568, *Sternfeld* 1610. OE pers.n. **Sterni*, genitive sing. **Sternes*, + **feld**. The specific may, however, be the unidentified element ***sterne** seen also in SEWESTERN Lei SK 8821, SYDERSTONE TF 8332 Norfolk and possibly Tansterne Humbs TA 2237, *Tanstern(e)* 1086–1828, although this name has been explained as 'Tann's pool', ON pers.n. *Tannr*, genitive sing. *Tanns*, + **tjǫrn**. DEPN, Baron, YE 60.

STERT Wilts SU 0359. 'The tail of land'. *Sterte* 1086–1242, *Stuerte* 1196, *Sto(e)rte* 1189–1388, *Steorte, St(e)urte* 14th cent. OE **steort**. The ME *ue, oe, eo, eu* sps indicate a pronunciation [stør t], the antecedent of modern local pr [stə:rt]. Wlt 314.

STERT FLATS Somer ST 2645. 'Stert mud flats'. Cf. *Stert poynt* 1610. P.n. Stert or STEART ST 2745 + ModE **flats**.

STETCHWORTH Cambs TL 6458. 'Tree-stump enclosure'. *Steuicheswrðe* [1042×66]13th S 1051, *Steuicesuuorde* [c.1050]12th LibEl, *Steueche(s)worð(e), Stuuicesworde, Sti(ui)cesuuorde* 1086, *Steu(e)chew(o)rde -a -w(o)rthe* [1017×35]12th S 1520–1341, *Steu(e)chw(o)rth(e)* [1017×35]14th–1340, *Stew(e)ch(e)wrda -w(o)rth(e)* 1216, 1223, 1456, *Stekworth* 1277, *Stech(e)worde -worth(e)* 1374–1567. OE ***styfiċ**, genitive sing. ***styfiċes**, genitive pl. ***styfiċa**, + **worth**. Ca 119.

STEVENAGE Herts TL 2325. '(Place at) the strong oak-tree'. *Stithenæce* c.1060, *-ache* 1204–1428 with variant *Sty-*, *Stigenace* 1086, *Sti- Styvenach(e)* 1201–1320, *Stevenach(e)* c.1295–1451, *Stevenage* 1511. OE **stīth**, definite form dative case **stīthan**, + **ǣċ**, dative sing. of **āc**. Confusion between fricatives [ð] and [v] led to association with pers.n. Stephen in the forms *Stephenhache* 1368, *Stephyn Hache* 1527. Hrt 137 gives prs [sti:vənidʒ, stivnidʒ], TC 175.

STEVENTON Hants SU 5447. Partly uncertain. *Stivetvne* 1086, *Stivinton(a) -en-* 1167–1341 with varaints *Sty-* and *-ing- -yng-*, *Steventhon* 1280, *Stephynton* 1535. Possibly the 'estate called after Stif(a)', OE pers.n. **Stīf* or **Stīfa* + **ing**^{4} + **tūn** with later influence from the pers.n. *Steven*. But the occurrence of the identical name form at STEVENTON Oxon SU 4691, STEVINGTON Beds SP 9853 and Stevington End Essex near Bartlow TL 5845, *Stauin- Steui- Staum- Steintuna* 1086, *Stev- Stivinton(e)* 1197–1285, has led to the suggestion of OE ***styficing** 'place where trees have been grubbed up' < ***styfic** 'stump' + **ing**^{2} as the specific. Ha 155, BdHu 46, Berks 417, Ess 507, PNE ii.166.

STEVENTON Oxon SU 4791. 'Stifa's estate'. *Stivetune* 1086, *Stiueton'* 1224–1337 with variants *Sty- -v-* and *-ton(a), Estiventona* [1122]1305, *Stiuinton' -en-* 1220–1369 with variants *Sty- -v-* and *-ton(a), Stiuingtona* [1230]c.1250, *Stevyngton*

1295–1380. OE pers.n. *_Stīfa_, genitive sing. _Stīfan_, + **tūn** possibly varying with *_Stīf(a)_ + **ing**[4] + **tūn**. The earliest reference is to the boundary of the people of Steventon, _(to) stifingc hæma ge mæra_ [964]c.1200 B 1142, _(to) stifingehæme gemære_ [964]c.1240 ibid. in which _stifinge hæme_ is short for *_stifingtun hæme_. Brk 417.

STEVINGTON Beds SP 9853. Partly uncertain. _Stiuentone_ 1086–1316, _Stiuington_ 1247–1350, _Stevyngton_ 1315–1515. The recorded forms seem to point to pers.n. *_Stīfa_ or *_Styfa_, genitive sing. *_Stīfan_, *_Styfan_, + **tūn**, 'Stifa or Styfa's estate' or ***styfing** 'place marked by tree-stumps' + **tūn** and so 'settlement at the tree-stumps' or 'place called or at _Styfing_'. Bd 46, DEPN, Årsskrift 1974.59.

STEVINGTON END Essex → STEVENTON Hants SU 5447.

STEWARTBY Beds TL 0242. A new village built for workers in the brick-making industry and named after Halley Stewart, chairman of a local brick company. Mills 1991.309.

STEWKLEY Bucks SP 8526. 'Clearing with a stump or stumps'. _Stivelai_ 1086–1197, _Stiuecle(ia) -lai_ [c.1135]c.1230, 1182–1241, _Stucle(y)_ 1394, 1446, _Stukeley_ 1485. OE ***styfic** + **lēah**. Bk 72. Jnl 2.33 gives local pr [stu:tli].

STEWTON Lincs TF 3686. 'The stub farm or village'. _Stivetona -u-_ 1086, _Stiuetuna_ c.1115, _-tun_ 1199. OE ***styfic** + **tūn**. DEPN.

STEYNE SYorks → River LOXLEY SK 2989.

STEYNING WSusx TQ 1811. Either '(the place) amongst the Staningas, the people of the stone or stony place', or 'the people called after Stan'. _(þone hám æt) Stæningum_ 'the homestead or estate at S' [873×88]11th S 1507, c.1030 Secgan, _Stænig_ 11th coins, _Estaninges_ 1086, _Staninges_ 1086–1301 with variant _-ing_ 1228, 1261, _Stening(e) -yng(e(_ 1274–1301, _Steyninge_ 1316, _Stenning_ 1641, 1726. OE folk-n. *_Stǣningas_ < **stān** 'a stone' or ***stǣne** 'a stony place' or pers.n. *_Stān_ + **ingas**, dative pl. *_Stǣningum_. Alternatively this might be the dative pl. of OE ***stāning**, ***stǣning** 'a stony place' < **stān** + **ing**[2]. As there is no clear geological feature here to explain the folk-n., derivation from pers.n. _Stān_ may be best. An exact parallel is found in _stæninga haga_ 'the enclosure of the people of _Staines_' TQ 0471. Sx 234 gives pr [steniŋ], ING 40.

STIBB Corn SS 2210. 'Tree-stump'. _Stibbe_ 1318 Gover n.d., _Stybbe_ 1327, _Stibb_ 1606. OE **stybb**. Cf. STIBB CROSS Devon SS 4314, STIBB GREEN Wilts SU 2262. PNCo 159, Gover n.d. 10.

STIBB CROSS Devon SS 4214. 'Stibb crossroads'. _Stibb Cross_ 1809 OS. P.n. Stibb + ModE **cross**. Stibb, _Stibbe_ 1330, is OE **stybb** 'a stub, a tree-stump'. D 96, PNE ii.165.

STIBB GREEN Wilts SU 2262. 'The green at Stibb, the tree-stump'. _Steep Green_ 1773, 1817 OS, _Stibb Green_ c.1840. Cf. the lost name _Stibmarshe_ 1613, _Stibbe Marsh_ 1626. P.n. *_Stibb_, OE **stybb** 'a tree-stump', + ModE **green**. The p.n. probably occurs in the surname of Alan _atte Stubbe_ 1264, 'Alan at the stub'. Wlt 338.

STIBBARD Norf TF 9282. Uncertain. _Estanbyrda, Stabyrda, Stabrige_ 1086, _Stiberde_ 1202–91, _Stibrede_ 1238, _Stibyrd_ 1270, _Stiburde_ 1316. The forms seem to point to three variant names, OE **stān** + **byrde** 'stone border, stone edge', **stan** + **brycg** 'stone bridge', and **stīġ** + **byrde** 'path or road edge, road side'. The relationship between these variants and their significance are not known. DEPN, Studies 1936.163, RES 5.75–6, PNE i.73.

STIBBINGTON Cambs TL 0898. 'Estate called after Styb(b)a'. _Stebintune -tone_ 1086, _Stibinctuna_ [c.1150]14th, _Sti- Stybington_ 1291–1428. OE pers.n. *_Styb(b)a_ + **ing**[4] + **tūn**. Hu 197.

STICKER Corn SW 9750. Probably 'tree-stumps'. _Stekyer_ 1319, 1398, 1416, (wood of) _Stekyer_ c.1395, _Stykker_ 1389. Co ***stekyer**, a plural formed from **stok** 'a tree-stump' (a loan word from English) + pl. ending **-yer**. PNCo 159, Gover n.d. 414.

STICKFORD Lincs TF 3559. 'The ford marked by a stick or sticks' or 'leading to the island called _Sticca_'. _Stichesforde_ 1086, _Sticceforda_ 1142, _Stikeford(e)_ 1185, [1188]14th, 1209×19. OE **sticca** + **ford**. The second explanation is suggested by the proximity of STICKNEY TF 3456. DEPN, Cameron 1998.

STICKLE 'A steep place'. OE ***sticele**, dial. **stickle**.

(1) ~ PIKE Cumbr SD 2193. _Stickle Pyke_ 1786, 1801. Dial. **stickle** + **pike** (OE _pīc_, ON _pík_) 'a mountain peak'. A steep pointed peak rising to 1231ft. La 193 f.n.2.

(2) ~ TARN Cumbr NY 2807. _Stickel Tarn_ 1800. P.n. Stickle as in Harrison STICKLE + dial. **tarn** 'a lake' (ON _tjǫrn_). We i.17.

(3) Harrison ~ Cumbr NY 2807. _Harrison Stickle_ 1865. Local family name Harrison + dial. **stickle**. The name of one of the Langdale Pikes rising to 2401ft. We i.206.

STICKLEPATH Devon SX 6494. 'Steep path'. _Stikelepethe_ 1280, _-pathe_ 1292. OE **sticol** + **pæth**. The old road from Exeter to Okehampton climbs steeply out of the village. The identical name occurs at SS 5532, _Styklepeth_ c.1280, and Somer ST 0436 and 3012 at the foot of vertiginous lanes. D 166, 27.

STICKNEY Lincs 3456. 'Sticca's island' or 'the island called _Sticca_, the Stick'. _Stichenai_ 1086, _Sticcenaia_ 1142, _Stikenei(a)_ 12th, _Stickenay_ 1202. OE pers.n. *_Sticca_ or **sticca**, genitive sing. *_Sticcan_, **sticcan**, + **ēġ**. It has been suggested that the site of the village, a long stretch of ground between parallel streams was called _Sticca_ 'the stick'; but an OE by-name _Sticca_ is also possible. DEPN, L 39.

STIFFKEY Norf TF 9742. 'Tree-shaped island' or 'island with tree-stumps'. _Stiuekai_ 1086, _Stiuekeia_ 1203, _Stivekeye_ 1242. OE ***styfic** + **ēġ**. DEPN gives pr [stju:ki].

River STIFFKEY Norf TF 9332. A back-formation from the p.n. STIFFKEY TF 9742. RN 377.

North STIFFORD Essex TQ 6080. ModE **north** + p.n. _Stifort -forda, Estinfort_ 1086, _(on) Stiþforde_ c.1090, _Sti- Styf(f)ord_ from 1225, 'lamb's cress' or 'difficult ford', OE **stīthe** or **stīth** + **ford**. Ess 128, DEPN, Jnl 2.42.

STIFFORD'S BRIDGE H&W SO 7348. _Stiffords Bridge_ 1832 OS. A bridge across Cradley Brook between Hereford and Worcester. Surname _Stifford_ + ModE **bridge**.

STILLINGFLEET NYorks SE 5941. 'Stretch of river called or at _Styfeling_, the place called after Styfel' or 'of the Styfelingas, the people called after Styfel or living at _Styfeling_'. _Steflingefled -feld, Steflinflet_ 1086, _Steuelingeflet_ 1190, 1191, _Stiuelingflet(e) -y- -v-_ 12th–1416, _Stilingflet(e) -y-_ 1330–1568. OE *_Stȳfeling_ < pers.n. _Stȳfel(a)_ + **ing**[2], locative–dative sing. *_Stȳfelinge_, varying with *_Stȳfelingas_ < pers.n. *_Stȳfel(a)_ + **ingas**, genitive pl. *_Stȳfelinga_, + **flēot**. The reference is to a creek draining into the river Ouse. YE 266, Jnl 29.81.

STILLINGTON Cleve NZ 3723. 'Estate called after Styfel'. _Stillingtune_ c.1190, _Stilington(e) -yng-_ c.1324–1415, _Stel(l)yngton -ing-_ 1382–1561, _Stillington_ from 1433. OE pers.n. _Stȳfel(a)_ + **ing**[4] + **tūn**, cf. STILLINGTON NYorks SE 5867, STILLINGFLEET NYorks SE 5941. The alternative suggestion that the specific is OE **still** 'a place for catching fish' seems unlikely on both topographical and phonological grounds (the expected dial. form is _stell_). Årsskrift 1974.55, Trans D&A n.s.6.92.

STILLINGTON NYorks SE 5867. 'Estate called after Styfel'. _Stiuelinctun_ 1086, _Stiuelinton'_ 1176, _Sti- Styvelington(e)_ 1280–1351, _Stillyngton_ 1371, 1442. OE pers.n. *_Stȳfel(a)_ + **ing**[4] + **tūn**. YN 27.

STILTON Cambs TL 1687. 'Settlement by the ascending path'. _Stic(h)iltone_ 1086, _Stichelton_ 1167, _Stikelton_ 1181, _Stilton_ from 1219, _Stig(h)elton_ 1227–8. OE **stiġel** + **tūn**. The reference is to the rising ground due west of the village; a road climbs from 50ft. to 170ft. on the way to Caldecote. Forms with _-ch-_ and _-k-_ suggest that the original specific was OE ***sticel(e)**. Hu 199.

STILTONS NYorks SE 5984. 'Til(l)i's estate or farm'. _Tilstun(e)_ 1086, _Thilleston', Thyllestonam_ [c.1180]16th, 1252, _Tilston'_ 1230. OE pers.n. _Til(l)i_, genitive sing. _Til(l)es_, + **tūn**. The strange development _Tilston_ > _Stiltons_ is unexplained. YN 73.

STINCHCOMBE Glos ST 7398. Possibly 'sandpiper valley'. *Sti- Styntescumb(e) -comb(e)* 1150×60–1413, *Sti- Stynch(e)comb(e)* c.1290–1651. OE ***stint** 'sandpiper, dunlin', genitive sing. ***stintes**, possibly used as a pers.n., + **cumb**. Gl ii.250, L 93, L 103.

STINSFORD Dorset SY 7191. 'Sandpiper ford'. *Sti(nc)teford* 1086, *Stinteford* 1236–88, *Sti- Styntesford* 1244–1446, *Stynsford* [1270]1372, 1492, *Stynford* 1575, 1591. OE ***stint** 'sandpiper, dunlin', genitive pl. ***stinta**, genitive sing. **stintes**, + **ford**. Do i.366, L 71.

STIPERSTONES Shrops SO 3698. *Stiperstones* 1836 OS. A range of hills rising to 1760ft.

STIRCHLEY Shrops SJ 7006. 'Stirk clearing or pasture'. *Styrcleage* 1002×4 Charters II, *Stirclege* 1177, *Stirchleia -le(g') -ley* from c.1200, *Stirchle(gh) -ley(e)* 1203–1560. OE **stirċ** 'a calf' + **lēah**. Sa i.280.

STISTED Essex TL 8024. Partly uncertain. *Stistede* [1042×53]13th S 1535, [1052×66]13th S 1519, *Stiestedā* 1086, *Stisted(e)* c.1095–1476 with variants *-steda -stude -stode -styde, Stidsted(e)* 1198, 1204, *Stithstede* 1298. The generic is OE **stede** 'place, site', the specific any of OE **stīġ** 'a path', **stigu** 'a sty, a pen', **stīthe** 'lamb's cress' or adj. **stīth** 'stiff' referring to the soil which here is loam and clay overlying clay. Ess 460 gives pr [staistid], Stead 206, Jnl 2.46.

STITHIANS Corn SW 7336. 'St Stithian's (church)'. *(ecclesia) Sancte Stethyane* 1268, *(rector) Sancte Stediane* 1282–1341, *Seint Stethyent* 1353, *Seynt Stedyan* 1478, *Stethyans* 1524, *Stithians* 1610. Saint's name *Stithian* of whom nothing is known apart from the dedication of the church. PNCo 159, Gover n.d. 529, DEPN.

STITHIANS RESERVOIR Corn SW 7136. P.n. STITHIANS SW 7336 + ModE **reservoir**. Constructed 1965. PNCo 159.

STITTENHAM NYorks SE 6767. '(Settlement at) the steep places'. *Stidnun* (for *Sticlum*) 1086, *Stitlum* 1185–1295, *Stiklum* [c.1260]13th, *Sti- Stytel(l)um* 13th cent., *Stitnum* 1310, *Stytnam* 1316–1615. OE **sticol** 'steep', ***sticel(e)** 'a steep place', dative pl. **sticlum**. The village stands at the top of a very steep hill overlooking Sheriff Hutton Carr. Early spellings with *t* may be due to *c/t* confusion in early handwriting or to sound substitution. YN 33, PNE ii.152.

STIVICHALL WMids SP 3376. 'The tree-stump nook'. *Stivichall* from [c.1144]1348, *-hale* 1274, 1280, *Stychehale* 1421, *Stychall* 1535, *Stichall, Steachell, Stichell* 17th. OE ***styfiċ** + **halh**. Wa 179 gives pr [staitʃəl], L 106.

STIXWOULD Lincs TF 1766. 'Stig's wold or woodland'. *Stigesuuald -walt -walde* 1086, *Sticheswald(a) -uuald* 1115–60, *-weld* c.1115, *Stichewald* 1150×60, *Stikeswald* 1212. ODan pers.n. *Stig*, genitive sing. *Stigs*, + OE **wald**. Stixwould is situated on raised ground in a low-lying area near the Witham. DEPN, SSNEM 222, L 226–7.

STOAK Ches SJ 4273. 'Outlying farms'. *Stok* 13th–1360, *Stoke* 1260–1620, *Stooke* 1666–1785, *Stoak* from 1724, *Stokes in Wirhale* 1284–1582. OE **stoc**, pl. OE ***stocu**, ME **stokes**. Che iv.181 gives pr [stouk], formerly [stu:ək].

STOBOROUGH Dorset SY 9286. 'Stone hill or barrow'. *Stanberge* 1086, 1284, *Stabergh* 1253–1319, *Stoburgh(e)* 1315–1477, *-bor(r)o(u)gh(e)* from 1512. OE **stān** + **beorg**, later replaced by ME **burgh** (OE *burh*). Do i.73.

STOBOROUGH GREEN Dorset SY 9285. *Stoborough Green* 1811. P.n. STOBOROUGH SY 9286 + ModE **green**. Do i.78.

STOCK Essex TQ 6998. 'The tree-stump'. *Stocke* 1337, 1583, *Stokk(e)* 1475, 1551. Short for 'Hereward's Stock or tree-stump', *Her(e)ward(e)stoc(ke) -stok(e)* 1234–1372, OE pers.n. *Hereweard*, genitive sing. *Hereweardes*, + **stocc** later taken as a p.n. There were many variations of this name in later times, *Hereforstok* 1373, *Harfoth Stoke, Hertford Stock, Stock Hereford* 16th, *Harvord Stook* 1604, *Harrard Stock* 1608. Ess 269.

STOCKBRIDGE Hants SU 3635. 'Log bridge'. *Stocbrugge* 1221, *Stoc- Stokbrigg(e) -brig' -brug(ge)* 1227–1316. OE **stocc** + **brycg**. This place is probably referred to in DB f.47b under the name *Sōmburne*; Stockbridge was also known as White Somborne. Ha 155, Gover 1958.186, Proc.Hants 32.93–101.

STOCKBURY Kent TQ 8461. 'Fortified place of the Stoke people'. *Stochingeberge* 1086, *Staca- Stocingabere* c.1100, *Stoking(e)bir' -beri* 1208–1242×3, *Stokyngbery, Stoke Ingeberi* 1253×4, *Stockebir'* 1233, *Stockbery* 1610. OE folk-n. **Stocingas* < **stoc** perhaps used as a p.n. referring to STOKE TQ 8275 or one of the other Kent *stokes*, + **byriġ**, Kentish **beriġ**, dative sing. of **burh**. The reference may have been to a manor house which preceeded the motte and bailey castle here. PNK 230, PNE ii.155.

STOCKCROSS Berks SU 4368. Probably 'cross-roads at a tree-stump'. *Stok(e)crosse* 1547, *Stockcross* 1817. ModE **stock** + **cross**. Brk 268.

STOCKDALEWATH Cumbr NY 3845. '*Stokker* ford'. *Stokerwath* 1340, *Stokkerwath* 1372, *Stokhalwath* 1440, *Stokelwath* 1488, *Stockellwath* 1608. P.n. **Stokker* 'stock or tree-trunk shieling', ON **stokkr** or OE **stocc** + ON **ǽrgi**, or OE occupational term ***stoccere** 'one who deals with tree-stumps, a tree feller' or the surname *Stocker* derived from it, + **vath**. A ford across the Roe Beck. The reference is to tree-felling, whether for timber or land clearance. Cu 246, SSNNW 71, DBS 334.

STOCKERSTON Leic SP 8397. 'The timber stronghold'. *Sto(c)tone* 1086, *Stocfaston' -k-* c.1130–1576, *Stokeston* c.1291, 15th cent., *Stokerston* 1535–1610, *Stockerson* 1572–3. OE **stocc** + **fæsten**. Lei 255.

STOCK GAYLARD HOUSE Dorset ST 7213. P.n. Stock Gaylard + ModE **house** referring to an 18th cent. mansion here. Stock Gaylard, 'Stoke belonging to the Coillard family', is *Stok Coillerd* 1299, *Stock Coylard* 1302, *Stoke Coil(l)ard* 1305–1431, *Stock Gaylard* from 1393. Earlier simply *Stoches* 1086, *Stok(e)* 1268–1575, 'outlying farm', OE **stoc**. Do iii.350, Studies 1936.20, Newman-Pevsner .399.

STOCK GREEN H&W SO 9959. *Stoke Green* 1558. P.n. Stock, *la Stolke* (sic) 1271, *Stoke* 1275, 1364, *(atte) Stocke* 1327, ME **stokke** (OE *stocc*) 'a tree-trunk', + ModE **green**. Wo 329.

STOCKINGFORD Warw SP 3491. 'The ford by or leading to the cleared ground'. *Stoccingford* 1157, *Stockingeford* 1221–1335, *Stok(k)- Stockinford -y-* 1221–1413, *Sto(c)kenford* 1504–1633. OE ***stoccing** 'a clearing of stumps, a piece of ground cleared of stumps' perhaps used as a p.n. + **ford**. Wa 91.

STOCKING PELHAM Herts TL 4529 → Stocking PELHAM.

STOCKLAND Devon ST 2404. 'Newly cultivated land at Stoke'. *Stokeland* [934]17th S 391, *Stocland* [998]18th S 895 and from 1202 with variants *Stok-* and *-lond(e) -laund(e)*. OE **stoc** used as a p.n. + **land**. The DB form is *Ertacomestoche* '*Ertacumb* outlying farm'. D 647, L 249.

STOCKLAND BRISTOL Somer ST 2443. 'S belonging to Bristol' sc. Corporation. *Stockland Bristol* 1809 OS. Earlier simply *Stocheland(e)* 1086, *Stoclande* 1166, *Stokeland* [1295] Buck, 'newly cultivated land at the dependent settlement', OE **stoc** + **land**. The reference is to reclamation of coastal marsh near the river Parrett estuary, the dependent settlement most likely being the *stoke* of STOGURSEY ST 2042. Also known as *Stocland Gaunt*. The estate was granted by Maurice de Gaunt to the Hospital of St Mark in Bristol (Gaunt's Hospital) and at the suppression in 1541 by the king to the Mayor and Corporation of Bristol. DEPN, L 248–9, VCH VI.126.

STOCKLEIGH 'Wood or clearing where stocks or logs are cut and obtained'. OE **stocc**, genitive pl. **stocca**, + **lēah**. Identical with STOCKLEY.

(1) ~ ENGLISH Devon SS 8506. 'S held by the English family'. *Stokeley Engles* 1268, *Stokleyenglisch* 1390. The manor was held in 1242 by Gilbertus Anglicus 'the Englishman'. Earlier simply *Stochelie* 1086, *Estocheleia* 1086 Exon, *Stockelegh* 1242. 'English' for distinction from STOCKLEIGH POMEROY SS 8703. D 418.

(2) ~ POMEROY Devon SS 8703. 'S held by the Pomeroy family'. *Stokelegh Pomeray* 1261. The manor was held by Henry de la Pumerai in 1200. Earlier simply *Stochelie* 1086, *Estocheleia* 1086 Exon, *Stochelega* 1187. 'Pomeroy' for distinction from STOCKLEIGH ENGLISH SS 8506. D 418.

STOCKLEY Wilts SU 0067. 'The tree-stump wood or clearing or where stocks are obtained'. *Stokele* 1281, 1289, *Stokeleye -legh* 1306, 1332, *Stoclygh* 1319, *Stockle* 1336, *Stokkele* 1351, 1372. OE **stocc**, genitive pl. **stocca**, + **lēah**. Wlt 258.

STOCKPORT GMan SJ 8990. 'Market-place at or belonging to a stock-farm or dependent hamlet'. *Stokeport* 1188–1550 with variants *-porte -pord -part*, *Stokport(e)* 1249, 1269, *Stockport* from c.1274. OE **stoc** + **port**. Che i.294, TC 175.

STOCKSBRIDGE SYorks SK 2698. A 19th cent. township created from Bradfield SK 2692. There are no early forms, but the name would mean 'bridge made of logs'. OE **stocc** + **brycg**. YW i.257.

STOCKSFIELD Northum NZ 0561. 'The open land belonging to a *stoc*, an outlying hamlet' or possibly 'a religious house'. *Stokesfeld'* 1242 BF, 1296 SR. OE **stoc**, genitive sing. **stoces**, + **feld**. If the sense 'religious house' is admissible for **stoc** the reference in this case would be to Hexham Abbey. NbDu 190, Ekwall Studies 1936.37.

STOCKS RESERVOIR Lancs SD 7356. P.n. Stocks + ModE **reservoir**. Stocks, also known as Stocks in Bowland, *Stokes* 1246, *Stocks* 1771, is ME plural **stokes, stockes** of either OE **stoc** 'an outlying dairy-farm' or **stocc** 'a tree-stump'. YW vi.203.

STOCKTON 'A settlement with or belonging to a *stoc*, an outlying farm', OE **stoc** + **tūn**; or 'a settlement built of logs or where logs are obtained', OE **stocc** + **tūn**. Studies 1931.33–4.

(1) ~ H&W SO 5161. Possibly 'farm belonging to Stoke Prior'. *Stoctune* 1086, *Stoc(k)tuna* 1123, 1186×98, *Stocton(e)* 1291–1326, *Stok(es)ton(')* 1324, 1332. Stoke Prior appears next to Stockton in the list of the constituents of Leominster manor in DB. He 112.

(2) ~ Norf TM 3894. *Stoutuna* 1086, *Stocton* 1180ff. DEPN.

(3) ~ Shrops SO 7299. *Stochetone* 1086, *Stoc- Stokton(')* 1244–1492, *Stockton* from 1577. Also known as *Body Stocton* 1392 from the family of Richard Body, recorded as owner of *Parva Stokton'* 'little Stockton' in 1284 and 1291×2. Sa i.282.

(4) ~ Warw SP 4363. *Stoc- Stokton* 1249–1535, *Stockton* 1290. The old name of the inhabitants of Stockton, OE *Stoc-hæme*, + OE **hyll**, survives in a field-name in the neighbouring parish of Southam, *Stochemehull* 1206. Wa 146, Studies 1931.33.

(5) ~ Wilts ST 9838. *Stottune* 1086, *Stoctun'* 1166, *-ton* 1189–1350, *Stocktun* c.1250. Either an enclosure made of stocks or where stocks were manufactured. Wlt 230.

(6) ~ HEATH Ches SJ 6186. *Stoaken Heath* 1682, *Stockton Heath* 1831. P.n. Stockton + ModE **heath**. Stockton, *Stoketon(a)* 1190×99, 13th, *Stocton* c.1200, *Stok-* 1220×40, *Stockton* from 1444. Che ii.145.

(7) ~ ON-TEES Cleve NZ 4419. *Stokton upon Tease* 1630. P.n. Stockton + r.n. TEES. Stockton, *Stoctun'* [1183]14th, *-ton* 1197–1331, *Stokton(e)* c.1245–1535, *Stoketon* 1195×1221–1339, *Stockton* from 1338, is probably 'the outlying farm'. Stockton, which later became the bishop of Durham's principal manor house in the south of the county, seems to have originated as an outlying settlement of an estate perhaps originally centred at Norton. But the specific might be OE **stocc** 'a tree-trunk'. Studies 1936.11–43.

(8) ~ ON TEME H&W SO 7167. *(into, of) stoctune* [?781×96]11th S 1185, *(into, of) stoctun* [n.d.]12th S 1595, *Stotvne* (sic) 1086, *Stocton* 1194 etc. Wo 80.

STOCKTON ON THE FOREST NYorks SE 6555. 'A stockaded enclosure'. P.n. *Stocthun, Stochetun* 1086, *Stocatuna* [1145×53]15th, *Stoke- Stoceton'* [1170×88]15th–1316, *Stoc(k)ton* 1218–1577. OE **stocc**, genitive pl. **stocca**, + **tūn**, + ModE **on the forest** referring to location within the old royal forest of GALTRES. YN 11.

West STOCKWITH Notts SK 7994. *Westokheth* 1348, *West Stokewythe -with* 1546–7. ME adj. **west** + p.n. *Stochith'* 1226, *Stockehithe* 1305, *Stokkyth* 1340–1400, *Stokheth(e)* 1266–1355, *Sto(c)ket(h), Stokketh* 14th, *Stokwith* 1373, the 'landing-place with a post', OE **stocc** + **hȳth** later confused with ON **vithr** 'wood'. 'West' for distinction from East STOCKWITH Lincs SK 7994 on the opposite bank of the Trent. Nt 39, L 77.

STOCK WOOD H&W SP 0058. *Stock Wood* 1831 OS. P.n. Stock as in STOCK GREEN SO 9959 + ModE **wood**.

STODMARSH Kent TR 2160. 'Stud marsh, horse marsh'. *(in marisco qui appellatur) Stodmerch* 'in the marsh called S.' [675]15th S 7, *Stodmersche* [686]15th S 9, *Stodmerse* 1198. OE **stōd** + **mersc**. KPN 9.

STOFORD Somer ST 5613. 'Stony ford'. *Stafford* 1225, *Stoford* 1274, *Stoforde* 1610. OE **stān** + **ford**. DEPN, L 68–9.

STOFORD Wilts SU 0835. 'The stone ford'. *Stanford* [943]14th S 492, *Staneford* 1196, *Stafford* 1242, *Stoford* from 1285, *Stovorde* 1559. Wlt 228, L 68–9.

STOGUMBER Somer ST 0937 → STOKE.

STOGURSEY Somer ST 2042 → STOKE.

STOKE 'A secondary settlement, an outlying farmstead, a dairy-farm'. OE **stoc**, dative sing. **stoce**, pl. ***stocu**. Studies 1936.11–43, PNE ii.153.

(1) ~ Devon SS 2324. Properly 'St Nectan's Stoke', *Nistenestoch* 1086, *Nectanestoke* [1189]1365, *Stoke Sancti Nectani* 1361. Irish saint's name *Nechtan* from the dedication of the church. D 75.

(2) ~ Hants SU 9851 → STOUGHTON SU 9851.

(3) ~ Hants SU 4051. *(æt) stoce be hysseburnan* 'S by Hurstbourne' [900]11th S 359, *Stokes* 1208, *Stoke* from 1237. This is possibly the Stoke referred to in DB f.40c as *Stoches*. Also known as *Crocker(e)stok'* 1255, 1297, 'potters' S' with surname *Crockere* for distinction from STOKE CHARITY etc. It was a grange of the monks of Abingdon or St Swithin's at different times. Ha 156, Gover 1958.155, Studies 1936.21.

(4) ~ Hants SU 7202. *Stoke* from 1327. Also known as *Northsto(c)ke* 1470 for distinction from East Stoke, *(æt) east stoce, stoccæ* [956]12th S 604, *Est(st)oke* 14th. Gover 1958.16, Studies 1936.21.

(5) ~ Kent TQ 8275. *De Stokes que antiquitus vocabatur Andscohesham* 'concerning S which was anciently called A' [738]12th S 27, *Sto(c)he* [c.975]12th B 1321–2, *(E)stoches* 1086. The old name *Andscohesham* is 'Hondscioh's homestead or estate', OE pers.n. *Hondscīoh*, genitive sing. *Hondscīohes*, + **hām**. KPN 36, ASE 2.29.

(6) ~ ABBOTT Dorset ST 4500. 'S belonging to the abbot' sc. of Sherborne. *Stok Abbatis* 1272, *Stoke Abbots* 1275, *~ Abbot* 1348. Earlier simply *Stoche* 1086, *Stokes* 1247, short for *Osanstok* and *Wulfheardigstok* [998]12th S 895, 'Osa's *stoc*' and '*stoc* called after Wulfheard', OE pers.n. *Wulfheard* + **ing**[4] + **stoc**. Do 137, Studies 1936.20.

(7) ~ ALBANY Northants SP 8087. 'S belonging to the Albini family'. *Stok(e) Daubeny* 1274–1388, *~ Aubeny* 1301, *Stoake Albane* 1626. Earlier simply *Stoche* from 1086 with variant *Stoke, Stoke* 1175–1242. Held by William de Albini in 1155. An outlying farm of Wilbarston SP 8188. Nth 171, Studies 1936.27.

(8) ~ ASH Suff TM 1170. 'S with the ash-tree'. *Stokeaysche* 1524, *Stoke Ash* 1837 OS. P.n. *Stoca* 1044×65, *Stoches* 1086, *Stoche* c.1095, *Stoke* from 1316, + ME **ash, aish** 'an ash-tree'. *Ash* looks like a manorial addition but no family of this name is known here. DEPN, Baron.

(9) ~ BARDOLPH Notts SK 6441. 'S held by the Bardolph family'. *Stokebardulf* 1269, *Stoke Bardolf* 1304. Earlier simply *Stoches* 1086, *Stoke(s)* 1197–1266 etc. The manor was held by Doun Bardolf in 1194. Nt 177.

(10) ~ BLISS H&W SO 6562. 'S held by the Blez family'. *Stoke de Blez* 1242. Earlier simply *Stoch* 1086. The manor was held in

1211×12 by William de Bledis, whose family came from Blay, Calvados, *Bleis* 1077, Blies or Blye, Ain, *Bleis* 1176, or Blé, Vienne. Subsequent spellings of the manorial addition are *Bles* 1277–1431, *Blys, Blisse* 1535, 1544. Wo 80.
(11) ~ BRUERNE Northants SP 7449. 'S held by the Briwerre family'. *Stok(e)bruere -brewere* 1275–1376, *Stokebruerne* 1428, *-bruin* 1657, *-brewing* 1710. P.n. *Stoche* 1086–1283 with variants *Stoke(s), Stok in Salcey forest* 1296, + manorial name of William de Briwerre, who held the manor in the late 12th cent. Nth 107, Studies 1936.27.
(12) ~ -BY-NAYLAND Suff TL 9836. *Stokeneylond* 1272. P.n. *Stoke* [946×51]13th S 1483, 1610, *Stoc* [970]12th S 780, 1288–9, *Stokes* 1086, + p.n. NAYLAND TL 9743. DEPN.
(13) ~ CANON Devon SX 9397. 'Stoke held by the canons' sc. of Exeter monastery, to whom it was given by King Athelstan. *Stoke Canonicorum* 1281, 1288 ('of the canons', Latin genitive pl. of **canonicus**), *Stoke Canon* 1535. Earlier simply *(æt) Stoce* 1033 S 971, *Stoche* 1086. Also known as *Hrócastoc* *[925×39]11th S 389, 'rooks' outlying farm', OE **hrōca**, genitive pl. of **hrōc**, and *(æt) Stoctune* [c.968×93]11th S 1452, 'the outlying farm enclosure'. D 447.
(14) ~ CHARITY Hants SU 4839. 'S held by the la Charité family'. *Stokecharite* c.1270–1341. Also known as Old Stoke, *Eledestok(e)* 1256–1354, *Eldestok(e)* 1276–1428, *Oldestoke* 1364, 1564. Here *Old* is a re-interpretation of OE **ǣlede** 'burnt' which became ME *elede, elde* and so was confused with southern *elde* < lOE **ēald** 'old'. A dependent farm of Micheldever held by Hyde abbey and in 1276 by Henry de la Charite. DB f.40c *Stoches* probably does not belong here but under STOKE SU 4051. Ha 156, Gover 1958.178, Studies 1936.22.
(15) ~ CLIMSLAND Corn SX 3674. *Stokesclymesland* 1282 Gover n.d., *Stok in Clymeslond* 1302. Earlier simply *(ecclesia de) Stoke* 1266, 1335. 'Climsland', for distinction from other Stokes in Devon, referring to the manorial centre at Climson SX 3774, *Clymes tun* mid 11th BCS 1247, *Clismestone* 1086, *Clemeston(a)* 1177, 1223, *Climeston* 1194, 'estate called *Climes*', unexplained p.n. *Climes* + OE **tūn**, from which the whole district was called *Climeslande* c.1215×20, *Clymeslaunde -lond* 1274, 1284, 'tract of land called or at *Climes*'. At SX 3775 there is also an Oldclims, *Oldeclynes* 1337, possibly the original manorial centre. PNCo 159, Gover n.d. 205, DEPN s.n. Climsland.
(16) ~ D'ABERNON Surrey TQ 1259. 'S held by the Abernon family'. *Stokes de Abernun, Stok(e) juxta Coveham* (Cobham TQ 1060), ~ *Daubernon* 13th cent., *Stookdabbron* 1493. P.n. *Stoche* 1086, *Stokes* 1178–86, 1190, + family n. *Abernun* from Abenon, Calvados, tenants here c.1135×54. Sr 95, OEB 66, DBSurrey 19, 32.
(17) ~ DOYLE Northants TL 0286. 'S held by the D'Oyly family'. *Stoke Doyly* 1344, 1431, ~ *Doyle* 1621. P.n. *(æt) Stoce* [c.908]c.1200 B 1130, *Stoche* from 1086 with variant *Stoke, Stokes* 1175–1301, ~ *by Undele* 'by Oundle' 1301, *Stokys juxta Pylketon* 'by Pilton' 1285, + family name of John de Oyly, who held the manor in 1286. An outlying farm of Oundle. Nth 218, Studies 1936.27.
(18) ~ DRY Leic SP 8596. 'Dry S'. P.n. *Stoche* 1086, *Stok(e)* 1179–1610 + ME prefix **dry**, *Dri-* 1205–1330, *Dry(e)-* 1220–1376, postfix *-drie* 1281–1436, *-dry* from 1492. The village is on a hill above the valley of Eye Brook which was probably marshy in former times. R 298, Studies 1936.28.
(19) ~ FERRY Norf TF 7000. 'Stoke with a ferry'. *Stokeferie* 1248. Earlier simply *Stoches* 1086. Named from a ferry over the Wissey. DEPN, Studies 1936.27.
(20) ~ FLEMING Devon SX 8648. 'Stoke held by the Fleming family'. *Stoke Flandrensis* 'Flemish Stoke' 1261, *Stokeflemeng* 1270. Earlier simply *Stoc* 1086, *Stokes* 1218. The family of le Fleming is first mentioned here in 1218. D 331.
(21) ~ GABRIEL Devon SX 8457. 'St Gabriel's Stoke'. *Stokegabriel* 1309, *Gabrielstok(e)* 1313, 1356, *Stoke Sancti Gabrielis* 1396, 1453. Saint's name *Gabriel* from the dedication of the church. Earlier simply *Stoke* 1307. D 522.
(22) ~ GIFFORD Avon ST 6280. 'S held by the Giffard family'. *Stok(e) Elye Giffardi* 1221, ~ *Gi- Gyffard* 1268–1398, ~ *Gi- Gyfford(e)* 1464–1619. Earlier simply *Stoche, Estoch* 1086, *Stok(e)* 1221–1373. The manor was held by Osbern Gifard in 1086, by Helyas Giffard in 1169. The family name is a nickname from OFr *giffard* 'chubby-cheeked, bloated'. Gl iii.140 gives pr ['stouk 'gifəd], DBS 144.
(23) ~ GOLDING Leic SP 3997. 'S held by the Goldington family'. Originally plural p.n. *Stochis* before 1175, *Stokes* 1200–93, *Stoke* from 1201, + affix *-goldington* 1316, *-golding(e)* from 1576. Stokes was held by Petrus de Goldinton in 1200. Lei 507, Studies 1936.28.
(24) ~ GOLDINGTON Bucks SP 8748. 'S held by Peter of Goldington'. *Stoch Petri de golð* 1167, *Stoke Goldington* 1262. Earlier simply *Stoches* 1086. Bk 13.
(25) STOGUMBER Somer ST 0937. *Stoke Gunner* 1225, 1248, *Stok Gomer, Stokegumer* 1249, *Stoke Cromer* 1610, *Stogumber* 1809 OS. P.n. Stoke, + manorial addition *Gummer* from CG *Gundmar*. The DB form is *Waverdinestoch* 1086, unidentified element or pers.n. + **stoc**. DEPN, Studies 1936.18, DBS 159.
(26) STOGURSEY Somer ST 2042. 'S held by the de Courcy family'. *Stoke Curci* before 1189, ~ *Curcy* [c.1224] Buck, 1241, *Stokeurcy* [1295] Buck, *Stokgussey* 1610. Earlier simply *Stoche* 1086, before 1171. Also known as *Suntinstoch* before 1107, *Sutinstoche* before 1160, 'outlying farm called or at *Sunting*', OE ***sunting** 'marsh place, pool place' < OE ***sunt** 'marsh, pool' + *ing*[2]. The manor was held c.1100×35 by William de Curci whose family came from from Courcy, Calvados, *Curceium* 1035. DEPN, Turner 1950.121–2.
(27) ~ HAMMOND Bucks SP 8829. 'S held by Hamon' son of Mainfelin, a 12th cent. descendant of the DB holder of the manor. *Stokes Hamund* 1242, *Stokehamon* 1535. Earlier simply *Stoches* 1086. The OFr pers.n. *Hamon* is of Frankish origin and corresponds to OHG *Haimo*. Bk 25 gives pr [stouk hæmən].
(28) ~ HEATH Shrops SJ 6529. 'Heath belonging to Stoke'. *Stoke Heath* 1833 OS. P.n. Stoke as in STOKE ON TERN SJ 6428 + ModE **heath**.
(29) ~ HOLY CROSS Norf TG 2301. 'S belonging to the Holy Cross'. *Crouchestoke* 1140×53. P.n. *Stoches* 1086 + **holy cross** from the dedication of the church, ME **crouche**. DEPN, Studies 1936.27.
(30) ~ LACY H&W SO 6249. 'S held by the Lacy family'. *Stokelacy* 1234×9. Earlier simply *Stoches* 1086, c.1174, *Stoka* 1143×8, *Stokes* 1160×70, c.1225. Possibly a dependent settlement of Much Cowarne. He 185.
(31) ~ LYNE Oxon SP 5628. '(William) Lynde's S'. *Stokelynde* 1526, *Stoke-lyne* 1658. P.n. *Stoches* 1086, *Stokes* 1198–1255, *Stoke* from 1209 + surname of William Lynde who bought the estate in the early 15th century. Also known as *Stoke Insula* 1316, ~ *del Isle* 1328, ~ *Lisle* from the earlier owners, the *del Isle* or *Insula* family. O 237.
(32) ~ MANDEVILLE Bucks SP 8310. 'S held by the Mandeville family'. *Stoke Mandeville* 1284, *Stoke Manfyld* 1552. Earlier simply *Stoches* 1086. Bk 156, Jnl 2.29.
(33) ~ NEWINGTON GLond TQ 3286 → Stoke NEWINGTON TQ 3286.
(34) ~ ON TERN Shrops SJ 6428. *Stoke upon Tirne* from 1309 with variants *super Tyrne, on Tirne* etc. P.n. *Stoche(s)* 1086, *Stoke* from 1228, *Stoak(e)* 18th cent., + r.n. TERN SJ 7037. Also known as *Nordestok* 1199, *North Stoke* 1346, 1428, 1431, for distinction from no.48 STOKESAY SO 4381. Sa i.283, Studies 1936.29.
(35) ~ -ON-TRENT Staffs SJ 8745. *Stoke Super Trent* 1686, *Stoke-upon-Trent* 1747. P.n. *Stoche* 1086, *Stoch* 1166, *Stokes* 1223, *Stoke* from 1224, + r.n. TRENT SJ 9330 for distinction from Stoke-by-Stone SJ 9133, *Stoca* 1086. DEPN, TC 176.

(36) ~ ORCHARD Glos SO 9228. Originally, 'Stoke which supplies an archer' sc. for the king's service, later 'Stoke held by the Archer family'. *Stok(e) le Archer* 1269, ~ *Archer* 1295–1561, ~ *Orchard* 1498. Earlier simply *(æt) Stoce* [967]11th S 1313, *Stoches* 1086, *Stok(e)* 13th cent. Several tenants from the 12th to 14th cents. are called *Archer*, Latin *Archerius*, but the land was originally held by the service to the king of supplying an archer equipped with bow and arrows for 40 days a year. Gl ii.93.

(37) ~ POGES Bucks SU 9884. 'S held by Imbert le Pugeis' in 1255. *Stoke Puges,* ~ *Pogeys* 1302. Earlier simply *Stoches* 1086. Imbert le Pugeis came to England with queen Eleanor of Provence who married Henry III in 1236. He came from Le Puy-en-Velay, Haute-Loire, and the surname means 'man from Puy' or 'dweller by the hill'. Bk 243 gives pr [poudʒis], a spelling pr which has ousted the expected [pju:dʒis], DBS s.n. Poggs.

(38) ~ POINT Devon SX 5645. *Stoke P^t* 1809 OS. P.n. Stoke SX 5646 + ModE **point**. Stoke is short for Revels Stoke, *Rawelestok* 1219, *Rewelstoke* 1418, *Revelstok(e), Rolstoke* 16th cent., 'Revel's Stoke' from the family name Revel, cf. Richard Revel, sheriff of Devon c.1189×99. D 257.

(39) ~ PRIOR H&W SO 9467. 'S held by the prior' sc. of Worcester. *Stok Prioris* 1275. Earlier simply *Stoke* *[770]11th S 60, *Stoche* 1086, *Stokes* [c.1086]1190, 1221, and possibly *(in) Stoce* [941]17th S 1842. Wo 359.

(40) ~ PRIOR H&W SO 5256. S held by the prior' sc. of Leominster. *Stoca* 1086 and possibly *(æt) Stoce* 1016×35 S 1462. DEPN.

(41) ~ RIVERS Devon SS 6335. 'Stoke held by the Rivers family'. *Stoke Ryvers* 1285–1339. Earlier simply *Stoche* 1086, *Stok(e)s* 1242, 1291. In the 12th cent. the manor was held by the Redvers or Riveres family, the first earls of Devon. D 69.

(42) ~ ROCHFORD Lincs SK 9127. 'S held by the Rochford family'. *Stoke Rocheforthe* 1545, ~ *Rochford* from 1623. P.n. *(æt) Stoce* [1066×8]c.1200 Wills, *Stoche* 1086, 1172×80, *Stoke* 1123–1428, + family n. of Ralph de Rocheford who held land here in 1303. Also known as *Sudstoches* 1086, *Southstoke* 1329–1576 for distinction from North Stoke SK 9127, *Nortstoches* 1086, *Northstoke* from 1248. Perrott 477, 479.

(43) ~ ROW Oxon SU 6884 → Stoke ROW SU 6884.

(44) ~ ST GREGORY Somer ST 3427. 'St Gregory's S'. (chapel of) *St Gregory of Stoke at Northcuri* 1233, *gregorytoke* (sic) 1610. P.n. *Stokes* 1225 + saint's name *Gregory* from the dedication of the church. DEPN.

(45) ~ ST MARY Somer ST 2622. 'St Mary's S'. *Mary Stok* 1610, *Stoke St Mary* 1809 OS. P.n. *Stoc* *[854]12th S 310, *(æt) Stoce* [882]12th S 345, *Stocha -e* 1086 + saint's name *Mary* from the dedication of the church. DEPN.

(46) ~ ST MICHAEL Somer ST 6646. 'St Michael's S'. *Mikelstok'* 1243, *Stok Michel* 1303, *Stoke Michaelis* 1428. P.n. Stoke + saint's name *Michael* from the dedication of the church. Studies 1936.18.

(47) ~ ST MILBOROUGH Shrops SO 5782. 'St Mildburh's S'. *Stok' Milburge* 1271×2, *Stoke St Milburgha* from 1272 with variants *Milbridg* and *Milburdge, Milborough* 1729. P.n. *Stok'* 1231 + saint's name *Mildburh* referring to the ownership of the manor by St Milborough's priory at Much Wenlock. Also known as *Godestoch* 1086, 'God's S', and *Stok' Prioris* 1271×2 'the prior's S'. Later forms include *Milburghstoke* 1389, *Stock Milburne* and *Mylbornstoke* 16th as if from *mill-burn* 'a mill-stream', and *Milverstoke* 1687–1700. Sa i.283, Studies 1936.29.

(48) STOKESAY Shrops SO 4381. 'S held by the de Say family'. *Stoke Say* from 1255. P.n. *Stoches* 1086, *Stoake* 1675, + manorial addition from the *de Say* family who held this place in the 12th and 13th cents. Also known as *Suthstoke* 1178×9–1428, 'south S', for distinction from STOKE ON TERN SJ 6428, *North Stoke* 1346 etc. Sa i.282, Studies 1936.29.

(49) ~ SUB HAMDON Somer ST 4717. *Stokes under Hamden* 1248. P.n. *Stoca, Stoche(t)* 1086, *Stoke* 1610† + Latin **sub** + p.n. Hamden as in HAM HILL ST 4816, *Hamedone* c.1100. Studies 1936.18.

(50) ~ TALMAGE Oxon SU 6799. 'S held by the Thalemalche family'. *Stokes Talemasche* 1219, 1222×3, *Stoketalemacch'* [c.1225]c.1280, *Stoke St. Almage* 1675. P.n. *Stoches* 1086, *Stokes* [1148×55]c.1200–1211, *Stoke* from 1220×2 + manorial affix from the family n. of Peter Thalemalche 1148×55. O 93.

(51) ~ TRISTER Somer ST 7428. 'S held by the Trister family'. *Tristrestok* 1265, *Stoke Tristre* 1284×5. P.n. *Stoca, Stoche* 1086, *Stoke* 1610, + family name *Trister*. This family name is not known in connection with this place but derives from ME *triste, tristur* 'a hunting station' and denoted the man in charge of the hounds and preparations for the hunt. Studies 1936.19, DBS 354–5.

(52) Chew ~ Avon ST 5561. 'S in Chew hundred'. *Stoche* 1086. DEPN.

(53) East ~ Dorset SY 8786. *East Stoke* c.1628. Earlier simply *Stoches* 1086‡, *Stokes* 1166–1358, *Stok(e)* 1284–1498. 'East' probably in relation to Bindon abbey SY 8586. Do i.145, Studies 1936.20.

(54) East ~ Notts SK 7549. *Eststoke* 1340, *East Stoake* 1693. ME **est** + p.n. *Stoches* 1086, *Stokes* 1176–1280, *Stok(e)* from 1256. Also known as *Stoke juxta Farnedon* and ~ *juxta Neuwark* 'S by Farndon' and 'S by Newark' 1330. A dependent settlement of Newark. 'East' in relation to STOKE BARDOLPH SK 6441. Nt 217.

(55) Itchen ~ Hants SU 5632. 'S dependent on Itchen' sc. Abbas *Ichenestoke* 1185–1497. P.n. Itchen as in ITCHEN ABBAS SU 5333 + p.n. *Stoche* 1086††. Ha 101, Gover 1958.80, Studies 1936.22.

(56) Limpley ~ Wilts ST 7860. *Lympley Stoke* 1585. Unknown p.n. or family n. *Limpley* + ModE **stoke** (OE *stoc*). Earlier known as *Hangyndestok* 1263, *Hanging Stoke* 1322, 'hanging Stoke' from its position below a steep hill-side, and simply *Stoke* 1333–1611, 'the outlying farm' sc. of Bradford on Avon. Wlt 121.

(57) Lower ~ Kent TQ 8376. ModE adj. **lower** + p.n. STOKE TQ 8275.

(58) North ~ Avon ST 7069. *Norþstoc* [808 for 757×8]12th S 265, 'northern outlying farm' sc. of St Peter's abbey, Bath. Cf. SOUTHSTOKE ST 7461. DEPN.

(59) North ~ Oxon SU 6186. *Northstok* 1218, *Northestoke* 1479. ME adj. **north** + p.n. *Stoches, Estoche* 1086, *Estoches* [1087]15th, *Stokes* c.1181–1276, *Stoke* from 1209×18. 'North' for distinction from South STOKE SU 5983.

(60) North ~ WSusx TQ 0210. *Norstok* 1253, *Northstoke* 1271. OE, ME adj. **north** + p.n. *Stoches* 1086. Situated in a bend N of the r. Avon. Cf. South STOKE TQ 0210. Sx 173.

(61) Purton ~ Wilts SU 0990. 'S belonging to Purton'. *Purytonstoke* 1422, *Pirton Stoke* 1509. P.n. PURTON SU 0987 + p.n. *Stoche* 1086, *Stoke* 1257. Wlt 39, Studies 1936.23.

(62) Rodney ~ Somer ST 4850. 'S held by the de Rodene family'. *Radnestoke* 1610. P.n. *Stoches* 1086, *Stoke* [1159]B, + family name of Richard de Rodene who acquired the manor before 1303 by marriage into the Giffard family. Also known as *Stokes Giffard* 1243. DEPN, Studies 1936.20.

(63) Severn ~ H&W SO 8544. 'S by the river Severn'. *Savarnestoke, Savernestok'* 1221–75. Earlier simply *(in) stoce* 972 S 786, *Stoche* 1086. 'Severn' for distinction from *suthstoc* 'south Stoke', modern Hawkesbury Glos ST 7687, another estate of Pershore abbey. Wo 227, Hooke 177.

†*Stoke* [924×39]13th S 1717, [979×1016]13th S 1779 possibly belong here.

‡The identification is not certain.

††The form *Ytingstoce* [960]12th S 683 is sometimes identified with Itchen Stoke. It is, however, a different name, to be identified with BISHOPSTOKE SU 4619.

(64) South ~ Oxon SU 5983. *Suthstok* 1281, 1318. ME **south** + p.n. *Stoch* 1086, *Stoches* 1109–60, *Stoke* from 1152. Also known as *Bishopestoke* 1213–20 and *Stoke Abbatis* 'the abbot's S' 1268–1428. The estate was part of the bishop of Lincoln's manor of Dorchester in 1086, given to Eynsham abbey in the early 12th cent. O 156.

(65) South ~ WSusx TQ 0210. *Sudstok* 1242, *Suth Stok* 1281. OE, ME adj. **sūth** + p.n. *Stoches* 1086. Situated in a bend S of the r. Avon. Cf. North STOKE TQ 0210. Sx 142.

(66) West ~ WSusx SU 8308. *West Stoke* 1585. ModE adj. **west** + p.n. *Stokes* 1205, 1291. Also known as *Stoke Juxta Cycestre* 'S by Chichester' 1288. W of Lavant SU 8508. Sx 61.

(67) Winterbourne ~ Wilts SU 0740. 'S on the Winterbourne'. *Wintreburnestoch* 1086, *-bornestoca* 1086 Exon, *Winterburnestoke* 1169, 1211. Also known as *Wynterbourne juxta Chitterne* 'W beside Chitterne'. P.n. WINTERBOURNE 'a stream that flows in winter' + **stoc** 'an outlying farm'. The stream is the river Till, a back-formation from Tilshead, formerly called *Winterbourne Water* c.1540. Wlt 237.

STOKEFORD Dorset SY 8687. 'Ford near or leading to Stoke'. *Stokford(e)* 1244–1463, *Stokeford* from 1355, *Stock(e)ford(e)* 1535, 1628. P.n. Stoke as in East STOKE SY 8786 + OE **ford**. Do i.148.

STOKEHAM Notts SK 7876. '(At) the Stokes or outlying settlements'. *Estoches* 1086, *Stokum* 1242–1497, *Stockham* 1542, *Stockham* 1542, *Stokeham* 1577. OE **stocum**, dative pl. of **stoc**. Nt 60, Studies 1936.28.

STOKEINTEIGNHEAD Devon SX 9170. 'Stoke in the Ten Hide(s)'. *Stokes in Tynhide* 1279, *Stokeyntynhede* 1492, *Stockinge Tynid* 1558×1601. Ten Hide(s) was a Domesday Book area of thirteen manors containing about ten hides. The name was subsequently associated with the river name Teign and so reformed with that name here and in COMBEINTEIGNHEAD SX 9071. OE **hīd** was a feminine noun, pl. **hīde** without *-s*. Stoke, *Stoches* 1086, *Stokes* 1242, is OE **stoc** 'an outlying farm'. The forms with *-s* are probably AN reinflected nominative singular rather than plural forms. D 460 gives pr [stoukəntini(d)].

STOKENCHURCH Bucks SU 7696. 'Log church'. *Stockenechurch* c.1200. OE **stoccen** + **ċiriċe**. Bk 194.

STOKENHAM Devon SX 8042. 'Stoke in Ham'. *Stok(e) in Hamme* 1276–1344, *Stokeinhamme* 1341. P.n. *Stokes* 1242, 1258, *Stoke* 1291, 'outlying farm', OE **stoc**, + **in** + p.n. Ham as in The South Hams, the name given to the fertile district lying between Plymouth and the Dart estuary and bounded by Dartmoor on the N, *Southammes* 1396, *the Southhammes* c.1550. This seems to have comprised at least two districts called Ham, 20 hides of land *on Homme* 847 S 298, an area bounded by the river Avon SX 6745, Sorley SX 7346 and Thurlstone SX 6742, and the *Hamme* of Stokenham. A *Northamme* 1242 is recorded in Loddiswell, now Ham SX 7249. The sense of OE **hamm** in these names is uncertain. D 331 gives pr [stoukən'hæm], 264.

STOKES BAY Hants SZ 5898. *Stokes Bay* 1810 OS. P.n. Stoke referring to ALVERSTOKE SZ 6099 + ModE **bay**. Studies 1936.21.

STOKESAY Shrops SO 4381 → STOKE no. 48.

STOKESBY Norf TG 4310. 'Farm village belonging to Stoke'. *Stokesbei, -bey* 1086, *Stokebi* 1152×8–1198, *Stokesbi* before 1164–1254×75, *-by* from 1164. P.n. *Stoches* 1086, OE **stoc**, genitive sing. **stoces**, + ON **bȳ**. The *Stoke* in question is probably this very place to which ON **bȳ** was subsequently added. Nf ii.22, Studies 1936.32.

STOKESLEY NYorks NZ 5208. 'Clearing belonging to *Stoke*'. *Stocheslag(e)* 1086, *Stokesley* from 1112×22. OE **stoc** 'an outlying farm, a religious place, a town, a place' probably used as a p.n., genitive sing. **stoces**, + **lēah**. It has been suggested that the *stoke* in question may have been Stockton on Tees in Durham. YN 169, DEPN.

STOLFORD Somer ST 2245. *Stolford* 1809 OS. Earlier forms are needed for this ancient settlement name. It could be 'ford with or by a stool', OE **stōl** + **ford**. PNE ii.157.

Great STONAR Kent TR 3359. *Great Stonar* 1819 OS. ModE **great** + p.n. *(E)stanores* 1203, *Stanore(s)* 1227–80, *Stonore* 1270, 'stone bank', OE **stān** + **ōra**, referring to a long shingle ridge. 'Great' for distinction from *Little Stonar* 1819 OS lost TR 3358. PNK 605, Jnl 21.18.

STONDON MASSEY Essex TL 5800. 'S held by the Marci family'. *Standon de Marcy* 1238, *Stondon Masse* 1542. Earlier simply *Staundune* *[1062]12th S 1036, *Staundon(e)* 1291–1374, 'stone hill', OE **stān** + **dūn**. Ralph de Marci held an estate in 1086 at nearby Kelvedon Hatch. Ess 81 gives pr [stoundən].

STONE 'The stone(s)'. OE **stān**, pl. **stānas**, dative pl. **stānum**.

(1) ~ Bucks SP 7812. 'The stones'. *Stanes* 1086–1346, *Stone* from 1307×27. The reference is unknown. Bk 164.

(2) Glos ST 6895. 'The stone'. *Stane* 1204–1439, *La Stane* 1261, *Stone* from c.1315. OE **stān**. Gl ii.225.

(3) ~ H&W SO 8675. 'The stones'. *Stanes* 1086–1327, *Stone* 1346. OE **stān**, pl. **stānas**. The reference is unknown. Wo 255.

(4) ~ Kent TQ 5774. *of stane* [c.975]12th B 1322, *Estanes* 1086, *Stanes* 13th cent. The reference may have been to the promontory of chalk cliff overlooking the Thames estuary, on which the church stands. KPN 304, Newman 1969W.524.

(5) ~ Staffs SJ 9033. 'The stone or stones'. *Stanes* c.1154–1293. OE **stān**, pl. **stānas**. The *-s* may either be pl. or show AN nominative sing. ending *-s*. The reference is unknown, but a legend of the martyrdom of Wulfhad and Ruffin preserved in a 12th cent. hagiography probably produced at the Augustinian priory founded here c.1135 associates them with this place, which became a Mercian cult centre. The name, here regarded as a pl., is explained as commemorating the practice of devotees bringing stones to the place for building. The story is a calque on the Bedan story of the Jutish princes martyred at Stone in Hants. DEPN, Rumble 1997.307–19.

(6) ~ IN OXNEY Kent TQ 9427. *Stane* c.1185, 1254, *Stanes* 1240, 1265, *Stone* from 1254. Possibly the reference is to the stone altar to the Roman god Mithras, now in the parish church. PNK 488, Newman 1969W.528.

The STONE Corn SW 5526. *The Stone* 1748. A rock in the sea in Mount's Bay. Cf. RUNNEL STONE SW 3620. PNCo 159.

STONEBROOM Derby SK 4159. 'Broom growing in stony ground'. *Stanbro(u)m* 1323. OE **stān** + **brōm**. Db 300.

STONEGATE ESusx TQ 6628. *Stone Gate* 1813 OS. ModE **stone** + **gate**.

STONEGRAVE NYorks SE 6577. 'Stone valley'. *Sta(i)ne-Steinegrif* 1086, *Steingrave* c.1150, *-greve* 13th cent., *Staingrive* 1190, *-greue* 1200×10–1301, *Stangreve* 15th cent. ON **steinn** + **gryfja**. However, ON **gryfja**, Yorkshire dial. *griff*, is normally used of a steep-sided valley or ravine as at SKINNINGROVE NZ 7119 and MULGRAVE NZ 8412, which does not fit the topography here. This is, therefore, a late Scandinavian rationalisation of the earlier OE name, *Staningagrave* [757×8]11th B 184, the site of an Anglo-Saxon monastery, meaning 'quarry of the people of (the) Stone', OE folk-n. **Stāningas* < **stān** + **ingas**, genitive pl. **Stāninga*, + **græf**. The reference is to a prominent outcrop of rock on the hill-side above the village, where stone was quarried for building. YN 54, SSNY 166, Jnl.17.14–19.

STONEHAUGH Northum NY 7976. Named after *Stonehaugh Shields* 1866 OS, 'the shielings at Stony Haugh'. ModE **stone** + dial. **haugh** 'land in the bend of a river' (OE *halh*).

STONEHENGE Wilts SU 1242. 'The stone hinges'. *Stanenges* c.1130, *Stanheng* 12th, *-henge* c.1200, *-henges* 13th, *Stonheng* c.1250, 1297 with variant *þe stonheng(e)*, *Stonehenge* from 1610, *the stone heng(l)es* 1470, *the stonege* 1547, *Stonage* 1668. OE **stān** + ***hencg** 'something hanging, a hinge'. ModE *henge, hodie* 'a stone circle' (since 1932), originally 'something hanging', is not evidenced before Stukeley 1740 (*Stonehenge* ii.8):

'Pendulous rocks we now call henges in Yorkshire, and I have been informed of another place there called Stonehenge, being natural rocks. So that I doubt not, Stonehenge in Saxon signifies the hanging stones'. It is probably a back-formation from the name Stonehenge, since OE ***hencg** normally gives ModE *hinge* although the variant *henge* is also found. The word (and its diminutives, **hengle**, **hingle**) is related to the verb 'to hang' and like its MHG cognate *hengel* means 'something hanging'. The reference is to the megalithic lintels of the great sarsen circle of Phase III which were actually called *the Hanging stones in Wilts* in 1617 (Thomas Middleton, *A Fair Quarrel* v.i.181) and *pières pendues* in 1155 (Wace, *Roman de Brut* 8178, translating Geoffrey of Monmouth's *Stanhenges*). An alternative name for this sort of monument was *the Giauntes carole* 'the giants' dance' 1470, Wace's *carole as gaianz* 1155, Layaman's *eotende ring* c.1200. The modern pr ['stoun'hendʒ] is an artificial one replacing earlier [stɔnidʒ]. Wlt 360, xli†.

STONE HOUSE Cumbr SD 7686. *Stone House* 1858. YW vi.259.

STONEHOUSE Glos SO 8005. '(The) stone house or houses'. *Stanhus(e)* 1086–1270, *Stonehouse* from 1316. OE **stān** + **hūs**, pl. **hūs**. Gl ii.202.

STONEHOUSE Northum NY 6958. 'The stone house'. *Stonehouse* 1866 OS. ModE **stone** + **house**.

STONELEIGH Warw SP 3372. 'The stony clearing'. *Stanlei* 1086, *-leia -lega -lege* 1153–1259, *Ston(e)ley(e)* 1275–1564. OE **stān** + **lēah**. According to Dugdale (1656) 'the soil where the Town stands is rocky'. Wa 180, L 205.

STONELY Cambs TL 1067. 'Stony wood or clearing'. *Stanlegh* 1260, *Stonle(gh)* 1260 etc., *Stonley* 1766. OE **stān** + **lēah**. Hu 244 gives former pr [stɔnli].

STONES GREEN Essex TM 1626. *Stones Green* 1644, *Stone Green* 1777, 1805 OS. Surname *Stone* + ModE **green**. Contrasts with nearby *Wickes Green*, *Bockings Green* and *Goose Green*, all 1805 OS. Ess 346.

STONESBY Leic SK 8224. Partly uncertain. *Stovenebi* 1086, *Stounesbi(a) -by -v-* c.1130–1276, *Stounesbi -by -is- -ys-* 1245–1449, *Stonesby* from 1317. The generic is ON **bȳ** 'village, farm', the specific either ON pers.n. **Stofn*, genitive sing. **Stofns*, or more likely ON **stofn**, 'a tree-stump', genitive sing. **stofns**, perhaps replacing OE **stofn** in the same sense, genitive sing. **stofnes**, and so the 'tree-stump village or farm'. OE, ON **stofn** may in fact have been an earlier p.n. here as in STOVEN Suff TM 4481, and so properly 'the **bȳ** at, called or of *Stofne*, (the settlement) at the tree-stump'. Lei 193, SSNEM 71.

STONESDALE MOOR NYorks NY 8904. P.n. *Sconesdale* (for *Stones-*) 1298, *Stonedale* 1577, the 'valley of the stone' OE **stān**, genitive sing. **stānes**, + ON **dalr** probably replacing OE **dæl**, + ModE **moor**. YN 273.

West STONESDALE NYorks NY 8802. OE **west** + p.n. Stonesdale as in STONESDALE MOOR NY 8904.

STONESFIELD Oxon SP 3917. 'Stunt's open land'. *Stuntesfeld* 1086, *Stuntesfeld(e) -feud* 1167–1406, *Stuntefeld(a)* 1130–1274, *Stonysfelde* 1526, *Stunsfeld* 1676. OE pers.n. **Stunt*, genitive sing. **Stuntes* varying with weak variant *Stunta*, + **feld**. A *Stuntescumb'* 1273×4, *Stuntesdon* 1246, *Stontesford* and *-ham* 1278×9 occur in the same parish. O 283 gives prs [stʌnz-] and [stounzfi:ld], 284.

STONESIDE HILL Cumbr SD 1489. *Stones Head Fell* 1783. Cu 419.

STONE STREET Kent TR 1348. *Stanstrete* 1240, *Stonstrete* 1278–1313, *Stonestreet* 1690. OE **stān** + **strǣt**. The Roman road from Canterbury to Lympne, Margary no. 12. PNK 547.

†The suggestion that the generic is rather OE **hencgen** 'an instrument for hanging, a gallows' and that the 'gaunt framework of the trilithons might well be imagined to resemble a series of great gallows' is phonologically possible (cf. the loss of final *-en* in *clew* 'a ball of thread' < OE **cliwen**, maid < OE **mægden**) but unecessary.

STONE STREET Suff TM 3686. *Stone Street* 1837 OS. ModE **stone** + **street**. The name of the Roman road from Halesworth to Woodton, Margary no. 36.

STONEY CROSS Hants SU 2611. *Stony Cross* 1670. A cross-roads possibly with reference to the metalling of the Roman roads that intersect here, Margary nos.422 and 424, or to the gravelly soil. ModE adj. **stony** + **cross** 'a crossroads'. Ha 157, Gover 1958.209.

STONEYCROFT Mers SJ 3991. 'Stony croft'. No early forms.

STONEYGATE Leic SK 6002. A suburb of Leicester begun in the early 19th cent., *Stoneygate* 1833 OS. The reference is to the Roman road from Leicester to Huntingdon, Margary no. 57a, *(le) Stangate* c.1250–early 14th, *Stanegate* early 14th, *(le) Stongate* early 13th, '(the) stone road'. OE **stān** + ON **gata**. Lei 215, 106.

STONEYHILLS Essex TQ 9497. No early forms. ModE **stony** + **hill(s)**.

STONHAM ASPALL Suff TM 1359. 'S held by the Aspal family'. *Aspalestonham* 1404, *Stonham aspoll* 1610. P.n. *Stonham* [1035×44]13th S 1521, *Stana- Sta(n)ham* 1086, *Stanham* 1177–1327, *Stonham* from 1285, 'the stone homestead', OE **stān** + **hām**, + manorial addition from Roger de Aspale of ASPALL TM 1765 who held the manor in 1292. The reference is to the stony gault soil here. DEPN, Baron.

Earl STONHAM Suff TM 1158. 'S held by the count'. *Stanham Comitis* 1254, *Erlestonham* 1610. P.n. Stonham as in STONHAM ASPALL TM 1359 + ModE **earl**, Latin **comes**, genitive sing. **comitis** referring to Roger Bigod mentioned in connection with this place in 1212. DEPN.

Little STONHAM Suff TM 1160. *Parva Stonham* 1219, *Stonham p'ua* 1610. ModE adj. **little**, Latin **parva**, + p.n. Stonham as in STONHAM ASPALL TM 1359. DEPN.

STONNALL Staffs SK 0704. 'The stone nook'. *Stanahala* 1143, *Stanhala* 1167, *Stonhal* 1216×72. OE **stān**, genitive pl. **stāna**, + **halh** in the sense 'a hollow, a small valley'. DEPN, L 105, 110.

STONOR Oxon SU 7388. 'The stony flat-topped ridge'. *Stanora* [late 10th]11th B 216, *Stanore* 1204–91, *Stonor(e)* from 1252. OE **stān** + **ōra**. O 84, Jnl 21.21.

STONTON WYVILLE Leic SP 7395. 'S held by the Wyville family'. P.n. Stonton + affix *Wyvile* 1265, *Wyvyll* 1340. Stonton, *Stantone* 1086, *Stanton(e)* c.1130–1610, *Staunton(e)* 1230–c.1545, *Stonton(')* from 1306, is the 'stone farm or village', OE **stān-tūn**, possibly referring to building in stone rather than stony ground. The manor was held by Robert de Wivele in 1230. Lei 256.

STONYHURST COLLEGE Lancs SD 6939. An ancient house which became the home of the dissolved English College at St Omer in 1794. Stonyhurst, *del Stanyhurst* 1358, *Stonyhirst -hurst* 1577, is the 'stony wooded hill', OE **stāniġ** + **hyrst**. It is in a commanding position. La 141, L 197.

STOODLEIGH Devon SS 9218. 'Stud clearing'. *Stodlei*, *Stollei* 1086, *Stodleg(he)* 1205–1373, *Studley* 1675, *Stoodlley* 1809 OS. OE **stōd** 'a stud, a herd of horses' + **lēah**. D 393.

STOODLEIGH BEACON Devon SS 8818. *Stoodley Beacon* 1809 OS P.n. STOODLEIGH SS 9218 + ModE **beacon**.

STOPHAM WSusx TQ 0219. 'Stoppa's river meadow'. *Stopeham* 1086–1478, *Stopham* from 1234. OE pers.n. *Stoppa* + **hamm** 1 or 3, or perhaps 7 'land on a shelf over a river'. The site is near the junction of the Rother and the Arun. Sx 120, ASE 2.46.

STOPSLEY Beds TL 1023. 'Stoppa's wood or clearing'. *Stoppelee* 1199, *Stop(p)ele(g)* 13th, *Stop(p)esle(gh)* 1207–1504. OE pers.n. **Stoppa* + **lēah**. For the pers.n. cf. STOPHAM WSusx TQ 0219 and *Stoppingas* [717×37]11th S 94 'the people called after Stoppa', the name of an ancient *provincia* in Warwickshire. Alternatively the specific might be the OE appellative **stoppa** 'a pail, a bucket' referring to wooden vessels made here. Bd 163.

STORETON Mers SJ 3084. Probably 'brushwood farm or

village'. *Stortone* 1086, *-ton(a)* 1202–1619, *Storeton* from 1305, *Stortton* 13th. ON **storth** + **tūn**. Che iv.254, SSNNW 189.

STORRIDGE H&W SO 7548. 'Stone ridge'. *Storug(g)e* 1219–1320. OE **stān** + **hrycg**. The reference is to a ridge of ground extending NW from the Malvern Hills. He 62.

STORRINGTON WSusx TQ 0814. 'The storks' farm'. *Estorchetone, Storgetune* 1086, *Storg(h)eton* 1242–1428, *Storketon* 1276–92, *Storghton* 1314–1493, *Storghton al. Storrington* 1583. OE **storc**, genitive pl. **storca**, + **tūn**. Sx 161, Ritter 109, SS 75.

STORRS Cumbr SD 4094. 'Plantations, lands overgrown with brushwood'. *(ye) Storthes* 1292, 1553, *Storr(e)s* from 1606. ON **storth** + secondary ME pl. **-es**. We i.186, SSNNW 166, L 59.

Bishop's STORTFORD Herts TL 4821. *Bysshops Stortford* 1587, *Bishop Stafford* 1710. ModE **bishop** + p.n. *Storteford* 1086–1428, *Sterte-* 1198, 1305, *Strat- Strotford* 1493–1619, 'Steorta's ford', OE pers.n. **Steorta* + **ford**. Alternatively this might be a compound name in OE **steort** 'a tail, a projecting piece of land', genitive pl. **steorta**, referring to the tongues of higher ground reaching down to the river N and S of the town. The r.n. Stort, *Stour* 1576, *Stort* 1586, 1598, is a back-formation from the town name. The bishop of London held the manor in 1086. Hrt 201 gives pr [stɔ:fəd], commonly [sta:fəd], 5.

STORTH Cumbr SD 4779. 'Plantation, land overgrown with brushwood'. *Storth* from 1446, *le Storth* 1534, *the Storthe* 1542. ON **storth**. We i.70, SSNNW 166, L 69.

STOTFOLD Beds TL 2136. 'Stud enclosure'. *Stodfald* 1007–1227, *Stotfalt* 1086, *Stotfold* from 1199. OE **stōd** + **fald**. Bd 178, IntroSurvey 150.

STOTTESDON Shrops SO 6782. 'Stud hill, horse hill, herdsman's hill'. *Stodesdone* 1086, *-don(') -dun* 1162–1717, *Stotesdona -dune -don(e)* c.1090–1804, *Stottesdun' -don(e)* from 1160, *Stoteresden* 1284×5, *Stot(t)eresdon* 1317, 1341, *Stotterton* 16th cent. OE **stōd** 'a stud, a herd of horses', genitive sing. **stōdes**, varying with **stott** 'a horse, an inferior horse', genitive sing. **stottes**, + **dūn**. The *-er-* forms point to a further variant of the name with OE, ME ***stottere** 'a herdsman', genitive sing. ***stotteres**, as specific. Sa i.285 suggests OE ***stōdere** 'a horse-herd, one who tends a stud' but this fits the recorded sps less well. The usual term for a horse-herd in OE seems to have been ***stott-hierde** to judge from the surname Stothard. Gelling 1998.79.

STOTTINGWEY Dorset SY 6684 → BROADWEY SY 6683.

STOUGHTON 'Outlying farm with an enclosure'. OE **stoc-tūn**. Studies 1936.33–4.

(1) ~ Leic SK 6402. *Stoctone* 1086, *Stoc(k)- Stokton(e)* 1202–1435, *Stoghton'* 1320–84, *Stoughton(')* from 1349, *Stouton* 1316–1464, *Staughton* 1610, *Stawton* 1631. Lei 257.

(2) ~ Surrey SU 9851. *Stoctune* 12th(p), *-ton(a)* 13th cent., *Stoghton* 1294, *Stoughtonesmore* 'the moor of S' 1356. P.n. *Stochæ* 1086, *Stok(e)(s)* 13th, + OE **tūn** referring to an outlying farm of Guildford 1 m N of the town in a bend of the river Wey. The 1356 form refers to associated marshland, OE **mōr**, used for grazing the animals kept at the Stoke. Sr 151 gives pr [stautən].

(3) ~ WSusx SU 8011. *Estone* (sic) 1086, *Stocton(a)* 1135–1279, *Stohtun* c.1200, *Stoghton* 1327, 1346. Sx 54 gives pr [stɔ:tən], PNE ii.155.

(4) West ~ Somer ST 4148. *West Stoughton* 1817 OS. ModE **west** + p.n. Stoughton, for which no early forms are known. It probably represents OE **stoc** + **tūn** 'dependent settlement' in this case on Wedmore ST 4347. Contrasts with Middle Stoughton ST 4249 and Stoughton Cross ST 4249, *Middle Stoughton, Stoughton Cross* 1817 OS.

STOULTON H&W SO 9049. 'Stool settlement'. *STOLTUN, Stoltun* [840]11th S 192, *Stoltun* 1086, *-ton(a)* [c.1086]1190-1454, *Stoulton* from 16th, *Stowton* 1577. OE **stōl** 'seat, throne' + **tūn**. The reference is either to the hill on which Stoulton stands perhaps called 'the stool' or to a seat of authority connected with the meeting-place of Oswaldslow hundred at Low Hill SO 9152 or to the manufacture of stools. Wo 166 gives pr [stoutən].

River STOUR 'Violent, fierce one'. Probably an OE r.n. *Stūre* < adj. ***stūr** from the root seen in *storm* 'tumult, onrush', *styrian* 'stir, rouse', and ME *stour*, MDu *stuur*, ON *styrr* 'tumult, strife' and the G r.n. Stör, a tributary of the lower Elbe, *Sturia* 9th, 11th, *Store* 1225. RN 378ff, LHEB 195 note 1, Berger 251.

(1) ~ ~ Dorset ST 7821, 8014, SY 9699. *(on, anlang) Sture* [944]15th S 502, *(of, on) Stoure* [968]14th S 764, *Sture* c.1150–1586, *Stoure* 1324–1575, *Stowre, Stour* c.1540. RN 379.

(2) ~ ~ Essex TL 9233. *Sture* 1000×2 S 1486–1296 including [894]c.1000 Asser and ASC(D, E) under year 885, *(le) Stoure* 1399, 1576. Ess 12 gives pr [stauə], RN 379.

(3) ~ ~ H&W SO 8083. *Stur* 736 S 89, *(on) Sture* [866]18th S 212, [951×55]c.1400 S 579, [964]17th S 726, [985]11th S 1351, [996]17th S 1380, *Stoure* 1280, 1300. The river gave its name to a lost Anglo-Saxon estate at Ismere near Kidderminster, *(de) Sture* [700 for 716×757]17th S 1826, *(æt) Sture* [757×775]11th S 1411, *(æt) sture* [781]11th S 1257, *Norðstur* c.1000 S 89 endorsement. RN 380, Hooke 27, 29, 32, 63.

(4) ~ ~ Kent TR 2763. *Stur* [814]13th S 176, 822 S 186, 839 S 287, 845 S 296, ?859 S 1196, 863 S 332, *Sture fluminis* 811 S 168, *flumine Sturæ* 839 S 287, *Stoure* 1264, 1575. RN 378.

(5) ~ ~ Warw SP 2248. *(on) Sture* [699×709]11th S 64, [922]17th S 1289, [972]10th S 786, [977]10th S 1330, *(innan, andlang) Stúre* [988]10th S 1356, *Stuur, (of) Sture, Stures (stream)* [757]11th S 55, *Stur* *[764×75]11th S 61, [817]11th S 182, *Sturae (fluminis)* *[964]12th S 731, *Stura* 1247, *Sture* 1262, *Stoure* 1296, 1501–1586, *þe Stowre* [c.1270]c.1460, *(le) Stower* 1460, 1669. Wa 6, RN 380 gives pr [stauə].

(6) ~ PROVOST Dorset ST 7921. 'The Stour estate held by the provost' sc. of King's College, Cambridge. *Stowr' Provost* 1549. R.n. STOUR + ModE **provost**.The manor was given to King's College in the 15th cent. 'Provost' is actually an adaptation of earlier *Stures Prewes* 1268, *Sture Preauus* 1270, *St(o)ure Prewes* 1288–1468, 'S held by Préaux abbey', Seine-et-Marne, France, Latin *Stures Pratellorum* 1243, *Sture Pratellis* 1285†. Earlier simply *Stur* 1086, *St(o)ure* 1324–1412. Do iii.73.

(7) ~ ROW Dorset ST 8221. 'Row of houses in Stour (Provost)'. Possibly identical with *le Woderew(e)* 1421, 1477, 'row of houses at the wood'. P.n. Stour as in STOUR PROVOST + ModE **row** (ME *rewe*, OE *rǣw*). Do iii.76.

(8) East ~ Dorset ST 7922. 'Eastern Stour estate'. *East Stower* 1664, 1811. ModE **east** + p.n. *Sture* 1086, 1304, *Stures* 1212–15th from r.n. STOUR. Also known as *Stur(e) Cosin, Cosyn* 1244–1314, 'S estate held by the family of William Cusin' (1216×72). 'East' for distinction from West STOUR ST 7822. Do iii.67.

(9) East ~ RIVER Kent TR 0837. ModE **east** + r.n. STOUR TR 2763.

(10) Great ~ Kent TR 1651. ModE **great** + r.n. STOUR TR 2763, the main branch of the river so called for distinction from the lesser branch or Little Stour TR 2424. RN 379.

STOURBRIDGE WMids SO 9084. 'The bridge over the r. Stour'. *Sturbrug, Sturesbrige* 1255. R.n. STOUR H&W SO 8278 + ME **brugge** (OE *brycg*). Wo 311 gives pr [stɔ:bridʒ].

STOURHEAD Wilts ST 7734. 'The source of the river Stour'. R.n. STOUR Dorset ST 7821, 8014, SY 9699 + ModE **head**. The reference of the modern name is to a mansion house built in 1721–4. Pevsner 1975.495.

East STOURMOUTH Kent TR 2662. ModE adj. **east** + p.n. *Sturmutha* 1089, *Sturmude* c.1100, *Sturmue* 1176×7, *Sturmuth* 1235, 'the mouth of the Stour', r.n. STOUR TR 2763 + OE **mūtha**. PNK 513, RN 379.

†Genitive and dative pl. respectively of Latin *pratellum* 'small meadow'. But the form of this name in France is *apud Perellos* 1169, 'at the peartrees', Latin *pirellus*. Dauzat-Rostaing 525.

West STOURMOUTH Kent TR 2562. ModE **west** + p.n. Stourmouth as in East STOURMOUTH TR 2662.

STOURPAINE Dorset ST 8609. 'Stour estate held by the Payn family'. *Sture(s) Paen* 1242×3, 1268, ~ *Payn* 1280, 1288, *Stoure(e) Payn -payn(e)* 1303 etc. Earlier simply *Sture* 1086, *Sture(s)* 1208–88, r.n. STOUR + family name *Payn* who held the manor in the 13th and 14th cents. (Pagan son of William 1226, Bartholomew Payn 1303, 1316 etc.). Do ii.116 gives pr ['stauəpein].

STOURPORT-ON-SEVERN H&W SO 8171. R.n. STOUR SO 8083 + ModE **port**. A new town built at the confluence of the Stour and Severn round the basin linking the river Stour with the Staffordshire and Worcestershire canal built 1766×71 to connect the Trent and Mersey canal at Great Heywood Junction, Staffs, to the river Severn. The original name of the settlement here was Mitton, *Mettune* 1086, *Mutton(e)* 1227~1359, *Mytton* 1420, OE **(ġe)mȳthe** + **tūn** 'settlement at the confluence of the rivers'. Room 1983.117, Wo 254.

STOURTON 'The settlement or estate on the r. Stour'. R.n. STOUR + OE **tūn**.

(1) ~Staffs SO 8585. *Sturton* 1227, 1255. R.n. STOUR H&W SO 8278 + OE **tūn**. DEPN, Duignan 145.

(2) ~Warw SP 2936, *Sturton* 1206–1549, *Stourton* 1316–1565, *Stowreton -er-* 1445, 1609–1807. R.n. STOUR SP 2248 + OE **tūn**. Wa 301.

(3) ~Wilts ST 7734. *Stortone* 1086, *-ton* 1279–1310, 1657, *Sturton(e)* 1182–1327, *Stourton* from 1332, *Stowerton* 1606. R.n. STOUR + OE **tūn**. Wlt 181.

STOURTON CAUNDLE Dorset ST 7115 → Stourton CAUNDLE.

STOVEN Suff TM 4481. 'The tree stump'. *Stou(o)ne* 1086, *Stovene* 1201–1346, *Stofne* 1254, *Stoven* 1524. OE, ON **stofn**. DEPN, Baron.

STOW 'A place, a place of assembly, a holy place'. OE **stōw**. PNE ii.158, Stōw 182–96.

(1) ~ Lincs SK 8882. Short for 'St Mary's Stow', *(æt, into) Sce̅ Marian stowe* [1053×5]12th S 1478, [1066×8]12th Wills, *S' Maria de Stou, Scae Mariae Stov* 1086, *ecclesia sancte Marie de Stou* 1090. There was a religious house there in the mid-10th cent. DEPN, Stōw 191, Cameron 1998.

(2) ~ BARDOLPH Norf TF 6205. 'S held by (William) Bardulf' in 1244. Earlier simply *Stou* 1086, *Stowe* 1244. DEPN.

(3) ~ BARDOLPH FEN Norf TF 5604. *Bardolph Fen* 1824 OS. P.n. STOW BARDOLPH + ModE **fen**.

(4) ~ BEDON Norf TL 9596. 'Stow helf by (John de) Bidun' in 1212. *Stouwebidun* 1287. Earlier simply *Stou* 1086. DEPN.

(5) ~ CUM QUY Cambs TL 5260. 'Stow with Quy'. *Stowe et Coye* 1279, *Stowe cum Quey* 1316. P.ns. Stow + Quy. Stow is *Stoua* 1086, *Stowe* 1189–1327, Quy (pronounced [kwai]), *Choeie, Coeia -e, Cuege* 1086, *Queya -e -eie* 1218–1324, *Cow(e)ye* 1268–71, *Quie, Quye* 1327–1434, 'cow island', OE **cū** + **ēġ**. Ca 132, 133.

(6) STOWLANGTOFT Suff TL 9668. 'S held by the Langtoft family'. *Stowelangetot* 13th, *Stowlangtoft* 1610. P.n. *Stou, stoua* 1086, *Stowe* 1206, + manorial addition from Richard de Langetot who held the manor in 1206. His family came from Languetot in Normandy, itself an ON p.n. *Langatoft* 'the long toft'. DEPN.

(7) ~ LONGA Cambs TL 1171. 'Long Stow'. *Longestowe* 1286, 1327. ME **long**, Latin **longa**, + p.n. *Estou, Estove* [1086]12th, *Stowe* 1219. Also known as Over Stow, *Oueristowe* 1248. The village is long and straggling and lies on high ground. Hu 248.

(8) ~ MARIES Essex TQ 8399. 'S held by the Mareys family'. *Stowe Mareys* 1412, *Stowe Mary(s) alias Marysh(e), Maris alias Marshe* 1601. Earlier simply *(la) Stowe* 1222–1366, 'the meeting place'. 'Maries' is from the family name of Robert de Marisco 'of the marsh' 1250, but as the church is dedicated to St Mary the name was easily transformed into Stow St Mary. Ess 228.

(9) STOWMARKET Suff TM 0558. 'S with the market'. *Stoumarket* [1173×87]1286, *Stowemarket* c.1190–1347. P.n. *Stou* 1086, *Stowe* 1254–1316, 1523, 1610, + ME **mercat**. The earliest reference to the market is *forum de la Stowe* 1253 in which the Fr definite article **la** implies that **stōw** in this name was still felt as an appellative, 'the meeting place'. It replaced an earlier name, *Tornei(a) -ai* 1086, *Thorneie -ey(e)* 1212–1346, 1563, 'thorn-tree island', OE **thorn** + **ēġ**. DEPN, Baron.

(10) ~ -ON-THE-WOLD Glos SP 1925. *Stow(e) super le Olde, ~ on the Olde, Wowld* 1557–77. Earlier simply *Stoua* 1213, *Stow(e)* 1221–1733. This was originally *Eduuardesstou* 1086, *Stone S(ci') Edwardi* 1260–1585, 'the holy place of St Edward', OE saint's name *Ēadweard*, genitive sing. *Ēadweardes*, + **stōw**. The church is dedicated to St Edward, king and martyr, 975–8. The village stands on a high exposed hill-top of the Cotswolds. Gl i.225, Stōw 191 no.1.

(11) STOWUPLAND Suff TM 0660. Either 'the upper land at Stow' or '(the part of) Stow in the country'. *Stow Uplande, Upland of Stow* 1524. P.n. Stow as in STOWMARKET TM 0458 + ME **upland** (< *uppe land*) 'a rural district'. Baron.

(12) West ~ Suff TL 8170. *Westowe* 1254, *W: Stow* 1610. ME adj. **west** + p.n. *Stowa* 1086. The exact sense of **stōw** here is uncertain. Roman pottery kilns have been found a mile upstream and an entire pagan Saxon village of the period c.400–650 has been excavated. The original site was an island in the marsh on the N side of the r. Lark. Evidence of weaving sheds has been found. 'West' for distinction from STOWLANGTOFT TL 9568 or STOWMARKET TM 0458. DEPN, Pevsner 1975.482.

STOWBRIDGE Norf TF 6007. 'Bridge leading to Stow' sc. Bardolph. *Stow Bridge* 1824 OS. P.n. STOW BARDOLPH + ModE **bridge**.

STOWE 'A place, a place of assembly, a holy place'. OE **stōw**. PNE ii.158, Stōw 182–96

(1) ~ Shrops SO 3073. *Stoe* c.1200, *La Stowe* 1255, *Stowe* from c.1291. OE **stōw** 'a venue for a particular activity'. The site of Stowe at the top of a cul-de-sac valley perhaps suggests a religious sense here. Sa i.285.

(2) ~ -BY-CHARTLEY Staffs SK 0027. P.n. *Stowea* 1199, *Stowe* 1242–1304, **stōw**, + p.n. Chartley SK 0128, *Certelie* 1086, *Certelea* 1192, *Cerdel'* 1231, *Scerteley* 1236, *Certeley* 12th–14th cents., 'Cearda's wood or clearing', OE pers.n. *Ċearda* + **lēah**, for distinction from Stowe SK 1210 in Lichfield, *Stowe* 1221, traditionally the site of St Chad's first church. DEPN, PNE i.91, Stow 191, TSSAHS 14.30–1, Horovitz.

(3) ~ SCHOOL Bucks SP 6737. An independent school founded in 1923 at Stowe House, an 18th cent. mansion. P.n. Stowe + ModE **school**. Stowe, *Stou* 1086, *Stowa* 1255. The exact significance here is unknown. Bk 48.

(4) Upper ~ Northants SP 6456. Adj. **upper** + p.n. Stowe as in Church STOWE SP 6357. Also known as Little Stowe, *Parva Stowe* 1576, and *Butter Stowe* 1791. Nth 30.

(5) Church ~ Northants SP 6357. 'S with the (new) church'. *Stowe Nichurche* 1386, *Stow with the Nyne Church* 1418, ~ ~ ~ ~ *Churches* 1439, ~ *cum novem ecclesiis* 1595. The head form means 'Stowe with the church' referring to St Michael's church with its Saxon tower. The form *Nichurche* looks like 'new church', OE **nīwe**, * **nīġe**, although ***nīġe** is a S dial. form and perhaps questionable here. *Nyne Church* is either a folk-etymology of this or possibly a late echo of the OE dative inflexional form **nīwan** in the prepositional construction *æt thām nīwan ċirċan* 'at the new church'. Earlier simply *(æt) Stowe* [956]16th S 615, *Stowe* from 1086. Nth 30, Stōw 190, Campbell §411, AB 28.295.

STOWELL Somer ST 6822. 'Stony spring'. *Stanwelle* 1086, *Stawell* 1225, 1243, *Stowell* 1610. **stān** + **wella**. DEPN.

West STOWELL Wilts SU 1362. *Weststouwelle* 1392, *West Towel* (sic) 1817 OS. ME adj.**west** + p.n. *Stawele* 1176, 1202, *Stowelle* 1229, 1249, 'the stone spring', possibly one with a steened channel, i.e. an edging of stone, though most examples of this name-type seem to refer to natural stones, OE **stān** + **wella**.

'West' for distinction from East Stowell SU 1462, *Eststowelle* 1327, *East Towel* (sic) 1817 OS. Wlt 326, L 31.

Nether STOWEY Somer ST 1939. *Nutherestoweye* 1276, *Nether Stowley* (sic) 1610. ME adj. **nether** (OE *neothera, nythera*) + p.n. *Stalvvei* 1086, *Stawaye* 1243, 'paved road', OE **stān** + **weġ**. The reference is to the stony track from the village to a pass on the Quantock ridgeway. 'Nether' for distinction from Over Stowey ST 1838, *Overstaweie* 1220. DEPN.

STOWFORD Devon SX 4386. Partly uncertain. *Staford* 1086, *Estatforda* Exon, *Stafford(e)* 1242–91, *Stoford* 1287, *Stouford* 1306. If the Exeter DB form is reliable this may be OE **stæth-ford** 'ford with a bank or shore'. But its subsequent development is more like East and West Stowford SS 5642 and 5342, *Staveford* 1086, *Eststaufford, Weststafford* 1289, *Stouford(e)* 1289, 1330, 'ford marked by staves', OE **stæf**, genitive pl. **stafa**, + **ford**, or Stowford Wilts ST 8157, *Stanford* [987]12th S 867, Stoford Wilts SU 0835, *Stanford* [943]14th S 492, or STOLFORD Somer ST 2245, *Stanford* [899×909]16th S 380, all from OE **stān** + **ford** 'stone ford'. The expected development is exhibited by Stafford Devon SS 5811, *Staford* 1086, *Stafort, Stadforda* 1086 Exon, *Stouford* 1333 and STAFFORD SJ 9223. D 208 gives pr [stoufəd], 41, 367.

STO(W)FORD Wilts ST 8157, SU 0835 → STOWFORD Devon SX 4386.

STOWLANGTOFT Suff TL 9668 → STOW.

STOWMARKET Suff TM 0558 → STOW.

STOWTING Kent TR 1242. 'The hill, the hill-place'. *Stuting* c.1045 S 1471, *Stotinges, Estotinghes* 1086, *Stu- Stotinge(s)* 11th–1240, *Scuting'* (for *St-*) 1244, *Stutynge* 1253×4, *Stouting* 1257. OE ***stūting** < ***stūt** + **ing**[2]. The reference is to the circular mound on which the motte of Stowting Castle was later built. The apparently plural forms are probably AN *-s*. cf. RUCKINGE TR 0233. PNK 429, KPN 325, ING 188, BzN 1967.347, 355, Cullen.

STOWUPLAND Suff TM 0660 → STOW.

STRADBROKE Suff TM 2374. Partly uncertain. *State- Stetebroc* 1086, *Stradebroc -brok(e)* 1100–1454, *Stradbrok* 1610. This has been taken to be OE **strǣte** + **brōc**, but there is no Roman road here and no brook. Formally the specific could be OE **strǣde** 'a pace, a step'; with **brōc** this might mean a 'brook that can be crossed at a pace or a stride'. DEPN, Baron.

STRADISHALL Suff TL 7452. 'The shelter on the (Roman) road'. *Stratesella* 1086, *Strateshell* 1203, *Stratezell* 1228, *Stradesele*. OE **strǣt** + **ġesell**. No Roman road is known here. DEPN.

STRADSETT Norf TF 6605. 'Fold, camp or place on the Roman road'. *Strateseta* 1086, *-sete* 1254, *Stradesete* 1242. OE **strǣt** + **(ġe)set**, pl. **setu**. On the Roman road from Smallburgh to Denver, Margary no. 38. DEPN.

STRAGGLETHORPE Lincs SK 9152. Possibly 'Straker's outlying farm'. *Stagarthorp* late 12th, *Stragerthorp* 1242–81, *Tragertorp -thorpe* 1212, 1242×3, *Stragelthorpe -il- -yl-* 1306–1610, *-le-* 1548–1629, *Stragglethorpe* from 1600. Pers.n. or surname *Straker* + **thorp**. Straker would be either a derivative of ME *strake* 'to stroke' < OE *stracian* or more likely identical with ModE *streaker* 'a hunting dog, a hound', ME *stracur* < OFr *estrac* 'track'. Perrott 361, DBS 336, OED s.v. *streaker*.

STRAMSHALL Staffs SK 0835. 'Strangric's hill'. *Stagrigesholle* 1086, *Stranricheshill* 1199, *Sterangricheshull* 1208, *Strangricheshall -hull* 1221, 1227, *Strongushull* 1274, *Strangsil* 1419, *Strongeshulf* 1209, *Stronsheff* 1415. OE pers.n. **Strangrīċ*, genitive sing. **Strangrīċes*, + **hyll** partly confused with **scylf** 'a shelf'. The village occupies a hill-top rising to 300ft. overlooking the r. Team. DEPN, Horovitz.

STRANGEWAYS GMan SJ 8499. '(Place subject to) strong flooding'. *Strangwas* 1322–56, *Strang(e)way(e)s* 1326–1577, *Strangwyshe* 1551. OE **strang** + **(ġe)wæsc** changed by popular etymology. Strangeways is in a tongue of land between the rivers Irk and Irwell. La 33, L 59.

STRASHLEIGH Devon SX 6055 → STRETCHOLT Somer ST 2944.

STRATFIELD 'Open land by a Roman road'. OE **strǣt** + **feld**. PNE i.166, ii.162, L 239, 243.

(1) ~ MORTIMER Berks SU 6664. 'S held by the Mortimer family'. *Stratfeld' Hugonis de Mortemer* 'Hugh de M's Stratfield' 1167, 1175, *Stratfeld Mortem'* 1275×6 etc. with variants *Mortimer, Mortymer, Mortymar*. Earlier simply *Stradfeld* 1086, *Strafeld'* 1190, *Straffeld* 1230, *Stretfeld* 1224×5. The three Stratfields (S Mortimer, S Saye, S Turgis) lie on either side of the Roman road from London to Silchester, Margary no. 4a. Ralf de Mortimer (from Mortemer, Seine-Maritime) held S Mortimer in 1086 and his descendants kept it until the accession of Edward IV (1461). Brk 216, L 239, 243, Byn 101–21.

(2) ~ SAYE Hants SU 6861. 'S held by the Say family'. *Stratfeldsay* 1263–1486. Earlier simply *(at) Stratfeld* [1053×66]14th S 1129, 1158–1227, *Stradfelle* 1086. The reference is to the road from Silchester to London, Margary no. 4a. The manor was held in 1227 by Robert de Say, whose family came from Sai in Normandy. The addition is for distinction from the other Stratfield estates, S MORTIMER Berks SU 6764 and S TURGIS SU 6960. Ha 157, Gover 1958.122, DEPN, L 239, 243.

(3) ~ TURGIS Hants SU 6960. 'S held by the Turgis family'. *Stratfeud Turgis* 1287, *Stratfeld Turgys* 1293–1331. Earlier simply *Stradfelle* 1086. P.n. Stratfield as in STRATFIELD SAYE SU 6861 + family name Turgis, a Normanised form of the Scand pers.n. *Thórgils*. Ha 157, Gover 1958.123.

STRATFORD 'Ford on a Roman road'. OE **strǣt-ford**. Cf. STRETFORD. PNE ii.162.

(1) ~ GLond TQ 3883. See Stratford atte BOW.

(2) ~ ST ANDREW Suff TM 3560. 'St Andrew's S'. *Straford St. Andrew* 1837 OS. P.n. *Straffort* 1086, *Strafford* 1254, *Stratford* from 1316, + saint's name *Andrew* from the dedication of the church. No Roman road is known here. DEPN, Baron.

(3) ~ ST MARY Suff TM 0434.'St Mary's S'. P.n. *Strœtford* [962×91]11th S 1494, 975×1016 S 1487, *Stredford* [962×91]13th S 1494, *Stratfort* 1086, *Stratford* from 1242, *Stretford* 1610, + saint's name *Mary* from the dedication of the church. Stratford lies on the Roman road from Colchester to Baylham, Margary no. 3c. DEPN, Baron.

(4) ~ TONY Wilts SU 0926. 'S held by the Tony family'. *Stratford To(u)ny* 1332, 1363. P.n. *(on) stretford* *[? 793×6]12th S 229, *(on) stre(a)t ford* [948]11th S 540, *Stradford* 1086, *Stratford* 1242–9, *Stretford* 1279–1338, + manorial addition from the family name of Ralph de Touny who held the manor in 1242. Wlt 224.

(5) Stony ~ Bucks SP 7940. *Stani Stratford* 1202, *Stony Stretteford* 1290. 'Stony' for distinction from Fenny STRATFORD SP 8834 and Old Stratford SP 7741, *Westratford* 1278, OE **west**, *Forstratford* 1330, ON **forn** 'old', *Oldstratford* 1498. The ford carries Watling Street, Margary no 1e, across the Ouse. Bk 18, Nth 97.

(6) ~ -UPON-AVON Warw SP 2055. *Ufera Stretford bi Eafene* 'upper Stratford by Avon' [845]11th S 198, *Stratford super Avene* 1247. P.n. *(œt) Stretfordœ* [691×9]18th S 76, *(œt) Stretforda* [781]11th S 1257, *Streatforda* [714]16th S 1250, *(on) Strœtford* [985]11th S 1350, *(innan) Strœtforda* [988]11th S 1356, *Stradforde* 1086, *Stratford* from 1221, *Stretford* 1251–1309, 1564, + r.n. AVON. A ford on the Roman road from Ealington to Droitwich, Margary no. 56b. There must have been two fords to judge from the first form cited above and *œt Uferanstrœtforda* [966]11th S 1310, 'at the upper street-ford'. Wa 236.

(7) ~ -UPON-AVON CANAL Warw SP 1765. P.n. STRATFORD-UPON-AVON + ModE **canal**.

(8) Fenny ~ Bucks SP 8834. *Fenni Stratford* 1252, *Fenny Stretford* 1338. ME adj. **fenni** + p.n. Stratford. The place lies on the Roman road from St Albans to Towcester (Watling Street), Margary no. 1e. Marshy ground lies at the bottom of the hill

on which the town sits high and dry. 'Fenny' for distinction from Stony STRATFORD SP 7940. Bk 26.

(9) Stony ~ Bucks SP 7940. *Stani Stratford* 1202, *Stony Stretteford* 1290. ME adj. **stani** + p.n. Stratford. The place lies on the Roman road from St Albans to Towcester (Watling Street), Margary no. 1e, at the point where it crosses the river Ouse. The bed of the river was filled with stones to make the ford. 'Stony' for distinction from Fenny STRATFORD SP 8834. Bk 18.

(10) Water ~ Bucks SP 6534. *Stratford ad Aquam* 1542. Also known as *West Watrestretford* 1383. The furthest W of the three Stratfords in Bucks lying on the Ouse on the Roman road from Bicester to Towcester, Margary no. 160a. Earlier simply *Stradford* 1086, *Stret- Stratford* 13th cent. Bk 49.

STRATTON 'Street settlement, settlement on or by a Roman road'. OE **strǣt-tūn**. Cf. STRETTON.

(1) ~ Dorset SY 6593. *Stratton* from 1212. The reference is to the Dorchester–Ilchester road, Margary no. 47, which is here joined by a branch from Stinsford, Margary no. 470. Do i.373.

(2) ~ Glos SP 0103. *Stratone* 1086, 1215, *Stretton* 1200–1221, 1600, *Stratton(e)* 1221–1597. The road is Ermine Street between Cirencester and Gloucester, Margary no. 41c. Gl i.65.

(3) ~ AUDLEY Oxon SP 6026. 'S held by the Audeley family'. ~ *Audeley* 1318. P.n. *Stratone* 1086 *Stratton(a)* from 1109 + manorial affix from the family n. of William de Alditheleg' 1244. O 239.

(4) ~ -ON-THE-FOSSE Somer ST 6550. *Stratton super la Fosse* 1347, ~ *in the Vorswey* 1610. P.n. *Stra- Stretone* 1086, *Stratton* [n.d.] Buck + road name Foss as in FOSS WAY ST 5021. DEPN.

(5) ~ ST MARGARET Wilts SU 1787. 'St Margaret's S'. *Stratton Sce Margarete* 1294, 1578, *St Margaret Stratton* 1630. P.n. *Stratone* 1086, *Strettuna -e* c.1150, 1216×72, *-ton* 1253–1446, *Stratton(e)* 1195 etc., + saint's name *Margaret* from the dedication of the church. Also known as *Netherstratton* 1268, 1578, *Lower Stratton* 1509, and *Nethetowne al. Throppe* 1630 for distinction from Upper Stratton SU 1687, *superior Stratton* 1316, *Upper Stratton* 1509. Both lie on Ermine Street, Margary no. 41b. Wlt 33.

(6) ~ ST MICHAEL Norf TM 2093. 'St Michael's S'. *Strattone Sancti Michaelis* 1254. Earlier simply *Estratuna, Stra- Stretuna* 1086. Saint's name *Michael* from the dedication of the church. On the Roman road from Baylham to Caistor St Edmund, Margary no. 3d. DEPN.

(7) ~ STRAWLESS Norf TG 2220. 'Strawless S'. *Strattone Streles* 1446, ME **strēles**. Earlier simply *Stratuna* 1086. This place lies on no known Roman road about 2 miles S of the Roman road from Smallburgh to Denver, Margary no. 38. DEPN.

(8) East ~ Hants SU 5439. *Est Strattone* 1316. ME adj. **est** + p.n. *strattone* [903]14th S 370, *Stratune* 1086. The reference is to the Roman road from London to Winchester, Margary no. 42a. 'East' for distinction from West Stratton SU 5240, *West Stratton* from 1250. Ha 157, Gover 1958.83, 82, DEPN.

(9) Long ~ Norf TM 1992. *Long Stratton* 1275. ME adj. **long** + p.n. Stratton as in STRATTON ST MICHAEL TM 2093. Also known as *Stratton Sancte Marie* 1291, 'St Mary's S', from the dedication of the church. DEPN.

(10) Stoney ~ Somer ST 6539. *Stoney Stratton* 1817 OS. Earlier simply *Strettun* *[1042]18th S 1042, *Stratton* 1262. The village does not lie on a known Roman road. DEPN.

STRATTON Corn SS 2206. Originally 'valley of the (River) Neet'. *Strætneat* [873×88]11th S 1507, *Stratone -a* 1086, *Stratton* from 1187, *Stretton* 1249. OCo ***strad** 'valley' + r.n. **Nēth* (*Neth, Neet, Neh(e)t* 13th), identical with the River Neath W Glam and Powys SN 9013, *Nido* [ablative] [c.300]8th AI, the Yorks r. NIDD, *Nid* c.715, and Corn Glasney, *Glasneyth* 1291, 1306, 'the blue *Neth*', Co **glas**, probably the old name of the Penryn River SW 7834, from a base form **Nidā* seen also in the Old European continental r.ns. Nidda, *Nitta* 782, *Nidda* 800, a tributary of the Main at Frankfurt with its own tributary, the Nidder, *Nitorna* 1016 < **Nidurna*, and the Roman settlement *Nida* 2nd AD near Heddernheim, the Nied, *Nita* 1018, *Neda* 1121, a tributary of the Saar in Lorraine, the Nethe, a tributary of the Weser near Höxter, the Belgian Nethe, *Hnita* 726, and the Nitja in Norway, all from the IE root **neid-/*nid-* 'flow'. The name was early assimilated to the pattern of STRATTON < OE **strǣt-tūn** 'settlement on a Roman road' from which a new r.n., the River Strat evolved by back-formation, although the r.n. Neet is still used, probably as a result of antiquarian revival. There is no Roman road at Stratton. PNCo 159, RN 301, BzN 1957.251–2, Berger 194, 198.

STREAT ESusx TQ 3515. 'The (Roman) road'. *Estrat* 1086, *Strete* 1271–15th, *Strates, Stretes* 13th cent. OE **strǣt**. The reference is to the Roman road from Barcombe Mills to Hardham, Margary no. 140. Sx 304.

STREATHAM GLond TQ 2972. 'Homestead or enclosure by the Roman road' sc. The Roman road from London to Brighton, Margary no. 150. *Estreham* 1086, *Stretham* 1225 etc. including *[727]13th S 1181 and *[933]13th S 420, *Stratham* 1175–1255 including *[1062]13th S 1035, *Streteham* 1247–1432, *Streetham* 1422, *Streatham* 1510. OE **stræt** + **hām** or **hamm**. Sr 33 gives pr [stretəm].

STREATLEY 'Wood or clearing by the street or Roman road'. OE **strǣt** + **lēah**.

(1) ~ Beds TL 0728. *Strœtlea* [c.1053]13th S 1517, *Stradl(e)i, Strailli* 1086, *Stratle(gh) -ley* 13th–14th cents., *Stredle(y)* 13th cent., *Stretle(ye)* 13th–18th cents. The street is the road from Luton to Bedford. It is not known to have been Roman. Bd 164, L 82, 206.

(2) ~ Berks SU 5980. *Stretlea* [688×90]12th S 252, *Stretleœ* [687]12th S 239, *Stretleg' -le(ye) -l(e)y -lee -ley* 1224–1535, *Estralei* 1086, *Streyt(e)ley* 16th cent., *Streatly* 1583. The reference is to the Roman road from Dorchester-on-Thames to Silchester, Margary no. 161c. Brk 531 gives pr [stretli], L 82, 206.

STREET Lancs SD 5252. No early forms. OE **strǣt, strēt** 'a Roman road' referring to the Roman road from Ribchester to Lancaster, Margary no. 704.

STREET NYorks NZ 7304. *Street* 1861 OS.

STREET Somer ST 4836. 'The Roman road'. *Stret* *[725]12th S 250, *Strete* *[971]13th S 783, 1330, *Streete* 1610. The reference is to the Street causeway which crosses the marsh between Street and Glastonbury, Margary no. 511. DEPN.

STREET END WSusx SZ 8599. 'The end of the street or hamlet'. *Streetend Farm* 1743. ModE **street** + **end**. Sx 85.

STREETHAY Staffs SK 1410. 'The enclosure by the Roman road'. *Strethay* before 1176, 1256, *Stretheye* 1262, 1286, *Streetahie* 1601. OE **strēt** + **(ġe)hæġ**. The reference is to Ryknild Street, Margary no. 18c. DEPN, Duignan 145, Horovitz.

STREETLY WMids SP 0998. 'The wood or clearing by the Roman road'. *(on) strœtléa, (into) strétlíe* [957]12th S 574. OE **strēt** + **lēah**. The place lies on Icknield Street, Margary no. 18b. DEPN.

STREFFORD Shrops SO 4485. 'Roman road ford'. *Straford* 1086, *Stratford* 1232, *Streford* 1255–1318. OE **strēt-ford**. Presumably named from Stretford Bridge SO 4385 where the Shropshire Watling Street, Margary no. 6b, crosses the river Onny. Bowcock 228.

STRENSALL NYorks SE 6360. Uncertain. *Strenshale* 1086, *Strensale* 1127×8–1319, *Strenes(h)ale* 1222–1302, *Strensall* from 1316. The earlier explanation, 'Streon's nook of land', OE pers.n. **Strēon*, genitive sing. **Strēones*, + **halh** has been questioned owing to the peculiarity that, apart from its appearance as a name-element in the OE pers.n. *Strēonbercht* in LVD and as a by-name in *Godric Strēona*, the inferred pers.n. **Strēon* only appears in p.ns. in composition with *halh*. This suggests that *Streoneshalh* is more likely to have been a compound appellative with OE **strēon** 'gain, acquisition, property, treasure, procreation', genitive sing. **strēones**, as its first element. It has been suggested that the sense may have been 'begetting corner, secluded spot used by lovers'. Identical with *Streaneshalh, Streonœshalch* [c.731]8th BHE, the earlier name of Whitby. YN 13, Jnl.13.50–3, L 109, Signposts 189.

STRENSHAM H&W SO 9140. 'Streng's promontory'. *Strenchesham* [c.1086]1190, 1275, *Strengesham* before 1198–1428, *Strensham* 16th cent. OE pers.n. **Streng*, genitive sing. **Strenges*, + **hamm** 2a. The earliest reference is *(in) strengesho* 972 S 786 'Streng's hill-spur' referring to the cliff above the Severn on which Strensham stands, OE pers.n. *Streng* + **hōh**. Wo 229 gives pr [strensəm].

STRETCHOLT Somer ST 2944. 'Stretch wood'. *Stretheholt* (for *Streche-*) 1242, *Streccholt* 1344. ME ***strecche** 'a stretch of land' + **holt**. Cf. STRASHLEIGH Devon SX 6055, *Strecchelegh* 1285. D 273.

STRETE Devon SX 8347. 'The road'. *Streta* 1194, *Strete* from 1244. OE **strǣt** 'a road, a Roman road'. The place lies on an ancient trackway. D 317, DEPN.

STRETFORD 'Roman road ford'. OE **strēt-ford**. L 62, 70, 82.

(1) ~ GMan SJ 7994. *Stretford* from 1212, *Stratford(e)* 1292, 1549. The ford carried the Roman road from Chester to Manchester, Margary no. 7a, across the Mersey. La 32.

(2) ~ COURT H&W SO 4455. P.n. Stretford + ModE **court**. Stretford, *Stratford* 1086, *Stretford* 1316, is the 'Roman-road ford', The reference is to a crossing of Stretford Brook by Watling Street, the Roman road from Leintwardine to Monmouth, Margary no. 6c. DEPN.

STRETHALL Essex TL 4939. 'Nook of land by the Roman road'. *Strathala* 1086, *-hale* 1235–1336, *Strethal(e) -hall* 1212–1317, *-all* 1571. OE **strǣt** + **halh**. The reference is to a nook of land at the county boundary and the Roman road from Braughing to Worsted Lodge, Margary no. 21b. Cf. CHRISHALL TL 4439. Ess 534 gives pr [stretəl].

STRETHAM Cambs TL 5174. 'Homestead on the Roman road' from Cambridge to Littleport (Akeman Street), Margary no. 23b. *(œt) Strœtham* c. 975, *Stratham* 1086–1303, *Stradham* 1086, *Stretham* from 1170. OE **strǣt** + **hām**. Ca 237.

STRETTINGTON WSusx SU 8907. 'Settlement of the dwellers at *Stratone*'. *Estretementona* [1100×35]1332, [1154×89]1387, *Estremeton* 1155, *Stretham(p)ton* 1288. OE folk-n. **Strǣt-hǣme*, short for **Strǣt-tun-hǣme* 'the inhabitants of *Stratone*', genitive pl. **Strǣt-hǣma*, + **tūn**. In 1086 it is simply *Stratone* 'street settlement', OE **strǣt** + **tūn**. The reference is to Stane Street, *Stan(e)stret* 1270, 1279, 'the stone street', the Roman road from Chichester to Pulborough, Margary no.15. Sx 68.

STRETTON 'Settlement on a Roman road'. OAngl **strēt** + **tūn**. Identical with STRATTON < WS **strǣt-tūn**.

(1) ~ Ches SJ 6182. *Stretton(a)* [1154×89, 1199×1216]17th, and from 1259. Margary no.70a. Che ii.121.

(2) ~ Ches SJ 4453. *Stretton* from 12th. Margary no.6a. Che iv.56.

(3) ~ Derby SK 3961. *(œt) Strœttune* [1002×4]c.1100 S 1536, *Stratune* 1086, *Strattun -ton* 1154×89–1415, *Stretton(e)* from 12th. The village lies on Ryknild Street, Margary no.18d. Db 307.

(4) ~ Leic SK 9415. *Stratune -tone* 1086, *Stratton'* 1107–1503, *Stretton(e)* 1176–1610. Also known as *Est-Stratton* 1306 and *Stretton in the Strete* 15th cent. for distinction from Great and Little STRETTON SK 6600. The village lies on Ermine Street, Margary no. 2c. R 37.

(5) ~ Staffs SJ 8811. *Estretone* 1086, *Stretton(a)* from 1166, *Stratton(a)* 1175–1375. The reference is to the Roman road from Stretton to Whitchurch, Margary no. 19. St i.178.

(6) ~ Staffs SK 2526. *Stretton* from [941]13th S 479, *Strœttun* [1002×4]11th S 1536, *Stratone* 1086. The reference is to Ryknild Street, Margary no. 18c. DEPN.

(7) ~ EN LE FIELD Leic SK 3011. P.n. *Streitun, Stretone* 1086, *Stretton(e)* c.1130–1610, + ME affix *in le, in the Feld(e)* 1412–1475, *in le Field* 1617, *en le Field* 1795, 'in the open country' for distinction from other Strettons. No Roman road is known, but the village is on the line of the road from Grantham to Barrow on Soar, Margary no. 58a, which was used as a Salt Way. Lei 571.

(8) ~ GRANDISON H&W SO 6344. 'S held by the Grandison family'. *Stretton Graundison* 1350. Earlier simply *Stratvne* 1086, *Strettona* c.1180. S lies on the Roman roads from Dymock and to Kenchester, Margary nos. 610, 63a. DEPN.

(9) ~ HEATH Shrops SJ 3610. 'Heath belonging to (Stoney) Stretton'. *Streton's Heath* 1624, *Stretton Heath* 1836 OS. P.n. Stoney STRETTON SJ 3809 + ModE **heath**. A squatter settlement. Sa ii.64.

(10) ~ -ON-DUNSMORE Warw SP 4172. *Stratton super Dunnesmor* 1262, *Stretton on Dounesmor* 14th cent. P.n. *Stratone* 1086, c.1126, *Stratton* 1159–1535, *Stretton(a)* c.1190–1444, + p.n. DUNSMORE. Stretton lies on the Fosse Way. Wa 146.

(11) ~ -ON-FOSSE Warw SP 2238. *Stretton super le Fosse* 1263, ~ *on the Fosse al. on the Force* 1651. P.n. *Stratone* 1086, *Stratton* c.1170–1291, *Stretton* from 1235, + road-n. FOSSE WAY. The village lies beside the Fosse Way. Wa 306.

(12) ~ SUGWAS H&W SO 4642. 'S near Sugwas'. *Strettone by Sugwas* 1347. Earlier simply *Stratone* 1086, *Strattone* 1285, *Strettone* 1340. Originally two separate manors, Stretton and Sugwas, *Svcwessen* 1086, *Sugwas* from 1160×70, 'alluvial land frequented by sparrows', OE **sucga** + **wæsse**. Stretton lies on the Roman road from Stretton Grandison to Kenchester, Margary no. 63a. Sugwas lies on the alluvial plain of the river Wye. He 186, L 59–60.

(13) ~ UNDER FOSSE Warw SP 4581. *Stretton subtus Fosse* 1656. P.n. *Stretton juxta Kirkeby monach.* 1303 'S beside Monks Kirby'. The village lies just off the Fosse Way. Wa 119.

(14) ~ WESTWOOD Shrops SO 5998. *Stretton Westwood* 1934. P.n. Stretton, here probably a recent creation for a row of houses along a track at right angles to the road along Wenlock Edge (the 'street') + p.n. *Westwud* 1235, cf. *Westwood Barn, Common* and *Cottage* 1833 OS, 'the west wood', OE **west** + **wudu**. Gelling.

(15) All ~ Shrops SO 4695. 'Alfred's S'. *Auredesstratton', Aluredesstretton'* 1261×2, *Aluestretton'* 1291×2, *Ould Stretton* c.1540, *Alstretton* 1577, *All-Stretton* 1662. OE pers.n. *Ælfrēd*, genitive sing. *Ælfrēdes*, for distinction from Church STRETTON. Sa i.286–7.

(16) Church ~ Shrops SO 4593. 'S with the church'. *Chirch' Stratton', Chirich -ech Stretton* 1261×2, *Chyrchestretton'* 1271×2, *Chirchestretton* 14th cent., *Churchstretton* from 1577. ME **chirche** + p.n. *Stratun(e)* 1086, *Strettun(a) -ton(')* from 1156. Also known as *Magna Stretton* 1261, *Great Stretton* c.1540, for distinction from Little STRETTON, and as *Stratton in Strettondale* 1227 etc., *Stretton Le Dale* 1666, ~ *en le Dale, in the Dale* 18th cent. The reference is to the Shropshire Watling Street, Margary no. 6b. Sa i.286.

(17) Little ~ Leic SK 6600. Latin prefix **parva**, *Parua* 1327–1368 (postfix 1303–1627), ModE **little**, *Little* 1610, + p.n. *Straton(e)* 1086–1368, *Stratton(e)* 1156–1435, *Stretton(e)* 1183–1610. 'Little' for distinction from Great Stretton SK 6500, *Magna* prefix 1283–1344, postfix 1319–1629, *Much* 1467×72, *Great* 1610. Lei 258.

(18) Little ~ Shrops SO 4491. *Parua Stretton'* 1261×2, ~ *Stratton* 1271×2, *Little Stretton* from c.1540. 'Little' for distinction from Church or Great STRETTON. Sa i.286.

(19) Stoney ~ Shrops SJ 3809. *Stoney Stretton* 1833 OS. ModE adj. **stony** + p.n. *Stretton(')* from 1255. Also known as *Stretton Parva* 'little S' 1612 for distinction fron Church STRETTON etc. The reference is to the Roman road from Wroxeter to Trefeglwys, Margary no. 64, which was known locally as *The Stoney Causeway* 1512. Sa ii.64.

Great STRICKLAND Cumbr NY 5522. *magna Sterkelangd* (sic) 1278, *Magna Stri(c)kland* 1540, *Greate Strickland(e)* 1580, 1643. Adj. **great**, Latin **magna**, + p.n. *Stirc- Stirk(e)- Styrkeland(a) -lond* 1193–1401, 'newly cultivated land where stirks (heifers) are kept', OE **stirc**, genitive pl. **stirca**, + **land**. 'Great' for distinction from Little STRICKLAND NY 5619. We ii.149, L 248–9.

Little STRICKLAND Cumbr NY 5619. *parva Sti- Styrk(e)land* c.1233–1415, *Li- Lyt(y)le Stri(c)k- Strykland* 1527–1632. Adj. **little** (Latin **parva**) + p.n. Strickland as in Great STRICKLAND NY 5522. We ii.152

STRINES RESERVOIR SYorks SK 2390. P.n. Strines + ModE **reservoir**. Strines, *Stryndes* 1591, is Mod dial. **strine** 'a ditch, a water channel' (OE *strind* 'a stream'). YW i.228.

STRINGSTON Somer ST 1742. 'Strang's farm or estate'. *Strangestona* 1084, *Strengestvne, Strenegestone* 1086, *Strengeston* [1295] Buck, *Strenixton* 1610, *Stringston* 1809. If the 1084 form is reliable this is AScand pers.n. **Strang* (ODan *Strangi)*, genitive sing. *Stranges*, + **tūn**. DEPN gives OE pers.n. *Strengi*.

STRIXTON Northants SP 9061. 'Strikr's farm or village'. *Strixton(e)* from 12th, *Strickson* 1658. ON pers.n. *Strikr*, genitive sing. *Striks*, + **tūn**. Strikr is probably identical with the *Stric* recorded in Domesday Book as holding land in neighbouring Wollaston and Bozeat in the time of King Edward the Confessor. Nth 197 gives pr [striksən], SSNEM 194.

STROAT Glos ST 5797. 'The Roman road'. *to Stræt* [1061×5]12th S 1555, *Strawte* 1575, *Stroote* 1587, *Stroat(e)* 1597, 1637, *Strete* 1624. The place is on the Roman road to S Wales, Margary no. 60a, and the earliest form in S 1555 points to OE **strǣt**. The later forms, however, except *Strete* 1624, cannot have the same derivation and must be related to the W name of nearby Tidenham ST 5695, *Istrat Hafren* 12th 'Severn valley', OW **Strat Habren*. The S 1555 form could itself, in fact, represent OW **strat** as OE *Strœtneat* [c.880]11th for STRATTON Corn SS 2306 'valley of the river Neet', represents OCo **strad**. Gl iii.265, L 82.

STROOD Kent TQ 7269. 'Marshy land overgrown with brushwood'. *strōdes* (genitive sing.) [889]late 9th S 1276, *Strodes* 1100×35–1219, *Strode* 1158×9–1205. OE **strōd**. KPN 228.

STROUD Glos SO 8505. 'Marshy land overgrown with brushwood'. *(la) Strode* 1200–1540, *Strowde* 1542–1698, *Stroud(e)* 1652, 1694, *Strood(e)* 1561, 1592. OE **strōd**. OE, ME **strōd** would normally give [stru:d] as in STROOD Kent TQ 7269. There was, however, a late ME variant with *ū* from *ō* in non-standard speech which gave dial. Stroud pronounced [straud]. Gl i.139, Jordan §53, Dobson §158, L 34, 58–9.

STROUD Hants SU 7223. 'The marsh'. Cf. the surname *(atte) Strode* 1327, *(atte) Stroude* 1333, OE **strōd**. Stroud lies beside one of the headwaters of the Rother. Ha 158, Gover 1958.63.

STRUBBY Lincs TF 4582. 'Village, farm on a promontory'. *Strobi* 1086, *Strubby* [1115]14th, *Stru(b)bi* [1125]13th, 1154×89. These forms may be supplemented by those for Strubby Hall at TF 1577, *Strubi* 1086, *Stru(te)bi* c.1115, *Strubi* [late 12th]13th. ON **strútr** 'a pointed hood' used topographically of something projecting + **bȳ**. Strubby stands on a promontory of raised ground projecting into marsh and Strubby Hall on a small pointed promontory. DEPN, SSNEM 71–2. Cameron 1998 prefers the pers.n. *Strútr* derived from **strútr**.

STRUMPSHAW Norf TG 3507. 'Scrubland with stumps'. *Stromessaga* 1086, *Trumeshah* 1204, *Strumeshag* 1212, *Strumpsawe* 1291, *Strumpeshache* 1295. OE ***strump** + **sceaga**. DEPN, PNE ii.164.

STUBBINGTON Hants SU 5503. 'Settlement at the *stubbing*, the cleared land with tree-stumps'. *Stvbitone* 1086, *Stub(b)inton(e) -yn-* 1202–1316. OE, ME **stubbing** < OE **stubb** + **ing**[2], + **tūn**. Ha 158, Gover 1958.29, DEPN, Årsskrift 1974.58.

STUBBINS Lancs SD 7918. 'The cleared land'. *Stubbyns Halle* 1559, *Stubbyng* 1563. ME **stubbing**, pl. **stubbinges**. La 64.

STUBHAMPTON Dorset ST 9113. 'Estate of the *Stybbhǣme*, the inhabitants of *Stybbham*, the tree-stump homestead'. *Stibemetune* 1086, *Stubehampton(e)* 1233, 1280, *Stubhampton(e)* from 1280. OE folk-n. **Stybbhǣme* 'dwellers at **Stybbhām*', genitive pl. **Stybbhǣma*, + **tūn**. **Stybbhām* would be 'tree-stump homestead or village', OE **stybb** + **hām**. Do ii.245.

STUBTON Lincs SK 8748. 'Tree-stump farm or village'. *Stvbetvne, Stobetun* 1086, *Stubeton(a)* 1206–1395, *Stubton* from 1210×23. OE **stubb**, genitive pl. **stubba**, + **tūn**. Pers.n. *Stubba* is a possible alternative. Perrott 385.

STUCKTON Hants SU 1613. Partly uncertain. *Stuketon -tune* c.1210–1256, *Stickton* 1536, *Stukton* 1541. A further form, *Styker(e)ton* 1307×27, may also belong here, although the identification is not certain and it does not fit the run of forms; it might represent OE **sticera-tūn* 'estate of the stickers or (pig) slaughterers'. OE **stūc** 'a heap, a stook', genitive pl. **stūca**, is possible, as is the unknown **stūca*, **stūce* of *stucan wise* [935]12th S 430 in the bounds of Havant. Gover 1958.215.

East STUDDAL Kent TR 3249. *Eststodewold', Eststodwolde* 1270. ME adj. **est** + p.n. *Stodwald(e)* 1240, 1254, *Stodwolde, Stotwold'* 1254, 'the horse-herd forest', OE **stōd** + **weald**. 'East' for distinction from West Studdal TR 3049, *Weststodwolde* 13th. PNK 569, L 227.

STUDHAM Beds TL 0215. 'Stud homestead' or 'enclosure where horses are bred'. *(æt) Stodham* [1053×66]13th S 1235–16th cent., *Estodham* 1086, *Studham* 1326. OE **stōd** + **hām** or **hamm**. Bd 132.

STUDLAND Dorset SZ 0382. 'Newly cultivated land by the horse-stud'. *Stollant* 1086, *Stodland(e) -lond(e)* 1210–1514, *Stoudlond* 1327–1543, *Studland* from 1512. OE **stōd** + **land**. Do i.43, L 127, 129.

STUDLAND BAY Dorset SZ 0383. *Studlandbaye* 1575. P.n. STUDLAND SZ 0382 + ModE **bay**. Do i.50.

STUDLEY 'Clearing used for a stud of horses, stud pasture'. OE **stōd** + **lēah**. L 206.

(1) ~ Warw SP 0763. *Stodlei* 1086, *-leia -lege -legh* 1130–1507, *Studeley* 1453, *Studlegh* 1548, *Stoodeley* 1555. Wa 225.

(2) ~ Wilts ST 9671. 'The stud pasture'. *Stodleia -lega* 1175–98, *-le(gh)* 13th, *Studley* 1653. OE **stōd** + **lēah**. Wlt 258, L 206.

(3) ~ ROGER NYorks SE 2970. 'Roger's S'. P.n. Studley + manorial addition *Roger* from 1288 referring to Roger de Mowbray. Studley is *Stollai -lei(a)* 1086, *Stodleia -le(e) -lay -ley* 1166–1428, *Studley* from 1285. Also known as *Nether* 1198, *Parva* 'little' c.1250 and *Sowth* Studley for distinction from STUDLEY ROYAL SE 2770 and North Studley (lost), *North Stodelay* 13th. YW v.190.

(4) ~ ROYAL NYorks SE 2770. P.n. Studley + manorial additions *rg'* (for Latin *regis*) 1589, *Royal* 1822. Studley is *(on) Stodlege* c.1030 YCh 7, *Estollaia* 1086, *Stodleia -lega -ley -lay* c.1132–1552. Also known as *Magna* 'great' 1297, *Studley Magna* 1481, and *Overstodelay* 1198. YW v.192.

Great STUKELEY Cambs TL 2275. *Magna Steuecle* c.1200. ModE adj. **great**, Latin **magna**, + p.n. *Stivecle* 1086–1306 with variants *Stiu- Styu-* and *-kle*, *Stucle(y)*, *Stuckle* 1362–1433, 'tree-stump clearing', OE ***styfic** + **lēah**. 'Great' for distinction from Little STUKELEY TL 2075. Hu 224.

Little STUKELEY Cambs TL 2075. *Parva Stiueclai* 1193. ModE **little**, Latin **parva**, + p.n. Stukeley as in Great STUKELEY TL 2275. Hu 224.

STUMP CROSS Cambs TL 5044. No early forms.

STUMP CROSS Essex TL 5044. No early forms. ModE **stump** + **cross**.

STUNTNEY Cambs TL 5578. '(At) the steep island'. *Stuntenei* 1086, *-(e)ie -eia -eya -ey(e)* 1086–1418, *Stumpney* 1541. OE adj. ***stunt** 'steep', oblique case definite form ***stuntan**, + **ēġ**. OE ***stunt** may be identical with **stunt** 'foolish' ultimately from the IE root *(s)tewd-* 'push, prick, thrust'. The island rises suddenly from the surrounding fen. Alternatively we may have a pers.n. **Stunta*, genitive sing. **Stuntan*, and so 'Stunta's island'. Ca 220, L 37, 39.

STURBRIDGE Staffs SJ 8330. A puzzle. No early forms and there is no bridge, river or stream here.

STURMER Essex TL 6944. 'Stour pool'. *Sturmere* from 1086 including [c.993]17th *Battle of Maldon*. R.n. STOUR TL 9233 + OE **mere**. According to Harrison's *Description of Britain* 1577, 1586, the 'Sture or Stoure ariseth at Stouremeere, which is a poole conteining twentie acres of ground at the least'. The pool is marked on the 1805 OS ½ mile E of Sturmer; the river flows through the pool but arises some 10 miles above it. Ess 462.

STURMINSTER MARSHALL Dorset SY 9499. 'S held by the Marshal family'. The affix *Marescal* first appears in 1268, *Mare(s)schal(l)* 1284–1465. Sturminster is *(æt) Sture minster* [873×88]11th S 1507, *Sturminstre* 1086, *-min(i)str(e) -minister -myn(i)stre -mynystre -mynster* 1204 etc., 'the church on the river Stour', r.n. STOUR + OE **mynster**. 'Marshall', referring to possession by the Marshal family, earls of Pembroke, of whom William Marescallus was there in 1204 etc., is for distinction from STURMINSTER NEWTON ST 7814. Do ii.45 gives pr ['stə:Rmistə R].

STURMINSTER NEWTON Dorset ST 7814. This is a combination of two originally separate names, Sturminster on the N side of the Stour and Newton on the S.

I. *Sturministr'* 1288 *-mynster* 1294, *-mynstre (Abbatis)* 1333–1444, 'church on the r. Stour', r.n. STOUR + OE **mynster**. *Abbatis* 'abbot's' refers to possession by Glastonbury abbey from 968.

II. *Nywetone* [968]14th S 764, *Newentone* 1086, *Neuton(e), Niwetun, Ny- Niw(e)ton(e)* 1196–c.1350, *New(e)ton(')* from 1268, 'the new settlement', OE **nīwe**, definite form **nīwa**, dative case **nīwan**, + **tūn**, dative case **tūne**. Also known as *Newetone Kastel* [1016]1727, *Ni- Nyw(e)ton(e)castel* 1297–1350 from the Iron Age promontory fort within which lay the medieval manor house of the abbot of Glastonbury.

III. The combined name is recorded as *Stur(e)m(n)inistr(e) Nyweton(e)* 1291, *Sturmynstre Neuton* 1407, *Sturmynstre Neutoncastell* 1437, *Sturmister Newton* 1666. Do iii.188.

STURRY Kent TR 1760. 'The Stour district'. *uillam nomine Sturigao* 'vill called S' *[605]12th S 4, *Sturia* [678]15th S 1648, *Sturgeh* *[690]13th S 10 (with variant *Sturegh*), *Sturige* [690]13th S 11, *Sturrie* [690]13th S 13 (with 15th cent. variant *Stureie*), *Estvrai* 1086. R.n. STOUR TR 2057 + OE ***ġē**. KPN 6, RN 379, Charters IV.14, 139, 147, 149.

STURTON 'Farm or village by the paved road'. OE **strǣt, strēt** + **tūn**.

(1) ~ Northum NU 2107. *Strattona* [c.1220]14th, *Stretton* [1241×8]14th. NbDu 191.

(2) ~ BY STOW Lincs SK 8880. P.n. *Stratone* 1086, *Strettuna* c.1115, *Stretona* 1150×60, *Straton* 1191 + p.n. STOW SK 8882. The road in question is Margary no. 28a. DEPN, Cameron 1998.

(3) ~ LE STEEPLE Notts SK 7884. 'S with the steeple'. *Sturton le Steeple* 1732, *~ in the Steeple* 1769. P.n. *Estretone* 1086, *Stretton(a)* 1215–1459, *Stratton'* 1166–1280, *Sty- Stirton* 1499–1512, *Sturton* from 1513, + Fr definite article **le** + ModE **steeple**. The church possesses a tall steeple which can be seen for miles. Also known as *Stretton, Styrton in le Clay* 1263–1505. Sturton lies on clay soil on the Roman road from Lincoln to Doncaster, Margary no. 28a. Nt 40.

(4) Great ~ Lincs TF 2176. *magna Stretton* 1216×72. Latin **magna**, ModE **great**, + p.n. *Stratone* 1086, *Strettuna* c.1115, *Stratton* 1199, [1209]1252. The road in question is Margary no. 27. 'Great' for distinction from Little Sturton TF 2175, *Stratone* 1086. DEPN, Cameron 1998.

STUSTON Suff TM 1377. 'Stut's estate'. *Estutes- Stutestuna* 1086, *Stuteston* 1189×99, *Stuston* from 1254, *Sturston* 1610. OE pers.n. **Stūt*, genitive sing. **Stūtes*, + **tūn**. DEPN.

STUTTON NYorks SE 4841. 'Stump farm or village'. *Stouetun, Stutun(e) -tone* 1086, *Stutton* from 13th. ON **stúfr** perhaps replacing OE ***stuf** which must lie behind the elements ***styfic** 'a stump', ***styfiht** and **styfecing**, + **tūn**. The reference is either to a prominent tree-stump or possibly to Wingate Hill which forms a marked peak above the village. YW iv.75, SSNY 115.

STUTTON Suff TM 1434. 'The hill settlement'. *Sto(t)- Stuttuna* 1086, *Stutton* from 1220. OE ***stūt** + **tūn**. Stutton lies on a prominent hump or hill overlooking Holbrook Bay on the r. Stour. DEPN, PNE ii.165.

STYAL Ches SJ 8383. Partly uncertain. *Styhale* c.1200–1337, *Stiale* 1331, 1337, *Styall* from 1364. It is impossible to say whether the specific is OE **stigu** 'a sty, a pen' or **stīġ** 'a path'; the generic is OE **halh** in the sense 'small valley': 'nook with a pen' or 'with a path'. Che iv.229, L 81, 106, 110. Addenda gives 19th cent. local pr [staiə], now [staiəl].

STY HEAD Cumbr NY 2209. 'Head of the path'. *the Stey heade* 1540, *the Stime (or Stye) head* 1578, *Stye Head* 1774. ModE **sty** (OE *stīġ*) + **head**. The reference is to the *sty* or path which makes a long ascent from Borrowdale to cross into Wasdale. Cu 357.

SUCKLEY H&W SO 7251. 'Sparrow wood or clearing'. *Svchelei -lie* 1086, *Succhele(ia)* 1174, 1222, *Suggelega* 1180–90, *Suckele, Sukkele(ya)* 1242–1401×5. OE ***succa** + **lēah**. Wo 81, PNE ii.166.

SUCKLEY HILLS H&W SO 7352. P.n. SUCKLEY SO 7251 + ModE **hills**.

SUDBOROUGH Northants SP 9682. 'The southern fortified place'. *Suthburhc* 1065, *-burg(a) -buri* 13th cent., *Sutburg* 1086, *Sudburc* 1168, *-burgh* 1293 etc. Possibly so named with reference to Brigstock SP 9485. Nth 187.

SUDBOURNE Suff TM 4153. 'The southern stream'. *Sutborne* [1042×66]13th S 1051, *Sutburna -e, Sudburnha* 1086, *Sudburne* 1160–1610, *-bourne* from 1316. OE **sūth** + **burna**. The name is problematic, as there is no obvious stream here now and no obvious feature which might be a **Northbourne*. DEPN.

SUDBROOKE Lincs TF 0375. 'The southern stream'. *Sutbroc, Sudborc* 1086, *Sudbroc* 1202, *Suthbroc(a)* 1166, 1209×19. OE **sūth** + **brōc**. S in relation to Scothern. DEPN, L 16, Cameron 1998.

SUDBURY Derby SK 1632. 'The south fortification'. *Sudberie* 1086, *Suberia* c.1141, *Sudburia -bury* 1178–1318 etc. OE **sūth** + **byriġ**, dative sing. of **burh**. 'South' in contrast to NORBURY SK 1242. Db 610.

SUDBURY Suff TL 8741. 'The southern fort or manor'. *Sudberi* c.1200 ASC(F) under year 798, *(into) Suðbyrig* 1000×2 S 1486, *Sutberie* 1086, *Sudburye* 1610. OE **sūth** + **byriġ**, dative sing. of **burh**. 'South' for distinction from the **byriġ** at Bury St Edmunds. DEPN.

SUDELEY CASTLE Glos SP 0327. *castle of Sudeley* 1606. P.n. Sudeley + ModE **castle**. A 15th cent. castle largely built by Ralph Boteler between 1398 and 1469. Sudeley, *Svdlege* 1086, *Sudlei(e) -leg(a) -ley(e) -le(e)* 1168–1590, *Sudelegh -ley(a)* 1175–1683, *Siudle(ye)* 1354, *Seude- Sewdley* 1468–1712, *Shudley* 1637, is 'south wood or clearing', OE **sūth**+ **lēah**, because of its position S of Winchcomb SP 0228. Such a name would normally be expected to become *Sudley [sʌdli]. Occasionally, however, ME *ū* became [iu:] after palatal sounds, possibly in this case aided by French influence on the name. An alternative possibility is that the specific is not **sūth** but OE **(ġe)syd** 'wallowing place' or **scydd** 'a shed' with AN [s] for [ʃ]. Gl ii.26 gives prs ['su:dli, 'siudli], Dobson §178, L 206, Verey 1070C.438.

SUDGROVE Glos SO 9308. 'The south wood'. *Sodgraue* 1248, *Suthgrave* 1307, *Sudgraue* 1494, *-grove* 1620. OE **sūth** + **grāf**. S in the parish of Miserden SO 9309. Gl i.130.

SUFFIELD Norf TG 2332. 'South open-land'. *Sudfelda* 1086, *-feld* 1168, 1191. OE **sūth** + **feld**. S of Roughton Heath or of the woodland of what became Gunton Park. DEPN, L 242, 244.

SUFFIELD NYorks SE 9890. 'South open land'. *Sudfelt -feld* 1086, *Suffeld, Suthfeld* [12th]15th. OE **sūth** + **feld**. YN 115, L 244.

SUFFOLK '[The territory of] the southern people'. *(innon) Suffolke* [1043×5]13th S 1531, *(in to) Suðfolce* [1047×65]12th S 1124, *Suðfolc* 11th ASC(D) under year 1076, *Suthfolc, Suthfulc, Sudfolc* 1086. OE **sūth** 'south' + **folc** 'folk, people'. Named in contrast to NORFOLK, the northern people of the East Anglian region. The form *in pago Suthfolchi*, ascribed to 895, is in a document forged after the Norman conquest (S 349). There is no certain evidence for either county-division before the mid-11th century.

SUGNALL Staffs SJ 8031. 'Sparrow hill'. *Sotehelle* (for *Soce-*) 1086, *Sugenhulle* 1222, *Sogenhul* 1242. OE ***sucga**, genitive sing. ***sucgan**, genitive pl. ***sucgena**, + **hyll**. The exact species of bird in uncertain, possibly a titlark or wagtail; OE **hegesugge** surviving as dial. *haysuck* meant 'hedge-sparrow'. DEPN, PNE ii.167.

SULBY RESERVOIR Northants SP 6581. P.n. Sulby + ModE **reservoir**. Sulby, *Solebi* 1086–1316 with variant *-by, Sulebi -by* 1158 etc., *Sulby* from 1310, is probably the 'farm in the gully', OE **sulh** + **bȳ**. Sulby lies in a shallow valley. But alternative possibilities are ON **súla** 'a pillar, a cleft, a fissure' or ON pers.n. *Súla*. Nth 121, SSNEM 72.

SULGRAVE Northants SP 5545. Probably the 'gully grove'. *Sulgrave* from 1086, *Sole- Sulegrave* 12th–1329, *Sowgrave* 1556. OE **sulh** + **grāfa**. Sulgrave lies on a low spur in a broad deep-cut valley. But 'Sula's grove' is also possible, OE pers.n. **Sula*. Nth 36, L 194.

SULGRAVE MANOR Northants SP 5645. P.n. SULGRAVE SP 5545 + ModE **manor**. An Elizabethan manor house. Pevsner 1973.421.

SULHAM Berks SU 6474. 'Homestead at a gully or furrow'. *Soleham* 1086–1322, *Suleham* [1142×84]12th–1276, *Sulham* from 1284. OE **sulh** 'furrow' used in a topographical sense of a narrow valley + **hām**. Brk 220, PNE ii.167.

SULHAMPSTEAD Berks SU 6368. 'Homestead at a furrow or narrow valley'. *Silamested'* 1197×8, *Silhamsted(e)* 1220–1429, *Selehamsted'* 1267×8, *Silhamstede Banastre* 1297–1342, ~ *Abbatis* [1317]14th, *Sulhampsted Banaster, ~ Abbatis* 1535, *Sulhamstead Banister als Michills* 1757. OE **sulh** 'furrow' used in a topographical sense of a narrow valley, genitive or dative sing. **sylh**, + **hāmstede**. The manor of Sulhampstead Abbots belonged to Reading abbey at the end of the 12th cent. S Bannister is named after the *Banastre* family: William Banastre, son of John Banaster of Sulhampstead, granted Reading abbey land here c. 1200×13. *Michills* in the 1757 form refers to the dedication of the church to St Michael. Brk 184, PNE i.232, ii.167, Stead 267.

SULLINGTON WSusx TQ 0913. 'The farm or estate called or at *Syling*, the willow-copse or the muddy place'. *Sillinctune* *[959]12th S 1293, *Sillingtune* *[1066]13th S 1043, *Sillintone, Semlintun* 1086, *Si- Syllyngton* 1291, 1409, 1722, *Sullyngton(e)* 1291 etc., *Sullington al. Sillington* 1641. OE ***syling** < **syle** 'a bog, a miry place' or ***syle, *siele** 'a willow-tree copse' + **ing**[2], possibly used as a p.n., + **tūn**. Another possibility might be OE ***sielling** 'a gift'. Sx 179, DEPN, SS 75.

SUMMERBRIDGE NYorks SE 2062. Partly uncertain. *Somerbrig(g)e* 1536–1615, *Sum(m)erbridge* 1658, 1677. Either 'a bridge used in summer', ME **summer** + **brigge** or the specific is the surname *Summer*. YW v.147, DBS 338.

SUMMERCOURT Corn SW 8856. Apparently 'court-yard used in summer'. *Somercourt* [1711]1748. There was similarly a 'summeryard' at Newton, St Neot, *Le Somer Yerde* 1516. There was an important September fair there in the Middle Ages, *Longaferia* 1227, *La Lunge feire* 1234, (fair of) *Langchepyng* 1351, 'the long fair', ME **lang, long** + **feire** and **chepyng** 'market', alluding to the length of its extent along the main road. Possibly this was the 'summer court' in question, although there is no evidence that 'court' was ever used in this way. PNCo 160, Gover n.d. 333.

SUMMER DOWN Wlts ST 9148. *St Maur's Down* 1830. Part of the 13th cent. manor of St Maur. Wlt 146.

SUMMERFIELD Norf TF 7438 → SOMERSHAM Cambs TL 3677.

SUMMERSEAT GMan SD 7914. 'Summer shieling'. *Sumersett* 1556, *Somerseat* 1618. The evidence is too late to decide between an English origin, OE **sumor** + **set** 'fold', and a Scandinavian one, ON **sumarr** + **sætr**. *Sommersæt* is a common p.n. in N Norway. La 62.

SUMMIT GMan SD 9418. The highest point on the road, rail and canal link between Littleborough and Todmorden.

SUNBIGGIN Cumbr NY 6508. 'Sunny' or 'south building'. *Sunnebyggyn(ge)* 1310, *Sunbiggin* from 1577. ON **sunna** or OE **sunne** 'sun', + or ON **sunn** 'south', + ME **bigging**. The place lies on the S side below High Pike NY 6509 (1220ft.), and contrasts with Friar Biggins NY 6309, buildings belonging to the brethren (*frere*) of Conishead Priory. We ii.44, 46, SSNNW 206.

SUNBURY Surrey TQ 0769. 'Sunna's fortified place or manor'. *Sunnabyri* [959]12th S 1293, *(æt, into) Sunnan byrg, Sunnan burges (bōc)* 'the S charter' c.950×68 S 1447, *(æt, to) Sunnanbyrig* 962 S 702, *Suñeberie* 1086, *Sunneberi* 1198, *Sonnebery* 1291, *Sunbury* 1535. OE pers.n. *Sunna* + **burh**, locative-dative sing. **byrig**. Mx 22.

SUNDERLAND 'Detached (part of an) estate'. OE **sundor-land**. PNE ii.168, L 249.

(1) ~ Cumbr NY 1835. *Sonderland in Blankrayk* 1278, *Sunderlond* 1279, *Sunderlande* 1299. For *Blankrayk* see BLINDCRAKE NY 1434. Cu 324, L 249.

(2) ~ T&W NZ 3957. *Sunderland* from 1168, often distinguished from SUNDERLAND BRIDGE Durham NZ 2637 by the additions *iuxta mare, nigh the sea* etc. 1388–1758. The name originally applied only to a small area of land at the mouth of the river Wear whose parish church was Holy Trinity. This was probably the *terra trium familiarum ad austrum Vuiri fluminis iuxta ostium* 'the three holdings south of the river Wear at the mouth' for which Benedict Biscop exchanged two silk robes on his return from his sixth visit to Rome c.683. It has often been claimed that Sunderland was the birthplace of the venerable Bede, on the grounds that his statement that he was born in the territory (*in territorio*) of the joint Monkwearmouth-Jarrow monastery is translated in the OE version of the *Ecclesiastical History* as *on sundurlonde þæs ylcan mynstres* 'on the *sundorlond* of the same minster'. However, OE **sundorlond** meant 'land set apart for a special purpose, private land' as well as 'detached land'. There is no certainty that Sunderland was the particular **sundorlond** on which Bede was born, even if the OE translation is reliable. In fact, one of the early lives of Bede records that he was born in a hamlet, by tradition Monkton, in the territory of the Gyrwe on the banks of the Tyne, i.e. at or near Jarrow. NbDu 192.

(3) ~ BRIDGE Durham NZ 2637. *Sonderlandbrigg* 1563, *Sunderland next (nigh, near) the Bridge* 1622–1717, *Sunderland Bridge* from 1675. P.n. Sunderland + ModE **bridge**. Sunderland, *Sunderland* c.1168, ~ *next Croxdale* 1420–1636, *Sonderland* 1620, *Sunderland* from 1675 (when it is described as 'a house or two'), was a detached portion of the parish of St Oswald, Durham, from which it was cut off by the river Wear. A bridge was already in existence by the 13th cent. The addition is for distinction from SUNDERLAND BY THE SEA T&W NZ 3957. NbDu 192.

(4) ~ POINT Lancs SD 4255. P.n. Sunderland + ModE **point**. A headland on the N bank of the Lune estuary. Sunderland, *Sinderlaund* 1246, *Sunderland* 1254, is the southernmost part of Overton township. The forms cited, however, come from surnames and may not belong here. La 175, Jnl 17.100.

North SUNDERLAND Northum NU 2131. ModE adj. **north** + p.n. *Suðlanda* 1176, *Sutherlannland* 12th cent., *Sunderland* 1187,

13th cent., the 'southern land' sc. of Bamburgh parish, OE **sūth(er)** + **land** early assimilated to OE **sundor-land** 'detached land'. 'North' for distinction from SUNDERLAND T&W NZ 3957. NbDu 192.

Lower SUNDON Beds TL 0526. ModE adj. **lower** + p.n. *Sunnandun* [c.1050]13th S 1517, *Sonedone* 1086–1286, *Sune(n)don* 1247, *Sonin(g)don -yng-* 1276–1373, *Sundon* 1390×2, 'Sunna's hill', OE pers.n. **Sunna*, genitive sing. **Sunnan*, + **dūn**. 'Lower' for distinction from Upper Sundon TL 0527, *Upper Sundon* 1834 OS. Bd 165, L 151.

SUNDON PARK Beds TL 0525. P.n. Sundon as in Lower SUNDON TL 0526 + ModE **park**.

SUNDRIDGE Kent TQ 4755. 'Detached ploughed field'. *Sunderhirse* [1072]n.d., *Sondresse* 1086, *Sunderhersce* c.1100, *Sund(e)resse* 1210×12, 1226. OE **sundor** + **ersc** later replaced by **ridge**. PNK 69.

SUNK ISLAND Humbs TA 2619. *Sunk Island* from 1678. Now part of the mainland since the filling up of the north area of the Humber, this patch of land developed from a small island rising out of a sand-bank in the Humber in the earlier 17th cent. It is situated on a former part of the mainland which had previously been washed away. Cf. *Sunk Sand* 1824 OS. YE 24.

SUNK SAND Essex TM 3000. 'The sunken sand-bank'. *The Sonke Sande* 1509×47. ModE **sunk** + **sand**.

SUNNINGDALE Berks SU 9567. A modern name short for *Sunning Hill Dale* 1800. P.n. SUNNINGHILL SU 9467 + ModE **dale**. The ecclesiastical parish of Sunningdale was formed in 1841 from parts of Old Windsor, Sunninghill, and Egham and Chobham, Surrey. It became a civil parish in 1894. Brk 87, Room 1983.119.

SUNNINGHILL Berks SU 9467. 'Hill of the Sunningas, the people called after Sunna'. *Sunigehill'* 1185, *-hull'* 1246, *Sunningehull'* 1190, *-hell'* 1221, *Sunninghull'* 1191, *So(u)nnynghulle* 1327, *Sondynghill* 1447. OE folk-n. **Sunningas* as in SONNING SU 7575 < pers.n. **Sunna* + **ingas**, genitive pl. **Sunninga*, + **hyll**. Brk 88, ING 45.

SUNNINGWELL Oxon SP 4900. 'The spring of the Sunningas, the people called after Sunna'. *(ad) Sunnigwellan* *[811]c.1200 S 166, *Sunningauuille* *[821]c.1200 S 183, *Suniggawelle* *[821]c.1240 ibid., *(on) sunningawylles broc, (andlang) sunninga wylle broces* [956]c.1200 S 605, *Soningeuuel* 1086, *Sunni(n)gewell(e)* [1066×87]c.1240, 1241–3, *Sunningwell(')* from 1242×3. OE folk-n. *Sunningas* < pers.n. *Sunna* + **ingas**, genitive pl. *Sunninga*, + **wella**, **wylle**. The Sunningas were a group whose territory covered much of E Berks including SONNING SU 7575 and SUNNINGHILL SU 9367. Sunningwell is 20 m. NW of the former and must represent a detached group of this people. Brk 459, 932–3.

SUNNISIDE Durham NZ 1438. *Old Sunnyside* 1865 OS. A modern pit village named from Sunniside Grange NZ 1438, *Sunneyside Homestall* 1839 *TA*, on the S side of Billy Hill and so distinguished from North Side NZ 1438, *North Side* 1865 OS.

SUNNISIDE T&W NZ 2058. 'The sunny hill-side'. *Sonnyside* 1322, 1342, *Sunnyside* [1728]19th, 1768 Armstrong. OE ***sunniġ** + **sīde**. Situated on the sunny side of Whickham Fell and contrasting with Fellside NZ 1959, *Fellside* 1768 Armstrong. NbDu 192.

SUNNYSIDE ESusx TQ 3937. No early forms.

SURBITON GLond TQ 1867. 'The south barton or corn farm' sc. of Kingston. *Suberton(e)* 1179–1253, *Sur- Sorbelton* 1241–1352, *Surbeton(e)* 1263–1765, *Surbiton* 1597, *Surbetton al. Sutton* 1626, *Surbiton al. Surton* 1673, 1718. OE **sūth** + **beretūn**. 'South' for distinction from Norbiton TQ 1969, 'the north barton', *Nor(t)berton* 1205, *Norbeton(e)* 1272–1323, *Norbiton* 1531. Sr 63.

SURFLEET Lincs TF 2528. 'The sour creek'. *Sverefelt* 1086, *Surflet(e)* 1172–1535 including [1133]14th, *-fleet* from 1281. OE **sūr** + **flēot**. Payling 1940.97, L 22, Jnl 29.81.

SURFLEET SEAS END Lincs TF 2729. *The Seas End* 1653. Cf. *Surfleteskore* 1352, 1364, *Score* 1375, *Shore* 1362, 'Surfleet shore'. This place marks the original position of the sea-bank before the 18th cent. drainage and reclamation of the Fens. Payling 1940.99.

SURLINGHAM Norf TG 3106. 'Homestead of the Sutherlingas, the southern people', i.e. S of the river Yare. *S(c)utherlingaham, Suterlingeham* 1086, *Surlingham* from 1197, *Surlingeham* 1250, *Suthrlingham* 1275. OE folk-n. **Sūtherlingas* < **sūthor** + **lingas**, genitive pl. *Sūtherlinga*, + **hām**. Possibly a variant of the folk-n. in SEETHING TM 3197. If, however, the form *Herlingaham* 1046 Thorpe belongs here the folk-n. might be the *Sūth Herlingas* 'the south Herlingas, the people called after Herela'. See East HARLING TL 9986. DEPN, ING 138–9.

SURREY 'The southern district'. *in regione Sudergeona* 731 Bede, on *Suþrige* c.900 ASC(A) under year 722. OE **sūther** + **ġē**. Bede, and some other Anglo-Saxon writers, use the county name as a folk-name: 'the people of Surrey'. Sr 1–2.

SURREY HILL Berks SU 8864. *Surry Hills* 1800, *Surry Hill* 1816. County name SURREY, *on Suþrige* 9th ASC(A), OE **sūther-ġē** 'southern district', + ModE **hill**. Surrey Hill is on the old county boundary with Surrey. Brk 41.

SUSSEX '(The land of the) south Saxons'. *Suþ Seaxe, Suðseaxnaland* 9th ASC(A). OE folk-n. *Sūthseaxe*, genitive pl. *Sūthseaxena*, + **land**. In Latin they are the *Australes* or *Meridiani Saxones* c.731 BHE. Nunna *rex Suthsaxonum* 'king of the south Saxons' attests a charter of 692 S 45. Sx 1.

SUSTEAD Norf TG 1936. 'The southern place'. *Sur- Sutstede* 1086, *Suthsted(e)* [1101×7]13th–1379, *Susted(e)* 1254–1471. OE **sūth** + **stede**. Stead 185.

SUSWORTH Lincs SE 8302. Partly uncertain. *Silkeswath, Sirke(s)wad'* 1202, *Susworth* 1824 OS. ON pers.n. *Silki*, secondary genitive sing. *Silkes*, or OE ***sīluc** 'a gully a drain', genitive sing. ***sīluces** + ON **vath** 'a ford'. Cameron 1998, SPN 240.

SUTCOMBE Devon SS 3411. 'Sutta's coomb'. *Svtecoma* 1086, *Suttecumb' -combe* 1242–1428, *Suthcombe* 1291. OE pers.n. **Sutta* + **cumb**. Occasionally reinterpreted as if 'south coomb'. D 168.

SUTON Norf TM 0998. *Suton* 1838 OS applied to a two mile stretch of the road from Wymondham to Attleborough between Gonville Hall and Decoy Common. Presumably it must represent OE **sūth-tūn** 'south settlement' with long vowel retained.

SUTTERTON Lincs TF 2835. 'Shoemaker farm or village'. *Sutterton* [810]lost S 1189 and from 1234 with variants *Suter-* 1177–1348, *Soter-* 1252–1405. OE **sūtere** + **tūn**. Payling 1940.100, *SelPap* 74, Cameron 1998.

SUTTON 'South settlement'. OE **sūth** + **tūn**. PNE ii.169.

(1) ~ Beds TL 2247. *Sudtone* 1086, *Sutton(e)* from 1086. S in relation to Potton TL 2249. Bd 110.

(2) ~ Cambs TL 0998. *suðtun* [948]12th S 533, *Sutton(a)* from 1189. S of Upton. Np 243.

(3) ~ Cambs TL 4479. *Sudtone* 1086, *Sutton(e)* from 1246. S of Mepal. Ca 239.

(4) ~ GLond TQ 2564. *Suþtone* *[727]13th S 1181, *Suðtone* *[1062]13th S 1035, *Sudtone* 1086, *Suthton(a)* 12th, 1177 etc. South-west of Carshalton. Sr 54.

(5) ~ Kent TR 3349. *Suttone* 1154×5, 1156×7, *Sutton'* from 1226. PNK 576.

(6) ~ Norf TG 3823. *Suttuna -e* 1086, *Sutton(e)* from 1185. S in relation to Stalham. Nf ii.127.

(7) ~ Notts SK 7637. *Suttun'* 1235, *Sotton* 1284, 1290. Also known as *Sutton juxta Granby* 'S by Granby' 1301–30. S in relation to Elton SK 7438. Nt 226.

(8) ~ Notts SK 6884. *(æt, to) Suttune* [958]14th S 679, *Suttuna*

†OE *hām*, normally feminine gender, must have been assimilated to the masculine/neuter declension.

1171×9. Also known as *Sutton in Loundale, ~ on Lounde* 1551, 1552. SW of Lound SK 6986. Nt 100.
(9) ~ Oxon SP 4106. *Sutton(')* from 1207. S of Eynsham. O 283.
(10) ~ Shrops SO 7286. *Sutton(')* from 1271×2. S of Chelmarsh SO 7288. Sa i.287 no.1.
(11) ~ Shrops SO 5183. *Sudtone* 1086, *Sutton* from 1208×9. In the S part of the parish of Diddlebury SO 5085. Sa i.288.
(12) ~ Shrops SJ 6631. *Sudtone* 1086, *Sutton(')* from 1255×6; also known as *Sutton by Drayton* 1661 and *Sutton upon Tern* 1833 OS. S of Market Drayton SJ 6734. Sa i.288.
(13) ~ Staffs SJ 7622. *Sutton* from 1203. S of NORBURY SJ 7823 and WESTON JONES SJ 7524. St i.147.
(14) ~ Suff TM 3145. *Sut(h)tuna* 1086, *Sutton(')* from 1166. S in relation to Woodbridge. DEPN, Baron.
(15) ~ Surrey TQ 1046. *Sudtone* 1086, *-ton(a)* [1135×54]1318, *Sutton* from 1461×83. S of Paddington TQ 1047. Sr 251.
(16) ~ WSusx SU 9715. *Suðtun* [c.880]c.1000 B 553, *Sudtone* 1086, *Sutton* 1331. SE of Barlavington SU 9176. Sx 120, SS 75.
(17) ~ AT HONE Kent TQ 5570. 'S at the boundary stone'. *Sutton' atte hone* 1240, *Sutton' Atte(n)hone* 1254. P.n. Sutton + ME **atten hone** < OE *æt thǣm hane*[†], OE **hān**. Earlier simply *Suttone* [1087]13th, *Suthtuna* c.1100, *Sutton(')* from 1199. PNK 49.
(18) ~ BASSETT Northants SP 7790. *Sutton Basset* from 1309. P.n. *Sutone* 1086, *Sutton* 1185, + family name of Richard Basset who held the manor in the 12th cent. 'South' for distinction from Weston SP 7791 and in relation to Ashley SP 7990. Nth 172.
(19) ~ BENGER Wilts ST 9478. 'S held by Berenger'. *Sutton(e) Berengeres* 1377, *~ Benger* 1488. P.n. *(at) Suttune* [854]12th S 305, *Sutton(e)* [956]13th S 1577, + manorial addition from Berenger, the under-tenant in 1086. S with reference to Hullavington. Wlt 74.
(20) ~ BINGHAM RESERVOIR Dorset ST 5410. P.n. SUTTON BINGHAM Somer ST 5411 + ModE **reservoir**. Sutton Bingham, 'S held by the Bingham family', is *Sutone* 1086. The manor was held by John de Bingham 1100×35. It lies S of East and West Coker. DEPN.
(21) ~ BRIDGE Lincs TF 4721. Formed in 1894 as a separate parish out of Long SUTTON TF 4322. The original bridge across the Nene was erected in 1831 replacing a 2 mile crossing of the Wash by ford or boat. It was replaced in 1851 and again in 1894×7. Wheeler 129 and Appendix i.37, Pevsner-Harris 1964.686.
(22) ~ CHENEY Leic SK 4100. 'S held by the Chaynel family'. P.n. *Svetone* 1086, *Sutton(')* from 1221 + manorial affix *Cheynell* 1411, 1540, *Cheyney* 1577, *Cheney* 1611. The manor was held by John Chaynel in 1293. Lei 546.
(23) ~ COLDFIELD WMids SP 1296. *Sutton (super, in) Col(l)efeld* 1269–1549, *~ Colfeld* 1289, *~ Coldefilde* 1605. P.n. *Sutone* 1086, *Sutton(e)* 1176–1487, + p.n. *le Colfeld* 1313, *the great wast called Colfield* 1656, 'the open land where charcoal is burnt', OE **col** + **feld**. Wa 49, 12, L 237, 240–1, 244.
(24) ~ COURTENAY Oxon SU 5093. 'S held by the Courtenay family'. *Suttone Curteney* 1284, *Sutton Curtenay* 1294. P.n. *Suthtun* [c.870]c.1240 S 539, *suðtune* [c.895]c.1200 S 355, *Suðtun* [983]c.1240 S 851, [1042]c.1200 S 993, *Suttun* [1000]c.1240 S 897, *Sudton(e), Suttone* 1086, *Suttun', Sutton(')* from 1157 + manorial affix from the family n. of Reynold de Courtenay c.1180. S of Abingdon. Brk 424.
(25) ~ CROSSES Lincs TF 4321. 'Cross roads at S', i.e. Sutton St Mary's or Long SUTTON TF 4322. *Sutton Cross End* 1824 OS.
(26) ~ GRANGE NYorks SE 2874. *grangiam de Sutton* 1156. P.n. *(on) Suðtune* c.1030 YCh 7, *Sudtunen -ton* 1086, *Sutton* from 1132×40, + **grange**, a grange of Fountains Abbey. S primarily in relation to North STAINLEY SE 2876. YW v.162.
(27) ~ GRANGE NYorks SE 7970. P.n. *Sudton(e)* 1086, *Sutton(a) - tuna* 1121×37 etc. + ModE **grange**. S in relation to NORTON SE 7971. YE 140.
(28) ~ HILL Shrops SJ 7003. Cf. *Sutton Hill House* 1833 OS. P.n. Sutton as in SUTTON Maddock + ModE **hill**.
(29) ~ HOO TUMULI Suff TM 2848. P.n. *Sutton Haugh* 1837 OS + Latin **tumulus** 'a burial mound', pl. **tumuli** referring to the royal cemetery of Rendlesham containing at least 11 barrows including the famous 7th cent. ship burial uncovered in 1939. P.n. SUTTON TM 3145 + ModE **haugh** 'land in a river bend' for earlier *Hou, hoi* 1086, *Hoo* 1442, *Howell* 1451, *Howhills* 1629, OE **hōh**, 'the hill-spur' with later addition of **hill**. Both terms are applicable to the site. Arnott 70, Baron.
(30) ~ HOWGRAVE NYorks SE 3179. P.n. *Sudton(e), Sutone* 1086, *Sutton* from 1157[‡], + p.n. HOWGRAVE SE 3179. Also called *Sutton Rugemond* 1280 from Ralph de Rougemond who held a third of a knight's fee here. YN 221.
(31) ~ IN ASHFIELD Notts SK 4959. *Sutton in Essefeld* 1276, *~ in Assefeld* 1287. P.n. *Sutone* 1086, *Sutton* 1213, + p.n. Ashfield as in KIRKBY IN ASHFIELD Notts SK 5056. S of Skegby SK 4961. Nt 134.
(32) ~ -IN-CRAVEN NYorks SE 0043. *Sutun* 1086, *Sutton* from 1184. Situated in the S of Kildwick parish. See also CRAVEN. YW vi.26.
(33) ~ LANE ENDS Ches SJ 9271. *Sutton Lane* occurs 1620 and *Lane Ends* 1831. Sutton is *Sutton* from [early 13th]1608 and 1246. S of Macclesfield. Che iv.155, 148.
(34) ~ LEACH Mers SJ 5292. *Sutton* from 1200. The modern development there is named after *Leech Hall* and *Toad Leech* 1842 OS, OE **leċe** 'a brook'. La 108, Jnl 17.61.
(35) ~ MADDOCK Shrops SJ 7201. *Sutton' Madok* 1255×6, *Sutton Maddock* from 1722. P.n. *Suptone* [949]13th S 549, *Suthtune* 1002×4, *Sudtone* 1086, *Sutton(')* from 1189, + W pers.n. *Madoc* referring to holders of the manor in the 12th and 13th cents. S of Brockton SJ 7203. Sa i.288.
(36) ~ MALLET Somer ST 3736. 'S held by the Malet family'. *Sutton Malet* 1280. Earlier simply *Sutone* 1086, *Sutton* 1610. S was held by Ralph Malet in 1200. It lies S of the Polden Hills. DEPN.
(37) ~ MANDEVILE Wilts ST 9828. 'S held by the Mandeville family'. *Sutton Maundeville* 1278, *~ Maundvild* 1535. P.n. *Sudtone* 1086 + manorial addition from the family name of Robert de Mandevile who held the manor in 1224. W 217.
(38) ~ MONTIS Somer ST 6224. 'S held by the Mons or Montacute family'. *Sutton Mountagu* 1335. Earlier simply *Svtone* 1086. S was held by Drogo de Monteacuto in 1086. It lies S of S Cadbury and Cadbury Castle. *Montis* is the genitive sing. of Latin **mons** 'hill', the short form of the family name. DEPN.
(39) ~ -ON-HULL Humbs TA 1232. P.n. *Sudton(e)* 1086–1206, *Sutone* 1086, *Sutton(a) -ton(a)* 1150×67–1531, + r.n. HULL TA 0646. Two miles S of Swine. YE 42.
(40) ~ ON SEA Lincs TF 5281. P.n. *Su(d)tune -tone* 1086, *Suttuna* c.1115, *Sutton* 1824 OS + ModE **on sea**. Also known as Sutton in the Marsh. DEPN.
(41) ~ ON THE HILL Derby SK 2333. *(in) Suðtone* [949]13th, *(æt) Suðtune* [c.1002]c.1100 S 1536, [1104]14th S 906, *Suttun* [c.1002]c.1100 S 1536, *-ton(e)* 1243–98 etc., *Sudtun* 1086. 'South town' for distinction from the 'church town' half a mile to the NE. Db 612.
(42) ~ -ON-THE-FOREST NYorks SE 5864. P.n. *(on) suþtune* 10th S 1660, *Su(d)tune, Suton* 1086, *Suttune* 1145×53, 1252, *Sutton* from 1166, + ModE *in the Forest* 1577 referring to location in the Forest of GALTRES whence the early addition *in Gatris* 1242. Also known as *Ouergate Suton'* 1231 with ME **over** 'upper' + ON **gata** 'road'. YN 19.

[†]OE **hān**, normally fem. gender, must have been assimilated to the masc./neuter declension.
[‡]The form *(on) suþtune* 10th S 1660 is assigned to Sutton Howgrave by YN 221 and DEPN, to Sutton-on-the-Forest by S.

(43) ~ ON TRENT Notts SK 7965. *Sutton super Trente* 1221×30, ~ *uppon Trent* 1651. P.n. *Sudtone* 1086, *Sutton* 1176, *Suttun* c.1200, + r.n. TRENT. S in relation either to Normanton SK 7968 or Weston SK 7767. Nt 196.

(44) ~ PARK WMids SP 1096. Cf. *Sutton chace* 'S hunting ground or park' 1477. P.n. Sutton as in SUTTON COLDFIELD SP 1296 + ME, ModE **park** 'an enclosed tract of land for beasts of the chase'. Wa 52.

(45) ~ ST EDMUND Lincs TF 3613. 'St Edmund's S'. *Sutton Sancti Edmundi* 1526. P.n. Sutton + saint's name *Edmund* from the dedication of the church. Payling 1940.51.

(46) ~ ST JAMES Lincs TF 3918. 'St James's S'. *Parish of St James in Sutton in Holand* 1525, *Sutton St James* 1530. P.n. Sutton + saint's name *James* from the dedication of the church. Payling 1940.51.

(47) ~ ST NICHOLAS H&W SO 5345. 'St Nicholas's S' from the dedication of the church. Originally simply *Sv(d)tvne* 1086, *Suttone* 1242. The estate lies in the south part of the district called Maund. DEPN, He 187.

(48) ~ SCOTNEY Hants SU 4439. 'S held by the Scotney family'. *Sutton Skoteny* 1287. Earlier simply *Svdtvne* 1086, *Suttun'* 1235. S in relation to Norton in the same parish. The manor was held by Walter de Scoteny in 1235 whose family came from Étocquigny in Seine-Maritime, Normandy. Ha 158, Gover 1958.179.

(49) ~ -UNDER-BRAILES Warw SP 3037. *Sutton juxta Brales* Latin 'S beside Brailes' 1268, ~ *subtus Brayles* 1573. P.n. *Svdtvne* 1086, *Sutton(e)* from 1204, + ModE **under**, Latin **subtus**, + p.n. Brailes as in Lower and Upper BRAILES SP 3139, 3039. Sutton lies 2 m. to the S. Wa 301.

(50) ~ -UNDER-WHITESTONECLIFFE NYorks SE 4882. *Sutton subtus Whitstanclif* [13th]15th. P.n. SUTTON + ModE **under** + p.n. Whitestone Cliff SE SE 5083, *Whitstanclyff* [13th]15th, *Whitstoncliffe* 1613, the 'white-stone cliff'. Earlier simply *Sudtune -tone* 1086. YN 199.

(51) ~ UPON DERWENT Humbs SE 7046. *Sutton on Der(e)went(e)* from 1233. P.n. *Sudton(e)* 1086, *Sutton* from 1230, + r.n. DERWENT SE 7035. Also known as *Quenersuttona -tun* 1164×72, 1172×9, which is unexplained. Lying a mile S of a moated site at SE 7148 and two miles from Newton upon Derwent. YE 189.

(52) ~ VALENCE Kent TQ 8149. 'S held by the Valence family' whose connection with the manor dates from 1265. Earlier simply *suðtune* (dative sing.) 814 S 173, *Svdtone* 1086, *Sutton(')* from 1219. K PNK 232, KPN 132.

(53) ~ VENY Wilts ST 8941. 'Fenny Sutton'. *Fennisutton* 1268, *Veny Sutton* 1535, *Sutton Veny* 1817 OS. ME adj. **fenny** + p.n. *Su(d)tone* 1086. Also known as *Magna Sutton* 'great Sutton' 12th. 'Fenny' from its marshy situation, and 'south' in contrast to NORTON BAVANT ST 9043 which may have influenced the change in word order from Veny Sutton to pseudo-manorial Sutton Veny. Wlt 154.

(54) ~ WALDRON Dorset ST 8615. 'Waleran's S'. *Sutton(e) Walerand, Walronde, Walra(u)nd* 1297–1451, *Sutton Walron* 1545, ~ *Waldron* 1664. Earlier simply *(at, of) suttune* [932]15th S 419, *Sudtone* 1086. The affix refers to Waleran Venator 'the huntsman' who held the manor in 1086 or one of his descendants such as Walter Walerand who occurs 1210×12. Do iii.80.

(55) ~ WALLS H&W SO 5246. *Sutton Walls* 1832. P.n. Sutton as in SUTTON ST NICHOLAS SO 5345 + ModE **walls**. The reference is to an Iron Age hill-fort built c.300 BC and occupied down to the 3rd cent. AD. It may have been the *caput* of the district known as the Maund. Thomas 1976.151, He 13.

(56) ~ WEAVER Ches SJ 5479. *Sutton* from late 12th. The S settlement of the peninsula between the river Weaver and the Mersey and in systemic relation to NORTON SJ 5582, WESTON SJ 5080 and ASTON SJ 5578. Che ii.180.

(57) Bishop ~ Avon ST 5859. Three *Suthtunes* are mentioned in *[1065]c.1500 S 1042 referring to Bishop's S, Sutton Wick and Sutton Court. Bishop S probably belonged to Chew Magna which was held by the Bishop of Wells in 1086. ECW 152.

(58) Bishop's ~ Hants SU 6131. *Suttona Episcopi* 1167. Earlier simply *sudtunam* [982]14th S 842, *Sv- Sudtone* 1086. The estate was acquired by the Bishop of Winchester in 1136. Ha 34, Gover 1958.88.

(59) Chart ~ Kent TQ 7950. Short for *Chert near Sutthon* 1280, *Chert by Sutton* 1305. Chart Sutton adjoins SUTTON VALENCE TQ 8149. Cf. CHART SUTTON TQ 7950. K 209.

(60) Full ~ Humbs SE 7455. 'Dirty Sutton'. ME prefix *Ful* (OE **fūl**) from 13th + p.n. *Sudtone* 1086, *Suttune -ton(a)* 1156×7–1587. YE 185.

(61) Guilden ~ Ches SJ 4468. *Guldenesutton* c.1200–1306, *Gulden-* 1329, c.1350, 1600, *Guilden-* from 1398. Earlier simply *Sudtone* 1086, *Suttona* [1154×60]1329, 1190×1211. 'Golden Sutton' meaning 'splendid, wealthy'. At the S extremity of the Domesday hundred of *Wilaveston*. Che iv.126.

(62) Kings ~ Northants SP 4936. *Sutton Regis* 1252, *Kinges Sutton* 1294. ME **king**, genitive sing. **kinges**, Latin **rex**, genitive sing. **regis**, + p.n. *Sudtone* 1086, *Sutton(e)* from 1155. 'South' in relationship to Middleton Cheney SP 4942 or Great Purston SP 5139. Cf. also CHARLTON SP 5235.

(63) Long ~ Hants SU 7347. *Longesuttone* c.1220. Earlier simply *suðtun* [979]12th S 835, *Sudtune* 1086. 'Long' for distinction from SUTTON SCOTNEY and Bishop's SUTTON. S of Odiham SU 7451. Gover 1958.111.

(64) Long ~ Lincs TF 4322. *Lange- Longsutton* 1386. ME adj. **lang, long** + p.n. *Sudtone* 1086, *Sutton* from 1177. Also known as *Est Sutton voc Seynt Mary* n.d., *Sutton St Mary's* 1824 OS. Payling 1940.51

(65) Long ~ Somer ST 4625. *Langesutton* 1312, 1610. OE adj. **lang**, definite form **langa**, + p.n. *Sudton* [852 for 878]17th S 343, *Sutvne -tone* 1086. S of Somerton. DEPN.

SWABY Lincs TF 3877. 'Svafi's village or farm'. *Suabi* 1086, c.1115–84, *Suauebi* 1160×75, *Swabi* after 1169. ON pers.n. *Sváfi*, genitive sing. *Sváfa*, + **bȳ**. The name is unlikely to be a Scandinavianisation of an earlier English p.n. referring, like SWAFFHAM Norf TF 8208 and SWAFFHAM BULBECK and PRIOR Cambs TL 5562 and 5764, to a settlement of Swabians during the migration period on account of the existence of the ON pers.n. *Sváfi* in the form *Swafa* as a moneyer of Harold Harefoot, Hathacnut and Edward the Confessor. Cf. SWAYTHORPE Humbs TA 0368. DEPN, SSNEM 72.

SWADLINCOTE Derby SK 3020. 'Swartling's cottage(s)'. *Sivardingescotes* 1086, *Su- Swartlin(g)cot(e) -yn(g)-* 1208–1481, *Swar(d)lincote* 13th, *Swadlin(g)cote* from 1528. Pers.n. OE **Sweartling* or ON **Svartlingr* + OE **cot**, pl. **cotu**. The DB form points rather to the common pers.n. *Siward* (OE *Sigeweard*). Db 663, SSNEM 381.

SWAFFHAM Norf TF 8208. 'The homestead of the Swabians'. *Suafham* 1086, *Suaffham* c.1130, *Swafham* 1230. OE folk-n. *Swǣfe, Swāfe*, genitive pl. *Swǣfa, Swāfa*, + **hām**. Alternatively the compound may be formed on the stem form *Swǣf-, Swāf-*. DEPN gives pr [swɔfəm].

SWAFFHAM BULBECK Cambs TL 5562. 'S held by the Bulbeck family'. *Swafham Bolebek* 1218. Earlier simply *Suafham* 1086–1312 including [1042×66]12th S 1051, *Suafam* 1086, *(altera) Suuafham* 1086 Inquisitio Eliensis, 'homestead of the Swabians', OE folk-n. *Swǣfe*, genitive pl. *Swāfa*, or stem-form *Swǣf-*, + **hām**. 'Bulbeck', referring to Hugo de Bolebec(h) who held the manor in 1086, is for distinction from SWAFFHAM PRIOR TL 5764. Ca 133.

SWAFFHAM PRIOR Cambs TL 5764. 'S held by the prior' sc. of Ely. *Swafham prioris Elyensis* 1232. P.n. Swaffham as in SWAFFHAM BULBECK TL 5562 + Latin, ModE **prior**, Latin genitive sing. **prioris**. Ca 133.

SWAFIELD Norf TG 2832. Possibly 'the open land through which there is a track'. *Suaf(f)elda, Suauelda* 1086, *Sw- Suadefeld* c.1145–c.1250, *Su- Swathefeld* c.1150–1362, *Swathfeld* 1195–1518, *Suafeld* 1208, *Swaiefeud* 1254, *Swayfeld* 1636. OE **swæth, swathu,** + **feld**. OE **swæth** meant 'the mark left by a moving body, a swathe'; extension of the meaning to 'track, road' is possible but unproven. DEPN gives pr [sweifi:ld], Nf ii.195.

SWAINBY NYorks NZ 4701. 'Farm or village of the young men'. *Suanebi* 1086, *Suenebi* [1111–22]15th, *Swayneby* [1184]15th–1560. OE **swān** probably replacing ON **sveinn**, genitive pl. **swāna**, + **bȳ**. YN 225, SSNY 38.

SWAINSHILL H&W SO 4641. *Swineshill* 1831 OS. Without earlier forms it is impossible to know the correct form of this name.

SWAINSTHORPE Norf TG 2200. 'Sveinn's outlying farm'. *Torp, Sueinestorp* 1086, *Sueinestorp* 1198, *Sweinestorp* 1202. ON pers.n. *Sveinn*, secondary genitive sing. *Sveines*, + **thorp**. DEPN, SPNN 356.

Upper SWAINSWICK Avon ST 7568. ModE **upper** + p.n. *Sweyneswyk* 1291, 1302, 'Swein's dairy-farm'. Earlier simply *Wiche* 1086. ME pers.n. *Swein* (< ON *Sveinn*), genitive sing. *Sweines*, + OE **wīċ** used here as a p.n. 'Upper' for distinction from Lower Swainswick ST 7666 a mile down valley. DEPN.

SWALCLIFFE Oxon SP 3737. 'Swallow cliff'. *Sualewclive* c.1166, *Sualewecliue* 1189×91, *Swaleweclyve* 1327, *Swaleclive -clyve* [c.1190]c.1280–1382, *Swalclyf -cliff* 1266 etc., *Swakeley* 1546. OE **swalwe** + **clif**. O 425 gives pr [sweiklif], formerly [sweikli].

The SWALE Kent TQ 9866. 'The rushing stream'. *Suuealuue fluminis* 812 S 169, *Sualuæ* (nominative case) 815 S 178, *Swale* 1361. OE **Swealwe*, identical with r.n. SWALE NYorks SE 2796. KPN 281, RN 383.

River SWALE NYorks SE 2796. 'The whirling, rushing river'. *Sualua, Swalwa* [c.731]8th BHE, *Swal(e)wan, Sweal(e)wan* [c.890]c.1000 OEBede, *Swale* from 1157(15th). OE r.n. **Swalwe* usually compared with MHG *swal(l)* 'a spring, a gush of water', *swalm* 'a whirlpool' and the r.ns. Schwalb in Franconia, *Swalawa* c.802, Schwale in Holstein, *Suala* 12th, Schwalbach, *Sualbahc* 893, Schwalm, *Sualmanaha* 782, and Zwalm, Brabant, *Sualma* 11th. More recently these names and the Lithuanian r.n. *Swale* have been related to IE **swel-* 'burn' with the sense 'shining stream'. If so the r.n. Swale belongs to the pre-Germanic Old European stratum of hydronymy. YN 6, RN 384, Berger 240, Li v.144–6, Laur s.v. Schwale.

SWALECLIFFE Kent TR 1367. 'The swallow cliffe'. *(æt) Spalepanclife, Spalepanclifes (land gemero)* 'bounds of S' [946×51]13th S 578, *Soaneclive* 1086, *Swalclive* [1087]13th, *Swal(u)ecliu(e)* 13th cent. OE **swealwe**, genitive sing. or pl. **swealwan**, + OE **clif**. KPN 281, Charters IV.108.

SWALEDALE NYorks SE 0298. 'The valley of the river Swale'. *Sualadala* 1128×32, *Swaledale* from 1159, *Swaldale* c.1180×5–1401, *Swawdall -dell* 16th. R.n. SWALE SE 2796 + ON **dalr**. YN 269 gives a pr [swɔ:dil].

SWALLOW Lincs TA 1730. Uncertain. *(in) Svalun -vn, (in) Sualun* 1086, *Sualwa* c.1115, *Swalwe* 1163–1327, *Su- Swalewe* 1175–1361, *Sualue, Sualowe, Swalewe* early 13th. The DB forms look like an OE dative pl. ***swal(w)um** of ***swalwe** 'a whirlpool, rushing water' as in the r.n. SWALE NYorks SE 2796. There is, however, no such feature here although it lies in a well-marked valley. An alternative possibility is the dative pl. of OE ***swalg** 'a pit, a pool' seen in The Swallow YE 189, *Swalewe* 1252, a place in marshy land near a field called The Dimple. The reference is to 'a mill called *Swalewe*', perhaps a nick-name meaning 'the devourer'. At Swallow, too, there is reference to the 'miller of *Swalwe*' in 1281. Li v.145 reports the existence of a stream here that rises from an underground source, flows E, and disappears again, possibly the 'swallow' in question. Comparison with the G r.ns. Schwalb, *Swalawa* c.802, and Schwale, *Suale* 12th, and the Lithuanian *Swale*, however, perhaps suggests a pre-Gmc Old European formation on the root **swel-* 'burn' in the sense 'shine'. L v.144, RN 384, YE 189, PNE ii.170, Flussnamen 26, Laur s.n. Schwale.

SWALLOWCLIFFE Wilts ST 9627. 'The swallow's cliff'. Latin *rupis irundinis, id est Swealewanclif* 'the rock of the swallow, that is, S' [940]14th S 468, *Svaloclive* 1086, *Swalweclive -klive -clyve* c.1155–1294, *Swaleclive -clyve* 1241–1316, *Swallowcliffe al. Swalclyff* 1546, *Swakley* 1618. OE **swealwe**, genitive sing. **swealwan**, + **clif**. Wlt 192 gives former pr [sweikli], L 131–2, 136.

SWALLOWFIELD Berks SU 7264. 'Open land by the river *Swalewe*'. *Sualefelle* 1086, *Sualewesfeld'* c.1160, *Su- Swal(e)wefeld(')* 1167–1361, *Swallowfelde* 1316. Lost r.n. *Swalewe* + OE **feld**. The reference is to the stream formed by the union of the Blackwater SU 7364 and the Hants Whitewater SU 7360. It appears as *Swalewe* 1272, 1300, *Swawe* [1272×1307]17th and is comparable with the G r.n. Schwalb, *Sualauua* 793, which is assumed to represent Gmc **Swalawja* from the root **swal-* 'to swell' or **swel-* 'burn'. Brk 7, 108, RN 384, Bach ON 325.3, 749, Ortsnamenwechsel 286.

SWAN GREEN Ches SJ 7373. *Swan Green* 1769. ModE **swan** + **green**. Che ii.221.

SWANAGE Dorset SZ 0378. 'Work-place of the herdsmen or peasants', or, 'swannery'. *(æt) Swanawic* late 9th ASC(A), 12th ASC(E) both under year 877, *Suanauuic, Suanewic* [893]11th Asser, *Swan- Sonwic* 1086, *Su- Swanewic -wik -wyk* 1210–80, *Swanewich(e) -wych(e)* 1270–1448, *Swanwych(e) -wich(e)* 1244–1603, *San(d)wich(e)* 16th, *Swanage* from 1795. Either OE **swān** 'herdsman, peasant', genitive pl. **swāna**, or **swan** 'a swan', genitive pl. **swana**, + **wīċ** 'dairy-farm, work-place'. Do i.52 gives pr [ˈswɔnidʒ].

SWANAGE BAY Dorset SZ 0379. *San(e)wich(e) baye* 1575, c.1586, *Swanage Bay* 1811. P.n. SWANAGE SZ 0378 + ModE **bay**. Do i.59.

SWANBOURNE Bucks SP 8027. 'The swans' stream'. *Suanaburna* *[795 for 792]13th S 138, *Soeneberno, Sueneberne -borne* 1086, *Swaneburne* 1160–1242. OE **swān**, genitive pl. **swāna**, + **burna**. Bk 73, L 18.

SWANLAND Humbs SE 9927. 'Svan's wood'. *Suenelund* 1181–91, *Swan(n)eslund* 1237–1333, *Swanlund* 1303, 1329, *-lond* 1285–1462, *-land* from 1302. ON pers.n. *Svanr*, secondary genitive sing. *Svanes* probably later confused with **swan** 'a swan' or **swān** 'a herdsman', + **lundr** later replaced by ME **lond, land**. YE 218, SPN 271.

SWANLEY Kent TQ 5168. 'Swine-herd wood or clearing'. *Swanleg* 1203, *Swanley* 1573. OE **swān** + **lēah**. PNK 51.

SWANMORE Hants SU 5716. 'Swan pool'. *Suanemere* 1205–33, *Swanemere* 1207–1327, *Swanmere* ´1316, 1496. OE **swan**, genitive pl. **swana**, + **mere**. There is no lake at this place today; it lies near the source of the Hamble. Ha 159, Gover 1958.48.

SWANNINGTON Leic SK 4116. 'Estate called after Swan'. *Swaneton'* 1199–1266, *-i- -y-* 1209–1349, *-in- -yn-* 1243–1502, *-ing- -yng-* 1274–1610, *Swannington* from 1507. OE pers.n. **Swan* + **ing**[4] + **tūn**.

SWANNINGTON Norf TG 1319. 'Estate of the Swaningas, the people called after Swan'. *Sueningatuna* 1086, *Suaneton* 1191, *Sueiningeton* 1192, *Suenninge- Sweinnigeton, Swenigtun(e)* 1198, *Sueington', Swaningeton'* 1202. OE folk-n. **Swāningas* < pers.n. **Swān* + **ingas**, partly influenced by ON *Sveinn*, genitive pl. *Swāninga*, + **tūn**. DEPN, SPNN 360.

SWANSCOMBE Kent TQ 6074. 'The herdsman's *camp*'. *Suanescamp* *[677]16th S 1246, *Svinescamp* 1086, *Swanescamp(e)* 1166–1239, *Swanescombe* 1292. OE **swān**, genitive sing. **swānes**, + **camp** 'uncultivated land' from Latin *campus*, later replaced by ME **combe** 'a coomb'. KPN 17, Signposts 76–7.

SWANTON 'The herdsmen's settlement'. OE **swān**, genitive pl. **swāna**, + **tūn**.

(1) ~ ABBOTT Norf TG 2625. 'S held by the abbot' of Holme Abbey. *Abbot Swanton* 1451. Earlier simply *Swaneton* *[1044×7]13th S 1055, *Suanetuna* 1086. DEPN.

(2) ~ MORLEY Norf TG 0116. 'Swanton held by (Robert de) Morle' in 1346. Earlier simply *Suanetuna* 1086, *Swaneton* 1212. DEPN.

(3) ~ NOVERS Norf TG 0232. 'Swanton held by (Milo de) Nuiers' in 1200 from Noyers-Bocage in Normandy, *Noers* 11th. Earlier simply *Suaneton* [1047×70]13th S 1499, 1200, *-tuna* 1086. DEPN, Dauzat-Rostaing 498.

SWANWICK Derby SK 4053. 'Dairy-farm of the herdsman'. *Swanwyk(e) -wick* 1272×1307–1501, *Swanick* 1721. OE **swān** + **wīc**. Db 189 gives pr [swɔnik].

SWANWICK Hants SU 5109. 'The herdsmen's farm'. *Suanewic(h)* 1185, 1210, *Swanewic -k -wyk(e) -wich -wyche-* 1231–1324. OE **swān**, genitive pl. **swāna**, + **wīċ**. Ha 159, Gover 1958.32.

Lower SWANWICK Hants SU 5009. ModE **lower** + p.n. SWANWICK SU 5109. Lower Swanwick is downhill from Swanwick SU 5109 beside the Hamble. In 1810 OS they are both simply *Swanwick*.

SWARBY Lincs TF 0440. 'Svarri's village or farm'. *Svarrebi* 1086–1202, *Swar(r)eby* 1202–1428, *Swarby* from 1271. ON pers.n. *Svarri*, genitive sing. *Svarra*, + **bȳ**. Perrott 50, SSNEM 73.

SWARDESTON Norf TG 2002. 'Sweord's estae'. *Suerdestuna* 1086, *Swerdeston(e)* 1202, 1254, *Suerdesdon* 1230. OE pers.n. *Sweord*, genitive sing. *Sweordes*, + **tūn**. DEPN gives pr [swɔrstn], PNE ii.172.

SWARKESTONE Derby SK 3628. 'Swerkir's farm or village'. *Suerchestune*, *Sorchestun* 1086, *Sworkeston* 1210, 1214, 1652, *Swerkeston(e) -is-* 1230–1481, *Swerston* late 13th–1431, *Swarkeston(e)* from 1330, *Swarston* 1594–1676, *Swaston*, *Swarson* 17th. ODan pers.n. *Swerkir*, genitive sing. *Swerkis*, + **tūn**. Db 664, SSNEM 195.

SWARLAND Northum NU 1601. 'Heavy newly-cultivated land'. *Swarland* 1242 BF, *Swar(e)- Swerla(u)nd* 13th cent., OE **swār**, **swǣr** + **land**. Sometimes known as Old Swarland for distinction from the new development at Swarland Estate NU 1603. NbDu 193, L 249.

SWARLAND ESTATE Northum NU 1603. A modern development named after SWARLAND or Old SWARLAND NU 1601 + ModE **estate**.

SWATON Lincs TF 1337. Possibly 'Swafa's village or estate'. *Svave- Suaui- Svavintone* 1086, *Sua- Swaveton(e)* 1123–1327, *Swa- Suaventon* 1278, 1322, *Swafton* 1230, *Swaton* from 1327. OE pers.n. **Swāfa*, genitive sing. **Swāfan*, + **tūn**. Alternatively the specific could be ON pers.n. *Sváfi*, genitive sing. *Sváfa*. Perrott, SSNEM 381.

SWAVESEY Cambs TL 3669. 'Swæf's landing place'. *Suauesheda* 1086, *-hed'* 1163, *-hide* 1203, *Suauishith* 1290, *Swaveshith -hyth -heth* 1212–1313, *Suauesy(e)* 1086, *Swavesey(e)* from 1228 with variants *-eia -eie -ay(e) -y(e)* etc. to 14th. OE pers.n. *Swǣf*, genitive sing. *Swǣfes*, + **hȳth**. Ca 172, L 62, 76, 78.

SWAY Hants SZ 2798. Unexplained. *Svei(a)* 1086, *Sweia* 1227, *Sweye* 1248–1331. Several suggestions have been made about this name, OE **swēġ** 'noise' or **swēġe** 'sounding' referring to the stream that runs past the village, OE **swæth** 'a track' referring to the path that opens out N from the village onto the heath, without carrying total conviction. Ha 159, Gover 1958.210, DEPN.

SWAYFIELD Lincs SK 9922. Probably the 'open land characterised by swathes'. *Suafeld* 1086, *Swafeld(e)* 1198–1576, *Swauefeld* 1202, *Swathefeld* 1319, 1321, *Swayfelde* 1549. OE **swæth, swathu** + **feld**. This explanation depends on the 14th cent. forms in *Swathe-* and the comparable p.n. SWAFIELD Norf TG 2832. The precise sense of **swæth** 'a track, a mark left by a moving body, a swathe' is uncertain. Perrott 202, L 243.

SWAYTHLING Hants SU 4315. Unexplained. *(be) swœðelinge* *[909]11th S 376, cf. *(to) swœðelingeforda, (of) swœðelingforda* [932]13th S 418, *(on) swœðeling wylle* [1045]12th S 1012, *Swatheling(e) -yng(e)* from 1256, *Swavelyng* 1527. In form this is apparently a stream name in **-(l)ing**[2] on OE **swæth** 'a track', **swathul** 'smoke' or **swethel, swæthel** 'swaddling band'. If **swathul** could mean mist, perhaps 'stream where mist gathers'; alternatively 'ribbon or ribbon-like stream' referring to its straight course. But all this is speculative. Ha 160, Gover 1958.41, ING 193.

SWAYTHORPE Humbs TA 0368. Probably 'Svafi, the Swabian's outlying farm'. *Suauetorp* 1086, *Swauet(h)orp* 12th–1276, *Swathorp(e)* 1240–1516, *Swai- Swaythorpe* 1530, 1579. ON pers.n. *Sváfi*, genitive sing. *Sváfa*, + **thorp**. Possibly a Scandinavianisation of an earlier English p.n. referring like SWAFFHAM Norf TF 8208 and SWAFFHAM BULBECK and PRIOR Cambs TL 5562 and 5764 to a settlement of Swabian foederati between c.360 and 450. Cf. also SWABY Lincs TF 3877. YE 97, SSNY 68.

SWEETHOPE LOUGHS Northum NY 9482. *Sweethope Loughs* 1868 OS. P.n. *Swethop'* 1242 BF, 1269 SR, *Swectoppe* (sic), *Suet- Swethopp(e)* 1269 SR, *Sweetup* 1663, 'the sweet valley', OE **swēte** + **hop**, + N dial. **lough** 'a lake' (OE *luh*). The sense is 'valley with good pasture', cf. Sweethope Hill Scotland NT 6939. NbDu 193, L 116.

SWEFFLING Suff TM 3463. Uncertain. *(in) Suestlingan, Sweflinga, Suestlingua* 1086, *(in) Suuetelincge* [1086]c.1180, *Sueftlinges* c.1150, *Swiftling* 1222, *Sweftling'* 1254, *Swiftlinghe* 13th, *Swiftlyngge* 1321, *Sweftlynge* 1342, *Swesflinge* 1610, *Swefflinge* 1674. This is either a sing. **ing**[2] or a pl. **ingas** formation on an OE pers.n. **Swiftel*, **Swiftla* or **Swǣftel*, either 'the place called after Swiftel' etc., or 'the Swiftlingas, the people called after Swiftel' etc. The sp *Suestlingan* with <s> for <f> looks like the reflex of dative pl. **Swiftlingum*. DEPN, Baron, ING 55.

Lower SWELL Glos SP 1725. *Nether(e) Swell(e)* 1274–1575, *Lower Swell* 1699. ME adj. **nether**, ModE **lower**, + p.n. *Swelle* *[706]12th S 1175, *(œt) Suuelle* [1055]12th S 1026, *Sv(v)elle* 1086, *Swell(e)* 1221–1597, OE **(ge)swell** or ***swelle** 'a swelling' referring to land rising steeply to Swell Hill. 'Lower' for distinction from Upper SWELL SP 1726. Gl i.226, Studies 1936.151, L 124.

Upper SWELL Glos SP 1726. *Over(e) Swell(e)* 1274–1536, *Upper Swell* 1611. ME adj. **over**, ModE **upper**, + p.n. Swell as in Lower SWELL SP 1725. Also known as Greater Swell, *(œt) Suella major* *[714]16th S 1250. Gl i.227.

SWEPSTONE Leic SK 3610. 'Sweppi's farm or village'. *Scopestone* (sic) 1086, *Swepeston' -is-* c.1130–1471, *Swepston(')* from 1234. OE pers.n. **Sweppi*, genitive sing. **Sweppes*, + **tūn**. Lei 403.

River SWERE Oxon SP 4733. *Sowar, Swere* 1577, 1586 Harrison. A back-formation from SWERFORD SP 3731. RN 386.

SWERFORD Oxon SP 3731. 'The ford by a col'. *Svrford* 1086, *Su- Swereford(')* 1122–1428, *Swerford(')* from 1230, *Swyreford* 1337, 1346. OE **sweora** + **ford**. O 382.

SWETTENHAM Ches SJ 8067. 'Sweta's homestead or village'. *Suetenham* late 12th, 1220×30, *Swetenham* early 13th–1621, *Swettenham* from late 13th, *Swetnam* 1557, *Swittnam* 1660. OE pers.n. *Swēta*, genitive sing. *Swētan*, + **hām**. Che ii.283 gives pr ['swetnəm].

River SWIFT Leic SP 5283. *Swift* from 1577, 1586. A modern name[†]. The original may have been OE **Hlūtre* 'the clean one,

[†]Ekwall's assumption RN 387 of an older origin, OE **Swifte* the 'sweeping, winding one', related to OE *swīfan*, is not supported by the evidence.

the pure one' from OE **hlūttor** 'clean, pure', preserved in LUTTERWORTH SP 5484. Lei 99.

SWILLAND Suff TM 1852. 'Newly cultivated land where pigs are kept'. *Suinlanda* 1086, *Suinelanda* 1185, *Swinelande -londe, Swynelond(e)* 1242–1336, 1568, *Swellond* 1524, *Swilland* 1610, 1674. OE **swīn**, genitive pl. **swīna**, + **land**. DEPN, Baron.

SWILLINGTON WYorks SE 3830. Possibly 'the settlement at *Swinling*, the swine-hill or swine-pasture place'. *Suillictun -igtune -intun* 1086, *Suin- Swin- Swynlington(a) -tun(a)* 12th–1342, *Swi- Swylinton(a) -yn-* 12th–1415, *Swillington* from 1196. OE p.n. **Swīnling* < **Swīn-hyll* 'pig hill' or *Swīn-lēah* 'swine pasture' + **ing**[2], + **tūn**. YW iv.93.

SWIMBRIDGE Devon SS 6229. Partly uncertain. Originally simply *Birige* 1086, 'the bridge', OE **brycg**, later *Swynbrigge* 1274, *-brug(ge)* 1334, 1422, *Swinebregge* 1286, *Svimbrige* 1225, *Swym- Sumbridge* 16th, *Som(e)- Sone- Symbridge* 17th. Formally the specific could be OE ***swin** 'a creek, a channel' though the other known occurrences of this element are all on the east coast. The owner of the manor in 1086 was Sawin (OE pers.n. *Sǣwine*) and this name may therefore be the explanation of the first element. The name was later associated with ME *swine* 'a pig'. D 350, PNE ii.172.

SWINBROOK Oxon SP 2812. 'Pig brook'. *Svinbroc* 1086, *Swi- Swynbroc(h) -brok(e)* 1196 etc., *Swynebrok(e)* 1278–1351. OE **swīn** + **brōc**. O 383.

Great SWINBURN Northum NY 9375. *Great Swinburn* 1868 OS. ModE adj. **great** + p.n. *Swineburn'* 1236, 1242 BF, *Swyn- Swymb(o)urn* 1346, originally a stream name, the 'stream where swine are watered or pastured', OE **swīn**, genitive pl. **swīna**, + **burna**. Also known as West Swinburn, *Swineburn' occidentalem* 1236 BF, *West Swyneburne* 1296 SR, for distinction from Little or East Swinburn NY 9577, *Swyneburne Est* 1296 SR. NbDu 193, L 18.

SWINDALE BECK Cumbr NY 5113. *Swyndellbeck* 1249, *Swyndale bek* 1473. P.n. Swindale + ON **bekkr**. Swindale, *Swin- Swyndal(e) -dall* c.1200–1777, is 'valley where swine are kept', ON **svín** or OE **swīn** + ON **dalr**. We ii.177, 171, SSNNW 167.

SWINDEN RESERVOIR Lancs SD 8933. P.n. Swinden + ModE **reservoir**. Swinden is 'swine valley', *Swyndene* 1562, OE **swīn** + **denu**, a narrow twisting valley opening into the river Brun. La 86.

SWINDERBY Lincs SK 8663. Partly uncertain. *Sunderby, Suindrebi* 1086, *Suinderbi* 1185, *Sunderby* 1209×35, *Squinderby* 1284×5, *Swinderby* from 1220×35. Possibly 'swine Derby', p.n. *Derby* 'the deer farm', ON **djúra-bȳ**, with *swine*, ON **svín**, prefixed for distinction from Darby, lost Lincs SE 8717, *Derbi* 1086, and DERBY Derby SK 3536. Woodland pasture is mentioned at Swinderby (DB 56.16, 367a). The forms do not support the earlier explanation from ODan **sundri, syndri** 'southern'. DEPN, SSNEM 73.

SWINDON 'Swine hill'. OE **swīn** + **dūn**. L 144–9, 157.

(1) ~ Glos SO 9375. *Svindone* 1086, *Swi- Swyndon(a)* 1220–1777. Gl ii.112.

(2) ~ Staffs SO 8690. *Swineduna* 1167, *Suindun* 1236, *Swyndon* 1271, 1332. Swindon was in Kinver Forest where pasturage for swine was an important privilege. DEPN, Duignan 146, Horovitz.

(3) ~ Wilts SU 1585. *Svindone -dune* 1086, *Swindon(a) -(e)* from 1156. The reference is to the high ground around the church at SU 1583 south of the later railway town. Wlt 276.

SWINE Humbs TA 1335. 'The creek'. *Suuine* 1086, *Swine -y-* from 1154×89. OE ***swin**. The reference is to an ancient channel or creek of the Humber now drained by Holderness Drain. YE 51, Studies 1931.88–9.

SWINEFLEET Humbs SE 7722. 'Tidal inlet called Swin'. *Swynefleth* 1154×89, c.1190×1207, *Swineflet(e) -y-* 1189×1207–1382, *Swinflet(e) -y-* 1266–1448, *-flett* 1557. OE ***swin** + **flēot**. The reference is to a former side channel of the r. Ouse. YW ii.10, Studies 1931.89–90, L 21, Jnl 29.81.

SWINESHEAD Beds TL 0565. 'Pig's headland'. *Suineshefet* 1086, *Swynesheued* 1209–1428, *Swynyshed* 1525, *Swaneshead al. Swinshead* 1595. OE **swīn**, genitive sing. **swīnes**, + **hēafod**. The sense is probably 'projecting snout of land' rather than 'headland where pigs are herded' or 'where wild pigs are seen'. The reference is to a long spur of higher ground N of the village. The theory that this class of name was literally meant and referred to the ritual exposure of animal heads in pagan times is no longer accepted. Bd 20 gives pr [swinzhed].

SWINESHEAD Lincs TF 2340. 'The head of the creek called *Swin*'. *Swines hæfed, (at) Swinesheafde* 1121 ASC(E) under years 675 and 777, *(æt) Suinesheabde* [786×96]12th S 1412, *Sui- Swi- Swyneshaved -heved -et -is-* 12th–1401, *Swyneshed(e)* 1327–1495, *Swynnesheved* 1353, 1380, *-hed(e)* 1526. R.n. **Swin* < OE ***swin** 'a creek, a channel', genitive sing. **Swines*, + **hēafod**. The reference is to Bicker Haven, formerly *aqua de Swyn'* 1276, at one time a tidal creek which reached as far as Swineshead and is probably referred to in the lost *Swyreflet* (for *Swyne-*) 1202, *Swynfleet -flete* 1337–1514, r.n. **Swin* + OE **flēot** 'an inlet, a river'. The lost name *Swin* is identical with *Het Zwin*, the old name of part of the Schelde estuary near Knokke-Heist on the Belgium-Holland border, MDu, Flem *zwin, zwen* 'an inlet, a creek'. The term is related to OHG *swīnan* 'decrease, diminish, dry up' and seems to refer to a channel that tends to dry up at low tide. Payling 1040.104, Studies 1931.90, YE 51, Jnl 29.81–2. L 160–1 takes the name to be one of a group of names meaning 'pig's head', OE **swīn**, genitive sing. **swīnes**, + **hēafod** with the meaning 'projecting snout of land'.

SWINESHEAD BRIDGE Lincs TF 2142. *Highbridge* 1824 OS, *(Swineshead) High Bridge* 1896 Wheeler. A road-bridge over the South Forty-Foot at Three Gibbet Hill. Later a halt on the Great Northern Railway line from Grantham to Boston. Wheeler 452 and Appendix i.38.

SWINFORD 'Swine ford'. OE **swīn**, genitive pl. **swīna**, + **ford**. L 5–6, 71.

(1) ~ Leic SP 5779. *Svin(e)- Svinesford* 1086, *Swy- Swi- Suineford* c.1130–1373, *Suinford'* 1176–93, *Swynford(e)* 1277–1555, *Swinford* 1576. Lei 466.

(2) ~ Oxon SP 4408. *Swinford* from *[931]c.1200 S 410, *Swynford* *[931]c.1240 ibid., 1327, *Swineforde* [c.1180]13th, *Swyneford'* 1242×3. The road from Oxford to Eynsham crosses the Thames here. Brk 446.

(3) Old ~ WMids SO 9083 → KINGSWINFORD SO 8888.

SWINGFIELD MINNIS Kent TR 2142. 'Swingfield common'. *Swingfield Minnis alias Folkstone Common* 1698. P.n. Swingfield as in SWINGFIELD STREET TR 2343 + ModE **minnis** (OE *(ġe)mǣnnes*). Cullen.

SWINGFIELD STREET Kent TR 2343. 'Swingfield hamlet'. P.n. Swingfield + Mod dial. **street**. Swingfield, *Sumafeld* (for *Suina-*) c.1100, *Swi- Swyne(s)feld'* 1212–1253×4, *Swynkfield* 1610, is the 'open land where pigs are kept', OE **swīn**, genitive pl. **swīna**, + **feld**. PNK 454.

SWINHOE Northum NU 2028. 'Swine headland'. *Swinhou* 1242 BF, *Swyneho* 1280, *Swinhow* 1296 SR 148, *Swynowe* 1315. OE **swīn** + **hōh**. For a traditional rhyme about Swinhoe, see BUCKTON Northum NU 0838. NbDu 193, L 168.

SWINHOPE Lincs TF 2196. 'Pig valley'. *Su- Svinhope* 1086, *Suin(ah)opa* c.1115, *Swinope* 1185, 1592, 1641, *Swi- Swynhop* 1226–1719, *-hoppe* 1593–1693, *Swinhope* from 1606. OE **swīn** + **hop**. Li iv.160, L 115, 117.

SWINITHWAITE NYorks SE 0489. 'Clearing made by burning'. *Swiningethwait, Swiningtweit* 1202, *Swynigt(h)wayt* 1295, *Swynythwayt* [1315]16th. ON **svithnungr** + **thveit**. YN 256.

SWINMOOR H&W SO 4240 → MOCCAS SO 3542.

SWINSCOE Staffs SK 1348. 'Swine wood'. *Swinescho -skou* 1203, *Swyneschouh -scou -skow, Svineskoch, Swineskoc -scow* etc. all 13th. ME **swin** (OE *swīn*, ON *svín*) + **skow** (ON *skógr*). Oakden, L 209.

SWINSTEAD Lincs TF 0122. 'The pig farm'. *Suinhā, Suinhā-Svinham- Suamestede* 1086, *Swi- Swynhamsted(e)* 1130–1280, *Swinsteda* c.1150, *Suinesteda* 1169, *Swi- Swyn(e)sted(e)* 1227–1610. OE **swīn** + **hāmstede**. Perrott 204, Stead 288.

SWINSTY RESERVOIR NYorks SE 1953. P.n. Swinsty + ModE **reservoir**. Swinsty is the 'pig-sty', ME, OE **swīn** + **sty** (OE *stigu*); forms are recorded in the p.ns. *Swynstye More* 1585, *Swinsty Hall* 1647. Another *Swinesti* 13th is recorded in Gwendale. YW v.129, 155.

SWINTON 'A pig-farm'. OE **swīn** + **tūn**.

(1) ~ GMan SD 7701. *Suinton* 1258, *Swynton* 1276–93 etc. La 41.

(2) ~ NYorks SE 7573. *Suintun(e)* 1086, *Swinton* from 1219. YN 47.

(3) ~ NYorks SE 2179. *Suinton* 1086. YN 234.

(4) ~ SYorks SK 4598. *Suinton(e)* 1086–1193, *Swintone* 1086, *Swin- Swynton(a)* from 12th. YW i.115.

SWITHLAND Leic SK 5413. 'Wood or grove by land cleared by burning'. *Swi- Swythelond(e)* 1196×1208–1341, *-lund(e)* 1224–1304, *Swi- Swythlund'* 1236, *-lond* 1363–1514, *-land* from 1239. ON **svitha** 'land cleared by burning' + **lundr**. Lei 406, SSNEM 173, L 208.

SWITHLAND RESERVOIR Leic SK 5513. P.n. SWITHLAND SK 5413 + ModE **reservoir**.

SWORTON HEATH Ches SJ 6884. *Sworton Heath* 1548, *Sor(e)ton Heath* 1842, 1849, *Sowton Heath* 1831. P.n. Sworton + ModE **heath**. Sworton, *Suerton* [12th]17th, *Swerton* late 13th, *Sworton* from 13th, *Soorton* late 13th, is the 'settlement at a neck of land', OE **swēora** + **tūn**. The place is on a ridge running W from High Legh SJ 7083. Che ii.46.

SWYNNERTON Staffs SJ 8535. The forms are complicated:

I. *Svlvertone* 1086, *Silverton* 1205, *Swiluerton', Soulverton* 1206, *Swilverston* 1236, *S(w)ilveston* 13th, *Silvereston* 1275.

II. *Swaneforton* c.1195, *Sonnerton(e)* 1264, 1372, *Swonnerton* 1320.

III. *Suin(n)erton* 1205, 1242, *Swynaferton, Swynforton, Swinefarton, Swinnerton* 13th, *Swynemerton* 1289, *Swineforton* 1355, *Swynarton* 1404.

IV. *Sumerverton* 1228.

There appear to be two forms at least of this name: I. 'the settlement at Swilford, the dirty ford', p.n. **Swilford* < OE ***swille** 'a sloppy or liquid mess' + **ford**, + **tūn**, III. 'the settlement at Swinford, the swine ford', OE p.n. **Swīnford* < **swīn** + **ford**, + **tūn**. In addition II. seems to be 'the settlement at Swanford, the herdsmen's or swans' ford', OE **swān** or **swan**, genitive pl. **swāna**, **swana**, + **ford**, IV. 'the settlement at Somerford, the summer ford', OE **somer** + **ford**. The village stands on a tributary of Meece Brook. DEPN, Horovitz.

SWYRE Dorset SY 5288. 'The neck of land'. *Suere* 1086, *Swere* 1196, *Suure* 1275, *Swyre* 1288. OE **swēora** varying with **swȳre**. The reference is to a col or pass in the coastal hills giving access to the sea. Do 140.

SYDE Glos SO 9511. 'The hill-side'. *Side* 1086–1698 with variant *Syde* from c.1220. OE **sīde**. Gl i.162, L 187.

SYDENHAM GLond TQ 3571. 'Cippa's homestead or enclosure'. *Chipeham* 1206, *Shippenham* 1315, *Sipeham* 14th, *Sypenham* 1560, *Sidenham* 1690, *Sidnum* c.1762. OE pers.n. *Cippa* + **hām** or **hamm**. DEPN, LPN 223.

SYDENHAM Oxon SP 7301. 'At the wide *hamm*'. *Sidreham* 1086, *Si- Sydeham* [1148–84]c.1200–1308, *Si- Sydenham* from 1216. OE **sīd**, definite form dative case **sīdan**, + **hamm** in the phrase *æt thǣm sīdan hamme*. O 114.

SYDENHAM DAMEREL Devon SX 4076. 'S held by the D'Albemarle family'. *Sidenham Albemarlie* 1297, ~ *Daumarll* 1303. Earlier simply *Sidelham* (sic) 1086, *Sideham* 1201, *Sidenham* 1283, '(at) the wide *hamm*', OE **sīd**, definite form oblique case, **sīdan**, + **hamm**, dative case **hamme**. Johannes de Alba Mara or de Albemarle is first mentioned in connection with the manor in 1242. Also known as *South Sydenham* 1809 OS. D 209, NoB 48.160.

SYDERSTONE Norf TF 8332. Unexplained. *Cide- Scidesterna* 1086, *Sidesterne* 1198, 1203. This appears to be a compound of OE **sīd** 'large, spacious, long' + ***sterne** of unknown meaning. It has been suggested that *sterne is a metathesised form of OE **(ġe)strēon** 'property', but this is very uncertain. An alternative suggestion is that the generic may be ON **tjǫrn** 'a tarn, a small lake' preceded by a pers.n. **Sīd*, short for names in *Sīde-* such as *Sīdeman, Sīdewine, Sīdelufu*. cf. SIDLESHAM Sussex SZ 8599. On the other hand ON **tjǫrn** seems not to be used outside Yorkshire, Cumbria and Lancashire. DEPN.

Up SYDLING Dorset ST 6201. 'Upper Sydling'. *Upsidelinch* 1230, *-ling* 1244, *-lyng* 1311. OE **upp** + p.n. Syding as in SYDLING ST NICHOLAS SY 6399. The place lies a mile further up the valley. Do 141.

SYDLING ST NICHOLAS Dorset SY 6399. 'St Nicholas's S'. P.n. Sydling + saint's name *Nicholas* from the dedication of the church. Sydling, *Sidelyng* [843 for 934]17th S 391, *Sidelince* 1086, *Sideling* 1212, is the 'broad ledge', OE **sīd**, definite form **sīda**, + **hlinċ**. Also known as *Brodesideling* 1288, 'great S', for distinction from Up SYDLING ST 6201. Do 140.

SYDMONTON Hants SU 4857. 'Sidemann's estate'. *Sidemanestone* 1086–1204, *Si- Sydemanton(')* 1169–1346, *Sidemonton* 1205, *Sidemountaine* 1586. OE pers.n. *Sīdemann*, genitive sing. *Sīdemannes*, + **tūn**. Sidemann is possibly to be identified with the *Sydeman minister* who signed the Ecchinswell charter of 931 (S 412). Ecchinswell is 1 mile N of Sydmonton. Ha 160, Gover 1958.149.

SYERSTON Notts SK 7447. Possibly 'Sighere's farm or village'. *Sirestun(e)* 1086, *-ton* 1203–1577 with variants *Syres- Siris-*, *Sierson* 1576, 1661. OE pers.n. *Siġhere*, genitive sing. *Siġheres*, + **tūn**. Nt 218.

SYKE GMan SD 8915. *Syke* 1843. N dial. **sike** 'a small stream' (OE *sīc*, ON *sík*).

SYKEHOUSE SYorks SE 6317. 'The house(s) by the stream'. *Sike- Sykehouse* from 1404. OE **sīc** + **hūs**. YW i.16.

SYKES Lancs SD 6351. 'The streams'. *Sykes* 1423, *Syckes* 1581. ME **sike** (OE *sīc* or ON *sík*), plural **sikes**. The reference is to rivulets descending from the fell. YW vi.214.

SYKES FELL Lancs SD 6150. *Sykes Fell* 1845. P.n. SYKES SD 6351 + ModE **fell** (ON *fjall*). YW vi.214.

SYLEHAM Suff TM 2178. 'The homestead at the furrow or at the miry place'. *Seilam, Seilanda* 1086, *Seleham* 1156, *Sileham* 1174, *Silham* 1275–1346 including [942×c.951]13th S 1526, *Syleham* 1568. OE **sulh**, genitive/dative sing. **sylh**, or **syle**, **sylu**, + **hām**. The 'furrow' would refer to a narrowing of the Waveney valley NE of the village. DEPN, Baron.

SYMONDSBURY Dorset SY 4493. 'Sigemund's hill or barrow'. *Simondesberge* 1086, *Simunisberge* 1212, *Symundesberg* 1237, *Symondesburgh* 1340. OE pers.n. *Siġemund*, genitive sing. *Siġemundes*, + **beorg** later confused with **bury** (OE *byriġ*, dative sing. of *burh*). There are round barrows in the parish, which is also situated between two round hills. Do 141, L 128.

SYMONDS YAT H&W SO 5516. 'Simund's gate or pass'. *Symons Yate* 1665, *Simmons Gate* 1831. Pers.n. *Simund* + ME **yatt** (OE *ġeat*) referring to the narrow passage of the Wye at this place between hills. Bannister 182.

SYREFORD Glos SP 0320. *Syreford* 1779, *Syersford* 1777, *Cyford* 1792. The evidence is far too late for certainty, but this could be 'Sigehere's ford', OE pers.n. *Siġehere*, genitive sing. *Siġeheres*, + **ford**. Gl i.185.

SYRESHAM Northants SP 6341. 'Sigehere's homestead'. *Sigre(s)ham* 1086, *Sig(e)resham* c.1147–1318, *S(c)hiresham* 1162–1361, *Sir- Syresham* from 12th, *Syseham* 1600. OE pers.n.

Siġehere, genitive sing. *Siġeheres*, + **hām**. Nth 59 gives prs [saisəm, səːrsəm] and [sairəsəm].

SYSTON Leic SK 6211. Partly uncertain. *Sitestone* 1086, *Sy- Sithes- Sy- Sideston(e) -is-* 1204–1368, *Sy- Sieston'* c.1130–1576, *Sy- Siston(e)* 1277 and from 1380, *Sison* 1622. The generic is OE **tūn**, the specific a shortened form of a pers.n. such as *Siġethrȳth* or *Siġehǣth*. Lei 322.

SYSTON Lincs SK 9240. 'The broad stone'. *Sidestan -sthā* 1086, *Si- Sydestan* 1205–76, *Si- Sythestan* 1154×89–1321, *-ston* 1265–1526, *Si- Systan* 1242–1338, *Syston* from 1338. OE **sīd**, definite form **sīda**, + **stān**. The reference is unknown. DEPN.

SYTCHAMPTON H&W SO 8466. 'Home farm by the water-course'. *Sychampton* 1575–89. Dial. **sitch** (OE *sīċ*) + **hampton** (OE *hām-tūn*). Wo 272 gives pr [sitʃən].

SYWELL Northants SP 8267. 'The seven springs'. *Snewelle* (sic) 1086, *Siwell(a)* from 1086 with variants *Sy-* and *welle*. OE **syfan** + **wella**. A common p.n. type, cf. SEWELL Beds SP 9922, Seven Springs Ches SJ 9884, Glos SO 9617, Northum NY 8249 and the German p.n. type Siebenbrünnen. Nth 139 gives pr [saiəl].

SYWELL RESERVOIR Northants SP 8365. P.n. SYWELL SP 8267 + ModE **reservoir**.

T

Bishop's TACHBROOK Warw SP 3161. '(The part of) T held by the bishop' sc. of Chester in 1086. *Bishops Tachebroke* 1511. ModE **bishop's** + p.n. *Tæcelesbroc* [1033]13th S 967, *Taschebroc* 1086, *Tache(s)brok(e) -broc* 1112–1298, *Tachelesbroc* c.1149–1227, *Thacklesbrok* c.1320, 'the boundary brook', OE ***tǣċels** 'a boundary mark, a boundary' + **brōc**. The stream here was anciently the boundary between the dioceses of Lichfield and Worcester. The same term occurs in the boundary clause of S 1323, *an tæcles broc, of tæcles broce* [969]11th, on the boundary between Little Witley H&W SO 7863 and Shrawley SO 8064. 'Bishop's' for distinction from Tachbrook Mallory SP 3162, *Tachebroke Malory* 1547, earlier *Tacesbroc* 1086, the other manor held by Henry Mallore in 1200. Wa 258, PNE ii.174, WASCB 278–80.

TACKLEY Oxon SP 4820. 'Tæcca's wood or clearing'. *Tachelie* 1086, *Tak(k)ele(a) -ley(e)* 1176–1525, *Tackele(gh) -lee -ley(e)* 1196–1347, *Tacle* 1278×9, *Takley* 1383, *Tackley* 1675. OE pers.n. **Tæcca* + **lēah**. O 285.

TACOLNESTON Norf TM 1395. 'Tatwulf's estate'. *Tacoluestuna* 1086, *Takolueston* 1185, *Tacolneston* 1203. OE pers.n. *Tātwulf*, genitive sing. *Tātwulfes*, + **tūn**. The change of *Tat-* to *Tac-* is a dissimilatory one, and that of *-v-* (*-u-*) to *-n-* is due to misreading. DEPN gives pr [tæklstn].

TADCASTER NYorks SE 4843. 'Tada or Tata's Roman fort'. *Tatecastr(e) -er* 1086–1311, *Tadecaster -re* [c.1150]copy–1428, *Tadcaster* from 1269. OE pers.n. **Tāda* or *Tāta*, + **cæster**. On the Sunday before the battle of Stamford Bridge near York (SE 7155) in 1066 King Harold's troops were at a place which ASC(C) calls *to Táda*. If this is meant to stand for Tadcaster as is usually supposed, the specific is OE pers.n. **Tāda*, a by-n. from OE *** tāde**'a toad', and the *-t-* spellings are due to assimilation to the following voiceless [k]; otherwise OE pers.n. *Tāta*. The name of the Roman settlement at Tadcaster, *Calcaria* 'lime-works, the limestone quarries', did not survive. YW iv.76, RBrit 288.

TADDEN Dorset ST 9901. 'Toad haven'. *Tadhavene* 1327, *Taddehauene* 1332, *Tadden* 1847. ME **tadde** (OE **tāde, tadde*) + **haven** (OE *hæfen*). The name may be jocular or contemptuous or refer literally to the marshy ground by the stream here. Do ii.167.

TADDINGTON Derby SK 1471. 'Estate called after Tada'. *Tadintun(e)* 1086, 1225, *-ynton(e)* 1194–1462, *Tadington -yng-* 1209–16th, *-enton* 1287, 1317, *Tatinton -yn-* 1200–1327, 1458, *Taddington -yng-* from 1487. OE pers.n. *Tāda* + **ing**[4] + **tūn**. The forms do not support the suggestion in Årsskrift 1974.43 that the specific is an OE ***tāte** 'a hill', but OE ***tāde** 'a toad' would be possible, perhaps **Tāding* 'toad place' + **tūn**. Db 169.

TADLEY Hants SU 6061. Probably 'Tada's wood or clearing'. *(æt) tad(d)anleáge* *[909]12th S 377, *Taddele(e) -leye* 1205–1316. OE pers.n. *Tāda* or *Tadda*, genitive sing. *Tad(d)an*, + **lēah**. The specific might alternatively be OE **tadde**, ***tāde**, genitive sing. **tad(d)an** 'toad's meadow or pasture'. Gover 1958.142, Ha 161, PNE ii.174.

TADLOW Cambs TL 2847. 'Tada's mound'. *Tadeslaue -lawe* 1086 ICC, 1230, 1332, *Tadelai* 1086, *-law(e)* 1199–1307, *-lowe* 1218–1399, *Tadlow(e)* from 1312. OE pers.n. **Tāda* + **hlāw**. Ca 66.

TADMARTON Oxon SP 3937. Possibly the 'frog pool settlement'. *Tademær tun, (æt) Tademærtune* 956 S 618, *Tad(e)mærton* [956]16th S 584, *Tademertun* [956]c.1200 S 617, *Tademertone* 1086–1285 with variants *-tun* and *-tona, Tadmertun, (to) Tadmertune* [956]18th S 611, *Tademarton(')* 1241–1316, *Tadmarton* 1428. OE ***tāde-mere** + **tūn**. The early *mær* spellings, however, point rather to OE **(ġe)mǣre** 'a boundary'. O 406, Jnl 25.47–8.

TADWORTH Surrey TQ 2356. 'Theodda's settlement'. *þeddewurthe* [727]13th S 1181, *-uuerþe* [933]13th S 420, *(æt) Ðæddeuurðe* [1062]13th S 1035, *Tadorne, Tadeorde* 1086, *Tadeswurthe, 'Taddewurth', Thadeworth* 13th cent., *Tadworth* from 1540. OE pers.n. **Thēodda* + **worth**. Sr 70–1.

TAKELEY Essex TL 5521. 'Tæca's wood or clearing'. *Tacheleiam -ā* 1086, 1123–33, *Takele(e) -leg -leia -leye* 1119–1306, *Tha(c)kelega -leiam -legh -ley* 1165–1389, *Tayclay* 1526. OE pers.n. **Tæca* + **lēah**. Ess 535.

TALATON Devon SY 0699. 'Settlement, estate on the river Tale'. *Taleton(e)* 1086 etc., *Talliton* 1587. R.n. TALE ST 0702 + OE **tūn**. D 571 gives pr [tælətən].

TALE Originally a r.n. Unexplained. RN 388.

(1) Higher ~ Devon ST 0601. *High Tail* 1809. ModE **higher** + p.n. *Tale* from 1086. Transferred from the r.n. TALE ST 0702. 'Higher' for distinction from Lower Tale ST 0602, *Lower Tail* 1809 OS. D 567, RN 388.

(2) River ~ Devon ST 0702. *(on) tælen* [1061]1227, *Tale* from 1238. The forms point to an OE **Tæle* which could be from the OE adj. **(ġe)tæl** 'quick, active, swift'. But the stream is a sluggish one. D 13, RN 388 give sp [teil].

TALKE Staffs SJ 8253. 'The forehead, brow or gable end'. *Talc* 1086, *Talk(e)* from 1252, *Talk on the hill* 1836 OS. OW **talcen** with ME loss of final *-n*. The name of the marked hill (559ft.) on which the village stands. DEPN, Coates 1988.33.

TALKIN Cumbr NY 5457. 'White brow'. *Talcan* c.1195, c.1225, *Talkan* c.1230–1490, *Talken* 1485, *Talkyn* 1597, *Taukin* 1610. PrW ***tal** 'brow, end' + **can** 'white'. Originally a hill name. Cu 88 gives pr [tɔːkin], GPC 407.

TALKIN TARN Cumbr NY 5458. *Talkaneterne* 1295, *talken tarne* 1589, *Tawkin tern* 1618. P.n. TALKIN NY 5457 + ME **tern** (ON *tjǫrn*) 'a lake'. Cu 35.

TALLAND BAY Corn SX 2257. *Talland Bay* 1813. P.n. Talland SX 2251 + ModE **bay**. Earlier referred to as *a crikket betwixt Poulpirrhe and Low* c.1540, 'a creek between Polperro and Looe'. Talland, *Tallan* [c.1205]14th, c.1250–1533, *Talland* 1291, 1610, *Tallant* 1440 is probably 'hill-brow church-site', Co **tal** + ***lann**, from which a fictitious saint's name was subsequently invented, *Sanctus Tallanus* 1452. PNCo 160, Gover n.d. 297, DEPN.

TALLANTIRE Cumbr NY 1035. 'End of the land'. *Talentir* c.1160–1584 with variants *-tire -tyr(e), Tallentyre* 1558, *Tallantire* 1580. PrW ***tal** 'brow, end' + definite article **in** + **tir** 'land'. Tallantire is situated at the west end of a ridge of high ground. Cu 324.

TALLINGTON Lincs TF 0908. Possibly the 'estate called after T(e)alla'. *Talintvne -tone -tune* 1086, *-inton(e)* 13th cent., *Talingtun -ton -yng-* 1212–1555, *Tallington -yng-* 1242–1607. OE pers.n. **T(e)alla* + **ing**[4] + **tūn**. Perrott 434.

TALSKIDDY Corn SW 9165. Possibly 'hill-brow of land-clearance'. *Talschedy -skydy* 1297, *-skedy* 1300, *-skidy -skythy -skithy* 1310–58. Co **tal** + an unrecorded word corresponding to Breton *skidiñ* 'to clear land'. An alternative possibility is 'brow of shadows', Co **tal** + **skeudi**, plural of **schus** 'shade, shelter' (OCo *scod*). Cf. Tolskithy SW 6841, *Tolskithey* 1748. PNCo 161, Gover n.d. 323.

River TAMAR Devon SX 2999–4266. *Ταμάρου (ποταμοῦ ἐκβολαί)* [c.150]13th Ptol, *Tamaris* [c.650]13th Rav, *Tamur* 980×8 Crawford, *Tamer* 1018–c.1870, *Tamar* [12th]c.1200, 1429 and from c.1540, *Tambra* c.1125, *Tambre(is)* [c.1220]13th, *Taumbre* 1242. An ancient European r.n. formation appearing as *Tamaros* in British, from the root ***tam** + suffix **-ar-** as also in the Tambre, Coruña, Spain, *Tamara* [c.150]11th Ptol, the Tammaro, a tributary of the Calore, near Benevento, Italy, *Tamarus* [3rd]7th AI, and the Demer, a tributary of the Dyle, Belgium, *Tamera* 908×15. The interpretation of ***tam-** as 'dark' is no longer held. Rather it is an *m* extension of the IE root **ta-* /**tə-* 'to flow'. Tamar thus belongs to an extensive family of ancient European names with three branches, those in *-m-*, THAME, THAMES, TEAM, TEME etc. and, with British lenition of *m* > *v*, TAF(F), TAVY, TEVIOT, TAWY; those in *-n-*, TAIN, TEAN, TONE etc.; and those in *-u-*, TAW, TAY, for all of which there are continental parallels. The modern spelling with *-ar* is an artificial revival of the classical form. PNCo 161 gives pr [teimə], RN 389, RBrit 465, BzN 1957.256–62, FT passim, ScotPN 190.

Lower TAMAR LAKE Devon SS 2911. A modern lake on the upper Tamar River.

Upper TAMAR LAKE Devon SS 2812. A modern lake on the upper Tamar River.

River TAME OBrit **Tamā*, a pre-Celtic Old European r.n. on the IE root **tā-*/**tə-* 'flow'. On this prolific root are names formed with an *m*-suffix, TAME, TEAM, THAME, TAVY, TAMAR, TEME, THAMES, an *n*-suffix, TONE, TEAN, TANAD, and a *u*-suffix TAW. BzN 1957.256–62.

(1) ~ GMan SJ 9092. *Tome* 1292, *Thame* [13th]14th, *Tame* [1322]15th, 1577. La 27, RN 390, BzN 1957.256.

(2) ~ NYorks NZ 5211. *Tame* [1129]15th, [12th, 13th]13th. YN 6, RN 390.

(3) ~ Warw SP 2091. Staffs SK 1807. *Tame* from c.1025, *Thame* 1282–1509. The boundary of the people who dwelt by the Tame is *Tomsetna gemœre* [849]11th S 199. St i.20, RN 389, BzN 1957.256.

TAMERTON 'Settlement, estate on the r. Tamar, R.n. TAMAR SX 3682 + **tūn**.

(1) ~ FOLIOT Devon SX 4761. 'T held by the Foliot family'. *Tamereton Foliot* 1262. Earlier simply *Tambretone* 1086, *Tamerton* 1212, 1242. The Foliot family held the manor from 1242; the addition is for distinction from North TAMERTON Corn SS 3197. Ptolemy's *Ταμάρα* (*Tamara*), a city on the Tamar, is more likely to be at Launceston SX 3384 than here. D 242, RBrit 464.

(2) North ~ Corn SX 3197. *Northtamerton alias Tamerton-Bridge* 1565. ModE adj. **north** for distinction from TAMERTON FOLIOT Devon SX 4761 + p.n. *Tamert'* c.[1162]15th, *Tamerton* from 1180, 'settlement on the river Tamar', r.n. TAMAR SX 3682 + OE **tūn**. PNCo 161.

TAMWORTH Staffs SK 2104. 'The enclosure by the r. Tame'. *Tamouuorðie -thige* [781]11th S 120–1, *Tome worðig* 799 S 155, 808 S 163, *Tomoworðig -in* [814]11th S 171–2, *(æt) Tomanworðie, (in) Tomeuuorthie* [841]11th S 195–6, *Tamweorthin* [855]11th S 207, *(æt) Tame worþige* 922 ASC(A), *Tamuuorde* 1086, *Thamworthe* 1313. R.n. TAME SK 1807 + OE **worthiġ** varying with **worthiġn** and **worth**. *Tomtun* [675×92]12th S 1804 is probably an earlier form of the name. RN 390, Horovitz.

TANDRIDGE Surrey TQ 3750. Possibly 'the swine-pasture ridge'. *Tenhric* [963×75]c.1150 S 815, *Tenrige* 1086–1316 with variants *-rigge -rugge -regge* etc., *Tanruge* c.1212–1539 with variants *- rygge -rich* etc., *Tunrug(g)e, Tanerig(ge)* 13th cent., *Tannerigge* 1263, *Tanridge alias Tandridge* 1612. Possibly OE **dænn** + **hrycg**. The change of initial *d* > *t* is due to preceding *æt*, cf. TIDENHAM Glos ST 5596, *æt Dyddanhamme* [956]12th S 610, and Trimworth Kent TR 0649, *Dreaman wyrð* 824 S 1434. Sr 335, L 169.

TANFIELD Durham NZ 1855. 'Open land on the river Team'. *Tainefeld* (for *Tame-*) c.1190, *Tamfeld* 1196×1215, 1444, *Tanfeld* 1309–1590, *Taun-* 1388–1528, *Tanfield* [1286]1332 and from 1612. R.n. TEAM + OE **feld**. NbDu 193, L 243.

TANFIELD LEITH Durham NZ 1855 → LEES GMan SD 6500.

East TANFIELD NYorks SE 2878. A deserted medieval village. *Estanfeld(e)* 1157, 1396, 1579, *Est Tanefeld* 1280, 1292. ME **est** + p.n. *Tanefeld* 1086–c.1291, *Danefeld -felt* 1086, *Tanne-* 1198, 1301, either 'open land where young shoots grow or are obtained', OE **tān**, genitive pl. **tāna**, or 'Tana's open land', OE pers.n. **Tāna* + **feld**. YN 221, L 243.

West TANFIELD NYorks SE 2678. *Westanfeld(e)* 1282, 1396. ME **west** + p.n. *Tanefeld(e)* 1086, as in East TANFIELD SE 2878. YN 222.

TANGLEY Hants SU 3252. 'Wood or clearing of the spits of land'. *Tangelea* 1174, *-lie -le(gh)* 1212–1327. OE **tang**, genitive pl. **tanga**, + **lēah**. The reference is to three projecting points of land on the middle of which between two small dry valleys stands the church of Tangley. Gover 1958.161, Ha 161, PNE ii.176.

TANGMERE WSusx SU 9006. Partly uncertain. *Tangmere* from *[680 for ? 685]10th S 230, *Tangemere* 1086–1327. The name is a compound of OE **tang** 'a tong, forceps' and **mere** 'a pond' but the significance is unclear. The pond survives beside the church. Sx 97.

TAN HILL Durham NY 8906. Uncertain. No early forms.

TAN HILL Wilts SU 0864. Originally *S. Anns Hill* 1610, *St Anns Hill* 1817 OS. The church of All Cannings in which parish Tan Hill lies is dedicated to St Anne. Site of the famous Tan Hill Fair held on St Anne's day, 6th August. Wlt 312.

TAN HOUSE H&W SO 2448 → STANSBATCH SO 3561.

TANKERSLEY SYorks SK 3499. 'Thancred's wood or clearing'. *Tancreslei(a)* 1086, *-ley* 1185×1215, *Tancredeslay* 1194, 1196, *Tankersley -lay* from 13th. Pers.n. OE or CG *Thancrēd*, genitive sing. *Thancrēdes*, + **lēah**. YW i.297, xi.

TANNINGTON Suff TM 2467. 'The estate called after Tata'. *Tatintuna* 1086, *Tat(t)ingeton(a) -tun* 1167–99, *Tat(t)ington(e)* 1224–1336, *Tadding(e)ton'* 1188, 1190, *Tanyngton* 1524, 1610. OE pers.n. *Tāta* + **ing**[2] + **tūn**. DEPN, Baron.

TANSHELF WYorks SE 4422. Possibly 'Tædden's shelf of land'. *Taddenesscylf* 11th ASC under year 947, *Tatessella, Tateshal(l)e -halla* 1086, *Tan(e)self* 1255×8, 1353, *Tans(c)helf(e)* 1256–1619. OE pers.n. **Tædden*, genitive sing. **Tæddenes*, + **scelf**. The spelling with **scylf** is standard WSax. Since the pers.n. **Tædden* is not on record, it has been suggested that the specific is OE p.n. **Tād-denu* 'toad valley'. Tanshelf was the original name of Pontefract as Symeon of Durham records: Taddenessclyf *erat tunc villa regia quae nunc vocatur* Puntfraite *Romane, Anglice vero* Kirkebi, '*Taddenessclyf* (erroneously for *-scylf*) was then a royal estate which is now called *Puntfraite* in French but *Kirkebi* (probably referring to the early twelfth cent. Cluniac priory) in English'. YW ii.83, 75.

TANSLEY Derby SK 3259. Probably 'wood or clearing where shoots are obtained'. *Tanes- Teneslege* 1086, *Taneslea* 12th cent., *Tanneslegh -ley(e) -leie -lay* 1276–1519, *Tansley* from 1577. OE **tān**, genitive sing. **tānes**, + **lēah**. The specific could, however, be an unrecorded OE pers.n. **Tān* parallel to OHG *Zeino*. Db 406.

TANSOR Northants TL 0590. 'Tan's promontory'. *Tanesovre* 1086, *-ouera -overe -owere* c.1152–1307, *Tanesores* 1205, *Tansore* 1346–1428. OE pers.n. **Tān*, genitive sing. **Tānes*, + **ofer**. Tansor is situated on the side of a rounded spur in the marshes beside the r. Nene. Nth 208, L 177.

TANSTERNE Humbs TA 2237 → STERNFIELD Suff TM 3961.

TANTOBIE Durham NZ 1754. Uncertain. *Tantovy* 1768 Armstrong, *Tantoby* 1857 Fordyce, 1865 OS. Tantobie is a modern colliery village close to TANFIELD NZ 1855. The specific may be the r.n. TEAM NZ 2455. Possible association with the ModE hunting and political word *tantivy* seems unlikely.

TANTON NYorks NZ 5210. 'Settlement on the r. Tame'. *Tametun -ton(a)* 1086, *Tameton* 1170–1224, *Tanton* from [1203×7]15th. R.n. TAME NZ 5211 + OE **tūn**. YN 170.

TANWORTH-IN-ARDEN Warw SP 1170. P.n. *Tanewrthe -word -worth* 1201–1315, *Tannewurth -worth* 1249, 1304, *T(h)onew(o)rth* 1275–1480, *Tonworth al. Tanworth* 1538, 'the enclosure of branches', OE **tān**, genitive pl. **tāna**, + **worth**, + forest name ARDEN WMids SP 1859. It is unnecessary to posit unrecorded pers.n. **Tanna* for this name. Wa 291.

TAPELEY Devon SS 4729. 'Wood where pegs are made'. *Tapeleia* 1086, 1195, *Tappeleg(he)* 1177–1314. OE **tæppe** + p.n. Leigh as in WESTLEIGH SS 4728 and Eastleigh SS 4827, *Est legh* 1242, all parcels of an area E of Bideford originally called *Leia* 1086, 'the wood or clearing', OE **lēah**. D 124.

East TAPHOUSE Corn SX 1863. *Ye East topp House* 1675, *Easter Taphouse* 1699. ModE adj. **east** + **taphouse** 'an ale-house'. 'East' for distinction from Middle and West Taphouse, SX 1763 and 1563, first recorded as *Taphouse* c.1533, isolated inns for travellers across the moors from Liskeard to Lostwithel, like Jamaica Inn on Bodmin Moor and INDIAN QUEENS SW 9158. PNCo 161, Gover n.d. 246.

TAPLOW Bucks SU 9182. 'Tæppa's barrow'. *Thapeslau* 1086, *Tapeslawe* 1186–8, *Tappelawe* 1189×99–1262, *Teppe-* 1227–97, *Toppelewe* 1247, *Toplo* 1571. OE pers.n. **Tæppa* + **hlāw**. The reference is to the tumulus in the churchyard which in 1883 yielded the Taplow Anglo-Saxon hoard of treasures now in the British Museum. Bk 231 gives occasional pr [təplou].

TARBOCK GREEN Mers SJ 4687. *Tarbock Green* 1842 OS. P.n. Tarbock + ModE **green**. Tarbock, *Torboc* 1086–1256, *-bok* 1257–1354 etc., *-broke* 1311, *Thorboc -k, Thorebok, Turbok* 13th, *Tarboche, Terbok* 1346, is probably 'thorn-tree brook', OE **thorn** + **brōc** with dissimilatory loss of *r*. Alternatively the specific might be ON pers.n. *Thor*. La 113, SSNNW 171, L 15, Jnl 17.63.

TARLETON Lancs SD 4521. *Tarelton* [c.1200]1268–1451, *-il-* [c.1212]1268, *Tarleton* from 1246. Usually explained as 'Tharald's farm or village' from a supposed side-form of the ON pers.n. *Thóraldr* with *a* instead of *ó*. Such forms, however, only become frequent in the late 14th cent. in Norway and other examples of the name in Lancashire retain the *ó*-vocalisation. An alternative possibility suggested by Dr Insley is 'settlement on the river **Tærla*' from a pre-Celtic IE r.n. **Tarla* on the root **ter-*/**tor-* 'quick, strong' as in the continental r.ns. Taro, a tributary of the Po, *Tarus* AD 77 Pliny NH, Tara, a river in Calabria, *Τάρας* c.AD 150 Pausanias, cf. the p.n. Taranto, *Τάρα* n.d. Kinnamos Hist.III.7, a river in Illyria, Thérain, a tributary of the Oise, Seine-Inférieure, *Thara* 879, *Tara* 968, Tharaux, Gard, *Taravua* 1192 etc. **Tarla* could be an earlier name of the Douglas although formation with the suffix *l* would be unique in our present knowledge of Old European hydronymy. La 138, SSNNW, Jnl 17.78, Krahe 1964.57, NoB 87.71–80.

TARLSCOUGH Lancs SD 4314. Possibly the 'wood by the river **Tærla*'. *Tharlescogh* c.1190, *Tarlescou -scogh'* 13th, *Terlesco wood* 1577. OE r.n. **Tærla* (Pre-Celtic **Tarla*) + ON **skógr**. The same issues arise with this name as with nearby TARLETON SD 4521. La 124, SSNNW 167, Jnl.17.69, NoB 87.71–80.

TARLTON Glos ST 9699. Partly uncertain. *Torentvne, Tornentone* 1086, *Torleton* 1204–1590, 1777, *Tarlton* 1830 OS. The DB spellings point to OE **thorn-tūn** 'settlement among' or 'fenced with thorns', the later spellings to an OE compound p.n. **Thorn-lēah* 'thorn wood or clearing' + **tūn**. The modern form shows SW dial. *a* for *o*. Gl i.106.

TARN BAY Cumbr SD 0790. *Tarn Bay* 1864 OS. P.n. The Tarn SD 0789, *Tarne* 1646, ModE **tarn** (ON *tjǫrn*) 'a lake', + **bay**. Cu 348.

TARNBROOK Lancs SD 5855. 'Tarn stream'. *Tyrn(e)brok* 1323–4. ON **tjǫrn** + OE **brōc**. There are two small tarns near the headwaters of Tarnbrook, one at Hare Syke SD 6057, another at Brown Syke SD 6158. *Terne-* rather than *Tyrne-* is the expected spelling; either this is a case of N dial. *i* for *e* or there has been confusion with OE **thyrne**, ON **thyrnir** 'a thorn-tree'. La 172, SSNNW 253.

TARNBROOK FELL Lancs SD 6057. P.n. TARNBROOK SD 5855 + ModE **fell** (ON *fjall*).

TARPORLEY Ches SJ 5562. Partly uncertain. *Torpelei* 1086, *-le(gh) -ley(e)* [c.1208]17, 1287–1549, *Torplei -le(gh) -ley* 1284–1653, *Torp(e)le(gh) -ley(e)* 1284–1653, *Torperley* 1198–1755 with variants *-or- -ur- -ir-* and *-legh(e) -le(e) -leigh(e)* etc., *Thorperleg(h) -le(y)* 1280–1349, *Tarporley* from 1394. The generic is OE **lēah** 'a wood, a clearing', the specific possibly OE ***thorpera**, genitive pl. of **thorpere** 'one who lives in a *thorp*', parallel to ON **thorpari** 'a peasant, a cottager'. Although there is no known suitable p.n. in **thorp** surviving nearby, this solution is formally satisfactory and less problematical than other attempts. Che iii.294 gives pr ['tarpəli], older local ['tarpli].

TARR Somer ST 1030. Possibly the 'tor or rocky peak'. *H.ʳ* and *L.ʳ Tarr* 1809 OS. Possibly OE **torr** with later south-western *a* for *o*. The Tarrs lie on a hill which rises to 405ft.

River TARRANT Dorset. 'Strongly flooding one'. *(to) Terrente* [1016]12th S 935, *Terente dene* 'the Tarrant valley' [956]15th S 630, *Tarente* 1253, *Tarent* 1449. Brit **Trisantonā*, PrW **Trihanton* and with metathesis **Tirhanton*. Identical with the r.n. TRENT. RN 416. A series of estates named from the river lies along its course.

(1) ~ CRAWFORD Dorset ST 9203. 'Tarrant estate beside Crawford', originally 'Crawford in Tarrant', cf. *Lordshipp of Lyttle Craford within Tarrant* 1508.

I. Tarrant, 'estate on the r. T', *Tarente* 1086–1403. Also known as *litle tarenta* 1372, *Nuns Tarent* 1377, *Tarent Monachorum als. Tarent Abbey* 1582, referring to the former abbey here.

II. Little Crawford, *Little, Parva Craweford* 1280, 1291, *Little Crauford* 1347–86. So called for distinction from the lost Great Crawford on the other bank of the r. Stour in Spetisbury ST 9002, *Craveford* 1086, *Crawe- Craueford* 1242–1367, 'crow ford', OE **crāwe** + **ford**. The combination *Crawford Tarrant* first occurs in 1795. Do ii.181, 64.

(2) ~ GUNVILLE Dorset ST 9212. 'T estate held by the Gundeville family'. *Tarente Gundeville* 1233–17th with variants *Tarrent(e), Tar(r)ant(t)* from 1316, *G(o)und(e)- Gun-* from 1475, *Gon-* 1503, and *-uil(l)(e) -vyl(l)(e) -fylde* (1504), *-field* (1610). P.n. *Tarente* 1086†–1320 + family name *Gundeville*. The manor, held by Robert de Gundevill' in 1180×1, was also known as *Gondevileston -vyles-* 1264–1405 and simply *Goundevill'* 1395–1494, *Gunville* 1563, *Gunfield* 1618. Do ii.242.

(3) ~ HINTON Dorset ST 9310. 'T estate called Hinton, the religious community's estate'. *Tarent(e) Hyneton(')* 1280–1428. P.n. Tarrant + p.n. *Hineton'* 1280, *Hynton* 1548, 'monks' estate', OE **hī(ġ)na**, genitive pl. of **hīwan** 'members of a (religious) household', + **tūn**. Earlier simply *(in) Tarente* [871×7]15th S 371(1,2), *(ad) Tar- Terentam* [935]15th S 429, *Tarente* 1086–1245. The estate was granted to Shaftesbury abbey by Kings Alfred and Athelstan. Do ii.119.

(4) ~ KEYNESTON Dorset ST 9204. 'T estate called Keyneston, the Keynes family's estate'. *Tarent(e) Keyneston, C- Kayneston* 1303–1469, *Tarrant Kayns- Kainston* 1617. Other forms of the name are simply *Tarente* 1086–1292, *Tarent(e) Kaaign(e)s, Kahaines, Kay(g)nes* etc. 1242–1432, *Kayneston(e)* 1278–1431, *Keinson al. Keinston* 1624, and *Keynes* 1280, 1364. The manor was held by William de Cahaignes in 1199. Do ii.122 gives pr ['keinstən].

†The DB identification is probable rather than certain.

(5) ~ LAUNCESTON Dorset ST 9409. 'T estate called Launceston, Leofwine or Lowin's estate'. *Tarente Loueweni(e)ston', Louinton'* 1280, ~ *de Lowyneston* 1285, ~ *Launston* 1397. Also known simply as *Tarente* 1086 and *Louineston', Lowyneston', Louynton'* 1280, *Launston* 1431, OE pers.n. *Lēofwine* or the ME surname *Lowin* derived from it, genitive sing. *Lēofwines, Lowines,* + **tūn**. Do ii.125.

(6) ~ MONKTON Dorset ST 9408. 'T estate called Monkton, the monks' estate'. *Tarent(e) Moneketon(e)* 1280–1366, ~ *Monketon(')* 1288–1402, ~ *Monkton* 1332. Earlier simply *Tarente* 1086 or *Tarent Monachorum* 'monks' T' 1291–1428. Monkton is ME **monketon** 'monks' estate' (OE *munuc*, genitive pl. *muneca*, + *tūn*) and refers to possession of the manor by Cranborne priory in 1086 or by Tewkesbury abbey in 1154×89. Do ii.292.

(7) ~ RAWSTON Dorset ST 9306. 'T estate called Rawston, Ralph's estate'. *Tarrant Rawston* 1535. The manor was held in 1086 by one *Radulfus* (Ralph). Also known simply as *Tarente* 1086 and *Tarente Willelmi de Antioche* 'William of Antioch's T estate' 1242×3, *Tarente Antyoche, A(u)ntioch(e)* 1288–1346 or *A(u)ntyocheston(')* 1268–1431 and *Tarent(e) Auntyocheston* 1399 ('T estate called Antioch's estate'). Do ii.127.

(8) ~ RUSHTON Dorset ST 9305. 'T estate called Rushton, the estate of the Russeaux family'. *Tarent(e) Russcheweston* 1307–1432 with variants *Russea(u)ston, Russh(es)ton, Rissheton* etc., *Tarrant Rushton* 1609. Earlier simply *Tarente* 1086†–1244, *Tarente Petri de Russell'* 'Peter de Russeaux's T' 1242×3, *Tarente Russe(a)us -eals -eaux* 1280–1399. The manor was granted to Peter de Rusceaus in 1216. Also known as *Tarente Vileres, Vylers, Villers* etc. 1291–1432 because Roger de Vilers held land here in 1227. Rushton, *Rus(s)cheuston(e)* 1283, 1327, *Russe(a)uston(e)* 1315, *Russ(c)heton(e)* 1326, *Rushton* 1588, properly 'the Russeaux estate', was misinterpreted as 'rush settlement'. Do ii.251.

TARRING NEVILLE ESusx TQ 4404. 'T held by the Neville family'. *Thoryng Nevell* 1339, *Tarringe Nevell* 1588. P.n. Tarring + family name Nevile first associated with this place in 1253. Tarring, *Toringes* 1086, 1194, *Torringes* 1215, 1251, *Tarringes* 1291, *Torryng(ge) -inge* 13th–16th cents., *Terring(es)* 1274–91, is 'the Teorringas, the people called after Teorra', OE folk-n. **Teorringas* < pers.n. **Teorra* + **ingas**. Cf. West Tarring TQ 1304, *Teorringas* [946]13th S 515, *Terringes* 1086–1312 including *[941]13th S 477, *Ter(r)inge* 1210–1316, *Tarring* 1253, *Torryng* 1280. Sx 339, 194, ING 40, SS 64–5.

TARRINGTON H&W SO 6140. 'Estate called after Tata'. *Tatintvne* 1086, *Tatin(g)ton -yn-* 1200–1447×8, *Tadintona -ynton* c.1135×44, 1160×70, 1306. OE pers.n. *Tāta* + **ing**[4] + **tūn**. For the subsequent sound substitution of [r] for [d] cf. DERRYTHORPE Lincs SE 8208, DORRINGTON Shrops SJ 4703. He 187.

TARVIN Ches SJ 4967. Transferred from *Tervin*, the old name of the river Gowy. *Terve* 1086, *Terne* 1152–1296, *Teruen -v-* 1185–1576, *Teruin -vyn(n) -vin* 1222–1719, *Taruen* 1220×40, *Tarvin* from 1287. The r.n., recorded as *Tervin* 1209, *Teruen* 1209–90, is 'the boundary (river)', PrW ***tervīn** (ultimately from Latin *terminus*). Che iii.281, i.26.

TASBURGH Norf TM 2095. 'The agreeable or pleasant fort'. *Taseburc* 1086, 1197, *-burg(h)* 1202, 1242, *Tasseburc* 1199. OE **(ġe)tǣse** + **burh**. Alternativey the specific might be an OE pers.n. **Tǣsa* and so 'Tæsa's fort'. The reference is to the Iron Age fort at Tasburgh Camp. DEPN gives pr [teiz-].

TASLEY Shrops SO 6994. Partly unexplained. *Tasseleya* c.1143–1443 with variants *-leg(h)(') -le(e)* and *-ley(e), Tassheley* 1416, *Tashley* 1728, 1803, *Tasley* from 1535, *Tastley, Tarsley* 18th. The generic is OE **lēah** 'a wood, a clearing', the specific unidentified; OE **tǣsel** 'teasel' does not fit the forms and only an element ***tassa** is consistent with the persistent *-ss-* spellings and absence of *e-* vocalism. This might be an OE pers.n. though none such is known. Cf. however, the German p.n. Zazenhausen near Stuttgart, *Zazenhausen* [788]12th, *Zazzenhusen* 1274, 'Zazo's settlement', in which pers.n. *Zazo* is explained as a pet form of **Tatto*, cf. OE *Tata, Tǣta*. Sa i.290, Reichert 1982.175.

TASTON Oxon SP 3621. 'Thor's stone'. *Thorstan* 1278×9, *-stane* 1316, *Torstone* 1492, *Taston* 1608×9. Divinity n. *Thor* for OE *Thunor* + **stān**. The reference is to a giant monolith. O 380.

TATENHILL Staffs SK 2022. 'Tata's hill'. *Tatenhyll* [942]13th S 479, *Tatenhala* 1188, *Tattenhull(e)* 1227, 1251. OE pers.n. *Tāta*, genitive sing. *Tātan*, + **hyll**. DEPN, L 170, Horovitz.

TATHAM Lancs SD 6069. 'Tata's homestead'. *Tathaim* 1086, 1215, *Tateham* 1202, 1463, *Tatham* from 1226, *Tatam* 1297. OE pers.n. *Tāta* + **hām** partly replaced with ON **heimr**. La 182, SSNNW 206, Jnl 17.102, 18.18.

TATHAM FELLS Lancs SD 6763. P.n. TATHAM SD 6069 + ModE **fell** (ON *fjall*).

TATHWELL Lincs TF 3282. Partly uncertain. *(æt) Taðawyllan* [1002×4]c.1100 S 1536, *taþa willan* [1004]late 11th S 906, *Tadewelle -uuelle* 1086, *-wella* c.1115, *Taddewell* 1156. Later forms show the name was understood to be the 'toad spring', OE ***tāde, tadde** + **wella**. The earlier forms, if they do in fact belong here, point to ON **tath** 'dung' as the specific, + OE **wella**, dative sing. **wellan**, pl. **wellum**. The earliest forms show the standard WSax written form **wylle** for Anglian **wella**. DEPN, SSNEM 222, L 31.

TATSFIELD Surrey TQ 4157. 'Tatel or Tætel's open land'. *Tatelefelle* 1086, *Thatle- Tat(e)les- Tetlesfeld, Tetesfeud* 13th cent., *Tattelesfeld* 1313, *Tattes-* 1392, *Tattys-* 1488, 1500. OE pers.n. *Tātel* or *Tǣtel*, genitive sing. *Tā- Tǣteles*, + **feld**. Sr 337, L 239, 243.

TATTENHALL Ches SJ 4858. 'Tata's nook'. *Tatenale* 1086, *-hale* 1157–1487, *Tattenhall* [1287]17th and from 1400, *Tat(t)nall* 1473–1729, *Tatna* 1690. OE pers.n. *Tāta*, genitive sing. *Tātan*, + **halh** in the sense 'small valley'. Che iv.97 gives pr ['tætnɔ:l], older local ['tætnə], L 106.

TATTENHAM CORNER STATION Surrey TQ 2258. P.n. Tattenham Corner + ModE **station**. Tattenham Corner is called after *Totnams* 1603×25, *Tattenham* 1841, originally a manor held by the family of Roger *de Tot(t)enham* 1294, 1332, which probably came from TOTTENHAM GLond TQ 3390. The generic is ME **corner** 'a corner, a nook'. Sr 71, PNE i.108.

TATTERFORD Norf TF 8628. 'Tathere's ford'. *Taterforda* 1086, *-ford* 1203, *Tateresford* 1207, *Tatersford* 1254. OE *Tāthere* + **ford**. The same pers.n. occurs also in nearby TATTERSETT TF 8429. DEPN.

TATTERSETT Norf TF 8429. 'Tathere's fold'. *Tatessete* 1086, *Tatersete* 1199–1254. OE p.n. *Tāthere* + **(ġe)set**. The same pers.n. occurs also in nearby TATTERFORD Norf TF 8628. DEPN.

TATTERSHALL Lincs TF 2157. 'Tathere's nook'. *Tateshale* 1086, 1170×5, *-ala* c.1115, *-hala* 1140×50, *Tatersala, Tatrehalla* 12th, *Tatersale* 1187, c.1200, *Tatersall* 1212. OE *Tāthere*, genitive sing. *Tātheres*, + **halh**, dative sing. **hale**, probably in the sense 'nook of dry ground in marsh'. DEPN, L 103, 110.

TATTERSHALL BRIDGE Lincs TF 1956. *Tattershall Bridge* 1824 OS. P.n. TATTERSHALL TF 2157 + ModE **bridge**.

TATTERSHALL THORPE Lincs TF 2159 → Tattershall THORPE.

TATTINGSTONE Suff TM 1337. 'Tating's estate' or 'the estate of Tating, the place called after Tata'. *Tati(s)tuna* 1086, *Tateston* 1254, *Tatinmgeston(e) -tun* 1219–1341, *Tattingstone* 1591×2. OE pers.n. **Tāting* or p.n. **Tāting* < pers.n. *Tāta* + **ing**[2], genitive sing. *Tātinges*, + **tūn**. DEPN, Baron.

TATTON HALL Ches SJ 7481. *Tatton Hall* 1831. P.n. Tatton + ModE **hall**, referring to the house built c.1794 to replace Old

†The DB identification is probable rather than certain.

Hall. Tatton, *Tatune* 1086, *Tatun(ia)* 12th, 13th, *Tatton* from early 13th, *Taitun* 1385, is 'Tata's estate', OE pers.n. *Tāta* + **tūn**. Che ii.64.

TATWORTH Somer ST 3205. 'Tata's enclosure'. *Tattewurthe* 1254, *Tateworth* 1315. OE pers.n. *Tāta* + **worth**. DEPN.

TAUNTON Somer ST 2224. 'Settlement on the river Tone'. *Tantun* [737]12th S 254, *[854]12th S 310, *938 S 443, 891 ASC(A), 1121 ASC(E) both under year 722, *Tantone* 1086, *Tanton(')* [1165–1380] Buck, *Taunton(ie)* [c.1152–1363] Buck, 1610. R.n. TONE ST 3127 + OE **tūn**. RN 411, TC 179.

TAVERHAM Norf TG 1614. 'Red-lead homestead'. *Taver- Tauresham* 1086, *Tauerham* 1168, 1191. OE **tēafor**+ **hām**. The reference is uncertain; either this was a place where the colour was made or it had a building so decorated. It is interesting to note that the word **tēafor** survives in the E Anglian dilalect as *tiver* 'red ochre for marking sheep'. DEPN gives pr [tei-].

TAVISTOCK Devon SX 4774. 'Outlying farm on the river Tavy'. *(at) Tauistoce* [981]13th S 838, *(æt) Tefing stoce* 1121 ASC(E) under year 997, *Tæfingstoc* c.1050 ASC(C, D) under year 997, [1046]12th S 1474, *Tæu- Tæfistoc* c.1000 Saints, *(on) Tæfingstoce* [1046]12th S 1474, *Tavestoc(h)* 1086, *Tauistoc(h)* c.1115–1160, *Tastocke* 1602. R.n. TAVY SX 4765 + OE **stoc**. The town grew up around an abbey here founded by Ordulf, earl of Devon. It is possible, therefore, that the sense of **stoc** in this name is 'religious place'. D 217, RN 393, PNE ii.154.

TAVY Originally a r.n. lOE **Tæfi* < Brit **Tamīos* on the same root as TAME, THAME(S), TEAM etc. RN 393, BzN 1957.257.

(1) Mary ~ Devon SX 5079. 'St Mary's Tavy estate'. *St Mary Tavy* 1290, *Marytauey* 1577. Earlier simply *Tavi* 1086, *Tavy* 1242. Saint's name *Mary* from the dedication of the church and for distinction from Peter TAVY SX 5177, + r.n. TAVY SX 4765. D 200, RN 393.

(2) Peter ~ Devon SX 5177. 'St Peter's Tavy estate'. *Petri Tavy* 1270, *Peterestavi* 1276, 1295, *Patryxtavy* 1535. Saint's name *Peter* (Latin genitive sing. *Petri*) from the dedication of the church and for distinction from Mary TAVY SX 5079, + p.n. *Tawi* 1086, *Taui* 1165, transferred from the r.n. TAVY SX 4765. D 231 gives pr [pi:- pitəteivi], RN 393.

(3) River ~ Devon SX 4765. *Taui* c.1125–1281, *Tavie* 1291. D 14, RN 393.

River TAW Devon SS 6614. *Eltabo* (for *Fl*, i.e. *flumen*, *Tabo*) [c.650]13th Rav, *Thau(e)* 1198, 1350, *Tau* 1244, *Taw(e)* from 1249, *Towe(e)* c.1280, 1394, 1398. The earliest English form refers to the mouth of the Taw, *Táwmuða* c.1100 ASC(D) under year 1068, the site of Appledore SS 4630. Identical with the r.ns. Tay, Scotland, NO 1138, 1221, *(ad) Taum* (accusative case) [98]9th Tacitus, *Ταούα* [c.150]13th Ptol, *Taba* [c.700]13th Rav, *Táu, Taú, Tau* n.d., *Taye* 1342, and La Thève, a tributary of the Oise, *Tavia* [c.300]8th AI. The Gaelic form for Tay, *Tatha* < earlier *Toē* presupposes a Brit **Tau̯i̯ā* ultimately from the IE root **ta-/*tə-* 'flow' as at the base of TAME, THAME(s), TEAM, Thève and other r.ns. in Italy, Dalmatia and Galatia. The older derivation from Brit **tau̯o-*, **tau̯ā* 'silent, peaceful' as in W *taw*, Ir *toi* is an unlikely description of these rivers. D 14, 102, RN 394, RBrit 470, BzN 1957.260–1.

TAW GREEN Devon SX 6596. 'Green beside the r. Taw'. *Tawegreene* 1635. D 451.

TAWSTOCK Devon SS 5529. 'Outlying farm on the river Taw'. *Tauestoch(e)* 1086, *Taustoche -stok(')* 1157×60–c.1400, *Toustok* 1227, *Tostock* 1594. R.n. TAW SS 6614 + OE **stoc**. D 121.

TAWTON 'Settlement, estate on the r. Taw'. R.n. TAW SS 6614 + **tūn**.

(1) Bishop's ~ Devon SS 5630. *Tautone Episcopi* 1284, *Bischopestautone* 1374. Earlier simply *Tautone* 1086. The estate belonged to the bishop of Exeter in 1086. D 352.

(2) North ~ Devon SS 6601. *Nortauton* 1262, *Northtauton* 1316. Earlier simply *Tau(u)etona* 1086, *Tauton* 1158. Also known as *Chepin(g)tauton* 1199–1244, 'market Tawton', ME **cheping**. 'North' and *Cheping* for distinction from South TAWTON SX 6954. D 370.

(3) South ~ Devon SX 6954. *Suthtauton* 1212. Earlier simply *Taue(s)tone* 1086. 'South' for distinction from North TAWTON SS 6601. D 448.

TAXAL Derby SK 0079. Possibly 'valley of the land-holding'. *Ta(c)keshale* c.1251–1527 with variants *Tach- -is-, -hal(l), Taxhale -hal(l)* 1274–1690, *Tackesal(e)* 1285, 1335, *Tacsal(e) -all* 1288–1493 with variants *Ta(c)k- Thac-, Taxall* 1502. The form *Tatkeshal* cited in DEPN is a mistake for *Tackes-* and cannot, therefore, support pers.n. *Tatuc* as the specific. Possibly ME **tak** 'a lease, a tenure, a revenue', genitive sing. **takes**, + **hale** (OE *halh*) 'a nook'. Che i.172, L 106.

TAYLORGILL FORCE Cumbr NY 2211. P.n. Taylorgill, family name *Taylor* + **gill** 'a ravine', + **force** 'a waterfall'. Cu 352.

TAYNTON Glos SO 7321. 'Farm, village or estate called after Tǣta'. *Te- Tatinton* 1086, *T(h)eþing- Thointone* 12th, *Tedinton(a)* 1154×89, 1240, *Tei- Teyntun(e) -ton(a)* 12th–1486, *Tai- Taynton* from 1253. OE pers.n. *Tǣta* + **ing**[4] + **tūn**. Gl iii.187.

TAYNTON Oxon SP 2313. 'The settlement on the r. *Teign*'. *(æt) Tengetune* [c.1055]11th S 1105, *Teintuna -ton(e)* [1059]11th S 1028–1676 with variant *Teyn-, Tenton, Teigtone* 1086, *Taynton* 1675. Unrecorded r.n. identical with TEIGN SX 7689 + OE **tūn**. O 385.

TEALBY Lincs TF 1590. Partly uncertain. *Taveles- Tauele(s)bi* 1086, *Tauelesbi* 1136–1218, *Teuelebi -by* 1219–20, *Teuelby -v-* 1242–1723, *Teilebi* 1210, *Tel(e)by* 1252–1529, *Teelby* 1375, *Tealby* from 1526. All earlier explanations are problematic. The specific can hardly be OE pers.n. *Theabul*, which stands for OE *Thēoful, Tēoful*, or OE **tæfl(e)**, **tefle** 'a chess-board', a feminine noun which would not have *-es* genitive sing. OE masculine ***tæfli**, invented to evade this difficulty, is compared to ODan **tafl** 'a square-shaped piece of land, a piece of raised ground' which is not, however, found in genitival composition in Dan p.ns. such as Tavlgaard, Tavlov etc. Possibly, therefore, OE folk-n. **Tāflas, *Tǣflas* < *Taifali*, an east Gmc tribe recorded in Britain in the early 5th cent., used as a simplex p.n. + later ON **bӯ**, 'the **bӯ** at or called *Taveles*'. Li iii.131–6, DEPN, PNE ii.174, SSNEM 74.

River TEAM Durham NZ 2455. *Tame* before 1128–1652, *Thame* 1223–1427, *Tayme* 1485, *Team River* 1768 Armstrong. *Tomemuthe* 'the mouth of the river Team' occurs c.1107. A British r.n. **Tamā* on the root **tā-/*tə-* 'to flow' found also in the r.ns. TAMAR, THAME, TEME etc. NbDu 194, RN 390, BzN 1957.256, RBrit 465.

TEAN Scilly SV 9016. Short for St Tean's island. (Island of) *Sancta Teona* [c.1160]15th, (island of) *Sancta Theona* [1193]16th, *Tyan* c.1540. Unknown female saint from the dedication of the chapel. PNCo 162 gives pr ['ti:ən].

Upper TEAN Staffs SK 0139. *Upper Tean* 1836 OS for distinction from *Lower Tean* ibid. SK 0139. ModE **upper** + p.n. *Tene* 1086–1355, *Thene* 1204, 1208, *Teyne* 13th, 14th, transferred from r.n. Tean, *Tene* 1389, *Tayne* 1577, [c.1600]1717, *Teine* 1577, 1586, of unknown origin and meaning. St i.20, RN 395, Horovitz.

TEBAY Cumbr NY 6104. 'Tiba's water-meadow or island'. *Ti- Tybei(a) -ai -ey(e) -ay* 1178–1588, *Ti- Tybbei(a)* etc. 1199–15th, *Tebay(e) -ey* 1292–1807. OE pers.n. *Tib(b)a* + **ēġ**, referring either to a small island in the Lune or to water-meadows in the bend of the river at Castle Green, Old Tebay, NY 6105. We ii.50, L 36, 38.

TEBWORTH Beds SP 9926. 'Teobba's enclosure'. *Teobbanwyrþe* [926]13th S 396, *T(h)eb(b)worth(e)* 13th cent. OE pers.n. **Teobba*, a pet-form of a name like *Thēodberht* or *Thēodbeald*, genitive sing. **Teobban*, + **wyrth** later replaced by **worth**. Bd 118.

TEDBURN ST MARY Devon SX 8194. 'St Mary's Tedburn'. *Sct. Marytedborne* 1577. Saint's name *Mary* from the dedication of the church + p.n. *Teteborna* 1086, *Tettaborna* c.1120, *Tette(s)burn(e)* 1234–66, 'Tette or Tetta's stream', OE pers.n. masc. **Tetta* or feminine *Tette*, + **burna**. 'St Mary' for distinction from Venny Tedburn SX 8297, *Fennytetteburn(e)* 1299, 1311, *Venytedborne* 1589, 'marshy Tedburn'. Tedburn is the ancient name of the stream now called the River Culvery, *Culver* 1630, a derivative of OE **culfre** 'a dove, a pigeon' with the same suffix as the river OTTER, earlier *Oteri*. D 451, 406, 4.

TEDDINGTON GLond TQ 1671. 'Estate called after Tuda'. *tudintún* [969]c.1100 S 774, *Tudingtune* [c.1000]13th, *Tu- Todinton* 1198–1336, *Tud(d)ing- Todington* 1274–1618, *Tedinton* 1294, *-yng-* 1428, *Teddington* 1754. OE pers.n. *Tūda* + **ing**^{4} + **tūn**. Another early form, *(æt) Tudincgatunæ* [c.968×71]12th S 1485, may belong here: if so it represents an alternative form of the name, 'estate of the Tudingas, the people called after Tuda', OE folk-n. **Tūdingas* < pers.n. *Tūda* + **ingas**, genitive pl. **Tūdinga* + **tūn**. For the unusual change *u* > *e* cf. Tedfold Sx 150, *Tuddefolde* 1296, Bedham Sx 126, *Budeham* 13th and Wlt xx, Nt xxii. Mx 24, GL 91.

TEDDINGTON Glos SO 9633. 'Farm, village or estate called after Teotta'. *Teottin(c)gtun* [780]11th S 116, [969]11th S 1326, *Teotingtun* [780]c.1000, *Teotintun* [964]12th S 731, 1086, *(to) Teodintune* 969, [977]11th S 1334, *Tedinton(a) -yn-* c.1086–1674, *-ing-* 1246, *Teddington* 1426, 1709. OE pers.n. **Teotta* varying with *Tēoda* + **ing**^{4} + **tūn**. Gl ii.45 suggests that *Teotta* or *Tēoda* is a pet form of the Thēodbald whose name is preserved in The Tibble Stone, a stone formerly standing beside the road from Tewkesbury to London and later erected in the forecourt of Teddington Hands Garage, which gave its name to Tibblestone Hundred, *Teoboldestan hundred* 11th. Gl ii.46, 41.

TEDSTONE DELAMERE H&E SO 6958. 'T held by the Delamare family'. *Thoddesthorne la Mare* 1243, *Teddesthorne Delamare* 1283, *Tedston Delamare* 1373. Earlier simply *Tedesthorne* 1086, 'Teod or Teodi's thorn-tree', OE pers.n. *Tēod* or **Tēodi*, genitive sing. *Tēodes*, + **thorn**. A series of aberrant spellings also probably belong here, *Chetestor* 1086, *Ketestorna* 1160×70, *Chedesthorn* c.1200. The first of these may stand for *Tetestorp* as in TEDSTONE WAFRE SO 6759. He 188.

TEDSTONE WAFRE H&W SO 6759. 'T held by the Wafre family'. *Tedestorna R. Walfr'* 'Robert le Wafre's T' 1160×70, *Tedston Wafre* 1373. Earlier simply *Tetistorp* 1086, *Thoddesthorne* 1243, 'Teod or Teodi's outlying farm' or 'thorn-tree', OE pers.n. *Tēod* or **Tēodi*, genitive sing. *Tēodes*, + OE **thorp** varying with **thorn**. He 189.

TEES BAY Cleve NZ 5429. *Tees Bay* 1861 OS. R.n. TEES + ModE **bay**.

River TEES Durham NY 7733, NZ 2711. Uncertain. *Tesa* 1026–1377, *Tese* [c.1040]12th–1520, *Teisa, Teysa* 1087×93–1382, *Teise, Teys(e)* 12th–1608, *Teisia* 1214, *Teese* 1442–1763, *Tease* 1515–1743, *Tees* from 1381. The various spellings point to an OE **Tēs* or **Tēse* which Ekwall related to W *tês* 'heat, sunshine', < Brit **Tess*, cf.Ir *teas* 'heat', OIr *tess*, < **tep-stu-* on the root *tep-* seen in Latin *tepeo, tepidus* 'warm, tepid', with the same suffix as Gothic **maihstus* 'dung-heap' < PIE **meyĝĝh-s-tu-* 'urination', ON *lǫstr* 'fault, flaw' < Gmc **lahstuz*, ModE *oast(-house)*, OE *āst*, < IE **aidh-* 'burn' + *-s-tu-*, *mist*, OE *mist*, < **mih-s-ta-*, cf. Gk 'ὀμίχλη, also from PIE **meyĝh-*. LHEB finds this 'not very convincing' and the derivation of OIr *tess* and W *tes* from *tepstu-* is not certain. Nevertheless it would certainly be apt if the reference could be to the 'boiling' surging water of so much of the upper river, especially at places like Cauldron Snout. NbDu 194, RN 395, MacBain 363, Kluge §133A, GS iii.128, Krahe i.40, 91, Thurneysen 227, LHEB 343.

TEES-SIDE AIRPORT Cleve NZ 3613. P.n. Teesside + ModE **airport**. Teesside is a modern creation for the short-lived County Borough of *Teesside* created in 1968.

TEETON Northants SP 6970. 'The beacon'. *Teche* 1086, *Tek(e)ne, Tecne* 1220–1386, *Tetene* 1316, *Teton* 1551. OE ***tǣcne**. Teeton is situated at the end of a prominent 500ft. ridge. Nth 88, DEPN, PNE ii.174, SSNEM 382.

TEFFONT EVIAS Wilts ST 9931. 'T belonging to (the barons of) Ewyas (Harold)'. *Hewyas Tefunte* 1242, *Teffunt Ewyas* 1275, *Teffont Evyas* 1304. P.n. *(be) Tefunte* [860]15th S 326, *(at) Teofunten* [964]15th S 730, *Tefonte* 1086, *Tefunte* 1241, 'the boundary spring', OE ***tēo** + ***funta**. Teffont lies on the border of two hundreds, but the name must be older than the formation of the OE hundreds. The boundary between Teffont and Wylye is referred to as *Teofuntinga gemære* 'the boundary of the people of Teffont' in [940]14th S 469. 'Evias' for distinction from TEFFONT MAGNA ST 9832. Wlt 193 gives pr [tefənt i:vaiəs], formerly [ju:əs], Signposts 74, 86.

TEFFONT MAGNA Wilts ST 9832. 'Great Teffont'. *Teffont Magna* 1817 OS. P.n. Teffont as in TEFFONT EVIAS ST 9931 + Latin **magna**. Also known as *Over Tefunte* 1268, *Upperteffont* 1584. Wlt 193.

TEIGH Leic SK 8616. 'The meeting-place, the court'. *Tie* 1086, *Ti* 1202, *Ty* 1254–1513, *Tygh(e)* 1434–1610, *Teigh(e)* 1605, 1629. OE **tīġ, tīh, tīġe**. Most authorities derive this name from OE **tēag** 'a small enclosure', primarily a SE word. The evidence seems rather to point to **tīġ**. R 48 gives pr [ti:].

River TEIGN Devon SX 7689. 'River, stream'. *on teng* [739]11th S 255, *(on, of) Teynge* [739]15th ibid., *Teine* c.1200, *Tyng* 1240, 1575, *Teyng(e)* 1244–15th, *Teigne* c.1540, 1577, *Ting* 1773. The mouth of the Teign is *on tenge muðan* 1044 S 1003. From Brit **tagnī*, **tagnō* cognate with W **taen** 'sprinkling' (< **tagnā*) and Latin *stagnum*. Dr Andrew Breeze has recently shown that the proper sense of the Welsh etymon of this r.n. is 'sweeper, scatterer, flooder', D&CHQ xxxviii, 1998, 101–3. D 14 gives pr [ti:n], RN 398.

TEIGNGRACE Devon SX 8473. 'Teign estate held by the Grace family'. *Teyngegras* 1331, *Graceteyng* 1543, *Tingrace* 1675. Earlier simply *Taigne* 1086, *Teng(ue)* 1277, 1292, and *Teyne Bruere* 1281. The name is transferred from the river TEIGN SX 7689. The manor was held by the family of de la Bruere 1277 and by 1352 had passed to Geoffrey Gras, a kinsman of 'John called Gras' i.e. the fat one, a canon of Torre abbey 1351. D 486 gives pr [tiŋgreis], RN 398.

TEIGNMOUTH Devon SX 9473. 'The mouth of the Teign'. *(on) tenge muðan* 1044 S 1003, *Teignemudan* 1148, *Tinemuth* 1213, *Teignemouth* 1253, *Tengmuth* 1301, *Tingmouth* 1675. R.n. TEIGN SX 7689 + OE **mūtha**. D 503 gives pr [tinməθ], RN 398.

River TEISE or TYSE → TICEHURST TQ 6930.

TELFORD Shrops SJ 6910. A happy adoption of the surname of Thomas Telford (1757–1834) by the Dawley Development Corporation in 1968 for the new town begun in 1963 incorporating Dawley SJ 6807, Oakengates SJ 7010 and Wellington SJ 6511. It commemorates the Scottish civil engineer who was appointed surveyor of Shropshire in 1786 and agent and engineer to the Ellesmere Canal Company in 1793 and went on to build the suspension bridge over the Menai Straits and the Caledonian Canal. Ironically, despite appearances, the surname Telford (recorded in this form since 1685) is not by origin a p.n. but represents OFr *taille fer* 'cut iron, iron-cleaver' used as a nick-name for a man who could cleave clean through the armour of his foe and was borne by the mounted *jongleur* who sang to Duke William of Charlemagne and Roland before the battle of Hastings. Telford's true name was Telfor and it is recorded that, thinking it had no meaning, he changed it to Telford. Room 1983.121, 1988.354. DBS 344, Raven 197.

TELLISFORD Somer ST 8055. Possibly 'ford of the table or plateau'. *Tefleford* [1001]15th S 899, *Tablesford* 1086, *Tevelesford* 1209×35, *Teflesford* 1232, *Telsford* 1610, *Telisford* 1817 OS. OE **tefl**, genitive sing. **tefles**, + **ford**. The ford across the Frome lies in a valley below the church which is situated at the edge of a plateau of higher ground stetching E towards

Norton St Philip. But cf. TEALBY Lincs TF 1590. DEPN, Charters V.116.

TELSCOMBE ESusx TQ 4003. 'Tytel's coomb'. *Titelescumb* [966]12th S 746, *Titteles- Tytelescumbe* 1248, 1268, *Titlescomb* 1509, *Tet(t)elescombe* 1272–91. OE pers.n. *Tytel*, genitive sing. *Tyteles*, + **cumb**. Sx 326 gives pr [tælz'kum], L 89, 92, SS 69.

River TEME Shrops SO 3273. A Celtic r.n. ultimately from the same root as r.n. THAMES.
I. English forms: *Tamede* (Latin oblique case *Tamedam*) [757×75]c.1100 S 142, [934]11th S 406, 1256, *Temede* (oblique case *Temedan*) [757×75]11th–1324, *Temethe* [c.1200]c.1260, *Temed* 1502, *Temd(e)*, *Tende* 16th, *Teame* 1515, *Teme* 1577.
II. Welsh forms:
(a) *Deueityat* late 14th, *Teveidiad*, *Tybhediad* 16th, *Tefeidiad* 1913,
(b) *Teueityawc* c.1380, *Tefedioc*, *Tyfeidioc* 16th,
(c) *Tavidiot* 15th.
This is explained as Celtic **Tam-íjā* whence OBrit **Tamiðā*, British **Tamið*, PrW **Taṽið*, while the dative form **Taṽïðē* became PrOE **Tamidu* and with i-mutation OE *Temede*; the two W forms (a) with suffix *-*ịad* < *-*ịati* and (b) with suffix *-*ịǭc* < *-*ịāko* were probably originally district names rather than applied to the river itself.
The ultimate origin is the large family of *m* extensions of the IE root **tā-*/**tə-* 'to flow' discussed under River THAMES. The name must have been borrowed into English about 600. RN 398, FT 410–55, BzN 1957.257–8, LHEB 487, 611, 629n.1.

TEMPLE CLOUD Avon ST 6257 → Temple CLOUD.

TEMPLE Corn SX 1473. From the ownership of the place by the Knights Templar. *(ad calceam de) Templo* 'at T causeway' [1241]15th, *Temple* from 1284. Latin **templum**, ME **temple**. The estate is mentioned but not named in 1185 as 'one land on *Fawimore*', Foy Moor, the original name of BODMIN MOOR SX 1876. PNCo 162, Gover n.d. 133.

TEMPLECOMBE Somer ST 7022. '(Estate called or at) Combe held by the Knights Templar'. *Cumbe Templer* 1291, *Templecombe* [1295] Buck, 1387, 1617. Earlier simply *Come* 1086, 'the coomb', OE **cumb**. 'Temple' commemorating possession of the manor before 1185 by the Knights Templar and for distinction from Abbas COMBE ST 7022 'the abbot's (estate at) Combe'. DEPN.

TEMPLETON Devon SS 8813. 'The Templars' estate'. *Templeton* from 1335. Earlier simply *Templum* 1206, *Temple* 1238. ME **temple**, Latin **templum**, + **toun**. A late parish formed round the Knights Templar's manor of Coombe SS 8815. D 394, DEPN.

TEMPSFORD Beds TL 1653. 'Ford (on the road) leading to the river Thames'. *Tæmeseford* c.950 ASC(A) under year 921, *Temesanford* c.1200 ASC(E) under year 1010, *Tamiseforde* 1086–1242, *Temes(e)ford* 1227–1428. OE r.n. *Temese*, genitive sing. *Temesan*, + **ford**. The ford lies at the confluence of the Ivel and the Great Ouse. Either one of these rivers had the alternative name *Temese* or the reference is to a road leading S to the Thames 50 miles distant. Bd 110, RN 403–4, L 72.

TENBURY WELLS H&W SO 5968. P.n. Tenbury + ModE **wells** referring to mineral springs discovered here in 1839. The place briefly became a spa. Earlier simply *Temedebyrig* 11th, *Tame(t)deberie* 1086, *Temedbury* 1275, 1327, *Tenbury* 1465, 'fortified place on the river Teme', OE r.n. *Temede* + **byriġ**, dative sing. of **burh**. Wo 83, Pevsner 1968.277.

TENDRING Essex TM 1424. Uncertain. *Tenderinge* 1086, 1294, *Tendringa -e* 1086, *Tendring(g)es* 1135×54–1309, *Tendring* from 1234. Possibly OE *Tyndringas*, Essex dial. form *Tendringas*, 'the people from Tündern' near Hameln, *Tundiriun* 1004, *Tundirin* 1025: or a folk-n. related to OE **tynder** 'tinder' but in what sense is unknown. Ess 351, ING 24, Bahlow 491.

TEN MILE BANK Norf TL 6096. *Ten Mile Bank* 1898 Gaz.

TENTERDEN Kent TQ 8833. 'Swine-pasture of the people of Thanet'. *Tentwardene* 1178×9, *Tentwarden(n')* 1226–51, *Tentyrden'* 1255 etc. OE folk-n. **Tenetware*, genitive pl. *Tenetwara*, + **denn**. The folk-n. also occurs in *Tenet wara brocas* 'the brooks of the people of Thanet' 968 S 1215, in the bounds of land at Heronden TQ 8127, four miles SW of Tenterden. Tenterden belonged to the manor of Minster in Thanet. PNK 355, KPN 296.

TERLING Essex TL 7715. 'The Terhtlingas, the people called after Terhtel'. *Terlinges* [1017×35, 1042×66]12th, [1042×66]13th S 1051, 1185–1255, *Terlinga(s)* 1086, *Tertlinces* [1086]c.1180, *Terdlinge* 1237, *Terling* from 1204, *Tarlinges* 1238, *-(e)* 1509×47, 1601, 1718. OE folk-n. **Terhtlingas* < pers.n. **Terhtel*, Essex dial. form of *Tyrhtel*, + **ingas**. Ess 296 gives prs [ta:liŋ, ta:lən], ING 25.

River TERN Shrops SJ 7037. 'The powerful one'.
I. Welsh forms: *Tren* [12th]c.1275, [c.1400].
II. English forms: *Tirne* c.1200–1360, *Terne* c.1200–1577, *Tyrne* c.1223–c.1540, *Teryn* 1341, 1439, *Tern* 1833 OS.
W **tren** 'strong, fierce'. Raven, however, describes the river as an 'unimpressive waterway which for most of its length is little more than a drainage ditch'. Raven 200, RN 400.

TERNHILL Shrops SJ 6332. 'Hill by the river Tern'. *Terynhyll* 1520, *Tirnhull* 1535, *Tern Hill* 1833 OS. R.n. TERN SJ 7037 + p.n. *Hulle* 1232, 1271×2, *Hull* 1284×5, 'the hill', OE **hyll**. A tree-covered rise just S of the river. The present settlement is a modern one that grew up to service the RAF station established here in 1916 and the subsequent army camp, although there was an earlier settlement. Raven 200, Gelling.

TERRICK Bucks SP 8303 → Kimble WICK SP 8007.

TERRINGTON NYorks SE 6770. Partly uncertain. *Teurin(c)tun(e)*, *Teurinton(e)* 1086, *Tiverington -yng-* 1226–1367, *Teverington* 1275, *Teryngton* 1495–6, 1545. Possibly OE **tēfrung** 'a picture' + **tūn**. The reference might be to mural decoration or possibly to magical practice; **tēfrung** is a derivative of the root seen in OE **tēafor** 'red colouring' and so cognate with OHG *zaubar*, ON *taufr* 'sorcery, magic' which seems to have been the original sense (the meaning 'red' developed from the practice of staining magic runes this colour). YN 34, DEPN.

TERRINGTON MARSH Norf TF 5523. P.n. Terrington as in TERRINGTON ST CLEMENT TF 5520 + ModE **marsh**.

TERRINGTON ST CLEMENT Norf TF 5520. 'St Clement's T'. *Terington St Clement* 1824 OS. P.n. Terrington + saint's name *Clement* from the dedication of the church. Terrington, *Tilinghetuna* 1086, *Terintona* 1121, *Tirintuna* 1103×31, 1133×69, *Tiringet'* 1205, is the 'settlement of the Tiringas, the people called after Tir or Tira', OE folk-n. **Tīringas* < pers.n. **Tīr(a)* + **ingas**, genitive pl. *Tīringa*, + **tūn**. 'St Clement' for distinction from TERRINGTON ST JOHN TF 5314. DEPN.

TERRINGTON ST JOHN Norf TF 5314. 'St John's T'. *Terrington St John* 1824 OS. P.n. Terrington as in TERRINGTON ST CLEMENT TF 5520 + saint's name *John* from the dedication of the church which dates back to the 13th century.

River TEST Hants SU 3637. Unexplained.
I. *(on, of) terstan* [877–985]12th S 1507, 359, 378, 465, 840, 857, *þan alde tersten*, *þar ealde terste* [971×5]14th S 812, *þære éá Tærstan* c.1025, *amnem Tærstan* [before 1085]12th, *Tærstan stream* 1045, *Terste* 1234, 1426, *Trest* 1357.
II. *(andlang) testan* [877]12th S 1277, *Test* 1425 and from 1575.
The OE name was clearly *Terste* of uncertain origin. It might be thought to represent IE **der-* 'run' + suffix *-st-*, but this would normally give Celtic *-ss-*. The name remains a problem. Gover 1958.5, Ha 162, RN 401.

TESTON Kent TQ 7053. 'Stone with a gap or hole'. *de terstane*, *of cærstane* (sic for *tær-*) [c.975]12th B 1322, *Testan* 1086, *terstane* [1087]13th, *Terstana*, *Testane* c.1100, *Terstan(e)* 13th cent. OE **tær**, **ter** + **stān**. PNK 166, KPN 305, PNE ii.175.

TESTWOOD Hants SU 3514. 'Wood by the river Test'. *Terstewode*

1174×99–1327, *Testewode* 13th, *Testwode* 1362. R.n. TEST SU 3637 + ME **wode** (OE *wudu*). Ha 162, Gover 1958.196, RN 401.

TETBURY Glos ST 8993. 'Tette's fortified place'. *(to) Tettan byrg* [c.903]11th, *Tettan byrig* [n.d.]11th B 1320, *Teteberie* 1086, 1165×77, *Tetteberia -bur(y) -bir(e) -biria -buri* c.1170–1499, *Tetbury* from 1438. OE feminine pers.n. *Tette*, genitive sing. *Tettan*, + **byriġ**, dative sing. of **burh**. Also known as 'Tette's monastery', *Tettan, Tectan monasterium* [681]late 14th S 73, 71. Tette may be identical with Tette, sister of King Ine of Wessex, who founded Wimborne Abbey c.700. Gl i.109.

TETCHILL Shrops SJ 3932. 'Tetin's hill'. *Teteshull', Tecneshull* for *Tetnes-* 1255×6, *Tetneshul* 1279×80, *Tetushull'* 1427, *Tetch Hill* 1577. OE pers.n. **Tetīn*, genitive sing. **Tetīnes*, + OE **hyll**. Identical with Tetsill SO 6671, *Tedenesolle* 1086, *Tetneshull(e)* 1211×2, 1287, *Tetis- Tetteshull* 1448. Gelling, Studies 1936.12.

TETCOTT Devon SX 3396. 'Tette or Tetta's cottage(s)'. *Tetecote* 1086, 1217, *Tettecot(e)'* 1238–1346, *Tetticot'* 1242. OE masc. pers.n. **Tetta* or feminine *Tette* + **cot** (pl. **cotu**). D 168.

TETFORD Lincs TF 3374. 'The public ford'. *Tes- Tedforde* 1086, *Tedforda* c.1115, *Thetford* 12th. OE **thēod**+ **ford**. OE **thēod** 'people, tribe, region' was probably used in the sense 'public, national' of features such as fords and roads not under individual control. Cf. THETFORD Norf TL 8783, Little THETFORD Cambs TL 5376, a lost Thetford in Baston TF 1114, *Tetford(e)* 1259–1577, *Thetford* 1281–1821, and the continental parallels Dietfurt on the Altmuhl, Bavaria, *Deituorten* 1144, *Dietfvrtth* 1158, *Dietfurt*, Dietfurt on the Thur, St Gallen, Switzerland, *Dietfurt* 1099, Dietfurth on the Bode NE of Quedlinburg, *Deotfurdi* 974, *Thietforde* 1155 and Bad Salzdetfurth, *salina apud Thetforde* 1195, all with OHG *diot* 'people' + *furt*. DEPN, PNE ii.203, Perrott 395, LbO 100, Berger 77, 233.

TETNEY Lincs TA 3101. 'Tæte's island'. *Tatenaya* [1085]16th, *Tatenai* 1086, c.1115, *-ay* 1194–5, 1250, *Tetenay -eia-ey* c.1170–1514, *Tetney* from 1381. OE feminine pers.n. *Tǣte*, genitive sing. *Tǣtan*, + **ēġ**. Li v.152, L 38.

TETNEY LOCK Lincs TA 3402. *Tetney Lock* 1779. P.n. TETNEY TA 3101 + ModE **lock**. The lock controls the outfall of the Louth Navigation into the Humber estuary. Li v.156.

TETSILL Shrops SO 6671 → TETCHILL SJ 3932.

TETSWORTH Oxon SP 6601. 'Tætel's enclosure'. *Tetteswrd'* [c.1146]c.1200–late 14th with variants *-wrde -wrda* and *-w(o)rth(e), Tetleswrthe* [1148–55]c.1200, *Tettleswrd(e)* [c.1197, c.1200]early 13th. OE pers.n. **Tætel*, genitive sing. **Tæteles*, + **worth**. O 143.

TETTENHALL WMids SJ 8800. 'Teotta's nook'. *(æt) Teotanheale* 11th ASC(C), c.1121 ASC(E) both under year 921, *(æt) Totanheale* c.1050 ASC(D) under same year, *Totehala, Totenhale* 1086, *Tettenhala* 1169. OE pers.n. **Tēotta*, genitive sing. **Tēottan*, + **halh** in the sense 'a slight hollow', dative sing. **hale**. DEPN, L 104–5.

TEVERSAL Notts SK 4861. Uncertain; possibly the 'painter or sorcerer's stronghold'. *Tevreshalt* 1086, *Tevers(h)alt -d* 1275–1393, *-hall* 1548, *-all* 1577, *Ti- Tyvresholt -er(e)s-* 1204–1387, *Tyvers(h)alt -aut* 13th–1428. Possibly OE ***tīefrere** 'painter' used in the sense 'sorcerer' as G *Zauberer*, genitive sing. ***tīefreres**, + **(ġe)heald** 'hold, shelter' later confused with **holt** 'wood'. But this remains very uncertain. Formerly pronounced [tə:rsəl, ti:əsə]. Nt 135.

TEVERSHAM Cambs TL 4958. Partly uncertain. *Teuuresham* [1042×66]12th S 1051, *Teuresham* 1086–1310 including [1042×66]13th ibid., *T(h)euresham* 1086–1392, *Teuersham* 1086–1553, *Taversham* 1130–1588. Uncertain element or name + **hām** 'homestead'. On formal grounds OE **tēafor** 'red, red lead, vermilion, purple' would fit here, but it is difficult to make sense of this or account for the genitival form. Probably we have an unrecorded pers.n. or occupational term, such as **Tēofer* or ***tēafrere** 'a painter'. The cognates of both *Tēofer* and **tēafrere** are associated in other Germanic languages with magic (ON *taufr*, OHG *zoubur*); possibly the 'sorcerer's homestead'. Ca 146, PNE ii.177.

Duns TEW Oxon SP 4528. 'Dunn's meeting-place or close'. *Donestiua* c.1210, *Dunnestywa* c.1233–1294×5 with variants *Dunes- Don(n)es- -tiwa -e* and *-tywe, Dunstew(e)* 1428. OE pers.n. *Dunn*, genitive sing. *Dunnes*, + p.n. *Teowe, Tewa(m), Tvvam* 1086, *Tiwe -a* 1192–1215, OE ***tīwe** 'the meeting place, the court'. Alternatively Tew might be an OE ***tiewe** 'a row, a lengthy object' referring to the ridge on which the Tews stand. O 287, DEPN.

Great TEW Oxon SP 3929. *Magna Tywe* 1195–1428 with variants *Tiwa -e, Tywa, Tewe, Great Tywe* 1336 etc. Latin **magna**, ME **grete** + p.n. *(æt) Tiwan* [1003×4]12th/13th S1488, *Tewam* 1086 as in Duns TEW SP 4528. O 289.

Little TEW Oxon SP 3828. *parua Tywa* 1200–1316 with variant *Tiwa, Little Tiwe* 1268. Latin **parva**, ME **litel** + p.n. *Te(o)we, Teova, Tewa* 1086 as in Duns TEW SP 4528. O 291.

TEWIN Herts TL 2714. 'The Tiwingas, the people called after Tiwa'. *(terram) Tiwingum* [944×6]13th S 1497, *Tywingam* 1015, *Tywhinham* [1015]12th S 1503, *Teuuinge, Theunge* 1086, *Ti- Tywinge -ynge* 1166–1307, *Tewing -yng* 1198, 1303, *Tu(w)yng* 1327, 1368, *Tuyn* 1596. OE folk-n. **Tīwingas* < pers.n. *Tīwa* + **ingas**, dative pl. **Tīwingum*. It is possible that the Tiwingas were the 'worshippers of Tiw' or the 'dwellers on or by the *Tiw*' as in Duns and Great TEW Oxon SP 4528, 3929. Hrt 231.

TEWKESBURY Glos SO 8932. 'Teodec's fortified place or manor house'. *Te(o)dekesberie -b'(r)ie -eches-* 1086, *Theotokesberia* c.1125, *Teodekesbyri* c.1150, *T(h)eokesbir(ia) -bur(y) -beria -bery* 1105–1337, *Teu- Tewkesbury(e) -bery* 1254–1583, *Teu- Tewxbury* 1316, 1475–1700. OE pers.n. **Tēodec* or **Tēoduc*, genitive sing.**Tēodeces*, + **byriġ**, dative sing. of **burh**. The form of c.1125 is from William of Malmesbury, who explains the p.n. as a Greek-English hybrid, *Dei genitricis curia*, 'manor-house of the mother of God', a piece of learned etymology alluding to the virgin Mary's Greek title *Theotokos* 'mother of God'. Gl ii.61 gives prs ['tiuksberi] and ['tʃuksbəRi], ii.xi.

Great TEY Essex TL 8925. *Magna Teye* 1231. ModE **great**, Latin **magna**, + p.n. *(at) Tygan, Tigan* [946×c.951]13th S 1483, *(into) Tigan* 1000×2 S 1486, *Teiam* 1086, *Teia, Teya, Teye* 1135–1556, *Teyen* 1285, *Tayn* 1768, '(at the) enclosures', OE **tīeġum**, dative pl. of **tīeġe**, perhaps referring to several settlements here and at Little Tey, *Parva Teye* 1321, and Marks TEY TL 9123. Great Tey was also known as *Teye al. Clocher* 1254, *They(e) a la steple* 1286, referring to the church tower, OFr **clocher**, ME **stepel**. Ess 400 gives pr [tei].

Marks TEY Essex TL 9123. *Merkysteyn* 1439, *Merkys Teye* 1475. Also simply *Merkys* 1412. Family name *Merk* from the descendants of Adolf de Merck 1267 + p.n. Tey as in Great TEY TL 8925. Also known as *Teye de Mandevill* 1238 from the DB tenant Geoffrey de Mandeville 1086 and also as *Teye atte (N)elmes* 1342, 1364, 'T at the elm-trees'. Ess 400.

TEYNHAM Kent TQ 9562. Probably 'Tena's homestead or estate'. *Tene- Tenaham* [798]13th and 16th S 1258, *Terhā* 1086, *Tenham* 1139×47–1226 including [791]12th S 1613. OE pers.n. **Tēna* + **hām**. The same name or element occurs in nearby Timbold Hill TQ 9156, *Tinfold Hill* 1819 OS, p.n. *Tinfold* + ModE **hill**. *Tinfold* is *(to) teninge faledun* [850]13th S 300, *Tengefald* 1196, *Teningefeld -fold, Tenin(g)fold* 13th, *Thenyngefaud'* 1226, 'the fold(s) of the Teningas, the people called after Tena', OE folk-n. **Tēningas* < OE pers.n. **Tēna* + **ingas**, genitive pl. **Tēninga*, + **falod**, dative pl. **falodum**. In this case, however, the folk-n. **Tēningas* might be the Kentish dial. form for **Tȳningas* 'the people who live in or by a **tȳning** or enclosure'. PNK 278, 227, KPN 83, 191.

THAKEHAM WSusx TQ 1017. 'The thatch homestead'. *Tacaham* 1073, 1330, *Taceham* 1086, *Tach(eh)am -k(e)-* 1167–1595, *Thakeham al. Fakeham* 1641, 1789, *Thackham* 1175–1708. OE

thæċ 'thatch' or **thaca** 'a thatched roof' + **hām**. Sx 180 gives pr [θækəm], ASE 2.33, SS 86.

River THAME Bucks SP 7713, Oxon SP 7613. *(on, andlong) Tame* 956 S 587, *(on) Tama, (andlang) Tamœ* [1004]1313 Frid, *Tame Strem* [1004]15th, *Tame* 1241, 1387, *Thame* from 1220. An Old European r.n. form with an *m*-extension on the IE root **tā-/tə-* 'to flow'. Identical with the r.ns. TEAM, TAME and **Taf(f)**, Wales. RN 390, BzN 1957.256.

THAME Oxon SP 7005. A transfer from the r.n. THAME Sp 7613. *(œt) Tame* c.1000 ASC(B) under year 971, 1086–1675, *Thamu* [675]13th S 1165, *Tham* 1086, 1235, 1278×9, *Thame* from 1230. O 146 gives pr [teim].

River THAMES Kent TQ 5576. An Old European r.n. on the IE base **ta-/*tə-* 'to flow'.

I. Latin and Greek forms (cited in nominative case form only);

(a) *Tamesis* [51 BC]9th/10th Caesar BG, [6th]11th Gildas, [c.800]11th HB, [894]c.1000 Asser, 12th, 1547, *Tamensis* [c.415] Orosius, [6th]11th Gildas, [680]12th S 72, [c.731]8th Bede HE, 12th.

(b) *Tamesa* [115×17]11th Tacitus *Annals*, [c.700]13th Rav, *Ταμήσα (Tamesa)* [c.150]13th Ptol, [c.200] Dio Cassius, *Tamisa* 704–13th including B 56, S 65, S 93, S 241, S 1171, *Tamisia* 10th–13th including S 132, S 537, S 1628, *Thamis(i)a* 1198.

II. English forms:

(a) *Tamis* [790 for ?795]10th S 132, [898 or 9]13th S 1628, *Tamese* [before 1085]12th, *Thamis* [672×4]13th S 1165.

(b) *Tœmese* (oblique cases) [693]11th S 1248, [957]12th S 645, 966 S 738, [966]11th, 1121 ASC(E) under years 1070, 1114.

(c) *Temes* (nominative), *(ðare), (ðœre) Temese* (oblique cases) 891–13th including ASC(A) under years 823 etc., Alfred's *Orosius*, OE Bede and S 353, 354, 896, 927, 1022, 1025, 1271.

III. ME, French and later forms: *Tamise* [c.1140]13th Gaimar, 1247, *Thamyse* 1244, [c.1300]c. 1450, *Tammes* 1386, *Temese* c.1300, *T(h)emse* [1377]c.1460, 1387, 1462, *Thames* from 1577.

An *m*- extension of the root **ta-/*tə-* as also in the r.ns.TEAM, TAME, THAME, TAVY, TAMAR, TEME, TOME. The classical sources record this name in both *i*-stem (Ia) and *ā*-stem (Ib) declensional forms, the latter both masculine and feminine gender. The late British forms seem to have been **Taμēsa* < **Tamēsa*, although the origin of the *s*- suffix is problematic. In the Anglo-Saxon dialects of southern Britain in the 5th century the sound *ē* did not exist and was replaced with OE *ī*, whence the earliest OE form **Tamīsu*, later *Tamis* (IIa), and with *i*-mutation, **Temis*, later *Temes* (II b,c). Forms with *Th*- are AN, and the modern spelling with *-a-* is due to antiquarian and learned influence from knowledge of the Latin forms. RBrit 466, RN 402, BzN 1957.258–9, LHEB 331, FT *passim*.

THAMES HAVEN Essex TQ 7581. R.n. THAMES + ModE **haven**. An industrial district of CORYTON TQ 7482, now an oil storage depot, but originally opened in the 19th cent., cf. *Thames Haven Dock, Railway and Station* 1884 OS. The name is paralleled by Shellhaven, another part of Coryton developed by the Shell Oil Company in 1912 at the estuary of Holehaven Creek. By happy coincidence the site was already called *Shell Haven* 1509×47, 1805 OS. Ess 16, Room 1983.27.

THAMESMEAD GLond TQ 4780. A new town developed from 1967 on Erith marshes. R.n. THAMES + ModE **mead**. Room 1983.121, LPN 225.

Isle of THANET Kent TR 3267. Uncertain.

I. Latin and Greek forms:

Τολίατις νῆσος (*TON*- misread as *ΤΟΛΙ*-) 'the island *Toliatis* (for *Tonatis*)' [c.150]13th Ptol (with variant *Τολίαπις, Toliapis*), *Tanatus* [3rd] Solinus, *Tanatos insula* 'the island T' [c.620]8th Isidore with variants *Tanathos, Tanatus Τολιάτις, Taniatide* (with variant *Tamatide*, oblique case of **Taniatis*) [c.700]13th Rav, *Tanatos insula* [c.731]8th BHE, *Taneti* (genitive), *(in) Taneto* [694]13th S 15, [748]15th S 91, [925]13th S 394, *Thœneti* (genitive) [694]15th S 15, *(in) Thaneto* [694]15th S 15, [696]15th S 17, [826]13th S 1267.

II. English forms: *tenid* 679 S 8, *Tenet* [689]13th S 10 (with variant *Thanet*), [690]13th S 14, [763 or 4]13th S 29 (with variants *Thanet* 13th, *Tanœt* 15th), [826]13th S 1267 (with variant *Thanet*), c.891 ASC(A) under year 853, 943 S 512, [c.890]c.1000 OE Bede, [894]c.1000 Asser, [1027×35]13th S 990 (with variant *Tœnœt*), c.1121 ASC(E) under years 851–2, 865, 1046, *Tenyt* 944 S 497, *tœnett* 949 S 546, *Tanet* [716 or 7 for ? 733]13th S 86 (with 15th cent. variant *Tœnet*), [c.800]11th HB, [925]13th S 394, 1086, *Tenœt* [c.761]15th S 29, *Tanatos* [1042×5]13th S 1048, *Tœnate* 1205, *Thœnet* [?733]15th S 86, *Thanet* from 13th including [689]13th S 10 and [925]13th S 394.

British **Tannēton* either from the root **tenos*, **tenet-* as in Irish *teine* 'fire' (OIr *tene*, genitive sing. *tened*), W *tân*, Co Breton *tan*; the reference may have been to the lighthouse or fire beacon: or Gaulish **tann* 'oak-tree' as in Breton *tann*, OFr *tan* (whence English *tan, tannin* etc.) + suffix *-ētum* and so 'oak-wood'. With normal early Anglo-Saxon substitution of *ī* for *ē* and consequent *i*-mutation this became OE *Tœnid*, *Tenid*, later *Tanet, Tenet*. The modern form with *Th*- seems to be due to a learned but false connection with Greek *thanatos* 'death', first proposed by Isidore of Seville following a hint in Solinus, *Etymologies* 14: 'Tanatos, an island of the ocean separated from Britain by a narrow channel... called *Tanatos* from the death of serpents; for while it has none of its own, soil taken from it to any place whatsoever kills snakes there.' Nennius HB 32, c.800, offers an alternative name for this island which Vortigern handed over to Hengest and Horsa, *insulam quae in lingua eorum vocatur Tenet, britannico sermone Ruoihm* 'the island called in their, sc. the Germans', tongue T, in British R', a tradition repeated by Asser c.894, who gives the name as Ruim. KPN 10, RBrit 468, 70,85, DEPN, FT 579, LHEB 331, BzN 1957.260.

THANINGTON Kent TR 1356. Uncertain. *Tan(n)ingtune* [833×58]13th S 323, 1623, *Tanintune* 1086×7, 1142×8, 1174×82, *-a* c.1090, *Tenitune* c.1100, *Tanintone* 1177, 1185×90, *Thanintune* 1177, *Thanington* 1251. Possibly the 'estate called after Tana', OE pers.n. **Tān(a)* + **ing**[4] + **tūn**. But this does not explain the modern form with *Th*-. Association with p.n. THANET TR 3267 is possible, either through false etymology, or because the original form may have been **Tœningatun* < **Tœnetingatūn*, the 'estate of the people of Thanet', OE folk-n. **Tœningas* for **Tœnetingas*, genitive pl. **Tœn(et)inga*, + **tūn**. PNK 501, Cullen.

THARSTON Norf TM 1894. Partly uncertain. *Thers- Ster(e)stuna* 1086, *Terston'* 1209, *Therestone* 1254, *Therston* 1286. Possibly OE **thyrs** 'giant' used, like ON **thurs**, as a by-name, + **tūn** and so 'Thyrs's estate'. Otherwise an ON pers.n. **Therir*, genitive sing. **Theris*, has been suggested. But there is no certain independent evidence for the existence of such a name. DEPN, SPNN 389.

THATCHAM Berks SU 5167. 'Thatched homestead' or 'river-meadow where thatching material is obtained'. *þœcham* [c.968×71]12th S 1485, *Taceham* 1086, *Tac(c)eham* [1202]1227, c.1225, *Thacheham* c.1225, *Thatcham* from 1224×5. OE **thæċċ** + **hām** or **hamm** 1. Brk 188.

THATTO HEATH Mers SJ 5093. *Thatto Heath* 1842 OS. P.n. Thatto + ModE **heath**. Thatto, *Thetwall* 12th, *Thotewell* 1246, is 'pipe spring', OE **thēote** 'a torrent, a fountain, a water conduit, a pipe' + **wælla, wella**. La 108, L 32.

THAXTED Essex TL 6131. 'Thatch place'. *Thacstede* [c.1004]12th, *Tachestedā* 1086, *Tac- Taxsted(e)* 1199–1548, *Thac(k)- Thaksted(e)* 1263–1517, *Thaxsted(e)* 1291–1548, *Thaxted(e)* from 1314. OE **thæċ** + **stede**. A place where reeds for thatching were obtained. Ess 496, YE lvii, Stead 206.

THEAKSTON NYorks SE 3085. 'Theofoc or Theoduc's estate'. *Eston* (sic) 1086, *Texton(e)* 1158×66–1270, *Thekeston* [1157]15th,

1285, *Thexton* [1184]15th–1409. OE pers.n. **Thēofoc* or **Thēoduc*, genitive sing. **Thēofoces*, **Thēoduces*, + **tūn**. Cf. TEWKESBURY Glos SO 8933, *Teodechesberie* 1086, *Thekesbury* 1233. YN 228.

THEALBY Humbs SE 8917. 'Thiódulfr's village or farm'. *Tedul(f)bi* 1086, *Tedolfbi* c.1115, *Tethelby* 1209×19. ON pers.n. *Thiódulfr* + **bȳ**. DEPN, SSNEM 74.

THEALE Berks SU 6471. 'The planks' or 'at the plank'. *Teile* 1208, *(La) Thele* 1220–[1272]14th, *Theale vulgo Dheal* 1675. OE **thel**, pl. **thelu**, dative sing. **thele**. The reference may have been to a plank bridge or causeway across the river Kennet. Brk 221, PNE ii.203, L 79.

THEALE Somer ST 4645. 'The planks'. *Thela* 1176, *la Thele* 1310. OE **thel**, pl. **thelu**. The reference must have been to a plank path across marshy ground like the prehistoric tracks leading from the Polden Hills to Westhay and Meare (Sweet Track, Walton Heath Track and Abbot's Way). DEPN, Archaeology 31–7.

THEARNE Humbs TA 0736. 'The thorn-tree'. *Thoren* 1297, 1298, 1303, *Thorn(e)* 1309–1566, *Thurne* 1536–1577, *Thearne* 1828. OE **thorn**. YE 201 gives prs [θən] and [θɔn].

THEBERTON Suff TM 4365. 'Theodbeorht's estate'. *Thewardetuna* 1086, *Tiberton* 1178–90, *Tiburton* 1198, *Teberton* 1200, 1254, 1302, *Thebertun* 1216×72, *-ton(')* from 1275. OE pers.n. *Thēodbeorht* + **tūn**. DEPN, Baron.

THEDDEN GRANGE Hants SU 6839. P.n. Thedden + ModE **grange**. Thedden, *Tedena* 1168, *Þuddene* 1203, *Theddene* 1207, *Thutdene* c.1270 is 'valley with a fountain or conduit'. OE **thēote** + **denu**. Possibly a reference to Roman water-works of some kind. Gover 1958.96, Ha 162.

THEDDINGWORTH Leic SP 6685. Possibly the 'enclosure of the Theodingas, the people called after Theoda'. *Tevlingorde, Tedi- Dedig- Tedingesworde* 1086, *Tedingewrth'* 1206, *Thedingew(o)rth' -yng-* 1207, 1379, *Thedingw(o)rth(e) -yng-* c.1130–1556, *Theddingworth -yng-* from 1316, *Taingwurda -uurda -wrda -wrde -wrth* 1140–late 12th, *Teinge- Thainge- Theingew(o)rth(a) -(e)* c.1200–1. OE **Thēodingas* < pers.n. *Thēoda* + **ingas**, genitive pl. **Thēodinga*, + **worth**. The specific may however be a singular OE **Thēoding*, the 'place called after Theoda', genitive sing. **Thēodinges*, 'the enclosure of *Theoding*', locative–dative sing. **Thēodinge* 'the enclosure at *Theoding*'. Lei 259.

THEDDLETHORPE ALL SAINTS Lincs TF 4688. *Thedelthorp' Omnium Sanctorum* 1254. P.n. *Te(d)lagestorp* 1086, *Tedolf- Dedloncstorp* c.1115, *Thedlac- Tedlaue- Torlaue- Teldes- Totlaue(s)torp* 12th, *Theloueftorp* early 13th, 'Theodlac's outlying settlement', OE pers.n. **Thēodlāc*, genitive sing. **Thēodlāces*, + **thorp**, + affix *All Saints* from the dedication of the church and for distinction from THEDDLETHORPE ST HELEN TF 4788. Also known as *West Theddlethorpe* 1824 OS. The Theddlethorpes lie in an area not fully settled until the construction of protective sea-banks in the early 11th cent. The element *Thēod-* is rare in early OE pers.ns, more frequent from 10th cent. owing to the influence of LG names in *Thiad-*. Cognate OHG *Theotleih* can be ruled out since it does not occur in the W Frankish and LG areas with which the Anglo-Saxons had direct links but belongs rather to central and southern G. Some of the early sps show confusion with Frankish *Theodulf* or ON *Thjóðúlfr*. Other nearby p.ns. containing CGmc pers.ns. include GRAINTHORPE TF 3897, MABLETHORPE TF 5084 and TRUSTHORPE TF 5183. SSNEM 118, Cameron 1998.

THEDDLETHORPE ST HELEN Lincs TF 4788. *Thedelthorp' Sancte Elene* 1254. P.n. Theddlethorpe as in THEDDLETHORPE ALL SAINTS TF 4688 + saint's name *Helen* from the dedication of the church. Also known as *East Theddlethorpe* 1824 OS. Cameron 1998.

THELBRIDGE BARTON Devon SS 7812. P.n. Thelbridge + ModE **barton** 'manor farm'. Thelbridge, *Talebrige* 1086, *Thelebrig(ge) -brug(ge)* 1242–1328, *Thelbrigge* 1291, *Thelbridge* 1809 OS, is 'plank bridge', OE **thel** + **brycg**. The bridge must have crossed the stream at Woodhouse Villa N of the village. A similar example of the late addition of *barton* is the neigbouring Pedley Barton SS 7712, *Piedelega* 1086, *Piddelegh'* 1242, *Peddelegh* 1285, *Pedlegh* 1391, *Pidley* 1809 OS, 'Pidda's clearing'. D 395, 401, PNE i.31.

THELNETHAM Suff TM 0178. Uncertain. *Theluete- Teluette- teolftham* 1086, *Thelfetham -uel-* c.1095, *Teluedham* 1196, *Elnetham* 1202, *Thelnetham* from 1254. This name must be taken with Great and Little Welnetham Suff TL 8859 and 8960, *telueteha', hvelfihā, Hvelfitham* 1086, *Weluetham* 1170, *Welfuetham* 1179, *Welueteham* 1198, *Welnetham* 1206–7, 1254, *Whelnetham* 1242, *Weltham magna* 1610. The current forms with *-net-* are clearly due to misreading of *-uet-* for *-fet-*. The name seems to be element + p.n. **Elfethamm* 'the swan river-meadow', OE **elfitu** + **hamm** 3. In the case of Thelnetham the suggested specific is OE **thel** 'a plank' and so 'the plank-bridge at *Elfethamm*' in the case of Welnetham OE **hwēol** and so 'the (water-)wheel at *Elfethamm*'. DEPN.

THELWALL Ches SJ 6487. Probably 'pool with or by a plank bridge'. *(to) þel wæle* 923 ASC(A), *Thealwæle* [920]c.1118, *Thelewella -well(e) -wall(e)* 1190–1499, *Thelwall* from 1289. OE **thel** + **wēl** 'a deep pool in a river' later confused with **wella**, **wælla** 'a well, a spring'. ASC refers to a *burh* established here in 922 presumably to guard the crossing of the Mersey at Latchford SJ 6187. The deep pool was probably a place in the Mersey. Che ii.138. Addenda gives 19th cent. local pr ['θelwəl], now ['θelwɔ:l].

THEMELTHORPE Norf TG 0523. Partly uncertain. *Timeltorp* 1203, *-thorp* 1219, *Thymelthorpe* 1248. Usually interpreted as 'Thymel or Thymli's outlying farm', OE pers.n. **Thȳmel* or ON pers.n. **Thymli* + **thorp**. But the absence of any genitival ending may suggest OE **thȳmel** 'a thimble' as the specific in this name, either with reference to the making of thimbles or to some thimble-like topographic feature or perhaps humorously to the size of the settlement. DEPN, SPNN 432.

THENFORD Northants SP 6970. 'Ford of the thegn or thegns'. *Teworde, Taneford* 1086, *Teyn- Tein(e)ford* 1175–1285, *Thenford* from 1242, *Fenford* 1567. OE **theġn**, genitive pl. **theġna**, + **ford**. The reason for the name is unknown. Nth 61, L 70.

THERFIELD Herts TL 3337. 'Dry open land'. *Đerefeld* [c.1062]14th S 1030, *Derevelde, Furreuuelde* 1086, *Ferefeld* 1114×30, 1235, *Therefeld -feud* 1161–1428, *Therfeld* 1259–1402, *Tharfelde* 1482, *Tarfel* 1700†. OE **thyrre** + **feld**. Hrt 166 gives pr [θa:rfəl], ECTV no. 145.

River THET Norf TL 9584. *Thet* from 1586. A back-formation first evidenced in Camden from the p.n. THETFORD TL 8783. RN 405.

Little THETFORD Cambs TL 5376. *Liteltedford* 1086, *-theotford -teodford -teoforð* 1086, *-tædford -tiedford* 1144. OE adj. **lȳtel** + p.n. *(æt) þiutforda* c.972 *ASC, Thedford'* c.1155, *Thetford(e)* from 1251, *Tefford* 1198–1285, 'people's ford', OE **thēod** + **ford**. The ford across the Ouse seems to have taken a track from Ely to Fordey Farm and Wicken skirting the edge of Soham Mere. 'Little' for distinction from THETFORD Norf TL 8783. Ca 242, 198–9.

THETFORD Norf TL 8787. 'The people's ford'. *Þeodford* 891 ASC(A) under year 870, *Þeot ford* c.1100 ASC(D) under year 952, *Tedforda -fort, Tetford* 1086. OE **thēod**+ **ford**. From the use of **thēod** in other compounds in OE the sense seems to be either 'public ford' (as opposed to all the fords with pers.ns. as their specific) or 'chief ford' (cf. OE *thēodloga, thēodfēond* 'arch-liar, arch-fiend'). Thetford was a major route centre and ford of the Little Ouse on the Roman road from London to

† *þyrefeld* *[769]13th S 150 belongs to TURVILLE Bucks SU 7691.

Caistor by Norwich (Icknield Way), Margary nos. 333, 37. DEPN.

THETFORD WARREN Norf TL 8383. *Thetford Rabbit Warren* 1836 OS. P.n. THETFORD TL 8783 + ME **wareine**. Thetford Warren Lodge in the middle of the warren is a 15th cent. building. Pevsner 1962NW.347.

THEYDON BOIS Essex TQ 4598. 'T held by the Boys family'. P.n. Theydon + manorial affix *de Bosco* 1235, *Boys* 1257, from the family of William de Bosco 1166. Theydon is *Teidanam, Taindenā* 1086, *Tei- Teydene* 1264–74, *They- Theiden(e)* 1236–1428, *Thayden* 1238, *Thoydon* 1311–1535. The earliest reference is to the bounds of Theydon, *þecdene gemære* *[1062]13th S 1036. This is a post-conquest compilation but the p.n. forms are authentic: *þecdene* is the 'valley where thatch is obtained', OE **þæć** + **denu**. The subsequent development *Thec-* > *Thei-* is unusual but paralleled by Braydon in Braydon Hook Wilts SU 2167, *brœcdene* [968]13th S 756, *Brayden* 1257, the 'valley with land broken up for cultivation', OE **bræc** + **denu**. 'Bois' for distinction from Theydon Garnon TQ 4799, *Teyden Gernuns* 1287, 'T held by the family of Ralph Gernon' (1202, from NFr *grenon* 'moustache') and Theydon Mount TQ 4999, p.n. Theydon + *ad Montem* 'by the hill' 1256, *(del) Munt* 1273, 1280, Latin **mons**, OFr **munt** referring to the hill on which the church stands. Ess 82 gives pr [boiz].

THICKWOOD Wilts ST 8272. 'The thick wood'. *Ticoode* 1086, *Thickwude* 13th. OE **thicce** + **wudu**. Wlt 94.

THIMBLEBY Possibly 'Thymill or Thymli's farm or village', ON pers.n. **Thymill* or **Thymli*, related to ON **thumall** 'thumb', genitive sing. **Thymla*, + **bȳ**. An alternative suggestion is OE **thymel** 'thimble' or by-n. **Thymel*.

(1) ~ Lincs TF 2369. *Stim- Stinblebi* 1086, *Timlebi* c.1115, *-el-* 1162, late 12th, *Thimelbi* 1196×8, *Thymelby* 1219. An alternative suggestion is OE **thȳmel** 'thimble, thumb-stall' referring to the long thin ridge on which the village stands. DEPN, SSNEM 74.

(2) ~ NYorks SE 4595. *Timbelbi -belli* 1086, *thémelebi* 1088, *Thimilby, T(h)imelebi* 12th cent., *Thi- Thymelby* 1233–1316, *Thimbelby* 1359. YN 214, SSNY 38.

THIRKLEBY NYorks SE 4778. 'Thorketill's farm or village'. *Turchilebi* 1086, *Thurkillebi -by* 13th cent., *Thirtleby* 1202. ON pers.n. *Thorketill* + **bȳ**. YN 189, SSNY 39.

THIRLBY NYorks SE 4884. 'The serfs' farm or village'. *Trillebi(a), Trylleby* [1189]16th–1285, *Thrilleby* [1273]16th, 1301, *Thirleby* 1271–1579. ON **thrǽll**, genitive pl. **thrǽlla**, + **bȳ**. Cf. TURSDALE Durham NZ 2035. YN 199 gives pr [θɔrlbi], SSNY 7.

THIRLEY COTES NYorks SE 9795. P.n. Thirley + ModE **cot** 'a cottage', pl. **cotes** (OE *cot*). Thirley is *Tornelai -lay, Torneslag* 1086, 1204, *Thornelay(e)* 1109×14–1314, *Thornelac* [1199]15th, *Thirley* 1619, the 'thorn-tree clearing', OE **thorn** + **lēah**. YN 114.

THIRLMERE Cumbr NY 3116. 'Lake in the hole or perforation'. *Thyrlmere* 1573, *Thirlmere Lake* 1787. OE **thyrel** + **mere**. Thirlmere is a long narrow lake piercing the mountains of Cumbria. Cu 35.

THIRLWALL CASTLE Northum NY 6666. *Thirlwall Castle* 1866 OS. P.n. Thirlwall + ModE **castle**, a ruined tower of uncertain date, possibly 14th cent. Thirlwall, *Thurlewall'* 1256, *Thirlewalle* 1279, *Murus Perforatus* 14th Fordun, *Thrilwall* 1479, is 'the pierced wall', OE **thyrel** + **wall**, referring to a gap in Hadrian's Wall where the Tipalt Burn breaks through the high ground on which the wall stands. NbDu 194, Pevsner-Grundy etc. 1992.582.

THIRLWALL COMMON Northum NY 6769. *Thirlwall Common* 1866 OS. P.n. Thirlwall as in THIRLWALL CASTLE NY 6666 + ModE **common**.

THIRN NYorks SE 2186. 'The thorn-bush'. *Thirn(e)* 1086–1551. OE **thyrne**. YN 235 gives pr [θɔrn].

THIRSK NYorks SE 4382. 'The lake or fen'. *Tresc* 1086–c.1160, *Tresch(e)* 1086–c.1285, *T(h)resk(e)* c.1200–1492, *Thirsk, Thyrsk* from 1403, *Thrusk(e)* 1580, 1733. ON ***thresk** corresponding to OSwed *thræsk*. YN 188 gives prs [θɔsk, θrusk], SSNY 105.

East THIRSTON Northum NZ 1999. *East Thirston* 1868 OS. ModE adj. **east** + p.n. *Thras- Th(r)afriston* 1242, *T(h)ra(s)terston -re-* 1257–1346, *Tresstereston* 1296 SR, *Trautreston* 1298, *Thrist- Thresterton* 15th cent., *Thruston* 1580, *Thriston* 1628, 'Thrasfrith's farm or village', OE pers.n. **Thræsfrith*, genitive sing. **Thræsfrithes*, + **tūn**. 'East' for distinction from West Thirston NZ 1899. NbDu 194, Studies 1931.91.

THISTLETON Leic SK 9117. 'Thistle settlement'. *Tister- Tisteltune* 1086, *Thy- Thisteltun -ton(e)* 1212–1610, *Thistleton* 1610. OE **thistel** + **tūn**. It has been suggested that the reference may have been to thistles growing on the large deserted Romano-British site ½ m. SW of the church. But some varieties of thistle (Holy Thistle, Milk Thistle, Carline Thistle) were valued and even cultivated for medicinal use and this may be the meaning here. R 52, Grieve 795, 797, 800.

THISTLEY GREEN Norf TL 6676. No early forms. ModE **thistley** + **green**.

THIXENDALE NYorks SE 8461. 'Sigsteinn's valley'. *Sixte(n)- Xistendale* 1086, *Si- Syxtendale -a* [a.1139]16th–1413, *Sixteendale* 1617, *Thyssyndalle* 1548, *Thixindale -en-* from 1566. ON pers.n. *Sigsteinn* + **dalr**. The specific was early associated with OE *sixtīene* 'sixteen'. YE 133, lx, SSNY 105.

THOCKRINGTON Northum NY 9579. 'The estate called after Thocer'. *Thokerinton* 1223, 1242 BF, 1256, *Thokerington* 1256–1296 SR, *Tokerington* 1269 Ass. Pers.n. **Thocer* related to OE ***thocor** 'unsteady', *thocerian* 'move to and fro' and **thoccere* 'a tramp, a vagabond' + **ing**[4] + **tūn**. Cf. Doggaport in Langtree, Devon SS 4515, *Thokirport* 1330, *-er-* 1333, 1346, 'the tramp's town'. NbDu 195, DEPN, D 95.

THOLOMAS DROVE Cambs TF 4006. 'Tholymer's cattle road'. *Tolymeresdrowe* 1275, *-draue* 1370, *Tholymessedrove* 1438, *Tholomas Drove* 1824 OS. ME surname *Tholymer* as in Walter Tholymer 1248 + **drove**. Ca 294.

THOLTHORPE NYorks SE 4766. 'Outlying farm belonging to *Thurulfestun*'. *Turulfes- Turoluestorp* 1086, *Turold' Torp* 1176, *Thoraldethorp* 1285–1328, *Thoralthorp'* 1295, *Tholthorp* 1505, *-thropp* 1614. P.n. *Thurulfestun* abbreviated to *Thurulfes* + p.n. *Thorp* [972×92]11th S 1453, ON **thorp**. *Thurulfestun* 'Thorulf's estate', ON pers.n. *Thórulfr*, genitive sing. *Thórulfs*, + **tūn**, is a lost p.n. mentioned in S 1453 in association with *Thorp* which seems to have been originally a secondary settlement dependent on it. It seems likely that the present Tholthorpe occupies the site of the lost *Thurulfestun*. The history of the name shows later substitution of ON pers.n. *Thóraldr* for *Thórulfr*. YN 21, SSNY 68, 130.

THOMPSON Norf TL 9296. 'Tumi's estate'. *Tomestuna* (4×) 1086[†], *Tomestun'* 1191, *Tomeston'* 1201, *Tumestone* 1242. ODan pers.n. *Tūmi, Tummi*, secondary genitive sing. *Tūmes, Tummes*, + **tūn**. DEPN, SPNN 385.

THONG Kent TQ 6770. Uncertain. *Thange* 1147×82, *Thuange, Twonge* 1185×1214, *Thonge* 1226×35, *Twonge* 1327. OE **thwang** 'thong' used in some transferred sense such as 'narrow strip of land, narrow place, passage, trap' or 'tethering place'. The exact sense here is uncertain; the topography is not decisive. KPN 291, PNE ii.221.

THORALBY NYorks SE 0086. 'Thorald's farm or village'. *Turo(l)desbi* 1086. ON pers.n. *Thóraldr*, secondary genitive sing. *Thóraldes*, + **bȳ**. YN 268, SSNY 39.

THORALDBY Clevel NZ 4907. 'Thorald's farm or village'. *Tur- Toroldesbi* 1086, *Thoroldeby* 1219, *Thorald(e)by* [c.1280×90]15th

[†]An alternative form *Tumersteda* is recorded once in 1086.

etc. ON pers.n. *Thóraldr*, genitive sing. *Thóralds*, + **bȳ**. YN 175, SSNY 39.

THORESBY NYorks SE 0290. 'Thorir's farm or village'. *Toresbi* 1086, *Thoresby* from [1184]15th. ON pers.n. *Thórir*, genitive sing. *Thóris*, + **bȳ**. YN 266, SSNY 39.

THORESBY HALL Notts SK 6371. P.n. Thoresby + ModE **hall**. A great country house built in 1864–75 to replace the earlier *Thoresby House* 1840 OS. Thoresby, *Yuresby* (sic for *þur-*) [958]14th S 679, *Turesbi* 1086, *Thuresby* 1234–1316, *Thoresby* from 1287, is 'Thurir's village or farm', EScand pers.n. *Thūrir*, genitive sing. *Thūris*, + **bȳ**. Nt 92, SSNEM 74.

North THORESBY Lincs TF 2998. *North Thoresby* from 1292. ME **north** + p.n. *Toresbi* 1086–1219, *Thorisby* 1242–92, *Thoresbi -by* from 1202, 'Thorir's village or farm', ON pers.n. *Thórir*, genitive sing. *Thóris*, + **bȳ**. 'North' for distinction from South THORESBY TF 4076. Li iv.165, SSNEM 74 no.3.

South THORESBY Lincs TF 4076. *South Thoresby* 1426. ME adj. **suth** + p.n. *Toresbi* 1086, *Thoresbi* after 1217, as in North THORESBY TF 2998. DEPN, SSNEM no.4, Cameron 1998.

THORESWAY Lincs TF 1696. Partly uncertain. *Toreswe* 1086, *Tore(s)weia* c.1115, 1212, *Thoreswaia* 1154×89, *-wey(e)* 1242–1549, *-way* from 1275. The usual explanation, 'Thorir's road', ODan pers.n. *Thōrir, Thori*, genitive sing. *Thōris*, + OE **weġ**, has recently been questioned and the DB spelling taken to represent ODan **Thorswǣ*, 'the (pagan) shrine of Thor' with ODan **wǣ** (ON *vē*) early replaced by ME **wey**. If this is correct the name is a direct parallel to the Swed p.n. Torsvi and this must be one of the earliest Scand coinages in England. Li iii.150, SSNEM 222.

THORGANBY Lincs TF 2197. 'Thorgrim's village or farm'. *Turgrī- Torgre(m)- Turgrebi* 1086, *Torgrim(e)bi* c.1115–1200, *Thorgrimbi* 1154×89, *Torgram(e)bi -by* 1202–43, *Thorgramby* 1226–1431, *Thorganby* from 1296. ON pers.n. *Thorgrímr* + **bȳ**. Li iii.157, SSNEM 74.

THORGANBY NYorks SE 6942. 'Thorgisl's farm or village'. *Turgisbi* 1086, *Turgrime(s)bi* 12th, *Thorgremby* c.1200, *Torgramebi, Turgraneby* 13th, *Thur- Thorgramby* 1271–1406, *Thorganby* from 1420. ON pers.n. *Thórgisl -gils* later replaced by *Thorgrímr*, + **bȳ**. YE 263, SSNY 39.

THORGILL NYorks SE 7096. *High, Low Thorgill* 1861 OS. Uncertain element or name + ModE **gill** (ON *gil*) 'a ravine'.

THORINGTON Suff TM 4274. 'The thorn-tree settlement'. *Tornin- Turnin- Torin- Torentuna* 1086, *Turritune* 1199×1216, *Thurintone* 1254, *Thurringeton* 1270, *Thurington* 1323, *Toriton'* 1275, *Thorynton(e)* 1303, 1327, *Thoryngton* 1357, *-ing-* 1610. OE **thorn** varying with **thyrne** + **tūn**. DEPN, Baron.

THORINGTON STREET Suff TM 0135. 'Thorington hamlet'. *Thrington Street* 1805 OS. P.n. Thorington, no early forms but cf. THORINGTON TM 4274, + ModE **street**.

THORLBY NYorks SD 9652. 'Thorald's farm or village'. *Toredere- Toreilderebi* 1086, *Thordelbi, Thoreldby* 13th, *Thorleby -bie* 1303–1779, *Thorlby(e)* from 1405. ON pers.n. *Thóraldr*, genitive sing. *Thóraldar*, + **bȳ**. The genitive sing. in *-ar* is unexpected and led Stenton to suggest the ON feminine pers.n. *Thórhildr*, genitive sing. *Thórhildar*. But DB spellings are not invariably reliable. YW vi.76, SSNY 39, TRHS 24.23, 28.12.

THORLEY Herts TL 4719. 'Thorn-tree wood or clearing'. *Torlei* 1086, *Thorlegh -leie -leye -le(e)* from 1230, *Thorneley* 1212, *Thornle* 1291. OE **thorn** + **lēah**. Hrt 204.

THORMANBY NYorks SE 4974. 'Thormoth's farm or village'. *Tur- Tormozbi* 1086, *T^r modesbi* 1167, *Thormodeby* 1193×1208, 1230, *Thormotheby* 1295, *Thormanby* from 1481. ON pers.n. *Thormóthr*, genitive sing. *Thormóths*, + **bȳ**. YN 26, SSNY 40.

THORNABY-ON-TEES Cleve NZ 4516. 'Thormoth's village or farm'. *Turmoz- Thormozbi, Tormozbi(a)* 1086, *Thormodebi -by* 1202–[1333]16th, *-mot(h)ebi -by* 13th cent., *-motby* 1301, 1369, *Thornaby* 1665. ON pers.n. *Thormóthr* + **bȳ**. YN 172, SSNY 40.

THORNAGE Norf TG 0536. 'Thorn-tree enlosure or pasture'. *Tornedis* 1086, 1166, *Thornedisch* 1254, *Thornege* 1291. OE **thorn** + **edisc**. DEPN.

THORNBOROUGH 'Thorn-tree hill'. OE **thorn** + **beorg**.

(1) ~ Bucks SP 7433. *Torneberge* 1086–1227, *Thorneberge* 1227, 1241, *T(h)orneburuwe* 1235. Bk 56, L 128, 221.

(2) ~ NYorks SE 2979. *Thorn(e)bergh* 1198–1399, *Thorn(e)bargh* 16th, *Thornbrough* 1654. YN 223.

THORNBURY 'Thorn-tree fortification or manor'. OE **thorn**, genitive pl. **thorna**, + **byriġ**, dative sing. of **burh**.

(1) ~ Avon ST 6490. *Tvrneberie* 1086, *Thorn(e)bir(i)* etc. from 1214†. The reference is unknown. Gl iii.14.

(2) ~ Devon SS 4008. *Torneberie* 1086, *-bury* 1193, *Thornebiria* 1224. D 169.

(3) ~ H&W SO 6259. *Torneberie* 1086, *Thorn(e)bir' -bury* 1216–1382. The reference may be to the Iron Age hill-fort called Wall Hills SO 6938 which is surrounded by a ring of thorn-trees. He 189, Thomas 1976.151.

THORNBY Northants SP 6775. 'Thorn-bush farm or village'. *Torneberie* 1086, *Thirnebe* c.1160, *-by* 1236–1379 with variant *Thyrne-, Thurnebi -by* 1189×99–1361, *Thornby* from 1418. ON **thyrnir** + **bȳ** probably replacing OE **thorn** + **byriġ**, dative sing. of **burh**, the 'fortification with a thorn-hedge'. Subsequently the name was refashioned under the influence of the common word *thorn*. Nth 74, SSNEM 14.

THORNCLIFFE Staffs SK 0158. *Thorn(e)cley -clay(e)* 1230×32, *Thorenteleye* 1279, *Thorntileg* 13th, *Thorn(e)teley* 1476, *Thorncliffe* c.1600. The earliest forms point to 'thorn-tree clay land', OE **thorn** + **clæġ**. But **clæġ** does not otherwise appear as a specific and it is more likely that they are bad sps for OE ***thornett** 'a thorn-tree copse' + **lēah** 'a wood or clearing', for which *cliff* is frequently substituted in weak stress. Horovitz.

THORNCOMBE Dorset ST 3703. 'Thorn-tree valley'. *Tornecoma* 1086, *-cumba* 12th, *Thorncumbe* 1236, *Thornecombe* 1399. OE **thorn** + **cumb**. Do 143, L 93, 221.

THORNCOMBE STREET Surrey TQ 0042. 'T hamlet'. *Thorncomestrete* 1518×29, *Marshall alias Thornecombe Strete* 1596. P.n. *Torncūba* 1206, *Thorn(e)cumbe* 13th cent., *Thorncombe* 1322, 'the thorn-tree valley', OE **thorn** + **cumb**, + ME **strete**. Robert *Marshall* held the manor in 1502. Possibly contrasts with Gate Street Farm TQ 0141, *Gate Street* 1816, cf. Juliana *Attegate* 1263. Sr 228.

THORNDON Suff TM 1369. 'The thorn-tree hill'. *Torn(e)duna* 1086, *Thorndune* c.1095, *Thorndon* from 1254. OE **thorn** + **dūn**. DEPN, Baron.

THORNDON CROSS Devon SX 5393. 'Thorndon crossroads'. P.n. Thorndon SX 5192 + ModE **cross**. Thorndon, *Thorndon* 1244, is 'thorn-tree hill', OE **thorn** + **dūn**. D 208.

THORNE SYorks SE 6913. 'The thorn-tree'. *Torne* 1086–1300, *Thorn(e)* from c.1147. OE **thorn**. YW i.2, L 220.

THORNE COFFIN Somer ST 5127 → THORNE ST MARGARET.

THORNE MOORS OR WASTE SYorks SE 7315. P.n. THORNE SE 6913 + OE **mōr**, ME **waste**. *mora versus Thorn* 1190×1202, *Thorne Waste* 1771. YW i.5.

THORNER WYorks SE 3840. 'Thorn-tree ridge'. *Torneure* 1086, *-oure -oura* 1086–1160×75, *Thornoure* 1174–1303, *-ouer* c.1200–1428, *Thornor(e)* 1251–1594, *Thorner* from 1441. OE **thorn** + **ofer**. YW iv.103, L 176.

THORNE ST MARGARET Somer ST 0921. 'St Margaret's Thorne'. *Thorn St Margaret* 1251, *Margretsthurne* 1610. P.n. Thorne + saint's name *Margaret* from the dedication of the church and for distinction from Thorne Coffin ST 5127 and THORNFALCON ST 2823. Earlier simply *Torne* 1086, 'the thorn-tree', OE **thorn**. DEPN.

†The form *to þornbyrig* [896]11th S 1441 does not belong here.

THORNE WASTE OR MOORS SYorks SE 7315 → THORNE MOORS OR WASTE SE 7315.

THORNESS BAY IoW SZ 4594. *Thornbeye* 1346, *Thornesbay* 1395, *Thorney bay* 1583, *Thorness Bay* 1781. P.n. Thorness + ME **baye**. Thorness represents the genitive sing. of p.n. Thornhay, *Torneyam* 1198×1216, *Thorneye* 1285–1398, *Thorn(h)ay* 1299, *Thorn(h)ey(e)* 1311–1803, the 'thorn-tree hedge', OE **thorn** + **heġe** or **hæġ**. Wt 191, Mills 1996.102.

THORNEY Notts SK 8573. 'The thorn-tree enclosure'. *Torneshaie* 1086, *Turnaie* 1200, *Thorney* from 1512, *Thornhay* 1572, *Thorn(e)hawe -hagh(e)* c.1230–1450. OE **thorn** + **hæġ**, **thorn** + **haga**. Nt 207. L 221.

THORNEY 'Thorn-tree island', OE **thorn** + **ēġ**.

(1) ~ HILL Hants SZ 2099. P.n. Thorney + ModE **hill**. Thorney, *Thorney* 1280, is 'thorn-tree island', OE **thorn** + **ēġ**. Gover 1958.223.

(2) ~ ISLAND WSusx SU 7503. P.n. *þorneg* c.1100 ASC(D) under year 1052, *Tornei* 1086, + pleonastic ModE **island**. Sx 62.

(3) West ~ WSusx SU 7602. *Westthorneye* 1291. ME adj. + p.n. Thorney as in THORNEY ISLAND SU 7503. 'West' for distinction from Thorney in E and W Wittering, *Torneia* [945]14th S 506, *Thorny* 1340. Sx 62, 89.

THORNFALCON Somer ST 2823. 'Thorn held by the Fagun family'. *Thorn fagun* 1265, 1268. P.n. Thorn + family name *Fagun* for distinction from Thorne Coffin ST 5127 and THORNE ST MARGARET ST 0921. Earlier simply *Torne* 1086, *Thorne* 1610, 'the thorn-tree', OE **thorn**. DEPN.

THORNFORD Dorset ST 6013. 'Thorn-tree ford'. *ðorn- Torneford* [903 for 946×51]12th S 516, *ðorford* [998]12th S 895, *Torneford* 1086–1212, *Thorneford* 1163–c.1500, *Thorn-* from 1249. OE **thorn** + **ford**. Do iii.376 gives pr ['ða:RnvəRd].

THORNGUMBALD Humbs TA 2026. 'Thorn held by the Gumbaud family'. *Thorn(e)gumbaud* 1305, *-gumbald* 1374, *Thorngumbald als. Gumberthorne* 1579. P.n. *Torn(e)* 1086–1260, *Thorn(e)* 1175×95–1490, 'the thorn-tree', OE **thorn** used as a p.n. + family name Gumbaud. YE 38 gives pr [gumbəθɔn].

THORNHAM Norf TF 7343. 'Thorn-tree homestead'. *Tornham* 1086, c.1140, 1197. OE **thorn** + **hām**. DEPN.

THORNHAM MAGNA Suff TM 1071. 'Great T'. *Marthorham* (sic) 'greater T' 1086, *Magna Thornham* 1235. OE adj. **māra**, Latin **magna** + p.n. *T(h)ornham* 1086, 'the thorn-tree homestead', OE **thorn** + **hām**. The reference may have been to a *hām* fenced with thorns. DEPN, Baron.

THORNHAM PARVA Suff TM 1172. 'Little T'. *(in) paruo Thornham* 1086. Latin adj. **parvus**, dative case **parvo**, + p.n. Thornham as in THORNHAM MAGNA TM 1071. DEPN.

THORNHAUGH Cambs TF 0600. 'Thorn enclosure'. *Thornhawe* 12th–1346, *Thornhagh* 1275, *Thorney* 1526. OE **thorn** + **haga**. Nth 243.

THORNHILL 'Thorn-tree hill'. OE **thorn** + **hyll**. L 171, 221.

(1) ~ Derby SK 1983. *Thorn(e)(h)ull(e)* 1200–1428, *Thornhill* from 1216×72. Db 171.

(2) ~ EDGE WYorks SE 2518. *Thornhill Edge* 1613. P.n. Thornhill SE 2518, *Torni(l)* 1086, *Thornhill -hyll* 1196×1202–1822, + ModE **edge**, *The edge* 1612, referring to the long escarpment S of Thornhill overlooking Howroyd Beck. YW ii.210, 215.

THORNHILL Hants SU 4712. This is *Town hill* 1810 OS, earlier *Tunhill(e) -hulle* 1256–2, *Tounhull(e)* 1289–1336, 'hill with a farm', OE **tūn** + **hyll**. Gover 1958.43.

THORNICOMBE Dorset ST 8703. 'Thorn-tree valley'. *Tornecome* 1086, *Thornacumba, Thornecumbe* 1199×1216, *Thornecombe* 1398 etc. OE **thorn**, genitive pl. **thorna**, + **cumb**. Do ii.73, L 93, 221.

THORNLEY Durham NZ 1137. 'Wood or clearing where thorn-trees grow'. *Thorniley* c.1280, *Thornle* 1377, *-ley* 1382. OE **thorn** + **lēah**. The first spelling may point to OE **thorniġ** 'growing with thorns' or even an OE ***thorning** 'thorny place'. NbDu 195, PNE ii.204, L 203–4, 221.

THORNLEY Durham NZ 3639. 'Thorn-tree hill'. *(æt) Þornhlawa* 1071×80, *Tornlauum* 1144×52, *Thorn(e)law(e)* 1144×52–1522, 1739, *Thorn(e)ley* from 1522, *Thornl(a)y* 16th OE **thorn** + **hlāw**. NbDu 195.

THORNS Suff TL 7455. 'The thorn-trees'. No early forms but contrasts with nearby *Little Oaks, Oaks Plantation* and *The Oaks*, all 1836 OS. ModE **thorn**.

THORNTHWAITE 'Thorn-tree clearing'. OE, ON **thorn** + ON **thveit**.

(1) ~ Cumbr NY 2225. *Thorn(e)thwayt -thweyt -thwait* 1230–1515, *Thornthat* 1605. Cu 371 gives pr [θɔ:nθət], SSNNW 169 no.3, L 211, 221.

(2) ~ NYorks SE 1758. *Tornthueit* 1230, *Thorn(e)thwait(e) -thwayt* 1301–1651. YW v.137.

THORNTON 'Thorn-tree settlement, settlement with a thorn hedge'. OE **thorn-tūn**. PNE ii.204.

(1) ~ Bucks SP 7535. *Ternitone* 1086, *Thornton(e)* from 1208. The DB form may point to OE **thyrne** or even **thyrning** as the specific. Bk 64.

(2) ~ Cleve NZ 4713. *Torentun* 1086. YN 171.

(3) ~ Humbs SE 7645. *Torte- Tornetun* 1086, *Thorn(e)ton* from 13th, ~ in *Spaldinggemor* 1290. YE 184.

(4) ~ Lancs SD 3442. *Torentun* 1086, *Thorinton* [1201]1268, *Torrenton* 1226, *Thornton(a)* from [1245]1268. La 157, Jnl 17.91.

(5) ~ Lincs TF 2467. *Torintvne* 1086, *Torentuna* c.1115. DEPN.

(6) ~ Mers SD 3300. *Torentun* 1086, *Thorinton* 1212, before 1250, *Thorn(e)ton* from 1246. La 118, Jnl 17.65.

(7) ~ Northum NT 9547. *Thorneton'* [1183]c.1320, *-tona* 1208×10 BF, *-ton* 1539. NbDu 195, DEPN.

(8) ~ WYorks SE 1033. *Torentune -ton(e)* 1086, *Thor(r)enton(a)* c.1150, 1276, *Thornetun -ton(a)* 1190×1220–1379, *Thornton* from 1300. YW iii.271.

(9) ~ BECK NYorks SE 8381. *Thorntonebech* [1167×79]16th. P.n. Thornton as in THORNTON DALE SE 8382 + ON **bekkr**.

(10) ~ BRIDGE NYorks SE 4370. P.n. Thornton + ModE **brigge** 1576. Thornton is *Torenton(e)* 1086, *Thorneton(a)* [13th] etc. Also known as *Thornenton on Swale* 1275. YN 24.

(11) ~ CURTIS Humbs TA 0817. 'T held by the Curtis family', *Thornton Curteys* 1430–55. Earlier simply *Torentune -one* 1086, *-tuna* c.1115, *Torneton* 1159–78, *Torinton(')* 1190–1234, *Thorneton(')* 1203–1576, *Thorenton(e)* [1139×47]1301–1548, *Thornton(')* from 1213. Li ii.279.

(12) ~ DALE NYorks SE 8382. *Thornton in vallem de Pykerynge* [1248]15th. P.n. *Torentōna -tun(e)* 1086, *Thornetun -ton* 1157×8 + ModE **dale** (ON *dalr*). Thornton stands at the mouth of a long narrow dale stretching into the North York Moors. YN 88.

(13) ~ FIELDS Cleve NZ 6118. P.n. *Tornetun* 1086 + ModE **fields**. DG 536.

(14) ~ HOUGH Mers SJ 3081. 'T held by the Hough family'. *Thorneton Hough* 1624. Earlier simply *Torintone* 1086, *Thornton* from 1260. Also known as *Matheue Thornton* 1287–1656 with variants *Thorneton Maheu* 1307, *Mayew, Mayo(w)*, and *Thorn(e)ton Gra(u)nge* 1415–1620. There was a grange of Basingwerk Abbey here. Mathew de Thornton was a tenant c.1252 and Hough is from the surname of the family of Richard del Hogh c.1329 'of the hill-spur' (OE *hōh*, ME *hogh*). Che iv.230 gives pr ['θɔrntən 'ʌf].

(15) ~ -IN-CRAVEN NYorks SD 9048. *Thorinton in Craven* 1201. P.n. *Torentun(e)* 1086, *Thorneton* 1276–1537, *Thornton* from 1300 + p.n. CRAVEN. YW vi.32.

(16) ~ IN LONSDALE NYorks SD 6873. P.n. *Tornetun* 1086, *Thorn(e)ton* 1293–1584, *Thorenton* 1295, + ME suffix *in Lonesdale* 1280 'the valley of the r. Lune' as in KIRKBY LONSDALE Lancs SD 6178. YW vi.250.

(17) ~ -LE-BEANS NYorks SE 3990. *Thornton in le Beynes, ~ in Fabis* 1534. P.n. Thornton + Fr definite article **le**, short for **in** + ModE **beans**, Latin **fabis**, dative pl. of **faba** after preposition **in** for distinction from THORNTON-LE-MOOR SE 3988. Earlier recorded as *Gristorentun* 1086, *Grisethorntune* 1088, ON pers.n. *Gríss* or appellative **gríss**, genitive pl. **grísa**, 'pigs' Thornton', and *Thorinton super vivarium* 'T on the fishpond' 1208, Lat **vivarium**. YN 208, SSNY 150.

(18) ~ -LE-CLAY NYorks SE 6865. 'T in the clay'. P.n. *Torentun(e)* 1086, *Thorneton* c.1100×15, 1301, + Fr definite article **le** + ModE **clay**. YN 39.

(19) ~ LE MOOR Lincs TF 0596. 'T on the moor'. *Thornton in Mora* 1291, *~ in La Mora* 1318, *~ in le More* 1506, *~ le Moor* 1824 OS. P.n. *Torentvn -tune -tone* 1086, *Torntuna* c.1115, *Thorneton* 1236–1606, *Thornton* from 1303, + Fr definite article **le** short for *en le* + ModE **moor**, Latin **mora**. Li iii.163.

(20) ~ -LE-MOOR NYorks SE 3988. 'T in the moor'. *Thorinton' in mora* 1208, *Thornton in the More* 1327. P.n. *Torentone -tune* 1086 + Fr definite article **le** + ME **mor** for distinction from THORNTON-LE-BEANS SE 3990 and THORNTON-LE-STREET SE 4086. YN 209, DG 536.

(21) ~ -LE-MOORS Ches SJ 4474. *Thorneton le Moors* 1291, *~ in le Moore* 1299, *~ in the Moors* 1477. Earlier simply *Torentune* 1086, *Torinton'* 1198×1216, 1272, *Thorneton* early 13th–1620, *Thornton* from early 13th. 'le Moors'. Fr definite article **le** (short for *en le*) + ME **more**, for distinction from other Thorntons and alluding to its location beside the marshes of the river Gowy. Che iii.258.

(22) ~ -LE-STREET NYorks SE 4086. 'T on the Roman road'. *Thorinton in via* 1208, *~ in le Strete* c.1291. P.n. *Torentun* 1086 + OFr definite article **le** + ME **strete** referring to the Roman road Margary no. 80a and for distinction from THORNTON-LE-MOOR SE 3988. YN 205.

(23) ~ MOOR (RESERVOIR) WYorks SE 0533. P.n. THORNTON SE 1033 + ModE **moor** + **reservoir**.

(24) ~ RESERVOIR Leic SK 4707. P.n. Thornton + ModE **reservoir**. Thornton is *Torenton' -in-* 1201–65, *Thorenton'* 1239–73, *Thorneton(')* 1276–1714, *Thornton(')* from c.1274. Lei 482.

(25) ~ RESERVOIR NYorks SE 1888. P.n. Thornton as in THORNTON STEWARD SE 1787 + ModE **reservoir**.

(26) ~ RISEBOROUGH NYorks SE 7583. Short for *Torneton sub Riseberg* 'T under Riseborough' c.1200, *Thornton under Isberg* 1406. P.n. *Tornitun, Tornentun* 1086, *Torinton'* 1167, + p.n. *Ri- Ryseberg(h)' -berch* c.1200–1318, 'brushwood hill', OE **hris** + **beorg**. Riseborough is a prominent hill (232ft.) in the brushwood-grown carrs W of Pickering. YN 77.

(27) ~ RUST NYorks SD 9788. 'Rossketill's T'. *Thorneton Ruske* 1153, *~ Rust* from [12th]15th. P.n. *Tore(n)ton* 1086 + manorial suffix *Rust*, a folk-etymological reformation of *Rusk* representing ON pers.n. *Rossketill*. YN 268, SPN 225.

(28) ~ STEWARD NYorks SE 1787. 'The steward's T'. P.n. Thornton, *Tornenton(e)* 1086 + manorial suffix *Steward* 1252. Also known as *Thorneton Dapifer* [1157]15th. Thornton was held by the steward of the earls of Richmond, Latin **dapifer**. YN 248.

(29) ~ WATLASS NYorks SE 2385. P.n. *Torreton -tun* 1086, *Thorn(e)ton* from [1184]15th, + p.n. *Wadles* 1086, *Watlass* from 1263, *Watlous* 1269–1555, *Wattelaws* 1376, *Watloose* 1576, originally a separate village, whose name means 'waterless', ON **vatn-lauss**. The site of Watless to the S is high and there is no stream nearby. YN 235, SSNY 107.

(30) Bishop ~ NYorks SE 2663. *Thorn(e)ton episcopi* 1198, *Bishop ~* 1362 etc. Earlier simply *(on) þorntune* c.1030 YCh 7, *Torentune -tone* 1086, *Thorna- Tornetunam* 1135×40, *Thorn(e)ton* 1220×46–1457. The estate lay in the archbishop of York's manor of Ripon. YW v.184.

(31) Childer ~ Ches SJ 3677. *Childrethornton* 1288, *Childerfrom* 1305. Earlier simply *Thorinthun* 1200×20, *Torinton* 1209×29, *Thornton* from 13th. 'Childer', ME **childre** (from OE *ċildra*, genitive pl. of *ċild* 'a child'), because the manor was allocated to the upkeep of, or was owned by, young men, probably the junior members of St Werburgh's abbey (Chester Cathedral) which possessed the estate. Che iv.196.

THORNWOOD COMMON Essex TL 4705. *Thornwood Common* 1805 OS. P.n. Thornwood + ModE **common**. Thornwood, *Thornenewod' -ine-* 1297, *Thornyngwod* 1323, *Thornewode* 1480, is the 'thorn-tree wood', OE ***thornen** 'growing with thorn-trees', < OE **thorn** + **-en**, definite form ***thornena**, + **wudu**. Ess 86.

THOROTON Notts SK 7642. 'Thorfrøthr's farm or village'. *Toruenton, Toruertune* 1086, *Turvertun -ton(e)* [1174×81]1337, 1176–99, *Thurverton* 1236–1594, *Thurwerton* 1287, *Thoroton* from 1552. ON pers.n. *Thorfrøthr* + **bȳ**. Nt 229 gives pr [θʌrətən], SSNEM 195.

THORP(E) 'Secondary settlement, dependent outlying farmstead or hamlet', ON **thorp**. PNE ii.205.12.

(1) ~ Cumbr NY 4926. *Thorp(e)* from 1333. We ii.208, SSNNW 59 no.5.

(2) ~ Derby SK 1550. *Torp* 1086–1250, *Thorp(e)* from late 12th, *~ iuxta Mapeltone, ~ iuxta Assheburn* 1330, *~ in (the) Clottes* 1323, 1387, 1585. *Clottes* is OE **clot(t)** 'lump' here with transferred sense 'hill'. Db 407, SSNEM 118.

(3) ~ Lincs TF 4982. *Thorpe* 1824 OS. Short for *Fugelestorp* 1210, *Fugletorp* 1218×9, *Fulestorp* c.1200, 'Fugl's outlying farm', ON pers.n. *Fugl*, OE *Fugol*, genitive sing. *Fugles*, + **thorp**. 'Fugl's' for distinction from TRUSTHORPE TF 5183.

(4) ~ Norf TM 4398. *Torpe* 1254, *Thorp cum Hadesco* 1316. An outlying farm of Haddiscoe. DEPN.

(5) ~ NYorks SE 0161. *Torp* 1086–1234, *Thorpe* from c.1202, *Torph juxta Brunsella* c.1190. An outlying farmstead dependent on Burnsall. YW vi.96, SSNY 68 no.5.

(6) ~ Notts SK 7649. *Torp* 1086–1235, *Thorp juxta Newerk* 'T by Newark' 1269, *Throppe near Newarke* 1564. An outlying settlement from Newark. Nt 218, SSNEM 119.

(7) ~ Surrey TQ 0268. *T(h)orpe* [672×4]13th S 1165, *Thorp* [871×99, 933, 967]13th S 353, 420, 752, *(æt) Đorpe* [1062]13th S 1035, *Torp* [1053×66]13th S 1094, 1086, [1154×89]1285 etc. with variant *Thorp(e), Throp* 1272. Sr 134.

(8) ~ ABBOTTS Norf TM 1979. 'Thorpe held by the abbot' sc. of Bury St Edmunds. *Torp Abbatis* 1254. P.n. *Thorp* 1086 + Latin **abbas**, genitive sing. **abbatis**. DEPN.

(9) ~ ACRE Leic SK 5120. 'T held by the hawker'. P.n *Torp'* 1086–late 13th, *Thorp(e)* from c.1130 + affix *Hauker(e)* 1343–1461, *Hawker* 1535, 1537, *Awker* 1612–17, *Aker* 1579, *Acre* from 1608, ME **hauker** 'a hawker, a falconer' (OE *hafocere*) or surname *Hawker*. Lei 382, SSNEM 118 no. 7.

(10) ~ ARCH WYorks SE 4345. 'T held by the de Arches family'. *Thorp(e)archis -ys, de Archis, Darches* 13th cent., *~ Arch(e)* from 1333. P.n. *Torp* 1086–1310, *Thorp(e)* 1254–1646, + manorial addition from the family name of Osbern de Arches who held the three manors of Thorp at the time of DB. The family came from Arques in Normandy, *Archis*, dative pl. of Latin *arcus* 'an arch of a bridge'. YW iv.244, SSNY 69.

(11) ~ ARNOLD Leic SK 7720. 'T held by Ernald' sc. de Bosco who held the manor in 1156 followed by three successors of the same name. P.n. *Torp* 1086–13th, *Thorp(e)* from c.1130 + manorial affix *-(h)ernald -old* 1238–1449, *Arnoldes-* 1214, *~ Arnald -old* from 1254. Lei 195, SSNEM 118 no.8.

(12) ~ AUDLIN WYorks SE 4716. 'T held by the Audlin family'. *Thorp(e) Aud-, Awd(e)lyn(e) -lin* from 13th. P.n. *Torp(e)* 1086–1216, *Thorp(e)* from 1154×9 + manorial addition from William son of Audlin (CG *Aldelin*) who held here in 1190. YW ii.97, SSNY 69.

(13) ~ BAY Essex TQ 9284. A modern name from Thorpe as in

Thorpehall, *Thorp* 1086, *S(o)uthorp(e)* 1275–1341, *Thorpehalle* 1434. OE **thorp** 'outlying farm' of Southchurch, + ModE **bay**. Ess 201.

(14) ~ BASSET NYorks SE 8673. *Thorp Basset(t)* 1267–1608. P.n. *Torp* 1086, 1200, *Thorp(e)* from 1204, + manorial addition *Basset*. Also known as *Thorpe juxta Wintringham* 13th, ~ *in Hauerfordlyth* 13th, and *Thorphelis -ys, Thorp El(e)ys, Elisthorpe* 13th cent. The manor was held by William Basset in 1204. YE 137, SSNY 69 no.15.

(15) ~ BY WATER Leic SP 8996. 'T by the water' sc. of the r. Welland. P.n. *Torp* 1086–1296, *Thorp(e)* from 1231, + affix *on Welond* 1358, *by the Watre* 1428, *juxta aquam (de Welland)* 1459, 1701. R 302, SSNEM 119 no.20.

(16) ~ CONSTANTINE Staffs SK 2609. 'T held by the Constantine family'. *Thorp Costentin* c.1245, *Thorpe Constantyn* 1318. P.n. *Torp* 1086 + manorial addition from Galfrid de Costentin, who held a fee here in 1212. DEPN, Horovitz.

(17) ~ END Norf TG 2811. A garden village built at the end of THORPE ST ANDREW TG 2609.

(18) ~ FENDYKES Lincs TF 4560 → Thorpe FENDYKES TF 4560.

(19) ~ GREEN Suff TL 9354. *Thorpe Green* 1837 OS. P.n. Thorpe as in THORPE MORIEUX + ModE **green**.

(20) ~ HALL NYorks NZ 1014. P.n. Thorpe + ModE **hall**. Named after a lost village, *Torp* 1086, *Thorp(e)* [1184]15th, *Thrope near Tease* 1574. YN 301, SSNY 68 no.3.

(21) ~ HALL NYorks SE 5731. *Thorp(e)hall* 1371, 1817. P.n. Thorpe + ME **hall**. Thorpe is one of the *twegen þorpas* 'two Thorpes' of c.1030 YCh 7, *Thorp(e)* 12th etc. Also called *Westhorp', Thorp' Seleby* 13th cent. The other Thorpe of c.1030 YCh 7 is THORPE WILLOUGHBY SE 5730. YW iv.34, SSNY 70 no.41.

(22) ~ HALL NYorks SE 5776. P.n. *Torp* 1086, *Thorp* from c.1142, + ModE **hall**. Also known as *Thorpe le Willows*. YN 193, SSNY 69 no.27.

(23) ~ HESLEY SYorks SK 3796. 'T by or belonging to Hesley'. *Thorpe(e) Hesteley* 1594, *-hesley* 1611. P.n. *Tor(p)* 1086, 1170×1200, *Thorp(e)* from 1229, + p.n. Hesley as in HESLEY HALL SK 6917 in Ecclesfield. YW i.187, SSNY 69.

(24) ~ HILL NYorks SE 5964. A lost village, *Torp* 1086, + ModE **hill**. DBY 1Y1, SSNY 68 no.7.

(25) ~ HILL NYorks SE 7159. A lost village. *Torp* 1086, + ModE **hill**. DG 536, DBY 23N29, SSNY 69 no.19.

(26) ~ IN BALNE SYorks SE 5910. *Thorp(e) in Balne* from 1339. P.n. *Thorp(e) in Balne* from 1339. P.n. *Thorp(e)* from 1150, + district name BALNE SE 5919. YW i.19.

(27) ~ LANGTON Leic SP 7492. 'T by Langton' as in Church and East LANGTON SP 7292–3. P.n. *Torp* 1086–1243, *Thorp(e)* from c.1130, + affix *iuxta Langeton'* c.1130, 1294, *Langton* 1610. Lei 260, SSNEM 119 no.13.

(28) ~ LARCHES Durham NZ 3826. A modern formation from p.n. Thorpe as in THORPE THEWLES Cleve NZ 4023 + ModE **larch** first recorded in 1548.

(29) ~ LE FALLOWS Lincs SK 9180. 'T in the ploughed land'. *Thorp en les Falous* 1325. P.n. *Torp* 1086–1185 + Fr definite article **le** short for *en le* + ME **falu** (OE *falh*) 'ploughed land'. Also recorded as *Turuluestorp* c.1115, *Thorelthorp* 1254, 'Thurulf's Thorp', ODan pers.n. *Thurulf*, secondary genitive sing. *Thurulfes*, + **thorp**. SSNEM 118 no.10, Cameron 1998.

(30) ~ -LE-SOKEN Essex TM 1822. 'T in the soke' sc. of The NAZE TM 2624. *Thorpe in ye Sooken* 1612, ~ *le Soaken* 1687. P.n. *Torp(eia)* [1119–1202]n.d., 1181×1222, *Thorp(e)* 1222, 1254, 1366, + Fr definite article **le** + ME **soken** (OE *sōcn*). Ess 352.

(31) ~ LE STREET Humbs SE 8543. 'T on the Roman road'. *Thorp in Strata* 1301–1434, ~ *in le Strete* 1413, *Thorp-le-Street* 1828. Earlier simply *Torp(i)* 1086, *Thorp* 1226–84. On the Roman road from York to Brough on Humber, Margary no. 2e. YE 229, SSNY 69.

(32) ~ LE WILLOWS NYorks → THORPE HALL SE 5776.

(33) ~ MALSOR Northants SP 8379. 'T held by the Malesoures family'. *Thorp Malesoures* 1220, *Thorpe Malsor al. Malsworth* 17th. P.n. *Thorp* from 12th + family name of Fucher Malesoures who held this place in the 12th cent. In DB it is *Alidetorp* 1086, 'Alida's thorp', possibly representing the OE feminine pers.n. *Æthelgyth*, genitive sing. *Æthelgythe*. Nth 122, SSNEM 120.

(34) ~ MANDEVILLE Northants SP 5344. 'T held by the Mandeville family'. *Throp Mondeville* 1300, *Thorp Mandevill* 1316, *Thrupmounfeld* 1539. P.n. *Torp* 1086, *Thorpe* from 12th, *Trop* 1220, + manorial addition from the family name *Amundeville* as in COATHAM MUNDEVILLE NZ 2820. Nth 61, SSNEM 119 no.15.

(35) ~ MARKET Norf TG 2435. 'Thorpe with a market'. *Torpmarket* 1251. P.n. *Torp* 1086 + ME **market**. DEPN.

(36) ~ MORIEUX Suff TL 9553. 'T held by the Morieux family'. *Thorp Morieux* 1330. P.n. *þorp* [962×91]11th S 1494, *Torp(a)* 1086, *Thorpe* 1610, + manorial addition from the family name of Roger de Murious who held the manor in 1201. His family came from Morieux near St Brieuc in Brittany. Also known as *Guvetorp* 1201, 'Gua's T' from Gua, Roger's mother. DEPN.

(37) THORPENESS Suff TM 4759. P.n. *torp* 1086, *Thorp* 1275–1610, + ModE **ness** 'a headland'. Baron.

(38) ~ ON THE HILL Lincs SK 9065. *Thorp(e) super Collem* 1277, 1279, 1557, 1572, *Torp sur le Tertre* 1281, *Thorp(e) on the Hill(e)* from 1325. Earlier simply *Torp* 1086–1277, *Thorp(e)* 1254–1349. 'On the hill' was Latinised as *super collem* (Latin *collis* 'a hill') and once translated into French (Fr *tertre* 'a hill'). Perrott 268, SSNEM 119 no.12.

(39) ~ PERROW NYorks SE 2685. P.n. Thorpe + manorial addition *Pirrow(e)* 1285. Thorpe is *Torp* 1086, *Thorp* [1184]15th and was held by the lords of *Pir(n)hou* in Ditchingham Nf. A lost village. YN 229 gives pr [θɔ:p pɔrə], SSNY 69 no.13.

(40) ~ ST ANDREW Norf TG 2609. 'St Andrew's T'. P.n. *Torp* 1086, *Thorp juxta Norwycum* 'T beside Norwich' (literally 'the north wicks') 1302, + saint's name *Andrew* from the dedication of the church. DEPN.

(41) ~ ST PETER Lincs TF 4860. 'St Peter's T'. P.n. *Torp* 1086–1185, *Thorpe* 1824 OS, + saint's name *Peter* from the dedication of the church. DEPN, Cameron.

(42) ~ SALVIN SYorks SK 5281. 'T held by the Salvin family'. *Thorp(e) Saluayn(e) -ey- -vayn* 1255–1454, ~ *Salvin(e) -yn* from 1588. P.n. *Torp* 1086–c.1200, *Thorp(e)* 13th cent. The manor was held by the Salvain family in the 13th cent. Also known as *Richenildtorp, Thorprikenil* from the ancient road name RIKENILD STREET. YW i.151, SSNY 69.

(43) ~ SATCHVILLE Leic SK 7311. 'T held by the Satchville family'. P.n. *Torp'* c.1141–1243, *Thorp(e)* from c.1130, + manorial affix *Sec(c)h(e)vill(e), Seg(g)evile, Segefeld(e), Sacheville -feild* 1262–1610. The manor was held by Ralph de Sechevill in 1234. Lei 330, SSNEM 119 no.18.

(44) ~ THEWLES Cleve NZ 4023. 'T held by the Thewles family'. *Thorpp' Thewles* 1265, *Thorpethewles* from 1312 with variants *-theules -theweles* and *-thewlesse* 1596×7–1684. Earlier simply *Torp(')* 1144×52–1382, *Thorp(')* c.1170–1821. *Thewles* is probably a manorial addition from the Y surname *Thewless, Thewlis*, OE **thēawlēas** 'wanton, dissolute'. Pr [θju:lz]. DBS 346.

(45) ~ UNDERWOOD NYorks SE 4659. P.n. *Thorp(e)* from 1175×99 etc. + suffix *underwod(e)* from 1198. There are woods on Thorpe Hill overlooking the village. The earliest recorded name is *Tuadestorp* 1086, 'Thorald's thorp', ON pers.n. *Thóraldr* possibly in the earlier form *Thorvaldr*, genitive sing. *Thorvalds*, + **thorp**. Also known as *Thorpe in*

Burghshire 1175×99 for which see ALDBOROUGH SE 4066. YW v.5, SSNY 68 s.n. Tharlesthorpe no.2, DG 509.

(46) ~ WATERVILLE Northants TL 0281. 'T held by the W family'. *Thorpe Watervile* 1300. P.n. *Torp(e)* 12th–1206, *Thorp(e)* from 1220, + family name of Ascelin de Waterville who held the manor in the 12th cent. Nth 219, SSNEM 119 no.21.

(47) ~ WILLOUGHBY NYorks SE 5730. P.n. Thorpe + manorial addition *Wyleby* 1276. Thorpe is one of the *twegen þorpas* 'two Thorpes' of c.1030 YCh 7, *Torp* 1086, 1110×30, *Thorp(e)* 1110×30 etc. and was held by the Willoughby family. The other *thorp* of c.1030 YCh 7 is probably THORPE HALL SE 5731. YW iv.30, SSNY 69 no.26.

(48) Burnham ~ Norf TF 8541. 'Burnham outlying farm'. *Brunhamtorp* 1199, *Burnhamtorpe* 1201. P.n. Burnham as in BURNHAM DEEPDALE TF 8043. DEPN.

(49) Gayton ~ Norf TF 7418. 'Gayton outlying farm'. *Geytonthorp* 1401×2. P.n. GAYTON TF 7219 + p.n. *Torp* 1086, *Thorp* 1302. Also known as *Aylswiththorp* 1316, *Aylswythorp* 1390, 'Æthelswiths's outlying farm', OE feminine pers.n. *Æthelswīth*. DEPN.

(50) Ixworth ~ Suff TL 9172. 'Ixworth outlying farm'. *Ixworth thorp* 1305. P.n. IXWORTH TL 9370 + p.n. *torp(a), Torp* 1086. DEPN.

(51) Tattershall ~ Lincs TF 2159. 'T belonging to Tattershall'. *Tattershall Thorpe* 1824 OS. P.n. TATTERSHALL TF 2157 + p.n. *Torp* 1086, c.1115, *Thorp* 1150×60. SSNEM 118 no.6, Cameron 1998.

THORPEFIELD 'Thorpe (open) field'. ON **thorp**, + ME **feld**.

(1) ~ NYorks SE 4179. *campo de Thorp* [1243]15th, *Thorpfeld* 1303. *Torp* 1086, 'the outlying farm'. Latin **campus**, dative case **campo**. Also known as *Petithorp juxta Thresk* c.1142, 'little T by Thirsk'. YN 187, DG 536, SSNY 69 no.16.

(2) ~ NYorks (lost near Scarborough). P.n. *Torp* 1086, 'the outlying farm'. DG 536, DBY 13N10, SSNY 69 no.17.

THORPENESS Suff TM 4759 → THORPE.

THORVERTON Devon SS 9202. 'Thurferth's estate'. *Torverton* 1182–13th, *Thorverton(e)* from 1263, *Thurfurton* n.d., 1340. Anglo-Scand pers.n. *Thurferth* + OE **tūn**. D 572.

THRANDESTON Suff TM 1176. 'Thrandr's estate'. *Thrandestone* [11th, probably before 1038]13th S 1527, *Thrandes- Thrundes- Frondestuna* 1086, *Throndestun* c.1095, *Thrandeston* from 1286. ON pers.n. *Thrándr*, secondary genitive sing. *Thrándes*, + **tūn**. DEPN, Baron.

THRAPSTON Northants SP 9978. Partly obscure. *Trapeston(e)* 1086–1247, *Strape(s)ton(a)* 1130–1275, *Thrapston* from 12th, *Thrapson* 1631. A name of great antiquity the specific of which may be an ancient Germanic pers.n. **Thræpst*, genitive sing. **Thræpstes*, + **tūn**. Nth 220.

THREAPWOOD Ches SJ 4445. 'Disputed wood'. *Threpewood* 1548, *Threapwoode* 1550. OE **thrēap** + **wudu**. So named from its location on the border of England and Wales and belonging wholly to neither. In 1773 it was reported that Threapwood was reputed to be in no county, parish, town or hamlet, that no land tax or rates were payable, that the sheriffs of Cheshire and Flint had no jurisdiction here, nor had the JPs, and that no civil court could deal with cases arising in the area. Che iv.61 gives pr [θri:pwud].

THREE BRIDGES WSusx TQ 2837. *three bridges between Worth and Crawley* 1598, *(the bridge called) Le three bridges* 1613. ModE **three** + **bridge**. The precise reference is unknown: the road crosses two small streams. Sx 282.

THREE HOLES Norf TF 5000. Short for *Three Holes Bridge* 1824 OS, a bridge with three arches.

The THREE HUNDREDS OF AYLESBURY Bucks SP 8607. The original 18 hundreds of Buckinghamshire were subsequently grouped into triple hundreds. The three hundreds of Aylesbury were Aylesbury, Risborough and Stone. Bk 3.

THREEKINGHAM Lincs TF 0836. Partly uncertain. *Triching(he)hā -inge-* 1086, *Tricingeham* [after 1131]early 13th, *Tri(c)k- Trykingham* 1093×1100–1538, *Thri- Thrykyngham* c.1128–1428, *Trikingeham -ynge-* 1219–1346, *Threkingham -yng-* 1359–1652. Possibly the 'homestead of the Tricingas, the people called after *Tric*', OE folk-n. **Tricingas* < unidentified element or name *Tric* + **ingas**, genitive pl. **Tricinga*, + **hām**. The element *Tric*, which occurs as a former name for Skegness, *Tric* 1086, is unexplained but such a form with *ĭ* would explain the later forms with *ē* by ME open syllable lengthening. Perrott 145.

THREE LEG CROSS ESusx TQ 6831. *Threlegged Crosse* 1556, *Threlegg' Crosse* 1562. Sx 455.

THREE LEGGED CROSS Dorset SU 0506. *Three Legged Crosse* 1591. Either a reference to a T-junction or an allusion to the gallows which was nicknamed *Three legged mare*. Local tradition, however, says that the name referred to a tripod surmounted by a wooden cross that was a direction beacon to persons crossing the heaths. Do ii.258.

THREE MILE CROSS Berks SU 7167. *Threemile Cross* 1761, 1846, 1973. A crossroads three miles from Reading SU 7173. Brk 106.

THREEMILESTONES Corn SW 7844. *Three Mile Stone* 1884. A 19th cent. hamlet named from its position three miles outside Truro. PNCo 162.

THREE PIKES Durham NY 8334. *Three Pikes* 1866 OS. N country dial. **pike** (< *pīc*) 'a pointed hill'.

THRELKELD Cumbr NY 3125. 'Thralls' spring'. *Trellekell'* 1197, *-keld(e)* 1278–1305, *Threlkeld(e)* from c.1240, *Thirkeld* 1278, *Thyrlkel'* 1392, *Threlket* 1575. ON **thrǽll**, genitive pl. **thrǽlla**, + **kelda**. Cu 252 gives prs [θrelkəl], formerly [θrelkət], locally [trɛlkət], SSNNW 170, L 22.

THRESHFIELD NYorks SD 9863. 'Open land where threshing takes place'. *Freschefelt* 1086, *Tresche- Threskefeld* 12th, *Treskefeld(e)* 1192–1314, *Thresfeld* a.1212–1535, *Threschefeld* 1316. OE ***thresc-feld** with AN *t* for *th* and ON *sk* for *sh* perhaps under the influence of ON **thresk* 'a fen, a lake'. YW vi.105, SSNY 166.

THRIGBY Norf TG 4612. 'Thrykki's farm or village'. *Tru- Trikebei* 1086, *Try- Trikeby* 1219–75, *Thir- Thyrkeby* 1219–1519, *Thryckeby* 1291, 1428, 1553, *Thrickby* 1303. ON pers.n. **Thrykki*, genitive sing. **Thrykka*, + **bȳ**. Nf ii.26, SPNN 431.

THRINGARTH Durham NY 9322. 'Thorn-bush enclosure'. *Thyrnegarth* 1251, 1301, *Thryngarth* 1561. OE **thyrne** or ON **thyrnir** + ME **garth**. YN 309.

THRINGSTONE Leic SK 4217. Partly uncertain. *Trangesbi -by* 1086, *Trengeston' -is-* c.1200–1396, *Threngeston(e) -is-* 1245–1444, *Thri- Thryng(g)eston(e) -is-* 1268–1547, *Thringstone* from 1389. Possibly 'Thræingr's village', ON pers.n. **Thræingr*, genitive sing. **Thræings*, + **bȳ** early replaced by OE **tūn**. Lei 358, SSNEM 75.

THRINTOFT NYorks SE 3293. 'Thorn-bush messuage'. *T(h)irnetoft(e)* 1086–1562, *Tirnetoste* 1086, *Thrumtoft* 1597. ON **thyrnir** or OE **thyrne** + **toft**. The specific might alternatively be ON pers.n. *Thyrnir*. YN 276, SSNY 89.

THRIPLOW Cambs TL 4446. Partly uncertain. *Tripelau* [1042×66]12th S 1051, *-laue -lawe* 1086–14th, *Triplaw(e) -low(e)* 1231–1553, *Trippelawe -a* c.1060–1338, *-laue* c.1169–1412, *Trepeslau -lai* 1086, *Thri- Thryppelowe* 1298–1409, *Thry- Thriplow* 1341, 1541–76. Unknown element or pers.n. + OE **hlāw** 'tumulus, mound, hill'. DEPN suggests a pers.n. **Tryppa* related to *treppan* 'to tread' or hypocoristic for **Thrȳthbeorht* or the like; perhaps a form **Thryppa* is more appropriate, possibly related to *thrēapian* 'to reprove, correct'. Ca 90, DEPN.

THROCKING Herts TL 3330. Uncertain. *Trochinge* 1086, *Trocking(e)* 1175–1243, *Throck- Throkkyng(e) -ing* 1279–1428, *Thorkyng' -ing* 1278, 1611, 1676. Probably an **ing**[2] formation on OE **throcc** 'a table, a piece of wood on which the ploughshare is fixed'. Perhaps this was a place where such objects were made. Cf. THROCKLEY T&W NZ 1566. Hrt 187, ING 26.

THROCKLEY T&W NZ 1566. Uncertain. *Trocchelai* 1160, *Trokelawa* 1176–1255, *Throclau* 1236 BF, *Thro(c)kelawe* 1242 BF–1479. If the 1160 form is trustworthy this would appear to be 'the post clearing, the clearing or wood where posts are obtained', OE **throcc**, genitive pl. **throcca**, + **lēah**. The *-law* forms are persistent, however, and this may rather be 'Throcca's tumulus', OE pers.n. **Throcca* (derived as a nick-n. from **throcc**) + **hlāw**. In northern names OE **hlāw** is regularly reduced to *-ley*, although not as early as 1160. Throckley is situated on high ground overlooking the Tyne valley, and barrows are known in this area in similar locations. NbDu 196, PNE ii.213, L 163.

THROCKMORTON H&W SO 9849. 'Farm, settlement called or at *Throcmere* or *Throcmor*'. *Throcmortune* [11th]18th, *Trochemerton -ke-* 1176–1254, *Throgmarton, Throkmorton* 1436–51, *Frogmorton* 1696. OE p.n. **Throcmere* 'pond with a drain', OE **throcc** + **mere** or **Throcmōr* 'marsh with a drain', OE **mōr**, + **tūn**. Wo 169.

THROPHILL Northum NZ 1385. 'Hamlet hill'. *Trophil* 1166, *Throp(p)hill'* 1242 BF, *Throppille* 1296 SR, *Troppil, Thorpill* 14th cent. Newm, *Thropple* 1663. OE **throp** + **hyll**. NbDu 197.

THROPTON Northum NU 0202. 'The estate with a *throp* or outlying hamlet'. *Tropton* 1176, *Troptone* [1248]14th Newm, *Thropton'* 1242, 1296 SR, *Thorpton* 1334. OE **throp** + **tūn**. NbDu 197, DEPN.

THROWLEIGH Devon SX 6690. 'Trough wood or clearing'. *Trula* 1086, *T(h)rulegh* 1243–86, *Throulegh'* 1242–1341, *Throughly* 1631. OE **thrūh** 'a trough, a water-pipe, a conduit, a coffin' + **lēah**. The reference is probably to the manufacture of troughs and the like. Cf. THROWLEY Staffs TQ 9955. D453.

THROWLEY Kent TQ 9955. 'Wood or clearing where troughs or conduits or coffins are obtained' or 'wood, clearing in a trough'. *Brulege, Trevelai* 1086, *Thniliga* (sic for *Thru-*) [1087]13th, *T(h)rulege* c.1100, *T(h)rullega -leg(e) -legh(e)* 1153–1278. OE **thrūh** + **lēah**. The reference may be to the long narrow valley due E of Throwley. PNK 298.

THROWLEY FORSTAL Kent TQ 9854. P.n. THROWLEY TQ 9955 + Kent-Sussex, dial. **fostal** 'paddock, a way leading to a farm-house'. PNE i.184.

THROXENBY NYorks TA 0189. 'Thurstan's farm or village'. *T'stanebi* 1167, *Thur- Thorstanby* 1276–1475, *Throssenby* 1537, *Frostenby* 1577. ON pers.n. *Thórsteinn* with OE *-stān* for *-steinn* + **bȳ**. For the development of *x* cf. ROXBY NZ 7616. YN 110, SSNY 7.

THRUMPTON Notts SK 5031. 'Thormothr's farm or village'. *Tvrmodestvn* 1086, *Thurmodeston(e)* c.1189–1227, *Thurmeston* 1242–1330, *Thurm(e)ton* 1297–1393, *Thrumpton* from 1409. ON pers.n. *Thormóthr* + **tūn**. Cf. THURMASTON Leics SK 6109. Nt 250, SSNEM 195.

THRUNTON Northum NU 0810. Partly uncertain. *Trowentona* [c.1180]14th, *Torhenton* 1199, *Throingtun* 1236 BF, *Throunton alias Trowynton* 1253, *T(h)row(i)nton* 13th cent. *Trowenton', Thorowinton'* 1279 Ass, *Thorneton* 1296 SR, *Thrunton* 1649. This might be 'Thurwine's farm or village', OE pers.n. *Thurwine* + **tūn**. Thurwine is recorded as *Thruwin* in LVD. The 1199 spelling, however, suggests a specific with OE *-rh-*, possibly OE **thrūhum**, dative pl. of **thrūh** 'a water-pipe, a conduit, a coffer', and so 'the settlement at the pipes or troughs'. Thrunton is close to a Roman site. NbDu 197.

THRUNTON WOOD Northum NU 0709. P.n. THRUNTON NU 0810 + ModE **wood**.

THRUPP Glos SO 8603. 'Outlying secondary settlement'. *Trop'* 1261, *Throp(e)* 1393–1594, *Throppe, Throupe* 16th cent., *the Thrup* 1779. OE **throp**. An outlying settlement of Chalford SO 9002. Gl i.140.

THRUPP Oxon SP 4815. 'The hamlet, the farm'. *Trop* 1086–1297, *Throp* 1247–1428, *Thorp* 1284–1306, 1797, *Thrope* 1306. OE **throp**. O 292.

THRUSCROSS RESERVOIR NYorks SE 1558. P.n. *Thorescros(se)* [c.1142]15th–1512, *Thurescrosse* 1379, *Thurscros* 1617, *Thruscros(se)* 1562–1643, 'Thorir's cross', ON pers.n. *Thórir*, genitive sing. *Thóris*, + ON **kross**, + ModE **reservoir**. YW v.126.

River THRUSHEL Devon SX 4789. A back-formation from THRUSHELTON SX 4487. *Frischel* (sic) 1244, *Thrushel* 1577. D 14, RN 406.

THRUSHELTON Devon SX 4487. 'Thrush farm or settlement'. *Tresetone* 1086, *Thrisselton'* 1242, 1303, *Thri- Thryssheton -sch-* 1242–1334, *Thrushelton* 1306. OE ***thryscele** + **tūn**. D 210.

THRUSHGILL Lancs SD 6562. 'Goblin ravine'. *Thursgill* 1631, *Thurskeale* 1672. ME, ModE **thurse**, **thrush** (ON *thurs* 'a giant') + N dial. **gill** (ON *gil*). Other examples are a lost *Thursgyll* c.1350 near Capernwray and THURSGILL NYorks SD 6893. La 182, Mills 140.

THRUXTON H&W SO 4334. 'Thurkil's estate'. *Torchestone* 1086, *Thurk(el)eston* 1265–91, *Throkeston* 1485. ODan pers.n. *Thurkil*, secondary genitive sing. *Thurkilles*, + **tūn**. He 189.

THRUXTON Hants SU 2945. 'Thurkill's estate'. *T'killeston* 1167, *Turkilleston* 1174, 1199, *Thur-* 1202, *Turcleston* 1236, *Trokeleston* 1269, 1316, *Thrukeleston* 1325, *Throkeston* 1442, *Thruxton* 1670, *Fruxon* 1675. ON pers.n. *Thurkil*, genitive sing. *Thurkilles*, + **tūn**. The original name of the manor was *Anne* 1086 as in ANDOVER SU 3645, Abbotts ANN SU 3243 and AMPORT SU 3044. Another *Anne* manor which developed an alternative name for distinction from its homonyms is Sarson SU 3042, *Anne* 1203, *Anna Savage* 1242, 'Anne held by the Savage family', *Sauvageton* 1269 'the Savage manor', *Savageston* 1367, *Saveston* 1491, *Sarston alias Savieston alias Ansavage* 1597. Ha 162, 144, Gover 1958.168, 163.

THRYBERGH SYorks SK 4695. 'The three hills'. *Triberge -berga* 1086, *Tri- Tryberg(e)* 13th, *Thri- Thryberg(h)* from 1297. OE **thrēo, thrī** masc., + **beorg**. Thrybergh lies in a depression between three hills. YW i.173, L 128.

THUNDERSLEY Essex TQ 7788. 'Thunor's wood'. *Thunresleam* 1086, *Thundreslee -legh -ley* 1257–1397. OE divinity n. *Thunor*, genitive sing. *Thunres*, + **lēah**. The name implies a pagan sacred grove dedicated to Thunor. Ess 172, ES 19.155, Evidence 107.

THURCASTON Leic SK 5610. 'Thorketill's farm or village'. *Tur- Tvrch(it)elestone* 1086, *Thurketleston' -is-* c.1130–1284, *Turkilles- Turkel(l)eston(e) -is-* 1175–1349 with variants *Thur-*, *Thurkeston(')* 1268–1528, *Thurcaston* from 1357. ON pers.n. *Thorketill*, ME genitive sing. *Thorketilles*, + OE **tūn**. Lei 406, SSNEM 195.

THURCROFT SYorks SK 4988. 'Thori's enclosure'. *Thurscroft* 1319, *Turcroft* 1327, *Thurcroft* 1822. ON pers.n. *Thórir*, genitive sing. *Thóris*, + OE **croft**. YW i.142.

THURGARTON Norf TG 1834. 'Thurgar's estate'. *Ðurgartun* *[1044×7]13th S 1055, *Turgartuna(m), Turgaitune* 1086, *Turgarton'* 1205–6. AScand pers.n. *Thurgār* < ON *Thorgeirr* + **tūn**. DEPN, SPNN 405.

THURGARTON Notts SK 6949. 'Thurgar's farm or village'. *Tvrgarstvne* 1086, *Turgareston* 1205, *T(h)urgarton(a)* c.1170–1279 etc. Pers.n. *Thurgar* < ON *Thorgeirr* + **tūn**. Nt 178 gives pr [θɔ:gətən, θɔ:gə'n], SSNEM 196.

THURGOLAND SYorks SE 2901. 'Thurgar's newly cultivated land'. *Turgesland* 1086, *Turgarland(a)* 1090–1202, *Thurgerland(e)* 1321–1565, *Thurgoland* from 1549. ON pers.n. *Thorgeirr* + OE **land**. YW i.314, SSNY 106, L 246, 249.

THURLASTON 'Thorlaf's village'. ON pers.n. *Thorleifr* with OE *-lāf* for *-leifr*, genitive sing. *Thorlafes* + OE **tūn**.

(1) ~ Leic SP 5099. *Lestone* (sic) 1086, *Turlaueston'* 1196–1230, *Thurlaveston* 1228, *Tur- Thurleston(e) -is-* 1200–1535, *Thurlaston* from 1292. Lei 549, SSNEM 196 no.1.

(2) ~ Warw SP 4670. *Torlavestone* 1086, *Thurlaveston(e)* 1154×89–c.1228, *Thorleston* 1251, *Thorlaston* 1316, 1386, *Thurlaston* from 1317. Wa 147, SSNEM 196.

THURLBY 'Thorulf's village or farm'. ON pers.n. *Thórúlfr*, ODan *Thorulf*, + **bӯ**.

(1) ~ Lincs SK 9061. *Tvrolf- Turulf- Turolue(s)bi* 1086, *Turlebi -by* before 1140–1243, *Thurleby* c.1224×31–1607, *Thurlby* from 1322. Perrott 269, SSNEM 75 no.2.

(2) ~ Lincs TF 0916. *Turlebi* [c.990]c.1150–1406 with variant *-by*, *Tvrolve- Tvrolde-* (with *d* corrected to *t*), *Torulf- Turoluebi* 1086, *Thurleby* 1146–1547, *Thurlby(e)* from c.1221. Perrott 436, SSNEM 75 no.3.

THURLEIGH Beds TL 0558. 'At the wood or clearing'. *Thyrley(e)* 1372–16th, *Thurleigh al. Raleigh al. Laleigh* 1641. Earlier *(La) Lega* 1086–13th, *(La) Legh, Leye* 13th cent., *(la) Rel(e)ye* 1287–1427. OE **æt thǣre lēage** falsely analysed as *at the releye* and *at therleye*. The stressing of the modern name on the second syllable derives from this phenomenon. Bd 47 gives pr [θə'lai].

THURLESTONE Devon SX 6742. 'The pierced rock'. *(fram) ðyrelan stan* 847 S 298, *Torlestan* 1086, *Thurlestan(e)* 1238, 1242, *Thrulston* 1546. OE **thyrel** 'having a hole', dative sing. definite form **thyrelan**, + **stān**. The reference is to a great coastal rock here pierced by a natural hole. D 312. See Little Guide to Devon, illustr. opp. p. 287.

Great THURLOW Suff TL 6750. *Trillawe Magna* 1254. ModE **great**, Latin **magna**, + p.n. *Tritlawa, Tridlauua, Thrillauura* 1086, *Thrillauue* c.1095, uncertain element + OE **hlāw** 'a hill, a burial mound'. The DB forms suggest *thrid-* possibly representing OE ***thride** 'deliberation' < *thridian* 'to deliberate'. This would make the hill or barrow an assembly site. Another suggestion is OE *thrӯth-hlāw* 'the warriors' burial mound' from OE **thrӯth** 'a troop, a host of warriors'. DEPN, PNE ii.217.

Little THURLOW Suff TL 6851. *Trillawe Parva* 1254, *Thirlow p'ua* 1610. ModE adj. **little**, Latin **parva**, + p.n. Thurlow as in Great THURLOW TL 6750.

THURLOXTON Somer ST 2730. 'Thurlac's estate'. *Turlakeston* 1195, *Thurlokeston* 1285, *Thurloxton* 1610. Pers.n. *Thurlāc* (ODan *Thurlāk*, ON *Thorleikr*), genitive sing. *Thurlāces*, + OE **tūn**. DEPN.

THURLSTONE SYorks SE 2303. 'Thurulf's farm or village'. *Turulfestune, Turoluestun* 1086, *Thurueleston* 1246, *Thurlestone(e)* from 13th cent., *Thurston* 1569. ON pers.n. *Thórulfr*, genitive sing. *Thórulfs*, + **tūn**.YW 1.339, SSNY 130.

THURLTON Norf TM 4198. 'Thurferth's esate'. *T(h)uruertuna* 1086, *Thurvertone* 1254. AScand pers.n. *Thurferth* < ON **Thorfrøthr* + **tūn**. DEPN, SPNN 402.

Lower THURLTON Norf TM 4299. *Low Thurlton* 1838. ModE adj. **low(er)** + p.n. THURLTON TM 4198. Lower down towards the marshland than the parent village.

THURMASTON Leic SK 6109. 'Thormothr's village'. *Tvrmodestone* 1086, *Turmodeston(e) -is-* 1175–[c.1250]1404, *Thur- Thormedeston'* c.1130, *Thurmodeston(e) -is-* 1191–1340, *Thurmeston' -as-* 1203–1610. ON pers.n. *Thormóthr*, ME genitive sing. *Thormóthes*, + OE **tūn**. Lei 325, SSNEM 195.no.2.

THURNE Norf TG 4015. 'The thorn-bush'. *Đirne* *[1044×7]13th S 1055, *Thura, T'na, Tnā* 1086, *Thirne* 1198 etc. OE **thyrne**. Nf ii.42, RN 406.

River THURNE Norf TG 4017. *Thurine* 1577, 1586, *Thrin* 1622, *Thyrn* 1724. A back-formation first recorded in Harrison from the p.n. THURNE TG 4015. RN 406.

THURNHAM Kent TQ 8057. 'Thorn-tree homestead or estate'. *Tvrneham* 1086, 1232, 1240, *T(h)orneham* c.1100, *Tho- Thurn(e)ham* 1174×84–1242×3. OE **thorn**, genitive pl. **thorna**, + **hām**. The original *u*-form of this name is remarkable. Occasional comparable forms occur in DB e.g. *Tvrneberie*, THORNBURY Avon ST 6490, but do not persist. The explanation is unknown. PNK 233.

THURNHAM Lancs SD 4554. '(Settlement) at the thorn-trees'. *Tiernun* 1086, *Thurnum* [before 1160–1230]1268, *Thirnum -om* 1301–32, *Thurneham* 1569. OE **thyrne** or ON **thyrnir**, dative pl. **thyrnum**. La 171, L 221, Jnl 17.98.

THURNING Norf TG 0829. 'Thorn-bush place, thorn-brake'. *Ty- Ti- Turninga* 1086, *T(h)erning* 1196, 1212, *Tiringes* 1203, *Thirning* 1244, *Tirning* 1254. OE ***thyrning** < **thyrne** + **ing**[2]. DEPN, ING 200, PNE ii.223.

THURNING Northants TL 0883. 'The place growing with thorns'. *Torninge* 1086, *Thirninge* [c.1140]14th, *Thorney* [12th]13th–1428 with variants *Thyrn-* and *-yng(e)*, *Thurning* from 1175. OE ***thyrning**. Nth 221, ING 200, PNE ii.223.

THURNSCOE SYorks SE 4505. 'Thorn-tree wood'. *Ternusc(h) -usche, Dermescop* 1086, *Thirnescoh* c.1090–1229, *Thi- Thyrnescogh* [1154×89]15th–1428, *Thrynesc(h)o* before 1196, 1428, *Thornescogh* 14th cent., *-skowe* 1607, *Thurnscoo* 1542, *Thrunscoe* 1786. ON **thyrni-skógr**. There was woodland for pannage here in 1086. YW i.91, SSNY 106, L 209, 221.

Little THURROCK Essex TQ 6477. *Lytelthroke* 1523. ME **litel**, Latin **parva** (1201) + p.n. *Thurroce* [1040×2]12th, *turroc* 1085×6, *Tur(r)oc, Turocham, Thurrucca, Turrucā* 1086, *T(h)urroc(h) -o(c)k* from 1195, *Furrok* 1485, OE **thurruc** 'ship's bottom' in the sense 'place where filthy water collects'. The reference is to the marshes at West Thurrock, perhaps especially to the inlet called *The Breach* 1805. 'Little' for distinction from West THURROCK TQ 5877 and GRAYS Thurrock TQ 6177. Ess 129, Jnl 2.42.

West THURROCK Essex TQ 5877. *West Thorruk* 1320, *West Hurrock* 1408, *West(t)horke* 1535, 1552. ME adj. **west** + p.n. Thurrock as in Little THURROCK TQ 6477. Ess 129.

THURSBY Cumbr NY 3250. 'Thori or Thuri's village or farm'. *Thoresby -bi* c.1165–1414, *Thursby* from 1277, *Thuresby vulg. Fearsby* 1675. ON pers.n. *Thórir* or ODan *Thūrir*, genitive sing. *Thóris, Thūris*, + **bӯ**. Cu 152, SSNNW 41.

THURSFORD Norf TF 9833. 'Giant ford'. *Tureforde, Turesfort* 1086, *Turesford* 1231, *Thirsford* 1291. OE **thyrs** perhaps used, like ON *Thurs*, as a by-name, + **ford**. Otherwise the reference must be to some piece of local folklore. DEPN.

THURSGILL NYorks SD 6893. 'Giant or spectre ravine'. *Thursegilemos* [1220×50]1268, *Thursgill* 1609. ON **thurs** + **gil**. Cf. THRUSHGILL Lancs SD 6562. YW vi.270.

THURSLEY Surrey SU 9039. Partly uncertain. *Thoresle* 1292–1373, *-l(e)y -l(y)e* 14th cent., *Thursle* 1329, 15th cent., *Thursley* 1332(p), *Thirs-* 1554, *Thursleystret* 1609. The generic is OE **lēah** 'a wood or woodland clearing', the specific possibly OE *Thunor*, the name of the heathen god, genitive sing. *Thunres, Thūres*. Association with OE **thyrs** 'giant' seems impossible in view of the early spellings with *-o-*. Cf. Thundersley Essex TQ 7988, *Thureslei, Thursle* 1226. Sr 211, Evidence 108, NQ n.s. 43.388.

THURSTASTON Mers SJ 2484. 'Thorsteinn's village, farm or estate'. *Turstanetone* 1086, *Thurstanton* 1119×38–1621, *Thurstaneston* 1121×9–1819 with variants *-stan(y)s-* and *-tun*, *Thurstaston* from 1553. ON pers.n. *Thorsteinn* with OE *-stān* replacing *-steinn*, genitive sing. *Thorstānes*, + **tūn**. Che iv.279, SSNNW 191.

THURSTON Suff TL 9256. 'Thur's estate'. *Thurs- torstuna* 1086, *Thurstune* c.1095, *Thurston* 1226×8, 1610. ON pers.n. *Thūr*, genitive sing. *Thūrs*, + **tūn**. DEPN.

THURSTONFIELD Cumbr NY 3156. 'Thorsteinn's newly

cultivated land'. *Turstanfeld* c.1210–1350, *Thurstaneffeld* 1279, *Thirstonfeild* 1589, *Thrustenfeld* 1504, *Thrustenfeild*, *Thristeenfield* 17th, *Threstonfield* 18th. ON pers.n. *Thorsteinn* (Latinised as *Turstanus*) + OE **feld**. Cu 128 gives pr [θrusənfi:ld], SSNNW 254, L 246, 249.

THURSTONLAND WYorks SE 1610. 'Thurstan's newly cultivated land'. *Tostenland* 1086, *Turstain(e)land(a)* 1184×91, 1202×10, *Thurstanland* 1202–1542. ON pers.n. *Thorsteinn*, ODan *Thorsten, Thursten*, + **land**. YW ii.251, L 246, 249.

THURTON Norf TG 3200. 'Thorn-bush settlement'. *Tortuna* 1086, *Thermtona* (for *Therni-*) c.1150, *Thuriton* 1248, *Thurnton* 1302. OE **thyrne** varying with ***thyrning** or ***thyrniġ** 'thorny' + **tūn**. DEPN.

THURVASTON Derby SK 2437. 'Thurferth's estate'. *Tor- Turverdestune* 1086, *Turverdeston(a)* 1190, 1197, 1199×1216, *-wardeston(a)* c.1141, 1226, *Thurwaston(e) -vas-* 1272–1331 etc. ON pers.n. *Thorfrøthr -frithr*, secondary genitive sing. *Thorfrøthes*, + **tūn**. Db 593, SSNEM 195.

THUXTON Norf TG 0307. 'Thurstan's estate'. *Turstanestuna, Tures- T(h)urs- Thus- Toruestuna* 1086. AScand pers.n. *Thurstān* < ON *Thorsteinn*, genitive sing. *Thurstānes*, + **tūn**. DEPN, SPNN 425.

THWAITE 'A woodland clearing, a meadow'. ON **thveit**.

(1) ~ Norf TG 1932 → THWAITE ST MARY.

(2) ~ NYorks SD 8998. 'The clearing'. *Thwaite* 1860 OS. ModE **thwaite** (ON *thveit*).

(3) ~ Suff TM 1168. *Theyt* 1228, *Thueyt* 1216×72, *Tweyt(e)* 1254–1336, 1568, *Thweyt(e)* 1316–44, *Thwayght* 1524, *Thwate* 1610. DEPN, Baron.

(4) ~ ST MARY Norf TM 3395. 'St Mary's T'. P.n. *Thweit* 1254 + saint's name *Mary* from the dedication of the church. 'St Mary' for distinction from Thwaite near Aylsham TG 1932, *Đweyt* *[1044×7]13th S 1055, *Tuit* 1086, *Thweit* 1254. DEPN.

THWING Humbs TA 0570. 'The strip of land'. *Tu(u)enc* 1086, *T(h)ueng, T(h)weng(e)* 12th–1402, *Twyng(e) -i-* 1268–1582, *Thwyng(e)* 13th–1516. ON **thvengr** 'a shoe-lace, a thong' or ONb **thweng** 'a thong, a strap' used in a topographical sense, perhaps here of the narrow dry valley near which it lies; cf. the Dan p.ns. *Tvœng, Tving*. YE 112 gives prs [(θ)wiŋ], SSNY 106.

TIBBERTON Glos SO 7622. 'Tidbriht's estate'. *Tebriston* 1086, *Tribrichtonia -thuna* c.1145, *Ti- Tybert(h)un(a) -t(h)on(a)* 12th–1592. OE pers.n. *Tīdbriht* + **tūn**. Gl iii.194.

TIBBERTON H&W SO 9057. 'Estate called after Tidbriht'. *(into) tidbrihtincgtune* [c.978]11th S 1369, *Tidbertvn* 1086, *Tibrittune, Thibrictun* 1240, *Tiburtone* 1248, *Tibrinton* 1283. OE pers.n. *Tīdbriht -beorht* + **ing**[4] + **tūn** varying with *Tīdbriht* + **tūn**. Wo 170.

TIBBERTON Shrops SJ 6820. 'Tidbeorht's estate'. *Tetbristone* 1086, *Tibrihtona* 1155, *Tibricton(')* 1203–1256 with variants *Th-* and *-briht-*, *Ti- Tybrighton -brygh-* 1286–1431, *Tybritun' -ton* 1236×7, 1271×2, *Tiberton'* 1255×6, *Tyb(b)erton* 1535, *Tibberton* from 1577. OE pers.n. *Tīdbeorht -briht* + **tūn**. Sa i.291.

TIBSHELF Derby SK 4360. 'Tibba's shelf'. *Tibecel* 1086, *Ti- Tybb(e)s(c)helf(e) -self* 1179–1441, *Ti- Tyb(e)s(c)helf(e) -s(c)elf -chelf* c.1200–43 etc. OE pers.n. *Tibba* + **scelf**. Db 313.

TIBTHORPE Humbs SE 9655. 'Tibbi or Tibba's outlying farm'. *Tipetorp* 1086, 1166, *Tibe-* 1086, *Ti- Tybethorp(e)* 1274–1362, *Tibthorpe* from 1285, *-throppe* 1565. Pers.n. ON *Tibbi*, genitive sing. *Tibba*, or OE *Tibba* + **thorp**. Cf. the Dan p.n. *Tibberup*. YE 167 gives pr [tibθrəp], SSNY 70.

TICEHURST ESusx TQ 6930. 'Wooded hill frequented by kids'. *Tycheherst* 1248, *T(h)ichesherst -hurst* 1263–14th, *Thycehurst* 1291, *Tys(e)herst -hurst* 1438–16th, *Ticehurst* 1590. OE **tiċċen** + **hyrst**, SE dial. **herst**. The expected modern form is **Titchurst*: [s] for [tʃ] is an Anglo-Norman sound substitution and the long vowel [ai] for [i] a spelling pronunciation. The river Teise or Tyse is a back-formation from this name. Sx 450.

TICHBOURNE Hants SU 5730. 'Kid stream'. *ticceburna* [909]12th S 385, *(be) ticceburnan* [938]12th S 444, *Ti- Tycheburne -born* 1235–1428. OE ***tiċċe** + **burna**. The stream itself is *(inon) ticceburnan* *[701]12th S 242. Ha 163, Gover 1958.75.

TICKENCOTE Leic SK 9809. 'Kid cottage(s) or shelter(s)' *Tichecote* 1086, *Tichencote* 1154×89, *Ty- Tikencot(e) -k- -in-* 1199–1550, *Tickencote* 1610. OE **tiċċen**, genitive pl. **ticna**, + **cot**, plural **cotu**. The meaning is probably '(young) goat farm'. R 165.

TICKENHAM Avon ST 4471. 'Tica's dry ground in marsh or homestead' or 'kid enclosure'. *Tichehā* 1086, *Tiche(s)ham* 1201. OE pers.n. *Tica*, genitive sing. *Tican*, or **tiċċen**, genitive pl. **ticcna**, + **hamm** 2a, 3, or 4, or **hām**. The site is low on the edge of Tickenham Moor beside the Land Yeo. DEPN, PNE ii.178.

TICK FEN Cambs TL 3384. *Thickffenn* 1603, *Tick Fenn* 1824 OS. This might be 'thick fen' referring to dense undergrowth or 'kid fen', i.e. a fen where young goats were pastured (OE *tiċ- ċen*), but the forms are too late for certainty. Ca 253.

TICKHILL SYorks SK 5893. Either 'Tica's hill' or 'kid hill'. *Ti- Tykehil(l)' -hyll* c.1115–1438, *-hull(e)* 1189–1315, *Ti- Tykhill -hil(e) -hyll* 1249–1494, *-hull* 1347–1540, *Tickhill* from 1559. OE pers.n. *Tica* or ***ticca** or **ticcen** 'a kid' with ONb loss of *-n* + **hyll**. The hill is the site of the castle. YW i.52, L 171.

TICKLERTON Shrops SO 4890. Probably 'Ticcela's enclosure'. *Tichelevorde* 1086, *Ticlewrthin, Ti- Tykelewordin -wrth(yn)'* etc., *Tikelesworth, Tykelesworthyn, Tykewarthin' -wurthin', Tigwardin, Tykelyngworthin, Tyklingwrth'* all 13th cent., *Tyklarden* 1550, *Tycklerton* 1587. OE pers.n. **Ticcela* + **worth** varying with **worthiġn**. The variety of 13th cent. forms is exceptional. Sa i.292.

TICKNALL Derby SK 3523. 'Kid nook'. *(on) Ticenheale* [c.1002]c.1100 S 1536, [1004]14th S 906, *Tychenhal'* [1004]14th ibid., *Tichenhalle* 1086, *-hale -hala* 1086–1185, *Ti- Tykenhala -hal(e)* 1178–1413, *Ticknall* from 1445. OE **tiċċen** + **halh**, dative sing. **hale**. Db 665, L 107, 111.

TICKTON Humbs TA 0641. 'The kid-farm' or 'Tica's farm or village'. *Tichetone* 1086, *Ti- Tyketon(a)* 1297–1402, *Tykton* 1310–1436, *Tickton* from 1566. OE **tiċċen** or pers.n. *Tica*, genitive sing. *Tican*, + **tūn**. YE 201.

TIDCOMBE Wilts SU 2958. 'Titta's coomb'. *Titicome* 1086, *Ti- Tytecumbe -combe* 1197–1332, *Tydecumbe* 1249–1316, *Tidecombe* 1286, *Titcombe al. Tidcombe* 1748. OE pers.n. **Titta* + **cumb**. Wlt 356.

TIDDINGTON Oxon SP 6504. 'Tytta's hill'. *Titendone* 1086, *Teten- To- Tatindon'* 1208, *Tudendon* 1240×1, *Tet(t)indon'* 1242–6, *Tudington* 1244, *Tidington* 1797. OE pers.n. **Tytta*, genitive sing. **Tyttan*, + **dūn**. O 191.

TIDDINGTON Warw SP 2255. 'The estate called after Tida'.

I. *(æt) Tidinctune* [969]11th S 1318, *Tidingtun* [1016]18th S 1388, *Tydington* 1306, 1440×91, *Tiddington* 1629.

II. *Tidantun* [985]11th S 1350, *Tidinton* 1200–1318, *Tiddenton* 1629.

OE pers.n. *Tīda* + **ing**[4] varying with genitive sing. *Tīdan*, + **tūn**. Wa 232.

River TIDDY Corn SX 3064. Uncertain. *Tudi* late 11th, *Tody* c.1317, *Tyddie* early 17th. The basis is an OCo *Tūti, Tūdi* with normal change of *u* > *i*, possibly cognate with Irish *tuath* 'north', OIr *túath* 'left, north' < **toutâ*, **touto-s* adjective, 'left hand, left, good', Gothic *thiuth* 'good', W *Tut* 'good' used as a title of Morgan le Fey, all from PIE *tew-* 'look on in a friendly manner' (cf. Latin *tueor, tūtus*). PNCo 161, RN 406, GED þ 45.

TIDEBROOK ESusx TQ 6130. 'Tida's stream'. *Tydebroke* 1439. OE pers.n. *Tīda* + **brōc**. The stream probably marked the boundary of Tida's land, although **brōc** may have had the SE sense here 'boggy land, meadow'. Sx 389.

TIDEFORD Corn SX 3459. 'Ford across the river Tiddy'. *Tutiford* 1201, *Tuddeford* 1284, *Tediford (brigge)* 1345, *Todiford* 13th, 1395, *Tidiford* 1813. R.n. TIDDY SX 3064 + OE **ford**. PNCo 163 gives pr [tidifəd], Gover n.d. 223, DEPN.

TIDENHAM CHASE Glos ST 5598. *Tidenham Chase* 1830. P.n. TIDENHAM ST 5596 + ModE **chase** 'unenclosed parkland'. Gl iii.267.

TIDENHAM Glos ST 5596. 'Dydda's water-meadow'. *(æt) Dyddanhame, (to) Dydan hamme, (of, on) Dyddanhamme* [956]12th S 610, [1061×5]12th S 1555, *(æt) dyddan hamme* [1061×5]12th S 1426, *Dyddenhamm* 1060×6 S 1426, *Tede(ne)ham* 1086, *Tude(n)ham* 12th–1468, *Tydnam* 1526. OE pers.n. **Dydda*, genitive sing. **Dyddan*, + **hamm**. The change of initial *D-* to *T-* is probably due to the preceeding preposition *æt* 'at'. Gl iii.264 gives pr ['tidnəm].

TIDESWELL Derby SK 1575. 'Tidi's spring'. *Ti- Tydeswell(e) -is- -us-* 1086–1205 etc., *-wall(e)* 1259–1758, *Ti- Tyddeswell(e) -is-* 1192–17th, *-wall(e)* 1301–1658. OE *Tīdi*, genitive sing. *Tīdes*, + **wælla**. In the same parish is Tides Low SK 1477, *Tidislawe -is-* 1251, 1216×72, 'Tidi's burial-mound'. Db 172, 173.

TIDMARSH Berks SU 6374. 'Common marsh'. *Tedmerse* 1196, *Thedmers(h) -merch -mersche -mersshe* 1213 etc., *Tydmers* 1284, *-merssch* 1412, *T(h)ud(e)mers(he)* 1300. OE **thēod** 'nation, people' used in p.ns. with the sense 'common, public' + **mersc**. Brk 223.

TIDMINGTON Warw SP 2638. 'The estate called after Tidhelm'. *(æt) Tidelminctune* [977]11th S 1330, *Tidelmintun* 1086, *Ti- Tydelminton, Tidelintun* 11th, *Tydaminton* 1252, *Tidmington* 16th, *Tidillmington* 1685. OE pers.n. *Tīdhelm* + **ing**[4] + **tūn**. Wo 171.

TIDPIT Hants SU 0719. *Tidpit* 1811 OS. This must be associated with *Tide Farm* and *Tite Down* ibid., but the forms are too late for explanation. Tidpit lies in a hollow and the generic must be OE **pytt** 'a pit, a natural hollow'.

South TIDWORTH Hants SU 2448. *South Tedeworth* 1362. ME adj. **south** + p.n. *(æt) tudanwyrðe* [979×1015]14th, *Tedorde* (1×), *Todeorde* (2×) 1086, *Tudewrth* 1203, *-worth* 1236, 'Tuda's enclosure', OE pers.n. *Tuda*, genitive sing. *Tudan*, + **wyrth**. 'South' for distinction from North TIDWORTH SU 2349 accross the county boundary in Wilts. Ha 163, DEPN.

North TIDWORTH Wilts SU 2349. *Northtudewrthe* 1280, *Northtodeworth* 1313, *North Tydworth* 1724. ME adj. **north** + p.n. *(æt) Tudanwyrðe* [977×82]14th S 1498, *Todew(o)rde, Todwrde* 1086, *Tudeworda -wurda -worth* 1186–1316, 'Tuda's enclosure', OE pers.n. *Tūda*, genitive sing. *Tūdan*, + **worth, wyrth**. 'North' for distinction from South TIDWORTH Hants SU 2448 on the other side of the river Bourne. Wlt 370 gives pr [tedwəθ].

TIFFIELD Northants SP 6951. Partly obscure. *Ti- Tyfeld(e)* 1086–1275, *Ti-Tyffeld* from 1182, *Tythefeld al. Tyghfeld* 1551, *Tighfeld al. Tiffeld* 1695. Obscure element *ti-* + **feld** 'open country'. *Ti-* might represent OE **tīġ, tih, tīġe** 'a meeting place, a court', ***tiġe** 'a goat', or possibly an unrecorded word for a bee or swarm of bees cognate with the first element of OHG *zīdalweide*, MHG *zīdelweide* 'woodland area for bee-keeping'. The difficulty of explaining this name testifies to its high antiquity. See also *tie* in the Netherlands, NGN vii 30–1. Nth 93, lii, DEPN, L 241, 245.

TIGLEY Devon SX 7560. Possibly 'Tigga's wood or clearing'. *Tiggele(gh)* 1399. OE pers.n. **Tigga* + **tēah**. The pers.n. *Tigga* is not on record and an OE **tigga** might be postulated beside ***tiġe** 'a goat' and **ticcen** 'a kid, a young goat'. The sense would then be 'goat pasture', cf. GATELEY Norf TF 9624. D 297.

TILBROOK Cambs TL 0769. 'Tilla's stream'. *Tilebroc* 1086, 1202, *Tillebroc* 1202–42, *Ti- Tyllebrok* 1227–76 etc., *Tylbroke* 1276. OE pers.n. *Tilla* + **brōc**. The original name of the River TILL Beds TL 0268 and River KYM Cambs TL 1066. Hu 248.

TILBURY Essex TQ 6376. 'Tila's fortified place'. *Tilaburg* [c.731]8th BHE with 8th cent. variants *tiila-, tilla-* and *-burug, Tilaburh* [c.890]c.1000 OEBede, *Tillabyri* 1066–85, *Tiliberia* 1086, *Ti- Tyllebir(e) -ber(y) -bury* 1199–1335. OE pers.n. *Tila* + **burh**, dative sing. **byriġ**. Modern Tilbury clusters round Tilbury Docks opened in 1896 and Tilbury Fort built in 1670–83 on Tilbury Marshes. The medieval settlements were at East and West TILBURY TQ 6878, 6677. Another ancient Tilbury occurs at TL 7540, *Tileberiam* 1086, *Ti- Tyl(l)eberia -bery -bury* 1142–1427, and a *Tilleberie* 1086 in West Wycombe, Bucks. The triple occurrence of this compound may suggest that the specific is not a pers.n. but the OE adj. **til** 'good, useful' perhaps used as a stream, n. *Tila* 'the useful one'. Ess 173, 463, Bk 208, TC 180, Anglia 1996.545–6.

East TILBURY Essex TQ 6878. *Esttillebire* 1201. ME adj. **est** + p.n. TILBURY TQ 6376. Ess 173.

West TILBURY Essex TQ 6677. *Westtillebire* 1202. ME adj. **west** + p.n. TILBURY TQ 6376. Ess 173.

TILE CROSS WMids SP 1686. *Tylecross* 1725, *Tile Cross* 1834 OS. A crossroads. Wa 48.

TILE HILL WMids SP 2878. *Tylhull* 13th, *Tile Hill* 1834 OS. Possibly a place where tiles were made. Wa 186.

TILEHURST Berks SU 6673. 'Wooded hill where tiles are made'. *Tigelherst* 1167–1342 with variants *Ti- Tyg(h)el- Thigel-* and *-hurst(')*, *Tylhurst* 1224×5, *Tyleherst* 1517, 1675, *Tilehurst* 1761. OE **tiġel** + **hyrst**. Brk 194, L 198.

TILFORD Surrey SU 8743. Either 'Til(l)a's ford' or 'the good ford'. *Tileford* [c.1140]1341, 13th cent., *Tyle-* 1258–1346, *Tille- Tylle-* 1210–1332, *Tolle-* 1372. OE pers.n. *Til(l)a* or adjective **til**, definite form dative sing. **tilan**, 'useful, good', + **ford**. Sr 173, PNE ii.179.

River TILL Beds TL 0268 Cambs TL 0869. A back-formation from Tilbrooke TL 0769, *Tilebroc* 1086, 1202, *Ti- Tyllebroc -brok* 1202–76 etc., 'Tilla's' or 'useful stream', OE pers.n. *Til(l)a* or adj. **til**, definite form **tila**, + **brōc**. Hu 248.

River TILL Northum NT 9732. *Till* from [c.1040]12th, *Tille, Tylle* 13th cent., *Tilne* 1256, 1560. A pre-English r.n. on the root **tei-/*ti-* 'dissolve, flow'. Identical with French Tille, a tributary of the Saône, *Tyla* 7th cent., *Tila* 830, and ultimately related to r.n. TYNE. NbDu 179, RN 407, BzN 1957.262.

River TILL Lincs SK 9078. Possibly 'good river'. OE **til** 'good, useful'. *Til* [c.1190]14th, *Tyl(le)* c.1250, [1216×72]14th, *the Til* 1577×80. On the Till is Till Bridge, *Tilbrigge* 1357, *Til(le)brigges* 1367. RN 407, Cameron 1998.

TILLINGHAM Essex TL 9903. 'Homestead of the Tillingas, the people called after Tilla or Tilli'. *Tillingaham* [c.1000]c.1125, *Tillingeham* *[604×16]13th S 5, 1181–1247, *Tilingham* [942×c.951]13th S 1526, *Tillingham* from 1086. OE folk-n. **Tillingas* < pers.n. *Tilla* or *Tilli* + **ingas**, genitive pl. *Tillinga*, + **hām**. Ess 229, ING 122.

River TILLINGHAM ESusx TQ 8720. Named from Tillingham Farm TQ 8920, *Ty- Tillingeham* 1296, 1327, *Ty- Tillingham -hame* 14th cent.–1428, 'promontory of the Til(l)ingas, the people called after Tila or Tilli'. OE folk-n. **Til(l)ingas* < pers.n. *Tila* or *Tilli*, + **ingas**, genitive pl. *Til(l)inga*, + **hamm** 2. Sx 532, ING 125, ASE 2.46, SS 86.

TILLINGTON 'Estate called after Tulla'. OE pers.n. *Tulla* + **ing**[4] + **tūn**.

(1) ~ H&W SO 4645. *Tul(l)inton(e)* c.1170×85–1391, *Tullington -yn(g)* 1229–1535. He 50.

(2) ~ WSusx SU 9621. *Tullingtun* 960 S 687, *Tul(l)- Tolin(g)ton -y(n)-* 1135–1626, *Ty- Tillington* from 1528, *Ti- Tylleton* 1558–1609. Sx 121.

(3) ~ COMMON H&W SO 4546. *Tillington Common* 1832 OS. P.n. TILLINGTON SO 4645 + ModE **common**.

TILLY WHIM CAVES Dorset SZ 0376. The remains of an ancient stone quarry in the face of the cliff, unworked since 1812. *Tilly Whim* 1811. Surname *Tilly* + **whim** 'a windlass'. Do i.59.

TILMANSTONE Kent TR 3051. 'Tilman's estate'. *Tilemanestun* [1072]n.d., *-tone* 1086, *Tile(s)mannestune* c.1100, *Ti- Tylemaneston(')* 1205–1242×3. OE pers.n. *Tilmann* or **tilmann** 'husbandman, farmer', genitive sing. *Tilmannes*, + **tūn**. PNK 584.

TILNEY ALL SAINTS Norf TF 5617. *Tilney All Saints* 1824 OS. P.n. Tilney + *all saints* from the dedication of the church. Tilney, *Tilnea* 1170, 1190, *Tillenee* 1197, *Tilneie* 1207, *Tilneye* 1242, is 'Tila's island', OE pers.n. *Tila*, genitive sing. *Tilan*, + **ēġ**. 'All Saints' for distinction from TILNEY ST LAWRENCE TF 5413. DEPN.

TILNEY HIGH END Norf TF 5613. This place is called *Tun Green* 1824 OS.

TILNEY ST LAWRENCE Norf TF 5413. *Tilney St Lawrence* 1824 OS. P.n. Tilney as in TILNEY ALL SAINTS TF 5617 + saint's name *Lawrence* from the dedication of the church.

TILSHEAD Wilts SU 0347. 'Theodulf's hide of land'. *Theodulveside, Tidulfhide* 1086, *Tidolfeshida* 1167, *Ty- Tidulfeshid -olf-* 1242–1332, *Tydolside* 1390, *Tyleshide* 1403, *Tyleshed* 1502, *Tilsett* 1639. OE pers.n. *Thēod(w)ulf*, genitive sing. *Thēod(w)ulfes*, + **hīd**. The pers.n. was early confused with *Tīd(w)ulf*. Later the whole name was wrongly interpreted as 'the head of the river Till', Till itself being a back-formation from the p.n. for the stream originally called *Winterbourne Water* c.1540. Wlt 236 gives pr [tilzhəd], formerly [tilzed], 10.

TILSTOCK Shrops SJ 5437. 'Tidhild's outlying farm'. *Tildestok* 1211, *Tyldestok* 1327. OE feminine pers.n. *Tīdhild*, genitive sing. *Tīdhilde*, + **stoc**. DEPN.

TILSTON Ches SJ 4651. 'Tilli or Tilla's stone'. *Tilleston* 1086, 1291, *Tilestan* 1217×72, *Ty-* 1260, c.1345, *Tilstan* 1217×72–1662 with variant *Ty-*, *Tilston* from 1287 with same variant to 1767. OE pers.n. *Tilli* or *Til(l)a*, + **stān**. Possibly a mile-stone on the Roman road here (Margary no.6a). Che iv.58. Addenda gives pr ['tilstən], older local ['tilssn].

TILSTON FEARNHALL Ches SJ 5560. *Tilston Farnhale* 1427, *Tidleston et Farnall* 1475, is a compound of two separate p.ns., Tilston, *Tidulstane* 1086, *Ti- Tydulstan -tona* early 13th–1310, *Tydelstan* 1311, *Ty- Tidleston* [1265×84]1640, 1347–1536, *Tilstan* 1440, *Tilston* from 1417, 'Tidwulf's stone', OE pers.n. *Tīdwulf*, genitive sing. *Tīdwulfes*, + **stān**, and the lost *Fornale* 1353, *Farn(h)ale* 1363–1473, *Farn(h)all* 1438–1553, *Fern(eh)all(e)* 1539–1842, *Fearnall* 1546, 'fern nook', OE **fearn** + **hale**, dative sing. of **halh**, which was within Tilston but a separate manor. Che iii.317–8.

TILSWORTH Beds SP 9724. 'Thyfel or Tyfel's enclosure'. *Pileworde* (sic for *þile-*) 1086, *Thuleswrthe -worth* 13th cent., *Tul(l)es- Til(l)esworth* 13th–15th cents. OE pers.n. **Thȳfel* or **Tyfel*, genitive sing. **Thȳfeles*, **Tyfeles*, + **worth**. Alternatively the specific might be OE **thȳfel** 'a bush, a thicket'. Bd 133.

TILTON ON THE HILL Leic SK 7405. *Tilton on the Hill* 1688. Earlier simply *Tile- Tillintone* 1086, *Ti- Tylton(e)* c.1130–1610, 'Tila's farm or village', OE pers.n. *Tila*, genitive sing. *Tilan*, + **tūn**. Lei 326.

TIMBERLAND DELPH Lincs TF 1560. P.n. TIMBERLAND TF 1258 + Mod dial. **delph** 'a drain' (OE *ġedelf*).

TIMBERLAND Lincs TF 1258. 'The grove where timber is obtained'. *Timbrelund, Timberlunt* 1086, *Ti- Tymberland -lond* 1185–1642, *-lound* 14th cent. ODan **timbær** or OE **timber** + **lundr**. Perrott 340, SSNEM 161, L 208.

TIMBERSBROOK Ches SJ 8962. 'Brook running through timber-trees'. *Timber Brook* 1831, *Timbers Browk* 1840. ModE **timber** + **brook**. Che ii.292.

TIMBERSCOMBE Somer SS 9542. 'Coomb where timber is obtained'. *Timbrecōbe* 1086, *Timbrescumba* 1176, *Ti- Tymbercumbe* 1227, [13th] Buck, *Timbercomb* 1610. OE **timber**, genitive sing. **timbres**, + **cumb**. DEPN, L 93.

TIMBLE NYorks SE 1853. Uncertain. *tun mel* (for *tim mel*) [972×92]11th S 1453, *(on) Timbel* c.1030 YCh 7, *Tinbel* 1173×85, *Ti- Tymble* 1086–1679, *Ti- Tymbel* 1155×70–1303. Possibly an OE ***tymbel**, ***timbel** 'that which falls' from *tumbian* 'to tumble, to fall'. The reference would be to the steep ridge on which the village lies or to a landslip or a tumbling watercourse. YW v.128.

TIMPERLEY GMan SJ 7888. 'Clearing or wood where timber is obtained or prepared'. *Timperleie* 1211×25, *Ti- Tymperleg(h) -legh(e) -le(e)* etc. 1293–1350, *Tympurley* 1419. OE **timber** + **lēah**. Che ii.31.

TIMSBURY Avon ST 6658. 'Timber grove'. *Temesbare, Timesberie* 1086, *Timberbarewe* 1200, *Timberesberwe* 1233. OE **timber**, genitive sing. **timberes**, + **bearu**. The later forms are unambiguous and refer to woodland where building timber was obtained. There is, however, an unnamed unrecorded camp a mile to the E on Tunley Hill at ST 6859 which might account for the alternative DB form in *-berie* and the modern forms in *-bury* (< OE *byrig*, dative sing. of *burh* 'a fortified place') and hence 'timbered fort' with DB *r/s* confusion. DEPN, cf. Domesday Studies 136.16.

TIMSBURY Hants SU 3424. 'Timber fort or manor house'. *Timbreberie* 1086, *Timberbury* 1252, *Timbresbury* 1236. OE **timber** + **byriġ**, dative sing. of **burh**. Ha 163.

TIMWORTH GREEN Suff TL 8669. *Timworth Green* 1836 OS. P.n. *timwrtha, timeworda* 1086, *Timuuorde* c.1095, *Timeworthe* 1166, *Tymworth* 1610, 'Tima's enclosure', OE pers.n. **Tīma* + **worth**, + ModE **green**. DEPN.

TINCLETON Dorset SY 7791. 'Valley of the small estates'. *Tincladene* 1086, *(Up, Hole) Tincle- Tynkelden(e)* 1260–1535, *Tynkelton* 1535. OE ***tȳninċel**, genitive pl. ***tȳn(in)cla**, + **denu**. The small estates were divisions of Tincleton subsequently known as *Hole, Up, Est* and *West Tincleden*. Do i.329.

TINDALE Cumbr NY 6159. 'Tyne valley'. *Ti- Tyndale* from late 12th. R.n. South TYNE + OE **dæl**. Cu 36, RN 426.

TINDALE TARN Cumbr NY 6058. 'Lake called or at *Tyniel*'. *Tymelterne* (for *Tyniel-*) 1295, *Tynyel-logh* 1485, *Tyniell, Tynnyell Tarn* 1589, *Tyndale Tarne* 1610. Cf. *Tynyelfell* 1485, *Tiniellfell* 1603 for Tindale Fells NY 6057. P.n. *Tyniel* of unknown meaning + ME **tern** (ON *tjǫrn*). The earlier explanation of *Tyniel* as 'fertile upland region by the river Tyne', r.n. (South) TYNE + **ial*, is based upon a ghost-word. Possibly identical with Co Tinnel SX 4263, **tyn-yel*, unknown element + adjective suffix **-yel*. Whatever its origin the form of *Tyniel* has been assimilated to that of TINDALE NY 6159. Cu 36, RN 426, CPNE 138–9.

TINGEWICK Bucks SP 6532. 'Dairy farm, trading place at Tiding, the place called after Tida'. *Tedinwiche* 1086, *Tinsuicȝ* 1088×94, *Tingwic(h')* 1163, c.1198, *Tengewich(i)a* 1167, *Tyngewik* c.1218, 1262, *Tynchewyk* 1281–1422, *Ti- Tyngwik -wyk* 1227–96. OE p.n. **Tīding* < pers.n. *Tīda* + **ing**², locative-dative sing. **Tīdinġe*, + **wīċ**. Bk 65 gives pr [tindȝik], BzN 1967.386.

TINGLEY WYorks SE 2826. 'Hill where the *thing* or council meets or met'. *Thing(e)- Thynglau -law(e)* early 13th–1411, *Tinglawe* 1608. OE **thing** + **hlāw**. Probably the meeting-place of Morley Wapentake. The site is a mound at SE 281261 on top of the hill just S of the crossing of the Wakefield-Bradford and Leeds-Huddersfield roads. YW ii.175.

TINGRITH Beds TL 0032. 'Assembly brook'. *Tingrei* 1086, *Tyngri* 1220–1365, *Tingrith* 1276–1509, *Tyngryff(e)* 15th–16th cents. OE **thing** + **rīth**. This was the meeting place of Manshead hundred. PNE ii.204 takes the specific to be 'probably OE', Bd 135 assumes English origin without question. There is probably no need to suggest ON **thing** here. Bd 134, L 29.

TINHAY Devon SX 3985. 'Between waters'. *Bituinia* 1194, *Tuneo* 1316, *Tenyowe* 1486×1515, *Twyneow* c. 1500, *Tinhay* 1809 OS. OE **betwēonan** + **ēa** 'a stream'. Tinhay occupies a tongue of land between the Thrushel and Lyd rivers. Other examples of this p.n. type in Devon are Tinney SS 3099 between the Tamar and Devril Water, *Tuneo* 1529×32, *Tyneowe* 1550, and Twinyeo SX 8476 between the Bovey and the Teign, *Betunia* 1086, *Betwynyo*

1303, *Twyneya* 1242, *Twyneyo* 1311. In these names the accusative pl. **ēa** seems to be used rather than the expected dative pl. **ēam**. D 190, 163, 479, Campbell §628.4 .

TINHEAD Wilts ST 9252. 'Ten hides'. *Tunheda* 1190, *-hyde -hid(e), Tynhide, Tenhyd(e)* 1216–1373, *Tynhedd* 1558×1603. OE **tīen** + **hīd**. Wlt 141, PN and History 22.

TINNEL Corn SX 4263 → TINDALE TARN Cumbr NY 6058.

TINNEY Devon SS 3099 → TINHAY SX 3985.

TINSHILL WYorks SE 2539. Possibly 'Tind's hill'. *Tinshill* 1639. Possibly ON pers.n. *Tindr*, genitive sing. *Tinds*, + **hyll**. YW iv.191, SPN 285.

TINSLEY SYorks SK 4090. 'Tynni's mound'. *Ti(r)neslauue* 1086, *Ti- Tyneslaw(e)* 1196–1379, *Ti- Tynneslawe* 1303–1418, *Ti- Tyneslowe* 1327–1612, *Ti- Tynsley* from 1525. OE pers.n. **Tynni*, genitive sing. **Tynnes*, + **hlāw**. YW i.190.

TINTAGEL Corn SX 0588. 'Fort of the constriction'. *Tintagol* [c.1137]mid12th, *Tyntagel* [c.1220]c.1275, [1233]c.1300, *Tintagel* 1212, *Tintaieol(e), Tintageol* [c.1220]c.1260, *Tynthagel* 1259, *Dyntagel* 1302, *Tyntagelle* 1427. Co ***dyn** + ***tagell** with AN [dʒ] for Co [g]. The reference is to the Dark Age occupation of the headland and the narrow neck of land joining the promontory to the mainland. PNCo 163, Gover n.d. 82, DEPN.

TINTAGEL HEAD Corn SX 0489. 'Tintagel headland'. *Tintagell Head* 1748. P.n. TINTAGEL SX 0588 + ModE **head**. PNCo 163.

TINTEN Corn SX 0675 → TINTINHULL Somer ST 4919.

TINTINHULL Somer ST 4919. 'Hill called or at *Tinten*'. *Tintanhulle* [939×46]18th S 1728, [959×75]18th ECW, *Tvtenelle, Tintehalle* 1086, *Tintenellae, Tintehella, Tuttehella* 1086 Exon, *Tintenell(e)* [1091×1106–1189×99]14th M, *-hell(e)* [1107×18, 1174×80]14th M, *Ty- Tintenhulle* [1107×18–1361×2]14th M, *Tintenhille* 1168, *Ti- Tyntehull(e)* 1219–1316, *Tinteshull* [1279×80]14th M. P.n. *Tinten* as in Tinten Corn SX 0675, *Thinten* 1086, *Tinten* 1086 Exon, + OE **hyll**. Turner suggests a Brit **tintinio-* of unknown meaning, cf. Tintigny, Belgium, *Tintiniacum*, though this is likely to be a doublet of the Fr p.n. Tinténiac containing pers.n. *Tintinius*. Probably, therefore, ***tin** for PrW ***din**, Co **dyn**, ***tyn** 'fort', as in Tenby, Dyfed SN 1300, 'little fort', *Dinbych* c.1275, TINTAGELL Corn SX 0588, Tintern Gwent SO 5200, 'royal fort', *Dindyrn* c.1150 (PrW **dinteryn* < Cetlic **tegerno-* 'lord'), TRENARREN Corn SX 0348, originally *Tyngharan* 1302 'crane fort' (Co *tyn-garan*) + unknown element. A W explanation seems preferable to unrecorded OE pers.n. **Tinta*, genitive sing. **Tintan*, + **hyll**. DEPN, Turner 1950.122, 1952–4.19, Holder ii.1854, LHEB 279–80, 320, 446–7, CPNE 84. The Wades (1907) record an obsolete pr [tiŋknəl].

TINTWISTLE Derby SK 0297. 'Fork of the river *Teign*'. *Tengestvisie* 1086, *-twysel -twissell* 1345, 1360, *T(h)enge- Tinge(t)twisel(e) -twy- -ell -il -ul* 1286–1494, *Tingtwisel* 1369, *-twistle* 1603, *Tintwistle* 1724. R.n. **Teign* + OE **twisla**. The reference is to the meeting of ARNFIELD BROOK and the river ETHEROW. The former is a modern name and it is suggested that its original form was Teign identical with r. TEIGN Devon SX 7689. Che i.320.

TINWELL Leic TF 0006. Partly uncertain. *Tedinwelle* 1086, *Tineguella* 1125×8, *Tiningewelle* c.1250, *Tineuuell(e) -well(e)* c.1150–1347, *Ty- Tinwell(e)* 1221–1553. The generic is OE **wella** 'a stream, a spring', the specific either OE pers.n. *Tīda*, genitive sing. *Tīdan*, or possibly **Tīdna* or **Tīdin*; or, if the DB form is erratic, OE ***tiġe** 'a goat', genitive pl. ***tiġna**, and so either 'Tida's spring' or 'the goat stream'. The single form *Tiningewelle* is probably insufficient to support the otherwise attractive explanation 'spring of the Tyningas, the people called after Tyni', OE folk-n. **Tȳningas* < pers.n. **Tȳni*, a diminutive of names in *Tūn-* such as *Tūnbeorth* etc., + **ingas**, genitive pl. **Tȳninga*, + **wella**. R 167, PNE ii.179.

TIPTON ST JOHN Devon SY 0991. P.n. Tipton + saint's name *John* from the dedication of the church (1840). Tipton, *Tipton* from 1381, *Tippeton* 1413, *Tuppeton* 1382, is either 'Tippa's farm or estate' or 'edge farm', OE pers.n. **Tippa* or ***tipp(a)** + **tūn**. Tipton lies beneath the end of a promontory of high ground; it is tempting to see here an early example of ModE *tip*, a word not certainly recorded in English before c.1440. Although its origin is unclear it is a good example of an apophonic formation in the series *tip-tap-top*, of which the last two do occur in OE, *tæppa, tæppian* and *topp*. D 606.

TIPTON WMids SO 9592. 'Tibba's estate'. *Tibintone* 1086, *-ton* 1242. OE pers.n. *Tibba* + **ing**[4] or genitive sing. *Tibban*, + **tūn**. DEPN.

TIPTREE Essex TL 8916. 'Tippa's tree'. *Tipentrie* 1154×89, *Ty- Tippetre* 1243–1309, *Tiptre(e)* 1401, 1521. OE pers.n. **Tippa*, genitive sing. **Tippan*, + **trēow**. Ess 307.

TIRLEY Glos SO 8328. 'Circular wood or clearing'. *Trineleie* 1086, *Tri- Trynlei(e) -leg(a) -legh -ley(e)* 1086–1539, *Trilleg(h)' -ley(e) -lee* 1221–1506, *Turley* 1551, *Tyrley* c.1560. OE ***trind** + **lēah**. Gl iii.149.

TIRRIL Cumbr NY 5026. 'Shieling where resinous wood is obtained'. *Ti- Tyrerghe* c.1189, *-erh(e) -ergh(a)* c.1200–1348, *Tyrrer* 1247–1346, *Tyregh(e)* 1256–1401, *Tyreth* 1377, 1401, *Ti- Tyrel(l)* 1265–1787, *Tirril* from 1619. ON **tyri** + **ǽrgi**. The change from *Tyrrer* to *Tyrel* is probably a case of dissimilation. We ii.208, SSNNW 72.

TISBURY Wilts ST 9429. 'Tyssi's fortified place'. *Ty- Dyssesburg* c.800 Boniface, *(to) Tyssebyrig* 899×924 S 1445, *(æt) Tissebiri* [983]15th S 850, *Tisseberie* 1086, *Tussebury* 1295–1309. OE pers.n. **Tyssi*, genitive sing. **Tysses*, + **byriġ**, dative sing. of **burh**. Wlt 194 gives pr [tizbəri].

TISSINGTON Derby SK 1752. 'Estate called after Tidsige'. *Tizinctun* 1086, *Ti- Tyssinton -yn-* late 11th–1610, *Ti- Tyscintun -ton(e)* c.1141–1297, *Tizinthon'* c.1215, *Ti- Tyssington(e) -yngton(e)* 1276–81 etc. OE pers.n. *Tīdsīġe* + **ing**[4] + **tūn**. Forms in *Tiz-, Tisc-* are AN spellings for *Tits-* < *Tids-*. Db 409.

East TISTED Hants SU 7032. *Est Tystede* 1263. ME adj. **est** + p.n. Tisted as in West TISTED SU 6529. Ha 164, Gover 1958.94.

West TISTED Hants SU 6529. *Westysted(e)* 1236–1462, *Westistede* 1346. ME adj. **west** + p.n. *ticces stede* [932]12th S 417, *west Ticces tede* (rubr.), *(æt) Ticcestede* [941]12th S 511, *ticce stedæ* (rubr.), *(æt) Ticcetesde, Ticceststede* (sic) [943]12th S 488, *Tistede* 1086–1346 with variant *Ty-*, 'place where kids are kept'. OE ***tiċċe**, genitive sing. **tiċċes**, + **styde**. Alternatively the specific might be a pers.n. **Tiċċa* from **tiċċe*. Ha 164, Gover 1958.89 which gives pr [tistid], Stead 273.

TITCHBERRY Devon SS 2427. 'Tetti's fortified place'. *Tetisbyr'* 1249, *Tettesbyry* 1308, *Tyttesbury* 1396, *Titsbury* 1546, *Tichbury* 1634. OE pers.n. **Tetti*, genitive sing. **Tettes*, + **byriġ**, dative sing. of **burh**. D 76.

TITCHFIELD Hants SU 5305. 'Kid open land'. *ticcefeld* [982]14th S 842, *Ticefelle* 1086, *Tichesfeld* 1182, *Tychefeld(e)* 1218–42. OE ***tiċċe** + **feld**. The form *ticcanfelde* also in S 842 does not belong here but on the Isle of Wight. It is possible that the specific is a pers.n. *Tiċċa* derived from *tiċċe*. Gover 1958.32, Ha 164, ECW 58.

TITCHMARSH Northants TL 0279. 'Marsh where kids are pastured'. *Tut(e)anmersc* [973]15th, *Ticanmersc(e), Ticceanmersce* [c.975]14th, *Tircemesse, Ticemerse* 1086, *Ti- Tychemers(he)* 1175–1428. OE **tiċċe**, genitive sing. **tiċċan**, + **mersc**. Nth 221 followed by L 53 prefers pers.n. *Tiċċea*, genitive sing. *Tiċċean*, as the specific.

TITCHWELL Norf TF 7643. 'Kid spring'. *(et) Ticeswelle* c.1045 S 1489, *Tigeswella, Tigeuuella* 1086, *Tichewell* 1206. OE **tiċċe**, genitive sing. **tiċċes**, + **wella**. DEPN.

TITHBY Notts → TYTHBY SK 6936

TITLEY H&W SO 3360. 'Tit wood or clearing'. *Titel(l)ege* 1086,

Titellega 1123, *Tyteleye* 1304. OE * **titel**, a diminutive of ***tit(e)** 'a titmouse', + **lēah**. Neither ***tit** nor its diminutive ***titel** are recorded in OE, the earliest occurrence being in the compound *titemose* c.1325. Cf. ON **titlingr** 'a tit, a sparrow'. The underlying sense (as of OE *titt* 'teat') is probably 'something small'. He 190.

TITLINGTON Northum NU 0915. 'The estate called after Tytel or Tyttla'. *Tedlintona* 1123×8, *Titlington(a)* 1197–1336, *Tithlington* 1166, *Tidlington* 1167, *Thitelittonam* [c.1150]14th. OE pers.n. *Tyt(t)el(a)* + **ing**[4] + **tūn**. NbDu 197, DEPN.

TITTENSOR Staffs SJ 8738. 'Tittin's ridge'. *Titesovre* 1086, *Tichesoura* 1167, *Tidesovre* 1200, *Tineshovere, Tinneshore, Tiddesor(e)* 1203, *Titneshovere* 1236, *-overe* 1242, *Tyntnesoure* 1351, *Tentenhall otherwise Tentenshale otherwise Tytenshall otherwise Tittensor otherwise Titensore* 1617. OE pers.n. **Tittīn*, genitive sing. **Tittīnes*, + **ofer**. The village occupies a shelf at the lower end of a long ridge of high ground. DEPN, L 175–6, Horovitz.

TITTERSTONE CLEE HILL Shrops SO 5978.'(That part of) Clee Hill called T'. *Theterston Cle* c.1540, *Stitterstones Hill* 1577, *Titterstone Hill* 1832 OS. P.n. *Tyderstone* c.1540, ModE **titterstone** 'a stone that see-saws', ModE dial. **titter** 'sway to and fro, see-saw', + p.n. CLEEHILL SO 5975. Gelling.

TITTESWORTH RESERVOIR Staffs SJ 9959. P.n. *Tetesword'* 1203, 1222×32, 1547, *-worþe* c.1300, *-worth(e)* 1302–1614, *Tettesworth(e)* 1216×72–1695, *Tetsworth* 1558×1603, *Tittesworth* from 1811, 'Tetti's enclosure', OE pers.n. *Tetti*, genitive sing. *Tettes*, + **worth**, + ModE **reservoir**. The site of Tittesworth, *Tittesworth House* 1842 OS, has been drowned by the lake. Oakden.

TITTLESHALL Norf TF 8921. 'Tyttel's nook'. *Titeshala* 1086, *Titleshal* 1200, 1205f, *Tetleshal* 1205f, *Tutleshal* 1275. OE *Tyttel*, genitive sing. *Tyttles*, + **halh**. DEPN.

TIVERTON Ches SJ 5560. 'Red-lead farm'. *Tevreton* 1086, *Teverton(a)* early 13th–1819, *Tiverton* from 1253, *Tar(u)- Tere- Tear- Teirton* 16th cent. OE **tēafor** + **tūn**. The name might denote a red-painted building, but more likely a place where red-lead was available. Che iii.320 gives pr ['tivərtn] and older local ['ti:ərtn].

TIVERTON Devon SS 9512. 'Settlement, estate at or called *Twyfyrd*, the double ford'. *Tuuertone* 1086, *Tovretona* 1086 Exon, *Tuiverton* 1141×55, *Twi- Twy- Tuyverton* 1228–1341, *Tivertun* 1219, *Tiverton al. Twyford Town* 1695. P.n. *Twyfyrd* + OE **tūn**. Tiverton lies in the junction of the Lowman and the Exe, both of which were fordable here. The original double ford is recorded in S 1507 as *(æt) Twyfyrde* [873×88]11th, OE **twī** + **fyrde**. D 541 gives pr [tivətən].

TIVERTON JUNCTION STATION Devon ST 0311. A station on the line from Taunton to Exeter, once the junction for Tiverton.

TIVETSHALL ST MARGARET Norf TM 1686. *Tivetshall St Margaret* 1838 OS. P.n. *(of) Tifteshale* 11th, *Teuetessalla, Tiuetessala, Teueteshala* 1086, *Tiueteshale* [1087×98]12th, *Tiftes- Tivetshale* 1254, apparently OE ***tifet** or ***tyfet** of unknown meaning, genitive sing. ***tiftes, *tyftes**, + OE **halh** 'a nook', + saint's name *Margaret* from the dedication of the church. The specific has been associated with 18th–19th cent. Northern English *tufit, tewfet* 'a lapwing', but this is very uncertain. 'St Margaret' for distinction from TIVETSHALL ST MARY TM 1686. DEPN.

TIVETSHALL ST MARY Norf TM 1686. *Tivetshall St Mary* 1837 OS. P.n. Tivetshall as in TIVETSHALL ST MARGARET + saint's name *Mary* from the dedication of the church.

TIXALL Staffs SJ 9722. 'The kid's nook'. *Ticheshale* 1086, *Tikeshala* 1167, *-hale* 1242, *Tixhaul* c.1540. OE **tiċċen**, genitive sing. **ticnes**, + **halh** in the sense 'a small valley'. DEPN, L 105, 111.

TIXOVER Leic SK 9700. 'The ridge where kids are grazed'. *Tichesovre* 1086, *Tic(h)esoure* 1104×6–1392 with variants *Tyk- Tik-* and *-our(a) -ovr(e) -ofre -hov(e)re, Tixover* from 1461, *Tichesora* c.1131, 1130×3, *Tixor* 1758. OE **tiċċen**, genitive sing. **ticnes**, + **ofer** partly replaced by **ōra**. R 304, L 177.

TOCKENHAM Wilts SU 0479. 'Tocca's homestead or water-meadow'. *Tockenham* from 1322 including [854]14th S 306, *Tochehā* 1086, *T(h)okeham* 1194–1294, *Tokenham* 1289. OE pers.n. *Tocca*, genitive sing. *Toccan*, + **hām** or **hamm**, possibly in the sense 'dry ground in marsh'. Wlt 272.

TOCKENHAM WICK Wilts SU 0381 → Tockenham WICK.

TOCKETTS Clevel NZ 6117. Partly uncertain. *Theoscota* [1043×60]12th, *Toscutun, Tocstune* 1086, *Tou- Tofcotes* 1187–1412, *Toscotes* 1202, *Thocotes* 13th cent., *Tokotes* 1301, *Tockets* 1665. The generic is OE **cot** 'a cottage', pl. **cotu**, dative pl. **cotum**, the specific uncertain. YN 153.

TOCKHOLES Lancs SD 6623. Partly uncertain. *Cokolles* (for *T-*) 1199, *Tocholis* [c.1200]1268, *-holes* 1246, 1497, *Tokhol, Thoc(h)ol* 13th, *Tockholes* from 1311. Pers.n. OE *Tocca* or ON *Tóki* or ON by-name **Tók* 'fool' + OE **hol(h)**, ON **hol** 'hollow, valley' with secondary ME plural in **-es**. Tockholes is situated on the slope of Winter Hill (905ft.) from which three shallow valleys descend to the W. La 75, Mills 141, SSNNW 170, Jnl 17.47.

TOCKINGTON Avon ST 6086. 'Estate called after Toca'. *Tochintvne* 1086, *Tokinton(a) -yn-* 1154×89–1540, *To(c)kington -yng-* 1308–1595. OE pers.n. *Toc(c)a* + **ing**[4] + **tūn**. Gl iii.121.

TOCKWITH NYorks SE 4752. 'Toki's wood'. *Tocvi* 1086, *Tocwid -wyd* [1109×35]13th–1305, *Tok- Tocwic -wik* [1119×35]13th–1259, *-with* 1121×7–1536. ON pers.n. *Tóki* + **vithr** possibly replacing or replaced by OE pers.n. *Toc(c)a* + **wīc**. YW iv.250, SSNY 106.

TODBER Dorset ST 7920. Uncertain. *Todeberie* 1086, *-bire* 1209, *-ber(e)* 1244–1362, *Toteberg(a) -berge* 1177×94, 1212, *-bire* c.1217, *-ber(e)* 1299–1495, *Todbur(y)* 1278, 1411, *-bere* 1644. The specific is either OE pers.n. *Tota* or **tōte** 'a look-out', the generic either **beorg** 'hill, barrow', **bearu** 'wood, grove' or **bǣr** '(woodland) pasture', with some confusion with *bury* (< OE *byriġ*, dative sing. of *burh*). A 'look-out hill' would suit the topography of the parish where land in the N rises to over 250ft. Do iii.82 gives pr ['tɔdbə].

TODDINGTON Beds TL 0128. 'Hill of the Tudingas, the people called after Tuda'. *Dodintone, Totingedone* 1086, *Tudingedon -ton* 1166–1243, *Tu- Todington -yng- -don* 1219–1526. OE folk-n. **Tudingas* < pers.n. *Tuda* + **ingas**, genitive pl. **Tudinga*, + **dūn** varying with **tūn** 'estate'. Bd 135, L 145, 151.

TODDINGTON Glos SP 0332. 'Estate called after Tuda'. *Todintvn* 1086, *-inton(a) -yn-* 1269–1342, *-ing- -yng-* 1287–1708, *Tudinton(a)* 12th–1248, *Tuddyngton -ing-* 16th cent. OE pers.n. *Tuda* + **ing**[4] + **tūn**. Gl ii.28 gives pr ['tɔdintən].

TODENHAM Glos SP 2436. 'Teoda's water-meadow'. *Todanhom* [804]11th S 1187, *Teode- Toteham* 1086, *Thuden- Teudenham* 1221, *Todeham* 1231–1300, *-hamme* 1327, *Todynham* 1287, 1298, *Todenham* from 1291. OE pers.n. *Tēoda*, genitive sing. *Tēodan*, + **hamm**. Gl i.258.

TODHILLS Cumbr NY 3663. 'Foxholes'. *Todholes* 1568. ME **todhole** 'fox-earth'. Cu 149.

TODMORDEN WYorks SD 9324. 'Totta's boundary valley'. *Tottemerden, Totmardene* 1246, *Todmereden* 1298, *-marden* c.1300, *Todmorden* from 1641. OE pers.n. *Totta* + **(ġe)mǣre** + **denu**. The reference is to the county boundary with Lancashire. YW iii.174 gives prs ['tɔdmədin, 'tɔdmɔ:dən].

TODWICK SYorks SK 4984. 'Tata's dairy-farm'. *Tateuuic* 1086, *-wich* 12th–1316, *Tatwic* 13th, *Tadwy(c)k* 1305, *Totewik(e) -wyk(e) -wick* 13th–1366, *Todewyk(e)* 1268–1428, *Todwik(e) -wyk* 1391–1822. OE pers.n. *Tata* + **wīc**. YW i.157.

TOFT 'A building plot, a curtilage, a deserted site'. ON **toft**. This element belongs to a younger stratum of name-giving than **bȳ** and **thorp**. Holmberg 1946, Hellberg 1967, KLMM s.v. *-toft*, SSNEM 137–8.

(1) ~ Cambs TL 3655. *Tofth* 1086, *Toft(e)* from 1086, *Toftes* 1086–1314. Ca 164.

(2) ~ Lincs TF 0617. *Toft* from 1086. Perrott 206, SSNEM 150 no.3.

(3) ~ HILL Durham NZ 1528. 'Hill at Toft, the building site'. *Toft Hill* 1647. ODan, late OE **toft** + **hyll**. *Toft* is often used of a deserted site. This one was earlier referred to as *les toftes de baronia* 'the tofts of the barony' (sc. of Evenwood NZ 1524) 1382, *le Toft* 1418, *Les Toftes next Raby* 1459.

(4) ~ MONKS Norf TM 4294. 'T belonging to the monks' sc. of Préaux abbey in Normandy. *Toft monachorum* 1386. P.n. *Toft* 1086 + Latin **monachorum**, genitive pl. of **monachus**. DEPN.

(5) ~ NEXT NEWTON Lincs TF 0488. *Toft iuxta Neuton* 1311, *Tofte next Newton* 1530. P.n. *Tofte* 1086–1292, *Toft* from c.1115, *Toftis* [1110]c.1200, + p.n. NEWTON TF 0587. Li iii.64, SSNEM 150 no.1.

(6) ~ SAND Lincs TF 4440. P.n. Toft as in FISHTOFT TF 3642 + ModE **sand**. A sand-bank in The Wash.

(7) Bircham ~s Norf TF 7732. 'T belonging to or beside Bircham'. *Brechamto[f]tes* 1272. P.n. Bircham as in Great BIRCHAM TF 7632 + p.n. *Toftes* 1205, 'the (deserted) buildings'. DEPN.

TOFTREES Norf TF 8927. 'Toft trees'. P.n. *Toftes* 1086, 1254, 'the house sites, the deserted village', lOE **toft** < ON **topt**, pl. **toftes**, + ModE **tree**, pl. **trees**. DEPN.

TOFTWOOD Norf TF 9911. 'Toft wood'. Cf. *Toftwood Mill* 1838 OS. ModE **toft** (lOE *toft*) + **wood** (OE *wudu*).

TOGSTON Northum NU 2401. 'Tocg's valley'. *Toggesdena* 1129, *Tockis-* 1176, *Tog(g)esden(e) -is-* 1236–1425, *Tokesden* 1250 BF, *Togston* 1663. OE pers.n. **Tocg*, a strong variant of *Tocga*, genitive sing. **Tocges*, + **denu**. NbDu 198.

TOKERS GREEN Oxon SU 6977. *Talkers Green* 1797, *Tokens Green* 1822. Probably another of the Chilterns squatter settlements; cf. SHEPHERD'S GREEN SU 7183. O 77.

TOLL BAR SYorks SE 5508. A toll-bar on the turnpike road from Doncaster to Selby.

TOLLAND Somer ST 1032. Partly uncertain. *Taalande* [1066]11th ECW, *Talanda, Talham* 1086, *Taland* [1174–c.1250] Buck, 1266, *-landa* [c.1175] Buck, *Tolonde* 1327, *Tolland* 1610. The generic is OE **land** 'newly cultivated land'. The specific has been taken to be the r.n. TONE ST 3127. Tolland is a remote place not on the r. Tone but by the upper reaches of one of its branches. All the forms point to OE **tā** 'a toe' perhaps used in some topographical way of the lower extremity or projection of one of the surrounding hills. DEPN.

TOLLARD ROYAL Wilts ST 9417. 'T held by the crown'. *Tollard Ryall* 1535, *Toller Riall* 1547, *Tollard Royall* 1620. P.n. *Tollard* from 1086, *Toulard* 1227–82, PrW ***tull** 'a hole, a hollow' + ***arð** 'a height', referring to the numerous dramatic hollows that characterise the Tollard massif, + ModE adj. **royal** referring to King John's possesion of a knight's fee here in right of his wife Isabella. Wlt 208.

TOLLER FRATRUM Dorset SY 5797. 'The brothers' T'. *Tolre Fratrum* 1340. Earlier simply *Tolre* 1086, *Tollre* 1244. R.n. *Toller* + Latin **fratrum**, genitive pl. of **frater**, referring to the order of Knights Hospitaller who held the estate. The place is named from the river on which it stands, formerly called *Tollor* [1035]12th S 975, now the river Hooke, *Owke* 16th cent., named from HOOKE ST 5300. *Tollor* possibly represents PrW ***tull** (Co *toll*) 'a hole' + ***duβr** (Co *dour*) 'water'. The sense would be either 'stream with holes in it' or 'stream in the deep valley'. Do 145, RN 410.

TOLLER PORCORUM Dorset SY 5697. 'Pigs' Toller'. A semi-facetious name for contrast with TOLLER FRATRUM SY 5797. *Tolre Porcorum* 1340. R.n. *Toller* as in TOLLER FRATRUM + Latin **porcorum**, genitive pl. of **porcus**. The place was famous for its herds of swine and was also known as *Swyne Tolre*. Do 145.

TOLLERTON Notts SK 6034. 'Thorlaf's farm or village'. *Troclavestvne* 1086, *Tur- Torlaueston(a)* 1165–c.1300, *Torlauetun -ton* c.1190–1221×30, *T(h)orlaton* 1280–1332, *Tollerton* from 1499, *Torlagheston -lawes-*, *Thorlaxton -las-* 13th cent. Pers.n. *Thorlaf* < ON *Thorleifr* + **tūn**. Some 13th cent. spellings show confusion with pers.n. *Thorlac* < ON *Thorleikr*. Nt 242, SSNEM 196.

TOLLERTON NYorks SE 5164. 'The tax-gatherer's estate'. *Toletun* [972×92]11th S 1453, *Tolentun, Tolletune* 1086, *Tolereton'* 1167, *Tollerton* from 1256, *Tolnertona, Tolnorton* 1293. OE **tolnere** + **tūn**. YN 22 gives pr [touləṭən].

TOLLESBURY Essex TL 9510. 'Toll's fortified place or manor house'. *Tolesberiam* 1086, *-bir(ia)* 1212, 1232, *Tollesbur(e) -bery -bury* 1230–1337, *Towlesburye* 1594. OE pers.n. **Toll* as also in nearby TOLLESHUNT D'ARCY TL 9312, genitive sing. **Tolles*, + **byriġ**, dative sing. of **burh**. Ess 304, 306 gives pr [toulzbəri].

TOLLESBY Clevel NZ 5015. 'Toli's farm or village'. *Tol(l)esbi* 1086, *Tol(l)esby* from 12th, *Toulesbi -by* 1166–1310, *Towsby* 1364. ON pers.n. *Tólir*, genitive sing. *Tólis*, + **bȳ**. YN 163, SSNY 40.

TOLLESHUNT D'ARCY Essex TL 9312. 'T held by the Darcy family'. *Tolshunt(e) Darcy* 1472. P.n. Tolleshunt + family name of Robert Darcy 1441 whose family succeeded that of Robert Gernon 1086, William de Tregoz 1141 and John de Boys all of whom are remembered in manorial additions before 1441. Tolleshunt, *(of) Tollesfuntan* [c.1100]c.1125, *Tolesfunte* [1068]1309, *-fonte* 1377, *Toleshunta(m)* 1086, *Tol(l)eshunt(e)* from 1196, *Tholsunt* 1220, *Tolshunt(e)* 1321–1609, *Tolson* 1553–1605, is 'Toll's spring', OE pers.n. **Toll* as also in nearby TOLLESBURY TL 9510, genitive sing. **Tolles*, + ***funta**. For the development *funt* > *hunt* cf. BOARHUNT Hants SU 6008, CHADSHUNT Warw SP 3453. Tolleshunt lies on the course of a Roman road and the spring may have been a Roman antiquity. 'D'Arcy' for distinction from TOLLESHUNT MAJOR TL 9011 and Tolleshunt Knights TL 9114, 'the knight's T', *Toleshunt(e) Chevalers* 1254, ~ *Militis* 1303, *Knights* 1608, also called *Little* 1547 and *Busshes* 1540 from the surname *Boys* (Fr *bois* 'wood, bush') as above. Ess 306 gives pr [toulzn(t)], Signposts 84.

TOLLESHUNT MAJOR Essex TL 9011. 'Malger's T' later misunderstood as 'greater T'. *Toleshunt(e) Maugar* 1253, *Tolsunt Major* 1536. P.n. Tolleshunt as in TOLLESHUNT D'ARCY TL 9312 + pers.n. *Malger*, the DB undertenant. Also known as Great T, *Magna* 1291 for distinction from Little or Knights T, perhaps influenced by the form *Major* < *Mauger* with ME *ā* < *au*, and as *Beckinghams* 1604 from tenure by the family of Stephen Beckingham 1543. Ess 306.

TOLPUDDLE Dorset SY 7994. 'Tola's Piddle estate or manor'. *Tol(l)epidele -pyd(e)le* 1210–1428, *Tolpud(e)l(l)e* [1270]1372–1626, *Towpiddle* 1575. Earlier simply *Pidele* 1086. ODan feminine pers.n. *Tola* + r.n. PIDDLE SY 8591. Tola is a known person, the widow of Edward the Confessor's *huscarl* Urc; she gave the estate to Abbotsbury abbey 1058×66 to whom it belonged in 1086. This is a good example of a donor's name becoming affixed to the p.n. of an estate given to a religious house. For a similar example, including the change *-piddle* to *-puddle* for æsthetic reasons, cf. AFFPUDDLE SY 8093. Do i.331 gives pr ['toupidl].

TOLSKITHY Corn SW 6841 → TALSKIDDY SW 9165.

TOLWORTH GLond TQ 1965. 'Tala's enclosure'. *Taleorde* 1086, *-worda -wurth(a) -wurða -worth* 1130–1314, *Talle-* 1179–1613, *Talworth* 1352–1866, *Tolworth* from 1601. OE pers.n. **Tala* + **worth**. **Tala* would be a derivative of OE **tæl** 'quick, prompt' (ModE *tall*). Sr 67.

TONBRIDGE Kent TQ 5947. 'Bridge by the town'. *Tonebrige* 1086, *-brigga* c.1100, *-bricge* c.1121 ASC(E) under year 1087, *Tunebrig -brug* 13th cent. OE **tūn** + **brycg**. The town grew up at an important crossing of the Medway, originally a ford and subsequently a bridge. The *o*-spellings have persisted for distinction from TUNBRIDGE WELLS TQ 5839. PNK 180.

River TONE Somer ST 3127. An Old European r.n. *Tan* [672 for 682]16th S 237, [705]n.d. S 248, [979 for 879]12th S 352, [899×909]16th S 380, 1280, *Táán* [854]12th S 307, *Tán* *[854]12th S 311, [882]12th S 345, [979 for 879]12th S 352, *Thon* 1243, c.1540, *Tone* from 1303. An *n*- formation on the IE root **tā-/*tə-* 'flow'. RN 411, BzN 1957.259.

TONG Shrops SJ 7907. Uncertain. There are several types:
I. *(into) Tweongan* c.975 S 1534, *(æt) Twongan* late 11th S 1536†, *Tuange* 1086, *Toenga* c.1090–1138, *Twanga* 1167, *Twenge* 1221×2, *Twonge* 1291×2.
II. *Tonga* c.1145, *Tonge* 1221–1733, *Tong* from 1255.
III. *Tang(e)* 1203–1236.
IV. *Thonke* 1212, *Thonge* [1217]1285, 1395.
V. *Tung(e)* 1220–c.1540, *Toong* 1401, *Toung(e)* 1689.
II and III represent OE **tong**, **tang** 'a tong, forceps' used like ON **tangi** of a spit of land, here referring to Tong Hill between Morning Brook and an unnamed stream to the N. IV and V represent OE **thwang** 'a thong' and **tunge** 'a tongue' both also used topographically, cf. NETHERTHONG SE 1309, UPPERTHONG SE 1308, TONGE Leics SK 4123; I has no known antecedent but clearly points to an OE ***twang(a)**, ***twong(a)**, *** twenġa**< **twang-j-on-* related to the verbs *twenġan* 'to pinch' < **twang-i-an* and *twingan* 'to press' and referring to the pinched or restricted nature of the site between streams, cf. G name Zwinger used of the restricted pinched area of land between the wall of a city and the ditch from G *zwingen*, OHG *dwingan*, an exact parallel to OE *twingan* < Gmc **thwengan*. The form *Tweongan* is difficult to account for. The rare word ***twang(a)**, ***twenġa** was subsequently replaced by the commoner **tang**, **tong**. Sa i.293.

TONGE Leic SK 4123. 'The tongue of land'. *Tvnge* 1086, *Tung(e)* 1276–[1335]15th, *Tong(e)* from 1267. The village is situated on a tongue of land between two streams. Lei 349.

TONGHAM Surrey SU 8849. Partly uncertain. *Tuangham* 1189, *Tuange- Twang(e)- T(h)wong- Tweng- Tang- Tong(e)-* 13th cent., *Thong-* 1509×47. OE ***t(w)ang** 'fork of a river', **tong** 'a spit of land' or **thwang** 'a thong, a narrow strip of land', + **hām**. The reference is uncertain but the topographical explanation in Sr 182 is impossible. Cu lxxii, PNE ii.176–7, ASE 2.32.

TONWELL Herts TL 3317. 'The town well, spring belonging to a *tun*'. *Tunwelle* 1220, 1294, *Tone(s)well* 1296, 1322. OE **tūn** + **wella**. Hrt 216.

TOOT BALDON Oxon SP 5600 → Toot BALDON SP 5600.

TOOT HILL Essex TL 5102. 'Look-out hill'. *Toote Hill* 1611. Ultimately from OE **tōt** + **hyll**. Ess 79.

TOOTHILL Hants SU 3818. 'Look-out hill'. *Totahulle* 1233, *Totehill* 1513, *Towte Hill* 1579, *Toot Hill* 1610. Ultimately from OE ***tōt-hyll**. Gover 1958.41.

TOOTING BEC GLond TQ 2972. 'Tooting held by Bec'. *Toting de Bek* 1255, ~ *Beck* 1316. The estate was held by the abbey of Bec-Hellouin, Normandy, cf. WEEDON BEC Northants SP 6359. Earlier simply *Totinge* [672×4]13th S 1165, [933]13th S 420, *Totinges* [1057×66]13th, 1086–1247, *Toting(e)* 1225–1595, *Tutyn* 1542, either the 'look-out place', OE ***tōting**, or the 'place called after Tōta', OE **Tōting*. The plural forms may represent OE **Tōtingas* 'the people of Tōta or of Tooting' or may refer to the separate manors at Tooting, Tooting Bec and Lower or South Tooting, later Tooting Graveney held by Richard de Gravenel in 1215. Sr 35, ING 30, BzN 1967.348.

Upper TOOTING GLond TQ 2772. *Upper Tooting* 1816. ModE **upper** + p.n. *Totinge* *[727]13th S 1181, *[933]13th S 420, *Totinges* 1086–1247 including [1057×66]13th S 1136,*Totingas* 1070×85, *Toting(e)* 1225–1316, *Tooting* 1486×93, *Tutin* 1685, 'the Totingas, the people called after Tota' or 'the people of the look-out place', OE folk-n. **Tōtingas* < OE pers.n. *Tōta* (from ***tōta** 'a look-out man') or ***tōt** + **ingas**, possibly varying with OE singular ***tōting** 'the look-out place' or 'place called after Tota', pers.n. *Tōta* + **ing**[4]. 'Upper', otherwise Tooting Bec, *Totinge de Bek* 1255, held in 1086 by the abbey of Bec-Hellouin in Normandy, for distinction from Lower or South Tooting, otherwise Tooting Graveney TQ 2771, *Toting' Grauel'* 1255, *Thoting Gravenel* 1272, held by Richard de Gravenel in 1215 whose family came from Graveney, Kent. Sr 35, ING 30, BzN 1967.348.

TOPCLIFFE NYorks SE 3976. 'The highest part of the cliff'. *Topeclive* 1086–1251, *Toppe-* 1154×81–1301, *Topclif* 1288–1371. OE **topp** + **clif**. The village lies on the upper edge of a steep and lofty bank overlooking the river Swale. YN 186.

TOPCROFT Norf TM 2692. 'Topi's enclosure'. *Topecroft -cropt* 1086, *Topecroft* [1087×98]12th–1208, *Topescroft* [1087×98]12th, *Toppecroft* 1206. ON pers.n. *Tōpi*, genitive sing. *Tōpa*, + **croft**. DEPN, SPNN 380.

TOPCROFT STREET Norf TM 2691. 'Topcroft village'. *Topcroft Street* 1838 OS. P.n. TOPCROFT TM 2692 + dial. **street** 'a hamlet'.

TOPPESFIELD Essex TL 7337. 'Open land of the hill-top'. *Topesfelda -e* 1086, *Top(p)esfeld(e)* 1196–1205 etc. OE **topp**, genitive sing. **toppes** + **feld**. This explanation fits the topography and is preferable to unrecorded pers.n. **Topp*. Ess 463 gives pr [topəzfl].

TOPPINGS GMan SD 7213. *Toppings* 1843. Probably the plural of dial. **topping** 'a hill top'. Cf. Roseberry Topping under NEWTON-UNDER-ROSEBERRY Cleve NZ 5613.

TOPSHAM Devon SX 9688. 'Hill-top promontory'. *Toppesham, æt Toppeshamme* [937]11th S 433, *(æt, to) Toppes hamme, (of) toppes hām land* 12th, *Topeshant* 1086. OE **topp**, genitive sing. **toppes**, + **hamm** 1, 2a. Topsham occupies a promontory of high ground between the rivers Exe and Clist. Some late spellings show dial. *a* for *o* and confusion arising from misdivision of *at Topsham, Tapsam* 1359, *Opsham* 1480×3, *Topsham vulg. Apsum* 1675. D 454 gives pr [tɔpsəm].

TOR BAY Devon SX 9259. *Torrebay* 1401. P.n. Torre as in TORQUAY SX 9164 + ME **baie**. D 20.

TORBAY Devon SX 8962. A modern seaside resort named from TOR BAY SX 9259.

TOR BROOK Devon SX 5558. See under River TORRIDGE SS 5509.

TORBRYAN Devon SX 8166. 'Torre held by the Brion(n)e family'. *Torre Briane* 1238. Earlier simply *Torre* 1086, 1242, OE **torr** 'a rock, a rocky outcrop' referring to the hill just W of the church which rises to 250ft. Wido de Brion(n)e is first mentioned here in 1238. 'Bryan' for distinction from Tormoham, the earlier name of Torquay, *Torre Moun* 1279, *Torremohun, Torre Mohoun* 1331, 'Torre held by the Mohun family', and simply *Torre* 1086, *T(h)orre* 1233, 1238, referring to the tor beneath the S side of which Torre Abbey was built. D 522–3.

TORCROSS Devon SX 8242. 'Cross by the tor'. *Tarcross* 1714. The tor is the hill rising to 323ft. S of the village on which there is a cross at SX 8141. D 332.

TORKSEY Lincs SK 8378. 'Turoc's island'. *(æt) Tureces iege* late 9th ASC(A) under year 873, *TOR, TVR(CE)* 975–1003 Coins, *(in) Turces ige* c.1000 Æthelweard, 1121 ASC(D, E), *(æt) Turkes ege* c.1000 ASC(B) under year 873, *Torchesey -yg -ig -iy, Dorchesyg* 1086, *Torchesi -eia* 1100–53, *Torkeseie* [1147]13th. OE pers.n. **Turoc*, genitive sing. **Turoces*, + **ēġ**. The earliest sp suggests **Turoc*; the rest would be compatible with pers.n. **Turc* < Brit **torco-* 'a boar', cf. Breton pers.n. *Turch*. DEPN, L 38, Cameron 1998.

TORMARTON Avon ST 7778. Perhaps 'Thormar's estate' or 'thorn-tree Marton'. *Tormentone* 1086, *Tormarton(e) -tun* 1166–1706, *Tormerton(a)* 1183–1584. ODan pers.n. *Thormar* + **tūn** or OE **thorn** + p.n. Marton 'pond settlement', OE **mere-tūn**. Like RODMARTON Glos ST 9397 and DIDMARTON Glos

†Probably wrongly assigned by Charters II xxviii to Tonge Leics, whose forms it does not match. See TONGE Leics SK 4123.

ST 8287 where ponds survive, this place too once possessed its pond. Gl iii.56, Jnl 25.46.

River TORNE SYorks SE 6502. A back-formation from the lost p.n. Torne Wath. *Thorn(e)* c.1160–early 14th, *Torne* from 1539. Torne Wath, a lost n. in Rossington is *Theornewat* 1088, *Tornewat* 12th, 'ford marked by a thorn-tree' or 'on the road to Thorne', OE **thorn** or p.n. THORNE SE 6913 + ON **vath**. YW i.50, DEPN.

TORPENHOW Cumbr NY 2039. 'Hill-spur called *Torpen*'. *Torpennev* c.1160, *Torpennoc'* 1163, *Torpenno(h')* 1222–79, *Torpenho(w) -hou* 1222–1312, *Torpenny* 1576. P.n. *Torpen* 'head, end of the high ground', PrW ***tor** + ***penn**, + pleonastic OE **hōh**. Cu 325, CPNE 221.

TORPOINT Corn SX 4355. 'Crag headland'. *Tar Point* 1736, *Torpoint* 1746, *Torr Point* 1748, earlier simply *The Torr* and *Tor Park* 1617. ModE **tor** + **point**. An 18th cent. new town named from the nearby point. The 1736 form shows dial. *a* for *o*. For an earlier name → START POINT SX 0486. PNCo 163, Gover n.d. 215.

TORQUAY Devon SX 9164. 'Quay or landing stage at Torre'. *Torre Key* 1509×47, *Torquey* 1577, *Torkay* 1668, *Fleete otherwise Torkey within the parish of Tormohun* 1670†, *Tor Quay* 1765, 1809 OS. P.n. *Torre* as in Tormoham, the earlier name of Torquay, discussed under TORBRYAN SX 8166, + ME **key**. Although Torquay is chiefly a 19th cent. development, there was a small village here in the 17th cent. The earliest reference to the quay here is *un lieu appelle le Getee de Torrebaie*, 'a place called the jetty of Tor Bay', where a captured French ship laden with wine was brought into Devon in 1412. D 524 gives pr [tɔ:'ki:].

River TORRIDGE Devon SS 5509. 'Violent, rough stream'. *Toric* [802]13th S 1693, *on toric stream* [925×39]11th S 388, *T(h)oriz* 1238, *Toryz* 1249, *Tor(r)ygg(e)* 1345–89. A British r.n. related to W *terig* 'ardent, severe, harsh'. The identical name occurs again at Tor Brook SX 5558, *Torygg* 13th, *Torey broke* c.1550. D 14,15, RN 413.

TORRINGTON 'Settlement, estate on the r. Torridge'. R.n. TORRIDGE SS 5509 + OE **tūn**.

(1) Black ~ Devon SS 4605. *Blachetoriton(a)* 1167×1316 with variant *Blake-, Blaketorrintun* 1219, *Blaketoryngtone* 1359. Earlier simply *Torintone* 1086. It is said that the river here has a blackish colour which it loses three miles downstream. D 170, RN 413.

(2) Great ~ Devon SS 4919. *Torytone Magna* 1366, *Mochel Torynton* 1443. Latin adj. **magna** ME **muchel** + p.n. *Tori(n)tona* 1086–1246 with variants *Tory- Thorin-*. Also known as *Chippingtoriton* 1284, 'market Torrington'. 'Great' and *Chipping* for distinction from Little TORRINGTON SS 4916. D 123.

(3) Little ~ Devon SS 4916. *Parva Toriton'* 1238, *Little Torygton* 1411. Earlier simply *Toritone* 1086. D 110, RN 413.

East TORRINGTON Lincs TF 1483. *Est Tyrington* 1232, *East Torrington* 1824 OS. ME **est** + p.n. *Terintone* 1086, *Tiringtuna* c.1115, *-tun -tona* 12th, 'the estate called after Tira', OE pers.n. ***Tira* + **ing**[4] + **tūn**. 'East' for distinction from West TORRINGTON TF 1382. DEPN.

West TORRINGTON Lincs TF 1382. *Westiringtun* 1195×1200, *West Torrington* 1824 OS. ME **west** + p.n. Torrington as in East TORRINGTON TF 1483. DEPN, Cameron 1998.

TORRISHOLME Lancs SD 4564. 'Thorald's island'. *Toredholme* 1086, *Thaurrandeshal'* 1201, *T(h)oraldes- Turoldesholm* 1201–12, *Thoredesholm* 1233, *Thorisholme* 1323, *Torisholm* 1322, 1327, *Torryshulme* 1557. ON pers.n. *Thorrøthr* later replaced by *Thóraldr*, genitive sing. *Thóralds*, + **holmr**. The reference is to an island of higher, dry ground in marsh. La 177, SSNNW 171, Nomina 10.174, Jnl 17.100.

†There is a Fleet Street in Torquay where a small stream (OE *flēot*) formerly entered the sea.

TORSIDE RESERVOIR Derby SK 0698. P.n. Torside + ModE **reservoir**. Torside as in Torside Clough SK 0797, *Torside Clough* 1843, Torside Naze SK 0797, *Torside Naze* 1843, and Torside Castle SK 0796, *Torside Castle* 1843, all named from *Tor Top* 1767, 1817, ModE **tor** + **top**. Db 71.

TORTINGTON WSusx TQ 0005. 'The estate called after Torhta'. *Tortinton* 1086, *Torti(nge)ton -y(ng)-* 1291–1397, *Torton* 1304×7, 1641, 1732. OE pers.n. ***Torhta* + **ing**[4] + **tūn**. Sx 143 gives pr [tŋ:tən].

TORTWORTH Avon ST 7093. 'Torhta's enclosure'. *Torteword* 1086, *-wurð -w(o)rth(e)* c.1100–1378, *Tortworth* from 1487 with variant *-worthy* 1555–1652, *Tortery* 1583. OE pers.n. ***Torhta* + **worth**. Gl iii.41.

TORVER Cumbr SD 2894. Partly uncertain. *Thoruergh* 1190×99, 1246, *T(h)oruerg(h)* 1246–1320, *Thorwerghe* 1202, *Thorfergh* 1246, *Torver* from 1537. Either ON pers.n. ***Thorfi* + **ǽrgi** 'Thorfi's shieling' or ON **torf** + **ǽrgi** 'peat shieling'. It is imposible to say whether the *Th-* or *T-* forms are primary since either could be a graphy for the other. La 215, SSNNW 72.

TORWORTH Notts SK 6586. Probably the 'dung enclosure'. *Turdeworde* 1086, *Thord(e)- Torthew(o)rth(e)* [1189×99]1232, 1209–1400, *Tord(e)w(o)rth(e)* 1280–1316, *Torword* 1274, *-worth* from 1327. OE **tord** + **worth**. Some spellings seem to show substitution of ODan **torth** 'dung' and confusion with the ON pers.n. *Thórthr*. Nt 100 gives pr [tɔrəθ], SSNEM 215.

TOSELAND Cambs TL 2362. 'Toglos's grove'. *Toleslund* 1086 DB, c.1180, 1255, *Touleslund* 1231, 1286, *Toulislond -es-* 1284–1362, *Towe(s)lond* 1364–1443, *Towes- Towseland* 16th cent. ON nickname *Tauglauss* 'ropeless' + **lundr**. The reference is to the Danish earl called *Toglos* ASC(A) under year 921 who was killed by King Edward's forces at Tempsford. He is also called Toli in the *Liber Eliensis*. This may account for the DB spelling though the *Tou- Tow-* spellings clearly point to Toglos, presumably his nickname. In full he must have been known as Toli Tauglauss. 'Ropeless' probably referred to a man who did not have the right ropes for his ship. Hu 272.

TOSSIDE Lancs SD 7756. 'Thórir's hill-side or hill-pasture'. *Torsyd(e)* 1581, 1598, *Tosset(t)* 1585, 1662, *Tossid(e) -syd(e)* from 1591. ON pers.n. *Thórir* + ME **side** (OE *sīde*) or **set** (ON *sǽtr*). YW vi.182.

TOSSIDE BECK Lancs SD 7853. *Tosside Beck* 1858. P.n. TOSSIDE SD 7756 + ModE **beck**. YW vi.182.

Great TOSSON Northum NU 0200. *Mangnam Tossin* (sic) 1236 BF, *Magnam Tossen* 1242 BF, ~ *Tossyn* 1296 SR, *Mykle Tosson* 1553 Dixon, *Great Tosson* c.1715 AA 13.13. ModE adj. **great**, Latin **magna**, ME **mikel** + p.n. *T(h)osse* 1150×62, *Thos(s)an* 1203, 1229, *Tossin* 1236, *Tossen -an* 13th cent., *Tosson* 1346, probably the 'look-out stone', OE **tōt** + **stān**, referring to the rocky Burgh Hill NU 0201 above Great Tosson with its ancient hill-fort commanding extensive views up and down the valley of the Coquet. 'Great' for distinction from Little Tosson NU 0101, *Parvam Tossen* 1242 BF, ~ *Tossan* 1296 SR, *Little Tosson* 1553 Dixon 331. DEPN, Dixon.

TOSSON HILL Northum NZ 0098. *Tosson Hill* 1868 OS. P.n. TOSSON NU 0200 + ModE **hill**. A hill rising to 1447ft.

TOSTOCK Suff TL 9563. 'The look-out's place'. *totestoc, totstocha* 1086, *Totstoche* c.1095, *Totestok* 1226×8, *Totstok* 1610. OE ***tōta** + **stoc**. DEPN, Studies 1936.27.

Great TOTHAM Essex TL 8511. *Magna Totham* 1255, *Great Totham Nevile* 1260, *Myche Tottam* 1552. ModE **great**, Latin **magna**, ME **mich**, + p.n. *Totham* from 1000×2 S 1486 including 1086 and [946×c.951]13th S 1483, *Toteham* 1086, *Totteham* 1255, 1276, 'look-out place homestead', OE ***tōt(e)** + **hām**. The original site is unknown but the parish contains high ground including Beacon Hill TL 8512. 'Great' for distinction from Little TOTHAM TL 8812. Ess 310 gives pr [tɔtəm].

Great TOTHAM Essex TL 8613. This place is *Bung Row* 1805 OS.

Little TOTHAM Essex TL 8812. *Parva Toteham* 1154×89. ModE **little**, Latin **parva**, + p.n. Totham as in Great TOTHAM TL 8613.

TOTLAND IoW SZ 3287. 'Newly cultivated land at the look-out'. *Toteland* c.1240, 1341, *Totteland* 1337, *Totland* 1608. OE ***tōt** + **land**. The site is close to a former beacon mentioned as early as 1324 on Headon Hill looking out to sea. Wt 227 gives pr ['tɔ:tlən], Mills 1996.102.

TOTLAND BAY IoW SZ 3186. *Totland Bay* from 1720, *Tollands Bay* 1810 OS, 1852. P.n. TOTLAND SZ 3287 + ModE **bay**. Mills 1996.102.

TOTLEY SYorks SK 3079. Partly uncertain. *Totingelei* 1086, *Totinley(e) -yn-* c.1250–1469, *Tot(e)le -lay -ley(e)* from 1275. The forms are compatible with several alternatives, OE p.n. **Tōtinge*, locative–dative sing. of ***tōting** 'a look-out place' + **lēah** 'a wood or clearing', or OE **Tōtinga*, genitive pl. of folk-n. **Tōtingas* 'the look-out place people' or 'the people called after Tota', OE ***tōt** or pers.n. **Tōta*, + **lēah**; and also simply 'Tota's' or 'the look-out wood or clearing', OE pers.n. *Tota* or ***tōt(e)** + **lēah**. Db 315.

TOTNES Devon SX 8060. 'Totta's headland'. *Totanæs* 979–1016 coins, *(to) Tottanesse* 11th, *Totenais, Totheneis* 1086, *Tot(t)enes(se) -as -ays -eys* 1167–1337, *Tottneis* 1364. OE pers.n. *Totta* + **næss** referring to the prominent point of land on which the castle stands. D 334 gives pr ['tɔtnəs].

TOTTENHAM GLond TQ 3390. 'Totta's homestead'. *Toteham* 1086, 12th cent., *Tot(t)enham* 1189–1543. OE pers.n. *Totta*, genitive sing. *Tottan*, + **hām**. The same pers.n. also occurs in Tottenham Court (as in Tottenham Court Road), *Totnalcourt* 1411, *Tottenham Court* 1741×5, 'Tottenham manor house', p.n. Tottenham + ME **court**. Tottenham, *(of) þottanheale* c.1000, *Totehele* 1086, *Thotte(n)hal(e)* 13th, *Totenhala -hal(e)* [1083]15th–1341, is 'Totta's nook', OE pers.n. *Totta, Tottan*, + **healh**. The modern form is due to the influence of Tottenham TQ 3390. Mx 78, 143.

TOTTENHILL Norf TF 6410. 'Totta's hill'. *Tottenhella* 1086, *Totehill* 1251. OE pers.n. *Totta*, genitive sing. *Tottan*, + **hyll**. DEPN.

TOTTERIDGE Bucks SU 8893. 'Tota's or look-out hill'. *Tuterugge* 1179, *Toterugge* 1179, 1416. OE pers.n. *Tōta* or **tōta** + **hrycg**. Bk 203.

TOTTERIDGE GLond TQ 2494. 'Tata's ridge'. *Taderege* 12th, *Taterige -rugg -regge -riche -reche* 1230–54, *Tatteridge al. Totteridge* 1608. OE pers.n. **Tāta* + **hrycg**. Dollis Brook, which forms the boundary between Totteridge and the county of Middlesex, is called *(of) tateburnan* and *(on) tataburnan* 'Tata's stream' [957]12th in the boundary clause of S 645. Hrt 149.

TOTTERNHOE Beds SP 9921. 'Look-out hill-spur'. *Totenehou* 1086, *Toternho(u)* 12th–14th cents., *Totternhoe* 1657. OE **tōt-ærn** + **hōh**. The reference is to the earthwork called Totternhoe Castle, which commands an extensive view of Watling Street from steep high ground. Bd 139, L 168.

TOTTIFORD RESERVOIR Devon SX 8183. P.n. Tottiford + ModE **reservoir**. Tottiford, *Toteworthi* 1333, is 'Totta's enclosure', OE pers.n. *Totta* + **worthiġ**. D 472.

TOTTINGTON GMan SD 7712. 'Settlement at or called *Tōting*, the look-out place, or called after Totta'. *Totinton* 1212, 1235, *Totington -yng-* 1233–1332 etc. with variants *Tod-* 1242, *-don* 1251, *Tottyngton* 1285. OE p.n. **Tōting* 'look-out place, look-out hill', OE ***tōt(e)** + **ing**[2], or pers.n. *Tot(t)a* + **ing**[4], + **tūn**. The hill beside which Tottington is situated is crossed by the Roman road from Manchester to Ribchester, Margary no. 7b. The 896ft. summit at SD 7504 commands extensive views and is marked by an old stone cross. Cf. Upper TOOTING GLon TQ 2772. La 63, Årsskrift 1974.44, Jnl 17.43, 18.22.

TOTTON Hants SU 3613. 'Look-out settlement'. *Dodintvne* 1086, *Totintone* 1086–1242, *Totyngton -ing-* 1174×99–1279, *Tottone* 1327, *Tatton* 1817. OE ***tōting** + **tūn**. This explanation well suits the site, which stands at the confluence of the Test and Southampton Water on the line of the Roman road from Dibden to Lepe, Margary No. 423. Alternatively 'estate called after Tot(t)a' or, if the form *Dodintvne* 1086 also belongs here, 'estate called after Doda'. Gover 1958.196, Ha 164, Årsskrift 1974.44.

River TOVE Northants SP 6647. 'The dilatory one, the laggard'. *Toue* 1219, *Tove* 1437, *Tea* 1622. OE ***tof**. The reference is to its winding course. For a possible earlier name see TOWCESTER SP 6948. Nth 4, RN 414.

TOVIL Kent TQ 7554. 'Tubba's open country'. *Tobbeffeld* 1218, *Topifield, Toppefeud* 1226, *Tog(h)feld'* 1304, *Toufeld* 1313, *Tof(f)eld(e)* 1327–1485. OE pers.n. **Tubba* + **feld**. Some later forms (1304, 1313) point to OE **tōh** 'tough, sticky, hard'. PNK 142, PNE ii.181.

TOW LAW Durham NZ 1139. 'Look-out hill'. *Tollaw(e)* 1423–59, *Towlawe, Tollow* 1647, *Towlaw* 1857. OE **tōt-hlāw**. In 1841 it was a solitary farm-house. The town owed its development to the establishment there of blast-furnaces by the Weardale Iron Company in 1845. NbDu 199.

TOWAN Corn SX 0149 → PENTEWAN SX 0147.

TOWAN HEAD Corn SW 7962. 'Towan headland'. *Towan Head* 1748, *Towan point* 1699. P.n. Towan, *Tewyn* 1289, *Towyn* 1504, 'sand-dunes', Co **towan**. (MCo **tewyn*), + ModE **head**. PNCo 163, Gover n.d. 328.

TOWCESTER Northants SP 6948. 'The Roman camp on the river Tove'. *Tofeceaster* c.925 ASC(A) under year 921, *Tovecestre* 1086–1404 with variant *Toue-, Towecestre* 1199×1216, *Touchestre* 1294, *Towchester* 1577, *Toceter* 1564, *Toster* 1675. R.n. TOVE Northants SP 6647 + **ċeaster**. Site of the RBrit settlement of *Lactodurum* in which *Lacto-* may preserve an older name of the r. Tove meaning 'milky water'. Nth 94 gives pr [toustə], RN [tɔustə], RBrit 382.

TOWEDNACK Corn SW 4838. 'To-Winnoc's (church)'. *(parochia) Sancti Tewennoch* 1327, ~ ~ *Tewennoci* 1333, ~ ~ *Tewynnoci* 1335, *Tewynnek* 1524, *Twyd- Tuidnac* 1584. OCo honorific prefix ***to** + saint's name *Winnoc*, a pet form of the saint's name *Winwaloe*, from the dedication of the church. See ST WINNOW SX 1156. Co *nn* became *dn* in the second half of the 16th cent. PNCo 164, Gover n.d. 663.

The TOWER OF LONDON GLond TQ 3380. First mentioned as *þone tur* c.1150 ASC(E) under year 1097, OFr **tur**. The original reference is to the White Tower erected by William Rufus, but the name *The Tower* was subsequently used of the whole fortress. Mx 154.

TOWERSEY Oxon SP 7305. 'Eye held by the Tours family'. *Turrisey* 1237–40, *Tur(e)seya* 1241, 1262, *Tour(e)sey(e)* 1302–1458, *Towresey* 1422, 1527. Family n. *Tours* who held land here 13th cent. + p.n. *Eie* 1086, *Eye* 1235, 'the island', OE **ēġ**. Also known as *Parva Eie* 'Little Eye' 1227. The manorial addition distinguishes this Eye from neighbouring KINGSEY SP 7406, 'the king's or great Eye'. Bk 128.

TOWN END Cambs TL 4195. 'Town-end' sc. of March. Cf. *Townesend Dike* 1636, *Towne end Greene* 1669, *Town End Common* 1824 OS. Ca 258.

TOWNELEY HALL Lancs SD 8530. P.n. Towneley + ME **hall**, a medieval courtyard house. Towneley, *Tunleia* c.1200, *-ley* 1243, *Touneley* 1296, *Tounley* 1322, is the 'clearing or pasture belonging to the town', OE **tūn** + **lēah**. The 'town' in question must have been Burnley. La 84, Pevsner 1969.82, Jnl 17.52.

TOWNHEAD Cumbr NY 6334. 'Thorn-tree headland'. *Thornheued* 1278–1331 with variants *T(h)orne-* and *-hefd*, *Thornhead* 1595. OE **thorn** + **hēafod**. Cu 229.

TOWNSHEND Corn SW 5932. From the family name Townshend. *Townsend* 1867. A 19th cent. village created by the Duke of Leeds (family name Townshend) who inherited the Godolphin lands here. Cf. LEEDSTOWN SW 6034. PNCo 164.

TOWTHORPE NYorks SE 6258. 'Tofi's outlying farm'. *Touetorp* 1086, *Touthorp(e)* c.1157×70–1310, *Towethorp* 1372, *Towthorpp* 1419, *-throppe* [1316]16th. ON pers.n. *Tófi*, genitive sing. *Tófa* + **thorp**. YN 13 gives pr [təuθrəp], SSNY 70.

TOWTON NYorks SE 4839. 'Tofi's estate'. *Touetun -ton* 1086–1341, *Towton* from 1379. ON pers.n. *Tófi*, genitive sing. *Tófa*, + **tūn**. YW iv.72, SSNY 130.

TOXTETH Mers SJ 3588. 'Toki or Tok's landing-place'. *Stochestede* 1086, *Tokestath* 1212, *Toxtath(e)* 1221–1504×5, *Toxstath* 1297, 1323, *-steth* 1446×7, *Tokstaffe* c.1540. ON pers.n. *Tóki* or ODan by-name **Tōk* 'fool' + **stọth**, (genitive sing. **stathar**), referring to a landing-place on the River Mersey or Dingle Brook. Alternatively the generic might be ON **stathir** 'a place, a site'. La 115, SSNNW 59, Jnl 17.64.

TOY'S HILL Kent TQ 4651. *Toys Hill* 1819 OS. Surname *Toy* + ModE **hill**. For the surname cf. DBS 353.

TOYNTON ALL SAINTS Lincs TF 3964. *Thoynton' Omnium Sanctorum* 1254, *Toynton All Saints* 1824 OS. P.n. *Totintun(e)* 1086, *Totingtun(a) -intona* 12th, 'the settlement at *Toting*, the look-out place', OE p.n. ***tōting** + **tūn**, + affix **All Saints**, Latin **omnes sancti**, genitive pl. **omnium sanctorum** from the dedication of the church. Toynton occupies a strategic site overlooking the one-time inundated East Fens. DEPN, BzN 1981.321.

TOYNTON FEN SIDE Lincs TF 3962. 'The side of T Fen'. *Toynton Fen Side* 1824 OS. P.n. Toynton as in TOYNTON ALL SAINTS TF 3964 + ModE **fen** and **side**.

TOYNTON ST PETER Lincs TF 4063. 'St Peter's T'. *Thoynton Sancti Petri* 1254, *Toynton St Peters* 1824 OS. P.n. *Totintun(e)* 1086 as in TOYNTON ALL SAINTS TF 3964 + saint's name *Peter* from the dedication of the church. DEPN, BzN 1981.321.

High TOYNTON Lincs TF 2869. *Tynton Superior* 1254, *Ouertincton'* 1279, *upper Toynton* 1661, *High Toynton* 1824 OS. ModE adjs. **high**, **upper**, ME **over**, Latin **superior** + p.n. *Todintune* with suprascript *l* (*vel* 'or') *Ti*, *Tedintone*, *Tedlintune* 1086, *Tidinton'* 1166, *Teinton'* 1199, *-tune* 1230, the 'estate called after Teoda', OE pers.n. *Tēoda* + **ing**[4] + **tūn**. There were two developments of this name, (1) *Tēodingtun* > *Tēding- Tēdin(g)-* > *Tēin-* > *Tinton*, (2) *Tēodingtun* > *Teóding-* > *Todin(g)-* > *Tointon*, or this form was influenced by TOYNTON ALL SAINTS. 'High' for distinction from Low Toynton TF 2770, *Toynton Inferior* 1254, *Lower Toynton* 1661. DEPN, BzN 1981.320.

TRABOE Corn SW 7421. 'Farm of Gorabo'. *Trefwurabo* [977]11th S 832, 1059 S 1027, *Treurabo* [c.1240]14th, *Trewerabo* 1291, *Trerabo* 1611. Co ***tref** + pers.n. *Gorabo*. PNCo 164, Gover n.d. 559.

TRAFFORD PARK GMan SJ 7896. *Trafford Park* 1843 OS. P.n. (Old) Trafford + ModE **park**. Trafford, *Trafford* from c.1200, *Stratford(e)* 1200–1212 in the surname of the *de Trafford* family, is a doublet of STRETFORD SJ 7994 of which it was a part until created a separate manor. The loss of initial *S-*, due to AN influence, was reinforced by the need to differentiate the two geographically distinct places. La 32, Mills 142.

Bridge TRAFFORD Ches SJ 4571. '(The part of) Trafford at the bridge'. *Bregetrouford* 13th, *Bridge Trafford* from 1553. ME **brugge**, **brigge** + p.n. Trafford, *Trosford* 1086 etc. as in Mickle TRAFFORD SJ 4469. Che iii.261.

Mickle TRAFFORD Ches SJ 4469. 'Great Trafford'. *Magna Trogthforde* 1290–16th with variant spellings *Tro(u)gh-Trafford* (from 1528), *Mekeltroghford* 1379, *Mickle Trafford* from 1616. ME **micel**, (Latin **magna**) + p.n. *Tra- Tro(s)ford* 1086, *Troc(h)- Trokford* 12th–1295, *Trog- Trogh(e)- Troughford* 1254–1521, *Trouford* 1267×8, 1327–8, *Trafford* from 1287, 'trough-ford', OE **trōg** + **ford**, alluding to some characteristic of the ford, possibly deeply worn approaches to it along the narrow ridges of firm ground across the Gowy marshes. 'Great' for distinction from Bridge TRAFFORD SJ 4571. Che iv.133.

TRANWELL Northum NZ 1883. Possibly 'the crane's spring'. *Trennewell* 1267, *Trenwell* 1270–1428, *Tran(e)well* 1288, 1296. ON **trani** + OE **wella**. NbDu 199, L 31.

TRAWDEN Lancs SD 9138. 'Trough-like valley, or, valley called *The Trough*'. *Trochdene* 1296, *Trouden(e)* 1305–56, *Troweden* 1311. OE **trōg** + **denu**. The village is situated in a broad trough-like valley. La 88 gives pr [trɔ:din], Jnl 17.53.

The Forest of TRAWDEN Lancs SD 9338. An area formerly under forest-law. See TRAWDEN SD 9138.

TRE ESSEY H&W SO 5021 → TRETIRE SO 5224.

TREADDOW H&W SO 5424 → TRETIRE SO 5224.

TREALES Lancs SD 4433. 'Court homestead or village'. *Treueles* 1086–1332 etc., *Treules* 1324, 1327, *Treeles* 1431, *Tra(y)les* 16th, *Treals* 1792, 1832. PrW ***treβ** + ***lïs** 'a hall, a court, the chief home in a district'. Identical with Trelease Corn SW 7017 and 7621 and Trellys-y-Criwc Wales SM 8935, Trellys-y-Coed SM 9034 and Treflys SN 5869. Treales is situated in a central position within Amounderness and may have been the original administrative centre of the whole district. La 152 gives pr [tre:lz], Jnl 1.49, 51, 17.88, CPNE 150, 223.

TREBARTHA Corn SX 2677. Partly unexplained. *Triberthan* 1086, *Trebartha* from 1284. Co **tre** 'estate, farmstead' + unidentified element as in *Polbartha* 1333, a lost place, Co **pol** 'pit, pool, stream, creek', POLBATHIK SX 3456, *Polbarthek* 1365, and the W r.n. Barthen, Dyfed. PNCo 164, Gover n.d. 168.

TREBERON H&W SO 5025 → TRETIRE SO 5224.

TREBETHERICK Corn SW 9377. 'Farm of Pedrek'. *Trebederich* 1284, *Trebedrek* 1302, *Trebethrick* 1657. Co **tre** + saint's name *Pedrek* (*Peder* from Latin *Petrus* + diminutive suffix *-ek*). The name is probably connected with the monastery of St Petrock (*Pedreck*) at Padstow, just across the estuary. PNCo 164, Gover n.d. 130.

TREBOROUGH Somer ST 0136. 'Tree hill'. *Traberge* 1086, *Trebergh* 1225, *Treberge* [14th, n.d.] Buck, *Treboro* 1610. OE **trēo** + **beorg**. Possibly the sense was 'tree tumulus' or a hill marked by a conspicuous tree. The village lies at the edge of a 1000ft. oval hill. DEPN, L 211.

TREBUDANNON Corn SW 8961. 'Farm of Pydannan'. *Trebedannan* 1279, *Trepydannen* 1355, *-pidannen* 1415. Co **tre** + pers.n. **Pydannan* with lenition of *p* to *b*. PNCo 164, Gover n.d. 323.

TREBUMFREY H&W SO 5222 → TRETIRE SO 5224.

TREBURLEY Corn SX 3477. 'Borlay's farm'. *Treburley (juxta Beaulyuford)* from 1370 Gover n.d. Co **tre** + family name Borlay as in William *Borlay* 1327. PNCo 164, Gover n.d. 159.

TRECILLA H&W SO 5321 → TRETIRE SO 5224.

TREDAVOE Corn SW 4528. 'Farm of Gorthavow'. *Trewordavo* 1298, *Treworthavou* 1328, *Tredavo alias Trewardavo* 1633. Co **tre** + pers.n. *Gorthavow*. PNCo 165, Gover n.d. 656.

TREDINGTON Warw SP 2543. 'The estate called after Treda'. *Tredincgtun* [757]11th S 55, *Tredinctun* [978]11th S 1337, 1086, *Tredintun* [991]11th S 1366, *-ton* [c.1086]1190–1275. OE pers.n. **Tre(o)da* + **ing**[4] + **tūn**. There is, however, an alternative form *Tyrdintun* [964]12th in the unreliable charter S 731, which can only be taken as containing the pers.n. *Tyrda*. According to S 55 Tredington had previously been held by *comes Tyrdda*. Wo 172.

TREDINNICK Corn SW 9270. 'Brackeny farm'. *Treredenek* 1286, *Tredenek* 1406, *Tredinnek* 1610. Co **tre** + **redenek** (Co *reden* 'bracken' + adjectival suffix *-ek*). PNCo 165, Gover n.d. 316.

TREDRIZZICK Corn SW 9576. 'Brambly farm'. *Tredreyseg* 1262, *-dreysek* 1409. Co **tre** + **dreysek** (Co *dreys* 'brambles' + adjectival suffix *-ek*). PNCo 165, Gover n.d. 131.

TREDUCHAN H&W SO 5209 → TRETIRE SO 5224.

TREDUNNOCK H&W SO 5221 → TRETIRE SO 5224.

TREEN Corn SW 3923. 'Farm of the fort'. *Trethyn -in* 1321–c.1480 including [1284]1446, *Tredine* c.1540, *Treen* 1699. Co **tre** + ***dyn** with lenition of *d* to *th*. The reference is to the Iron Age promontory fort of Treen Dinas SW 3922, (castle of) *Trethyn* 1478, Co ***dynas**. PNCo 165, Gover n.d. 641, Thomas 1976.65.

TREETON SYorks SK 4387. 'The settlement at the trees'. *Tre(c)tone* 1086, *Tretun -ton* 1154×89–1531, *Treeton* from 1564. OE **trēow** + **tūn**. YW i.162.

TREFONEN Shrops SJ 2527. 'The ash-tree farm'. *Tre- Travenen* 1272, *Trefennen* 1302, *Trefonnen* 1307. W **tref** + **onnen**. O'Donnell 100, Gelling.

TREGADILLETT Corn SX 2983. Probably 'farm of Cadyled'. *Tregadylet* [1076]15th, *Tregadilet* [c.1212]15th. Co **tre** + pers.n. **Cadyled*. PNCo 165, Gover n.d. 150.

TREGATHERALL Corn SX 1189 → CHICKERELL Dorset SY 6480.

TREGEARE Corn SX 2486. 'Farm by the fort'. *Tre(n)gyer* 1284. Co **tre** + definite article **an** + **ker** with lenition of *k* to *g*. The reference is to the earthwork on Tregeare Down at SX 2586. PNCo 165, Gover n.d. 144.

TREGEARE ROUNDS Corn SX 0380. P.n. Tregeare SX 0379 + dial. **rounds** 'hill-fort'. Also known as *Tregeare Castle* 1876, an Iron Age hill-slope fort of the 2nd-1st cents. BC. Tregeare, *Tregayr* 1425, is 'farm by the fort', Co **tre** + **ker** with lenition of *k* to *g*. Sometimes called Dameliock Castle in 18th cent. antiquarian writings and later. But this is a wrong identificaion of Geoffrey of Monmouth's *Dimilihoc* (*History of the Kings of Britain* vii.19) which actually refers to Domellick SW 9458. PNCo 165, Thomas 1976.66.

TREGIDDEN Corn SW 7522. Partly unexplained. *Tregudyn* c.1190, 1331. Co **tre** + unknown element, possibly Co *cudin* 'lock of hair' or a pers.n. *Cudynn*. PNCo 165, Gover n.d. 560.

TREGOLE Corn SX 1998. Unexplained. *Tvrgoil* 1086, *Tregoul* 1270, *Tregaul* 1293, *Tregowel* 1306, *Tregawle* c.1550. The generic may be Co **tre** 'farmstead' if the 1086 form is a bad spelling; the second word is unidentified. PNCo 106, Gover n.d. 80.

TREGONETHA Corn SW 9563. 'Farm of Kenhetho'. *Tregenhetha* 1341, *Treganeytha* 1357, *Treganetha* 1449. Co **tre** + pers.n. *Kenhetho* with lenition of *k* to *g*. PNCo 106, Gover n.d. 359.

TREGONNING HILL Corn SW 6029. *Tregonning Hill* 1687. P.n. Tregonning SW 6030 + ModE **hill**. Tregonning, *Tregonan* 1341, is 'farm of Conan', Co **tre** + pers.n. *Conan* wih lenition of *c* to *g*. PNCo 166, Gover n.d. 495.

TREGONY Corn SW 9244. Possibly 'farm of Rigni'. *Tref hrigoni* 1049 S 1019, *Trelingan, Treligani* (with *l* for *r*) 1086, *Trigoni* 1201, 1260, *Tregeny* 1214, *Treguni* 1229, *Tregony* from 1267, *Tregny* c.1540, *Tregnye* 1610. Co **tre** + pers.n. **Rigni*. PNCo 166, Gover n.d. 443, DEPN.

TREGURRIAN Corn SW 8565. Partly unknown. *Tregurien* 1232 Gover, *Tregurrian -on* 1606. Co **tre** + unknown element. PNCo 166, Gover n.d. 350.

TREKNOW Corn SX 0586. 'Farm of the valley'. *Tretdeno* 1086, *Trefnou* c.1245, *Trenou* 1337–99, *Trenow* 1590. Co ***tref** + ***tnou**. Situated at the head of a short side-valley to the N of the main valley running down to the coast. PNCo 166, Gover n.d. 84.

TREL EVAN H&W SO 5222 → TRETIRE SO 5224.

TRELAN Corn SW 7418. Probably 'farm at the church-site'. *Tre(t)land* 1086, *Trelanmur* c.1260, *Trelan* from 1288. Co **tre** + ***lann**. No early church-site is known here, but the reference might be to an Iron Age cemetery between Trelan and Trelanvean SW 7519. *Trelanmur* is 'great Trelan' for distinction from Trelanvean 'little Trelan', *Trelanbigan* c.1250 Gover n.d., Co *byghan*. PNCo 167, Gover n.d. 560.

TRELANVEAN Corn SW 7519 → TRELAN SW 7418.

TRELASDEE H&W SO 5023 → TRETIRE SO 5224.

TRELASH Corn SX 1890 → TRELIGHTS SW 9979.

TRELIGGA Corn SX 0584. Partly unexplained. *Trelvge* 1086, *Treluga* 1086 Exon –1427, *Trelygy* 1569. Co **tre** + unidentified element, possibly a pers.n. **Luga* (? < Latin *Lucas*). PNCo 167, Gover n.d. 137.

TRELIGHTS Corn SW 9979. Partly unexplained. *Trefflectos* 1302, *Treleghtres* 1425. Co ***tref** + unidentified element identical to that in Trelash SX 1890, *Trefleghtres* 1355. This cannot be Co ***legh-res** 'slab ford' parallel to the W p.n. type Llechryd for topographical reasons. PNCo 167, Gover n.d. 114.

TRELILL Corn SX 0478. 'Farm of Lulla'. *Trelulla* 1262–1317, *Trelille* c.1540. Co **tre** + OE pers.n. *Lulla*. PNCo 167.

TRELISSICK Corn SW 8339. Probably 'farm of Ledik'. *Trelesyk* 1275, *Trelesyc* 1327. Co **tre** + pers.n. **Ledik* with later Co *d* > *z*. PNCo 167, Gover n.d. 121.

TRELOWARREN Corn SW 7124. Partly unexplained. *Trellewaret* 1086, *Treluueren* 1086 Exon, *Trelewarent* 1227, *-leweren* 1290, *-laweren* 1610. Co **tre** + unidentified element. PNCo 168, Gover n.d. 573.

TRELUBBAS Corn SW 6629 → RELUBBUS SW 5631.

TREMAIL Corn SX 1686. 'Farm of Mel'. *Tremail* from 1086 with variant *-mayl*. Co **tre** + pers.n. *Mel*. PNCo 168, Gover n.d. 56.

TREMAINE Corn SX 2388. 'Farm of the stone'. (Chapel of) *Tremen* [c.1230]15th, *Tremene* c.1450–1610, *Tremeen* 1523–9, *Tremaine* 1582. Co **tre** + **men**. PNCo 168, Gover n.d. 174.

TREMAR Corn SX 2568. 'Farm of Margh'. *Tremargh* 1284, 1305, 1413. Co **tre** + pers.n. *Margh* (from Latin *Marcus*). Alternatively this could be 'horse-farm', Co **tre** + **margh**. PNCo 168, Gover n.d. 256.

TREMATON Corn SX 3959. 'Estate, farm called Tref meu'. *Tref meu tun* mid 11th B 1247, *Tremetone* 1086, *-ton* 1201–1322, *Trematon* from 1337. Co p.n. *Tref meu*, Co ***tref** 'farm' + unidentified element or name, + OE **tūn**. PNCo 168, Gover n.d. 241.

TRENANCE Corn SW 8567. 'Farm in the valley'. *Trenans* 1327, 1414. Co **tre** + **nans**. The original settlement was at the head of a short valley running down to the coast. PNCo 168, Gover n.d. 350.

TRENARREN Corn SX 0348. 'Fort of a crane'. *Tyngharan* 1302, *Tingaran* 1307, *Trenyaren* 1556. Co ***tyn** (later replaced by **tre** and definite article **an**) + **garan**. The reference is to Cliff Castle on Black Head SX 0448. PNCo 168, Gover n.d. 388.

TRENCH Shrops SJ 6913. 'The woodland road'. *le Trenche iuxta Wombrugge* 1306×7, *Trenche Way* c.1630, *The Trench* 1692–1701. OFr, ME **trenche**, 'a woodland road with clearings on either side to make the route safe for travellers'. The modern settlement grew up in the 17th–19th cents. along the Wellington–Newport road. Gelling.

TRENCHFORD RESERVOIR Devon SX 8082. P.n. Trenchford + ModE **reservoir**. Trenchford, *Trenchford* 1558×1603, *Transfer* 1809 OS, is probably 'ford across the trench', ME **trenche** referring to a tributary of the Teign that flows in a deep valley, + **ford**. D 424.

TRENDRINE HILL Corn SW 4738. *Trendrine Hill* 1888. P.n. Trendrine SW 4739 + ModE **hill**. Trendrine, *Trendreyn* 1302, *Trendryn* 1327, is 'farm of the thorn-bushes', Co **tre** + definite article **an** + **dreyn** 'thorns, thorn-bushes'. PNCo 168, Gover n.d. 688.

TRENEGLOS Corn SX 2088. 'Farm of the church'. *(ecclesia Beati Gregorii de) Treneglos* from 1269. Co **tre** + definite article **an** + **eglos**. PNCo 169, Gover n.d. 86.

TRENEWAN Corn SX 1753. Partly unexplained. *Trenewien* 1207, *Trenewyen* 1424, *Trenewen* c.1523. Co **tre** 'farm' + unidentified element, possibly a pers.n. **Nowyen*. PNCo 169, Gover n.d. 269.

TRENT Dorset ST 5918. Originally a Brit r.n. identical with TRENT Derby SK 2826. *Trente* 1086–1409, *Trent* 1575. Do iii.393.

TRENT AND MERSEY CANAL Derby SK 3529. Staffs SK 2018. R.ns. MERSEY Mers SJ 4081 and TRENT Derby SK 2826. Otherwise known as the Grand Trunk Canal, it begins in the Bridgwater Canal at Preston Brook Chesh and joins the Trent at Cavendish Bridge in Derby. It is about 93 miles long and was cut in 1777. Gaz 1896 s.n.

River TRENT Derby SK 2826 etc. Possibly 'great wanderer' or 'great flooder'. *Treenta, Treanta* 731 BHE, *Treontan (stream)* [c.890]c.1000 OE Bede, 924 ASC(A), 1009, *(on) Tr(a)entan* [956]13th S 602, S 659, *Treonte* c.1025, *aqua Trentae* 1086, *Treante* [before 1085]12th, *(be, andlang) Trentan* 1121 ASC(E) under year 679, 1013, *Trente* 1156–1595, *Trent* from 1267. The Welsh forms are *Trahan(n)oni fluminis* [c.800]11th HB, *Tranon* c.1071, *Taranhon* [12th]c.1275. The forms are consistent with derivation from British intensive prefix ***tri-** + ***santŏn-** as in the Gaulish folk-n. *Santones*. The meaning of ***santŏn-** is uncertain. Attempts have been made to link it with Brit **sento-* (PrW **hint*, W *hynt* 'a way, a path') and a British divinity name *Sentona* 'goddess of the way'. Such a r.n. would mean 'great wanderer' referring both to the great length of the Trent and to its regular flooding. The river was famous for its tidal wave or *eagre* described in 1839 as 'a white curling wave varying from one to four feet in perpendicular height' which, like the Severn *bore*, was regarded as a manifestation of the river divinity. Unfortunately, however, the vowel of **sento-* is difficult to reconcile with that of **santŏn-*. An alternative proposal is the root of Latin *sentina* 'bilge water' < IE **sem-* 'to pour'. The r.n. would then mean 'flooding strongly, draining thoroughly'. British *Trisantonā*, if correct, would produce PrW **Trihanton* borrowed into OE as *Treante*, which together with variant *Treonte* would give ME *Trente*. The W spellings represent OW *Trahannan* < PrW **Trihanton* with vowel harmony *i - a* > *a - a* and PrW *nt* > OW *nn*. Identical with the Devon TARRANT and Sussex *Tarente*, the earlier name of the river ARUN. Db 18, Nt 9, St 21, RN 415, RBrit 476–8, 510, LHEB 524–5.

River TRENT OR PIDDLE Dorset SY 8591 → River PIDDLE OR TRENT.

TRENT VALLEY Staffs SJ 9922. R.n. TRENT SJ 9330 + ModE **valley**.

TRENTHAM Staffs SJ 8741. 'The homestead or estate on the r. Trent'. *Trenham* 1086, *Trentham* from c.1145. R.n. TRENT SJ 9330 + OE **hām**. DEPN, RN 416, Horovitz.

TRENTISHOE Devon SS 6448. 'Hill-spur of Trendel, the circular hill'. *Trendesholt* 1086, *Trendelesho(u) -hoo* 1242–1326, *Tryndsho* 1441, *Trenshoo, Trynshow* 16th. OE **trendel** possibly used as the name of the circular hill of Trentishoe Down (1059ft. at SS 6347), genitive sing. **trendeles**, + **hōh**. The church is situated on a spur of high ground extending NE from Trentishoe Down. D 56.

TREREECE H&W SO 5220 → TRETIRE SO 5224.

TRERIBBLE H&W SO 5122 → TRETIRE SO 5224.

TRERICE Corn SW 8458. 'Farm by the ford'. *Trereys* 1302, *Treres* 1326–74, *Trerees* 1359, *Trerys* c.1400, *Trerise* 1610. Co **tre** + ***res** (OCo *rid*). PNCo 169, Gover n.d. 373.

TRESCO Scilly SV 8914. 'Elder-bush farm'. *Trescau* 1086, c.1210. Co **tre** + ***scaw**, plural of **scawen** 'an elder tree'. Identical with TRESCOWE SW 5730. PNCo 169, CPNE 205.

TRESCOTT Staffs SO 8497. 'The cottages by the r. *Tresel*'. *(æt) Treselcotum* [985]12th S 860, *Trescote* 1190×9, 1271, *Tressecot* 1259. R.n. Tresel as in TRYSULL SO 8594 + OE **cot**, dative pl. **cotum**. RN 419.

TRESCOWE Corn SW 5730. 'Farm of elder-bushes'. *Trescav* 1086, *-scau* c.1210–83. Co **tre** + ***scaw**, pl. of **scawen** 'an elder-tree'. PNCo 169, Gover n.d. 495.

TRESHAM Avon ST 7991. Partly uncertain. *Tresham* from 972 S 786, *Tressam* 1549–1701. Perhaps OE ***træs**, ***tres** 'brushwood' + **hām** 'homestead'. However, Tresham stands at the head of a deep hemmed-in valley at the mouth of which stands Hammouth Hill ST 7889, *Homoth Hill* 1830 OS, 'mouth of the *ham*': perhaps, therefore, this is rather 'brushwood valley bottom' with **hamm** 6. Gl iii.31

TRESILLIAN Corn SW 8646. 'Farm of Sulyen'. *Tresulyan* 1325, 1592, *-ian* 1336, *Tresilian* 1451, cf. *Tresiliam bridg* 1610. Co **tre** + pers.n. *Sulyen*. PNCo 169, Gover n.d. 472.

TRESMEER Corn SX 2387. Probably 'farm of Gwasmeur'. *Treguasmer* [1076]15th, *Tre(w)asmur* [1185]15th, *Tresmur* [1275]15th, 1291, *Tresmere* 1284, 1610. Co **tre** + pers.n. **Gwasmeur*. PNCo 169, Gover n.d. 174.

TRESWELL Notts SK 7879. 'Tir's spring'. *Tireswell(e)* 1086–1520, *Tressewelle* 1428, *Truswell* 1511. OE pers.n. **Tīr*, genitive sing. **Tīres*, + **wella**. Nt 61 gives local pr [trʌsw(ə)l].

TRETHEVY QUOIT Corn SX 2568. *Trethevy Quoit* 1885. P.n. Trethevy + ModE **quoit** 'a ring of iron used in the game of quoits'. This is a 19th cent. antiquarian use of *quoit* in the sense 'discus, giant's play-thing', cf. the Devil's Quoit, Dyfed, SM 8800 and SR 9696, referring to a burial chamber and a standing stone respectively. Also known as *Trethevystones* 'called in Latin *Casa gigantis* ("the giant's house"), a litle howse raysed of mightie stones', c.1605 (PNCo 169), *Trevethy* (sic) *Stone* 1842. The reference is to an outstanding Neolithic tomb c.3200–2500 BC consisting of seven upright stones, (one fallen) and a massive capstone. Trethevy, *Trethewy* 1284, is 'farm of Dewi', Co **tre** + pers.n. *Dewi* with lenition of *d* to *th*. The form of 1842 is a deliberate corruption to make the name appear to contain Co *beth* 'grave' with reference to the Quoit. PNCo 169, Gover n.d. 256, Pevsner 1951.144, Thomas 1976.58.

TRETHEWEY Corn SW 3823. 'Farm of Dewi'. *Trethewy* 1320, 1435. Co **tre** + pers.n. *Dewi* with lenition of *d* to *th*. PNCo 170, Gover n.d. 642.

TRETHURGY Corn SX 0355. 'Farm of Devergi'. *Tretheverki* c.1230, *Trethevergy* 1251, *Trethergy* 1360. Co **tre** + pers.n. *Devergi*. Alternatively 'farm of the otter', OCo **doferghi**. PNCo 170, Gover n.d. 388.

TRETIRE H&W SO 5224. 'Long ford'. *Rythir* 1210×12, *Retir, Ret(t)yr* 1276–1314, *Tretire* 1831 OS. OW **rit** + **hir**. The modern form has been influenced by neighbouring p.ns.in W **tre** 'a homestead', Treberon SO 5025, *Trebereth* 1302×3, *Treberyn* 1334, 'Peren or Beren's farm', Trevase SO 5125, *Trevays* 1334, 'field homestead', W **tre** + **maes**, Trelasdee SO 5023, Tre Essey SO 5021, *Treosseth* 1334, Trereece SO 5220, *Trerease* 1587, 'Rhys's farm', Treaddow SO 5424, *Trerado(u)* 1227–1334, possibly 'settlement at the gap', W *tre'r adwy*, Trecilla SO 5321, *Tresele* 1302×3, 'Sely's farm', Treduchan SO 5209, *Tradraghaun* 1499, possibly 'Trychen's farm', Tredunnock SO 5221, *Treredennok* 1302×3, 'ferny farm', W **tre** + **rhedynog**, Trea Evan SO 5222, *Trezevan* 1334, 'Ieuan's farm', Treribble SO 5122, *Trerevel* 1207, probably 'smithy farm', W *tre'e efail*. Another instance of the spread of **tre** is Trebumfrey SO 5222, *Trebumfrey* 1722, a late translation of *Humfreston* 1224, 'Humphrey's estate'. He 132–3, Bannister 187, Celtic Voices 311-3.

TREVALGA Corn SX 0890. Partly uncertain. *Trevalga* from 1238, *Treualgy* 1610. Co **tre** + unknown element, possibly pers.n. *Melgi* with lenition of *m-* to *v-* if reduction of *-i* to *-a* is possible as early as the 13th cent., or a pers.n. **Melga* if not. PNCo 170, Gover n.d. 88.

TREVANSON Corn SW 9772. 'Farm of Antun'. *Trevansun* 1259, *Trevanson* 1284. Co ***tref** + pers.n. *Antun* (from Latin *Antonius*) with later Cornish *ns* for *nt*. PNCo 170, Gover n.d. 317.

TREVARREN Corn SW 9160. Partly uncertain. *Treverran* 1201, *-veran* 1244, 1302, *Trevarran* 1336. Co **tre** 'farm' + unidentified element or name, possibly a pers.n. *Meren* corresponding to OBret *Meren*, with lenition of *m-* to *v-*. PNCo 170, Gover n.d. 324.

TREVARRICK Corn SW 9843. Partly unexplained. *Trevarek* 1327, 1332, *Trevarrek* 1428. Co **tre** 'farm' + unknown element or name. PNCo 170, Gover n.d. 401.

†The form *Trisantona* cited from Tacitus *Annals* 12.31 ([115×7]11th) is an emendation by Henry Bradley of MS *castris antonam* and must, however brilliant, be left aside.

TREVASE H&W SO 5125 → TRETIRE SO 5224

TREVELLAS Corn SW 7452. Partly uncertain. *Trevelles* 1306–76, *-as* 1341. Co **tre** + unidentified element or name, possibly a pers.n. *Melyd* corresponding to the W saint's name *Melyd* with lenition of *m-* to *v-*. PNCo 170, Gover n.d. 366.

TREVERVA Corn SW 7531. Partly uncertain. *Trewruvo* 1327, *Trefurvo* 1358, *Trevyrvo* 1407, *Treverve* 1500. Co ***tref** 'farm' + unknown element or name, possibly a pers.n. **Urvo* if the 1327 spelling is a mistake for **Trewurvo*. PNCo 171, Gover n.d. 500.

TREVIGRO Corn SX 3369. Partly unexplained. *Trevigora* [c.1230]1348, *Trefigerow* 1327, *Trevigro* 1622. Co ***tref** or **tre** + unexplained element **igra*, **migra* or **bigra*. PNCo 171, Gover n.d. 205.

TREVISCOE Corn SW 9456. 'Farm of Otker'. *Tref otcere* 1049 S 1019, *Trevyscar* 1333, *Treviscoe* 1748. Co **tref** + OCo pers.n. *Otcer* with later change of *t* to *s*. PNCo 171, Gover n.d. 426.

TREVONE Corn SW 8975. Partly unexplained. *Treavon* 1302, 1333, *Trevone* 1427. Co **tre** 'farm' + unidentified element or name. This cannot be Cornish **auon** 'river' unless an exceptional stress-shift from the first to the second syllable of **auon** is admitted as a possibility. PNCo 171, Gover n.d. 357.

TREVOSE HEAD Corn SW 8576. *Trevose Head* 1748, earlier ~ *Point* 1699. P.n. Trevose SW 8675 + ModE **head** 'a headland'. Trevose, *Trenfos* 1302, *Trefos* 1373, is 'farm by the bank', Co **tre** + definite article **an** + **fos** 'a dyke, a ditch'. The name suggests the existence of a promontory fort on the headland, which is further suggested by the name Dinas Head (Co ***dynas** 'fort') for part of the headland. There is a tumulus at Dinas Head but no known fort. Alternatively the reference may have been to the (undated) earthwork 500 yards E of the farm. PNCo 171, Gover n.d. 353.

TREWARMETT Corn SX 0686. Partly uncertain. *Trewerman* 1302, *-warman -worman* 1337, *-warmett* 1599. Co **tre** 'farm' + unidentified element or name, possibly a pers.n. **Gorman* with lenition of *g-* to *w-* and unexplained 16th cent. *-ett*. PNCo 172, Gover n.d. 85.

TREWARTHENICK Corn SW 9044. 'Farm of Gwethenek'. *Trewythynek* 1284. Co **tre** + pers.n. *Gwethenek* with lenition of *gw-* to *w-*. PNCo 172.

TREWASSA Corn SX 1486. 'Farm of Gwasso'. *Trewas(s)a* 1284, 1305. Co **tre** + pers.n. *Gwasso* (OCo *Was(s)o*). PNCo 172, Gover n.d. 57.

TREWAVAS HEAD Corn SW 5926. *Trewavas Head* 1813. P.n. Trewavas + ModE **head** 'a headland'. Trewavas, *Trewaevos* 1289, *Trewaves* 1504, is 'farm of the winter-home', Co **tre** + ***gwavos** 'winter-dwelling' (OCo *goyf* 'winter' + **bod*). PNCo 172, Gover n.d. 496.

TREWELLARD Corn SW 3733. Partly uncertain. *Trewyllard* 1307, *-wylard* 1327, *-welard* 1376. Co **tre** + uncertain element, possibly a pers.n. **Gwylarth* with lenition of *gw-* to *w-*, or better ME pers.n. or surname *Wil(l)ard* (< ? OE *Wilheard*). PNCo 172, Gover n.d. 634, DBS 385.

TREWEN Corn SX 2583. Probably 'white farm'. *Trewen* [c.1293]15th. Co **tre** + **guyn** with lenition of *gw-* to *w-*. PNCo 172, Gover n.d. 175.

High TREWHITT Northum NU 0105. *High Trewhitt* 1868 OS. ModE adj. **high** + p.n. *Tirwit* 1150×62, *Tyr(e)wyt -wit* 13th cent., *Tirwit(h)* 14th cent., *Trewhytt* 1542 of uncertain meaning and origin. It has been taken to be the 'fire-wood bend', ON **tyri**, **tyrfi** 'a resinous wood for fire-making' + OE ***wiht** 'a bend' sc. in Wreigh Burn. Derivation from OE ***thwīt**, ON **thveit** 'a clearing' would be better but the absence of spellings in *thweit* is a difficulty and the element **thveit** is not otherwise known in Northum. 'High' for distinction from Low Trewhitt NU 0004, *Low Trewhitt* 1868 OS. NbDu gives pr [trufit], DEPN.

TREWIDLAND Corn SX 2559. Possibly 'farm of Gwydhelan' or 'farm by a cemetery'. *Trewithelon* 1297, *-wythelan* 1298. Co **tre** + pers.n. *Gwydhelan* parallel to W *Gwyddelan* seen in Dolwyddelan, Gwynedd SH 7352, and Llanwyddelan, Powys SJ 0801. Alternatively, in view of the different stress pattern in the Co name, the second element might be a common noun **gwyth-lann* 'cemetery' parallel to W *gwyddlan*. But no such feature is recorded here. PNCo 172, Gover n.d. 277.

TREWINT Corn SX 1897. 'Windy farm'. *Trewynt* 1303, 1342, *Trewent* 1336, *Overa- Nideratrewynt* c.1520. Co **tre** + **guyns** (older **gwynt*) with lenition of *gw-* to *w-*. PNCo 173, Gover n.d. 81.

TREWITHIAN Corn SW 8737. Partly uncertain. *Trewythyan* [c.1270]14th, *-ian* from 1302. Co **tre** 'a farm' + pers.n. *Gwethyen* or *Gwythyen*. PNCo 173, Gover n.d. 453.

TREWOON Corn SW 9952. 'Farm on the downs'. *Tregoin* 1086, *Trewoen* 1284–1428, *Trewone* c.1520, *Trewoone alias Troone* 1680. Co **tre** + **goon** with lenition of *g-* to *w-*. Cf. TROON SW 6638. PNCo 173 gives pr [tru:ən], Gover n.d. 414.

TREYARNON Corn SW 8673. Probably 'farm of Yarnenn'. *Trearvan* (sic for *-nan*) c.1210, *Treyarnen -an* c.1240–1464, *Trearnen* c.1245. Co **tre** + pers.n. **Yarnenn* or *Yarnan*. PNCo 173, Gover n.d. 354.

TREYFORD WSusx SU 8218. 'The tree ford'. *Treverde* 1086, 1428, *Treferd* 1256–1332, *Treford* 1272, 1279, 1428, *la Threferde* 1330, *Treaford* 1585, *Trayford* 1641. OE **trēo** + ***fyrde**. The exact form of the generic is uncertain, either an adj. 'forded' or the locative-dative case of **ford** or a *ja*-stem ***fyrde**. The reference may have been to a ford marked by a tree or laid with or crossed by tree-trunks. Sx 43, PNE i.180.

TRIANGULAR LODGE Northants SP 8382. A small late-Elizabethan building built by the Catholic convert Sir Thomas Tresham in 1597. Its design is entirely directed by the number three and it allegorises the Trinity in stone. ModE **triangular** + **lodge**. Pevsner 1973.400.

TRICKETT'S CROSS Dorset SU 0901. Surname *Trickett* + ModE **cross** 'a crossroads'.

TRIGON HILL Dorset SY 8989. 'Triangle hill'. *Trigdon* 1811, *Trigon Hill* 1842. ModE **trigon** referring to the shape of the hill or the farm there. The 1811 OS form *Trigdon* is due to mistaken association with names in *-don* < OE **dūn** 'a hill'. The earlier name of the hill was *Hungerhill* 1774, *Hungerhull* 1318, a well-known derogatory name for poor land, OE **hungor** + **hyll**. Do i.165.

TRIMDON Durham NZ 3633. 'High ground with a wooden cross'. *Tremeldon* 1196, *Tremedun* [1183]13th, [1183]c.1382, *-dona -done* c.1230–1311×6, *Tremdon* 1234–1421, *Trendon'* 1314, 1330, *Trimdon* from 1539. OE **trēo-mǣl** + **dūn**. Probably referred to an early preaching cross on the site where the parish church was later built. Cf. MELDON Northum NZ 1183. NbDu 200.

TRIMDON COLLIERY Durham NZ 3835. *Trimdon Colliery* 1863 OS. P.n. TRIMDON NZ 3633 + ModE **colliery**.

TRIMDON GRANGE Durham NZ 3635. *Trimdon Grange* 1863 OS. P.n. TRIMDON NZ 3633 + ME **grange**.

TRIMINGHAM Norf TG 2738. 'Homestead of the Trymingas, the people called after Trymma or Trymi'. *Trimingeham* 1185, *Trimigham* 1254, 1291, *Tremingham* 1276. OE folk-n. **Trym(m)ingas* < OE pers.n. **Trymma* or **Trymi*, genitive pl. **Trym(m)inga*, + **hām**. DEPN, ING 139.

TRIMLEY ST MARTIN Suff TM 2738. 'St Martin's T'. *Tremele Sancti Martini* 1254. P.n. *Tremlega, Tremelaia* 1086, *Tremelye* 1221, *Trem(e)ley(e)* 1270–1558, *Tremle* 1345, *Trym(e)ley* 1505–96, *Trimley* 1674, 'Trymma's wood or clearing', OE pers.n. *Trymma* + **lēah**, + saint's name *Martin* from the dedication of the church. The churches of the two parishes, T St Martin and T St Mary, lie on the edge of their respective parishes and occupy the same churchyard. DEPN, Arnott 35, Baron.

TRIMLEY ST MARY Suff TM 2836. 'St Mary's T'. *Tremle Beate Marie* 1254. P.n. Trimley as in TRIMLEY ST MARTIN TM 2738 + saint's name *Mary* from the dedication of the church. DEPN, Baron.

TRIMPLEY H&W SO 7978. 'Trympa's wood or clearing'. *Trinpelei* 1086, *Try- Trimpelege -ley(e)* 1255–1550, *Trimpley* 17th cent. OE pers.n. **Trympa*, an *i*-mutation weak variant of *Trump* as in TRUMPINGTON Cambs TL 4454, + **lēah**. Wo 252.

TRIMPLEY RESERVOIR H&W SO 7778. P.n. TRIMPLEY SO 7978 + ModE **reservoir**.

TRIMSTONE Devon SS 5043. Partly unexplained. *Trempelstan* 1238, *Trympeston* 1330, *Trimpestone* 1351. Unknown element **trempel** + OE **stān**. Possibly OE ***trempel-stān** 'a stone used as a stile' related to the verb *tramp*, Gothic *ana-trimpan*. D 42.

TRIMWORTH Kent TR 0649 → TANDRIDGE Surrey TQ 3750.

TRING Herts SP 9211. 'The wooded slope'. *Trevinga -ung(e), Tredvnge -unge -unga -ung', Trunga* 1086, *Trawinge* 1176, *Treing* 1200, *Treange* 1208, *Treinges* 1221, *Tre(h)eng(e)* 1222, 1254, *Trynge* 1367, *Trungla* 1135×54, 1154×89, *Triangre* 1199×1216, *Trehangr', Threhangre* 1199, *Trehanger* 1265. On the strength of the 1265 form this is OE **trēow** + **hangra** 'tree wooded-slope', the 1086 forms perhaps pointing to a side form with OE **trēowede** 'covered with trees'. The development was *Trēow-hangra* > *Trehangre* > *Treange, Treunge, Treenge* > *Treinge* > *Tring*. This is the preferred explanation of ING 228 and EHN 3.29 to Hrt's tentative suggestion of ON **thrithjungr** 'a third part'. Hrt 25, 51.

TRISPEN Corn SW 8450. Partly obscure. *Tredespan* 1325, *Trethespan* 1382, *Trevisprin* 1462, *Trethispen* 1513, *Trispan* 1695. Co **tre** 'farm' + unknown word or name *despan* with lenition of *d-* to *dh-* and with *v* substituted for [ð]. The earlier form of the name survives in Trevispian Vean SW 8450, 'little Trispen' (Co **byghan**). PNCo 173 gives pr ['trispən], Gover n.d. 447.

TRITLINGTON Northum NZ 2092. 'The estate called after Tyrhtel'. *Turthlyngton* c.1170, *Tirlington* 1210, *Tierclinton* 1212, *Ti- Tyrt(e)lington* 1242 BF-1346, *Trit(e)lin(g)ton* 1256, 1279. OE pers.n. *Tyrhtel* + **ing**[4] + **tūn**. NbDu 200, DEPN.

TROON Corn SW 6638. 'Farm on the downs'. *Trewoen* 1327, 1442, *Trewon* 1430, *Troon* 1768. Co **tre** + **goon** with lenition of *g-* to *w-*. Cf. TREWOON SW 9952. PNCo 173, Gover n.d. 586.

TROSTON Suff TL 8972. 'The estate called after Trost or Trosta'. *Trostingtun* 975×1016 S 1487, *Trostuna* 1086, *Troston* 1610. OE pers.n. **Trost* or **Trosta* + **ing**[4] + **tūn**. DEPN.

TROTTISCLIFFE Kent TQ 6460. 'Trott's cliff'. *Trottes clyva, Trottesclib* [788]12th S 129, *Trotescliua -e* [c.975]12th B 1321–2, *Totesclive* (sic) 1086, *Trot(t)e(s)cliue* 1218–68, *Trosclyffe* 1610. OE pers.n. **Trott* as in TROTTSWORTH Surrey SU 9967, genitive sing. **Trottes*, + **clif**. A variant form Trosley is given by Bartholomew's 1971 ½ in. map. KPN 70, L 130, 136.

TROTTON WSusx SU 8322. Possibly 'the settlement by the stepping stones'. *Traitone* 1086, *Tratin(g)- Tradin(g)ton* 13th cent., *Tratton* 1316, 1421–56, *Trotton* 1545. Possibly OE ***trædding** + **tūn**. The reference may have been to a crossing of the Rother. Alternatively 'the estate called after Trætt', OE pers.n. **Trætt* + **ing**[4] + **tūn**. Sx 44.

TROTTSWORTH Surrey SU 9967. 'Trott's enclosure'. *Troteswrthe* 1166, *Trottesw(o)rth(e)* 1166–1332. OE pers.n. **Trott* as in TROTTISCLIFFE Kent TQ 6460, genitive sing. **Trottes*, + **worth**. Sr 125.

TROUGHEND COMMON Northum NY 8591. *Troughend Common* 1869 OS. P.n. Troughend NY 8692 + ModE **common**. Troughend, *Trocquen* 1242 BF, *Tre(h)quen(ne)* 1279, 1292, *Tr(o)uwhen* 1336 SR, 1460, 1590, *Troughwen* 1399, 1618, *Trocken* 1584, *Trough End* 1663, is obscure. It is probably identical with Torquhan Scotland NT 4447, *Torquhene* 1593, Troquhain Scotland NS 3709, *Treu(e)chane* 1371, *Troquhan* 1511, *Torquhane* 1506 and NX 6879, *Trechanis* 1467, *Torquhane* 1590, PrW, PrCumb ***treβ** 'a farmstead, a homestead, a hamlet' + an unknown element, possibly PrW ***wïnn** 'white', if the name was borrowed after the sound change *w* > *gw*. This is not normal with English p.ns., the first datable example being the personal name *Cwœspatrik* BCS 1254 < W *Gwaspatric* 'servant of St. Patrick', which seems to show the same English sound substitution for PrW *gw* as the early forms for Troughend. The stage *gw* is dated to the late 6th, early 7th cent. The p.n. type ***treβ** + ***wïnn** is well evidenced in Co, e.g. TREWEN SX 2583 and four other examples, and also in Drefwen, the W name of WHITTINGTON Shrops SJ 3231, *Trefwen* 1254. NbDu gives pr [trufend], ScotPN 168, PNE ii.185, 269, Jnl 1.51, LHEB 385–94, CPNE 231.

TROUTBECK Cumbr NY 4003. 'Trout stream'. *Trutebek* 1272–1345, *Trutbek* 1444, *Trout(e)bek -beck(e)* 1324–1656, *Trought(e)bek* 1437–51. ME **trought** (OE *truht*) + **beck** (ON *bekkr*). We i.188, SSNNW 255.

TROUTBECK BRIDGE Cumbr NY 4000. *Trowtbe(c)kebrigge* 1454. R.n. Trout Beck as in TROUTBECK NY 4003 + ME **brigge** (OE *brycg*). We i.197.

TROUTS DALE NYorks SE 9287. 'Trout pool'. *Truzstal* 1086, *Trucedal(e)* [1314]n.d.–1619, *Trowt(t)es- Troutesdale* 1497–1665. OE **truht** + **stall**. This replaces the earlier explanation 'Trut's valley', ON by-n. *Trútr*, genitive sing. *Trúts*, + **dalr**. YN 98 gives pr [tru:tsdil], SSNY 259.

TROWBRIDGE Wilts ST 8557. 'The tree bridge', i.e. made of tree-trunks. *Straburg* (sic) 1086, *Trobrig(g)e* 1184–1242, *Trou- Treu-* 1211–1332, *Trowbrugg(e)* 1230, *Troughbrigge* 1358, 1395. OE **trēow** + **brycg**. Some later sps have been influenced by the word 'trough'. Wlt 133 gives pr [troubridʒ].

TROW GREEN Glos SO 5706. 'Trollop green'. *Trolley grene* 1618, cf. *Trowfeild* 1635. This is thought to be early ModE **trowle, trulle** 'a loose woman, a trollop' + **green**; but the specific could equally well be ModE **trolley** 'a cart'. Gl iii.239.

TROWLE COMMON Wilts ST 8458. *Trowl Common* 1817 OS. P.n. *Trole* 1086, *Trol(le)* 1154×89–1384, *Trul(le)* 1154×89, 1242, *Troulle* 1571, *Trowle* 1611, unidentified element or name ***trūl** of unknown meaning, possibly a stream name, + ModE **common**. Wlt 117.

TROWSE NEWTON Norf TG 2406. *Trowse Newton* 1838 OS. Short for two separate names, *Trowes cum Newtone* 'Trowes with Newton' 1316. P.n. *Treus(sa)* 1086, 'the tree house' referring to a wooden building, OE **trēo** + **hūs** or ON **tré-hús**, + p.n. Newton as in Trowse NEWTON, *Newotona* 1086, 'the new settlement', OE **nīwe** + **tūn**. DEPN.

TRUDOXHILL Somer ST 7443. *Truddoxhill* 1817 OS. Uncertain element + ModE **hill**.

TRULL Somer ST 2122. 'The ring'. *Trendle* 1225, 1314, *Trull* from 1483. OE **trendel**. This word or its variants *trindle, trundle* is used of pre-historic earthworks, cf. Trundle or Trendle Ring Camp at Bicknoller, ST 1139. An alternative possibility is reference to gearing associated with a water-wheel, OED s.v. *trindle* sb.1, *trundle* sb.2. The precise reference here is unknown. DEPN.

TRUMPET H&W SO 6539. *Trumpet* 1831 OS. An inn name, also recorded in London at the time of Pepys. DPN s.n., ETN 100, 102.

TRUMPINGTON Cambs TL 4455. 'Estate called after Trump'. *Trumpintune* 1086, 1170, [1042×66]13th S 1051, *-tona -tone* 1086, *Trumpi(n)tone -y(n)- -e-* 1202–1470, *Trumpington(e) -yng-* from 1198. OE pers.n. *Trump* + **ing**[4] + **tūn**. Ca 91.

TRUNCH Norf TG 2834. Uncertain. *Trunchet* 1086, 11th, *Tru(n)ch* 1203, 1254. *Trunchet* may be a transferred name from one of the French Tronchets, Le Tronchet, Ille-et-Vilaine, *Trunchetum* [1154×89]1291, *Tronchetum* 1159×78, Tronchet, Sarthe, *Trunchet* 1097, Tronchoy, Haut-Marne, *Trunchetum* 1249×54, or Tronchoy, Somme, *Truncetum* 1130, all meaning 'wood', Latin **truncetum** < **truncus** 'a tree-trunk'. An interesting alternative possibility suggested by Ekwall is the PrW equivalent of *Tronget* in Restronguet Corn SW 8316, *Restonget* 1322, 'hill-spur called Tronget, the nose wood', Co. ***ros** + p.n. Tronget < Co. ***tron-gos** < Co **tron** 'nose, promontory', W

trwyn, + **cos** 'wood', W **coed**, PrW **cẹ̄d*. In PrW the same name would have been **troncẹ̄d*. DEPN, Dauzat-Rostaing s.n. Troncq (Le), CPNE 235. For a different W etymology see now Celtic Voices 173.

The TRUNDLE WSusx SU 8711. A name now used of the whole of St Roche's Hill, *St Rokeshill* 1579, *Rooks Hill* 1813 OS, but originally applied only to the Iron Age earthwork on it. It represents OE **tryndel** 'a cricle' found frequently in minor names. Sx 54, PNE ii.185, Thomas 1980.212.

TRURO Corn SW 8244. Uncertain. *Triueru* c.1173–1214, *Treueru* 1194–1299 Gover n.d., *Triu(e)reu* [12th]1285, *Triwereu* 1201, *Truru* c.1280–1390, *Trufru* 1289–1362, *Truro* 1610. There are two possibilities; either Co ***try-** 'triple' or 'very' + **berow** 'a boiling' possibly meaning '(place of) great water-turbulence' referring to winter flooding where the two fast rivers Allen and Kenwyn meet to form the Truro river; or the name is to be compared with that of Trier, *Treveri* 1st BC–7th AD, *Triera* 10th, from the Gaulish tribal name *Treveri* thought to mean 'the people of the river-crossing', i.e. the people who transport goods across the Moselle, Celtic *trē* 'through' < **trei-* + **uer* 'water, wet', cf. OIr *treóir* 'water crossing'. There are two river-crossings at Truro. However, the ending *-(e)u* remains unexplained unless regarded as parallel to the suffix in *Cornowii* 'horn people' (Co **corn**, ModCo *Kernow*) as in CORNWALL. PNCo 174, Gover n.d. 483, Berger 259, DAG 753–4.

TRUSHAM Devon SX 8582. Uncertain. *Trisma* 1086, 1370, *Trisme, Trysme* 1259–1431, *Tryssam* 1535, *Trusham* 1577. Perhaps Co **dreys** 'brambles' + p.n. forming suffix **-ma** 'place'. D 503, CPNE 88, 155.

TRUSLEY Derby SK 2535. Possibly 'brushwood clearing, clearing where brushwood is obtained'. *Trusselai -leia -ley(e) -legh* etc. c.1141–1466, *Truslegh(e)* 1224×48, *-ley(e)* from 1331. OE **trūs** + **lēah**. The almost universal *-ss-* in early forms, however, casts some doubt on this explanation. Db 613.

TRUSTHORPE Lincs TF 5183. Partly uncertain. *Dr(e)uistorp* 1086, *Turstorp, Trustorp'* 1189–1212, *Truscetorp* 1226, *Strutt(h)orp'* 1196–1231. The DB forms point to 'Drew's outlying settlement', CGmc pers.n. *Drogo*, OFr *Dreus, Drues* (nominative), *Dru, Drui, Dreu* (accusative), the later forms to replacement of Drew by respectively ON pers.n. *Thórir*, ME **trouse** (OE *trūs*) 'brushwood' and either ON ***strútr** 'a pointed hood, a promontory' as in STRUBBY TF 4582 or OE ***strūt** 'strife, dispute'. Trusthorpe is situated on the coast on land which may have formed a promontory before the present shore-line was established. For other p.ns. in this area with CGmc pers.ns. as specific cf. GRAINTHORPE TF 3897, GRIMOLDBY TF 3988, MABLETHORPE TF 5084 and THEDDLETHORPE TF 4688. DEPN, SSNEM 120.

TRYSULL Staffs SO 8594. Transferred from the r.n. *Tresel*, the old name of Smestow Brook. *Treslei* 1086, *Tresel* 1176, 1295, *Trisel* 1236, *Treasle* 1686. The r.n. is *(on) Tresel* [985]12th S 860, *(into) Tresel* [996 for 994]17th S 1380, *Tresel* 13th cent., *Tressul* 1307, probably a compound of OW **tres** 'uproar, turmoil, commotion' and stream-n. forming suffix **ell** (cf. G r.n. Mosel). St i.22, CPNE 93, Staffordshire Studies 10.77–8, Horovitz. RN 419 gives pr [tr:izl].

TUBNEY Oxon SU 4398. 'Tubba's island'. *Tobenie* 1086, *Tubbeneia* [1066×87]c.1240, [1166]13th with variants *-ai -eya -ey(e)*. OE pers.n. *Tubba*, genitive sing. *Tubban*, + **ēġ**. The same pers.n. occurs in Tubworth Barn in the same parish, *(on) tubban forda* 'Tubba's ford' [942]c.1240 S 480, *Tubford* 1535, *Tubberd* 1553, *Tubworth* 1828. Brk 425–6.

TUCKENHAY Devon SX 8156. Possibly 'at the oak enclosure'. *Tokenhey* 1550. ME **atte oken hay** (OE *æt thæm ācenan (ġe)hæġe*). D 315.

TUCKHILL Shrops SO 7888. Possibly 'Tucga's hill'. *Tug(g)- Tuck- Tokehill* 1617, *Tuchill* 1617, *Tugehill* 1649, *Tuckhill* 1833 OS. OE pers.n. **Tucga* as in TUGFORD SO 5587 + **hyll**. But the evidence is rather late for certainty. For a different suggestion ('a hill where loads are dragged' ME **tuggen** 'to pull violently') cf. TOW LAW Durham NZ 1139. Gelling.

River TUD Norf TG 0912. A modern back-formation from the p.n. North TUDDENHAM TG 0414.

TUDDENHAM 'Tudda's homestead'. OE pers.n. *Tudda*, genitive sing. *Tuddan*, + **hām**.

(1) ~ Suff TL 7371. *Todenhā, todeha', Totenhā* 1086, *Tudeham* 1154×89, *Tudenham* 1235. DEPN.

(2) ~ Suff TM 1948. *Todden- Totden- Tude(n)ham* 1086, *Tudenham* c.1236–1610, *Tuddenham* from 1524. DEPN, Baron.

(3) East ~ Norf TG 0711. *East Tudenham* 1086. OE adj. **ēast** + p.n. *Toddenham* 1086, *Tudenham* 1198, *Tuddeham* 1199. 'East' for distinction from North TUDDENHAM TG 0414. DEPN.

(4) North ~ Norf TG 0414. *Nord Tudenham* 1086. OE adj. **north** + p.n. Tuddenham as in East TUDDENHAM TG 0711. DEPN.

TUDELEY Kent TQ 6245. Probably 'at the wood or clearing growing with ivy'. *Tivedele* 1086, *-u-* c.1100, *Theodelei* c.1100, *Teodele, Tywedeleg'* 1254, *Teudele(ye)* 1261×2–1327. There is no known element to account for the spellings *Tivede- Tywede- Teude-*. Possibly initial *T-* is due to misdivision of OE **æt** + ***īfed**, oblique case definite form **īfedan**, + **lēah** 'at ivy wood'. PNK 174, DEPN.

TUDHOE Durham NZ 2635. 'Tuda's hill-spur'. *thudoue* early 13th, *Tudhou* 1242×3, *-how(e)* 1296–1522, *-ow(e)* 1368–77, *Tod(e)how* 1288–1372, *Tod(d)ow(e)* 1335–86, *Tuddo(e)* 1558–1638, *Tudhoe* from 1717. Pers.n. *Tud(d)a* + OE **hōh**. *Tuda* is the short form of a name of Welsh origin such as *Tudwal*. Pronounced locally [tudə]. NbDu 201, L 168.

TUE BROOK Mers SJ 3892. *Tue Brook* 1842 OS.

TUFFLEY Glos SO 8315. 'Tuffa's wood or clearing'. *Tuffelege -le(ya) -ley(e)* 1086–1540, *Tuffley* from 1535. OE pers.n. *Tuffa* + **lēah**. Gl ii.142.

TUGBY Leic SK 7600. 'Toki's village or farm'. *Tochebi* 1086–1184, *Tokebi -by* 1176–1551, *Tokby(e)* 1275–1535, *Tugby* from 1519. ON pers.n. *Tóki*, genitive sing. *Tóka*, + **bȳ**. Lei 329, SSNEM 75.

TUGFORD Shrops SO 5587. 'Tucga's ford'. *Dodefort* (sic) 1086, *Tugafort* c.1090, 1138, *Tuggeford(')* *-fort* c.1143–late 14th, *Tugford* from 1500. OE pers.n. **Tucga* + **ford**. Sa i.294.

TUMBY Lincs TF 2359. 'Village, farm by or with a *tūn*'. *Tvnbi* 1086, *Tūbi* c.1115, *Tumbi -by* from 1170×5, *Tuneby* 1223. OE or ODan **tūn** 'an enclosure, ? a fence' + **bȳ**. Alternatively perhaps **tūn** is used in this name like **tūn-stede** 'village site' referring to a deserted settlement. SSNEM 75, Cameron 1998.

TUMBY WOODSIDE Lincs TF 2657. P.n. TUMBY TF 2359 + p.n. *Wood Side* 1824 OS so named from *Tumby Wood* ibid. at TF 2458.

TUMMER HILL SCAR Cumbr SD 1867. *Tummer Hill Scar* 1852 OS. A coastal reef. P.n. Tummer Hill + **scar** 'a rocky outcrop' (ON *sker*). Cf. *Tummer Hill Marsh, Tummer Hill House* 1852 OS.

Royal TUNBRIDGE WELLS Kent TQ 5839. Originally simply *Tunbridge Wells* 1819 OS, 'springs near Tonbridge', p.n. TONBRIDGE TQ 5947 + ModE **wells** referring to mineral springs discovered here in 1606 by Baron North when staying at Lord Abergavenny's hunting seat at Ebridge. Passing through a wood on his way back to London he noticed a clear spring of water which bore on its surface a shining mineral scum and left a ruddy ochreous substance on the bottom. He tasted the water and sent one of his servants for bottles to take a sample to his London doctor who gave a favourable judgment of the quality of the chalybeate water. The following year he came back for a full course of the waters and in 1608 seven springs were found and enclosed. They became famous and Queen Henrietta Maria spent six weeks taking the waters in 1630 after the birth of Prince Charles. The first public building came in the 1660s and the main expansion in the 1830s. The prefix 'Royal' was granted in 1909 to commemorate the many royal visitors. Room 1983.126, Newman 1969W.553.

TUNDRY GREEN Hants SU 7752 → DOGMERSFIELD SU 7852.

TUNSTALL 'A farm-site, a farmstead'. OE ***tūn-stall**. PNE ii.198.

(1) ~ Humbs TA 3031. *Tunestal(e)* 1086–1228, *Donestal(l)* 1098×1102–1160×2, *Tunstall* from 1189×99. YE 57 gives pr [tunstəl].

(2) ~ Kent TQ 8961. *Tvnestelle* 1086, *Tunsteal, Tunestele* c.1100, *Tunstall(')* from 1208. PNK 271.

(3) ~ Lancs SD 6073. *Tunestalle* 1086, *Tunstall* from 1235. La 183, Jnl 17.102.

(4) ~ Norf TG 4108. *Tunestalle* 1086, *Tunstal* 1196. DEPN.

(5) ~ NYorks NZ 5312. *Ton(n)estale* 1086, *Tunstall'* 1189. YN 166.

(6) ~ NYorks SE 2196. *Tunestale* 1086, *Tunstale* [1157]15th etc. YN 246.

(7) ~ Staffs SJ 8651. *Dunstall'* 1162, *Tunstal* 1212, *-stall* 1227. DEPN, Horovitz.

(8) ~ Staffs SJ 7727. *Tunestal* 1086, *Tunstall* 1267. DEPN.

(9) ~ Suff TM 3655. *Tunestal* 1086, *Tunstall* 1242. DEPN.

(10) ~ FOREST Suff TM 3854. P.n. Tunstall, no early forms, + ModE **forest**.

(11) ~ RESERVOIR Durham NZ 0641. P.n. Tunstall + ModE **reservoir**. Tunstall, *Tunstall* 1768 Armstrong, replaces earlier *Tounstedhows* 1382, *Tunstead House* 1647, a vaccary in Wolsingham Park. The park was originally a hunting reserve of the Bishop of Durham. OE **tūn-stede** 'enclosure place'. *Tunstead* and *Tunstall* frequently interchange in p.ns. Stead 90–2.

TUNSTEAD Norf TG 2921. 'Village, farmstead'. OE **tūn-stede**. *Tune- Ton(e)- Consteda, Stunetada* 1086, *Tunsted(e)* 1101×7–1535 including *[1044×7]13th S 1055. Nf ii.134, 197, Stead 76, 185.

TUNWORTH Hants SU 6748. 'Tunna's enclosure'. *Tvneworde* 1086, *Tun(n)ewrthe -worth* 1193–1331, *Tunworth* 1579. OE pers.n. *Tunna* + **worth**. Ha 165, Gover 1958.129.

TUPSLEY H&W SO 5340. Probably 'ram wood or clearing'. *Topeslage* 1086, *-le* 1241, 1265, *T(h)opesley(e)* 1292–1506. OE * **tōp**(ME *toup*), genitive sing. ***tōpes**, + **lēah**. Alternatively the specific might be an OE pers.n. derived from ***tōp** or ODan *Topi*. He 95.

TUPTON Derby SK 3966. Either 'Tupi's farm' or 'ram farm'. *Top(e)tune* 1086, *Topton* 1216×72, 1272×1307, 1438, *Tupton(e)* from 1185. Either ON pers.n. *Túpi* or ME **tup(p)e** + **tūn**. Db 317, SSNEM 196.

TUR LANGTON Leic SP 7194. Possibly the 'estate called after Tyrli'. *Cher- Terlintone* 1086, *Terlin(g)ton(e)* 1205–6, *Tirlinton' -yn-* 1165–1316, *-ing- -yng-* 1209×35–1518, *Turlinton(e)* 1165–1205, *-ing- -yng-* 1253–1617, *Tir- T(h)urlangton* 1526–1688, *Tur Langton* 1835 OS. OE pers.n. **Tyrli* + **ing**[4] + **tūn**. Lei 262.

TURGIS GREEN Hants SU 6959. *Turges Green* 1817 OS. Short for Stratfield Turgis Green. Cf. Turgis Court, *Turges Court* 1817 OS. For the family name Turgis see STRATFIELD TURGIS SU 6960.

TURKDEAN Glos SP 1017. 'Valley of the river *Turce*'. *(on) Turcandene* [737×40]11th S 99, [949]12th S 550, *(vallis qui dicitur) turcadenu* 'valley called T' 779 S 114, *(ofer) turcendene* [816]11th S 179, *Tvrchedene* 1086, *Turkeden(a) -dene* 1151–1535, *Turkdean* 1737. R.n. *Turce*, genitive sing. *Turcan*, + **denu**. *Turce* is paralleled by the W r.n. type Twrch, Dyfed SN 6446, *Turghe* 1578, and Dyfed SN 7117, *Turc(h)* c.1150, *Turch* 1577, Gwyn SH 8828, Gwyn SH 9024, Powys SH 9714, *Turgh* 1578, which is said to be W *twrch* 'boar' and so 'boar river', apparently in the metaphorical sense 'a river forming deep channels or holes into which it sinks into the earth and is lost for a short distance'. The stream that runs through Turkdean disappears underground at SP 0919. Gl i.183,xii, RN 420, AfcL 3.45, L 97, 99.

TURNASTONE H&W SO 3536. 'The Tournai family's estate'. *Tornerieton'* 1210, *Tho- T(h)urneston(e)* 1242–1356, *To- Turnaston(e)* from 1311. Family n. *Tournai* from Tournai-sur-Dive, Orne, + ME **toun** (OE *tūn*). This estate is probably identical with *Wlvetone* 1086 'Wulfa's estate' which was held in 1160×70 by Robert Turuei (sic for Turnei), probably son of Ralph de Tornai who was associated with this part of Hereford c.1132. Cf. WALTERSTONE SO 3425. He 192.

TURNDITCH Derby SK 2946. 'Winding ditch'. *Turnyndedyche* 1346, *Turn(e)dich(e) -dyche* 1387–96 etc., *-ditch(e)* 1560, *Turndyke* 1490. ME **turnende** varying with **trun**, **turn**, + **dīċ**. Db 614.

TURNERS HILL WSusx TQ 3135. *Turnoureshill* 1427, *Turner Hill* 1816 OS. Surname *Turner* + ME **hill**. A Galfridus le Turnur occurs in 1296. Sx 283.

TURNERS PUDDLE Dorset SY 8393 → Turners PUDDLE.

TURNWORTH Dorset SY 8207. 'Thorn-bush enclosure'. *Torneworde* 1086, *-worth(e)* 1280–1327, *Turnewrd(a) -wurth(') -worth(')* 1199–1530, *Turnet* 1412, *Turnwood* 1575, 1795, *Turnworth* 1795. OE **thyrne** varying with **thorn**, genitive pl. **thorna**, + **worth**. Probably an enclosure formed of thorns. Do ii.254.

TURSDALE Durham NZ 2035. 'The serf's valley'. *Trellesdene* c.1150, *Trillesdene* c.1200, *Tursdaile* 1649. ME **threll** (ON *thrǣll*), genitive sing. **threlles**, + **dene** (OE *denu*) later replaced by **dale**.

TURTON BOTTOMS Lancs SD 7415. 'Flat, wet valley-floor by Turton'. P.n. Turton as in Turton Tower SD 7315 + **bottom** (OE *botm*) referring to its situation in the narrow valley of Bradshaw Brook. Turton, *Turton* from 1212, *Torton* 1246, 1282, *Thurton* 1257, 1303, is 'Thorr or Thori's farm or village', AScand pers.n. *Thórr* or ODan *Thóri* + **tūn**. La 47, SSNNW 191.

TURTON MOOR Lancs SD 6918. P.n. Turton as in TURTON BOTTOMS SD 7415 + **moor** (OE *mōr*).

TURVES Cambs TL 3396. *The Turves* 1668. ModE **turf**. The reference is probably to peat cutting. Ca 265.

TURVEY Beds SP 9452. 'Turf island'. *Torueie, Torveia* 1086, *Turueia -eie -ey(e)* from 1138×47. OE **turf** + **ēġ**. The reference is to a patch of land with good grass. Bd 48, YE xlvi, L 39.

TURVILLE Bucks SU 7691. 'Dry open land'. *þyrefeld* *[796]13th S 150, *Tilleberie* 1086, *Ti- Tyrefeld* 1175–1339, *Turfeld* 1445–1548, *Turville or Turfield* 1826. OE **thyrre** + **feld**. Bk 196, ECTV no.145, L 241–2. Jnl 2.33 gives local pr [turvul].

TURVILLE HEATH Bucks SU 7491. Cf. *Turville heath Wood* 1830 OS. P.n. TURVILLE SU 7691 + ModE **heath**.

TURWESTON Bucks SP 6037. 'Thurwe's estate'. *Turveston(e)* 1086–1296, 1634, *Turueston(a)* c.1200–1394, *Turlestone* 1227, *Turwestone* 1243, 1707, *Turreston'* 1242, *Turston* 1732. AScand pers.n. **Thurwe* (ODan *Thorwīf*), secondary genitive sing. **Thurwes*, + **tūn**. The exact form of the pers.n. here is uncertain: alternative suggestions are OE *Thurferth* (ON *Thorfrøthr*), *Thurfæst* (ODan *Thurfastr*) or *Thurulf* (ON *Thórúlfr*). Bk 49, xxxii, DEPN. Jnl 2.33 gives local pr [turstən].

TUTBURY Staffs SK 2128. Partly uncertain.

I. *Toteberia* 1086, *Totesbery* 1140×50, *-berie* 1141, *Tuttebury* 1200.

II. *Stutesberia* 1139×60, *Stuteberia* 1176.

The generic is OE **byriġ**, dative sing. of **burh** 'a fortified place, a manor'. The forms are consistent with either pers.n. *Tutta* or pers.n. *Stūt*, genitive sing. *Stūtes*. Either the latter is original with loss of initial *S-* as in NOTTINGHAM for Snottingham (OE pers.n. *Snot*), or there were two separate names for this place which subsequently became partly confused with each other. DEPN.

TUTNALL H&W SO 9970. 'Tota or the look-out's hill'. *Tothehel*

1086, *Tot(t)enhull* 1262–1535, *Toutnell* 1675. OE pers.n. *Tōta* or ***tōta** 'a look-out', genitive sing. *Tōtan*, ***tōtan**, + **hyll**. Tutnall occupies a good vantage point on a hill-ridge. Wo 363.

TUTSHILL Glos ST 5494. 'Tot or Tutt's hill'. *Tutteshill* 1635, *Tutshill* 1655. OE pers.n. **Tōt* or **Tutt*, genitive sing. **Tōtes* or **Tuttes*, + **hyll**. **Tōt* and **Tutt* would be strong forms corresponding to the recorded names *Tot(t)a* and *Tutta*. Alternatively the specific might be ***tōt** 'a look-out' although this is not otherwise found in the genitive sing. The site is a hill overlooking the rivers Wye and Severn. Gl iii.267.

TUTTINGTON Norf TG 2227. 'The settlement of the Tuttingas, the people called after Tutta'. *Totington, Tutintune* *[1044×7]13th S 1055, *Tutincghetune* 1086, *Tuttington* 1198, *Tutingeton* 1200. OE folk-n. **Tuttingas* < pers.n. *Tutta* + **ingas**, genitive pl. **Tuttinga*, + **tūn**, perhaps varying with **Tuttingtūn* 'the settlement called after Tutta', pers.n. *Tutta* + **ing**[4] + **tūn**. DEPN.

TUXFORD Notts SK 7370. Partly uncertain. *Tuxfarne* 1086, *-forne* 1273, *Tukesford* 1212–1420 including [1154×89]1316, *Tuxford* from 1231×9. This seems to be pers.n. **Tuk*, genitive sing. **Tukes*, + **ford**. But even if this pers.n. is acceptable, the 1273 and the DB spellings remain unexplained. Nt 62.

River TWEED Northum NT 9452. Unexplained. *Tuidi fluminis* [c.731]8th BHE, *Tuidon streames, Tuede streames* [c.890]c.1000 OEBede, *Twīode* c.1030 Secgan, [before 1085]12th, *Tuidam, Tweoda* [c.1040]12th, *Tuyda* 12th cent., *Twida* 1204, *Twydy -i -e, Tvidi, Twide* 14th cent., *Tweda* c.1107–13th cent., *Twede* 1245–1576, *Tweid* 15th cent. Scots sources. There seem to have been two OE forms of this name, (1) **Twīd(e)* from which develop the long series of *i/y* spellings with [i:], (2) **Twīude, *Twīode, *Twēode* from which develop the ME *e* and Scots *ei* spellings with [e:]. These forms have never been satisfactorily related and no acceptable explanation of the name has yet been proposed. It may be pre-Celtic and non-Indo-European. Identification of Ptolemy's *Τούεσις (Touesis)*, itself unexplained, with the Tweed has not been accepted. RN 421–3, RBrit 400–1.

TWEEDMOUTH Northum NT 9952. 'The mouth of the river Tweed'. *Tuedemue* 1208×10 BF, *Tuedemuthe* 1228 FPD, *Twedmouth* c.1410, *Tweidis mowth* c.1500, *Twedmoith* 1539 FPD. R.n. TWEED NT 9452 + OE **mūtha**. NbDu 202, RN 422.

TWELVEHEADS Corn SX 0854. *The Twelfe Heades* early 17th. ModE **head** in the sense 'hammer in a set of tin-stamps'. The place is at a confluence of streams where there must have been a set of water-driven tin-stamps. PNCo 174, 461.

TWELVEWOOD Corn SX 2065 → DOUBLEBOIS SX 1964.

TWEMLOW GREEN Ches SJ 7868. *Twemlow Green* 1831. P.n. Twemlow + ModE **green**. Twemlow, *Tu- Twam(e)lou(e) -law(e -a) -low(e) -lau* 12th–1299, *Twemlowe* c.1200–1640, *-low* from 1408, *-loe* 1697, is 'by, at the two mounds', OE **tweġen**, dative **twǣm**, + **hlāw**. In 1819 it was reported that there were five tumuli in the township, Jodrell Hall SJ 7970 (originally Twemlow Hall) standing between the second and third, which may have been the particular mounds that gave rise to the name and the original site of Twemlow. Che ii.230, v.xxxi.

TWENTY Lincs TF 1520. No early forms: a lost hamlet possibly meaning 'land having an area of 20 acres'. Perrott 112.

TWENTY FOOT RIVER Cambs TL 3397. *Twenty Foot River* 1824 OS.

TWERTON Avon ST 7264. 'Settlement at, estate called *Twifyrde*'. *Tvvertone* 1086, *Twyuerton* 1225, *Twiverton* 1236. OE **twi** + **fyrde** 'double ford' + **tūn**. Identical with TIVERTON Devon SS 9512. DEPN.

TWICKENHAM GLond TQ 1473. 'Twicca's river-bend land'. *Tuican hom, Tuiccanham* [704]8th S 65, *(in) Tuicanhamme* [790 for ?795]10th S 132, *Twicknem* 1651, *Twitnam(e)* 1644, 1721. OE pers.n. **Twic(c)a*, genitive sing **Twic(c)an*, + **hamm** 1. On phonological grounds this seems more likely than the alternative suggestion '*hamm* of the river fork', OE ***twiċċe**, genitive sing. **twiċċan**, + **hamm**. Mx 29 gives former pr [twitnəm], GL 94.

TWIGWORTH Glos SO 8522. 'Twigga's enclosure'. *Tuiggewrthe* 1216, *Twi- Twyggeworth(e)* 1251–1546, *Twi- Twygworth(e)* 1248–1605. OE pers.n. *Twigga* + **worth**. Gl ii.157.

TWINEHAM WSusx TQ 2519. '(The place) between two rivers'. *Twienen, Tuineam, Twyenen* 1066×87, *Twynem* 1242–1405, *Twynham* 1280. OE **(be)twēonan** + **ēam**, dative pl. of **ēa**. The place lies between two streams. Sx 279, PNE i.32.

TWINHOE Avon ST 7459. Partly uncertain. *Tynho* 1253 Ass, *Twyniho* [1267]late 15th Hung, *Twiniho* late 15th ibid., *Twinhoe* 1817 OS. Uncertain element + **hōh** 'a hill-spur'. Twinhoe lies on a spur of land between Cam Brook and Wellow Brook towards their confluence and it is tempting to equate the name with Twinyeo Devon SX 8476, *Betunia* 1086, *Bitweneya* 1263, *Betwynyo* 1303, *Twyneya* 1242, *Twynya* 1303, *Twyneyo* 1311, 'between the waters', OE **betwēonan** + **ēam**, dative pl. of **ēa** 'river'. If this is right, at some time the position of Twinhoe on a hill-spur has led to modification of the element **ēa** (subsequently *eá* and *yeo*, cf. SW dial. *yeo* 'river') to **hōh** 'hoe, hill-spur'. D 479.

TWINSTEAD Essex TL 8637. 'Double place'. *Tumesteda* (for *Tuine-*) 1086, *Twi- Twynsted(e)* 1203–1412. OE **twinn** + **stede**. A possible alternative specific might be OE ***twiġen** 'abounding in twigs or brushwood'. Ess 465, Stead 206.

TWINYEO Devon SX 8476 → TINHAY SX 3985.

TWISLEBROOK NYorks. (Lost) in Swinton SE 2179. 'Fork stream'. *Twislebroc* 1086, *Tuisebrok* [1184]15th. OE **twisla** + **brōc**. YN 234.

TWITCHEN Devon SS 7830. 'The cross-roads'. *Twechon* 1442, *Twycchyn* 1524, *Twitchen* 1609. ME **twichen** (OE *twiċen*). D 353.

TWITCHEN Shrops SO 3779. 'The road junction'. *Twitching* 1691, *Twichin* 1682, *The Twitchen -inn* 1727×8, 1751. ME **twichen** (OE *twiċen*), OED s.v. *twichel*, Gelling.

TWO BRIDGES Devon SX 6074. There is only one bridge here – ignoring the modern road bridge – over the West Dart River so this must be Devon dial. **to** + **bridge** 'at the bridge'. *To brygge* 1422×61, *(terr' vocat') too bridge* 1573, *Two Bridges* 1659. D 197.

TWO DALES Derby SK 2863. 'Fox earths'. *Todeholes* 1590, *Todhole* 1646, *Toadehole* 1668. ME **tod-hole** twice reformed by popular etymology. Db 83.

TWO GATES Warw SK 2101. *Two Gates* 1770. ModE **two** + **gate**. The place is a crossing of the road from Tamworth to Kingsbury and Watling Street. Wa 28.

TWYCROSS Leic SK 3305. Either 'the two crosses' or 'the double cross' perhaps referring to a four-armed signpost. *Tvicros* 1086, *Twy- Twicros -cross(e)* 1209×35–1576. OE **twī** + **cross**. Lei 557.

TWYFORD 'The double ford'. OE **twi-fyrde**, **twi-ford**. PNE i.50, ii.199.

(1) ~ Berks SU 7975. *Tuiford'* 1170, *Twiford'* 1224×5, *Twyvorde* 1327, *Twyford* from 1332. The road from Reading has to cross two branches of the Loddon here. Brk 135, L 68–9.

(2) ~ Bucks SP 6626. *Tueverde, Tuiforde* 1086, *Twyforde* 1224. The old road to Cowley crosses two streams at Twyford Mill. Bk 57, L 68–9. Jnl 2.33 gives local pr [twaivurd].

(3) ~ Hants SU 4825. *tuifyrde* [963×75]12th S 827, *Tviforde* 1086, *Twyforde* 1144, 1291, 1327. The Itchen still runs in two main channels here. Ha 165, Gover 1958.76.

(4) ~ Leic SK 7210. *Tuiuorde, Taiworde* 1086, *Tuy- Tui- Twy- Twiford(e)* c.1130–1610. There were two fords of the stream that flows through the village. Lei 330.

(5) ~ Norf TG 0124. *Twyford* 1254. The road here crosses two small streams. DEPN.

(6) ~ COMMON H&W SO 5135. P.n. Twyford + ModE **com-**

mon. Twyford, *Twiford* 1281, is the 'double ford' across Red Brook. He 52.

TWYNING Glos SO 9037. 'Between rivers'. *Tweoneaum* [c.740]12th, *Bituinœum* [814]11th S 172, *Tv(e)ninge* 1086, *Tu(u)e- Tweninga- -ing(e) -yng(e)* 1095×1122–1404, *Twin(n)ing* 1221–1777. An elliptical name for 'land between rivers', OE preposition **betwēonan** + ***ēum**, dative pl. of **ēa**. The parish is an isolated one N of the Avon and lying between the Avon and the Severn. The post-Conquest form of the name is a late OE folk-n. derived from a contracted form of the original p.n. + **ingas**, 'folk living in or at *Bituinœum*'. Gl ii.71 gives pr ['twiniŋ].

TWYNING GREEN Glos SO 9036. *Twyning Green* 1830. P.n. TWYNING SO 9037 + **green**. Gl ii.72.

TWYWELL Northants SP 9578. 'The two streams'. *Twiwel* [1013]14th S 931, *Twi- Twywell(e)* from [1017×25]14th, *Tuiwella, Tevwelle* 1086. OE **twī** + **wella**. The village is situated in the fork of two streams. Nth 188.

TYBERTON H&W SO 3839. 'Estate called after Tidbriht'. *Tibrintintvne* 1086, *Tibritona* 1160×70, *Tyberton(e)* from 1279. OE pers.n. *Tīdbriht -beorht* + **ing**^{4} + **tūn**. He 192.

TYBURN GLond SP1490 MARYLEBONE TQ 2881.

TYBURN WMids SO 1391. *Tyburn* 1834 OS. A transfer c.1730 from the London r.n. Tyburn, *(andlang) teoburnan* 951 for ?959 S 670, *Tyburn* 13th, 'the boundary stream', OE ***tēo**, genitive sing. ***tēon**, + **burna**. Wa 75, Mx 6, RN 424.

TYBY Norf TG 0827. 'Tythi's farm or village'. *Tytheby* 1086. ON pers.n. **Tythi*, genitive sing. **Tytha*, + **bȳ**. DEPN compares Tiby in Sweden, *Tidhœby* 1309, 'Tidhe's *by*'. Nt 242.

TYDD GOTE Lincs TF 4517. *Tyddegot* 1316, *Tiddegote* 1362, *Tydgote* 1365, 1428. P.n. Tydd as in TYDD ST MARY TF 4418 + ME, ModE **gote** 'a water-course, a channel, a stream' (OE **gotu*). According to Wheeler 133 'the Hamlet of Tydd Gote is named from the fact of the outfall gote or sluice being built here. The earliest recorded sluice is mentioned in 1293, the second in 1551, the third and present – called Hill's Sluice, or Tydd Gote Bridge – in 1632'. Payling 1940.60.

TYDD ST GILES Cambs TF 4216. 'St Giles's T'. *Tydde Sancti Egidii* 1250, ~ *Seynt Gilles* 1504. Earlier simply *Tit* c.1165, *Tid(d)(e)* 1170–1559. The meaning and origin of the name are unknown. OE **titt** 'a teat' in the transferred sense 'small hill' has been suggested perhaps in connection with salt-making, but no such feature is now evident here or at TYDD ST MARY Lincs TF 4418. An OE ***tydd** has also been proposed in the sense 'brushwood' or 'hillock' cognate with G *Tudden* 'hump, hummock', Icel *toddi* 'small wood'. But this is pure guesswork. Ca 283, SN 5.3, PNE ii.200.

TYDD ST MARY Lincs TF 4418. 'St Mary's T'. *Tyd(d) St Mary* from 1263, *Seintemaritidde* 1391, *Tid Mary* 1525, *Tidd St Maries* 1616, *Tydd St Mary's* 1824 OS. P.n. *Tite, Stith, Tid* 1086, *Tit* 1094, *Tid, Tyd* 1168–1348, *Tydd(e)* from 1200, *Tidde* 1285–1330, either OE ***tydd** 'brushwood, shrubs', or **titt** 'a teat, a breast' used topographically of a small hill very likely with reference to the salthill on which the settlement stands. Payling 1940.56, Cameron 1998.

TYLDESLEY GMan SD 7001. 'Tilwald's wood or clearing'. *Tildesleia* c.1200, *Ti- Tyldesleg(h) -le(ge) -ley(e)* 1212–1346 etc., *Tylys- Tillisley* 1451–61. OE pers.n. *Tīlwald*, genitive sing. *Tīlwaldes*, + **lēah**. The contraction of the second element of the pers.n. is due to weak stress; it is not necessary to posit a shortened form **Tild* or **Tildi*. La 101.

TYLER HILL Kent TR 1460. 'Kiln slope'. *Tylerhelde* 1304, *Teghelere- Tegularynhelde* 1363–1516, *Tylo^{r}hill* 1535. OE ***tiġel-ærn** or **tiġlere** 'a tiler' + **helde**, later replaced by **hill**. PNK 498, PNE ii.179.

TYLERS GREEN Bucks SU 9094. No early forms.

River TYNE Northum NZ 0261. *Tinea* [c.700]14th Rav, *Tino* (abblative case), *Tini* (genitive), *Tinam* (accusative) [c.731]8th BHE, *Tiine seo éa, (be) Tínan þære ea* [c.890]c.1000 OEBede, *(be, into) Tinan, (to) Tine* 1121 ASC(E), *Tina* [c.1040]12th, c.1107–1256, *Tine* [c.1130]12th–c.1540, *Tyne* from [c.1130]12th. A pre-English r.n. on the root **tei-/*ti-* 'to melt, to flow'. Ptolemy's *Τίνα (Tina)* does not belong here; it is a corrupt reading for **Ituna*, the river Eden, Fife. NbDu 202, RN 425, RBrit 473, 380.

River South TYNE Northum NY 6854. *Tinam Australem* c.1130, *South Tyne* 1866 OS. Adj. **south** + r.n. TYNE NZ 0261. RN 425.

TYNEHAM Dorset SY 8880. 'Goat enclosure'.
I. *Tigeham* 1086, 1185, *Tiham* 1194, 1244.
II. *Tingeham* 1086, *Est(t)ingham* 'east T' 1288.
III. *Tyn(h)am* 1244–1464, *Tyneham* from [1445]16th.
OE * **tiġe**, genitive sing. ***tiġan** or pl. ***tiġena**, + **hamm** (or **hām** 'homestead'). Type II may point to an OE variant **tiġinge-hamm* in which ***tiġinge** might be the locative–dative sing. of ***tiġing** 'goat place, goat farm', OE ***tiġe** + **ing**^{2}. Do i.101 gives pr ['tainəm].

TYNEMOUTH T&W NZ 3669. '(The settlement at the) mouth of the Tyne'. *(œt) Tinan muþe* c.1121 ASC(E) under year 792, *(œt) Tine muðan* c.1121 ASC(E) under year 1095, *Tinemutha* c.1107–c.1170×74 DEC, *Ti- Tynemue* 1235×6–1260, *Tynemuwe* 1242 BF, *-mewe* 1296 SR, *Tynnemouth* 1485, *Tinmouth* 1637, 1763, 1768 Armstrong, *Tynemouth* 1768 ibid. R.n. TYNE Northum NZ 0361 + OE **mūtha**. NbDu 202 gives pr [tinməθ].

TYRINGHAM Bucks SP 8547. Partly uncertain. *Te(d)lingham* 1086, *Ti- Tyringeham* 1185–1242, *Tyringham* from 1227, *Tirincham* 13th. The DB form possibly points to OE pers.n. *Tīdhere* as the specific with AN *l/r* confusion, the rest to a short form of a name in *Tīr-* such as *Tīrweald* or *Tīrwulf*. The single 13th cent. spelling with *-ch-* also possibly points to a locative-dative form with assibilation: perhaps, therefore, 'homestead of the Tiringas, the people called after Tira', OE folk-n. **Tīringas* < pers.n. *Tīra* + **ingas**, genitive pl. **Tīringa*, or p.n. *Tīring* 'the place called or at Tiring, Tira's place' < *Tīra* + **ing**^{4}, locative-dative sing. *Tīrinġe*, + **hām**. Bk 14 gives pr [tiriŋəm].

Lower TYSOE Warw SP 3445. *Nethertyseho* 1451, *Nether Tysho al. Temple Tysho* 1570, *Churche Tyshoe al. Nether Tyshoe* 1606, *Temple or Lower Tysoe* 1801. ModE adj. **lower**, ME **nether**, + p.n. *Tiheshoche* 1086, *Tys- Tisho* 1185–1549, 'the hill-spur of Tiw', OE pagan deity name *Tīw, Tīġ*, genitive sing. *Tīġes*, + **hōh**. The Templars held land here in 1185. Wa 284, L 3–4, 168, Settlements 103, West Mids 92, Gelling 1998.83.

Middle TYSOE Warw SP 3444. ModE adj. **middle** + p.n. Tysoe as in Lower TYSOE SP 3445. Identical with *Chyrchetisho* 1299, *Church Tysoe* 1796. Also known as *Churcheton* 'the church town' 1529. Wa 284.

Upper TYSOE Warw SP 3343. *Overtyso* 1546. ModE adj. **over** + p.n. Tyshoe as in Lower TYSOE SP 3445. Also known as *Overton* 'the upper town' 1529. Wa 284.

TYTHBY Notts SK 6936. 'Tithi's village or farm'. *Tied(e)bi* 1086, *Titheby* c.1190–1428 with variants *Tythe- Tydhe-* and *-bi, Tyby* 1338. ODan pers.n. **Tīthi*, genitive sing. **Tītha*, + **bȳ**. Cf. TYBY Norf TG 0827 and Tiby, Sweden, *Tidhœby* 1309. Nt 242 gives pr [tiðbi], SSNEM 75.

TYTHERINGTON Avon ST 6888. 'Estate called after Tidhere'. *Tidrentvne* 1086, *Ti- Tyd(e)rin(g)ton(a) -yn(g)-* 1167–1592 with variant *Tu-* 1272×8, 1287, 1378, *Titherington -yng-* 1592, 1594. OE pers.n. *Tīdhere* + **ing**^{4} + **tūn**. Gl iii.19, xi.

TYTHERINGTON Ches SJ 9175. Probably 'cattle-farm' or 'piggery'. *Tidderington* c.1245, 1573, *Ti- Tyderinton* 1249, 1260, *Tidrinton(a)* c.1280, *Tyd(d)rinton -yn(g)-* 1316–97, *Tythrenton* 1480, *Tytherington* from 1620. OE **tȳd(d)rung, tȳd(d)ring** 'production, propagation' in a concrete sense, 'stock-breeding', + **tūn**. Otherwise 'estate called after Tydre', OE pers.n. **Tȳdre* + **ing**^{4} + **tūn**. Che i.214, v.I.i.xx, SN 43.574–5, 44.270. Addenda gives pr ['tiðrin- 'tiðriŋtən], 19th cent. local ['tiθitn].

TYTHERINGTON Somer ST 7645. *Tytherington* 1817 OS. Early forms are needed.

TYTHERINGTON Wilts ST 9141. Partly uncertain. *Tuderinton -yn(g)-* 1242–1332, *Toderington* 1279, *Ted(e)ryngton* 1459, 1489. An -**ingtūn** formation on OE **tūdor** 'offspring, child, young animal' or pers.n. *Tūder* derived from it, or the related adj. **tīedre** 'weak, frail, infirm' or pers.n. **Tīedre*, **Tȳdre*. The senses would be 'farm where young animals are reared' or 'estate called after Tuder, Tiedre or Tydre'. A less likely possibility is that **Tiedring* was the name of a weak or intermittent stream. Wlt 168, Årsskrift 1974.32.

TYTHERLEIGH Devon ST 3103. 'The thin or tender woodland'. *Tiderlege* 1154×9, *Ty- Ti- Tuderlegh* 1247–1342, *Tydderley* 1581. OE **tīedre** + **lēah**. D 654.

East TYTHERLEY Hants SU 2928. *Estuderlegh* 1291, *Est Tyderley* 1414, *Est Tetherley* 1579. ME adj. **est** + p.n. *Tiderlege -lei, Tederleg* 1086, *Ti- Te- Tuderle(gh) -lee* 1218–1352, 'tender, fragile wood', OE **tīedre** + **lēah**. 'East' for distinction from West TYTHERLEY SU 2729, *Westuterlie* 1212. Gover 1958.191 gives pr [tiðəli], Ha 165, DEPN.

West TYTHERLEY Hants SU 2729. *Westuterlie* 1212, *Westiderleg(a) -ley* 1219–54, *West Tetherley* 1579. ME adj. **west** + p.n. Tytherley as in East TYTHERLEY SU 2928. In 1086 simply *alia Tiderlege -lei, Tederlee* 'the other Tytherley'. Gover 1958.191.

East TYTHERTON Wilts ST 9675. ModE adj. **east** + p.n. *Tedelin- Te(d)rintone* 1086, *Tid(e)rinton(e)* c.1155–1242, *Tuderington -yng-* 1202–91, *Titherton* 1603, identical with TYTHERINGTON ST 9141. Also known as *Tytherton Kellaways* 1828 OS, 'Tytherton beside Kellaways' ST 9795, *Keylewayes* 1585, *Kellawayes* 1637, earlier *Tuderinton(e) Kaylewai* 1257, *Tuderyngton Kaylewey* 1289, '(William de) Cailleway's part of Tytherton'. 'East' for distinction from West TYTHERTON ST 9574. Wlt 91, 99, Årsskrift 1974.32.

West TYTHERTON Wilts ST 9574. *West Tytherton* 1721. ModE adj. **west** + p.n. Tytherton as in East TYTHERTON ST 9675. Also known as *Tuderyngton Lucas* 1289, 1428, *Tethryngton Lucas* 1573, *Titherton Lucas* 1603, from Adam Lucas who held the manor in 1249. Wlt 91.

TYTTENHANGER Herts TL 1805. 'Tida's wooded slope'. *Tydenhangre -er* 1199×1216–1334, *Tytenhangre* 13th–1427, *Tytnangre -er* 1542, 1561, *Tinnanger* 1669. OE pers.n. *Tīda*, genitive sing. *Tīdan*, + **hangra**. Hrt 85 gives former prs [titnæŋə, tinæŋə].

TYWARDREATH Corn SX 0854. 'House on the strand'. *Tiwardrai* 1086, *Tywardrait* [c.1150]13th, *Tiwardraith* 1235, *Trewardreth* 1367, 1610. OCo **ti** partly later replaced with **tre** + **war** + **trait** (OCo-t = [θ]). PNCo 175 gives pr [tauə 'dreθ], Gover n.d. 426.

UBBESTON GREEN Suff TM 3273. P.n. *Upbestuna* 1086, *Ubbestun* 1154×89, *-ton* from 1206, *Vp- Upston* 1336, 1344, 1568, *Vppeston* 1610, 'Ubbi's estate', ODan pers.n. *Ubbi*, secondary genitive sing. *Ubbes*, + **tūn**, + ModE **green**. DEPN, Baron.

UBLEY Avon ST 5258. 'Ubba's wood or clearing'. *Hulban-* or *Hubbanlege* [959×75]13th S 1771, *Tvmbeli* (sic) 1086, *Ubbele(ia)* 1213, 1223. OE pers.n. *Ubba* + **lēah**, dative sing. **lēaġe**. The DB form may be explained as showing *T-* from the OE preposition in the phrase *æt Ubbanleage* and confusion of capital *B* and capital *M*. DEPN, Domesday Studies 136.21.

UCKERBY NYorks NZ 2404. Perhaps 'Utkari's farm or village'. *Ukerby* 1198, *Huckerby* [c.1250]13th, *Ukkerby* 1285 etc. ON pers.n. **Útkári*, genitive sing. **Útkára*, + **bȳ**. YN 278, SSNY 7.

UCKFIELD ESusx TQ 4721. 'Ucca's open land'. *Ukke- Uckefeld* 1220–1428, *Uckfeud(e)* 1296, 1316. OE pers.n. *Ucca* + **feld**. Sx 396 gives prs [ʌkfəl] and [ukfəl], L 243.

UCKINGTON Glos SO 9124. 'Estate called after Ucca'. *Hochinton* 1086, *Okin(g)ton -yn(g)-* 1227–1560, *Ukinton'* 1248. OE pers.n. *Ucca* + **ing**[4] + **tūn**. Gl ii.85.

UDIMORE ESusx TQ 8718. 'Uda's lake'. *Dodimere* (sic) 1086, *Odimer(e)* 1249–1535, *Ude- Udimere* 1291–1675, *Udimore* 1295×7. OE pers.n. *Uda* + **mere**. The persistence of spellings with *-i-* suggests that this name was originally a compound in **ing**[4], OE **Udingmere*. Sx 516.

UFFCOTT Wilts SU 1277. 'Uffa's cottage(s)'. *U(l)fecote* 1086, *Uffecot(e) -kote* 1115–1282, *Uphcot* 1687. OE pers.n. *Uffa* + **cot** (pl. **cotu**). Wlt 296.

UFFCULME Devon ST 0612. 'Uffa's Culm'. *Offaculum* [839×55]13th S 1697, *Offecoma* 1086, *Uffe Culum, Offeculum* 1175–1249. OE pers.n. *Uffa* + r.n. CULM used as the name of an estate. D 537.

UFFINGTON 'The estate called after Uffa'. OE pers.n. *Uffa*, + **ing**[4] + **tūn**.

(1) ~ Lincs TF 0607. *Offinton(e) -tvne* 1086–1285, *Uffinton'* 1219, 1243, *Offington -yng-* 1219–1453, *Uffington* from 1231. Perrott 438.

(2) ~ Shrops SJ 5213. *Ofitone* 1086, *Offinton(a) -yn-* 1155×62–1462, *Uffitun' -tona* 1177, 1203×10, *Uffeton* c.1291, *Uffinton(e) -yn-* 1195–1668, *Uffington* from 1535. The *ing-* spellings appear rather late, however, so this might be genitive sing. *Uffan* + **tūn** 'Uffa's farm etc.' Sa i.295.

UFFINGTON Oxon SU 3089. 'Uffa's farm'. *Uffentune* [c.931]c.1200 S 1208, *Uffinton' -yn-* 1197×8 etc., *Uffyngton'* 1248, *Offentone* 1086, *Offinton' -yn-* 1221–1370. OE pers.n. *Uffa*, genitive sing. *Uffan* + **tūn**. Before Uffa's time the estate, which originally also included Woolstone SU 2987, was known as *(in to) æscæsbyriges (suðgeat)* 'the south gate of Ashbury' [c.931]c.1200 S 1208, *(to) æscesbyrig* [953]13th S 561, 'Æsc's fort', OE pers.n. *Æsc*, genitive sing. *Æsces*, + **byriġ**, dative sing. of **burh** referring to Uffington Castle Iron Age hill-fort. Brk 379, 380, Thomas 1976.181

UFFINGTON CASTLE Oxon SU 3086. *Uffington Castle* 1830. P.n. UFFINGTON SU 3089 + ModE **castle** referring to the Iron Age hill-fort there. Brk 380, Thomas 1976.181.

UFFORD Cambs TF 0904. 'Uffa's enclosure'. *Uffewurda -w(u)rth* 1178–1227, *Upford* 1202, 1206, *Ufford(e)* from 1209. The earliest reference is to the bounds of Ufford, *(to) uffawyrða gemære* [948]12th S 533. OE pers.n. *Uffa* + **worth**. Np 244.

UFFORD Suff TM 2952. 'Uffa's enclosure'. *U- Offeworda, Uf(fe)forda* 1086, *Ufford* from 1195. OE pers.n. *Uffa* + **worth** early replaced by **ford**. DEPN, Baron.

UFTON NERVET Berks SU 6367. 'U held by the Neyrnut family'. *Offeton' Nernut, Uffington' Nermyt* 1284, *Ufton Nermyte* 1552. P.n. Ufton + affix Nervet from an OFr surname *Neirenuit* 'black night': Richer Neyrnut held the manor here in 1242. Ufton, *Offetvne* 1086, *-ton'* 1178, *Vffetona* 1179, *Uffinton' -en-* [1225×6]13th, 1275, *Ufton(e)* from 1316, is 'Uffa's estate', OE pers.n. *Uffa*, genitive sing. *Uffan*, + **tūn**. Brk 224, DEPN.

UFTON Warw SP 3762. Possibly 'Wulfwiht's estate'.

I. *Vlchetone* 1086, *Ul(u)ghton* 1284–1315, *Oulghetone* 1306.

II. *Ol(o)uton* 1262–97, *Ol(o)ughton* 1278–1549.

III. *Olefton* 1257, *Ulfton* 1267, 1352, *Ufton* from 1535, *Olufton* 17th. These forms must be taken together with *wulluht graf* 1001 S 898, the name of a point on the bounds of Long Itchington corresponding to Ufton Wood SP 3862, in which *wulluht* is possibly a late form of pers.n. *Wulfwiht* and **graf** is 'grove'. Type III shows normal substitution of fricative [f] for the peripheral phoneme [x]. The alternative suggestion, OE ***hul(u)c** 'a shed, a hut', takes no account of the evidence of S 898. Wa 186.

UGBOROUGH Devon SX 6755. 'Ucga's hill'. *Ulgeberge* 1086, *Uggeberge -bergh* 1200–44, *-byri -burga -burghe* 1266–1306, *Ugborough al. Ubbourowe* 1570. OE pers.n. **Ucga* + **beorg**. D 284.

UGBOROUGH BEACON Devon SX 6659. *Ugborough Beacon* 1809 OS. P.n. UGBOROUGH SX 6755 + ModE **beacon**. This is properly the Eastern Beacon 1232ft., so distinguished from Western Beacon SX 6557, *Western Beacon* 1809 OS.

UGBOROUGH MOOR Devon SX 6462. *Uggbroghe Moor* 1557. P.n. UGBOROUGH SX 6755 + ModE **moor**. D 288.

UGGESHALL Suff TM 4480. 'Uggeca's nook'. *Uggecehala, Uggiceheala, Ugghecala, Wggessala* 1086, *Uggecala* c.1095, *Huggechale* 1242, *Ugec- Ugeshale* 1254, *Vggeshall* 1610. OE pers.n. **Uggeca* + **halh** in the sense 'valley'. DEPN.

UGGLEBARNBY NYorks NZ 8807. 'Uglubarthr's farm or village'. *Ugle- Ulgeberdesbi* 1086, *Ugel- Uglebardeby* 1100×c.1115–[1222×7]15th, *Ugglebarnby* 1613. ON pers.n. **Uglubarthr* 'Owl-beard', partially anglicised secondary genitive sing. *Ugluberdes*, + **bȳ**. The late change of *-bardby* to *-barnby* is due to the proximity of BARNBY NZ 8112. YN 121, SSNY 40.

UGLEY Essex TL 5128. 'Ucga's wood or clearing'. *Uggele, Ugeleiam* [c.1041]c.1300, *Uggheleam* 1086, *Uggele(gh)* 1232–1428. OE pers.n. *Ucga* + **lēah**. Ess 553, Jnl 2.48.

UGTHORPE NYorks NZ 7911. 'Uggi's dependent farmstead'. *Ug(h)etorp* 1086, *Uggethorpe* 1161–1242, *Ugthorpe* from [c.1180]n.d. ON pers.n. *Uggi*, genitive sing. *Ugga*, + **thorp**. YN 183 gives pr [ugθrəp], SSNY 70.

ULCEBY Humbs TA 1215. 'Ulf's village or farm'. *U- Vlvesbi* 1086, *Ulesbi -by* c.1115–1343, *Ulseby* 1163–1431, *Ulceby* from 1270. ON pers.n. *Ulfr*, genitive sing. *Ulfs*, + **bȳ**. A purely Scandinavian name preserving ON genitive sing. [s] instead of expected [z], cf. the Lincs p.ns. BRACEBY TF 0135, HACEBY TF 0236, LACEBY TA 2106, RAUCEBY TF 0146. Li ii.291, SSNEM 62.

ULCEBY Lincs TF 4272. 'Ulf's village or farm'. *Vlesbi* 1086–[1154×89]13th, *Vlsebi* c.1115, *Ulseby* 1201. ODan pers.n. *Ulf*, genitive sing. *Ulfs*, + **bȳ**. The pr with [s] preserves the original OScand genitival inflection and shows that the name must have been created by OScand speakers. DEPN, SSNEM 62.

ULCOMBE Kent TQ 8449. 'The owls' coomb'. *Uulacumb* [941]13th S 477, *Vlan- Ulecumbe, Vlecumb* [946]14th S 515, *Olecūbe* 1086, 1210×12, *Wula- Vle- Hulecumba* c.1100, *Ul(l)e- Hule- Wolcumbe, Wullecumb'* 13th. OE **ūle**, genitive pl. **ūla(n)**, + **cumb**. PNK 236, KPN 252, L 89,93, and Brunner §276 A.5 for forms of the genitive pl.

ULDALE Cumbr NY 2537. 'Ulf's' or 'wolf valley'. *Ulvesdal'* 1216, 1279, *Uluedal(e)* 1228–1399, *Ulledale* 1332, *Uldale* from 1391. ON pers.n. *Úlfr*, secondary genitive sing. *Úlfes*, or ON **úlfr**, secondary genitive sing. **úlfes**, genitive pl. **úlfa**, + **dalr**. It is impossible to say which is the correct explanation. Cu 327 gives pr [uldəl], SSNNW 122, L 94.

ULDALE HEAD Cumbr NY 6400. 'Head of Uldale'. *Uldale Head* 1865. P.n. Uldale, *Uluedalebank* 13th, *Uldale als. Ulnedale* (for *Ulue-*) 1411, is 'wolves' valley', ON **úlfr**, genitive pl. **úlfa**, + **dalr**. We ii.48, 53.

ULDALE HOUSE Cumbr SD 7396. P.n. Uldale SD 7597 + ModE **house**. Uldale, *Ulnedale* (for *Ulue-*) [1154×89]1645, *Uluesdaile* [1224]1651, *Uldall* 1585–1730, is 'wolves' valley', ON **úlfr**, genitive pl. **úlfa**, + **dalr**, although the 1224 form shows the same ambiguity as ULDALE NY 2537. We ii.33.

ULEY Glos ST 7998. 'Yew-tree wood or clearing'. *Euuelege* 1086, *Eweleg(h)* 13th cent., *I- Yweleg(e) -l(e) -legh -ley(e)* 1166–1400, *Ule -ley(e)* 1388–1571, *Yewley -leigh* 16th. OE **īw** + **lēah**. Gl ii.253 gives prs ['ju:li] and ['jiuli], L 222.

ULGHAM Northum NZ 2392. 'Owl nook'. *Wlacam, Hulgam* [1139]14th, *Ucham* 1226, 1251, 1316, *Ulgh(h)am* 1290, 1316, *U- Wlweham* 1242 BF, 1296 SR, *Vlwham'* 1547 Newm, *Howgham* 1570, *Ougham* 1663, *Uffham* 1812. OE **ūle** + **hwamm**. NbDu 203 gives pr [ufəm].

ULLENHALL Warw SP 1267. Probably 'the owl's nook'. *Holehale* 1086, *Hulehale* c.1185, c.1210, *Vlenhale* 1187, *Ullenhale* 1221, c.1330, 1375, *-hall* 1375 and from 1618, *Ullnal al. Ownall* 1566. OE **ūle**, genitive sing. **ūlan**, + **hale**, dative sing. of **halh**. Alternatively the specific might be an OE pers.n. *Ulla*, genitive sing. *Ullan*. Cf. nearby OLDBERROW SP 1266. Wa 245, L 105.

ULLENWOOD Glos SO 9416. Possibly 'owls' wood'. *Ullen Wood* 1830. Cf. *Ullen Fm.* 1777. Possibly originally OE **ūle**, genitive pl. **ūlena**, + **wudu**. Another possibility would be 'Ulla's wood', OE pers.n. *Ulla*, genitive sing. *Ullan*, + **wudu**. Gl i.153.

ULLESKELF NYorks SE 5240. 'Ulf's shelf of land'. *Oleschel, Oleslec* 1086, *Ulfskelf(f)* 1170×7–1428, *Ulskelf(e)* 1177–1641, *Ulleskelf(e)* from 1284. ON pers.n. *Úlfr*, genitive sing. *Úlfs*, + ON **skjalf** or OE **scelf** with OScand *sk* for *sh*. The reference is to the flat bank of the r. Wharfe. YW iv.67, SSNY 107, L 187.

ULLESTHORPE Leic SP 5087. 'Ulfr's outlying farm'. *Vlestorp* 1086, *Olestorp' -thorp(e) -is-* 1190–1440, *Ulvesthorp(e)* 1285, 1311, *Ullesthorp(e)* 1278–1622, *-thropp* 1631. ON pers.n. *Úlfr*, ME genitive sing. *Ulfes*, + **thorp**. Lei 467, SSNEM 120.

ULLEY SYorks SK 4687. 'Owl wood or clearing'. *Ollei(e)* 1086, *Ullay -ley* from 13th cent. OE **ūle** + **lēah**. YW i.163.

ULLINGSWICK H&W SO 5950. 'Dairy farm called after Ulla'. *Vllingwic* 1086, *Ol(l)ingewiche' -wike* c.1127, 1188×1205, *Ullyngwick* 1292, *Ullingswick* 1832 OS. OE pers.n. *Ulla* + **ing**[4] + **wīc**. He 193.

ULLOCK Cumbr NY 0724. 'Place where wolves play'. *Ulnelaike* (for *Ulue-*) 1248, *Uluelaykes* c.1265, *-laik* 1279, *Ulleyk -layk(e)* 1295–1367, *Ullocke* 1570. ON **úlfr**, genitive pl. **úlfa**, + **leikr**. Cu 367, SSNNW 172.

ULLSCARF Cumbr NY 2912. *Ullscarf* 1868 OS. A peak in the Borrowdale Fells rising to 2370ft. ON **úlfr** 'wolf', genitive sing. **úlfs**, + **skarth** 'a notch, a cleft, a mountain pass' with *f* for *th* as frequently with this element in field-names. Cu xliii.

ULLSWATER Cumbr NY 4220. 'Ulf's lake'. *Ulueswater* 1220, *Ulveswatre* 1327, *Ulleswat(e)r* 1357–1671, *Eeleswater* 1657. ON pers.n. *Úlfr*, secondary genitive sing. *Úlfes*, + OE **wæter**. Cu 36, We i.17.

ULPHA Cumbr SD 1993. 'Wolf hill'. *Wolfhou* 1279, *Ulfhou* 1337, *Ulpho* 1449, *Ulpha* from 1625, *Ouffa* 1777. ON **úlfr** (replaced by OE **wulf** in 1279) + **haugr**. Cu 437 gives pr [u:(l)fə], SSNNW 256.

ULROME Humbs TA 1656. 'Wulfhere or Wulfwaru's homestead'. *Ulfram, Vlreham* 1086, *Uleram* 12th, *Ulram* 12th–1573, *Ulrome als Owram* 1604. OE pers.n. masculine *Wulfhere* or feminine *Wulfwaru*, + **hām**. Loss of initial *W-* is due to Scand influence. YE 84 gives prs [u:ərəm, ulrəm], SSNY 150.

ULVERSTON Cumbr SD 2877. 'Ulfarr's farm or village'. *Vlurestun* 1086, *Olueston(am)* 1127–96, *Ulverston* from 1180×4, *Ulreston* 1246, 1336, *Ullerston* 1327, *U'ston* 1867. ON pers.n. *Úlfarr* (possibly replacing OE *Wulfhere*), secondary genitive sing. *Úlfares*, + **tūn**. La 211, SSNNW 191.

UMBERLEIGH Devon SS 6023. Partly uncertain. *Umberlei* 1086–1310 with variants *Umbre-* and *-leg(h), Wu- Wo- Wymberlegh* 1270–1440. Possibly stream name Umber identical with Wimborne in WIMBORNE MINSTER SZ 0199 + **lēah** 'a wood or clearing'. D 357.

UNDERBARROW Cumbr SD 4691. '(Place) under the hill'. *U- Vnderbarra -barro(e)* from 1517 with variants *-barr(e)y* 16th, *-barrow(e)* from 1558. ME **under** + **barrow** (OE *beorg*) referring to Helsington Barrows to the E at SD 4990, *Helsington barrey* 1170×84, *(le) Berghes* 1301, *le Bergh de Helsington* 1332. We i.100, 109.

The UNDERCLIFF 'Land below a cliff'. OE, ModE **under** + **cliff.**

(1) ~ IoW SZ 3882. A modern name. Wt 75.

(2) ~ IoW SZ 5376. *sub falasia de Newetona* 'under the cliff of Niton' [13th]1781, *under le Clif(f)e* 1608, *under Clift* 1771. Wt 184, Mills 1996.102.

UNDERRIVER Kent TQ 5525. 'Beneath River Hill'. *Great, Lit. Under River* 1819 OS. ModE **under** + p.n. River as in River Hill TQ 5352, *River Hill* 1819 OS.

UNDERWOOD Notts SK 4750. '(The settlement) under the wood'. OE L 229. *Underwode* 1287, *-wood* 1490. **under-wuda**. Nt 132.

UNSTONE Derby SK 3777. 'On's farm or village'. *(H)onestune* 1086, *On(e)ston(e) -is- -ys-* 1263–96 etc., *Honeston -is-* 13th, *Honston* 1519, *Ouns- Ownston(e)* 1368–1757. OE pers.n. **Ōn*, genitive sing. **Ōnes*, + **tūn**. Db 318.

UNTHANK Cumbr NY 4536. *(H)unthanc* 1274, *Unthank* 1332. A common local name from OE **unthanc** 'ill-will, ingratitude, displeasure', used of land held against the will or without the consent of its lawful owner; a squatter's holding. Cu 241.

UPAVON Wilts SU 1355. 'The upper part of Avon'. *Oppavrene* (sic) 1086, *Upeavena* 1172, *Upavene* 1211. OE **upp** + r.n. AVON. 'Up' in contrast to NETHERAVON SU 1449. Wlt 324.

UPCHURCH Kent TQ 8467. 'High church'. *Vpcyrcean* c.1100, *Upechereche* 12th, *Vpcherche* 1219, 1238. OE **up(pe)** + **ċiriċe**. The church and village are situated on a hill. PNK 272.

UPCOT Devon SS 6118 → NORTHCOTE MANOR SS 6218.

UPCOTT H&W SO 3251. 'The higher cottage(s)'. *Up(pe)cote* 1160×70, *Uppekote* 1243. OE **upp** + **cot**, pl. **cotu**. He 26.

UPEND Cambs TL 7058. 'Upper end' sc. of Kirtling. *Upend* from 1612, *Upyng* 1669, *Upping* 1821. These forms are probably reformations of *Upheme* 'the up-dwellers' 13th, 1483, *Upyeme* 1477. OE **upp-hǣme**. Upend is on high ground NE of Kirtling. Ca 126.

UPHAM Devon SS 8808. No early forms. Probably the 'upper settlement' of Cheriton Fitzpaine SS 8606.

UPHAM Hants SU 5320. 'Upper estate'. *Upham* from c.1170, *Uppham* 1291. OE **upp** + **hām**. Ha 166, Gover 1958.48, DEPN.

Lower UPHAM Hants SU 5219. *Upham Low.r End* 1810 OS. ModE adj. **lower** + p.n. UPHAM SU 5320.

Upper UPHAM Wilts SU 2277. *Upper Upham* 1828 OS. ModE adj. **upper** + p.n. *(to) Uphammere* 'to the Upham boundary' [955]14th S 568, *Upham* from 1201, 'the high homestead', OE **upp** + **hām**. 'Upper' for distinction from Lower Upham SU 2077, *Lower Upham* 1828. Upper Upham is situated above the

800ft. contour in the middle of Aldbourne Chase, Lower Upham beneath the escarpment of the high ground. Wlt 293.

UPHILL Avon ST 3158. '(Settlement) above the hill or creek'. *Opopille* 1086, *Uppepull* 1197, *Uppehill* 1176. OE **uppan** + **pylle**, dative sing. of **pyll**. The place stands on a creek at the mouth of the river Axe. DEPN, L 28.

UPLEADON Glos SO 7527. 'Estate futher up the river Leadon' sc. than HIGHLEADON SO 7723. *Up- Vpleden(e)* 1253–1398, *Upledon(e)* 1316–1659, *-leadon* 1605. Earlier simply *Ledene* 1086–1327. OE **upp** + r.n. LEADON SO 7628. Gl iii.189.

UPLEATHAM Cleve NZ 6319. 'Upper Leatham'. *Upelider* 1086, *Upli(th)um* [1119]15th–1310, *Lyum* 13th cent., *Uplethum* 1407, *Up-Leatham* 1665. ON **hlíth**, nom.pl. **hlíthir**, dative pl. **hlíthum**. *Up* in relation to KIRKLEATHAM NZ 5921. YN 153.

UPLEES Kent TQ 9964. *Up^r, Lower Lees* 1819 OS. ModE **up** + **lease** (OE *lǣs*) 'pasture, meadow-land'.

UPLODERS Dorset SY 5093. 'Loders higher up-stream'. *Uppelodres* 1445, *Uplodre* 1446, *Uplodres* 1467. Earlier simply *Lodre* 1086. OE **upp** + p.n. LODERS SY 4994. Do 148.

UPLOWMAN Devon ST 0115. 'Settlement, estate up the river Lowman'. *Oppelaume* 1086, *Oplomia, Oppaluma* 1086 Exon, *Uplomene* 1303, *Uppelomyn* 1489. OE **upp** + r.n. Lowman, *Loman* 1563. Also known simply as *Lu- Lomene* 1242–1327 and as *Richardeslomene* 1317, 'Richard's Lowman' from the feudal tenant Ricardus of 1242 and for distinction from Chieflowman ST 0015, *Lonmine* 1086, *Childelomene* c.1200, *Chill(el)omene* 1281–1316, *Chiffeloman* 1548, 'Lowman estate of the youths', OE **ċild**, genitive pl. **ċilda**, + r.n. Lowman, and Craze Lowman SS 9814, *Lonmele* 1086, *Lomene Clavile* 1285–1318, *Clavylys Lomyn* 1456, 'Loman estate held by the Clavile family' from John de Clavile who held the manor in 1284. Uplowman was probably first so named by the people of Tiverton. D 552, 542, 8, RN 265.

UPLYME Devon SY 3293. 'Settlement, estate up the r. Lyme'. *Uplim* 1238, *Up Lym* 1282. OE **upp** + r.n. Lyme, *Lim* [774]12th S 263, *Lym* [938]14th S 442, *Lyme* 1322 as in LYME REGIS Dorset SY 3492 which is called *Nytherlym* 1310 'lower Lyme'. Also known simply as *Lim* 1086, *Lym* 1284. D 649, 8.

UPMARDEN East MARDEN WSusx SU 8014.

UPMINSTER GLond TQ 5685. 'Upper minster'. *Upmonstrā -munstrā -munstre* 1086, *-min(i)str(e)* 1212–35, *Upmynstre* *[1062]13th S 1036, 1329–80, *Upmestre* 13th, *-mister* 1535, 1543. OE **upp(e)** + **mynster**. The reference is to the site on slightly raised ground above the r. Ingrebourne. Ess 131.

UPNOR Kent TQ 7570. 'Higher Nore'. *Upnore* 1374. ME **up(pe)** + p.n. *atte Nore* 1292, a misdivision of ME *atten Ore* < OE *æt thæm oran*, 'at the flat-topped hill', OE **ōra**. The **ōra** seems to be the long bank which runs behind the shore to the N of Lower Upnor. 'Up' probably refers to the S part of the settlement which is higher up the Medway. PNK 116, Jnl 22.27.

UPOTTERY Devon ST 2007. 'Settlement, estate up the r. Otter'. *Upoteri* [1005]17th S 911, 1291. Also simply *Otri* 1086. OE **upp** + r.n. OTTER SY 0996. 'Up' in relation to Honiton and for distinction from Mohun's Ottery, [mu:nz otəri], the manor of the Mohun family at ST 1905†, *Mounesotery* 1453, earlier *Otermoun* 1285 and simply *Otri* 1086, and OTTERY ST MARY SY 0995. D 650, 642.

UPPARK WSusx SU 7717. 'The upper park'. *Uppark* 1427. ME **up** + **park**. Earlier *le Overpark* 1350 for distinction from Down Park SU 7822, *le Netherpark* 1350 and later *Dunpark* 1427. Sx 38.

UPPER END Derby SK 0976. 'Upper end' sc. of Wormhill township. *Overend(e)* 1461×83, 1573, 1580. ModE **upper**, ME **over** + **ende**.

UPPER GREEN Berks SU 3763. ModE adj. **upper** + **green**. This is a hamlet about a mile SE of Lower Green, itself situated at the N end of Inkpen. Formerly known as *Haslewick Green* from the lost p.n. Haslewick, 'dairy-farm at the hazel-tree', *Haselwyk* 1241–1375, OE **hæsel** + **wīċ**. Brk 311, 309.

UPPERMILL GMan SD 9905. *Upper Mill* 1730. YW ii.317.

UPPER STREET Hants SU 1518. No early forms, but nearby are Flood Street SU 1417, *Flood Street* 1811 OS, and North Street SU 1518, *North Street* ibid. ModE **upper** + dial. **street** 'a hamlet'. Gover 1958.217.

UPPER STREET Norf TG 3517. 'The upper hamlet' sc. of Horning. *Horning Upper Street* 1838 OS. Contrasts with *Horning Lower Street* 1830 ibid., which is now simply HORNING TG 3417. There is another Upper and Lower Street at TG 3217 and 3116, *Upper Street, Lower Street* 1838 OS, related to Hoveton St John TG 3018. Nf ii.165.

UPPERTHONG WYorks SE 1308. 'The upper part of Thong'. *Uverthwong(e)* 1297, *(Hover)thoung(e)* 1275, *Overthonge* 16th. Adj. **upper**, ME **over**, + p.n. *Thwnge* 1274, *Thwong(e)* 1274–1618, *Thong* 16th, OE **thwang** 'a narrow strip of land'. 'Upper' for distinction from NETHERTHONG SE 1309. YW ii.288.

UPPERTON WSusx SU 9522. 'The upper settlement' sc. of Tillington. *Upperton* 1191. ME **upper** + **tūn**. Sx 124.

UPPINGHAM Leic SP 8699. 'The homestead of the Yppingas, the upland people'. *Yppingeham* 1067, 1066×1087, *Ippingeham* 1157, *Uppingeham* 1080×7–1245, *Uppingham* from 1198. OE folk-name **Yppingas* < OE **yppe** 'upper place, hill' + **ingas**, genitive pl. **Yppinga*, + **ham**. Uppingham stands on a long high ridge. The specific might alternatively be an OE p.n. **Ypping* 'the upland place' < **yppe** + **ing**[2], locative-dative sing. **Yppinge*, and so the 'homestead at or called *Ypping*'. R 210.

UPPINGTON Shrops SJ 5909. Probably 'the estate called after Uppa'. *Opetone* 1086, *Opyton -i-* late 13th, *Upetuna* c.1140, *Upton'* 1251, *Op(p)inton(') -yn-* 1195–1375, *Uppinton(') -yn-* 1208–1768, *Uppington -yng-* from 1284. The boundary of this place is referred to in 975 (S 802) as *uppinghæma gemære* 'the boundary of the dwellers at Uppington' in which *hǣme* implies an OE p.n. **Uppingtūn* just as *stifingc hæma gemære* [964]12th S 724 refers to Steventon Berks, *Stivetune* 1086, *Stivinton* 1220, OE **Stīfingtūn* and *dræg hæme* 1021×3 S 977 refers to Drayton Northants. Such a name would normally mean 'the estate called after Uppa', OE pers.n. **Uppa* + **ing**[4] + **tūn**. But there was clearly an alternative understanding of the name as **uppe-tūn** 'the up settlement' although this does not fit the topography particularly well. It is unlikely, therefore, that the true explanation is OE ***upping** < **uppe** + **ing**[2] 'the up place' + **tūn**. Sa i.295, PNE i.217, Årsskrift 1974.57.

UPSALL 'The high dwellings, high huts'. ON pl. **upsalir**. Possibly a name-type transferred from p.n. Uppsala in Sweden. See also UPSLAND SE 3080. SSNY 89.

(1) ~ Cleve NZ 5415. *Upes(h)ale* 1086, *Uppesale* [1155×65]15th–1412, *Upsall* from 1443. YN 158, SSNY 89 no.1.

(2) ~ NYorks SE 4587. *Upsale* 1086, *Uppesale* 1185×95. Situated on the slope of a steep hill. YN 200, SSNY 89 no.2.

UPSHIRE Essex TL 4100. 'Upper district' sc. of the shire or liberty of Waltham. *Up(e)scire* 1274, *-shire* 1423. OE **uppe** + **scīr**. Waltham shire is *Waltamscire* 1108×18. Ess 31, 27.

UPSLAND NYorks SE 3080. 'The high dwellings, high huts'. *Opsala, Upsale* 1086, *Oppeslund* [1184]15th, *Uppeslunde* 1285, *-lounde* 1301, *-lond* 1406, *Uppislande* 1556. ON pl. **upsalir**, possibly transferred from p.n. Uppsala in Sweden, to which was later added ON **lundr** 'a wood, a grove, a sacred grove or sanctuary'. See UPSALL. YN 221, SSNY 89 no.3.

UPSTREET Kent TR 2263. 'The higher Roman road'. *Upstreet* 1690. Ultimately from OE **up(pe)** + **strǣt**. The reference is to the upper part of the Roman road from Canterbury to Thanet, Margary no. 11, where it rises out of the low ground between the mainland and Thanet. PNK 508.

†Also known as *Otery Flandrensis* 1247, *Ottery Flemeng'* 1279 from the family of William le Flemmeng who had an interest here in 1219–1244.

UPTON 'Higher settlement', OE **upp(e)-tūn**. PNE ii.227, L227.

(1) ~ Berks SU 9879. *Optone* 1086, *Vpton'* 1176, *Upton(')* from 1220. Brk 534.

(2) ~ Bucks SP 7711. *U- Opetone* 1086, *Upton* from 1204. Situated higher up than Dinton in WESTLINGTON SP 7610. Bk 160.

(3) ~ Cambs TL 1778. *Opetune* 1086, *Upton* 1285. Upton is on high ground overlooking Alconbury Brook. Hu 249.

(4) ~ Cambs TF 1000. *Upton* [948]12th S 533, *(on) uptune, (on) Optune* [972]c.1200 B 1130, *Upton* from 1179. Nth 245.

(5) ~ Ches SJ 4169. *Huptun* [958]14th S 667, *Optone* 1086, *Uppetuna* [1096×110]1280, [1121×9]150, *Uptuna* [1121×9]1285, 13th, *Upton* from 1260. A manor of Chester Abbey. Che iv.142.

(6) ~ Ches SJ 5087 → DITTON Ches SJ 4985.

(7) ~ Dorset SY 9893. *Upton* from 1463. 'Up' in relation to the lower settlement at either Hamworthy or Lytchett Minster. Do ii.16.

(8) ~ Hants SU 3655. *Optvne* 1086, *Upton* from 1312. The hamlet is higher up the river Test from Hurstbourne Tarrant SU 3853. Gover 1958.160.

(9) ~ Hants SU 3717. *Upton* 1410. Gover 1958.183.

(10) ~ Leic SP 3699. *V- Upton(e)* c.1130–1576. Lei 542.

(11) ~ Lincs SK 8686. *Opetune* 1086, *Uppetune, Uptuna* c.1115. The village is situated on a hill overlooking the r. Till. DEPN.

(12) ~ Mers SJ 2788. *Optone* 1086, *Upton* from 1265. The village is on a hill and contrasts with MORETON SJ 2690. Che iv.305.

(13) ~ Norf TG 3912. *Uptune* 1086, *Uppeton* 1165. DEPN.

(14) ~ Northants SP 7160. *Opton(e)* 1086–1317 with variant *Up-, Up(p)eton* 1158–75. The reference is probably to its situation higher up the Nene from Northampton. Nth 88.

(15) ~ Notts SK 7476. *Upetun(e) -tone* 1086, *Upton* from 1280. Nt 52.

(16) ~ Notts SK 7354. *Uptune* [958]14th S 679, *Upetun* 1086, *Upton(')* from 1185. Also known as ~ *Archiepiscopi* 'the archbishop's U' 1185, ~ *juxta Suwell* 'U by Southwell' 1305, and ~ *on Hill* 1586. The centre of the village is at the top of a low hill. The estate was granted with Southwell to the Archbishop of York by King Eadwig in 956. Nt 179.

(17) ~ Oxon SU 5186. *Optone* 1086, *Upton(')* from 1220. U stands higher than Blewbury and West Hagbourne. Brk 534.

(18) ~ Somer SS 9929. *Upton* 1225, 1610. DEPN.

(19) ~ WYorks SE 4713. *Uptun -ton* 1086. YW ii.98.

(20) ~ BISHOP H&W SO 6427. 'U held by the bishop' sc. of Hereford. *Opton Episcopi* 1291, 1459. Earlier simply *Vptvne* 1086, *Uptona -ton(e)* c.1200–1334. 'Up' in relation to Ross-on-Wye which, together with Walford, was part of a grant of land at Ross made by Edmund Ironside to the bishop of Hereford in 1016. He 193, ECWM 145.

(21) ~ CHEYNEY Avon ST 6969. 'U held by the Cheyney family'. *Upton(e) Chaun(e)* 1325, 1482, ~ *Cheyney* 1570. Earlier simply *Vppeton* 1190. *Upton(e)* 1208–1584. Upton lies on the slope of Hanging Hill above Bitton ST 6869. Gl iii.75.

(22) ~ CRESSETT Shrops SO 6592. 'U held by the Cressett family'. *Upton Cressett* from 1796, *Hopton Cressett* 1535. P.n. *Ultone* (sic) 1086, *Opeton'* 1167, *Upton(')* from 1201 + manorial addition from Thomas Cressett who married into the *de Upton* family c.1250. Sa i.296.

(23) ~ CROSS Corn SX 2872. *Upton Cross* 1870, 'crossroads near Upton'. A 19th cent. village. Upton SX 2772, *Oppeton* 1311 G, *Upton* 1378 G, *Uppeton* 1474, is 'higher farm'. PNCo 175, Gover n.d. 165.

(24) ~ GREY Hants SU 6948. 'U held by the Grey family'. *Upton Grey* 1281. Earlier simply *Huppeton* 1100×35, *Upton* 1202. The site is at the top of a rise on the Roman road from Chichester to Silchester, Margary no. 155. The manor was acquired c.1260 by John de Grey whose family came from Graye in Normandy. Ha 166, Gover 1958.134.

(25) ~ HELLIONS Devon SS 8403. 'U held by the Hellion family'. *Uppetone Hyliun* 1270, *Upton Hellyng* 1557. The manor was held by William de Helihun, a Breton name, in 1242. Upton here means the upper of the two manors called Creedy, *(be) cridian* [1016×20, probably 1018]11th S 1387, *Cridia, Creda* 1086, *Cridie* 1242, later *Crydihelyhun* 1242 (i.e. Upton) and *Cridie Peyteveyn* 1305, held by Robert le Peytevin (i.e. of Poitou) in 1242, now Lower Creedy SS 8402. Creedy is a r.n. as in CREDITON SS 8300 and CREEDY PARK SS 8301. D 419.

(26) ~ MAGNA Shrops SJ 5512. 'Great U'. *Great Upton under Haghmon* 1408, *Upton Magna* from 1535. ModE **great**, Latin **magna**, + p.n. *Uptune* 1086, *Upton(')* from c.1145. 'Magna' or 'Great' for distinction from Waters UPTON SJ 6319. *Haghmon* is Haughmond Hill SJ 5414, *Hag(h)eman* 1156–1320, *Haghmon(')* c.1160–c.1540, *Hamon* 1203×10–1695, *Haughmon* c.1225, 1398, 1627, *Hawe- Hauman* 1241–1329, *Haumond* 1346, *Haughmond* from 1553, OE ***hagaman** of unknown meaning, pronounced locally [heimən] and assimilated in spelling to Fr **mont** 'a hill'. Sa i.297, 148–50.

(27) ~ NOBLE Somer ST 7139. 'U held by the Le Noble family'. *Upton le Noble* 1291, *Vpton noble* 1610. Earlier simply *Opetone* 1086. DEPN.

(28) ~ PYNE Devon SX 9197. 'U held by the Pyne family'. *Uppeton Pyn* 1283, *Uppen Pyne* 1624. Earlier simply *Opetone* 1264, *Uppeton(e)* 1275–1316. Herbert de Pyn held the manor in the 13th cent. 'Up' with reference to Brampford Speke SX 9298.

(29) ~ ST LEONARDS Glos SO 8615. 'St Leonard's U'. *Upton(e) Sancti Leonardi* 1287–1610, *Seynt Leonards* ~ 1486, ~ *St Leonardes* 1593. Earlier simply *Optvne* 1086, *V- Upton(e)* 1195–1535. The reference is to the dedication of the church. Upton lies on the lower slopes of the Cotswolds. Gl ii.170.

(30) ~ SCUDAMORE Wilts ST 8647. 'U held by the Scudamore family'. *Upton Squydemor* 1275, ~ *Escudemor'* 1281, ~ *Skydemour* 1301, *Skydemorysupton* 1439. P.n. *(æt) Uptune* [after 987]14th S 1505, *Uptone* 1086, + manorial addition from the family name of Peter de Skydemore who held the manor in 1216 and Geoffrey de Escudamor in 1242. Wlt 156.

(31) ~ SNODSBURY H&W SO 9454. Originally two separate vills, *Upton juxta Snodebure* 'U beside S' 1280, *Upton Snodesbury* 1327. Earlier simply *Upton Stephani* 'Stephen's U' 1212. Snodsbury, *(in) snoddesbyri* 972 S 786, *Snodesbyrie* 1086, *Snodesbury* 13th, *Snodgbury* 1700, is 'Snodd's fortified place', OE pers.n. **Snodd*, genitive sing. **Snoddes*, + **byriġ**, dative sing. of **burh**. The same person gave his name to *snoddes lea* 'Snodd's wood' in the bounds of the adjacent parish of Crowle [n.d.]11th S 1591. The reference is to the Iron Age hill-fort on Castle Hill SO 9355. The church-town of Snodsbury is at 157ft. at Upton on ground overlooking most of the low-lying parish. Wo 230 gives pr [snodʒberi].

(32) ~ UPON SEVERN H&W SO 8540. *Upton super Sabrinam* (Latin form) 1327. Earlier simply *Uptun* [889]17th ECWM 267, [962]11th S 1300, 1086. Upton is further up the Severn in relation to Ripple of which it was once a portion. Wo 174, Hooke 102, 244.

(33) ~ WARREN H&W SO 9367. 'Warin's U'. *Opton Warini* 1290 from the tenure of the father of William fitz Warin who was here in 1254. Warin is ONFr *Warin*, a pers.n. of Frankish origin. Earlier simply *Uptona* [716]12th S 83, *Vptvne* 1086. Also known as *Shirreue(s) Upton* 1300, 1319, from the Beauchamps, the hereditary sheriffs of Worcester who were the overlords here. 'Up' from its position further up the Salwarpe from Droitwich. Wo 311.

(34) Hawkesbury ~ Avon ST 7887. *Haukesbury Upton* 1439, *Hawkesburyesupton* 1599. Earlier simply *Upton* 972 S 786–1601. P.n. HAWKESBURY ST 7687 + *Upton*. Upton stands about 250ft. higher than Hawkesbury. Gl iii.32.

(35) Tetbury ~ Glos ST 8895. 'U belonging to Tetbury'. *Upton(e) -tune* 1086–1587. Upton lies on higher ground NW of Tetbury. Gl i.111.

(36) Waters ~ Shrops SJ 6319. 'Walter's U'. *Upton Waters* 1346,

Wateres Upton 1431, *Waters Upton* from 1517. Pers.n. *Walter* from Walter de Opton', tenant in 1242, + p.n. *Uptone* 1086. Also known as *Upton' Parva* late 14th, 'little Upton' for distinction from UPTON MAGNA SJ 5512, 'Great Upton'. Sa i.297.

UPTON LOVELL Wilts ST 9440. 'U held by the Lovell family'. *Ubbedon Lovell* 1476, *Lovells Upton* 1526, *Upton Lovell* 1597. P.n. *Ubbantun* [957]14th S 642, *Ub(b)eton* 1199–1430, *Ubton* 1624, 'Ubba's estate', OE pers.n. *Ubba*, genitive sing. *Ubban*, + **tūn**. The bounds of the people of Upton are *Ubbantuninga gemære* [957]14th S 642. Wlt 171 gives pr [lʌvəl], DEPN.

UPWALTHAM WSusx SU 9413 → Up WALTHAM SU 9413.

UPWARE Cambs TL 5370. 'Upper fishing weir'. *Upwere* 1170–1419. OE **ūp** + **wer**. Ca 204.

UPWELL Norf TF 5002. 'Upper Well'. *Upwell(e)* from 1221 with variants *Up(p)ewell(e), Opwell(e)* 1269–1480. OE **up**, **uppe** + p.n. *(æt) Wellan* [? 963]12th S 1448, *(æt) wyllan* [970]c.1100 S 779, [973]14th, *(et) Willan* [970]13th S 779, *Welles* [1077×1130]12th–1368 including [974]13th S 798[†], *Welle* 1086–1549 including [1021×3]12th S 980, '(the settlement) at the springs', OE **æt** + **wellum**, dative pl.of **wella**. 'Up' for contrast with OUTWELL TF 5104 with which it originally formed a unit. The form *Welles* is a new ME pl. Ca 288.

UPWELL FEN Cambs TL 4795. P.n. UPWELL TF 5002 + ModE **fen**.

UPWELL FEN Norf TL 5599. Called *Outwell Low Fen* 1824 OS. P.n. UPWELL TF 5002 + ModE **fen**.

UPWEY Dorset SY 6684. 'Higher settlement on the r. Wey'. *Uppeweie* 1241, *Upway -wey(e)* from 1327. Earlier simply *Wai* 1086[‡], *Wai(e)* 1194–1259, *Way(e)* 1237–1363. Also known as *Way(e) Ba(y)(h)(o)us(e)* 1237–1428, *Waye Pi- Pygace* 1243, 1392, *Waye Raba(y)ne* 1288, *Way(e) Hamundevill -vyle* 1249, 1343×5. ME **uppe** (OE *upp*) + p.n. Wey as in River WEY SY 6681. The alternative forms, alluding to the Baieux family from Bayeux, Normandy (e.g. Alan de Bayocis c.1200, John Bayouse 1244)[††], the Rabayne family (Maud de Rabayn held land here 1259–60) and the Pigace and Amundevill families (no individuals are known for these names), are for distinction from the other Wey manors enumerated under BROADWEY SY 6683. Do i.245, 196.

UPWOOD Cambs TL 2582. 'Upper wood'. *Upehude* 1086, *Upwude* 1253, *-wode* 1303. OE **uppe** + **wudu**. 'Upper' in contrast to lower woodland on the edge of the fens W of the village. Hu 225.

URCHFONT Wilts SU 0457. Probably 'Eohric's spring'. *Ierchesfonte* 1086, *Erchesfont(e)* 1175–1332, *Archesfunte* 1179, *Urichesfunte* 1242, *O- Urchesfunte* 1259, 1289, *Archfounte al. Urshent* 1564, *Earchfount* 1605, *Urchefount al. Urshent* 1611. OE pers.n. **Eohrīċ*, genitive sing. **Eohrīċes*, + **funta**. Wlt 315 gives prs [ʌʃənt, ərʃənt], Signposts 84, 86, L 22.

URDIMARSH H&W SO 5249. Partly uncertain. *Urdimarsh* 1832 OS. The generic is ModE **marsh**, but the evidence is too late to explain the specific.

River URE NYorks SE 2085. Probably the 'strong or swift river'. *Earp* 11th (for *Earp*, i.e. *Earw*), *Jor(e), Yor(e)* c.1150–1415, *Your(e)* 1295–1751, *Ure* from c.1540. The root is IE **is-*/**eis-* as in r.n. AIRE SE 6723, with a different suffix, Brit **Isurā* from which the Roman town of ISURIUM is named. This would give PrW **Ior* borrowed into OE as **Īor, Ēor*. If the form *Earp* is a mistake for *Earp, Earw*, i.e. *Ear wæter* 'Ear water', it shows ONb *ea* for *eo*. YN 7, RN 427, LHEB 362, 523, BzN 1953.239, Krahe 1964.56.

URISHAY COMMON H&W SO 3137. *Uris-hay Common* 1832 OS. P.n. Urishay + ModE **common**. Urishay, *Haya (H)urri* c.1200, *Urysay* 1307, is 'Urry's hay', pers.n. *Urri*, possibly a Norman form of OE *Wulfrīċ* or CG *Udalric*, + p.n. *Haia* 1166, c.1212×17, 'the enclosure', ME **haye** (OE *(ġe)hæġ*). A forest enclosure held by Urri or Ulric de la Hay c.1136×48. Probably a renaming reflecting the settlement policy of William fitz Osbern of the lost *Alcamestvne* 1086, *Alch Hemestona* 1160×70, 'Alhelm's estate', OE pers.n. *Alhhelm*, genitive sing. *Alhhelmes*, + **tūn**. Cf. WALTERSTONE SO 3425. He 165, DB Herefordshire 29.9–10, DBS s.n. Hurry.

URLAY NOOK Cleve NZ 4014. *Early Nooke* 1739, *Urlay Nooke* 1826. P.n. *Lurlehou* c.1220, *Lur(e)lau* c.1264, *Lurlaw* 1347, *Urlawe* 1509–27 + ModE **nook**. Possibly OE pers.n. **Lurla* + **hōh** 'a hill-spur' or **hlāw** 'a hill' (the earliest form possibly represents **Lurlan hlāw* 'Lurla's hill or tumulus' + **hōh**). For the pers.n. **Lurla* cf. LARLING Norfolk TL 9889. The subsequent history of the name seems to be the result of false analysis as if it were *l'urlawe* with the French definite article *le* elided. The suffix *law* regularly becomes *ley* in Northern p.ns.

URMSTON GMan SJ 7594. 'Wyrm's farm or village'. *Wermeston* 1194, *Wurmeston* 1219, *Urmeston(e)* 1212–1341 etc., *Ormeston* 1284. OE pers.n. **Wyrm* with ME loss of *w* before [u], genitive sing. **Wyrmes*, + **tūn**. La 37, SSNNW 191, Dobson §419, Jnl 17.33, Årsskrift 1987.49.

URRA NYorks NZ 5701. 'The dirty hill'. *Horhowe* 1301, *Orrow(e)* 1377. Probably a ME coinage ultimately from OE **horh** + ON **haugr**. YN 70.

URSWICK Cumbr SD 2673. 'Farm at *Urse*, the bison lake'. *Ursewica* c.1150, *(H)ursewic, Wrsewik, Vrs(e)wic(h)* 12th, *Urs(e)wik -wyk* 1269–1327, comprising Great Urswick SD 2774, *Magna Urswic* 1180×90, *Great Urswyk* 1277, and Little Urswick SD 2673, *Parva Urswik* 1257, *Little Ursewyk* 1299. P.n. **Urse*, OE **ūr** 'bison, aurochs' + **sǣ** 'lake' referring to Urswick Tarn SD 2774, + **wīċ**. Mills 1976.144.

USHAW MOOR Durham NZ 2242. A modern colliery village. The moor from which it was named is *mora de Ulshawe* 1420. P.n. Ushaw + ME **mōr**. Ushaw, *Vlueskahe* 1180×96, *Ulueschawe* 1382, *Ulshaw(e)* 1393–1623, *Uuesshawe* 1312, *Vssawe* 1349, *Usshawe* 1567, is 'wolf wood', OE **wulf**, genitive pl. **wulfa**, + **sceaga**. The loss of initial *W-* is due to the influence of ON **úlfr**. Wolves were native to Durham in the Middle Ages; there was a lost *Vlshawe* 1418 in Bishop Auckland at NZ 2178 and a farm at NZ 1739 is still called Wooley, *Wolleys* 1349–53, *Wollyhal* 1425, *Wolly* 1768, 'wolves' clearing', OE **wulfa** + **lēah**. NbDu 204, 220, Nomina 12.36.

USSELBY Lincs TF 0993. 'Oswulf's village or farm'. *Osoluabi, Osoluebi* c.1115, *Os(s)elby* [1154×89]1269–1622, *Osolfby* 1209×35, c.1221, *Usselby* from 1324. OE pers.n. *Ōswulf*, probably an Anglicised version of ON *Ásulfr*, + **bȳ**. The absence of genitival *-es* suggests that medial *-e-* represents OScand genitive sing. *-a* < *-ar*. Li iii.168, DEPN, SSNEM 76.

USWAY BURN Northum NT 8713. *Osweiburne* [1153×95]14th, *Oswaiburne* [n.d.]14th. P.n. Usway + **burna**. Usway, *Usway* 1817 Dixon 29, appears to be 'Osa's road', OE pers.n. *Ōsa* + **weġ**. On it is Uswayford NT 8814, *Useyfoord* 1743. NbDu 204.

UTKINTON Ches SJ 5465. 'Estate called after Uttoc'. *Utkinton* from 1188 with variant *-yn-*, *Utkyngton -ing-* 1296–1671, *Utkuton* 1324. OE pers.n. **Uttoc* + **ing**[4] + **tūn**. Ch iii.298 gives pr ['utkintən], older local ['utkitn].

UTLEY WYorks SE 0542. 'Utta's wood or clearing'. *Utelai* 1086, *Vtteleie* 1152×62, *Utley* from 1497. OE pers.n. *Utta* + **lēah**. YW vi.4.

UTON Devon SX 8298. 'Settlement, estate on the r. Yeo'. *Yeweton* 1285, 1435, *Iuweton* 1296, *Uton* 1486. R.n. YEO SX 8199 + OE **tūn**. D 406, 17, RN 480, NoB 14.54.

UTTERBY Lincs TF 3093. 'Uhtred or Uhthere's village or farm'. *Vthterbi* late 12th, *Uthterby* [1150×60]1409, *Utterby* from c.1221. OE pers.n. *Ūhtrēd* or *Ūhthere* + ON **bȳ**. Li iv.35, SSNEM 82.

[†]This is the identification in Ca 288 though S 798 refers it to WELLS NEXT THE SEA TF 9143. In the list of places in the charter *Welles* follows Hilgay TL 6298 and Walsoken TF 4710.

[‡]The identification is not certain.

[††]The name survives in Wabyhouse, a liberty in Culliford Tree Hundred 1664–1863 and Wabey House, a private residence in Upwey.

UTTOXETER Staffs SK 0933. 'Wuttoc's heath'. *Wotocheshede* (for *-hedere*) 1086, *Uttokishedere* 1175, *Wittokeshather* 1242, *Uttoxatre* 14th, *Utseter* 16th. OE pers.n. **Wi(u)ttuc*, **Wuttuc*, genitive sing. **Wi(u)- Wuttoces*, + ***hæddre**. DEPN, Duignan 157, PNE i.214.

UXBRIDGE GLond TQ 0583. 'Bridge (or causeway) leading to the Wixan'. *Wixebrug' -brigge* c.1145–1294, *Oxebruge* c.1145, *Uxebrigg(e) -bridge* 1200–1560, *Wu- Woxebruge -brigg(e) -bridge* 1219–1493. OE folk-n. *Wixan* + **brycg**. The East and West Wixan are mentioned in the 7th cent. Tribal Hidage as a people apparently settled in or near the Lincolnshire fens. The name seems to mean little more than 'the village people' being a form of **Wihsan*, the exact equivalent of OHG nominative pl. *Wihsa* for the modern Bavarian p.n. Weichs. The origin is IE **u̯eikos*, whence Gothic *weichs* 'village', Latin *villa* (< **ueiks-lā*), cognate with ablaut variant **u̯oikos*, the source of Greek *οἶκος* 'home', Latin *vīcus*, OE *wīc*. Tribal migration certainly took place, but the meaning of the name or nick-name appellation suggests that it could have arisen independently in more than one place. Hence perhaps its occurrence in a third location in the H&W stream-name Whitsun Brook in Flyford Flavel SQ 9754, *pixenabroc* [972]10th S 786 (with genitive pl. *Wixena*) 'the brook on the boundary) of the Wixan'. The territory of the GLond Wixan seems to have extended to both Harrow, where the road names Uxendon Hill and Crescent preserve the memory of Uxendon Farm, *Woxindon* 1257, *Wxindon* 1275, *Uxendon* 1593, 'hill of the Wixan' (OE *Wixena* + **dūn**), and Norwood, where there was once a Waxlow Farm, *Buxle* (sic for *Wuxle*) 1249, *Woxeleye* 1294, 'woodland clearing of the Wixan', OE *Wixan* + **lēah**. The sound development *Wix-* > *Wux-* > *Ux-* is paralleled by the history of the word *woman*, OE *wīfman* > lOE *wimman* > ME *wumman* > dial. *'ooman*. Mx 48, 45, 54, Wo 16, DEPN, Bach 2 §601, Frings I.37 f.n., I.85 f.n., II 502, Campbell §§304, 318, 484, Jordan §§36 A.2, 162.2, Luick §774, Dobson §420.ii, 421, Hooke 190, 192.

VALE OF BELVOIR Leic SK 7838. *(de) valle Beauver* 1250, *the Vale of Beauer* 1449. ME **vale** + **of** + p.n. BELVOIR SK 8233.

VALE OF BELVOIR Notts SK 7838. *Vallem de Bello Vero* 1289, *vale of Beauver* 1399, *vallis de Bevor* 1491, *vale of Beaver* 1657. ME **vale**, Latin **vallis**, + p.n. BELVOIR Leic SK 8233. Nt 12.

VALE OF BERKELEY Glos SO 6900. ModE **vale** + p.n. BERKELEY ST 6899.

VALE OF CATMOSE Leic SK 8709. *The Vale of Catmouse* 1576, *Vale of Catmus* 1613, ~ *Catmose* 1684, 1695, ~ *Catmoss* 1809. ModE **vale** + p.n. Catmose, probably the 'wild-cat marsh', OE **catt** + **mos**. R 3.

VALE OF EVESHAM H&W SP 0942. ModE **vale** + **of** + p.n. EVESHAM SP 0344.

VALE OF GLOUCESTER Glos SO 8320. *vallis Gloecestræ* c.1125, *valle de Gloccestria* 1316. Latin **vallis**, ME **vale**, + p.n. GLOUCESTER SO 8318. Gl i.2.

VALE OF MAWGAN or LANHERNE Corn SW 8964. See LANHERNE or Vale of MAWGAN.

VALE OF PEWSEY Wilts SU 1158. ModE **vale** + p.n. PEWSEY SU 1660.

VALE OF TAUNTON DEANE Somer ST 1727. *Taunton Deane* 1894 OED. P.n. TAUNTON ST 2224 + ModE **dean** 'a vale, a wide valley' (OE *denu*).

VALE OF WHITE HORSE Oxon SU 2689. (The vale of) *Whithors* 1368, *the fruteful vale of White-Horse* 1542. ModE **vale** + White Horse as in White Horse Hill SU 3086. White Horse Hill is *mons ubi ad album equum scanditur, locus qui vulgo mons albi equi nuncupatur* 'the hill where you climb up to the white horse, the place called in the vernacular the Hill of the White Horse' [before 1170]c.1200, *Whytehorse* 1322, ME **white** + **horse**, referring to Uffington White Horse, an Iron Age chalk-cut figure $\frac{1}{4}$ m. S of Uffington Castle, the oldest 'white horse' in Britain, possibly the oldest chalk-cut hill figure. It was probably cut in the 1st cent. BC as a tribal emblem ultimately related to animal worship. Brk 4, 380, Thomas 1976.181.

VANGE Essex TQ 7287. 'Marsh district'. *(æt) Fengge, (to) fænge* 963 S 717, *Fenge* [963]13th S 1634–1313, *Phenge* 1086, *Fange(s)* 1203–1339, *Vahnge* 1216×72, *Vange* from 1395. OE **fenn**, Essex dial. form **fænn**, + **ġē**. Ess 174 gives pr [vændʒ], formerly [va:ndʒ].

The VAULD H&W SO 5349. 'The fold'. *(La) Falde* 13th cent., 1470×4, *The Valde* 1614. OFr definite article **la** + ModE **folde** (OE *falod*) with west-country voicing of initial *f* to *v*. He 144.

VELLAN HEAD Corn SW 6614. *Velland Point* 1841. P.n. Vellan + ModE **head** 'headland'. Vellan might be Co **melin** 'a mill, a windmill' with lenition of *m* to *v* after a lost definite article **an**. But no windmill is known at this spot. PNCo 175, Gover n.d. 578.

VENFORD RESERVOIR Devon SX 6870. P.n. Venford + ModE **reservoir**. Venford is *Wenford* 1355, *Wendford* 1358, *Wandeford* 1452, *Wendford lake* (i.e. stream) 1609. The evidence is insufficient to provide a secure explanation. D15.

VENN OTTERY Devon SY 0791 → Venn OTTERY.

VENNINGTON Shrops SJ 3409. 'The marsh settlement'. *Feniton* 1256, *Fen(n)yton* 1305, 1396, *Veniton* 1760, *Venington alias Fenington* 1618. Probably ***fenning** 'fen place' + **tūn**. Alternatively the specific might be adj. **feniġ** 'marshy'. Sa ii.66.

VENNS GREEN H&W SO 5348 → WALKER'S GREEN SO 5247.

VENNY TEDBURN Devon SX 8299 → TEDBURN ST MARY SX 8194.

VENTA Hants SU 4729. The RBrit name for the Roman city of Winchester. Οὐέντα [c.150]c.1200 Ptol, *Venta Belgarum* [c.300]8th AI, *Venta Velgarom* [c.700]13th Rav, 'Venta of the Belgae' whose capital it was. The origin of the element **venta** is disputed; there is no evidence that it meant 'market', 'field' or 'hill'. Most likely it is a pre-British element meaning 'place, chief place' cognate with Albanian *vend* 'place'. RBrit 492, 262–5, Rostaing 295–6, Jnl.16.1–24.

VENTNOR IoW SZ 5677. Probably a manorial name from the family of William *le Vyntener* c.1340. *Vintner* 1591, 1617, *Vyntnor* 1607, *messuage or tenement called Vintner or Vintners* 1633, *Ventnor* 1769. The surname represents OFr *vintenier* 'an officer in command'. In the Middle Ages the whole island was divided into nine districts for military defence, each district commanded by a *vintenier*. William le Vyntener was one of four jurors appointed in 1341 to supervise the collection of taxes in Bonchurch 1 mile E of Ventnor. This was probably his estate. Wt 231–2, Mills 1996.103.

River VER Herts TL 1209. *VVer, Verus* 1572. An antiquarian invention by Humphrey Lhuyd from the RBrit name VERULAMIUM TL 1307. An earlier name was *Redburne* [1284]15th 'reed stream', OE **hrēod** + **burna**. RN 429.

VERCOVICIUM Northum NY 7968. 'The place of the *Vercovices*'. *Germani Cives Tuihanti Cunei Frisiorum Vercovicianorum* before 225 Inscr, *Velurtion* [c.700]14th Rav, *Borcovicio* [c.400]15th ND. The Roman name of the fort at Housesteads. Folk-n. **Vercovices* means 'the effective fighters' < Brit ***u̯erco-** 'work' + ***u̯ic-** 'fight'. RBrit 493–4.

VERNHAM DEAN Hants SU 3456. 'Vernham valley'. *Farnhamsdene* 1410, *Vernhams Deane* 1558×1603. P.n. Vernham + ME **dene** (OE *denu*). It is uncertain whether the forms *Farnhams/Vernhams* represent a genitive, a pseudo-genitive, or a plural formation. The 1817 OS shows a series of places *Vernhams, Vernhams Dean, Vernhams Row, Vernhams Street*. Today Vernhams Dean is in use as the parish name, but the others together with Vernham Bank SU 3356 and Vernham Manor SU 3556 are recorded in forms without *-s*. Vernham, *Ferneham* 1210, 1219, *Fernham* 1232, 1316, is 'bracken estate' or 'enclosure', OE **fearn** + **hām** or **hamm**. Ha 167, Gover 1958.161, DEPN.

VERNHAM STREET Hants SU 3557. 'Vernham hamlet'. *Vernhams Street* 1817 OS. P.n. Vernham as in VERNHAM DEAN SU 3456 + ME **strete**, *la Strete* 1324. Ha 167, Gover 1958.161.

VERNOLDS COMMON Shrops SO 4780. *Vernalls Common* 1729, 1832 OS, *Vernolls Common* 1783, *Varnit(t)s Common* 1770, *Varnil's Common* 1772. Surname *Vernall* (*Farnell*) + ModE **common**.

VERTERAE Cumbr NY 7914. The RBrit name of the fort at Brough Castle meaning 'at the summits'. *Verteris* [2–3rd]8th AI, *Valteris* [c.700]13th Rav, *Verteris* [c.408]15th ND. RBrit *Uerteris*, locative-dative pl. of Brit **Uertera̅* from ***u̯ertero-** < **u̯er-* 'over' + comparative formant **-tero*, nom.pl. **Uerterae*. There is no evidence of survival of this name, which would have become **Werther*, cf. W *gwarther* 'summit'. RBrit 496.

VERULAMIUM Herts TL 1307. The RBrit name of the Roman town which preceded St Alban's. *VER, VERL, VERO, VERLAMIO, VIR* 20 BC–AD 10 coins, *Verulamio* (dative case) [50]11th Tacitus, Οὐρολάνιον (*Urolanium*) [c.150]13th Ptol,

Verolami(o), Virolamo [4th]8th AI, *Virolanium* [c.650]13th Rav. The RBrit name is remembered by Gildas who calls St Alban *Verolamiensem* 'Verulamian' and by Bede, *civitatem Uerolamium quae nunc a gente Anglorum Uerlamacaestir sive Uaeclingacaestir appellatur* 'the city V now called by the English *Uerlamacaestir* or *Uaeclingacaestir*, the Roman town of the Verlame or Wæclingas'. The name has not been satisfactorily explained. RBrit 497, Hrt 86.

VERWOOD Dorset SU 0908. 'Beautiful wood'. *Fairwod(e)* 1329–1436, *-wood* 1553, *Verwood* 1774. ME **faire** (OE *fæġer*) with southern voicing of initial *f* > *v* + **wode** (OE *wudu*). The earliest occurrence is the French form *Beubo(y)s* 1288, OFr **beau** + **bois**. Do ii.256 gives prs ['vɔ:wud, 'vɔ:rud].

VERYAN Corn SW 9139. 'St Symphorian's (church)'. *(Parochia) Sci Simphoriani* 1278, *Sanctus Symphorianus* 1281, *Severian* 1525, *Seyntveryan* 1534, *Verian* 1607, 1610, *Veryan* 1617. Saint's name *Symphorian* from the dedication of the church, a Gaulish saint of the 2nd or 3rd century martyred at Autun. The modern form of the name is due to the change of *Symphorian* to *Severian* and false analysis as *Saint-Verian*. The DB name of the manor and parish was *Elerchi* 1086, *Elerky* 1231, 'swan-stream, swan-place', Co **elerhc** 'swans' + name suffix ***-i**, surviving in Elerkey SW 913395. PNCo 175–6, Gover n.d. 484, DEPN.

VERYAN BAY Corn SW 9640. *Veryan Bay* 1813. P.n. VERYAN SW 9139 + ModE **bay**. PNCo 176.

Lower VEXFORD Somer ST 1135. *Low Vexford* 1809 OS. ModE **low(er)** + p.n. (Higher) Vexford ST 1035, *Fescheforde* 1086, *Vexford* 1750 map, presumably for *Frescheforde* 'fresh-water ford', OE **fersc**, ***fresc**, definite form ***fresca**, + **ford**. There is, however, no possibility of contrast between fresh and salt water here as at FRESHFORD Avon ST 7860. DEPN.

VICARAGE Devon SY 2088. *Vicarage* 1809 OS. The reference is to Branscombe vicarage.

VICKERSTOWN Cumbr SD 1868. A company town built in 1901–4 for the workmen of Vickers Sons and Co. who bought the Barrow shipyard in 1896. Mills 1976.144, Room 1983.129.

VICTORIA Corn SW 9861. *Victoria* 1888. From the Victoria Inn here. PNCo 176.

VICTORIA PARK GMan SJ 8595. *Victoria Park* 1843 OS. An estate of large villas laid out in 1836, built mostly c.1850–60, surrounded by walls and gates. Pevsner 1969.325.

VICTORIA STATION GLond TQ 2979. The station, opened in 1862, was built at the end of Victoria Road which was cut through the slums of *Duck Lane*, *Thieving Lane* and *The Ambry* in the 1850s. Mx 184, Pevsner 1973L.518, 661.

VIEWING HILL Durham NY 7833. *Viewing Hill* 1866 OS. ModE **viewing** + **hill**. A common name for a hill commanding an extensive view.

The VILLAGE WMids SO 8989. A modern development with no early forms. ModE **village**.

VINE'S CROSS ESusx TQ 5917. Surname *Vine* + ModE **cross**. The family name Vine or Vyne is recorded here from the 16th cent. Sx 467.

VINEHALL STREET ESusx TQ 7520. 'Vinehall hamlet'. P.n. Vinehall + Mod dial. **street**. Vinehall, found in surnames *Fynhage* c.1310, *-hawe* 1327, then *Vynawes* 1566, *Vine Hall* 1813 OS, is 'enclosure by the wood-heap' or 'on the hill', OE **fīn** 'heap', sometimes applied to a hill, + **haga**. Voicing of initial [f] to [v] is responsible for the later refashioning of the name. Sx 476.

VIRGINIA WATER Surrey SU 9967. An artificial lake created in 1748 by William Augustus, Duke of Cumberland, the newly appointed Ranger of Windsor Great Park, so called in memory of his governorship of Virginia in the United States. The American state was so named in 1584 in honour of the Virgin Queen, Elizabeth I. Sr 125, Room 1983.129.

VIRGINSTOW Devon SX 3792. 'The virgin's holy place'. *Virginestowe* 1174×83, *Virstawe* 1463. ME **virgine** + **stōw**. The reference is to St Bridget the Virgin to whom the church is dedicated. The earliest form here antedates the first recorded occurrence of the word *virgin* in OED (c.1200). D 212.

VIRLEY CHANNEL Essex TM 0011. P.n. Virley TL 9413 + ModE **channel**. Virley, *Virley* 1501, is short for *Salcote Verly, Virly* 1291, 1323, 'Salcott held by the Verli family', p.n. SALCOTT TL 9413 + family name of Robert de Verli, the tenant in 1086. Ess 323.

VIROCONIUM Shrops SJ 5608. Short for *Viroconium Cornoviorum* 'V of the Cornovii', the RBrit name of the Roman city at Wroxeter, capital of the Cornovii, of uncertain meaning. *Οὐιροκόνιον* (Viroconium) [c.150]13th Ptol, *Urioconio -cunio, Uiro- Uiriconio* [4th]8th AI, *Utriconion Cornoviorum* [c.700]13th Rav. Possibly 'the town of Virico', pers.n. **Uiricō* + suffix **-on-io-*. The original reference was to the Iron Age hill-fort on the Wrekin SJ 6308, the former tribal capital of the Cornovii before they were resettled at Wroxeter. Sa i.330, RBrit 505, Thomas 1976.185.

VOBSTER Somer ST 7049. 'Fobb's tor'. *Fobbestor* 1234, *-ter* 1243, *Vobster* 1817 OS. OE pers.n. **Fobb*, genitive sing. **Fobbes*, + **torr**. DEPN.

VOREDA Cumbr NY 4938. The Roman fort at Old Penrith, Plumpton Wall. *Voreda* [2nd-3rd]8th AI, *Bereda* [c.700]13th Rav. British **Uorēdā* 'horse-stream' from **u̯orēdo-* 'horse'. The river here is the Petteril of unknown meaning. RBrit 508.

VOWCHURCH H&W SO 3636. 'Multi-coloured church'. *Fow(e)chirch(e)* 1291–1316, *Vowechurche* 1397, 1508. ME **fawe** (OE *fāh, fāge*) + **chirche**. Identical with Falkirk Scotland NS 8880, *la Faukirk* 1298, a translation of *Egglesbreth* (for *-brech*) 1080, Gaelic *eaglais b(h)rec* 'speckled church'. He 194, ScotPN 7–16.

The VYNE Hants SU 6357. *The Vyne* 1531. OFr, ME **vine** ' vine'. A Tudor mansion built between 1518 and 1527. The name may have been created for the house since the original name of the estate is Sherborne Cowdray 'S held by the Cowdray family', *Syrburne Coudray* 1272, *Manor of Vyne alias Sherborne Cowdray* 1550, p.n. Sherborne as in SHERBORNE ST JOHN SU 6255 + family name of Fulk de Coudray who held the manor in 1251. Ha 167, Gover 1958.129.

WABERTHWAITE Cumbr SD 1093. 'Clearing of the hunting or fishing booth'. *Waybyrthwayt'* c.1210, *Wayburthwayt* c.1215–1392 with many variants *Wey-*, *-byr-* *-bir-* *-bur(g)-* *-burgh-* and *-thwait* *-thweyt* etc., *Waibut(h)wait* c.1225, *Waythebuthwayt* c.1250, *Wawburthwaite* 1576. ON ***veithibúth** 'hunting or fishing booth', genitive sing. ***veithibúthar** possibly alternating with ***veithibúr** 'hunting or fishing store-house', + **thveit**. Cu 439 gives pr [wɔ:bərθwət], SSNNW 173, L 210–1.

WACKERFIELD Durham NZ 1522. Partly uncertain. *Wacarfeld* [c.1040]12th, c.1107, *Wakerfeld* with occasional variant spellings *-ir-*, *-yr-* 1283–1580. The specific might be the late OE pers.n. *Wacer* < OE **wacor** 'watchful', but this would normally occur in the genitive sing. form *Wacres*. The sense might be 'open land with a look-out': it is a good look-out site for observing movement in the Tees valley. Alternatively it may have had a significance similar to that of WAKEFIELD WYorks SE 3320, 'open land of the annual wake or festival'. Wackerfield was centrally sited in the shire of Staindrop. The alternative suggestion, 'open ground where osiers grow', depends on an unrecorded OE ***wācor** related by ablaut to OSwed *vīker*, ModE *wicker*. NbDu 204, PNE ii.234.

WACTON Norf TM 1791. 'Waca's estate'. *Waketuna* 1086, *Waketone* [1101×7]13th, *-tun* 1198. OE pers.n. *Waca* + **tūn**. DEPN.

WADBOROUGH H&W SO 9047. 'Woad barrows or hills'. *(in) uuadbeorhan, (on) ƿad beorgas, (of) ƿad beorgan* 972 S 786, *Wadberge -œ* 1086, *Wadbarewe* 1454, *Wadborough* 1628. OE **wād** + **beorg**, pl. **beorgas**, dative pl. **beorgum**. This may be evidence of the cultivation of woad in Anglo-Saxon times. Wo 220.

WADDESDON Bucks SP 7416. 'Weott's hill'. *Votesdone* 1086, *Wottesdune* 1156, *Wot(t)esdon* 1211–1327, *Wettesdon(')* 1167–1337, *Wod(d)esdon* 1327–1477, *Wotesdon or Waddesdon* 1755. OE pers.n. **Weott*, genitive sing. **Weottes*, + **dūn**. The same pers.n. occurs in *(on, of) Wottesbroce* 'Weott's brook' [1004]14th ECTV p.183, the name of a stream which flows SW from SP 7415 to Watbridge Farm at 7214. Bk 137 gives pr [wɔdzdən], YE addenda, DEPN, L 150.

WADDINGHAM Lincs SK 9896. Either 'the homestead of the Wadingas, the people called after Wada' or 'the homestead at or called *Wading*, the place called after Wada'. *Wading(e)-* *Widingehā* 1086, *Wadingheheim*.c.1115, *Wad(d)ingeham* 1168, 1200, *Wadincham* 1212. OE folk-n. **Wadingas* < pers.n. *Wada* + **ingas**, genitive pl. **Wadinga*, or p.n. **Wading* < pers.n. *Wada* + **ing**[2], locative–dative sing. **Wadinge*, + **hām**. DEPN, ING 145.

WADDINGTON 'The estate called after Wada'. OE pers.n. *Wada* + **ing**[4] + **tūn**.

(1) ~ Lancs SD 7243. *Widitun* (sic) 1086, *Wadingtun -ton, -yng-* c.1231–1483, *Waddington* from 1303. YW vi.199.

(2) ~ Lincs SK 9764. *Wadintun(e) -tone* 1086–1327 with variant *-ton(')* from 1169, *Wadingtun(a) -ton(')* 1185–1576, *Waddington* from 1289. Li i.216.

WADEBRIDGE Corn SW 9872. 'Bridge at Wade'. *Wadebrygge* 1478, *Wadebridge* c.1540. P.n. Wade, *Wade* 1358–1478, is 'the ford', OE **wæd**, + ME **brigge**. A bridge was built here – one of the best medieval bridges in England – replacing the earlier ford about 1460, on packs of wool to prevent the sixteen piers from sinking into the quick sand. PNCo 176, Gover n.d. 314.

WADEFORD Somer ST 3110. *Wadford* 1809 OS. Without earlier forms it is impossible to decide between 'Wada's ford', OE pers.n. *Wada* + **ford**, 'woad ford', OE **wād**, or 'ford that can be waded'.

WADENHOE Northants TL 0183. 'Wada's hill-spur'. *Wadenho* 1086–1428, *Wadnoe* 1603×25, 1730. OE pers.n. *Wada*, genitive sing. *Wadan*, + **hōh**. Nth 222 gives pr [wɔdnou], L 168.

WADESMILL Herts TL 3517. 'Wade's mill'. *Wadesmeln* 1294, *-myle* 1397, *Wadgemill* 1674. Surname *Wade* as William Wade 1287 + ME **miln** (OE *myln*), Herts dial. form **meln**. Hrt 206.

WADHURST ESusx TQ 6431. 'Wada's wooded hill'. *Wadehurst* 1253–1455, *Wad(e)herst* 1272–1613, *Woodhurst* 1633. OE pers.n. *Wada* + **hyrst**, SE dial. **herst**. Sx 385, SS 68.

WADSHELF Derby SK 3171. 'Wada's shelf of land'. *Wadescel* 1086, *-s(c)elf* *-s(c)helf(f)* *-chelf* 1199×1216–1290 etc. OE pers.n. *Wada* + **scelf**. A *Wade* is recorded in DB as holding land in Brampton and Wadshelf at the time of Edward the Confessor. Db 222.

WADSWORTH MOOR WYorks SD 9833. *Wadsworth Moor* 1817. P.n.*Wadesuurde* 1086, *-worth(e)* 14th cent., *Waddesw(o)rth(e)* 13th–1536, *Wadsworth* from 1577, *Wodsworthe* 1572, 'Wæddi's enclosure', OE pers.n. **Wæddi*, genitive sing. **Wæddes*, + **worth**, + ModE **moor**. There is no village and the original site is unknown. It may have been at Old Town SD 9928, *(le) Oldtowne* 1533. YW iii.199.

WADWORTH SYorks SK 5797. 'Wad(d)a's enclosure'. *Wadeword(e)* *-u(u)rde* 1086–1208, *-worth* 1218, c.1280, *Wathewurthe* 1166, *Waddewurth'* *-wrth(e)* *-worth(e)* 1200–1456, *Wadword* early 13th, *-worth* from 1257. OE pers.n. *Wad(d)a* + **worth**. YW i.59.

WAGBEACH Shrops SJ 3602 → SNAILBEACH SJ 3702.

WAINFLEET ALL SAINTS Lincs TF 4958. 'All saints' W'. *Weynflet Omnium Sanctorum* 1229, *All Hallows* 1661. P.n. *Wen-* *Wemflet* (for *Wein-*) 1086, *Wein-* c.1115, *Waineflet* c.1165, 'the wagon creek', i.e. one that can be crossed by wagons, OE **wæġn** + **flēot** + ModE affix **all saints, all hallows**, Latin **omnes sancti**, genitive pl. **omnium sanctorum**. Also known as High Wainfleet for distinction from Low Wainfleet or Wainfleet St Mary TF 4958, *Weynfled Beate Marie* 1254, *Wainfleet St Mary* 1824 OS. DEPN, Jnl 29.80, 82.

WAINFLEET BANK Lincs TF 4759. P.n. Wainfleet as in WAINFLEET ALL SAINTS TF 4958 + ModE **bank**. The site of the original church of All Saints.

WAINFLEET SAND Lincs TF 5445. *Wainfleet Sand* 1824 OS. P.n. Wainfleet as in WAINFLEET ALL SAINTS TF 4958 + ModE **sand**. A sand-bank in The Wash.

WAINHOPE Northum NY 6790. 'The wagon valley'. *Waynhoppe* 1279, *-hop* 1325, *Wayneshopp* 1376. OE **wæġn** + **hop**. NbDu 205.

WAINHOUSE CORNER Corn SX 1895. *Wainhouse Corner* 1748, named from a lost wine-house or inn, *Winhouse* 1417, *Wynehous* 1440, ME **wīnhouse**. PNCo 176, Gover n.d. 6.

WAINSCOTT Kent TQ 7471. 'Wagon shed'. *Wainscot* 1819 OS. ModE **wain** + **cot**.

WAINSTALLS WYorks SE 0428. 'The wain or wagon sheds'. *Wayne-* *Wainestalles* 1525–1733. ModE **wain** (OE *wæġn*) + **stall**. YW iii.125.

WAITBY Cumbr NZ 7508. Probably 'wet farmstead'. *Watheby* 12th, 1203×27, *Watebi -by* c.1170–1647, *Wai- Waytby(e)* 1522–1625. ON **vátr** + **bȳ**. We ii.24 gives pr ['we:tbi], SSNNW 42.

WAKEFIELD WYorks SE 3320. Either 'Waca's open land' or 'the open land of the annual wake or festival'. *Wachefeld -felt* 1086, *Wakefeld(a)* 1091×7–1489, *-field* from 1509. OE pers.n. *Waca* or **wacu** + **feld**. In the latter case the reference would be to an annual wake or festival, perhaps the origin of the celebrations during which in later times the well-known Towneley cycle of mystery plays was presented. Wakefield is the traditional capital of the West Riding of Yorkshire, the centre of a very extensive manor, and must have been of considerable importance in pre-Conquest times. YW ii.163, TC 187, L 244.

Great WAKERING Essex TQ 9487. *Magna Wakering(e)* 1251 etc., *Muche Wakeryng* 1431–42, *Great Wakelin* 1695. ModE adj. **great**, ME **much**, Latin **magna**, + p.n. *Wacheringa -elingam* 1086, *Wakeringa -yng(ge) -ing(e)* 1181–1552, *Wakeringes* 1182×98, 1195×1215, 1216×72, usually taken to represent OE folk-n. **Waceringas* or **Wæceringas* 'the people called after Wacor or Wæcer', OE pers.n. *Wacor, Wæcer* + **ingas**. The place, however, is situated on a hill overlooking the Thames estuary and may alternatively be an **ing**[2] formation on the adjective **wacor** 'awake, watchful' or on its definite form **wacra** 'the watchful one' referring to and meaning a 'look-out place' or 'the people of the look-out place', cf. WACKERFIELD Durham NZ 1522. 'Great' for distinction from Little WAKERING TQ 9384. Ess 203, ING 25.

Little WAKERING Essex TQ 9384. ME **little** (1282, 1297 etc.), Latin **parva** (12th) + p.n. Wakering as in Great WAKERING TQ 9487. Ess 203.

WAKERLEY Northants SP 9599. Partly uncertain. *Wacherlei* 1086, *Wakerlea -le(ga) -ley* from 1184. The generic is OE **lēah** 'a wood, a clearing'. The specific presents the same problems as WACKERFIELD Durham NZ 1522. Either OE ***wācor** 'osier' or, since the situation overlooks a crossing of the Welland on the county boundary, perhaps 'clearing of the watchful ones', OE adj. **wacor**, genitive pl. **wacra**, + **lēah**. Nth 172 gives pr [weikəli], PNE ii.234.

WALBERSWICK Suff TM 4974. 'Walbert's trading place'. *Walberdeswike -wyke* [1199]1319, *Walbereswic* 1235, *Walberdeswyk* 1275, 1286. CG pers.n. *Walbert (Waldbert)*, genitive sing. *Walbertes*, + **wīċ**. Walberswick was an important port in early times. DEPN.

WALBERTON WSusx SU 9705. 'Wealdburg's farm or estate'. *Walburgetone* 1086, c.1200, *Wauburguetone* 1162, *Walberton* from 1203. OE feminine pers.n. *Wealdburh*, genitive sing. *Wealdburge*, + **tūn**. Sx 143.

WALBURY HILL Berks SU 3761. P.n. Walbury + ModE **hill**. Walbury is a hill-fort referred to as 'a square camp called Wallborough, or bury, by Aubrey Corn hill' in Gough's edition of Camden's *Britannia* (1789). The absence of early forms makes any attempt at an etymology futile although the generic seems to be OE **burh**, dative sing. **byriġ**. Brk 317, Thomas 52.

WALCOT 'The cottage(s) of the Welshmen'. OE **wala-cot(u)**. Jnl 12.18–20.

(1) ~ Lincs TF 0635. *Walecote* 1086–1281 with variants *-cot(a)*, *Walcot(e)* from [c.1020]c.1150. Perrott 149, Jnl 12.44.

(2) ~ Shrops SO 3485. *Walecot(e)* 1255–1307, *Walcote* 1316. A constituent memeber of the North Lydbury estate. Sa i.298, DEPN.

(3) ~ Shrops SJ 5912. *Walecote* c.1138, 1230, *-cota* 1160. Sa i.298, DEPN, Jnl 12.44, 50.

(4) ~ Warw SP 1258. *Walecote* 1235, 1279, *Walcote* 1445, *Wolcote* 1596. Wa 212, Jnl 12.44, 50.

WALCOTE Leic SP 5683. 'The cottages of the Welshmen'. *Walecote* 1086, *-cot(e)* 1166–1322, *Walcot(e) -Kot(e)* 1288–1610. OE **walh**, genitive pl. **wala**, + **cot**, pl. **cotu**. Lei 458, Jnl.12.44

WALCOTT 'The Welshmen's cottages'. OE **walh**, genitive pl. **wala**, + **cot**, pl **cotu**.

(1) ~ Lincs TF 1356. *Walecote* 1086–1281, *Walecot(')* 1199, 1200. Perrott 342, Jnl 12.44, Cameron 1998.

(2) ~ Norf TG 3632. *Walecota, Wealchota* 1086, *Walecot(e) -kote* 1198–1316, *Walcot(e)* 1225–1535. This was a village of Welshmen or servants in the manor of Happisburgh. Nf ii.128, Jnl. 12.49.

WALDEN NYorks SE 0083. 'Valley of the Welshmen'. *Walden(e)* 1270–1536, *Waledene* 1321, *Wawden* 1574. OE **walh**, genitive pl. **wala** + **denu**. YN 265 gives pr [wɔ:dən], Jnl 12.42, 15.

WALDEN HEAD NYorks SD 9880. P.n. WALDEN SE 0083 + ModE **head** referring to the head of the valley.

WALDEN STUBBS NYorks SE 5516. 'Walding's S'. *Stubbes Walding* 1280, *Walden Stubes* 1587. Pers.n. *Walding* (OG *Waldin* or OE patronymic *Walding* 'son of *Walda*') + p.n. *Eistop, Istop* 1086, *Stubbis -ys* 1175×83–1501, *Stubbes* c.1210–1607, *Stubbs* from 1281, 'the tree-stumps', OE **stubb**, pl. **stubbas**. YW ii.53.

King's WALDEN Herts TL 1623. *Waldon(e) Regis* 1190, 1296, *Waledon Regis* 1294, *Kinges Waldene* 1222. ME **king**, Latin **rex**, genitive sing. **kinges, regis**, + p.n. Walden as in St Paul's WALDEN TL 1922.

Little WALDEN Essex TL 5438. *Parva Walden(e)* 1248. ModE adj. **little**, Latin **parva**, + p.n. Walden as in Saffron WALDEN TL 5438. Ess 537.

Saffron WALDEN Essex TL 5438. 'W where saffron grows'. *Saffornewalden* 1582, *Safron Waldon* 1594, ME **safron** + p.n. *Wealadene* c.1000 K 1354, *Waledanā* 1086, *-den(e)* 1119–1387, *Walden(a)* from 1141, 'the valley of the Welshmen', OE **wealh**, genitive pl. **weala**, + **denu**. Also known as Walden St Mary (1295), saint's name *Mary* from the dedication of the church, *Castel Walden* 1285, *Chepyng Walden* 1328 and Great Walden (*Magna*) 1254, all for distinction from Little WALDEN TL 5441. The cultivation of saffron was introduced into England c.1340. The earliest reference to saffron here is 1545, but there is mention of a *Safrongardyn* in nearby Widdington in 1467. Ess 537, Jnl 2.48.

St Paul's WALDEN Herts TL 1922. *Pawles Walden* 1558×1603. Earlier *Waldene Abbatis* 'abbot's W' c.1275. St Paul's Walden was given to St Alban's abbey in 888 and after the dissolution was granted to St Paul's cathedral, London. Walden, *Waleden(e)* 1158–1285 including *[888]13th S 220, *(on) Wealadene* [11th]12th K 1354, *Waldene(i)* 1086, is the 'valley of the Welshmen', OE **wealh**, genitive pl. **weala**, + **denu**. Hrt 22.

WALDERSEY Cambs TF 4304. 'Waldhere's island'. *Walderse* 1248–1420, *-sey* 1368, *-sea* 1658, *Walters(h)e* 1251–1480. OE pers.n. *Waldhere*, genitive sing. *Waldheres*, + **ēġ**. Ca 269.

WALDERSHARE HOUSE Kent TR 2848. A Queen Anne mansion built 1705–12. P.n. Waldershare + ModE **house**. Waldershare, *Waldmeres scora* [824]?10th S 1266, *Walwalesere* 1086, *Wald(e)wares(c)hare* 1253×4–78, is the 'share, district or boundary of the *Waldware*, the forest-dwellers', OE folk-n. ***wealdware**, genitive pl. ***wealdwara**, + **scearu**. The earliest spelling is probably a corruption of this, although an OE pers.n. *Wealdmǣr* and ***scora** 'a slope' or **scoru** 'a boundary mark' are possible alternatives. Adjacent to Waldershare is SIBERTSWOLD TR 2647, another **weald** name. PNK 585, KPN 163, L 1, 226.

WALDERSLADE Kent TQ 7663. 'Forest valley'. *Waldeslade* 1190–1346, *Walderslade* from 1278. OE **weald** + **slæd**. The *er* forms may point to a specific **Wealdware* as in WALDERSLADE HOUSE Kent TR 2848 and so 'valley of the forest dwellers'. PNK 128, L 123.

WALDERTON WSusx SU 7810. 'The estate called after Wealdhere'. *Walderton(e) -re-* from 1167, *Waldriton* 1291, *-yngton* 1331. OE pers.n. *Wealdhere* + **ing**[4] + **tūn**. Sx 55.

Great WALDINGFIELD Suff TL 9043. *Waldingfeud Magna* 1254.

ModE **great**, Latin **magna**, + p.n. *Wœaldinga fœld, Wealdinga feld* [962×91]11th and 13th S 1494, *Waldingefeldā, Wal(d)ingafella* 1086, 'the open land of the Waldingas, the *Wald* or woodland people', OE folk-n. **Waldingas* < **wald** + **ingas**, genitive pl. *Waldinga*, + **feld**. 'Great' for distinction from Little WALDINGFIELD TL 9254. DEPN.

Little WALDINGFIELD Suff TL 9254. *Waudingefeud Parva* 1254. ModE **little**, Latin **parva**, + p.n. *(altera) Walingafella* 'the other W' 1086 as in Great WALLINGFIELD TL 9043. DEPN.

WALDITCH Dorset SY 4892. 'Ditch with a wall'. *Waldic* 1086, *Waudich* 1236, *Waldich* 1268, *-dysche* 1416. OE **weall** + **dīċ**. Do 150.

River WALDON Devon SS 3610. *Waldon river* 1765. Named from Waldon Farm 3610, *Waldon(e)* 1330, 1492, possibly the 'hill of the Welshmen', OE **weala** + **dūn**, cf. Walland SS 4011, *Waleland* 1249, 'land of the Welshmen'; but 'wall hill' is also possible, OE **weall** + **dūn**. D 16, 153, Jnl 12.43.

WALDRIDGE Durham NZ 2550. Partly uncertain. *Walrigge* [1286]1332, *-rig(e) -ryg(g)e* 1382–1723, *-rege* 1485–1512, *-ridge* 1573, 1647, *Waldridge* from 1647. The forms are too late for certainty: the *d* is a late intrusion so that this could be 'ridge with a wall', OE **wall** + **hrycg**. Forms with *Wale-* would be needed to establish the alternative possibility 'ridge of the Welshmen', OE **wala** + **hrycg**. NbDu 205, L 169, Nomina 12.36.

WALDRINGFIELD Suff TM 2844. 'The open land of the Waldringas, the people called after Waldhere'. *Waldringfeld* [942×c.951]13th S 1526, *Waldringafelda* 1086, *Waldryngfeld* 1227, 1327, *Waudringfeud* 1289, *Waldringfylde* 1568. OE folk-n. **Waldringas* < pers.n. *Waldhere* + **ingas**, genitive pl. **Waldringa*, + **feld**. DEPN, Arnott 16.

WALDRON ESusx TQ 5419. 'Forest house'. *Waldere, Waldrene* 1086, *Waldern(e)* 1242–1610, *Waldron* from 1336, *Waudern(e)* 1230–94. OE **weald** + **ærn**. Sx 405.

WALES SYorks SK 4882. 'The Welshmen'. *Wales, Walis(e)* 1086. OE **walas**, from **walh**. The form *æt Þaleshó* 'at Walh's hill-spur' [1002×4]11th S 1536 may belong here. The name designated a surviving settlement of Welshmen isolated on the boundary between Northumbria and Mercia and probably between the *Pēcsǣtan* and the Welsh enclave of Elmet. YW i.155, Jnl 12.13–4.

WALESBY 'Valr's village or farm'. ON pers.n. *Valr*, secondary genitive sing. *Vales*, + **bȳ**.

(1) ~ Lincs TF 1392. *Walesbi* 1086–c.1210, *-by* from 1223. Li iii.172, SSNEM 76.

(2) ~ Notts SK 6870. *Walesbi* 1086, c.1190 etc. with variant spelling *-by*. Nt 63, SSNEM 76.

WALFORD H&W SO 3972. Partly uncertain. *Waliforde* 1086, *Wal(l)eford* 1292, 1305, *Walford* 1249. The generic is OE **ford** 'a ford' referring to a N–S crossing of the river Teme. The specific might be OE **wælla** 'a spring' or perhaps **weala**, genitive pl. of **wealas** 'the Welshmen'. He 127.

WALFORD H&W SO 5820. 'Welsh ford'. *Wal(ec)ford* 1086, *Wal(e)ford(e)* 1166–1342. OE *Wealh* 'Welsh, Welshman' varying with **weala**, genitive pl. of **wealas** 'the Welshmen', + **ford**. A crossing of the Wye linking English and Welsh territory. He 197, Jnl 12.10, 19, 41, 48.

WALFORD Shrops SJ 4320. 'The ford by the spring'. *Waleford* 1086, 1255×6, *Walleford(e)* 1201–1428, *Walford* from 1203. OE (Mercian) **wælla** + **ford**. Sa i.298.

WALGHERTON Ches SJ 6949. 'Walhere's farm or village'. *Walcretune* 1086, *Walcerton(a)* 1260–1429 with variants *-quer- -ker-*, *Walgherton* from 1295. OE pers.n. *Walhere* + **tūn**. Che iii.73 gives pr ['wɔldʒərtən], older ['wɔːkər-].

WALGRAVE Northants SP 8072. 'Grove belonging to Old'. *Wald-Wold(e)grave* 1086, *Waldegrauia* 1185–1410 with variants *Waude-, Wald-* (from 1241) and *-grave, Walgrava* 1195, *Walgrave* 1576. P.n. OLD SP 7873, *Walda* 1086, + OE **grāfa**. Nth 131, L 194.

WALKDEN GMan SD 7403. 'Walca's valley'. *Walkeden* 1325, 1514, *-dene* 1408. OE pers.n. **Walca* + **denu**. La 41.

WALKER T&W NZ 2964. 'The marsh by the (Roman) wall'. *Waucre* 1242 BF, *Walkyr* 1267, *Walker* from 1296. OE, ME **wall** + ME **ker** (ON *kjarr*). NbDu 205.

WALKER FOLD Lancs SD 6741. No early forms. Probably surname *Walker* + **fold** (OE *fald* 'an enclosure for animals').

WALKERINGHAM Notts SK 7692. 'Homestead called or at *Walkering* or of the Walceringas, the people called after Wealhhere'. *Wachering(e)hā* 1086, *Walcringham(a)* 1212, [1154×89]1316, [1100×35]17th, *Walcringeham* 1205–69, *Walkeringham* from 1242, *Wauk-* 1262. OE p.n. **Wealhhering* < pers.n. *Wealhhere* + **ing**[2], locative–dative sing. **Wealhherinġe*, or folk.n. **Wealhheringas* < *Wealhhere* + **ingas**, genitive pl. **Wealhheringa*, + **hām**. Nt 41 gives pr [wɔːkriŋəm], ING 149.

WALKERITH Lincs SK 7893. 'The landing-place of the fullers'. *Walkerez* 13th, *Walkreth -creth* 1300–32. OE **walcere**, genitive pl. **walcera**, + **hȳth**. The place lies close to WALKERINGHAM Notts SK 7692 but it is difficult to see how the two names can be linked. DEPN, SelPap 76, ING 149, Cameron 1998.

WALKERN Herts TL 2926. 'Fulling mill'. *Walchra* 1086, 1119, *Walcra -e -kre* 1248, 1281, *Walkern(e)* from 1222, *Waukerne* 1279. OE **wealc-ærn**. Hrt 141 gives prs [waːk- wɔːkən].

WALKER'S GREEN H&W SO 5247. *Walkers Green* 1832 OS. Surname *Walker* + ModE **green**. Nearby are Paradise Green SO 5247, *Paradise Green* 1832 OS, Venn's Green SO 5348, *Venns Green* 1832 OS, p.n. Venn SO 5449, *Fenne* 1086, *La Venne* 1381, 'the fen', OE **fenn**, + ModE **green**, and *Layfield Green* 1832 OS. He 39.

WALKERWOOD RESERVOIR GMan SJ 9898. P.n. Walkerwood + ModE **reservoir**. Walkerwood is *Walkerwood* 1842, possibly surname Walker + **wood**, but the evidence is too late for certainty. Che i.320.

River WALKHAM Devon SX 4870, 5576. A back-formation from WALKHAMPTON SX 5369. *Walkam(p) Walkham* [1291]1408. For a similar back-formation (from Okehampton) cf. the form *Okem* 1244 for the river OKEMENT Devon SX 6000. Walkhampton itself seems to be named after this river, whose original OE name was probably **Wealce* 'the rolling one', related to OE *wealcan* 'to roll', *wealca* 'a roller, a wave'. Also recorded as *Stour(e)* 1586–1797, which is either a genuine survival or a late antiquarian invention. Devon 16, 244, RN 430.

WALKHAMPTON Devon SX 5369. Partly uncertain.

I. *Walchentone* 1084, *Wachetone* 1086, *Walchinton* 1187, *Walkyngton -ing-* 1548, 1733.

II. *Walcom(e)- cam(e)ton, Walkamton(e) -cam-* 13th cent., *Walkham(p)ton(e)* from 1270.

Form I may be 'settlement on the river *Wealce*', OE r.n. **Wealce* 'the rolling one', genitive sing. **Wealcan*, + **tūn**; form II 'settlement of the people of *Wealcham*, the homestead on the *Wealce*', OE **Wealchǣme*, genitive pl. **Wealchǣma* + **tūn**. **Wealce* on this interpretation must have been an early name of the river Walkham, *aqua de Walkamp* [1291]1408, *Walkham* from 1684, itself a back-formation from Walkhampton. It was also known as *the Stoure* 1586, probably one of Harrison's antiquarian inventions. D 243 gives prs [wækiŋtən] and [wækətən], 16, RN 430–1.

WALKINGTON Humbs SE 9937. 'Estate called after Walca'. *Walkintun(a) -ton(a) -yn-* 1080×6–1374, *-yng- -ing-* from 1279, *Walchinton(e)* 1086, *Waukinton* 1251. OE pers.n. **Walca* + **ing**[4] + **tūn**. YE 203 gives pr [wɔːkitən].

WALK MILL Lancs SD 8630. 'Fulling mill'. ModE **walk-mill** (ME *walkmiln*, from *walken* 'to beat and press woollen cloth to cause felting of the fibres, with consequent shrinking and thickening').

WALL 'The wall'. OE **wall**.

WALL Northum NY 9169. *Wal* 1165. The reference is to Hadrian's Wall. NbDu 205.

WALL Staffs SK 1006. *Walla* 1167, *La Wal* 1242. This is the site of the Roman town of LETOCETUM substantial ruins of which are recorded by 18th cent. antiquarians. DEPN.

WALL 'A spring'. OE Mercian **wælla**.

(1) ~ UNDER HEYWOOD Shrops SO 5092. *Walle sub Eywode* 1255, *Wall under Heywood* 1833 OS. P.n. Wall as in East WALL SO 5293 + p.n. *Heywode* 1250, 'the wood with an enclosure', OE **(ġe)hæġ** + **wudu**, for distinction from East WALL SO 5293. DEPN.

(2) East ~ Shrops SO 5293. *East Wall* 1833 OS. ModE adj. **east** + p.n. *Walle, Welle* 1200. DEPN.

WALLAND MARSH Kent TQ 9923. *Walland Marsh* 1565, 1583, *Walling Marsh* 1819 OS. P.n. Walland + ModE **marsh**. Walland, *Walland -lond* 1471, 1475, is the 'newly reclaimed land by the wall', ME **wall** + **land**. PNK 477, Cullen.

WALLASEA ISLAND Essex TQ 9693 → River ROACH TQ 9592.

WALLASEY Mers SJ 2992. 'The island of *Waleye*'. *Waleyesegh* 1351, *Waleseye* 1377, *Wallasegh* 1418, *Wallasey* from 1545. P.n. *Waleye*, genitive sing. *Waleyes*, + pleonastic ME **eye** (OE *ēġ*). Waleye, *Walea* 1086, *Waleie* [1096×1101]1150, 1150, *Waley(e)* [1096×1101]1280–1555, *Walley(e)* 1259–1621, is the 'Welshmen's island', OE folk-n. *Walas*, genitive pl. *Wala*, + **ēġ**. In former times *Waleye* was cut off at high tide by Wallasey Pool (now West and East Float) and The Birket or Main Fender (ModE dial. **fender** 'a drainage ditch') which formed an arm of the Mersey estuary. Wallasey is properly the name of the whole island so formed, but was transferred to the village otherwise known as *Kirkeby in Waleya* c.1180–1245, the 'church village in Wallasey'. Che iv.323, 332, SSNNW 34.

WALLBURY Essex TL 4918. 'Wall camp, the camp with walls'. *Wal(le)bury -bery* 1324–96. P.n. *Wallā* 1086, *Walla* 1212, *la Walle (in Hallinggeberi)* 1240–67, 'the wall', OE **wall**, + **byriġ**, dative sing. of **burh**. The reference is to Wallbury Camp Iron Age hill-fort. Cf. Great HALLINGBURY TL 5119. Ess 35, Thomas 1976.123.

WALLER'S HAVEN ESusx TQ 6607. *Wallereshaven* 1455. Family name *Waller* recorded here as early as 1327 + ME **haven** (OE *hæfen*). Sx 484.

WALL HILLS H&W SO 6359. *Wall hills* 1832 OS. ModE **wall** + **hill(s)**. The reference is to *Wall Hill Camp*, an Iron Age hill-fort occupying the top of the steep-sided hill. Thomas 1976.151.

WALLING FEN Humbs SE 8829. 'Fen of the Welshmen or called or at *Walling*, the place called after Wealh or the Welshman'. *Walefen* 1228, *Walingfen* 1228, 1300, *Wallyngfen* 1556. OE **w(e)alh**, genitive pl. **wala**, varying with p.n. **Waling* < *W(e)alh* or **w(e)alh** + **ing**[2], + **fenn**. Three miles to the NW at SE 8433 lies Wholsea, *Walsay* 1285, 1338, *Whalsey* 16th cent., *Wholsea* 1828 which could be the 'Welshman or Welshmen's lake', OE **w(e)alh**, genitive pl. **wala** or sing. **wales**, + **sǣ** or the 'Welshman's island', OE **wales** + **ēġ**. YE 234, 248.

WALLINGFORD Oxon SU 6089. 'The ford of the Wealhingas, the people called after Wealh'. *Welengaford* [c. 895]10th Alfred's *Orosius*, *(æt) Welingaforda* [879×99]12th S 354, *(to) Wælingforda* [c.915]16th, *(æt, to) Wealingaforda* [1003×4]13th S 1488, c.1050 ASC(C) under year 1013, c.1100 ASC(F) under year 1006, *(to) Wealungaforda* c.1100 ibid. under year 1013, *(on) Wallingeforde* [1065×6]13th S 1148, *Walinge- Walenge- Warengeforde, Walengefort* 1086, *Warengefort -ford(ia)* c.1100–late 12th, *Walengeford(ia)* 1152–14th, *Wallingford* from 1208×9. OE folk-n. **Wealhingas* < pers.n. *Wealh* + **ingas**, genitive pl. **Wealhinga*, + **ford**. Brk 535.

WALLINGTON 'Estate of the Welshmen'. OE **wealh**, genitive pl. **weala** + **tūn**.

(1) ~ GLond TQ 2863. *Waletona* 1076×84, *Waleton(e)* 1086–1373, *Wallyngton* 1377, *Wallington al.Waleton* 1713. Sr 39, 55, Jnl 12.45, 49.

(2) ~ Hants SU 5806. *Waletune* 1233, *Waleton* 1307–1453, *Walintone -yn-* 1288–1333, *Walyngton* 1544. Ha 168, Gover 1958.29, DEPN, Jnl 12.24, 45.

WALLINGTON Herts TL 2933. 'Estate called after Wændel'. *Wallington(e)* from 1086, *Walingeton* 1199, *Wandelington(a)* 1154×89, 1280, *Wandelingetuna* c.1180, *Wendling(g)etone* 1199, 1201. OE pers.n. **Wændel* + **ing**[4] + **tūn**. Forms with medial *-e-* may point to an alternative formation, 'settlement of the Wændelingas, the people called after Wændel', OE folk-n. **Wændelingas* < pers.n. **Wændel* + **ingas**, genitive pl. **Wændelinga*, + **tūn**. Hrt 168.

WALLINGTON HALL Norf TF 6208. P.n. Wallington + ModE **hall**, referring to an early 16th cent. mansion house. Wallington, *Wal(l)inghetuna* 1086, *Wallingtone* 1216×72, may be the 'settlement of the Wallingas, the folk who live by the wall' referring to an embankment along the river Ouse, OE folk-n. **Wallingas* < **wall** + **ingas**, genitive pl. **Wallinga*, + **tūn**. DEPN.

WALLISWOOD Surrey TQ 1238. Perhaps 'the wood called after the Waley family'. *Wallis Wood* 1619. Cf. Cristina *Waley* 1332. Sr 262.

Middle WALLOP Hants SU 2937. *Middle Wallop* 1817 OS. ModE **middle** + p.n. Wallop as in Nether WALLOP SU 3036.

Nether WALLOP Hants SU 3036. 'Lower Wallop'. *Wallop inferior* c.1270, *Netherwellop* 1271, *Netherwallop* 1316, *Lower Wallop* 1817 OS. ME adj. **nether**, Latin **inferior**, + p.n. *Wallop(e)* 1086, *Wallop* 1130–1295, *Wal(le)- Wel- Wolhop(e) Well(e)- Wollop* all 13th, a name which cannot satisfactorily be explained from OE elements. Probably it is to be identified with the form *Guolopp(um)* 9th HB from PrW **Wolop* of unknown meaning. 'Nether' because lower down the valley of Wallop Brook than Over Wallop. Ha 168, Gover 1958.192, DEPN, Antiquity 13.105–6, L 112.

Over WALLOP Hants SU 2838. *Overwellop* 1280, *Wallop superior* 1283, *Ouuere Wallop* 1283, *Upwalhope* 1338, *Upper Wallop* 1817 OS. ME adj. **over**, Latin **superior**, + p.n. Wallop as in Nether WALLOP SU 3036. Ha 169, DEPN, Gover 1958.192.

The WALLS Shrops SO 7896. *The Walls* 1833 OS. The reference is to an Iron Age enclosure. Raven 228.

WALLSEND T&W NZ 2966. '(The place at the) end of the (Roman) wall'. *Wallesende* c.1107–c.1170×4 DEC, *Wal(l)eshend* 1154×66–1539. OE **wall**, genitive sing. **walles** + **ende**. NbDu 205.

WALMER Kent TR 3750. 'Lake or sea of the Welshmen'. *Walemere* [1087]13th, 1227–53×4, *Wealemere* c.1100. OE **wealh**, genitive pl. **weala**, + **mere**. PNK 577, KPN 298, Jnl 12.42.

WALMER BRIDGE Lancs SD 4824. 'Bridge near *Waldmire*'. *Waldemurebruge* [before 1251]1268. P.n. **Waldmire* + OE **brycg**. Waldmire, *Walde(s)mure* [before 1251]1268, is either 'Walda's mire' or 'forest mire', OE pers.n. *Walda* + ON **mýrr**, or OE **wald** + **mýrr** 'bog, swamp'. The bridge crosses a small stream draining through marshland into the river Asland. La 137, SSNNW 256.

WALMERSLEY GMan SD 8013. Possibly 'Waldmer or Walhmer's wood or clearing'. *Walmeresley* 1262, *-legh* 1318, *Womersley* 1522, *Wamessley* 1555, *Walmsley* 19th. OE pers.n. **Waldmǣr* or *Walhmǣr*, genitive sing. **Wald- Walhmǣres*, + **lēah**. The incidence of pers.ns. compounded with OE **lēah** is high, but the evidence of the spellings of the specific in this name is ambiguous. It could be a lost p.n. **Wald-mere* 'lake by a wood' and so 'wood, clearing of the forest lake', or **Wald-ġemǣre* 'wood boundary' and so 'wood, clearing of the wood-boundary'. There is no obvious topographical support for these possibilities; but there are mill-pools at Walmersley, and a nearby p.n. *Lumb* 1843 may represent dial. *lum* 'a pool'. L 62, Jnl 17.43.

WALMLEY WMids SP 1393 'The warm wood or clearing'. *Warmelegh* 1232, *-leye* 1332, *Warmley* 1656, 1666, 1834 OS. OE **wearm**, definite form **wearma**, + **lēah**. Wa 50.

WALNEY ISLAND Cumbr SD 1768. P.n. Walney + pleonastic ModE **island**. Walney, *Wagneia(m)* 1127–90, *Wageneia* 1155–94, *Wagneya* 1200, *Waghenay* 1336, *Wannegia* (for *Waun-*) 1246, *Wawenay* 1404, *Walney* 1577, is a compound in ON **ey** or OE **ēġ** 'an island'. The specific is either OE ***wagen** 'quagmire, quaking sands' or ON **vǫgn**, 'a grampus or killer whale', genitive pl. **vagna**. In the first case the reference would be to the sands between the island and the mainland, in the second to the sighting of killer whales off the coast of the island or to the supposed similarity of its long thin shape to that of the whale. In any case OE **wagen-ēġ* would have been re-interpreted by the Vikings as **vagna-ey*. The modern spelling is hypercorrect owing to the pronunciation [wɔ:] for *Wal-* as well as *Waw-* < *Wag-*. La 205, SSNNW 173, PNE ii.232, 239.

WALPOLE Suff TM 3674. 'The pool of the Welshmen'. *Walepole* 1086, *Walepol(e)* 1254–86, *Walpole* from 1291. OE **walh**, genitive pl. **wala**, + **pōl**. DEPN, Baron.

WALPOLE HIGHWAY Norf TF 5113. *Walpole Highway* 1824 OS. P.n. Walpole as in WALPOLE ST ANDREW TF 5017 + ModE **highway**. A drove road from Walpole to the fens, cf. WALTON HIGHWAY TF 4912.

WALPOLE ST ANDREW Norf TF 5017. 'St Andrew's W'. *Walpole St Andrew* 1824 OS. P.n. Walpole + saint's name *Andrew* from the dedication of the church. Walpole, *Walepol* [1042×66]12th S 1051, *Walpola* 1086, *-pol(e)* 1121–1200, is partly uncertain. The S 1051 form points to 'pool of the Welshmen', OE **walh**, genitive pl. **wala**, + **pōl**, but Walpole stands on a line of the Roman sea-wall and is probably, therefore, 'pool by the (sea-)wall', OE **wall** + **pōl**. Cf. WALSOKEN TF 4710. 'St Andrew' for distinction from WALPOLE ST PETER TF 5016. DEPN, Schram 142.

WALPOLE ST PETER Norf TF 5016. 'St Peter's W'. *Walpole St Peter* 1824 OS. P.n. Walpole as in WALPOLE ST ANDREW TF 5017 + saint's name *Peter* from the dedication of the church.

Higher WALREDDON Devon SX 4871. Simply called *H^r Town* 1809 OS. ModE adj. **higher** + p.n. *Walradene* 1329, *-reden* 1421, *-redon* 1809 OS, *Waldron* 1719, possibly 'community, administrative district of the Britons', OE folk-n. *Wēala*, genitive pl. of *Wealh*, + **rǣden**. The site is an isolated corner of ground S of Tavistock and close to the Cornish border. The forms, however, are not conclusive. D 248 gives pr [wɔ:ldərn], Jnl 12.41.

WALSALL WMids SP 0198. 'Walh's nook'. *Ƿaleshale* c.1100 Charters II.xxxvii, *Waleshale* 1163, *-hal* 1201†. OE pers.n. *Walh*, genitive sing. *Wales*, + **halh** in the sense 'small valley, hollow', dative sing. **hale**. DEPN.

WALSALL WOOD WMids SK 0503. *Walsall Wood* 1834 OS. P.n. WALSALL SP 0198 + ModE **wood**.

WALSDEN WYorks SD 9322. 'Welshman's or Walsa's valley'. *Walseden* 1235. OE **walh**, genitive sing. **wales**, or pers.n. **Walsa* + **denu**. YW iii.186, L 98.

WALSGRAVE ON SOWE WMids SP 3881. Kelly's Directory for 1855 states that this is the original name of the village known otherwise as *Sowe* [1043]17th S 1000, *Sowa* 1086, *Sow* 1293–1834 OS including [c.1043]17th S 1226, *Sogh* 1200, 1232, *Sohe* 1203, *Souhe* 1274, which is simply the r.n. SOWE SP 3777. Walsgrave, *Woldegrove* 1411, *Walgrove* 1576, is 'the grove in or near a forest or near *Wold*', ME **wold** (OE *wald*) possibly used as a p.n. + **grove** (OE *grāf(a)*). Wa 188, L 194.

WALSHAM LE WILLOWS Suff TM 0071. 'W in the willows'. *Walsham-le-Willows* 1836 OS. P.n. *Wal(e)sam, Washā* 1086, *Walesham* c.1095, 1203, 'Walh's homestead', OE pers.n. *Walh*, genitive sing. *Wales*, + **hām**, + Fr definite article **le** short for *en le* + ModE **willows**. DEPN.

North WALSHAM Norf TG 2830. *Norðwalsham* *[1044×7]13th S 1055, *Norwalesham* 1169. OE adj. **north** + p.n. *Walsam* 1086, *Walesham* 1203, probably 'Walh's homestead', OE pers.n. *Walh*, genitive sing. *Wales*, + **hām**. Ekwall suggested that the S 1055 spelling *-walsham* might point to 'Wæls's homestead', but this is a forged charter in a late copy. 'North' for distinction from South WALSHAM TG 3613. DEPN.

South WALSHAM Norf TG 3613. *Suðwalsham* *[1044×7]13th S 1055. OE adj. **sūth** + p.n. *Walessam, Walesham, Walsam* 1086, *Walesham* 1190ff, as in North WALSHAM TG 2830. DEPN.

WALSHAW DEAN (RESERVOIR) WYorks SD 9633. *Walchedene* 1309. P.n. Walshaw + ME **dene** 'a valley' (OE *denu*). Walshaw, *Wallesheyes* 1277, *-schaghes* 1323, *Walschagh* 1379, *-shay(e), shawe* 16th, is 'the Welshmen's copse', OE **walh**, genitive pl. **wala**, + **sceaga**. YW iii.202, Jnl 12.15, 43.

Great WALSINGHAM Norf TF 9437. *(in) Walsingeham magna* 1199, *Magna Walsingham* 1200, 1226×8, 1242×3, *Old Walsingham* 1838 OS. ModE adj. **great**, Latin **magna**, + p.n. *Walsingaham* 1023×38 S 1489, 1086, *Galsingham* (sic representing AN [g^w] for [w]) 1086, 'the homestead of the Wælsingas, the people called after Wæls', OE folk-n. **Wælsingas* < pers.n. *Wæls* + **ingas**, genitive pl. **Wælsinga*, + **hām**. Wæls is a heroic pers.n. found in *Beowulf*. 'Great' for distinction from Little WALSINGHAM TF 9336. There is another lost Walsingham in Earl Carlton at TG 1802, *(at) Walsingham* [1042×53]13th S 1535, [1052×66]13th S 1519, *Wa(l)sincham* 1086. DEPN gives pr -[s]-, ING 139.

Little WALSINGHAM Norf TF 9336. *Parva Walsingham* 1226×8, *New Walsingham* 1838 OS. ModE adj. **little**, Latin **parva**, + p.n. Walsingham as in Great WALSINGHAM TF 9437. DEPN, ING 139.

WALSOKEN Norf TF 4710. 'The wall soke'. *Walsocna -e* *[974]13th and 14th S 798, *Wallsocne* *[1060]13th S 1030, *Walsoca* 1086. OE **wall** + **sōcn** 'a district over which a legal right is exercised, an estate'. The reference is to the Roman sea-wall here. DEPN.

WALTER'S ASH Bucks SU 8398. *Walters Ash* 1822 OS.

WALTERSTONE H&W SO 3425. 'Walter's estate'. *Walterestun* 1249, *Walterston* 1292–1355. ME pers.n. *Walter*, genitive sing. *Walteres*, + **toun**. The manor was held post-Conquest by Walter de Lacy. One of several estates in this part of west Herefordshire re-named after Anglo-Norman tenants, Chanstone SO 3635, *Chenestun'* 1243, 'the Cheyney estate' from the family of Lawrence Chanu 1207 (OFr *chanu, chenu* 'white haired') replacing earlier *Elnodestune* 1086, *Elnotheston* 1206,'Ægelnoth's estate', ROWLESTONE SO 3727, 'Rolf's estate' and TURNASTONE SO 3536, 'the Tournai estate', earlier 'Wulfa's estate', Walterstone (lost at ? SO 3636), *Waltereston'* 1224. Most of these places are 11th cent. renamings of earlier settlements under the rule of Gilbert fitz Thorold, the representative of William fitz Osbern who was earl-palatine in Herefordshire and sheriff 1071×86 and whose policy was one of consolidating the Norman hold on this important border area. For other early renamings cf. Mynyddbrydd SO 2841 'speckled mountain', earlier *Rvvenore* 1086 'at the rough ridge', OE **rūh**, dative case definite form **rūwan**, + **ōra**, URISHAY SO 3137, 'Urry's Hay' replacing *Alcamestvne* 1086, 'Alhhelm's estate', Godway SO 3540, *God(e)way(e)* 1232 'good road', earlier *Beltrov* 1086 'well found', OFr **bel trouvé**, and PETERCHURCH SO 3438, earlier *Almvndestvne* 1086, 'Alhmund's estate'. DEPN, He 74, 162–3, 194, 196–7, DB Herefordshire 23.4, Nomina 10.72–4.

WALTHAM 'A forest estate centre'. OE **wald** + **hām**. The name-type seems frequently to have been used of royal administrative centres in forest areas and to have been current during the early Anglo-Saxon period from c.450 to c.550. Wealdhām 198–201, L 226.

(1) ~ Humbs TA 2603. *Waltham* from 1086. Li iv.182.

(2) ~ Kent TR 1048. *Waltham* from 1086×7, *Wealtham* c.1100, *Wautham* 1240, 1254. PNK 548, Cullen.

(3) ~ ABBEY Essex TL 3800. *Waltham Abbey* 1805 OS. P.n.

†The form *æt Ƿaleshó* 'at Walh's hill-spur' [1002×4]11th S 1536 probably belongs to WALES SYorks SK 4882 rather than here, cf. Charters II.xxvi.

Waltham from *[1062]13th S 1036 + ModE **abbey**. Also known as *Holyrode Waltham* 1285, 1495, *Waltham Sce Crucis* 'holy cross W' 1293, from a miraculous cross said to have been entrusted to the church by its founder Tofig in 1030. Ess 27 gives pr [wɔ:ltəm], MA 19.200–1.

(4) ~ CHASE Hants SU 5615. *Waltham forest* 1579, *Waltham Chase* 1810 OS. P.n. Waltham as in Bishop's WALTHAM SU 5517 + ModE **chace**. Earlier called *(la) Hordareswode* 1301, 1350, *Hordereswode* 1307, 'horder's wood', OE **hordere** 'treasurer, keeper of provisions', genitive sing. **horderes**, + **wudu**. The reference may have been to an officer of the bishop of Winchester. Ha 169, Gover 1958.8.

(5) ~ ON THE WOLDS Lincs SK 8025. *Waltham on the Wolds* 1824 OS. Earlier simply *Waltham* from 1086. DEPN.

(6) ~ ST LAWRENCE Berks SU 8276. 'St Lawrence's W'. *Waltamia Sancti Laurencii* 1224×5, *Wautham* ~ ~ 1284, *Waltham Seynt Laurens* 1400. P.n. Waltham + saint's name *Lawrence* from the dedication of the church. Waltham is *(æt) Wuealtham, (to) Wealtham* [940]13th S 461, *(æt) Wealtham* [1007]13th S 915, *parva Waltham, Walt(h)am* 1086, *Wautham* 1212, *Lutelwaltham* 1315. 'Little' (*parva, Lutel-*) for distinction from White WALTHAM SU 8577. Brk 112, L 226.

(7) Bishop's ~ Hants SU 5517. 'Waltham held by the bishop' sc. of Winchester. *Waltham Episcopi* 1237, *Bisshoppeswaltham* 1505. Earlier simply *Waltham* [904]12th S 372, *(æt) wealtham* [963×75]12th S 816, *Waltham* from 1086. The boundary of the men of Waltham is mentioned as *(to) wealthæminga mearce* 'to the *Wealthæmingas*' mark' *[11th] S 376. The estate was granted to the see of Winchester by Edward the Elder in 904. Also known as *Suthwaltham* 1281 for distinction from North WALTHAM SU 5646. Ha 169, Gover 1958.48, MA 19.201.

(8) Cold ~ WSusx TQ 0216. *Cold Waltham* 1340. ME adj. **cold** + p.n. *Waltham* from [957]14th S 1291. 'Cold' referring to its bleak position on a heath (cf. *Waltham on the hethe* 1539) and for distinction from UPWALTHAM SU 9413. The forest alluded to is that of the Weald or *Andredesleage* 'the wood of Anderidos' i.e. Pevensey, known from the 7th cent. as *desertum Ondred* 'the wild area called Andred' or simply as *se micla wudu* 'the great wood', lying between the North and South Downs covering much of west Kent, nearly all of Sussex and parts of Hants. The Walthams were probably settlements projected into the Weald from places not in the Weald. Sx 126, 1, ASE 2.33, SS 86–7, Wealdham 198–201.

(9) Great ~ Essex TL 6913. *Magna Waltham* 1236×8, *Much* ~ 1484. ModE **great**, Latin **magna**, ME **much** + p.n. *Waldham* 1086, 13th, *Waltham* from 1065 including 1086. 'Great' for distinction from Little WALTHAM TL 7012. Ess 270, MA 19.200.

(10) Little ~ Essex TL 7012. *Little Waltam* 1198. ME **litel** + p.n. Waltham as in Great WALTHAM TL 6913. Ess 270, MA 19.200–1.

(11) New ~ Humbs TA 2804. ModE adj. **new** + p.n. WALTHAM Humbs TA 2603. A civil parish created out of WALTHAM TA 2603 in 1961. Li iv.183.

(12) North ~ Hants SU 5646. *Norwaltham* 1208, *Northwautham* 1256, 1280. Earlier simply *wealtham* *[909]12th S 377, *Waltham* 1167. 'North' for distinction from Bishop's WALTHAM SU 5517 also called *Suthwaltham*. Gover 1958.139, Ha 169, MA 19.200, 201.

(13) Up ~ WSusx SU 9413. 'Upper W'. *Uppwaltome* 1641. ModE **up** + p.n. *Waltham* from 1086. 'Up' for distinction from Cold ~ TQ 0216. Both were forest vills in the Weald. Sx 77, 126, Wealdhām 200.

(14) White ~ Berks SU 8577. *Wytewaltham* 1242×3, *Blaunche Wautham,* ~ *Wauttam, Alba Wautham* 1284, *Qwyt Waltham* 1346, *Whyte* ~ 1381, *White* ~ 1557. ME adj. **white**, OFr **blaunche**, Latin **alba**, + p.n. *(on) Waltham, (into) Weltham* [1052×66]13th S 1477, *Waltham* from 1086, *Wautham* 1214. 'White' probably with reference to chalky soil and for distinction from WALTHAM ST LAWRENCE SU 8276. Brk 70, L 226.

WALTHAMSTOW GLond TQ 3788. Possibly 'Wilcume's meeting or holy place'. *Wilcumestowe* 1066×87, *-stouue* 1067, *-stou* 1086, *Wylkomestowe* 1285, *Welcom(e)stow(e) -stowa -stou* 1107–1529, *Walcumstowe* 1398–1507, *Walthamstow(e)* 1446 and from 1536×9. OE femine pers.n. *Wilcume* + **stōw**, but early understood as 'place where strangers are welcome', OE **wilcuma** 'one whose coming is pleasing, a welcome arrival', ME **welcome**. Ess 103 gives pr [wo:təmstou].

WALTON 'Settlement of the Welshmen'. OE **w(e)alh**, genitive pl. **w(e)ala**, + **tūn**. Jnl 12.1–53.

(1) ~ Derby SK 3569. *Waletune* 1086, *-tona -ton(e)* 1183–1323, *Walton(e)* from 1243. Jnl 12.45, Db 321.

(2) ~ Leic SP 5986. *Waltone* 1086, *-ton(e)* 1160×1200–1573, *Waleton'* 1199(2×), 1209×35(2×), *Wauton'* 1231–68. Lei 451, Jnl 12.45.

(3) ~ Mers SJ 3694. *Waletone* 1086, *Waleton* 1094–1252, *Walton(a)* from 1219×20. La 115, Jnl 12.45, 17.64.

(4) ~ NYorks. (Lost) in Welburn SE 6784. *Waleton -tun(e)* 1086. Jnl 12.45.

(5) ~ Suff TM 2935. *Wealtune* [975×1016]11th S 1487, *Waletuna* 1086, *-ton* 1159–1322, *Walton* from 1160. Some doubt about this name must remain. The S 1487 form, if it belongs here, points rather to OE **weall-tūn** '*tūn* with or by the wall'. DEPN, Arnott 42, Baron, Jnl 12.45, 47.

(6) ~ WYorks SE 3517. *Waleton -tun(a)* 1086–1307, *Walton* from 1252. YW ii.112, Jnl 12.46.

(7) ~ WYorks SE 4447. *Walitone* 1086, *Waletone -tune* 1086–1310, *Walton* from 1251. YW iv.246, Jnl 12.46.

(8) ~ -LE-DALE Lancs SD 5527. *Walton in La Dale* 1304, 1332, *Walton in Le Dale* 1318 etc. Earlier simply *Waletune* 1086, *Walatun* [1190–1212]1268, *Waleton* [1213]1268, 1246. *Le Dale*, 'in the valley', OFr definite article **le** short for *en le* + ME **dale**, refers to the situation in the valley of the river Darwen and is for distinction from Higher WALTON SD 5727. The site is associated with a Roman fortlet and was a dependent settlement of the large estate of Leyland. La 68, Jnl 12.45, 47, 17.44, 21.34.

(9) ~ -ON-THAMES Surrey TQ 1066. 'The W on the river Thames', as distinct from WALTON ON THE HILL TQ 2254. *Waleton super Thamse* 1279, 1354, ~ *by Tamys* 1314, *Walton uppon Thames* 1569. P.n. *Waleton* 1086, *Waletona* 1167, *-tun* 1156–1428 with variant *-ton(e), Walleton* 1272, 1324, + r.n. THAMES. Sr 96, Jnl 12.45.

(10) ~ -ON-THE-HILL Staffs SJ 9521. *Walton on the hill* 1812. P.n. *Waletone* 1086, *-ton(a)* c.1166–1236, *Walton* from 1235 + ModE **on the hill**. Also known as *Walton super Canoke* 'on Cannock Chase' 1326. St i.31, Jnl 12.45.

(11) ~ -ON-THE-NAZE Essex TM 2521. *Walton at the Naase* 1545, ~ *-on-the-Nase* 1714. P.n. *Walentonie* 11th, *Waletun(a) -ton(a)* 12th–c.1300, *Woulton* 1686, + ModE prep. **on** + p.n. The NAZE TM 2624. The name does not mean 'wall-enclosure' referring to a failed sea-embankment, and Naze originally referred to the whole manor or soke of that name, cf. *Walton in the Sooke* 1558×1603, *Wolton le Soken* 1805 OS. While the name Walton is ancient, today's Walton-on-the-Naze is a modern development begun about 1825. There is evidence of Roman salt-workings in the area. Ess 354, MM 16.71, Pevsner 1965.410, Jnl 2.45, 12.22.

(12) ~ ON THE WOLDS Leic SK 5919. P.n. Walton, *Waletone* 1086, *-ton(e)* 1195–1254, *Walton(e)* 1209×35–1610, *Wauton'* 1247, + affix *super Waldas* 1354, + *on le Wold(e)* 1415–1506, *on the Wo(u)lds* 1604, 1631, *super Olds* 1678. Lei 331, Jnl 12.45, 49.

(13) ~ -ON-TRENT Derby SK 2118. *Walton super Trente* 1275 etc. Earlier simply *(æt) Waletune* [942]13th, 1086, *-ton* 1243–1316, *Walton* from 1274. Db 667, Jnl 12.45, 50.

(14) Isley ~ Leic SK 4225. Probably short for Isley and Walton. *Isly Walton'* 1327, *Isley Walton* from 1543. Earlier simply *Waletona* 1185, *-ton* 1208, *Walton* 1325–59. 'Isley' does not appear to be a manorial or feudal affix since the manor belonged to the Knights Templar from the twelfth century. It probably represents a lost p.n. 'Isa or Ielf's wood or clearing', OE pers.n. **Isa* or *Ielf*, genitive sing. *Ielfes*, + **lēah**. Lei 366, Jnl 12.23, 45.

(15) Higher ~ Ches SJ 5985. *Superior Walton'* late 13th, *Over Walton* 1307×27, *Upper* ~ 1498, *High* ~ 1774. Adj. **higher**, Latin **superior**, ME **over, upper**, + p.n. *Waletun* 12th, *-ton'* 1191–1331, *Walton* from 1190. 'Higher' for distinction from Lower Walton SJ 6085, *Walton inferior* 1270×90, *Nether Walton* 1295, *Lower Walton* 1642. Che ii.158,157, Jnl 12.45.

(16) Higher ~ Lancs SD 5727. Adj. **higher** + p.n. Walton as in WALTON-LE-DALE SD 5527. Higher Walton is situated higher up the river Darwen from Walton-le-Dale.

WALTON 'Wall settlement'. OE **wall** + **tūn**. PNE ii.244.

(1) ~ Cumbr NY 5264. *Walton* from 1169, *Wauton* 1590×1, *the Waltowne* 1599. The reference is to Hadrian's Wall. Cu 114.

(2) ~ Warw SP 2853. *Waltone* 1086, *-ton(e)* from 1123, *Waleton(a)* 1176, 1200, [1154×89]1478, *Wauton* 1239–85. Wa 287.

(3) ~ CARDIFF Glos SO 9032. 'W held by the Cardiff family'. *Walton(e) Kardif, Kerdef, Kerdif(f)e -yf(f)* 1292–1487, ~ *Cardiff* 1345. Earlier simply *Walton(e)* 1086–1535. There is a small bank of uncertain age 300 yards S of the church. Robert de Cardif occurs 1182×1202, William de Kardif 1249. Gl ii.73.

(4) ~ HIGHWAY Norf TF 4912. *Walton Highway* 1824 OS. P.n. Walton as in West WALTON TF 4713 + ModE **highway**. The reference is to a drove road leading from West Walton to Walton Fen. Cf. WALPOLE HIGHWAY TF 5114.

(5) ~ ON THE HILL Surrey TQ 2254. 'The W on the hill', as distinct from WALTON-ON-THAMES TQ 1066. It lies on the N Downs. *Wauton on the Hull* 1398, *Waulton on the Hill* 1471. P.n. *Waltone* 1086, *Walton(e)* 13th cent., *Waleton(e)* 13th cent., 1428, *Walleton* 1272, *Wauton* 1225–1359, *Waweton* 1288, 1389, + preposition **on** + ME **hull, hill**. Sr 83.

(6) East ~ Norf TF 7416. *Est Waleton* 1252. ME adj. **est** + p.n. *Waltuna* 1086. 'East' for distinction from West WALTON TF 4713. DEPN.

(7) West ~ Norf TF 4713. *Westwaletone* 1254. ME adj. **west** + p.n. *Waltuna* 1086, *-tona* 1081×7. The reference is to the Roman sea-wall here. 'West' for distinction from East WALTON TF 7416. DEPN.

WALTON 'Wood settlement'. OE **w(e)ald** + **tūn**. PNE ii.241.

(1) ~ Bucks SP 8836. *Walton* [1159]14th, *Waldone* c.1218, c.1225, *Walton(e)* c.1225, 1284. Bk 39, Jnl 2.25.

(2) ~ Somer ST 4636. Partly uncertain. *Waltone* 1086, *Walton* from 1196. The generic is OE **tūn** 'village, farm, estate', the specific either **weald** 'forest' or **weall** 'a wall'. DEPN.

(3) ~ IN-GORDANO Avon ST 4273. *Waltone in Gordano* 1333. Earlier simply *Waltona* 1086, *Wauton* 1252, probably 'settlement in high woodland', OE **weald** + **tūn**. For the suffix see CLAPTON-IN-GORDANO ST 4774. DEPN.

WALTON Shrops SJ 5918. 'The spring ' or 'the wall settlement'. *Walton(')* from 1255×6. OE (Mercian) **wælla** or **wall** + **tūn**. Sa i.299.

WALWORTH Durham NZ 2318. 'Enclosure of the Welshmen'. *Walewrth* 1207, 1305×6, *-worth* 1291, 1382, *Walleword' -worth* 1305–1384, *Walworth* from 1313. OE **walh**, genitive pl. **wala**, + **worth**. NbDu 206, Jnl 12.43.

WAMBROOK Somer ST 2908. Partly uncertain. *Awambruth* (sic) [802×39]17th ECW, *Wambrok* 1280, 1306, *-brock* 1610, *Wrambroke* 1291. This is usually explained as '(at the) crooked brook', OE *(æt thæm) wōn brōce* from **wōh**, definite form dative case **wōn**, + **brōc**. ½ mile N of the village the brook bends right then sharply left. If the explanation is correct the form *Wam-* must be due to shortening of *ō* before the consonant cluster *-nbr-* and assimilation of *n* > *m* before *b*. It is difficult to see, however, why *ō* should have become *ă*. Perhaps the specific is rather OE **wamb** 'womb' in some topographical sense or even the rare **hwamm** 'corner, angle' usually found in NCy but also apparently once in the bounds of a charter dealing with the hundred of Taunton Deane which, although spurious probably contains genuine material (S 311). DEPN.

River WAMPOOL Cumbr NY 2553. Partially obscure. *poll Waðœn* [11th]13th, *Wathenpoll* 1190, *-pol(e)* 1291–1540, *Wampoole or Wathenpoole* 1578, *Wathepol(')* 1189, 1201, *Wathelpol* [1189×99]1307, 13th cent. Cf. *Overwampolbrygge* 1501. The first form is in Celtic word-order, OIr **poll** 'pool, stream' or PrCumbr **poll** 'pool, creek', + unidentified element. Subsequent loss of medial *-n-* in the form *Wathenpol* allowed partial re-interpretation as a compound with ON **vathill** 'a ford'. Cu 29, SSNNW 424, Jnl 2.56.

WANBOROUGH Wilts SU 2183. 'Wen barrow(s)'. *(æt) Wenbeorgan, Wænbeorgon* [854]12th S 312 and 313, *Wemberge* 1086, *Wam- Wanberg(a) -(e) -berghe -borge* 1091 etc., *Wanbrow* 1553. OE **wenn, wænn** 'a wen, a tumour' + **beorg**, dative pl. **beorgum**, lOE **beorgon, beorgan**. The reference is to a group of conspicuous Bronze-Age bell- and bowl-barrows on Suger Hill SU 2477. Wlt 283, Thomas 1976.234.

WANDLEBURY Cambs TL 4953. 'Wændel's fort'. *Wendlesbiri* [10th]17th, *Wandlebiria* c.1211, *Wendelbiri* c.1225, *Wandlesbury* 1594, *Wandlebury* 1719, *Vandelbury Camp* 1808. OE pers.n. *Wændel*, genitive sing. *Wænd(e)les*, + **byriġ**, dative sing. of **burh**. Wandlebury is an Iron Age hill-fort, possibly built by the Norfolk Iceni against the expanding Essex Belgæ. Wændel is probably a mythological name applied to a great structure. Ca 88, Bd 114, Thomas 1976.54.

WANDSWORTH GLond TQ 2676. 'Wændel's enclosure'. *(to) Wendles wurðe* [693]11th S 1248, *Wendleswurthe* 1067, *Wendelesorde* 1086, *Wend(e)leswurda -w(u)rth -worth* 1184–1428, *Wand(el)esorde* 1086, *Wand(e)leswurde* 1195–1328, *Wandesworth* 1200, *Wannesworth* 1393, *Wandlesworth vulg. Wansworth* 1675. OE pers.n. **Wændel*, genitive sing. **Wænd(e)les*, + **worth**. Sr 36.

WANGFORD Suff TM 4679. 'The ford by the open fields'. *Wankeforda* 1086, *Wangeford* 1238–1327, *Wanford* 1231, 1291, 1674, *Wangford(e)* from 1327. OE **wang**, genitive pl. **wanga**, + **ford**. Wangford Common remained open common until 1817. DEPN, Baron, L 72.

WANGFORD FEN Suff TL 7484. *Wangford Fen* 1836 OS. P.n. *Waine- Wanne- Wamforda(m)* 1086, *Weinford* c.1095, *Waineford* 1190, *Wainford* 1197–1254, *Waynforth* 1491, 'the wagon ford', OE **wæġn**, genitive pl. **wæġna**, + **ford**, + ModE **fen**. The reference is to a crossing of the Little Ouse. DEPN, L 72.

WANGFORD WARREN Suff TL 7782. *Wangford Warren* 1836 OS. P.n. Wangford as in WANGFORD FEN TL 7484 + ModE **warren**.

WANLIP Leic SK 5910. 'The solitary one'. *Anelepe* 1086, *An(e)lep(e)* 1205–1381, *Onelep(e)* 1316–1541, *Wonlop* 1439, *Wanlep* 1449, *-leape* 1597, *-lip(p) -y-* 1576–1729. OE adjective **ānliepe** used perhaps of an isolated tree. Lei 412.

River WANSBECK Northum NZ 1185. Unexplained. *Wenespic* [1137]14th, *Wenspik(e)* [12th cent.]14th, *Wanspic* [12th]14th, 1271, *-spik(e) -spyk(')* [12th, 13th]14th, 13th cent., *Wansbeke* c.1540, *Wanspek* c.1576, *-becke* 1577. Possible association with OE **spic** 'brushwood' seems unlikely as the latter is a South-Eastern word. The name has been assimilated to ModE **beck** 'a stream'. NbDu 206, RN 432, PNE ii.138.

WANSDYKE Avon ST 6763, Wilts SU 1264. 'Woden's dike'. *wodnes dic* [903]14th S 368, 939 S 449, [961]12th S 694, *wondesdich* [936]14th B 710, *Wodenesdich -dik* 1259–60, *Wannysdiche*

1499, *Wansdiche* 1563, *Wensditch* 1670, *Wansditch* 1819, *Wansdyke* 1817 OS. OE pagan divinity name *Wōden*, genitive sing. *Wod(e)nes*, + **dīċ**. The reference is to the massive frontier earthwork stretching from Morgan's Hill SU 0367 to Savernake Forest SU 2266 thrown up to bar incursions along the Ridgeway which it crosses at SU 1265. The precise historical context of this earthwork is still uncertain, recent suggestions being either the late Romano-Britons of S Wiltshire for protection against the early Saxon settlers of Thames Valley in the late 5th cent., or the Wiltshire Saxons against the Thames Valley Saxons in the troubled times of the late 6th cent. Wlt 17 gives pr [wɔnzdaik], Pevsner 1975.54, Myres 1986.155–6 referring to Arch Jnl 115.1–48, Antiquity 2.89–96, Costen 71, Cunliffe 294.

WANSFORD Cambs TL 0799. 'Ford of the spring'. *Wylmesford* [983×85]12th S 1448a *Walmesford* 1224–1589 including [1185]15th, *Wanesford* 1346. OE **wielm**, Anglian **wælm**, genitive sing. **wielmes**, **wælmes**, + **ford**. The reference is to the stream which flows into the Nene immediately upstream of the site of the ford. Hu 198.

WANSFORD Humbs TA 0656. 'Wændel's ford'. *Wandesford* 1199×1216–1514, *Wande-* 1218–c.1310, *-forth* 1276, 1279×81, *Wandlesford* [1234], *Wonsforth* 1695. OE pers.n. *Wændel*, genitive sing. *Wændles*, + **ford**. YE 95 gives prs [wanzfəd, wanzwəθ].

WANSTEAD GLond TQ 4087. Probably the 'place by the wen or tumour-shaped rise'. *Wænstede* 1042×66, *[1066]13th S 1043, *Wansted(e)* 1196 etc., *Wenestedā* 1086, *Wenstude -sted(e)* 1212–91. OE **wæn** 'lump, tumour, wen' + **stede**. The place is on rising ground above the Roding. Ess 109.

WANSTROW Somer ST 7141. 'Wændel's tree'. *Wandestreow* *[1065]18th S 1042, *Wandestrev* 1086, *-tre* 1182, 1201, *Wandelestr'* 1225, *Wandestre, Walestrena* [13th] Buck, *Wanstrawe* 1610. OE pers.n. *Wændel*, genitive sing. *Wændeles*, + **trēow**. DEPN, L 212–3.

WANSWELL Glos SO 6801. Possibly 'Wægn's spring'. *Weneswell(a)* 1170×90–1339, *Waneswell(e)* early 13th–1573, *Wanneswill -well -wyll* 15th cent., *Wanswell* from 1428. OE pers.n. **Wæġn*, genitive sing. **Wæġnes*, + **wella**. The reference is to Holywell Spring 300 yards N of Wanswell Court. Gl ii.230.

WANTAGE Oxon SU 4088. Transferred from an original stream n. *(æt, to) Waneting* [c.880]11th, [951×5]16th S 1515, [c.954]15th EHD, *Uuanating* [893]11th Asser, *Wáneting* [c.995]12th, *Wanetinz* 1086, *Waneting(a) -e -ynge* 1176–1345, *Wanetench'* 1267×8, *Wanting' -yng* 1255–1380, *Wantynche* 1429, *Wantage* from 1517. The stream-n. occurs as *(on) wanotingc broc* [956]13th S 597, *wanotingc* [956]12th ibid., of uncertain origin and meaning, perhaps related to OE *wanian*, **wanotian* 'to wane, decrease'. Cf. ON *vanta*, ME *want*, referring to an intermittent stream. Brk 491, 17, RN 433.

WAPLEY Avon ST 7179. 'Spring wood or clearing'. *Wapelie -lei(a) -lai -lay(a)* 1086–1221, 1522, *Wappelei(a) -legh' -ley(e)* 1164–1497, *Wap(p)ley* 1519–1705. OE **wapol** 'bubble, spring' possibly used as a stream name + **lēah**. Gl iii.57.

WAPPENBURY Warw SP 3769. 'Wæppa's fortified place'. *Wapeberie* 1086, *Wap(p)enbiri -beri -bure -bury* from c.1200, *Wappyngbury* 1279, 1325. OE pers.n. *Wæppa*, genitive sing. *Wæppan*, + **byriġ**, dative sing. of **burh**. The reference is to the large first cent. BC rectangular entrenchment which encloses the village. Wa 148, Thomas 1980.214.

WAPPENHAM Northants SP 6245. 'Wæppa's homestead'. *Wapeham* 1086–1284, *Wappenham* from 12th, *W(h)apnam* 1542. OE pers.n. **Wæppa*, genitive sing. **Wæppan*, + **hām**. One spelling, *Wappehamm'* 1220, suggests that the generic might be OE **hamm** 'land hemmed in by water or marsh; river meadow; cultivated plot on the edge of woodland or moor'. Nth 62 gives pr [wɔpnəm].

WARBLETON ESusx TQ 6018. 'Wærburg's estate'. *Warborgetone* 1086, *Warberton(e)* 1166, 1262, *Warbelton -le-* 1273–1404, *Warblinton -yng-* 1242–1309. OE feminine pers.n. *Wǣrburh*, genitive sing. *Wǣrburge*, + **tūn**. The *ing* forms are secondary in this name. Sx 468, SS 76.

WARBOROUGH Oxon SU 5993. 'Watch hill'. *Wardeberg* 1200, *Warberge -burgh -burwe -borowe* 1200–1304. OE **weard**, ***wearda** or ***wearde** + **beorg**. Cf. WARDINGTON SP 4946. O 138.

WARBOYS Cambs TL 3080. 'Wearda's bush'. *Wærdebusc* [1077]17th, *Wardebusc* 1086–1253, *-busche* [1123×30]14th, 1167, *-boys -bois* 1148×50–1378. OE pers.n. *Wearda* + ***busc**. Hu 226, xli, PNE i.64.

WARBSTOW Corn SX 2090. 'Holy place of St Wærburg'. *Capella Sancte Werburge* c.1180, *Warberstowe* 1309, *Warbestow(e)* 1342, 1377, *Warpstow* 1553, *Warpestow* 1610. OE saint's name *Wærburg* (Latin *Werburga*) + **stōw**. She was an Anglo-Saxon princess, daughter of King Wulfhere of Mercia, buried in Chester and an object of pilgrimage in the Middle Ages. PNCo 176, Gover n.d. 89, DEPN.

WARBURTON GMan SJ 7089. 'Wærburg's farm or village'. *War(e)burgetune -tone* 1086, *Werberton(a)* 1190×1211, *Warburton(a)* from late 12th with variants *-ber- -byr-*, *Werburgtuna* c.1311. OE feminine pers.n. *Wǣrburg*, genitive sing. *Wǣrburge*, + **tūn**. The dedication of the church is derived from the p.n. rather than vice versa. It has been argued that Warburton is the *æt Weard byrig* (dative case of *Weard-burh* the 'look-out fort') of ASC(C) 915 fortified with Runcorn by Æthelflæd, lady of the Mercians, but the forms do not support this suggestion. Che ii.34 gives pr ['wɔːrbərtən].

WARCOP Cumbr NY 7415. Either 'hill with a cairn' or 'look-out hill'. *Wardecop(p)* 12th–15th, *Warthecop(p) -coppe* 1199–1392, *Warthcop(p)* 1301–80, *Warcop -copp(e)* from 1370. Either ON **vartha** 'cairn, heap of stones' or **vǫrthr** 'watch, look-out', genitive sing. **varthar**, + OE **copp** 'a hill, a summit'. Either explanation is suitable. The reference is to one of the hills near the village which would offer extensive views up and down the valley of the river Eden, a main thoroughfare. We ii.82 gives pr ['waːkəp], SSNNW 256, L 137.

WARCOP FELL Cumbr NY 7820. *Warcupp Fell* 1633, *Warcop Fell* 1777. P.n. WARCOP NY 7415 + ModE **fell** (ON *fjall*).

WARDEN 'Watch hill'. OE **weard** + **dūn**. L 145–6, 153, 158.

(1) ~ Kent TR 0271. *Wardoñ -n(')* 1207–92, *Wardeñ* 1215. PNK 274, L 145–6, 153, 158.

(2) ~ POINT Kent TR 0272. *Warden Pt.* 1819 OS. P.n. WARDEN TR 0271 + ModE **point**.

(3) Chipping ~ Northants SP 4948. 'W with the market'. *Chepyng Wardoun* 1389, *Chepingwarden* 1483, OE **ċēping** + p.n. *Waredon(e)* 1086, 1205, *Wardon* from 1163. Also known as *Westwardon* for distinction from Old WARDEN TL 1348. Chipping Warden stands on high ground overlooking the Cherwell and a tributary of the Cherwell about a mile SW of the hill now called Warden Hill SP 5150 which overlooks Welsh Road. Nth 36, L 153.

(4) Old ~ Beds TL 1343. *Old Wardon* 1495. ME **old** + p.n. *Wardone* from 1086, *Warden* 1359. 'Old' for distinction from Warden Street TL 1244, *Wardon in le Strete* 1549. Bd 97–8.

WARD GREEN Suff TM 0463. *Ward Green* 1837 OS.

WARDINGTON Oxon SP 4946. 'The estate called after Wearda'. *Wardinton* c.1180–1370 with variants *-tona -tone* and *-yn-*, *Wardington* from 1268. OE pers.n. **Wearda* + **ing**[4] + **tūn**. Alternatively the specific may have been an OE ***wearde**, ***wearda** related to ON **vartha -i** 'a beacon, a cairn' and so the 'estate called or at *Wearding*, the beacon or watch place', OE **wearding* < ***wearde** + **ing**[2]. Cf. WARBOROUGH SP 5993. O 427, DEPN.

Upper WARDINGTON Oxon SP 4945. ModE **upper** + p.n. WARDINGTON SP 4946.

WARDLE 'Watch hill'. OE **weard** + **hyll**.

(1) ~ Ches SJ 6157. *Warhelle* 1086, *Wardle* from 1184, *Ward(h)ul(l)* early 13th–1681, *Word(e)hull(e)* 1272–1602. Che iii.322.

(2) ~ GMan SD 9116. *Wardhul* before 1193, *-hull* 13th, 1329, *-hil* 1190×8. The reference is to Brown Wardle Hill SD 8918, *Brown Wardle* 1580, which rises to 1314ft. and commands views of the road between Rochdale and Bacup. Modern Wardle is properly *Little Wardle* 1786 and there were formerly other settlements at SD 9017, *Brown Wardle, Higher Wardle* 1843 OS. La 57.

WARDLEY Leic SK 8300. Partly uncertain. *Werlea* 1067, *-leia(m)* 1080×7, 1066×87, *Warleie* c.1125–1311 with variants *-lei(a) -le(a) -lee -leg' -leye* etc., *Wardeley(e)* 1263–1535, *Wardleyh* 1235, *-ley* from 1535. The form *Wardeley(e)* points to OE *wearda-lēah* 'wood or clearing of the watchmen', OE **weard**, genitive pl. **wearda**, + **lēah**, which would be appropriate to the elevated site on the road from Peterborough to Leicester and overlooking the valley of Eye Brook or might be associated with the keeping of the Forest of Rutland or Leighfield. This appears, however, to be a reformation of an earlier *Wær-lēah* with either **wær** 'a weir' referring to a device in Eye Brook, or **wǣr** 'an agreement, compact, treaty'. R 126.

WARDLOW Derby SK 1874. 'Look-out hill'. *Wardelawe* 1258, *-low(e)* 1292–1516, *Wardlow(e)* from 1275. OE **weard** + **hlāw**. The reference is to Wardlow Hay Cop, *Wardeley hey* c.1550, *Wardlowe hey* 1617, OE **(ġe)hæġ** 'a fence, an enclosure', and *Wardlow Copp* c.1620, ModE **cop** (OE *copp*)'a summit, a peak', which rises to 1214ft. at SK 1773. Db 175.

WARDLOW Staffs SK 0947 → WEAVER HILLS SK 0946.

WAR DOWN Hants SU 7219. *Var Down* 1810 OS. The evidence is too late for explanation.

WARD'S STONE Lancs SD 5858. Surname *Ward* + ModE **stone**. The stone marks a summit of 1839ft.

WARDY HILL Cambs TL 4782. *Wardie Hill* 1589, *Waudy Hill* 1821, *Wardow Hill* c.1825. P.n. Wardy + ModE **hill**. Wardy, *Wardey(e)* 1251–1423×34, is 'look-out island', OE **weard** + **ēġ**. Ca 230.

WARE Herts TL 3614. 'The weirs'. *Wara(s)* 1086, *Wares* 1190–1240, *Ware* from 1200. OE **wær**, a by-form of **wer**. Cf. Fuller 1662, '*Weare* is the propper name of that *Town* (so *called* anciently from the *Stoppages*, which there about obstruct the River)'. Hrt 206.

WAREHAM Dorset SY 9287. 'Hemmed-in land at the weir or fish-trap'. *(æt, into, from, fram, to, on) Werham* late 9th–1405 including ASC(A) under years 784, 876, 877, ASC(C) under year 982, ASDC(E) under year 877, [893] early 11th Asser, [914×9]16th BH and coins 979–1066, *(æt, to) Wærham* c.930, 12th ASC(E) under years 784, 876, 979, 980, 1113, *(on, æt) Werhamme* 11th ASC(D) under years 979, 980, *(at, æt) Warham* 1086–1431 including early 12th ASC(F) under years 979, 980, *Wareham* from 1476. OE **wer, wær** probably + **hamm** 'hemmed-in land, land surrounded by water' later replaced by **hām**. Do i.152 gives prs ['wɛːəRəm, 'wɛːrəm] and ['wɔrəm], SN 6.120, BH 108.

WAREHAM FOREST Dorset SY 8792. P.n. WAREHAM SY 9287 + ModE **forest**.

WAREHORNE Kent TQ 9832. 'Guard or look-out promontory'. *Werahorna, (æt) Worahornan* [845]11th S 282, *(æt) Werhornan* [1032]13th S 1465, *Werhornas* 1042×66 S 1047, *Werahorne* 1086, *Werehorn(e)* c.1100–1210×2. OE ***weru** + ***horna**. An OE ***weru** would be related to the verb *werian* 'guard, keep, defend'. Warehorn occupies a horn of land jutting into Romney Marsh. The compound seems to be established by other occurrences at Warehorn TR 2467 in St Nicholas at Wade and Warehorne Farm TR 2860 in Ash, both overlooking the once tidal Wantsum, and Warehorne TR 2458 in Preston in a commanding spot overlooking major rivers and roads. PNK 474, KPN 166, Cullen.

WAREN BURN Northum NU 1630. 'The alder stream'. *Pharned* [c.1040]12th, *Warned* 12th, *Warnet* [1157]14th, 1212, 1370, *Warne* 1550, 1577, 1580. PrW ***werned** (Brit **u̯erno-* + suffix *-eto-*), identical with the common French p.n. Vernet, Latin *Vernetum.* Cf. W *gwern* 'alder, thicket, marsh'. OE **Weorned* by breaking of *ĕ* before *rn* became *Wearned* in ONb whence ME spellings *Warned* by contrast with Werneth GMan SD 9104, *Wornyth* c.1200, *Vernet* c.1224, *Wyrnith* 1323, and Werneth GMan SJ 9592, *Warnet* 1086, *Wernyt(h) -th'* 1285–1819, *Werneth* 1415 where OE *eor* > ME *er* and only later *ar*. RN 435, PNE ii.230, Jnl.1.52, La 51, Che i.302.

WARENFORD Northum NU 1328. *Warneford'* 1256, *Warinford, Warneford'* 1296 SR. R.n. Waren as in WAREN BURN NU 1630 + **ford**. RN 435.

WAREN MILL Northum NU 1434. *molend de Warnet* [1157]14th, *molendinum de Warnet* 1212 BF. R.n. Waren as in WAREN BURN NU 1630 + ModE **mill**, Latin **molendinum**. This was the last water-mill in the region to be involved in the grain trade, closing c.1980. RN 435, Pevsner-Grundy etc. 1992.609.

WARENTON Northum NU 1030. 'The homestead on Waren Burn'. *Warnetham* 1208–1242 BF, *Warent- Warende- Warondham* 1256, *Warneham* 1269, *Waryndham* 1296 SR, *Warnd(h)am* 14th cent. R.n. Waren as in WAREN BURN NU 1630 + OE **hām**. NbDu 207.

WARESIDE Herts TL 3915. No early forms. Possibly 'beside WARE' TL 3614.

WARESLEY Cambs TL 2454. 'Weder's wood or clearing'. *Wedreslei(e), Wederesle* 1086, *Weresle(a) -lai* 1169–1535, *Waresle(g)* 1199–1316. OE pers.n. *Weder*, genitive sing. *Wederes*, + **lēah**. Doubt has been cast on the DB forms and on the early loss of *-d-* (or *-th-* if that is the original consonant here). But the name *Weder* is on record. Otherwise 'Wær's wood or clearing', OE pers.n. **Wǣr*, genitive sing. **Wǣres*. Ca 273 gives pr [weizli].

WARFIELD Berks SU 8872. 'Open land by the river dam'. *Warwelt* 1086, *Warefeld(')* 1171–1347, *Werre-* 1186, 1228, *Warfeld* 1263. OE **waru** 'shelter, defence, guard' used in a transferred sense of a river-dam + **feld**. The persistent *-a-* spellings speak for OE **waru** rather than **wer, wær**. The existence of OE **wær** as a side-form of **wer** calls for renewed investigation since we are concerned here with etymological *e* from the IE base **u̯er-*. The forms of 1186 and 1228 show confusion with ME **werre**. Brk 115, PNE ii.255.

WARGRAVE Berks SU 7878. 'Grove by the weir'. *Weregrauæ* [1042×65]12th S 1062, *Weregrave -graua* 1086–1316, *Werregraua* 1167–1241, *Wergraua -grave* 1157–1402, *Waregrava* [1144×84]12th, *Wargraue* 1517. OE **wer**, dative sing. **were**, + **grāf(a)**. DEPN, Brk 119, L 320 wrongly assume **wer** to be plural here.

WARHAM Norf TF 9441. 'The weir homestead'. *Warham* from 1086 with variant *Guarham* 1086 (with AN [gʷ] for [w]). OE **wær** + **hām**. DEPN.

WARING'S GREEN WMids SP 1274 → CHESWICK GREEN SP 1376.

WARK 'A fortification'. OE **(ġe)weorc**. PNE ii.254.

(1) ~ Northum NT 8238. *Werch* 1157, *Werk(e)* 13th cent. An important border castle was raised here in the early 12th cent. by Walter Espec. NbDu 207, HN XI 44.

(2) ~ Northum NY 8677. *Werke* 1279, *Wark* 1294. Wark was the capital of the regality of Tynedale and formerly possessed a motte and bailey castle destroyed before 1538. NbDu 207, HN XV 282.

(3) ~ FOREST Northum NY 7377. P.n. WARK NY 8677 + ModE **forest**. A modern Forestry Commission forest.

WARKLEIGH Devon SS 6422. 'Spider wood or clearing'. *Wauerkelegh'* 1242, *Warkeleye* 1277, *-le(gh)* 1326, 1381, *Wortl(e)y* 1675, 1737. OE **wǣferce** + **lēah**. D 349.

WARKSBURN Northum NY 8176. 'Wark's stream'. *Werkesburn* 1293, *Warksburn* before 1769 Map. P.n. WARK NY 8677, genitive sing. *Warks* (ME *Werkes*) + ME **burne**. NbDu 207.

WARKTON Northants SP 8979. 'The estate called after Weorca or Weorce'. *Wurcingtun* [946]copy S 520, *Werchintone* 1086, *-kin-* 1176, *-kene-* 1228, *Wercheton -ke* 1166–1386, *Warton* 1449. OE pers.n. *Weorca* masculine or *Weorce* feminine, + **ing**[4] + **tūn**. Nth 188 gives pr [wo:ktən].

WARKWORTH Northum NU 2406. 'Werca's enclosure'. *Werceworthe* [c.1040] 12th cent., *-worde* c.1107, *Werkewurt, Werkworth* 1139×52, *Wercworth* 1212 BF, *Warkeworth* 1428. OE pers.n. *We(o)rca* or *We(o)rce* feminine + **worth**. Bede mentions a Verca who was abbess of Tynemouth in the 7th cent. NbDu 207.

WARLABY NYorks SE 3591. 'Wærlaf or the warlock's farm or village'. *Warlaues- Werleges- Wergelesbi* 1086, *Warlauby* 1283, 1328, *Warlow(e)by* 14th cent., *Worleybye* 1550. OE pers.n. *Wǣrlāf*, genitive sing. *Wǣrlāfes*, or **wǣrloga**, + **bȳ**. YN 276, SSNY 40.

WARLAND WYorks SD 9420. 'Tenant land, taxable land'. *Warland* 1783. ME **warland**. YW iii.186.

WARLEGGAN Corn SX 1569. Uncertain. *Wrlegan* c.1250, *Worlegan* c.1260–1378, *Warlaygan* 1377, *Warlegen* 1380, *Warlegan* 1610. Possibly a compound in Co ***gor-** 'over' meaning 'watch-place', related to the early W verb *gorllwg* 'watches over, guards'. If so the reference might be to Carburrow Tor SX 1570 at a height of 1240ft. PNCo 177, Gover n.d. 304.

Great WARLEY Essex TQ 5890. *Great Warley alias Walette* 1536, *Great Warley alias Warley Waylett alias Abasse Warley* 1604. ModE **great** + p.n. *Werle* [1035×44]13th S 1521, *War(e)leia* 1086, *Wearlea* c.1090, *Warle(e) -le(y)gh -ley* from 1212, 'weir wood or clearing', OE **wer, wær** + **lēah**. *Waylett* is ME **weylate** (OE *weġ-ġelǣtu*) 'crossroads' as in DUNTON WAYLETTS TQ 6590, and *Abasse* refers to possession by the abbess of Barking. Also known as *West Warley* 1356. 'Great' for distinction from Little WARLEY TQ 6090. Ess 133 gives pr [wɔli], DEPN, Jnl 2.42.

Little WARLEY Essex TQ 6090. *Little Warley* 1548. ModE adj. **little** + p.n. Warley as in Great WARLEY YQ 5890. Also known as *Est* Warley 1276 and Warley *Moeles* 1276, *Septem Molarum* 1301, Warley *Setmel* 1251, ~ *set miles rectius Semeles* 1216×72, all from the family name of William *Setmell, Setmeles, de Septem Mol(endin)is* 'seven mills' c.1212 from Sept Meules S of Eu in NE Normandy. Ess 133.

WARLEY MOOR (RESERVOIR) WYorks SE 0331. *Warley Moor* 1843. P.n. Warley SE 0525, *Werlafeslei* 1086, *Werloweley* 1274–1309, *Werlo- Werlu- Werleley* 14th cent., *Warley* from 1416, 'Werlaf's wood or clearing', OE pers.n. *Wērlāf*, genitive sing. *Wērlāfes*, + **lēah**, + ModE **moor**. Some spellings seem to point to popular association with OE **wǣrloga**, ME **werlaw** 'oath-breaker, traitor' and one form, *Warwolfley* 1323, to association with OE **werewulf**, ME **warwolf** 'a werewolf'. Cf. WARLABY SE 3591. YW iii.130, 122.

WARLINGHAM Surrey TQ 3458. 'The homestead of the Wærlingas, the people called after Wærla'. *Warlyngham* 1144, *-ing-* 1198–1272, *-ing(g)e, Werlingham* 13th cent., *Wor-* 1291, 1318. OE folk-n. **Wǣrlingas* < pers.n. *Wǣrla*, a diminutive of names such as *Wǣrlāf*, + **ingas**, genitive pl. **Wǣrlinga*, **hām**. Sr 339, ING 122–3, ASE 2.32.

WARMFIELD WYorks SE 3720. Partly uncertain. *Warnesfeld* 1086, *Warnefeld(e)* 1119–1486, *Warmefeld(e)* 12th–1562, *Warmfeld* 1532, *-field* 1641. Either 'open land frequented by stallions', OE ***wræna**, ***wǣrna** + **feld**, or 'open land frequented by wrens', OE **wrenna, wrænna, wærna** + **feld**. YW ii.117, L 244.

WARMINGHAM Ches SJ 7161. 'Homestead, village called or at *Wœrming*'. *Warmincham* 1259–1879, *Wermingham -yng-* 1260–1336, *Warmengeham* 1306, *Warmyngcham* 1492, *Warmingeham* 1574. P.n. **Wœrming*, locative–dative sing. **Wœrmingė* < OE pers.n. **Wœrma* or *Wǣrm(und)* + **ing**[2] 'place associated with or called after *Wœrma* or *Wǣrmund*, + **hām**. Che ii.262 gives pr [-iŋəm], older local [-intʃəm].

WARMINGTON Northants TL 0791. 'The estate called after Wyrma'. *Wyr- Wermingtun* [c.980]c.1200 S 1448a, *War- Wermintone* 1086, *Wermin(g)tone -yng-* 12th–1428. OE pers.n. **Wyrma* + **ing**[4] + **tūn**. Nth 215.

WARMINGTON Warw SP 4147. 'The estate called after Wærma'. *Warmintone* 1086, *-ton(e)* 1138–1327, *Warmington* from 1416. OE pers.n. **Wǣrma* + **ing**[4] + **tūn**. Wa 274.

WARMINSTER Wilts ST 8644. 'The minster on the river Were'. *Worgemynster* 899×924 S 1445, *Wori(men)* 979×1016 Coins, *Wori(m), Wor* 1016×40 Coins, *Worem'* 1253, *Gverminstre* 1086, *Werminister -menistra -e -munstre* 1115–1207, *Wereministre -mynstre* 1253–1343, *-mestre* 1357, *Warmenistre -ministre* 1155, 1158, *Warmestre* 1448, *Warmister* 1558×1603. R.n. WERE + OE **mynster**. Wlt 157.

WARMSWORTH SYorks SE 5400. 'Wermi's enclosure'. *We(r)mesford(e)* 1086, *Wermedeswithe* (sic) c.1200, *Wermesw(o)rth(e)* c.1110–1412, *Wermundesworth* 1267, *Warmesworth(e)* c.1195–1577. OE pers.n. *Wermi*, a characteristically Mercian name, derived from OE *Wēr- Wǣrmund*, genitive sing. *Wermes, Wǣrmundes*, + **worth**. YW i.62.

WARMWELL Dorset SY 7585. 'The warm spring'. *Warm(e)welle, Warmemoille* (sic) 1086, *Werm(e)well(e)* 1152–1353, 1518, *-wull(e)* c.1165, 1242×3, *Warm(e)well(e)* from 1166, *Wermell* 1244, 1288, *Warmell* 1440–1682. OE **wearm**, definite form **wearme**, + **wylle**. Do i.170 gives pr ['wɔ:məl].

WARNATBY Leic SK 7123. 'Wærcnoth's village or farm'. *Worcnodebie* 1086, *Wargnodebi* 1102×6, *Warcnodbi* c.1200, *Warc- Warknothebi* early 13th, *Warkeneth(e)by* 1285, 1355, *Warnot(e)by* 14th cent., *Wark(e)naby -eby* 1262–1523, *Warnaby(e)* 16th cent. OE pers.n. **Wrœc-* **Wœrcnōth*, Scandinavianised genitive sing. **Wœrcnōtha*, + **bȳ**. Lei 173, SSNEM 76.

North WARNBOROUGH Hants SU 7351. *Northwargheburne* 1276, *Northwarneburgh* 1379. ME adj. **north** + p.n. Warnborough as in South WARNBOROUGH SU 7247. Ha 170, Gover 1958.116.

South WARNBOROUGH Hants SU 7247. *Suthwargheborgh alias Suthwargheburn* 1291, *Suthwarneburgh* 1431. ME adj. **suth** + p.n. *(œt) weargeburnan* [973×4]12th S 820, *Warg(h)eburne* 1207, 1219, *Warweburne* 1236, *Warneburne* 1183–1235, the 'felons' stream', OE **wearg**, genitive pl. **wearga**, + **burna**. The boundary of the men of (? South) Warnborough is *(on) weargeburninga gemœre* [1046]12th S 1013. The reference is believed to be to the ancient custom of executing felons by drowning them, hands tied beneath their knees, in a stream. There is no stream at South Warnborough to which the charter forms are held to refer. The reference must therefore be to North Warnborough on the Whitewater River. Dial. *werg* 'willow' suggested in Ha 170 is not in question here: that seems to be a derivative of OE **withiġas** 'willows' to judge from the forms for The Wergs near Wolverhampton, *Withegas* 1202, *Wyrges* 14th, *Wythegus* 1418. DEPN, RN 473–4, Gover 1958.135, Brk 917.

WARNDON H&W SO 8857. 'Wærma's hill'. *Wermedvn* 1086, *Warmendone* [c.1086]1190, 1240, *Warmedon(e)* 1208–1374, *Warndon* from 16th. OE pers.n. **Wǣrma*, genitive sing. **Wǣrman*, + **dūn**. Wo 175.

WARNFORD Hants SU 6223. 'Wærna's ford'. *(œt) wernœforda* [12th] S 1476, *Warneford* 1086–1341. OE pers.n. **Wǣrna* + **ford**. Alternatively the specific might be either OE ***wrǣna**,

****wǣrna*** 'a stallion' or **wrenna**, **wrænna** or **wærna** 'a wren'. Ha 171, L 71, Gover 1958.51, Studies 1936. 67–8, PNE ii.277–8.

WARNHAM WSusx TQ 1523. 'The stallion paddock'. *Werneham* 1166, 1316, *Wernham* 1256–1327, *Warnham* from 1329. OE ***wǣrna** + **hamm** 5b. Cf. nearby HORSHAM TQ 1730. Sx 238, ASE 2.46–7, SS 82.

WARNINGLID WSusx TQ 2526. Partly uncertain. *Warthynglithe* 13th, *-lythe* 1373, *Warlingelide -lithe* 1260×72, *Wardinge-Wardlinglithe* 1279, *Warnynglyth* 1456, *Warninglead* 1629. Possibly 'Wearda's hollow slope', OE pers.n. *Wearda* + **ing**[4] + **hlith**. There is a narrow tree-filled cleft in the hill (Gelling). Sx 278.

WARREN Ches SJ 8970. *The Warren* 1842. ModE **warren** (ME, OFr *warene*). Che i.69.

WARREN ROW Berks SU 8180. *Warren Row* 18th cent. ModE (rabbit) **warren** + **row**. Brk 64.

WARREN STREET Kent TQ 9253. 'Warren hamlet'. *Warren Street* 1819 OS. ModE **warren** + **street**. Nearby are *West Street* TQ 9054, and Payden Street TQ 9254, *Peyton Street* 1819 OS.

WARRINGTON Bucks SP 8953. Possibly 'settlement called or at Wearding, the look-out place' or 'estate called after Wearda or the look-out'. *Wardintone* c.1175, *Wardington -yng-* 1343–1403, 1604, *Waryngton* 1474, 1545. Either OE p.n. **Wearding* < **weard** + **ing**[2] or pers.n. *Wearda* or **weard** + **ing**[4], + **tūn**. Warrington lies at the foot of higher ground which rises to 354ft. on the Bucks/Northants boundary where there may have been a look-out place. Bk 15.

WARRINGTON Ches SJ 6088. 'Settlement by a river-dam'. *Walintune* 1086, *Werineton* 1228, *Werington* 1246–1332 etc., with variant *-yng-*, *Warryngton* 1332. OE ***wering** (< OE *wer* + *ing*[1]) + **tūn**. La 96, Mills 1976.146, PNE ii.255.

WARSASH Hants SU 4906. 'Wær's' or 'fish-weir ash-tree'. *Weresasse* 1272, *-as(c)he* 1272×1307, 1308, cf. *Warishassefeld* 1537. Either OE pers.n. *Wǣr*, genitive sing. *Wǣres*, or OE **wer**, genitive sing. **weres**, + **æsc**. Ha 171, Gover 1958.31.

WARSLOW Staffs SK 0858. Probably 'the look-out tumulus'. *Wereslei* 1086, *Werselaw(e)* 1198–9, *Werselow(e)* 1301–1564, *Warcelowe* 1477, *Wars(e)lowe* 1566–1607, *Waslowe* 1666. Probably a reduction of OE **weardsetl** 'a look-out place, a watch-tower' which would suit the topography, + **hlāw** 'a tumulus'. There is an ancient barrow here. Oakden.

WARSOP Notts SK 5667. 'Wær's or fish-weir valley'. *Wares(h)ope* 1086, *Warshop(e)* c.1150–1321×4, *Warsope* 1086 etc., *Worsop* 1569. OE pers.n. *Wǣr* or **wær**, genitive sing. *Wǣres*, **wæres**, + **hop**. Also known as *Warsopp Markett towne* 1653 and *Markett Warsop* 1716 for distinction from Church WARSOP SK 5668. The reference is to the well-marked valley between the settlement here and Church WARSOP. Nt 101 gives pr [wa:səp], L 115, 116.

Church WARSOP Notts SK 5668. 'W with the church, the church-town of W'. *Warsopp Church towne* 1653, *Church Warsopp* 1716. ModE **church** + p.n. WARSOP SK 5667. Nt 101.

WARTER Humbs SE 8750. 'The gnarled tree'. *Warte* 1086, *Wartre* 1086–1542, *Warter* from 1338. OE **w(e)arr** 'callosity, a warre, a knot in a tree' + **trēow**. An alternative possibility is OE **w(e)argtrēow** 'gallows'. YE 168 gives pr [wa:θə], Studies 1931.91, L 212, 214.

WARTH HILL Cumbr SD 5684. *Warth Hill* 1857. Rises to 892ft. half a mile E of Warth SD 5584. P.n. *Warth* 1721, ON **vartha** 'a cairn, a heap of stones' + ModE **hill**. We i.64.

WARTHILL NYorks SE 6755. 'The look-out hill'. *Wardhilla -(h)ille* 1086, *Warthill(e) -hil -hyll* 1194×8–1416. OE ***weard** + **hyll**. ON **vartha** may have influenced the 1295 spelling *Warthehill*. The topography, however, does not favours this explanation unless the original site was on higher ground to the E overlooking the Roman road from York to Stamford Bridge, Margary no. 81a. YN 11 gives pr [wa:til], SSNY 259.

WARTLING ESusx TQ 6509. 'The Wertelingas, the people called after Wertel'. *Werlinges* 1086, *Wertlinges* 12th–1244, *Wortling* 1199–1407, *Wartling'* 1230–1483, *Wertling* 1232–1429. Metathesised forms *Wrotlyng(ge), Wret(e)lyng, Wryt(e)ling* are also found. OE folk-n. **Wertelingas* < pers.n. **Wertel* or **Wertla*, SE dial. forms for **Wyrtel*, **Wyrtla*. The same pers.n. occurs in Worsham TQ 7509, *Wyrtlesham* [772]13th S 108, *Wertelsham* 1279–1332, *Wersham* 1595, *Worseham* 1795, 'Wyrtel's promontory, river meadow or hemmed in land', pers.n. **Wyrtel*, genitive sing **Wyrtles*, + **hamm** 2, 3 or 6. Worsham is 6 miles E in Bexhill. Both places were probably named after the same individual and Wartling was most likely originally a district name for part of Bexhill hundred. If **Wyrtel* represents Gmc **Wurtila-* its etymology is unclear; if it represents **Wiertel* it could be a derivative of OE *wearte* 'a wart'. Sx 483, 494, ING 41, SS 65, 83.

WARTON 'Look-out enclosure or settlement'. OE **weard** + **tūn**.

(1) ~ Lancs SD 4028. *Wartun* 1086, *-tuna* 1153×60, *Warton* from [1199]1268. OE **waroth** 'shore' is also a possible specific for this instance since the village lies close to the Ribble estuary. La 151, Jnl 17.87.

(2) ~ Lancs SD 5072. *Wartun* 1086, *Warton* from 1246. The village stands at the foot of Warton Crag (534ft), an old beacon hill, with remains of ancient earthworks. La 188.

(3) ~ Northum NU 0002. *Wartun* 1236, *Warton* from 1242. NbDu 208.

(4) ~ SANDS Lancs SD 4472. P.n. WARTON SD 5072 + ModE **sand(s)**.

WARTON Warw SK 2803. Possibly 'the settlement on the river Waver'. *Wavertune* c.1155, *-ton* [1242]1398–1606, *Warton* from 1343. OE **wæfre** 'unstable, restless, wandering' or ***wæfer** 'that which wanders' + **tūn**. *Wǣfer* may have been the name of the stream here, or a reference to swampy ground. Wa 23.

WARWICK Cumbr NY 4656. 'Shore farm'. *Warthwic* 1131–c.1225 with variants *-wi(c)k -wyk(e)*, *Warthewic* [c.1155]c.1200–1413 with same variants, *Warwick(e)* from c.1225. OE **waroth** + **wīċ**. Warwick is situated on the banks of the river Eden. Cu 157 gives pr [warik].

WARWICK Warw SP 2865. 'The dwellings by the weir'. *(into) Wǣringc wican* [716×37]11th S 94, *(in) Wǣrinc wicum* 1001 S 898, *(on) Wǣrincwican* [1016]18th S 1388, *(æt) Wǣring wicum* c.1050 ASC(C) under year 914, *Wǣrinc wic* c.1050 ASC(D) under year 915, *Warwic -uuic -uic* 1086, *Wǣrewic* c.1150, *Warwick* from 1268. OE ***wæring** + **wīc**, dative pl. **wīcum**. Wa 259.

WARWICK BRIDGE Cumbr NY 4756. *pontem de Warthwyc* c.1170, *Warthwikbrig'* 1393, *Warwikkebrig* 1470. P.n. WARWICK NY 4656 + OE **brycg**. Cu 157.

Nether WASDALE Cumbr NY 1204. *Netherwacedal -wasdale* 1338, *-washedale* 1540, *-wasdaill* 1552. Adj. **nether** + p.n. Wasdale as in WASDALE HEAD NY 1808. 'Nether' because at the lower end of the valley from Wasdale Head. Cu 440.

WASDALE HEAD Cumbr NY 1808. 'The head of Wasdale'. *Wascedaleheved* 1334, *Wasdale Hede* 1448, *Wasshedaylehed* 1519, *Wastedale heade* 1540, *Wastellhead* 1628. P.n. Wasdale + ME **heved** (OE *hēafod*). Wasdale, *Wastedal(e)* 1279–1323, *Wasse- Wa(s)cedale* 1279–1363, *Wast Dale, Wasdell, Wastall* 17th, is 'lake valley', ON **vatn**, genitive sing. **vatns**, or ON **vatz** 'water, a body of flowing or still water', + **dalr**, a name type paralleled in Icelandic *Vatnsdalr* and *Vazdalr*. The reference is to WAST WATER NY 1606. Cu 390 gives prs [wɔsəl], occasionally [wɔʃdəl], SSNNW 174, L 30, 94.

WASHAWAY Corn SX 0369. *Washaway* 1699, *Washway* 1732. Probably ModE **washway**, 'part of a road crossed by a shallow stream', 'a concave road deeper in the middle than at the sides' or 'a place where soil has been washed away'. PNCo 177, Gover n.d. 102.

WASHBOURNE Devon SX 7954. 'Stream used for washing'. *Waseborne* 1086, *Wa(y)sseburn(e)* 1238–84, *Wayssheborn* 1333, *Wassh(e)born* 1303. OE **wæsce** + **burna**. The reference is to a sheep-wash or to the washing of clothes. D 324, L 19, 59.

Great WASHBOURNE Glos SO 9834. *Great Washbourne* 1659. Earlier simply *Waseborne* 1086, *Wasseburn(a) -bo(u)rn(e)* 1105–1327, *Was(s)he- Washburn(e) -b(o)urn(e)* 1456–1682, 'stream with alluvial land subject to flooding'. 'Great' for distinction from Little Washbourne SO 9933, *Little Washborne* 1642. Earlier simply *(æt) Uassanburnan* [780]11th S 116, *Uuassanburna* [840]11th S 192, *(æt) Wasseburne* [780]c.1000 S 116, [977]11th S 1336, 1203–1221, *-bourne* 1286–1428. OE **wæsse**, genitive sing. **wæssan**, + **burna**. Gl ii.51, 46, L 18, 60.

River WASHBURN NYorks SE 1261. 'Walc's stream'. *Walkesburn(e)* 1130×40–1330, *Walsh- Walchesburn* 14th, *Washe burne* c.1540, *Washburn* from 1577. OE pers.n. **Walc*, genitive sing. **Walces*, + **burna**. YW vii.141, RN 438.

WASHFIELD Devon SS 9315. Either 'open land by a sheep-wash' or 'beside land subject to flooding'. *Wasfelle* 1086, *Was(se)feld(e)* 1166–1317, *Was(s)hfelde* 1 316, 1342, *Waysh-* 1329, 1354. OE **wæsce** or **(ġe)wæsc** + **feld**. In the second case the reference is to the flat marshy land beside the river Exe. D 419, L 59, 243.

WASHFORD Somer ST 0441. 'Ford leading to Watchet'. *Wecetforda* [963×75]12th S 825, *Wecedforda* [n.d.]12th S 1572, *Wach(e)ford* [1142×66]Buck, *Wecheford* 1243, *Wetheford* [n.d.] Buck, *Wachetford* 1367. P.n. WATCHET ST 0743 + OE **ford**. DEPN, L 72.

WASHFORD PYNE Devon SS 8111. 'W held by the Pyne family'. Herbert de Pinu held the manor in 1219, cf. Pyne Fm SS 8010, *Pyne* 1650. Washford, *Wa- Wesford, Wasforde* 1086, *Was(se)ford(e)* 1242–1280, *Wassheford* 1292, *Waysh(e)ford* 1316, 1340, is probably the 'sheep-wash ford', OE **wæsce** + **ford**. D 397, L 59.

WASHINGBOROUGH Lincs TF 0270. Partly uncertain. *Washingeburg* 1086, *Wassingburche* [1042×55]c.1150, *Wassing(e)burc -burg* 1156–1229, *Washyngburg* 1212, *Washingburgh* 1585, *-borow* 1666×7, *Whassingburg(h) -yng-* 1229–1305, *Whassingburgh -yng-* 1328–1458, *Quassing- -yng-* 1291–1379. The generic is OE **burh** 'a fortified place, a manor house', the specific possibly an **ing**[2] derivative of OE **(ġe)wæsc** 'a washing, a flood' alluding to land by a river which is subject to flooding and draining quickly. Washingborough is situated on rising ground overlooking Car Dyke and the Witham flood plain. Forms with medial *-e-* point to OE locative–dative sing. ***wæscinge** 'at the washing'. Alternatively 'the fortified place of the Wassingas, the people called after Wassa'. Perrott 345, Cameron 1998.

WASHINGTON T&W NZ 3056. 'The estate called after Hwæssa'. The forms fall into three types:

I. *Wessint'-ton* c.1170×80–1371, *Wessington(') -yng-* [1183]c.1320, 1196×1215–1473, *Wesshyngton -ing-* 1411–1556.

II. *Wassinton'* 1211, *Wassington -yng-* 1382, 1418, *Wasshin(g)ton* 1406, *Washington* from 1581.

III. *Quessigton'* 1280, *Quessigton'* c.1310, *Whessyngton* 1475, 1548, *Qwassyngton* 1388×1406, *Whassington -yng-* 1350–70.

OE pers.n. **Hwæssa* + **ing**[4] + **tūn**.

The name seems to have been influenced by other OE p.n. elements, viz OE **wæsse** and **gewæsse**, neither of which are appropriate to the topography of Washington. For the pers.n. *Hwæssa* cf. Whessoe Durham NZ 2718, *Wessehou* 1304, 'Hwæssa's hill-spur'. NbDu 208, 212.

WASHINGTON WSusx TQ 1212. 'The settlement of the Wassingas, the people called after Wassa'. *Wessingatun* [946×7]12th S 1504, *(æt) Wassingatune, Wasingatun* [947]13th S 525, *(æt) Wasingatune* [963]13th S 714, *Wassingatune* before 1080, *Wasingetune* 1086, *Wassington* 1261–1439, *Washington* 1397. OE folk-n. **Wassingas* < pers.n. *Wassa* + **ingas**, genitive pl. **Wassinga*, + **tūn**. Sx 240, SS 75.

WASING Berks SU 5764. Obscure. *Walsince* 1086, *Wawesing' -enge* 1186–1236, *Wausynge* 1316, *Waghesing(')* 1220, *Wasinges* 1235×6, *Wasinge* 1242×3, *Wasing* 1547×53. ING postulates an OE **wagōsa*, **wagusa* corresponding to OHG *waganso*, ON *vagsni* 'ploughshare' and suggests an OE sing. p.n. **Wagesing* 'place by a tongue of land'. For topographical and phonological reasons this is untenable. G. Kristensson suggests an OE **Wagesing* based on a verb **wagesian*, a derivative of OE *wagian* 'move, shake' used to describe the characteristic movement of water in a stream. Wasing lies S and E of the rivers Kennet and Enborne. Brk 271, ING 46, NoB 61.52–4, Anglia 102.415–8.

WASKERLEY Durham NZ 0545. *Waskerley* 1863 OS. Probably identical with High, Low Waskerley Northum NZ 0853, *Waskyrleye* 15th, *Waskerley* 1624×5, '*Wasker* wood or clearing', p.n. *Wasker* + **lēah**. *Wasker* is ON **wasi** + **kjarr**, 'marsh with a brushwood path'. NbDu 208, Nomina 12.30.

WASKERLEY RESERVOIR Durham NZ 0244. P.n. WASKERLEY NZ 0545 + ModE **reservoir**. The reservoir lies at the head of Waskerley Beck, *Walkerhopburne* (sic for *Wasker-*) 1242×3, *Wescrow river* 1768, *Wascrow Beck* 1861 OS, in Waskerley Park, *parcum de Wascroppe* 1311. The beck descends to the Wear at Wolsingham through a side-valley called *Wascropp'* 1283, p.n. *Wasker* as in WASKERLEY NZ 0545 + OE **hop**. NbDu 208, Nomina 12.30.

WASPERTON Warw SP 2658. 'The pear orchard by the land that floods'. *Wasmertone* 1086, *Waspreton* 1139, *Waspurton* early 13th–1349, *Washparton* 1514. The charter forms *Waspertune* S 1000 and *Wasperton* S 1226 purporting to belong to 1043 are in fact forgeries surviving only in 17th cent. copies. OE ***wæsse** 'land by a meandering river which floods and drains quickly' + **per-tūn**. Wasperton lies in a meander of the Avon. Wa 266, L 60.

WASS NYorks SE 5579. 'The fords'. *Wasse* 1541. Probably a development of ME **wathes** similar to that in SMALLWAYS NZ 1111, from ON **vath**. The village lies at the junction of several streams. YN 195 gives pr [wæs].

WAST WATER Cumbr NY 1606. Short for Wasdale Water. *Wassewater* 1294, *Wastwater* 1338. *Was-* from *Wasdale* as in WASDALE HEAD NY 1808 + ME **water** (OE *wæter*). Cu 36.

WATCH HILL Cumbr NY 6246. *Watch Hill* 1866 OS. A summit of 1999ft near the county boundary with Northumberland.

WATCHET Somer ST 0743. 'Lower wood'. *(to) peced* [914×9]12th–16th BH, *Wæced* 918 ASC(A), *Weced* 12th ASC(D) under year 915, *Wecedport* c.1121 ASC(E) under year 987, c.1050 ASC(C) under same year, *Wæcet* [962]12th S 701, *Wacet* 1086, *Wechet* 1243, *Watchet* 1610. PrW ***gwo** + ***cēd**. The name is identical with *montem Vocetium* (Bözberg in East Jura, Switzerland), Tacitus *Histories* I.68, Gaulish **vo-ceto-n* < **vŏ-* 'lower' + **cēto-* 'wood'. DEPN, Holder ii.425, BH 112.

WATCHFIELD Oxon SU 2590. 'Wæcel's open land'.

I. *Wacenesfeld* *[726×37]12th S 93, *Uuacenesfeld* *[821]12th S 183, *Wæthenesfeld* *[821]13th ibid., *Wachenesfeld* 1086–1428 with variants *-feud'* and *-felde*, *Wachenefeld* 1224×5, 1327, *Watchyngfeld* 1547, *Watchfeld* 1585, *Watchfield alias Watchingfield* 1737.

II. *(æt) Wæclesfeld, (ed) Weclesfeld* [931]12th S 413, *(æt) Wæclesfeld* [931]13th ibid., *Wakelesfeld* 1187.

OE pers.n. **Wæċel* as in WATLINGTON SU 6894 and *Wæclingacæstir*, the ancient name of VERULAMIUM TL 1307 (ST ALBANS TL 1507), genitive sing. **Wæcles*, + **feld**. This rather than **Wæċċin* seems to be the pers.n. in question; the *n*-forms, current only from DB and late unreliable charters, are due to AN sound substitution but gave rise to the folk-etymologies 'watching field' and 'watch field'. Brk 382.

WATCHFIELD Somer ST 3446. *Watchfield or Watchwell* 1791 Collinson, *Watchfield* 1809 OS. Possibly 'watch field, look-out field', but early forms are needed.

WATCHGATE Cumbr SD 5299. 'Look-out gate'. *Watchgate* 1793. A one-time gate on the Kendal-Penrith road – possibly referred to in the surname of Alan *de la Watce* ('of the watch') 1332 of nearby Skelsmergh. We i.150.

WATCHINGWELL IoW SZ 4488 → WATTISFIELD Suff TM 0174.

WATENDLATH Cumbr NY 2716. Possibly 'rivulet at Watend'. *Wattendlane* 1189×99, *-lan* 1199×1216, 1213, *Wathendeland* c.1250, *Wa(y)teleth, Waidenleth, Wat(t)endleth* 16th, *Watanlath* 18th. ON ***vatn-endi** 'lake end' referring to Watendlath Tarn + OE **lanu** 'lane; hollow course of a large rivulet in meadowland; slowly moving section of a river'. This name gave scribes much trouble as the full range of spellings shows. Cu 352, SSNNW 257, L 30.

WATER END ModE **water** + **end**.

(1) ~ ~ Herts TL 0310. *Waterende* 1559 sc. of Great Gaddesden. Contrast with POTTEN END TL 0108 of Great Berkhamstead and PICCOTTS END TL 0509 of Hemel Hempstead. Hrt 36.

(2) ~ ~ Herts TL 2304. *Water ende* 1545 sc. of North Mimms parish. Hrt 66.

WATER Lancs SD 8425. No early forms. A reference to Whitewell Water.

WATERBEACH Cambs TL 4965. Properly Water Beach, 'Beach by the water'. *Waterbech(e)* 1237–1395, *-beach* 1337. ME **water** + p.n. *Bechia* [before 1086]c.1280, *Beche* 1169×70–1412, '(at) the ridge', OE **bæċ**, locative–dative sing. **beċe**. 'Water' for distinction from LANDBEACH TL 4765: both places are situated on a ridge of higher ground N of Cambridge, Waterbeach overlooking the Granta. Also known as 'Out Beach', *Vtbech* 1086, *Udbec(h)e -becce, Vtbeche* 1086 InqEl. Cf. the form *Inbeche* for LANDBEACH. Ca 184, L 12, 30, 126.

WATERDEN Norf TF 8836. 'The stream valley'. *Waterdenna* 1086, *-dene* 1188, *-dena* 1191. OE **wæter** + **denu**. DEPN.

WATERFALL Staffs SK 0851. 'The waterfall'. *Waterfal* 1201, *-fall* 1272. OE **wæter-(ġe)fall**. The sense here is 'a place where water disappears underground' referring to the disappearance of the r. Hamps here due to the limestone formations in the area. DEPN.

WATERFOOT Lancs SD 8321. No early forms. The place is situated in Rossendale Valley at the foot of Whitewell Water.

WATERFORD Herts TL 3114. *Waterford* from 1214. OE **wæter** + **ford**. The significance of the name is not clear. Hrt 216.

WATERGATE BAY Corn SW 8264. *Watergate Bay* 1813. Lost p.n. *Watergate* 'sluice-gate' (e.g. for a mill-stream) + ModE **bay**. PNCo 177, Gover n.d. 350.

WATERGROVE RESERVOIR GMan SD 9017. P.n. Watergrove + ModE **reservoir**.

WATERHEAD Cumbr NY 3703. 'Head of the lake' sc. Windermere. *Watterhead* 1597, *(ye) Waterhead* 1604, 1714. ModE **water** in the sense 'lake' + **head** 'top end'. We i.184.

WATERHOUSES Durham NZ 1841. 'House(s) beside the river'. *Waterhouse* 1596, *Water House* 1768 Armstrong. ModE **water** + **house**. The hamlet lies beside the river Deerness.

WATERHOUSES Staffs SK 0850. 'The houses by the stream'. *Over water house* c.1571, *le Upper or Over Waterhowses* 1580, *Waterhouses* 1836 OS. ModE **water** + **house**. The place lies on the r. Hamps. Horovitz.

WATERINGBURY Kent TQ 6853. Uncertain. *(of) Woðringaberan, Woðringabyras, Uuotringeberia, Uuotryngebyri* [975×87]12th S 1511, *Oteringaberiga, Wohringa byran* [c.975]12th B 1321–2, *Otrin(ge)berie* 1086, *Wotringaberia* c.1100, *Watringbury* 1610. Neither the specific nor generic are clear in this name. The former may be a formation on OE **wōth** 'noise', a word related to the Germanic vocabulary of song , poetry and divine inspiration, hence perhaps a folk-n. *Wōthringas*; the latter is unclear as between OE **bǣr** 'woodland pasture', **byriġ** the dative sing. of **burh** 'fortification' and **bȳre** 'a shed'. Similar formations are *(æt) Woþhringe* [932]13th S 418, *Wot(h)eryng* ibid., the name of a meadow in North Stoneham Hants SU 4317 (which may be a formation in OE **hring** 'a ring, circle') and *Wotring* 1285 for Otteridge in Berstead WSusx SU 9300. Possibly we have to do with a r.n. **Wother*, an *r*-extension of **wōth** as in the r.n. Wooth Dorset SY 4796, *Woth* 1207. PNK 167, KPN 250, ING 232, RN 469.

WATERLIP Somer ST 6544. *Waterlip* 1817 OS. Possibly a compound name in OE ***hlēp**, WS **hlīep, hlȳp** 'a leap, a jump, a leaping place, a steep declivity', but early forms are needed. The reference is to a water-filled hollow in the hill-side (?) formed by quarrying.

WATERLOO Dorset SZ 0094. Named after the *Waterloo Iron Foundry* c.1850. Do ii.7.

WATERLOO Mers SJ 3198. A district of Liverpool that grew up as a resort around the Waterloo Hotel (originally the Royal Waterloo Hotel) founded in 1815, the year of the victory at Waterloo. Room 1983.131, Pevsner 1969.419.

WATERLOO Norf TG 2219. *Waterloo* 1838 OS. Transferred from Waterloo, Belgium, site of the last battle of the Napoleonic wars in 1815.

WATERLOO STATION GLond TQ 3179. Opened in 1848 and named from the original Waterloo Bridge opened on 18th June 1817, the second anniversary of the Battle of Waterloo, a village near Brussels. Encyc 959.

WATERLOOVILLE Hants SU 6809. *Waterloo Ville* 1898. Earlier simply *Waterloo* mid 19th. A suburban satellite of Portsmouth developed c.1830 and named from an inn variously recorded as the Waterloo Hotel or The Heroes of Waterloo, today represented by a later building simply called The Heroes. The place was earlier called *Whateland End* 1759, *Waitland End* 1817, *Wheatland End* c.1840 apparently after a farmhouse called Wait Lane End, a lane named after the Wait or Wayte family recorded in this area from the 16th cent. Gover 1958.23, Ha 171, Pevsner-Lloyd 644, Room 1983.131.

WATERMILLOCK Cumbr NY 4422. 'Wether Mellock'. *Weþermeloc* early 13th, *Weyermelok'* (for *Weþer-*) c.1250, *Wethermelok(e) -meloch -melock(e)* 1253–1682, *Waltermelocke* 1568, *Watermelocke* 1572, *Watermillock* 1678. OE **wether** 'a castrated ram' + p.n. Mellock, PrW ***mę̄lǭg** 'little bare hill', the W form of Little Mell Fell NY 4224. The prefix **wether** was subsequently replaced by the pers.n. *Walter* and ModE **water** on account of the situation of the place on Ullswater. Cu 254.

WATERPERRY Oxon SP 6206. *Waterperi* 1186×91 etc. with variants *-pyrye -pury* and *-per(r)y*. ME **water** + p.n. *Perie* from 1086† with variants *Pery(e), Peri(a)* to 1320, 'the pear-tree'. 'Water', for distinction from WOODPERRY SP 5710, alludes to the liability of the land here to flooding. O 192.

WATERROW Somer ST 0525. *Waterwood Spars* 1809 OS.

WATERSFIELD WSusx TQ 0115. 'The open land of the water'. *Watresfeld* before 1226, *Wateresfeld* 1256–1418. OE **wæter**, genitive sing. **wæteres**, + **feld**. The hamlet lies low on a small stream. Sx 126, SS 70.

WATERSTOCK Oxon SP 6305. 'Water Stoke'. *Waterstokes* [1208×13]c.1300, *Waterstok(e) -sto(c)k* from 1247. ME **water** + p.n. *Stoch* 1086, *Stoches* c.1188×91, 'the outlying farm', OE **stoc**. O 148.

WATERTHORPE SYorks SK 4382. 'Walter's outlying farmstead'. *Waltert(h)orp(e)* 1276–1356, *Waterthorpe* from 1571. A p.n. of post-conquest origin, pers.n. CG *Walter* + ME **thorpe**. Db 210.

WATER YEAT Cumbr SD 2889. *Water Yeat* 1864 OS. ModE **water** + dial. **yate** 'a gate, an opening' (OE **ġeat**). Situated at the S end of Coniston Water.

†Another DB spelling is *Pereivn*.

WATFORD Herts TQ 1196. 'Hunting ford'. *Watford* from [990×1001]13th S 1497, *(of) wat forda* 1007, *Wathford(a)* c.1180–13th. OE **wāth** + **ford**. Hrt 103.

WATFORD Northants SP 6068. Probably 'the ford that can be crossed by wading'. *Watford* from 1086, *Wad-* 1176, *Wath-* 1238, 1290. OE **wæd** + **ford**. Most fords would have been crossed on horse-back. An alternative possibility is OE **wāth** + **ford** 'hunting ford'. Nth 75, L 67, 72, SSNEM 382.

WATH 'A ford' ON **vath**.

(1) ~ NYorks SE 3276. *Wat* 1086, 13th cent., *Wath* from [1184]15th. YN 219, SSNY 107.

(2) ~ NYorks SE 1467. Short for *Acchewath -wad* 'Aki's ford' 1154×91, *Ackewath* 1170×9, or *Youdenwath* 'Yeaden ford' 1307, *Ewdinwath* 1651, ODan pers.n. *Áki*, genitive sing. *Áka*, and p.n. YEADEN SE 1467. YW v.150.

(3) ~ NYorks SE 6775. *Wad* 1086, *Wath* 1224×30. The ford carries the supposed Roman road from Mallon to Horingham, Margary no. 814, over Wath Beck. YN 52.

(4) ~ UPON DEARNE SYorks SE 4300. P.n. *Wade, Wat(e)* 1086, *Wad* 1153, *Wat* 1164×81, *Wath(e)* from 1208, + r.n. DEARNE SE 3408. YW i.118, SSNY 107.

WATLASS → THORNTON WATLASS NYorks SE 2385.

WATLING STREET Beds Herts Leic TL1110–SP4490. 'The Road of the Wæclingas, the people called after Wæcel', originally the OE name for the Roman road from London to St Alban's, Margary no. 1 f-g. *Wætlinga stræt* c.880×90, *(andlang, be norðan) Wætlinga stræte* [956]16th S 615, 1121 ASC(E) under year 1013, *(in) wæclinga stræte, wæxling(g)a stræte* [926]13th S 396, 944 S 495, [975]12th S 802, [986]c.1000, 11th ASC(C) under year 1013, *Weatlinga-strætæ, id est strata quam filii Weatlæ regis per Angliam straverant* 'W S, that is the road made through England by the sons of King Weatla' before 1118 FW, *Watlingestrate -strete -ynge-* before 1154–1337. OE folk-n. *Wæclingas* < pers.n. **Wæċel* + **ingas**, genitive pl. *Wæclinga*, + **strǣt**. *Wæclingas* appears to have been the name of the Anglian settlers of ST ALBANS, Bede's *Uaeclingacæstir* c.737, 'Roman camp of the Wæclingas'. Pers.n. **Wæċel* occurs also in WATLINGTON Oxon SU 6894 and WATCHFIELD Oxon SU 2590. The name was subsequently applied to the extension of the road to Wroxeter, and then by antiquarian writers of the 12th cent. and later to Roman roads stretching SE from London and N or W from Wroxeter and elsewhere, especially the road through Yorkshire to Corbridge and Scotland. According to Chaucer's eagle, in a typical gossipy aside, it was also a name of *the Galaxie*

> Which men clepeth the Milky Wey
> For hit ys white - and somme, parfey,
> Kallen hyt Watlynge Strete,
> *House of Fame 936–9,*

'a famous old road', F. N. Robinson notes *The Works of Geoffrey Chaucer*, 2nd ed. 1957, *ad loc.*, 'which probably ran from Kent to the Firth of Forth'(!). Bd 7, OED s.v. *Watling-street.*

WATLINGTON Norf TF 6110. Partly uncertain. *Watlingetun* 11th, *-tone* 1166, *Watlingtone* 1254. If the specific is identical with that of WALTING STREET, *Wæclingastret* 944 S 495, *Wætlingastræt* [956]16th S 615, and the early name of St Albans, *Uaeclingacæstir* c.731 BHE, *Wætlingaceaster* [1005]13th S 912, this is the 'settlement of the Wæclingas, the people called after Wacol', OE folk-n. **Wæclingas* < pers.n. *Wacol* + **ingas**, genitive pl. **Wæclinga*, + **tūn**. Alternatively this might be the 'settlement at the hurdle or wattle place', OE ***wateling** < **watel** + **ing**[2], + **tūn**. Watlington is at the edge of the fens and might well have been a place for the manufacture of wattle or hurdles. DEPN.

WATLINGTON Oxon SU 6894. 'The estate called after Wæcel' and 'of the Wæclingas, the people called after Wæcel'. *(æt) Wæclinctune* [880 for 887]11th S 217, *(æt) Wætlinctune, Wæcling tun, Huuætlingatune* 11th Heming, *Watelintone -tune* 1086, *Watlinton(a) -tuna -tone* 1129–1320, *Watlingtun' -ton(')* from 1218. OE pers.n. **Wæċel* + **ing**[4] + **tūn** and folk-n. **Wæclingas* < **Wæċel* + **ingas**, genitive pl. **Wæclinga*, + **tūn**. O 94.

WATNALL Notts SK 5045. 'Wata's hill-spur'. *Watenot* 1086, *Wat(t)enho(u)* 1199–1275, *Watenowe* 1280–1330, *Watnow(e)* 1292–1550, *Watnall* from 1586. OE pers.n. **Wata*, genitive sing. **Watan*, + **hōh**. Also known as *Watnow(e) Chaworthe* 1392 and ~ *Cauntcliffe* 1541 referring to division of the estate into two separate manors which passed through heiresses to Thomas de Chaworth and Nicholas de Cantelupe in the 13th cent. Nt 146 gives local pr [wɔtnə:], L 168.

WATTISFIELD Suff TM 0174. Partly uncertain. *Wat(l)es-Watefelda* 1086, *Watlesfeld* c.1150, 1197, *Uueatlesfeld* c.1095. Uncertain element or pers.n. + **feld** 'open land'. Elsewhere there are instances of later *Watl-* for original *Wacl-*, e.g. WATLING STREET SJ 8311 etc., WATLINGTON Oxon SU 6994. This name could, therefore, be 'Wacol or Wæcel's open land'. But other possible pers.ns. are **Wætel*, a diminutive of **Wata*, cf. OHG *Wazo, Wezilo*, or **Hwætel*, a diminutive of **Hwæt* as in Watchingwell IoW, *(to) Hwætincgle* [968]14th S 766, *Watingewelle* 1086, *Whatingewelle* 1287×90, *Whatlyngwelle* 1316, 'the spring of the Hwætingas, the people called after Hwæt', or even OE **watol** 'wattle', and so 'the open land where wattle is collected'. Cf. WATTISHAM TM 0151. DEPN.

WATTISHAM Suff TM 0151. 'Wæcci's homestead'. *Weceshā* 1086, *Wechesham* 1182, *Wachesham* 1184. OE pers.n. **Wæċċi*, genitive sing. **Wæċċes*, + **hām**. DEPN.

WATTON Humbs TA 0150. 'Hill at the wet place'. *Ueta dun* [c.731]8th BHE, *Weatadun* [c.890]c.1000 OEBede, *wǽta dún, wetadun -dún* [c.890]11th ibid., *Watun -ton* 1086–1351, *Watton* from 12th. ONb **wēt** replaced by ON **vátr**, + **dūn**. The reference is to the marshlands of Watton Carrs which lie to the W along the r. Hull. YE 158 gives prs [watən, waʔn].

WATTON Norf TF 9100. 'Wada's estate'. *Wadetuna* 1086, *-ton* 1203, *Waditone* 1254. OE pers.n. *Wada* + **tūn** possibly varying with *Wada* + **ing**[4] + **tūn**. DEPN.

WATTON AT STONE Herts TL 3019. *Watton Stone* 1304, ~ *at Stone* 1311. P.n. Watton + ME **stone** referring to an old stone formerly under the horse-trough at the Waggon and Horses inn. Watton, *Wattun* [969]11th Wills, *Wattune* [1049]14th, *Wattone* [1049]12th S 1123, *(æt) Wadtune* 11th K 1354, *Uuattune* 1065, *Watone, Wodtone* 1086, is the 'woad farm', OE **wād** + **tūn**. Hrt 142.

WATTON BECK Humbs TA 0448. P.n. WATTON TA 0150 + ModE **beck** (ON *bekkr*).

WATTY BELL'S CAIRN Northum NT 8902. A boundary marker at an elevation of 1166ft. between Coquetdale and Redesdale. Pers.n. *Watty Bell* + ModE **cairn**. *Watty* is a pet form for *Walter*.

WAVENDON Bucks SP 9037. 'Wafa's hill'. *Wavendone, Wawendene* 1086, *Wauendon* 1186–13th cent., *Wavyn(g)don* 1300–75, 1512, *Waunden* 1405×20, 1482, 1535, *Wondon* 1509. The earliest reference is *gemæru wafanduninga* 'the boundary of the people of Wavendon' 969 S 772. OE pers.n. *Wafa*, genitive sing. *Wafan*, + **dūn**. Bk 150 gives prs [wɔndən, wɔ:ndən] and [wa:ndən], L 150.

River WAVENEY Suff TM 2381, 4691. 'The quagmire river'. *Wahenhe* 1275, *Wagenho* 1286, *Wawneye* 1485, *Waueney* 1575, 1577, *Waveney* 1724. OE **Wagen-ēa* < **Wagen* as in WAWNE Humbs TA 0937 + **ēa**. Alternatively the specific might be OE ***waga** 'that which moves, a quagmire' < *wagian*. The subsequent development is regular, **Wagen-ēa* > **Waɣʷene* > *Wawene*. The present form is due either to misreading of <u> = [w] in *Waueney* as [v] or to earlier sound substitution of peripheral spirant phoneme [ɣ] by neighbouring spirant [v]. RN 439.

River WAVER Cumbr NY 1950. 'Wandering river'. *Wafyr* [11th]13th, *Waver* from 1189×99. OE **wæfre** 'wandering'. Cu 30, RN 440.

WAVERLEY ABBEY Surrey SU 8645. P.n. *Waverleiacum, Vaverliacum* [c.1140]1318, *Waverleia -legh(e) -ley(e) -lai* etc. 1147 etc., *Waferlewe* 1265, 'the wood or woodland clearing' either 'in swampy ground' or 'with swaying trees', OE **wæfre** + **lēah**, + ModE **abbey** (OFr *abbeie*). The first two forms are pseudo-Gaulish and reflect Fr influence on this Cistercian monastery founded 1128. Sr 174, PNE ii.235–6.

WAVERTON Ches SJ 4564. Partly uncertain. *Wavretone* 1086, *Waverton(e)* from 1188×91, *Warton* 1363–1690. The specific is OE **wæfre** 'unstable, restless' or rather a noun derived from it, ***wæfer** 'that which is unstable, restless' + **tūn**; and so 'farm at a waving tree' or 'farm at a quaking bog' or 'where brushwood is obtained'. Cf. WHARTON SJ 6666, WAVERTREE Mers SJ 3889, WAVERLEY ABBEY Surrey SU 8645. Che iv.103, V(I.i.)xi, PNE ii.235.

WAVERTON Cumbr NY 2247. 'Settlement by the river Waver'. *Wauerton -v-* from 1183, *Warton* 1452–1656. R.n. WAVER NY 1950 + OE **tūn**. Cu 159 gives pr [wa:rtən].

WAVERTREE Mers SJ 3889. 'Swaying tree'. *Wauretreu* 1086, *Wavretre* 1196–1251 etc., *Wavertree* from 1201, *Wartre* 1577. OE **wæfre** + **trēow**. The reference might be specifically to aspen-trees which still flourish in the area. Dial. *waver*, however, simply has the sense 'an isolated young tree, a young tree left standing by itself in a felled wood'. La 112, Mills 1976.147, PNE ii.235.

WAWNE Humbs TA 0937. 'The quagmire, the quaking fen'. *Wag(h)ene* 1086, *Waghen* 1150×3–1478, *Wawene* 1223, *Waune* 1228×31, *Wawne* from 1371. OE ***wagen**. YE 44, RN 440, PNE ii.239.

WAXHAM Norf TG 4326. Partly uncertain. *Waxtonesham* *[1044×7]13th S 1055, *Wacstanest, Wacstanes- stenesham* 1086, *Waxstonesham* 1127–1595, *Wachesham* 1244, *Waxham* from 1440. This might be an OE ***wac** + **stān** 'a watch stone' or a pers.n. **Wǣġstān*, genitive sing. ***wac-stānes**, **Wǣġstanes*, + **hām**, 'the homestead of or by the watch-stone, the stone where a watch is kept' or 'belonging to Wæġstan'. Nf ii.131.

WAXHOLME Humbs TA 3229. 'Homestead where wax is produced, bee farm'. *Was(s)ham, Wassum* 1086, *Waxham* 1086–1512, *Waxsome* 1549, *Washolme* 1650. OE **weax** + **hām**. YE 29.

WAY VILLAGE Devon SS 8810. *Way* 1809 OS. In the same parish (Cruwys Morchard) are West Way SS 8710, *West Way* 1809 OS, and Hookway SS 8910, *Hookeway* 1657. D 381.

WAYFORD Somer ST 4006. 'Way ford'. *Waiford* 1206–25. OE **weġ** + **ford**. The significance of this name is not clear, but the reference is to a ford at 400064 on the now disused track from Wayford to Ashcombe and Winsham. DEPN, L 70, 83, VCH iv.

WAYLAND'S SMITHY Oxon SU 2885. 'Weland's smithy'. *(be eastan) welandes smiððan* [955]13th S 564, *Wayland Smiths Forge* 1828. OE pers.n. *Wēland*, the famous smith of Germanic legend, genitive sing. *Wēlandes*, + OE **smiththe**. The reference is to a Neolithic chambered long barrow of the 3rd millennium BC. Local legend held that the smith would shoe horses in return for a groat laid on the roof slab of one of the three chambers. Brk 347, Thomas 1976.178.

WEALD 'Upland forest'. OE **weald**.

(1) North ~ BASSET Essex TL 4904. 'North W held by the Basset family'. *Welde Basset* 1291, *Northwell Basset* 1555. Earlier *Northwolde* 1244, *Northwelde* 1299, and simply *Walda -ā, Wallam* 1086, *Walda -e* 1227–55, *Weld(e)* 1251–1354. Manorial affix from a 13th cent. owner, Sir Philip Basset; 'North' for distinction from South WEALD TQ 5793. Ess 86, YE lvi.

(2) South ~ Essex TQ 5793. *S(o)uthweld* 1262–1347 with variants *-walde* 1281, *-wealde* 1316, *-wold* 1347, *Southwell* 1447–1551. ME **south** + p.n. *Welda* 1086, *Weld(e)* *[1062]13th S 1036–1346, *Walde* *[1062]13th ibid., 1224–8. 'South' for distinction from North WEALD BASSET TL 4904. Ess 135.

(3) The ~ ESusx TQ 6035. *þe welde* 1290, *le Walde* 1330, *the Welde* c.1480, *the Wilde* 1667, *the Weilde* 1675. The forest is referred to in ASC under the year 893 as *ðæs miclan wudu eastende ðe we Andred hataþ ... seo ea lið ut of ðæm walda* 'the east end of that great wood which we call Andred... the river flows out of that wood'. Under the year 477 the wood is called *Andredesleage* 'the wood of Andred', in 755 simply *Andred* and in 1018 *Andredesweald* 'the high forest of Andred'. In Eddi's *Life of St Wilfrid* [c.700]c.1000 it is *desertum Ondred* 'the wilderness Ondred'. Andred represents *Anderidos -tos* [c.410]15th ND, the Romano-British name of Pevensey, also called *Andredesceaster* 'the Roman fort of Andred' ASC(A) under year 491. The meaning is 'great ford', Brit ***ande-** as in Gaulish *Anderitum*, Javols, Lozère, France, + ***ritu-** as in *Camboritum*, Lackford, Suff, *Durolitum*, Little London, Chigwell, Essex, and PENRITH Cumbr NY 5130. The reference must have been to a crossing of some former inlet at Pevensey. Sx 1f, RBrit 250f.

(4) Upper ~ Bucks SP 8037. *Upper Weal* (sic) 1834 05. ModE **upper** + p.n. *Wald(e)* 1199–1324, *Waude* 1241, 1324, *(la, atte) Welde* 1291–1353, *Weale* 1766–1826, part of Whaddon Chase. 'Upper' for distinction from Lower and Middle Weald SP 7838, 7938, *Lower, Middle Weal* 1834 OS. Bk 18, L 275–6.

WEALDSTONE GLond TQ 1689. *Weald Stone* 1754. P.n. *Weald* as in Harrow Weald + ModE **stone**. Harrow Weald, *Harewewelde* 1388, *Harrow Weelde* 1553, earlier simply *Weldewode* 1282, *Weld(e)* 1294–1453, is 'Harrow woodland', p.n. Harrow as in HARROW ON THE HILL + OE **weald**. The reference is to a sarsen stone embedded in the pavement outside the Red Lion at Harrow Weald, probably a boundary stone separating Harrow Weald from the rest of Harrow parish. The develpment of Wealdstone dates from the opening of the London and Birmingham railway in 1838. Mx 54, Encyc 961.

River WEAR Durham NZ 1134. *Uiurus, Uuirus* [c.730]8th BHE, *Uuirus, Wirus, Wyrus, Uiurus* [c.716]10th BHA, *Wiire* [c.890]c.1000, *Wirus* c.1107–c.1190×5, *Uuira, Wira* [c.1125]c.1150, *Weirus* [c.716]12th, *Weor* [c.1040]12th, [1072×83]15th, c.1107, *Wer* c.1148–1243, *Were* c.1250–1621, *Weere* 1285, 1418, *Weir* 15th cent., *Weare* 1572–1647, *Wear* from 1438. Identical with the Lancs r. WYRE SD 4341, 5553, *Wir* 1170×84, and the German river Weser, OHG *Wesera, Wisara, Wisura* 'flowing water', from the IE root **wis-/*weis-* 'liquid, flow'. The best elucidation of the various forms is that of Nicolaisen BzN 1957.236: intervocalic *s* was lost in British so that British **Uisuriā* > PrW **Wiir* > OE **Wīr* latinised as *Wirus, Wira*, while the alternative British form **Uisurā* > PrW **Wior* latinised as *Wiurus* > OE *Wēor* > ME *Wēr*, ModE *Wear*. Another name of the river is preserved in Ptolemy's *Geography*, *Οὐέδρα (Vedra)* [c.150]13th. This appears to have survived in the Welsh name of the river, *Gweir*, and in the p.n. *Kaer Weir* 14th (for ?Durham, ?Wearmouth). *Vedra* is cognate with the German river name Wetter, *Wetteraha* 772, French Vézère and Vesdre, and also with the Scots Quair Water and Traquair, *Treverquyrd* c.1124, *Trauerquayr* c.1150–1242 from IE **wedōr-* 'water' on the base **wed-*. It is not unusual for rivers to possess more than one recorded name, either of different date, or applied to different stretches of their course. But for a more recent association of *Gweir* with the meaning 'bend' (highly appropriate for the Wear) see now DAJ 13.87. NbDu 209, RN 441, 475, LHEB 362, BzN 1957.236, RBrit 489, Berger 274, 275.

WEARE Somer ST 4152. 'The weir'. *Werre* 1086, *Wera* 1169, *Were* 1242, 1610. OE **wer**. Named from a fishing weir on the river Axe. DEPN.

WEARE GIFFARD Devon SS 4721. 'W held by the Giffard family'. *Weregiffarde* 1328. The Giffard family is first mentioned in connection with this place in 1219. Weare, *Were* 1086–1290, is OE **wer**, referring to a weir (no doubt with fish-traps) in the river Torridge. D 111 gives pr [dʒifəd].

Lower WEARE Somer ST 4053. *Lower Weare* 1817 OS. Probably the lower part of WEARE ST 4152 rather than 'the lower weir'. It lies downstream from Weare.

WEARHEAD Durham NY 8539. 'The source of the river Wear'. *Wereheved* 1372, *Wear- Warehead* 1647, *Weares Head* 1685. R.n. WEAR NZ 1134 + OE **hēafod**. NbDu 209.

WEASDALE Cumbr NY 6903. Perhaps 'weasel valley'. *Wels(s)dall* 1581–1607, *Weelsdaill* 1628, *Weesdal(l) -daile* 1643–1747, *Weasdale* 1655. The evidence is late but this might be ME **wesele** + **dale**. We ii.33.

WEASENHAM ALL SAINTS Norf TF 8421. 'All saints' W'. *Wesinham Omnium Sanctorum* 1291. P.n. Weasenham + ModE **all saints**, Latin **omnes sancti**, genitive pl. **omnium sanctorum**, from the dedication of the church. Weasenham, *Wesenham* 1086–1242, *Weseham* 1205, is probably 'Weosa's homestead', OE pers.n. **Wisa*, **Weosa*, genitive sing. **Weosan*, + **hām**. 'All Saints' for distinction from WEASENHAM ST PETER TF 8522. DEPN.

WEASENHAM ST PETER Norf TF 8522. 'St Peter's W'. *Weasenham St Peter* 1824 OS. P.n. Weasenham as in WEASENHAM ALL SAINTS TF 8421 + saint's name *Peter* from the dedication of the church.

River WEAVER Ches SJ 5477. 'Winding stream'. *Weuer* [c.1130]1479, *Wever(e)* c.1230–1719, *Weever* [1133]n.d., 1358–1819, *Weaver* 1341, 1656, *Wy- Wiver(e)* 1300–1586. OE **wēfer(e)**. The Weaver is a very winding river: Harrison in his *Description of Britain* (1577) knows no other 'that fetcheth more or halfe so many windlesses and crinklings'. Che i.38, RN 443, PNE ii.248.

WEAVER HILLS Staffs SK 0946. *Wever Hills* 1682, 1795. Apparently a r.n. *Wevere in Stanton* 1315, + ModE **hill**. The r.n. must have applied to one of the streams draining S from the hills and means 'the winding stream', OE ***wefer(e)** as in r. WEAVER Chesh SJ 5877. The hills are Ribden SK 0747, *Wrybbedon* 1327, *Ribden* 1608, 'Wrybba's hill', Wredon SK 0846, *Reeden* 1686, *Raydon Hill* 1836, 'rye hill', Cauldon Low SK 0848, *Caldelawe* 1203, *-lowe* 1329, *Caldon Low* 1836, 'Cauldon tumulus', perhaps originally 'the cold tumulus', Rue Hill SK 0847, *Rowlow* 1686, 'rough hill', Wardlow SK 0947, *Wardlow* 1836 OS, 'the look-out hill', Queen Low, *Queen Low* 1686, *Queens Knowl* 1836 OS, Gallows Knoll, *Gallows Knoll* 1686, and Castlow Cross, *Castlow Cross* 1836 OS. St i.23, Horovitz.

WEAVERHAM Ches SJ 6174. 'Village by the river Weaver'. *Wivreham* 1086, 1581, *Weverham* [1096×1101]1284, 1150–1724, *Weeverham* 1270, 1656, *Weaverham* from 1558, *Wereham* 1420–1656, *Wearham* 1695. R.n. WEAVER SJ 5477 + OE **hām**. Che iii.205 gives pr ['wi:vərəm], locally ['wɛ:rəm, 'wɔ:rəm].

WEAVERTHORPE NYorks SE 9670. 'Vithfari's dependent farm'. *Wifretorp* 1086, *Wi- Wyuertorp -thorp(e)* [1114×21]13th–1541, *Wyrthorp(e)* 14th cent., *Werthorp(e)* 1355–1553, *War-* 1543, *Weuerthorp(e)* 1419–1582. ON pers.n. *Víthfari*, genitive sing. *Víthfara*, + **thorp**. An alternative suggestion for the pers.n. is OE *Wīgferth*. YE 122 gives pr ['wi:əθrəp], SSNY 71.

WEBHEATH H&W SP 0226. *Web Heath* 1831 OS.

WEDDINGTON Warw SP 3693. Probably the 'estate called after Hwæt'. *Watitune* 1086, *Whetinton* 1221, *Wetinton(e) -yn(g)-* 1235–85, *Wedinton -yn-* 1291–1332, *-ing-* 1331, *Weddyngton* 1535. OE pers.n. **Hwæt* + **ing**[4] + **tūn**. Wa 93.

WEDHAMPTON Wilts SU 0657. 'The weed farm'. *Wedhampton* from 1249 with variants *Wede-* and *Weyd-*, *Weddington* 1773. OE **wēod** + **hām-tūn**. Wlt 316 gives pr [wediŋtən].

WEDHOLME FLOW Cumbr NY 2253. P.n. Wedholme + ModE **flow** in the sense 'watery moss or bog'. Wedholme, *Waytheholm'* 1189, *Waitheholm* 1201, *Wedholm(e)* from 1570, is 'hunting island', referring to a patch of higher ground surrounded by bog, ON **veithr** 'bag (in hunting)' + **holmr**. Cu 308 gives pr [wedəm flau], SSNNW 174.

WEDMORE Somer ST 4347. Partly obscure.

I. *Wethmor* [676×85]13th S 1667, [680×709]13th S 1674, *Weþmor* 892 ASC(A) under year 878.

II. *Wedmor* 873×88[11th] S 1507, 11th–12th ASC(C,D,E) under year 878, *Weddmor* c.1000 ASC(B) under same year, *[1065]18th S 1042, *(æt) Weodmor, Wedmor'* [1061×66]13th S 1115, *Wedmore* from 1086 including [1066×75]13th S 1241, *Wetmor* [n.d.]18th ASC(W).

III. *Wædmor* [c.894]c.1000 Asser, *Wadmor* 1201.

Uncertain specific + **mōr** 'marsh'. Type III points to OE ***wǣthe**, a mutated side form of OE **wāth** 'hunting': this would not be expected to give spellings with *e* in the WS dial. as in types I and II but the sense 'hunting marsh' would be suitable. Later folk-etymology associated the name with OE **wædd, wedd** 'a pledge' on account of the baptism of the Danish King Guthrum here in 878. Cf. WEMBDON ST 2837. DEPN, Turner 1951.158, L 54–5.

WEDNESBURY WMids SO 9995. 'Woden's fort'. *Wadnesberie* 1086, *Wodnesberia* 1166, 1190, *Wednesbiri* 1227. OE heathen divinity-n. *Wōden*, genitive sing. *Wōd(e)nes*, + **byriġ**, dative sing of **burh**. Wednesbury may have been a site of heathen worship unless the allusion is to an antiquity loosely named after Woden like the use of *devil* in p.ns. DEPN gives prs [wei-] and [wensbri]. Evidence 109, Signposts 161, L 2.

WEDNESFIELD WMids SJ 9500. 'Woden's open land'. *Wodnesfeld* [996 for 994]17th S 1380, 1227, *-felde* 1086, *Wednesfeld* 1251. OE heathen divinity-n. *Wōden*, genitive sing. *Wōd(e)nes*, + **feld**. DEPN gives prs [wei-] and [wenzfi:ld]. Evidence 109, L 240, 244.

WEEDON Bucks SP 8118. '(Heathen) shrine hill'. *(æt) Weodune* *[1065]14th S 1040, *[1066]12th S 1043, *Wedone* 1220–1328, *Weodone* 1346, 1376. OE **wēoh** + **dūn**. Bk 85, Evidence 102, 109, Signposts 158, L 144, 150, 153.

WEEDON BEC Northants SP 6359. 'W held by Bec'. *Wedon Beke* 1379. The estate was held by the abbey of Bec-Hellouin, Normandy, cf. TOOTING BEC GLond TQ 2972. Weedon, *Wedon(e)* 1086–1401 with variants *-dun -down*, is the 'hill with a heathen temple', OE **wīġ, wēoh** + **dūn**. The bounds of the people of Weedon are mentioned 944 S 495 as *weoduninga gemære*. 'Bec' for distinction from WEEDON LOIS SP 6047. The most likely site for the temple is the hill on which Weedon Hill Farm lies, 1 mile SW at SP 6158. Nth 30, Evidence 102, 109, Signposts 161, L 144.

WEEDON LOIS Northants SP 6047. 'Weedon with the well of St Loys or St Lewis'. *Leyes Weedon* 1475, *Loyeswedon* 1524. Earlier *Wedune Pynkeny* 1282, *Weedon Pynkenye* 'Weedon held by the Pinkney family', and simply *Wedon(e)* 1086–1316, the 'hill with a heathen temple', OE **wīġ, wēoh** + **dūn**. The manor was the head of the Pinkney honour. Nth 45 gives prs [wi:dənloi] and [loiwi:dən], Evidence 109.

Upper WEEDON Northants SP 6258. Adj. *upper* + p.n. Weedon as in WEEDON BEC SP 6359. 'Upper' for distinction from Lower Weedon SP 6259.

WEEFORD Staffs SK 1404. 'The heathen temple ford'. *Weforde* 1086, *-ford* 1200–93, *Wyford* 1200, *Weoford* 1291, *Wifford* c.1540. OE **wēoh** + **ford**. The 1291 sp must go back to OE **wēoford** with loss of *h* before Anglian smoothing; forms with *We-* show monophthongisation of *ēo* > *ø* > *ē* since where smoothing occurred before loss of *h* the resultant monphthong was *ī*. DEPN, Evidence 101, L 72, Campbell §230.

WEEK 'Farm, village'. OE **wīċ**.

(1) ~ Devon SS 7316. *Wik* 1242, *Est- Westwyke* 1378. D 379.

(2) ~ ORCHARD Corn SS 2300 → WEEK ST MARY SX 2397.

(3) ~ ST MARY Corn SX 2397. 'St Mary's Week'. *Wyk S. Marie* 1291, *Seintemarywyk* 1321. Otherwise simply *Wich* 1086, c.1170, *Wyke* 1327, *Wike* 1610. P.n. WEEK + saint's name *Mary* from the dedication of the church and for distinction from WEEK

ORCHARD SS 2300, 'orchard by Week', *Orcert* 1086, *Orchard* from 1202, OE **ortġeard**, GERMANSWEEK SX 4394 and PANCRASWEEK SS 2905. PNCo 177, Gover n.d. 33, 35, DEPN.

WEEKLEY Northants SP 8880. 'The wood or clearing by the *wīc*'. *(to) wiclea (forde)* [956]12th S 592, *Wiclei* 1086–1382 with variants *Wyk- Wik-* and *-le(e) -ley, Wi- Wykele(a) -ley* etc. 1172–1527. OE **wīc** + **lēah**. OE **wīc** as a specific in p.ns. may sometimes reflect Latin **vicus** and be related to Roman settlement, here the RBrit settlements at SP 8780 or 8881, the former a Roman industrial centre exploiting local iron-stone deposits. Nth 173, Pevsner 1973.26, 276.

WEELEY Essex TM 1422. 'Willow wood or clearing'. *Wilgelea al. Wiglea* [1042×66]c.1090, *Wigeleia -lai* c.1100, 1181, *Wileiam -leia -lega -lege -ley(e) -leya* 1086–1346, *Willeye* 1119, *Wyle(e) -leia -leya -leye -legh(e)* 1235–1362, *Wel(e)y* 1512×50, *Weighleigh alias Wyghleigh* 1548, *Weeley or Wiley* 1768, *Wheeley* 1805 OS. OE ***wiliġ, weliġ** + **lēah**. The 11th cent. form *Wilgelea* confirms this explanation and shows that the name cannot contain OE **wīġ, wēoh** 'a pagan shrine'. Ess 355, DEPN, Settlements 110, Jnl 2.46.

WEELEY HEATH Essex TM 1520. *Wheeley Heath* 1805 OS. P.n. WEELEY TM 1422 + ModE **heath**.

WEEPING CROSS Staffs SJ 9421. *Weeping Cross, so stiled from its vicinity to the antient place of execution* 1787 Pennant cited in OED s.v. ModE **weeping** + **cross**. A wooden cross stood here in the 16th cent. and according to Duignan 169 it was once frequented by penitents. The name-type occurs elsewhere, the earliest example being *Crucem Lacrymantem* before 1500 near Bury St Edmunds. The proverbial phrase *To come home by Weeping Cross*, i.e. to suffer grievous loss or disappointment, is recorded from 1579. OED s.v. Weeping Cross, O 396.

WEETHLEY Warw SP 0555 → WITHLEIGH Devon SS 9012.

WEETING Norf TL 7788. 'The wet place'. *Wetinge* [c.975]12th, 1086, 1275, 1610, *Watinge* [1042×66]12th S 1051, *Wetinga* 1086, *Watinga* [1086]c.1180, *Weting'* 1185–6. OE ***wǣting** < **wǣt** + **ing**^{2}. DEPN, ING 200.

WEETON 'Willow farm or settlement'. OE **withiġ** + **tūn**.

(1) ~ Lancs SD 3834. *Widetun* 1086, *With[.]tun'* 1219, *Wi- Wytheton* 1236–1327, *Wythington* 1286, *Wetheton* 1324–46 etc., *Weton* 1341 etc. La 153, Jnl 17.89.

(2) ~ NYorks SE 2847. *Widetone -tun(e)* 1086, *Wi- Wytheton* 1170×80–1348, *Wi- Wyton(a)* 12th, 13th, *Weton* 1343–1615, *Weeton* from 1539. YW v.51.

(3) ~ STATION NYorks SE 2747. P.n. WEETON SE 2847 + ModE **station** 'a railway station'.

WEETS HILL Lancs SD 8544. *Weets Hill* 1857. P.n. Weets, *Waytes* 1693, *Weets* 1817, probably **wēt** 'wet' in the sense 'wet place', + ModE **hill**. YW vi.36, 172.

WEETWOOD HALL Northum NU 0129. *Weetwood Hall* 1866 OS. P.n. *Wetewude* 1196, 1256, *Wetwod(e)* 1242 BF–1279, *Wetewod'* 1296 SR-1542, *Wheitwod* 1579, *Weetwood* 1628, 'the wet wood', OE **wēt**, definite form **wēta**, + **wudu**, + ModE **hall**, a late 18th cent. house. NbDu 210, Pevsner-Grundy etc. 1992.620.

Little WEIGHTON Humbs SE 9833. *Little Weighton* 1828. ModE **little** + p.n. *Wideton(e)* 1086, *Wi- Wytheton(a)* c.1207–1428, *Weton* 1410–1568, *Weeton* 1593, 'the willow farm', OE **withiġ** + **tūn**. The modern form has been influenced by nearby Market WEIGHTON SE 8841 from which the place then came to be distinguished as 'Little W'. YE 205 gives pr [la:tl wi:tn].

Market WEIGHTON Humbs SE 8841. *Market-Weighton* 1828. ModE **market** + p.n. *Wicstun* 1086, *Wi- Wychton(a)* 1165–1301, *Wi- Wycton(a) -tun* c.1150×60–1279×81, *Wi- Wyhtun(a) -ton* 1156–1285, *Wi- Wyghton* 1257–before 1678, OE **wīc-tūn** of uncertain significance, possibly 'settlement, village near the Roman *vicus*'. There was a market here since 1252 but the addition of *Market* is comparatively modern for distinction from Little WEIGHTON SE 9833. Previously it was also known as *Wighton subtus Olde als. under le Olde* 'under Wold' 1553 and *Wighton on the Woulde* 1569. Market Weighton lies between the Roman roads from York and Malton to Brough on Humber, Margary nos. 2e and 2g. YE 229 gives pr [ma:kitwi:tən].

WEIR Lancs SD 8725. No early forms. The place is situated on the Irwell, a mile from its source.

WEIR DYKE Humbs SE 9714. ModE **weir** + **dyke**.

WEIR WOOD RESERVOIR ESusx TQ 3934. P.n. Weir Wood + ModE **reservoir**. Weir Wood is the 'wood by the weir' ultimately from OE **wer** + **wudu**. The weir is probably referred to in the name of William atte Ware 1296. Sx 330.

WELBECK ABBEY Notts SK 5574. P.n. Welbeck + ModE **abbaye**, a Premonstratensian house founded in 1153 in the royal forest of Sherwood and subsequently a stately home dating from the 17th–19th cents. Welbeck, *Wellebech* c.1161–86, *-bec -be(c)k(e)* 1187–1324, is the 'well stream', ME **welle** (OE *wella*) + ME **bek** (ON *bekkr*) referring to the stream that rises at Cresswell Derby SK 5274 and feeds the Great Lake, *rivulus qui dicitur Wellebec* [1154×89]1327, *Wilebek-water, Wilebek streme* 1541. Nt 103.

WELBORNE Norf TG 0609. 'The spring stream'. *Walebruna* 1086, *Welebrun* 1203, *Wellebrunne* 1254. OE **wella** + **burna** influenced by ON **brunnr**. DEPN.

WELBOURN Lincs SK 9654. 'The spring stream'. *Weleburn(e)* [c.1080]13th, 1202–81, *Wellebrvne* 1086, *-burn(e)* 1212–1327, *Welburn(e)* 1276–1642, *-bourn(e)* from 1341. OE **wella** + **burna**. The village is situated on the spring line where the limestone ridge overlies the upper lias clay. Perrott 236, L 18, 30.

WELBURN 'Stream with a spring'. OE **wella** + **burna**.

(1) ~ NYorks SE 6784. *Wellebrune* 1086. The form is influenced by ON **brunnr**. YN 66, SSNY 167 no.2.

(2) ~ NYorks SE 7268. *Wellebrune* 1086, *Welbrun* 1243, *Welleburn(e)* 1167–1310, *Welburn(e)* from 1301. Forms with *-brun* are due to the influence of ON **brunnr**. YN 40, SSNY 167 no.1.

WELBURY NYorks NZ 3902. 'Spring hill'. *Welberga, Welleberg(e)* 1086, *Welleberyg -byry* 14th cent., *Welbery* 1400–1508. OE **wella** + **beorg**. The village lies on a hill-side with a spring known as *Hali Well*. YN 216.

WELBY Lincs SK 9738. 'The spring village or farm'. *Wellebi* 1086, *Wellebi -by* 1202–1467, *Welby* from 1346. OE **wella** + **bȳ**. Perrott 513, SSNEM 77.

WELCHES DAM Cambs TL 4786. *Welshes Dam* 1651, *Welsh his dam* 1652. Named after a dam across the Old Bedford River built by Edmund Welsh, an overseer to the Adventurers, the speculators who undertook to drain the fens in the 17th cent. Ca 242.

WELCOMBE Devon SS 2218. 'Spring valley'. *Walcome* 1086, *Welcuma -cumb(e)* [1189]1365, 1244, 1316, *Welcombe* 1356. OE **wella** + **cumb**. There are several springs in the valley, including a St Nectan's Well in the village and Well nearby. D 79.

WELDON Northants SP 9289. 'The hill with a spring'. *Wale(s)done, Weledene* 1086, *Weledone* 1086–1300, *Welledon* 1166–1324, *Weldon* from 1355. OE **wella** + **dūn**. Nth 174, L 30,145.

WELFORD Berks SU 4173. 'Willow ford'. *Weliford* [821]12th S 183, *Wœlingford* ibid. 13th, *(æt) Welig forda, (to) Weligforda* 949 S 552, *(æt) Weligforda* [956]12th S 622, *Waliford* 1086, *Waleford* 1284, *Weli- Weleford(e)* 1167–1316, *Welford* 1386. OE **weliġ** + **ford**. Brk 272, L 70, 221.

WELFORD Northants SP 6480. 'The spring ford'. *Wellesford* 1086, *Welleford(ia) -e* [1155×8]1329–1483, *Welford* 1347. OE **wella** + **ford**. Nth 76, L 30, 70.

WELFORD-ON-AVON Warw SP 1552. *Welford super Auene* 1433. P.n. *Welleford* 1086–1325, *Welneford* 1177–c.1560, *Welford* from 1314, 'the ford near the spring or springs', OE **wella**, genitive sing. **wellan**, genitive pl. **wellena**, + **ford**, + r.n. AVON. Gl i.259.

WELHAM GREEN Herts TL 2305. *Whethelhom grene* 1377×99. P.n. Welham + ME **grene**. Welham, *W(h)ethyngham* 1307×27, *Wethenham(feld)* 1333, *W(h)ethelham* 1387, is probably the 'withy enclosure', OE adj. ***withiġen** 'growing with willows' or ***withiġn** 'a willow copse' + **hamm**. Hrt 66.

WELHAM Leic SP 7692. Probably 'Weola's homestead'. *Wal(end)e- Welehā* 1086, *Weleham* c.1125–1328, *Welle-* 12th–1362, *Wel(l)am* c.1291–1550, *Welham* from 1208. OE pers.n. *Wēola* + **hām**. The many spellings with *Welle-* probably point to reinterpretation of this name as if containing OE **wella** 'spring, stream' rather than to its original form. The DB foms probably show AN *a* for *e* and possible influence of the r.n. WELLAND SP 8995. Lei 263.

WELL 'The spring'. OE, ME **wella**.

(1) ~ Hants SU 7646. *(la) Welle* 1237–1528. Ha 171, Gover 1958.112.

(2) ~ Lincs TF 4473. *Welle* 1086–1234, *Well'* 1197. DEPN, L 30, Cameron 1998.

(3) ~ NYorks SE 2681. *Welle* 1086 etc. The reference is to springs in the township now known as The Springs, St Michael's Well and Whitwell. YN 229.

WELLAND H&W SO 7940. Partly uncertain. *Wenelond* [1182]18th, *-land* [1190]1335, 1233, *Wenland -londe* [c.1197]18th, 1326–1649. Unidentified element + OE **land** 'newly cultivated land'. It has been suggested that the specific is a Welsh r.n. **Wen*, PrW **wïnn** 'white', referring to the stream which runs through the village and forms a tributary and branch of the Wyndbrook, earlier Wen brook, *(to) ƿen broce, (ond long) ƿen broces* [967]11th S 1314 in the bounds of Pendock, and that the Welland branch of Wyndbrook shared its name. More recently a fully Welsh etymology has been propounded, PrW **winn-lann* 'white enclosure' as in the common W p.n. Gwenllan. Wo 17, 177, Hooke 264–5, Celtic Voices 215.

River WELLAND Lincs TF 2727. Uncertain. *Weolud* 921 ASC(A), *Ueoolod* c.1000 Æthelweard, *Welund* [before 1118]12th, *Wey- Weiland* 1200–15th, *Weilound'* 1247, *Weland(')* 1218–1488 including [1135×54]14th, *Welland* from 15th. The OE form was *Wēolud* with *ēo* rather than *ĕo* because of the ME spellings *ei/ey* in which *i/y* is taken to be a graphy denoting vowel length. Suffix *-ud* may represent the Gmc noun suffix *-uth* but *wēol-* remains unexplained. In ME the name was variously assimilated, (1) to *lound* (ON *lundr*) 'a grove', (2) to *land* (OE *land*) 'newly cultivated land', and (3) to the present participle ending *-ande* as if it were *wellande* 'the welling, bubbling stream' from ME *wellen* 'to spring'. Perrott 40, RN 445, Kluge §99. LHEB 221 doubts a Celtic origin.

WELLESBOURNE Warw SP 2855. Possibly 'the pool stream'. *(in, æt) Weles burnan* [840]11th S 192, [862]c.1400 S 209, [969]17th S 773, *(æt) Walesburnan* [862]14th S 209, *Waleborne* 1086, *Walesborna -burn(a)* 1100×35–1246, *Welesburna -e* 1100×35–1235, *Wellesburn(e) -born(e) -bourne* from 1267. Forms from 1267 have been influenced by ME **welle** (OE *wella*) 'a spring, a stream'. The earliest spellings seem to point to OE **wǣl, wēl** 'a deep pool, a deep place in a river', genitive sing. **wǣles, wēles**, although pers.n. *Wealh*, genitive sing. *Wēales*, is perhaps not impossible. Wellesbourne is on the river Dene. Wa 286, PNE ii. 249.

WELL HILL Kent TQ 4963. *Well Hill* 1819 OS. ModE **well** + **hill**.

WELLING GLond TQ 4775. Uncertain. *Wellyngs* 1362, *Wellynges* 1367, *Wellyng* 1370, *Well end* c.1762. This might be a manorial name short for 'Welling's place or manor' or it might represent ME **welling** in the pl. There are springs in the vicinity. LPN 243.

WELLINGBOROUGH Northants SP 8967. 'Wendel's fortified place'. *Wendle(s)- Wedlingeberie* 1086, *Wendlesburg* 1221, *Wendlingburch -burgh -borwe* 1178–1389, *Wellingburg(h) -yng-* 1167, 1316. OE pers.n. **Wendel*, genitive sing. **Wendles*, varying with **Wendling* < **Wendel* + **ing**[4], + **burh**, dative sing. **byriġ**. Nth 140.

WELLINGHAM Norf TF 8722. 'Homestead at or called Welling, the place of springs, or of the folk who live at the springs'. *Walnccham* (sic) 1086, *Uuelingheham* c.1190, *Wellingeham* 1198, *Welingham* c.1200, 1242×3, *Wyllingham* 1232×3. OE **welling* < **wella** + **ing**[2] perhaps used as a p.n., locative-dative sing. **wellinge*, or folk-n **Wellingas* < **wella** + **ingas**, genitive pl. **Wellinga*, + **hām**. DEPN, ING 139.

WELLINGORE Lincs SK 9856. 'The flat-topped hill at or called *Welling*, the spring'. *Wallingour(e)* 1070×87, 1207×10, *Walingoure* 1096, *Wel(l)ingoure* 1086–1526, *Welling(h)or(e)* 1205–1650. OE **welling* < **wella** + **ing**[2] + **ofer**. The reference is to the rounded promontory on which the village is sited jutting out from the high ground of the Lincoln Cliff. The specific might alternatively be OE **Wellinga*, genitive pl. of an OE folk-n. **Wellingas* 'the spring people', and so 'the promontory of the Wellingas, the people of the spring'. A spring rises just S of the village and flows into the Brant at SK 9456. Perrott 239, L 175.

WELLINGTON H&W SO 4948. 'Spring settlement'. *Walintone* 1086, *-tona(m)* 1123×7–c.1227, *Welintun'* 1243, *-ynton(e)* 1247–1334. OE WMids form **wælling** + **tūn**. He 199, Sa 301–2.

WELLINGTON HEATH H&W SO 7140. *Wellington Heath* 1831 OS. P.n. Wellington + ModE **heath**. Wellington is *Weolintun* 1016×35 S 1462, *Walynton(e)* 1200×15–1305, *Wel(l)in(g)tone* 1210×12, 1295, a difficult name: the earliest form suggests 'estate called after Weola', OE pers.n. *Weola* + **ing**[4] + **tūn** or possibly 'farm called or at *Weoling*' from OE ***wēoling** 'a mechanical device' < OE ***weol** + **ing**[2]. The persistent spelling *Wal-* is difficult and might be due to interference from the p.n. Wellington in WELLINGTON SO 4948. He 200.

WELLINGTON Shrops SJ 6511. Partly uncertain. *Walitone* 1086, *Walintona -yn-* 1121, 1138, *Welintun -ton(a) -yn-* c.1145–1455, *Welington(') -yng-* 1177–1729, *Wolinton'* 1195–1214, *Wellington* 1244 and from 1535. The generic is OE **tūn** 'an estate', the specific an OE **wœling* or **weling* of unknown meaning and origin, but possibly an **-ing**[2] derivative of OE **walu** 'a ridge, a bank' referring to some kind of ditch or embankment, perhaps to the *agger* of Watling Street which runs through Wellington, Margary No 1h. The *Wol-* sps remain, however, a difficulty, as do the early *Wal-* spellings for explanations from OE **wēoling* 'a device, an artefact' or a putative earlier p.n. **Wēoh-lēah* 'the grove with a heathen temple'. Sa i.301, Årsskrift 1974.34.

WELLINGTON Somer ST 1320. Partly obscure. *We(o)lingtun* [899×909]16th S 380, *Welingtun* *[1065]18th S 1042, *Walintone* 1086, *-ton* 1178, *Wellinton* 1225, *Wellington* 1610. Unknown element + OE **tūn**, cf. WELLINGTON H&W SO 4948, Shrops SJ 6511. The specific might be an OE **Wēoling*, OE pers.n. **Wēola* + **ing**[4], and so 'estate called after W', or an unattested OE ***wēoling** 'a trap' < OE ***wēol** 'artifice' + *ing*[2] and so 'settlement, farm, village where a trap was used'. Neither explanation is very satisfactory and the name must remain partly unexplained. DEPN, NQ 1974.124, Sa i.301–2.

WELLOE Corn SW 5825. Uncertain. *The Welloe* 1748, 1826. The evidence for the name of this rock in the sea is too late for certainty; possibly Co **gwelow**, pl. of **gwel** 'sight', **guella** 'best' (i.e. best fishing) or a word cognate with W **gwäell** 'knitting needle, skewer, splinter'. Cf. also a rock in St Just in Penwith called *Guely breteny* c.1300, Co **guly** 'bed' + unknown element. PNCo 178, Gover n.d. 496.

WELLOW Avon ST 7358. Originally a stream name probably meaning 'winding stream'. *Welewe* [873×88]11th S 1507, *(on) Weleweheia* [984]12th S 854, *Weleuue* 1084, *Wel(l)ewe* 1233–1324, *Welowe* 1276. The stream is *Weluue, Welwe* [766 for ?774]c.1500 S 262, and on it lies RADSTOCK ST 6854, formerly *Welewe stoce* [984]12th S 854. OE **Welewe* from Brit *u̯elu̯o- related to Latin *volvo* and derived words in English such as *convolution*. RN 446, LHEB 281, 387.

WELLOW IoW SZ 3888. '(Settlement at) the willow-tree'. *(æt) Welig* [873×88]11th S 1507, *Welige* 1086, *Welegh(e)* 1293–1329, *Wel(e)we* 1185–1398, *Welowe* 1439–1507, *Wellow* 1559. OE(WS) **weliġ**. Wt 213 gives prs ['weɫə] and ['weɫou], Mills 1996.106.

East WELLOW Hants SU 3020. *Estwelewe* 1310. ME adj. **est** + p.n. *(æt) welewe* [873×88]11th S 1507, *(æt) Welowe* [931]13th S 1604, *Welle, Weleve* 1086, *Welewe* 1246, 1251, *Wellowe* 1428, by origin a river name, *(on) welewe* *[before 672 for ?793×6]12th S 229, *[825]12th S 275, *[905 for 931×4]12th S 393, *[948]11th S 540, [997]12th S 891, identical with WELLOW Somer ST 7358, PrW **welw* < Brit *u̯elu̯o- 'turned, turning'. 'East' for distinction from West WELLOW SU 2919. Ha 172, Gover 1958.193, RN 446–7, LHEB 387.

West WELLOW Hants SU 2919. *Westwelwe* 1461. ME adj. **west** + p.n. Wellow as in East WELLOW SU 3020. Gover 1958.193.

WELLS Somer ST 5445. '(At) the springs'. *(æt) Wyllan* (OE), *(in) Well'* (Latin) [1061×66]13th S 1115, [1066×75]13th S 1241, *Welle* 1086, *Welles* 1212, 1225, *Wells* 1610. OE **wylle**, dative pl. **wyllum**. An early Latin form of the name is *Fontinetum* [725]12th S 250. The reference is to the freshwater springs that still flow SE of the cathedral and which are referred to as *fontem magnum quem vocitant Wielea* (for *Wiella*) 'the great spring called *Wiella*' [767 for ? 774]c.1500 S 262. DEPN, L 30.

WELLS-NEXT-THE-SEA Norf TF 9143. *Wells Next the Sea* 1838 OS. Earlier simply *Guelle -a* 1086, *Wellis* 1291. The earliest form probably represents OE dative pl. **wellum**, '(the settlement) at the springs'. DEPN.

WELLSBOROUGH Leic SK 3602. 'The hill of Wheel'. There are four types:

I. *Wethelesberg(e)* 1181, 1208, *Wethelesberue -we* 1185, 1266, 1289.

II. *Quelesberge* 1210, *Whelesbergh(e) -is-* 1272–1377, *Wel(l)esber(e)we* 1293–1338, *W(h)ellesburgh* 1328–1542, *-borough* 1413.

III. *Weulesberg* 1285–1336, *Whewelesbergh* 1301, *Wheulesberwe -is-* 1304, 1327, *-bergh(e)* 1330–61.

IV. *Wheglesbergh* 1354.

They can be reconciled by assuming an OE **Hweogoles-berg*, OE **hweogol** 'a wheel', gentive sing. **hweog(o)les**, used either in a transferred topographical sense of the low rounded hill on which the village stands or as a p.n. **Hweogol* 'the wheel', + **beorg**. This may survive in type IV. Type I shows substitution of fricative [ð] for the peripheral phoneme [γ] of **hweogol**, type III substitution of [w] for [γw] and type II loss of medial [w] or [γ]. Lei 543.

WELLWICK Bucks SP 8057 → Kimble WICK SP 8007.

Great WELNETHAM Suff TL 8859. *Weltham magna* 1610, *Great Welnetham* 1837 OS. ModE adj. **great**, Latin **magna**, + p.n. *telueteha'* (< *at eluetham*), *hvelfihã* 1086, *Weluetham* 1170, *Welfuetham* 1179, *Welueteham* 1198, *Welnetham* 1206×7, 1254, 'wheel *Elfethamm*, the swan river-meadow', OE **hwēol** perhaps referring to a water-wheel here + p.n. *Elfethamm* < **elfitu** + **hamm** 3. The prefix is for distinction from THELNETHAM TM 0178, which also contains OE p.n. *Elfethamm*. 'Great' for distinction from Little WELNETHAM TL 8960.

Little WELNETHAM Suff TL 8960. *Wheltham p'ua* 1610, *Little Welnetham* 1837 OS. ModE adj. **little**, Latin **parva**, + p.n. Welnetham as in Great WELNETHAM TL 8859.

WELNEY Norf TL 5293. 'The river Well'. *Wellenhe* [n.d.]c.1350, *Welney* 1608. Latin forms *aqua de Welle* 'water of W' 1250, 1257. OE **wella** apparently used as a r.n., genitive sing. **wellan**, + **ēa**. Probably identical with *Wellestream* [1077]1444, *Wellestrem* c.1235, 1358, in the bounds of West Walton, OE **wella** used as a r.n. + **strēam**. DEPN, RN 447.

WELSH END Shrops SJ 5136. Short for *Whixall Welsh End* 1833 OS, 'the Welsh end of Whixal parish'.

WELSHAMPTON Shrops SJ 4335. 'Welsh Hampton', referring to proximity to a detached part of Flintshire. *Welch Hampton* 1649, *Welsh Hampton* 1726. ModE **welsh** + p.n. *Hantone* 1086, *-ton* 1255, *Henton* 1272, 1274, *Hampton* from 1316, 'at the high settlement', OE **hēah**, definite form dative case **hēan**, + **tūn**. Sa i.302.

Lower WELSON H&W SO 2949. *Lower Welson* 1833 OS. ModE adj. **lower** for distinction from Upper Welson SO 2951, *Upper Welson* ibid., + p.n. *Walston(e)* 1262, 1327, *Welston(e)* 1393, 1663, possibly 'the Welshman's estate', OE **wealh**, genitive sing. **weales**, + **tūn**. An alternative possibility is 'stone at the spring', OE Mercian dial. form **wælla** + **stān**. He 78.

WELTON 'Spring or stream settlement'. OE **wella** + **tūn**. L 30.

(1) ~ Cumbr NY 3544. *Welleton* 1272–1305, *Welton* from 1292. Cu 152.

(2) ~ Humbs SE 9527. *Wealletune* 1080×6, *Welletuna -e -ton(e)* 1080×6, 1086–1379, *Welton* from 1212, *Walleton* 1279×81. There are several springs, including St Anne's Well, in or near the village. YE 219.

(3) ~ Lincs TF 0079. *Welletonam* 1070×87, *-tone* 1086–1147×53, *Wellatuna* c.1115. DEPN, Cameron.

(4) ~ Northants SP 5866. *Welin- Waleton* 1086, *Weletone* 1086–1324, *Welton* from 1167. Nth 31.

(5) ~ LE MARSH Lincs TF 4768. 'W in the marsh'. *Welton in le Marshe* 1546. P.n. *Waletone -tune* 1086, *Welletuna* c.1115, *-ton'* 1185–1203, + Fr definite article **le** short for *en le* + ModE **marsh**, a modern addition for distinction from WELTON LE WOLD TF 2787. DEPN, Cameron 1998.

(6) ~ LE WOLD Lincs TF 2787. 'W on the wold'. P.n. *Welletone -tvne -tune* 1086, *-tuna* c.1115, *-ton'* 1156×8, after 1175, + Fr definite article **le** short for *en le* + ModE **wold**, a modern addition for distinction from WELTON LE MARSH TF 4768. DEPN, Cameron 1998.

WELWICK Humbs TA 3421. 'Dairy farm near the spring'. *Weluuic* 1086, *-wike -wyk(e)* 1199×1216–1486, *Wellewick* 1217, *-wic -wik(e) -wyk(e)* 1219–1417. OE **wella** + **wīc**. YE 22 gives pr [welik].

WELWYN Herts TL 2316. '(Place at) the willow-trees'. *(ad) Welingum* [944×6]13th S 1497, *(on) Welugun* [11th]12th K 1354, *We- Wilga -e* 1086, *Welewe(s)* 1198–1324, *Welewen(e)* 1220, 1307, *Welwyn* 1545, *Welwyn al. Welwys* 1626, *Wi- Wellen* 1675. OE **weliġ, wiliġ**, pl. **wel(i)gas**, dative pl. **wel(i)gum**. The modern form is probably a new ME pl. *Welwen* beside *Welwes* rather than a continuation of the dative pl. form *Welingum* (which seems to have been influenced by names in *-ingum*, dative pl. of *-ingas*). Hrt 144 gives pr [welin].

WELWYN GARDEN CITY Herts TL 2412. P.n. WELWYN TL 2316 + ModE **garden city**. A modern planned town built according to a scheme of Ebenezer Howard and designated a New Town in 1948. Room 1983.133.

WEM Shrops SJ 5129. 'Dirty ground'. *Weme* 1086, 1367, 1500–1638, *Wemme* 1228–1727, *Wem* from 1577. OE **wemm** 'spot, blemish, filth' referring to the marshy nature of the surrounding country. Cf. G field names *Wemme, Wemmewiesen*. Sa i.303, Bahlow 529.

WEMBDON Somer ST 2837. 'Huntsman hill'. *Wadmendvne* 1086, *Wemedon* 1227–57, *Wemdon* 1610. OE ***wǣthemann** + **dūn**. DEPN, PNE ii.238, L 143, 145, 147–8.

WEMBLEY GLond TQ 1985. 'Wemba or Wæmba's wood or clearing'. *(æt) Wemba lea* 825 S 1436, *Wambeleg' -le(ye)* 1294–1401, *Wembele* 1282, 1368, *Wembley* 1535, *Wemlee* 1387. OE pers.n. **Wæmba* or **Wemba*, + **lēah**. Mx 55.

WEMBURY Devon SX 5148. Partly uncertain. *Weybiria* [1100×35]1329, *Wenbir(ia) -bury* 1154×89–1318, *Wembury* 1359, *Waynbury* 1365. The generic is OE **byriġ**, dative sing. of **burh** 'fortification, manor', the specific probably OE **wenn** 'a wen, a tumour' referring to a local barrow or perhaps to one of the hills here; or possibly OE **wæġn** 'a wagon', which fits the spellings well but seems an odd compound with **burh**†. D 260, PNE ii.254.

WEMBURY BAY Devon SX 5147. P.n. WEMBURY SX 5148 + ModE **bay**.

WEMBWORTHY Devon SS 6609. 'Wemma or Wemba's enclosure'. *Mameorda* (sic) 1086, *Wemmewrth' -worthe -worthi -y* 1242–1333, *Wembworthy* 1420, *Wemry* 1629×51. OE pers.n. **Wemma* or **Wemba* + **worthiġ**. D 372 gives pr [wemǝri].

WENDENS AMBO Essex TL 5136. 'The two Wendens'. P.n. Wenden + Latin **ambo**, a modern addition referring to Great and Little Wenden parishes united in 1662. Wenden, *Wendena* 1086, *-den* from 1251, *Wandenne* 1206, is 'wen or bending valley', OE **wenn, wænn** possibly used as a p.n. referring to a well-marked round hill round which the valley bends, or ***wende**, ***wænde** 'a bend', + **denu**. Ess 542.

WENDLEBURY Oxon SP 5619. 'Wændla's fortified place'. *Wandesberie* 1086, *Wendelberi* 1163–1428 with variants *-bur(ia) -bury* and *-bir(y), Wendlebur'* from 1231 with similar variants. OE pers.n. **Wændla* + **byriġ**, dative sing. of **burh**. O 241 gives pr [windǝlberi].

WENDLING Norf TF 9312. Uncertain. *Wenlinga[m]* 1086, *(ad) Uuenlinge* [1087×98]12th, *Wenlingauilla* 'Wendling vill' 1166, *Wenling(e)* 1167–1275, *Wentlingg* 1254. It is unclear whether this is by origin a folk-n. **Wenlingas* 'the people called after Wennel', dative pl. **Wenlingum*, which would give ME *Wenling(e)*, or a singular p.n. **Wenling* < **Wennel* + **ing**² 'the place called after Wennel'. Nor is the pers.n. in question clear. The absence of early forms with *-d-* points to an OE **Wennel* corresponding to OHG *Wanilo*. On the other hand the consonant cluster *-ndl-* in the pers.n. *Wendel* in composition is often simplified, e.g. *Wenlingeburg* 1197 as opposed to *Wendle(s)berie, Wedlingeberie* 1086 for Wellingborough Northants. DEPN, ING 62.

WENDOVER Bucks SP 8607. Originally a stream name 'bright waters'. *(æt) Wændofran* (dative pl.) [c.968×71]12th S 1485, *Wendoure* 1086–1199, *Wandoura* 1189×99, *Wendovere* 1231. PrW ***wïnn** + **duβr**. Bk 157, RN 448.

WENDRON Corn SW 6731. 'St Wendron's (church)'. *(Ecclesia) Sancte Wendrone* 1291–1428, *Seynt Wendron* 1384, ~ *Gwendurne* 1514, *Wendron*, 1522, *Gwendron* 1569, *Gwendern* 1610. Saint's name *Wendern*, of whom nothing is known, from the dedication of the church. A Co form of the name with **eglos** 'a church' is found 1513, *Egloswendron*. PNCo 178, Gover n.d. 533, DEPN.

WENDY Cambs TL 3247. 'Island at the river bend'. *Wendeie -a -ey(e)* 1086–1457, *Wendy(e)* from 1272, *Wandrie, Wandei* 1086. OE ***wende** possibly used as a p.n. **Wende* 'watercourse with a bend, the Bend', + **ēġ**. Ca 67, L 37, 39.

Great WENHAM Suff TM 0738. *magna Wenham* 1327, *Great Wenham* 1805 OS. ModE adj. **great**, Latin **magna**, + p.n. *Wenham* from 1086, uncertain element + OE **hām** 'a homestead, an estate'. A possible specific is OE **wenn** 'a tumor' used as a hill-n., or ***wynne** 'a pasture'. There are several hill-spurs here. 'Great' for distinction from Little Wenham TM 0839, *Parva Wenham* 1242×3. DEPN, Baron.

WENHASTON Suff TM 4275. 'Wynhæth's estate'. *Wenadestuna* 1086, *Wenhaestun* [1199]1319, *Wenhaueston* 1197, 1230, *Wenhaston* from 1254. OE pers.n. **Wynhæth*, SE form **Wenhæth*, genitive sing. **Wenhathes*, + **tūn**. DEPN, Baron.

WENLOCK EDGE Shrops SO 5190. *Wenlock yitch* 1675, *Wenlock Edge* 1833 OS. P.n. Wenlock as in Much WENLOCK SO 6299 + p.n. *The Egge* c.1540, ModE **edge** 'an escarpment'. Gelling.

Little WENLOCK Shrops SJ 6407. *Parva Wenlok'* [1190]15th, ~ *Wenlac* 1232, ~ *Wenloc'* 1334, *Littell Wenloke* 1609. ModE **little**, Latin **parva**, + p.n. *Wenloch* 1086 as in Much WENLOCK SJ 6299 whence it was transferred to this new settlement on the NE edge of its territory. Sa i.305.

Much WENLOCK Shrops SJ 6299. *Magna Wenlak'* 1291×2, *Moche, Much Wenlok'* 1550×1, *Great Wenlock* 1749–56. ME **moche**, **much**, Latin **magna**, ModE **great** + p.n. *Wen-Wynlocan* c.1030 Secgan, *Wenloch* 1086, *-loc -lok(k)(e) -lock(e)* 1200 etc., *Wenlac -lake* 1200–1262, *Winloc'* c.1147, *Wynnlok'* 1291×2, *Weneloc(h) -lok -lac(h)* 1167–1290, *Wentlok* 1236, *Wendlok* [1281]1348, of uncertain origin. The earliest references are *Wininicas* [675×90]13th S 1798 and 901 S 221 the first syllable of which may contain PrW **wïnn** 'white' perhaps with reference to the limestone of Wenlock Edge. The name may be a compound of this syllable *Win-* with OE **loca** 'an enclosed place' possibly referring to a monastery. However, the name remains puzzling and even the form *Wininicas* is normally read as *Wimnicas*. St Mildburg founded a monastery here in c.675. Sa i.303, ECWM 197ff, esp.201.

River WENNING Lancs SD 6169, NYorks SD 7367. 'The river of the Wenningas, the people called after Wenna'. *Wenninga(m)* 1165×77, before 1177, *Wenningga* [c.1220]1412, *W(e)nnyng'* [1165×77–c1220]1412, *Wenning* from [1240×60]1268, *Wenn(e)y* 1577. OE folk-n. **Wenningas* < pers.n. **Wenna* + **ingas**, genitive pl. **Wenninga*, + **ēa**. The pers.n. **Wenna* would be a regular short form for OE names such as *Wer(e)nbeorht* etc. The alternative explanation, 'dark one', OE **Wenning* < **wann** + **ing**² depends upon *i*-mutation, but the settlement of this area took place after the period of this sound change. La 169 gives pr [wenin], DEPN, RN 448, Anglia 1993.75–81.

River WENNING NYorks SD 7167. 'The dark one'. *Wenning* from 1165×77. OE r.n. **Wenning* from OE **wann** 'dark' + **ing**². YW vii.142, RN 448.

WENNINGTON Cambs TL 2379. 'Estate called after Wenna'. *Weninton, Wennitona* [10th]14th S 1808, *Weninton* 1167, *Weny(ng)ton* 1293, 1322, *Wen(n)yngton* 1555, *Wy- Wempton* 1286. OE pers.n. *Wenna* + **ing**⁴ + **tūn**. Hu 221.

WENNINGTON GLond TQ 5381. 'Estate called after Wynna'. *Winintune* *[969]?12th S 774, [1042×44]13th S 1117, *Wini(g)tune* [1042×66]14th, *Wemtunā* (for *Weni-*) 1086, *Weninton* 1190–1253, *Wen(n)igton* 1235, 1247, *Weninge- Weington'* 1248, *Wenyngton* 1324, 1328, 1408, *Wynnyngton* 1553, *Wannyngton -ing- alias Wallyngton -ing-* 1569×70. OE pers.n. *Wynna*, SE form *Wenna*, + **ing**⁴ + **tūn**. Ess 139.

WENNINGTON Lancs SD 6170. Probably 'settlement on the river Wenning'. *Wennigetun, Wininctune* 1086, *Weni(n)gton* 1212–1271, *Weninton* 1229–43, *Wenyngton* 1332–46 etc. R.n. WENNING + OE **tūn**. Explanation is complicated by the existence of a place called Old Wennington, *Old Wenigton* 1227, *Old Weninton* 1229, to the NE on the river Greta, probably the original site. Greta is an ON r.n. presumably replacing an earlier name which, however, can hardly also have been Wenning. The name might be identical with WENNINGTON Cambs TL 2379 but it seems unlikely that the r.n. Wenning can be a back formation from the settlement name, in view of the earliness of its forms. Old Wennington is possibly, therefore, the 'settlement of the people who came from the neighbourhood of the Wenning', OE **Wenningas*, genitive pl. *Wenninga*, + **tūn**. La 181, Mills 147, YW vii.142, JNl 17.102, 18.23.

WENSLEY Derby SK 2661. 'Sacred grove of Woden'. *Wodnesleie* 1086, *-leg(a) -le(i)a -le(gh)* 1167–1240, *Wednesleg(a) -ley(e)* etc. 1180–1463, *Wendesley(e) -lay* c.1270–1587, *Wen(n)esley -le(e)* 1414–1573, *Wensley* 1446. OE *Wōden*, genitive sing. *Wōd(e)nes*, + **lēah**. Db 411, Settlements 110.

WENSLEY NYorks SE 0989. 'Wændel's forest clearing'. *(Alia) Wendreslaga* 'the second W', *(in duabus) Wentreslage* 'the two Ws' 1086, *Wandesle(i) -ley -legh* 1199 etc., *Wendesle(y) -lay* 1201–1396, *Wenselawe* 1363, *Wenslaugh* 1536. OE pers.n. *Wændel*, genitive sing. *Wændles*, + **lēah**. YN 257.

†The form *Wiganbeorth* late 9th ASC(A) under year 851 can only be OE pers.n. *Wicga* + **beorg** and does not belong here.

WENSLEYDALE NYorks SD 9988. 'Wensley valley'. *Wandesleydale* 1142, *Wandelesdale* [c.1146]14th, *Wendesleileyley- Wendeslacdale* 13th cent., *Wenslawdale* 14th. P.n. WENSLEY SE 0989 + ON **dalr**. Wensleydale is the valley of the river Ure. YN 246.

River WENSUM Norf TG 0518. 'The winding river'. *Wenson* [1096, 1119]c.1500, 1444, 1543, *Wensum* from 1250. OE adj. ***wendsum**. This name occurs also in Kent, *Uantsumu* [c.730]8th BHE, *Wantsumo Stream* [c.890]c.1000 OEBede, *(on þa ea, andlang) Wantsume* 944 S 497, 1586, OE ***wændsum**, the ancient name of the channel that formerly separated the Isle of Thanet from mainland Kent. RN 433f, E&S i.34, PNE ii.254.

River WENT NYorks SE 6117. 'The pleasant stream'. *Wenet(am)* [1154×89]13th–1235, *Wente* 1190–1559, *Went* from c.1220. PrW ***wïnet** or ***wïnēd** from Brit **ueneto-* or **uenētio* related to W *gŵen* 'smile, mirth, favour' with suffix *-eto* or *-ētion*, cognate with ON *yndi* 'charm, delight, joy' < **wunethia*, OE *wunōdsam* 'pleasant', ModE *winsome* and many other similar words all from PIE **wen-* 'strive, wish, love, content' seen in Lat *venus* 'love, pleasure' etc. **wïnet*/**wïnēd* would normally give OE **winet* as in NYMET SS 7108, 7200, but **wenet* would also be possible if the borrowing took place after the reduction of PrW pre-tonic *ï* to *ı* or ə. There is no ground for linking Went with the r.n. *Uinuaed* [c.731]8th BHE, *Winwede, Wunwœde stream* [c.890]10th OEBede, the site of Oswiu's defeat of Penda in 654. This means 'battle ford', OE **(ge)winn** 'fight' + **wæd** 'ford', and is either a complete invention or an etiological rationalisation of **winet*. YW vii.142, 35 f.n.l, RN 449, LHEB 279, 286, 672–81, GPC 1634.

WENTBRIDGE WYorks SE 4817. 'Bridge across the river Went'. *pontem de Wente* 1190, ~ *Wenet* 1190×1210, *Wentbrig(g) -bryg* 1302–1638, *-bridge* from 1545. R.n. WENT SE 5017 + **brycg**. It carried the Roman road from Doncaster to Tadcaster, Margary no. 28b. YW ii.51.

WENTNOR Shrops SO 3892. 'Wenta's flat-topped ridge'. *Wantenoure* 1086, *-ouura -overa* 1121, 1155, *Wontenour(e) -owr* [c.1200]1292–1305, *Wentenour(e) -overe* 1251–1333, *Wentnore* 1385. OE pers.n. *Wenta*, genitive sing. *Wentan*, + **ofer**. The variation between *a*/*o* forms and *e* forms is problematic; the former may represent a French sp for nasalised *e* before *nt*. Sa i. 306, Gelling 1998.84.

WENTWORTH 'Vineyard enclosure', 'Wintra's enclosure' or 'enclosure used in winter'. OE ***winter** 'a vineyard' or pers.n. *wintra* or **winter** + **worth**.

(1) ~ Cambs TL 4878. *Winteuuorde* 1086, *Wi- Wynteworða -uurð -word(a) -w(o)rth(e)* 1086–1428, *Wynterw(o)rd(')* c.1150, *-wrth* 1285, *Went(e)worth(e)* 1335–1573. Ca 243.

(2) ~ SYorks SK 3898. *Wint(r)euuorde* 1086, *Winterwurd* 1196, *Winteword(a)* 1152×5, 1202×10, *Wi- Wyntew(u)rth(e) -worth(e)* 1169×80–1432, *Wyntworth* 13th–1455, *Went-* from 1498. YW i.120.

WENTWORTH CASTLE SYorks SE 3203. *Wentworth Castle* 1771. Surname *Wentworth* as in p.n. WENTWORTH SK 3898 + ModE **castle**. A house built by Thomas, Earl of Strafford, one of the Wentworth family, in 1730. YW i.313.

WEOBLEY H&W SO 4051. 'Wibba's wood or clearing'. *Wibelai* 1086–1187, *Webbele(ya) -ley(e)* 1101×2–c.1540, *Weobley* 1832 OS. OE pers.n. *Wibba* + **lēah**. It seems unlikely that the modern form goes back to the OE back-mutated form of the pers.n., **Wiobba*, **Weobba*; more likely it is an antiquarian spelling under the influence of *eo* in Leominster – both are pronounced [e]. He 200.

WEOBLEY MARSH H&W SO 4151. *Weobley Marsh* 1832 OS. Already a hamlet in 1832. P.n. WEOBLEY SO 4051 + ModE **marsh**.

River WERE Wilts ST 8644. *Were* 1695, 1822. Earlier forms are to be found in the p.n. WARMINSTER ST 8644, *Worge* 899×924 S 1445, *Wer(e)* 1115–1357, *Wori* 10th cent. coins. OE ***wōriġ** 'wandering, unstable'. Wlt 10, RN 449.

WEREHAM Norf TF 6801. 'Homestead on the river Wigore'. *Wigorham* [c.1050]13th S 1529, *Wigreham* 1086, *Wir(e)ham* 1203, *Werham* 1251. OE r.n. **Wigore* + **hām**. This r.n., representing Brit **Uigora*, occurs in the French r.ns. Vière and Voire < Gaulish *Vigora* of unknown origin. Cf. also WORCESTER H&W SO 8555. DEPN, RN 476.

The WERGS WMids SJ 8701. 'The willow-trees'. *Witheges* 1202, *Wy-* 1306, *Withegis* 1327, *The Wergs* 1834 OS. OE **wīthiġ**, pl. **wīthigas**. DEPN, L 221.

WERNETH GMan SD 9104 → WAREN BURN Northum NU 1630.

WERNETH GMan SJ 9592 → WAREN BURN Northum NU 1630.

WERRINGTON Cambs TF 1703. 'Estate called after Wither'. *Witheringtun, Wiðringtun* [972]12th S 787, *Witherin(g)ton* 1199–1284, *Widerington(e)* 1086–1227, *Wirrintona* 1125×8, *Wy- Wirrington* 1563, 1712, *Weteryngton* 1285, *Weryngton* 1428, 1525. OE pers.n. *Wither* + **ing**[4] + **tūn**. The same pers.n. occurs in nearby WITTERING TF 0502 and may refer to the same individual. Nth 246.

WERRINGTON Corn SX 2492. 'Farm, village, estate called after Wulfred'. *Ulvredintone* 1086, *Wolvrinton* [1171]13th, *Wulfrinton* 1249, *Worryngton* 1324, *Werington* 1593, *Waryngton* 1610. OE pers.n. *Wulfrǣd* + **ing**[4] + **tūn**. PNCo 178.

WERRINGTON Staffs SJ 9447. 'The estate called after Wær'. *Werinton* 1259, *-yn-* 1272, *-ing-* 1279, *-yng-* 1330–1508, *Werrington* 1775. OE pers.n. **Wǣr*, short for *Wǣrheard* or the like, + **ing**[4] + **tūn**. Oakden.

WERVIN Ches SJ 4272. Uncertain. *Wivrevene, Wivevrene* 1086, *Weruenam* (accusative) [1096×1101]1280, *Werven* 1695, *Wervyn* 1351–1579, *-vin* from 1509, *Wy- Wirvyn -in* 1157–1724. Of the various attempts to solve this difficult name, perhaps the best are 'shaking fen', OE ***wifer** + **fenn** or 'cattle fen', OE **weorf**, genitive pl. **weorfa**, + **fenn**. Che iv.137, v(I.i)xli, L 41.

WESHAM Lancs SD 4232. '(Settlement, place) at the western houses'. *West(h)usum* 1189, 1246, *Westhus* 1204, *West(e)som* 1292, *Westsum* 1327, 1332, *Wessum* 1371, *Weshame* 1610. ON **vestr** or OE **west** + **hūsum**, locative-dative pl. of ON **hús** or OE **hūs**. The reference is to the situation NW of Kirkham. La 153, SSNNW 60, Jnl 17.89.

WESSINGTON Derby SK 3757. 'Wigstan's farm, village or estate'. *Wistanestune* 1086, *-ton* c.1210, *Wi- Wystanton(e)* 1154×9–1330, *Wi- Wyssington -yng-* 1309–1539, *Wes(s)yngton* 1321, 1349, *Was(s)hington* 1569–19th cent. OE pers.n. *Wīġstān*, genitive sing. *Wīġstānes*, + **tūn**. Db 322.

WEST BAY ModE **west** + **bay**.

(1) ~ ~ Dorset SY 6773. *The West Bay* 1710. W of Portland. Do i.225.

(2) ~ ~ Dorset SY 4690. A modern name.

WESTBERE Kent TR 1961. 'West byre or cowshed'. *Westbere* from 1212, *Westbyr'* 1278, *Westbir'* 1299. OE **west** + **bȳre**. K 515, Cullen.

WESTBOROUGH Lincs SK 8544. 'The western fortified place or manor'. *Westbvrg -burg* 1086–1316, *-burgh(e)* 1242–1603, *-borugh* 1303, *-borow(e)* 1405–1675. OE **west** + **burh**. Situated 2½ m. SW of Hougham and Marston. Perrott 387.

WESTBOURNE Dorset SZ 0791. Like Northbourne SZ 0895 and SOUTHBOURNE SZ 1491, a modern name for a district of Bournemouth. Do 152.

WESTBOURNE WSusx SU 7507. *Westbo(u)rne* 1302×7. ME **west** + p.n. *Borne, Burne* 1086, 'the stream', OE **burna**. Still referred to locally as Bourne in 1929, *west* was added for distinction from EASTBOURNE TV 6199. Sx 55, SS 72.

WESTBURY '(At the) west fortified place or manor house'. OE **west-byriġ** < **west** + **byriġ**, dative sing. of **burh**.

(1) ~ Bucks SP 6235. *Westberie* 1086, *Westbiry* [1154×89]1269, *Westburi* 1302. Bk 50.

(2) ~ Shrops SJ 3509. *Wesberie* 1086, *Wesbury* 1346, 1625–1811, *Westbir' -bur' -biry -buri(e) -bury(e)* from 1203×4. W of Pontesbury SJ 4006. Sa i.306.

(3) ~ Wilts ST 8751. *Wes(t)berie* 1086, *Westbiri* from 1115 with variants *-bury -buri -bery* etc. 'West' perhaps because near the western border of the county. Wlt 149.

(4) ~ LEIGH Wilts ST 8650 → Westbury LEIGH ST 8650.

(5) ~ -ON-SEVERN Glos SO 7214. *Westbury(e) on Sevarne* 1297. Earlier simply *(wið) Wæst byrig* 11th ASC(A) under year 1053, *Wes(t)berie* 1086–1167, *-buri(a) -bury(e) -bir' -byry* 1154×89–1595, *Wes(e)bury(e)* 1438–1592. The place lies W of Gloucester and the Severn on the Roman road to Wales, Margary no. 60a. Gl iii.201.

(6) ~ -SUB-MENDIP Somer ST 5048. *Sub Mendip* 'under Mendip' is a modern addition to *Westbyrig* *[1065]c.1500 S 1042, *Westberie* 1086, *Westbury -bery* [1159, 1174×91], *Westbury* 1610. The bishop of Bath had a park here and this was his 'western manor'. DEPN.

WESTBY Lancs SD 3831. 'The west village or farm'. *Westbi* 1086, *Westby* from 1226. ON **vestr** or OE **west** + **bȳ**. W of Kirkham. La 151, SSNNW 43.

WESTCLIFF-ON-SEA Essex TQ 8685. A modern seaside resort at Milton, *Mildentunā* 1086, *Mid(d)elton(a) -e* 1155–1548, *Mi-Mylton* from 1342, the 'middle settlement' between Leigh TQ 8385 and Southchurch TQ 9186. The name is entirely modern and does not appear in the 1805 OS. Ess 191.

WESTCOMBE Somer ST 6739. *West Combe* 1610. ModE **west** + **combe**. The hamlet lies W of BATCOMBE SY 6838.

WESTCOTE Glos SP 2220. 'West cottage(s)'. *Westcote* from 1315, *Wescote* 1457. OE **west** + **cot**, pl. **cotu**. Gl i.228.

WESTCOTT 'The western cottage(s)'. OE **west** + **cot**, pl. **cotu** (ME *cote*).

(1) ~ Bucks SP 7117. *Westcote* c.1200. A hamlet two miles W of Waddesdon. Bk 141.

(2) ~ Devon ST 0204. *Westcote* 1439. The place lies W of Bulealler ST 0204. D 562.

(3) ~ Surrey TQ 1448. *Wescote* 1086, *Westcot'* 1202, *-cotes* 1210×12, *Westcote juxta Dorkyng* 1342. ME secondary pl. **cotes**. W of DORKING TQ 1649. Sr 273.

WESTCOTT BARTON Oxon SP 4225 → Westcott BARTON SP 4225.

WESTDEAN ESusx TV 5299. 'West valley'. *Wes(t)den(e)* [1189]14th–1389. OE **west** + **denu**. 'West' for distinction from East DEAN TV 5598. In DB Westdean is simply *Dene* 1086, 'the valley'. Sx 419.

WEST END ModE **west** + **end**.

(1) ~ Avon ST 4469. 'West end' sc. of Nailsea parish'. *West End* 1817 OS.Contrasts with EAST END ST 4870.

(2) ~ Hants SU 4614. *Westend* 1607. But this is not the west end of any parish. Possibly it represents OE **wǣstenne** 'waste, wilderness' reshaped under the influence of HEDGE END SU 4912. Ha 172, Gover 1958.42.

(3) ~ Humbs SE 9130. 'The west end' of the village of South Cave. *West End* 1824 OS. ModE **west** + **end**.

(4) ~ Norf TG 4911. The 'west end' of Caister-on-Sea. Cf. *(Great, Little, Further) West End Marsh* 1841 TA. Nf ii.5.

(5) ~ NYorks SE 1457. *West End* 1733 sc. of Thruscross Township. **west** + **end**. YW v.127.

(6) ~ Oxon SP 4204. *Westende* 1278×9, *West End Village* 1774. The west end of Stanton Harcourt. O 283.

(7) ~ Surrey SU 9460. *West End* 1680. The 'west end' of Chobham parish. Sr 119.

(8) ~ GREEN Hants SU 6661. *West End Green* 1759. The west end of Stratfield Say and so named for distinction from Fair Oak Green SU 6660, *Fayer Oake* 1622, *Fair Oak Green* 1759. Gover 1958.123, 122.

WESTERDALE NYorks NZ 6605. 'Western valley'. *Westerdale -dala* [1154×81]16th etc. ON **vest(a)ri** + **dalr**. One of the more westerly valleys of Eskdale. YN 134.

WESTERDALE MOOR NYorks NZ 6602. P.n. WESTERDALE NZ 6605 + ModE **moor**.

WESTERFIELD Suff TM 1747. 'The western open land'. *Westrefelda* 1086, *Westerfeld* 1206, 1568, *Westerfeild* 1610. OE **westerra** + **feld**. 'Western' for distinction from WALDRINGFIELD TM 2844. DEPN, Baron.

WESTERGATE WSusx SU 9305. 'The western gate'. *Westgate* 1230, 1288, *Westregate* 1271. ME *west, wester* + **gate**. *West* for distinction from EASTERGATE SU 9405. Cf. also WOODGATE SU 9304. Possibly all were gates onto common land. Sx 64.

WESTERHAM Kent TQ 4454. 'Western homestead or estate'. *Westarham* 871×89 S 1508, *Westerham* from [c.975]12th B 1321, *O(i)streham* 1086. OE **westra** + **hām**. The DB spelling is odd, cf. *Oistreham* 1086 for Ouistreham, Calvados, France, 'eastern village', Gmc *ooster*. KPN 227.

WESTERLEIGH Avon ST 7079. 'More westerley wood or clearing'. *Westerlega -le(i)gh(e) -ley(e)* 1176–1595. OE **westerra** + **lēah**. Gl iii.68.

WESTERNHOPE MOOR Durham NY 9133. *Westernhope Moor* 1866 OS. P.n. Westernhope + **mōr**. Westernhope, *Whestanhope* 1418, *Westenhope* 1768, and Westernhope Burn, *Westanburnshele* 1457, *Whesnup bourn in Wardaill* 1598, derive from OE **hwetstān** + **hop** and **hwetstān** + **burna**, 'valley, stream where whet-stones are found or manufactured'. NbDu 211.

WESTERN ROCKS Scilly SV 8306. *The Western Rocks* 1792. PNCo 178.

WESTERTON Durham NZ 2331. 'Western settlement' sc. of the three Merringtons. *Westerton* 1647. OE **westra** + **tūn**. Otherwise known as West Merrington, *West Merington -yng-* 1296–1580, *West Merrington* 1647, adj. **west** + p.n. Merrington as in Kirk MERRINGTON NZ 2631.

WEST FELL Cumbr NY 6601. *West Fell* 1865. We ii.48.

WEST FEN 'West fenn'. ME, ModE **west** + **fen**.

(1) ~ Cambs TL 3698. *Westfen* 1251. Like WESTRY TL 3998 W of March. Ca 258.

(2) ~ Cambs TL 5182. *Westfen* 1251. W of Ely. Ca 225.

(3) ~ Lincs TF 3154. *The Weste Fenne* 1661, *West Fen* 1824 OS. The western part of the Fens between the Witham and the seacoast drained and enclosed 1801–18. Wheeler 197ff.

WESTFIELD ESusx TQ 8115. 'Western' or possibly 'waste open land'. *Westewelle* 1086, *-felde* 12th, *-feud* 1291, *Westerfeuld* 1248. OE **west** or **weste** + **feld**. W in relation to Guestling. Sx 505, SS 70.

WESTFIELD Norf TF 9909. 'West open land'. *Westfeld* *[1042×66]12th S 1051. OE **west** + **feld**. W of Whinburgh and Yaxham. DEPN.

WESTGATE Durham NY 9038. *(the) Westgate (in Stanhope Park)* 1647, cf. *Westyatshele* 1457 'Westgate shieling'. OE **west** + **ġeat**, later replaced by ModE **gate**. The reference is to the west gate of the Bishop of Durham's hunting park at Stanhope in Weardale. NbDu 211, Trans D&N n.s.4.31.

WESTGATE Humbs SE 7707. *Westgate* 1824 OS. ModE **west** + **gate**.

WESTGATE Norf TF 9740. 'The west hamlet'. *Westgate* 1838 OS. ModE **west** + **gate** < ON *gata*.

WESTGATE ON SEA Kent TR 3270. 'The western gates' leading to the sea. *Westgata* 1168, *-gate* from c.1300, OE **west** + **ġeat**, pl. **gatu**, + ModE **on sea**. 'West' of MARGATE TR 3670. The 'gate' is a gap in the hills leading to the sea. Cullen.

WEST GREEN Hants SU 7456. *West Green* 1817 OS. W of Hartley Wintney SU 7756 and contrasting with PHOENIX GREEN SU 7655. Gover 1958.115.

WEST HALL Cumbr NY 5667. *West Hall* 1589. Cu 98.

WESTHALL Suff TM 4181. 'The west nook'. *Westhala* 1169–94, *-hal(e)* 1177–1344, 1524, *-hall* 1195 and from 1568. OE **west** + **halh**, dative sing. **hale**, in the sense 'valley'. 'West' in contrast to Sotherton TM 4479, 'the southern estate', *Sudretuna*

1086, *Sutherton* 1229–1316, *Sotherton* from 1275, OE **sūtherra** + **tūn**. DEPN, Baron.

WESTHAM ESusx TQ 6404. 'West promontory'. *Westham(me)* 1222–1428. OE **west** + **hamm** 2b, 4. The place lies W of Pevensey. Sx 446, ASE 2.28, SS 83.

WESTHAM Somer ST 4046. 'West pasture'. *West Ham* 1817 OS. ModE **west** + **ham** (OE *hamm*). The hamlet lies at the W end of the old west field of Wedmore ST 4347.

WESTHAMPNETT WSusx SU 8806. *Westhamton'* c.1200, *Westhamptonette* 1279. ME **west** + p.n. *Hentone* 1086, *Hamptoneta* c.1187, '(the place) at the high settlement', OE **hēan-tūne** < **hēah**, definite form dative case **hēan**, + **tūn**, + OFr diminutive suffix **-ette**. 'West' for distinction from East Hampnett SU 9106, *Esthamton* 1288, *Esthamptonette* 1347, earlier simply *Antone* 1086. The earliest reference is to the land of Hampton, *terram heantunensem* (Latin) *[680 for ?685]10th S 230. Sx 78, 67, SS 74.

WESTHAY Somer ST 4342. *Westhay* 1817 OS. Earlier forms are needed for this name, which may represent OE **west īeġ** 'west island'. It lies at the end of a low ridge W of Meare and Godney. ECW 109.

WESTHEAD Lancs SD 4307. 'The west headland'. *Westhefd* c.1190, *Le Westheued* 1366. OE **west** + **hēafod**. La 123.

WEST HEATH Hants SU 8556. *West Heath* 1817 OS. W of Farnborough SU 8753, part of Aldershot Heath.

WESTHIDE H&W SO 5844. 'West Hide'. *Westhyde* 1243, 1292. ME adj. **west** + p.n. *Hide* 1086, *(La) hide* 1157×62–1196, *Hyde* 1291, 'the hide of land', an estate so assessed for taxation. 'West' for distinction from Monkhide SO 6143, which belonged to Gloucester abbey, *Hida Monachorum* Latin 'H of the monks' 1243, *Monkshide* 1321, 1413, earlier simply *(la) Hyda -e* 1113×20, 1273, and Little Hide, now Monkesbury Court SO 6143, *Little Hyde* 1329, earlier Latin *Parua Hyda* c.1220. He 201, 215.

WEST HILL Devon SY 0693. Called *Ottery Hill* 1809 OS. Possibly identical with *Westdowne* 16th. W of the river Otter. D 616.

WESTHOPE H&W SO 4651. *Westhope* 1832. There are no early forms but this must be the 'western secluded valley' in relation to HOPE UNDER DINMORE SO 5052, ultimately from OE **hop**.

WESTHOPE Shrops SO 4786. 'The west valley'. *Weshope* 1086, *Westhop' -hope* from c.1200. OE **west** + **hop**. Cf. EASTHOPE SO 5695. Sa i.307.

WESTHORPE Lincs TF 2131. 'The western thorp or outlying settlement'. *Westthorpe* 1318, *Westhorp(e)* 1394–1458, *Westrop* 1571. ME **west** + **thorp**. W of Gosberton TF 2331. Payling 1940.88, Cameron.

WESTHORPE Suff TM 0569. 'The west outlying settlement'. *West(t)orp, Westurp* 1086, *Westhorpe* from 1285. OE **west** + **thorp**. DEPN, Baron.

WESTHOUGHTON GMan SD 6505. *Westhalcton* c.1240, *-hal(gh)ton* 14th, *-houghton* 16th, *Westhowftun* 1864. ME adj. **west** + p.n. *Halcton* c.1210, 1258, *Halcghton* 1246, *Halghton* 1332, the 'nook settlement', OE **halh** + **tūn**. The reference is probably to one of the valleys nearby, perhaps that at Water's Nook SD 6605. 'West' for distinction from Little Houghton. La 43, L 107–8, Jnl 17.36.

WESTHOUSE NYorks SD 6774. 'The west houses'. *West(e)hous(e)* 1387, *-houses* 1392, *Westus* 1609. OE **west** + sing./pl. **hūs**. The place lies W in the parish of Thornton in Lonsdale. YW vi.250.

WESTHOUSES Derby SK 4258. A modern development named after West House Farm, *West House* 1840. 'West' in relation to Blackwell SK 4458, in which township it is situated. Db 213.

WESTHUMBLE Surrey TQ 1651. Partly uncertain. *Wy- Wistumble-* 13th cent., *Westhumble* from 1536, *Wystomble, Whistumble* 17th cent. The generic is OE ***stumbel** 'a tree stump'. The specific may be OE **wiċe** 'a wych-elm' or **wisce** 'a marshy meadow'. Sr 82.

WESTLAKE Devon SX 6253. *Westlake* 1813 OS. The place is possibly referred to in the surname of Richard *Bywestelake* 1333, 'west of the streamlet', OE **west** + **lacu**. D 274.

WESTLAND Devon SS 6542 → WISTLANDPOUND RESERVOIR SS 6441.

WESTLEIGH 'West wood or clearing'. OE **west** + **lēah**.

(1) ~ Devon SS 4728. *Weslege* 1086, *Westlegh* 1238. 'West' for distinction from Eastleigh SS 4827, *Estlegh* 1242. Earlier simply *Leia* 1086. These two places together with TAPELEY SS 4729 seem to have been parcels of an area originally simply called Leigh. D 124.

(2) ~ Devon ST 0616. *Westlegh* [c.1200]15th. This is the W division of a wood of which the other part is Canonsleigh ST 0616, where an abbey was founded c.1170, *Leghe Canonicorum* 1282, *Canounleye* 1403. Originally simply *Lei* 1086, *Lega* [1173×5]1329, 'the wood', which must have referred to the whole area. D 548, 547.

WESTLETON Suff TM 4469. 'Vestlithi's estate'. *Westledes- Westlens- Weslestuna* 1086, *Westleton* from 1202, *Westleveton(')* 1268–1323. ON pers.n. *Vestlithi*, genitive sing. *-litha*, secondary genitive sing. *-lithes*, + **tūn**. DEPN, Baron.

WESTLEY 'West wood or clearing'. OE **west** + **lēah**.

(1) ~ Shrops SJ 3607. *Westlegh'* 1291×2, *-ley* from 1562, *Woosley* 1672. 'West' in contrast to ASTERLEY Shrops SJ 3707, 'the eastern wood or clearing'. Sa ii.71.

(2) ~ Suff TL 8264. *Westlea* 1086, *Uuestlea* c.1095. W of Bury St Edmunds. DEPN.

(3) ~ WATERLESS Cambs TL 6256. 'Westley pastures'. *Westle Waterles* 1285, *Westeley Waterlesse al. Westley Waterleyes* 1556. P.n. Westley + ME **water-leyes**. There is no lack of water here. Westley, *(at) Westle* [1043×5]13th S 1531, *Westlai* [1042×66]13th S 1051, *Weslai* 1086, *Westle(e)* 1220–1441, is 'west wood or clearing'. W of Dullingham Ley and SW of Stetchworth Ley. The whole district was once well wooded and probably called *Lēah*. Ca 120.

WESTLINGTON Bucks SP 7610. Probably the 'place west in the township'. *Westinton* 1384, *Westlington* 1696, *Weslin(g)ton* 1709–10, *West Dinton* c.1825. The present form points to ME **west** + p.n. DINTON 7610 but this is probably a popular etymology for an earlier name. ME **west in toune** 'the place west in the township', a formula normally confined to western counties, would account for the 1384 form. Bk 160.

WESTLINTON Cumbr NY 3964. 'West Linton'. *Westlevington* c.1200–1682 with variant *-yng-*, *West Linton* 1583. OE **west** + p.n. Linton 'settlement on the river Lyne' (*Leuen* 1259). 'West' for distinction from KIRKLINTON NY 4366. See further River LYNE NY 4972. Cu 117.

WESTMARSH Kent TR 2761. *Westmersh* 14th, *Westmarsh* from 1484×5. ME **west** + **mersh**. PNK 532, Cullen.

WESTMESTON ESusx TQ 3313. 'The most westerly settlement'. *Westmæstun* [c.765]14th S 50, *Westmestun(e)* 1086, *-ton* from 1291. OE **westmest** + **tūn**. So called by the people of Lewes on account of its position beyond Plumpton. Sx 304, SS 75.

WEST MILL Herts TL 3627. *Wes(t)mele* 1086, *Westmella -melna -melle -muln -mull -milne* 1130–1303, *Westmyll* 1590. OE **west** + **myln**. W possibly in relation to a lost mill on the opposite bank of the Rib. Hrt 209 gives common pr [wesməl].

WESTMINSTER GLond TQ 2979. The 'western monastery' sc. in relation to the city of London. *Westmunster* [785]12th S 124, *-minster* [959]c.1100 S 1293, *-mynster* 972×8 S 1451. OE **west** + **mynster**. According to tradition the abbey was originally known as *Torneie* [969]c.1100, *(in) loco terribili que ab incolis nuncupatur Thorney* 'in a terrible place called by the inhabitants T' 972×8 S 1451, 'thorn-tree island', OE **thorn** + **ēġ**. The reference is to the inhospitability of the original site, an island between two branches of the Tyburn at its marshy outfall into the Thames. The name is remembered in Thorney Street of c.1931. Mx 165, Encyc 887.

WEST MOORS Dorset SU 0802. *West Moors* 1591. Earlier simply *La More* 1310, *Moures* 1407, '(the) moors', OE **mōr**. The sense here is 'marshy ground' beside the Moors river. 'West' for distinction from East Moors SU 1002, *East Moors* 1811 OS, and *Little Moors* ibid. Do 152.

WESTMORLAND. Former county absorbed in Cumbria in 1974. 'District of those living west of the moors'. *Westmoringa land* c.1150 ASC(E, F) under year 966, *Westmerland(e)* 1129–1707, *Westmarilond -meriland(e)* 1150–1357, *Westmorland* from 1257. OE folk-n. *Westmōringas* 'the people west of the moors', genitive pl. *Westmōringa*, + **land**. ME *-meri-* spellings may show influence of OE **(ġe)mǣre** 'boundary' or reflect a Scandinavian form *Vestmøringa-*. We i.1.

WESTNEWTON Cumbr NY 1344. *West Newton* 1777. ModE adj. **west** + p.n. *Neutona* c.1187, *Neuton* before 1195–1399, OE **nīwa-tūn** 'the new settlement'. 'West' apparently for distinction from NEWTON ARLOSH NY 1955. Cu 328.

WESTOE T&W NZ 2366 → WESTOW NYorks SE 7565.

WESTON 'West settlement, farm, village, estate'. OE **west** + **tūn**.

(1) ~ Avon ST 7266. *(æt) Westtune* [946]12th S 508, *(æt) Westune* [961 for ?956]12th S 661, *Westone* 1086. W of Bath; the estate belonged to Bath abbey. Contrasts with BATHEASTON ST 7867, another estate of the same abbey. DEPN.

(2) ~ Berks SU 4073. *Westun* 1086, *Westuna -ton(')* from 12th. W of Welford SU 4173. Brk 274.

(3) ~ Ches SJ 5080. *Westone* 1086, *-ton* from 1260. At the W tip of the Runcorn peninsula and contrasting with NORTON SJ 5582, SUTTON WEAVER SJ 5479 and ASTON SJ 5578. Che ii.182. Addenda gives 19th cent. local pr ['wesn], now ['westən, 'wessn].

(4) ~ Ches SJ 7352. *Westun* 13th, *Weston* from 1260. W of Barthomley SJ 7652. Che iii.75 gives pr ['westn], older local ['wessn].

(5) ~ Dorset SY 6871. *Weston* from 1324, *Wesson* 1608, *West Town* 1774. On the opposite side of the Isle of Portland to EASTON SY 6971. Do i.220.

(6) ~ Herts TL 2630. *Westone* 1086. Hrt 146.

(7) ~ Lincs TF 2925. *Westune* 1086, *Weston* from 1176. W of Moulton. Payling 1940.60, Cameron 1998.

(8) ~ Northants SP 5847. *Weston(e)* from 1162, *Weston Pynkney* 1311. The place is situated at the W end of Weedon Lois parish; Weedon Lois was the head of the Pinkney Honour. Nth 45.

(9) ~ NYorks SE 1747. *Weston(e) -tun(a)* 1086–1689. W of Otley. YW v.64.

(10) ~ Notts SK 7767. *Westone* 1086, *-ton(a)* 1130 etc. 'West' in relation to Normanton SK 7968. Nt 197.

(11) ~ Shrops SJ 5629. *Westune* 1086, *-ton(')* from 1195, *Wesson* 1666, 1693. The W part of the ancient parish of Hodnet. Sa i.310.

(12) ~ Shrops SO 5993. *Weston* 1255. Also known as *Weston Monachorum* 'the monks' W' 1291, *Moncke Weston* 1569. In the W of Monkhopton parish which belonged to Wenlock Priory. Sa i.308.

(13) ~ Staffs SJ 9727. *Westone* 1086. DEPN.

(14) ~ BAMPFYLDE Somer ST 6124. 'W held by the Baumfeld family'. *Weston Bampfield* 1811 OS. Earlier simply *Westone* 1086. The manor was held by John de Baumfeld in 1316. It lies W of Cadbury Castle and contrasts with SUTTON MONTIS ST 6224, the settlement S of Cadbury Castle. DEPN.

(15) ~ BAY Avon ST 3060. P.n. Weston as in WESTON-SUPER-MARE ST 3261 + ModE **bay**.

(16) ~ BEGGARD H&W SO 5841. 'W held by the Beggard family'. *Weston Bagard* 1831 OS. Earlier simply *Westvne* 1086. West of Yarkhill SO 6042. He 202.

(17) ~ BY WELLAND Northants SP 7791. ~ *super Wylond* 1377, *Weston Bassett al. Weston by Wolland* 1609. P.n. *Weston(e)* from 1086 + r.n. WELLAND SP 8995. W in relation to Ashley SP 7971. The manor was held by the Basset family in 1191. Nth 175.

(18) ~ COLEVILLE Cambs TL 6153. 'W held by the Coleville family'. *Westone Coluyle* 1236, *Westkoleulle -colvill* 1324, 1327. Earlier simply *Westone* [1043×5]13th S 1531, 1086, 1176 etc. 'West' because W of Carlton. Held by William de Coleville 1202. The boundary of the folk of Weston is *(oð) West tuniga gemæra* [974]11th S 794. Ca 121, Jnl 2.55.

(19) ~ CORBETT Hants SU 6847 → WESTON PATRICK SU 6946.

(20) ~ GREEN Cambs TL 6252. *Weston Green* c.1825. P.n. Weston as in WESTON COLEVILLE TL 6153 + ModE **green**.

(21) ~ HEATH Shrops SJ 7713. P.n. Weston as in WESTON-UNDER-LIZARD Staffs SJ 8010 + ModE **heath**, cf. HEATH HILL SJ 7614 and *Heath Birches* 1833 OS.

(22) ~ HILL Shrops SO 5582. *Weston Hill* 1833 OS. An outlier of the Clee group of hills, rising to 1052ft. P.n. Weston as in Cold Weston SO 5583, *Coldeweston(')* 1255–1563 *Cold Weston* from 1395, earlier simply *Westona* c.1090. W either in relation to Stoke St Milborough SO 5782 or to the Clee Hills SO 5975. Sa i.309.

(23) ~ HILLS Lincs TF 2720. *Weston Hill* 1824 OS. P.n. WESTON TF 2925 + ModE **hill(s)**.

(24) ~ -IN-GORDANO Avon ST 4474. *Weston in Gordenlond* 1271, ~ *in Gordene* 1343. Earlier simply *Westone* 1086. The suffix is as in CLAPTON-IN-GORDANO ST 4774. DEPN.

(25) ~ JONES Staffs SJ 7624. 'John's W'. *Weston Johannis* 1236, *Weston J(h)ones* from 1245. P.n. *Weston* 1242 + manorial addition pers.n. *John*. The village lies in the most westerly part of Norbury parish. St i.176, Horovitz.

(26) ~ LONGVILLE Norf TG 1115. P.n. *Westuna* 1086, *Weston* 1201, 1838 OS, + p.n. Longville as in NEWTON LONGVILLE Bucks SP 8431. DEPN.

(27) ~ LULLINGFIELDS Shrops SJ 4224. *Weston' super Lullingfeld* 1291×2, *Westone in Lollyngfyld'* 1516. P.n. *Weston* 1255 + p.n. Lullingfeld, the 'open country of the *Lullingas*, the folk called after Lulla'. Cf. STANWARDINE IN THE FIELDS Shrops SJ 4124. Sa i.308.

(28) ~ -ON-THE-GREEN Oxon SP 5328. P.n. *Westone* 1086 etc. + ModE **on the green**. O 243.

(29) ~ -ON-TRENT Derby SK 4028. *(into) Westun(e)* 1009, *-ton(e) -tona* from 1086. W in relation to the adjoining parish ASTON ON TRENT SK 4129. 'On Trent' for distinction from WESTON UNDERWOOD SK 2942.

(30) ~ PATRICK Hants SU 6946. 'W held by Patrick' sc. de Chaworth, *Paterik de Chaworces* 1257. *Weston Patrik* 1314. Earlier simply *Westone* 1086, *Weston* 1212. 'Patrick' for distinction from Weston Corbett SU 6847 held by the Corbet family in 1203, *Westone Corbet* c.1270, earlier simply *Weston* 1203. Ha 173, Gover 1958.135, DEPN.

(31) ~ RHYN Shrops SJ 2835. *Weston Ryn* 1302, ~ *Rhyn* 1670. P.n. *Westone* 1086, 1272, + p.n. Rhyn, referring to a ridge of high ground known as Y Rhyn 'the hill or promontory', *Ryn* 1302, Welsh def. article **y** + **rhyn** 'a point, a peak', for distinction from Weston Coton in Oswestry. Sa i.309, O'Donnell 101.

(32) ~ -SUB-EDGE Glos SP 1241. 'W beneath the edge' sc. of the Cotswolds. *Weston(e) sub Egge* 1255–1535, ~ *under Egge* 1303–1478. Earlier simply *Weston(e)* 1086–1248. 'West' in relation to ASTON SUBEDGE SP 1341, the 'east settlement'. The boundary of the people of Weston is referred to in [1005]12th S 911 as *Wæsðæma gemære*, i.e. *Westhæma*, genitive pl. of *Westhæme* 'the west dwellers'. Gl i.261.

(33) ~ -SUPER-MARE Avon ST 3261. 'W on sea'. *Weston super Mare* 1349. P.n. *Weston* 1266 + Latin **super** 'on' **mare** 'sea' for distinction from WESTON ZOYLAND ST 3534. DEPN, TC 191.

(34) ~ TURVILLE Bucks SP 8511. 'W held by the Turville family'. *Westone Turvile* 1302, *Weston Turwyld -vyld* 1552. Earlier simply *Weston* 1086. In 1146 Geoffrey de Turville granted John of the Lee one hide of his demesne in Weston. 'West' in relation to ASTON CLINTON SP 8712. Bk 166, Jnl 2.30.

(35) ~ -UNDER-LIZARD Staffs SJ 8010. *Weston subtus Lus(e)yord -yerd* 1349–1420, ~ *under Lizard* 1833. P.n. *Guestona* 1081, *Westone* 1086, *Weston* c.1150–1428 + ModE **under**, Latin **subtus**, + p.n. *Lusgerde* [664]13th S 68, *Lusgeard* [680]12th S 72, *Lu- Lisgarde* 1291, 'the hall by the hill', PrW ***līs** + **garth**. The parish lies in the W part of Cuttlestone Hundred NE of Lizard Hill in Shrops. St i.180, DEPN.

(36) ~ UNDER PENYARD H&W SO 6323. *Weston sub Penyard*. Earlier simply *Westvne* 1086. Penyard SO 6122, *Penþard* (with <þ> for <y>) 1216×72, *Penyerd -iard -(i)erd* 1227–1230, *Penyard* 1280, is PrW ***penn** 'top, head'+ ***garth** < ******garto-* 'hill' referring to Penyard Hill which rises to 621ft. a mile W of the village. Weston is the *west tūn* contrasting with the *ēast tūn* at ASTON INGHAM SO 6823. DEPN, He 202, 174, Turner 1950–2.115.

(37) ~ UNDER WETHERLEY Warw SP 3669. *Weston juxta Wethele(ye)* 1327, ~ *sub(tus) Wethele(y)* 1354, ~ *under Wetherly* 1695. P.n. *Westone* 1086, *Weston(e)* from 1173, + ModE **under** + p.n. Waverley SP 3471, *Wethele* 1204–84, 1656, *(bruille de* 'forest of ~ 1204, *bruera de* 'heath of ~' 1279), *Waveley* c.1830, 'the hunting wood or clearing', OE ***wǣthe** + **lēah**. Wa 189, 185.

(38) ~ UNDERWOOD Bucks 8650. 'W under the forest' sc. of Yardley Chase. *Westone Underwode* 1363. Earlier simply *Westone* 1086. Bk 16.

(39) ~ UNDERWOOD Derby SK 2942. *Weston Underwode* 1301–1415. Earlier simply *Westune* 1086, *-ton(e)* 1154×9–1346. 'Underwood' for distinction from WESTON-ON-TRENT SK 4028. Db 508.

(40) ~ ZOYLAND Somer ST 3534. 'W in land belonging to *Sowi*'. *Westonzoyland* 1809 OS. Earlier *Westsowi* c.1245 'west *Sowi*' and *Weston* 1263, 1610. In DB the estate is included in MIDDLEZOY ST 3733, *Sowi* 1086. DEPN.

(41) Alconbury ~ Cambs TL 1777. *Alkmundebir Weston* 1227, *Aukingbury Wiston* 1675. P.n. ALCONBURY TL 1875 + p.n. *Westune* 1086. W of Alconbury. Also known as *Wodeweston* 1260, 1286, 'wood Weston' in contrast to Old WESTON TL 0977, 'wold Weston'. Hu 249.

(42) Buckhorn ~ Dorset ST 7524. 'Bleachers' W'. *Boukeresweston* 1275–1403, *Bokernes-* 1344, *Bo(u)kerne Weston* 1346–1431, *Weston Bukkehorne* 1535, *Buchorne Weston* 1664. Earlier simply *Weston(e)* 1086–1428, referring to the situation in relation to Gillingham ST 8026. ME **bouker** 'a buckwasher, a bleacher' (< *bouken*, OE **bucian* 'steep in lye, bleach'), genitive sing. **boukeres**, + p.n. Weston. *Bouker* may be used here as an occupational term (perhaps referring to a special feudal duty) or as a surname. The genitive sing. *-es* was identical with the genitive pl. in ME and the spellings with *-n-* probably derive from the alternative SW genitive pl. ending *-ene* < OE *-ena*; **boukerene* was subsequently changed to *buck-horn* 'the horn of a buck' (not recorded before 1447×8) as a result of popular etymology. Do iii.84.

(43) Coney ~ Suff TL 9577. The modern forms, *Conyweston* 1610, *Coney Weston* 1837 OS, point to ModE **cony** (ME *cuni*) 'a rabbit' + p.n. Weston as in Market WESTON TL 9871. These are probably reinterpretations of earlier sps *(at) Cunegestone* [1051×2 or 1053×7]13th S 1080, *Cunegestuna* 1086, 'the king's manor', OE **cyning** influenced by ODan **kunung**, genitive sing. **cy- kuninges**, + **tūn**, with the same development *-inge-* > *-ige-*, *-ege-* > *-ewe-* as in HONINGTON TL 9174, MONEWDEN TM 2458 etc. DEPN.

(44) Edith ~ Leic SK 9305. 'W belonging to Eadgyth'. Prefix *Edi- Edy-* 1263–1610, *Edith(e)* 1309–1634 + p.n. *Weston(e)* 1114–1610. Ēadgȳth, wife of Edward the Confessor, held large possessions in Rutland in 1066. The village lies W of Ketton SK 9084. The prefix is for distinction from COLLYWESTON Northants SK 9902. R 221.

(45) Hail ~ Cambs TL 1662. 'W on the river Hail'. *Heilweston* 1199, *Hai- Hayleweston* 1219–1376. R.n. Haile + p.n. Weston. W of St Neots. Hail, *Haile* [c.1180]13th, *Hayle* 1256, 1276, is an ancient name of the river Kym. It may represent British **Saliā* on the root *sal-* 'grey' as in OIr *salach* 'dirty'. 'Dirty' is an accurate description of this clogged, sluggish river. BdHu 7, 275, RN 188.

(46) Market ~ Suff TL 9871. 'W with the market'. *Market Weston* 1837 OS. ModE **market** + p.n. *Westuna* 1086, *Weston* 1202. Contrasts with Blo NORTON Norf TM 0179. DEPN.

(47) Old ~ Cambs TL 0977. *Oldweston* 1535, *Owld Wessen* 1594, is really a form of *Wald Weston* 1227, *Wold(e)weston* 1249–1346, 'Weston on the wold' sc. of Bromswold (as in LEIGHTON BROMSWOLD TL 1115). OE **wald** (ME *wold*, dial. *old*) + p.n. *Westune* 1086. 'Wold' for distinction from Alconbury or Wood WESTON TL 1777. 'West' in relation to EASTON TL 1471. Hu 250, Dobson 421.

(48) Priest ~ Shrops SO 2997. 'The priests' W'. *Presteweston', Prestes Weston', Weston' Prustes* 1291×2, *Prest(es)weston* 1317–1421, *Priest Weston* from 1637, *Preece* ~ 1672. OE **prēost**, genitive pl. **prēosta**, + p.n. *Westune* 1086, *Weston* 1228 etc. W in the county and held by Chirbury Priory. Sa i.309.

(49) South ~ Oxon SU 7098. *Southwestone* 1526. ModE **south** + p.n. *Westone -tune* 1086 etc. O 99.

(50) Stalbridge ~ Dorset ST 7216. *Stalbridge Weston* from early 17th. Earlier simply *(æt) Westtune, (apud) Westonam* (Latin version) [933]12th S 423, *Westun* [998]12th S 895, *Westone* 1086. W of Stalbridge. Also known as *Weston Abbots* 1334 referring to possession by Sherborne abbey from the 10th cent. Do iii.283.

WESTON Hants SU 7221. '(Place) west of the village' or 'west in the estate'. *Westeton* 12th–1314, *Westreton* 1248, 1304, *Westin(g)ton -yn- -i-* 1278–1347. The later form suggests OE **west in tūne** 'west in the village or estate' although **(be) westan tūne** 'west of the village' and **westerra tūn** 'the western settlement' are also possible. Ha 173, Gover 1958.56.

WESTONBIRT Glos ST 8589. 'W held by the le Bret family'. *Bretteweston* 1309, *Weston(e) Brut(t)* 1322–92, ~ *Britt(e) -y-* 1324–1501, ~ *Birt(e) -y-* 1535–1717, earlier simply *Weston(e)* 1086–1749, 'west settlement', OE **wes** + **tūn**. Westonbirt lies west of Tetbury. It was held in 1242 by Richard le Bret. Gl i.114.

WESTONING Beds TL 0332. 'W held by the Ing family'. *Westone Ynge* 1365. Earlier simply *Westone* 1086–1365, 'the west settlement', OE **west** + **tūn**. The manor was acquired in 1303 by William Inge, chief justice under Edward II (1314–16). Bd 141.

WESTOW NYorks SE 7565. Probably the 'women's meeting place'. *Wi- Wyueston -v- -w* [12th]15th–1451 with occasional *n* for *u*, *Wystowe* 1365, *Westow(e)* from 15th. OE **wīf**, genitive pl. **wīfa**, + **stōw**. Alternatively the specific may be OE pers.n. **Wifa* or **Wife*. Cf. Westoe Tyne & Wear NZ 2366, *Wiuestoue* c.1125, *-stowe* 1228, *Westowe* 1539. YE 145, NbDu 211.

WESTPORT Somer ST 3820. No early forms.

WEST ROAD ESusx TR 0016. ME **road** 'Sheltered water suitable for offshore anchorage', West of Hastings towards Dungness.

WEST ROW Suff TL 6775. *West Row* 1836 OS. ModE **west** + **row** 'a row of houses'. Contrasts with BECK ROW TL 6577 and HOLYWELL ROW TL 7077. W of Mildenhall.

WESTRY Cambs TL 3998. 'Western island'. *West(e)rey(e)* 1221–c.1450, *Westry* 1597. OE **westerre** + **ēġ**. The sense is probably 'island W of March', in contrast to Estover Farm TL 4298, *Estiw(o)rth(e)* 1221–1331, *Estworth* 1251–1331, *Easterforth, Estwith, Eastover* 17th, the 'eastern enclosure', OE **ēasterra** + **worth**. Cf. EASTREA TL 2997. Ca 257, 256.

WEST SCAR Cleve NZ 6026. *West Scar* 1861 OS. ModE **scar** 'a sunken rock, a reef' < ON **sker**.

WEST STREET Kent TQ 9054. 'West hamlet'. *West Street* 1819 OS. ModE **west** + Kentish dial. **street**. Nearby is WARREN STREET TQ 9253 and Payden Street TQ 9254, *Peyton Street* 1819 OS.

WEST TOWN Avon ST 4868. 'West settlement' sc. of Backwell ST 4968. *West Town* 1817 OS.

WESTWARD Cumbr NY 2744. 'West ward'. *le Westwarde in Allerdale* 1354. OE **west** + **weard**. The west ward or division of the forest of Inglewood. Cu 329.

WESTWARD HO! Devon SS 4329. A modern settlement named from the title of Charles Kingsley's novel (1855), many of the scenes of which are laid in this neighbourhood. D 103.

WESTWEEK Devon SX 4293 → GERMANSWEEK SX 4293.

WESTWELL 'The west spring or stream'. OE **west** + **wella**.

(1) ~ Kent TQ 9947. *Westwell(')* from 1226, *-welles* 1247–70. Earlier simply *Welle* 1086, 'the spring', OE **wella**. 'West' for distinction from Eastwell in EASTWELL PARK TR 0147. KPN 198.

(2) ~ Oxon SP 2210. *Westwelle* 1086 etc. O 332 gives pr [westəl].

(3) ~ LEACON Kent TQ 9647. 'Westwell common'. *Westwell Leacon* 1798, ~ *Leaton* 1819 OS. P.n. WESTWELL TQ 9947 + Kent dial. **leacon** 'a wet swampy common'. PNK 402, Cullen.

WESTWICK 'The west dairy-farm'. OE **west** + **wīċ**.

(1) ~ Cambs TL 4265. *Westuuiche* 1086, *-wich'* 1196–1201, *-wic(a) -wik(e) -wyk(e)* 1130–1428. Ca 154.

(2) ~ Norf TG 2726. *Westuuic* 1086, *-wyc -wik -wyk -wic(h)* [1101×7]13th–1534. Nf ii.199.

WESTWOOD Devon SY 0198. No early forms. W of Clyst St Lawrence SY 0299. ModE **west** + **wood**.

WESTWOOD Wilts ST 8059. 'The west wood'. *(to) Westwuda* [987]12th S 867, *Westwode* 1086–1363 with variant *-wude*. OE **west** + **wuda**. Wlt 122.

WEST WOODLANDS Somer ST 7743. Short for *Frome West Woodlands* 1817 OS. Contrasts with East Woodlands ST 7844, *Frome East Woodlands* 1817 ibid.

WESTWOODSIDE Humbs SK 7499. ModE **west** + p.n. *Woodside*.

WETHERAL Cumbr NY 4654. 'Wether nook'. *Wetherhala* c.1100, *Wetherhal(e)* from 1229, *Weder- -ir-* 1131–1539. OE **wether** 'castrated lamb' + **halh**, dative sing. **hale**, here in the sense 'raised ground in marsh'. Cu 160, L 109, 110.

WETHERBY WYorks SE 4048. 'Wether-sheep farmstead'. *Wedrebi* 1086, *Werrebi -by* c.1150–1310, *Wetherby* from 1221. ON **vethr**, genitive sing. **vethrar**, + **bȳ**. YW v.38, SSNY 41.

WETHER CAIRN Northum NT 9411. *Wether Cairn* 1869 OS. ModE **wether** + **cairn**. A boundary cairn at 1834ft. where the parish boundaries of Alnham, Biddlestone and Alwinton meet. It lay on the boundary of the medieval Kidland estate of the monks of Newminster Abbey, and is therefore to be identified with the boundary cross which the monks set up at the head of Allerhope Burn. Newm 76, Dixon 38–9.

WETHERDEN Suff TM 0062. 'Wether valley'. *Wederdena* 1086, *Wetherden(n)* from 1197 with variants *-dene -don(e)*. OE **wether** + **denu**. DEPN, Baron.

WETHER FELL NYorks SD 8787. *Wether Fell Side, Wether Peat Ground* 1860 OS. ModE **wether** + **fell**.

WETHERINGSETT Suff TM 1266. Probably 'the fold(s) of the Wetheringas, the people of Wetherden'. *Weddreringesete* [1017×35]12th S 1520, *Wetheringsete* [1043×5]13th S 1531, *Wederingesete* [1042×66]12th S 1051, *Weringhe- Wederingaseta* 1086, *Wetheringeset* 1201, *Wetheringset(e) -yng-* 1261–1346, 1568, *Wetheringsett* 1338. OE folk-n. **Wetheringas* < *Wether-* as in p.n. WETHERDEN TM 0062 + **ingas**, genitive pl. **Wetheringa*, + **set**, pl. **setu**. DEPN, Baron.

WETHER LAIR Northum NY 7096. *Wether Lair* 1869 OS. A hill rising to 1622ft. on the boundary between Kielder and Emblehope. ModE **wether** + **lair**.

WETHERSFIELD Essex TL 7131. 'Wihthere's open country'. *Witheresfelda, Westrefeldā* 1086, *Wihterefeld* 1135×54, *Westerfeld* 1216×72, *Weðres- Wether(e)sfeld -is- -feud* 1177–1430, *Were(s)feld -feud* 1185–1281. OE pers.n. *Wihthere* later replaced by **wether** 'a wether', genitive sing. *Wihtheres*, **wetheres**, + **feld**. The first four forms, two with AN <s> for [ç], point to *Wihthere* which was later replaced by **wether**. Ess 456 gives pr [wʌðəzfl].

WETHERUP STREET Suff TM 1464. *Wetherup Street* 1837 OS. The name means 'Wetherup hamlet' and is presumably related to nearby p.n. WETHERINGSETT TM 1266.

WETLEY ROCKS Staffs SJ 9649. *Wetley Rocks* 1784. Cf. *ye Rocks* 1734, 1795. P.n. Wetley as in *Wetley Moor* 1529, *Watteley Moor* 1535, *Wyttley more* 16th, *Whitle moore* c.1540, 'the wet woodland glade', ultimately OE **wēt** + **lēah**, + ModE **rocks**. Oakden. Horovitz.

WETTENHALL Ches SJ 6261. 'Wet nook'. *Watenhale* 1086–1360, *-hall* 1581, [17th]1724, *Wetenhale* 1121×9–1671, *-hall(e)* 1238–1724, *Wettenhall* from 1240×6, with variant *Wetnall*. OE **wēt** + **halh**, in the sense 'dry ground in marsh', dative sing. definite form **(æt thǣm) wētan hale** 'at the wet nook'. Che iii.166 gives pr ['wetnɔ:l], older local ['wetnə], L 106, 110.

WETTON Staffs SK 1155. '(The settlement at) the wet hill'. *Wetindona* 1188×94, *Wettindun* 1252, *Wetton* 1327. OE **wēt**, definite form dative sing. **wētan**, + **dūn**. DEPN, Horovitz.

WETWANG Humbs SE 9359. 'Field of summons for the trial of a dispute, trial field'. *Wetuu- Wetwangha'* 1086, *Wetewang(e)* c.1155–1376, *Wetwang* from 1297. ON **véttvangr**. YE 128 gives pr [wetwan] beside [wetwang], SSNY 108, Saga-book iv.102.

WETWOOD Staffs SJ 7733. 'The wet wood'. *Wetewode* 1291, *Wetwode* 1312. OE **wēt**, definite form **wēta**, + **wudu**. DEPN, L 228, Horovitz.

WEXCOMBE Wilts SU 2759. 'Wax valley'.

I. *Wexcumbe* 1167–1332, *Waxcombe* 1426, 1817 OS;

II. *West Cumbe* 1155, *Westcumba -e* 1156–1275.

OE **weax** + **cumb**. The reference is to a valley where bees' wax was obtained or produced. Wexcombe lies W of Tidcombe, whence the alternative name 'west coomb'. Wlt 347, L 93.

River WEY An Old European r.n. **u̯aisā* < IE **u̯oisā* on the root **vis-*/**veis-* 'water, liquid'. See also MEDWAY, WYE. RN 451, BzN 1957.237.

(1) ~ Dorset SY 6681. *Waye* 1244, 1288, *Weye* 1367, *Way or Wile* c.1540, *Wey* 1586. For other forms see BROADWEY SY 6683, UPWEY SY 6684 and WEYMOUTH SY 6779. RN 451, BzN 1957.237.

(2) ~ Surrey TQ 0557. *(andlang) Wœgan, Wegan* [956]13th S 621, *Waie* 12th cent., *Waie(muþe)* [672×4]13th S 1165, *Waye* 13th cent., *Weye* 1342, *le Wey(mouth)* 16th cent., *Wye* 17th cent. There are two views about the origin of this r.n. type. The OE form of the name, **Wœġe*, is either a reflex of Brit **U̯aisā* < IE **weis-ā* on the root **vis-*/**veis-* 'water, flow' as in r.ns. WEAR, WYRE, WYE etc., or possibly to be derived from IE **wegh-* 'move, shake' seen in Latin *vehere*, OE *wǣġ* 'a wave (in the sea)'. Sr 7, RN 451, BzN 1957.237–8, LHEB 45.

WEYBOURNE Norf TG 1143. Partly uncertain. *Wabrunna -brune* 1086, *Walbruna* 1158, *Wabrun(n)* 1177, 1228. The generic is ON **brunnr** possibly replacing OE **burna** 'a stream', the specific unknown. DEPN.

WEYBREAD Suff TM 2480. 'The expanse of land by the road'. *Weibrada -e* 1086, *-breda -e* 1187–1346, *Weiebred* c.1200, *-brade* c.1205, *Waybred(e)* 1269–1568. OE **weġ** + **brǣdu**. The road in question is the Roman road from Pulham to Peasenhall, Margary no. 35. DEPN, Baron.

WEYBRIDGE Surrey TQ 0764. 'The bridge over the river Wey'. *Wai- Weigebrugge* [672×4, 727]13th S 1165, 1181, *Webruge -brige* 1086, *Waibrigge* 12th–1439 with variants *Way- -brugge* and *-bregge*, *Wabrigge* 1177, *-s* c.1270, *Weybrugge* [933]13th S 420, *-brigge* 1294, 1300, *-brigga* [1062]13th S 1035, *Waybrigga* [967]13th S752. R.n. WEY TQ 0557 + OE **brycg**. Sr 98, L 65.

WEYHILL Hants SU 3146. 'Hill at Wey'. *Wayehyll* 1571, *Woo al. Way al. Weyhill* 1700. P.n. Wey + ModE **hill**. Wey, *la Wou* c.1270, *(la) Woe* 1299–1341, *(la) Weo* 1318, 1379, *(la) Wee* 1379–1543, *Weye* 1412, is of uncertain meaning. It has frequently been taken to represent OE **wēoh** 'a heathen shrine', but this word did not survive into ME and could hardly have been used with the definite article. Possibly, therefore, it is OE ***weġ-hōh** 'hill-spur climbed by a road'. Gover 1958.169, Evidence 100–1, 111, Ha 174.

WEYMOUTH Dorset SY 6779. 'Mouth of the r. Wey'. *(of) Waimouþe, (on ðan) Waymouþe* [843 for 934]17th S 391, *(apud) Waimudā* 1130, *Waymue* 1152–1293, *-muth(e)* 1252–1382, *Weymue* 1225–1308, *-muth(e)* 1248–1408, *-mouth(e)* from 1290. R.n. WEY SY 6681 + OE **mūtha**. Do i.250.

WEYMOUTH BAY Dorset SY 6980. *Weymouth Bay* 1811 OS. Earlier *the Roade of Waymouth* 1587, *Weymouth Road* c.1825, ModE **road** 'an anchorage for ships'. In S 938 the sea here is referred to as 'the east sea', *in þare æst Sæ* [978×84]14th, as opposed to the 'west sea' of Lyme Bay. Do i.257.

WHADDON 'Wheat hill'. OE **hwǣte** + **dūn**. L 144, 149–53.

(1) ~ Bucks SP 8034. *Hwætædunæ* *[966×75]12th S 1484, *Wadone* 1086, *Waddon* 1152–1333, *Whaddon* from 1238. Bk 74 gives pr [wɔdən], ECTV no.152, L 144, 149–50, 153.

(2) ~ Glos SO 8313. *Wadvne* 1086, *Waddona - don(e) - dun* 1148×79–1631, *Whaddon* from 1252. Gl ii.173.

(3) ~ CHASE Bucks SP 8032. *chace of Whaddon* 13th. P.n. WHADDON SP 8034 + ME **chace** 'a hunting reserve'.

WHADDON Wilts SU 1926. 'The wheat valley'. *Watedene* 1086, 1242, *Hwatedena* 1109×20, *W(h)addene* 1273–4, *Whaddon* 1332, *Waddon* 1811 OS. OE **hwǣte** + **denu**. Wlt 375, L 98, 144.

WHADDON Wilts ST 8761. 'Woad hill'. *Wadone* 1086, *Waddune -don* 1234–1394, *Whadone* 1363, 1428. OE **wād + dūn**. This name has been assimilated to the form of WHADDON. Wlt 144, L 144.

WHALE Cumbr NY 5221. 'Isolated hill'. *Vwal* 1178, *Vhala* c.1230, *Hvale* [before 1239]1294, *Whale* from 1234×6. ON **hváll** 'an isolated round hill' referring to a hill rising to 856ft at NY 5321. We ii.183 gives pr [we:l], SSNNW 175, L 169.

WHALE CHINE IoW SZ 4678. *Whale Chine* 1810 OS. ModE **whale** + Hants/IoW dial. **chine** (OE *ċinu*) 'a gorge'. Probably so called from the stranding of a whale here, or possibly from the local family name Wavell [weiəl], who owned nearby Atherfield farm 1557–1636. Wt gives pr [wiəltʃain]. Mills 1996.107.

WHALE ISLAND Hants SU 6302. *Whale Island* 1722. Earlier simply *Whalle* 1535, 1506, perhaps representing OE **hwæl-ēġ** 'whale island'. Gover 1958.10.

WHALEY Derby SK 0081. 'Wood, clearing or pasture at the road'. *Wal'* c.1211×25, *Walegh* 1290, *-ley* 1411, *Weile, Weyle(g'), Wayle(y) -legh* 1216×72–1357, *Waley* 1399. OE **weġ** + **lēah**. The early *Wa-* spellings reflect Mercian **wæġ** for **weġ**, Campbell §328. The reference is to the road from Buxton to Stockport. Che i.176.

WHALEY Derby SK 5171. Uncertain. *Wallie -ley(e) -leg'* 1231–17th cent., *Whal(e)y* from 1540. The generic is OE **lēah** 'wood, clearing', the specific any of OE **wall** 'a wall', **wala**, genitive pl. of **walh**, 'of the Welshmen' or **wælla** 'a spring'. Db 215.

WHALEY BRIDGE Derby SK 0181. *Whaley-bridge* c.1620, *Wely Bridge* 17th cent. P.n. WHALEY SK 0081 + ModE **bridge**. Refers to a bridge near the r. Goyt. Che i.178.

WHALEY THORNS Derby SK 5371. *Whaley Thorns* c.1840. P.n. WHALEY SK 5171 + ModE **thorns** 'thorn-trees'. The place is a modern development. Nt 85.

WHALLEY Lancs SD 7336. 'Clearing by the hill'. *Hwælleage* c.1050 ASC(D) under year 798, *Hweallæge* c.1121 ASC(E), *Wallei* 1086, *Walalege* c.1130, *Walleya -lega -leye* 1124–1284, *Whalley* 1246, *Qwalley* 1346, *Whaulley* c.1540. OE ***hwæl** + **lēah**, dative sing. **lēaġe**. The reference is to the great whale-like headland of Whalley Nab SD 7335 which rises to 558ft. overlooking the r. Calder. La 76, PNE i.271, Jnl 17.48.

WHALLEY RANGE GMan SJ 8294. *Whalley Range* 1896.

WHALTON Northum NZ 1381. 'The settlement by the rounded hill'. *Walton(a)* 1203–91, *Wauton* 1218, *Whalton* from 1205, *Qu- Qwalton* 1296 SR 1312, 1424, *Whawton* 1638. OE ***hwæl** + **tūn**. The reference is to Whalton Hill and hill-fort at NZ 1482. NbDu 211 gives pr [wa:tən], PNE i.271.

WHAM NYorks SD 7762. 'The valley, the marshy hollow'. *Quane* (sic) [13th]n.d., *Wham* 1771. OE **hwamm**. YW vi.146.

WHAPLODE Lincs TF 3224. 'The eelpout drainage channel'. *Cope- Copolade* 1086, *Quappelad(e) -a* 1170–1403, *Whappelod(e)* 1333–1404, *Woplade* 1311, *Quaplode* 1335, 1388, 1497, 1645, *Whaplode* from 1386. OE ***cwappa** + **lād**. *Lade* or *lode* is one of the terms used in ME for the sewers or drains in the Lincolnshire fens. These sewers were freshwater water-courses rich in fish which were taken by means of weirs or nets. Already c.1000 Ælfric mentions the eelpout or burbot as one of the river fish taken by his *piscator*. The *pout* of eelpout, OE **pūte**, like ***cwappa** itself, means 'bag, something swollen', cf. OHG *quappa, kape* 'bag'. The purported 9th and 10th cent. form *Cappelade* occurs in late copies of charters of dubious authenticity (S 189, 213, 538). Payling 1940.63, Studies 1936.100, L 24, Cameron 1998.

WHAPLODE RIVER Lincs TF 3429. *Whaplode River* 1824 OS. P.n. WHAPLODE TF 3224 + ModE **river**, replacing earlier dial. **eau** (OE *ēa*) 'a river' as in *Whaplo(e) Ee, ~ Eau* 1572, 1812. Payling 1940.69.

WHAPLODE DROVE Lincs TF 3113. *magnam drauuam de Quappel'* late 12th, before 1235, *Whapplod drowe* 1272×1307, *Whaplod(e)drove* c.1520. P.n. WHAPLODE TF 3224 + ME **drove** 'a road along which horses or cattle are driven; a channel for drainage'. Created as a parish in 1812. Payling 1940.69.

WHAPLODE FEN Lincs TF 3320. *Whaplode Fen* 1824 OS. P.n. WHAPLODE TF 3224 + p.n. *Fen* 1665, ModE **fen**. Payling 1940.69.

WHARFE NYorks SD 7869. 'The nook'. *Warf* 1224, *Qwarf(e)* 1297, 1383, *Qwerff'* 1379, *Wharfe* from 1596. ON **hvarf** 'a bend, a nook, a corner'. Wharfe lies in a nook of land in the hills N of Austwick. YW vi.230 gives pr [wa:f].

River WHARFE NYorks SE 0262. 'The winding river'. *ye orf* (for *peorf*, i.e. *weorf*) [963]14th S 712, *Werf* [1158]14th–1280, *Wherf* 12th–1469, *Querf(f)* 1239×42–1436, *Wharfe* from 1269. Brit **u̯erbeiā*, a formation from **u̯erb-* on PIE *wer-* 'turn, twist' found also in continental names, e.g. *Verbannos*, the Gaulish name of Lago Maggiore, and cognate with E *warp* (OE *weorpan*). A Roman altar set up at Ilkley carries the inscription (RIB 635) *VERBEIAE SACRUM* 'sacred to Verbeia' the goddess of the river Wharfe whose name was identical with that of the river but has also been associated with OIr *ferb* 'cattle' as a goddess of cattle. The subsequent history of the name shows influence by ON *hwerfi* 'a bend, crook' and the normal ME sound change *er* > *ar*. Identical with r.n. WORFE SO 7698. YW 143, RN 454, RBrit 493, LHEB 282, Holder iii.181, Ross 279 etc.

WHARFEDALE WYorks SE 2646. 'Valley of the river Wharfe'. *Hwerverdale* 12th, *Wervedal'* 1204, *Querfdale* 1198, *Werfedalle* 1272×1307, *Wherfedale* 1393, *Wharl- Qhwardale* 1269–82, *Querdale* 1401, *Whardayll* 1457. R.n. WHARFE SE 0262 + OE **dæl**, ON **dalr**. The *Wharl-* forms may be due to popular association with OE **hwerfel** 'something circular' as the spelling *Wherveldale* suggests. Origin of the surname *Wardle*. YW v.2.

WHARLES Lancs SD 4435. Uncertain. *Quarlous* 1249, *Werlows, Warlawes* 1286, *Wharlowes* 1617. Possibly 'the round-topped hills, the mounds', OE **hwerfel**, nominative pl. **hwerf(e)las**, or **hwerfel** + **hlāw**, nominative pl. **hwerfel-hlāwas**. La 152.

WHARNCLIFFE SYorks SK 3095. 'Cliff where querns or mill-stones are obtained'. *Querncliffe* 1406, *Wharnecliffe* 1598. OE **cweorn** + **clif**. The reference is to a long steep cliff and rocky edge overlooking the Don valley. YW i.299, L 132,136.

WHARNCLIFFE SIDE SYorks SK 2994. 'W hill-side'. *Wharnetliffe Side* (sic with dial. *tl* for *cl*) 1634. P.n. WHARNCLIFFE + OE **sīde**. YW i.240.

WHARRAM LE STREET NYorks SE 8666. P.n. *Warham, Warran*

1086, *Warrum* [1119×35]13th–1276, *Warram* [1154–13th]15th, *Wharrom* 1197×1210–1376, *Wharrum* 1234–1560, of uncertain origin + suffix *in the Strete* 1333, 'on the Roman road', ME **strete** (OE *strǣt*). The best explanation on topographical grounds is ON **hvarf** 'a bend', dative pl. ***hvarfum**, 'at the bends' referring to the S-bends of the valley in which both Wharrams lie. But the absence of spellings with *-rf- -rv-* is problematical. The easiest source linguistically is ONb ***hwær**, OE **hwer**, 'a kettle, a pot, a cauldron', dative pl. ***hwarum**, but this does not suit the topography in any obvious way. 'Le Street', Fr definite article **le** + ModE **street**, refers to location on the Roman road from Malton to Wetwang, Margary no. 813. YE 135, SSNY 259.

WHARRAM PERCY NYorks SE 8564. 'W held by the Percy family'. P.n.*Warran -on* 1086, *Wharrom* 1150×60–1368, *-um* 1279×81–1387, *Qwharrum* 1311, as in WHARRAM LE STREET SE 8666, + manorial suffix *Percy* 1351. A deserted medieval village held by the Percy family in the 13th cent. YE 134 gives pr [warəm piəsi], SSNY 259.

WHARTON Ches SJ 6666. Partly uncertain. *Wanetune* (sic for *Wau-*) 1086, *Waverton* 1216–1628, *War(e)ton* 1288–1633, *Wharton* 1398 and from 1728. Like WAVERTON SJ 4564 this is OE ***wæfer** + **tūn**, either 'farm at a waving tree' or 'settlement at a quaking bog' or 'where brushwood is obtained'. Che ii.213 givs pr ['ʌ-, 'wɔːr-, 'waːrtən].

WHARTON H&W SO 5055. 'Settlement on the Waver'. *Wavertvne* 1086, *Waverton(a) -u-* 1103×4–1401, *Warton* 1599. R.n. **Waver* + OE **tūn**. Wharton actually lies on the river Lugg which makes a number of meanders here; it must have had an Anglo-Saxon nick-name 'wanderer, winding stream', OE **wæfre** as in the r.n. WAVER NY 1850. W is thus identical with WAVERTON Cumbr NY 2247. He 126.

WHASHTON NYorks NZ 1406. Partly uncertain. *Whasinga-Whassingetun* [1154×69]copy, *Wassingtun -ton* 1208–1316, *Quassyngton -ing-* 1285, 1301, *Qwaston* 1492, *Whassheton* 1574. Possibly OE **Hwassinga-tūn* 'settlement, village of the Hwassingas, the people called after Hwassa', or 'village, settlement called **Hwassing*', OE folk-n. **Hwassingas* < pers.n. **Hwassa* + **ingas**, genitive pl. **Hwassinga*, + **tūn**, or p.n. **Hwassing* < unknown element, possibly OE **hwæss** 'sharp', + **ing**[2], the 'sharp or pointed place', + **tūn**. The reference would be to the steepness of the slope on which Whashton stands. YN 292, BzN 1968.178.

WHATCOMBE Dorset ST 8301. Probably 'the wet coomb'. *Watecumbe* 1288, 1316, *-combe* 1332, *Whatecumbe* 1288, *-combe* 1412, *Whatcombe* from 1510. OE(WSax) **wǣt**, definite form **wǣta, wāta**, + **cumb**. The Winterborne flows here. Formally the specific could be OE **hwǣte** 'wheat'. Do ii.84, L 93.

WHATCOTE Warw SP 3044. 'The wheat cottage(s)'. *Quatercote* 1086, *Quatcote* c.1186, *Whatcote* from 1206. OE **hwǣte** + **cot**, pl. **cotu**. Wa 288.

WHATFIELD Suff TM 0246. 'The open land where wheat grows'. *Wate(s)felda, Watefella* 1086, *Whatefeld*. OE **hwǣte** + **feld**. DEPN.

WHATLEY Somer ST 7347. 'Glade where wheat grows'. *Watelea* [939×46]13th S 1726, *Watelei* 1086, *-leg* 1225, *Whatelay* 1610. OE **hwǣte** + **lēah**. DEPN.

WHATLINGTON ESusx TQ 7618. 'Settlement of the Hwætlingas, the people called after Hwætel'. *Watlingetone* 1086, *-lin(g)ton* [c.1150]15th, 1291, *Wetlingetun* 12th, *Whatlingetune* 1195, *Whatlington* 1724. OE folk-n. **Hwætlingas* < pers.n. *Hwætel* + **ingas**, genitive pl. **Hwætlinga*, + **tūn**. Sx 500, MA 10.24, SS 76.

WHATSTANDWELL Derby SK 3354. Short for Whatstandwell Ford. *Wattestanwell ford* 1390, *Watstandwell* 1444. The 1390 form comes from a charter of agreement between the Abbot of Darley and John Stepul in which the latter intends to build a bridge over the river Derwent next to the house which *Walter Stonewell* alias *Stondewell* 1347, *Standewell'* 1350, had held as tenant of Darley Abbey. Hence the name of the ford, *Watte* (short for *Walter*) *Stanwell ford*. Db 438.

WHATTON Notts SK 7439. 'The wheat farm'. *Watone* 1086, *Watton(a)* c.1160–1291, *Whatton(e)* from 1189. OE **hwǣte** + **tūn**. Nt 230.

Long WHATTON Leic SK 4823. *Long(e)whatton(e)* 1337–1610. ME adj. **long** + p.n. *Wact(h)on'* c.1130, *Wahton* 1240, *Watton(e)* 1190–1406, *Whatton(e)* 1210–1610, of uncertain origin, possibly 'the watch settlement', OE **wacu** + **tūn**, later reinterpreted as 'the wheat farm', OE **hwǣte** + **tūn**. Lei 412.

WHAW NYorks NY 9804. 'The enclosure near the fold'. *Kiwawe* 1280, 1283, *le Kuawe* 1285, *Quagh* 1342. ON **kví** + **hagi**. A **kví** was a pen or fold where sheep were milked. YN 296.

WHEATACRE Norf TM 4693. 'Newly cultivated land where wheat is grown'. *Hwateaker* 1086, *Qwet- Whetacre* 1254. OE **hwǣte** + **æcer**. DEPN, L 232–3.

WHEATENHURST Glos SO 7609 → WHITMINSTER SO 7708.

WHEATHAMPSTEAD Herts TL 1713. 'Wheat homestead'. *Wathemestede* [c.960]17th S 1807[†], *hwathamstede* [c.1060]14th, *Huuœthamstede* [1065]n.d., *Huuœthampstede* [1065]18th S 1042, *Watamestede* 1086, *W(h)athamsted(e)* 1170–1505, *W(h)ethamsted(e)* 1226–1580. OE **hwǣte** + **hāmstede**. Hrt 54 gives pr [wetəmstid], Stead 242.

WHEATHILL Shrops SO 6282. 'Wheat hill'. *Wethulla* c.1188–1409 with variants *-hull'* and *-hul(l), Hwethill'* c.1200, *Hwet-Whethull(')* 1255–1690 with variant *-hill(e), Wheathill* from 1618, *Whettill* 1495, *Whettle* 1577–1695. OE **hwǣte** + **hyll**. The DB form, *Waltham* 1086, is unusually inaccurate. Sa i.310.

WHEATLEY 'Clearing where wheat is grown'. OE **hwǣte** + **lēah**.

(1) ~ Hants SU 7840. *Wateligh* 12th, *-legh* 1269, *Hwathelee* 12th, *Whatele(y)* 1316, 1350. Gover 1958.98, L 206.

(2) ~ Oxon SP 5905. *Hwatelega* 1163, *Watelega -leie -le(e) -legh -leye* 1176–c.1425, *Wheatly* 1797. O 193.

(3) ~ LANE Lancs SD 8338. 'The lane leading to Wheatley'. P.n. Wheatley + ModE **lane**. Wheatley, *Whetley* 1423, *Whitley* 1502, *le Wheyteley* 1516, *Witley* 1526, was originally a district of considerable extent: Carr Hill SD 8438 is named from *Wheteleycarre* 1464, 'Wheatley marsh', ON **kjarr**. La 81, SSNNW 257, Jnl 17.50.

(4) North ~ Notts SK 7585. *Northwetley* 1280. ME adj. **north** + p.n. *Wateleia -e -laie* 1086, *-leia -lega -le(gh)* 1088–1234, *Hwetele(y) -lay* 1154×89–1331, *Whetley* 1504×15. Also known as *Northbekwetelay* 'north-beck W' 1327. 'North' for distinction from South Wheatley SK 7685, *Wetelegsudbek'* 'W south-beck' 1242, *Suthwetley* 1280. The two villages lie N and S of a stream. Nt 42.

South WHEATLEY Corn SX 2492. *South Wheatley* 1748. ModE adjective **south** + p.n. Whiteley, *Whytele(ye)* 1249, *Whiteleye* 1332, 'the bright clearing', OE **hwīt**, definite form **hwīta**, + **lēah**. 'South' for distinction from other divisions of Whiteley, Middle Whiteley SX 2494, *Wheatleywist* 1699, *North Whetley* 1813, and Higher Whiteleigh SX 2492, *Wyteleye* 1327, *Wheatlyweeke* 1699 (referring to its location in Week St Mary). PNCo 178.

WHEATLEY HILL Durham NZ 3738. *Wheitleyhill* 1515, *Whettley on the Hill, Whetelehill* 1529, *Wheatley hyll* 1568, *~ Hill* 1625. P.n. Wheatley + ModE **hill**. Wheatley, *Wuatlaue* 1180, *Qwetelawe* 1259×60, *Quetlau -lawe* 14th cent., *Whetlawe* 1349–1474, *Whetlay* 1388, *Wheitley* 1558×9, is 'wheat hill', OE **hwǣte** + **hlāw**. NbDu 212.

WHEDDON CROSS Somer SS 9238. 'Wheddon crossroads'. *Wheddon Cross* 1809 OS. P.n. Wheddon + ModE **cross**. Wheddon, *Wheteden* 1243, 1253, *-don* 1253, is 'wheat valley' or 'hill', OE **hwǣte** + **denu** or **dūn**. DEPN.

WHEELDALE MOOR NYorks SE 7898. P.n. Wheeldale + ModE

[†]So B 1012. K 966 prints *Wathamestede* and Hrt 54 *Wathemstede*.

moor. Wheeldale, *Wheel- Welledale* 1252, *Whele-* 1335, is the 'wheel valley', OE **hwēol** + **dæl**, so named because its course forms a large arc of a circle. Identical with WHELDALE SE 4526. YN 130.

WHEELER END Bucks SU 8093 → CADMORE END SU 7892.

WHEELERSTREET Surrey SU 9440. 'The row of houses or hamlet called after the Wheeler family'. *Whelerstret(e)* 16th cent. Family name *Wheeler* + ME **strete**. Cf. Nicholas *Whelere* 1373. Sr 218.

WHEELOCK Ches SJ 7759. Transferred from the river WHEELOCK SJ 7062. *Hoiloch* 1086, *Weloc(k)* 13th, *Que- Qwelo(c)k* late 13th–1321, *Whelok -o(c)ke* 1288 etc., *Wheelock* from 1387. Che ii.273 gives pr ['wi:lɔk], Addenda 19th cent. local ['wilək].

River WHEELOCK Ches SJ 7062. 'Winding river'. *Quelok* c.1300, *Qwelok* 1321, *Whelok -ocke* 1577, *Wheelock* 1619. W ***chwylog** (W *chwyl* 'a turn' + adjective suffix *-og*) < Brit **Suilāco-*. A 7th cent. borrowing from OW into OE. Che i.38, RN 455.

WHEELTON Lancs SD 6021. 'Wheel settlement'. *Weltonam* c.1160, *Whelton* [c.1200]14th–1332 etc., *Wyl- Wel- Quelton* 1250–88. OE **hwēol** + **tūn**. The reference is uncertain: the 13th cent. name *Whelcroft* is found in Wheelton. It might have been a water-wheel in the Lostock, a circle of stones, a place where wheels were manufactured, or possibly the small detached hill immediately N of the village on which Prospect House stands. La 132, Jnl 17.75.

WHELDALE WYorks SE 4526. 'Wheel valley'. *Weldale* 1086, 1252, *Queldale -dal(am)* 1086–1604, *Wheldale* from 1419. OE **hwēol** + **dæl** referring to a bend in the river Aire here. Identical with p.n. Wheeldale in WHEELDALE MOOR SE 7898. YW ii.67, L 96.

WHELDRAKE NYorks SE 6845. Probably 'death *Ric*, the strip of land or water where a death took place'. *Coldrid* 1086, *Coldric* 1167–1285, *Qu- Qweldrik(e) -ric -y-* 12th–1400, *Weldrik* [13th]copy, *Wheldrake* from 1535. OE **cweld** + ***riċ** with Scand [k] for English [tʃ], perhaps already used as a p.n. The reference is to a narrow ridge of land extending from Stillingfleet via ESCRICK SE 6342, 'ash-tree *Ric* or strip of land', to Wheldrake. YE 269, SSNY 260, L 184.

WHELFORD Glos SU 1799. 'Ford by the deep river-pool'. *Welford(e)* 12th–1779, *Whelford(e)* 1535–1822. OE **wēl** + **ford**. The *Whel-* spellings are too late to be significant. Gl i.39, L 30, 70.

WHELPLEY HILL Bucks SP 9904. *Whelpley Hill* 1822 OS P.n. Whelpley + ModE **hill**. Whelpley, *Welpelie* c.1200, *Whelplege* 1206, is 'whelps' wood or clearing', OE **hwelp**, genitive pl. **hwelpa**, + **lēah**. The reference is to a clearing where young animals played. Bk 213.

WHENBY NYorks SE 6369. 'The women's farm or village'. *Quennebi* 1086, *Quenebi -by* 1202–1333, *Whenby* from 1454. ON **kona** 'woman', genitive pl. **kvenna**, + **bȳ**. Cf. Swedish p.n. Kvinneby, OSwed *Quinnæby*. YN 30, SSNY 41.

WHEPSTEAD Suff TL 8358. 'The brushwood place'. *Wepstede, Whepste* [942×c.951]13th, 14th S 1526, *Hwipstede* 975×1016 S 1487, *Hwip- Hwepstede* 11th, *Huepestede* 1086, *Huepstede* 1087×98, *Qwep- Quepsted(e)* 1254, 1291, *Wepsted(e)* 1156–1420, *Whepsted(e)* 1202–1568. OE ***hwip(pe)** 'a whip, a switch'. The sense is 'place where material for switches is obtained'. Stead 193.

WHERNSIDE 'Hill-side where millstones are found'. OE **cweorn** + **sīde**.

(1) ~ NYorks SD 7381. *Querneside -syde* 1202×8–1346, *Quernsyd* 1204, 1343, *Whornesyde* 1592. YW vi.254.

(2) Great ~ NYorks SE 0174. *Great Wharnside* 1771. ModE **great** + p.n. *Querneside* 1214, 1348, 1609. YW v.218.

(3) Little ~ NYorks SE 0277. *Little Wharnside* 1771. ModE **little** + p.n. *Querneside* 1214, 1348, 1609. YW v.218.

WHERSTEAD Suff TM 1540. 'The shore place'. *Weruesteda(m)* 1086, *Warvestede* 1207, *Wersted(e)* 1254–1338, *Qu- Qwersted(e)* 1254–91, *Whersted(e)* 1296–1524, *Wested* 1506. OE **hwearf** 'a wharf' + **stede**. Wherstead must have been a landing-place on the upper Orewell river just S of Ipswich. Stead 194.

WHERWELL Hants SU 3840. 'Cauldron streams'. *Hwerwyl* [951×5]15th S 1515, *(æt) werewelle* [959]14th, [1002]14th S 904, 1201–1316, *(to) hwerwillom* c.1121 ASC(E) under year 1052, *hwaerwellan* 11th ASC(D) under same year, *Warwelle* 1086, *Wherewell(e)* 1217–1327, *Whar(e)well* 1445, 1539, *Whorwel(l)* 1535, 1579, *Horwell* 1636. OE **hwer** + **wylle**, dative pl. **wyllum**. The merging rivers Test and Dever here run in multiple channels; the sense is 'bubbling streams'. Ha 174, Gover 1958.172 gives pr [(h)ʌrəl].

WHESSOE Durham NZ 2718 → WASHINGTON T&W NZ 3056.

WHESTON Derby SK 1376. 'The whetstone'. *Weston* 1225–1305, *Whetstan* 1251, *Whestan* 1231, *Wheston* from 1254. OE **hwet-stān**. Probably refers to a place where stone suitable for whetstones was found or to a hill shaped like a whetstone. Db 176.

WHETSTED Kent TQ 6545. Partly uncertain. *Hwætonstede* [838]10th S 280 with later variants *Hwætan- hwetenstede, Whetstede* 1226, 1327, *We(t)stede* 1226–1323. This name can be interpreted either as a compound of OE **hwǣten** 'growing with wheat' + **stede** 'place, locality', or OE **hwǣte** 'wheat' + either **hām-stede** 'homestead' or ***hamm-stede** 'enclosure site'. PNK 176, 179, PNE i.271, Stead 228.

WHETSTONE Leic SP 5597. 'The whetstone'. *Westhā* (sic) 1086, *Whet(e)stan* 1154×89–1245, *-ston(e)* 1370–1680. OE **hwet-stān**. This may have referred to an ancient standing stone no longer extant. Lei 470.

WHICHAM Cumbr SD 1382. 'Homestead, village called or at *Hwitinge*'. *Witingham* 1086–1291 with variants *Wyt-* and *-yng-*, *Wintinghaham* c.1130, *Hwithingham* c.1175, *Whittingham* 1279, *Quytyngham* 1396, *Wycheham* 1550, *Whycham* 1573. OE p.n. **Hwīting* 'place associated with, called after *Hwīta*' < pers.n. *Hwīta* + **ing**[2], locative-dative sing. **Hwītinġe*, + **hām**. Cu 443, BzN 1967.391.

WHICHFORD Warw SP 3134. Either 'the wych-elm ford' or 'the ford leading to the Hwicce'. *Wicford* 1086, *Wicheford(a)* c.1130–1316, *Wyc(c)heford* 1253–1549, *Whichford* 1263, *Whiccheford* 1322. Probably OE **wiċe**+ **ford**. The overwhelming majority of forms in *W-* favours **wiċe** but the modern form and the occasional earlier *Wh-* sp may point to the alternative explanation, 'the ford leading to the Hwicce', as the origin. The Hwicce were an Anglo-Saxon people settled in Gloucestershire, Worcestershire and the western half of Warwickshire. Wa 301, L 70.

WHICKHAM T&W NZ 2061. 'The homestead with a quickset hedge'. *Quicham* [1183]c.1320, 1197–1438×9, *Quykham* 1311–1498, *Quickam* 1580, *Wicham* 1354, *Whyk- Whikham* 1380–1547, *Whickham* from 1567. OE **cwic** + **hām**. NbDu 212.

WHIDDON DOWN Devon SX 6992. *Whyddon Doune* 1565, *Whiddon Downe* 1661. P.n. Whiddon + ModE **down**. Whiddon is probably 'white hill', ultimately from OE **hwīt** and **dūn**. D 451.

North WHILBOROUGH Devon SX 8766. *N^th^. Whilborough* 1809 OS. ModE adj. **north** + p.n. *Weghelburgh* 1292, *Whelberewe* 1294, 'circular hill', OE **hwēo(g)l** + **beorg**, referring to one of the circular hills here. 'North' for distinction from South Whilborough SX 8766, *Sutwhegelbergh* 1294, *South Welbergh* 1342. D 515.

WHILLAN BECK Cumbr NY 1802. *Whillon* 1587, *Whillan Beck* 1794. Cu 30.

WHILTON Northants SP 6364. 'Wheel farm or village'. *Woltone* 1086, *Whelton* 12th–1401, *Wheleton* 1404, 1428. OE **hwēol** + **tūn**. The reference is either to the circular hill on which the village stands or the circular course of the stream which flows round it. Nth 89.

WHIMPLE Devon SY 0497. Originally a stream name, 'white stream'. *Winple* 1086, *Wimpoll* 1218, *Win- Wun-* or *Wimpol, Wimple* 1238, *Wympol* 1296, *Whympel* 1391. PrW ***wïnn** + **poll**.

The place where the stream rises in Talaton is referred to as *Wympelwell in parochia de Taleton* 1281. D 579, RN 456, CPNE 120.

WHIMPWELL GREEN Norf TG 3829. *Whimpwell Green* 1838 OS. P.n. Whimpwell + ModE **green**. A mile N lies *Whimpwell Street* 1838 ibid., 'Whimpwell hamlet'. Whimpwell, *Hwamp- Hwimpwella* 1086, *(H)wimp(e)well(e)* 1240–1400, is possibly the 'nook spring', OE **hwamm** 'nook, corner' or ON **hvammr** 'grassy slope or vale', cf. dial. *wham* 'a swamp, a marshy hollow', + **wella**. Nf ii.93–4.

WHINBURGH Norf TG 0009. 'Whin hill'. *Wineberga* 1086, *Quyneberge* 1254. ME **whin** (ON **hvin*) + **bergh** (OE *beorg*). DEPN.

WHINFELL BEACON Cumbr NY 5700. *Whinfell Beacon* 1777. P.n. Whinfell + ModE **beacon**. Whinfell, *Quin- Quyn- Qwynfel(l)* 1154×89–1593, *Whin- Whynfel(l)* from c.1200, is 'hill overgrown with whins or gorse', ON ***hvin** + **fjall**. We i.142, SSNNW 176 no.2, L 159.

WHINLATTER PASS Cumbr NY 1924. *Passes of Whinlate* 1794. P.n. Whinlatter, *Whynlater* 1505, *Whinlatter* 16th, 'furze-covered slope', ME **whin** (ON **hvin*) + **later** (OIr *lettir* 'hill, slope'), + ModE **pass**. Cu 409.

WHINS BROW Lancs SD 6353. A precipitous hill-top rising to 1561ft. ModE **whin** 'whin, gorse' (ON **hvin*) + **brow** 'the projecting edge of a hill' (OE *brū*).

WHIPPINGHAM IoW SZ 5193. 'Homestead of the Wippingas, the people called after Wippa'. *Wippingeham* [735]c.1300, 1155×8–66 with variant *Wyppinge-*, *Wit- Wipingeham* 1086, *Wipingeham* 1235, *Wippingaham* 1193×1217, *Wi- Wyppingham* 1213–1428, *Whippingham* from 1284. OE folk-n. **Wippingas* < pers.n. **Wippa* + **ingas**, genitive pl. **Wippinga*, + **hām**. Wt 234 gives pr ['wipnəm], Mills 1996.108.

WHIPSNADE Beds TL 0117. 'Wibba's detached ground'. *Wilbesnede* (sic) 1202, *Wi- Wybsnede, Wib(b)esnade* 1202–1350, *Wyp(pe)snade* 1227–1491. OE pers.n. *Wibba* + **snǣd**. Bd 142.

WHIPTON Devon SX 9493. 'Wippa's estate'. *Wipletone* 1086, *Wippeton* 1341, *Whipeton* 1393. OE pers.n. **Wippa* + **tūn**. If the 1086 form is not an error, it points to a diminutive **Wipela*. D 441.

WHISSENDINE Leic SK 8214. 'The valley of the Hwiccingas, the people called after Hwicca, or of the Hwicce'. *Wichingedene* 1086, *Wisingheden* 1295, *Wy- Wissingden(e)* 1265–1343, *Wy- Wissenden(e)* 1203–1610, *Whittsonden* 1491, *Whytsondyne* 1506, *Whitsondine -sen- -sun-* 17th, *Why- Whissendyne* 1539, *Whissendine* 1695. OE folk-n. **Hwiċċingas* < pers.n. *Hwiċċa* + **ingas**, genitive pl. **Hwiċċinga*, + **denu**. Alternatively the specific might be OE folk-n. *Hwiċċe*, genitive pl. *Hwiċċena*. Evidence of a possible presence of the Hwicce in Rutland is supplied by the hundred n. Witchley, *Hwiccleslea* [before 1075]12th, *Wice(s)lea* 1086, 'Hwicci' or 'the Hwicce's woodland', OE pers.n. *Hwiċċi*, genitive sing. *Hwiċċes*, or folk-n. *Hwiċċe*, genitive pl. *Hwiċċa*, + **lēah**. Earlier explanations of the specific from OE **wīċinga**, genitive pl. of **wīċing** 'a pirate', or **Wiċinga*, genitive pl. of **Wiċingas* 'the people called after Wic', OE pers.n. **Wiċ*, are unsatisfactory. R 221–2.

WHISSONSETT Norf TF 9123. 'The fold of the Wicingas, the people of Wic'. *Witcingkeseta* 1086, *Wichingseta* 1191, *Wicingesete* 1196. OE folk-n. **Wīċingas* < **wīċ** + **ingas**, genitive pl. **Wīċingas*, + **(ġe)set**. The same folk-n. occurs in Great & Little Witchingham TG 1020, 1220, *Wicingha- Wit(t)cinge- Witcincham* 1086, *Wichingeham* 1130, 1198, *Wichingham* 1155, 1199, *Wikingeham* 1180ff. The *Wicingas* are probably the people who dwell at or come from a place called *Wic*; the Wicingas who occur in the OE poem *Widsith* at line 47 are thought by one scholar to be the people from Bardowick near Lüneburg, *Bardaenowic* 805 'the settlement of the Lombards'. DEPN, ING 140, Berger 48.

WHISTLEY GREEN Berks SU 7974. *Whistley Green* 1761. P.n. Whistley + ModE **green**. Whistley, *(æt) Wisclea* [968]12th S 769, *(æt) Uuiscelea* ibid. 13th, *Wiscelea -leie* [before 1170]12th, *Wiselei* 1086, *Wiselega, Wistle* 1185, *Whisseley* 1378, *Whistley* 1659, is 'marshy-meadow pasture', OE **wisc** + **lēah** here in the sense 'pasture'. Brk 100, L 199, 250.

WHISTON 'The white stone'. OE **hwīt** + **stān**, or definite form **hwīta**, + **stān**.

(1) ~ Mers SJ 4791. *Quistan* [1190]1268–1332, *Quicstan* (for *Quit-*) 1246, *Wytstan -ston* 1252, *Whistan* 1272. The reference is unknown. La 108, Jnl 17.60.

(2) ~ Staffs SK 0347. *Witestone* 1086, *Whytston* 1306, *Wytston* 1328, 1331, *Wystan* 1307, 1348, *Wy- Wiston* 13th–1659, *Whiston* from 1294. The township is noted for large amounts of exposed rock which deposit a fine white sand used in the manufacture of cosmetics. Oakden, Horovitz.

(3) ~ SYorks SK 4590. *Witestan, Widestha' -stan* 1086, *Witstan(e), Wytstan* 12th–1428, *Wi- Wystan(e)* 12th–1323, *Wize Wizestan* 1195, 1196, *Hwitstan* 13th, 1347, *Whyt- Whitstan* 1301–1420, *-ston* 1407, *Whi- Whyston* from 1386. The allusion is no longer known but may have been connected with the site of the meeting place of the district of MORTHEN. YW i.167.

WHISTON Northants SP 8460. Either 'the estate called after Hwicca' or 'the settlement of the Hwicce'. *Hwiccingtune* [974]13th S 798, *Hwiccintunæ* [1066×87]13th, *Wice(n)tone* 1086, *Wichenton(a) -in-* 1175–1229, *Hwi- Hwychetone* 1231, *Whiston* from 12th with variant *Whyston*. Either pers.n. **Hwiċċa* + **ing**[4] or genitive sing. ***Hwiċċan**, + **tūn**, or folk-n. *Hwiċċe*, genitive pl. *Hwiċċena*, + **tūn**. **Hwiċċa* would be an individual from the tribe called *Hwiċċe*, whose homeland was the lower Severn valley. Nth 152 gives pr [wiʃtən].

WHISTON Staffs SJ 8914. 'Witi's estate'. *Witestun* [988]11th S 1356[†], [1004]11th S 906, *-tone* 1086, *Wi- Wyston* c.1255–1546, *Why- Whiston* from 1257. OE pers.n. **Witi*, genitive sing. **Wites*, + **tūn**. St i.103.

Nether WHITACRE Warw SP 2393. *Nethere Whittacre* 1262, *Nethere Whitacre* 1330. ME adj. **nether** + p.n. *Witecore* 1086, *Wytacre* 1166–1275, *Hwitacre* 1206, 1221, *Whit(e)acre* 1330, 1356, 1545, 'the white cultivated land', OE **hwīt** + **æcer**. 'Nether' for distinction from Over WHITACRE SP 2591. Wa 94, L 232–3.

Over WHITACRE Warw SP 2591. *Over Wytacre* 1275, *Overewhitacre* 1310. ME adj. **over** 'upper' + p.n. *Witacre* 1086, as in Nether WHITACRE SP 2393. Wa 94, L 232–3.

WHITBECK Cumbr SD 1184. 'White beck'. *Witebec* [c.1160]12th–1292 with variants *Wyte-* and *-be(c)k*, *Whitebec* c.1200, *Quitebec* c.1250, *Whitbeck* 1568. ON **hvítr** or OE **hwīt** + ON **bekkr**. Cu 447, SSNNW 176.

WHITBOURNE H&W SO 7256. Originally a stream name, 'the white stream'. *Witebern* 1163, *W(h)yteburn(e)* 1219–1356, *Wyteborne* 1412. OE **hwīt**, definite form **hwīta**, + **burna**. The reference is to an efflorescence of lime in the bank of the stream. He 203, L 18.

WHITBURN T&W NZ 4062. 'The white barn'. The forms fall into three types:

I. *W(h)itebern(e)* [1183]c.1320, 1342–1418, *Hwiteberne* c.1190, *Wyteberne* 1292, c.1400, *Qwyt- Qwit(e)bern'* [1293]c.1400, 1303, 1364, *Whit- Whytbern(')* 1382–1591, *Whittebarn'* 1438×9, *Whitbarne* 1536;

II. *Qwytberme* [1293]c.1400, *Wyteb'ine -berme* 1296, 1304, *Whit(he)berm'* 1316–50;

III. *Witeburne* 1242×3, *Whitborn'* 1303, *Whitburn* from 1382;

OE **hwīt**, definite form **hwīta**, + **bern**. Type II has probably been affected by the spellings of HEBBURN NZ 3164. Type

[†]This form is more reliable than that in the earlier copy, *Witestan* [1002×4]11th S 1536.

III shows suffix confusion with OE **burna** 'a stream'. NbDu 213.

WHITBY 'White village'. ON **hvítr** + **bȳ**, probably referring to white-washed stone buildings. SSNNW 43.

(1) ~ Ches SJ 3975. *Witeberia* [1096×1101]1150, 1150, *Witebia* [1096×1101]1280, *Wy- Witebi -by* 1188–1316, *Whiteby* 1241–1547 with variants *Whyte- Quite- Qwyte-*, *Whitby* from 1402. The earliest form shows that the original name was OE 'white manor or village', OE **hwīte**, dative case **hwītan**, + **byriġ**, dative sing. of OE **burh** later reformed under Scand influence. Che iv.198, SSNNW 43.

WHITBY Partly uncertain. Probably 'Hviti's farmstead', ON pers.n. *Hvíti*, genitive sing. *Hvíta*, + **bȳ**. But the specific could in fact be any of pers.ns. *Hvíti*, OE *Hwīta* or *Hwīte* feminine, or ON **hvítr** or OE **hwīt** 'white'.

(2) ~ NYorks NZ 8910. *Wi- Wytebi -by* 1086–1298, *Whitby* from 1138, *Whi- Whyteby* c.1150×60–1361, *Quiteby* 13th cent. In the 13th cent. ON *Morkinskinna* the form *(við) hvita by* [c.1220]c.1275 is clearly understood as 'Hviti's farmstead'. For the earlier name see STRENSALL SE 6360. YN 126 gives pr [widbi], SSNY 41, Townend 42.

WHITCHURCH 'The white church'. OE **hwīt**, definite form **hwīte**, + **ċiriċe**. The reference is either to a church built of stone as opposed to wood, or to whitewashing.

(1) ~ Avon ST 6167. *Hwite circe* *[1065]c.1500 S 1042, *Wytchirche* 1230. DEPN.

(2) ~ Bucks SP 8020. *Wicherce* 1086, *Hwitchirche* 1185. Bk 86.

(3) ~ Devon SX 4972. *Wicerce* 1086, *Wi(t)cherche* 1166–7, *Whit(e)chirche* 1238, 1290. D 247.

(4) ~ Hants SU 4648. *hwitan cyrice* [909]12th S 378, *(to) hwitcyrcan* [n.d.]12th S 1821, *(æt) hwitciricean* 1001 ASC(A), *Witcerce* 1086, *Whitchurche* 1188. Probably a white-washed church, but it may have been built of stone. The present building is 13th, 18th and 19th cents., but the mid-9th cent. gravestone of an Anglo-Saxon female Frithburg survives. Ha 174, Gover 1958.156, Pevsner-Lloyd 651.

(5) ~ H&W SO 5517. *Wytechirche* 1320, *Whitchirch* 1325, *~ Church* 1535. Earlier Latin forms are *Albi Monasterii* (genitive case) 1148×63, *Album Monasterium* 1313, Latin adj. **albus** 'white', neuter gender **album**, + **monasterium**. W forms of the name refer to the saints venerated here, *Lann Tiuinauc* [1054×1104]c.1130, 'St Winnoc's church', W saint's name *Winnoc* + hypocoristic prefix **ty-**, *Fendenerac* (sic for *-deuerac*) 1277, *Landeuenok* (sic for *-deuerok*) 1334, 'St Dyfrig's church', W saint's name *Dyfrig*, the present dedication of the church. He 205.

(6) ~ Oxon SU 6377. *(æt) Hwitecyrcan* 990×2, *Hwitcyrce* [1012]13th S 927, *Witecerce* 1086–1285 with variants *Wy-* and *-cherch(e) -church(e) -chirch(e), Witcherch(e) -chirch(e) -chirch(e)* 1167–1342, *Whitchirche* 1241, *Whytchurch* 1285. O 62.

(7) ~ Shrops SJ 5441. *Whytchyrche* 1271×2 etc. with variants *Whit-* and *-chirch(e), -church(e), Whitechirche* 1307–1750 with variants *Whyte-* and *-church(e) -cherche*. Also known in French and Latin forms as *Blancmustier* c.1200, *Blancmostiers* c.1320 and *Blancum Monasterium*. The earliest name is *Westune* 1086, 'the west settlement', but the reason is unknown. Sa i.310.

(8) ~ CANONICORUM Dorset SY 3995. 'W belonging to the canons' sc. of Salisbury cathedral. *Whitchurch Canonicorum* from 1262. Earlier simply *Witcerce* 1086, *Witechurch* 1231, *Whytchyrche* 1241. The dedication to St Candida 'White' is probably derived from the p.n. rather than vice-versa. There are Roman tiles in the external walls of the church. Do 153, Newman-Pevsner 458.

(9) ~ HILL Oxon SU 6478. *la Hulle de Witchurche* 1254×5. O 63.

(10) ~ MAUND H&W SO 5649 → MAUND BRYAN SO 5650.

WHITCOTT KEYSETT Shrops SO 2782. 'Keysett's W'. *Hodecote Keyset* 1284, *Hottecote Keyset* 1399, *Hudcotte Ekysed* 1513, *Yutcott Kysset* early 16th, *Whitcote Kysset* late 16th, *Whitcott Keysett* 1672. P.n. **Hodcott* 'Hoda's cottage(s) or sheep shelter(s)', OE pers.n. *Hoda* + **cot**, pl. **cotu**, identical with Hodcott Berks SU 4781, *Hodicote -y-* 1086–1412, *Hod(d)ecote* 1291–1428, *Hodcott* 1657 (possibly with **ing**[4]), later influenced by ME *white* as if 'the white cottage(s)', + manorial addition from surname *Keysot, Kesyat* 1328, 1336, W *ceisiad* 'a sergeant of the peace', for distinction from Whitcott Evan SO 2781, 'Iefan's W', *Hodecote Yevan* 1284, *Hutcot Jenii* 1542, *Hudcote Jevan* 1561, *Whytcott Jevan* early 16th, *Whitcott Evan* 1601×2, p.n. *Hodicote -y-* 1302, 1371, *Hodecote* 1344, + W pers.n. *Ieuan*. Morgan 1997.57, Berks 505.

WHITECHAPEL Lancs SD 5541. No early forms. The chapel dates from 1738. Pevsner 1969.149.

WHITECLIFF BAY IoW SZ 6486. *Whitecliff Bay* 1769. P.n. Whitecliff + ModE **bay**. Whitecliff is *The White Clyffe* 1611, a translation of earlier *la Blaunche Faloyse* 1322. ModE **white** + **cliff**, OFr **blanc**, feminine **blanche**, + **faloise**. Wt 40, Mills 1996.108.

WHITE COPPICE Lancs SD 6119. ModE adj. **white** + **coppice** (ME *copis*).

WHITECROFT Glos SO 6206. 'White croft or messuage'. *Whitecroft (forge)* 1669. Gl iii.231.

WHITE FEN Cambs TL 3492. *Whites Fen* 1636. Surname *White* + ModE **fen**. Ca 247.

WHITEFIELD GMan SD 8105. '(The) white open land'. *Whitefeld* 1292, *Whitefield or Stand* 1843 OS. OE **hwīt**, definite form **hwīta**, + **feld**. Possibly refers to an area where snow lay long. L 242. La 49.

WHITEGATE Ches SJ 6369. 'White gate'. *Whytegate* 1540, *Whitegate* 1542. Named from the outer gate of Vale Royal abbey. Che iii.178.

WHITEHAVEN Cumbr NX 9717. Short for Whitehead Haven. *Qwithofhavene* c.1135, *Witehovedhafne* c.1140, *Whytehauene* 1279, *Whithaven* 1535. P.n. *Withoue* c.1180 'white head(land)', ON **hvít** + **hǫfuth**, + ON **hǫfn** 'harbour' (whence late OE *hæfen*). The second element **hǫfuth** was early lost. The reference is to the headland S of the harbour where there is 'a great rock or quarry of white hard stone which gives name to the village and haven' (Thomas Denton 1687×8). It contrasts with another headland not now identified but once called 'black head(land)', *Swartahof* c.1125, *Suuartahoft* [c.1135]12th, ON **svart** + **hǫfuth**. Cu 450 records former prs [ʍitən] and [ʍitheivən], SSNNW 176, L 161.

WHITE HILL Lancs SD 6758. No early forms. ModE adj. **white** + **hill**. A hill rising to 1786ft. on the Lancs-Yorks border, a place where snow lies long. YW vi.214.

WHITEHILL Hants SU 7934. *White Hill* 1816 OS. Possibly a reference to the chalk soil. Ha 174, Gover 1958.94.

WHITE HORSE Dorset SY 7184. This is not an ancient example but a figure cut in 1815 to represent George III on horseback. It was his visits to Weymouth in 1780 and again in 1789–1805 which made the place famous as a seaside resort. Do i.214, Newman-Pevsner 308, 450.

WHITEHORSE HILL Oxon SU 3086 → VALE OF WHITE HORSE Oxon SU 2689.

WHITE ISLAND Scilly SV 9217. *Whites Iland* 1652, *White Island* 1689. It is uncertain whether this is surname *White* or adj. **white**. Co 179.

WHITELAKE Somer ST 5240. 'White stream'. *White Lake* 1817 OS. Adj. **white** + SW dial. **lake** 'a water-channel, a small stream' (OE *lacu*). A drainage channel in East Sedgemoor.

WHITE LAW Northum NT 8526. A hill on the Scottish–English boundary, no doubt where snow lies long.

WHITELEY VILLAGE Surrey TQ 0942. A model village laid out in 1911 and built 1914–21 by William Whiteley of Whiteley's Stores, London. Room 1983.135.

WHITEMANS GREEN WSusx TQ 3025. *Whitman's Greene* 1604. Surname *Whit(e)man* + ModE **green**. Sx 268.

WHITEMOOR Corn SW 9757. 'White marsh'. *Whitemoor* 1748. ModE **white** + **moor**. Named in deliberate contrast to Blackmoor, the name of the tinning district in which it lies. PNCo 179.

WHITEMOOR RESERVOIR Lancs SD 8743. P.n. White Moor SD 8644 + ModE **reservoir**. White Moor, *White Moor* 1771, rises to over 1180ft. and is often snow-covered. YW vi.37.

WHITEPARISH Wilts SU 2423. Short for 'White Church parish'. *Whyteparosshe* 1289, *la Whiteparosse* 1301, *la Whyteparyshe* 1307×27. Earlier *la Whytechyrch* 1278, *Whit- Whytechirche* 1289. ME **white** + **church** replaced by **paroche, parishe**. The reference must have originally been to the colour of the church, the oldest parts of which are Norman. Wlt 387, Pevsner 1975.571.

WHITE ROCKS H&W SO 4424. No early forms, but cf. *Rocks Bottom* 1831 OS.

WHITESAND BAY Corn SW 3527. *Whitson Bay* [1580]18th, *Whitsande Baye* 1582, *Whitland Baya* (sic) 1610. ModE **white** + **sand** + **bay**. PNCo 179, Gover n.d. 663.

WHITE SHEET HILL Wilts ST 8034. Cf. *White Sheet Castle* 1817 OS, an Iron Age hill-fort on the hill. Identical with WHITESHEET HILL ST 9524.

WHITESHEET HILL Wilts ST 9524. 'Hill called Whitesheet'. *Whiteshete Hille* c.1540. P.n. *White Sheet* 1608, probably a nickname descriptive of the hill-side when covered with snow, + ModE **hill**. Alternatively *Sheet* might be a survival of OE ***scēot** 'a slope, a steep place', but the evidence is too late for certainty. Wlt 184, 175.

WHITESHILL Glos SO 8407. A modern name, *Whitehill* 1830, from the surname White. Gl ii.193, Room 1983.135.

WHITESHOOT HILL Hants SU 2833. Cf. *White Shoot Down* 1757. ModE **shoot** in the sense 'steep slope' (OE **sceōt*). Gover 1958.188.

WHITESIDE Northum NY 7069. 'White hill-side'. *Whiteside* 1866 OS. ModE adj. **white** + **side**.

WHITESMITH ESusx TQ 5214. *Wythesmyth* 15th cent. The generic is OE **smiððe** 'a smithy, a metal worker's shop'. A whitesmith is a worker in 'white iron', a tinsmith, or one who polishes and finishes metal goods as opposed to forging them. Sx 403.

WHITESTAUNTON Somer ST 2810. *Whitestaunton* 1337, *Whitstaunton* 1610. ME adj. **white** + p.n. *Stantvne* 1086, 'stone settlement', OE **stān-tūn**. Blue lias occurs in the parish and old workings here point to stone quarrying. DEPN, Wade 274.

WHITESTONE Devon SX 8693. 'The white stone'. *Witestan* 1086, *(on) hwita stane* c.1100, *Whitestan* 1238, *Whyteston* 1303, *Whiston* 1333. OE **hwīt**, definite form **hwīta**, + **stān**. D 456.

High WHITE STONES Cumbr NY 2809. *High White Stones* 1865 OS. A summit of 2500ft.

WHITEWAY Glos SO 9210. Probably the name of an old salt-way from Tewkesbury to Rodmarton. Gl i.131,20.

WHITEWAY HOUSE Devon SX 8782. P.n. Whiteway + ModE **house**. Whiteway is *Whiteweye* 1330. D 491.

WHITEWELL Lancs SD 6547. 'The clear spring'. *Whitewell* 1343. OE **hwīt**, definite form **hwīta**, + **wella**. YW vi.210.

WHITEWORKS Devon SX 6071. *White Works* 1809 OS. A china-clay mine.

WHITFELL Cumbr SD 1593. *Whitfell* 1865 OS. 'White fell', ultimately from ON **hvítr** + **fjall** 'a mountain'. A hill rising to 1881ft.

WHITFIELD 'The white open land'. OE **hwīt**, definite form **hwīta**, + **feld**.

(1) ~ Avon ST 6791. *Whit- Wythfeld* 1497, *Whitefeld* 1509, 1533, *Whitfield* 1638. The original tract of land extended to Whitfield House ST 6891, *Whitefeld* 1320. 'White' is sometimes used of places where snow lay long. Gl iii.7, 4, L 241–2.

(2) ~ Kent TR 3045. *Wytefeld'* 1228, *W(h)y- Whit(e)feld(e)* 1228–93. KPN 52, L 242.

(3) ~ Northants SP 6039. *Witefelle* 1086–1293 with variants *Wyte-* and *-feld -feud, Hw- Whi- Whytefeld* 1185–1307, *Whitfeld* 1381, *Qui- Qwytefeld* c.1240, 1285. Nth 64, L 241–2.

(4) ~ Northum NY 7858. *Witefeld* before 1274, *W(h)itefeld* 1279. The reference is to open land where snow lies long. NbDu 213, L 241–2.

(5) ~ MOOR Northum NY 7453. *Whitfield Moor* 1866 OS. P.n. WHITFIELD NY 7858 + **moor**.

WHITFORD Devon SY 2595. 'The white ford'. *Witefort* 1086, *-ford* 1167, *Whytford* 1228. OE **hwīt**, definite form **hwīta**, + **ford**. D 630.

WHITGIFT Humbs SE 8122. 'Hwīta, Hwīte or Hvíti's dowry land'. *Wite- Wytegift -gyft* [c.1070], 1078×85–1304, *White- Whytegift -gyft* 1232–1372, *Whitgift* from 1412. OE masculine pers.n. *Hwīta* or feminine *Hwīte* or ON *Hvíti*, genitive sing. *Hwíta*, + **gipt**. Morgay Farm in Ewhurst WSusx is a similar name, *Morgtheve* c.1240, *Morgheyve* c.1248, from OE **morgen-gifu** 'morning-gift', the gift given by the husband to the wife on the morning after the marriage. YW ii.11, Sx 519.

WHITGREAVE Staffs SJ 8928. 'The white grove'. *Witegraue* 1193, *-greve* 1251. OE **hwīt**, definite form **hwīta**, + **grāf(a)** or **grǣfe**. DEPN.

WHITLEY 'The white wood or clearing'. OE **hwīt(a)** + **lēah**, referring to either the bark or the blossom of the trees, L 205.

(1) ~ Berks SU 7170. *Witelei* 1086, *W(h)iteleia* 1198, *W(h)ytele(y) -legh, Wytteleyg'* 1284, *Whitley* 1539×40. Brk 177, L 205.

(2) ~ Ches SJ 6179. *Witelei* 1086, *Wit(t)eleia -le(g)a -leth -le(i)gh -ley* 1182–1465, *Whete-* 1284–1361, *Whit(t)e-* 1295–1367, *Qwe- Qwy- Qwite-* 1316–57, *Whitley* from 1386. Che ii.124.

(3) ~ NYorks SE 5621. *Wi- Wytelai(e) -lay -legh* 1086–1280, *Whitley -lay* 1316–1591, *Whiteley, Qwytlay* 14th cent. YW ii.60.

(4) ~ Wilts ST 8866. *Witlege* [1001]15th S 899, *Witelie* 1086, *Whitlee* 1286. Wlt 129, L 205.

(5) ~ BAY T&W NZ 3572. P.n. *Wyteleya* 1154×89, *Witeleya* 1203, *(H)wyteleya* [1198]1271, + ModE **bay**. NbDu 213, DEPN.

(6) ~ CHAPEL Northum NY 9257. *Whitley chapell* c.1715 AA 13.9. P.n. *Whiteley* 1349 + ME **chapel**. NbDu 213.

WHITLEY ROW Kent TQ 4952. 'Row (of buildings) at Whitley, the white hill'. *Whitley Row* 1819 OS. P.n *Whitehell'* 1313, *Whytclyffe* 1535, *Whitley al. Whitcliff* 1555, 1574, 'the white hill', OE **hwīt**, definite form **hwīta**, + **hyll**, Kentish **hell**, later varying with ModE **cliff**, + ModE **row**. K 55.

WHITLOCK'S END WMids SP 1077. *Whitloxend* 1579, *Whitlocks End* 1834 OS. P.n. Whitlock, short for *Witlakeffeld* 1292, *Whitlakesfeld* 1340, *Whitlokkesfeld* 1528×9, 'Wihtlac's open land', OE pers.n. *Wihtlāc*, genitive sing. *Wihtlāces*, + **feld**, + Mod **end**. Pers.n. *Wihtlāc* has been reformed as if the surname *Whitelock* 'white hair'. Wa 72.

WHITMINSTER Glos SO 7708. *Whitnester* 1535, *Whitmyster* 1577, 1695, *Whitminster* 1667, 1675, *Upper Whitminster* 1830 OS. This name has nothing to do with the word *minster* but is a late adaptation of nearby Wheatenhurst SO 7609, *Witenherst* 1086, *Wytenherste* 1195, *-hurst(e)* c.1275–1428, *Whi- Whytenherst -hurst(e)* 1248–1425, *Whetenhurst(e) als. Whi- Whynester* 1579–1623, ~ *als. Whitmyster* 1612, *Wheatenhurste als. Whitminster* 1645, 1830. Wheatenhurst is either '(at the) white wooded hill', OE **hwīt**, dative sing. definite form **hwītan**, in the sense 'bright', + **hyrst**, or the specific might be the OE pers.n. *Hwīta*, genitive sing. *Hwītan*, 'Hwita's wooded hill'. Gl ii.204.

WHIT MOOR Lancs SD 5964. Adj. **white** (OE **hwīt**) + **moor**. This is the northern side of Caton Moor SD 5663 rising to 1184ft. where snow lies long.

WHITMORE Staffs SJ 8141. 'The white moor'. *Witemore* 1086,

Whytemore 1227. OE **hwīt**, definite form **hwīta**, + **mōr**. The reference may be to a place where snow lies long. DEPN.

WHITNAGE Devon ST 0215. 'At the white oak-tree'. *Witenes* 1086, *Witenech* c.1200, *Hwytenych* 1316. OE **hwīt**, dative sing. definite form **hwītan**, + **ǣċ**, dative sing. of **āc** 'an oak-tree'. Cf. BRADNINCH SS 9903, RADNAGE Berks SU 7897 and WHITNASH Warw SP 3263. D 553, L 219.

WHITNAL Hants SU 4851 → East WOODHAY SU 4061.

WHITNASH Warw SP 3263. '(The settlement) at the white ash-tree'. *Witenas* 1086, *-asche -asse -ash* 1221–1316, *Whitenasshe* 1327–1458, *Whitnash* 1535. OE **(æt thǣm) hwitan æsce**. Wa 190.

WHITNEY-ON-WYE H&W SO 2747. P.n. Whitney + r.n. WYE SO 4341. Whitney, *Witenie* 1086, *Wi- Wytteney(e)* c.1130–1230, *Hwytene* 1272×1307, *Whytney* 1412, is either 'Hwita's island' or '(at the) white island', OE pers.n. *Hwīta*, genitive sing. *Hwītan*, or adj. **hwīt**, definite form oblique case **hwītan**, + **īeġ** in the sense 'dry land in marsh'; the site is on the Wye and the surrounding land must have been subject to inundation. He 206, L 39.

WHITRIGG 'White ridge'. ON **hvítr** + **hryggr**. SSNNW 177.

(1) ~ Cumbr NY 2038. *Whyterigg'* 1278, *Whitrig* 1399. Cu 326, SSNNW 177 no.3.

(2) ~ Cumbr NY 2257. *Wyterik'* 1249, *Whyterigg, Quyterig, Wythirgg'* 1279. Cu 126, SSNNW no.1.

WHITSAND BAY Corn SX 3751. *Whitesand Bay* 1813. Identical with WHITESAND BAY SW 3527. PNCo 179.

WHITSBURY Hants SU 1219. 'Wych-elm fort'. *Wiccheberia* [1132×5]c.1195, *Wi- Wyc(c)heburi -y- -beri -biri* 1207–1428, *Whichbury* 1812, *Whittisbury* 1536, *Whitsbury* 1611. OE **wiċe** + **byriġ**, dative sing. of **burh**. The reference is to Whitsbury Hill Iron Age hill-fort. Wlt 224, Thomas 1976.143.

WHITSTABLE Kent TR 1166. '(Place at) the white post'. *Wit(en)estaple* 1086, *W(h)i- W(h)ytstapel* 1178×9–1240, *Whitstable* 1610. OE **hwīt**, definite form oblique case **hwītan**, + **stapol**. The DB forms refer to the hundred; Whitstable itself is *Nortone* 1086, 'north settlement', a name subsequently superseded by that of the hundred which met there, presumably at the white post or pillar. The hundred included Blean and Swalecliffe as well as Whitstable. PNK 493.

WHITSTONE Corn SX 2698. 'The white stone'. *Witestan* 1086, *Wi- Wyteston(e)* 1201–1319, *Wytestane* 1263, *Whyteston* 1333, *Whyston* 1525, *Whitston* 1610. OE **hwīt**, definite form **hwīta**, + **stān**. PNCo 179, Gover n.d. 37, DEPN.

WHITSUN BROOK H&W SO 9951 → CONDERTON SO 9637.

WHITTINGHAM Northum NU 0611. 'The homestead or estate called or at **Hwiting*, the place called after Hwita'. *Hwitincham* [c.1040]12th, *Hwittingaham* c.1107, *Witing(e)ham* 1160–1236 BF, *W(h)yt(t)incham -yngeham* 1253–1327. OE p.n. **Hwīting* < OE pers.n. *Hwīta* + **ing**[2], locative–dative sing. **Hwītinġe*, + **hām**. Still pronounced [witindʒəm]. NbDu 214, BzN 1967.392.

Little WHITTINGHAM GREEN Suff TM 2876. *Little Whittingham Green* 1836 OS. Adj. **little** + p.n. *Wy- Witingham* [11th, probably before 1038]13th S 1527, *Wettingaham* 1086, *Witingeham* 1212×12, *Witincham* 1236, probably 'the homestead of the Wittingas, the people called after Witta, or the homestead at or called Wittingham, the place called after Witta', OE folk-n. **Wittingas* < pers.n. *Witta* + **ingas**, genitive pl. **Wittinga*, or p.n. **Witting* < *Witta* + **ing**[2], locative–dative sing. **Wittinġe*, + **hām**, + ModE **green**. Baron, ING 131.

WHITTINGSLOW Shrops SO 4389. 'Hwittuc's burial-mound'. *Witecheslawe* 1086, *Witi- Wi- Wytekeslowe* 1208×9, *Wi- Wyttokeslawe -lowe* 1242–72, *Wittingeslaw* 1200, *Whi- Whyttingeslowe -yng-* 1301–1502, *Whittingslowe* 1601. OE pers.n. *Hwittuc*, genitive sing. *Hwittuces*, + **hlāw**. A *hwittuces hlæwe* [955]13th S 564 occurs also in the bounds of Compton Beauchamp, Berks. Sa i.311.

WHITTINGTON 'Hwita's estate'. OE pers.n. *Hwīta*, genitive sing. *Hwītan*, + **tūn**.

(1) ~ Staffs SO 8682. *Quitenton* 1203, *Whitinton, Whytynton, Whitenton* 13th, *Wytyndon, Wytinton* 1286, *Whittington* 1686 OS. Duignan 171, Horovitz.

(2) ~ Staffs SK 1608. *(æt) Hwituntune* [925]13th S 395, *Witinton* 1182, *Whytynton* 14th, *Whittington* 1686. DEPN, Duignan 171, Horovitz.

WHITTINGTON 'Estate called after Hwita'. OE pers.n. *Hwīta* + **ing**[4] + **tūn**.

(1) ~ Derby SK 3874. *Witintune* 1086, *Wi- Wytin(g)ton(e) -yng-* 1171–1361, *-tt-* 1167–1350, *Witenton(a)* 1185×95, early 13th, *Whi- Whytington(e) -yng-* 1197–1201 etc., *-tt-* 16th. Db 324.

(2) ~ Glos SP 0121. *Witetvne* 1086, *Wi- Wytinton'* 1205–1309, *-ing- -yng-* 1285, 1287, *Whitinton -yn-* 1278–1373, *-ing- -yng-* 1303–1620, *Whittington -yng-* 1488, 16th cent. Gl i.184.

(3) ~ H&W SO 8752. *HUITINGTUN* [816]11th S 180, *Hpitintun, æt hpitintune* [989 for 983×5]11th S 1361, *Widintvn* 1086, *Wi- Wytinton(a)* 1190–c.1255, *Whitenton, Whytinton -yntone* 1227–1365. The same pers.n., *Hwīta*, may occur in the boundary point *hpitan dene* 'Hwita's valley' or 'white valley' [n.d.]11th S 1601 in the bounds of the neighbouring estate of Stoulton. Wo 178 gives pr [hwitəntən], Hooke 113, 115, 315, 420.

(4) ~ Lancs SD 6076. *Witetvne* 1086, *Witington* 1212, *Quit- Wyttin(g)ton -yng-* [before 1219]1268–1333, *Whi- Whytington -yn-* 1246–1399, *Whittyngton* 16th. La 184, Jnl 17.102, 18.23.

(5) ~ Norf TL 7199. *Whittington* 1824 OS.

(6) ~ Shrops SJ 3231. *Wititone* 1086, *Quitentona* c.1127, *Whi- Whytenton* 1264–97, *Wi- Wytinton(') -yn-* 1160–1380, *Whi- Why-* 1198–1427, *Whi- Whytington -yng-* 1304–1730, *Whyttyngton* 1544, *Whittington* from 1711. OE pers.n. *Hwīta* + **ing**[4] + **tūn**, varying with genitive sing. *Hwītan* + **tūn**. The latter form allowed confusion with dative sing. definite form **hwītan** of adj. **hwīt** 'white' as in the phrase *æt thǣm hwītan tūne* '(at the) white farm', which is clearly how the name was understood in the 13th and 14th cents. A W form occurs, *Trefwen* 1254, 'white farm', W **tref** + **gwyn**, and a French form *Blauncheville e englois Whytyntone* c.1320. Sa i.312.

WHITTLE-LE-WOODS Lancs SD 5821. 'Whitehill in the woods'. *Wythill* or *Whithull in the Wode* 1304, *Whithull in bosco* 1327, 1332, *Whitle in le Woods* 1565. Earlier simply *Witul* c.1160, *Withull'* 1220, *Quythull* 1292, *Qwytyll* 15th, 16th cent., 'white hill', OE **hwīt** + **hyll**, + OFr preposition **en** + definte article **le** + ME **wode**. La 133, Jnl 17.75.

WHITTLEBURY Northants SP 6943. 'Witla's fortified place'. *Witlanbyrig* [c.930]c.1100, *Witleberia -y* 1185–1309, *Whyttlebiry* 1269, *Whittelbur'* 1285, *Whittlebury* from 1316. OE pers.n. **Witla*, genitive sing. **Witlan*, + **byriġ**, dative sing. of **burh**. The form has been influenced by the *Wh-* of nearby WHITTLEWOOD FOREST SP 7242. Nth 4.

WHITTLESEY Cambs TL 2797. 'Witel's island'. *Witlesig* c.972, [973]15th S 792, *Wi- Wytleseia -eie -eya -ey(e) -e* c.1120–1553, *Witesie* 1086. OE pers.n. *Witel*, genitive sing. *Witles*, + **ēġ**. Ca 258.

WHITTLESEY MERE Cambs TL 2290. 'Witel's mere'. *Witlesmere* [?963]12th S 1448, [1146×53]14th, 1270, *Witelesmere* [1020×23]12th S 1463, [c.1150]14th, *Witelesmare* 1086, *Whittelsmere* 1535. Pers.n. *Witel* as in WHITTLESEY TL 2797, genitive sing. *Witles*, + **mere**. Hu 191.

WHITTLESFORD Cambs TL 4748. 'Witel's ford'. *Witelesford* 1086, *Wi- Wyt(t)(e)lesford(e)* 1170–1492, *Willesford* 1427, *Wittesford* 1490, *Witzer* 1722. OE pers.n. *Witel*, genitive sing. *Witles*, + **ford**. The same man also gave his name to *Witlisoo* 1504, 'Witel's hill-spur' (OE **hōh**) in this parish. Ca 98 gives former pr [witsə].

WHITTLEWOOD FOREST Northants SP 7242. P.n. Whittlewood

+ ModE **forest**. Whittlewood, *Whitlewuda -e* 1100×1135, 1247, *Wytlewod* [1154×89]1383, *Witlewuda* 1196, is possibly p.n. *Whitley 'the white clearing', OE **hwīt-lēah**, + **wudu**. Nearby is WHITTLEBURY SP 6943. If the forms are to be trusted, despite their similarity the two names are of different origin. Nth 2.

WHITTON Cleve NZ 3822. 'Hwita's farm or village'. *Witenton* 1235, *Wyttun* 1245×69, *Qwitton'* 1304, *Quytton* 1382, *Whitton(')* from 1356. OE pers.n. *Hwīta*, genitive sing. *Hwītan*, + **tūn**.

WHITTON Humbs SE 9024. 'Hwīta's island'. *Witenai* 1086, *Witena* c.1115, *Wihitene* [1179]1328, *Whiten* 1276. OE pers.n. *Hwīta*, genitive sing. *Hwītan*, + **ēġ**. DEPN.

WHITTON Northum NU 0501. 'Hwita's' or 'the white settlement'. *Witton* 1228, *Wytton* 1256, *W(h)itton* 1275. OE pers.n. *Hwīta* or adj. **hwīt** + **tūn**. NbDu 213.

WHITTON Shrops SO 5772. Probably 'the white estate'. *Witet'* c.1174(p), *Witinton* c.1180(p), *-tone* 1210×12(p), *W(h)ytinton* 1255×6, *Whittington* 1233(p), *Whitenton* 1237, *Wi- Wytton(e)* 1236–1334, *Why- Whitton* from 1255×6. OE **hwīt**, definite form dative sing. **hwītan**, + **tūn**. Forms that point to OE *Hwītingtūn*, 'estate called after *Hwīta*', occur mostly in surnames perhaps deriving from one of the genuine *Hwitingtuns* rather than belonging here. Sa i.313.

WHITTON Suff TM 1447. 'The estate called after Hwita'. *Widituna* 1086, *Witton* 1212, *Wytenton* 1216×72, *Qiting- W(h)ytyngton(e)* 1254–1346, *Whytenton* 1295, *Whitton* 1568. OE pers.n. *Hwīta* + **ing**[4] + **tūn**. DEPN, Baron.

WHITTONDITCH Wilts SU 2972. '(The settlement) at the white ditch'. *Whitedic* 1249, *Wytendyche* 1268, *Whytendich* 1289. OE **hwīt** + **dīċ**, dative case definite form **(æt thǣm) hwītan dīċe**. Wlt 289.

WHITTONSTALL Northum NZ 0757. 'The farmstead with a quickset hedge'. *Quictunstal* c.1150–1242 BF, *Quikcumstal* (sic) 1296 SR, *Whi- Whyttonstal(l)* 1256–1307. OE adj. **cwic** + **tūnstall**. NbDu 214.

WHITWELL 'The white spring'. OE **hwīt**, definite form **hwīta**, + **wella**. L 31.

(1) ~ Derby SK 5276. *Hwitan wylles geat* 'Whitwell gate' 10th ASC(A) under year 942, *(æt) Hwitewylle* [c.1002]c.1100 S 1536, *Witewelle* 1086–1316 with variant *Wyte-*, *White- Whytewell(e)* 1258–1535, *Whitwell(e)* from 1330. The first form is 'the gap of Whitwell', referring to the valley in which the village lies. Db 327.

(2) ~ Herts TL 1821. *Wy- Wetewelle* 1278, *Whitewelle* 1321, *Whytwell* 1539. Hrt 116.

(3) ~ IoW SZ 5278. *Quitewell'* 1212, *Why- Whitewelle* 1255–1462, *Whitwell* from 1428. Wt 251 gives pr ['witəɫ], Mills 1996.109.

(4) ~ Leic SK 9208. *Wy- Witewell(e)* 1086–1326, *Why-Whitewell(e)* 1197–1553. R 62.

(5) ~ NYorks SE 2899. *Witeuuella* 1086, *Whitwell(e)* from 1285. YN 279.

(6) ~ -ON-THE-HILL NYorks SE 7265. P.n. *Witeu(u)elle* 1086, *Whytte- Qwyttwell* [12th cent.]copy + ModE **on the hill**. YN 39.

WHITWICK Leic SK 4316. 'The white' or 'Hwita's farm'. *Witewic* 1086, *Wi- Wytewic(a) -wyc -wich -wik(e) -wyk(e)* 1152–1535, *Whi- Whytewik(e) -wyk(e) -wich -wick* after 1204–1512, *Whi- Whytwi(c)k(e) -wy(c)k(e)* 1270–1551. OE **hwīt**, definite form **hwīte**, or pers.n. *Hwīta*, + **wīċ**. The reference may have been to the outcrop of white sandstone here. Lei 360.

WHITWOOD WYorks SE 4024. 'The white wood'. *Witeuude -wde* 1086, *Witewde* c.1090–1240, *White- Whytew(u)d(e) -wode* 1211–1497, *-wood* 1230, 1588. OE **hwīt**, definite form **hwīta** + **wudu**. YW ii.124, L 228.

WHITWORTH Lancs SD 8818. 'Hwita's enclosure'. *Whiteword -worth* [13th]14th, *Wytewurth(e)* 1246. OE pers.n. *Hwīta* + **worth**. The specific could also be adj. **hwīt** 'white'. La 61.

WHIXALL Shrops SJ 5134. 'Hwittuc's nook'. *Witehala* 1086, *Withekisall, Witcheshill* 1203×4, *Whitekeshal* 1240, *Wickeshal(l) -is-* 1255–1316 with variants *Wy(c)kes- Wyches-* and *-hale*, *Qui(c)keshale* 1255×6, 1362, *Quyxhale* 1334, *Quik- Quixsale* 14th, *Wixall* 1615–1742, *Wixhall otherwise Quixhall* 1665. OE pers.n. *Hwittuc*, genitive sing. *Hwittuces*, + **halh** in the sense 'firm ground in a wet area', dative sing. **hale**. Sa i.314.

WHIXLEY NYorks SE 4458. 'Cwic's forest glade or clearing'. *C(r)ucheslage, Cuselade* 1086, *Qui- Quyxele(a) -lai -lay -ley* 12th–1301, *Qui- Quy- Qwyxle(y) -lay* 12th–1502, *Whi- Whyxley -lay* 1353–1685. OE pers.n. **Cwic*, genitive sing. **Cwices*, + **lēah**. YW v.9.

WHOLSEA Humbs SE 8433 → WALLING FEN SE 8829.

WHORLTON Durham NZ 1014. 'Settlement on the *Cweorning*'. *Queorningtun* [c.1040]12th, *Cueornington* [1072×83]15th, *-tun* c.1107, *Quer(n)ington -yng-* 1306–60, *Whanton* 1512, 1524, *Quarrington alias Whor(e)l(e)ton* 1624, 1637, *Walton* 1627, *Whorlton* from 1516. R.n. **Cweorning* 'stream where millstones are found', OE **cweorn** + **ing**[2], + **tūn**. The forms of the name show two developments:

I. *cweorningtun* > *cwernington* > *cwerrington* > *cwarrington* (cf. QUARRINGTON HILL NZ 3337);

II. *cweorningtun* > *cwernington* > *cwerlington* > *hwerlton* > *hwarlton* > *worlton* Cf. DARLINGTON NZ 2914. NbDu 215.

WHORLTON NYorks NZ 4802. 'Settlement by (the hill called) Whorl (*Wirvel*)'. *Wirveltun(e)* 1086, *Weruelthun -ton* 1189×99–1294, *Qwerlton* 1198, *Wheruelton* 1202, *Wherleton* 1299–1354, *Whorl(e)ton* 1399–1575. OE **hwerfel** 'a circular or round-topped hill' + **tūn**. Whorl hill is a prominent circular hill nearby. YN 177.

WHORLTON MOOR NYorks SE 5098. *Whorlton Moor* 1857 OS. P.n. WHORLTON NZ 4802 + ModE **moor**.

WHYGATE Northum NY 7776. *Whygate* 1866 OS. Possibly ON **kví** 'a pen, a fold' + **gata** 'road, gate'.

WHYLE H&W SO 5661. Unexplained. *Hvilech* 1086, *Whiale* 1123, *Whilai* c.1150, *Wielai* 1160×70, *Wihale* 1158×61, *Wyley* 1220, *Wi- Wyle* 1211×12–1307, *Whi- Whyle* from 1341 including *Le While* 1431, *The Whyle* 1832 OS. This name remains unexplained; the spellings suggest that the generic was taken to be OE **lēah** 'a wood, a clearing, a pasture' but this may have been an attempt to make sense of an already obscure name. The specific appears to have been disyllabic but remains unidentified. He 171.

WHYTELEAFE GLond TQ 3458. A residential district and railway station named from a field called *White Leaf Field* 1839 from the aspen-trees growing there. Sr 313.

WIBDON Glos ST 5697. Unexplained. *Wibdon* 1624, *Webdon* 1694. The evidence is too late, but the generic is probably from OE **dūn** 'a hill'. Gl iii.267.

WIBTOFT Warw SP 4787. 'Wibba or Vibbi's curtilage'. *Wibbetofte* [1002]c.1100, 1227, 1468, *Wibetot* 1086, *Wibtoft* from 1271. OE pers.n. *Wibba* or OEScand *Vibbi*, genitive sing. *Vibba*, + **toft**. Wa 120.

WICHENFORD H&W SO 7860. 'Wych-elm ford'. *Wiceneford* 11th, *Wi- Wychenford* 1208–94, *W(h)ichingford* 1594, 1675. OE **wiċe**, genitive pl. **wiċena**, + **ford**. Wo 179, PNE ii.263.

WICHLING Kent TQ 9256. Partly uncertain. There are two forms associated with this place;

I. *Winchelesmere* 1086, *Wincelesmere* c.1100, *W(h)icclesmer(e)* 1242×43.

II. *Winchelinge(s)* 13th, *Wy- Wicheling(e)* 1236×7–78.

The first appears to be 'the pool (or boundary) of *Wincel*' in which *Wincel* is OE ***winċel** 'a nook, a corner' referring to the small valley here, genitive sing. **winċeles**, + **mere** or **(ġe)mǣre**; the second either a p.n., OE ***winċeling** 'a nook-like place', or a folk-n. **Winċelingas* 'the people of the *winċel*, the folk living at the nook or corner'. PNK 238, ING 16, BzN 1967.349, PNE ii.268.

WICK 'Collection of buildings, farm, workshop'. OE **wīċ**, dative pl. **wīcum**.

(1) ~ Avon ST 7073. *Wik(e), Wika, Wyke, Wica* 1189–1587, *We(e)ke* 1588, 1595. Also known as *Berde(s)wyke* 1222 with the ME by-n. *Berde* 'beard'. Gl iii.71.

(2) ~ Dorset SZ 1591. *la Wych* 12th, *Wyke* 1263, *la Wyk* late 13th. Do 155.

(3) ~ H&W SO 9645. *Wiche* 1086, *la Wike* 1261, *Wyke* 1275. Wo 231.

(4) ~ WSusx TQ 0203. *Wyke* 1261, 1288, 1718, *Weeke* 1677, 1718. A dairy-farm attached to Lyminster. Sx 171.

(5) ~ Wilts SU 1621. *Wicha* 1166, *Wyke* 1268, 1279, *Weke* 1524, 1585, *Weeke* 1642. An outlying farm belonging to Downtown SU 1820. Wlt 395.

(6) ~ DOWN SU 1321. 'Down belonging to Wick'. *Wick Down* OS 1810. P.n. WICK + ModE **down**.

(7) ~ HILL Berks SU 8064. Probably short for Upwick Hill, *Upwicks Hill* 1816. *Upwick(s)* is perhaps 'upper dairy farm' or possibly a surname. The form is late and the name need not be of any great antiquity. ModE **up** + **wick**. Brk 97.

(8) ~ ST LAWRENCE Avon ST 3665. 'St Lawrence's W'. *Wike* 1225, *la Wyk* 1243, + saint's name *Lawrence* from the dedication of the church. The use of the French definite article in the 1243 form shows that **wīc** was still felt to be a common noun, 'the dairy-farm'. DEPN.

(9) Hannington ~ Wilts SU 1795. 'W belonging to Hannington'. *Hanyndoneswyk* 1364, *Hanyngdon Weke* 1547. Earlier simply *Wyk* 1289. P.n. HANNINGTON SU 1793 + WICK. Wlt 25.

(10) Haydon ~ Wilts SU 1387. 'W belonging to Haydon'. *Haydonwyk* 1249, *Haydoneswyk* 1321, *Heydon Week* 1595, also known simply as *Wyke* 1332. P.n. Haydon SU 1288, *Haydon(e)* 1242–1428, either 'the enclosure hill', OE **ġehæġ** + **dūn**, or 'hay hill', OE **hēġ** + **dūn**, + WICK. Wlt 32.

(11) Kimble ~ Bucks SP 8007. 'Kimble dairy-farm'. *Kimblewick* 1834 OS. Earlier simply *Wyk(a)* [1154×89]1313, 1227. P.n. KIMBLE SP 8206 + OE **wīċ**. One of a series of wicks or dairy-farms in the low-lying ground N of Kimble, cf. LONGWICK SP 7805, OWLSWICK SP 7806, Terrick SP 8308, Wellwick SP 8057. Bk 164.

(12) Milborne ~ Somer ST 6620. 'W belonging to Milborne'. *Milborne Week* 1811. P.n. Milborne as in MILBORNE PORT ST 6718 + p.n. Wick, Week. Earlier simply *Weke* 1610. DEPN.

(13) North ~ Avon ST 5865. 'North dairy-farm'. *Northwick* 1817 OS. N of Norton Hawkfield ST 5964.

(14) Potterne ~ Wilts SU 0057. 'Dairy-farm belonging to Potterne'. *Poternewike* 1203. P.n. POTTERNE ST 9958 + **wīc**. Wlt 246.

(15) Stanton ~ Avon ST 6162. 'W belonging to Stanton', sc. Stanton Drew ST 5962. *Stanton Wick* 1817.

(16) Tockenham ~ Wilts SU 0381. 'Dairy farm belonging to Tockenham'. *Tockenhamweke* 1559×79, *Tokenham Wike* 1561, *Tottenham Weeke* 1558×1603. P.n. TOCKENHAM SU 0479 + **wīc**. Wlt 272.

(17) West ~ Avon ST 3661. *West Wick* 1809 OS. Contrasts with Way Wick ST 3862, *Waywick* 1817 OS.

WICKEN Cambs TL 5770. '(At) the dairy-farms'. *Wi- Wyken* c.1200–1570. OE **wīcum**, dative pl. of **wīċ**. Singular forms *Wich, Wicha* 1086, *Wyk(e)* 1227–1353 are also recorded, along with pl. forms *Wicrena* [1124×35]1337, *Wicre(s)* 1145×8, [1160×71], 1337, *Wi- Wykes* 1208–1428. Ca 203.

WICKEN Northants SP 7439. 'The Wicks'. *Wicha -e* 1086, *Wika -e, Wyke* [1100×35]1267–1284, *Wykes* 1209–18, *Wicne* 1235, *Wyken* from 1457. OE **wīċ**, ME pls. **wikes**, **wiken**. There were two separate manors called Wick, 'the dairy farm, the work or trading place'. Nth 107.

WICKEN BONHUNT Essex TL 5033. Originally two separate names, *Wicam* 1086, *Wykes* 1239–1381, *Wykyn, Wiken* 1412–75, OE **wīcum** 'at the dwellings', dative pl. of **wīċ**, and *Banhuntā* 1086, *-hunt(e)* 1141, 12th, *Bonhunt(e)* 1236–1412, 'helpful' or 'bone spring', OE **bān**+ **funta** with early substitution of *h* for *f* as in TOLLESHUNT TL 9011, 9312. The first solution, preferred by Dr Insley, rests on an unattested OE ***bān** cognate with ON *beinn*, the origin of the r.n. BAIN Lincs TF 2473. But since **funta** might refer to a Roman antiquity the reference here might be to a cultic spring where sacrificed animal bones were deposited. The forms do not support derivation from OE pers.n. *Bana* or **bana**. Ess 544, Anglia 1995.207, 1996.544.

WICKENBY Lincs TF 0881. 'The village or farm of Viking or the Vikings'. *Wichinge- Wighingesbi* 1086, *Vichen- Vichinghebi* c.1115. ODan pers.n. *Vīking*, secondary genitive sing. *Vīkinges* or ON **víkingr**, genitive pl. **víkinga**, + **bȳ**. DEPN, SSNEM 77.

WICKERSLEY SYorks SK 4891. 'Vikar's wood or clearing'. *Wicresleia, Wincreslei* 1086, *Wi- Wykereslai -ley(e)* 1185–1449, *Wi- Wykersleg -ley -lay* [1199]1232–1525. ON pers.n. *Víkarr*, genitive sing. *Víkars* + **lēah**. YW i.171, SSNY 167.

WICKFORD Essex TQ 7593. 'Ford at or called Wick, or by the dairy-farm'. *(æt) Wicforda* 962×91 S 1494, *Wi(n)cfort* 1086, *Wic- Wyc- Wik(e)- Wyk(e)ford* 1194–1330. OE **wīċ** possibly used as a p.n. + **ford**. Ess 176, Signposts 1988.70, 247.

WICKHAM 'Homestead associated with a Roman or Romano-British settlement'. OE **wīc-hām**. Evidence 8–26, Signposts 67–74.

(1) ~ Berks SU 3971. *Wicham* [821]12th S 183, 1167 etc. with variants *Wick- Wycham, Wyke- Wickham* 1550. Brk 274, 802–4, L 323.

(2) ~ Hants SU 5711. *(æt) wicham* [955×8]14th S 1491, *Wicheham* 1086, *Wic- Wik- Wykham* 1167–1330, *Wickham* 1341. Ha 175, Gover 1958.34, Evidence 10.

(3) ~ BISHOPS Essex TL 8412. 'W held by the bishop' sc. of London. *Wycham Episcopi* 1291, *Wykham Bishops* 1313. P.n. Wickham + ME **bishop**, genitive sing. **bishopes**, Latin **episcopus**, genitive sing. **episcopi**. Wickham is *Wicham* 1086–1297 including [c.940]13th S 453, *Wicam* 1086, *Wyc- Wykham* 1221–1458. The area is rich in Roman finds. 'Bishops' for distinction from WICKHAM ST PAUL TL 8336. Ess 313, Settlements 9, 25, Signposts 72.

(4) ~ MARKET Suff TM 3055. *Wickham Markett* 1674. P.n. *Wik(h)am* 1086, *Wicham* 1254, *Wi- Wykham* 1286–1524, + ModE **market**. There is no known association with Roman settlement here but the place lies on a Roman road, Margary no. 340. DEPN, Baron Evidence 12.

(5) ~ ST PAUL Essex TL 8336. 'St Paul's W'. *Wyk(h)am Sancti Pauli* 1254, 1428, *Wikehampaule, Wykkam Pawley, Wickam St Paul alias Wickam Pole* 16th cent., *Wickham poole* 1607. P.n. *Wic(h)am* 1086, + saint's name *Paul* from St Paul's cathedral, London, to the canons of which the estate belonged, cf. *Canouns Wykham* 1322. Roman remains have been discovered at and near Wickham. Jnl 2.47, however, cites a form *Hinawicum* [c.1000]c.1125 for this name which points to OE **wīcum**, dative pl. of **wīċ**, as its source instead of **wīc-hām**. *Hina* is OE **hīġna**, genitive pl. of **hīwan** 'members of a religious house', here referring to the convent of St Paul's. Ess 467, Settlements 9, 25.

(6) ~ SKEITH Suff TM 0969. 'W with the racecourse'. *Wicham Skeyth* 1368, *Wickham Skeithe* 1568, *Wickham Skithe* 1610. P.n. *Wic(c)hamm* 1086, *Wic- Wyc- Wykham* 1236–1367 including [1189]1253, 'the water-meadow with a dwelling, OE **wīċ** + **hamm**, + ON **skeith** or a family name derived from it. The place is 1 m. W of a Roman road, Margary no.3d, on which there is a small Roman settlement. DEPN, Baron, Evidence 12.

(7) ~ STREET Suff TL 7554. P.n. Wickham as in WICKHAMBROOK TL 7554 + Mod dial. **street** 'a hamlet'.

(8) ~ STREET Suff TM 0969. *Wickham Street* 1836 OS. P.n. Wickham as in WICKHAM SKEITH TM 0969 + Mod dial. **street** 'a hamlet'.

(9) East ~ GLond TQ 4283. *Estwycham* 1284, *Est Wycham* 1292, *East Wickham* c.1762. ME **est** + p.n. *Wikam* 1242. 'East' for distinction from West WICKHAM TQ 3866. DEPN, LPN 73.

(10) West ~ Cambs TL 6149. *West Wikham* 1266–1428. Earlier simply *Wicham* 1086–1318, *Wicheham, uuichehâm* 1086, *Wi- Wykham* 1218–1426. The earliest mention is in the bounds of the neighbouring parish of West Wratting, which ran *oð wichammes gemære* 'to the bounds of W' [974]11th S 794. The *-mm-* in this spelling suggests a p.n. in **hamm** rather than **hām** but no other form supports this. The site is not notably appropriate for a **hamm**. There was a minor Romano-British settlement at Horseheath 1 mile S of Wickham on the line of the Roman road from Cambridge to Sible Hedingham, Margary no. 24. 'West' for distinction from WICKHAM BROOK Suff TL 7454. Ca 112, Evidence 9.

(11) West ~ GLond TQ 3866. *Westwycham* 1284. ME **west** + p.n. *Wichamm* 955 for ?973 S 671, *Wicheham* 1086, *Wicham* 1231. On the Roman road from Lewes to London, Margary no.14; there are Roman settlement remains. The boundary of the people of Wickham is recorded as *Wichema mearcæ* 862 S 331 and *Wic hammes gemæru* 955 for ?973 S 671. 'West' for distinction from East WICKHAM (12) TQ 4576. DEPN, Signposts 72, 76, 81, LPN 246.

WICKHAMBREAUX Kent TR 2259. 'Wickham held by the Brewse family'. *Wykham Breuhuse, Wyckham Breuse* 1270 etc. Earlier simply *wichám* 948 S 535, *Wichehā* 1086, *Wicham* [1087]13th, 1236, 'village near a Roman *vicus*', OE **wīc-hām**. The manor was held by William de Brayhuse in 1265; the family name is from Briouze in Normandy. KPN 277, DEPN, Settlements 10, Signposts 72.

WICKHAMBROOK Suff TL 7554. 'W with the stream'. *Wichambrok* 1254. P.n. *Wichā* 1086, *-ham* 13th, OE **wīċ-hām** 'the homestead, the dwelling place', + **brōc**. There is no known association with any Roman settlement here. DEPN, Evidence 12.

WICKHAMFORD H&W SP 0641. 'Wigwenne ford'. *Wike Waneford* 1255, *Wycanford* 1550, *Wic(k)hamford* from 1593. P.n. Wigwenne + OE **ford**. Medieval scribes had difficulty with the specific of this name, which was obscure to them. It occurs again in nearby CHILDSWICKHAM Glos SP 0738. The forms for Wickhamford are *wigorne* [714]16th S 1250, *wicwona* *[709]12th S 80, *Wiqvene* 1086, *Wikkewan* 1251. The boundary of people of Wicwona is mentioned in S 226, *Wycweoniga gemære* [c.860]c.1200. Wo 273, Hooke 23, 46.

WICKHAMPTON Norf TG 4205. 'Wickham estate or village' or 'Hampton wick or dairy-farm'. *Wichamtuna* 1086, *-ton* 1206. This is either p.n. **wīċ-hām** 'dwelling place, manor' + **tūn** or **wīċ** + **hām-tūn** '(dairy) farm called or at Hampton'. DEPN.

WICKLEWOOD Norf TG 0702. 'Wickley wood'. *Wickelewuda* 1086, *Wicklewuda* 1168, *-wode* 1242, 1254. P.n. *Wikele* probably representing OE *wīċe-lēah* 'wych elm wood' < OE **wiċe** + **lēah**, + pleonastic **wudu**. DEPN.

WICKMERE Norf TG 1733. 'The lake or pond by the farm'. *Wicmara -mera* 1086, *Wikemere* 1166. OE **wīċ** + **mere** . There is no lake here today. DEPN.

WICKWAR Avon ST 7288. 'Wick held by the la Warre family'. *Wi- Wykewarre* 1216×72–1535, *Wickwar* 1552. Earlier simply *Wichen* 1086, 'at the dwellings', OE dative pl. **wīcum** of **wīċ**, and *Wy(c)ke, Wike* 1220–81. The manor was confirmed to John la Warre by King John. Gl iii.42.

North WIDCOMBE Avon ST 5758. *North Widcombe* 1817 OS. ModE adj. **north** + p.n. *Widecomb* 1303, *Wydecomb* 1321, 'the wide coomb', OE adj. **wīd**, definite form **wīda**, + **cumb**. The reference of **cumb** here is to an indentation in the escarpment between Burledge Hill ST 5858 and White Hill St 5857. 'North' for distinction from South WIDCOMBE ST 5856. DEPN, L 93.

South WIDCOMBE Avon ST 5856. *South Widcombe* 1817 OS. ModE adj. **south** + p.n. Widcombe as in North WIDCOMBE ST 5758.

WIDDALE NYorks SD 8287. 'Wood valley'. *Withdale* 1217, *Wyddale* 1307. ON **vithr** + **dalr**. YN 267.

WIDDALE FELL NYorks SD 8088. P.n. WIDDALE SD 8287 + ModE **fell** (ON *fjall*).

WIDDINGTON Essex TL 5331. 'Willow-tree settlement'. *Widi(n)tuna* 1086, *Wi- Wyditon* 1204–1328,*Wi- Wythiton(e)* 1123–1254, *Wythington* 1327, *Widdi(ng)ton* 1594. OE **withiġ(n)** + **tūn**. Ess 545, PNE ii.271.

WIDDOP RESERVOIR WYorks SD 9333. P.n. Widdop + ModE **reservoir**. Widdop, *Wyd- Widhop(e)* 1368, 1695, *Wi- Wydop(e)* 1548, 1709, is the 'wide valley', OE **wīd** + **hop**. The upper part of Hebden Dale opens out into a wide basin which now forms the reservoir. YW iii.202.

WIDDRINGTON Northum NZ 2595. 'The estate called after Widuhere'. *Vuderintuna* c.1160, *Wud(e)rinton* 1170, 1177, *Wod(e)ring(a)ton(e)* 1166–1346, *Widerintune -ton(e)* c.1180–1346, *Wydrington* 1356. OE pers.n. **Widuhere* + **ing**[4] + **tūn**. NbDu 216.

WIDDRINGTON STATION Northum NZ 2494. P.n. WIDDRINGTON NZ 2595 + ModE **station**.

WIDDYBANK FELL Durham NY 8330. *Widdybank Fell* 1866 OS. P.n. Widdybank + ModE **fell**. Widdybank, *Widdybank* 1768, 'wooded hill-side', is OE **widiġ**+ ME **banke**. **widiġ** is the earlier form of **wudiġ**, cf. WITTON.

WIDE OPEN T&W NZ 2372. *Wide Open* 1863 OS. A 19th cent. colliery settlement; the name alludes to its exposed position.

WIDECOMBE IN THE MOOR Devon SX 7176. *Whithecombe in the More* 1362, *Wydecomb yn the More* 1461. Earlier simply *Widecumb(a)* [1100×35]1270–1453 with variants *Wyde-* and *-comb(e)*, 'the wide coomb', OE adj **wīd**, definite form **wīda**, + **cumb**. Occasional spellings with *-th-* for *-d-* may point to OE **withiġ** 'a withy, a willow', but the overwhelming majority of forms has *-d-*. D 526.

WIDEGATES Corn SX 2857. *Wide-gates* 1673. ModE **wide** + **gate** in the sense 'road leading on to the downs'. PNCo 179.

WIDEMOUTH BAY Corn SS 2002. A holiday village called *Widemouth* 1969, *Widemouth Bay* 1971, from Widemouth SS 2001, *Wide- Witemot* 1086, *Widemutha* 1181, *Wydemouthe* 1337, 'wide gap', OE **wīd** + **mūtha**. The bay itself is *Widemouth Bay* 1813. The reference is probably to the mile-long gap in the cliffs here. PNCo 179 gives pr ['widməθ], Gover n.d. 81.

WIDFORD Essex TL 6905. 'Withy ford'. *Witford* 1202–47, *Wiliford* 1254, *Wydiford* 1254, 1295, *Wydford* 1428. OE **withiġ** + **ford**. Ess 275.

WIDFORD Herts TL 4115. 'Withy ford'. *Wideford(e)* 1086–1402 with variant *Wyde-*, *Widi- Wydiford* 1205–1315, *Wydforde* 1428. OE **withiġ** + **ford**. This might seem to be the 'wide ford', but the normal word for this aspect is OE **brād** rather than **wīd**. Hrt 210.

WIDMERPOOL Notts SK 6328. Probably 'the pool called *Widmere*, the wide pond'. *Wimarspol(d)* 1086, *Widmerepol* 1180–3, *Widmerpol* 1235–1327 with variants *Wyd-* and *-poll*. P.n. **Widmere* 'the wide pool' < OE **wīd**, definite form **wīda**, + **mere** + OE **pōl**. There is still a pond in the village today. Nt 257, L 26.

WIDNES Ches SJ 5185. 'The wide promontory'. *Wydnes* c.1200–1271 etc., *Wydenes* 1242, *Widnesse* 1271. OE **wīd** + **ness**. A headland jutting out into the Mersey at a narrow point called Runcorn Gap. La 106.

Lower WIELD Hants SU 6340. *Lower Weild* 1657, *Lower Weald* 1817 OS. ModE adj. **lower** + p.n. *Walde* 1086, 1236, *Welde* 1256–1341, *la Wolde* 1309, *Wylde* 1316, *Weelde* 1428, *Weild* 1621, 1657, 'forest', OE **weald**. 'Lower' for distinction from Upper WIELD SU 6238. Also known as *Estwild* 'east Wield' 1337. Ha 176, Gover 1958.80.

Upper WIELD Hants SU 6238. ModE adj. **upper** + p.n. Wield as in Lower WIELD SU 6340. In OS 1817 it is simply *Weald*.

Probably the same place as *Westwelde* 'west Wield' 1369. Ha 176, Gover 1958.80.

WIGAN GMan SD 5805. Probably 'homestead of Wigan'. *Wigan* from 1199 with variants *Wygan* c.1215–1387, *Wyan* 1420, *Wi- Wygayn -ain -eyn* 13th cent. PrW ***treβ** + W pers.n. *Wigan* (OW *Uicant*). For the loss of ***treβ**, cf. Wigan Anglesey and for the pers.n., cf. the lost *Bodewygan* 1294 'dwelling of Wigan', Llanddeusant, Gwynedd SH 3485, also in Anglesey. The alternative view, that Wigan derives from OE dative pl. **wīcum** 'at the dwellings' with reference to Wigan's Romano-British origins, is not supported by the forms of the name or by the assumed parallel of Wigan Farm Cambs SU TL 3176, *Wyken* 13th, 1405–1538, *Wiggin* c.1750. La 103, TC 193, Jnl 17.58 gives pr [wigin], 21.46, BdHu 229.

Great WIGBOROUGH Essex TL 9615. *Wykeberwe Magna* before 1272, *Great Wigborough* 1805 OS. ModE adj. **great**, Latin **magna**, + p.n. *Wicghebergā, Wigheberga(m)* 1086, *Wi- Wyg(h)eberg(h)e* 1206–1347, *Wi- Wyggeberig' -beregh' -berwe* 1248, 'Wicga's hill', OE pers.n. *Wicga* + **beorg**. Great W is situated on a marked round hill. 'Great' for distinction from Little Wigborough TL 9715, *Little Wygbar(r)owe* 1433. Ess 323.

WIGGATON Devon SY 1093. 'Wicga's estate'. *Wigaton* 1281, *Wyggeton* 1333. OE pers.n. *Wicga* + **tūn**. The same pers.n. and same owner occurs in the form *wicgincland* [1061]1227 'land called after Wicga'. D 607 gives pr [wikətən].

WIGGENHALL ST GERMANS Norf TF 5914. 'St German's W'. *Wygenhale Sancti Germani* 1254. P.n. Wiggenhall + saint's name *Germanus* from the dedication of the church. Earlier simply *Wigrehala* (sic) 1086, *Wiggehal* 1160, *Wiggenhal* 1196, 'Wicga's nook', OE pers.n. *Wicga*, genitive sing. *Wicgan*, + **halh**. 'St Germans' for distinction from WIGGENHALL ST MARY MAGDALEN TF 5911, WIGGENHALL ST MARY THE VIRGIN TF 5813 and Wiggenhall St Peter TF 6013, *Wygenhale Sancti Petri* 1254. DEPN.

WIGGENHALL ST MARY MAGDALEN Norf TF 5911. 'St Mary Magdalen's W'. *Wygenhale Magdalene* 1254. P.n. Wiggenhall as in WIGGENHALL ST GERMANS TF 5914 + Saint's name *Mary Magdalene* from the dedication of the church. DEPN.

WIGGENHALL ST MARY THE VIRGIN Norf TF 5813. 'St Mary the Virgin's W'. *Wygenhale Matris Christi* 'the mother of Christ's W' 1254. P.n. Wiggenhall as in WIGGENHALL ST GERMANS TF 5914 + saint's name *Mary* from the dedication of the church. DEPN.

WIGGINTON NYorks SE 5958. 'Viking or Wicga's estate'. *Wichis- Wichintun* 1086, *Wi- Wyginton'* 1231–1337, *-ing-* 1291–1330, *Wiggenton* 1579. ON pers.n. *Víkingr* or OE pers.n. *Wicga* + **ing**[4] + **tūn**. YN 14, SSNY 130.

WIGGINTON 'Wicqa's farm or estate'. OE pers.n. *Wicqa*, genitive sing. *Wicqan*, + **tūn**.

(1) ~ Herts SP 9410. *Wigentone* 1086, *Wigeton* 1200, 1202, *Wi- Wyginton(e)* 1201–1371. Hrt 59.

(2) ~ Oxon SP 3833. *Wigentone* 1086–1364 with variant *-tona, Wigintun', Wi- Wygenton(a) -tone -yn-* 1220–1389, *Wigingtone* 1225×6. *Wyggeleuam* c.1250, 'Wicga's tumulus', is a field-n. in neighbouring Hook Norton. O 408, 357.

(3) ~ Staffs SK 2106. 'The estate called after Wicga'. *Wicgintun* 11th, *Wigetone* 1086, *Wyggenton* 1230. OE pers.n. *Wicga* + **ing**[4] + **tūn** or genitive sing. *Wicgan* + **tūn**. DEPN.

WIGGLESWORTH NYorks SD 8157. 'Wincel's enclosure'. *Wiclesforde, Winchelesuu(o)rde* 1086, *Wic- Wyck- Wikleswrthe -worth* 12th–1303, *Wickelesword, Wi- Wykelesworth* 1217–1540, *Wi- Wyglesworth* 1198–1617. OE pers.n. *Wincel*, genitive sing. *Wincles*, + **worth**. For loss of *n* in this name cf. WICHLING TQ 9256. YW vi.162

WIGGONBY Cumbr NY 3053. 'Wigan's village or farm'. *Wy- Wigayneby* 1278, 1279, *Wyganby* 1285–1485 with variant *Wi-, Wygganby* 1292. OFr pers.n. *Wigan, Wigayn* (from OBreton *Uuicon, Uuecon*) + **bȳ**. One of a group of late *by*-names compounded with OFr pers.s.n in the Carlisle area, cf. ALLONBY NY 0842, JOHNBY NY 4333, LAMONBY NY 4135, MORESBY NX 9921 and SSNNW 15, 21ff. Cu 120, SSNNW 43.

WIGGONHOLT WSusx TQ 0616. 'The wych-elm wood'. *Wikeolte* 1195, *Wi- Wykeholt* 1212–88, *Wyken(h)olt* 1279–1332, *Wygeholt* 1230, 1255, *Wygenholt* 1316–1510. OE **wiċe**, genitive pl. **wicna**, + **holt**. Sx 163, PNE ii.263, L 196.

WIGHILL NYorks SE 4748. Probably 'nook of land with a dairy farm'. *duas Wicheles* 'the two Ws' 1086, *Wic- Wyc- Wik- Wykhal(e)* 13th cent., *Wi- Wyghal(e)* 1303–1535, *Wighill* 1538. OE **wīċ**+ **halh**. It has been suggested that **wīc** in this name might be used in its earliest sense of 'Romano-British settlement'. Wighill lies 2 m. N of Tadcaster and 1 m. E of Rudgate, the Roman road from Tadcaster to Whixley, Margary no. 280. YW iv.242, L 111.

Isle of WIGHT SZ 4985. 'The fork' or 'watershed'. *Ἴκτιν* (accusative case, *Ictim*) [c.30]11th BC Diodorus Siculus v.22†, *Vectis* [AD 77]9th Pliny NH iv.103, *Vectem* (accusative case) [c.120]9th Suetonius *Vespasian* 4, *Οὐηκτίς* (*Vectis*) [c.150]13th Ptol with variant *Οὐικτίς* (*Victis*), *Vecta* [c.300]8th MI, Panegyric VII(v), [683]c.1300, [731]8th BHE–1537 including S 274, 281, 543, 821 all in 12th–13th cent. copies, *Βέκτην* (accusative *Bectem*) 4th Eutropius-Paeanius, *Wi(e)hte ealond, Wiht(land)* 9th–11th ASC under years 449, 530, 534 etc., ?1071–1249 including S 274 and 543 in 12th–13th cent. copies, *Wight* from 1199 with many variants. W forms include *Gueith* c.800 HB and the two spellings cited below. Brit **Ueχta* < **u̯ekto-* the sense of which is derived from 13th cent. interpretations of the name, *quam Britones insulam Gueid vel Guith quod latine divorcium dici potest* 'the Britons call the island Gueid or Guith which can be translated *divorcium* in Latin'. Latin *divorcium* means 'fork' in a road and 'watershed'. The reference is to the position of the island in the fork of the Solent forming a division of the waters. Wt 1, 281, RBrit 487, FT 118, Mills 1996.109.

WIGHTON Norf TF 9439. 'Dwelling place, farm with a dwelling place'. *Wistune* 1086, *-tona* 1130, *Wihton* 1161, *Wichton* 1165, *Wigton* 1194, *Wicton* 1212. OE **wīċ-tūn**. DEPN.

WIGMORE H&W SO 4169. 'Quaking marsh'. *Wig(h)emore* 1086, *Wyg(g)emore* 1262–1535. OE **wicga** 'unstable ground' + **mōr**. He 207, L 55–6, Nomina 9.104.

WIGMORE Kent TQ 8064. 'The wide pool'. *Wydemer(e)* 1270–1347, *Wygem'* 1347, *Wigmer* 1610, *-more* 1819 OS. OE **wīd**, definite form **wīda**, + **mere**. PNK 130.

WIGSLEY Notts SK 8670. 'Wicg's wood or clearing'. *Wigesleie* 1086, *Wi- Wyggesley(e)* c.1160–1428. OE pers.n. **Wicg*, genitive sing. **Wicges*, + **lēah**. Nt 208.

WIGSTHORPE Northants TL 0482. 'Vikingr's outlying farm'. *Wykingethorp* 1232, *Wykingestorp* 1247, *-thorp* 1253, 1330, *Wygingestorp* 1337, *Wigisthorp* 1428, *Wigstroppe* 1624. ON *Víkingr*, either a pers.n. or appellative, genitive sing. *Víkings*, + **thorp**. An outlying settlement of Lilford TL 0383. Nth 185, SSNEM 134.

WIGSTON Leic SP 6099. 'Vikingr's farm or village'. *Wichingestone* 1086, *Wy- Wiking(g)estone -yng- -is-* 1191–1509, *Wy- Wigingeston' -yng-* before 1250–1385, *Wy- Wikinston(e) -yn-* c.1247×60–1503, *Wy- Wigston* from 1453. ON pers.n. *Víkingr*, ME genitive sing. *Víkinges*, + **tūn**. Lei 471, SSNEM 197.

South WIGSTON Leic SP 5898. A modern suburb of Leicester. ModE adj. **south** + p.n. Wigston as in WIGSTON SP 6099.

WIGTOFT Lincs TF 2636. 'The creek curtilage'. *Wigetoft(e)* 1180–1332, *Wichetoft* 1185, 1202, *Wi(c)ke- Wyketoft* 1200–1391, *Wyge-* 1234–1362, *Wigtoft(e)* since 1305. ON **vík** + **toft**. The ref-

†Sometimes said to be St Michael's Mount rather than Wight.

erence is to Bicker Haven, an arm of the sea which formerly nearly reached as far as the village. Payling 1940.108, SSNEM 165, Wheeler 93 and Appendix I.41.

WIGTON Cumbr NY 2548. 'Wicga's farm or village'. *Wi- Wyggeton'* 1163–1349, *Wi- Wygeton* 1191–1385, *Wy- Wiginton* 1232, 1261×72, *Wigenton* 1306, *Wygton* 1241–1426, *-don* 1539. OE pers.n. *Wicga*, genitive sing. *Wicgan*, + **tūn**. Cu 166.

WILBARSTON Northants SP 8188. 'Wilbeorht's farm or village'. *Wilbertes- Wilberdeston(e)* 1086–1285, *-stoch -stoke* 1196–1220, *Wilberston* 1301, *Wybbarston* 1602. OE pers.n. *Wilbeorht*, genitive sing. *Wilbeorhtes*, + **tūn** occasionally replaced with **stoc**. Nth 175 gives pr [wibəstən].

WILBERFOSS Humbs SE 7351. 'Wilburg's ditch'. *Wilburcfos(s)a* 1148, 1156, *Wilburfos(se)* 1170×80–1402. OE feminine pers.n. *Wilburg* + **foss**. YE 188.

Great WILBRAHAM Cambs TL 5557. ModE **great**, **much** (*Moche* 1515), Latin **magna** (c.1250) + p.n. *(æt) Wilburgeham* 975×1016 S 1487–1428, *Witborham, Wiborg(e)ham -burge* 1086, *Wi- Wylbur(g)ham* c.1169–1553, *Wi- Wylbram* 1360–1550, *Wy- Wilbraham* from 1509×47, 'Wilburg's homestead or estate', OE feminine pers.n. *Wilburh*, genitive sing. *Wilburge*, + **hām**. 'Great' for distinction from Little WILBRAHAM Cambs TL 5458. Ca 137.

Little WILBRAHAM Cambs TL 5458. ModE **little** (*Litle* 1500), Latin **parva** (*Parua* 12th) + p.n. Wilbraham as in Great WILBRAHAM TL 5557. Ca 137.

WILBURTON Cambs TL 4875. 'Wilburg's estate'. *(on) Wilburhtune* [970]13th S 780, *Wilburton* from 1251 with variant *Wy- Wilbertone* 1086–1553. OE feminine pers.n. *Wilburh* + **tūn**. Ca 243.

WILBY Norf TM 0389. 'Willow-tree farm or village'. *Wilgeby, Willebeih* 1086, *Wileby* 1220, *Willobi* 1254. OE ***wiliġ**, genitive pl. **wilga**, + **bȳ**. DEPN.

WILBY Northants SP 8666. 'Villi's village or farm'. *Willabyg* [c.1067]12th, *Wilebi* 1086–1334 with variants *Wyle-* and *-by*, *Wi- Wyllebi -by* 1186–1343, *Wyli- Wyly- Wylu(gh)by* 1230–1394, *Wylby* 1321. ON pers.n. *Villi*, genitive sing. *Villa*, or OE pers.n. *Willa*, + **bȳ**. 13th and 14th cent. forms show assimilation to OE **wiliġ** 'a willow-tree'. Nth 141, SSNEM 78.

WILBY Suffolk TM 2471. Probably 'the circle of willow-trees'. *Wilebey -bi* 1086, *Wyleb(e)ye -begh -bey(ghe)* 1254–1346, *Wylbey(gh)* 1524, 1568. OE **wiliġ-bēag**, dative sing. **wiliġ-bēaġe**. Although **bȳ** is sometimes spelled *-bei -bey* the sps are persistent and not occasional in this n. and point to **beaġe**, dative sing of **bēag** 'a ring'. DEPN, Baron.

WILCOT Wilts SU 1460. 'The spring cottages'. *(æt) Wilcotum* 940 S 470, *Wilcote* 1086–1341, *Wy- Wilecote* 1268–1327, *Welcot(e)* 1279–89. OE **wylle** + **cot**, dative pl. **cotum**. Wlt 325.

WILDBOARCLOUGH Ches SJ 9868. 'Wild boar ravine'. *Wildeborclogh'* 1357, *Wildboar(s) Clough* 1724. ME **wilde-bor** (OE *wīlde-bār*) + **clogh** (OE *clōh*). Che i.159. Che ii.vii gives pr ['waildbɔ:r'kluf], ['wilbərklʌf], ['wil(d)bər 'kluf], 19th cent. local ['wilbərtluf], ['wilbərklʌf].

WILDEN Beds TL 0955. Partly uncertain. *Wilden(e)* from 1086, *Wilding* 1798. This could be any of **wīl** 'a trap', **wild** 'wild', **wilġ** 'willow' or pers.n. *Wil(l)a* + **denu** 'valley'. Bd 66, L 98.

WILDEN H&W SO 8272. 'Beetle hill'. *Wineladuna* (sic for *Wiuela-*) [1182]17th, *Wineldune* (for *Wiuel-*) [1182]18th, *Wiveldon* 1299, *Wildon* 1480. OE **wifel**, genitive pl. **wifela**, + **dūn**. Wo 247.

WILDHERN Hants SU 3550. Uncertain. *Wildherne* 1635. The evidence is too late to decide between ME **wilde herne** 'wild corner or recess of land', OE **wilde** + **hyrne**, or OE, ME **wilderne** 'wasteland, wilderness'. Ha 176.

WILDMORE FEN Lincs TF 2552. *Wildmore Fen* 1801, cf. *Wildemore Commō* 1661. P.n. *Wildamora* c.1140, *Wildemore* [1198]1328, *la Wildemore* 1206, 'the uncultivated marshland', OE **wilde** + **mōr**, + ModE **fen**. The fen was drained and allotted in the early 19th cent. and created a parish in 1880. DEPN, Wheeler 222, 228, Cameron 1998.

WILDSWORTH Lincs SK 8097. 'Wifel's enclosure'. *Winelesworth* (for *Wiueles-*) [1199]1232, *Wyveleswurth* 1280, *Wylessworth* 1316. OE pers.n. *Wifel* from **wifel** 'a beetle', genitive sing. *Wifeles*, + **worth**. DEPN.

WILEY SIKE Cumbr NY 6470. Cf. *Wylysikefoot* 1742. Wiley is possibly 'willow stream', ultimately OE **wiliġ** 'willow' + OE **sīc**, ON **sík** 'a small stream'. Cu 30.

WILFORD Notts SK 5637. Partly uncertain. *Wilesforde* 1086, *-ford'* 1168, *Wylesford* 1280, *Wi- Wyleford* c.1190–1287, *Wi- Wylleford -forth* c.1240–1371, *Wilford* from 1302. Probably OE pers.n. *Willa* + **ford**. But the specific might alternatively be OE **wiliġ** 'a willow-tree' or **wīle** 'a weel, a basket, a fish-trap'. Nt 251.

WILKESLEY Ches SJ 6241. 'Beetle's claw (of land)'. *Wiuelesde* (sic for *-cle*) 1086, *Wivelescle* [1230, 1282]1331, *Wy-* 1253, *Willescle* [12th]14th, *Willesley* 1437, *Wilskelegh* 1404, *Wilkesley* from 1535. OE **wifel**, genitive sing. **wifeles**, + **clēa** 'a claw' referring to the tongue of land between the river Ducklow and the stream forming the county boundary near Shavington Shrops at SJ 6439. The generic has been replaced with *-ley* from OE **lēah**. Che iii.93.

WILLAND Devon ST 0310. 'Waste, uncultivated land', or 'newly cultivated land by the waste'. *Willelande* 1086, *Wildelanda* [1155×8]1334, *(la) Wy- Wildelonde* 1244–1438. OE **wilde** + **land**. The reference is either to newly cultivated land or perhaps to cultivated land that has reverted to waste. D 553, L 248–9.

WILLASTON 'Wiglaf's farm, village or estate'. OE pers.n. *Wīġlāf*, genitive sing. *Wīġlāfes*, + **tūn**.

(1) ~ Ches SJ 3377. *Wylaveston* [1230]1580, *Wi- Wylaston* 1286–1512, *Willaston* from 1481, *Wollaston* 1546–1860 with variants *Wollos- Wool(l)as-*. The earliest form occurs in the hundred name *Wilaveston Hundred* 1086. Che iv.232.

(2) ~ Ches SJ 6752. *Wilavestune* 1086, *Wylaveston -tun* c.1250, *Willawestun* c.1250×60, *Wi- Wylaston(e)* c.1180–1393, *Willaston* from 1466, *Wolaston* 1216×72,1295, *Wollaston* 1550–90. Che iii.78.

WILLEN Bucks SP 8741. 'At the willow-trees'. *Wyle, Wily* 1189, *Wilinges* 12th, *Wilie(s)* 1208–42, *Wylien(e) -yen(e)* 1235–1490, *Wyllyn* 1517. OE **wiliġ**, dative pl. **wiliġum**. Alternatively the *n*-forms possibly represent OE **wiliġn** 'a willow, a willow copse', as in Church and Great Wilne Derbys SK 4431, 4430, *Wilne* from 1096, *Wi- Wylene* 1236–1330. Bk 26, PNE ii.266, L 221.

WILLENHALL WMids SO 9799. 'Willa's nook'. *Willanhalch* [733]15th S 86, *Willenhale* [996 for 994]17th S 1380, *Þilinhale* c.1100 Charters II.xxxxvii, *Winenhale* 1086. OE pers.n. *Willa*, genitive sing. *Willan*, + **halh** in the sense 'small valley, hollow'. DEPN, L 105.

WILLENHALL WMids SP 3676. 'Willow-tree nook'. *Wilenhala* 1100×35, *Wi- Wylenhal(e) -hall* 1221–79, *Wilihale* 1195, *Wi- Wyln(h)al(e)* 1236–1369, *Wynhale* 1291, *Wy- Winnall* 1535–1675. OE **wiliġn** 'a willow copse' or ***wiliġen** 'growing with willows' + **halh** in the sense 'a small valley', dative sing. **hale**. Wa 190 gives pr [winəl], L 105–6, 110.

WILLERBY Humbs TA 0230. 'Wilheard's village or farm'. *Wilgardi* 1086, *Wil(l)ardebi* 1196, 1206, 1208, *Wollerby* 1566. OE pers.n. *Wilheard* + ON **bȳ**. YE 218, SSNY 41.

WILLERBY NYorks TA 0079. 'Wilheard's farm or village'. *Wi- Wyllardebi -by* [12th]c.1300–1446, *Wil(l)ardby* [12th]c.1300–1416, *Wi- Wyllarby* 1290–1583, *-er-* 1548. OE pers.n. *Wilheard* + **bȳ**. YE 117, SSNY 7.

WILLERSEY Glos SP 1039. 'Wilhere or Wilheard's island'. *Willersey(e) -ei(a)* [709]12th S 80, [714]16th S 1250, *-ei* 1086, *(æt) Willereseie*, *(into, on) Wyllerese(i)ge*, *of þere ege* ('from the island') [840×52]12th S 203, *(on) Wyllersege*, *(of) Wylleresegge* [n.d]12th S 1599, *Wi- Wyl(l)ardeseye* c.1220–1303, *Wi-*

Wyllarsey(e) 1269–1577. OE pers.n. *Wilhere*, genitive sing. *Wilheres*, + **ēġ**. Some spellings point to the alternative pers.n. *Wilheard*. Gl i.263, L 38.

WILLERSLEY H&W SO 3147. Partly uncertain. *Willaveslege* 1086, *Willelmesle* 1142, *Wylardesl(eye)* 1291–1327. Uncertain pers.n. + OE **lēah** 'wood or clearing'. The 1086 form points to OE *Wīġlāf*, but this was early replaced by CG *Willelm* and subsequently by *Willard* < CG *Widelard* or OE *Wilheard*. He 208.

WILLESBOROUGH LEES Kent TR 0324. 'Willesborough pasture'. *Willesborough Leeze* 1797, *Wilsborough Lees* 1798. P.n. Willesborough TR 0241 + ModE **leaze** (OE *lǣs*, Kentish *lēs*). Willesborough, *wifeles berge* 863 S 332, *Wifeles beorge* [996]15th S 877, *Wyuel(l)esberg(he)* 1243–70, *-berwe* 1270, *Willesborow* 1610, is 'Wifel's hill', OE pers.n. **Wifel*, genitive sing. *Wifeles*, + **beorg**. PNK 419, KPN 220, Cullen.

WILLESDEN GLond TQ 2284. 'Spring hill'. *Wellesdune* [939]13th S 453, *Wellesdone* 1086, *Willesdone* [924×39]12th S 453, *Willesden -don(a)* 1181–1535, *Wylsdon* 1563. OE **wylle**, secondary genitive sing. **wylles**, + **dūn**. The archaic spelling *Willesden* was adopted by the London and Birmingham railway c.1840 and subsequently replaced the historically correct form *Wilsdon*. Mx 160, TC 215, GL 99.

WILLETT Somer ST 1033. Transferred from the lost r.n. Willett. *Willet* 1086, *Wellet* 1238×58, *Wellyt* 1285. R.n. *Willite (rivulus)* *[854]12th S 311, possibly 'gushing stream', OE(WS) **wylle** + ***ġīete**, ***ġīte**. RN 460, DEPN.

WILLEY 'Willow wood or clearing'. OE ***wiliġ** + **lēah**.

(1) ~ Shrops SO 6799. *Wilit* 1086, *Welileia* [1120]copy, *Wililega* c.1180–1406 with variants *Wyli- Wyly-* and *-leia -leg(h)' -lee -ley(e), Wilegh'* 1209, *Wyleye* 1416, *Willey* from 1453. Sa i.316.

(2) ~ Warw SP 4984. *Welei* 1086, *Wele* 1372, *Wilee -lega* 1130–1401 with variants *Wy- -le -ley(e) -lei(e) -lie* and *-lye, Wi- Wylley(e)* 1230, 1262, 1535. Wa 121, L 203, 221.

High WILLHAYS Devon SX 5789. *Hight Wyll* 1532, *High Willhayse* 1809 OS, *High Willows* 1827. The highest point on Dartmoor rising to 2038ft. *Hight Wyll* may be 'height spring', ultimately from OE **wylle**; the later forms have been assimilated to **hay** 'an enclosure'. D 203 gives pr [wiliz].

WILLIAMSCOT Oxon SP 4745. 'William's cottages'. *Williamescote* [1166]c.1203–1428 with variants *Wylliames- Willames-* and *-kote, Walmescote* [1166]c.1280, [1208–13]c.1300, *Willescot(a) -e* c.1240, c.1270. ME pers.n. *William*, genitive sing. *Williames, Willes*, + **cote** (OE *cot*, pl. *cotu*). O 427 gives pr [wilzkɔt].

WILLIAN Herts TL 2230. '(Place) at the willows'. *Wilie* 1086–1428 with variants *Wyl-* and *-ye, Wilian* 1212, *Wylien* 1216–1523 with variants *Wylyene, Wilyen, Wyllyne*. OE **wiliġ**, dative pl. **wiliġum**. Hrt 147, PNE ii.266.

WILLIMONTSWICK Northum NY 7763. 'Willimot's farm'. *Willimoteswike* 1279, *Willymounteswyke* 1542. OFr diminutive pers.n. *Wil(li)mot*, secondary genitive sing. *Willimotes*, + OE **wīc**. A fortified manor of the Ridleys. NbDu 217.

WILLINGALE Essex TL 5907. 'The nook of the Willingas, the people called after Willa'. *Ulinghe- Willingehalam* 1086, *Wi- Wylling(e)hal(e)* 1198–1428. OE folk-n. **Willingas* < pers.n. *Willa* + **ingas**, genitive pl. **Willinga*, + **halh**. Ess 500 gives pr [winigl].

WILLINGDON ESusx TQ 5902. 'Willa's hill'. *Wil(l)e(n)done* 1086, *Wy- Wil(l)indon(e)* 1229–1340, *Wy- Wil(l)ingdon* 1290–1500. OE pers.n. *Willa*, genitive sing. *Willan*, + **dūn**. Sx 424, L 143, 146, SS 74.

WILLINGDON HILL ESusx TQ 5600. P.n. WILLINGDON TQ 5902 + (pleonastic) ModE **hill**.

WILLINGHAM 'Homestead of the Wifelingas, the people called after Wifel'. OE folk-n. **Wifelingas* < pers.n. **Wifel* + **ingas**, genitive pl. **Wif(e)linga* + **hām**.

(1) ~ Cambs TL 4070. *Uuinlingeham* (sic for *Uuiulinge-*) [1042×66]13th S 1051, *Vuivlingeham* [1042×66]13th ibid., *Wiuelingeham* 1086 ICC, InqEl, 1221, *Wiuelincga- Wuiuelinge- Wiuelincgeham* 1086 InqEl, *Wyvelingeham* 1238, 1244, *Wi- Wyuelingham -yng-* [c.1060]14th–1638 including 1086, *Wil(l)y(ng)ham -ing-* 1272–1553. Ca 173, ING 129.

(2) ~ BY STOW Lincs SK 8784. *Wivelingeham et Stowe* 1210×12. P.n. *We- Wilingehā* 1086, *Viflinghe- W- Uiflingeheim, Wiflingham* c.1115, *Wiueling(e)ham* 1202–1218, *Wiflingeham* 1233, *Wiflincham* 1212, + ModE **by** + p.n. STOW SK 8882. 'By Stow' for distinction from North WILLINGHAM TF 1688 and South WILLINGHAM TF 1983. DEPN, ING 146, SSNEM 383, Cameron 1998.

(3) Cherry ~ Lincs TF 0372. *Cherwellyngham* 1373, *Chyry Wylynham* 1386, *Cherry Willingham* 1824 OS. ME **chiri** 'a cherry-tree' + p.n. *Vlinge- VVlinge- Wilingheham, Gullingham* 1086, *Wlling(e)heim* c.1115, *Willingham* 1163, 'the homestead of the Willingas, the people called after Willa', OE folk-n. **Willingas* < pers.n. *Willa* + **ingas**, genitive sing. *Willinga*, + **hām**. ING 145.

(4) North ~ Lincs TF 1688. *North Willyngham* 1502. ModE **north** + p.n. *Wiuilingeham* 1086, *-el-* 1193, c.1221, *Wiflingeham* c.1115, *Wiflingham* 1086–before 1224, *Wyflyngham -ing-* c.1150–1560, *Wylingham* 1242–1375, *Willingham* from 1526, as in WILLINGHAM BY STOW SK 8784. Li iii.181, ING 146, SSNEM 383.

(5) South ~ Lincs TF 1983. *South Willingham* 1824 OS. ModE adj. **south** + p.n. *Vlingeham -hā* 1086, *Wllingheham -heim* c.1115, *Wel(l)ingeham* 1121×3, c.1160, either the 'homestead at or called *Welling*, the spring place', OE p.n. **Welling* < **wella** + **ing**², locative–dative sing. **Wellinge*, + **hām**, or the 'homestead of the Wellingas, the people of the spring, or of the Willingas, the people called after Willa', OE folk-n. **Wellingas* < **wella** + **ingas**, genitive pl. **Wellinga*, or **Willingas* < pers.n. *Willa* + **ingas**, genitive pl. **Willinga*, + **hām**. There is a series of springs and streams in the valley S and E of the village. 'South' for distinction from North WILLINGHAM TF 1688. ING 145.

WILLINGTON Beds TL 1049. 'Willow-tree farm, willow settlement'. *Welitone* 1086, *Wil(l)itona -e* c.1150–14th cent., *Wyl(l)ington* 1276–1539. OE **weliġ** and ***wiliġen** or ***wiliġn** + **tūn**. Bd 99.

WILLINGTON CORNER Ches SJ 5367. *Willington Corner* 1831. A corner of the boundary of Delamere Forest. P.n. Willington + ModE **corner**. Willington, *Winfletone* 1086, *Wynfleton* [1233×7]1347–1499, *Win- Wynlaton* 1208×29–1475, *Wi- Wy- Wol(l)aton* 1304–1819, *Willyngton* 1526, *-ing-* from 1547, is 'Wynflæd's estate', OE feminine pers.n. *Wynflǣd* + **tūn**. Che iii.286, 285.

WILLINGTON Derby SK 2928. 'Willow farm or village'. *Willetune* 1086, *Wi- Wyl(l)intun(e) -ton(a) -yn-* c.1100–1610, *Wilenton(a)* 1114–1216×72, *Wi- Wyl(l)ington(e) -yng-* 1198–1269 etc. OE ***wiliġn** + **tūn**. Db 513.

WILLINGTON Durham NZ 1935. 'Farm, village called after Wifel'. *Wyvelintun* c.1190, *Wyflington* 1285, *Wilyngton* 1384–1441, *Willyngton* 1421–1524, *-ington* from 1421. OE pers.n. *Wifel* + **ing**⁴ + **tūn**. Alternatively the specific might be OE **wifel** 'a beetle', 'settlement, farm infested with beetles'. Pronounced locally [wilitən]. NbDu 217, Årsskrift 1974.52.

WILLINGTON T&W NZ 3167. 'The estate called after Wifel'. *Wiflintun* c.1107×23–c.1170×4 DEC, *Wiflinctun* before 1195 ibid., *Wiuelintonam* 1154×66, *Wiuelington'* 1204, *Wiflingtun* [c.1160×65]c.1220×30, [before 1195]c.1220×30, *Willyngton* 1430, 1539, *-toune* 1539. OE pers.n. *Wifel* + **ing**⁴ + **tūn**. The pers.n. is derived from OE **wifel** 'a weevil, a beetle' and an alternative suggestion is that the name means 'farm or village at the beetle-infested place', OE ***wifeling** + **tūn**. NbDu 217, Årsskrift 1974.52.

WILLINGTON Warw SP 2639. 'The estate called after Wulflaf'.

Vllavintone 1086, *Wolaui(g)tona* 1224, *Wu- Wollavynton* 1232–88, *Wollavyngtone* 1441, *Wolington* 1372, 1403, *Wyllington* 1546. OE pers.n. *Wulflāf* + **ing**[4] + **tūn**. The modern form has been assimilated to the name of the *Willington* family who bought nearby Barcheston in 1507. Wa 297.

WILLITOFT Humbs SE 7435. 'Willow curtilage'. *Wilgetot* 1086, *-toft* 1190×1, 1246, *Wi- Wylegh- Wi- Wyl(o)ughtoft* 14th cent., *Wyllytoft -i-* from 1534. OE **wiliġ** + **topt**. YE 242, SSNY 150.

WILLITON Somer ST 0741. 'Settlement on the r. Willett'. *Willettun* *[904]12th S 373, *Willetone* 1086, *Wiliton* 1273, *Williton* 1610. R.n. Willett as in WILLETT ST 1033 + OE **tūn**. DEPN.

WILLOUGHBY 'Willow-tree farm or village'. OE **wiliġ**, genitive pl. **wiliga**, + **bȳ**. Some of the Willoughby names might be Scandinavian versions of OE **Wiliga-bēag* 'a circle of willow-trees' or partial reformations of OE **Wiliġ-tūn*. SSNEM 78.

(1) ~ Lincs TF 4771. *Wilgebi* 1086–1199, *Wilghebi* 1191. DEPN, SSNEM 78 no.4, Cameron 1998.

(2) ~ Warw SP 5167. *Wiliabyg* [956]11th S 623, *Wilebere -bene -bei -bec* 1086, *Wilegebi* 1176, *Wi- Wylebi -by* 12th–1400, *Wylewby* 1279, *Wilouby* 1316, *Wylluby* 1345, *Wylughby* 1333–1434, *Wylloughby* 1535. Wa 148, SSNEM 78 no. 3.

(3) ~ -ON-THE-WOLDS Notts SK 6325. P.n. *Willebi* 1086, *Wi- Wyl(l)eby* c.1190–1316, *Wilgebi* 1086–1212, *Wilghebi* c.1180, *Wylgheby* 1280, *Wilwebi -by* c.1190, c.1240, *Wy- Wil(l)ughby* 1342–1517, *Willoughbye* 1633, + affix *super le Wolde, super Waldas* 13th cent., *on le Wold* 1343, *on Oldes* 1584, *uppon the Would* 1652. Nt 258.

(4) ~ WATERLEYS Leic SP 5892. 'W water-meadows'. P.n. *Wile(che)bi* 1086, *Wy- Wilubi -by* c.1130–1428, *Wy- Wilebi -by -ll-* 1183–1330, *Wy- Wilug(h)by -ll-* 1292–1430, *Willoughby* from 1507 + affix *Waterles* 1420–1579, *Waterlies -leas -leys* 17th, 'water-meadows', ME **water** + **leas**, pl. of OE **lēah**. The modern form *Waterless* is a corruption of *Waterleys*. Lei 476, SSNEM 78 no.6.

(5) Silk ~ Lincs TF 0543. *Sylkwylowby* 1477, *Silkwilloughby* from 1604, a contraction of *Wylughby cum Silkeby* 'W with Silkby' 1316. Earlier simply *Wilgebi* 1086, c.1150, *Wi- Wyl(h)geby -gh-* 1210–1303, *Wilugh(e)by* 1242×3, 1332, *Wyluby* 1247, 1291, 1428, *Willoughby* from 1278. Silkby, a former hamlet at TF 055430, *Silkebi* [1189]1341, 1212, *-by* 1231–1496, *Selkeby* 1211×2, 1234, 1357, *Silkby* 1375 is possibly 'Silki's village or farm', ON pers.n. *Silki*, genitive sing. *Silka*, + **bȳ**. But, whether a by-n. from *silki* 'silk' or *selki* 'a young seal', Silki is an unusual pers.n. and the specific might rather be an OE appellative ***sīluc**, ***sīoluc**, ***sēoluc** 'a gulley, a drain', as postulated for SILKMORE Staffs SJ 9021. Perrott 91, SSNEM 78 no.8, 82.

WILLOUGHTON Lincs SK 9393. 'The willow-tree settlement'. *Wilchetone* 1086, *Wilgatuna* c.1115, *Wilgheton* 1220. OE **wiliġ**, genitive pl. **wil(i)ga**, + **tūn**. DEPN, L 221.

WILLY HOWE Humbs TA 0672. *Willy Howe* 1856 OS. One of the many barrows on the Yorks Wolds. ON **haugr** 'a burial mound'. Thomas 1976.154.

WILMCOTE Warw SP 1658. '(The settlement) at the cottage(s) called after Wilmund'. *Wilmundigcotan* [1016]18th S 1388, *Wilmundecot* 1274, 1285, *Wilmecote* 1086, 1228–1316, *Wilmecoate al. Wimcott* 1657. OE pers.n. *Wilmund* + **ing**[4] + **cot**. The ending *-cotan* of the earliest form probably represents OE dative pl. **cotum**. Wa 198 gives pr [wimkət].

WILMINGTON Devon SY 2499. 'Wilhelm's estate'. *Wilelmi- Willelmatona* 1086, *Wilhamtyn* 1332, *Willem'ton* 1391. OE pers.n. *Wilhelm* + **ing**[4] + **tūn**. D 629.

WILMINGTON ESusx TQ 5404. 'Wilma's estate'. *Wineltone, Wilminte* 1086, *Wilmentun -ton* c.1150, 1287, *Will- Wylmingtone* 1314–1401. OE pers.n. *Wilma*, genitive sing. *Wilman* + **tūn**. The *ing* forms are secondary in this name. Sx 412, SS 76.

WILMINGTON Kent TQ 5372. 'Estate called after Wighelm'. *Wilmintuna* 1089, 1154×89, *Wilmentuna* c.1100, *Wy- Wilmington(')* from 1226, *Wi- Wylmenton'* 1254, 1270. OE pers.n. *Wīġhelm* + **ing**[4] + **tūn**. PNK 35.

WILMSLOW Ches SJ 8481. 'Wighelm's mound'. *Wilmesloe* mid 13th, *Wi- Wylmeslow(e) -lawe -is-* 1287–1354, *Wimmislowe* 1286, *Wymslowe* 1498, *Wimslow* 1696, *Wilmslowe* 1286, *-low* from 1620. OE pers.n. *Wīġhelm*, genitive sing. *Wīġhelmes*, + **hlāw**. Che i.219, Che ii.vii gives 19th cent. local pr ['wimzlə], now ['wi(l)mzlou].

Church WILNE Derbys SK 4431 → WILLEN Bucks SP 8741.

Great WILNE Derbys SK 4430 → WILLEN Bucks SP 8741.

WILNECOTE Staffs SK 2201. 'Wilmund's cottages'. *Wilmundecote* 1086–1332, *Wilmecota -e* 1217–1606 with variant *-en-*, *Wilnecote* 1317 and from 1607, *Wincote* 1656, 1694. OE pers.n. *Wilmund* + **cot**, pl. **cotu**. Wa 27 gives pr [winkət].

WILPSHIRE Lancs SD 6832. Partly uncertain. *Wlyp- Wlipschyre, Wlyp- Wlipsire* 1246, *Wlipschire* 1258, *Wypsire* 1272, *Wilp- Wylps(c)hire* 1311, 1396, *Lip- Whilpshire* 1589. Possibly 'estate of Wlips', OE by-name **Wlips* (from OE **wlips**, **wlisp** 'lisping') + **scīr** 'a division, a shire, an estate'. But this is very uncertain. La 72, Jnl 17.46.

WILSDEN WYorks SE 0936. 'Wilsige's valley'. *Wilsedene* 1086, *W(i)l- Wylsi(n)den(e)* 1190×1220–1298, *Wilsden* from 1340. OE pers.n. *Wilsiġe*, genitive sing. *Wilsiġes*, + **denu** alternating with *Wilsiġe* + **ing**[4] + **denu**. YW iii.274, L 98.

WILSFORD 'Wifel's ford'. The generic is OE **ford**, the specific pers.n. *Wifel*, genitive sing. *Wifeles*.

(1) ~ Lincs TF 0043. *Wivelesford -u-* 1086, *Wi- Wyuelesford(e)* 1086–1290, *Wi- Wyllesford* 1180–1535, *Wilsford* 1339. Perrott 515.

(2) ~ Wilts SU 1057. *Wifeles ford* [892]14th S 348, [933]14th S 424, *Wivlesford* 1086, *Wi- Wyvelesford* 1185–1332, *Wylesford* 1292. Wlt 326.

(3) ~ Wilts SU 1339. *Wiflesford(e)* 1086, *Wivelesford* c.1200, *Wylesford* 1279. Wlt 372.

WILSILL NYorks SE 1864. 'Vifill, Wifel or beetle's nook of land'. *Wifeles healh* c.1030 YCh 7, *Wifleshale* 1086, *Wiueles-* 1086–12th cent., *Wilsell'* 1535, *Wilsill* 1645. ON pers.n. *Vífill*, genitive sing. *Vífills*, or OE pers.n. **Wifel* or appellative **wifel** 'a beetle', genitive sing. *Wif(e)les*, **wif(e)les**, + **halh**. The existence of an OE pers.n. **Wifel* has been rejected on the grounds that its frequency in the names for boundary markers in Anglo-Saxon charter boundary clauses links the element with animal terms rather than pers.ns. However, **halh** in major names seems to have had a quasi-habitative significance and the most frequent category of specific compounded with it is pers.ns. The reference is to a small valley in the steep hill-side overlooking the river Nidd. YW v.150, SSNY 167, L 107–8, 110, Kitson 1994.75–6.

WILSON Leis SK 4024. 'Vifill's farm or village'. *Wy- Wiueleston' -is-* 1203–[1329]14th, *Wy- Wivelston'* 1248–1428, *Wy- Willeston'* 1345–1553, *Wy- Wilston* 1429–1610, *Wy- Wil(l)son* 1553–16112. ON pers.n. *Vífill*, genitive sing. *Vífils*, + **tūn**. Lei 350.

WILSTHORPE Lincs TF 0913. Partly uncertain. *Wiuelestorp* 1086–1338 with variants *Wy-* and *-is-*, *Wi- Wyllestorp -thorpe* 1180–1551, *Wilsthorpe* from 1428. The generic is OE, ON **thorp**, the specifc the genitive sing. of either ON pers.n. *Vífill* or OE **Wifel*, or of OE ***wifel** or ODan ***wivæl** 'a pointed piece of ground (cf. OE *wifel* 'a javelin, a dart'). Neither pers.n. is recorded independently in L or Y. The village stands on raised ground in a marshy area. Perrott 398, SSNEM 121.

WILSTONE Herts SP 9014. 'The beetle's thorn-tree'. *Wivelestorn(e)* 1220, 1233, *Wyvelesthorne* 1278, 1385, *Wylstorn* 1459, *Wilsterne* 1656, OE **wifel**, genitive sing. **wifeles**, + **thorn**. For the argument against OE pers.n. **Wifel* in p.ns. see FLH 14.35, 75. Hrt 53.

WILSTONE RESERVOIR Herts SP 9013. P.n. WILSTONE SP 9014 + ModE **reservoir**.

WILTON 'Wild, uncultivated settlement'. OE **wild** + **tūn**.

(1) ~ Cleve NZ 5819. *Widtune* 1086, *Wiltune* 1086, [1155×65]15th. YN 159.

(2) ~ NYorks SE 8682. *Wiltun(e)* 1086–1247, *Wi- Wylton(am)* 1167 etc. YN 90.

WILTON Norf TL 7288 as in Hockwold cum Wilton. 'Willow-tree farm'. *Wiltuna* 1086, *-tona* 1121, *-ton* 1242. OE **wiliġ** + **tūn**. DEPN.

WILTON Wilts SU 0931. 'the settlement on the river Wylye'. *Uuiltún* 838 S 1438, *Wiltun* [854]?11th S 308, [894]c.1000 Asser, 959×75, 1016×35 Coins, *(æt, to, in) Wiltune* 9th ASC(A) under year 871, 1086, 14th cent. copies of 10th cent. charters (S 424, 438, 767, 1515), *Wiltun -tone* 1086 etc. R.n. WYLYE + OE **tūn**. Wlt 218.

WILTON Wilts SU 2661. 'The wool farm'. *Wulton* 1227–1386, *Wolton(e)* 1289–1403, *Wilton* from 1402. OE **wull** + **tūn**. Wlt 347.

Bishop WILTON Humbs SE 7955. 'W held by the (arch)bishop' sc. of York. *Bis(s)hop Wilton* from 14th cent. ME **bishop** + p.n. *Widton, Wiltone* 1086, *Wilton* from 12th, probably the 'willow farm', OE **wiliġ** + **tūn** rather than the 'wild, uncultivated enclosure or farmstead', OE **wild** + **tūn**. Held by the Archbishops of York from the time of Edward the Confessor. YE 175.

WILTON HOUSE Wilts SU 0931. *Wilton Ho.* 1811 OS. P.n. WILTON SU 0931 + ModE **house**. The first earl of Pembroke built a house here in the mid 16th cent. in the former nunnery's estate. Pevsner 1975.580.

WILTSHIRE 'The district dependent upon Wilton'. *Wiltunscir* 9th ASC(A) under year 898, [951×5]15th S 1515, *Wiltescire* 1086, *-a* 1130–1367, *county called le Wiltshire* 1447, *Wilsher'* 1523, 1606. P.n. WILTON SU 0931 + OE **scīr**. The county name is a late one and belongs to a period when a considerable degree of centralised local government had evolved. Before that the people of Wiltshire were known as the *Wilsætan* 9th ASC(A) under years 800, 878, *Wilsæte* 12th ASC(E), 'the dwellers by the river Wylye', r.n. WYLYE + OE **sǣte**. Wlt 1 gives local pr [wilʃər], xvi.

WIMBISH Essex TL 5936. Possibly 'Wina's copse of bushes'. *Wimbisc* 1042×3 S 1530, [1043×5]13th S 1531, *Winebisc* [1042×3]12th, *Wimbeis* 1086, *Wy- Wimbis(se) -bysse* 1227–60, *Wymbys(s)h -bis(sc)h -bussh* 1265–1467, *Winbiss(e)* 1235, *Wynebys -bych* 1254, 1285. OE pers.n. *Wina* + **bysc**. But DEPN rejects this on the ground that in Essex OE **bysc** should give ME *besch* and suggests alternatively OE ***winn** 'a meadow, a pasture' + ***biosiċ** 'a reedy place'. Ess 546, PNE i.74, Jnl 2.48.

WIMBISH GREEN Essex TL 6035. *Wimbish Greens* 1805 OS. P.n. WIMBISH TL 5936 + ModE **green**.

WIMBLEBALL LAKE Somer SS 9730. A modern reservoir. No early forms.

WIMBLEDON GLond TQ 2470. Probably 'Wynnman's hill'. *Wunemannedune* [942–c.951]13th S 1526, *Wymmendona* 1154–61, *Wymendon, Wimedon* 13th, *Wimeldon* 1202, *Wimmeldun'* 1212, *Wi- Wymbeldon(a)* 1211–94. The boundary of the people of Wimbledon is referred to in a ME copy of an Anglo-Saxon charter as *(bi) wimbedounyngemaerke* [967]14th S 747. OE pers.n. **Wynnman* + **dūn**. The development of the name must have been *Wimendon* > *Wimeldon*, a common Anglo- Norman sound substitution, > *Wimbeldon* with instrusive *b* in the cluster *ml*. Sr 38, LPN 250.

WIMBLEDON PARK GLond TQ 2472. *Wimbledon Park* 1816. P.n. WIMBLEDON TQ 2470 + ModE **park**. LPN 251.

WIMBLINGTON Cambs TL 4192. 'Estate called after Wimbel'. *Wimblingetune* [c.975]12th, *Wimblingtune* 12th, 1170, *Wymb(e)lington -yng-* 1251–1617. OE pers.n. **Wimbel*, probably a pet form of *Wynnbeald*, + **ing**[4] + **tūn**. Ca 265.

WIMBLINGTON FEN Cambs TL 4589. *Wimblington Fen* 1824 OS.

WIMBORNE MINSTER Dorset SU 0100. 'W with the monastery church'. *Winburnan monasterium* [893]early 11th Asser, *(on, æt) Winburnan mynstre* c.1000, 12th ASC(E) under year 871, *Wy- Winbo(u)rn(e) Mynstre, Mynster* 14th–1547, *Wi- Wymbo(u)rn(e) Min(i)stre, Mynster* from 1268. R.n. Wimborne + Latin **monasterium**, OE **mynster**, originally referring to a nunnery here founded before 705. Wimborne, *(æt) Winburnan* late 9th ASC(A) under years 718, 871, 12th ASC(E) under year 718, *Wimburnan* late 10th ASC(A) under year 961, *Winburne -borne* 1086, *Wi-Wymburn(e)* c.1165 etc., was the original name of the r. Allen (itself a back-formation from *Aldewynebrig* 1268, *Aldwynesbrigg* 1280, *Aleynsbrydge* 1543, properly 'Ealdwine's bridge, later taken as 'bridge of the Aleyn or Allen'). The sense is 'meadow stream', OE ***winn**, ***wynn**, + **burna**. 'Minster' for distinction from Wimborne St Giles SU 0312, 'St Giles' W', *Wymbourn St Giles* 1394, earlier simply *Winburne* 1086 or *Up Wi- Wymburn(e)* [1154×89]1496, [c.1183]14th, 1213 etc. 'settlement higher up the Wimborne', + saint's name *Giles* from the dedication of the church. Do ii.183, 263.

WIMBOTSHAM Norf TF 6205. Probably 'Winebaud's homestead'. *Winebodesham* *[1060]13th S 1030, c.1140, *Winebotesham* 1086, *Winebadisham* 1195. OG pers.n. *Winebaud*, genitive sing. *Winebaudes*, + **hām**. *Winebaud* may, however, conceal OE pers.n. *Winebald*. DEPN.

New WIMPOLE Cambs TL 3450. No early forms. ModE adj. **new** + p.n. Wimpole as in WIMPOLE HALL TL 3351.

WIMPOLE HALL Cambs TL 3351. *Wimple Hall* 1695. P.n. Wimpole + ModE **hall**. A mansion house built c.1632 for Sir Thomas Chicheley and subsequently altered. Wimpole, *Winepol(e)* 1086–1242, *Wimpol(e)* from 1195, *Wymple* 1553, is 'Wina's pool', OE pers.n. *Wina* + **pōl**. Ca 81 gives former pr [wimpl], 82, Pevsner 1954.402.

WIMPSTONE Warw SP 2149. Partly uncertain. *Wylmestone* 1313, *Wilm(e)ston* 1498–1614, *Wymston* 1605, *Wimpston* 1656. The evidence is too late for certainty, but the generic is OE **tūn**, the specific probably a pers.n., perhaps *Wilhelm* or *Wīġhelm*, genitive sing. *Wilhelmes, Wīġhelmes*. Wa 307.

WINCANTON Somer ST 7128. 'Settlement on the river *Wincawel*'. *Wincainie- Wincautone* 1084, *Wincaletone* 1086, *Winchaulton* [13th]Buck, *Wynkauelton* 1243, *Wincaneton* [1275×92]Buck, *Wincaulton* 1291, *Winecaunton* 1610. R.n. *Wincawel* [956 for 953×5]14th S 570 + OE **tūn**. *Wincawel* is the 'white Cawel', PrW ***wïnn** + r.n. CALE ST 7126, of which the white Cawel is an arm. RN 63, DEPN, Turner 1952–4.13.

East WINCH Norf TF 6916. *Eastuuininc, Estwinic -uuinc* 1086, *Estweniz* 1242, *Estwinch* 1254. OE adj. **ēast** + p.n. Winch, 'the meadow farm', OE ***winn** + **wīċ**. 'East' for distinction from West WINCH TF 6315. DEPN.

West WINCH Norf TF 6315. *Wesuuenic -uuinic* 1086, *Westweniz -winic* 1198, *Westweniz* 1242, *West Weniz* 1203, *Westwinch* 1254. OE adj. **west** + p.n. Winch as in East WINCH TF 6916. DEPN.

WINCHAM Ches SJ 6775. 'Wigmund's homestead'. *Wimundisham* 1086, *Wymundsham* 1306, *-ysham* 1581, *Wy- Wimingham* late 12th, 1209, *Wymmincham* c.1270, *Wymincham -yn-* 1306–1641, *Wyncham* 1435, *Wi-* from 1687. OE pers.n. *Wīġmund*, genitive sing. *Wīġmundes*, + **hām**, later reformed as if a formation in **-ing**[2] with palatalisation (**-inġ**), an example of popular etymology arising from the devoicing of *-d* before genitival *s* in forms like **Wimundsham* and the subsequent development of [ts] to [tʃ], **Wimuncham*. Che ii.136 gives pr ['winʃəm].

WINCHCOMBE Glos SP 0228. 'Valley with a bend in it'. *(æt, on) Wincelcumbe* 796×821 S 1861, c.815, 1014×16, [1042]11th, 11th ASC(D) under year 1053, *Wi(n)celcvmbe -cumbe, Wicecombe* 1086, *Winchelcumb(e) -comb(a)* 1130–1586, *Wi- Wychecumb(e) -comb(e)* 1059–1509, *Wi- Wynchcumb -comb(e)* 1234, 1316, 1586. OE **winċel** + **cumb**. Winchcombe was an important centre in Anglo-Saxon times and was the centre of a shire. Only a selection of OE forms is given. Gl ii.30, L 91, 93.

WINCHELSEA ESusx TQ 9017. 'River-bend island'. *Wencles* 973×5 coins, *Winceleseia* 1130, *Winchenesel* 1130–91,

Winchelese(e) 1165–1342. OE ***winċel**, genitive sing. ***winċeles**, + **ēġ**. The name is descriptive of the site of Old Winchelsea rather than its successor at Iham TQ 9017, *Iham(me), Ihomme* 1200–1509, 'island' or 'yew-tree promontory', OE **īeġ** or **īġ**, **īw** + **hamm** 1, 2, 5, the cliff on which the new town of Winchelsea was built c.1280. Sx 537, L 39, SS 83.

WINCHELSEA BEACH ESusx TQ 9115. P.n. WINCHELSEA TQ 9017 + ModE **beach**.

Nether or Lower WINCHENDON Bucks SP 7312. *Winchendone Inferior* 1242, *Nether Wynchedon* 1262, ~ *Witchingdon* 1627, ~ *Winchendon* 1833 OS. ME **nether**, Latin **inferior** + p.n. *yincandum* (sic for *pin-* i.e. *wincandun*), *(into) wincandone* [1004]14th S 909, *Wichendone* 1086–1227[†], *Winchendon* from 1176, 'hill of or at the bend or winch', OE **winċe**, genitive and dative sing. **winċan**, + **dūn**. The reference is to a bend in the river Thame, a nook or corner in the hill, or to a winch. Earlier explanations from the OE pers.n. **Wineca*, genitive sing. **Winecan*, do not work because the absence of palatalisation in this name. 'Nether' for distinction from Upper WINCHENDON SP 7414. Bk 111 gives pr [witʃəndən], PNE ii.267, L 150.

Upper WINCHENDON Bucks SP 7414. *Wychendon Superior* 1262, *Over Wynchyndone* 1375. ModE adj. **upper**, ME **over**, Latin **superior**, + p.n. Winchendon as in Nether WINCHENDON SP 7312. Bk 112.

WINCHESTER Hants SU 4729. 'The Roman town Venta'. *Uintan cæstir* [c.730]8th BHE, *Wi- Wentancestre* 11th ASC(F) under year 731, *(on, to, æt) wintan ceastre* 9th ASC(A) under years 744, 855, 860–12th including 933, 963, 982 ASC(A) and S 463, 1513, 1560, *(on) Winte ceaster* 897, 909 ASC(A), [900]11th S 360, *Winte caester* 934 S 425, *(on) Winceastre* [938]12th S 444, c.1025, [c.1060]1317, and 11th and 12th cent. MSS of ASC, *Wincestre* 1086. RBrit p.n. *Venta* + OE **ċeaster**. Venta, *Οὐέντα* [150]13th Ptol, *Venta Belgarum* 'Venta of the Belgæ' [c.300]8th AI, *Venta Velgarom* [c.700]13th Rav, *(a civitate) Uenta* [c.730]8th BHE, *Wæntan* (genitive sing.) 1121 ASC(E) under year 731, is a pre-British, possibly IE, name meaning something like '(chief) place' of a tribe. Also found in Latin texts in the adjectival form Wintonia, *in Wentana civitate* [877]12th S 1277, *in Wintonia civitate* [900]12th S 1284, *in Wintonia urbe* [947]15th S 526. Ha 176, Gover 1958.10, RBrit 492 and discussion under Bannaventa 262–5, Jnl 16.1–7.

WINCHFIELD Hants SU 7654. 'Open land by the nook'. *Winchelefeld* 1229, *Wi- Wynchefeld* 1249–91. OE ***winċel** + **feld**. The reference is to a small indentation in the hills. Ha 177, Gover 1958.118.

WINCHMORE HILL Bucks SU 9395. *Winȝemerehull* 1270, *Winsmore hill* 1611, 1639, *Winchmore Hill* 1639, 1706, *Winshmore hill* 1674. P.n. Winchmore + ME **hull** (OE *hyll*). Winchmore probably represents OE **winċe**, ***winċel** 'nook, angle, corner' + **(ġe)mǣre** 'boundary'. The reference is to the county boundary which makes a sharp turn here. Bk 228, Jnl 22.52.

WINCHMORE HILL GLond TQ 3195. Partly uncertain. *Wynsemerhull* 1319, *Wynsmorehyll* 1509×40, *Winchmore Hill* 1586. The second part of this name may represent OE **(ġe)mǣre** + **hyll** 'boundary hill' referring to the southern boundary of the ancient parish of Edmonton. The specific is unclear, probably an OE pers.n. such as *Wine*, genitive sing. *Wines*, or *Wynsiġe*. The evidence is too late for certainty. Mx 70.

WINCLE Ches SJ 9666. Uncertain. *Winchul* c.1190, *Wi- Wynchul(l) -k-* [c.1190]1285, 1237–1471, *Winkel* early 13th, *Wynkell -ull* 1357–1499, *Wincle* from 1531 with variants *Wyncle, Win(c)kle* to 1724. The simplest explanation is OE ***winċel** 'a nook, a corner' referring to the hamlet's location in the bend of a valley, with confusion of *-el* with ME **hull** 'a hill' (< OE *hyll*). The other possibilities are 'winch hill' or 'hill at or in a bend', OE **winċe** 'winch, pulley, ?angle, ?bend' + **hyll** referring, if the topographical sense is accepted, to the hill-spur that bends round the hamlet from N to W; and 'Wineca's hill', OE pers.n. **Wineca* + **hyll**. Che i.164, ii.vii which gives 19th cent. local pr ['winkə].

WINDBURY POINT Devon SS 2826. Cf. *Windbury Head* 1809 OS. The reference is to earthworks on the headland (ultimately from OE **wind** + **byriġ**, dative sing. of **burh**, 'windy fort').

WINDERMERE Cumbr SD 3995 (the lake). 'Winand's lake'. *Wi- Wynendermer(e), Winendermer* [1154×89]1412, *Wi- Wynandremer(e) -er-* 1157–1681×4, *Wi- Wyndermer(e)* 16th–1738. CG pers.n. *Winand* + OScand genitive sing. *-ar* + OE **mere**. The addition of OScand genitive sing. *-ar* to a CG pers.n. cannot have taken place before the 12th cent. and shows the continued use of Norse as a living language. We i.18, SSNNW 258, SPN 7017.

WINDERMERE Cumbr SD 4198 (the town). Named from WINDERMERE SD 3995 (the lake). *Wi- Wynandermer(e)* 1154×89–1607, *Windermer(e)* from 1560. We i.192, SSNNW 258.

WINDERTON Warw SP 3240. 'The winter settlement'. *Winterton(e)* 1166, 1334, *Wynterton* 1235–1547, *Wynderton* 1262–1568, *Winderton* 1618. OE **winter-tūn**, a farmstead used in winter. Wa 277.

WINDLESHAM Surrey SU 9363. Partly uncertain. *Windesham* 1178(p), *Wy- Wind(e)les-* 1223–1535, *Wendes-* 1401, *Wyns- Wens-* 16th cent., *Winsom(e)* 1749. Possibly 'the settlement with or by a windlass', OE ***windels** + **hām**. Sr 152 gives a former pr [winsəm] which matches the 1749 sp. ASE 2.32.

WINDLEY Derby SK 3045. 'Pasture clearing'. *Wi- Wynleg(a) -leia -legh -lee -ley(e)* 1138×48–1559, *Wynd(e)ley(e)* 1417–1635, *Wind(e)ley* 1607. OE **winn** + **lēah**. Later forms and the pronunciation have been influenced by **wind** 'wind'. Db 617 gives pr [waindli].

WINDMILL HILL ESusx TQ 6412. *Windmill Hill* 1587. ModE **windmill** + **hill**. Sx 482.

WINDMILL HILL Somer ST 3116. *Wind-mill Hill* 1809 OS.

WINDRUSH Glos SP 1913. Named from the river WINDRUSH SP 1816. *Wenric(a)* 1086, 1175, 1201×3, *Wenrich(e)* 12th–1398, *Wyrych(e) -riche* 1295–1577, *-rys(s)he -risshe* 1409–1501, *Wyndriche* 1562, *-rishe* 1599, *Windrush* from 1675. Gl i.209.

River WINDRUSH Glos SP 1816. Probably 'white fen'. *Uuenuuœnrisc* 779 S 114, *Wenrisc* [969]12th S 771, 11th, *Wenris, Wœnric* [949]c.1600 S 550, *Wenric* [1016]12th S 935, [1044]12th S 1001, c.1055, *Wen(e)rich(e) -rych'* 13th cent., *Wynrysshe* 1500×15, *Winruche, Wynderusch* c.1540, *Windrush* 1577. PrW ***wïnn** (Brit **u̯indo-*) + ***rēsc** < IE **reisko-*, corresponding to OW *ruisc*, Ir *riasg* 'moor, fen', Gaelic *riasg* 'dirk-grass, land covered with sedge or dirk-grass', Scots dial. *reesk* 'coarse grass, marshy land'. **reisko-* is probably a derivative of a root **rei-* 'to flow' related to Greek *ῥέω* (rheo) < IE **sr-ew-*). The PrW name has been assimilated to OE **risc** 'rush' and possibly ***riċ** 'a narrow strip'. The upper parts of the river were also called *Gytingbroc* [780]11th S 116 as in GUITING POWER SP 9214 and *Theodningc* 779 S 114, the 'prince stream', OE **thēoden** + **ing**[2]. Gl i.14, RN 461.

WINDSOR Berks SU 9676. 'Ridge with a windlass'. *Windlesoran* [1042×66]12th S 1141, 1121 ASC(E) under year 1096 etc., *Windlesore(s)* 1087–[c.1265]c1450 with variants *Wyndles-* and *-or(') -our, Vuindisor* 1072, *Windesore* 1086–1600 with variants *Wyndes-* and *-or(is), Windsor* 1618. OE ***windels** + **ōra**. The lost names Underore, *Underore* 12th, and Upnor, *Vpenore* 1172 'upon the ridge', suggest that there was a tract of land here called *Ōra* 'the Ore, the ridge', of which Windsor, 'The windlass Ore', was part. Brk 26, 29, 30, TC (1986) 194, L 181–2.

[†]Another DB spelling is *Witchende*: *Wich-* is for *Winch-* by omission of the nasal suspension.

WINDSOR FOREST Berks SU 9373. *Foresta de Windesores* 1086, *Foresta de Windleshora* [1109×20]13th. P.n. WINDSOR SU 9676 + MLat **foresta**. Here **foresta** is used in its legal sense denoting an area of woodland in royal hands devoted to the hunting and preservation of game and subject to the forest laws. In 1300 the royal forest of Windsor included all the ancient county of Berkshire east of the Loddon. Brk 4.

WINDSOR GREAT PARK Berks SU 9572. *parcum de Windes'* 1243, *Wyndesore Park* 1239, *The Greate Park* 1601. P.n. WINDSOR SU 9676 + ModE **great** + **park**. 'Great' for contrast with *The Litle Parke* (in New Windsor) 1607 and *Littelparke* 1526. Brk 32.

Old WINDSOR Berks SU 9874. *Vet'i Windesores* 1212×13, *Old Windeshour* 1251. ME **old**, Latin **vetus, veterem** etc., + p.n. WINDSOR SU 9676. 'Old' for distinction from New Windsor, *(on tham) niwan windlesoran* 1121 ASC(E) under year 1110, *Nova Wyndsore* 1316, the name given to the settlement which grew up near Windsor Castle, a fortress built in the late 11th cent. Brk 26.

WINDY CRAG Northum NT 7705. *Windy Crag* 1869 OS. ModE **windy** + **crag**. A rocky edge rising to 1607ft.

WINDY GYLE Northum NT 8515. *The wyndy gole* 1541 Dixon, *the Windy-gyle* 1831 AA 2.290. Possibly ModE **windy** + **gyle** from ON **geil** 'a narrow ravine, a way, a narrow lane'. Windy Gyle is one of the highest peaks on the Scottish-English border at 2032ft. and is traversed by Clenell Street, one of the ancient track-ways connecting the two countries.

WINEHAM WSusx TQ 2320. Probably 'the river meadow by the winding stream'. *Wyndeham* 1248–1332, *-hamme* 1288, 1316, *Windham* 1810 OS. OE **(ġe)wind** + **hamm** 3. Sx 213 gives pr [wainəm], ASE 2.28, SS 84 s.n. Wyndham.

WINESTEAD Humbs TA 3024. 'Women's place'. *yinestede, yiuestode, yifesta* (for *Wiue- Wifestede*) [1033]14th, *Wife- Wiuestad, Wi(fe)stede* 1086, *Wi- Wyvested* 1268, *Wi- Wynestede* 1256–1527, *Wysted(e)* 1429–1610. OE **wīf**, genitive pl. **wīfa**, or masc. pers.n. *Wīfa* or feminine *Wīfe*, + **stede**. Difficulty in reading medial *-u-* for [v] and confusion with *n* accounts for the modern form. The earliest forms show influence of ON **stathr**. YE 29, SSNY 150, Stead 295.

WINFARTHING Norf TM 1086. 'Wina's quarter'. *Wineferthinc* 1086, *-ferding* 1165, *-ferðing* 1168. OE pers.n. *Wina* + **feorthing** 'a fourth part of an estate'. DEPN.

WINFORD Avon ST 5465. Originally a stream name, *Wunfrod* [984×1016]12th S 1538, *Wenfre* 1086, *Wenfrod* 1086 Exon, *Wi- Wynforð -frod -fred -fryth* 1169–1491, *Wineford* 1274, 'white, fair, holy stream', PrW ***wïnn-frud**. Identical with Winford in WYNFORD EAGLE Dorset SY 5895, WINFRITH NEWBURGH SY 8084, and the Welsh names Gwenffrwd Dyfed SN 5960, 7746, Powys SO 0331, *Guenfrut* 12th, and *Wenferð* [866]11th S 211, a lost stream name in Kidderminster. DEPN, RN 463, Wo 16, PNE ii.269, Jnl i.47, 52.

WINFORTON H&W SO 2947. Originally 'Widfrith's estate'. *Widferdestvne* 1086, *W(i)lfreton(e)* 1219–78, *Wynfreton(e)* 1219×34, 1265, *Wynforton* 1535. OE pers.n. **Wīdfrith*, genitive sing. **Wīdfrithes*, + **tūn**. Later spellings suggest substitution by OE *Wilfrith* and *Winfrith*. He 208.

WINFRITH NEWBURGH Dorset SY 8084. 'W held by the Newburgh family'. *Wynfrod(e) Newburgh, Neuburg* 1210–88, *Winfred(e) Neuburgh -burth, Neburg(h)* 1288–1447, *Wunfreth Neuburgh* 1431, *Winfrith Newborough* 1705. Earlier simply *Winfrode* 1086, *-ford(e)* 1210×12, *Wynfryth* 1399. Originally a stream-n., 'white stream', PrW ***wïnn** + **frud**, identical with WINFORD Somer ST 5465 and WYNFORD EAGLE SY 5895. The *de Neuburh* or *De Novo burgo* family from Le Neubourg near Rouen, Normandy (*Novus Burgus* 1089 'the new fortification'), was here from 1100×35. Do i.174.

WING Bucks SP 8822. '(Among) the Weowungas, the people called after Wiwa'. *(æt) Weowungum* [966×75]c.1150 S 1484, *Witehunge* 1086, *Wienge* 1181, *Wiungua* 1154×89, *Weng(e)* c.1215–1400, *Wyeng(e)* 1325–97, *Wyng(e)* 1349, 1512. OE folk-n. **Weowungas* < pers.n. **Wiwa* + **ingas**, dative pl. **Weowungum*. This explanation assumes that the DB form is due to a misreading of runic ƿ (w) as þ (th). The mistake may, however, have been the reverse and *Weowungum* a misreading for *Weoþungum*, OE folk-n. **Withungas*, dative pl. **Weothungum*. The territory of the Weowungas or Withungas must have been considerable since it contained both Wing and WINGRAVE SP 8619 three miles to the S. Bk 86, ING 47.

WING Leic SK 8903. Uncertain. *Weng(e)* 1136×9–1428, *Wieng'* 1203–4, *Whe(e)nge* 1315–1407, *Wyng(e)* 1294–1543, *Wing* 1516. Either ON **vengi** 'in-field' used in Denmark of the three large fields in a village, or ODan ***wæng** postulated in the same sense for the Dan p.n. Veng and the Sw p.n. Vänge, OSwed *Vængia*. The earliest reference is *(into) wenge forde* 'the ford on the road to Wing' [1046]12th S 1014, a lost p.n. located at SK 868014. R 228, DEPN, SSNEM 162.

WINGATE Durham NZ 4037. '(At the) wind gates'. *æt Winde gatum* 1071×80, *Winde- Wyndegate* 1144–1312, *-gates* 1307–1456, *Windgate* 1180, 1303, *Wyngatte* 1428, *Wyngait, Wingayt, Win(d)gait* 16th, *Wingate* from 1605×6. OE ***wind** + **ġeat**, dative pl. ***wind-gatum**. A wind-gate is a place where the wind is funneled through a valley or a trough; the reference is to the gap in the hills between Wheatley Hill and Deaf Hill. NbDu 217.

WINGATES 'Wind gates'. OE ***wind** + **ġeat** ME secondary pl. **windyates** or **windgates**. The reference is to a gap or valley through which the wind drives.

(1) ~ GMan SD 6507. *Windyates* 1272, *lee Wyndzates* 1451, *Win Yates* 1843 OS. Cf. WINGATE Durham NZ 4037. La 44.

(2) ~ Northum NZ 0995. *Wyndegates* [1208]14th, *Windegatis -es* 1236, 1242 BF, *Wi- Wyndegat(e)* 1242 BF–1296 SR. NbDu 217.

WINGERWORTH Derby SK 3767. 'Winegar's enclosure'. *Wingreurde* 1086, *Wi- Wyngerwurth -worth(e) -ir- -ur-* 1238–1307 etc. OE pers.n. *Winegār* + **worth**. Db 330.

WINGFIELD Beds SP 9926. Partly uncertain. *Wi- Wynfeld* c.1200 and 13th cent., *Wintfeld* 13th, *Wynchefeld* 1276, *Wyndeselde* (sic for *-felde*) 1535, *Winfield* 1675. The generic is OE **feld** 'open land', the specific either OE pers.n. *Winta*, genitive sing. *Wintan* with later strong variant *Wintes*, or **winċe** 'a winch, roller, pulley' or ***winċel** 'a nook, angle, corner'. The earliest forms, however, point to a variant with OE ***winn** 'pasture'. Cf. WINCHFIELD Hants SU 7654, WINGFIELD Wilts ST 8256, Derby SK 3755, 4165. Bd 118, DEPN, L 245.

WINGFIELD Suff TM 2376. 'The open land of the Wigingas, the people called after Wiga'. *Wingefeld* [11th, probably before 1038]13th S 1527, *Wighefelda* 1086, *Wihingefeld* 1185–6, *Wyng(e)feld -feud* 1254–1346. OE folk-n. **Wiġingas* < pers.n. *Wiga* + **ingas**, genitive pl. **Wiġinga*, + **feld**. An alternative less likely suggestion is that the *Wigingas* might be 'the people of the heathen temple' < OE *wīġ, wēoh* + **ingas**. DEPN, Baron, Evidence 111, L 244.

WINGFIELD Wilts ST 8256. 'Wina's open land'. *(on) Wuntfeld* [964]14th S 727, *Winefel* 1086, *Wi- Wynefeld* 1241–1428, *Wynfeld* 1446–1541, *Wynkefeld(e)* 1535, 1556, *Wyngfeld* 1542, *Winkfeild al. Wingfeild* 1707. OE pers.n. *Wina* + **feld**. There is no evidence for the unexplained intrusive [k] or [g] before 1535. The sp in S 727 is unreliable. Wlt 122 (under Winkfield).

North WINGFIELD Derby SK 4165. *North Wynfeld* 1439, 1503, *Northwingfield(e)* 1556, 1607. Adj. **north** + p.n. *Wy- Winnef(eld)* [c.1002]c.1100 S 1536, [c.1002]13th, [c.1004]14th S 609, *Winnefelt* 1086, *Wi- Wynefeld(e)* 1243–1330, *Wyn-* 1384–1438, either 'open land used for pasture', OE **winn** + **feld**, or 'disputed open land', OE **(ġe)winn** + **feld**. 'North' for distinction from South WINGFIELD SK 3755. Db 333, L 244.

South WINGFIELD Derby SK 3755. *Suwinnefeld'* 1198×1208, *Suthwynnefeld* 1280, 1299, *-wynfeld* 1330, *Southwynfeld*

1392–1610, *Southwingfield* 1652. ME adj. **sūth** + p.n. *Winefeld* 1086, *-feud* 1200–1351, *Wi- Wynnefeld(e) -feud* 1154×9–1336, *Wi- Wynfeld(e)* 1247–1535×43, with same meaning and etymology as Wingfield in North WINGFIELD. 'South' for distinction from North WINGFIELD SK 4165. Db 335.

WINGHAM Kent TR 2457. 'Homestead of the Wigingas'. *Uuigincgga ham* 825×32 S 1268, *Winganham* [941]13th S 477, *Wuung- Wyngeham* [939×46]13th S 515, *Wing(h)eham* 1086, *Wi- Wynge(e)ham* 1164×5–1229. OE folk-n. **Wiġingas* < pers.n. *Wiga* + **ingas**, genitive pl. **Wiġinga*, + **hām**. A possible alternative origin for the folk-n. is 'the people of the shrine', OE **wēoh, wīh, wīġ** + **ingas**. PNK 537, 384, KPN 158, ING 120, ASE 2.30, Rumble 1997.383–9.

WINGRAVE Bucks SP 8619. 'Grove belonging to Wing or to the Weowungas or Withungas'. *Wit(h)ungrave* 1086, *Wiengraua* 1175, 1198, *Wengrave* 1189–1399, *Wi- Wyngrave* from 1262, *Wingrove* 1365. P.n. WING SP 8822 or folk-n. **Weowungas* or **Withungas*, genitive pl. **Weowunga*, **Withunga*, + **grāf(a)**. Bk 88 gives pr [wingruv], ING 47.

WIN GREEN Wilts ST 9220. *Win Green* 1811 OS.

WINKBURN Notts SK 7158. Originally a stream-n., 'the stream with a bend'. *Wicheburne* (for *Wīche-*) 1086, *Winkeburn(a)* c.1150–1330 with variants *Wynke-* and *-burne*, *Winkerburn'* 1188, *Winkelburn* 13th. Stream name Wink ultimately < OE ***winċel** 'a nook, a corner'. The Wink has corners N and S of the village. Nt 197, L 17, 18.

WINKELBURY Wilts ST 9521. *Winkelbury* 1811 OS. Cf. *Winkelbury Hill* 1773. An Iron Age promontory fort with a pagan Anglo-Saxon cemetery immediately to the S. Another Iron Age hill-fort at SU 6152 near Basingstoke Hants is also called Winklebury, *Winklebury* 1817 OS. Although the evidence is very late this might be OE ***winċel** 'a nook, a corner' referring to a bend in the hill + **byriġ**, dative sing. of **burh**. On the other hand the boundary clause of S 582 ([955]14th) refers to the Wilts Winkelbury as the Winter Camp, *to Winterburge geat* 'to the gate of *Winterburh*'. The similarity between Winkelbury and *Winterburh* makes it likely but not certain that the latter derives from the former. Wlt 202, Grundy 1920.33, PNE ii. 268, Thomas 1980.239, 143.

WINKFIELD Berks SU 9072. 'Wineca's open land'. *Winecanfeld* [942]12th S 482, *Wenesfelle* 1086, *Wineche(s)feld'* 1176, *Winekefeld* [before 1170]12th, *Wynekfeld* 1316, *Wi- Wynkefeld* 1224–1517, *Winkfield* 1685. OE pers.n. **Wineca*, genitive sing. **Winecan*, + **feld**. Brk 36, 835, L 243, Signposts 195.

WINKFIELD ROW Berks SU 8970. *Winkfield Row* 1761. P.n. WINKFIELD SU 9072 + ModE **row**. Brk 41.

WINKHILL Staffs SK 0651. 'The nook hill'. *Wycleshull* 1278, *Wykynghull* 1307, *Wynkeshull* 1329, *Wynkyll* c.1538, *Winkle-hill* 1686. OE ***winċel**, genitive sing. ***wincles**, + **hyll**. Cf. WINCLE Ches SJ 9666. Che i.164–5, Horovitz.

WINKLEIGH Devon SS 6308. 'Wineca's wood or clearing'. *Wincheleia* 1086–1426 with variants *Winke- Wynke-* and *-legh(e)*. OE pers.n. *Wineca* + **lēah**. Alternatively the specific might be ***winċel** 'a nook, a corner' referring to Winkleigh's site on a hill between streams. D 373.

WINKSLEY NYorks SE 2571. 'Winuc's forest glade or clearing'. *Wichingeslei* 1086, *Winchesl(aie)* 1086, 12th, *Wi- Wynkeslay -le(y) -legh* 12th–1598, *Wincksley* 1623. OE pers.n. *Winuc*, genitive sing. *Win(u)ces*, + **lēah**. *Wichingeslei* looks like an alternative form 'Viking's forest clearing', possibly for OE *Winucinges* + **lēah** 'the forest clearance of *Winucing*, the place called after Winuc'. YW v.195.

WINMARLEIGH Lancs SD 4647. 'Winemær's wood, clearing or pasture'. *Winemerleie* [1190×1200]1268, *Wynemerislega*, *Wynermerisle* 1212, *Winmerleie* [c.1220]1268, *Wymerley* [1216×35]1268, *Wimmerlee* [1220×50]1268, *Wynmerlegh* 1343, *Wimmerlaw* c.1540. OE pers.n. *Winemǣr* + **lēah**. La 164 gives pr [winma:li], Jnl 17.96.

WINNERSH Berks SU 7870. 'Ploughed field by the meadow'. *Wenesse* 1190, 1194, *Weners(sh)* 1247×8, 1397, *Winnersh* from 1617. OE ***wynn** + **ersc**. Brk 136, PNE ii.269, L 235.

WINNIANTON Corn SW 6620 → GUNWALLOE FISHING COVE SW 6522.

WINSCALES Cumbr NY 0226. 'Windy shielings'. *Wy- Windscales* 1227–1334, *Wy- Winskales* from 1254, *Wynscales* c.1258. ME **wind** (ON *vindr*) + pl. **skales** (ON *skáli*). This is not the Winscales after which the nuclear power station was originally named. That is Winscales NY 0209, *Windscales* 1294, of identical origin. Cf. also WINSKILL NY 5835. Cu 341. 454, SSNNW 73 no.1.

WINSCAR RESERVOIR SYorks SE 1502. No early forms: cf. Winscar Holes, YW i.343. P.n. Winscar, possibly ModE **whin** 'whinstone' + **scar** 'rock' (ON *sker*), + ModE **reservoir**.

WINSCOMBE Avon ST 4257. 'Wine's coomb'. *Wynescumbe* [959×75]13th S 1762, *Winescome* 1086, *-cumb* 1196. OE pers.n. *Wine*, genitive sing. *Wines*, + **cumb**. DEPN, L 92.

WINSCOT Devon SS 4912 → WINSWELL SS 4913.

WINSFORD Ches SJ 6566. 'Wine's ford'. *Wynisford* 1216×72, *Wi- Wynesford* 1255, *Winsford* from c.1350. OE pers.n. *Wine*, genitive sing. *Wines*, + **ford**. Che iii.173.

WINSFORD Somer SS 9034. 'Wine's ford'. *Winesford* 1086, 1610, *Wynesford* 1251. OE pers.n. *Wine*, genitive sing. *Wines*, + **ford**. DEPN.

WINSFORD HILL Somer SS 8734. *Winsford Hill* 1809 OS. P.n. WINSFORD SS 9034 + ModE **hill**.

WINSHAM Somer ST 3706. 'Wine's homestead or estate'. *Winesham* [1046]12th S 1474, *[1065]18th S 1042, 1086, *Winsham* 1610. OE pers.n. *Wine*, genitive sing. *Wines*, + **hām**. DEPN.

WINSHILL Staffs SK 2623. 'Wini's hill'. *(on) Wineshylle* [1002×4]11th S 1536, [1004]11th S 960, *-halle* 1086, *-hulla* 1113, *-hill* 1150×9, *Wyncehulle* 1316, *Wynsul* 1322, *Wynsell* 1521. OE pers.n. *Wini*, genitive sing. *Wines*, + **hyll**. DEPN, L 170, Horovitz.

WINSKILL Cumbr NY 5835. 'Windy shielings'. *Wyndscales* 1292, *Wynskales* 1541, *Winskill* 1564. ME **wind** (ON *vindr*) + pl. **skales** (ON *skáli*). Identical with WINSCALES NY 0226. Cu 208, SSNNW 73.

WINSLADE Hants SU 6548. 'Wine's watercourse'. *Winesflot* 1086, *-flote* 1219, *Wi- Wynesflod(e)* 1154×89–1348, *Winslode* 1399, *Winslade* 1535. OE pers.n. *Wine*, genitive sing. *Wines*, + **flōde**. Ha 178, Gover 1958.129.

WINSLEY Wilts ST 8061. 'Wine's wood or clearing'. *Winesleg'* 1242, *Wi- Wyneslege -legh -leye* 1249–1332, *Winslegh* 1363. OE pers.n. *Wine*, genitive sing. *Wines*, + **lēah**. Wlt 124.

WINSLOW Bucks SP 7627. 'Wine's barrow'. *(et) Wineshlauue* *[795 for 792]13th S 138, *Weneslai* 1086, *Wineslawe* [1154×89]1301, 1247, *Wyneslowe* 1247, *Winslawe* 1262, *Wynslow* 1492. OE pers.n. *Wine*, genitive sing. *Wines*, + **hlāw**. The village stands on the brow of a hill where there may have been a barrow. Bk 75, Sx xli.

WINSON Glos SP 0908. 'Wine's estate'. *Winestvne* 1086, *Wi- Wyneston(a)* 1224–1327, *Wi- Wynston* 1220–1587, *Winson* from 1622. OE pers.n. *Wine*, genitive sing. *Wines*, + **tūn**. Gl i.185.

River WINSTER Cumbr SD 4185. Uncertain. *Wi- Wynster* from 1170. This could be ON **Vinstra* 'the left one' (ON **vinstri**), i.e. westerly river compared with the river Gilpin which runs on a parallel course 3 miles to the E. It may, however, be identical with the W r.n. *Gwensteri* [6th]c.1100, the site of a battle in one of the Taliesin poems. If so it would be 'white stream', PrW ***wïnn** (Brit **uindo-*) + ***ster** (related to Breton *ster*, pl. *steri*, 'river, stream'). Whitish clay has been dredged from the river. We i.15, RN 463, SSNNW 425, Williams 1975.126.

WINSTER Cumbr SD 4193. Named from the River WINSTER SD 4185. *Wi- Wynster(e)* 13th–1777. We i.178.

WINSTER Derby SK 2460. Possibly 'Wine's thorn-bush'. *Winsterne* 1086, *Winesterna* 1121×6, 1155, *Win(e)s- Wynster* late

12th–1260 etc., *Wi- Wynstre* c.1200–1758. OE pers.n. *Wine*, genitive sing. *Wines*, + **thyrne**. Db 177.

WINSTON 'Wine's farm or village'. OE pers.n. *Wine*, genitive sing. *Wines* + **tūn**.

(1) ~ Durham NZ 1416. *Winestona* [1091×2]12th, [1183]c.1382, *Wynston* 1244×9–1624, *Winston* from 1586. NbDu 217.

(2) ~ Suff TM 1861. *Winestuna* 1086, 1109×31, *Wy- Wineston(e)* 1252–91, *Wynston* 1316, 1338, 1524, *Winston* 1674. Occasional sps *Winerdeston(a)* [1156×62, 1189×99]1396 show either that *Wine* was short for *Wineheard* or confusion with WYVERSTONE TM 0467. DEPN, Baron.

WINSTONE Glos SO 9609. 'Wynna's stone'. *Winestan(e)* 1086, 1178, *Wunne- Wone- Wynestan* 1211×13–1303, *Wunne- Wi- Wyneston(e)* 1264–1327, *Wi- Wynston* 1291–1742. OE pers.n. *Wyn(n)a* + **stān**. Gl i.142.

WINSWELL Devon SS 4913. 'Beetle spring'. *Wifleswille* 1086, *Wyveliswille* 1321, *Wyllyswell* 1492, *Winswell* 1809 OS. OE **wifel**, genitive sing. **wifeles**, + **wylle**. The form of the name has been influenced by nearby Winscot SS 4912, *Winescote* 1086 'Wine's cottage(s)'. The original form survives in Willeswell Moor SS 4913. D 98.

WINTER HILL Lancs SD 6615. No early forms. 'Hill used in winter'. Rises to 1498ft. Mills 151.

WINTERBORNE 'Winter stream', i.e. a stream that flows in winter. OE **winter-burna**. PNE ii.270, L 19.

(1) ~ CLENSTON Dorset ST 8303. 'W estate called Clenston, the estate of the Clench family'. *Wynterborn Cleyngestone* 1273, *Wy- Winterburn -bo(u)rn(e) Clencheston(e)* 1303–1510, ~ *Clenston* from 1535. Earlier simply *Wintreburne* 1086†, *Winterborn' Clench* 1242×3 or *Clenchton'* 1268, *Clenges-* 1316, *Clenches-* 1327, 'Clench's estate', from Robert Clencg 1232. Do ii.79.

(2) ~ HERRINGSTON Dorset SY 6888. 'W estate called Herringston, the estate of the Herring family'. *Wynterborn(e) -burne Heringeston* 1288, *Winterborn-Herringston or Winterborn-Herring* 1774. Earlier simply *Wintreburne* 1086†, *Wi- Wynterborn(e) -burn(e) Harang* 1242–85, *Wynterbo(u)rn(e) Heryng(e)* 1327–1435 and *Heryng(e)ston, Her(r)ing(e)ston* 1464–1615, 1863, 'Herring's estate' referring to the *Harang* family which was here 1243–1336 (*Heryng* from 1327). Do i.264.

(3) ~ HOUGHTON Dorset ST 8104. 'W estate called Houghton, Hugh's estate'. *Wi- Wynterbo(u)rne Hueton(e)* 1246, 1302, ~ *Hout(t)on(e), Hugheton, Huweton* etc. from 1279. Earlier simply *Wintreburne* 1086, *Wi- Wynterburn(') -born(e) Fercles* 1208–90, and *How(e)ton* from 1303 with variants *Hugh(e)- Ho(u)gh(e)-* etc., 'Hugh's estate', probably referring to the DB tenant Hugh de Boscherbert, although the first member of the Fercles family was also called Hugh, late 12th cent. Do ii.128 gives pr ['hautən].

(4) ~ KINGSTON Dorset SY 8697. 'The king's W estate'. *Wynterb(o)urn(e) -bron(e) Kyng(e)ston* 1280–1616. Earlier simply *Wintreburne* 1086†, *Winterburn* 1196, *Kingeswinterburn(e)* 1194–1346 with variants *Kyng(g)es, Kingis-* and *-wynterburn(e) -born(e)*, and *Kingston* 1244–1577, 1811. The estate was held by the crown as early as the time of John, 1199×1216. Do i.283.

(5) ~ MONKTON Dorset SY 6787. 'W estate called Monkton, the monks' estate'. *Wy- Winterb(o)urn(e) Moneketon(e)* from 1268 with variants *Mu- Monketon*. Earlier simply *Wintreburne* 1086, *Wi- Wynterburn(e)* 1212–90 and *Moneketon(e)* 1285–1332, *Monke-* 15th cent., *Munck-* 16th cent., *Monkton* 1811, 'monks' estate', OE **muneca**, genitive pl. of **munuc**, + **tūn**. Also known as *Wy- Winterburn(e) -bo(u)rn(e) Waste* 1244–1591. The estate was held by the Cluniac priory of Le Wast near Boulogne from the early 13th cent. Do i.266.

(6) ~ STICKLAND Dorset ST 8304. 'W estate called Stickland'. *Wi- Wynterburn(e) -bo(u)rn(e) Sti- Stykel(l)ane* 1203–1385, ~ *Stikeland* 1205, 1316, 1379. Earlier simply *Winterborna* 1068×84, *-burne* 1086 and *Sti- Stykelane* 1268–1548, *Stickland* 1577, 'steep lane', OE **sticol** + **lanu**. Lanes run into the hills E and W out of the deep valley of the r. Winterborne in which the village lies. Do ii.131.

(7) ~ WHITECHURCH Dorset ST 8300. 'W estate called White Church'. *Winterburn Albi Monasterii* (Latin genitive sing. of **album monasterium** 'white church') 1201–99 with variants *Wy-* and *-born, Winterburn' Blancmustier* (French version) 1212, ~ (and) *Bla(u)ncmuster, Blaun(c)mynstre -min-* 1249–80, *Wynterborn' (et) Wytecherch(e)* from 1268 with variants *W(h)y- Whitechurch(e)* etc. Earlier simply *Wintreburne* 1086† and *Wytchirch* 1291, *W(h)y- Whit(e)church(e)* 1316–1575. The affix, OE **hwīt**, definite form **hwīte** + **ċiriċe**, probably referred to a stone church. Do ii.82.

(8) ~ ZELSTON Dorset SY 8997. 'W estate called Zelston, the de Seles manor'. *Wy- Winterbo(u)rne Selyston* 1350–1514, ~ *Seleston'* 1398–1468, ~ *Selston* 1535, ~ *Zelston* 1626. Earlier simply *Wintreburne* 1086†, *Wi- Wynterburn(') -born(e)* 1214–1546. Also known as *Wy- Winterburn(e) -bo(u)rn(e) Mal- Maureward(e)* 1230–1620 and *Marlewardeston* 1275. The Malreward family was here 12th–16th cents. Zelston is from the *de Seles* family probably from Zeals, Wilts, who had association with this place in the 13th cent. Do ii.68.

WINTERBOURNE 'A stream that flows in winter', i.e. one that is dry in summer. OE **winter-burna**. L 19, 97. Identical with WINTERBORNE.

(1) ~ Avon ST 6580. *Wintreborne* 1086, *Wi- Wybterburn(a) -e* 12th–1440, *-bourn(e)* from 1291. Gl iii.123.

(2) ~ Berks SU 4572. *Wintrebvrne -borne* 1086, *Wy- Winterburn(e) -bourn* from 1178. Also known as *Wynterborn Gray, Wynterborn Mayn* 1476 and *Winterbourne Davers* 1752 from the family of Sir Robert Grey of Rotherfield, Oxon, who married into the family holding this estate in the early 14th cent., from a man called *Mayn*, who died seized of the manor in 1260×1, and from the Danvers family, who held here in the 13th cent. Before the Conquest there were three manors here but they seem to have been united by the end of the 15th cent. The boundary of the people of Winterbourne is mentioned as *Winterburninga gemære* [951]16th S 558. The name of the stream here has been transferred to the settlements on its bank. Brk 277.

(3) ~ ABBAS Dorset SY 6190. 'W estate held by the abbot' sc. of Cerne. *Wynterburn Abbatis* (Latin genitive sing. of **abbas**) 1244, ~ *Abbots* 1297. Earlier simply *Winceburnan* (sic for *Winter-*) [987]17th S 1217, *Wintreburne* 1086. Also known as *Watreleswyntreburn* 'waterless W' from the dryness of the South Winterborne river in some seasons. The estate was given to Cerne abbey by the founder Æthelmær in 987. Do 159.

(4) ~ BASSETT Wilts SU 1075. 'W held by the Basset family'. *Wynterburn' Basset* 1249. P.n. *Wintreborne -burne* [869]14th S 341, 1086, *-burn(a)* 1114–1220, + manorial addition from the family name of Alan Basset to whom the manor was confirmed of the gift of his uncle Water de Dunstanville by Richard I in 1194. Winterbourne here is the name of the upper part of the river Kennet. W 309.

(5) ~ DAUNTSEY Wilts SU 1734. 'W held by the Dauntsey family'. *Wynterburne Dauntesie* 1268, *Winterborne Dansey* 1641. P.n. *Winterburne* 1086 + manorial addition from the family name of Roger Danteseye who held the manor in 1242. Winterbourne here is the name of the river Bourne. Wlt 383.

(6) ~ EARLS Wilts SU 1734. 'W belonging to the earl' sc. of Salisbury. *Winterburn(e) Comitis Sar'* 1198, ~ *le Cunte* 1250, *Heorleswynterbourne* 1324, *Erles-* 1401, *Winterbourne Errells* 1553. The same stream as in WINTERBOURNE DAUNTSEY SU 1734. W 383.

(7) ~ GUNNER Wilts SU 1834. 'Gunnora's W'. *Winterburn*

†It is not certain to which of the Winterbornes this spelling applies.

Gonor 1267, *Wynterburne Gunnore* 1268. Gunnora de la Mare held the manor in 1249. The same stream as in WINTERBOURNE DAUNTSEY SU 1734. W 383.

(8) ~ MONKTON Wilts SU 1072. 'The monks' W'. *Winterburne Monachorum* 1249, *Moneke Wynterburn* 1251. Also simply *Monketon* 1348, *Monkton* 1571 'the monks' estate'. The manor was held by Glastonbury Abbey in 1086. The same stream as in WINTERBOURNE BASSETT SU 1075. Wlt 309.

(9) ~ STEEPLETON Dorset SY 6289. 'W estate called Steepleton, the village with a church steeple'. *Wynterburn Stepilton* 1244, earlier *Stipelwinterburn* 1199 and simply *Wintreburne* 1086 or *Stepelton* 1219, ME **stepel** (OE *stīepel*) + **toun** (OE *tūn*). Do 159.

(10) ~ STOKE Wilts SU 0740 → Winterbourne STOKE SU 0740.

WINTERBURN 'Stream that flows or becomes a torrent only in winter'. OE **winter-burna**.

(1) ~ NYorks SD 9358. *Witrebvrne* 1086, *Wi- Wynterburn(a) -e* [late 11th]copy–1567. YW vi.50.

(2) ~ RESERVOIR NYorks SD 9460. P.n. WINTERBURN SD 9358 + ModE **reservoir**.

WINTERINGHAM Humbs SE 9222. 'Homestead of the Winteringas, the people called after Winter or Wintra, or called or at *Wint(e)ring*'. *Wintringeham* 1086, c.1115, *Uuintrigham* 12th, *Wintrincham* 1212. OE folk-n. **Winteringas* < pers.n. *Wintra* or **Winter* + **ingas**, genitive pl. *Winteringa*, or p.n. **Wintering* < pers.n. *Wintra* or **Winter* + **ing**[2], locative–dative **Winteringe*, + **hām**. DEPN, ING 146.

WINTERLEY Ches SJ 7457. Partly uncertain. *Wyntan(e)legh* 1329, 1344, *Wyntenley* 1435, *Wi- Wynteley* 1570–1621, *Winterley* 1840. Unidentified first element + OE **lēah** 'a wood, a clearing'. One suggestion is an OE ***wīn-tān** 'vine-shoot', referring to some wild climbing plant. Che iii.14.

WINTERSETT WYorks SE 3815. 'Fold used in winter'. *duabus Wintersetis* 'the two Ws' 12th, *Wy- Winterset(am)* 1119×35–1534, *-sett* from 1546. OE **winter** + **(ġe)set**. YW i.262.

WINTERSHILL Hants SU 5217. 'Winter's hill'. *Wintereshull* 1272, *Wynt(e)reshull* 1280–1350, *Wyntershyll* c.1460. OE pers.n. or ME surname *Winter*, genitive sing. *Winteres*, + OE **hyll**, ME **hull**. Ha 179, Gover 1958.48.

Middle WINTERSLOW Wilts SU 2432. 'The middle settlement at W'. *Midlewinterslewe* 1397. Earlier *Wynterslewe Middelton* 1345, also simply *Midelton* 1347. ME **midel** + p.n. *Wintreslev -lei* 1086, *Wintereslawe -lewe* 1215–70, *Wyntreslowe* 1301, 'Winter's burial mound', OE pers.n. *Winter*, genitive sing. *Wint(e)res*, + **hlǣw, hlāw**. There is a group of barrows including two very large bell-barrows a mile N of Middle Winterslow. 'Middle' for distinction from West WINTERSLOW SU 2332. Wlt 385, Pevsner 1975.593, Thomas 1980.236.

West WINTERSLOW Wilts SU 2332. *West Winterslewe* 1262. ME adj. **west** + p.n. Winterslow as in Middle WINTERSLOW SU 2423. Also called *Wyntreslewe Uppinton* 1351, OE **upp-in-tūne** '(land) higher up in the village'. 'West' for distinction from East Winterslow SU 2332, *Estwinterslewe* 1257 (now Roche Court), *East Winterslowe* 1632, also *Wynterslowe Eston* 1361, 'the east settlement of W'. Wlt 385, PNE ii.227.

WINTERTON Humbs SE 9218. 'Farm or village of the Winteringas, the people called after Wintra or Winter, or the settlement called or at *Wint(e)ring*'. *Wintringatun* c.1067, *Wintrintune* 1086, *-ingtuna* c.1115, *-ingeton* 1228. OE folk-n. **Winteringas* < pers.n. *Wintra* or **Winter* + **ingas**, genitive pl. **Winteringa* varying with p.n. **Wintering* < pers.n. *Wintra* or **Winter* + **ing**[2], locative–dative sing. **Winteringe*, + **tūn**. This may have been a place dependent upon Winteringham SE 9222. DEPN, ING 146.

WINTERTON-ON-SEA Norf TG 4919. P.n. *Winttertonne* *[1044×7]13th S 1055, *Wintretuna* 1086, *Wy- Winterton(')* from [1094×5]13th, 'the settlement used in winter', OE **winter** + **tun**, + ModE **on sea**, an addition consequent upon the development of Winterton as a resort. The name contrasts with nearby SOMERTON TG 4719 further inland. DEPN.

WINTHORPE Lincs TF 5665. 'Wina's outlying settlement'. *Winetorp* 1154×89–1203, *Wintorp* 1175×81–1212, *Winthorp* 1209×35. OE pers.n. *Wina* + **thorp**. DEPN, SSNEM 143, Cameron 1998.

WINTHORPE Notts SK 8156. 'Wigmund or Vigmund's outlying farm'. *Wimuntorp* 1086, *Wimethorp* 12th, *Winetorp* c.1135, 1228, *Wintorp* 1175×81, *Wynthorp'* 1280–1342, *Winthropp* 1615, *Wimeltorp'* 1215, *Wymbilthorp -el-* 1286, 1325, *Wym(e)thorp(e)* 1280–1335. Pers.n. OE *Wīgmund* or ON *Vigmundr*, + **thorp**. Nt 208, SSNEM 121.

WINTNEY → HARTLEY WINTNEY.

WINTON Cumbr NY 7810. 'Wind-swept' or 'pasture farm or village'. *Win- Wynton(ia) -tuna* 1090–1663. OE **wind** or **winn** + **tūn**. We ii.28.

WINTON Dorset SZ 0893. A modern coinage for a district of Bournemouth built on part of an estate owned by the Talbot family. It either represents *Winton*, the abbreviated form of the Latin title *Wintoniensis* of the bishops of Winchester, or the name of the earl of Winton, a title created in 1859 for Archibald William Montgomerie, 13th earl of Eglinton, a relative of the Talbots. Room 1983.136, Ha 179.

WINTON FELL Cumbr NY 8307. *Winton Fell* 1859. P.n. WINTON NY 7810 + ModE **fell** (ON *fjall*). We ii.29.

WINTRINGHAM NYorks SE 8873. 'Homestead of the Winteringas, the people called after Winter'. *Wentrigha' -igeha'* 1086, *Wintringham* from [1169]13th, *Winteringeheim* 1190, 1191. OE **Winteringas* < pers.n. *Winter* + **ingas**, genitive pl. **Winteringa*, + **hām**. The rarity of forms with *-inge-*, however, suggests a possible alternative, OE p.n. **Wintering* < ***wintering** 'the vineyard place' < ***winter** 'a vineyard' + **ing**[2], locative-dative sing. **Winteringe*. But this is very uncertain. YE 136.

WINWICK Cambs TL 1080. 'Wina's dairy-farm'. *Wineuuiche* 1086, *Winewic -wik* 1195–1428, *Wynwyk* 1348, 1428, *Winnick* 1641. OE pers.n. *Wina* + **wīċ**. Hu 251.

WINWICK Northants SP 6273. 'Wina's farm or trading-place'. *Winewican* [1043]15th S 1000, *Winewincle* 1086, [1189]1332, *Winewiche* 1086–1443 with variants *-wich -wik(e) -wyk(e)*, *Winwik* 1284, 1346, *Wynnycke* 1627. OE pers.n. *Wina* + **wīċ**. The *Winewincle* spellings point either to an alternative generic ***winċel** 'a nook or corner', or to an OE diminutive of **wīċ**, ***wīċinċel**. Nth 77 gives pr [winik], DEPN.

WINYATES H&W SP 0767. No early forms but probably identical with WINGATE Durham NZ 4036, 'wind gates', a place through which the wind blows, ultimately from OE **wind** + **ġeat**.

WIRKSWORTH Derby SK 2854. 'Wyrc's enclosure'.

I. *Wyrcesuuyrthe* [835]13th S 1624, *Wy- Wirkeswurth(e) -wurt -wrth -word(e) -worth(e) -is- -us-* after 1180–*Wirksworth(e)* 1536, 1549 etc.

II. *Werchesuorde -worde* 1086, *Werkeswurda -w(o)rthe* etc. 1169–1540, *Work(e)sworth(e) -ys-* 1329–1631.

OE pers.n. *Wyrc*, varying with *We(o)rc*, genitive sing. *Wyrces*, *We(o)rces*, + **worth**. Cf. WORSOP Notts SK 5879. Db 413 gives prs [wɛ:sə, wusə].

WIRRAL Ches SJ 3187. 'Bog-myrtle nook(s)'. *(on) Wirhealum* 894 ASC(A), *-halum* [1002]11th, *(of) Wirheale* 895 ASC(A), *Wirhale* [1096×1101]1280–1397 with variants *Wyr- Uyr-* and *-hal*, *Wiral* 1260–c.1390, *Wirral* 1278, *Wirrall* from 1291. OE **wīr** + **halh**, dative pl. **halum**. The use of **halh** is not clear. In the plural it may mean 'remote, secluded places', but it is not obvious which part of the peninsula may be meant; in the singular it refers to the whole peninsula in the sense 'peninsula, corner of land, land between rivers (sc. Mersey and Dee) where bog-myrtle grows'. In the 14th cent. romance *Sir Gawain and the Green Knight* part of the hero's itinerary from North Wales to the Green Chapel traverses *the wyldernesse of Wyrhale* where *wonde . . . but lyte That auther God other gome wyth goud hert louied*, 'where few lived that loved either God or man with good heart' (701–2). Che i.7, L 106,109, addenda gives 19th cent. local pr ['wurəl], now ['wirəl].

WIRSWALL Ches SJ 5444. 'Wighere's spring'. *Wireswelle -uelle* 1086 *Wyriswall* c.1180, *Wyreswell(e)* 1276–1324, *-wall(e)* 1288–1499 with variant *Wires-*, *Wirswall* from 1321. OE pers.n. *Wīġhere*, genitive sing. *Wīġheres*, + **wella**, **wælla**. Che iii.112 gives pr ['wə:zwɔ:l], locally [-(w)əl], L 31.

WISBECH Cambs TF 4609. Either 'valley' or 'ridge of the marshy river'. *Wisbece* 1086, *Wi- Wysbech(e)* from 1284 with variants *-bich(e) -bych(e) -bitch* 1436–1594, *Wisebache* 1086 InqEl, *Wysbach* 1470, *Wi- Wysebec(e) -bec(c)he* c.1150 ASC(E) under year 655, 1173–1494, *Wisbeach* 1824 OS. The problems here are two, whether the specific is OE **wisce** 'a marshy meadow' or ***wise** 'a river, a swamp' as in the r.n. WISSEY and whether the generic is OE **beċe** 'a stream, or **bæċ** 'a ridge'. Wisbech lies on the Nene and occupies a low ridge in marshy ground. Ca 291 gives prs [wizbitʃ] and [wisbidʒ], L 12, 126, 250.

WISBECH ST MARY Cambs TF 4208. 'St Mary's Wisbech'. P.n. WISBECH TF 4609 + saint's name *Mary* from the dedication of the church and for distinction from Wisbech St Peter.

WISBOROUGH GREEN WSusx TQ 0426. *Wysebrough al. Grene* 1604, *Westborow Green* 1614. Really two separate names, *Wisebregh* 1227, *Wyseberg(h)(e)* 1227–1397, *Wys(e)ber(u)we* 1261–79, *Whish- Wyshbergh* 1307, 1366, *Wyseborowe* 1509, 'the marsh hill', OE ***wise** varying with **wisce** + **beorg**, and *Grene* 1520, 1740, 'the green', ModE **green**. The village lies on a small hill above a winding tributary of the Arun. Sx 130, L 128, 250.

WISETON Notts SK 7189. Either the 'farm or village beside the wet meadowland' or 'Wisa's farm or village'. *Wisetone* 1086, *Wyseton* 1219, 1226, *Wi- Wyston* 1217–1334. OE **wisce** or pers.n. **Wīsa*, + **tūn**. Wiseton lies on a slight rise beside the water-meadows and marshland of the river Idle. Nt 43 gives pr [wistən].

WISHAW Warw SP 1794. Partly uncertain. *Witscaga* 1086, *Wiðshada* 1166, *Wittes(s)hage* 1182, 1183×93, *Wys(c)hawe* 1262–1375, *Wis(s)haw'e* 1272–1401. The generic is OE **sceaga** 'a copse', the specific possibly **withiġ** 'a willow-tree', **withthe** 'a withe, a tie, a thong, an osier' ('copse where withes are obtained') or **wiht** 'a bend' referring to a large curving hollow E of the village or a smaller one to the W. Wa 52, YE lviii, DEPN, L 209.

Great WISHFORD Wilts SU 0835. *Wykford Majori* c.1190, *Majori Wicheford* 1208, *Muchelewychford* 1332, *Magna Wyccheford* 1428, *Wyssheford Magna* 1513. ModE **great**, Latin **major, magna**, ME **micel**, + p.n. *Wicheford* 1086–1242, 'the elm-tree ford', OE **wiċe** 'a wych-elm' + **ford**. 'Great' for distinction from Little Wishford SU 0736, *Litel Wycford* 1324, *Lytelwychford* 1332, *Parva Whicheford* 1316, *Lytlewisheford* c.1570, earlier simply *Wicheford* 1086. Wlt 230, 228, L 70, 222.

River WISKE NYorks NZ 4300, SE 3497. 'The marshy stream'. *Wisca* 1088, *Wi- Wysk, Wisc* [1157]15th, 1210–1483, *Wiske* from late 13th. OE **wisce** 'marshy place', dial. **wish** 'marshy meadowland, river-land liable to floods' with Scandinavian [sk] for [ʃ]. Identical with Swed r.n. Viskan, OSwed *Visk*, and cognate with OHG *wisa*, G *wiese* 'meadow', OE *wāse* 'ooze, mud' and ultimately with the r.ns. WEAR NZ 1134, Weser etc. from the root **u̯is-*/**u̯eis-*. YN 8, RN 464, BzN 1957.236, Krahe 1964.50–1.

WISPINGTON Lincs TF 2071. Probably the 'brushwood settlement'. *Wispinton* 1060, 1253, *Wispinctune* 1086, *-ingtuna* c.1115. OE ***wisping** 'a place growing with brushwood, a thicket' < ***wips**, ***wisp** 'a wisp, a twig' + **ing**[2], + **tūn**. DEPN, PNE ii.270, Årsskrift 1974.49.

WISSETT Suff TM 3679. Uncertain. *Uuitsede, Wis(s)eta* 1086, *Wicsota* 1162, *Witseta* 1165, *Wytsett* 1235, *Wiseta, Wi- Wys(s)et(e)* 1163–1430. The forms for this name are really indecisive. The generic might be OE **(ġe)set**, pl. **(ġe)setu**, or **sǣte** 'settlers, dwellers', the specific possibly pers.n. *Wit(t)a* and so 'Witta's folds', or an unknown element or name, perhaps the name of the stream on which the village stands. DEPN, Baron.

River WISSEY Norf TL 6797. 'The marsh river'. *Wusan* 905 ASC(A), *Wúsan* c.1100 ASC(D) under year 905, *Wissene* 1257, *Wi- Wyssenhe* [1277]13th, *Wise* 1314–16, *Wisse* 1314, 1575, *Wissey* 1662. Since the drainage of the fens the original course of the Wissey is unknown but it probably flowed through WISBECH TF 4609 [wizbitʃ, wisbidʒ] the forms of which therefore become relevant, *Wisbec(c)e -bech(a)* 1291–1469 etc., *Wisebache -bec(e) -bec(c)h(e)* 1086–1484, *Wissebech(e)* 1291–1469, 'the stream or valley of the Wisse'. There is also an unidentified *Wi- Wysem(o)uth(e)* c.1250–1603 in Wisbech Fen near Whittlesey TL 2797 which, however, seems too far W to be relevant here. The OE forms must have been **Wise*, later *Wuse*, and **Wisse*, genitive sing. **Wis(s)an, Wusan*, so that Wissey is the 'stream of the Wis(s)e', **Wis(s)an* + **ēa**. **Wise* is probably cognate with OE *wāse* 'mud' < Gmc **waisōn*, ultimately from IE **u̯eis-/u̯is-* 'flow' as in the r.ns. WEAR, Weser etc. The reduced grade of this, Gmc **wisōn*, would give OE *wise* and *wuse* with w-umlaut as in *wudu* beside original *widu*. **Wisse* represents a declensional variant **wisjōn* as in OE **wisse* 'a meadow, a marsh'. RN 465, Ca 12,291, PNE ii.270.

WISTANSTOW Shrops SO 4385. Probably 'the holy place of St Wigstan'. *Wistanestou* 1086, *-stowa -stowe* 1177–1443 with variant *Wy-, Wy- Wistanstow(e) -stou* from 1255. OE saint's name *Wīġstān*, genitive sing. *Wīġstānes*, + **stōw**. Wigstan was a member of the Mercian royal family murdered in 849 or 50 and buried at Repton Derbyshire. The place of the murder is given as *Wistanstow* which would however suit Wistow Leics SP 6496, *Wistanestou* 1086 better geographically. Sa i.317–8, *Stōw* 190.

WISTANSWICK Shrops SJ 6629. 'Wigstan's farm or work-place'. *Wistaneswick* 1274, *Wystaneswyk* 1285. OE pers.n. *Wīġstān*, genitive sing. *Wīġstānes*, + **wīc**. DEPN.

WISTASTON Ches SJ 6853. 'Wigstan's farm or village'. *Wistanestune* 1086, *Wistaniston* 12th–1535 with variants *Wy-*, *-stan(e)s- -stams-* (for *-stanis-*?) *-stann(e)s-*, *Wistastun* 1314, *-ton* from 1421, *Wistason* 16th cent. OE pers.n. *Wīġstān*, genitive sing. *Wīġstānes*, + **tūn**. Che iii.45 gives pr ['wistastən], older local ['wistisn].

WISTLANDPOUND RESERVOIR Devon SS 6441. P.n. Wistlandpound + ModE **reservoir**. Wistlandpound SS 6342, *Westland Pound* 1809 OS, 'pound belonging to Westland', p.n. Westland SS 6542 + ModE **pound**. Westland is *Wistland* 1678. It lies on the W of Challacombe parish, but another Wistland occurs in Dolton, *le Wystlond* 1522. Possibly these names contain OE **wist** in some such sense as 'resting place, lair, dwelling'. Ḋ 61, 367, PNE ii.270.

WISTON WSusx TQ 1413. 'Wigstan or Winestan's estate'. *Wistanestun* 1086, *Winestaneston'* 1190, 1193, *Wysteneston* c.1230–1357, *Wysteston* 1296, *Wytston* 1509, *Wiston* 1552. OE pers.n. *Wīġstān* varying with *Winestān*, genitive sing. *Wīġ- Winestānes*, + **tūn**. Sx 243 gives pr [wisən].

WISTON PARK WSusx TQ 1512. P.n. WISTON TQ 1413 + ModE **park**.

WISTOW Cambs TL 2781. 'Dwelling place'. *Wistov* 1086, 1114×33, *Wyrstowe* (sic for *Wyc-*) 1227, *Wyn(e)stowe* 1270, *W(h)ytstowe* 1286, *Wistowe* 1321. The earliest reference is *Kingestune id est Wicstoue* *[974]14th S 798. The place is called *Kingeston* 'the king's manor' again in 1253. It is suggested that **wīc-stōw** may have meant 'site of the royal house'. Hu 228.

WISTOW NYorks SE 5935. 'The dwelling place, the camp'. *Wicstow, (on) Wic-stowe* c.1030 YCh 7, *Wi- Wystow(e)* [1109×14]copy–1641, *Wike- Wykestow(e)* 1154×63, 13th cent. OE **wīc-stōw**. YW iv.36.

WISWELL Lancs SD 7437. 'Spring of the river *Wise*'. *Wisewell* 1207, *-wall(e)* 1243–1332 etc., *Wysewell* 1246, 1272, *Wyswell* 1281. R.n. **Wise* + **wella**. For the r.n. **Wise* cf. WISBECH Cambs TF 4609, River WISSEY Norf TF 8502, TL 6797. La 77, L 32, PNE ii.270, Jnl 17.48.

WITCHAM Cambs TL 4680. 'Wych-elm promontory'. *(on) Wichamme* [970]13th S 780, *Wiceham* 1086, *Wi- Wyc(c)ham* c.1120–1418. OE **wiċe** + **hamm** 2a 'promontory of dry land into marsh'. Witcham is 2 miles W of Witchford. Ca 244.

WITCHAMPTON Dorset ST 9806. 'Settlement of the *Wichæme*, the dwellers at *Wicham*'. *Wichemetune* 1086, *Wichamatuna* 1086 Exon, *Wichamton(')* 1216–88, *Wi- Wych(h)ampton(e)* 1242–1456. OE **wīchǣme**, genitive pl. **wīchǣma**, + **tūn**. The **wīchǣme** were either the inhabitants of a **wīc**, a dairy-farm or work-place of some kind, or more likely the inhabitants of a place called **Wīchām* 'village associated with a Romano-British settlement or *vicus*' referring to the extensive Roman remains and villa at nearby Hemsworth ST 9605. Do ii.260 gives pr [witʃ 'æmtən], Settlements 25.

WITCHFORD Cambs TL 5078. 'Wych-elm ford'. *Wiceford(e)* 1086, *Wi- Wycheforde* c.1120–1467, *Wi- Wych(e)ford(e)* 1138–1499. OE **wiċe** + **ford**. Witchford is 2 miles E of Witcham. Ca 245.

Great & Little WITCHINGHAM Norf TG 1020, 1220 → WHISSONSETT TF 9123.

Great WITCOMBE Glos SO 9114. *Magna Wydecombe* 1397. Earlier simply *Wi- Wydecumbe -comb(e)* 1220–1461, *Wydcombe* 1535, 1543, *Wyttecombe* 1512, *Witcombe* 1535, 'the wide valley', OE **wīd**, definite form **wīda**, + **cumb**. 'Great', Latin **magna**, for distinction from Little Witcombe SO 9115, *Wyd(d)ecombe* c.1300–1544, *Parva-* 1316, *Little Wittcomb* 1720. The earliest occurrence is *Widecomesege* 1121, 'Witcombe's edge', OE **ecg**, referring to the edge or scarp of the Cotswolds into which the two Witcombes form a wide semi-circular valley. Gl ii.158, 116, L 93.

WITHAM Essex TL 8114. Possibly 'homestead by the bend' sc. in the river Brain. *Wit ham* from c.925 ASC(A) under year 913 including 1086, *Witanham* c.1050 ASC(D) under same year, *Hwitham* c.1150, *Wi- Wyham* 1199–1346. OE **wiht** + **hām**. In the absence of any *Wiht-* spellings this explanation remains insecure: the forms hardly, however, support OE adj. **hwīt** or pers.n. *Hwīta*. Ess 299 gives pr [witəm].

WITHAM FRIARY Somer ST 7440. *The Friarye* 1610, *Witham Friary* 1817 OS. P.n. Witham + ModE **friary**. The site of the first English Carthusian monastery founded in 1178×9, and therefore inaccurately called a friary. Also known properly as *Witteham Charterhouse* [1238]Buck. Witham, *Witeham* 1086–1212, *Withe- Wutteham* 1212, *Witham* 1160, is probably 'Wit(t)a or the councillor's homestead or estate', OE pers.n. *Wit(t)a* or **wīta**, + **hām**. DEPN, Pevsner 1958N.341.

WITHAM ON THE HILL Lincs TF 0516. *Wytham super Montem* 1719, *Witham-on-the-Hill* 1821. P.n. *Withā -ham* 1086–1719 with variant *Wy-, Wi- Wyham* 1106–1303, the 'village by the bend' sc. in the r. Glen, OE **wiht** + **hām**, + ModE **on the hill**, Latin **super montem** for distinction from North and South WITHAM. Perrott 210.

North WITHAM Lincs SK 9221. *Nortuuine* 1086, *Nort Widhem* before 1160, *Norwime -wyme* 1154×89–1314, *North ~* 1226–1428, *North Wytheme -wythme -withom -wythom* 1259–1490, *-witham* from 1364. OE **north** + p.n. *Wime* 1086, *Wi- Wyme -a* c.1160–1261, *Widme* 1185, 1202, *Widham* 1235 transferred from the r.n. WITHAM SK 9363, TF 2548 used as a p.n. 'North' for distinction from South WITHAM SK 9219. Perrott 212.

River WITHAM Lincs SK 9363, TF 2548. Unexplained. *Wiðma* [c.1030]11th Secgan, [before 1085]12th, *Wi- Wydme* c.1150–14th, *Wi- Wytham* from [1115]13th, *Withme* 14th cent., *Wyme* 1224–1323. Ekwall's identification with Ptolemy's *Εἰδυμανίος* is almost certainly wrong and no etymology can be based on this form. Possibly a Gmc r.n. with the concrete forming suffix *-ma-* as in OE *strēam* < IE **sru-/*sreu-/*srou-* 'flow', but no obvious verbal root offers itself in Gmc or Celtic. Possibly, therefore, a pre-Celtic name. Perrott 41, RN 467 gives pr [wiðəm].

South WITHAM Lincs SK 9219. *Suthwyme* 1231–1428, *Suthwithme -wythme* 1265–1346, *Southwitham* from 1284. ME **suth** + p.n. *Wim(m)e, Widme* 1086 from the r.n. WITHAM SK 9363. 'South' for distinction from North WITHAM SK 9221.

WITHENS CLOUGH (RESERVOIR) WYorks SD 9823. *Withens Clough* 1843. P.n. *(le) Withens* 1314, 1642, 1843, 'the willow-trees', OE ***wīthiġn**, + ModE **clough** (OE *clōh*) 'a dell'. YW iii.163.

WITHERENDEN HILL ESusx TQ 6426. 'Wither's swine-pasture'. *Wytherenden(ñ) -in-* 1180–1296, *Wytheringden(ne)* 1279, 1307. Pers.n. **Wither(a)* + **ing**[4] + **denn**. Sx 454.

WITHERIDGE Devon SS 8014. 'Wether ridge'. *Wirige* 1086, *Wyrig'* 1242, *Wetherigge* 1249, *Wytherigge -rugge* 1256, 1262 etc. OE ***wither**, genitive pl. ***withra**, varying with **wether**, + **hrycg**. D 397, PNE ii.271.

WITHERLEY Leic SP 3297. 'Wigthryth's wood or clearing'. *Widredele* 1154×89, *-legh'* 1224, *Widredesly* before 1173, *Wyd- Wid- Wyth- Witherdele(e) -ley(e) -legh* c.1204–1403, *Wy- Witherle(e) -ley(e)* 1259–1576, *W(h)etherley* 1426–c.1555. OE feminine pers.n. *Wīġthryth*, genitive sing. *Wīġthrythe*, + **lēah**. Lei 55.

WITHERN Lincs TF 4382. Probably the 'wood house'. *Widerne* 1086, *-erna* c.1115, *Wierne* 1154×89, *Wihernia* 1199×1216. Probably OE **widu** 'wood' replaced by ON **vithr** + **ærn**. Alternatively this might be an *-ern* derivative of OE **wīthiġ**, ***wīthern** 'a clump of willows'. DEPN, SN 6.90ff.

WITHERNSEA Humbs TA 3427. 'Withorn lake'. *Witfornes, Widfornessei* 1086, *Wi- Wythorn(e)se* 13th–15th cent. Lost p.n. *Withorn* + **sǣ**. The lake has disappeared but is frequently referred to as *(lacus vocat') Wi- Wythornse Mar(re)* 'the pool of Withorn lake or fen' 1260 etc., p.n. *Withorn* + ON **marr** 'a pool, a fen, a marsh'. *Withorn* is recorded independently as *Wi- Wythorn* 12th–14th, probably the '(settlement) by the thorn-tree', OE **with** or ON **vithr** + **thorn**. YE 26 gives pr [wiðrənsi].

WITHERNWICK Humbs TA 1940. '*Withorn* dairy-farm'. *With- Wid- Witforneuuinc, Widforneuuic* 1086, *Wi- Wythorn(e)wik -wyk* 1190–1521, *Withernwyk* 1316, *Withrinwick* 17th. Lost p.n. *Withorn* as in WITHERNSEA TA 3427 + **wīc**. YE 69 gives pr [wiðrənwig].

WITHERSDALE STREET Suff TM 2681. 'W hamlet'. *Withersdale Street* 1836 OS. P.n. *Weresdel* 1086, *Wideresdal(a) -dale* 1184–1286, *Wi- Wytheresdal(e)* 1201–92, *Wetheresdale* 1254, *Wethersdale* 1523, 'Vitharr's valley', ON pers.n. *Vítharr*, lOE *Wither*, genitive sing. *Víthars, Witheres*, + ON **dalr**, OE **dæl**, + Mod dial. **street**. Alternatively the specific might be OE **wether** 'a wether'. DEPN, Baron, SPNN 442–3, L 95–6.

WITHERSFIELD Suff TL 6547. 'The wether's open land'. *Vrdres-* for *Vedres-*, *Wedresfelda* 1086, *Wetherisfeud* 1254, *Witheresfeud* 1235, *Wytheresfeld* 1242, *Withersfeilde* 1610. OE **wether**, genitive sing. **wetheres**, + **feld**. DEPN.

WITHERSLACK Cumbr SD 4383. Partly unncertain. *Wi- Wytherslak(e) -slack(e)* 1186–1672. This is either ON **vithr** 'a wood', genitive sing. **vithar**, or ON **vith** 'willow', genitive sing. **vithjar**, + **slakki** 'a hollow': either 'hollow of the wood' or 'hollow of the willow-tree'. We i.77, SSNNW 178, L 123, 222.

WITHERSLACK HALL Cumbr SD 4386. *A mansion house commonly called Wither Slacke* 1654, *Witherslack Hall* 1777. P.n. WITHERSLACK SD 4383 + ModE **hall**. We i.79.

WITHIEL Corn SW 9965. 'Wooded district, forest'. *Widie* 1086, *Widel* 1201, *Wythiel* 1274–1377, *Guythiel* 1355. Co **guyth** 'trees' + adjectival suffix **-yel**. Cf. LOSTWITHIEL SX 1059 which may have been part of the same district. PNCo 180, Gover n.d. 360, DEPN.

WITHIEL FLOREY Somer SS 9833. 'W held by the de Flury family'. *Wythele Flory* 1305. Earlier simply *Wiðiglea* [737]12th S 254, *938 S 443, *Withiclea* [956]12th S 596, *Withiglea* [961]10th S 697, *Wythel* 1237, *Wethihill* 1610, 'the willow-tree wood', OE **wīthiġ** + **lēah**. The manor was held by Randulfus de Flury in 1237. DEPN, L 203.

WITHINGTON 'Willow farm or settlement'. OE **wīthiġn** + **tūn**.
(1) ~ Ches SJ 8169. *Widinton -en-* 1185–6, *Withinton -en-* 1210, *-ing-* from 1249. Che i.88. Former pr [wiθitn].
(2) ~ Glos SP 0315. 'Widia's hill'. *Wudiandun* [736×7]11th S 1429, [774]11th S 1255, *Uuidiandun* [774]11th ibid., *Widiandune* [c.800]11th S 1556, *Widindvne* 1086, *Wi- Wydindon(e)* 1221–1328, *Wi- Wythindon(a) -yn- -done* 12th–1547, *Wythingdon -yng-* 1248–1439, *Wi- Wythinton -yn-* 1211×13–1314, *-ing- -yng-* 1573–1654. OE pers.n. *Widia, Wudġa*, genitive sing. *Widian, Wudġan*, + **dūn**. Gl i.186, L 149.
(3) ~ GMan SJ 8592. *Wy- Withington -yn-* 1212–1409, *Witheton* 1219, 1222, *Wythington -yng-* 1246, 1416. La 30, Jnl 17.30.
(4) ~ H&W SO 5643. *Widingtvne* 1086, *Wi- Wyt(h)inton(e)* 1228–1355. He 211.
(5) ~ Shrops SJ 5713. *Wientone* 1086, *Withentunie* c.1160×72, *Wi- Wythinton(') -yn-* 1242–1743, *Wythington* 1271×2, 1535, 1768, *Wi-* from 1646. Sa i.318.
(6) ~ GREEN Ches SJ 8071. *Withington Green* 1842. P.n. WITHINGTON SJ 8169 + ModE **green**. Che i.90.

WITHLEIGH Devon SS 9012. 'Withy wood or clearing'. *Witheleg(h)* 1219, 1316. OE ***wīthiġ** + **lēah**. Identical with Weethley Warw SP 0555 and WITHIEL Somer SW 9965. D 546.

WITHNELL Lancs SD 6422. 'Willow-tree hill'. *Withinhull* c.1160, *Withenhull* 1246, *Wythinhull* 1332, *Wynnell* 1580. Either OE ***wīthiġen** 'growing with willows' or ***wīthiġn** 'a willow, a willow copse', + **hyll**. The reference is to Pike Lowe SD 6222, a prominent hill rising to 720′. La 132, PNE ii.271, Jnl 17.75 gives local pr 'Winnell'.

WITHYBROOK Warw SP 4384. 'The willow-tree brook'. *Wythibroc* 12th–1316, *Withibroc* 1205–91. OE **wīthiġ** + **brōc**. Wa 121, L 15.

WITHYCOMBE Somer ST 0141. 'Willow-tree coomb'. *Hwithicumb* *[1065]c.1500 ECW (S 1042)[†], *Widicumbe* 1086, *Withicumbe* 1225, *Wethicomb* 1610. OE **wīthiġ** + **cumb**. DEPN.

WITHYHAM ESusx TQ 4935. 'Withy enclosure'. *Wideham* c.1095, *Withiham, Wydy- Wythyhamme* 13th cent. OE **wīthiġ** + **hamm** 2a 'promontory into lower land' or, if the original site is TQ 4936, **hamm** 3 or 6a 'river meadow, valley-bottom land'. Sx 370, ASE 2.28, SS 82.

WITHYPOOL Somer SS 8435. 'Willow-tree pool'. *Widepolle* 1086, *Widipol* 1185, *Wethipole* 1610. OE **wīthiġ** + **pōl**. DEPN.

WITHYPOOL COMMON Somer SS 8234. P.n. WITHYPOOL SS 8435 + ModE **common**.

WITLEY Surrey SU 9440. 'Witta's wood or clearing'. *Witlei* 1086, *-lea* 1174, *-le(e)* 1204–1370, *-lege -legh* 1291, *Wittelega* 1185, *Wytle(g) -lye* 1255–1320, *Wyt(t)e- Whit(e)- Whyt(e)- Wite-* 13th cent. with variants *-le(gh) -ley(e) -leg(he)* etc. OE pers.n. *Witta* + **lēah**. Sr 215.

WITLEY COURT H&W SO 7664. *Witley Court* 1832 OS. A Jacobean and Victorian mansion, since 1937 a burnt-out shell. P.n. Witley as in Great WITLEY SO 7566 + ModE **court** 'a large house, a manor house'.

Great WITLEY H&W SO 7566. *Whitele Major* 1275, *Great Witley* 1832 OS. ModE adj. **great**, Latin **major** 'greater', + p.n. *Wytleye* 1290, 1307, *Wyttel(ie) -eleye* 1316–1428, as in Little W SO 7863. Wo 86.

Little WITLEY H&W SO 7863. *Parva Wyttelege* 1249, *Litelwytele* 1388. ME adj. **litel**, Latin **parva**, + p.n. *Ƿittlæge* [964]12th S 731, *ƿitleah -g* [969]11th S 1323, *(æt) ƿitlea(ge)* [975]11th S 1372, *Wihtlega* 11th Heming, *Witlege* 1086, 'wood or clearing in the recess', OE ***wiht** + **lēah**. The reference is to the recess in the hills to the W between Woodbury Hill SO 7464 and Abberley Hill SO 7667. Wo 183, Hooke 145, 278, 299.

WITNESHAM Suff TM 1850. 'Wittin's homestead'. *Witdesham* 1086, *Witlesham* 1195–1286, *Witnesham* from 1254 with variant *Wy-*. OE pers.n. **Wittīn*, genitive sing. **Wittīnes*, + **hām**. Sps with *-les-* are either due to AN *l/n* liquid confusion or evidence of an alternitive diminutive of *Witta*, **Wittel*. DEPN, Baron.

WITNEY Oxon SP 3509. 'Witta's island'. *(æt) Wyttanige* [969]12th S 771, *(to) Wittannige* [1044]12th S 1001, *Witenie* 1086–1285 with variants *-(e)ia -ega* and *-e(ye), Witten(eia) -e(ga) -ay -eie -eya* etc. 1151–1471, *Wytney* 1316. OE pers.n. *Witta*, genitive sing. *Wittan*, + **(ī)eġ**. There is mention of *Wittan mor* 'Witta's marsh' in the bounds of S 771 [969]12th. O 333.

Little WITTENHAM Oxon SU 5693. *Little Wittenham* 1830. ModE adj. **little** + p.n. *Wittanham* [862]12th, *Withennam* [862]13th S 335, *Wittanhamme* [892×99]12th S 355, *Witeham* 1086, *Wittenham* from c.1180, 'Witta's river-bend land', OE pers.n. *Witta*, genitive sing. *Wittan*, + **hamm** 1. Also known as *Estwittenham* 'east W' 1338, 1343, for distinction from Long WITTENHAM SU 5493, and as *Wytte(n)ham Abbatis* 'abbot's W' 1284 as a possession of the abbot of Abingdon. Brk 427, L 43, 49.

Long WITTENHAM Oxon SU 5493. *Longe Witnam al. Witnam Counts* 1548×9. ME **long** + p.n. Wittenham as in Little WITTENHAM SU 5693. Also known as *Witteham comitis Giffardi* 1177, *West Wittenham Earls* 1301, *Wittnam Comity* 1517 from its possession by the Giffard family in the 11th and 12th cents., and later by the earls of Pembroke and Gloucester. Brk 427.

WITTERING Cambs TF 0502. 'The Witheringas, the people called after Wither'. *Witering(a)* 1166–1428, *Witeringes* 1202×3, *Wythering* 1227, *Wittering* 1284, 1346. OE **Witheringas* < pers.n. **Wither*, a short form of a name like *Witherġild*, + **ingas**. The earliest reference to the Witheringas is to the 600 hides in the Tribal Hidage called *Widerigga land* 7th. In DB the place is called *Wit(h)eringaham* 1086 'the homestead of the Witheringas', and elsewhere 'the island of the Witheringas', *(æt) Wiðering ige, (on) Wiðeringa eige, (wið) Wyðeringa ige* B 1130, *Widringaiȝ, Wiðringaig* [c.973]14th S 792, genitive pl. **Witheringa* + **ēġ**. The same people are recorded in the nearby name WERRINGTON Northants TF 1703. Nth xiv, 247, ING 69.

East WITTERING WSusx SZ 7996. *Estwightryng* 1320, 1400. ME **est** + p.n. *Wihttringes* *[683]14th S 232, *Wystrings* [733×54]14th S 46, *Westringes* 1086, 1166, *Witteringes* 1227×33, *Wytering(e)* 1241, 1261, 'the Wihtheringas, the people called after Wihthere', OE folk-n. **Wihtheringas* < pers.n *Wihthere* + **ingas**. 'East' for distinction from West WITTERING SZ 7898. Sx 87, ING 41, ASE 10.23, SS 64.

West WITTERING WSusx SZ 7898. *Westwyghtryngg'* 1292. ME adj. **west** + p.n. Wittering as in East WITTERING SZ 7996. Sx 87.

WITTERSHAM Kent TQ 8927. 'Wihtric's promontory'. *(æt) Wihtriceshamme* [1032]13th S 1465, *Wihtricesham* c.1090, *Wit(t)ricesham* 1226, 1233. OE pers.n. *Wihtrīċ*, genitive sing. *Wihtrīċes*, + **hamm** 2. KPN 319, ASE 2.22, Cullen.

WITTON H&W SO 8962 → NORTHWICH Ches SJ 6573. 'Enclosure, settlement at Wich or at a *wīċ*'. OE **wīċ** + **tūn**.

WITTON 'Wood settlement'. OE **widu** + **tūn**, often referring to a specialised settlement concerned with wood-felling, woodworking, and the supply of timber and fuel, and playing a specialised role in the economy of a manor or estate. **widu** is the earlier form of OE **wudu**. Cf. WOTTON, WOOTTON.
(1) ~ BRIDGE Norf TG 3431. Simply called *Stone Brigg* 1838. OS P.n. Witton + ModE **bridge**. Witton, *Wittuna, Widituna* 1086, *Witton* from 1194, is the 'wood settlement', OE **widu-tūn**. Nf ii.201.

[†]This is a better reading from MS 3 than DEPN's *Hwidigcum*.

(2) ~ GILBERT Durham NZ 2345. 'W held by Gilbert'. *Witton Gilbert* from 1382. The estate was granted to Durham Abbey by Gilbert the Sheriff before c.1185. Earlier simply *Wyttone iuxta Dunelmum* [c.1185×95]13th, *Witton' -y-* [1183]c.1320–1524, ~ *next the Beyrpark* 1524). Pronounced [witən dʒilbət]. NbDu 218.

(3) ~ -LE-WEAR Durham NZ 1431. 'W in Weardale or on the Wear'. *Wotton' in Werdale* 1348–1422, *Wy- Witton in Werdale* 1408, ~ *in Wardalle* 1416, *Witton-uppon-Weire* 1587, ~ *upon Weare* 1590, ~ *le Wear* 1862 OS. Earlier simply *Wudutun* [c.1040]12th, *Wudetun* [1072×83]15th, c.1107, *W(u)ttun* 1163×80, c.1200, *Wotton'* [c.1294]18th–1422, *Wittun* c.1261, *Wy- Witton* 1303–1587. NbDu 218.

(4) ~ PARK Durham NZ 1730. *Witton Parke* 1647. P.n. WITTON + ModE **park** A modern industrial settlement named from the park of Witton Castle.

(5) East ~ NYorks SE 1486. *E(a)stwitton* 1156, *Est Wotton* 1316. ME adj. **est** + p.n. *Witun(e) -tone* 1086, 1204, 1396, *Wittun -ton* c.1150–1298. YN 249.

(6) West ~ NYorks SE 0688. *West Witton* [12th]15th. ME adj. **west** + p.n. *Witun* 1086, *Widtona* 1166. 'West' for distinction from East WITTON SE 1486. YN 255.

WIVELISCOMBE Somer ST 0827. 'Beetle or Wifel's coomb'. *Wifelescumb* *[854]12th S 311, *Wifelescombe* *[1065]c.1500 ECW (S 1042)†, *Wivelescome* 1086, *-comba* [13th] Buck, *Wyvelescomb(e)* [13th–1397] Buck, *Wiuelscomb* 1610. OE **wifel**, genitive sing. **wifeles**, or pers.n. *Wifel*, genitive sing. *Wifeles*, + **cumb**. DEPN.

WIVELSFIELD ESusx TQ 3420 → WIVELSFIELD STATION TQ 3219.

WIVELSFIELD GREEN ESusx TQ 3519. *Wivelsfield Green* 1813 OS. P.n. WIVELSFIELD TQ 3420 + ModE **green**.

WIVELSFIELD STATION WSusx TQ 3219. P.n. *Wifelesfeld* [c.765]13th S 50, *Wiu- Wyvelesfeud -feld(e)* 1235–1580, *Wylsfelde* 1580, *Weevelsfield* 1637, 1651, 'the weevil's open land', OE **wifel** or pers.n. *Wifel* derived from it, genitive sing. **wifeles**, *Wifeles*, + **feld**. Sx 305, SS 70.

WIVENHOE Essex TM 0321. 'Wifa's hill-spur'. *Wiunhov* 1086, *Wyvenho(o) -howe* 1238–1548. OE pers.n. **Wifa*, genitive sing. **Wifan*, + **hōh**. Ess 403 gives pr [wivnou].

WIVENHOE CROSS Essex TM 0324. Probably 'W crossroads'. *Wivenhoe Cross* 1805 OS.

WIVETON Norf TG 0443. 'Wife or Wifa's estate'. *Wiuentona, Wiuetuna* 1086, *Wiventone* 1254, *Wyveton* 1242. OE pers.n. *Wife* or *Wifa*, genitive sing. *Wifan*, + **tūn**. DEPN.

WIX Essex TM 1628. 'The dwelling places or dairy farms'. *Wykes* 1238–1309 including [1022×43]13th S 1537, *Wicā* 1086, *Wika, Wyke* 1195–1254, *Wikes* 1200–1421, *Weekes* 1720, *Wickes* 1805 OS. OE **wīċ**, ME pl. **wikes**. A weak pl. also occurs, *Wykyn* 1504. Ess 357.

WIXFORD Warw SP 0954. 'Wihtlac's ford'. *Wihtlachesforde* *[962]12th S 1214, *Witelavesford* 1086, *Wichtlakesford* [c.1086]1190, [1155]1340, *Wihtlaxford* 1287, *Witlakesford* 1200–1251, *Wykesford* 1262, *Wicksford* 1656. OE pers.n. *Wihtlāc*, genitive sing. *Wihtlāces*, + **ford**. Wa 227, L 69.

WIXOE Suff TL 7142. 'Widuc's hill-spur'. *Wlteskeou* for *Witkes-* 1086, *Widekeshoo* 1205, *Wydekesho* 1219, 1326, *Wicksoo* 1610. OE pers.n. *Widuc*, genitive sing. *Widuces*, + **hōh**. DEPN.

WOBURN Beds SP 9433. 'Winding stream'. *Woberne -burne* 1086, *Wuburn* 13th cent. OE **wōh** + **burna**. The boundary of the people of Woburn is mentioned [969]11th S 772 as *land gemœru Þōburninga*. Bd 143 gives pr [wu:bən], L 18.

WOBURN ABBEY Beds SP 9632. *Woburn Abbey* 1834 OS. P.n. WOBURN SP 9433 + ModE **abbey**. The seat of the duke of Bedford dating from the 17th and 18th cents.

WOBURN SANDS Bucks SP 9235. *Woburn Sands* 1834 OS. P.n. WOBURN Beds SP 9433 + ModE **sand(s)**.

†This is the reading of MS 3 compared with DEPN's later *Wyfelescumbe*.

WOKEFIELD PARK Berks SU 6765. Cf. *Oakfield House* and *Green* 1817 OS. P.n. Wokefield + ModE **park**. Wokefield, *(æt) Weonfelda* [946×51]13th S 578, *Hocfelle, Offelle* 1086, *Weke- Wechefelda* 1167, *Wegh(e)- Wog(h)efeld(')* 13th cent., *Wo(g)hfeld* 1325–74, *Wocfeld(')* 1220–55, *Wokefeld* 1367, *Wookefeld* 1552, *Wokefield* 1846, is probably '(at) the holy field', OE ***wēoh** 'holy', dative sing. definite form **wēon**, + **feld**, dative sing. **felda**. Alternative suggestions, OE pers.ns. **Wocca* or **Weohha*, present major phonological difficulties. The early forms suggest the solution offered with later *-o(o)-* spellings resulting from accent shift *éo* > *eó*. The 19th cent. forms in *Oak-* show dial. loss of *w* before [u:] < ME *ō* parallel to the *Ok-* forms of WOKINGHAM SU 8068, followed by popular association with ModE *oak* 'an oak-tree'. Brk 227, DEPN, L 324, Dobson 421.

WOKING Surrey TQ 0058. 'The Wocingas, the people called after Wocc or Woca'. *W- Uuocchingas* [708×15]c.1200 B 133, *Wocingas* 1121 ASC(E) under year 777, *Wocc-* [757×96]c.1150 S 144, *Wochinges* 1086–1156, *Wok-* 1155–1233, *Wokk-* 1166, *Wockinges -ynges* 13th, *Wokinge -y-* 1154–1540, *Wockinga -e* 1172–1499, *Ok(k)yng* 1474, 16th cent., *Oking* 1541, 1675, *Woakeing* 1693. OE folk-n. **Woc(c)ingas* < pers.n. **Wocc(a)* + **ingas**. Sr 156, ING 30.

WOKINGHAM Berks SU 8068. 'Homestead of the Wocingas, the people called after Woca'. *Wokingeham* [1146]15th–1228, *Wockingeham* [c.1179×80]c.1280, *Wokkyngham* 1284, *Wokingham* from [1146]copy, *Okynham* 1517, *Okingham* 1600. OE folk-n. **Wocingas* < pers.n. **Woc(a)* + **ingas**, genitive pl. **Wocinga*, + **hām**. The *Wocingas* of Wokingham are probably identical with those of WOKING Surrey TQ 0058, some 14 miles away. The 1517 and 1600 spellings in *Ok-* show the same dial. loss of initial *W* before ME *ō* as the *Oak-* spellings of WOKEFIELD. Brk 139, 815, 840, ING 127, DEPN, Dobson 421.

WOLD FELL Cumbr SD 7885. *Wold Fell* 1858, cf. *Wouldfoote* 1592. ModE **wold** 'open high ground' + **fell** (ON *fjall*). YW vi.259.

WOLDINGHAM Surrey TQ 3756. 'The homestead of the Wealdingas, the people called after Wealda'. *Wallingehā* 1086, *Walding(e)ham* 13th cent. with variant *-yng(e)-*, *Walryngham* 1285, *Woulding-* 1593, 1674, *Wold-* from 1608. OE folk-n. **Wealdingas* < pers.n. *Wealda* + **ingas**, genitive pl. **Wealdinga*, + **hām**. Sr 339–40, ING 123, ASE 2.32.

The WOLDS 'The forest land, especially the high forest land'. OE (Angl) **wald**. Later used of any wasteland in lofty country.

(1) ~ Humbs SE 9763. *Waldas* 1303, *Waldo* 1322, *Waldam* c.1400, *the Woolde* c.1540, *the Woolds* 1652, *York(e)s Wold* 1551, 1695. Cf. HOLME ON THE WOLDS SE 9646, MIDDLETON-ON-THE-WOLDS SE 9449, and Wold NEWTON TA 0473, TF 2498.

(2) ~ Lincs TF 2583 → CUXWOLD TA 1701, STIXWOULD TF 1766.

River WOLF Devon SX 4290. No early forms.

WOLF FELL Lancs SD 6045. ModE **wolf** + **fell** (ON *fjall*). Wolf Fell is half a mile N of Wolfen Hall SD 6044, *Wolffehall* 1600, *Woolfhall* 17th. La 143.

WOLF ROCK Corn SW 2711. 'Whirlpool rock'. *une basse qui sappelle la roussee et en bretō goulff* c.1485, translated (and wrongly associated with RUNNEL STONE SW 3620) as *a Rocke called Reynolde stone called golfe in brytysshe'* 1528, *gofre* 1548, *The Gulf(e)* 1564–1801, *De Wolff* 1584, *the Gulf or Woolfe* c.1698, *The Gulf, by some call'd the Wolf Rock* 1744 Gover n.d., *the Wolf* 1817. lME **gulf** + **rock**. According to popular etymology the modern form is said to have been so called 'from the continued and melancholy howling which the waves make in breaking round it' 1817 (PNCO 180); it appears to have been a Dutch sailors' substitution finally established through the standardising influence of British Admiralty charts. Alternatively the name may allude to the rock's dangerous nature. The forms with *G-* may, however, point to OCo **guillua** 'look-out place' as the ultimate origin, parallel to W *gwylfa* in Wylfa Head Anglesey SH 3594. Also known as *losei -ey* 1327–1552, *petra lusia* 1490, 1562, possibly 'loss, destruction island', OE **los** + **ēġ**, referring to shipwrecks here; and as *la (basse) rossée* 'wave-beaten reef', probably a phonological alteration of *losey*. PNCo 180, Gover n.d. 642, Jnl 24.13–29.

WOLFERLOW H&W SO 6761. 'Wulfhere's tumulus'. *Vlferlav* 1086, *Wulferlaw'* 1160×70, *Wolfrelowe* 1303. OE pers.n. *Wulfhere* + **hlāw**. He 212.

WOLFERTON Norf TF 6528. 'The Wulfhere estate'. *Wulferton* 1166f, 1196. OE pers.n. *Wulfhere* + **tūn**. DEPN.

WOLFHOLE CRAG Lancs SD 6357. 'Crag at Wolf's Lair'. *Wolfalcrag, Wlffalcragge* c.1350, *Wulfo crag, Wulfcragge* 1577, *Westwofa Cragg* 1652, *Wool Fell Craig* 1771. P.n. **Wulf-halh* 'wolf's nook' or **Wulf-hol* 'wolf's lair' + ME **cragge** 'a crag, a rock' (Ir *creag*). The reference is to a rock outcrop on Tarnbrook Fell at a height of 1731ft. on the old boundary between Lancashire and Yorkshire. La 170, YW vi.215.

Great WOLFORD Warw SP 2534. *Magna Woluuarde* 1247, *Wolforde magna* 1327, *Much, Mychell Wulford* 1518, 1544, *Greate Woollford* 1695. ModE adj. **great**, Latin **magna**, ME **moche, michel**, + p.n. *Vlware, Vlwarda, Vol- Worwarde* 1086, *Wlfuuardia* c.1140, *W(u)lfward* 1138×47, 1212, *Wlfarde, Wlford* 1199, *Wolford* from 1351, 'the wolf watch, the watching place for wolves', OE **wulf** + **weard** early assimilated to OE **ford** 'a ford'. 'Great' for distinction from Little WOLFORD SP 2635. Wa 302.

Little WOLFORD Warw SP 2635. *Parva Wlward* 1227, *Wolfarde parva* 1279. ModE adj. **little**, Latin **parva**, + p.n. Wolford as in Great WOLFORD SP 2534. Wa 302.

WOLLASTON Northants SP 9063. 'Wiglaf's farm or village'. *Wilavestone* 1086, *Willeveston* 1219, *Wullaueston(ia)* c.1150–1274 with variants *Wol(l)- Wul-*, *Wollaston* from 1279. OE pers.n. *Wiġlāf*, genitive sing. *Wiġlāfes*, + **tūn**. Later spellings with *Wu- Wo-* may point to the alternative pers.n. *Wulflāf* or to rounding of shortened *i* after *w*. Nth 197, DEPN.

WOLLASTON Shrops SJ 3312. 'Wiglaf's estate'. *Willavestune* 1086, *Willaueston'* c.1200–55, *Wa- Wullaueston(')* 1203×4, *Wi- Wylaston(') -ll-* 1242–1695, *Wollaston(')* from 1305×6, *Woolleston -a-* 17th cent. OE pers.n. *Wīġlāf*, genitive sing. *Wīġlāfes*, + **tūn**. Sa i.319.

WOLLERTON Shrops SJ 6230. 'Wulfrun's estate'. *Uluretone* 1086, *Wluretona -ton(e)* 1121–c.1275, *Wluruntona -ton'* c.1135, c.1144, *Wulfrinton(')* 1200, 1240, *Wolverton(')* c.1291–late 14th, *Wollerton* from 1535. OE feminine pers.n. *Wulfrūn*, genitive sing. *Wulfrūne*, + **tūn**. Sa i.319.

WOLSINGHAM Durham NZ 0737. 'Homestead called after Wulfsige'. *Wlsingham* c.1148–1309, *Wulsingeham* 1197, *Wulsingham -yng-* [1183]c.1320–1558, *Wolsingham* from 1311, *Wals-* 1313–1472, *Wools-* 1558, 1573, 1612×3. OE pers.n. *Wulfsiġe* + **ing**[4] + **hām**. Pronounced [wulsiŋəm]. NbDu 218.

WOLSINGHAM PARK MOOR Durham NZ 3040. *Wolsingham Park Moor* 1862 OS. P.n. Wolsingham Park + **mōr**. The Bishop of Durham had a hunting park at Wolsingham in the Middle Ages, *Wolsingham Park(e)* 1582, 1647. The park is mentioned in Bishop Hatfield's survey of episcopal estates made in 1382.

WOLSTON Warw SP 4175. 'Wulfric's estate'. *Vlvrice- Vlvestone* 1086, *Wlfricheston(e)* 1086×94, 1100×35, *Wulfriceston(e)* 1227, 1235, *Wolfricheston* 1309–1487, *Wlfreston* 1252, *Wolfreston* 1395, *Wul(f)ston* 1535, *Woolston* 1673, *Woolson* 1698. OE pers.n. *Wulfrīċ*, genitive sing. *Wulfrīċes*, + **tūn**. Wa 151.

WOLVERCOTE Oxon SP 4909. 'Wulfgar's cottages'. *Vlfgarcote* 1086, *Wlgaricote* c.1130–1389 with variants *-cot(a), Wolgarcote* 14th–15th cents., *Wulvercot* c.1185, *Woolvercote* 1675. OE pers.n. *Wulfgār* (probably) (+ **ing**[4] + **cot**, pl. **cotu**. O 33–4.)

WOLVERHAMPTON WMids SO 9198. 'Wulfrun's Hampton'. *Wolvrenehamptonia* 1074×85, *Wulfrunehanton* 1169. OE feminine pers.n. *Wulfrūn*, genitive sing. *Wulfrūne*, + p.n. HAMPTON SO 9198, *(æt, into) Heantune* [985]12th S 860, c.1000 S 1534, 'the high settlement'. Wulfrun is the name of the lady who received the estate from King Æthelred in 985. In 996 she gave it and other estates to endow a monastery she was founding in Wolverhampton. When OE *Hēan-tūn* became *Hampton* her name was prefixed for distinction from other Hamptons. DEPN, TC 195.

WOLVERLEY H&W SO 8379. 'Wulfweard's wood or clearing'. *Ƿluardele* [700 for 716×57]1705 S 1826, *Ƿlfardileia* [864 for 716×57]1705 S 1827, *Ƿulfferdinleh* [866]18th S 212, *DE UULFUUARILEA, uulfordilea, (æt) pulfpeardig lea, (æt) Ƿulfferdinlea* *[866]11th S 211, *(æt) Wulfweardiglea* 1052×7 S 1232, *Vlwardelei* 1086, *Wolffardeleye* 1279, *Wolvardelegh -ley* 1292–1411, *Wolverley* 1535. OE pers.n. *Wulfweard* + **ing**[4] + **lēah**. S 212 is a grant of land by Burgred king of Mercia to Wulfred; it would be tempting to believe that this is the man whose name is preserved in the p.n., but the forms and dates of S 1826–7 disprove this. Wo 256, Hooke 27, 120–3, ECWM 91.

WOLVERLEY Shrops SJ 4731. 'Wulfweard's clearing'. *Ulwardelege* 1086, *Woluardesl'* 1255×6, *Wluardele* 1271×2, *Wolfardeleye* 1380, *Wulverley* 1446, *Wolverley* from 1586×7. OE pers.n. *Wulfweard* + **lēah**. The pers.n. means literally 'wolf guard, wolf watchman'. Sa i.320.

WOLVERTON 'Estate called after Wulfhere'. OE pers.n. *Wulfhere* + **ing**[4] + **tūn**.

(1) ~ Bucks SP 8140. *Wlverintone* 1086, *Wulfrinton* 1195, *Wulver- Wulvre- Woluerington* 1227, *Wolvertone* 1302. Bk 27.

(2) ~ Hants SU 5558. *Ulvretune* 1086, *Ulferton(a)* 1159–62, *Wu- Wolferton(a)* 1167–1389, *Wlfrin(c)ton* 1248, 1272, *Wolverton* from 1316. OE pers.n. *Wulfhere* + **tūn** varying with *Wulfhere* + **ing**[4] + **tūn**. Ha 179, Gover 1958.150.

WOLVERTON Warw SP 2062. 'The estate called after Wulfweard'. *Vlwarditone* 1086, *Wlfwardinton'* c.1200, *Wolwardyn(g)ton* 1273–1335, *Wolverdyngton* 1336–1632, *Wolverton* 1545. OE pers.n. *Wulfweard* + **ing**[4] + **tūn**. In the same area was the lost p.n. *Wolwardinghul, Wuluardinghulle* 13th with OE **hyll** 'a hill'. Wa 228.

WOLVEY Warw SP 4387. Probably 'wolf island'.

I. *Vlveia* 1086, *Wulfeie* 1195, *Wu- Wolvey -eie* from 1242.

II. *Wulfheia* 1195–1221, *Wu- Wolfheye* 1230–1346.

Probably OE **wulf** + **ēġ**, but type II may point to **wulf** + **(ġe)hæġ** 'enclosure'. Wa 122.

WOLVISTON Cleve NZ 4525. 'Wulf's farm or village'. *oluestona* [1091×2]12th, *wlfestuna* 1114×28, 1153×95, *Wluestun(a)* 1128–c.1350, *-ton'* 1218×34–1326, *Wolueston(')* c.1250–1682, *Wolviston* from 1296, *Wlestu'* early 13th, *W(u)lleston'* 1256–1370, *Wlston* 1322–71, *Wolston* c.1340–1668, *Woulston(e)* 1573–1614, *Wuston* 1576, *Woo(w)ston* 1579, *Woston(e)* 1580, 1695. OE pers.n. *Wulf*, genitive sing. *Wulfes*, + **tūn**. This name, for which there is an exceptionally rich run of forms illustrating the local pronunciation, commemorates the unidentified donor of the estate to Durham Priory.

WOMBLETON NYorks SE 6683. 'Wynn- or Winebald's settlement'. *Winbel- Wilbetun* 1086, *Wimbeltun* [c.1159]16th–[c.1250]13th, *Wi- Wymbelton(a)* 1231–1385, *Wimbleton* 1301. OE pers.n. *Wynn-* or *Winebald*, + **tūn**. YN 67.

WOMBOURNE Staffs SO 8793. '(The settlement) at the winding stream'. *Wambvrne* 1086, 1271, *-burn(a)* 1167, 1175, *Womburne -borne* 1236, 1242. OE **wōh**, definite form dative case **wōn**, + **burna**. The development must have been *Wōnburne* > *Wŏn-* > *Wanburn* beside *Wōnburne* > *Wōm-* > *Wŏm-* > *Wamburn* with same change *ōm* > *ǎm* as in OE **brōm**-names. Alternatively the specific might be OE **wamb** 'a womb, a belly' in the transferred sense 'a hollow' which would fit the topography here. DEPN, Duignan 175, PNE ii.272, L 17, 18., Horovitz.

WOMBRIDGE Shrops SJ 6911 → WOMBWELL SYorks SE 3902.

WOMBWELL SYorks SE 3902. Possibly 'Wamba's spring'. *Wanbella, -buelle* 1086, *Wambewell(e)* 12th–1316, *Wamb-* 13th–1550, *Wam-* 13th–1428, *Wombe-* 13th–1550, *Womb-* from

13th. OE pers.n. **Wamba* + **wella**. Alternatively the specific could be OE **wamb**, 'a womb, a belly', used in some topographical sense such as 'hollow' or 'lake': cf. WAMBROOK Somer ST 2908, Wombridge Shrops SJ 6911, *Wombrugga* 1181, *Wambrigg* 1207, which may have referred to a small lake here, and the Swed lake n. *Våmbsjön*. YW i.102 gives a pronunciation [wumwel].

WOMENSWOLD Kent TR 2250. 'Forest of the Wimlingas'. *Wimlincga wald* [824]?10th S 1266, *Wimlingweald* c.1090, *Wymelingwaud -wald' -wold'* 1254, 1265, *Wymingswold* 1610. OE folk-n. **Wimlingas* 'the people called after Wimel(a)' < OE pers.n. **Wimel(a)* + **ingas**, genitive pl. **Wimlinga*, + **weald**. PNK 539, KPN 160, Cullen.

WOMERSLEY NYorks SE 5319. 'Wilmer's forest clearing'. *Wilmereslege* 1086, *Wi- Wylmersley(e) -lay* 1286–1428, *Wolmer(s)ley* 1311, 1313, *Wymers-* 1316, *Wommers-* 1501, *Womersley* from 1556. OE pers.n. **Wilmēr*, genitive sing. **Wilmēres*, + **lēah**. YW ii.54 gives pr ['wumǝzlǝ].

WONERSH Surrey TQ 0245. '(The seetlement) at the crooked ploughland'. *Woghenhers* 1199, *-ers* 13th cent., *-ersh* 1305, *Wunhers* 1224, *Uners* 1249, *Wonerse* 1258, *-ersh* from 1334 with variants *-erche -arshe* etc., *Wowenersch* 1279, *Onerssh* 1430, *Ognershe* 1571. OE **wōh**, dative sing. definite form **wōgan**, **wōn**, + **ersc**. Sr 253, PNE ii.272, L 235.

WONSON Devon SX 6789. Possibly 'at the crooked stone'. *Wnston* 1238, *Woneston* 1244, 1254, *atte Wonston* 1374. Possibly OE *æt thæm wō(ga)n stāne*, from **wōh**, dative sing. definite form, **wōgan**, **wōn**, + **stān**. D 453.

WONSTON Dorset ST 7408. 'Wulfmær's estate'. *Wolmerston'* 1280, *Wom(e)ston* 1580, 1607, *Wonston* 1774. OE pers.n. *Wulfmǣr*, genitive sing. *Wulfmǣres*, + **tūn**. Do ii.102.

WONSTON Hants SU 4739. 'Wynsige's estate'. *(on) wynsiges tune* [900]11th S 360, *Wenesistune* 1086, *Wensi(e)ston* 1205–27, *Won(s)sinton(e) -yn-* c.1124–1331, *Wonston alias Wonsyngton* 1443, *Wonson al. Wonsyngton* 1753. OE pers.n. *Wynsiġe*, genitive sing. *Wynsiġes* varying with *Wynsiġe* + **ing**[4] + **tūn**. Ha 179, Gover 1958.178 gives pr [wʌnstǝn].

South WONSTON Hants SU 4635. A modern development not recorded by 1817 OS. ModE adj. **south** + p.n. WONSTON SU 4739.

WOOBURN Bucks SU 9087. 'Stream with walled banks'. *Waburna* c.1075, *Waborne* 1086, *Wauburn* 1200, *Wo(u)burn(e)*, *Wuburn*, *Wyburne* 13th cent., *Woghburne* 1331, *Oburn* 1487, 1490, *Obornes* 1675. OE **wāg** + **burna**. The early spellings without *-gh-* are surprising: otherwise the normal Southumbrian rounding of *ā* > *ō* and vocalisation of the group *ōg* > *ou* are evidenced. Cf. WEYBOURNE Norf TG 1143. The later forms may be compared with WOBURN Beds SP 9433 and OBORNE Dorset ST 6518, 'crooked stream', OE **wōh** + **burna**. Bk 196 gives pr [u:bǝn], PNE ii.238–9, L 18.

WOOBURN GREEN Bucks SU 9189. *Woburn Green* 1822 OS. P.n. WOOBURN SU 9087 + ModE **green**.

WOODALE NYorks SE 0279. 'The wolves' valley'. *W(u)lvedale* 1223. OE **wulf**, genitive pl. **wulfa**, + **dæl**. YN 254.

WOODBASTWICK Norf TG 3315. 'Wood Bastwick'. *Wodbastwyk* 1253. ME **wode** (OE *wudu*) + p.n. *Bastwik* *[1044×7]13th S 1055, *Bastuuic* 1086, 'the bast farm', OE **bæst** + **wīć**. Bast is the bark of lime-trees used for rope-making, which is probably the reference here, although **bæst** might simply mean 'lime-tree'. 'Wood' for distinction from BASTWICK TG 4217. DEPN.

WOODBECK Notts SK 7777. 'The wood stream'. *Wodebec* c.1300. OE **wudu** + ON **bekkr**. Nt 58.

WOODBOROUGH Notts SK 6247. 'The fortified place by the wood'. *Ude(s)burg* 1086, *Wudeburc* 1168, *Wodeburg(h)* 1211–1302, *Woodburgh or Wodborow* 1490. OE **wudu** + **burh**. The reference is to the fortification at Fox Wood SK 6148. Nt 180.

WOODBOROUGH Wilts SU 1159. 'The wood hill'. *Wideberghe* 1208, 1242, *Wodeberg(e) -bore -burgh -berwe -bergh -borgh* c.1220–1344, *Woodborowe* 1594. OE **widu** varying with **wudu** + **beorg**. Wlt 327, L 128, 228.

WOODBRIDGE Suff TM 2747. 'The wood bridge', probably 'the bridge by or leading to the wood' rather than 'the wooden bridge'. *Oddebruge* [1042×66]12th S 1051, *Wudebrige -bryge*, *Udebriga -e -bryge*, *Wdebrige* 1086, *Wodebreg(g)e -brugge -brig(g) -brigge -brygge* 1254–1463, *Woodbrygge* 1523, *Woodbridge* 1610. OE **wudu** + **brycg**. DEPN, Baron.

WOODBURY Devon SY 0187. 'Wood fort'. *(on) Wudebirig* 1072×1103, *Wodeberie* 1086, *-byr(e) -biri* 1242–86. OE **wudu** + **byriġ**, dative sing. of **burh**. The reference is to Woodbury Camp Iron Age hill-fort SY 0387. D 601, Thomas 1976.97.

WOODBURY HILL H&W SO 7564. *Woodbury Hill* 1832 OS. P.n. Woodbury + ModE **hill**. Woodbury, *Oldbury* 1275, 1327, is the 'ancient fort' referring to the Iron Age hill-fort that occupies the summit of this 904ft. hill. OE **ald** + **byriġ**, dative sing. of **burh**. The modern form is either a new name or a continuation of the medieval name showing the dial. development of *w* before ME *ǫ* as in ModE [wûn] for 'one', and dial. forms such as *wuts* for 'oats' and 15th cent. *wold* for 'old'. Dobson §431 and notes, EDG §236, OED s.v. old, Thomas 1976.152.

WOODBURY SALTERTON Devon SY 0188 → Woodbury SALTERTON.

WOODCHESTER Glos SO 8402. 'Roman settlement in the wood'. *Uuduceastir* [716×45]11th with variant spelling *-cester* S 103, *Wuduceaster* [896]11th S 1441, *Widecestre* 1086, *Wodecestr(e)* 1220–1376, *-chestr(e)* 1297–1436, *Woodcestre* 1287, *-chestre -er* 1465, 1535, *Woodchester* from 1584. OE **wudu** + **ċeaster**. Remains of a Roman villa and tessellated pavement have been found just N of the village. The DB form *Wide-* is a vestige of OE **widu**, the archaic form of **wudu**. Gl i.115.

WOODCHURCH Kent TQ 9434. 'Wood church'. *Wudecirce* c.1100, *Wo- Wudechirche* 1211×12–1261×2. OE **wudu** + **ċiriċe**. The sense is the 'church by the wood', referring to the S edge of the High Weald. PNK 364, Gelling 1981.8.

WOODCOTE 'The cottages at the wood'. OE **wudu** + **cot**, pl. **cotu**.

(1) ~ Oxon SU 6481. *Wdecote* 1109–1399 with variants *Wude-* and *-cota -kote*, *Woodcott* 1639. O 157.

(2) ~ Shrops SJ 7615. *Udecote* 1086, *Wudecota -cote* 1177–1204, *Wodecot(e)' -kote* 1200–1466, *Woodcote* from 1577. Sa i.321.

WOODCROFT Glos ST 5495. 'Woodland messuage'. *Wodecroft* 1360. OE **wudu** + **croft**. Gl iii.267.

WOODDITTON Cambs TL 6559. 'Wood Ditton'. *Dittune siluatica* 1086, *Dittone Silvestre* 1170, *Wodeditton* 1227. Latin adjectives **silvaticus -a**, **silvestris -e**, ME **wode** + p.n. *Dictune* 1022–86, *Dic(h)ton* 1218–98, *Ditton(e)* from 1169×71, 'settlement by (the Devil's) dyke', OE **dīć** + **tūn**. Also known as *(Wode)ditton Valoy(g)nes* 1218–17th with variants *Valence*, *Valeynce* etc. from the family of Robert de Valoines. "'Wood' to make a distinction from ye other Ditton (i.e. Fen DITTON TL 4860) beinge well wooded and in the wood country as wee call it" (MS c.1640). Ca 127.

WOODEATON Oxon SP 5312. 'Eaton by the wood'. *Wode Eton'* 1204–1316 with variant *Eton(e)*. ME **wode** (OE *wudu*) + p.n. *Etone* 1086 'the river settlement', OE **ēa** + **tūn**. 'Wood', for distinction from Water EATON SP 5112. O 194.

WOOD END ModE **wood** + **end**.

(1) ~ Herts TL 3225. *Wodend* 1474. ME **wode** (OE *wudu*) + **end** sc. of Ardeley parish. Hrt 152.

(2) ~ Warw SP 1071. Short for Wood End of Tanworth, where John *Atte Wode* probably lived in 1317. Wa 295.

(3) ~ Warw SP 2498. Short for Wood End of Kingsbury, where William atte Wode probably lived in 1327. Wa 19.

WOODEND Cumbr SD 1696. *Woodend* 1606, possibly identical with *the Woodhend* c.1441. ME **wode** (OE *wudu*) 'wood' + **end**. Unclear whether this means 'the end of the wood' or 'the wood end of Ulpha township'. Cu 438.

WOODEND Northants SP 6149. 'The wood end' sc. of Blakesley SP 6250. *Wodende* 1316, 1371, *Wodendeblacovesley* 1420, *Woodende al. Woodblakesley* 1522. Earlier known as Little Blakesley, *Little Blacolvesle* 12th, *Parva Blakoluesl'* 1275 and Wood Blakesley, *Wodeblakolesle* 1247. OE **wudu** + **ende**. The reference is to that side of Blakesley parish close to Plumpton Wood, in contrast to Quinbury End SP 6250, *Quinbury End* 1761. Nth 46, 40.

WOODEND WSusx SU 8108. 'The wood end' sc. of Funtington. *Woodend* 1810 OS. ModE **wood** + **end**. The wood may be mentioned as *in bosco de Funtintone* 'Funtington wood' 1273. Sx 61.

Lower WOODEND Bucks SU 8187. *L^r Wood End* 1822 OS. ModE adj. **lower** + p.n. Woodend as in Woodend Fm SU 8187, *Wood End* 1822. ModE **wood** + **end**.

WOODFALLS Wilts SU 1920. 'The wood fold(s)'. *Wudefolde* 1258, *Wu- Wodefald -faud* 1268–1400, *Wodefolde* 1309, *Woodfall* 1721, *Woodfield* 1811 OS. ME **wode** + **fald, fold**. An enclosure near the edge of the New Forest. Wlt 398.

WOODFORD 'The ford by the wood'. OE **wudu** + **ford**. L 70,228.

(1) ~ Corn SS 2113. *Wdeford* 1197, *Wodeford* 1302. PNCo 180, Gover n.d. 24, DEPN.

(2) ~ Devon SX 7950. *Woodford* 1809 OS.

(3) ~ GLond TQ 4090. *Wdefort* 1086, *-ford* 1177, *Wu- Wodeford(e)* 13th cent. including *[1062]13th S 1036, *Widefordefrith* 1264, 1285, 1369, *Woodworth -forth* 1554. Ess 110.

(4) ~ GMan SJ 8982. *Wid(e)ford(e)* 1248, 1249 etc., *Wyd(e)ford* 1276–1369, *Wodeford* 1296, 1376, *Wood(e)ford(e)* from 1430. Che i.217, L 70, 228.

(5) ~ Northants SP 9677. *Wodeford* 1086–1353 with variant *W(u)de-*. A ford across the Nene leading to a wood. Nearby is Woodwell SP 9577. Nth 189.

(6) ~ BRIDGE GLond TQ 4291. *Woodfordbrigge* 1429. P.n. WOODFORD TQ 4090 + ME **brigge**. Ess 110.

(7) ~ GREEN GLond TQ 4192. *Woodford Green* 1883. Earlier *Woodford Row* 1805. P.n. WOODFORD TQ 4090 + ModE **green**. LPN 252.

(8) ~ HALSE Northants SP 5452. 'W belonging to the manor of Halse'. *Wodeford* from 12th with variant *Wude-* + p.n. HALSE Northants SP 5640. Nth 37.

(9) Low ~ Wilts SU 1235. *Netherwodeford* 1279, *Lower Woodford* 1817 OS. ModE **low, lower**, ME **nether** + p.n. *(toþæm ealdan) wuduforda* [972]14th S 789, *Wo- Wudeford* 1120–1332. 'Low' in the sense 'lower down the river', for distinction from Middle and Upper WOODFORD SU 1136, SU 1237. Wlt 373, L 70, 228.

(10) Middle ~ Wilts SU 1136. *Middle Woodford* 1817 OS. ModE adj. **middle** + p.n. Woodford as in Low WOODFORD SU 1235.

(11) Upper ~ Wilts SU 1237. *Upwodford* 1524, *Upper Woodford* 1817 OS. ModE adj. **up(per)** + p.n. Woodford as in Low WOODFORD SU 1235. Wlt 373.

WOODGATE H&W SO 9766. *Woodgate* 1831 OS. ModE **wood** + **gate**.

WOODGATE Norf TG 0216. 'Wood hamlet'. *Woodgate* 1838 OS. Contrasts with Greengate TG 0116, *Greengate* 1388 ibid., both hamlets near Swanton Morley. ModE **wood** + **gate** ultimately < ON **gata**.

WOODGATE WMids SO 9982. No early forms. ModE **wood** + **gate**.

WOODGATE WSusx SU 9304. 'The gate at or by the wood'. *Wodegate* 1327. ME **wode** (OE *wudu*) + **gate** (OE *ġeat*). Cf. EASTERGATE SU 9405 and WESTERGATE SU 9305.

WOOD GREEN Gr Lond TQ 3191. 'Village green by the wood'. *Wodegrene* 1502. ME **wode** + **grene**. Originally a hamlet on the edge of Enfield Chase. Mx 80, TC 215, Encyc 995.

WOODGREEN Hants SU 1717. *Woodgreen(e)* from 1667. A small extra-parochial area next to Densom Wood from which presumably it is named. ModE **green** is frequently used in SCy in the names of secondary settlements. Ha 179, Gover 1958.217.

WOODHALL NYorks SD 9790. 'The hall at the wood'. *Le (La) Wodehall(e)* c.1281. OE **wudu** + **hall**. The reference is probably to a building housing people with specialised wood functions within a larger administrative unit. YN 262, Sa i.322–3.

WOODHALL SPA Lincs TF 1963. P.n. Woodhall TF 2268 + ModE **spa**. A spring beneficial to sufferers from gout or rheumatism was discovered here in 1824. Woodhall, *Wudehalle -a* 12th, *Wudehall* 1212, is the 'hall in the wood', OE **wudu** + **hall**, possibly the hall for the forest court. DEPN, Pevsner-Harris 1964.429.

WOODHAM 'Settlement in or by a wood'. OE **wudu** + **hām**.

(1) ~ Surrey TQ 0362. *Wodeham* [672×4]13th S 1165, [933]13th S 420, 1199, *Wude-* [871×99]13th S 353, 1202. Sr 112, ASE 2.32.

(2) ~ FERRERS Essex TQ 8097. A modern development at a place originally called Saltcoats, *le Saltcote* 1332, *Saltecotes* 1469, the 'salt-making cottages', ME **salt** + **cot**, pl. **cotes**, S of Woodham at TQ 7999. Ess 277.

(3) ~ FERRERS Essex TQ 7999. 'W held by the Ferrers family'. *Wudeham de Ferers, Wodeham Ferrers* 1230, *Woodham Ferris* 1805 OS. P.n. Woodham + family name of Henry de Ferreriis 1086. Woodham, *(æt) Wudaham* [962×91]11th S 1494, *Wuduham* 1000×2 S 1486, *U- O- Wdeham* 1086, *Wod(e)ham* 1203–1428, is the 'wood estate'. 'Ferrers' for distinction from the neighbouring WOODHAM MORTIMER TL 8205, WOODHAM WALTER TL 8006 and South WOODHAM FERRERS TQ 8097, all members of the wooded W end of Dengie Hundred. Ess 231.

(4) ~ MORTIMER Essex TL 8205. 'W held by the Mortimer family'. *Wudeham de Mortimer* 1230, *Wodeham Mortimer* 1255. P.n. Woodham as in WOODHAM FERRERS TQ 7999 + family name of Robert de Mortimer 1154×89. Ess 232.

(5) ~ WALTER Essex TL 8006. 'W held by Walter', the founder of the family of Robert Fitzwalter 1154×89. *Wod(e)ham Walte'i* 1248, ~ *Roberti* 1255, ~ *Water* 1303. P.n. Woodham as in WOODHAM FERRERS TQ 7999 + pers.n. *Walter*. Ess 231.

WOODHAM Durham NZ 2826. '(Settlement at the) woods'. *Wdum* 1154×66, early 13th, *Wudū* [1199×1216]1300, 1203×4, *Wodon* [1091×2] early 12th, 1444, *-um* 1243–1323, *-om(')* 1348–1424, *Wodeham* 1311, *Woodham* from 1611. OE **wudum**, dative pl. of **wudu**. NbDu 219, L 227.

East WOODHAY Hants SU 4061. Partly uncertain. *Estwydehay* 1350, 1382, *-wodehay* 1379. ME adj. **est** + p.n. *Wodeheya* 1189×99, *-haye* 1280, 1291, 1428, 'wood enclosure', OE **wudu** + **(ġe)hæġ**. Another run of spellings, *Wideheia* c.1150, *Wy- Widehaye -haie* 1171–1346 can be seen as containing OE **widu**, the original form of **wudu**; but if so this was early confused with OE adj. **wīd**, as if **æt thǣm wīdan hæġe* 'at the wide enclosure', *Wydenhaya* 1144, 1189×99, *Widenhai* 1189, *Wydenaye* 1316. Whether *Windenaie* 1086 belongs here or under Whitnal SU 4851, *(on) whitan leas heal, (þurh) hwitan leashal* 'white wood nook' [909]12th S 378, is uncertain. 'East' for distinction from West WOODHAY Berks SU 3963 across the county boundary, *Wydehaye* 1220 etc. 'the wide enclosure', *Westwydehaye* 1284, *Westwodehaye* 1428. It seems most likely that the *wudu-* forms of East Woodhay are primary and that West Woodhay was influenced and reshaped by them in the 15th cent. Ha 180, Gover 1958.157, Brk 328.

West WOODHAY Berks SU 3963. *Westwydehaye* 1284, *West Wodehay* 1428. ME adj. **west** + p.n. *Widehieie* 1203, *Wy- Widehaie, -hay(e)* 1220–1327, '(at) the wide enclosure', OE **wīd**, definite form dative case **wīdan**, + **(ġe)hæġ**. Also known as *Wydehaye Osevill* 1221 from Sewal de Osevile who held the manor in the early 13th cent. 'West' for distinction from East WOODHAY Hants SU 4061. Brk 328.

WOODHENGE Wilts SU 1543. *Woodhenge* 1929. A modern coinage on the pattern of STONEHENGE for the wooden henge monument discovered by air photography in 1925. Cunnington 1929.

WOODHILL Shrops SO 7384. *Wood Hill* 1833 OS.

WOODHORN Northum NZ 2989. 'The horn of land by the wood'

or 'the wooded horn of land'. *Wudehorn* 1177, *Wodehorn'* 1242 BF–1296 SR. OE **wudu** + **horn**. The reference is either to Beacon Point NZ 3189 or to the headland at the N end of Newbiggin Bay NZ 3188.

WOODHOUSE 'The house(s) at the wood'. OE **wudu** + **hūs**. L 227.

(1) ~ Leic SK 5415. *(les) Wodehuses* 1209×35, *-houses* 1277–1365, *Woodhouse* 1514. Lei 417.

(2) ~ SYorks SK 4185. *Wdehus* 12th, *Wodehousis* 13th cent., *-house* 14th cent., *Wodhouse(s)* 13th–1488. Later *Handsworth Woodhouse* 1608, p.n. HANDSWORTH SK 4186 + *Woodhouse* for distinction from Hatfield WOODHOUSE SE 6808. YW i.166.

(3) ~ EAVES Leic SK 5314. *Woodhowse Eves* c.1570, *Woodhouse Eves -Eaves* 1605–75. Earlier simply *les Eves* 1481, 'the edge or border' sc. of Woodhouse Woods, ME **eves** (OE *efes*) wrongly interpreted as a plural. Lei 418.

(4) Annesley ~ Notts SK 4953. *Annesley Wod(e)house* 13th–1362. P.n. ANNESLEY SK 5053 + p.n. *Wdehus* c.1190. Nt 112 gives local pr [wudəs]. Nt 112.

(5) Dronfield ~ Derby SK 3378. 'W belonging to Dronfield'. *Dranfeild Woodhouse* 1633. Earlier simply *Wodh(o)us* 1216×72, 1272×1307, *Wod(e)hous(e) -h(o)uses* before 1290–1546. P.n. DRONFIELD SK 3578 + WOODHOUSE. Db 244.

(6) Hatfield ~ SYorks SE 6808. P.n. *Haitefeild* 1591 as in HATFIELD SE 6509 + p.n. *Wodhous* 1404, *Woodhouse -howse* 1590–1697. YW i.8.

(7) Horsley ~ Derby SK 3945. 'W belonging to Horsley'. *Horselewodehus* 1303, *Hors(e)ley Wod(e)hous(e)* 14th cent. Earlier simply *Wudehus* 1225, *Wod(e)hous* 1317, 1330, *-hows* 1431, *-houses* 1318. P.n. HORSLEY SK 3744 + WOODHOUSE. Db 472.

(8) Mansfield ~ Notts SK 5463. *Mamesfeud Wodehus* 1280, *Mammesfeld Wodehuses* 1305. P.n. MANSFIELD SK 5361 + p.n. *Wodehuse* 1230, *Wudehus* 1242, *Nortwodehus* 'north W' c.1250, *le Wodehouse* 1316. Nt 127 gives local pr [wudəs].

(9) Norwell ~ Notts SK 7462. *Northwelle Wodehouse* 1300, *Wodehouses by Northwell* 1310, *le Wodhouse juxta Northwell* 1327. P.n. NORWELL SK 7761 + ME **wodehous**. Nt 194.

WOODHURST Cambs TL 3176. 'Wooded hill'. *Wdeherst* 1209, *-hirst* 1234, *Wodehyrst* 1252. OE **wudu** + **hyrst**, which may originally have been used as a p.n. as in OLDHURST TL 3077, 'Wold Hurst', from which it is thus distinguished. *Hurst* may originally have been the name for both places before division. Hu 229, L 198.

WOODINGDEAN ESusx TQ 3605. *Woodendean Farm* 1813 OS.

WOODKIRK WYorks SE 2724. 'The church in the wood'. *Wdekirk(a), Wudechirch(a)* 12th–13th cent., *Woodkirk(e)* from 1539. OE **wudu** + **ċiriċe** replaced by ON **kirkja**. Also known as *West Ardsley or Woodkirk* 1843 OS and possibly also as *Kirkham* as in KIRKHAMGATE SE 2922. YW ii.176.

WOODLAND Devon SX 7968. Probably 'newly cultivated land by the wood'. *Wodeland* 1424–73, *Woodland* 1535. ME **wode** (OE *wudu*) + **land**. D 525.

WOODLAND Durham NZ 0726. *Wodland* 1382–1564, *Woodlande* 1563. OE **wuduland**, which has recently been taken to refer to 'land cleared for cultivation in or near a wood'. This may be the sense here, but it is equally possible that the reference is to a particular district with a specialised function within the rural economy of the ancient multiple estate of Staindropshire. It is noteworthy that in at least one OE text *wudulond* is explicitly contrasted with *erythlond* 'arable' as well as with *etelond* 'pasture' and *medlond* 'meadow' (charter of Burgred king of Mercia [869]10th S 214). L 247.

WOODLAND FELL Cumbr SD 2689. P.n. Woodland SD 2489 + ModE **fell** (ON *fjall*). Woodland, *Kirkeby wodelands* 1544, *Wodland* 1577, is the woodland of Kirkby-in-Furness Lancs SD 2282. ModE **woodland** (OE *wudu-land*). La 221.

WOODLAND FELL Durham NZ 0325. *Woodland Fell* 1861 OS. P.n. WOODLAND NZ 0726 + ModE **fell**.

WOODLANDS 'Newly cultivated land at the wood'. ME **wode** + **land** (OE *wudu, land*).

(1) ~ Dorset SU 0509. *Wodelande* 1244, 1345, *(La) Wodelond(e)* 1268–1664, *Woodlands* 1774. Woodlands is situated near Verwood SU 0908 on the edge of Ringwood Forest SU 1008. Do ii.284 gives prs ['wudlændz] and ['udlənz], L 228, 247, 249.

(2) ~ Hants SU 3211. *Wodelonde* 1379. Woodlands is on the edge of the New Forest. Gover 1958.200, L 228, 247, 249.

WOODLANDS PARK Berks SU 8578. No early forms. P.n. Woodlands + ModE **park**. Here the name would have the ModE sense 'woodland' rather than the medieval sense 'land cleared for cultivation in or by a wood'. L 247.

WOODLANDS ST MARY Berks SU 3474. P.n. Woodlands + saint's name *Mary* from the dedication of the church. A modern settlement in *East Garston Woodlands* 1761, part of a series of woodland areas between Lambourn Woodlands, *Wodland in Chepynglambourne* 1451 and Shefford Woodlands. As in WOODLANDS PARK SU 8578 the sense is ModE 'woodland'. Brk 331, 339.

WOOD LANES Ches SJ 9381. *Wood Lane* 1831. The reference is to the *boscus de Adelinton* 1286, *Adlyngton Wode* 1462, *common or wastes called Adlington Wood* 1680. Che i.182.

WOODLEIGH Devon SX 7348. 'Wood clearing'. *(æt) Wudeleage* 1008×12, *Odelia* 1086, *Wodelegh(')* 1242, 1338–9. OE **wudu** + **lēah**. D 312.

WOODLESFORD WYorks SE 3629. The 'ford by the thicket'. *Wri- Wryd(e)lesford(e)* 12th–1425, *Widlesford* 1188×1202, *Wodelesford* [1252]17th, *Wudelesford* 1425, *Woodlesford -forth* 1626, *Wriglesforth* 16th. OE ***wrīdels** + **ford**. The *r* was lost by assimilation as in the 1188×1202 form, and the modern form developed directly from this. The ford carried the Wakefield-Tadcaster road across the r. Aire. YW ii.141, xi.

WOODLEY Berks SU 7673. 'Meadow by a wood'. *Wodlegh'* 1241, *Wodleyhe* 1341, *Woodley* 1761. OE **wudu** + **lēah**. Brk 146, L 228.

WOODMANCOTE 'The woodman's or woodmen's cottage(s)'. OE **wuduman**, genitive sing. **-mannes**, pl. **-manna**, + **cot**, pl. **cotu**.

(1) ~ Glos SP 0009. *Wodemancote -kot* 12th–1429, *Wod-* 1497–1533, *Wodemanecota -e* 1220, *Wudemone- -manne-* 1232–1311, *Woodmacote* 1632. Gl i.148.

(2) ~ Glos SO 9727. *W(u)de- Wod(e)mon(e)cot(e) -kot(e)* c.1200–1385, *-man-* 13th–1547, *Woodmancote* 1535, *Woodmacott* 1612, *Woodmecoat* 1704. Gl ii.94 gives pr ['wudməkət].

(3) WOODMANCOTTE Hants SU 5642. *woedemancote* (sic) [903]16th S 370, *Vdemanecote* 1086, *W(o)deman(n)ecote* 1242–1327, *Wydemancott'* 1256. The settlement probably began as a temporary residence for workmen performing special tasks connected with exploiting the woodlands of the Candover estates. Ha 181, Gover 1958.86.

(4) ~ WSusx TQ 2314. *Odemanscote* 1086, *Wodemane(s)cot(e)* 1240, 1271, *Wudemanekote* 1242. Sx 220.

(5) ~ WSusx SU 7707. *Wodemancot* 1332. Sx 57.

WOODMANSEY Humbs TA 0537. Probably 'the woodman's lake'. *Wod(e)manse* 1289–1512, *Woodmancy* 1573, *-sey* 1577. OE **wuduman** + **sǣ**. Woodmansey is near a farm called Sicey, *Si- Sydese* 1297, 'the broad lake', another name in **sǣ** in a former area of marsh and alluvium; but the generic could be OE **ēa** 'a stream' after genitive sing. **wudumannes**. YE 202.

WOODMANSTERNE Surrey TQ 2760. Partly uncertain. *Odemerestor* 1086, *Wudumaresthorne* c.1192, *-mare-* 1222, *Wudemeresthorn* 1205–1450 with variants *Wode-* and *-mers-*, *Wudemaunestthorne* 1253, *Woodmansterne alias Woodmersterne* 1559. The generic is OE **thorn** 'a thorn-tree'. The specific may be OE pers.n. **Wudumǣr*, later reinterpreted as the ME pers.n. or by-n. *Wudemann*, or alternatively a compound p.n. **wudu-(ġe)mǣre* 'the wood boundary'. The parish lies between the two important Middle Saxon estate-centres of BANSTEAD TQ 2559 and Coulsdon Surrey TQ 3058. Sr 56, Rumble 1976, 173–5.

WOODMINTON Wilts SU 0022. 'The woodman's settlement'. *Wodmanton* 1254, *Wu- Wodeman(e)ton* 1255–1456, *Woodmington* 1773. ME **wodeman** 'huntsman, one who cuts trees' + **toun**. Wlt 204.

WOODNESBOROUGH Kent TR 3056. 'Woden's hill'. *Golles-Wanesberge* 1086, *Wodnesbeorge -berga, Wanneberga* c.1100, *Wodnesberg(h)e* 1198, 1253×4, *-burue* 1211×12, *Woodnesborough* 1610. OE divinity-n. *Wōden*, genitive sing. *Wōd(e)nes*, + **beorg**. There was a mound by the church at the end of the 18th cent. and traditions of grave-goods being found there but the hill here is a perfect **beorg**. The DB spellings are difficult: *Golles-* may represent OFr initial *gu-* [g] or [gw] for OE *w* and AN *l* for *n* in the group *-dn-* with subsequent assimilation of *-dl-* > *-ll-*. *Wanes-* may have evolved from **Wœnnes-* < **Wennes-* < *Wednes-* from the OE variant *Wēden* for *Wōden*. But **Wœnnes-beorg* 'hill of the mound' from OE **wenn, wænn** 'a wen, a blister, a mound' is perfectly possible to explain one of the DB spellings. PNK 586, Settlements 112, Domesday Studies 133.

WOODNEWTON Northants TL 0394. 'Newton at the wood'. *Wodeneuton* 1255–1324, *Wood Newton* 1599. ME **wode** (OE *wudu*) + p.n. *Niwetone* 1086, *Neweton* 12th, *Newenton* 1196–1384, 'the new settlement'. The prefix *Wood-* refers to its situation within the forest of Cliffe and distinguishes it from other Northamptonshire Newtons. Nth 208.

WOODPERRY Oxon SP 5710. 'Perry at the wood'. *W(o)deperi(e)* 1220, 1242×3. ME **wode** (OE *wudu*) + p.n. *Perie* 1086 as in WATERPERRY SP 6206. O 188.

WOODPLUMPTON Lancs SD 5034. 'Plumpton at the wood'. *Wodeplumpton* 1327–69 etc. Earlier simply *Pluntun* 1086, *Plumpton* 1256, *Plumton* 1287. ME **wode** (OE *wudu*) + p.n. Plumpton, OE **plūme** + **tūn** 'plum-tree farm'. 'Wood' for distinction from Fieldplumpton, now PLUMPTON SD 3833. La 161, Jnl 17.94.

WOODRISING Norf TF 9803. 'Wood Rising'. *Woderisingg* 1291. ME **wode** (OE *wudu*) + p.n. *Risinga* 1086, *Resinges* 1121, 1206, *Risinges* 1185, 1226, as in Castle RISING Norf TF 6624. The DB entry mentions woodland for 200 swine here. DEPN, ING 60.

WOODSEAVES 'The border of the wood'. ModE **wood** + **eaves**.
(1) ~ Shrops SJ 6830. *Wodseves* 1548, *Woodseaves* 1833 OS. Cf. Haywood Fm SJ 7030, *Heywood* 1833 OS. Gelling.
(2) ~ Staffs SJ 8025. *Wood- Woddease* 1594, *Woodeseves* 1612. Originally the name of a single farm. Horovitz.

WOODSEND Wilts SU 2276. *Wood End* 1773, 1828 OS. ModE **wood** + **end**.

WOODSETTS SYorks SK 5583. 'The folds in the wood'. *Wodesete(s)* 1324–1439, *-setts* 1496, *Woodsetts* from 1771. OE **wudu** + **(ġe)set**. YW i.150, L 228.

WOODSFORD Dorset SY 7690. Probably 'Wigheard's ford'. *Werdesford* 1086–1457, *Werdeford* 1196–1388, *Wardesford* 1086, *Warde(s)ford* 1221–1428, *Wi- Wyrde(s)ford(e)* 1280–1498, *Wodesford* 1280, *Woodford* 1575, *Woodsford* 1682. OE pers.n. *Wīġheard*, genitive sing. *Wīġheardes*, + **ford**. An alternative specific might be OE ***wierde** 'a beacon', and so 'ford of the beacon'. Do i.186.

WOODSIDE Berks SU 9371. Probably 'the side of the wood'. *Woodside* 1839. ModE **wood** + **side**. Brk 41.

WOODSIDE Herts TL 2506. '(Place) beside the wood'. *Woodside* 1538. ME **wode** (OE *wudu*) + **side**. Hrt 130.

WOODSTOCK Oxon SP 4416. 'The outlying farm at the wood'. *Wudestoce* c.1000, *Wodestoch* 1086–1464 with variants *Wude-* and *-stoc(ha) -stok(e)*. OE **wudu** + **stoc**. The name was explained by Symeon of Durham as *silvarum locus* 'the place of the woods'. O 292.

WOOD STREET Surrey SU 9551. 'The wood hamlet'. *Woodstrete* 1544, *Wodstrate* 1609. ME **wode** + **strete**. Sr 165.

WOODTHORPE 'Outlying settlement at the wood'. OE **wudu** + ON **thorp**. L 228.
(1) ~ Derby SK 4574. *Wdesthorp* 1154×89, *Wodesthorp* 1264, *Wod(e)thorp(e)* 1268–1496. The place belonged to Staveley and was also known as *Staueley Woddethorp* 1299, *Staueley Wod(e)thorp* 1387. Db 303.
(2) ~ Leic SK 5417. *Wudethorp* 1253, *Wodethorp(e)* 1277–1456. ME prefix **wode** (OE *wudu*) + p.n. *Torp* 1236, *Thorp(e)* 1243–1445. Lei 384.

WOODTON Norf TM 2993. 'Wood settlement'. *W(o)detuna, Videtun* 1086, *Wudeton* 1254. OE **wudu, widu** + **tūn**. DEPN.

WOODTOWN Devon SS 4925. *Woodtown* 1809 OS. A settlement at the wood referred to as *Le Wode* in 1294. There are other examples in Devon of the late addition of **town** to existing place-names. D 86, 676.

WOODVILLE Derby SK 3119. This was originally the *Wooden Box Station* 1836. The name was changed to *Woodville* in 1845 and the place made a chapelry by order in Council in June 1847. *Wooden Box* was described in 1856 as having been 'a populous village of potters' that sprang up in the neighbourhood of Butt House and was called *Wooden Box* after a port wine butt from Drakelow Hall which was used as a hut for the collection of tolls at the turnpike. Db 670.

WOODWALTON Cambs TL 2180. *Wodewalton* 1300, *Woodwallton* 1567. ME **wode** + p.n. *Waltune* 1086, 1236, *Walton(a)* from 1155, 'wall settlement, farm or estate', OE **wall** + **tūn**. The specific could alternatively be OE **wald** 'wood'.

WOODYATES Dorset SU 0219. '(At) the gate or gap in the wood'. *in publico loco qui dicitur at Wdegate* 'in the open place called *At Wdegate*' [859 for 870]15th S 334, *in þare stowe þat is inemned at wudegate* 'in the place called *at wudegate*' [870]15th S 342, *(to) wideyate* [944×6]14th S 513, *Widamgate* (for *Widian-*) 959×75 S 1772, *(besuðan) wudigan gœte* c.970, *Odiete* 1086, *Wo- Wud(e)iat(e)* 1208–33, *Wodeyate* 1244–1428, *-gate* 1199–1325, *Wodeyats* 1535, *East- West-Woodyates* 1774. **wudu** + **ġeat** varying with OE adj. **wudiġ**, definite form oblique case **wudigan**, + **ġeat**. Do ii. 271 gives prs ['wudiəts] and ['udiəts].

WOODY BAY Devon SS 6849. *Woodabay* 1809 OS.

WOOFFERTON Shrops SO 5268. Probably 'Wulfhere's estate'. *Wulfreton* 1221–1304, *Wulferton* 1221, *Wolferton* 1304. OE pers.n. *Wufhere* + **tūn**. The evidence does not allow certainty about the precise pers.n. in question, and OE *Wulffrith* is a possible alternative. Bowcock 259, DEPN.

WOOKEY Somer ST 5145. 'Snare, noose, trap'. *Woky* *[1065]18th S 1042, 1231, [1240–70]Buck, *Wochi* 1178, *Wokey* 1610. OE **wōciġ**. DEPN.

WOOKEY HOLE Somer ST 5347. 'Wookey cave'. *Wokyhole* *[1065]18th S 1042, 1249, *Wokey hole* 1610. P.n. WOOKEY ST 5145 + OE **hol**. The oldest known cave in England entered through a large natural archway from which the river Axe issues. DEPN.

WOOL Dorset SY 8486. '(At) the springs'. *(œt) Wyllon* 1002×12[†], *Wille* 1086, *Welle* 1086–1344, *Woll(e)* 1249–1663, *Wool* from 1575. If *œt Wyllon* belongs here this is OE **æt wyllum** 'at the springs', dative pl. of **wylle**. The reference is to abundant springs S of the village. Do i.188.

WOOLACOMBE Devon SS 4543. Probably 'spring valley'. *Wolne- Wellecome* 1086, *Wellecomb* 1301, *Wollecumb' -comb(e)* 1242–1379. OE **wylle** + **cumb**. The 1086 spelling *Wolnecome*, however, if not a mistake, might stand for *Wolue-* from OE **wulfa**, genitive pl. of **wulf** 'a wolf'. D 54.

[†]Identification not certain.

WOOLAGE GREEN Kent TR 2349. P.n. Woolage + ModE **green**. Woolage, *Wuleheth'* 1254, *W(o)lfeth(e)* 1270, 1304, *Wulfhecch'* 1313, *Wollwych* 1535, *Wollwich* 1819 OS, is 'wolf hatch', OE **wulf** + **hæċċ**, perhaps in the sense 'wolf-trap'. PNK 540.

WOOLASTON Glos ST 5899. 'Wulflaf's estate'. *Odelaveston* (sic) 1086, *Wul(l)aveston(a)* [1154×89]1307, *Wol(l)aueston(e) -v-* 13th cent., *Wu -Wolaston(e)* c.1270–1535, *Woollaston* 1793. OE pers.n. *Wulflāf*, genitive sing. *Wulflāfes*, + **tūn**. Gl iii.268 gives pr ['wuləstən].

WOOLAVINGTON Somer ST 3441. There are two different forms of the name:
I. *Hvnlauintone* 1086.
II. *Wilaveton* 1201, *Wulavinton* 1248, *Wollavyn-*, *Wilavinton* 1276, *Wollauington* 1610.
Form I is 'estate called after Hunlaf', OE pers.n. *Hūnlāf* + **ing**[4] + **tūn**, form II 'estate called after Wiglaf or Wulflaf', OE pers.n. *Wīġlāf* or *Wulflāf*. DEPN.

WOOLBEDING WSusx SU 8722. 'The Wulfbædingas, the people called after Wulfbæd'. *Welbedlinge* 1086, *Wolbed(d)ing(e) -yng(e)-* 1191–1641, *Wulfbeding* c.1230, *Wlbetinges* 1248. OE folk-n. **Wulfbædingas* < pers.n. *Wulfbæd* + **ingas**, dative pl. **Wulfbædingum* (source of the *-inge* forms). The DB sp has been thought to point to pers.n. *Wulfbeald* but this is not necessary. Sx 31 gives pr [wulbəiŋ], ING 41, SS 64, MA 10.23.

WOOLER Northum NT 9928. 'The spring promontory'. *Wullovre* 1187, *Welloure* 1196, 1203, 1346, *Wilour* 1249, *W(u)ll- Wollovera -(e)* 1199–1637, *W(u)ll- Wolloure* 1256–1334, *Wooler* 1663. OE **wella** + **ofer**. There is an ancient wishing-well on Horsdean Hill above Wooler at NT 9827. Rounding of *e* > *o* after *w* is well evidenced in ONb. NbDu 219, DEPN, Campbell 319, L 177, Tomlinson 476, Jnl 22.37.

WOOLEY Durham NZ 1739 → USHAW MOOR NZ 2242.

WOOLFARDISWORTHY 'Wulfheard's enclosure'. OE pers.n. *Wulfheard*, genitive sing. *Wulfheardes*, + **worthiġ**.
(1) ~ Devon SS 8208. *Vlfaldeshodes* 1086, *Wolfaresworth'* 1242, *Wolfres- Wolveresworthy* 1281, *Wulfardeswurth -is- -worthe -worthy* 1244–1378, *Woolfardisworthy als. Woolsworthy* 1825. D 399 gives pr [wulzəri] and the spelling *Woolsery* occurs on local signposts.
(2) ~ Devon SS 3321. *Olvereword* 1086, *Wlfereswrthi* 1238, *Wolfarysworth'* 1242, *Wolfardesworthi* 1263, *Wollesworthye* 1550, *Woolsry* 1629×57, *Woolsery* local signpost. D 80 gives prs [wu-] and [ulsəri].

WOOLHAMPTON Berks SU 5766. 'Estate called after Wulflaf'. *Ollavintone* 1086, *Ullavinton* 1155×62, *Wul(l)auinton'* 1242×3, *Wolauinton'* 1294, *Wolavyngton* 1341, *Wullaminton* 1275×6, *Wolhamton* 1327–1517, *Wol(h)ampton(e)* 1341–1428, *Woolhampton vulgo Wollington* 1675. OE pers.n. *Wulflāf* + **ing**[4] + **tūn** with later confusion with **hām-tūn**. Brk 229.

WOOLHOPE H&W SO 6135. 'Wulfgifu's Hope'. *Wolvythehope* 1219×27, *Wulveve Hope* 1243, *Wulhope* 1526, *Hope Wolnith -yth* (sic for *-uith*) 1221, 1407. OE feminine pers.n. *Wulfġifu*, genitive sing. *Wulfġife* + p.n. *Hopam* before 1066, *Hope* 1086–1273, OE **hop** the 'secluded valley'. The place was a gift to the canons of Hereford by two women, *Wulfeva* (OE *Wulfgifu*) and Godiva (*Godgifu*) before 1066. He 213, DB Herefordshire 2.13.

WOOLLAND Dorset ST 7707. Probably 'estate with a high proportion of meadow'. *Wennland* [833]15th S 277, *(æt) Wonlonde* (probably for *Wen-*) [843 for 934]17th S 391, *Winland* 1086, *Wynlond(')* 1288, *Wunlanda -e* 1170×1, 1231, *Wuland'* 1212, *Wollond(e)* 1268–1480, *Woulond(e)* 1285, 1311, 16th. OE ***winn**, ***wynn**, + **land**. Do iii.232, L 248–9.

WOOLLEY 'Wolves' wood or clearing'. OE **wulf**, genitive pl. **wulfa**, + **lēah**. PNE ii.281.
(1) ~ Cambs TL 1474. *Ciluelai* (sic for *Olue-*) 1086, *Wulueleia* 1158, *Wolvele* 1220–1315, *Wullegh* 1260, *Wolle(y)* 1276–1627.
(2) ~ WYorks SE 3213. *Wiluelai* 1086, *Wulwineleya* 1201, *Wluineleys* 1200×18, *Wlve- Wluele(y) -lay* 12th–1350, *Wolue- Wolvelay -ley* 1296–1487, *Wollay -ley* 1330–1541, *Woolley* from 1381. OE **wulf**, genitive pl. **wulfa**, + **lēah** varying with **wylfen** 'she-wolf', genitive pl. **wylfena**, + **lēah**. YW i.286.
(3) ~ HOUSE Berks SU 4179. P.n. Woolley + ModE **house**. Wooley, *Olvelei* 1086, *Wulvelye* 1205, *Wulueleye* 1284, *Wolley* 1517, *Woolley* 1761, is 'wolves' wood or clearing'. Brk 291.

WOOLMER GREEN Herts TL 2518. *Woolmer Grene* 1605. Surname Woolmer, or p.n. *Woluemere* 1296, 'wolves' pond', OE **wulf**, genitive pl. **wulfa** + **mere**, + ModE **green**. Hrt 145.

WOOLPIT Suff TL 9762. 'The wolf pit(s)'. *Wlpit* [?978×1016]13th S 1219, *Wulpettas* 11th, *Wlfpetā* 1086, *Uulfpet* c.1095, *Wulpet* 1610. OE **wulf-pytt** 'a pit for trapping wolves', SE form **-pett**, pl. **-pettas**. DEPN.

WOOLSCOTT Warw SP 4967. 'Wulf's cottage(s)'. *Wulscote* c.1235, 1452, *Wulvescot'* 1336, *Wo(o)scote* 1533, 1569, *Woolscott* 1649. OE pers.n. *Wulf*, genitive sing. *Wulfes*, + **cot**, pl. **cotu**. Wa 131.

WOOLSINGTON T&W NZ 1970. 'The estate called after Wulfsige'. *Wulsinton* 1203, *Wolsyngton* 1360, *Wissington* 1663, 1798. OE pers.n. *Wulfsige* + **ing**[4] + **tūn**. NbDu 220 gives pr [wisintən].

WOOLSTASTON Shrops SO 4598. 'Wulfstan's estate'. *Ulestanestune* 1086, *Wolstaneston(')* 1204×10–1301, *Woolstanston(e)'* 1204×10–1721, *Wusaston* 1577, *Wolstaston* 1601, *Woolstaston* from 1749. OE pers.n. *Wulfstān*, genitive sing. *Wulfstānes*, + **tūn**. 'Wolstaston alias Wollaston, usually written Woolstaston and pronounced Woosasun', Plymley 1803. Sa i.323–4.

WOOLSTHORPE Lincs SK 8333. 'Wulfstan's outlying settlement'. *Vlestane(s)torp* 1086, *Uulstanes- Wistanestorp* [1106×23]1333, *Wolestorp* 1202, *-thorp(e)* 1299–1462, *-troppe* 1506, *Wolsthorp(e)* 1282–1475, *-tropp(e)* 1535–1602, *Woolstrop(e)* 1590–1603. OE pers.n. *Wulfstān*, genitive sing. *Wulfstānes*, + **thorp**. Perrott 481, SSNEM 122.

WOOLSTON 'Wulf's estate'. OE pers.n. *Wulf*, genitive sing. *Wulfes*, + **tūn**.
(1) ~ Ches SJ 6489. *Oscitonam* 1094, *Ocsitonam* 1122, *Oxsitonum* 1155, *Ulfitonam* 1142 (all Latin accusatives), *Wolues- Wulueston* 1246, *Wlston* 1257, *Wolston* 1327–89 etc. La 95.
(2) ~ Hants SU 4310. *Olvestune* 1086, *Ulveston(e)* 1212, 1272, *Wolveston* 1280–1424. There is also a form *Wolverichestone* 1284 belonging here, 'Wulfric's estate'. The relationship of the two names is not understood. Ha 181, Gover 1958.40.
(3) ~ GREEN Devon SX 7766. *Wolston greene* 1662. P.n. Woolston + **green**. Woolston, *Holveston* 1353, *(W)olveston* 1384, *Ulston* 1431, *Wolston* 1467. D 521.

WOOLSTON Devon SX 7141. 'Wulfsige's estate'. *Vlsistone* 1086, *Wols(t)ingthon'* 1242, *Wolseton* 1285. OE pers.n. *Wulfsiġe*, genitive sing. *Wulfsiġes*, + **tūn**. D 289.

WOOLSTON Shrops SJ 3224. 'Oswulf's estate'. *Osulvestune* 1086, *-ton* c.1230, *Osselton* 1272, *Oseleton* 1397, *Osselston* 17th, *Wolston* 1577, *Woolston* from 1708, *Wooston* 1672, 1808. OE pers.n. *Ōswulf*, genitive sing. *Ōswulfes*, + **tūn**. Sa i.324.

WOOLSTON Shrops SO 4287. 'Wulfhere's estate'. *Wlureston'* 1208×9, *Wulfre(s)- Wol(e)ures- Wolfres- Woluers- Wolfers- Wylfris- Wlfereston(e)* 13th cent., *Wolston* 1431–1745, *Wulston* 1577, *Woolston* from 1705, *Woson* 1558×1603, *Wooson* 1717, *Wooston* 1703. OE pers.n. *Wulfhere*, genitive sing. *Wulfheres*, + **tūn**. The DB form *Wistanestune* 1086 is an error for the parish Wistanstow in which Woolston lies. Sa i.324.

WOOLSTONE Bucks SP 8638. 'Wulfsige's estate'. *Ul- Wlsiestone* 1086, *Wulfsieton* 1186, *Wolse(s)ton* c.1218, *Wulston*, *Wooston* 16th. OE pers.n. *Wulfsiġe*, genitive sing. *Wulfsiġes*, + **tūn**. Bk 27 gives pr [wulsən].

WOOLSTONE Oxon SU 2987. 'Wulfric's estate'. *Olvricestone* 1086, *W(u)lvriches- Wolfricheston(e)* 13th cent., *Wolriston*

[1325]14th, *Woolstone* 1830. OE pers.n. *Wulfrīč*, genitive sing. *Wulfrīčes*, + **tūn**. The two estates of East and West W, formerly known as Ashbury, the OE name of UFFINGTON SU 3089, were given to the thegn Wulfric in 944 and c.955. Brk 383.

WOOLTON Mers SJ 4286. 'Wulfa's farm or village'. *Vluentune, Vuetone* 1086, *Wlueton* 1187, 1258, *Wlvinton* 1188, *Wolve(n)ton* 1322–3, *Wolueton* [c.1180]1270, 1200–1341, *Wolleton* 1403. OE pers.n. *Wulfa*, genitive sing. *Wulfan*, + **tūn**. La 111, Jnl 17.62.

WOOLTON HILL Hants SU 4361. *Woolton Hill* 1817 OS. P.n. Woolton, *Woulthorne* 1604, + ModE **hill**. The evidence is too late to offer an explanation. Ha 182, Gover 1958.158.

WOOLVERSTONE Suff TM 1838. 'Wulfhere's estate'. *Ulueres-Hulferestuna* 1086, *Wolferston(e)* 1196–1505 with variant *-res-*, *Wolverston* 1524, *-stone* 1656. OE pers.n. *Wulfhere*, genitive sing. *Wulfheres*, + **tūn**. DEPN, Baron.

WOOLVERTON Somer ST 7953. 'The estate called after Wulfhere'. *Wulfrinton* 1196, *Wolfrington* 1291, *Wuluerton* 1610. OE pers.n. *Wulfhere* + **ing**[4] + **tūn**. DEPN.

WOOLWICH GLond TQ 4478. 'Trading place or harbour for exporting wool'. *Uuluuich* [918]13th S 1205a, *Wulleuic* [964]14th S 728, *Vulwic* 1016, *Hulviz* 1086, *Wolewic* 1089, *Wulewic* 1227. OE **wull** + **wīc**. DEPN.

WOOPERTON Northum NU 0420. Partly uncertain. *Wepredane -den* 1180, *Weperden* 1242 BF–1428, *-don* 1255, 1346, 1586, *Woperdon* 1596, *Wop(p)erton* 1663–74, *Wooperton* 1746, *Wap(p)erton* 1637, 1811. Unknown element or pers.n. + OE **denu** 'valley' later confused with **dūn** 'a hill'. NbDu compares W p.n. Wepre near Ewloe, Clwyd SJ 2969, *Wepre* 1086, 1335, *Wapir* 1281, 1351, *Weper* 1285, *Weppra* 1452, 1612, *Gwep(p)ra* 1489–1699, which, taking a clue from DEPN, Hywel Wyn Owen takes to be OE *weoh-beorg* 'idol or pagan temple hill'. But this is very uncertain. NbDu 220 gives prs [wo- wapətən], Nomina 1987.105, EFlint 168–9, 177, DEPN.

WOORE Shrops SJ 7342. Uncertain. *Wavre* 1086, *Wau(e)re, Wafre* 1255×6, 1284×5, *Wour(r)e* 1291–1334, *Wouere -v-* 1292×5–1431, *Wore* 1429, *Woore* 1703, *Owre* 16th, *Oare, Woer* 17th, *Wooer* 1701, 1706. OE **wæfre** 'unstable, restless, wandering', ***wæfer** 'that which wanders'. Elsewhere this occurs as a r.n. (WAVER Cumbr NY 1950) or in composition with the element 'tree' (WAVERTREE Mers SJ 3889). Perhaps the sense here is 'an isolated young tree, a young tree left standing by itself in a felled wood' as for dial. *waver*, or the like. Sa i.325, OED s.v. *waver*.

WOOTH Dorset SY 4795 → River BRIT SY 4795.

WOOTON 'The settlement at or by the wood'. OE **widu**, later **wudu**, + **tūn**. This name type often implies special function as a constituent member of a larger economic unit.

(1) ~ Humbs TA 0916. *Udetune -tone, Vdetone* 1086, *Wit(t)una, Wttuna* c.1115, *Witton(')* 1166–13th, 1550, 1609–1711, *Wutton(')* 1200–55, *Wotton(')* 1185–1642, *Wooton* from 1609. Li ii.298.

(2) ~ Northants SP 7656. *Witone* 1086, *Witton'* 1235, *Wutton(a)* 1162–1271, *Wotton* 12th–1324, *Wootton* 1346. W lies W of Yardley Chase and Salcey Forest. Nth 153.

(3) ~ Oxon SP 4702. *Uudetun* *[821]12th S 183, *Wudtun* *[821]13th ibid., *(æt) Wuttune* [985]13th S 858, *Wutton'* 1234×5, *Wotton(e)* 1224–1548×53, *Wooton* 1830. Brk 462.

WOOTTON 'Wood settlement or estate'. OE **widu**, **wudu** +**tūn**. L 227.

(1) ~ Beds TL 0045. *Otone* 1086, *Wutton* 1197–1397, *Wotton* 1223 etc. Bd 86.

(2) ~ Hants SZ 2498. *Odetvne* 1086, *Wodeton* 1256–1331, *Wotton* 1429, *Wootton* 1670. The settlement began as a place in Milton parish for exploiting New Forest woodland. The forms *Wodyton(e) -i-, -yn-* 1307–1327 point to OE **wudiġ** 'wooded' and ***wuding** 'wood place' as alternative specifics. Ha 182, Gover 1958.229.

(3) ~ Kent TR 2246. *Uudetun* [687]13th S 1610, *Wudu tun* [799]10th S 156, *(on) otþunes hyldan* 'to Wootton hill' [944]13th S 501, *Wodetone* 1210×12. PNK 15, KPN 263, Cullen.

(4) ~ Oxon SP 4702 → WOOTON SP 4702.

(5) ~ Staffs SK 1045. *Wodetone* 1086, *Wotton* 1274, 1322. DEPN, Horovitz.

(6) ~ Staffs SJ 8227. *Wodestone* 1086, *Wotton* 1253. DEPN.

(7) ~ BASSETT Wilts SU 0783. 'W held by the Basset Family'. *Wotton Basset* 1272, ~ *Bassett* 1547, *Bassetteswotton* 1315. P.n. *Wdetun* *[680]12th S 1166, [745]13th S 256, [937]12th S 436, *Wodetone* 1086, *Wotton(e)* 13th cent., + manorial addition from the family name of Alan Basset, who acquired the manor before 1212 in right of his wife. The boundary of the people of Wooton is *Wudetunningca gemære* [983]12th S 848. Wlt 272.

(8) ~ BRIDGE IoW SZ 5491. *Wotton bridge* 1608. P.n. Wootton + ModE **bridge**. Wootton is *Odetone* 1086, *Wudeton* 1189×1204, *Wodi(n)tone -y(n)-* 1248–1519, *Wdington* 13th, *Wodyngton(e)* 1332–1428, *Wotton* 1378, *Wutton* 1559, *Wootton* 1608. The spellings show that there were alternative forms of the specific **wudiġ** and **wuding**. Wt 249 gives pr [utn], Mills 1996.111.

(9) ~ COMMON IoW SZ 5391. *Wootton Common* 1769. P.n. Wootton as in WOOTON BRIDGE SZ 5491 + ModE **common**.

(10) ~ COURTENAY Somer SS 9343. 'W held by the Courtenay family'. *Wotton Courtenay* 1408, ~ *Courtnay* 1610. Earlier simply *Otone* 1086, *Wotton* 1274. The manor was held by John de Curtenay in 1274. DEPN.

(11) ~ FITZPAINE Dorset SY 3695. 'W held by the Fitz Payn family'. *Wodeton Roberti filii Pagani* 'Robert son of Pagan's W' 1316, *Wotton Fitz Payn* 1392. Earlier simply *Wodetone, Odetun* 1086, *Wudeton* 1244. Do 162.

(12) ~ RIVERS Wilts SU 1963. 'W held by the Rivers family'. *Wotton Ryver* 1428, *Otten Rivers* 1743. P.n. *Wdutun* [801×5]12th S 1263, *Otone* 1086, *Wotton(a)* 1212–1428, + manorial addition from the family name of Walter de Riperia who held the manor in 1212 and William de la Rivere in 1222. Wlt 357.

(13) ~ ST LAWRENCE Hants SU 5953. 'St Lawrence's W'. *Laurens Wotton* 1580. Saint's name *Lawrence* from the dedication of the church + p.n. *(æt) wudatune* [990]12th S 874, *Odetone* 1086, *Wu- Wotton(a) -(e)* 1167–1341. Ha 182, Gover 1958.142.

(14) ~ WAWEN Warw SP 1563. 'Wagene's W'. *Wageneswitona* 1138×47, *Waghnes Wotton* 1285, *Wau(n)es Wotton* 1285, 1308, *Wotton Wawen* 1517, 1578, *Wawens Wotton* 1578. ODan pers.n. *Vagn* as in *Wagene de Wotton* mentioned in the spurious charter S 1226 (supposedly c.1043, surviving in 15th–17th cent. copies), genitive sing. *Wagenes*, + p.n. *Uuidutuun* [716×37]11th S 94, *Wotone* 1086, *Wotton(e)* 12th–1320. Wa 242, DEPN.

(15) Butleigh ~ Somer ST 5034. 'W near or belonging to Butleigh'. *Butleigh Wooton* 1817 OS. P.n. BUTLEIGH ST 5233 + p.n. Wootton.

(16) Glanvilles ~ Dorset ST 6708. *Glamvileswotton* 1396, *Glanvile-* 1428, *Glanvilles Wooton* 1811. Also known as *Wotton' Clauyle* (sic for *Glan-*) 1280, ~ *Gla(u)nuill' -vyle -vyll* 1288–1558. Earlier simply *Widetone* 1086, *Wotton(')* 1268–1547. The manorial affix is from the Glanville family (Geoffrey de Gla(u)nvile 1258 etc.) from Glanville in Calvados, Normandy, *Glandevilla* 1079, 'Glanda's vill'. Also known as *Wolfrenewotton, Wolvern Wotton* 14th cent., 'Wulfrun's W', OE feminine pers.n. *Wulfrūn*, genitive sing. *Wulfrūne*. Do iii.267.

(17) Leek ~ Warw SP 2969. 'Leek W'. *Lecwotton* 1285, *Leekwotton* 1306. OE **lēac** 'a leek' + p.n. *Wttuna* 1100×35, *Wottone* 1203–1459. For the affix cf. LECKHAMPSTEAD Berks SU 4375, Bucks SP 7237, LECKHAMPTON SO 9419. Wa 190.

(18) North ~ Dorset ST 6514. *North(e) Wotton* 1569×74, 1575, *North Wooton* 1811. Earlier simply *Wotton'* c.1180–1617 and *Wotton Episcopi* 1316, *Wotton Bishops* 1393 from its possession by the bishops of Salisbury. 'North' for distinction from Glanvilles WOOTTON. Do iii.380.

(19) North ~ Norf TF 6423. *Nordwitton* 1166, *Nort Wottone* 1254. Earlier simply *Wdetuna* 1086. 'North' for distinction from South WOOTTON TF 6422. DEPN.

(20) North ~ Somer ST 5641. *North Wotton* 1610. Earlier simply *Wudetone* [760]13th S 1684† *Wodetone* [855×60]13th S 1700, [946]14th S 509, *Vtone* 1086. 'North' for distinction from Butleigh WOOTTON ST 5035. DEPN.

(21) South ~ Norf TF 6422. *Sudwutton* 1182, *Suth Wottone* 1254. ME **sūth** + p.n. Wootton as in North WOOTTON TF 6424. DEPN.

WORBARROW BAY Dorset SY 8679. *Worthbar(r)ow bay(e)* 1575, c.1586, 1861, *Worbarrow Bay* from 1579, *Warbarrow Bay* 1773. P.n. *Wyrebarowe* 1462, *Wyrbarow* c.1500, 'watch hill', OE ***wierde** + **beorg**, + ModE **bay**. In 1462 this is referred to as a place where watches are set, presumably at Worbarrow Tout SY 8679 (OE *tōte* 'look-out hill'), 'a little rocky conical hill ... almost environed by the sea, being joined to the continent by a neck of land' (J. Hutchins, *The History and Antiquities of the County of Dorset*, 3rd ed. 1861–70, I.619). Do i.103.

WORCESTER H&W SO 8555. 'Roman town of the Weogoran'. There are over 70 OE charter forms for Worcester of which the most important spellings are: *Uueogorna civitate* [691×99]17th S 77, [778]11th S 113, *Uueogorna cœstre, ceastre* [789]11th S 1430, [803]11th S 1260, [899×904]16th S 1283, [c.903]11th S 1446, *Wigorna ce(a)stre* *[777×9]11th S 118, [804]11th S 1187, *[851]11th S 201, [883]11th S 218, [967]11th S 751, *[972]17th S 788, *Weogra ceastre* [793×6]11th S 146, *Wi- Wygraceaster -cœstre -cestre* [904]11th S 1280, [961×84]11th S 1375, c.1050 ASC(D) under years 1033, 1047, c.1130 ASC(H) under year 1114, with later spellings *Wirecestre* 1086, *Wircestre* 1350–c.1400, *Worcetre* 1396, *Worcester al. Wurcestre* 1487, *Wysseter* 1538×40. OE folk-n. *Weogoran*, genitive pl. *Weogorena*, + **ċeaster**. The folk-n., which occurs again in the name of a lost estate immediately W of Worcester Hallow SO 8258, *(in) þeogorena leage* [816]17th S 180 with variant *þeogorna leage* 11th, 'wood or clearing of the Weogoran', and in WYRE FOREST SO 7576 and Wyre PIDDLE SO 9647, is unexplained. One suggestion is that it derives from a lost r.n. parallel to Gaulish *Vigora*. It does not derive from the RBrit name of Worcester which seems to have been *Vertis*. Wo 19 gives a pr [ustə], Hooke 113, TC 195, RBrit 496.

WORCESTER AND BIRMINGHAM CANAL H&W SO 9465. *Worcester and Birmingham Canal* 1831 OS.

WORCESTER PARK Surrey TQ 2266. *Worcester Park* 1816 OS. P.n. WORCESTER + ModE **park**. The residence of the Earls of Worcester in the early 17th cent., cf. *Worcester House* 1680. Sr 74.

WORDSLEY WMids SO 8987. 'The wolf-guard' or 'Wulfweard's wood or clearing'. *Wuluardeslea* 12th, *Wordesley* 1834 OS. OE ***wulfweard** or pers.n. *Wulfweard*, genitive sing. ***wulfweardes**, *Wulfweardes*, + **lēah**. Mills 1998.

River WORFE Shrops SO 7696. Uncertain. *Wrhe* c.1211, *Wurgh* [1227]1603×25, *Wrgh* 1247, *Wornh* (for *Woruh*) 1248, 1285, *Worgh* 1300, *Wo(r)th* 13th, *Worfe* from 1577. The forms point to an OE **Wurh* or **Weorh* of uncertain origin, with subsequent substitution of [f] for the peripheral phoneme [x]. Perhaps ultimately to be associated with IE **wer-* 'turn', *wer-gh* 'press, strangle'; the river is very winding and mostly runs between high banks. Cf. Gothic **wraiqs* 'crooked' < **wr-ey-ĝ-*. Sa i.327, RN 470, GED W 91.

WORFIELD Shrops SO 7597. 'The open land by the r. Worfe'. *Wrfeld* 1086, 1265, *Wur(e)feld(') -feud(')* 1177–1449, *Woresfeld* 1177, *Worfeld(e) -feud -fyld -field* from 1230×1, *Wer(e)feld(')* 1196–1214, *Wirfeld* 1205, 1265, *Worveld(e)* 1334, 1570, *-vill* 1577, 1765. R.n. WORFE SO 7696 + OE **feld**. Sa i.326.

WORKINGTON Cumbr NX 9928. 'Farm, village, estate called after Wyrc'. *Wi- Wyrkynton -in- -en-* c.1125–c.1540, *Wi- Wyrkyngton* c.1130–1564, *Workington* 1564, *Wirketon'* 1211†. OE pers.n. *Wyrc* + **ing**[4] + **tūn**. Cu 454 gives pr [wərkitən].

WORKSOP Notts SK 5879. 'Werc or Wyrc's valley'.

I. *Warchesoppe, Werchesope* 1086, *Werckessop -hope, Werkessop(e) -hop'* 1183–1202, *Werk- Warksop* 1268, *Worksopp* 1332, *Worsop* 1462, *Wurksoppe* 1535.

II. *Wi- Wyrkesop(e)* 1154×89–1332, *Wi- Wyrkeshop(e)* 1237–1301, *Wirksop* 1301–2, *Wyrsop* 1354.

The two forms correspond to OE pers.n. *Weorc*, genitive sing. *Weorces*, and *Wyrc*, genitive sing. *Wyrces*, + **hop**. Cf. WIRKSWORTH Derby SK 2854. Nt 105 records three pronunciations, [wɔ:ksəp, wusəp] (local), and [wɔ:səp] (obsolete).

WORLABY Humbs TA 0113. 'Wulfric's village or farm'. *U- Vlurice- Wirichebi* 1086, *Wlfriches- Wlfrechebi* c.1115, *Wolurikesbi* 1202, *W(u)lfrikeby* 13th, *Wolri(c)kby* 1314–87, *Wolriby* 1354–81, *Worleby* 1395–1664, *Worlaby* from 1478. OE pers.n. *Wulfrīċ* + **bȳ**. Li ii.302, SSNEM 79.

WORLD'S END Berks SU 4876. *The Worlds End* 1830. Probably refers to a distant field like the *California, Botany Bay* type of field name. Brk 233.

East WORLDHAM Hants SU 7438. *Estwerldham* 1254, *Est Wordleham* 1327×77. ME adj. **est** + p.n. *Werilde- Wardham* 1086, *Wer(i)ld- Worildham* 1200–36, *Werldham* 1204–70, *Worldham* 1258, *Wordeleham* 1280, *Ward-le-ham* 1789, 'Wærhild's estate', OE feminine pers.n. **Wǣrhild*, genitive sing. **Wǣrhilde*, + **hām**. 'East' for distinction from West WORLDHAM SU 7436. Ha 182, Gover 1958.106.

West WORLDHAM Hants SU 7436. *Westwerldham* 1236, *West Worldham* 1258. ME adj. **west** + p.n. Worldham as in East WORLDHAM SU 7438. Ha 182, Gover 1958.106.

WORLE Avon ST 3562. 'Wood-grouse wood or clearing'. *Worle* from 1086, *Wurle* 1226, 1257. OE ***wōr** + **lēah**, dative sing. **lēa**. DEPN, Studies 1936.95.

WORLEBURY Avon ST 3162. 'Worle camp'. *Worle Berry* 1809 OS. P.n. WORLE ST 3562 + dial. **berry, bury** 'a camp, a fort' < OE **byriġ**, dative sing. of **burh**. The reference is to the Iron Age fortress on Worlebury Hill. Pevsner 1958N.337.

WORLESTON Ches SJ 6556. 'Werulf's farm or village'. *Werblestune* 1086, *Uerulestane* [1096×1101]1150, 1150, *Werleston* 1216×50–1367 with variants *-is- -as- -us-*, *Worleston* from 1282 with same variants. OE pers.n. *Wēr(w)ulf*, genitive sing. *Wēr(w)ulfes*, + **tūn**. DEPN, Che iii.151 gives pr ['wɔ:rlstn].

WORLINGHAM Suff TM 4489. Partly uncertain. *Wer- Warlingaham* 1086, *(ad) Uuerlinge* [1087×98]12th, *Werlingeham* 1168×74, 1212, *Werlingham* 1173–1327, *Wir- Wyr-* 1221–91, *Wur-* 1568, *Worlingham* 1674. Possibly 'the homestead of the Werlingas, the people called after Werel', OE folk-n. **Werlingas* < pers.n. *Werel* + **ingas**, genitive pl. **Werlinga*, + **hām**. The pers.n. is uncertain and some other n. in *Wēr- Wǣr-* may be in question. Baron, ING 132.

WORLINGTON Devon SS 7713. 'Estate called after Wulfræd'. *Vlfredintvne, O- Vlvrintone* 1086, *Wolvringt(h)on* 1242, *Wulrynton* 1249, *Worlington* 1581. OE pers.n. *Wulfrǣd* + **ing**[4] + **tūn**. D 401.

WORLINGWORTH Suff TM 2268. 'The enclosure of the Wilringas, the people called after Wilhere'. *(et) Wilrincgawerþa* 1023×38 S 1489, *Wirlingaweorð* 11th, *Wyrlingwortha* 1086, *-wurðe -wrth -worth(e)* 1065–1339, *Wirlingworth* 1275–1468 with variant *Wilryng-* 1319–1403, *Wo- Wurlyngworth(e) -ing-* 1483–1592, *Worlingworth* from 1521, *Warlingworth* 1610. OE folk-n. **Wilringas* < pers.n. *Wilhere*, genitive pl. **Wilringa*, + **worth**. DEPN, Baron.

WORMBRIDGE H&W SO 4331. 'Bridge over Worm Brook'. *Winebruge* [1199]17th, *Wermebrige* 1207, *Worm(e)brigg(e)*

†Probable identification.

†The form *Wurcingtun* [946]copy S 520 belongs not here but to WARKTON Northants SP 8979.

1292–1338. R.n. Worm + ME **brigge** (OE *brycg*). Worm, *Guormui -muy, Gurmuy* c.1130, *Worme* c.1540, is 'the dark stream', OW ***wurm** (W *gwrm*) 'dark brown, dark blue'. For the W suffix cf. the W forms of the river LUGG SO 5251. He 214, RN 471.

WORMEGAY Norf TF 6611. 'Wyrm's island'. *Wermegai* 1086, 1162, *Wirmegeie* c.1150, *Wurmegai* 1159, *Wirmingai* 1173. OE pers.n. *Wyrm* or **Wyrma* + **ing**[4] + **ēġ**. DEPN.

WORMELOW TUMP H&W SO 4930. *Wormelow Tump* 1831 OS. P.n. Wormelow + pleonastic ModE **tump** 'a small hill, a mound, a tumulus'. Wormelow, *Vrmelauia, Wermelav* 1086, *Wurmelawe* 1227, is 'hill by Worm Brook', r.n. Worm as in WORMBRIDGE SO 4331 + OE **hlāw**. The hill was the meeting place of one of the Herefordshire hundreds. Bannister 214, RN 471.

WORMHILL Derby SK 1274. 'Wyrma's hill'. *Wruenele* 1086, *Wermenhull -hil(l)* 1205–86, *Wurmen- Wyrmenhull -hille* 1230–72, *Wurmehull(e) -hill* 1204–1304, *Wurmhull(e) -hill* 1185–1330, *Wormhill* from 1278. OE pers.n. *Wyrma*, genitive sing. *Wyrman*, + **hyll**. The form *Wyrman* cannot easily be linked to OE **wyrm** 'a snake, a serpent, a dragon', genitive sing. **wyrmes**, genitive pl. **wyrma**, and must, therefore, represent a pers.n. Db 179.

WORMINGFORD Essex TL 9332. 'Withermund's ford'. *Widemondefort* 1086, *Wiðer- Wythermunde(s)ford* 1186–1290 with variants *Wi- Wyder- Wi- Wydre-, Wy- Wir(e)munde(es)ford* 1254–90, *Wormand(e)ford* 1254–1333, *Wa- We- Wormyngford* 1459–97, *Wormingford(e)* 1535. OE pers.n. **Withermund*, genitive sing. **Withermundes*, + **ford**. Ess 403.

WORMINGHALL Bucks SP 6408. Either 'nook of land called after Wyrma' or 'snake nook'. *Wermelle* 1086, *W(u)rme- Wi- Wyrmehal(e)*, 1163–1302, *Wormenhale* 1280–1382, *Worminghall* from c.1450, *Wurnall* 1535. OE pers.n. *Wyrma* or **wyrm** + **ing**[4] or **ing**[2] + **halh**. Bk 129 gives pr [wɔ:nəl] showing assimilation of *-mn-* to *-n-* as in the 1535 spelling, PNE ii.282, L 102, 111.

WORMINGTON Glos SP 0436. 'Wyrma's estate'. *Wermetvne -tone* 1086, 1221, *Werminton'* 1201, *-i-* 1221, *-ing-* 1248, *Wrmitune* 1185, *Wormin(g)ton(e) -yn(g)-* 1220–1654. OE pers.n. **Wyrma*, genitive sing. **Wyrman*, varying with **Wyrma* + **ing**[4] + **tūn**. Gl ii.40.

WORMINSTER Somer ST 5742. 'Wyrm's or the dragon's tor'. *Wormester* (sic for *-tor*) [946]14th S 509, *Wuormestorr* *[1065]18th S 1042, *Weremestorre* 1176. OE pers.n. *Wyrm* or **wyrm**, genitive sing. *Wyrmes*, **wyrmes**, + **torr**. The reference is to *Currington* 1817 OS (? 'the quern-stone hill'), a hill rising steeply to 416ft. NE of the village.

WORMLEIGHTON Warw SP 4453. 'Wilma's herb garden'. *Wilman lehttune* [956]13th S 588, *Wilmaleahtun* [990×1006]13th S 937, *Wimeres- Wimenes- Wimelestone* 1086, *Wilmelec(h)ton(a)* 1100×35–1200, *-leghton* 1256–1354, *Wermelistone* 1199, *Wirmelegton -lech-* 1221, *Wurmeleghton* 1232, *Worme-* 1316, 1497, *Worme Leighton* 1397. OE pers.n. *Wilma*, genitive sing. *Wilman*, + **lēac-tūn**. The same man is named in *Wilmanforda* 'Wilma's ford' in the bounds of Wormleighton [956]13th S 588 and *Wilmanbroc* 'Wilma's stream' in the bounds of Ladbroke and Radbourne ibid. Wa 275.

WORMLEY Herts TL 3605. 'Wyrma's or snake infested wood or clearing'. *Wrmeleia* c.1060, *(æt) Wurmeleá* [1053×66]12th S 1134, *Uurmelea* *[1066]12th S 1043, *Wermelai* 1086–1317 with variants *-le(ye) -legh*. OE pers.n. *Wyrma* or **wyrm**, genitive pl. **wyrma**, + **lēah**. Hrt 233.

WORMLEY WEST END Herts TL 3306. *Wormley West End* 1805 OS. P.n. WORMLEY TL 3605 + ModE **west, end**.

WORMSHILL Kent TQ 8857. Uncertain. There are three types:
I. *Godeselle* 1086, *Godeshelle* c.1100.
II. *Wotnesell* c. 1225, *Wodnes(s)ell'* 1232, 1242×3, *Wodneshill' -helle* 1270, 1275.
III. *Worneshell(e) -hull(e) -hill* 1253×4–70, *Wormeshille -hull* 1270, 1316, 1346 etc.
Type I may be OE **god** 'a pagan god', genitive sing. **godes**, or may be due to OFr *gu-* [g] or [gw] for OE *w*. Type II is apparently 'Woden's hill', OE pagan godn. *Wōden*, genitive sing. *Wōd(e)nes*, + **hyll**, Kentish **hell**. Type III is possibly 'the (?) flock's hill', OE **w(e)orn** 'troop, company, multitude crowd', later assimilated to ME **worm** 'a worm, a snake', genitive sing. **w(e)ornes**, + **hyll**, Kentish **hell**. Type I also suggests OE **(ġe)sell** 'shed, a shelter' as the original generic. No certainty is possible. PNK 240, Settlements 112.

WORMSLEY H&W SO 4247. 'The snake's wood or clearing'. *Wrmeslev, Wermeslai* 1086, *Wu- Wormesley(e)* late 13th–1397. OE **wyrm**, genitive sing. **wyrmes**, + **lēah**. He 215.

WORPLESDON Surrey SU 9753. 'The hill with a path'. *Werpesdune* 1086, *Werplesdon* 1215–1396, *Werplindon -en-, Wereplesdune -don* 13th cent., *Werpelesdon* 1313, *Warples-* 1288–1478, *Worpelesdene* 1428, *Worplesden* 1509. OE ***werpels** + **dūn**. The reference may be to the steep track leading up the hill on which the church stands. Sr 161–2, L 146.

WORRALL SYorks SK 3092. 'The nook of land where bog-myrtle grows'. *Wihale -a* (sic) 1086, *Wir- Wyrhal(le)* 1218–1362, *Wir- Wyral -all* 13th–1441, *Werall* 13th, 1442, *Worhall* 1461, *Worall* 1557–1604. OE **wīr** + **halh**. Probably a secluded bit of land overlooking the r. Don. YW i.230, L 106.

High WORSALL NYorks NZ 3809. ModE **high** + p.n. *Wircheshel, Wercesel* 1086, 'Weorc's nook of land', OE pers.n. *Weorc*, genitive sing. *Weorces*, + **halh**. The reference is to land in a bend of the river Tees. YN 280, 173.

Low WORSALL NYorks NZ 3909. ModE **low** + p.n. *Wercesal, Wirceshel, Wercheshal(e)* 1086, *Wi- Wyrkeshale* [1158]15th–1335, *Wi- Wyrkesale* 1285–1367, *Wirsal* [1316]16th, 1369, *Parva Worsall* 'little W' 1483, as in High WORSALL NZ 3809. YN 280, 173.

WORSBROUGH SYorks SE 3503. 'Wirc's fortification'. *Wircesburg* 1086, *Wirkeburc -burg(a)* 1154×89–1260×80, *Wi- Wyrkesburg(h) -burc* 1170×84–1382, *Workesburg(h -e)* 1449–1524, *Worsburgh* 1458–1612, *Worspur* 1556, 1695, *Wosper* 1669, 1765. OE pers.n. **Wirc* or **Wyrc*, genitive sing. **Wirces*, + **burh**. One of a series of fortifications in the Don–Dearne valleys. YW i.292.

WORSHAM ESusx TQ 7509 → WARTLING TQ 6509.

WORSLEY GMan SD 7501. Partly uncertain. The forms fall into three types:
I. *Werkesleia* 1196, *Worke(s)ley(e)* 1300, *Workeslegh* 1332 etc., *Worsley* 1843 OS.
II. *Wyrkedele* 1212, *Wirkedley* 1219, *Wirkidele* 1246, *Wi- Wyrkithele(ye) -lege* 1246, *W(o)rkedele(y)e* c.1225–99.
III. *Wurkythesle* 1246, *Workedesle(gh)* 1259, 1278.
Type I could be 'Weorc's wood or clearing', OE pers.n. *Weorc*, genitive sing. *Weorces*, + **lēah**, but this cannot explain types II and III. They might contain a British p.n. consisting of an unknown element + PrW ***cę̄d** 'a wood', endingless in II, with the ME secondary genitive sing. **-es** in III. An alternative possibility is that II and III contain an OE feminine pers.n. **Weorcgȳth*, genitive sing. **Weorcgȳthe*, secondary genitive sing. **Weorcgȳthes*, but no such name is on record. La 40, Mills 152.

WORSTEAD Norf TG 3026. 'Enclosure site, enclosure, farmstead'. *Wrðestede* *[1044×7]13th S 1055, *Wrde- Ordesteda(m), Vrdestada* 1086, *W(u)rthsted(e)* 1135×54–1431 with variants *W(o)rde- Worth(e)-* etc., *Wursted(e)* 1256, 1320, 1518, *Worsted(e)* 1291–1477. OE **worth** + **stede**. The exact sense of this term is uncertain. Nf ii.204, Stead 186. DEPN gives pr [wusted].

WORSTHORNE Lancs SD 8732. 'Weorth's thorn-tree'. *Worthesthorn, Wrdestorn* 1202, *Wurthes- Worthesthorn(e)* 1246, 1285, *Worstorn* [1296]14th, *Worsthorne* 1496. OE pers.n. **Weorth*, genitive sing. **Weorthes*, + **thorn**. OE pers.n. **Weorth* occurs in London in the 12th cent. La 84 gives pr [wɔ:sthɔ:n], Jnl 17.52.

WORSTON Lancs SD 7642. 'Weorth's farm or village'. *Wrtheston* 1242, *-is-* 1258, *Wurtheston* 1285, *Worston(e)* from 1296. OE pers.n. **Weorth*, genitive sing. **Weorthes*, + **tūn**. Half a mile to the NE is a hill rising to 725ft. called Worsaw Hill SD 7743, *Worsow* 1529, *Worsaw* 1538, 'Weorth's hill', OE pers.n. **Weorth*, genitive sing. **Weorthes*, + **hōh** or ON **haugr**. La 78, Jnl 17.48.

WORTH 'An enclosure'. OE **worth**.

(1) ~ Kent TR 3356. *Wurth(e)* 1226–59, *Wrth'* 1240, *Worth(e)* 1247. c.1295. PNK 590, Cullen.

(2) ~ WSusx TQ 2936. *Orde* 1086, *Wurða -e* 1175–6, *Worth(e)* from 1268, *Woore* 1568, *Woord* 1641, 1716. Sx 280, SS 79.

(3) ~ ABBEY WSusx TQ 3134. P.n. WORTH TQ 2936 + ModE **abbey**.

(4) ~ MATRAVERS Dorset SY 9777. 'W held by the Matravers family'. *Worth Matrauers* 1664. Earlier simply *Orde, Wirde* 1086, *Orda, Urda, Wirda* 1086 Exon, *Wrde* 1086–c.1170, *W(u)rth(e)* 13th cent., *Worth(e)* 1230–1575, and *Wurth Fitzpayne* 1544. Robert Fitz Pain held land here in 1212, John Mautravers in 1335. Do i.63.

(5) River ~ WYorks SE 0137. *The Worthe* 1577. A back-formation from HAWORTH SE 0337 or OAKWORTH SE 0338 through both of which it runs. YW vii.144.

WORTHAM Suff TM 0877. Perhaps 'the enclosed homestead'. *Wrtham* 1241–86 including [942×c.951]13th S 1526, *Wortham, Word(h)am* 1086, *Wurtham* c.1200. OE **worth** + **hām**. DEPN, Baron.

WORTHEN Shrops SJ 3204. 'The enclosed settlement'. *Wrdine* 1086, *Worthyn -in* 1246–1787, *Worthen* from 1501. OE **worthiġn**. Sa i.328.

WORTHING Norf TF 9919. 'The enclosure'. *Worthing* 1282, *Worthene* 1355. OE **worthiġn**. DEPN, ING 225.

WORTHING WSusx TQ 1402. 'The Weorthingas, the people called after Weorth'. *(M)ordinges* 1086, *Wording(e) -yng* 1240–1327, *Worthing(e)* from 1244, *-inges* 1288. OE folk-n. **Weorthingas* < pers.n. **Weorth* + **ingas**. Sx 194, ING 42, SS 64.

WORTHINGTON Leic SK 4020. Possibly the 'estate called after Weorth'. *Werditone* 1086, *Wrd- W(u)rthinton' -yn-* c.1130–1340, *Wrþ- W(u)rthington(e)* c.1130–1340, *Worthington(e) -yng-* 1260–1610. OE pers.n. **Weorth* + **ing**[4] + **tūn**. Lei 421.

WORTHY 'The enclosure'. OE **worthiġ**.

(1) ~ DOWN Hants SU 4534. *Worthy Down* 1817 OS. P.n. Worthy as in Headbourne WORTHY SU 4932 and King's WORTHY SU 4933 + ModE **down**. Gover 1958.77.

(2) Abbots ~ Hants SU 4932 → Headbourne WORTHY SU 4932.

(3) Headbourne ~ Hants SU 4932. 'Hydeburne Worthy'. *Hideburnewurthy* 1236, *Hyd(e)burn(e) Wordy -worthy* 1291, 1341, *Hedborneworthy* 1579. R.n. Hydeburne, *(into) hydiburnan* *[854]12th S 309, + p.n. *(æt, to) worðige* *[854]12th ibid., *Worðig* 1001 ASC(A), *Ordie* 1086, *Worthy* 1285. 'Headbourne', apparently the 'hide stream', OE **hīd** + **burna**, a small tributary here of the Itchen, for distinction from Abbots Worthy SU 4932, *Abboteswrth'* 1248, granted to the abbot of Hyde in 909. Earlier simply *(with) easton worðige* 'east Worthy' or 'Easton Worthy' (from EASTON SU 5132) [909]11th S 376, *Wurthige* [1026]12th S 962, *Ordie* 1086, Kings WORTHY SU 4933 and Martyr Worthy SU 5132, *Wordi Lamartre* 1243, *Martrewordi* c.1296, held by Henry le Martre 'the marten or weasel' in 1201, earlier simply *(æt) worþige, worðige* [825]12th S 273, *Ordie* 1086: subdivisions of what was originally probably a single estate, although there were already two distinct manors by the middle of the 10th cent. as is shown by the phrase *æt þan twan worþigum* 'at the two Worthies' (? King's and Martyr Worthy) [955×8]14th S 1491. Headbourne Worthy was also known as *Wordy Comitis* 'the count's Worthy' 1291, *Wordy Mortimer* 1303 and *Hydeburnemortemer* 1408 referring to its tenure by Roger de Mortuo Mari (i.e. Mortimer) in 1212. The boundary of the people of Headbourne Worthy is referred to as *(to) worthigsætena mearc, (andlang) worthihæma mearc* [904]16th S 374, 'the boundary of the *Worthiġsǣte* or *Worthiġhǣme*, the people of (Headbourne) Worthy', and *(be) hide burninga gemære* [909]11th S 376, 'the boundary of the *Hideburningas*, the people of Headbourne'. Among the puzzling features of these names is why OE **worthiġ** 'enclosure, curtilage', which normally refers to minor settlements, was used in the name of so important a royal estate, why **worthiġ**, a characteristically SW word, appears here so far east, and what the meaning is of the r.n. Headbourne, 'the hide stream, stream of the hides'. Ha 182–4, Gover 1958.76–7.

(4) King's ~ Hants SU 4933. *Chinges Ordia* 1155, 1159, *Kingeswurthy -worthy* 1227, 1382, *Worthya Regis* c.1270. Earlier simply *Ordie* 1086. lOE **cing**, genitive sing. **cinges**, + p.n. Worthy as in Headbourne WORTHY SU 4932. Held by the king in 1066. Ha 182–3, Gover 1958.77.

(5) Martyr ~ Hants SU 5132 → Headbourne WORTHY SU 4932.

WORTLEY SYorks SK 3099. 'Clearing used for growing vegetables'. *Wirtleie, W(i)rlei(a)* 1086, *Wrt(e)ley -lay* 1185×1215–1251, OE **wyrt** + **lēah**. YW i.298.

WORTON Oxon SP 4329. 'The settlement by the ridge'. *Ortune* [1050×52]13th S 1425, *Orton(a)* 13th cent., *Ouer- Uuer- Over- Netherorton(e)* 1200–1337, *Hortone* 1086, *Horton'* 1199, *Nother- Nether- Overhorton* 1241–3, *Wurtton'* 1194, *Wrtton'* 1220, *Over- Netherwortone* 1526 and simply *Overton(e)* 1275–1341. OE **ōra** varying with ME **over** + **tūn**. O 295.

WORTON Wilts ST 1173–1305. 'The herb or vegetable garden'. *Wrton(a)* 1173, *Worton(e)* from 1220, *Wourton* 1558×1603, *Woorton* 1622. OE **wyrt** + **tūn**. Wlt 248 gives pr [wɛːtən].

Nether WORTON Oxon SP 4230. *Nitherortun'* 1220–1428 with variants *Netherorton' -ton(a), Nutheroverton'* 1273–6, *Nether-* 1284×5, *Netherwortone* 1526. ME **nether** + p.n. *Ortune* [1050×2]13th S 1425–1246 with variants *-ton' -ton(a), Overton(')* *-tone* [1236]c.1300–1341, 'the settlement at the ridge', OE **ōra** + **tūn**. 'Nether' for distinction from Over Worton *Overorton'* 1200 etc., *Overwortone* 1526. Forms in *Over-* show substitution of **ōra** by **ofer** 'a flat-topped ridge'. O 295, Jnl 22.35.

WORTWELL Norf TM 2784. *Wortwell* 1837. This might be 'herb well or spring', ModE **wort** (OE *wyrt*) + **well**. Early forms are needed.

WOTHERTON Shrops SJ 2800. 'The settlement at *Wudaford*, the wood ford'. *Udevertune* 1086, *W(u)deverton(')* 1206–1228, *Wode-* 1242–1326, *Woderton(')* 1274–1623, *Wooderton* 17th cent., *Wotherton* from 1649. OE p.n. **Wudaford*, OE **wudu**, genitive sing. **wuda**, + **ford**, + **tūn**. Sa i.328.

WOTTER Devon SX 5561. 'Wood tor'. *Wodetorre* 1263, *Wottorre* 1345, *Wottor* 1445. OE **wudu** + **torr**. D 260.

WOTTON 'Wood settlement'. OE **wudu** + **tūn**. Probably often a specialised place within a land-unit or multiple estate where wood for fuel or building was obtained or worked. See also WITTON, WOOTTON. PNE ii.280.

(1) ~ Surrey TQ 1247. *Odetone* 1086, *Wodeton* 1235–1355, *Wude- Woden- Wodi-* 13th cent., *Wodyng-* 1292–1456, *Wotton* from 1548, *Wutton* 1610. Sr 278 gives pr [wutən], L 227.

(2) ~ -UNDER-EDGE Glos ST 7593. *Wotton under Egge* 1359–1592. Earlier simply *(æt, to) Wude- Wudutune* [940]12th S 467, *Vutune* 1086, *Wottun -tona* 1154–1532. Wotton is situated immediately beneath the Cotswold escarpment. Gl ii.255 gives pr ['wutn 'undereʒ].

(3) ~ UNDERWOOD Bucks SP 6816. *Wotton-under-Bernewode* 1415. Earlier simply *(in) wudotune* 844×5 S 204, *Oltone* 1086, *Wutton'* 1167. For *Bernewode* see GRENDON UNDERWOOD SP 6820. Bk 113.

WOUGHTON ON THE GREEN Bucks SP 8737. *Woughton on the green* 1834 OS, earlier simply *Ulchetune* 1086, *Woche- Wo- Wuketon(e)* 1167–1362, *Wokton* 1318, *Woughton* from 1459, *Wafton al. Woveton* 1700. OE pers.n. **Wēoca* + **tūn**. **Wēoca* (from *Wēoh-* in names like *Wēohstān*) will acount for the predominant *Wo-* spellings and also for *Weketune* 1199, *Wicheton*

1200 from the smoothed form *Wēca*. Aspiration of [k] > [x] seems to be a later development which, with sound substitution of [f] for the peripheral phoneme [x], produced the pr [wu:ftən] recorded in Bk 28 and the 1700 spellings. The village has grown up round a large central green. Jnl 2.33 gives local prs [u:fun] and [wu:ftun].

WOULDHAM Kent TQ 7164. 'Wulda's homestead or estate'. *uuldaham* 811 S 165, *Wulda ham* 960×88 S 1458, *Uulde -Uuldaham* [960×88]12th ibid., *Wuldaham* [995]12th S 885, *Oldeham* 1086, *Wu- Wold(e)ham* 1089–1240 etc. OE pers.n. **Wulda* + **hām**. PNK 152, KPN 123, 360, ASE 2.29.

WRABNESS Essex TM 1731. 'Wrabba's headland'. *Wrabenasā* 1086, *-nase -nese -nas(se) -nose* 1121–1428, *Wravnes, Wrabnish* 17th. OE pers.n. **Wrabba* + **næss, ness, nōse** or **nosu**. This is a marked headland protruding into the Stour. Since **Wrabba* is a formation on an apophonic variant of the root of *wry* 'bent, twisted' it is conceivable that there was an OE adj. ***wrabb** with the same sense. The headland itself may have been called *Wrabbe* 'the bent one', cf. Archivum Linguisticum 1954.74–80. Ess 358, Jnl 2.46.

WRAFTON Devon SS 4935. Possibly 'farmstead built on piles'. *Wratheton* 1238–1340, *Wrathton* 1412, *Wraghton* 1284, *Wroughton* 1738, *Wrafton* 1809 OS. OE **wrathu** 'prop, support' + **tūn**. The reference may be to an estate or farm where props were manufactured or obtained, or rather to a building erected on piles – Wrafton lies low near the Taw estuary. The history of the name shows substitution of the phoneme [f] for the peripheral phoneme [x] which itself was substituted for [θ]. D 45.

WRAGBY Lincs TF 1378. Probably 'Vragi's village or farm'. *Waragebi* 1086, *Wrag(h)ebi* c.1115, *Wraggebi* 1154×89, *Wrackebi* 1212, *Wraghby* 1353, *Wraweby* 1268, *Wrauby alias Wragby* 1539. The development of this name with $g > g^w > w$ points to a specific with medial spirant [γ], probably ON pers.n. *Vragi*, genitive sing. *Vraga*, + **bȳ**. The pers.n. *Vragi*, however, is not independently recorded in Lincs or Yorks, and the specific may possibly be ON ***vragi** 'a bollard', genitive sing. ***vraga**, referring to the high ground on which the village stands. It is situated in Wraggoe wapentake, *Waragehou Wapentac* 1086, *Wraghehov* c.1115, *Wragehou* 1168, *Wraggeho* 1169–1202, *Wraghou* 1206, probably 'Vragi's burial mound', ON pers.n. *Vragi*, genitive sing. *Vraga*, + **haugr**; but again the specific might be ***vragi**. SSNEM 79, 163, Cameron.

WRAKENDIKE T&W NZ 3262. 'The ditch or dike of the exiles'. *Vrakendic* 1144×52, [1144×52]c.1220×30, *Wrakendyk* c.1225, *Wrackendike* 1647, *Wreken Dike* 1863 OS, cf. *Wracennhegge* c.1190, 'the hedge of the exiles'. OE **wreċċa, wræċċa**, genitive pl. **wræccna**, + **dīċ**. The reference is to the Roman road from Wrekenton to South Shields, Margary No. 809. It may have been so called as a route used by fugitives from the law seeking exile under the terms of sanctuary, or, the monks of Wearmouth-Jarrow may themselves be referred to. Men who in Anglo-Saxon times adopted the monastic or eremitical life were regarded as *peregrini*, pilgrims for the love of God and exiles from the world. *Peregrinus* in Psalm 68.9 is translated *wrecca* in the OE version and at least one hermit is described in OE as *Godes wrœcca* 'God's exile'. NbDu 220.

WRAMPLINGHAM Norf TG 1106. Uncertain. *Wranplincham, Waranpli(n)cham* 1086, *Wramplingham* from c.1185. This is not obviously an *-ingaham* folk-n. Perhaps 'homestead' or 'promontory called or at Wrampling, the bend place', OE ***wrampling** + **hām** or **hamm** 2b. **Wrampling* is either an *ing* derivative of the root seen in ME *wrimple* 'wrinkle' or a *ling* derivative of the root seen in 17th cent. *wramp* 'a twist'. Possibly **Wrampling* was the name of the little twisting stream on which Wramplingham lies. DEPN, ING 140.

WRANGLE Lincs TF 4250. Originally a r.n., 'the crooked stream'. *Weranghe -angle* 1086, *Wrangle* from 1202, *Wrengle* 1191–1250. A *-ula* derivative of ON *vrangr* 'bent, crooked' or possibly of an OE **wrang*. Cf. the Norw r.n. Rangla and Swed Vrangel, the name of a spring. In Denmark Vrangel is the name of an irregular shaped piece of land in S Schleswig. Payling 1940.132, SSNEM 163.

WRANGWAY Somer ST 1217. The name means 'crooked road' ultimately < OE **wrang** + **weġ**, but early forms are needed.

WRANTAGE Somer ST 3022. Uncertain. *Wrentis* 1199, 1201, *Wrentis(s)e* 1227, 1246, *Wrentyssh* 13th. Possibly a compound of OE ***wrǣna** + ***etisc** 'stallion pasture'. DEPN.

West WRATTING Cambs TL 6052. *Westwratting(e) -yng(e)-* from 1272, *Westwrotyng(e)* 1285, 1405. ME **west** + p.n. *(œt) Wre(a)ttinge* [974]11th S 794–1273, *(œt) Wrœttincge* 975×1016 S 1487, *Wratinge* [1042×66]12th S 1051, *Waratinge* 1086, *Wratting(e)* from 1218 including [c.1050]14th, *Wrotting(e) -yng(e)* 1199–1450, 'place where madder is obtained' OE **wrætt** + **ing**[2]. In glosses **wrætt** has the sense *Rubia tinctorum*: its roots were used as a source of dye. Ca 121, ING 198.

Great WRATTING Suff TL 6848. *Magna Wrothing* 1269. ModE **great**, Latin **magna**, + p.n. *Waracatinge, Wr- Vratinga* 1086, *(ad) Uurartinge* [1087×98]12th, *Wratig* c.1200, *Wrotting, Wretting'* 1206, *Wrettinge* 1242×3, 'the place where madder grows', OE ***wrætting** < **wrætt** + **ing**[2]. 'Great' for distinction from Little Wratting TL 6847, *Parva Wrottyng* 1296×1375. ING 198.

WRAWBY Humbs TA 0108. 'Wraghi's village or farm'. *Waragebi* 1086, *Wragebi* c.1115, *Wragheby* c.1200–1327, *Wra(u)ghby* 1328–1441, *Wraweby* 1276–1438, *Wrawby* from 1375. ON pers.n. *Wraghi*, genitive sing. *Wragha*, + **bȳ**. Identical with WRAGBY Lincs TF 1378. Li ii.307, SSNEM 79.

WRAXALL 'Buzzard's nook'. OE ***wrocc, *wrōc**, genitive sing. ***wrocces, *wrōces**, + **healh**. The reference is to a type of secluded hollow harbouring prey which attracted buzzards and over which they were seen to circle. Studies 1936.96ff., PNE ii.279, L 110.

(1) ~ Avon ST 4972. *Werocosale, Worocosala* 1086, *Wrokeshall* 1227, *-hale* 1327. The reference is to a recess in the hills N of the Yeo river. DEPN, Studies 1936.96, L 101, 109–10.

(2) ~ Somer ST 6036. No early forms.

(3) ~ Wilts ST 8364. *Wroxhal* 1227, *Wrokeshal'* 1242. Also known as *Suthwroxhall* 'South W' 1468 for distinction from North WRAXALL ST 8175. Wlt 125.

(4) North ~ Wilts ST 8175. *Northwroxhall* 1468, *Northe Wraxhale* 1506, *North Wraxall* 1582. ME **north** + p.n. *Werocheshalle* 1086, *Wro(c)keshal(e)* 1229–1314, *Wraxhal'* 1229, 'North' for distinction from WRAXALL ST 8364. Wlt 113.

WRAY Lancs SD 6067. 'The nook, the corner'. *Wra* 1227–71 etc., *Wraa* 1327, 1332. ON **vrá**. The village lies in a corner of the river Hindburn on the opposite side to the rest of the township. Cf. WREA GREEN. La 182, SSNNW 178.

WRAY CASTLE Cumbr NY 3701. P.n. Wray as in High WRAY SD 3799 + ModE **castle**.

High WRAY Cumbr SD 3799. *the Heywray* 1619. Adj. **high** + p.n. *Wraye* c.1535, 1537, ME, early ModE **wra(y)** (ON *vrá*) 'a nook, a corner' referring to a remote place. 'High' for distinction from Low Wray SD 3702, *Lowrey* 1656. La 219.

WRAYSBURY, formerly WYRARDISBURY, Bucks TQ 0073. '(At) Wigræd's fortified place'. *Wirecesberie* 1086, *Wiredesbur'* 1195, *Wyredesburi* 1284, *Wyredeberia -biry* 13th cent., *Wyrardesbury -ys-* 13th–1489, *Wyrardisbury* 1925, *Wraysbury* from 1536. OE pers.n. *Wīgrœd*, genitive sing. *Wīgrœdes*, + **byriġ**, dative sing. of **burh**. Bk 244 gives pr [wreizbəri].

WREA GREEN Lancs SD 3931. 'Green at Wrea'. P.n. Wrea + ModE **green**. Wrea, *Wra* 1201, 1226, *le Wra(a)* 1323, 1327, *Wraa* 1324, 1380, *Wro* 1322, is 'the nook, the corner', ON **vrá**. Identical with WRAY SD 6067. La 152, SSNNW 178.

River WREAKE Leic SK 6616. 'The crooked, twisted river'. *Werc* 1154×89, c.1235, 1237, *Wreþech* 1224×30, *Wrethec -k(e)* c.1225–1501, *Wrek* 1272–1492, *Wreic -k, Wreyk(e)* 1243–1486, *Wreak* 1576. ON r.n. **Wreithk* < ON *(v)reithr* 'crooked, twisted' + r.n. suffix *-k*. **Wreithk* was simplified to *Wrethk*, a spelling occurring once in 1320, which with intrusive vowel became *Wrethek*. A metathesised form **Werthk* probably lies behind the earliest spelling and its 16th cent. descendants *Warke, Urke* 1577, 1586. RN 472.

WREAY '(The) nook'. ME **wra(y)** (ON *vrá*).

(1) ~ Cumbr NY 4423. *Wra* 1487, *the Wraye* 1581. Situated at the edge of Watermillock township. Cu 258.

(2) ~ Cumbr NY 4349. *the Wrey* 1600, *Wraye* 1619, *Wrea* 1777. Short for *Petrelwra* 1272–1650 with variants *Peterel- Paytrel- Petrille-* etc., 'nook or corner of land by the river Petteril', river n. PETTERIL NY 4543 + WREAY. Situated at the SE edge of St Cuthbert's parish, Carlisle, to which it was added in 1934. The forms *Patterdaylewrye* and *Peterdaylewraye* 1541 are due to folk-etymology. Cu 167, SSNNW 258.

WREDON Staffs SK 0846 → WEAVER HILLS SK 0946.

WREIGHILL Northumb NT 9701 → WRELTON NYorks SE 7686.

WREKENTON T&W NZ 2759. *Wrekenton* 1863 OS. It is odd that no earlier forms are recorded. The name is clearly related to WRAKENDIKE NZ 3262 and means either 'the settlement of the exiles' or more likely 'the farm or village on Wrakendike'.

The WREKIN Shrops SJ 6308. An outcrop of pre-Cambrian volcanic rock forming a conspicuous hill of 1334ft. with an Iron Age hill-fort, probably the original tribal capital of the Cornovii who were later settled by the Romans at *Viroconium*, the 'town of Virico'. For the Latin forms of the name see VIROCONIUM SJ 5608. Subsequent forms are:

I. English: *(on) Wrocene* [975]12th S 802, *La Wrokene* 1284, *La Wrekene* 1278.

II. Welsh: *Cair Guricon* [c.800]9th HB, *(G)ureconn* [9th]14th Canu Llywarch Hen.

The people dwelling near the Wrekin were called the *Wreocensǣtan* in OE charters, *Wocensætna land* 7th BCS 297, *(in) Wreocensetun* [855]11th S 206, *provincia Wrocensetna* [963]12th S 723.

The original p.n. must have existed in two forms, **Uriconion* (Latin *Viroconium*) and **Uricono-* which is needed to explain OW *Guricon* and PrOE **Wricun* > **Wriocon* > *Wreocen* from which ME forms *Wrok-* (representing dial. [wrø:k]) and *Wrek-* (representing standard [wre:k]) are normal. DEPN, LHEB 601–2, Thomas 1976.185.

WRELTON NYorks SE 7686. Partly uncertain. *Wereltun* 1086, *Wrelton* from 1282, *Wherlton* 1316. Unidentified specific + OE **tūn** 'farm, village, settlement'. Suggestions for the specific are OE p.n. *Wearg-hyll* 'felon hill' found in Horrell Corn SX 3088 and Wreighill Northumb NT 9701, *Werghill* 1292, or *Hwerwella* 'well with a kettle' as in WHERWELL SU 3840. YN 81, NbDu 220, PNE ii.247.

WRENBURY Ches SJ 5947. Probably 'Wrenna's fortified place'. *Wareneberie* 1086, *Wrennebury(e)* 1284–1510 with variants *-buri(e), Wrenbury* from 1327 with similar variants. OE pers.n. **Wrenna* + **byriġ**, dative sing. of **burh**. Alternatively the specific might be OE **wrenna** 'a wrenn'. The suggestion **wrǣna** 'a stallion' is less likely in view of the rarity of forms with *Wrene-* (*Wrenebur(y) -ber* c.1327, 1328, 1488). Che iii.119, Studies 1936.67.

WRENINGHAM Norf TM 1698. Either 'homestead of the Wrenningas' or possibly 'homestead' or 'promontory called or at Wrenning, the wren place'. *Wreningham* [c.1050]13th S 1516, *Vrnincham* 1086, *Wreningeham* 1197, *Wrenningham* 1254. OE folk-n. **Wrenningas* < **wrenna** 'a wren' used as a pers.n. + **ingas**, genitive pl. **Wrenninga*, + **hām**, or p.n. **Wrenning* < **wrenna** + **ing**[2], + **hām** or **hamm** 2b. DEPN, ING 140.

WRENTHAM Suff TM 4982. 'Wrenta's homestead'. *Wret- Uuereteham* 1086, *Wrentham* from 1228, *Wrantham* 1272, 1327. OE pers.n. **Wrenta* + **hām**. For the pers.n. cf. MDu *wrant* 'quarrelsome', OFris *wranten* 'grumble'. DEPN, Baron.

WRESSLE Humbs SE 7131. Uncertain. *Weresa* 1086, *Wresel(l)* 1183–1468 with variants *-ill(e) -ull, Wressell* 1285–1609. Possibly an OE ***wrǣsel** from *wrāse* 'knot, lump', referring to the old site of the castle. However, the underlying sense 'something twisted or knotted' may refer to the twist or bend of the r. Derwent here. YE 242 gives pr [ræzl].

WRESTLINGWORTH Beds TL 2547. 'Wrestling enclosure'. *Wrastlingewrd* c.1150, *Wrestlingewurda* 1194, *Wrestlingwurth* 1234. OE ***wræstling** + **worth**. The two early spellings with medial *-e-* have been taken to point to an OE folk-n. **Wrǣstlingas* < pers.n. **Wrǣstel*, **Wrǣstla*, + **ingas**, genitive pl. **Wrǣstlinga*, but the later spelling *Wraxlingworth* 1434 shows association with ME **wraxle** 'wrestle'. Cf. PLAWSWORTH Durham etc. Bd 111 gives pr [resliŋwərθ].

WRETHAM Norf TL 9290. 'Homestead or promontory where crosswort or hellebore grows'. *W(e)retham* 1086, *Wretham* from 1177, *Wrot(t)ham* 1199, 1244. OE **wrætt(e)** + **hām** or **hamm** 2b. Both plants had important medicinal uses. Cf WRETTON TL 6899. DEPN gives pr [retəm], PNE ii.278.

WRETTON Norf TL 6899. 'Crosswort farm'. *Wretton* from 1198, *Wrottun* 1251. OE **wrætt** + **tūn**. Cf. WRETHAM TL 9290.

WRIBBENHALL H&W SO 7975. 'Wrybba's nook'. *Gurberhale* (sic) 1086, *Wrbenhala* [n.d.]11th, *Wrubbe(n)hale* [c.1160]c.1240, [c.1200]c.1250, *(W)rignall* 1565, 1581, *Ri- Rybbenhall* 1633, 1678. OE pers.n. **Wrybba*, genitive sing. **Wrybban*, + **hēalh** in the sense 'small valley'. Wo 253, L 103.

WRIGHTINGTON BAR Lancs SD 5313. P.n. Wrightington + ModE **bar** 'a toll-house gate or barrier'. Wrightington, *Wrstincton* 1195, *Wrichtington* 1202, *Wrytinton* 1256, *Wrightyngton -ing-* from 1314, is 'the settlement of the wrights', OE **wyrhta**, genitive pl. **wyrhtena**, + **tūn**. La 130, Jnl 17.73.

WRINEHILL Staffs SJ 7547. 'Wrime hill'. *Wrinehull* 1225, *le Wrimehull* 1486. P.n. *Wrime* 1278, *Wryme* 1299, 1332, + ME **hull, hill** (OE *hyll*). *Wrime*, of unknown origin and meaning, was probably the name either of the hill on which the village stands or of Checkley Brook which it overlooks. One possibility is the root **u̯reik-* 'bend' seen in ModE *wry* 'twisted', OE *wrīġian* 'to turn, twist, bend', with the concrete-forming suffix *ma-*: but this is very uncertain. DEPN, PNE ii.278, Kluge §88, Horovitz.

WRINGTON Avon ST 4762. 'Settlement on the river Wring'. *Wring'* [904]14th S 371, *Wringtone* [939×46]13th ECW, *Weritone* 1086, *Wringeton* 1243, *Wri- Wryngton* 1225, 1274, 1434. Lost r.n. *Wring* + OE **tūn**. The r.n. is recorded as *(on) Wring', (on, endlang) Wryng* [904]14th S 371, *Wrynge* 1276, possibly representing an OE **Wrīoing* 'winding stream' from OE ***wrīo, wrēo** 'twisting'. The modern name of the river is YEO ST 4662 < OE **ēa** 'river'. DEPN, RN 474, PNE ii.278.

WRITTLE Essex TL 6606. By origin a r.n., the 'chattering stream'. *Writelam, Writa* 1086, *Writele(a) -(a)* 1138–1241 etc., *Wrutel(e)* 1246, 1274, *Writ(t)l(e)* from 1218, *Wrickell* 1498, *Ritle* 1657. OE ***writol** 'chirping, chattering' sometimes compounded with **ēa** 'a stream'. The r.n. proper, *Writolaburna* [685×94]8th S 1171, was later replaced by *Wid River* 1848, a back-formation from WIDFORD TL 6905. Ess 13, 277, Jnl 2.39, 44.

WROCKWARDINE Shrops SJ 6212. 'Enclosure by the Wrekin'. *Recordine* 1086, *Rocardina* 1138, *Rockardyne* 1316, *-erdine -ardine* 17th cent., *Wrocwordina* from c.1155 with variants *Wro(c)k-* and *-wurthin(') -wurdin(') - wardin(e)* etc., *Rockwardine, Rockadine* 17th cent., *Wrockwardine* from 1616. *Wrok- Wrek-* from The WREKIN SJ 6308 + OE **worthiġn**. Sa i.329.

WROOT Humbs SE 7103. 'The snout, the snout-like spur of land'. *(insula de) Wroth* 1157, *Wrot* 1193, 1212. OE **wrōt** 'snout' used in a topographical sense. The village and church lie at either end of a ridge of high ground in Hatfield Moor. DEPN.

WROTHAM Kent TQ 6059. 'Wrota's homestead or estate'. *Uurotaham* [788]12th S 129, *Wroteham* c.1100, 1177, 1184×90, *Broteham* (sic for *Wrote-*) 1086, *Wrotham* from 1197. OE pers.n. **Wrōta* + **hām**. PNK 155, KPN 71, 360, ASE 2.29.

WROTHAM HEATH Kent TQ 6358. *Wrotham Heath* 1819 OS. P.n. WROTHAM TQ 6059 + ModE **heath**.

WROTHAM PARK Herts TQ 2599. *Wrotham Park* 1754. A house built for the Byng family in 1750 and named after their ancient seat WROTHAM Kent TQ 6259. Mx 77.

WROUGHTON Wilts SU 1480. 'The settlement on the river *Worf*'. *Wertune, Wervetone* 1086, *Worfton* 1196–1385, *Werf- Warfton* 13th cent., *Warghton* 1365, *Wrofton* 1418, *Wroghton* 1428, *Wroughton* 1466, *Roahtun* (sic) 1655. R.n. *Worf* discussed under River RAY SU 1191 + **tūn**. The earlier name was *Ellendun* c.890 ASC(A), c.1000 ASC(B) c.1000 Æthelweard, c.1100 ASC(C) all under year 823, [956]12th S 585, *Ellandun* c.1100 ASC(D), c.1121 ASC(E), *Ællandun* c.1100 ASC(F) all under year 823, *Elendune* 1086, *Elendone quod est Worftone* 'Elendone that is W' c.1270, *Ellyngdon, Elington al. Wroughton* 16th cent., either 'the elder-tree hill', OE **ellen** + **dūn**, or, 'Ella's hill', OE pers.n. *Ella*, genitive sing. *Ellan*, + **dūn**. This was the understanding of the chronicler Florence of Worcester c.1120, *in Ellandune, id est in monte Ealle* 'in *Ellandun*, that is, in the hill of *Ealla* (Latin genitive sing. *Ealle*)' and of the other early *-an* sps, but whether it was the original form of the name or a piece of folk-etymology it is now impossible to say. However, in the bounds of *Ellendun* occurs the marker *to þæm ællen stybbe* 'to the elder-tree stump' [956]12th S 585. Elcombe SU 1380 in the same parish, *Elecombe* 1086, *Ellecumba -e* 1168–1281, presents the same problem, either 'the elder-tree coomb' or 'Ella's coomb'. *Ellendun* was the site of a battle between Beornwulf king of Mercia and Egbert king of the West Saxons in 823. Wlt 278 gives pr [rɔ:tən], Grundy 1970.55, L 149.

WROXALL 'The buzzard's nook'. OE ***wrocc, *wrōc**, genitive sing. ***wrocces, *wrōces**, + **h(e)alh**, dative sing. **h(e)ale**.
(1) ~ IoW SZ 5579. *(æt) Wroccesheale* [1043×4]12th S 1391, *Warochesselle* 1086, *Wroxala* 1155×60, *Wrokeshal(e)* 1184×5–1333, *Wroxhall* 1267, *Wroxall* 1368, *Wraxsale, Wraxhall* 15th. Wt 256 gives prs ['raksəɫ] and ['rɔkəɫ sər], Mills 1996.111.
(2) ~ Warw SP 2271. *Wroches(s)ale* 1157–67, *Wro(c)keshale* 1178–1275, *Wroxhale* 1227×35–1316, *-hall* 1535. Wa 228, Studies 1936.96.

WROXETER Shrops SJ 5608. 'The Roman town (called) *Uricon*'. *Rochecestre* 1086, *Wrochecestria* 1155, *Wrokecestr'* 1255, *Wroxcestr' -cestre -cester* 1155–1737, *Wroxeter* from 1636, *Wroxtor* 1535, *Roxiter* 1588–1669. *Wrok-* from *Uricon* as in VIROCONIUM SJ 5608 and The WREKIN SJ 6308 + OE **ċeaster** with AN initial [s] for English [tʃ]. The commonest sp before 18th cent. was *Wrox(c)ester* with <x> for [ks]. The loss of the second *s* appears to be a late dissimilatory change in the sequence *s–s*, [roksestə] > [roksetə]. Sa i.330–1.

WROXHAM Norf TG 3017. 'Wroc's homestead or river bend land'. *Vrocs- Vrosc- Wross- Vrochesham* 1086, *Wrokesham* c.1220, 1254. OE pers.n. *Wroc*, genitive sing. *Wroces*, + **hām** or **hamm** 1. An alternative suggestion here is that *Wroc* is not a pers.n. but represents OE ***wrocc, *wrōc** 'a buzzard'. DEPN, Studies 1936.96, PNE ii.279.

WROXTON Oxon SP 4141. 'Buzzards' or buzzard's stone'. *Werochestan* 1086, *Wrokeston' -stan(e) -stone* 1203–38, *Wrox(s)tan -ton(e)* 1204–1428, *Wraxton* 1285, *Raxston* 1675. OE ***wrocc, *wrōc**, genitive pl. ***wrocca, *wrōca**, genitive sing. ***wrocces, *wrōces**, + **stān**. O 409 gives pr [rɔkstən], occasionally [ra:kstən].

WRYDE CROFT Cambs TF 3107. *Wyrde Croft* 1824 OS. P.n. Wryde + ModE **croft** 'messuage'. Wryde, *(le) Wride, Wryde* c.1250, c.1270 etc., *Ryde* 1674, was originally the name of a stream near Thorney abbey, *Wridelake* 1586, *Wrydestreame* 1597, *Wryde River* 1720, perhaps representing OE **wride-ēa* 'twisting stream' from ***wride** 'a winding, a twist, a bend' + **ēa**. Ca 282.

WRYNOSE PASS Cumbr NY 2702. Mountain name Wrynose NY 2602, 2704 + ModE **pass**. Wrynose, 'twisted nose', *Wrenose* 1576, *the mountain Wrynose* 1610, is a folk-etymological adaptation of an earlier *Wreines* seen in the forms *Wraines-Wreineshals*, *Wrenhalse* 1157×63, *Wranishals* c.1180, *Wreneshals* 1196, with OE **hals** 'a neck, a pass'. *Wreines* is of uncertain origin; it might represent ON **vrein-nes* 'stallion headland', ON **vreini** + **nes**, or ME **wreye-nes* 'twisted headland', OE ***wrēo, wrīo**, + **nes**. The reference is to the convolution of the massive hill-side of Wrynose Fell on the N or S side between which the road passes. We i.205, Cu 437, La 194, PNE i.226, ii.278.

WYASTON Derby SK 1842. 'Wigheard or Widheard's farm or village'. *Widerdestune* 1086, *Wyardeston(e) -toun -is-* 1244–1403, *Wi- Wyaston* 1363–1452 etc., *Wireson, Wierson* 1743. OE pers.n. *Wīġheard*, genitive sing. *Wīġheardes*, + **tūn**. In this interpretation the DB form is regarded as erroneous. It might, however, point to a CG name, OHG *Wīdheard*, as the specific. Db 557.

WYBERTON Lincs TF 3141. 'Wigbeorht, Wigburg or Wibert's farm or village'. *Wibertune -tone* 1086, *-ton* 1180–1524, *Wybertun -ton* from 1154×89. OE pers.n. *Wīġbeorht, Wīburh* feminine, genitive sing. *Wīburge*, or CG *Wī(g)bert* + **tūn**. Payling 1940.110, Cameron 1998.

WYBOSTON Beds TL 1656. 'Wigbald's farm or estate'. *Wiboldestone* 1086, *Wyboston* 1297, *Wiberson* c.1750. OE pers.n. *Wīġbald*, genitive sing. *Wīġbaldes*, + **tūn**. Bd 59 gives pr [waibəsən].

WYBUNBURY Ches SJ 6949. Possibly 'Wigbeorn's fortified place or manor house'. *Wimeberie* 1086, *Wybbunberi* from 1199×1216 with variants *W(h)yb- Wib(b)- -en- -in- -yn- -on- -em-* (1291) *-um-* (1452) *-ym-* (1452, 1464) and *-bur(e) -buri(e) -bury(e) -bir(y)*, *Widen- Wyd(d)enbury* 16th cent. OE pers.n. *Wīġbeorn* + **byriġ**, dative sing. of **burh**. It is, however, surprising that no forms with medial *-r-* or *-rn-* appear apart from the genitival forms *Wylbrisbur' -bury -biry, Wyb(b)risbur(y), Wybresbur(y), Wib(b)ris- Wybrysbir(y)* 1288–90 all from local Court or Eyre rolls of Chester. Che iii.80 gives pr ['wibmbri, 'winbəri, 'wibənbəri] and older local ['widnbəri], ['wimbəri].

High WYCH Herts TL 4614. *Highwick* 1676–c.1825. ModE adj. **high** + p.n. *Wyches* 1540, 'the farms, the enclosures', ultimately from OE **wīċ**. Hrt 196 gives pr [waitʃ].

Higher WYCH Ches SJ 4943. 'The higher factory'. *Upper Wych* 1208×49, *le Overe Wych* c.1305, *the Hygher Wyche* 1530. ModE adj. **higher** (ME **upper**, **over**) + **wīċ**. 'Higher' for distinction from Lower Wych in Iscoyd, Clwyd, SJ 4844, *Layerwyche* 1526, *the Loour W(h)iche* 1528. These were salt-wiches also known as *Wyz Maupas* 1257, *the wiches hard by le Malpas* 1357, *Fulewic* [1096–1101]1150, 1150, 'dirty works', *Nether(e)- & Over(e)folwich -wych* 1328, and with the same meaning *Dritwyche* 1482, *Dirtwich* 1485, *Netherdretwyche* 1524, *the Hygher, the Lower Drayte- Droytewhiche* 1530. Che iii.51.

WYCH CROSS ESusx TQ 4231. *Wyggecrouche* 1356, *Wygecrouche* 1407, *Wiggecross* 1564, *Witchcrosse* 1579. P.n. *la Wigge* 1274 + ME **cruche**, ModE **cross**. The tradition that this place marks the resting place of the body of Bishop Richard de la Wych on its way from Chichester to Kent is not supported by the early forms of the name. The piece of ground where the cross later stood seems to have been called 'the Widge', implying OE ***wicga**, an ideophonic formation on the root seen in *wag, wiggle* meaning perhaps 'unstable, swampy ground'. Cf WIGMORE H&W SO 4169. Sx 351.

WYCHBOLD H&W SO 9266. 'Building by or belonging to Wich' i.e. Droitwich. *(DE) UUICBOLDA, uuicbold* [692]11th S 75, *Wicelbold* 1086, *Wichebald -baud* 1160–1308. OE p.n. *Wic* referring to DROITWICH SO 8963 + **bold**. Wo 285.

WYCHE H&W SO 7744. 'The farm'. *The Wich* 1831 OS. Ultimately from OE *wīc*.

WYCHWOOD Oxon SP 3317. 'The forest of the Hwicce'. *Huiccewudu* [841]11th S 196, *Hucheuuode* 1086, *Wichewude* 1185–1362 with variants *W(h)ic(c)he- W(h)yc(c)he-* and *-wud(')* - *wod(e)*. OE folk-n. *Hwiċċe*, genitive pl. *Hwiċċa*, + **wudu** referring to an extensive forest on the border of the tribal territory of the Hwicce, i.e. Gloucestershire, Worcestershire and the W half of Warwickshire. Cf. ASCOT- SP 2918, MILTON- SP 2618 and SHIPTON-UNDER-WYCHWOOD SP 2717. O 335, 386.

WYCK Hants SU 7539. 'The dairy farm'. *Wikes* 1236, 1298, *(la) Wyk(e)* 1250–1316. OE **wīċ**. Ha 184, Gover 1958.98.

WYCK RISSINGTON SP 1921 → RISSINGTON.

High WYCOMBE Bucks SU 8593. *High Wycombe* 1822 OS. ModE adj. **high** + p.n. *(æt) Wicumun* [c.968×71]12th S 1485, *Wicumbe* 1086–1212, *Wycumbe* c.1220, 1235, of uncertain formation†. *Wicumun* is a dative pl. perhaps representing r.n. Wye + **cumbum**, dative pl. of **cumb** referring to the two Wycombes, High and West W. TC 109, however, explains *Wicumun* as OE **wīcum** 'at the dwellings', dative pl. of **wīċ**, with the addition of a second dative plural ending *-um* (*-un*) because there were two settlements here. Also known as East, Great and Chipping Wycombe, *Chepingwycomb* 1478, 'W with the market' ME **cheping** (OE *ċēping*), *Estwicomb* 1509, *Magna Wykeham* 1545, for distinction from West WYCOMBE SU 8394. Bk 200, DEPN, Signposts 67–8, L 91.

West WYCOMBE Bucks SU 8394. *West Wicumbe* 1195. ME adj. **west** + p.n. Wycombe as in High WYCOMBE SU 8593. Bk 200.

WYCOMBE MARSH Bucks SU 8891. *Wycombe Marsh* 1822 OS. Earlier simply *Merhs* 13th. P.n. Wycombe as in High WYCOMBE SU 8593 + ME **mersh** (OE *mersc*). Bk 204.

WYDDIAL Herts TL 3731. 'Willow nook'. *Widihale* 1086–1324 with variants *Wydi- Wydy-* and *-hall*, *Withial* 1257, *Wethyhall al. Wedyall* 1527, *Widjel* 1662. OE **withiġ** + **halh**. Hrt 188 gives prs [widjəl, widʒəl], L 110.

River WYE An Old European r.n. ***u̯aisā** < IE **u̯oisā* from the root **vis-/*veis-* 'water'. Cf. r.n. WEY. BzN 1957.237–238.

(1) ~ Derby SK 2169. *Waya* 1242, *Weya, Wey(e)* 1235–18th, *Wee* 1335–1631, *Wy(e), Wie* from c.1460. The forms are too late for certainty but this name probably belongs here. Db 19, RN 451, BzN 1957.237–238.

(2) ~ Wales, H&W SO 1054, 1940, 3046, 4242, 5928, ST 5398.
I. *Guoy* [c.800]11th HB, *Gui, Guy* 12th cent., *Gwy, Wy* from 12th, *Guai* c.1150.
II. *Guaia* 1154×1216, *Gwaia* 13th cent.
III. *(on, neah, æt þære ēá) Wæge* [956]c.1200 S 1555, c.1025, *Wege* [before 1085]12th, *Waia* 1086–1266, *Waie* 1086–[c1200]13th, *Waya* [1066×87]13th, 1150×4–1412, *Waye* 1227–1330, *Weye* 1200–1324, *Wye* from 15th.
Type I is Welsh forms, type II Latinised Welsh forms, and type III English forms. RN 451–2, LHEB 452, BzN 1957.237–8.

WYE Kent TR 0546. 'The pagan shrine'. *(in villa regia) An' Uuiæ* 'in the royal vill On W' 839 S 287, *(uillam regalem que nominatur) With* (?for *Wich*) 'the royal vill called W' [762]13th S 25 (with 15th cent. variant *Wyth*), *(in illa famosa uilla que dicitur æt) Uuie* 'in the famous vill called at W' [845]13th S 297, *Wii* 858 S 328, 1044×8 S 1473, *Wi* 1086‡, *Wi, Wy* c.1121–1226 etc., *Wyc(h')* 1254–92. The inhabitants of Wye, the *Wiware*, genitive pl. *Wiwara*, are referred to in *on Weowera wealde* 'to the forest of the Wiware' [724]15th S 1180, *Wiwarawic (quæ ante subjecta erat to Wii)* 'the wick of the Wiware, formerly subject to W' 858 S 328, *Wiwarlet, Wiuuartlest, Wiwart lest* 'the lathe of the Wiware' 1086 and Worthgate, one of the gates of Canterbury, 'the gate of the Wiware', *uueowera get* 845 S 296, *Wur- Worgate* 13th cent. Usually explained as a derivation of OE **wīġ**, **wēoh** 'a (pagan) temple or shrine' with later influence of OE **wīc** 'a farm'. However, Wye lies at the centre of a rich farming district on the upper Stour, a r.n. of English origin. It seems as likely, therefore, that this important early estate preserves a British name of the river. If so, this must have been **U̯īsa* and so provides another example of the family of Old European r.ns. derived from the IE root *wīs-* 'water' to add to the rivers WEAR Durham NZ 1134, WEY Hants SU 7742 and Surrey TQ 0557, and WYE H&W SO 3045 and Derby SK 2069. K 384, KPN 182–3, Settlements 112, BzN 1957.237, Charters iv.44, 80.

†Another early form which may belong here is *Wichama* 764 for 767 S 106.
‡The name of the hundred is once given as *Wit* 1086 with the *-t* of preceding *Wiwarlet* erroneously added.

WYKE 'Farm, dairy-farm, trading place'. OE **wīċ**, dative pl. **wīcum** 'at the dairies'. Sa i.332.

(1) ~ Dorset ST 7926. *Wyke* from 1244. Do iii.18.

(2) ~ Shrops SJ 6402. *Wike* 1221, 1255, *Wyk'* 1261×2. Gelling.

(3) ~ WYorks SE 1526. *Wich(e)* 1086, *Wik(e), Wyk(e)* 1276–1783. YW iii.33.

(4) ~ REGIS Dorset SY 6677. 'The king's Wyke'. *Kingeswik* 1242, *Kyngeswyk(e)* 1309, 1365, *Wyke Regis* (Latin genitive sing. of **rex** 'king') from 1407. Earlier simply *Uuike, (to) Wike* [978×84]14th S 938, *Wi- Wyk(e)* 1212–1566. The sense here may be either 'dependent farm' or 'harbour, fishery'. It was an ancient royal demesne. Do i.267.

(5) The ~ Shrops SJ 7306. *Wy- Wiches* 1219, *Wike* 1285 etc., *The Wyke, The Wike* 1736–42. Gelling.

The WYKE NYorks TA 1082. 'The bay'. *The Wyke* c.1855 OS. ON **vík**.

WYKEHAM NYorks SE 9683. 'Roman habitation'. *Wicam* 1086, *Wic- Wi- Wykham* 1086–1408. OE **wīc-hām**. YN 99 gives pr [waikəm], Evidence 12, Signposts 72.

WYKEHAM NYorks SE 8175. '(The settlement) at the buildings with a specialised function'. *Wich, Wic(h)um* 1086, *Wicum* [1268]14th, 1301, *Wycom(b)* 14th. OE **wīċ**, dative pl. **wīcum**. YN 45.

WYKEY Shrops SJ 3925. '(The settlement) at the dairies'. *Wiche* 1086, *Wi- Wykes* 1254, *Wikey* 1577–1810, *Wykey* from 1672, *Weykie -ey* 1656, 1672. OE **wīċ**, dative pl. **wīcum**. The development to Wykey seems to be unparalleled. Sa i.331–2.

WYLAM Northum NZ 1164. '(The settlement) at the fish-traps'. *Wylum* c.1120, 13th cent., *Wi- Wylom(e)* 1326, 1380, 1428. OE **wīle** < **wiliġe** 'a willy, an osier basket', locative–dative pl. **wīlum**. NbDu 220, Nomina 7.41.

WYLDE GREEN WMids SP 1294. Cf. *Will Green House* 1834 OS. P.n. *le Wylde* 1537, *le Wyle* 1544×7, *the Wild* 1667, 'the wild, uncultivated land', OE **wilde**, + ModE **green**. Wa 50.

WYLYE VALLEY Wilts SU 0137. 'The valley of the river Wylye'. R.n. WYLYE SU 0536 + ModE **valley**.

River WYLYE Wilts SU 0536. Uncertain.
I. *Wileo* *[688]13th S 234, *Wilig* [860]15th S 326–[988]14th S 870, *(on) wili stream* [1045]14th S 1010, *Wy- Wily, Wili, Wylie* 13th cent.;
II. *Guilou* [c.894] c.1000 Asser.
PrW **Wīlou* < **wīl* of unknown origin and meaning + suffix *ou̯i̯ā* probably reformed as OE **Wīl-ēa*. Identical with the W r.ns. Gwili Dyfed SN 4223 (a tributary of the Tywi) and SN 5707 (a tributary of the Loughor), *Camguili* c.1150 'the winding or false Gwili'; and possibly with the continental r.ns. Wile, *Vilia* 1st, Weil, a tributary of the Lahn, Germany, *Hwilina* 821, *Wilena* 849, *Wile* 1276, Wilster, Germany, *Wilstria* 1221, unexplained r.n. **u̯il-* + suffix *-str-*, and Vileika, Russia, and Vilnius, Lithuania, both on the r. Vilia (Wilja). Wlt 11 gives pr [waili], RN 457, LHEB 341, 352, FT 287–96, Berger 270.

WYLYE Wilts SU 0037. '(The settlement beside the river) Wylye'. *(æt) Wilig* [977]12th S 831, *Wil(g)i* 1086, *Wili* 1154×89, 1180, *Wy- Wily* c.1190–1817 OS. Originally *Biwilig* 'beside *Wilig*'

[901]12th S 901. Transferred from the r.n. WYLYE SU 0536. Wlt 231 gives pr [wɑili].

WYMERING Hants SU 6506. 'The *Wigmœringas*, the people called after Wigmær'. *Wimeringes* 1086–1229, *Wymering(g)es* 1242, 1270, *Wi- Wymering(g) -(e)* from 1215. OE folk-n. **Wīġmǣringas* < pers.n. *Wīġmǣr* + **ingas**. Ha 184, Gover 1958.20, ING 43.

WYMESWOLD Leic SK 6023. 'Wigmund's part of the Wold'. *Wimvndewalle, Wimvndeswald -wale* 1086, *Wy- Wimundeswald(e)* 1205–1316, *-wold(e)* 1212–1428, *Wy- Wim(m)eswold(e) -ys-* 1397–1610. OE pers.n. *Wīġmund*, genitive sing. *Wīġmundes*, + **wald**. Lei 332, L 226–7.

WYMINGTON Beds SP 9564. 'Wimma's estate'. *Wimenton(e)* 1086 etc., *Widminton* 1195, 1211, *Wym(m)ington* from 1284. OE pers.n. **Wimma*, genitive sing. **Wimman*, + **tūn**. *Wimma* would be a pet form for a name such as *Wīdmund* which would explain the spelling *Widminton*. Bd 49 gives pr [wimiŋtən].

WYMONDHAM Leic SK 8518. 'Wigmund's homestead'. *Witme- Wimvndeshā* 1086, *Wi- Wym(m)undeham* 1094×1123–1409, *Wi- Wymundham* c.1130–1460, *Wymondham* from 1269. OE pers.n. *Wīġmund* + **hām**. Lei 199.

WYMONDHAM Norf TG 1101. Partly uncertain. *Wimund(e)ham* 1086, *Wimundehamia* c.1150, *-ham* 1168, 1200, *Wimundham* 1202, 1212. Influenced by the absence of genitival *-s-* and the presence of medial *-e-* Ekwall believed this to be a reduced form of OE *Wīġmudingahām* 'the homestead of the people called after Wigmund'. This is possible but hazardous and it seems best to regard the Norfolk Wymondham as identical with the Leicesteshire WYMONDHAM SK 8518, viz. 'Wigmund's or the Wigmund homestead', OE pers.n. *Wīġmund* + **hām**. DEPN, ING 140.

WYMONDLEY Herts TL 2128. 'Wilmund's wood or clearing'. *(œt) Wilmundeslea* [11th]12th K 1354, *Wilmundele(ye)* 1198–1402, *Wimunde(s)lai* 1086, *Wymesley* 1525, *Wymley* 1535–98. OE pers.n. *Wilmund*, genitive sing. *Wilmundes*, + **lēah**. Hrt 148 gives pr [wimli].

WYNFORD EAGLE Dorset SY 5895. 'W held by the del Egle family'. *Winfrot Gileberti de Aquila* 'Gilbert of A's W' 1204, *Wynford Aquile* 1275, *Wynfrod Egle* 1288. The manorial addition commemorates Gilbert del Egle whose family came from Laigle, Orne, France, *Aquila* 'eagle' 1136. Wynford, *Wenfrot* 1086, is an ancient Celtic r.n., PrW ***wïnn** + **frud** 'white stream' as in WINFRITH NEWBURGH SY 8084 and WINFORD Avon ST 5465. Do 164, Dauzat-Rostaing 378.

WYRE FOREST H&W SO 7576. *foresta de Wira* 1177. Wyre is identical with the first part of the name WORCESTER SO 8555.

River WYRE Lancs SD 4440, SD 5454. Identical with River WEAR Durham NZ 2436–2851. *Wir* [1170×84, 1200×17 etc.]1268, 1194×9, [1227×36]1412, *Wer* 1226×8, *Wyr* [1190×1213]1268, 1292, *Wyre* from [c.1250]15th. La 139, RN 475, BzN 1957.236, Jnl 17.79.

Great WYRLEY Staffs SJ 9907. *Great Wyrleg* 1222–93, ~ *Wy- Wirley(e)* from 1279. ME **great** + p.n. *Wirlega* 1170, 1176, *Wi- Wyrley(e)* 1228–1406, as in Little WYRLEY SK 0106. St i.71.

Little WYRLEY Staffs SK 0106. *Little Wyrle* 1293. ME **litel** + p.n. *Wereleia* 1086, 'bog-myrtle wood or clearing', OE **wīr** + **lēah**. 'Little' for distinction from Great WYRLEY SJ 9907. DEPN.

WYSALL Notts SK 6027. Partly uncertain. *Wisoc* 1086, *Wi- Wyshou* c.1130–1340, *Wisou -ow* 1235–1316, *Wy- Wisho(u)w(e)* 1280–1330, *Wyshawe* 1327, *Wysall* from 1553. The generic is OE **hōh** 'a hill-spur', the specific possibly OE **wīġ**, **wīh**, **wēoh** 'a heathen temple' in the genitive sing. form **wīges**, ***wīs**, and so perhaps the 'hill-spur of the heathen temple'. Nt 259 gives pr [waisə], Evidence 113, L 168.

WYTCH HEATH Dorset SY 9784. *Wych Heath* 1811, *Wytch Heath* 1844. P.n. Wych + ModE **heath**. Wych, *Wicha* [1189×99]1372, *Wych(e)* 1498, 1611, OE **wiċe** 'wych-elm', is transferred from the original name of the Corfe river. Do 164 and i.20.

WYTHALL H&W SP 0875. Uncertain. *Warthuil* 1086, *Wyhtehalle* 1283, *Witho* 1577, *Withall* 1650, 1672, *Withorn* 1763. The forms are few, late and inconsistent; no explanation is possible. Wo 358.

WYTHAM Oxon SP 4708. 'The homestead at the bend' sc. in the r. Thames. *(ad, œt) Wihtham* [955×7]12th, 13th S 663, *Winteham* 1086, *Wi(c)htham, With(h)am* 13th., *Wightham* 1316, *Wi- Wytham* from 1221. OE **wiht** + **hām**. Brk 464.

WYTHBURN FELLS Cumbr NY 3112. P.n. Wythburn NY 3213 + ModE **fell(s)** (ON *fjall*). Wythburn, *Withebotine* c.1280, *Wytheboten* 1345, *Wythbottom* 1552, *Withbone* 1554, *Wyeborne* 1564, *Withburn or Wyburn* 1703, is 'withy valley', ME **withy** (ON *vithir* 'willow-tree' or OE *withiġ* 'withy, willow') + ME, ON **botn** 'valley' later replaced by English **bottom** and **burn** 'a stream'. The reference is to silted-up land at the S end of Thirlmere producing a 'bottom' or area of flat damp land since drowned by the late 19th cent. damming of the lake. Cu 315 gives pr [waibərən], SSNNW 179, Jnl 20.40, L 87.

Great WYTHEFORD Shrops SJ 5719. *Magna Wydiford* 1255×6. ModE **great**, Latin **magna**, + p.n. *Wic- Wideford* 1086, *Wideford* 1257, *Withiforde* 1155×62–mid 19th with variants *Wi- Wythy-* and *-ford, Wytheford'* 1242, *Witheforde* 1255×63, 'the willow ford', OE **withiġ** + **ford**. A crossing of the r. Roden. 'Great' for distinction from Little Wytheford SJ 5619 on the other side of the river, first noted 1293 (*Parva* ~). Sa i.332.

WYTHENSHAWE GMan SJ 8287. 'Willow copse'. *Witenscawe* 13th, *Wythensache* [late 13th]18, *-s(c)hagh -schawe* 1306–54, *Wythenshawe* 1548. OE **wīthiġn** + **sceaga**. Che i.236.

WYVERSTONE STREET Suff TM 0367. 'W hamlet'. P.n. WYVERSTONE TM 0467 + Mod dial. **street**.

WYVERSTONE Suff TM 0467. 'Wigferth's estate'. *Wiuerthes(s)tuna, Wiuertes- Wiuerdes-* 1086, *Wi- Wyverdeston* 1203, 1231, *Wyveriston* 1254, *Wyverston(e)* from 1291. OE pers.n. *Wiġferth*, genitive sing. *Wiġferthes*, + **tūn**. DEPN, Baron.

YADDLETHORPE Humbs SE 8807. 'Eadwulf's outlying farm'. *Iadulf(es)torp* 1086, *Edoluestorp* c.1115. OE per.n. *Ēadwulf*, genitive sing. *Ēadwulfes*, + **thorp**. DEPN, SSNEM 122.

YAFFORD IoW SZ 4481. 'Hatch ford'.
I. *Heceford* 1086, *Hecceford'* 1251, *Hachford* 1255.
II. *Egeford* 1235, *Haygford'* 1287×90, *Ecford* 1299.
III. *Eb(be)ford(e)* 1220–64.
IV. *Jakeforde* 1235, *Yauford* 1292, *Yagheford* 1468, *Yafford* from 1651.
V. *Ac(ke)ford* 1255.
VI. *Efford(e)* 1273–99.
OE **hæċċ(e)** + **ford**. The reference is probably to a fence-like construction probably used for trapping fish. The forms are complex: OE *hæċċe-ford* would normally give *Hatchford, but *hæċċ-ford* might give *Hackford and with prosthetic [j] *Yackford and Yafford. Confusion with OE *heġe-ford, ebba-ford* and *ēa-ford* is apparent. Wt 224, Mills 1996.112.

YAFFORTH NYorks SE 3494. 'The river ford'. *Eiford* 1086, 13th, *Iaforde -be* 1086, *Jaf(f)ord(e)* 1198–1316, *Yafford* 1283–1530, *Yafforthe* 1574. OE **ēa** + **ford**. The early appearance of initial [j] from OE **ēa** with accent shift to *ēá* is paralleled in e.g. the DB forms for Yaddlethorpe Lincs, *Iadulfestorp* (pers.n. *Ēadwulf*), Yarborough Lincs, *Gereburg* (OE *eorth*), Yarnfield Somer, *Gernefelle* (OE *earn*) and Yealand Lancs, *Jalant* (OE *hēa-land*). The river in question is the Wiske. YN 277.

YALDING Kent TQ 7050. 'The Ealdingas, the people called after Ealda'. *Uuestaldingis* 'west Y' 1087×9, *Hallinges* 1086, c.1100, *Ellinges* 1086, *Ealdinga* c.1100, *Alding'* 1162–1242×3, *Ealdynges -inges* 1191, 1226, *Yeldyng* 1450, *Yaldyng* 1451. OE folk-n. **Ealdingas* < pers.n. *Ealda* + **ingas**. PNK 168, ING 17.

YANWORTH Glos SP 0713. Partly uncertain. *Janew(o)rth(e) -wurth'* 1043×66–after 1412, *Teneurde* (for *Iene-*) 1086, *ȝeneworþe* 1154×89, *þaneword(i)am -worthe -worþe* (sic for *ȝane-*) 12th cent., *Ianeword(am) -worþia -orþe -w(o)rth(e)* c.1162–1384, *ȝanew(o)rth(e) -wrþe* 1211–1428, *Yanew(o)rth(e)* 1251–1413, *Yanworth* 1316, and from 1540. The generic is OE **worth** 'an enclosure', the specific either OE ***ēan** 'lamb' with prosthetic [j] or pers.n. **Ġeana*, **Ġæna*, a shortened form of pers.ns. such as *Ġænbald*, *Iaenbeorht* etc. Gl i.190.

YAPHAM Humbs SE 7852. '(The settlement) at the slopes'. *Iapun* 1086, *Yapum* 1150×61–1350, *Yapam* 1276, *-ham* 1451. OE **ġēap**, dative pl. **ġēapum**. The village lies high up on a fairly steep slope. YE 182 gives pr [japəm].

YAPTON WSusx SU 9702. 'The estate called after Eabba'. *Abyn(g)ton* 12th–1279, *Ab(b)iton(a) -y-* c.1187–1285, *ȝapeton* 12th, *Yapeton* 1295, *Yapton* 1329, *Yaby(n)ton* 1288, 1315. OE pers.n. **Eabba* + **ing**^{4} + **tūn**. Sx 144.

River YAR IoW SZ 6186. A back-formation from Yarbridge SZ 6086, *Yarnbrigge* 1462, *Yarsbridge* 1769, *Yarbridge* 1781. Nearby was *Yarneforde* 1324, possibly identical with *Arneford* 1248, either '(sea-)eagles' ford', OE **earn**, genitive pl. **earna**, + **ford**, 'ford fit for riding', OE ***ærneford**, or 'gravelly ford', OE ***ēaren** + **ford**. Wt 62, Mills 1996.112.

YARBRIDGE IoW SZ 6086 → River YAR SZ 6186.

YARBURGH Lincs TF 3593. The 'earth fortification'. *Gereburg* 1086, *Ierburc* c.1115, *Jerdeburc(h)* 12th, *Jerdburg* 1212, *Yerburg* 1156×8, 1203, *Yardbury* 1209×19. OE **eorth-burh**. Probably a late Roman fortified site obliterated by Anglo-Saxon building and cultivation. DEPN, Jnl 28.52, Cameron 1998.

YARCOMBE Devon ST 2408. 'Valley of the river Yarty'. *Ercecombe* (for *Erte-*) [934]17th S 391, *Herticome, Ertacomestoche* 1086, *Erticoma, Ertacomestoca* 1086 Exon, *Erti(n)cumb(e)* 1155–1249, *Artecumbe* 1269, *Yartecumbe* 1278, *Yorkham* 1648. R.n. YARTY ST 2505 + OE **cumb**. D 651.

North YARDHOPE Northum NT 9201. *North Yardhope* 1869 OS. ModE adj. **north** + p.n. *Yerdhopp* 1324, *-hope* 1331, *Yardhope* 1604, the 'valley where rods are obtained', OE **ġerd** + **hop**. 'North' for distinction from South Yardhope NT 9200. NbDu 221.

YARDLEY 'The wood where rods are obtained'. OE **ġerd** + **lēah**.
(1) ~ WMids SP 1386. *Gyrd leah* 972 S 786, *Gerlei* 1086, *Ierdele* 1220–1305, *Yerdeleye* c.1300, 1478, *Yardeley* 1541. Wo 231, L 205.
(2) ~ CHASE Northants SP 8455. 'Y hunting preserve'. *The chase of Jordele* 1277, *bosco de Jerdele* 1285, *chace of Yerdeleye* 1302. P.n. Yardley as in YARDLEY HASTINGS SP 8656 + ME **chace**. Nth 154.
(3) ~ GOBION Northants SP 7644. 'Y held by the Gobion family'. *Yerdele(ie) Gobioun* 1353, *Yardeley Gubbyn* 1580, *Yardley Gubbins* 1702. P.n. *Gerdeslai* 1166, *Jerdelai* 1166–1316 with variants *-le(gh) -leye*, + family name of Henry Gubyun who held land here in 1228. Nth 108 gives pr [gʌbinz].
(4) ~ HASTINGS Northants SP 8656. 'Y held by the Hastings family'. P.n. *Gerdelai* 1086, [1155×8]1329, *Jerde- Yerdele* 1220–1348, *Yardele* 1314, + family name *Hastinges* 1316, holders of the manor from c.1250. Nth 153.

River YARE Norf TG 1108, 3604. Uncertain. *Γαριέννου ποταμοῦ ἐκβολαί* (*Gariennou potamou ekbolai*) 'the mouths of the river Gariennus' [c.150]13th Ptol with variant *Γαρυέννου*, *Gerne* c.1150, 1324, 1586, *Hierus, Hiere, Jere, Gare* 16th cent., *Yare* from 1577. Cf. also the RBrit name of BURGH CASTLE TG 4804, *Gariannum* c.425 ND, and Great YARMOUTH TG 5207. This r.n. has been associated with the root **gar-/ger-* 'shout, talk' as in Latin *garrio* 'chatter' (English garrulous) as if this were the 'babbling river'. The Yare, however, is a slow river in flat country and on the face of it this meaning is not appropriate. Possibly, therefore, to be compared with the French River Garonne, Latin *Garumna*, the etymology of which is unknown. RN 477, RBrit 366.

YARKHILL H&W SO 6042. 'Kiln with a yard'. *Geardcylle* 811 S 1264, *Archel* 1086, *Yard(h)ull(e)* c.1158–1350, *Archil(l)a, Arkil* 1158×63–c.1250, *Iarculn'* 1243. OE **ġeard** + **cyln**, lOE ***cyll**. He 215.

YARLET Staffs SJ 9128. Partly uncertain. *Erlide* 1086, 13th, *-lida* 1167, *-lyde* 13th, *Erlid* 15th, *Yarlett* from 1566. Probably 'the gravel slope', OE **ēar** + **hlid**, or 'the eagle slope', OE **earn** + **hlid**. The settlement lies on the slope of Yarley Hill which rises to 463ft. at Peasley Bank SJ 9030. DEPN, Duignan 177, Horovitz.

YARLINGTON Somer ST 6529. Partly uncertain. *Gyrdlingatone* [939×46]13th S 1731, *Gerlintvne* 1086, *Gerlincgetuna* 1086 Exon, *Gerlingatune* [1107×22]14th M, *Gerlintune* [1091×1106]14th M, *Gerlin(g)tone* [1100×18–1152×8]14th M, *Gerlingstone* [1174×80]14th M, *Gerlingeton* 1212, *Yarlington* 1610. If the form in S 1731 is reliable this is the 'settlement of the *Gyrdlingas*, the people called after Gyrdel or Gyrdla'. OE folk-n. **Ġyrdlingas* < pers.n. **Ġyrdel* or **Ġyrdla*, genitive pl. **Ġyrdlinga* + **tūn**. Otherwise 'settlement at or called *Gerling* or of the *Gerlingas*', p.n. **Ġerling* 'place associated with

Gerla' (OE pers.n. *Ġerla* + **ing**[2]) or OE folk-n. **Ġerlingas* < pers.n. *Ġerla* + **ingas**, genitive pl. **Ġerlinga*, + **tūn**. DEPN.

YARM Cleve NZ 4111. '(Settlement at) the fish-weirs'. *Iarun, Gerou* (sic for *Geron*) 1086, *J- Yarum* 1198–1436, *Yarom* 1285–1470, *Yarm(e)* from 1300. OE **ġear**, dative pl. **ġearum**. OE **ġear** means 'a yair, a weir, a pool formed by a weir for catching fish'. YN 172, Nomina 7.35, 37.

YARMOUTH IoW SZ 3589. 'Gravelly, muddy estuary'. *Ermud* 1086, *Hernemue* c.1180, *Ernem(o)uth(e)* 1223–1488×9, *Eremuth(e)* 1224–1351, *Aremuthe* 1299, *Yarmouth* from 1293. OE **ēar** or adj. ***ēaren** + **mūtha**. Wt 260, Mills 1996.00.

Great YARMOUTH Norf TG 5207. *Magna Gernemue* 1252. ModE adj. **great**, Latin **magna**, + p.n. *Gernemwa -mutha* 1086, *Gernemuda* [1121–45]n.d., *-muth(a) -muða* 1136×45–1168, *-mue* [c.1140]13th, 1196, *(við) Járnamóðu* 13th, *ȝarne- ȝerne- ȝarmouþe* c.1300, the 'mouth of the YARE'. 'Great' for distinction from Little Yarmouth Suff, *Parva Gernamuta* 1219, earlier simply *Gernemutha* 1086. DEPN, RN 477.

YARNBURY CASTLE Wilts SU 0340. *Yarneberie castell* 1591. P.n. Yarnbury + ModE **castle**. Yarnbury is probably 'the eagles' fort', OE **earn**, genitive pl. **earna**, + **byriġ**, dative sing. of **burh**. One of the finest Iron Age hill-forts in Wiltshire with massive multivallate defences. Wlt 228, Pevsner 1975.108, Thomas 198-.239.

YARNFIELD Staffs SJ 8632. 'The eagles' open land'. *Ernefeld* 1266, *Yarnefylde* 16th. OE **earn**, genitive pl. **earna**, + **feld**. DEPN, Duignan 177, L 244.

YARNSCOMBE Devon SS 5623. 'Eagle's coomb'. *Hernes- Herlescome* 1086, *Ernescumbe -comb(e)* 1238–1346, *Jernescom* 1275, *Yernesco(m)be* 1300, 1303, *Yerscombe* 1413, *Yarescombe* 1558×1603. OE **earn**, genitive sing. **earnes**, + **cumb**. D 82.

YARNTON Oxon SP 4712. 'The estate called after Earda'. *(æt) Ærdintune* [1005]12th S 911, *Harb- Hardintone* 1086, *Aerdintona* 1091–1198, *Erdinton' -tun' -ton(e) -yn-* 1206–1390, *Erdington(e) -yng-* 1273–1592, *Eardyngton* 1285, *Yarnton* from 1529. OE pers.n. *Earda* + **ing**[4] + **tūn**. O 296.

YARPOLE H&W SO 4764. 'Fish-trap pool'. *Larpol(e)* (for *Iar-*) 1086, *Yarpol(e)* 1173×86–1328. OE **ġear** + **pōl**. He 216.

River YARROW Lancs SD 5117. Uncertain. *Earwe* 1203, *Yarwe* [1184×99]1268, 1292, *Yaruhe* 1199×1215, *Yarewe* 1246, *Yarowe* [13th]14th, 1577, *Yaro* c.1540. This could be identical with the River ARROW Warw SP 0861, Old European **Arvā*, the River ARROW H&W SO 2451, 3560, 4058, Old European **argu-*, or the Scottish Yarrow Water NT 3257, *Gierua, Gierwa* 12th, and Welsh Afon Garw SS 9088, *Garewe* 1207, *Garwe* 1314 from W **garw** 'rough, tempestuous' (Brit **garŏ-*) with reference to rapids in the upper reaches of the river. La 127, RN 478, BzN 1957.231, Jnl 17.71, GPC 1383.

YARSOP H&W SO 4147. Possibly 'Eadred's remote valley'. *Ardeshope, Edreshope, Erdeshop, Erdesope* 1086, *Erdeshop(e)* 1137×1397. OE pers.n. *Ēadrēd*, genitive sing. *Ēadrēdes*, + **hop**. He 217.

YARTLETON H&W SO 6821 → YORKLEY Glos SO 6307.

River YARTY Devon ST 2505. Uncertain. *Jerti* 1238, *Yearte* 1467, 1479, *Yerty* 1511, *Yartey* c.1550, *Artey* 1577. For earlier forms see under YARCOMBE ST 2408. A suggested source is OE **Ēart-*, later **Yart-*, possibly from Brit **arto-* 'a bear' (OIr *art*, W *arth*), as in Afon Arth Dyfed SN 4862, 5462, *Aber Arthe* 1425, and the Fr r.n. Arce, a tributary of the Seine, *Artia* 1263. In view of the Gallo-Latin *Artio* 'bear goddess' such a r.n. might have meant 'river of Artio, the bear goddess'. Alternatively this might be OE **earte** 'a wagtail'. D 17, RN 479, Anglia Beiblatt 46.17.

YARWELL Northants TL 0797. Partly uncertain. *Jarewelle* 1166–1289, *Yarewelle* 12th, *Iaruwelle* c.1220, *Yarwelle* 1318. Possibly 'spring by the yairs or fish-traps', OE **ġear**, genitive pl. **ġeara**, or 'yarrow-grass spring', OE **ġearuwe** + **wella**. The first explanation would have to refer to a site on the Nene in which case **wella** might be a replacement of original **wǣl** 'a deep place in a river', as in some Northumbrian river fishery names. Nth 209 gives pr [jærəl], DAJ 2.56, 13.89.

YATE Avon ST 7182. 'The gate'. *(æt) Ge(a)te* [777×9]11th S 147, *Giet(e)* 1086, 1167, 1287, *Yate* from 1207. OE **ġeat**. The reference is probably to the gap in a low range of hills through which the headwaters of the Frome flow. Gl iii.44 gives pr ['jæt].

YATELEY Hants SU 8160. Partly uncertain. *Yat(t)elegh* 1236, *Yatele(ye)* 1248–1387. This appears to be 'wood or clearing of the gates', OE **ġeat**, analogical genitive pl. ***ġeata** (instead of *gata*), + **lēah**, but the significance is unknown. A possible alternative explanation is 'Geata's wood or clearing', OE pers.n. **Ġeata* + **lēah**. Ha 185, Gover 1958.112.

YATESBURY Wilts SU 0671. Possibly 'Geat's fort'. *Etesberie* 1086, *Yatebur(y)* 1279, 1297, *Yetesbur(y)* 1239, 1306, *Yatesbur(y)* from 1242, *Yeates- Yatsbury* 17th. OE pers.n. *Ġēat*, genitive sing. *Ġēates*, + **byriġ**, dative sing. of **burh**, probably referring to the Neolithic camp on Windmill Hill SU 0871. Wlt 264.

YATTENDON Berks SU 5574. 'Valley of the Geatingas, the people called after Geat'. *Etingedene* 1086, *Yetingeden(e)* 13th cent., *Jetingedon'* 1223, *Yetingden* 1242×3, *Jatingden* 1232, *Yat(t)in(g)den(')* 1241–1348 with variants *-yn(g)-* and *-don*, *Yatendone* 1517, *Yattington* 1574, *Yaddinton* 1813. OE folk-n. **Ġēatingas* < pers.n. *Ġēat* + **ingas**, genitive pl. **Ġēatinga*, + **denu**. The alternative suggestion, based on the DB form and a handful of spellings in *(H)et-*, that the pers.n. involved is *Ēata*, requires the assumption of accent shift in the diphthong *éa* > *eá* and development of prosthetic [j], normally a late ME development. Brk 279, Dobson 429, L 326.

YATTON Avon ST 4265. Partly uncertain. *Latvne* (sic) 1086, *Jatton* 1178–1243, *Jactun -ton* 1227–76. If the *-c-* forms are reliable this could be 'cuckoo settlement', OE **ġēac** + **tūn**. Alternatively the specific might be OE **ġeat** 'a gate'. DEPN.

YATTON H&W SO 4366. 'Settlement at the gate or pass in the hills'. *Getvne* 1086, *Giet(t)ona* 1137×9, 1160×70, *Yatton* from 1307. OE **ġeat** + **tūn**. The reference is to the col in the hills through which Walting Street, Margary No. 6c, passes on its way from Leintwardine to Hereford. He 217.

YATTON KEYNELL Wilts ST 8676. 'Y held by the Keynall family'. *Yatton Kaynel* 1289, *Yatton Keynell* 1618. P.n. *Getone* 1086, *Yeton* 1247, 1289, *Yatton(e)* 1242–1346, 'the gate or gap village', OE **ġeat** + **tūn**, + manorial addition from the family name of Henry Caynel who had a holding here in 1242. The 'gate' is the head of the well-marked valley to the W of the village. Wlt 114.

YAVERLAND IoW SZ 6185. 'Newly cultivated land where boars are kept'. *Ewerelande* [683]c.1300, *Everelant, Evreland* 1086, *Iwerland* 1189×1204, *Euerlaund(e) -lond(e)* 1279–1462, *Yoverland* 1311–2, *Yaverlond(e)* 1324–1535, *Yaverland* from 1403. OE **eofor**, genitive pl. **eofora**, + **land**. Wt 262, L 248–9.

YAXHAM Norf TG 0010. 'Cuckoo promontory'. *Jaches- Jakesham* 1086, *Iakesham* 1254. OE **ġēac**, genitive sing. **ġēaces**, + **hamm** 2. DEPN.

YAXLEY Cambs TL 1892. 'The cuckoo's clearing'. *(æt) Geakeslea* [956]14th S 595, *Geaceslea* [?963]12th S 1448, c.1000, *Iaceslea* [963×75]12th S 1377, *Geaceslea, oþer geakeslea* [973]14th S 792, *Iacheslei* 1086, *Iakesle(e)* 1203–1347, *Iakele(ye)* 1253–1327, *Yaxlee* 1389. OE **ġēac**, genitive sing. **ġēaces**, + **lēaġe**, dative sing. of **lēah**. Hu 201.

YAXLEY Suff TM 1273. 'Cuckoo wood or clearing'. *Jaches- Iacheslea, Lachele(i)a* 1086, *Iakesle(a)* 1170, 1254, *Iachesle* 1198, *Yakesle(y)* 1336–1419, *Yaxley* from 1523. OE **ġēac**, genitive sing. **ġēaces**, + **lēah**. DEPN, Baron.

YAZOR H&W SO 4046. 'Iago's ridge'. *Lavesovre* (for *Ia-*) 1086, *Iagesoure* 1160×70, 1243, *Iawesore* 1198, *Eausore* 1243, *Iaveso(u)re* 1277, 1303, *Yausore* 1313, *Yasoure* 1334, *Yazore* 1397. W pers.n. *Iago* + OE, ME genitival **-es**, + OE **ofer**. He 217.

YEADEN NYorks SE 1467. 'Yew-tree valley'. *Iweden* 1154×91,

Youden 1307, *Yowden* 16th cent., *Yudding* 1771. OE **īw** + **denu**. YW v.150 gives pr ['jiədən].

YEADING GLond TQ 1182. 'The Geddingas, the people called after Geddi'. *Geddinges* [716×757]12th S 100, *Geddingas* [790 for ?795]10th S 132, *(æt) Geddincggum* 825 S 1436, *Geddings* 1210×12, *Yeddinggs* 1325, *Yeddyng* 1331. OE folk-n. **Ġeddingas* < pers.n. *Ġeddi* + **ingas**. In the 757 charter *Geddinges* is described as a *regio*. Mx 40, ING 27–8.

YEADON WYorks SE 2041. 'The steep hill'. *Iadon -dun* 1086, 1167, *Yadun* 1185×1215, *Yedon* 1285–1556, *Yeadon* from 1446. OE ***ġǣh** + **dūn**. The place stands on the upper slope of a lofty hill. YW iv.155, xi, L 157.

YEALAND 'Newly-cultivated high land'. OE **hēa** + **land** < adj. **hēah**, definite form **hēa**, + **land**. L 248–9.

(1) ~ CONYERS Lancs SD 5074. *Yeland Coygners* 1301, 1341, ~ *Conyers* 1353. Earlier simply *Jalant* 1086, *Yaland* [1206–25]1268, *Yeland* [1190]1268, 1243 etc., *Hieland(e)* 1202–12. Situated on the E slope of a ridge. The development was *hēa* > *heā*, [hja] > [ja]. The original manor was divided after the Conquest and this part was held by Robert de Conyers in 1242. La 188, DEPN, Jnl 17.104.

(2) ~ REDMAYNE Lancs SD 5075. *Yeland Redman* 1341, 1395. Earlier forms as for YEALAND CONYERS. This part of the manor was held by the Redman or Redmayne family from Redmain Cumbr NY 1333. La 188, DEPN, Jnl 17.104.

River YEALM Devon SX 6056. *Yhalam* 1310, *Yalme* 1414, *Yaulme* c.1550. For earlier spellings see YEALMPTON SX 5751. OE r.n. **Ealme* from Brit **Almā*, a *-mā* formation on the IE root **al-* seen also in the r.ns. Alma, Italy, *Alma* [3rd]8th AI, Alme, Lithuania, Westphalia, Lom, Bulgaria, *Almus*. This would give ME **Elme* or with accent shift *éa* > *eá Yalme*. D 17, RN 479, BzN 1957.225.

YEALMPTON Devon SX 5751. 'Settlement, estate on the river Yealm'. *Elintone* (for *Elm-*) 1086, *Alentone* 1205, *Almenton* 1249, *Yalminton* 1244, *Yalmetun* 1275, *Yalmton* 1297, *Yalampton* 1303, *Yampton al. Yealmpton* 1670. R.n. YEALM SX 6056 + OE **tūn**. D 261 gives pr [jæmptən].

YEARSLEY NYorks SE 5874. 'Wild boar wood or clearing'. *Eureslage* 1086, *Euereslei -ley(a)* 1176–1303, *Euersle(gh) -ley* 1204–1304, *Yever(e)sley(e)* 14th cent., *Yearsley* 1577. OE **eofor**, genitive sing. **eofores**, + **lēah**. YN 193 gives pr [ja:zlə], Anglia 34.293.

YEATON Shrops SJ 4319. 'The river settlement'. *Aitone* 1086, *Eton* 1271–1335, *Eaton* 1687, *Yeaton* from 1702. OE **ēa** + **tūn**. The modern form cannot be due to OE accent shift *éa* > *eá* > *iá* as this should have produced **Yatton*, cf. YAFFORTH NYorks SE 3495, OE *ēa-ford* and possibly YATTON Avon ST 4265. It must be an example of sporadic ME and ModE excrescent [j] before *ē*, e.g. 17th cent. *yeast* for *east*. Sa i.333 gives pr [jeton], Dobson §428–30.

YEAVELEY Derby SK 1840. 'Geofa's wood or clearing'. *Gheveli* 1086, *Yv- Yuele(e) -legh -leie -ley(e)* 1251–1272×1307, *Yev- Yeueleg' -le(e) -ley(e)* 1275–1640 *Yeeveley*. OE pers.n. *Ġeofa* + **lēah**. Db 619.

YEAVERING BELL Northum NT 9229. P.n. Yeavering NZ 9330 + ModE **bell** 'a bell-shaped hill'. Yeavering, *Ad Gebrin, Ad Gefrin* [c.731]8th BHE, *æt Gefrin* [c.890]c.1000 OEBede, *Yev(e)r(e)* 1242 BF, 1316, 1359, *Yverine* 1296 SR, *Yevern* 1404, 1442, *Yeverin* 1663, *Yevering* 1796, is the 'goats' hill', PrW ***gavr**, pl. ***gevr** + ***vrïnn**, lenited form of ***brïnn**. The reference is to the hill of Yeavering Bell. PrW *Gevrïnn* may in fact have been the name of the Celtic *oppidum* which occupied the summit. It was transferred to the Anglo-Saxon royal complex at Old Yeavering, NT 9230. Pronounced [jivrin]. NbDu 221, Yeavering 15.

YEDINGHAM NYorks SE 8979. 'Homestead of the Eadingas, the people called after Eada'. *Edingham* 1170×5–1333, *Edengeham* 1168, *Yedingeham* 1246, *-ingham* from 1185×95. OE **Ēadingas* < pers.n. *Ēada* + **ingas**, genitive pl. **Ēadinga*, + **hām**. YE 121, ING 154, Anglia 34.293.

YELDEN Beds TL 0167. 'Valley of the river Ivel or of the Gifle'. *Giveldene* 1086–13th, *Guuelden(e)* 13th, *Yevelden(e)* 14th–1660, *Yelden* c.1390 and from 1780. R.n. *Ġifle*, the original name of the Till identical with r.n. IVEL TL 1938, or folk-n. *Ġifle*, genitive pl. *Ġifla*, + **denu**. Bd 21 gives pr [jeldən].

Great YELDHAM Essex TL 7638. *Magna Geldham* 1248, *Moche Yeldham* 1494. ModE adj. **great**, Latin **magna**, ME **much**, + p.n. *Ger- Geld(e)ham* 1086, *Geldham* 1166–1254, *Yeldham* from 1314, possibly 'homestead liable to pay a tax', OE **ġeld** + **hām**. Ess 468.

Little YELDHAM Essex TL 7738. *Parva Gehelham* 1303, *Over alias Littell Yeldham* 1562. ModE **little**, Latin **parva**, + p.n. Yeldham, as in Great YELDHAM TL 7638. Ess 468.

YELFORD Oxon SP 3604. 'Ægel's ford'. *Aieleforde* 1086, *Eleford(e)* 1200–1428, *Ei- Eylesford* 1245×6, 1304. OE pers.n. *Æġel*, genitive sing. *Æġeles*, + **ford**. The same pers.n. occurs in *(to) Æglesuuillan broce* 'the brook of Ægel's spring' *[958]13th S 1632 on the boundary between Ducklington and Yelford. O 324.

YELLAND Devon SS 4932. 'Ancient tilled land'. *Yollelonde* 1330. OE **eald** + **land**. D 115.

YELLING Cambs TL 2562. 'The Gellingas, the people called after Gella'. *Gy- Gillinges* [974]12th S 798, *Gilling' -inge* [974]14th ibid., *Gillinge* [c.1062]13th, 14th S 1030, *G(h)ellinge, Gelinge* 1086, *Gellinches* 1135×54, *Gillinges* 1210×12, 1220, 1228, *Gi- Gylling(e)* 1218–1535, *Yilling* 1344, 1385, *Illyng* 1507, 1540, *Yellyng -ing* from 1535, *Yealding al. Yealing* 1601. OE folk-n. *Ġellingas* < pers.n. **Ġella* + **ingas**. Hu 275, ING 68.

YELVERTOFT Northants SP 5975. 'Geldfrith's cottage(s)'. *Celvrecot, Gelvrecote, Givertost* 1086, *G(h)elvertoft(e)* 12th–1253, *Jelvertoft* 1235, *Yelvertoft* from 1276, *Yellowtoft* 1517. OE pers.n. **Ġeldfrith* + **cot**, pl. **cotu**, later replaced with ON **toft** 'curtilage'. Nth 77, SSNEM 215.

YELVERTON Devon SX 5267. 'Settlement at Ella's ford'. *Elverton* 1765. Earlier *Elleford* [1291]1408. This is an example of the late addition of **ton** to an existing p.n., 'Ella's ford', OE pers.n. *Ella* + **ford**. The farm is still called Elfordtown; Yelverton is the local dial. form with prosthetic *y* adopted by the Great Western Railway when the station was built in 1859. D 225.

YELVERTON Norf TG 2902. 'The Geldfrith estate'. *Ailuertuna* 1086, *Ielverton* 1198, *Gelvertone* 1254. Yelverton is not near a ford. The specific cannot, therefore, be OE **gēol-ford* and must be a pers.n. The DB form points to OE *Æthelfrith* which appears in DB as *Aeluer(t)*, the later forms to **ġeldfrith* + **tūn**. DEPN.

YENSTON Somer ST 7121. Unexplained. *Yenston* 1610, *Yeanston* 1811 OS. The evidence is too late to offer an explanation.

River YEO. Yeo is a SW dial. form simply meaning 'river' from OE **ēa** with shift of accent *éa* > *eá* and rounding of *ā* to *ō*. It frequently replaces an earlier name often of Celtic origin. Many streams in Devon and Somerset have this name e.g. Land Yeo ST 4271, Lox Yeo ST 3856, earlier simply Lox as in LOXTON ST 3755, Yeo SS 4322, *Yoe* c.1400, Yeo SS 5734, *North Yeo* 1797, YEO SS 7306, earlier the *Nymet*, Yeo SS 7626, 8326, *Yoe 1499*, also earlier *Nymet*, Yeo SX 8399, Yeo ST 4723, 5913, earlier the *Ivel* at YEOVIL ST 5515 and YEOVILYON ST 5422 (where it may well, in fact, be a late back-formation from the p.ns. themselves influenced in spelling by the r.n. type Yeo).

(1) ~ Avon ST 4662. A modern name replacing the old name *Wring* as in WRINGTON ST 4762.

(2) ~ Devon SS 7306. *Yeo* 1558×1603. An alternative name was *Lovebrook -broke* 1590, 'Leofa's brook', OE pers.n. *Lēofa* + **brōc**. The earliest name of the river survives in the p.n. NYMET TRACEY SS 7200 and in the DB form for ZEAL MONACHORUM SS 7103. D 17.

(3) ~ Devon SS 7726. *Yeo River* 1809 OS. D 337.

River YEO Devon SX 8199. Possibly 'yew stream'. *(on) eowan* [739]11th S 255, *Iouwe* 1238, *Iwe* 1216×72, *Iou* 1244, *Yowe* 1361, *Yew* 1630, *Yeo* 1809 OS. OE *Ēowe* from **ēow**. The name has been assimilated to the form of the other Yeo rivers. On the Yeo are YEOFORD SX 7898 and UTON SX 8298. The alternative name Fortin, *Forten* 1586, is one of Harrison's inventions, a back-formation from FORDTON SX 8499, *Fordeton* 1249, 1289, *Forteton* 1256. D 406, 17, RN 480, NoB 14.54.

River YEO Dorset ST 6214, Somer ST 4723. Originally the *Yevel* as in r. IVEL Beds. *(on) Gifle* [903 for 946×51]12th S 516, *Yevel* [852 for 878]17th S 343, *Givell, Gi- Gyuele, Geule* 13th cent., *Ivel(le)* c.1540, *Iuell* 1577, 1586, *Euill* 1577, 'forking stream', OE **Ġifle, Ġefl* < PrW ***gefl**, Brit **gabliā*, **gablio-* cognate with W *gafl*, Breton *gaol* 'fork'. The Yeo has two main arms, the Yeo itself which rises near Sherborne, and the Cam which rises near Yarlington in Somerset. The modern form *Yeo* 1811 is probably a back-formation from the p.n. YEOVIL Somer ST 5515 aided by the existence of the r.n. YEO Somer ultimately from OE **ēa** 'river'. RN 221, CPNE 103, GPC 1370.

YEOFORD Devon SX 7898. 'Ford across the river Yeo'. *Ioweford* 1242, *You(u)eford* 1281, 1334, *Yewford* 1754. R.n YEO SX 8199 + OE **ford**. D 408.

YEOLMBRIDGE Corn SX 3187. Partly uncertain. *Yambrigge* 1216×72, *Yhombregge, Yombrigge* 1308, *Yalme bridge* c.1540. The generic is ME **brigge** (OE *brycg*) 'a bridge', the specific possibly OE **ēa** + **hamm** 'river meadow' with shift of accent from *éa* to *eá* > *ya*. The *l* is not original and probably owes its intrusion to Yealmbridge Devon SX 5952 on the River Yealm. The river here is the Ottery. The present bridge is a 14th cent. construction. PNCo 181, Pevsner 1851.220.

YEOMAN WHARF Cumbr SD 3564. Cf. *Yeomans Bank* 1852 OS. ModE **yeoman** + **wharf** in the sense 'embankment, sand-bank' (OE *hwearf*).

YEOVIL Somer ST 5515. Transferred from the r.n. Ivel, the old name of the river YEO ST 4723. *(æt) Gifle* 873×88 S 1507, *(to) Gyfle* ?c.950 S 1539, *Givele, Ivle* 1086, *Gi- Gevele, Gywele* [13th]14th M, *Gyvele* 14th M, *Yeuele* 1414, *Euyll* 1610, *Yeovil* 1811 OS, a spelling influenced by the common SW stream-name Yeo from OE **ēa** 'river'. RN 221.

YEOVIL JUNCTION STATION Dorset ST 5714. A halt on the railway from Salisbury to Exeter with a branch to Yeovil. P.n. YEOVIL Somer ST 5515 + ModE **junction** and **station**.

YEOVIL MARSH Somer ST 5418. P.n. YEOVIL ST 5515 + ModE **marsh**.

YEO MILL Devon SS 8426. *Yeo Mill* 1809 OS. R.n. YEO SS 7726 + ModE **mill**.

YEOVILTON Somer ST 5422. 'Settlement on the river Ivel'. *Giffeltone* [899×925]13th S 1708, [959×9]13th S 1754, *Geveltone* 1086, *Gi- Geueltona* 1086 Exon, *Givelton* 1202, [1251]Buck, *Geveltone* [1269×84]14th M, *Ivelton* [1346]Buck, *Euilton* 1610. R.n. Ivel as in YEOVIL ST 5515 + OE **tūn**. RN 221.

YES TOR Devon SX 5890. 'Eagle's tor'. *Ernestorre, Yernestorr* [1240]copy, *Yestor* 1765. OE **earn**, genitive sing. **earnes**, + **torr**. D 205.

YETLINGTON Northum NU 0209. 'The estate called after Geatela'. *Yetlinton* 1186, *Yetlingtun* 1236 BF, *-ton* from 1247. OE pers.n. **Ġēatela* + **ing**[4] + **tūn**. NbDu 221, Jnl.8.29.

YETMINSTER Dorset ST 5910. 'Eata's church'. *Etiminstre* 1086, *Eteministr* 1214, *Yateminstre* 1226, *Yeteministr* 1252. OE pers.n. *Ēata* + **mynster**. Do 165.

YETTINGTON Devon SY 0585. 'Settlement of the dwellers at the Gate'. *Yetematon', Yethemeton'* 1242, *Yet(t)emeton(e)* 1260, 1292, *Yeadmeton* 1316. OE folk-n. **Ġeat-hǣme*, genitive pl. **Ġeat-hǣma*, + **tūn**. The **Ġeat-hǣme* were the people who lived at **Ġeat* 'the gate', OE **ġeat**, referring to the narrow valley at this point. D 582.

YIEWSLEY GLond TQ 0680. 'Wife's wood or clearing'. *Wiuesleg'* 1235, *Wyvesle* 1406, *Wewesley* 1593, *Yewsley* 1819. OE pers.n. **Wife*, genitive sing. *Wifes*, + **lēah**. Vocalisation of medial [v] > [w] is well established in ME, as in *gewgaw* < *giuegoue*, dial. *deul* < *devil*. Subsequently initial [w] was lost from the awkward combination [wju:zli]. Mx 42, Jordan §216.1.

YOCKENTHWAITE NYorks SD 9079. 'Eogan's clearing'. *Yoghannesthweit* [1241]copy, *Yokenthwaite* 1499, *Yoke-and-white* 1745. OIr pers.n. *Eogan*, secondary genitive sing. *Eoganes*, + ON **thveit**. YW vi.117.

YOCKLETON Shrops SJ 4010. Unexplained.

I. *Ioclehuile* 1086, *-hulle* 1121, *Jokelhull* 1322, *Yokelcul* 1255.

II. *Iochehulle* 1138, *Yukehull'* 1268(p).

III. *Thokethul* c.1143, *Lokethull'* c.1155, *Yokethil -hull(e)* 1246–1333.

IV. *Yokelthul* 1274, *-ul* 1291×2, *-hull* 1301.

V. *Yekelton* 1316, 1357, *Yokelton* 1349–1430, *Yockleton* 1577.

VI. *Yokton* 1347.

If type II represents the original form of the name this could be OE **ġeoc-hyll** 'yoke hill', form I *Yoke-hill* with pleonastic OE **hyġel** 'a hillock' or **hyll** 'a hill', form V 'the **tūn** or settlement at *Yoke-hill*'. Form III could be a variant 'yoked hill', OE **(ġe)ġeocod-hyll**. The reference would have to be to some feature of the topography of Yockleton, which sits opposite a marked small hill bounded by streams. Association with OE *ġeocled -let* 'a measure of land, a small manor' chiefly found in Kent seems unlikely on both formal and semantic grounds. Too little is known of 19th cent. dial. *yokel* 'a green-wood-pecker, a yellow-hammer' to associate it safely with this name. Sa i.333.

YOKEFLEET Humbs SE 8224. Probably 'Joki's tidal inlet or stretch of water'. *Iugu- Iucufled* 1086, *Yoke- Jukeflet(e)* 12th cent., *Jokesflied, Yokesfliet* 1199, *Yukesflet* 1231, *Yuc- Yukflet(e)* 1246–1498, *Yowk- Yewk(e)flete* 16th cent. ON pers.n. *Jóki* + OE **flēot**. The stretch of water may have been a fishery. YE 255 gives prs [jukflit] and [jo:kflit], SSNY 168.

YORK NYorks SE 6052. 'Place of the yew-tree(s)'.

I. *Ἐβόρακον (Eborakon)* [c.150]c.1200 Ptol, *Eburacum -i -o* [4th]8th, [c.650]13th, [c.731]8th BHE, *Eboracum -i -o* 210–1428.

II. Welsh forms. *Cair Ebrauc* [c.800]11th HB, *Urbs Ebrauc* [866]12th, *Kaer Efrawc* 14th, ModW *Kaer Efrog*.

III. Anglo-Saxon forms. *Evoraca urbs* [10th]17th, *Eoforwic(ceaster)* c.895–c.1150, *Euerwic(h) -wik -wyk* [1019]12th S 956–1415, *Euruic* 1086.

IV. ME forms. *Ȝerk* 14th, *Ȝarke* 1619.

V. ON forms. *Jórvík* [10th]14th, [c.1040]14th, 13th, *Jórk* c.1230, *Y(e)orc* 13th, *York(e)* c.1330–1421, *York(e)* from late 13th.

Brit **eburos** 'a yew-tree' + **-aco-**. Alternatively the specific could be Brit pers.n. *Eburos* and the p.n. mean 'estate of Eburos'. Brit *Eburācon* developed to something like **Evorōg* by the time of the Anglian settlement and this was associated with OE **eofor** 'a boar' and **wīc** probably in the sense 'Roman town'. This in turn was adapted by the Scandinavian settlers as **Éorvík* which with shift of accent from *e* to *o* became *Eórvík, Jórvík* and finally with loss of medial *v Jórk, Yeork* and *York*. There was also a native dial. development of *Eoforwic* to *Yerk* and, with the change of *er* to *ar*, *Yarke*, parallel to that in JERVAULX SE 1785 and YEARSLEY SE 5874. YE 275, Anglia N.F.36.291–6, RBrit 355, LSE 18.141.

YORKLETTS Kent TR 0963. 'The half-hide'. *Yoclete* 1254–1332. OE **ġeocled -t** with later pseudo-manorial *-s*. KPN 105, PNE i.199.

YORKLEY Glos SO 6307. Partly uncertain. *Yark(e)leye* 1338, *Yorkeley* 1628, *Yarkle* 1669. This name is to be compared with Yartleton H&W SO 6821, *Y- Iarclesduna -done* 1154×89, c.1275, *Yarkelton* 1345, 1441, and Yarkhill H&W SO 6042, *(æt) geard-cylle* 811 S 1264, *Archel -il(l)* 1086–1216×72, *Iarculn, Yarchulle* 1243. The latter is 'enclosure kiln', OE **ġeard** + **cyln**, but Yartleton may contain the RBrit p.n. *Ariconium*, PrW **Argin*, PrOE **Arc- *Ærc- *Earc-* as in the district name

ARCHENFIELD 'open land at *Ercing*'. Yartleton may, therefore, be 'hill of Yarkley', OE p.n. **Earc-lēah*, genitive sing. **Earc-lēas* + **dūn**. Yorkley may contain the same elements as this hypothetical OE p.n. **Earc-lēah* meaning 'wood or clearing called or at **Earc*'. Gl iii.228, 192.

YORKSHIRE 'The province of the city of York'. *Eoferwicscire* [1060×5]16th S 1067, 1065 ASC(C), *Eoforwicscire* 1065 ASC(D), *Evrvicscire* 1086, *Yorkshire* 14th Chaucer *Summoner's Tale* 1. P.n. YORK SE 6052 + OE **scīr** 'shire, county'. YW vii.117.

YORTON Shrops SJ 5023. 'The settlement with a yard'. *Iartune* 1086, *Iyarton* 1255, *Yorton* from 1255×63. OE **ġeard** + **tūn**. Sa i.334.

YOULGREAVE Derby SK 2164. 'Yellow or Geola's grove'. *Giolgrave* 1086, *Yolegreue -greve* 1159×6–1312, *Yolgreue -greve* 1255–1490 with variants *-graue -grave* 1252–18th cent. OE **ġeolu** or pers.n. *Ġeola* + **grǣfe**. Db 182, L 193.

YOULSTONE Devon SS 2715. 'Geoloc's hill'. *Yulkesdon* 1353. This is properly East Youlstone; across the county border in Cornwall is West Youlstone, *Yulkesdon* 1302, *Yolkesdon* 1476. OE pers.n. **Ġeoloc*, genitive sing. **Ġeoloces*, + **dūn**. D 134.

YOULTHORPE Humbs SE 7655. 'Eyjulfr's', later 'Yole's, outlying farm'. *Aiul(f)torp* 1086, *Hiel- Hioltorp* 12th cent., *Yolt(h)orp(e)* 12th–1359, *Youlthorp(e)* from 1372. ON pers.n. *Eyjulfr* + **thorp**. From the 12th cent. this name contains a different pers.n., ME *Yole* from ON *Jól, Jóli*. YE 175 gives pr [jɔuθrəp], SSNY 71, SPN 157.

YOULTON NYorks SE 4963. 'Joli's estate'. *Loletun(e)* (for *Iole-*) 1086, *Yolton'* 1295–1508, *Youlton* from 1574. ON pers.n. *Jóli*, genitive sing. *Jóla*, + **tūn**. YN 22, SSNY 130.

YOUNG'S END Essex TL 7319. *Youngs End* 1805 OS, sc. of Great Leighs. Surname *Young* + ModE **end**. Ess 257.

YOXALL Staffs SK 1419. 'The nook of the yoke' sc. of land. *Locheshale* (for *Ioches-*) 1086, *Yoxhal* 1222, *J- Iokeshale* 1236, 1242, *Yoxsall* 1589×90. OE **ġeoc**, genitive sing. **ġeoces**, + **halh**, dative sing. **hale**, in the sense 'small valley'. Yoxall lies in the valley of the r. Swarbourn which forms a narrow side-valley opening off the Trent. The sense of **ġeoc** in this name is not clear. DEPN, L 105, 110.

YOXFORD Suff TM 3968. 'The yoke ford'. *Gokesford, Iokesfort* 1086, *Jokeford* 1203, *J- Iokesford* 1254–1311, *Yoxford* from 1316. OE **ġeoc**, genitive sing. **ġeoces**, + **ford**. The sense is probably 'the ford passable by a yoke of oxen'. DEPN, Baron.

Z

ZEAL 'Hall, manor house'. OE **sele** with Southern initial [z] for [s]. This is the simplest explanation.

(1) ~ MONACHORUM Devon SS 7103. 'The monks' Zeal'. *Sele Monaco'* 1275, *Monekenesele* 1346. Latin genitive pl. **monacorum** translating ME **monekene** 'of the monks', + p.n. *Sele* 1228–76[†]. The manor belonged to the monks of Buckfast Abbey, the manorial addition being for distinction from other examples of the common SW p.n. Zeal. The DB name of the manor was *Limet* 1086, the common DB form for *Nymet* as in NYMET TRACEY SS 7200, the original name of the river YEO SS 7306. D 375.

(2) South ~ Devon SX 6593. *Southsele* 1544. ME adj. **south** + p.n. *(la) Sele* 1167–1322. Also known as *Zele Tony* 1299, from the family of Constancia de Tony which held the manor of South Tawton in 1212 and for distinction from other examples of the common SW p.n. Zeal. D 450.

ZEALS Wilts ST 7831. 'The willow-tree(s)'. *(at) Seale* [956]14th S 637[†], *Sele -a* 1086, *Seles* 1176–1377, *Sayles* 1629, *Zailes* 1637, *Zeales* 1665. OE **sealh**, dative sing. **sēale**, pl. **sēalas**. Wlt 182, xl.

ZELAH Corn SW 8151. 'The hall'. *Sele* 1311, *Zela* 1613. OE **sele** with southern voicing of *s* to *z* as in South ZEAL Devon SX 6593 and ZEAL MONACHORUM SS 7103. PNCo 181, Gover n.d. 432.

ZENNOR Corn SW 4538. 'St Senar's (church)'. *(ecclesia) Sancti Sinari* c.1170, *(apud) Sanctam Sinaram* 1235, *(ecclesia) Sancte Senare* 1270–1300, *Senar* 1522, *Sener* 1562, *Zenar* 1582. Saint's name *Senar* from the dedication of the church. Nothing is known of this saint, not even his or her gender. The modern form shows S dial. voicing of *S-* to *Z-*. PNCo 181, Gover n.d. 666, DEPN.

ZONE POINT Corn SW 8431. 'Long cleft point'. *Savenheer or the long coved point* 1597, *The Zone Point* 1813. P.n. *Savenheer*, Co **sawn** 'cleft, gully, a hole in a cliff through which the sea passes' + **hyr** 'long', + ModE **point**. Co **sawn** has been Anglicised as *zone*. On some OS maps this was misprinted with a sideways N misread as Z and giving rise to a form Zoze Point. The cleft has now nearly fallen in. PNCo 181, Gover n.d. 434.

[†]The DEPN explanation, OE **seale**, dative sing of **sealh** 'a sallow-tree', depends on the form *at Seale* [956]14th S 639, but this belongs rather to ZEALS Wilts ST 7831. D 375n., Wlt xl.

[†]Identification uncertain.

Lightning Source UK Ltd.
Milton Keynes UK
UKOW040149140112

185366UK00002B/17/P